Encyclopedia of
ROSE SCIENCE

ENCYCLOPEDIA OF
ROSE SCIENCE

Editor-in-Chief
ANDREW V ROBERTS

Editors
THOMAS DEBENER
SERGE GUDIN

ELSEVIER
ACADEMIC
PRESS

Amsterdam Boston Heidelberg London New York Oxford
Paris San Diego San Francisco Singapore Sydney Tokyo

Elsevier Ltd., The Boulevard, Langford Lane, Kidlington, Oxford, OX5 1GB, UK

Elsevier Inc., 525 B Street, Suite 1900, San Diego, CA 92101-4495, USA

First edition 2003

Library of Congress Control Number: 2003112413

A catalogue record for this book is available from the British Library

ISBN: 0-12-227620-5 (set)

This book is printed on acid-free paper

Printed in Spain

EDITORIAL ADVISORY BOARD

FOREWORD

No other flower is as universally loved and grown or has a more illustrious history than the rose. Both reality and symbol, the rose has been a constant source of inspiration throughout history in such different fields as religion, architecture, literature, art, the perfume industry, fashion, gastronomy and, of course, horticulture and gardening. Throughout the centuries the rose has been regarded as the queen of flowers, and it is a fact that there is very little to surpass the sight of well-grown roses in full bloom. Whether grown by themselves or among other plants, roses bring to bear many years of development in an unparalleled array of functionality and beauty. However, the incredible history of roses starts in the hands of passionate, meticulous and attentive breeders, who imagine and develop colours, fragrances and shapes for the queen of the garden. What are the secrets of their alchemy? How is a rose born? The *Encyclopedia of Rose Science* attempts to answer these mysterious questions.

The research methods used by rose breeders to create and screen new varieties at this stage are mainly based on classical processes. Rose-breeding efficiency is still very dependent on the breeder's experience, despite recent scientific advances that have contributed to plant improvements. Moreover, science plays an increasingly important role in the production of this flower which is at the centre of one of the world's biggest floricultural industries. Rose breeders such as myself recognise this and have long backed scientific research, beginning in the 1970s and continuing to the present day. It has always seemed to me to be important to drive the organisation of international scientific symposia, to develop internal programs, and facilitate national and international cooperation in order to improve and speed up the whole creation and selection process.

The *Encyclopedia of Rose Science* draws together the whole range of rose techniques and scientific knowledge. For breeders, rose growers and avid amateurs, as well as researchers and students, this book will be an invaluable reference guide which will supply a treasury of information.

Alain Meilland
Le Cannet-des-Maures

September 2003

INTRODUCTION

Roses have been admired for their beauty and fragrance since the dawn of civilization. Over the centuries, poets, artists, historians, gardeners, apothecaries and scientists, each in their own way, have contributed to our appreciation, enjoyment and knowledge of roses. The *Encyclopedia of Rose Science* focuses on scientific contributions but also contains articles, including those on the history of cultivated roses, pruning and intellectual property rights, which provide background information and advice for an international readership of scientists, breeders, growers and knowledgeable rose enthusiasts.

Approximately 8 billion rose stems, 80 million potted plants and 220 garden roses are sold annually. This flourishing horticultural trade is served by many scientists in commercial organizations, research institutes and universities throughout the world, who have made substantial but sometimes under-appreciated contributions to knowledge of the rose. The *Encyclopedia of Rose Science* brings together in a single publication and on an unprecedented scale, overviews, explanations and interpretations that no recent publication has attempted. Whilst exploring their areas of special interest, readers will be able to follow cross-links (manually or electronically) to less familiar but related topics.

The editors have been helped in their tasks of identifying topics for inclusion in the *Encyclopedia*, choosing the best authors and reviewing the completed articles, by an Editorial Advisory Board that consists of 20 distinguished scientists and practitioners from many countries. Many topics, including those concerned with breeding, propagation, production, *in vitro* culture, physiology, fragrance, pests and pathogens, are well subscribed to in the scientific literature, but particular care was needed to provide essential information about some relatively neglected topics. For example, information about the distribution and ecology of the rose is difficult to obtain because few rose scientists presently specialize in these areas and literature sources are scattered in diverse, often inaccessible publications. To cover these areas, information was sought from representatives of the national rose societies affiliated to the World Federation of Rose Societies. Their knowledge of the local conditions, the distribution of wild species and the cultivars that grow best in their localities, provides unique information that will provide a valuable reference for formal ecological studies in the future.

The description of species of roses is an activity that was once at the very centre of scientific investigation and exploration. However, there is no recent work that is as comprehensive and detailed as the chapter on *Rosa*, in Rehder's *Manual of Cultivated Trees and Shrubs Hardy in North America* (2nd edition, 1960). As an important reference source, the chapter concerning *Rosa* has been reproduced in this *Encyclopedia*. For descriptions of species and classification, schemes are also given in some articles on Distribution and Ecology and Classification. Clearly, a monograph covering recent perspectives on the taxonomy, classification and description of the genus *Rosa* is urgently needed. Readers should note that there have been name changes, including the change from *R. wichuriana* Crép. to *R. wichurana* Crép. One name that appears in many scientific papers, but not in taxonomic works, is the name *R. hybrida* L. or (*R.* × *hybrida* L.). The origin of this name is mysterious! Searches by professional taxonomists have failed to identify any such species named by Linnaeus. It is, therefore, unclear as to what the name signifies. The origin of this name needs to be clarified or a new name introduced to describe an unambiguously defined category of hybrid roses.

It is likely that readers of the *Encyclopedia* will wish to refer repeatedly to descriptions of cultivated roses that are cited in the articles. To facilitate this, the reader can refer manually or electronically to *Modern Roses XI* which is reproduced in the *Encyclopedia*. This is the 11th edition of a periodically revised publication owned by the American Rose Society (the official registration body of cultivar names), edited by Tommy Cairns and published by Academic Press (an Elsevier imprint). It lists 24 000 cultivars. The reader will find that some cultivars referred to in the articles are not listed, often because they are old cultivars that are rarely seen, new cultivars that will appear in the next edition of *Modern Roses*, or rootstock varieties.

Most articles in the *Encyclopedia of Rose Science* include a statement of future developments anticipated by the authors. Perhaps an additional editorial comment is needed on one area of development that may have a significant impact. Biotechnology has arrived and is likely to be an area of much activity and discussion! It has been said, disparagingly, that modern cultivars are made by man but wild roses were made by God. However, most of us applaud the achievements of rose breeders. Over the centuries, the processes of genetic modification and selection that fashioned the wild roses have been extended by human intervention. Early introductions to rose gardens, including roses with multipetalled flowers and the recurrent-flowering habit, resulted from selections of naturally occurring mutants. Hybridization that enabled breeders to introduce new combinations of characters was a later development. Now modern rose breeders are able to plan breeding programmes, carry out difficult crosses and select seedlings with greater skill than ever before, to provide us with ever more attractive and healthy roses. However, the exploitation of new techniques in molecular biology may significantly challenge old practices. The possibilities are, of course, enormous and some of these are regarded with apprehension. The use of genetic probes that will enable selection to be carried out with infinitely greater efficiency than before will not change the essential genetic nature of rose crops. More controversially, genetic manipulation, already carried out in some laboratories, allows genes to be introduced that do not naturally occur in the genus *Rosa*. The verdict on genetic manipulation will be delivered by a discriminating public that is wary of innovation but also recognizes the environmental benefits, for example, of not spraying roses for infectious diseases. We hope that the articles on Genetics in this *Encyclopedia* will assist the reader in understanding the techniques involved and potential applications of the new approaches.

The editors gratefully acknowledge the expert contribution of the Major Reference Works unit at Elsevier, who have ensured that the *Encyclopedia of Rose Science* is presented to a high standard. We particularly appreciate the contribution of the Senior Developmental Editor, Mark Knowles: his patient advice, friendly support and, when necessary, tactful coercion of editors and authors alike have ensured that publication has kept to its schedule. The support of the Editorial Advisory Board has been invaluable and the distinguished contributions of the authors of the articles themselves will be apparent to all readers. We hope that the success of the *Encyclopedia* will be a fitting tribute to all who have made its production possible.

Andrew V Roberts
Thomas Debener
Serge Gudin

September 2003

GUIDE TO USE OF THE ENCYCLOPEDIA

Structure of the Encyclopedia

The material in the Encyclopedia is arranged as a series of entries in alphabetical order. Most entries consist of several articles that deal with various aspects of a topic and are arranged in a logical sequence within an entry. Some entries comprise a single article.

To help you realize the full potential of the material in the Encyclopedia we have provided three features to help you find the topic of your choice: a Contents List, Cross-References and an Index.

I. Contents List

Your first point of reference will probably be the contents list. The complete contents list, which appears at the front of volumes one and two will provide you with both the volume number and the page number of the entry. On the opening page of an entry a contents list is provided so that the full details of the articles within the entry are immediately available.

Alternatively you may choose to browse through a volume using the alphabetical order of the entries as your guide. To assist you in identifying your location within the Encyclopedia a running headline indicates the current entry and the current article within that entry.

You will find 'dummy entries' where obvious synonyms exist for entries or where we have grouped together related topics. Dummy entries appear in both the contents list and the body of the text.

Example

If you were attempting to locate material on the molecular biology of flavonoids via the contents list:

MOLECULAR BIOLOGY *see* BIOTECHNOLOGIES FOR BREEDING: Genetic Transformation. CLASSIFICATION: Chemotaxonomy and Molecular Taxonomy. FRAGRANCE AND PIGMENTS: Anthocyanins and other Flavonoid Pigments; Flavonoid Molecular Biology; Functional Genomics. GENETICS: Nuclear DNA Amounts; Inheritance in the Dogrose; DNA Fingerprinting; Gene Mapping; Molecular Markers; Cloned Genes; Marker-Assisted Selection; Research Trends. GROWTH REGULATION: Floral Induction. POSTHARVEST PHYSIOLOGY: Ethylene.

The dummy entry directs you to the Flavonoid Molecular Biology article, in the FRAGRANCE AND PIGMENTS entry. At the appropriate location in the contents list, the page numbers for articles under Fragrance and Pigments are given.

If you were trying to locate the material by browsing through the text and you looked up Molecular Biology then the following information would be provided in the dummy entry:

Molecular Biology *see* **Biotechnologies for Breeding**: Genetic Transformation. **Classification**: Chemotaxonomy and Molecular Taxonomy. **Fragrance and Pigments**: Anthocyanins and other Flavonoid Pigments; Flavonoid Molecular Biology; Functional Genomics. **Genetics**: Nuclear DNA Amounts; Inheritance in the Dogrose; DNA Fingerprinting; Gene Mapping; Molecular Markers; Cloned Genes; Marker-Assisted Selection; Research Trends. **Growth Regulation**: Floral Induction. **Postharvest Physiology**: Ethylene.

Alternatively, if you were looking up **Fragrance and Pigments** the following information would be provided:

FRAGRANCE AND PIGMENTS

Contents

2. Cross-References

All of the articles in the Encyclopedia have been extensively cross-referenced.

The cross-references, which appear at the end of an article, serve three different functions. For example, at the end of the GENETICS/DNA Fingerprinting article, cross-references are used:

i. To indicate if a topic is discussed in greater detail elsewhere.

> GENETICS/DNA Fingerprinting.
> *See also:* **Classification**: Conventional Taxonomy (Wild Roses); Horticultural Classification Schemes; Cultivar Identification by Image Analysis; Chemotaxonomy and Molecular Taxonomy. **Genetics**: Inheritance in the Dogrose; Species Crosses; Mutation; Gene Mapping; Molecular Markers; Marker-Assisted Selection. **Intellectual Property**: Plant Patents and Trademarks; Technical Examination and Conventional Characteristics. **Rose Collections and Trials**: Rose Collections and Herbaria.

ii. To draw the reader's attention to parallel discussions in other articles.

> GENETICS/DNA Fingerprinting.
> *See also:* **Classification**: Conventional Taxonomy (Wild Roses); Horticultural Classification Schemes; Cultivar Identification by Image Analysis; Chemotaxonomy and Molecular Taxonomy. **Genetics**: Inheritance in the Dogrose; Species Crosses; Mutation; Gene Mapping; Molecular Markers; Marker-Assisted Selection. **Intellectual Property**: Plant Patents and Trademarks; Technical Examination and Conventional Characteristics. **Rose Collections and Trials**: Rose Collections and Herbaria.

iii. To indicate material that broadens the discussion.

> GENETICS/DNA Fingerprinting.
> *See also:* **Classification**: Conventional Taxonomy (Wild Roses); Horticultural Classification Schemes; Cultivar Identification by Image Analysis; Chemotaxonomy and Molecular Taxonomy. **Genetics**: Inheritance in the Dogrose; Species Crosses; Mutation; Gene Mapping; Molecular Markers; Marker-Assisted Selection. **Intellectual Property**: Plant Patents and Trademarks; Technical Examination and Conventional Characteristics. **Rose Collections and Trials**: Rose Collections and Herbaria.

3. Index

The index will provide you with the page number where the material is located, and the index entries differentiate between material that is a whole article, is part of an article or is data presented in a figure or table. Detailed notes are provided on the opening page of the index.

4. Appendices

In addition to the articles that form the main body of the Encyclopedia, there are a number of appendices, which provide an historical list of Rose Awards, extensive lists of refereed papers relevant to articles, Rehder's classification scheme for *Rosa*, reproduced from the *Manual of Trees and Shrubs, Hardy in North America* (Second Edition), and a listing of rose names, reproduced from the American Rose Society's *Modern Roses XI*.

The appendices are located in volume 2, before the index, and in volume 3.

5. Contributors

A full list of contributors appears at the beginning of volumes one and two.

CONTRIBUTORS

A J J Akkerman
DLV Plant BV
Wageningen, The Netherlands

J P André
INRA
Antibes, France

J O Becker
University of California
Riverside, CA, USA

H Ben-Sade
The Hebrew University of Jerusalem
Rehovot, Israel

J-P Biolley
Université Claude Bernard
Lyon, France

J Blaas
Business Unit Biosciences
Plant Research International
Wageningen, The Netherlands

P S Blake
Horticulture Research International
East Malling,
Kent, UK

T J Blom
University of Guelph
Guelph, ON, Canada

W A Brandenburg
Business Unit Biosciences
Plant Research International
Wageningen, The Netherlands

N Bredmose
Danish Institute of Agricultural Sciences
Aarslev, Denmark

H Brichet
World Federation of Rose Societies
San Terenziano, Italy

D H Byrne
Texas A&M University
College Station, TX, USA

R I Cabrera
Texas A&M University
Dallas, TX, USA

T Cairns
American Rose Society
Studio City, CA, USA

C Casey
North Carolina State University
Raleigh, NC, USA

A Chaanin
W. Kordes Soehne
Klein Offenseth-Sparneshoop, Germany

D Chimonidou
Agricultural Research Institute
Nicosia, Cyprus

A C M Clerkx
Enschedé, The Netherlands

H A Collin
University of Liverpool
Liverpool, UK

J M Costa
Wageningen University
Wageningen, The Netherlands
Fundação para a Ciência e Tecnologia
Lisbon, Portugal

Y M Crane
Purdue University
West Lafayette, IN, USA

L Crespel
Meilland Star Rose
Le Luc en Provence, France

R R Cutler
University of East London
London, UK

D P de Vries
Plant Research International
Wageningen, The Netherlands

Present address: Foundation Sub Rosa
Ede, The Netherlands

T Debener
Federal Centre for Breeding Research on Cultivated Plants
Institute for Ornamental Plant Breeding
Ahrensburg, Germany

A Dohm
Federal Centre for Breeding Research on Cultivated Plants
Institute of Ornamental Plant Breeding
Ahrensburg, Germany

R Drewes-Alvarez
University of Applied Sciences Dresden
Dresden, Germany

L A M Dubois
Plant Research International
Wageningen, The Netherlands

D Esmenjaud
INRA
Antibes, France

J Ferare
The Conard–Pyle Co.
Hollister, CA, USA

F Ferrero
INRA
Saint Aygulf, France

B Feuerhahn
Berlin, Germany

J-L Fiasson
Université Claude Bernard
Lyon, France

K Fiasson
Université Claude Bernard
Lyon, France

R G Foottit
Agriculture and Agri-Food Canada
Ottawa, ON, Canada

G Forkmann
Technical University of Munich
Freising-Weihenstephan, Germany

H W M Fuchs
University of Delft
Delft, The Netherlands

M Fukuchi-Mizutani
Institute for Fundamental Research
Suntory Ltd
Osaka, Japan

M-H Gandelin
Geves Cavaillon
Les Vignères, France

N García Victoria
PPO Glasshouse Horticulture
Aalsmeer, The Netherlands

M L Gleason
Iowa State University
Ames, IA, USA

J-F Gonnet
Université Claude Bernard
Lyon, France

C Grossi
Université Claude Bernard
Lyon, France

S Gudin
Université d'Aix-Marseille III,
Marseille, France

W Guoliang
Jiangsu Forestry Academy
Dongshanqiao, Nanjing, China

S J Helland
Iowa State University
Ames, IA, USA

A P Hert
University of Florida
Gainesville, FL, USA

R K Horst
Cornell University
Ithaca, NY, USA

Y Jacob
INRA
Saint Aygulf, France

M Jay
Université Claude Bernard
Lyon, France

J B Jones
University of Florida
Gainesville, FL, USA

F Joyaux
Roseraie de la Cour de Commer
Commer, France

J F Karlik
University of California
Bakersfield, CA, USA

J W Kelly
Clemson University
Clemson, SC, USA

P G Kevan
University of Guelph
Guelph, ON, Canada

M Kilian
Bayer Environmental Science
Monheim, Germany

T Kwakkenbos
Community Plant Variety Office (CPVO)
Angers, France

M Le Bris
Université d'Aix-Marseille III
Marseille, France

R Lewis
University of East London
London, UK

M Linde
Federal Centre for Breeding Research on Cultivated Plants
Ahrensburg, Germany

P T Lynch
University of Derby
Derby, UK

N Marissen
PPO Glasshouse Horticulture
Aalsmeer, The Netherlands

M Marriott
David Austin Roses Ltd
Wolverhampton, UK

J G Mason
Florigene Ltd
Collingwood, Victoria, Australia

J Mattock
Oxford, UK

H E L Maw
Agriculture and Agri-Food Canada
Ottawa, ON, Canada

J Meynet
INRA
Saint Aygulf, France

J Mottley
University of East London
London, UK

J Mouchotte
Meilland Star Rose
Le Luc en Provence, France

R Müller
Royal Veterinary and Agricultural University
Frederiksberg, Denmark

B C Murphy
Cal Agri Products
Davis, CA, USA

T M Murphy
University of California
Davis, CA, USA

R Noack
Guetersloh, Germany

H Nybom
Swedish University of Agricultural Sciences
Kristianstad, Sweden

C O'Donnell
University of California
Davis, CA, USA

G Otto
Dresden, Germany

M P Parrella
University of California
Davis, CA, USA

H B Pemberton
Texas A&M University
Overton, TX, USA

T Pettitt
Horticulture Research International
Wellesbourne, Warwick, UK

F Piola
Université Claude Bernard
Lyon, France

M Podwyszynska
Research Institute of Pomology and Floriculture
Skierniewice, Poland

H M C Put
Cut Flower Hygiene
Deventer, The Netherlands

S Rajapakse
Clemson University
Clemson, SC, USA

O Raymond
Université Claude Bernard
Lyon, France

M Reid
University of California
Davis, CA, USA

A V Roberts
University of East London
London, UK

A Samach
The Hebrew University of Jerusalem
Rehovot, Israel

C Sanlaville-Boisson
Université Claude Bernard
Lyon, France

U K Schuch
University of Arizona
Tucson, AZ, USA

H Schulz
Federal Centre for Breeding Research on Cultivated Plants
Institute for Plant Analysis
Quedlinburg, Germany

A Schum
Federal Centre for Breeding Research on Cultivated Plants
Institute of Ornamental Plant Breeding
Ahrensburg, Germany

R C Shattock
University of Wales
Bangor, UK

N Shishkoff
Cornell University
Long Island Horticultural Research and Extension Center
Riverhead, NY, USA

J D Shorthouse
Laurentian University
Sudbury, ON, Canada

A Skytt Andersen
Royal Veterinary and Agricultural University
Frederiksberg, Denmark

W Spethmann
University of Hanover
Hanover, Germany

U Steiner
Rheinische Friedrich-Wilhelms-Universität Bonn
Bonn, Germany

B M Stummann
Royal Veterinary and Agricultural University
Frederiksberg, Denmark

M S Szyndel
Warsaw Agricultural University
Warsaw, Poland

Y Tanaka
Institute for Fundamental Research, Suntory Ltd
Osaka, Japan

S A Tjosvold
University of California Cooperative Extension
Watsonville, CA, USA

S Torre
Agricultural University of Norway
Aas, Norway

M J Tsujita
University of Guelph
Guelph, ON, Canada

Y Ueda
Chiba University
Matsudo, Japan

L Urban
INRA/CIRAD
Saint-Pierre de la Réunion, France

A Vainstein
The Hebrew University of Jerusalem
Rehovot, Israel

P A van de Pol
Wageningen University
Wageningen, The Netherlands

U van Meeteren
Wageningen University
Wageningen, The Netherlands

H A Verhoeven
Plant Research International
Wageningen, The Netherlands

M-R Viricel
Université Claude Bernard
Lyon, France

D Weiss
The Hebrew University of Jerusalem
Rehovot, Israel

G Werlemark
Swedish University of Agricultural Sciences
Kristianstad, Sweden

V Wissemann
Friedrich-Schiller-Universität Jena
Jena, Germany

X Xu
Horticulture Research International
East Malling, West Malling, UK

D Zamir
The Hebrew University of Jerusalem
Rehovot, Israel

D Zhang
Geves Le Magneraud F-17700
Surgères, France

CONTENTS

VOLUME 1

VOLUME 2

I

J

M

N

O

P

A

Anthocyanins *see* **Fragrance and Pigments**: Odoriferous Substances and Pigments; Anthocyanins and other Flavonoid Pigments; Flavonoid Molecular Biology.

B

Bacteria *see* **Disease**: Overview; Crown Gall.

Bactericide *see* **Disease**: Bactericides and Fungicides.

BIOTECHNOLOGIES FOR BREEDING

Contents

Early Embryo Rescue

R Drewes-Alvarez, University of Applied Sciences Dresden, Dresden, Germany

Introduction

Embryo culture is a technique for cultivating an embryo under aseptic conditions on a nutrient medium. The method can be divided into two applications. One is performed with mature embryos and helps mainly in shortening the period of germination by overcoming seed dormancy. The other is performed with immature embryos and is called early embryo rescue. The purpose of this application is to grow embryos that would otherwise have been aborted (*see* **Propagation**: Seed Maturation and Germination). In some crosses, e.g. interspecific or intergeneric, the nutrient tissue (endosperm) fails to develop and causes the abortion of the embryo. These embryos can be rescued by culturing them on artificial nutrient medium until they are fully developed and able to germinate. Early embryo rescue has recently been used to obtain hybrids from the interspecific cross *Rosa rugosa* Thunb. × *R. foetida* Herrm. and from crosses between English rose varieties, e.g. *R.* × *hybrida* 'Heritage'® × *R.* × *hybrida* 'Lucetta'®, and to obtain dihaploid plants from *R.* × *hybrida* 'Sonia'.

History

Embryo rescue is one of the oldest techniques of *in vitro* culture. As early as 1904 artificial cultures of embryos were used in Brassicaceae. In the first half of the nineteenth century several experiments were done with seeds from interspecific crosses. The authors excised the embryos during the early seed development and succeeded in growing them into plants by culturing them on aseptic media. The first published results of early embryo rescue in Rosaceae go back to 1933 and were done with cherry and peach. In *Rosa* L. embryo rescue with mature achenes was done in 1946 for the first time (**Table 1**), but results using immature embryos were only published in 1994.

Table 1 Embryo rescue in rose

Purpose	*Author(s) and year*
Immediate germination of seeds	Lammerts (1946), Asen (1948), Lauzer and Laberge (1996)
Investigations on seed dormancy	Semeniuk *et al.* (1963), Von Abrams and Hand (1956)
Early embryo rescue	Gudin (1994), Marchant *et al.* (1994)

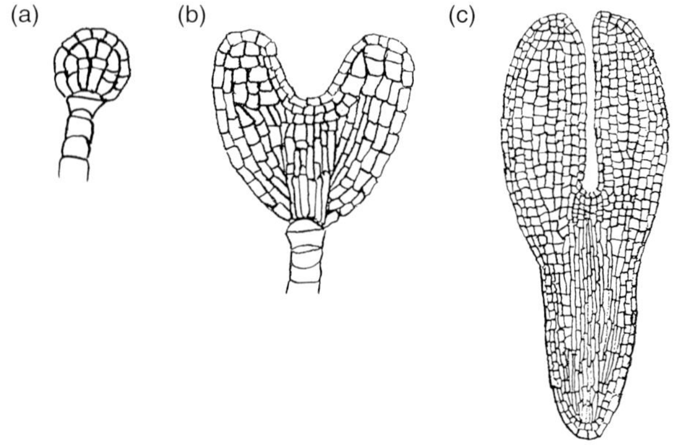

Figure 1 Development of an embryo. (a) globular stage; (b) heart-shaped stage; (c) torpedo-shaped stage. (Redrawn after Sitte *et al.* (2002).)

Embryo Development

In angiosperm plants the embryo develops after fertilization first into a round globular stage. During further development the dividing cells will form the cotyledons, the shoot tip and the root tip. The embryo appears first and is heart-shaped, followed by torpedo-shaped (**Figure 1**) and finally the cotyledonary stage is reached, when the cotyledons have reached their mature size in the seed. The mature embryo in *Rosa* L. is covered by a seed-coat (testa) and a woody pericarp, which is composed of three layers, of which the inner one is the most impermeable. **Figure 2** shows a longitudinal section through a rose achene. A cotyledon, the hypocotyl and the radicle are visible.

The achenes are normally considered to be mature after 3.5–4 months.

Techniques of Embryo Rescue

In *Rosa* the embryo is well-protected by the surrounding pericarp so surface sterilization procedures can be rather harsh. Sterilization of the hips includes either 10–15 min in sodium hypochlorite or 15 min in $HgCl_2$ followed by washing in sterile water. After this the achenes can be extracted from the hips and kept in sterile water. Very young embryos, before reaching the heart-shaped stage, do not develop on artificial media. Embryos that have reached the heart-shaped stage should be cultured *in ovulo* and only be isolated after some further development. Torpedo-shaped embryos can be isolated directly.

Starting from approximately day 35, the pericarp starts to become woody and about 50 days after pollination the pericarp is very hard and difficult to remove. So in achenes that are harvested from day 50 onwards the problem is to get rid of the pericarp without damaging the embryo. In some varieties soaking the achenes in water or enzymatic solutions will make dissection easier. In almost mature achenes treatment with acid (e.g. a mixture of zinc chloride and concentrated hydrochloric acid) could give good results but requires constant checking, otherwise the acid may damage the embryo. After removing the pericarp, the testa must also be taken out because it has inhibiting effects on development and germination. These embryos can now be placed on nutrient medium. **Figure 3** shows germinating embryos 2 weeks after isolation.

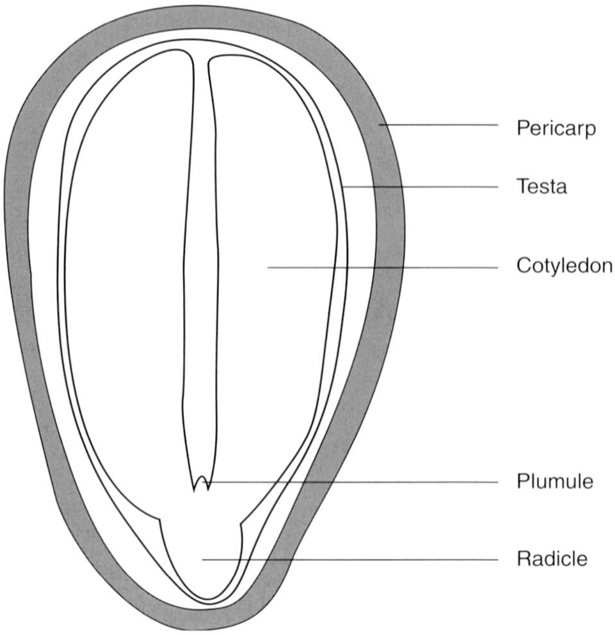

Figure 2 Longitudinal section through a rose achene.

Media for Embryo Culture

While mature embryos easily germinate on a nutrient medium with mineral salts and 2.5–5% sucrose, it is more difficult to find a suitable medium composition for immature embryos. The younger embryo depends a lot on nutrition through the maternal tissue and the endosperm. In this stage a more complex medium is required.

The mineral salts that are required follow a formulation of Murashige and Skoog or in other experiments a formulation of Asen and Larson and contain

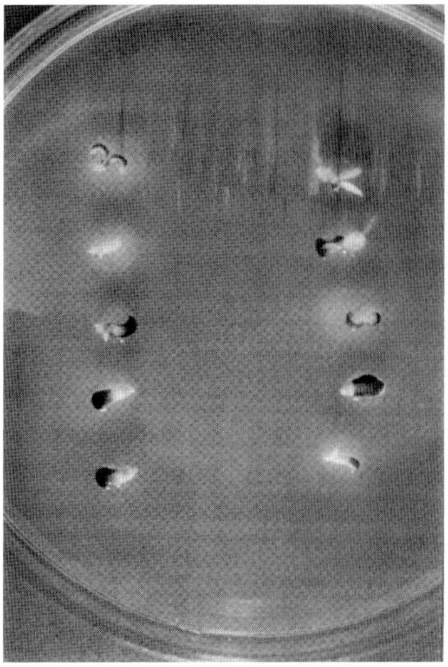

Figure 3 Germinating rose embryos, 2 weeks old, on culture medium.

macro- and microelements (*see* **Cell, Tissue and Organ Culture**: Micropropagation). No significant differences in embryo growth were recorded on media with either mineral salt formulation. Sometimes the strength of the mineral salts is diluted to one-half or one-quarter of the original recipe. Different sources of carbohydrates can be used, e.g. glucose, sucrose and fructose. Sometimes vitamins and amino acids were added to the medium. Adding the liquid endosperm of coconut to the culture medium also had growth-promoting effects on the embryos. The incorporation of 6-benzylaminopurine, known to break seed dormancy in *R. arvensis* Huds., had no significant effects. The pH of the culture medium is adjusted to 5.5–5.8 and the media are solidified with 0.5–0.6% agar.

Culture Conditions

Different light and temperature regimes can be applied. Most of the time the cultures were kept in the dark for an initial period of 2–6 weeks and were then transferred into the light with a photoperiod of 16 h. The temperatures in the culture room were kept between 22 and 23 °C. Cold treatment of 4 weeks at 4 °C in the dark can promote growth and germination of immature rose embryos.

Further Treatment of Germinated Embryos

The germinated embryos can be kept in the culture vessels until they have grown into young seedlings. The plantlets can be micropropagated like any other microshoots that have developed from shoot tips or nodal segments (*see* **Cell, Tissue and Organ Culture**: Micropropagation; Rooting of Micropropagated Shoots) or they can be transferred to soil directly when they reach about 5 cm height. Seedlings that have germinated on agar require acclimatization to conditions of lower humidity after transfer to soil.

See also: **Cell, Tissue and Organ Culture**: Micropropagation; Rooting of Micropropagated Shoots. **Morphology and Anatomy**: Pollen Grains and Tubes. **Propagation**: Seed Maturation and Germination.

Further Reading

Asen S (1948) Embryo culture of rose seeds. *American Rose Annual* 33: 151–153.

Bridgen MP (1994) A review of plant embryo culture. *HortScience* 29: 1243–1245.

Gudin S (1994) Embryo rescue in *Rosa hybrida* L. *Euphytica* 72: 205–212.

Gudin S (1995) Rose improvement, a breeder's experience. *Acta Horticulturae* 20: 125–127.

Lammerts WE (1946) Use of embryo culture in rose breeding. *Plants and Garden* 2: 111.

Lauzer D and Laberge C (1996) Establishment of a collection of *Rosa* species through *in vitro* embryo culture. *HortScience* 31(3): 458–459.

Marchant R, Power JB, Davey MR and Chartier-Hollis J (1994) Embryo rescue, for the production of F_1 hybrids, in English rose. *Euphytica* 74: 187–193.

Semeniuk P, Stewart RN and Uhring J (1963) Induced secondary dormancy of rose embryos. *Proceedings of the American Society for Horicultural Science* 83: 825–828.

Sitte P, Weiler EW, Kadereit JW, Bresinsky A and Körner C (2002) *Strasburger–Lehrbuch der Botanik für Hochschulen*. Heidelberg, Berlin: Spektrum.

VonAbrams GJ and Hand ME (1956) Seed dormancy in *Rosa* as a function of climate. *American Journal of Botany* 43: 7–12.

Manipulation of Ploidy Level

L Crespel, Meilland Star Rose, Le Luc en Provence, France
J Meynet, INRA, Saint Aygulf, France

Introduction

Ploidy level is a term referring to the number of chromosome sets in somatic cells of the diplophase ($2n$) or gametophytic cells of the haplophase ($1n$). It is indicated by a number followed by the *x* letter. Diploid cells have two sets of chromosomes and are indicated by $2x$. Cells with more than two sets of chromosomes are termed polyploid and are indicated by $n\ (>2)\ x$.

Chromosome numbers in the genus *Rosa* are based on multiples of seven and range from $2n = 2x = 14$ to $2n = 8x = 56$. Aneuploids are rare. The subgenera *Hulthemia*, *Platyrhodon* and *Hesperodos* include only one species each, in which $2n = 2x$. The fourth subgenus, *Rosa*, contains over 120 species grouped into 10 sections. In the sections *Banksiae*, *Bracteatae*, *Indicae*, *Laevigatae* and *Synstylae* $2n = 2x$, in the *Carolinae* and *Gallicae* $2n = 4x$, in the *Pimpinellifoliae* $2n = 2x$ and $4x$, in the *Caninae* $2n = 4x$, $5x$ and $6x$, and in the *Cinnamomeae* $2n = 2x$, $4x$, $6x$ and $8x$.

Modern roses are grouped into horticultural classes that include, for example, Polyanthas ($2n = 2x$), Hybrid Teas ($2n = 3x, 4x$), Floribundas ($2n = 3x, 4x$) and Miniatures ($2n = 2x, 3x, 4x$).

Biotechnological techniques, such as haploidization or mitotic, meiotic and somatic polyploidization, are available to reduce or increase ploidy level, respectively. It has long been known that some wild species, mainly diploids, represent potentially valuable genetic resources, especially concerning resistance to diseases. To access these wild diploid resources two different strategies based on the manipulation of ploidy level can be envisaged: the haploidization of tetraploid cultivars and polyploidization of diploid wild species or diploid interspecific hybrids.

Haploidization

Possible Methods

Since 1922, when the first haploid plant was observed in *Datura stramonium*, several methods have resulted in plants with the chromosome structure of a sexual cell.

In some species, haploidization can spontaneously occur *in situ* especially as a result of polyembryony: one synergid develops without fertilization (apogamy) at the same time as the fertilized egg, so twin embryos appear, one is haploid, the other is diploid. In some other species, the male and female nuclei divide separately without a true fusion (semigamy), and produce a mosaic of tissues of male and female origin, with one of these two tissues being progressively eliminated.

The best-studied *in situ* haploidization methods result from interspecific crosses in *Solanum* and *Hordeum* genera. The cross between *Solanum tuberosum* and *Solanum phureja* can lead to the formation of haploid embryos of *S. tuberosum*. The male gametes of *S. phureja* are grouped and exclusively merge with the secondary nuclei of the embryo sac, and the unfertilized egg parthenogenetically develops to form a haploid embryo. The cross between *Hordeum vulgare* and *Hordeum bulbosum* results in a classical double fertilization but the chromosomes of *H. bulbosum* are progressively eliminated from both the embryo and the albumen. In this case, the albumen degenerates. Hence, embryo rescue by *in vitro* culture is necessary.

However, the success of haploidization in a great number of species requires the use of *in vitro* culture, either in order to induce a diversion of the male or female gametophytes to an embryogenetic development and the formation of viable plants, or to secure embryo rescue.

Androgenesis by *in vitro* culture of anthers remains the most commonly used haplomethod since the pioneering works in Solanaceae (*Datura innoxia* and *Nicotiana*). One of the main limits of that method is the great specificity of the androgenetic abilities according to genetic origin of the donors. Some botanical families are often very responsive (Solanaceae, Ranunculaceae), while others are not (Compositae), some being quite refractory (Chenopodiaceae). In any case, the anthers must be cultured at a very precise stage (generally at the first mitotic division of the pollen, when the nucleus of the microspore is in a central position).

Androgenesis by *in vitro* culture of isolated microspores should permit the best control of the technique since the contingent effects of the sporophyte tissues contained in the anther are avoided. That method is more recent than anther culture; it is mainly used in some cereal species.

Gynogenesis consists in inducing the development of unpollinated female gametophytes in embryos and plantlets through *in vitro* culture. The first successful trials in *Gingko biloba* were confirmed in *Hordeum vulgare*, then in a great number of species which were sometimes refractory to androgenesis, such as some Compositae (*Gerbera jamesonii* or *Helianthus annuus*).

Finally, pollen fertility inactivation, especially by irradiation, can lead to *in situ* parthenogenetic development of eggs (pseudogamy). This method often requires *in vitro* embryo rescue. It was successfully applied in a great number of different species, especially in some fruiting members of the family Rosaceae such as apple.

Methods Tested in Roses

The first attempts at haploidization in roses used *in vitro* culture of anthers of tetraploid roses, *Rosa damascena* Miller and *Rosa* cultivars. They allowed some partially diploid calluses to be obtained; presumably they originated from pollen but they never resulted in dihaploid plants.

Isolated microspores cultured in a liquid medium produced an enlargement of the cells and, at best, a few cell divisions.

Gynogenesis attempts by *in vitro* culture of unfertilized eggs were carried out in our laboratory; they only produced calluses, probably from placenta origin.

Up to now, the only successful haploidization method has resulted from *in situ* induction of parthenogenesis using irradiated pollen and subsequent *in vitro* culture of immature seed.

In Situ Parthenogenesis in Roses

Haploidization is obtained by *in situ* parthenogenesis after pollination with irradiated pollen and subsequently embryo rescue. Such pollination is believed to stimulate parthenogenetic egg cell development without fertilization.

The female parents must be chosen among highly fertile varieties because there is a positive correlation between competence for haploidization and the female fertility of the tetraploid donors.

Anthers are collected from flower buds of the male parents 1 to 2 days prior to flower opening. The anthers are dried for 2 days in an incubator at 30 °C to induce dehiscence. The pollen is sifted and stored in a desiccator at 4 °C for the duration of the hybridization period, e.g. from April to June in European countries. All the flower buds on the maternal plants are emasculated 2 days prior to anthesis by removing the calyx, petals and anthers with a forceps. After pollination, the flowers are protected with paper bags.

Dry pollen samples are exposed to gamma irradiation from a cobalt-60 source for a total exposure of 600 Gy. Pollinations with irradiated pollen are made on the day of irradiation or the day after.

Eight weeks after pollination, the hips that form are collected and achenes are extracted from each hip. The enlarged achenes are plunged into water. The probability that achenes contain an embryo is approximately 15 times higher in achenes of density >1 than in floating ones. The achenes that sink are disinfected in 30 g l^{-1} $CaCl_2$ for 20 min, and subsequently rinsed three times for 5 min in sterile distilled water. Embryos are aseptically removed from the endocarp.

Embryos are placed on a solid culture medium (**Table 1**) in darkness at 4 °C for 4 weeks, and then transferred to a 16-h photoperiod provided by daylight fluorescent illumination (90 μmol m^{-2} s^{-1}) and at 22 °C for 2 weeks to obtain rooted plantlets.

After incubation, the embryos germinate and the rooted plantlets are planted into a classical horticultural medium and grown under greenhouse conditions. The ploidy level of the plants obtained can then be determined by flow cytometry or vegetative meristem chromosome counts.

Table 1 Composition of the medium used for embryo rescue

Component	*Concentration (mg l^{-1})*
KNO_3	950
KH_2PO_4	85
$MgSO_47H_2O$	185
NH_4NO_3	200
$CaCl_22H_2O$	220
$MnSO_44H_2O$	22.3
H_3BO_3	6.2
$CuSO_45H_2O$	0.025
$CoCl_26H_2O$	0.025
$ZnSO_47H_2O$	8.2
KI	8.2
$Na_2MoO_42H_2O$	0.25
FeEDTA	5
Inositol	100
Thiamine	10
Pyridoxin	1
Nicotinic acid	1
Calcium panthotenate	1
Benzylamino purine	0.2
Agar	7500
Sucrose	30000
pH	5.8

Characteristics of Dihaploids

Characteristics such as the guard cell length, chloroplast number, stem length, leaf and flower size are reduced in dihaploids as compared with their tetraploid donors (**Figure 1**).

The male fertility of the dihaploids is generally very weak. The pollen viability determined by staining with Alexander's solution varies from 0 to 35% in dihaploids derived from different rose cultivars with an average of 9%, and is near to 0% in haploids of *R. canina*.

The female fertility of the dihaploids is extremely variable. The dihaploids often display proliferation phenomena (secondary flowers from ovaries) which perturb the female reproduction function of the flower. However, many progeny have been obtained from a hybridization programme between dihaploids of rose cultivars, used as female parents, and diploid wild species (e.g. *R. gigantea* Colett ex Crép., *R. wichurana* Crép., *R. rugosa* Thunb., *R. roxburghii* Thratt.), used as male parents.

Dihaploids and hybrids derived from crosses between dihaploid cultivars and 2*x* wild species, produce some 2*n* eggs and 2*n* pollen, i.e. gametes with a somatic chromosome number. The analysis of the diameter of pollen produced by dihaploids generally

Figure 1 Two dihaploids compared with their tetraploid donors. (a) H_3, a dihaploid characterized by a proliferation phenomenon and (b) 'Sweet Promise' ('Sonia Meilland') its 4*x* donor and (c) H190 a fertile dihaploid derived from (d) 4*x* 'Meirilocra' ('Zambra').

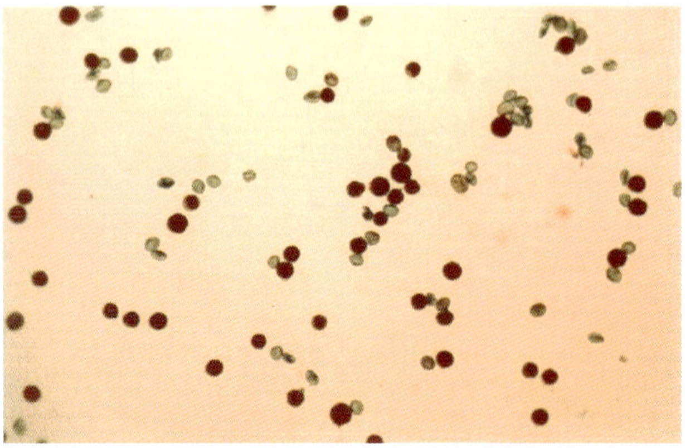

Figure 2 A sample of pollen from a dihaploid (derived from 4*x* FJV6) × *R. rugosa* hybrid. The small, irregular and clear grains are sterile; the size of the others is variable; the largest ones correspond to 2*n* pollen.

reveals two subpopulations corresponding to *n* (nearly 32 ± 7 μm) and 2*n* (nearly 43 ± 5 μm) (**Figure 2**).

Unreduced (2*n*) gametes are formed in two primary ways: (1) an incomplete first meiotic division (first division restitution; FDR), or (2) an incomplete second meiotic division (second division restitution; SDR). Unreduced gametes via FDR are comprised mainly of the nonsister chromatids of each homologous pair of chromosomes, whereas in SDR the sister chromatids are included in the same gametes. As a result, 2*n* gametes formed by FDR transmit more of the parental heterozygosity into progenies than those formed by SDR.

Meiotic abnormalities lead to 2*n* gametes of different types, second division restitution without crossing-over (SDR-NCO), second division restitution with crossing-over (SDR-CO), first division restitution without crossing-over (FDR-NCO) and first division restitution with crossing-over (FDR-CO). It has been estimated that the 2*n* gametes of SDR-NCO, SDR-CO, FDR-NCO and FDR-CO types transfer 0%, ±40%, 100% and ±80% of the parental heterozygosity, respectively. The proportions of the different gamete types produced are genotype-dependent. The formation of 2*n* gametes is often accompanied by *n*, 3*n* and 4*n* gametes. This ability to produce 2*n* gametes is transmitted to the progenies, but the genetic control of this trait is still unknown.

In haploids of *R. canina* (originally $2n = 5x = 35$), vegetative meristem chromosome counts revealed a mixoploid character. The chromosome number varied from 25 to 32 with a majority of tetraploid cells ($2n = 4x = 28$).

Polyploidization

In roses, spontaneous polyploidization is of rare occurrence and it has been never described in detail. However, the case of 'Max Graf', a sterile diploid *R. rugosa* × *R. wichurana* hybrid is well known. Spontaneously, it produced a fertile tetraploid seedling, *R.* × *kordesii*, which founded a very important group of roses.

Chromosome doubling can be obtained via the three following experimental methods: mitotic, meiotic and somatic polyploidization.

Mitotic Polyploidization

Mitotic polyploidization consists in doubling the somatic chromosome number using chemicals with antimitotic activity such as colchicines (*see* **Biotechnologies for Breeding**: Amphidiploidy).

Colchicine has been used for doubling the number of chromosomes of many crop plants over a period of more than 50 years. This alkaloid, obtained from the wild *Colchicum autumnale*, blocks the cell at metaphase of mitosis. It binds to one tubulin end and prevents its polymerization and causes disappearance of the mitotic spindle.

Other chemicals with antimitotic activity similar to colchicine such as colcemid, or more recently oryzalin, are equally used.

Colchicine-Induced Polyploidization in Roses

Attempts were made to double the chromosome numbers of the diploid species by either of two methods.

1. Colchicine treatments of seedling shoots An aqueous solution containing 0.5% colchicine was applied three times at 2-day intervals on the top three lateral buds in active growth (located at the four- to eight-leaf stage) of seedlings of diploid species. The three laterals were then cut and rooted under mist. The success of this system was erratic, but it has been successfully performed with different genotypes worldwide. The tetraploids obtained by this system are characterized by:

- flowers, leaves and roots of larger size than in the diploid original
- larger stomata and pollen grains than in the diploid original.

2. Growing seedlings *in vitro* in a liquid medium after a culture phase on a medium containing colchicine Unbranched shoots of a diploid species, *R. wichurana*, were formed on a multiplication medium and nodal sections were subcultured. Shoots with five to six leaves were transferred to 50 ml rooting medium containing 0.05% colchicine for 12 h, then to rooting medium without colchicine for 7 days in Erlenmeyer flasks (250 ml) on an orbital shaker (100 rpm). Both multiplication and rooting media contained Murashige and Skoog (MS) salts and vitamins and 40 g l^{-1} sucrose. The pH was adjusted to 5.6. The multiplication medium was supplemented with 1 mg l^{-1} benzylamino purine, 0.005 mg l^{-1} naphthaleneacetic acid (NAA) and 8 g l^{-1} agar, and the rooting medium was only supplemented with 0.1 mg l^{-1} NAA (no agar). The culture media were sterilized by autoclaving at 121 °C for 15 min. All the cultures were maintained at 23 °C in a 16-h photoperiod supplied by Grolux fluorescent lights giving 90 μmol $m^{-2} s^{-1}$. Cytological studies were then carried out on cells of root tips which were excised after the treatments. Seven days after exposure to 0.05% colchicine for 12 h gave (1) complete spindle arrest; (2) the highest frequency of tetraploid cells in the root meristem (32.7%).

Amphidiploidy

Recently, amphidiploidy was used for the development of black-spot resistant rose germplasm. From five interspecific diploid hybrids involving the black-spot resistant diploid species *R. wichurana*, *R. roxburghii*, *R. banksiae* Ait., *R. rugosa rubra* Hort. and *R. setigera* Michaux, amphidiploids were induced using two colchicine treatment protocols through agitation of excised nodes in a colchicine solution or tissue culture of shoots on a medium supplemented with colchicine.

Protocol 1: Single nodes cut from greenhouse-grown plants The excised nodes were surface sterilized, and then shaken orbitally at 90 rpm in quarter-strength MS salts solution with 2.5 mmol colchicine for 48 h in darkness at 25 °C. After this treatment, the nodes were transferred onto full-strength, solid MS medium with 3% sucrose, 0.7% agar, 8.9 mmol 6-benzyladenine (BA) and 0.5 mmol NAA and held for 2 weeks to initiate shoots. These shoots were then transplanted onto solid MS medium (one-fourth strength) with 3% sucrose, 0.7% agar, and indoleacetic acid (IAA) at 0.6 mmol for rooting. Culture conditions for shoot initiation and rooting were continuous cool-white fluorescent lighting (90 μmol $m^{-2} s^{-1}$) at 25 °C.

Protocol 2: Axillary shoots (≈1.5 cm long) regenerated *in vitro* from stock plants These shoots were directly transferred onto solid MS medium (half-strength) with 3% sucrose, 0.7% agar and 1.25 mmol colchicine. The shoots were cultured in darkness at 10 °C for 120 h, and were then transferred onto MS medium (one-fourth strength) with 0.6 mmol IAA for rooting at 25 °C under continuous cool-white fluorescent lighting (90 μmol $m^{-2} s^{-1}$).

At the end of the 2 weeks' incubation on the rooting medium, the rooted plantlets were planted into a medium of 2 pine-bark mulch : 1 peat : 5 sand (by volume) and grown under ambient daylight and relative humidity in an unshaded greenhouse with a day/night cycle of 25/22 °C ± 5. Chromosome counts confirmed amphidiploids among the plants that survived colchicine treatment.

The average frequency of induced amphidiploids varied from 8% to 13% for protocol 1 and from 2% to 3% for protocol 2, depending on the genotype.

Characteristics of amphidiploid roses No mixoploid was detected in the chromosome counts of the plants that survived colchicine treatment. The amphidiploids were visually distinguishable from the diploid plants (**Figure 3**):

- leaflets of the amphidiploids have a greater width-to-length ratio and are more crinkled along the midrib than in their diploid original hybrids
- the fertility of the amphidiploids was much higher than that of the original diploid hybrids.

Meiotic Polyploidization

Sexual polyploidization is the process by which a polyploid zygote is formed by natural fertilization involving $2n$ gametes. The case of interploidy crosses (e.g. $2x$–$4x$), where one $2n$ gamete fertilizes a normally reduced gamete has been termed 'unilateral polyploidization', whereas 'bilateral polyploidization' is characterized by the fusion of two $2n$ gametes (as might occur in a $2x$–$2x$ cross). Depending on its origin, a single $2n$ gamete may convey little or nearly all of the heterozygosity present in somatic cells. In order to maximize heterozygosity in the polyploid progeny produced by sexual polyploidization, the use of FDR type gametes is preferable to the use of SDR gametes.

In roses, $2n$ gametes produced by dihaploid rose cultivars and their diploid hybrids are used to obtain new tetraploid genotypes through unilateral and bilateral polyploidization. At this stage, only crosses between dihaploids of rose and rose cultivars have been realized. The percentage of tetraploid progenies obtained was varied from 14.2% to 100% according to the dihaploid parent used. A strong gametic selection is observed, in favour of $2n$ gametes, especially when a tetraploid partner is involved.

Figure 3 Morphological characteristics of a polyploidized plant, *Rosa rugosa* (right) as compared to its diploid origin (left). (Photo courtesy of Andrea Dohm; from work performed at the Institute of Ornamental Plant Breeding in Ahrensburg.)

Somatic Polyploidization

In this case, protoplast fusion is used (*see* **Cell, Tissue and Organ Culture**: Protoplast Culture). In roses, the fusion of callus or cell suspension-derived protoplasts from various cultivars was induced using polyethylene glycol. These protoplasts were used either with protoplasts from cherry or blackberry. Putative somatic intergeneric hybrids were regenerated. However, results of cytological and random amplified polymorphic DNA (RAPD) analyses indicated that these plants rather corresponded to genetic variants of the rose cultivar. Protoplasts of the diploid hybrid *R. persica* Michx. × *R. xanthina* Lindl. have been self-fused and tetraploid plants were obtained. These tetraploids exhibited unusual morphology, and had significantly increased guard cell lengths and chloroplast numbers.

Prospects for Manipulation of Ploidy

An important pool of variation is still available in *Rosa*. Although close to 200 rose species have been described worldwide, only about 10 have been used to create the enormous present cultivar diversity. Tremendous improvement potential is still available through traditional hybridizing but new techniques may be required to better exploit this variation. Various biotechnology techniques, such as mitotic, meiotic and somatic polyploidization and haploidization, are currently available for rose breeding. However, some of these new techniques have only been applied recently to *Rosa* and therefore need further research before they can be used with full efficiency.

In any case, increased knowledge in *Rosa* basic genetics is urgently needed. Ongoing research such as that on haploidization has great potential in that respect.

See also: **Biotechnologies for Breeding**: Early Embryo Rescue; Amphidiploidy. **Cell, Tissue and Organ Culture**: Protoplast Culture. **Genetics**: Karyology.

Further Reading

Bretagnolle F and Thompson JD (1995) Gametes with the somatic chromosome number: mechanisms of their formation and role in the evolution of autopolyploid plants. *New Phytologist* 129: 1–22.

Byrne DH, Black W, Ma Y and Pemberton HB (1996) The use of amphidiploidy in the development of blackspot resistant rose germplasm. *Acta Horticulturae* 424: 269–272.

Drewes-Alvarez R (1992) *Untersuchungen am Pathosystem Sternrusstau* (Marssonina rosae) – *Rose* (Rosa *L.) zur Klärung der Frage nach pilzlichen Rassen und zur Übertragung der Resistenz aus der diploiden* Rosa multiflora *Thunb. auf tetraploide Gartenrosen*, PhD Thesis, University of Hamburg, Germany.

El Mokadem H, Meynet J and Gudin S (2000) Utilization of parthenogenetic diploid plants of *Rosa hybrida* L. *Acta Horticulturae* 508: 185–190.

El Mokadem H, Crespel L, Meynet J, Gudin S and Jacob Y (2001) Gametic behaviour of parthenogenetic plants of *Rosa canina* L. and *Rosa hybrida* L. *Acta Horticulturae* 547: 289–296.

El Mokadem H (2001) *Haploïdisation des Rosiers, Comportement des Dihaploïdes de* Rosa hybrida *L. et Utilisations à des Fins de Sélection et d'Analyses Génétiques*. PhD Thesis. University of Toulouse, France.

Mottley J, Yokoya K, Matthews D, Squirrell and Wentworth JE (1996) Protoplast fusion and its potential role in the genetic improvement of roses. *Acta Horticulturae* 424: 393–397.

Veilleux R (1985) Diploid and polyploid gametes in crop plants: mechanisms of formation and utilization in plant breeding. *Plant Breeding Reviews* 3: 253–288.

Amphidiploidy

D H Byrne, Texas A&M University, College Station, TX, USA

Y M Crane, Purdue University, West Lafayette, IN, USA

Introduction

Given that there is a polyploid series within *Rosa* that extends from diploid to octoploid, it is natural that the evolution and breeding of this genus have involved the transfer of genes among these ploidy levels. Amphidiploidy is both a natural and an induced phenomenon that combines two diploid species into a fertile tetraploid – a potentially useful approach in the transfer of genes from the diploid to the tetraploid level.

Types of Polyploidy among Plants

Classically, polyploidy is classified into two extreme situations: autopolyploidy and allopolyploidy. An autotetraploid plant has four copies of the same chromosome set. In contrast, an allotetraploid or amphidiploid plant has two diploid sets of chromosomes derived from distinct species. This distinction has important genetic consequences (*see* **Genetics**: Meiosis). Autopolyploids generally have no novel alleles and therefore little genetic advantage over their progenitor diploid species. The fertility of autotetraploids is frequently reduced because of multivalent chromosome association at meiosis I. Allotetraploids combine the diverse genetic composition of two distinct species, and their chromosomes regularly pair as bivalents at meiosis I. The resulting hybrid vigour and high fertility often give allotetraploids a competitive advantage over their progenitor diploid species. Among plants, there is a continuum from autopolyploid to allopolyploid situations.

Frequency among Plants

More than half of all angiosperm species are polyploid, and polyploidy has been an important force in their evolution. The majority of these polyploids are allopolyploids that have arisen through fusion of unreduced gametes or chromosome doubling in diploid hybrids. The unreduced gametes can develop through occasional failure of either meiotic division in the parental diploid species, or because of irregular meiosis in diploid hybrids. Sometimes the sequence is diploid to triploid to tetraploid, with the fusions involving one unreduced, diploid gamete at the first stage and two diploid gametes at the next stage. Complicated hybrid situations become possible in cultivation, where hybrids can cross with or without human intervention, and with or without the production of unreduced gametes. Often in such hybrids the chromosomes will pair preferentially if like partners are present, and remain capable of pairing across species differences if like partners are not present. In this case, an advanced-generation tetraploid hybrid can contain chromosome segments from more than four parental diploid species.

Polyploid Series among Rose Species and Cultivars

The large, diverse genus *Rosa* has about 200 species. Taxonomic opinion varies as to the exact number, in part because the general cross-compatibility within and across subsections and subgenera has allowed natural hybridization and introgression to occur. This pattern of species relationships is common in woody plant genera. Although most rose species are diploid, ploidies range up to octoploid. Odd-numbered and even-numbered polyploids in subgenus *Rosa* sect. *Caninae* are maintained by a peculiar meiotic mechanism that transmits different chromosome sets through the male and female gametes (*see* **Genetics**: Inheritance in the Dogrose). Most rose cultivars are tetraploid and trace their origin to a set of 8–10 diploid and a few tetraploid progenitor species.

Incorporation of Diploid Species into Tetraploid Roses via Triploids

Ornamental roses have been cultivated in Europe and China for over two millennia. Before the nineteenth century, most European roses were tetraploids that flowered once or twice per growing season. The introduction of the ever-blooming diploid rose from China permitted the development of modern free-blooming roses. Much has been written about the four stud roses that brought about this dramatic change (*see* **History of Roses in Cultivation**: Modern (Post-1800)). It should be noted that this involved hybridization between a diploid rose (*Rosa chinensis*) and tetraploid roses (various species hybrids centred on *R. damascena*). The initial crosses resulted in triploid hybrids with generally low fertility. These produced low numbers of tetraploids, probably through diploid egg cells and pollen from the surrounding tetraploid plants in the breeding garden. The resulting tetraploid hybrids (founders of the Hybrid Perpetuals) were used in breeding, with the most fertile contributing the most. This process has been repeated many times with other diploid species such as *R. wichurana*, *R. multiflora*, *R. brunonii*, *R. moschata* and *R. persica*.

In this approach, the primary triploid hybrid is usually sterile and thus a difficult bottleneck in breeding. An alternative approach is the induction and use of tetraploids from diploid species. Although autotetraploids have been used, as in the case of *Rosa rugosa* in the creation of the Spotless cultivars, the production of amphidiploids is more promising.

Induction of Amphidiploids

There are two practical ways to produce amphidiploids from diploid rose species:

1. Double the diploid species and cross the resulting autotetraploids.
2. Cross the diploid species and double the resulting diploid hybrid.

The second approach is easier, because it avoids the typical infertility of primary autotetraploids. In addition, since diploid interspecific hybrids are produced relatively easily, these diploid hybrids can be selected for horticultural traits before doubling the chromosome number.

The general rule is that the less fertile the interspecific diploid hybrid, the more fertile the amphidiploid derived from it. Thus the parents used in the process need to be as distantly related as possible as well as have all the characteristics desired in the final amphidiploid. This is illustrated by the various amphidiploids produced from hybrids of *R. wichurana* with *R. rugosa* developed in Germany (*R. kordesii*), in Canada (G12), and in the USA (86-7). These diploid hybrids are easily made and are sterile. Nevertheless, meiotic studies of the diploid hybrid 86-7 indicated that there was principally bivalent pairing with occasional univalents, as was seen in the diploid parents. In all these cases, fertility was greatly improved in the amphidiploid. In the amphidiploid 86-7, meiotic studies indicated that most of the chromosomes formed bivalents, and about 30% were involved in multivalent or univalent formation (**Figure 1**). Thus the evidence suggests that the chromosomes of these species are still quite similar. In a strict sense, these amphidiploids are intermediate between autotetraploids and allotetraploids.

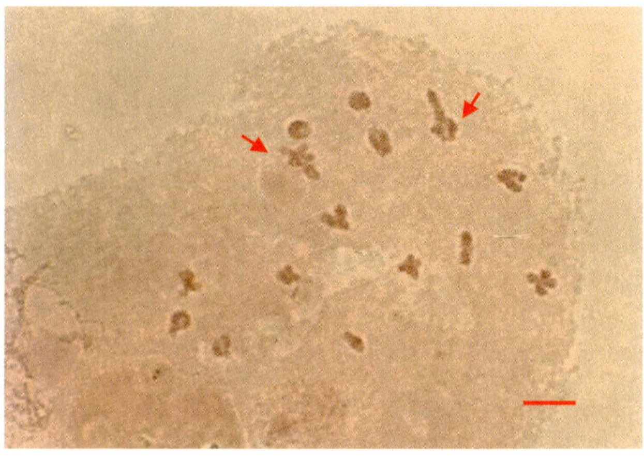

Figure 1 Meiotic chromosomes of the tetraploid rose (Cyt 66) showing bivalents, univalents and two trivalents (arrowed). Bar = 2.5 μm.

Colchicine application is commonly used to double the chromosome number of a rose, although oryzalin is also effective at micromolar concentrations in disrupting the spindle mechanism of plant cells. These chemicals prevent the spindle from pulling daughter chromosomes apart and cause the chromosomes to remain clumped together. Thus various higher ploidy levels can result as the chromosomes decondense into the next interphase. Colchicine is quite toxic, and the effective dose is scarcely less than the lethal dose. Therefore, it is not surprising that success has been erratic, with the best treatments giving 2–14% chromosome doubling. Effectiveness depends upon concentration, duration, genotype and condition of the exposed tissue. Both *in vivo* and *in vitro* applications have been tried with roses.

Dr Robert Basye used an *in vivo* approach. He applied 0.06% colchicine in water to the first young bud of a rose seedling for four consecutive days (one dose per day) while maintaining the plants in high humidity. He obtained two amphidiploids (86-7 (*R. rugosa* var. *rubra* × *R. wichurana*) and 86-3 (*R. laevigata* × *R. banksiae*)), one mixoploid (84-1000 (*R. laevigata* × *R. roxburghii*)) and autotetraploids of *R. bracteata, R. laevigata, R. roxburghii* and *R. setigera*.

Two *in vitro* approaches have been tried:

1. Soaking nodal sections in 2.5 mmol l^{-1} colchicine solution for 48 h before planting into gelled medium.
2. Growing explants at least 5 days on medium containing 1.25 mmol l^{-1} colchicine.

Both *in vitro* approaches have worked reasonably well, but they are limited to rose genotypes that grow easily *in vitro*. In addition, genotype affects the optimal concentration, duration and culture conditions. This work has resulted in additional amphidiploids from the hybrids *R. wichurana* × *R. roxburghii* and *R. wichurana* × *R. setigera*.

Spontaneous doubling has been opportunistically exploited, as in the origin of *R. kordesii* from cultivar 'Max Graf'. It is likely that *R. kordesii* arose from fusion of unreduced gametes during the repeated self-pollination of 'Max Graf', although some authorities have suggested that *R. kordesii* arose from an outcross. The genotype G12 is known to have come from an outcrossing on 'Max Graf'. A third example is Dr Basye's amphidiploid 67-305, obtained from the diploid hybrid of *R. abyssinica* with *R. rugosa*. All three of these spontaneous tetraploids are fertile enough for continued breeding.

Induced tetraploidy is visually detected by typical 'gigas' characteristics: slower growth, thicker leaflets with a greater width-to-length ratio, tissue crinkling near the leaf midrib and greater overlapping of the leaflets (**Figures 2** and **3**). Other indicators are larger guard cell size, and, once the plants begin to flower, larger pollen grain (**Table 1**) and flower size. Ultimately a chromosome count is needed (**Figure 4**) (*see* **Genetics:** Karyology). The procedures for pollen measurements and guard cell measurements are provided in the appendix to this article. Chromosome count procedures are detailed elsewhere (*see* **Genetics:** Karyology).

Figure 2 Leaflets (adaxial and abaxial sides) of diploid (bottom) and tetraploid (top) forms of a *Rosa rugosa* var. *rubra* × *R. wichurana* hybrid showing the broader and thicker leaflets of the tetraploid as compared to the diploid form.

Since colchicine (or any other antimitotic agent) works on a single cell level, the original plant is likely to be a chimera – a plant with sectors of diploid and tetraploid tissue. Thus care needs to be taken to examine the plant thoroughly and isolate the doubled sector.

Fertility of the Initial Amphidiploids

Diploid hybrids and their derived amphidiploids are inversely correlated for fertility, because genomic dissimilarity promotes preferential pairing of homologous chromosomes at meiosis I. Thus a sterile diploid usually produces a more fertile amphidiploid, and a fertile diploid usually produces a less fertile amphidiploid. Many diploid hybrids are partially fertile, and consequently so are their induced amphidiploids. In the diploid hybrid incomplete chromosome pairing causes this partial fertility whereas in the induced tetraploid it is due to multivalent formation.

In reality, amphidiploid roses vary tremendously in their initial fertility, even when the initial diploid hybrids are sterile. Amphidiploid 86-7 had abundant pollen, whereas 86-3 and 84-1000 produced only limited quantities of pollen. Their pollen fertility affected the results of hybridization trials (**Table 2**). The pollen of 86-7 gave at least twice as much hip set and seed per cross as did 86-3 and 84-1000. Nevertheless, all the amphidiploids were more fertile than the diploid interspecific hybrids from which they were obtained,

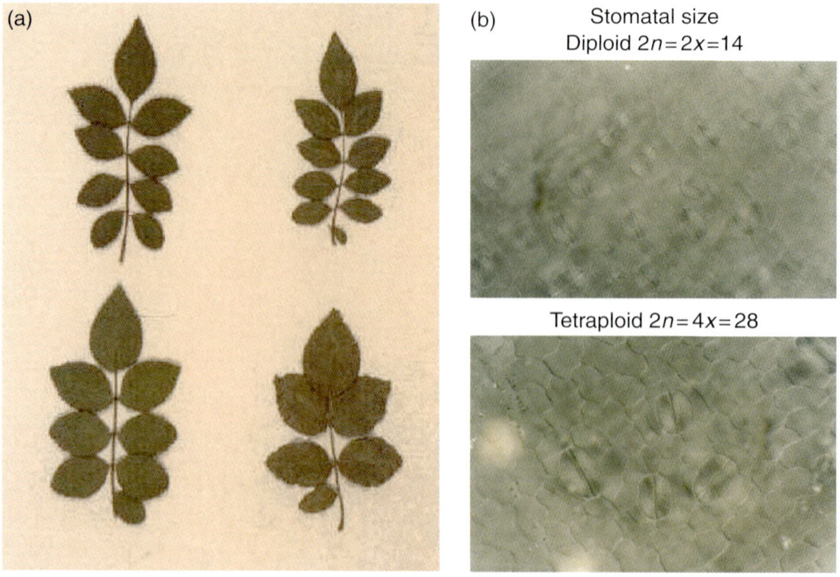

Figure 3 (a) Leaf and (b) stomata size of diploid (top) and tetraploid (bottom) hybrid interspecific roses from a *Rosa wichurana* × *R. roxburghii* cross.

Table 1 Comparison of leaflet, stomatal and pollen size of the amphidiploids 86-3 (*R. laevigata* × *R. banksiae*) and 86-7 (*R. wichurana* × *R. rugosa* var. *rubra*) with their progenitor diploid parents

Trait	R. banksiae	R. laevigata	*86-3*	R. rugosa	R. wichurana	*86-7*
No. of leaflets	5–7	3	5	9	9–11	5–9
Leaflet length (cm)	3.5	3.3	2.6	2.7	1.5	2.6
Leaflet width (cm)	1.1	1.6	1.4	2.0	1.1	2.6
Ratio of leaflet width to length	0.32	0.49	0.55	0.73	0.74	0.98
Stomatal length (μm)	5.7	5.4	7.7	5.1	6.1	8.1
Stomatal width (μm)	3.8	4.0	5.0	3.3	3.5	4.4
Pollen diameter (μm)	9.7	9.5	11.1	10.3	10.4	12.4

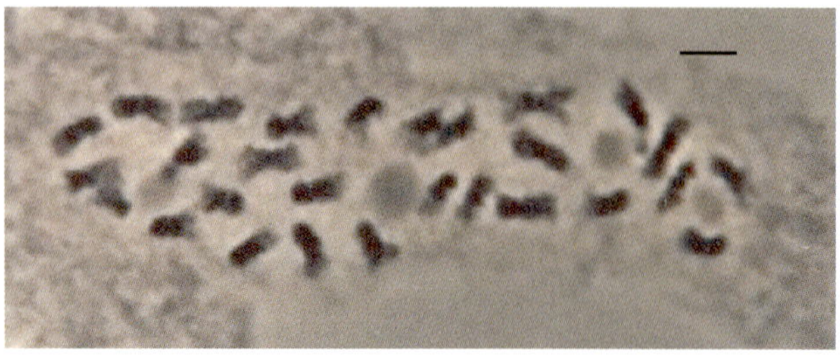

Figure 4 Mitotic chromosomes of 86-7 showing 28 chromosomes. Bar = 2.5 μm.

Table 2 Effectiveness of two rose amphidiploids as male parents

	86-3	*86-7*
No. of crosses	427	755
Stainable pollen (%)	85	74
Percentage of hips set	19%	38%
Seed/hip	4.5	5.7
Seed/cross	0.8	2.1

and all are sufficiently fertile for use in breeding. It is possible to select for increased fertility in hybrid populations derived from the amphidiploids.

Use of Amphidiploids

The most widely used amphidiploid in modern rose breeding is *R. kordesii*, which gave rise to the Kordesii hybrid roses in European and Canadian breeding programmes. Basye's amphidiploids 67-305, 86-3 and 86-7 are being hybridized with commercial rose germplasm in order to raise the general black spot resistance (*see* **Breeding:** Selection Strategies for Disease and Pest Resistance. **Disease:** Overview; Black Spot) of the breeding population. Other interspecific hybrids are being actively created to assess their usefulness in breeding.

Conclusions

There is a need to transfer traits efficiently from wild diploid species to tetraploid cultivars. These traits include disease resistance, cold-hardiness, growth habit and novel flower types, such as the red centre from *R. persica*. Traditionally this has been done via triploids, but the active induction of new interspecific tetraploids is more efficient and would expand the gene pool available to rose breeders.

Appendix

Pollen Fertility and Size

Pollen is collected from opening or freshly opened flowers, dried for 3 days at 25 °C, dispersed in 0.1% aniline blue in lactophenol on a slide, and covered with a cover slip. Stainability is visually assessed for 200–300 randomly chosen pollen grains per genotype, and the fully stained pollen grains are measured with an ocular micrometer.

Guard Cell Measurement

Collect four to seven leaves at approximately equal angles around each branch. Transversely slice the terminal leaflet from each leaf and clear the slices to measure guard cells in a whole mount. To do this, heat the slices in boiling 2 : 1 dimethyl sulfoxide (DMSO) : chloroform ($CHCl_3$) to extract the chlorophyll. Then incubate the slices 30 min or more in the following solutions at 25 °C: 2 : 1 : 1 DMSO : $CHCl_3$: benzyl benzoate (BB), 1 : 1 : 2 DMSO : $CHCl_3$: BB, 1 : 3 $CHCl_3$: BB, and 1 : 3 dibutyl phthalate (DBP) : BB. The last solution is an effective clearing agent in which leaf tissue may be stored indefinitely. Mount the cleared leaflets in 3 : 1 BB : DBP for examination under differential interference-contrast optics (**Figure 3**). Measure the length and width of approximately 100 guard cells per genotype with an optical micrometer.

See also: **Biotechnologies for Breeding**: Manipulation of Ploidy Level. **Genetics**: Karyology; Meiosis; Inheritance in the Dogrose.

Further Reading

Byrne DH, Black W and Ma Y (1996) The use of amphidiploidy in the development of blackspot resistant rose germplasm. *Acta Horticulturae* 424: 269–272.

Eigsti OJ and Dustin P Jr (1957) *Colchicine – in Agriculture, Medicine, Biology, and Chemistry*. Ames, IA: Iowa State College Press.

Grant V (1980) *Plant Speciation*, 2nd edn. New York: Columbia University Press.

Lewis WH (1980). *Polyploidy: Biological Relevance*. New York: Plenum.

Genetic Transformation

A Dohm, Federal Centre for Breeding Research on Cultivated Plants, Institute of Ornamental Plant Breeding, Ahrensburg, Germany

Introduction

The term 'genetic transformation' describes the integration of foreign genetic information into a particular genotype via genetic engineering. The first successful transformation event in higher plants was described for tobacco in 1983. With respect to the huge opportunities offered by transformation technology, numerous sophisticated techniques for gene transfer have been developed during the last decades, including nearly all plant species of economic or scientific relevance. Simultaneously, more and more genes encoding important plant characteristics have been identified and isolated and are now available for gene transfer. In addition to the scientific information which has been obtained by the use of genetic transformation, genetically modified plants also offer great potential in terms of commercialization. Up to now, transgenic plants have been generated in all important agricultural crops and several have also been brought on to the market. In ornamentals, however, until now only genetically engineered carnation cultivars have been commercialized. For roses, different transformation protocols have been established, and there are several examples of promising modifications of important plant characteristics. In the following article, the principles of genetic transformation will be outlined. Subsequently, transformation techniques used in roses will be described and, finally, some examples of genetically modified traits in roses will be presented.

Transformation Technology – A Brief Introduction

Since the first description of successful plant transformation, considerable progress has been made concerning the development of efficient transformation systems. Indirect gene transfer systems make use of the microorganisms *Agrobacterium tumefaciens* or *A. rhizogenes* and viral vectors for gene transfer into plant cells. In addition, different techniques for direct DNA import have been established. In such cases the DNA is physically delivered into plant cells, for example by particle bombardment, by transformation of protoplasts mediated by polyethylene glycol (PEG) or electric pulses as well as by microinjection. The appropriate technique must be chosen with regard to the respective plant species and the gene of interest. With the exception of microinjection, which is rarely applied in plant transformation, all techniques for direct gene transfer require *in vitro* regeneration systems for the recovery of nonchimeral transgenic plants from single transformed cells. The indirect gene transfer via *A. tumefaciens*, however, has also been successfully applied to transformation of germ line cells *in vivo*. All gene transfer systems currently used have particular shortcomings limiting their universal application. In roses, up to now *Agrobacterium*-mediated gene transfer and particle bombardment have been successfully applied. Therefore, both techniques will be described in more detail.

Agrobacterium-mediated Transformation

Agrobacterium tumefaciens is a Gram-negative soil bacterium, which in many plant species induces tumours by introducing a segment of oncogenic DNA into plant cells at wounded sites, resulting in the so-called crown gall disease (*see* **Disease**: Crown Gall). The transferred (t) DNA, which is located on a plasmid called Ti-plasmid, is integrated into the plant genome via illegitimate recombination. In order to obtain transgenic plants the transformed plant cells must be regenerated *in vitro*. As the transfer of the t-DNA is only dependent on the so-called border repeats and independent of the sequence information in between, the oncogenes at both sides can be replaced by nearly every gene of interest and, subsequently, such disarmed bacteria can be used as 'DNA-ferries'. Usually the gene of interest is located on so-called binary vectors with the border repeats at both sides. The remaining Ti-plasmid harbours the virulence genes necessary for transformation. During the last decades protocols for *A. tumefaciens*-mediated gene transfer have been established for many different plant species, including monocotyledonous plants, which do not belong to the natural host range of agrobacteria. Furthermore, different plant tissues have been used as targets, although not every cell type is accessible to agrobacteria. For example, up to now there have been only a few reports on the successful transformation of meristematic tissue via *A. tumefaciens*.

Similarly to *A. tumefaciens*, *A. rhizogenes* naturally transfers DNA into plant cells, but instead of inducing a well-developed tumour at the crown level, it induces smaller tumours bearing adventitious roots transformed by so-called *ROL* genes. Disarmed strains of *A. rhizogenes* can be used as 'gene ferries' as described for *A. tumefaciens*. However, an alternative strategy is the induction of transformed adventitious shoots mediated by the *ROL* genes and the subsequent regeneration of transgenic shoots from these roots.

Biolistic Transformation

The biolistic transformation approach requires particular technical equipment: naked DNA is bound to gold or tungsten microparticles of only a few micrometres in diameter and shot with a particle gun into living plant tissue. A certain number of cells survive the bombardment, integrate the foreign DNA into their own genome via illegitimate recombination and express the incorporated genetic information. Whole plants may be regenerated from those transformed cells. The most popular set-up of a particle gun consists of a vacuum chamber, in which the DNA-coated particles are accelerated down the barrel of the gun towards the plant tissue by helium overpressure (**Figure 1**). There are several factors influencing the efficiency of gene transfer associated with the technical equipment, such as distances between rupture disc, macrocarrier, stopping screen and target tissue, or the applied vacuum and rupture disc pressure, respectively. Further critical factors concern the preparation of the t-DNA as well as the treated plant material. All these parameters must be optimized for each respective plant genotype, plant tissue and genes considered. In contrast to *Agrobacterium*-mediated gene transfer, the biolistic approach is theoretically applicable to every plant species as well as to every cell type. For this reason, particle bombardment has been used as a priority in monocotyledons which were not accessible to *Agrobacterium* transformation for many years. One further advantage of the biolistic approach is the possibility of transforming plastids and mitochondria.

Genetic Transformation of Roses

Rose Transformation via *Agrobacterium*

According to the examples of successful transformation listed in **Table 1**, in roses *Agrobacterium*-mediated gene transfer has been the preferred transformation method. In addition to primary explants, embryogenic tissue and protoplasts are suitable targets.

Embryogenic tissue Somatic embryos as well as embryogenic calli have proven to be appropriate tissue for agroinfection (**Table 1**). The origin of embryogenic tissue varies from vegetative parts such as leaves or internodes to floral plant tissue. Almost all plant parts can be used as starting material for somatic embryogenesis (*see* **Cell, Tissue and Organ Culture**: *In Vitro* Regeneration Techniques). For transformation, the embryogenic tissue is briefly immersed in an *Agrobacterium* suspension. Different disarmed bacterial strains of *A. tumefaciens* have been used successfully, for example LBA 4404, GV

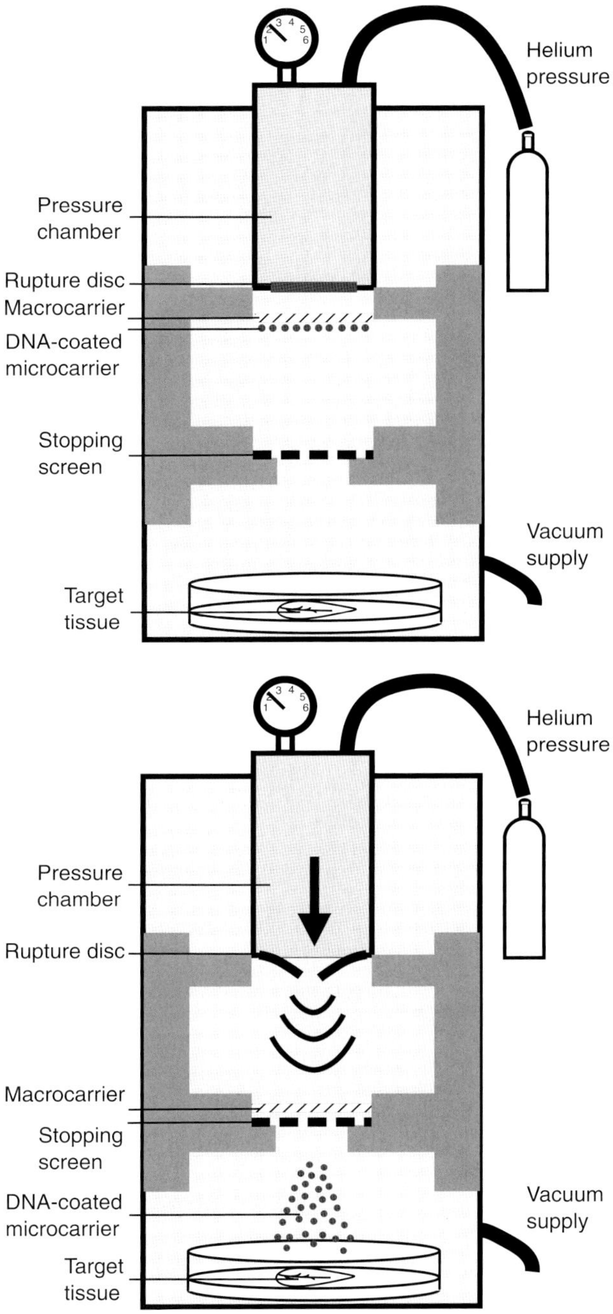

Figure 1 Scheme of biolistic transformation. Top: before bombardment; bottom: after bombardment.

3101, GV 2260 or EHA 105. For transformation of friable embryogenic callus, disarmed strains of *A. rhizogenes* also proved to be efficient. Transformation efficiencies vary with the rose genotype and the kind of plant tissue. The *Agrobacterium* suspension usually starts from a single colony, which is inoculated overnight in especially rich culture media. Further preparation of the bacterial suspension strongly influences the overall transformation efficiency. There is no clear evidence regarding the optimal developmental stage of the immersed plant tissue, which most likely also determines the transformation success, as has been demonstrated on other plant species. Wounding of the plant tissue prior to bacterial immersion in order to induce exudation of phenolic compounds enhances the transformation frequency, but is not compulsory, either for embryogenic calli or for somatic embryos. Wounding may be performed mechanically as well as by ultrasonic treatment or by particle bombardment. Likewise, the addition of the phenolic compound acetosyringone is not compulsory for successful transformation.

After immersion for about 1 h the *Agrobacterium* suspension is removed from the inoculated plant material, which is subsequently incubated on a regeneration medium for a cocultivation period of 2–6 days, depending on the *Agrobacterium* strain. Unlike in other plant species cocultivation at lower temperatures (about 20 °C) has not increased the transformation frequency in roses.

Cocultivation is terminated by transfer of the plant tissue on to a medium containing antibiotics to kill the agrobacteria. These antibiotics have to be bactericidal but not phytotoxic. In roses in particular, cefotaxime or carbenicillin in the range of 50–500 mg l^{-1} was efficiently used, as well as vancomycin (100–400 mg l^{-1}) and Timentin® (150 mg l^{-1}). Furthermore, combinations of several antibiotics as well as alternating treatments were also efficient. In some systems the supply of antibiotics for only 2 months of culture is sufficient to remove bacteria completely.

Subsequently, the transformed embryogenic tissue regenerates either via maturation and germination or via adventitious shoot formation (**Figure 2**) (*see* **Cell, Tissue and Organ Culture:** *In Vitro* Regeneration Techniques). The final outgrowth of shoots requires culture media with comparably high 6-benzylaminopurine (BAP) concentrations. Usually, the regeneration process is initiated under selection pressure in order to prevent the outgrowth of nontransformed shoots. In the case of transformation of somatic embryos and subsequent adventitious shoot formation, regeneration from single cells is absolutely necessary to avoid the formation of chimeras. Concerning *A. tumefaciens*-mediated gene transfer into somatic embryos, the overall transformation frequency for transient as well as for stable events is up to 100%. However, the percentage of transgenic plants referred to the originally treated embryos reaches a maximum of only 3%. **Figure 3** summarizes the process of *A. tumefaciens* transformation of embryogenic rose tissue.

Stem discs The inoculation of primary explants in *Agrobacterium* suspension and their subsequent

Table 1 Selected reports on rose transformation

Reference rose genotype	*Modified traits*	*Integrated genes encoding for*	*Transformation procedure*		*Target tissue*
			Transformation technique	*Mode of regeneration*	
Firoozabady *et al.* (1991, 1994) 'Royalty'	Flower colour	Chalcone synthase	*Agrobacterium tumefaciens* *A. rhizogenes*	Somatic embryogenesis	Embryogenic callus
Matthews *et al.* (1994) *R. persica* × *xanthina*	Marker gene	β-glucuronidase (intron)	*A. tumefaciens*	Somatic embryogenesis	Protoplasts
Souq *et al.* (1996) 'Deladel'	Plant habit Flower colour	*Rol C* Chalcone synthase	*A. tumefaciens*	Somatic embryogenesis	Embryogenic callus, somatic embryos
Marchant *et al.* (1998) 'Glad Tidings'	Disease resistance	Chitinase	Particle bombardment	Somatic embryogenesis	Embryogenic callus
Van der Salm *et al.* (1998) 'Moneyway'	Root characteristics	*Rol B*, *Rol A + B + C*	*A. tumefaciens*	Adventitious root formation, somatic embryogenesis and adventitious shoot formation	*In vitro* stem discs
Dohm *et al.* (2001) 'Heckenzauber' 'Pariser Charme'	Disease resistance	Chitinase, glucanase, ribosome-inactivating protein, lysozyme	*A. tumefaciens*	Somatic embryogenesis and adventitious shoot formation	Somatic embryos

Figure 2 Regeneration of transgenic rose shoots from somatic embryos.

regeneration to plants is the most straightforward strategy for transformation. However, up to now, almost all attempts made to regenerate inoculated explants have resulted in the production of transformed callus, which did not develop into plants. Nevertheless, a method based on the formation of roots from transformed stem discs followed by subsequent initiation of embryogenic callus has successfully been applied. In a first step transformed roots are produced upon immersion of *in vitro* stem slices in a suspension of *A. tumefaciens* and subsequent culture on medium containing indolebutyric acid (IBA). Exertion of selection pressure results in the regeneration of only transformed roots, from which embryogenic callus is initiated on a culture medium supplied with up to 11 mg l^{-1} 2,4-dichlorphenoxy acetic acid (2,4-D). The development of somatic embryos occurs on a

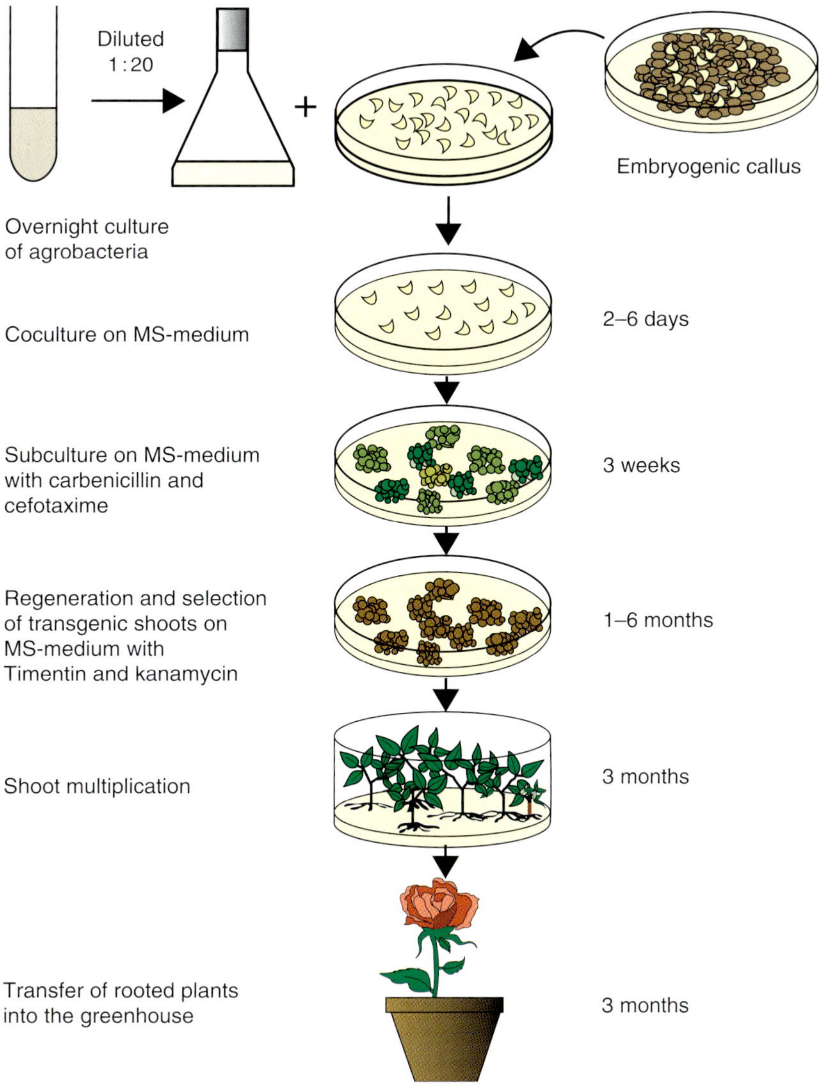

Figure 3 *Agrobacterium* transformation of somatic rose embryos. MS, Murashige and Skoog.

growth-regulator-free medium. The embryos are finally regenerated to plants by the addition of BAP to the culture medium. The elimination of agrobacteria is achieved as described for embryogenic tissue.

Protoplasts Compared to the transformation methods described previously, the use of protoplasts as targets for agrobacteria is the most time-consuming procedure, and it requires the largest amount of technical knowhow. However, it allows both efficient transformation and rigorous selection and it has the most promising potential for isolating nonchimeral transformants. Protoplasts of *Rosa persica* × *xanthina* can be isolated by enzyme digestion from callus as well as from suspensions (*see* **Cell, Tissue and Organ Culture**: Protoplast Culture). Following a preculture period of 24 h, protoplasts are cocultivated for another 24 h with the *A. tumefaciens* strain LBA 4404 harbouring disarmed binary vectors. Bacteria are supplemented in a ratio of approximately 400 per protoplast. Following cocultivation the agrobacteria are removed from the cell culture by the addition of the antibiotics carbenicillin and vancomycin at 250 mg l^{-1} each. The recovery of transgenic plants occurs via callus initiation and subsequent regeneration by somatic embryogenesis (*see* **Cell, Tissue and Organ Culture**: Protoplast Culture). The transgenic status of plants resulting from this procedure was confirmed by measuring the expression of the β-glucuronidase (GUS) reporter gene.

Rose Transformation via Particle Bombardment

The biolistic approach of rose transformation has been established for embryogenic callus of 'Glad Tidings'. Embryogenic callus is initiated from petioles of *in vitro*-grown shoots by successive culture on media with up to 5 mg l^{-1} 2,4-D. Pieces of these embryogenic calli with white, globular proembryos are subjected to biolistic gene delivery. The t-DNA is coated on to gold microprojectiles. Bombardment is done with a helium-driven apparatus with the following parameters: a 30 mm distance between the rupture disc and macrocarrier and a 10 mm distance between the macrocarrier and stopping screen, and a vacuum of 736 mm Hg. Results of experiments with GUS(Int) as the target gene indicated that a firing distance of 70 mm (distance between the stopping screen and the bombarded plant tissue) combined with 9000 kPa rupture disc pressure is best maintained for a maximum number of transformation events. Furthermore, the conditioning of plant cells by adding an osmoticum to the culture medium contributes to successful transformation. The supply of 0.25 mol l^{-1} myoinositol 4 h before as well as 24 h after bombardment proved to be most efficient. The recovery of transformed plants from the treated tissue occurs via maturation and germination (*see* **Cell, Tissue and Organ Culture**: *In Vitro* Regeneration Techniques).

Selection of Transformed Cells and Tissues

During regeneration of plants from treated tissues, transformed plant cells compete with nontransformed cells. In order to promote the regeneration frequency of transgenic cells, different strategies have been developed. These are mainly based on the inhibition of nontransformed cells by supplying the culture medium with special compounds disturbing the regeneration process or even leading to cell death. The transformed cells, however, possess corresponding resistance genes resulting from transformation with a vector harbouring such a gene next to the gene of interest. The compounds preferably used as selective agents are antibiotics. In roses, independent of the applied transformation and regeneration system the antibiotic kanamycin has been successfully used (**Figure 4**). Resistance to kanamycin is conferred by the *nptII* gene (neomycin phosphotransferase) from the transposon Tn5, which is found in various Gram-negative bacteria. The gene is under the control of the nopalin synthase promoter from *A. tumefaciens*. Appropriate selective kanamycin concentrations vary with the type of plant tissue. For example, in callus cultures, concentrations of up to 250 mg l^{-1} kanamycin sulphate are supplied, which are 5–10 times higher than those used for embryos or shoots. For the transformation of rose protoplasts, resistance to the antibiotic hygromycin has been used as a selection marker, which is mediated by the *aphIV* gene (aminoglycoside phosphotransferase type IV) from *Klebsiella* driven by the constitutive 35S-promoter of the cauliflower mosaic virus (CaMV). Hygromycin concentrations close to 30 mg l^{-1} proved to be sufficient for selection.

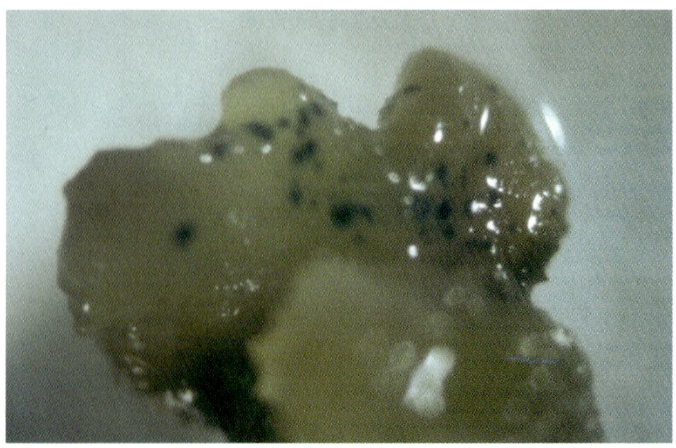

Figure 4 Transient β-glucuronidase expression of embryogenic rose callus.

As one of the major arguments against the field release of transgenic plants is the horizontal transfer of genes conferring resistances to antibiotics, many efforts have been made to develop alternative selection strategies. For example, the use of markers, which are not based on resistances to antibiotics, has been discussed. In *A. tumefaciens*-mediated transformation of the rose cultivar 'Heckenzauber', the *bar* gene (phosphinotricin acetyltransferase) from *Streptomyces hygroscopicus* conferring resistance to the herbicide phosphinotricin has been successfully applied for selection. Another strategy is based on the resistance to 2-deoxyglucose (2-DOG) conferred by the 2-DOG-6-P-phosphatase gene from yeast. In rose *in vitro* regeneration 2-DOG works efficiently as a selective agent; however, the corresponding resistance gene has not yet been introduced into roses. Further promising approaches are based on different enzymes, which are essential for sugar metabolization like phosphomannose isomerase or xylose isomerase. It could be shown that *in vitro* regeneration of roses responds to different carbon sources. However, the efficient use of the corresponding genes as selection markers, already demonstrated on other plant species, has still to be proven. Furthermore, different strategies for the development of transformation systems resulting in marker-free transgenic plants have been described. These approaches use cotransformation strategies, site-specific recombination systems, transposable elements or intrachromosomal recombination in order to eliminate the selectable marker during the subsequent breeding process. For roses, such strategies have not been developed yet.

Ideal markers for the selection of transformed plant tissue during regeneration are so-called visible markers like the firefly luciferase or the green fluorescent protein (GFP) from *Aequoria victoriae*. Expression of the corresponding genes creates a strong visible fluorescence when supplied with the appropriate substrate (luciferase) or excited with ultraviolet or blue light. Roses have been successfully transformed with both the luciferase and the GFP gene, but reliable selection proved to be difficult due to weak fluorescence as well as to disturbing autofluorescence from the rose tissue.

One further visible marker efficiently used in roses is the GUS gene from *Escherichia coli*. GUS is a hydrolase that catalyses the cleavage of a wide variety of β-glucuronides, many of which are commercially available as spectrophotometric, fluorometric and histochemical substrates. This marker system allows the quantitative measurement of gene expression as well as its histochemical localization. The GUS assay, however, usually destroys the analysed plant material. For this reason the GUS gene is not applicable for continuous selection during the regeneration process, but it provides a valuable tool for the establishment of a transformation system. **Figures** 5–7 show GUS expressing rose tissue.

Analysis of Transgenic Roses

Detection of the Transgenes

For the confirmation of transformation, several molecular and biochemical techniques are available, indicating both the presence of the respective transgene and the number of integrated copies, as well as its expression.

PCR analysis Generally, polymerase chain reaction (PCR) analysis is the easiest method for the molecular detection of the transgene. First of all, only small amounts of crudely purified DNA must be isolated from the putative transformant (*see* **Genetics**: DNA

Figure 5 β-glucuronidase expressing adventitious shoot on a somatic rose embryo.

Figure 6 β-glucuronidase expressing leaves from *in vitro* rose shoots.

Fingerprinting; Molecular Markers). Total DNA is subjected to PCR analysis with primer pairs specific for the respective transgene. Subsequently the amplified sequences are separated on agarose gels and visualized by staining with ethidium bromide. The bands can be identified by their size compared to fragments of standardized size. One major disadvantage of this comparably simple method is the possibility of obtaining false-positive signals due to agrobacteria which have survived in the inoculated plant tissue. Bacterial contamination, however, can be detected by further PCR analysis with primer pairs specific for bacterial chromosomal DNA sequences. Furthermore, the integration of plasmid DNA located outside the border sequences following *Agrobacterium*-mediated gene transfer can be detected with specific primers.

Southern blot analysis In addition to the identification of putative transformants, Southern blot analysis provides further information on the number of integrated gene copies. This detection method requires the isolation of larger amounts of highly purified total DNA. Due to secondary plant metabolites, this might be difficult in roses. However, during the last years several protocols developed for DNA isolation from woody plants have been adapted to roses. Ten to 50 μg of total DNA from putative transformants is restricted with specific enzymes, separated electrophoretically on an agarose gel, transferred to a nylon membrane and hybridized to labelled probes specific for the respective transgene. Labelling is performed either by ^{32}P random priming or by nonradioactive procedures according to standard protocols (*see* **Genetics**: DNA Fingerprinting; Molecular Markers). Following hybridization, the filter is exposed to an X-ray film and DNA fragments containing the transgene produce a positive signal. The number of such positive signals indicates the number of copies of the transgene incorporated into the genome. Furthermore, Southern blot analysis offers the possibility of identifying identical transformants as the analysis of plants originating from independent transformation events should produce different integration patterns.

Figure 7 Kanamycin selection of transformed rose shoots. True transgenic shoots stay green (left), whereas escapes turn pale (right).

Northern blot analysis For Northern blot analysis total RNA is isolated according to standard protocols or by the use of commercially available kits. Close to 20 μg of total RNA is separated electrophoretically on a denaturing-agarose gel, blotted to a nylon membrane and hybridized to labelled probes specific for the integrated transgenes. Labelling as well as the further handling of the membranes are mainly performed as described for Southern blot analysis. After exposure to an X-ray film, signals indicate the expression of the respective transgenes (**Figure 8**). Moreover, quantitative measurements of the signal intensity allow conclusions on the intensity of gene expression. **Figure 8** shows an example of Northern blot analysis in roses.

Protein analysis The final confirmation of a transgenic plant requires the detection of the gene product encoded by the incorporated transgene. Suitable assays depend on the type of protein and must be adapted to the given circumstances. Such assays are often based on substrate specificity and many include

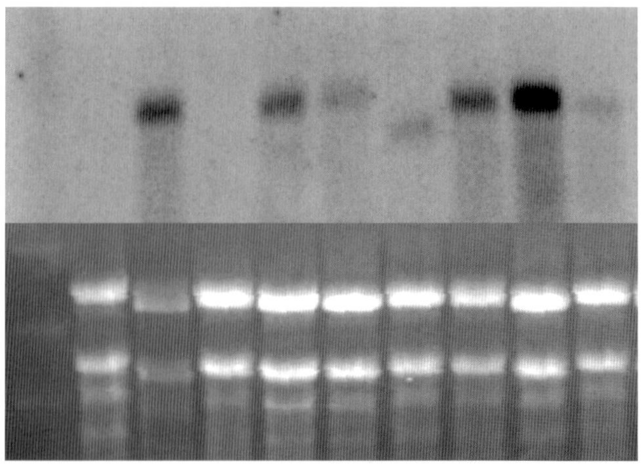

Figure 8 RNA analysis of transformed rose plants: total RNA is separated electrophoretically on a denaturing gel (bottom), blotted and hybridized to the corresponding DNA probe (top). Expressing plants give a signal. Lane 1, RNA-standard; lane 2, nontransformed control plant; lanes 3–10, independent transgenic plants.

staining reactions or quantitative determination of particular reaction products. In a different strategy total protein is separated electrophoretically and transferred to a membrane by Western blotting. Subsequently, the proteins encoded by the incorporated transgenes are immunodetected with the corresponding antibodies.

Genetic Stability of Transgenic Roses

Assessment of genetic stability has to consider two aspects: (1) stability of expression of the incorporated transgene and (2) somaclonal stability of the respective rose genotype. Ideally, plants derived from transformed cells are identical to the mother plant except for modifications conferred by the integrated transgene. However, further deviations from the original phenotype or genotype may occur. This phenomenon corresponds to the expression of somaclonal variation. Somaclonal variation is induced by the *in vitro* regeneration process. It is due to mutations during *in vitro* cultivation or resulting from chimeral plant tissue (*see* **Cell, Tissue and Organ Culture:** *In Vitro* Regeneration Techniques).

Furthermore, additional variations may be caused by transgene integration, a phenomenon described as gene silencing. On the one hand, expression of the transgene itself can be inhibited; on the other hand plant genes homologous to the integrated gene may be concerned. Two general types of gene-silencing phenomena have been distinguished: in the first case positional effects are responsible in such a way that the flanking plant DNA or the chromosomal location negatively affect the expression of a single-copy transgene; in the second case gene expression is disturbed by either multiple copies of the integrated transgene or by similarities of the transgene sequence to plant genes. Since this phenomenon is based on sequence homology it has been termed homology-dependent gene silencing. Such homology-dependent gene silencing may occur transcriptionally or post-transcriptionally. Whereas transcriptional gene silencing is characterized by altered methylation patterns or chromatin structure, posttranscriptional gene silencing describes the subsequent degradation of RNA from homologous sequences.

Gene silencing has not yet been described for transgenic rose plants. Available plant transformation techniques do not allow control of either the number of integrated gene copies or the site of integration. Especially, higher copy numbers impair the expression of other plant genes and therefore reduce the overall plant quality. Furthermore, they complicate the subsequent breeding process. In most cases the occurrence of somaclonal variation is readily recognized during plant development to flowering. The stability of transgene expression, however, must be confirmed for several years. In roses, several investigations on somaclonal stability have been performed (*see* **Cell, Tissue and Organ Culture:** *In Vitro* Regeneration Techniques), but up to now there has been no detailed information on the genetic stability of transgenes, especially during the subsequent breeding process. However, intensive studies on GUS-transformed rose plants will hopefully provide further insights on the genetic stability of transgenic roses.

Genetically Engineered Traits in Roses

In roses, resistance to diseases and pests, modified flower colours as well as enhanced plant quality are considered worth modifying via genetic engineering (**Table 1**). Realization depends in part on the availability of the corresponding gene encoding the particular trait.

Disease Resistance

Roses are attacked by different pathogens and insects, causing economically relevant damage (*see* **Disease:** Overview). In garden roses, black spot (*see* **Disease:** Black Spot) resulting from infection with the fungus *Diplocarpon rosae* is the most important disease. The expression of different antifungal proteins introduced via genetic engineering in several rose cultivars successfully reduced their susceptibility. For example, the integration of a basic (class I) chitinase gene from rice into the genome of 'Glad Tidings' by particle bombardment resulted in enhanced resistance. Bioassays on transformed leaf discs demonstrated that expression of the chitinase gene reduced the severity of black spot development by 13–43%. In a different strategy a barley gene encoding a ribosome-inactivating protein (RIP) fused to a potato gene coding for a signal peptide was transferred into the rose 'Pariser Charme'®. Due to transport of the RIP into the apoplast, in bioassays on leaves or plants the transgenic genotypes showed an overall susceptibility reduction of 60%.

Flower Colour

One of the most desired flower colours in roses, unknown within the natural gene pool, is blue. By integration of a flavonoid 3′5′hydroxylase gene, transgenic plants with bluish flowers have been obtained. However, they have not been introduced into the market.

An alternative approach for the modification of flower colour is the manipulation of the rose chalcone synthase gene by transformation (*see* **Fragrance and Pigments:** Flavonoid Molecular Biology). However, neither the integration of the full-length DNA sequence in antisense orientation nor the incorporation of an incomplete sequence in sense orientation under control of the 35S-promoter resulted in total suppression of chalcone synthase synthesis. The transgenic plants showed reduced intensity of their red flower colour, but did not display completely white petals.

Growth Habit

Plant growth characteristics as axillary bud release or root formation are primarily controlled by the plant hormones auxins and cytokinins. Potential genes, which proved to interfere with plant hormone metabolism, are the *ROL A*, *B* and *C* genes from *A. rhizogenes*. Transformation of the rootstock 'Moneyway' with *ROL B* driven by the strong CaMV 35S promoter or with a combination of *ROL A + B + C* in combination, each under control of its own promoter, increased root formation on stem discs *in vitro*. Due to a higher sensitivity to auxins, explants transformed with a combination of *ROL A + B + C* showed an optimum rooting response at lower auxin concentrations as compared to nontransformed control explants or explants transformed with a GUS(Int) gene. In contrast to the control, they were even able to regenerate roots without exogenous auxin supply. Furthermore, micropropagated shoots as well as greenhouse cuttings of transgenic plants carrying *ROL A + B + C* or *ROL B* alone under influence of their own promoters showed improved rooting characteristics. One *ROL A + B + C* transformant, which was proven to express *ROL C*, was grafted with 'Madelon' as a scion. This experiment resulted in both a stimulation of root development of the transformed rootstock as well as in the enhanced branching of the grafted cultivar. The frequency of basal shoot formation was enhanced about four times as compared to control plants. Since basal shoots are an important component of flower production, the use of this *ROL A + B + C* transformed rootstock might increase flower productivity. However, *ROL* gene expression in the currently available transgenic plants resulted in altered phenotypes such as decreased apical dominance, modified leaf and flower shape and reduced plant weight. Furthermore, some of the genetically modified plants exhibited reduced fertility as well as increased susceptibility to diseases and pests.

Conclusions

During the last years different transformation protocols have been established for several rose cultivars and have been successfully applied for the modification of particular traits. However, the available transformation protocols, based on *A. tumefaciens*-mediated gene transfer and particle bombardment (**Table 1**), are restricted to few genotypes, which are not of economic importance but may be used as progenitors for the development of cultivars. The restriction of the described transformation strategies to particular genotypes mainly results from the requirement of *in vitro* regeneration systems for the recovery of transgenic plants. Up to now, in roses regeneration of transformed plant cells has only been successful via somatic embryogenesis. Somatic embryogenesis in roses is a long-term process and, accordingly, the production of transgenic plants takes 1 year at least, which makes genetic engineering an expensive breeding tool. Furthermore, compared to many model plant species, transformation frequencies are low. However, a regeneration frequency of 3% transgenic plants from the treated plant material is considered sufficient for routine production of genetically modified roses. As *in vitro* regeneration of roses bears the risk of somaclonal instability (*see* **Cell, Tissue and Organ Culture:** *In Vitro* Regeneration Techniques), a transformation strategy independent of *in vitro* regeneration would be desirable. Therefore, efforts are made to establish an *in planta* transformation system using germ line cells as target for *Agrobacterium*-mediated gene transfer. Such systems have been realized for several Brassicaceae. Furthermore, early experiments with *R. canina* rootstocks demonstrated the feasibility of obtaining shoots from *A. rhizogenes*-mediated transgenic roots. This strategy might also have the potential for an *in vivo* transformation approach. A major challenge for optimization of the available transformation systems is the development of alternative selection strategies, which are not based on resistance to antibiotics.

In summary, in spite of certain limitations, efficient transformation methods are available for roses, which are continuously being improved. Furthermore, there are promising reports on valuable modifications of different traits, such as resistance to black spot. However, up to now genetically modified roses have not been introduced into the market. There are several reasons for this: first of all, for many traits of economical relevance the respective genes are missing, for instance, genes mediating insect resistance. Additionally, nearly all genes which could be interesting candidates for rose transformation are protected by patents. This also applies for many techniques which are used for the production or analysis of transgenic plants. As a consequence, licence fees must be paid in the case of commercial use and, in addition to these economic considerations, the owner of a patent may even prohibit the use of the respective gene or technique. Furthermore, at present in most countries consumers predominantly disapprove of genetically transformed plants. And finally, especially in Europe, the legislation on genetically modified plants has become increasingly restrictive. Therefore, as for other species, the success of transgenic roses will be determined not only by technical progress, but also by consumers' acceptance and by political decisions.

See also: **Cell, Tissue and Organ Culture**: *In Vitro* Regeneration Techniques; Protoplast Culture. **Disease**: Overview; Crown Gall; Black Spot. **Fragrance and Pigments**: Flavonoid Molecular Biology. **Genetics**: DNA Fingerprinting; Gene Mapping; Molecular Markers; Marker-Assisted Selection; Research Trends.

Further Reading

Deroles SC, Boase MR and Konczak I (1997) Transformation protocols for ornamental plants. In: Geneve RL, Preece JE and Merkle SA (eds.) *Biotechnology of Ornamental Plants*, vol. 2, pp. 13–34. Wallingford, UK: CAB International.

Gudin S (2000) Rose: genetics and breeding. *Plant Breeding Reviews* 17: 159–189.

Potrykus I and Spangenberg G (eds.) (1995) *Gene Transfer to Plants*. Berlin: Springer-Verlag.

Rout GR, Samantaray S, Mottley J and Das P (1999) Biotechnology of the rose: a review of recent progress. *Scientia Horticulturae* 81: 201–228.

BREEDING

Contents

Overview

S Gudin Université d'Aix-Marseille III, Marseille, France

Introduction

Plant breeding can broadly be defined as the improvement of the genetic patterns of domestic plant populations. Based on the importance of plants for human life, humans have been involved in manipulating and improving plants for their needs for more than 11 000 years. Though merely intuitive in the beginning and without a scientific basis for most of the time, scientific knowledge has revolutionized plant breeding over the last decade. These changes, though first effective in the breeding of agricultural crops, have now also reached the field of ornamental plant breeding and in particular rose breeding.

Rose cultivars grown around the world are probably more numerous than any other plant cultivars. As roses have been grown and admired since the origins of civilization, this has surely been true since the beginning of their domestication. There is evidence that they were already cultivated 5000 years ago by the ancient civilizations of China, western Asia and northern Africa. According to the different geographical regions and periods of history, rose cultivars were grown for different main uses; hence, they resulted from different needs that obviously had an influence on the selected forms. Whereas, if they are nowadays bred essentially for ornamental purposes, and to a lesser extent for their use for perfumes, medicine and cooking (*see* **History of Roses in Cultivation**: History of the Perfume Industry. **Secondary Metabolites**: Culinary Uses and Nutritional Value; Medicinal and Pharmaceutical Uses), they were essentially grown to be appreciated in the form of separated petals in ancient Rome, where they were also largely used to make remedies derived from flowers and hips, and healing scent. To give another example of changes in fashion, much later, at the beginning of the nineteenth century, mossy cultivars were extremely popular in western Europe but the moss character nowadays rarely corresponds to a breeding objective.

It is likely that the first acts of rose domestication corresponded to the cultivation and multiplication of species collected in the wild. The proximity of these species in cultivated areas led to the spontaneous occurrence of interspecific hybrids that have long been considered as original species. Thus *Rosa damascena* Mill., which is now thought to have arisen from *R. gallica* L. × *R. moschata* Herrm. or *R. gallica* × *R. phoenicea* Boiss., *R. alba* L. from *R. canina* L. × *R. gallica* and *R. centifolia* L. from (*R. rubra* Blackw. × *R. moschata*) × *R. alba*, respectively. Some of these spontaneous hybridizations were only possible because of human activity as the original species involved were collected in distant parts of the world and cultivated together in gardens. *R. foetida* Herrm., which is probably the origin of the yellow flower colour of most modern cultivars, and originates from Asia, was introduced at the end of the sixteenth century in Holland and England. *R. chinensis* Jacq., which brought the recurrent-flowering character to modern cultivars (together with *R.* × *odorata* (Andr.) Sweat) and also originates from Asia, was probably introduced in Italy at the beginning of the sixteenth century before it arrived in England at the end of the eighteenth century.

Together with the huge popularity of specialized rose gardens, rose breeding is a feature of the nineteenth and twentieth centuries. Since the beginning of the nineteenth century, there are records of controlled crosses have been known. Hence, cross-breeding corresponds to the most traditional breeding practice. It is nowadays still widely used and its efficiency actually benefits from a recent increase in rose genetic knowledge that started 30 years ago and has developed with increasing speed over the last decade with the advent of molecular markers. Marker-assisted selection of important traits should also be available very soon for rose breeders (*see* **Genetics:** Marker-Assisted Selection). Complex interspecific cross-breeding is carried out and made possible by the application of recently developed biotechnologies, such as manipulation of ploidy levels and embryo rescue. Mutation breeding has also been used in the last decades. Other technologies more recently applied on roses, such as genetic transformation, protoplast fusion or *in vitro* somaclonal variation, will probably represent new breeding tools in the near future (*see* **Biotechnologies for Breeding:** Manipulation of Ploidy Level; Genetic Transformation. **Cell, Tissue and Organ Culture:** Protoplast Culture. **Genetics:** Mutation).

Breeding objectives during the twentieth century continuously became more specific to the classical or more recently appeared rose product lines and they are quite different when the cultivars are selected as garden, cut-flower-producing, landscape, pot or rootstock roses. For example, apart from ornamental characters, disease resistance became one of the most important breeding goals in garden roses whereas ornamental characters, productivity and long vase-life are more important breeding goals in cut roses.

Compared to some other crop species, genetic knowledge concerning roses (although it seems to be rapidly increasing) is still scarce. The inheritance of only a few agronomic traits is known (*see* **Genetics:** Inheritance of Characteristics), and this makes the rose breeder's task to combine particular characteristics in one genotype of rose very difficult. Different features may explain this status. Although roses are among the economically most important ornamental crop, they are still much less investigated than some of the more important agricultural species such as corn, rice and tomato. Moreover, as woody plants are mainly polyploids, and highly heterozygous due to their essentially allogamic sexual reproduction, with low fructification, few seeds per hip and low germination, they correspond to a difficult model for the geneticists to work with. However with their short time from seed germination to first blooming, recurrent-flowering habit (for many) and comparatively small genome, they are nowadays increasingly used in genetic research, essentially based on molecular approaches. Despite some recently acquired genetic data that can contribute to plant improvement, rose-breeding efficiency is still very much dependent on the breeder's experience. Indeed, in such an ornamental crop, aesthetical traits that correspond to subjective selection criteria are essential.

Breeding Objectives

A discussion of the breeding objectives in rose breeding faces the problem that, due to a number of different rose types and different trait complexes, every scheme must be somewhat artificial. However, as the rose types and trait combinations are the most obvious parameters which have to be considered in this respect, the following overview will be divided into breeding objectives according to rose type and breeding objectives according to trait combinations.

Different Rose Types

Cut roses Ornamental characters still form the baseline for all breeding activities in cut roses (*see* **Breeding:** Selection Strategies for Cut Roses). However, breeding for cut roses is nowadays very much influenced by the production region for which a selected variety is intended. Thus, if such a traditional breeding objective as increased productivity is still favoured, it will be much more important when a variety is bred for a region that is typified by intensive cultivation with high investment prices and low economical margins per sold unit, such as northern Europe, than when it is bred for a region with a more extensive strategy corresponding to lower investment cost and higher margins per unit, such as Latin America. In the same way, the needs for tolerances to pests and diseases are not equal in the different production structures that correspond to the different producing areas, as the rose parasites that have the major detrimental effect are not the same in a highly heated, artificially illuminated, CO_2-supplemented and computerized climatically controlled glasshouse, with soilless and individual drip fertilization systems, located in The Netherlands, and a nonheated plastic-house, with soil cultivation, hand fertilization and sprinkling irrigation, located in the savannah of Bogotá, Colombia. Furthermore, especially as cut-rose production moves increasingly to developing countries of the subtropics, far from the rich northern countries, postharvest longevity and adaptation to complex flower transport networks are increasingly important. Moreover, because of easier manipulation and bundling by mechanical bunching, such as are now widespread in northern Europe, thornless stems are highly desirable. There are other desirable traits that have appeared with the evolution of the production techniques. Thus, some varieties are nowadays selected for their ability to perform well on their own roots and can, therefore, be efficiently propagated by cuttings at a lower cost than by grafts, and are better adapted to soilless cultivation systems. Others that are characterized by reliable and frequent bottom-break sprouting are favoured when the popular bending technique is used, as it is associated with low plant height management.

New traits, that correspond to new consumer tastes, developing into new fashions, are now under selection by breeders. Fragrant cut roses are thus highly desired by consumers. However, there is some indication that high and early expression of 'real rose' fragrance is physiologically incompatible with good postharvest life. This may explain why lately some newly released fragrant cultivars are characterized by scents such as aniseed or lemon, that do not appear to be linked to poor vase-life. Some newly released cultivars display a change in corolla colour throughout flower opening, which originally accompanies the shape of evolution of the opening flower bud. Some others have different-colour striped petals, such as only a few isolated old garden cultivars previously had. More than two decades ago fashion led to the appearance of new lines such as the spray rose which inherited the multiflora habit from polyantha progenitors, whereas only hybrid teas were formerly used by breeders to create the more traditional single-flowered cut roses. Lately, foliage quality, in terms of glossiness and depth of the green colour, has become an important selection feature in cut roses.

Garden roses As in cut roses, there is a growing demand for fragrant cultivars in garden roses. Here too, the 'real rose' fragrance appears to be problematic as it could be genetically linked to susceptible diseases, such as black spot and botrytis blight, while more recent rose fragrances such as aniseed, lemon, apple or raspberry characterizing some newly bred cultivars do not seem to be linked to such susceptibilities.

Because of the continuously growing market for landscape roses and of increasing consumer environmental sensitivity, especially in northern Europe, genotypes with exceptional tolerance to diseases and pests are selected for, as they require less pesticide applications than more traditional garden roses (*see* **Breeding**: Selection Strategies for Disease and Pest Resistance). As the landscape rose concept is gaining in popularity, all the different flower shapes and petal numbers as well as plant growth habits that are available in *Rosa* are selected for. The same concept, as it is linked to the idea of 'carefree', has led to the selection of genotypes that can grow vigorously without pruning. Moreover an all-season decorative effect, such as an important and aesthetic hip set, is selected for.

In the near future, multiuse cultivars could constitute a new line that could be grown in the garden, producing cut flowers with attractive hips in the winter season, and eventually be vigorous on their own roots.

Miniature roses/pot roses The miniature cultivars are mainly used as pot plants. Hence, they are selected for their compact growth habit. In the last decades, strong selection pressure has led to ever-smaller cultivars, with good rooting ability, which can be propagated by cuttings, in order to satisfy the growing industry of pot roses mainly in northern Europe and North America (*see* **Breeding**: Selection Strategies for Pot Roses). The continuous decrease in plant size has been accompanied by a proportional decrease in flower bud size. Selection efforts are made in order to keep the miniaturized plant sizes and increase the flower sizes and flower numbers. At the same time, more traditional rose bud shapes with an elongated and turbinated corolla, as can be found in classical garden varieties, are sought. Some newly introduced traits such as simple corollas (with only five petals), striped petals or mossy habit have recently appeared in this product line. High selection pressure is also put on pot rose postharvest quality, which is especially strongly affected during long-distance truck transportation by darkness and ethylene accumulation, or reduced light and humidity accompanied by high temperature during distribution in a typical container.

Miniature roses, as they are more and more planted in gardens of reduced-size and balconies, are nowadays selected with an increased concern for tolerance to pests and diseases.

Rootstocks In contrast to the other types of roses, which mainly result from cross-breeding, those used as rootstocks essentially derive from clonal selection and are much closer to the original wild species than any other cultivated type. Only the compatibility with the main fashionable cultivars is tested in the fields. The rootstocks are selected for rapid development, in order to produce large numbers of cuttings when they are used as mother plants, resistance to soil diseases such as crown gall, and, when they are propagated by cuttings, for high rootability. Inermism (stems lacking prickles), facilitating grafting manipulation, is also appreciated.

Highly Desirable Traits

As mentioned earlier, basic ornamental characters, such as flower shape, flower colour, inflorescence structure and overall plant habit still form the basis of rose breeding for all rose types except rootstocks. To pursue these characteristics, efficient selection pressures have to be organized by breeders, as little is known about the genetic determination of these traits. An example of nonornamental characters, which are important in present-day rose breeding, is described.

Disease and pest resistance Until recently, improvement in disease and pest resistance could only result by selecting for new progenitors, discarding the most sensitive genotypes, as a result of field observations.

Newly described *in vitro* and *in vivo* selection procedures are now available that may be used to identify sources of resistance in rose germplasm, especially in the case of cryptogamic diseases such as black spot, stem canker, botrytis blight, powdery and downy mildew. Susceptibility tests to *Agrobacterium tumefaciens* (responsible for crown gall disease) and *Meloidogyne hapla* (a root-knot nematode) have also been described.

Flower productivity Flower productivity, which typically corresponds to a quantitative character, is improved by additive effect: the breeders, through selection on this criterion of the progenitors they use, accumulate genes associated with the best productivities available at a given time. This selection is carried out by comparing the productivity of the candidate progenitors to that of highly productive reference cultivars, grown in the same cultivation area.

Some correlation has been established between the time to first blooming following seedling emergence and further flower productivity. Hence, some early selection may be practised at the seedling stage. A major determinant of flower yield, blind shoot occurrence, may also be screened at the same early stage under low irradiance conditions.

Postharvest longevity Postharvest longevity depends on numerous factors that involve many physiological determinants: the best-known comprise water absorption capacity of the cut stem, resistance to bacterial invasion (for the cut roses), stomate sensitivity and resistance to ethylene. To select for that complex criterion, according to the type of varieties, cut or pot roses, the breeders usually utilize vase-life and simulated transportation tests, respectively, in standardized conditions. The vase-life tests are of primary importance not only to appreciate longevity as such but also slow evolution and aesthetic opening of the flower bud, which is a highly desirable trait.

Early longevity indicator tests, based on the measurement of some physicochemical characteristics of the petal cell sap, may be used in cut roses. The flower bud opening speed can also be estimated at an early stage during the selection process (before enough flowers are available to make vase-life tests), as openings *in planta* and in the vase are highly correlated.

Adaptation to different growth conditions Cut roses are produced in very different climatic conditions, from northern Europe or North America to the subtropics. These different regions correspond to extremely variable production systems that greatly influence the cultivars' growth conditions, with artificial and sophisticated techniques or, on the contrary, basic ones. The main breeders have thus organized in the last two to three decades world testing networks. Hence, multilocation trials are done so that the adaptability of potential cultivars to different systems can be tested at an early stage of selection. At the same time, adaptability to cutting or graft propagation on different rootstocks can be observed as these characteristics vary greatly from one production region to another.

Breeding Technologies

Cross-Breeding

Intraspecific crosses The explosion of rose breeding since the beginning of the nineteenth century and the formerly numerous spontaneous hybridizations exploited by human beings have led to the modern rose cultivars that correspond to a complex artificial species, mainly tetraploid, often referred to as *R. hybrida*. Admittedly only 10 wild species mainly contributed to the merging of this artificial species and it is within that species that twentieth-century breeders made almost all their crosses. Intraspecific cross-breeding in that sense has thus been by far the most exploited breeding approach lately. This has been made possible by both the high heterozygosity of the artificial species, that implies high variability available in the germplasm, as attested by the tremendous diversity of the thousands of bred varieties, and the obviously highly appreciated tetraploid level that characterizes a large majority of the most popular cultivars (*see* **Breeding:** Methods of Cross Breeding).

Interspecific crosses As stated earlier, interspecific hybridization, either spontaneous or artificial, is at the origin of today's roses. Among the 10 original wild species that have constituted the artificial species *R. hybrida*, most are diploid, but some are tetraploid or pentaploid. The modern rose is mostly represented by tetraploid varieties. This demonstrates that nature or humanity has overcome ploidy compatibility problems. It is thus known that the Bourbon roses resulted from natural hybridizations between the tetraploid autumn Damask rose and the diploid 'Old Blush'. The first Bourbon roses were logically triploid. Later, tetraploid Bourbon roses appeared, probably resulting from unreduced gamete formation. The same development occurred in the hybrid China group, which also resulted from crosses between tetraploid and diploid species. Moreover, the nowadays most popular hybrid tea group, today essentially represented by tetraploid forms, originated from the triploid interspecific hybrid 'La France', obtained by L. Guillot in 1867.

As the great majority of the 120 species belonging to the genus *Rosa* (according to Rehder's classification) have not been involved in modern rose varietal creation, and as some are known to be potential sources of high resistance to diseases or unique growth behaviours, there is considerable interest in interspecific breeding. This is also to be linked with the new technologies of ploidy level manipulation and early embryo rescue that can indeed overcome some interspecific incompatibilities. Crosses between initially diploid and tetraploid species or cultivars have thus recently been made possible by the use of amphidiploidy or haploidization. Chromosome doubling attempts have apparently been less successful until now. Early embryo rescue, of which the first results in roses were only published in 1994, also enabled the production of original interspecific hybrids. It was also part of the only haploidization method that was successfully applied to roses, following parthenogenesis induced using irradiated pollen.

The production of tetraploid unreduced gametes, which are assumed to be produced by some tetraploid varieties, could in the near future, if it is verified, lead to attempts to intercross these cultivated forms with wild species of high-ploidy, such as the octoploid *R. acicularis* Lindl.

Possible sexual barriers between species may in the future be overcome by using protoplast fusion, as entire plant regeneration from rose protoplasts has recently been reported.

Mutation Breeding

Using radiation and chemical mutagens Many popular rose cultivars are spontaneous sports of pre-existing cultivars. Based on this observation, many attempts have been made, mainly in the 1970s and 1980s, in some leading breeding companies to induce mutations in roses, essentially through the use of ionizing radiation, sometimes combined with *in vitro* culture (*see* **Genetics:** Mutation).

Although many induced sports resulted from this work, very few cultivars that issued from the technique were registered, unlike in other flower crops like alstroemerias, dahlias and chrysanthemums. This probably explains why mutation breeding is now infrequently used. However, as all the attempts were carried out to induce colour variants of commercialized cultivars, such as those traditionally obtained by spontaneous mutation, it is possible that the technique could be better applied in the future, in attempts to increase physiological variability.

Exploiting somaclonal variation Large proportions of variants have sometimes been obtained as a result of regenerating callus grown *in vitro*. If the method is combined with strategic and efficient *in vitro* selection pressures, it could become a new breeding tool to obtain mutants.

Marker-Assisted Selection

Unlike in other crops, morphological and biochemical markers have been used very rarely, if at all, in rose selection.

Interestingly, recently, a few rose genetic maps have been drawn up using molecular markers, such as randomly amplified polymorphism DNA, amplified fragment length polymorphism and DNA microsatellites. On these maps, some morphological and agrophysiological traits have been located, such as petal quantity or resistance to black spot disease. This represents the first step towards future marker-assisted selection in roses (*see* **Genetics:** Marker-Assisted Selection).

Genetic Transformation

In the last decade, very few successful transformations in *Rosa* have been reported. Among them, only one transformant was reported to have a selective advantage compared to the original variety, in the form of increased tolerance to black spot. Transformation feasibility is thus now finally fully demonstrated in roses, although it does not appear to be as easy to realize as in other crops. However, the largest obstacle for the utilization of transgenic roses is public concern related to the use of transgenic plants in Europe and the resulting legal restrictions in the use of transgenic plants and their release into the environment.

Conclusion

Rose breeding is a dynamic force; 25–30 highly competitive international companies and many amateur breeders practise it. The selection criteria are increasingly numerous and specific: today, apart from more traditional traits, new parameters such as flower opening shape and colour evolution, hip aesthetic value, foliage glossiness, increased fragrance and good development on own roots receive special attention. The breeding objectives also grow in number and are increasingly defined. Careful identification of the improvement strategy, according to the objective, is of primary importance, especially as different methods are nowadays available. However, there is little doubt that cross-breeding will retain its dominant role for a long period, surely accompanied and reinforced in efficiency by technologies such as embryo rescue, ploidy level manipulation and molecular marker-assisted selection. Indeed, the rose germplasm, as a source of variability, especially the part that corresponds to most of the wild species and old cultivars, is far from being fully exploited by humanity. In such a context, and as genetic transformation is not as easy to obtain in roses as it is in other crops, its economic value will probably still be efficiently challenged for some time by highly competing and well-established methods such as cross-breeding, which has benefited from a recent increase in the knowledge of rose genetics. The use of genetic transformation in rose improvement will depend, probably more than in other crops, on the genuineness and economical value of the traits to be introduced, especially when one is aware that such a complex and subjective criterion as aesthetic aspect is first selected by the breeders.

See also: **Biotechnologies for Breeding**: Manipulation of Ploidy Level; Genetic Transformation. **Breeding**: Methods of Cross Breeding; Selection Strategies for Cut Roses; Selection Strategies for Pot Roses; Selection Strategies for Disease and Pest Resistance. **Cell, Tissue and Organ Culture**: Protoplast Culture. **Fragrance and Pigments**: Functional Genomics. **Genetics**: Inheritance of Characteristics; Mutation; Marker-Assisted Selection; Research Trends. **History of Roses in Cultivation**: History of the Perfume Industry. **Secondary Metabolites**: Culinary uses and Nutritional Value; Medicinal and Pharmaceutical uses.

Further Reading

Berninger E (1992) Le rosier de serre pour fleurs à crouper. In: Gallais A and Bannerot H (eds.) *Amélioration des Espèces Végétales Cultivées*, pp. 490–504. Paris: INRA.

Broertjes C and Van Harten AM (1978) *Application of Mutation Breeding Methods on the Improvement of Vegetatively Propagated Crops*. Amsterdam: Elsevier.

Courtney-Gutterson N (1994) The biologist's palette: genetic engineering of anthocyanin biosynthesis and flower colour. In: Ellis BE (ed.) *Genetic Engineering of Secondary Metabolism*, pp. 93–124. New York: Plenum Press.

Crane MB and Lawrence WJC (1956) *The Genetics of Garden Plants*. New York: Macmillan St Martin's Press.

Darlington CD and Wylie AP (1955) *Chromosome Atlas of Flowering Plants*. London: George Allen & Unwin.

Dohm A, Ludwig C, Schilling D and Debener T (2001) Tranformation of roses with genes for antifungal proteins. *Acta Horticulturae* 547: 27–33.

Rehder A (1940) *Manual of Cultivated Trees and Shrubs* New York: Macmillan.

Roberts AV, Yokoya K, Walker S and Mottley J (1995) Somatic embryogenesis in *Rosa* spp. In: Jain S, Gupta P and Newton R (eds.) *Somatic Embryogenesis in Woody Plants*, vol. 2, pp. 278–289. Amsterdam: Kluwer Academic.

Shepherd RE (1954) *History of the Rose*. New York: Macmillan.

Testu C (1984) *Les Roses Anciennes*. Paris: Flammarion, La Maison Rustique.

Methods of Cross-Breeding

L Crespel and J Mouchotte, Meilland Star Rose, Le Luc en Provence, France

Introduction

The genus *Rosa* comprises more than 100 botanical species:

- the subgenera *Hulthemia*, *Platyrhodon* and *Hesperodos*, which have only one species each
- the subgenus *Eurosa*, which contains over 120 species grouped into 10 sections (*Banksianae*, *Bracteatae*, *Indicae*, *Laevigatae*, *Synstylae*,

Gallicanae, Carolinae, Pimpinellifoliae, Caninae and Cinnamomeae) (*see* **Classification**: Conventional Taxonomy (Wild Roses)).

Cultivated roses are popular as plants for house and garden, and cut flowers are sold in substantial numbers; more than 18 000 cultivars are registered and almost all are issued from conventional crossing procedures.

Numerous crossings and hybridizations were performed between rose founder species of European origin on the one hand and of Chinese origin on the other. This domestication has allowed the introgression into modern roses of important horticultural characters such as winter-hardiness, resistance to pests, floral complexity and flower doubling brought by European roses, while recurrent flowering or perpetuity as well as colour brightness came from their Chinese counterparts.

Various taxonomical and cytological studies have concluded that only 11 species were used to create the modern rose we know today: *Rosa canina* L., *R. chinensis* Jacq., *R. foetida* Herrm., *R. gallica* L., *R. gigantea* Colett ex Crép., *R. moschata* Herrm., *R. multiflora* Thunb. ex Murr., *R. phoenicea* Boiss., *R. rugosa* Thunb., *R. wichurana* Crép. and *R. rubra* Blackw. A few more species have also been used in the past but they have not been used for breeding for many decades, such as *R. banksiae* Ait., *R. laevigata* Michx, *R. bracteata* Wendl. *R. roxburghii* Tratt., *R. indica odorata* Andr., *R. clinophylla* Wendl. and *R. persica* (Michx.) Bornm, except for the recent reuse of *R. banksiae* Ait. and *R. laevigata* Michx. (*see* **History of Roses in Cultivation**: European (Pre-1800); Modern (Post-1800)).

Methods of Cross-Breeding: Practice

The mechanics of rose breeding, from pollination to seedling care, consist of several operations.

Choice of Parents

The parents are chosen for:

- their complementary agronomic and ornamental value
- their female and male fertility and hip production (previous test crosses are carried out to evaluate characteristics).

Pollen and Pollination

Cross-pollination is the process of applying pollen from one flower to the pistils of another flower. Pollination occurs in nature with the help of insects and wind. This process can also be done by hand to produce offspring with desired traits, such as colour, size or pest resistance.

Figure 1 The stamens are removed from the rose after petals and sepals. (Courtesy of Doc. Meilland, Photo E Ulzega.)

To cross-pollinate it is necessary to prepare two different roses that are at a similar stage of development. In general, in contrast to diploid roses which are mostly self-incompatible, tetraploid roses are self-compatible. To prevent self-pollination of the selected seed parent, the anthers (pollen sacks) must be removed from the blooms before any pollen is released (**Figure 1**). This is best accomplished at sunrise when blooms are in the one-third to one-half open stage. First, all of the petals are removed. This procedure allows better access to the centre of the flower. Next, with tweezers or small scissors, the anthers are removed. If the variety selected as a seed parent will also be used as a pollen parent with other roses, the anthers may be collected in a cup and placed uncovered in a dry place where they will mature and release pollen by the next day (**Figure 2**). After this operation of emasculation has been performed, the remaining portion of the bud is covered with a white paper bag to prevent unwanted pollen from reaching the stigmas.

Blooms that have been appropriately emasculated are immediately ready to receive pollen from another variety. There is a difference of opinion as to the best time of day to pollinate. Some insist it should be done in the morning, others at noon, while a third group insists that best results are obtained if the pollen is applied in late afternoon; and the proper conditions may exist for an hour or so to a day or more and are dependent upon

Figure 2 Matured pollen grains. (Courtesy of Doc. Meilland, Photo E Ulzega.)

both the variety of rose and weather conditions. The time of year in which pollination must be performed is far more important than the time of day, May–June proved to be optimal in the south of France.

The pistils are located in the very centre of the bloom. On a rose this structure is shaped like a gourd with a thin elongated neck, topped with a knob called the stigma. Pollen tubes enter the top of the pistil through the stigma and travel down the style to the ovules which are contained in the ovary at the base of the pistil. If the ovules are fertilized they develop into seeds.

The success of pollinating depends on placing the pollen on the stigmas at the proper time. This is determined by looking through a magnifying glass. On each stigma a sticky fluid exudes. When the sticky exudate appears on the stigmas, it is time to apply the pollen. Pollen which has the appearance of yellow or tan powder may be applied with a paint brush to the stigma of the seed parent (**Figure 3**). The paint brush must be cleaned each time a different pollen is used. In order to remember what was crossed with what, a label should be gently tied to the stem below the bloom which has just been pollinated. Information written on the tag should include the cross and the date. Proper notation for a cross always lists the seed parent first, followed by the pollen parent.

After hybridizing, the crosses are covered with a white paper bag to protect against water, sun and insects (**Figure 4**). After 4 days, the cover is removed. If the cross is successful, a rose hip will begin to form in 2–3 weeks.

Figure 3 The cross-pollination itself: the pollen is applied smoothly on the stigmas. (Courtesy of Doc. Meilland, Photo E Ulzega.)

Figure 4 The cross-pollinated flowers are protected by a white paper bag. (Courtesy of Doc. Meilland, Photo E Ulzega.)

Improvements Based on Increased Scientific Knowledge

Fertility and germination being low and variable in roses, breeding technology has mainly progressed in the last decade by understanding better rose sexual reproduction physiology. New attention has been given to pollen care and pollination.

Storage of pollen The cryoconservation of pollen for several months, without loss of germinative quality and fertilizing power, has been shown to be successful.

This is very important for some crosses where both partners do not completely overlap in flowering time.

Optimal germination conditions Pollen germination is improved at 23–30 °C with 60–65% relative humidity, by the choice of pollination season (May–June in the south of the France), and by choosing female parental cultivars with a stigmatic exudate of low pH – approximatively 5.

Pollination accompanying treatments A 1-month cold treatment (4 °C) of rose plants just after cutback increased hip and seed sets after self- or cross-pollination as a result of improved pollen germination and female fertility.

The repeated pollination of the same flower at least twice between 1 and 4 days after emasculation increased the number of seeds per fruit, fruit weight and weight of seeds.

A putrescine treatment of the stigmas increased pollen germination and female fertility.

Seed Maturation and Germination

The hips begin ripening about the middle of September in the south of France and are harvested as soon as they begin to lose their normal 'growing-green' colour or the stem begins to darken. Extremely premature fall usually denotes that the seeds are hollow and therefore valueless. Germination is somewhat more rapid, and better, if the seeds are not permitted to become too ripe and are not dried unnecessarily between removal from the hip and planting; both of these apparently cause the seed shell to become extremely hard, and delay or prevent the emergence of the cotyledons (*see* **Propagation**: Seed Maturation and Germination).

The pericarp (which is a well-known obstacle to germination) is composed of three layers that can be interpreted as epi-, meso- and endocarp. The inner layer was demonstrated to be the most impermeable one and its thickness, determining germinability, was influenced by genetic origin and, through environment conditions, by the speed of embryo development.

Seeds from hips are extracted by hand by cutting into each hip and digging out the seeds. Seeds are dried on absorbent paper for 1–2 days and are mixed into a plastic bag containing moistened, clean, washed quartz sand or vermiculite. Very clean sand and vermiculite are essential to avoid disease. The bag is sealed and placed in darkness at 23 °C for 1 month and then at 4 °C for 2 months of cold stratification. After stratification, the seeds are removed from the sand or vermiculite by hand and immediately sown in good potting soil. After up to 2 months' germination in a cold greenhouse, the new seedling population is ready for selection.

Prospects

Conventional crossing procedures should remain preponderant for a long time as:

- they have proved the diversity they can create, starting from only a few original species;
- there are so many species and botanical varieties as yet unexploited to be included in future hybridization programmes;
- they can still benefit from new knowledge of sexual physiology, and from 'sophisticated' techniques, such as embryo rescue, protoplast fusion and ploidy level manipulation, to exploit genetic diversity better.

See also: **Biotechnologies for Breeding**: Early Embryo Rescue; Manipulation of Ploidy Level. **Classification**: Conventional Taxonomy (Wild Roses). **Genetics**: Karyology; Species Crosses; Mutation. **History of Roses in Cultivation**: European (Pre-1800); Modern (Post-1800). **Propagation**: Seed Maturation and Germination.

Further Reading

Gudin S (1989) *Quelques Aspects de la Reproduction Sexuée chez le Rosier, de la Pollinisation à la Germination.* Thesis. Nice: University of Nice.

Gudin S (1995) Rose improvement, a breeder's experience. *Acta Horticulturae* 420: 125–128.

Leroy A (1955) *Les Rosiers.* Paris: Baillière JB.

Rondeau AS (1998) *La Grande Famille des Roses.* Paris: Editions Rustica.

Vries de DP and Dubois LAM (1988) Factors affecting fruit and seed set in the Hybrid Tea-rose 'Sonia'. *Acta Horticulturae* 226: 223–230.

Selection Strategies for Cut Roses

A Chaanin, W. Kordes Soehne, Klein Offenseth-Sparneshoop, Germany

Introduction

As for all ornamental plants, aesthetic value stands out as the centre of interest in the field of rose breeding. A number of further characteristics such as yield, vase-life or stem quality are tested in the different stages of selection. Due to the expansion of planting areas to South America and Africa new aspects have had to be added to take into account the climatic and cultural conditions in these areas. Therefore, it has become necessary to test the selected elite clones under different climatic conditions, thus creating a quantity of varieties which are especially suitable for one planting area or another. Many of the more successful varieties in Ecuador, for example, are of minor importance in Europe or are even not cultivated there. The development of the market in the southern hemisphere has led breeders recently to revise their breeding activities or to start a separate breeding and selection programme in these new areas.

Given the immense range of varieties of different colours and sizes breeders do their best to improve the existing assortment by producing a large diversity of several thousands of seedlings and by trying to find the few elite plants which are worth being tested to create one new variety. This is of great importance, because breeders in most countries generally operate with the most successful existing varieties. And this mode of operation has turned out to be successful. Cross-breeding has resulted in new hybrids, although the material has been unified to a high degree due to constant hybridization. It can therefore be assumed that the kind of breeding and selection we have used until now will come to its limits, if for example breeding is focused on physiological attributes like resistance to plant diseases or high production at low light or temperature. For these aims, crossing with suitable *Rosa* species is a possibility as a way of introducing new properties in cut roses and also of transferring new 'blood' into the breeding material. Using biotechnological methods offers other strategies for overcoming problematic breeding issues such as blue colour or fragrance in connection with good vase-life.

A comparison of the current varieties with those varieties cultivated for cutting 100 years ago, like for example 'Ulrich Brünner Fils' or 'Maréchal Niel', shows nonetheless how much progress breeders have achieved in this period of time with regard to vase-life, production and colour diversity.

Crossing Strategy

Based on our own results over the last years and the current developments in the world market for roses, crossing strategies have to be changed every year. Because usually space capacity for crossing parents is limited, the choice must be made very carefully. Varieties with high production and good bud size in different colours are normally used. Thornless and fragrant varieties are also involved in the crossing design programme. Recently classical forms with globular shapes such as the 'English Roses' have gained more and more attention.

The parents usually originate from existing varieties on the market. Test material from the last years which include some desirable characteristics are also taken to make backcrosses with one of the parents. In the crossing programme combinations of the same colour are planned, but such crosses are made between for example small and large-sized varieties so as to achieve more production in the large-sized roses.

Large-scale crosses are carried out to be sure of obtaining useful results in the progenies. For refreshing the genetic basis and for the creation of new characters like classical forms, crosses with different species must be made; but this is time-consuming and not always successful.

Another strategy, for example to get large buds, is the utilization of the heterosis effect using inbred parents. This strategy has not been tested yet. It has to be kept to a small scale in order to reduce risks. In a commercial setting the major aim is to get good and saleable varieties in the shortest possible time. This is one of the reasons why such efforts have not been practised in the commercial sector at this time.

Crossing Procedure

Crosses are usually carried out from May until the end of June (in Europe, that is) but may also be made at other times. Petals and anthers of closed but fully developed flowers are discarded in the morning. Anthers collected from planned pollen parents are kept to dry in Petri dishes at room temperature overnight. Pollen received after drying by shaking the Petri dishes will be transferred to the stigma of the castrated flowers. Pollinated flowers of the seed parents will be labelled with the name or reference number of the pollen parents. Mature hips are harvested in October and registered in a crossing-book. After cleaning, disinfected seeds are sown in a mixture of sand and soil in seed benches or in artificial substrates such as peat or cocos fibres. They are kept at a moderate moisture level at 5 °C for about 2 months. If sowing was done in the winter months a special technique for cool treatment (stratification) is not necessary. However, crosses that are done in the winter in order to get ripe hips in the summer need a cool treatment to force the germination.

The number of crosses made depends on the space capacity of the breeder (**Table 1**).

Stages of Breeding

The conventional breeding of cut roses proceeds, as with other plants, in three principal stages:

1. Creating new genetic variability by crossing.
2. Selection and tests of the elite plants.

3. Propagation of the selected plants and introduction onto the market.

This course does not differ essentially between most breeding companies. In different stages of tests certain characteristics are taken as the main selection criteria for keeping or eliminating one seedling, but the general impression of a plant, the subjective evaluation, is the most important criterion in all different stages. The time needed to introduce a novelty is nearly 5 years (**Table 2**).

Seedling Phase

The type of selection at the seedling stage is based on negative criteria, i.e. the elimination of unsuitable candidates. To this end, seeds are sown directly in rows on seed benches with a capacity of about 15 000 seeds each bench (**Figure 1**).

For more efficient selection the population size is quickly reduced by removing those individuals that are clearly not useful. Each decision is coupled with a short comparison with the existing varieties on the market. Many of the seedlings are single or semi-doubled and can therefore be eliminated at an early stage from the population. Furthermore some characteristics like thorns or poor length of stems are tolerated at this stage if the flowers are well doubled. Mostly two flushes are evaluated at the seedling stage. This occurs within 7 months from sowing to the transfer in the next stage of selection. Seedlings that do not flower in this time are nonrecurrent and are removed. By this kind of selection the capacity for recurrent blooming (perpetual flowering) has been established in roses. Crossing with nonrecurrent flowering species in order to transfer certain characteristics such as intense colour or fragrance is unfortunately partly combined with the transfer of this undesirable characteristic. These types are nonrecurrent bloomers, and to change this characteristic takes a long time.

Table 1 Efficiency of breeding in one commercial rose breeding operation (approximate numbers)

Number of parent plants	2000
Crossing combinations	1000 (seed parents × pollen parents)
Pollinated flowers	25 000
Harvested hips	20 000
Sown seeds	200 000
Germinated seeds	120 000
Seedling phase	
Step 1	3000 individuals
Clonal phase	
Step 2	400 elite clones
Step 3	50 elite clones
End selection	4–6 novelties

Figure 1 Seedling phase: 5 months after germination. (Reproduced with permission from W. Kordes' Söhne.)

Table 2 Time-scale and basic scheme of the breeding of cut roses

Stage of programme	*Year of programme*	*Targets*	*Selection criteria*
Production of seeds	1	Realization of several thousand crosses. Creation of genetic variability	—
Seedling phase (stage 1)	2	Evaluation of nearly 200 000 seedlings	Colour, form, double flowers and decorative effect
Clonal phase (stage 2) (group of three plants)	2–3	Selection of approx. 1000–3000 elite plants; reduction to approx. 2–3% of the quantity of seedlings by negative selection	Colour, form, double flowers, growth habit, quality of stems
Clonal phase (stage 3) (group of 50 plants)	3–4	Selection of approx. 50–100 favourites and setting up of a small group (50–100 plants) of each candidate	Aesthetic value as well as 'inner qualities', like vase-life, production, characteristics of production under different cultural conditions, propagation capacity
Introduction on the market	5	Introduction of four to six novelties	Suitability for commerce, acceptance by consumers, licensed production

Besides colour and flower size, stem quality and stability of the neck can be evaluated positively. Plants with nicking necks are eliminated as well as those with a small number of petals. Taking into account the available space in the greenhouse for further cultivation of selected elite plants, some 1000–3000 seedlings (about 2–3% of the seedlings) are chosen for the next test stage.

Clonal Phase

Groups of three plants Selected seedlings are mostly propagated on their own roots by cuttings or by budding on a rootstock to create a group of three or more plants per clone depending on greenhouse capacity. At this stage seedlings are evaluated under normal cultural conditions using a technique from Japan with partial bending of the stem, cultivation on a suitable substrate using cocos fibres, perlite or others with fertilizing irrigation and partly artificial supplementary lighting. The aesthetic value of the plant here again is the most important characteristic. Furthermore the quality of the stems after bending can be evaluated. The stems arising from the buds at the base of the plant are strong and long and provide the best quality in any variety. Therefore the cultivation of cut roses is aimed at stimulating these stems by the bending system or by cutting the plant down during the harvest. As a rule five flushes in the course of about 1 year are evaluated on the plants in order to get a first impression of the time needed until next flush as well as the reaction of the clones during winter in poor light conditions. In the course of selection in this group candidates with good characteristics are selected and continually propagated for the next step. An uncertainty will remain concerning the selection for the next test phase, because the decision in favour of one rather than another elite clone has usually been done by evaluating only one to three plants of an individual clone. Therefore it is recommended to build an intermediate phase of around 200–400 clones with 6–10 plants from each favourite clone to get early information about their suitability for propagation and to produce more plant material for the next propagation (**Figure 2**).

Figure 2 Clonal phase: the group of 200–400 clones. (Reproduced with permission from W. Kordes' Söhne.)

Groups of 50 plants In this stage of selection there is more material available, so that one group of 50 to 100 plants for each favourite clone is formed. Propagation is done in different ways, as by cuttings or budding on *Rosa* 'Natal Briar', *Rosa canina* 'Inermis', *Rosa manettii* (Krüssmann, 1986) or *Rosa indica* 'Major'. In this test phase it is very reasonable to introduce different types of propagation in the trials, because intensity of colour, size of flower or stem length as well as production characteristics will depend on it. Selected elite plants are evaluated the most intensely in this stage. Apart from the morphological characteristics of the plants, data on the production of the single clones and on the vase-life of the flowers are gathered. Additionally data about the flowering habits of the plants, such as the ability to open fully or the reaction of the flowers to being cooled, are collected. Further opening when the flowers are stocked cool or the blueing of red varieties, are for example, undesirable attributes for future varieties. Of great importance also is the stem length in different seasons, especially in summer because varieties tend to produce short stems in that season. Tests are made at this stage over a whole year to estimate all morphological and physiological characteristics under the different climatic conditions that are found in Africa and South America.

Disease resistance of the plants is evaluated too. Testing of the elite clones is carried out under normal conditions. Seedlings with a high sensitivity for powdery or downy mildew or other plant diseases are normally rejected at the earlier stages of selection. In later stages this will also be observed because any such future varieties will have problems in production and the expense of keeping them healthy will be high and not justified. Problems with pests like spider mites or thrips will also be observed. Here however there is need for research about the sensitivity of different genotypes to these pests, because effective chemicals (especially against spider mites) do not exist or those existing are not satisfactory. In addition, the number of available plant-protecting agents has decreased steeply over recent years in many countries due to new regulations and laws. The aim of the producers to use few chemicals and to produce plants under integrated pest management conditions has increased in recent years, too.

Out of this group candidates will be chosen for which Plant Breeder's Rights will be applied for and which will go ahead to introduction on the market.

These are the novelties combining an attractive colour and good flower production with a good size of flower, good vase-life and reliable opening.

Introduction on the Market

Rose breeding companies use different strategies. Many of them have their own independent production facilities, and as most of the breeding companies have their headquarters in The Netherlands or at least a partner or representative there they offer the new varieties at the rose auction in Holland. This is usually how a new variety is introduced on the market. Another possibility is to grant exclusive production rights to certain producers. In this way the breeder can assure a constantly good quality of flowers while introducing the variety, so that demand will be initiated.

Many of the cut-rose producers take personal care of the choice of their varieties. While visiting the elite plants at the breeding companies at stage 3 (the groups of 50 plants) they choose the varieties for their next planting. The producers are interested in cultivating the newest varieties even before their introduction on the market, before the evaluation has been done at the breeders and before enough data and information on these plants have been collected. As a consequence some of the varieties have a short life on the market, as the faults in these varieties will show up in the course of production. Then there will be no alternative but to uproot these plants.

Main Aims in Breeding Cut Roses

Crossing aims generally do not differ worldwide, though of course characteristics will differ depending on the climatic and cultural conditions. However, the interest of breeders is focused on some targets which can be summarized as follows: colour, size and shape of flower, leaves, neck, thorns, stem length, vase-life, tolerance of transport and cooling treatment, fragrance, disease resistance, yield and winter growth.

Beautiful and Attractive Colour

The classic colours of red, yellow, pink or white will always have great importance worldwide; red ranks first with more than 30% share of the market. Therefore efforts of breeders concentrate on establishing a good red variety on the market. Besides the classic colours, fashion colours have a certain importance. For a long period the sunshine colours like orange and apricot dominated the market in fashion colours, so the breeders reacted with an abundance of varieties in these intonations. In the last 4 years pink has been praised as the most important colour in the different fashion magazines. But this trend has only been successful for a short time. The breeders reacted with proper offers. Other extraordinary colours like brown, purple or green are interesting, but they only have a small market share, so that these colours cannot be planted as mass varieties. Bicoloured and striped varieties also have a certain demand belonging to the extra-colours (**Figure 3**). Besides the colour itself the lastingness of the colour is important. A variation of the colour on the plant is a negative characteristic and is to the disadvantage of the future variety.

Figure 3 Outstanding forms like the novelty Abracadabra® create a high demand on the market. (Reproduced with permission from W. Kordes' Söhne.)

As the time-scale for breeding cut roses is approximately 5 years, all colours are considered in the different stages of selection. The colour is the first impression that the breeder will gain from the seedling in the first stage of the selection. Nevertheless some intonations of the colour depend on intensity and spectrum of light, as for example the colours yellow, yellow/red and white/red. Here there are great differences in the intonations in the same varieties whether planted in South America, Africa or Europe.

Tough Petals and Double Flowers

These characteristics can be evaluated even in the early stages on the plants and the selection will mainly be done in the seedling phase. Individuals with double flowers normally have a longer vase-life, whereas seedlings with fragrance normally have softer petals and their vase-life will be shorter than those with tougher petals. But the seedlings with fragrance will be observed further and will get throughout the stages of selection the so-called 'fragrance bonus'. Because of the selection which was and is still based in this earlier stages on tough petals, fragrance in cut roses has been neglected and has practically been lost in most of the present-day varieties. Although fragrant varieties can gain good prices the sales are limited due to the fact that they have a limited vase-life.

The number of petals is genetically determined and does not depend on the nutritive conditions of the plant. In unpublished tests the petals of several semi-doubled seedlings were counted in the first stage (seedling phase). The number did not change during the following flushes in this stage. After budding on *Rosa canina* 'Inermis' there was no change in the number of petals in comparison with those of the origin seedling.

Quality of Stems

Strength and length of stems must be in proportion to the size of flowers. The flower neck has to possess a certain stability. While length and strength of stems cannot be evaluated definitively in the seedling phase, this is possible for the flower neck. Observations of breeders and producers have showed that seedlings with soft or nicking necks never lead to plants with straight necks in later stages of propagation. This is also true for garden roses; they preserve this attribute under outdoor climatic conditions after transfer to field tests.

The desired length of the stem depends on the size of the flower. The demand alters from country to country. Whereas in The Netherlands or Italy long stems up to 100 cm are mostly demanded, the main interest in Germany or Kenya concentrates on a medium stem length of up to 50 cm with a large flower.

Another aspect for evaluating the stem length is the absence of thorns or a reasonably small number of thorns at least in the upper half of the stems. Strong thorns can cause damages to the stems during transport leading to reduction of quality and decrease of vase-life. The work of florists is also complicated by thorns so they will probably have an aversion to this variety which will therefore be doomed to failure.

Size of Flowers

An important selection criterion is the size of flowers. Even in the first stage of selection (seedling phase) this characteristic can be evaluated positively. The seedlings are however not selected very rigorously on this attribute here because in next stages (clonal phases) the size of the flowers and their relation to the stem can be evaluated better and unmistakably.

Figure 4 Cherry Girl®: a large-flowered variety. (Reproduced with permission from W. Kordes' Söhne.)

In cut-rose production the following types of roses are offered:

- Large-flowered or long-stemmed roses
- Intermediate or medium-stemmed roses
- Sweetheart roses or short-stemmed roses
- Spray roses.

Large-flowered varieties These have a stem length of 50–100 cm and a flower diameter of approx. 14 cm (**Figure 4**). This group has gained importance in recent years and the breeders have introduced a number of new varieties on the market.

Intermediate roses The intermediate roses (**Figure 5**) are a relatively new group. The stem length is approx. 50–70 cm and the size of flower is nearly as large as in the long-stemmed roses. It is their stem length that leads to their classification as medium-sized. Intermediate roses are very popular in Kenya and Zimbabwe. They are cultivated there with great success and prepared for export to Europe and Japan.

Small-flowered or sweetheart roses These reach a stem length between 35 and 55 cm (**Figure 6**). As the harvesting costs amount to 80% of the costs of cut-rose production there is no commercial benefit in this group in countries with high labour costs.

Spray roses Spray roses (**Figure 7**) produce five or more flowers per stem. They are not very important in Europe but are popular in Japan.

Figure 5 Sunbeam®: an intermediate rose. (Reproduced with permission from W. Kordes' Söhne.)

Figure 6 Frisco®: the top-selling sweetheart rose in The Netherlands. (Reproduced with permission from W. Kordes' Söhne.)

Durability and Transporting Qualities

While the selection is very subjective in the first two stages, characteristics of the so-called 'inner quality' are evaluated at stage 3. Because of the long periods of transportation from the cultivation areas the demand for a long vase-life is very high. Normally the varieties should have a vase-life of at least 12–14 days, but there are varieties on the market with 18 days

Figure 7 Taifun®: a spray rose type. (Reproduced with permission from W. Kordes' Söhne.)

vase-life and more. Varieties with fragrance, however, have a vase-life of 8–10 days. The tests for vase-life at the breeders are done under uniform conditions with lighting of $3\,\mathrm{W\,m^{-2}}$ for 12 h, temperature of $20\,^\circ\mathrm{C} \pm 1\,^\circ\mathrm{C}$ and a relative air humidity of $60 \pm 5\%$. After the cutting for testing purposes flowers will be stocked in a cool room at 2–4 °C in water for 1 day, and after that a simulation of transport in normal boxes without water will be made for another day at room temperature. On the third day a few centimetres at the end of stem are cut and the roses are put into vases with water under the above-mentioned conditions in the 'vase-life testing room'. During this testing phase the opening of the varieties and their sensitivity to botrytis are evaluated. To find out the best time for cutting, flowers in different opening stages are involved in the trials. The different types of roses react differently in their flowering attitude. Sweetheart roses can be cut relatively early, in the bud stage, and are still able to develop to a fully open bloom, whereas large-flowered varieties have to be cut half-open, otherwise they will not open fully because of the double flowers. The characteristics of the groups have to be found out by tests of the novelties and confirmed under normal conditions.

Production

Together with the increasing costs for salaries and investment the yield for one variety per square metre per year as well as the required time until the development of a flush are important aspects while evaluating a novel variety. At the stage of the 50-plants groups data on production parameters are investigated. In relation to the size of flowers the possible yield is lower in large-flowered than in sweetheart roses. It is 90–140 stems m^{-2} $year^{-1}$ for the large-flowered varieties. They require 60–80 days until harvesting under glasshouse conditions in northern areas between April and September. In Columbia and Ecuador with an altitude around 2500 m and low temperatures this phase lasts 85–100 days. Yet this longer period for the development of flowers leads to larger flowers and more intensive colour than roses grown under Northern European climatic conditions.

The production of medium-flowered roses is approx. 220 flowers m^{-2} $year^{-1}$. This group takes the main share of the mass production of cut roses. Sweethearts or small-flowered roses produce 200–300 stems $m^{-2}year^{-1}$ with a length between 35 and 55 cm, but varieties with production of 600 stems under advantageous growth conditions exist as e.g. the variety Frisco® (**Figure 6**). The production of spray roses is normally low with approx. 120 stems $m^{-2}year^{-1}$. Even the required time between the flushes is comparable with those of the large-flowered varieties. The price of large sprays is nearly the same as that of long-stemmed roses and justifies therefore the cultivation of this group.

When testing new varieties in regard to their suitability for the market, especially in the red range, a comparison with the existing varieties regarding their production qualities cannot be avoided. For a long period the variety First Red® was the most planted red variety. When a new red variety was being considered for introduction, First Red's production as well as stem length was constantly compared with the new red candidate under different cultural conditions. Novelties with lower production or shorter stems have a poor chance of being established on the market. This comparison is not done all the time in other colours as in the red ones.

Conclusion

Breeding roses in the conventional way has a large element of luck in the discovery of good types. Each seedling obtained by crossing will be unique. The careful choice of parents is a very important step to reduce risks. However, thousands of seedlings are raised every year in the search for desired characteristics. As long as our knowledge about the inheritance of certain characteristics of cut roses with their tetraploid number of chromosomes ($2n = 4x = 28$) is still small, we need to raise a seedling population of an increasing size to find out the desired combinations of characteristics. Tests to transfer the characteristic 'fragrance' onto other nonscented varieties have not so far led to better varieties in respect of vase-life and quality of flowers than those already existing on the market. Crossing varieties with and without fragrance in red coloured roses resulted in a progeny of scented seedlings, but these did not differ in colour and flower shape from the scented cross-bred variety. Self-pollination of scented varieties has resulted in great losses in the progeny due to inbreeding. In the few remaining seedlings the characteristic 'fragrance' was found, but it was connected to other characteristics of the mother variety like colour, shape and soft petals. By crossing varieties with different colours, like white and red, scented types in white with red edges were found in the progeny. But the vase-life of these varieties was very limited. In the early selection process (seedling phase) the toughness of the petals give a clue about the vase-life: flowers with soft petals are usually fragrant. Recent experiments have discovered a correlation between petal structure and the suppression of scent emission of petals.

While looking at the inheritance of colour, intonations have been found that are situated between the colours of both parents. Therefore breeders try to build up homogeneous groups in colour, because their aim will in most cases be to breed a better red or yellow rose or other varieties with higher production or longer vase-life. However, the possibilities presented by conventional breeding of achieving some breeding aims such as the combination of long-vase life with fragrant varieties or the creation of blue colour are limited and have not advanced up to the present time. Biotechnology may help to overcome such breeding problems. The use of embryo rescue or protoplast fusion to break crossing barriers offers the opportunity to expand the genetic variability in the actual breeding material. Because of the need for skilled labour and specialized equipment and also the relatively long time taken to establish them, biotechnological methods are not widespread in the practice of rose breeding.

Despite all these uncertainties the success up till now of the classic breeding techniques can be assessed. A visit to the greatest exhibition of flowers – the 'Horti Fair' in The Netherlands – will produce the evidence for that.

See also: **Breeding**: Overview. **Fragrance and Pigments**: Functional Genomics. **Genetics**: Mutation. **Production and Marketing**: Cut Rose Production.

Further Reading

Dohm A, Debener T and Grunewaldt J (1996) Regeneration und Transformation bei Rosen. *GPZ-Tagung Vortr. Pflanzenzüchtung* 32: 172–174.

Glasau F (1961) *Rosen im Garten*. Berlin: Paul Parey Verlag.
Haenchen E and Haenchen F (1974) *Das neue Rosenbuch*. Berlin: VEB Deutscher Landwirtschaftsverlag.
Holton AT and Tanaka Y (1994) Blue roses: a pigment of our imagination? *Trends in Biotechnology* 12: 40–42.
Krüssmann G (1986) *Rosen, Rosen, Rosen*. Berlin: Paul Parey Verlag.
Marshall HH, Campbell CG and Collicutt LM (1983) Breeding for anthocyanin colors in *Rosa*. *Euphytica* 32: 205–216.
Pertwee J (2000) *Production and Marketing of Roses*. Dordrecht: Elsevier International Business Information.
Rupprecht H (1970) *Rosen unter Glas*. Basel: Verlag J. Neumann.
Vries DP de (1976) Selection of hybrid tea-rose seedlings for cut flower yield. *Euphytica* 25: 361–363.
Vries DP de and Dubois L (1977) Early selection in hybrid tea-rose seedlings for cut stem length. *Euphytica* 26: 761–764.
Walther F and Sauer A (1986) Analysis of radiosensitivity: a basic requirement for *in vitro* somatic mutagenesis. III. Rose cultivars. *Gartenbauwissenschaften* 51: 40–43.

Selection Strategies for Pot Roses

D P De Vries, Plant Research International, Wageningen, The Netherlands
Present address: Foundation Sub Rosa, Ede, The Netherlands

Introduction

As a rule, pot roses are cultivars that remain ‘small’. Small in this sense means that even when fully grown, plants are conspicuously less tall, have thinner shoots, and bear smaller flowers and leaves than any variety from the current Floribunda, Hybrid Tea or Grandiflora classes. A general term for these small roses is ‘dwarf roses’, often incorrectly named ‘Miniatures’. While the natural height of dwarf roses does not exceed 40–50 cm, that of marketable pot roses is limited to about 25 cm. This height may be attained by application of length-reducing chemicals. In course of the last 200 years, dwarf varieties have occurred in several classes of roses. Numerous of their plants have been grown in pots, with varying success. Classes of which varieties have been used are: ‘Koster roses’, ‘Dwarf Polyanthas’ and ‘Compactas’, but particularly ‘Miniatures’. In addition to these accepted varieties, the B-quality of 1-year-old field budded ‘large’ garden rose varieties (*see* **Rootstocks:** Usage of Rootstocks) have also been potted to compensate for otherwise lost profits of nurserymen. Such plants commonly attain a disappointingly large size, thus confirming that ‘not any rose in a pot is a pot rose’. Although considerable quantities of pot roses in northwest European nurseries are still being produced in 1-year cycles, at the turn of the century the bulk of pot roses, the so-called ‘floriculture’ pot roses, is produced in the glasshouse all year round, in 14–18-week cycles. Statistics about production and area of pot roses are not available, but the estimated production in the United States (mainly California, New Mexico and Florida) is 50×10^6 pots annually. In Europe the numbers are about 30×10^6 for Denmark, 10×10^6 for The Netherlands and 0.5×10^6 pots in France and Italy together. Where initially pot roses were used to decorate balconies, walls or terraces, and were given great care to ensure their survival, current pot roses have become just one of many indoor plants with a relatively short shelf-life. In addition to indoor use, dwarf roses are increasingly grown for patios and small gardens, particularly in favourable climates. Prior to the times of plant patent or plant breeders' rights (1930 and 1941, respectively), pot rose breeding was a hobby rather than a profession. Even today, professional pot rose breeders are rare. In Europe De Ruiter (The Netherlands), Poulsen (Denmark), Kordes (Germany) and Meilland (France) control more than 80% of the market. In Canada and the United States, the breeders Laver and Hill as well as a number of amateur breeders provide the market with new varieties. Commonly, royalties amount to about €0.10 per 8–10-cm pot, but there is a tendency to raise royalties as pots are larger or contain more cuttings.

Despite the small number of active breeders, considerable progress has been made as to plant habit, flower number, flower size, hardiness and shelf-life since the first dwarf roses were marketed. Besides the common dwarf factor, any character of the Hybrid Tea or Floribunda classes has now been incorporated in the current categories of dwarf roses. Presently, breeders are aiming for larger flowers, preferably in combination with hardiness, disease resistance and prolonged shelf-life. Research in pot roses has been devoted to inheritance studies of the dwarf factor, comparison of the performance of micropropagated plants with those from cuttings, variation for rooting capacity, selection criteria for pot plants in segregating populations and, currently, the search for prolonged shelf-life.

History of Dwarf Roses

Literature as to the origin and descent of dwarf roses is often unclear and controversial. Confusion is mainly caused by poor historical records, the occurrence of dwarf roses in several species and classes of cultivars, and the fact that some authors seem to have copied each other. In chronological order of origin, the classes of dwarf roses are:

1. Miniatures.
2. Dwarf Polyanthas.
3. Koster roses.
4. Compacta roses.

Miniatures

Authors generally agree on *Rosa chinensis* var. *minima* (Sims) Voss as the common ancestor of the major Miniature varieties. This diploid ($2n = 2x = 14$) recurrent-flowering species and several of its cultivars, which represent the tiniest roses known, have been widely used as pot plants since the nineteenth century. The origin of Miniatures is a colourful story. Their direct import in Europe has not been recorded, but plants described under this species are not indigenous to China. In 1810 the British botanist Robert Sweet obtained a dwarf rose from the isle of Mauritius (or Réunion), which was a British colony and an important intermediate station for sailing ships on their way to and fro the Far East. Sweet's plant was disseminated as *R. lawranceana* Sweet from 1818 onwards. However, John Sims had already in 1815 described a dwarf rose under the name *R. semperflorens* var. *minima*, which he had obtained from seed. According to the late British breeder Jack Harkness this was a dwarf seedling selected from a segregating population of *R. chinensis* var. *semperflorens*. Because *R. chinensis* var. *semperflorens* 'Slater's Crimson China' was already known in France prior to the year 1800, it is not improbable that the story of the Miniatures goes back to before 1810. In 1894 the botanist Andreas Voss described and denominated a dwarf rose as *R. chinensis* var. *minima*, which was probably the same as *R. semperflorens* var. *minima*. *Rosa chinensis* var. *minima* (Sims) Voss would have been introduced and cultivated on the continent prior to import in England, although this has not been confirmed. The Swiss botanist Pyramus de Candolle also reports having obtained a dwarf rose from Mauritius, which he disseminated in Europe under the name *R. indica* var. *humulus*. Several authors mention that another *R. chinensis* var. *semperflorens* cultivar 'Parsons' Pink China', would have yielded one or more dwarf seedlings. One of them was named 'Dwarf Pink China' or 'Fairy Rose' (not 'The Fairy') as a cultivar of *R. lawranceana*. From 1825 till about 1870 Miniatures were desirable pot plants in France. Varieties like 'Pompon de Paris', 'Caprice des Dames', 'Gloire des Lawranceanas', 'Jenny', 'Nigra', 'Pallida' or 'Retour de Printemps' were sold on a large scale by Paris nurseries. However, from 1870 these tiny roses became out of style and were gradually replaced by the somewhat larger and more winter-hardy Polyanthas. They had sunk so far into oblivion that a dwarf rose that was newly 'discovered' by the military physician Roulet in the Swiss Jura in 1920 was not recognized, and was denominated as *R. rouletii* (also *R. roulettii*) by the botanist Correvon in 1922. Later, the German breeders Peter Lambert (Trier) and Wilhelm Kordes Sr (Sparrieshoop) confirmed that *R. rouletii* was synonymous with 'Pompon de Paris'. *Rosa rouletii* has become the major ancestor of the present class of Miniatures. Today, only 'Pompon de Paris' (**Figure 1**) and 'Climbing Pompon de Paris' are available. Miniatures or their descendants form the major part of modern pot roses all over the world.

In early crosses Miniature varieties remained diploids ($2n = 2x = 14$), but owing to breeding with varieties from tetraploid ($2n = 4x = 28$) Floribunda or Polyantha classes, a range of triploid dwarfs with hitherto unknown forms and flower colours was soon created. Because triploid roses readily produce viable diploid pollen, the new generation of Miniatures, like practically all modern large-flowered rose varieties, became tetraploids within several decades. Owing to their somewhat sturdier growth, tetraploid Miniature varieties have become attractive pot plants. Famous Miniature breeders were Jan de Vink (Boskoop, The Netherlands), Pedro Dot (Barcelona, Spain) and Ralph Moore (Visalia, USA). Free exchange of plant material among these breeders accelerated the development of new varieties. The current tendency to name all dwarf roses 'Miniatures' is incorrect as the term only applies to descendants of *R. chinensis* var. *minima* (Sims) Voss.

Figure 1 *Rosa chinensis minima* 'Pompon de Paris': plant raised from softwood cuttings. Plant Research International, Wageningen, The Netherlands.

Dwarf Polyanthas

In 1865, the plant collector Robert Fortune shipped a dwarfing, nonrecurrent pink-flowered spray form of *R. multiflora* from Japan to Britain. The French nurseryman and breeder Jean Sisley at Lyon, France, seized upon a number of those plants, and succeeded in raising a number of seedlings. Most of them would have been nonrecurrent-flowering Multiflora-like ramblers. However, in the next generation the French nurseryman Guillot selected several recurrent-flowering dwarfing plants. Guillot introduced the first diploid ($2n=2x=14$) Dwarf Polyanthas: 'La Pâquerette' (1875) and 'Mignonette' (1880). At the beginning of the twentieth century Dwarf Polyanthas were popular as pot plants, but they gave way to the more manageable Miniatures. The British geneticist Hurst concluded that the original cross of these Dwarf Polyanthas must have been *R. multiflora* × *R. chinensis* var. *minima*. According to *Modern Roses*, the recurrent-flowering *R. multiflora* var. *nana* Hort. would have a similar descent, which seems questionable because populations do not segregate in 'giants' and 'dwarfs'. All later Dwarf Polyanthas appear to descent from Guillot's initial varieties. Important varieties have been: 'Mlle Cécile Brünner' (Ducher, 1880), 'Katharina Zeimet' (Lambert, 1901), 'Gloria Mundi' (De Ruiter, 1929) and 'The Fairy' (Bentall, 1932). The originally diploid ($2n=2x=14$) Dwarf Polyanthas have been extensively used to create new forms in the Miniature class.

Koster Roses

The basis of all Koster roses is just one natural dwarf mutant, a sport of the deep-pink-flowered rambler 'Tausendschön' ('Thousand Beauties'), discovered in 1914 by Peter Lambert in Trier, Germany. Probably because 'Tausendschön' descends from the multifloras 'Daniel Lacombe' × 'Weisser Herumstreicher', the pink-flowered dwarf mutant 'Baby Tausendschön' (syn. 'Echo') strongly resembles a Dwarf Polyantha. For that reason *Modern Roses* classifies them incorrectly as Polyanthas. No other rose than 'Baby Tausendschön' has been the starting point of so large a family of mutants from mutants, mainly discovered between 1914 and 1960 by the Dutch nurseryman Koster. Presently, the family of the Koster roses consists of several dozens of named varieties, all with a similar plant habit, but with flower colours that range from white via pink to dark pink and red, and even orange. Only yellow is lacking. The plant habit of Koster roses is larger than that of the Miniatures. Contrary to the expectation, the diploid ($2n=2x=14$) Koster roses have seldom been used to breed other dwarf roses. Only the American breeder Morey selected a number of new varieties of which 'Blushing Jewel' and 'Ruby Jewel' are best known. Between 1930 and 1990 Koster roses were beloved pot plants that were mainly produced in Dutch nurseries and exported to Scandinavia. At the present time, they have given way to new Miniature varieties.

Compacta Roses

Compacta roses owe their exclusive origin to the Dutch breeder De Ruiter who, starting in 1954, initially crossed Dwarf Polyanthas with Hybrid Teas, and in later years also with Miniatures. The selected plants grow in an attractively 'compact' habit and are seldom taller than 45 cm. Despite the fact that *Modern Roses* does not recognize Compactas as an independent class and lists them under Polyanthas, the plants can be easily distinguished from either Dwarf Polyanthas or Miniatures. By definition, the early Compactas were triploid ($2n=3x=21$). Famous Compactas, which have been extensively used for pot culture, are the so-called 'seven dwarfs', viz. 'Happy', 'Doc', etc. Compactas have greatly contributed to the breeding of modern dwarf roses by De Ruiter nurseries.

Breeding Dwarf Roses

Inheritance of the Dwarf Gene

Phenotypically, dwarf roses of each class have a similar small plant habit in common. The genetic base for 'dwarfness', however, is not necessarily the same. When Dwarf Polyanthas, Koster roses or Compactas are crossed with Floribunda or Hybrid Tea varieties, a clear segregation between 'small' and 'large' plants does usually not occur; in hybrid populations plants of all sizes may be seen. Continuous variation for plant size suggests quantitative inheritance of that character in these classes. This means that in populations this kind of dwarf roses always form a minority of a few percent only. Quantitative inheritance does not explain the apparent ease by which breeders like Jan de Vink or Pedro Dot selected their first Miniature varieties from crosses with *R. rouletii*. Initially, the mode of inheritance of the dwarf character in *R. rouletii* was assumed to be recessive. Not until 1987, when Dutch scientists at Plant Research International, Wageningen, studied populations of *R. rouletii* × (diploid) Polyanthas, *R. rouletii* × *R. multiflora* var. *nana*, and selfed a number of dwarf varieties, was it demonstrated that dwarfness in Miniatures is controlled by a single dominant gene, that was named *D*. In the cultivars tested, *R. rouletii* and its dwarf descendants were heterozygous for *D*, while all Polyanthas were homozygous for *d*.

Tendencies in Breeding

There is an increasing interest in 'small roses with large flowers', preferably with some resistance to black spot (*Diplocarpon rosae* Wolf) and powdery mildew (*Podosphaera pannosa* (Wallr.) Lev.), so that they may meet current demands for use as patio plants. Although some improvement of flower size has been attained since 1930, flowers have not by a long way the size of Floribundas or Hybrid Teas. To solve problems temporarily, pot rose varieties that are too tall are chemically dwarfed by e.g. repeated Bonzi sprayings. Because small roses generally have small flowers and vice versa, only long-term breeding programmes with repeated use of suitable Floribundas or Hybrid Teas may truly solve the problem. Since dwarfness is inherited in a monogenic manner, Miniature varieties are an ideal starting-point for breeding programmes. Because early Miniatures originated from direct crosses between *R. rouletii* and larger-flowered varieties, populations initially showed gigantic variation. Since the middle of the twentieth century, however, new varieties mainly originate from Miniature × Miniature crosses. Continuous use of the same gene pool, however, has its restrictions as to new, desirable characters, while even mild inbreeding affects seed and fruit set. For that reason, breeders like Moore and Tantau have used species roses that have not yet previously been incorporated in Miniatures, e.g. *R. wichurana* Crép. and *R. sempervirens* L. Also the 'moss' character, available in nonrecurrent-flowering Moss rose varieties (*R. centifolia* var. *muscosa* L.), has been introduced in Miniatures. Besides crosses between Miniatures and large roses, two other possibilities may be applied to obtain 'large-flowered dwarfs'. First, breeders of either garden or glasshouse roses have noticed that a kind of 'dwarf' rose seedling accidentally occurs in several segregating populations. Commonly these plants are discarded because they do not meet the breeding aim. Retaining them, however, may be a new source of large-flowered dwarfs. Another possibility has recently been offered by biotechnology. The so-called IPT gene, when introduced into various crops has a natural shortening effect on internodes while maintaining flower size. Its introduction in roses has been frustrated by general difficulties in genetic modification of the crop and by restrictive European legislation. Other improvements of dwarf roses may be obtained in using evergreen species such as *R. bracteata* Wendland ($2n = 2x = 14$) in crosses.

Selection of Dwarf Seedlings

In practical breeding, the segregation of 'dwarfs' and 'giants' in Miniature seedling populations is very conspicuous only a few weeks after seed germination (**Figure 2**). This means that in aiming at Miniature varieties, giant seedlings may be discarded in an early stage, thus retaining the desired dwarfs. Selection is facilitated when seedlings are cultivated in pots in the controlled environment of the glasshouse.

Figure 2 Seedling population no. 330 'Meitrisical' × *R. rouletii*: segregation for (left) giants and (right) dwarfs. Plant Research International, Wageningen, The Netherlands.

Commonly, seedling plants will flower at 50–70 days from germination. Apart from the common dwarfing gene, a population of dwarf seedlings is highly heterozygous for all characters. Hence, according to the aim of the cross and the preference of the breeder, first selection may be carried out for:

- number of flowers per stem
- flower colour
- flower size
- number of petals
- plant habitus (attractiveness, number of shoots etc.).

At first selection commonly 5–10% of the dwarfs are retained. These plants are transplanted into larger pots, where they may grow for several months before being multiplied by cuttings. The relation between the growth and development of Miniature seedlings and their clonal plants has been studied at Plant Research International, Wageningen. In principle seedling and clonal plants perform similarly, but the level of expression of seven characters studied may differ significantly. Hence, in seedlings the flower bud occurs earlier, but actual flowering is later than that of clonal plants. At anthesis seedlings have shorter shoots, fewer leaves, more flower buds (absolute and per lateral), a higher percentage of sprouted axillary buds and fewer petals than their clonal plants. Owing to these results the process of selection in both seedling and clonal plants has been considerably facilitated.

Clonal Selection in Dwarf Seedlings

Retained dwarf rose seedlings may be multiplied in various ways. Commonly, the first small clones of these seedlings consist of own-rooted plants raised by softwood cuttings. As in floricultural practice, cuttings are struck at three to five cuttings per genotype in 9–15-cm pots, and rooted under water vapour at high temperatures. Since there is substantial genetic variation for the formation of adventitious roots, speed of rooting and number of roots have become important selection criteria for automated propagation. Thus, even beautiful young clonal plants may be discarded on account of their poor rooting performance. After rooting and weaning, the small clones of 5–10 pots each are placed on tables in the glasshouse. When the first flower buds appear, plants are pruned down to stimulate formation of laterals. In addition to the characters already selected for in the seedlings, important criteria for the first clonal plants are:

- number of days from pruning to flowering
- plant height at flowering (without chemical dwarfing)
- number of flowers (buds).

Depending on the breeder's policy, clonal plants may be tested at this stage for (simulated) dark and/or cooled transportability and shelf-life under (simulated) room conditions. It is not uncommon that fewer than 5% of the clones are suitable for a next, larger, clonal propagation. In order to further speed up the selection process, breeders tend to board out large-scale clonal testing with specialized pot-rose growers. It is notable that, although in both first seedling and clonal selection 5% of the plants are retained, only 1 out of 400 ($1/20 \times 1/20 = 1/400$) seedling plants of a population may produce a desirable clone. But even when young clones have been thoroughly tested, large-scale production and general introduction into the market may be disappointing. The marketing policy of many breeders is therefore to distinguish between the consumer preferences of various countries.

Propagation of Dwarf Roses

Floriculture by Softwood Cuttings

In floriculture, pot roses from softwood cuttings are grown year-round in 14–18-week cycles in heated glasshouses. This method, which has been in vogue since about the 1980s, is the fastest way to produce potted plants. Because of the expense of labour, in most concerns propagation has been completely or almost completely automated. Even the selection and striking of cuttings, as well as the preparation of plants for auctioning is now (partly) robotized.

Figure 3 Striking softwood cuttings of pot roses is often still done by hand. Messrs. Pouw Rozen, De Kwakel, The Netherlands.

Prior to striking, most softwood cuttings are soaked in solutions of fungicide and/of acaricide. After grading for size and number of leaflets, cuttings are struck (by hand) in pots that sit on conveyer belts in week 0 (**Figure 3**). Commonly, rooting hormones (e.g. 0.25% indolebutyric acid) mixed with fungicide are applied at the proximal end of cuttings, although current varieties may root also without. Then 8–10 pots with three to five cuttings per variety are simultaneously placed on rolling tables that, when complete, are automatically conveyed to the rooting glasshouse (**Figure 4**). Under supplementary light, high temperature and high air humidity, cuttings will have rooted by week 2. After weaning, in the cultivation glasshouse shoots of the young plants are pruned back for the first time in week 5, to stimulate lateral formation. In weeks 9–10 the plants are pruned back for the second time. These prunings are the starting material for the next generation of softwood cuttings. Presently, specialized enterprises have been founded that only handle the pots up till the rooted stage, after which they are transported to growers who grow the plants on until flowering (**Figure 5**). Depending on the variety and the probable destination of the plants, three to four sprayings with a shoot-length inhibitor, e.g. Bonzi, are applied after the second pruning. Commonly, plants are wrapped in coloured plastic sleeves before being auctioned in weeks 14–16 (**Figure 6**). There is a trend in pot rose production to adapt the flower colour of varieties to the mood of the customer at the particular time of the year,

Figure 4 Just-rooted pot rose cuttings lifted from conveyor belt, on their way to rolling tables. Messrs. Pouw Rozen, De Kwakel, The Netherlands.

Figure 6 Pot roses in 'sleeves' on a trolley ready for auctioning. Messrs. Pouw Rozen, De Kwakel, The Netherlands.

Figure 5 A sea of floriculture pot roses about 10 weeks old. Messrs. Pouw Rozen, De Kwakel, The Netherlands.

Figure 7 Flowering 'Margo Koster', raised by (left) micropropagation or (right) softwood cuttings. Plant Research International, Wageningen, The Netherlands.

viz. white in winter, yellow or orange in spring and warm red in autumn.

Micropropagation

Besides softwood cuttings, in floriculture *in vitro* (micro)propagation is sometimes applied (*see* **Cell, Tissue and Organ Culture**: Micropropagation). Although several cultivars may readily multiply *in vitro*, the method is laborious and rather expensive. At Plant Research International, the performance of plants of 36 dwarf rose varieties, obtained from both *in vitro* propagation and softwood cuttings, has been compared (**Figure 7**). *Ex vitro* plants flowered significantly earlier, had shorter shoots, fewer and shorter internodes, and more and longer laterals than those from softwood cuttings. Differences for the number of bottom-breaks, number of flower buds, or number of petals were not significant. However, the growth rate of *ex vitro* plants was twice that of cuttings. In general, differences between the two types were in favour of the *in vitro* plants. Despite the fact that *ex vitro* plants had more shoots and were somewhat bushier than plants from cuttings, they look rather thin when planted solitary in a pot. Three or more plantlets per pot, however, give an attractive impression. Because the price of one *in vitro* plantlet is about the same as three cuttings, micropropagation in pot rose culture is considered too expensive for general use.

Traditional Culture by Softwood Cuttings

Contrary to the practice in floriculture, in traditional or 'nursery' practice, pot roses are propagated only once a year, which is in late summer. Softwood cuttings are taken from mother plants that are cultivated in unheated glasshouses for that purpose. Softwood cuttings, consisting of parts of the shoot with two leaves, are rooted in the glasshouse under double glass or sometimes mist or fog. Rooted plants are struck in wider pots than in floriculture, e.g. 15 cm. From late summer to spring plants are cultivated under glass, commonly with little heating. Constant pruning and slow growth make very attractive plants for Valentine's Day or Mother's Day. Pots are mainly sold in garden centres.

Traditional Culture by Grafting

Another way to produce attractive pot roses is by grafting varieties onto 6–8 mm *R. multiflora* seedling stocks (**Figure 8**). This trade is mainly practised in Dutch nurseries around Boskoop. Grafting, which is carried out in the glasshouse during winter, is comparable to the propagation of glasshouse varieties on seedling 'Inermis' stocks (*see* **Production and Marketing**: Cut Rose Production. **Rootstocks**: Usage of Rootstocks). When scion and stock have united, plants are potted, likewise in 15-cm pots, and grown in the glasshouse until auctioning, again for Valentine's Day or Mother's Day. Although somewhat thinner, the final product is comparable to the softwood cuttings of traditional culture, rather than to those of floriculture.

Figure 8 Dwarf roses in traditional culture raised by grafting on *R. multiflora* stocks. Messrs. Kuyf, Boskoop, The Netherlands.

Traditional Culture by Cutting–Grafting

Still another way to produce potted dwarf roses is by cutting–grafting (*see* **Propagation**: Stenting). As a stock, the shoots of thornless *R. multiflora* which have been harvested in the previous autumn are used. In late winter these shoots are cut into pieces of 30, 60 or even 90 cm length, which are subsequently rind-grafted (crown-grafted) on the distal side with two leafy scions of the variety. The proximal side is dipped in auxin (powder or solution), after which the cutting–graft is placed in a glasshouse under plastic at high temperature and high air humidity. Growing together of scion and the stock and rooting of the stock occurs within 3 weeks. Rooted plants are potted in 15–18-cm pots and further grown in the glasshouse until auctioned, again mainly for Mother's Day or early summer. Suitable varieties on so-called 'foot stems' (30 cm) or taller make highly attractive plants for small gardens. Because this method of propagation has been practised over 100 years in the Boskoop area of The Netherlands, nurserymen were quite sceptical about the 'new' method of stenting glasshouse roses (*see* **Propagation**: Stenting), it being an exact copy of their line of work.

Shelf-Life of Pot Roses

For continuous flowering, the performance of pot roses is good only when the plants are kept in full daylight, either indoors or outdoors. When well taken care of, potted dwarf roses may prosper for several years. However, when the plants are placed in centrally heated, badly aerated living rooms at low light intensities, shelf-life is seriously shortened. Under those conditions flowers will fade within a week and like the leaves, unopened flower buds will yellow and drop prematurely. Compared to most indoor plants, pot roses show poor shelf-life, and commonly are a disappointment for the consumer. Although dwarf roses do show genetic variation for shelf-life, this is hardly sufficient to meet the hardships of the living room. Current research aims at prolonging their shelf-life: rather than using genetics, experiments are under way to improve keeping quality by means of temperature regimes during cultivation and transport, the levels and kinds of macro elements in the nutrition, and various exogenously applied chemicals or phytohormones. As to temperature effects, it was demonstrated that compared to 24 °C day–18 °C night (which regime is not uncommon for floriculture pot roses), high production temperatures (29 °C day–24 °C night) increased postproduction flower longevity and decreased bud drop. In general, however, it is known that for roses high temperatures lead to smaller flowers and less vigorous plants. Postcultivation storage at low (4 °C) temperature for 4 days or longer improved shelf-life performance at 21 °C. A general rule seems to be that flower respiration correlates negatively with flower longevity in spring and summer, but positively in autumn and winter. Several independent experiments have shown that the postharvest life and decorative value of pot roses increases with a decreasing potassium : calcium ratio in the nutrition. Thus increasing calcium levels decreases the number of wilted flowers, particularly in the presence of low concentrations of NH^{4+} ions. Because pot roses are highly sensitive to the ageing hormone ethylene, which is one of the main causes of premature senescence, an important part of shelf-life research has been dedicated to ethylene effects, prevention and/or reduction. Ethylene is produced by the rose flowers themselves, but endogenous production is stimulated by exposure to exogenous ethylene. The ethylene inhibitor 1-methylcyclopropene (1-MCP) appears to be more effective in young than in senescing flowers. Likewise the application of 1 mmol l^{-1} silver thiosulphate (STS) significantly promotes flower display, but it does not reduce leaf drop. In the same experiment, the cytokinin benzyladenine (BA) alone only decreased leaf yellowing, but in combination with STS it improved postproduction quality of rose plants. Endogenous ethylene production of individual varieties was negatively correlated to period of shelf-life. Hence, the variety 'Vanilla', which was a low ethylene producer, showed longer shelf life than 'Bronze' which produced more ethylene. Other experiments showed that exogenous applications of the ageing hormone abscisic acid (ABA) promoted senescence by increasing endogenous ethylene production. The results strongly suggest that natural ABA level is raised by ethylene, either exogenous or endogenous. Remarkable effects have been observed from ethanol (30% + nutrients) sprayings or exposition to gaseous ethanol. Ethanol increased the number of flowers, as well as the fresh and dry weights of plants, and prolonged the flowering time of single flowers by 2 days.

Future Developments

As in cut or garden roses, novel developments in dwarf rose breeding are handicapped by the use of the same gene pool by all breeders. The creation of dwarf plants with large flowers remains an important aim. Since this can hardly be attained by breeding among existing dwarf varieties, crosses with large-flowered varieties seem to be obvious. Likewise, accidentally occurring 'large-flowered dwarfs' in for instance garden rose seedling populations may be used in subsequent crosses with true dwarf varieties. Further, when future European legislation on genetically modified organisms (GMOs) may become more liberal, dwarfing gene constructs could be inserted in Hybrid Tea varieties to obtain suitable patio plants and progenitors. In aiming at increased number of flowers per shoot, old spray varieties or even botanical species may be successfully used. Improvement of shelf-life, i.e. sustained flowering of pot rose plants under room conditions, seems difficult to attain just by breeding, unless perhaps species roses could be involved. In that context evergreen rose species from the *Bracteatae* (Thory) Rehd. section might be of interest. It may be anticipated, however, that chemistry rather than genetics will substantially contribute to future improvements in shelf-life. As to the latter, ideal growing substances like cytokinins or ethylene inhibitors, as well as specialized fertilizers, will doubtless be developed. Full automation and even robotization of rose culture, aiming at still lower costs of labour, will increase the demand for more uniformly rooting and flowering varieties, thus allowing exactly the same growing cycles in one glasshouse. Further adaptation of the flower colour and plant habit of varieties to the desires of the consumers is expected.

See also: **Cell, Tissue and Organ Culture**: Micropropagation. **Genetics:** Inheritance of Characteristics. **Production and Marketing:** Pot Rose Production; Cut Rose Production. **Propagation:** Stenting. **Rootstocks:** Usage of Rootstocks.

Further Reading

Cairns T (2000) *Modern Roses*, 11th edn. San Diego, CA: Academic Press.

Fitch CM (1977) *The Complete Book of Miniature Roses*. New York: Hawthorn/Dutton.

Garner RJ (1979) *The Grafter's Handbook*, 4th edn. London: Faber & Faber.

Harkness J (1978) *Roses*. London: J. M. Dent & Sons.

Hartmann DT, Koster DE and Geneve RL (1997) *Plant Propagation: Principles and Practices*. Englewood Cliffs, New Jersey: Prentice-Hall.

Jäger A (1936) *Rosenlexikon*. Reprint of the original edition by Zentralantiquariat der deutschen demokratischen Republik, Leipzig.

Krüssmann G (1986) *Rosen, Rosen, Rosen*. Berlin: Paul Parey-Verlag.

Le Grice EB (1965) *Rose Growing Complete*. London: Faber & Faber.

Pinney ME (1964) *The Miniature Rose Book, for Outdoor and Indoor Culture*. New York: van Nostrand.

Shepherd RE (1954) *History of the Rose*. New York: Macmillan.

Selection Strategies for Disease and Pest Resistance

R Noack, Guetersloh, Germany

Introduction

Among the rose diseases, black spot (*Diplocarpon rosae*), botrytis (*Botrytis* spp.), downy mildew (*Peronospora sparsa*), powdery mildew (*Podosphaera pannosa* var. *rosae*), rust (*Phragmidium mucronatum*) and crown-gall disease (*Agrobacterium tumefaciens*) have to be mentioned, with a differing importance for cut and garden roses. Thus powdery mildew is mainly important in cut roses whereas black spot and downy mildew are limiting factors in garden rose cultivation. Differing between growing regions and influenced by the number of plants grown in the same place, lack of high tolerance or resistance results in a severe reduction of plant material sold.

Regarding pests, nematodes, cyniped wasps (*Diplolepsis sinosa* and *D. radicum*), different aphids, different mites and thrips (mainly *Frankliniella* and *Thrips tabaci*) have to be taken into account. Although genetic variability for pest resistance and tolerance is available even among cultivated roses, actual selection strategies are concentrated on disease resistance. Reasons for this are quite different: a pronounced host alternation of most of the parasites, a linkage of pest susceptibility with desired rose variety characters and, as far as investigated, a complex genetic background of pest resistance. Specific selection strategies for pest resistance will therefore not be included in this contribution.

Due to changing plant protection legislation and the ecological awareness of consumers, the need to breed rose genotypes with at least high tolerance against the diseases and pests mentioned is a prerequisite to withstand customer rejection and to regain the economic basis to breed and produce roses.

Cultivated roses are mainly vegetatively propagated polyploid (tetraploid) plants. In general they exhibit cross-fertility that allows the combination of desired traits, although crossing barriers, production of unreduced gametes, apomixis and other mechanisms may significantly reduce hybridization rates. The resulting hybrids are highly heterozygous and selfing as well as out-crossing will result in polymorphic progenies, allowing the selection of desired types. But generative reproduction also destroys a once-selected genetic constitution, e.g. combinations of favourable characters.

An easy systematic combination of traits as in diploids cannot be expected in tetraploid rose progenies. Selection strategies in general are based on the generation of populations suitable for selection. As mentioned before, the majority of cultivated roses are highly heterozygous. The directed combination of desired traits and the minimization of undesired background effects are both time-consuming and laborious because very large populations have to be created.

Most of the economically important characters, like vigour, fragrance, vase-life and number of flower heads per plant, are polyfactorially controlled. As far as is known, disease resistances are inherited by only few, in some cases only single genes. This makes the strategy of combining resistance with other traits easier, at least according to the resistance situation.

This general approach has to be differentiated, as genotypic variability within populations of pathogens occurs. For example, populations of black spot include at least six races, and the composition of the population varies from area to area. The single-gene resistance reaction is therefore more or less an interaction between a rose and the local race mixture of the pathogen and may differ distinctly from one growing area to the next. The pathogen populations develop continuously, producing new races with lower or even higher virulence. Thus, in most cases, a race-specific resistance will last only a limited period of time.

Populations for Selection

The simplest population for selection consists of actual and recent rose varieties. As a consequence of the high heterozygosity and the inclusion of numerous, genetically divergent rose genotypes in past breeding programmes, a high tolerance or even resistance can be selected between existing varieties.

As breeders and growers are well informed about the resistance level of varieties, they can select those with a pronounced, long-lasting resistance reaction over years and, if given, a high ornamental value. The combination of these types results in populations with an increased resistance in some of the offspring candidates. As breeders may use different resistant varieties as parents and a high heterozygosity is given, the disadvantage of this approach is rather limited: the resistance basis will not be reduced as fast as for example in breeding of diploid cereals. In contrast to seed-propagated species, the selection pressure in vegetatively propagated roses is much lower, assisted by the heterozygous constitution.

As an extension of producing useful population for selection the back-cross method can be applied. This enables a breeder to accumulate resistance in one genetic background and simultaneously reduce undesired features within a system of recurrent selection.

If resistance has to be imported from outside the variety level or existing breeding material, distant crosses between adapted and nonadapted genotypes have to be performed. As an example, the introduction of black spot resistance from *Rosa multiflora* is described. *Rosa multiflora* is a diploid species which can only be crossed with the tetraploid garden roses after producing tetraploid *R. multiflora*. The scheme is as follows.

Colchicine treatment of *R. multiflora* plantlets *in vitro* resulted in tetraploid shoots, which were raised to flowering plants (**Figure 1**). One of them was crossed with the tetraploid variety 'Caramba'®. To avoid expected inbreeding depression the F_1 plants were crossed with the tetraploid variety 'Heckenzauber', and resulting individuals back-crossed with either 'Heckenzauber' or 'Caramba'. This modified back-cross strategy yielded donor lines including black spot resistance against a variety of races and, due to selection on morphological traits, pronounced a variety of characters. This material can be used as donor lines to transmit black spot resistance to different rose classes.

In addition to the combination of gametes in conventional *in vivo* systems the fusion of isolated somatic cells, protoplasts, opens an additional possibility to combine desired traits. This procedure is mainly of interest when crossing barriers cannot be overcome by applying other methods. For example, experiments have been performed to transfer black spot resistance from *R. wichurana* Crép. into cultivated roses. However, plants from such fusions could not be regenerated

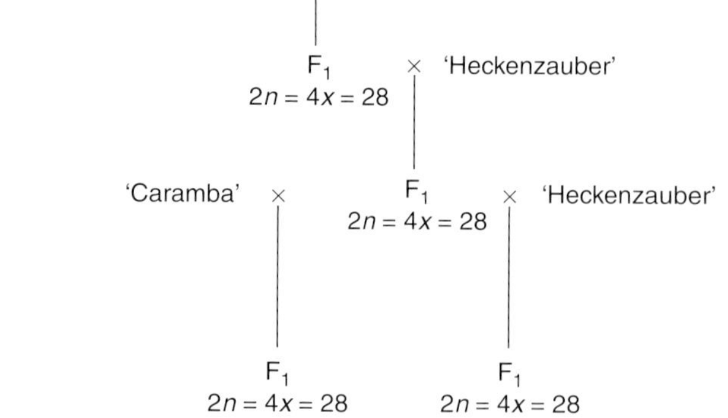

Figure 1 Introgression of black spot resistance from a diploid *Rosa multiflora* into tetraploid garden roses.

yet from hybrid callus. As a side aspect, protoplasts may be used for direct gene transfer (*see* **Biotechnologies for Breeding**: Genetic Transformation. **Cell, Tissue and Organ Culture**: Protoplast Culture).

As an alternative to the integration of resistance genes via sexual crosses, genetic engineering may be used. One advantage of plant transformation over conventional cross-breeding is the introduction of only desired resistance gene(s) without the often undesired gene background of the donor genotype. Another advantage is the possibility of transferring nonrace-specific genes from plant taxa other than *Rosa*, and even from other kingdoms. These genes are expected to ensure a long-lasting resistance because the interaction between pathogen and host plant completely differs compared with that of race specificity.

Potential candidates for nonrace-specific resistance genes against black spot and powdery mildew are, for example, those for ribosome-inhibiting proteins, chitinase and glucanases isolated from barley, the T_4-lysozyme gene from the T_4 phage, and an antisense construct of the flavonol-hydroxyl-transferase (FHT) gene. More examples and details can be found elsewhere in this Encyclopedia (*see* **Biotechnologies for Breeding**: Genetic Transformation. **Disease**: Black Spot).

As a result, in all transgenics produced in the black spot resistance experiments, 80% showed expression of the transgene. The expected reduction of black spot susceptibility was only detected after secretion of the ribosome-inhibiting protein into the apoplast, and the infection rate was as low as 60% compared with the susceptible control. The transgenics obtained partly expressed a reduced number of leaflets per leaf, a slightly increased number of petals and a reduction in male fertility. Thus these transgenics can only be useful as resistance donors. Calculating the costs of producing transgenics, their benefit in general will not be direct use as selected resistant varieties. They are only useful in cross-breeding programmes.

The transformation may be performed via *Agrobacterium* or in direct transfer. Regeneration of transgenic rose plants is possible by somatic embryogenesis from transformed leaf tissue or even after infiltration of rose flowers with a wild strain of *A. tumefaciens*, isolated from roses. There is evidence that harvested seed from these flowers results in transgenic roses.

Selection Strategies

There exist a variety of strategies to select resistant rose material. The following strategy depends on the segregating population produced, and the individual approach of the breeder. In general, one may distinguish between direct and marker-assisted selection strategies. Both are artificial and include several cycles which are not performed within one vegetation period.

Direct selection is based on visually detectable characters in plants. For example, one uses a seedling reaction on a spontaneous or induced mildew infection to select a resistant seedling. An indirect or marker-assisted selection is performed, when for example a weak petal is used as an indicator of short flower sustainability. Molecular markers represent a new generation of indicators to select indirectly desired traits. Up to now there have existed in roses rather few of these markers. As the efficiency to use them is obvious, their number will increase rapidly. Breeders are very interested to use early detectable characteristics as tools to reduce the large seedling populations grown as soon as possible.

However, marker-assisted selection has not been used in commercial rose breeding to date.

As a whole bundle of characters have to be combined in a potential variety candidate, neither direct nor indirect selection procedures on seedlings, viz. a single plant, can result directly in a variety. The selection cycles have to be extended to the official variety testing or patenting to release a variety.

General Selection Procedure

A selection procedure for garden-rose varieties is shown in **Table 1**. The whole selection cycle covers

Table 1 Scheme for garden-rose breeding

Year	*No. of plants and clones*	*Targets*	*Selection criteria*[b]
0		Crosses between selected parents, mainly firm-owned breeding material. Less than 5% varieties from other firms	High field resistance combined with superior garden-rose characters
1	About 100 000 seedlings	Selection of about 3000 seedlings in greenhouse	General positive seedling behaviour and types underrepresented in released varieties
2	Three per clone: about 3000 clones	Field test in breeding station	General behaviour of plant development; stability of resistance reaction; frost tolerance; flowering time and duration; flower colour and sustainability; fragrance; behaviour of cuttings versus ovulations in ground-cover roses, landscape roses and ramblers. Within rose groups: *Bed roses*: number of flowers per inflorescence; uniform blooming; first and following blooms flourishing at the same height *Hybrid teas*: flower shape and keeping quality; stability of flower colour; heat tolerance *Miniature roses*: compact growth; small plants and flowers *Landscape roses*: shoot length; solitary habitus *Climbing roses*: number and length of shoots; number of ground and side shoots *Ground-cover roses*: high number and length of branching shoots; flowering
3	15–20 per clone: about 300 clones	Field test in breeding station	
4	300 per clone: About 20–25 clones	Field trial in breeding station and other national stations (ADR)	As in years 2 and 3. In addition, productivity and adaptability to differing environments
5	300 and more per clone:[a] About 20–25 clones	Field trial in breeding station, national stations (ADR) and abroad. Selection of 3–5 candidates for official variety testing and US patenting	
6	300 and more per clone:[a] About 20–25 clones	Field trial in breeding station, other national stations (ADR) and abroad	
7	300 and more per clone:[a] About 20–25 clones		

[a] According to market demand.
[b] Strategies for selection of disease resistance are omitted for restrictions in space and are indicated in the text.
ADR, Allgemeine Deutsche Rosenprüfung – The General German Examination for New Roses.

seven vegetation periods, and in order to select for resistance no plant protection should be applied.

The cross parents included should be selected after an intensive observation over 10–20 years under different environmental conditions. As cross parents, either only firm-owned variability is used, excluding released varieties, or an extensive import from other breeders is performed. As a consequence of these different approaches, the genetic diversity between the gene pools of different breeders will either increase (in the first case) or decrease rapidly.

In addition to general variety characteristics, such as frost tolerance, flowering time and duration, vase-life, fragrance, performance and compatibility after grafting on rootstocks (**Table 1**), the main emphasis should be given to female and male fertility and the resistance behaviour during the whole time of observation.

Crosses are performed in year 0 in the greenhouse. To enable easy handling the selected parents are grown in pots. Emasculation of the female parent (**Figure 2**) will be performed 1 or 2 days before pollination with either freshly harvested or stored pollen. The emasculated and pollinated flowers are isolated against contamination with paper bags. The developed seed (**Figure 3**) is harvested during late autumn and after mechanical separation from the remaining hip it is sown in sterilized sand on tables in the greenhouse. During winter stratification takes place and heating of the greenhouse starts in early spring of year 1 (**Figure 4**, **Table 1**) to allow germination and seedling development (**Figure 5**). The germination rate of seeds varies from progeny to progeny, between 20% and 60%, indicating the genetic similarity of cross parents combined and the viability of seedlings expected. Ungerminated seeds are discarded.

During year 1 (**Figure 6**) a selection of about 3000 seedlings from a total of 100 000 is performed, i.e. about 3% of the seeds sown. Selection parameters are either direct ones, such as resistance reaction,

Figure 2 Emasculated seed parent.

Figure 3 Developing hips. Labels on the stems give details about the parents.

Figure 4 Seedlings in the greenhouse in February.

Figure 5 Seedlings in the greenhouse in March.

Figure 6 Seedlings in the greenhouse with first blooms in May.

flower colour and leaf shape (**Table 1**), or indirect ones, such as strong petals, number of leaves or molecular markers. In general, candidates will be favoured that display character combinations underrepresented in their own or other breeders' registered varieties.

From each selected seedling three buds are taken, inoculated on *R. canina* 'Pfänders' and planted in rows in the field of the breeding station. In some cases the remaining selected original seedling is raised on its own root to evaluate the interaction of rootstock and inoculated seedling genotype and to test for the production of unwanted suckers. These original seedlings are also planted in the field together with the respective clone. During year 2 a negative mass selection is performed, reducing the number of clones from 3000 to 300. Besides the general selection criteria (**Table 1**) there are rose group-specific characteristics. Thus potential bed roses are selected for the number of flowers per inflorescence, uniformity of blooming and height of first and second flourishing; hybrid teas are selected for flower shape, keeping quality, stability of flower colour and heat tolerance; while miniature roses should have compact growth, small leaves and flowers and long shelf-life. Landscape roses are selected for shoot length and solitary habitus; climbing roses should represent long and numerous shoots and sufficient branching near the ground; ground-cover roses should produce a large number of branching shoots and bright flowering above the leaves. At the same time the remaining clones are multiplied to obtain 15–20 plants per clone. The multiplication is performed by inoculation on to rootstocks of *R. canina* 'Pfänders' and for ground-cover roses, landscape roses and ramblers through cuttings. This gives information on the productivity of genotypes on its own roots versus grafted material and in addition winter-hardiness can be compared between both groups.

Year 3 ends with the selection of 20–25 clones, multiplied up to 300 plants per clone. They are inoculated on seedling rootstock *R. corymbifera* 'Laxa' and grown in the fourth year in field trials in the breeding station and, for example in Germany, at different locations for the General German Examination for New Roses (Allgemeine Deutsche Rosenprüfung or ADR roses).

As a first result of the field trials, some of the most promising clones are further multiplied on *R. corymbifera* 'Laxa' to be grown in the fifth year in local field trials as well as in national and international regions. About three to five potential variety candidates are selected from these trials and, for example in Germany, sent for official variety testing in Germany and patenting abroad.

In years 6 and 7 field trialling and variety testing are continued. The selection cycle ends up with (hopefully) one or more new released varieties and one or more patents. If a long-lasting, healthy variety is the breeding goal, the resistance should be of type 'field resistance'. Following experiences with varieties of other important crops, this is the most promising genetic constitution to withstand fast specialization of pathogens.

As a final calculation of the demonstrated general selection procedure, it can be concluded, that, from starting material of about 100 000 seedlings after a 7-year selection cycle, up to three varieties may be obtained (**Table 1**).

Selection for Resistance to Diseases and Pests

Selection for disease and pest resistance in roses can be performed either under laboratory conditions or under field conditions. For an efficient selection under lab conditions the following prerequisites have to be fulfilled:

- isolates of the major pathogenic organisms must be available
- infection assays for major pathogens and pests that make it possible to screen a very large number of seedlings have to be manageable within small and medium-sized companies.

As only a selection of pathogenic isolates for only a few pathogens are available for research purposes, a screening for broad resistance against all the relevant pathogens and pests is not possible under laboratory conditions with the given financial resources of most rose breeders. Therefore, screens for resistance are almost exclusively performed in the field. For garden roses the main diseases to be considered are black spot, powdery mildew, rust and *Spacaeloma* (*see* **Disease:** Black Spot; Downy Mildew; Powdery

Mildew; Rust). For pests screens are mainly done for thrips, as spider mites are less problematic in the field then in the greenhouse (*see* **Insects and other Animals**: Thrips). A major prerequisite for resistance screens in the field is cultivation under pesticide-free conditions. Therefore several breeders have started to test their promising genotypes on fields which have been left untreated for a minimum of 3 years. Practical experience has shown that this time is required for the infection pressure of the various pathogens to build up. In addition to the omission of fungicides, one should cultivate clones in different soils, including suboptimal nutrition. This ensures that resistance or tolerance to the important pathogens and pests remains stable over a wider range of conditions for cultivation. After 3 years, promising candidate genotypes will be further tested at other sites, including the General German Examination for New Roses (ADR), and this is generally performed under pesticide-free conditions.

Marker-assisted Selection

Marker-assisted selection may be performed starting with the first segregating population. While for a very long time only morphological and/or physiological characters were used for selection, more recently molecular markers have been developed. These have the advantage of being detectable at a very early stage of organism development, and in plants only some milligrams of fresh material are needed to detect them. Thus, one of the first pair of leaflets can be used for fingerprinting and detection of the markers, closely linked to the desired character(s).

As resistance is often found to be controlled by a single gene or major genes together with modifying genes, this character possesses ideal prerequisites for a marker-assisted selection. It allows all seedlings lacking a desired resistance gene at the two-leaf stage to be eliminated.

At the moment relatively few molecular markers linked to resistance genes in roses are known; examples are resistance to black spot and powdery mildew (*see* **Disease**: Black Spot; Powdery Mildew. **Genetics**: Molecular Markers). This situation may rapidly change when breeders become interested in this tool for early selection.

If desired characters are inherited quantitatively, such as the number of shoots and flowers, more than one gene is responsible for the character expression. A great deal of effort is being put into finding these so-called QTLs (quantitative trait loci) and genotypic markers. Unfortunately the success rate, as with plants other than roses, is very poor.

If desired genetic variability has to be established after an interspecies cross, for example *R. multiflora* × garden-rose variety, a series of back-crosses with garden-rose varieties is needed to eliminate undesired traits of the wild species parent. An elegant method of selecting individuals with the lowest amount of the donor species genome is offered by the molecular quantification of *R. multiflora* and the garden-rose genome portion in back-cross individuals. This can be achieved by developing molecular markers specific for only one of the cross parents, respectively. Calculating the amount of *R. multiflora* specific markers in a second back-cross generation with a garden rose, resistant individuals with only about 10% of the *R. multiflora* markers were found. As these were not detected by visual inspection of breeders, the advantage of marker-assisted selection is obvious for continuing back-crossing.

Conclusion

Decades ago the market favoured exclusively garden-rose varieties developed and grown under intensive chemical protection from diseases and pests. Enhanced ecological awareness and the drastic reduction of plant protection as a consequence of changing regulations now offer the basis for a broad acceptance of healthy roses in the market.

The breeding concept to produce adequate, highly resistant varieties follows a very traditional procedure with overwhelming success. The integration of modern techniques to produce and select desired genetic constitutions has only been sporadically applied, although its value to increase breeding efficiency is obvious and demonstrated. Probably it needs more time to calculate expenditure and returns before investing in these techniques. Concerning gene technology, however, a broad rejection in developed countries is the main reason for rose breeders to avoid this technique.

See also: **Biotechnologies for Breeding**: Genetic Transformation. **Cell, Tissue and Organ Culture**: Protoplast Culture. **Disease**: Black Spot; Downy Mildew; Powdery Mildew; Rust. **Genetics**: Inheritance of Characteristics; Molecular Markers; Marker-Assisted Selection; Research Trends. **Insects and other Animals**: Thrips.

Further Reading

Debener T (2001) Molecular tools for modern ornamental plant breeding and selection. *Acta Horticulturae* 552: 121–127.

Debener T, Malek Bv, Mattiesch L and Kaufmann H (2001) Marker assisted selection for blackspot resistance in roses. *Acta Horticulturae* 547: 349–352.

Debener T, Mattiesch L and Vosman B (2001) A molecular marker map for roses. *Acta Horticulturae* 547: 283–287.

Dohm A, Ludwig C, Nehring K and Debener T (2001) Somatic embryogenesis in roses. *Acta Horticulturae* 547: 341–347.

Ferrero F, Cadour-Marvaldi P, Guilloteau E *et al.* (2001) Evaluation of the resistance to powdery mildew, *Sphaerotheca pannosa* var. *rosae*, of rose-tree species and hybrids. *Acta Horticulturae* 547: 377.

Poncet C, Bonnet G, Pionnat S *et al.* (2001) Spread of crown gall disease in rose cultures. *Acta Horticulturae* 547: 75–81.

Schum A and Hofmann K (2001) Use of isolated protoplasts in rose breeding. *Acta Horticulturae* 547: 35–44.

Shorthouse JD (2001) Galls induced by cynipid wasps of the genus *Diplolepis* (Cynipidae, Hymnoptera) on cultivated shrub roses in Canada. *Acta Horticulturae* 547: 83–92.

Zieslin N and Agbaria H (eds.) (2001) Proceedings of the Third International Symposium on Rose Research and Cultivation. *Acta Horticulturae* 547.

C

CELL, TISSUE AND ORGAN CULTURE

Contents

Micropropagation

A V Roberts, School of Health and Bioscience, University of East London, London, UK
A Schum, Federal Centre for Breeding Research on Cultivated Plants, Institute of Ornamental Plant Breeding, Ahrensburg, Germany

Introduction

Micropropagation is the clonal propagation of plants in closed vessels under aseptic conditions. Inside the vessels, the plants are grown on culture media that contain nutrients and growth regulators, and are described as *in vitro*, which means 'in glass'. In contrast, soil-grown plants are described as *in vivo*. Plants that are propagated *in vitro* are smaller than *in vivo* plants, hence the term 'micropropagation'. Rose cultivars are clonally propagated and are, therefore, suitable subjects for micropropagation. The most commonly used method of micropropagating roses is based on shoot multiplication, which is the *in vitro* equivalent of *in vivo* propagation from cuttings. It is also possible to produce plants from cells and tissues through the formation of adventitious shoots and somatic embryos (*see* **Cell, Tissue and Organ Culture**: *In Vitro* Regeneration Techniques). However, apprehensions of genetic instability have discouraged the widespread use of these methods in rose micropropagation. In this article, culture methods, characteristics of micropropagated plants and prospects for rose micropropagation by shoot multiplication are discussed.

Culture Media and Aseptic Technique

Culture media include formulations of salts and vitamins, sugar and growth regulators. The pH is adjusted to 5.6–5.8 and, if a gelling agent is to be used, it is added after the pH has been adjusted. The media are then sterilized by heating in an autoclave to 121°C under pressure (106 kPa) for 15 min. Various general-purpose formulations of salts and vitamins can be used but a satisfactory medium that was adapted specifically for micropropagation of roses is shown in **Table 1**.

The routine inclusion of calcium gluconate (6 mmol l^{-1}) in culture media is strongly recommended because it has markedly beneficial effects on some types of rose (as described, below, under stage 2 and stage 3) and has no detrimental effect on others. Sucrose, at a concentration of 30–40 g l^{-1}, is the most commonly used sugar.

The term 'growth regulator' includes naturally occurring hormones and synthetic substances with similar properties to hormones. The main categories of growth regulators used in rose micropropagation are the cytokinins and auxins. The cytokinins

Table 1 A formulation of salts and vitamins for the micropropagation of roses

Salts containing macroelements ($mg\,l^{-1}$)	
KNO_3	1900
KH_2PO_4	170
NH_4NO_3	1650
$CaCl_2 \cdot 2H_2O$	440
$MgSO_4 \cdot 7H_2O$	370
Salts containing microelements ($mg\,l^{-1}$)	
$MnSO_4 \cdot 4H_2O$	22.3
H_3BO_3	6.2
$CuSO_4 \cdot 5H_2O$	0.025
$CoCl_2 \cdot 6H_2O$	0.025
$ZnSO_4 \cdot 7H_2O$	8.6
KI	0.83
$Na_2MoO_4 \cdot 2H_2O$	0.25
FeEDDHA	96.0
Vitamins and amino acids ($mg\,l^{-1}$)	
Nicotinic acid (B_3)	0.5
Pyridoxine HCl (B_6)	0.5
Glycine	2.0
Inositol	100.0
Thiamine HCl (B_1)	0.1

Based on a modification (by van der Salm TPM, van der Toorn CJG, ten Cate CHH *et al.* (1994) Importance of the iron chelate formula for micropropagation of *Rosa hybrida* L. 'Moneyway'. *Plant Cell, Tissue, Organ Culture* 37: 73–77) of the formulation of Murashige T and Skoog F (1962) *Physiologia Plantarum* 15: 473–497.

stimulate shoot multiplication, whereas auxins promote stem elongation and rooting. Synthetic growth regulators, such as the cytokinin 6-benzyladenine purine (BAP) and the auxins α-naphthylene acetic acid (NAA) and indole-3-butyric acid (IBA) are often preferred to naturally occurring hormones because they are more stable during heat sterilization and the culture of plants at room temperature. The gelling agents agar ($8\,g\,l^{-1}$) or phytagel (also known as gelrite: $2.5\,g\,l^{-1}$) are included in the medium to provide a semisolid medium that supports the plant so that its shoots do not become submerged. Agar is a natural product derived from seaweed and gives a semiopaque medium. Phytagel is a product of bacterial fermentation and gives a much clearer medium. Phytagel has a higher matric potential than agar and, in some plant species, tissues become hyperhydrated and shoot growth is distorted. However, in roses, it gives an appropriate degree of hydration and it is recommended in preference to agar.

If heat-resistant culture vessels are used, the culture medium is heated to melt the phytagel, dispensed into the culture vessels, then autoclaved. Otherwise, the medium is first autoclaved in a heat-resistant vessel then dispensed, whilst still molten, into sterile disposable culture vessels. If heat-labile hormones or other additives are used, they must be filter-sterilized and added after autoclaving. One type of culture vessel used for the micropropagation of roses is a honey jar of 300 ml capacity (Freeman & Harding, Erith, UK) with a screw-on lid of heat-resistant plastic (Roberts' Capsule Stopper, London), to which 60 ml of culture medium is added. Honey jars have the advantage of low cost, resulting from their mass production, and they are reusable.

All manipulations of the plant material are carried out with sterile instruments in a laminar air-flow cabinet that blows sterile air over the working surface. Instruments such as scalpels and forceps can be wrapped in foil and autoclaved or heated in a hot-air oven at 180 °C. However, instruments should be repeatedly sterilized during cutting operations and the use of freshly sterilized autoclaved instruments is impractical. It is, therefore, usual to flame instruments intermittently and to rest them, briefly, on a sterile rack to cool. There is some uncertainty as to the best method of carrying out this intermittent sterilization. It was once common practice to dip the instruments into absolute alcohol before flaming, but this operation can be hazardous. Spillages occur and there have been incidents in which laminar air-flow cabinets and clothing have been set on fire! A safer procedure is to flame the instruments over a Bunsen burner, periodically scraping off encrusted plant remains. The cutting surface should be sterile and frequently replaced to avoid the carry-over of contamination from one plant to another. In small-scale operations, it is convenient to use sterile Petri dishes as cutting surfaces. In commercial operations, costs are reduced by a variety of ingenious methods, including the use of autoclaved tiles and cardboard.

Culture Conditions

The cultures are best maintained in culture rooms that are illuminated for 16 h per day. A temperature of 20 °C is suitable for rose micropropagation. The most efficient use of space in the culture room is achieved by stacking the shelves and arranging fluorescent tubes over each shelf. However, heat from the tubes rising to the shelf above creates problems. Water evaporates from the culture medium, condenses on the lid of the vessel and drips back on to the medium. This can cause variations in the concentration of substances in the culture medium and contaminating organisms spread rapidly. Care should, therefore, be taken to minimize this transfer of heat. The problem can be overcome by circulating cool water in pipes laid within the shelves. Bottom cooling, adjusted so that the medium is 2 °C lower than the temperature of the air in the culture vessel, results in a lowering of the relative humidity in the culture vessel from 100% to about 96%. This leads to an improvement in the hardiness of plants. If sucrose is included in the medium, the primary requirement of light is to prevent etiolation of the shoots, for which a photosynthetic photon flux density (PPFD) of $10\,\mu mol\,m^{-1}\,s^{-1}$ is adequate. If sucrose is excluded, with the objective of obtaining autotrophic plants (see stage 3, below), a higher PPFD is needed.

Stages of Micropropagation

There are five main stages of micropropagation by shoot multiplication. Stage 0 is the preparation of plants for introduction into culture, stage 1 is the introduction of plant material into culture, stage 2 is multiplication of shoots, stage 3 is the preparation of plants for transfer to soil and stage 4 is the acclimatization of plants after transfer to soil.

Stage 0

Stem pieces (explants) that are to be introduced into culture should be taken from vigorously growing shoots of healthy plants. The plants should be inspected to ensure that they are good representatives of the species or cultivar and that there is no evidence of infection. The required level of plant care is best achieved in a well-managed greenhouse, in which preventive action is taken against pests and pathogens.

Stage 1

Explants consist of shoot tips and nodal sections. These are trimmed, open leaves are removed from shoot tips and nodal sections are prepared by removing the leaf at the node and cutting the stem above and below the axillary bud. The explants are then surface-sterilized. One protocol is to wash the explants in running tap water for 1 h, dip them for 10 s in 70% ethanol, immerse them for 15 min in solution of sodium hypochlorite (0.5% available chlorine), then thoroughly wash in three changes of sterile (autoclaved) distilled water, each of 15 min duration. The initial washing in water reduces the number of external microorganisms and the soaking improves the killing rate during treatment with alcohol and sodium hypochlorite. The household bleach, Domestos, which contains 2% available chlorine, is a convenient source of sodium hypochlorite and includes a surfactant. Explants are then inoculated on to the surface of a multiplication medium.

The appearance of fungal or bacterial colonies juxtaposed to the explant indicates that surface sterilization has not been completely effective. If there are several explants per culture vessel, uncontaminated explants should be transferred to a fresh vessel and

Figure 1 Stage 1: initiation of rose explants in a honey jar, showing axillary shoots developing from nodal sections.

contaminated explants discarded. Cross-contamination can be avoided by the use of disposable Dilu vials with translucent plastic clip-on lids (LIP, Shipley, UK). They are of suitable size (30 ml capacity, 10 ml culture medium) to accommodate a single explant and the cost per vial is low. Growth of shoot tips or axillary buds should be apparent (**Figure 1**) 2–4 weeks after initiation.

Surface sterilization does not eliminate endogenous contaminants that may be present in the internal tissues. Often these contaminants are inconspicuous but can be detected as stains and discolourations in the culture medium emanating from the cut surfaces of the explants. They are deleterious and the use of endogenously contaminated explants should be avoided. Phytagel forms a clearer gel than agar and its use facilitates detection of endogenous contaminants. A sensible strategy is to introduce many explants into culture and discard all those with endogenous contaminants. If the number of explants is limited because of the rarity of a type of plant, endogenous contaminants can be eliminated by treatment with antibiotics (**Table 2**). These should be filter-sterilized and added to culture medium that has already been autoclaved. It may be necessary to expose shoots to this medium for 14–21 days to eliminate the contaminants.

If a viral infection is suspected, cultures should be initiated from shoot meristems. These can be excised using a stereo microscope. Viruses do not usually invade plant meristems and unpublished studies have shown that explants consisting of shoot meristems and one to two leaf primordia are virus-free.

Stage 2

Multiplication of shoot cultures involves the stimulation of axillary buds so that they develop into shoots. This can be done by including BAP in the culture medium at concentrations of $1\text{–}4\,\mu mol\,l^{-1}$. High

Table 2 A combination of antibiotics for the treatment of endogenous bacterial contaminants of *in vitro* plants

Antibiotic	*Concentration ($mg\,l^{-1}$)*
Tetracycline	25
Polymyxin B	25
Rifampicin	6
Cefotaxime	6

Data from Young PM, Hutchins AS and Cranfield ML (1984) Use of antibiotics to control bacteria in shoot cultures of woody plants. *Plant Science Letters* 34: 203–209.

concentrations of BAP may give faster multiplication rates but may adversely affect rooting at stage 3. One cultural approach, the subdivision method (**Figure 2a**), is to subdivide the plants into their constituent shoots, reculture each shoot on fresh medium and discard the residual pieces. This procedure can give a multiplication two- to fivefold every 4–6 weeks. If 100 honey jars, occupying 0.74 m^2 of bench space, are used and each honey jar contains five shoots, a steady-state production of up to 24 000 shoots per annum can be maintained. An alternative approach, the mother-plant method, is to excise shoot tips of defined size (two to three opened leaves), retain the basal part of the plant for further rounds of culture and use the shoot tips to initiate stage 3 cultures. The shoot tips are removed at intervals of 2–3 weeks. The basal part grows into a mother-plant (**Figure 2b**) that becomes progressively more productive until, after about 6 months, its size indicates that it should be subdivided into smaller units. The basal part can be transferred to a fresh culture medium every 6–8 weeks or the same medium can be reused and the plantlet revitalized by the addition of liquid medium. A single mother-plant can produce, on average, 28 or more shoot tips per month and 100 mother-plants (one per honey jar), occupying a bench surface of 0.74 m^2, gives a steady-state production of 34 000 cuttings per annum. The removal of shoot tips releases axillary buds in the mother-plant from the repressive effects of apical dominance. This enables the development of axillary shoots to occur at low concentrations of BAP; low concentrations of BAP at stage 2 facilitate rooting at stage 3.

The concentration of cytokinin needed to ensure adequate development of axillary shoots varies between cultivars and species. For introductions into culture (stage 1) and stage 2 multiplication by the subdivision method, BAP at a concentration of 2 $\mu mol\,l^{-1}$ is adequate for most species and cultivars of rose. For stage 2 multiplication by the mother-plant method, 1 $\mu mol\,l^{-1}$ is adequate. An auxin, usually NAA (0.2–0.5 $\mu mol\,l^{-1}$), is sometimes included in the multiplication medium but this may be unnecessary and is positively deleterious for some roses, including *Rosa canina*.

Figure 2 Stage 2: shoot multiplication. (a) The method in which plantlets are subdivided into constituent shoots and the remaining pieces discarded; (b) the mother-plant method in which shoot tips are excised and subcultured and the remaining plant is recultured.

Some types of rose require more calcium salt than is provided in the medium described in **Table 1**. Calcium deficiency results in shoot-tip necrosis and hyperhydration of stem and leaf tissues. These effects can be overcome by the addition of calcium chloride but, as this salt has toxic side-effects, the use of calcium gluconate is preferred. In those roses that express calcium deficiency, the inclusion of calcium gluconate (6 $\mu mol\,l^{-1}$) in the multiplication medium significantly increases shoot multiplication rates and stem length at stage 2 and increases the number of roots formed at stage 3. It has no negative effects on roses that do not suffer from calcium deficiency, so its routine inclusion in the multiplication medium is recommended.

Cultures can be stored, without subculture, for periods of 6 months in an incubator at 4 °C. Shoots grow, although slowly, at this temperature, so lighting is needed to prevent etiolation.

Stage 3

Shoot-tips consisting of the terminal bud and two to three opened leaves can be induced to form roots (**Figure 3a**) on a medium containing salts and vitamins at half the strength used in multiplication medium, and a combination of the auxins NAA (0.25 $\mu mol\,l^{-1}$) and IBA (0.25 $\mu mol\,l^{-1}$). The proportion of shoots that form roots is influenced by the culture medium at stage 2; in particular, it is inversely related to the concentration of cytokinin and, in some roses, it is promoted by the presence of calcium gluconate (6 $mmol\,l^{-1}$). Rooting can be induced in semisolid medium but can also occur in liquid medium after shoots have been inserted into Sorbarod plugs (Baumgartner Papiers, Lausanne; **Figure 3b**).

Sucrose can be excluded from the culture medium at stage 3, provided that there is sufficient light and CO_2 to promote photosynthesis. In a sealed container, the concentration of CO_2 falls to a low level during the photoperiod and photosynthesis is negligible. In order to maintain adequate concentrations of CO_2, the culture vessel must be designed to allow CO_2 to enter by diffusion or filtered air must be pumped in. Culture vessels with air-permeable filters that are commercially available include Magenta B-caps (Sigma-Aldrich, Poole) that are designed as closures for baby-food jars. In order to obtain sufficient light at low cost, culture vessels can be maintained in a partially shaded greenhouse rather than a culture room. In experiments in which plants benefited from enhanced PPFD and CO_2, plants were found to form larger leaves and better roots on media without rather than with sucrose. Relative humidity in the culture vessel is reduced if there is gaseous exchange and may be further reduced if the culture vessels are bottom-cooled. This conditions the plant to resist desiccation at stage 4.

(a)

(b)

Figure 3 Stage 3: stem elongation and rooting of plantlets. (a) On phytagel-solidified medium; (b) in Sorbarod plugs (roots obscured).

Stage 4

Micropropagated plants are vulnerable at the time of transfer to soil. If plants are maintained in a water-saturated atmosphere at stage 3, they lose water rapidly from the leaves at stage 4. This is because the stomata are unable to close in response to water deficits, the cuticle is thin and the surface wax that is needed to prevent loss of water is poorly developed. Uptake of water is limited if the roots are damaged during transfer to soil. If sugar is supplied at stage 3, the plant is unable to fix sufficient CO_2 to maintain itself and depends heavily on stored food reserves. Restoration of normal photosynthetic ability only occurs in the new leaves that are formed after the plantlet has been removed from the culture vessel. Thus, plantlets that have been rooted in agar, without stage 3 treatments to promote hardiness, require acclimatization (weaning) in a misting or fogging unit after transfer to soil.

Plantlets that have been rooted in Sorbarod plugs can be transferred to soil in the plugs. They are protected during transfer to soil and, in the short term, the plug provides the plant with a reservoir of water. If the plantlets have also been exposed to a relative humidity as low as 96% at stage 3, they may be sufficiently vigorous and resistant to desiccation to be transferred to soil in a greenhouse or garden without misting or fogging (**Figure 4**). After 2 months in soil, a saleable plant is obtained (**Figure 5**).

Performance of Micropropagated Roses

Traditionally, rose plants have been propagated on rootstocks which affect traits such as plant architecture, vigour, resistance to adverse environmental conditions, flower yield and quality (*see* **Rootstocks**: Scion–Rootstock Relationships; Seedling Stocks). However, knowledge of the influence of rootstocks on the characters of rose scions is sparse compared with fruit trees. Due to advancing glasshouse technology and the development of new culture methods, the interest in own-rooted rose plants has steadily increased during the last three decades. Numerous rose cultivars and species are amenable to propagation by cuttings. This is especially true of pot roses and container-grown plants of bush and ground-cover roses, which are sold during the growing season for use in private gardens, public parks and landscaping. Such roses, growing on their own roots, require less maintenance, as removal of suckers from the rootstock becomes unnecessary. In the case of hybrid garden roses, most breeders and producers still favour grafted material, which is said to be more vigorous and free-flowering. However, some nurseries successfully grow all rose classes, including the Hybrid Teas, on their own roots. With regard to greenhouse-grown roses for cut-flower production, it was argued that grafted plants are longer-lasting and more productive in winter. However, with the development of soil-less culture systems, increasing numbers of plants are grown on their own roots.

As micropropagated roses are own-rooted, they have to compete with material produced from cuttings. The decision of the producer depends on the assessment of the suitability of a cultivar for a particular propagation method, production costs and plant quality. There are few published reports on experiments which compare *in vitro*-produced roses with cuttings and grafted plants, respectively. In addition, in many cases the collected data must be carefully interpreted, as the plant material was not of comparable developmental state at the beginning of greenhouse or field trials. Furthermore, uniform trial conditions may have favoured one or the

Figure 4 Stage 4: rooting of plantlets in a multiplug tray (Baumgartner vessel). The lid has been removed to display the plantlets in Sorbarods. On each side of the lid are five holes covered with a bacterial filter (Tyvek). Air exchange through the filter results in reduced humidity in the vessel and the raised CO_2 concentration permits photosynthesis.

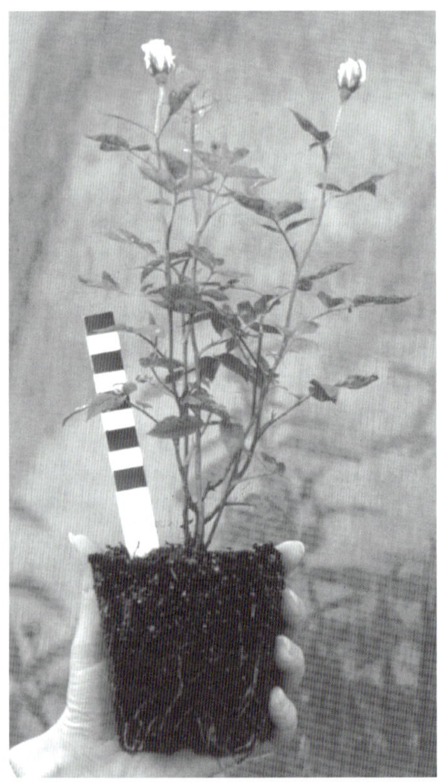

Figure 5 Plant of 'Clarissa'® 3 months after weaning (tag marked in segments of 10 mm).

other material. **Table 3** summarizes the available information.

Plant Architecture

Plants derived from *in vitro* culture (*ex vitro* plants) are frequently described as being more vigorous and more compact than cuttings or budded material (**Table 3**). Typically, shoots are initially shorter due to fewer and shorter internodes but there are many more axillary bud-breaks. In some cases, this higher branching potential was associated with a reduced diameter of the lateral shoots. Such characters, typical for micropropagated roses may, at least partly, result from the aftereffects of cytokinins that were included in the culture media. The extent of continuing axillary shoot formation depends on the type and concentration of the phytohormones applied, and on the sensitivity of the given cultivar. Adjustment of the cytokinin concentration to the specific demand of a given cultivar may allow production of plants of optimal quality.

Roots

It has been claimed that *ex vitro* plants develop more branched root systems, similar to those of seedlings (**Table 3**). Furthermore, the rooting capacity of cuttings derived from micropropagated stock plants was found to be generally superior to that of conventionally propagated ones. In general, the positive effects of micropropagation were smaller in cultivars of superior rooting ability. Improved rooting competence, often associated with rejuvenation, has frequently been observed in woody plant species after repeated *in vitro* subculture on multiplication media. Interestingly, the sprouting of the scions was promoted by rootstock cuttings taken from *in vitro*-derived plants.

Flower Yield and Quality

Observations on flower yield are inconsistent. Reduced flower bud setting, considerably increased flower yields as well as no significant differences between micropropagated plants, cuttings and budded material have been reported (**Table 3**). In some cases enhancement or diminution of flowering of micropropagated plants was found to be dependent on the cultivar.

In several experimental designs, *in vitro*-produced plantlets were too small at the time of flower bud setting to be fairly compared with grafted plants. In such cases flower yields eventually became equal after establishment of plants. In general, micropropagated rose plants appear to have flower yield potentials similar or superior to those of grafted plants or cuttings.

In several investigations, no significant differences in flower quality were observed with regard to the size of buds or number of petals per flower (**Table 3**). However, differences in length of flower stalks were reported between *in vitro*-derived and conventionally produced plants used for cut-flower production. This phenomenon is attributed to the high frequency of lateral bud-breaks of the *ex vitro* plants, as the length of flower stalks correlates inversely with their quantity. Appropriate disbudding adjusted to the developmental characteristics of *in vitro*-derived plants may result in production of cut flowers of first-class quality.

Physiological Traits

Earlier flowering of micropropagated rose plants was observed as well as delayed flower bud setting (**Table 3**). The latter can be partly explained by differences in the developmental state of plants obtained by different methods of propagation at the time of forcing. Increased growth rates of *ex vitro* plants were noted in several cases. No problems have been reported in respect of winter-hardiness of well-established micropropagated plants. Furthermore, in some cases increased resistance to low temperatures was observed. When plants die back to the ground in severe winters, typically own-rooted roses come back in the spring true to type, whereas grafted material forms suckers from the rootstocks.

Prospects for Rose Micropropagation

Micropropagated roses have the potential to be commercialized in all market sectors, including production of pot, garden and landscape plants as well as for cut-flower production. Furthermore, as micropropagation is a very rapid, all-year-round method of propagation, it can be used to build up numbers rapidly before the launch of new cultivars. Thus, substantial investment by a breeder in the development of a new cultivar can be turned to profit with minimal delay. International travel is facilitated by the small size of the plants *in vitro* and their disease-free condition minimizes problems with phyto-sanitary inspectors.

The quality and cost of weaned plants are factors that will determine the market share of micropropagated roses. In several studies, *ex vitro* plants proved to be superior in vigour, plant architecture and flower production. The best results may be seen with modern cultivars that have been selected, in part, for their performance on their own roots, rather than with the older cultivars that were selected only

Table 3 Selected reports on performance of *in vitro*-propagated plants

Plant category (number of cultivars investigated) Comparisons (where specified)	*Characters of* ex vitro *plants*			*Source*
	Plant architecture/root system	*Flower yield/quality*	*Physiological traits*	
Pot roses (1 cv.) *Ex vitro* vs budded material	After 1 year, *ex vitro* plants smaller and diameter of shoots reduced	Reduced flower bud setting	—	Tweddle D *et al.* (1984) In: Novak FJ, Havel L and Dolezel J (eds.) *Plant Tissue and Cell Culture Application to Crop Improvement*, pp. 529–530. Prague: Czech Academy of Science
Pot and tub roses (4 cvs) General observations	More compact	—	—	Onesto JP *et al.* (1985) *Agronomie* 3: 303–306
Pot roses (36 cvs) *Ex vitro* vs single-node softwood cuttings	Shorter shoots with fewer and shorter internodes; more axillary bud-breaks and longer laterals; similar number of bottom-breaks	Number of flower buds and petals per flower similar	Growth rate almost double that of cuttings; earlier-lowering	Dubois LAM *et al.* (1988) *Scientia Horticulturae* 35: 293–299
Garden roses (many cvs) *Ex vitro* vs budded material	Superior vigour; more laterals	Flower yield increased by 10–20%	Faster growth; earlier-flowering; more resistant to low temperatures	Anonymous report on experiments at Delbard nursery *Zierpflanzenbau* (1984) 589–590
Garden roses; greenhouse roses (cut-flower production) General observations	More compact; superior vigour; more laterals; well-branched root system of seedling type	—	—	Angiboust A (1985) *Revue Horticole* 258: 43–44
Garden roses General observations	Well-branched root system extending deeper into soil	—	Increased resistance to low temperatures and frost	Martin C *et al.* (1983) *Agronomie* 3: 303–306
Garden roses General observations	—	—	Good performance over several years, including two severe winter periods	Horn WAH (1992) (In: Bajaj YPS ed.) *Biotechnology in Agriculture and Forestry*, vol. 20, pp. 320–342. Berlin: Springer-Verlag
Garden roses (7 cvs) *Ex vitro* vs grafted material	Vigour less in the first year, but equal or superior later; more compact (depending on cultivar)	Lower flower yield in the first year, however eventually equal when plants become well established in soil	No difference in disease susceptibility and frost tolerance	Reist A and Lê LC (1989) *Revue Suisse de Viticulture, d'Arboriculture et d'Horticulture*, 21: 19–23
Garden roses (1 cv.) *Ex vitro* vs stem cuttings and budded material	More laterals but of reduced diameter; more bottom-breaks	Flower yield considerably increased	—	Waldenmaier S and Spethmann W (personal communication)

Ground-cover roses (4 cvs) *Ex vitro* vs stem cuttings and budded material	Low habit due to horizontal growth of laterals; shoots shorter and fewer in the first 2 years of culture but eventually equal or superior (depending on the cultivar).	—	—	Wisniewska-Grzeszkiewicz H and Podwyszynska M (2001) *Acta Horticulturae* 547: 371–376
Greenhouse roses for cut-flower production *Ex vitro* vs grafted material	—	Flower yield increased by 10% in the first year and by 20% in the second and third years of culture	—	Martin C *et al.* (1981) *Comptes Rendus de l'Académie de Science Paris* 293: 175–177
Greenhouse roses for cut-flower production (2 cvs) *Ex vitro* vs grafted material	—	Productivity equal or slightly inferior; quality (length of flower stalks) superior at the beginning, inferior at the end of the trial period	Flowering delayed by 3 weeks	Reist A (1985) *Revue Suisse de Viticulture, d'Arboriculture et d'Horticulture* 17: 173–176
Greenhouse roses for cut-flower production (2 cvs) *Ex vitro* vs cuttings	Comparable in vigour; increased branching (depending on cultivar)	Productivity equal or superior (by 20%) depending on cultivar; inferior quality (shorter flower stalks)	Flowering delayed by 2 weeks	Reist A (1985) *Revue Suisse de Viticulture, d'Arboriculture et d'Horticulture* 17: 361–364
Greenhouse roses for cut-flower production (3 cvs) *Ex vitro* vs cuttings and grafted material	More bottom-breaks	High flower yield Inferior quality (length of flower stalks and fresh weight)	—	De Dood J (1986) *Vakblad voor de Bloemisterij* 41: 40–41
Greenhouse roses for cut-flower production (3 cvs) *Ex vitro* vs cuttings and budded material	—	Flower yield higher, lower or equal (depending on the cultivar); after establishment of plants, the size and quality of flowers did not differ between the propagation methods	—	Wisniewska-Grzeszkiewicz H and Rejman S (1992) *Acta Horticulturae* 314: 227–231
Clonal rootstocks for cut roses (6 genotypes) Comparison of softwood cuttings of *ex vitro* vs *in vivo* origin	Improved sprouting of the scion	—	Rooting of cuttings from micropropagated stock plants generally superior	De Vries DP and Dubois LAM (1994) *Gartenbauwissenschaft* 59: 81–85

for their performance on rootstocks. The age at which micropropagated plants are sold needs careful thought; plants should be comparable in vigour to those propagated by other methods but the time between weaning and sale should not be delayed unnecessarily. In scientific investigations, precisely defined media and culture conditions are preferred in order to ensure repeatability of measurements but this adds to the cost of propagation. For commercial purposes, cost is a more important criterion and savings may be made by greater use of commercial-grade chemicals and natural lighting.

If production costs were to be reduced, there can be little doubt of a large market demand for micropropagated plants. But to what level must the price be reduced to achieve the necessary threshold and how might the necessary savings be achieved? The future prospects of micropropagation depend on satisfactory answers to these questions.

See also: **Cell, Tissue and Organ Culture**: Rooting of Micropropagated Shoots; *In Vitro* Regeneration Techniques; Storage and Cryopreservation. **Rootstocks**: Scion–Rootstock Relationships; Seedling Stocks.

Further Reading

Chu IYE (1992) Perspective of the micropropagation industry. In: Kurata K and Kosai T (eds.) *Transplant Production Systems*, pp. 137–150. Dordrecht: Kluwer Academic.

Horn WAH (1992) Micropropagation of rose (*Rosa* L.). In: Bajaj YPS (ed.) *Biotechnology in Agriculture and Forestry*, vol. 20, *High Tech and Micropropagation IV*, pp. 320–342. Berlin: Springer-Verlag.

Kozai T, Fujiwara K, Hayashi M and Aitken-Christie J (1992) The *in vitro* environment and its control in micropropagation. In: Kurata K and Kosai T (eds.) *Transplant Production Systems*, pp. 247–282. Dordrecht: Kluwer Academic.

Pierik RLM (1987) *In Vitro Culture of Higher Plants*. The Hague: Martinus Nijhof.

Roberts AV, Smith EF, Horan I *et al.* (1994) Stage III techniques for improving water relations and autotrophy in micropropagated plants. In: Lumsden PJ, Nicholas JR and Davies WJ (eds.) *Physiology, Growth and Development of Plants in Culture*, pp. 314–322. Dordrecht: Kluwer Academic.

Short KC and Roberts AV (1991) Roses (*Rosa: in vitro* culture micropropagation and secondary products). In: Bajaj YPS (ed.) *Biotechnology in Agriculture and Forestry*, vol. 15, *Medicinal and Aromatic Plants III*, pp. 376–397. Berlin: Springer-Verlag.

Skirvin RM, Chu MC and Young HJ (1990) Rose. In: Ammirato PV, Evans DR, Sharp WR and Bajaj YPS (eds.) *Handbook of Plant Cell Culture*, vol. 5, pp. 716–743. New York: McGraw Hill.

Rooting of Micropropagated Shoots

M Podwyszynska, Research Institute of Pomology and Floriculture, Skierniewice, Poland

Introduction

Rooting of shoots propagated *in vitro* (microcuttings) is stage 3 of micropropagation. It follows stage 1, introduction into culture, and stage 2, shoot multiplication (*see* **Cell, Tissue and Organ Culture**: Micropropagation). Shoots can be rooted *in vitro* either on media solidified with gelling agent or in liquid medium with, for example, cellulose plugs used as a supporting material. The *in vitro* rooting system is followed by acclimatization of plants after transplanting to soil – stage 4. Microcuttings can also be rooted *ex vitro* in nonaseptic conditions, e.g. in peat substrate in a greenhouse, thus combining stages 3 and 4 into a single stage. Rooting of cuttings is an essential process in horticultural mass production in both traditional vegetative propagation and in micropropagation. The production of large numbers of microcuttings with low ability for rooting is associated with a risk of serious plant material losses. Therefore, the knowledge of the morphological and physiological phases of adventitious root formation and factors influencing this process is necessary. Rhizogenesis is controlled by external and internal factors. External factors like temperature, light and medium composition can be easily regulated. Internal factors, e.g. endogenous production of phytohormones, carbohydrates and phenolic compounds, all result from genetic characteristics. The physiological status and quality of microcuttings are also affected by the conditions of stage 2. Within all these factors, auxins either produced endogenously or applied exogenously play the key role in the rooting of cuttings *in vitro*. The auxin effect is modulated by a complex of several other factors. Among them, the mineral salt concentration in the rooting medium and the genotype-dependent capacity to synthesize and respond to ethylene are considered the most important for rooting of rose.

Phases of Adventitious Root Formation

Adventitious root formation *in vitro* has been studied in detail in many species, e.g. in microcuttings of grapevine (*Vitis vinifera* L.), apple (*Malus* spp.) and mung bean (*Vigna radiata* L.). Several investigations indicate that this process proceeds similarly in rose. Generally, adventitious root formation both *in vitro* and *in vivo* can be divided into four phases:

1. Induction, also called the dedifferentiation phase, begins after taking the cuttings and lasts approximately 24 h. Certain cells close to the vascular tissue turn meristematic and become sensitive to auxin-inducing root formation. Such competent cells accumulate starch. During the first hours ethylene production is induced by the wounded tissue resulting from the excision of microcuttings, followed by an accumulation of endogenous auxin at the base of cuttings. Presumably, ethylene enhances tissue sensitivity to auxins. A second peak of ethylene production occurs usually 24 h later in response to auxin accumulation. The auxin–ethylene interaction affects the complex of enzymatic activities involved in cell dedifferentiation and division.
2. Initiation (subsequent 2–4 days) is the second phase. The competent cells are able to respond to auxins (produced endogenously or applied externally) that induce cell division, resulting in the formation of meristematic centres named meristemoids or root initials.
3. Formation of root primordia from meristemoids occurs 4–5 days after placing microcuttings on the rooting medium. At this time auxin is no longer required.
4. In rose, root elongation is apparent after 5–6 days, as root tips emerge through the tissue at the base of the microcuttings. In this phase exogenously supplied auxins might act in an inhibitory way. Roots with root caps usually become visible 6–8 days after placing shoots on a rooting medium.

Influence of Microenvironment During Shoot Multiplication on the Rooting Capacity of Microcuttings

Quality of Microcuttings

Physiological status and morphological characteristics constitute the quality of microcuttings. It is well known that shoots grown *in vitro* differ from plants grown *ex vitro* due to the specific conditions creating a microenvironment of *in vitro* culture: high relative humidity, low light intensity, limited gas exchange, high level of mineral and organic nutrients and application of synthetic growth regulators, mainly BAP (6-benzylaminopurine), GA_3 (gibberellic acid) and auxins (*see* **Biotechnologies For Breeding:** Early Embryo Rescue). Shoots grown *in vitro* have a thin cuticle, poorly developed vascular system and abnormally functioning stomata. Also hydathodes, apparatus discharging excess water, are not able to close. Moreover, *in vitro*-formed roots usually have weak vascular connections with the shoot, and water uptake is relatively poor. *In vitro*-grown rose shoots possess a much higher ability to enhance photosynthesis in response to increased light intensity and CO_2 levels in comparison with, for example, gerbera. However, under standard conditions photosynthetic activity is often limited, mainly due to low photosynthetic photon flux density (PPFD), the presence of sucrose in the culture medium and inappropriate CO_2 concentrations in the tightly closed containers – too low during the light period and too high in the darkness. Such plants are sensitive to water loss after the transfer to soil. Hence, even under optimal conditions during *ex vitro* acclimatization, there is a risk of plant material loss. The efficiency of stages 3 and 4 of micropropagation may be enhanced by the improvement of microcutting quality. Resistance to desiccation and rooting ability is influenced by the chemical (composition of the medium) and physical (humidity, temperature and light) microenvironment of the multiplication – stage 2.

Microcuttings consisting of shoot tips with two to three unfolded leaves are most frequently used for rooting (**Figures 1** and **2**). Shoots should be more than 10 mm long and must not show any symptoms of hyperhydration or chlorosis. The proper size and best quality of microcuttings are very important, especially in recalcitrant cultivars that are difficult to root or show early symptoms of senescence (leaf yellowing and abscission often associated with shoot tip necrosis, beginning in the third or fourth week of a multiplication cycle). In many cases, shoots transferred to root induction medium directly after a 4–5-week multiplication subculture period produce roots only sporadically. Satisfactory rooting may be obtained by just prolonging the duration of the last subculture period to 6, 8 or even 10 weeks, depending on genotype, resulting in shoot elongation. Longer shoots have an increased rooting capacity due to the higher rate of tissue lignification, and a lower level of endogenous

Figure 1 Microcuttings 3–4 cm in length suitable for *ex vitro* rooting.

Figure 2 Microcuttings (2–3 cm long shoot tips and triple-node shoot segments) rooted *in vitro* for 7 days on modified agar Murashige and Skoog (MS) medium containing 0.5 μmol l^{-1} of indolebutyric acid (IBA) and 1.7 μmol l^{-1} of paclobutrazol, suitable for planting in a greenhouse.

Table 1 The macroelement compositions of MS and two alternative media (QL and WPM) used in rose micropropagation

Salts containing macroelements	*Concentration (mg l^{-1})*		
	MS	*QL*	*WPM*
KNO_3	1900	1800	
K_2SO_4			990
KH_2PO_4	170	272	170
$MgSO_47H_2O$	370	370	370
NH_4NO_3	1650	400	400
$CaCl_22H_2O$	440		96
$Ca(NO_3)_24H_2O$		1204	556

MS: the medium of Murashige T and Skoog F (1962) A revised medium for rapid growth and bioassays with tobacco tissue culture. *Physiologia Plantarum* 15: 473–497.
QL: also called Lepoivre's medium, based on the medium of Quoirin M, Lepoivre P and Boxus Ph (1977) Un premier bilan de 10 années de recherches sur les cultures de méristèmes et la multiplication *in vitro* de fruitiers ligneux. pp. 93–117 in *CR Rech. 1976–1977 et Rapports de Synthèse, Stat. Cult. Fruit. et Maraîch.*, Gembloux, Belgium, described by Zimmerman RH (in Sharp WR, Evans DA, Ammirato PV and Yamada Y (eds.) (1984) *Handbook of Plant Cell Culture*, vol. 2. *Crop Species*, pp. 369–395. New York: Macmillan.
WPM: the medium of Lloyd G and McCown B (1981) Commercially-feasible micropropagation of Mountain laurel, *Kalmia latifolia*, by use of shoot tips culture. *International Plant Propagators Society Proceedings* 30: 421–427.

cytokinin and nitrogen. High-quality microcuttings that root easily can be harvested from the cost-saving system of pruned mother plants (*see* **Cell, Tissue and Organ Culture:** Micropropagation). In rose genotypes with strong apical dominance that form robust and long shoots *in vitro*, single-, double- or triple-noded microcuttings without apical bud (**Figure 2**) can also be used (some cultivars of hybrid teas, climbing and rambler ones). The best quality is particularly important in rooting *ex vitro*, for which only robust 2–3-cm-long shoots with high rhizogenic potential can be used (**Figure 1**).

Mineral Composition of Medium

In general, rose shoots are successfully multiplied on the Murashige and Skoog (MS) modified medium (*see* **Cell, Tissue and Organ Culture:** Micropropagation). However, microcuttings of some cultivars manifest poor growth or chlorosis on this medium. Growth improvement may be obtained by replacing MS macroelements with those of QL medium (**Table 1**) that contains a quarter of the level of NH_4^+, 40% higher Ca^{2+} concentration and a very low amount of Cl^-. In some genotypes of *Rosa chinensis* Jack. and *R. nitida* Willd., positive results are obtained using WPM macroelements, recommended for multiplication of woody plants, containing a quarter of the level of both ammonium and nitrate in MS medium. Sometimes, an improvement of shoot growth is noted by diluting the MS macroelements to half. Application of a different iron form, viz. sequestrene Fe 330 at a concentration of 40 mg l^{-1} or an increase in magnesium level by 50%, may counteract chlorosis. Raising the calcium concentration by adding calcium gluconate (6 mmol l^{-1}) helps to prevent symptoms of early leaf senescence in sensitive rose cultivars and generally results in increased shoot quality, chlorophyll content and rooting ability (**Figure 3**). Shoot-tip necrosis is classified as a calcium-related physiological disorder, resulting from deficiency or rather its inefficient distribution, due to the reduced transpiration rate under high humidity in culture containers. It is possible that the calcium-related physiological disorder is also

Figure 3 Shoot growth and multiplication of the rose 'Starina'® sensitive to ethylene on a medium with various calcium (Ca^{2+}) levels (½, half, 1, full and 2, double) in relation to Murashige and Skoog (MS) medium; 6 weeks after subculturing. Early leaf senescence is visible at the half and full Ca^{2+} levels.

influenced by the auxins and auxin-induced ethylene, which induce the release of this element from cell membranes. Thus, the increased calcium level reduces membrane permeability and cell wall flexibility as well as extensibility, making plants more resistant to the harmful effect of excessive ethylene and auxin.

Growth Regulators

One of the major factors influencing the morphology of microcuttings and rhizogenic competence is the cytokinin concentration (e.g. BAP) used in the multiplication medium. This growth regulator stimulates the development of axillary shoots and simultaneously inhibits root formation. If applied at the supraoptimal level of 8–13 $\mu mol\,l^{-1}$, depending on genotype, BAP can accumulate in plant tissue at high concentrations in the form of conjugates, inactive glucosides, as well as physiologically active ribosides or ribotides. In contrast to other cytokinins (kinetin, 2iP, *meta*-topolin or zeatin), BAP conjugates are very stable, being slowly degraded, and exert a prolonged inhibitory effect on rooting. But the replacement of BAP by another cytokinin, in order to improve rooting ability, adversely affects shoot growth. Another negative side-effect of supraoptimal BAP levels manifests as hyperhydration of the shoot tissue which occurs in a higher degree under calcium deficiency. Shoots formed under high BAP levels are tiny and have small leaves, while the rooting ability correlates positively with shoot length and quality. Therefore, in the multiplication system of cyclic shoot subdivision, lowering the BAP concentration to half or even quarter strength in the last subculture prior to rooting is recommended. Another growth regulator, GA_3 (0.3–3 $\mu mol\,l^{-1}$), in the presence of BAP, enhances the multiplication of several rose genotypes, but it may cause tissue hyperhydration and an excessive shoot elongation. Gibberellins are also known to inhibit rooting *per se*. In turn, auxins, usually applied at a relatively low concentration during stage 2, may stimulate early leaf senescence in some cultivars sensitive to ethylene. Thus, improvement of shoot quality can be obtained by excluding GA_3 and/or auxins from the multiplication medium.

Relative Humidity, Gas Exchange and Light

In the last subculture prior to rooting, the ability of shoots to resist desiccation and to perform photosynthesis may be enhanced by increased PPFD and culture container ventilation. This reduces humidity, enables the diffusion of ethylene and other detrimental gases and provides an appropriate CO_2 level. The selection of plant culture vessels and closures depends on the rooting system (sterile or unsterile, mixotrophic or photoautotrophic) and is frequently based on cost and availability. Several vessels and lids with bacteriological filters (0.22 μm pore membrane), providing ventilation through natural diffusion, were developed especially for plant tissue culture. These include Vitro Vent Containers (Duchefa Biochemie BV, The Netherlands); food jars covered with different closures, e.g. lids with a notch on the screw, transparent polypropylene film or aluminium foil with or without filter, caps with a hole filled with cotton wool or Magenta B-Cap-Vented (Sigma-Aldrich Co.); Magenta GA-7 vessels with vented closure (Sigma-Aldrich Co.); and polypropylene vessels closed with vented LifeGuard lids (Osmotek Ltd, Rehovot, Israel). A certain humidity reduction of the atmosphere from close to 100% to 96% can be obtained by bottom cooling of culture vessels. Robust and hardy shoots with high rooting capacity are also produced in photomixotrophic (with sucrose concentration reduced to

$20\,g\,l^{-1}$) or photoautotrophic systems in which sucrose is excluded from the medium, under increased PPFD (100–180 $\mu mol\,m^{-2}\,s^{-1}$) and CO_2 levels (0.05–1%). For these techniques larger vessels providing the proper level of CO_2 are used. Such systems are often concomitant with a forced ventilation, robotization and automation.

Factors Influencing Rooting *in vitro*

Auxins

From the production point of view, indole-3-acetic acid (IAA), the auxin which is endogenously produced by rose shoot apex and leaves, is often insufficient for effective rooting of microcuttings. Therefore, in addition to IAA, synthetically produced auxins such as 3-indolebutyric acid (IBA), 1-naphthylacetic acid (NAA) and 2,4-dichlorophenoxyacetic acid (2,4-D) or their combinations are used. They are usually applied to the base of microcuttings which take them up through the cut surface by the vascular system. Auxin penetration through the cuticle is poor. These hormones are only needed during the formation of root initials. Therefore, optimal auxin treatment ought to be short, 1–10 days; if applied for a longer time at supraoptimal concentrations they may retard elongation of roots and cause their thickening. These plant growth regulators can be applied in several ways. In pulse treatment, which corresponds to traditional vegetative propagation, shoots are dipped for a few seconds or minutes in aqueous or alcoholic solutions or in talcum powder containing auxins at relatively high concentrations (millimolar range). Sometimes, this method is also used for rooting *ex vitro* long and robust microcuttings. In another approach, microcuttings are incubated overnight on a gelled medium containing 1000 $\mu mol\,l^{-1}$ of IAA or 100 $\mu mol\,l^{-1}$ IBA (the latter is recommended for *R. damascena* Mill.) and subsequently transplanted to liquid or gelled media without auxins, where the shoots become more robust and preinitiated root primordia develop into roots. But this rooting method is time-consuming. Usually, microcuttings are exposed to a medium containing lower amounts of auxin for a period of 5–10 days and subsequently planted *ex vitro*. In such a system, microcuttings possess only root primordia and very short roots suitable for planting. In the case of rose genotypes, producing thin or short shoots, the period of rooting *in vitro* on gelled medium or in Sorbarod plugs (Baumgartner Papiers, Lausanne, Switzerland) can be prolonged to 14–30 days. For such long-term root induction, recommended auxin treatments include IAA at a concentration of 0.57–5.7 $\mu mol\,l^{-1}$ for miniature roses, IBA at 0.5–10 $\mu mol\,l^{-1}$ for tea roses and NAA 0.05–1.5 $\mu mol\,l^{-1}$ for *R. canina* L. Combinations of synergistically acting auxins are also used, particularly for difficult-to-root roses: IAA (0.57–5.7 $\mu mol\,l^{-1}$) with IBA (0.5–5 $\mu mol\,l^{-1}$) or NAA (0.05–0.5 $\mu mol\,l^{-1}$). Combinations of IAA at low concentrations with other auxins usually improve root development, reduce undesirable ethylene production and promote plant growth after transplanting to soil, compared with the application of IBA or NAA alone. Regarding large varietal differences, an examination of rhizogenic potential of each micropropagated genotype 2–3 months before the date of scheduled mass rooting is recommended. In easy-to-root roses, auxin can be excluded from the medium. It is worth noting that auxins differ in their rhizogenic activity. IBA and NAA, generally preferred to IBA alone, are more active than IAA and, therefore, are commonly used for rooting (**Figure 4**).

In order to obtain a similar rooting efficiency, NAA ought to be applied at concentrations of between a quarter and a tenth of IBA. These auxins, applied at a supraoptimal concentration, inhibit root elongation, often cause leaf senescence and stimulate excessive

Figure 4 Microcuttings of difficult-to-root 'White Gem' rooted *in vitro* for 3 weeks on medium without auxin (0) and in the presence of 1 $mg\,l^{-1}$ (5.7 $\mu mol\,l^{-1}$) indole-3-acetic acid (IAA) or 1 $mg\,l^{-1}$ (4.9 $\mu mol\,l^{-1}$) indolebutyric acid (IBA). There are few and long roots in the presence of IAA, whereas in the presence of IBA roots are numerous and short.

callus growth at the shoot base. After their uptake, auxins are rapidly metabolized by conjugation and oxidation, the latter being an irreversible degradation. Thus, in the plant tissue, only very small amounts (less than 1%) of the auxin taken up occurs in a free form. These auxin conversions constitute a control mechanism for the hormone levels. Auxins are conjugated with amino acids, myoinositol as well as oligo- and polysaccharides, and are accumulated in plant tissue in these physiologically inactive forms. Such an auxin conversion is reversible and the hormone may be released from conjugates to the free form. Rhizogenesis is only induced by free auxins. Differences in rhizogenic activity of auxins result from differences in uptake, transport and metabolism as well as from genotype-dependent tissue sensitivity to a particular auxin. Thus, the stronger activity of IBA may result from its faster absorption and hydrolysis of the free form as well as from slower degradation, compared with IAA. IBA also significantly enhances the level of endogenous free IAA. As auxins are only required for root induction and subsequently inhibit root elongation, unstable IAA is usually recommended for prolonged culture periods on rooting media. The more stable IBA and NAA – the latter is not oxidized – are used for pulse and short (1–7 days) auxin treatments. Differences in rhizogenic activity between auxins can also result from their various effects on ethylene production. It is well documented that ethylene is involved in rhizogenesis. Studies on the role of auxin–ethylene interactions in rooting of difficult-to-root rose cultivars have shown that IBA and NAA strongly stimulate both ethylene production and rhizogenesis, whereas IAA only slightly affects both these processes.

Ethylene

This gaseous plant hormone is produced by all plant tissues and induces leaf senescence, fruit ripening and developmental processes such as callus or root formations (*see* **Postharvest Physiology:** Ethylene). The ethylene biosynthesis is stimulated, among other factors, by auxins and wounding, such as the excision of microcuttings. This gaseous hormone may promote, inhibit or not influence rhizogenesis. The proposed physiological mechanism by which ethylene contributes to rooting is based on auxin–ethylene interactions in which ethylene may inhibit the auxin transport, reduce its metabolism and change tissue sensitivity to auxins. One of the known mechanisms, at a biochemical level, is that ethylene affects the complex of enzymatic activities involved in cell dedifferentiation and division. Thus, auxin stimulates acid cellulase activity which is involved in root primordium initiation (cell division and xylem differentiation), whereas ethylene increases the activity of basic cellulase responsible for cortical degradation, enabling root emergence. Effects of endogenous ethylene depend on its concentration, phase of root development and plant genotype. Thus, in apple and mung bean, the stimulatory effect of ethylene on rooting (induced by auxin and wounding) occurs during the first induction phase, only at an auxin suboptimal level. In turn, the retarding effect of ethylene is evident in the case of auxin excess during the second phase of root meristemoid formation. In some rose cultivars which are sensitive to ethylene, manifesting early senescence symptoms, the wounding and auxin-induced ethylene is responsible for the inhibition of root formation and elongation, as well as for shoot decay during the rooting stage. This effect may result from ethylene accumulation in the basal part of a microcutting submerged in the gelled or liquid medium. In plant tissue, the ethylene level is reduced mainly by its release into the atmosphere. As ethylene diffusion in water is very poor (10 000-fold less than that in air), its escape from the rooting zone of a microcutting immersed in a medium is very limited. Therefore, providing high aeration of the rooting zone by placing microcuttings in an air-porous supporting material, e.g. Sorbarod plugs or rock wool, allows ethylene escape that markedly improves the rooting efficiency and morphology of roots, which develop root hairs. However, the latter process requires a certain level of ethylene, but lower or higher concentrations are inhibitory. Ethylene may also decrease the quality of microcuttings. This gas may accumulate to high physiological concentrations in tightly closed vessels, causing leaf senescence and shoot tip necrosis. This problem can be partially overcome by container ventilation. But in some recalcitrant rose cultivars it does not prevent damage because their tissue has a high ability both to produce and to bind to ethylene.

Supplement of chemicals inhibiting ethylene biosynthesis or action could be another approach towards reducing ethylene damage. Several compounds have been tested: inhibitors of ethylene biosynthesis like aminoethoxyvinylglycine (AVG), aminooxyacetic acid (AOA), $CoCl_2$, acetyl salicylic acid (ASA) and 1-aminoisobutyric acid (AIB), as well as inhibitors of ethylene action. Chemicals containing silver ions (Ag^+) are used the most frequently due to their high antisenescence activity, $AgNO_3$ and silver thiosulphate (STS) (5–100 $\mu mol\, l^{-1}$). Ag^+ acts by attaching to ethylene receptor sites, thereby preventing binding of ethylene. However, due to their detrimental side-effects, none of these chemicals is routinely applied in commercial rose propagation. AIB (250 $mg\, l^{-1}$) and $AgNO_3$ (2.5–10 $mg\, l^{-1}$) reduce senescence symptoms and stimulate root emergence but inhibit root hair development and influence negatively further plant growth in greenhouse. AIB causes chlorosis and retards plant growth, and Ag^+, as a toxic heavy metal, causes plant decay. Compounds containing silver ions, however, are used in the regeneration of rose adventitious shoots and embryos which are associated with biotechnology and breeding (*see* **Cell, Tissue and Organ Culture:** *In Vitro* Regeneration Techniques). All of the inhibitors mentioned above have also been used in studies distinguishing the role of ethylene and auxin in the rooting process. For this purpose ethylene precursor 1-aminocyclopropane-1-carboxylic acid (ACC) and ethylene-releasing compound, etephon (2-chloroethylphosphonoic acid) were applied, but they have not been utilized in rose micropropagation to improve rooting. The most beneficial effect of delaying the senescence process stimulated by ethylene is attributed to a proper calcium level in plant tissue.

Mineral Salts

Lowering the level of macroelements in the medium during *in vitro* rooting promotes rhizogenesis of microcuttings of most plants. In rose, these mineral salts are usually diluted to a half, sometimes to a third or even a quarter strength of the MS macroelements used in the multiplication medium. Further salts dilution does not give any more satisfactory results. The beneficial effect of low-mineral-salt concentration on rooting is attributed to nitrogen. Lowering its level to one-eighth (7.5 mmol l^{-1}) of the MS medium (60 mmol l^{-1}) and diluting the remaining macroelements to half gives efficient rooting. Nitrogen is required in the medium in the form of NO_3^- and NH_4^+. For rose, the optimal NO_3^- : NH_4^+ ratio is 3 : 1.

Carbohydrates

The initial carbohydrate level in microcutting tissue must be sufficient to supply energy and structural materials needed for root formation. The need for sucrose in the medium as a carbon source depends on the rooting method. In rooting *in vitro* at a relatively low PPFD (10–60 μmol m^{-2} s^{-1}), the addition of sucrose is necessary at a concentration of 20–40 g l^{-1}, with higher concentrations at lower PPFD. Under such low PPFD, the photosynthetic activity is insufficient to provide the optimal amount of carbohydrate. Besides mostly recommended sucrose, glucose is also utilized at a concentration of 20–45 g l^{-1}, especially for recalcitrant rose genotypes, such as *R. rugosa* Thunb. and *R. villosa* L. It is known that cuttings with a high carbon-to-nitrogen ratio, which is correlated with low N level in the medium, root better than those with a low C : N ratio. The molar ratio of sucrose to nitrogen in the medium beneficial for rooting of rose microcuttings is above 10.

Other Medium Ingredients

If shoots are rooted on gelled media, it is recommended to apply the same brand and concentration of gelling agent (agar or phytagel) as has been used during multiplication.

Among amino acids examined in several species belonging to the Rosaceae family, aspartic acid, glutamic acid, tryptophane, arginine and proline (400–800 μmol l^{-1}) favourably affected shoot rooting in apple rootstock and *Cydonia oblonga* L. In rose, arginine and proline increase rooting efficiency.

In addition to routinely applied vitamins, simultaneous inclusion of riboflavin (10–30 μmol l^{-1}) and auxin (e.g. IBA 10–20 μmol l^{-1}) can have a pronounced effect on rooting (*see* **Cell, Tissue and Organ Culture:** Micropropagation). This vitamin is involved in photooxidation of the auxin. The strategy of this rooting method is the initial maintenance of shoots in the darkness for 2–5 days, while the auxin induces root formation, and their subsequent exposure to a 16-h photoperiod where the auxin (no longer required) is oxidized by light. Such a rooting method is applied in *Carica papaya* L., apple and, sporadically, in rose.

Activated charcoal, known to absorb toxic substances excreted by plantlets, may enhance rooting of some rose cultivars, e.g. the miniature rose. This compound is added to the medium at a concentration of 0.5–1 g l^{-1} but it should be applied carefully because it can delay root formation by absorbing promoting substances.

Phenolic compounds are known to act as rooting cofactors. Among them, phloroglucinol, phloridzin and ferulic acid (600–1000 μmol l^{-1}), used in the presence of auxin, favourably influence rooting of *Malus* spp. The stimulatory rooting effect of phenols is attributed to the protection of auxins from oxidation. These compounds are also considered to function as antioxidants in preventing tissue damage caused by the wounding stress. A positive effect of phloroglucinol in the presence of NAA has been noted in the rooting of *R. damascena* Mill.

Polyamines (spermidine, spermine and putrescine) occurring ubiquitously in plants are involved in many developmental processes. In the presence of auxin, they play a regulatory role during the initiation of the process of root primordium formation. Within these compounds, putrescine added to a medium (30–100 μmol l^{-1}), stimulates root growth in several plant species, e.g. *Olea europea sativa* L. and *Vigna radiata* L. But to date, there is no information about polyamine effects on rose shoot rooting *in vitro*.

Other Factors

Rose genotypes differ significantly in their requirements and responses to culture conditions as well as in rooting capacity. Thus, the rooting method developed for one particular cultivar is not always suitable for another. The genotype-based ease of rooting is very important from an economic point of view and should be considered by breeders during their evaluation of germplasm. Generally, cultivar-dependent rooting capacity observed in traditional propagation by cutting is highly correlated with the *in vitro* rooting capacity. On the other hand, difficulties with rooting can be overcome using *in vitro* propagation. Problems with root formation of some rose genotypes may result from inadequate endogenous levels of auxins and other growth regulators, phenolics or enzymes involved in their turnover and biosynthesis. For instance, in *Asparagus* spp., rooting was inhibited by high levels of gibberellins, that was overcome by including growth retardants, inhibitors of gibberellin biosynthesis from the triazol group, as ancimidol and paclobutrazol. In rose, the addition of paclobutrazol (0.17–3.4 μmol l^{-1}) to the rooting medium promotes rooting (**Figure 2**) and improves leaf structure as well as stomata function, resulting in better plant survival in the greenhouse.

Temperature and light also influence the rooting of microcuttings. In the majority of standard procedures, shoots are rooted at a temperature of 21–23 °C in a 16-h photoperiod. However, with particular rose genotypes, optimal temperatures for rooting were determined between 15 and 18 °C. Sometimes, incubation in darkness for the first 3–5 days improves root formation.

Ex vitro Rooting and Acclimatization

Currently, in many commercial laboratories, microcuttings of several plant species are rooted directly in nonsterile conditions, skipping the phase of rooting *in vitro* (**Figure 5**). Root formation proceeds simultaneously with plantlet acclimatization. The costs of *in vitro* rooting, depending on plant species, are calculated at 35–50% of the total micropropagation costs. *Ex vitro* rooting shortens the micropropagation cycle by 2–4 weeks and reduces its costs. In such a system, roots develop properly and possess root hairs. In rose, direct rooting is, however, only recommended for genotypes producing *in vitro* long and robust shoots of a high rooting capacity (**Figure 1**). Microcuttings 2–4 cm long can be planted in different growing substrates composed of peat with an addition of perlite (**Figure 6**), sand or vermiculite (4 : 1 or 3 : 1 v/v), supplemented with complete fertilizer at 0.5–1 g l^{-1}, pH 6. Jiffy pots (**Figure 7**), oasis or Florialite (a mixture of cellulose fibres and vermiculite) can also be used. High percentages of rooting and plant survival can be obtained in rock wool blocks (Grodan BS 25/50) soaked with aqueous solutions of a complete fertilizer (e.g. 0.05% Peters Professional 10 + 52 + 10 plant starter), with electrical conductivity about 1.0 mS cm^{-1} and pH 5.6–6.0 adjusted with nitric acid (**Figure 8**). Such plant material is particularly suited for further cultivation on rock wool. In the mezoclimatic zone, the efficiency of *ex vitro* rooting depends on the season and is usually higher during the spring and summer to early autumn, reaching up to 100%. In winter and early spring it can drop to 80% and requires supplementary lighting provided for at least a 13-h photoperiod (PPFD 100 μmol m^{-2} s^{-1}). The proper temperature, ranging from 20 to 36 °C with an optimum at 23/26 °C, night/day, may be maintained by bottom heating. In most cases, in order to stimulate rhizogenesis, microcuttings are treated with auxins by dipping in liquid or powdered rooting compounds. Alternatively, shoots may be sprayed 2–3 times with an aqueous solution containing auxins at concentrations comparable to those used for *in vitro* rooting, e.g. 57 μmol l^{-1} IAA or IBA in combination with IAA, both at 5 μmol l^{-1} (**Figure 6**). Such treatment can be combined with fungicides against soilborne diseases and grey mould (*Botrytis cinerea*).

The efficiency of rose direct rooting varies with genotypes. Usually, in difficult-to-root ones, root formation and emergence are tardy. In such cases, the directly rooted microcuttings grow markedly slower than those rooted *in vitro*. This growth inhibition results from the water stress that inhibits photosynthesis and causes deficiency of carbohydrates which are necessary for root development. Thus, the presence of roots, even poorly developed *in vitro*, permits one to compensate high transpiration rates during the first few critical days following transfer to *ex vitro* conditions.

Microcuttings transplanted to soil, both rootless and rooted *in vitro*, require shading and acclimatization in high relative humidity, provided by misting and fogging chambers or polyethylene tents (**Figure 5**). After 1–3 weeks (depending on the rooting system, season and climate), as new leaves have occurred, relative humidity and shading are gradually reduced. Preliminary hardening of shoots *in vitro* (e.g. bottom cooling or container ventilation) as well as a short-term root induction or rooting in Sorbarod plugs provide fast and successful adaptation to greenhouse conditions. In order to obtain sufficient light at

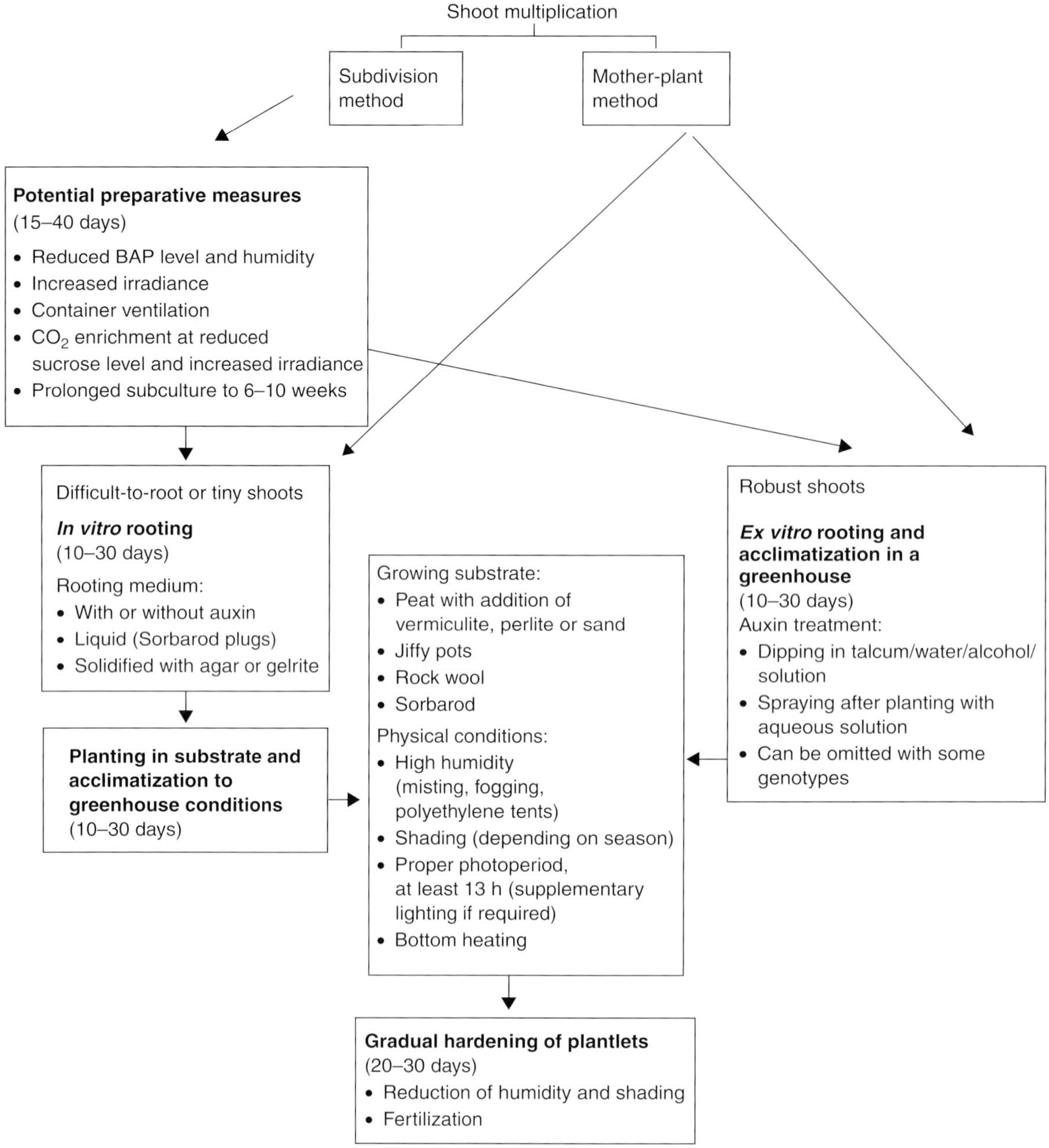

Figure 5 Scheme of rooting and acclimatization of micropropagated shoots of rose.

low costs, culture vessels can be maintained in a partially shaded greenhouse instead of in culture rooms. Plantlets rooted in a Sorbarod system can be transferred to soil in the plugs, which protect roots and serve as water reservoirs for a short time. Usually, after 2 months in soil, a saleable plant is obtained.

Prospects of Rose Rooting

Rooting and acclimatization are critical stages in the micropropagation of many plants, including roses. Currently, several studies to identify markers or indicators of rooting capability of microcuttings have been undertaken at the physiological, biochemical and molecular level. Endogenous factors specific for rhizogenic competence, like hormones, enzymes and genes, have been determined. Further attempts aim at enhancing rooting efficiency through the timing of treatments with auxins and other substances regarded as rooting cofactors (polyamines, phenolic substances). Certain studies, also including roses, were devoted to the use of environmentally friendly (ecologically compatible) methods which might improve rooting and acclimatization. Among them, the potential for a red light treatment to stimulate root formation *in vitro* or *ex vitro* has been demonstrated as a

Figure 6 Stage 3 and 4: *ex vitro* rooting of the microcuttings of easy-to-root rose 'Starina'® in a medium composed of peat and perlite (4 : 1 v/v), 5 weeks after planting in a greenhouse. Left: untreated microcuttings; right: sprayed with aqueous solution of indolebutyric acid at 57 $\mu mol\ l^{-1}$.

Figure 7 Micropropagated plant of 'Sabrina' 10 weeks after *ex vitro* acclimatization in a Jiffy-7 pot; root induction *in vitro* for 7 days in the presence of 4.9 $\mu mol\ l^{-1}$ indole-3-acetic acid.

Figure 8 Micropropagated plant of 'Sonia' 5 weeks after *ex vitro* rooting in a block of rock wool (Grodan SBS). In order to stimulate root formation, microcuttings were sprayed with aqueous solution of 4.9 $\mu mol\ l^{-1}$ indolebutyric acid and 5.7 $\mu mol\ l^{-1}$ indole-3-acetic acid.

substitute for auxin application. Another approach, earning special attention, is the infection of microcuttings with mycorrhizal fungi and beneficial bacteria during rooting *in vitro* or acclimatization. Since the rhizospheric microorganisms are absent during propagation *in vitro*, there is a risk of infection with pathogens causing plant decay after planting in soil. A protective role is attributed to inoculation with symbiotic microflora. It is worth noting that mycorrhization has been shown to promote plant growth in soil through the improvement of water and nutrient uptake in several micropropagated species such as strawberry, rhododendron, potato, grapevine and raspberry.

Promising results concerning the enhancement of growth, quality and rooting of rose shoots have recently been obtained by the use of fluorocarbon, an inert, liquid organic compound having the ability to dissolve large amount of gases, e.g. CO_2. Fluorocarbon, enriched with CO_2, can be distributed on a semisolid medium where it will subsequently release CO_2, resulting in atmospheric enrichment. Spraying microcuttings with glycine betaine seems to be another promising approach for increasing growth after transplanting. This amino acid, extracted from sugar-beet molasses, is known to protect plants against stresses caused by water deficiency during transplantation to soil. Research on alternative methods for the stimulation of rooting and improvement of plant growth *ex vitro* may in future not only decrease the micropropagation costs but also reduce the use of chemicals harmful to the environment.

See also: **Biotechnologies for Breeding:** Early Embryo Rescue. **Cell, Tissue and Organ Culture:** Micropropagation. **Growth Regulation:** Hormones in Growth and Development. **Postharvest Physiology:** Ethylene.

Further Reading

Collet GF and Le CL (1987) Role of auxin during *in vitro* rhizogensis of rose and apple-trees. *Acta Horticulturae* 212: 273–280.

De Klerk GJ, Van Der Krieken W and De Jong JC (1999) The formation of adventitious roots: new concepts, new possibilities (review). *In Vitro Cellular and Developmental Biology – Plant* 35: 189–199.

Fujiwara K and Kozai T (1995) Physical microenvironment and its effects. In: Aitken-Christie J, Kozai T and Smith MAL (eds.) *Automation and Environmental Control in Plant Tissue Culture*, pp. 319–370. Dordrecht: Kluwer Academic.

George EF, Puttock DJM and George HJ (1987) *Plant Culture Media*, 1st edn, vol. 1. pp. 1–567. Edington, Westbury, UK: Exegetics.

Hartman HT, Kester DE and Davies FT (1993) *Plant Propagation: Principles and Practices*, 5th edn, pp. 199–255: Englewood Cliffs, NJ: Prentice-Hall International.

Moncousin Ch (1991a) Rooting of microcuttings: general aspects. *Acta Horticulturae* 289: 301–310.

Moncousin Ch (1991b) Rooting of microcuttings: fundamental aspects. *Acta Horticulturae* 289: 311–317.

Moncousin Ch (1991c) Rooting of microcuttings: unmanipulated factors. *Acta Horticulturae* 289: 319–327.

Podwyszynska M and Hempel M (1988) Factors influencing acclimatization of *Rosa hybrida* plants multiplied *in vitro* to greenhouse conditions. *Acta Horticulturae* 226(II), 639–642.

Preece JE and Sutter EG (1991) Acclimatization of micropropagated plants to the greenhouse and field. In: Debergh PC and Zimmerman RH (eds.) *Micropropagation: Technology and Application*, pp. 71–94. Dordrecht: Kluwer Academic.

Roberts AV, Walker S, Horan I, Smith EF and Mottley J (1992) The effect of growth retardants, humidity and light at stage II on stage IV of micropropagation in chrysanthemum and rose. *Acta Horticulturae* 319: 153–158.

Roberts AV, Smith EF and Horan I (1994) Stage III techniques for improving water relations and autotrophy in micropropagated plants. In: Lumsden PT, Nicholas JR and Davis WJ (eds.) *Physiology, Growth and Development in Plants in Culture*, pp. 314–322. Dordrecht: Kluwer Academic.

Rout GR, Samantaray J, Mottley J and Das P (1999) Biotechnology of the rose: a review of recent progress. *Scientia Horticulturae* 81: 201–228.

Uosukainen M, Rantala S, Manninen A and Vestberg M (2000) Improvement of microplant establishment through *in vitro* and *ex vitro* chemical applications. *Acta Horticulturae* 530: 325–331.

Williams RR (1995) The chemical microenvironment. In: Aitken-Christie J, Kozai T and Smith MAL (eds.) *Automation and Environmental Control in Plant Tissue Culture*, pp. 405–440. Dordrecht: Kluwer Academic.

In Vitro Regeneration Techniques

A Schum and A Dohm, Institute of Ornamental Plant Breeding, Ahrensburg, Germany

Introduction

The term ‘regeneration’ includes the replacement of organs either by development from preformed dormant buds or by new formation of structures originating from previously differentiated cells. While the former process is the common mechanism of spermatophyta for the substitution of lost organs, under natural conditions the latter is restricted to particular species. In contrast, *in vitro* culture systems allow the exploitation of plant cells' totipotency by induction of morphogenetic processes leading to the formation of adventitious organs or somatic embryos in theoretically any species.

Mechanisms of *In Vitro* Regeneration

Figure 1 summarizes the different modes of *in vitro* regeneration which in principle follow three distinct patterns: (1) bud regeneration, (2) adventitious regeneration of organs and (3) somatic embryogenesis. In the first case, explants containing meristems are taken into *in vitro* culture while the latter two systems generally start with explants devoid of such structures. All three patterns of regeneration can be realized in roses.

Bud Regeneration

The most simple process of *in vitro* shoot regeneration is initiated from isolated meristems, apical or axillary buds included in bigger explants like nodal segments with dormant eyes. By manipulation of the phytohormone balance such preformed shoot initials are forced to develop according to their predetermined pattern. The process of bud regeneration is closest to the *in vivo* situation and comprises simple growth and development. Axillary buds of regenerated shoots can be subjected to continuous regeneration cycles by subculture on appropriate media with increased levels of cytokinins (*see* **Cell, Tissue and Organ Culture:** Micropropagation).

Corresponding developmental processes are realized in root cultures starting from root initials and in germination of zygotic embryos (*see* **Biotechnologies for Breeding:** Early Embryo Rescue).

Adventitious Regeneration

Adventitious organs regenerate from *de novo* formed meristems originating from cells that were once differentiated, i.e. specialized for particular functions. In roses, adventitious regeneration *in vivo* occurs in the form of rooting of cuttings or spontaneous formation of suckers on roots, but under *in vitro* conditions it comprises additional possibilities. Regeneration of adventitious organs requires an initial step in which specialized cells of the sporophyte or gametophyte dedifferentiate, a process which is generally connected with a reduction of the vacuole, reversion of chloroplasts and proliferation of cytoplasm. Dedifferentiation is stimulated by wounding of organized tissues which leads to altered endogenous phytohormone balances. *In vitro* cultures promote such a process,

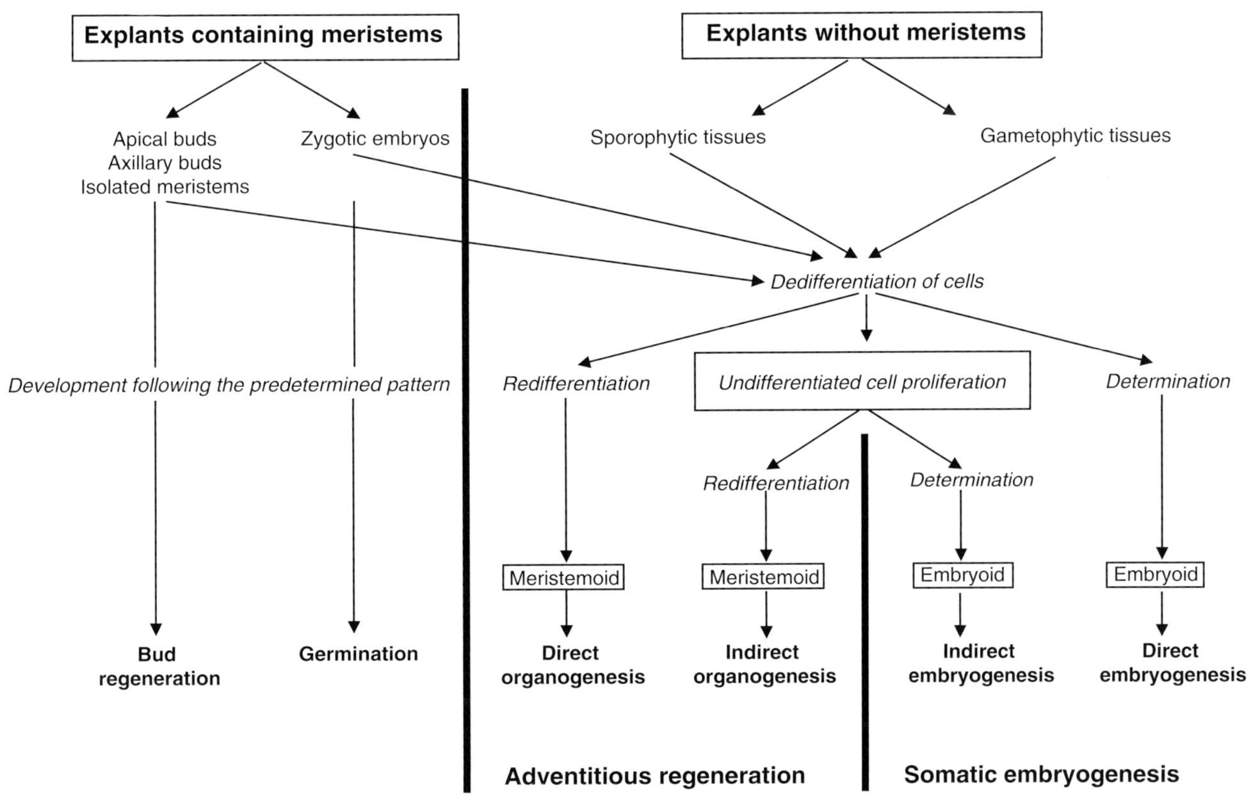

Figure 1 Scheme of *in vitro* regeneration patterns.

as taking of explants in itself constitutes a disruption of the organized structure of the whole plant. It can be further stimulated by providing appropriate exogeneous growth regulators in the culture medium.

Dedifferentiated cells regain meristematic potential. Affected cells can be redifferentiated immediately, resulting in the formation of structured unipolar meristemoids. According to their determination these organ initials will develop into shoots or roots, respectively. This mode of regeneration is known as direct organogenesis. Alternatively, dedifferentiation is followed by a phase of undifferentiated cell proliferation resulting in callus formation. Only at later stages, generally induced by a change of the incubation conditions, cells of the callus redifferentiate and develop meristemoids which subsequently grow out to organs. This mode of regeneration is called indirect organogenesis.

Somatic Embryogenesis

The initial stages of the third regeneration pattern, i.e. somatic embryogenesis, are analogous to those of organogenesis in dedifferentiation of specialized cells of the sporophyte or gametophyte. Subsequently, such cells are embryogenically determined, with or without an intermediate callus phase. The following cell divisions lead to the formation of bipolar embryoids which morphologically resemble zygotic embryos. Ideally, such somatic embryoids germinate into whole plants by simultaneous development of shoot and root poles. In case of embryoid initiation directly from cells of the explant the process is known as direct embryogenesis, in case of an intermediate callus phase as indirect embryogenesis. In roses, somatic embryogenesis is unknown to occur *in vivo*.

Both adventitious organs as well as somatic embryos may originate from either single cells or from small groups of cells.

Adventitious Regeneration and Somatic Embryogenesis in Roses

Explants and Culture Conditions

Theoretically any part of the rose plant can be used as explant for the regeneration of adventitious organs as well as for induction of somatic embryos. In practice, due to their overall availability vegetative parts of the shoot such as internode or leaf segments are preferred rather than floral parts or roots. However, in spite of the generally accepted concept of plant cell totipotency, it must be kept in mind that different types of tissue might differ in their amenability for induction of dedifferentiation and redifferentiation processes. In this respect the use of floral parts may be of advantage for induction of somatic embryogenesis in particular cases. Explants may be taken from *in vitro* stock cultures with the advantage that surface sterilization is not needed. Furthermore, once the absence of endogeneous contaminants has been confirmed, any problems associated with concomitant microorganisms are bypassed. However, interactions with preculture conditions have to be considered, which especially applies to the phytohormone regime of the media used for stock cultures.

If *in vitro* cultures of a specific genotype are not available, the plant parts of interest must be harvested from greenhouse or outdoor grown plants. These need to be chemically surface sterilized (*see* **Cell, Tissue and Organ Culture:** Micropropagation). Depending on the cultural condition of the mother plants, a varying percentage of contaminated explants has to be expected. Presence of endophytes will pose special problems and may require the application of antibiotics. Endogenous phytohormone levels vary with the season and climatic conditions and will interact with the *in vitro* performance of explants. Standardization of preculture conditions is hardly possible as compared to the use of *in vitro* grown stock cultures. On the other hand, an abundant number of different genotypes might be more easily available as *in vivo* grown plants. In any case, special attention has to be paid to plant nutrition as well as to plant protection.

All handling of the plant material must be performed with caution in order to avoid drying out or squashing of explants. One problem, which mainly emerges with the *in vitro* culture of woody plants, is the secretion of phenolic compounds by wounded plant cells which may disturb the regeneration process. Short-term incubation of explants in liquid medium prior to transfer onto solid culture medium may help to eliminate such interfering substances. Furthermore, the incubation of explants in the dark during the first days of culture helps to prevent browning of tissue and media caused by polymerization of phenolic substances.

For nearly all modes of regeneration the cultivation of the plant material at 25 °C has proved to be suitable. Although the temporary incubation in complete darkness seems to be favourable for particular stages during the regeneration process, the cultivation in a 16-h photoperiod is best maintained for most processes and a large number of different rose genotypes.

Indirect Shoot Organogenesis

In general, indirect organogenesis comprises four distinct stages, i.e. (1) the induction of callus, (2) the callus proliferation, (3) the induction of meristemoids and (4) the outgrowth of organs. Each of these steps requires specific cultural conditions which, however, are interdependent. Furthermore, the environmental conditions during preculture of source material as well as the given genotype decisively influence the regeneration success.

Callus induction and proliferation Induction of callus with competence for adventitious shoot regeneration can be accomplished from different types of explants, including segments of internodes, leaves, roots and petals (**Table 1**). If whole leaflets are used as starting material, callus induction is promoted by transversal cuts through the main veins. The explants are placed on agar solidified culture medium. Suitable, for example, is the one developed in 1962 by Murashige and Skoog in either full or half concentration. Induction of morphogenetic callus is highly dependent on appropriate phytohormone regimes. Type, quantity and ratio of exogenous growth regulators added to the culture medium have to be determined mostly by 'trial-and-error' approaches in each specific case. Suitable regimes depend on the given genotype, type and age of explants, preculture conditions and interacting endogenous phytohormone levels. In general, roses most readily respond to the auxins naphthalene acetic acid (NAA) or 2,4-dichlorophenoxyacetic acid (2,4-D) and, in case of cytokinins, to 6-benzylaminopurine (BAP) or thidiazuron (TDZ). In most cases, auxins are essential for the induction of callus, while simultaneous supplement of media with cytokinins promotes its organogenic character. It has to be considered that the type of auxin applied in the callus induction phase might restrict organogenesis in subsequent cultural phases. For this first stage of indirect organogenesis in roses, screening of NAA in the range of 0.005 to 5 $mg\,l^{-1}$ or 2,4-D in concentrations between 0.2 and 2 $mg\,l^{-1}$, respectively, in combination with BAP (0 to 3 $mg\,l^{-1}$) is convenient. Higher growth-regulator levels are rarely found necessary to induce callus capable of plant regeneration. Examples of appropriate exogeneous phytohormone doses in the induction medium are shown in **Table 1**.

Callus formation starts at the wounded parts of the explants and eventually overgrows the entire primary tissue. Subculture of induced callus for further propagation can often be performed on the identical culture media. In other cases lowering of the phytohormone levels promotes subsequent regeneration of adventitious meristemoids. However, callus should not be propagated and maintained for too long, as this is usually connected with an eventual loss of morphogenetic capacity. Furthermore, prolonged phases of undifferentiated cell proliferation are known to affect adversely the genetic stability.

Induction of meristemoids and regeneration of shoots Early organogenesis can sometimes be observed in the course of callus proliferation, but mostly is promoted by a shift of the auxin /cytokinin ratio in the culture medium. As a rule, regeneration is achieved by either lowering the auxin level or raising the cytokinin level or both (**Table 1**). Again, one has to keep in mind that there are interactions between a multitude of environmental and tissue-related factors affecting the response of callus cells to specific stimuli.

The induction of meristemoids becomes apparent by emergence of protuberances on the callus surface. These consist of shoot primordia which under appropriate cultural conditions further develop into shoots. The outgrowth of shoots may be stimulated by lowering the cytokinin concentration in the culture medium or by addition of gibberellic acid (GA_3) (**Figure 2**).

Bud regeneration was obtained in up to 98% of the induced calluses upon optimal adjustment of phytohormone supplements to the given genotype. Regenerated adventitious shoots are amenable for cloning by stimulation of axillary branching (*see* **Cell, Tissue and Organ Culture:** Micropropagation) and finally can be rooted (*see* **Cell, Tissue and Organ Culture:** Rooting of Micropropagated Shoots) and transferred to the greenhouse.

Direct Shoot Organogenesis

In direct organogenesis two main cultural stages can be distinguished, i.e. (1) the induction of meristemoids and (2) the outgrowth of organs. In roses, root and especially leaf segments have proved to be the most promising explant types for direct regeneration of adventitious shoots without an intervening callus phase (**Table 2**). Just as in the case of indirect organogenesis, agar-solidified media according to the formulation of Murashige and Skoog in either full or half concentration are well suited to promote direct adventitious regeneration. However, in this case, cytokinins prove to be the principal exogeneous type of hormone required for the initiation of cell divisions leading to differentiation of shoot initials. The cytokinins BAP or TDZ are especially effective with roses. Addition of auxins to the culture media is not obligatory, but may promote induction of meristemoids depending on the physiological state of the explants. Examples of effective phytohormone regimes are given in **Table 2**.

Under appropriate exogenous growth regulator conditions meristemoids appear within 2 to 4 weeks after initiation of cultures as green protuberances with smooth surfaces. The formation of such shoot initials generally has been observed from subepidermal cell

Table 1 Selected reports on indirect regeneration of adventitious shoots

Explant type	Genotype	Induction culture medium		Regeneration culture medium		Reference
		Basic medium	Phytohormone supplements	Basic medium	Phytohormone supplements	
Internodal segments						
From *in vitro*-grown shoots	*Rosa persica* × *R. xanthina*	MS	0.1–0.3 mg l^{-1} NAA + 1.0–2.0 mg l^{-1} BAP or 0.005 mg l^{-1} NAA + 0.5 mg l^{-1} BAP	MS	0.1–0.3 mg l^{-1} NAA + 3.0 mg l^{-1} BAP or 0.005 mg l^{-1} NAA + 1 mg l^{-1} BAP	Lloyd *et al.* (1988)
From *in vivo*-grown plants	*R. damascena*	MS	1.9 mg l^{-1} NAA	MS without NH_4NO_3	0.19 mg l^{-1} NAA + 2.25 mg l^{-1} BAP	Ishioka and Tanimoto (1990)
From greenhouse-grown plants	*R. hybrida* 'Landora'	1/2 MS (full-strength vitamins)	5.0 mg l^{-1} NAA + 2.0 mg l^{-1} BAP	1/2 MS (full strength vitamins)	0.01 mg l^{-1} NAA + 0.2 mg l^{-1} BAP + 20 mg l^{-1} GA_3 (+600 mg l^{-1} L-proline or L-glutamine) or 0.01 mg l^{-1} NAA + 0.2 mg l^{-1} BAP + 0.01 mg l^{-1} GA_3 (+20 mg l^{-1} adenine sulphate)	Rout *et al.* (1992)
From *in vitro*-grown shoots	*R. hybrida* 'Meirutral'	MS	0.0006 mg l^{-1} NAA + 1.0 mg l^{-1} BAP	MS →MS + vitamins according to Morel and Martin (1955)	0.2 mg l^{-1} NAA + 3.0 mg l^{-1} BAP 0.5 mg l^{-1} IAA + 1.0 mg l^{-1} BAP	Arene *et al.* 1993
From *in vitro*-grown shoots	*R. hybrida* 'Carefree Beauty'™ *R. chinensis minima* 'Meirutral', 'Baby Katie'™	1/2 MS	2.05–21.1 mg l^{-1} NAA	1/2 MS	5.1 mg l^{-1} TDZ + 1.04 mg l^{-1} GA_3	Hsia and Korban (1996)
Leaf discs						
From *in vitro*-grown shoots	*R. hybrida* 'Landora'	1/2 MS (full strength vitamins)	0.1 mg l^{-1} NAA + 0.5 mg l^{-1} BAP	1/2 MS (full strength vitamins)	0.01 mg l^{-1} NAA + 0.2 mg l^{-1} BAP + 10 mg l^{-1} GA_3 (+600 mg l^{-1} L-proline or L-glutamine) or 0.01 mg l^{-1} NAA + 0.2 mg l^{-1} BAP + 0.01 mg l^{-1} GA_3 (+20 mg l^{-1} adenine sulphate)	Rout *et al.* (1992)

Leaf segments						
From *in vitro*-grown shoots	*R. hybrida* 'Meirutral'	MS	0.2 mg l^{-1} NAA + 3.0 mg l^{-1} BAP	MS →MS + vitamins according to Morel and Martin (1955)	0.2 mg l^{-1} NAA + 3.0 mg l^{-1} BAP 0.5 mg l^{-1} IAA + 1.0 mg l^{-1} BAP	Arene *et al.* (1993)
From *in vitro*-grown shoots	*R. hybrida* 'Carefree Beauty'™	MS	0.22–44.21 mg l^{-1} 2,4-D or 2.05–21.1 mg l^{-1} NAA	1/2 MS	5.1 mg l^{-1} TDZ + 1.04 mg l^{-1} GA_3	Hsia and Korban (1996)
	R. chinensis minima 'Meirutral', 'Baby Katie'™	1/2 MS				
Root segments						
From *in vitro* cultured plantlets	*R. hybrida* 'Meirutral'	MS	0.2 mg l^{-1} NAA + 3.0 mg l^{-1} BAP	MS →MS + vitamins according to Morel and Martin (1955)	0.2 mg l^{-1} NAA + 3.0 mg l^{-1} BAP 0.5 mg l^{-1} IAA + 1.0 mg l^{-1} BAP	Arene *et al.* (1993)
Immature embryos	*R. hybrida* 'Bridal Pink'™ × 'Amorous', 'Bridal Pink'™ × selection 83-5052, 'Bridal Pink'™ × 'Haute Pink'	1/2 MS (full-strength vitamins, modified)	0.01 mg l^{-1} NAA + 0.23 mg l^{-1} BAP	1/2 MS (full-strength vitamins, modified)	0.01 mg l^{-1} NAA + 0.23 mg l^{-1} BAP	Burger *et al.* (1990)

→ shows that explants were transferred to the indicated medium after the first culture period.
MS, culture medium based on the formulation of Murashige and Skoog (1962) *Physiologia Plantarum* 15: 473–497; NAA, 1-naphthaleneacetic acid; IAA, indole-3-acetic acid; 2,4-D, 2,4-dichlorophenoxy acetic acid; BAP, 6-benzylaminopurine; TDZ, thidiazuron; GA_3, gibberellic acid.
Sources: Arene *et al.* (1993) *Euphytica* 71: 83–90; Burger *et al.* (1990) *Plant Cell, Tissue and Organ Culture* 21: 147–152; Hsia and Korban (1996) *Plant Cell, Tissue and Organ Culture* 44: 1–6; Ishioka and Tanimoto (1990) *Plant Cell, Tissue and Organ Culture* 22: 197–199; Lloyd *et al.* (1988) *Euphytica* 37: 31–36; Morel and Martia (1955) *C.R. Acad. Sci. Paris* 41: 472–475; Rout *et al.* (1992) *Indian Journal of Experimental Biology* 30: 15–18.

Figure 2 Indirect adventitious shoot regeneration from a peduncle explant. (Photograph by Felten.)

layers in the region of the vascular bundles of petioles, leaf veins and pericycle of roots.

Subsequent outgrowth of differentiated adventitious buds (phase 2) is stimulated by lowering the activity of available exogenous cytokinins. This can be accomplished by a shift to physiologically weaker types of phytohormones and/or by lowering their concentration. **Figure 3** shows direct adventitious shoot regeneration from a peduncle explant.

Very high regeneration frequencies have been obtained if leaf explants consisting of the basal part of a leaflet with its subtending petiole from greenhouse-grown plants are used as explants (**Table 2**). Depending on the cultivar 70–100% of the explants formed adventitious shoots.

Indirect and Direct Root Organogenesis

The *in vitro* regeneration of adventitious organs in roses is not restricted to shoots. Corresponding indirect and direct meristemoid formation patterns can also be used for the recovery of roots. This process is referred to as rhizogenesis. The induction of adventitious roots is generally more feasible and works more readily with a wider range of genotypes as compared to adventitious shoots.

Regeneration of adventitious roots generally is controlled by presence of auxins in the culture medium. In addition to NAA and 2,4-D, indolebutyric acid (IBA) was found to be especially promotive in case of roses. Induction of adventitious roots from isolated explants may be of interest, as such organs have been discussed as target tissue for genetic transformation with subsequent regeneration of potential transgenic adventitious shoots or somatic embryos (*see* **Biotechnologies for Breeding:** Genetic Transformation). Induction of adventitious roots from *in vitro* microshoots is indispensable for the transfer of plants to *in vivo* conditions (*see* **Cell, Tissue and Organ Culture:** Micropropagation; Rooting of Micropropagated Shoots).

Figure 3 Direct adventitious shoot regeneration from a leaf explant. (Photograph by Felten.)

Somatic Embryogenesis in Roses

In vitro regeneration via somatic embryogenesis was first described for the carrot (*Daucus carota*) in 1958. Ideally, this regeneration process comprises four main developmental stages: (1) the induction of embryogenic callus, (2) callus proliferation, (3) the induction and maturation of somatic embryos and (4) the germination of somatic embryos into plants. In roses, however, regeneration via somatic embryogenesis does not always correspond to all these stages, which are regularly observed in ideal systems like carrots. In some cases developmental stages of somatic embryogenesis are combined with adventitious shoot formation. The germination of embryos into plants, for example, may be replaced by adventitious shoot formation and subsequent rooting. But in spite of the occurrence of organogenesis during somatic embryogenesis in roses, a clear distinction can be made between the two processes: in the case of somatic embryogenesis the regenerating cells follow developmental stages, which are typical for zygotic embryogenesis. Furthermore, in contrast to callus competent of adventitious shoot formation, embryogenic callus can be maintained and propagated for several years without losing its morphogenic capacity. To our knowledge, somatic embryogenesis in roses is always

Table 2 Selected reports on direct regeneration of adventitious shoots

Explant type	*Genotype*	*Induction culture medium*		*Regeneration culture medium*		*Reference*
		Basic medium	*Phytohormone supplements*	*Basic medium*	*Phytohormone supplements*	
Root segments						
From *in vitro*-cultured plantlets	*Rosa persica* × *R. xanthina*	MS	2.0 mg l^{-1} BAP	MS	2.0 mg l^{-1} BAP	Lloyd *et al.* (1988)
From *in vitro*-cultured plantlets or from rhizogenic calli derived from different floral parts	*R. hybrida* 'Meirutral'	MS	2.0 mg l^{-1} BAP	MS	2.0 mg l^{-1} BAP	Arene *et al.* (1993)
Leaves						
From *in vitro*-cultured plantlets	*Rosa persica* × *R. xanthina* *R. wichurana* *R. laevigata*	MS	0.5 mg l^{-1} BAP	MS	0.5 mg l^{-1} BAP	Lloyd *et al.* (1988)
Leaf segments						
From *in vitro*-cultured plantlets	*Rosa hybrida* 'Meirutral'	MS	0.5 mg l^{-1} BAP	MS	0.5 mg l^{-1} BAP	Arene *et al.* (1993)
Leaf explants						
Consisting of the basal half of a leaflet with its subtending petiole from greenhouse-grown plants	24 *R. hybrida* cut-rose cultivars (including 'Madelon', 'Only Love', 'Presto', 'Sonia', 'Tineke') and *R. indica* rootstocks	1/2 MS (full strength vitamins)	1.5 mg l^{-1} TDZ + 0.05 mg l^{-1} NAA	MS	0.5 mg l^{-1} BAP + 0.01 mg l^{-1} NAA or 0.5 mg l^{-1} BAP + 0.01 mg l^{-1} NAA + 0.1 mg l^{-1} GA_3	Dubois and de Vries (1995) Dubois and de Vries (1997)
Receptacles						
From immature flower buds	Unspecified	MS	1–7.5 mg l^{-1} TDZ + 0.25 mg l^{-1} NAA	MS	1–7.5 mg l^{-1} TDZ + 0.25 mg l^{-1} NAA	Lu (1993)

MS, culture medium based on the formulation of Murashige and Skoog (1962) *Physiologia Plantarum* 15: 473–497; NAA, 1-naphthaleneacetic acid; BAP, 6-benzylaminopurine; TDZ, thidiazuron; GA_3, Gibberellic acid.
Sources: Arene L *et al.* (1993) *Euphytica* 71: 83–90; Dubois and de Vries (1995) *Gartenbauwissenschaft* 60: 249–253; Dubois and de Vries (1997) *Acta Horticulturae* 447: 79–85; Lloyd *et al.* (1988) *Euphytica* 37: 31–36; Lu (1993) *In Vitro Cellular and Developmental Biology* 29: 92–96.

preceded by callus formation; direct embryogenesis has not been described so far.

In the following, the four main stages of somatic embryogenesis are described for roses.

Induction of embryogenic callus Nearly all parts of the plant can be used as starting material for somatic embryogenesis (**Table 3**), but the success with particular explant types strongly depends on the genotype. In order to induce callus formation the explants must be wounded, for example, by transversely cutting the midribs of intact leaves or by using stem slices. The basic culture medium is mainly based on Murashige and Skoog (MS) medium with varying vitamin supplements. However, for some rose genotypes it might be of advantage to use other formulations, for example the one according to Schenk and Hildebrandt (SH). The main effect of the SH medium results from the reduced nitrogen supply. Of further importance is the choice of the optimal carbon source. Particular rose cultivars respond better to glucose, fructose or maltose rather than to sucrose. Furthermore, several rose genotypes exhibit enhanced frequencies of somatic embryogenesis, if the natural gelling agent agar-agar, which is routinely used for solidification of the culture medium, is replaced by agarose or by the synthetic compound gelrite. Of fundamental significance for the regeneration process, however, is the supply with adequate phytohormones. These again must be adapted to the respective rose genotype. A formulation which has proved to induce somatic embryogenesis in about 50 different rose genotypes belonging to different species is the supplement of NAA (1 to 2 $mg\,l^{-1}$) or 2,4-D (0.5 to 2 $mg\,l^{-1}$) for callus induction followed by further subcultures on medium with either zeatine (2 to 4 $mg\,l^{-1}$) or TDZ (0.5 to 2 $mg\,l^{-1}$) (**Table 3**). As a first response to exogeneous auxin the explants form a hard type of callus, which is non-embryogenic. The development of true embryogenic callus is initiated upon transfer onto culture medium with cytokinins (**Figure 4**). Simultaneous supply with both plant hormone types, however, often results in the prompt formation of embryogenic callus. Depending on the rose genotype, the frequency of explants exhibiting somatic embryogenesis may vary from below 1% to nearly 100%.

Callus proliferation The second stage of somatic embryogenesis is initiated with the separation of the embryogenic callus from the original explant and its transfer onto fresh medium. Callus proliferation is usually performed on media supplied with a combination of auxins and cytokinins. With some genotypes, however, the formulation of the culture medium used for callus induction can be maintained (**Table 3**). Again, the exact formulation has to be adapted to the genotype of interest as well as to the explant type.

The preferred type of embryogenic rose callus represents a fast-growing, smooth type with proembryogenic and embryogenic structures (**Figure 5**). Its colour may vary between white, yellow, green, beige and light or dark brown. In order to maintain this callus type its handling during subculture is of particular importance. Only callus showing proembryogenic or embryogenic structures should be selected and transferred to fresh medium. Callus pieces without embryogenic structures as well as shootlike structures should be aborted during the subcultivation process. Selection of the 'wrong' callus type may result in the loss of the embryogenic capacity. Furthermore, both the multiplication rate as well as a proper callus development are enhanced by spreading the cells to a very thin layer.

Development of somatic embryos The two developmental stages of callus proliferation and induction of somatic embryos, respectively, in most cases are difficult to differentiate clearly due to continuous development of proembryogenic structures within the calluses into mature embryos (**Figure 6**). However, the embryogenic capacity of the calluses can be enhanced to a certain degree by raising the sucrose content of the culture medium to 9% or by addition of organic compounds like L-proline (**Table 3**).

The appearance of the somatic embryos varies markedly with regard to their size, colour and number of cotyledons. In contrast to ideal systems in dicotyledonous plants, rose embryos often form more than two cotyledons. Furthermore, secondary embryogenesis at the base of the primary embryos may result in the formation of clusters.

Embryo germination and plant development The further development of somatic rose embryos into plantlets may occur as true germination, but in many cases adventitious shoot formation mostly originating from cotyledons or a mixture of both regeneration processes can be observed. The germination usually is induced on MS-based culture medium devoid of phytohormones. However, in other cases the medium used for the induction of somatic embryos will also promote the regeneration process. The germination can be enhanced by addition of abscissic acid (ABA) (**Table 3**) or methyl laurate as well as by treatment with low temperatures. Adventitious shoot formation is induced by transfer of the somatic embryos onto MS-based culture medium with higher concentrations of cytokinines. For example, 2 $mg\,l^{-1}$ BAP proved to work for a broad range of different rose genotypes (**Table 3**). The embryos respond with indirect shoot organogenesis (**Figure 7**). The addition of BAP also promotes the further development of germinated embryos into plantlets. These can be transferred into soil immediately, whereas adventitious shoots must be further subcultured for rooting (*see* **Cell, Tissue And Organ Culture:** Micropropagation; Rooting of Micropropagated Shoots). **Figure 8** shows greenhouse-grown plants of the cultivar 'Heckenzauber' derived from somatic embryogenesis.

Regeneration potentials In addition to the media composition, light conditions were found to have a crucial effect on somatic embryogenesis. While low light intensities promote callus induction and early stages of somatic embryogenesis, higher light intensities were observed to be essential for embryo development and shoot regeneration. Accordingly, some protocols recommend to incubate explants initially in the dark for 6–8 weeks and subsequently transfer cultures to a 16-h photoperiod. The exposure of somatic embryos to lower temperatures enhanced germination in some cases but had no promotive effects in other protocols.

Regeneration percentages vary greatly between species or cultivars. Most protocols compiled in **Table 3** were applied to only one or a very limited number of genotypes. In responsive cultivars generally 5–14% of the explants formed embryogenic callus and regeneration of shoots from somatic embryos could be achieved in frequencies between 12% and 60%. In contrast, the procedure developed in our laboratory (**Table 3**) worked with 69% of 50 different rose genotypes tested with a maximum regeneration frequency of 95% referring to the number of originally cultivated leaf explants.

Somaclonal Variation

Plants obtained by bud regeneration from preexisting meristems are most likely to be true to type with the source genotype. In contrast, adventitious regeneration as well as somatic embryogenesis are recognized as bearing a certain risk of somaclonal variation. This phenomenon was first described in 1981 by Larkin and Scowcroft. Such genetic alterations can be due to mutations which were already present in cells of the original explant. Especially older plant parts are known to carry somatic mutations. In addition, genetic alterations may be induced in the course of *in vitro* culture especially during phases of undifferentiated cell proliferation. Therefore, the processes of indirect organogenesis or indirect embryogenesis are apt to create sometimes high degrees of genetic aberration. The reasons for these alterations are still mostly unknown. Possible mechanisms are the methylation of particular nucleic acids as well as transposon activity.

Corresponding observations have been made in roses. Somaclonal variation does not usually arise following axillary bud regeneration and, up to now, it has not been described for plants regenerated via direct adventitious shoot formation. In plants regenerated after a preceding callus phase via indirect organogenesis or somatic embryogenesis, however, somaclonal instability is a serious problem. Phenotypic variation is observed both in vegetative and in floral characters. Affected plants may show deviating leaf morphology (**Figure 9**) including size, colour, shape and number of leaflets. Variation is also found in plant habitus, as for example a reduction of internodal length. Aberrations observed in floral characters affect size, colour, shape and number of petals (**Figure 10**). Frequently an increased number of petals coincides with reduced male and female fertility. In addition to deviating phenotypes further alterations of the genotype can be detected with molecular markers. Application of the amplified fragment length polymorphism (AFLP®) technology have revealed extensive DNA polymorphisms in plants regenerated by somatic embryogenesis as compared to the original plants. In roses, somaclonal variation seems to be restricted to alterations of single genes. Instability of the ploidy level or aneuploidy has not been described, although intensive cytological investigations have been performed.

The degree of somaclonal variation is significantly determined by the rose genotype and the explant type as well as by the culture conditions during *in vitro* regeneration. The comparative analysis of plants regenerated from somatic embryos of three garden rose cultivars demonstrated tremendous differences between genotypes with regard to alterations of both the phenotype and the genotype. Furthermore, in addition to the plant material long-term maintenance of callus cultures as well as high amounts of synthetic auxins like 2,4-D promoted somaclonal variation.

Conclusions

Practical Application of Different *In Vitro* Regeneration Techniques

The different modes of *in vitro* regeneration were introduced into practice mainly in the 1960s. Since then they have become established tools for micropropagation and a multitude of aspects in plant breeding (**Table 4**). To date, several of the theoretically possible applications have also become well established in roses.

Micropropagation of roses is routinely being performed on a commercial scale by application of

Table 3 Selected reports on somatic embryogenesis of roses

Rose genotype	*Explant type*	*Culture*	*Media used to induce*[a,b]	*Reference*
'Domingo', 'Vicky Brown'	*In vitro* leaves	Callus induction and proliferation and embryo development	1/2 MS + 2% sucrose + kinetin (0.1–0.2 mg l^{-1}) + NAA (0.01–0.05 mg l^{-1}) or NOA (0.05–0.1 mg l^{-1})	De Wit *et al.* (1990)
		Embryo germination	MS + 2% sucrose + 2 mg l^{-1} BAP + 1 mg l^{-1} IAA	
R. persica × *R. xanthina*	*In vitro* roots	Callus induction and proliferation	SH + 3 mg l^{-1} 2,4-D	Roberts *et al.* (1990)
		Embryo development	SH + 1 mg l^{-1} 2,4-D + 0.3 mg l^{-1} GA_3 + 1 mg l^{-1} ABA	
		Embryo germination	SH + 0.1 mg l^{-1} IBA + 16 mg l^{-1} phloroglucinol	
'Royalty'	*In vivo* stamens	Callus induction	B5 + 329 mg l^{-1} NH_4SO_4 + 1,5 mg l^{-1} zeatine + 2 mg l^{-1} 2,4-D → MS + 1.5 mg l^{-1} zeatine + 0.25 mg l^{-1} NAA + 1 mg l^{-1} GA_3	Noriega and Söndahl (1991)
		Callus proliferation	N6 + 1 mg l^{-1} zeatine + 5 mg l^{-1} 2,4-D + 1 mg l^{-1} GA_3	
		Embryo development	MS + 1 mg l^{-1} GA_3 + 2 mg l^{-1} ABA → MS + 10% coconut water + 1 mg l^{-1} GA_3 + 0.2 mg l^{-1} ABA	
		Embryo germination	1/2 N6 without growth regulators → MS + 3 mg l^{-1} BAP + 0.3 mg l^{-1} IAA	
'Landora'	*In vivo* leaves and internodes	Callus induction	1/2 MS + 0.5 mg l^{-1} BAP + 1 mg l^{-1} NAA + 2,4-D (0.5–2 mg l^{-1})	Rout *et al.* (1991)
		Embryo development and proliferation of embryogenic callus	1/2 MS + 0.5 mg l^{-1} BAP + 0.01 mg l^{-1} NAA + 0.1 mg l^{-1} GA_3 + L-proline (200–800 mg l^{-1}) → 1/2 MS + 0.5 mg l^{-1} BAP + 0.01 mg l^{-1} NAA + 0.1 mg l^{-1} GA_3	
		Embryo germination	1/2 MS + 0.5 mg l^{-1} BAP + 0.1 mg l^{-1} GA_3 + 4.5 mg l^{-1} adenine sulphate	
'Meirutal'	*In vitro* leaves, roots and internodes, zygotic embryos	Callus induction and proliferation, embryo development and germination	MS + 4% sucrose + BAP (0.5–3.0 mg l^{-1}) + NAA (0.0005–0.2 mg l^{-1}) or 1 mg l^{-1} 2,4-D	Arene *et al.* (1993)
R. rugosa	Zygotic embryos	Callus induction	MS without growth regulators	Kunitake *et al.* (1993)
		Callus proliferation	MS + galactose (1.8%–3.6%) without growth regulators	
		Embryo development	MS + 1.8% fructose or 3.4% sucrose without growth regulators	
		Embryo germination	MS + 1.8% sorbitol without growth regulators	
'Carefree Beauty', 'Baby Katie'	*In vitro* leaves and internodes	Callus induction and proliferation	MS + 3% sucrose + 2,4-D (0.2–2.2 mg l^{-1})	Hsia and Korban (1996)
		Embryo development	1/2 MS + sucrose (2%/4%) or glucose (2%/4%) + 5 mg l^{-1} TDZ + 1 mg l^{-1} GA_3	
		Embryo germination	MS without growth regulators	
'Trumpeter'®, 'Glad Tidings'	*In vitro* petioles and roots	Callus induction and proliferation	MS or SH + 5 mg l^{-1} 2,4-D → MS or SH + 2,4-D (1–4 mg l^{-1}) + 300 mg l^{-1} L-proline	Marchant *et al.* (1996)
		Embryo development and germination	MS or SH + 1 mg l^{-1} 2,4-D + 0.3 mg l^{-1} GA_3 + 1 mg l^{-1} ABA + L-proline (0–600 mg l^{-1}) → MS or SH + 3% maltose + BAP (0–1 mg l^{-1}) + 0.1 mg l^{-1} IBA	

'Arizona'	*In vivo* petals	Callus induction and proliferation	MS + BAP (0.5–2.0 mg l^{-1}) + kinetin (0.5–2.0 mg l^{-1}) + dicamba (0.5–1.0 mg l^{-1}) → MS + 0.022 mg l^{-1} 2,4-D + 2.45 mg l^{-1} 2iP	Murali *et al.* (1996)
		Embryo development and germination	1/2 MS + ABA (0–0.5 mg l^{-1}) → 1/2 MS + 0.5 mg l^{-1} NAA + 100 mg l^{-1} phloroglucinol	
'Moneyway'	*In vitro* roots	Callus induction and proliferation	SH + 0.3% gelrite + 11 mg l^{-1} 2,4-D	Van der Salm *et al.* (1996)
		Embryo development	SH without growth regulators	
		Outgrowth of shoots	SH + 500 mg l^{-1} casein hydrolysate + 0.3% gelrite + BAP (mg l^{-1}) + IBA (mg l^{-1})	
'Soraya'®	*In vivo* leaves	Callus induction and proliferation and embryo development	MS + 1 mg l^{-1} kinetin + 10 mg l^{-1} *p*-CPA	Kintzios *et al.* (1999)
		Embryo germination	MS + 1.2 mg l^{-1} BAP + 1 mg l^{-1} IAA	
50 different genotypes	*In vitro* leaves and internodes	Callus induction	MS + 2,4-D (0.5–2.0 mg l^{-1}) or NAA (1.0–2.0 mg l^{-1}) → MS + zeatine (1.0–4.0 mg l^{-1}) or TDZ (1.0–2.0 mg l^{-1})	Dohm *et al.* (2001)
		Callus proliferation and embryo development	MS + 1.5 mg l^{-1} zeatine + 0.25 mg l^{-1} NAA + 1 mg l^{-1} GA_3	
		Regeneration to plants	MS + 2 mg l^{-1} BAP + 0.05 mg l^{-1} IBA + 1 mg l^{-1} GA_3	

→ shows that explants were transferred to the indicated medium after the first culture period.

[a] Mineral salts according to: MS, Murashige and Skoog (1972) *Physiologia Plantarum* 15: 473–497; SH, Schenk and Hildebrandt (1972) *Canadian Journal of Botany* 50: 199–204; B5, Gamborg *et al.* (1968) *Experimental Cell Research* 50: 151–158; N6, Chu *et al.* (1975) *Scientia Sinica* 18: 659–668.

[b] Growth regulators: NAA, 1-Naphthaleneacetic acid; NOA, 2-naphthoxyacetic acid; BAP, 6-benzylaminopurine; IAA, indole-3-acetic acid; 2,4-D, 2,4-dichlorophenoxyacetic acid; GA_3, gibberellic acid; ABA, abscisic acid; IBA, indolebutyric acid; TDZ, thidiazuron; 2iP, 6-γ-γ-(dimethylallylamino)-purin; *p*-CPA, *p*-chlorophenoxyacetic acid.

Sources: Arene *et al.* (1993) *Euphytica* 71: 83–90; De Wit *et al.* (1990) *Plant Cell Reports* 9: 456–458; Dohm *et al.* (2001) *Acta Horticulturae* 547: 341–348; Hsia and Korban (1996) *Plant Cell, Tissue and Organ Culture* 44: 1–6; Kintzios *et al.* (1999) *Plant Cell Reports* 18: 467–472; Kunitake *et al.* (1993) *Plant Science* 90: 187–194; Marchant *et al.* (1996) *Plant Science* 120: 95–105; Murali *et al.* (1996) *Euphytica* 91: 271–275; Noriega and Söndahl (1991) *BioTechnology* 9: 991–993; Roberts *et al.* (1990) In: J de Jongh (ed.) *Integration of* In Vitro *Techniques in Ornamental Plant Breeding*, pp. 100–115; Rout *et al.* (1991) *Plant Cell, Tissue and Organ Culture* 27: 65–69; Van der Salm *et al.* (1996) *Plant Cell Reports* 15: 522–526.

Figure 4 Somatic embryogenesis from a leaf explant. (Photograph by Felten.)

Figure 7 Adventitious shoot regeneration from a somatic embryo. (Photograph by Felten.)

Figure 5 Embryogenic rose callus. (Photograph by Felten.)

Figure 8 Plants of the garden rose 'Heckenzauber' derived from somatic embryogenesis. (Photograph by Felten.)

Figure 6 Embryogenic rose callus with mature embryos. (Photograph by Felten.)

Figure 9 Somaclonal variation in leaves of somatic embryo derived plants of 'Pariser Charme'® (left: control). (Photograph by Felten.)

axillary bud regeneration techniques (*see* **Cell, Tissue and Organ Culture:** Micropropagation). For this purpose, adventitious shoot regeneration as well as somatic embryogenesis generally are not favoured due to their inherent high risk of generating somaclonal variants. However, there is interest from commercial laboratories in making use of somatic embryogenesis in future, as it allows a certain automation of micropropagation. A limited degree of somaclonal variation is thought to be tolerable if it is outweighed by a drastic reduction in labour costs.

For mutation induction, polyploidization and regeneration of plants with potentially increased horticultural value, *in vitro* bud regeneration techniques rather than systems making use of adventitious shoots or somatic embryos have been applied. As described before, the latter techniques *per se* create a potentially high degree of genetic variation. However, as far as we are aware, exploitation of somaclonal variation has only been a side aspect and has never been the main purpose of breeding programmes in roses (*see* **Genetics:** Mutation).

Both adventitious shoot regeneration and somatic embryogenesis techniques have been applied in rose breeding, for example for segregation of chimeras, genetic transformation or regeneration of protoplasts (**Table 4**) (*see* **Biotechnologies for Breeding:** Genetic Transformation; **Cell, Tissue and Organ Culture:** Protoplast Culture).

Figure 10 Somaclonal variation in flowers of somatic embryo derived plants of 'Heckenzauber'. (Photograph by Felten.)

Table 4 Practical applications of different *in vitro* regeneration modes

Regeneration mode	*Possible fields of application*	*Realized applications in roses*
Bud regeneration	Micropropagation	+
	Mutagenesis	+
	Polyploidization	+
	Embryo rescue	+
Adventitious regeneration	Micropropagation	–
	Mutagenesis	(+)
	Generation of somaclonal variants	(+)
	Segregation of chimeras	+
	Reduction of ploidy level	–
	Protoplast regeneration	–
	Transformation	+
Somatic embryogenesis	Automatic handling of micropropagation	–
	Mutagenesis	–
	Generation of somaclonal variants	(+)
	Segregation of chimeras	(+)
	Reduction of ploidy level	–
	Protoplast regeneration	+
	Transformation	+

+, established in practice; (+), reported; –, not reported.

Outlook

The general application of *in vitro* regeneration techniques is hampered by the fact that protocols are often restricted to particular genotypes and must be adapted for specific purposes. Furthermore, in many cases low frequencies of regeneration and the sometimes protracted regeneration process prevent an efficient use of *in vitro* culture techniques in rose breeding. This especially applies for organogenesis and somatic embryogenesis, but also the comparably simple process of bud regeneration can pose problems with specific genotypes.

In spite of increasing experience with *in vitro* regeneration techniques, the development of regeneration protocols is still based on trial-and-error approaches, which are time-consuming and will not necessarily lead to success with the genotype of interest. Therefore, there is an urgent need for more basic knowledge of the processes underlying and controlling adventitious regeneration and somatic embryogenesis as well as interactions with environmental and inherent plant factors.

Further Reading

Bouman H and de Klerk GJ (1997) Somaclonal variation. In: Geneve RL, Preece JE and Merkle SA (eds.) *Biotechnology of Ornamental Plants*, vol. 10, 165–184. Wallingford, UK: CAB International.

Dohm A, Ludwig C, Nehring K and Debener T (2001) Somatic embryogenesis in roses. *Acta Horticulturae* 547: 341–348.

Larkin PJ and Scowcroft WR (1981) Somaclonal variation: a novel source of variability from cell cultures for plant improvement. *Theoretical and Applied Genetics* 60: 197–214.

Merkle SA (1997) Somatic embryogenesis in ornamentals. In: Geneve RL, Preece JE and Merkle SA (eds.) *Biotechnology of Ornamental Plants*, vol. 2, pp. 13–34. Wallingford, UK: CAB International.

Roberts AV, Yokoya S, Walker S and Mottley J (1995) Somatic embryogenesis in *Rosa spp.* In: Jain S, Gupta P and Newton R (eds.) *Somatic Embryogenesis in Woody Plants*, vol. 2, pp. 277–289. Dordrecht, The Netherlands: Kluwer Acadamic Publishers.

Rout GR, Samantaray S, Mottley J and Das P (1999) Biotechnology of the rose: a review of recent progress. *Scientia Horticulturae* 81: 201–228.

Thorpe AT (ed.) (1999) *In Vitro Embryogenesis in Plants*. Dordrecht, The Netherlands: Kluwer Academic Publishers.

Tripepi RR (1997) Adventitious shoot regeneration. In: Geneve RL, Preece JE and Merkle SA (eds.) *Biotechnology of Ornamental Plants*, vol. 4, pp. 45–72. Wallingford, UK: CAB International.

Protoplast Culture

J Mottley, University of East London, London, UK

Introduction

Plant protoplasts are cells without cell walls. Exposure of the naked protoplast membrane allows us to investigate more easily fundamental aspects of cell physiology and cell surface interactions with toxins and pathogens. It also allows isolated genes to be directly introduced into the genome in the form of DNA and protoplasts from different plants to be fused. The *in vitro* culture of these protoplasts can lead to the synthesis of new cell walls and, under suitable culture conditions, cell division is possible. As protoplasts separate into individual units when the cell wall is removed, cells and plants derived from protoplasts are derived from single cells and protoplasts can be used for regenerating nonchimaeric plants.

The most promising use of protoplasts for plant improvement is cell hybridization. Using appropriate techniques it is possible to induce fusion between protoplasts from plants that do not normally hybridize sexually. Subsequent regeneration of plants from such somatic hybrid cells can lead to the production of somatic hybrid plants that exhibit novel nuclear and cytoplasmic genetic combinations. Protoplast fusion allows us to produce hybrids with combined genomes and increased ploidy levels, as opposed to the combination of meiotically reduced genomes found in sexual hybridization or the limited gene transfer that is achieved with genetic transformation. In addition, protoplast fusion facilitates the combination of polygenetically inherited traits due to the absence of genetic recombination that occurs during gamete formation. Somatic hybridization therefore provides us with a unique tool for studying and manipulating plant genomes.

In this article, methods are described for isolating, culturing, fusing and genetically transforming protoplasts, and regenerating plants from protoplasts.

Isolation and Culture of Protoplasts

Protoplasts of roses can be isolated from several differentiated and undifferentiated tissues, including leaves, from plantlets cultured *in vitro*, exponential-phase suspension cultures and fast-growing callus. However, successful protoplast isolation and culture and the regeneration of plants from protoplasts very much depends on genotype and the tissue of origin. For example, plants have only been successfully regenerated from protoplasts obtained from cultured rose tissues (callus and cell suspensions). Leaves have been found in several laboratories to yield protoplasts that are only capable of cell division to form non-embryogenic callus.

Cell walls are removed from the cells using commercial preparations of hydrolytic enzymes derived from certain plant pathogenic fungi. An acid pH is required for optimal activity of cell wall degrading enzymes during protoplast isolation. During the enzymic digestion, a suitable osmoticum must be present in the suspending buffer at an appropriate concentration to prevent bursting of the isolated protoplasts. Cell membrane stability during subsequent purification of the protoplasts is also aided by calcium ions and alkaline pH.

The successful culture of protoplasts requires optimal conditions for cell wall regeneration and subsequently for sustained cell division. Critical conditions are osmoticum concentration, the presence or absence of light and the culture medium, especially plant growth regulators and nutrients that may become limiting during the lengthy culture process. For most plant species, usually only a low proportion (<10%) of viable protoplasts eventually forms viable colonies that can be further cultured and regenerated to plants.

The following protocols are the ones used for the isolation, culture and fusion of protoplasts from rose cell cultures in the author's laboratory. They are based on those of Matthews *et al.* (1991) and Mottley *et al.* (1996) (see **Table 2** below for further details).

Buffers and Enzymes

The protoplasts are suspended in cell and protoplast washing (CPW) salts (**Table 1**), supplemented with either mannitol (13% w/v) for CPW13M, or sucrose

Table 1 (A) Inorganic and (B) organic constituents of culture media used for the isolation and culture of rose protoplasts

A Inorganic constituents

Constituent	*Concentration (mg l^{-1})*		
	KM8p[a]	*SH*[b]	*CPW*[c]
KNO_3	1900	2500	101.0
NH_4NO_3	600		
$NH_4H_2PO_4$		300	
$CaCl_2.2H_2O$		200	1480.0
$CaCl_2$	453		
KCl	300		
KH_2PO_4	170		27.2
$MgSO_4.7H_2O$	146.55	400	246.0
Na_2EDTA	37.25	20	
$FeSO_4.7H_2O$	27.85	15	
$MnSO_4.H_2O$	10	10	
H_3BO_3	3	5	
$ZnSO_4.7H_2O$	2	1	
KI	0.75	1	0.16
$Na_2MoO_4.2H_2O$	0.25	0.1	
$CuSO_4$		0.2	
$CuSO_4.5H_2O$	0.025		0.025
$CoCl_2.6H_2O$	0.025	0.1	

B Organic constituents

Constituent[d]	*Concentration (mg l^{-1})*	
	KM8p	*SH*
Mannitol	45000	
Glucose	34200	
Sucrose	250	
Sorbitol	250	
Fructose	250	
Mannose	250	
Rhamnose	250	
Ribose	250	
Xylose	250	
Casamino acids	250	
Cellobiose	250	
Myo-inositol	100	1000
Citric acid	40	
Fumaric acid	40	
L-malic acid	40	
Pyruvic acid	20	
Ascorbic acid	2.0	
Thiamine-HCl	1.0	5
Pyridoxine-HCl	1.0	0.5
Ca pantothenate	1.0	
Choline chloride	1.0	
Nicotinamide	1.0	
Nicotinic acid		5
Folic acid	0.4	
Riboflavin	0.2	
p-aminobenzoic acid	0.02	
Vitamin B_{12}	0.02	
Biotin	0.01	
Vitamins A, D (each)	0.01	

[a] The KM8p formulation is modified from Kao KN and Michayluk MR (*Planta* 1975, 16: 105–110).
[b] The SH medium was made according to Schenk RU and Hildebrandt AC (*Canadian Journal of Botany* 1972, 50: 199–204).
[c] The CPW medium is that of Frearson EM, Power JB and Cocking EC (*Developmental Biology* 1973, 33: 130–137).
[d] Coconut water is also included at a concentration of 20 ml l^{-1}.

Table 2 Examples of publications in which isolated protoplasts were regenerated into callus or shoots

Culture characteristics	*Authors*[a]					
	Pearce and Cocking (1973)	*Krishnamurthy* et al. *(1979)*	*Strauss and Potrykus (1980)*	*Matthews* et al. *(1991) and Mottley* et al. *(1996)*	*Marchant* et al. *(1997)*	*Schum* et al. *(2001)*
Source tissue	NE suspension cultures	NE callus	NE suspension cultures	NE and E callus	*In vitro* leaves	NE and E suspension cultures, *in vitro* shoots
Protoplast isolation						
Salts and organics[b]	Sucrose	White's salts, dextran sulphate and mannitol	Mannitol and $CaCl_2$	CPW13M	CPW13M	CPW11M + MES
Enzymes	Cellulase Onozuka 3000	Cellulase and Macerozyme R10	Cellulase Onozuka R10, hemicellulase, pectinase Pectinol fest	Cellulase Onozuka R10, Pectolyase Y-23 and hemicellulase	Hemicellulase, macerozyme, Cellulase RS and Pectolyase Y-23	Cellulysin, driselase and macerase
Maximum protoplast yield per g FW^{-1}	Not reported	2.4×10^6	Not reported	3.6×10^7	1.87×10^6	4×10^6
Protoplast culture						
Salts and organics[b]	Modified MX_1	Modified MS	Modified MX_1	Modified KM8p	Modified KM8p	Pelletier salts and KM8p organics
Growth regulators	2,4-D + kinetin	NAA	2,4-D + kinetin	NAA and BA	NAA and BA	2,4-D + NAA + BA or 2-iP or TDZ
Gelling agent	Agar or liquid culture	Agar or liquid culture	Agar	SeaPlaque agarose	SeaPlaque agarose	Alginate
Plant regeneration Medium[b]	None	None	None	SH + 2,4-D	SH + 2,4-D	MS + 2,4-D ↓ MS + NAA, zeatin, GA_3 ↓ MS + GA_3, ABA ↓ MS + NAA, BA, GA_3
Final product[c]	Callus	Callus	Callus	Callus, somatic embryos, adventitious shoots and plants	Callus with roots	Plants or callus with roots

NE, nonembryogenic; E, embryogenic; CPW, cell and protoplast washing; MES, 2[*N*-morpholino]ethanesulfonic acid; 2,4-D, 2,4-dichlorophenoxyacetic acid; NAA, α-naphthaleneacetic acid; BA, 6-benzylamino purine; 2-iP, *N*-isopentenylamino purine; TDZ, thidiazuron.

[a] Citations to articles on rose protoplasts: Pearce RS and Cocking EC (1973) *Protoplasma* 77: 165–180. Krishnamurthy KV *et al.* (1979) *Plant Science Letters* 15: 135–137. Strauss A and Potrykus I (1980) *Physiologia Plantarum* 48: 15–20. Matthews D *et al.* (1991) *Plant Cell, Tissue and Organ Culture* 24: 173–180. Mottley J *et al.* (1995) *Acta Horticulturae* 424: 393–397. Marchant *et al.* (1997) *Plant Cell Tissue and Organ Culture* 50: 131–134. Schum *et al.* (2001) *Gartenbauwissenschaft* 66: 115–122.

[b] Citations to articles on culture media: White's salts: White PR (1963) *The Cultivation of Animal and Plant Cells*. New York: Ronald Press. CPW13M: Frearson EM *et al.* (1973) *Developmental Biology* 33: 130–137. CPW11M + MES: Schum A *et al.* (2001) *Gartenbauwissenschaft* 66: 115–122. MX_1: Nash DT and Davies ME (1972) *Journal of Experimental Botany* 23: 75–91. MS: Murashige T and Skoog F (1962) *Physiologia Plantarum* 15: 473–497. KM8p: Kao KN and Michayluk MR (1975) *Planta* 126: 105–110. SH: Schenk RU and Hildebrandt AC (1972) *Canadian Journal of Botany* 50: 199–204. Pelletier salts: Pelletier GC *et al.* (1983) *Molecular Genetics* 191: 244–250.

[c] Species, hybrids and cultivars in which regeneration of adventitious shoots or somatic embryos were obtained include: Matthews *et al.* (1991): *R. persica* × *xanthina*; Mottley *et al.* (1996): tetraploid *R. persica* × *xanthina* following protoplast fusion, the putative somatic hybrids *R.* 'Frensham' + *Prunus* 'Colt' and *R.* 'Frensham' + *Rubus* 'Thornless Oregon'; Schum *et al.* (2001): *R. persica* × *xanthina*, 'Heckenzauber', 'Pariser Charme', the putative somatic hybrids *R. wichurana* + 'Heckenzauber' and *R. wichurana* + 'Pariser Charme'.

(21% w/v) for CPW21S. The pH of the CPW13M solution may be stabilized with 5 $mmol\,l^{-1}$ (2[*N*-morpholino]ethanesulphonic acid) and adjusted to 5.8. The CPW13S is adjusted to pH 7.5. The solutions are sterilized by autoclaving for 15 min at a temperature of 121 °C and pressure of 106 kPa. Freeze-dried enzymes are dissolved in CPW13M and the pH is readjusted to 5.8. A number of suitable enzyme mixtures have been found by different authors for general use with most rosaceous genotypes (**Table 2**). The enzyme solutions need to be clarified by centrifugation (15 min at 600 **g**) and should then be sterilized using a filter with a 0.2 μm pore size.

Isolation and Purification of Protoplasts

Callus that is 3–4 weeks old is preplasmolyzed in CPW13M for 30 min initially to separate the cell membranes from their associated cell wall. The CPW13M is then replaced with CPW13M enzyme solution to provide 10 ml for every 2–4 g FW of callus. The suspension is incubated at 24 °C in the dark for 16 h with gentle agitation on a rotary shaker. The digested plant material is filtered through a 100 μm nylon sieve, and the filtrate is centrifuged for 10 min at 100 **g**. The pellet is resuspended in 10 ml of CPW21S and recentrifuged. The resulting floating layer of protoplasts is made up to 5 ml with CPW21S and centrifuged for 5 min at 100 **g**. The floating protoplasts are washed twice in CPW13M, and finally resuspended in CPW13M (**Figure 1a**). Protoplast yield is assessed using a haemocytometer and expressed as the number of protoplasts per g FW of callus used. This yield can vary greatly according to the condition of the callus, its culture conditions prior to use and the genotype, but averages about 6×10^6 protoplasts g FW^{-1}. In most cases, protoplast viability exceeds 90% when determined using fluorescein diacetate (FDA). A drop of FDA (0.1% w/v in CPW13M, diluted with CPW13M from 1% w/v FDA in acetone) is added to a drop of protoplast suspension on a cavity slide and the proportion of brightly fluorescing protoplasts counted with a UV fluorescence microscope after 10 min.

Culture of Rose Protoplasts

Rose protoplasts are cultured in filter-sterilized modified KM8p medium (**Tables 1** and **2**) containing 4.5 $\mu mol\,l^{-1}$ 2,4-dichlorophenoxyacetic acid (2,4-D), 5.4 $\mu mol\,l^{-1}$ 1-naphthaleneacetic acid (NAA) and

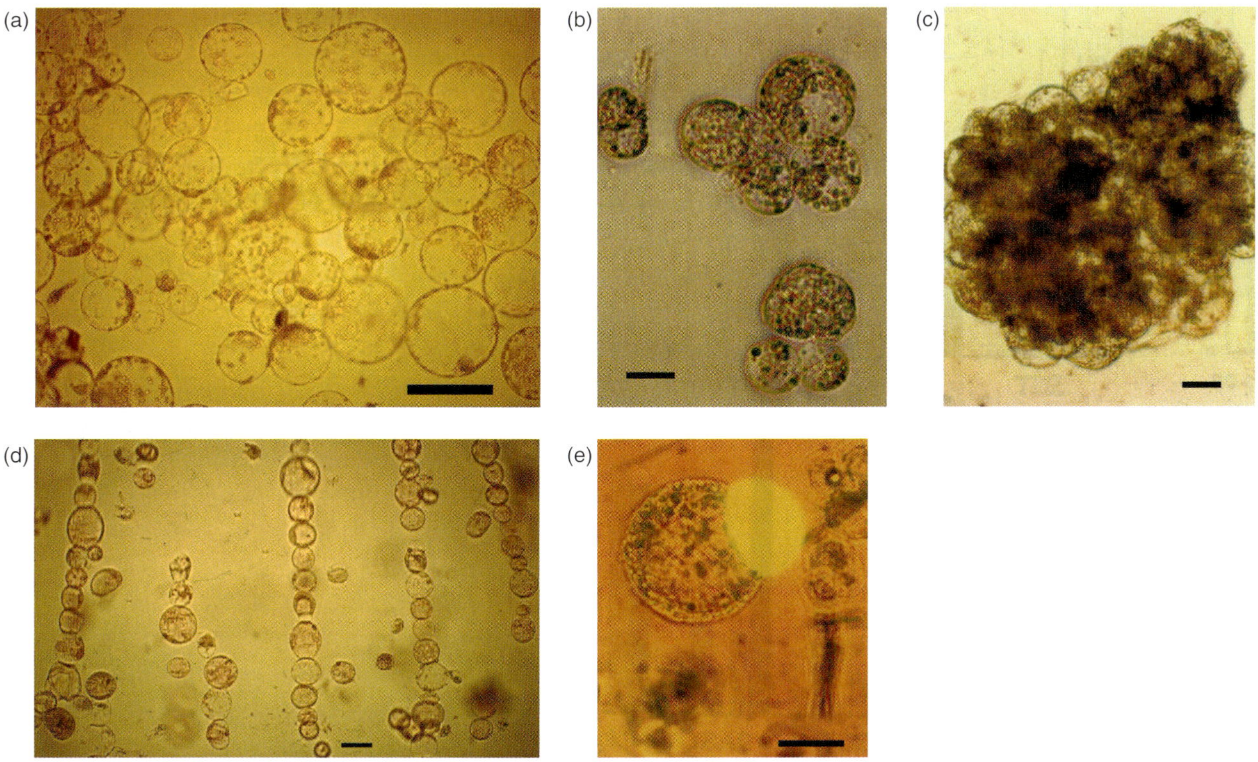

Figure 1 Fusion of rose protoplasts. Each bar line represents 100 μm. (a) Population of freshly isolated and purified protoplasts of *Rosa* 'Frensham'; (b) cultured protoplasts of *R.* 'Frensham' after 10 days of culture (courtesy of K. Yokoya); (c) cultured protoplasts of *R.* 'Frensham' after 28 days of culture (courtesy of K. Yokoya); (d) protoplasts of the same variety align along electromagnetic field lines in an electrofusion chamber prior to fusion; (e) protoplasts of *R.* 'Frensham' fuse together. The brightly glowing protoplasts were stained with fluorescein diacetate prior to mixing with unstained ones and then polyethylene glycol was added to cause adhesion. The protoplasts were viewed with a mixture of white light and ultraviolet light so that both types of protoplast could be seen.

2.2 μmol l^{-1} 6-benzylaminopurine (BAP). One ml of protoplasts suspended in 9% mannitol solution (DW9M) at a density of 5.0×10^4 protoplasts ml^{-1} is transferred to a six-well dish with 35-mm diameter wells (Flow Laboratories Inc., Rickmansworth, UK) containing 1 ml of double-strength modified KM8p medium. The dish is gently swirled to distribute the contents evenly and then incubated in the dark at 24 °C. Some reports indicate that the use of alginate films to immobilize cultured protoplasts can be beneficial if liquid culture proves unsuccessful. During the next few weeks, three main processes are initiated sequentially in the cultures: cell wall regeneration, cell division and colony formation. New cell wall formation can be detected after a few days using an ultraviolet microscope after staining samples of the culture with calcofluorowhite (0.1 ml protoplasts mixed with 5.0 μl of a 0.1% solution in water), but can be more readily assessed by observing slight changes in the spherical shape of the protoplast surface.

Cell division usually occurs within a few days (**Figure 1b**) under optimal conditions and small colonies containing 10 cells or more can be observed microscopically within 4 weeks (**Figure 1c**). Microcolonies of approximately 1 mm diameter can be seen with the naked eye within 8 weeks. After 4 weeks, 1 ml of mannitol-free modified KM8p medium is added to the cultures to reduce the level of osmoticum and replenish the available nutrients. After a further 4 weeks, the culture containing a multitude of microcolonies is poured into 9-cm Petri dishes containing 25 ml of Phytagel – solidified callus induction medium (CIM), consisting of Schenk and Hildebrandt (SH) salts and vitamins (**Table 1**) with 2,4-D (13.6 μmol l^{-1}). Colonies that form are subcultured on to fresh CIM every 4 weeks. The proportion of the plated rose protoplasts which survives and finally grows to form callus after several weeks (termed the final plating efficiency) is usually less than 1% but has been reported to be as high as about 15% by some authors. The protoplast isolation and culture conditions described above were originally developed for *R. persica* × *xanthina* callus and subsequently used for other genotypes, but alternative ones have been published for protoplasts derived from suspension cultures and leaves (**Table 2**).

Plant Regeneration from Rose Protoplasts

To date, there have been only a few reports on the regeneration of plants from protoplasts of roses (**Table 2**). Plants are more commonly regenerated through somatic embryogenesis (*see* **Cell, Tissue and Organ Culture:** *In Vitro* Regeneration Techniques). Somatic embryos can be initiated by transfer of embryogenic callus on to a medium consisting of SH salts and vitamins (**Table 1**), 30 g l^{-1} sucrose and 13.6 μmol l^{-1} 2,4-D, pH 5.8. The plates are incubated at 23 °C in a 16-h photoperiod under Grolux lights giving a light intensity of 40 μEm^{-2} s^{-1}. Globular and later-stage embryos are produced after two to three subcultures, each of 4 weeks, duration, on this medium. Somatic embryos are germinated on SH medium without growth regulators under the same environmental conditions. In one study, approximately 30% of the embryos of *R. persica* × *xanthina* developed into plantlets after 4 weeks under these conditions. Regenerated plantlets can be rooted on Sorbarod plugs (Baumgartner Papiers SA, Switzerland) with liquid rooting medium (*see* **Cell, Tissue and Organ Culture:** Micropropagation) and finally transferred to soil in a greenhouse. Somatic embryo production and plant regeneration from protoplasts have so far proved unsuccessful with some rose genotypes, including 'Albéric Barbier', *R. canina*, *R. caudata*, *R. corymbifera* 'Laxa', *R. multiflora* 'Wuci Fenhong', *R. multiflora* and *R. spinosissima*.

Protoplast Fusion

Protoplast fusion may be carried out to combine whole genomes from different parents (symmetric fusion) or to transfer partial genomes from one parent to the other (asymmetric fusion). The latter process usually involves the elimination of the nucleus of one of the parents using X-rays. Fusion of such protoplasts containing only active cytoplasmic genomes with protoplasts of another genotype containing functional nuclei leads to the formation of fusion products containing mixed cytoplasmic components, termed cybrids.

Protoplast fusion usually generates only a small number of fusion products that must be separated from the background of unfused parental protoplasts and the products of self (homologous) fusions. This can be achieved in various ways, such as the visual identification and mechanical isolation of fused protoplasts and the use of genetic complementation of auxotrophic cell lines. However, the most convenient universally applicable method involves the use of a different specific inhibitor for each of the two parental populations. When the inactivated protoplasts are fused, biochemical complementation occurs in the fusion products, so that they are able to grow further, whilst any nonfused parental protoplasts and homologous fusion products remain blocked.

Various methods have been devised to induce the fusion of protoplasts, including chemical and electrical (electrofusion) methods. Electrofusion involves the use of a fusion chamber holding electrodes between which a small volume of protoplast suspension is placed. An AC current is then applied and this causes the protoplasts to associate in chains (**Figure 1d**). The application of a high-voltage pulse of DC current for a few milliseconds then produces perforations in the plasma membranes at the point of contact of the adhering protoplasts. The fusion of these membranes, followed by the mixing of the cytoplasm and nuclei of the two protoplasts, continues during subsequent incubation of the protoplasts. Electrofusion has not, however, been used extensively with rose protoplasts.

Chemical methods are the simplest ways to fuse protoplasts. The most widely used chemical method involves the use of polyethylene glycol to bind and fuse cell membranes (**Figure 1e**). This method usually results in very low fusion frequencies and survival rates for the fusion products compared to electrofusion, but it does not require sophisticated skills or apparatus, and so is the preferred technique in many laboratories.

Chemical Inactivation of Rose Protoplasts

Protoplasts can be inactivated using rhodamine 6G (R6G) or iodoacetic acid (IOA), both of which block different parts of the energy-yielding respiratory chain and hence indirectly inhibit cell division. IOA blocks the initial biochemical process of glycolysis, whilst R6G inactivates the terminal section of aerobic respiration by attacking mitochondrial membranes. Protoplasts are treated with R6G or IOA for 25 min at the lowest concentration required to prevent cell division in each genotype. This concentration should be determined empirically for each genotype because it has been found to vary from 50 to 90 $mg\,l^{-1}$ for R6G and 80–150 $mg\,l^{-1}$ for IOA. The protoplasts are then washed three times in CPW13M and finally resuspended in DW9M at a density of 10^5 protoplasts per ml. There is no need to treat fusion partners that are unable to divide and would automatically be eliminated during subsequent culture.

Chemical Fusion of Rose Protoplasts

Protoplast suspensions of the two fusion parents, each at a density of $7 \times 10^5\,ml^{-1}$, are mixed in equal proportions and one drop (50 μl) of the mixture is placed in the centre of a Petri dish. A drop of filter-sterilized polyethylene glycol fusion medium (30 $g\,l^{-1}$ PEG 4000, 40 $g\,l^{-1}$ sucrose and 10 $mmol\,l^{-1}$ $CaCl_2$) is then added to induce fusion (**Figure 1e**). The mixture is left undisturbed for 5 min and then gradually diluted with 2 ml of modified KM8p medium. For each fusion experiment, two controls are set up consisting of nonfused R6G- and IOA-treated parental protoplasts cultured separately.

The production of callus and regeneration of plants following protoplast fusion are carried out using the same procedures as for protoplast-derived cultures described above.

Confirmation of Hybridity

Selection methods for somatic hybrids are rarely completely reliable so it is necessary to confirm, or at least provide further evidence, that plantlets regenerated from putative fusion products are true somatic hybrids before they are further propagated. There are a number of approaches to this, ranging from the morphological to the molecular and some provide more reliable evidence than others. The ones described here have been used to characterize putative somatic hybrids in roses.

Plant Morphology

The observation and measurement of various vegetative and floral characteristics provide clues to hybridity. However, the high incidence of culture-induced (somaclonal) variation often found in culture-derived rose plants means that only visible traits derived from both parental genotypes, rather than new types of variation, found in putative hybrids should be considered as preliminary indicators of hybridity. Various morphological traits have been compared in plants regenerated from fusion products involving protoplasts of *Rosa* 'Frensham' with those of *Rubus* 'Thornless Oregon' and *Prunus* 'Colt'. The fusion-derived plants were characterized by their significant variability compared to the 'Frensham' parent in features such as altered flower colour and leaf shape; occasionally more leaves developed prior to flowering, and there were smaller flowers with more petals (**Figure 2**), and shorter leaves, internodes and pedicels.

Assessment of Ploidy

The number of chromosomes in a putative hybrid can provide evidence that nuclear fusion has taken place. The morphology of chromosomes can provide further evidence of the origin of the fused nuclei. However, if the chromosomes from two genotypes are visually indistinguishable, as with most rose + rose fusions, it is possible that increased ploidy could have arisen from self-fusion of the protoplasts of a single parent. A number of methods that have been used for determining ploidy levels in rose tissues are described below.

(a)
(b)
(c)

Figure 2 Flowers of *Rosa* 'Frensham' (a), a putative somatic hybrid between *R.* 'Frensham' + *Rubus* 'Thornless Oregon' (b) and between *Prunus* 'Colt' + *R.* 'Frensham' (c). Note the effect of the fusion process on the difference in leaf shape, flower colour and number of petals between (a) and (b), and the shortened, dark red petals, leaf-like sepals and the foliar development of carpels in (c).

Chromosome counts Counting the number of metaphase chromosomes in mitotic cells is a technically simple and accurate method for determining ploidy levels, but it is time-consuming if large numbers of hybrids are to be screened. Chromosome counts can be obtained from shoot tips (*see* **Genetics**: Karyology), or by fixing 10-mm excised root tips in 3 : 1 glacial acetic acid and 5 mol l^{-1} HCl for 20 min and 10 min, respectively, at 23 °C. The root tips are then washed three times in distilled water, transferred to 70% ethanol, dissected with tungsten needles and stained with aceto-orcein. A recent investigation of products of fusions between *Rosa* 'Frensham' and *Prunus* 'Colt' and *Rubus* 'Thornless Oregon' showed identical chromosome numbers to those of the rose parent. This indicated that somatic incompatibility between the combined genomes led to the elimination of the nuclear component of the nonrose parent. Possible reasons for this are discussed in the section on DNA analysis below.

Flow cytometry Flow cytometry can be used as a rapid and accurate method of assessing ploidy levels in most soft tissues of roses, but requires expensive equipment and skilled operators. Briefly, this method involves the release of nuclei from tissues into buffer by chopping tissues with razor blades. The nuclear suspension is then filtered and treated with DNA-specific fluorescent dyes (fluorogens) that bind stoichiometrically to DNA. The suspension of nuclei is then passed through a flow cytometer (*see* **Genetics**: Nuclear DNA Amounts). This instrument irradiates the streaming suspension of stained nuclei with light of defined wavelength and detects emitted longer-wavelength radiation. The number of pulses of radiation detected is related to the number of stained nuclei and the intensity of emitted radiation from each nucleus is a function of the amount of DNA in the nucleus. Using this method, it is possible to determine the nuclear DNA amount in a putative hybrid within minutes. The value obtained for true somatic hybrids should be equal to the sum of the nuclear DNA amounts of both parents. It must be stressed, however, that flow cytometry does not directly determine chromosome size or numbers, but the latter can often be deduced from the values obtained for the nuclear DNA amounts. Most studies so far show that rose callus produced from protoplasts show the parental nuclear DNA level or its doubled equivalent, whilst plants regenerated from such calli usually have only the parental nuclear DNA level. Callus derived from fused protoplasts exhibits a wide variation in nuclear DNA levels, whilst plants regenerated from such callus often show DNA levels corresponding more closely to only one of the parents. This indicates that the fusion process may lead to genomic stress resulting in anomalous DNA levels and that the regeneration process tends to reduce the level of anomalous DNA in the nuclear genomes.

Guard cell characteristics The size and chloroplast numbers of guard cells correlate well with ploidy levels in many plant species. The size of guard cells of fully expanded leaves can be achieved by pressing the lower leaf surface on to transparent adhesive tape (e.g. Sellotape). Most of the leaf tissue is then scraped away with a scalpel, leaving the lower epidermis attached to the tape. The lengths of guard cells are measured with a microscope. Guard cell chloroplasts are best visualized by exploiting endogenous red fluorescence produced with an ultraviolet microscope fitted with a 480-nm filter. In the only published account where self-fusions were carried out to produced tetraploids of *R. persica* × *xanthina*, the chromosome number of the tetraploid was exactly double that of the diploid. The mean guard cell lengths and chloroplast numbers were also significantly increased in the tetraploid.

DNA fingerprinting The analysis of DNA fingerprints (*see* **Genetics:** DNA Fingerprinting) from regenerated protoplasts in fusion experiments is the only secure method for confirming the hybrid status of somatic hybrids. However, care has to be taken in choosing suitable DNA primers that identify sections of DNA (bands) from both parents in the hybrid nuclear DNA. If not, then it is possible for hybrids to be mistaken for one of the parental genotypes.

Total genomic DNA is extracted from 100 mg of young, fresh leaf material. Contaminating proteins are removed with two chloroform extractions and DNA is then selectively precipitated from contaminating carbohydrates in the presence of 2.5 mol l^{-1} NaCl. The extracted DNA is pure enough to be amplified by the polymerase chain reaction. Total DNA of the different plants has been compared by random amplification of polymorphic DNA (RAPD), using various primers. Electrophoresis of samples can be carried out on a horizontal slab of agarose gel (200 × 180 mm, 10 g l^{-1} agarose, 1 × TAE (tris-acetate-EDTA) buffer, 100 V for 3–4 h). DNA bands are visualized by adding ethidium bromide (50 μg l^{-1}) to the TAE buffer and viewed under a UV transilluminator.

The results of RAPD analysis show that fusion products involving parents that are both embryogenic usually show bands from both parents. If only one of the parents is embryogenic, e.g. *R.* 'Frensham' or *R.* 'Abraham Darby'™, it is possible that only bands from the embryogenic parent will appear in the fusion products because strong selection pressures favour the genomes of the regenerating partner. Novel bands belonging to neither parents, indicating that mutational events have occurred, are often found in fusion-derived cell lines. These mutations probably result from genomic stress in the fused protoplasts. This genomic stress could cause the elimination of the chromosomes of the nonembryogenic parent, leaving the chromosome complement of only the embryogenic parent. An additional factor in fusion combinations involving a rose parent and a nonrose parent could be that the culture conditions in these experiments are optimized for rose genotypes, which could lead to reduced cell vigour in nonrose genotypes. To date there has been only one report of the positive identification of somatic hybrid rose callus tissue using RAPDs and another involving a more sensitive molecular technique termed amplified fragment-length polymorphism (AFLP). In the case where RAPD analysis was used, three hybrid cell cultures were produced involving fusions between the protoplasts of *R.* 'Albéric Barbier' and *R.* 'Frensham', *R.* 'Albéric Barbier' and *R.* 'Abraham Darby', and *R.* 'Abraham Darby' and *Prunus* 'Colt'. The AFLP technique was used with putative hybrids involving 'Pariser Charme'® and *R. wichurana*. In all these cases, plants regenerated from cultures did not exhibit hybrid status and it seems likely that, during morphogenesis, the chromosomes of only one of the parents were retained.

Genetic Transformation of Rose Protoplasts

There have been only two reports so far of attempts to use protoplasts for the genetic transformation of roses. In one case, protoplasts of *R. persica* × *xanthina* and *R.* 'Allgold'® were cocultivated with *A. tumefaciens* carrying a gene for β-glucuronidase (GUS). In another report, rose protoplasts were subjected to polyethylene glycol-induced direct transformation with a plasmid carrying the green fluorescent protein (GFP). The GFP-transformed cells showed both transient and stable expression of the introduced genes, indicating successful transformation, but only the GUS-transformed protoplasts were successfully regenerated to transformed plants.

Prospects for Protoplast Technology in Roses

Commercial rose breeding relies on the introduction of novel traits into existing varieties. However, the gene pool of modern roses is narrowly based on about eight species (*see* **Genetics:** Species Crosses) and the availability of genes from some other species is limited by F_1 sterility. Protoplast fusion could widen this gene pool by the introduction of single or multigene traits through the transfer of nuclear and/or cytoplasmic genes between sexually incompatible genotypes. This transfer can be achieved without the complex procedures involved in the molecular characterization or isolation of the genes that would be required for genetic transformation. Novel genetic combinations can be brought about through protoplast fusion in three main ways:

1. The combination of two complete genomes from sexually incompatible genotypes (symmetric hybridization).
2. Partial genome transfer (asymmetric hybridization) from one genotype to another.
3. Transfer of cytoplasmic genes contained in the chloroplasts and/or mitochondria (cybridization).

Important traits that may be conferred with genome transfers include disease resistance, and novel flower and vegetative characteristics from species of *Rosa* or other genera within the family Rosaceae. Even if whole genome fusions between members of different genera lead to infertility, the resulting somatic hybrids may still be vegetatively propagated and used as ornamental plants or as rootstock varieties. In any case, such infertility may be avoided by employing asymmetric fusion, where part of the protoplast genome of one species is inactivated by irradiation prior to fusion, leading to limited gene transfer from one of the parents. This approach could be utilized, for example, where the vegetative and floral characters of one variety are combined with disease resistance from an unrelated genotype. Somatic hybridization can also generate new combinations of both nuclear and cytoplasmic genes that are difficult or impossible to achieve by conventional means. This is particularly important where valuable hybrids can only be propagated vegetatively. Protoplast fusion offers a unique way of increasing cytoplasmic variability that can be used for the analysis of the functioning, genetics and modification of cytoplasmic genes. Such genes are known to control important traits in many plant species, such as chlorophyll content, photosynthetic efficiency, resistance to certain herbicides and male sterility. In addition, the doubling of ploidy levels has already been achieved by protoplast fusion in the infertile hybrid *R. persica* × *xanthina* and so it may represent an additional technique for improving fertility in some sexual hybrids.

Protoplast technology has only been used with roses in a few laboratories and so it is not surprising that it has, so far, been unsuccessful in producing rose genotypes that are of direct proven benefit to rose breeders. Factors that have limited the implementation of protoplast fusion methods include the requirement for regenerating somatic hybrid plants from protoplasts that, even in amendable genotypes, can take several months to achieve. In addition, the use of nonspecific selection techniques that do not involve selectable genetic markers coupled with chemical fusion techniques that produce low fusion frequencies and cell viability also limits the success rate. Nonspecific culture-induced variation in ploidy levels and plant phenotype are known to occur in roses and this variation could increase the variability of the phenotypes recovered after fusion experiments. This decreases the predictability of the outcome. There is some limited evidence, as presented in this article, that this somaclonal variation may even be enhanced by the fusion process and may be used by breeders as an additional source of variation. If the variant characters are undesirable, they could be bred out of somatic hybrids or reduced by using protoplasts from differentiated rose tissues, such as leaves, thus removing the need for cell culture prior to protoplast isolation. However, protoplasts of these origins have not so far been regenerated to plants.

If the technology proves successful in the future, the source of new contributions to the rose gene pool is likely to be limited to rosaceous species. This would still, however, represent a quantum leap for rose breeding. It is also less likely to engender negative reactions from the buying public than gene transfer techniques that are used to introduce genes from unrelated species.

See also: **Cell, Tissue and Organ culture**: Micropropagation; *In Vitro* Regeneration Techniques; Protoplast Culture. **Classification**: Conventional Taxonomy (Wild Roses). **Genetics**: Karyology; Nuclear DNA Amounts; Species Crosses; DNA Fingerprinting.

Further Reading

Bajaj YPS (1994) Somatic hybridization: a rich source of genetic variability. In: Bajaj YPS (ed.) *Somatic Hybridization in Crop Improvement 1, Biotechnology in Agriculture and Forestry*, vol. 27, pp. 3–32. Berlin: Springer-Verlag.

Earle ED (1994) Analysis of somatic hybrids and cybrids obtained by fusion of *Brassica rapa* and *B. oleracea*. In: Bajaj YPS (ed.) *Somatic Hybridization in Crop Improvement 1, Biotechnology in Agriculture and Forestry*, vol. 27, pp. 305–319. Berlin: Springer-Verlag.

Galun E (1995) Protoplast-fusion derived cybrids in Solanaceae. In: Terz M, Cella R and Falavigna A (eds.) *Current Issues in Plant Molecular and Cellular Biology*, pp. 161–169. Dordrecht: Kluwer Academic.

Hinnisdaels S, Jacobs M and Negrutiu I (1994) Asymmetric somatic hybrids. In: Bajaj YPS (ed.) *Somatic Hybridization in Crop Improvement 1, Biotechnology in Agriculture and Forestry*, vol. 27, pp. 57–71. Berlin: Springer-Verlag.

Matthews D, Mottley J, Yokoya K and Roberts AV (1994) Regeneration of plants from protoplasts of *Rosa* species. In: Bajaj YPS (ed.) *Plant Protoplasts and Genetic Engineering V, Biotechnology in Agriculture and Forestry*, vol. 29, pp. 146–160. Berlin: Springer-Verlag.

Mottley J, Yokoya K, Matthews D, Squirrell J and Wentworth JE (1996) Protoplast fusion and its potential role in the genetic improvement of roses. *Acta Horticulturae* 424: 393–397.

Mottley J, Yokoya K and Roberts AV (2000) 'D-Flowering' – the flow cytometry of plant DNA. In: DiMaggio S and Diamond R (eds.) *In Living Colour: Protocols in Flow Cytometry and Cell Sorting*, pp. 391–420. Berlin: Springer-Verlag.

Roest S and Gilissen LJW (1989) Plant regeneration from protoplasts: a literature review. *Acta Botanica Neerlandica* 38: 1–23.

Roest S and Gilissen LJW (1993) Regeneration from protoplasts: a supplementary literature review. *Acta Botanica Neerlandica* 42: 1–23.

Rout GR, Samantaray S, Mottley J and Das P (1999) Biotechnology of the rose: a review of recent progress. *Scientia Horticulturae* 81: 201–228.

Schum A and Hofmann K (2001) Use of isolated protoplasts in rose breeding. *Acta Horticulturae* 547: 35–44.

Schum A, Hofmann K and Felten R (2002) Fundamentals for integration of somatic hybridization in rose breeding. *Acta Horticulturae* 572: 29–36.

Yokoya K, Walker S and Sarasan V (1996) Regeneration of rose plants from cell and tissue cultures. *Acta Horticulturae* 424: 393–397.

Cell Culture and Secondary Metabolites

T M Murphy, Section of Plant Biology, University of California, Davis, CA, USA

Introduction

The technique of culturing plant cells and tissues in defined chemical media and under aseptic conditions has been highly developed since its first successful attempt with tomato roots by White in 1934. Within a few years, the discovery and recognition of the importance of the growth regulator auxin allowed the proliferation of unorganized cells from carrot, tobacco and many other plant species. Rose tissues have been cultured *in vitro* since at least 1946. The development of culture techniques has often been spurred by the desire to study physiology and biochemistry with simplified experimental materials. Conversely, the desire to proliferate plant cells and to stimulate the development of tissues and organs has led to key discoveries in physiology, for instance the isolation of the cytokinin kinetin by Miller and Skoog in 1953.

This article will focus on culture systems producing rose cells and cell aggregates with minimal differentiation. It will first describe the types of systems used to culture plant cells. It will describe the advantages and disadvantages of various culture systems, relative to each other and to whole plants, in the study of plant and plant cell physiology and biochemistry. Finally, it will describe examples of studies using rose cells and discuss the potential of culture in the production of rose secondary products.

Culture Systems

Suspensions

Under the appropriate conditions, including agitation or aeration in liquid medium containing a balanced combination of auxin and cytokinin, rose cells will proliferate, synthesizing protein, nucleic acids and other metabolites, and dividing approximately once a day (**Figures 1** and **2**). In general, in early stages of the growth period, the cells adhere to each other, forming aggregates of 10 to hundreds of cells. As the culture matures, cell division slows and stops, the cells expand and the aggregates break apart, maintaining their relatively small size. There is no sharp demarcation

Figure 1 Suspension cultures of *Rosa damascena* Mill. Top, left to right: flasks showing the increase in turbidity from day 0 (inoculation) to day 7 and day 14. Bottom left: cell aggregate from 7-day culture. Bottom right: cell aggregate from 14-day culture. Scale marker represents 100 μm and applies to both micrographs.

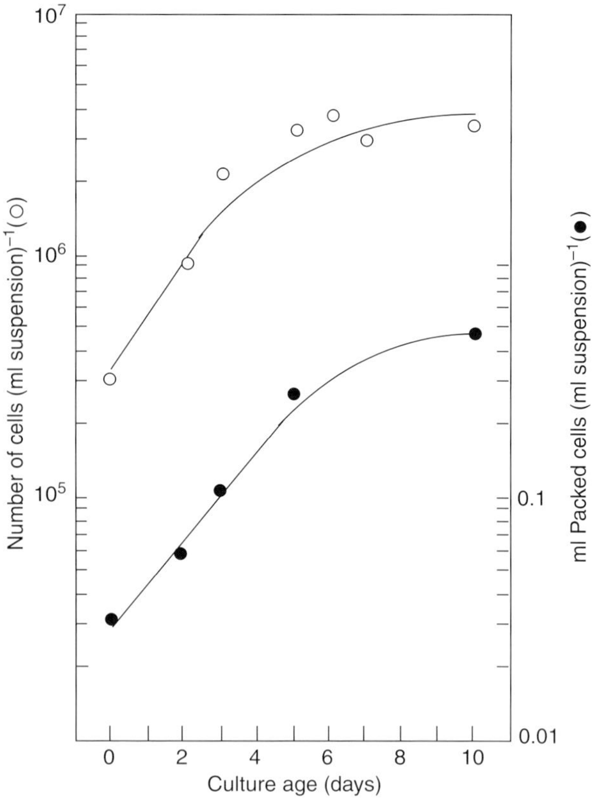

Figure 2 Growth of *Rosa damascena* Mill. in suspension culture. Five ml of 7-day-old suspension was inoculated into 50 ml of MX_1 medium (**Table 1**) in a 250-ml Erlenmeyer flask on day 0. The suspension was swirled at 100 rpm at 25 °C. Cells occurred in small aggregates: there were 90–100 cells per aggregate after 3–4 days of culture and 40–60 cells per aggregate after 10 days of culture. Reproduced with permission from Murphy TM *et al.* (1979) *Plant Physiology* 64: 936–941 © American Society of Plant Biologists. The volume of packed cells is a convenient measure of growth but overestimates the true protoplast volume by approximately a factor of 3 (Lorences EP and Fry SC (1991) *Plant Cell, Tissue and Organ Culture* 24: 211–215).

between cell division and expansion, however. The agitation and the relatively small size of the aggregates lead to the description of this system as suspension culture. Whereas with many species (e.g. tobacco) suspension cultures will grow only if inoculated at sufficiently high concentrations, at least some cultures of rose cells do not seem to be limited in this way. The cells of even a single aggregate can divide and grow.

Callus

Tissue samples growing on solid medium – nutrient medium solidified, for example, with agar or agarose and with a balanced concentration of auxin and cytokinin – will grow in size through cell division and expansion without differentiation. Such cultures are described as callus cultures. They may be further characterized as primary cultures (initiated from explants of whole plants) or as secondary or established cultures (initiated by subculturing from previously cultured material). Rose suspension cultures will form callus when they are mixed with 0.7% purified agar in MX_1 medium and plated. Because single aggregates of rose cells will form calluses with high efficiency, plating can be used to test their viability, just as with bacteria and yeast.

Differentiated Tissues and Organs

With many plant species, it is possible with appropriate combinations of hormones and other medium components to induce the initiation and growth of differentiated cells and organs. Among examples of differentiation can be the production of secondary cell walls (modelling the development of tracheids) and the production of anthocyanins and other secondary compounds. More complex development, as in the formation of embryos or the induction of shoots and roots, is generally only successful with primary cultures or ones that have been cultured over relatively few transfers. Callus from established cultures is often refractory to organogenesis; this is true of rose cultures (*see* **Cell, Tissue and Organ Culture**: *In Vitro* Regeneration Techniques). This may occur because of genetic or epigenetic changes in the DNA, or it may simply represent the loss of unidentified inductive substances present in the whole plant but lost during repeated subculturing.

Media

Most media for plant tissue culture contain the elements comprising the basic structures of proteins, nucleic acids, lipids and carbohydrates (C, H, O, N, S, P), elements needed for osmotic regulation (K, Cl), elements important for the activities of certain enzymes (Mg, Ca, Fe, Mn, B, Zn, I, Cu, Co, Mo), an energy source (a sugar), certain vitamins and the growth regulators auxin and cytokinin. The composition of a medium used to maintain a rose suspension culture for over 25 years is given in **Table 1**.

Advantages and Disadvantages

Axenicity

Plant tissue cultures exclude other organisms, including bacteria and fungi that are either pathogenic or compete for nutrients. This is an advantage for producing healthy plant tissue and for determining the characteristics of plant cells in the absence of biological stress.

Table 1 Medium for rose suspension and callus cultures

Inorganic macroelements	*mmol l^{-1}*
$MgSO_4 \cdot 7H_2O$	1.0
KCl	10.1
$CaCl_2 \cdot 2H_2O$	0.5
KH_2PO_4	1.0
Inorganic microelements	*μmol l^{-1}*
$FeSO_4 \cdot 7H_2O$	15.0
H_3BO_3	3.2
$MnSO_4 \cdot 4H_2O$	4.3
$ZnSO_4 \cdot 7H_2O$	1.7
KI	0.6
$CuSO_4 \cdot H_2O$	0.08
$CoSO_4 \cdot 6H_2O$	0.04
$NaMoO_4 \cdot 2H_2O$	0.08
Organic nutrients	*mmol l^{-1}*
Sucrose	56.0
Na glutamate	5.0
Vitamins	*μmol l^{-1}*
Myo-inositol	560
Ca pantothenate	2.1
Pyridoxine	2.4
Thiamine HCl	1.5
Nicotinic acid	8.1
Growth regulators	*μmol l^{-1}*
2,4-Dichlorophenoxyaceticacid	5.0
Kinetin	2.3

Based on MX_1 of Nash DT and Davies ME (1972) *Journal of Experimental Botany* 23: 75–81, with the substitution of Na glutamate for $NaNO_3$. See also Nesius KK *et al.* (1973) *Planta* 106: 173–176 for media used successfully with rose suspension cultures and Murashige T (1974) *Annual Review of Plant Physiology* 25: 135–166 for media used to support the development of differentiated organs from shoot tips.

Homogeneity

It is relatively easy to design culture systems that subject all the cells to the same nutrients and conditions of light, temperature and aeration. This tends to put the cells of undifferentiated cultures in the same physiological state and is particularly important for experiments on cell function or metabolism. However, an experimenter should be aware that, because of genetic or epigenetic changes, even long-term established cultures of apparently undifferentiated cell aggregates or callus may not be entirely homogeneous.

Ease of Treatments

Suspension cultures are useful for treating cells with chemicals, for instance, nutritional intermediates, growth regulators or toxins, and with ultraviolet radiation. Callus cultures are similar, but without direct evidence an experimenter cannot be certain that compounds diffuse or are transported uniformly throughout a large callus.

Ease of Extraction

Rose cells in suspension and calluses are thin-walled and tend to contain few tannins. Thus it is easy to extract metabolites, enzymes and nucleic acids from them. Plant cells secrete some metabolites and proteins, potentially allowing the collection of products from continuous suspension cultures – cultures in which growth medium is removed and fresh medium added continuously or periodically.

(Lack of) Differentiation

While the lack of differentiation and relative homogeneity of suspension and callus cultures is an advantage for experiments on basic cell functions and metabolism, it naturally mediates against experiments on the control of differentiation and complex development. Even in cultures in which development can be induced, it is not necessarily possible to conclude that the regulation of development is the same *in vitro* as in whole plants.

Genetic Instability

It has often been noted that over long periods of time and repeated subculturing, plant cell cultures develop changes in phenotypic or chromosomal characteristics. Over long periods of time, one may expect cultured strains to adapt to the conditions of culture, and this may involve the loss of genes important in the reproduction and development of intact plants. This is a disadvantage relative to the use of plants, in which mutations may be selected against, either naturally or by an experimenter.

Maintenance

Intact plants can be preserved as seed or, as in the case of rose varieties, propagated as long-lived clones. In contrast, the preservation of cultures is more difficult. Cells in suspension cultures begin to die after as little as 1–2 weeks. One method to maintain such cultures is to subculture them (dilute them into fresh media) at regular intervals. Alternatively, it is possible to minimize the work, as well as the danger of genetic changes that are inherent in the maintenance of cultures through continuous subculture, by the use of cold temperatures to slow metabolism. Suspension cultures of rose cells, which are relatively tolerant to cold and anaerobic conditions, can be stored in a refrigerator for several weeks. Undifferentiated plant cells of many plant species, including rose, can be stored for long periods under cryogenic conditions by treating them first with protective agents, such as glycerol and dimethylsulphoxide (*see* **Cell, Tissue and Organ Culture:** Storage and Cryopreservation).

Physiological Studies

Undifferentiated cultured plant cells can be used to study many aspects of cell physiology, including the regulation of metabolic pathways, processes of mitosis, cytoplasmic streaming, gene expression, the genesis of many organelles and certain responses to the environment. Below are described two examples of research in which rose suspension cultures have played important roles.

Cell Wall Composition and Synthesis

The wall is a key to understanding much about the structure and growth of plant cells. The primary cell wall, even though thin, is exceptionally strong, resisting turgor pressure that typically amounts to 0.7 MPa (7 atm, 7 kg cm^{-2}, approximately 100 lb in^{-2}) and can be much greater. Breaking the chemical bonds that hold components of the wall together, for instance with cellulase, polygalacturonase or other hydrolytic enzymes, destroys the wall and allows unrestricted swelling of the protoplast (leading to death of the cell). On the other hand, the wall must be able to expand slowly and plastically in order to allow growth of the cell. The expansion of the wall depends on: (1) sufficient turgor pressure; (2) active metabolism, including the synthesis of new wall material; (3) acidification of the solution that permeates the wall. During growth, chemical bonds that hold the preexisting wall components together are broken and new components with new bonds are inserted. The question of how bonds that are strong enough to contain high pressure can be broken without destroying the integrity of the wall has been a focus of research for many years.

Suspension-cultured rose cells – specifically, an established culture of cells from 'Paul's Scarlet Climber' rose – have been the subject of several investigations into the nature and synthesis of the cell wall. The walls of these cells contain the basic components found in the walls of other dicotyledonous plants: cellulose, arabinogalactans, xyloglucans, pectins and proteins. They tend to lack the phenolic molecules that are often found in other plant cell walls and that, through cross-linking, can confound the extraction and analysis of other components.

In the mechanism of wall extension, xyloglucans may play an important role. Xyloglucans are polysaccharides composed of xylose, glucose, galactose and fucose residues. They are synthesized inside the cell and secreted through the plasma membrane by way of Golgi vesicles. In rose cell suspension cultures, xyloglucans can be rapidly labelled with radioactivity by the addition of [^{3}H]arabinose (a portion of which is converted to xylose). Some of the radioactivity found in newly synthesized xyloglucan is soluble or weakly bound to the cell wall, but much is bound very strongly.

The strong binding of the new xyloglucans to the rose cell walls can be ascribed to transglycosylation, and this process may also help explain wall extension. Transglycosylation is the transfer of a sugar residue or a polymer of sugar residues from one polysaccharide (donor) to the nonreducing end of another polysaccharide (acceptor). Acting either as donor or acceptor, a newly synthesized xyloglucan can attach covalently to a preexisting xyloglucan that is bound strongly to cellulose or other wall components. By growing cells in medium containing glucose labelled with heavy isotopes (^{2}H, ^{13}C) followed by glucose labelled with light isotopes (^{1}H, ^{12}C), S.C. Fry and coworkers could distinguish between preexisting and newly synthesized xyloglucans. They provided a pulse of radioactively labelled sugar after the transition to light isotopes and found that the radioactive xyloglucan had a density intermediate between heavy and light, indicating that some of the newly synthesized xyloglucan had been covalently attached to preexisting, heavy xyloglucans. Over time (hours), the radioactive xyloglucan became progressively heavier, reflecting a continued change in connections between old and new molecules. Over longer times (days), the radioactive xyloglucan became lighter, as new light xyloglucan was added to the wall. A fraction of the xyloglucans, about one-third, is covalently attached to acidic pectins. How these attachments occur, and what significance they have to growth is not known.

The model of cell wall metabolism that emerges from these studies suggests a continual rearrangement of connections between wall components that allows growth when the wall is subjected to pressure and new material is available (**Figure 3**). Although similar studies are being conducted with other plant materials, the rose suspension cultures have been particularly useful because of their rapid growth, the lack of differentiation of their cell walls and the facility with which experimenters can change medium components.

'Paul's Scarlet' rose cultures have also been useful in obtaining and characterizing arabinogalactan-proteins (AGPs). AGPs are complex molecules found in cell walls of all plants. Each AGP contains a polypeptide chain coded by one of several genes (there are at least eight in *Arabidopsis*) that is covalently connected to large, branched polysaccharide chains. At least some AGPs have an attached glycosylphosphatidyl-inositol lipid, which serves to anchor them in the plasma membrane, although phospholipase C activity can break the attachment. Using cultured rose cells, E. A. Nothnagel and his coworkers have isolated AGPs from the plasma membrane, cell walls and culture medium. Although the functions of AGPs have

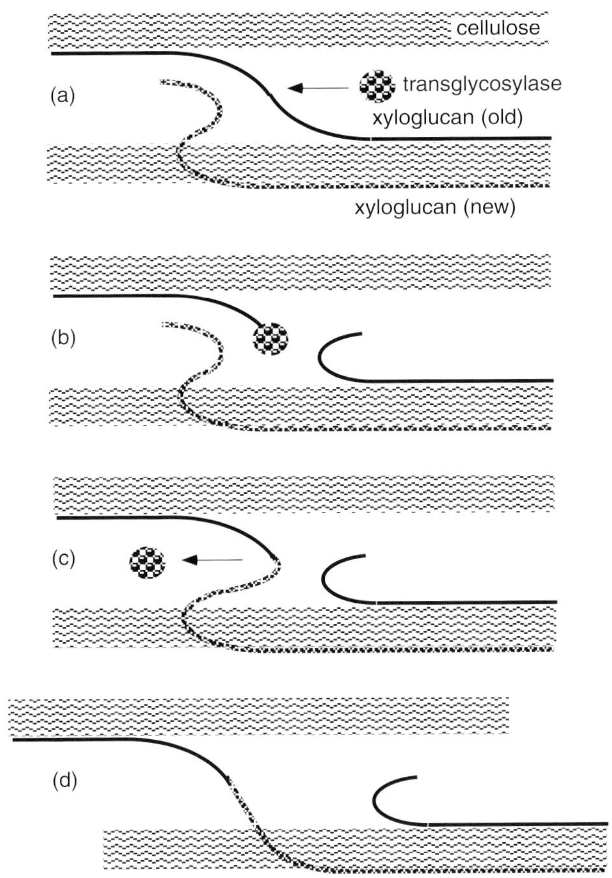

Figure 3 Transglycosylation leading to expansion of a cell wall. (a) Preexisting ('old') and newly secreted ('new') xyloglucans are both tightly bound to cellulose microfibrils through hydrogen bonds. The new xyloglucan has a free nonreducing end. (b) Xyloglucan transglycosylase attacks the old xyloglucan, cleaving it but staying attached to the reducing end. (c) The transglycosylase attaches the reducing end of the old xyloglucan to the nonreducing end of the new, and is released. (d) Under turgor pressure, the new arrangement of xyloglucans allows microfibrils to slip relative to each other. Adapted from Fry SC *et al.* (1990). Oligosaccharins from xyloglucan and cellulose: Modulators of the action of auxin and H^+ on plant growth. In Roberts J, Kirk C and Venis M (eds) *Hormone Perception and Signal Transduction in Animals and Plants*, pp. 285–298. Cambridge: Company of Biologists, Ltd © Society for Experimental Biology with permission.

not been identified, they may connect the plasma membrane and the cell wall. Furthermore, their complex structures and locations suggest that they may serve as signals (or receptors) in the regulation of growth and/or development of plant organs. For instance, the presence of specific AGPs marks developing xylem tissue in dicot roots. The development of xylem tracheids involves autolysis of the protoplast, a type of programmed cell death. Thus, the observation that reagents that bind to and aggregate AGPs inhibit growth and stimulate programmed cell death in rose cell cultures suggests that these cultures may provide a useful system for elucidating steps in the signal pathway leading to this developmental process.

Oxidative Burst

Plants respond to various physical and biological stresses – especially the presence of certain pathogens – with an oxidative burst, a rapid production of reactive oxygen compounds, including superoxide (O_2^-) and hydrogen peroxide (H_2O_2). The significance of this response is unclear. Originally, it was thought to protect against pathogen infection, either by killing the pathogens or killing the host cells near the point of infection. (This would be analogous to a process by which white blood cells in mammals inactivate bacterial invaders that they ingest by phagocytosis.) More recently it has been suggested that hydrogen peroxide is a signalling compound that in intact plants travels to and induces defences in cells at a distance from the area of infection (*see* **Breeding:** Selection Strategies for Disease and Pest Resistance). The ability to respond is a property of individual cells, making suspension cultures a useful tool in investigating the mechanism of perception of the stress and regulation and synthesis of the reactive oxygen.

Suspension cultures of *Rosa damascena* have been shown to produce superoxide and hydrogen peroxide when they are irradiated with far-ultraviolet radiation (254 nm, produced by a mercury vapour lamp). The addition of a chemical elicitor, contained in the partially hydrolysed components of cell walls from a *Phytophthora* species isolated from a rose plant, has the same effect. The superoxide is apparently formed outside the plasma membrane. The mechanism is unconfirmed, but probably involves the transfer of electrons from cytoplasmic NAD(P)H to extracellular O_2 by a reductase intrinsic to the plasma membrane. Hydrogen peroxide is produced by the dismutation of superoxide, either through the agency of superoxide dismutase, or nonenzymatically in the mildly acidic conditions of the cell wall.

The oxidative burst in rose cells may be related to changes in the flow of ions across the plasma membrane (**Figure 4**). Both far-ultraviolet radiation and *Phytophthora* elicitor induce a rapid loss of K^+ ion from the cells. This occurs with about the same kinetics as the production of reactive oxygen, that is, it starts about 15 min after the treatment and continues for 1–2 h. An initial effect of ultraviolet radiation is an inhibition of the plasma membrane adenosine triphosphatase that is responsible for the production of the trans-plasma membrane electrical potential through proton efflux. There is a virtually immediate reduction of the plasma membrane potential. This may be followed by an influx of Ca^{2+}: an external

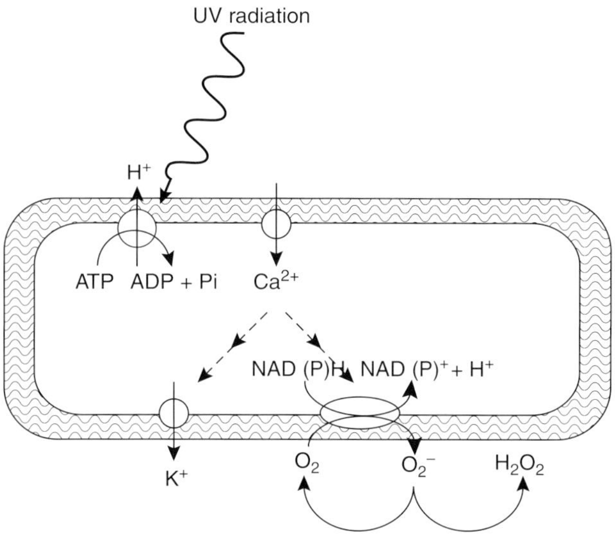

Figure 4 Plasma membrane events following far-ultraviolet (UV) irradiation of rose cells. UV radiation inactivates the H^+ pumping adenosine triphosphatase by oxidation of trp residues in the enzyme (Imbrie CW and Murphy TM (1984) *Photochemistry and Photobiology* 40: 243–248). In the absence of the proton pump, membrane potential drops (Huerta AJ and Murphy TM (1989) *Physiologia Plantarum* 77: 525–530), possibly leading to the opening of a Ca^{2+} channel. K^+ efflux and the reduction of O_2 to form superoxide (O_2^-) both depend on Ca^{2+} influx (Murphy TM (1988) *Physiologia Plantarum* 74: 537–543), probably mediated by a protein kinase, but the mechanisms are still undetermined.

concentration of at least 10^{-4} mol l^{-1} Ca^{2+} is essential for the efflux of K^+. The efflux of K^+ is accompanied by an efflux of HCO_3^- and consequent acidification of the cytoplasm by 0.2–0.3 pH units. In other plant systems, there is a coincidence between ion fluxes and an oxidative burst, and an influx of Ca^{2+} has been shown to be required for the induction of an oxidative burst. However, the functional connection between ion fluxes and the oxidative burst, if any, has yet to be discovered. Many other questions remain, including how the *Phytophthora* elicitor induces K^+ efflux and an oxidative burst and why inhibitors of protein synthesis have a similar effect. The rose cell system has been useful in these studies because of its homogeneity and small aggregate size, which allows rapid transport of materials between the cell apoplast and the bulk medium.

Secondary Compound Synthesis

Examples from Other Plants

Plants produce a wide variety of compounds, many of which are considered secondary in the sense that they are not intermediates in energy metabolism or the synthesis of compounds essential to the life of any cell. However, the secondary compounds have important functions in the physiological ecology of a plant species. Some provide protection against physical stress (such as flavonoids, which shield cells from solar ultraviolet radiation). Some, such as alkaloids and cyanogenic glycosides, protect against herbivores or pathogens. Aromatic terpenoids and colourful anthocyanins and carotenoids serve to attract pollinators and seed-dispersing animals. Many of these compounds are useful commercially, for example as food additives or pharmaceuticals (*see* **Secondary Metabolites**: Culinary Uses and Nutritional Value; Medicinal and Pharmaceutical Uses).

Although many – perhaps most – secondary compounds produced by plants are associated with differentiated cells and tissues, it has been possible to coax cultured cells to produce measurable amounts of some interesting chemicals. Compounds that have been identified in plant cultures include saponins from *Panax ginseng*, digitoxin from *Digitalis purpurea*, nicotine from tobacco, morphine and codeine from *Papaver* species, caffeine from *Coffea arabica*, scopoletin from *Nicotiana tabacum* and serpentine from *Catharanthus roseus*. The levels of such compounds are generally lower in cultured cells than in intact plant organs, but can sometimes be increased markedly by manipulating the medium or selecting high-yielding cell lines. One approach is biotransformation, feeding cell cultures late intermediates of a biochemical pathway with the intent that the cells will provide one or more enzymatic steps to make a final product (or at least a useful intermediate that is difficult to form by synthetic chemistry).

Rose Phenolics

For rose lovers (and many animal pollinators), the colours of petals represent an important visual attraction. These colours are the result of the accumulation of phenolic compounds – yellow flavones and flanonols and red anthocyanins – in the vacuoles of the cells.

Under certain conditions, rose cultures are able to produce and store phenolic compounds. In a 1972 classic paper, M.E. Davies described the phenolic compounds in a suspension culture of 'Paul's Scarlet' rose. Phenolic compounds tend to accumulate in the later stages of culture growth. The total amount of phenolics depends on the concentrations of auxin (being reduced by higher concentrations of 2,4-dichlorophenoxyacetic acid and naphthaleneacetic acid), of carbon source (increased by higher concentrations of sucrose), of nitrate (reduced by higher concentration) and light (stimulated relative to growth in dark). In the light, anthocyanins are formed, giving Davies's cultures a deep purple-red colour late in the growth cycle. The light-stimulated induction of

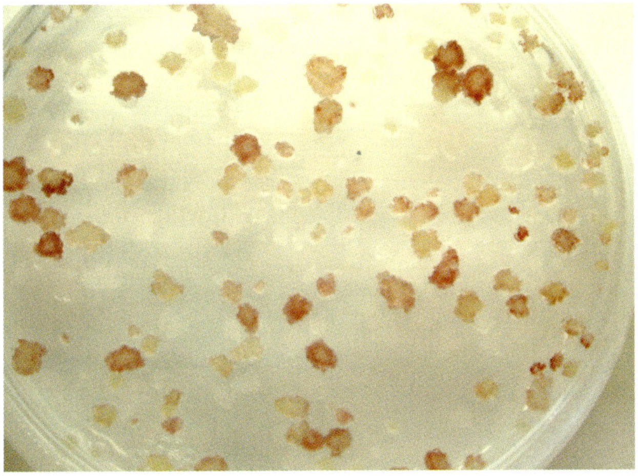

Figure 5 Callus cultures of *Rosa damascena* Mill. Cell aggregates in suspension culture, normally pale yellow, were plated in agar, then subjected to refrigerator temperature followed by high light to induce anthocyanin synthesis.

phenolics is associated with the induction of phenylalanine ammonia lyase, the first enzyme in the biochemical pathway leading to phenolics. Although since that time little has been done in this field with rose cultures, this system could be used to study the molecular aspects of the regulation of phenolic metabolism in rose cells.

The accumulation of anthocyanins in rose cell aggregates provides a clear representation of the heterogeneity of the differentiation states of cells in culture (**Figure 5**).

Rose Oil (Potential)

Another substance produced by roses, and one with substantial commercial value, is rose oil, a complex mixture of monoterpene-based compounds, including linalool, citronellol, nerol, geraniol and β-phenylethyl alcohol (*see* **Fragrance and Pigments**: Odoriferous Substances and Pigments; Fragrance Profiles of Wild and Cultivated Roses. **Secondary Metabolites**: Extraction and Industrial Processes). Attempts to obtain rose oil from cultures have been made since the 1970s. Several unsuccessful attempts to produce monoterpenes in rose cell cultures have been reported, but in one successful procedure mixtures of monoterpenoids were found in callus cultures from petals, sepals and leaves of Damask rose. In this report, the accumulation of terpenes was said to be slow, requiring several weeks of culture, but eventually a mixture similar to that of rose oil from an intact plant was achieved. In another type of study, a rose culture was shown to convert geraniol and its pyrophosphate to a mixture of nerol, citral-a and citral-b and citronellal. 'Paul's Scarlet' rose cultures have been reported to form more complex terpenoids, including four different carotenoids (tetraterpene-derived) and the triterpenoid, β-amyrin.

The available information thus suggests that under the right conditions rose tissue cultures can produce rose oil compounds. At higher concentrations, the components of rose oil can be toxic. In plants, they are often confined to specialized cells and organs. Undifferentiated cells may protect themselves through biochemistry: cultured cells of *Citrus limon* have been shown to metabolize geraniol, and this may also occur in cultured rose cells. A better understanding of the control of monoterpene catabolism could lead to more success in using rose cultures for the production of rose oil components.

See also: **Cell, Tissue and Organ Culture**: *In Vitro* Regeneration Techniques; Storage and Cryopreservation. **Fragrance and Pigments**: Odoriferous Substances and Pigments; Fragrance Profiles of Wild and Cultivated Roses. **Secondary Metabolites**: Culinary Uses and Nutritional Value; Medicinal and Pharmaceutical Uses; Extraction and Industrial Processes. **Selection Strategies**: Resistance to Diseases and Pests.

Further Reading

Arnott T and Murphy TM (1991) A comparison of the effects of a fungal elicitor and ultraviolet radiation on ion transport and hydrogen peroxide synthesis by rose cells. *Environmental and Experimental Botany* 31: 209–216.

Charlwood BV and Rhodes MJC (eds.) (1990) *Secondary Products from Plant Tissue Culture*. Oxford: Clarendon Press.

Davies ME (1972) Polyphenol synthesis in cell suspension cultures of Paul's Scarlet rose. *Planta* 104: 50–65.

Fry SC, McDougall GJ and Lorences EP *et al.* (1990) Oligosaccharins from xyloglucan and cellulose: modulators of the action of auxin and H^+ on plant growth. In: Roberts J, Kirk C and Venis M (eds.) *Hormone Perception and Signal Transduction in Animals and Plants*, pp. 285–298. Cambridge: Company of Biologists.

Fu T-J, Singh G and Curtis WR (eds.) (1999) *Plant Cell and Tissue Culture for the Production of Food Ingredients*. New York: Kluwer Academic/Plenum.

Huerta AJ and Murphy TM (1989) The kinetics of efflux of K^+ from cultured rose cells treated with ultraviolet light. *Physiologia Plantarum* 77: 525–530.

Imbrie CW and Murphy TM (1984) Mechanism of photoinactivation of plant plasma membrane ATPase. *Photochemistry and Photobiology* 40: 243–248.

Kireeva SA, Bugorskii PS and Reznikova SA (1977) Cultivation of Damask rose tissue and accumulation of terpenoids in them. *Soviet Plant Physiology* 24: 673–679.

Majewska-Sawka A and Nothnagel EA (2000) The multiple roles of arabinogalactan proteins in plant development. *Plant Physiology* 122: 3–9.

Murphy TM (1988) Ca^{2+} dependence and La^{3+} interference of ultraviolet radiation-induced K^+ efflux from rose cells. *Physiologia Plantarum* 74: 537–543.

Murphy TM and Auh C-K (1996) The superoxide synthases of rose cell plasma membrane. *Plant Physiology* 110: 621–629.
Murphy TM and Huerta AJ (1990) Hydrogen peroxide formation in cultured rose cells treated with ultraviolet light. *Physiologia Plantarum* 78: 247–253.
Nash DT and Davies ME (1972) Some aspects of growth and metabolism of Paul's Scarlet rose cell suspensions. *Journal of Experimental Botany* 23: 75–81.
Nesius KK, Uchytil LE and Fletcher JS (1972) Minimal organic medium for suspension cultures of Paul's Scarlet rose. *Planta* 106: 173–176.
Staba EJ (ed.) (1980) *Plant Tissue Culture as a Source of Biochemicals*. Boca Raton, FL: CRC Press.
Street HE (ed.) (1977) *Plant Tissue and Cell Culture*, 2nd edn. Berkeley: University of California Press.
Thompson JE and Fry SC (2001) Restructuring of wall-bound xyloglucan by transglycosylation in living plant cells. *Plant Journal* 26: 23–34.
Thompson JE, Smith RC and Fry SC (1997) Xyloglucan undergoes inter-polymeric transglycosylation during binding to the plant cell wall *in vivo*: evidence from $^{13}C/^{3}H$ dual labelling and isopycnic centrifugation in caesium trifluoroacetate. *Biochemical Journal* 327: 699–708.

Storage and Cryopreservation

P T Lynch, Division of Biological Sciences, University of Derby, Derby, UK

Introduction

The usual method of choice for conserving plant germplasm is by collection and storage of seeds. However, this approach is generally not suitable for roses as elite varieties do not come true from seed and seed germination rates can be low. Many rose varieties are maintained in gardens worldwide, often in association with rose and horticultural societies. Rose breeders also maintain collections to support their breeding programmes, the characteristics of which tend to be determined by commercial pressures (*see* **Rose Collections and Trials**: Rose Collections and Herbaria; Rose Societies; Awards). However, plants preserved under glasshouse or field conditions are vulnerable to loss due to pests, diseases and environmental factors, such as extreme weather conditions, and are expensive to maintain in terms of labour and land use. Biotechnological approaches provide an important alternative and have the advantage of allowing the storage and distribution of disease-free material and the potential to allow the rapid multiplication of specific accessions when required.

The ability to maintain a diversity of rose tissue types (e.g. shoot cultures, somatic embryos, callus and cell suspension cultures) under *in vitro* conditions is an essential requirement for the exploitation, by the rose industry, of biotechnological techniques, such as clonal propagation (micropropagation) and genetic manipulation (e.g. somatic hybridization). However, the maintenance of *in vitro* cultures in an active growth state can be expensive, in terms of the need for specialist laboratory facilities (e.g. environmentally controlled rooms), laboratory consumables (e.g. culture media) and skilled personnel. Valuable cultures are vulnerable to loss due to equipment failure, microbial contamination, human error and the effects of time-related morphological and genetic changes which are the basis of somaclonal variation. Hence approaches to store plant germplasm under secure *in vitro* conditions have been developed to support biotechnology programmes; additionally such approaches have an important role in the conservation of plant germplasm.

Reducing the growth rate, hence increasing the period between subculture cycles, of *in vitro* plant cultures can provide a convenient option for medium-term germplasm storage. Medium-term storage under reduced (slow or minimal) growth conditions provides a cost-effective option and is now used, for example, by major international germplasm centres for vegetatively propagated crops and commercial micropropagation companies. However, it is not suited to long-term storage, because of risks of selection due to stresses imposed on the cultures during storage. The only option for the long-term storage of *in vitro* propagules is cryopreservation. Cryopreservation is the preservation of cells, tissues and organs at ultralow temperatures, normally in liquid nitrogen (−196 °C), in such a manner that viability is maintained after thawing. Although *in vitro* storage is increasingly used to conserve plant germplasm, *in vitro* rose cultures have received relatively little attention. In this article the practical approaches for the medium-term storage (slow growth) and long-term storage (cryopreservation) of rose *in vitro* cultures will be discussed.

It is important to remember that a good understanding of the tissue culture response of germplasm to be stored *in vitro* is vital to the success of medium- or long-term storage. Although rose tissue can be difficult to manipulate under *in vitro* conditions, the extensive studies on rose micropropagation by shoot multiplication support *in vitro* storage since such shoots provide appropriate tissue for germplasm storage.

Medium-Term Storage (Slow Growth)

A variety of approaches have been used separately and in combination to reduce the growth rates of *in vitro* cultures, ranging from temperature reduction or environmental manipulation to chemical additions in the growth media. Although this approach to *in vitro* storage is used in a number of laboratories worldwide, there is relatively little literature on the subject. The most generally successful strategies involve reducing the culture incubation temperature. Cold-tolerant species, including members of the Rosaceae, have been successfully maintained at 0–4 °C for over a year between subcultures. The duration tissue can be stored under reduced temperature and slow growth conditions can be enhanced by reducing the light intensity. Modifications to the culture medium, including the addition of osmotically active compounds, such as mannitol, and the reduction of the medium's nutrient status permit maintenance of cultures in slow growth. Overlaying cultures with refined mineral oil, which reduces the available oxygen, has also been successfully used to extend subculture periods. The simplicity and cost of this particular approach make it particularly useful in 'low-tech' situations. For members of the Rosaceae, such as *Malus* and *Prunus* species, the most effective approaches have involved maintenance at reduced incubation temperatures (generally 1–4 °C) in the dark or with a 16-h photoperiod. However, it must be remembered that response to slow-growth storage conditions varies significantly between species and even varieties. Where slow-growth storage involved media modification, particularly of growth regulators, the response is even more genotype-dependent.

Latent microbial contaminants often become evident during storage, which can lead to the loss of valuable germplasm. This problem highlights the need to index *in vitro* cultures for latent bacterial and fungal contaminants. Indexing using bacteriological media to detect microbial contaminants is more effective than visual examination of cultures. Such disease indexing will also provide safer germplasm for distribution. Using heat-sealed tissue culture bags can reduce microbial contamination during storage.

Medium-term storage has been applied to callus cultures and shoot cultures. However, most success in terms of post-storage recovery of phenotypic stability has been achieved with shoot cultures. Callus-cultures have generally exhibited significant morphological and biochemical changes after relatively short periods of slow-growth storage as a result of somaclonal variation. This observation is probably a reflection of the genetically less stable nature of callus and suspension cultures. Although slow growth is an important approach for plant germplasm storage, even with increased time between subculture, the maintenance of large *in vitro* collections can be problematic.

Cryopreservation

To date, the only approach which has been successfully used for the cryopreservation of rose shoot tips is encapsulation/dehydration, which is based on artificial seed technology.

Preculture and Preparation of Tissue for Cryopreservation

Shoot tips derived from micropropagated shoot cultures are a suitable tissue for cryopreservation, since under *in vitro* conditions they can regrow directly into shoots (*see* **Cell, Tissue and Organ Culture:** Micropropagation; Rooting of Micropropagated Shoots). Avoiding the production of plants through the formation of adventitious shoots and regeneration from callus is important since plants derived from these *in vitro* propagation routes are susceptible to somaclonal variation.

Preculture treatments are usually applied to tissue before cryopreservation, to enhance post-thaw survival. Typical treatments include cold acclimation/hardening, the application of osmotic agents which reduce the water content of tissue prior to freezing and treatment of tissue with chemicals known to be associated with cold hardening, for example proline and abscissic acid. The specific details of preculture treatments and their significance to postthaw regrowth are highly genotype-dependent. Interestingly, the protocol reported for the cryopreservation of rose shoot tips does not include a preculture treatment. However, the assessment of the influence of preculture treatments on the postthaw regrowth of rose requires further investigation.

Shoots tips of rose for cryopreservation should be excised from healthy, proliferating shoot cultures maintained under normal growth conditions, that are approximately halfway through their normal subculture cycle. Terminal buds of shoots, with 2 mm long shoot tips, comprising the meristematic dome with one pair of leaf primordia, are dissected aseptically under a stereomicroscope using a scalpel and fine forceps (**Figure 1**). Commonly, when dissecting rose shoot tips the tissue rapidly browns, as a result of oxidative browning. This can be overcome by dissecting shoot tips in a Petri dish containing an antioxidant solution (0.2 mol l^{-1} phosphate buffer, pH 5.7, containing 50 g l^{-1} ascorbic acid and 15 g l^{-1} sodium borate), but particular care is required to ensure that the sterility of the excised shoot tips is maintained.

To protect the excised rose shoot tips from damage during the cryopreservation process, shoot tips are individually encapsulated in calcium alginate beads. To achieve this, excised shoot tips are suspended in liquid, calcium-free shoot multiplication medium

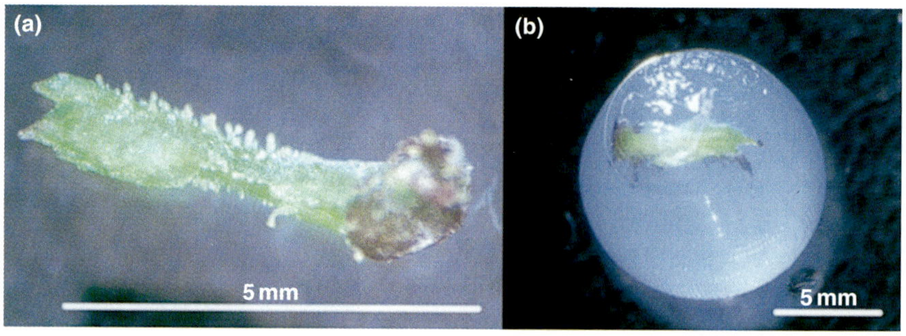

Figure 1 (a) Excised rose shoot tip; (b) non-frozen shoot tip encapsulated in calcium alginate bead.

containing 30 g l^{-1} sodium alginate (*see* **Cell, Tissue and Organ Culture**: Micropropagation). The shoot tips are drawn up into sterile, wide-bore 10 cm^3 disposable pipettes and dripped into liquid shoot multiplication medium containing 14.7 g l^{-1} calcium chloride. The sodium alginate surrounding the shoot tips polymerizes in the presence of calcium to form calcium alginate; thus, each shoot tip is encapsulated in a bead of calcium alginate. The beads are incubated in the calcium-enriched medium for between 15 and 30 min to ensure the complete polymerization of the sodium alginate around the excised shoots. The beads are then collected on a sterile nylon mesh filter prior to cryopreservation preculture treatment.

Osmoprotection and Dehydration Treatments

Osmoprotection and dehydration treatments are required to ensure that the tissue to be cryopreserved is able to survive the dehydration and chilling stresses associated with cryopreservation. Osmoprotection is achieved by subjecting encapsulated tissue to osmotic dehydration by incubation in sucrose solutions, normally 0.5–0.75 mol l^{-1}. As sucrose cannot penetrate cell membranes it acts as an external osmoticum capable of inducing passive water efflux and hence cell plasmolysis. This treatment results in cellular dehydration, but several other properties of sucrose assist postthaw survival and regrowth, including its ability to stabilize cell membranes and proteins during desiccation and promote non-crystalline ice formation in the cellular matrix, an important factor in freeze tolerance. After sucrose incubation the encapsulated tissue is physically desiccated, normally to a water content of approximately 20% (fresh-weight basis). This can be achieved by placing the beads in the sterile air current of a laminar air-flow cabinet or over silica gel in a sealed jar. These treatments ensure that the moisture content of the tissue is sufficiently low so that there is no water available for ice formation and the viscosity of the cell becomes so high that water freezes to a noncrystalline amorphous 'glassy state'.

After encapsulation, rose shoot tips are transferred to a liquid shoot multiplication medium containing 0.5 mol l^{-1} sucrose (10 ml of medium per 5-cm Petri dish containing five beads). The dishes are sealed with wax laboratory film (e.g. Nescofilm, Nippon Schoji Kaisha Ltd, Osaka, Japan) and incubated at 18 °C for 24 h in the dark with gentle shaking (40 rpm) on an orbital shaker. After osmoprotection the beads are removed from the sucrose-enriched culture medium and individually rapidly surface-dried by rolling them on sterile filter paper (Whatman no. 1). The beads, in groups of 10, are then transferred on to 5.5-cm filter papers in the lids of 5-cm Petri dishes which are aseptically placed in 250-ml glass powder jars containing 80 g of silica gel (the jars were sterilized and the silica gel dried by incubation at 160 °C for 6 h). The jars are sealed with a screw lid and wax laboratory film. The beads are desiccated over silica gel for 2 h in the dark.

Freezing

Precultured/cryoprotected cells and tissue may be frozen and placed into cryogenic storage using one of two approaches, involving either direct rapid immersion into liquid nitrogen or controlled-rate (programmable) cooling. Rapid freezing is normally used to induce tissue vitrification. Vitrification is the process by which water undergoes a phase transition from a liquid to an amorphous 'glassy state'. In this form water does not possess a crystalline structure, whereas when plant tissue is cooled at a controlled rate in the presence of colligative cryoprotectants (e.g. glycerol and dimethyl sulphoxide) ice crystal formation occurs extra-cellularly. This results in the movement of unfrozen water from inside the cells moving extra-cellularly, resulting in cell dehydration and an increase in intra-cellular solute concentration high enough to prevent intra-cellular ice formation when the samples are plunged into liquid nitrogen. Vitrification has become the main approach for plant germplasm cryopreservation, particularly of organized tissue, while controlled-rate freezing, regarded as the traditional cryopreservation approach, tends to be used for *in vitro* cell cultures (callus and cell suspensions).

After sucrose treatment and silica gel dehydration, encapsulated rose shoot tips are transferred into 2-ml polypropylene cryovials (Starsted Ltd, Boston Road, Leicester, UK), 10 beads per vial. The vials are then clipped on to aluminium canes (**Figure 2**) and plunged into liquid nitrogen in a small benchtop liquid nitrogen dewar and then transferred into the long-term storage dewar.

Cryogenic Storage

Storage dewers (sometimes known as liquid nitrogen refrigerators) are twin-walled vacuum vessels. The two walls are made of metal and are joined at the neck by some insulating material. Between the two walls is an ultra-high vacuum. The neck plug has to be impervious to gas, capable of insulating a 200 °C temperature gradient and supporting the weight of the inner vessel full of samples, inventory system and liquid nitrogen. There is a diversity of dewers available on the commercial market (**Figure 3**). An important characteristic of a dewer is its holding time, i.e. the maximum time between 'topping up' the liquid nitrogen reservoir to maintain cryogenic temperature. Sample inventory systems within the dewers are an important part of the long-term management of cryopreserved tissue samples as an inefficient system can lead to the loss or misidentification of samples. Generally, for small, regularly accessed collections an inventory system based on cane-mounted vials is adequate. However, for larger collections a drawer-based inventory system provides a more efficient option. The evidence for a wide variety of plant species and tissue types is that as long as the samples are maintained at stable ultralow temperatures, the duration of storage does not affect postthaw recovery.

Thawing

After storage, samples are generally thawed rapidly to prevent the damaging formation of ice crystals. For frozen encapsulated rose shoot tips, thawing is achieved by plunging the vials containing the beads into sterile water at 40 °C for approximately 4 min. The vials are carefully removed from the water, dried and the beads transferred on to semi-solid shoot multiplication medium (five beads per 5-cm Petri dish) and incubated for 24 h at 18 °C in the dark.

Postthaw Recovery

On thawing, plant tissue inevitably suffers damage and stress. Therefore specialist postthaw culture conditions have been developed to support initial recovery and regrowth of thawed samples. Modification of culture media, for example the removal of ammonium ions, or the addition of activated charcoal and antioxidants, has been shown to enhance postthaw cell regrowth. Modifying the subculture regime and incubation conditions, e.g. maintaining tissue in the dark or under reduced light intensity, can also benefit the frequency of postthaw regrowth.

After 24 h incubation in the dark the thawed rose shoot tips are excised from the alginate beads and placed on to semi-solid shoot multiplication medium (four shoot tips per 5-cm Petri dish) and incubated in the dark at 18 °C for a further 10 days. After this, the samples are transferred to a 16 h photoperiod and light intensity of 3–7 $\mu mol\ m^{-2}\ s^{-1}$ photosynthetic active radiation. The recovering shoot cultures are transferred to fresh micropropagation medium at 30-day intervals. Regrowth of shoot tips is first observed approximately 40 days after thawing; approximately 25% of thawed shoots exhibit regrowth. Callus growth is observed around the base of shoot tips after thawing, a not uncommon indication of cryoinjury (**Figure 4**). No shoot regeneration is observed from the callus that develops after thawing, but occasionally the growing shoot may be overgrown by this callus. Generally the shoots develop to a stage where they can be excised from the callus approximately 60–80 days after thawing and maintained as

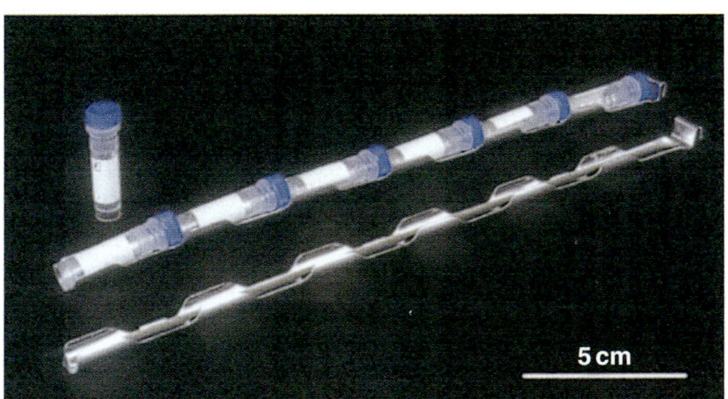

Figure 2 Polypropylene cryovials (2 ml) with aluminium cane.

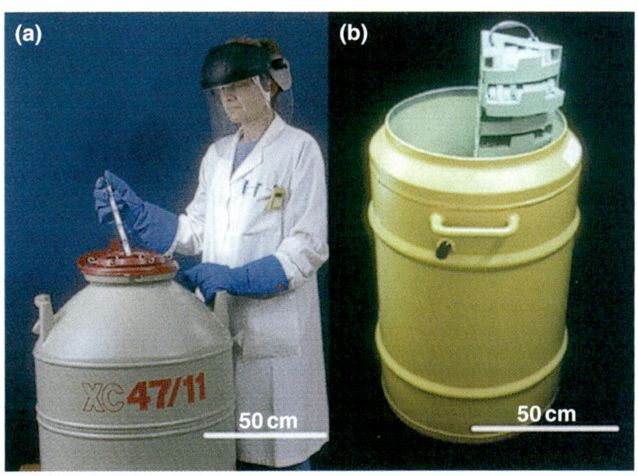

Figure 3 Storage dewers: (a) narrow neck with cane inventory system; (b) wide neck with draw inventory system.

Figure 4 Rose shoot tips regrowing 50 days after thawing: (a) little callus growth at base of shoot; (b) healthy shoot growth, but with significant callus proliferation at the base of shoot; (c) poor shoot growth, overgrown by proliferating callus.

micropropagated shoot cultures prior to transfer to compost (*see* **Cell, Tissue and Organ Culture**: Micropropagation; Rooting of Micropropagated Shoots).

Prospects for Rose *In Vitro* Conservation

Although a large number of roses are produced worldwide by micropropagation each year (*see* **Cell, Tissue and Organ Culture**: Micropropagation; Rooting of Micropropagated Shoots), there is a lack of reports concerning the *in vitro* conservation of rose germplasm. Reported methods require further development, for example an assessment of different preculture treatments, to improve the frequency of poststorage regrowth and applicability to a wide range of *Rosa* species and cultivars. As for plants derived from micropropagation, the genetic stability of germplasm after storage requires further assessment. The ability to cryopreserve rose germplasm may also become commercially more significant. When a patent involving a biological culture is filed, a sample must be lodged with a laboratory accredited with International Depository Authority, awarded by the World Intellectual Property Organisation, Geneva. This intellectual property protection may become of increasing importance for genetically modified plants. However, to be effective and valuable, biotechnological methods need to be carefully targeted and utilized as complementary and enabling technologies; they must be applied in context.

See also: **Cell, Tissue and Organ Culture**: Micropropagation; Rooting of Micropropagated Shoots; Cell Culture and Secondary Metabolites. **Rose Collections and Trials**: Rose Collections and Herbaria; Rose Societies; Awards.

Further Reading

Anon (1995) *Guide to the Deposit of Micro-organisms under the Budapest Treaty*. Geneva: World Intellectual Property Organisation.

Ashmore SE (1997) *Status Report on the Development and Application of* In Vitro *Techniques for the Cryopreservation and Use of Plant Genetic Resources*. Rome: International Plant Genetic Resources Institute.

Benson EE (1995) Cryopreservation of shoot-tips and meristems. In: Day JG and McLellan MR (eds.) *Methods in Molecular Biology*, vol 38: *Cryopreservation and Freeze-Drying Protocols*, pp. 121–132. Totowa, NJ: Human Press.

Benson EE (1999) Cryopreservation. In: Benson EE (ed.) *Plant Conservation Biotechnology*, pp. 83–96. London: Taylor & Francis.

Constabel F and Shyluk JP (1994) Initiation and micropropagation of plant cell and tissue cultures. In: Vasil IK and Thorpe TA (eds.) *Plant Tissue Culture*, pp. 3–15. Dordrecht: Kluwer Academic Publishers.

Engelmann F and Engels JMM (2002) Technologies and strategies for *ex situ* conservation. In: Engels JMM, Ramanatha Rao V, Brown AHD and Jackson MT (eds.) *Managing Plant Genetic Diversity*, pp. 89–104. Wallingford, UK: CAB International.

Harding K (1999) Stability assessments of conserved plant germplasm. In: Benson EE (ed.) *Plant Conservation Biotechnology*, pp. 97–107. London: Taylor & Francis.

Kartha KK and Engelmann F (1994) Cryopreservation and germplasm storage. In: Vasil IK and Thorpe TA (eds.) *Plant Cell and Tissue Culture*, pp. 195–230. Dordrecht: Kluwer Academic Publishers.

Lynch PT (1999) Tissue culture techniques in *in vitro* plant conservation. In: Benson EE (ed.) *Plant Conservation Biotechnology*, pp. 41–62. London: Taylor & Francis.

Lynch PT (2002) Cryopreservation of *Rosa* (rose). In: Towill LE and Bajaj YPS (eds.) *Biotechnology in Agriculture and Forestry*, vol 50: *Cryopreservation of Plant Germplasm II*, pp. 244–353. Berlin: Springer-Verlag.

Reed B (2002) Implementing cryopreservation for long-term germplasm preservation in vegetatively propagated species. In: Towill LE and Bajaj YPS (eds.) *Biotechnology in Agriculture and Forestry*, vol 50: *Cryopreservation of Plant Germplasm II*, pp. 22–33. Berlin: Springer-Verlag.

Reed BM and Chang Y (1997) Medium- and long-term storage of *in vitro* cultures of temperate fruit and nut crops. In: Razdan MK and Cocking EC (eds.) *Conservation of Plant Genetic Resources* In Vitro, vol 1: *General Aspects*, pp. 67–105. Enfield, NH: Science Publishers.

Sakai A (2000) Development of cryopreservation techniques. In: Engelmann F and Takagi H (eds.) *Cryopreservation of Tropical Plant Germplasm*, pp. 1–7. Rome: International Plant Genetic Resources Institute.

Stacey GN, Lynch PT and Benson EE (1999) Plant gene banking: agriculture, biotechnology and conservation. *Agro Food Industry Hi-Tech* 10: 9–14.

Staritsky G (1997) Background and principles of *in vitro* conservation of plant genetic resources. In: Razdan MK and Cocking EC (eds.) *Conservation of Plant Genetic Resources* In Vitro, vol 1, *General Aspects*, pp. 26–49. Enfield, NH: Science Publishers.

Towill LE (2002) Cryopreservation of plant germplasm: introduction and some observations. In: Towill LE and Bajaj YPS (eds.) *Biotechnology in Agriculture and Forestry*, vol 50: *Cryopreservation of Plant Germplasm II*, pp. 3–21. Berlin: Springer Verlag.

CLASSIFICATION

Contents

Conventional Taxonomy (Wild Roses)

V Wissemann, Friedrich-Schiller-Universität Jena, Jena, Germany

Introduction

Taxonomy of wild roses is notoriously difficult, for a variety of reasons. On the one hand the enormous phenotypic variability is based on genetic complexicity, while on the other hand the close relatedness of the genomes and features of the life cycle have begun to be clarified in recent years. However, traditional taxonomy based on morphological and anatomical data is in itself subject either to biological phenomena such as reproductive biology and ecological variability, or to theoretical problems like What is a species? and How can biological diversity be translated into taxonomically and nomenclaturally valid studies? The most common system in use is the classification by Alfred Rehder (see vol. 3 of this Encyclopedia): nevertheless, there is an urgent need for an update of that system, for both nomenclatural and biological reasons. In this article, methods of rose classification are discussed and the Rehder classification is updated.

Classification of the Genus *Rosa*

Roses have influenced human cultural history in various ways and the enormous interest in roses led to several attempts to classify the genus. During the Renaissance, in Baroque times and up to the early eighteenth century, rose systems were just divided into wild and gentle species, followed by subdivisions based on petal colour. The recognition of hybridization in roses was first remarked by Linnaeus, who wrote in the *Species Plantarum* of 1753 that he had the impression that *Rosa* species are difficult to distinguish, more difficult to determine, and that nature just for fun mixes species to form new ones. Also Herrmann claimed in his thesis from Strasbourg in 1762 that species are very difficult to determine on the one hand because of the lack of characters, and on the other hand because horticulture had merged the species so that recognition of 'pure' species was no longer possible.

The first attempt to create a real system of *Rosa* was made by Linnaeus. He regarded the shape of the hip to be useful for classification, and this character was accepted nearly exclusively until the beginning of the nineteenth century. Willdenow remarked in 1811 that form and presence of prickles as well as hairs and glands can serve as species-specific characters. This remark led to several new systems which were based more on personal preference rather than on observations in nature. The application of various characters to treat the genus systematically was the first sign of the enormous increase in names up to the beginning of the twentieth century (**Table 1**). The number of different studies on the genus *Rosa* at this time is not a sign of real insight; it is an expression of uncertainty facing the bewildering diversity of the genus.

Table 1 Number of species recognized from the sixteenth to the nineteenth century

Gessner (1561): 10 species
Dodonaeus (1569, 1616): 10 species
Lobelius (1581): 11 species
Gerard (1597): 16 species
Spigelius (1633): 16 species
Elsholz (1663) 15 species
Welsch (1697): 18 species
Salmon (1710): 12 species and 32 cultivars
Linnaeus (1735): 12 species
Linnaeus (1772): 10 species
Forsyth (1794): 28 species
Laicharding (1794): 31 species
Willdenow (1811): 34 species
Smith (1819): 57 species
Trattinnick (1823): 24 series with > 200 species
Seringe (1825): 146 species
Lindley (1830): 101 species, about 300 synonyms
Reichenbach (1832): 77 species, about 200 synonyms
Döll (1855): 114 species, about 200 synonyms
Déséglise (1877): 15 sections, section *Caninae* alone with 329 species

However, not all members of the genus *Rosa* faced the same problem. Mainly the dogroses, section *Caninae*, were subject to intensive species splitting.This culminated in 1886 in the article on 'buissonnomanie' by Crépin, who vehemently accused the rhodologists of naming each rosebush with a new name.

The change in understanding of the biological diversity of European wild roses started in 1873 with the study of the Swiss rhodologist Christ, who reduced the inflation of names to about 30 species by applying his synthetic classification system. This synthetic method was based on classification considering combinations of preferable correlated characters. In contrast to the analytical methods applied before Christ, his approach circumscribed taxa which he recognized as being natural groups, rather than artificial entities. However, molecular data clearly indicate (*see* **Classification:** Chemotaxonomy and Molecular Taxonomy) that these groups have to be revised again. Going back in the history of rhodology it is obvious that the taxonomic treatment of the genus *Rosa* is influenced by a European view of the genus. Systems available today show that, based on the increase in knowledge of biological diversity, there is a strong need for systematic and taxonomic studies in roses worldwide. Whereas the knowledge of middle European roses is quite advanced, next to nothing is known about eastern European species, and Asian species and even north American species are quite unknown from a systematic and ecological point of view. Thus morphological classification of the genus results in at least some unnatural groups.

In this article the key for identification of Rehder's main categories is replaced by a key oriented on the whole species diversity. Nevertheless, morphological classification is critical in sections like *Cinnamomeae* versus *Pimpinellifoliae*, where unique characters for the sections are lacking, and where molecular data from both ribosomal and chloroplast DNA indicate that morphological partition is artificial. Another example is the circumscription of section *Carolinae*. These roses are generally described as not stoutly built, but they do not show any significant specific characters for a sectional classification. Looking at these roses in their natural habitat, one can instantly see that their life in the shadow under trees in the woods forces such a fragrant growth with single flowers. However, we know from cultivation experiments in other sections that these characters are strongly dependent on ecological features like sun intensity, temperature and humidity, but there is no knowledge on the plasticity of these phenotypic characters in section *Carolinae*, so section and species delimitation may not be natural.

Changes in the Rehder System

From a nomenclature point of view, the subgenus *Eurosa* must be replaced by the subgenus *Rosa*. Based on generic type, the subgenus *Rosa* contains section *Rosa*. Section *Rosa* replaces section *Gallicanae*, because *R. centifolia* L. is the generic type of *Rosa*. Following the rules of the International Code of Botanical Nomenclature, the first lectotypification by Britton & Brown in 1913 on *R. centifolia* as the generic type is still a valid choice, but is disputed. The proposal to replace this choice by *R. cinnamomeae* was rejected by the general committee for nomenclature at the Tokyo Botanical Congress in 1995. The proposal was again not accepted in the Saint Louis Code of Botanical Nomenclature in 2000. A revised proposal will be discussed at the Vienna Congress 2005 (**Table 2**).

Addenda and Corrections to the System of Rehder 1940

Subgenus *Hulthemia* (Dumort.) Focke 1888

Monotypic subgenus: ***Rosa persica*** Michx. ex Juss. 1789 (=*R. simplicifolia* Salisb., *R. berberifolia* Pall., *Hulthemia berberifolia* (Pall.) Dumort., *Hulthemia persica* Bornmüller, *Lowea berberifolia* Lindl.) ($2n = 2x = 14$).

Subgenus *Rosa*

Sect. 1. *Pimpinellifoliae* (DC.) **Ser. 1825** About 15 species from Asia and Europe. Taxonomic subdivisions uncertain. Molecular data support the

Table 2 Key to the sections of *Rosa*

1	Leaves entire (1 leaflet), without stipules	Subgenus *Hulthemia*
1*	Leaves imparipinnate (> 1 leaflet), with or without stipules	2
2	Receptacle smooth or ± gland-tipped acicles or bristles	4, subgenus *Rosa*
2*	Receptacle with prickles like a chestnut, sepals pinnate and erect	3
3	Stipules adnate, with divergent, rounded or broadened auricles, leaflet usually 3 or 5, rarely 7	Subgenus *Hesperhodos*
3*	Stipules adnate, with subulate auricles, leaflets usually > 7 (–15)	Subgenus *Platyrhodon*
4	Stipules persistent and half-adnate with auricles	7
4*	Stipules free and ± deciduous	5
5	Receptacle and younger branchlets tomentose or pubescent	Sect. 10: *Bracteatae*
5*	Not this combination	6
6	Sepals erect and persistent after anthesis, receptacle bristly	Sect. 9: *Laevigatae*
6*	Sepals reflexed and deciduous, receptacle smooth	Sect. 8: *Banksianae*
7	Sepals conspicuously pinnate (the two sepals outside pinnate on both sides, the next one only on one side and all of the two inner sepals	8
7*	Sepals entire or only the two outer ones with slight appendices	9
8	Prickles glabrous, straight and mixed with gland-tipped bristles	Sect. 2: *Rosa*
8*	Prickles straight or curved, not mixed with bristles	15 (sect. 3: *Caninae*)
9	Sepals much longer than petals, younger stems with a purplish bloom, leaflets with purple pigmentation, prickles straight or curved	Sect. 3: subsect. *Rubrifoliae*
9*	Not this combination	10
10	Sepals short, erect, leaflets glandular and ± hairy, low and compact shrub	Sect. 3: subsect. *Vestitae*
10*	Not this combination	11
11	Styles agglutinated after anthesis like a column	Sect. 6: *Synstylae*
11*	Styles free	12
12	Sepals after anthesis erect and persistent	13
12*	Sepals after anthesis reflexed or spreading, deciduous	14
13	Flowers mostly solitary, solitary flowers without bracts	Sect. 1: *Pimpinellifoliae*
13*	Flowers solitary or corymbose; if solitary, then with bracts	Sect. 5: *Cinnamomeae*
14	Styles not exserted, covering the orifice	Sect. 4: *Carolinae*
14*	Styles exserted, about half the length of the inner stamens	Sect. 7: *Indicae*
15	Sepals entire or only with very few appendices, sepals conspicuously longer than petals, leaflets not glandular and glabrous	Subsect. *Rubrifoliae*
15*	Outer sepals significantly pinnate, if entire then leaflets glandular and ± hairy	16
16	Prickles straight or curved (like a sickle), homoacanth	17
16*	Prickles hooked, sometimes heteracanth (hooked and straight mixed)	18
17	Leaflets glabrous on the surface, underneath glabrous or hairy, but not pannose. Veins often prominent like a relief and glandular. Rhachis with glands. In all, glands more noticeable than hairs. Pinnate sepals deciduous after anthesis, pedicel glandular or with gland-tipped bristles. Leaves multiserrate with glands, fresh leaves not smelling like resin or turpentine	Subsect. *Trachyphyllae*
17*	Leaflets conspicuously hairy on both sides. Leaflets underneath with glands hidden in the hairs. Hairs more noticeable than glands. Fresh leaves smelling like resin or turpentine. Sepals sometimes ± entire	Subsect. *Vestitae*
18	Leaflets sticky, at least on the underside, with numerous glands smelling like apples or vine. Leaflets glabrous or hairy, glands more noticeable than hairs, hooked prickles sometimes subheteracanth-heteracanth, mixed with gland-tipped bristles	Subsect. *Rubigineae*
18*	Leaflets not sticky-glandular, leaflets glabrous or hairy, aglandular or with non-smelling glands	19
19	Leaflets hairy, at least always on the glandular midrib, mostly also the rhachis, the veins, the margins of the stipules and the multiserrate leaflet-margin with glands. ± equal impression of glands and hairs	Subsect. *Tomentellae*
19*	Leaflets glabrous or hairy, aglandular or sparely glandular on rhachis, veins and margins of the stipules. Leaflets uniserrate–multiserrate. If hairy, then hairs more prominent than glands. Confusion possible between glandular forms of *R. corymbifera* and subsect. *Tomentellae*	Subsect. *Caninae*

existence of three major groups which are mostly merged together: subsect. *Pimpinellifoliae*, *Luteae* Crép. and *Sericeae* Crép. Chromosome numbers known with $2n = 2x$, $4x = 14$, 28, balanced. ***R. ecae*** Aitch 1880 (=*R. mogoltavica* Juz.), ***R. foetida*** J. Herrm. 1762 (=*R. lutea* Mill.) ($2n = 4x = 28$), ***R. hemisphaerica*** J. Herrm. 1762. ***R. hugonis*** Hemsl. 1905 ($2n = 2x = 14$), ***R. koreana*** Komar. 1901

($2n = 2x = 14$), ***R. myriacantha*** DC 1805 ($2n = 4x = 28$), ***R. omeiensis*** Rolfe 1912 ($2n = 2x = 14$), ***R. primula*** Boul. 1936 ($2n = 2x = 14$) (=*R. ecae* Kanitz, non Ait.), ***R. sericea*** Lindl. 1820 (=*R. tetrapetala* Royle, *R. wallichii* Tratt.) ($2n = 2x = 14$). The morphological difference between *R. omeiensis* and *R. sericea* is the thickened stalk of *R. omeiensis*. Usually *R. sericea* is attributed as having only four sepals and petals, but also forms with five exist, ***R. spinosissima*** L. 1753 (=*R. pimpinellifolia* L.): see Rehder no. 2 ($2n = 4x = 28$). There is still a great deal of uncertainty about the subdivision of the species. Especially under the name *R.* sp. var. *altaica* (Willd.) Rehd. or *R. altaica* Willd., several gene pools exist which are used for horticultural purposes. ***R. xanthina*** Lindl. 1820 ($2n = 2x = 14$).

Sect. 2. *Rosa* (=sect. *Gallicanae* (DC.) Ser. 1825) One species in Europe and W. Asia. ***R. gallica*** L. 1759 (nomen ambiguum *R. gallica* L. 1753) ($2n = 4x = 28$) (=*R. provincialis* Herrm., *R. austriaca* Crantz, *R. pumila* Scopoli non Jacq, nom. illegit.). Section *Rosa* harbours a number of taxa given species rank. These taxa are of hybrid origin, long cultivated, the status and knowledge of natural occurrence are uncertain. All species have ($2n = 4x = 28$): ***R. alba*** L. 1753. ***R. centifolia*** L. 1753 (=*R. gallica* var. *centifolia* (L.) Regel). ***R. damascena*** Mill. 1768 (=*R. gallica* var. *damascena* Voss). ***R. francofurtana*** Münchhausen 1774 (=*R. turbinata* Ait.). ***R. polliniana*** Spreng. 1813.

Sect. 3. *Caninae* (DC) Ser. 1825 Stems upright or arching with hooked, slender or straight prickles, outer sepals lobed (three exceptions), sepals persistent and erect, reflexed-deciduous or patent after anthesis. All members characterized by the *Caninae* meiosis with unbalanced heterogamous fully sexual reproduction. About 50 species in Europe.

Subsect. Trachyphyllae *H. Christ 1873* ***R. jundzillii*** Besser 1815 (=*R. marginata* Wallr., *R. trachyphylla* A. Rau): see Rehder no. 19 (*R. marginata* Wallr.) ($2n = 6x = 42$, unbalanced, heterogamous). Of hybrid origin, maternal parent from sect. *Caninae*, paternal parent *R. gallica* L.

Subsect. Rubrifoliae *Crép. 1892* *R. glauca* Pourr. 1788 (=*R. ferruginea* Vill., *R. rubrifolia* Vill.; non *R. glauca* Vill.): see Rehder no. 20 (*R. rubrifolia* Vill.) ($2n = 4x = 28$, unbalanced, heterogamous). The smooth or glandular entire sepals sometimes with lateral appendages. Morphology places the species into *Cinnamomeae*, but meiosis is of *Caninae* type.

Subsect. Vestitae *H. Christ 1873* ***R. mollis*** Sm. 1812 (=*R. mollissima* Fr., *R. pomifera* subsp. *mollis* (Sm.) Schwertschl.): see Rehder no. 13 ($2n = 4x, 5x, 6x = 28, 35, 42$, unbalanced heterogamous) flowers pink-red, white. ***R. pseudoscabriuscula*** (R. Keller 1931: 276) Henker & G. Schulze 1993 (=*R. tomentosa* subsp. *pseudoscabriuscula* R. Keller) ($2n = 5x = 35$, unbalanced, heterogamous). Shrub up to 2 m, prickles slender and curved, homoacanth: leaflets (5)-7, multiserrate and glandular, underneath with glands hidden in the pubescence covering both sides of the leaf. Rachis pubescent and glandular, sepals spreading but often erect at hip ripening. Flowers pink. ***R. sherardii*** Davies 1813 (=*R. omissa* Déségl.): see Rehder no. 13 ($2n = 4x, 5x, 6x = 28, 35, 42$, unbalanced, heterogamous) flowers dark pink–red. ***R. tomentosa*** Sm. 1800 (=*R. cinerascens* Dumort.): see Rehder no. 14 ($2n = 5x = 35$, unbalanced, heterogamous). Prickles slender and curved, not straight. Leaflets mostly uniserrate with conspicuous wide and short teeth; if biserrate than a slight occurrence of glands, sepals reflexed and deciduous, flowers slightly pink–white. ***R. villosa*** L. 1753 (=*R. pomifera* Herrm.): see Rehder no. 13 ($2n = 4x, 8x = 28, 56$, unbalanced, heterogamous), flowers pink–red, white nail.

Subsect. Rubigineae *H. Christ 1873* ***R. agrestis*** Savi 1798 (=*R. sepium* Thuill.): see Rehder no. 16 ($2n = 5x, 6x = 35, 42$, unbalanced heterogamous) sepals reflex and deciduous after anthesis. Back of sepals without glands. R. elliptica Tausch 1819 (=*R. graveolens* Gren. et Godr.): see Rehder no. 15 (synonym of *R. inodora*) ($2n = 5x, 6x = 35, 42$, unbalanced heterogamous), sepals erect, persistent after anthesis. Flowers pinkish. ***R. inodora*** Fr. 1814 (=*R. elliptica* subsp. *inodora* Fries), see Rehder no. 15 ($2n = 5x, 6x = 35, 42$, unbalanced heterogamous), somewhat of an intermediate species between *R. agrestis* and *R. elliptica*, sepals patent and deciduous. ***R. micrantha*** Borrer ex. Sm. 1812: see Rehder no. 16 ($2n = 4x, 5x, 6x = 28, 35, 42$, unbalanced, heterogamous), usually small hips with reflexed and deciduous sepals early after anthesis. ***R. rubiginosa*** L. 1771 (=*R. eglanteria* L., *R. umbellata* Leers): see Rehder 15 (synonym of *R. eglanteria*) ($2n = 5x = 35$, unbalanced, heterogamous), aromatic smelling leaf (apple), sepals persistent and erect. ***R. rubiginosa*** subsp. ***columnifera*** Schwertschl. 1910 (=*R. columnifera* (Schwertschlager) Henker & G. Schulze 2000 (homonym illeg., non Fries 1814) ($2n = 5x = 35$ also $6x$?, unbalanced, heterogamous), intermediate species combining characters of *R. rubiginosa* and *R. micrantha*. Sepals patent and partly deciduous, flowers pinkish-white.

Subsect. Tomentellae H. *Christ 1873* ***R. tomentella*** Léman in Cassini 1818 (*R. obtusifolia* auct. (mult.) non Desv.) ($2n = 5x = 35$, unbalanced, heterogamous). Shrub up to 2 m, homoacanth, prickles curved. Leaflets 5–7, mostly pubescent on both sides, veins underneath the leaflets densely covered with red glands which do not smell, increase of glands towards the margin of the leaflets. Rhachis pubescent and with glands. Sepals reflexed and deciduous after anthesis. Receptacle usually smooth as well as the pedicel, flowers white. ***R. abietina*** Gren. ex Christ 1873 (=*R. obtusifolia* subsp. *abietina* (Gren. ex Christ) F. Herrmann) ($2n = ?$, unbalanced, heterogamous). Shrub about 1.5 m high, prickles slender or hooked. Leaflets (5)-7, both sides or only underneath pubescent, always with glands. Pedicels stipitate-glandular, back of the sepals covered with glands. Sepals reflex and deciduous. Flowers reddish-pink.

Subsect. Caninae *R. caesia* Sm. 1812 (=*R. coriifolia* Fr.): see Rehder no. 18 (synonym *R. coriifolia*) ($2n = 5x$, $6x = 35$, 42, unbalanced heterogamous). Sepals first patent after anthesis, than erect and persistent. Flowers intensely pink. ***R. canina*** L. 1753 (=*R. nitidula* Besser, *R. afzeliana* Fr., *R. lutetiana* Léman, *R. squarrosa* (A. Rau) Boreau, *R. scabrata* Crép., *R. canina* agg. Ehrendorfer): see Rehder no. 19 ($2n = 5x = 35$, unbalanced, heterogamous). Leaflets and rhachis not pubescent. Sepals reflex and deciduous after anthesis. Flowers pinkish-white. ***R. corymbifera*** Borkh. (*R. dumetorum* Thuill., *R. obtusifolia* Desv., *R. deseglisei* Boreau, *R. brilonensis* G.H. Loos): see Rehder no. 18 ($2n = 5x = 35$, unbalanced, heterogamous). Leaflets pubescent on both sides or only underneath or even only the rhachis. Sepals reflex and deciduous after anthesis. Flowers white, sometimes pinkish. ***R. dumalis*** Bechst. 1819 (=*R. vosagiaca* Déségl., *R. glauca* Vill. ex Loisel, *R. reuteri* (Godet) Reuter): see Rehder no. 19 ($2n = 5x$, $6x = 35$, 42, unbalanced heterogamous). Sepals first patent after anthesis, than erect and persistent. ***R. montana*** Chaix in Villars 1786: see Rehder no. 19 ($2n = 5x = 35$, unbalanced, heterogamous). ***R. stylosa*** Desvaux 1809: see Rehder no. 17 ($2n = 5x$, $6x = 35$, 42, unbalanced, heterogamous). Leaflets 5–7, the single one often formed like a flame of a candle. Pedicels glandular and very much longer than the eglandular receptacle. Sepals on the back without glands, reflexed and deciduous after anthesis. Flowers pinkish-white. Styles agglutinated but not connate as in section 6: Synstylae. Disc formed like a cone. ***R. subcanina*** (H. Christ) R. Keller 1891 (=*R. reuteri* f. *subcanina* H. Christ, *R. dumalis* subsp. *subcanina* (H. Christ) Soó) ($2n = 5x = 35$, unbalanced, heterogamous). Shrub up to 2.5 m, intermediate species between *R. canina* and *R. dumalis*, sepals patent and deciduous. Flowers pale pink. ***R. subcollina*** (H. Christ) R. Keller 1891 (=*R. coriifolia* f. *subcollina* H. Christ; *R. caesia* subsp. *subcollina* (H. Christ) Soó) ($2n = 5x = 35$, unbalanced, heterogamous). Shrub up to 3 m, intermediate species between *R. corymbifera* and *R. caesia*, sepals patent and deciduous. Flowers pink.

Sect. 4. *Carolinae* Crép. 1891 5(?) species from North America. ***R. carolina*** L. 1753 ($2n = 4x = 28$) (=*R. humilis* Marsh., *R. lyonii* Pursh, *R. serrulata* Rafinesque). ***R. foliolosa*** Nutt. ex. Torr. & A. Gray 1840 ($2n = 2x = 14$). ***R. nitida*** Willd. 1809. ***R. palustris*** Marsh 1785 ($2n = 2x = 14$) (=*R. carolina* Gray non L.) ***R. virginiana*** Herrm. 1762 ($2n = 4x = 28$) (=*R. virginiana* Mill. 1768, in most cases the citation of the author for this species is wrong; the name by Miller is a younger homonym).

Sect. 5. *Cinnamomeae* (DC.) Ser. 1825 Uncertain number of species (about 80) in Asia, North America, Europe. ***R. acicularis*** Lindl. 1820 ($2n = 4x$, $8x = 28$, 56) (=*R. alpina* Pall. non L., *R. granulosa* Keller, *R. gmelinii* Bunge). ***R. amblyotis*** C.A. Mey. 1847 ($2n = 2x = 14$) (=*R. davurica* Hulten non Pall.). ***R. arkansana*** Porter ex I.M. Coult. 1874 ($2n = 4x = 28$) (=*R. rydbergii* Greene). ***R. banksiopsis*** Baker 1914 ($2n = 2x = 14$). ***R. beggeriana*** Schrenk 1841 ($2n = 2x = 14$). ***R. bella*** Rehd. & E.H. Wilson 1915 ($2n = 4x = 28$). ***R. blanda*** Ait. 1789 ($2n = 2x = 14$). ***R. davurica*** Pall. 1788 ($2n = 2x = 14$) (=*R. cinnamomea* Ldb. non L., *R. wildenowii* Sprengel). ***R. californica*** Cham. et Schltdl. 1827 ($2n = 4x = 28$). ***R. caudata*** Baker 1914 ($2n = 2x$, $4x = 14$, 28). ***R. corymbulosa*** Rolfe 1914 ($2n = 2x = 14$). ***R. davidii*** Crép. 1874 ($2n = 4x = 28$). ***R. fedtschenkoana*** Regel 1878 ($2n = 4x = 28$). (=*R. caraganifolia* Sumn., *R. coeruleifolia* Sumn., *R. epipslia* Sumn., *R. laurenkoi* Sumn., *R. lipschitzii* Sumn., *R. minusculifolia* Sumn., *R. oligosperma* Sumn.). ***R. forrestiana*** Boulenger 1936. ($2n = 4x = 28$), ***R. gymnocarpa*** Nutt. ex Torr. et A. Gray 1840 ($2n = 2x = 14$). ***R. laxa*** Retz 1803 ($2n = 2x = 14$) (=*R. soongarica* Bge., *R. alpina* Ldb. non L.). ***R. majalis*** Herrm. 1762 (wild: $2n = 2x = 14$; hort.: $4x$, $8x = 28$, 56) (=*R. cinnamomea* L. 1759 non 1753, *R. spinosissima* Rydb.). ***R. marretii*** H. Lév. 1910 ($2n = 2x = 14$). ***R. moyesii*** Hemsl. & E.H. Wilson 1906 ($2n = 4x = 28$). ***R. multibracteata*** Hemsl. et E.H. Wilson 1906 ($2n = 4x = 28$) (=*R. latibracteata* Boul., *R. orbicularis* Baker, *R. reducta* Baker, *R. rotundibracteata* Cardot). ***R. nanothamnus*** Boulenger 1935 ($2n = 4x = 28$) (*R. beschnauensis* Sumn., *R. botryoides* Sumn., *R. vitaminifera*

Sumn.). ***R. nutkana*** C. Presl 1851 ($2n = 6x = 42$) (=*R. fraxinifolia* Boor. ex Hook. non Borkh., *R. woodsii* Rgl. non Lindl.). ***R. pendulina*** L. 1753 ($2n = 4x = 28$) (=*R. alpina* L. 1762, *R. cinnamomea* L. 1753, *R. rupestris* Crantz). ***R. pisocarpa*** A. Gray 1872 ($2n = 2x$, $4x = 14$, 28). ***R. prattii*** Hemsl. 1892 ($2n = 2x = 14$). ***R. rugosa*** Thunb. 1784 ($2n = 2x = 14$) (=*R. kamtchatica* Red., *R. ferox* Lawrance, *R. pubescens* Baker). ***R. sertata*** Rolfe 1913 ($2n = 2x = 14$). ***R. setipoda*** Hemsl. et E. H. Wilson 1906 ($2n = 4x = 28$) (*R. hemsleyana* Täckholm, *R. macrophylla* Crép. non Lindl., *R. macrophylla* var. *crasseaculeata* Vilmorin, *R. setipoda* Rolfe non Hemsl. et E.H. Wilson). ***R. suffulta*** Greene 1899 (=*R. arkansana* var. *suffulta* (Greene) Cockerell). ***R. sweginzowii*** Koehne 1910 ($2n = 6x = 42$). ***R. webbiana*** Wall. ex Royle 1835 ($2n = 2x = 14$). ***R. willmottiae*** Hemsl. 1907 ($2n = 2x = 14$). ***R. woodsii*** Lindl. 1820 ($2n = 2x = 14$).

Sect. 6. *Synstylae* DC. 1813 About 25 species from Asia, North America, Africa. ***R. abyssinica*** R. Br. ex Lindl. (?). ***R. anemoniflora*** Fortune ex. Lindl. 1847, see Rehder no. 53 ($2n = 2x = 14$) (=*R. triphylla* Roxb. ex Hemsl, *R. triphylla* Roxb. ex Lindl. 1820, *R. sempervirens* var. *anemoniflora* Regel). ***R. arvensis*** Huds. 1762 ($2n = 2x = 14$) (=*R. repens* Scopoli, *R. silvestris* Herrm., *R. serpens* Wibel, *R. sempervirens* subsp. *arvensis* (Huds.) Malagarriga). ***R. brunonii*** Lindl. 1820 ($2n = 2x = 14$) (=*R. brownii* Tratt, *R. moschata* var. *nepalensis* Lindl., *R. clavigera* H. Lév., *R. pubescens* Roxb.). ***R. filipes*** Rehd. & Wils. 1915 ($2n = 2x = 14$). ***R. helenae*** Rehd. & Wils. 1915 ($2n = 2x = 14$). ***R. henryi*** Boulenger 1933 ($2n = 2x = 14$) (=*R. moschata* var. *densa* Vilm., *R. paucispinosa* H.L. Li). ***R. longicuspis*** Bertol. 1861 ($2n = 2x = 14$) (=*R. lucens* Rolfe, *R. willmottiana* Lév., *R. irridens* Focke) ***R. l.*** var. ***sinowilsonii*** (Hemsl.) T.T. Yu et T.C. Ku 1981 ($2n = 2x = 14$) (=*R. sinowilsonii* Hemsl. 1906, see Rehder no. 57, published as *R. sinowilsoni* by Hemsl.). ***R. luciae*** Franch. & Rochebr. 1871: uncertain status and relation to *R. wichurana*. Differs by a reduced size of flowers and fruits as well as a reduced number of leaflets ($2n = 2x = 14$). ***R. maximowicziana*** Regel 1878 ($2n = 2x = 14$) (=*R. luciae* var. *aculeatissima* Crép. ex Regel, *R. fauriei* H. Lév., *R. granulosa* Keller, *R. spinosissima* var. *mandschurica* Yabe). ***R. moschata*** Herrm. 1762 ($2n = 2x = 14$): hybrids with *R. multiflora* form the Lambertiana roses (? = ***R. phoenicia*** Boiss.). ***R. multiflora*** Thunb. ex Murr. 1784 ($2n = 2x$, $3x = 14$, 21) (=*R. thunbergii* Tratt., *R. wichurae* K. Koch, *R. polyantha* Sieb et Zucc., non hort., *R. intermedia* Carr.). ***R. rubus*** H. Lév. et Vaniot 1908 ($2n = 2x = 14$) (=*R. ernestii* Stapf ex Bean, *R. moschata* var. *hupehensis* Pamp., *R. henryi* var. *puberula* (Hand.-Mazz.) F.P. Metcalf). ***R. sempervirens*** L. 1735 ($2n = 2x = 14$) (=*R. balearica* Persoon, *R. alba* Allioni non L., *R. atrovirens* Viviani). ***R. setigera*** Michx. 1803 ($2n = 2x = 14$) (=*R. trifoliata* Rafinesque, *R. fenestrata* Tratt.). ***R. soulieana*** Crép. 1896 ($2n = 2x = 14$). ***R. wichurana*** Crép. 1887 ($2n = 2x = 14$) (uncertain status and relation to *R. luciae*).

Sect. 7. *Indicae* Thory 1820 Three species from China. ***R. odorata*** (Andrews) Sweet 1818 (=*R. indica* var. *odorata* Andrews). ***R. gigantea*** Collet ex Crép. 1888 (=*R. odorata* var. *gigantea* (Collet ex Crép.) Rehder et E.H. Wilson 1915). ***R. chinensis*** Jacq. 1768 ($2n = 2x, 3x, 4x = 14, 21, 28$) (=*R. montezumae* Bertol., *R. indica* Lindl. non L., *R. sinica* L.)

Sect. 8. *Banksianae* Lindl. 1820 Two (?) species from China ***R. banksiae*** Ait. 1811. ($2n = 2x$, $4x = 14$, 28 balanced). Uncertain status of ***R. cymosa*** Tratt. 1823.

Sect. 9. *Laevigatae* Thory 1820 One species from China ($2n = 2x = 14$, balanced). ***R. laevigata*** Michx. 1803 (=*R. cherokeensis* Donn., *R. triphylla* Roxb. ex Lindl., *R. ternata* Poirr., *R. hystrix* Lindl., *R. nivea* DC., *R. camelia* Sieb., *R. sinica* Ait. non L.)

Sect. 10. *Bracteatae* Thory 1820 One or two species in South-East Asia ($2n = 2x = 14$, balanced) ***R. bracteata*** Wendl 1798 (=*R. macartnea* Dum. de Courset). Uncertain status of ***R. clinophylla*** Thory.

Subgenus *Platyrhodon* (Hurst) Rehder 1940 One species from Japan and China. ***R. roxburghii*** Tratt. 1823 (=*R. forrestii* Focke, *R. microphylla* Roxb. non Desf., *R. microphylla* var. *glabra* Regel) ($2n = 2x = 14$, balanced).

Subgenus *Hesperhodos* Cockerell 1913 Two species in North America ($2n = 2x = 14$, balanced). ***R. stellata*** Wooton 1898 (=*R. vernonii* Greene, *Hesperhodos stellatus* Boulenger), ***R. stellata*** var. ***mirifica*** (Greene) Cockerell, ***R. minutifolia*** Engelmann 1882.

See also: **Classification**: Horticultural Classification Schemes; Chemotaxonomy and Molecular Taxonomy; **Distribution and Ecology**: South Africa; Australia and New Zealand; Continental Asia and Japan; Europe; Israel; North America; South America. **History of Roses in Cultivation**: Ancient Chinese Roses; European (Pre-1800); Modern (Post-1800).

Further Reading

Beales P (1997) *Classic Roses*, 2nd edn. London: Harvill Press.

Christ H (1873) *Die Rosen der Schweiz mit Berücksichtigung der umliegenden Gebiete Mittel- und Süd-Europas*. Basel: H. Georg.

Darlington CD and Wylie A (1961) *Chromosome Atlas of Flowering Plants*. London: Allan & Unwin.

Graham GG and Primavesi AL (1993) *Roses of Great Britain and Ireland*. BSBI Handbook no. 7. London: Botanical Society of the British Isles.

Henker H (2000) *Rosa*. In: Conert EI, Jäger J, Kadereit J, Schultze-Motel W, Wagenitz G and Weber HE (eds.) *Illustrierte Flora von Mitteleuropa*, vol. IV/2C, LFg. A. Berlin: Verlag Paul Parey.

Keller R (1931) *Synopsis Rosarum spontanearum Europae mediae*. Vol. 65. Denkschriften der Schweizerischen Naturforschenden *Gesellschaft, Zuerich*.

Krüssmann G (1974) *Rosen, Rosen, Rosen. Unser Wissen über die Rose*. Berlin: Paul Parey.

Saakov SG (1976) *Wild- und Gartenrosen*. Berlin: VEB Deutscher Landwirtschaftsverlag.

Thomas GS (1994) *The Graham Stuart Thomas Rose Book*. London: John Murray.

Horticultural Classification Schemes

T Cairns, American Rose Society, Studio City, CA, USA

Historical Background

By definition, a species is a population of wild plants similar in characteristics. It is generally recognized that the genus *Rosa* contains more than 140 different species (*see* **Classification:** Conventional Taxonomy (Wild Roses)), typically having five sepals and petals (four of each in one species), abundant stamens and, in the centre of the bloom, numerous styles, each topped by a sticky stigma. Furthermore, when crossed together or when self-pollinated, it produces harmonious progeny. It is the welcome task of humanity to bring order and logic to such a grouping of plants to understand their development and natural evolution. Taxonomy is the scientific discipline which concentrates on devising systems of classification. In the botanical classification scheme for the genus *Rosa* as originally defined by Rehder (see vol. 3 of this Encyclopedia) and later modified (*see* **Classification:** Conventional Taxonomy (Wild Roses)), there are four subgenera – three very small and anomalous, namely *Hulthemia, Hesperrhodos* and *Platyrhodon*, and one major, *Rosa*, containing the roses proper. While *Rosa* is divided into 10 sections, three can be regarded as very small and of little importance as far as garden roses are involved. This subdivision within *Rosa* is based on certain physical parameters (stipules, sepals, blooms, styles, leaves, thorns, etc.). However, the number of chromosomes in the cell nucleus does also provide an important characteristic that can be used in classification. Most wild roses carry 14 chromosomes (i.e. two sets of seven) and are therefore called 'diploids'. Roses containing 28 chromosomes are termed as 'tetraploid'. Roses with 42 chromosomes are 'hexaploid'. Only a few roses have eight sets of chromosomes (i.e. 56) and are called 'octaploid'. Hybrids between diploids and tetraploids are triploids with 21 chromosomes (*see* **Genetics**: Meiosis). Roses with this odd-numbered set of chromosomes usually lack fertility and do not survive in nature. However, dog roses (section *Caninae*), which have an unusual mode of inheritance (*see* **Genetics**: Inheritance in the Dogrose), include fertile pentaploids.

The sections of the subgenus *Rosa* are listed below (*see* also **Classification**: Conventional Taxonomy (Wild Roses)):

- *Pimpinellifoliae* – about 8–10 species compose this division, with all except two being diploid
- *Rosa* – one tetraploid species only, *R. gallica*
- *Banksianae* – one diploid species only, from China, *R. banksiae*
- *Bracteata* – southern North America, *R. bracteata* and *R. clinophylla* which are diploids
- *Caninae* – generally referred to as the dog roses which include tetraploids, pentaploids and hexaploids
- *Carolinae* – sometimes defined as a subsection within the *Cinnamomeae*
- *Indicae* – two climbing diploids only, *R. chinensis* and *R. gigantea*
- *Cinnamomeae* – largest section comprising more than 50 diploid, tetraploid, hexaploid and octaploid species from throughout the northern hemisphere, e.g. *R. macrophylla, R. moyesii, R. acicularis*
- *Synstylae* – predominantly a large number of white-flowered, diploid climbers
- *Laevigatae* – one diploid species from east Asia which has become naturalized in the southern states of the USA.

Subsequent development and evolution of the species into today's modern garden roses from their primeval ancestors demanded a taxonomic system that incorporated both the old and the new, partially botanically based but with growth habit and plant structure playing an important role. This evolutionary construction is still very much a work in progress. However, some day the family relationships among members of the genus will be disclosed by DNA analysis (*see* **Genetics**: Research Trends; Gene Mapping). Until that day, the classification system adopted should reflect some scientific logic but be clearly overruled by the need to communicate in language suitable to describing the modern progeny in simple easy-to-understand terms.

The ideal system for classifying may well be found in a detailed description of the anatomical parts of the rose such as corolla, stigma, style, ovary (petal, disc, stamens, bract, prickle, calyx tube, leaf, leaflet, auricle, stipule, peduncle, stem, prickles, etc.). Within each of these characteristics a system could be set up with definitions of well-recognized shapes. For instance, for the petal shape the following criteria could exist: rounded smooth edges, notched indented or heart-shaped, abrupt point at apex, slightly lobed edges, fringed or ruffled edges, blunt, etc. Until such times as there is a marriage between emerging scientific data and recognition of plant characteristics, an absolute classification system is beyond our construction, for it will certainly have faults. Evolution of the rose is constantly taking place and the genus *Rosa* should be capable of accommodating new classes within that framework. For the moment, tolerance must be exercised in the shortcomings of our knowledge base and the hybrid classification schemes adopted to attempt to move forward towards the ultimate goal.

Classification of Cultivated Roses

During the last half of the twentieth century an evolution of rose classification schemes took place in an attempt to apply basic taxonomic principles. Such a synthetic separation of similar plants has its limitations since the family tree may be driven not purely by botanical concepts, but by commercial ones. The World Federation of Rose Societies (WFRS), for instance, formulated a family tree based on three distinct groups – wild roses, old garden roses and modern roses. This system of classification was rudimentary to understand, with major emphasis on recurrent versus nonrecurrent and climbing versus nonclimbing. The recurrent nonclimbing roses represent the majority of modern roses with a subdivision into large, cluster and dwarf flowered varieties. However, the application of such a low-resolving-power approach avoided botanical emphasis and depended on human perceptions.

This system, initiated and embraced by the Royal National Rose Society of Great Britain (RNRS), was employed as a fundamental mechanism to describe roses and their growing habit as well as serving as the main criteria for exhibiting roses. However, the terms hybrid tea, floribunda and miniature did exist solely for commercial purposes and communication to the public, but the fundamental concept of grouping on the basis of growing habit dominated. This system was acceptable to most countries of the world and received widespread adoption by breeders and introducers alike throughout the rose world, with the sole exception of the American Rose Society (ARS).

This disagreement stemmed from the fact that, in 1955, the ARS was formally designated as the International Registration Authority for Roses (IRAR) and chose to adopt a scheme based on a combination of botanical ancestry and commercial designations for sales. Rose registrations were approved according to this defined classification scheme, with 54 different classes and 18 different colour classes.

Then in 1989, Jack Harkness was commissioned by the British Association of Rose Breeders (BARB) to propose a classification scheme with improved utilization for hybridizers. The consensus of BARB members was that the existing system was too restrictive and did not allow for expansion to accommodate the latest evolutionary steps in the history of the rose. Jack Harkness developed such an elegant and innovative classification system, based on 30 distinct classes that he had used with great success since 1977 (**Table 1**). Within this scheme, several new classifications appeared that encompassed both classical and modern developments in rose breeding, e.g. Florishrub, Patio, Wichurana Ramber, Wichurana Carpet, Wichurana Shrub, English, Scotch and Sweet Briar. Alas, little attention was paid to this pioneering effort by the nomenclature official bodies to bring common sense to a complex problem. Clearly the success of this system was based upon the improved increased resolving power utilizing ancestry whenever possible, and yet this system was still rejected by the ARS.

However, the final consensus chapter in devising a universally acceptable classification during the twentieth century took place in Houston, Texas at the 12th World Rose Convention in 2000. It was at the Classification Committee Meeting chaired by David Ruston of Australia that the new ARS classification scheme was embraced by both the WFRS and the RNRS. This acceptance had been brought about by the recent overtures by ARS towards the WFRS system and the recepetiveness of WFRS towards the main

Table 1 Scheme proposed by Jack Harkness (1989)

Number	*Class*	*Botanical affinity/characteristics/similarity*
1	Species	Wild roses
2	China	'Old Blush' or other early Chinas
3	Noisette	'Blush Noisette' or a tea-influenced relative thereof
4	Tea	Both an early Tea-scented China and a China
5	Hybrid tea	Tea, but with improved habit and hardiness
6	Floribunda	Hybrid tea, usually with the influence of polyantha
7	Florishrub	Floribunda, but of more spreading growth
8	Miniature	China, but on a miniature scale
9	Patio	Both a miniature and a floribunda
10	Climbing hybrid tea	Hybrid tea, but of climbing habit
11	Climbing floribunda	Floribunda, but of climbing habit
12	Climbing miniature	Miniature or patio, but of climbing habit
13	Polyantha	*R. multiflora* as the dominant factor among any other classes present
14	Climbing polyantha	*R. multiflora* as the dominant factor among any other classes present
15	Hybrid musk	Both a polyantha following the style of Pemberton's hybrid musks
16	Wichurana rambler	*R. wichurana* or similar species as the dominant factor
17	Wichurana carpet	*R. wichurana* or similar species as the dominant factor
18	Wichurana shrub	*R. wichurana* or similar species as the dominant factor
19	Gallica	*R. gallica*
20	Damask	Gallica, but of more open growth and clusters
21	Centifolia	Gallica and damask, but of more lax growth and more pendulous leaflets
22	Moss	Centifolias, noticeable glandular growths on the sepals, seed pods
23	Portland	Gallica, damask or centifolia on the one side, and to a China on the other
24	Bourbon	Gallica, damask or centifolia on the one side, and to a China or tea on the other
25	Hybrid perpetual	Bourbons, representing a progress from them towards hybrid teas
26	English	Old garden rose on the one side and to a modern rose on the other
27	Scotch	*R. spinosissima* (*R. pimpinellifolia*)
28	Alba	*R. canina* or a similar species on the one side and to an old garden rose
29	Sweet briar	*R. eglanteria* (*R. rubiginosa*)
30	Rugosa	*R. rugosa*

ARS reasons for a family tree based on botanical as well as commercial classification systems to blend with the rose registration process.

The ARS Classification Scheme 2000 (officially adopted by WFRS for international registration of roses)

In 1999 the ARS foresaw the urgent need to bring harmonization to the rose classification scheme and adopted a new family tree incorporating the best ideas from the RNRS/WFRS scheme. As IRAR they approved a new classification scheme that reflected both the botanical and evolutionary commercial progress of the rose. There were three main groupings: species (i.e. wild roses), old garden roses (classes in existence before 1867) and modern roses (classes not in existence before 1867). However, several major changes were accepted by ARS. For instance, many of the old garden roses were reclassified as hybrids rather than species, e.g. China to Hybrid China. Within the modern roses, certain ramblers were given a more appropriate family name, Hybrid Wichuranas (HWich), with a spelling correction committed decades ago. Moreover, the new class of Mini-Flora was introduced to accommodate those plants equidistant from miniatures and floribundas in flower size and growth habit.

Then, at the very end of the twentieth century at the WFRS World Convention in Houston, Texas, USA, it was agreed to adopt this new scheme set out by ARS (**Figure 1**). As always, it was stressed that it was a work in progress.

Colour classification The ARS, as the custodian of the registration process, has devised a classification scheme for flower colours that divides the spectrum into 18 distinct colour classes. These colours and their abbreviations, defined in **Table 2**, are also used to describe roses in *Modern Rose XI*, reproduced in vol. 3 of this Encyclopedia.

Species Roses

Often referred to as 'wild roses', they are usually single-petalled (four to five petals), once-blooming and have a size range from 60 cm to 6 m. They are

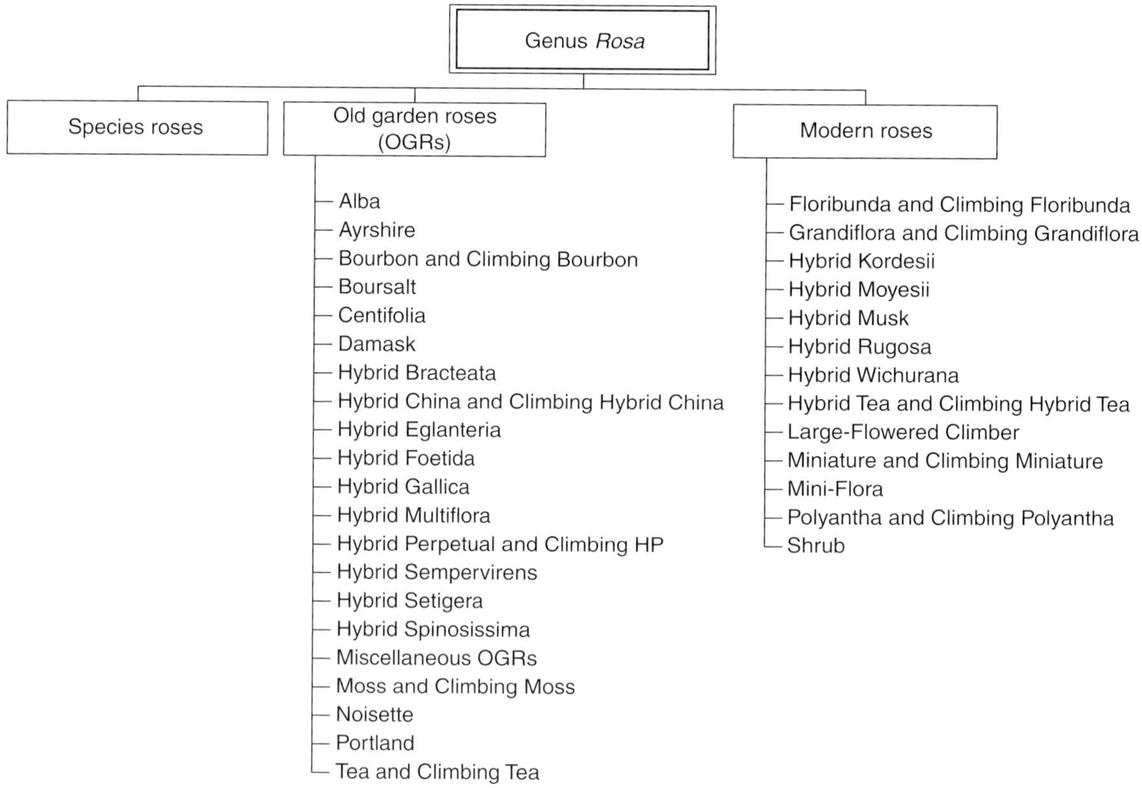

Figure 1 The American Rose Society classification scheme (2000). Reproduced from *Modern Roses XI* with permission.

Table 2 The American Rose Society classification of rose flower colour

Abbreviation	*Description*
w	White, near white and white blend
ly	Light yellow
my	Medium yellow
dy	Deep yellow
yb	Yellow blend
ab	Apricot and apricot blend
ob	Orange and orange blend
op	Orange-pink and orange-pink blend
or	Orange-red and orange-red blend
lp	Light pink
mp	Medium pink
dp	Deep pink
pb	Pink blend
mr	Medium red
dr	Dark red
rb	Red blend
m	Mauve and mauve blend
r	Russet

Reproduced from *Modern Roses XI* with permission.

listed according to their Latin name, beginning with *R.* for *Rosa*, and can have common synonyms, e.g. *R. foetida* var. *bicolor* is also known as 'Austrian Copper' (*see* **Classification**: Conventional Taxonomy (Wild Roses)).

Old Garden Roses

In 1966 ARS adopted the definition of old garden roses as referring to those types that existed prior to 1867, the year of introduction of the very first hybrid tea, 'La France'. Within this generic definition a number of popular subdivisions exist based on natural historical developments and characteristics. The flower form can be quartered, cupped, imbricated or expanded, reflexed, globular or compact. After an initial spring crop of blooms, some varieties may produce no more flowers for the rest of the year but their hip production does add a different kind of beauty to the garden. The beauty of the old garden roses often lies in the heavy fragrance they can impart to the garden. The classifications recognized (for botanical affinities see Harkness descriptions, **Table 1**) are:

Alba Affinity to *R. canina* and *R. damascena* or *R. gallica*. Once-blooming. Known as 'white roses', these plants are upright, often climbing, dense blue-green foliage and disease-resistant. Highest-rated member is 'White Rose of York'.

Ayrshire Developed from *R. arvensis*. A group of climbing, sprawling roses that originated in Scotland. Once-blooming.

Bourbon First repeat-flowering roses developed from the Hybrid Chinas. They derive their name from the location of the first members of the class, the Ile de Bourbon in the Indian ocean. Plant size can range from 60 cm to 4.5 m tall. Repeat-blooming.

Boursault A small group of thornless rambling-type roses developed by the French gardener who gave his name to the class. Once-blooming.

Centifolia (Provence Roses) Not a true species as once, but a complex hybrid of *R. gallica, R. phoenicia, R. moschata, R. canina* and *R. damascena*. These Dutch hybridized roses derive their class name from the fact that the flowers often contain over 100 petals. Plants are 120–240 cm tall and winter-hardy. They are also referred to as 'cabbage roses', featured in Dutch flower paintings of the seventeenth century. Developed mainly by the Dutch in selecting improved strains of *R. centifolia*. Once-blooming.

Damask These roses brought by crusaders around the thirteenth century from the Middle East are best known for their intense heavy fragrance and recurrent-blooming and prized for their fragrance. 'Autumn Damask' was highly characteristic of this class. Plants are generally in the size range of 1–2 m. Some varieties are repeat-blooming.

Hybrid China Plants are generally small, ranging from 60 cm to 1 m tall. Stems are often too weak to support fully the clusters of blooms which have a spicy fragrance. Plants are not hardy and require winter protection in cold climates. Repeat-blooming.

Hybrid Gallica Affinity to *R. gallica*. Plants are small (1–1.3 m tall), fragrant, winter-hardy and come in brilliant colours. Once-blooming.

Hybrid Perpetual Popular during the nineteenth century, this group of roses is characterized by repeat blooming, size (about 2 m tall upright plant), fragrance and colour range (mostly pinks and reds).

Moss Serendipitous breeding from a sport of a *R. centifolia*. Named for the mossy thorn growth on the peduncle just below the bloom and sepals, this group releases a pine-scented oleoresin when the moss is rubbed between the fingers. Plants are generally winter-hardy and 1–2 m tall. Some varieties are repeat-blooming.

Noisette This classification originated in the USA by Philippe Noisette of Charleston, South Carolina, who later introduced them in France when he moved there in 1817. Plants are large and sprawling, often reaching up to 6 m tall. Blooms are produced in fragrant clusters.

Portland This small group of plants derived from crosses involving hybrid Gallica, Damask, Centifolia and Hybrid China are small in stature (usually 1.3 m tall), repeat-blooming and have very short peduncles. Class named in honour of the Duchess of Portland.

Tea Characterized as small to medium-sized plants (1.3 m tall), they have large blooms on weak stems, which cause the flowers to droop. Flower form is symmetrical and the forerunner of modern hybrid teas.

Modern Roses

The era of modern roses was established in 1867 with the introduction of the first hybrid tea, 'La France' by the French breeder, Guillot. This variety was considered unique for a number of important horticultural reasons. One, it possessed the general habit of a Hybrid Perpetual as well as the elegant shaped buds and free-flowering character of a tea rose. By the late twentieth century, more than 10 000 hybrid teas has been bred with great success. Introduction of 'La France' heralded the era of modern roses. Breeders were quick to recognize that planned parenthood could evolve new flower forms and size, growth habit and colours. Therefore, the following new classifications based on growth habit evolved.

Floribunda and Polyantha Second only to the hybrid tea and Grandiflora in popularity, the floribunda is characterized by its profuse ability to bear flowers in large clusters or trusses with more than one bloom in flower at any one time. This class is unrivalled for providing massive colourful displays which are long-lasting for garden display. The distinct advantage of the floribunda is its ability to bloom continually, whereas the hybrid tea exhibits a bloom cycle every 6–7 weeks. The floribunda as a class are hardier, easier to care for and more reliable in wet weather than their hybrid tea counterparts.

Polyanthas are generally smaller but sturdy plants with large clusters of small 25-cm-diameter blooms, often used for massing, edging and hedges.

Hybrid Tea and Grandiflora Perhaps the most popular class of modern roses is the hybrid tea, easily recognized by the large, shapely blooms containing

30–50 petals. Flowers are borne on long stems, either singly or with several side buds. In 1945, the 'Peace' rose heralded the modern era of the elegantly formed hybrid teas. So dramatic was the overwhelming public acceptance and praise accorded this variety that its place in history was instantaneous. Since 1945, many thousands of new hybrid teas have been bred and introduced.

In 1954, the introduction of a rose bred from crossing the hybrid tea, 'Charlotte Armstrong', with the floribunda, 'Floradora', resulted in a carmine-rose and dawn-pink variety displaying not only the characteristics of a hybrid tea but also the ability to bear clusters or trusses and grow to a commanding height of 2–3 m or more. To accommodate this variety, the class of Grandiflora was born and 'Queen Elizabeth'® had the distinction of being the very first member of this class.

Large-Flowered Climber and Hybrid Wichurana These varieties are dominated by their growth habit, long-arching canes with the ability to climb (if properly trained and tied) up fences, over walls, through trellises, arbors and pergolas. These varieties offer a wide range of flower form, shape and colour.

Shrub (Classic and Modern) Shrubs are easily characterized by their sprawling habit. There are five popular subdivisions within the class: Hybrid Kordesii, Hybrid Moyesii, Hybrid Musk, Hybrid Rugosa and Shrub. They can grow from 1.5 to 4.5 m or more in every direction given the correct climate and growing conditions. Noted for their hardiness, they are usually vigorous and produce large quantities of clusters of flowers.

The unique group of roses hybridized by David Austin (often called English roses) belongs to this class. They resemble old garden roses in shape and form but are recurrent-bloomers and have fragrance.

Miniature and Mini-Flora These classes have increased in popularity due to their novelty and versatility. They can be used for edging beds, growing in containers and rockeries or even for taking indoors as temporary pot plants for decoration. The height of the average plant is about 35–75 cm and flower form and foliage are indeed miniature versions of both hybrid tea and Floribundas.

Mini-flora roses are a new classification adopted by the ARS in 1999 to recognize another step in the evolution of the rose, i.e. intermediate in bloom size and foliage between miniatures and Floribundas.

Rose Registration Process

Pivotal to developing an appropriate classification scheme is the intelligent collection of detailed descriptions of characteristics such as flower colour, petal number, bloom size, fragrance, blooming habit, prickles, foliage, growth habit, and, most of all, accurate parentage. That process has already been begun by ARS in the rose registration process (see *Modern Roses XI*, pp. xviii–xix). Moreover, the widespread use of DNA analysis may reveal more definitive relationships which are as yet undiscovered.

Today, however, there exists some confusion in the minds of average rose growers throughout the world as to why two different rose varieties with exactly the same name can be sold. The reasons are simple. Most breeders prefer to register their roses officially using a unique generic variety denomination. That denomination is often a name, difficult to pronounce, constructed according to the code system supported by Communauté Internationale des Obtenteurs de Plantes Ornementales et Fruitières de Reproduction Asexuée (CIOPORA). At the same time the breeder makes a formal announcement of the intended fancy commercial names under which the variety will be sold. Unless those fancy names are trademarked, there is no legal protection against using such a fancy name over and over again. For instance, Kordes in Germany has introduced 'Valencia'® and 'Liebeszauber' several times decades apart, as is their right!

In the twenty-first century, however, there is comfort in the knowledge that the naming and protecting of variety denominations has quickly evolved by taking a few giant steps in the right direction. While the solution to preventing two roses with the same fancy name from occurring again is by no means completely resolved, the designation of the variety denomination for registration purposes should help proper identification. To understand this tale of nomenclature, the following brief explanation should assist the reader in the complexities of the process of naming to preserve valuable data for future generations of rose growers.

History of rose registration In 1913 the ARS pioneered the process of recording new rose varieties from all over the world as a means of harmonizing nomenclature. By 1916 it had successfully promulgated registration guidelines and began the long tradition of publishing such registrations. In 1921 the *American Rose Annual* contained a feature section called 'New Roses of The World' which has continuously appeared every year into the modern era. Then in 1930 the late Dr J. Horace McFarland, then Editor of the Society's publications, began to compile lists of rose names and descriptions which he published under the title of *Modern Roses*. Each subsequent edition of *Modern Roses* built on the initial policies and procedures of including only those varieties currently in cultivation. In the 1958 edition, *Modern Roses V*, the goals had been enlarged to include all the names of species and varieties presently in cultivation or of current interest, together with many varieties of historical and botanical importance.

However, all editions up to and including *Modern Roses X* have been published strictly using the fancy name for intended use by American rose growers. Names were always directly applicable to the USA while the rose might enjoy other names in other countries. For instance, 'Peace' was the fancy name in the USA while in France it was called 'Mme A. Meilland', in Germany it was sold under the name 'Gloria Dei' and in Spain 'Gioia'. These other names are often referred to as synonyms. Unless protected by trademark or registered, the names could in theory be used by another grower or introducer. So today many breeders rely heavily on trademarks to protect the assigned fancy names of their roses. Without such protection the names could be reused in the commercial environment. A good example of this situation is the name 'Lady in Red'. Originally registered as a miniature rose with IRAR, the name has recently been reused to describe a new red hybrid tea by Jackson & Perkins for sale in New Zealand.

With the growing popularity of roses throughout the world, the 14th International Horticultural Congress (IHC) meeting in Scheveningen, The Netherlands, in 1955 appointed the American Rose Society as IRAR. This act by the IHC was a solid declaration of the need for a single central focus for the registration system to prevent duplication of variety denominations. ARS subsequently accepted the role and, more importantly, the responsibility and implementation. It was clearly stated for 'the good of the industry and botanical and horticultural science that all hybridizers or nurserymen register the names of their roses with one of the many National Registration Centers' set up in association with various national rose societies. During the years since 1955 the ARS has continued to approve the registration of a variety denomination, ensuring that no duplication was permitted. However, as the regulatory authority, ARS had no compliance or enforcement policy should an independent grower wish to market a rose under a name already registered but unprotected.

Around 1958 it became apparent that regulating the name of the rose was a much more complex process. In spite of guidelines and controls, there still existed the possibility that different roses with the same name could be sold to the public. On this occasion, the breeders took matters into their own hands and developed a simple system of unique coding to avoid duplication, at least in the registration process. Code names were simply a combination of the first three letters of their name or company in capital letters followed by some additional letters, usually derived from the slang name given the rose during its evaluation before introduction. In this way it became easy to identify the roses from a specific breeder, for example KOR for Kordes in Germany, MAC for McGredy, HAR for Harkness in the UK, MEI for Meilland in France and SIM for Nola Simpson from New Zealand. While such code names were included in registration submissions for approval by ARS, they were not included with the descriptions of roses in *Modern Roses* until the edition published in 1986.

In 1978 the Convention of the International Union for the Protection of New Plant Varieties in Geneva, Switzerland voted for the notion that each new rose variety registered be given a unique descriptor, the code name, to identify clearly the plant whatever the final name given the rose for sale to the public. This code name was intended to become the designated variety denomination within the formal registration process. In 1981 the ARS was appointed to serve as the Registrar for CIOPORA. Then, in 1985, the US Patent and Trademark Office proposed rule making in the *Federal Register* to implement the consensus reached by the Convention in Geneva. However, they fully intended to adopt the registered names listed in the latest edition of *Modern Roses* which, in many cases, were popular fancy commercial names and not the unique variety denomination (i.e. usually the code name). To correct this situation, the ARS in combination with American growers and CIOPORA moved to change the registration strategy by declaring the code name as the variety denomination. Within this agreement the breeders agreed that they would indicate on the registration application which cultivars were patented or trademarked under a fancy name. The new registration form was approved for use in January 1998.

For the future In the registration of a new variety, the unique code name will automatically become the variety denomination. For most registrations, this guideline will be directly applicable. In most instances, when an amateur breeder registers a rose, the lack of a code name places the given fancy name as the variety denomination. Trademarks under which the variety enters commerce will be annotated only at the request of the breeder. However, it is routine collection of exact data on each cultivar that will serve the purpose of improving the current classification system.

Conclusions

Classification of roses has long been a topic for discussion as to how best to preserve the botanical significance and yet fulfil a commercial benefit to the public. On the international scene, there is no doubt that common sense prevails in assigning classes based on several basic principles, such as climbing versus nonclimbing, recurrent versus nonrecurrent. This grand design accommodates cultivated roses in a clean subdivision that leaves no doubt as to the growth habit of the cultivar. For commercial purposes, however, the public need an identifiable class to connect with. Introducers and hybridizers have been innovative in introducing new classes to describe their efforts. For instance, Floribunda and Grandiflora were American inventions to capture the growing habits of new types of roses. This system becomes a necessity within the trade to communicate easily with the potential buyer.

In the registration process delegated to ARS, introducers have the option to use such classes, as approved by ARS, in their submissions to describe in simple terms what type of rose they have hybridized, irrespective of its parents and heritage. At the present time, however, there exists some confusion in that the registration data may not reflect how the rose is actually being sold. Any classification system must recognize the importance of the hybridizer/introducer and the rose-growing community. Hopefully, the recent consensus reached to bring common sense and logic to this important aspect of roses is practical. The evolution of the rose into many new flower forms and growing habits should be accommodated by a classification system that is flexible and is a reflection of reality.

See also: **Classification**: Conventional Taxonomy (Wild Roses). **Genetics**: Meiosis; Species Crosses; Inheritance in the Dogrose; Gene Mapping; Research Trends. **History of Roses in Cultivation**: European (Pre-1800); Modern (Post-1800).

Further Reading

Allen EF (1972) *A Simplified Rose Classification for Gardeners*. St. Albans, UK: Royal National Rose Society Annual.

Cairns T (2000) Modern Roses XI. London: Academic Press.

Lammerts WE (1945) *Scientific Basis of Rose Breeding*. Shreveport, LA: American Rose Society.

Rehder A (1940) *Manual of Cultivated Trees and Shrubs Hardy in North America*. 2nd edn. New York.

Cultivar Identification by Image Analysis

D Zhang, Geves Le Magneraud, F-17700 Surgères, France

M-H Gandelin, Geves Cavaillon, Les Vignères, France

Introduction

Rose breeders need to protect their intellectual property rights for their new varieties. This requires an accurate description of varieties. In the process of protecting new plant varieties, all applicant varieties must undergo tests of distinctness, uniformity and stability (DUS) according to the Union pour la Protection des Obtentions Végétales (UPOV) convention. The assessment of the distinctness of a new variety is based on the investigation of 46 phenotypic characteristics listed in the *Guidelines for the Conduct of Tests for Distinctness, Homogeneity and Stability* of roses established by UPOV. All the observations made on these 46 characteristics fall into three categories: behaviour (development type), morphology (size, form) and flower colour. The principal weakness of this system is the lack of precision and subjectivity of the observations. For example, the size of petal should be assessed as 'very small, small, medium, large and very large' according to the UPOV guidelines for roses. With the increasing number of protected varieties of roses in the world, this becomes insufficient for accurate cultivar identification. This is why new techniques and descriptors are being explored for cultivar identification in roses. Image analysis is one of the new tools which can allow an automated numeric description of different organs of roses; this may make the description of rose morphology and flower colour more precise, objective and reliable, adding to the human investigations in the field or greenhouse.

Materials Necessary for Image Analysis

Flowers, petals, leaves and leaflets can all be used for image analysis in rose. In the present article, only flowers and petals will be considered. The sampling can be made on plants cultured in a greenhouse or in the fields. In order to assess the variability within a cultivar, the study should be performed on several plants per cultivar, several flowers per plant and several petals per flower, over several years and several seasons, including flowerings in greenhouse and field.

In Geves Sophia-Antipolis, we used a colour scanner Sharp JX450, with a resolution of 100 dpi (digit per inch = 4 pixels per millimetre), to acquire images of flowers and petals. The scanner, via an interface GPIB (General Purpose Interface Bus), is piloted by a Sun station and the software Cameleo®. The images are scanned in a Visilog® format; each colour image is made up of three files corresponding to the three channels, red, green and blue.

Image Acquisition

After removing the sepals if necessary, flowers are put directly on the glass window of the scanner. The petals are stuck on a white or black background in order to obtain maximum contrast between the background and the object to be analysed (**Figures 1** and **2**).

The images are analysed using a Sun station with 8 bits and 16 Mb of memory. The programs for image analysis are developed in language C; some of the basic routines are included in the package Visilog®.

Image Analysis

Image processing consists of two principal steps: (1) pretreatments leading to the isolation of objects to be analysed (flowers or petals), inside the images; and (2) adaptive description according to the objects under consideration.

Isolation of the Objects of Interest

This phase is fundamental. It consists of isolating the flower or petal from the background images which may contain different foreign bodies, including pollen and dust. This stage includes different steps:

- creation of a grey-level image from the three files of colour images (red, green and blue channels); this allows simplification of calculations and reduction in the time of image processing
- construction of a histogram of the grey level on the whole image
- detection on the histogram obtained of peaks corresponding to objects of interest and the background
- computation of a threshold value corresponding to the transition background object
- creation of a binary image where the background noise is set to 0 and the object of interest set to 1 (**Figure 3**)
- elimination of particles or residual point effects (which have the same grey level as the object of interest but with a negligible area) by an opening of order 3
- formation of a mask of the objects (flowers or petals) by the multiplication of the original image by the binary image (**Figure 4**).

After this preprocessing, the background of the image becomes really black (0 grey level) and only the object of interest will be taken into account in the subsequent computations. Now it is ready to describe the object of interest, whether flower or petal.

Figure 1 Original colour images of flowers.

Figure 2 Original colour images of petals.

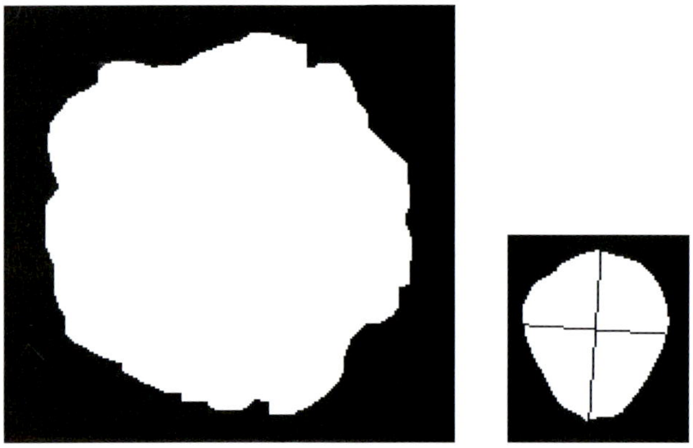

Figure 3 Isolation of object of interest: binary image creation with determination of an axial reference for petals.

Figure 4 Grey-level mask of the objects (multiplication of binary mask and original image).

Description of the Object of Interest

Flower The global morphological parameters, including area, perimeter and compactness, are computed on the binary image.

For a fine description of flower shape, the radius as well as the diameter are computed at every 2° angle from the barycentre to the contour. These measurements quantitatively define the subcategories of the flower shape (viewed from above) in the class 'irregularly rounded'. It is very useful to determine whether the flower is round or star-shaped and to quantify the irregularity of flowers that are irregularly rounded.

On the flower colour mask, the global colorimetric parameters, such as hue, intensity and saturation, are computed.

Petal On petals, as for flowers, the global morphological parameters such as area, perimeter and compactness are computed on the binary image.

To describe the shape of petals precisely, an axial reference is established from the global morphology

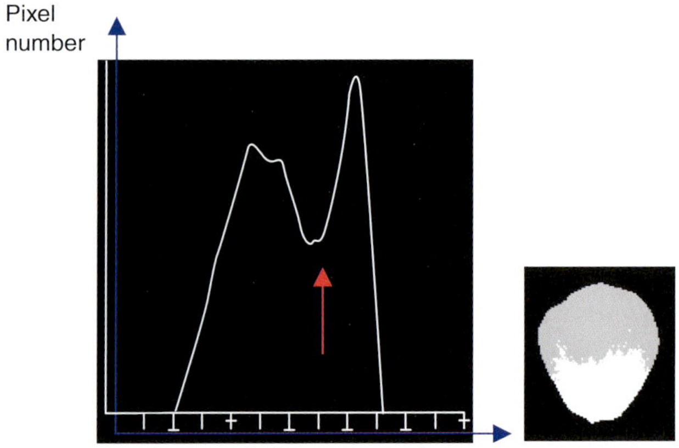

Figure 5 Image analysis on a petal: building the grey-level histogram on the mask image; computing threshold value (arrow), using different colour areas for isolation on the petal.

of petals. The axial reference is determined by the intersection of the vertical and the horizontal axes (**Figure 3**).

On this axial reference, the following computations and measurements are made:

- axial width of petal
- axial height of petal
- position of the maximum width in relation to the height
- area of the four different sectors of the petal: right–left and top–bottom

The last measurements allow the observation of symmetry and asymmetry of petals. A fine analysis of the petal shape can be realized on the profiles of distance contour-axis of the four different sectors of the petal.

The distribution of colours along each axis is analysed in order to characterize and separate the different zones on the petal. Densitometric profiles of each axis are constructed with the average of colours measured on 3 pixels along each axis. The detection of breaks in slopes or peaks on the profiles indicates the existence of different colour zones.

If different colour zones are detected, a more accurate analysis is made. For this, a histogram of colours is built by considering the totality of points making up the petal (**Figure 5**). From peaks or nonnegligible detectable zones of transition on the histogram, one or several threshold values are computed and one given colour is assigned to the different zones detected (**Figure 5**). Once isolated, the different zones are quantitatively characterized by their area, shape, hue, saturation and position. Their colorimetric characteristics can be quantitatively established by building the colour images of different zones by multipliying the image of zones and the colour mask of the petal.

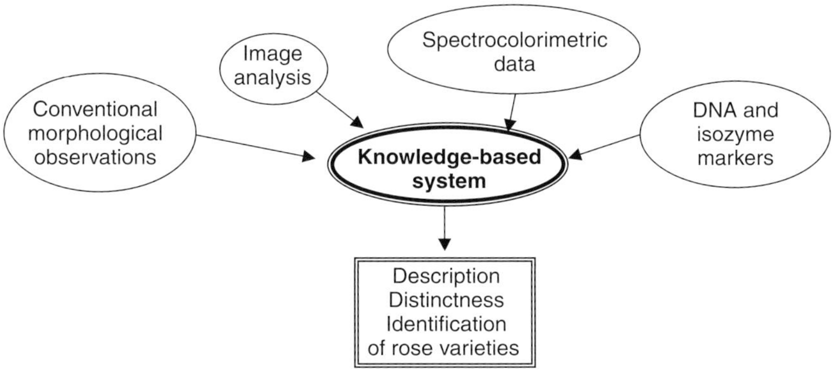

Figure 6 Prospect: a knowledge-based system imaged for rose varieties: description and identification integrating data generated by different tools.

Variability of Measurement

For flowers, as for petals, samples can present important morphological variations within a particular year, place and flowering, even for the flowers sampled from a same plant and the petals sampled from the same flower. The principal difficulty of image analysis concerns the quantification of this 'normal' variability of a given cultivar and how this may change with time.

Prospects of Image Analysis for Rose Cultivar Description

Image analysis can generate useful information and data for the description of roses that can help the identification of rose cultivars. The big challenge using this tool for rose variety description is how to quantify reliably the existing morphological variations that can be observed on roses, and to generate acceptable data by image analysis and processing that can be used for cultivar identification. It is clear that using only one single tool, like image analysis alone, will not be sufficient to make an accurate description of rose varieties and to allow a reliable decision on distinctness. It would be very interesting to build a knowledge-based system for rose variety description and identification, integrating data generated by conventional morphological observations, image analysis, spectrocolorimetric analysis and DNA and isozyme markers (**Figure 6**).

See also: **Breeding**: Selection Strategies for Cut Roses. **Classification**: Conventional Taxonomy (Wild Roses). **Genetics**: Mutation. **Intellectual Property**: Plant Patents and Trademarks; Technical Examination and Conventional Characteristics.

Further Reading

Ballard DH and Brown CM (1982) *Computer Vision*. Englewood Cliffs, NJ: Prentice-Hall.

Gandelin MH, Mistou MN and Poetto L (1995) Protection et recensement des variétés de rosier: nouvelles méthodes d'étude. Les dossiers n°11 INRA mensuel n° 81. *La rose, le rosier et recherches* 81: 43–45.

Kender JR (1976) *Saturation, Hue and Normalized Color: Calculation, Digitalization Effects, and Use*. Pittsburgh: Department of Computer Science, Carnegie-Mellon University.

Zhang D, Besse C, Cao MQ and Gandelin MH (2001) Evaluation of AFLPs for variety identification in modern rose (*Rosa hybrida* L.). *Acta Horticulturae* 546: 351–357.

Chemotaxonomy and Molecular Taxonomy

J-L Fiasson, O Raymond, F Piola, C Sanlaville-Boisson and M Jay, Université Claude Bernard, Lyon, France

Introduction

As in all plant groups, the systematics of the genus *Rosa* was based at first on morphological features, an approach reaching its pinnacle with the classification proposed by Rehder (*see* **Classification:** Conventional Taxonomy (Wild Roses)). Then, taxonomists tried to take advantage of other kinds of descriptors, increasingly more quantifiable, subtle and 'closer to the genome'. The aim was initially to take into account characters that are both more objectively measurable and less sensitive to developmental and environmental variations. As soon as it was understood that chromosomes were the material vector of heredity, systematicians sought to 'read' them more directly. Indeed, phytochemicals, such as colorants and fragrance molecules, reflect directly the activity of enzymatic proteins translated from DNA; diversity of isozymes echoes directly that of alleles; while the ultimate step is to decipher the blueprint of DNA itself (**Figure 1**).

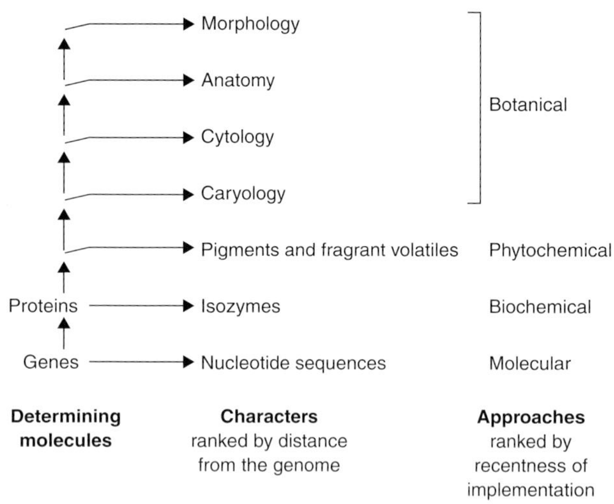

Figure 1 Historical evolution of the distance between taxonomical approach and genetic information. The translation from genotype to phenotype is likely to be blurred at each intermediate step by the interference of endogenous as well as exogenous factors. Methodological improvements, which allow more and more subtle features to be taken into account, reduced the number of these steps and thus the amount of 'noise' in taxonomical characters.

Since the widespread acceptance of Darwinian theory, it is generally accepted that classification should reflect the genealogical relationships between species. The implicit model of organism evolution is the phylogenetic tree, reflecting successive ramifications and divergence of lineages. Actually, this branching pattern does not take into account the possibility of hybridization, which is peculiarly frequent in the plant kingdom and even more so during plant breeding. Therefore, the real phylogeny is often reticulate rather than simply branched (**Figure 2**). Concerning the genus *Rosa*, it is unclear whether its accepted classification reflects the evolutionary process. It is all the more important that hybridization, sometimes leading to chromosome doubling – a phenomenon called allopolyploidization (**Figure 3**) – seems to be more frequent in this genus than previously acknowledged.

The Use of Phytochemical and Molecular Markers in Rose Systematics

Chemical Analyses: Interpreting the Organoleptic Phenotype

Most phytochemical analyses of roses were done with an agronomic aim, from selection of parents to

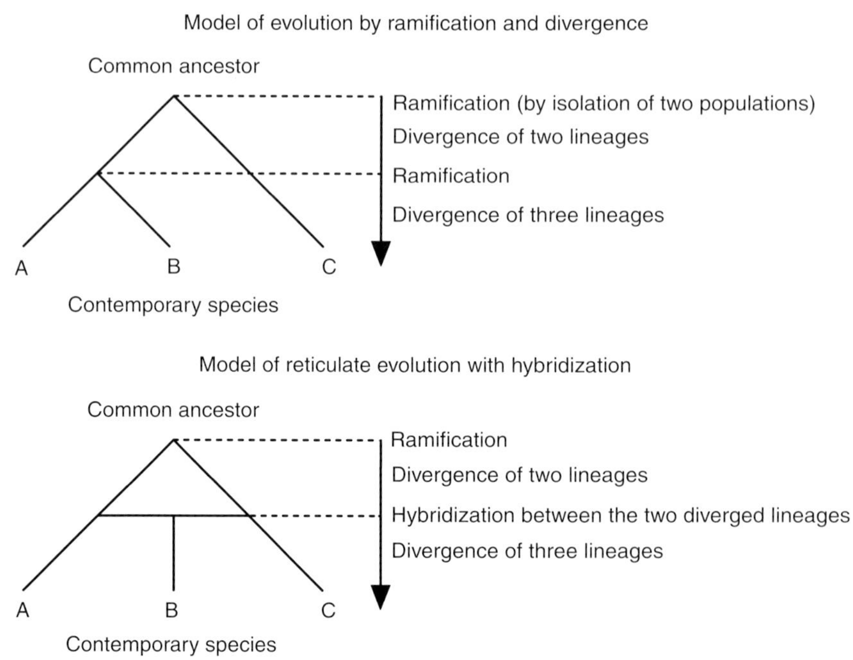

Figure 2 Phylogenetic relationships between taxa under divergent and reticulate evolution. Classical models of Darwinian evolution imply that species appear after isolated populations diverge. Arborescent genealogical trees describe this process, but they are misleading when populations hybridize, which is a frequent phenomenon in the plant kingdom. In this case, the paradigmatic assumption that sharing derived characters (apomorphies) tags a single (monophyletic) lineage is no longer valid, since a hybrid taxon harbours derived characters inherited from multiple ancestors. Reticulate trees (or networks) more accurately reflect this evolutionary scenario.

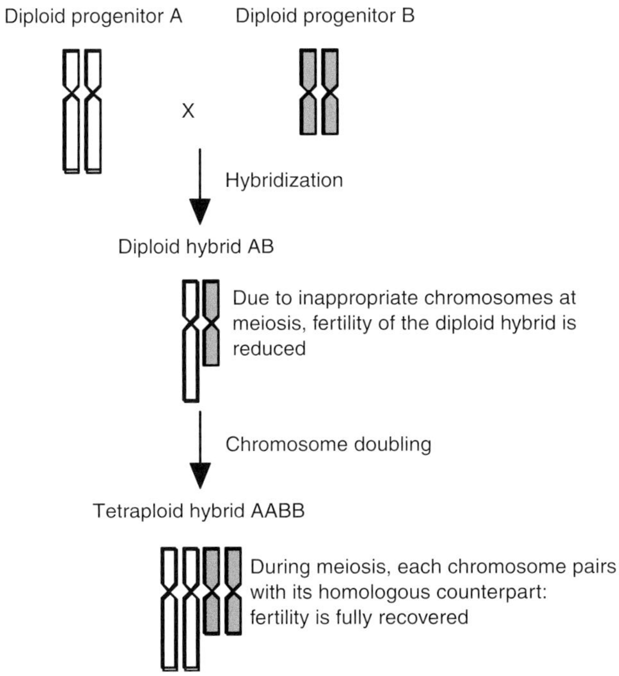

Figure 3 Allotetraploid formation by hybridization between diploid species followed by chromosome doubling. In higher plants, as in animals, each cell of each individual usually bears two sets of chromosomes, one inherited from each parent. During the formation of pollen and egg (meiosis), homologous chromosomes pair, and sometimes exchange pieces of hereditary material. Pairing of parental chromosomes is hampered by too much dissimilarity between them, resulting in reduced fertility. It can occur when two different diploid species hybridize: the diploid progeny is often infertile. Normal chromosomal pairing in the hybrid can be restored through chromosomal doubling: after this rare event each parental chromosome has a twin counterpart, with which pairing is possible. Each cell of the doubled hybrid organism, an allotetraploid, possess twice as many chromosomes as its diploid ancestors. The parental subsets of chromosomes in an allopoplyploid are termed 'homologous'.

characterization of patented cultivars. Consequently, sampling is both limited and discontinuous, preventing rigorous conclusions to be drawn (**Table 1**).

The fragrance of the rose (Figure 4) The perfume industry was at the origin of the elucidation of the chemical basis of rose scent: most work was done on the Damask rose, mainly on essential oils. Several tea-scented China roses were also investigated for the origin of the soft and refreshing notes in their fragrance, and the composition of volatiles of several cultivars was analysed in a search for new compounds. Many different molecules make the odoriferous bouquet; perhaps the most striking result is that characteristic odours are due to minor compounds: for the essence of the Damask rose, the β-damascenone accounts for most of the scent but only 0.1% of the weight of volatiles. *Rosa gigantea* and its progeny are characterized instead by the production of 1,3-dimethoxy-5-methyl-benzene but, in *Rosa* as a whole, such discriminative chemicals are only part of a common background, mainly of monoterpenoids such as citronellol and geraniol, and aromatic compounds such as 2-phenyl ethanol. This sprinkling of analyses of rose fragrant volatiles is all the less usable in a chemotaxonomical perspective since results are highly variable depending on physiological state, environmental conditions and experimental processes.

The colour of the rose (Figure 5) Flower colour has long been used as a taxonomic character; since different combinations of pigments can produce the same hue, elucidation of the chemical structures was a prerequisite to turn this macroscopic feature into a reliable taxonomic character. Chemically, rose petal colour results from the accumulation of pigmentary molecules in the epidermis – yellow carotenoids dissolved in oil droplets in the plastids, and red water-soluble anthocyanins in the vacuoles, derived from mevalonic and shikimic metabolism, respectively.

Carotenoids in rose petals Carotenoids are responsible for yellow colour and, in combination with anthocyanins, for the bronze and orange hues. While such pigments are universally part of the photosynthetic apparatus of green plants, no yellow shade was present in the petals of the old roses of Mediterranean origin. The pale yellow colour displayed by several early noisette and tea roses was inherited from 'Park's Yellow Tea-Scented China', but the intense, warm and bright-yellow pigmentation of modern roses was inherited later from *R. foetida*. Indeed, the biosynthesis of carotenoids in petals is suppressed in old western roses, blocked at the step of pale-yellow carotene precursors in 'Parks' Yellow Tea-Scented China', and fully expressed, up to the brightly coloured oxygenated xanthophylls, in *R. foetida*. As a whole, the occurrence of carotenoids in rose petals is far too uncommon to be used in chemotaxonomy.

Flavonoids in rose petals (and leaves) All shades of red in rose petals are due to polyphenolic, heterosidic pigments belonging to the anthocyanin family, part of the biosynthetic group of flavonoids: indeed, the colourless flavonol glycosides (mainly of kaempferol and quercetin), which always co-occur with them in vacuolar sap, also play a part in the petal hue by acting as copigments. The systematic exploration of this metabolism in the genus *Rosa* led to the recognition of only three flavonol chemotypes for the 10 botanical sections, and the distribution of anthocyanins does not bring much additional information: in particular, the progenitors of old domesticated roses (*Gallicanae*,

Table 1 Distribution of relevant secondary metabolites among subdivisions of *Rosa*

Subgenus	*Section*	*Phytochemistry*			
		Fragrance	*Colour*		
		Volatile compounds	*Carotenoids*	*Polyphenols*	
			Petals	*Petals*	*Leaves*
Hulthemia					
Rosa		2-phenyl ethanol, citronellol and geraniol generally predominant in subgenus *Rosa*			Mainly Q heterosides
	Pimpinellifoliae		Complete carotenoid metabolism in *R. foetida*	4′-glucosides of Q and K	
	Gallicanae	β-damascenone in *R. damascena*		3-glucosides of Q and K	
	Caninae			3-diglucosides of Q and K	
	Carolinae			3-diglucosides of Q and K	
	Cinnamomeae			3-diglucosides of Q and K; *pelargonidin*	
	Synstylae			3-glucosides of Q and K	
	Chinenses	1,3-dimethoxy-5-methyl-benzene in *R. gigantea*	ζ-carotene in *R. gigantea*	3-glucosides of Q and K; paeonidin rare	
	Banksianea			3-glucosides of Q and K	
	Laevigatae			3-glucosides of Q and K	
	Bracteatae			3-glucosides of Q and K	
Plathyrhodon					Mainly K heterosides
Hesperhodos					

Q, quecertin; K, kaempferol (see **Figure 4**).

Synstylae, *Indicae*) share the same flavonoid pattern except that paeonidin is rare in the *Indicae*. Using a refined and quantitative approach, taking statistical differences between old representatives of these three sections, hybrids do not display a simple additivity of ancestral features; furthermore, biogenetic equilibria between flavonols have been distorted by selection for high yields of anthocyanins. The uncommon pelargonidin, which is found in modern cultivars, probably comes from *Cinnamomeae* ancestors. Overall, floral flavonoid patterns, as they are currently known, turn out to be poorly suited for genealogical reconstruction within the domesticated roses complex, and for taxonomy of the subgenus *Eurosa* as a whole.

Leaf flavonoids, acknowledged as chemotaxonomical markers for wild plants, were until recently neglected in rose, as in most other plants grown for their flowers. In the leaves, flavonoid patterns are far less complex and varied than in petals: in the subgenus *Rosa* most molecules correspond to monoglycosides of the simple flavonol quercetin, while in *R. roxburghii* glycosides of kaempferol are predominant. While variations of quercetin/kaempferol between sections may be significant, it must be stressed that no single chromatographic system allows a satisfactory separation of their glycosides, while all the information about their substitution is lost by hydrolysis. In the rose, leaf polyphenol metabolism, as currently understood, is too constant to mark lineages. Botanical sections cannot be defined by their chemotypes; uncommon flavonoid balances seem rather to reflect recent hybridizations.

Conclusion In most plant taxa, the chemistry of the secondary metabolites (those small molecules accumulated by plants that we perceive as colours, smells, tastes) has turned out to be a powerful taxonomical tool; but, currently, the chemotaxonomic study of *Rosa* appears disappointing; the observed variations in accumulated secondary metabolites are either very limited, or reflect too many unknown factors. Improvement of the analytical techniques combined with a wider sampling would none the less provide the taxonomist with a robust framework of data, which could, at the very least, be useful to test the conclusions reached through other approaches.

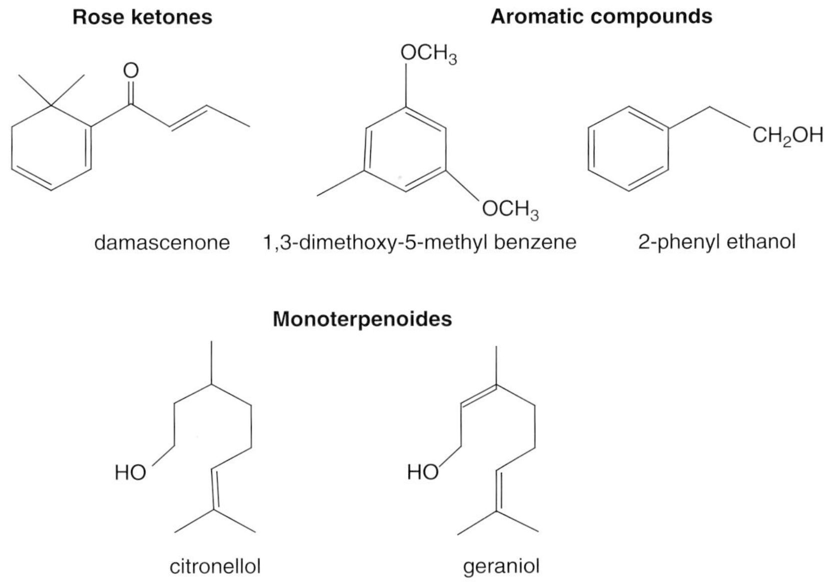

Figure 4 Fragrant molecules.

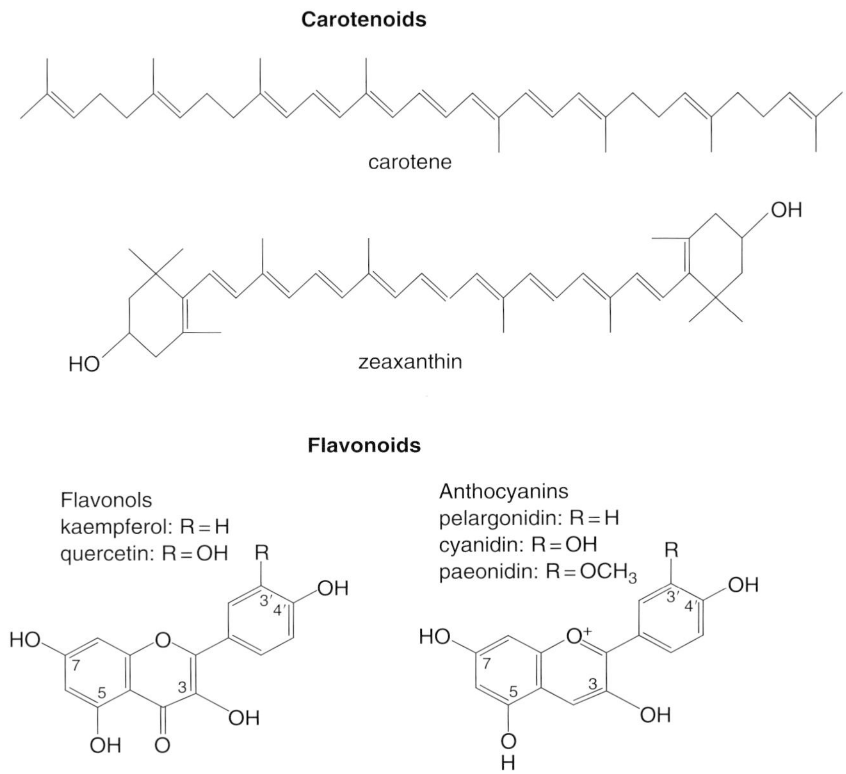

Figure 5 Colouring molecules.

Molecular Analyses: Chemically Deciphering the Genotype

The genus *Rosa* is large and complex, composed of wild and domesticated species, grouped on the basis of morphological characters. Hybridizations and allopolyploidization have occurred frequently in this genus and make difficult the classification and the search for relationships between species and varieties. The molecular approach seems to be the more effective because it allows direct access to the hereditary material (the genome) and makes it possible to understand the relationships between plants.

DNA is the repository of the genetic information and is present in all cells in the chromosomes. The chromosomes include genes which directly determine the various characters of a plant. In closely related organisms, such as those in the genus *Rosa*, all individuals possess the same broad gene patterns which control the same fundamental biological functions. However, all individuals differ from each other by minute variations, called allelic variations; they are all the more important in genetically distant taxa. The genes run into tens of thousands and yet they have to be transmitted from the parents to the progeny in a precise and balanced manner. Consequently, the knowledge of the genetic structure of rose varieties allows us to identify the individuals and to understand their genealogy.

DNA Fingerprints

Polymerase chain reaction (PCR) and controlled digestion (restriction) make the development of molecular markers possible (*see* **Genetics:** Gene Mapping; Research Trends), popularized by their use in forensic science. Different types of these markers have recently been used to throw new light on *Rosa* taxonomy.

DNA fingerprints are obtained after a random but highly specific sampling of the genome by PCR, resulting in a collection of DNA fragments of various sizes. After separation by electrophoresis, these DNA fragments constitute a kind of 'barcode' characteristic of the variety analysed. The fact that two plants share common PCR fragments indicates some degree of relationship between them. Using computer programs permits the pairwise comparison of the PCR patterns (or 'barcodes') and allows representation of the genetic relatedness between all the plants analysed. In rose studies, arbitrarily primed PCR (AP-PCR) as well as random amplified polymorphic DNA (RAPD) markers have demonstrated their effectiveness for the identification of varieties, the study of genealogical relationships between cultivars, the differentiation of different genomes and the phylogenetic relationship between species and varieties. Recently, a method named amplified fragment length polymorphism (AFLP), based on the selective amplification of restriction fragments – obtained after enzymatic digestion of genomic DNA (*see* **Genetics:** Molecular Markers) – has been employed for the genetic mapping of two modern rose cultivars. This method could also be used for classification purposes, though the results of this application to this field remain to be published.

Methods based on pairwise comparisons of banding patterns have been variously applied in classification at both supra and infraspecific levels. Generally, they reveal a good agreement between the botanical sections and the identified clusters of related DNA fingerprints. When applied to the comparison of domesticated gene pools and their ancestral botanical species, these methods brought to light particular features of the domestication process in *Rosa*. For example, the European (mostly *Gallicanae*) and Chinese ancestors (mostly *Indicae* and *Synstylae*) of our modern cultivars were consistently retrieved by these analyses; they were also used to quantify the relative contributions of each gene pool to the hereditary material of early and recent hybrids. The important conclusion is that domestication in *Rosa* entailed a reduction of genetic diversity, a commonly observed part of the so-called domestication syndrome. Finally the infraspecific classification of cultivars in races was unequally supported, probably due to wide and recurrent interracial hybridizations.

Nucleotide Sequences

Other molecular analyses focus on the local architecture of DNA, described by the sequence of its elementary constituents, namely its nucleotides. They aim at assessing genome similarities on a finer scale than fingerprinting methods. These markers are under recent development for the rose: among the numerous sequences retrieved from *Rosa* DNA, chloroplastic genes and nuclear internal transcribed spacer (ITS) offered new hypotheses on phylogeny and possible hybridization events. However, these two types of sequences are prone to documented caveats: the chloroplastic genome on the one hand has a uniparental inheritance and a low rate of evolution and, therefore, is partially informative, especially for studying recently diverged taxa. On the other hand, ITS, though widely used for inferring past hybridization, has been demonstrated to undergo concerted evolution, i.e. homogenization of parental sequences in hybrid taxa. All in all, the conclusions drawn from these methods are either disappointing (chloroplastic genome led to poorly resolved phylogeny) or possibly misleading (the triparental origin of *R. damascena* remains to be supported by more evidence than partial sequence ITS additivity).

At the present time, none of the published molecular rose phylogenies is fully resolved, either because of an incomplete scanning of the range of taxonomic diversity or because the evolution rate detectable by the chosen markers is inappropriate for the study of recent divergence events. So, what type of sequences could be harnessed to reconstruct more efficiently the phylogeny of *Rosa* species?

First, rapidly evolving introns of low-copy nuclear genes were successfully used in other genera (*Paeonia*, *Oryza*, *Gossypium*) to understand the origin of allopolyploid species. Given the frequency of such allopolyploidization events in *Rosa*, such sequences could provide us with new insights on the hybrid origin of many rose species and early cultivars. Although a prerequisite to retrieving nucleotide sequences from low-copy nuclear genes in allopolyploid organisms is the separation of homologous genes by means of cloning, which could be time-consuming, future phylogenies of *Rosa* species and horticultural groups should include evidence from such DNA sequences.

Then, microsatellites, short sequences of tandemly repeated nucleotides, exhibit high evolutionary rates and could therefore be used to address those problems of plant classification that come from short-time scale evolution. Their use is currently under development in *Rosa*.

Chromosome Painting

Finally, molecular techniques provide a new way of staining chromosomes. These methods of *in situ* hybridization rely on the affinity of known DNA fragments – probes that are experimentally labelled with a fluorescent dye – for specific genomic regions. The probes can then be visualized in chromosomes at metaphase. Some probes have a high affinity for nuclear organizer regions (NOR) that are helpful to identify species of hybrid origin. In *Rosa*, this approach has been successfully employed twice, to demonstrate the allopolyploid origin of *R. gallica* and of modern cultivars. More sophisticated techniques use the DNA of the whole genome as a labelled probe (genome *in situ* hybridization or GISH) and can help to identify, by differential staining of homologous chromosomal subsets, the progenitors of allopolyploid species. It requires that the progenitors have sufficiently divergent genomes to avoid crossed *in situ* hybridization of probes with the analysed subsets of chromosomes. This method has not yet been applied to *Rosa*, but it may be promising to screen for and confirm the identification of progenitors of known allopolyploids such as *R. gallica*.

Molecular studies will allow us to understand better the history of the genus *Rosa* and could lead to a new classification of this complex genus. Indeed, it has already been demonstrated using different types of molecular markers that the Rehder's sections *Indicae* and *Synstylae* are very closely related. This result shows that the taxonomy of the rose is very complicated and it will be necessary to make use of all the available descriptors (morphological, chemical, biochemical and molecular) to propose a new classification reflecting the real history of roses.

Conclusion and Perspectives

Classification within the genus *Rosa* is a daunting challenge. Recognition of wild specimens is not easy due to the imprecise nature, and, sometimes, high variability, of diagnostic morphological criteria. As an example of this difficulty, five different forms – sometimes even termed species – of *R. gallica* were described. This may be the consequence of an important genetic diversity in populations, as adaptations to local conditions, and even of phenotypic plasticity.

As a result of these classification issues, several attempts were made to establish species definition on the basis of allegedly objective biometric measures. A convincing illustration of this effort comes from nordic dog roses (sect. *Caninae*), among which five species (*R. canina*, *R. dumalis*, *R. rubiginosa*, *R. sherardii*, *R. villosa*) were successfully distinguished by elliptic Fourier analysis of leaflet shape. However, the accuracy of the morphometric delineations obtained can be misleading. Although the leaflets of these five dog roses harbour distinctive patterns of elliptic Fourier harmonics, the underlying genetic stucture, assessed by means of RAPD similarity, is organized around three subgroups only. Unlike the earlier example of *R. gallica*, dog roses show that morphologically distinct taxa can be more genetically related than expected. This may reflect common ancestry or gene flow between pentaploid taxa that exhibit similar reproductive features involving unbalanced meiosis.

The examples of *R. gallica* and dog roses emphasizes that phenotypic diagnoses of taxa are not necessarily congruent with genetic relatedness. These discrepancies between phenotype and genotype classifications are only meaningful when interpreted as symptoms of the relevant evolutionary processes. It seems self-evident that taxonomy should be used as a tool to unravel genealogical relationships between species and sometimes shed light on peculiar and original evolutionary pathways. However, with few notable exceptions, this has not been the case in the genus *Rosa*. One of these exceptions is the continuous focus on the section *Caninae*, since the early 1920s, when an original meiotic behaviour, referred to as '*canina*-meiosis', was discovered. In the pentaploid *Caninae*, for example, only the haploid microspores are viable, whereas fertile macrospores are tetraploid (*see* **Genetics:** Inheritance in the Dogrose). As a consequence of this unbalanced mechanism, heredity in the section *Caninae* was expected to be predominantly maternal, and this prediction has been formally demonstrated recently with the help of RAPD markers.

Despite the interest in *canina*-meiosis, it cannot be denied that it is extremely rare in the plant kingdom. Though its evolutionary consequences deserve to be thoroughly described *per se*, its significance is likely to be marginal. However, the section *Caninae* illustrates the potential of combining complementary methods: high-tech morphometry; translating leaflet shape into harmonics; phytochemistry, revealing possible links between *Caninae* species and some *Cinnamomeae* species through the sharing of rare quercetin and kaempferol sophorosides; caryology, identifying unusual patterns of chromosome segregation; and finally, genetics, using reciprocal crosses and molecular markers to confirm the prediction of matroclinal inheritance. This integration of methodologies, in association with recent refinements of the phylogenetic rationale, will contribute to a renewed classification of *Rosa*, translating the current distribution patterns of characters into a historical evolutionary process. We are convinced that more general insights on plant evolution can be obtained from this integrative approach of *Rosa* taxonomy. So what general evolutionary concepts could benefit from studies of genus *Rosa*?

Polyploidization is now acknowledged as a driving evolutionary force, with as much as 2–4% of speciation events in angiosperms resulting from chromosome doubling. Polyploidization can arise from spontaneous chromosome doubling of an initially diploid ancestor, resulting in an autotetraploid neospecies. Though this phenomenon is known in numerous plant genera (including *Dactylis*, *Lotus*, *Epilobium* and *Centaurea*), it may be less frequent in *Rosa* than the alternative process in which an allotetraploid neospecies originates from an initial hybridization between two diploid progenitors (**Figure 3**). Indeed, the recent physical mapping of ribosomal DNA of *R. gallica* supported an allotetraploid origin of this species. The range of ploidy levels in genus *Rosa* (2× – 8 ×) and the extreme diversity of species are strong incentives to investigate its phylogeny looking for allopolyploidization events.

Genomic methods in general and nucleotide sequences in particular can be used to infer reticulation events entailing allopolyploidization. However, the choice of a methodological approach must be carefully reasoned and sampling issues must be correctly handled. Once again, the reconstruction of phylogenetic relationships between *Rosa* species was often poorly supported. Either the molecular probes were unsuitable – because they represented parts of the genome that were evolving too fast or too slowly (submitted to concerted evolution, for example) – or because the coverage of the taxonomic range was poor (sometimes both!). Despite these concerns, recent phylogenetic studies suggest some interesting directions for future research. For instance, the allopolyploid origin of section *Gallicanae*, suspected by Hurst on the basis of karyological and ecodevelopmental characteristics, now seems well established. More precisely and though this statement deserves cautious consideration, for the reasons mentioned above, one might be struck by the possible implication of *Synstylae* species in the origin of the *Gallicanae*, as suggested by ITS sequence additivity in *R. gallica* and *R. damascena*. This last study, however, only focused on the so-called triparental origin of *R. damascena* and completely overlooked the essential point that *R. gallica* could also have a hybrid origin! Whatever the final interpretation of these controversial results, it seems likely that molecular tools will lead to reappraisal of the position of section *Gallicanae* with respect to section *Synstylae*.

The possibility of hybridizations is known to create taxonomic ambiguities that are not taken into account in Rehder's classification. The ease with which *Rosa* species hybridize has been widely exploited to create new cultivars, and hybridization also occurs spontaneously between wild species. The extent of natural hybridization – between species with the same chromosome number (homoploid hybridization) – has not been quantified, probably because homoploid hybrids are difficult to identify. Nevertheless, preliminary results obtained with the help of RAPD markers indicate that diploid species of the section *Indicae* could be of hybrid origin. Spontaneous hybridization between species of different ploidy levels is well documented in *Rosa*. European populations of triploid hybrids between tetraploid *R. gallica* and diploid *R. arvensis* (*Synstylae*) have been reported and recent literature has mentioned the existence of natural triploid hybrids between *R. gallica* and diploid *R. moschata* (*Synstylae*). These observations give some support to the idea of a close relationship between *Gallicanae* and *Synstylae*, but also suggest future research: triploid hybrids are known to be of reduced fertility – a phenomenon known as triploid block – however, they could act as bridges for gene exchange between tetraploid and diploid species. Their contribution to gene flow and to the population dynamics of hybridizing tetraploid and diploid species remains to be quantified; *Rosa* is a promising model for examining these currently disputed issues.

See also: **Classification:** Conventional Taxonomy (Wild Roses); **Distribution and Ecology** Overview. **Genetics:** Karyology; Inheritance in the Dogrose; Gene Mapping; Molecular Markers; Reserach Trends.

Further Reading

Britton G (1983) *The Biochemistry of Natural Pigments*. Cambridge: Cambridge University Press.

Goodwin TW (ed.) (1988) *Plant Pigments*. London: Academic Press.

Grazzini F, Carrer C and Wise T (1996) *Flower, Why Do You Smell So Nice?* New York: Kane/Miller.

Harborne JB (ed.) (1996) *Introduction to Ecological Biochemistry*, 4th edn. London: Academic Press.

Harborne JB and Tomas-Barberan FA (eds.) (1991) *Ecological Chemistry and Biochemistry of the Plant Terpernoids*. Oxford: Oxford University Press.

Hillis DM, Moritz C and Mable BK (1991) *Molecular Systematics*. Sunderland, MA: Sinauer Associates.

Krinsky NI (ed.) (1990) *Carotenoids. Chemistry and Biology*. New York: Plenum Press.

Lea PJ and Leegood RC (eds.) (1999) *Plant Biochemistry and Molecular Biology*. New York: John Wiley.

Leitch IJ and Bennett MD (1997) Polyploidy in angiosperms. *Trends in Plant Science* 2: 470–476.

Mulder-Krieger T and Verpoorte R (1993) *Anthocyanins as Flower Pigments: Feasibilities for Flower Colour Modification*. Dordrecht: Kluwer Academic.

Nei M and Kumar S (2000) *Molecular Evolution and Phylogenetics*. Oxford: Oxford University Press.

Sondheimer E and Simeone JB (eds.) (1970) *Chemical Ecology*. London: Academic Press.

Wilder LB (1974) *The Fragrant Garden: A Book About Sweet-Scented Flowers and Leaves*. Dover: Dover Publications.

Cross Breeding *see* **Breeding**: Methods of Cross Breeding.

Cytogenetics *see* **Genetics**: Karyology; Meiosis.

D

DISEASE

Contents

Overview

R K Horst, Cornell University, Ithaca, NY, USA

Infectious diseases are caused by fungi, bacteria, viruses and nematodes. Fungi are organisms which have no chlorophyll, reproduce by sexual and asexual spores and typically possess mycelia or a mass of interwoven threads (hyphae). There are about 70 000 species living as parasites or saprophytes on other organisms or residues and more than 8000 species causing plant disease. The most serious and widespread fungal diseases of rose are powdery mildew caused by *Podosphaera pannosa* (Wallr.: Fr.) de Bary and black spot caused by *Diplocarpon rosae* Wolf.

Plant pathogenic bacteria are generally unicellular, nonspore-forming rods. There are about 80 species of bacteria which cause plant disease and many of them consist of numerous pathovars. The identification of bacteria can be very difficult; it involves special staining techniques, biochemical tests and serology. Pathogenic bacteria enter plants through insects or other wounds, through stomata, hydathodes, lenticels and flower nectaries. The most widespread bacterial disease of rose is crown gall caused by *Agrobacterium tumefaciens* (E. F. Sm. & Town.) Conn.

Viruses contain two major components: nucleic acid surrounded by a protein coat. There are over 850 described plant virus species. Some attack a large number of different plants and are of great economic importance: others are confined to a single host. Viruses are transmitted from plant to plant by insects, mites, fungi, nematodes, water, and by rubbing, abrasion or other mechanical means. Viruses are differentiated and identified by host specificity, physical properties, purification, electron microscopy, electrophoresis and serology. The most widespread viral disease of rose is rose mosaic. Rose mosaic is generally associated with isolates of *Prunus* necrotic ringspot virus (PNRSV) or with mixed infections with PNRSV, apple mosaic virus and *Arabis* mosaic virus.

Nematodes are unsegmented roundworms that inhabit fresh and salt water, decaying organic matter, soil, plants and animals throughout the world. More than 2000 species are parasitic on plants. Most plant nematodes are root parasites, but some live in stems, leaves or buds. Some cause galls and others produce general yellowing, stunting or dieback that is often ascribed to other causes such as nutrient imbalance.

Commercial production of plants free of disease requires frequent observations and intelligent strategic planning. Prevention is essential for good disease control. Although control can vary from partial to economic to perfect, the best disease control is when it is profitable to crop production. Even before the discovery of Bordeaux mixture, it was determined by trial and error that chemicals could be useful for pest control. For example, sulphur was known to avert pests by Homer in 1000 BC and was probably generally used by 1800. In 1824 Robertson recommended sulphur for control of powdery mildew of peach. Lime-sulphur found its way into disease control by 1906 and copper sulphate was recommended for roses in 1862.

These early observations and discoveries led to mixing lime and copper sulphate. This mixture was first used on grapes to deter pilfering in Médoc, France. However, Millardet in 1882 observed that the mixture (called Bordeaux mixture, after the location where it was first used) controlled the severe downy mildew disease of grapes. This discovery marked the true beginning of the modern era of chemicals to control plant disease. Bordeaux mixture is still used today; its greatest deterrent is a tendency to be phytotoxic. Sulphur, lime-sulphur and Bordeaux mixture were the major tools for disease control in ornamentals from 1900 to 1943.

The horizons for chemical controls were broadened measurably in 1934 with the announced development of alkyl dithiocarbamate fungicides. However, technical difficulties in development delayed by 10 years making them available to the public. Then in 1943 the dilemma was solved and nabam, zineb and maneb were introduced (the bisdithiocarbamate family of chemicals). These fungicides were broad-spectrum in control of fungal pathogens; production yields in agriculture increased 65% from 1901 to 1945 (a period of 44 years) and by 90% from 1945 to 1954 (a period of 9 years). The bisdithiocarbamates are still used today; however, their availability may be in peril due to the Food Quality Protection Act passed in 1996 by the US Congress.

Chemical soil fumigation, beginning with chloropicrin in 1948 and with methyl bromide in 1951, has been used in indoor and outdoor production of ornamental plants. Excellent control of soil-borne pathogens has been achieved with their use; however, alternatives are being sought due to toxicities associated with their use and their likely elimination for use by 2006.

The use of chemicals to control fungal, nematode and bacterial pathogens from 1945 to 1975 was exciting to plant pathologists and the industry since most disease could be controlled and better and more flowers grown and sold. Prior to 1945 disease control amounted to 'putting out fires' and after 1945 it was preventive. Preventive disease control programmes could be designed and successfully performed with proper use of pesticides. Exceptions to this success story were systemic disease and viral-caused diseases. Systemic fungicides and viricides were difficult to develop since the genomes of pathogens and host plants shared some similarities; host and pathogen physiology shares some similar functional responses to some chemical compounds. Thus, pesticides that would control vascular wilts caused by fungi, bacteria and viral pathogens were also probably phytotoxic to the host. For example, prior to 1942 the control of *Verticillium* wilt of chrysanthemums and other vegetatively propagated ornamentals was dependent on selection of stock plants which were visually free from disease, on use of short terminal cuttings and the application of best preventive sprays then available. This programme enabled the small grower to avoid disastrous losses, but few were ever free from *Verticillium*. However, during the 1940s and 1950s propagation became concentrated in the hands of a few specialists.

In 1942 A.W. Dimock showed that laboratory procedures for determining the presence or absence of pathogenic fungi and bacteria in the vascular tissues of plants had potential for commercial adaptation. The culture-index was enlarged to include virus-indexing, i.e. culture virus-indexing. The programme also included thermotherapy and meristemtip culture methods for obtaining plants free of virus when detection methods showed them to be infected. The success of the chrysanthemum programme led to similar culture virus-indexing programmes for carnations, geraniums, roses and orchids. These programmes are being used commercially for ornamentals on the European continent as well as in North and South America. These classic programmes have also been modified for use in agronomic crops. The employment of the culture virus-index programme has been so effective that many young growers do not recognize some of the diseases which in 1940 caused costly losses to every chrysanthemum grower in the business.

In 1961 a landmark book was published by Rachel Carson, entitled *Silent Spring*. This book had a major impact on the indiscriminate and successful use of pesticides. The main focus was on DDT and toxic environmental influences of its use. Public concerns raised by this book caught the attention of the US government which took action to establish the US Environmental Protection Agency (EPA) in 1970. The EPA's responsibility was to register and regulate the use of pesticides. These restrictions required exceptional planning in ornamental production.

During the 1980s there was increased interest in the reduced use of toxic pesticides in the environment. These concerns generated more interest in the use of integrated pest management (IPM). The main objective of IPM is to reduce the use of toxic chemicals which contaminate the environment and still control diseases. An important component of a successful IPM programme is the forecast of conditions which are favourable for disease epidemics and to spray only when necessary. IPM strategies which can be used include the following:

- Apply pesticides only when necessary.
- Make use of application methods that apply less pesticide or use a more efficient spray system.
- Use biocompatible chemicals as they become available.
- Use biological controls when available and when appropriate.
- Use cultural practices which are favourable to healthy plant growth.

A successful IPM programme depends on four basic techniques:

1. Scouting: regular and random visual observations provide early warning to disease problems.
2. Disease identification: the first and most important step is to identify the problem; misdiagnosis results in improper control.
3. Timing: improper timing of control measures will result in disease control failure; the control measure must be timed correctly to the stage of disease development.
4. Records: brief accurate records are a good tool for disease control decisions.

A grower's use of IPM strategies can reduce pesticide use by 50%. During the mid-1980s and into the 1990s there has been intense interest in the development and use of biological controls, genetically engineered disease-resistant plants, biocompatible chemicals and more efficient spray delivery systems. Biological controls have been developed from species of *Trichoderma*, *Streptomyces* and *Gliocladium* for control of fungal-caused diseases. Biocompatible chemicals are 'soft' chemicals and/or 'natural' substances that may bear profound antimicrobial properties and that exhibit low mammalian and environmental toxicities. Some examples are bicarbonates (used in baking), neem from the neem tree (*Melia azedarach*), horticultural oils and strobulins (from mushrooms). There has also been interest in improved efficiency in spray delivery systems, such as air-assisted aerosol generators and air-assisted electrostatic sprayers. Fungicides are dispersed effectively and at reduced dosages.

On 6 August 1996 the US Congress passed the Food Quality Protection Act. This Act requires EPA to reassess tolerances of all 9600 pesticides available to the industry. Thirty-three per cent of these pesticides must be reevaluated by August 1999, 66% by 2002 and all of them by 2006. The impact this process will have on the ornamental industry will be great since it is likely that many commonly used pesticides will be eliminated for commercial use. Alternative disease control measures will become more important and in some instances disease control will be difficult. Growers and plant pathologists will need to develop creative disease control strategies to be successful in a very competitive ornamental market.

See also: **Breeding**: Selection Strategies for Disease and Pest Resistance. **Disease**: Crown Gall; *Botrytis*; Black Spot; Downy Mildew; Powdery Mildew; Rust; Replant Problems and Soil Sickness; Bactericides and Fungicides. **Insects and other Animals**: Overview of Insects; Aphids; Mites; Thrips; Nematodes; Galls Induced By Cynipid Wasps; Insecticides. **Integrated Pest Management**.

Further Reading

Binns AN and Thomashow MF (1988) Cell biology of *Agrobacterium* infection and transformation of plants. *Annual Review of Microbiology* 42: 575–606.

Braun V (1995) The powdery mildews (*Erysiphales*) of Europe. Jena: G. Fischer.

Brierley P and Olson CJ (1956) Development and production of virus-free chrysanthemum propagative material. *US Department of Agriculture Plant Disease Reporter Supplement* 238: 63–67.

Carson R (1962) *Silent Spring*. Boston: Houghton Mifflin.

Debener T, Drewes-Alvarez R and Rockstroh K (1998) Identification of five physiological races of black spot (*Diplocarpon rosae* Wolf) on roses. *Plant Breeding* 117: 267–270.

Dimock AW (1943) A method of establishing *Verticillium*-free clones of ornamental plants. *Phytopathology* 33: 3.

Dimock AW, Williamson CE and Nelson PE (1964) Diseases. In: *Chrysanthemums. A Manual of the Culture, Diseases and Insects and Economics of Chrysanthemums*, pp. 116–150. New York State Extension Service and New York Flower Growers Association.

Horst RK (1983) *Compendium of Rose Diseases*. St Paul, MN: APS Press.

Horst RK (1993) Current trends, ornamentals. In: Naegly SK (ed.) *Plant Health Guide*, pp. 138–139. Willoughby, OH: Meister.

Horst RK (1997) Fungicides for the future. *Grower Notes* 2(2): 1–3.

Horst RK (1999) Technology on production: defeating disease. In: *History of US Floriculture, Greenhouse Growers*, pp. 97–98, Willoughby, OH: Meister.

Horst RK (2001) Integrated pest management. In: *Westcott's Plant Disease Handbook*, 6th edn, pp. 30–31. Boston: Kluwer.
McCallan SEA (1956) Rediscovery of sulfur as a fungicide. *Phytopathology* 46: 582.
Moury B, Cardin L, Onesto JP, Candresse T and Ponpet A (2001) Survey of *Prunus* necrotic ringspot virus in rose and the variability in rose and *Prunus* species. *Phytopathology* 91: 84–91.
Wong SM, Horst RK and Langhans RW (1988) Symptomatology and occurrence of apple mosaic and *Prunus* necrotic ringspot on rose in New York. *Acta Horticulturae* 234: 437–447.

Crown Gall

A P Hert and J B Jones, University of Florida, Gainesville, FL, USA

Introduction

Bacterial crown gall is caused by the bacterium *Agrobacterium tumefaciens* (Smith and Townsend 1907). This bacterium is commonly found on members of the rose family (apples, rose, raspberry, peach, etc.) and is worldwide in distribution. The organism is a Gram-negative, rod-shaped, non-spore-forming, motile, aerobic bacterium. Upon infection of plants, *A. tumefaciens* incites abnormal cell proliferation (hyperplasia) which results in tumour formation. *Agrobacterium* is closely related to *Rhizobium* which forms nitrogen-fixing nodules on clover and other leguminous plants. Recently a proposal has been made to transfer the group *Agrobacterium* to *Rhizobium*.

Disease Cycle

The bacterium gains entry into the host plant through fresh wounds. Wounds that commonly serve as infection sites are those made during cultivation, pruning, harvesting or machinery operations or are a result of freezing injury, growth cracks, soil insects or any other factor that causes injury to plant tissues. The bacterium is most commonly introduced into a planting site on or in planting material.

The disease was given the name crown gall because of the large tumour-like swellings, or galls, which typically develop at the crown as a result of grafting (**Figure 1**). Galls can also form on stems and shoots, although aerial galls are not common. *Agrobacterium tumefaciens* is abundant in the outer portions of primary galls, which is often sloughed off into the soil.

Figure 1 Crown gall on rose. (Photograph courtesy of Dr Powelson, Oregon State University.)

When the sloughed tissue comes in contact with wounded healthy tissue of a susceptible host, the bacterium enters the plant and induces gall formation, thus completing the disease cycle. Tissue surrounding the gall may appear to be healthy; however, it has been shown that the bacterium can spread systemically throughout the whole plant.

In early stages of development, the galls appear as tumour-like swellings that are more or less spherical, white or flesh-coloured, rough, spongy (soft) and wart-like. As galls age, they become dark brown to black, hard, rough and woody. If infection is severe, plants may be stunted or show various deficiency symptoms due to impaired uptake and transport of nutrients and water. The bacterium is a soilborne pathogen that can persist in soil or plant debris for up to 3 years. The bacterium survives best in moist alkaline soils and is especially severe in loose sandy or loam soils.

Mode of Action

Agrobacterium tumefaciens has a unique mode of action (**Figure 2**). There are chromosomal virulence genes (*chv* genes) that are involved in pathogenicity in disease development; however, most of the genes involved in virulence (*vir* genes) and crown gall disease (T-DNA) are located on a large plasmid, termed the tumour-inducing plasmid (Ti plasmid). Pathogenic *Agrobacterium* species are able to sense their host by recognizing acetosyringones, or wound phenolics released by the plant when wounded. These phenolics activate the virulence genes (*chv* and *vir* genes) as well as endonucleases which excise the T-DNA. *Chv* genes are responsible for attachment of the bacterium to plant cells at the wound site. *Vir* genes are responsible for preparation of the T-DNA, and the exact genes

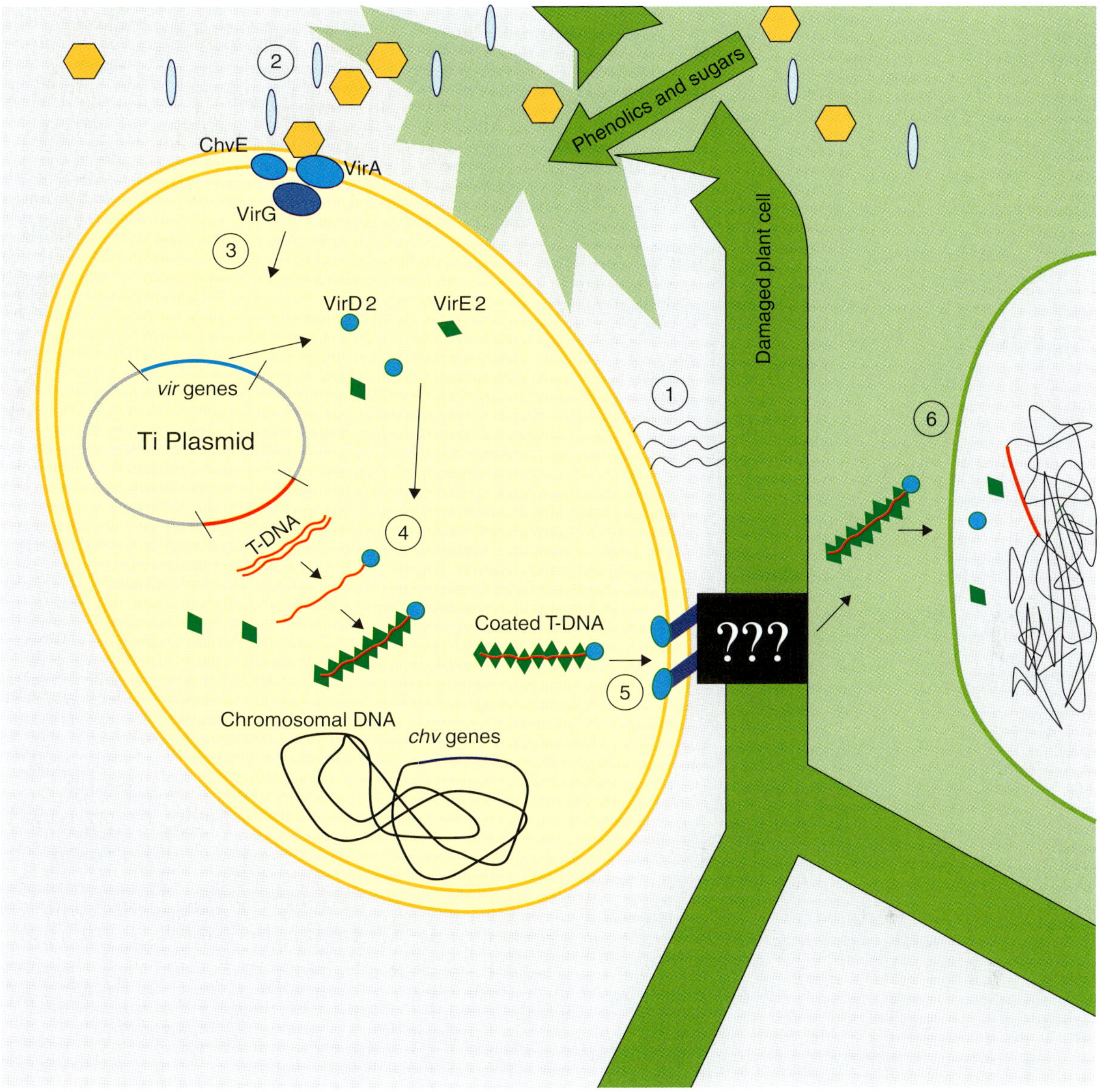

Figure 2 *Agrobacterium*–plant cell interactions. This diagram illustrates the interactions between *Agrobacterium* and a damaged host plant cell. *Agrobacterium* and the plant cell are represented in yellow and green, respectively. There are six basic steps: (1) attachment of the bacterium to the host cell; (2) recognition of plant phenolics and sugar; (3) activation of *vir* genes and production of T-DNA; (4) coating of T-DNA by VirD2 and VirE2; (5) delivery into the host cell; and (6) transport into the host cell nucleus and integration of T-DNA into the host chromosomal DNA. Black Box – delivery of the T-DNA across the plant cell wall has still not been determined definitively.

responsible for delivery of the T-DNA into the host plant is still unknown; however, the *virB* operon is thought to be responsible.

Once inside the host cell the T-DNA must enter the nucleus and integrate into the plant chromosomal DNA for the gene(s) on it to be expressed. Two *vir* proteins, VirD2 and VirE2, are responsible for signalling the T-DNA for transport across the nuclear membrane of the plant cell. Both VirD2 and VirE2 have been shown to contain nuclear localization signals, which signal transport of the T-DNA into the cell nucleus.

Once the T-DNA is inside the nucleus it must be expressed; however, in order to be expressed the T-DNA must be integrated into the chromosome of the plant cell. It is thought that the VirD2 and VirE2 are also responsible for helping with the integration of the T-DNA into the chromosomal DNA. When the T-DNA is successfully inserted into the chromosomal DNA of the host plant cell the proteins encoded on it are expressed.

The T-DNA encodes a number of proteins including those involved in auxin and cytokinin synthesis. Synthesis of auxins and cytokinins (plant hormones) results in an unregulated and disorganized growth of plant cells to the extent that galls develop. The galls in turn produce novel carbon compounds, called opines, that are compounds almost exclusively utilized by *Agrobacterium*. The Ti plasmid encodes the necessary enzymes to allow the bacteria to utilize the opines. The opines produced by the tumours are exuded and utilized by the *Agrobacterium* as carbon sources as a result of the catabolic gene systems encoded on the Ti plasmid. Interestingly, the T-DNA itself has no effect on this whole process up to the expression stage of the transformation. It has been found that the T-DNA can be modified significantly without affecting transfer of the T-DNA and integration. Consequently, much of the T-DNA can be replaced by 'your favourite gene' (YFG) of interest and in this way theT-DNA acts as a vector for genetically transforming of plants with YFG.

Genetic Engineering

Agrobacterium expressing a modified Ti plasmid can transfer genes of interest and that DNA will be integrated into the genome of the host plant. This technique has a high frequency of transformation, with low copy number for correct expression of the introduced genes. *Agrobacterium*-mediated transformation is commonly utilized for transformation of a number of plant hosts.

This technology has been utilized in disease resistance for roses. The black spot disease caused by the ascomycete fungus *Diplocarpon rosae* Wolf is the most common and damaging disease of rose. Breeding for resistance had been relatively unsuccessful, and fungicides are costly and undesirable. These problems have led to the need for development of a transgenic approach to disease resistance in roses. Researchers determined that the Type I ribosome inhibiting protein from barley was an excellent candidate (for disease control) to be used for *Agrobacterium*-mediated transformation of rose. They engineered *in planta* expression of the Type I protein. It was shown that secretion of the ribosome inhibiting protein into the extracellular space significantly reduced susceptibility to the black spot fungus.

Diagnosis

A good way to prevent crown gall is to exclude the pathogen from plant material by determining if you have the bacterium before the symptoms appear. Therefore, a good prerequisite for the control of crown gall is to have a reliable detection method. Polymerase chain reaction (PCR) is an extremely sensitive technique. Furthermore, primers specific to the Ti plasmid are available for laboratory use for detection of pathogenic *Agrobacterium* spp. Unfortunately PCR is not an at-home diagnostic test. Therefore isolation of the bacterium from infected host material or from infested soil is the method of choice.

Agrobacterium can be isolated from galls and vascular fluid, and from symptomless plants. For optimal isolation, cream-coloured gall tissue from young, actively growing galls is best. Wash the gall surface with running water and surface sterilize for 10–20 min in a 1% sodium hypochlorite solution (or 20% household bleach) prior to isolation. Rinse the tissue several times with sterile water before proceeding. Remove two or three small pieces of the gall tissue and place in a culture tube with sterile distilled water or buffer, vortex and let stand for 30 min or longer to allow the bacteria to diffuse into the liquid. Streak a loopful of the tissue suspension onto agar media.

The bacterium can also be isolated from soil and water. The use of semiselective media is especially helpful for isolation of *Agrobacterium* from soil and water. Even on selective media a few other bacterial species can grow. Sensitivity of the selective media is limited to about 10^4 colony forming units per gram of soil. Incubate plates at 25–27 °C.

Once isolated from plant material, soil, or water, *Agrobacterium* strains can be differentiated into one of four species: *A. tumefaciens*, *A. rhizogenes*, *A. rubi* and *A. vitis*. There are several diagnostic tests for identification of *Agrobacterium* to species; however, these diagnostic tests do not differentiate between pathogenic and nonpathogenic strains.

Several molecular techniques, including PCR, are used to identify *Agrobacterium* to the species level. Serological techniques using monoclonal antibodies are currently available for identifying *A. vitis* strains. Fatty acid analysis has also proven to provide differentiation between members of *A. tumefaciens*, *A. rhizogenes*, *A. rubi* and *A. vitis*. Carbon source utilization provides another useful molecular tool for identification of *Agrobacterium* to the species level. The Biolog system (Biolog, Inc., Hayward, CA, USA) is one of the most widely used systems for carbon source utilization analysis. This system allows for a metabolic fingerprint based on utilization patterns of the different biolog carbon compounds. The patterns are characteristic at both the *Agrobacterium* genus level as well as the species level.

Control

The best control measure is prevention through the use of cultural practices. There are no chemical pesiticides currently available for control of this pathogen; however, biological control methods are commercially available.

Cultural Practices

To prevent crown gall, plant only certified, disease-free roses, and take care not to wound plants, especially the root systems, at planting time. Plant only in disease-free soils without a prior history of the disease, or wait 3–5 years before replanting. Fumigation of new or renovated rose beds is suggested before planting. Steam the soil at least to 50 °C for 1 h, especially if the soil was infested with *A. tumefaciens*. Crown gall control starts with continuous monitoring to detect infected plants. Promote growth of young rose plants and make sure environmental factors such as fertilization, temperature, humidity and light are optimal. Disinfect all cutting and pruning tools that have been used near crown gall infected plants by dipping them in a solution of 0.5% sodium hypochlorite. Cuttings taken from healthy tissue of plants containing galls may also be infected, because, as mentioned previously, the bacterium has been shown to move systemically into healthy tissues. Avoid infestation of soil insects and nematodes, which are able to damage the roots and stems of rose plants and thus provide an entry for the bacterium. Remove infected plants as soon as galls are observed. If possible, remove and discard the soil from areas where infected plants are located.

Biological Control

Treatment of plants before planting with a commercially available biological control agent may prevent infection by *A. tumefaciens*. The root system of roses may be protected from crown gall with a preplant root-dip of *A. radiobacter* strain K84. Strain K84 is antagonistic to most strains of *A. tumefaciens* due to the production of the bacteriocin agrocin 84. Bacteriocins are compounds produced by a bacterium that are inhibitory or harmful only to closely related bacteria. As is the case for most bacteriocins, agrocin 84 is expressed on a transmissible plasmid. Therefore, the *A. radiobacter* strain K84 utilized as a biological control agent has been modified to delete the transfer genes (*tra* genes) in order to reduce the possibility of transferring the bacteriocin-producing plasmid into pathogenic strains of *Agrobacterium*. The genetically engineered transfer deficient plasmid-containing strain is strain K1026. Agrocin 84 is deleterious to *A. tumefacien* strains, acting by blocking DNA synthesis. Galltrol-A and Norbac 84C are commercially available products utilizing strain K1026.

Nomenclature

Agrobacterium was originally divided into five species: *A. tumefaciens*, *A. radiobacter* (Beijerinck and van Delden 1902), *A. rubi* (Hildebrand 1940), *A. vitis* (Ophel and Kerr 1990), and *A. rhizogenes* (Riker *et al.* 1930) (**Table 1**). Pathogenicity symptoms expressed in infected plants caused by the particular strain of *Agrobacterium* were originally used as a tool to determine nomenclature: *Agrobacterium tumefaciens* strains caused characteristic crown gall symptoms; *A. radiobacter* strains were the nonpathogenic species of *Agrobacterium*; *A. rubi* strains were limited to cause crown gall on *Rubus* species of plants of the Rosaceae family; *A. vitis* strains caused crown gall on *Vitis* species of plants (Vitaceae or grape family); and *A. rhizogenes* strains caused hairy root which induces abundant root proliferation at the site of infection.

As knowledge about the bacterium increased, it was determined that pathogenicity by *Agrobacterium* species is mediated by the Ti or Ri plasmids as mentioned previously. The discovery that pathogenicity genes were carried on a plasmid and were transferable between strains was an unreliable determinant of species designation. More recently chromosomally encoded phenotypic and genotypic characteristics were used to place *Agrobacterium* into three chromosomal groups or biovars: biovar 1, biovar 2 and biovar 3 (**Table 2**). However, in 1994 the biovar again changed to biovar species status with the nomenclature being as follows: *A. tumefaciens* (biovar 1), *A. rhizogenes* (biovar 2)

Table 1 Original nomenclature of *Agrobacterium* species

Nomenclature	*Phytopathogenic physiology*
Agrobacterium tumefaciens	Causal agent of crown gall
Agrobacterium radiobacter	Nonpathogenic
Agrobacterium rubi	Causes gall on *Rubus* spp.
Agrobacterium vitis	Crown gall on *Vitis* spp.
Agrobacterium rhizogenes	Causal agent of hairy root

Table 2 Comparison of old and new nomenclature for species of *Agrobacterium*

Proposed taxonomy	*Current taxonomy*	*Previous taxonomy*
Rhizobium radiobacter	*A. tumefaciens*	*A. tumefaciens* biovar 1 *A. radiobacter* biovar 1 *A. rhizogenes* biovar 1
Rhizobium rhizogenes	*A. rhizogenes*	*A. tumefaciens* biovar 2 *A. radiobacter* biovar 2 *A. rhizogenes* biovar 2
Rhizobium vitis	*A. vitis*	*A. tumefaciens* biovar 3 *A. radiobacter* biovar 3
Rhizobium rubi	*A. rubi*	*A. rubi*

and *A. vitis* (biovar 3), regardless of pathogenicity (**Table 2**). Finally in 2001 the nomenclature was again proposed to place all species currently in the genus *Agrobacterium* in the closely related genus *Rhizobium*.

See also: **Breeding**: Selection Strategies for Disease and Pest Resistance. **Disease**: Bactericides and Fungicides. **Integrated Pest Management**.

Further Reading

Binns AN and Thomashow MF (1988) Cell biology of *Agrobacterium* infection and transformation of plants. *Annual Review of Microbiology* 42: 575–606.

Bouzar H, Jones JB and Bishop JB (1995) Simple cultural tests for identification of *Agrobacterium* biovars. In: Gartland KMA and Davey MR (eds.) *Methods in Molecular Biology*, vol. 44, pp. 9–13. Totowa, NJ: Humana Press.

Bradbury JF (1986) *Guide to Plant Pathogenic Bacteria*. Kew, UK: Commonwealth Agricultural Bureaux.

Broembsen S (1994) *Extension Facts: Diseases of Roses*. Stillwater, OK: Oklahoma State University.

Dohm A, Ludwig C, Schilling D and Debener T (2001) Transformation of roses with genes from antifungal proteins. *Acta Horticulturae* 547: 27–33.

Gelvin SB (2000) *Agrobacterium* and plant genes involved in T-DNA transfer and integration. *Annual Review of Plant Physiology and Plant Molecular Biology* 51: 223–256.

Horst RK (1995) *Compendium of Rose Diseases*, pp. 23–25. St Paul, MN: American Phytopathological Society.

Kersters K and De Ley J (1984) Genus III. *Agrobacterium*. In: Krieg NR and Holt JG (eds.) *Bergey's Manual of Systematic Bacteriology*, vol. 1, pp. 244–254. Baltimore, MD: Williams & Wilkins.

Klee H, Horsch R and Rogers S (1987) *Agrobacterium*-mediated plant transformation and its further applications to plant biology. *Annual Review of Plant Physiology and Plant Molecular Biology* 38: 467–486.

Moore LW, Bouzar H and Burr T (2001) *Agrobacterium*. In: Schaad NW, Jones JB and Chun W (eds.) *Laboratory Guide for Identification of Plant Pathogenic Bacteria*, 3rd edn, pp. 17–35. St Paul, MN: American Phytopathological Society.

Nutter RC, Thomashow MF, Gordon MP and Nester EW (1981) *Agrobacterium*: nature's genetic engineer. In: Panopoulos NJ (ed.) *Genetic Engineering in the Plant Sciences*, pp. 17–35. New York: Praeger Press.

Pionnat S, Dessaux Y, Nesme X and Poncet C (1996) Detection and determination of pathogenic *Agrobacterium* of roses with PCR. *Acta Horticulturae* 424: 227–232.

Sheng J and Citovsky V (1996) *Agrobacterium*–plant cell DNA transport: have virulence proteins, will travel. *Plant Cell* 8: 1699–1710.

Botrytis

M L Gleason and S J Helland, Department of Plant Pathology, Iowa State University, Ames, IA, USA

Introduction

Grey mould, caused by the fungus *Botrytis cinerea* Pers.:Fr., the imperfect stage of *Botryotinia fuckeliana*, is among the most damaging diseases of cultivated roses worldwide, attacking both outdoor (**Figure 1**) and greenhouse (**Figure 2**) plantings as well as cut roses during storage and transit. Common names of the disease on roses include botrytis blight and botrytis rot. The pathogen is ubiquitous in greenhouses, sporulates abundantly on senescing and dead plant tissues, and is pathogenic on scores of ornamental and vegetable crops in addition to roses. Symptoms on rose flowers include small flecks that develop into brown, necrotic spots that can ultimately spread to engulf the entire flower (**Figures 3–5**). Stem lesions are elongated, sunken, and black and brown in colour. High humidity and cool to mild temperatures are conducive to abundant conidia production; free water on plant surfaces is required for infection. Infections are frequently latent when roses are harvested, but symptoms may develop during storage and shipment. Small, rounded, black sclerotia are the most common overseasoning stage of *B. cinerea*. Valuable cultural management techniques include: use of partially resistant cultivars; rigorous sanitation; suppression of high humidity in greenhouses by forced-air circulation, venting and heating, and in shipment by adequately ventilated containers; and watering techniques that minimize the time that plants are wet. Fungicides are widely used to suppress grey mould of roses, but resistance and other factors can compromise their effectiveness. Numerous experimental treatments have the potential to supplement or replace conventional fungicides for grey mould management, but these treatments are not yet available for commercial use.

Symptoms

On petals of greenhouse roses, symptoms typically appear as small flecks that expand into brown, necrotic spots. These spots can ultimately engulf entire petals and the receptacle, finally causing collapse of the flower head and abscission of petals. On susceptible roses, one to three lesions per flower are sufficient to render a flower unmarketable. *Botrytis cinerea* petal infections on greenhouse roses are frequently

Figure 1 Target-like canker that developed on the cane of a field-grown rose after prolonged humid weather. (Courtesy of RK Jones, North Carolina State University.)

Figure 2 Lesion developing on the petal of a greenhouse-grown rose. (Courtesy of Iowa State University.)

Figure 3 Flower blighted by *B. cinerea* infection. (Courtesy of Iowa State University.)

Figure 4 Base of flower blighted by *B. cinerea* infection. (Courtesy of Iowa State University.)

latent; that is, they are not visible when flowers are harvested, but develop under humid postharvest storage conditions or as flowers mature. Severe economic losses can occur when symptoms appear on flowers in storage or during shipment.

Lesions on field-grown roses, and sometimes those grown in greenhouses, may develop woolly masses of tan-brown mycelium and greyish *B. cinerea* conidia under humid conditions. Young twigs and cut stubs die back and large, sunken, elongated, target-like, black and brown blotches develop on canes. Small, black, oval to hemispheric sclerotia frequently form on or just below the cuticle or epidermis in these lesions, and are firmly attached.

Disease Cycle and Ecology

Botrytis cinerea persists year-round in greenhouses as conidia, mycelia or sclerotia on living or dead plants, and as sclerotia or conidia in infested soil. Outdoors, the fungus overwinters in infested plant debris or soil. Sclerotia are the most important structures for field survival, although conidia may also overseason in the field. The survival stages can be spread long distances

Figure 5 Blight of entire bloom, caused by advanced *B. cinerea* infection. (Courtesy of Iowa State University.)

through movement of infested plant debris or soil. Sclerotia commonly germinate by producing conidia, occasionally by producing hyphae, and rarely by producing apothecia and ascospores.

Conidia developing from sclerotia, infected plants or infested plant debris are dispersed in large numbers by air currents. Sporulation is favoured when air circulation within the plant canopy is reduced or absent. *Botrytis cinerea* spores in a greenhouse can originate within the crop itself or can enter the house from external sources through vents. Spore germination and infection are most rapid at temperatures of 22–25 °C with free moisture or high relative humidity (90–100%). *Botrytis cinerea* is also active at relatively low temperatures, however, and can cause considerable damage to cut-rose flowers and plants stored to 0–10 °C. When roses in shipping cartons are transferred to room temperature during transit or at delivery sites, condensation of water on the cold flowers triggers germination of *B. cinerea* conidia that were deposited during the flower production phase, resulting in lesions.

Although germinating conidia seldom penetrate intact, actively growing tissue, penetration is common through wounds such as cut stubs of stems and tipburned leaves. Ageing flower petals and senescing leaves are readily colonized. Levels of ethylene, a compound produced during flower senescence, were positively correlated with grey mould severity in roses.

Susceptibility of roses to grey mould varies with the growing environment. Susceptibility of a greenhouse-grown cultivar may differ seasonally; for example, 'Golden Wave' flowers grown in a California greenhouse during winter were more susceptible than the same cultivar grown in the autumn, presumably because increased humidity, lower temperatures and lower light conditions during the winter reduced cuticle thickness and thus facilitated invasion by *B. cinerea*. Seasonal changes in the greenhouse environment, including relative humidity and vapour pressure deficit, can also affect the amount of airborne conidia and conduciveness to conidial germination on plant tissues. Greenhouses that did not rigorously remove senescing and dead plant tissue, on which *B. cinerea* can sporulate copiously, were more vulnerable to severe grey mould outbreaks than greenhouses with adequate sanitation practices.

Management

Successful management of grey mould requires integration of the best available strategies, since no single measure is adequate alone.

Cultural Strategies

Some rose cultivars are moderately resistant to *B. cinerea*, but the degree of resistance has not been clearly determined for most cultivars. In greenhouses and field plantings, strict sanitation is a critically important cultural practice. All spent blossoms, fallen leaves and other plant debris, as well as any symptomatic buds, flowers and canes, should be removed, from both plant beds and beneath benches, hauled away from the vicinity and destroyed. When soil or recycled potting media are used for pots and plant beds, they should be steamed at 82 °C for 30 min or 71 °C for 1 h in order to destroy *B. cinerea* inoculum.

Greenhouse relative humidity should be held below 85–90% by forced-air circulation, as well as venting and heating at night. Plant spacing should be adequate to allow air movement within the crop canopy. The goal of these practices is to minimize the duration of periods when moist, still air remains around the plants. Adequate spacing is also important for outdoor rose plantings, and selection of sites with adequate soil and air drainage. Effective weed suppression also speeds drying and minimizes the duration of wet periods. Watering at the base of plants is preferable to overhead irrigation. Where overhead irrigation is unavoidable, however, watering during the morning is advised, since it allows time for the plants to dry before nightfall.

Since *B. cinerea* readily invades wounds, it is helpful to avoid wounding plants unnecessarily.

Improvements in shipping practices can help suppress development of grey mould. Using cardboard boxes perforated with numerous, large ventilation holes reduced germination of *B. cinerea* conidia in comparison to boxes without holes. Promptly unpacking flowers and arranging them in vases also suppressed disease by accelerating drying of flower buds and petals.

Fungicides and Fungicide Alternatives

Fungicides are widely used to suppress *B. cinerea* in rose production, both in greenhouses and in the field. Effectiveness of fungicides can be limited in several ways, however. First, it is nearly impossible to protect all flowers even when fungicides are sprayed weekly, since flowers that are closed during treatment may develop and be harvested before the next spray, and are thus vulnerable to *B. cinerea*. Second, *B. cinerea* has developed widespread resistance to several fungicides that were formerly mainstays of grey mould control, including those in the benzimidazole and dicarboximide classes. Third, since the presence of visible fungicide residues on harvested roses hampers marketability, many growers prefer to use products that leave little or no residue.

Several treatments have shown promise as alternatives or supplements to conventional fungicide sprays. Dipping of harvested flowers into tap water at 50 °C for 20–40 s reduced grey mould severity by 60% without reducing vase-life of the flowers. Spraying detached rose petals with a solution of GA_3, a gibberellic acid, suppressed development of grey mould, possibly by delaying petal senescence and stimulating the formation of endogenous compounds inhibitory to *B. cinerea*. Suppression of grey mould by GA_3 was enhanced by addition of the growth regulator paclobutrazol. However, postharvest spray applications of GA_3 to intact rose flowers did not provide significant suppression of *B. cinerea*. Grey mould was reduced on cut roses that had taken up methyl jasmonate, a naturally occurring compound in many plants, because methyl jasmonate apparently induced systemic acquired resistance mechanisms; flower quality was not impaired by this treatment. Dipping rose buds in the antibiotic pyrrolnitrin, which is derived from the bacterium *Burkholderia cepacia*, and subsequently inoculating with *Botrytis cinerea* conidia, suppressed grey mould in cold storage as effectively as a conventional fungicide, and also enhanced poststorage fresh weight gain.

Inhibitors of ethylene can help to delay flower senescence and thus suppress grey mould. Incubating cut roses in solutions of $CaSO_4$, $CaCl_2$, or sodium thiosulphate reduced the severity of grey mould symptoms by 60%, probably by inhibiting increases in ethylene production and membrane permeability associated with natural senescence.

The biological control fungus *Clonostachys rosea* (Link:Fr.) Schroers, Samuels, Siefert, and W. Gams, formerly known as *Gliocladium roseum*, markedly suppressed sporulation of *B. cinerea* in rose petals and leaves when applied before the arrival of *B. cinerea* spores, and rapidly colonized dead leaves and petals as well. This finding suggested that *C. rosea* has the potential to effectively suppress the production of *B. cinerea* conidia in rose production systems.

Although none of these alternative tactics is currently registered for use against *B. cinerea* on roses, they may be useful in the future as supplementary treatments to decrease reliance on conventional fungicides that are resistance-prone or environmentally toxic.

See also: **Breeding**: Selection Strategies for Disease and Pest Resistance. **Disease**: Bactericides and Fungicides. **Integrated Pest Management**.

Further Reading

Elad Y (1988) Involvement of ethylene in the disease caused by *Botrytis cinerea* on rose and carnation flowers and the possibility of control. *Annals of Applied Biology* 113: 589–598.

Elad Y (1988) Latent infection of *Botrytis cinerea* in rose flowers and combined chemical and physiological control of the disease. *Crop Protection* 7: 361–366.

Elad Y and Volpin H (1991) Heat treatment for the control of rose and carnation grey mould (*Botrytis cinerea*). *Plant Pathology* 40: 278–286.

Hammer P and Evenson K (1994) Differences between rose cultivars in susceptibility to infection by *Botrytis cinerea*. *Phytopathology* 84: 1305–1312.

Hammer P and Evensen K (1996) Effects of the production environment on the susceptibility of rose flowers to postharvest infection by *Botrytis cinerea*. *Journal of the American Society of Horticultural Science* 121: 314–320.

Hazendonk A, ten Hoope M and van der Wurff T (1995) Method to test rose cultivars on their susceptibility to *Botrytis cinerea* during the post-harvest stage. *Acta Horticulturae* 405: 39–45.

Horst R (1983) *Compendium of Rose Diseases*. St Paul, MN: APS Press, American Phytopathological Society.

Marois J, Redmond J and MacDonald J (1988) Quantification of the impact of environment on the susceptibility of *Rosa hybrida* flowers to *Botrytis cinerea*. *Journal of the American Society of Horticultural Science* 113: 842–845.

Meir S, Droby S, Davidson H *et al.* (1998) Suppression of botrytis rot in cut rose flowers by postharvest application of methyl jasmonate. *Postharvest Biology and Technology* 13: 235–243.

Morandi A, Sutton J and Maffia L (2000) Effects of host and microbial factors on development of *Clonostachys rosea* and control of *Botrytis cinerea* in rose. *European Journal of Plant Pathology* 106: 439–448.

Nameth S (2002) *Botrytis Gray Mold in Greenhouse Floral Crops*. Ohio State University Extension FactSheet HYG-3070-96.

Rodriguez A, deVis R and Arbelaez G (1999) Effect of a photoselective plastic and a climatic screen on the disease caused by *Botrytis cinerea* pers. in a rose farm. *Acta Horticulturae* 482: 227–233.

Shaul O, Elad Y and Zieslin N (1995) Suppression of botrytis blight in cut rose flowers with gibberellic acid: effects of postharvest timing of the gibberellin treatment, conidial inoculation, and cold storage period. *Postharvest Biology and Technology* 6: 331–339.

Volpin H and Elad Y (1991) Influence of calcium nutrition on susceptibility of rose flowers to botrytis blight. *Phytopathology* 81: 1390–1394.

Black Spot

R Drewes-Alvarez, University of Applied Sciences, Dresden, Germany

Black spot is a common disease of roses caused by the fungus *Diplocarpon rosae* Wolf (conidial stage: *Marssonina rosae* (Lib.) Died.). It causes early defoliation and weakening of plants. All rose varieties are more or less susceptible to the disease. In this article facts about the history, the symptoms and the lifecycle are given. As the fungus is not a strictly biotrophe parasite, it can be grown on artificial media. Isolates of the fungus can be differentiated by morphological traits or according to their interaction with different roses. Breeding for black spot resistance requires the existence of resistance genes. After evaluation of species and hybrid roses, resistance genes were found in some species of roses, such as *Rosa multiflora* Thunb. ex Murray and *R. rugosa* Thunb. and breeders are trying to transfer these genes into cultivars by cross-breeding or to incorporate resistance genes from other plants into transgenic roses.

History of the Disease

The conidial stage of the fungus was first reported in 1815 in Sweden by Fries. After that there were reports from Belgium and France and in the second half of the nineteenth century the disease was also reported from Germany and the UK. Reports from North America and Australia followed at the end of the nineteenth century. The disease was reported from China in 1910 and in 1914 from Japan. More details can be found in **Table 1**. All these papers report the fungus in its conidial stage. The first report of the perfect stage was by Wolf in 1912. He found apothecia on overwintered leaves in Ithaca (NY, USA). Nowadays the fungus is distributed worldwide and is even found on islands like the Philippines and Hawaii.

Symptoms of the Disease and Lifecycle of the Fungus

In spring the first infections are caused by spores that overwintered in fallen leaf-material, or in structures called acervuli, on stems. Acervuli are fruiting body structures that the fungus develops in leaves, forming conidia, initially subarticularly or subepidermally. In **Figure 1** the life cycle of black spot is shown. The spores stick to the leaf cuticle, germinate and penetrate into epidermal cells. The fungus forms strands of subcuticular hyphae and grows deeper into the leaf, with intercellular hyphae sending nutrient-absorbing structures called haustoria into the plant cells. Small brown spots appear on the upper surface of the leaves and develop into irregular spots with fringed margins. In **Figure 2** part of a rose leaf with black spot symptoms is shown. While the brown spots develop on the leaves, small black acervuli with new spores (conidia) are often formed under the cuticle in concentric circles but are also distributed irregularly. The diameter of the acervuli can vary from 50 to 400 μm. In **Figure 3** a photograph made under an electron microscope shows the leaf surface with hyphae and three acervuli. Inside each acervulus there are masses of two-celled conidia. In **Figure 4** another photograph made under the electron microscope shows a leaf surface with some ruptured acervuli and surrounding two-celled conidia. After the rupture of the cuticle, the conidia are visible as a white, slimy mass in and around the acervuli (**Figure 2**). The conidia are spread by rain

Table 1 First reports of black spot in different countries

Country	*Author and year*
Sweden	Fries (1815)
Belgium	Libert (1826)
France	Duby (1830)
Germany	Bonorden (1853)
England	Berkeley (1860)
Scotland	Stevenson (1879)
Italy	Saccardo (1884)
USA	Scribner (1887)
Australia	Cooke (1892)
The Netherlands	Oudemans (1904)
Hungary	Laszlo (1910)

Data from Aronescu (1934) and Frick (1943).

water, insects or arachnids and cause infection on new rose leaves during the entire summer if humidity is sufficiently high.

The leaf tissue around the spots turns yellow and the leaflets will fall at the slightest touch. Inside the yellow tissue the metabolic activity is very high. A high enzymatic activity can be measured in the yellow tissue and phenolics and amino acids are accumulated. Ethylene production is also associated with yellowing and abscission of leaflets. The ethylene production in rose leaves with black spot is very high at the beginning of the infection process but decreases as leaves turn yellow.

In autumn the fungus invades deeper into the leaf tissue and new acervuli are developed under the epidermis. Sometimes one-celled spores (spermatia) are formed together with the two-celled conidia. The subepidermal acervuli are capable of overwintering (**Figure 1**).

If conditions are favourable for the fungus, the perfect stage of the lifecycle will be formed in autumn. Ascopores (which are haploid and require mating (karyogamy) and meiosis) can be found the next spring; these are formed by the fungus on leaves, in structures called apothecia. The perfect stage was seldom reported in Europe but has occasionally been found in the United States.

Pure Culture of Black Spot

Black spot pathogen is not a strictly obligate biotrophe parasite. The fungus is also capable of utilizing artificial media for its nutrition. It can be cultured on

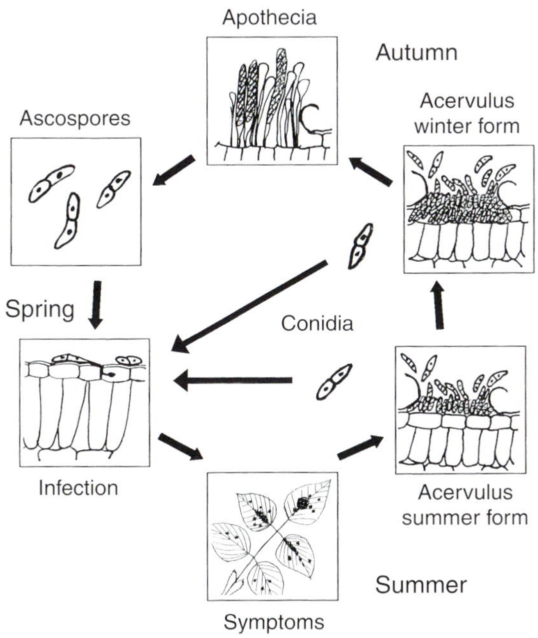

Figure 1 Life cycle of black spot.

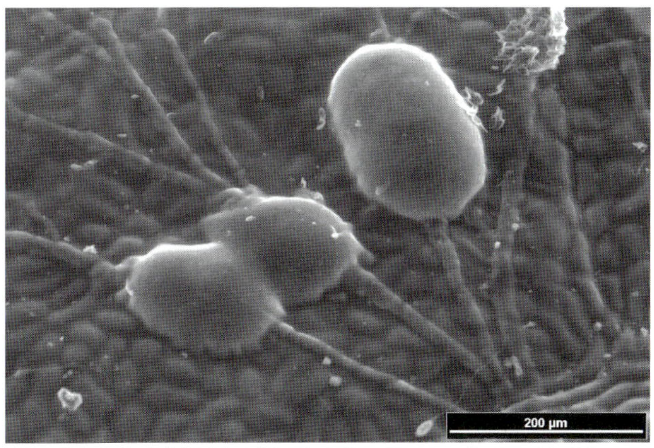

Figure 3 Acervuli and subcuticular hyphae under the electron microscope. (Photograph courtesy of T. Debener.)

Figure 2 Black spot symptoms on a rose leaf. (Photograph courtesy of T. Debener.)

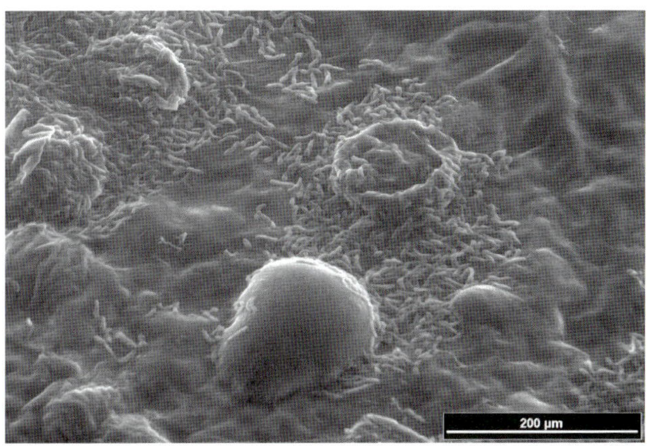

Figure 4 Ruptured acervuli with conidia under the electron microscope. (Photograph courtesy of T. Debener.)

Table 2 Media for pure culture

Medium used	*Author*
4% malt, 2% agar	Frick (1943)
Canned green pea infusion agar	Jenkins (1954)
Oatmeal	Shirakawa (1955)
Pea extract-agar	Stewart and Semeniuk (1965)
3% malt, 2% agar	Drewes-Alvarez (1992)
Quarter-strength potato dextrose agar	Ali *et al.* (2000)

Table 3 Conidia sizes

Measured sizes	*Author and year*
13.58–25.24 × 3.92–6.74 μm	Frick (1943)
18–25 × 5–6 μm	Sivanesan and Gibson (1976)
22.1–25.3 × 5.3–7.1 μm	Wenefrida and Spencer (1993)

different media (**Table 2**). The best results for sporulation were obtained on malt agar or malt–yeast agar.

Single-conidia isolates

Isolates made of single conidia are necessary for investigations if the fungus can be divided into different morphotypes (morphological races) and pathotypes (physiological races). Infected leaves with black spot symptoms should be placed in a Petri dish with wet paper to guarantee high humidity. After 1–2 days fresh acervuli are formed and conidia can be scratched off under 50× magnification. The conidia should be spread on 2% water-agar and can be isolated after germination. These single germinated conidia can then be transferred to the different multiplication media, such as malt extract or potato dextrose agar.

Morphotypes

The black spot fungus grows very slowly on artificial media. Depending on the culture medium, the mycelium diameter never exceeded 20 mm in 4 weeks. Morphological differences can be found in the size of conidia, in colony colour or in mycelium growth. The first investigations in pure culture were done with single-spore isolates on 4% malt agar. In these experiments colony colour was not a constant trait in parallel cultures. Measurement of mycelium diameter and differences in conidia sizes were revealed but the isolates were not subdivided into distinct morphotypes.

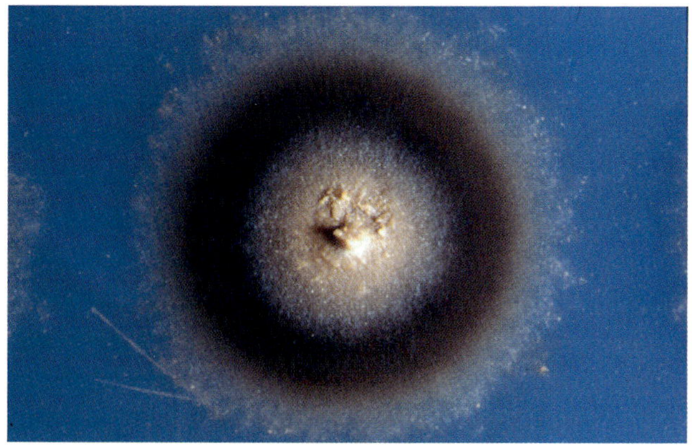

Figure 5 Black spot morphotype with compact colony.

Figure 6 Black spot morphotype with star-shaped colony.

Other workers also found morphological differences. In some investigations the differences were not considered large enough to identify morphotypes but more recent reports described morphotypes. In German isolates one compact colony (**Figure 5**) and one star-shaped colony (**Figure 6**) were found. In American isolates colour groups were described.

Beside the different colony colours, several workers have measured conidial size. Conidia length varied from 13.6 to 25.3 μm and conidia width from 3.9 to 7.1 μm. In **Table 3** details of different conidial size measurements are given. There is great variation depending on isolate and infected rose variety. In recent experiments a relation was found between colony colour and conidial size. The longer conidia were associated with lighter yellow colonies and smaller conidia with darker colonies.

Genetic Diversity of the Fungus

Fungal DNA can be extracted for use in molecular studies to look at the variability of black spot isolates. For example, mycelium and conidia from liquid or solid culture can be used to extract DNA. Several protocols were tested to optimize the extraction process. In German investigations, isolates from different geographic regions showed clear DNA polymorphism after random amplified polymorphic DNA (RAPD) analysis and could be clearly differentiated from one another. Isolates from one geographic region showed few differences.

Genetic diversity was found in British investigations by simple sequence repeat (SSR) and amplified fragment length polymorphism (AFLP). The AFLP technique proved more reproducible and sensitive. Preliminary results show that there are probably two main clusters of strains.

Loss of Pathogenicity and Maintaining Isolates

By cultivating the fungus on artificial media it will ultimately lose its pathogenicity. Depending on the medium, the loss of virulence is measurable after 3 months (4% malt agar) or more (3% malt agar) in culture. Investigations at Beltsville, Maryland showed that the decline of virulence was not linear in pure cultures stored at 1–2 °C but showed peaks with higher infection rates.

This loss of pathogenicity makes it necessary to inoculate roses with the different isolates at least every 4–5 months followed by reisolation of the fungus. As this is very laborious and time-consuming, it is important to develop a method for maintaining the virulence of the different isolates. A possible solution is freezing the isolates either on leaves or in suspensions. The vitality of the conidia can be estimated after staining with phenosafranin. Vital conidia are stained red and dead conidia remain unstained.

Freezing the leaves at −20 °C was used in several experiments and proved to be a simple solution. The vitality of the conidia was still about 60% after 2 years' storage. Freezing conidia suspension at −20 °C resulted in vitality rates about 10% lower. Suspensions stored at −80 °C showed about the same vitality as conidia frozen on leaves. These preliminary experiments indicate that all three techniques can be used to maintain isolates.

Rose–Black Spot Interaction

Inoculation Methods

Inoculations could be done on whole plants or on detached leaflets. Suspensions of conidia could be prepared from black spot-infected leaves or artificial cultures by washing off the conidia with sterile distilled, tap or rain water. For controlled inoculations the conidial density must be adjusted by spore counts. Different concentrations have been used in the past: the lowest was 30 000 conidia ml^{-1} and the highest 800 000 conidia ml^{-1}. The suspensions could be applied in different ways – either spraying the whole plant with an atomizer or applying single drops on to the leaflets.

Plants should be placed under high humidity after inoculation for at least 2 days to ensure good germination of the fungus conidia. Black spot conidia germinate within a range of 0 °C and 33 °C, but the optimal temperature is approximately 18 °C. Under favourable conditions germination should be complete after 9–18 h. Optimal temperature for disease development is 24 °C, according to investigations at Cornell University, New York.

Early Stages of Invasion

Several workers used light microscopy or scanning electron microscopy (SEM) techniques to reveal the plant–fungus interaction from germination to conidia formation.

Conidia stick to the cuticle with a mucilaginous substance. By using SEM it was proved that the conidia were unornamented. The conidia normally germinate from one of the two conidial cells and generally from the tip or the middle of that cell. The germ tubes can vary in their length but are seldom long. Long germ tubes failed to form appressoria (which are specialized structures that the fungus forms before the penetration of the leaf). The fungus can penetrate the cuticle with or without forming appressoria but was never found to enter the leaf through stomata. Germination is often complete after 9 h. Conidia can germinate on susceptible and resistant roses.

Under the cuticle the fungus grows with subcuticular hyphae and sends haustoria into the epidermal cells. Haustoria were first observed 15 h after inoculation. After 2 days parallel strands of hyphae start to develop, sending haustoria into new epidermal cells. At several places along the parallel strands, new hyphae develop in a circular shape. Conidia are formed on these hyphae until the cuticle ruptures under the pressure of the conidial masses. Haustoria can also be observed in black spot-resistant species, e.g. *R. roxburghii* Tratt. and *R. wichurana* Crép., but the affected epidermal cells responded with a hypersensitive reaction.

Pathotypes

The existence of physiological races was investigated with American, Canadian, Swiss, German and British black spot isolates. Pathotypes can be described by qualitative criteria (infected or without any symptoms) or by considering quantitative criteria such as size of lesion. However, since roses can vary in their susceptibility to black spot during the year, lesion size is not only influenced by pathotypes.

In order to characterize black spot races, rose cultivars and species of roses with different reactions to black spot infections are needed. These roses are called a differential set. After inoculation with conidial suspensions from single-spore isolates, roses of that differential set should be infected by some of the isolates, depending on the genetic background of the rose, and thus allow a conclusion about the existence of different physiological races of the fungus. In recent experiments four pathotypes were described in the United Kingdom and six pathotypes in Germany. Since the differential sets that were used are composed of completely different roses, it is not possible to decide whether the British and German pathotypes are identical.

Control of Black Spot

Cultural Practice

As black spot conidia need to be in contact with water before they germinate, rose leaves should not be allowed to remain wet for several hours. If roses are watered by sprinkling, it should only be done on bright mornings. Good air circulation should be guaranteed by avoiding dense planting so that the leaves of the rose bushes will dry more quickly. The overwintering potential of the fungus can be reduced by removing fallen leaves and by pruning canes with black spot lesions.

Chemical Control

Once the pruning is done in spring, a protectant fungicide should be applied to kill the newly spread conidia. The treatment should be repeated at 10–14-day intervals when new leaves appear. If black spot becomes established on rose plants it is difficult to control. The following substances are currently used as fungicides against black spot in Germany: bitertanol, dichlofluanid, fenarimol, kresoxim-methyl, tebuconazole and sulphur. In field trials in 1999 and 2000 in Alabama the protectant fungicide chlorothalonil provided better control than other fungicides, including myclobutanil and triforine, when applied at 14-day intervals.

Some experiments to control black spot were done with sulphur dioxide. Concentrations of $100\,\mu g\,m^{-3}$ reduced black spot infection after 2 days' exposure.

Breeding for Black Spot Resistance

Breeding rose cultivars which are resistant to black spot is a time-consuming process due to the complex genetic constitution of roses. Resistance to black spot seems to operate at two levels – mechanical and/or metabolic. A thick cuticle may prevent the penetration of the fungal germ tube. On the other hand, substances can be produced by the plant to inhibit germination of conidia or to stop the growth of the fungus after it has invaded the plant tissue. In breeding for resistance the first step is to find the sources of black spot resistance. For this purpose species and hybrid roses must be evaluated. The resistant roses must then be crossed with other roses and the seedling must be tested as early as possible using one of the inoculation techniques, followed by further selection for other traits, such as flower shape, growth habit and scent.

Evaluation

Many workers have evaluated roses for black spot resistance either under natural conditions or after inoculation in the greenhouse or in the laboratory. Different evaluating schemata were used. Evaluation can be done by determining the percentage of affected leaves on roses grown in the field. Another method of determining black spot resistance in roses is to inoculate species or hybrid roses in the laboratory on detached leaflets.

In recent Canadian investigations the incubation period, the lesion length, the number of lesions, the leaf area with symptoms and the sporulation capacity were evaluated as components of partial resistance to black spot.

Most modern roses are susceptible to black spot butthere are several diploid and tetraploid species of roses that show no sign of the disease. High resistance to black spot was found in genotypes of *R. banksiae* Ait., *R. carolina* L., *R. laevigata* Michx., *R. multiflora* Thunb. ex Murray, *R. rugosa* Thunb., *R. roxburghii* Tratt., *R. virginiana* Herrm. and *R. wichurana* Crép.

Resistance Genes

Relatively little is known about the inheritance of important traits in roses. Detailed genetic studies have been performed for recurrent flowering and for winter-hardiness (*see* **Genetics**: Inheritance of Characteristics; Species Crosses; Mutation; Research Trends).

At present several groups are working on identifying black spot resistance genes. The *Rdr 1* gene is the first resistance gene described in the genus *Rosa*. For this resistance gene a 'gene-for-gene' type of interaction in the pathosystem *Diplocarpon*/*Rosa* was suggested. As breeding for black spot resistance is a time-consuming process, molecular markers that are tightly linked to the resistance genes can be used in marker-assisted selection. Markers for the resistance gene *Rdr 1* with tight linkage were recently found.

Cross-breeding

Most of the rose cultivars are tetraploid (28 chromosomes in four sets of seven). Breeding on the diploid level (14 chromosomes in two sets of seven) is much easier than on the tetraploid level due to the splitting of genotypes (*see* **Genetics:** Karyology; Inheritance of Characteristics; Mutation). Selections for resistant genotypes had been done in offspring of the diploid species *R. multiflora* and *R. rugosa*. But if the selection is done on the diploid level, techniques for chromosome doubling are required before crosses with cultivars should be performed. Different substances, like amiprophos-methyl, colchicine, oryzalin, pronamide, can be used for that purpose. The treatment can be done in the greenhouse or, for better success, the substances should be applied to roses cultured *in vitro*.

Another approach is to combine two distantly related resistant diploid species like *R. wichurana* and *R. rugosa*. The offspring of that cross are sterile diploids and need a chromosome doubling to become fertile. The resulting tetraploids are called amphidiploids. An example of this is *R. kordesii*, which is a *R. rugosa* × *R. wichurana* amphidiploid.

Transgenic Plants

The integration of resistance via cross-breeding with wild species will in most cases result in integration of a lot of undesired traits. It is possible to increase resistance against the pathogen via transformation by using elements of plant-specific defence. These elements can be pathogenesis-related proteins (PR proteins), ribosome inactivating proteins (RIPs) or phytoalexins. The most important PR proteins used in genetic engineering are chitinases, glucanases and lysozymes. These enzymes inhibit fungal growth by attacking the cell walls. RIPs are the best-known plant toxins. They block protein biosynthesis either in the pathogen or in the infected plant cells, resulting in local cell death. Phytoalexins, e.g. stilbenes, are substances with antimicrobial effects that accumulate in plant tissue after pathogen infection.

Several groups are working on methods to develop transgenic roses with increased resistance against black spot. Different resistance genes with nonrace-specific antifungal effects were integrated into rose genomes by *Agrobacterium*-mediated transformation. For example, genes for RIPs, for chitinases and glucanases from barley were used. Transgenic plants which secrete the antifungal proteins into the extracellular space were reduced in their susceptibility to black spot. The severity of black spot development was reduced by 13–43% in a black spot-susceptible rose, by transferring a rice gene for chintinase into embryogenic tissue, called callus, which has undifferentiated cells.

Introducing disease defence genes provides a method of producing black spot-resistant rose cultivars.

See also: **Disease:** Overview; *Botrytis*; Downy Mildew; Powdery Mildew; Rust. **Genetics:** Karyology; Inheritance of Characteristics; Species Crosses; Mutation; Research Trends. **Selection Strategies:** Resistance to Diseases and Pests.

Further Reading

Ali A, Hall AM, Aquino de Muro M and Cannon P (2000) The investigation of the biology of the rose blackspot fungus, *Diplocarpon rosae* leading to improved methods of control. *BCPC Conference in Brighton*, 13th–16th November.

Aronescu A (1934) *Diplocarpon rosae*: from spore germination to haustorium formation. *Bulletin of the Torrey Botanical Club* 61: 291–329.

Debener T, Drewes-Alvarez R and Rockstroh K (1998) Identification of five physiological races of blackspot (*Diplocarpon rosae* Wolf) on roses. *Plant Breeding* 117: 267–270.

Drewes-Alvarez R (1992) Untersuchungen am Pathosystem Sternrusstau (*Marssonina rosae* (Lib.) Died.) – Rose (*Rosa* L.) zur Klaerung der Frage nach pilzlichen Rassen und zur Uebertragung der Resistenz aus der diploiden *Rosa multiflora* Thunb. auf tetraploide Gartenrosen. PhD Thesis, University of Hamburg, Germany.

Frick L (1943) Untersuchungen über Biologie und Pathogenität von *Diplocarpon rosae* (Lib.) Wolf. *Phytopathologische Zeitschrift* 14(6): 525–591.

Horst RK (1983) *Compendium of Rose Diseases*. St Paul, MN: American Phytopathological Society.

Jenkins WR (1954) Variation in pathogenicity and physiology of *Diplocarpon rosae* (Lib.) Wolf, the rose blackspot pathogen. PhD Thesis, University of Maryland, USA, 47 pp.

Melching JS (1966) *Effects of Environment and of Selected Chemicals upon the Development of Black Spot of Rose and of the Pathogen,* Diplocarpon rosae *Wolf*. Memoir 397. Cornell: Cornell University, Agricultural Experiment Station.

Shirakawa HS (1955) The nutrition of *Diplocarpon rosae*. *American Journal of Botany* 42: 379–384.

Sivanesan A and Gibson IAS (1976) *Diplocarpon rosae*. *CMI Descriptions of Pathogenic Fungi and Bacteria* 49, No. 485, 2pp.

Stewart RN and Semeniuk P (1965) Report on Rose Research in the US Department of Agriculture. *American Rose Annual* 50: 99–105.

Wenefida I and Spencer JA (1993) *Marssonina rosae* variants in Mississippi and their virulence on selected rose cultivars. *Plant Disease* 77: 246–248.

Yokoya K, Kandasamy KI, Walker S, Mandegaran Z and Roberts AV (2000) Resistance of roses to pathotypes of *Diplocarpon rosae*Annals of Applied Biology 136: 15–20.

Downy Mildew

X Xu, Horticulture Research International, East Malling, West Malling, UK
T Pettitt, Horticulture Research International, Wellesbourne, Warwick, UK

Overview

Downy mildew of rose caused by *Peronospora sparsa* Berk. is a serious foliar disease. The main effect of the disease is short-term disfigurement caused by foliar lesions and extensive premature leaf abscission or 'leaf drop'. Whilst it has long been speculated that *P. sparsa* overwinters in rose as resting mycelium in the woody stem tissues, recent work has shown little evidence for this, and oospores appear more likely to be the main overwintering phase of this disease. As with other downy mildews, *P. sparsa* is favoured by cool, humid conditions. Infection is markedly increased by long periods of leaf wetness. However, sporulation does not require the presence of free water, and high humidity is optimal. The disease can be controlled using chemicals that are effective against other downy mildews, but applications need to be made at regular intervals to maintain this control. Possible areas for future research leading to sustainable disease management are discussed in this article.

Introduction

Downy mildew on rose is found virtually everywhere roses are cultivated. It is more commonly seen on glasshouse-grown roses than on plants growing outdoors. The disease was first described in 1862 in England and from then into the early 1900s was reported throughout Europe. Downy mildew on roses had spread to the midwestern USA by 1880s and has more recently been seen in all parts of the USA. It also occurs on roses in Canada, New Zealand and Australia. In recent years its geographical distribution has expanded further with reports on its occurrence in both China and South Africa.

The disease can be very destructive but tends to be sporadic in its occurrence. Although not as common as powdery mildew, downy mildew is still considered in the UK to rank in the 'top four' most important foliar diseases of cultivated roses along with rust, powdery mildew and black spot.

The Downy Mildew Pathogen

Downy mildew is caused by *P. sparsa* Berk. Mycelia are intercellular in host tissues. Erect conidiophores on lower leaf surfaces are dichotomously branched (**Figure 1**). Under humid conditions, these conidiophores are generally about 350 μm long, bearing subelliptical sporangia 17–22 × 14–15 μm, and emerge through stomata on the lower surfaces of infected leaves. Oospores may be found in infected leaves, sepals, flower buds and stems. The pathogen can also infect hosts other than roses, including arctic brambles, raspberry, dryberry and boysenberry. The downy mildew of raspberry (*P. rubi*) can also infect rose leaves.

(a)
(b)

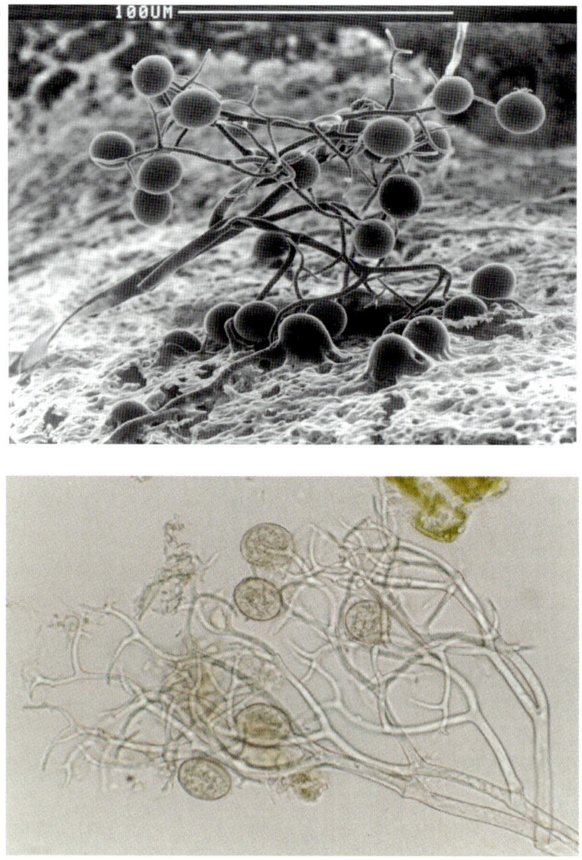

Figure 1 (a) Scanning electron microscope and (b) light microscope images of rose downy mildew conidiophores (*Peronospora sparsa*).

Symptoms

Whilst downy mildew can infect leaves, stems, peduncles, calyces and petals, infection is generally restricted to young, apical plant growth. Leaves develop purplish-red to dark-brown irregular spots (**Figure 2a**) and may turn yellow. Downy mildew lesions in rose normally have clearly defined edges that are often angular in appearance as a result of the lesion ending at a major leaf vessel. Foliar symptoms can often resemble the scorch symptoms typical of pesticide/herbicide phytotoxicity. When inoculated

(a)

(b)

Figure 2 Symptoms of rose downy mildew (*Peronospora sparsa*): (a) typical severe field symptoms and (b) symptoms on detached leaves artificially inoculated *in vitro*.

in vitro, the colour of lesions is sometimes more greyish and their margins are less well-defined and angular (**Figure 2b**). These symptoms (sometimes also seen in the field) can easily be confused with those of rose black spot (*Diplocarpon rosae*). Whilst purplish to reddish-brown spots develop more clearly on the upper surfaces of leaves, sporulation normally occurs on the lower surfaces. The undersides of infected leaves bear the delicate dichotomously branched conidiophores carrying infective conidia (**Figure 1**). During the later stages of infection, leaves may become necrotic. Occasionally, the flowers and flower buds are malformed and retarded even when infection of the leaves is slight. Rose downy mildew can be distinguished from rose powdery mildew as the latter appears drier and more powdery due to the abundant production of conidia; also with young powdery mildew lesions or 'pustules' this white mycelium can be rubbed away with the fingers to reveal green leaf surfaces below. Furthermore, powdery mildew mycelia can develop on both upper and lower surfaces of leaves, whereas downy mildew mycelium only appears on the lower surface.

A typical symptom of rose downy mildew is the occurrence of severe leaf drop caused by the abscision of infected leaves. This symptom is one of the most damaging impacts of the disease and sometimes apparently healthy leaves may be abscised in the process. It is also another symptom that can be confused with rose black spot which can also cause leaf drop. However, in downy mildew leaf drop is normally more severe than in black spot and in artificial inoculation experiments, most inoculated leaves are abscised soon after inoculation (generally about a week after inoculation), often without the development of obvious leaf spot symptoms.

Epidemiology

Most published research on rose downy mildew has been concerned with its distribution and on attempts at its control. Research on the epidemiology of the disease is very sparse and often based on casual observations, rather than systematic studies. Unless otherwise specified, most of this section on epidemiology is from the recent work conducted by the authors at Horticulture Research International (HRI), funded by the British Department of Environment, Food and Rural Affairs.

Primary Inoculum/Overwintering Inoculum

It has been speculated that rose downy mildew may overwinter as mycelia on wood or as systemic infections in host tissues, inferred from other downy mildew diseases. However, this speculation is not supported by recent studies.

The occurrence of downy mildew was monitored on whole diseased plants; these plants were either artificially inoculated or collected from natural epidemics on both protected and outdoor crops, and when they entered dormancy all dropped leaves were collected and disposed of. Once overwintered, these plants were placed under controlled environment (CE) conditions to grow out of dormancy. Plants were treated with frequent misting to encourage severe downy mildew symptoms to develop. Downy mildew symptoms have not been observed in any new leaves developing on any of these previously diseased plants. Similar observations were made on a larger scale with plants maintained in polythene tunnels despite provision of frequent long wetting periods. These observations clearly indicated that, if all the infected leaves were removed from diseased plants, new leaves arising from nodes previously bearing infected leaves did not develop any downy mildew symptoms. These results also suggest that mycelia did not spread from infected leaves to stems to act as an inoculum source. This is further supported by histological observations of rose downy mildew mycelium development inside leaf tissues; this was often locally restricted and rarely seen to invade even minor vascular tissues, and certainly not seen in petiolar vascular tissues, thus indicating that invasion of stems through petioles was unlikely.

Oogonia are frequently observed in cleared diseased leaf tissue taken from field infected plants. In inoculation experiments, oogonia are often observed inside the mesophyll soon after infection, and their formation is recorded as early as 1 week after inoculation of detached leaves *in vitro*, using light microscopy and scanning electron microscopy (**Figure 3**). In detailed studies of their distributions in infected leaves, oogonia were usually seen within lesions and were rarely seen in the green parts of affected leaves, although their possible formation in green tissues of abscised leaves still needs to be determined. These observations, combined with those reported above, suggest that oospores are likely to be the primary overwintering inoculum of *P. sparsa* in rose. Further research is needed to study the viability and pathogenicity of oospores in relation to environmental conditions.

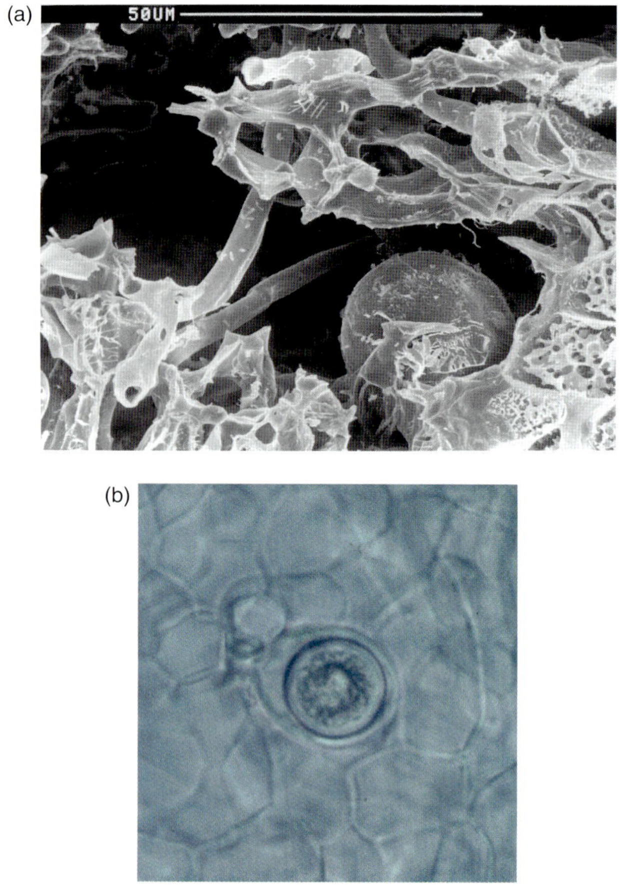

Figure 3 (a) Scanning electron microscope and (b) light microscope images of rose downy mildew (*Peronospora sparsa*) oogonia.

Germination and Infection

Downy mildew generally develops under comparatively cool and humid conditions. *In vitro* inoculation of detached leaves indicated that free water was essential for infection. Germination of conidia has been studied *in vitro* on deionized water agar (1.2% w/v) in relation to temperature. Germination of two *P. sparsa* isolates under different temperatures is shown in **Table 1**. Overall germination remained very high, near the maximum, between 2–14 and 2–18 °C for the two isolates. Germination declined sharply beyond these temperature ranges, and was virtually zero at 26 °C. Interestingly, the pattern of germination with temperature was very similar for two isolates of *P. rubi*, that were also included in this study. The effects of temperature can be well described using a Gompertz model, but the model parameter estimates differ significantly between the two *P. sparsa* isolates.

The effects of temperature and wetness of duration on infection of rose leaves by *P. sparsa* have been investigated in CE cabinets (**Figure 4**). The effects of temperature in the range of 9–18° C are generally small. The incidence of disease increased gradually with increasing duration of wetness from 16 to 96 h, and then increased sharply from 96 to 120 h.

Table 1 Percentage germination of two *Peronospora sparsa* isolates on deionized water agar at various temperatures

Temperature (°C)	*Isolate 1*	*Isolate 2*
2	92	93
5	90	94
9	90	93
11	88	94
14	86	87
17	71	90
19	43	88
22	38	82
23	26	73
24	19	40
26	1	0
29	0	0

Data from Breese *et al.* (1994).

Sporulation

Rose downy mildew is renowned for its generally sparse asexual sporulation, hence its species epithet *sparsa*. However, recent work at HRI East Malling

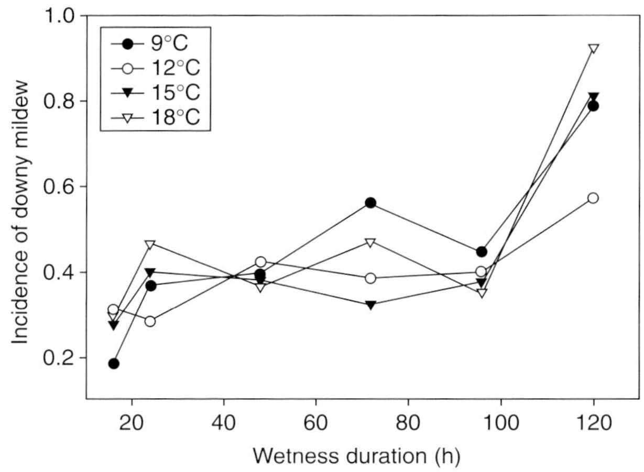

Figure 4 Incidence of rose downy mildew (*Peronospora sparsa*) on leaves inoculated with conidia of *Peronospora sparsa* in relation to the duration of wet periods and the wet period temperature.

showed that abundant sporulation is often observed under prolonged humid conditions. Production of conidiophores can often take place in infected leaves before the appearance of lesions, thus at the height of an epidemic apparently healthy leaves can be infective. *In vitro* studies show that there is a very large variation in the sporulation intensity of individual lesions between leaves even within the same Petri dishes. Observations on inoculated plants in CE cabinets indicate that sporulation appears not to require free water and high humidity may be sufficient.

The spatial distribution of oogonia and conidia on artificially inoculated and naturally infected plants has been investigated to understand the dynamics of spore production. Each sampled leaf was photographed and then cleared for microscopic investigation. The presence and positions of oogonia and conidiophores were recorded directly on to enlarged printouts of each photographed leaf. As outlined above, oospores were frequently found in large numbers ($>30\,cm^{-2}$) in infected leaves. However, the formation of oospores showed great variation from leaf to leaf, even on the same plant. Oogonia were normally seen within necrotic lesions and rarely seen in the green parts of affected leaves. This observation was in contrast to the distribution of conidiophores, which were largely present at the margins of lesions or in the green areas of leaves. Under optimum conditions, sporulation on the green parts of infected leaves tends to be sparser than on the necrotic areas.

Detection

Although the presence of systemic infections in propagation material seems unlikely as a primary source for the majority of downy mildew epidemics in rose, it may still be important in some circumstances to detect low-level infections, for instance in material being selected for tissue culture. Sequences of the internal transcribed spacer (ITS) regions 1 and 2 of rDNA genes have been determined for several *P. sparsa* isolates, obtained from arctic bramble (*Rubus arcticus*), where systemic infection can often occur. Little sequence variability was detected. Two pairs of primers were designed to the ITS regions that could be used to detect *P. sparsa* by polymerase chain reaction using two different types of thermal cycles ('heated block' mini cycler and 'hot air flow' rapid cycler). These two primers are 5′-GGCTGGCTGCTACTGGGCA-3′ and 5′-GCCGACTGGCCACGCGGA-3′. As little as 0.25 pg of *P. sparsa* DNA could be detected in the reaction mix containing 3 ng of total DNA of the host (arctic bramble) DNA. However, these two primers have not been tested against *P. sparsa* isolates from other hosts.

Disease Management

Currently, control of downy mildew in roses relies almost totally on fungicide applications, especially in crops grown outdoors. Successful control with fungicides usually only results from the implementation of routine programmes with 10–14-day intervals between applications. In UK container-grown crops this can mean 10–15 sprays applied over the period from leaf emergence in spring to the onset of dormancy in autumn. Surprisingly little work has been carried out directly on roses, but recent work by Dr O'Neill of ADAS (UK) has shown that compounds for control of potato blight that are effective against *Peronospora* species infecting other host species can give good control of rose downy mildew. Phenylamide fungicides (e.g. metalaxyl and oxadixyl) generally give good control. However, resistance to these compounds is widely seen in the oomycetes and especially in other downy mildews (e.g. *P. parasitica*). These compounds are generally only available formulated with other active ingredients to combat this, but it is still a wise precaution when using these products in spray regimes to alternate them with other compounds. Protectant chemicals such as chlorothalonil can give good control, but it is important to achieve complete coverage of the plant as sporulation predominantly occurs on the undersurfaces of leaves. The dithiocarbamates are another group of protectant chemicals that give downy mildew control, although recent studies have shown that mancozeb works better against *P. sparsa* when combined with metalaxyl. The approach to application is also apparently important with fosetyl-aluminium, which can give good control of downy mildew when applied as a high-volume spray but almost no control when applied as a drench. This may be due to poor upward movement of the chemical in rose as drench applications, whereas when applied to other host species, this compound can give good mildew control. One group of compounds that have not been studied on roses are the strobilurins (e.g. azoxystrobin). These have been shown to give reasonable control of other downy mildews and also have a broad spectrum of activity, opening up the possibility of combined downy and powdery mildew control. The downside to this group of chemicals again is the danger of fungicide resistance developing in the pathogen population.

Apart from the use of chemicals it is possible to achieve some degree of downy mildew control using cultural measures. Direct measures such as removal and burning of all debris resulting from disease outbreaks have long been recommended and work already described here emphasizes the validity of this approach. Another strategy is to manipulate the growing environment, which can be especially effective when growing plants under protection if facilities for managing temperature and humidity are available. Although often difficult to achieve under UK conditions, US growers find that reducing the relative humidity to 80% by heating and venting can virtually eliminate infection. Recent UK studies have shown that the irrigation regime can strongly influence the rate of disease build-up, with significant reductions being achieved using subirrigation as opposed to overhead sprinklers. This effect is strongest in protected crops but can still be achieved in outdoor crops. Whilst these irrigation treatments can greatly slow the rate of disease spread, if applied alone they cannot halt the eventual development of severe disease. However, it is important to note that, even with overhead irrigation, the disease only occurs when the weather conditions are conducive to disease, with appropriate temperatures and long periods of leaf wetness. Current research is using data on conditions conducive to infection and sporulation to generate predictive models to identify times of high disease risk and thereby produce a more rational strategy for disease control. This would identify appropriate times for applying sprays and/or the rates of active ingredient needed for optimal control. In years when the disease pressure is low, such a strategy would greatly reduce the amounts of chemicals applied compared to a routine spray programme.

Future Research

Current research on the epidemiology of the disease is limited. Further research is needed before a fully integrated disease management strategy can be developed and implemented.

- The effect of drying on potential oospore formation in dropped green leaves needs to be studied (oospores of other oomycetes are often formed under drying conditions).
- The viability and pathogenicity of oospores in relation to environmental conditions are unknown but are important for practical disease management.
- Research is needed to identify effective biological treatment of infected leaves to destroy oospores or prevent oospores from forming.
- For breeding durable resistance, more research is needed to identify resistance sources and to elucidate the genetics of host–pathogen interactions.
- The current heavy use of fungicides for downy mildew control needs to be greatly reduced. The need for free water on plant surfaces and high humidities for infection and sporulation means that it should be possible to develop models to help predict periods of high disease risk. With the development of such models a more rational and possibly integrated approach to disease control may be possible in the near future.
- More studies are needed to determine whether *P. rubi* and *P. sparsa* belong to the same species.
- It would be useful to investigate the efficacy of strobilurin compounds against rose downy mildew.

See also: **Disease**: Overview; Black Spot; Powdery Mildew; Bactericides and Fungicides.

Further Reading

Aegerter BJ (2001) *Epidemiology and Control of Rose Downy Mildew in a Bare-Root Production System.* Dissertation, University of California, Davis.

Wheeler BEJ (1981) Downy mildew of ornamentals. In: Spencer DM (ed.) *The Downy Mildews*, pp. 473–485. London: Academic Press.

Powdery Mildew

M Linde, Federal Centre for Breeding Research on Cultivated Plants, Ahrensburg, Germany

N Shishkoff, Cornell University, Long Island Horticultural Research and Extension Center, Riverhead, NY, USA

Introduction

Powdery mildew of roses, a disease first described by Theophrastus in 300 BC, is a problem worldwide, in greenhouses or outdoors, wherever roses are grown.

Figure 1 Powdery mildew on a highly susceptible cultivar.

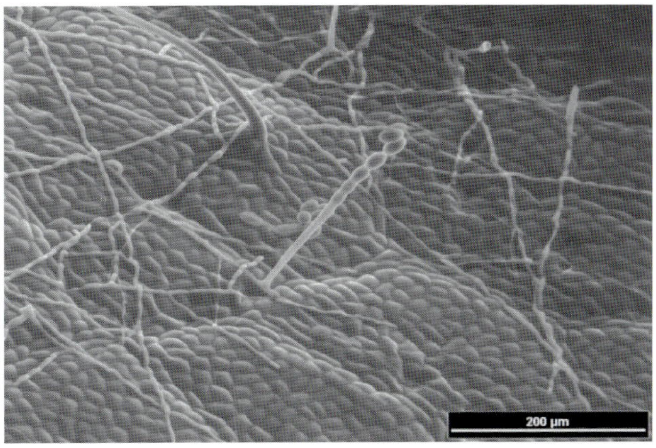

Figure 2 Mycelium and a conidiophore of *Podosphaera pannosa* on the surface of epidermal rose cells (raster electron microscope).

Nearly 40% of the fungicide sprayed on roses is to control powdery mildew. The disease can cause distortion and death of leaves and shoots, but even a mild case makes plants unsightly (**Figure 1**). Powdery mildew reduces the quality of cut flowers and makes nursery stock less saleable to consumers. Its ubiquity causes gardeners and landscapers to think twice about planting roses. In a time when many parks are cutting maintenance budgets and limiting the use of pesticides in public places and when the use of many pesticides is restricted or not allowed for many breeders and growers of cut and garden roses in Europe, this disease continues to be one of the major scourges of roses. In this article we will describe the pathogen causing powdery mildew, its hosts, symptoms of the disease and how it spreads and the means of controlling it.

The Pathogen

Powdery mildews belong to the family Erysiphaceae in the class Ascomycetes. The Erysiphaceae include 13 genera and probably about 400 species. Powdery mildews are obligate parasites; they can't survive without a living host. Unlike most plant pathogen fungi, which grow within plant tissue, powdery mildews live epiphytically on the outer surface of plants in mats of whitish hyphae. The hyphal strands are anchored to the leaf surface by pegs that penetrate epidermal cell walls, form haustoria and allow the absorption of nutrients from leaf tissue. Vegetative hyphae called 'conidiophores' grow away from the leaf surface and produce spores in chains from their tips, giving the colony its characteristic powdery appearance (**Figure 2**). The fungus also produces spherical structures called 'ascomata' that contain the sexual spores. These are occasionally visible in colonies as reddish-brown dots.

The species causing powdery mildew of rose had long been known as *Sphaerotheca pannosa* var. *rosae*; recent changes in nomenclature result in the pathogen now being called *Podosphaera pannosa* (Wallr.: Fr.) de Bary. The 'var. *rosae*' is not necessary, since the isolates found on rose are not morphologically distinct from isolates found on *Prunus* species (such as peach and apricot). There is some degree of specialization among fungal isolates, so that isolates from rose tend to be less damaging on *Prunus*, and vice versa.

Infection Process

The dispersal of *P. pannosa* happens almost exclusively by the distribution of conidia by wind. The conidia germinate with one or two thick-walled germ tubes 2–6 h after deposition on the leaf surface. This germ tube subsequently develops into mycelium and is the starting point for the development of the fungal colony.

Optimal conditions for germination are around 22 °C and nearly 100% humidity. Temperatures above 30 °C prevent germination. Incubated in moist conditions, conidia can withstand long periods of 0 °C without loss of viability; survival at 0–3 °C for up to 3 months has been described. Germination under lower humidity is also possible down to 50%, but with lower frequency. The direct contact of conidia with large drops of free water in the first 6–8 h after infection reduces the number of germinated conidia which grew to a sporulating colony, possibly by a water layer between the host surface and the conidia. Leaf wetness seems not to inhibit the production of germ tubes in the first 6 h.

After infection, a nipple-shaped unlobed appressorium is built at the end of the germ tube. Appressoria attach the mycelium to the host surface and initiate the formation of the haustoria. The appressorium appears to be closely fixed to the cuticle of a host cell by a fine slime layer. With enzymatic degradation and mechanical penetration of the cuticle and the epidermis cell wall, a very fine hypha called an infection peg emerges through a pore of the appressorium and enters the epidermal cell. In the lumen of the epidermal cell the penetration peg enlarges to an elongated structure, the haustorial neck.

About 1 day after infection, mature haustoria can be observed, originating from the centre of the attachment of each appressoria. These globose, multilobed structures are the feeding organs of the fungi which establish contact with the nutrients of the host. They are formed at regular intervals as hyphae continue to grow along the surface of the leaf. The intense activity of the haustorium complex is indicated by the occurrence of many elongated mitochondria and a large nucleus in the dense ribosome-rich matrix. The haustoria are surrounded by an extrahaustorial membrane, which may be derived from the host cell wall, or from *de novo* synthesis as a reaction to the development of the haustorium. The matrix seems not to contain chitin (commonly found in fungal cell walls) or cellulose and pectin (components of the plant cell wall). The water, minerals and metabolites from the host leaf are exported to the hyphae, where they are used for mycelial growth and spore formation.

Figure 3 Inoculation tower with rose leaves in Petri dishes in front.

Races of the Pathogen

Physiological races of plant pathogenic fungi are classically defined by differences in virulence of individual isolates of the pathogen species on a differential set of host genotypes. For this, it is essential to use isolates derived from single conidia (and therefore of a single genotype) and to test these isolates on appropriate host plants under identical conditions.

Single spore isolates of powdery mildew fungi can be derived in various ways: one possibility is to transfer single conidia or conidiophores from naturally infected leaves to mildew-free leaves (leaves from *in vitro* plants or leaves from the greenhouse wiped off with 70% ethanol) from a susceptible genotype. After about 2–3 weeks single conidia from colonies can be transferred to new leaves. This process should be repeated several times because of possible contamination by conidia from more than one conidiophore. Another method is to blow a few conidia from naturally infected leaves on to the susceptible leaf material using an inoculation tower (**Figure 3**). After about 5–10 days, leaves with only single mildew colonies are used as the conidia source for a second inoculation of mildew-free leaves in the tower. The resulting single colonies are assumed to be single-spore isolates.

Separate forms of *P. pannosa* were reported in the mid twentieth century based on disease frequency in peach orchards. In the 1960s inoculation experiments with a differential set of host genotypes were made, but not using monoconidial isolates. First reports of races of *P. pannosa* using the method mentioned above are from 1984. In a collection of nine isolates from different rose genotypes, five races of *P. pannosa* were identified by their significant variation in virulence. Different races of *P. pannosa* were also identified from isolates collected from a very limited area. Five isolates were taken from naturally infected genotypes from the plantations of the Federal Centre for Breeding Research on Cultivated Plants in Ahrensburg (Germany) and one from the Risoe National Laboratory in Denmark. Within these six monoconidial isolates, five races could be distinguished, including the Danish isolate, using a differential set of eight rose genotypes (**Table 1**), suggesting a differentiation of *P. pannosa* races in a small geographic area on different rose genotypes.

Susceptibility of Rose Cultivars

Each year several new rose cultivars are brought into the market, many of them bred to have a high degree of resistance against *P. pannosa* as well as for the horticulturally important characters desired by consumers, like flower colour and architecture or the vase-life of cut flowers. However, only a few display a high level of

Table 1 Differentiation of races of *Podosphaera pannosa*

Rose genotype	*Single-spore isolate*[a]				
	1	*2*	*3*	*4*	*5*
95/13-39	+[b]	+	+	+	+
Spalier 3	+	+	−	−	+
93/1-119	−[c]	+	−	−	−
'Rebell kordana'®	−	+	+	+	+
'Queen Elizabeth'®	−	+	−	+	−
Rosa multiflora	+	+	+	−	+
'Elina'®	−	+	−	+	+
'Pariser Charme'®	+	+	+	−	+

[a]Origin of the isolates: 1–4: Federal Centre for Breeding Research on Cultivated Plants, Ahrensburg, Germany; 5: Risoe National Laboratory, Denmark.
[b]Resistant rose genotype (genotypes with only single conidiophores were treated as resistant).
[c]Susceptible genotypes: mean values from 3–5 replicated experiments 10 days after inoculation with ~2 conidia mm^{-2}.

disease resistance. This may be due to the development of new genetic races of *P. pannosa* that can overcome this resistance. Several publications have rated differences in susceptibility of various cultivars. Among these cultivars few possess complete resistance. And the situation is even worse if we look not only at the natural infection of cultivars in the field, but also take into account the results from artificial inoculation experiments. Reports concerning a different degree of susceptibility between the different horticultural classes are not consistent in the literature. Whereas some authors tend to say that climbing roses, ramblers and hybrid teas are more susceptible than wichurianas, others conclude that floribunda and polyantha cultivars are generally more susceptible than hybrid teas. In a survey of 120 rose varieties commonly cultivated in Poland, miniature cultivars like 'Zwergkönig 78', 'Fresh Pink', 'Eleanor' or 'Degenhard' were far less susceptible than hybrid teas or floribunda and polyantha cultivars.

In the wild species there seems to be no clear distinction in powdery mildew resistance among the taxonomical sections within the subgenus *Eurosa*. Whereas some members of the sections *Caninae* (*R. agrestis, R. glutinosa*) and *Pimpinellifoliae* (*R. foetida* var. *persiana, R. omeiensis* var. *pteracantha*) seem to be highly resistant in field observations with natural infections and artificial inoculation experiments, there are some species in these sections which are highly susceptible (*R. britzensis, R. vosagiaca* = *R. caesia* subsp. *glauca* in the *Caninae* and *R. pimpinellifolia* in the *Pimpinellifoliae*).

Disease Symptoms

The appearance of infected tissue depends on many factors, including when it was infected. Rose leaves are most susceptible in the 3 days following unfurling. When young, rapidly expanding tissue is infected, it

Figure 4 Rose leaves showing distortion caused by infection by powdery mildew.

may become stunted or distorted (**Figure 4**). Fungal colonies on young tissue expand quickly, mostly starting on the upper-leaf epidermis. Sometimes they coalesce to cover the entire surface of a leaf, giving it a whitish appearance. Badly infected tissue will frequently shrivel and die. When older, less susceptible tissue is infected, colony expansion is slower, and leaves will appear normal except for discrete circular fungal colonies. As colonies age, they stop producing conidia and become thicker and feltier. Primary (young) mycelium of *P. pannosa* shows a cell wall structure similar to that of the conidia (it is colourless, septate and thin-walled); the persistent secondary mycelium is dense and pannose, forming a greyish felt of sparsely branched, thick-walled hyphae.

Symptoms may vary from cultivar to cultivar. In some cultivars only leaves and young shoots are infected; in others the stems may also become felted. Buds can become infected, and even opened flowers, especially in some cultivars of miniature roses. In resistant cultivars infection may be limited to a few cells that quickly die, killing the fungus with them, causing a brown flecking of tissue, called hypersensitive response.

Environmental conditions also affect the appearance and severity of the disease. Optimum conditions are found on warm, humid days without rain and cool nights with high humidity. Closely planted roses, roses grown in shade or cultivars with dense foliage are also susceptible because of poorer air circulation, allowing moist air to remain around plant surfaces for longer amounts of time.

Spread of Disease Outdoors

The life cycle of *P. pannosa* is shown in **Figure 5**. In autumn, *P. pannosa* infects dormant buds and overwinters in them. When the buds unfurl in early spring, the fungus is already present on the young leaves and begins to produce asexual spores called conidia, whose genetic makeup is identical to the parent colony. In dry weather mature conidia are captured by gusts of wind and blown away. Spores that land on young rose tissue form additional colonies. Depending on temperature, a conidium can germinate, form a colony and produce new conidia in 4–10 days. Thus the disease spreads with amazing speed. The production of asexual spores continues throughout the summer as long as there is young tissue to infect. Raindrops knock conidia out of the air and off leaves, reducing inoculum levels, and large droplets of water on leaves inhibit conidial germination, so long periods of rainy weather will inhibit powdery mildew (although it will promote the spread of most other fungal pathogens).

In late summer, if both mating types of the fungus are present, the sexual state may form. The fruit bodies of the powdery mildews are spherical or flattened ascocarps (e in **Figure 5**). They are formed by the sexual organs of the fungi, the gametangium and the ascogonium, which develop on lateral branchlets of the mycelium (d in **Figure 5**). They contain 4–8 ascospores in one ascus. The formation of ascocarps by *P. pannosa* does not occur regularly and is rather erratic. In field observations ascocarps were only found on less than 5% of over 700 rose species and cultivars investigated over several seasons in the UK and the USA. More ascocarps were detected on old bush roses and ramblers than on modern hybrids, suggesting the presence of host factors influencing their formation. In addition, many powdery mildews require cool temperatures in autumn to trigger ascocarp formation, and this may also be the case with the rose powdery mildew. Therefore, in roses, overwintering of the sexual stage seems to be of little importance. However, if it occurs it increases the chances of genetic recombination.

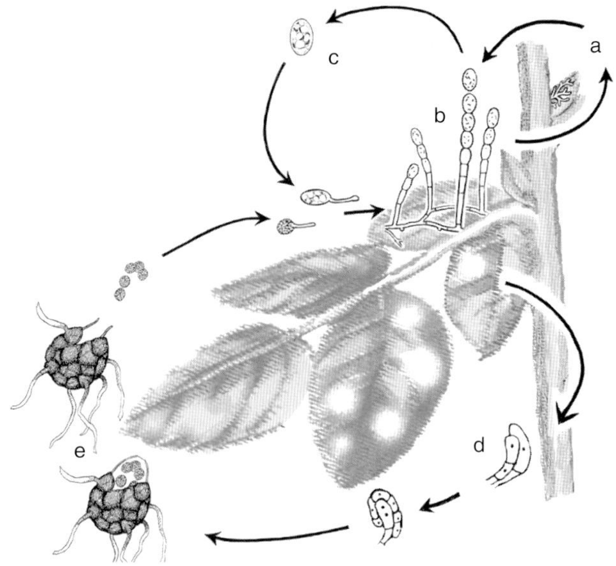

Figure 5 The life cycle of *Podosphaera pannosa*. In autumn, *P. pannosa* infects dormant buds and overwinters in them (a). When the buds unfurl in early spring, the fungus is already present on the young leaves and quickly forms colonies that produce asexual spores (b). In dry weather the spores are captured by gusts of wind (c); spores that land on young rose tissue form additional colonies. In late summer, if both mating types of the fungus are present, the sexual state will form (d). Although this step has not been reproduced in the laboratory, it is likely that in spring the sexual spores are released on to young rose tissue (e).

Weather-forecasting models have been developed to identify periods when environmental conditions favour disease development, based on measurements of temperature, leaf wetness and relative humidity. It may be possible to use such models to reduce the number of fungicide treatments while achieving the same level of control.

Spread in the Greenhouse

Epidemics in greenhouses begin with the introduction of infected plant material or by conidia blown in from outdoors. Greenhouse conditions are ideal for spread of conidia; its humid and air circulation is good enough to disseminate spores but not enough to keep humid air from accumulating around plant surfaces. Conidia can also be spread on the hands and clothing of greenhouse workers, and possibly by gnats and whiteflies. Once a greenhouse is infested with mildew, the only sure way to clean it is to remove all roses from it, wait long enough for any conidia to settle from the air and die, and then reintroduce clean material. It can be difficult to be sure that a plant is 'clean', particularly in the early stages of infection, or in a resistant cultivar with few symptoms. Even in a clean greenhouse, infection can occur from conidia produced elsewhere entering through air vents.

Control by the Plant's Own Defence Mechanisms

After the deposition of the conidia on the leaf, the first barrier to overcome is the cuticle of the leaf and the epidermal cell wall. Differences in resistance against powdery mildew can not be explained by cuticle thickness, and penetration of the cells is apparently not prevented by any morphological barrier.

The first active reaction of the rose leaves against the powdery mildew attack and to the formation of the appressoria seems to be the development of a papilla. Papillae are cell wall appositions located very near to the fungal penetration peg between the plasma membrane and the outer epidermal cell wall of the host. Cytoplasmic aggregates of organelles and vesicles seem to play a major role in the deposition of the papilla. Callose, phenolics, proteins, lignin and other compounds have been detected in the papilla. In roses, papillae are found about 1 h after the wounding of the epidermal cell wall from the appressoria, often associated with a restricted development of the penetration peg. The detailed working mechanisms of the papilla have not been clear up to now. However, the penetrating hyphae are sometimes stopped in incompatible taxa due to papillae preventing successful colonization.

The development of a collar from fibrillar-granular material around and along the haustorial neck is another feature of the host reaction to the fungal pathogen. The host cell wall modifications consist of an amorphous area with relatively low cellulose content surrounded by cellulose-rich fibrillar layers. There seems to be a strong correlation between the occurrence of a collar and the poor development of haustoria, maybe due to preventing a close association between the extrahaustorial membrane and the neck wall. If powdery mildew germlings have penetrated the papillae and were not stopped by a collar, the fungus can be arrested from the host by the hypersensitive response (HR). The HR is a type of programmed cell death of a single host cell or a group of host cells due to the mildew attack. In barley this happens only if a race-specific resistance gene in the host matches a corresponding avirulence gene in the fungus. The activation of the pathway to the HR requires the presence of a haustorium in the epidermal cells. Then there is probably a contact between the avirulence gene products of the pathogen, so-called elicitors, and the resistance gene products, which serve as the sensory apparatus of the plant.

Epidermal cells of roses and barley which died hypersensitively by an incompatible mildew race exhibit a characteristic fluorescence, due to callose incorporation into the cell wall and accumulation of phenolic compounds in the cell (**Figure 6**).

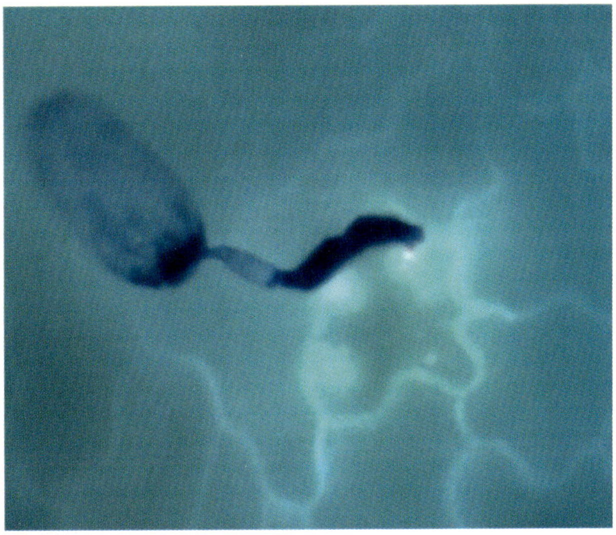

Figure 6 Conidia with appressoria and the penetration peg on an epidermal rose cell showing a fluorescence of the cell wall indicating a hypersensitive response.

Single cell death occurs rapidly after penetration, when the fungus had only contact with one or two plant cells. In contrast, multicell HR starts after limited fungal growth with several haustoria; it is visible by eye and is called a necrotic lesion (as shown in **Figure 6**). The occurrence of single or multicell HRs seems to depend on a particular combination of avirulence and resistance genes in fungus and host. The damage of the plant cell membrane seems to be an early event in the HR, followed by the accumulation of phenolics and a halt to cytoplasmic streaming after 15–18 h. The final stage is cellular collapse at 18–26 h after inoculation in barley. Since the fungus is dependent on living host cells, it is likely that cell death due to the HR can kill powdery mildew.

Control by Chemicals

The asexual conidia are 'clones' which are identical to the parent. However, they are produced in such abundance that chance mutations may occasionally occur, in perhaps 1 in 1 000 000 spores. Of those mutated spores, perhaps one in a million has an advantageous mutation, such as a reduced sensitivity to a given fungicide. Of those spores, perhaps one in a thousand lands on a rose leaf and infects. This may seem like poor odds for pesticide resistance to develop, but over the course of a season so many spores are produced that resistance will almost invariably develop in a few years unless growers alternate between different control methods to make sure that a spore that survives one method will be killed by the subsequent one.

Therefore it is important to understand how each type of control measure works.

Many modern fungicides work by interfering with a single site in a metabolic pathway of the pathogen. If one uses a single fungicide to control powdery mildew, spraying frequently, the vast majority of spores will be killed. However, a single mutation altering the shape of one enzyme might be enough to endow a conidium with resistance. The mutant pathogen will have no competition for space on the now pristine rose leaves and will spread rapidly. Alternating between two pesticides in the same class is not much better than using a single fungicide, because the two fungicides have similar modes of action. Fungicides like strobilurins (Compass, Heritage and Cygnus) are not only less toxic to nontarget organisms than older fungicides, they have a different mode of action from older classes of chemicals like triazoles. Because they are also single-mode-of-action fungicides, they should not be used alone; they should be incorporated into a management scheme. One should alternate single-mode-of-action fungicides (like strobilurans and triazoles) with those having broader modes of action: copper compounds, sulphur compounds, bicarbonate salts, horticultural oils, antifungal plant extracts (such as neem extract, milsana and cinnamite), biocontrol agents (*Ampelomyces quisqualis* Cesati ex Schlechtendahl (AQ10 Biofungicide), *Pseudozyma flocculosa* (Traquair, LA Shaw & Jarvis) Boekhout & Traquair, *Trichoderma harzianum* Rifai (Topshield)), and activated-resistance compounds. Some of these may be phytotoxic, others are less effective than traditional fungicides, but many can be used effectively as one part of a management scheme. Biocontrol agents generally need higher levels of moisture to infect than the powdery mildews they parasitize, but under greenhouse conditions it may be possible to manipulate the environment to make them effective.

Control by Breeding Resistant Cultivars

One of the most environmentally safe ways to control rose powdery mildew is to cultivate resistant varieties. Identifying molecular markers closely linked to resistance genes could be a key factor for resistance breeding (*see* **Breeding:** Overview). Several studies have shown that disease resistance may be the combination of one or two major factors and a few minor ones. These can be tagged by a quantitative analysis and quantitative trait loci (QTL) mapping or by the cosegregation of genetic markers and qualitative resistance data. A putative resistance gene against *P. pannosa* (Rsp-1) could be mapped using a bulked segregant analysis. A backcross population of a resistant and a susceptible diploid parent was analysed for the resistance against a single-spore isolate. The data from the segregating population were compared with the distribution of genetic markers (amplified fragment length polymorphism or AFLP) for about 150 different individuals. AFLP-marker linked to Rsp-1 in a distance between 4 and 16 cM could be mapped on one side of the putative gene (**Figure** 7).

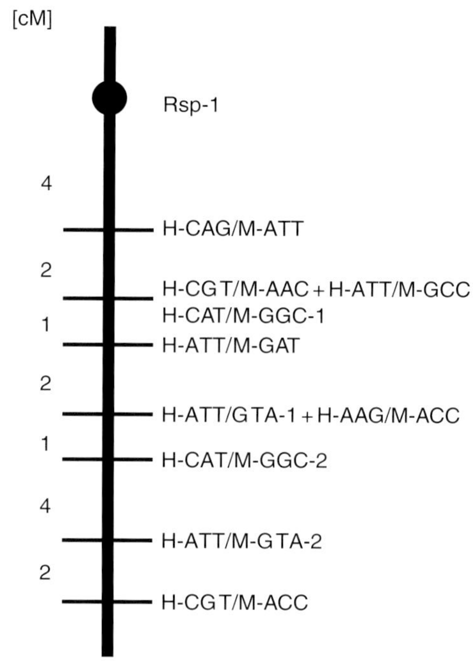

Figure 7 Genetic map of the putative resistance gene Rsp-1 and the linked amplified fragment length polymorphism (AFLP) marker.

Integrated Control

In summary, powdery mildew can be reduced by selecting and breeding resistant cultivars (by testing wild species and introducing their resistance genes by backcrossing with cultivars with traditional methods or using molecular markers: marker-assisted selection (MAS); phytopathological tests of cultivars under strict conditions), growing them in a site with full sun and good air circulation to eliminate conditions favouring conidium germination, and providing adequate but not excessive fertilization. Plants should be inspected regularly to catch the onset of disease and treatment should consist of an alternation of several methods of chemical control. Single-mode-of-action fungicides should be used with those having broader modes of action in order to reduce the chance of resistance developing in the pathogen. The usage of weather-forecasting models based on measurements of temperature, leaf wetness and relative humidity to reduce the number of fungicide treatments while achieving the same level of control is also constructive (*see* **Integrated Pest Management**).

See also: **Disease**: Overview; Downy Mildew; Bactericides and Fungicides; **Insects and other Animals**: Overview of Insects. **Integrated Pest Management.**

Further Reading

Belanger RR, Labbe C and Jarvis WR (1994) Commercial-scale control of rose powdery mildew with a fungal antagonist. *Plant Disease* 78: 420–424.

Bender CL and Coyier DL (1984) Isolation and identification of races of *Sphaerotheca pannosa* var. *rosae. Phytopathology* 74: 100–103.

Braun U (1995) The powdery mildews (Erysiphales) of Europe. Jena: G. Fischer.

Cobb GS, Hanan JJ and Baker R (1978) Environmental factors affecting rose powdery mildew in greenhouses. *HortScience* 13: 464–466.

Coyier DL (1985) Powdery mildew. In: Strider DL (ed.) *Diseases of Floral Crops*, vol. 2, pp. 405–417. New York: Praeger Scientific.

Hajlaoui MR, Benhamou N and Bélanger RR (1991) Cytochemical aspects of fungal penetration, haustorium formation and interfacial material in rose leaves infected by *Sphaerotheca pannosa* var. *rosae. Physiological and Molecular Plant Pathology* 39: 341–355.

Horst RK (1983) *Compendium of Rose Diseases*. St Paul, MN: APS Press.

Kunoh H (1995) Host–parasite specificity in powdery mildews. In: Kohmoto K, Singh US and Singh RP (eds.) *Pathogenesis and Host Specificity in Plant Diseases: Histopathology, Biochemistry, Genetic and Molecular Bases*, vol. 2, pp. 239–250. Oxford: Elsevier.

Mence MJ and Hildebrandt AC (1966) Resistance to powdery mildew in rose. *Annals of Applied Biology* 58: 309–320.

Price TV (1970) Epidemiology and control of powdery mildew (*Sphaerotheca pannosa*) on roses. *Annals of Applied Biology* 65: 231–248.

Tjosvold SA and Koike ST (2001) Evaluation of reduced risk and other biorational fungicides on the control of powdery mildew on greenhouse roses. *Acta Horticulturae* 547: 59–67.

Rust

R C Shattock, University of Wales, Bangor, UK

Introduction

Rust is an infectious fungal disease of roses principally attacking leaves, but in some cases stems and inflorescences are affected too. Rusted plants are physiologically affected in much the same way as those attacked by powdery and downy mildew and blackspot, i.e. photosynthesis is reduced, transpiration is increased and the pathogen acts as a sink for nutrients and photosynthates and thereby weakens the host. As in the case of blackspot and downy mildew, rust-affected leaves of some varieties and species of *Rosa* may fall prematurely further affecting plant vigour and winter hardiness.

The Pathogen

In North America nine species of the rust genus *Phragmidium* occur on native roses with *Phragmidium mucronatum* (Pers.) Schlecht described as the most common species on hybrid tea and floribunda varieties. In contrast, only four species are recognized in the United Kingdom and *P. tuberculatum* Mull. is cited as the principal species on nonspecies bush roses. It is clear, however, that both *P. mucronatum* and *P. tuberculatum* have been repeatedly confused with each other. This situation comes about because host ranges of the two species probably overlap as do characteristics associated with size and shape of spores (there are five such spores in the life cycle of *Phragmidium* subsp.). Other factors, not so far investigated, could be due to hybridization between the closely related *Phragmidium* subsp. Alternatively, both *P. mucronatum* and *P. tuberculatum* could be conspecific. Conspecificity describes variants of the same species. Such a situation has been shown recently in the case of the downy mildew pathogens of *Rubus* hybrids such as tayberry, and rose. Downy mildew of *Rubus*, *Peronospora rubi* Rabh., is now recognized as a variant of the older named species *Peronospora sparsa* Berk., responsible for rose downy mildew. Clearly, *Phragmidium* subsp. on *Rosa* subsp. require critical revision and modern methods of molecular taxonomy will undoubtedly be useful in this respect.

In the absence of a taxonomic re-evaluation, *P. tuberculatum* in the United Kingdom is thought to be restricted to *Rosa* cultivars, *R. dumalis* Bechst. *R. rubiginosa* L. and *R. rugosa* Thunb. *Phragmidium mucronatum*, however, has been reported on the latter two species but essentially attacks the Albi group and roses and species of *Bracteatae* and *Caninae* including *R. corymbifera* 'Laxa', the important understock species for bush roses generally. Two other species of *Phragmidium* are recorded in Britain: *P. fusiforme* Schrot, on *R. pendulina* L. and *R. glauca* Pourr., and *P. rosae-pimpinellifoliae* (Rabh.) Diet on *R. pimpinellifolia* L., the Burnet rose, and probably some of its hybrids.

Figure 1 Leaflets of *Rosa* 'Albertine' affected by *Phragmidium tuberculatum*. Left, symptoms on upper epidermis; right, lower epidermis with yellow–orange uredinia and black telia.

Life Cycle

Rust fungi have the most complex of life cycles of all fungal plant pathogens with up to five different spore stages infecting, in some cases, two taxonomically different plant hosts. Rusts of the genus *Phragmidium* normally produce all five different spores, but all on the same host.

Rose growers are most familiar with symptoms expressed during mid to late summer (**Figure 1**). On the upper epidermis of rust-affected leaflets these usually resemble a scattering of red–purple and yellow–brown irregular spots and lesions. On the lower side of affected leaflets orange–yellow pustules, uredinia (**Figures 1** and **2**) producing urediniospores appear first followed later and intermixed with black telia (**Figure 1**), which give rise to the overwintering brown–black teliospores. The urediniospore, which is single-celled, is the repeating spore stage, and is produced throughout summer and allows rust to develop rapidly on susceptible genotypes, if not adequately controlled. Teliospores produced later in the season are multicelled (*P. tuberculatum*, 5–7 cells; *P. mucronatum*, 6–9 cells) with a stalk or pedicel 70–140 μm or 80–150 μm, respectively, in length (**Figure 3**).

Figure 2 Close-up of uredinia on lower epidermis.

Figure 3 Stalked, multicelled, brown–black, thick-walled overwintered teliospore producing four thin-walled short-lived basidiospores (products of sexual reproduction) from one of its seven cells.

Whilst urediniospores are wind- and rain-dispersed and can travel some distance, teliospores usually remain close to their source on fallen leaves, stems, soil or adjacent debris. This is facilitated by the pedicel which is hygroscopic and becomes sticky when wet permitting detached teliospores to stick to stems of roses and adjacent plants and other garden structures such as fencing. Teliospore cells are both thick-walled and pigmented allowing them to remain dormant throughout winter before germinating the following spring to produce short-lived, thin-walled basidiospores (**Figure 3**). The basidiospores, products of sexual reproduction, germinate and in the case of *P. tuberculatum* infect newly emerging leaflets by direct penetration of the epidermis. Several days later yellow spots 0.4–0.6 mm in diameter are produced on leaves. These spots (spermagonia) produce another thin-walled spore, the spermatium, which upon transfer by rain or insects to other spermagonia, re-establishes the binucleate condition required for formation of the next spore stage, the aeciospore. Aeciospores arise from concentric yellow rings of aecia around the fertilized spermagonia. Aeciospores are wind- and rain-splash dispersed and upon infection lead to urediniospores produced in the yellow pustule-like uredinia. Both germinating aeciospores and urediniospores enter rose leaflets via natural openings on leaves, namely stomata (**Figure 4**). The other British rusts on roses produce the same sequence of spores, but in the case of *P. mucronatum* and *P. rosae-pimpinellifoliae* aecia can also be found on veins of leaves, petioles, parts of the inflorescence and stems. On the latter, aecia may be large and confluent (**Figure 5**). In some cases, stem- and petiole-borne aecia arise not from basidiospore infection but from long-lived, perennating systemic mycelial infection of stems. Such confluent aecia usually produce swollen and distorted buds (galls) (**Figure 6**) from which are

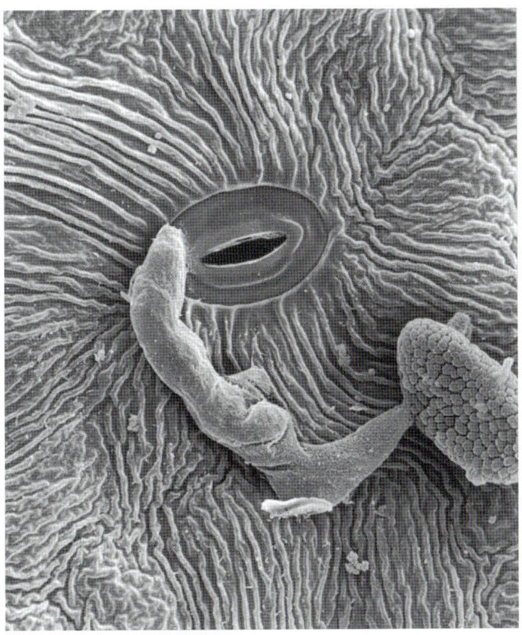

Figure 4 Germ tube from aeciospore growing towards a stomatal pore of the leaf. Both aeciospores and urediniospores germinate and infect the host by this route.

Figure 5 Orange aecial galls produced by perennating mycelium of *P. rosae-pimpinellifoliae* on *R. pimpinellifolia*.

Figure 6 Aecial galls producing distorted buds from perennating mycelium of *P. mucronatum* on *R. corymbifera* 'Laxa'.

released vast numbers of aeciospores. This difference in symptomatology helps to distinguish these two species and *P. tuberculatum*. However, the main difference is microscopical, with aeciospores of *P. tuberculatum* being verrucose (warty) whereas those of *P. mucronatum* and *P. rosae-pimpinellifoliae* are echinulate (spiny). Because there is little morphological difference between urediniospores and teliospores (the most commonly observed stages of the rust life cycle) of all *Phragmidium* subsp. it is easy to understand the confusion and uncertainty over the different species, namely *P. tuberculatum* vs. *P. mucronatum* repeatedly reported in the literature.

Free water is necessary over several hours for both aeciospores and urediniospores of all species to germinate and infect susceptible roses. Temperature exceeding 25 °C during the growing season limits germination and thereby restricts rust normally to areas with cool wet summers such as northwestern regions of both North America and Europe rather than districts with hotter, drier summers. The overwintering teliospores may be killed by very cold winters although the normal events of wetting and drying, freezing and thawing of winter in temperate areas are usually required to break inherent dormancy and thereby trigger teliospore germination leading to basidiospore production in spring.

Effect of Rust on Roses

Because rust affects photosynthesis, transpiration and other physiological components of both vegetative and reproductive growth of roses, very susceptible genotypes can be seriously weakened and killed by heavy and repeated attacks of rust: new growth, in particular, can be severely affected by rust.

Whilst selection criteria by breeders are unlikely to promote host genotypes susceptible to rust for commercial development it is by no means certain that resistance expressed by a particular genotype will be permanent over time or in all regions. This is because rusts, like powdery and downy mildews, are both ecologically biotrophic plant pathogens (i.e. they only grow and reproduce on living host tissue) and have high host specificity. Mutants with novel pathogenicity towards host genotypes previously considered resistant can arise and be selected for in the pathogen population. Such 'boom-and-bust' scenarios, with loss of host resistance as the frequency of novel host-specific pathotypes increases, are well documented in rust diseases of arable crops, e.g. cereals, and have been recorded in the related species *Phragmidium rubi-idaei* causing yellow rust of red raspberry in the United Kingdom.

Since roses are long-lived perennials, and with several thousand genotypes available, it has to be accepted that those sold as resistant to rust may succumb to the disease at any time and thereby adding to the long list of older but popular and widely planted rust-susceptible varieties, such as the rambler 'Albertine', the climber 'New Dawn', the floribunda 'Queen Elizabeth'® and many others.

The damaging effect of rust on roses was quantified in a study of severely affected *R. corymbifera* 'Laxa' infected with *P. mucronatum* compared to plants kept free of the disease with regular spraying with systemic fungicides. Over two growing seasons the fresh weight of shoots and roots was halved, and the number of flowers reduced by a quarter by rust. Similar reductions in growth are likely to occur with bush roses attacked by *P. tuberculatum*. In a related experiment the commercial quality of four bush rose varieties, 'Elizabeth of Glamis'®, 'Gold Crown', 'Josephine Bruce' and 'Christian Dior', budded into rusted and non-rusted 'Laxa' stock was measured when maiden plants were harvested. The quality of the four cultivars bud-grafted into 'Laxa' rusted in year 1 of the 2-year production cycle, i.e. before 'heading back' of the 'Laxa' stems to permit growth of the budded scion, was significantly lower when compared with lifted plants budded into rust-free 'Laxa'. Bud-take was not affected by presence or absence of rust on 'Laxa', but on lifting 29 (90%) out of a total of 32 plants classified as first-quality saleable plants were on understocks of rust-free 'Laxa', as were 41 (56%) of 73 second-quality plants. In contrast, the majority, 63 (71%) of a total of 89 unsaleable-quality bush roses were from rust-affected budded 'Laxa' understocks. Clearly, rust caused by *P. mucronatum* on foliage of 'Laxa' in year 1 of the 2-year production cycle severely reduced plant growth, especially root development of the understock, mainly due to earlier defoliation of rusted plants. This reduction in growth could not be compensated for by the growth of the scion of budded maidens in year 2 of the cycle even though the maiden bush roses remained free of rust infection, i.e. *P. tuberculatum*, in year 2 and were not susceptible to *P. mucronatum* throughout the experiment (**Figure** 7).

Disease Control

As with other principal fungal diseases of rose, namely blackspot and powdery mildew, inoculum of the rust fungi must be assumed to be ever present, especially as aeciospores and urediniospores are wind-borne and therefore widely dispersed. Cultural control by removing inoculum by removing rust-affected leaves or

Figure 7 Susceptibility and resistance to *P. mucronatum*. The compound leaf of *R. corymbifera* 'Laxa' (left) is heavily rusted, whereas *Rosa* 'Pascali'® (right) growing as a scion on the 'Laxa' stock, and which has not been 'headed back', is unaffected.

collecting fallen rusted leaves is unlikely to be successful. Furthermore, later in the year teliospores are also dispersed, albeit shorter distances, and stick, as stated earlier, to stems and other permanent substrates. Removal of rust-forming galls in the case of *P. mucronatum* and *P. rosae-pimpinellifoliae*, however, may help to reduce aeciospore inoculum arising from perennating systemic mycelial infection.

Chemical control can be effective and a number of systemic fungicides, e.g. benodanil, oxycarboxin, penconazole, propiconazole and triforine, have given excellent control over the past couple of decades. The latter two have been available to the amateur gardener but have become outclassed or withdrawn for a variety of reasons. Myclobutanil is a readily available systemic fungicide providing protective, curative and eradicant control and, of course, is active against powdery mildew, too. Myclobutanil is an azole compound and this family of chemicals inhibits sterol biosynthesis and therefore disrupts cell membranes of target fungal pathogens. Commercial growers, in the United Kingdom at least, have a wider choice of compounds to choose from including new-generation azoles developed for the cereal market and other compounds too, such as the strobilurins which inhibit cellular respiration in pathogens. In the United Kingdom, subject to the usual restrictions, fungicides approved for use on any growing edible crop, e.g. cereal or top fruit, can be used, at present, on commercial agricultural or horticultural holdings, where no part of the plant is to be consumed by humans or animals. This applies to hardy ornamental nursery stock, and therefore includes roses.

See also: **Disease**: Overview; *Botrytis*; Black Spot; Downy Mildew; Powdery Mildew. **Breeding**: Selection Strategies for Disease and Pest Resistance.

Further Reading

Henderson DM (2000) *A Check List of the Rust Fungi of the British Isles*. Kew, UK: British Mycological Society.

Horst RK (1983) *Compendium of Rose Diseases*. St Paul, MN, USA: American Phytopathological Society.

Laundon GF and Rainbow AK (1969a) *Phragmidium tuberculatum*, CMI Descriptions of Pathogenic Fungi and Bacteria no. 204. Kew, UK: Commonwealth Agricultural Bureaux.

Laundon GF and Rainbow AK (1969b) *Phragmidium mucronatum*, CMI Descriptions of Pathogenic Fungi and Bacteria no. 205. Kew, UK: Commonwealth Agricultural Bureaux.

Laundon GF and Rainbow AK (1969c) *Phragmidium rosae-pimpinellifoliae*, CMI Descriptions of Pathogenic Fungi and Bacteria no. 208. Kew, UK: Commonwealth Agricultural Bureaux.

Losing H (1988) Bekampfung von Rosenrost. *Deutsche Baumschule* 40: 518–519.

Shattock RC and Rahbar Bhatti MH (1983) Rust on Laxa stocks. In: Harkness J (ed.) *The Rose Annual 1983*, pp. 150–162. St Albans, UK: Royal National Rose Society.

Replant Problems and Soil Sickness

W Spethmann, University of Hanover, Germany
G Otto, Dresden, Germany

Introduction

Declines in growth and yield have been observed in roses and several other genera when the same genus has repeatedly been planted in the same location. The degree of decline depends on the plant species, various soil factors, and pests present in the soil. To avoid such replant problems growers tend to move to a new field, fumigate, or cure with hot steam (**Figure 1**). Crop rotation schedules, allowing planting intervals for the species in question as well as fallow periods, are recommended.

The two main causes of replant problems are first, nematodes and second, specific replant disease or soil sickness. Both factors will be discussed in this article. At the present time it is not known whether nematodes are connected directly with the specific replant disease. The organisms causing specific replant disease in roses have not yet been identified, although economic losses are very high. After many years of intensive research in apples, the causal organisms associated with specific

Figure 1 Disinfected soil plots with optimum growth of *Rosa corymbifera* 'Laxa' beside plots with depressed growth caused by soil sickness and nematodes in Bönningstedt (Schleswig–Holstein). (Photograph by W. Spethmann.)

replant disease were identified as *Actinomycetes*. It is possible that these microorganisms could have a similar effect in roses as in apples; therefore the mode of action of these microorganisms in *Malus* are discussed in detail.

Besides nematodes and specific replant disease, replant problems could be caused or increased by fungi, lack of mineral substances, trace elements, or soil compaction.

Characteristics and Terminology

Some plant species suffer severe damage when grown repeatedly in the same field. This cannot be avoided even by having a long interval before returning with the same crop, especially where species of the Rosaceae family are concerned. The terms 'replant disease', 'soil sickness' and 'soil exhaustion' are often used to describe this problem. The problem is said to be species-specific (for apples, for roses, etc.). The persistence of the cause of replant disease can be observed in the soil even decades later, but other factors should also be taken into account. The phenomenon appears to be strictly limited to the field in which the particular crop was grown previously, as especially noted in nurseries which changed the direction in which rows of roses or apple liners were planted. The disease occurs uniformly within entire rows of plants, not in nestlike pockets. Furthermore, it is known that replant disease is reversible. Plants from affected soil which are transplanted into virgin soil will soon recover and grow normally. Typically, replant disease occurs with varying severity in different locations. In some areas serious problems rarely occur or only after several repetitions of the same crop, whereas in other areas a single, short-term planting produces immediate damage for the following crop. Of course, this particularly applies to nurseries. Steaming or chemical fumigation of the soil will improve the problem.

Numerous investigations have suggested a range of theories on the cause of replant problem. They include macro- or micronutrient deficiencies, nematodes, structural changes in the soil, toxin accumulation and an imbalance in microorganism population. A conclusive explanation of the cause of replant problems has not yet been presented. It has sometimes facetiously been said that "soil sickness or replant disease is when nobody knows the cause". The replant problem in roses often results from a combination of soil sickness or replant disease and nematode damage.

Many factors may play a role when the same species is planted successively, and this multiplicity of causes has made it difficult to diagnose replant problems. Therefore definitions need to be precise. Replant problems include all those factors which cause reduced growth and yield when the same species or close relatives are planted repeatedly. One primary cause is nonspecific or more or less specific pests, such as the nematode *Pratylenchus penetrans*. Its hosts include over 200 plant species, and therefore it cannot be considered to be an exclusive agent in the context of replant disease.

Physical soil factors may cause replant problems, such as soil compaction or waterlogging. Intentional monocropping of the same crop instead of a proper crop rotation programme may give rise to replant problems. Deficiencies in macro- or micronutrients may develop under such conditions. However, this should not be assumed to be the cause of the replant disease, because other plant species will react in a similar way. Neither can the long-term persistence of the problem in one location be explained in such cases. Similar arguments ignore the effect of root excretions or toxic substances formed when the remaining roots decompose after grubbing. Such substances are not expected to remain in the soil for long periods of time. Neither is it expected that the after effects of herbicide applications are a causal agent of replant disease.

All these factors have generated various theories on the cause of replant disease. Replant disease or soil sickness is but one factor in this range of considerations. It differs from other factors by having a specific effect on one plant species and persisting in the long term. Savory coined the term 'specific replant disease' and distinguished between 'specific apple replant disease' and 'specific peach replant disease' etc. emphasizing the biological species-specificity. Practical experience provides evidence to support this concept, since pome fruit trees can be grown after stone fruit and vice versa with no replant problems. It also correlates with Klaus's definition of 'plant specificity' which is not necessarily associated with any species in the taxonomic sense. A *Malus* rootstock taxonomic species may well show different symptoms from a previously cropped *Malus* of a different species, and other Rosaceae may show reductions in growth when planted after a crop of apple trees. Thus, plant or fruit tree specificity cannot be considered a true characteristic of replant disease.

In spite of these endeavours to define 'specific replant disease' clearly, the lines of difference today are not clearly drawn. Depending on respective authors' scientific fields, factors are often cited that are not necessarily associated with the characteristics of replant disease. Opinions on the cause of replant disease are divided at this time. Investigations must continue in order to resolve this problem, as virgin soils in suitable climates become scarcer and scarcer. Fumigation is not an acceptable proposition for environmental reasons. New methods of dealing with this problem cannot be developed until we know its true cause.

Soil Sickness/Replant Disease with Roses

Compared with fruit trees, much less is known on replant disease in roses. There is no evidence on whether the phenomenon only occurs during the first few years after replanting, and whether plants can overcome it after a short time on site. It had been assumed that apples could recover from replant disease; however it was found that symptoms continued until the replanted trees were 11 years old. Obvious yield reductions in 36% of the compared fields were measured in replant trees during the entire life of these trees compared with those planted on virgin soil. No definitive figures exist in roses on a 'weary' soil. The growth performance of four rose rootstocks in a replant plot was only 21–78% compared with steam-treated soil. In the same experimental study the same four rose stocks replanted on an 'apple-weary' soil only attained 41–68% performance compared with steamed soil.

Reports on the frequency of replant problems are also quite rare. In a survey of 166 German rose nurseries, about 75% of growers' replies mentioned replant problems, and 60% observed reduced growth. The same order of magnitude was found in 300 replies to a survey on apples. In this survey 85% of respondents did not recommend replanting apples without adequate treatment. Investigations are hampered by the fact that nurseries either avoid replanting or fumigate their soil regularly after one or two crops. Usually fields are assumed to be at risk of replant problems, and are given prophylactic treatment.

Little is known about persistence and reversibility of replant disease in roses. Moreover there is no evidence that transplanting roses from a 'sick' soil to a virgin soil will improve the symptoms. One observation prevails for both roses and fruit trees: effective control is only possible with soil fumigation. This indicates that the microorganisms involved are hard to control.

There is no conclusive evidence that roses planted after other woody species or following a reverse planting sequence will have reduced performance. It has been observed that apples planted in a former rose nursery plot produced less growth than on virgin soil, and that several Rosaceae species are unsuitable crops when pome or stone fruits follow. These observations, as well as similarities in the replant disease in apples and roses, justify a more detailed description of the research results on the cause of specific apple replant disease. The intention is to clarify the factors associated with the replant disease in roses. It may also aid in designing further studies on replant disease in roses.

Investigations on the Cause of Specific Replant Disease in Apples

As there were different opinions about the cause of replant disease in apples, suitable methods were required to exclude or confirm theories. Attempts were first made to transfer replant disease within the root ball of an apple seedling from a sick soil to virgin soil. The roots, including the fibrous roots, were freed from attached soil particles. Successful transfer was only possible when the fibrous roots were present. This indicates that there was no causal association with nutrient deficiencies, soil physical factors or toxins in the soil. Further experiments revealed successful transfer when sick soil was freed from visible fibrous roots. A greater percentage of sick soil mixed with virgin soil resulted in increased replant disease. These results suggested a causal association of microorganisms in soil particles and in the fibrous root system.

To identify these microorganisms, the temperature levels needed to remove replant disease in steaming practices were varied and compared with the temperature tolerance of groups of microorganisms. This procedure eliminated microscopic fungal microorganisms as a possible cause. Nematodes have a lower temperature tolerance than fungi and can therefore also be excluded. *Actinomycetes* and other bacteria could not be excluded.

Actinomycetes are organisms of the family *Actinomycetaceae* in the order *Actinomycetales*. They show a vegetative mycelium, whose hyphae have a diameter lower than 1 μm. In the mycelial hyphae rod-shaped or claviform segmentations were found. They are mostly Gram-positive, aerobic and prefer neutral pH. The temperature tolerance of some strains of *Actinomycetes* isolated from soils with apple replant disease is very close to the critical temperature for replant disease to be removed and therefore it has become the target for further investigations. There are several modes of action whereby *Actinomycetes* may produce replant disease. An accumulation of *Actinomycetes* in the rhizosphere may have negative physiologically influence on replanted trees by metabolic substances. Pathogenic *Actinomycetes* may cause damage to the fine roots. *Actinomycetes* were found in longitudinal sections of fibrous roots from sick soil (**Figure 2**). These *Actinomycetes* grow into young light fibrous roots, accumulate in the root cortex, and functionally damage the root. This observation correlates with the appearance of root systems from sick soils. The fibrous roots appear to be brittle, they are brownish, and the total root mass is considerably reduced. Electron micrograph research has confirmed that the *Actinomycetes* were indeed phytopathogenic.

This was the first plausible explanation of the cause of apple replant disease. However, it was difficult to comprehend the failure to find comparable amounts of *Actinomycetes* in the fine roots. Infection occurred as early as 10 days after the formation of fine roots, and reached considerable dimensions subsequently.

With severe replant disease, *Actinomycetes* was found to be associated with up to 80% of the fibrous roots. Surprisingly low *Actinomycetes* counts were found at the end of the growing season, unrelated to the degree of replant disease. There is a connection of the infection rate with host plant metabolic activities. The infection intensity is promoted by auxins and cytokinins. If their flow into the roots is interrupted, the attack on the fibrous roots is reduced, which means that during a phase of intensive synthesis of growth regulators during the main growth phase, there is a vigorous attack on the fibrous roots. During the second half of the growth period the attack is reduced.

These observations help to explain a further phenomenon of replant disease in apple trees. On strongly affected fields apple trees display pronounced abatement of growth and yield. Seedlings are reported to produce nothing but rosette shoots. In spite of such impairment, apple trees do not die more frequently than in plantings on virgin soil. Apple tree performance is reduced as a result of damage to the fibrous roots during the main season such that they produce much less shoot growth and fruit yield. In the second half of the season the fibrous roots are only infected by *Actinomycetes* on a very small scale, so root regeneration takes place and the trees will survive.

In longitudinal cuttings of fibrous roots development of sexual stages could be observed only occasionally. Apparently it only started in the weakened cortex of the fibrous roots and proceeded by fragmentation of *Actinomycetes* hyphae (**Figure 3**). As damaged root tissue decomposes, the permanent form of *Actinomycetes* gets into the soil and increases the number of germs locally. Fibrous roots entering these parts of the soil are attacked again and the enrichment of parasitic organisms in the rhizosphere leads to replant disease. This cyclic process is repeated year after year. It is independent of grubbing and this explains why it occurs in existing orchards. This ties in with the fact that replant disease is strictly limited to the extent of a previous orchard or nursery plot. The extent of accumulation of root pathogenic *Actinomycetes* is a deciding factor for the level of replant disease (**Figure 4**), which is also said to depend on pH and O_2 content.

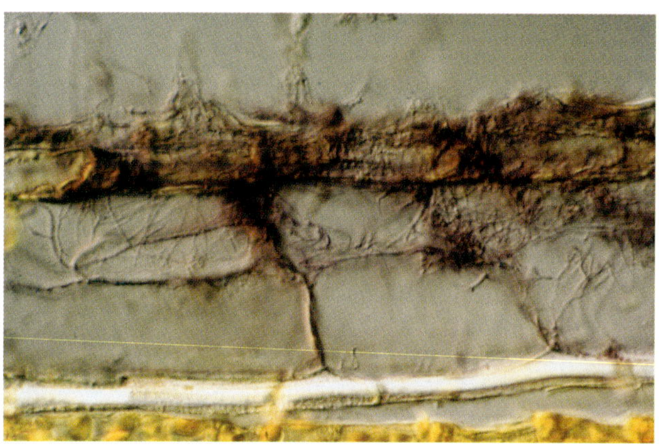

Figure 2 *Actinomycetes* in the root cortex of fibrous roots of an apple seedling from a sick soil with the beginnings of damage of the root tissue. (Photograph by G. Otto.)

Figure 3 Permanent forms in catenulate order formed by fragmentation of *Actinomycetes* hyphae. (Photograph by G. Otto.)

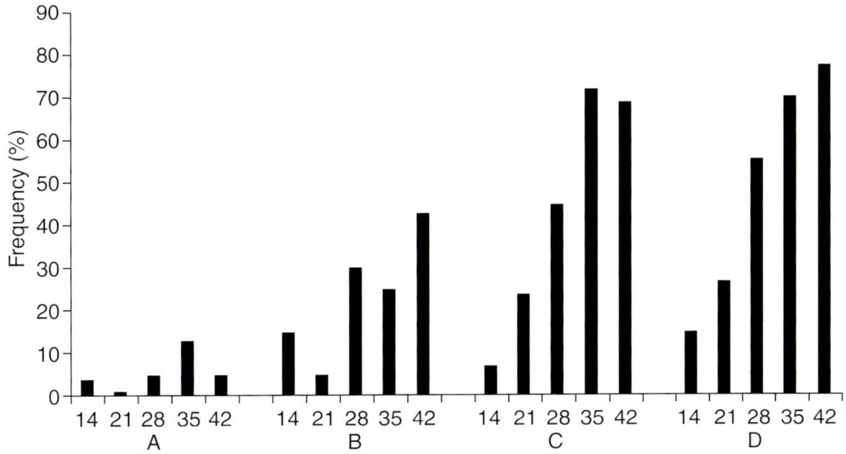

Figure 4 Frequency of *Actinomycetes* in rootlets of apple seedlings from soils with increasing degrees (A to D) of soil sickness 2–6 weeks after planting. (Data after Otto G and Winkler H (1993) *Acta Horticulturae* 324: 53–59.)

The permanent forms of root pathogenic *Actinomycetes* which cause replant disease with apples are probably ubiquitous in the soil at low microbial levels. Furthermore they can be transferred with their roots from nursery fields when planted in another location. As development begins, some growth regulators or other substances get into the soil from root excretions. They act as signal substances on permanent forms of the root pathogenic *Actinomycetes* and reactivate them. The hyphae of germinating permanent forms either penetrate the epidermis of the fibrous roots or enter across root hairs into the cortex (**Figure 5**). They injure it to such an extent that entire parts of it disintegrate or die.

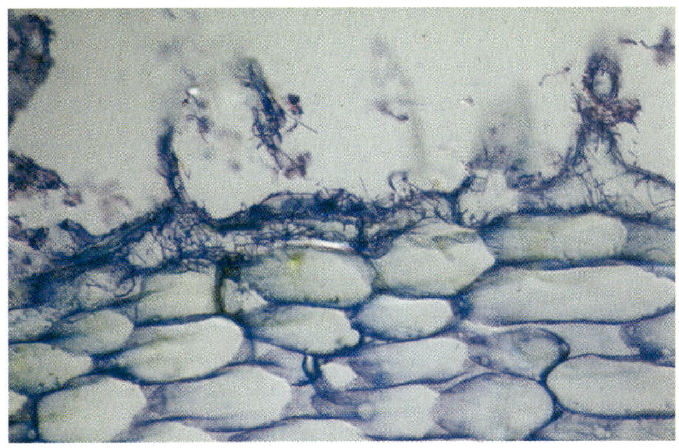

Figure 5 Hair roots and epidermal cells of fibrous roots of an apple seedling damaged by *Actinomycetes*. (Photograph by G. Otto.)

Only when new fine roots could not grow into soil areas without accumulation of *Actinomycetes*, growth and yield start being reduced. This may explain why high-density plantings which have a greater number of trees per hectare show an effect sooner than extensive orchards which only have a few hundred trees per hectare.

It has been suggested that exchanging the soil of the planting hole may overcome replant disease. Any positive results were lost the smaller the volume of exchanged soil had been. Improved performance could only be sustained as long as only the exchanged soil was filled with roots. When new root growth reached the sick soil, the enriched *Actinomycetes* produced a typical growth depression.

These findings have led to a new view on replant disease and the need for new discussions. Replant disease in apple is no longer a 'specific replant disease', but rather a phytopathological problem in the rhizosphere of apple trees. But up to now no conclusive proof has been supplied. Not all authors agree with the *Actinomycetes* theory, sometimes preferring a multifactor complex of fungi like *Phythium* or *Rhizoctonia* with bacteria like pseudomonads and *Pratylenchus*, but investigations could not support this clearly. Investigations over some decades support the *Actinomycetes* theory for central Europe and the United States, but in other continents different organisms could be prevalent.

Soil Sickness/Replant Disease with Various Rosaceae Species

Since it has been established that replant disease with apples is a phytopathological problem, the question arises as to what is the range of hosts for the disease. This question is important because other species of the Rosaceae family are also afflicted by replant disease. Additionally, negative effects have been seen when apples are grown after pears, or rowan after apples. Therefore an experiment was carried out to find out whether fibrous roots of woody Rosaceae plants on an 'apple-sick' soil were carrying *Actinomycetes*. It was noted that fibrous roots of rowan were just as infected as the roots of apple seedlings. Only in pears was the frequency slightly lower than with apples 11 weeks after planting. The appearance in root cortex portions was the same as in apple cortex. In fibrous roots of cherry, plum and rose (*Rosa canina*) no *Actinomycetes* were found in these investigations. This led to the conclusion that replant disease cannot be caused by the same pathogen in these species as with apple, pear and rowan. This confirmed growers' observations that successive planting of pome and stone fruits or vice versa does not result in specific replant disease.

Further investigations were carried out on ornamental woody plants of the Rosaceae family regarding infestation by *Actinomycetes*. These trees or shrubs were planted either on previous apple plots or on plots of the same species. *Actinomycetes* was found in the fibrous roots of all six ornamentals from previous apple plots. The species tested were hawthorn (*Crataegus monogyna*), *Cotoneaster*, Japanese quince (*Chaenomeles* spp.), firethorn (*Pyracantha* spp.), *Spiraea* and rose (*R. glauca*) (**Figure 6**). The frequency of infestation varied, but in *Pyracantha* it was as high as in apple. With the exception of *Spiraea* spp., all other woody ornamental Rosaceae carried *Actinomycetes* when grown in repetition. The frequency of infestation was greater than in fibrous roots in previous apple plantings, except in *Chaenomeles*, whose symptoms corresponded to those in fibrous apple roots. The trials were not large enough to enable safe conclusions to be made about the infestation of fibrous roots of these species and also to correlate the infestation directly with growth reduction.

At any rate, this was the first confirmation of involvement in Rosaceae of *Actinomycetes*, which are known to cause replant disease. The occurrence of *Actinomycetes* in fibrous roots of roses is of particular interest here. In *R. canina* grown on previous apple soil no *Actinomycetes* were found, whereas *R. glauca* carried them on both apple soil and in previous rose plots. The *Actinomycetes* infestation was generally not very great, but it was clearly higher in previous rose soil than in apple plots. This may indicate that different rose rootstock species are afflicted differently, and they may develop resistance to root pathogenic *Actinomycetes*. On the other hand the *Actinomycetes* on different Rosaceae may be different species of this causative agent. Root pathogenic *Actinomycetes* may have preferred host plants – infestation is roughly similar in apple, rowan and *Pyracantha*, and lower in pear and other Rosaceae on previous apple plots. It is not yet possible to state this categorically, as only a limited number of cases have been investigated to date. However, it appears safe to claim that the *Actinomycetes* causing apple replant disease have a host range beyond any *Malus* species, but which does not include all other Rosaceae.

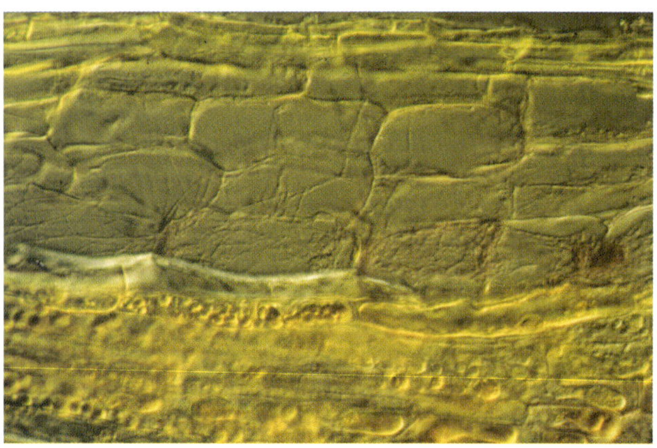

Figure 6 *Actinomycetes* in the root cortex of a fibrous root of *Rosa glauca* from a sick soil. (Reproduced with permission from Otto G. *et al.* (1995) *Zeitschrift für Pflanzenkrankheiten und Pflanzenschutz* 102(6): 599–605.)

What Are the Future Expectations Regarding Replanting of Roses?

There are similarities between apple and rose specific replant disease, although significant proof has not yet been supplied. The existence of *Actinomycetes* in the fibrous roots of *R. glauca* grown on both previous apple and rose soil is a first clue. Because trials have so far been limited, it cannot be ascertained whether infestation of fibrous roots depends on the growth rhythm during the vegetation period in roses as it does in apples. Thus it has not been definitively established whether infestation of fibrous roots occurs in *R. canina* grown on previous rose or apple plots as it does that in one case, in *R. glauca*. The investigation date must be carefully checked. This is also true for microscopic checks of four rose rootstocks: it was only in *R. multiflora* that a certain amount of infestation of fibrous roots by *Actinomycetes* could be established. Perhaps, as with apples, the infection is limited to certain growth periods and therefore would not be detected at every time of investigation. Furthermore, microscopic analysis of the fine structures of *Actinomycetes* in the fibrous roots requires a great deal of knowledge and experience.

Of course, the question whether only one or several species of *Actinomycetes* are involved cannot be resolved. This requires isolation of the *Actinomycetes*, but this has not yet been accomplished even with apples. So postulates could not be proved until now. Presumably *Actinomycetes* are obligatory biotrophic organisms whose isolation and pure culture are very difficult.

The literature does not present definitive statements on the persistence of replant disease and its reversibility. There are no indications about the soil factors which promote the disease or its occurrence even before grubbing in the range of still-developing roots. Finally there has not been any conclusive evidence about the extent of reactions of rootstocks to replanting after rose cultivation. Tests have indicated that there is more than just nematode damage when roses are cultivated repeatedly. Only steam sterilization or fumigation is effective at preventing this with roses – as with apples. This indicates that the cause must also be microorganisms in roses as well. The occasional occurrence of *Actinomycetes* in fibrous roots of roses is an indication of this. However, replant disease in roses possibly may be caused by other factors, such as have been proposed with stone fruit trees. *Actinomycetes* have not been detected in *Prunus mahaleb*, although repeated cultivation results in severe growth depression.

The development of a prognostic system for replant problems would be very important for the commercial rose grower. If root pathogenic *Actinomycetes* are the main cause of replant disease in rose fields, it will be crucial to develop new methods of controlling this organism. Strategies may include the application of antagonistic microorganisms. Using antagonistic plant species for intercropping may also be valuable.

Nematodes as Important Causal Agents of Replant Problems

In most nursery soils different species of nematodes are present (**Figure 7**). In rose nursery plots with replant problems, 50–1500 nematodes, mainly *Pratylenchus penetrans*, were counted in 250 cm^3 soil. Other species observed were *P. crenatus* and *P. neglectus*. The damaging effects of the latter two species, as well as those of the genera *Tylenchorhynchus*, *Paratylenchus*, *Trichodorus* and *Rotylenchus*, have not yet been well documented. A large number of *Paratylenchus* were counted only on aeolic loess soil. Differences between locations give rise to variations in the range of genera. Investigations on a loess field showed differences between rose species regarding the ratios of *Pratylenchus* and *Paratylenchus*. Whereas in *R. canina* 'Pfänders' and *R. corymbifera* 'Laxa' approximately the same proportions of both genera were found, in *R. rugosa* and *R. acicularis* the genus *Paratylenchus* was prevalent (**Table 1**).

One- or two-times rose cropping resulted in increasing *Pratylenchus* populations, while three- or four-times cropping reduced the number of *Pratylenchus* in favour of other nematode genera. The optimum conditions for *Pratylenchus* are found at pH 5.5–5.8, but at pH 6.7 the population decreases considerably. High soil temperatures ($>18\,°C$) or high precipitation levels are reported to have a negative effect on *Pratylenchus* development. The largest proportion of nematodes (about 80%) is found in the upper 30-cm layer, and during the growing season they are found to a depth of 60 cm (**Figure 8**). From April until the summer season varying populations of *Pratylenchus* were counted in the soil surface layer (30 cm): from October they decreased, but at a depth of 60–90 cm nematode density increases, indicating that part of the population retreats to deeper soil layers in the autumn.

In the plant roots *Pratylenchus* could only be found from July onwards. Until October the figures rose to

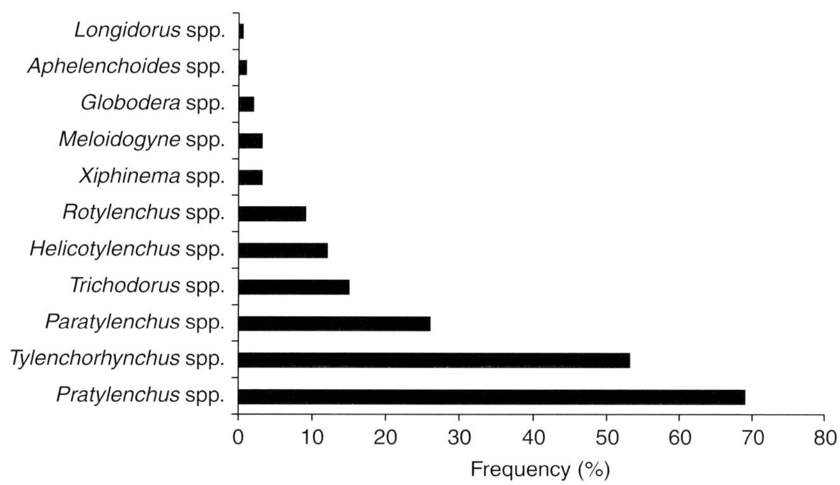

Figure 7 Frequency of nematodes found in nursery soils in the Schleswig–Holstein area (Germany) 1986–1994 ($n = 638$). (Data after Lösing H (1995) PhD thesis, University of Berlin, Germany.)

Table 1 Number of *Pratylenchus* and *Paratylenchus* nematodes (in 250 cm^3) in sandy soil plots in Bönningstedt (Bö) (Schleswig–Holstein) after a culture of *R. corymbifera* 'Laxa' and in loamy loess soil plots in Ruthe (Hanover) after a 3-year culture of some *Rosa* species (Kloczko and Spethmann)

Plot	Pratylenchus *spp.*					Paratylenchus *spp.*				
	7 May	*23 Jun*	*4 Aug*	*10 Sep*	*27 Oct*	*7 May*	*23 Jun*	*4 Aug*	*10 Sep*	*27 Oct*
Bö 1	24	120	30	69	159	0	0	0	0	0
Bö 3	12	30	18	107	54	0	0	0	3	0
Bö 4	6	141	42	27	—	0	0	0	0	0
Bö 5	6	189	48	54	33	0	0	0	0	0
Bö 6	24	0	0	3	3	0	0	0	0	0
Bö 9	72	102	21	237	504	0	0	0	0	0
Bö 10	0	27	21	45	39	6	0	0	0	0
Bö 12	24	246	9	231	63	0	0	0	0	0
Bö 13	6	12	45	3	3	0	0	0	0	0
Bö 15	24	51	39	120	30	6	0	0	51	27
R. canina 'Pfänders'	0	48	3	45	24	0	8	0	3	52
R. corymbifera 'Laxa'	30	60	0	24	36	48	39	0	18	—
R. acicularis	24	9	15	12	3	24	24	0	18	27
R. micrantha	24	18	3	3	36	18	3	0	0	9
R. rugosa	30	3	6	9	3	48	18	66	81	45
R. nitida	30	0	0	—	6	0	18	0	—	0
R. arvensis	0	0	3	0	—	6	0	0	0	—
R. canina	6	15	3	9	—	0	3	0	3	—
R. rubiginosa	18	6	0	6	6	0	3	0	0	6
R. glauca	18	12	3	6	24	0	0	0	0	0

Data from Kloczko M and Spethmann W (1994) Untersuchungen über Schäden durch Nematoden an Rosen in Baumschulen. *Taspo Gartenbaumagazin* 3(5): 10–11.

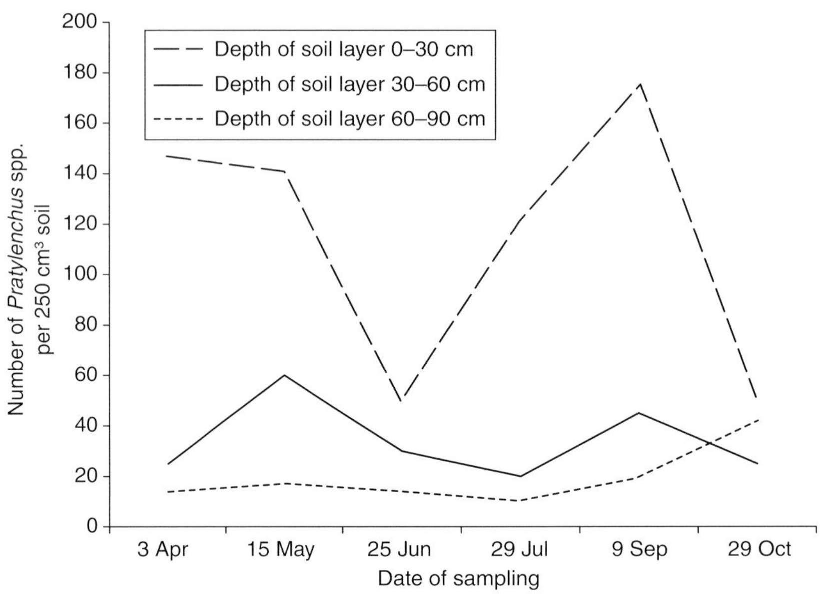

Figure 8 Number of *Pratylenchus* spp. per 250 cm^3 soil dependent on date of sampling and depth of the soil layer. (Unpublished data of M. Kloczko and W. Spethmann.)

a maximum of 4500 nematodes per 5 g roots. As roses are suitable host plants, the number of *Pratylenchus* increased in the successive year. Even after one rose crop, the number may increase to such an extent that damage becomes evident in the following rose crop. The number of nematodes is considerably influenced by the preceding crop (**Figure 9**). Cereals, maize and legumes as a preceding crop increase the number of *Pratylenchus* nematodes. A preceding crop of conifers, deciduous trees or shrubs, e.g., fruit plants, slightly

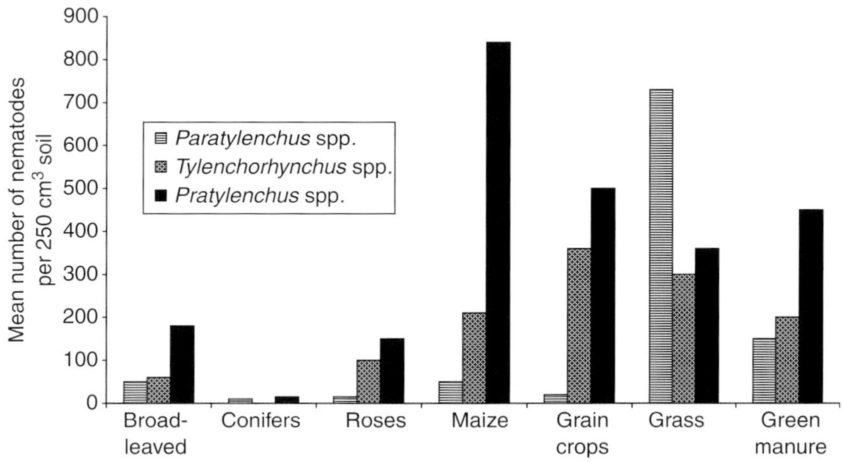

Figure 9 Influence of the preceding crop on the number of *Pratylenchus* spp., *Tylenchorhynchus* spp. and *Paratylenchus* spp. in the Schleswig–Holstein nursery area (Germany) ($n = 506$). (Data after Lösing H (1995) PhD thesis, University of Berlin, Germany.)

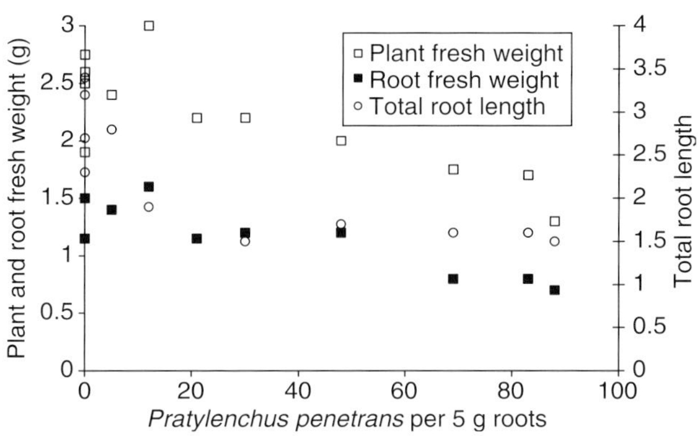

Figure 10 Plant and root fresh weight and total root length of *R. corymbifera* 'Laxa' dependent on concentration of *Pratylenchus penetrans* per 5 g roots. (Unpublished data of M. Kloczko and W. Spethmann.)

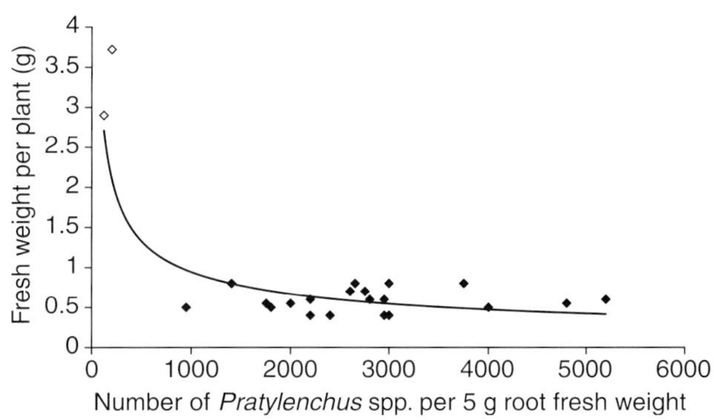

Figure 11 Rose plant fresh weight dependent on the number of *Pratylenchus* in the roots per 5 g root fresh weight (open marks = disinfected plots). (Unpublished data of M. Kloczko and W. Spethmann.)

increases the number. Reduced *Pratylenchus* numbers are observed after a fallow period and a *Tagetes* crop, particularly *T. patula* and *T. erecta*. These species are specifically used to control nematode pests (*see* **Insects and other Animals**: Nematodes).

Visual Nematode Damage

Nematode damage on roses is caused not only by *Pratylenchus penetrans*, but also by *P. vulnus*, *P. thornei* and *Meloidogyne hapla*, and possibly also by *P. crenatus* and *P. neglecta* (in greenhouses). Experimental data on their damaging effects are only available on *P. penetrans*.

A comparison of roses on nematode locations with roses from 'clean' plots shows reduced growth. Localized, nest-like reduced growth or blanks are usually typical of nematodes. Rosette shoot formation, i.e. decreasing length of internodes, reduced leaf number, reduction in root length and fresh weight of plants, are typical characteristics (**Figure 10**). The germination of rose seed is often inhibited. The extent of the damage increases with increasing numbers of nematodes: however, damage remains constant with 1000 nematodes per 5 g roots (**Figure 11**).

It remains to be seen whether other growth-inhibiting factors supplement the nematode effects. To find out whether that is the case, inoculations were carried out in a greenhouse. The materials chosen were *Pratylenchus* isolated from *R. corymbifera* 'Laxa' roots which were mixed into the TKS substrate (peat and sand mixture) at rates of 0–1500 nematodes in 250 g substrate. Two-week-old seedlings of *R. micrantha*, *R. obtusifolia* and *R. tomentosa* were planted, and the presence of *Pratylenchus* resulted in an infection in all cases. Increasing inoculum densities caused increasing root infection, depending on the species. In *R. micrantha* almost 800 nematodes per 5 g roots were counted, but in *R. obtusifolia* only 200 could be found (**Figure 12**).

The damage depended on inoculum density, as could be shown by inoculation of *R. corymbifera* 'Laxa' (**Figure 13**). Growth reductions became more severe between 30 and 300 nematodes per 5 g, but above that level, only slightly more damage was observed. As regards the number of nematodes in the substrate, damage only increased at rates between 50 and 750 nematodes per 250 g soil. This confirms results obtained in the field. The threshold of definitive damage is 250 nematodes in 250 g soil; the critical threshold is seen to be 700–800 nematodes. Depending on where the investigation is carried out, these figures are lower or, in a few cases, higher.

The results confirm that different rose species have different susceptibility to nematodes. The search for resistant rose species or rootstocks therefore appears promising. In the survey on replant problems the most serious damage was observed with *R. corymbifera* 'Laxa' and *R. canina* 'Pfänders', while minimal damage was seen with *R. canina* 'Inermis', 'Pollmers' and *R. multiflora*.

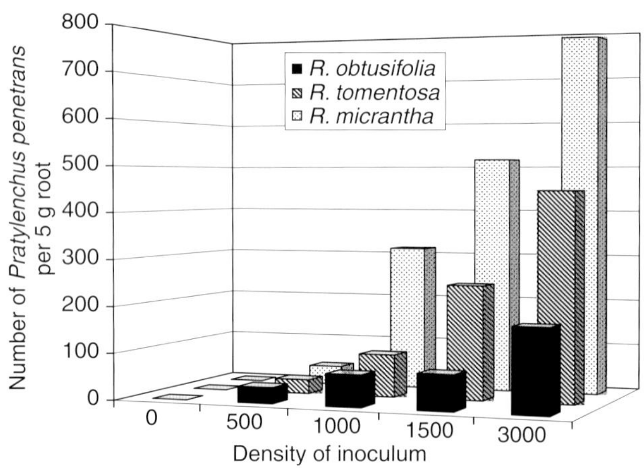

Figure 12 Number of *Pratylenchus* per 5 g root fresh weight dependent on the density of the inoculum (*Pratylenchus penetrans*) per 500 g substrate. (Data after Spethmann W (1993) *Rosenjahrbuch Verein Deutscher Rosenfreunde* 1993: 67–82.)

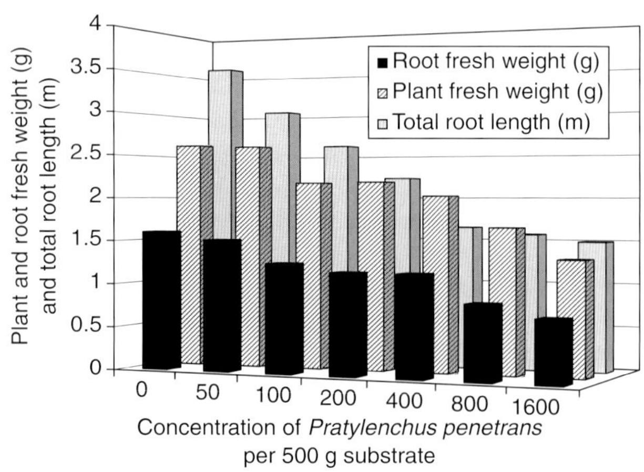

Figure 13 Plant and root fresh weight and total root length of *R. corymbifera* 'Laxa' dependent on concentration of *Pratylenchus penetrans* per 500 g substrate. (Unpublished data of M. Kloczko and W. Spethmann.)

Demarcation of Replant Disease and Nematode Damage

Since for places showing replant problems it could not been determined whether the replant problems are caused by replant disease, nematodes, or both, various tests have been developed. One such test, developed for apples, is direct evidence comparing growth in treated and untreated soil. Shoot growth in untreated soil was compared with that in steamed soil or in soil fumigated with Dazomet and Metham-sodium compounds. Evidence of exclusive nematode damage can be obtained by comparison with an aldicarb nematicide-treated soil. The growth difference indicates how serious the replant disease is.

Based on the tests developed for apple, a test for *Rosa* was introduced which helps to estimate the proportions of the two causes – replant disease and nematodes. In this test *R. corymbifera* 'Laxa' seedling growth is assessed in untreated soil, soil partially heat-treated at 50 °C, and soil fully heat-treated at 100 °C. Since nematodes are destroyed at 45–50 °C, the comparison of a partially treated soil and an untreated soil demonstrates how much damage is caused by factors other than nematodes. On the average of several tests, the 50 °C steaming results in a growth increment of 45–90% compared to untreated soil; after full treatment at 100 °C the biomass increment is around 300% higher (**Figure 14** and **Table 2**). This demonstrates that replant disease is an important factor in addition to nematode injury in the development

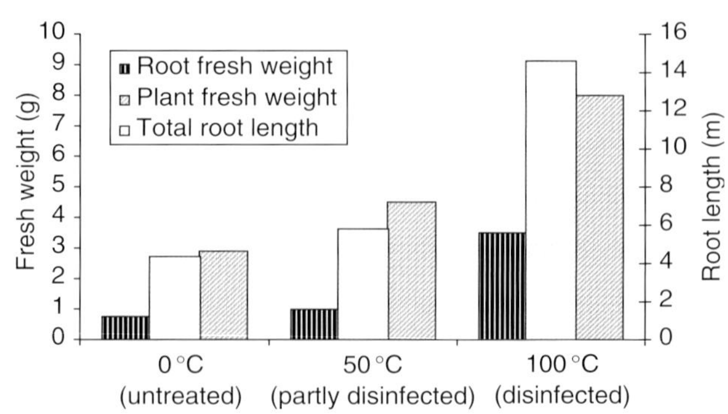

Figure 14 Plant and root fresh weight and total root length of *R. corymbifera* 'Laxa' dependent on part and total disinfection (mean of 15 plots). (Data after Kloczko M and Spethmann W (1994) *Taspo Gartenbaumagazin* 3(5): 10–11.)

of replant problems. Another test, carried out with *Rosa* raised *in vitro* ('The Fairy' and 'Komet') is based on the same principle. The *in vitro* plants are grown in the respective substrates and growth is compared after 4 weeks.

Measures to Control Nematodes

Treatments with nematicides (e.g. Temik 5 G) or fumigants (Basamid) result in a clear reduction in the nematode population (**Figure 15**). The increasing withdrawal of pesticides has led to a search for alternative methods. Inserting a complete fallow period in the crop rotation could considerably reduce the nematode count. Steaming at >50 °C is quite effective, but is expensive. As mentioned above, inserting a crop of *Tagetes patula* and *T. erecta* (10 kg ha^{-1} of seed) reduces the nematode population and is already practiced over large surface areas (**Figure 16**). The problem here is the slow initial development which allows weeds to flourish in competition with the *Tagetes*. Herbicides or mechanical weeding is indispensable. With *T. erecta* pre- and postemergence treatments with Afalon and Goltix WG are compatible with roses; in pre-emergence Tenoran, Bandur or Stomp SC is effective. For *T. patula* Goltix WG is considered to be compatible. The cultivar *T. minuta* 'Nemanon', however, produces only minimal nematicidal effect, and *Phacelia* is likewise unsuitable as a crop before roses. Some authors claim that sunflowers have a positive effect, while others refute this.

It makes sense to use rootstocks that suffer less damage. Of the two main rootstocks, 'Laxa' and 'Inermis', the latter shows less damage, particularly its small-fruited selection. 'Heinsohns Rekord' also produces minimal damage symptoms. Growth reduction has not been found with *R. multiflora* 'Stachellos' and *R. carolina*. The search for resistant species needs to be intensified to provide basic material for breeding nematode-resistant rootstocks.

Regarding the cause of specific replant disease in roses many questions remain. Will further investigations give more evidence for the contribution of *Actinomycetes*? Are other organisms like fungi concerned? Are nematodes a separate phytopathological problem, or are they a contributing part of

Table 2 Percentage of reduction of plant and root fresh weight and total root length by nematodes and specific soil sickness (100% = total reduction compared with plants growing in disinfected soil)

Soil of nursery plots	*Reduction of plant fresh weight (%)*		*Reduction of root fresh weight (%)*		*Reduction of total root length (%)*	
	Nematodes	*Soil sickness*	*Nematodes*	*Soil sickness*	*Nematodes*	*Soil sickness*
Bönningstedt	39	61	30	70	16	84
Klei	43	57	45	54	38	62
Mehlen	32	68	47	53	40	60
Ruthe (loess)	46	54	39	61	49	51

Data from Kloczko M and Spethmann W (1994) Untersuchungen über Schäden durch Nematoden an Rosen in Baumschulen. *Taspo Gartenbaumagazin* 3(5): 10–11.

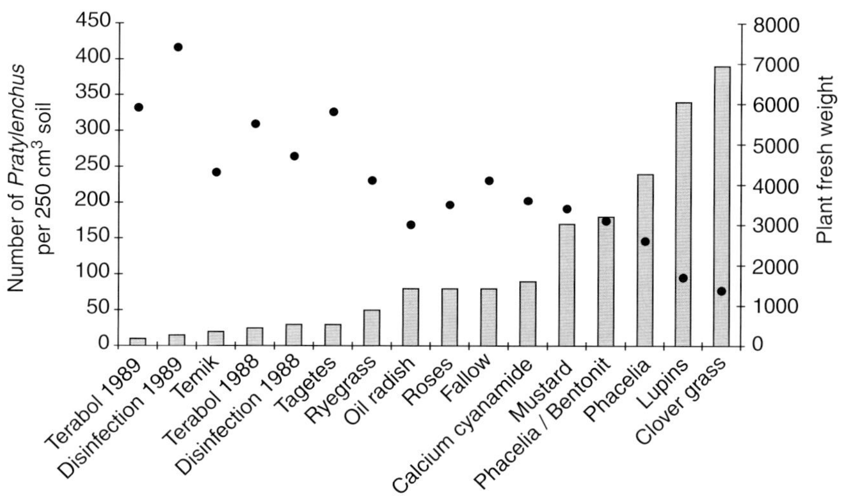

Figure 15 Number of *Pratylenchus* per 250 cm^3 soil (shaded bars) and rose plants fresh weight (g) (solid circles) of the total plot after different disinfection treatments or cash crops. (Data modified after Strassburger T (1992) PhD thesis, University of Kiel, Germany.)

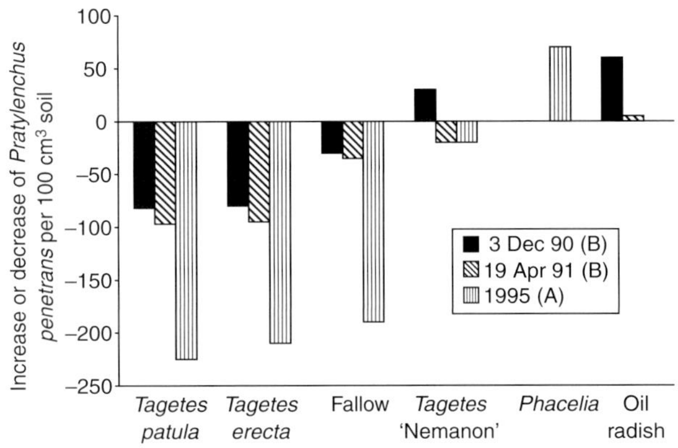

Figure 16 Increase or decrease of *Pratylenchus penetrans* (per 100 cm^3 soil) in percent after different cultures of fallow. (Data after Lösing H (1995) PhD thesis, University of Berlin, Germany (A) and unpublished data of M. Kloczko and W. Spethmann (B).)

a multifactor complex together with bacteria or fungi, maybe as vectors? Many investigations must be undertaken to clear up the problem.

See also: **Breeding**: Selection Strategies for Disease and Pest Resistance. **Disease**: Bactericides and Fungicides. **Insects and other Animals**: Nematodes. **Integrated Pest Management**. **Morphology and Anatomy**: Roots. **Rootstocks**: Usage of Rootstocks.

Further Reading

Dauck H and Spethmann W (1991) Der Bodenmüdigkeit auf der Spur: Eine bundesweite Umfrage zu Nachbauproblemen bei Rosen. *Deutsche Baumschule* 43(4): 164–167.

Dressler H (1997) *Untersuchungen zur Besiedlung von Rosenwurzeln durch Bakterien im Hinblick auf Bodenmüdigkeit.* PhD Thesis, University of Hanover, Germany.

Hoestra H (1968) Replant diseases of apple in the Netherlands. *Mededelingen Landbouwhogeschool* 68: 13.

Krüssmann G (1997) *Die Baumschule.* Berlin: Paul Parey Verlag.

Lösing H (1995) *Bedeutung und integrierte Bekämpfung wandernder Nematoden* (Pratylenchus *spp.*) *als eine Ursache von Nachbauschäden bei Rosen in der Baumschulproduktion.* PhD Thesis, University of Berlin, Germany.

Maeseneer de J (1967) *Biologie en Bestrijding van* Pratylenchus penetrans *en* P. crenatus. PhD Thesis, Rijksfaculteit der Landbouwwetenschapen, Ghent, Belgium.

Mazzola M and Gu Yu Huan (2000) Phyto-management of microbial community structure to enhance growth of apple in replant soils. *Acta Horticulturae* 532: 73–78.

Oostenbrink M (1961) Nematode damage and 'specific sickness' in *Rosa, Malus* and *Laburnum. Tijdschrift over Planteziekten* 67: 264–272.

Otto G (2000) Bodenmüdigkeit. In: Friedrich G and Fischer M (eds.) *Physiologische Grundlagen des Obstbaues*, 3rd edn, pp. 281–296. Stuttgart, Germany: Eugen Ulmer.

Otto G and Winkler H (1993) Soil fumigation for controlling replant problems in apple orchards. *Acta Horticulturae* 324: 53–59.

Otto G and Winkler H (1998) Influence of root pathogenic actinomycetes on the trimming of the rootlets of some species of *Rosaceae* with root hairs. *Acta Horticulturae* 477: 49–54.

Otto G, Winkler H and Szabó K (1993a) Influence of growth regulators on the infection of rootlets of apple seedlings in SARD soils by actinomycetes. *Acta Horticulturae* 363: 101–107.

Otto G, Winkler H and Szabó K (1994) Proof of actinomycetes in rootlets of species of *Rosaceae* from a SARD soil: a contribution to the specificity of replant diseases. *Acta Horticulturae* 363: 43–48.

Savory BM (1966) *Specific Replant Diseases.* Farnham, UK: Commonwealth Agricultural Bureaux.

Spethmann W and Dauck H (1992) Forschungskonzepte zum Nachbauproblem bei Rosen. *Gartenbaumagazin* 1(7): 18–20.

Strassburger T (1992) *Alternative Verfahren zur Beseitigung von Nachbauproblemen in Baumschulen.* PhD thesis, University of Keil, Germany.

Szabó K, Winkler H, Petzold H and Marwitz R (1998) Evidence for the pathogenicity of *Actinomycetes* in rootlets of apple seedlings from soils conductive to specific apple replant disease. *Acta Horticulturae* 477: 55–65.

Tagliavini M, Utkhede R and Zucconi F (eds.) (1990) Symposium on soil sickness and replant diseases in fruit trees. *Acta Horticulturae* 324: 1–109.

Viruses

M S Szyndel, Warsaw Agricultural University, Poland

Introduction

Plant viruses are infectious, intracellular, submicroscopic, spherical or rod-shaped particles. They are composed of only a single type of nucleic acid surrounded by protein coat. The parasitism of viruses occurs at the molecular, intracellular level, where virus particles copy themselves using the replicative system of a plant cell. Most plant viruses may induce a wide variety of disease symptoms although infection may sometimes be symptomless. Viruses, as infectious agents, can be transmitted from diseased to healthy plants in a variety of ways, mainly during vegetative propagation or grafting of host plants and/or can be transmitted by natural vectors such as aphids, mites and nematodes. Like all vegetatively propagated plants, roses are excellent suscepts to infection by viruses and viral diseases of roses occur throughout the world wherever roses are grown.

Many virus and 'virus-like' diseases of roses were described during the last 80 years. Many different viruses have been suspected to be associated with the pathogenesis of those diseases. Up to the early 1970s investigations mainly pertained to the symptomatology of rose viral diseases. Since the technology was not readily available to purify and maintain the viral pathogens, available viral specimens are sparse today for comparison study of infectious agents. Moreover, differences in symptom expression by rose plants to virus infection may depend on numerous factors, including virus species, virulence of virus strain, presence of unrecognized virus, rose cultivars and natural or greenhouse environment (i.e. temperature, light). For this reason an educated or guess speculation can aid researchers with the classification placement of some virus-like disorders of roses. The results obtained with using very specific, sensitive and efficient virological techniques have indicated that currently most common viruses infecting *Rosa* spp. belong to Ilarviruses: *Prunus* necrotic ringspot virus, apple mosaic virus and Nepoviruses: arabis mosaic virus, strawberry latent ringspot virus. This article does not provide detailed characterization of viruses infecting roses because they are not specific to *Rosa* spp. alone, and these complex viral characterizations are found in numerous plant virus descriptions or virus taxonomy books. The main purpose of this article is to present the rose virosis (the viral diseases) and their significance in rose gardening or rose production.

Representative virus and virus-like rose diseases, described in the literature, are listed in **Table 1**.

Rose Mosaic Disease

Although symptoms caused by many plant viruses can be specifically associated with the viral causal agent, mosaics, including rose mosaic, are complex and highly variable in symptom expression. The most characteristic symptoms are prominent chlorotic areas commonly adjacent to the midrib of the leaflet or leaf, greyish-yellow areas with green islands, mottles, chlorotic line patterns (**Figure 1**) consisting of defined or partly defined rings (**Figure 2**), erratic wavy lines (narrow or broad bands), watermarks (fine greyish lines in the leaf surface) and oak-leaf patterns. Occasionally chlorosis is confined to one side of the leaf and the unequal growth of leaf margins results in leaf distortion. Typical symptoms include yellow mosaic with very bright and yellow chlorotic areas, chlorotic patching, vein banding (yellow bands along the vein) and yellow-net (**Figure 3**) symptoms, when the entire vein network may be affected. Infected roses can be dwarfed with bleached and damaged buds. They may suffer more severe winter dieback than plants free of rose mosaic. Not all leaflets or leaves on an infected plant express symptoms.

Rose mosaic is an old disease, which occurs in roses planted in landscapes, rose gardens, nurseries, commercial fields and even in greenhouse production. This disease was first described by R.P. White in 1928 as an infectious chlorosis of rose. The unknown causal agent, transmissible only by shoot or buds grafts, was designated as *Rosa* virus 1. The yellow variety of White's 'ordinary rose mosaic' was characterized as rose yellow mosaic caused by *Rosa* virus 2. The disease named rose ring pattern (RRP) on hybrid tea roses exhibits symptoms which are very similar to mosaic virus: ring pattern, fine line pattern, chlorotic flecking on the leaves and colour-break rings in petals. However, there was no necrotic reaction and gumming around the graft side after graft transmission on *Prunus serrulata* 'Shirofugen' (flowering cherry) used as an indicator plant for indexing rose mosaic. A very strong symptomatic reaction occurred however on *Rosa multiflora* Burr. This thornless rootstock is a reliable indicator for RRP. Symptoms in greenhouse-maintained plants remain constant and include severe stunting, deformity, rugosity and mottling on leaflets. Other rootstocks remain symptomless. RRP disease was designated a component of the rose mosaic complex. Similar results were reported with rose X disease on commercial roses in California and Oregon in 1977. The disease appears to be identical to RRP and is presently recognized as part of the RRP syndrome.

Another disease considered to be a form of rose mosaic is rose line pattern. Rose line pattern virus was associated with yellow banding of leaves of greenhouse-grown rose bushes. Clearly defined wavy yellow bands sometimes conforming to oak-leaf patterns were observed on young red leaves. Stem necrosis occurred directly beneath the developing flower buds, causing their death before flower opening. Excessive growth of lateral buds strongly reduced the vigour of affected plants. Some differences in host range, symptomatology and effectiveness of graft transmission yielded the name yellow vein mosaic on the rootstock *Rosa bourbiana*. On leaf veins and veinlets different symptoms occurred, including light clearing to strong vein yellowing or finally yellow-net vein system. Disease symptoms on 'Talisman' also included a slight reduction in flower size followed

Table 1 Virus and virus-like diseases of roses

Disease or Virus	*Main disease symptoms*	*Geographical distribution*	*Transmission*	*References (see footnote)*
Apple mosaic virus	Distortion of leaflets with chlorotic mottles, mosaics, bright chlorotic rings, patches, puckering leaflets, shoots stunting	Worldwide	Grafting	Fulton (1968, 1972); Thomas (1981, 1984b); Wong *et al.* (1988); Wong and Horst (1993)
Arabis mosaic virus	Chlorotic vein banding, vein netting, vein mottling, chlorotic ringspots, reduction in plant vigour	UK	Grafting, nematodes	Cammack (1966); Thomas (1981, 1982b, 1984a)
Frisure, curly head	Proliferation of short shoots, distortion of leaflets (rosette), chlorosis, dieback of plants	France	Grafting	Cardin and Devergne (1971); Devergne and Goujon (1975)
Infectious chlorosis (*Rosa* virus 1)	Yellowing, vein clearing, dwarfing, distortion of leaflets	Worldwide	Grafting	White (1928, 1930, 1932); Smith (1937)
Little leaf	Bud and shoot proliferation, shoot dwarfing, severe leaf size reduction, shoot dieback, recovery possible	South Africa	Grafting	Meyer (1960)
Prunus necrotic ringspot virus	Fine or broad chlorotic rings, lines, patterns, mottles, occasionally leaf distortion	Worldwide	Grafting	Cochran (1950); Kirkpatrick *et al.* (1962); Traylor *et al.* (1966); Thomas (1981, 1982a, 1984a, b); Moury *et al.* (2001)
Rose bud proliferation, curly head (Kroeskoppen)	Bud, shoot proliferation into tiny witches' brooms, plant deformations, plants deteriorate, desiccate, seldom plants recover	Netherlands	Grafting	Bos and Perguin (1975)
Rose colour break	Flower fails to open normally, flower distortion, crimpling of petals, colour breaking	New Zealand, UK	Budding	Hunter (1966); Farrar and Frost (1972)
Rose cryptic virus	Symptomless	USA	Seeds	Handley (1986)
Rose dieback (*Rosa* virus 3)	Recurved appearance of the leaflets forming a ball, defoliation of pale green or yellowish leaflets, dieback, sometimes temporary recovery occurred	USA, UK	Grafting	Grieve (1931); Cheo (1970); Hollings *et al.* (1974); Thomas (1975, 1976, 1980)
Rose flower break	Flower breaking, marginal petal distortion	New Zealand, UK	Grafting	Ikin and Frost (1974)
Rose flower proliferation	Flower proliferation, phyllody	Italy, UK	No data	Gualaccini (1963)
Rose leaf curl	Epinasty, size reduction of young leaves, yellow flecking of veins, premature defoliation in spring	USA	Grafting	Cheo (1970); Gumpf and Weathers (1974); Slack *et al.* (1976a)
Rose line pattern virus	Yellow bands, oak-leaf patterns, stem necrosis, premature bud death	USA	No data	Fletcher and Kingham (1962)
Rose mosaic	Foliar mosaics, mottling, yellow line patterns, vein banding, yellow nets, ring spots, chlorotic patches	Worldwide	Grafting	Weiss and McWhorter (1930); McWhorter (1931); Brierley and Smith (1940); Fulton (1952); Hunter (1965); Bos (1976); Thomas (1984b); Wong *et al.* (1988)
Rose mosaic virus (*Rosa* virus 1)	Complex of viruses caused rose mosaic diseases	Worldwide	Grafting	Smith (1937); Fulton (1952, 1967); Fry and Hunter (1956); Hunter (1965); Basit and Francki (1970)

Table 1 Continued

Rose mottle mosaic virus	Green mottling on *Rosa rugosa* plants	USA	Grafting	Kirkpatrick *et al.* (1968)
Rose ring pattern	Rings, yellowing, fine line patterns, chlorotic flecking on the leaves, colour-break rings in petals	USA	Grafting	Secor and Nyland (1978)
Rose rosette	Redding and crinkling of leaflets, witches' brooming, phyllody	USA	Grafting, mites	Thomas and Scott (1953); Amrine and Hindal (1988); Epstein and Hill (1995)
Rose spring dwarf	Growth of balled or rosetted leaves in spring, vein clearing, reduced size	USA	Grafting	Traylor *et al.* (1971); Slack *et al.* (1976b); Secor *et al.* (1977)
Rose streak (rose streak virus, *Rosa* virus 4)	Green-brown ring patterns, brown vein banding, defoliation	USA, UK, Germany, South Africa	Grafting	Brierley (1935); Smith (1937); Brierley and Smith (1940); Schmelzer (1967); Klesser (1967); Ikin and Frost (1974)
Rose stunt	Proliferation of small stunted shoots, shoots 'blinding', growth ceasing	UK	Grafting	Ikin and Frost (1974); Hollings *et al.* (1973, 1974); Thomas (1975)
Rose tobamovirus	Colour-break of petals	UK	Mechanical	Hicks and Frost (1984)
Rose wilt (*Rosa* virus 3)	Downward leafroll, vein clearing, shoot proliferation, premature defoliation, terminal dieback	Australia, UK, New Zealand, South Africa	Grafting	Grieve (1931); Smith (1937); Meyer (1960); Fry and Hammett (1971); Hammett (1971); Hollings *et al.* (1973, 1974); Gumpf and Weathers (1974); Thomas (1975, 1976, 1980)
Rose X	Stunting, leaflet deformation, green mottling, fine line patterns, ring spots	USA	Grafting	Secor *et al.* (1977)
Strawberry latent ringspot virus	Reduction in size of shoots and leaves; loss of vigour, leaf malformation	Europe – UK	Grafting, nematodes	Cammack (1966); Sweet (1975a); Ikin and Frost (1976); Thomas (1982b, 1984a)
Tobacco necrosis virus	Virus found in rose nursery soils and isolated from roses with 'frisure' disorder	France, UK	Fungi	Cardin and Devergne (1971); Devergne and Goujon (1975); Sweet (1975b)
Tobacco ringspot virus	Pale green or yellow wavy lines, broad band line, ringspots and blotches	USA	Grafting, mechanical	McDaniel *et al.* (1971)
Tobacco streak virus	Symptomless or chlorotic patches, chlorosis, twisting of leaves, shoot stunting	USA	Grafting, mechanical	Fulton (1970); Converse and Bartlett (1979)
Tomato ringspot virus	Mosaics on the leaves	USA	Grafting, nematodes	Halliwell and Milbrath (1962)
Witches' broom of rose	Stem elongation, breaking axillary buds, leaflet deformation, wrinkling, bright red pigmentation, phyllody, increasing succulent thorniness	USA	Budding	Viehmeyer (1961); Crowe (1984)

Continued

Table 1 Continued

Disease or Virus	*Main disease symptoms*	*Geographical distribution*	*Transmission*	*References (see footnote)*
Yellow mosaic virus (*Rosa* virus 2)	Chlorosis, brighter and lighter yellow than the ordinary rose mosaic	USA, Germany, Italy	No data	Brierley (1935); Smith (1937); Gualaccini (1963); Schmelzer (1966)
Yellow vein mosaic	Yellow net, vein yellowing	India	Cutting	Sastry (1966)

Amrine JW Jr and Hindal DF (1988) Rose rosette: a fatal disease of multiflora rose. *West Virginia University Agricultural Forestry Experimental Station Morgantown*, Circular 147, August; Basit AA and Francki RJB (1970) Some properties of rose mosaic virus from South Australia. *Australian Journal of Biological Science* 23: 1197–1206; Bos L (1976) Symptom expression and variation of rose mosaic. *Netherlands Journal of Plant Pathology* 82: 239–249; Bos L and Perguin FW (1975) Rose bud proliferation, a disorder of still unknown aetiology. *Netherlands Journal of Plant Pathology* 81: 187–198; Brierley P (1935) Symptoms of rose mosaic. *Phytopathology* 25: 8; Brierley P and Smith FF (1940) Mosaic and streak disease of rose. *Journal of Agricultural Research* 61: 625–660; Cammack RH (1966) Soil-borne viruses in rose. *Plant Pathology* 15: 47–48; Cardin L and Devergne JC (1971) The isolation of prunus necrotic ringspot and tobacco necrosis virus from roses. *Annales de Phytopathologie* 3: 539–540; Cheo P (1970) Rose wilt or dieback. *Lasca Leaves* 20: 88–89; Cochran LC (1950) Infection of apple and rose with the ringspot virus. *Phytopathology* 40: 964; Converse RH and Bartlett AB (1979) Occurrence of viruses in some wild *Rubus* and *Rosa* species in Oregon USA. *Plant Disease Reporter* 63: 441–444; Crowe FJ (1984) Witches' broom: a new outbreak in several central states. *American Rose Annals* 69: 83–87; Devergne JC and Goujon C (1975) Studies on growth abnormalities on roses grown in glasshouses. *Annales de Phytopathologie* 7: 71–79; Epstein AH and Hill JH (1995) The biology of rose rosette disease: a mite-associated disease of uncertain aetiology. *Journal of Phytopathology* 143: 353–360; Farrar EH and Frost RR (1972) A new disease of roses in Britain. *Plant Pathology* 21: 97; Fletcher JT and Kingham HG (1962) Rose line pattern virus. *Plant Pathology* 11: 92; Fry PR and Hammett KRW (1971) Rose wilt virus in New Zealand. *New Zealand Journal of Agricultural Research* 14: 735–743; Fry PR and Hunter JA (1956) Rose mosaic virus in New Zealand. *New Zealand Journal of Science and Technology Section A* 37A: 478–482; Fulton RO (1952) Mechanical transmission and properties of rose mosaic virus. *Phytopathology* 42: 413–416; Fulton RO (1967) Purification and serology of rose mosaic virus. *Phytopathology* 57: 1197–1201; Fulton RO (1968) Serology of some viruses isolated from cherry necrotic ringspot, plum line pattern, rose mosaic and apple mosaic. *Phytopathology* 58: 635–638; Fulton RO (1970) A disease of rose caused by tobacco streak virus. *Plant Disease Reporter* 54: 949–951; Fulton RO (1972) Apple mosaic virus. *C.M.I./A.A.B. Description of Plant Viruses* no. 83; Grieve BJ (1931) 'Rose wilt' and 'dieback'. A virus disease of roses occurring in Australia. *Australian Journal of Experimental Biology Medical Science* 8: 107–121; Gualaccini F (1963) Singolare vinescenza, prolungamento dell'asse fiorale e anomalie forliari su Rosa ottenuta da seme di pianta affetta da'mosdaico giallo ('rose yellow mosaic'). *Boll. Staz. Pat. Veg. Roma.* 21: 45–46; Gumpf DJ and Weathers LG (1974) Detection of a new infectious disease of roses in California. *Acta Horticulturae* 36: 53–57; Halliwell RS and Milbrath JA (1962) Isolation and identification of tomato ringspot virus associated with rose plants and rose mosaic virus. *Plant Disease Reporter* 46: 555–557; Hammett KRW (1971) Symptom differences between rose wilt virus and *Verticillium* wilt of rose. *Plant Disease Reporter* 55: 916–920; Handley MK (1986) *Identification and Partial Characterization of Two Cryptic Viruses in Rose by Analysis of Double-stranded RNA Content.* PhD thesis, University of California, Davis; Hicks RGT and Frost RR (1984) Isolation of tobamovirus from rose. *Plant Pathology* 33: 581–587; Hollings M, Stone OM, Brunt AA, Atkey PT and Fry PR (1973) Rose *Rosa* spp.; rose wilt. *Glasshouse Crops Research Institute Annual Report for 1972*: 105; Hollings M, Stone OM, Atkey PT and Pawley RR (1974) Rose *Rosa* spp.; Rose dieback (rose wilt). *Glasshouse Crops Research Institute Annual Report for 1973*: 118; Hunter JA (1965) Further studies of rose mosaic virus in New Zealand. *New Zealand Journal of Agricultural Research* 8: 327–335; Hunter JA (1966) Rose colour break: a virus disease of roses. *New Zealand Journal of Agricultural Research* 9: 1070–1072; Ikin R and Frost RR (1974) Virus disease of roses. Part I. Their occurrence in the United Kingdom. *Phytopathologische Zeitschrift* 79: 160–168; Ikin R and Frost RR (1976) Virus diseases of roses. Part II. Strawberry latent ringspot virus R/1 : 2.6 or 2 × 1.6/ 38 : S/S : S/Ne. *Phytopathologische Zeitschrift* 87: 205–223; Kirkpatrick HC, Lindner RC, Cheney PW and Graham SO (1962) Rose as a source of prunus ringspot virus. *Plant Disease Reporter* 46: 722–723; Kirkpatrick HC, Lindner RC and Cheney PW (1968) Symptomless infection of stone fruits with mechanically transmissible virus from rose. *Plant Disease Reporter* 52: 449–502; Klesser PJ (1967) Occurrence of rose streak virus in South Africa. *South African Journal of Agricultural Science* 10: 849–856; McDaniel GL, Buck GJ and Ford RE (1971) Isolation of tobacco ringspot virus from rose. *Phytopathology* 61: 45–49; McWhorter FP (1931) Further report on rose mosaic in Oregon. *Plant Disease Reporter* 15: 1–3; Meyer MA (1960) Virus diseases of roses in South Africa. *South African Journal of Agricultural Science* 3: 467–471; Moury B, Cardin L, Onesto JP, Candresse T and Poupet A (2001) Survey of prunus necrotic ringspot virus in rose and the variability in rose and *Prunus* species. *Phytopathology* 91: 84–91; Sastry KSM (1966) Yellow vein mosaic of *Rosa bourbiana. Indian Phytopathology* 19: 316–317; Schmelzer K (1966) Zur Atiologie des Rosengelbmosaiks. *Zentrblatt Bakt. Parasitkde Abt.* 120: 233–236; Schmelzer K (1967) Die Strichelkrankheit der Rose (rose streak) in Europa. *Phytopathologische Zeitschrift* 58: 92–95; Secor GH and Nyland G (1978) Rose ring pattern. A component of the rose mosaic complex. *Phytopathology* 68: 1005–1010; Secor GA, Kong M and Nyland G (1977) Rose virus and virus-like diseases. *California Agriculture* 31 (3): 4–7; Slack SA, Traylor JA, Williams HE and Nyland G (1976a) Rose leaf curl, a distinct component of a disease complex which resembles rose wilt. *Plant Disease Reporter* 60: 178–182; Slack SA, Traylor JA, Nyland G and Williams HE (1976b) Symptoms, indexing and transmission of rose spring dwarf disease. *Plant Disease Reporter* 60: 183–187; Smith KM (1937) Rosa virus 1–4. In: *A Texbook of Plant Virus Diseases*, pp. 152–156. P Blakiston's; Sweet JB (1975a) Strawberry latent ringspot virus in glasshouse roses. *Plant Pathology* 24: 93–96; Sweet JB (1975b) Soil-borne viruses occurring in nursery soils and infecting some ornamental species of Rosaceae. *Annals of Applied Biology* 79: 49–54; Thomas BJ (1975) Rose *Rosa* spp.; rose dieback (rose wilt). *Glasshouse Crops Res. Inst. Ann. Rep. for 1974*: 118–119; Thomas BJ (1976) Rose *Rosa* spp.; rose dieback. *Glasshouse Crops Res. Inst. Ann. Rep. for 1975*: 123; Thomas BJ (1980) Some degeneration and dieback diseases of the rose. *Glasshouse Crops Res. Inst. Ann. Rep. for 1979*: 178–190; Thomas BJ (1981) Studies on rose mosaic disease in field grown roses produced in United Kingdom. *Annals of Applied Biology* 98: 419–429; Thomas BJ (1982a) The effects of prunus necrotic ringspot virus on field grown roses. *Annals of Applied Biology* 100: 129–134; Thomas BJ (1982b) Rose – *Rosa* spp. The effect of arabis mosaic and strawberry latent ringspot viruses on growth and flower quality. Seed transmission in *Rosa* spp. Detection viruses in new varieties. *Glasshouse Crops Res. Inst. Ann. Rep. for 1981*: 143–145; Thomas BJ (1984a) Epidemiology of three viruses infecting the rose in the United Kingdom. *Annals of Applied Biology* 105: 213–222; Thomas BJ (1984b) Rose mosaic disease: symptoms induced in roses by graft inoculation with both prunus necrotic ringspot virus and apple mosaic virus. *Plant Pathology* 33:155–160; Thomas HE and Scott CE (1953) Rosette of rose. *Phytopathology* 43: 218–219; Traylor JA, Williams HE and Nyland G (1966) Symptoms caused by strains of prunus necrotic ringspot virus in rose resemble typical rose mosaic. *Phytopathology* 56: 152; Traylor JA, Wagnon HK and Williams E (1971) Rose spring dwarf, a graft-transmissible disease. *Plant Disease Reporter* 55: 294; Viehmeyer G (1961) A 'new' disease of roses. *American Rose Annals* 46: 98–101; Weiss F and McWhorter FP (1930) Pacific coast survey for rose mosaic. *Plant Disease Reporter* 14: 203–205; White RP (1928) An infectious chlorosis of roses. *Plant Disease Reporter* 12: 33–34; White RP (1930) An infectious chlorosis of rose. *Phytopathology* 20: 130; White RP (1932) Chloroses of the rose. *Phytopathology* 22: 53–69; Wong SM and Horst RK (1993) Purification and characterization of an isolate of apple mosaic virus from rose in the USA. *Journal of Phytopathology* 139: 33–47; Wong SM, Horst RK and Langhans RW (1988) Symptomatology and occurrence of apple mosaic and prunus necrotic ringspot on rose in New York. *Acta Horticulturae* 234: 437–447.

Figure 1 Rose mosaic – line pattern and mottle symptom on leaf. Courtesy of R. K. Horst, Cornell University, with permission from Kluwer Publishing.

Figure 2 Rose mosaic – ringspot symptom on leaf. Courtesy of R. K. Horst, Cornell University, with permission from Kluwer Publishing.

by mild deformations. The aetiology of causal agents for diseases described above, as different forms of rose mosaic disease, has not been conclusively established. Nevertheless, several known viruses were isolated from roses showing mosaic symptoms. Infected plants harboured single virus species or very often mixed infections. *Prunus* necrotic ringspot virus (PNRSV) is currently the most prevalent rose virus in Europe. PNRSV alone or together with apple mosaic virus (ApMV) are the most common causal agents of rose mosaic in the USA and probably throughout the world.

PNRSV, with spherical particles about 23 nm, is only disseminated by vegetative propagation in rose production. PNRSV infection alone induced fine or broad foliar chlorotic ringspots, chlorotic mottle, line and oak-leaf pattern, and chlorotic vein netting. Leaves were occasionally distorted. During prolonged periods of high temperature chlorotic vein banding occurred. Since rose mosaic is not a lethal disease, there is an unsubstantiated opinion that it does no significant damage to rose. PNRSV, in rose field production, delayed the onset of flowering, reduced flower production (the size and number of blooms), caused poorer flower quality by increased proportion of deformed flowers, caused earlier autumnal leaf-fall, reduced plant survival, cold-hardiness and reduced rates of budding success. Healthy greenhouse rose 'Sonia' produced 25% more flowers per plant than those infected. The average stem length was longer and the proportion of first-class flowers was greater in healthy 'Mercedes' than in those infected by PNRSV.

Figure 3 Rose mosaic – yellow-net symptom on leaf. Courtesy of R. K. Horst, Cornell University, with permission from Kluwer Publishing.

ApMV has isometric virus particles about 26 nm in diameter, and, similar to PNRSV, is only transmitted by vegetative propagation in rose production. Plants infected by ApMV were stunted and lacked vitality;

leaflets were distorted with severe chlorotic mottle. In *Rosa multiflora* chlorosis was severe, almost white; chlorosis of the midrib caused leaf puckering and vein banding. During summer months slightly chlorotic rings and lines occurred on leaves. Plants infected with this virus produced significantly fewer flowers each spring. Other symptoms caused by ApMV are similar to those for PNRSV.

Apple mosaic virus is a synonym for rose mosaic virus (RMV), described by RW Fulton. For many years this special rose virus was thought to be the major causal agent of mosaic diseases. RMV was identical serologically with ApMV and was only distantly serologically related to PNRSV. The presence of a continuous serological spectrum of rose viral isolates was noted, extending from PNRSV to the ApMV serotype.

Arabis mosaic virus (ArMV) has spherical (30 nm diameter) virus particles and is transmitted in rose production by budding and grafting and by the nematode *Xiphinema diversicaudatum*. It caused chlorotic vein banding on leaves produced after the temperate summer months of May–June, and sporadic vein mottling. ArMV slightly reduced the vigour of plants and increased production of deformed flowers. This virus, together with PNRSV, caused chlorotic rings and line pattern in leaves produced in temperate spring and summer, as well as vein banding and vein netting.

The following viruses were also isolated from roses showing mosaic symptoms:

- Tomato ringspot virus causing typical mosaics was serologically related to Fulton's RMV.
- Tobacco ringspot virus, on *R. multiflora*, caused symptoms such as pale green, creamy white or yellow wavy lines, broad bands, ringspots and blotches which persisted throughout the growing season.
- Rose mottle mosaic virus (RMMV), which induced green mottled leaves in *R. rugosa*, was transmitted mechanically to herbaceous indicators, mahaleb cherry and Lovell peach seedlings. It was graft-transmitted from mahaleb cherry to stone-fruit trees, where it remained symptomless. RMMV caused a necrotic reaction in 'Shirofugen', but there were differences in host range and symptoms induced on them compared to those induced by PNRSV.

Visual inspections of rose are inaccurate for virus detection and even less so for virus identification. It is not possible to determine the presence of single or mixed infection by symptomatology alone. Scientifically it is desirable to name the virus/viruses associated with each report of rose mosaic. From a practical point of view, the economic loss will be the same for rose mosaic regardless of the associated virus species. However, specific viral pathogen identification can help to design an effective control programme.

Control of Rose Mosaic Virus Complex

The traditional statement that 'prevention is better than cure' is certainly true for virus diseases. In practice, the infected plants remain infected and often symptomless throughout their life. There are no chemicals (pesticides) to control viral diseases and growers can only remove and destroy diseased plants or use only virus-free plants for vegetative propagation. Diseased plants are infected systemically and virus can occur in every plant tissue, even those without symptoms. Viruses causing rose mosaic can be eliminated from infected plants by heat treatment and meristem tip culture.

The procedures involve subjecting infected plants to high temperatures (35–39 °C) for several weeks followed by dissection of the apical meristem with some meristematic leaves to propagate them on tissue culture media. The newly regenerated plants are used to create virus-free rose stock, which must then be indexed for the presence of RMV. Virus indexing includes bioassays with woody indicators such as 'Shirofugen' or 'Burr' rootstocks, very sensitive serological tests such as an enzymatic test enzyme-linked immunosorbent assay (ELISA), immunodiffusion in gels, latex flocculation tests, immunoelectron microscopy and the polymerase chain reaction (PCR) technique. Tested material can then receive a certification of 'virus-free' or 'virus-tested' according to a European and Mediterranean certification scheme for rose pathogens including rose viruses. It is noteworthy that heat-treated and virus-tested rose clones yielded 13.5% more blooms per plant per year than commercial untreated clones.

Growers as well as most nurseries lack the facilities to free plants of virus diseases, but there are official certification services or programmes available which offer commercially virus-free propagative material. While the mosaic disease is not contagious, except through budding or grafting, the use of indexed rootstocks and buds guarantees that plants remain free of rose mosaic disease. There is no legitimate excuse today for the continued existence of rose mosaic in rose nurseries or gardens.

Strawberry Latent Ringspot Virus (SLRV)

This virus belongs, along with ArMV, to the genus *Nepovirus*. Its properties are very similar to ArMV, although the viruses are not serologically related. SLRV occurs naturally in many cultivated and wild plant species. SLRV is transmitted in rose production by vegetative propagation and by nematode vectors *Xiphinema diversicaudatum* and *X. coxi*. Rapid infection of roses during the 2-year production cycle can occur when viruliferous nematodes are present. SLRV causes a reduction in size of shoots and leaves, a leathery texture and distortion of leaf shape; leaves also show yellow angular flecks. There is a high frequency of scion bud death on plants infected with this virus. SLRV causes a marked reduction in the number and diameter of shoots produced at the bud union, reduced number of flowers per bush, increased proportion of deformed flowers and strongly decreased vigour of diseased bushes. The size and vigour of infected rootstocks are also reduced.

Rose Rosette Disease (RRD) – Witches' Broom of Rose

Rosette or witches' broom of rose is an indigenous disease in North America and was reported in the 1940s, 1960s and 1970s from central and eastern states of the USA. This disease has been found on hybrid teas, floribundas, miniatures, climbers, 'antique rose' as well as *R. arkansana, R. canina, R. corymbifera, R. eglanteria, R. hugonis, R. montezumae, R. multiflora, R. nutkana, R. odorata, R. pomifera, R. rubrifolia and R. woodsii*. The disease remained obscure until 1976 when the increased incidence of RRD was as high as 90% and mainly in multiflora rose hedges.

Rosette is a graft-transmissible disease and the only known natural means of transfer is by *Phyllocoptes fructiphilus*, an eriophyid mite, which is wingless and possibly windborne. There are three stages of symptom development: early symptoms, early rosetting and late rosette. New infection starts to appear in mid-July. The underside of a leaf vein and leaf blades become deep red in colour; shoots elongate rapidly and develop a deep magenta coloration. The shoots are more succulent with strong proliferation of thorns. Leaves become deformed with a development of mosaic, red pigmentation, crinkling and rugosity. Second-stage leaves are more distorted, many lateral buds start to grow and petioles shorten, resulting in a rosette appearance. Flowers are rarely formed. In late rosette stage leaves are greatly reduced in size and are intensely red. Petioles are very short and all lateral buds produce short, intense red shoots. Plants produce very few rootlets, and apical growth is weak. Plants do not survive winter conditions. Symptoms on commercial ornamental cultivars are very similar to those in multiflora rose, with very strong bud distortion and thorn proliferation.

The aetiology of rose rosette remains unknown. Four double-stranded RNAs were associated with the tissues of symptomatic multiflora rose. Moreover, quasispherical, double-membrane-bound particles 120–200 nm in diameter and two types of inclusions were found in the cytoplasm of all cell types in symptomatic leaves, but their pathogenicity is still not confirmed. RRD is lethal to *R. multiflora*, a vigorous, thorny shrub, introduced to North America. Multiflora rose has been used as a rootstock, as an ornamental, as an aid to soil conservation, as a wildlife cover and as a 'living fence'. It is presently recognized as a pest species and a noxious weed. There are attempts to use RRD for biological control of *R. multiflora* by augmentation, which is defined as intensification of a native agent to achieve biological control of an exotic, invasive weed.

Control of rose rosette includes removing infected ornamental roses when the first symptoms are observed. Pruning infected canes sometimes result in eliminating the disease, but if not, the entire plant with rootstock should be destroyed. New rose cultivars should be isolated from known stands of *R. multiflora*. Chemical control of the eriophyid mites is very difficult and only succeeds if vectors are killed before feeding initiation.

Rose Colour Break

Symptoms of colour break in the rose flowers are rare. Affected flowers of affected rose 'Queen Elizabeth' fail to open normally and the petals are crimped at their outer margin. Moreover, white, pink and deep rose areas occurred on normally dawn-pink-coloured petals. A green tinge was visible on the outer petals. Mild chlorotic leaf symptoms included chlorosis near the leaf margin. These leaf symptoms could not be recognized at the end of the growing season. The causing agent was transmitted by budding to *R. multiflora*. Petals were crimped at the margin with streak symptoms made apparent by petal veins which were prominently darker in pigmentation than the rest of the petal. These symptoms were observed on affected cultivars 'Fragrant Cloud', 'Highlight' and 'Super Star'. Although electron microscopy revealed spherical virus-like particles, neither electron microscopy nor bioassays confirmed a viral nature of a causal agent. Tobamovirus, serologically related to tobacco mosaic virus, was isolated from climbing rose cultivar 'Danse du Feu', showing flower break symptoms. The virus was mechanically transmitted to herbaceous plants but attempts to transmit it back to roses failed.

Tobacco Streak Virus

On naturally infected *R. setigera* tobacco streak virus (TSV) caused irregular chlorotic patches and vein chlorosis. Leaves tended to be twisted and infected shoots were stunted. Symptoms were so severe that the virus is not considered a part of the rose mosaic complex. There is no information about natural transmission of TSV in rose production. Severe symptoms helped to eliminate infected plants and they were not used for propagation.

Rose Streak Disease

Similar to rose mosaic disease, rose streak is an old disease discovered on more than 60 species and cultivars of roses in the 1930s. Infection was recognized on teas, hybrid teas, hybrid rugosas, hybrid multifloras, perpetuals, polyanthas and many others. Disease symptoms were exhibited on leaves as brownish or reddish patterns, rings and irregular pale green areas or chlorotic streaks along the veins. The streaks were etched by necrotic lines. Necrotic rings, line patterns and yellowing were visible on both surfaces of the leaves. Defoliation occurred earlier than normal. Half-ripened fruits showed waterish, brownish discoloured spots and rings. Symptoms on rose canes included brownish or greenish and often water-soaked ring patterns and light green areas which became purple and deep purple. Disease was only transmitted by budding and on hybrid teas 'Madame Butterfly', 'Ophelia' or 'Briarcliff' black, necrotic areas developed in the budding sites and the distal parts died. The uncharacterized, graft-transmissible agent was named *Rosa* Virus 4 by K.M. Smith in 1937. This virosis can be controlled in the same manner as the rose mosaic complex.

Rose Dieback (Rose Wilt, Stunt) Disease

Dieback, stunt and wilt symptoms have been observed on many rose cultivars wherever roses are grown. They have been described as separate virosis by different researchers, but comparative studies have revealed that these were symptoms of one and possibly a complex of diseases and were named rose dieback or rose wilt. Infected plants exhibited a range of symptoms which could have been influenced by cultivars, development stage of plants during infection or climate and general environmental conditions. Symptoms first appear during the spring months, when the new growth is initiated after pruning. Infected scion shoots were reported to be stunted, broad at the base and tapering to the tip and new shoots exhibited dieback from the tip of the shoots. Leaves had a curved appearance, were brittle and crowded together forming balls. Sometimes leaves turned pale green or yellowish followed by defoliation leaving bare branches. Symptoms on well-established plants were associated with extensive dieback and loss of apical dominance on one or more shoots. Apart from terminal buds many lateral buds produced rosette growth of shoots which curved downward and balling of lower leaves. Dieback from terminal shoots led to the death of the whole plant.

Early descriptions of rose dieback (wilt) symptoms were complicated by the presence of a fungus *Verticillium albo-atrum* which also caused rose wilt. Symptom differences between rose wilt 'virus' and *Verticillium* wilt of rose were characterized by Hammett in 1971. Vegetative propagation was the recognized means of transmitting the causal agent, but its aetiology is still unknown and confusion still exists concerning the primary causal agent.

Rose dieback disease has appeared infrequently and there has been no evidence of its spread. Maintaining the highest plant quality by using only indexed plant material for propagation and eliminating any affected plants are the only ways of controlling this disease.

Rose leaf curl disease, which was described in Californian roses, could be a distinct component of the rose dieback complex disease. This graft-transmissible disease has only been found in public rose gardens. The nature of the infectious agent has not been determined. Affected established plants exhibited loss of vigour and significant shoot dieback. Leaf epinasty was observed on young leaves in the spring. Vein clearing, flecking of leaflets and even vein necrosis occurred. Roses exhibited defoliation, shoot tip necrosis and lateral bud break. Necrotic vein splitting occurred on leaflets, and mature canes split, revealing longitudinal areas of rough, corky bark. 'Queen Elizabeth' was used as an indicator plant, on which bark necrosis occurred in mature canes 8–9 months after inoculation. It was suspected that a natural means of spreading could have occurred, although the disease has never been identified in nurseries. Because symptomless or mild infection can occur, nursery stock should be considered a possible infection source.

Rose Spring Dwarf Disease (RSD)

RSD is another graft-transmissible disease which was discovered in nursery stock and landscape roses in California and Oregon. Although the causal agent is unknown, experiments have indicated that it is very heat-stable and remains in plants after thermotherapy

(26 weeks at 38 °C). Based on symptomatology, RSD is a distinct disease from rose leaf curl. Typical symptoms occurred only in early spring and consisted of delayed or dwarf shoots with downward-curved, rolled leaves and leaflets. Growth following bud break was 'balled' or 'rosetted'. Initial leaflets were generally smaller than normal and with distinct vein clearing or vein netting. Defoliation did not readily occur as in rose leaf curl infection. Since shoot elongation was often delayed, symptoms become less obvious and the summer growth was symptomless. In late autumn new shoot growth developed mild leaf epinasty. Burr multiflora may be used as indicator plants and exhibit characteristic balling and vein netting on infected plants. Fortunately, the incidence of RSD is very low and no natural spread of the disease has been observed. Control is achieved through the use of indexed propagation material.

Other diseases similar to rose dieback disease and worth mentioning are:

- 'Frisure' which occurs in France in glasshouse production on hybrid tea cultivars grafted on *R. indica major* rootstocks. Short, numerous shoots with chlorotic rosetted leaflets were observed in spring or after pruning. Dieback of plants or, more frequently, partial recovery occurred. The unidentified causal agent was sensitive to heat treatments.
- Little leaf disease was observed on hybrid tea and hybrid polyantha (floribunda) in South Africa in 1960. Symptoms appeared after pruning. Proliferation of buds occurred; these developed two or more short, deformed shoots with small pale leaves. Growth may be permanently retarded and affected shoots died back from the tips. Plants sometimes recover but at the end of the season symptoms reappeared. The causal agent of the disease has not been determined.
- In the Netherlands a disease named rose bud proliferation was observed. Infected buds proliferated into curly heads (Dutch: *kroeskoppen*) and tiny witches' brooms which usually died prematurely. Plants which occasionally recovered never showed a recurrence of the previous abnormalities. A similar disease called rose stunt was described in the UK. This disorder consisted of a proliferation of small stunted shoots around the scion, or one or more longer tapering shoots which became 'blind' (failure of buds to open). In both cases bud graft transmission experiments did not demonstrate an infectious causal agent.
- There is little information about rose flower proliferation disorder which was described in 1963 in Italy. The centre of the flower grew a short distance as a stem and then produced one or more flowers.

Conclusions

Many of these diseases are only suspected to be caused by plant viruses and are traditionally called 'virus-like diseases'. They were described many years ago and today it is impossible to verify them. The system of rose propagation from virus-indexed plant material which is commonly used in rose production may have helped to eliminate many 'unidentified virus-like causal agents'. Important rose viruses are well-defined today (except for rose rosette) and in recent years most laboratories have improved virus detection and molecular characterization technology.

See also: **Disease**: Overview. **Production and Marketing**: Overview of Roses and Culture.

Further Reading

Brown LG and Bliss W (2001) Viruses. In: Jones RK and Benson DM (eds.) *Diseases of Woody Ornamentals and Trees in Nurseries*, pp. 30–35. St Paul, MN: APS Press The American Phytopathological Society.

Brunt AA, Crabtree K, Dallwitz MJ, Gibbs AJ and Watson L (1996) *Virus of Plants. Descriptions and List from the VIDE Database*, p. 1484. Cambridge: CAB International University Press.

Clark MF and Martelli GP (1988) Viruses. In: Smith IM, Dunez J, Phillips DH, Lelliot RA and Archer SA (eds.) *European Handbook of Plant Diseases*, pp. 13–115. Oxford: Blackwell Scientific.

Cooper JI (1993) *Rosa* – Rosaceae. In: *Virus Diseases of Trees and Shrubs*, 2nd edn, pp. 135–142. London: Chapman & Hall.

Coyier DL (1985) Roses – virus diseases. In: Strider DL (ed.) *Diseases of Floral Crops*, vol. 2, pp. 473–488. New York: Praeger Scientific.

Horst RK (1983) Diseases caused by viruses. In: *Compendium of* Rose *Disease*, 3rd edn, pp. 26–30. St Paul, MN: APS Press, The American Phytopathological Society.

Horst RK (2001) Pathogens and Diseases – Virus, Viroid, Phytoplasma. *Rosa*. In: *Westcott's Plant Disease Handbook*, 6th edn, pp. 475–516, 829–832. Boston: Kluwer Academic.

Philley G, Hagen AK and Chase RC (2001) Rose diseases. In: Jones RK and Benson DM (eds.) *Diseases of Woody Ornamentals and Trees in Nurseries*, pp. 342–348. St Paul, MN: APS Press The American Phytopathological Society.

van Regenmortel MHV, Fauquet CM, Bishop DHL *et al.* (2000) *Virus Taxonomy. Classification and Nomenclature of Viruses. 7th Report of the International Committee on Taxonomy of Viruses*, p. 1121. San Diego: Academic Press.

Bactericides and Fungicides

M Kilian, Bayer Environmental Science, Monheim, Germany
U Steiner, Rheinische Friedrich-Wilhelms-Universität Bonn, Bonn, Germany

Introduction

Powdery mildew (*Podosphaera pannosa* var. *rosae*) is one of the best-known disease of roses and is widely distributed in glasshouses, garden- and field-grown plants. A number of other diseases, including rust diseases (*Phragmidium* spp.), black spot (*Diplocarpon rosae*), botrytis blight (*Botrytis cinerea*), downy mildew (*Peronospora sparsa*), stem canker (*Coniothyrium fuckelii*) and dieback (*Verticillium dahliae*) may affect roses where appropriate conditions prevail (*see* **Disease:** *Botrytis*; Black Spot; Downy Mildew; Powdery Mildew; Rust). As the acreage of this crop has expanded worldwide and new cultivation techniques have been adopted, some diseases have increased in severity. Therefore, the application of plant protection agents may be necessary and fungicides are availble in a number of countries. A disease of economic importance in roses caused by a bacterium is crown gall (*Agrobacterium tumefaciens*) (*see* **Disease:** Crown Gall). Plant protection products registered to protect roses against bacteria are not on the market.

Economic Importance of Fungicide Use

Roses are one of the economically most important ornamental plants and form a considerable segment within the fungicide market of ornamentals. Cut and pot roses are produced worldwide over an estimated area of 16 000 ha in glasshouses or plastic tunnels and on around 3000 ha in the open field. Roses produced in nurseries and for fragrance production are not included in these numbers.

The intensity of plant protection varies depending on production conditions and the intended use of the roses. Customers normally expect cut and pot roses to be completely free of disease symptoms. Therefore, the frequency of fungicide applications in this market segment is relatively high. Glasshouse conditions favour the development of some diseases, mainly powdery mildew. On average up to 20 applications per year are necessary to protect the roses against this pathogen. This corresponds to up to 40% of fungicide treatments in glasshouses. The annual cost invested by professional rose growers for fungicide treatments can reach 0.8 € m^{-2}. In the open field the number of sprayings is normally lower due to the fact that the vegetation period is shorter and environmental conditions are less favourable for the pathogen. On the other hand the possibility of influencing the incidence of disease due to changes in environmental conditions is only small.

Development and Registration of Fungicides

The costs of developing and registering a new active ingredient as a plant protection product are extremely high – more than €125 million. From an economic point of view a new rose fungicide can only be developed on the basis of active ingredients which have already been registered in other major agricultural or horticultural crops.

The market structure (cultivated area and frequency of sprays) must be considered separately for each country to calculate whether the costs can be justified for the local development of a product tailored for application in roses. In consequence, economic problems for commercial growers emerge as a result of the registration rules and the fact that the availability of products varies in different countries from year to year. In most countries the application of unregistered plant protection products for use in roses is strictly forbidden and penalties are very high if local laws are not complied with.

Like other plant protection products, fungicides and bactericides need to pass an official registration process before being commercialized and applied legally. In general, the producing companies or local suppliers apply for registration. The applicant has to prove by scientific studies following strict guidelines that the active ingredient and formulated product are not harmful to humans, nontarget organisms and the environment. Products foreseen for use in domestic gardens must fulfil stricter safety requirements, because in most cases users of consumer products are not trained and educated to handle plant protection products, unlike professional rose growers.

Furthermore, the European registration rules require detailed efficacy data on target diseases and plant safety as well as a management strategy to avoid the quick emergence of resistance in the pathogen. New products must show a level of efficacy which is at least as good as that of products already on the market. In general the registration of a product is limited to a period of 5–10 years. Re-registration of established products is possible if the applicant can prove that the product fulfils the current regulation rules.

Classes of Fungicides and their Mode of Action

There are two types of fungicides available to growers: immobile and systemic fungicides. Immobile or contact fungicides remain on the outside of the plant and prohibit the invasion of fungal pathogens. In growing plants, the expansion of foliage surface breaks the protective fungicide layer, allowing a fungal pathogen to invade. Consequently, immobile fungicides need frequent applications to be effective. Because they are surface-bound, immobile fungicides can be multisite inhibitors, and have a corresponding broad spectrum of activity.

Systemic fungicides penetrate the cuticle of leaves and stems and enter the apoplastic space as well as the epidermal or vascular tissues of the plant. This passive transport depends on the physical properties of the product, such as diffusion characteristics and solubility in different compartments of the plant tissue. Once systemic fungicides enter the water stream within the plant they are transported to the leaf margins. These products can be applied as soil drench or foliar spray, which is the main technique in roses. Only a few fungicides are able to move within the phloem-mediated transport system. An important factor in the redistribution of fungicides is their level of vapour-phase activity. Compounds that are relatively immobile in the plant may be extensively redistributed through the vapour phase and effect commercially acceptable disease control.

Systemic fungicides are generally compounds with a single site of action; they are very effective at low rates of application and provide disease control for extended periods. Disease control may be achieved after symptom expression, as in the eradication of visible powdery mildew colonies, or prior to disease expression. The difference is embraced in the use of the terms eradicant, curative and protectant. Strictly, eradicant fungicides are only active against the visible stages of the fungal life cycle. Curative fungicides are active against early but postpenetrative events in the infection process. Protectant fungicides prevent infection. They are active against spore germination, germ tube development and growth and appressorium formation. Traditionally, protectants were synonymous with immobile fungicides.

Several biochemical processes have been validated as targets for fungicide activity (**Table 1**). They include the inhibition of different fungal developmental stages and biosynthetic sequences. The oldest and still the most commercially valuable group of fungicides in roses are the general cell toxicants.

Nonsystemic Multisite Inhibitors

This group includes heavy metals, inorganic compounds and organic complexes. Although some have been implicated as inhibitors of respiration and have been shown to precipitate protein at high concentrations, others have a more general action, including binding to enzymes containing sulphydryl or amino groups, resulting in their inactivation and a general loss of cell integrity.

Inorganic compounds Sulphur products have been used to manage powdery mildew for centuries as a protectant acting through inhibition of conidial germination. They are highly effective when applied before disease symptoms appear, but have no eradicative action. If applied incorrectly, residues on the rose leaf surface may cause plant damage. Copper, as sulphate in the Bordeaux mixture, is mainly used in the control of *Oomycetes*. Fixed copper is a general-purpose fungicide and bactericide most often used as a protectant against various leaf-spot pathogens, *Botrytis* spp. and anthracnose. However, overall growth of some plants may be reduced after application. Therefore, label directions have to be followed carefully to reduce the risk of phytotoxicity.

Organic compounds Organic multisite inhibitors include two major groups, dithiocarbamates and phthalimides. Both groups display a broad spectrum of antifungal effects, but lack activity against the powdery mildews. Maneb, mancozeb, propineb, ziram and methiram belong to dithiocarbamates. In ziram the anion is the basis for activity mediated through complexes with several enzymes containing Cu^{2+} or sulphydryl groups. Maneb and mancozeb exert their action though conversion into ethylene diisothiocyanate and subsequent reaction with enzyme sulphydryl groups. The phthalimide captan preferentially reacts with enzyme sulphydryl groups but may also attack amio groups and inhibit enzymes that do not contain sulphydryl groups. Dichlofluanide and tolylfluanide both belong to the sulphamides; they are close relatives of the phthalimides and have the same mode of action. Replacing the chlorine atom by fluorine in these substances introduces activity against powdery mildew. The fungicidal activity is focused on *Botrytis cinerea*. Additional activity in the vapour phase is important for its efficacy.

Chlorothalonil, introduced in the mid-1960s, is still a major protectant fungicide in roses. It is effective for the control of *B. cinerea* and *D. rosae* as well as other fungi. It reacts with glutathione, coenzyme A,

Table 1 Classification of fungicides according to chemical class, mode of action and registration for application in roses (for status of registration, see local registration list)

Mode of action	*Chemical class*	*Common names*	*Main application*
Multisite inhibitors	Copper-containing products	Copper sulphate	Downy mildew, powdery mildew, black spot
		Penthahydrate	
		Copper octanoate	
		Copper oxychloride	
	Sulphur-containing compounds	Sulphur	Powdery mildew
	Dithio-carbamates	Mancozeb	Black spot, powdery mildew, downy mildew, rust
		Metiram, maneb	Downy mildew, rust
		Propineb	Rust
		Ziram	Black spot
	Phtalimides	Captan	Powdery mildew
	Others	Clorthalonil	Black spot, powdery mildew, downy mildew
		Diclofluanid/methylfluanid	Black spot, powdery mildew, rust
Sterol biosynthesis inhibitors:	Piperazine	Triforine	Black spot, powdery mildew, rust
	Pyridines	Pyrifenox	Powdery mildew
D14-α demethylation inhibition	Imidazole	Fenarimol	Black spot, powdery mildew, rust
		Imazalil	Powdery mildew
	Triazoles	Bitertanol	Black spot, powdery mildew, rust
		Difenconazol	Black spot, powdery mildew
		Flusilaziole	
		Myclobutanil	Black spot, powdery mildew, rust
		Penconazole	Powdery mildew
		Propiconazole	Black spot, powdery mildew, rust
		Tebuconazole	
		Triadimefon	Black spot, powdery mildew, rust
$\Delta 8/\Delta 7$ iso-merase, $\Delta 14$-reductase inhibition	Morpholines	Dodemorph	Powdery mildew
	Spiroketalamines	Spiroxamine	
3-ketoreductase inhibition	Hydroxyanilide	Fenhexamide	Botrytis blight
Respiration inhibitors	Strobilurines	Azoxystrobin	Rust
		Kresoxyim-methyl	Black spot
			Powdery mildew
		Trifloxystrobin	Black spot, powdery mildew, rust
Compounds interfering with cell membranes	Guanidines	Dodine	Black spot
Tubulin biosynthesis inhibitors	Benzimidazoles	Carbendazim	Botrytis blight, powdery mildew
		Thiophanate-methyl	Black spot, powdery mildew
Inhibitor of appressorium formation	Aminopyrimidines	Bupirimate	Powdery mildew
Compounds with undefined mode of action	Organic phosphor compounds	Fosetyl-aluminium	Downy mildew
	Carbamates	Propamocarb	Downy mildew
	Dicarboxamide	Iprodione	Botrytis blight
		Cymoxanil	Downy mildew
Biological fungicidal agents		Vegetable, mineral oils, lecithin	Powdery mildew
		Ampelomyces quisqualis	Powdery mildew
		Bacillus subtilis	Powdery mildew
			Botrytis blight

2-mercaptoethanol and other compounds, forming *S*-derivatives in *in vitro* experiments. In addition it inhibits a direct interaction with SH-groups of enzymes that can occur, leading to an irreversible destruction of enzyme activity.

Sterol Biosynthesis Inhibitors

Sterols are localized in cell membranes, conferring stability and controlling permeability and membrane function. In all fungi except the Phycomycetes, ergosterol is a main sterol and a reduction in its availability disrupts membrane integrity. Sterol biosynthesis inhibitors (SBIs) are unsuccessful at preventing spore germination, which is independent of ergosterol biosynthesis, but impair elongation of germ tubes after consumption of sterol supply (**Figure 1**). SBI fungicides are effective at controlling rose diseases caused by Ascomycetes, Basidiomycetes and Fungi imperfecti but they are not effective against Oomycetes. In general they are mobile within plants; an exception is bitertanol with its high lipophilicity, which is only locosystemic. In some cases SBIs provide vapour-phase activity. The SBIs have been divided into the C14α-sterol inhibitors (demethylation inhibitors or DMIs) and the Δ8/Δ7 isomerase and/or Δ14 reductase inhibitors.

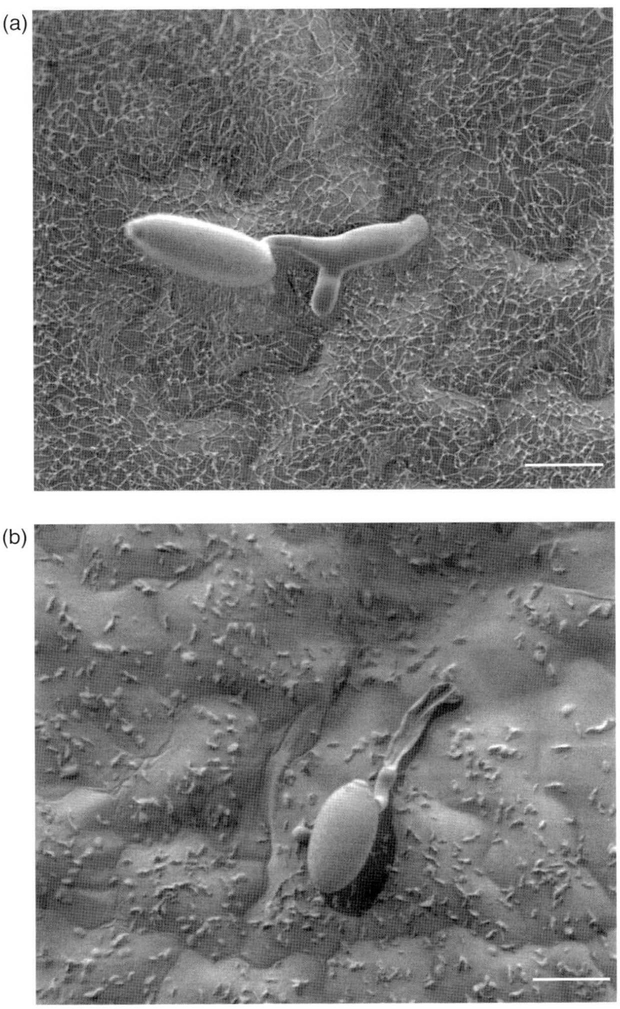

Figure 1 Effect of a protective application of a sterol biosynthesis inhibitor (tebuconazole) on germination and germ tube elongation of *Podosphaera pannosa* on rose leaves. (a) Untreated leaf: conidia germinate to form an appressorium; (b) fungicide-treated leaf: conidia germinate and the germ tube has collapsed. Bar 10 μm. Courtesy of Dr G. Haensler, Bayer Crop Science AG.

The C14α-demethylation inhibitors comprise a diverse group of chemistries, including the azoles, comprising 1,2,4-triazoles, imidazoles and the pyridines and piperazines. The principal action of the DMI fungicides is to inhibit the action of the cytochrome P450-dependent removal of the C14α-methyl group from 24-methylene dihydrolanosterol, obtusifoliol and 14α-methylfecosterol, together with a consequent reduction in ergosterol synthesis. These reactions disrupt the normal functioning of cell membranes, being the basis of DMI activity. Differences in performance between DMIs may be due to their binding affinities to the cytochrome P450-dependent demethylase enzyme of different pathogens.

Some highly effective SBI fungicides such as tebuconazole have side-effects as regards the length of developing shoots. This can be favourable for the production of small pot roses and also results in more bushy plants in nursery and domestic garden situations, which is often seen as an improvement in plant quality. However, in greenhouse-grown cut roses the length of flower shoots acts as a quality criterion and therefore SBIs have to be used carefully and in rotation with other products.

Like the DMIs, morpholines and spiroketalamines affect fungal sterol biosynthesis, but they inhibit the step between the Δ8 and Δ7 isomerase as well as the sterol Δ14 reductase. Isomerization seems to be a common site of action of morpholine fungicides and their analogues. The new fungicide spiroxamine is an example of the complexity of possible enzyme inhibition. With this fungicide the primary target site is the Δ14-reductase. Additionally other sites in the pathway, including squalene epoxidase, epoxidosqualene cyclase and sterol Δ8/Δ7-isomerase, are inhibited. The action on multiple sites may be advantageous in combating resistance.

The hydroxyanilide fungicide fenhexamide is a novel botryticide described as a sterol biosynthesis inhibitor which has been identified as being highly effective against *Botrytis cinerea* and the related pathogen *Sclerotinia* sp. It exhibits only limited locosystemicity, but its protective effectiveness is associated with outstanding long-lasting activity. It does not affect *in vitro* the germination of the conidia of *B. cinerea*, whereas germ-tube elongation and mycelium growth are inhibited at low concentrations. The germ tubes are distorted and bulge. These symptoms are similar to those induced by various sterol biosynthesis inhibitors. Fenhexamide treatments result in an accumulation of 3-keto-sterols. Therefore the target site seems to be the 3-ketoreductase involved in the C4 demethylation of 4,4-dimethylfecosterol and 4α-methylfecosterol.

Respiration Inhibitors

The strobilurins are an important class of fungicides. Their discovery was inspired by a group of natural fungicidal derivatives of methoxyacrylic acid, one of which is strobilurin A. These natural products are produced by a range of basidiomycete wood-rotting fungi, for example *Strobilurus tenacellus* (Pers. ex Fr.) Singer.

The mode of action of these fungicides was identified as inhibition of the mitochondrial respiratory chain in fungi by blocking the electron transport at the so-called Q_o site of cytochrome *b*. Cytochrome *b* is part of the cytochrome bc_1 complex, located in the inner mitochondrial membrane of fungi and other eukaryotes. When one of the inhibitors binds, it blocks electron transfer between cytochrome *b* and cytochrome c_1, which, in turn, disrupts the energy cycle within the fungus by halting the production of adenosine triphosphate. Despite having the same mode of action, their spectra of activity against fungal pathogens and their systemicity in plants vary widely between strobilurines.

The commercialized strobilurins in roses, azoxystrobin, kresoxim-methyl and trifloxystrobin, have a broad spectrum of activity. Of particular importance is their broad activity, but not all give high levels of control of all four major groups of plant pathogenic fungi. For example, kresoxim-methyl and trifloxystrobin only moderately protect plants against diseases caused by Oomycetes, such as downy mildew, and give only poor or moderate control of a number of basidiomycete diseases, such as rust. The special strength of trifloxystrobin is the long-lasting protection against powdery mildew of roses. Studies with azoxystrobin, kresoxim-methyl and trifloxystrobin demonstrated that spore germination and zoospore motility are stages of fungal development that are particularly sensitive. This can be explained by their biochemical mode of action, namely disruption of energy production, with the consequence that they are particularly effective against these highly energy-demanding stages of fungal development.

Strobilurins are best applied prior to infection or in the early stages of disease development in order to capitalize on their potent effects against spore germination (**Figure 2**). Although a number of strobilurins are active in the vapour phase, it should be emphasized that none of the strobilurins is highly volatile. Azoxystrobin is xylem-systemic; in contrast, kresoxim-methyl and trifloxystrobin are nonsystemic.

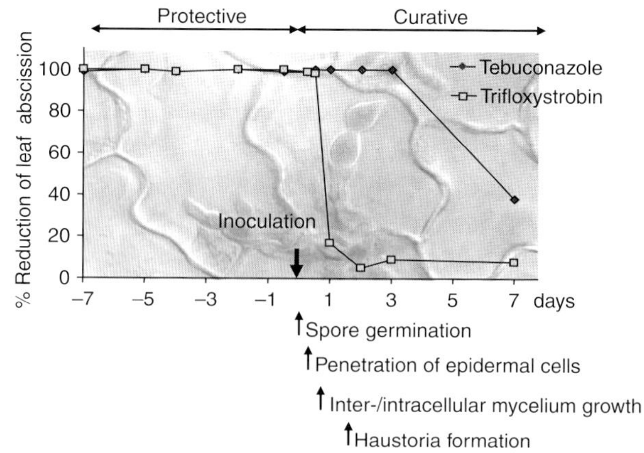

Figure 2 Effects of treatments before and after inoculation of *Diplocarpon rosae* conidia (protective and curative treatments dependent on different development stages) on leaf abscission of roses with an azole (tebuconazole, 0.01% a.i.) and a strobilurin (trifloxystrobin, 0.01% a.i.) fungicide. Background picture shows haustoria of *D. rosae*.

Compounds Interfering with Cell Membrane Structure

Cell membranes are bilayers containing phospholipids, sterols and globular proteins. The structure is an essential requirement for stability. The guanidines have long-chain alkyl groups and act as nonspecific detergents. They belong to the multisite-acting fungicides. The alteration of permeability of cell membranes and interaction with mitochondrial membranes are considered to be primary actions. The lipophilic alkyl chain attaches to the lipid fraction of membranes whilst the polar guanidino portion remains in the aqueous phase. Dodine is used as a protectant fungicide in roses, although curative and eradicant actions are described. Among the side-effects of dodine, the bactericidal action of the fungicide may be mentioned.

Tubulin Biosynthesis Inhibitors

The benzimidazole fungicides have a high affinity for tubulin proteins and disrupt mitosis by attacking the mitotic spindle. They are highly selective, depending on single structural differences of the microtubule binding site of the pathogens. Benomyl is used for its protective and eradicant activity against several fungi like *Botrytis* spp. and *Verticillium* spp. The fungicide is less toxic against rust fungi belonging to the basidiomycetes. Like thiophanate-methyl, it converts in plants and soil into methylbenzimidazol-2-yl carbamate, carbendazim, which has a similar spectrum of activity.

Inhibitors of Appressorium Formation

The 2-aminopyrimidine bupirimate is the sulphamate ester of ethirimol and its spectrum of activity includes powdery mildews like *Podosphaera pannosa* on roses. One reason for its remarkable protective and eradicative efficacy may be its good vapour-phase activity. It remains within veins when applied as a root drench, but is sufficiently volatile to be redistributed as vapour across leaf surface. Powdery mildews enter their hosts through the cuticle. An essential first step involves the formation of an appressorium, and its inhibition by 2-aminopyrimidine fungicides is described as the main fungicidal effect seen microscopically. Other steps may be inhibited, although these may be of limited practical significance. At a biochemical level aminopyrimidines seem to interfere with purine metabolism, which is why they are very effective against powdery mildews. The auxotrophy of *Erysiphe graminis* for purine may be an indicator for the selectivity of the fungicide.

Compounds with Undefined Modes of Action

Propamocarb is selectively active against Oomycetes. The fungicide is xylem-mobile and can be used as drench, soil incorporation, dip or foliar spray. The action is related to membrane function, causing an efflux of cell compounds. Leakage out of the mycelium can be inhibited by the addition of sterols. Production of an unusual fatty acid was later reported in mycelial cells of Oomycetes.

Aluminium tris-O-ethylphosphonate (fosethyl-Al), known as Aliette, is a systemic organic phosphate compound used to control diseases caused by Peronosporales. It readily degrades to phosphoric acid and carbon dioxide in water, soil and plants. Phosphoric acid is considered to be the active and systemic principle through disruption of phospolipid metabolism. The fungicide controls diseases at concentrations which have little or no effect on growth of the pathogens on artifical media. This low *in vitro* toxicity has been interpreted as an indication that antifungal activity is indirect and host defence reactions are stimulated. The application enhances the accumulation of phenols and antifungal stilbenes and flavonoids, resulting in a reduced symptom expression.

Cymoxanil is a systemic product with protectant and curative activity against Oomycetes. Mycelial growth is more sensitive to cymoxanil than early growth phases, including the release of zoospores from sporangia and their germination. The fungicide inhibits nucleic acid and protein synthesis.

Dicarboximides are used in ornamental plant cultivation, in bulb flowers and other flower crops for disinfection against *Botrytis* spp. They represent a group of highly active and selective fungicides. The compounds inhibit spore germination, but less strongly than mycelia development. Germinated spores developed short and stumpy germ tubes; the cell wall structure is altered in the region of the hyphal tips that are still growing. Disturbances in cell wall synthesis occur, associated with an outflow of cytoplasm. Effects on cell division have been reported but the main metabolic pathways are not distrubed, or only to a small degree. Recent studies have suggested that their mode of action is via their effect on lipid peroxidation and the inhibition of transport-associated phosphorylation of glucose. Signal transduction, exchange of ions and substances through membranes, cell cycle and biogenesis of cell structures and other functions of organisms are controlled in this way. Additionally the glutathione system is suggested to be a target of the dicarboximides because *in vitro* the fungicides inhibit the glutathione synthetase. The dicarboximides are classed among the contact fungicides which are generally applied prophylactically.

Biological Fungicidal Agents

Biological fungicidal agents are natural compounds or beneficial microorganisms formulated into a product that, when sprayed on the plant, impaire fungal pathogen development. *Ampelomyces quisqualis* is a parasitic fungus that actively attacks and destroys powdery mildew fungus. Instead of powdery mildew conidia the parasite develops pycnospores in pycnidia on the condiophores of *P. pannosa*. The bacterium *Bacillus subtilis* is the active ingredient in products that help prevent powdery mildew from infecting the plant. The bacteria are characterized by their ability to form highly resistant endospores and to produce a wide range of antifungal products, including subtilin and iturins. Other possible modes of antagonism could be competition for nutrients or space on leaves and the induction of resistance. Reports are available on the production of biosurfactant that affects the surface tension of the droplets. They can reduce periods of surface wetness and have an indirect effect on germination by limiting water availability to other pathogens, like *Botrytis*. These products have some effect in controlling the powdery mildew organism but are not as effective as the oils or sulphur. Particularly interesting is the success encountered under different cultural and climatic conditions by applying *Agrobacterium radiobacter* as a biocontrol agent against the pathogenic bacterium *A. tumefaciens*. *Agrobacterium radiobacter* strain K84 acts by competing with the pathogen at the infection site. It is commercialized in many countries and applied by dipping at transplant.

Natural products such as sodium bicarbonate and oils, including horticultural oils, neem oil and jojoba oil, proved effective at reducing powdery mildew, also showing efficacy against black spot. Oils work best as eradicants but also have some protectant activity. Based on their lipophilic nature they are believed to accumulate in the lipid layer and alter membrane structure and function. Neem oil is a broad-spectrum botanical pesticide derived from the neem tree that is effective against various fungal diseases, including black spot on roses, powdery mildew, downy mildew, anthracnose and leaf spot. Another natural product is lecithin, which, applied as a direct spray, inhibits spore germination and reduces hyphal elongation of powdery mildew fungi. The precise mechanism by which it impairs fungal development is not known, but it appears to be due to the influence on membrane composition on the synthesis of the hyphal wall polymer chitin.

Formulations of Fungicides

Fungicides do not consist of the active ingredients alone. The preparation in a form suitable for practical applications is referred to as a formulation and includes a sophisticated mixture of compounds. It influences the bioavailability, which is the basis for initial and long-lasting efficacy as well as plant safety. Crop plants differ significantly in composition and the structures of their cuticles and epicuticular waxes, which are the first, if not the most important, barrier to penetration of plant protection agents. Therefore, it is necessary to develop specific formulations for certain crops like roses to achieve optimal protection by a product. The hydrophobic nature of the waxes makes it difficult to obtain good wetting of leaf surfaces with aqueous sprays. Deposition, distribution and retention of spray droplets are especially affected by the leaf surface microroughness caused by waxes, cell surface contours, leaf venation and trichomes. On leaf surfaces with a contact angle of above 90°, like that of roses occurring with a smooth layer of superficial wax, wettability is significantly affected. Adjuvants affect the physical properties of spray liquids and may serve as emulsifiers, wetting agents, spreaders, sticker, antifoaming agents or buffers in these situations. Additionally, application properties like size of spray droplets, retention on the plant surface and rainfastness of the product are defined by the specific formulation.

Fungicides for professional use in roses are formulated as emulsion concentrates (EC) or suspension concentrates (SC) or as water soluble (WG) or water dispersible granules (WDG) or powders (WP). What components are essential for the formulation depends on the chemical and physical properties of the active ingredient and the intended mode of application. Inert compounds are needed to adjust the concentration of the active ingredient. Organic solvents are used to prepare a liquid formulation if the active ingredient is solid and water cannot be used due to insufficient solubility or hydrolytic stability. Emulsifiers or dispersants ensure a homogeneous distribution of the active ingredient in the aqueous spray broth and stabilize a dispersion of water with insoluble particles or hydrophobic droplets, and are therefore essential to make a homogeneous spray broth consisting of insoluble actives. Crystallization inhibitors help to prevent an active ingredient from rapid crystallization either within the spray broth or at the surface of plant material. There are specific penetration enhancers which improve the penetration of systemic acting compounds through the plant's cuticle.

In the preparation of dry formulations the solid actives must be ground to small particles of a few micrometres. The particle size affects the distribution of the active ingredient on the leaf surface, which may result in a modified long-lasting efficacy and penetration into the leaf. Wettable powders have a high proportion of particles less than 5 μm, and all particles should pass through a 44 μm screen. Microencapsulated formulations have been developed for the controlled release of pesticides. While in the past the user toxicity profile was a important reason to encapsulate active ingredients, today's targets are also related to the initial and also the long-lasting efficacy. Varying the wall thickness and type of polymer as well as the size of the microcapsule can control the persistence of a deposit. Beside the above-mentioned main components, formulations contain components like antifreeze, preservatives and dyestuffs. In particular, water-based products such as those used in ready-to-use pump sprays need to be preserved against mould.

Fungicide Application Strategies

Typical plant pathogens have basically four phases – prepenetration, penetration into the plant, post-invasion and finally sporulation to disperse the pathogen. The best control of fungal infection is still obtained with preventive spray programmes. Variations in onset of disease between seasons and areas make it difficult to time the application of a fungicide, although minimeteorological stations can provide local weather data to help decide whether treatment is required. The addition of adjuvants to increase the rainfastness of a spray deposit may be beneficial in roses. Plant tissue damaged by a pathogen can only be removed by pruning.

A protectant fungicide must be applied several times to limit the spread of disease. Systemic fungicides are easier to apply, as deficiencies in application can be compensated to some extent by redistribution after treatment. Curative or eradicative spray strategies can only be applied if the intended use of the plant tolerates a certain level of damage. The visible effect of a curative fungicide is the same as a protectant fungicide but the flexibility, that is, the period during which control can be effected, is much greater with a curative. On the other hand protectants like copper with their broad-spectrum activity have remarkable side-effects on minor diseases. The decrease in use of this type of product has in some cases increased the incidence of diseases like stem canker or dieback. The development of biological fungicidal agents for use in roses enlarges the means of plant protection; however, the level of control is not always comparable to that of chemical products.

The optimal use of fungicides requires a basic knowledge of the biology of pathogens, the mode of action of fungicides and a careful monitoring of the rose plants to detect and identify disease as early as possible. On outdoor roses, powdery mildew caused by *P. pannosa* can be expected to occur when rainfall is low or absent, and humidity is high at night and low during the day. Protective products must be applied at short intervals of 4–7 days. On glasshouse roses *P. pannosa* develops when the temperature range is 20–25 °C and humidity conditions are similar to those cited above. Under these environmental conditions protective sprays should be applied and repeated on a 7-day schedule. Also fumigation of protectives like sulphur is often used. Systemic products give longer protection but only on new growth with that proportion of the fungicides sprayed on lower plant parts. In dense stands translocation over the vapour phase helps to protect new plant parts. *D. rosae* causing black spot requires 14–35 days to develop visible colonies. For successful infection the pathogen needs several hours of permanent leaf wetness; its optimal developmental temperature is 20 °C. First sprays in spring, when leaves emerge, prevent an early disease spread and can delay an outbreak for weeks. Protective products do not stop inter- and intracellular mycelium growth and spore formation within the rose leaves, which is possible by applying systemic fungicides like tebuconazole (**Figure 2**). Under wet climatic conditions and high disease, it is recommended to use pressure sprays with an interval of 7 days for protective and of 7–14 days for systemic fungicides, as this system is more flexible in respect to accurate timing of sprays. Against botrytis blight, protective fungicide applications should be used to cover wounds. Stored roses should be sprayed or dipped when they are brought into storage. However the level of residue left by the chemicals is not always acceptable.

Fungicide Resistance Management

A requirement for modern fungicides is their specific mode of action. They usually inhibit a specific metabolic process (targets) in the pathogen and are therefore referred to as single-site inhibitors. Almost all single-site fungicides are considered to be high-risk fungicides because single mutations in the genes encoding for the target molecule may prevent binding of the active fungicidal ingredient. The extended use of single-site inhibitors could lead to the occurrence of either fungicide resistance or insensitivity. If only one or a few genes are involved, fungicide treatments lead to a selection of fungal subpopulations of either sensitive or extremely highly resistant strains. Examples of this are the benzimidazoles and carboximides with resistance factors greater than 50. If the involved mode of resistance is more complex and a number of gene mutations are necessary, genotypes with different levels of resistance occur, resulting in resistance development in slow steps; this is called shifting. This is the mode of resistance development described for DMI fungicides, e.g. for some powdery mildews in cereals. Reduced dosages of fungicides can increase selection for resistance of DMI fungicides.

Against strobilurins, which account for a large proportion of the agricultural fungicide market, several fungal pathogens developed resistance soon after introduction. This indicates the enormous flexibility of fungi and the need for an appropriate fungicide resistance management which should also be used to protect against rose disease. The large number of generations of rose powdery mildew and, to some extent also, of black spot makes it likely that fungicide resistance can also occur for this disease after intensive use of a single product.

Resistance management can be defined as the use of all available means of plant protection to reduce the selection pressure on fungal pathogens. This can be done by reducing the number of fungal generations and disease pressure wherever possible by changing environmental conditions, sanitary circumstances, use of resistant cultivars and precise spray techniques and avoiding sublethal doses of the fungicide (*see* **Integrated Pest Management**). It is most important to apply products with different modes of action in a rotation scheme. In practice this is narrowed due to the limited number of registered products with different modes of actions in roses. For example, applications of protective fungicides including fumigation with sulphur can alternate with single-site inhibitors like azoles, morpholines or strobilurines. Biological products offer the possibility of introducing new modes of action in a spray scheme. Commercial formulations of neem oil often show remarkable effect on powdery mildew. In general, rotation of products is preferred over combinations. However, for application in home and garden situations, combination products offer greater convenience and achieve acceptable levels of protection.

See also: **Breeding**: Selection Strategies for Disease and Pest Resistance. **Disease**: Crown Gall; *Botrytis*; Black Spot; Downy Mildew; Powdery Mildew; Rust. **Integrated Pest Management**.

Further Reading

Copping LG and Hewitt HG (1998) *Chemistry and Mode of Action of Crop Protection Agents*. Cambridge, UK. Royal Society of Chemistry.

Dehne HW, Gisi U, Kuck KH *et al.* (2001) *Modern Fungicides and Antifungal compounds III*. 13th International Reinhardsbrunn Symposium. Bonn: AgroConcept.

Delp CJ (ed.) (1988) *Fungicide Resistance in North America*. St Paul, MN: APS Press.

Gullino ML and Garibaldi A (2001) Diseases of roses: evolution of problems and new approches for their control. In: Morisot A and Ricci P (eds.) *Second International Symposium on Roses. Acta Horticulturae* 424: 195–201.

Hock B and Elstner EF (1995) *Schadwirkungen auf Pflanzen*. Heidelberg: Spektrum Akad. Verlag.

Horst K (1983) *Compendium of Rose Diseases*. St. Paul, MN: American Phytopathological Society.

Horst L (1995) *Modern Selective Fungicides: Properties, Applications, Mechanisms of Action*. Jena: Gustav Fischer Verlag.

Lyr H, Russell PE, Dehne HW and Sisler HD (1998) *Modern Fungicides and Antifungal compounds II*. 12th International Reinhardsbrunn Symposium:. Andover: Intercept.

Matthews GA (2000) *Pesticide Application Methods*. Oxford, UK: Blackwell Science.

Disease Resistance *see* **Breeding**: Selection Strategies for Disease and Pest Resistance.

DISTRIBUTION AND ECOLOGY

Contents

Overview

H Brichet, World Federation of Rose Societies, San Terenziano, Italy

It is a well-known fact that originally the natural distribution of the genus *Rosa* lay in the northern hemisphere. In North America roses were present in today's Canada, the United States and Mexico; in Europe they were spread across the entire continent as far as the Arctic Circle. The Mediterranean basin was fertile ground for the rose, which spread all along its coastline. The Near East is home to numerous rose species, while in Asia, with the exception of the desert and tropical areas, the rose is at home from India to the Japanese islands. However the greatest treasure house of roses was undoubtedly China, as it still is, also with much other plant material.

Thus was the 'paradisiacal' situation, but the rose long ago started to migrate (*see* **History of Roses in Cultivation:** Ancient Chinese Roses; European (Pre-1800); Modern (Post-1800)), and particular impetus was given following the voyages of discovery from Europe to the Far East and the Americas. Today, large collections in specialized gardens harbour innumerable cultivars, as well as many species roses of diverse origin. These gardens are a joy to the eyes of amateur gardeners, but are also a treasure trove to professional breeders who have in them an immense gene bank. These gardens may also be seen as historical mirrors of centuries of man's cultural development. The most renowned in Europe are the collections at L'Haÿ-les-Roses in France, Sangerhausen in Germany and Cavriglia in Italy. In North America distinguished collections are numerous, while in the southern hemisphere man has not been slow in redressing nature's distraction: at Renmark in Southern Australia and at Elgin in Southern Africa, but also in New Zealand and Uruguay remarkable collections may be visited. The southern hemisphere too has become the preferred grounds for large-scale cultivation of modern commercial cut-flower varieties – great tracts in Chile, Zimbabwe and the East African states, to name only a few, present favourable climatic conditions for open-air production aimed, above all but not exclusively, at northern markets during their winter season.

In many countries some species and early hybrids have 'escaped', as it were, and become naturalized in the wild or in local countryside on both sides of the equator. Few scientific studies have been made on the ecology of the rose but descriptions of species are often accompanied by descriptions of soils, localities and other ecologically relevant information. There are also repositories of local knowledge of the climatic and soil conditions under which wild and cultivated roses grow best. In order to tap this valuable source of information, use has been made of the network of rose societies that are affiliated to the World Federation of Rose Societies. These societies exist in practically all countries where roses grow. For the purpose of the survey presented in this article members of the World Federation of Rose Societies were asked to assist in gathering information in their countries. Most were happy to do so. They were requested to report on the climatic and soil conditions in their geographical locations, the occurrence of wild roses (indigenous and feral), cultivated varieties which do best and the uses to which roses are put. Descriptions of indigenous species and their distributions are given but readers should also refer to descriptions of species roses in the reproductions of Modern Roses (11th edn) and Rehder's classification of roses in Volume 3 of this Encyclopaedia. Those interested in venturing more deeply into the individual rose situations are encouraged to take advantage of the references quoted at the end of the article and, naturally, the magazines, annuals and other publications of the National Rose Societies of the contributors.

See also: **Distribution and Ecology**: South Africa; Australia and New Zealand; Continental Asia and Japan; Europe; Israel; North America; South America. **History of Roses in Cultivation**: Ancient Chinese Roses; European (Pre-1800); Modern (Post-1800). **Rose Collections and Trials**: Rose Collections and Herbaria; Rose Societies.

Further Reading

Austin D (1988) *The Heritage of the Rose*. Woodbridge, UK: Antique Collectors' Club.
Beales P (1985) *Classic Roses*. London: Collins Harvill.
Dickerson BC (1992) *The Old Rose Advisor*. Portland, OR: Timber Press.
Dickerson BC (1999) *The Old Rose Adventurer*. Portland, OR: Timber Press.
Genders R (1965) *The Rose: A Complete Handbook*. London: Robert Hale.
Harkness J (1978) *Roses*. London: J. M. Dent & Sons Ltd.
Krüssmann G (1987) *Roses*. London: Batsford.
Paul W (1978) *The Rose Garden*. London: Heydon.
Thomas GS (1955) *The Old Shrub Roses*. London: J. M. Dent & Sons Ltd.
Thomas GS (1962) *Shrub Roses of Today*. London: J. M. Dent & Sons Ltd.
Thomas GS (1965) *Climbing Roses Old and New*. London: Phoenix House.

South Africa

H Brichet, World Federation of Rose Societies, San Terenziano, Italy

The author gratefully acknowledges the contribution of Ludwig Taschner (Federation of Rose Societies of South Africa) in the preparation of this article.

South Africa

Climate and Soils

The vastly differing climatic regions in South Africa can be categorized as follows:

1. The subtropical region has temperatures ranging between 15 °C and 40 °C and generally frequent rainfall throughout the year. Roses flower continually and do not experience winter dormancy in the northern and easternmost parts of the country.
2. There is summer rainfall in the vast central plateau and interior of the country at an altitude between 1300 and 1800 m with temperatures ranging between 10 and 40 °C during summer from October to May, and winter temperatures from −10 °C at night to 25 °C at midday with virtually no rainfall from May until October.
3. Winter rainfall occurs in the coastal region of the Cape Province with temperatures varying in summer between 15 °C minimum and 35 °C maximum (November to March) with hardly any rain, and winter temperatures between 3 °C and 18 °C. The roses start flowering by mid-October and extend into May.
4. Semidesert conditions prevail in the Karoo and the Kalahari Desert with extremely hot summers and very cold winters and a minimum of rainfall. Roses perform well in this region when irrigated regularly, flowering from October into May.

Various soil types are found within individual gardens in all regions. The prevalent soils are red sandy loam and a black turf with varying percentages of sand and clay. The irrigation water obtained from wells and irrigation dams is often of a poor quality for rose growing as it contains a high percentage of sodium.

Wild Roses

There are, of course, no indigenous species of rose in South Africa. *Rosa rubiginosa* is naturalized in the Drakensburg mountainous areas at an altitude between 2000 and 3000 m. The winters are sufficiently cold in that region for the seeds to germinate after having been distributed by birds. This rose grows widely in the region, along roadsides and in grazing fields. It has now been declared a noxious weed in South Africa.

'Natal Briar', which is thought to be the result of natural hybridization between a *R.* × *multiflora* and a *R.* × *damascena*, grows wild in the subtropical climate of the Natal Province. It can be seen on embankments and fences as well as along roadsides.

Cultivated Roses

The first rose plants were brought to South Africa from The Netherlands by the first Governor of the naval refreshment station and later Cape Colony, Jan van Riebeeck, in 1657. Subsequently settlers from various countries brought European varieties into the country while ships returning from the Far East seem to have left species and varieties from those distant and exotic lands.

Nurserymen have discovered that rooted cuttings of 'Natal Briar' make excellent understock for budding roses and distribution of this rose has taken place in the form of suckers from budded roses and also from rooted runners. The canes are frost tender, but it is evergreen in regions with warm winters. Thus, it is the preferred rootstock for commercial cut roses for year-round production in The Netherlands, Kenya, Colombia and Ecuador. Millions of cuttings are being exported annually to these countries from South Africa.

At present about 1000 rose cultivars are grown throughout the country. Rose nurseries in South Africa represent most well-known rose breeders and their roses are grown in trial grounds, the most suitable varieties being released commercially.

The favourite rose variety over the past two decades has been 'Iceberg'. Commercial rose growers report that above 5% of total rose production is that of 'Iceberg' produced either as a bush, budded onto a standard stem or as the climbing sport. Mature bushes of 'Iceberg' easily grow to dimensions of 2 × 2 m.

Floribundas are popular for public places, large gardens and landscaped area. Hybrid Tea varieties are still the most popular roses for the home owner. Cutting blooms for the home is an important reason for planting roses. David Austin's 'English Roses' and other fragrant nostalgia roses are planted selectively, mostly for their fragrance. Here they do have the stigma of being overly vigorous and shy bloomers. Groundcover, Climbing and Miniature roses are also widely grown.

Virtually all garden roses are propagated by skin budding onto a hardwood cutting of the understock *R.* × *multiflora* 'Brooks' 48'. This rootstock, amongst others, was introduced into South Africa in the 1950s from Texas. It was made virus free through the use of heat therapy in the 1960s by the Horticultural Research Institute in Pretoria, indexed and made available in large quantities to all rose nurseries. About 2 million garden roses are sold and planted in South Africa annually.

About 7 million cut-flower rose blooms are purchased every month by consumers in South Africa. In addition about 2 million blooms per month are exported to various countries. Most of the cut roses are produced in modern polysheet-roofed greenhouses. These flowers are cut from about 10 million rose plants.

The commercial production of cut roses has become one of the prime export commodities in other African countries. Estimated numbers of plants grown for cut-flower production are given in **Table 1**.

Table 1 Annual production of cut roses in Africa

Country	*Number of plants grown (million)*
Kenya	80
Malawi	2
Tanzania	4
Uganda	4
Zambia	6
Zimbabwe	30

See also: **Distribution and Ecology**: Overview. **History of Roses in Cultivation**: Modern (Post-1800). **Production and Marketing**: Cut Rose Production. **Rootstocks**: Rootstock Breeding. **Rose Collections and Trials**: Rose Collections and Herbaria.

Further Reading

Carter G (1914) *The South African Home Garden*. Maritzburg, South Africa.
Fagan G (1988) *Roses in the Cape of Good Hope*. Cape Town, South Africa: Bree Street Publishers.
Fairbridge D (1924) *Gardens of South Africa*. Cape Town, South Africa: Maskew Miller.
Pertwee J (2003) Production and marketing of roses. Elsevier International Business Information, int@ebi.nl.
Readers' Digest (1971) *Complete Guide to Gardening in South Africa*, 2 vols. Readers' Digest: London.
Smith CW (1953) *Rose Growing in South Africa*. Cape Town, South Africa: Central News Agency.
Taschner L (date) *Roses*. Cape Town, South Africa: Struik Publishers.
Taschner L (date) *Living with Roses*. Pretoria, South Africa: Ludwig's Roses Pty Ltd.
Taschner L (date) *Retail Catalogue*. Pretoria, South Africa: Ludwig's Roses Pty Ltd.
Van der Spuy KR (1969) *Old Nectar and Roses*. Cape Town, South Africa: Books of Africa Pty Ltd.

Australia and New Zealand

H Brichet, World Federation of Rose Societies, San Terenziano, Italy

The author gratefully acknowledges the contribution of Margaret Macgregor, David Ruston, Malcom Watson and Richard Walsh (National Rose Society of Australia), and Doug Grant (The National Rose Society of New Zealand) in the preparation of this article.

Australia

Climate and Soils

Large central tracts of this vast continent are deserts and thus inhospitable to humans and roses. The northern coastal areas have a tropical climate not suitable for rose growing as too are the mountain ranges of the southeast and west as well as that of Tasmania. The rose thrives in areas that range from hot and dry, provided adequate water is available, to the cooler coastal regions where water is adequate, but where fungal diseases can be a problem. The choice of understock is also determined by climatic conditions. In Western Australia *Rosa fortuniana* is used because its roots penetrate deeper into the light soils than others. This is an advantage in the sandy soils and hot summer climate of Perth. *R. multiflora* is widely used in the cool to mild climates of the southern and eastern coastal areas of New South Wales, Tasmania, Victoria and South Australia, where the soils are neutral and slightly acid. *R. indica* var. *major* adapts to a wide range of climates and soils and is eminently suitable to the hot, dry, alkaline conditions found in much of inland Australia. 'Dr Huey' has become popular in some of the hot, dry areas, with conditions similar to those areas where it is favoured in North America.

Wild Roses

Of course there are no native wild species in Australia, but over the years a number of species, particularly those imported originally as understock, *R. indica* var. *major*, *R. multiflora*, *R. fortuniana*, *R. canina* and *R. rubiginosa* have become naturalized in those parts of the country that most suit their climatic and soil preferences.

Cultivated Roses

The first settlers from England are presumed to have brought roses as early as 1788. By the middle of the nineteenth century, roses abounded in the gardens surrounding the more prosperous settlers' homes. In 1893 the first rose specialist nursery was established in Melbourne, and in 1920 the first rose catalogues appeared and many growers were licensed to grow roses from the northern hemisphere nurseries after trial under quarantine conditions. This system is still in place today and roses are imported from practically all countries. New varieties are tested to see how they stand up to the Australian conditions before being released to the public.

Roses are used as groundcover, miniature pot plants, shrubs, formal standards, hedges, ramblers, climbers and as ornamental weepers in private gardens as well as in a large number of public parks and city gardens. Most rosarians choose to grow roses of many types in their gardens. Unfortunately in Australia the rose suffers the same diseases and is plagued by the same pests as in the northern hemisphere. Black spot is, in most areas, the most damaging of the diseases (*see* **Disease**: Black Spot). The southern coastal areas are the worst affected where warm days and heavy overnight dews provide ideal conditions for fungal diseases. Pests, in particular thrips, are a major problem in spring, but numbers vary according to the weather (*see* **Insects and other Animals**: Thrips). In Melbourne the advent of hot northerly winds heralds their arrival. Environmental concerns are now being addressed seriously and many rosarians are taking a responsible position with respect to the use of sprays. Most now concentrate on keeping their roses healthy and growing vigorously in the belief that the healthier the plant, the more resistant it will be to disease.

With respect to cut flowers in Australia, roses make up 30–40% of cut flowers sold and include the full range of colours available.

The success of the rose in Australia is well demonstrated by the establishment of rose societies, the earliest being the National Rose Society of Victoria in 1899. A year later shows were held in Melbourne. Membership of rose societies grew steadily, especially in the decades following the Second World War, as home ownership increased. The National Rose Society of Australia was formed in 1972 by representatives from all the Australian state rose societies and today has approximately 3500 members. The National Rose Trial Garden of Australia was established during 1995 and is housed within the Adelaide Botanic Gardens Precinct. The Heritage Rose Society was founded in 1979 in Adelaide and its current membership is around 1500.

Of particular interest are the roses bred in Australia with the country's specific conditions in mind. By far the most important breeder was Alister Clark (1864–1949), who raised more than 120 varieties between 1912 and his death. He was the first hybridizer to use *R. gigantea* in his breeding programme. Its early flowering habit was inherited by its offspring, which herald the start of the rose season. Amongst these are 'Jessie Clark', 'Flying Colours', 'Squatter's Dream', 'Tonner's Fancy', 'Traverser', and the best known of this strain, 'Nancy Hayward'. One of Australia's best-loved roses is 'Lorraine Lee' (1924). The rather loose flowers are at their best in the winter months. To get them at this time, bushes were pruned at Easter. In the 1930s and 1940s hundreds of hedges of this rose were to be seen in most suburbs in Australia. It is still popular for winter blooms in warmer areas. By far the largest group of Clark's roses was the repeat-blooming climbers. These roses were bred for the cooler areas of Australia and are not so spectacular for the hot inland. 'Black Boy' was released in 1919 and is still Australia's most popular red climber. Other notable breeders include Mrs. H.C. Fitzhardinge,

Frank Riethmuller, George Dawson, Bill Allender, Ron Bell and Eric Welsh. All Australian breeders are amateurs who have added a number of really worthwhile varieties to our gardens. It is indeed unfortunate that, due to very strict quarantine rules, they may not be admired in other countries.

New Zealand

Climate and Soils

New Zealand is located in the South Pacific southeast of Australia. It comprises two main islands, North and South Island, and numerous smaller ones. The country is generally mountainous with several large regions of plains. Two-thirds of the area is between 200 and 1100 m above sea level. The climate is mild, and seasonal differences are not great. The north end of the Auckland Peninsula is predominantly subtropical with the warmest climate; the coldest climate occurs on the southwestern slopes of the Southern Alps. Rainfall is generally moderate to abundant and, except in a small area in the south central part of South Island, exceeds 500 mm annually.

New Zealand has a wide range of soils. North Island soils have generally been developed from volcanic ash. In the centre and along the east coast the ash is of rhyolitic origin while to the north, west and south, the soils are derived from andesitic ash. Other soils of the North Island beyond the volcanic belt have been formed from a great variety of rocks mainly mudstone, sandstone, greywacke and limestone.

In the South Island most of the parent materials of the cultivated or farmed soils are derived from the mountain ranges and consist largely of greywacke and mica schist. This material, carried down from the mountains and foothills by rivers and wind and aided by glacier action, has given rise to the bulk of the South Island soils.

Wild Roses

There are no indigenous roses in New Zealand but some roses have become naturalized. *Rosa rubiginosa* is recorded as being planted in the Bay of Islands sometime between 1815 and 1820. It was used for hedging purposes and flourished under New Zealand's conditions. It spread to places where it was not welcome, particularly in the South Island hill country. Its hips were a source of vitamin C for the prospectors during the gold rush of the 1860s to combat the dangers of scurvy. The sweet briar was listed as an outlawed plant in the Noxious Weeds Act of 1900, but it is still common in the Central Otago region of the South Island where birds, rabbits and possums are thought to be responsible for its spread.

The dog rose, *R. canina*, was grown in the country as early as the 1820s. It was more common around the Auckland area than further north. However it is still found growing wild in parts of the South Island. It was also used as a source of vitamin C for the gold prospectors.

The 'Chestnut Rose', *R. roxburghii*, is common in the Bay of Islands.

Cultivated Roses

The first rose to be introduced into New Zealand was *R. chinensis* 'Slater's Crimson China' in 1814 by Christian missionaries from England. During the period of settlement between the 1850s and 1870s the new colours and forms of roses being created in Europe were much in demand, and many of roses that are now popular, such as 'Félicité et Perpétue' and 'Anaïs Segalas' were introduced.

Rose growing and breeding flourishes in New Zealand and many Hybrid Teas, Floribundas, miniatures and patio types, climbers, and shrubs and ground-cover types are popular in the gardens of New Zealand. The types of rose that have been successfully bred in New Zealand include spray roses for the cut-flower trade, miniature, patio, groundcover types and roses for exhibitors. The National Rose Society of New Zealand was founded in 1931, the first National Rose Show was held in 1943 in Auckland and the National Rose Society Trial Grounds opened at Palmerston North in 1969. The New Zealand Rose Breeders' Association was formed with the aim of bringing together all rose hybridizers and assisting them in raising worthwhile roses. The Rose Introducers of New Zealand is an organization of professional rose growers who represent in New Zealand the majority of the world's leading rose hybridizers. It acts as the agency for the collection of royalties for plant breeders' rights. With regards the plant breeders' rights, 'Matangi'®, by Sam McGredy, who had come to the country from Ireland in 1972, was the first variety of any species to receive protection in New Zealand.

See also: **History of Roses in Cultivation**: European (Pre-1800); Modern (Post-1800). **Intellectual Property**: Plant Patents and Trademarks. **Rose Collections and Trials**: Rose Collections and Herbaria.

Further Reading

Griffiths T (1984) *The Book of Old Roses*. London: Michael Joseph.

Griffiths T (1986) *My World of Old Roses*, 2 vols. Christchurch, New Zealand: Whitcoulls.

Nottle T (1983) *Growing Old-Fashioned Roses in Australia and New Zealand*. Kenthurst, Australia: Kangaroo Press.
Ross D (1985) *Rose Growing for Pleasure*. Melbourne, Australia: Lothan Publishing Co.
Steen N (1966) *The Charm of Old Roses*. Wellington, New Zealand: A.H. & A.W. Reed.
Thomas AS (1950) *Better Roses*. London: Angus & Robertson.
Thomas AS (1983) *Knowing, Growing and Showing Roses in Australia*. London: Macmillan.

Continental Asia and Japan

H Brichet, World Federation of Rose Societies, San Terenziano, Italy

The author gratefully acknowledges the contribution of Dr and Mrs Viru Viraraghavan (Indian Rose Federation), Dr Mahmooda Hashmi (Pakistan National Rose Society), and Professor Akira Ogawa (Japan Rose Society) in the preparation of this article.

China

Climate and Soils

China is located in Central and Eastern Asia and covers an area of 9 596 961 km^2. Over 66% of China is upland hill, mountains and plateaux, while the highest mountains and plateaux are found in the west. To the north and east of the Tibetan Plateau the altitude decreases to the desert or semidesert areas of Sinkiang and Inner Mongolia. To the northeast the broad fertile Manchurian Plains are separated from North Korea by the densely forested upland of Changpai Shan. East of the Tibetan Plateau and south of Inner Mongolia is the Sichuan Basin, which is drained by the Yangtze River flowing east across the southern plains to the East China Sea. The southern plains along the east coast of China have rich, fertile soils and are protected from the northern winds.

The climate of China may be divided into seven climatic zones.

1. Northeastern China has cold winters brought about by strong northerly continental winds, while summers are warm and humid with unreliable rainfall.
2. Central China has warm humid summers with the coastal regions occasionally subject to cyclones and typhoons.
3. South China has hot and humid summers with heavy rainfall between April and September.
4. Southwestern China is mountainous with the summer temperatures consequently moderated by the altitude, while the wet winters are mild.
5. The Tibetan region is a high plateau where winters are severe with frequent light snow and frost. While summers are warm during the day, temperatures drop to extremes during the night. Rainfall is also heaviest in summer.
6. The western interior zone has an arid desert climate with cold winters, and rainfall is distributed evenly throughout the year.
7. Inner Mongolia comprises the mountain ranges and semidesert lowlands and has an extreme continental climate with cold winters and warm summers. Rainfall is high and strong winds in winter and spring further reduce the temperatures.

The pattern of China's climate is an alternating exchange of air masses that form in a great high-pressure area over Siberia, Mongolia and the China seas. During the summer months the monsoon moves in from the tropical South China Sea bringing warm and humid weather, but air masses also invade China from the north during summer and from the ocean the year around. Hence, few countries have such contrasts in pressure and temperature. Another characteristic is the cyclonic storm, the typhoon, which is caused by the conflict of southwest and southeast trade winds in the China seas. Typhoons occur regularly every summer along the coast south of the Yangtze River and account for much of the precipitation in that area.

China has a diversity of soil types that follow, in general, the physiographic regions. The soils of northern China include the grey desert soil of the Tarim Basin, the chestnut and fertile loess soils of the Loess Plateau, the brown forest soil of Manchuria, and the productive alluvial soil of the Great Plain of northern China. In central China the Red Basin has fertile grey–brown forest soil, and the flood plain and delta of the Yangtze River have fertile alluvium. The soil of southern China is red soil, at times with lateritic materials. Tibet has stony soil similar to the mountain soils of the tundra zone.

Most of the natural vegetation of China proper disappeared many centuries ago through intensive cultivation of the land. Every available square metre of space of the usable land is cultivated and great care and painstaking attention to detail are employed. It is generally thought that China once had a discontinuous mixed temperate forest, including deciduous and coniferous varieties.

Wild Species

By far the most important contribution which this country has made to the knowledge and development of this flower is by way of the number of species which have been found, and hopefully will still be found in the future, in the wild.

Subgenus *Hulthemia*

R. persica Michx. The petals are yellow with a purple-crimson blotch on the base. This species is found in the Xinjiang region and grows on mountainsides, waste lands or by roadsides in dry areas.

Subgenus *Rosa*

Section Pimpinellifoliae

R. farreri Stapf ex Cox are small shrubs with cylindrical branchlets densely covered with acicles and sparsely with prickles. They are to be found in the provinces of Gansu and Sichuan, mostly in thickets at altitudes 1460–2800 m.

R. foetida Herrm. f. *foetida* are shrubs from 1.5 to 3 m high with slender branchlets and erect prickles. The flowers are dark yellow and solitary or, rarely, a few together, and have a foetid odour. The cultivar *R. foetida* Herrm. f. *persiana* has double yellow flowers and is cultivated in Xinjiang.

R. graciliflora Rehd. et Wils. has pink or dark crimson flowers and numerous stamens and fruit with erect and persistent sepals. It is found in Sichuan, Yunnan and Xizang (Tibet).

R. hugonis Hemsl. has strong branches which always droop. The prickles are flat and mixed with small acicles. The yellow flowers have numerous stamens. This rose is found in the provinces of Shanxi, Gansu, Qinghai and Sichuan on sunny mountainsides.

R. kokanica Regel ex Juzep. are small shrubs with straight and spreading branches, the surface of which is coarse and densely armed with straight acicles and glandular hairs. The leaflets' margins are sharply biserrate, the tips of the serrations are always glandular. The milky white or yellow flowers are solitary, while the fruit is spherical. In China this rose is found in the Xinjiang region mainly on mountainsides and notably as undergrowth in larch forests at altitudes between 1500 and 2000 m. It is also distributed in the central Asian area of the former Soviet Union, Iran and Afghanistan.

R. koreana Kom. var. *koreana* has dense branches closely covered with acicles, while the current year's branchlets are rather sparsely armed. The flowers are solitary, borne in the axils and the bracts are absent. It is found in Liaoning, Jilin and Heilongjiang at the edge of woods or on stony mountainsides. It is also found in Korea. *R. koreana* Kom. var. glandulosa differs from the original type in that the margins of its leaflets are biserrate and the undersides of its leaves are densely covered with glands.

R. mairei Lévl. is a small shrub with thick arched branches which are covered with long villi when young. The leaflets too are furnished with threadlike villi. This rose is found in Sichuan, Yunnan, Guizhou and Xizang (Tibet).

R. morrisonensis Hayata has erect prickles which always occur in pairs and can reach 1 cm in length, as well as sharply serrated leaflets. The four white petals are slightly longer than the sepals. It is found in Taiwan at altitudes of 3200–4100 m.

R. omeiensis Rolfe is a large upright shrub with nine to 13, and sometimes, 17 leaflets. The four white petals are obtriangular–ovate. This rose is found in Yunnan, Sichuan, Hubei, Shanxi, Ningxia, Gansu, Qinghai and Xizang (Tibet). It commonly grows on mountainsides. Bark of the roots contains 16% of tannin and glue may also be extracted. The dried and ground hips are mixed with flour and cooked. It is also used for medicinal purposes and alcoholic drinks are brewed from extracts. There are several varieties.

R. pimpinellifolia var. *pimpinellifolia*. This is a small shrub with spreading branches. The current year's branchlets are purplish-brown or reddish-brown and armed with erect prickles and densely covered with acicles. It is found in the Xinjiang region and grows in mountainous areas, grassy slopes, in woods and on river banks.

R. platyacantha Schrenk has leathery, roundish leaflets. The yellow, obovate flowers are solitary or two or three together and the dark crimson to purplish brown, lustrous fruit is spherical to ovoid. It is found in the Xinjiang region in thickets and amongst shrubs on the dry mountainous areas, in wasteland or in damp areas near rivers and lakes.

R. primula Bouleng. is a small shrub with flat prickles whose bases are swollen; the leaflets are nine to 15 in number and in rare cases seven. The flowers are pale yellow to yellowish-white, and the pedicels can be as long as 1.5 cm. It is found in Hubei, Shanxi, Gansu and Sichuan at altitudes of 1400–3450 m. It closely resembles *R. ecae* Ait. but it has more leaflets, the margins of which have denser serrations and the flowers are a paler yellow.

R. sericea Lindl. is an upright shrub with strong, drooping branches with sparse prickles. The flowers are white. It is found in Yunnan, Sichuan, Guizhou and Xizang (Tibet), but also found in India, Myanmar and Bhutan. There are several different varieties.

R. sikangensis Yu et Ku has branchlets which are almost glabrous, and sparse prickles mixed with acicles tipped with glands when young. The four petals are white. This species is found in Sichuan, Yunnan and Xizang (Tibet). It closely resembles *R. sericea* Lindl. There are two forms.

R. spinosissima L. var. *altaica* (Wild.) has a comparatively small number of bristles and flowers are comparatively large. It is found in the Xinjiang, Altai regions and also Siberia.

R. taronensis Yu et Ku has cylindrical branchlets armed with large prickles with thick bases and minute acicles. The yellowish white flowers have four petals measuring 3.5–4 cm across. The erect sepals are always persistent even when the fruit is ripe and the fruit stalks thicken.

R. transmorrisonensis Hayata is a small evergreen shrub which is distributed only in Taiwan in the mountains at over 2400 m.

R. tsinglingensis Pax. et Hoffm. closely resembles *R. graciliflora*; however in this rose the number of leaflets is always 13 and the flowers are white and slightly smaller. It is found in Shanxi and Gansu often growing in birch forests at altitudes of 2800–3700 m.

R. xanthina Lindl. f. *xanthina* is a tall, upright shrub with untidy, spreading branches. The yellow flowers are solitary and double or semidouble. This rose is often seen growing in gardens in the north and northeastern areas of the country presenting a wonderful sight in early spring. *R. xanthina* Lindl. f. *normalis* is considered to be the original type.

Section Cinnamomeae

R. acicularis Lindl. is widely distributed in the temperate zone of the northern hemisphere and there are considerable variations in its characteristics. However the branches are always armed with straight and slender prickles. The peduncles are always covered with glandular hairs. This rose closely resembles *R. davurica* Pall. It is found in numerous provinces of China.

R. albertii Regel resembles *R. beggeriana* Schrenk, however, in the latter prickles are broad, flat and hooked, the margins of the leaflets are always uniserrate. The leaflets have biserrate margins with occasional glands present at the tips of the serrations. The stipule margins are furnished with glandular hairs. The flowers are white, petals broadly obovate and the styles are covered with long villi, very many shorter than the stamens. The fruit is pear-shaped or ellipsoidal; the sepals and the top of the calyx tube fall off altogether.

R. banksiopsis Baker has many pink or crimson flowers borne in corymbs and grows in forests or mountainsides in Hubei, Sichuan, Jiangxi, Shanxi and other provinces.

R. beggeriana Schrenk has purplish-brown branches armed with pale yellow sickle-shaped prickles with swollen bases. The upper surface of the leaflets is dark green and glabrous, the under surface greyish green and covered with villi. The flowers are white or, rarely, pink. It is found in Xinjiang and Gansu.

R. bella Rehd. et Wils. bears solitary or two to three large flowers with pink petals. The fruit is vermilion with glandular hairs and stalks as long as 1.8 cm. It is found in Jilin, Hubei, Shanxi and other provinces. *R. bella* Rehd. et Wils. var. *nuda* is found in Shanxi and Hunan. The flowers are used for perfume oils and jams as well as for medicinal purposes. In Hubei and Shanxi the hips are used for medicinal purposes as substitutes for those of the *R. laevigata*.

R. calyptopoda Card. This small species has three to five foliate bracts under the inflorescences, and thus closely resembles *R. multibracteata* Hemsl. et Wils. However it has single flowers in most cases, the number of leaflets is smaller and the peduncles are very short.

R. caudata Baker can be as high as 4 m with thick triangular prickles. The numerous crimson flowers are borne in corymbs and the sepals are as long as 3 cm. It is found in Hubei, Sichuan and Shanxi. *R. caudata* Baker var. *maxima* Yu et Ku varies in that its leaves and flowers are larger than those of the original type.

R. chengkouensis Yu et Ku resembles *R. corymbulosa* Rolfe, but the latter rarely bears flowers singly and the leaves are larger, the undersurface of which is glabrous, but not furnished with distinct, glandular hairs. It is found in Sichuan.

R. corymbulosa Rolfe is a small shrub with crimson flowers in corymbs. It closely resembles *R. setipoda* Hemsl. et Wils. and is found in Hubei, Sichuan, Shanxi and Gansu.

R. davidii Crép. resembles *R. corymbulosa* Rolfe but there are always seven to nine leaflets, their margins are uniserrate, the bracts are large and the flowers are slightly larger. It is found in Sichuan, Shanxi, Gansu and Ningxia. The variety *R. davidii* Crép. var. *elongata* has longer and larger leaflets.

R. davurica Pall. is an upright, rather rampant shrub with paired, slightly hooked prickles. The pink flowers are solitary or two to three together and borne in axils. It is found in Hubei, Shanxi, Jilin, Neimenggu and other provinces and also grows in Korea, East Siberia and the southern areas of Mongolia. The fruits and roots are used for medicinal purposes. *R. davurica* Pall. var. *setacea* differs in that its branches are densely armed with prickles of uneven size. The leaflets of the *R. davurica* Pall. var. *glabra* are slightly larger. It has fragrant pink flowers and is grown as an ornamental plant. It is used as a rootstock while the flowers and hips are used for medicinal purposes. Glue may be obtained from the roots, bark and leaves.

R. fedtschenkoana Regel has oblong or ovoid fruit, 1.5–2 cm across, dark crimson and densely covered with glandular hairs. It is exceedingly subject to variation. In China it is found in Xinjiang and also in central Asia, Tianshan and Pamirs.

R. forrestiana Bouleng. has roundish leaflets and glands always present at the tips of serration and styles densely covered with yellowish white, long villi, almost as long as the stamens. This rose is found in Yunnan.

R. giraldii Crép. resembles *R. sertata* Rolfe whose leaflets are larger and both surfaces always glabrous. There are a number of varieties. It is found in Shanxi, Hunan, Gansu and other provinces.

R. laxa Retz. has pale yellow, hooked prickles and in this respect closely resembles *R. beggeriana* Schrenk.; however the inflorescences generally have fewer flowers, or in rare cases, the flowers are solitary. The fruit is oblong and, when ripe, the sepals are erect and retained. It is found in the Xinjiang region. *R. laxa* Retz. var. *mollis* Yu et Ku grows in flooded areas, swampy grassland and poplar woods.

R. macrophylla Lindl. is a shrub between 1.5 and 3 m high. It has leaves 7–15 cm long, crimson flowers 3.5–5 cm across and large, lustrous crimson fruit. It is found Xizang (Tibet), Yunnan and also in India, Sikkim and Kashmir. *R. macrophylla* Lindl. var. *glandulifera* is found in Tibet where the hips are used for medicinal purposes in place of those of *R. laevigata*.

R. moyesii Hemsl. has from seven to 13 leaflets that are 1–5 cm long. The dark crimson flowers are solitary or two to three together. The oblong, crimson fruit has glandular hairs and the sepals are erect and persistent. It is found in Yunnan, Sichuan and Shanxi. *R. moyesii* Hemsl. et Wils. var. *pubescens* has leaflets that are densely covered with villi.

R. multibracteata Hemsl. et Wils. resembles *R. willmottiae* Hemsl., but the latter has leaflets normally uniserrated or biserrated, its flowers are borne singly, the number of bracts is from one to two and the sepals fall off after the fruit is ripe. This species is exceedingly subject to mutation. On the basis of the different inflorescences and the shapes of bracts, some people separate the different types and consider them not to belong to the same species.

R. murielae Rehd. et Wils. resembles *R. prattii* Hemsl. and *R. sertata* Rolfe. All three are distributed in approximately the same areas and sometimes intermediate types are to be found. It is very difficult to identify a specimen when complete flowers or ripe hips are not available. They are found in Sichuan and Yunnan.

R. oxyacantha R. Bieb. closely resembles *R. acicularis* Lindl. which has however larger leaflets with uniserrated margins and a villous surface, peduncles longer and fruit larger. It is found in Xinjiang as well as in Siberia and Mongolia.

R. persetosa Rolfe has branches covered with waxy powder and densely armed with acicles and closely resembles *R. sertata* Rolfe. It is found in Sichuan.

R. prattii Hemsl. is a shrub whose pink flowers, normally two to seven, are borne in corymb-like inflorescences. It is found in Gansu, Sichuan, Yunnan and other provinces.

R. pseudobanksiae Yu et Ku is a small climbing shrub with three to five flowers arranged in corymb-like inflorescences, the petals of which are white or pink before the flower opens or when it has just opened. It is found in Yunnan and closely resembles the Banksian roses. However the stipule is almost wholly connate to the petiole and inseparable from it, styles count five to six, and are pubescent and exerted. It also resembles the *R. webbiana* Royle, but this has leaves with five to nine leaflets and flowers are pink and always borne singly.

R. rugosa Thunb. is an upright shrub armed with pale yellow prickles, surfaces of the prickles covered with downy hairs. The dark green surface of the leaflets is rugose and sharply serrated. The purplish-crimson to white, double to semidouble, fragrant and solitary flowers are borne at axils. This rose originates in the northern areas of China and also in Japan and Korea. It is cultivated in many areas of the country and there are a large number of garden cultivars. The flowers are used to distil oils, and to make sweets and wine. The dried petals are made into rose tea and the buds are used for medicinal purposes.

R. saturata Baker has large (3.5–5 cm) flowers of a crimson colour and closely resembles in *R. banksiopsis* Baker which, however has numerous, smaller, pink flowers in corymbs and seven to nine oblong leaflets. It is found in Hubei, Sichuan and Zhejiang. *R. saturata* Baker var. *glandulosa* Yu et Ku had particularly biserrate leaflets, the tips of which are furnished with glands.

R. sertata Rolfe, known as the blunt-leaved rose, is found in Gansu, Shanxi, Henan and other provinces. It is exceedingly subject to variation.

R. setipoda Hemsl. et Wils. can be as high as 3 m and is sparsely armed with broad, flat prickles. The biserrate leaflets have distinct glands on the tips of the serrations and are densely covered with glands, while the sepals always have a pinnate cleft. The pink flowers are borne in sparse corymbs at the base of which two to three bracts are present. It is found in Hubei and Sichuan.

R. sweginzowii Koehne closely resembles *R. setipoda* Hemsl et Wils. which has fewer prickles and larger leaflets and a large number of flowers borne in corymbs. It is found in Yunnan, Sichuan, Hubei and other provinces. *R. sweginzowii* Koehne var. *glandulosa* is found in Gansu, Sichuan, Yunnan and Xizang (Tibet).

R. tibetica Yu et Ku closely resembles *R. webbiana* Wall. et Royle but has white flowers, the leaflets' margins are biserrate and the undersurface is densely covered with glandular hairs. It is found in Xizang (Tibet) in pine and cedar forests or in secondary forests at altitudes of 3800–4000 m.

R. webbiana Wall. et Royle is found in Xizang (Tibet), growing on mountainsides, grasslands, in thickets or near paddy fields at altitudes of 2000–4500 m. It is also found in central Asia, the northern areas of India, Kashmir and Afghanistan.

R. willmottiae Hemsl. has weak branches and leaflets, from seven to nine in number and rarely 11, the whole length of the leaf measuring between 2–4 cm. The pink flowers are solitary. It is found in Sichuan, Shanxi, Gansu, Qinghai and other provinces. The variety *R. willmottiae* Hemsl. var. *glandulifera* Yu et Ku is found in high mountains at altitudes of 2500–3800 m.

Section Indicae

R. chinensis Jacq. is an upright shrub with three to five leaflets. The double to semidouble crimson, pink to white flowers are borne in small numbers. The flowering period is from April to September. It is cultivated in many parts of the country. Garden cultivars are numerous. The flowers, roots and leaves are used for medicinal purposes. *R. chinensis* Jacq. var. *spontanea* Yu et Ku has cylindrical branches, flat, broad prickles and crimson simple flowers. It is found in Hubei, Sichuan and Guizhou. It is regarded as the original type of 'Monthly Flower'. *R. chinensis* Jacq. var. *semperflorens* Koene has thin branches and five to seven leaflets which are reddish-purple. The flowers are mostly borne singly and are double and dark crimson with long, thin peduncles.

R. lucidissima Lévl. is an evergreen or semi-evergreen climber with normally three leaflets. The flowers are crimson and solitary. It is found in Hubei, Sichuan and Guizhou.

R. odorata Sweet is a rampant, evergreen climber with single, white, fragrant flowers 5–8 cm across. It is found in Yunnan and cultivated in the provinces of Jiangsu, Zhejiang, Sichuan and Yunnan. *R. odorata* Sweet var. *gigantea* Rehd. et Wils. has flowers 8–10 cm across and is found in Yunnan. *R. odorata* Sweet var. *pseudoindica* has double, yellow or orange flowers and is found in Yunnan. *R. odorata* Sweet var. *erubescens* Yu et Ku has double, pink flowers 3–6 cm across and is also found in Yunnan.

Section Synstylae

R. anemoniflora Fort. ex Lindl. is a climbing shrub with thin, glabrous branches sparsely furnished with hooked prickles and glandular hairs. The three leaflets are sharply serrulated. The pink flowers are solitary or in corymbs. It is found in Fujian. There exists a double-flowered variety which makes a good ornamental plant, bearing pale pink flowers; it is cultivated in gardens in the eastern parts of China.

R. brunonii Lindl. Is a climbing shrub from 4–6 m high with many white flowers arranged in compound corymb-like inflorescences. It is found in Xizang (Tibet) and Yunnan and grows mostly in forests or thickets in river valleys at altitudes of 2600–2750 m.

R. duplicata Yu et Ku is a small shrub with leaflets with biserrations on the margins and glands on the undersurfaces. The pale yellow to white flowers are borne singly or, rarely, two or three together. It is found in Xizang (Tibet) growing on footpaths in rice paddies or by the roadsides.

R. filipes Rehd. et Wils. is a 3–5 m shrub with long creeping, glabrous branches armed with short hooked prickles. The inflorescences in some cases contain more than 20 flowers. It is found in Gansu, Shanxi, Sichuan, Yunnan and Xizang (Tibet).

R. glomerata Rehd. et Wils. is a spreading shrub with long and creeping branches. The numerous flowers are densely borne in corymbs. It is produced in Hubei, Sichuan, Yunnan, Guizhou and other provinces.

R. helenae Rehd. et Wils. is a spreading shrub with long, thick creeping branches and short and thick prickles. The white flowers form corymbs, the fruits are ovoid and the stalks about 2 cm long and densely covered with glandular hairs. It is found in Hubei, Shanxi, Ggansu, Sichuan and other provinces.

R. henryi Bouleng. is very similar to *R. sambucina* Koidz. with small differences in the calyx tubes and the outer surface of the sepals. It is found in numerous provinces and regions.

R. kwangtungensis Yu et Tsai is a small climbing shrub with long creeping branches; from four to 15 white flowers are united in corymbs. It is found in Guandong, Guangxi and Fujian. *R. kwangtungensis* Yu et Tsai var. *mollis* has peduncles and sepals, leaflets and raches densely covered with villi and has double flowers. *R. kwangtungensis* Yu et Tsai var. *plena* gives double flowers and is possibly derived from a cultivar.

R. lasiosepala Metc. is a 10-m high climbing shrub with numerous flowers borne in compound corymb-like inflorescences. The sepals are lanceolate and 1.5–2 cm long. The stamens are numerous. It is found in the Guangxi region.

R. lichiangensis Yu et Ku is a climbing shrub with three to five leaflets, the upper surface being glabrous. The pink flowers are in groups of two to four and have numerous stamens. It is found at Lijiang in Yunnan and grows in thickets.

R. longicuspis Bertol. is a large climbing shrub with bowed branches and coriaceous, glossy leaflets and numerous corymb-like inflorescences; the outer surface of the petals is covered with silky hairs. It is found in Yunnan, Sichuan and Guizhou in forests and also in the northern areas of India. *R. longicuspis* Bertol. var. *sinowilsonii* Yu et Ku has larger leaves, more flowers and the leaflets are always five in number.

R. maximowicziana Regel. is a small shrub with long, creeping branches sparsely armed with small hooked prickles. The white or white-shaded pink flowers are borne in corymb-like inflorescences. It is found in Liaoning, Shandong and other provinces as well as in Korea and the far eastern areas of the former Soviet Union. The fruit is used for medicinal purposes, and the young stems are edible. It is also used as a rootstock.

R. multiflora Thunb. is a climbing shrub with many white flowers arranged in panicle-like inflorescences. The fruit is nearly spherical and reddish brown. It is found in the provinces of Jiangsu, Shandong and Hunan and is also commonly seen in Japan and Korea. It is exceedingly subject to variation. Cultivated varieties are often seen in gardens. There are numerous varieties. The roots, leaves, flowers and seeds are used for medicinal purposes, while glue is extracted from the roots and oils from the flowers for the cosmetic industry. In the northern areas of China this rose is often cultivated and used to make hedges, planted to consolidate slopes and on decorative trellises.

R. rubus Lévl. et Vant. is a creeping shrub as high as 5–6 m with from 10 to 25 white flowers in corymbs arranged like panicles. This species is widely distributed in the western, eastern and southern areas of China. It is easily identified by its densely villous branches, numerous fragrant flowers and spherical, red hips. The roots are used to make glue and the flowers for perfume oils. *R. rubus* Lévl. et Vant. f. *glandulifera* is found in Guangxi in mountainous areas.

R. soulieana Crép. is a large, upright spreading shrub with whitish, pale green branches and small obtuse leaflets. The yellowish-white flowers are numerous and borne in short and dense corymbs or, rarely, solitary. This species is widely distributed in the mountainous areas of southwestern China. There are a number of varieties.

R. uniflora Yu et Ku is a small shrub with thin branches having distinct stripes and patterns. The white flowers are solitary and borne at the axils. The styles are united. It grows in sunny places along the sea shores in Zhejiang.

R. weissiensis Yu et Ku is a small climbing shrub with branchlets of the current year furnished with glandular hairs. It resembles *R. rubus* Lévl. et Vant. but its leaflets and flowers are smaller, the margins of the leaflets are biserrate and the undersurface is furnished with glands.

R. wichurana Crép. is a climbing shrub 3–5 m high. The prickles are small, the five to seven leaflets are dark green on the upper surface and lustrous. Numerous white, fragrant flowers are borne in corymb-like inflorescences. It spreads long creeping canes flatly. There are a large number of cultivars and it is used to cover fences, stony slopes and trellises. It is found in Zhejiang, Guangdon, Guangxi and Fujiang and also in Taiwan, in Japan and Korea.

Section Banksianae

R. banksiae Ait. var. *banksiae* is a climbing shrub which can reach 6 m with the older branches armed with comparatively large, hardened prickles. The double to semidouble, small, white flowers are borne in umbels. It is found in Sichuan and Yunnan and is cultivated in many parts of the country as an ornamental plant, but is not winter-hardy; thus in northern and northeastern China it is grown in pots. *R. banksiae* Ait. var. *normalis* has single, white, fragrant flowers and is regarded as the original wild type of *R. banksiae*. It is found in Hunan, Gansu, Shanxi, Hubei, Sichuan, Yunnan and Guizhou. The roots are used to produce glue and also for medicinal purposes. *R. banksiae* Ait. f. *lutea* Rehd. has yellow, double flowers, but no fragrance. *R. banksiae* f. *lutescens* Voss. has single yellow flowers.

R. cymosa Tratt. is a climbing shrub closely resembles the single-flowered *R. banksiae* Ait., but it bears its flowers in compound corymbs and its sepals always have pinnate lobes. There is considerable variation in its characteristics and numerous intermediate types. It is found in numerous provinces. *R. cymosa* Tratt. var. *puberula* is found in Shanxi, Jiangsu, Hubei and Anhui.

R. fortuneana Lindl. has solitary, white, double flowers and is cultivated in Fuzhou in the Fujian province. There is doubt that this rose is really a hybrid between *R. banksiae* and *R. laevigata*.

Section Laevigatae

R. laevigata Michx. is an evergreen, climbing shrub which may reach 5 m in height. The leaflets are coriaceous and normally three in number. The large, white flowers are solitary and borne at the axils. The fruit is pear-shaped and densely armed with bristles. It is found in numerous provinces and likes sunny fields and areas near paddies. The roots are used to make glue, the hips are boiled for sugar and fermented for alcoholic beverages. The roots, leaves and hips are all used for medicinal purposes. *R. laevigata* Michx. f. *semiplena* Yu et Ku bears large semidouble flowers and is found in Jiangxi.

Section Bracteatae
R. bracteata Wendl. is an evergreen spreading shrub with long, creeping branches. The prickles are flat, hooked and always paired. The leaflets are coriaceous, dark green, roundish and lustrous. It has large, broadly ovate bracts and white flowers which are solitary or two to three together. It is found in Jiangsu, Zhejian, Taiwan, Fujian, Jiangxi, Hunan, Guizhou and Yunnan and is also found in Ryuchu in Japan. Its hips and roots are used for medicinal purposes. It is grown as evergreen hedges to contain animals. It has perhaps been crossed with *R. laevigata* to give a garden cultivar with double flowers of a fruit-like fragrance. *R. bracteata* Wendl. var. *scabriacaulis* has branches densely covered with acicles and glandular hairs. It is found in Shejiang, Fujian and Taiwan growing by valley streams or in coppices.

Subgenus *Platyrhodon*
R. kweichowensis Yu et Ku is closely related to *R. roxburghii* Tratt., but the petals are white and numerous flowers are borne in compound corymbs; the leaflets number seven to nine. It is found in Guizhou in half-shady places.

R. praelucens Byhouwer grows in Yunnan. Comparing the leaflet, the colours and the size of its flowers, it is clear that resembles *R. rugosa* Thunb. to some extent and might be regarded as an intermediate type between *Microphyllae* and *Cinnamomeae*.

R. roxburghii Tratt. is a spreading shrub with greyish-brown exfoliating bark and from nine to 15 leaflets. The double to semidouble, pale crimson or pink, rarely white, slightly fragrant flowers are borne at the top of short branchlets. The fruit is spheroidal, 3–4 cm across and densely armed with acicles. This rose is found in Shanxi, Gansu, Jiangxi, Anhui and many other provinces. It is also found in Japan. The fruit is used for sugar or brewed into alcoholic beverages, and is also used for medicinal purposes. It is also cultivated as an ornamental plant and for hedging purposes. *R. roxburghii* Tratt. f. *normalis* Rehd. et Wils. has single pink flowers and is regarded as the original type.

Cultivated Roses

Rosa gallica and *R. centifolia* are grown in many areas of China today. In recent years *R.* × *damascena* has been grown as an ornamental plant in China and also for extracting essential oils. *R.* × *alba* is cultivated as an ornamental plant.

The rose is a commonly cultivated flower in private gardens and public parks. There are many much-visited rose gardens. Of particular note is the extensive collection of species and old cultivars and the impressive display of modern roses at the Botanical Gardens in Beijing, headquarters of the Rose Society of China. There are also a number of amateur rose breeders.

India

Climate and Soils

India covers 3.28 million km^2. It extends from the 8° N to the 40° N, a length of over 2500 km, and from latitude 68° E to 97° E, a width of 3000 km. Climatically, there are two moderating factors: first, the monsoon (the southwest monsoon), which brings rain from June for 4 to 5 months, the rest of the year being dry; second, the Himalayas – the greatest of the world's mountain chains, stretching for nearly 2500 km, from the western ice deserts in Ladakh to the astonishingly moist regions in the east, in Assam. In fact, one area has over 12 500 mm of rainfall per year – the world's highest figure. The Himalayas, being well over 6000 m for much of the range, effectively prevent the movement of cold air from the highlands of central Asia and Siberia entering India. So here the climates are quite different from other places in the same latitudes. Such a great diversity of climates supports a correspondingly great variety of plant life. Certain other features which affect the climate deserve mention. The first is the mountain range called the Western Ghats, stretching for about 800 km along the western coast along the Arabian Sea. This mountain range, which at its highest point is more than 2700 m in altitude, is responsible for the southwest monsoon rains being very heavy on the western side of these mountains and comparatively lighter on the eastern side. The eastern side of peninsular India has two main divisions: the Deccan Plateau, which lies at an average elevation of about 600 m, and the east coast. Rainfall, particularly in the southern parts of the east coast adjoining the Bay of Bengal, is comparatively low during the monsoon period as the area lies in the rain shadow of the Western Ghats. However the same area receives rainfall later in the year, in November and December, which is the period of the withdrawal of the southwest monsoon. This is referred to as the northeast monsoon period. The southwest monsoon, as moderated by the Himalayas, the Western Ghats and other lesser mountain systems, is responsible for the variety of climates found in India.

The soils of India have been broadly classified into as many as 25 different types, but in this brief survey we are concerned with the five most widespread, which are:

1. Red soils which occupy 500 000 km^2, lying chiefly in the southern, eastern and northeastern parts of the country. The main features of these soils are their lighter texture and porous, friable nature,

an absence of limestone or free carbonates, and the presence of soluble salts in small quantities. These soils are neutral to acid in reaction.

2. Laterite soils occupy 130 000 km^2 and are composed essentially of a mixture of the hydrated oxides of aluminium and iron, and are found mainly in the hilly regions and on the western coast, especially where the rainfall is heavy and humid conditions prevail. These soils are highly acidic, owing to the leaching of calcium with heavy rainfall.
3. Black soils extend over large areas of the Deccan Plateau and adjoining areas. As distinct from red soils, these are generally much deeper and contain a fair percentage of limestone and carbonates, resulting in a high pH of 8.5–9.
4. Alluvial soils (including calcareous, saline, alkaline and coastal soils) occur mainly in the southern, northwestern and northeastern parts, and, most importantly, in the Indo–Gangetic Plain (i.e. the areas served by the rivers Indus and Ganges), and occupy 800 000 km^2). These soils are amongst the most fertile in the country and are very deep with the common feature of the presence of limestone. The pH is generally neutral to alkaline.
5. Desert soils occupy large areas in the arid parts of western and northwestern India, lying between the river Indus and the Aravalli mountain range. The Rajasthan or Thar Desert extends to about 100 000 km^2. These areas receive little or no rainfall and the soils are characterized by a high percentage of soluble salts and a high pH.

Wild Roses

The distribution of species roses in India is restricted to the mountains, with the notable exception of *R. clinophylla (R. involucrata)* which is perhaps the world's only tropical rose species. The species are listed below and further comments on some of their distributions and notable features follow.

Subgenus *Rosa*
Section Pimpinellifoliae
R. *ecae* Boul.
R. *foetida* Herrm.
R. *sericea* Lindl.

Section Cinnamomeae
R. *beggeriana* Shrenk.
R. *macrophylla* Lindl.
R. *webbiana* Royle

Section Indicae
R. *gigantea* Coll. (= R. *odorata* var. *gigantea* (Coll.) Rehd. et Wils.)

Section Synstylae
R. *brunonii* Lindl.
R. *leschinaultiana* Wight et Arnold (R. *moschata* Herrm. var. *leschinaultiana* Wight et Arnold)
R. *longicuspis* Bertol.
R. *moschata* Herrm.

Section Bracteatae
R. *clinophylla* Thory (= R. *involucrata* Roxb.)
R. *clinophylla* Thory var. *parviflora*
R. *lyelli* Lindl.

Distribution of Wild Roses Almost all of India's species are found in the Himalayas. In the extreme west, in the very dry climate of Ladakh, and in the upper reaches of Himachal Pradesh State (altitudes between 3000 and 4000 m) we have R. *ecae* and R. *foetida*, both with bright yellow flowers. Moving slightly east we have the lilac–pink R. *webbiana* on the rocky slopes of the western Himalayas. As we progress further east, with an increase in rainfall, R. *macrophylla* with bright pink flowers and very large hips is to be found at altitudes of 2100 and 3800 m.

A typical rose of the lower Himalayas found almost throughout its length is the white-flowering climber, R. *brunonii*, the Himalayan Musk Rose. The climber R. *moschata* is also reported to occur in the extreme west of the Himalayas, as also is R. *beggeriana* which is common in the Kashmir Valley. Another species that occurs throughout the length of the Himalayas is the shrubby, four-petalled R. sericea, with cream flowers; it occurs at altitudes of 2100–4500 m.

In the extreme east of the Himalayan range and, more particularly, on the adjoining Kashi Hills of Meghalaya State, is found R. *longicuspis*, another member of the musk rose complex with white flowers. This is a region with very heavy rainfall. In the adjacent hills of Manipur and Nagaland States of northeast India, near the border with Myanmar R. *gigantea*, a very vigorous climber with very large creamy yellow flowers and large globular hips, is found. This form of R. *gigantea* is distinct from the form found in Myanmar and southern China, which generally has white flowers. In addition to R. *brunonii* and R. *longicuspis*, a third member of the musk rose complex, R. *leschinaultiana* (R. *moschata* var. *leschinaultiana*) is found in the Nilgiri and Palni Hills of the Western Ghats in southern India. This is a climbing shrub with white flowers.

Apart from the species found in the mountains, we have R. *clinophylla (R. involucrata)*. This exists in three forms. The first form *(R. clinophylla parviflora)* is found in the flood plains of the Rivers Ganges and

Brahmaputra – on their banks and islands, in West Bengal State and in the adjoining Bangladesh. This form has the interesting capacity of withstanding prolonged submersion in floodwaters with only the tops of its branches visible above water. The second form is found in the Chota-Nagpur Plateau of Bihar State at an altitude of around 700 m. An interesting feature is that this form also occurs in many other parts of India – northern Orissa State, northern Andhra Pradesh State and the mountains of Coorg in Karnataka State. The third form, also called *R. lyelli*, is found in Mount Abu, which rises out of the Thar Desert in Rajasthan to an altitude of 1500 m. All three forms are shrubs with white flowers, which are solitary or paired and borne in corymbs in the third form.

Cultivated Roses

The diversity of climates and soil types results in substantial differences in methods of cultivation of roses. Perhaps the most important result is the requirement of different rootstocks for different regions. Broadly speaking, *R. multiflora* is the preferred rootstock for regions with acid soil and moderate summer temperatures, while *R. indica* var. *odorata* is preferred in areas with alkaline soil and high summer temperatures.

The plains of India are generally too hot to grow species roses, though some like *R. laevigata* and *R. banksia* are grown, especially in northern India. But some of the heritage roses, particularly the Teas, Chinas, and the sturdier Bourbons and Hybrid Perpetuals, are widely grown. A wide range of modern roses is also grown.

There are three important commercial features of roses in India:

1. The cut-flower industry, which has progressed rapidly in recent years, with over 50 big export-oriented enterprises, mostly based on the Deccan Plateau region, around Poona, Bangalore and Nashik. Exports of the red roses 'Grand Gala' and 'First Red'® for Valentine's Day are very profitable. The latest cut-flower varieties like 'Leonidas', 'Papillon' and 'Konfetti' are grown.
2. Field-scale growing of traditional varieties like 'Rose Edouard', 'Grüss an Teplitz' and 'Souvenir de la Malmaison' for use for garlands, temple worship and festivals.
3. Huge numbers of 'R. Edward' and to a limited extent *R.* × *damascena* are grown widely for the extraction of rose oil, rose-water and other products used for cosmetic, medicinal and culinary purposes, such as 'gulkhand' (rose preserve), 'gul roghan' (hair oil) and 'pankhuri' (dried rose petals).

Pakistan

Climate and Soils

The climate of Pakistan is subtropical to subtemperate. Therefore the conditions under which roses are cultivated in this country are very different from those of European countries. Only the conditions at hill stations like Murree, Quetta and some high valleys in the north are such that some comparison might be made. In Pakistan there is no truly dormant period and roses flower continuously throughout the year in a succession of flushes of blooms. In the northern and central parts of the country where some of the best roses are grown, summer is very harsh. The finest blooms are produced in the spring months of March and April. In terms of colour good blooms are produced during the winter, when there is hardly any rain, but the night temperatures may descend far enough to damage the open blooms or, in some cases, to prevent them from opening at all. In Islamabad low temperatures during January and February are problematic unless some resistant varieties are selected, and a similar situation is also found in Peshawar. However Karachi is humid all the year round and good blooming cannot be obtained. On the other hand in the area of Quetta the flowering period of the winter months could be extended to the entire summer period from April to September. Thus the irrigated northern and southern plains of the country provide the requisite agro-ecology for the production of roses during the winter months. For the domestic market, summer is a lean period in which the production in Quetta plays an important role.

Roses do well in soils having a pH between 6 and 7.5, that is, soils that are slightly acidic, neutral or slightly alkaline. The rose plant is sensitive to saline and alkaline soil. It is quite likely therefore that only occasionally would one come across a soil which would be found inherently ideal for rose growing. Few plants respond so well as roses to soil management, and this is the key to success in rose culture. Since most of the soils in Pakistan are alkaline in nature with pH values ranging from 7.5 to 8.5 a large effort is needed to improve the situation through a regular programme of organic manure application. In Islamabad the use of well-rotted pine needles is useful in lowering the pH range. Most of the soils are deficient in nitrogen and organic matter while a high soil pH range restricts the availability of critical trace elements.

Wild Roses

Records show that wild roses were once widespread in Pakistan but, due to the loss of habitats and human overpopulation, most of the species roses are now only found above 1500 m and some are threatened and require protection. The number of species roses mentioned in literature cannot in reality be observed in great numbers in the wild. A new assessment of the present state of roses is called for. The species listed below species are to be found in Pakistan, and further comments on some of their distributions and notable features follow.

Subgenus *Rosa*
Section Pimpinellifoliae
R. ecae Boul.
R. foetida Herrm.
R. kokanica Regel ex Jusep.
R. sericea Lindl.

Section Caninae
R. canina L.
R. nanothamnus Boul.

Section Cinnamomeae
R. beggeriana Shrenk.
R. lacerans (Boiss et Buhse) R. Keller (= *R. beggeriana* var. *lacerans* (Boiss et Buhse) R. Keller)
R. macrophylla Lindl.
R. webbiana Royle

Section Synstylae
R. brunonii Lindl.

Distribution of wild roses *Rosa foetida* and *R. hemisphaerica* are both found in Pakistan. The latter has hooked prickles and a pleasant smell as opposed to the straight prickles and unpleasant odour of *R. foetida*. Their flowering time is June and July. *R. ecae* is common in hedges around Kargil in Ladakh. It is a small shrub with small leaves and small solitary golden yellow flowers. *R. kokanica* flowers in May and June and is found in Baluchistan and the northern areas. *R. sericea* is also distributed in the northwest Frontier Province. *R.* × *damascena* is cultivated in the Chakwal District of Punjab as a small local industry. *R. canina* is found in Azad Kashmir, Mir Pur District, the northwest Frontier Province and Baluchistan. *R. beggeriana* is found in the northwest Frontier Province where it flowers in July and August; the same is true of its close relative *R. lacerans*. *R. macrophylla* is a very common and variable pink rose with large red fruit which are delicious and are utilized commercially for jams, etc. *R. nanothamus* flowers in May and June and is closely related to *R. webbiana* which it has replaced in the dry places of the inner valleys. *R. brunonii* is a great tree-climber and a wonderful sight in spring in the Murree Hills above Islamabad. *R. multiflora* is an escaped species in Pakistan where a double pink form is widely found. *R. laevigata*, also native to China, flowers in March and April, is a climber with large white flowers, rarely pink ones, and is to be found in the plains and lower hills.

Cultivated Roses

A wide range of modern cultivars grow well under the local climatic conditions and *R. banksiae* is cultivated as an ornamental plant in various areas.

In Pakistan all parts of the rose plant are widely used medicinally both as a single ingredient or as a compound with other ingredients. In addition jams, rose-water and attar of roses is widely prepared. The flowers are popularly employed in the making of garlands, potpourris and as cut flowers for floral decoration.

Japan

Climate and Soils

Japan is a mountainous island country. It consists of four major islands and dozens of smaller ones stretching for more than 2400 km. The total area is about 378 000 km^2 – smaller than the state of California in the USA – and 67% of the total surface is covered with mountains and hills. Thus the usable land is extremely limited. The southeastern side of the Japanese archipelago faces the Pacific Ocean and enjoys four distinct seasons of similar duration. Except in the northern and mountainous areas, where the weather in winter is severe, the average temperatures and humidity are high compared with Europe. The summer is extremely hot and humid, while in winter the temperature seldom falls below freezing. In the middle of June the rainy and humid season begins, lasting about a month. From June to September a number of typhoons occur in the southern Pacific Ocean and may cause serious damage across the Japanese islands.

About 80 mountains are active or dormant volcanoes, and geological features in this country are largely related to their activities. Thick layers of volcanic ash cover most of the gentle mountain slopes on which various types of forests have grown and added rich humus, turning them into brown or yellow forest soil. This is the most commonly observed type of soil in the mountain regions, and it accounts for about 50% of the total land area, which supports many orchards, tea plantations and vegetable fields in these regions.

A unique type of soil often referred to as 'kuro-boku' (meaning bland and crumbling) is widely observed in large areas of flat lands and gentle mountain slopes, about 16% of the total area. This soil, originating in volcanic ash and basically clayish, contains much of humus and forms a highly porous structure, advantageous for growing plants. In spite of this good physical characteristic, it has a rather high acidity, and often tends to fix phosphates in insoluble forms, thus preventing plants from taking them in. This poses some problems in growing roses too.

Since the country is small in area and mountainous, most rivers are short and fast flowing by continental standards. Even so, there are several alluvial plains, though small in area again by comparison to those of other continents. The soil there can be regarded basically as an enriched type of the above two.

The Japanese people have grown rice in paddy fields for more than a millennium. The soil in paddy fields, improved by human hands over generations, has many unique features ideal for growing agricultural crops. Other types of soils, such as podzol, limey soil, or peaty soil, also occur in limited localities in Japan.

Wild Roses

Subgenus *Rosa*

Section Cinnamomeae

R. acicularis Lind. is found on Hokkaido and Honshu in the high mountains on the Sea of Japan side of the central to northern areas, Sakhalin, the Korean Peninsula, China, Siberia and the Kamchatka Peninsula. In fact it is distributed widely in most of the areas of high latitude in the northern hemisphere – it is one of the 'circumboreal plants'.

R. marretii Lévl. grows on Hokkaido, Honshu, Sakhalin and the northern areas of the Korean Peninsula.

R. nipponensis Crép. is found on the high mountains on Honshu and amongst the conifer zones and dwarf Siberian pine zones of Mount Tsurugi on Shikoku.

R. rugosa Thunb. ex Murray is distributed on Hokkaido and Honshu along the coastal areas.

Section Synstylae

R. fujisanensis Makino is frequently seen in the high mountains in central Honshu, while it is sparsely distributed in large areas from the Chichibu Mountains to Mount Omine on Honshu Island and in the mountains in central Shikoku.

R. jasminoides Koidz grows in mountainous areas, mainly chestnut ranges in Honshu, Shikoku and Kyushu; it is not to be found on the Sea of Japan side of the archipelago.

R. luciae Franch. et Rochebr. grows on the downs and lower mountains in Honshu, and near Tokyo.

R. multiflora Thunb. ex Murray is found in the southwestern parts of Hokkaido, the lowlands of Honshu, Shikoku and Kyushu as well as the Korean Peninsula. This rose grows naturally in many parts of Japan with a considerable number of local variations, all with great fragrance. *R. multiflora* var. *adenochaeta* (Koidz) Ohwi is distributed on Honshu and Kyushu. Although generally considered to be a variety of *R. multiflora*, it could be a hybrid of *R. multiflora* and some other roses. The flowers are pink in varying degrees, the large bush is spreading, and its canes are very vigorous. There is a widely known colony of this rose on the banks of the Kuma River in Hitoyoshi City, Kumamoto Prefecture, Kyushu.

R. onoei Makino is also found in the mountainous areas of Honshu, Shikoku and Kyushu.

R. paniculigera Makino is distributed on Honshu, the northern parts of Shikoku and the downs and lower mountains in the north of Kyushu.

R. sambucina Koidz is to be found on Honshu, Shikoku and Kyushu.

R. wichurana Crép. is found on Honshu, Shikoku, Kyushu, Ryukyu, in the Korean Peninsula and China. *R. wichurana* var. *glandulifera* (Koidz.) Honda is found on southern Kyushu and Ryukyu.

Section Bracteatae

R. bracteata Wendl. is found on the Sakishima Islands, Formosa and the shores of southern China.

Subgenus *Platyrhodon*

R. hirtula (Regel) Nakai is found on Honshu.

Distribution of wild roses The 14 roses listed above are indigenous to Japan. Their hybrids are also to be seen growing in various parts of the country. Further, there are those species roses that were imported in earlier years, such as *R. laevigata* and *R. banksiae*.

Cultivated Roses

Most roses can grow and bloom in Japan, located as it is in a temperate zone and ordinarily blessed with a warm climate. Except in the northernmost island where winters are severe, even those roses regarded as sensitive to cold weather survive in Japan without special protection. Seasonal winds bring in humidity from the seas and precipitation is high. However, the high temperature in summer, almost comparable to that in tropical regions, is a grave disadvantage to some varieties. Hybrid Tea roses give flowers of such poor quality in hot weather that some rosarians prefer to remove the buds and a considerable part of new growth in late summer, considering that it is better to give a rest to the bushes so that they will give better flowers in the autumn.

The high temperature and humidity during the long growing season also provide good conditions for the spread of fungal diseases. Thus the most appropriate roses for gardens in Japan are those highly resistant to fungal diseases. Both old garden roses and modern varieties tend to suffer from them if left unprotected. Regular spraying of pesticides is also necessary, or rather a necessary evil in recent years when more and more people are oriented to environmentally friendly rose culture. Only a limited number of roses can be grown without chemical fungicides. *Rosa rugosa* and most of its hybrids and a few *R. wichurana* hybrids are known to show good resistance to black spot.

Until a decade ago Hybrid Tea roses of exhibition type were generally preferred in Japan, but recently Floribundas and 'English Roses' have acquired popularity. However gardens are mostly very small and rosarians are making every effort to grow as many roses as possible in limited space. Shrub roses are not yet established as a group while climbers of compact habit are gaining popularity. On the other hand Miniature roses are widely grown in Japan, in harmony with the local taste for something small but perfect, and also because they suit terrace gardening.

See also: **History of Roses in Cultivation:** Ancient Chinese Roses; European (Pre-1800); Modern (Post-1800). **Rose Collections and Trials:** Rose Collections and Herbaria.

Further Reading

Brandis D (1874) *The Forest Flora of North West and Central India.*

Flora Republicae Popularis Sinicae (1985) Vol. 37. Beijing: Science Press. (Presently being translated into English in cooperation with the University of Missouri. Regarding the genus *Rosa*, a provisional version may be found at the site http://hua.huh.harvard.edu/china/mss/volume09/Rosaceae-MO-Rosoidea_edited.htm)

Fortune R (1847) *Three Years Wanderings in the Northern Provinces of China.*

Fortune R (1853–7) *A Residence Among the Chinese.*

Fortune R (1863) *A Narrative of a Journey to the Capitals of Japan and China.*

Hickey M and King C (1997) *Common Families of Flowering Plants.* Cambridge, UK: Cambridge University Press.

Hooker JD (1879) *The Flora of British India.* London: L. Reeve & Co. Ltd.

Hutchinson J (1980) *The Genera of Flowering Plants.* Otto Koeltz.

Krüssmann G (1982) *Roses.* London: Batsford.

Le Rougetel H (1988) *A Heritage of Roses.* London: Allen & Unwin.

Moody M (1992) *The Illustrated Encyclopaedia of Roses.* Portland, OR: Timber Press.

Phillips R and Rix M (1988) *Roses.* London: Pan Books.

Europe

H Brichet, World Federation of Rose Societies, San Terenziano, Italy

The author gratefully acknowledges the contribution of Frans Thomas (Société Royale Nationale "Les Amis de la Rose", Belgium), Josef Thomas (Czech Rosa Club), of Klaus-Jürgen Strobel and Georg Timmermann (Verein Deutscher Rosenfreunde), Petra Meyer (Hellenic Rose Society), Stefano Marzullo and Guiseppe Fabrini (Associazione Italiana della Rosa), Ghislaine de Schorlemer (Letzeburger Rousefrenn), Beatrijis Arts van Overbeek (Nederlandse Rozenvereniging), Arthur Dixon (Rose Society of Northern Ireland), Rein Lae Solberg (Norwegian Rose Society), Irene Golebiowska (Polish Society of Rose Fanciers), Stefan Wagner (Asociatia Amicii Rozelor din Romania), Milly Weyermann (Gesellschaft Schweizerischer Rosenfreunde) and Andy Roberts (Royal National Rose Society, United Kingdom) in the preparation of this article.

Europe

Climate and Soils

The conventional limits of Europe are normally taken as the Arctic, Mediterranean, the North Sea and the Black Sea, with the line of the Ural Mountains, continued by the Ural River to the Caspian Sea, as the eastern boundary. Though in a sense it can be regarded as a peninsula of the much larger continent of Asia, Europe is itself a series of peripheral peninsulas – Iberia, Italy, Greece and Scandinavia – all having mountainous backbones. On the mainland mass of Europe, the Alpine arcs form a high central core.

The main northeast to southwest ridge of Scandinavia is of great climatic importance as it places part of Sweden in a rain shadow. The absence of any considerable relief over the rest of northwestern Europe permits the deep penetration of sea influences, which affect the climate of the area. The plain of northern Germany and Poland is flanked by the central uplands on its southern side, the outer periphery being the Jura Mountains. The Alpine chains tend to extend in long arcs around interior basins, such as the Po basin in Italy and the Hungarian plain. It is only in the Mediterranean regions that anything approaching a dry season – with its associated effects on the river regimes – is encountered; elsewhere the effect of snowmelt influences not only the regime but also the usefulness of the rivers.

Europe may be divided into four main climatic regions:

1. The western coastal margin enjoys to the greatest extent the warm North Atlantic Drift, the influence of which extends well inland, as well as the Gulf Stream. The region has mild temperatures and rain falls throughout the year with slight autumn and winter maxima. Northern Ireland is a clear example of this: 70% of the country lies between sea level and 150 m and is generally drier in the east and wetter in the west with pronounced rain shadow effects in the lee of mountainous areas. The annual rainfall lies between 800 and 1200 mm with the wettest period of the year from August to January.
2. The eastern European climatic region encompasses Hungary, Romania, much of Russia and north to Scandanavia. Winter temperatures remain well below freezing for about 3 months which cause damage to roses so that sensitive varieties, especially Hybrid Teas and some Floribundas, need protection. Summers are quite hot and springs can be capricious with late frosts. Rainfall is normally concentrated in summer and spring.
3. Between the maritime fringe and the continental interior is an area of transition. With no precise limits, it adopts the characteristics of the nearby climates. For example, the Czech Republic lies in the temperate zone of Europe with a regular weather pattern and no great extremes during winter or summer. It occupies a key position in central Europe having formed, in the glacial era, a narrow passage between the north European and Alpine areas of glaciation. In general, the area has temperatures of around or just below freezing in January, warm summers, with increasing temperature range, decreasing rainfall, and an increasing tendency toward summer maxima of rainfall towards the east. The Alps have a dominating influence over the formation of warm wind during the cold season on their northern slopes, which permit certain forms of Mediterranean vegetation to flourish far beyond their normal geographical limits in favourably situated valleys.
4. The southern part of Europe has a very distinctive climate delineated by the mountain barriers that mark its northern limit, holding back the extreme continental influences, and the seas which act as a continuation, in some ways, of the Atlantic, so that the narrow coastal belt enjoys mild winters and hot summers. The Mediterranean is the warmest area at all seasons and is also the area of winter rain and relative summer drought, enjoying many hours of sunshine. In Greece, as in Israel, this may mean too many hours of sun for roses, for often the leaves burn while the flowers take a siesta during the summer months, as they do in all countries around the Mediterranean. Italy, with its very long coastline and sun-baked islands, falls to a great extent into this group; however, its northern frontier is in the shadow of the Alps. Traditionally the Ligurian coast and the French Riveria furnished northern Europe with fresh flowers, and particularly roses, during the winter months. The Spanish Mediterranean coast too, as opposed to that country's dry central plateau, is very favourable to rose cultivation.

The greater part of Europe was covered by a zone of mixed forest and although the forest has largely been cleared, the characteristic brown earth soils remain. These merge with the podsols to the north and the Mediterranean soils with a reddish element in their colouring, especially those developed on calcareous rock. The soil types are dependent on underlying parent rock, glaciation, and local climatic and vegetational differences. Within each country there is a great variety of soil types. In a country as small as The Netherlands these range from poor sand, sandy soil, light loam, medium loam, black peat to light and heavy clay. In Switzerland, the Jura area has rightly been named the most privileged rose garden of Europe. No other area produces such a wide variety of forms, such a wealth of types and varieties. The reason for this extremely rich flora is the enormous diversity of subsoil on a small area. Switzerland claims of 0.4% of European territory but hosts more than 20% of the native European flora.

Wild Species

Subgenus *Rosa*

Section Synstylae

R. arvensis Huds. is deciduous with weak and trailing stems and protruding and fused styles, found in south, west and central Europe.

R. moschata J. Herrm. is a vigorous evergreen climber up to 12 m which is cultivated for its ornamental value, mainly in south and western Europe, and naturalized in the Mediterranean region.

R. phoenicia Boiss. is confined to north eastern Greece.

R. sempervirens L., evergreen, with long creeping stems, is found in the Mediterranean region as well as southwestern Europe and in western France.

Section Pimpinellifoliae

R. foetida J. Herrm. has stems of up to 4 m and curved prickles and yellow flowers. Although cultivated for ornamental purposes, it is locally naturalized in south and central Europe.

R. pimpinellifolia L. is of erect growth with abundant prickles and acicles, the flowers occur singly. It is found in all parts of Europe except the northeast, most parts of Fennoscandia, the extreme southwest and many of the islands. There are many hybrids between *R. pimpinellifolia* and members of the *Caninae* section for example, *R.* × *hibernica* (The Irish Rose).

Section Cinnamomeae

R. acicularis Lindl. has stems of up to 1 m with numerous slender acicles and long straight prickles. The purplish-pink flowers are solitary and scented. It is located in northeastern Europe and central and eastern Russia.

R. blanda Ait. has unarmed stems, the leaflets being dull and glabrous and paler beneath. It is cultivated for ornamental value and is naturalized in central Europe.

R. glauca Pourret has erect bluish-green stems when young which later become brown. The leaflets are bluish-green or purplish. Flowers are deep pink. It is native to the mountains of central Europe, extending southwards to the Pyrenees, central Italy and northern Albania.

R. majalis J. Herrm. has stems of up to 2 m, forming large patches and reddish-brown bark. It is confined to north and central Europe and the former Soviet Union except for the southwest.

R. pendulina L. has prickles usually absent and flowers solitary, the red fruit pendent. It is found in the mountains of central and southern Europe.

Section Rosa

R. gallica L. is a deciduous shrub forming large patches. The deep pink flowers are solitary and strongly scented. Its distribution extends from south and central Europe as far as Belgium and central France.

Section Caninae

All species in this section have a system of inheritance in which more chromosomes are received from the mother than from the father (*see* **Genetics**: Inheritance in the Dogrose). The resulting 'matroclinal' inheritance creates taxonomic difficulties. Consequently, the status of some species is uncertain, and the geographical distribution of many of the species is imperfectly known and can only be given in general terms.

R. abietina Gren. is native to the Alps.

R. agrestis Savi has smooth hips and pedicels. It is found in most of Europe but is rare in the north and east.

R. andegavensis Bast. is found in west, south and central Europe.

R. caesia Sm. in Sowerby is found in most of Europe eastwards to Estonia and the central Ukraine.

R. canina L. has green stems and pink to white flowers and is very widely found throughout Europe. It is distinguished from *R. caesia* by its smaller hips with reflexed sepals.

R. caryophyllaca Bess. is found in eastern and central Europe as far as the Balkan peninsula and the Ukraine.

R. corymbifera Borkh. is probably widespread throughout Europe but rarer in the north and northwest.

R. deseglisei Boreau is found mainly in central Europe, but the distribution is not fully known.

R. elliptica Tausch is present in west and central Europe, extending southeastwards to Albania and the western Ukraine.

R. glutinosa Sibth. and Sm., as indicated by the name, has leaflets more densely glandular. It is found in the eastern and central Mediterranean area and the Balkan peninsula.

R. heckeliana Tratt. is present on the mountains of the eastern Mediterranean region and Sicily.

R. jundzillii Bess. is native to central and eastern Europe, extending westwards to central France and northwestern Italy.

R. micrantha Borrer ex Sm. in Sowerby is a slightly larger plant and is current in western, southern and central Europe as far as northern Ukraine.

R. mollis Sm. in Sowerby has fairy leaflets and glandular pedicels. It is mainly in north and western Europe, but extends locally to south and central Russia.

R. montana Chaix in Vill. is confined to south and parts of south-central Europe.

R. nitidula Besser is found from Britain and northern Portugal eastwards to southern Sweden, the Carpathians and Greece.

R. obtusifolia Desv. is present in central, south and northwestern Europe.

R. orientalis Dupont ex Ser. is found in southern parts of the former Yugoslavia, northern Albania and Greece.

R. pouzinii Tratt. is confined to the Mediterranean region and Portugal.

R. rhaetica Gremli is present in the Alps.

R. rubiginosa L. has stout prickles usually mixed with acicles: it is common in most of Europe northwards to 61°N.

R. scabriuscula Sm. in Sowerby is probably widespread in most of Europe, except the northern and eastern margins.

R. serafinii Viv. occurs in the central Mediterranean region, Bulgaria and southern Yugoslavia.

R. sicula Tratt. is typical of the Mediterranean region.

R. sherardii Davies has erect sepals and large stylar orifices. It is found in north, west and central Europe, eastwards towards southwestern Finland and extending southwards to Bulgaria.

R. squarrosa Boreau is found mainly in central Europe, but its distribution is not fully known.

R. stylosa Desv. has hooked prickles, some with very stout bases. The styles are connate in a column which is shorter than the inner stamens. Its distribution is from Ireland and western Germany to southern Spain and Bulgaria, but also with localized appearances.

R. subcanina Dalla Torre and Sarnth. is much like the above species or subspecies, and is probably widespread in Europe, but rarer in the west.

R. subcollina Dalla Torre and Sarnth. is probably widespread in Europe except in the extreme north.

R. tomentosa Smk. receives its name from the densely pubescent or tomentose leaflets and is found in most of Europe except the extreme north.

R. turcica Rouy is also present in southeastern Europe.

R. villosa L. is present in central and southern Europe extending northwards to The Netherlands.

R. vosagiaca Desp. is found in most of Europe, northwards to Iceland and Fennoscandia, but rarely in the southwest of the continent.

Exotic species that are cultivated for ornamental purposes and also locally naturalized include *R. virginiana* and *R. rugosa*. Species that occur naturally within Europe but are naturalized outside their normal range include *R. moschata*.

Cultivated Roses

As has already been said with regard to the North American continent, considering the variety of soils and climatic conditions present across Europe, the natural home of so many species, cultivated roses of all types can be grown well somewhere and to some purpose. The history of rose cultivation in Europe dates from Greek and Roman times. Through the Dark and Middle Ages, the rose was practically confined to the gardens of monasteries and convents, until its rediscovery during the Renaissance period and later its supremacy as a garden plant with multiple uses. The modern era of rose cultivation in Europe can be said to have initiated with the importation of recurrent blooming varieties from the Orient.

Roses are grown extensively both in municipal and in private gardens. Most of the cities and towns and some villages have formal plantings of roses, these days mostly Floribundas for greater effect. In recent years the trend for municipal planting of groundcover and shrub varieties has increased due to their low maintenance costs. Some species such as *R. rugosa* are planted along the roadways where the soil is suitable, making it difficult to distinguish garden escapes from populations that may have been planted. In parks, fruit-bearing bushes are used to attract birds.

Roses are grown in many private gardens with a trend towards Patio and Miniature roses, although Floribundas are still widely grown. The climbing forms of Hybrid Teas are popular but have a limited winter hardiness. There is also great enthusiasm amongst rose lovers who have the space to grow old roses as well as species and early hybrids. Private gardeners are becoming more interested in disease-resistant roses and those with low maintenance costs. It is of interest that in the countries which were until 1989 behind the Iron Curtain in eastern Europe, the number of varieties now available to the public has greatly diminished. Previously budwood was imported illegally from the west, now however growers are obliged to pay patent rights which result in a huge increase of the price of the individual plants.

Many city municipalities and rose societies in Europe annually host prestigious International Rose Trials for new varieties in order to test roses from around the world under local conditions. The public is encouraged to visit these trial grounds and see which new varieties are likely to be most successful in their own gardens.

See also: **Classification**: Conventional Taxonomy (Wild Roses). **Distribution and Ecology**: Overview. **History of Roses in Cultivation**: European (Pre-1800); Modern (Post-1800). **Rose Collections and Trials**: Rose Collections and Herbaria.

Further Reading

Anonymous (1968) *Flora Europaea*, vol. 2. Cambridge, UK: Cambridge University Press.

Anonymous (1985–2001) *Flora of the Czech Republic*. Prague: Academia.

Anonymous (1996) *Flora Helvetica*. Bern: Paul Haupt Verlag.

Anonymous (1998) *Flora van België, Luxemburg, Noord-Frankrijk en de aangrenzende Gebieden*. Meise, Belgium: National Plantentuin van België.

Beales P (1985) *Classic Roses*. London: Collins Harvill.

Cappelletti C (1975) *Trattato di Botanica*, vols. 1 and 2, Third Edition, Torino: UTET.

Genders R (1965) *The Rose: A Complete Handbook*. London: Robert Hale.

Graham GC and Primavesi AL (1993) *Roses of Great Britain and Ireland*. London: Botanical Society of the British Isles.

Hole SR (1869) *A Book of Roses*. Edinburgh, UK: William Blackwood.
Krüssmann R (1982) *Roses*. London: Batsford.
Pignatti S (1982) *Flora d'Italia*, vol. 1. Bologna: Edagricole.
Shepherd R (1954) *History of the Rose*. New York: Macmillan.
Stace C (2001) *New Flora of the British Isles*. Cambridge, UK: Cambridge University Press.
Testu C (1984) *Les Roses Anciennes*. Paris: Flammarion.

Israel

H Brichet, World Federation of Rose Societies, San Terenziano, Italy

The author gratefully acknowledges the contribution of Dalit Kaslassi (Israel Rose Society) in the preparation of this article.

Israel

Climate and Soils

Israel has a predominantly Mediterranean climate with two principal seasons: a rainy winter and a dry summer. The climate is temperate with no extreme temperatures. Generally, the annual precipitation decreases steeply the further south or further east one goes. The average annual temperatures are very much higher in the south than in the north. Likewise, the Jordan River valley and the coastal plain are much hotter than the mountainous areas. The climatic areas in Israel listed below are largely characterized by their average temperatures and amount of precipitation:

1. Mediterranean climate: The Galilee, the mountainous areas and the coastal plain.
2. Semiarid climate: the northern Negev, Jordan River valley and the Beit She'an valley.
3. Desert climate: the southern Negev, the Dead Sea area and the Araa Desert.

The strong level of solar radiation throughout the year in Israel is exceptional. There is a total of 3300 h of sunlight a year, which is one of the highest rates in the world. Due to the angle of the radiation, a large amount of energy is released during the hours of sunlight. The high level of radiation causes a relatively high rate of evaporation, but, on the other hand, is very beneficial to agriculture.

There are some 20 different soil types in Israel. The main ones being the following:

- Terra Rosa, which is reddish-brown soil with a medium-heavy texture and contains little calcium. It is found in the mountain areas.
- Rendzina, which is greyish soil with a medium texture and a high calcium content. It is found in the rift valleys and the Shefela (foothills above the coastal plain).
- Basalt-based soil (formed by crumbled basalt rock). It is dark brown soil with a heavy texture and is commonly found in the eastern Galilee.
- Red sand, which is reddish in colour, with a medium to light texture. It is found in most parts of the central coastal plain.
- Loess, which is a fine dust, borne on the wind from the southern desert and swept further when the riverbeds flood. It has a yellow–brown colour and a medium to light texture, and is found mainly in the southern Negev.

Wild Roses

While over the years many species have been introduced into the country and are grown in public parks or amateur gardeners, there are only three species of wild roses native to Israel.

Subgenus Rosa

Section Caninae

Rosa canina L. which grows in the mountains, in moist Mediterranean woodland on north-facing slopes. Israel is the southern border of the distribution of this species. *R. pulverulenta* Bieb. is a mountainous species. The only place it is found in Israel is on Mount Hermon, where it grows even at a height of 2800 m. This is the southern border of the distribution of the species.

Section Synstylae

R. phoenicia Boiss. grows on the banks of watercourses and on the edge of swamps. It can now only be found in the area of the sources of the Jordan River and along watercourses that do not dry up in summer. In the past it could be found throughout the coastal plain, the Galilee, the Carmel and other places.

Cultivated Roses

The land of Israel has frequently served as an important transit station along the paths of conquest and wanderings of many nations on their way to and from the Orient. As a result a diversity of roses reached the region in ancient times and were used for the perfume, pharmaceutical and incense industries. *Rosa* × *damascena* was apparently brought to the region in the sixth century BC and planted extensively around Jerusalem and Jericho. During the Greek and Roman rule the rose became one of the most popular and profitable forms of cultivation. Later under the British Mandate (1920–48) the approach to rose growing changed from the cultivation for industrial purposes to growing for trade as a decorative garden flower. Towards the end of the 1920s work also began on producing grafted roses in Israel. The first rootstock was *R. canina* followed by 'Manetti' and others. Since the 1930s and up to the present the principal rootstock used in this country is *R. indica* var. *major* which easily produces shoots and remains green during the winter.

During the early 1960s initial attempts were made to export roses to Europe. The main areas of cultivation were the region of the Jordan River valley, where roses were picked in winter and pruned in the autumn, and the coastal plain where they were picked in summer and pruned in the spring. Israel pioneered growing cut roses in subtropical climates and became the main supplier of cut flowers to Europe during the winter months. In recent years 500 million rose flowers have been exported every year as well as 8 million grafting shoots. Rose nurseries are located in most parts of the country, even in the Arava Desert. The total rose growing area is about 250 ha.

Today, 1 million rose plants are produced every year for gardening purposes. There are dozens of sturdy garden varieties that perform very well in the local climate. Around 50% are Hybrid Teas, 30% cluster-flowered varieties (Floribundas and Miniatures) and the others are shrubs and climbers. The main blooming season lasts from April to November. In the Ein Gedi area (400 m below sea level, on the coast of the Dead Sea) roses bloom chiefly in the winter and spring and are almost dormant in the summer due to the very high temperatures. Hybrid Teas and Floribundas in that area are pruned at the end of summer. Garden roses grow in almost every part of the country. They have to be watered throughout the dry summer months, usually by means of a permanent drip-irrigation system.

In the Mediterranean areas of the country, *R. indica* var. *major* is commonly used to make a border for fruit and citrus orchards. It is also used to ornament the outside edges of forests, and along roadsides and intersections. *Rosa canina* is used in similar settings. Hybrid Teas and Floribundas are used to control disease and pests in young orchards and vineyards in the northern part of the country. The habit is perhaps more decorative today than in earlier years when, above all, the China variety 'Old Blush' was widely used in many Mediterranean countries as a early indicator of mildew. Druze living in the Golan Heights make a potion from the fruit of *R. canina* to cure sore throats and stomach aches and to treat colds and flu.

See also: **Disease**: Rust. **History of Roses in Cultivation**: European (Pre-1800); Modern (Post-1800). **Production and Marketing**: Cut Rose Production.

Further Reading

Goor A (1981) *The History of the Rose in the Holy Land throughout the Ages*. Tel Aviv: Massada & Am Hassefer.

Palevitch D and Yaniv Z (1991) *Medicinal Plants of the Holy Land*. Tel Aviv: Tammuz Publishers Ltd.

Pines N (1999) *Via Rosa: A Professional Guide for Cut Rose Production in Israel*. Tel Aviv.

Kadmon J (1980) *Israel*. Tel Aviv: Yavneh Publishing House Ltd.

Zieslin N (2001) Greenhouse rose industry in Israel: past, present, future. *Acta Horticulturae* 547: 17–19.

Zohary M (1982) *Vegetal Landscapes of Israel*. Tel Aviv: Am Oved Publishers Ltd.

North America

H Brichet, World Federation of Rose Societies, San Terenziano, Italy

The author gratefully acknowledges the contribution of Marily Young (American Rose Society) and George Pagowski (Canadian Rose Society) in the preparation of this article.

United States of America and Canada

Climate and Soils

Given the very large territory of the United States and Canada, it is not surprising that almost all climatic conditions and soil types can be found here. From the long winters and true tundra of northern Alaska and Canada to the baking summers in the desert of Arizona, from the temperate climate and shifting sand dunes of the east coast to the perpetual summers and rainforests of Hawaii, from the temperature extremes and rich alluvial soil of the Midwest to the dry, windy plains and 'caliche' clay of Texas, it is possible to find any imaginable ecological niche somewhere in this country.

For the purpose of discussing geology and soil types, the United States and Canada can be divided very roughly into large zones. In the northeast and along the Appalachian Mountains ridge, one finds a base of primarily plutonic rock. In some coastal areas, this has been eroded to sand. Travelling south from the terminus of the Appalachians, the underlying layer is sedimentary, having been deposited as the bottom of the 'Great Inland Sea' of the Quaternary Period. In many areas, the soil has a very high iron content and one hears frequently of the 'red clay of Georgia' and its surrounding states. In the Midwest and Mid-south, one finds, again, primarily sedimentary rock, but of an older date. Much of this was formed during the Palaeozoic period, with the exception of an extension of younger material along the western edge of the area. In the northern parts of the Midwest, the soil overlaying the rock was deposited with the retreat of the Wisconsin glacier and tends to be some of the most rich, fertile soil in the world. The western third of the United States is dominated by a much younger geology, consisting of the volcanic rocks and soils built during the formation of the Rocky Mountains. The mountainous area themselves are still sparsely populated, little cultivated, and subjected to extreme climatic conditions. Western coastal areas, however, are very temperate, ranging from the relative dryness of southern California to near-rainforest conditions in the Pacific Northwest. Southern California is considered by many to provide almost ideal conditions for growing roses.

Wild Species

With the wide range of climatic conditions and soil types found in the United States and Canada, it is not surprising to discover that over 60 species and forms of the genus *Rosa* are native, with approximately 10 additional species that have become naturalized.

Subgenus Rosa
Section Carolinae
This section includes species that are native to the eastern and central portions of the United States. Their most distinguishing feature is that the seeds are located only at the base of the hips. Some species tend to re-bloom under ideal conditions.

R. carolina is a highly variable group, and many variations have been named and described. It can be found along the entire eastern seaboard, from Maine to Florida, and west to Iowa, Missouri, Oklahoma and Texas. In Canada, it is delightful to see along country roads in Prince Edward Island, Nova Scotia, etc. It was in cultivation prior to 1826.

R. foliolosa Nutt. is nearly thornless. It is not common, but is distributed across parts of Oklahoma and Texas. It has been in cultivation since about 1880.

R. nitida Willd. Its most distinguishing characteristics are its generally dwarf habit, its small, glossy, almost fernlike foliage and its enlarged prickles on the stems. It is found primarily in the extreme northeastern states of Maine, Vermont, New Hampshire and Massachusetts, and in eastern Canada. One of its few cultivars is 'Metis', 1967.

R. palustris Marsh. is known as the 'Swamp Rose'. It is well adapted to very moist areas such as the edges of lakes, ponds and small rivers. It is very similar to *R. carolina* and is found from Maine to Florida and west to Minnesota, Iowa and Missouri and Nova Scotia.

R. rudiuscula Greene is distinguished by its very bristly stems and its distribution is concentrated in the central part of the United Stares, from Wisconsin to northern Oklahoma.

R. subserrulata Rydb. is sometimes considered to be a form of *R. carolina* and differs from it only in the serration of its leaflets. It is found in the lower Midwest, in the states of Missouri, Arkansas, Oklahoma and Texas.

R. virginiana Mill. is distinguished by its very glossy foliage. It is widely distributed, ranging from Maine to Georgia, and west to Iowa and Missouri. In Canada it is grown by some in harsh-climate zone gardens.

Section Cinnamomeae
Species of this section are both widely distributed and wide-ranging in their characteristics. Its species can be found throughout the northern hemisphere.

R. acicularis Lindl. The so-called 'Arctic Rose' is extremely hardy. It is found in the most northerly border areas of the United States, from Maine to New York, west to Colorado and north into Alaska and Canada. In Canada it has been used in breeding work by Skinner and Bugnet, including crosses with *R. rugosa*, in their search for hardy cultivars.

R. × *englemannii* S. Wats. is apparently a naturally occurring hybrid between *R. acicularis* and *R. nutkana* in Canada.

R. amblyotis Meyer. The blooms of this hardy species are red and borne either solitary or in small clusters. It is also known as the 'Kamchatka Rose' and has a very limited range in western Alaska.

R. arkansana Porter. There is some dispute as to whether this species is, in fact, distinct from *R. suffulta*. It can be found in the north–central states, from Wisconsin to Kansas and Colorado. In Canada this species has been hybridized with garden cultivars to give rise to the 'Parkland' series of hardy, repeat-blooming cultivars. The double-flowered *R. arkansana* var. *plena* is found in the wild.

R. blanda Ait. is another very hardy species and is also known as the 'Hudson's Bay Rose' or the 'Labrador Rose'. It is distributed from Maine, through New York and Pennsylvania to Illinois, Missouri, Wisconsin and Minnesota, and Canada. It was introduced into cultivation as early as 1773. It is also found in Canada and its best known hybrid is 'Betty Bland' by Skinner, 1925.

R. californica Cham. & Schlecht. is a highly variable species and is closely allied to *R. spithamea* and *R. pinetorum*. Each of these may be forms of a single species. It is found only along the western coast from British Columbia to Washington and to southern California.

R. gymnocarpa Nutt. is shade tolerant and grows naturally in forest-edge habitats. It is found in British Columbia to the northwestern United States from Washington and Idaho to central California.

R. macounii Greene is very hardy, occurring in the states of North and South Dakota and Montana and north as far as northern Alberta where the cultivar 'Athabasca' was found growing spontaneously in 1930.

R. melina Greene is a little-known species found in Colorado and Utah.

R. mohavensis Parish is confined to an area in southern Utah.

R. nutkana Presl. is distributed from southern Alaska to northern California and eastward to Oregon, Wyoming and Utah.

R. pisocarpa Gray is distributed in inland areas from southern Alaska to British Columbia to California and Utah.

R. pryfera Rydb. is very similar to *R. woodsii* and is limited to Colorado, Wyoming, Utah and other areas in the Rocky Mountains.

R. salictorum Rydb. is perhaps a form of *R. woodsii* and is native to Idaho and Nevada.

R. sonomensis Greene is found only in California and is not in cultivation.

R. spaldingii Crép. is very closely related to *R. nutkana*. Its distribution has not been documented.

R. suffulta Greene has the unusual characteristic of being somewhat recurrent and can produce multiple flushes of bloom: once on the previous year's wood, and then on the new year's lateral growth. It is well adapted to the dryness of the Great Plains region, and is found in Canada, southward to Texas and from Indiana to Idaho.

R. ultramontana Heller is probably a form of *R. woodsii*. It is distributed in the north and northwest parts of the United States and Canada.

R. woodsii Lindl. has a wide natural distribution, occurring from Canada to Texas and across the central part of the United States from Iowa to Oregon. It has been in cultivation since about 1880.

Section Synstylae

R. setigera Michx. is probably the most widely distributed species in the United States, being found in practically every state east of the Rocky Mountains. It is also present in Canada. *R. setigera* f. *inermis* and *R. setigera* var. *serena* are virtually thornless.

Subgenus Hesperhodos

R. minutifolia Englem. The range of this species is extremely limited, as it occurs naturally only in southern California.

R. stellata Woot. is very similar to *R. minutifolia* but produces deep pink blooms with a distinct purple cast. The foliage is noted for its covering of fine hairs. Its range includes the lower mountains of southern New Mexico, western Texas and eastern Arizona.

Feral Populations

Many species that are native to other parts of the world have succeeded in establishing feral populations in North America. *Rosa spinosissima, R. bracteata, R. canina, R. rubiginosa, R. micrantha, R. cinnamomea, R. rugosa, R. rugosa plena, R. laevigata* and *R. multiflora* are naturalized in many areas of the United States. In Canada, feral populations of *R. canina* can be found in hedgerows, especially on old farm properties, sometimes with isolated cultivars of garden roses. Other species, including *R. glauca* in southern Ontario, have colonized roadside wastelands, some old orchards and conservation areas. *Rosa rugosa* has invaded tracts along the St Lawrence River and pockets in other parts of the maritime provinces. *Rosa nitida* is planted in gardens in the unpredictable climate of Newfoundland, Canada with *R. blanda, R. carolina* and *R. virginiana.*

Unfortunately, as a control for the rose rosette disease, a few roses, including *R. multiflora* and *R. bracteata*, have become agricultural pests, and have been listed as noxious weeds by many state agriculture departments in the United States.

Cultivated Roses of the United States of America

With the variety of soils and conditions present in the United States cultivated roses of all types can be grown well somewhere, to some purpose. Since William Penn, George Washington and Thomas Jefferson first brought European hybrids to America in the late eighteenth century, representatives from every subgenus and section have been imported, and at least one new class – the Noisettes – has been developed.

The American Rose Society, the third oldest in the world, is by far the largest and has been for many years the official international registrar for new rose varieties. In addition, it publishes, every 5 years or so, updated editions of *Modern Roses* (there are now 11 numbers), which publication not only includes all new registered varieties but now also aims at including all roses existent in the world today. This latter task is undertaken in coordination with the World Federation of Rose Societies Conservation Committee.

Cultivated Roses of Canada

Roses are popular throughout Canada, with Hybrid Teas being the most desired for home gardens. The cultivation of these tender roses is not confined to the milder climatic areas of the country, and enthusiasts are prepared to go to great trouble in harsh-winter and short-summer areas in order to secure their prize blooms, particularly for exhibition purposes. Indeed, except for a few locations in southern Ontario and south-coastal British Columbia, winter protection of such roses is essential.

It is recognized that Hybrid Tea and Floribunda rose cultivars cannot withstand prolonged periods of temperature below −15 °C. Canada experiences periods below −35 °C, even in south–central Ontario. Therefore, hybridizers continue in their efforts to produce roses comparable to Hybrid Teas with the added characteristic of being able to withstand long periods of very low temperatures. To this end, they have employed both indigenous and introduced hardy species and, happily, there are now many choice cultivars available, although not yet popular with many home gardeners and commercial producers. These cultivars are offered by several nursery firms in Canada, Germany and the United States.

The use of indigenous roses in home gardens is not widespread, but there are plantings, particularly on the northern vacation properties and in harsh climatic areas. Reports speak enthusiastically about the results, often mentioning the low maintenance requirements of such plantings. Native species can be found in botanical gardens and teaching collections, but the general public is drawn to the more colourful displays of cultivated hybrid roses. There has apparently been no Canadian government agency promoting the use of exotic species for hedgerow or roadside planting, as was the case in parts of the United States.

Showy exotics and relatively hardy, older hybrids are frequently grown in private gardens, particularly in urban centres. Occasionally one's attention is drawn to arbours or trellises clothed in *R. multiflora*: presumably these have grown from rootstocks, the scion of the original climbing rose having been winter-killed. As far as garden varieties are concerned, we are seeing increasing numbers of Shrub roses including David Austin's 'English Roses' and the 'Explorer' and 'Parkland' series as well as introductions from other hybridizers.

Bermuda

Climate and Soils

Roses are cultivated under rather unusual conditions. The climate is subtropical, enjoying a temperature of between 21 °C and 17 °C between November and March, and between 26 °C and 22 °C from April to October.

Bermuda's soil is not ideal for the growing of roses because of the rather high pH (pH 8 and higher). In addition, the soil lacks sufficient depth and water restrictions in the dry months add to the complications. In the autumn, rain produces the first flush of blooms, usually in October, and the plants continue to produce flowers during the whole winter period until the temperature becomes too hot. This does not however afford the plants a period of respite.

The most common diseases are black spot, powdery mildew and rose cankers, while pests are caterpillars, worms, aphids and thrips.

Wild Roses

Although located in the northern hemisphere, the islands have no native roses. However, settlers from many lands as well as ships coming directly from the Far East soon brought in a number of varieties, some of which became completely naturalized. Among these, *R. bracteata*, *R. banksiae*, *R. fortuniana*, *R. indica* var. *major*, *R. laevigata* and *R. roxburghii* var. *plena* are common.

Cultivated Roses

There are many roses whose original names or provenance is unknown and which constitute the unique class of 'Mystery' roses. Some may be sports or seedlings of roses long established, while others may have been imported and their proper names forgotten. In most cases the roses have been given the names of the owner of the garden or the location where they were found.

Generally speaking, modern Hybrid Teas and Floribundas do less well than Old Garden Roses because they are imported from countries that provide a dormant period and are grafted onto rootstock that is suited to cooler climates and different soils. Far greater success has been had with a number of species and own-root varieties belonging to the China, Noisette, Bourbon, Hybrid Musk and Polyantha classes, as well as several climbers dating from around the beginning of the twentieth century.

See also: **Classification**: Horticultural Classification Schemes. **Distribution and Ecology**: Overview. **History of Roses in Cultivation:** European (Pre-1800); Modern (Post-1800). **Rose Collections and Trials:** Rose Collections and Herbaria.

Further Reading

Allen RC (1948) *Roses for Every Garden*. New York: M. Barrows.

Bailey LH (1914–17) *The Standard Encyclopaedia of Horticulture*. New York: Macmillan.

Beales P (1985) *Classic Roses*. London: Harvill.

Bermuda Rose Society (1984) *Old Garden Roses in Bermuda*. Bermuda: Bermudian Publishing Co.

Bermuda Rose Society (1997) *Roses in Bermuda*. Bermuda: Bermudian Publishing Co.

Cairns T (ed.) (1993) *Modern Roses*, 10th edn. Shreveport, LA: American Rose Society.

Cairns T (ed.) (2000) *Modern Roses*, 11th edn. San Diego, CA: Academic Press.

Little T (1942) The distribution of north American rose species. *American Rose Annual* 37–49.

Redouté RJ (1991) *Roses*. London: Wellfleet.

Shepherd, Roy E (1954) *History of the Rose*. New York: Macmillan.

Thomas GC (1920) *The Practical Book of Outdoor Rose Growing, U.S.A.* Philadelphia, PA: J.B. Lippincott.

Willmott E (1911–1914) *The Genus* Rosa. London: John Murray.

South America

H Brichet, World Federation of Rose Societies, San Terenziano, Italy

The author gratefully acknowledges the contribution of Mercedes Villar (Rose Society of Uruguay), and Isa Maria Bozzolo (Rose Society of Chile), and Lyda Chevalier Boutell, Marta P. de Pertile, Adelina Perz Larranaga, Amelia Ciccarelli, Nilda Sanchez and Maria Ines Penafort (Rose Society of Argentina), in the preparation of this article.

Uruguay

Climate and Soils

Uruguay is situated in the southeastern temperate zone between latitudes 30° and 35°S. Its official name is 'Oriental Republic of Uruguay' because of its geographic position on the eastern margin of the Uruguay River. It lies between the two largest South American countries, Brazil and Argentina. It is strategically placed on the River Plate estuary and has an uninterrupted coastline of beautiful sandy beaches on the Atlantic Ocean. It is the smallest country in South America and covers 484 138 km^2. Montevideo is the capital city.

Uruguay has a temperate, healthy climate. Average winter temperature is 12 °C (June to August) and 22 °C in summer (December to February). The temperature very seldom falls below zero and there is no snow. Average rainfall is 1000 mm per annum. However, the Uruguayan climate is well known for its unexpected variations, and it can sometimes have the conditions of all four seasons in one day.

Uruguay has good clayey soil adequate for the production of cereals, including wheat and maize. There are sandy soils all along the southern coastline.

Wild Roses

The genus *Rosa* never managed to cross the equator so there are no wild roses in South America and therefore none in Uruguay of local origin. Wild roses were introduced for use as rootstocks, mainly *Rosa canina*, in the late 1800s and *R. indica* var. *major* in the early 1900s; *R. multiflora* var. *inermis* is a more recent introduction for the same purpose. However, there is no written evidence of these dates. None of these species has become naturalized in the wild.

Cultivated Roses

Montevideo's Municipal Rose Garden was inaugurated in 1912 and for this occasion over 1000 rose varieties were introduced from France. Cuttings were handed out to the general public at pruning time and for this reason the varieties most suited to our climate still grow in all kinds of urban, suburban and rural gardens throughout the country. Uruguayans have love and respect for the rose and there are varieties that have been handed down in the family from one generation to another, without its proper name, only known as 'grandmother's rose'. With the creation of the Uruguayan Rose Society in 1983 these varieties are gradually being identified and, together with more recent introductions, it could be said that the following roses do best in this country, specifically in gardens:

R. banksiae var. *lutea* and var. *alba*
R. rugosa
R. roxburghii
Hybrid Alba: 'Mme Plantier'
Hybrid Multiflora: 'Veilchenblau'

Hybrid Wichurana: 'Albéric Barbier', 'François Juranville', 'Dorothy Perkins', 'American Pillar', 'Silver Moon', 'Excelsa'
Teas and Climbing Teas: 'Lady Hillingdon', 'Safrano', 'Souvenir de Mme Léonie Viennot', 'Général Galliéni'
Chinas: 'Louis Philippe, Cl.', *R. chinensis* var. *viridiflora*
Hybrid Perpetuals: 'Frau Karl Druschki'
Hybrid Polyanthas: 'Perle d'Or', 'Mlle Cécile Brünner'
Floribundas: 'Iceberg', 'Centenaire de Lourdes'®
Hybrid Teas: 'La France', 'Crimson Glory'.

Our Society, Asociación Uruguaya de la Rosa, continues to increase knowledge and enthusiasm for the rose, but the few existing rose nurseries must continue to improve the quality of their plants and increase the number of varieties for sale. Uruguay is a very small market which makes it even more of a challenge for anyone venturing in the rose trade. And it is up to our Society, through its activities, to encourage, create and attract interest for our beloved flower, the rose.

In November 1998 the Uruguayan Rose Society hosted the World Federation of Rose Societies' First Regional South American Rose Conference in Punta del Este. Two hundred and thirty-seven participants from 16 countries attended. Particularly significant was the presence of delegates from the newly created Rose Society of Chile as well as rosarians from Brazil, where interest in roses as garden plants is on the increase. A visit to the restored rose garden in Santa Teresa National Park solicited much comment.

Chile

Climate and Soils

Chile is extremely long (*c*. 5000 km) and narrow lying between the mountain range of the Cordillera de Los Andes and the Pacific Ocean. The climate and soil are therefore very variable but may be divided into three agricultural zones as follows:

1. The northern area, which extends from the city of Arica to Serena, is characterized by a mild climate along the coast, but is very hot during the day and very cold at night in the internal areas. This is a great mining area, particularly for copper and saltpetre, but also of vast deserts, e.g. the Desierto de Atacama. In the cities the soil is generally sandy, very permeable but too rich in salt, as too is the water. Rains are very scarce and the area certainly is not suitable for growing roses.
2. The central area, which extends from the city of Serena to the city of Chillan, including the capital, Santiago, is characterized by heavy soil, rich in organic material, which is not very permeable. Rains are generally regular during the winter season. This is the richest agricultural part of the country. Here exist numerous commercial cut-flower farms in isolated areas with their own microclimates. In Santiago and its surroundings, roses bloom very well during the spring and the autumn; however, during the summer the enormous differences between daytime and nighttime temperatures, as much as 25 °C, result in numerous fungal diseases weakening the plants.
3. The southern area extends from the city of Chillan to the city of Aysen. Here also the soil is very variable, but is predominantly very permeable, particularly in the zone of the lakes where the soil is sandy and of volcanic origin. Here the rains are very intense and also frequent in summer.

Wild Roses

In Chile there are of course no wild roses, but curiously we have a wild rose which 'escaped', and which traditionally was eradicated from agricultural land, but is today cultivated for the fruits or hips. This rose is locally called 'Rosa Mosqueta' or 'Coral' and 'Coralillo', or 'Rosa Silvestre' in Chile. It is extremely common in the fields of the central area as far as the province of Aysen and at altitudes as high as 2000 m. The 'Rosa Mosqueta', which is widely found in the central area and around Santiago, is a mixture of three species roses: mainly *Rosa eglanteria (R. rubiginosa)*, but also with some influence of *R. canina* and *R. moschata*. All three of these species were introduced into Chile by Spanish immigrants and missionaries.

Cultivated Roses

The hips of 'Rosa Mosqueta' are of great commercial value for the country and today about 1550 ha are cultivated for this purpose. The hips are picked manually in the autumn and dried before being exported to European countries for pharmaceutical and culinary use (*see* **Secondary Metabolites**: Culinary Uses and Nutritional Value; Medicinal and Pharmaceutical Uses), but also mixed into pig and chicken feed, having a great nutritive value due to the high vitamin C content. Reproduction of the plants is undertaken in greenhouses under strict quality control to ensure selection of the best plants before their plantation in the fields.

In Chile there are only three specialized rose nurseries, two in the area of Santiago and one in the central area near the city of Temuco. All other producers are occupied with the multiplication of cut-flower varieties, and only one of these exports a small quantity. Thanks to the tireless work of our young but enthusiastic Chilean Rose Society, by way of conferences and publications, we are now seeing the use of roses in public squares and on municipal walls. The diversified possibilities of these beautiful flowers are beginning to be appreciated by the private gardener but also by the public in general.

Argentina

Climate and Soils

The account which follows describes the conditions in the main rose-growing areas of the country.

In the province of Buenos Aires the soil is sandy but rich, although there are differences from zone to zone. Most of the area has regular winter frosts and very hot summers with long hours of daylight. In most places, the soil has a pH of around 7.

The region of Bariloche lies at 800 m above sea level and about 1600 km southwest of Buenos Aires where the great Andes Mountains spread into a landlocked lake area. It is exposed to the dominant winds coming from the Pacific Ocean and originating in Antarctica. Rains and temperatures are unpredictable with great differences between maximum and minimum even over very short periods. Winters are very cold and snow falls from June to September. The geological origin of the soils is volcanic and sandy with very little organic matter and has a pH of around 6.5.

In the region of Mendosa, the land is part of the Andean Desert, the climate is dry and rain is scarce with extreme temperatures. Water from the Mendoza River irrigates the northern oasis of the province. A manmade river known as the Cacique Guaymallen Trench constructed by the Huarpes Indians, the original inhabitants of the province, is still in use today.

The province of San Juan is a desert corner of Argentina with brilliant suns and luminous skies. The capital is located in the central western part of Argentina at the foot of the Andes where the valley of Tulub is the main oasis of the province. In the high ranges, rock is predominant while on the slopes a rough texture is characteristic with abundant gravel. On the plains and foothills, it is possible to locate soils that are somewhat more evolved. Rainfall is scarce and normally occurs during the summer months. Water is, however, adequate for irrigating roses and comes mostly from the San Juan River and drilled wells. The short temperate winters guarantee a well-deserved rest for the rose plants which do not completely lose their leaves.

Wild Roses

There are no indigenous roses in Argentina. *Rosa sicula* has become naturalized in the area of Mendosa. It is thought to have been introduced by the colonial immigrants and has been spread throughout the mountain canyons by the wild animals that feed on the hips. In the area of Villavicencio, a white variety is to be found next to the traditional pink bushes. By observing the areas where these bushes do well, the general rules for rose growing in Mendoza can be determined.

Garden escapes are established in the foothills of Barreal, in the region of San Juan.

Cultivated Roses

According to some sources, the first rose plants might have come to Argentina with John Thwaites during the eighteenth century. He was a Scottish gardener who emigrated with his countrymen and settled in the provinces of Buenos Aires and Entre Rios.

The first Rose Society in South America was formed in 1951 in Buenos Aires, where 50 years later, in 2001, the World Federation of Rose Societies' Second South American Regional Conference helped celebrate the event.

In the first years the Society was very much a male affair and English was the spoken language. Later Spanish was permitted and women began to take over the day-to-day organization of the Society, which began to flourish. The Rose Garden, which is situated in the 3 de Febrero Park and had been inaugurated as early as 1914, was a great inspiration. Only recently this garden has been restored by the Buenos Aires government and a private enterprise, adhering scrupulously to the original plan of borders and paths. The gates of the Rosedal were reopened in June, 1996. Then in the spring of 2000 the first International Rose Trials for New Varieties of Roses were staged in this park organized by the government of the city. These were the first rose trials on the South American continent. In the gardens and parks varieties of *R. rugosa* do very well with some shade. Most roses are used for gardens, fences and the ones that excel include: 'Elina'®, 'Savoy Hotel', 'Clair Matin'®, 'Versailles'®, 'The Fairy', 'Leonardo da Vinci'®, 'Jardins de Bagatelle'®, 'Kronenbourg' (flowers best in autumn), 'Altissimo'® and 'Nozomi'.

The Rose Society has continued to grow and, in a country as large as Argentina, it was natural that local branches be inaugurated. These are the Cordoba Branch, the Mendoza Branch, the Bariloche Branch and the San Juan Branch.

The Cordoba Branch celebrated its 15th anniversary in 2000, the year in which it started cooperation with the local municipality in taking care of the roses in the Rosedal of Parque Sarmiento. This rose garden, very popular with the citizens of the city, was laid out many years ago and some of the varieties are of great age.

Members meet once a month to exchange information, learn and teach rose growing as a hobby but also as a matter of concern for the city.

In Bariloche most of the roses do extremely well. Pests and diseases are rarely a problem although mildew does arise when the humidity reaches 60% with high temperatures. There are many popular varieties of Hybrid Teas, Floribundas, Climbers, Old Garden Roses and, most particularly, Miniatures, of which a great number are available.

The Mendoza Rose Society started in 1990 and had the advantage of a great rosarian as an ally: don Luis Correas, whose mother started growing and collecting roses for her garden in Las Acequias, Medrano in 1923. This interesting collection of over 700 varieties may still be visited today. By 1992, the Society had 30 members who took over the care of rose bushes for the local golf club and a home for the elderly. Recently, members have run successful pruning courses for municipal gardeners and the general public. Monthly meetings focus on practical instruction regarding growing roses and also the treatment of pests and diseases.

The San Juan branch was opened in 1995. In 1997 they presented the authorities with a project for a rose garden for the San Juan Bank, a garden which was inaugurated in December 1998. Today this garden is an important attraction in the city centre, the maintenance of which is supervised by the Society. It is a service-oriented organization, undertaking the pruning of public areas and inviting specialists to give lectures on a variety of subjects in order spread the knowledge of and love for the rose.

See also: **Distribution and Ecology**: Overview. **History of Roses in Cultivation**: Modern (Post-1800). **Rose Collections and Trials**: Rose Collections and Herbaria. **Secondary Metabolites**: Culinary Uses and Nutritional Value; Medicinal and Pharmaceutical Uses.

Further Reading

Annual magazines published by the Asociacion Argentina de Rosicultura.

Annual magazines published by the Asociacion Uruguaya de la Rosa.

Rose hip shells. *Cascarilla de Rosa Mosqueta.* Coesam Group, Santiago, Chile, 1999.

DNA *see* **Genetics**: Nuclear DNA Amounts; DNA Fingerprinting; Cloned Genes.

E

Ecology *see* **Distribution and Ecology**: Overview; South Africa; Australia and New Zealand; Continental Asia and Japan; Europe; Israel; North America; South America.

F

Flavonoids *see* **Fragrance and Pigments**: Odoriferous Substances and Pigments; Anthocyanins and other Flavonoid Pigments; Flavonoid Molecular Biology.

FRAGRANCE AND PIGMENTS

Contents

Odoriferous Substances and Pigments

H Schulz, Federal Centre for Breeding Research on Cultivated Plants, Institute for Plant Analysis, Quedlinburg, Germany

Introduction

Rose flowers provide one of the most important essential oils which are indispensable in perfumery production. Up to now more than 400 volatile substances have been identified by gas chromatography–mass spectrometry (GC-MS) and it has been found that most of these substances belong to the following three major biosynthetic pathways: phenylpropanoids, fatty acid derivatives and terpenoids. However, most modern cut-rose cultivars lack a distinct scent, because selection for improved disease resistance, long vase-life as well as the outer appearance of the blossom dominated the breeding activities for many years. Nevertheless, specific reasons explaining the loss of typical smell have not yet been found. About 20% of all rose species known today are classified as 'fragrant', *c.* 50% present only low smell intensity and the other species do not possess any scent. Odour profiles range from flowery aspects (reminiscent of lily of the valley or violet) to fresh citrus notes, spicy aspects and a typical musk scent. In the past some attempts have been made to classify the different rose cultivars according to their typical odours (**Table 1**). The three species *Rosa damascena* (pink damask rose) *R. centifolia* (light pink cottage rose) and *R. alba* (white cottage rose) are predominantly used in the production of rose oils (attar of roses) and rose extracts, whereas the modern rose cultivars (e.g. tea hybrids obtained by crossing tea roses and remontant roses) are exclusively cultivated for the floriculture and nursery industries.

Production of Rose Oil, Rose Concrete and Absolute

Conventional production of rose oil extracted from flowers is performed in two steps. The first step consists of solvent extraction, which yields an intermediate

Table 1 Classification of smell of rose species and cultivars

Type of smell	*Rose species or cultivars*
Damask (classic rose oil)	*R. damascena, R. centifolia*
Damask (modern)	'Papa Meilland'®
Tea	'Lady Hillingdon'
Fruity	'Harmonie', 'Double Delight'™
Blue	'Blue Perfume', 'Blue Moon'®
Spicy	*R. moschata, R. rugosa*
Miscellaneous	*R. multiflora*, 'Diana'

product called 'rose concrete'. It is mainly composed of fragrance-related substances, but also contains large quantities of paraffins, fatty acids, fatty acid methyl esters, di- and tri-terpenic compounds and pigments. In the second step, the concrete is postprocessed by hydrodistillation to obtain a volatile oil containing the most important rose fragrance components. Nowadays there are also some attempts to apply supercritical CO_2 extraction (e.g. at 80 bar and 40 °C) to postprocess the rose concrete in order to reduce thermal degradation and incomplete elimination of nonvolatile compounds such as paraffins.

For the production of essential rose oil as well as rose extracts the following two rose species are mainly used: *R. damascena* Mill. (so-called Bulgarian rose oil) and *R. centifolia* L., which is usually described as 'Moroccan rose oil'. Whereas *R. damascena*, cultivated in Bulgaria and Turkey, is predominantly used for rose oil production, *R. centifolia*, cultivated in southern France and Morocco, is mainly extracted with solvents such as petroleum ether or *n*-hexane in order to obtain the so-called *essence concrète de rose* (rose concrete).

The extractors used, having a capacity of about 1200 l, are arranged in rows of three or four and are connected so the solvent passes from one to the other. The flowers rest on perforated plates so thorough contact can be made by the solvent. The solvent is forced through by a pump and fresh solvent is added from time to time, usually for three washings. The combined washings are filtered, partially evaporated at 60 °C and the condensed solution is transferred to a vacuum still to remove the solvent residues. The resulting extract is a light brown to green-brown mass, which contains phenylethyl alcohol more or less in the same ratio as the blossoms. Rose concrete samples presenting a beeswax colour usually contain undesired substances, therefore compared with rose oil its market value is *c.* 8–10 times less. Rose absolute is obtained by extracting the concrete with aqueous ethanol. The resulting product is a reddish liquid with a typical rose odour, which is mainly referred to the high phenylethyl alcohol content of 60–75% in the volatile fraction. From 400–500 kg rose flowers 1 kg rose concrete and consequently 500–600 g rose absolute can be obtained.

Figure 1 Production hall with copper stills for the hydrodistillation of essential flower oils.

In addition to these two species *R. moschata* is important for rose oil production; it is mainly grown in southern Europe, North Africa and in Asia. In some regions of Turkey *R. alba* is also used for oil production. It was found that this species is more productive per hectare than *R. damascena*, but its oil content is about 50% less. Since the quality of this oil is comparatively low, the production of this product is not of great importance. The use of the three wild species *R. rugosa, R. bourboniana* and *R. gallica* is also limited and production can only be found at some local areas such as Turkey, Bulgaria or Russia relating to the traditional production of these special rose oil types.

Harvesting of the rose blossoms begins at the end of spring and beginning of summer just before sunrise and ceases before midday. During that time the roses are still sparkling with dewdrops and less likely to be bruised. Rose flowers have the highest amount of essential oils from 5 to 9 o'clock in the morning; by 12 o'clock losses are between 30 and 40% and in the evening losses of 70% have been observed. In order to obtain high yields of essential rose oil, the distillation process should be performed directly after hand-picking as the oil content quickly decreases.

Whereas in former times only small stills containing 120–150 l were available, today steam jacket, tinned copper or stainless steel stills are used; these can be charged with 400–500 kg flowers and 1500–2000 l water (**Figure 1**). The distillation process is usually carried out for 1.5–2.0 h; during this time the condenser temperature is kept at 35 °C to avoid solidification of waxes. The oil spreading from the distillate is collected in 200-l steel Florentine flasks; the overflow is passed into big-bottomed tanks ('first water'). Subsequently, the first water from several stills is combined and redistilled (cohobated). This fraction is allowed to stand until the oil rises to the top, when it is skimmed off. This crude oil is separated by decantation and is therefore also known as 'decanted oil', 'first oil' or 'direct oil'. The remaining water is subjected to further steam distillation using a more efficient still. The oil obtained from this process is known as 'recovered oil', 'second oil', 'redistilled oil' or 'indirect oil'. Today, a commercial rose oil is usually a blend of a decanted and a recovered oil. The finished product is kept in a darkened storehouse air-conditioned at a temperature between 23 and 25 °C for a period of 1 year. At that time the ageing process is nearly complete and the oil is ready for purchase.

Comparative studies of rose oil constituents extracted by dichloromethane from rose water and redistillation of rose water showed that the solvent extracts contain significantly higher amounts of phenylethyl alcohol and lower contents of citronellol, nerol, geraniol and linalool. It was also observed that total alcohols occur at higher levels in oil extracted and redistilled from rose water than in rose oil distilled directly from the flowers.

Bulgarian rose oil should meet the following specification data:

- Melting point: 21–25 °C
- Density at 30 °C: 0.844–0.868
- Refractive index at 25 °C: 1.452–1.466
- Solidification point: +18–23.5 °
- Acid number: 0.92–3.75
- Ester number: 7.2–17.1
- Ester number after acetylation: 197.0–233.3
- Stearoptine content: 17–21%.

There are enormous differences regarding the quality of rose oils offered in the market and sometimes products may be adulterated with cheaper synthetic chemicals or oils reconstructed from natural and synthetic components may be found. In order to identify these adulterations it is necessary to apply highly sophisticated analytical methods such as GC-MS and gas chromatagraphy–infrared (GC-IR). Furthermore, it is important to have a wide knowledge of different authentic rose oils produced worldwide. A representative gas chromatogram obtained from an authentic Bulgarian rose oil is shown in **Figure 2**. The molecular structures of important olfactory rose oil constituents are presented in **Figure 3**.

In this context, it should be considered that plant-derived essential oils such as rose oil do not represent the authentic profile of the living plant, but they contain complex artefact products developed during the hydrodistillation process, due to the influence of higher temperature and pH change inside the distillation stills.

For the production of 1 kg rose oil it is necessary to collect *c*. 5000 kg blossoms; this is the reason why the price of pure rose oil is extremely high (*c*. 2000 € kg^{-1} for Bulgarian rose oil up to more than 7000 € kg^{-1} for special qualities mostly originating from southern Europe). The annual world production of rose oil and rose concrete together is estimated at *c*. 20 tons, with Bulgaria and Turkey being the two largest producers. Also all rose products mentioned above are important for the creation of high-class perfume as well as for the production of valuable perfume oils used in the soap industry. Furthermore, rose oil is used to flavour sweets, chewing-tobacco and liqueurs.

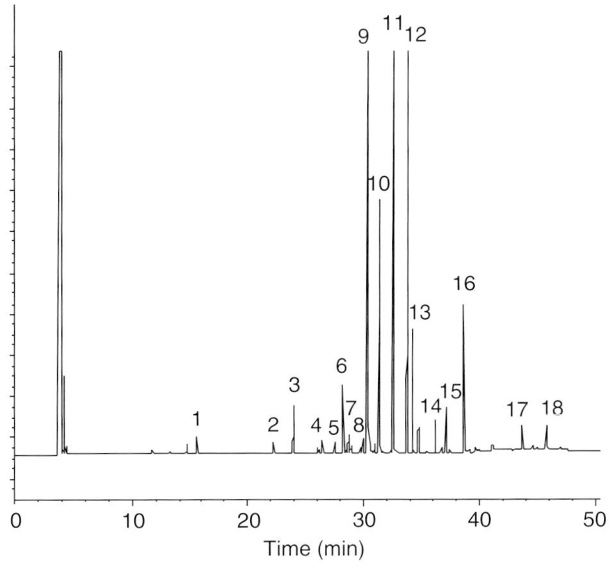

Figure 2 Gas chromatograph of authentic Bulgarian rose oil obtained with a polyethylene glycol column (HP-Innowax, 60 m × 0.25 mm × 0.5 μm) and flame ionization detection. 1, *p*-cymene; 2, pentacecane; 3, linalool; 4, β-caryophyllene; 5, citronellyl acetate; 6, hexadecane; 7, neral; 8, geranial; 9, citronellol; 10, nerol; 11, geraniol; 12, nonadecane; 13, nonadecene; 14, eicosane; 15, methyleugenol; 16, heneicosane; 17, farnesol; 18, docosane.

Analytical Characterization of Rose Flowers and Rose Oils

For chromatographic separation and identification of the volatile, odoriferous rose substances it is first necessary to apply optimal clean-up steps. For isolation of the volatile fraction, the following methods have mainly been used:

- solvent extraction
- steam distillation
- various headspace techniques.

The major limitation of solvent extraction is that it is useful only on samples that do not contain any lipids. Furthermore, solvent extraction and steam distillation are liable to produce artefacts by isolating nonvolatile decompositions. Therefore, headspace sampling, as a

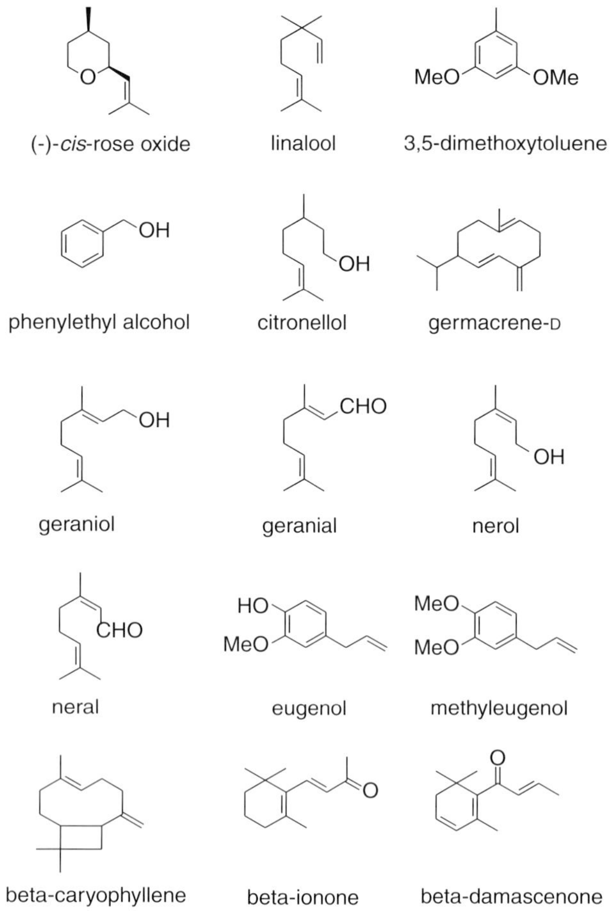

Figure 3 Molecular structures of some important olfactory rose volatiles.

nondestructive technique, which can be used on living plants either in the laboratory or in the field, has been found to be the method of choice. Generally, there exist four distinct methods of different headspace techniques (vacuum headspace technique, closed-loop stripping method, dynamic headspace technique and solid phase microextraction) that are commonly used today for analysing the rose volatile substances. The first one is the 'vacuum degassing' or 'vacuum headspace' technique in which the volatile constituents are separated from the plant matrix at a high vacuum and frozen out into traps cooled down by liquid nitrogen at $-196\,^{\circ}C$. The vacuum headspace apparatus consists of a round-bottomed flask filled with the fresh rose flowers, connected to two cold traps via a glass tube with a spherical ground joint. When the air is removed by a vacuum oil pump, the fragrance substances vaporise, together with the water present in the flowers, and are frozen out in the first cold trap. A second cold trap is installed in series with the first one, in connection with an active charcoal tube, to prevent the possibility that impurities from the vacuum pump may adulterate the analyte concentrate. Finally, the collected fragrance substances are passed into an organic solvent and after separation of the condensed water the mixture is separated and identified by coupled gas chromatography–mass spectrometry (GC-MS).

Although the vacuum headspace technique may provide a more or less authentic reproduction of the fragrance profile of an individual rose cultivar, it has the crucial disadvantage that flowers can only be analysed after they have been picked. In most cases enzymatic processes, initiated by the damage of plant cells, dramatically change the profile of the volatile fraction and in such cases it is very difficult to study the biochemical pathways occurring in the plant. Therefore, in recent times another technique, the so-called closed-loop stripping method, is usually applied in this context. Here, the fragrance substances released from the rose flower are carried along in a stream of carrier gas and are finally concentrated on a suitable adsorbent, for example, active charcoal or an organic polymer. To prevent impurities from the pump getting into the vapour phase surrounding the flower, a second charcoal tube is placed in series.

In order to perform on-site measurements, the rose flowers are placed in a suitable-sized conical flask and the opening of the flask is carefully closed off with a plastic film, taking care not to damage the flower stalks. A small suction pressure pump transports the volatiles in the stripping apparatus. The development of this dynamic headspace technique allows enrichment and reproducible identification of the most relevant volatile rose substances without damaging the flower being analysed. Subsequently, the loaded molecules to be analysed are thermally desorbed from the adsorbent cartridge by slow heating (2 min at $200\,^{\circ}C$) and it is possible to collect them in a loop cooled with liquid nitrogen. Finally, this loop is rinsed in helium steam, so that the volatile rose substances can be completely injected on a gas chromatography column. Active charcoal has been found to be an extremely powerful and nonselective adsorbent which can be applied very efficiently to a wide range of organic substances. However, because the adsorption is so strong, it may sometimes be a problem desorbing the molecules to be analysed; furthermore, often numerous artefacts are generated during thermal desorption. Porous synthetic polymers like Tenax(TM) (poly [2,6-diphenyl-*p*-phenylene oxide]) also exhibit a high adsorptive capacity towards volatile and semivolatile plant substances but they clearly show fewer problems when the analytes are desorbed.

First primitive GC-headspace techniques still implied the treatment of a relatively large quantity of raw material and the use of solvents which could produce artefacts. In comparison to that, dynamic GC headspace collection is much more effective and allows efficient enrichment of volatile, medium-volatile, and even low-volatile minor plant constituents.

During recent years, solid-phase microextraction (SPME) using various porous polymers as fibre materials (e.g. polydimethylsiloxane (PDMS)) has been increasingly applied in the analysis of plant volatiles. Compared with conventional liquid-liquid extraction, this method is convenient, easy to use, less time-consuming and requires no solvents. In the general SPME procedure, the coated fused silica fibre is exposed to the sample or its headspace and the volatile analytes present in the sample are adsorbed in the fibre coating. After that, the fibre is directly introduced into the injection port of a gas chromatograph; the extracted volatile substances are thermally desorbed and subsequently separated on an appropriate GC column. For the analysis of living rose flowers, one single blossom still attached to the plant is placed into a suitable glass chamber in such a way that the petals are not injured. This chamber contains sidearms closed with GC septa; the hole at the bottom of the glass flask is sealed with plastic film (**Figure 4**). Depending on the individual ambient temperature and the rose cultivar, the PDMS fibre is placed inside the glass chamber for 15–30 min. If the fragrant intensity of the rose blossom is high enough, this time is sufficient to extract most of the volatiles released by the plant material to be analysed.

Figure 5 shows the results of SPME-GC analyses performed on five different rose cultivars grown in the Europa-Rosarium in Sangerhausen, Germany. The cultivars 'Ave Maria' and 'Lavender Lassie' contain as main volatile constituents phenylethyl alcohol and phenylethyl acetate. Other characteristic rose fragrance substances such as hexyl acetate, citronellol, citronellyl acetate, methyleugenol, germacrene-D and heneicosane were also identified in the headspace of the flower. In contrast, the remontant rose reveals a completely different pattern of volatile components; here the main components are nerol, geraniol, citronellol and the corresponding acetates which significantly contribute to the fresh citrus fragrance of this rose cultivar. The individual smell of cultivars 'Roundelay' and 'Kore' is mainly influenced by the intensely odorous substances β-ionone (fruity character) and 3,5-dimethoxytoluene.

Figure 6 presents the SPME-GC separation of volatile rose substances collected from the flower of the old cultivar 'Pariser Charme'. The main components in the chromatogram, which have been detected by mass spectrometry, are germacrene-D, hexyl acetate, *cis*-3-hexenyl acetate, citronellol, citronellyl acetate, nerol and geraniol.

Several rose cultivars have been analysed by different GC-headspace techniques presenting clearly differing fragrance types. The volatile fraction of 'Othello', a modern English hybrid developed in 1986 by David Austin, contains as its main constituents citronellol, nerol and geraniol; the other fragrances identified by GC-MS occur in amounts of <3% (e.g. *p*-cymene, *cis*-hexenyl acetate, β-caryophyllene, neral, germacrene-D, geranial, nonadecene). The fragrance of the dark red blooms of this rose cultivar can be described as strongly rosy and narcotic.

Figure 4 Solid-phase microextraction field sampling of a rose blossom for subsequent gas chromatography analysis of the volatile compounds collected.

The famous old French rose cultivar 'Duchesse de Montebello', which has a fresh-floral, typically rosy fragrance, shows a similar pattern of volatile substances. One exception is the relatively high level of β-phenylethyl alcohol (approx. 30%) and β-phenylethyl acetate (approx. 7%). Other constituents in the volatile fraction are only present in lower amounts (e.g. α-pinene, sabinene, myrcene, *cis*-3-hexenyl acetate, β-caryophyllene, citronellol, nerol). In contrast, the flowers of the bush rose 'Lichtkönigin Lucia'® give off a floral-green scent (reminiscent to fresh cut grass) with spicy and fruity aspects. The main constituent of the headspace was found to be β-phenylethyl acetate (approx. 37%), but comparatively high concentrations of *cis*-3-hexenyl acetate, *trans*-2-hexenyl acetate and *trans*-2-hexenol were also detected. The last three secondary metabolites in particular are responsible for the characteristic fresh-green fragrance of this rose cultivar. The fruity note of this hybrid is mainly related to the occurrence of the intensely odorous β-ionone. This fragrance

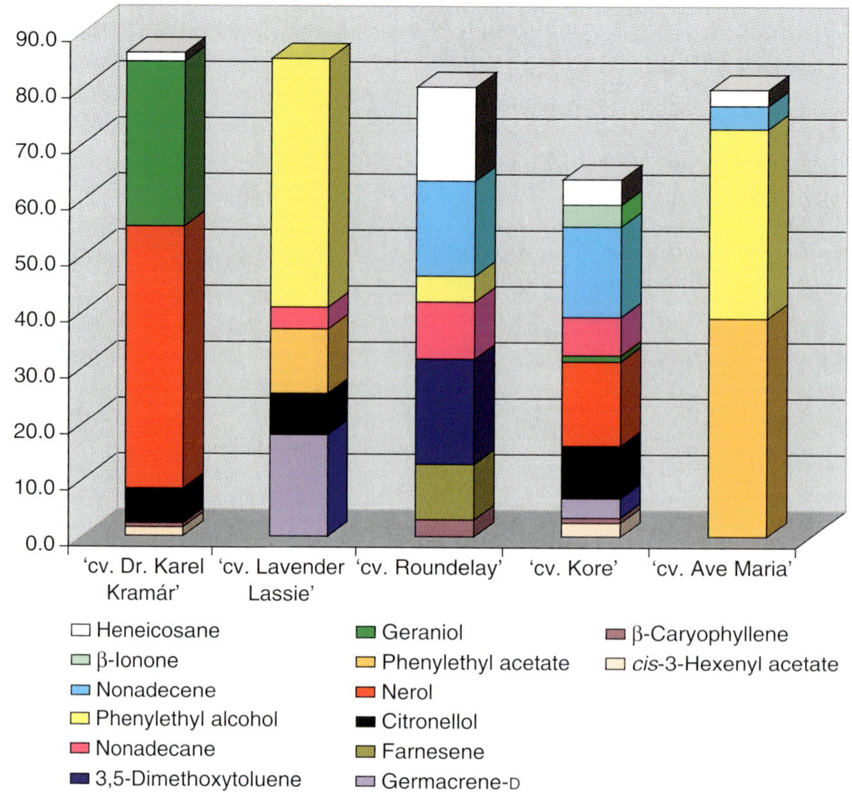

Figure 5 Results of nondestructive and rapid solid-phase microextraction gas chromatography on-site measurements on rose flowers.

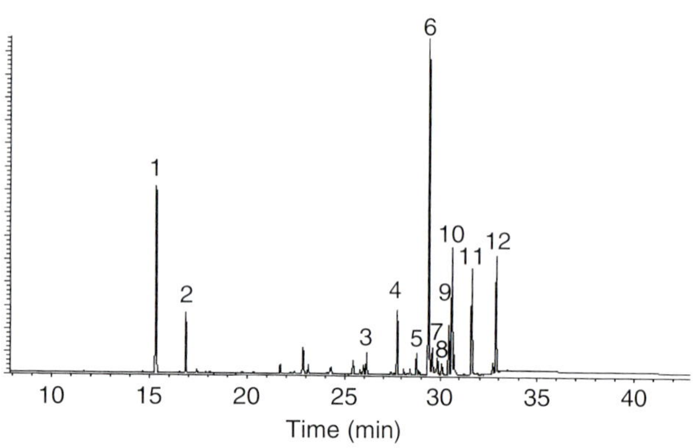

Figure 6 Solid-phase microextraction gas chromatography measurement of volatile rose substances collected from the flower of 'Pariser Charme'. Gas chromatographic separation was performed with a polyethylene glycol column (HP-Innowax, 30 m × 0.25 mm × 0.5 μm) and mass spectrometry detection. 1, hexyl acetate; 2, *cis*-3-hexenyl acetate; 3, β-caryophyllene; 4, citronellyl acetate; 5, nerol; 6, germacrene-D; 7, neryl acetate; 8, geranial; 9, geranyl acetate; 10, citronellol; 11, nerol; 12, geraniol.

material only occurs in amounts of about 0.4% in the volatile fraction but with regard to the very low odour threshold it is important for this individual rose type.

Generally, it has to be mentioned here that compounds to be detected in high concentrations are not necessarily the most efficient ones to express the scent; sometimes only a small number of minor compounds have a high impact on the whole fragrance perception.

The bright-pink flowers of the cultivar 'Queen Elizabeth'® ('Charlotte Armstrong' × 'Floradora') also possesses a very specific fragrance profile. The GC is dominated by 3,5-dimethoxytoluene. Other principal constituents are dihydro-β-ionone, dihydro-β-ionol as well as *cis*- and *trans*-theaspirane. Especially, dihydro-β-ionone was found to be responsible for the characteristic fruity and raspberry-like odour of the rose flower. Generally, ionone derivatives and theaspirane substances are known to be produced by enzymatic decomposition of various carotinoids which are predominantly found in hybrid roses in addition to other plant pigments such as flavonoids or anthocyanes.

Rose Scent and Composition of Essential Rose Oils Derived from Different Cultivars/Types

During the last 30 years our knowledge and experience regarding the chemistry and sensory properties of rose oil components have increased considerably. While alcohols like citronellol, geraniol, nerol and phenylethyl alcohol as well as some other oxygenated components have been known for half a century or more, most of the minor but olfactively essential rose components were identified during the last three decades. Among these characteristic substances are rose oxide and rose furan, the damascones and damascenone. The group of aldehydes also represents some constituents such as nonanal and decanal which are important for the typical scent of roses. Nonanal is described as having a very powerful and diffusive fatty-floral and waxy odour, becoming more pleasant in solution. Decanal has a penetrating, sweet-waxy, orange-peel-like odour which becomes refreshing in extreme dilution.

The content of volatile ester compounds also differs from one variety to another. *Cis*-3-hexenyl acetate, which is a main volatile substance of the rose variety 'Baronne Edmond de Rothschild', is described as being intensely green and sharp-fruity, whereas neryl acetate has a sweet, fruity floral odour with a typical raspberry character. Phenylethyl acetate has fruity notes, reminiscent of peaches but also of the smell of gardenia.

Today more than 400 volatile components have been identified in different rose species and varieties. This high number of analysed substances is mainly related to the dramatic increase of GC detection sensitivity which today gives the opportunity to identify many minor characteristic constituents in the volatile fraction. **Figure 7** presents some rose varieties in which one component exceeds 19–20%.

It is generally accepted that the group of rose ketones is very important for rose fragrance. In this context it should be stressed that β-damascone, β-damascenone and rose oxide possess their characteristic organoleptic properties in rose oils as trace components. It has been reported that neither β-damascenone nor the isomeric rose oxides occur in the aroma isolate of rose flowers. Consequently, both volatiles appear to be rearrangement products (artefacts) that are formed during the steam distillation process from extremely labile progenitors (3,7-dimethyl-5-octene-1,7-diol and other hydroxylated monoterpenoids).

In rose oil absolute two monoterpene butenolides occur, namely 3-(4-methyl-3-pentenyl)-2-buten-4-olide and 3-methyl-4-(3-methyl-2-butenyl)-2-buten-4-olide, which are structurally related to perillene and rose furan. These two furans have also been detected in rose absolute together with *cis*- and *trans*-2-(3-methyl-2-butenyl)-3-methyltetrahydrofuran.

In comparison to the roses used in perfumery, hybrid tea roses have a different scent which is reminiscent of tea notes. The headspace GC results show that the content of phenylethyl alcohol, citronellol and

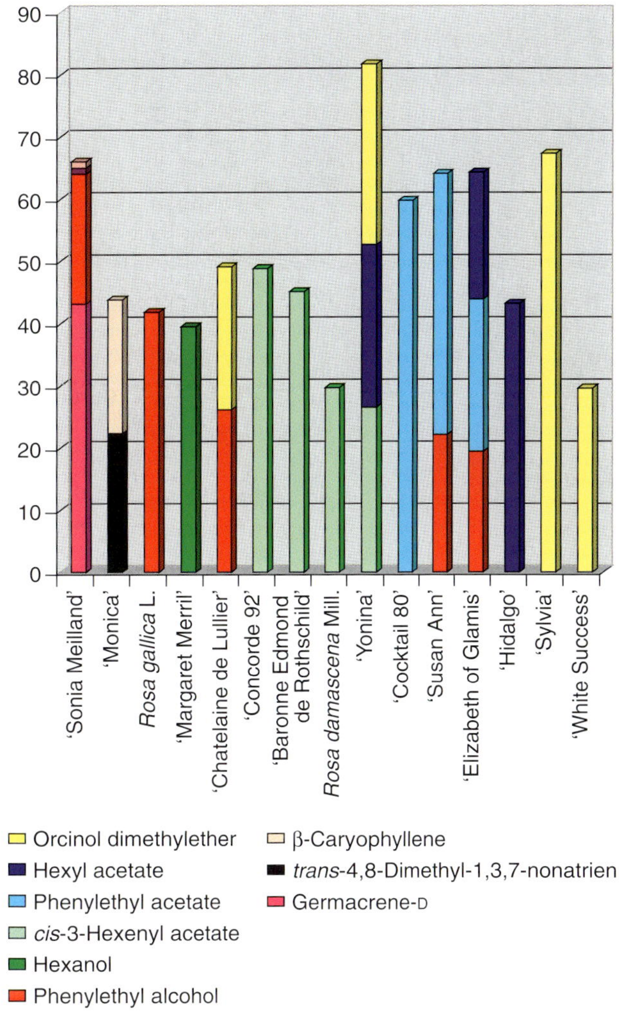

Figure 7 Main constituents of some selected rose varieties.

geraniol differs considerably from cultivar to cultivar. The volatile substance 1,3-dimethoxy-5-methyl benzene was found to occur in *R. gigantea*, in some tea rose cultivars and in hybrid tea roses, but it was not detected in *R. damascena* and *R. centifolia*, which are used for rose oil production. It has a fresh green, humid and spicy character reminiscent of the typical smell of a flower shop. It appears that this fragrance compound is a key component for the scent of tea rose cultivars and hybrid tea roses.

In comparison to *R. centifolia* or *R. damascena*, Chinese rose oil as well as rose absolute obtained from *R. chinensis* has a different odour. Citronellol, geraniol, phenylethyl alcohol, methyleugenol and eugenol are present as main components but rose oxide, linalool, geranial, neral, acetates of geraniol, nerol and citronellol have also been detected as minor constituents in these products. The profile of the substances mentioned is more or less similar to those of other commercial rose oils (e.g. oils from Bulgaria or Turkey) but geranic acid (3,7-dimethyl-2,6-octadienoic acid) was found to be a specific compound occurring only in Chinese rose oil in higher amounts.

It has been reported that the petals from modern ornamental hybrid tea roses also contain some characteristic sulphur components such as dibenzothiophene and its methyl derivatives, possessing a persistent, spicy and powdery note. In commercial rose oil dimethyl sulphide and dimethyl disulphide have also been found in traces. Relating to the very low odour threshold, traces of these substances may contribute to the naturalness of the individual rose scent.

Eugenol generally imparts a spicy note to the rose oil odour, whereas methyl eugenol has no relevant effect on the rosaceous character.

The hydrocarbon compounds occurring at amounts of *c.* 20% in the rose oil mainly consist of paraffin derivatives. These saturated or unsaturated substances are water-insoluble, nonvolatile and odourless; they are called the 'stearoptine fraction' because they solidify when the oil is chilled. In this context it is also worth mentioning that in authentic commercial rose oil the odourless component *cis*-9-nonadecane is present in a relatively constant concentration of about 3%; if the level of this paraffin substance is significantly lower, it can be assumed that the oil has been adulterated by synthetic materials.

Biosynthesis of Fragrance Substances

During recent years, detailed chemical analysis of volatile composition together with the identification of secondary metabolism-related genes whose expression coincidated with fragrance production in rose blossoms led to the discovery of several scent-related candidate genes in *R. chinensis* and some selected rose hybrids. In this context an annotated petal expressed sequence tag (EST) database was created and DNA chips were prepared and used for expression analyses of selected clones. The function of the isolated genes was biochemically determined using an *Escherichia coli* expression system. Although the individual biosynthetic pathways are not understood in detail, common biochemical modifications such as hydroxylation, acetylation and methylation have already been described.

Some investigations have also been performed to gain a deeper insight into the metabolism of the formation of odoriferous constituents occurring in various rose cultivars. In this context the time course of volatile emission by flowers of *R. hybrida* 'Honesty' was analysed by GC headspace. From bud opening to visibility of pollen, the flowers emitted 13 volatile compounds with a diurnal 24-h rhythm. On the day of maximal emission the major volatile compounds were citronellol, nerol, geraniol, 3,4-dimethoxytoluene and hexyl acetate. Minor headspace substances were citral, methyl geranylate, β-cubebene, *trans*-caryophyllene, 2-phenylethyl alcohol and 2-phenylethyl acetate. The regulatory role of light, which was observed during the studies performed, is most clearly shown by the absence of rhythmicity in flowers that were continuously kept in darkness before and during bud opening.

It has been found that ionone derivatives and theaspiranes are compounds which can arise through the oxidative or enzymatic decomposition of carotenoid plant colorants, and they are therefore only detected in those roses which contain these tetraterpene substances as pigments in addition to flavonoids and anthocyanins.

Petal Pigments

Pigments occurring in wild roses as well as garden rose cultivars can be related to the following three different chemical groups: chlorophylls, flavonoids (including anthocyanins) and carotenoids. The molecular structures of some selected anthocyanins occurring in rose blossoms are presented in **Figure 8**. The blossoms of old European roses such as *R. gallica, R. alba,*

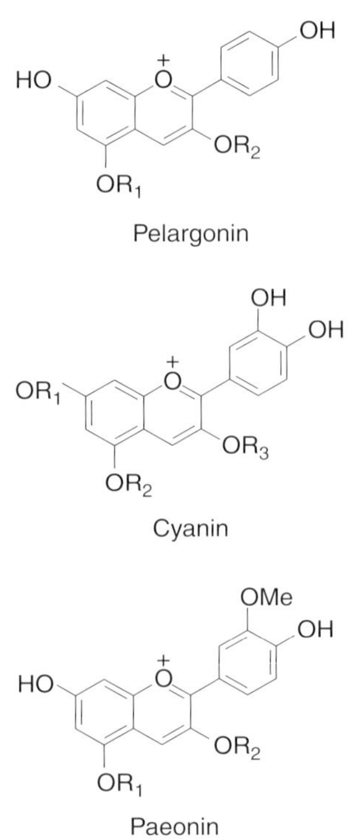

Figure 8 Molecular structures of some anthocyanins occurring in rose blossoms ($R_1 = R_2 = R_3 =$ β-D-glucose).

R. centifolia, *R. damascena* and moss roses (cultivated before *c.* 1830) presented only white, pink and red colours with all mixed shades. The exceptions are a few botanic roses which were already imported into Europe some centuries earlier (e.g. *R. hemisphaerica* and *R. foetida*). The petals of white roses reflect most of the incident light; nevertheless, they contain some flavonol glycosides, absorbing mainly in the ultraviolet region (e.g. quercetin, rutin, isoquercetin). Also some carotenoids have been detected in nonnegligible amounts (e.g. in *R. pimpinellifolia*, 'Mme Plantier', 'Yvonne Rabier' and 'Schneewittchen'). It is also worth mentioning that some of the apo-carotenoids such as rosafluin possess fluorescence properties and may therefore serve as optical brightening agents in the petals.

Roses with green blossoms are very rare; the best-known example of this group is the wild type *R. chinensis* var. *viridiflora*; sometimes some modern white roses also show a clear tinge of green. According to the latest investigations, the colour is mainly influenced by chlorophylls. Old yellow roses were only obtained when Chinese cultivars, which contain predominantly carotinoids as pigments, were used for breeding experiments. These rose cultivars predominantly contain carotenoids produced in early biosynthesis steps and they present a strong catabolic activity resulting in high amounts of decomposed carotenoid substances. It has been proved that catalytic reactions of the individual tetraterpene substances produce C-13 and C-27 compounds, which serve as fragrances or as other bioactive substances in the physiology of the plant. In contrast to that, new yellow garden roses, which have mainly been obtained by crossings with *R. foetida* var. *persiana*, possess a complete sequence in the carotenoid biosynthesis and catabolic reactions do not play an important role.

Cyanin and pelargonin, belonging to the group of anthocyanins, are among the most widespread pigments occurring in red roses (e.g. in 'Gloria Mundi'). Beside this, cyanin is not only responsible for the red colour of several rose cultivars, but it is also the most important pigment in violet cultivars such as 'Mainzer Fastnacht' or 'Veilchenblau'. It has been found that the different colours of this anthocyanin compound are related to the individual pH value in the epidermis cells of the plant. The stabilization of the anthocyanin chromophors in the individual pH range is reached by co-pigmentation with flavonoid glycosides (e.g. rutin, quercitrin, isoquercitrin). During this process a slack association between the flavylium ion (red colour, high pH value) or the chinone methide (violet colour, low pH value) and the flavonol (or a similar substance) is formed. In some rose cultivars also paeonin contributes to the specific violet colour of the blossoms. Generally, only glucose is used for glycosidation of the anthocyanins and it has been proved that the individual 3,5-diglucosides are those pigments with the highest colour intensity. Up to now no delphinidin derivatives have been detected in petals of roses, therefore it seems unlikely that in future a blue-coloured rose will be obtained by conventional breeding.

Several ionone derivatives and theaspiranes are produced through oxidative or enzymatic decomposition of carotenoid plant colourants, and should be therefore found only in those roses which contain these pigments in addition to flavonoids or anthocyanins. Contrary to that relationships between anthocyanins and particular volatiles produced by decomposition reactions in the rose flower have not yet been reported.

See also: **Fragrance and Pigments**: Anthocyanins and other Flavonoid Pigments; Flavonoid Molecular Biology; Functional Genomics; **History of Roses in Cultivation**: European (Pre-1800); Modern (Post-1800); History of the Perfume Industry. **Secondary Metabolites**: Medicinal and Pharmaceutical Uses; Extraction and Industrial Processes.

Further Reading

Agaoglu YS (2000) Rose oil industry and the production of oil rose (*Rosa damascena* Mill.) in Turkey. *Biotechnology and Biotechnological Equipment* 14: 8–15.

Croteau R and Karp F (1991) Origin of natural odorants. In: Müller PM and Lamparsky D (eds.) *Perfumes, Art, Science and Technology*, pp. 101–126. London: Elsevier Applied Science.

Dudareva N (2002) Molecular control of floral fragrance. In: Vainstain A (ed.) *Breeding for Ornamentals: Classical and Molecular Approaches*, pp. 295–309. Dordrecht, The Netherlands: Kluwer Academic.

Eugster H and Märki-Fischer E (1991) Chemie der Rosenfarbstoffe. *Angewandte Chemie* 103: 671–689.

Flament I, Debonneville C and Furrer A (1993) Volatile constituents of roses. Characterisation of cultivars based on the headspace analysis of living flower emissions. In: Teranishi R, Buttery RG and Sugisawa H (eds.) *Bioactive Volatile Compounds from Plants*, pp. 269–281. ACS Symposium Series no. 525. Washington, DC: American Chemical Society.

Kaiser R (1991) Trapping, investigation and reconstruction of flower scents. In: Müller PM and Lamparsky D (eds.) *Perfumes, Art, Science and Technology*, pp. 213–250. London: Elsevier Applied Science.

Knudsen JT, Tollsten L and Bergstrom G (1993) Floral scent: a checklist of volatile compounds isolated by head-space techniques. *Phytochemistry* 33: 252–280.

Lawrence BM (1991) Progress in essential oils: rose oil and extracts. *Perfumer Flavorist* 16: 43–77.
Ohloff G (1990) *Riechstoffe und Geruchssinn, die molekulare Welt der Düfte*. Berlin: Springer Verlag.

Fragrance Profiles of Wild and Cultivated Roses

H A Verhoeven, J Blaas and W A Brandenburg, Business Unit Biosciences, Plant Research International, Wageningen, The Netherlands

Introduction

In everyday life, scent plays an important role in many biological functions. Examples of particular functions are related to interactions between different organisms, but also within the same species, scent plays a role in attraction and repellence. Particularly in plants, which are bound to the same location due to their inability to move, defence against feeding insects and other animals is of paramount importance, as is the attraction of pollinators and the dispersal of seeds, packaged in fragrant fruits. Flowers have been known for a long time to be sources of sweet fragrances. For centuries, roses have played an important role in human society. Their ornamental value was not merely a matter of flower shape and size: the abundance of flowering and their fragrance have attracted attention.

As plant breeding has evolved over the last centuries into a science rather than an art, and plant production became more and more an industrial process, the focus was changed to plant production parameters rather than flower fragrance. This became even more pronounced after the concept was introduced that flower fragrance may show a negative correlation with the average vase-life of cut flowers. In the case of roses, this led to the development of many beautiful roses in the sense of flower shape and colour with an enhanced vase-life, but with only a fraction of the fragrance of earlier cultivars or wild species. It is only in roses cultivated for their essential oil, that the fragrance has received systematic scientific interest. A large number of publications on the chemical composition of rose oils is available, together with details of the influence of the extraction procedure on the composition and changes of oils due to thermal and chemical interactions. Also information is available with regard to culture conditions, soil and fertilizer use and pest control.

Recently, studies have been performed to elucidate the regulation of flower fragrance production in cultivated roses, especially in relation to rhythmicity of the fragrance release. These studies have shown that the daily emission pattern depends on the type of component, e.g. aromatic components show a different release pattern compared to esters. Many fragrance components are shown to be stored as glycosylated precursors in the rose petals, e.g. phenyl ethyl alcohol and the rose oxides. Their release is the consequence of the stimulation of the activity of glycosidase enzymes in a light-triggered reaction. All together, this suggests a complex interaction of physiological factors (e.g. light) with genetic regulation of metabolic activity.

Recent research has shown that many volatile components are the result of the activity of a limited number of genes, which encode for enzymes involved in particular biochemical conversions. In this case, the activity of these genes can be evaluated by the biochemical analysis of the volatile profile of flowers. As the evaluation of fragrance profiles requires a trained person, many plant-breeders are facing a problem. This problem can however be solved (to some extent) by the introduction of analytical chemical analyses. These analyses are often quite expensive and not comprehensive, as they need to be selective for a particular class of volatiles.

For sampling of flower fragrance, several different techniques are available. The oldest and most widely used technique relies on the absorption of the released fragrances of flowers, enclosed in a sealed container (preferably made of glass). This container is placed over the flower of interest, and air flow enters this container after it has passed a charcoal and/or tenax filter to remove contaminating volatiles. The excess air passes a cartridge, containing a small amount of absorbing matter, e.g. charcoal, tenax or wax. The volatiles, emitted by the flower(s) are trapped on to this material. After the desired exposure time, the absorbing substances are removed and extracted with a suitable solvent, such as hexane or pentane in the case of apolar volatiles (e.g. terpenes), or with a more polar solvent, such as ethyl-ether in the case of more polar volatiles, like alcohols. The largest amount of solvent may be removed by evaporation in order to concentrate the volatiles, often by bubbling the extract with nitrogen or argon, to avoid decomposition of labile components. The sample can then be analysed by gaschromatographic (GC) separation. Alternatively, thermodesorption of the absorbed volatiles can be performed, using dedicated equipment which allows tubes to be heated with absorbent inside the injection port.

In this article, a rapid and inexpensive method is described for the detailed analysis of flower fragrance, as well as how it can be used to classify different fragrances and how this information can be used for directed plant breeding in roses. Furthermore, an example is presented of a multivariate analysis of complex fragrance profiles by principal component analysis, resulting in the identification of key factors governing fragrance profiles. In contrast to most published chemical analysis, which rely on mass spectroscopic identification of individual peaks, this example shows the power of a more holistic approach using the entire chromatographic information in a principal component analysis. Principal component analysis is a statistical analysis tool which maps the sources of variation between samples in a newly defined multidimensional space. As such, it allows for correlation of differences in fragrance to the absence or presence of particular volatile components in samples.

Analytical Techniques

To distinguish differences due to physiological variation from these caused by genetic background, the flower material should be as uniform as possible. A number of scientific papers deals with the differences in volatile composition due to culture conditions, developmental stages, day/night rhythmicity and availability of organic carbon and energy sources. Most roses in this investigation were grown in greenhouses under standard growing conditions for roses, while some cultivars were bought from a local flowershop. Flowers were selected to be homogeneous with regard to their developmental stage (start opening of the flower buds, which is stage 4 according to Gutermann), since most roses produce the most intense fragrance at this stage. **Table 1** lists the names of the roses used in this study, and their properties, as they were described in the literature.

Sampling

Solid-phase microextraction A recently developed technique for sample preparation is solid-phase microextraction (SPME: Supelco, Bellmonte). This device has been designed to perform sampling, concentration and sample injection in one action. It consists of a silicone fibre, coated with a 100 μm thick layer of a strongly absorbing polymer, such as polydimethyl silane (PDMS). This coated fibre is mounted inside a stainless-steel hollow needle, and can be moved outwards to expose the absorbing layer to the volatiles to be sampled (**Figures 1–3**).

Volatiles are trapped inside this layer by an equilibrium mechanism, resulting in a concentrated sample of the headspace. The entire fibre can be inserted into the injection port of a GC, and quickly desorbed by the heat of the port, resulting in a rapid transfer of all absorbed components on to the GC column. In order to obtain a good chromatographic separation, it is advisable to use a column with a thick stationary phase, and to start temperature programming at low temperatures, such as 40–80 °C. After thermal desorption, the fibre is clean and ready for the next sampling. In this way, fibres can be used to sample flower fragrances for more than 100 times without loss of efficiency. Depending on the nature of the volatiles, it can be necessary to use a different fibre coating with an affinity which is more selective for more polar or apolar components. The method is suitable for mass screening using robotized injectors, e.g. the Combi PAL, without the need for extraction procedures.

The flowers to be sampled can be used in a number of ways. To determine emission patterns related to diurnal rhythm developmental stages, it is best to sample a flower that is still attached to the plant. The flower can be enclosed in a glass funnel of the appropriate size, which is closed with aluminium foil. The SPME needle is punctured through this foil and its coated surface is exposed during the sampling time – in this case 30 min. For screening activities, two methods can be used. In both approaches, rhythmicity is ignored, and results can only be used for screening. The first uses detached flowers, and the set-up shown in **Figure 1**. In the second approach, a detached flower petal is inserted in a glass vial which is closed with a PTFE-lined septum and cap. A small droplet of water is added to the vial in order to prevent rapid wilting of the petal (**Figure 2**). The production of volatiles by the petal continues for several hours, and all volatiles are accumulated in the headspace of the petal. The vial can be placed in the sampling tray of a GC/mass spectrometry (MS) system with a Combi PAL robotic injector (**Figure 3**). In this way, it is possible to screen the patterns of emitted volatiles of a large number of plant samples at high-throughput level.

Separation

The sample is introduced into the injection port of a GC. Depending on the type of column, the amount to be injected varies. Nowadays, most GCs are equipped with capillary columns, which can be used with microlitres of sample. By temperature programming, a suitable separation of the different components in the sample can be achieved, in combination with parameters such as linear gas velocity, type of stationary phase and film thickness. This separation is important, since most biological samples, including fragrance samplings, can contain from 10 to hundreds of

Table 1 Roses and their properties, as used in this study

Rose name	*Synonyms/trade name*	*Cultivar group*	*Flower colour*	*1*	*2*	*3*	*4*	*5*	*6*	*7*	*8*	*9*	*10*	*11*	*12*	*Fragrance*
'Ruirovingt'	Prophyta™	HT	Apricot blend	1.6	—	0.2	1.0	—	—	4.6	—	—	3.1	11.9	5.9	—
'Cara Mia'	Danina	HT	Medium red	0.2	—	—	—	—	—	4.7	—	—	13.6	31.6	8.7	Moderate
'Corso'®	—	HT	Orange	—	—	—	0.1	—	—	38.0	—	—	1.2	20.7	6.5	Slight
'Hocus-Pocus'	—	GF	Orange	0.1	—	—	—	—	—	15.9	—	—	5.6	0.4	40.4	Slight
'INTasprint'	Sprinter®	HT	Deep yellow	3.1	—	—	—	—	0.4	41.1	—	—	10.1	19.8	6.4	—
'KORcilmo'	Eskimo®	FL	White	0.3	—	—	0.2	—	—	40.0	—	—	2.3	2.6	5.6	—
'KORlingo'	Kardinal™	HT	Medium red	—	—	—	0.6	—	—	2.5	—	—	16.2	0.9	19.3	Slight
'KORp'	Prominent®	GF	Orange	—	—	—	—	—	—	30.7	—	—	17.4	12.6	9.3	Slight
'Meihelvet'	Sonia	GF	Pink blend	—	—	—	—	—	—	1.0	—	—	8.7	42.1	13.2	Intense
'Memory'	—	HT	Pink blend	0.3	—	—	0.1	—	—	10.2	—	—	0.5	0.3	24.5	Intense
'Olympia'	—	HT	Medium red	—	—	—	—	—	—	0.7	—	—	19.9	0.2	21.1	Moderate
'PEKcoujenny'	First Red®	HT	Dark red	1.0	—	—	—	—	—	54.2	—	—	30.5	1.5	2.0	None
'RUIco'	Fresco®	FL	Orange blend	0.1	—	—	—	—	—	20.0	—	—	4.5	0.1	22.4	Slight
'RUIdriko'	Vivaldi™	HT	White	—	—	—	—	—	—	35.4	—	—	6.7	2.7	16.9	—
'Selcarbonium'	Honesty™	HT	White	—	4.1	0.2	2.7	2.2	3.3	—	14.5	2.2	5.6	2.1	0.5	—
'Starlight'	—	HT	Medium yellow	0.6	—	—	—	—	0.4	60.8	—	—	1.7	0.4	10.7	Slight
'Sunblest'	Landora	HT	Deep yellow	0.6	—	—	0.1	—	—	15.6	—	—	23.3	3.0	15.5	Slight
'TANatesil'	Ilseta®	FL	Medium pink	—	—	0.2	—	—	—	2.6	—	—	1.1	3.4	32.8	None
'TANito'	Akito	FL	White	0.3	—	—	0.1	—	—	32.1	—	—	0.5	0.2	0.1	Slight
Rosa centifolia	—	SP	White	—	—	1.6	8.5	9.4	3.5	—	2.4	5.9	2.3	0.8	5.7	Moderate
Rosa spinosissima (L.)		SP	White	—	—	0.1	9.1	1.3	1.1	—	2.4	17.6	0.8	0.4	1.7	—

HT, hybrid tea; GF, grandiflora; FL, floribunda; SP, species.
Fragrance components are numbered: 1, 3-hexen-1-ol acetate; 2, acetic acid, hexyl ester; 3, D-limonene; 4, 2-phenyl ethyl alcohol; 5, geraniol; 6, 2-phenyl ethyl acetate; 7, 3, 5 dimethoxy toluene; 8, citral; 9, geranyl acetate; 10, caryophyllene; 11, butylated hydroxytoluene; 12, nonadecane.
The amounts of the individual fragrance components are expressed as percentages of total ion count per chromatogram. Name, synonyms, cultivar group, colour and fragrance conform to the descriptions in *Modern Roses XI* (see vol. 3).

Figure 1 The experimental set-up, with a rose flower enclosed by a glass funnel and aluminium foil. The solid-phase microextraction fibre is pierced through the foil, and samples the fragrances released by the flower.

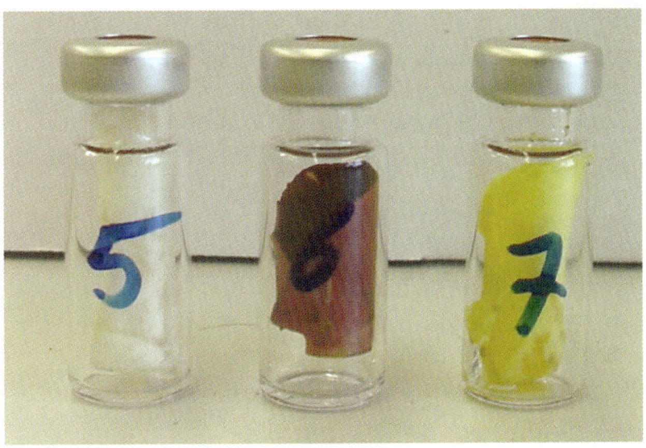

Figure 2 Three 1.5-ml vials with a small piece of rose petal inserted. The vial contains a small amount of water to prevent the samples from drying. This simple set-up can be used in automated sampling systems for large screening programmes.

components, which would be impossible to identify without proper separation. The GC, in our studies type GC8000 (Interscience Breda), was equipped with a 50 m HP-5 column, ID 0.32 mm and film thickness 1.05 μm. In our case, since SPME was used with

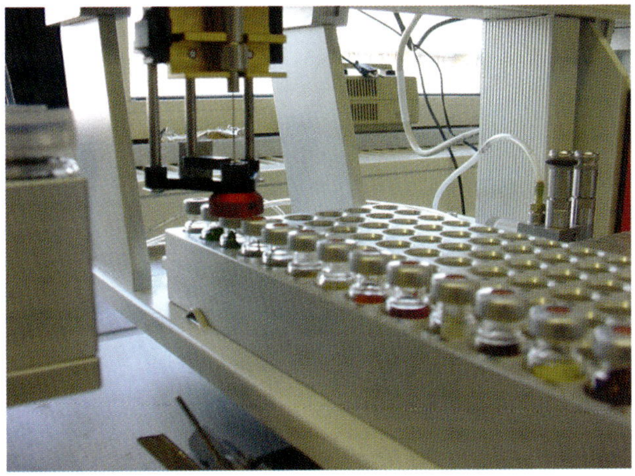

Figure 3 Overview of a sample tray of an automated sampling system, the Combi PAL, with the solid-phase microextraction device sampling the headspace of a vial.

PDMS as absorbents, the sample was introduced by thermal desorption of the trapped volatiles by heating the SPME fibre to 250 °C in the injection port. The fibre was desorbed during a period of 1 min. GC temperature was held at 80 °C for 2 min then increased to 250 °C with 8 °C min^{-1} ramp, and finally the column was cleaned at 5 min at 250 °C. A GC transfer line to the MS was set at 260 °C to prevent condensation of high-boiling components at this point. Helium with a purity of 99.999% at 47 kPa was used as carrier gas and further purified using a charcoal prefilter and an OMI-4 (Supelco) purifier. The injection port was equipped with a 0.5 ml glass liner for splitless injection in order to optimize transfer of heat to the exposed fibre for rapid desorption.

Identification

The eluting compounds can be detected using various detectors. The most common detector in GC is the flame ionization detector, which combines high sensitivity with a high degree of ruggedness. Its limitation is the lack of identification methods, so that one must rely on the stability of the chromatographic separation in combination with reference components to calculate the retention index. The retention index can then be used to identify the different components. Detailed information can be found in the reference books listed in the further reading section.

A more versatile and sensitive detector is the MS. In this device, the molecules eluting from the analytical column are bombarded with electrons with energy of 70 eV, resulting in ionization and eventually fragmentation of a part of the molecules. The fragmentation patterns can be used to identify the compound. Nowadays, large libraries are available with more than 200 000 different spectra. In this case, the NIST 98 library (National Institute of Standards and Technology, USA) was used with Excalibur software (Thermo Electron, Breda) system for identification. Once the library identified a component with a certain fit (usually when forward-fit parameters are above 600), a reference component was run to verify retention time and authentic spectrum. Since not all naturally occurring components can be purchased, this is not always possible. In these cases, literature was searched to validate the occurrence of that particular component in the plant species of interest.

Analytical Results

A typical chromatogram is shown in **Figure 4**. It contains more than 100 individual peaks, some of which have been identified by their mass spectrum, retention index, whether in combination with comparisons using an authentic standard. In total, 134 different components were detected during this investigation of two wild rose species, *Rosa centifolia* and *R. spinosissima*, and 18 cultivars of *R. hybrida*. The different components, as far as they could be identified, can be grouped into several classes. The fragrance of roses is composed of four major classes of components: aliphatic compounds, monoterpenes, benzoids and sequiterpenes. In addition, many minor components have been described for some species of roses or accessions, most notably in the case of roses for the production of rose oil (*see* **Secondary Metabolites:** Extraction and Industrial Processes).

The problem in the comparison of these complex patterns is the fact that many components only occur in some analyses, due to day/night rhythmicity, developmental stage, physiological status or genetic influences. This problem can be avoided by prolonged exposure time.

The most prominent feature of these analyses is the observation that most modern rose cultivars only produce minor amounts of a limited number of components. They all share emission of 3,5 dimethoxy toluene at RT (retention time) = 17.13, butylated hydroxytoluene at RT = 21.50 and nonadecane at RT = 28.14. Apart from these compounds, there is a large variation in the number and type of components per cultivar, as shown in **Table 1**. In order to analyse this in more detail, data were handled as follows. In each chromatogram, a peak list was composed after smoothing and baseline correction using the appropriate functions of the chromatographic software package. These peak lists were exported to an Excel spreadsheet, and all peaks were manually aligned according to the information of the mass spectra of each peak. After alignment, empty values (i.e. compounds which were not detectable in that particular sample) were filled with the baseline level, in this case intensity of 10 000 TIC (total ion count). Peak intensities were log-transformed to reduce the influence of abundant components on the principal component analysis. Peak lists were transformed to tab-delimited text files, and imported into Gene Maths version 1.0, a statistical package for the analysis of DNA microarray data (Applied Maths BVBA Kortrijk). Using the Pearson statistics to compare the row data, and the Euclidian distance for the column (showing the cultivar dependence), a graph was drawn up which reflected the correlation between the occurrence of the individual components and their distribution among the different cultivars (**Figure 5**).

A phylogenetic tree indicates the degree of correlation between the observed patterns of intensities of related components among the different cultivars. A similar tree is computed for the column data. It is striking to see that the four classes mentioned earlier were all present in the analysis of the wild rose species, whereas not a single cultivar of this investigation showed the presence of all four groups.The analysis of the above-mentioned cultivars and wild species reveals a number of striking differences. First of all is the apparent lack of variation between cultivars of the hybrid tea group. They all share a number of components, with very little differences, most of them belonging to the group of sesquiterpenes. The wild species, on the other hand, show a remarkable richness in the composition of the volatiles, with many monoterpenoids, aliphatic compounds, along with esters, benzoids and sesquiterpenes.

The principal component analysis plots (**Figure 6**) reveal the very close relationship between most rose cultivars, and the large differences with the two wild rose species tested. It is remarkable to see that 'Honesty' is positioned between *R. centifolia* and the group of hybrid tea and grandiflora cultivars. No striking differences were found between grandiflora, hybrid tea and floribunda cultivars, except for 'Ilseta'®, which shows more fragrance components belonging to the class of monoterpenes than most other cultivars.

New Developments in Genetics

The above-mentioned results have serious consequences for breeding programmes aimed at the (re)introduction of fragrance in future rose cultivars. Data obtained with 'Honesty' clearly illustrate that it is feasible to obtain the presence of important fragrance compounds such as phenyl ethyl alcohol and

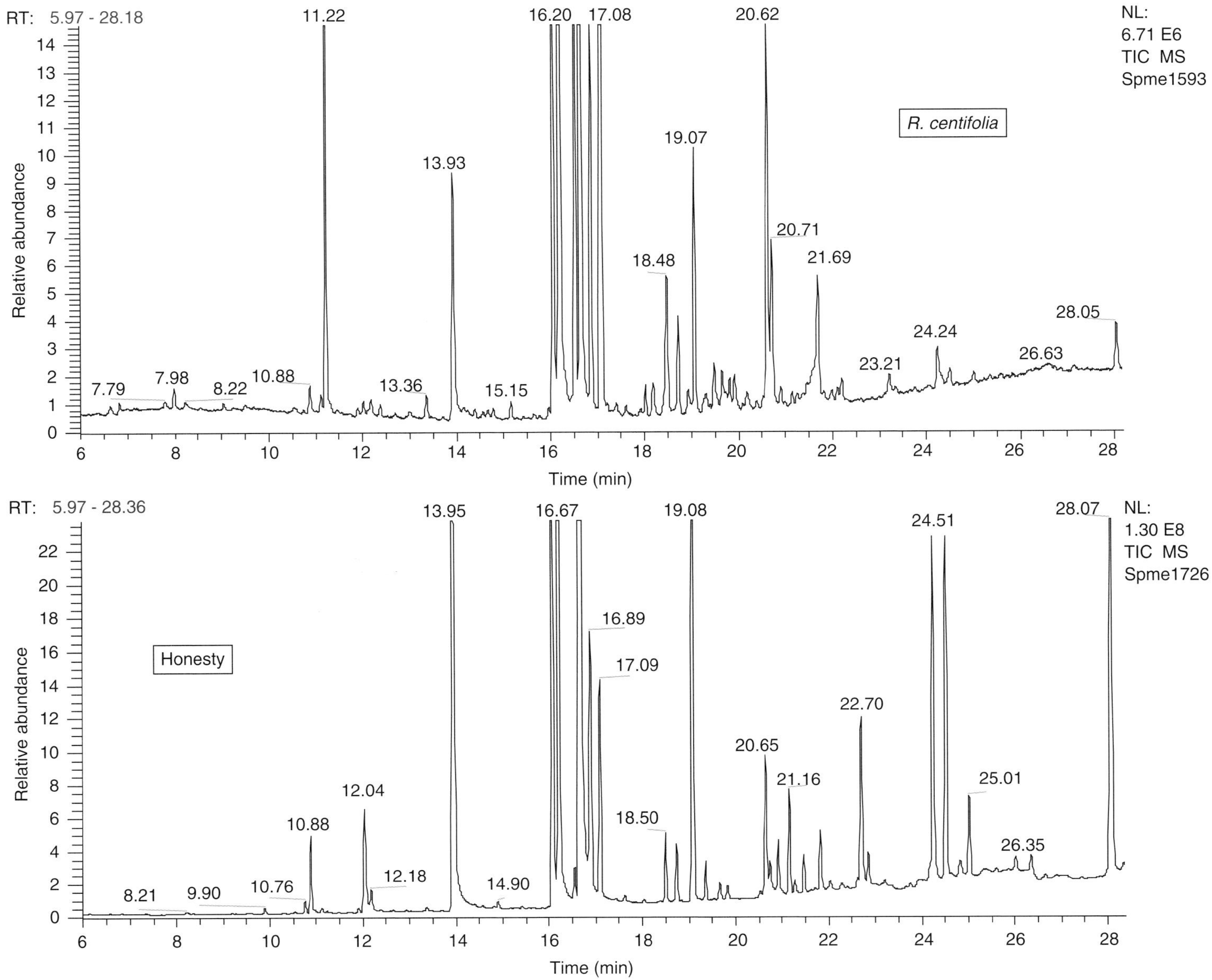

Figure 4 Typical chromatogram of a rose cultivar 'Honesty' (lower trace), compared with the chromatogram of *Rosa centifolia* (upper trace). The large difference in the number of peaks between these two roses indicates the loss of most rose fragrance components in the cultivated roses.

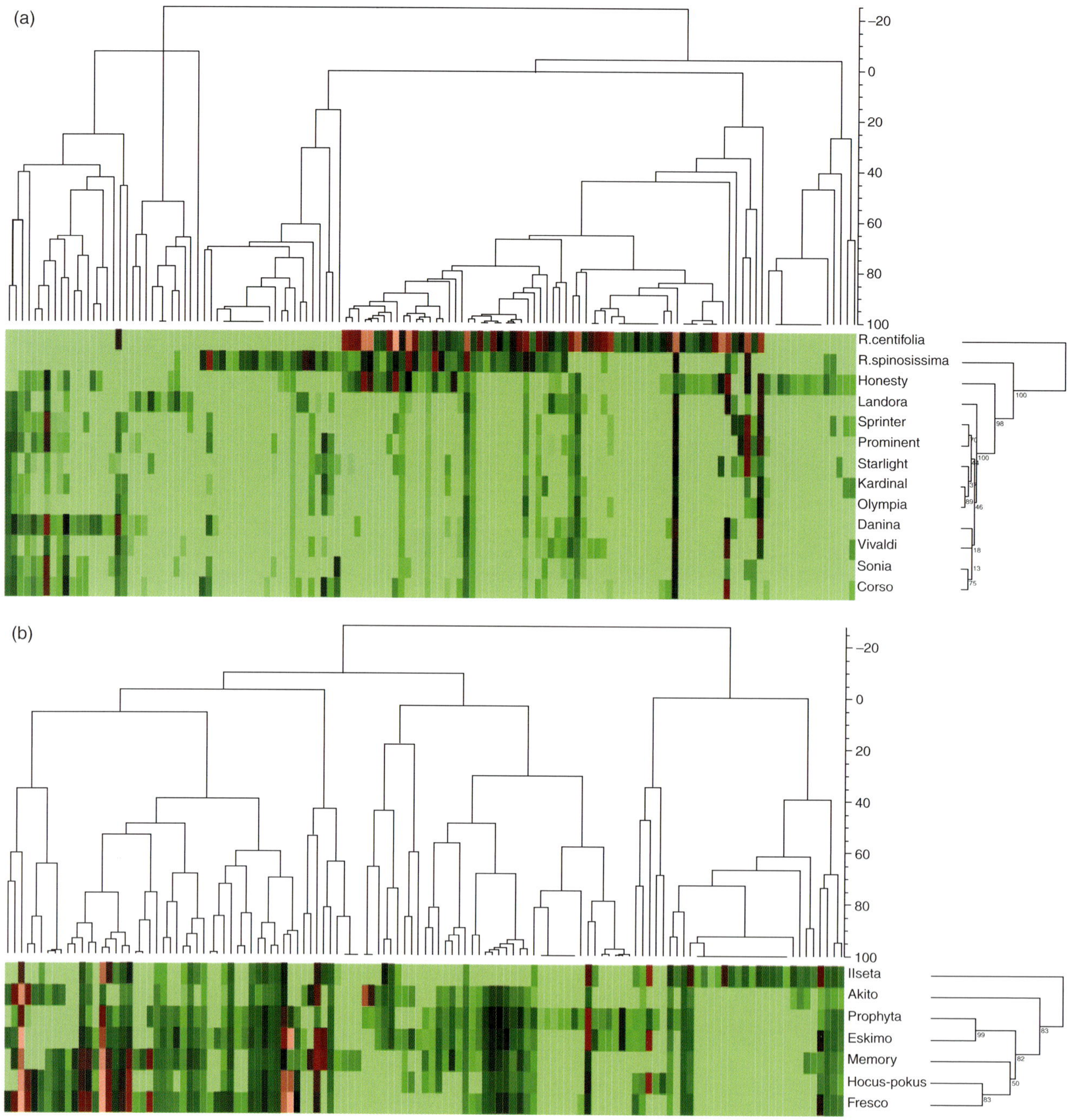

Figure 5 Phylogenetic tree of the different rose cultivars measured on (a) cut flowers and (b) pieces of rose petals. The two methods give slightly different results due to the presence of aldehydes as a consequence of tissue wounding caused by cutting the petals.

phenyl ethyl acetate. The biosynthesis of these compounds is probably related to the phenyl propanoid pathway, which is also involved in the production of flavonoids and anthocyanins. In particular, the desired presence of monoterpenoids, which contribute strongly to a rose-type fragrance, together with phenyl ethyl alcohol and its acetate derivative, offers new and exciting possibilities.

Research on a large number of monoterpene synthase genes has shown that most of the genes studied so far in many different plant species encode for enzymes that lead to a high production of a very limited number of monoterpenes. As a consequence, and using SPME for high-throughput analysis, it is possible to screen a crossing population for the presence of a particular product, thus allowing genes to be added from different

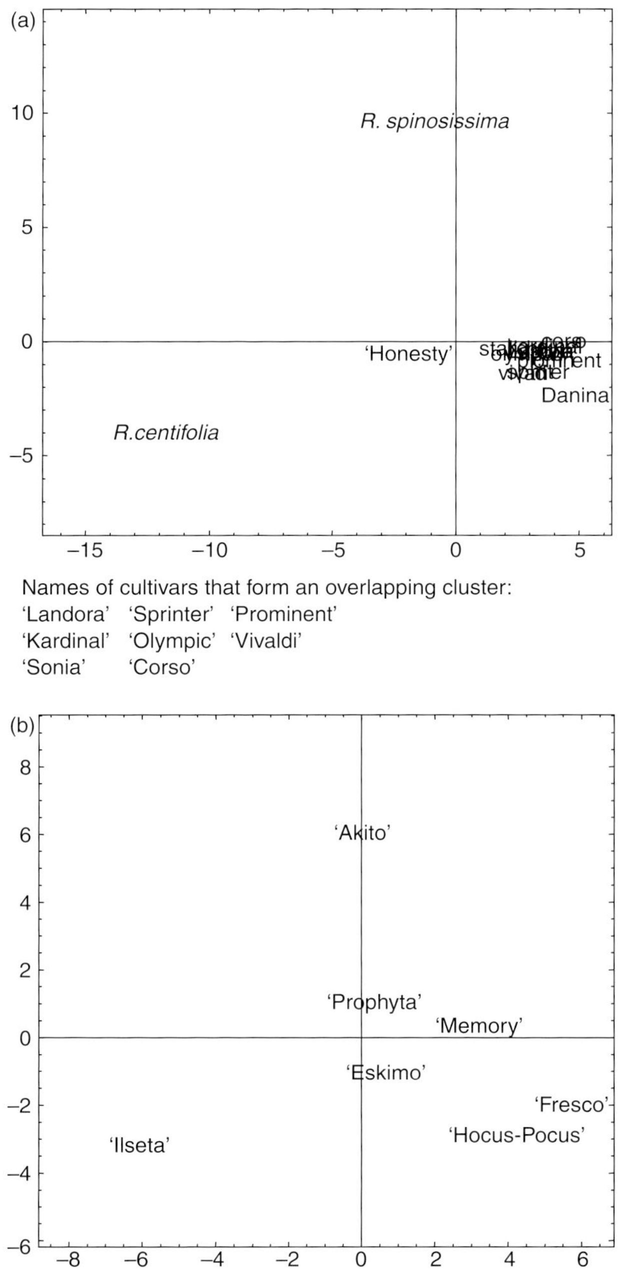

Figure 6 Principal component analysis of the fragrance composition of the samples of **Figure 5**. The high degree of similarity between most cultivars is indicated by the clustering of these cultivars in the right part of the diagram, whereas the wild species show remarkable differenes. Honesty is positioned between *R. centifolia* and the other cultivars. The upper panel (a) shows the analysis of intact cut roses, and the lower panel (b) the analysis of the flower parts, which give slightly different results due to the formation of breakdown products, and as such should be analysed separately.

species into a complex blend. The feasibility of this approach has been shown in studies with transgenic tobacco lines, each carrying a different monoterpene synthase gene, in this case limonene synthase, γ-terpinene synthase and β-pinene synthase. A plant was obtained that produced all the three monoterpenes by crossing. This finding makes it obvious that such a programme can only be successful when all desired genes can be found in the parental lines, and a careful selection should be made to combine the desired traits.

A similar approach can be performed for sesquiterpenes, which show a biosynthetic selectivity comparable to the monoterpene synthases. Since the substrates for these categories of components are present in most tissues of a flowering plant, the supply of precursors should not pose a limitation. A clear dependence of internal regulation has been observed in transgenic plants with regard to day/night regulation of endogenous precursor production. It appeared that the production of new monoterpenes by the introduced monoterpene synthase activity showed the same day/night rhythmicity as that of an endogenous terpene synthase, although they were under the control of different promotor signals.

With this information, comprehensive methods for fragrance analysis will make it possible to trace the origin of particular fragrance components to the genetic level. In scientific literature, papers can be found which specifically deal with the elucidation of the genetic background of fragrance production in roses. Combined with new emerging techniques, such as those described here, e.g. high-throughput headspace analysis using SPME and principal component analysis, but also those like metabolomics, more powerful tools for directed breeding will become available.

See also: **Fragrance and Pigments**: Odoriferous Substances and Pigments; Functional Genomics. **History of Roses in Cultivation**: History of the Perfume Industry. **Secondary Metabolites**: Culinary Uses and Nutritional Value; Extraction and Industrial Processes.

Further Reading

Croteau R and Karp F (1991) Origin of natural odorants. In: Muller PM and Lamparsky D (eds.) *Perfumes. Art, Science and Technology*, pp. 101–126. Glasgow: Blackie Academic and Professional.

Flament I, Sauvegrain P and Sauberli U (1992) De la science et de l'art structure et odeur des composants volatils de la rose. *Quadrimestrale dell Instituto Tetrahedron* 3: 1–13.

Flament I, Debonville C and Furrer A (1993) Volatile constituents of roses. Characterization of cultivars based on the headspace analysis of living flower emissions. *ACS Symposium Series* 525: 269–281.

Hay RM and Waterman PG (eds.) (1993) *Volatile Oil Crops: Their Biology, Biochemistry and Production*. Harlow: Longman Scientific & Technical.
Kaiser R (1991) Trapping, investigation and reconstitution of flower scents. In: Muller PM and Lamparsky D (eds.) *Perfumes. Art, Science and Technology*. Glasgow: Blackie Academic and Professional.
Kaiser R (1993) *The Scent of Orchids. Olfactory and Chemical Investigations*. Amsterdam: Elsevier.
Lucker J (2002) *Metabolic Engineering of Monoterpene Biosynthesis in Plants*. PhD thesis. The Netherlands: Wageningen University.
Weiss EA (1997) *Essential Oil Crops*. New York: CAB International.

Anthocyanins and other Flavonoid Pigments

M Jay, J-P Biolley, J-L Fiasson, K Fiasson, J-F Gonnet, C Grossi, O Raymond and M-R Viricel, Université Claude Bernard, Lyon, France

Introduction

Because of the ornamental value of roses, many studies have been devoted to the distribution of anthocyanins in flowers of rose species and cultivars. These molecules are also called pigments since they absorb the visible light in a specific manner which depends essentially on their chemical structure and also on the chemical and physical environment; they are mainly located in the vacuole of epidermal cells so that they are responsible for flower colours, providing most of the pink, salmon, orange, red colours, and also the more-or-less bluish mauve colour. Other phenolic compounds also contribute to the flower colour of roses. Flavonols and their glycosides are ultraviolet light-absorbing compounds: they are colourless to the human eye, but are able to act as copigments towards anthocyanins, and consequently to modify their visible light absorbance properties, and to change the basic colour of anthocyanins.

Anthocyanins and flavonols belong to the same biosynthetic pathway of the flavonoids (**Figure 1**). They derive their carbon skeletons from two basic compounds, malonyl-CoA and CoenzymeA ester of a hydroxycinnamic acid; the central step in their biosynthesis is the condensation of three units of malonyl-CoA with a hydroxycinnamic acid CoA ester to a C15 chalcone intermediate, catalysed by chalcone synthase (EC 2.3.1.74); a stereospecific action of the chalcone isomerase (EC 5.5.1.6) provides the first flavonoid, a flavanone. Furthermore, various reaction steps of oxidation and reduction will lead toflavonols and anthocyanidins; glycosylation of these aglycones gives rise to natural anthocyanins and flavonol-glycosides.

Minor flavonoids such as chalcones are also responsible for yellow colours in several plant species, but in roses, these compounds are not accumulated in the epidermal cells of petals. The carotenoids belong to a different metabolic pathway, the terpenoids, and are the main contributors to the yellow and orange colours in roses.

Finally, a white colour or yellowish hues are found in roses without visible light-absorbing molecules in their petals. Moreover, the visual perception of white requires the reflection of the most visible radiations; in white roses, air-filled intercellular spaces contribute to this complete reflection.

Structures of Anthocyanins Isolated from Roses

Anthocyanins result from the association of an anthocyanidin chromophore (C15 nucleus with two C6 rings, identified respectively as the A and B rings, and one C3 central oxygenated heterocycle) and a sugar moiety: the chromophore is positively charged and so is named the flavylium cation. Anthocyanins are the most important group of water-soluble plant pigments. They are usually in solution in the vacuole of epidermal cells, but occasionally they may crystallize or form spherical vesicles called anthocyanin vacuolar inclusions (AVIs). In roses, these natural pigments occur in flower petals, and sometimes in stems, leaves and hips.

The anthocyanins are usually classified into three groups according to the number of substitutions on the B ring (**Figure 2**): monosubstituted B ring or pelargonidin class, disubstituted or cyanidin class and trisubstituted or delphinidin class. In the last two classes, the original hydroxyl (–OH) substitutions on the B ring can be replaced by methoxyl (–OCH_3) groups. The six main compounds occurring in the plant kingdom are: pelargonidin (3,5,7,4′–OH), cyanidin (3,5,7,3′,4′–OH), paeonidin (3,5,7,4′– OH, 3′–OCH_3), delphinidin (3,5,7,3′,4′,5′–OH), petunidin (3,5,7,4′,5′–OH, 3′–OCH_3) and malvidin (3,5,7,4′–OH, 3′,5′–OCH_3).

In roses, the anthocyanin first identified (1915) was cyanidin 3,5-diglucoside (cyanin); it was retrieved from petals of *Rosa gallica*. Twenty-three years later, the cyanidin 3-monoglucoside (chrysanthemin) was discovered in the *chinenses* section of the subgenus *Rosa*. The black hips of *R. pimpinellifolia* also contain almost pure chrysanthemin.

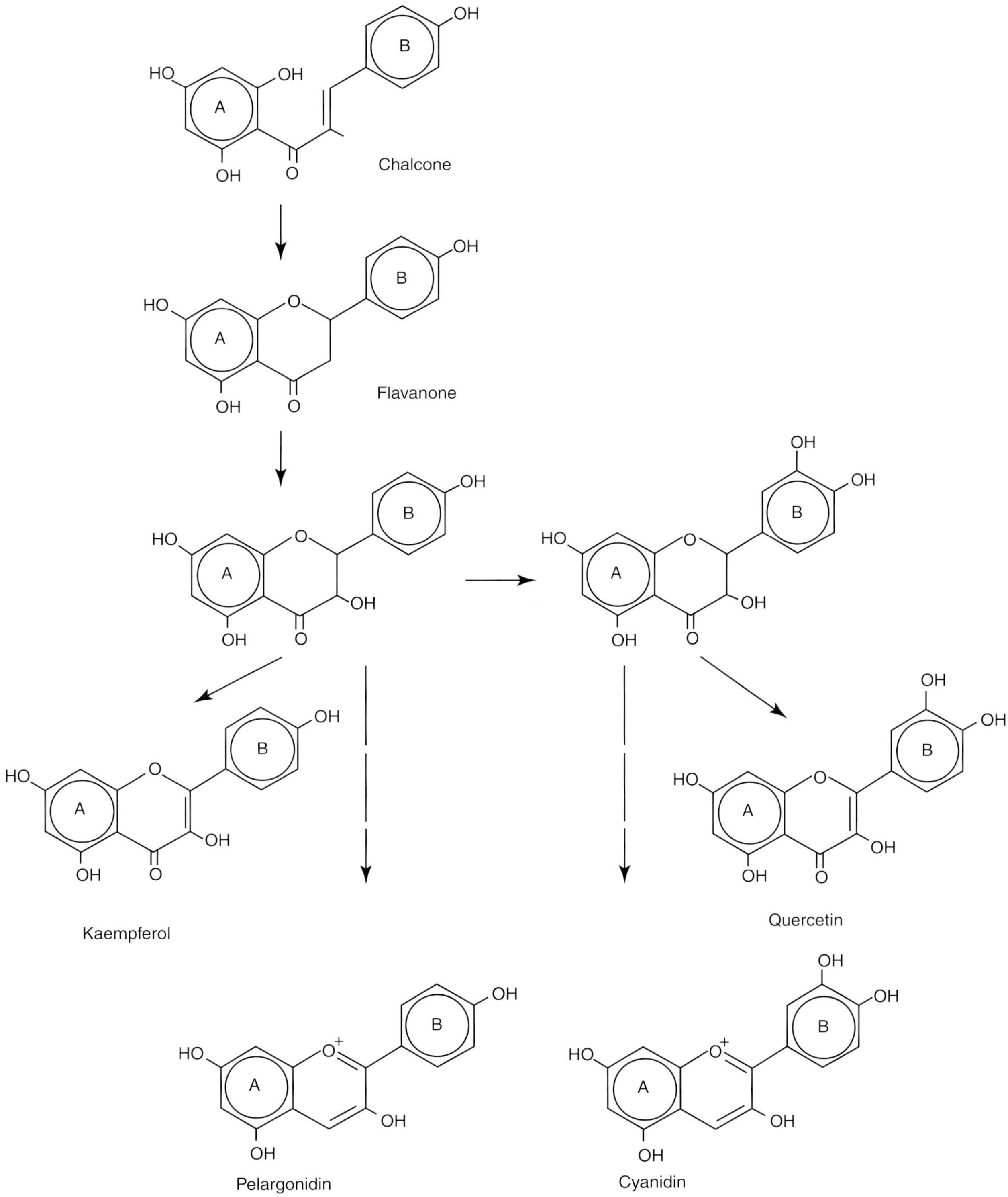

Figure 1 Biosynthetic organization of flavonoids.

Pelargonidin was reported for the first time in roses in 1934 as pelargonidin 3,5-diglucoside (pelargonin) from Polyantha cultivars such as 'Gloria Mundi' (de Ruiter), and in 1963 as pelargonidin 3-glucoside (callistephin). More recent analyses show that even old varieties contain significant amounts of pelargonidin glycosides: *R. centifolia* var. *muscosa* (Moss rose), 'Général Jacqueminot' (Hybrid Perpetual) (Roussel), 'Soleil d'Or' (Hybrid Tea) (Pernet-Ducher), as well as the botanical roses such as

Trivial name	*Characteristic substitutions*			
	R1	*R2*	*R3*	*R4*
Cyanidin type				
Cyanidin 3,5-diglucoside (cyanin)	gl	gl	OH	H
Cyanidin 3-glucoside (chrysanthemin)	gl	H	OH	H
Cyanidin 5-glucoside	H	gl	OH	H
Cyanidin 3-rutinoside	rhgl	H	OH	H
Cyanidin 3-sophoroside	glgl	H	OH	H
Cyanidin 3-(*para*-coumaroyl)glucosyl, 5-glucoside	Coum.gl	gl	OH	H
Paeonidin type				
Paeonidin 3,5-diglucoside (paeonin)	gl	gl	OCH_3	H
Paeonidin 3-glucoside (oxycoccicyanin)	gl	H	OCH_3	H
Paeonidin 5-glucoside	H	gl	OCH_3	H
Paeonidin 3-rutinoside	rhgl	H	OCH_3	H
Paeonidin 3-(*para*-coumaroyl)glucosyl, 5-glucoside	Coum.gl	gl	OCH_3	H
Pelargonidin type				
Pelargonidin 3,5-diglucoside (pelargonin)	gl	gl	H	H
Pelargonidin 3-glucoside (callistephin)	gl	H	H	H
Pelargonidin 5-glucoside	H	gl	H	H

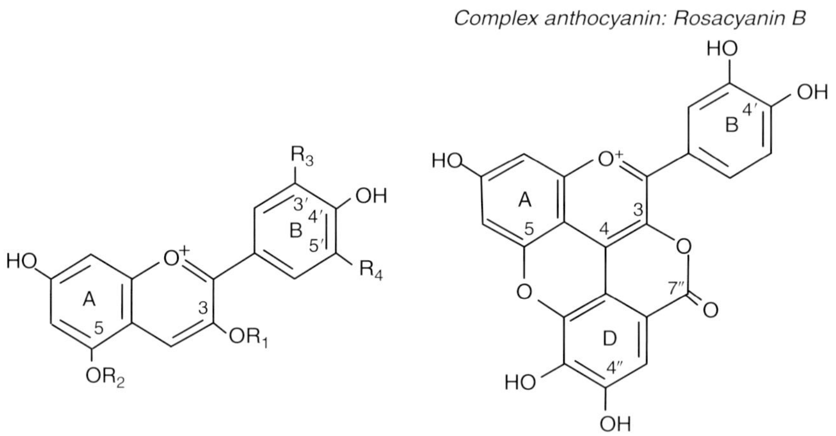

Figure 2 Structures of anthocyanins from rose petals.

R. gallica versicolor (Gallicanae) and *R. pendulina (Cinnamomae)*.

Among the methylated derivatives, paeonidin was first described in 1961 as paeonidin 3,5-diglucoside (paeonin) from petals of *R. rugosa* and derived varieties, e.g. 'Roseraie de l'Hay' (Cochet-Cochet). More recent studies show a broad distribution of this anthocyanin in the sections *Cinnamomae*, *Caninae* and *Pimpinellifoliae*. The old shrub rose 'Hansa' (Schaum & Van Pol) with mauve–red flowers seemingly represents a rich source of paeonin; this is also the case in *R. foetida* var. *bicolor (Pimpinellifoliae)*. The paeonidin 3-glucoside (oxycoccicyanin) has only exceptionally been found in roses.

For the third anthocyanidin class, delphinidin was reported in 1957 in several Hybrid Tea varieties with purple or mauve petals. This finding was invalidated three years later when it was shown that purple or mauve colours in roses are not produced by delphinidin–glycoside accumulation, but rather by copigmentation of cyanin or paeonin. This is the case with 'Tuscany' (Gallica rose), 'Reine des Violettes' (Millet-Malet), 'Generál Stefánik' (hybrid perpetual) (Böhm) (hybrid perpetual) and 'Veilchenblau' (climbing rose) (J.C. Schmidt).

Until recently, six anthocyanins based on pelargonidin, cyanidin and paeonidin were considered to be characteristics of roses: three major compounds for diglycosylated derivatives and three minor compounds for monoglycosylated forms. The proportion and intensity of all three main anthocyanins varied widely within a large rose collection. More recently, however, several new anthocyanin compounds have been identified in roses:

1. Three anthocyanidins bearing a sugar bond at the C5 position of the nucleus: pelargonidin 5-glucoside, cyanidin 5-glucoside and paeonidin 5-glucoside. These compounds are apparently rare.
2. Three molecules with a diglycosyl residue attached to the C3 position: cyanidin 3-rutinoside, paeonidin 3-rutinoside and cyanidin 3-sophoroside. These compounds have been identified in various groups (e.g. *Chinenses*, Tea roses) but rarely as major compounds.
3. Two acylated anthocyanins: cyanidin 3-*para*-coumaroylglucoside-5-glucoside and paeonidin 3-*para*-coumaroylglucoside-5-glucoside. These pigments have been found in the *Cinnamomae* section and more rarely in the *Chinenses* section.
4. In 2002 the last new rose anthocyanin was isolated, from the mauve flowers of the variety 'Madame Violet', and was named rosacyanin B. It is the first C4 substituted anthocyanin isolated from plants: a C–C bond and two ether linkages join a gallic acid moiety to a cyanidin chromophore. This pigment would be responsible for the violet colour of this rose.

Distribution of Anthocyanins within Rose Groups

The comparative distribution of major glycosides of pelargonidin, cyanidin and paeonidin within the main botanical taxa and biological groups of roses is shown in **Table 1**. In all cyanic taxa, the cyanidin 3,5-diglucoside was found as a major anthocyanin. As a consequence, pelargonidin 3,5-diglucoside, paeonidin 3,5-diglucoside and cyanidin 3-glucoside could be widely distributed but almost always as minor compounds. The paeonidin 3,5-diglucoside was found more sporadically, either as a minor (nine groups) or major (six groups) compound. Finally pelargonidin 3-glucoside seemed rather rare, except in Hybrid Tea roses. Only one group (*Banksianae*) was acyanic, and three groups (*Caninae*, *Pimpinellifoliae* and Hybrid Perpetuals) showed a low frequency of cyanic species or varieties.

The quantitative correlations between pelargonidin and cyanidin glycosides within modern roses (Hybrid Tea) are shown in **Figure 3**. Depending on the variety, the total anthocyanin amounts ranged from 1 to 120 mg per gram dry weight. All varieties accumulating more than 60 mg per gram petal dry weight were characterized by a cyanidin content greater

Table 1 Distribution of anthocyanins within sections and biological groups of roses. Citation number: 3, more than 80% of the species or varieties of group; 2, 20% to 80%; 1, less than 20%

Rose groups	*Cyanidin 3,5-diglucoside*	*Pelargonidin 3,5-diglucoside*	*Paeonidin 3,5-diglucoside*	*Cyanidin 3-glucoside*	*Pelargonidin 3-glucoside*
Section					
Banksianae	0	—	—	0	—
Caninae	2[a]	0	2[a]	0	0
Carolinae	3[a]	—	2[a]	—	0
Chinenses	2[a]	—	1	2[a]	0
Cinnamomae	3[a]	2	2[a]	1	1
Laevigatae	2[a]	—	0	0	—
Pimpinellifoliae	1[a]	0	0	0	0
Synstylae	2[a]	1	2[a]	1	0
Biological groups					
Alba	3[a]	1	1[a]	2	—
Bourbon	3[a]	2	0	2	—
Centifolia	3[a]	1	1	1[a]	—
China	2[a]	1	2	3	—
Floribunda	3[a]	2[a]	0	2	—
Gallica	3[a]	2	0	2	0
Hybrid Perpetual	1[a]	2	2	2	—
Hybrid × *rugosa*	3[a]	2	2[a]	2	—
Hybrid Tea	3[a]	2	0	3	2
Hybrid × *wichurana*	3[a]	2	1	3	—
Moss	3[a]	1	1	2	—
Noisette	3[a]	2	1	2	—
Polyantha	3[a]	2	2	2	—
Portland	3[a]	2	0	2	—
Tea	3[a]	1	1	2[a]	—

[a] Major compound (more than 10% of the total anthocyanin content).

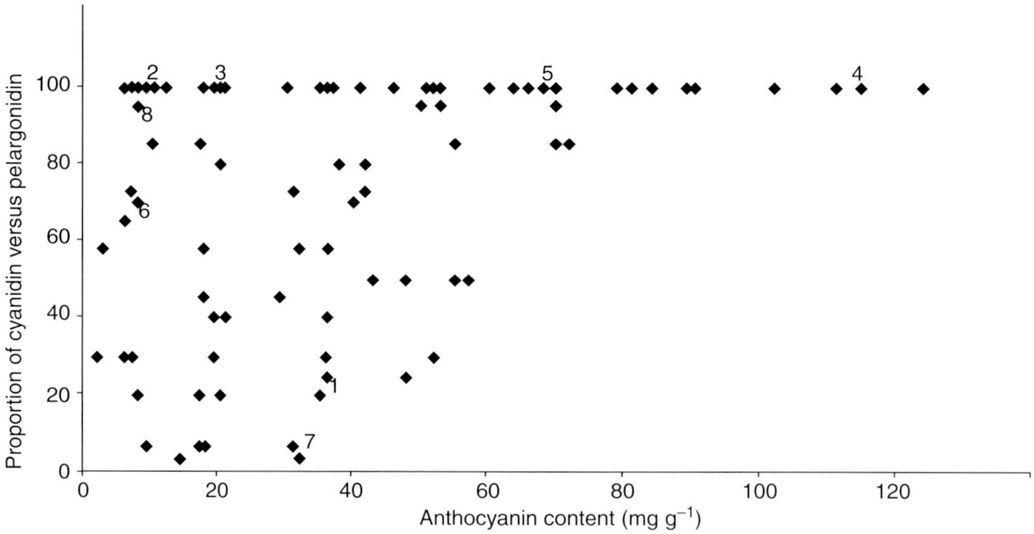

Figure 3 Distribution of about 100 modern roses (exclusively pigmented by cyanidin and pelargonidin glycosides) according to their petal anthocyanin concentrations, and the ratio between cyanidin and pelargonidin. The varieties are identified by a numerical code; some of them are keyed to the name of the variety: 1, 'Joconde'; 2, 'Mamy Blue'; 3, 'Lancôme'; 4, 'Perle Noire'; 5, 'Mme Georges Delbard'; 6, 'Eterna'®; 7, 'Mercedes'®; 8, 'Centenaire de Lourdes'®.

than 80% of the total anthocyanins; other varieties showed a continuous range of mixtures from 1% to 100% of pelargonidin or cyanidin reciprocally. While 92 cultivars exhibited a pure cyanidin glycosides content, only one variety ('Gloria Mundi') was a pure pelargonidin type.

Structures of Flavonol Glycosides Isolated from Roses

The first reports of flavonols in rose petals described the occurrence of quercetin and kaempferol in glycosidic form. In 1957, myricetin was reported as a new compound in flowers of 20 Hybrid Tea varieties, but this result was invalidated three years later. It is now clear that petal cells synthesize only two flavonol types: kaempferol with a monosubstituted B ring, and quercetin with a disubstituted B ring.

Numerous glycosides of flavonols have been isolated and formally identified from petals and more recently from leaves; **Figure 4** describes about 30 different chemical structures and their source organs. Several chemical families have been characterized on the base of the substitution types: flavonol 3-monohexosides, flavonol 3-dihexosides, flavonol 3-pentosides and flavonol-3-(acylated)-glycosides. The sugar residues are usually glucose, rhamnose, arabinose, xylose and (more rarely) galactose, glucuronic acid; the diosides on C3 are rutinose (rhamnosyl-glucose), sophorose (glucosyl-glucose) and rhamnosyl-galactose; the acylation is brought by aliphatic acids (acetic acid, hydroxymethyl-glutaric acid) and by aromatic acids (gallic acid, *para*-coumaroyl acid). Most of the compounds were reported in the petals as well as in leaves.

Distribution of Flavonol Glycosides within Rose Groups

The distribution of some important flavonol glycosides within the main botanical taxa and the major hybrid groups of roses is shown in **Table 2**. Globally, the amounts of kaempferol and quercetin are variously shared in rose petals according to the taxa; kaempferol is dominant in *Banksianae*, Centifolia roses, Noisette roses, Polyantha, Tea roses and Hybrid *Wichurana*, while quercetin is dominant in China roses, Hybrid Perpetuals, Hybrid *Rugosa*, some Floribunda and some Hybrid Tea roses. The balance is more or less even for all other groups.

Large amounts of flavonol 3-monoglycosides (kaempferol and quercetin bases) were detected in all groups of roses, except Hybrid *Rugosa*. On the contrary, kaempferol and quercetin 3-sophorosides seem quantitatively very characteristic of *Caninae, Carolinae, Cinnamomae*, Hybrid *Rugosa*, and to a lesser extent *Synstylae* and Polyantha. Similarly, kaempferol and quercetin 4′-glucosides are accumulated in large amounts in the section *Pimpinellifoliae*, and to a lesser extent as minor compounds in Bourbon and Floribunda roses.

Comparative Distribution of Flavonols and Anthocyanins in Hybrid Tea Roses

Figure 5 shows a strong positive correlation between kaempferol and pelargonidin contents, on the basis of

	Occurrence	
	Petals	*Leaves*
Monohexosyl-flavonols		
Quercetin 3-glucoside	+	+
Quercetin 3-galactoside	+	+
Quercetin 3-rhamnoside	+	+
Quercetin 3-glucuronide	+	
Quercetin 7-glucoside	+	
Quercetin 4′-glucoside	+	
Kaempferol 3-glucoside	+	+
Kaempferol 3-galactoside		+
Kaempferol 3-rhamnoside	+	+
Kaempferol 3-glucuronide	+	+
Kaempferol 7-glucoside	+	
Kaempferol 4′-glucoside	+	
Dihexosyl-flavonols		
Quercetin 3-rhamnoglucoside	+	+
Quercetin 3-rhamnogalactoside		+
Quercetin 3-sophoroside	+	
Kaempferol 3-rhamnoglucoside		+
Kaempferol 3-rhamnogalactoside		+
Kaempferol 3-sophoroside	+	
Pentosyl-flavonols		
Quercetin 3-arabinoside	+	+
Quercetin 3-xyloside		+
Kaempferol 3-arabinoside	+	+
Kaempferol 3-xyloside	+	+
Acylated glycosyl-flavonols		
Quercetin 3-acetyl-rhamnoglucoside		+
Quercetin 3-hydroxymethyl-glutaryl-galactoside		+
Quercetin 3-galloyl-glucoside	+	+
Quercetin 3-galloyl-rhamnoside		+
Quercetin 3-galloyl-rhamnoglucoside		+
Kaempferol 3-galloyl-rhamnoglucoside		+
Kaempferol 3-galloyl-rhamnoside		+
Kaempferol 3-*para*-coumaroyl-glucoside	+	+

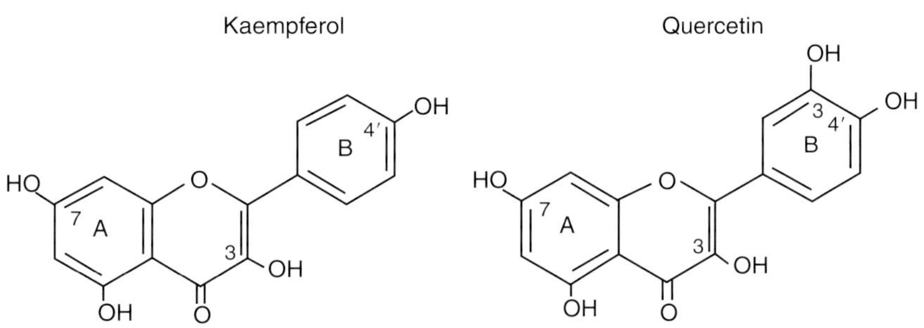

Figure 4 Structures of flavonol glycosides from rose petals and leaves.

phenolic composition of about 100 modern roses (Hybrid Tea). Thus varieties defined by less than 5% pelargonidin (i.e. containing more than 95% cyanidin) may contain from 5% to 50% of kaempferol derivatives. When pelargonidin exceeds 20%, regardless of cultivar, the kaempferol level is always more than 85% of the total flavonols. Obviously, a significant proportion of pelargonidin is never associated with a large amount of quercetin derivatives.

Contribution of Anthocyanins and Flavonols to the Colour of Rose Flowers

The wide colour range in rose flowers is among the characters that have contributed to the great popularity of rose plants: red, orange, and yellow colours appear alone or in various combinations to give many shades and colour patterns. The main molecules that directly contribute to petal colour are cyanidin

Table 2 Distribution of flavonol glycosides within section and biological groups of roses. Citation number: 3, more than 80% of the species or varieties of group; 2, 20% to 80%; 1, less than 20%

Rose groups	*Kaempferol*			*Ratio Q/K*	*Quercetin*		
	3-Monoglycoside	*3-Sophoroside*	*4′-Monoglycoside*		*3-Monoglycoside*	*3-Sophoroside*	*4′-Monoglycoside*
Section							
Banksianae	3[a]	0	0	<1	1	0	0
Caninae	3[a]	3[a]	0	—	3[a]	3[a]	0
Carolinae	3[a]	3[a]	0	—	3[a]	3[a]	0
Chinenses	3[a]	1	0	—	3[a]	2	0
Cinnamomeae	3[a]	3[a]	0	—	3	3[a]	0
Pimpinellifoliae	3	2	3[a]	—	2	2[a]	3[a]
Synstylae	3[a]	3	0	—	3[a]	3	0
Biological group							
Alba	3[a]	2	2	1	3[a]	2[a]	0
Bourbon	3[a]	2	3	1	3[a]	0	1
Centifolia	3[a]	0	2	<1	3[a]	0	1
China	3[a]	2	2	>1	3[a]	0	1
Floribunda	3[a]	2	3	variable	3	2	2
Gallica	3[a]	2	2	1	3[a]	0	1
Hybrid Perpetual	3[a]	2	2	>1	3[a]	2	1
Hybrid × *rugosa*	2	3[a]	2	>1	2	3[a]	1
Hybrid Tea	3[a]	2	2	variable	3	2	2
Hybrid × *wichurana*	3[a]	0	1	<1	3	2	1
Moss	3[a]	0	1	1	3[a]	1	1
Noisette	3[a]	2	2	<1	3	0	1
Polyantha	3[a]	2	2	<1	3	3	2
Portland	3[a]	1	2	—	3[a]	0	0
Tea	3[a]	2	2	<1	3[a]	0	0

[a] Major compound (more than 10% of the total flavonol glycoside content).

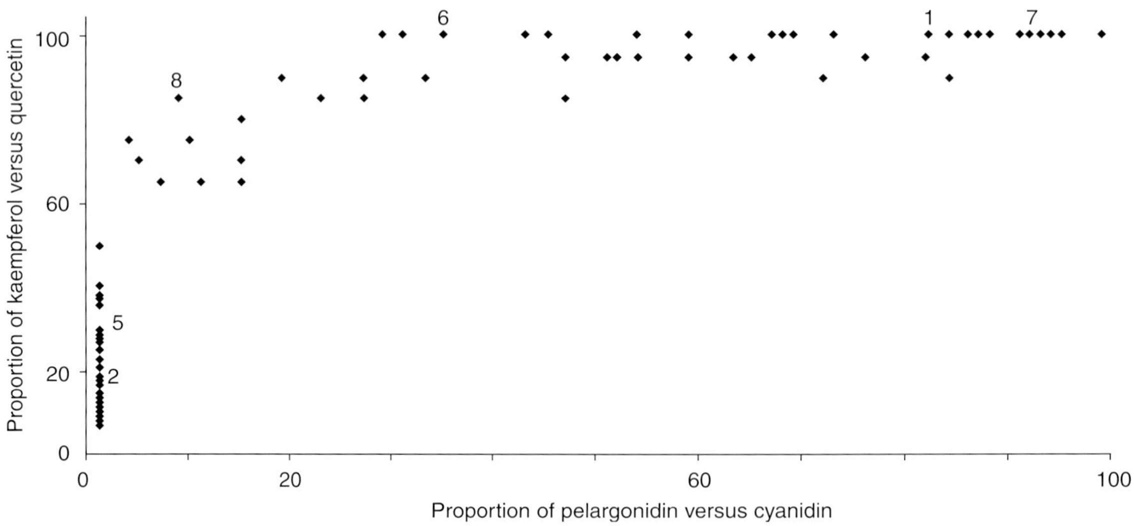

Figure 5 Quantitative correlation between anthocyanins and flavonol glycosides within a collection of about 100 varieties representative of modern roses (Hybrid Tea). The varieties are identified by a numerical code; some of them are keyed to the name of the variety: 1, 'Joconde'; 2, 'Mamy Blue'; 5, 'Mme Georges Delbard'; 6, 'Eterna'®; 7, 'Mercedes'®; 8, 'Centenaire de Lourdes'®.

3,5-diglucoside, pelargonidin 3,5-diglucoside and paeonidin 3,5-diglucoside. In addition, some flavonol glycosides act as copigments.

In vivo colour expression of anthocyanins is a complex issue because of their high reactivity to the chemical environment in the vacuolar compartment, specially regarding the pH. The typical red and magenta (purple–red) tonalities of the pelargonidin and cyanidin types respectively are those of the stable and coloured form of the flavylium cation, obtained only in an acidic medium (pH below 3). When, *in vitro*, the pH of the aqueous solution increases (up to 6–6.5 approximately), the anthocyanin nucleus is converted into multiple and poorly or even colourless forms; at pH above 6.5, coloured forms are present again, but they feature bluer tonalities than those of the basic flavylium cation.

At the most frequent pH values (between 3.5 and 5.5) reported for the vacuolar sap in epidermal cells of rose petals, the pigments would presumably be present as stable but colourless forms. In fact, the intense colours observed in fresh petals result from the existence of mechanisms preventing colour loss or even enhancing pigment colour expression and tonalities. These mechanisms are known as copigmentation and consist in the association, in a sandwich-type molecular stacking, of the pigment itself with other molecules present in the vacuolar solution, called copigments; in roses, the most efficient copigments are colourless flavonol glycosides and /or simpler phenolic acids. Three different types of copigmentation can coexist and compete in rose epidermal cells:

1. Intermolecular copigmentation, in which the copigment is not chemically (covalently) bound to the anthocyanin and its copigment. This occurs mainly with flavonol glycosides.
2. Intramolecular copigmentation, in which the copigment is part of the pigment structure as it is observed in anthocyanins acylated with cinnamic acid derivatives.
3. When anthocyanins are accumulated at high concentrations they interact as copigments due to the formation of chiral spiral stackings.

The main effect of copigmentation is to stabilize the pigment in its coloured forms in the vacuolar sap. But in addition, copigmentation also enhances the colour range of each basic pigment, resulting in two effects: a bathochromic effect currently reported as having a blueing effect on tonalities, and a hyperchromic effect producing more 'intense' colours. Recent colorimetric studies have revealed that the spectral variations originated by copigmentation throughout the visible light extent are producing more complex and subtle colour effects.

The global efficiency of copigmentation, and its subsequent effects on the colour, results from the multiple factors such as: (1) the vacuolar pH (most pronounced effects occur at pH between 3.5 and 4.5), (2) the pigment concentration (stronger effects at high concentrations) and (3) the nature of the pigment and the copigment(s) and their ratio (spectacular effects are obtained by very high – up to 40–100/1 – copigment/pigment ratios). In addition, each single pigment can display considerable variation of basic tonalities, simply by concentration effects. In acidic medium, the basic hues of solutions of cyanin (cyanidin-3-5-diglucoside, the most frequent anthocyanin present in roses) cover one-sixth of the complete colour circle: from orange (at very high concentrations) to red (medium concentrations) and finally magenta and purple–magenta (lowest levels).

At the macroscopic level (petals) the combined effects of all these chemical parameters create an extraordinary diversity of colours, even when relatively simple pigment/copigment patterns are involved: cyanin is virtually the only basic anthocyanin in cultivars 'Baccará'® (Meilland) or 'Mme Georges Delbard' (Delbard) (red/dark red colours), in 'Lancôme' (Delbard) or 'Maria Callas' (Meilland) ('fuschia' or 'cyclamen', i.e. magenta hues) but also in 'Mamy Blue' (Delbard) (light dull purple) too. Its colour palette can be even extended up to vivid blue–purple colours when it is included into particular vacuolar structures (AVIs) in the flowers of the newly released 'Rhapsody in Blue' (Cowlishaw) but also in those of ancient roses such as 'Bleu Magenta' (unknown breeder) or 'Cardinal de Richelieu' (Laffay).

See also: **Breeding**: Selection Strategies for Pot Roses. **Classification**: Cultivar Identification by Image Analysis. **Fragrance and Pigments**: Flavonoid Molecular Biology. **Genetics**: Inheritance of Characteristics.

Further Reading

Biolley JP and Jay M (1993) Anthocyanins in modern roses: chemical and colorimetric features in relation to the colour range. *Journal of Experimental Botany* 44: 1725–1734.

Biolley JP, Jay M and Viricel MR (1994) Flavonoid diversity and metabolism in 100 *Rosa* × *hybrida* cultivars. *Phytochemistry* 35: 413–419.

Brouillard R (1988) Flavonoids and flower colour. In: Harborne JB (ed.) *The Flavonoids*, pp. 525–537. London: Chapman & Hall.

Eugster CH and Märki-Fischer E (1991) The chemistry of rose pigments. *Angewandte Chemie International – English Edition* 30: 654–672.

Mikanagi Y, Yokoi M, Ueda Y and Saito N (1995) Flower flavonol and anthocyanin distribution in subgenus *Rosa*. *Biochemical Systematics and Ecology* 23: 183–200.

Raymond O, Biolley JP and Jay M (1995) Fingerprinting the selection process of ancient roses by means of floral phenolic metabolism. *Biochemical Systematics and Ecology* 23: 555–565.

Flavonoid Molecular Biology

G Forkmann, Technical University of Munich, Freising-Weihenstephan, Germany

Introduction

Flavonoids represent an important class of natural plant metabolites. Their wide occurrence, complex diversity and manifold functions, including the conspicuous and eye-catching colouration of flower petals, fruits and other plant organs, have made these compounds an attractive system for genetic, analytical and biochemical studies as well as for research on a molecular biological level. To date, flavonoids represent the best understood group of secondary plant products.

During the last 25 years, chemicogenetic, enzymological and molecular studies on flowers of many different ornamental plant species have not only greatly contributed to a detailed knowledge of flavonoids as flower pigments, but have also provided the tools and know-how for the successful creation of new and attractive flower colours by conventional and, in particular, molecular breeding methods. In spite of the economic importance of roses, however, the biosynthetic pathway leading to synthesis of flavonoid pigments in petals, fruits and leaves of roses has not yet been thoroughly elucidated.

In this article I will give a general overview of the biosynthetic pathway to the different flavonoid classes and of the regulation of biosynthesis. Then I will report on the very recent progress in elucidating flavonoid biosynthesis in rose petals and leaves. This is followed by a compilation of the flavonoid genes, which have been cloned from roses and partially characterized. Finally, the strategies used so far for the manipulation of flower colouration in general and respective approaches, particularly for roses, will be discussed.

General Overview of Flavonoid Biosynthesis and Regulation

Since the mid-1970s rapid and substantial progress has been made in elucidating the pathways to the flavonoid precursors and the individual steps to the flavonoid classes. The respective reaction steps are outlined in **Figure 1**.

The flavonoid precursors, malonyl-coenzyme A (CoA) and a hydroxycinnamic acid CoA ester, ordinarily 4-coumaroyl-CoA, are both derived from carbohydrates. Malonyl-CoA is formed from acetyl-CoA and CO_2 by acetyl-CoA carboxylase. 4-Coumaroyl-CoA and related hydroxycinnamic acid esters are supplied by the first steps of the general phenylpropanoid pathway. The respective reactions are catalysed by the enzymes phenylalanin ammonia-lyase, cinnamate 4-hydroxylase and 4-coumarate: CoA ligase. Hydroxylation of 4-coumaroyl-CoA in 3 position leads to caffeoyl-CoA, which may also serve as substrate for chalcone formation. 3-Hydroxylation may alternatively be attained at the level of unconjugated 4-coumarate.

The key reaction in flavonoid biosynthesis is the stepwise condensation of three acetate units from malonyl-CoA with 4-coumaroyl-CoA to naringenin chalcone, which is rapidly isomerized to the flavanone naringenin. The reactions are catalysed by chalcone synthase (CHS) and chalcone isomerase (CHI), respectively (**Figure 1**). The further reactions to the anthocyanins, the most important flavonoid class, are catalysed by flavanone 3-hydroxylase (FHT), dihydroflavonol 4-reductase (DFR), anthocyanidin synthase (ANS) and flavonoid 3-O glucosyltransferase (FGT). Other flavonoid classes are derived from different intermediates of the anthocyanin pathway. The most important enzymes involved are flavone synthase I and II (FNS I, FNS II), flavonol synthase (FLS), leucoanthocyanidin reductase (LAR) and anthocyanidin reductase (ANR). Proanthocyanidins are most probably synthesized by condensing leucoanthocyanidins with catechins, and isoflavones are formed from flavanones by the action of isoflavone synthase. The formation of 3-deoxyflavonoids from flavanones is most probably catalysed by DFR additionally expressing flavanone 4-reductase (FNR) activity. With the exception of ANS and the condensing enzyme, all the enzymes mentioned in **Figure 1** have already been demonstrated in protein preparations from tissues of many different plant species.

The CHS reaction with malonyl-CoA and 4-coumaroyl-CoA as the main physiological substrate leads to flavonoids with a 4′,5,7-trihydroxylation pattern. These simple flavonoids are substrates for extensive modification by hydroxylation, methylation of hydroxyl groups, glycosylation, acylation and a number of other reactions. Numerous enzymes catalysing these modifications have been described. Of particular interest are the enzymes flavonoid 3′-hydroxylase (F3′H) and flavonoid 3′,5′-hydroxylase (F3′5′H). These enzymes are responsible for B-ring hydroxylation at the appropriate positions. Hydroxylation of naringenin and dihydrokaempferol in 3′- or 3′5′-positions provides the precursors to luteolin and tricetin (flavones), quercetin (Qu) and myricetin (flavonols) and cyanidin (Cy) and delphinidin (Dp: anthocyanidins) (**Figure 1**). As a rule, pelargonidin (Pg)

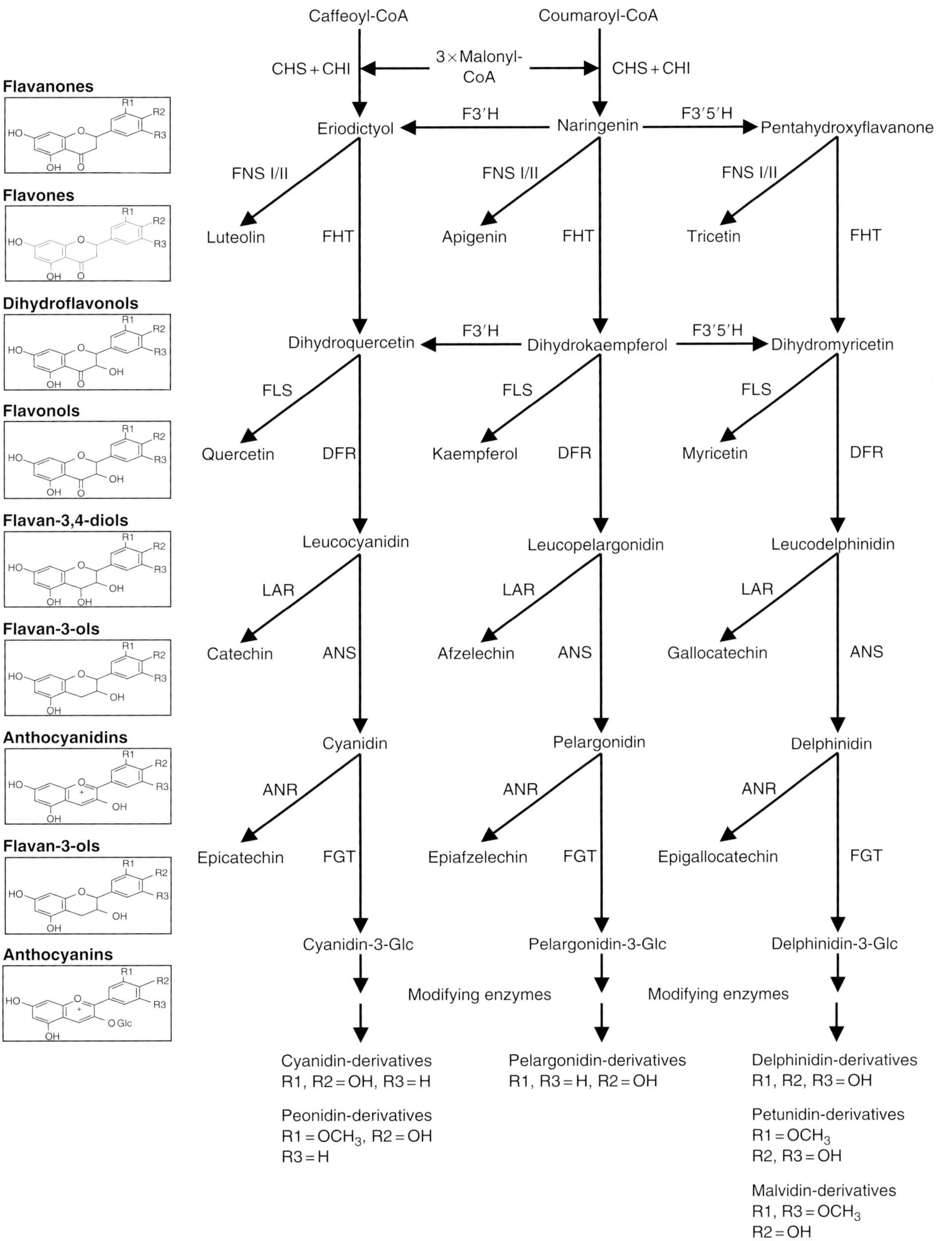

Figure 1 The biosynthetic pathways leading to various flavonoid classes of flowering plants. CHS, chalcone synthase; CHI, chalcone isomerase; FNS, flavone synthase; FHT, flavanone 3β-hydroxylase; F3′H, flavonoid 3′-hydroxylase; F3′5′H, flavonoid 3′5′-hydroxylase; FLS, flavonol synthase; DFR, dihydroflavonol 4-reductase; LAR, leucoanthocyanidin 4-reductase; ANS, anthocyanidin synthase; ANR, anthocyanidin reductase; FGT, flavonoid 3-*O*-glucosyltransferase.

derivatives are responsible for orange, pink and bright red colours, whereas Cy derivatives are the basis of red and magenta colours, and Dp derivatives for violet and blue colours, respectively.

The activity of the flavonoid enzymes is determined by the transcription rate of the respective structural genes, which in turn is controlled by specific regulatory genes. Some of these genes regulate the expression of a subset of structural genes, in particular of those involved in the late steps of anthocyanin formation; other genes are responsible for structural gene activation in dependence on plant development and/or environmental parameters. Molecular analyses of the regulatory genes in maize, snapdragon, *Arabidopsis*, petunia and some other plant species revealed that the regulatory genes of the flavonoid pathway belong to two gene families encoding transcription factors, either with homology to Myb oncoproteins (Myb-type) or with the basic helix-loop-helix domain of the Myc oncoproteins (Myc/bHLH-type).

Flavonoid Biosynthesis in Roses

Despite the manifold colours of rose petals, the pattern of the flavonoid pigments formed is rather simple. Up to now, the presence of only two flavonoid classes was demonstrated; glycosides of the flavonols kaempferol (Km) and Qu, respectively, and glucosides of the anthocyanidins Pg, Cy and peonidin (Pn), respectively (**Figure 1**). White or pure yellow flowering mutants lack the anthocyanins but flavonol glycosides are still present, indicating that the anthocyanin pathway is blocked at a late step. The typically yellow pigmentation of roses is not caused by flavonoids but by carotenoids.

The flavonoid pattern found in young shoots and leaves of roses is different from that of the petals. Besides few Km glycosides, mainly Qu glycosides are present and, with regard to anthocyanins, exclusively Cy/Pn-derivatives are formed. Moreover, as a rule, the presence of catechin and epicatechin can be demonstrated in leaves.

Over a long time, all efforts to demonstrate the activity of flavonoid enzymes in protein extracts prepared from rose petals or leaves with the methods successfully applied for many other plant species failed. Neither activity of soluble enzymes, such as CHS, FHT and DFR, nor of membrane-bound enzymes like F3′H could be demonstrated. The main reason was the high concentration of tannins and other secondary plant products in rose tissues. In the mid-1990s, the development of improved procedures for enzyme extractions from plant materials containing interfering compounds also allowed the demonstration of some flavonoid enzymes in extracts from rose tissues. Because of the differences in the flavonoid patterns of petals and leaves, the results of the enzymological studies will be presented separately.

Flavonoid Enzymes in Rose Petals

From the enzymes involved in flavonol and anthocyanin synthesis in rose petals CHS including CHI, FHT and DFR activities could be demonstrated so far in respective enzyme preparations. CHS provides the basic C15 chalcone intermediate from which all flavonoids originate. Because most CHS assays also include CHI activity, not chalcones but the respective 2S-flavanones are found as reaction products (**Figure 1**). Incubation of malonyl-CoA and 4-coumaroyl-CoA with enzyme extracts from rose petals led to the formation of appreciable amounts of naringenin. In rose cultivars mainly producing Qu- and Cy-derivatives (Cy-type) in the petals, caffeoyl-CoA was found to be accepted as substrate besides 4-coumaroyl-CoA, leading to the formation of eriodictyol as the product. 4-Coumaroyl-CoA was preferred at pH 7.5 and caffeoyl-CoA at pH 6.0 of the enzyme assay. However, in cultivars exclusively accumulating Km- and Pg-derivatives (Pg-type), caffeoyl-CoA was not accepted as substrate. In competition experiments with both substrates and enzyme preparations from petals of the Cy-type, 4-coumaroyl-CoA was found to be the preferred substrate independently of pH. These results indicate that in rose petals 4-coumaroyl-CoA is the main physiological substrate for the condensation reaction catalysed by CHS, and that the hydroxyl group in position 3′ of the B-ring (Cy-type) is introduced on the flavonoid level by the action of F3′H. Thus, the formation of the Cy-type of roses is achieved in the same way as in many other plant species. But, because only few cultivars have been investigated so far, it cannot be excluded that at least in petals of cultivars of the Cy-type CHS enzymes exist which use both 4-coumaroyl-CoA and caffeoyl-CoA.

FHT catalyses the stereospecific 3β-hydroxylation of 2S-flavanones to the respective dihydroflavonols (**Figure 1**). According to the cofactor requirement of 2-oxoglutarate, ferrous iron and ascorbate, the enzyme was classified as a 2-oxoglutarate-dependent dioxygenase. Incubation of naringenin or eriodictyol with enzyme extracts from rose petals in the presence of the required cofactors led to the formation of dihydrokaempferol or dihydroquercetin, respectively. As expected, FHT activity was clearly dependent on the presence of 2-oxoglutarate. There were no significant differences in specificity for naringenin and eriodictyol, respectively, indicating that FHT is not involved in the selection of a special flavonoid pattern.

DFR catalyses the stereospecific reduction of dihydroflavonols to the respective flavan 3,4-diols (leucoanthocyanidins) with NADPH as cofactor (**Figure 1**). As has been found for some other plant species, the DFR of rose petals showed a striking substrate specificity. Enzyme preparations from petals of a cultivar belonging to the Pg-type catalysed the reduction of both dihydrokaempferol to leucopelargonidin and dihydroquercetin to leucocyanidin, whereas enzyme extracts of a cultivar exclusively synthesizing Cy-derivatives did not reduce dihydrokaempferol at all, but used dihydroquercetin as substrate with high efficiency. Thus, at least two DFR enzymes with different substrate specificity may be expressed in petals of rose cultivars. As expected from the absence of catechins, LAR activity could not be demonstrated either with leucoanthocyanidins as substrate or in combination with DFR starting with dihydroflavonols as substrate in a two-step reaction.

In agreement with the presence of flavonols, CHS and FHT activity was demonstrated in pure white and yellow flowering cultivars, and DFR activity was also present. This result indicated that the lack of anthocyanins is due to a late block in the biosynthetic pathway.

All efforts to demonstrate enzyme activity of the soluble enzymes FLS and the membrane-bound F3′H have so far failed, although FLS should be present in all cultivars and F3′H should be highly active at least in petal extracts from Cy-types. The glycosyltransferases responsible for the glycosylation of flavonols and anthocyanidins and the methyltransferase catalysing methylation of Cy-glycosides to Pn-glycosides (**Figure 1**) have not yet been investigated.

Flavonoid Enzymes in Rose Leaves

From the enzymes involved in flavonol, catechin and anthocyanin formation in rose leaves CHS, including CHI, FHT, DFR and LAR, have so far been demonstrated.

The CHS/CHI activity expressed in rose leaves catalysed both the synthesis of naringenin from 4-coumaroyl-CoA and the synthesis of eriodictyol from caffeoyl-CoA, even in competition experiments with both substrates in the enzyme assay and at different pH values. Moreover, there were no differences in substrate specificity between CHS/CHI enzyme preparations from leaves of cultivars accumulating Pg- or Cy-derivatives in the petals. These results suggest that the CHS enzyme of rose leaves is different from the respective enzyme(s) expressed in rose petals.

As in the petals, the FHT of leaf extracts hydroxylated naringenin to dihydrokaempferol and eriodictyol to dihydroquercetin. Both substrates were converted at comparable rates. Thus, the FHT enzyme from leaves seems not to be different from that from petals. As expected, FHT activity was clearly dependent on 2-oxoglutarate as cofactor. No FHT activity could be demonstrated in enzyme assays in the presence of prohexadione-Ca, a structural analogue of 2-oxoglutarate, which competitively inhibits 2-oxoglutarate-dependent dioxygenases.

In agreement with the exclusive presence of Cy-derivatives and catechins (both 3′4′-hydroxylated), the DFR present in enzyme extracts from rose leaves was found to use dihydroquercetin but not dihydrokaempferol as substrate for the reduction reaction. Thus, the DFR enzyme in leaves may be the same enzyme as found to be present in petal extracts of the Cy-Typ.

Surprisingly, the enzyme extracts exhibiting DFR activity also expressed FNR activity, reducing eriodictyol to the respective flavan 4-ol, luteoforol (**Figure 2**). FNR is the key enzyme in the formation of 3-deoxyflavonoids. By heterologous expression of DFR cDNAs, it was recently shown that the DFR enzymes of several plant species, including roses, also possess FNR activity. 3-Deoxyflavonoids, including luteoliflavan (3-deoxycatechin), are minor compounds in rose leaves, if at all formed under natural conditions. But luteoliflavan is strongly accumulated after treatment of young shoots of roses with prohexadione-Ca. The inhibition of FHT activity by prohexadione-Ca leads to the accumulation of eriodictyol, which is then reduced to luteoforol and further to luteoliflavan by DFR/FNR and LAR, respectively (**Figure 2**). The fact that eriodictyol and not naringenin is accumulated *in vivo* after prohexadione-Ca treatment indicates that naringenin is a good substrate for F3′H.

LAR catalyses the NADPH-dependent reduction of flavan 3,4-diols (leucoanthocyanidins) to catechins. In agreement with the presence of catechins, LAR activity could be demonstrated in leaf extracts of roses, although this enzyme activity was not expressed in protein preparations from petals of the same plant. The enzyme extracts converted not only leucocyanidin to catechin but also luteoforol to luteoliflavan, indicating that the LAR activity of roses is also involved in the 3-deoxyflavonoid pathway. In enzyme assays containing both DFR/FNR and LAR activity, catechin or luteoliflavan was directly obtained in a two-step reaction from dihydroquercetin and eriodictyol, respectively.

As in petal extracts, the activity of FLS and F3′H has not yet been demonstrated. Studies on ANS, ANR (epicatechin formation), the glycosyltransferases and methyltransferase leading to Pn formation have so far not been performed.

From our general knowledge of flavonoid biosynthesis and the results of the enzymatic studies in roses, a putative biosynthetic pathway to the flavonoid compounds formed in petals and leaves can be established (**Figure 2**).

In petals, the pure Km/Pg-type is formed in the absence of F3′H activity and in the presence of CHS, which exclusively uses 4-coumaroyl-CoA as substrate,

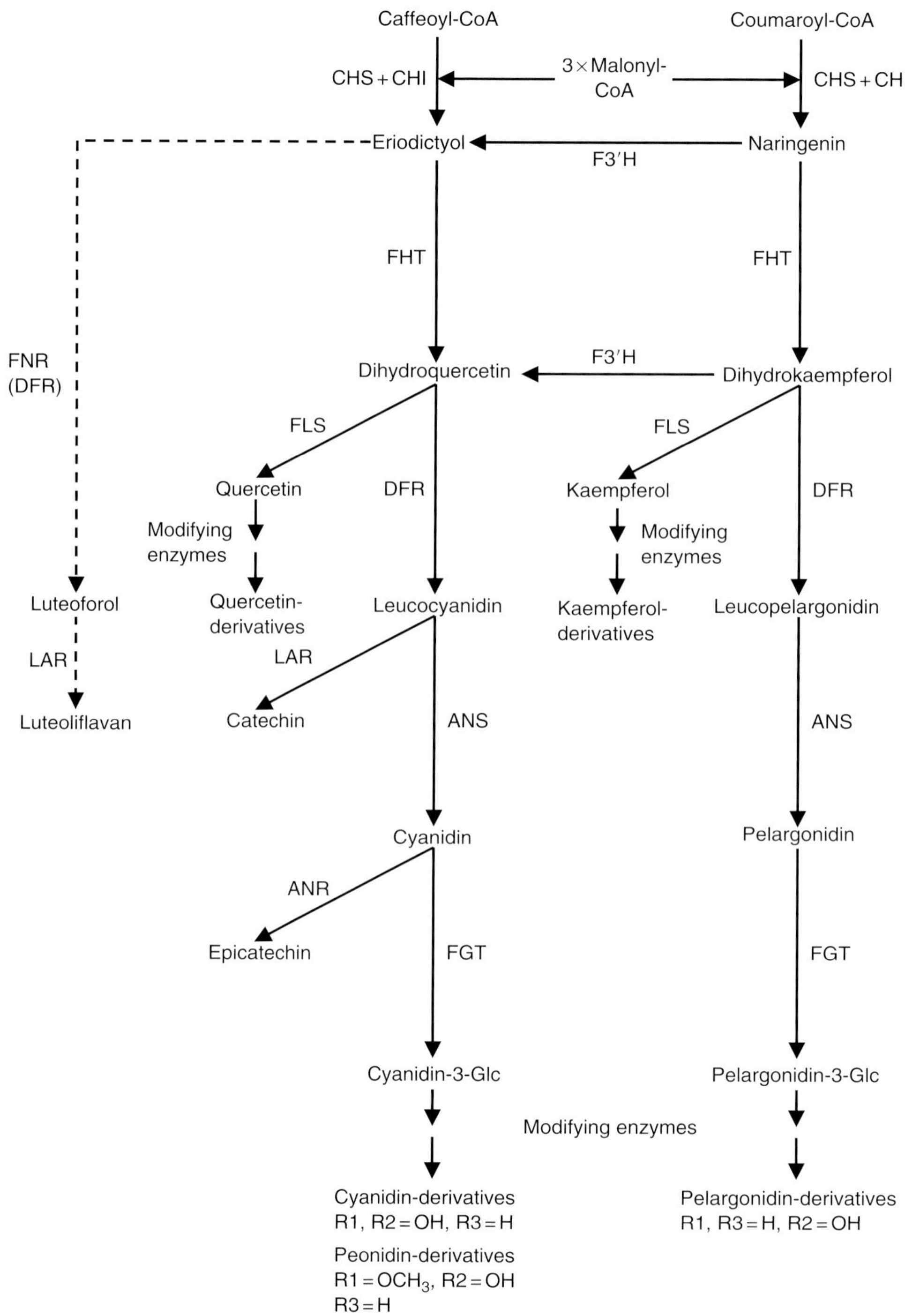

Figure 2 The putative biosynthetic pathways to the flavonoid compounds formed in petals and/or leaves of roses. CHS, chalcone synthase; CHI, chalcone isomerase; FHT, flavanone 3β-hydroxylase; F3′H, flavonoid 3′-hydroxylase; FLS, flavonol synthase; DFR, dihydroflavonol 4-reductase; FNR, flavanon 4-reductase; LAR, leucoanthocyanidin 4-reductase; ANS, anthocyanidin synthase; ANR, anthocyanidin reductase; FGT, flavonoid 3-*O*-glucosyltransferase.

and of DFR with specificity for dihydrokaempferol as substrate; the pure Cy-type is formed in the presence of F3′H activity, of CHS enzymes, which may use 4-coumaroyl-CoA as well as caffeoyl-CoA and of DFR with specificity for dihydroquercetin but not for dihydrokaempferol.

The different ratios of Km/Qu- and Pg/Cy-derivatives are probably due to the presence of higher or lower F3′H activity. The formation of mainly Qu-glycosides and exclusively Cy-derivatives and catechins in leaves can readily be explained by the presence of F3′H activity, a CHS enzyme with substrate specificity for both 4-coumaroyl-CoA and caffeoyl-CoA and DFR, using dihydroquercetin but not dihydrokaempferol as substrate (**Figure 2**).

Molecular Biology of Flavonoids

During the last 20 years a large number of flavonoid genes have been cloned from many different plant species. With the exception of the gene encoding the 'condensing enzyme' for proanthocyanidin formation, the structural genes of all of the enzymes mentioned in **Figure 1** have been isolated and further characterized. In addition, many structural genes encoding enzymes for flavonoid modification and some regulatory genes involved in flavonoid gene expression have been cloned. Moreover, the actions and interactions of structural and regulatory genes have been studied in detail.

Cloning of flavonoid genes from roses started in 1990. Screening of a cDNA library prepared from mRNA of a mixture of rose buds and petals with heterologous probes of known flavonoid genes led to the isolation of full-length cDNA clones for CHS, CHI, FHT and ANS, and of an incomplete DFR clone. All clones were identified by their high homology to the ortholog DNA and amino acid sequences of other plant species.

In the case of CHS, besides the full-length cDNA, a partial CHS clone with a slightly different DNA sequence was isolated, indicating that, in agreement with the enzymatic data, at least two different CHS genes were expressed in rose petals. Southern blots confirmed the presence of a small CHS gene family in roses.

Northern blot experiments with the CHS and ANS full-length clones and the DFR fragment as probes revealed that the three yellow and white flowering cultivars tested so far are exclusively blocked in ANS gene expression. This result is in agreement with the enzymatic studies on acyanic mutants of rose cultivars.

In 1995, a full-length DFR clone was isolated from a cDNA library of rose petals and identified by sequence comparison and transgenic expression in a suitable petunia line. Northern blot studies revealed a developmentally regulated expression of DFR mRNA in rose petals that paralleled anthocyanin formation. The same DFR clone was isolated from young leaves of a different cultivar by end-to-end polymerase chain reaction. Heterologous expression in yeast led to a funtional DFR enzyme catalysing the reduction of dihydroquercetin and even dihydromyricetin to the respective leucoanthocyanidins, whereas dihydrokaempferol was not accepted as substrate. This distinct substrate specificity is in agreement with the enzymatic data.

More recently, full-length clones of F3′H and FLS have also been isolated from a cDNA library of rose petals and further characterized, and within an analysis of expressed sequence tags (EST) of petals of *Rosa chinensis* a partial ANS clone could be identified.

Metabolic Engineering of the Flavonoid Pathway

Flower colour novelty is one of the major driving forces in ornamental plant breeding. In combination with classical breeding methods, metabolic engineering of the flavonoid pathway proved to be a powerful tool to meet this goal. Since the successful generation of Pg synthesis in flowers of petunia by expressing the DFR gene from maize in 1987, metabolic engineering of the flavonoid pathway has rapidly become an successful research area for many plant species.

Two main approaches have been used: first, the introduction of genes encoding missing enzymes and second, the specific inhibition of certain steps or of part of the flavonoid pathway. During the last 15 years, both strategies have been successfully used in the manipulation of flower colour. There are some essential prerequisites for the generation of desired flower colours by metabolic engineering of the flavonoid pathway. First, a detailed knowledge of the expression of genes, the activity of enzymes and the influence of other factors involved; second, the availibility of suitable genes and regulatory sequences; third, a well-characterized receptor line; fourth, an efficient transformation system for the respective receptor line; and fifth, stable expression and inheritance of the transgene.

As presented and discussed above, there is still a limited knowledge of biosynthesis and molecular biology of rose flavonoids. Moreover, transformation of roses is only established for a few cultivars. Approaches are based on the generation of embryogenic calli or on suitable shoots, their transformation using

Agrobacterium tumefaciens, and the regeneration of the transformed tissue to plantlets. But the existing methods are often inefficient and still time-consuming. Thus, despite their great commercial importance, the prerequisites for successful metabolic engineering are only partly fulfilled for roses. Nevertheless, a few approaches have been tested. However, detailed information is often inaccessible and the transformed rose plants have not yet been released to the market.

One example concerned the generation of white flowering rose plants by inhibiting the CHS reaction of a cultivar originally with deep red petals. Although this approach has already been successfully applied to a number of other floriculture crops, the recovered transgenic rose plants did not exhibit complete suppression of anthocyanin formation. Thus, pink instead of white petals were formed. This may be caused by the expression of several CHS genes in the receptor line.

Another example concerned the realization of an old dream of rose breeders and rose lovers, the creation of a truly blue flowering rose. Blue flower colouration mainly demands the occurrence of Dp-derivatives, which are not naturally formed in rose petals because of the lack of F3′5′H activity, the key enzyme for Dp synthesis. The respective gene is obviously not present in the gene pool of roses. Feeding petals of a light pink flowering cultivar with the Dp precursor dihydromyricetin (**Figure 1**) actually led to the formation of Dp-derivatives, proving that the internal enzyme set (DFR, ANS, FGT, etc.) of roses can convert the 3′,4′,5′-hydroxylated intermediates to the desired end-product. Thus, introduction of a DNA sequence encoding the F3′5′H enzyme in a suitable receptor line of roses should lead to hydroxylation of dihydrokaempferol to dihydromyricetin, which is converted to Dp-derivatives by the endogenous DFR, ANS and FGT enzymes of roses.

To date, F3′5′H genes are available from several plant species and the transformation of some rose cultivars has already been achieved. Nevertheless, using this approach for roses, there are at least two major obstacles: first, the competition of DFR, FLS, F3′H and the introduced F3′5′H for dihydrokaempferol as a common substrate and second, inappropriate conditions in vacuolar pH and/or copigmentation, impeding a truly blue colouration of the Dp-derivatives formed in petals of the transgenic plant. In a respective approach with *Dianthus*, the first problem could be circumvented by using a DFR and F3′H mutant as receptor line, which normally accumulates dihydrokaempferol. Introducing and expressing both the F3′5′H and the DFR reducing dihydromyricetin but not dihydrokaempferol from petunia resulted in the formation of exclusively Dp-derivatives. But violet flowers rather than truly blue flowers were formed, probably because of inappropriate vacuolar conditions. For roses, however, no DFR and F3′H mutant line has been identified so far. Therefore, the F3′5′H gene had to be introduced and expressed in cyanic rose cultivars. As expected, this approach led to the formation of Dp-derivatives, but anthocyanins based on Pg and/or Cy were still present in appreciable amounts, and, therefore, the colour of the petals was far from truly blue.

Prospects

For successful metabolic engineering of flavonoids in roses, more detailed information is necessary on the flavonoid genes and their expression and regulation in petals and leaves. This also requires further enzymological work, including studies on substrate specificity and the demonstration of further enzyme activities, in particular FLS, ANS and F3′H. This knowledge, together with feeding experiments and the application of specific enzyme inhibitors, will provide the basis for a more careful characterization of the target plants for metabolic engineering of the flavonoid pathway and may allow a prediction to be made on the outcome of a planned transformation experiment. Then the generation of truly blue flowering roses by genetic engineering might finally come true and other alterations of the flavonoid pathway of rose petals and leaves in a highly desired fashion will be possible.

See also: **Biotechnologies for Breeding**: Genetic Transformation. **Breeding**: Overview; Selection Strategies for Cut Roses. **Fragrance and Pigments**: Odoriferous Substances and Pigments; Anthocyanins and other Flavonoid Pigments. **Genetics**: Cloned Genes.

Further Reading

Bohm B (1998) *Introduction to Flavonoids. Chemistry and Biochemistry of Organic Natural Compounds*, vol. 2, Amsterdam: Harwood Academic Publishers.

Davies K (2000) Plant colour and fragrance. In: Verpoorten R and Alfermann AW (eds.) *Metabolic Engineering of Plant Secondary Metabolism*, pp. 127–163. Amsterdam: Kluwer Academic.

Dixon R and Steele C (1999) Flavonoids and isoflavonoids – a gold mine for metabolic engineering. *Trends in Plant Science* 4: 394–400.

Forkmann G (1993) Control of pigmentation in natural and transgenic plants. *Current Opinion in Biotechnology* 4: 159–165.

Forkmann G and Heller W (1999) Biosynthesis of flavonoids. In: Sankawa U (ed.) *Comprehensive Natural*

Products Chemistry, pp. 713–748. Amsterdam: Elsevier.

Forkmann G and Martens S (2001) Metabolic engineering and applications of flavonoids. *Current Opinion in Biotechnology* 12: 155–160.

Holten TA and Cornish EC (1995) Genetics and biochemistry of anthocyanin biosynthesis. *Plant Cell* 7: 1071–1083.

Mol J, Cornish EC, Mason J and Koes R (1999) Novel coloured flowers. *Current Opinion in Biotechnology* 10: 198–201.

Souq F, Coutos-Thevenot P, Yean H *et al.* (1996) Genetic transformation of roses, 2 examples: one on morphogenesis, the other on anthocyanin biosynthetic pathway. *Acta Horticulturae* 424: 381–388.

Functional Genomics

A Vainstein, D Zamir and D Weiss, The Hebrew University of Jerusalem, Rehovot, Israel

Flower fragrance is a composite character determined by a complex mixture of low-molecular-mass volatile molecules. Although all floral organs can emit fragrance compounds, petals are the main source of scent in most plants, including rose. For many years, research into flower fragrance focused on its chemical elucidation, and as a result, hundreds of compounds were identified (*see* **Fragrance and Pigments**: Fragrance Profiles of Wild and Cultivated Roses). These compounds are biosynthesized by a relatively small number of often overlapping metabolic pathways. In general, most plant volatiles are derived from three main classes of compounds – terpenoids, phenylpropanoids/benzenoids and fatty acid derivatives – which are often greatly modified (oxidized, esterified, methylated, etc.). Due to the invisibility of scent as a character, to the shortcomings of humans' sense of smell, and to the highly variable nature of the trait (in part because of strong environmental influences), no simple, efficient and reliable methods to screen for genetic variation have been developed. Moreover, to date, no convenient plant model systems that would enable biochemical and genetic studies of flower scent are available.

For centuries, the rose has been the most important crop in the floriculture industry. Its economic importance also lies in the use of rose petals as a source of natural fragrances and flavourings. Nevertheless, many modern cut-flower rose cultivars were selected for long vase-life, flower shape and colour, but not scent. As a result almost all cut-flower varieties lack a distinct scent, in contrast to some garden cultivars with an intense 'rose' scent. Studies of floral scent in numerous rose varieties have led to the identification of over 400 volatile compounds that can be classified into five major groups: (1) hydrocarbons (mostly sesquiterpenes such as β-caryophyllene), (2) alcohols (mostly monoterpenes such as geraniol, nerol and citronellol, or aromatics such as phenethyl alcohol), (3) esters (mostly acetates such as hexyl acetate, geranyl acetate and phenethyl acetate), (4) aromatic ethers (mostly 3,5-dimethoxytoluene [orcinol dimethyl ether], benzyl methyl ether, estragole and methyl eugenol), and (5) others (including aldehydes such as the monoterpene geranial, the aliphatic chain nonanal, and decanal; the rose oxides; and norisoprenes such as β-ionone). The floral scents of most rose cultivars are dominated by representatives of one or more of these five groups. However, note that the phenol ether orcinol dimethyl ether, for example, is emitted at some level from the flowers of most rose varieties.

The novel technologies of genomics allow quick access to plants which are poorly characterized genetically, enabling the choice of an existing model plant system based on the trait of interest (e.g. roses with copious scent emission), rather than being limited to established model systems. The National Center for Biotechnology Information (NCBI) stores 22 000 entries for Rosaceae nucleotides and 2000 protein records (*see* **Genetics**: Cloned Genes). Most of the DNA sequences were obtained for *Prunus*, from peach developing fruit mesocarp (14 000) and almond developing seed (3500). In the past year, there has been a surge in the number of rose petal sequences in the taxonomy browser of GenBank. *Rosa hybrida* cultivars are represented by 2931 expressed sequence tags (ESTs) and 31 protein sequences, while *Rosa chinensis* is represented by 1800 ESTs and five protein sequences. Nearly 20% of these rose sequences have no homologues in the database (**Figure 1**) and thus represent novel sequences that may encode rose-specific proteins. Indeed, more than a quarter of these novel genes have been identified in both *R. chinensis* and *R. hybrida*.

Almost all annotated *R. hybrida* ESTs come from two tetraploid varieties, 'Fragrant Cloud' and 'Golden Gate', with contrasting phenotypes. The large red flowers of the garden variety 'Fragrant Cloud' possess a strong scent, accumulate anthocyanins and have a short vase-life, whereas the small yellow flowers of the cut-flower variety 'Golden Gate' accumulate carotenoids, have a long vase-life, and lack a distinct odour. Detailed gas chromatography-mass spectrometry (GC-MS) analyses of 'Fragrant Cloud' flower

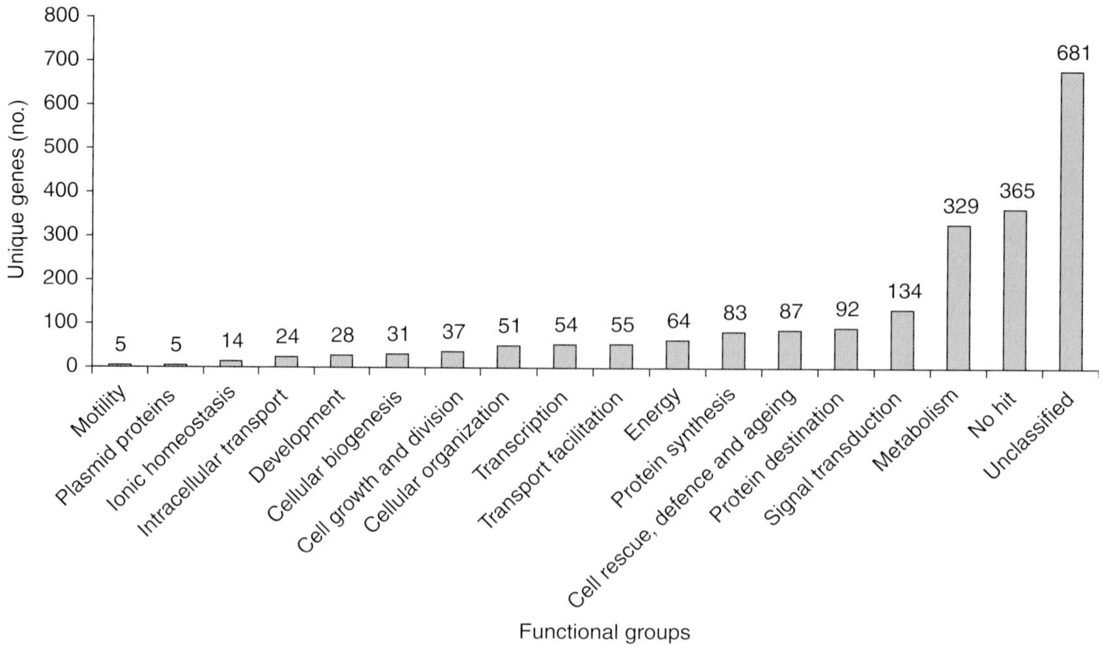

Figure 1 Functional classification of rose petal genes. Unique *R. hybrida* genes were classified into functional groups according to the Munich Information Center for Protein Sequences (MIPS).

Table 1 Functional characterization of *Rosa hybrida* genes involved in scent production

Clone (no.)	*Expressed sequence tags*		*Enzymatic activity*	*Substrate*	*Product*
	'Fragrant Cloud'	*'Golden Gate'*			
FC592	17	0	Sesquiterpene synthase	Farnesyl diphosphate	Germacrene D
FC1018	1	0	Alcohol acetyltransferase	Geraniol	Geranyl acetate
FC496	26	14	*O*-Methyltransferase	Orcinol	Orcinol methylether
FC895	12	20	*O*-Methyltransferase	Orcinol methylether	Orcinol dimethylether

headspace reveal emission of aromatic alcohols, monoterpenes, sesquiterpenes and various esters. Flowers of 'Golden Gate' emit, almost exclusively, methoxylated phenols (99% of total volatiles), which are hardly detected by the human nose.

To create the infrastructure to explore the relationships between DNA sequences and phenotypes in rose, expression profiling was performed using microarray analyses. Since most scent compounds are produced by mature petals but not by young green petals of 'Fragrant Cloud', and not by mature 'Golden Gate' petals, ESTs whose expression was upregulated during petal development in 'Fragrant Cloud' were overlapped with those that show higher expression in mature 'Fragrant Cloud' vs. mature 'Golden Gate' petals. By using a twofold change in gene expression as the threshold in the microarray analyses, 80 genes with higher expression in 'Fragrant Cloud' vs. 'Golden Gate' petals were identified. The analyses of gene expression during 'Fragrant Cloud' petal development allowed identification of ca. 100 genes that were upregulated in mature petals. Overlapping the resultant sequence information from these comparisons led to the identification of novel fragrance-related genes.

To prove the function of the candidate fragrance genes, lysates from *Escherichia coli* expressing these putative scent genes were employed. For example, to prove the function of a putative rose sesquiterpene synthase and to identify the type of sesquiterpene produced by this enzyme, the gene was cloned into an expression vector to obtain a catalytically active recombinant enzyme. *In vitro* assays using lysates from bacteria overproducing the gene product and the substrate farnesyl diphosphate resulted in the synthesis of a sesquiterpene, identified by GC-MS as germacrene D (**Table 1**). It is worth noting that the major sesquiterpene identified in the headspace of 'Fragrant Cloud' flowers was germacrene D, amounting to ca. 60% of the emitted sesquiterpenes. Similarly, functions of the putative alcohol acetyltransferase and of two O-methyltransferases (OMTs) were confirmed *in vitro*. The putative alcohol acetyltransferase was found to catalyse the transfer of the acyl group of acetyl-CoA to several alcoholic acceptor substrates, having the highest catalytic efficiency toward geraniol, and leading to the production of geranyl acetate. Both rose OMTs could methylate a variety of phenolic substrates using *S*-adenosylmethionine as the methyl donor. OMT1 exhibited the highest catalytic efficiency toward orcinol and produced orcinol methyl ether, and OMT2 used orcinol methyl ether most efficiently to produce orcinol dimethyl ether. It should be noted that geranyl acetate and orcinol methyl and dimethyl ethers are found in the headspace of rose flowers. To prove the function of rose scent genes in planta, transgenic petunia and tobacco plants expressing these rose genes were generated and the volatile composition of the transgenes was analysed by GC-MS. In this way, the production of germacrene D in transgenic plants expressing rose germacrene D synthase was obtained.

The elucidation of the molecular complexity of odour generation in general, and flower fragrance in particular, is still in its infancy. Using advanced metabolomics coupled with microarray techniques, it is possible, as shown here, to screen for genetic variation in fragrance production/emission in nonmodel plants such as rose and to identify relevant genes. Future application of these high-throughput analyses to an array of plants should make possible the discovery not only of genes coding for new fragrance biosynthetic enzymes, but also of those regulating these pathways.

See also: **Fragrance and Pigments**: Fragrance Profiles of Wild and Cultivated Roses. **Genetics**: Molecular Markers; Cloned Genes; Marker-Assisted Selection; Research Trends.

Further Reading

Croteau R and Karp F (1991) Origin of natural odorants. In: Muller P and Lamparsky D (eds.) *Perfume: Art, Science and Technology*, pp. 101–126. New York: Elsevier Applied Sciences.

Croteau R, Kutchan TM and Lewis NG (2000) Natural products (secondary metabolites). In: Buchanan B, Gruissem W and Jones R (eds.) *Biochemistry and Molecular Biology of Plants*, pp. 1250–1318. Rockville, MD: American Society of Plant Physiologists.

Dudareva N (2002) Molecular control of floral fragrance. In: Vainstein A (ed.) *Breeding for Ornamentals: Classical and Molecular Approaches*, pp. 295–309. Dordrecht, The Netherlands: Kluwer Academic Press.

Dudareva N and Pichersky E (2000) Biochemical and molecular genetic aspects of floral scents. *Plant Physiology* 122: 627–633.

Flament I, Debonneville C and Furrer A (1993) Volatile constituents of roses. In: Teranishi R, Buttery RG and Sugisawa H (eds.) *Bioactive Volatile Compounds from Plants*, pp. 269–281. Washington, DC: American Chemical Society.

GenBank (2003) Taxonomy browser. http://www.ncbi.nlm.nih.gov/Taxonomy/Browser/wwwtax.cgi.

National Center for Biotechnology Information (2003) DNA sequences of Rosaceae. http://www.genome.clemson.edu/projects/

Pichersky E and Gang DR (2000) Genetics and biochemistry of specialized metabolites in plants: an evolutionary perspective. *Trends in Plant Science* 5: 439–445.

Vainstein A (ed.) (2002) *Breeding for Ornamentals: Classical and Molecular Approaches*. Dordrecht, The Netherlands: Kluwer Academic.

Vainstein A, Lewinsohn E, Pichersky E and Weiss D (2001) Floral fragrance: new inroads into an old commodity. *Plant Physiology* 127: 1383–1389.

Weiss EA (1997) Rosaceae. In: Weiss EA (ed.) *Essential Oil Crops*, pp. 393–416. Wallingford, UK: CAB International.

Zuker A, Tzfira T and Vainstein A (1998) Cut-flower improvement using genetic engineering. *Biotechnology Advances* 16: 33–79.

Fungi *see* **Disease**: Overview; *Botrytis*; Blackspot; Downy Mildew; Powdery Mildew; Rust.

Fungicide *see* **Disease**: Bactericides and Fungicides.

G

GENETICS

Contents

Karyology

Y M Crane, Purdue University, West Lafayette, IN, USA
D H Byrne, Texas A&M University, College Station, TX, USA

Introduction

Karyotype analysis, the comparison of chromosome measurements, is a traditional cytogenetic step in the comparison of genomes among related species. Karyotypes can be described by the number of chromosomes, the mean and dispersion of chromosome length and arm ratio and the number and position of nucleolar organizer regions (NORs). Karyotypic changes are frequent in the evolution of higher plants and have frequently served as a basis of phylogenetic inference, since karyotypes can be compared without making interspecific hybrids. An alternative, more specific approach to comparing chromosome sets is meiotic analysis in hybrids (*see* **Genetics**: Meiosis). More recently, banding techniques and *in situ* DNA hybridization (ISH) have been used to compare chromosomes with increasing detail and specificity, but, with the exception of ribosomal probes, results with these methods in roses have not been published. With the rapid development of DNA length polymorphism markers and software for their phylogenetic analysis, karyotype analysis is no longer a major source of phylogenetic information, but it remains valuable for predicting the ability of related species to exchange genes in nature and in breeding programmes. The karyotype is less exposed to natural selection than are most morphological characters, which interact directly with the environment. Therefore, the karyotype is likely to be more stable than morphology in evolution at the sectional and generic level. In reality, the evolutionary rate of karyotypes varies greatly among plant families. In the Rosaceae, the subfamily Rosoideae is relatively stable, since the base chromosome number is 7 in most genera.

The following section discusses how karyotypes are observed and analysed. It is necessary to spread the chromosomes in a single layer on a microscope slide, take a picture of them and match them into groups on the basis of maximum similarity. As might be expected, improvements in microscopy and image analysis have made these tasks easier and more reliable since the first cytotaxonomists drew chromosomes with a camera lucida.

Methods of Preparing Rose Chromosomes

Plant chromosomes are best compared at mitotic metaphase, when the arms are contracted and the chromosomes can be more readily separated from one another. Squash methods have been popular since the 1930s, but enzymatic digestion-spreading methods have come to the fore during the past 10 years, because they can yield cleaner, better spread preparations suitable for ISH. The spreading force in the latter methods comes from the surface tension of evaporating alcohols and acetic acid, and the methods allow much of the cytoplasm to be washed off as the chromosomes adhere to the slide. Although the following protocol is for spreading chromosomes, digested shoot tips can also be squashed in traditional stains like acetocarmine or carbol fuchsin. As in many woody plants, it is difficult to obtain large numbers of usable metaphases in rose, even with pretreatment in antispindle agents. The cytoplasm of fixed shoot tips in rose is much less easily spread than that of root tips in most herbaceous species.

Collection and Pretreatment

Although most karyotypic studies have been performed on root tips, the most accessible, mitotically active material from established rose plants is the terminal 2–4 mm of a healthy, actively growing shoot, collected during the springtime burst of growth. Cell division is most frequent in 10–15 mm shoot tips collected in the morning. The shoot tips can be transported to the laboratory in ice water (0 °C); five to ten shoot tips fit easily in one 5-ml culture tube plunged in crushed ice. Shoot tips should be put into pretreatment solution within 60 min of collection. In the laboratory, carefully trim off the young outside leaves, and cut the shoot tip 2–4 mm below the apex. Immediately place this terminal 2–4 mm portion in the pretreatment solution. A 4-h incubation in the dark at 25 °C in a mixture of 8-hydroxyquinoline (0.002 mol l^{-1}) and 0.1% colchicine frequently gives well-spread chromosomes.

Fixation

Fixation stops the progress of cell division and preserves the contracted chromosomes for later examination. While 3 : 1 (v/v) absolute ethanol : acetic acid is a common, classical fixative, the cells are more easily spread after fixation in 2 : 1 (v/v) acetone : acetic acid +2% (w/v) polyvinylpyrrolidone (PVP) of molecular weight 40 000. Either fixative should be freshly prepared before use, and fixation should run about 24 h. Fixed shoot tips can be rinsed and stored in 70% ethanol at −20 °C, but ISH works better with recently fixed material.

Preparation for Spreading

This step transforms the shoot tips into protoplasts suitable for spreading by surface tension. After fixation or storage, soak the shoot tips in several changes of distilled water for 1 h to several hours to remove the fixative and soften the tissue. Optionally, the shoot tips can be further softened in 1 mol l^{-1} HCl for 20 min at 25 °C, followed by rinsing twice in distilled water. Soak the shoot tips 10 min in distilled water, then in 0.01 mol l^{-1} sodium citrate at pH 4.6 for 30 min.

Enzymatic Digestion

Incubate the shoot tips in 20–50 µl (twice to thrice their own volume) of cell-wall lytic enzymes in 0.5 ml microcentrifuge tubes. Optimum digestion occurs with 5.5% cellulase R10 (Yakult Honsha Pharmaceutical, Tokyo, Japan) and 1.25% pectolyase Y23 (Seishin, Tokyo, Japan) in 0.01 mol l^{-1} sodium citrate at pH 4.6 for 3 or 4 h at 37 °C. This duration is far longer than is needed in most other plant species, but shorter digestion in rose shoots results in worse spreading and greater retention of cytoplasm.

Slide Preparation

The slides themselves must be scrupulously clean, for example washed in ethanol acidified with a tenth volume of acetic acid, and dried with a clean cotton cloth. There is significant variation in cell retention among brands of slides; Gold Seal slides, which are cut from blocks of glass and then polished, work better than float glass slides.

At the end of incubation, carefully add water to the tubes, and remove the diluted enzyme solution with a fine-tipped pipette, made by flaming and pulling a glass Pasteur pipette. The shoot tips are very fragile at this stage, and care is needed not to dissociate them prematurely in the tube. Suck up a single shoot tip in a regular Pasteur pipette and release the tip in a drop of water on to the centre of the slide. Using folded absorbent paper or a fine-tipped Pasteur pipette, blot or suck water away from the tip without touching it. What comes next is largely an art that must be learned through practice. Place a drop of fresh 3:1

ethanol : acetic acid on the shoot tip, and quickly macerate it with fine-tipped forceps. Examine the slide under phase contrast at *c*. 160× until most of the cells have settled on to the glass. Remove the slide from the microscope and wash the macerated area with one to several drops of the 3 : 1 ethanol : acetic acid. Do this by releasing the fixative from a fine-tipped Pasteur pipette near the label end of the slide and allowing the fixative to flow over the cells. Wipe the edges of the slide with absorbent paper, and return the slide to the microscope to observe the flattening of the cells. The spreading is sensitive to relative humidity, and often it helps to breathe on the slide or humidify the air in the room. There is a stage in flattening where the cytoplasm appears to dissolve and disperse widely; as the cells approach this stage, remove the slide from the microscope and rapidly dry it in a stream of air. Otherwise all the metaphases will lose chromosomes. The slides can be stored at this point in a desiccator for days or weeks, to improve chromosome adhesion to the slide.

Staining

Carbol fuchsin is more effective than acetocarmine or Giemsa stain for rose chromosomes. Carbol fuchsin-stained chromosomes darken with time. Place a drop of the stain on the dried slide, apply a cover slip, squash between filter papers and seal the preparation with depilatory wax using a flat-bladed wood-burning tool (as can be obtained from a craft shop). The slides are now ready for examination (**Figure 1**).

Karyotype Analysis

Photograph several high-quality, intact metaphase cells per plant. If a digital camera is available, use the image directly; otherwise scan the print into a TIFF image. Bring the image into an editor (such as CorelDraw) that allows recording of the coordinates of telomeres and centromeres. When chromatids are separately visible, obtain coordinates from both of them. When the NOR has stretched, record coordinates for its boundaries, so that the intact chromosome length can be obtained by subtraction. If a chromosome arm has bent, record the coordinates of the bend in addition to the coordinates of each end of the arm. Calculate the length of each region between coordinates by use of the Pythagorean theorem:

$$\text{length} = \sqrt{\left((x_r - x_l)^2 + (y_r - y_l)^2\right)}$$

where (x_l, y_l) are the coordinates of the left end and (x_r, y_r) are the coordinates of the right end. Add together segments in bent or satellited arms to reconstruct the original arm length.

Accurately grouping the chromosomes into homologous groups is relatively difficult and imprecise in roses, where many of the chromosomes differ only slightly in length and arm ratio, because even minor measurement errors can obscure the matching. In a diploid, tetraploid or hexaploid, with respectively two, four or six copies of each chromosome per cell, one minimizes the following expression for chromosomal similarity to obtain the most likely correct pairing of homologous chromosomes:

$$F(i,j) = \text{sum over all chromosome pairs } i,j \text{ of } \left(w_1(le_i - le_j)^2 + w_2(lo_i - lo_j)^2 + w_2(sh_i - sh_j)^2 + w_3(n_i - n_j)^2\right)$$

where *le* is the total length of a chromosome, *lo* is the length of its long arm, *sh* is the length of its short arm, n is the coded presence ($n = 1$) or absence ($n = 0$) of an NOR, and w_1, w_2 and w_3 are arbitrary weighting factors by which the relative importance of lengths and NORs can be regulated in the analysis. For a triploid, the formula for $F(i, j, k)$ is more complicated, since the chromosomes must be matched into triplets, but the principle is the same: minimize the sum of squared differences in total length, arm lengths and the score for the presence of an NOR.

Because the total length and the two arms are represented in separate terms in calculating F, the analysis considers both total length and arm ratio whenever w_1 and w_2 exceed zero. In minimizing F, the number of chromosome pairing arrangements to be considered is $(2n)!/(n! \times 2^n)$, where ! is the factorial operator and n is half the chromosome number of the shoot tip. For a diploid rose with 14 chromosomes per cell, this number is 135 135, and a simple computer program can look at every possible matching of chromosomes. For a tetraploid rose with 28 chromosomes per cell, this number is 213 458 046 676 875, and it is infeasible to calculate a value for every single matching. Instead, a more elaborate program can obtain a reasonable – though not necessarily correct – matching by sorting the chromosomes by total length and then exchanging adjacent chromosomes in the pairing pattern until F is minimal.

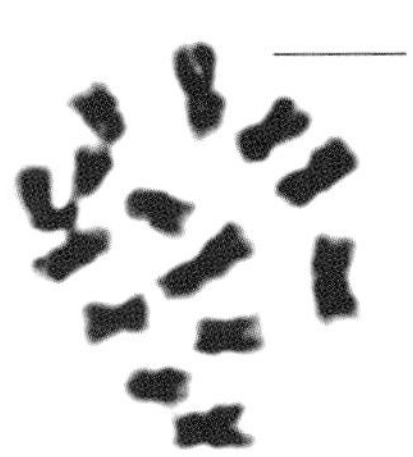

Figure 1 Mitotic chromosomes of *Rosa moschata* Miller, $2n = 14$. The bar indicates 2.5 μm.

Classical chromosome measurements are limited to arm lengths and the position of NORs. Various statistics and indices can be calculated from these data to express the similarity of the chromosomes to one another, but the most usual treatment is to classify the chromosomes according to arm ratio, the long arm length/short arm length for each individual chromosome. Metacentric chromosomes have arm ratio <1.7, and submetacentric chromosomes have arm ratio between 1.7 and 3.0. Acrocentric chromosomes have arm ratio >3.0, and telocentric chromosomes have only one arm, but these latter two types do not occur in normal rose karyotypes. The codes m and sm have been used to express the number of metacentric and submetacentric chromosomes in a basic set of chromosomes (subgenome). Thus 5 m + 2 sm designates a set of seven chromosomes, where five are metacentric and two are submetacentric, as defined above by the 1.7 threshold arm ratio.

The chromosome lengths and arm ratios are independent variables that can be used to infer relationships. Average linkage cluster analysis has been used to produce a phenogram (a 'tree' of similarities) on this basis for representatives from all the subgenera and sections of the genus *Rosa* (**Figure 2**).

Chromosome Numbers and Polyploidy in *Rosa*

Cytotaxonomists have repeatedly studied *Rosa* because of its horticultural value and the accessibility of its distribution in the north temperate zone. Since the first reports by Heslop-Harrison in 1921 and Täckholm in 1922, chromosome numbers have been reported for 147 of its 184 species, including representatives of all subgenera and sections. The largest remaining gaps in coverage are in Asiatic members of sections *Synstylae* (12 uncounted species) and *Pimpinellifoliae* (10 uncounted species, **Table 1**). Except for a few scattered, questionable reports of $2n = 16$, the reported counts are multiples of 7.

Polyploidy (the condition of having three or more chromosome sets per cell) is frequent in wild and cultivated roses. The great majority of modern rose cultivars are tetraploid, with 28 chromosomes, or triploid, with 21 chromosomes per cell. **Table 1** lists the range of known somatic chromosome numbers by subgenus and section. In section *Caninae*, there exists a peculiar genetic system (*see* **Genetics**: Inheritance in the Dogrose) that transmits one set of chromosomes through the pollen and the remaining sets through the ovule, so that pentaploids ($5n$) can reproduce stably by seed. This system prevents recombination of several sets of chromosomes, and results in the formation of relatively stable microspecies. For this reason, *Rosa canina* was once thought to be apomictic (asexually reproducing by seed), but it is not. Triploids in section *Chinenses* are associated with interspecific hybridity; most are cultivars arising from human effort or from spontaneous hybridization in gardens. Polyploids in sections *Carolinae* and *Cinnamomeae* are less well-documented, but most or all presumably belong to

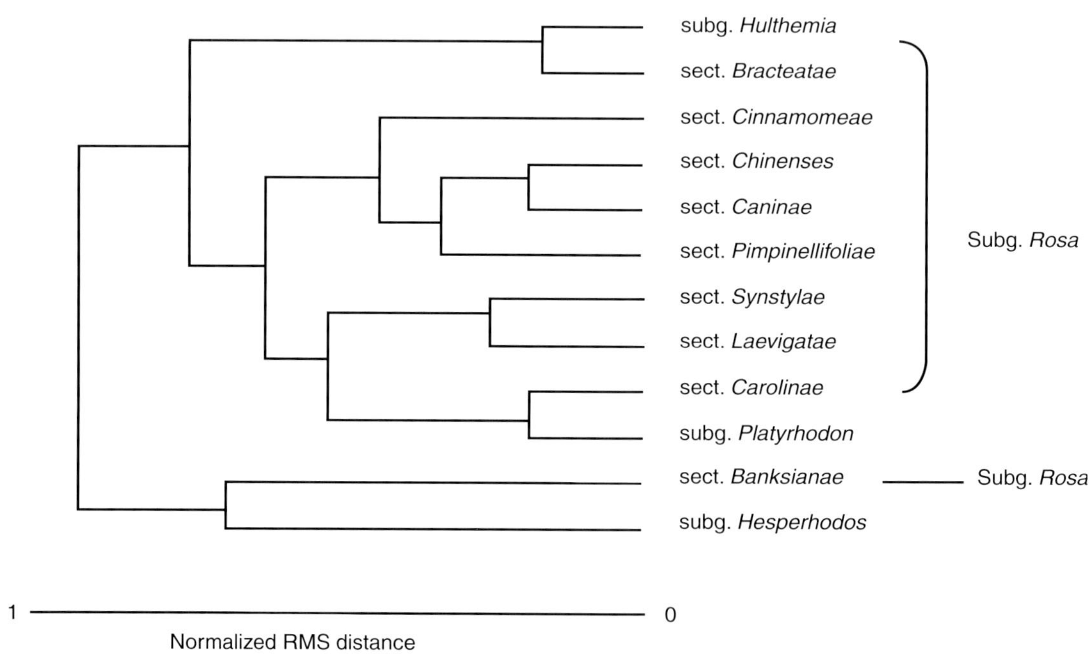

Figure 2 Phenogram depicting similarities among subgenera and sections of *Rosa* on the basis of karyotypes. RMS, root mean square. Adapted with permission from *Caryologia* 50 (3–4): 317–326 (1997).

ordinary polyploid complexes that have resulted from interspecific hybridization in nature.

The Rose Karyotype

Although chromosome counts are commonly reported, there have been far fewer measurements of rose chromosomes under standardized conditions. The base number is uniformly 7 in reliable reports. Rose chromosomes are generally small (1–3 μm long) and nearly metacentric, with arm ratio <1.7. Such chromosomes are classified as m in the usual karyotype formula; any chromosomes with arm ratio slightly over 1.7 are classified as sm. As shown in **Table 2**, most rose species have 6–7 m and 0–1 sm chromosomes per chromosome set. In a modern cultivar and five diploid species of subgenera *Rosa* and *Platyrhodon*, ISH with an 18S–26S ribosomal RNA probe has revealed only a single subterminal NOR per chromosome set (**Figure 3**). There is subtle variation in arm ratio, and chromosome size varies about twofold within the genus. Karyotypes are more uniform within sections than between sections, which is consistent with the extensive intrasectional hybrid fertility observed in cultivated roses. It also suggests that the currently recognized subgenera and sections are natural groupings based on phylogenetic relationships. **Table 2** presents quantitative information about a sample of karyotypes from six of the sections in subgenus *Rosa*, and three of the other subgenera.

Karyotypic similarity suggests a somewhat different pattern of relationships than is implied by the current sectional classification of *Rosa*. **Figure 2** shows this pattern as revealed by cluster analysis of arm lengths and the presence of an NOR. Most notably, the karyotype of *R. roxburghii* (subgenus *Platyrhodon*) is very similar to that of several species in subgenus. *Rosa*, particularly *R. carolina*. This implies that *Platyrhodon* should be merged into subgenus *Rosa*, in agreement with cross-compatibility between *R. roxburghii* and species in sections *Bracteatae* and *Cinnamomeae*. An isozyme study, comparing enzymes on the basis of their movement in an electrical field, also found similarity between these two sections and *R. roxburghii*.

On the other hand, the karyotype *of R. banksiae* has more variation in chromosome length than those of other species in subgenus *Rosa*, in accordance with isozyme evidence of its distinctiveness. On this basis it can be argued that the *Banksianae* merit subgeneric status.

The karyotype implies a relationship between the *Bracteatae* and the anomalous *R. persica*, the only rose to have simple leaves. Such a possible relationship needs to be investigated with other types of characters, since these species differ markedly in growth habit and morphology.

Modern cultivated roses trace their ancestry to approximately 10 wild species, mostly members of sections *Chinenses*, *Gallicanae* and *Synstylae*. Like diploids in these sections, the karyotypes of tetraploid modern roses range from 7 m to 5 m + 2 sm, times 4. ISH of tetraploid 'Angel Face' revealed a single NOR per chromosome set, with two of the NORs consistently larger than the other two. This is consistent with two NOR-bearing chromosomes from one section and the other two from another section, although NOR size remains to be systematically quantified in the different sections of *Rosa*.

Supernumerary chromosomes (B chromosomes) have been reported in *R. spinosissima* and *R.* 'Dearest' and 'Aunty Dora', although the great majority of rose species and cultivars lack B chromosomes. These Bs

Table 1 Distribution of chromosome counts among sections and subgenera in *Rosa*

Taxon (total species, counted)	*Percentage of section counted*	*2*n *chromosome number*
Subgen. *Hesperhodos* 2/2	100	14
Subgen. *Hulthemia* 1/1	100	14
Subgen. *Platyrhodon* 3/1	33.3	14
Subgen. *Rosa* 178/143	80.3	14, 21, 28, 35, 42, 56
Sect. *Banksianae* 4/3	75.0	14, 28
Sect. *Bracteatae* 1/1	100	14
Sect. *Caninae* 41/41	100	28, 35, 42
Sect. *Carolinae* 7/7	100	14, 28, 42
Sect. *Chinenses* 5/4	80	14, 21, 28
Sect. *Cinnamomeae* 53/45	84.9	14, 21, 28, 42, 56
Sect. *Gallicanae* 9/6	66.7	21, 28, 35, 42
Sect. *Laevigatae* 1/1	100	14
Sect. *Pimpinellifoliae* 21/10	47.6	14, 28
Sect. *Synstylae* 36/24	66.7	14, 21, 28, 35, 42

Table 2 Karyotype formulas of representative species roses

Subgenus	*Section*	*Example species*	*Karyotype formula*	*Mean arm ratio*
Hesperhodos	—	*R. minutifolia* Engelm.	6 m + 1 sm	1.60
Hulthemia	—	*R. persica* Michx. ex Juss.	5 m + 2 sm	1.42
Platyrhodon	—	*R. roxburghii* Tratt.	6 m + 1 sm	1.37
Rosa	*Carolinae*	*R. carolina* L.	6 m + 1 sm	1.35
—	*Chinenses*	*R. chinensis* Jacq.	6 m + 1 sm	1.37
—	*Cinnamomeae*	*R. rugosa* Thunb.	6 m + 1 sm	1.41
—	*Laevigatae*	*R. laevigata* Michx.	5 m + 2 sm	1.32
—	*Pimpinellifoliae*	*R. spinosissima* L.	7 m	1.36
	Synstylae	*R. wichuraiana* Crép.	7 m	1.33

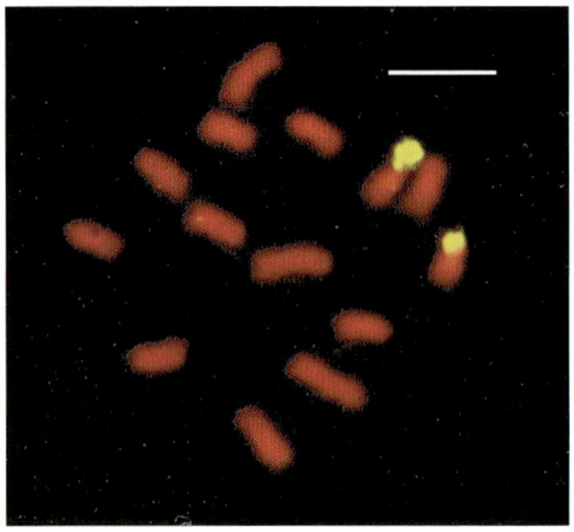

Figure 3 *In situ* DNA hybridization of *Rosa banksiae* R.Br. chromosomes. The two yellow spots are the nucleolar organizer regions, where the fluorescein isothiocyanate (FITC)-labelled probe has bound to the highly repeated genes that produce ribosomal RNA. Bar = 5 μm. Adapted with permission from *Caryologia* 50 (3–4): 320 (1997).

are about a quarter the size of the larger normal chromosomes, do not stain any differently from the normal (A) chromosomes, and are not preferentially transmitted to the progeny. In this regard, the Bs behave like centromeric fragments of normal chromosomes. However, they do not pair with any of the normal chromosomes at meiosis. The presence of two Bs reduces vigour, so it is questionable that these Bs persist for long periods of time in rose populations.

The Future of Karyotype Analysis in *Rosa*

It is unlikely that further karyotype analysis will change the major sectional alignments in *Rosa*, for two reasons: the karyotype is relatively conserved across the genus, and it is much easier to obtain and analyse DNA length polymorphisms in several different ways as biosystematic characters. Ironically, it may prove feasible to analyse karyotypes in the future without a microscope, at least from the standpoint of detecting structural rearrangements. The enabling technology is the DNA microarray, the experimental approach is deletion mapping and the objective is to follow structural rearrangements like inversions and translocations, whose breakpoints are unlikely to recur precisely during parallel evolution. Such an investigation requires a way of producing large numbers of nonchimeric plants after irradiation of an individual genotype, and production and fingerprint-screening of a library of a few thousand distinct single-copy DNA clones. The clones are spotted on the microarray and competitively hybridized with labelled probes from each mutant plant and its nonirradiated parent. Deletion is detected as a colour shift in the spot for that locus on the array, and a computer program can reconstruct the entire map from the frequencies with which loci are simultaneously deleted in pairs. Although this experiment is conceptually feasible today, progress in micropropagation and microarray preparation is needed to make it economically feasible.

See also: **Biotechnologies for Breeding**: Manipulation of Ploidy Level; Amphidiploidy; Genetic Transformation. **Genetics**: Meiosis; Nuclear DNA Amounts; DNA Fingerprinting; Cloned Genes.

Further Reading

Cave MS (ed.) (1956) *Index to Plant Chromosome Numbers (previous to 1956)*, vol. 1. Chapel Hill, NC: University of North Carolina Press.

Cave MS (ed.) (1964) *Index to Plant Chromosome Numbers (1956–1964)*, vol. 2. Chapel Hill, NC: University of North Carolina Press.

Darlington CD and Wylie AP (1956) *Chromosome Atlas of Flowering Plants*. New York: Macmillan.

Goldblatt P and Johnson DE (eds.) *Index to Plant Chromosome Numbers 1975–2001 Series*. St Louis, MO: Missouri Botanical Garden.

Jewell DC and Islam-Faridi MN (1994) A technique for somatic chromosome preparation and C-banding of maize. In: Freeling M and Walbot V (eds.) *The Maize Handbook*, pp. 484–493. New York: Springer-Verlag.

Krüssmann G (1981) *The Complete Book of Roses.* Portland, OR: Timber Press.
Moore RJ (ed.) *Index to Plant Chromosome Numbers 1967–1971.* Utrecht: Bohn, Scheltema & Holkema for the International Bureau for Plant Taxonomy and Nomenclature.
Moore RJ (ed.) *Index to Plant Chromosome Numbers 1972–1974.* Utrecht: Bohn, Scheltema & Holkema for the International Bureau for Plant Taxonomy and Nomenclature.
Stebbins GL (1971) *Chromosomal Evolution in Higher Plants.* London: Edward Arnold.

Meiosis

D H Byrne, Texas A&M University, College Station, TX, USA
Y M Crane, Purdue University, West Lafayette, IN, USA

Introduction

Plant and animal cell nuclei can divide in two ways, mitosis and meiosis, which are described at length in the Further Reading by Russell and Schulz-Schaeffer. At some point in the interphase before either type of division, the chromosomes replicate, i.e. an identical strand is copied from each chromosome. Thus each chromosome consists of two strands, termed chromatids, as it enters division. In mitosis, the two chromatids of each chromosome condense (shorten and thicken) during prophase. Then the chromosomes congregate in one plane (metaphase), before each chromosome's chromatids separate and migrate to opposite ends (poles) of the cell at anaphase. The chromosomes decondense, a nuclear membrane reforms and the nuclei enter interphase, having completed mitosis. Each daughter cell receives an identical copy of the genetic information in the dividing cell. Repeated mitotic divisions lead to the growth of an organism.

Meiosis consists of two divisions, meiosis I and meiosis II, separated by a short interphase during which there is no DNA replication. The divisions are basically similar to mitosis but differ in detail. In meiosis I, chromosomes pair, exchange segments of individual chromatids (cross over) and condense at prophase (**Figure 1**), align in one plane at metaphase (**Figure 2**) and separate as whole chromosomes rather than chromatids at anaphase. As meiosis I finishes, the paired chromosomes separate to opposite parts of the cell and decondense, and each of the two daughter nuclei has half the number of chromosomes that existed before meiosis I. At meiosis II, the recombined chromatids separate. Therefore, the two meiotic divisions transform the single premeiotic nucleus into four genetically distinct spore nuclei that have half the previous chromosome number (**Figures 4** and **5**). Meiosis is the reductional part of a sexual cycle; gametic fusion restores the somatic chromosome number later on. The meiotic reassortment of parental genes allows natural selection to operate on many more possible combinations of variant genes (genotypes) than would be possible without sex. This accelerates the rate of evolution and increases the chance of long-term survival of the lineage in a changing environment.

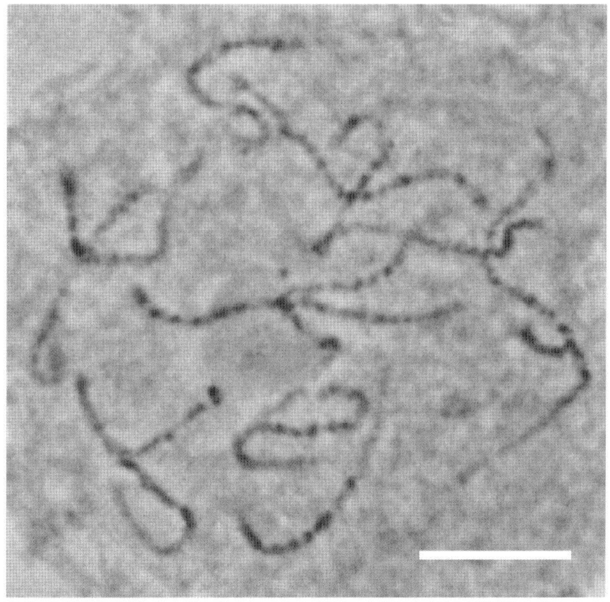

Figure 1 Pachytene in *Rosa spinosissima* (bar indicates 10 μm). This stage is mid-prophase and is significant because the chromosomes are now intimately associated, which allows the possibility of reciprocal exchanges between homologous chromosomes in an event known as crossing over. Courtesy of A. V. Roberts.

Meiosis is interesting for two practical reasons: hybrid fertility and the determination of species relationships. If chromosomes pair incompletely, they might not separate evenly at the end of meiosis I, and the spore nuclei arising from meiosis II will likely have different numbers of each parental type of chromosome (homologues). This genetic imbalance will prove fatal for many of the spores, resulting in partial or complete hybrid sterility. Similarly, if the homologous chromosomes of a hybrid differ in gene order (inversion heterozygosity) or gene content (translocation heterozygosity), many of the spore nuclei will have unbalanced genotypes, again resulting in partial to complete sterility.

Chromosome pairing is thought to reflect similarity in the DNA sequence along the chromosome, for at least a subset of the DNA. Absence of pairing therefore indicates dissimilarity and lesser relationship.

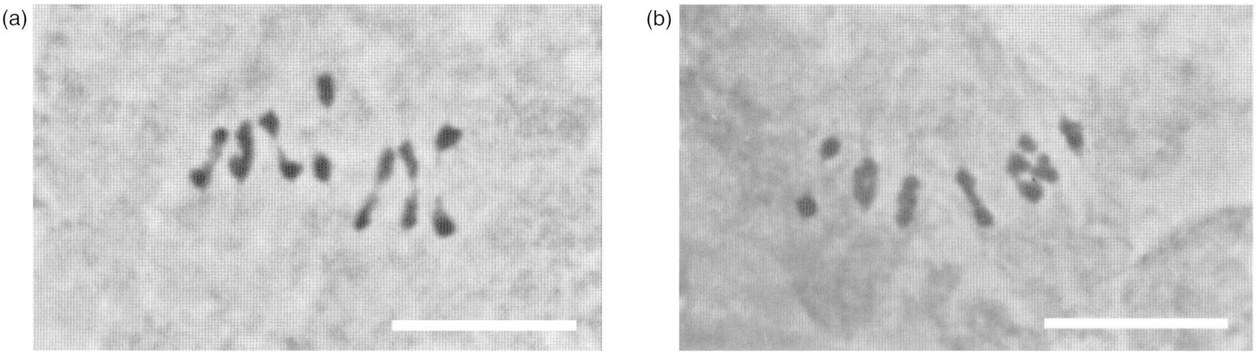

Figure 2 Metaphase of meiosis I: (a) *Rosa farreri* × *R. willmottiae*, showing seven bivalents with one chiasma each; (b) *R. forrestiana*, showing one quadrivalent with three chiasmata, two bivalents with two chiasmata, two bivalents with one chiasma, and two univalents recently separated from a bivalent association (bar indicates 10 μm). Courtesy of A. V. Roberts.

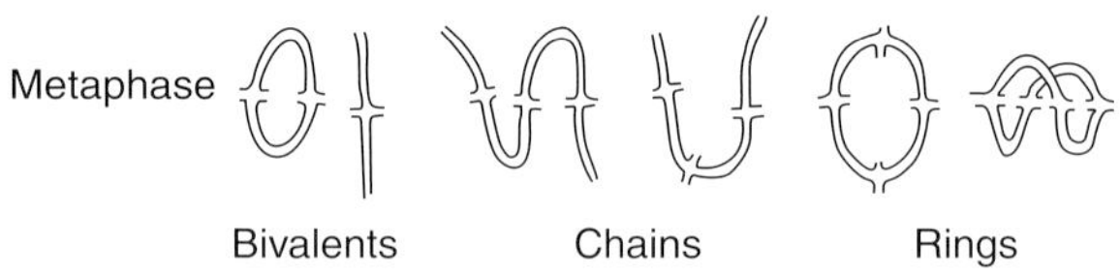

Figure 3 Diagrammatic representation of possible metaphase I configurations of a quadrivalent. Reproduced from Allard RW (1960) *Principles of Plant Breeding*. New York: John Wiley.

Different patterns of relationship produce different frequencies of chromosome associations in mid-meiosis I. For example, if all the chromosomes are highly dissimilar, none of them will pair, and all the chromosomes will be univalent (single) at mid-meiosis I. (Late prophase and metaphase of meiosis I are the best stages for seeing the chromosome associations clearly.) If the chromosomes are very similar and form pairs of homologues, as is the usual case in fertile, stable species, then there will be n bivalent associations (pairs) in meiosis I, where $2n$ is the somatic (body cell) chromosome number (**Figure 2a**). An intermediate degree of similarity produces intermediate frequencies of univalents (single chromosomes) and bivalents (chromosome pairs). More complex associations are possible when there are more than two sets of chromosomes per cell (polyploidy). If there are three similar sets of chromosomes per cell (i.e. the plant is triploid), the chromosomes can pair as trivalents (groups of three), because the two arms of each chromosome can pair more or less independently of each other. If there are three sets and two are similar to the exclusion of the third set, then only bivalents and univalents will be seen, at identical frequencies. If the plant has four sets of chromosomes, quadrivalents (associations of four chromosomes) are possible, in addition to trivalents, bivalents and univalents (**Figure 3**).

Each set of chromosomes is commonly termed a genome. Genomes are traditionally designated by single capital letters, like A, B or D. This tradition recognizes three similarity patterns (AAA, AAB, ABC) for triploids, and five patterns (AAAA, AAAB, AABB, AABC and ABCD) for tetraploids, with distinct expected association frequencies for each in its pure form. For example, ABC would form all univalents, AAB would form equal numbers of bivalents and univalents and AAA would form two trivalents for every bivalent. In the tetraploid, ABCD would form only univalents, AABC would form twice as many univalents as bivalents, AABB would form only bivalents, AAAB would form twice as many trivalents as bivalents and would have twice as many univalents as trivalents and AAAA would form equal numbers of quadrivalents and bivalents. The detailed mathematics of chromosome pairing depends on a number of assumptions, most prominently, independence of two chromosome arms and equivalence of all the homologous groups. It is well beyond the scope of this article and leads to the recognition of additional patterns of similarity that cannot be easily communicated as sets of capital letters.

Meiotic recombination is the main, most general source of exploitable variation in plant breeding. Mutation, induced polyploidy and transformation can also contribute in specific instances (*see* **Biotechnologies for Breeding:** Manipulation of Ploidy Level; Genetic Transformation. **Genetics:** Mutation).

Preparation of Meiotic Chromosomes

Squashes of developing anthers remain the primary method to prepare meiotic chromosome slides in flowering plants. Squash preparations of female meiosis require the dissection of the ovary and are, therefore, more difficult. Care must be taken to collect the correct bud size at the correct time of day. In many species, metaphase I is most frequent at early to mid-morning.

Flower buds generally have a thicker cuticle than root or shoot tips, because meiosis occurs only after the

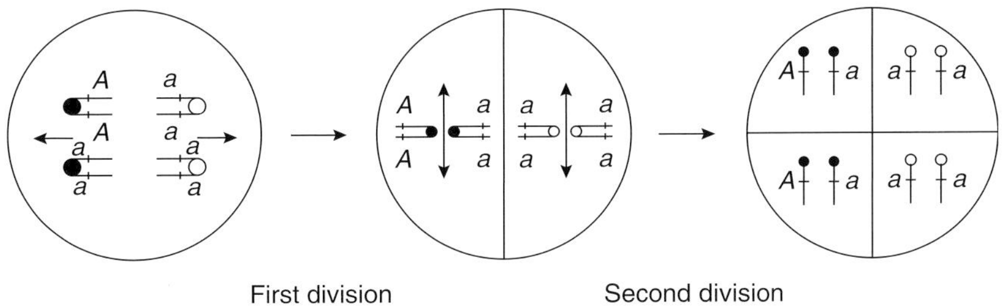

Figure 4 Gamete formation in a quadrivalent of a tetraploid in which the first division is reductional. Chromatids joined by the same centromere stay together in the first division since crossing over has not taken place between the centromere and the locus. This is an example of independent chromosome assortment. Reproduced from Allard R. W. (1960) *Principles of Plant Breeding*. New York: John Wiley.

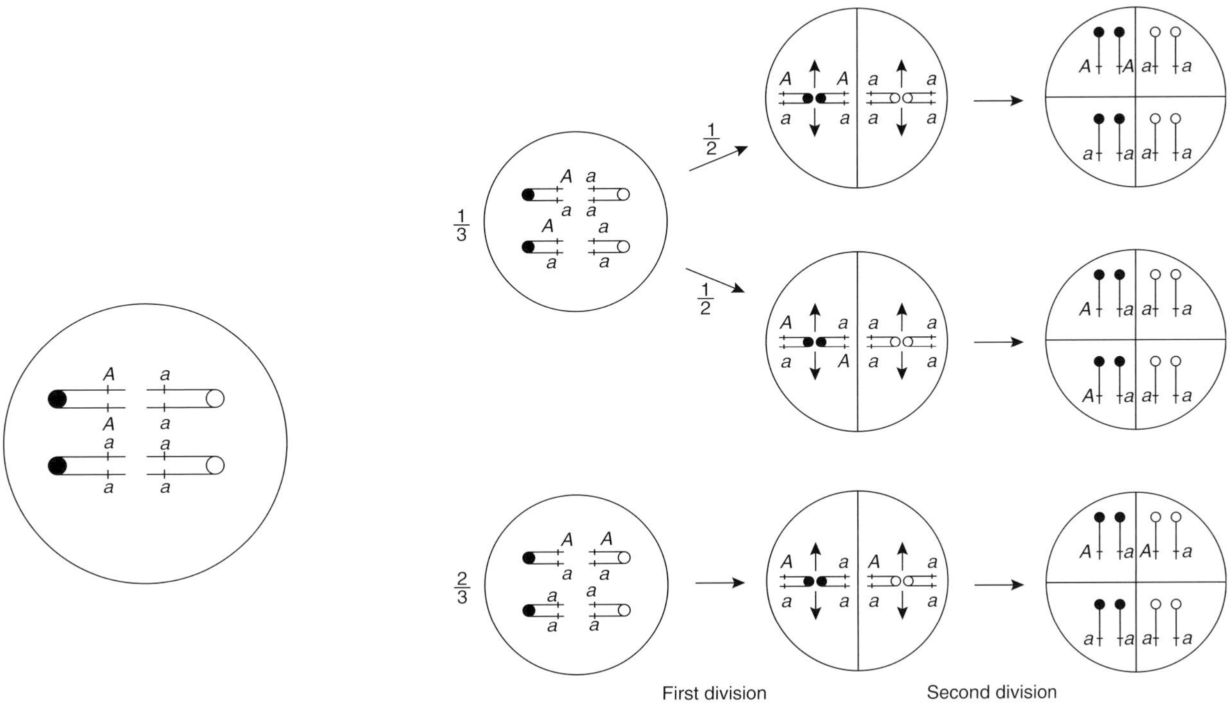

Figure 5 Gamete formation in a quadrivalent in which crossing over occurs between two chromosomes to give two *Aa* chromosomes from one *AA* and one *aa* chromosome. The *Aa* chromosomes go to the same pole (upper) one-third of the time and to opposite poles (lower) two-thirds of the time. When the *Aa* chromosomes pass to the same pole in the first division, it is possible for two *A* chromatids or two *a* chromatids to pass to the same pole in the second meiotic division, resulting in homozygous gametes. This is called double reduction and is possible with independent chromatid assortment. Reproduced from Allard R. W. (1960) *Principles of Plant Breeding*. New York: John Wiley.

enclosing structures (anthers, ovules) have differentiated significantly. Fixatives must penetrate the plant tissues rapidly enough to stop and preserve the chromosome associations as they are in life and disperse the cytoplasmic organelles and eliminate cytoplasmic staining. To ensure rapid penetration of the fixative, the bud is sliced longitudinally from the pedicel (stem) end before placing it in the fixative solution.

Fixatives have included freshly prepared 3 : 1 (v/v) ethanol : acetic acid, 6 : 3 : 1 (v/v/v) ethanol : chloroform : acetic acid, and 2 : 1 (v/v) acetone : acetic acid. Polyvinylpyrrolidone (2% w/v) is a useful additive to the first and third of these, to bind endogenous phenolic compounds that otherwise would harden the cytoplasm during fixation. Half-sliced buds are placed in the freshly prepared fixative at room temperature, and the fixative is replaced with more fixative after the first hour or so. Fixation can continue for 1–3 days at room temperature. Thereafter, the buds should be rinsed in 70% ethanol. The buds can be preserved in 70% ethanol at −20 °C until use. Even so, the buds will slowly deteriorate in storage, as indicated by loss of chromosome staining in the final preparations. Stored buds are best used within the first few months of storage.

Rose anthers are more difficult to squash than those of most plant species. The softening methods include soaking the fixed buds in water for several hours, and/or exposure to $0.1\,mol\,l^{-1}$ HCl for several min at room temperature. If soaking the buds in water, it is best to remove the fixative or ethanol, and use sterile distilled water. This avoids exposing the tissue to any nucleases. Nevertheless, the chromosomes lose DNA upon prolonged soaking. Equilibration with water also has the advantage of ensuring complete, immediate miscibility with the usual staining solutions, which tend to bead up and migrate away if the anthers are full of fixative.

The slides used for chromosome squashes must be free of oil, lint or glass fragments. The most frequent cause of poor squashing is material left on the slide, not insufficient pressure. Thus it is important to remove the anther walls and any other bud parts from the slide with fine-tipped forceps before squashing. The following protocol is but one of several workable ways to prepare meiotic squashes of rose for routine observation. It assumes the availability of a microscope for screening slides before squashing them.

1. Soak the fixed or stored buds in sterile distilled water for 10–15 min at room temperature. Sterile water is used to prevent exposure to DNAase that might be present in tap or bottled water.
2. Remove about one-quarter to one-third of the staminal mass from the bud, and place it in a drop of acetocarmine (1% carmine in 45% acetic acid) or carbol fuchsin on a scrupulously clean slide. Pop open the anthers with a half-spear. Quickly look at the slide under the 10× or 25× objective of a microscope to determine the range of stages present. If the first meiotic metaphase is present, use a needle or fine-tipped forceps to remove all debris from the slide.
3. Apply a scrupulously clean cover slip, lowering one edge first to avoid trapping bubbles. If the stain beads up outside the cover slip, use less next time.
4. Heat the slide. The classical method is to pass the slide over an alcohol lamp to the verge of boiling, without actually boiling the slide. A temperature-controlled slide-warming tray at 80 °C is more reliable. Place the slide cell-side up on the slide warmer for 30–60 s, then quickly put it on prefolded filter paper on the warming tray. Fold the filter paper over the slide and firmly squash down with the thumb once or twice. Excessive pressure or slippage of the cover slip will roll the cells up.
5. Seal the cover slip. The best material for this is depilatory wax, applied with a hot wood-burning blade that looks roughly like a paint scraper. Make sure to seal the edges of the cover slip completely, but do not leave large beads of the wax around the edge of the cover slip, as these can hit and damage the microscope objective during observation.
6. Label the slide and store it on glass rods in a humidified 15 cm Petri dish, which can be left refrigerated for several days if observation must be postponed.
7. Observe the slide. Viewing reddish stains like acetocarmine or carbol fuchsin with green-filtered light enhances contrast. Scan the slide in a raster pattern and record the association frequencies for each usable cell. Many roses exhibit meiotic abnormalities (stickiness, asynchronous contraction) that make accurate observation of association frequencies very difficult.

Analysis of Association Frequencies

The chromosome associations of interest are termed configurations. Unlike casual association, chromosomes in a configuration are bound together by chiasmata (sites of crossing-over), which withstand the tension of the metaphase spindle. The centromeres are pulled toward the poles of the spindle, and the similar orientation of centromeres among the associations usually remains evident after squashing the three-dimensional spindle into two dimensions on a slide. This co-orientation aids the interpretation of situations where associations touch or overlap one another.

Classical meiotic analysis usually stopped with the tabulation of numbers of configurations whereas modern models of meiotic chromosome association extend the analysis to various association subtypes on the basis of the number and relationship of the chromosome arms bound in them. Thus where the chromosome arm is the fundamental pairing entity, configurations can be classified into rods or chains and rings. Only one arm is chiasmatically bound in a rod bivalent, while both arms are bound in a ring bivalent. Thus ring bivalents are a stronger indicator of pairing preferentiality than are rod bivalents, given that the two arms of each chromosome are pairing independently. When odd numbers of chromosomes associate into multivalents and only one pairing per arm can occur, only chain configurations are possible. A configuration with an even number of chromosomes can be either a chain or a ring. A ring configuration occurs when every arm is bound. Ring multivalents are strong evidence of pairing randomness within groups of like chromosomes (**Figure 5**).

'Frying pan' configurations can occur when there are two or more cross-overs in the same chromosome arm, and each flanks a switch of pairing partners. Most numerical models of meiotic chromosome association do not take account of 'frying pans', treating them instead as rings if an even number of chromosomes is involved, and chains if the multivalent contains an odd number of chromosomes. The existence of 'frying pan' configurations indicates more independent chiasmata than either classical analysis or current numerical meiotic models are prepared to handle. Under these conditions, the current numerical models will overestimate pairing randomness.

Pairing Observations in Roses

The genus *Rosa* includes diploid, tetraploid, pentaploid, hexaploid and octoploid species. Generally, the number of species described decreases with increasing ploidy level. Cytological observations of chromosome pairing in rose have shown that diploid rose species typically exhibit complete chromosome pairing, with mostly ring bivalents, which have had a cross-over in each chromosome arm. Many tetraploid rose species also exhibit primarily bivalent pairing. Translocations have occurred frequently in the evolution of rose genomes, as evidenced by the frequent presence of one or two quadrivalent associations in diploid hybrids, and occasional translocation heterozygosity within some diploid species (**Figure 2b**).

One of the most unusual meiotic mechanisms in the plant kingdom occurs in the dog rose, *R. canina*. In this pentaploid species, there are seven bivalents and 21 univalents at metaphase I. The bivalents disjoin normally and all the univalents go to the same daughter cell. In the ovule, this daughter megaspore gives rise to the egg. In contrast, the surviving microspore cells have seven chromosomes. Therefore, the pollen transmits seven chromosomes, the egg transmits 28 chromosomes and the somatic number of 35 is restored in the zygote. The same mechanism operates in tetraploid and hexaploid species of the section Caninae. Because the univalents do not undergo crossing over, their gene content is passed intact to the next generation. Genes on the bivalent-pairing chromosomes recombine normally, but overall *R. canina* behaves as a collection of microspecies with predominantly maternal inheritance. The entire *R. canina* complex is probably of hybrid origin (*see* **Genetics:** Inheritance in the Dogrose).

Throughout the history of rose breeding, many triploids have been produced by crossing the basic tetraploid European-derived rose germplasm (including *R. damascena*) with diploid Asian species (*R. chinensis*, *R. multiflora* and *R. wichurana*) (*see* **History of Roses in Cultivation:** Modern (Post-1800)). These triploids typically have low fertility, as expected upon irregular chromosome segregation. However, some triploid genotypes occasionally produce viable haploid or diploid gametes, and a very few are consistently fertile, leading to fertile diploid and tetraploid plants in crosses with diploid and tetraploid plants, respectively.

Most modern commercial roses are tetraploids derived from intercrossing various tetraploid and diploid species. Consequently their meiotic behaviour can be quite complex. Translocation heterozygosity undoubtedly has contributed to the relatively high frequencies of trivalents, quadrivalents and occasionally higher associations seen in tetraploid cultivated roses, although most of these multivalent configurations reflect the overall high similarity of the parental genomes. Univalents are not common, and totally nonsynaptic hybrids have not been reported.

The existing numerical models can predict the meiotic configuration frequencies exactly in diploid rose species, but they do not give good fits of expected to observed frequencies in tetraploid rose hybrids, because the models predict too many chain quadrivalents and not enough ring quadrivalents. Nevertheless, one of the models predicts about 80% of the pairing to be preferential (i.e. genome A with genome A and genome B with genome B) in tetraploid hybrids involving *R. laevigata*. Meiotic studies of a synthetic tetraploid, its parental diploids and a series of hybrid progeny indicated that these roses were segmental allopolyploids, which represent a meiotic situation between an autopolyploid (AAAA or AAA) and an allopolyploid (AABB or AAB). Thus the analysis showed that anywhere from 4% to 74% of the chromosomes were involved in multivalent associations, which indicates that the genomes among the various rose sections are structurally similar, as is frequently seen among woody perennial plant species.

Genetic Consequences of Polyploidy

Polyploidy is a major evolutionary pathway in plants. Allopolyploidy (for example, AABB) adds new genetic diversity and gives the highest fertility amongst interspecific hybrids. In plant breeding, higher ploidy levels are sometimes induced to serve as a genetic bridge, create unique genetic combinations, increase hybrid fertility or to increase the size of vegetative and reproductive organs. In spite of these advantages, breeding with polyploids has certain genetic consequences related to the greater number of alleles that can interact for any given trait. In general, tetraploids, as compared to diploids, require more progeny and/or generations to fix a given trait.

The genetic outcome depends on the chromosome pairing behaviour of the polyploid. In tetraploid roses, the three extreme cases are as follows.

1. Disomic inheritance is expected in an allotetraploid (AABB) produced from two distantly related rose

species. This plant would exhibit completely preferential pairing. Genome A would pair only with genome A and genome B would pair only with genome B, so there is no genetic recombination between genomes. The genes segregate in a dihybrid pattern (**Figure 4**).
2. Tetrasomic inheritance with random chromosome assortment is expected from a diploidized autotetraploid in which all four chromosomes of each set are homologous but during meiosis only bivalent pairing occurs, perhaps because the chromosome arms do not pair independently. Since all the sets are structurally similar, the members within each set of homologous chromosomes pair at random (**Figure 4**).
3. An autotetraploid in which the chromosomes associate in groups of four (as quadrivalents) would exhibit tetrasomic inheritance but with assortment of the chromatids. In the extreme case, where quadrivalent formation is complete and there is a 50% cross-over rate, the chromatids are distributed into gametes completely at random. This situation is called random chromatid assortment (**Figure 5**).

This can be illustrated by segregation of the recessive gene for recurrent flowering. The phenotypic segregation ratios for the recurrent flowering gene upon self-pollination and selected cross-combinations for these three extreme tetraploid cases, as compared to a diploid, are shown in **Table 1**. The selfed progeny from a diploid nonrecurrent flowering heterozygote (*Rr*) would have 25% of the plants homozygous recessive and recurrent flowering, for a 1 : 3 recurrent-flowering : nonrecurrent-flowering phenotypic ratio.

In the allotetraploid situation, a nonrecurrent-flowering duplex heterozygote (*RRrr*) will produce all nonrecurrent-flowering offspring, if both *R* alleles reside on genome A and both *r* alleles reside on genome B. The ratio becomes 1 recurrent-flowering (*rrrr*) : 15 nonrecurrent (*R*– – –) if each genome of the allotetraploid carries one *R* and one *r* allele.

In the diploidized autotetraploid situation without crossing over, random chromosome assortment would yield the phenotypic ratio of 1 recurrent-flowering (*rrrr*) : 35 nonrecurrent-flowering (*R*– – –). Finally, if the autotetraploid has quadrivalent pairing and maximum crossing over, the segregation pattern would give 1 recurrent-flowering (*rrrr*) : 20.8 nonrecurrent-flowering. The most restrictive situation for recovering homozygous recessives in this example of the recurrent-flowering trait, is the allotetraploid, and the most permissive is the autotetraploid with random chromatid segregation. Even then, the recovery of recurrent-flowering recessive homozygotes is less than 20% of

Table 1 Phenotypic segregation ratios for various cross-combinations for diploid and tetraploid plants for a flowering habit in rose. Recurrent flowering (*r*) is recessive to nonrecurrent or seasonal flowering (*R*) habit

Diploid mating: genotype	*Phenotypic segregation*	*Tetraploid mating: genotype*[a]	*Phenotypic segregation*[b]		
			Disomic inheritance[c]	*Tetrasomic – random chromosome assortment*[d]	*Tetrasomic – random chromatid assortment*[e]
RR selfed	All NR	*RR-RR* selfed	All NR	All NR	All NR
		RR-Rr selfed	All NR	All NR	783NR : 1Rec
Rr selfed	3NR : 1Rec	*RR-rr* selfed	All NR	35NR : 1Rec	20.78NR : 1Rec
		Rr-Rr selfed	15NR : 1Rec	35NR : 1Rec	20.78NR : 1Rec
		Rr-rr selfed	3NR : 1Rec	3NR : 1Rec	2.48NR : 1Rec
rr selfed	All Rec	*rr-rr* selfed	All Rec	All Rec	All Rec
RR × *Rr*	All NR	*RR-RR* x *RR-rr*	All NR	All NR	All NR
		RR-Rr x *RR-rr*	All NR	All NR	129.67NR : 1Rec
		RR-Rr x *Rr-rr*	All NR	All NR	51.27NR : 1Rec
		RR-Rr x *rr-rr*	All NR	All NR	27NR : 1Rec
		RR-rr x *Rr-rr*	All NR	11NR : 1Rec	7.71NR : 1Rec
		Rr-Rr x *Rr-rr*	7NR : 1Rec	11NR : 1Rec	7.71NR : 1Rec
Rr × *rr*	1NR : 1Rec	*RR-rr* x *rr-rr*	All NR	5NR : 1Rec	3.67NR : 1Rec
		Rr-Rr x *rr-rr*	3NR : 1Rec	5NR : 1Rec	3.67NR : 1Rec
		Rr-rr x *rr-rr*	1NR : 1Rec	1NR : 1Rec	0.87NR : 1Rec

[a] Hyphen separates genotype into genomic composition for the allotetraploid (AABB) case. For example, *RR-rr* is $R^AR^Ar^Br^B$ and *Rr-Rr* is $R^Ar^AR^Br^B$.
[b] Phenotypic segregation: 'NR' or nonrecurrent phenotype = *RR* or *Rr*; *Rrrr*, *RRrr*, *RRRr*, or *RRRR*: 'Rec' or 'Recurrent' phenotype = *rr* or *rrrr*.
[c] Disomic inheritance: allotetraploid with preferential bivalent pairing, so traits are inherited in a diploid fashion without genetic recombination between the two genomes.
[d] Tetrasomic inheritance with random chromosome assortment: autotetraploid with random pairing of whole chromosomes within homologous sets and no crossing over between the locus and the centromere.
[e] Tetrasomic inheritance with random chromatid assortment: autotetraploid with quadrivalent formation and maximum crossing-over frequency.
Adapted from Allard R. W. (1960) *Principles of Plant Breeding*. New York: John Wiley.

that in a diploid. Whatever the pattern of genomic similarity, it will be more difficult to manipulate traits in the tetraploid than in the diploid. Cytological data for tetraploid roses indicate that tetrasomic inheritance (random chromosome or chromatid segregation) is possible, although confirmation with codominant molecular markers is required with a sufficiently large population to test statistical significance. On the basis of typical meiotic chromosome association frequencies, intermediates between allotetraploid and autotetraploid segregation are particularly likely.

See also: **Biotechnologies for Breeding**: Manipulation of Ploidy Level; Genetic Transformation. **Genetics:** Karyology; Inheritance of Characteristics; Inheritance in the Dogrose; Gene Mapping. **History of Roses in Cultivation:** Modern (Post-1800).

Further Reading

Allard RW (1960) *Principles of Plant Breeding*. New York: John Wiley.

Calloway DJ and Calloway MB (2000) *Breeding Ornamental Plants*. Portland, OR: Timber Press.

Erlanson WE (1933) Chromosome pairing, structural hybridity, and fragments in *Rosa*. *Botanical Gazette* 94: 551–565.

Grant V (1981) *Plant Speciation*, 2nd edn. New York: Columbia University Press.

Hurst HH (1929) Differential polyploidy in the genus *Rosa* L. *Zeitschrift für Induktiv Abstammungs- und Vererbungslehre* 2: 866–906.

Krüssmann G (1981) *The Complete Book of Roses*. Portland, OR: Timber Press.

Russell PJ (1992) *Genetics*, 3rd edn. New York: Harper Collins.

Schulz-Schaeffer J (1980) *Cytogenetics*. New York: Springer-Verlag.

Shepherd RE (1954) *History of the Rose*. New York: Macmillan.

Nuclear DNA Amounts

H Ben-Sade and A Samach, The Hebrew University of Jerusalem, Rehovot, Israel

Introduction

Crossing two different rose plants can often lead to surprising offspring. One phenomenon is an increase in size of all organs called 'gigantism'. What causes this and other breeding phenomena and how can we anticipate them? Here we will show that these phenomena are due to abnormal ploidy, a common situation in roses. Indeed, the surprising offspring could have been anticipated if the nuclear DNA content of the parents were available to the breeder.

This article will try to provide a background to understanding abnormal ploidy, the methods, difficulties and achievements reached in measuring DNA content of rose cells, and the uses this information provides.

Polyploidy

Almost every cell within an organism contains a nucleus and within that nucleus reside the DNA-containing chromosomes. Organisms that sexually reproduce normally have two sets of chromosomes, one from each parent, and are called diploids. The number of chromosomes in each set, and the size of each chromosome, is different for each species. Chromosomes contain genes and are made of double-stranded DNA. For example, humans, a species familiar to most of us, have around 3.5 billion basepairs of DNA distributed in 23 pairs of chromosomes. *Arabidopsis thaliana*, a plant whose DNA sequence has recently been deciphered, has 120 million basepairs of DNA distributed in five pairs of chromosomes.

A diploid plant of the genus *Rosa* contains seven pairs of chromosomes. Yet some plants in this genus contain more than two sets of chromosomes. This phenomenon, called 'polyploidy', is relatively common in all plant species. Plants in this genus can be diploids (14 chromosomes) up to octaploids (56 chromosomes). Teas, Chinas and Noisettes are mostly diploids, while Gallicas, Centifolias and Damasks are commonly tetraploids (with four sets of chromosomes). Crosses between the two groups usually give triploids that should be genetically equivalent. However, where the diploid types are the seed parents, the offspring tend to be larger and more vigorous. As mentioned above, a common result of an increase in chromosome number is the phenomena called 'gigantism', an increase in size of all organs. Indeed, a triploid hybrid from a diploid seed parent is more likely to exhibit gigantism than an autotriploid, such as the giant triploid form of *Rosa blanda*. The amount of nuclear DNA influences a plant's phenotype in two distinct ways, first by expression of its genetic content and second by the physical effects of its mass and volume. It is thought that the increase in nuclear material due to polyploidy disrupts the balance between nucleus and cytoplasm. For this, and many other reasons, it is very important for plant breeders to know the ploidy level of individuals crossed. Indeed, information about the ploidy levels of breeding stocks is widely used by rose breeders and there is interest in deducing this information from rapid estimates of DNA amount.

How do duplicated chromosome sets originate? The origin of the duplicated chromosome sets could be derived from the same species (autopolyploids) or from different species (allopolyploids). Natural causes of polyploidy are irregularities in cell divisions. Polyploidy could be caused in reproductive cells during meiosis (irregularities at meiosis may yield unreduced gametes) or in somatic cells during mitosis (spontaneous doubling of chromosomes due to replication without cell division). In the latter case, tetraploid shoots might give rise to diploid gametes in their flowers. Polyploidy can also be induced experimentally by various treatments (for example colchicin) which interfere with spindle formation during mitosis. After replication, rather than going to opposite poles the replicated chromosomes remain together in a single nucleus. As a result, the cell contains a double quantity of chromosomes.

DNA Amount and the C-Value

The DNA content is usually constant between cells in an individual, and relatively constant between individuals of the same species. The DNA amount in a cell will depend on both the state of ploidy and on the basic amount of DNA per one set of chromosomes. In 1950, Swift introduced the term 'C' (constant) as a value that states the DNA amount in the unreplicated haploid or gametic nucleus of an organism, irrespective of the ploidy level of the taxon. The amount of DNA per haploid cell is measured by weight in picograms. In general, 1 picogram (pg) of nuclear DNA corresponds to about 1 billion (10^9) basepairs (bp) of double-stranded DNA.

Nuclear DNA C-values differ by approx. 1000-fold among angiosperms, ranging from about 0.1 pg to about 125 pg, and tend to be characteristic for a taxon. For example *Arabidopsis thaliana*, a species with one of the smallest C-values known for an angiosperm, was estimated as 1 C = 0.2 pg or 164 Mbp. The 1 C value in the genus *Rosa* range from 0.6 pg to 3 pg (depending on the species/cultivars) somewhere between *Arabidopsis* and species with large C value like members of the Iridaceae family, for example *Iris zenaidae* with 1 C = 18.6 pg or 18 228 Mbp.

What information can we obtain from a plant C-value? The most obvious information is the plants level of ploidy and genome size. For example, polyploids with odd numbers of chromosome sets are completely sterile (or nearly so) because they produce gametes with incomplete sets of chromosomes. Allopolyploids are more fertile than autopolyploids, because only the homologous chromosomes of the same species pair with each other. DNA C-value reference lists provide data used for comparative studies of genome organization (intron size), taxonomy and systematics, genome evolution and phylogeny, and plant breeding. Since the employment of *in vitro* techniques in plant regeneration, the chance of obtaining polyploids has increased. Obtaining C-values enables detection of different ploidy levels in *in vitro* cultures. DNA amounts help detect aneuploid plants, in order to select individuals for breeding or to monitor the effect of a mutagenic treatment like radiation or chemicals.

Methods

The traditional method to determine ploidy level of plants is counting chromosomes during cell division, since they are enlarged and arranged at this stage. This method is slow and expensive where numerous plants are to be studied. To estimate ploidy levels in large numbers of samples it was necessary to develop high-throughput methods such as Feulgen microdensitometry and flow cytometry.

Feulgen Microdensitometry

The Feulgen reaction, proposed by Feulgen and Rossenbeck 75 years ago, is commonly regarded as one of the oldest and most specific reactions in histochemistry. Today, one of the most important applications of the Feulgen reaction is quantifying DNA amounts. Another area is the study of the structural organization of DNA *in situ*.

When DNA is exposed to a mild acidic environment purine bases are detached from the deoxyribose sugar exposing free aldehyde groups and leaving intact the DNA backbone. The DNA molecule becomes 'apurinic'. In this form, the free aldehyde groups can bind a dye such as the Schiff reagent acquiring a magenta color. Staining intensity (colour development) is proportional to the DNA concentration. DNA ploidy can be measured with the Feulgen reaction in nuclear or cellular suspensions spread onto a slide, in cell-cultured plates, in biological fluids, and in histological sections. Isolated intact single cells are by far the most commonly used targets to determine DNA ploidy with the Feulgen reaction.

Quantifying the stained Feulegen reaction was originally achieved using cytophotometry. Later it was found that DNA ploidy could also be measured by a microscope cytofluorometer. This method measures the fluorescence emitted from a nucleus stained with Schiff or Schiff-like reagents. Today, DNA amounts are usually measured using an electronic image cytometer.

For the Feulgen reaction to be specific, tissue fixation must be appropriate. Plant tannins can interfere with the Feulgen reaction to an extent, which makes it worthless as a quantitative method. Formaldehyde is the preferred fixative, because it binds to cellular tannins and prevents them from reacting with nuclear chromatin.

Flow Cytometry

In a flow cytometer a suspension of stained single cells passes singly through an apparatus which can measure the fluorescence of each cell individually. The most important feature of flow cytometric analysis is that large numbers, up to 100 000 particles or more, can be analysed one after the other in about 1 minute. The detection limit is as low as 100 fluorescent molecules per cell.

In order to use flow cytometry to measure DNA content, it is necessary to develop: (1) a method to separate cells or nuclei in suspension; and (2) a fluorochrome that stains nuclear DNA in a manner proportional to its content. Flow cytometry is now the method of choice for measuring DNA amounts. Statistical data like mean fluorescent intensity (and their shifts in time or their dependence on cell function) have proved to be very reliable. Since this method does not require dividing cells to estimate DNA amount, almost any plant material is suitable for DNA analysis. Indeed, nuclei from plant tissues usually emit two distinct peaks of fluorescence, one corresponding to nondividing cells (with a 2 C DNA amount) and the other to cells in mitosis (with a 4 C DNA amount).

The application of flow cytometry for calculating amounts of DNA was relatively delayed in plants compared to animal cells. Plant cells usually exist in complex tissues, and methods had to be developed for its preparation of such suspensions. Flow cytometric analysis of plant cells is also hindered by the presence of inclusion bodies such as starch grains and especially by the occasionally very high content of natural chromophores. Direct analysis of cell nuclei might overcome this problem. This would require osmotic lysis of protoplasts or a mechanical method.

Fluorochromes currently used for flow cytometric estimation of DNA content can be broadly classified into two groups:

1. Stains that intercalate with DNA with no base preference, such as ethidium bromide and propidium iodide.
2. Dyes and drugs that show a base preference and include Hoechst 33258 (H33258), 4′, 6-diamidino-2-phenylindole (DAPI), mithramycin, chromomycin A3 and olivomycin.

Flow cytometry measures DNA-related fluorescence and not DNA amount. Thus, we are provided with a relative value that needs to be compared to a reference standard with a known DNA content (**Figure 1**). Fluorochromes showing base preference might not be reliable if the ratio between overall molar AT and GC is different between the reference species and the measured species. It has been shown that use of base preference fluorochromes might lead to serious errors in the estimation of genome size in plants by flow cytometry. Even the use of a reference standard having the same AT/GC ratio as the sample cannot guarantee reliable results. For those reasons the Kew Angiosperm Genome Size Discussion Meeting chose propidium iodide as the recommended fluorochrome of choice. In addition, the same discussion group strongly recommended that material measured by flow cytometry should also be tested by traditional microscopic chromosome counts so that possible variations in C-value estimates could be directly attributed to ploidy level and chromosome aberrations.

Calibration Standards

The choice of the standard species used for calibration is of fundamental importance for accurate plant DNA C-value estimations. Indeed, many discrepancies in C-values reported for the same species were felt probably to reflect problems associated with the choice and/or use of calibration standards rather than genuine intraspecific variation. Flow cytometry was first developed using an animal standard; the most

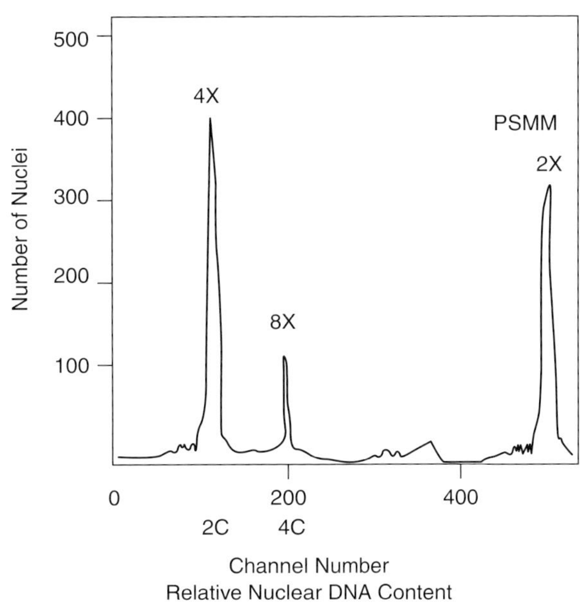

Figure 1 Illustration of a possible flow cytometer graph result showing the level of propidium iodide fluorescence staining in cell nuclei of a tetraploid rose line compared to an internal standard (*Pisum sativum* cv. Minerva Maple).

commonly used was chicken red blood cells (CRBC). Later, C-values for CRBC were reported to vary between sex and breeds. In view of this and some other problems the Kew Angiosperm Genome Size workshop recommended to stop using CRBC as the calibration standard, and pick a plant standard instead. The characteristics of an ideal plant calibration standard for both flow cytometry and Feulgen microdensitometry were that it should be a single cultivar from a diploid genetically stable species that is easily available from more than one source. Following discussions, three plant species conforming to these criteria were recommended as calibration standards: *Allium cepa* cv. Ailsa Craig, *Hordeum vulgare* cv. Sultan, *Pisum sativum* cv. Minerva Maple. In addition, the workshop recommended processing of the calibration standard tissue and the unknown tissue in the same tube since use of an external standard caused error. Therefore, DNA amount is calculated from the ratio of fluorescence intensities of standard to sample (**Figure 1**).

Calculating DNA Amounts in Roses

The diversity and the complexity of the genus *Rosa* are the consequence of the existence of numerous species (more than 120) and numerous interspecific hybrids. Modern roses are grouped into horticultural classes, the genomic origins of which have been partially obscured by intercrossing. Because of this complexity, it was necessary to estimate the ploidy levels on a large scale, in order to use plant material in breeding programmes or before undertaking any genetic improvement. Most data on DNA amounts in roses available today was obtained by flow cytometry.

High levels of tannins and other phenolics in roses have been shown to interfere with flourochrome binding to DNA. For example, parsley cells stained with DAPI are less fluorescent if chopped together with rose cells. It is assumed that phenolic compounds from the leaves of the rose enter the nuclei of both rose and parsley and interfere with the staining of DNA by DAPI. Polyvinylpyrrolidone (PVP) is known to bind to phenolics and was found to be useful in decreasing the influence of phenolics on fluorescence intensity of rose extracts. Yokoya and colleagues showed that the inclusion of PVP-40 in the nuclei isolation buffer did not affect the fluorescence intensity of DAPI-stained nuclei of parsley. However, when leaves of rose and parsley were chopped together, the fluorescence intensity of the nuclei of both species increased with PVP-40 concentration. There are results suggesting the PVP-40 is also useful in maintaining the integrity of the nuclear envelope. Its inclusion reduced nuclear debris peaks in flow cytometry histograms of propidium iodide-stained nuclei.

DNA Amounts and Ploidy Level in Roses

In general, we expect the C-value to increase with the ploidy level. All else being equal, the DNA amounts for polyploids are expected to increase in direct proportion to ploidy level. Tetraploids and hexaploids are expected to show double and treble the mean C-value for diploids. On the basis of the small sample of cultivars listed in **Table 1** it seems likely that diploid, triploid and tetraploid cultivars obey this rule. The C-value increases as the ploidy level (the number of chromosomes) increases. However, at higher ploidy levels, the C-value is smaller then expected. For example, the tetraploid *R. spinosissima* L. (section *Pimpinellifoliae*) has a similar amount of DNA to the triploid 'Frensham'. The pentaploid *R. canina* (section *Caninae*) has a higher DNA content than the hexaploid *R. moyesii* Hemsl. & Wils. (section *Cinnamomeae*). One reason for getting abnormal results might be due to predicting ploidy based on comparisons of roses from dissimilar taxonomic origin.

The literature abounds with examples where genome size in polyploids is smaller than expected. While some of these are probably technical artifacts, others seem real. Moreover, analysis of 2452 angiosperm taxa of known ploidy showed that mean C-values for diploids and polyploids were more similar than expected. DNA amount did not increase in direct proportion with ploidy level, and mean DNA amount per basic genome actually decreased with increasing ploidy in many cases. Polyploidy is often associated with selection and adaptation for rapid cell development, which in turn is correlated with small C-values and genome size, as in ephemeral weeds such as *Arabidopsis thaliana*. If so, DNA loss may commonly occur from constituent genomes in many polyploids after their formation. The alternative explanation, that C-value and genome size increase in diploids but not in their derived polyploids, seems unlikely. While the case for reduced genome size in many polyploids seems strong, more work is needed to confirm this at the molecular level and to describe the sequences involved.

Comparisons of Data on Rose DNA Amounts Obtained from Different Sources

As we have seen, the DNA amount data obtained can depend on the method, fluorochrome, internal reference and the contents of the nuclei isolation buffer. There are differences in data obtained by two groups, Bennett and Smith (1991) compared to

Table 1 Chromosome number (2*n*) and 2C DNA amount of rose species and cultivars

Group	*Section*	*Species/cultivar*	*2n=*	*2 C DNA amount (pg ± s.d.)*	*Method*	*Reference*
Subgenus *Eurosa*	*Banksianae*	*R. banksiae* Ait. var. *lutea* Lindl.	2*x*=14	1.04±0.08	Flow cytometry	Yokoya *et al.* (2000)
	Bracteatae	*R. bracteata* Wendl.	2*x*=14	1.15±0.08	Flow cytometry	Yokoya *et al.* (2000)
	Caninae	*R. canina* L.	5*x*=35	2.91±0.08	Flow cytometry	Yokoya *et al.* (2000)
		R. canina L.	5*x*=35	2.86±0.34	Feulgen microdensitometry	Greilhuber (1988)
	Carolinae	*R. foliolosa* Nutt.	2*x*=14	0.94±0.08	Flow cytometry	Yokoya *et al.* (2000)
		R. nitida Willd.	2*x*=14	0.95±0.08	Flow cytometry	Yokoya *et al.* (2000)
		R. virginiana Mill.	4*x*=28	1.95±0.08	Flow cytometry	Yokoya *et al.* (2000)
	Cinnamomeae	*R. iliensis* Chrshan.	2*x*=14	0.95±0.08	Flow cytometry	Yokoya *et al.* (2000)
		R. rugosa Thunb. var. *alba* W. Robins	2*x*=14	0.98±0.08	Flow cytometry	Yokoya *et al.* (2000)
		R. rugosa	2*x*=14	1.33±0.22	Flow cytometry	Moyne *et al.* (1993)
		R. willmottiae Hemsl.	2*x*=14	0.92±0.08	Flow cytometry	Yokoya *et al.* (2000)
		R. bella Rehd. & Wils.	4*x*=28	1.94±0.08	Flow cytometry	Yokoya *et al.* (2000)
		R. fedtschenkoana Reg.	4*x*=28	1.94±0.08	Flow cytometry	Yokoya *et al.* (2000)
		R. latibracteata Boul.	4*x*=28	1.91±0.08	Flow cytometry	Yokoya *et al.* (2000)
		R. moyesii Hemsl. & Wils.	6*x*=42	2.88±0.08	Flow cytometry	Yokoya *et al.* (2000)
		R. blanda Aiton	2*x*=14	0.6	Flow cytometry	Dickson *et al.* (1992)
		R. acicularis Lindl. var. *engelmannii* S. Wats.	6*x*=42	3.04±0.10	Flow cytometry	Yokoya *et al.* (2000)
		R. acicularis Lindl.	2*x*=14	1.3	Feulgen microdensitometry	Bennett and Smith (1991)
	Gallicanae	*R. gallica* L. var *officinalis* Thory	4*x*=28	2.20±0.08	Flow cytometry	Yokoya *et al.* (2000)
		R. damascena Mill. var. *versicolor* West.	4*x*=28	2.16±0.08	Flow cytometry	Yokoya *et al.* (2000)
		R. centifolia L.	4*x*=28	2.23±0.08	Flow cytometry	Yokoya *et al.* (2000)
	Indicae	*R. chinensis* Jacq.	2*x*=14	1.16±0.08	Flow cytometry	Yokoya *et al.* (2000)
	Laevigatae	*R. laevigata* Michx.	2*x*=14	1.14±0.08	Flow cytometry	Yokoya *et al.* (2000)
	Pimpinellifoliae	*R. xanthina* Lindl. 'Canary Bird'	2*x*=14	0.78±0.08	Flow cytometry	Yokoya *et al.* (2000)
		R. sericea Lindl. f. *pteracantha* (Franch.) Rehd. & Wils.	2*x*=14	0.78±0.08	Flow cytometry	Yokoya *et al.* (2000)
		R. foetida Herrm.	4*x*=28	1.95±0.08	Flow cytometry	Yokoya *et al.* (2000)
		R. spinosissima L. var. *hispida* (Sims) Koene	4*x*=28	1.85±0.08	Flow cytometry	Yokoya *et al.* (2000)
	Synstylae	*R. arvensis* Huds.	2*x*=14	1.12±0.08	Flow cytometry	Yokoya *et al.* (2000)
		R. multiflora Thunb. ex. J. Murr.	2*x*=14	1.65±0.11	Flow cytometry	Dickson *et al.* (1992)
		R. sempervirens L.	2*x*=14	1.14±0.08	Flow cytometry	Yokoya *et al.* (2000)
		R. wichurana Crép.	2*x*=14	1.13±0.08	Flow cytometry	Yokoya *et al.* (2000)
		R. wichurana Crép.	2*x*=14	0.2	Feulgen microdensitometry	Bennett and Smith (1991)
Subgenus *Hesperhodos*		*R. stellata* Woot. var. *mirifica* (Greene) Cockerell	2*x*=14	0.86±0.08	Flow cytometry	Yokoya *et al.* (2000)
Subgenus *Hulthemia*		*R. persica* Michx.	2*x*=14	0.84±0.08	Flow cytometry	Yokoya *et al.* (2000)
Subgenus *Platyrhodon*		*R. roxburghii* Tratt. var. *hirtula* (Reg.) Rehd. & Wils.	2*x*=14	0.95±0.08	Flow cytometry	Yokoya *et al.* (2000)
Polyantha		*R.* 'Ballerina'	2*x*=14	1.18±0.08	Flow cytometry	Yokoya *et al.* (2000)
Hybrid Sempervirens		*R.* 'Félicité et Perpétue'	2*x*=14	1.29±0.08	Flow cytometry	Yokoya *et al.* (2000)
Large-Flowered Climber		*R.* 'New Dawn'	3*x*=21	1.76±0.08	Flow cytometry	Yokoya *et al.* (2000)

Table 1 Continued

Group	*Section*	*Species/cultivar*	*2n=*	*2 C DNA amount (pg ± s.d.)*	*Method*	*Reference*
Floribunda		*R.* 'Frensham'	3*x*=21	1.79±0.08	Flow cytometry	Yokoya *et al.* (2000)
		R. 'Mountbatten'	4*x*=28	2.3±0.08	Flow cytometry	Yokoya *et al.* (2000)

Data from: Dickson EE, Arumuganathan K, Kresovich S and Doyle JJ (1992) Nuclear DNA content variation within the Rosaceae. *American Journal of Botany* 79: 1081–1086; Bennett MD and Leitch IJ (1998) Nuclear DNA amounts and genome size in angiosperm. *Annals of Botany* 82 (Supplement) 1; Greilhuber J (1988) 'Self-tanning': a new and important source of stoichiometric error in cytophotometric determination of nuclear DNA content in plants. *Plant Systematics and Evolution* 158: 87–96; Moyne AL, Souq F, Yean LH *et al.* (1993) Relationship between cell ploidy and regeneration capacity of long-term *Rosa* hybrida cultures. *Plant Science* 93: 159–168; Yokoya K, Roberts AV, Mottley J, Lewis R and Brandham PE (2000). Nuclear DNA amount in roses. *Annals of Botany* 85: 557–561.

Yokaya *et al.* (2000), in the same two rose sections (*R. acicularis* Lindl. and *R. wichuraiana* Crép.). The 2 C DNA amount measured by the latter group is larger. Bennet and Smith used Feulgen microdensitometry and did not account for self-tanning, while Yokaya and colleagues used flow cytometry and included PVP-40 in the nuclei isolation buffer. Also, in the case of *R. acicularis* the plants were not of the same clone. *R. rugosa* was estimated by both Moyne and colleagues (1993) and Yokoya and colleagues (2000). In this case the latter estimate was lower. Here the differences can be explained by the use of different fluorochromes and calibration standards. Moyne and colleagues used Hoechst 33258, a fluorochrome that has a preference for AT-rich regions and may have been influenced by the AT : GC ratio of the rose relative to their calibration standard (which was CRBC). In other more positive examples, different methods employed by different groups on the same line gave very similar results, for example, the DNA amount of *R. canina* estimated by both Greilhuber (1988) and Yokoya *et al.* (2000).

Jacob *et al.* (1996) used flow cytometry to assay the ploidy level of more than 90 rose genotypes (**Table 2**). They calculated the ploidy levels by comparing DNA amounts with the cultivar 'Sonia', whose ploidy level is known. From 90 genotypes analysed by Jacob *et al.*, 80% were in agreement with previous information. When we compare the data obtained in two relatively recent papers (Jacob *et al.* (1996) and Yokoya *et al.* (2000)), we find that of the 10 species/cultivars that were assayed by both, 80% were in agreement.

Hopefully disagreements in data will be reduced in the future by using a standard protocol, making no mistakes in clone identification, and avoiding the use of low-quality plant material.

Table 2 Ploidy level of some *Rosa* accessions, assessed by flow cytometry

Section	*Genotype*	*Ploidy level*
Cinnamomeae	*R. alpina stenodonta*	4
	R. cinnamomea	4
	R. oxyodon	4
	R. rugosa complicata	4
	R. prattii	4[a]
	'Amadis'	2
	R. beggeriana	2
	R. caudata	2
	R. davurica	2
	R. rugosa rubra	2
	R. woodsii	2
Gallicanae	*R. damascena*	4
	R. gallica	4
	'Rose à Parfum de l'Hay'	3
	R. centifolia major	3
	R. sylvatica	3[a]
	R. semperflorens	3
	KASANLIK	2[a]
Laevigatae	*Anemone Rose*	2
	R. laevigata	2
Synstylae	*R. brunonii*	5/6[a]
	R. setigera	4/5[a]
	R. triphylla	4/5[a]
	R. multiflora cathayensis	3[a]
	R. arvensis	2
	R. helenae	2
	R. longicuspis	2
	R. moschata	2
	R. moschata floribunda	2
	R. multiflora	2
	R. platyphylla	2
	R. polyantha grandiflora	2
	'Princess Marie'	2
	'ROSE MARIE VIAUD'	2
	R. sempervirens	2
	R. soulieana	2
	R. watsoniana	2
	R. wichurana	2
Banksianae	*R. banksiae*	2
	R. cymosa	2
	R. microcarpa	2
Caninae	*R. glutinosa*	6
	R. inodora	5
	R. canina	5

Present and Future Applications of DNA Estimations in Roses

The main use of measurements of DNA amounts in rose is in the determination of the ploidy of parents

Table 2 Continued

Section	*Genotype*	*Ploidy level*
	R. caryophyllacea	5
	R. coriifolia	5
	R. dumetorum	5
	R. stylosa	4/5
	R. ferruginea	4
	R. pomifera	4
	R. gypsicola	2[a]
	R. heterophylla	2[a]
	R. micrantha	2[a]
	R. serafinii	2[a]
	R. sherardii	2[a]
	R. stylosa	2[a]
Chinensis	'Soleil d' Or'	4
	R. × *noisettiana manettii*	4
	'La France'	3
	'Duchesse d'Auerstädt'	3
	'Mme Alfred Carrière'	3
	'China Rose'	3
	R. chinesis mutabilis	2
	R. gigantea	2
	R. indica	2
	'Old Blush China'	2
	'Pompon de Paris' climbing	2
	'Souvenir de Pierre Notting'	2
	R. viridiflora	2
	R. × *noisettiana*	2
	R. × *odorata*	2
Pimpinellifoliae	*R. lutea punicea*	4
	'Persian Yellow'	4
	R. spinosissima[a]	4
	R. foetida persiana	3[a]
	R. omeiensis	2
	R. swenginzowii	2
	R. sulphurea	2[a]
Carolinae	*R. nitida*[b]	6[a]
	R. nitida[b]	2
	R. lucida	2
	R. palustris nuttalliana	2[a]
Subgenus *Hulthemia*	*Hulthemosa hardii*	2
Subgenus *Hesperhodos*	*R. minutifolia*	2
Subgenus *Platyrhodon*	*R. roxburghii*	2
	'Conrad Ferdinand Meyer'	4
	R. oleifolia	2
	'Shafter Dr Huey'	3
	R. × *spaethiana*	2[a]

[a] Disagreement with other published data or supposed botanical section of classification.
[b] Two specimens present in the collection.
Data from Jacob Y, Teyssier C, Brown SC and Reynders-Aloisi S (1996) Use of flow cytometry for the rapid determination of ploidy level in the genus *Rosa*. *Acta Horticulturae* 424: 273–278.

and hybrids in breeding programmes and the investigation of ploidy in wild populations. In addition, measuring DNA amounts can help in the selection of haploids in anther cell cultures, and polyploids after treatment with spindle inhibitors. Cell cultures are a breeding ground for cytological abnormalities. Such abnormalities can lead to lack of regeneration.

The diversity and the complexity of the genus *Rosa* are the consequence of the existence of numerous species (more than 120) and numerous interspecific hybrids. Modern roses are grouped into horticultural classes, the genomic origins of which have been partially obscured by intercrossing. More than 18 000 registered cultivars are now listed. Those and others are preserved in botanical conservations, rose gardens and public or private collections. Mistakes caused by mislabelling can easily be propagated with time. Determining DNA amounts can serve as a rough 'fingerprint' of cultivars, since abnormal results would suggest mistaken identities.

Acknowledgments

Work in the Institute of Plant Sciences and Genetics in Agriculture is supported by the Israel Science Foundation–The Charles H. Revson Foundation (grant no. 436/00-1) and by the Chief Scientist of the Ministry of Agriculture and Rural Development (grant no. 812-0360-01).

See also: **Genetics**: Karyology; Meiosis; DNA Fingerprinting; Gene Mapping; Research Trends.

Further Reading

Bennett MD and Leitch IJ (1995) Nuclear DNA amounts in angiosperms. *Annals of Botany* 76: 113–176.

Bennett MD and Smith JB (1991) Nuclear DNA amounts in angiosperms. *Philosophical Transactions of the Royal Society of London B* 334: 309–345.

Cairns T (ed.) (1993) *Modern Roses*, 11th edn. Shreveport: American Rose Society.

Chieco P and Derenzini M (1999) The Feulgen reaction 75 years on. *Histochemistry and Cell Biology* 111: 345–358.

Dolezel J, Sgorbati S and Lucretti S (1992) Comparison of three DNA fluorochromes for flow cytometric estimation of nuclear DNA content in plants. *Physiologia Plantarum* 85: 625–631.

El-Lakany MH (1972) Quantitative variation in DNA as related to ploidy level and species in some wild roses. *Canadian Journal of Genetics and Cytology* 14: 347–351.

Griffith AJF, Gelbart WM, Miller JH and Lewontin RC (1999) *Modern Genetic Analysis*. New York: W.H. Freeman.

Muirhead KA, Horan PK and Poste G (1985) Flow cytometry: present and future. *Biotechnology* 3: 337–356.

Rehder A (1940) *Manual of Cultivated Trees and Shrubs*, 2nd edn. New York: Macmillan.

Swift H (1950) The constancy of deoxyribose nucleic acid in plant nuclei. *Proceedings of the National Academy of Sciences of the USA* 36: 643–654.

Inheritance of Characteristics

T Debener, Federal Centre for Breeding Research on Cultivated Plants, Institute for Ornamental Plant Breeding, Ahrensburg, Germany

Introduction

Since the rediscovery of Mendel's laws in 1901 by Correns, Tschernak and DeVries, genetic analyses have greatly increased our understanding of the basic principles of plant biology. Starting with traits that could be screened very easily, including flower colour and structural characters, both the methods of analysis and the genetic strategies were constantly refined, finally leading to the complete analysis of a plant genome in the *Arabidopsis* sequencing project. A molecular analysis alone rarely identifies the function of a given DNA sequence without the proper genetic experiments. Therefore, the classical genetic analysis of a particular trait is often the key step in the analysis of gene structure and function.

The Genetic System of Roses

Amongst 150 or more different species of the genus *Rosa*, the genetic system is not uniform. Ploidy levels can range from the diploid to the octoploid level. Due to the polyploid nature of most cultivated roses, the partial apomictic reproduction in some species of the section *Caninae* (*see* Genetics: Inheritance in the Dog-rose), self-incompatibility and fertility barriers between some species, thorough genetic analyses have been an exception in roses to date. Therefore, information on the inheritance of important rose characters is scarce. The complexity of genetic segregation patterns increases exponentially with the ploidy level of the segregating populations.

The simpler cases are found in segregating progeny derived within diploid roses. Nevertheless, genetic analysis is restricted by the self-incompatibility of many diploid species and varieties, which is an obstacle to the use of self-fertilization to create F_2 progenies. Thus the alternative 'double-pseudo-test-cross' strategy is used, in which unrelated parents with a high degree of heterozygosity are crossed with up to four alleles segregating in the progeny. In addition, backcrosses are also possible which will segregate for up to three alleles. In these cases segregation distortion has to be expected, as the presence of self-incompatibility factors similar to the gametophytic self-incompatibility system of *Prunus* would exclude the S-alleles shared and all alleles linked to them. Another disadvantage to genetic model systems is the small number of seeds produced by most of the diploid species which make the generation of sufficiently large populations very laborious. Furthermore, most diploid species and many diploid hybrids are not recurrent-flowering. This means that backcrosses can often be made only in the second year after sowing the F_1 seedlings.

The segregation of traits in tetraploids is much more complex than in diploids. Theoretically, up to eight different alleles per locus may segregate in the progeny of crosses between unrelated parents. In contrast to diploids, self-fertilization is possible in many tetraploid genotypes. However, even for characters determined by single dominant genes, much larger populations are required for tetraploids than for diploids to obtain the same genetic resolution (**Table 1**).

If a particular character is determined by more than one gene and dominance is incomplete, genes have additive effects and a huge number of possible genotypes can be expected in the progeny (**Table 2**). This tremendous difference between the possible number of genetic combinations in diploids and tetraploids also explains the difficulties of genetic analysis in tetraploids as compared to diploids.

A feature independent of the ploidy level was observed in a number of progeny from interspecific crosses (*see* **Genetics:** Species Crosses). Here, several characters, such as the shape and number of prickles

Table 1 Phenotypic ratios expected for the segregation of single loci in the tetraploid population

Mating	*Segregation of phenotypes*
AAaa selfed	35*A* : 1*a*
Aaaa selfed	3*A* : 1*a*
AAaa × *Aaaa*	11*A* : 1*a*
AAaa × *aaaa*	5*A* : 1*a*
Aaaa × *aaaa*	1*A* : 1*a*
Aa × *Aa*	3*A* : 1*a*
Aa × *aa*	1*A* : 1*a*

Table 2 Maximum number of different gametes that diploid and tetraploid roses may produce if all loci under consideration are heterozygous and multiallelic (for example, a^1a^2 for diploids and $a^1a^2a^3a^4$ for tetraploids)

Number of loci	*Number of different types of gametes*	
	In diploids	*In tetraploids*
1	2	6
2	4	36
4	16	1296
8	256	1 679 616
n	2^n	6^n

and the presence of glandular trichomes, are not inherited in an intermediate way but are rather inherited only through the female line, i.e. the offspring resemble the phenotype of the maternal plant. In some cases these F_1 hybrids are morphologically indistinguishable from their mother and only molecular markers may reveal their hybrid nature. Up to now, the biological basis of this phenomenon is unclear but one may speculate that processes connected to genetic imprinting and selective gene silencing may be involved. The most simple explanation, cytoplasmic inheritance, can largely be excluded because of the large number of characters involved, some of which have been shown to be nuclear-inherited (for example, the presence of prickles) and the observation that these complexes of characters will segregate in later generations.

Many speculations about the inheritance of morphological and physiological parameters have been published over the last century. However, most of these were based on observations of very few progeny and, thus, were not amenable to statistical analysis. On the other hand, the knowledge of commercial rose breeders based on decades of experience with genetic crosses and selection within segregating populations probably comprises a large amount of information about the inheritance of particular characters. But this information remains a company secret and is not available to the general public.

In the following article, only those characters will be considered which were investigated in controlled crosses and for which the methods of analysis meet the basic requirements for genetic experiments (Table 3).

Inheritance of Structural Characters

Structural characters and flower colours are easily observed and, therefore, were the first targets for genetic analysis of rose characters. Many observations have been made on leaf shape, moss character, whole-plant architecture, flower and inflorescence morphology, prickles and several other morphological features. However, the modes of inheritance have only been elucidated for flower morphology, moss character, the dwarf phenotype and prickles.

Flower Structure

Flower morphology consists of a large number of parameters, including the number and shape of petals, number of stamens, petal size and the number and arrangement of styles and ovaries. Some of these characters, for example, the size of floral organs, seem to be controlled by several genes, whereas single (five petals) versus double (>10 petals) was shown to be inherited by a single gene.

Wild rose species (with the exception of *Rosa sericea* and *R. omeiensis*) normally have five petals but a variable number of stamens and styles. In genotypes with double (multipetalled) flowers which have been selected several times during the early history of rose breeding, a certain number of stamens seems to have undergone homeotic transitions to petals, with some intermediate forms between both organ types. This is consistent with other plant species where these homeotic transformations have been observed for a long time. In roses these intermediate organ morphologies are common but the indefinite number of

Table 3 Rose characters that were analysed in genetic experiments

Character	*Mode of inheritance*	*Reference*
Recurrent flowering	Monogenic recessive	Semeniuk (1971a,b), De Vries and Dubois (1984), Debener (1999)
Yellow flower colour	Monogenic dominant	De Vries and Dubois (1984)
Pink flower colour	Monogenic codominant	Debener (1999)
Double flowers	Monogenic dominant	Debener (1999)
Prickles on stems	Monogenic dominant	Debener (1999), Rajapakse *et al.* (2001)
Prickles on petioles	Monogenic recessive	Rajapakse *et al.* (2001)
Dwarf phenotype	Monogenic dominant	Dubois and De Vries (1987)
Moss phenotype	Monogenic dominant	De Vries and Dubois (1984)
Resistance to black spot	Monogenic dominant	Malek and Debener (1998)
Resistance to powdery mildew	Monogenic dominant	Linde and Debener (2003)
Winter hardiness	Quantitative	Svejda (1977a, 1977b, 1979)

De Vries DP and Dubois LAM (1984) Inheritance of the recurrent flowering and moss characters in F_1 and F_2 Hybrid Tea × *Rosa centifolia muscosa* (Aiton) Seringe populations. Gartenbauwissenschaften 49: 97–100; Dubois LAM and De Vries DP (1987) On the inheritance of the dwarf character in Polyantha × *Rosa chinensis minima* (Sims) Voss F_1 populations. Euphytica 36: 535–539; Linde M and Debener T (2003) Isolation and identification of eight races of powdery mildew of roses *Podosphaera pannosa* (Wallr.: Fr.) de Bary and the genetic analysis of the resistance gene *Rppl. Theoretical and Applied Genetics* 107: 256–262. Malek B v and Debener T (1998) Genetic analysis of resistance to blackspot (*Diplocarpon rosae*) in tetraploid roses. Theoretical and Applied Genetics. 96: 228–231; Rajapakse S, Byrne D, Zhang L *et al.* (2001) Two genetic linkage maps of tetraploid roses. Theoretical and Applied Genetics 103: 575–583; Semeniuk P (1971a) Inheritance of recurrent blooming in *R.wichuraiana*. Journal of Heredity 62: 203–204; Semeniuk P (1971b) Inheritance of recurrent and non-recurrent blooming in "Goldilocks" × *Rosa wichuraiana* progeny. Journal of Heredity 62: 319–320; Svejda F (1977a) Breeding for improvement of flowering attributes of winterhardy *Rosa rugosa* hybrids. *Euphytica* 26: 697–701; Svejda F (1977b) Breeding for improvement of flowering attributes of winterhardy *Rosa kordesii* Wulf hybrids. *Euphytica* 26: 703–708; Svejda F (1979) Inheritance of winterhardiness in roses. *Euphytica* 28: 309–314.

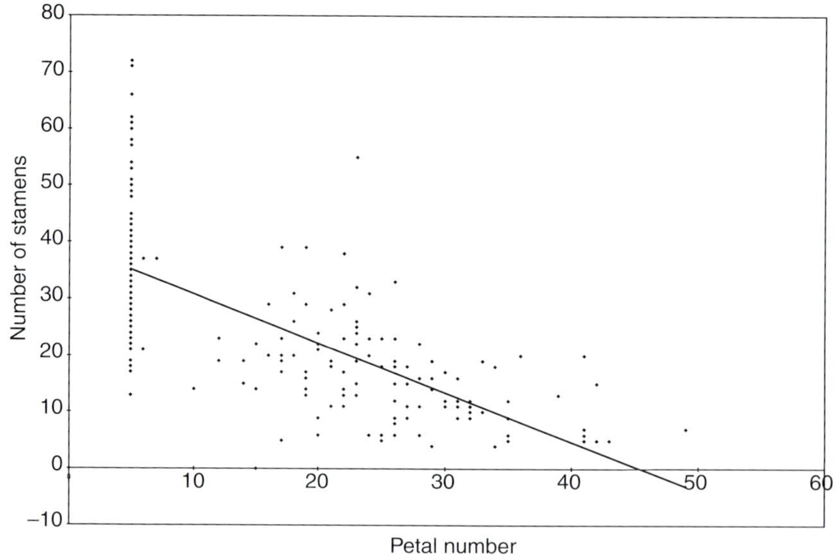

Figure 1 Regression curve demonstrating the negative correlation between the number of petals and the number of stamens in a diploid population segregating for a single dominant gene for double flowers and quantitative factors influencing the number of petals. Note that the distribution of stamen numbers for flowers with five petals indicates the variable number of organs.

stamens makes it difficult to correlate stamen and petal numbers. However, diploid crosses between double (>10 petals) and single (<6 petals) flowered genotypes revealed negative correlations between the number of stamens and petals (**Figure 1**) supporting the concept of homeotic transformation.

Whether or not this concept is accepted, it has been shown both in large diploid and tetraploid segregating populations that the double-flower phenotype is inherited as a single dominant gene (**Figure 2**).

In these populations where double flowers are caused by a single dominant gene, the number of petals in double-flowered genotypes displays a large variation, exceeding the values of the parental genotypes (so-called transgressive variation). By means of nonparametric statistical tests it could be shown that, in double flowers, differences between genotypes with respect to the mean number of petals per flower are not only due to environmental conditions but are also modulated by additional genes. These seem to be present in the genetic background of both double- and single-flowered genotypes.

Figure 2 Segregation of double flowers and flower colour in a diploid population. Parental genotypes = 93/1-117 (pink, single) and 93/1-119 (pink, double), 94/1 = individual plants from the segregating progeny.

The Moss Phenotype

The moss phenotype is characterized by the presence of a dense covering of glandular mossy protuberances on calyx tubes, petals, stems, petioles and pedicels. The character originated in the late seventeenth century as a sport of the cabbage rose *R. centifolia* L. Crosses between the tetraploid moss roses and tetraploid hybrid teas, originally performed to introduce recurrent flowering into the genetic background of the moss roses, revealed that the moss character is inherited as a single dominant gene.

Prickles

As for many other morphological characters of roses, prickles display an enormous variability in shape as well as in quantity and distribution. They are, therefore, used as a key character for the taxonomic distinction of some species and varieties. Despite the large and continuous variability, which implies a complex genetic basis, genotypes without prickles have been discovered relatively frequently, especially in diploid genotypes. Recent results based on the analysis of diploid segregating populations from crosses between prickled and unprickled plants (**Figure 3**) demonstrate

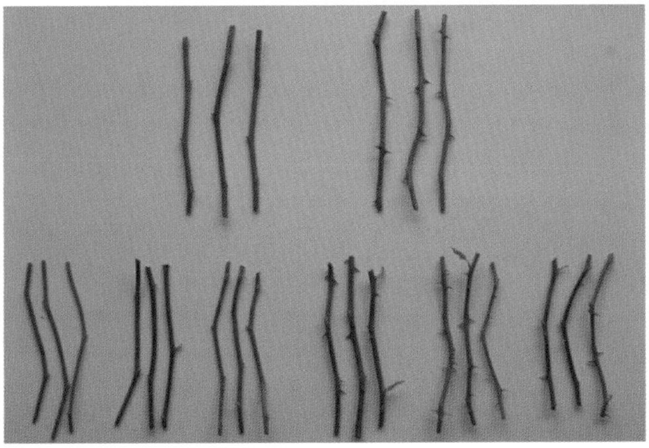

Figure 3 Segregation of the presence of prickles in a diploid population. Upper row, prickled and unprickled parents. Lower row, segregating progeny. Each group of three canes represents a single plant.

the inheritance of prickles on the stems as a single dominant gene at least in these genotypes.

As the prickled parental genotypes used for this study represent a very limited gene pool, it cannot be excluded that additional genes may be involved in other genetic backgrounds. Furthermore, genetic analyses conducted in diploid and tetraploid populations indicate that the presence of prickles on the stem and on the petiole are inherited independently.

Dwarf Phenotypes

Dwarf phenotypes mainly derived from the species *R. chinensis minima* (Sims) Voss. have been extensively used to breed miniature roses. The ease with which dwarf genotypes have been selected from crosses with diploid and tetraploid plants in the past and the observation that single-gene mutations cause dwarf phenotypes in plant models, such as *Arabidopsis thaliana* (L.) Heynh., would already imply simple inheritance of dwarfing in roses. Detailed genetic analysis of rose diploid populations indicated that the dwarf character is inherited as monogenic-dominant. Due to the different number of genes found in plant model systems, it cannot be excluded, as for the inheritance of prickles, that mutations in several different genes may lead to dwarf phenotypes.

Inheritance of Physiological Characters

Flower Colour

Rose petals contain a number of different pigments that determine their colour (*see* **Fragrance and Pigments**: Odoriferous Substances and Pigments; Anthocyanins and other Flavonoid Pigments). In the 1970s, paper chromatography analyses of tetraploid rose cultivars revealed differences in the amounts of carotenoids, anthocyanidins (pelargonidin, cyanidin) and flavonols (kaempherol, quercetin) and the inheritance of petal colour was postulated to be multigenic. However, the inheritance of yellow flower colour in tetraploid as well as pink flower colour in diploid populations was later shown to be controlled by major dominant and codominant genes, respectively (**Table 3**, **Figure 2**). Although the pigment composition of the latter populations may have been very simple, refined methods which distinguish not only general pigment classes but also their various glycosides may resolve even complex petal colorations into simple segregation patterns of individual pigment components.

Recurrent Flowering

Recurrent flowering was probably introduced into rose cultivars by introgression from *R. chinensis* Jaquin. and *R. odorata* (Crép.) Rehd. & Wils. in the early nineteeth century. As it affects the requirement for vernalization of rose shoots to induce flowering and significantly increases the flowering period, it is an extremely important character which greatly influenced modern rose breeding. It is relatively easy to obtain recurrent-flowering progeny from crosses between recurrent and nonrecurrent cultivars. Therefore, theories about the inheritance of this character as a recessive gene date back to the early 1940s and were favoured, for example, by Hurst. But it was only since 1971 that the first clear-cut genetic analysis in diploid *R. wichurana* Crép. crosses demonstrated that a single recessive gene was responsible for the observed variation. Since then, several other studies in both tetraploid and diploid populations confirmed these results. It is assumed that most genes for recurrent flowering were derived from the few original sources.

Winter Hardiness

Although frequently used in the rose literature, the term 'winter hardiness' is difficult to define precisely. The ability of roses to survive winter periods depends not only on the lowest temperatures, including frost tolerance (the ability to survive temperatures below freezing point), but also on the interplay of several other factors. These comprise the length of the acclimation period, humidity, periods of warm temperatures prevailing, frost factors and stress factors, including damage by pathogens and pests. Therefore, analyses of the percentage of winter kill in progeny derived from crosses between winter-hardy and nonhardy parents carried out in Canada in the 1970s displayed a broad variability, indicating an influence of multiple genes. However, estimates of heritability for different populations varied between 51 and 92%, implying the involvement of only a few or several closely linked loci.

Male Sterility

In contrast to statements that evidence for an inheritance of male sterility as a single gene had been obtained, no conclusive genetic experiments have been reported. First results were reported in the early 1960s on a population that resulted from a cross between a male sterile genotype of *R. setigera* Michx. with a male fertile genotype of *R. brunonii* Lindl. About 96% of the 70 plants from the F_1 progeny analysed were male sterile. However, to elucidate the genetic basis of this trait, backcrosses and further crosses between the male sterile line and other genotypes would have been needed. Therefore, the genetic basis of male sterility remains obscure.

Inheritance of Resistance to Pathogens

The study of plant resistance to pathogens involves higher levels of complexity than analyses of structural and physiological characters. Apart from genetic variation present in the host plant, the genetic constitution of the pathogen has to be considered as well. As was shown for numerous plant pathogen systems, the particular interaction between host and pathogen is often controlled by matching pairs of genes, following the so called 'gene-for-gene' interaction. These gene-for-gene interactions are also termed 'vertical resistance', whereas resistance that is independent of the pathogen genotypes and controlled by many genes with additive effects is often termed 'horizontal resistance'. Gene-for-gene interactions have been demonstrated for rose black spot and rose powdery mildew. Different pathogenic races may cause completely different reactions by the host plant and, therefore, polysporous isolates (i.e. mixtures of genetically diverse spores) often lead to host responses caused by several resistance genes simultaneously. As a consequence, genetic analyses focusing on these race-specific vertical resistance genes have to be performed under strictly controlled conditions with defined, genetically uniform strains of the pathogen. Apart from vertical resistance genes, quantitative or horizontal resistance genes acting on all pathogenic races may be present as well. This type of resistance is often less effective than vertical resistance in that it does not provide 100% protection against the pathogen but is generally expected to be more stable. It is also more difficult to manipulate in the breeding process because in most cases it is under polygenic control with different contributions of individual genes to the whole level of resistance.

Resistance to black spot disease The first genetic analysis of resistance to rose pathogens was performed for black spot caused by the hemibiotrophic ascomycete *Diplocarpon rosae* Wolf. Black spot is the most severe fungal disease of field-grown roses, especially in areas with high annual precipitation.

Screening of tetraploid as well as diploid populations carrying resistance from *R. multiflora* Thunb. with spores derived from a single conidial isolate revealed the presence of two single dominant genes *Rdr1* and *Rdr2* (**Figure 4** *see* **Disease:** Black Spot).

The data for *Rdr1* have been confirmed, with several molecular markers cosegregating with the gene. These markers are currently being used to isolate and analyse the gene. In addition to the vertical race-specific resistance to black spot, horizontal resistance may be present as well. Observations in different laboratories show continuous variation in susceptibility between rose genotypes and even within segregating populations. However, quantitative genetic data on the transmission of partial resistance are not yet available.

Resistance to powdery mildew Powdery mildew, caused by the biotrophic ascomycete *Podosphaera pannosa* (Wallr.:Fr.) de Bary, formerly known as *Sphaerotheca pannosa* var. *rosae*, is the most important fungal disease of roses grown in greenhouses and also a serious disease in the field (*see* **Disease:** Powdery Mildew). The first genetic analysis of powdery mildew resistance was based on roses of the same genetic background that were used for the genetic analysis of black spot resistance. A segregating population was screened with spores derived from a single conidial isolate and revealed the presence of a single major

Figure 4 Genotypes from a diploid population segregating for black spot resistance. Left, susceptible genotype displaying the typical black spot symptoms, like black spots and chlorosis. Right, resistant genotype without visible macroscopic symptoms. All plants were inoculated with spores from the single conidial *Diplocarpon rosae* isolate Dort E4, representing black spot race no. 5, at a density of 10^5 conidia ml^{-1}, and incubated for 10 days.

dominant gene. However, in contrast to black spot resistance, a large transgressive variation in the degree of susceptibility was observed in the susceptible fraction of the progeny. This implies the presence of additional factors involved in partial resistance in both parents which could be interpreted as a weak horizontal resistance.

Resistance to crown gall disease The spread of crown gall disease, caused by *Agrobacterium tumefaciens*, is a serious problem in nursery stocks of southern European countries (*see* **Disease:** Crown Gall). Therefore, first analyses concerning the degree of susceptibility were made in a collection of 28 different genotypes and pronounced differences in susceptibility were found for some of the genotypes. However, no segregating progeny was analysed and, therefore, no conclusions can be drawn concerning the inheritance of this character.

Implications for Rose Biology

Even the limited amount of information about the inheritance of rose characters is an important first step for the understanding of biological functions. One example is the analysis of disease resistance showing that rose genes for black spot and powdery mildew resistance (both pathogens are obligate parasites) follow the scheme of the so-called gene-for-gene interactions. The demonstration of single-gene segregation has important implications for both the biology of the rose–pathogen interactions and future breeding strategies. Furthermore, this genetic information is the prerequisite for further analyses of the characters under investigation. For the rose disease resistance genes, for example, efforts to isolate and analyse the DNA sequences that are under way rely on the original genetic data.

Another example are the genes known to influence the number of petals. The present data suggest that, apart from main switches deciding about double and single flowers, additional regulators determine the number of petals in double flowers. Utilizing the information about flower development in model plant species with several of the regulatory genes already cloned and analysed, the information available for roses may be used to unravel the regulation of flower development in roses down to the molecular level. This approach would rely on the homology of the main regulatory networks, as demonstrated in comparative analysis between *Arabidopsis*, *Antirrhinum* and several other plant species.

Another aspect under which genetic data from roses may facilitate structural and functional analyses of important genes is the concept of synteny. Synteny describes a conservation of gene order on chromosomal segments of different species from the same genus and even between different families. As extensive analyses on the genome of *Prunus* and *Malus* species, both belonging to the Rosaceae family, are under way, this information may soon help to detect and isolate rose genes.

Implications for Rose Breeding

Rose breeding has mainly relied on very simple strategies. In order to develop more targeted approaches, information about the inheritance of the key traits is indispensable. This is of particular importance in cases where genes have to be introgressed from wild species to highly developed varieties via wide crosses. As an example, the mode of inheritance of a particular resistance gene is important to calculate the minimum population size necessary for obtaining progeny with both resistance and the appropriate combination of morphological characters. The more characters that are known in respect to their genetic basis, the more precise strategies considering the number of progeny and the number of generations can be developed.

For traits that are difficult to screen, marker-assisted selection may be an alternative to the selection for the trait itself (*see* **Genetics:** Marker-Assisted Selection). Here again, markers can only be developed in populations segregating for a particular trait and in which precise knowledge about the inheritance is available.

Conclusions

Our knowledge about the inheritance of important rose characters is still limited if compared to well-investigated plant species. However, the availability of diploid populations which have been used to analyse several important traits will increase the speed with which we can obtain more genetic information in the near future. In this process a genetic analysis will often remain a first necessary step which, often in connection with molecular methods, will lead to new insights into biological processes. Due to several unique characters of roses that cannot be analysed in any model plant, detailed genetic analyses will remain indispensable over the next years. Apart from broadening our knowledge about principles of rose biology, information on the inheritance of key characters will be the prerequisite for the development of more advanced breeding strategies.

See also: **Breeding:** Overview; Methods of Cross Breeding; Selection Strategies for Disease and Pest Resistance. **Disease:** Black Spot; Downy Mildew; Powdery Mildew. **Genetics:** Karyology; Meiosis; Inheritance in the Dogrose;

Species Crosses; Mutation; Gene Mapping; Molecular Markers; Marker-Assisted Selection; Research Trends.

Further Reading

Debener T (1999) Genetic analyses of important morphological and physiological characters in diploid roses. *Gartenbauwissenschaft* 64: 14–20.
Debener T and Mattiesch L (1996) Genetic analyses of molecular markers in crosses between diploid roses. *Acta Horticulturae* 224: 249–252.
De Vries DP and Dubois LAM (1996) Rose breeding: past, present, prospects. *Acta Horticulturae* 424: 241–248.
Gudin S (2000) Rose: genetics and breeding. *Plant Breeding Reviews* 17: 159–189.
Krüssmann G (1974) *Rosen, Rosen, Rosen*. Berlin: Verlag Paul Parey.
Lindstrom EW (1936) Genetics of polyploidy. *Botanical Review* 2: 197–215.

Inheritance in the Dogrose

G Werlemark, Swedish University of Agricultural Sciences, Kristianstad, Sweden

Introduction

The dog roses, *Rosa* section *Caninae*, are native to Europe, western Asia and north Africa, and most of the taxa have their northern range margins in central Scandinavia and Finland (**Figure 1**). The Latin name comes from *canis*, meaning dog, and gives rise to the common name, dog rose. These species are long-lived perennial shrubs, growing in dry woodland margins and disturbed areas like roadsides and open pastures. Species within this section easily colonize new areas. One of them, *Rosa rubiginosa* L., eglantine or the briar rose, was introduced as an ornamental in Australia by the end of the nineteenth century and some 20 years later it had become one of the worst weeds in New South Wales.

Figure 1 *R. dumalis* × *R. rubiginosa* hybrid (*Rosa* section *Caninae*), in full bloom by the end of June. © Balsgård with permission.

The dog roses have been known since the Middle Ages, when they were planted as hedges, and rose hips, both the seeds and the fruit flesh, were used for medicinal purposes (*see* **History of Roses in Cultivation:** European (Pre-1800)). At the end of the eighteenth century, budding of ornamental roses was developed, and it was found that species within the section *Caninae* were very well suited as rootstocks, especially *R. canina* L., and the dog rose has retained this function to the present day (*see* **Rootstocks:** Usage of Rootstocks). Examples of this are the well-known rootstocks 'Inermis', 'Pfänder's' and 'Laxa', which bud early and produce few suckers. Rootstocks are usually propagated by seed, which germinates best in the second year (*see* **Rootstocks:** Seedling Stocks). The seedlings have been shown to be morphologically uniform, even if different genotypes of *R. canina* grow together in seed gardens. Propagation by cuttings from seedling selections have been tried, but the rooting behaviour has been variable, indicating genotypic differences (*see* **Rootstocks:** Clonal Rootstocks).

Not many of the dog rose species have ever been involved in breeding programmes for ornamental purposes: *R. rubiginosa* is one of the few exceptions. Lord Penzance introduced several hybrids with this species as one of the parents in the late nineteenth century, of which some still exist, e.g. 'Lady Penzance' and 'Lord Penzance', and the German rose breeder Kordes also used *R. rubiginosa* in crosses with Hybrid Tea roses. The result was new vigorous varieties, which lacked the characteristic apple scent of the briar rose, and had large flowers with a modern colouration. Recently, it has also been shown that *R. canina* is probably the seed parent to the *R.* × *alba* roses, together with a member of *R. gallica* as pollen parent.

The numerous rose hips of the dog rose plants can be used for tea, marmelade, wine and sherry. In Sweden, rose hip soup has been a well-known dessert for more than 100 years, and it is mainly the dog rose species that have the particular aroma suitable for rose hip soup. A plant-breeding programme was initiated at the Swedish University of Agricultural Sciences in the 1980s with the aim of developing new cultivars of dog roses which are, suitable for industrial cultivation. The world market price of rose hips has since declined, and it is presently more profitable to import the raw material (**Figure 2**). Annually, more than 3000 tonnes of dried rose hips are imported to Sweden, mostly from South America. The dog rose rose hips have a high vitamin C content – in some taxa up to 2 g $100\,g^{-1}$ fruit flesh (*see* **Secondary Metabolites:** Culinary Uses and Nutritional Value). This compares with strawberry (0.07 g $100\,g^{-1}$ fruit flesh), blackcurrant (0.12 g $100\,g^{-1}$ fruit flesh) and oranges (0.2 g $100\,g^{-1}$ fruit flesh). Recently, new valuable traits have been discovered in dog rose hips, including anti-inflammatory properties, that may become useful for therapy in patients with arthritis (*see* **Secondary Metabolites:** Medicinal and Pharmaceutical Uses).

The section *Caninae* is considered a very polymorphic group within the genus *Rosa*, and classification based on morphological characters has been difficult. At the beginning of the twentieth century, dog roses were exposed to an extensive nomenclatural proliferation, mainly by some Swedish rhodologists, and several hundreds of different forms were given species rank based on morphological characters. Today, a more conservative view is taken, and only 20–30 species are recognized (*see* **Classification:** Conventional Taxonomy (Wild Roses)).

The dog rose species are all polyploids ($2n = 28$, 35 or 42 chromosomes) with pentaploidy ($2n = 35$) as the most common ploidy level. In contrast to most other roses, they are self-compatible, but outcrossing is probably common, since several insects are attracted to the multitude of large showy flowers with abundant pollen production. Dog roses are also able to form hybrids with other *Rosa* species, both within the section and with species from other sections, but to what extent this happens in nature is not clear (*see* **Genetics:** Species Crosses). Dog rose species are very diverse, but there are no unique morphological traits for the section. However, all species belonging to this section have one thing in common and that is their peculiar meiosis with uneven distribution of chromosomes in male and female gametes. This article will try to shed some light on the odd meiosis and its consequences for the inheritance of morphological characters and molecular markers.

Figure 2 Harvesting dog rose hips in south Sweden, with a modified blackcurrant harvester. © Balsgård with permission.

Meiosis

Female Meiosis

In most plant species, the seed parent and the pollen parent transmit approximately the same amount of genetic material to their offspring, if both parents have the same ploidy level. Thus a diploid, tetraploid or hexaploid seed parent plant contributes half of its genetic material to its offspring, whereas the other half is donated from the pollen parent plant (*see* **Genetics:** Meiosis). However, a species group where pentaploidy is the most common ploidy level must solve this problem in another way if it is to continue with sexual reproduction, and dog roses have evolved a unique way of doing this.

At female meiosis, only two of the four, five or six genomes (depending on ploidy level) line up at the equator of the spindle for the bivalent formation. The other chromosomes remain unpaired as univalents at one of the spindle poles. At the first division, the bivalent chromosomes separate as usual and the daughter chromosomes move towards the poles, giving rise to two cells. The one closest to the micropylar end contains the seven chromosomes from the bivalent formation together with all the univalents, whereas the cell closest to the chalazar end only contains the seven chromosomes from the bivalents. At the second division, the univalents divide along with the chromosomes originating from the bivalents and this results in two cells with 21, 28 or 35 chromosomes, depending on ploidy level, and two cells with only seven chromosomes which are eventually lost. Of the two remaining cells, the one closest to the micropylar end develops into an embryo sac.

Male Meiosis

At the male meiosis, the univalents stay rather close around the equator when the 14 chromosomes line up for the bivalent formation. When the bivalent chromosomes divide, the univalents move to the equator where they split into chromatids. Most of the univalent daughters manage to reach the poles in time to be included in the daughter cells. However, at the second division, when the bivalent chromosomes divide normally, the chromatids from the univalents lag behind and the resulting tetrad will consist of four cells with seven chromosomes from the bivalent formation, often together with some micronuclei from the univalents. However, only pollen cells with seven chromosomes are fertile and, therefore, pollen viability within the section *Caninae* is rather low (*see* **Insects and other Animals:** Pollination). In the pure species the viability, as estimated with cotton blue stainability, is between 20% and 30% and in interspecific hybrids it is only 0–10%. On fertilization, the triploid/tetraploid/pentaploid egg cell fuses with the haploid pollen cell and restores the original chromosomal number.

Separation of the Sister Chromatids

The regulatory pathway, which governs the separation of the sister chromatids at meiosis, has received much interest lately. In the univalents of the dog roses, this separation takes place at different times in male and female meiosis, anaphase I and anaphase II, respectively. At male meiosis the sister chromatids of the univalent-forming chromosomes separate at anaphase I, whereas the chromatids of the bivalent-forming chromosomes stay together and separate at anaphase II, as is normal. In contrast, in female meiosis, the sister chromatids of the univalents stay together in anaphase I and separate, together with the chromatids of the bivalent-forming chromosomes, at anaphase II. There are two mechanisms which can explain how the univalent sister chromatids in the male meiosis separate, while at the same time they stay together in the bivalents: (1) the chromatid cohesion complex – the centromeres – which keeps the sister chromatids together, is placed in such a way in the bivalents that it is inaccessible to cleavage; (2) the bivalent-forming chromosomes have a chemically modified cohesion complex which is insensitive to cleavage. Without this complex, the chromatids would separate, as happens in the univalents. Either mechanism relies on a genetic control in both time and space over the chromatid cohesion.

Apomixis

Meiosis in the *Caninae* results in a pronounced influence of the maternal parent on its offspring, as demonstrated in several studies since the 1940s. The progeny plants, in most cases, resembled their seed parent in all details, which led to the conclusion that they originated from apomixis, whereas frequently occurring spontaneous interspecific hybridizations could lead to extremely polymorphic progeny plants. The dog roses were also believed to be pseudogamous, i.e. pollination is necessary for the development of the endosperm, but the embryo remains unfertilized just as in many other apomictic species in the Rosaceae family. However, a cytological study in the 1940s failed to find any unreduced embryo sacs, which for a while put a stop to speculations about apomixis in dog roses.

Recent studies have been performed on pairs of reciprocal crosses between Nordic dog rose species (*see* **Genetics:** Species Crosses). In one of these studies, pollen viability, seed and fruit characters and molecular markers in the progeny plants were compared to corresponding values for the parental plants, *R. dumalis* Bechst. and *R. rubiginosa*, respectively (*see* **Genetics:** DNA Fingerprinting). Approximately 10% of 70 offspring plants did not receive any of the molecular randomly amplified polymorphic DNA (RAPD) markers specific for the pollen donor. Furthermore, these plants had the same pollen viability as the parental plants, i.e. 20–30%, compared to the true hybrid plants, which had a pollen viability of 0–10% as expected. When the studied seed and fruit characters (i.e. seed weight, weight of fruit flesh and number of seeds per hip) and pollen viability values were combined in a multivariate analysis, four different progeny groups could be discerned; two hybrid progeny groups ascertained with specific molecular markers of the pollen parent, and with *R. dumalis* and *R. rubiginosa* as seed parents, respectively, and two progeny groups with the same two seed parents, respectively, which had not received any markers of the pollen parent. Plants which had not received any markers from the pollen donor, and had a higher pollen viability than their hybrid siblings, were assumed to have originated through apomixis. These plants were later also studied with microsatellite primer pairs, which also failed to detect any markers from the supposed pollen parent, and the result thus supported the conclusion that these plants were of apomictic origin.

The above-mentioned study was made on controlled crosses, performed in a greenhouse with optimal temperature and humidity. It is impossible to know to what degree these optimal conditions could have influenced the production of apomictically derived progeny plants as compared to under natural conditions. It is not known how common apomictically derived dog rose plants are in nature. Since dog roses are self-compatible and intraspecific variation is low as a consequence of the odd meiosis, apomictically derived plants are almost impossible to separate from those derived from self-pollination when using morphological characters only. But with the new molecular techniques that are now being developed, further studies in this area will be possible (*see* **Genetics:** DNA Fingerprinting).

Preferential Pairing of the Bivalents

In previous cytological studies, it was noted that, apart from the univalents, bivalents were mostly formed, whereas multivalents were rarely observed. This gave rise to two questions: (1) why were there only bivalent formations and no multivalents?; (2) why should this bivalent pairing be restricted to only two sets of chromosomes?

Studies performed with molecular RAPD markers on interspecific crosses of Nordic dog roses revealed that approximately half of the markers of the pollen parent are transferred to the offspring plants (**Figure 3**). Of these, each offspring plant inherits only on average one-third. The non-transferred markers specific for the pollen parent probably resides on the univalents and therefore never reaches the offspring plants. This is in pronounced contrast to the markers of the seed parent, since all or almost all of these markers are transmitted to the offspring. A similar result was achieved with microsatellite markers (sequenced-tagged microsatellite sites) (STMS) in a study made on some of these progeny plants. In addition to 23 shared alleles between *R. dumalis* and *R. rubiginosa*, all the 16 microsatellite DNA alleles specific for the seed parent were transmitted to the progeny plants, whereas only one to three alleles specific for the pollen parent were transmitted. It is noteworthy that the same pollen-specific markers (RAPDs as well as microsatellites) appear in all the offspring plants, not a random selection among all the markers available in the pollen parents (**Figure 3**). This points towards a strongly preferential pairing of the bivalent-forming genomes. If bivalents were instead formed at random from all the present genomes, all pollen-specific markers would appear in the offspring plants at one time or another.

In a study using microsatellite markers on three different pentaploid dog rose species, four different alleles could be found at several loci, but never five. Correspondingly, in a tetraploid dog rose species, three alleles were found in some loci, but never four. In a pentaploid species, three of the genomes can therefore be regarded as haploid, whereas the fourth is diploid. The pairing takes place mainly between the two members of the diploid genome, which leads to a very low level of recombination, as these two genomes are highly homologous and usually share

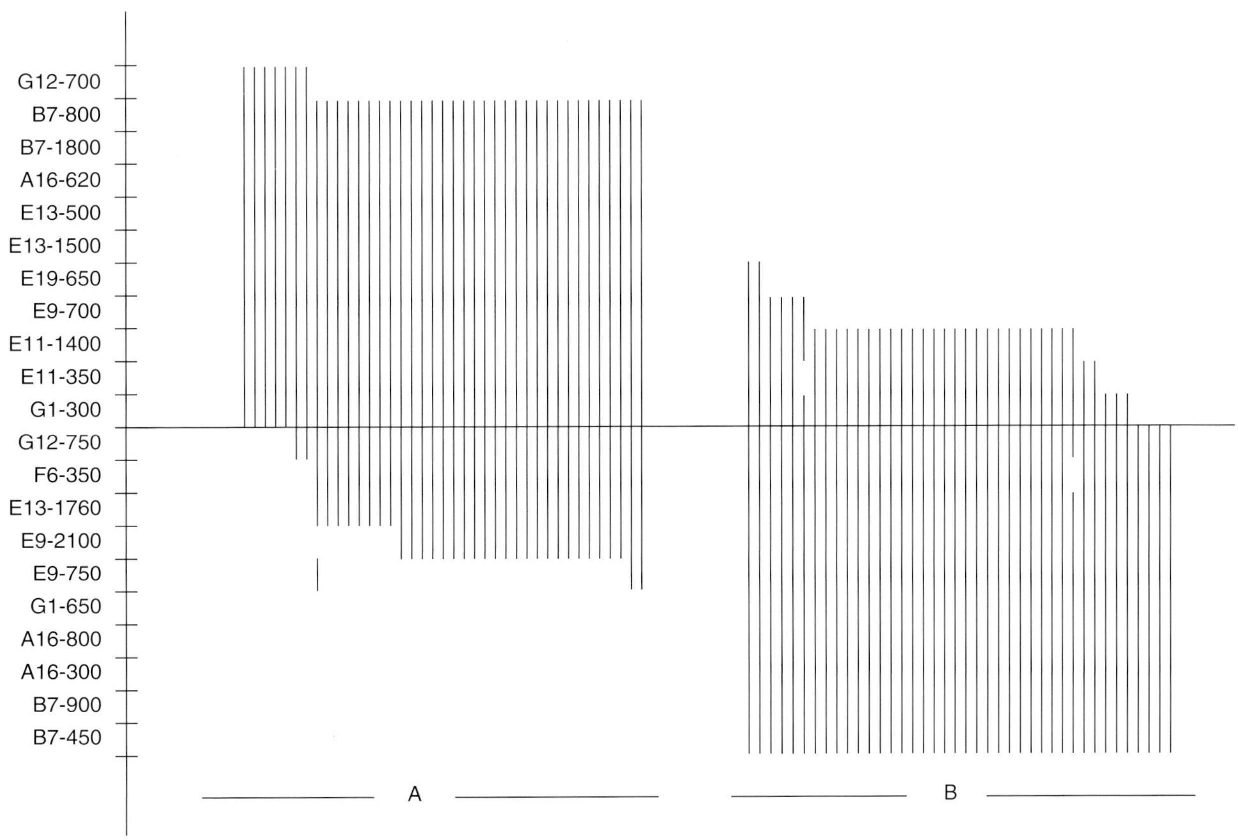

Figure 3 Distribution of randomly amplified polymorphic DNA bands in interspecific progeny plants between *Rosa dumalis* and *R. rubiginosa*. The bands above the horizontal line are specific for *R. dumalis* and the bands below the line are specific for *R. rubiginosa*. Each vertical line represents one progeny plant. A, *R. dumalis* × *R. rubiginosa*; B, *R. rubiginosa* × *R. dumalis*. Reproduced from Werlemark G, Uggla M and Nybom H (1999) Morphological and RAPD markers show a highly skewed distribution in a pair of reciprocal crosses between hemisexual dogroses species, *Rosa* sect. *Caninae*. *Theoretical and Applied Genetics* 98: 557–563, with permission.

identical microsatellite DNA alleles. The same applies in interspecific crosses, where bivalent pairing also appears to take place mainly among those chromosomes which have the most similar alleles. When studied with RAPD markers, the transmitted paternal markers appeared in all or almost all of the offspring, or in very few of the offspring, which also suggests that the marker most often occurs in both of the bivalent-forming genomes or in none (**Figure** 3). The more species-specific alleles appear to be placed on the univalents and are thus transferred solely through the seed parent. Since the recombination takes place mainly among the genomes, which strongly resemble each other, and more than half of the genetic material is transferred solely through the seed parent, the result is a reproductive system somewhere between selfing and apomixis (*see* **Genetics:** Species Crosses).

Studies of the nucleolar organization regions (NOR sites) with fluorescence *in situ* hybridization (FISH) have shown that each genome contains one NOR site, just as has been shown in other *Rosa* species. In dog roses, there is a marked size polymorphism among these sites. Two deviate by being either very small or very large, whereas the remaining three are of intermediate size and cannot as yet be separated from each other. The bivalent formation always takes place between two of these three intermediate-sized loci. Sequencing of the internal transcribed spaces (ITS) regions (the regions surrounding the NOR sites) has revealed four different sequences in a pentaploid *R. canina*, with one of the sequences occurring in more or less double the amount compared to the other three. This suggests that this sequence occurs in a diploid genome as opposed to the other three, which occur in one haploid genome each.

Originally, it was supposed that the univalent-forming chromosomes differed so much from each other that pairing between them was impossible. But later studies have shown that the meiosis mechanism is strictly regulated and probably under genetic control by restricting the pairing of the chromosomes, rather than preventing crossing overs. This has been demonstrated in other species as well, e.g. wheat (*Triticum aestivum*) and different dahlia (*Dahlia*) species. The control breaks down in interspecific hybrids. Crossing experiments between different dog rose species as the female parent and *R. rugosa* Thunb. (diploid, sect. *Cinnamomeae* = *Cassiorhodon* Dumort.) as the male parent have resulted in an elevated number of bivalents in the offspring. Even assuming that all seven chromosomes from *R. rugosa* pair with '*canina*' chromosomes, the excess of bivalents can only come from the univalents, which never pair in the '*canina*' parent itself.

A few successful attempts have been made to develop a tetraploid plant from a pentaploid dog rose species either through spontaneous parthenogenesis of a reduced egg cell, or through pollination with irradiated pollen and embryo rescue (*see* **Biotechnologies for Breeding:** Early Embryo Rescue). However, in these plants regular bivalent formation is destroyed and only one to three bivalents are formed. This lack of bivalent formation could reflect an absence of the male genome and, if this is so, the pollen parent genome is necessary for normal bivalent formation. This suggests that the control mechanisms responsible for the formation are located on both of the genomes which normally form the bivalents and are fully active only in double dose.

The described preferential pairing of the bivalents and the different times of division of the univalent chromosomes into chromatids in the male and female gametes suggest a strongly controlled meiosis. The combined data from the molecular studies also suggest that the genomes forming the bivalents are in some way very distinct. Epigenetic imprinting in the form of DNA methylation or histone acetylation has been proposed.

Inheritance of Characters

When comparing morphological characters in offspring plants from a reciprocal cross between two dog rose species, the maternal influence is clearly visible, in that there are two different offspring groups, each one resembling its respective seed parent (**Figure** 4). This has been demonstrated in several studies and, as previously mentioned, on several occasions has led to the conclusion that dog roses propagate by apomixis. Several interspecific crossing experiments, both within section *Caninae* and between *Caninae* species and species from the *Cinnamomeae* section (mostly *R. rugosa*) have demonstrated a stronger influence of the maternal morphological characters in the progeny plants. However, the inheritance of single characters seems to be dependent on what species combination is used. A reciprocal cross was performed between *R. canina* and *R. rubiginosa* (both $2n = 35$). Both progeny groups showed a strong resemblance to their respective seed parents, but in the cross *R. canina* × *R. rubiginosa* six morphologically different progeny plants were obtained, whereas in the reciprocal cross, *R. rubiginosa* × *R. canina*, 58 morphologically uniform progeny plants were obtained. There was also a difference in fertility, measured as seed set, in the two progeny groups: the first group was almost sterile and the second group was quite fertile (*see* **Genetics:** Inheritance of Characteristics).

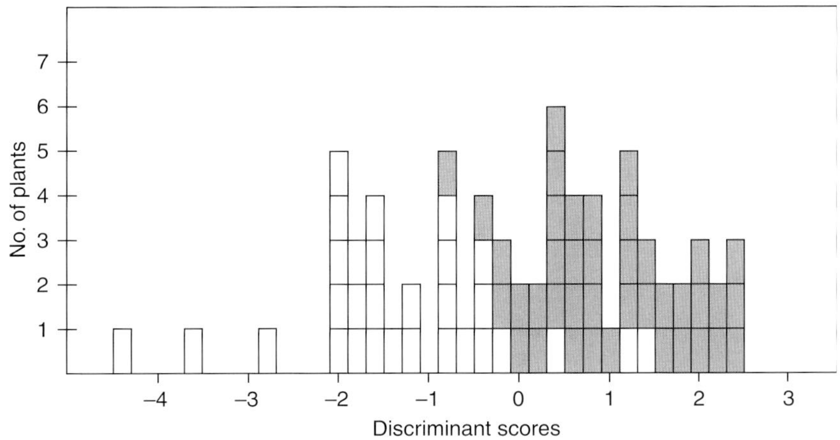

Figure 4 Discriminant analysis calculated on seven morphological characters for discrimination between a pair of reciprocal crosses. Open columns, *Rosa dumalis* × *R. rubiginosa*; shaded columns, *R. rubiginosa* × *R. dumalis*. Reproduced from Werlemark G, Uggla M and Nybom H (1999) Morphological and RAPD markers show a highly skewed distribution in a pair of reciprocal crosses between hemisexual dogroses species, *Rosa* sect. *Caninae*. *Theoretical and Applied Genetics* 98: 557–563, with permission.

Several studies have shown similar result; most of the time the progeny plants strongly resemble their seed parent, but sometimes more unique individuals appear. Leaf shape, shape of hip and pedicel length have been reported to be mostly matroclinally inherited, whereas shape of prickles has been reported as patroclinally inherited and shape and colour of the flower to be an intermediate inherited character.

Matroclinal inheritance is explained by the univalents being transmitted unpaired and thereby also unchanged to the progeny plants. Also, as previously mentioned, the pairing of the bivalents occurs between chromosomes with the most similar allelic configuration, so even progenies from quite distant crosses should strongly resemble that of their seed parent. In studies made on other species, hybrid plants often display a mixture of parental traits, depending upon dominance effect, and this could also be the reason why some characters in different dog rose crosses appear to be paternally inherited. Incomplete dominance or codominance could also lead to intermediate inheritance, as is often seen in the colouration of flowers.

Dog rose hybrids also often display characters that are extreme relative to those of either parent in the form of an increased amount of glandular hair and prickles. Some studies also indicate larger flowers and larger hips in intersectional hybrids then either of the parental species. The cause for this could be a complementary action of alleles or recessive genes present in a heterozygous form in the parents that become homozygous in the hybrid, or maybe a combination of both.

This extreme matroclinal inheritance has been used in the production of rootstocks for ornamental roses. The dog rose species used for this purpose can be propagated by seed instead of the more time-consuming and labour-intense propagation by soft- or hardwood cuttings (*see* **Rootstocks:** Clonal Rootstocks; Seedling Stocks). However, the matroclinal inheritance also complicates plant breeding. Very little of the genetic material from the pollen parent will ever reach the progeny plants. In a dog rose breeding programme for hip production, creating large rose hips with fewer than average seeds is one of the main aims, and this can often be achieved by making crosses beween species in the section *Caninae* as seed parents and species from the section *Cinnamomeae* as the pollen parent (*see* **Breeding:** Methods of Cross Breeding).

Population Genetics

Several studies on population genetics have been performed on Nordic dog roses, involving some of the most common taxa in the Nordic countries, e.g. *R. canina*, *R. dumalis* subsp. *dumalis, R. rubiginosa* and *R. villosa* subsp. *mollis* (Sm.) Kell. and Gams. (soft downy rose). Seedlings of these taxa, resulting from a huge Nordic seed collection, have been planted in experimental fields and the plants have been studied using both reproductive and vegetative morphological characters, including leaflet shape assessed with image analysis, as well as molecular RAPD markers. A canonical variates analysis (CVA) using seven morphological characters could reclassify only 55% of the plants into the proper taxa, which demonstrates the difficulties involved in discrimination among dog rose taxa. Hierarchical CVA revealed that 80% of the total morphological variation occurs between taxa and only 20% within taxa. Both morphological characters and molecular markers have demonstrated extremely low intraspecific variation in Nordic dog roses. Even so, the level of intraspecific variation varies between the examined taxa. In all studies *R. rubiginosa* has emerged as the most well-defined and homogeneous taxon, with very little variability between or within populations. Analysis of hierarchical partition of diversity using leaflet shape data showed that 28% of the total variability occurs between populations and the remainder within populations. Due to the overall low intraspecific variation in this taxon, interpretations of diversity partitions are, however, difficult. Molecular markers have shown hardly any variation at all among populations, and more or less the same RAPD phenotype was encountered in 90% of investigated individuals. The present Nordic *R. rubiginosa* population thus appears to be the result of only a few sexual recombination events. The resulting genotypes may have spread vegetatively or by dispersal of apomictic seeds, which would lead to populations consisting of only one or a few genotypes. *R. rubiginosa* also has a somewhat later flowering period than do the other dog roses, which may prevent introgression from other taxa.

In contrast, *R. dumalis* and *R. canina* have, in all studies, proved to be the most heterogeneous of the dog rose species investigated. They are, however, often difficult to separate from each other with morphological characters, and with molecular markers it has so far been impossible. Both taxa show heterogeneity between populations as well as within populations. A closer look at the partition of diversity within *R. dumalis*, using leaflet shape data, demonstrated that 57% of the variation occurred as a between-populations component. The high level of variation in *R. dumalis* and an overlap between populations demonstrated with the leaflet shape data suggest considerable gene flow among populations.

Somewhere in between these two extremes falls *R. villosa*, which shows some morphological variation between populations but very little within populations. A CVA based on leaflet shape data showed that 63% of the variability occurs between populations, and the remainder within populations. The overlap of the leaflet shape characters between populations also suggests considerable gene flow among populations.

Microsatellite studies have shown *R. dumalis* to have slightly more alleles per locus compared to both *R. rubiginosa* and *R. villosa*. This higher variability is also indicated in interspecific crosses. When *R. dumalis* acts as seed parent in interspecific crosses with *R. rubiginosa*, the offspring plants are more variable in both morphological and molecular characters compared to when *R. rubiginosa* is the seed parent. The morphological appearance of progeny plants was recorded from the reciprocal cross between *R. canina* and *R. rubiginosa*, and it was noted that, also in these combinations, progenies with *R. rubiginosa* as seed parent were more homogeneous than progenies with *R. canina* as seed parent. Since these two progeny groups also differed in fertility, it is likely that gene flow between dog rose taxa is more common in some directions than in others.

The partitioning of variability with relatively high between-population components in *R. dumalis* and *R. villosa* is consistent with reports for selfing or apomictic taxa. However, even if all dog roses were highly outcrossing, a similar partitioning of variability would be expected in the account of '*canina*' meiosis, which causes most genetic material to be inherited from the seed parent.

Evolution

It is highly unlikely that the peculiar '*canina*' meiosis has evolved more than once, and several theories have been developed to explain its evolution. The earliest was the so called septet theory of Hurst (1925). According to this theory, the genus *Rosa* evolved from a common northern decaploid ancestor which had five different diploid septets of chromosomes and each gave rise to specific morphological characters. The septets segregated as complete units, with the bivalent formation taking place between similar septets. When the species started to spread southward and the environment became easier to live in, there was no need to keep all the septets and, consequently, some were lost. The species within the *Caninae* section would have resulted from hybridization between species with only one septet in common, thereby giving rise to the odd meiosis.

Another theory is that this unique meiosis originated in a small isolated population of *R. canina*, or rather a predecessor thereof. Long-lasting inbreeding in this population first led to an increased homozygosity and eventually to the appearance of individuals with a recessive '*canina*' gene. With time, individuals carrying the recessive gene came together and produced one individual, which was homozygous in two genomes, which led to the '*canina*' meiosis. The occurrence of self-pollination in this individual kept it isolated from the rest of the population and its descendants were able to continue their existence when the original population degenerated due to inbreeding. As the new form started to expand, it also hybridized with other normal *Rosa* species, giving rise to the hybrid swarm we see today. The polymorphism of the NOR sites together with the preferential pairing of genomes and lack of multivalent formations are indicative of an allopolyploid origin of the section, a hypothesis that has been mentioned in several studies.

Most of the modern dog rose species occupy areas in Europe which were covered with ice during the ice age. The diploid ancestors of this section probably became extinct during this period and the hybrid complex we have presently has evolved since that time. No matter what the origin was, the meiosis must have conferred some sort of selective advantage in order for it to survive and proliferate for such a long time. It has been called a 'blind alley', but being a 'facultative sexual plant' it also has the benefit of two selective strategies: (1) the apomictic transmission of fit and vigorous alleles in the univalents together with selfing, making it easy to colonize new areas without being dependent upon separate pollen donors; and (2) the possibility to cross-pollinate, change the genetic constitution and lose disadvantageous alleles, should the external conditions change.

See also: **Biotechnologies for Breeding:** Early Embryo Rescue. **Breeding:** Methods of Cross Breeding. **Classification:** Conventional Taxonomy (Wild Roses). **Genetics:** Meiosis; Inheritance of Characteristics; Species Crosses; DNA Fingerprinting. **History of Roses in Cultivation:** European (Pre-1800). **Insects and other Animals**: Pollination. **Rootstocks:** Usage of Rootstocks; Clonal Rootstocks; Seedling Stocks. **Secondary Metabolites:** Culinary Uses and Nutritional Value; Medicinal and Pharmaceutical Uses.

Further Reading

Asker S and Jerling L (1992) *Apomixis in Plants*. Boca Raton, FL: CRC Press.

Graham GG and Primavesi AL (1993) *Roses of Great Britain and Ireland*, BSBI Handbook no 7. London: Botanical Society of the British Isles.

Grant V (1971) *Plant Speciation*. New York: Columbia University Press.

Hurst CC (1925) Chromosomes and characters in *Rosa* and their significance in the origin of species. *Experiments in Genetics* 38: 534–550.

Krüssmann G (1981) *Roses*. Portland, OR: Timber Press.

Olsson Å (1999) Morphometric and molecular variation in the Nordic dog roses (*Rosa* L. Sect. *Caninae*, Rosaceae). Doctoral thesis. Sweden: Department of Systematic Botany, Lund University.

Werlemark G (2001) Genetic variablity and reproductive strategies in Nordic dogroses, *Rosa* section *Caninae*. Doctoral thesis. Department of Horticultural Plant Breeding, Swedish University of Agricultural Sciences, Sweden.

Wylie AP (1976) Why the *Caninae* roses are different. *New Zealand Rose Annual* 67–79.

Species Crosses

W Spethmann, Section Tree Nursery Science, University of Hanover, Germany
B Feuerhahn, Berlin, Germany

Introduction

Every year rose breeders carry out hundreds of thousands of crosses, mainly amongst improved garden roses, to get new cultivars with improved character traits. These crosses are not difficult because, as most garden cultivars are tetraploid, the crosses are compatible. Usually, the hybrid seed of these garden cultivars germinate in the following year without showing seed dormancy. Furthermore, the seedlings develop flowers very early, often in the first year.

In the main lines of rose breeding, only 5–10% of the existing rose species have been included. To breed a new cultivar hundreds or thousands of seedlings have to be tested. Nevertheless, differences from existing cultivars are often small because the gene pool of the cultivars has become uniform, innovative character traits are rare and genotypes with dependable resistance to disease are not found at all.

The following species crosses were involved in the origin of some of the old rose classes: *Rosa gallica* × *R. moschata* led to the Damask roses, *R. chinensis* × *R. moschata* led to the Noisette roses, *R. rugosa* × *R. wichurana* Crép. were the parents of one of the first ground-cover roses and *R. multiflora* was used in breeding the Polyantha roses.

More than 100 rose species have yet to be integrated into breeding programmes. Resistance to fungal diseases is the most important breeding aim, but there are also other innovations that could be realized by breeding with species. With potatoes, cucumbers and grain, resistance character traits were introduced from wild species. In the genus *Rhododendron*, new, important hybrids were developed by including single species (*Rh. yakushimanum*, *Rh. forrestii* var. *repens*) in the breeding programmes.

Why do rose breeders not venture to breed with species? The arguments are manifold: rose species range from diploid to octoploid, therefore, crosses of cultivars to species of differing ploidy are rarely successful; hybrid seeds show seed dormancy, therefore, germination starts in the second year; the time of the first flowering is much later than in cultivar crosses, often after 5–6 years. To breed new successful cultivars, the breeding programmes would have to be continued over many years. Therefore, breeding programmes with species are usually enormously time-consuming and expensive.

Species naturally cross only where the areas of two species overlap. Many species in gardens and parks, which do not grow together in nature, produce many hybrids (for example, *R. rugosa* × *R. canina*). In the UK alone, 81 hybrids were named. Probably many species hybrids are not recognized, because many hybrids, especially with species of the *Caninae*, look like the seed parent. Spontaneous hybrids with garden cultivars are rare.

In this article, the experiences of crosses between roses species are described, difficulties are discussed and methods are described in detail.

Species Plants

Cross-fertility amongst roses is complicated by considerations of taxonomic distance and differences in ploidy (*see* **Biotechnologies for Breeding**: Manipulation of Ploidy Level). For each breeding trial an extensive knowledge of the material is needed, including the morphological and cytological characteristics of the parents as mother or as father plants. It is recommended to label each parent plant so that it is possible to trace the progeny to the parentage. Tolerance of and resistance to black spot disease (caused by *Diplocarpon rosae*) (*see* **Disease**: Black Spot) and powdery mildew (*Podosphaera pannosa*) (*see* **Disease**: Powdery Mildew) play an important role in rose breeding. Plants included in the breeding programme must be characterized for resistance to disease.

Pollen

To obtain the pollen it is necessary to harvest closed flower buds to prevent contamination with pollen from other species. The petals must be closed but show colour and the sepals must be partly opened. At this point the pollen is not completely ripe and must be afterripened by dehydration. All petals must be removed and all stamens taken from the flower. The stamens can be stored in Petri dishes and dehydrated in a desiccator over silica gel at room temperature. The humidity should be kept below 70% to prevent fungal attack. The freshest pollen is recommended for pollination. The pollen of several flowers from one plant can be used in the crossing experiments.

The time of flowering varies between rose species. For crossing an early-flowering with a late-flowering species it is necessary to store pollen. The harvested pollen must be sifted and placed in glass bottles which are impermeable to light. Pollen can be kept viable by storage near 0 °C. The duration of storage can be extended by deep freezing (−20 °C). However, regardless of storage conditions, the germination capacity of pollen declines with time. Before using the pollen in crossing experiments its viability should be tested by germination tests. Frozen pollen must be thawed and fresh pollen can be used immediately. Pollen grains are spread on a culture medium which contains 0.7% agar, 10% sucrose and 100 p.p.m. boric acid at pH 5.6. The germinated pollen grains can be counted after 24 h. All pollen tubes that are three times longer than the diameter of the pollen grain count as germinated. The pollen germination is complete after a further 3 h.

In the literature on roses, information about germination ability is little and varied. Not all breeders test for germination before crossing because it is expensive and time-consuming. Alternatives are possible, which give slightly different information. For example, it is possible to assess the viability of pollen as the proportion of grains that stain in tritetrazoliumchloride. But many scientists prefer the previously mentioned germination test.

Pollen germination rates differ from species to species and sometimes even from plant to plant within species, so the results of different breeders vary (**Table 1**). Breeders all agree that the quality of pollen varies between years and within the year between different harvesting points.

Species of the section *Caninae* have a lower germination ability than species of other sections and breeders often classified pollen of these species as 'bad pollen'. The low germination rate can be explained by the complicated cytology of the species of this section and the unusual events leading to gamete formation (*see* **Genetics**: Inheritance in the Dogrose).

A relationship between the number of chromosome sets and the fertility of pollen has been shown, among others, by experiments at the University of Hanover (Germany). The diploid species show the highest average pollen germination and the pentaploid species the lowest.

The quality of pollen is one key to the development of fruits after pollination. There is a positive correlation between the pollen fertility of the pollination parent and the growth of rose hips at the seed parent. The level of the correlation depends on the species included. In the literature, pollen with a germination ability of over 25% is characterized as fertile. Negligible development of rose hips will be achieved if the pollen germination rate is below 0.5%. Experiments at the University of Hanover show that pollen with germination ability below 25% will also give good fruit sets. So there must be other factors that influence crossing success, as discussed below.

Table 1 Records of pollen germination or fertility

Ploidy	*Species*	*Germination/viability (%)*	*Reference*
Diploid	*R. arvensis*	75	Cole (1917)
		75–95	Keller and Gams (1923)
		98–99	Keller (1931)
		55	Feuerhahn (1994)
	R. multiflora	80	von Rathlef (1937)
		85	Feuerhahn (1994)
	R. nitida	77	Cole and Melton (1986)
		73–87	Feuerhahn (1994)
	R. rugosa	100	Kroon and Zeilinger (1974)
		75	Cole and Melton (1986)
	R. rugosa 'Alba'	51–65	Feuerhahn (1994)
		60	Cole (1917)
		55–76	Feuerhahn (1994)
Tetraploid	*R. gallica*	50	Keller (1931)
		19–23	Feuerhahn (1994)
	R. mollis	40	Cole (1917)
		40	von Rathlef (1937)
		20	Feuerhahn (1994)
	R. pimpinellifolia	58–90	Cole (1917)
		25–95	Keller and Gams (1923)
		95–98	Keller (1931)
		98	Kroon and Zeilinger (1974)
		51–85	Roberts (1977)
		87–94	Pearson and Harney (1984)
		52	Feuerhahn (1994)
Pentaploid	*R. agrestis*	37	Keller (1931)
		31	Feuerhahn (1994)
	R. canina	37	Keller (1931)
		8	Feuerhahn (1994)
	R. corymbifera	36	Keller (1931)
		50–70	von Rathlef (1937)
		14	Feuerhahn (1994)
	R. micrantha	40	Keller (1931)
		12	Feuerhahn (1994)
	R. rubiginosa	16	Täckholm (1920)
		60	Keller (1931)
		0	von Rathlef (1937)
		6	Cole and Melton (1986)
		44–59	Feuerhahn (1994)
Hexaploid	*R. acicularis*	85–90	Cole (1917)
		83–100	Erlanson (1930)
		70–100	von Rathlef (1937)
		58	Feuerhahn (1994)
	R. jundzillii	50	Keller (1931)
		7	Feuerhahn and Spethmann (1995)
	R. moyesii	90	von Rathlef (1937)
		47	Feuerhahn (1994)

Cole RD (1917) *Botanical Gazette* 63: 110–123.
Cole P and Melton B (1986) *Journal of the American Society for Horticultural Science (USA)* 111(1): 122–125.
Erlanson EW (1930) *Genetics* 16: 76–95.
Feuerhahn B (1994) *Diplomarbeit.* Hanover: Institut für Obstbau und Baumschule der Universität Hanover.
Feuerhahn B and Spethmann W (1995) *Gehölzforschung*, vol. 3. Universität Hanover: Institut für Obstbau und Baumschule, Fachbereich Gartenbau.
Keller R (1931) *Denkschriften der Schweizerischen Naturforschenden Gesellschaft* 65: 23–27.
Keller R and Gams H (1923) *Illustrierte Flora von Mitteleuropa*, vol. IV/2, pp. 976–1052. München: HEGI.
Kroon GH and Zeilinger AE (1974) *Euphytica* 23: 345–352.
Pearson HM and Harney PM (1984) *HortScience* 19(5): 710–711.
Roberts AV (1977) *Botanical Journal of the Linnean Society* 74: 309–328.
Täckholm G (1920) *Svensk Botanisk Tidsskrift* 14(2–3): 301–311.
von Rathlef H (1937) *Die Rose als Objekt der Züchtung.* Jena: Fischer.

Crosses

Methods of Crossing

Among other things, the success of crossing depends on pollination and fertilization at the right time. The stigma is ready for pollination if it shows a shiny secretion. At this time the petals are slightly coloured, the sepals are spread out and the flower's own pollen has no pollination ability. Exceptions are *R. nitida* and *R. rugosa* which can self-pollinate at this stage. To avoid self-pollination it is necessary to prepare the stigma for cross-fertilization when the flower buds are still green, before the petals develop colour and before the sepals have spread.

The anthers must be removed, so the sepals, petals and filaments must be carefully separated. Pollination can be carried out with a pencil. It is recommended to pollinate several flowers with the same pollen. To increase the crossing success and the fruit set it is possible to repeat the pollination of the same flower. When the crosses are carried out outside, it is necessary to put paper bags over the pollinated stigmas to protect them from insects. Every cross must be labelled showing the seed and pollen parent. It is possible to emasculate, pollinate and isolate on the same day. When there is no secretion on the stigma, emasculation and isolation take place on one day and pollination on the following.

To test for self-compatibility, closed flower buds must be isolated. Self-compatibility is indicated if fruits are set. Apomixis (reproduction without fertilization) is indicated if fruit set occurs after emasculation (*see* **Genetics**: Inheritance in the Dogrose).

Success of Crossing

In the literature, various criteria of crossing success are used. One criterion is that crossing is successful if the progeny shows characteristics of both parents. Another is the fertility of the hybrids. Only hybrids which can produce a progeny can signify a successful crossing. The confirmation of hybridity, particularly by vegetative characteristics, may be difficult if the parent species are similar. For example, the hybrids *R. micrantha* × *R. rubiginosa* and *R. mollis* × *R. villosa* are difficult to distinguish from the parents. Floral, cytological and molecular characteristics are usually more dependable. In hybrids between some species, flower production takes place in the second or later years of life. It may therefore be necessary to observe the progeny for more than 5 years.

The recognition of hybrid characters is rendered more difficult if one of the parent species belongs to the section *Caninae*. In species of this section, more chromosomes are transmitted by the female than the male parent and apomixis sometimes occurs (*see* **Genetics**: Inheritance in the Dogrose).

Hybrids between two species of the section *Caninae* more closely resemble the seed parent, from which they inherited more chromosomes (matroclinal inheritance). This is most apparent in the first years of life.

Another pecularity which complicates the recognition of hybrids is apomixis (reproduction without fertilization). Until recently, the occurrence of apomixis in roses has been disputed but there is now convincing evidence of its occurrence (*see* **Genetics**: Inheritance in the Dogrose). One test that could be carried out for apomixis in a seed parent is to pollinate with mentor pollen. This is usually irradiated pollen of the mother species that is capable of pollination but not fertilization (*see* **Biotechnologies for Breeding**: Manipulation of Ploidy Level). Some species need pollination for the development of embryos but not unconditional fertilization (pseudogamy).

That fertility is not the only criterion for successful crossing can be shown by crossing of *R. rugosa* and *R. wichurana* Crép. The offspring *R. hybrida* 'Max Graf' was diploid and sterile. After repeated self-pollination, a fertile tetraploid seedling was produced, called *R.* × *kordesii*. Another fertile tetraploid species was obtained from crossing the diploid species *R. abyssinica* and *R. rugosa*. Spontaneous chromosome doubling makes it possible to get fertile species hybrids from formerly sterile ones. This can also be carried out artificially (*see* **Biotechnologies for Breeding**: Amphidiploidy). Thus, induced tetraploids can be crossed with garden roses, leading to a new diversity of shape.

Definition of an unsuccessful cross needs proof of repeatability. Because there are differences between the years in the success of crosses, it is recommended to repeat the cross several times. Different years and locations affect osmotic relations and, thus, the different activity of pollen tubes and embryos. Several years of observation are necessary to make concrete statements about cross-compatibility.

The literature gives many examples of successful and unsuccessful crosses between rose species. **Tables 2–4** give a survey about successful crosses that does not pretend to be comprehensive.

Ploidy level and crossing success Reports on the relationship between the ploidy levels of seed and pollen parents on crossing success differ. Some experiments showed an influence of ploidy level caused by the quantitative genome differences between the crossing partner, while others did not. It is difficult to compare all results because they were not all carried out with the same parents. The influence of ploidy is

Table 2 Published artificial interspecific hybrids and their origins

Seed parent[a]	*Pollen parent*	*Progeny denotation*	*Reference*
R. acicularis (28, 42, 56)	*R. glauca* (28)	—	Fagerlind (1944)
R. arvensis (14)	*R. arvensis* (14)	—	Fagerlind (1944)
	R. chinensis (14, 21, 28)	*R.* × *ruga* Lindl.	Krüssmann (1986)
	R. gallica (28)	*R.* × *polloniana* Spreng.	Fagerlind (1944)
	R. moyesii (42)	—	Fagerlind (1944)
	R. multiflora (14, 28)	—	Ratzek *et al.* (1939)
	R. rugosa (14)	*R.* × *paulii* Rehd.	Uilenburg (1985)
R. banksiae (14)	*R. laevigata* (14)	*R.* × *fortuniana* Lindl.	Krüssmann (1986)
R. blanda (14)	*R. chinensis* (14, 21, 28)	*R.* × *aschersoniana* Graebn.	Krüssmann (1986)
	R. rugosa (14)	*R.* × *warleyensis* Willm.	Krüssmann (1986)
R. californica (28)	*R. nitida* (14)	*R.* × *scharkeana* Graebn.	Krüssmann (1986)
R. canina (35, 42)	*R. gallica* (28)	*R.* × *waitziana* Tratt.	Fagerlind (1944)
	R. glauca (28)	*R.* × *pokornyana* Borb.	Persiel (1974)
	R. moyesii (42)	—	Kroon and Zeilinger (1974)
	R. multiflora (14, 28)	—	Wulff (1952)
	R. nitida (14)	—	Kroon and Zeilinger (1974)
	R. pimpinellifolia (28)	*R.* × *hibernica* Templeton	Fagerlind (1944)
	R. rubiginosa (35, 42)	*R.* × *nitidula* Besser	Gustaffson (1944)
	R. rugosa (14)	*R.* × *praegeri* W.-Dod	Gustaffson (1944)
R. carolina (28)	*R. rugosa* (14)	*R.* × *koehneana* Rehd.	Krüssmann (1986)
R. chinensis (14, 21, 28)	*R. damascena* (28)	*R.* × *borboniana* Desp.	Krüssmann (1986)
	R. gigantea (14)	*R. odorata*	Krüssmann (1986)
	R. moschata (14)	*R. noisettiana* Thory	Krüssmann (1986)
	R. multiflora (14, 28)	*R. polyantha* Carr.	Krüssmann (1986)
	R. rugosa (14)	*R.* × *calocarpa* (André) Willm.	Krüssmann (1986)
R. corymbifera (35)	*R. gallica* (28)	*R.* × *collina* Jasq.	Krüssmann (1986)
R. davorica (14)	*R. rugosa* (14)	*R.* × *kamtchatica* Vent.	Krüssmann (1986)
R. gallica (28)	*R. agrestis* (35, 42)	—	Fagerlind (1944)
	R. arvensis (14)	*R. axmanni* Gmelin	Fagerlind (1944)
	R. gallica (28)	—	Fagerlind (1944)
	R. glauca (28)	—	Fagerlind (1944)
	R. glutinosa (35, 42)	—	Fagerlind (1944)
	R. majalis (28)	*R.* × *francofurtana* Muenchh.	Täckholm (1920)
	R. moschata (14)	*R.* × *bifera* Pers.	Krüssmann (1986)
	R. multiflora (14, 28)	—	Fagerlind (1948)
	R. pendulina (28)	—	Fagerlind (1944)
	R. phoenicia (14)	*R.* × *richardii* Rehd.	Krüssmann (1986)
	—	*R.* × *damascena* Mill.	Krüssmann (1986)
	R. pimpinellifolia (28)	—	Fagerlind (1944)
	R. rugosa (14)	—	Hurst (1928)
	R. villosa (28)	—	Fagerlind (1944)
R. glauca (28)	*R. acicularis* (28, 42, 56)	—	Fagerlind (1944)
	R. gallica (28)	—	Fagerlind (1944)
	R. glauca (28)	—	Fagerlind (1944)
	R. glutinosa (35, 42)	—	Fagerlind (1944)
	R. multiflora (14, 28)	—	Fagerlind (1944)
	R. pimpinellifolia (28)	—	Fagerlind (1944)
	R. rubiginosa (35, 42)	—	Gustaffson (1942)
	R. rugosa (14)	*R.* × *rubrosa* Preston	Fagerlind (1944)
R. hugonis (14)	*R. sericea* (14)	*R.* × *pteragonis* Krause	Krüssmann (1986)
	R. xanthina (14)	*R. hybrid* 'Canary Bird'	Krüssmann (1986)
R. jundzillii (42)	*R. gallica* (28)	—	Klasterska and Natarajan (1974)
	R. jundzillii (42)	—	Jicinska (1975)
	R. rugosa (14)	—	Fagerlind (1944)
R. laevigata (14)	*R. odorata* (14)	*R.* × *anemonoides* Rehd.	Krüssmann (1986)
R. macrophylla (14)	*R. roxburghii* (14)	*R. coryana* Hurst	Krüssmann (1986)

Table 2 Continued

Seed parent[a]	*Pollen parent*	*Progeny denotation*	*Reference*
R. majalis (28)	*R. gallica* (28)	*R. hybrid* 'Francofurtana'	Krüssmann (1986)
	R. majalis (28)	—	Jicinska (1975)
	R. multiflora (14, 28)		Anonymous (1988)
R. mollis (28)	*R. villosa* (28)	—	Gustaffson (1942)
R. moschata (14)	*R. gallica* (28)	*R. bifera* Pers.	Krüssmann (1986)
	R. multiflora (14, 28)	*R. hybrid*, 'The Garland' (Wells)	Krüssmann (1986)
R. moyesii (42)	*R. acicularis* (28, 42, 56)	—	Fagerlind (1948)
	R. moyesii (42)	—	Fagerlind (1948)
	R. multiflora (14, 28)	—	Fagerlind (1944)
	R. pimpinellifolia (28)	—	Fagerlind (1948)
	R. rugosa (14)	—	Fagerlind (1944)
	R. setipoda (28)	*R.* × *wintonensis* Hillier	Krüssmann (1986)
	R. willmottiae (14)	*R.* × *pruhoniciana* Kriechbaum	Krüssmann (1986)
R. multiflora (14, 28)	*R. acicularis* (28, 42, 56)	—	Fagerlind (1948)
	R. arvensis (14)	—	Fagerlind (1948)
	R. gallica (28)	—	Fagerlind (1948)
	R. glauca (28)	—	Persiel (1974)
	R. moyesii (42)	—	Fagerlind (1944)
	R. multiflora (14, 28)	—	Fagerlind (1948)
	R. rugosa (14)	—	Paris and Maney (1939)
R. nitida (14)	*R. gallica* (28)	—	Ratzek *et al.* (1941)
	R. rugosa (14)	*R.* × *rugotida* Darthuis.	Ratzek *et al.* (1941)
R. palustris (14, 28)	*R. rugosa* (14)	*R.* × *spaethiana* Graebn.	Krüssmann (1986)
	R. virginiana (28)	*R.* × *mariae-graebnerae* Aschers. & Graebn.	Krüssmann (1986)
R. pendulina (28)	*R. pimpinellifolia* (28)	*R.* × *reversa* Waldst. & Kit	Krüssmann (1986)
	R. tomentosa (35)	*R.* × *spinulifolia* Dematra	Krüssmann (1986)
R. pimpinellifolia (28)	*R. acicularis* (28, 42, 56)	*R. oxyacantha*	Vogel (1937)
	R. carolina (28)	*R.* × *kochiana* Koehne	Krüssmann (1986)
	R. micrantha (35, 42)	—	Fagerlind (1948)
	R. mollis (28)	*R.* × *sabinii* Woods	Fagerlind (1948)
	R. pimpinellifolia (28)	—	Fagerlind (1948)
R. pratii (14)	*R. rugosa* (14)	*R. hybrid* 'Prattigosa'	Krüssmann (1986)
R. roxburghii (14)	*R. rugosa* (14)	*R.* × *micrugosa* Henkel	Krüssmann (1986)
R. rubiginosa (35, 42)	*R. glauca* (28)	—	Gustaffson (1942)
	R. harisonii (28)	*R. hybrid* 'Lucy Ashton'	Krüssmann (1986)
	R. multiflora (14, 28)	—	Blackhurst (1948)
	R. rugosa (14)	—	Gustaffson (1942)
R. rugosa (14)	*R. acicularis* (28, 42, 56)	—	Fagerlind (1944)
	R. arvensis (14)	*R.* × *paulii* Rehd.	Fagerlind (1944)
	R. damascena (28)	*R. hybrid* 'Atropurpurea'	Krüssmann (1986)
	R. gallica (28)	—	Fagerlind (1944)
	R. glauca (28)	—	Fagerlind (1944)
	R. glutinosa (35, 42)	—	Fagerlind (1944)
	R. harisonii (28)	*R. hybrid* 'Agnes Emily Carman'	Krüssmann (1986)
	R. jundzillii (42)	—	Fagerlind (1944)
	R. multiflora (14, 28)	—	Fagerlind (1944)
	R. nitida (14)	*R.* × *rugotida* Darthuis	Noack (1989)
	R. pimpinellifolia (28)	—	Fagerlind (1944)
	R. rubiginosa (35, 42)	—	Fagerlind (1944)
	R. rugosa (14)	—	Fagerlind (1944)
	R. sherardii (28, 35)	—	Jicinska (1975)
	R. tomentosa (35)	—	Fagerlind (1944)
	R. villosa (28)	—	Fagerlind (1948)
	R. wichurana (14)	*R.* × *jacksonii* Willm.	Krüssmann (1986)

Table 2 Continued

Seed parent[a]	*Pollen parent*	*Progeny denotation*	*Reference*
R. stylosa (28, 35, 42)	*R. acicularis* (28, 42, 56)	—	Fagerlind (1944)
	R. rugosa (14)	—	Fagerlind (1944)
R. villosa (28)	*R. acicularis* (28, 42, 56)		Fagerlind (1944)

[a] Chromosome numbers (in parentheses) are based on the published $2n$ value for each species.

Anonymous (1988) *Rosenverzeichnis Rosarium Sangerhausen*, 4th edn. Sangerhausen: Mitteldeutsches Druckhaus.
Blackhurst HF (1948) *Proceedings of the American Society of Horticultural Science* 52: 510.
Fagerlind F (1944) *Acta Horti Bergiani* 13: 247–302.
Fagerlind F (1948) *Acta Horti Bergiani* 15: 1–37.
Gustaffson A (1942) *Hereditas* 28: 235–238.
Gustaffson A (1944) *Hereditas* 30: 405–428.
Hurst CC (1928) Verhandlungen des V. internationalen Kongresses für Vererbungswissenschaft in Berlin 1927. Supplementband II: 866–906.
Jicinska D (1975) *Preslia* 47: 267–274.
Klasterska AJ and Natarajan AT (1974) *Hereditas* 76(1): 97–108.
Kroon GH and Zeilinger AE (1974) *Euphytica* 23: 345–352.
Krüssmann G (1986) *Rosen, Rosen Rosen*, 2nd edn. Berlin: Paul Parey.
Noack H (1989) *Wild- und Parkrosen.* Melsungen: Neumann-Neudamm.
Paris CD and Maney TJ (1939) *Journal of the Iowa Academy of Science* 46: 149–160.
Persiel F (1974) *Acta Pruhoniciana* 32: 58–66.
Ratzek JC, Yarnell SH and Flory WS (1939) *Proceedings of the American Society of Horticultural Science* 37: 983–992.
Ratzek JC, Yarnell SH and Flory WS (1941) *Proceedings of the American Society of Horticultural Science* 38: 637–654.
Täckholm G (1920) *Svensk Botanisk Tidsskrift* 14(2–3): 301–311.
Uilenburg NG (1985) *Groen* 41(3): 26–29.
Vogel M (1937) *Rosenjahrbuch* (*Verein Deutscher Rosentreunde*) 1937: 23–25.
Wulff HD (1952) *Der Züchter* 22: 233–244.

apparent if you look at the number of seedlings in the progeny.

Except for crosses in which the seed parent is diploid, the highest percentages of seed germination are obtained in crosses between parents of the same ploidy (**Table 5**). The cause is the so-called eu-incompatibility. This is a kind of gametophytic incompatibility, which is determined genetically and controlled by multiple alleles. The growth of the pollen tube is stopped and there is no formation of a zygote. A good hip set but fewer seedlings result from pseudoincompatibility. The egg cell is fertilized but the embryo degenerates. The cause of degeneration are differences in ploidy between the crossing partners. The mortality increases with higher ploidy differences and is caused by developmental disorders of embryo and endosperm.

The pentaploid species are good seed parents. Most belong to the section *Caninae*, which forms haploid pollen grains (*see* **Genetics:** Inheritance in the Dog-rose). These species behave like diploids with regard to compatibility. Sometimes they are superior to diploids. This leads to the assumption that the choice of individuals is more important in diploid than pentaploid species.

The formation of hybrids depends on the ploidy level of the parents. After crosses between two diploid species, the progeny can be up to 100% hybrid. If there are differences between the ploidy level, hybrids will be formed only after some crosses. But diploid species are more cross- than self-compatible; polyploids incline to self-fertilization (**Table 6**).

Taxonomic section and crossing success It is relatively simple to cross species which belong to the same section. Crosses between species from different sections are possible but not always successful. This can be attributed to qualitative genome differences and depend on the species.

The species of the section *Synstylae* are better pollen parents than seed parents, as confirmed by reciprocal crosses. In spite of the specialized cytology of species of the section *Caninae*, it is possible to cross them successfully with species of other sections. So, for example, *R. canina*, *R. rubiginosa* and *R. sherardii* can be crossed as female parents with *R. rugosa*. However, the progeny look more like the *Caninae* seed parent. The taxonomic relationship between species plays an important role in crossing ability.

Autogamy Some species of roses incline towards autogamy (self-pollination). In some roses, pollination can take place even if the flower is closed. In an autogamy trial, wild roses were divided into three classes. The first class, consisting of *R. multiflora*, *R. nitida* and *R. rugosa*, was self-compatible. The second class, consisting of *R. jundzillii*, *R. rubiginosa* and *R. glauca*, was both self- and cross-compatible. The third class, consisting of *R. acicularis*, *R. canina* and *R. pimpinellifolia*, was only cross-compatible. It is possible to encourage self-fertilization in species of the first and second classes by isolating closed flowers. Experiments at the University of Hanover showed that this is possible with *R. agrestis*, *R. glauca*, *R. glutinosa*, *R. jundzillii*, *R. mollis*, *R. obtusifolia*, *R. rubiginosa*,

Table 3 Interspecific hybrids raised by Feuerhahn (1994) and their origins of showing resistance to *Diplocarpon rosae* (black spot disease), *Podosphaera pannosa* (powdery mildew) and cold. Resistance is shown on a scale of 1 (susceptible) to 4 (resistance) by Menke and Unger (1999)

Seed parent[a]	*Pollen parent*	*Progeny classification*[b]	*Resistance*		
			D. rosae	P. pannosa	*Cold*
R. acicularis (28, 42, 56)	*R. majalis* (28)	Seed	2	3	4
	R. pimpinellifolia (28)	Intermediate	2	3	4
R. agrestis (35, 42)	*R. gallica* (28)	Intermediate	1	3	4
R. arvensis (14)	*R. gallica* (28)	Seed	2	2	4
	R. rugosa (14)	Intermediate	2	2	2
R. canina (35, 42)	*R. acicularis* (28, 42, 56)	Seed	2	3	4
	R. glutinosa (35, 42)	Seed	3	2	4
	R. jundzillii (42)	Pollen	3	2	4
	R. mollis (28)	Pollen	2	2	4
	R. pimpinellifolia (28)	Intermediate	2	3	4
	R. rubiginosa (35, 42)	Pollen	2	2	4
	R. rugosa (14)	Intermediate	2	2	4
	R. stylosa (28, 35, 42)	Pollen	2	2	3
R. gallica (28)	*R. gallica* (28)	—	2	2	4
	R. majalis (28)	Intermediate	3	1	2
	R. multiflora (14, 28)	Intermediate	3	2	3
	R. nitida (14)	Intermediate	4	1	4
	R. villosa (28)	Intermediate, pollen	3	2	2
R. glauca (28)	*R. acicularis* (28, 42, 56)	Intermediate	2	3	4
	R. nitida (14)	Intermediate	2	3	4
	R. pimpinellifolia (28)	Intermediate	2	2	4
	R. rubiginosa (35, 42)	Seed	1	3	4
	R. rugosa (14)	Intermediate	2	2	4
R. glutinosa (35, 42)	*R. pimpinellifolia* (28)	Seed	1	2	3
R. jundzillii (42)	*R. acicularis* (28, 42, 56)	Intermediate	2	4	4
	R. majalis (28)	Intermediate	2	2	3
	R. pimpinellifolia (28)	Seed	2	3	4
R. majalis (28)	*R. acicularis* (28, 42, 56)	Pollen	1	3	3
	R. pimpinellifolia (28)	Intermediate	2	3	2
	R. rugosa (14)	Intermediate	3	4	4
	R. sherardii (28, 35)	Seed	1	4	3
R. micrantha (35, 42)	*R. arvensis* (14)	Intermediate	1	3	3
	R. gallica (28)	Seed	2	2	3
	R. glutinosa (35, 42)	Seed	2	2	3
	R. micrantha (35, 42)	Seed	2	2	3
	R. mollis (28)	Seed	2	2	3
	R. multiflora (14, 28)	Seed	2	2	3
	R. rubiginosa (35, 42)	Pollen	2	3	3
	R. tomentosa (35)	Intermediate	2	2	4
R. mollis (28)	*R. acicularis* (28, 42, 56)	Intermediate	Dead	—	—
	R. arvensis (14)	Intermediate	2	3	3
	R. gallica (28)	Intermediate	1	2	4
	R. glauca (28)	Seed	1	3	4
	R. glutinosa (35, 42)	Pollen	2	4	4
	R. majalis (28)	Intermediate	1	3	4
	R. multiflora (14, 28)	Intermediate	Dead	—	—
	R. nitida (14)	Intermediate	1	3	1
	R. pimpinellifolia (28)	Seed	1	2	4
	R. rubiginosa (35, 42)	Seed	2	3	4
	R. rugosa (14)	Intermediate	2	3	4
	R. villosa (28)	Intermediate	2	1	4
R. multiflora (14, 28)	*R. arvensis* (14)	Seed	2	3	3
	R. multiflora (14, 28)	—	2	3	3
R. nitida (14)	*R. corymbifera* (35)	Seed	2	2	1
	R. glauca (28)	Intermediate	2	3	4
	R. nitida (14)	—	3	3	4
	R. pimpinellifolia (28)	Seed	3	3	4
	R. rubiginosa (35, 42)	Seed	2	3	4

Table 3 Continued

Seed parent[a]	*Pollen parent*	*Progeny classification*[b]	*Resistance*		
			D. rosae	P. pannosa	*Cold*
	R. sherardii (28, 35)	Seed	1	3	4
	R. villosa (28)	Seed	2	3	3
R. obtusifolia (35)	*R. gallica* (28)	Seed	1	2	4
	R. majalis (28)	Intermediate	1	2	3
	R. sherardii (28, 35)	Seed	1	2	4
R. pimpinellifolia (28)	*R. acicularis* (28, 42, 56)	Intermediate	1	3	4
	R. pimpinellifolia (28)	—	2	2	4
R. rubiginosa (35, 42)	*R. gallica* (28)	Seed	2	2	4
	R. glauca (28)	Intermediate	2	3	3
	R. glutinosa (35, 42)	Seed	2	2	2
	R. jundzillii (42)	Seed	2	3	3
	R. majalis (28)	Seed	2	4	2
	R. mollis (28)	Seed	2	2	4
	R. moyesii (42)	Seed	2	2	4
	R. multiflora (14, 28)	Intermediate	1	2	3
	R. nitida (14)	Intermediate	1	4	2
	R. rubiginosa (35, 42)	—	1	3	4
	R. rugosa (14)	Intermediate	2	3	3
	R. sherardii (28, 35)	Intermediate	2	2	2
	R. tomentosa (35)	Intermediate	1	3	3
R. rugosa (14)	*R. arvensis* (14)	Intermediate	2	3	2
	R. gallica (28)	Intermediate	2	3	4
	R. glauca (28)	Intermediate	2	3	4
	R. glutinosa (35, 42)	Intermediate	3	3	4
	R. mollis (28)	Seed	2	3	4
	R. multiflora (14, 28)	Intermediate	2	4	4
	R. nitida (14)	Intermediate	2	4	4
	R. obtusifolia (35)	Intermediate	3	3	2
R. sherardii (28, 35)	*R. gallica* (28)	Seed	1	2	4
	R. glauca (28)	Intermediate	Dead	—	—
	R. jundzillii (42)	Intermediate	1	1	4
	R. majalis (28)	Intermediate	1	3	3
	R. nitida (14)	Intermediate	1	2	1
	R. obtusifolia (35)	Seed	1	2	4
	R. rubiginosa (35, 42)	Intermediate	2	2	4
R. stylosa (28, 35, 42)	*R. gallica* (28)	Intermediate	Dead	—	—
	R. nitida (14)	Intermediate	2	4	3
	R. obtusifolia (35)	Seed	3	3	3
	R. rubiginosa (35, 42)	Pollen	3	3	3
	R. villosa (28)	Intermediate	3	3	4
R. villosa (28)	*R. acicularis* (28, 42, 56)	Intermediate	2	2	4
	R. majalis (28)	Seed	2	3	3
	R. nitida (14)	Intermediate	1	2	2

[a] Chromosome numbers (in parentheses) are based on published records for the species.
[b] Progenies are classified as being most similar to the pollen parent, seed parent or intermediate.
Feuerhahn, B (1994) *Diplomarbeit*. Hanover: Institut für Obstbau und Baumschule der Universität Hanover.
Menke O and Unger T (1999) *Diplomarbeit*. Hanover: Institut für Obstbau und Baumschule der Universität Hanover.

R. sherardii, *R. villosa*, *R. gallica* and *R. acicularis*. The results can differ because the choice of individuals is very important, and artificial selfing of two individuals of the same species can give different results. The negative results after artificial selfing can be explained by gametophytic and sporophytic incompatibility. Finally the pollen cannot germinate when an enzyme needed to break through the cuticle is absent.

Gametophytic and sporophytic incompatibility is mentioned by Stougaard after selfing of various species of roses. So it is possible to use this term in connection with pollen ploidy and selfing.

Harvesting of Seed and Seed Treatment

To avoid a hold-over of rose seed, it is recommended to harvest the hips of most of the species before full ripeness. Only in species of the section *Synstylae* must harvesting be delayed until full ripeness.

Table 4 Published records of naturally occurring interspecific hybrids and their origins

Seed parent[a]	*Pollen parent*	*Denotation*	*Reference*
R. acicularis (28, 42, 56)	*R. nutkana* (28)	*R.* × *engelmannii* S. Wats.	Krüssmann (1986)
R. agrestis (35, 42)	*R. canina* (35, 42)	*R.* × *beinensis* Ozan	Graham and Primavesi (1993)
	R. micrantha (35, 42)	*R.* × *bishopii* W.-Dod	Graham and Primavesi (1993)
	R. sherardii (28, 35)	—	Graham and Primavesi (1993)
	R. stylosa (28, 35, 42)	—	Graham and Primavesi (1993)
R. arvensis (14)	*R. canina* (35, 42)	*R.* × *verticillacantha* Mérat	Graham and Primavesi (1993)
	R. micrantha (35, 42)	*R.* × *inelegans*	Graham and Primavesi (1993)
	R. obtusifolia (35)	—	Keller (1931)
	R. rubiginosa (35, 42)	*R.* × *consanguinea* Gren.	Keller (1931)
	R. sherardii (28, 35)	—	Graham and Primavesi (1993)
	R. stylosa (28, 35, 42)	*R.* × *pseudorusticana* Crép. ex Rogers	Graham and Primavesi (1993)
	R. tomentosa (35)	—	Graham and Primavesi (1993)
R. canina (35, 42)	*R. agrestis* (35, 42)	—	Müller (1957)
	R. arvensis (14)	*R.* × *verticillacantha* Mérat	Graham and Primavesi (1993)
	R. damascena (28)	*R.* × *alba* L.	Krüssmann (1986)
	R. jundzillii (42)	—	Schwertschlager (1926)
	R. mollis (28)	*R.* × *molletorum* H.-Harr.	Graham and Primavesi (1993)
	R. obtusifolia (35)	*R.* × *dumetorum* Thuill.	Graham and Primavesi (1993)
	R. rugosa (14)	*R.* × *praegeri* W.-Dod	Gustaffson (1944)
	R. sherardii (28, 35)	*R.* × *rothschildii* Druce	Graham and Primavesi (1993)
	R. stylosa (28, 35, 42)	*R.* × *andegavensis* Bast.	Graham and Primavesi (1993)
	R. tomentosa (35)	*R.* × *scabriuscula* Sm.	Graham and Primavesi (1993)
R. gallica (28)	*R. agrestis* (35, 42)	—	Keller and Gams (1923)
	R. glauca (28)	—	Schnetz (1910)
	R. jundzillii (42)	—	Schnetz (1910)
	R. micrantha (35, 42)	*R.* × *sylvicola* Désélg. et Rip.	Keller and Gams (1923)
	R. obtusifolia (35)	—	Mayer (1904)
	R. rubiginosa (35, 42)	—	Keller and Gams (1923)
	R. tomentosa (35)	*R.* × *marlayana* Boulln	Keller and Gams (1923)
R. majalis (28)	*R. rugosa* (14)	*R.* × *acantha* Waitz	Anonymous (1988)
R. micrantha (35, 42)	*R. agrestis* (35, 42)	*R.* × *bishopii* W.-Dod	Keller (1931)
	R. arvensis (14)	*R.* × *inelegans* W.-Dod	Graham and Primavesi (1993)
	R. canina (35, 42)	*R.* × *toddiae* W.-Dod	Graham and Primavesi (1993)
	R. obtusifolia (35)	—	Graham and Primavesi (1993)
	R. rubiginosa (35, 42)	*R.* × *bigeneris* Duffort ex Rouy	Schwertschlager (1926)
	R. stylosa (28, 35, 42)	—	Graham and Primavesi (1993)
R. mollis (28)	*R. canina* (35, 42)	*R.* × *molletorum* H.-Harr.	Graham and Primavesi (1993)
	R. pimpinellifolia (28)	*R.* × *sabinii* Woods	Täckholm (1922)
	R. rubiginosa (35, 42)	*R.* × *molliformis* W.-Dod	Graham and Primavesi (1993)
	R. sherardii (28, 35)	*R.* × *shoolbredii* W.-Dod	Graham and Primavesi (1993)
R. obtusifolia (35)	*R. arvensis* (14)	*R.* × *rouyana* Duffort ex Rouy	Graham and Primavesi (1993)
	R. canina (35, 42)	*R.* × *dumetorum* Thuill.	Graham and Primavesi (1993)
	R. rubiginosa (35, 42)	*R.* × *tomentelliformis* W.-Dod	Graham and Primavesi (1993)
	R. stylosa (28, 35, 42)	—	Graham and Primavesi (1993)
	R. tomentosa (35)	—	Graham and Primavesi (1993)
R. pimpinellifolia (28)	*R. canina* (35, 42)	*R.* × *hibernica* Templ.	Graham and Primavesi (1993)
	R. glauca (28)	—	Keller (1931)
	R. rubiginosa (35, 42)	*R.* × *cantiana* W.-Dod	Graham and Primavesi (1993)
	R. sherardii (28, 35)	*R.* × *involuta* Sm.	Graham and Primavesi (1993)
	R. tomentosa (35)	*R.* × *coronata* Crép. ex Reuter	Graham and Primavesi (1993)
R. rubiginosa (35, 42)	*R. arvensis* (14)	*R.* × *consanguinea* Gren.	Graham and Primavesi (1993)
	R. canina (35, 42)	*R.* × *nitidula* Besser	Graham and Primavesi (1993)
	R. gallica (28)	—	Täckholm (1922)
	R. micrantha (35, 42)	—	Fagerlind (1940)
	R. mollis (28)	*R.* × *molliformis* W.-Dod	Hurst (1929)
	R. obtusifolia (35)	—	Keller (1931)
	R. pimpinellifolia (28)	*R.* × *cantiana* W.-Dod	Graham and Primavesi (1993)
	R. sherardii (28, 35)	*R.* × *suberecta* (Woods) Ley	Malmgren (1986)
	R. stylosa (28, 35, 42)	—	Graham and Primavesi (1993)
	R. tomentosa (35)	*R.* × *avrayensis* Rouy	Schwertschlager (1926)
R. rugosa (14)	*R. canina* (35, 42)	*R.* × *praegeri* W.-Dod	Graham and Primavesi (1993)
	R. majalis (28)	—	Anonymous (1988)

Table 4 Continued

Seed parent[a]	*Pollen parent*	*Denotation*	*Reference*
R. sherardii (28, 35)	*R. agrestis* (35, 42)	—	Graham and Primavesi (1993)
	R. arvensis (14)	—	Graham and Primavesi (1993)
	R. canina (35, 42)	*R.* × *rothschildii* Druce	Graham and Primavesi (1993)
	R. micrantha (35, 42)	—	Graham and Primavesi (1993)
	R. mollis (28)	*R.* × *shoolbredii* W.-Dod	Graham and Primavesi (1993)
	R. pimpinellifolia (28)	*R.* × *involuta* Sm.	Graham and Primavesi (1993)
	R. rubiginosa (35, 42)	*R.* × *suberecta* (Woods) Ley	Graham and Primavesi (1993)
	R. tomentosa (35)	*R.* × *suberectiformis* W.-Dod	Graham and Primavesi (1993)
R. stylosa (28, 35, 42)	*R. agrestis* (35, 42)	—	Graham and Primavesi (1993)
	R. arvensis (14)	*R.* × *pseudorusticana* Crép. ex Rogers	Graham and Primavesi (1993)
	R. canina (35, 42)	*R.* × *andegavensis* Bast	Graham and Primavesi (1993)
R. tomentosa (35)	*R. agrestis* (35, 42)	—	Graham and Primavesi (1993)
	R. canina (35, 42)	—	Graham and Primavesi (1993)
	R. micrantha (35, 42)	—	Graham and Primavesi (1993)
	R. mollis (28)	—	Graham and Primavesi (1993)
	R. obtusifolia (35)	—	Graham and Primavesi (1993)
	R. pimpinellifolia (28)	*R.* × *coronata* Crép. ex Reuter	Graham and Primavesi (1993)
	R. rubiginosa (35, 42)	*R.* × *avrayensis* Rouy	Graham and Primavesi (1993)
	R. sherardii (28, 35)	*R.* × *suberectiformis* W.-Dod	Graham and Primavesi (1993)

[a]Chromosome numbers (in parentheses) are based on the published $2n$ value for each species.

Anonymous (1988) Rosenverzeichnis Rosarium Sangerhausen, 4th edn. Sangerhausen: Mitteldeutsches Druckhaus.
Fagerlind F (1940) *Svensk Botanisk Tidsskrift* 34(4): 334–354.
Graham GG and Primavesi AL (1993) *Roses of Great Britain and Ireland.* Botanical Society of the British Islands.
Gustaffson A (1944) *Hereditas* 30: 405–428.
Hurst CC (1929) *Rose Annual* 23: 37–64.
Keller R (1931) *Denkschriften der Schweizerischen Naturforschenden Gesellschaft* 65: 23–27.
Keller R and Gams H (1923) *Illustrierte Flora von Mitteleuropa*, vol. IV/2, pp. 976–1052. München: HEGI.
Krüssmann G (1986) *Rosen, Rosen, Rosen.* 2nd edn. Berlin: Paul Parey.
Malmgren U (1986) *Svensk Botanisk Tidsskrift* 80(4): 209–227.
Mayer A (1904) *Flora von Tübingen und Umgebung.* Schwäbische Alb vom Plettenberg bis zur Teck, Balingen, Hechingen, Reutlingen, Urach Rottenburg, Herrenberg, Böblingen. Tübingen.
Müller K (1957) *Mitteilungen des Vereins für Naturwissenschaft und Mathematik Ulm* 25: 105–109.
Schnetz J (1910) *Mitteilungen der Bayerischen Botanischen Gesellschaft zur Erforschung der heimischen Flora* 2(14): 229–237.
Schwertschlager J (1926) *Berichte der Bayerischen Botanischen Gesellschaft München* 18: 1–126.
Täckholm G (1922) *Acta Horti Bergiani* 7: 97–381.

Most achenes of rose species do not germinate in the first year, even in good germination conditions. There are various methods for breaking dormancy, one of which is described here. A change of temperature during stratification gives germination results for *R. rugosa* 'Frau Dagmar Hastrup' of 60%. Achenes should be stratified directly after harvest. After a warm stratification (10–12 weeks, 20–25 °C) in a sand and vermiculite mixture with the addition of compost starter followed by cold stratification (10–12 weeks, 4 °C), up to 95% achenes of most rose species germinate within 6 months. Hybrid seed can be successfully germinated by the same treatment.

Confirmation of Hybridity

After crossing, it is necessary to confirm the hybridity of the progeny. With the help of only vegetative markers, this is difficult in some crosses, but floral markers are easier. Complicated cytological conditions, as in species of the section *Caninae*, render identification of hybrids more difficult. Even if it is possible to count chromosome numbers, molecular markers (*see* **Genetics:** Molecular Markers) will give the best results. Spontaneous hybrids were identified when *R. canina* was crossed with *R. jundzillii*, *R. mollis* and *R. stylosa*. Artificial crosses between these species give progenies that look more like the pollen parent (**Table 3**). Differences of classification are also noted for the progenies after crossing *R. micrantha* × *R. rubiginosa* and *R. mollis* × *R. pimpinellifolia*.

It seems that crossing between species of one section gives progenies which look like the seed or pollen parent. Most crossings between species of different sections give hybrids. As mentioned, hybrids of the *Caninae*, independent of ploidy, often look like the seed parent. But other combinations of ploidy levels also show hybrids similar to the seed parent, when the ovule contributes more chromosomes than the pollen. So the verification of the hybrid status is difficult. The use of random amplified polymorphic DNA (RAPD) was investigated in five plants each of the interspecific crosses *R. acicularis* × *R. majalis*, *R. obtusifolia* × *R. sherardii* and *R. stylosa* × *R. gallica*. Eighty-nine primer and primer combinations were used to identify markers specific for the parental genomes; 24–42

Table 5 Hips set and seedling germination in relation to ploidy in crosses by Feuerhahn (1994)

Seed parent	*Pollen parent*							
	Diploid (2x)		*Tetraploid (4x)*		*Pentaploid (5x)*		*Hexaploid (6x)*	
	SH%	*SP%*	*SH%*	*SP%*	*SH%*	*SP%*	*SH%*	*SP%*
Diploid (2*x*)	32.9	32.4	26.4	16.8	11.1	25.4	29	0
Tetraploid (4*x*)	36.8	14.8	39.1	21.4	23.7	18.2	40	0
Pentaploid (5*x*)	46.6	11.7	44.2	18.1	52.5	24.2	32.6	2.5
Hexaploid (6*x*)	0	—	10.6	48.4	0.8	0	0	—

SH%, set of hips (%); SP%, seedling percentage (%).
Feuerhahn B (1994) *Diplomarbeit.* Hanover: Institut für Obstbau und Baumschule der Universität Hanover.

Table 6 Hips set and seedling germination in relation to ploidy in selfing and outcrosses by Feuerhahn (1994)

	Crossing		*Self-fertilization*	
	SH%	*SP%*	*SH%*	*SP%*
Diploid (2*x*)	21.4	22.1	2.8	0
Tetraploid (4*x*)	37.5	18.9	44.8	20.5
Pentaploid (5*x*)	45.1	16.0	64.6	27.4
Hexaploid (6*x*)	10.6	48.4	44.2	40.4

SH%, set of hips (%); SP%, seedling percentage (%).
Feuerhahn B (1994) *Diplomarbeit.* Hanover: Institut für Obstbau und Baumschule der Universität Hanover.

specific markers could be found in the species. Between 39 and 67% of these fragments could be found in the putative hybrids. So all F_1 plants could be unambiguously identified as true hybrids.

Disease Resistance

Resistance or tolerance to disease, especially black spot and powdery mildew, is very important in rose breeding. Examples of resistance hybrids are shown in **Table 3**. Experiments at the University of Hanover showed that the progeny of crosses between *R. gallica* × *R. nitida*, *R. majalis* × *R. rugosa*, *R. majalis* × *R. sherardii*, *R. rubiginosa* × *R. majalis*, *R. rubiginosa* × *R. nitida*, *R. rugosa* × *R. multiflora*, *R. rugosa* × *R. nitida* and *R. stylosa* × *R. nitida* have a high tolerance of powdery mildew. With regard to black spot the trials showed that there were only a few hybrids with tolerance to this disease (*R. majalis* × *R. rugosa*, *R. nitida* × *R. pimpinellifolia*, *R. rugosa* × *R. glutinosa*, *R. rugosa* × *R. obtusifolia*, *R. stylosa* × *R. obtusifolia*, *R. stylosa* × *R. rubiginosa*, *R. stylosa* × *R. villosa*). In the case of powdery mildew the resistance of both the seed parent and pollen parent probably influence the resistance of the progeny. The progeny obtained when *R. acicularis*, *R. gallica*, *R. jundzillii*, *R. majalis*, *R. pimpinellifolia*, *R. rugosa* or *R. stylosa* were used as seed parent were often tolerant or highly tolerant. Tolerant or highly tolerant offspring were obtained when *R. acicularis*, *R. majalis* or *R. rugosa* were used as pollen parents. In contrast, most of the progenies of *R. micrantha* and *R. canina* 'Pfänders' as seed parent were highly susceptible to powdery mildew.

Reimann-Philipp tried to breed resistance against black spot from *R. multiflora* into garden cultivars. The triploid offspring were resistant but sterile. Later on, seeds developed spontaneously on the plants. An analysis of the plants show that they are substitution hybrids ($2n$) and once more crossable with *R. multiflora*. At the same institute some years later, a method was developed to produce tetraploid shoots by colchicine treatment *in vitro* (0.05–0.1% in the medium). These tetraploid plants were the basis for successful breeding work for resistance against black spot in the following years (*see* **Disease**: Black Spot. **Genetics**: Inheritance of Characteristics). These breeding programmes to develop rose cultivars resistant to black spot are time-consuming because a lot of back-crosses are necessary to reduce undesired character traits of the species. To reduce this time period, molecular markers closely linked to the resistance gene were used for marker-assisted selection procedures. *R. wichurana* Crép. and *R. rugosa* are other species with resistance to black spot that have been included in the breeding work.

Importance of Species Crosses for Rose Breeding

The primary importance of breeding with species is the introduction of resistance to fungal diseases or insect pests into the garden roses. Some successful results are referred to above. But other character traits could also be improved by species crosses. Breeding with *R. foetida* (*R. lutea* var. *persiana*) resulted in yellow rose cultivars for the first time. The other yellow species (*R. hugonis*, *R. ecae*, *R. primula*) could be included in breeding programmes to improve yellow flower colour combined with winter hardiness.

In roses, yellow flower colour is caused by lipophilic carotenoids of the chromoplasts. These are inherited only maternally, therefore only crosses with yellow seed parents are successful. The other colour looked for is pure blue. This colour only occurs if the anthocyanidin delphinidin is present alone. This anthocyanidin is not present in the genus *Rosa*. Therefore, species crosses could not be successful. Breeding of a blue rose will only be possible by gene transfer from a genus that can synthesize delphinidin. Another anthocyanidin in roses, cyanidin, gives colours of lilac to violet when it is present as diglycoside and red flowers when it is present as a monoglycoside.

Prickless genotypes occur in populations of *R. blanda* that may be used to breed this character but they are often sensitive to powdery mildew. Large ornamental prickles are characteristic of *R. sericea* subsp. *omeiensis* var. *pteracantha*. This trait could give roses an attractiveness outside the flowering season.

Rosa laxa from central Siberia is extremely winter-hardy, resistant against drought and powdery mildew. *Rosa suffulta*, from areas which are very dry in the summer, is very drought-resistant and shows a second flower in autumn. *Rosa arvensis* is a species with a special trait of shade tolerance. To get cultivars with tolerance to salt, species crosses are important, too. Often growers have problems in continuous rose cultivation by replant problems. Replant problems are often caused by nematodes like *Pratylenchus penetrans* and on the other hand by specific soil sickness whose origin and cause are not yet clear. Experiences of nurseries show that there are differences in sensitivity to replant problems between species. For example, *R. multiflora* seems to be less sensitive than *R. canina*, and *R. canina* 'Inermis' is less sensitive than *R. corymbifera* 'Laxa'. This species-specific resistance means that breeding for this character trait should be possible. Our own investigations with controlled inoculation of *Pratylenchus* have resulted in least damage in *R. obtusifolia*, and the greatest damage in *R. micrantha*. Investigations into differences between rose species in their tolerance to soil sickness and nematodes could lead to the identification of species which should be included in breeding projects to overcome the replant problem (*see* **Disease**: Replant Problems and Soil Sickness).

One of the biggest faults in breeding work with roses species is the neglect of the variability in populations or provenances of the species. Usually only the genotypes that accidentally happen to be present are investigated. Often the suitability attributed to the whole species is concluded from the character trait of the single plant. But in species of other genera, a great variability of morphology and the special character trait resistance is known. To select the optimal genotype regarding a specific character trait, populations and provenances from the original area should be examined before a species is included in a breeding project. Often the selection of the optimal provenance is responsible for success.

Many species have been excluded from all breeding projects by their lack of availability. In one of the greatest breeding programmes in Ahrensburg, on resistance to black spot disease, 60 species were tested for resistance (*see* **Disease**: Black Spot). But in this case again, mainly single plants from rose collections or botanical gardens were taken as representative of the whole species. The remaining 60 species that were not tested could also contain suitable sources for resistance.

See also: **Biotechnologies for Breeding**: Manipulation of Ploidy Level; Amphidiploidy. **Classification**: Conventional Taxonomy (Wild Roses). **Disease**: Black Spot; Powdery Mildew; Replant Problems and Soil Sickness. **Genetics**: Inheritance of Characteristics; Inheritance in the Dogrose; Molecular Markers. **History of Roses in Cultivation**: European (Pre-1800); Modern (Post-1800).

Further Reading

Cole P and Melton B (1986) Self- and cross-compatibility relationships among genotypes and between ploidy of the rose. *Journal of the American Society for Horticultural Science (USA)* 111(1): 122–125.

Erlanson EW (1930) Sterility in wild roses and some species hybrids. *Genetics* 16: 76–95.

Fagerlind F (1948) Compatibility, eu- and pseudo-incompatibility in the genus *Rosa*. *Acta Horti Bergiani* 15: 1–37.

Feuerhahn B and Spethmann W (1995) Kreuzungen bei Wildrosenarten. Gehölzforschung, vol. 3. Hanover: Institut für Obstbau u. Baumschule der Universität Hanover.

Graham GG and Primavesi AL (1993) *Roses of Great Britain and Ireland*. BSSI Handbook, vol. 7. London: Botanical Society of the British Islands.

Gustaffson Ä (1942) Cross experiments in the genus *Rosa*. *Hereditas* 28: 235–238.

Hurst CC (1929) The genetics of the rose. *Rose Annual* 23: 37–64.

Kroon GH and Zeilinger AE (1974) Apomixis and heterogamy in rose rootstocks (*R. canina* L.). *Euphytica* 23: 345–352.

Krüssmann G (1986) *Rosen, Rosen, Rosen* (2nd edn). Berlin: Paul Parey.

Ratzek JC, Yarnell SH and Flory WS (1941) Crossing relation of some diploid and polyploid species of roses. *Proceedings of the American Society of Horticultural Science* 38: 637–654.

Saakov SG (1976) *Wild- und Gartenrosen*. Berlin: VEB Deutscher Landwirtschaftsverlag.

Spethmann W (1993) Neue Methoden der Vermehrung und Züchtung von Rosen. *Rosenjahrbuch (Verein Deutscher Rosenfreunde)*, 1993: 67–82.
Stougaard B (1984) Analysis of variation in progeny from *Rosa multiflora* crosses. *Tidsskrift Planteavl* 88: 287–297.
Svedja F (1976) Breeding winter hardy and everblooming roses. *American Rose Annual* 61: 16–22.
Täckholm G (1920) On the cytology of the genus *Rosa*. *Svensk Botanisk Tidsskrift* 14(2–3): 301–311.

Mutation

A V Roberts and R Lewis, School of Health and Biosciences, University of East London, London, UK

Introduction

Every year, rose breeders raise more than one million seedlings, each with a unique genetic identity. Those few that are sufficiently novel and of good quality to be worth registering as new cultivars are listed in the periodically updated *Modern Roses*, the 11th edition of which, *Modern Roses XI*, is reproduced in vol. 3 of this Encyclopedia. Amongst the 30 000 or so cultivars included in *Modern Roses XI*, nearly 8% arose, not as seedlings, but by mutations of previously registered cultivars, from which they differ in only one or just a few characters. Most of these mutants arose spontaneously and were detected by chance but a few were deliberately induced with mutagenic agents.

A mutation is an event that results in a change in the hereditary material. A distinction is made between gene and chromosome mutations. Gene mutations are localized changes typically arising from errors in DNA replication. Chromosome mutations are larger-scale changes that involve many genes. They arise from breakage-rejoining events and can often be detected with a microscope. A chromosome mutation may involve the deletion or duplication of part of a chromosome, or a structural rearrangement involving either the inversion of a sequence of genes within a chromosome or an interchange of gene sequences between two chromosomes. The term 'mutant' is used to describe an organism that uniformly and stably expresses a character that arose as a result of mutation and the term 'sport' is sometimes used to describe a mutant that arose spontaneously. Bud mutations arise during the vegetative, as opposed to the reproductive, phase of plant development. 'Mutation breeding' is a term used to describe the artificial induction of mutations and the exploitation of mutations to obtain new varieties. Some authors regard a doubling in the number of chromosomes as a type of mutational event but in this Encyclopedia this type of event is discussed separately (*see* **Biotechnologies for Breeding**: Manipulation of Ploidy Level; Amphidiploidy).

Mutations, as the source of most genetic diversity, are an important component of evolutionary change. They occur spontaneously in low frequency but can also be induced by mutagenic agents that include certain radiations and chemicals. In roses and many other genera, spontaneous mutants are frequently found amongst plants regenerated *in vitro* via adventitious shoots and somatic embryos (*see* **Breeding**: Overview. **Cell, Tissue and Organ Culture**: *In Vitro* Regeneration Techniques). They are referred to as somaclonal variants. The components of the culture media, which apparently are responsible for these variants, are not mutagenic agents in the normal sense and the events that lead to the frequent appearance of somaclonal variants are not fully understood.

This article describes the spontaneous occurrence and artificial induction of mutants.

Naturally Occurring Mutations

Several mutant characters, including recurrent-flowering and multipetalled flowers, are of central importance in rose breeding (*see* **Growth Regulation**: Floral Induction. **Genetics**: Inheritance of Characteristics. **History of Roses in Cultivation**: Ancient Chinese Roses; European (Pre-1800); Modern (Post-1800)). However, most recorded mutants are merely useful variants of popular cultivars. Some genetic variants that repeatedly give rise to variant sectors, within a single petal, may result from the activity of transposons. These are mobile DNA elements, possibly of viral origin, that can move from place to place in the genome and also amplify their copy number within the genome during transposition.

New Cultivars Arising from Mutation

Mutants that arise by chance during routine cultivation are described as natural mutants or sports. The chance occurrence of a sport has often been the cause of celebration, as on the occasion when 'Kronenbourg' was discovered as a sport of 'Peace' (**Table 1**). Sam McGredy described its discovery in his nursery, in the 1960s, in the following terms: 'This sport was found by two of my staff who brought it to me. I gave them £50 apiece for their percipience

Table 1 Characters of 'Peace' and a selection of sports of 'Peace'

Cultivar name (date of introduction)	*Characters*
Progenitor	
'Peace' (1945)	Tall bush, dark foliage, high-centred flower, rose pink petals, yellow-edged, 40–45 petals
Sports	
'Climbing Peace' (1950)	Climber
'Lady Dallas Brooks' (1955)	Pink flowers
'Peaceport' (1960)	56–75 petals
'Baby Peace' (1962)	Miniature
'Jean McGregor Reid' (1962)	Cream flowers
'Dallas' (1963)	Crimson flowers
'Tony Peace' (1964)	Light-green foliage
'Kronenbourg' (1965)	Crimson/yellow bicolour flowers
'New Orleans' (1966)	Globular flowers

and one of them used the money to buy a motorbike, which was thereafter known as Kronenbourg.'

In an analysis of cultivars of roses described in *Modern Roses IX*, we found that, of the 18 000 cultivars listed, 1561 originated as sports (*see* **Intellectual Property:** Plant Patents and Trademarks). Some cultivars have given rise to many sports: 17 cultivars have given rise to 10 or more sports, the most prolific of which, 'Columbia', gave rise to 74 sports. The more bushes of a particular rose that are grown and the longer the period of cultivation, the greater the likelihood that cultivars will be found. Thus 'Peace', a very popular cultivar introduced in 1945, has produced 21 sports (**Table 1**). However, there is also evidence of an inherited propensity for producing mutants. This evidence comes from the high frequency of sports amongst closely related cultivars. The best example is that of 'Orléans', which produced 39 sports and, through crosses to other roses, gave rise to a family of high sporting cultivars (**Figure 1**). One reason for the high frequency of sporting in this group of related roses might be an inherited deficiency in a DNA repair enzyme. Enzymes that can repair damage to DNA immediately after a mistake in DNA replication has occurred are less well studied in plants than many other groups of organisms. In humans, for example, there is a mechanism for the repair of mutagenic damage induced by ultraviolet (UV) light immediately after the mutagenic event. However, individuals that inherit the recessive genetic disorder xeroderma pigmentosum are susceptible to multiple skin cancers because they lack the ability to repair these mutations.

Climbing roses are the single most common type of sport. Of the 1561 sports described in *Modern Roses IX*, 411 (26%) were climbing sports. Furthermore, in roses that have given rise to several sports, the first sport to be recorded has almost invariably been a climbing sport. As the records indicate only the first

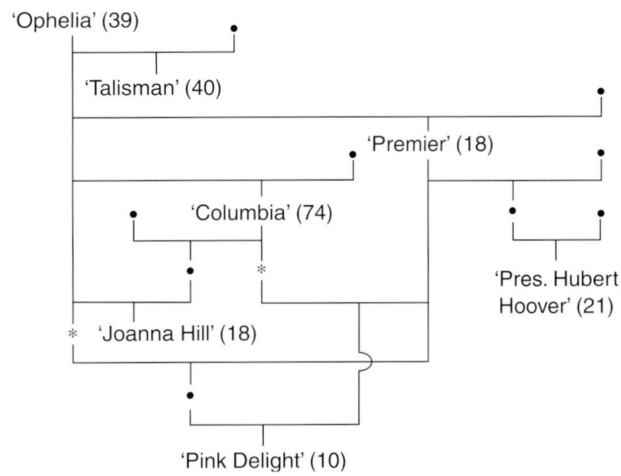

Figure 1 The genealogy of high-sporting cultivars derived from 'Ophelia' based on an analysis of the origins of cultivars listed in Hareing PA (1986) *Modern Roses IX*. Shreveport: American Rose Society. Horizontal lines represent crosses between the parents shown to left and right, above the lines; the offspring are shown below the lines. Parents (•) including sports (*) not named in this scheme can be identified by referring to information about the offspring in *Modern Roses XI* in volume 3 of this *Encyclopedia*.

appearance of a climbing sport, the possibility that climbing mutations have occurred repeatedly in the same cultivar is concealed. These climbing sports are recurrent-flowering mutants of recurrent-flowering bush roses. Therefore, unlike a mutation from a recurrent-flowering to a seasonal-flowering habit, the climbing sports differ from the bush progenitors specifically in the larger number of internodes that must be produced before a flower is initiated (*see* **Growth Regulation:** Floral Induction). The high frequency with which this particular type of mutation occurs is an interesting subject for speculation. One possibility is that shoots with the climbing character have a strong selective advantage over those with the bush habit. It might be advantageous that more energy-producing leaves are produced before a new flush of flowers depletes the resources of the plant. Another possibility is that the climbing habit may arise from the mutation of any one of several different genes, so that the net chances of a climbing mutation occurring in any one of these genes is greater than that of a mutation affecting a character determined by a single gene.

Transposons as a Possible Source of Genetic Instability

In some cultivars, of which 'Harry Wheatcroft' and 'Earthquake'™ are well-known examples, the petals are streaked and dotted with sectors that differ from the background colour (**Figure 2a**). Similar variegations are also found in the leaves of some roses, where yellow or white speckling is seen against a dark-green background (**Figure 2b**). One likely explanation for these unusual flower or leaf patterns is a genetic instability arising from the activities of transposons. Transposons have been described in several types of plant following the Nobel Prize-winning discovery, by Barbara McClintock, that they are responsible for the irregular pattern of streaks and spots in maize seeds. When they enter a gene, as they transpose from one location to another, they prevent the normal expression of that gene but when they spontaneously leave, the gene can resume its normal activity. They may remain stably integrated into a gene in meristematic tissues, so that an affected cultivar can be propagated true to type, but become unstable in other situations, for example, during the differentiation of petals and leaves. If, for example, a transposon is inserted into a gene responsible for anthocyanin pigmentation, it may remain stably incorporated in cells of a shoot meristem but randomly excised during cell division in a petal primordium. After excision, further cell division leads to the formation of groups of similar cells, the size of which varies according to how many cell divisions occur after the excision. The result would be the appearance of spots and streaks of pigmented cells against a pale background.

Another possible reason for erratic variations in flower colour, particularly where large sectors of flowers (**Figure 3**) or entire flowers are affected, may be that a cultivar harbours a mutant cell line in one or more of its different histogenic layers. A mutation affecting anthocyanin pigmentation that is present in the cortex may occasionally be destabilized and become established in the epidermis, the layer in which petals produce anthocyanin pigments.

(a)

(b)

Figure 2 Two unnamed varieties showing patterns of streaks and dots contrasting with the background colour of petals (a) and leaves (b). This type of variegation may result from the activities of transposable elements.

Induced Mutation

The main interest in inducing mutants in roses, as with many other vegetatively propagated crops, is to modify one or just a few characters in excellent cultivars. Adequate diversity can be generated by cross-breeding, so induction of mutations in seeds is of relatively little interest. The type of mutagen and the method of exposure are important practical considerations, and the published literature provides useful guidelines. X-rays and γ-rays are the most frequently used

Figure 3 'Mothersday' showing typical dark-red flowers and, on the same bush, pink flowers of the type found on 'Dick Koster', the variety from which 'Mothersday' arose as a sport. The frequent appearance of pink flowers suggests that 'Mothersday' is a periclinal chimera that harbours cells of 'Dick Koster' in the L2 or L3 layers.

mutagenic agents. Most attempts to induce mutants have been targeted at terminal or axillary buds on irradiated cuttings, bud wood or *in vitro* shoots. These give rise to shoots in which mutant tissues can be recognized, selected and established as mutant plants. Recently, leaves have been exposed to ionizing radiation and adventitious shoots (*see* **Cell, Tissue and Organ Culture**: *In Vitro* Regeneration Techniques) have been regenerated *in vitro*.

Mutagenic Treatments

Certain mutagenic chemicals and radiations are mutagenic agents that have the ability to induce mistakes during the replication of DNA and/or induce chromosome breaks. Both ultraviolet (UV) light and ionizing radiations are mutagenic. UV light has poor penetrating power and is rarely chosen as an agent for inducing mutations in plants. Ionizing radiations are so called because they release energy in quanta that are sufficiently energetic to ionize water. The ionization (radiolysis) of water molecules inside the plant leads to a series of reactions that result in the formation of hydrogen peroxide, peroxyl radicals and hydroxyl radicals which initiate mutagenic events. The ionizing radiations most frequently used to induce mutations in plants are γ-rays (produced by the radioactive decomposition of ^{60}Co) and X-rays (produced by X-ray machines). Chemical mutagens vary in their mode of action but those most commonly used to induce mutation in plants are the alkylating agents, including ethyl methanesulphonate (EMS). The mode of action of alkylating agents is to add alkyl groups (CH_3, C_2H_5, etc.) to the DNA molecule. The products resulting from radiolysis of water and the addition of alkyl groups to DNA cause both mistakes in DNA replication and chromosome breaks.

Ionizing radiations and alkylating agents give a similar spectrum of mutation and the choice of mutagenic agent is often driven by the availability of facilities. For those who have access to facilities for X-rays or γ-rays, these are often preferred to chemical mutagens. Safety is an important consideration because mutagenic agents are also mutagenic, teratogenic and carcinogenic to humans. Ionizing radiation requires routine shielding but an irradiated plant itself does not present a hazard. Safe usage of chemical mutagens, which should be fully discussed with a safety consultant, requires appropriate procedures for treatment of spillages and eventual disposal of the mutagen; furthermore, care must be taken to avoid exposure to the residual amounts of mutagen in treated plants.

DNA and chromosomes may be changed in innumerable different ways by mutagenic treatment and no two cells are likely to change in the same way. Some cells will die because the mutation load is too high, some are genetically changed but survive and others remain unaffected. The chances of obtaining mutant plants will be low if the exposure to the mutagen is either so high that the survival rate of exposed tissues is low, or so low that mutational events are rare. Treatments with X-rays and γ-rays that have been used to induce mutation are mainly in the range of 10–90 Gy. The optimum exposure is determined by several variables. It has been shown in many genera that polyploids survive higher exposures than diploids but this cannot readily be confirmed from the available literature on roses. If the subject of irradiation is cuttings, the proportion of survivors is increased by shielding the region where roots are formed with lead. In several *in vivo* studies and one *in vitro* study, the optimal exposure of ionizing radiation was 40–70 Gy. In one study (presently unpublished) where adventitious shoots were regenerated *in vitro* from leaves that were irradiated *in vivo*, the recommended optimum was 20–30 Gy. Effective exposures to EMS are likely to vary because uptake of the EMS into the target tissue depends on to the method of application, but published treatments include immersion in 0.2–2.0% EMS for 1–24 h.

Chimeras and the Retrieval of Stable Mutant Plants

As cells may be changed by mutation in an infinite number of ways, no two cells are likely to be changed in precisely the same way. Therefore, mutation is described as a single cell event. In most published work on roses, the target of mutagenic treatment has been shoot meristems. The units exposed to irradiation include cuttings, bud wood and *in vitro* plants. Within the shoot meristem of roses, the cells are organized into histogenic layers. It seems likely that, as in many dicotyledonous species, there are three such layers in roses: the L1 layer that gives rise to the epidermis, the L2 layer that gives rise to the subepidermal layers corresponding to the cortex of the stems and the L3 layer that gives rise to the innermost layers corresponding to the stele of the stems. A mutation may become established, for example, in the L3 layer, while the L1 and L2 layers remain of the original type. This arrangement remains stable throughout the plant body and in plants that are subsequently propagated by budding or from cuttings. Flower colour is determined by the epidermis (L1) which shows the anthocyanin (pink/red) pigments and carotenoid (yellow) pigments and the subepidermal layer (L2) which expresses the carotenoid but not the anthocyanin pigments. The pollen and ovules arise from the L2 layer and if a mutant character is to be passed on sexually, the mutation must occur in this layer. Adventitious roots that are formed by cuttings come from the stele (L3) and any shoots that arise from the roots as suckers will express the genetic character of that layer. Experimental evidence of the stability of the L3 layer was provided by Zimmerman and Hitchcock who found that root suckers of various sports they investigated showed the characters of the cultivars from which they were derived. Examples included 'Better Times', which reverted to 'Briarcliff', and 'Souvenir', which reverted to 'Talisman'. The genotype of the original cultivars had been retained in the L3 but was concealed in the mutant plant.

After mutagenic treatment, axillary or terminal meristems initially contain a mixture of cell lines which become attenuated as growth proceeds. A plant body which contains cells of different types is known as a chimera. Each cell line is likely to remain within the same histogenic layer in which it originated. However, the rates of cell division in different cell lines may vary, creating a turbulence within the meristem that may result in some cell lines penetrating other histogenic layers. The complex mixture of cell types with changed phenotype and differing growth rates can be appreciated in the first flowers to be formed following irradiation (**Figure 4**). In order that a stable and uniform mutant plant can be obtained and eventually registered as a new cultivar, it is essential that a uniform periclinal chimera or a pure mutant plant is established. This can be achieved most rapidly by repeatedly pruning the shoots so that growth is channelled through a series of axillary buds until, eventually, shoots of uniform morphology are obtained that can be propagated true to type.

Cells and tissues of many roses can be induced to form adventitious shoots and/or somatic embryos (*see* Cell, Tissue and Organ Culture: *In Vitro* Regeneration Techniques). However, there are few published accounts of these regeneration methods being used in association with mutagenic treatment. The distinctive feature of adventitious shoots or somatic embryos is that they arise either from single cells or small groups of cells that are not arranged in histogenic layers. They can thus be expected to lead to the formation of pure mutant plantlets without chimerism.

(a)

(b)

Figure 4 Flower of a control plant of 'Escapade'® (a) and a chimeric flower (b) which was the first flower to be formed after exposure of a cutting of 'Escapade' to X-rays (60 Gy). Note the irregular colouration of the chimeric flower and the tessellated margin resulting from different growth rates amongst the different cell lines.

Characteristics of Induced Mutants

Most mutants that have been induced are described in terms of changes in petal colour. This is at least partly because these variants are most easily recognized and selected after the mutagenic treatment. The examples shown in **Table 2** of mutant roses that were obtained by mutagenic treatment and registered as new cultivars are typical in this respect.

The type of variation that might be expected to arise from mutagenic treatments can be predicted from

Table 2 Examples of changed flower colour in registered rose cultivars obtained by mutagenic treatment

Progenitor's name and description of petals	*Mutant's name and description of petals*	*Mutagenic treatment*
'America's Junior Miss', coral pink	'Sukumari', almost white	X-rays (30 Gy)
'Contempo', orange blending to gold	'Tangerine Contempo', tangerine orange	X-rays (30 Gy)
	'Yellow Contempo', yellow	X-rays (30 Gy)
'Kordes' Perfecta', yellow with cream edges, later flushing crimson	'Permoser', light yellow with carmine edges	X-rays (15 Gy)
'Peace', rose pink yellow-edged	'Desi', golden with red stripes	X-rays (30 Gy)
'Queen Elizabeth'®, light pink	'Flamingo Queen', deep pink	γ-rays (70–80 Gy)
	'Paula', dusty salmon	γ-rays
	'Saroda', very light pink	X-rays (30 Gy)

Sources of information include: Broerties C and van Harten AM (1988) *Applied Mutation Breeding for Vegetatively propagated crops*. Amsterdam: Elsevier and *Modern Roses XI*, in volume 3 of this *Encyclopedia*.

(a)
(b)

Figure 5 Flowers of a control plant of 'Dame of Sark' (a) and a yellow mutant (b) after exposing a cutting to X-rays (60 Gy).

experiments reported in the literature. Comparisons between natural sports and the cultivars from which they arose can also be useful. There is certainly a strong element of similarity with regard to flower characters but, surprisingly, although climbing roses are the largest group of spontaneous mutants amongst roses, none has been induced by treatments with mutagens. The strikingly high frequency of spontaneous climbing sports, together with disparity between the frequency that they occur spontaneously and in response to mutagenic treatment, suggests that the switch from the bush to climbing habit may not involve a typical mutation of nuclear genes and justifies further investigation.

Most descriptions of induced mutations refer to conspicuous features affecting flower form or colour. Typical changes include increases or decreases in the number of petals and changes in petal colour. The most commonly reported flower colour mutants involve the loss of red or pink anthocyanin pigments and changes in shades of colour. When the ability to produce an anthocyanin pigment is lost, flowers that were initially red or pink are thus changed to white, and orange-coloured flowers, which are a blend of anthocyanin and carotenoid pigments, change to yellow (**Figure 5**). The acquisition of anthocyanin pigments and the loss of yellow pigment are uncommon. Increases in the amount of rose oil in Damask roses have also been reported. Generally, it is to be expected that structural rearrangements of chromosomes will lead to the loss of fertility after mutagenic treatment. However, the effects of mutagenesis can be unpredictable; Smilansky *et al.* found that a mutant rose obtained by -irradiation of the cultivar 'Mercedes', which is infertile as a female parent, was more fertile than its progenitor and showed 40% germination of seed.

Further Prospects

Before 1991, plant breeders' rights could easily be obtained for any sport or induced mutant that conformed to the requirements of distinctness, uniformity and stability. Sports and induced mutants could thus be 'easy pickings' for those wishing to commercialize new variants of successful cultivars. After the Union pour la Protection des Obtentions Végétales (UPOV) convention of 1991, it became more difficult for a third party other than the original breeder of a cultivar to obtain intellectual property rights over a mutant (*see* **Intellectual Property**: Plant Patents and Trademarks). It seems likely that new developments in mutation breeding will be initiated by breeders, as it is this group of people who stand to gain most from their exploitation. It seems probable that there will be more interest in mutating cut-flower roses than garden roses because of the higher income that can be made from royalties.

See also: **Biotechnologies for Breeding**: Manipulation of Ploidy Level; Amphidiploidy. **Breeding**: Overview; Selection Strategies for Cut Roses; Selection Strategies for Disease and Pest Resistance. **Cell, Tissue and Organ Culture**: *In Vitro* Regeneration Techniques. **Genetics**: Inheritance of Characteristics; Research Trends. **Growth Regulation**: Floral Induction. **History of Roses in Cultivation**: Ancient Chinese Roses. **Intellectual Property**: Plant Patents and Trademarks. **Secondary Metabolites**: Culinary Uses and Nutritional Value.

Further Reading

Bennetzen JL (2000) Transposable element contributions to plant gene and genome evolution. *Plant Molecular Biology* 42: 251–269.

Broertjes C and van Harten AM (1988) *Applied Mutation Breeding for Vegetatively Propagated Crops*. Amsterdam: Elsevier.

Klug WS and Cummings MR (2003) *Genetics: A Molecular Approach*. Upper Saddle River, NJ: Pearson Educational.

Schum A and Preil W (1998) Induced mutations in ornamental plants. In: Jain SM, Brar DS and Ahloowalia BS (eds.) *Somaclonal Variation and Induced Mutations in Crop Improvement*, pp. 333–366. Dordrecht: Kluwer Academic.

Soyfer VN (1987) DNA damage, repair and mutagenesis in higher plants. *Israeli Journal of Botany* 36: 1–14.

Van Harten AM (1998) *Mutation Breeding, Theory and Practical Applications*. Cambridge: Cambridge University Press.

DNA Fingerprinting

H Nybom, Swedish University of Agricultural Sciences, Kristianstad, Sweden

Introduction

In the last two decades, various molecular marker-based methods have been applied for discrimination among *Rosa* genotypes and assessment of genetic variability. These include DNA markers, isozymes and some secondary substances (*see* **Genetics:** Molecular Markers). Among the latter, polyphenolic compounds (e.g. flavonols and anthocyanins) as well as hydrolysable tannins have some discriminatory power but sensitivity to environmental influence is a serious drawback (*see* **Classification:** Chemotaxonomy and Molecular Taxonomy). Isozyme analysis has unfortunately proved to be less useful in roses than in many other plant groups, probably due mainly to the small number of consistently resolvable loci. Lately, very promising results have instead been reported with various DNA-based marker methods (*see* **Genetics:** Molecular Markers). This success stems mainly from the practically unlimited number of markers available and the fact that they are not influenced by environmental conditions. The concept of 'DNA fingerprinting' was introduced in 1985 by Alec Jeffreys, who developed what, at that time, amounted to a Holy Grail of biology, i.e. a method that could provide genotype-specific DNA fragment patterns. Numerous segments of the genome could be analysed simultaneously, resulting in a 'bar code' with much higher discriminatory power than any other available marker system. Although it was first applied to humans and various animal species, it was soon realized that this method would also be highly useful in plants.

Jeffreys' method involved the hybridization of restriction enzyme-digested DNA fragments to a minisatellite DNA probe and resulted in the simultaneous detection of many highly variable DNA loci. A number of other DNA-based and 'barcode'-providing methods have since been developed, and are now regularly employed to differentiate among cultivars, breeding selections and gene bank accessions in all kinds of plants, including the genus *Rosa*. The DNA fragments provided by these methods show Mendelian inheritance, and have therefore also been used to analyse genetic relationships and to quantify levels of diversity in cultivated, naturalized and spontaneous material.

Methods

The quality of the DNA samples is always a crucial point when working with marker systems that produce complex band profiles. Insufficient and uneven quality can interfere with enzyme digestion and primer annealing, and thus produce considerable artifacts that are not always evident because of the complexity of the band profiles. Unfortunately, considerable problems are often reported with obtaining sufficient DNA quality from rose samples. It is generally advisable only to use young leaves. Numerous DNA isolation protocols have been described in the literature. Most of these are based on CTAB (hexadecyltrimethylammonium bromide), but other isolation buffers may be less sensitive to problems with polysaccharides. Various pretreatments have been suggested, e.g. using large volumes of 2-mercaptoethanol, and adding PVPP (polyvinylpolypyrrolidone) or Triton X100. Purification steps include, for example, phenol extraction and gel filtration, or use of the Qiagen DNeasy plant mini kit.

Hybridization-Based Methods

Species-specific nuclear DNA probes RFLP (restriction fragment length polymorphism) techniques were introduced in the early 1980s. In the original and most simple version, restriction-enzyme digested DNA samples are size-separated with agarose gel electrophoresis and Southern blotted to a nylon membrane which is hybridized to a ^{32}P-labelled DNA probe. The resulting band patterns are detected by autoradiography and show varying levels of complexity depending on the probe used. The probes are usually developed from a genomic library containing random DNA sequences of the studied species (or a close relative). Single-locus RFLP probes provide codominantly inherited bands that are easily interpreted, at least in diploid organisms. In roses, a set of 15 multilocus RFLP probes have been developed using a genomic library from the cv 'Confection'. These probes yield rather complex patterns when applied to other cultivars, with up to nine different and unambiguously scoreable bands in a single probe/enzyme combination (**Figure 1**).

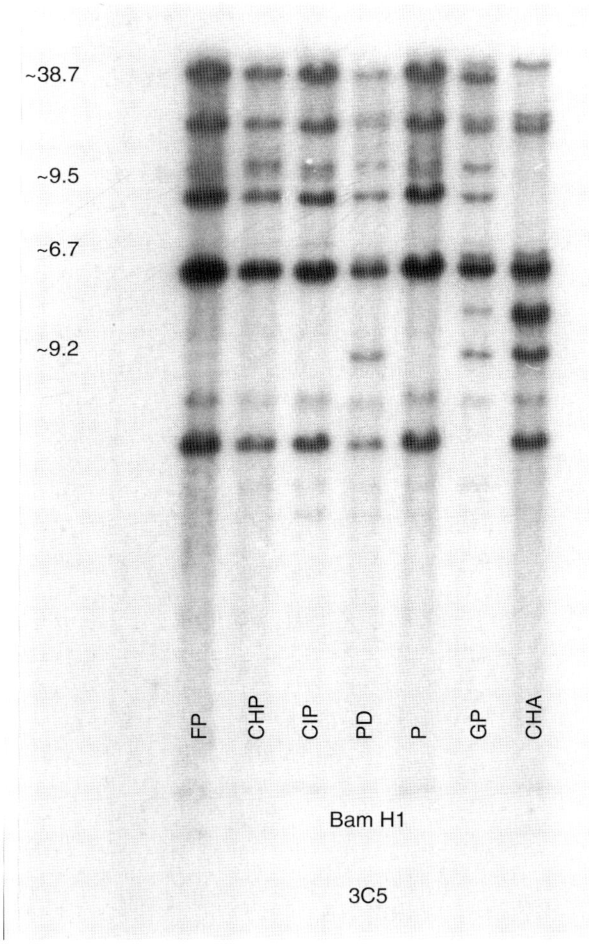

Figure 1 Band profiles obtained from hybridizing the RFLP probe 3C5 with *Bam*HI-digested DNA samples from a set of hybrid tea cultivars. FP 'Flaming Peace', CHP 'Chicago Peace', CLP 'Climbing Peace', PD 'Perfume Delight', P 'Peace', GP 'Garden Party', CHA 'Charlotte Armstrong'. FP, CHP and CLP are all sports of 'Peace', and show the same band profile. Numbers to the left of the figure indicate fragment lengths in kilobases. (Photograph courtesy of S. Rajapakse.)

Nonspecific nuclear DNA probes The introduction of Jeffreys' DNA fingerprint probes obviated the need for developing species-specific probes. Jeffreys' probes are cloned DNA fragments, originally isolated from human DNA, that are complementary to so-called minisatellite DNA. Similar probes have been obtained from various sources, for example, a sequence in the M13 bacteriophage. Another set of useful probes consist of artificially manufactured oligonucleotides that are complementary to so-called simple sequences or microsatellites. The high levels of polymorphism derives from differences in the number of tandem repeats of a short core sequence, 1–5 basepairs (bp) in simple sequence probes and approximately 10–60 bp in minisatellite probes. Consequently, these probes are also known as VNTR (variable number of tandem repeats) probes. The obtained band patterns are quite complex, making it impossible to identify individual loci or to discriminate between heterozygotes and homozygotes.

Three well-known DNA fingerprint probes – M13, Jeffreys' 33.6 and $(GACA)_4$ – have been used in studies of the genus *Rosa*. Each sample/enzyme combination typically yields 15–25 polymorphic bands, resulting in considerable discriminatory power. Individual genotypes usually contain their own, unique DNA fingerprints. The method is therefore very useful for identification, and also allows crude estimations of genetic diversity. Drawbacks are the high sensitivity to experimental conditions, and difficulties in scoring the complicated band patterns. Comparisons are therefore frequently made only among samples that have been loaded on adjacent lanes in the gel. This, of course, seriously hampers the throughput of large series of plant material.

Initial attempts to use both species-specific and nonspecific multilocus nuclear RFLP probes for cultivar discrimination were quite successful in the early 1990s, but the method has now been more or less abandoned and replaced with PCR-based DNA marker systems (described below).

Organelle DNA probes Whereas most DNA analyses are conducted on biparentally inherited nuclear DNA, nonrecombinant uniparentally inherited organelle genomes are very useful for analyses of interspecific relatedness and evolution. In comparison with nuclear DNA, both chloroplast DNA (cpDNA) and mitochondrial DNA (mtDNA) are characterized by low levels of substitution. Consequently, a number of nonspecific probes have been developed that hybridize with organelle DNA, especially cpDNA.

In roses, few reports have been presented on the application of organelle DNA probes. In one study, cpDNA probes developed in tobacco and mtDNA probes developed in sugar beet were employed for analysis of interspecific variation. However, the levels of polymorphism encountered were quite low, and identical band profiles were reported in several cases even for species that belong to different sections.

PCR-Based Methods

Almost all DNA marker applications today make use of the polymerase chain reaction (PCR) in some way. Among the different methods available, random amplified polymorphic DNA (RAPD) and amplified fragment length polymorphism (AFLP) have become the most commonly used in roses. Both these methods produce multilocus and highly polymorphic band patterns, mostly of nuclear DNA origin. A more locus-specific approach, with PCR amplification of repeated DNA sequences, so-called microsatellites, is gaining increased interest and has been applied both to nuclear and cpDNA in roses.

Random amplified polymorphic DNA In the RAPD method originally described in 1990, genomic DNA (template) is incubated together with an arbitrarily chosen primer that binds to complementary sites on the template DNA, an enzyme, nucleotides, and a suitable buffer system. Depending on where the primer binds to the sample DNA, various stretches of the sample DNA are copied by the action of the polymerase enzyme when submitted to a temperature cycle in a thermocycler. The resulting DNA amplification products are then size-separated by electrophoresis in agarose gels, and the bands are detected using ethidium bromide and ultraviolet (UV) illumination.

Usually short (10 bp), commercially available primers are used. Longer primers (up to 20 bp) have been reported to favour PCR amplification of nonrepeated sequences whereas repeated sequences are amplified preferentially by shorter primers. In roses, primers of different length (10, 15 or 20 bp) have all produced approximately the same number of bands in the resulting profiles. Band profiles obtained with longer primers (15 or 20 bp) may be more reproducible since they are used with a higher DNA annealing temperature (55 °C instead of the 35–40 °C used with shorter primers) which could result in more precise primer site pairing.

In roses, as in other plant groups, the number of scored polymorphic bands usually ranges from 5 to 15 in each primer/sample combination if the compared samples are genetically diverse. In order to obtain sufficient bands for an analysis, results from several primers frequently have to be pooled. Sometimes primers are used in pairs in the hope of thus obtaining a set of new bands, not detected when the same primers are used singly. In roses, levels of band pattern complexity are usually very similar regardless of whether the PCR primers are used pairwise or singly. Using the longer primers (15 or 20 bp) in pairs has, however, resulted in a higher proportion of new bands compared to using shorter primers in pairs. The fact that band complexity is not affected by length or number of primers suggests that primer site pairing is not perfect, and that competition occurs between primer sites.

Anonymous RAPD bands are often used to quantify genetic variation, and to aid in plant breeding, ecology, population genetics and systematics. In spite of the dominant inheritance of bands, RAPD has proved to be just as useful as codominantly inherited isozymes for detecting relationships between population genetics parameters and different life history traits (**Figure 2**).

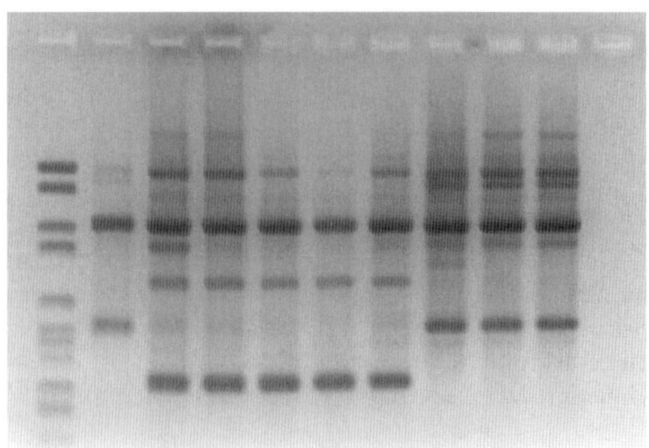

Figure 2 RAPD bands produced by primer OPA-16. From left to right: lane 1 molecular weight marker no. IV (Boehringer and Mannheim), lane 2 *R. dumalis*, lane 3 *R. rubiginosa*, lane 4–7 *R. rubiginosa* × *R. dumalis* hybrid plants, lanes 8–10 *R. dumalis* × *R. rubiginosa* hybrid plants. The three lowermost bands were scored and proved useful for discrimination between the parental plants as well as the two progeny groups. (Photograph courtesy of G. Werlemark.)

Amplified fragment length polymorphism The AFLP method incorporates elements of both RFLP and RAPD. Genomic DNA is digested into suitable fragments using two restriction enzymes. Adaptors (defined short nucleotide sequences) are ligated to both ends of the resulting fragments. PCR is then performed using these fragments as templates, and employing primers designed so that the 5′ end is complementary to one of the adaptor sequences and thus to one end of all the fragments. The 3′ end (from which the actual amplification takes place) is complementary to the other end of the fragments (i.e. to the other adaptor sequence); however, the primer 'sticks in' a little further (usually 1–4 bp) on the fragment. And thus only fragments that happen to have a sequence corresponding to this 1–4 bp part of the primer are used as templates for the amplification.

Usually there are two rounds of amplification: first, one with primers that will amplify numerous fragments, and then a second with primers designed to be more selective. In this second amplification, one of the primers may be labelled with a radioisotope or a fluorescent dye. Amplification products are electrophoresed on polyacrylamide gels and detected with autoradiography or, increasingly commonly, analysed using automated sequencing equipment. Numerous bands are produced for each primer pair. These bands are dominantly inherited as are the RAPDs.

The main difference between AFLP and RAPD is that AFLP yields more bands for each primer/sample combination, but it is also more technically demanding. Initially, AFLP may also be more time-consuming since there are more steps for which the experimental conditions must to be optimized. Thus AFLP is cost-effective mostly in situations where a large number of analyses need to be performed. AFLP has also been reported to yield more highly reproducible band patterns. To provide good reproducibility, DNA sample quality is however crucial since poor quality may seriously interfere with the activity of DNA restriction enzymes.

Finally, it should be mentioned that markers derived by AFLP and RAPD (and to a lesser extent RFLP) are prime candidates for genetic mapping (*see* **Genetics**: Gene Mapping) and marker-assisted selection (*see* **Genetics**: Marker-Assisted Selection).

Microsatellite DNA Another PCR-based method, which is gaining rapidly in interest, is sequence tagged microsatellite sites (STMS) using sequencing information to develop pairs of primers that amplify hypervariable microsatellite loci (also known as simple sequence repeats, SSR) in the nuclear DNA genome. The results are usually evaluated as alleles belonging to defined loci, and the method is therefore generally not included among the DNA fingerprinting techniques. Occasionally the band patterns obtained are treated instead as multilocus data, especially when analysing polyploid plants.

So far, there are very few published reports on the application of microsatellite DNA markers in rose research but a number of projects using this method have recently been initiated. The method has several features (individual loci are identified, allelic configuration can be ascertained, and bands are codominantly inherited) that make it more useful in certain areas than the more typical DNA fingerprinting methods. Promising results have thus been reported for the application of STMS to cultivar identification (*see* **Classification**: Cultivar Identification by Image Analysis), in-depth genetic analyses (*see* **Genetics**: Inheritance in the Dogrose) and genome mapping (*see* **Genetics**: Gene Mapping).

Microsatellite regions occur also in cpDNA and mtDNA, where they consist mainly of short poly(A) or poly(G) sequences. Variation in the number of repeat units can be studied using appropriate primers. Species-specific cpDNA microsatellite primers have proved to yield high levels of polymorphism between and within species, but their development requires sequencing information. Using a different approach, conserved primers for analysis of microsatellite motifs in cpDNA have recently been developed in tobacco and have proved to be informative in many other plant genera and even families (**Figure 3**). Levels of intraspecific variation using the conserved primers are limited. Some of these primers have, however, been successfully applied to the differentiation of rose species.

Additional methods Several additional methods have been applied to obtain polymorphic DNA band patterns in plants (*see* **Genetics**: Molecular Markers) but have not yet been much employed in rose research. One of the better-known methods for analysis of nuclear DNA is inter simple sequence repeat (ISSR) which is quite similar to RAPD but uses primers that have been designed to guide amplification mainly to regions with repetitive DNA. Consequently, ISSR markers are usually characterized by very high levels of polymorphism.

A series of so-called universal PCR primers has been developed which allow the amplification of chloroplast introns and intergenic spacers in a wide array of plant species. These amplification products may show some length variation but are usually treated with restriction enzymes to reveal nucleotide substitutions as well. No data have been reported for roses using this method, and only very low levels of both intra- and interspecific variation in other rosaceous

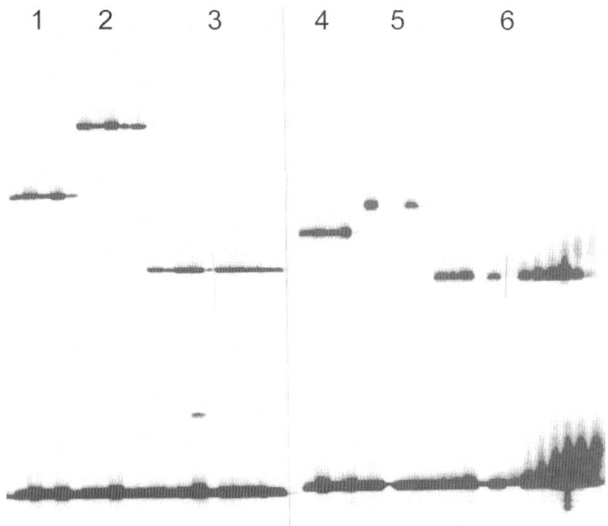

Figure 3 Amplification products obtained from two *Rosa* and two *Dahlia* genotypes, using two different conserved cpDNA microsatellite primer pairs. Each genotype is represented by five lanes in which the same DNA was subjected to slightly different reaction conditions. (1) *R. multiflora* hybrid and primer pair ccmp6; (2) *R. rugosa*, ccmp6; (3) two different *Dahlia* genotypes (still with five lanes each), ccmp6; (4) *R. multiflora* hybrid, ccmp10; (5) *R. rugosa*, ccmp10; (6) two *Dahlia* genotypes, ccmp10. (Photograph courtesy of T. Debener and L. Mattiesch.)

genera. Consequently the method appears, on the whole, to be more useful for studies at higher taxonomic levels.

Band Scoring and Statistics

DNA fingerprint bands, whether obtained with multilocus hybridization probes or with the PCR-based AFLP and RAPD methods, are usually scored as biallelic loci with bands being either present or absent. Linkage analyses have, in general, found very little association between the scored bands, indicating that primer sites are distributed over the entire genome instead of confined to restricted areas. Bands that are monomorphic (i.e. always present) across all samples studied are usually discarded and therefore not included in subsequent statistical treatments. Bands obtained in different samples using the same probe or primer and having the same size are usually treated as homologous. Studies in other plant genera have demonstrated that this assumption is quite valid if analyses are carried out within the same species or between closely related species.

Various similarity indices are calculated among pairs of samples. Some of these, like Jaccard or Dice's similarity indices, use only positive matches (shared band presences) whereas Apostol's simple matching index also includes negative matches (shared band absences). Comparisons among matrices derived with different similarity indices, however, show that they yield highly correlated results, although they differ in absolute values.

Analyses of larger sets of samples are usually based on matrices obtained with one of the similarity indices, and are conducted using cluster analyses, commonly unweighted pair group method with arithmetic mean (UPGMA) and principal coordinate analysis (PCO). These analyses represent relative similarities between single samples and /or groups of samples in a graphic form.

For estimating the level of variability within populations, expected heterozygosity, H_e, is commonly calculated. When based on polymorphic loci only, this parameter is identical to Nei's gene diversity, H_S (provided that sample size is identical for all populations) and can be calculated as $H = 1 - p^2 - q^2$, where p is the frequency of the allele for marker distribution and q is the frequency for the null allele. Unfortunately, p^2 cannot be determined directly from band profiles when using dominant markers. In roses, which are typically outcrossing or at least have mixed breeding, we must assume Hardy–Weinberg equilibrium for each locus in order to obtain the more practical formula: $H = 2q(1-q)/2$ where q is derived from x, with x being the frequency of individuals that lack the marker. Estimates of H_S based on AFLPs and RAPDs are usually almost identical when obtained on the same plant material, but isozyme analyses also generally yield similar values. By contrast, microsatellite DNA analysis usually yields values for H_e that are 2–3 times higher. In order to avoid making assumptions about Hardy–Weinberg equilibrium, another parameter is occasionally used for estimating within-population variability, namely Shannon's index. This can be calculated in different ways and the results are therefore less readily compared across studies.

Partitioning of genetic variability at different levels is usually conducted using analysis of molecular variance (AMOVA), Nei's G_{st} or Shannon's diversity index. The former two parameters yield very similar estimates for between-population differentiation when based on AFLP or RAPD, whereas the corresponding estimates obtained with isozymes and microsatellites are considerably lower. Shannon's index yields estimates that are correlated with the others but differ considerably in absolute values.

Applications

Differentiation of Cultivars

Cultivar identification is a major field of interest since infringement of patents constitutes a considerable problem in floriculture (*see* **Intellectual Property**: Plant

Patents and Trademarks; Technical Examination and Conventional Characteristics). Many rose cultivars have been obtained through separate events of genetic recombination and are therefore genetically quite different from each other (*see* **Classification**: Cultivar Identification by Image Analysis). By contrast, sports usually differ from the original cultivar only in ornamental characters like flower colour or flower shape (*see* **Genetics**: Mutation). Although DNA markers are usually more discriminatory than morphological traits, the ability to identify mutants remains very low since observed mutations usually involve only one locus or a few loci. Chances of finding a marker for exactly that locus (or those loci) are quite small, especially since marker polymorphism often originates from noncoding sequences.

Different kinds of molecular analyses have been employed in efforts to distinguish rose cultivars (*see* **Classification**: Cultivar Identification by Image Analysis). The level of success depends mainly on the amount of genetic differentiation among the cultivars examined and the number of polymorphic bands produced in the analysis. Hybridization-based DNA fingerprinting, using multilocus RFLP probes or some of the VNTR probes (M13, 33.6 and $(GACA)_4$), was applied in the early and mid 1990s to limited sets of genetically diverse cultivars. Complete discrimination of all the investigated cultivars (ranging between 13 and 19) was generally achieved using only one or two of the probes.

More recently, several studies have shown that RAPD and AFLP are similarly successful in cultivar discrimination. In general, the percentage of bands that are polymorphic is somewhat larger when using RAPD compared to AFLP, but the total number of bands analysed is usually much higher in AFLP studies. With RAPD, mostly smaller sets of cultivars (10–25) have been analysed, typically using 5–15 primers. By careful prescreening of RAPD primers, superior primers can be chosen that produce up to 15–20 polymorphic bands, thus enabling efficient discrimination with even fewer primers. Occasionally, RAPD primers are able to discriminate also between sports and their original cultivar but in other cases no polymorphism is found. In one example, a set of RAPD markers was not able to find any differences among four of the earliest Damask roses, which is taken as evidence that they have been derived from the same ancestor through vegetative propagation only. For production of very large numbers of markers, AFLP is usually more convenient and has been used for analyses of sets of 100 or more cultivars. However, even very large numbers of AFLP-generated markers (up to 400 polymorphic bands) can still separate only some of the sports from their original cultivars.

Differentiation of Gene Bank Accessions

The increased awareness about importance of preserving genetic resources has renewed the interest in various kinds of rose collections, from larger gene banks to smaller rosaries in botanical gardens or at plant breeding institutes (*see* **Rose Collections and Trials**: Rose Collections and Herbaria). Many of these collections have numerous problems with misidentification, mislabelling and incomplete definition of taxa. So far, DNA markers have not been as extensively utilized in the management of gene banks for roses as for many other plant groups. Still, some efforts to use RAPD, for example, for characterizing different accessions have been quite successful. Some accessions have proved to be identical in spite of being acquired from different sources, indicating their derivation by vegetative propagation from a common origin. High levels of similarity between other accessions have indicated that they most likely represent the same species in spite of sometimes being labelled differently.

Assessment of Relatedness

Since DNA fingerprinting bands appear to be mostly nuclear in origin and show Mendelian inheritance, relatedness among genotypes can be indicated from estimations of band pattern similarity. Thus, several studies have demonstrated that levels of DNA band pattern similarity among cultivars are closely associated with their known levels of relatedness according to pedigree information.

For many cultivars, exact pedigree information is, however, lacking. Instead, suggested parentage is often only speculative, based on recollection and morphological characteristics. By comparison of DNA band patterns obtained from the cultivar in question and its putative parents, some suggested relationships have been refuted and others appear highly likely.

In plant breeding programmes, knowledge about the exact origination of new genotypes is often essential. Since the regulations for plant breeders' rights to sports (essentially derived varieties) differ from those of other cultivars, it is important to distinguish progeny obtained from vegetative propagation from those obtained by sexual propagation (*see* **Intellectual Property**: Plant Patents and Trademarks). RAPD and AFLP analyses have proven very efficient in differentiating true hybrid progenies from their seed parents, indicating that sports (which generally differ by one or two bands, at the most) can be told apart from sexually derived cultivars (which differ by a large percentage of the analysed bands). Similarly, RAPD analysis has proved useful in differentiating between apomictically and sexually derived offspring in the section *Caninae* (*see* **Genetics**: Inheritance in the

Dogrose), where these two modes of reproduction appear to exist not only in the same species but also in the same genotype. Analysis of paternity can also help to distinguish seedlings emanating from successful cross-fertilizations and those resulting from unwanted fertilization events, e.g. selfing or fertilization by other parents.

The number of primers and bands needed to achieve sufficiently discriminatory marker profiles depends mainly on the level of relatedness among the putative parental genotypes and the amount of marker heterozygosity. The latter parameter can be estimated from analysis of offspring families and has been shown to vary considerably between genotypes, e.g. between 41 and 94% in one study using AFLP markers.

Cultivar Diversity

The strong selection for particular horticulturally important characteristics has resulted in a plant material where many cultivars show high levels of similarity (*see* **Classification**: Cultivar Identification by Image Analysis). Molecular markers have proved very useful for evaluation of the 'true' levels of genetic diversity in this material. RAPD-based dendrograms demonstrate that classical horticultural groups (*see* **Classification**: Horticultural Classification Schemes) consisting mainly of the old, founding cultivars (e.g. Gallicas, Damascenas, Chinenses) constitute relatively well-delimited entities, with a major division between Chinese founders and European founders. By contrast, representatives of the hybrid groups (e.g. Floribundas and Hybrid Teas) are more scattered in the dendrograms, most likely because they have been derived from interspecific hybridization events and now represent mosaics of the genomes of different species (*see* **Genetics**: Species Crosses). Nevertheless, estimates of genetic diversity obtained from RAPD band frequencies show an increasing reduction when initial founder groups are compared with groups containing the most modern cultivars.

Population Genetics of Wild or Naturalized Germplasm

Analyses of wild, or potentially wild germplasm have been carried out mainly to facilitate future collections and to decide on conservation strategies. Consequently, efforts have been made to estimate the proportions of ramets versus genets, and to quantify within- and between-population variation. An interesting case history has been provided by the recent use of RAPD in an effort to differentiate among 27 accessions of Cherokee rose, *R. laevigata* Michx., a species previously thought to be native in the USA but later shown to be introduced from China. In the southern states of the USA, this species occurs persistently at dwellings and other sites of deliberate planting but appears not to have reseeded spontaneously. Most accessions showed identical RAPD profiles suggesting that they had been derived from vegetative propagation, which according to historical records probably took place in 1820–21 and was followed by the distribution of material to a huge area from California in the west to Georgia in the east. This propagation appears to have involved mainly two clones, which differ not only in RAPD profile, but also in one isozyme band, and in the amount of bristles on the laterals.

Several RAPD-based studies have been conducted in wild Scandinavian populations of species belonging to the very homogeneous section *Caninae*. These species do not propagate vegetatively, except over very short distances. Nevertheless, there is exceptionally little intraspecific variation owing to severely restricted genetic recombination (*see* **Genetics**: Inheritance in the Dogrose). Consequently, levels of RAPD marker polymorphism are very low, and a considerable number of primers must be screened in order to obtain sufficiently polymorphic markers. In spite of these limitations, the RAPD-based estimates of intraspecific variability have been closely correlated with previous estimates obtained with morphological characterization. Partitioning of RAPD-generated variability at different levels (population, between maternal family, within maternal family) has also yielded informative results which demonstrate considerable differences between species.

So far, organelle DNA-based analyses have not been much used in roses. However, cpDNA microsatellites have yielded much valuable information in several other plant groups. Since the chloroplast genome does not recombine genetically, all the cpDNA microsatellite data for a certain individudal can be combined to form a 'haplotype', and genetic diversity measures at various population genetics and taxonomic levels can then be derived from haplotype frequency data. In the future, the application of genus-specific as well as conserved microsatellite DNA primers in roses may become a useful tool for various research areas, such as the detection of genetic bottlenecks during the domestication process, and for tracing the maternal genome after hybridization events.

Taxonomic Relationships and Phenetics

Several DNA-based approaches have been used to investigate taxonomic relationships at various levels, and the results have then been compared with more traditionally derived taxonomic treatments (*see*

Classification: Conventional Taxonomy (Wild Roses)). For phylogenetic and phylogeographic analyses, DNA sequencing information from carefully identified genes (especially their introns) is usually the preferred method (*see* **Classification**: Chemotaxonomy and Molecular Taxonomy). By contrast, phenetic analyses, i.e. a representation of present-day patterns of similarity among taxa, are very amenable to DNA fingerprinting methods.

Several smaller studies have been performed using RAPD, in which dendrograms constructed for intra- and interspecific analyses of 5–10 species have generally showed a good correlation with previous classifications based on morphology and karyology. In these analyses, samples belonging to the same species have usually been grouped together. Samples of the taxonomically rather isolated section *Caninae* have constituted a close-knit group with overall low levels of diversity.

The largest study to date is a phenetic analysis of 119 accessions, representing 36 species (comprising one species each from the subgenera *Platyrhodon* and *Hesperhodos*, and 34 species from a total of eight sections of subgenus *Eurosa*). A set of 10 preselected primers was used, which together yielded 213 polymorphic RAPD markers. All accessions could be differentitated, and almost all species were grouped into their respective sections when an UPGMA analysis was applied to a matrix based on Dice's similarity index. Apart from the surprisingly isolated section *Bracteatae*, a major division occurred between Asian sections on the one hand and primarily North American sections on the other hand. It was suggested that the two subgenera *Platyrhodon* and *Hesperhodos* should be placed within subgenus *Eurosa*, since they grouped together with the North American *Eurosa* sections *Carolinae* and *Cassiorhodon*.

See also: **Classification**: Conventional Taxonomy (Wild Roses); Horticultural Classification Schemes; Cultivar Identification by Image Analysis; Chemotaxonomy and Molecular Taxonomy. **Genetics**: Inheritance in the Dogrose; Species Crosses; Mutation; Gene Mapping; Molecular Markers; Marker-Assisted Selection. **Intellectual Property**: Plant Patents and Trademarks; Technical Examination and Conventional Characteristics. **Rose Collections and Trials**: Rose Collections and Herbaria.

Further Reading

Avise JC (1994) *Molecular Markers, Natural History and Evolution*. New York: Chapman & Hall.

Bachmann K (1997) Nuclear DNA markers in plant biosystematic research. *Opera Botanica* 132: 137–148.

Ballard RE, Rajapakse S, Abbott A and Byrne DH (1995) DNA markers in rose and their use for cultivar identification and genome mapping. *Acta Horticulturae* 424: 265–268.

Camlin MS (2001) Possible future roles for molecular techniques in the identification and registration of new plant cultivars. *Acta Horticulturae* 546: 289–296.

Crespel L, Zhang D, Meynet J and Gudin S (2001) Estimation of heterozygosity in two botanic rose species using AFLP markers. *Acta Horticulturae* 546: 187–191.

Cubero JL, Millán T, Osuna F, Torres AM and Cobos S (1995) Varietal identification in *Rosa* by using isozyme and RAPD markers. *Acta Horticulturae* 424: 261–264.

Debener T and Mattiesch L (1995) Genetic analyses of molecular markers in crosses between diploid roses. *Acta Horticulturae* 424: 249–252.

Gudin S (2000) Rose: genetics and breeding. *Plant Breeding Reviews* 17: 159–189.

Henry RJ (ed.) (2001) *Plant Genotyping: The DNA Fingerprinting of Plants*. Wallingford: CAB International.

Hodgkin T, Roviglioni R, De Vicente MC and Dudnik N (2001) Molecular methods in the conservation and use of plant genetic resources. *Acta Horticulturae* 546: 107–118.

Jeffreys AJ, Wilson V and Thein SL (1985) Hypervariable 'mini-satellite' regions in human DNA. *Nature* 314: 67–73.

Nybom H (2001) DNA markers for different aspects of plant breeding research and its applications. *Acta Horticulturae* 560: 63–67.

Nybom H and Bartish IV (2000) Effects of life history traits and sampling strategies on genetic diversity estimates obtained with RAPD markers in plants. *Perspectives in Plant Evolution, Ecology and Systematics* 3: 93–114.

Provan J, Powell W and Hollingsworth PM (2001) Chloroplast microsatellites: new tools for studies in plant ecology and evolution. *Trends in Ecology and Evolution* 16: 142–147.

Reynders-Aloisi S and Bollereau P (1995) Characterization of genetic diversity in genus *Rosa* by random amplified polymorphic DNA. *Acta Horticulturae* 424: 253–260.

Vos P, Hogers R, Bleeker M *et al.* (1995) AFLP: a new technique for DNA fingerprinting. *Nucleic Acids Research* 21: 4407–4414.

Weising K and Gardner RC (1999) A set of conserved PCR primers for the analysis of simple sequence repeat polymorphisms in the chloroplast genomes of dicotyledonous plants. *Genome* 42: 9–19.

Weising K, Nybom H, Wolff K and Meyer W (1995) *DNA Fingerprinting in Plants and Fungi*. Boca Raton: CRC Press.

Williams JGK, Kubelik AR, Livak KJ, Rafalski JA and Tingey SV (1991) DNA polymorphisms amplified by arbitrary primers are useful as genetic markers. *Nucleic Acids Research* 18: 6231–6235.

Gene Mapping

S Rajapakse, Clemson University, Clemson, SC, USA

Introduction

The term 'genome' combines both genes and chromosomes, the genetic material in cells that provides the basis for transmission of traits from parents to offspring. Genes, located linearly on chromosomes in the nucleus of cells, are the basic hereditary units that control morphological and physiological traits. In rose cell nucleus, chromosomes occur in sets of seven with some roses having two sets of seven (diploid), while others have three (triploid), four (tetraploid) or five (pentaploid), six (hexaploid) or even eight (octoploid). Some wild rose species are diploids, most hybrid tea roses are tetraploids and certain dog roses are pentaploids. In addition to the genes located on the chromosomes in the cell nucleus, rose cytoplasm also contains genes in mitochondria and chloroplasts. These cytoplasmic genomes are small compared to the large nuclear genome and have inheritance patterns that are different from the nuclear genome.

Rose genome analysis primarily involves understanding the structural organization of genes on the seven basic chromosomes and the function of these genes in regulating morphological and physiological characters. Understanding the structural organization of genes controlling important rose traits will provide useful information that can improve the efficiency of conventional rose breeding programmes. For example, DNA markers located close to important genes can be used to select breeding progeny for that trait through a marker-assisted selection programme (*see* **Genetics**: Molecular Markers; Marker-Assisted Selection). Furthermore, markers linked to traits, coupled with genetic maps, could be used to track the introgression of beneficial genes such as those that control disease resistance into roses lacking disease resistance. Genomic information also provides a systematic means of combining valuable characters found in different roses into one. Isolation of genes using map information, known as map-based gene cloning, could lead to the transfer of important genes to other roses lacking these traits through genetic engineering methods.

The structural organization of the rose genome can be fathomed by several means. The primary methods include constructing genetic maps and physical maps. In constructing genetic maps, the organization of genes on chromosomes is deduced from patterns of inheritance of various traits, whereas in constructing physical maps, the order of fragments of chromosomal DNA is determined. Comparing the gene organization with other related plants that have been studied in detail is another powerful method of understanding genomic organization. This method is known as comparative mapping. Until now all these methods have been used to study the rose genome only to a limited extent. However, these methods can be used to generate valuable genomic information in the future. Compared to crops like wheat (*Triticum aestivum*) and maize (*Zea mays*) that have large genomes, the small size of the rose genome makes it more amenable to analysis. But the high level of heterozygosity generally found in roses and the presence of more than two chromosome sets in polyploid roses complicate genomic analysis. The common methods used to study the genomic organization and our current understanding of the rose genome are presented below.

Genetic Mapping

Genetic maps provide insight into the organization of the rose genome via gene linkage and recombination. Certain genes located on the same chromosome are inherited together as a unit and are said to be linked, while genes located on separate chromosomes are unlinked and assort independently from each other. The linked gene units are termed linkage groups, equal in number to the paired chromosomes. During gamete formation in meiosis, linked genes are sometimes separated when sister chromatids exchange pieces. This gene shuffling breaks the linkage between genes. How often a pair of genes is separated during recombination provides a measure of relative distance between them. Two genes with a recombination rate of 1% (i.e. linkage is broken in 1 out of 100 meiotic events) are described as being 1 map unit or 1 centiMorgan (cM) apart. This genetic map unit honours the pioneer geneticist TH Morgan who discovered linkage in the early 1900s. Formulae have been developed to convert the rate of recombination to map distances. It is possible that a second recombination event can occur between the loci and link the genes again. These double crossovers have to be considered in computing map distances. The principles underlying genetic mapping apply to genes as well as to DNA markers. Thus, analysis of rate of recombination between genes (and/or markers) allows us first to assign a gene to a linkage group, and then to determine the order of the genes within the group.

Genetic linkage maps are constructed to span either all chromosomes or a localized region in a specific chromosome that harbours the gene of interest.

Maps covering all chromosomes provide a structural framework of the entire rose genome. Local linkage maps focused on one area of the genome often provide markers tightly linked to the gene useful for progeny selection in the breeding programmes and map-based cloning of the gene. Local maps will eventually be incorporated into main chromosomal maps. Maps produced by various researchers with different marker systems (*see* **Genetics**: Molecular Markers) can be combined to produce a single gene map for rose, provided that a sufficient number of anchor points exists among the maps. This consolidated gene map can serve as a genomic information resource base for understanding the genetic control of rose traits and the development of molecular breeding methods for the improvement of the rose germplasm.

A necessary prerequisite for studying genes controlling important traits is the development of test progeny that segregate for the traits of interest by controlled crosses such as hybridizing, backcrossing, selfing or other methods. Test families with a large number of offspring lead to more accurate estimates of rate of recombination between markers and genes, and hence map distances. The development of such large progenies remains one of the most challenging aspects of gene mapping in roses. Therefore, it is important first to identify the traits to be mapped and then carefully to select or develop test progeny informative for the trait. To construct a genetic map of the entire genome, initially a framework map of all chromosomes is made with DNA markers developed from the test population. These DNA markers provide landmarks on the rose chromosomes where genes controlling traits will be mapped. To identify genes controlling a specific trait such as resistance to black spot, the test progeny is first screened for actual disease resistance and then analysed for DNA markers that cosegregate with the resistance. Instead of screening the DNA of a large number of progeny plants individually, DNA from 10–15 resistant plants is pooled or 'bulked' and compared with a similar bulk made of DNA from susceptible plants. This approach expedites the discovery of markers (or pieces of DNA) linked to the resistance.

Traits that are known to be controlled by one or a few nuclear genes through genetic studies are the most common candidates selected for genetic mapping. Examples for such qualitative traits in rose include recurrent blooming, dwarfness and double corolla. Due to their simple genetic control and ease in phenotypic evaluation, these traits are studied and mapped more often. In contrast, quantitative traits such as fragrance, petal size and prickle size are thought to be controlled by a number of gene loci. Individually these genes may have a small influence on the trait, but collectively they have a strong effect on the trait. Mapping these quantitative trait loci (QTL) requires a large number of progeny segregating for the trait, a fairly saturated genetic map, precise quantification of the trait of interest and special mapping software. However, among the QTL regulating a trait, there could be few genes or chromosomal regions that contribute quite heavily to the genetic variation of the trait than the other loci. Markers in these major QTL regions could also be useful in selecting progenies in the breeding programmes, similar to the manner in which markers linked to qualitative traits are used.

Compared to crops in which inbred lines can be easily developed, the high level of heterozygosity found within the diploid rose genome can complicate the genetic mapping process. Polyploidy, or the presence of more than two chromosome sets in the cell nucleus of many roses, adds yet another level of challenge to rose genetic mapping. Diploid roses that have the desired gene to be mapped are sometimes hard to find and polyploids will have to be used. Because of the large number of genetic variants that can be found in a given locus, genetic mapping in polyploids is complex and requires large progeny sets and special data analytical methods. On the other hand, synthetic doubled haploid roses that do not have any allelic variation can be produced via anther or ovule culture and would simplify the genetic mapping process. However, such progenies are currently not available.

Rose Genetic Maps

Genome-wide genetic maps have been constructed for diploid and tetraploid roses. Two pairs of maps are presently available for diploid roses and one pair for the tetraploids. Genetic maps for diploid *R. multiflora* Thunb. ex Murr. have been constructed with F_1 hybrids generated from hybridization between two related roses. Two maps constructed, one for each parent of the cross, consist of seven linkage groups each (**Figure 1**). The maps cover 326 and 370 cM of the genomes with an average distance between markers of 2.4 and 2.6 cM, respectively. Due to the dominant nature of the randomly amplified polymorphic DNA (RAPD) and amplified fragment length polymorphism (AFLP) markers used (*see* **Genetics**: Molecular Markers), the maps are specific to each parent and are not combined. But many anchoring points between the two maps have been identified using markers common between maps. Two genes, one controlling double corolla (*Blfo*) and another for the pink flower colour (*Blfa*), are located on these maps. Many DNA markers tightly linked to these two genes located on separate linkage groups have been identified. Subsequently, two additional genes were

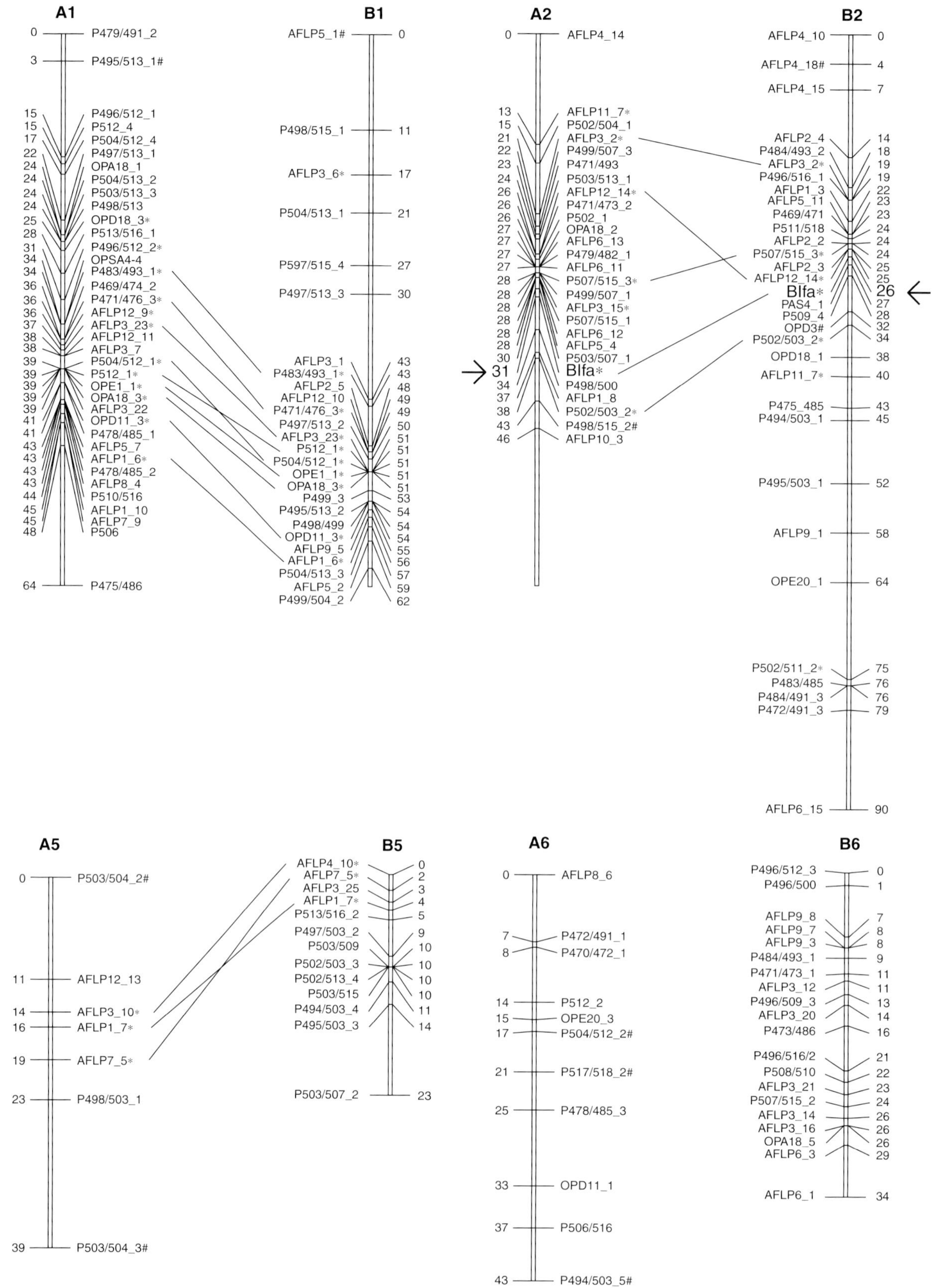

Figure 1 Linkage maps of the diploid *Rosa multiflora* rose genotypes 93/1117 and 93/1–119. A1–A7 linkage groups based on markers of the core map inherited from 93/1–117. B1–B7 linkage groups based on markers inherited from 93/1–119. Markers with skewed segregation ratios added at the second round of map construction are marked with a #. Markers inherited from both parents are marked with a *. Distances between markers are given in centiMorgans. Markers inherited by both parents and common to linkage groups are connected by solid lines. The two genes controlling the phenotypic characters doubleflowers, *Blfo*, and pink flower colour, *Blfa*, are denoted by arrows. Reproduced from Debener and Mattiesch (1999) Construction of a genetic linkage map for roses using RAPD and AFLP markers. *Theoretical and Applied Genetics* 99: 891–899 with permission of Springer-Verlag.

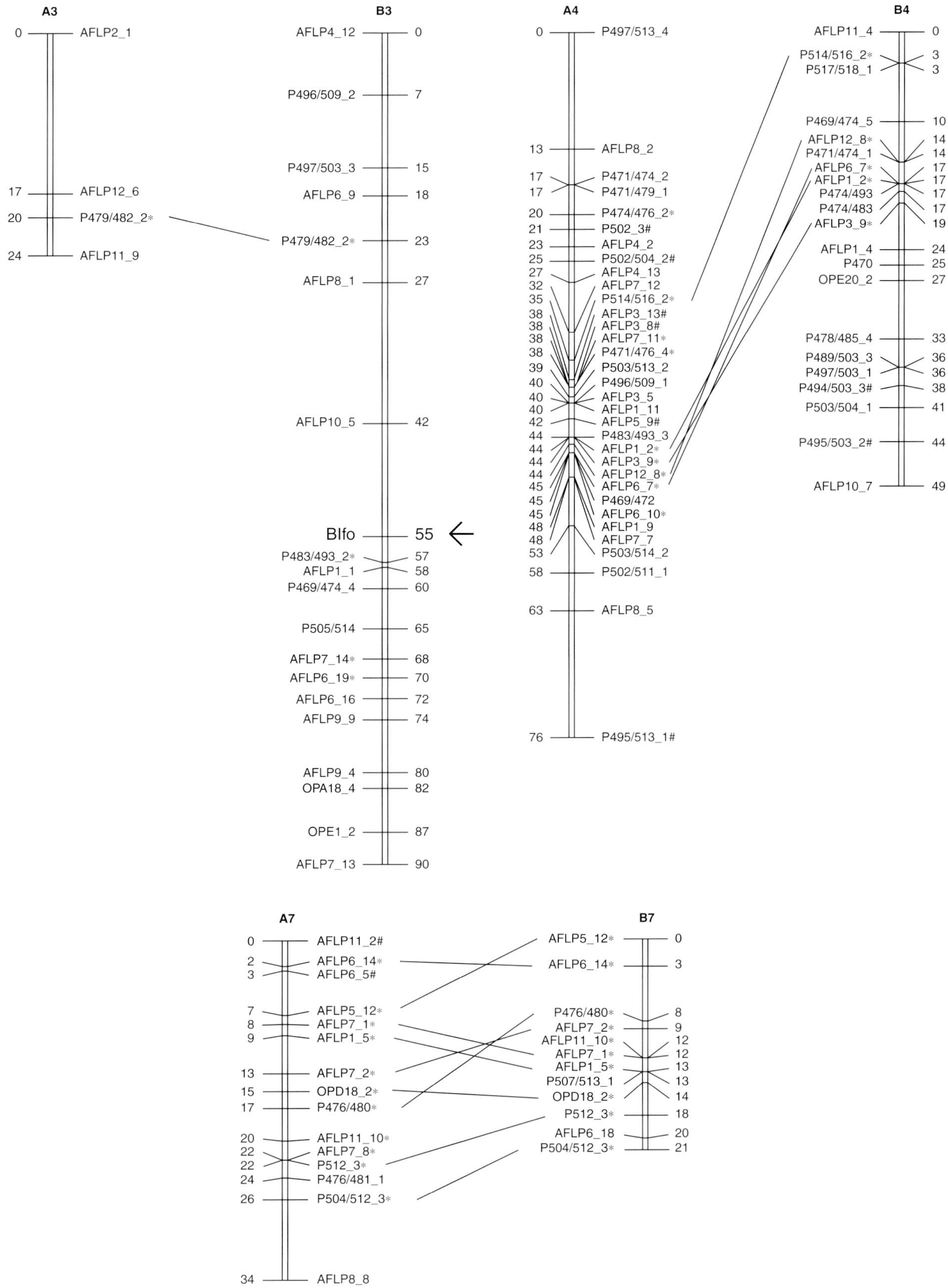

Figure 1 Continued.

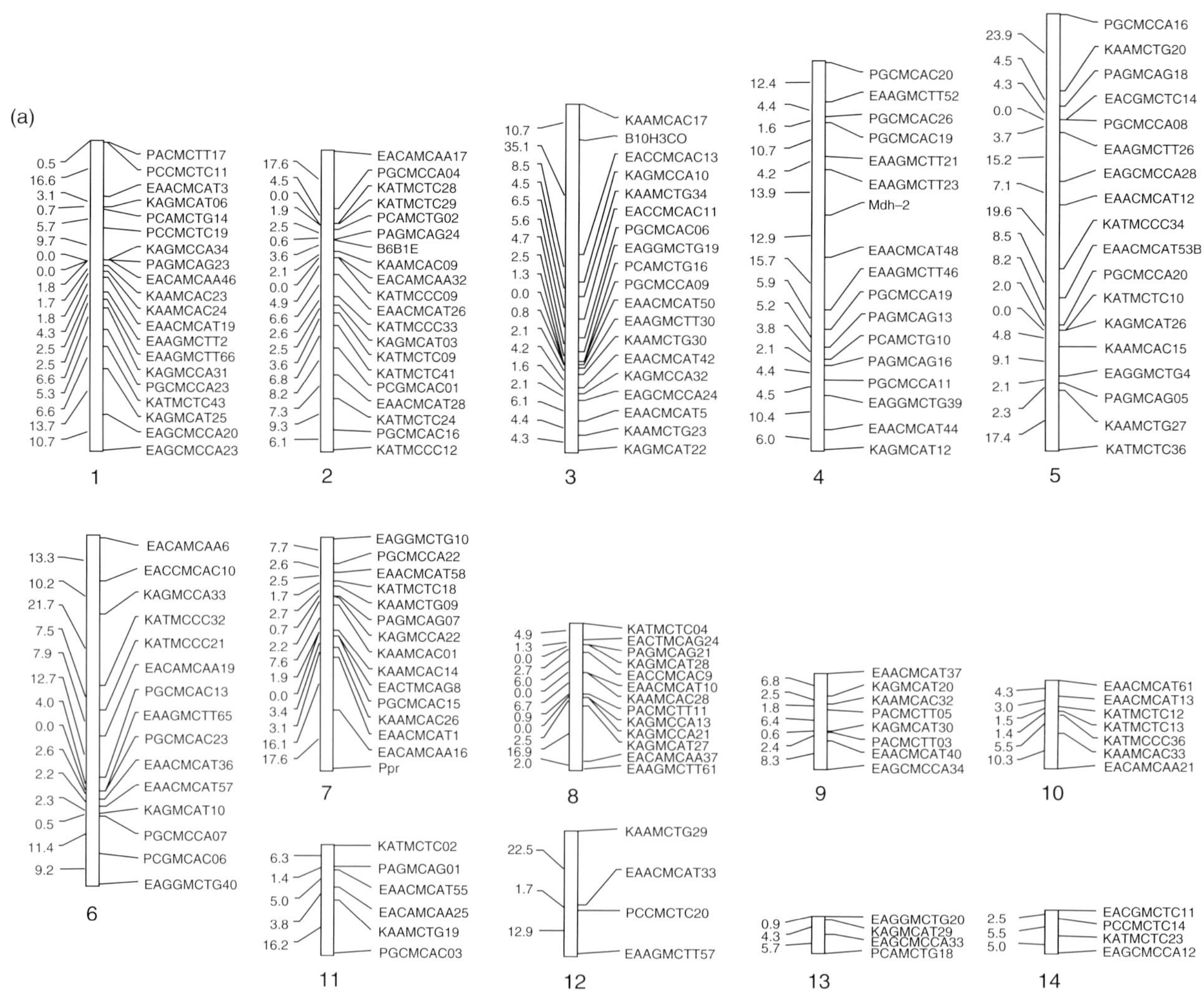

Figure 2 Genetic maps of tetraploid roses constructed with single-dose markers. (a) Linkage map of amphidiploid rose 86–7. (b) Linkage map of tetraploid rose 82–1134. Linkage groups are numbered arbitrarily. Markers are indicated to the right of the linkage groups and genetic distances (in centiMorgans) are given to the left of the group. Linkage groups with fewer than four markers were not drawn. A gene controlling the presence of prickles on the petiole (*Ppr*) was located at the end of linkage group 7 in the 86–7 map. Another gene coding for the production of one of the subunits of isozyme malate dehydrogenase (*Mdh-2*) is located on linkage group 4 of the 86–7 map. Reproduced from Rajapakse *et al.* (2001) *Theoretical and Applied Genetics* 103: 575–583 with permission of Springer-Verlag.

placed on these maps – a gene conferring resistance to black spot, identified as *Rdr1*, was located on linkage group 1, and a gene controlling resistance to powdery mildew, *Rsp1*, was placed on linkage group 3. These two genes were incorporated into maps by indirect means with the aid of markers linked to the genes in other segregating populations. The placement of *Rdr1* was confirmed by reciprocal mapping of a marker from the map on to the progeny segregating for *Rdr1*, from which the markers linked to *Rdr1* were first obtained.

The second set of diploid genetic maps (not drawn) have been constructed with hybrid progeny between a dihaploid rose, derived from a modern tetraploid *Rosa hybrida* cultivar, and diploid *R. wichuraiana* Crép. AFLPs were used as molecular markers in developing two separate maps for the parental genomes. The map of *R. wichuraiana* comprises six linkage groups covering 287 cM of its genome, while the map for dihaploid parent comprises eight linkage groups spanning 238 cM. No markers were currently identified to anchor the linkage groups in *R. wichuraiana* to the corresponding linkage groups in the dihaploid parent. Genes controlling two qualitative traits, recurrent blooming and double corolla, known to be inherited as single recessive and dominant genes, respectively, are located on separate maps. Furthermore, a quantitative trait, prickle density of the shoots, was also mapped in this population. It appears that prickle density is controlled by a major and a minor QTL, both of

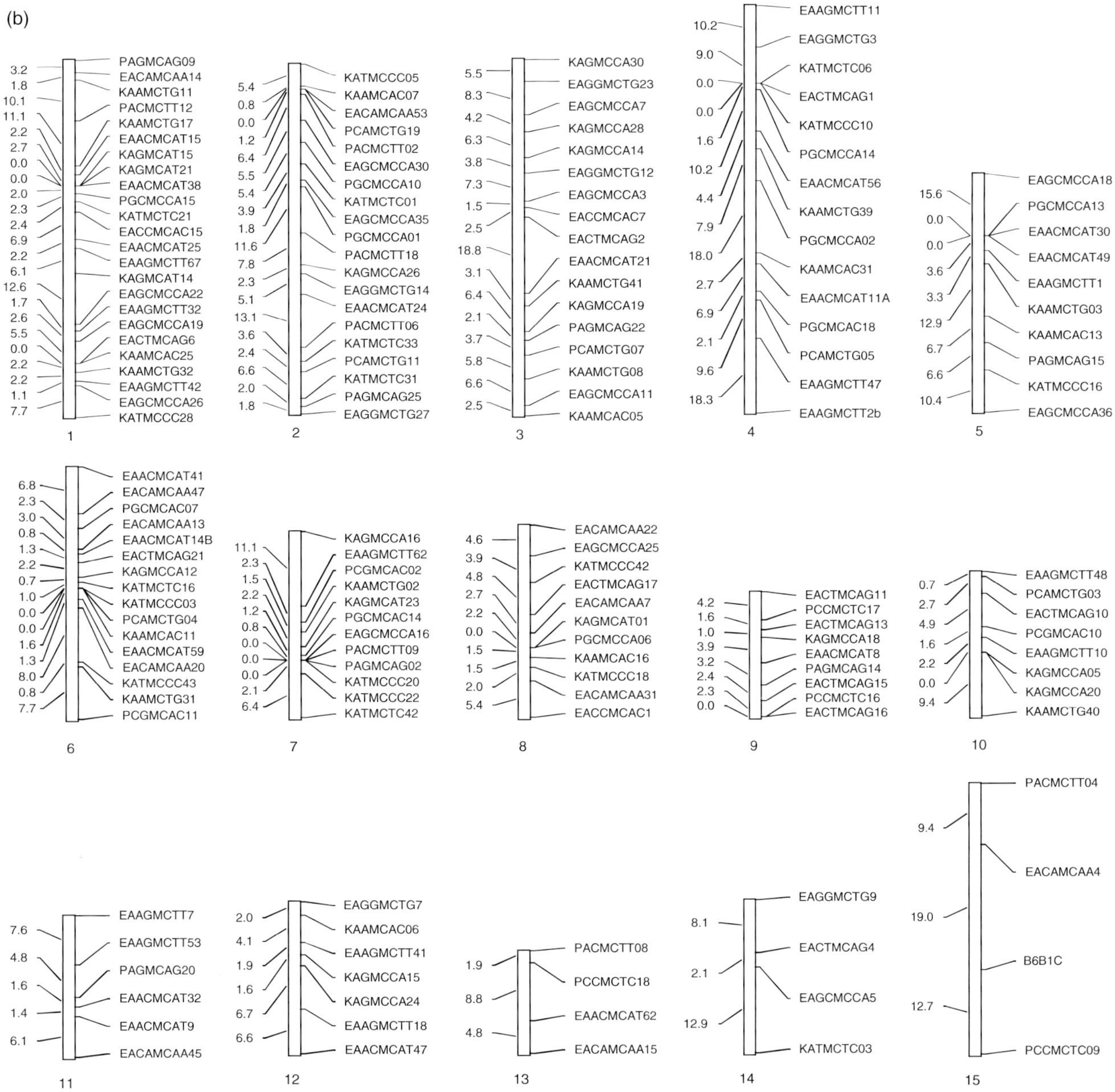

Figure 2 Continued.

which are located on the same linkage group at 36.5 and 3.2 cM from the seasonal/recurrent blooming gene, respectively, on the *R. wichuraiana* map. The single gene trait, double corolla, mapped in both sets of diploid maps (i.e. diphaplid rose in this and *R. multiflora* map) identifies the corresponding linkage groups in the two maps.

The tetraploid rose maps have been developed with F_2 progeny of a cross between an amphidiploid rose (combining the diploid genomes of *R. wichuraiana* cv. 'Basye's Thornless' and *R. rugosa* Thunb. var. *rubra*) and a hybrid known as 'Basye's Blueberry'. The map of amphidiploid consists of markers assigned to 15 linkage groups and covering more than 902 cM of the genome, while the map of 'Basye's Blueberry' consists of markers assigned to 14 linkage groups and covering more than 682 cM of the genome (**Figure 2**). It is estimated that both these maps cover less than half of the genomes in each parent. Two genes, one controlling the prickles on the leaf petiole (*Ppr*) and another coding for the production of one of the subunits of isozyme malate dehydrogenase (*Mdh-2*), are located on the amphidiploid 86–7 map. These tetraploid maps were initially constructed with dominant

AFLP markers, and no anchoring points were observed between the two parent-specific maps. Microsatellite or simple sequence repeat (SSR) markers developed subsequently in the progeny have since identified some of the corresponding groups. For example, linkage groups 1 and 10 in the 86–7 map correspond to groups 4 and 1 in the 82–1134 map, respectively. Development of additional SSR markers will eventually help to identify all of the corresponding linkage groups in both maps and produce a unified map for the tetraploids. In addition, SSR markers have already been used to identify one set of homeologous linkage groups within one map. It appears that linkage group 4 and another unnumbered small group of the 82–1134 map (not drawn) are homologous and arise from the two genome sets. Further refinement of tetraploid maps will help identify all homeologous groups and produce a combined map for the seven basic chromosomes.

All maps mentioned above are first-generation linkage maps in rose providing only a fragmental representation of respective genomes. There are many gaps and unmapped areas in all maps. Many of the linkage groups are small, reflecting the low number of markers used. However, these maps provide the basic genomic structures for incorporation of additional markers. Once saturated with markers, each map should consist of seven linkage groups, representing the seven haploid chromosomes in rose. At present, no common markers are available to combine these three sets of maps. This is primarily due to the nature of markers used in the construction of maps (*see* **Genetics**: Molecular Markers). Although AFLPs are the choice of markers for quick map construction, their limitations in mapping abound. One of the major limitations is their inability to be directly portable between mapping progenies. In contrast, markers such as restriction fragment length polymorphism (RFLP), SSR and sequence tagged site (STS) provide the means for direct screening across mapping populations. Efforts are currently under way to develop and screen microsatellite markers that can help merge separate maps into one and produce a marker-rich unified map for the rose genome. Common markers among the maps will also provide an opportunity to compare genome organization within the genus *Rosa*.

In addition to the above-mentioned maps covering the entire rose genome, a detailed localized map for a region of rose chromosome containing *Rdr1*, a gene for resistance to black spot, has also been constructed. This gene was found to be located at the telomeric end of linkage group 1 in the *R. multiflora* map (**Figure 1**). Many markers tightly linked to *Rdr1* have been identified and research towards cloning the *Rdr1* gene has begun. In addition to *Rdr1*, there may well be additional genes that control resistance to black spot that have not yet been identified. Combining several genes controlling resistance will help to build up stable disease resistance that is not easily overcome by the pathogen.

Physical Mapping

In contrast to genetic maps that represent the rose genome based on gene linkage and recombination, physical maps illustrate the physical organization of the genes along chromosomes irrespective of trait inheritance. While distance between genes or markers in genetic maps is measured by genetic map units (or cM), distances in physical maps are based on actual nucleotide sequences and are commonly expressed in kilo basepairs (kbp) or mega basepairs (Mbp). Construction of a physical map involves cutting DNA extracted from all chromosomes of one individual so as to obtain overlapping large pieces, cloning the pieces and arranging the pieces in the order they are located along the length of the chromosome. Thus, a physical map constructed provides a scaffold of large chromosomal pieces ready for sequencing the whole genome. The nucleotide sequence of each piece could be determined separately and then combined to reveal the entire genetic blueprint for that individual.

Cloned collections of fairly large chunks of chromosomes, known as large-insert DNA libraries, are crucial to the development of physical maps. Bacterial artificial chromosomes (BACs) provide most common vectors used in cloning as they can accommodate such large pieces, usually about 100 kbp. Collections of overlapping pieces of these large clones are known as contigs. A physical map consists of series of such contigs spanning chromosomes. Although the linear order between genetic maps and physical maps should be the same, the spacing between markers in corresponding genetic and physical maps may not be consistent due to differences in the rate of recombination along the chromosomes. Some areas of the chromosomes are known to be 'hot spots' for genetic recombination more than other areas. Therefore, it is important to have an idea of the physical distance between genes and markers in the specific area of interest in order to facilitate chromosome walking and gene cloning.

Construction of a physical map is a major undertaking that requires large amount of resources to develop large-insert DNA libraries and to fingerprint the clones in order to identify those that overlap. Therefore, it is unlikely that the whole rose genome will be physically mapped soon. However, physical maps are more often constructed for localized gene regions. One such region around *Rdr1*, controlling resistance to the black spot fungus, has been studied at the physical level. Markers tightly linked to this gene, obtained from genetic mapping, have been used to identify BAC clones containing the gene region and a contig has been assembled. This contig was found to carry several putative genes resembling disease resistance genes identified in other plant species. Further refinements in the region will produce *Rdr1* gene clones that can potentially transform species susceptible to black spot into resistant types. The ability to transform roses that lack the specific resistant variant of the gene is the ultimate test of the effectiveness of the genes isolated.

Comparative Mapping

Genetic maps constructed for related plants, such as wheat, maize and rice, have revealed remarkable conservation of gene content and gene order over about 60 million years of evolution in the grass family Poaceae. This discovery has led to a new way of understanding genomic organization and identifying genes by comparing with a genome of a related species that has been well studied at the genetic or physical level. Known as comparative mapping, this method can speed up the gene identification process considerably. Conservation of gene order, or synteny, has also been demonstrated in groups of other related plants, such as those within Solanaceae and Brassicaceae. Some level of colinearity of genes has been observed even between distantly related species that span across family borders such as *Arabidopsis thaliana* and tomato (*Lycopersicon esculentum*) as well as those that span dicotyledon and monocotyledon divergence, such as *Arabidopsis* and rice (*Oryza sativa*). *Arabidopsis* and rice have the best-studied genomes in the plant kingdom. Therefore, these two model species are often used in comparative studies.

For roses too, comparison with other plant species, both within and outside the family Rosaceae, could be a valuable source of genomic information in the future. Species most appropriate for comparative analysis within the family include peach, apple, pear, cherry, plum, almond and strawberry. Among these, peach (*Prunus persica* L. Batsch.) is the best genetically and genomically studied species in the family. Because of its small genome size, peach is emerging as a model species for indepth genomic analysis of Rosaceae plants. As a first step towards comparative analysis of rose and peach genomes, a few microsatellite markers developed and mapped in peach genetic maps have been tested in rose and a portion of these were found to be useful in rose and placed on rose tetraploid maps. However, there are not sufficient markers in common between peach and rose genetic maps for a meaningful comparison at this time.

Chloroplast and Mitochondrial Genomes

Located in the chloroplast, the chloroplast genome is also referred to as the 'plastid genome' or 'plastome' while the mitochondrial genome located in the mitochondria is referred to as the 'chondriome'. These cytoplasmic genomes are small compared to the large nuclear genome. Furthermore, these genomes are circular structures while the nuclear genome consists of linear chromosomes. Chloroplast genomes are unique to the organisms containing green photosynthetic pigments and mitochondrial genomes are only found in eukaryotic organisms. Unlike the nuclear genome, in which copies of both male and female parents are inherited through sexual reproduction, these genomes are usually maternally inherited, with exceptions in some plant species.

From a collection of rose BAC clones, four clones have been identified that collectively cover the entire rose chloroplast genome. Microsatellite markers of the rose chloroplast genome have also been identified. These markers appear to be polymorphic among roses, providing the means for understanding the evolutionary relationships among roses, since chloroplast genes are thought to evolve slower than the nuclear genes. Indeed, chloroplast DNA probes and gene sequences have been used in the evolutionary analysis of rose species.

Prospects of Rose Genome Analysis

Genomic analysis of major crop plants in the world has been ongoing for the last 15–20 years. Fairly saturated genetic maps have now been constructed for many of our crop plants, and the entire genome of the model species *A. thaliana* has been sequenced. One of the important staple foods in the world, rice, has also been the subject of major genomic studies throughout the world and the sequencing of its 12 chromosomes is now nearing completion. Rice has been chosen as a model species in the Poaceae to assist with the study of genomic organization of other cereals in the family that have much larger genomes. Compared to these model species as well as numerous other major food crops, our understanding of the rose genome is still in its infancy. As of yet, a highly saturated genetic map is unavailable and only a few genes are located on maps. On the other hand, if one considers the progress being made among the ornamental and floricultural plant species, roses are in the forefront of genomic analysis.

Of the many methods available today for genomic analysis, genetic mapping has been used to construct three separate pairs of medium-density maps providing random coverage of the rose genome. Physical mapping has been employed to study a single gene region. Comparative mapping has been explored only on a limited basis. Despite having one of the small genomes among the flowering plants, no method has been adequately employed to obtain a deep understanding of the rose genome. Consequently, our knowledge of important rose genes and their chromosomal localization remains sketchy at best. However, in the near future, fairly saturated marker maps will become available and we will have a consolidated genetic map covering the seven haploid chromosomes. While it is not anticipated that the entire rose genome will be sequenced soon, a number of gene regions will be studied at the physical level towards map-based cloning of valuable rose genes. What is most practical for understanding the genomic organization and identification of rose genes in the future is comparative mapping with related species, mainly with peach. Comparative analysis with related species can also shed light on the chromosomal rearrangements, duplications and other genomic changes associated with the evolution of the family Rosaceae. Furthermore, comparative analysis with *Arabidopsis* and other species such as lettuce (*Lactuca sativa*) and tomato, in which disease resistance genes have been studied thoroughly, would also provide the means for more efficient isolation of rose disease-resistance genes. Hence, comparative mapping methods hold the most promise for detailed rose genomics research in the future.

See also: **Fragrance and Pigments**: Functional Genomics. **Genetics:** Karyology; Molecular Markers; DNA Fingerprinting; Cloned Genes; Marker-Assisted Selection.

Further Reading

Baird WV, Ballard RE, Rajapakse S and Abbott AG (1997) Progress in *Prunus* mapping and application of molecular markers to germplasm improvement. *HortScience* 31(7): 1099–1106.

Crespel L, Chirollet M, Durel CE *et al.* (2002) Mapping of qualitative and quantitative phenotypic traits in *Rosa* using AFLP markers. *Theoretical and Applied Genetics* 105: 1207–1214.

Debener T and Mattiesch L (1999) Construction of a genetic linkage map for roses using RAPD and AFLP markers. *Theoretical and Applied Genetics* 99: 891–899.

Debener T, Mattiesch L and Vosman B (2001) A molecular marker map for roses. *Acta Horticulturae* 547: 283–287.

Delseny M, Salses J, Cooke R *et al.* (2001) Rice genomics: present and future. *Plant Physiology and Biochemistry* 39: 323–334.

Gale MD and Devos KM (1998) Comparative genetics in the grasses. *Proceedings of the National Academy of Sciences USA* 95: 1971–1974.

Gillham NW (1994) *Organelle Genes and Genomes.* New York: Oxford University Press.

Jauhar PP (1996) *Methods of Genome Analysis in Plants.* Boca Raton, FL: CRC Press.

Rajapakse S, Byrne DH, Zhang L *et al.* (2001) Two genetic linkage maps of tetraploid roses. *Theoretical and Applied Genetics* 103: 575–583.

Zhang H-B, Woo S-S and Wing RA (1996) BAC, YAC and cosmid library construction. In: Foster G and Twell D (eds.) *Plant Gene Isolation: Principles and Practice*, pp. 75–99. London: John Wiley.

Molecular Markers

S Rajapakse, Clemson University, Clemson, SC, USA

Introduction

Molecular markers are short segments of DNA that usually comprise 100–1000 basepairs, and are located somewhere along the long stretches of the genome of an organism that typically consists of several billions of basepairs of DNA. Because of the vast amount of resources needed to study the entire genomes of various organisms, more often, markers are used as 'windows to the genome' to extract various kinds of information regarding the structure and the composition of a genome. For example, two molecular markers mapped close to and on either side of a particular gene of interest indicate that the gene is located between the two markers. These markers linked to the gene provide valuable information about the gene's location because it is more difficult to pinpoint the precise location of a gene initially. Location of all kinds of markers are not known unless they have been mapped and not all markers represent genes since they can be located in nongene regions of the genome as well.

Other kinds of markers can provide information regarding the genetic composition. Those markers revealing specific basepair sequences unique to one or a group of organisms are valuable in their identification. This kind of marker can also be used to study genetic variation among individuals, offspring, populations, species and other taxonomic groups.

Molecular markers are broadly applicable to all organisms having DNA as their genetic material, including animals, plants, bacteria, fungi and roses are no exception. The advent of methods to generate DNA-based markers in plants during the early 1980s transformed the way scientists have been studying genetic differences in plants. Originally spun off from human DNA fingerprinting methods and later expanded by the development of the polymerase chain reaction (PCR), molecular markers have made rapid advancements possible in a wide range of applications including plant identification, taxonomic analysis, ecology, breeding, genetics, genome analysis, gene identification and gene cloning.

Prior to the development of methods to generate DNA markers, genetic analyses of plants relied heavily on morphology. Morphological characters that can be used to distinguish plants are limited in number. Their expression can also vary depending on the environmental conditions in which they grow. In contrast, molecular markers offer many advantages over conventional phenotypic markers. Molecular markers are developmentally stable, detectable in all tissues and unaffected by environmental conditions. Characterization of isozymes and secondary metabolites such as flavonoids has also been used for plant genetic analyses. While these methods have proven useful to some degree, they too suffer from the low number of markers available for generation and hence have a limited discriminatory power. In contrast, a practically unlimited number of markers can be generated with the wide range of DNA-based marker types now available. Molecular markers serve as powerful research tools for a broad spectrum of scientists analysing plants at the genetic level, from providing more efficient methods of evaluating progeny of crosses for desirable genotypes for the interest of plant breeders, to discovering and cloning new genes for the interest of biologists.

This article will provide a brief review of the main types of DNA markers and their potential applications, along with their limitations. Specific applications of molecular markers in roses for cultivar identification, DNA fingerprinting, marker-assisted selection and genome analysis are discussed in separate entries.

Types of Molecular Markers

RFLP

Following their successful application in humans and animals, restriction fragment length polymorphisms (RFLPs) were the first DNA-based markers applied to plants. These markers detect variation in the length of DNA segments after digestion of DNA with a restriction endonuclease enzyme. A large number of these enzymes is available, each recognizing a specific short (4–8) DNA sequence and cleaving the DNA wherever its target sequence occurs. DNA sequence alterations in a restriction endonuclease site or insertions and deletions between sites can give rise to detectable differences in the length of DNA fragments. Therefore, these genetic markers are referred to as RFLPs. To develop these markers, DNA is cut with a restriction enzyme and subjected to size separation by agarose gel electrophoresis. Due to the enormous amount and complexity of the DNA in any higher organism, this separation results in a more or less uniform smear of fragments, large and small. But due to the fact that DNA is composed of two complementary strands, distinct segments of DNA can be highlighted in this smear by using a method known as Southern hybridization. In order to perform this procedure, alkaline solutions are used to denature the double-stranded DNA in the gel. The resulting single-stranded DNA in the gel is then transferred to a nylon membrane and hybridized with a DNA fragment labelled using either radioactive or nonradioactive methods. The most common method of labelling the probe DNA is with radioactive ^{32}P. In this method, the hybridized membrane is placed on X-ray film to detect the areas of the membrane hybridized with the labelled DNA. Probes for hybridization can be selected from a variety of sources, including cloned DNA of the same species or from another species. The probe could also be a synthetic oligonucleotide. Human and bacteriophage-derived minisatellite probes (e.g. M13) have also proven quite useful in detecting RFLPs in plants. Minisatellite probes contain a repeated core sequence and were first used to detect highly polymorphic bands in humans.

The complexity of the resulting band pattern depends mainly on the probe. Probes specific to a single locus of the genome are preferred in genetic mapping. On the other hand, probes giving rise to RFLP patterns consisting of about 5–20 bands could constitute a universal product code (UPC)-type 'barcode' profile that would be useful in DNA fingerprinting. RFLPs generated with minisatellite probes or moderately repeated genomic clones are usually of this type.

Over the past 20 years, RFLP markers have brought tremendous progress to the field, such as genetic characterization of plant germplasm, genetic mapping and gene cloning. In roses, RFLPs have been developed as a means of cultivar identification by DNA fingerprinting and classification of wild roses. The major disadvantage of these widely used markers is the long and laborious procedure that is not well amenable to automation. Consequently, the entire process usually takes several days to accomplish. RFLPs also require fairly high-quality, undegraded DNA in amounts usually measured in micrograms, about 5–10 μg. However, once made, membranes carrying bound denatured DNA could be probed, stripped of the probe and reprobed. RFLP marker management and dissemination are also tedious. Because of these drawbacks, this method has largely been replaced by new classes of efficient PCR-based molecular markers. However, RFLPs remain one of the most reliable molecular markers.

RAPD, AP-PCR and DAF

A major leap in the development of molecular markers occurred with the discovery of PCR. PCR is a process for generating many copies of a specific target DNA sequence *in vitro*, using DNA polymerase enzyme and two short (typically <24 base) synthetic DNA fragments, called primers, that anneal to either end of the target sequence and enable the polymerase to copy the DNA between them. Subsequent heating separates the templates and the newly synthesized DNA strands, both of which can anneal to fresh primers when brought to a lower temperature, to begin a new cycle of DNA synthesis. After each cycle, the number of copies of the target region doubles and these copies serve as templates for the next cycle. Due to the exponential increase in the number of copies in each cycle, millions of copies are produced at the end of the PCR process that typically consist of 35–40 cycles. The use of thermostable DNA polymerase permits multiple cycles of primer annealing, DNA copying and strand separation to be performed using automated heating blocks, called thermal cyclers.

Several kinds of molecular markers based on PCR were developed initially for the analysis of plant genomes. These methods, termed random amplified polymorphic DNA (RAPD), arbitrary-primed PCR (AP-PCR) or DNA amplification fingerprinting (DAF), employ a single short (e.g. ≤ 10 base) oligonucleotide as a primer to amplify multiple regions in the genome. They produce a DNA banding pattern that depends on the substrate DNA and the primer used. Because of the use of a short primer, simply by chance it can anneal to sites in the DNA that are close enough together in the right orientation to permit PCR amplification of the intervening DNA segment. For all three methods, no prior knowledge of the target DNA sequence is necessary.

Theoretically, variations in primer recognition sites and insertions and deletions between the priming sites provide the basis for polymorphisms. However, in practice, the formation of secondary structures of initial PCR products is also known to produce variations in band profiles. Among these three types, RAPD and AP-PCR are very similar methods developed simultaneously by two independent laboratories. In these two methods, primers consist of 10 random bases. In RAPDs, PCR products are separated by size in agarose gels and visualized by staining with ethidium bromide. AP-PCR uses polyacrylamide gels coupled with autoradiography to analyse products. In the DAF method, the primers consist of only 5–8 bases and the PCR products are separated by polyacrylamide gel electrophoresis and visualized by silver staining. Due to the use of short primers, the DAF method gives rise to more variation than RAPDs and is more suitable for analysis of genetically very similar organisms.

These initial PCR-based methods offered many advantages over RFLPs and soon became the markers of choice for plant scientists. Among the three types, RAPD markers were used extensively in germplasm characterization, taxonomic assessments and gene mapping. The technical simplicity and the fast rate at which the markers can be developed are the most attractive features of these methods. These markers can be generated within just a few hours as opposed to the many days that it used to take for RFLPs. Another important advantage was the elimination of the requirement of a probe and associated steps in probe labelling and hybridization. Being PCR-based, only a small amount of DNA, usually a few nanograms per reaction, was needed for these methods. Therefore, mini DNA extraction methods from plant tissue often sufficed. Moreover, the procedures were amenable to automation to increase the throughput.

Despite these attractions, initially, questions were raised regarding the reproducibility of band profiles between different DNA extraction methods, PCR runs or laboratories. Amplification of genomic DNA using short random primers is sensitive to reaction conditions. Therefore, PCR reactions and temperature cycles had to be optimized carefully to obtain consistent results. However, with time, RAPDs became widely used and most scientists using this method were able to separate artifacts from true amplification products and relied only on bands that could be well reproduced. Informative RAPD markers were often converted to specific markers such as sequence characterized or amplified regions or SCARs to make screening for polymorphisms more efficient.

RAPD markers generally detect fairly high levels of variation, making them very useful in genetic analyses. But because of the high level of polymorphism revealed by this method, they are most useful in studies at or below the species level, even though at times they have been used at higher taxonomic levels. In roses, RAPDs have been developed to study phylogenetic relationships between species and sections, identify and fingerprint cultivars, and analyse wild populations. RAPDs have also been used in constructing genetic maps and marker-assisted screening of rose germplasm for resistance to black spot disease. In addition, they have been employed to test for apomictic origin of dog rose progeny and triparental origin of Damask roses. AP-PCR markers have been employed to study the domestication process of the modern roses.

A major drawback of these PCR-based markers is that they are dominant markers scored only as present or absent. In other words, the dominant allele, which gives rise to the PCR product, masks the presence of the recessive allele that does not give rise to a PCR product. Therefore, homozygotes of the genotype with two dominant alleles are not distinguishable from heterozygotes having one dominant allele and one recessive allele. However, a small percentage of RAPD markers show codominance at some loci, where different alleles of a marker can be seen. Development of RAPDs also accelerated the process of identifying markers linked to phenotypic characters through bulked segregant analysis (BSA). Instead of screening individual plants of a segregating progeny in search of markers linked to a trait, BSA allowed creation of bulks consisting of plants with similar phenotype and the rapid screening of the bulks with a large number of primers. Many markers linked to economically important traits have been discovered by this now-widespread method, including markers linked to a gene conferring resistance to black spot in roses.

AFLP

More recently, a novel marker system that scans the genome more rapidly has been developed. These markers, known as amplified fragment length polymorphism (AFLP), are generated by PCR amplification of a selection of restriction-digested genomic DNA. First, genomic DNA is cut with two restriction enzymes and short synthetic DNA adapters are ligated to the restriction fragments. PCR amplification is performed on the restriction fragments with primers based on adapter sequence along with two to four additional bases chosen at random. These additional bases select fragments to be amplified in each run from the initial restriction fragment pool, reducing the number of amplified fragments to allow them to be clearly resolved by electrophoresis. Then, denaturing polyacrylamide gel electrophoresis is used to separate amplified products. The AFLP marker system combines both PCR and RFLP technology, involves many steps in development and is technically fairly sophisticated.

The patented AFLP procedure is a powerful method that allows simultaneous identification of polymorphism at a large number of places in the genome. The method is easily adapted for bulked segregant analysis. Informative markers are routinely converted to sequence tagged sites (STS) or SCARs. As most markers are dominant, AFLPs share the drawbacks of the PCR-based methods described above. Because of the large number of markers that can be developed in a relatively short period of time, marker scoring is perhaps the most tedious aspect of developing AFLPs. Software for automated scoring is available when fluorescent dye-labelled primers are used in conjunction with sequencing machines.

AFLP has been the primary marker system used to construct two tetraploid and two diploid rose maps. Other rose maps used both RAPD and AFLP markers. An example of AFLP markers developed by denaturing polyacrylamide gel electrophoresis for map construction is given in **Figure 1**. AFLP markers linked to single-gene rose traits and quantitative trait loci have also been identified.

STS and SCAR

Once an informative marker has been detected among the bands in any one of the PCR-based methods described above, that particular marker is separated from the rest for effective genotype scoring. The specific band is cut from the gel and DNA-cloned. Then its DNA sequence is determined. Based on the information of the DNA sequence, primers specific to this product are designed and used in the amplification of that particular region in the genome. These markers, termed STS or SCAR, use a pair of unique primers, each about 20 bases long, to amplify specific short target sites that have a single occurrence in the genome. Conversion from random primers to specific primers makes screening of the DNA samples efficient. Another advantage of this is the conversion of dominant markers such as RAPD and AFLP to codominant types where multiple alleles can be visualized.

In practice, when RAPD or AFLP markers are converted to STS, sometimes the original polymorphism is lost. In this situation, additional steps such as restriction digestion of the PCR products are necessary to visualize polymorphisms again. Markers developed by cleaving the specific PCR products with restriction enzymes are termed cleaved amplified products (CAPs). STS markers derived from expressed genes are known as expressed sequence tags (ESTs).

SSR (STMS)

A special class of molecular markers based on simple sequence repeats (SSRs) has lately become very useful in observing genetic variability in plants. Tandem arrays of short nucleotide repeats, e.g. $(AT)_n$, $(CAT)_n$ or $(GATA)_n$, are ubiquitous in regions known as microsatellites in eukaryotic genomes. The copy number (n) of these repeats can vary among alleles and provides the

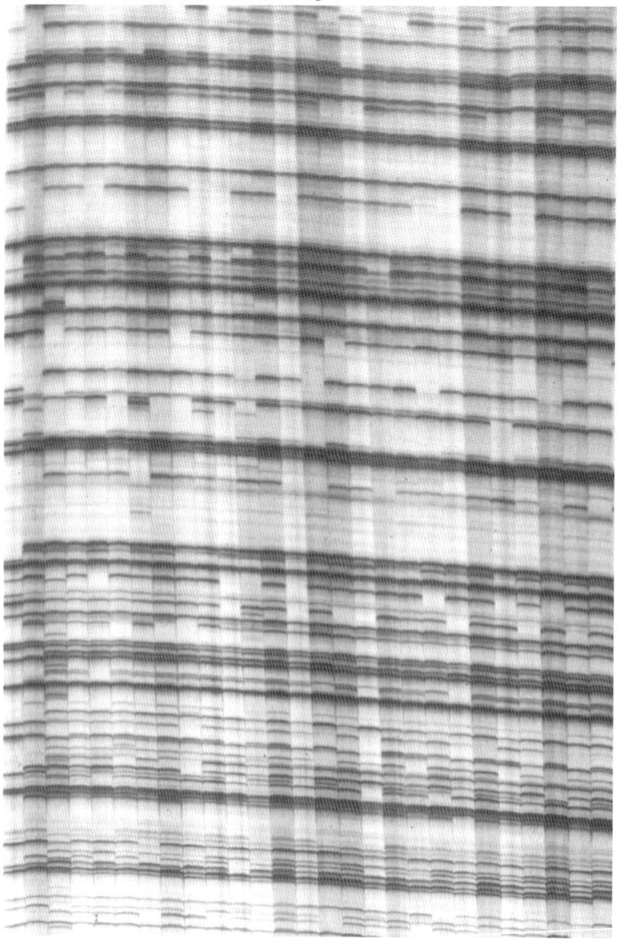

Figure 1 Denaturing polyacrylamide gel exhibiting amplified fragment length polymorphism (AFLP) markers generated in rose for genetic map construction. DNA from a tetraploid test progeny set identified as 90–69 was cut with restriction enzymes EcoRI and MseI. Bases used for further selection of cut pieces were ACC and CAT, for EcoRI and MseI, respectively (see text for details of the method). Lane representations are as below. Lane 1, GP_1, one of the grandparents of the progeny, *Rosa wichurana* Crép.$_{cv}$. 'Basye's Thornless'; lane 2, P_1, the pollen parent of the progeny, identified as 86–7; lane 3, P_2, the seed parent of the progeny, known as 'Basye's Blueberry'; lane 4, the F_1, 90–69. The rest of the lanes represent various F_2 progeny. Of all roses that produced these markers, only GP_1 is a diploid; all others are tetraploids. Note the large number of bands amplified from multiple loci in the genome. No allelic relationships can be directly inferred from these markers. Compare with **Figure 2**, that shows allelic variation arising from a single locus.

basis for polymorphism. In the classical method of generating these markers, a random collection of short cloned DNA fragments (or library) is screened with a number of microsatellite probes to identify clones containing SSRs. The DNA sequence is determined for SSR-containing clones and primers are designed based on unique sequences flanking SSR. The primers are used in amplifying genomic DNA to reveal polymorphisms based on number of repetitions present in each copy of the SSR. When a specific primer pair is used to amplify a single SSR-containing region, the marker is known as a sequence tagged microsatellite site (STMS).

SSRs are abundant throughout plant genomes and exhibit high levels of polymorphism. The codominant markers allow all alleles in a given locus to be observed. This is an advantage of microsatellite markers, especially in genetic analysis and mapping of polyploid species, where multiple alleles are present. SSRs are currently being used to map tetraploid roses. A polyacrylamide gel displaying SSR markers developed from a rose-mapping progeny set is shown in **Figure 2**.

The main shortcoming of microsatellite markers is the large amount of resources needed to generate the markers. Once established, they are simple PCR-based markers that are highly informative. While being effective in the species in which they were developed, SSRs can also be effective in species that are closely related to it. Indeed, several SSR primers derived from peach, apple and sour cherry, all of which are in the family Rosaceae, have been effective in amplifying rose DNA. Some of the peach SSR markers have been placed on rose maps. Both nuclear and chloroplast-derived SSRs have also been developed in rose. Nuclear-derived SSR markers have been employed in the genetic analysis of dog rose progenies and to elucidate the genomic composition in the polyploids. SSR markers are also being used to consolidate various genetic maps now available into one core map for rose. These markers will be extensively used in the future for diverse genetic studies of roses, including tracking of gene flow, population analysis and establishment of a core marker set for rose cultivar identification.

ISSR

Because of the initial expenses in developing SSR markers, variations of the above-mentioned original SSR method have come into use. In one such modification, single primers consisting of simple sequence repetitions are used in the PCR to amplify the region in between matching primer sites. These intersimple sequence repeats (ISSRs) are quite similar to RAPDs in the overall method of marker generation and visualization, simultaneous amplification of multiple loci and dominance of markers at most loci. However, ISSR primers are typically longer, consisting of about 15–25 bases, and stringent PCR reaction conditions are used. These differences result in more reproducible markers compared to RAPDs. ISSR primers have been claimed to be more likely to amplify unique regions in the genome compared to short random primers that tend to amplify repeated regions. Long RAPD primers (15–20 bases) are also known to amplify a low amount of repetitive

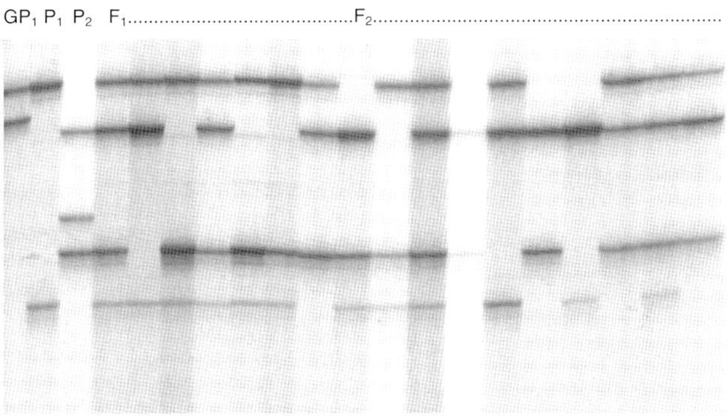

Figure 2 Denaturing polyacrylamide gel exhibiting microsatellite markers generated in rose for genetic map construction. Genomic DNA of a tetraploid test progeny set 90–69 was polymerase chain reaction (PCR)-amplified with specific primers to amplify DNA from a single locus (Rw5D11). Lane representations are as below. Lane 1, GP_1, one of the grandparents of the progeny *Rosa wichurana* Crép.$_{cv}$. 'Basye's Thornless'; lane 2, P_1, the pollen parent of the progeny, identified as 86–7; lane 3, P_2, the seed parent of the progeny, known as 'Basye's Blueberry'; lane 4, the F_1, 90–69. The rest of the lanes represent various F_2 progeny. Of all roses that were used to produce these markers, only GP_1 is a diploid; all others are tetraploids. Note that the few bands produced show variation of the four alleles of a single locus in the tetraploid genomes. Compare with **Figure 1**, that shows genetic variation at many loci.

DNA. There are no reports available on the use of ISSR markers in roses.

Applications of Molecular Markers

The main types of markers used in plants are described above. A plethora of terms are available to describe slight variations in these different kinds of markers. Hence, terms not widely used have been omitted from the above discussion. These markers are applicable in diverse fields, from plant breeding to gene cloning. In each field, specific features of markers become important for successful application. Some of the important features of various types of markers are summarized in Table 1.

Marker-assisted Selection

By using DNA-based markers, many problems associated with the phenotypic evaluation of traits can be avoided in breeding programmes. DNA markers are especially useful in selecting for polygenic traits that prove difficult to select based on phenotypic assessment alone. When breeding programmes are assisted by the use of molecular markers, the process is commonly known as marker-assisted breeding or molecular breeding. For assessing genetic variation in the germplasm, RAPD and AFLP markers that scan the genome quickly would be useful, although technical sophistication of the AFLP method limits its use with the breeder. Markers that can be effectively applied in selecting progeny should be technically simple methods that can be performed in breeders' setting as opposed to methods used in a research laboratory. Markers should also be cost-effective to be performed. Data handling and management should be easy. For screening large numbers of progeny for marker-assisted selections, simple PCR-based markers such as STS or SCARS are most appropriate. Allele-specific markers that would allow PCR-based allele identification are also quite useful for the breeder. Markers tightly linked to single-gene traits as well as quantitative trait loci regions controlling polygenic traits will first have to be established and then applied in the selection of progeny in breeding programmes. To track gene introgression, in addition to markers linked to the trait of interest in the nonrecurrent parent, markers distributed throughout the genome in the recurrent parent, in the form of a genetic map, are useful. The breeder and the molecular technologist must cooperate for the effective use of molecular markers as breeders trained in traditional ways alone are unable to use these methods due to the lack of skill and experience.

Ecological Studies

Molecular markers can provide highly detailed assessments of the dynamics of genetic variation in natural populations in relation to ecological, geographical and temporal distributions. They could also be used to identify clonal populations, hybrids and offspring of apomictic origin. Molecular markers can shed light on the origin of domesticated plants from wild relatives. RFLP and RAPD markers have been used in diverse ecological studies, but AFLPs are now more efficiently employed in whole-genome scanning while SSR and STS markers are used in the detailed analysis of selected regions in the genomes. Genetic marker loci could be followed through in dispersal and gene-flow studies. Seed and pollen dispersal could be distinguished by using maternally derived (cytoplasmic) and paternally derived (pollen parent-specific) genetic markers. Markers derived from the chloroplast, which are often maternally inherited with exceptions in some plant species, are used to trace the cytoplasmic lineage. Molecular markers could also be used to trace the presence and flow of transgenes arising from genetically modified plants. For this purpose, STS markers with specific primers designed for the transgene can be used.

Taxonomic/Phylogenetic Studies

Mutations in DNA provide a window to the evolutionary changes over millennia. Thus, molecular markers could be highly informative in uncovering

Table 1 Comparison of some of the key features of main types of DNA markers

Marker type	*Development based on*	*Primer specificity*	*Mode of marker action*	*Loci specificity*	*Ability to screen across progenies*
RFLP[a]	Restriction digestion	(Not applied)	Codominant	Single locus to few loci	Direct
RAPD AP-PCR DAF	PCR	Random primer	Dominant	Multiple loci	After conversion to specific markers
AFLP	Restriction digestion and PCR	Random primer	Dominant	Multiple loci	After conversion to specific markers
SSR (STMS)	PCR	Specific primer	Codominant	Single locus to few loci	Direct
ISSR	PCR	Random primer	Dominant	Multiple loci	After conversion to specific markers
STS EST	PCR	Specific primer	Codominant	Single locus	Direct

[a] See text for the definition of abbreviations.

the changes in genetic blueprint not visible in the phenotype. Markers shared among plant species and genera can provide a measure of their genetic distance which could be used to determine their phylogenetic relationships. Chloroplast genes are thought to evolve slower than the nuclear-genes. Therefore, chloroplast and nuclear-derived markers are used when studying evolution at different taxonomic levels.

Genetic Mapping

In genetic mapping, molecular markers consist of short segments of DNA that provide landmarks along the chromosomes. The markers provide a scaffold of the entire genome where genes are located. In order to appreciate the contribution of genetic markers to mapping, it is interesting to note the progress of mapping over the decades. Initial maps of plants were constructed based on morphology and were dependent on finding phenotypic mutations. Therefore, the process was extremely slow and only a few traits were mapped by following this method. Then, the discovery of RFLPs enabled the use of genetic markers that are not visible to the naked eye, to be used in constructing maps, while it was still a time-consuming process. With the eventual development of RAPD markers, it was possible to construct molecular maps of many plants in a few months – a relatively short time compared to what it used to be with the RFLPs. AFLPs further expedited the process, now routinely used, and maps developed in weeks. However, compared to co-dominant markers such as RFLP, STS and SSR, dominant markers such as RAPD and AFLP have a limited use in genetic analysis and mapping. Linkage maps specific to one parent are often constructed when using RAPDs or AFLPs with progeny derived from highly heterozygous parents, because a large number of progeny are required to find linkage relationships between markers arising from the two parents, which are said to be in repulsion. The ability to utilize markers in different progenies is best when using STS or SSR markers. RAPD and AFLP markers have to be converted to STS markers before they could be tested effectively in progeny sets other than the one from which they were first derived. STS and SSR markers are also useful in combining maps constructed from different progenies. They also serve as landmarks on physical maps.

Certain regions in the genomes of related plants are conserved in the course of evolution. Therefore, some of the markers that are developed from species related to rose, such as peach, apple and strawberry, can be useful in comparing the genomic structure between these species. Due to the similarity in the composition and order of genes among the related species, marker and map information derived from economically important crops in Rosaceae that have been studied extensively, such as apple and peach, can aid in rose map construction and saturation.

Gene Identification and Cloning

RAPD and AFLP markers are often used to identify markers linked to phenotypic traits using the bulked segregant analysis. These markers are very efficient when scanning the genome and identifying many markers linked to a gene. This is usually the first step in many of the gene identification and cloning experiments. The specific markers found linked to traits of interest will then be converted to PCR-based markers for efficient screening.

Conclusion

All genetic markers represent a variation in DNA sequence. The usefulness of a marker depends on its characteristics. Some markers detect variation at a single locus while others detect variation at multiple loci. Those that detect variation at a single locus are more descriptive and informative, while the latter kind examines more places in the genome for variation, albeit not providing as much genetic information as the single locus types. There are many other tradeoffs among the various markers in terms of simplicity of the method, efficiency of screening, cost and transferability to other studies. The choice of markers depends mainly on the purpose for which they will be used.

Not all genetic markers define a gene. Only those that have been developed from expressed genes or demonstrated to be located in a gene would. ESTs and RFLPs developed with cDNA probes are such gene markers. Molecular markers are a rapidly developing field and simpler, less expensive and more efficient methods are constantly being developed. New transposon-based markers known as miniature inverted repeat transposable elements (MITES) for large genome species are currently available, though not widely identified among plant species. A powerful new technique to detect single nucleotide polymorphisms (SNPs) has also been developed in humans and is now being applied to plants.

While many studies have already been aided by the use of molecular markers, the full potential of various molecular markers has yet to be realized in roses. Undoubtedly, the future will bring the application of molecular markers to more studies in diverse fields to understand the world of roses.

See also: **Classification**: Cultivar Identification by Image Analysis. **Fragrance and Pigments**: Functional Genomics. **Genetics**: DNA Fingerprinting; Gene Mapping; Molecular Markers; Cloned Genes; Marker-Assisted Selection.

Further Reading

Bachmann K (1994) Tansley review no. 63. Molecular markers in plant ecology. *New Phytologist* 126: 403–418.

Beyermann B, Nurnberg P, Weihe A *et al.* (1992) Fingerprinting plant genomes with oligonucleotide probes specific for simple repetitive DNA sequences. *Theoretical and Applied Genetics* 83: 691–694.

Botstein D, White R, Skolnick M and Davis R (1980) Construction of a genetic map in man using restriction fragment length polymorphisms. *American Journal of Human Genetics* 32: 314–331.

Caetano-Anolles G, Bassam BJ and Gresshoff PM (1991) DNA amplification fingerprinting using very short arbitrary oligonucleotide primers. *BioTechnology* 9: 553–557.

Jefferys AJ, Wilson V and Thein SL (1985) Individual specific 'fingerprints' of human DNA. *Nature* 316: 76–79.

Lander ES and Botstein D (1989) Mapping Mendelian factors underlying quantitative traits using RFLP linkage maps. *Genetics* 121: 185–199.

Michelmore RW, Paran I and Kesseli RV (1991) Identification of DNA markers linked to disease resistance genes by bulked segregant analysis: a rapid method to detect markers in specific genomic regions using segregation populations. *Proceedings of the National Academy of Sciences of the USA* 88: 9828–9832.

Mullis KB and Faloona FA (1987) Specific synthesis of DNA *in vitro* via a polymerase catalysed reaction. *Methods in Enzymology* 255: 335–350.

Mullis KB, Faloona FA, Scharf S *et al.* (1986) Specific enzymatic amplification of DNA *in vitro*: the polymerase chain reaction. *Cold Spring Harbor Symposia on Quantitative Biology* 51: 263–273.

Paran I and Michelmore RW (1993) Development of reliable PCR-based markers linked to downy mildew resistance genes in lettuce. *Theoretical and Applied Genetics* 85: 985–993.

Phillips RL and Vasil IK (2001) *DNA-based Markers in Plants*. Dordrecht: Kluwer Academic.

Sambrook J, Fritsch EF and Maniatis T (1989) *Molecular Cloning: A Laboratory Manual*, 2nd edn. Plainview, NY: Cold Spring Harbor Laboratory Press.

Vos P, Hogers R, Bleeker M *et al.* (1995) AFLP: a new technique for DNA fingerprinting. *Nucleic Acids Research* 23: 4407–4414.

Welsh J and McClelland M (1991) Fingerprinting genomes using PCR with arbitrary primers. *Nucleic Acids Research* 18: 7213–7218.

Williams JGK, Kubelik AR, Livak KJ, Rafalski JA and Tingy SV (1990) DNA polymorphism amplified by arbitrary primers as useful genetic markers. *Nucleic Acids Research* 18: 6531–6535.

Cloned Genes

Y Tanaka and M Fukuchi-Mizutani, Institute for Fundamental Research, Suntory Ltd, Osaka, Japan
J G Mason, Florigene Ltd, Collingwood, Victoria, Australia

Introduction

Recent rapid progress in genomics has made it possible to understand and discuss plant species in term of genes.

The genomic sequence of *Arabidopsis thaliana*, a model species of higher plants, is complete. The draft sequences of *Oriza sativa* L. subsp. *japonica* and *indica* (rice) were published in 2001 and a more accurate genome sequence was released in 2002. Rice is the most important crop and a model plant of the monocot. The genome project is in progress in *Lotus japonicus* and *Medicago truncatula*. Large-scale cDNA sequences are available for many plant species including *Triticum aestivum* (wheat), *Zea mays* (maize), *Glycine max* (soya bean) and *Solanum tuberosum* (potato). These approaches are changing not only plant biology but also plant breeding. Thus rose breeding will be accelerated by the information gained in plant genome projects.

In the case of roses, there were two milestones in 2002. Two groups (Israel/USA and France) published the sequences of a few thousand cDNAs from rose petals. The Israel/USA group published the sequences of petal-specific cDNA libraries of two hybrid tea roses and identified the genes involved in fragrance. The French group also conducted large-scale sequencing of rose petal cDNA clones and published the results. They identified the genes encoding enzymes that synthesize fragrance compounds and flavonoids.

In this article, we describe the cloned genes from the *Rosa* species in the following DNA databases: DDBJ, Genbank and EMBL, focusing on the two large-scale sequence projects especially fragrance biosynthesis, flavonoid biosynthesis and senescence, because these characters are commercially important. We also refer to the MADS-box (a highly conserved sequence motif found in a family of transcription factors) genes, which regulate floral organs, and the genes used for molecular phylogenetic analysis. Actual DNA sequences can be obtained from these DNA databases, as the accession number is shown for each sequence.

Large-scale Sequence Projects and Floral Scent Genes

Israeli and US scientists have sequenced a few thousand cDNA clones derived from rose petals to reveal rose floral fragrance genes and have published these results recently. They sequenced clones from petal cDNA libraries of two tetraploid *Rosa hybrida* cultivars, 'Fragrant Cloud' and 'Golden Gate'. 'Fragrant Cloud' has large red flowers possessing strong scent and a short vase-life while 'Golden Gate' has small yellow flowers with a long vase-life and no distinct odour. In all, 2873 high-quality sequences (1834 from 'Fragrant Cloud' and 1039 from 'Golden Gate') were obtained (the sequences have been deposited in the DNA databases with accession numbers BQ103855–BQ106728). A high number of unique sequences were found. Among these, 1288 and 746 sequences were unique to 'Fragrant Cloud' and 'Golden Gate', respectively. The largest group of sequences (about 32%) showed homology with genes coding for predicted proteins with unknown function. The second largest group (17%) contained unique sequences that have no apparent homologues. The third largest group (15%) encoded proteins involved in metabolism.

Microarray analyses of 350 unique sequences from 'Fragrant Cloud' revealed that 77 genes showed higher (more than two fold) expression in 'Fragrant Cloud' petals than in 'Golden Gate' petals. Two hundred and seventy sequences showed similar expression and three sequences showed higher expression in 'Golden Gate' petals. It was also revealed that 65 genes were upregulated and 14 genes were repressed during petal development of 'Fragrant Cloud'. A comparison of mature petals of the two cultivars showed that 40 genes were upregulated during 'Fragrant Cloud' petal maturation and exhibited higher expression in 'Fragrant Cloud' than in 'Golden Gate' flowers. Some of them appeared to be rose homologues of glutamate decarboxylase (BQ104391), methyglutaryl-coenzyme A (CoA) reductase (BQ105103), sesquiterpene synthase (BQ106271), the sequence was most similar to cotton (+)-δ-cardinene synthase among sesquiterpene synthases and shown to encode sesquiterpene synthase activity in recombinant *Escherichia coli*, two cytochromes P450 (BQ105028, BQ104118), tyrosine decarboxylase (BQ105113), xyloglucan endotransglycosylase (BQ104195), GAST (a tomato gene induced by gibberellin)-like protein (BQ105983) and aquaporin-like protein (BQ104371).

Some scent-related genes were found on further analysis. These were alcohol acetyl transferase (BQ106456; the activity was verified and geraniol-producing geranyl acetate was the most efficient acceptor substrate) and orcinol *O*-methyltransferases (OOMT, BQ106549, BQ106672; their functions were shown in recombinant *Escherichia coli* and their full-length sequences were AF502433 and AF502434, respectively).

A French group adopted a similar approach and sequenced numerous rose petal cDNA clones. They obtained a total of 1794 cDNA sequences derived from the diploid *Rosa chinensis* cultivar 'Old Blush' (also called 'Parsons' Pink China'). The sequence accession numbers are from BI977235 to BI979028. Among the sequences, 1159 were included in 242 clusters and the remaining 635 expressed sequence tags (ESTs) were classified as singletons. Some 26.6% did not show any homology to sequences in the database, while 9.2% were homologous to the proteins of unknown sequence. It was predicted that 15.1% had a defence or stress function and 10.1% encoded protein associated with the cell membranes. About 9.2% were similar to the genes encoding enzymes in primary and secondary metabolism. The sequences possibly related to floral scents were geranygeranyl pyrophosphate synthase (BI977759), isopentenyl diphosphate isomerase (BI977796), phytoene synthase (BI979026) and sesquiterpene cyclase (BI977742).

The same French group identified two OOMT sequences among the genes expressed in the petals of *Rosa chinensis* cv. Old Blush (AJ439731 and AJ439742) and obtained two additional OOMT sequences in the petals of *Rosa hybrida* 'Lady Hillingdon' (AJ439743, AJ439744). Each of the four sequences was closely related (over 96% amino acid identities). The group also cloned a caffeic acid *O*-methyltransferase coding sequence from petals of *R. chinensis* 'Old Blush' (AJ439740). The enzymatic activities encoded by the two OOMT sequences (AJ439731 and AJ439743) and caffeic acid *O*-methyltransferase sequence were exhibited in recombinant *E. coli*.

Flavonoid Genes

Flavonoid

Flavonoids are major constituents of rose flower colours, with the exception of yellow. The yellow colour in roses is derived from carotenoids, which are a different class of compounds from flavonoids. Colour is one of the most important characteristics that consumers focus on to select flowers at the point of purchase. Among flavonoids, anthocyanins, a coloured class of flavonoids, have direct roles in determining the colour of roses. Flavonols coexisting with anthocyanins also affect flower colour. The biosynthetic pathway leading to anthocyanidin 3-glucosides has been well characterized and is conserved among higher plants (**Figure 1**). The genes encoding enzymes involved in the biosynthesis of flavonoids that affect flower colour and transcriptional factors regulating the flavonoid biosynthetic pathway have recently been studied in some depth.

Anthocyanidin 3-glucosides are further modified with sugar groups (usually glucose), aliphatic acyl groups or aromatic acyl groups in a species- and variety-specific manner. Blue and violet flowers tend to produce highly modified delphinidin-type anthocyanins, which rose petals do not produce. Wild *Rosa* species flowers mainly contain 3-glucosides and 3,5-glucosides of cyanidin and peonidin. Some wild roses produce pelargonidin 3-glucoside and pelargonidin 3,5-diglucoside, which are now commonly found in commercial cultivars. A recent study has shown that cyanidin 3-rutinoside, peonidin 3-rutinoside, peonidin 3-*p*-coumaroylglucoside-5-glucoside and cyanidin 3-sophoroside are found in *Rosa* flowers. No delphinidin-type anthocyanins have been detected in rose petals. This is assumed to be the reason why roses do not produce any violet to blue variety. However, interestingly, some varieties produce delphinidin in their leaves.

Chalcone Synthase Gene

Chalcone synthase (CHS) is the first committed enzyme in flavonoid biosynthesis (**Figure 1**) and catalyses the stepwise condensation of three acetate units from malonyl-CoA using coumaroyl-CoA as a starter CoA to yield 2′,4′,6′,4-tetrahydroxychalone. The genes of CHS and its homologues have been cloned from more than 100 plant species. We cloned a CHS cDNA from the petal of a red rose of the cultivar 'Kardinal'® (AB038246). Among the published plant CHS sequences, rose CHS is homologous to *Camellia sinensis* (tea) CHS (94%), *Juglans nigra* × *Juglans regia* (walnut) CHS (92%) and *Citrus sinensis* CHS (92%).

A rose CHS cDNA has been successfully used to modify flower colour. The rose cultivar 'Royalty', which produces dark-red flowers, was transformed with a binary vector containing a rose CHS fragment under the regulation of a constitutive promoter in the sense direction. Fifteen of the 110 transgenic plants had uniformly reduced anthocyanins and produced pink flowers. A transgenic plant with completely white flowers was not obtained.

Flavonoid 3′-hydroxylase gene

Flower colour depends heavily on the hydroxylation pattern of the B-ring of anthocyanidin. The colour of anthocyanidins shifts towards blue with an increase in the number of hydroxyl groups on the B-ring. Flavonoid 3′-hydroxylase (F3′H) catalyses the hydroxylation of naringenin (a flavanone), dihydrokaempferol (a dihydroflavonol), kaempferol (a flavonol) and apigenin (a flavone) in position 3′. Dihydrokaempferol is modified to cyanidin and F3′H is thus essential to produce cyanidin. The F3′H gene was first cloned from petunia followed by many other species, such as rose, *Arabidopsis*, torenia and carnation. The rose F3′H gene was obtained from the cultivar 'Kardinal'®. Rose F3′H is 71% homologous to carnation and chrysanthemum F3′H.

Dihydroflavonol 4-reductase gene

Dihydroflavonol 4-reductase (DFR) catalyses a stereospecific reduction of (+) dihydroflavonols to 3,4-*cis* leucoanthocyanidins, which are the substrates of anthocyanidin synthase and the direct precursors of the anthocyanidins. The cDNA-encoding rose DFR (D85102) has been obtained from the cultivar

'Kardinal'®. Among the published DFRs, rose DFR showed 82% identity to apple DFR. A DFR-like gene was also obtained from EST analysis of *R. hybrida* cultivar 'Fragrant Cloud' (BQ105675).

The substrate-specificity of DFR enzymes sometimes contributes to the accumulation of selected anthocyanidins in plants. Petunia DFR is known to reduce dihydromyricetin efficiently and

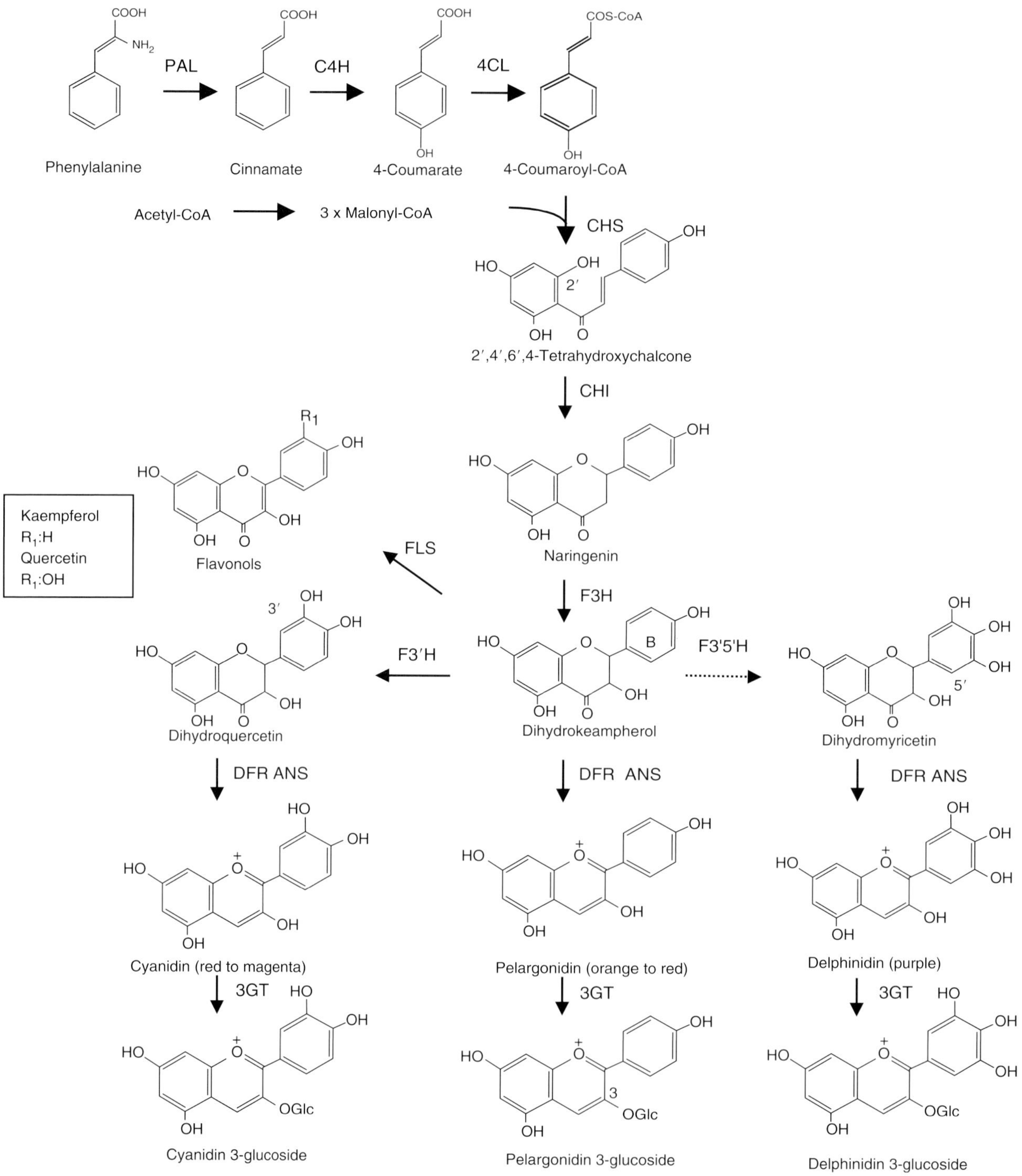

Figure 1 The flavonoid biosynthetic pathway. Rose petals do not accumulate delphinidin-type anthocyanins due to a deficiency in F3′5′H (shown with a dotted arrow). CoA, coenzyme A; PAL, phenylalanine ammonia-lyaze; C4H, cinnamate 4-hydroxylase; 4CL, hydroxycinnamate: CoA ligase; CHS, chalcone synthase; CHI, chalcone isomerase; F3H, flavanone 3-hydroxylase; F3′H, flavonoid 3′-hydroxylase; F3′5′H, flavonoid 3′5′-hydroxylase; FLS, flavonol synthase; DFR, dihydroflavonol 4-reductase; ANS, anthocyanidin synthase; 3GT, 3-glucosyltransferase.

dihydroquercetin reasonably well, but not dihydrokaempferol. This is the reason why petunia does not accumulate pelargonidin and exhibits no orange flower colours. A rose DFR cDNA was successfully expressed in petunia to produce pelargonidin and, thus, orange flower colour. Rose DFR can reduce dihydrokaempferol and dihydroquercetin as rose petals accumulate pelargonidin and cyanidin.

Flavonol Synthase Gene

Flavonols are synthesized from dihydroflavonols through the action of flavonol synthase (FLS) (**Figure 1**). FLS is one of the dioxygenases that require 2-oxyoglutarate, ferric ion and ascorbate for its activity. The flavonoid biosynthesis pathway contains some more dioxygenases: flavanone 3-hydroxylase (F3H), anthocyanidin synthase (ANS), flavone synthase I (**Figure 1**) and flavonoid 6-hydroxylase. Among them, FLS, F3H and ANS are widely distributed in flowering plants. Plants contain many dioxygenases which play important roles in plant metabolism, such as 1-aminocyclopropane-1-carboxylic acid (ACC) oxidase in ethylene biosynthesis, gibberellin 3-hydroxylase (G3H) in gibberellin biosynthesis, and hyoscyamine 6-hydroxylase (H6H) in alkaloid biosynthesis.

The FLS gene was first isolated from petunia followed by a few other species, such as arabidopsis and potato. A rose FLS cDNA has been cloned from 'Kardinal'® (AB038247). A rose ANS homologue was found in an EST of *R. chinensis* 'Old Blush' (BI977949).

A phylogenetic tree of the dioxygenase sequences, including FLS, ANS, F3′H, ACC oxidase, G3H and H6H, is shown in **Figure 2**. Proteins with the same enzyme activities form distinctive groups, which indicates that these genes diversified before the speciation of these plants. It is clear that these genes have evolved divergently, although convergent evolution of the enzymes involved in secondary metabolism is often observed. **Figure 2** also shows that FLS is closely related to ANS, and the genes of FLS and ANS diversified after their common ancestral gene had diversified from other dioxygenase genes.

Transcriptional Regulation of Flavonoid Genes in a Rose

We quantified the amount of flavonols and anthocyanins of petals of four developmental stages:

1. Stage 1, closed buds: petals are not pigmented (5 mm in diameter).
2. Stage 2, closed buds: petals are pale pink (12 mm in diameter).
3. Stage 3, sepals are cracking, and petals are pink (20 mm in diameter).
4. Stage 4, opened flower of *R. hybrida* 'Saphir', which produced more flavonols than anthocyanidins (**Figure 3a**).

Accumulation of flavonols preceded that of anthocyanidins. Anthocyanidins accumulated most rapidly from stage 2 to 3, which is consistent with the results obtained for the rose cultivar 'Kardinal'®, which predominantly contains anthocyanins. The expression profiles of FLS, CHS and DFR were examined in petals from each of the four developmental stages (**Figure 3b**). CHS and DFR transcripts were most abundant at stages 2 and 3, when anthocyanin biosynthesis was most active. FLS transcripts were most abundant at stages 1 and 2. The results agree with the early accumulation of flavonols (**Figure 3a**) and indicate that flavonol and anthocyanin biosynthesis is transcriptionally regulated in rose petals as in many plants. It is noteworthy that such early biosynthesis of flavonols compared with anthocyanins has been observed in the petals of *Matthiola incana*, petunia, and carnation. It is generally accepted that the flavonoid biosynthesis structural genes are regulated by *Myc* and *Myb* transcriptional factors, and these factors regulate a different subset of the genes

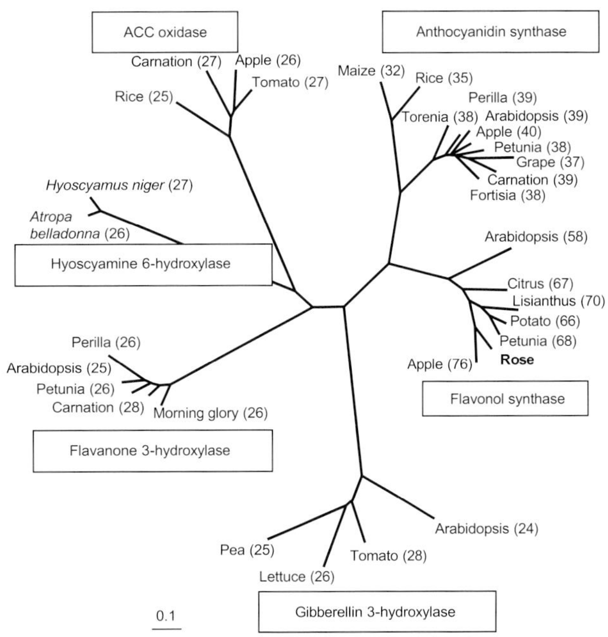

Figure 2 Unrooted phylogenetic tree inferred from amino acid sequences of various 2-oxoglutarate-dependent dioxygenases. The phylogenetic tree was constructed with the deduced amino acid sequences from the public. The figures in parentheses indicate the amino acid sequence identity to rose flavonol synthase. The scale bar indicates a divergence of 0.1 amino acid substitutions per site. Molecular phylogenetic distance usually agrees with the classical one. ACC, 1-aminocyclopropane-1-carboxylic acid.

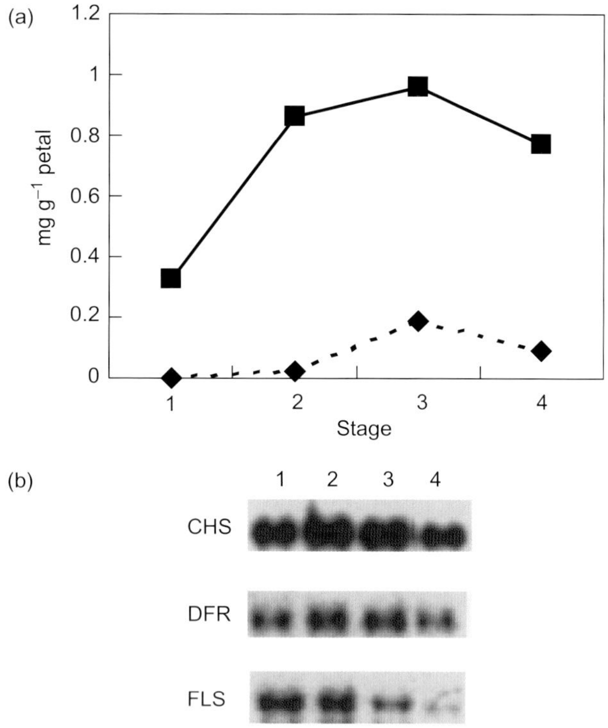

Figure 3 Accumulation of flavonols and anthocyanidins and RNA blot analysis of their biosynthetic genes in *Rosa hybrida* 'Saphir'. (a) Flavonols and anthocyanidins of the petals at the four developmental stages of the flower are quantified with high-performance liquid chromotography. Flavonol accumulation (solid line) precedes anthocyanidin accumulation (dotted line). (b) Total RNA derived from the petals at the four developmental stages of the flower was subjected to RNA blot analysis using ^{32}P-labelled rose chalcone synthase (CHS), dihydroflavonol 4-reductase (DFR) and flavonol synthase (FLS) cDNA fragments. FLS transcript levels peak earlier than those for CHS and DFR.

depending on the plant species. Two *Myb*-like genes (BQ106609, BQ106505) that are possibly involved in this regulation were found in petal ESTs of *R. hybrida* 'Fragrant Cloud' and 'Golden Gate'.

Senescence-related Genes

Ethylene-related Genes

Rose flowers, especially cut flowers, have a limited life. Longer flower life is beneficial to consumers, retailers and growers. Flower life is partly controlled by a gaseous plant hormone, ethylene. Ethylene is not a critical factor in regulating the vase-life of rose cut flowers. Practically speaking, blockage of the flower stem by microbial growth and subsequent inhibition of water uptake shorten the vase-life. The ethylene biosynthetic pathway and its signal transduction triggering senescence have been well-characterized (**Figure 4**). Suppression of ethylene biosynthesis genes (ACC synthase and ACC oxidase) successfully extends the vase-life of carnation and the shelf-life of some fruits, including tomato, melon and banana.

The ethylene signal is transduced to various physiological responses via a two-component regulatory-signal transduction system, which includes a kinase cascade. Ethylene receptors negatively regulate ethylene signal transduction (**Figure 4**). There are at least five ethylene receptors in *Arabidopsis*: *ETR1* (ethylene response 1), *ETR2*, *ERS1* (ethylene response sensor), *ERS2* and *EIN4* (ethylene insensitive). Similar ethylene receptor genes have been obtained from many plants, such as tomato and melon. The *CTR1* (constant triple response) gene, which is a negative regulator of ethylene signal transduction and is negatively regulated by ethylene receptors, has also been cloned from *Arabidopsis* and some other plants. *EIN2*, *EIN3*, *EIN5* and *EIN6* are known to be positive regulators of the ethylene response in *Arabidopsis*. Among them, *EIN3* is a transcription factor that binds to the promoters of ethylene-response genes. Four rose homologues of ethylene receptor genes (partial cDNAs, ETR1 (AF394914, AF127221), ETR2 (AF127220), ETR3 (AF154119), ETR4 (AF159172)) and a rose homologue of a *CTR1*-like protein kinase consisting of 847 amino acid residues (AY032953) have been registered in the DNA databases. A partial sequence from *R. hybrida* encoding an *EIN3*-like transcriptional factor (AY025825) has also been registered.

The utility of these genes in strategies aimed at prolonging flower life is untested in rose. Not all rose varieties are sensitive to the application of ethylene, as measured by petal drop. However, the genes detailed above have been used in related strategies in other flower crops, as mentioned above. Two broad strategies could be considered for application in roses. The first involves blocking endogenous ethylene production in the flower, and the second involves interference with ethylene perception in the flower, as a result of which endogenous and exogenous ethylene has no impact on senescence. Blocking ethylene production in the flower can be achieved via sense- or antisense-mediated suppression of ACC oxidase or ACC synthase synthesis. These enzymes are involved in the final steps leading to ethylene biosynthesis. *S*-adenosyl-L-methionine (SAM) is converted by ACC synthase to ACC, which is converted to ethylene, carbon dioxide and cyanide by ACC oxidase (**Figure 4**).

Through the introduction of an *Etr1* gene carrying, for example, the *Etr1–1* mutation of *Arabidopsis*, rose plants that would lack the ability to perceive ethylene could be generated and, thus, would not respond to endogenous or exogenous ethylene, a drawback of

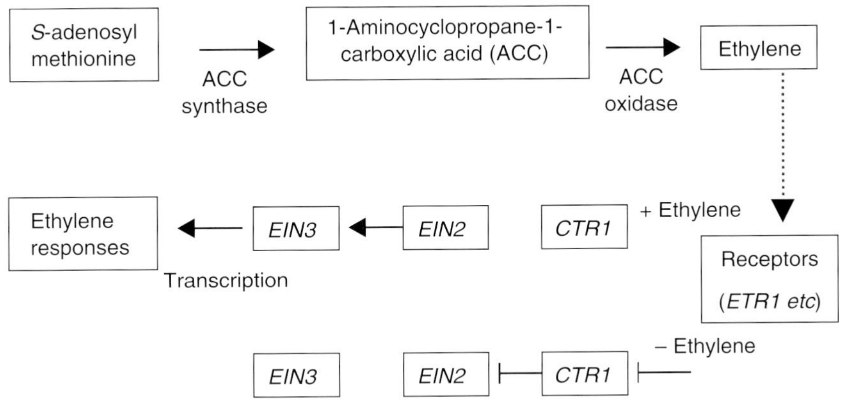

Figure 4 A simplified ethylene biosynthesis and signalling model. Ethylene is biosynthesized from *S*-adenosyl-methionine by 1-aminocyclopropane-1-carboxylic acid (ACC) synthase and ACC oxidase. When ethylene binds a receptor, negative regulation of *CTR1* is released, and ethylene responses then start.

the first strategy. If the expression of such a mutation were confined to the flower, then interference with ethylene transduction, which is important to other plant processes, such as the defence response, would be avoided. Such a strategy has been successful in carnation. In petunia, a similar approach utilizing *ERS* has been used to a similar effect.

Δ9 Fatty Acid Desaturase Homologue

Senescence-induced genes in petals of rose flowers have been cloned using differential hybridization in order to understand the molecular mechanism of the senescence of rose flowers. Among these cDNAs, the amino acid sequences deduced from three related cDNAs exhibited significant homology to that of Δ5 fatty acid desaturase from meadow-foam and those of membrane-bound Δ9 fatty acid desaturases of cyanobacteria, a yeast and mammals (D49383–D49385). Northern blot analysis indicated that the transcripts of the cDNAs were specifically expressed in petals at late developmental stages and during senescence, and no transcripts were detected in other vegetative and floral organs analysed so far. The biological roles of the temporal and petal-specific expression of this putative fatty acid desaturase are unclear. However, it is suggested in carnation petals that microsomal phospholipids with polyunsaturated fatty acids were selectively degraded during senescence and that fatty acid desaturases have roles in generating polyunsaturated fatty acids, which are more susceptible to catabolism than saturated fatty acids. Thus, the rose fatty acid desaturase homologue may also play an important role in converting saturated fatty acids to unsaturated ones to promote the degradation of membrane lipids during petal senescence.

Lipoxygenase

A cDNA encoding lipoxygenase homologue (Rlox1) has been isolated from senescing rose petals. By expressing the gene in *E. coli*, Rlox1 was shown to encode a soluble lipoxygenase with an acidic optimal pH. The transcript of Rlox1 was increased in senescent petals or following ethylene treatment. It was suggested that this lipoxygenase might be involved in lipid metabolism in the senescence of rose flowers or the biosynthesis of jasmonates to accelerate the process of rose flower senescence. The transcript was also detected in sepals, ovaries and anthers.

MADS-box Genes Regulating Floral Identity

MADS-box proteins containing a highly conserved DNA-binding domain called the 'MADS box' are transcriptional factors. They also contain a weakly conserved K-box domain that is involved in the dimerization of transcriptional factors. MADS-box genes are classified into distinct subfamilies on the basis of their sequence, expression patterns and function. The best-known function of the MADS-box protein is to control flower development, and the A, B and C genes in the ABC model of floral development MADS-box genes. The ABC model for the determination of floral organ identity has generally been applicable to flowers of most plant species. Sepals are made from the function of the A gene, petals by the A and B genes, stamens by the B and C genes and carpels by the C genes. The A and C functions antagonize each other.

The *Agamous* gene in *Arabidopsis* is a C gene. Mutation of the *Agamous* gene cause the homeotic conversion of stamens to petals and carpels to indeterminate flowers. In order to understand the molecular mechanisms related to the homeotic changes that occur in the development of the rose flower, an *Agamous* homologue (*RAG*, U43372) was isolated from *R. hybrida* 'Motrea'. *RAG* transcripts are detected only in stamen and carpel tissues after the formation of flower parts in the floral meristem. Two types of *Agamous*-like genes were isolated from *R. rugosa* Thunb. Ex Murray.

One of them, *MASAKO C1* (AB025644) showed high similarity with *R. hybrida RAG1* (98% in amino acid sequence), *Populus balsamifera PTAG1* (73%), *Corylus avellana CaMADS1* (72%), *Silene latifolia SLM1* (69%) and *Agamous* (63%). Interestingly, six mRNA species (*Masako C1–C6*) arose via alternative splicing. The position of the introns of *MASAKO C* is identical to those of *Agamous* and *Agamous*-like genes. The *MASAKO C* gene is specifically expressed in stamens and carpels. These results suggest that *MASAKO C1* is a *C* function gene of rose.

The second homologue, *MASAKO D1* (AB025643), exhibited 65% similarity to *MASAKO C1* and about 70% similarity to *Pan ax ginseng GAG2*, *Solana lycopersicum TAG1* and *Arabidopsis AGL1*. The *MASAKO D1* gene is expressed in stamens and carpels and at low levels in petals.

Phylogenetic Analysis

Molecular phylogeny using molecular biological techniques often gives a more accurate phylogenetic relationship of these species/varieties than classical taxonomy. Classification based on the direct sequence comparison of cloned genes from different species/varieties is thus a powerful tool. Some studies have been published to clarify the original species of modern commercial roses. Only seven species (*R. foetida*, *R. gallica*, *R. moschata*, *R. multiflora*, *R. wichuraiana*, *R. chinensis* and *R. gigantea*) from about 150 *Rosa* species have been found to contribute to modern commercial roses.

The *mat*K gene

Since the nucleotide substitution rate depends on each gene, a gene with a higher substitution rate is suitable to analyse on a subspecies level. The *mat*K gene in the chloroplast DNA had evolved at a higher rate than several other genes currently used in systematic studies. The DNA sequences of the *mat*K gene are determined for *R. persica* Michx. (accession number AB011974), *R. foetida* Herrm. var. *bicolor* (Jacq.) Willmott (AB011975), *R. spinosissima* L. var. *pimpinellifolia* HOOK. (AB011976), *R. sericea* Lindl. (AB011977), *R. gallica* L. (AB011978), *Rosa* × *macrantha* Desp. (AB011979), *R. canina* L. (AB011980), *R. eglanteria* L. (AB011981), *R. glauca* Pourr. (AB011982), *R. carolina* L. (AB011983), *R. virginiana* Mill. (AB011984), *R. blanda* Ait. (AB011985), *R. californica* Chem. & Schlecht. 'Plena' (AB011986), *R. cinnomomea* L. (AB011987), *R. moyesii* Hemsl. & Wils. 'Geranium' (AB011988), *R. rugosa* Thunb. ex Murray (AB011989), *R. moschata* Herrm. (AB011990), *R. multiflora* Thunb. ex Murray (AB011991), *R. wichuraiana* Crép. (AB011992), *R. chinensis* Jacq. var. *spontanea* (AB011993), *R. gigantea* Collett ex Crép. (AB011994), *R. banksiae* Ait. F. *lutea* (Lindl.) Rehd. (AB011995), *R. banksiae* var. *normalis* Regal (AB011996), *R. laevigata* Michx. (AB011997), *R. bracteata* Wendl., *R. roxburghii* Tratt (AB011999), *R. stellata* Woot. var. *mirifica*. Cockerell (AB012000) and *Agrimonia pilosa* Ledeb. var. *japonica* (AB012001). About 7.0% nucleotide and 12.5% amino acid substitution rates are found in these genes. The molecular classification of these species based on these sequences is consistent with traditional classification. Six out of seven species that contributed to modern roses belong to only one clade. This confirms the narrow genetic background of the modern rose.

Fifteen Japanese *Rosa*, 17 additional Asian and two American taxa were further analysed for phylogenetic relationship by determining the *mat*K DNA sequences (AB039289–AB039323). Seventy-two base substitutions in 1518 nucleotides were observed, and 17 substitutions were parsimony-informative characters. Sections *Synstylae* and *Indicae* are confirmed as being closely related. *Rosa brunonii* and *R. moschata* in the section *Synstylae* formed an inner clade. *Rosa acicularis* formed one single clade that was separated from the other section *Rosa* species. In the section *Rosa*, *R. rugosa* and *R. marretii* formed one clade, while *R. acicularis* var. *nipponensis* and *R. willmottiae* formed another clade. *Rosa acicularis* var. *nipponensis* is a primitive taxon and should be reassigned to *R. nipponensis*.

The *mat*K sequences of *R. farreri*, *R. hugonis*, *R. nanothammus* and *R. omeiensis* have also been determined.

The Internal Transcribed Spacers Region

A similar analysis to that described above performed on the basis of the DNA sequences of the internal transcribed spacers (ITS) region (638 nucleotides) of 18–26S nuclear ribosomal DNA (AB035657–AB035659, AB038440–AB038461). All 13 tested taxa in the section *Synstylae* formed one clade, while the section *Rosa* was separated into several different clades. The traditional subgenus *Rosa* and the relationship between the subgenus *Rosa* and the subgenus *Platyrhodon* should be reexamined.

The origin of the oldest Damask varieties was examined using the tools of biology. Other than randomly amplified polymorphic DNA (RAPD), the ITS sequences of nuclear ribosomal DNA are widely used for phylogenetic analysis at lower taxonomic levels because they evolve rapidly. The sequences from the following have been determined: three Damask roses (AB043830–AB043838), a variety of *R. gallica officinalis* (AB043824, AB043835), three varieties of *R. phoenica* (AB043826–AB043829), two varieties of *R. moschata* (AB043000, AB043003, AB043004), two varieties of *R. moschata nepalensis* (AB043001, AB043002), a variety of *R. moschata plena* (AB043005), a variety of *R. fedtschenkoana* (AB043822, AB043823) and a variety of *R. stellata mirifica* (AB043212). On the basis of the sequence analysis, *R. moschata*, *R. gallica* and *R. fedtschenkoana* are parental species of the original hybridization that contributed to forming the four oldest Damask varieties. DNA sequences of chloroplast *psb*A–*trn*H spacer sequences from these rose varieties (AB048211, AB043945–AB043952) have also been determined to reveal that all of the oldest Damask varieties had chloroplasts derived from *R. moschetta*, which indicates that this species is the universally maternal species of Damask roses. The spacer *psb*A–*trn*H region is known to evolve faster than the *mat*K coding region.

The nucleotide sequence of the *R.* × *fortuniana* ITS has also been determined (AB034194). *R.* × *fortuniana* was confirmed to be a hybrid between a seed parent, *R. laevigata*, and a pollen parent, *R. banksiae*.

The region between the *rbc*L and *atp*B genes in chloroplast and a part of the *rbc*L gene from *Rosa damascena* Mill (X69753) were determined. The rbcL sequence from *R. woodsii* was also determined (U06824).

The 5′ end-portion of the granule-bound starch synthase gene from 18 species in 14 genera in Rosaceae, including *R. mutiflora* (AF285993), was determined to understand the evolution within Rosaceae.

Concluding Remarks

It is obvious that the number of genes cloned from roses is still limited, although two groups undertook large-scale sequence projects of petals. Further molecular analysis such as large-scale cDNA sequence analysis of genes expressed in organs and/or a rose genome project are necessary to boost our knowledge of roses in terms of molecular biology. This will contribute to both the traditional and molecular breeding of roses.

See also: **Biotechnologies for Breeding:** Genetic Transformation. **Fragrance and Pigments:** Anthocyanins and Other Flavonoid Pigments; Flavonoid Molecular Biology. **Genetics:** DNA Fingerprinting. **Intellectual Property:** Plant Patents and Trademarks.

Further Reading

Bewley JD, Hempel FD, McCormick S and Zambryski P (2000) Reproductive development. In: Buchanan BB, Gruissem W and Jones RL (eds.) *Biochemistry and Molecular Biology of Plants*, pp. 988–1043. Rockville, MD: American Society of Plant Physiologists.

Croteau R, Kutchan TM and Lewis NG (2000) Natural products (secondary metabolites). In: Buchanan BB, Gruissem W and Jones RL (eds.) *Biochemistry and Molecular Biology of Plants*, pp. 1250–1318. Rockville, MD: American Society of Plant Physiologists.

Davies K (2000) Plant colour and fragrance. In: Verpoorte R and Alfermann AW (eds.) *Metabolic Engineering of Plant Secondary Metabolism*, pp. 127–163. Amsterdam: Kluwer Academic.

Forkmann G and Hellwer W (1999) Biosynthesis of flavonoids. In: Sankawa U (ed.) *Comprehensive Natural Products Chemistry*, pp. 713–748. Amsterdam: Elsevier.

Gutterson N (1995) Anthocyanin biosynthetic genes and their application to flower colour modification through sense suppression. *HortScience* 30: 964–966.

Mikanagi Y, Saito N, Yokoi M and Tatsuzawa F (2000) Anthocyanins in flowers of genus *Rosa*, section *Cinnamomeae*(= *Rosa*), *Chinenses*, *Gallicanae*, and some modern garden roses. *Biochemistry and Systematics Ecology* 28: 887–902.

Mol J, Grotewold E and Koes R (1998) How genes paint flowers and seeds. *Trends in Plant Science* 3: 212–217.

Mol J, Cornish E, Mason J and Koes R (1999) Novel coloured flowers. *Current Opinion in Biotechnology* 10: 198–201.

Rout GR, Samantaray S, Mottley J and Das P (1999) Biotechnology of the roses: a review of recent progress. *Scientia Horticulturae* 81: 201–228.

Stepanova AN and Ecker JR (2000) Ethylene signaling: from mutants to molecules. *Current Opinion in Plant Biology* 3: 353–360.

Tanaka Y, Tsuda S and Kusumi T (1998) Application of recombinant DNA to floriculture. In: Chopra VL, Malik VS and Bhat SR (eds.) *Applied Plant Biotechnology*, pp. 181–235. New Delhi: Oxford & IBH.

Tanaka Y, Shinzo T and Takaaki K (1998) Metabolic engineering to modify flower colour. *Plant Cell Physiology* 39: 1119–1126.

Trewaves A (2000) Signal perception and transduction. In: Buchanan BB, Gruissem W and Jones RL (eds.)

Biochemistry and Molecular Biology of Plants, pp. 930–987. Rockville, MD: American Society of Plant Physiologists.
Vainstein A, Lewinsohn E, Pichersky E and Weiss D (2001) Floral fragrance. New inroads into an old commodity. *Plant Physiology* 127: 1383–1389.
Winkel-Shirley B (2001) Flavonoid biosynthesis. A colourful model for genetics, biochemistry, cell biology, and biotechnology. *Plant Physiology* 126: 485–493.

DNA Databases

DDBJ http://www.ddbj.nig.ac.jp/
EMBL http://www.ebi.ac.uk/embl/
Genbank http://www.ncbi.nlm.nih.gov/genbank/index.html
French group http://www.ens-lyon.fr/RDP/

Marker-Assisted Selection

D H Byrne, Texas A & M University, College Station, TX, USA

Introduction

Most rose breeding programmes perform a two-stage phenotypic recurrent selection programme. In the first stage, seedlings are selected for a few essential traits on unreplicated plots. Frequently, this initial selection occurs in the greenhouse and only the best 1% are selected and propagated for the second-stage evaluation in clonally replicated plots in the field for landscape roses or in the greenhouse for miniature and cut roses. Since in most breeding the germplasm is everblooming, the rose seedlings flower within 6–8 weeks of germination so selection can be practised on flower characteristics, leaf and stem morphology, plant growth habit, and if the conditions are appropriate, resistance to powdery mildew in the greenhouse. This is an efficient system that allows the screening of large number of seedlings in a relatively small greenhouse space. A common planting density of the seed is $1550\,m^{-2}$ which if germination is 40% results in a seedling population of about $624\,m^{-2}$ seedlings. After selection with a retention rate of 1% there is about 1 seedling per $5\,m^2$. The other case is when environmental adaptability is of primary importance or in the evaluation of nonrecurrent flowering seedling populations, as in the development of disease-resistant roses where the source of resistance is not recurrent flowering and consequently the hybrids with recurrent flowering parents do not flower in the first year. In these programmes the first cycle of selection is done in the field with up to $6\,m^2$ of field space for 2 years, a much more costly procedure.

Over the last several decades, research and development of molecular markers has generated a plethora of genetic markers that have been used for variety/genotype identification, germplasm analyses and management, phylogenetic analyses, and for map construction. With this seemingly unlimited number of DNA-based markers available derived from a diverse number of molecular approaches (*see* **Genetics:** Molecular Markers), it became possible to mark the presence of specific genes. In addition, molecular markers distributed throughout the genome can help identify the parental contribution to the chromosomal segments in the progeny. These are the possibilities that have led to an enthusiastic research push in the area of marker-assisted selection (MAS) in rose breeding.

One of the challenges of plant breeding is to select the best genotype from phenotypic data. Thus the ability to select for the genes that confer specific phenotypic traits using markers tightly linked to genes controlling these traits through MAS is attractive to a plant breeder. This allows for selection for a genotype and avoids problems with the interaction of genetic and environmental factors on the phenotypic expression of a trait.

A major consideration in any plant breeding programme is the cost of the procedure since most are operated on a fixed budget. Among the traits in which MAS in a breeding programme is most likely to be cost efficient are:

1. Traits that require difficult and expensive evaluation protocols. In rose this might include pest and pathogen evaluations that frequently involve multisite and/or multiyear evaluations or labour-intensive greenhouse or laboratory techniques which are highly dependent on both environmental conditions and the genotypes of the pest/pathogens. A programme to pyramid resistance genes into one genotype would benefit from markers for each race-specific resistance gene. These markers would avoid the need to do race-specific pathogen testing on every progeny to identify the genes present. Evaluating traits such as postharvest performance of the flower or plant or cold hardiness generally require replicated trials (replicated in terms of plants and/or years) and thus can only be carried out once enough plant material is available. All these traits are expensive to measure in time, labour or space.

2. Traits for which progeny testing is needed to confirm its genetic presence. This would include recessive genes for any trait.
3. Introgression of simply inherited traits via backcrossing. In this case, selection could be for a desired donor allele and against markers unique to the donor rose. Again the most cited trait would be disease resistance but this could also be useful in the rapid incorporation of major genes for postharvest longevity, cold hardiness, fragrance, and other traits. The obvious limitation of this in practice would be the susceptibility of the recurrent rose parent to inbreeding depression.

Considerations of MAS

Rose germplasm is a multispecies complex and thus has a high degree of allelic diversity. Although many cultivated roses can be inbred, outcrossing is common and much of the breeding emphasizes the avoidance of inbreeding by using a diverse array of parents. This approach also tends to promote less consistency in marker-gene linkages on a population scale because among the highly heterozygous rose breeding populations there is a greater probability that any given marker is linked to different allelic forms of the gene of interest in different germplasm as compared to highly developed inbred crops such as rice, barley and wheat. Consequently, these linkages will tend to be cross or germplasm specific. Thus to use MAS in breeding, the marker-gene linkages may need to be confirmed for each cross and /or recombination cycle. This adds a significant cost to the practice of MAS. One approach to avoid this problem is to develop allele specific markers which are part of the gene and therefore not dependent on linkage stability. Unfortunately, these markers require more extensive research to develop than do linkage-based markers.

Selection for Juvenile Traits

If MAS is used to select for one or more genes on the juvenile stage, the trait chosen for selection must be essential to have in all subsequent breeding populations. Such selection will reduce the variability for other traits available for selection in the second stage of selection due to genetic drift and linkage drag. Thus it may be necessary to increase the size of the base population to avoid this problem.

The Ease of Using the Markers as a Major Constraint of MAS Implementation

At present, most markers require a lengthy DNA extraction and purification protocol followed by gel-based electrophoretic separation and identification procedures. This represents a major investment in labour, materials and laboratory equipment which most breeders are not able to make. The marker screening procedures need to be simplified to be able to perform the analysis on plant tissue with minimum preparation, such as crushed tissue, and incorporate non-gel-based rapid allele detection systems. To be effective, markers should be more broadly applicable to most germplasm.

The Effectiveness of MAS is Dependent on a Firm Understanding of Genetics of the Trait

At present, few traits (qualitative or quantitative) have been well studied in rose (*see* **Genetics**: Inheritance of Characteristics). Recent work at identifying genes for resistance to black spot and powdery mildew has created an opportunity to develop markers in a MAS procedure to develop roses with greater disease resistance. Although the manipulation of qualitative traits is useful in plant breeding, most important commercial characteristics are inherited by multiple quantitative trait loci (QTL). The major limitation of MAS for QTL selection is the extensive work it takes to reliably identify these QTL. Only a well-replicated genetic study with hundreds of progeny run over several years and/or sites will give the necessary information to separate the genetic and environmental effects of the phenotype. These are expensive studies to complete and if such studies need to be run for each cross and/or recombination cycle, it would be unlikely such an approach would be cost effective.

The feasibility of MAS is dependent on the relative costs of plant maintenance and evaluation with and without MAS. The two major cost advantages of MAS would be: (1) the elimination of plants at a juvenile stage and thus saving the cost of conventional maintenance and evaluation for the early eliminated plants, and (2) the increase in selection efficiency so that smaller populations can be used or a given objective can be achieved in fewer generations.

Backcross Programmes

A frequently suggested MAS programme is the incorporation of a dominant disease-resistance gene(s) into a commercial background using a backcross programme. A typical backcross programme (**Figures 1** and **2**) would begin with a cross between a susceptible commercial genotype (*aa* or *aaaa*: recurrent parent) and a resistant noncommercial genotype (*AA* or *AAAA*: donor). The F_1 hybrids will be heterozygous resistant plants (*Aa* or *AAaa*) and consequently no selection for resistance need be done. All subsequent backcross generations would involve a cross between

	Commercial cultivar	Progeny expected	Donor species	Percentage of donor genome
	aa	×	*AA*	100%
F_1		*Aa*	All resistant Select for horticultural traits	50%
BC_1	*aa* × *Aa*	1 *aa*:1 *Aa*	50% resistant Select resistant genotypes Evaluate these for horticultural traits	25%
BC_2	*aa* × *Aa*	1 *aa*:1 *Aa*	50% resistant Select resistant genotypes Evaluate these for horticultural traits	12.5%
BC_3	*aa* × *Aa*	1 *aa*:1 *Aa*	50% resistant Select resistant genotypes Evaluate these for horticultural traits	6.25%
BC_4	*aa* × *Aa*	1 *aa*:1 *Aa*	50% resistant Select resistant genotypes Evaluate these for horticultural traits	3.125%

Figure 1 Backcross programme to introgress a single dominant gene from a wild rose species into a commercial cultivar. Example with diploid roses, or tetraploid with disomic inheritance.

the susceptible commercial parent (*aa* or *aaaa*) and the resistant heterozygous selections (*Aa* or *Aaaa* or *AAaa*). Thus with a diploid situation each backcross generation will have about 50% of the seedlings that would be resistant. In the tetraploid case, the first backcross generation is expected to have 83% of the progeny resistant. This percentage decreases towards 50% as the generations progress. In such a backcross programme, four backcross generations are needed to recover 95% of the commercial genome combined with the donated gene of interest.

Cost Effectiveness

The question that needs to be answered is whether this is a cost-effective approach to breeding for resistance. A recent study analysed the costs and benefits of MAS for fruit crops, which gives a good framework to assess this. A typical backcross generation would begin in the spring and early summer with a cross between a heterozygous resistant selection and the commercial susceptible parent. The hips are harvested in the autumn, seed extracted and stratified, and in the following spring (year 2) seed are planted. If one assumes that 2000 seedlings are produced, in a conventional programme these would be planted at low density to allow rapid seedling growth so cuttings could be taken for a disease-resistance screen in the autumn. All the 2000 seedlings would be evaluated for their horticultural characteristics during the summer and autumn. Once the 2000 seedlings have been evaluated for disease resistance, the best of the resistant seedlings are retained for the subsequent backcross generation. In a backcross programme, juvenile MAS would save costs by eliminating susceptible genotypes as small seedlings and not by decreasing the length of the breeding cycle. Thus the saving is dependent on the percentage of seedlings that are eliminated early. In the case of a diploid, if one dominant resistant gene was involved, 50% of the progeny should be eliminated and only 1000 of the original seedlings need to be evaluated for their horticultural

	Commercial cultivar	Progeny expected	Donor species	Percentage of donor genome
	aaaa	×	*AAAA*	100%
F_1		*AAaa*	All resistant Select for horticultural traits	50%
BC_1	*aaaa* × *AAaa*	1 *AAaa* 4 *Aaaa* 1 *aaaa*	83% resistant Select resistant genotypes Evaluate these for horticultural traits	25%
BC_2	*aaaa* × *AAaa*	1 *AAaa* 4 *Aaaa* 1 *aaaa*	57% resistant Select resistant genotypes	12.5%
	aaaa × 4 *Aaaa*	12 *Aaaa* 12 *aaaa*	Evaluate these for horticultural traits	
BC_3	*aaaa* × *AAaa*	1 *AAaa* 4 *Aaaa* 1 *aaaa*	52% resistant Select resistant genotypes	6.25%
	aaaa × 16 *Aaaa*	48 *Aaaa* 48 *aaaa*	Evaluate these for horticultural traits	
BC_4	*aaaa* × *AAaa*	1 *AAaa* 4 *Aaaa* 1 *aaaa*	51% resistant Select resistant genotypes	3.125%
	aaaa × 52 *Aaaa*	156 *Aaaa* 156 *aaaa*	Evaluate these for horticultural traits	

Figure 2 Backcross programme to introgress a single dominant gene from a wild rose species into a commercial cultivar. Example in a tetraploid rose with tetrasomic inheritance with independent chromosome assortment.

traits. Following this further, if two or three dominant disease-resistance genes were being incorporated, the proportion of the population being eliminated every generation would be 75% (leaving 500 for subsequent evaluation) and 87.5% (leaving 250 for subsequent evaluation) respectively (**Table 1**). The break-even point, i.e. the increased amount that could be spent per selected seedling over conventional costs, would be about 45% of the conventional cost for the one-gene case, 65% for the two-gene case, and 73% for the three-gene case. Thus if your conventional cost of maintenance and selection is US$2.00 then you could spend US$2.90 per seedling for a one-gene MAS programme and ultimately have the same cost for the programme. As the number of genes increases and the cost of conventional selection increases, there would be more financial resources available for the costs of developing and using a MAS system (**Table 2**).

Progeny Testing

The incorporation of a recessive gene into the germplasm from a noncommercial germplasm source would require progeny testing to determine which of the backcross individuals had the recessive gene(s) of interest. In both the conventional and the MAS approach, the backcrosses are made the first year and the seed are planted the spring of the second year. Taking 2000 backcross seedlings in the greenhouse, the conventional approach would require wide spacing for the seedlings to allow a test-cross (test parent would be homozygous recessive) to be done the second year as compared to a MAS approach which would allow a high-density planting and assaying the young seedlings with markers to determine if they have the proper genotype. No test-crosses would be needed and 1000, 500 or 250 of the seedlings would be retained for a one-, two- or three-gene

Table 1 One backcross cycle for a conventional compared to MAS programme to introgress a dominant disease-resistance gene into the diploid rose. It is assumed that this trait is the primary selection focus and it cannot be easily evaluated on a small seedling plant. This would be the case if several genes for resistance to the same disease were being pyramided (combined in the same plant)

Year	*Conventional programme*		*MAS programme*			
	Operation	*Seedling number*	*Operation*	*Seedling number*		
				Percentage of seedlings selected[a]		
				50	*25*	*12.5*
1	Cross heterozygous resistant plant with commercial susceptible parent Harvest seed in fall Stratify seed		Cross heterozygous resistant plant with commercial susceptible parent Harvest seed in fall Stratify seed			
2 Spring	Germinate seedlings in greenhouse bed	2000	Germinate seedlings in greenhouse bed Perform MAS on seedlings in bed	2000	2000	2000
2 Summer	Root three cuttings of each seedling and grow for disease resistance evaluation	2000	Horticultural evaluation	1000	500	250
2 Autumn	Greenhouse disease resistance evaluation	250–1000	Horticultural evaluation			
	Horticultural evaluation	20	Select best 1% of plants as parents for next backcross cycle	20	20	20
	Propagate best 20		Propagate best 20			
3 Spring and summer	Use best 1% of plants as parents for the next backcross generation Evaluate these in the field	20	Use best 1% of plants as parents for the next backcross generation Evaluate these in the field	20	20	20

[a] Seedling selected categories of 50%, 25% and 12.5% correspond to the incorporation of one, two or three dominant genes respectively.

Table 2 Mean break-even cost (US$) per selected seedling for juvenile phase (year 1) marker-assisted selection (MAS) with no selection, 50% selection, 25% selection and 12.5% selection for a plant breeding programme

Selection cost without MAS	*Percentage of plants selected by MAS*[a]		
	50	*25*	*12.5*
1.00	0.45	0.65	0.73
2.00	0.90	1.30	1.46
5.00	2.25	3.25	3.65
10.00	4.50	6.50	7.30

[a] In a backcross programme with a diploid plant, 50%, 25% and 12.5% selection would correspond to selection for one, two and three genes respectively. After Luby JJ and Shaw DV (2001) Does marker-assisted selection make dollars and sense in a fruit-breeding program? *HortScience* 35: 872–879, with permission of ASHS.

selection scheme respectively. The seedlings are evaluated for horticultural traits for the rest of this year. In the third year, in the conventional approach the test-cross seedlings are planted in the greenhouse. For a one-gene scheme only five seedlings are needed to be able to ensure at a 95% confidence level that the recessive heterozygote will be detected. For two- and three-gene schemes, 11 and 23 seedlings are required for the same confidence. Thus assuming a 40% germination rate, about 13, 29 and 59 seed need to be sown for each test-cross for a total of 26 000, 58 000 and 118 000 seed for the one-, two- and three-gene case respectively. If the test-cross progeny evaluation is not done by the time the backcross progeny are ready to pollinate, to keep the cycle to two years, the next set of backcrosses needs to be made on the 2000 plants now. Once the evaluation is done, only the seed from the selected plants are collected and used in the following generation. The other alternative is to wait until the following year to do the next set of backcrosses only on the 20 selected plants. In contrast, the third year of the MAS case would be to do the next set of backcrosses on the selected 20 plants (**Table 3**). Consequently the use of MAS would eliminate the need to do the test-crosses or to evaluate

Table 3 One backcross generation for a conventional compared to MAS programme to introgress a recessive gene into the diploid rose. It is assumed that this trait is the primary selection focus and that it needs to be evaluated by a progeny test

Year	*Conventional programme*		*MAS programme*			
	Operation	*Seedling number*	*Operation*	*Seedling number*		
				Percentage of seedlings selected[a]		
				50	*25*	*12.5*
1	Cross heterozygous resistant plant with commercial susceptible parent Harvest seed in fall Stratify seed		Cross heterozygous resistant plant with commercial susceptible parent Harvest seed in fall Stratify seed			
2 Spring	Germinate seedlings in greenhouse at wide spacing	2000	Germinate seedlings in greenhouse bed at high density Perform MAS on seedlings in bed	2000	2000	2000
2 Summer	Do test-cross with homozygous recessive tester on all seedlings Grow all backcross seedlings	2000	Grow the remaining seedlings for horticultural evaluation	1000	500	250
2 Autumn	Harvest test-cross, need 13 to 59 seed per test-cross[b]	2000	Horticultural evaluation	1000	500	250
	(if this is not possible the cycle will require another year)		Keep best 1% of the plants as parents for the next generation	20	20	20
3 Spring	Grow out 5–23 test-cross seedlings for each backcross plant[b] Make crosses for next backcross on all backcross seedlings (if this is not possible the cycle will require another year)	10 000 to 46 000 test-cross seedlings 2000 backcross progeny	Make backcross to 20 selected plants	20	20	20
3 Summer	Evaluate test-cross seedlings Eliminate backcross seedlings without gene of interest Select to best 1% on horticultural traits	Discard test-cross seedlings 1000 plants with gene of interest 20 backcross progeny	Grow crossed plants for seed	20	20	20
3 Autumn	Collect backcross seed from selected plants	20	Collect backcross seed from the selected plants	20	20	20

[a] Seedling selected categories of 50%, 25% and 12.5% correspond to the incorporation of one, two or three genes respectively.
[b] To ensure a 95% confidence level of detecting the backcross seedlings carrying the desired recessive gene(s), the number of seedlings needed for analysis is 5, 11 and 23 for the one-, two- and three-gene case respectively. Assuming a 40% germination rate, the minimum number of test-cross seed that would be needed is 13, 29 and 59 seed for the one-, two- and three-gene case, respectively.

test-cross progeny evaluation, and would most likely cut a year off the backcross cycle.

Elimination of Noncommercial Genes

Beyond incorporating the gene of interest into a population, a backcross programme also is designed to eliminate noncommercial genes from the population as quickly as possible. Theoretically the proportion of these genes decrease by 50% each generation (**Figures 1** and **2**) so it takes four backcrosses after the initial cross to lower this to less than 5%. Work with other crops, and preliminary work with rose, indicates that this process can be accelerated if selection is done directly to eliminate the donor parent genome with molecular markers. Consequently, there is a possibility of completing a backcross programme with one or two fewer generations of work. This improves the attractiveness of MAS. For example, if a conventional program cost US$2.00 per seedling per generation, a four-generation backcross programme would cost US$8.00 per seedling (ignoring the cost of producing the hybrid seed each generation that remains fixed). Given a one-gene MAS program, the break-even cost would be US$3.60 for the MAS protocol. If you could do this program in two backcross generations instead of four, your break-even cost for MAS would be US$1.80 for the two selections of MAS plus US$4.00 per seedling for the two generations that did not happen, for a total of US$5.80 per seedling. The additional advantage is that the rose with the new trait could be introduced to market 4–6 years earlier. This is very attractive from a commercial standpoint and shows the potential power and savings of MAS.

Unfortunately, most traits of economic importance are not simply inherited but are genetically conditioned by multiple loci and alleles. These traits are further complicated by the great influence that the environment has on the expression of a given genotype on the resultant phenotype. This, of course, is the reason markers would be an effective tool since these are not influenced by the environment. There have been many studies to find molecular markers linked to QTL in a range of plants.

Response to Selection

MAS is basically an indirect selection protocol and as such its effectiveness in terms of response to selection depends on several factors: (1) selection intensity, (2) heritability of the trait and the markers, (3) the strength of the genetic association or linkage between the trait of interest and the markers, and (4) genetic variance for the trait. If we assume that there is constant selection intensity and that the genetic variance for the trait is high, the ultimate success depends on the heritability (additive genetic component) of a trait and the strength of the genetic association between the traits. Thus the effectiveness of the markers alone for selection will generally be less than using phenotypic selection, for two main reasons. The first is that the strength of the genetic association as measured by a correlation will be less than 1 (maximum value) and in many cases considerably less than 1. Second, the heritability (h^2) of the markers is usually less than h^2 of the trait. In theory, this approach is most attractive when dealing with traits with low h^2 but in practice as the heritability for a trait decreases so does the ability to detect useful markers.

Another approach would be to use index-MAS which combines phenotypic and marker-assisted selection at the same time to improve the efficiency of selection. For quantitative traits, this approach would save costs by decreasing the population size needed to obtain the same objective (**Table 4**). The table indicates that the most savings are realized for traits with low heritability (0.2 and less), strong genetic association, and at high selection pressure. Unfortunately, the power to detect QTL markers for traits with low heritability in unreplicated genetic studies with 100–300 progeny that are generally reported in the literature is insufficiently low. Well-planned genetic studies with larger numbers of seedlings that are replicated at several sites and over several years are required to separate the environmental and genetic variance. Only with this type of data can useful QTL be reliably detected. These studies are expensive to do and consequently the development costs to use MAS on these types of traits are also high. The price increases if the genetic linkages are not stable, as might be expected in a highly heterozygous germplasm. This would require repeating the marker identification studies every generation and in different genetic backgrounds.

Summary

Marker-assisted selection works given the necessary investment into understanding the genetics of the trait and identifying markers on or closely linked to it. The rose community is just beginning to tackle this challenge.

The pertinent question is whether it is a cost-effective approach. If all of the costs of developing the markers are borne by one breeding programme, it would be prohibitively expensive for many, especially the smaller breeding programmes. However, with proper industry-wide and governmental funding for this developmental work, the approach may become economically feasible for simply inherited traits that are expensive to evaluate and/or need a progeny test to detect recessive traits. They could also be used to select against a donor parent genetic background in a backcross breeding programme. In all these situations, the savings are accomplished by not having to grow as many seedlings during a given season because they are eliminated sooner by juvenile screening or by cutting a year or more off a breeding programme. In the case of traits inherited quantitatively, it is doubtful that MAS would be economically feasible given the expense required to develop quality markers for QTL for traits with low heritability.

Nevertheless, as the protocols of marker screening and development become less expensive to run and the rose community develops a strong body of

Table 4 Estimated progeny number required from a family that a breeder needs to test and mean internal break-even cost for phenotypic selection or index marker-assisted selection programmes for a normally distributed quantitative trait as related to narrow sense heritability (0.1, 0.2 and 0.5) if markers account for 20%, 40% or 60% of the genetic variance ($P = 0.2, 0.4$ or 0.6) to be 99% certain of selecting one or more progeny that are among the top 1% for genotypic value

h^2	*Number of progeny to be tested using:*				*Mean internal break-even cost per US$1.00 per seedling spent on evaluation for implementing MAS*		
	Phenotypic selection	*Index-MAS*					
		P = *0.2*	*0.4*	*0.6*	P = *0.2*	*0.4*	*0.6*
		10% selected, 1% target					
0.1	1424	468	458	458	$2.72	$2.80	$2.80
0.2	998	495	459	458	$1.35	$1.56	$1.57
0.5	579	503	468	459	$0.20	$0.31	$0.35
		5% selected, 1% target					
0.1	2298	507	460	458	$4.71	$5.32	$5.35
0.2	1492	573	473	459	$2.14	$2.87	$3.00
0.5	734	590	506	466	$0.32	$0.60	$0.77
		1% selected, 1% target					
0.1	7677	916	655	554	$9.83	$14.28	$17.12
0.2	4678	1151	762	605	$4.08	$6.85	$8.97
0.5	1682	1208	914	714	$0.52	$1.12	$1.81

Adapted from Luby JJ and Shaw DV (2001) Does marker-assisted selection make dollars and sense in a fruit-breeding program? *HortScience* **35**: 872–879, with permission of ASHS.

information on the genetics of the major economically important traits of the rose, the economic feasibility of this technique will become more favourable.

See also: **Genetics**: Meiosis; Inheritance of Characteristics; Gene Mapping; Molecular Markers; Research Trends.

Further Reading

Debener T and Mattiesch L (1999) Construction of a genetic linkage map for roses using RAPD and AFLP markers. *Theoretical and Applied Genetics* 99: 891–899.

Debener T, Linde M and Dohm A (2003) The utilization of molecular tools for rose breeding and genetics. *Acta Horticulturae* (in press).

Gudin S (2000) Rose: genetics and breeding. *Plant Breeding Reviews* 17: 159–189.

Luby JJ and Shaw DV (2001) Does marker-assisted selection make dollars and sense in a fruit-breeding program? *HortScience* 35: 872–879.

Malek V and Debener T (2000) Identification of molecular markers linked to *Rdr1*, a gene conferring resistance to black spot in rose. *Theoretical and Applied Genetics* 101: 977–983.

Rajapakse S, Byrne DH, Zhang L *et al.* (2001) Two genetic linkage maps of tetraploid roses. *Theoretical and Applied Genetics* 103: 575–583.

Ribaut JM and Hoisington D (1998) Marker-assisted selection: new tools and strategies. *Trends in Plant Science* 3: 236–239.

Young ND (1999) A cautiously optimistic vision for marker-assisted breeding. *Molecular Breeding* 5: 505–510.

Research Trends

S Rajapakse, Clemson University, Clemson, SC, USA

Introduction

Owing to its inherent nature, analysing the genetics of rose traits is a challenge. The high level of heterozygosity in the diploid rose genome, polyploidy, self- and cross-incompatibility, reduced seed set and poor seed germination contribute to this challenge. In spite of these difficulties, some progress has been made in the last decade towards understanding genetics of rose traits. The mode of inheritance is now known for traits such as double corolla, recurrent-blooming ability, pink flower colour, moss character, male sterility, dwarfness, spreading growth habit, prickles on the stem and resistance to black spot and powdery mildew. Chromosome maps representing the rose genome organization based on the rate of recombination have been produced with at least three different sets of progenies. Single genes controlling double corolla, pink flower colour, absence of prickles in the petiole, recurrent-blooming ability and production of enzyme malate dehydrogenase as well as several chromosome regions controlling petal number and the density of prickles on the stem have been discovered and located on these maps. Furthermore, two genes known to confer resistance to common diseases, black spot and powdery mildew, have been discovered and mapped. The gene controlling resistance to a single race of the fungus causing black spot (*Rdr1*) has been studied further with fine mapping and isolation of bacterial artificial chromosome (BAC) clone collections that contain the gene region. This gene was the first disease-resistance gene discovered from rose and it appears to contain several features common to many of the disease-resistant genes in plants.

There has been a heightened interest in the past 5 years in understanding the genetic basis of floral scent production. The fragrance of rose is unique and valuable to the perfume industry. Thus, roses are especially suitable to study fragrance compounds, corresponding genes and proteins. Fragrance is a result of thousands of volatile chemicals and their interactions, and it is difficult to screen for and quantify. For these reasons, genetic methods that begin with trait evaluation, which are used to isolate genes controlling most other traits, are not practical for cloning genes controlling floral fragrance. Instead, new genomic techniques have been applied to discover genes normally expressed in rose petals, the site of the scent production. Among thousands of sequences expressed in petals, a few that appear to be related to fragrance have been identified by expression analysis using gene-chips.

Progress has also been made in our ability to transform and regenerate roses through genetic engineering. Both *Agrobacterium* and gene gun-mediated methods have been developed and show promising results for the future transformation of rose. Past research has also revealed some encouraging facts for continued genetics and genomics studies in rose. We now know that the rose genome is relatively small, comprising about 600 mega basepairs (Mbp). A small genome usually means that it has a relatively reduced level of nongene, repetitive DNA sequences, which comprise the bulk of the genomes in the large-genome species.

We now possess a wealth of knowledge to manipulate the rose characters by traditional breeding methods, aided with new genomic and molecular marker information, and by genetic engineering. However, we have only begun to unlock the secrets hidden within the rose genome and much work remains to be done in the future. The pace of rose genetics and biotechnology research is expected to accelerate in the next 5 years as we move closer to producing a consensus rose genetic map, cloning major genes which have already been identified, isolating new genes and studying their function. Though it is difficult to predict the direction of future research, the trends and prospects of main avenues of genetics-related research are discussed below.

Construction of a Consensus Chromosome Map

A high-density chromosome map that can be applied to all roses in general is an invaluable tool for many facets of future rose research. This map can serve as a permanent genetic and genomic resource that locates all rose genes discovered as well as undefined expressed sequence tags (ESTs), provides a basis for future map-based gene isolation, facilitates physical mapping and assists in progeny selection in rose breeding. Such a widely applicable genetic map is currently not available for rose, although there are three separate pairs of linkage maps, each providing partial coverage of rose chromosomes and largely specific to the cross from which it was developed. A few important rose genes, both Mendelian type and quantitative trait loci, have been discovered and located on these maps.

At present, no common markers exist to combine these three sets of maps. Efforts are currently under way to join these maps using markers known as microsatellite or simple sequence repeats (SSRs). SSR markers are polymerase chain reaction (PCR)-based, highly polymorphic and applicable to all roses in general. Locating the same SSR markers in separate maps will provide anchoring points to join the maps to produce a unified consensus rose genetic map in the near future. In order to locate a SSR marker every 1 Mbp in the rose genome consisting of 600 Mbp, about 600 SSR markers are needed. Development of SSR markers is time-consuming and laborious; therefore it is unlikely that this many SSRs will be developed in the near future. Consequently, even after joining the maps, we will probably still have a fairly low-saturation level of widely applicable markers on the consolidated map. A set of about 40 SSR markers has already been identified by screening a rose genomic library and is used in map consolidation. In the future, new SSR markers will instead be identified by screening BAC libraries (collections of large-insert BAC clones).

Developing new SSR markers by screening BAC libraries has the advantage of simultaneous identification of an overlapping set of BAC clones assembled along the lengths of chromosomes by the mapped SSR markers. These BAC clones can provide a scaffold of the rose genome for future physical mapping.

Gene Identification and Isolation

The rose genome harbours many valuable genes. Isolation of these genes will help in understanding the mechanisms behind trait regulation as well as provide gene clones for transformation of roses lacking the trait. Moreover, certain rose genes isolated will have applications far beyond roses in enhancing the germplasm of other species. Elucidating the inheritance pattern of rose traits is usually the first step towards discovering the genes controlling the traits. Because of the complexities of genetic analysis of modern polyploid roses, inheritance and subsequent analyses have been made possible by the development of segregating populations in diploid backgrounds. Genes that control floral traits and disease resistance have received particular attention for isolation in the past and this trend is expected to continue in the future. Single genes controlling floral traits such as double corolla, pink petal colour and recurrent-blooming ability have been identified, located on maps and markers linked to the genes identified. In addition, quantitative trait loci controlling petal number and prickle density have been mapped. The next phase of rose research will see attempts to zero-in on these genes for isolation.

There is a special interest in cloning the recessive gene controlling the recurrent-blooming ability. The recurrent-blooming rose, one that blooms from spring until autumn in temperate regions, appears to overcome the seasonal barriers to which the spring-only blooming roses are subjected. A natural mutation first discovered in China over a thousand years ago and bred into other roses, recurrent or perpetual blooming ability is one of the main traits that made it possible for roses to become popular garden plants. Most of the hybrid tea roses have this gene variant that makes it possible for them to bloom throughout the warm season. This rose gene (identified as *r*) is the only single gene identified so far in the woody species that appears to overcome seasonal barriers for repeated blooming. This rose gene provides an excellent opportunity to understand the molecular basis of flowering in woody ornamental and fruit crops and subsequently manipulate this agriculturally important trait. Thus, the recurrent-blooming gene is one rose gene that will have implications for the improvement of other woody species.

Research is now being conducted to clone the *r* gene via genetic map-based methods and candidate gene approaches. The map-based approach, also known as positional cloning, makes use of the markers located close to and flanking the gene in the genetic maps to identify BAC clones containing the markers. Additional markers developed from BAC clones are used to saturate the gene region in the map further with additional markers that help identify BAC clones closer to the gene in the chromosome and eventually identify specific BAC clones containing the gene. This process, commonly known as chromosome walking, can be very slow, laborious and inefficient. In contrast, the candidate gene approach is a particularly efficient method that can eliminate a lot of steps involved in map-based gene cloning. In this method, genes with similar functions cloned in other plants are checked to investigate whether the one being sought in rose is similar to those candidate genes in basepair sequence. For example, to clone the recurrent-blooming gene by the candidate gene approach, other genes related to flowering and dormancy that have been cloned from *Arabidopsis*, apple, rice, peach, etc., are checked for similarity with the rose gene. PCR primers are designed in the conserved regions at the two ends of such candidate genes and used to amplify rose DNA. The resulting PCR products are used as markers and mapped on rose maps. The location of candidate genes in rose maps can quickly eliminate those that are not related to the rose gene.

A major area of genetic research interest is in the isolation of disease-resistance genes. Absence of natural resistance to common diseases in most of our modern cultivated roses has prompted this research. Identification of rose disease-resistance genes will allow us to study their mode of action and transfer to susceptible roses through genetic engineering and marker-assisted progeny selection. One gene conferring resistance to black spot, *Rdr1*, has already been identified, mapped and studied at the physical level. For this, a BAC library has been constructed from highly disease-resistant *Rosa rugosa* Thunb. var. *rubra*. Markers tightly linked to *Rdr1*, obtained from genetic mapping, have been used to isolate BAC clones. Chromosome walking initiated from BAC clones containing the markers led to an assembly of a contig (collection of continuous overlapping large DNA fragments) of six BAC clones spanning the *Rdr1* locus. This rose contig was found to carry several putative genes resembling disease-resistance genes identified in other plant species. Future research in this project will focus on further refinements in the gene region to produce smaller subclones of the contig and identify specific subclones harbouring the *Rdr1* gene and have the potential to transform species susceptible to black spot into resistant types. The ability to transform susceptible rose genotypes by genetic complementation with the cloned DNA fragments is the ultimate test of the effectiveness of the genes isolated. The identity of *Rdr1* will then be revealed by complete sequencing of the specific subclones.

Successful transformation is necessary to verify the nature of isolated gene sequences. Therefore, our ability to demonstrate the successful cloning of rose genes also depends on our ability to transform roses genetically. For this purpose, future small-insert libraries that will be made in this and other gene-cloning projects will most likely be made from transformation-ready vectors.

In addition to *Rdr1*, a gene conferring resistance to powdery mildew gene, *Rsp1*, has also been discovered and positioned on maps. In the future, attempts will be made to isolate this gene by the map-based methods as well. As with most other species studied in detail, disease resistance in roses is probably controlled by many genes. Therefore, additional rose genes conferring disease resistance will undoubtedly be discovered in the course of the next few years. Disease-resistance genes discovered in other species appear to have some features in common, regardless of the pathogen type involved. Proteins synthesized from these resistance genes are known to have conserved domains such as the *Toll*-interleukin receptor (TIR), nucleotide binding site (NBS) and leucine-rich repeats (LRR). Degenerate primers (primers with less stringency) made in these conserved regions of the resistance genes could be used to amplify resistance gene analogues (RGAs) from practically any plant. This method has been widely practised to discover many RGAs in other crops. Undoubtedly, RGAs will be identified from rose by this method. However, RGAs discovered by this PCR method need to be studied further and verified by other methods such as gene mapping and complementation. As more and more disease-resistance genes are identified from rose, the future will see efforts to pyramid the genes by breeding and marker-assisted selection methods to achieve more durable disease resistance.

Apart from floral and disease-resistance genes, other genes under investigation include those that control the presence of prickles, growth habit, dwarfness and genetic aspects in transformation. Research on these genes is expected to continue as well, although with lower priority.

Synteny with Other Crops

Besides selecting genes with similar function isolated in other crops, the candidate gene approach can also make use of the synteny or the conservation of gene order observed among plants for gene identification and cloning. In this method, rose chromosome maps are compared to maps of other model species that have been studied in great detail to identify the chromosomal segments that show conservation or colinearity. Genes located in these conserved regions in the model species can yield valuable information on candidate genes to discover genes in the corresponding chromosomal segments of rose. In our example of isolating the recurrent-blooming gene by this method, first it has to be established whether the rose chromosomal region carrying this gene is conserved with any other well-studied species. Genetically related species show the most synteny. Thus, comparative mapping with other species in the Rosaceae holds the most potential to provide conserved regions with rose and expedite the process of gene discovery in rose. Species most appropriate for comparative analysis within the family include peach, apple, pear, cherry, plum, almond and strawberry. Of these, peach is the most genomically studied species in the family, and future rose genetics research will involve comparative mapping between rose and peach to investigate their synteny. Apart from comparisons with peach and other Rosaceae crops, roses can benefit from the vast amount of information generated from model plants *Arabidopsis* and rice. The genome of *Arabidopsis* comprising about 25 000 genes is completely sequenced and a draft sequence of the rice genome is complete. A good proportion of *Arabidopsis* genes has known functions. Although no research is currently being conducted to examine the synteny of roses with model species in a systematic manner, all DNA sequences generated from other rose projects are routinely checked against existing gene sequences in the databanks to find matches that can aid in discovering rose genes.

Expressed Sequences

Isolating a rose gene that matches with a phenotype is a complex process that takes a long time and vast amount of resources. In another high-throughput strategy to identify and catalogue rose genes, random cDNA clones developed from specific rose organs (such as leaf, petal and bud) are sequenced. This gives rise to a collection of ESTs. ESTs sequence-generated are compared to sequences of known genes from other species in data banks or used in expression analysis to study their function. The EST approach is especially helpful in discovering genes for traits for which the other gene-cloning methods are not suitable. One such trait is the floral fragrance. Fragrance is difficult to quantify and mutations are hard to detect. Model plant genomes *Arabidopsis* and rice are poorly suited to study floral fragrance. Two rose petal EST databases have been developed with an aim of identifying genes controlling floral fragrance. One of these petal databases consists of about 2100 unique sequences; for about 350 of these there is no match to existing Genbank sequences. These ESTs are valuable resources for the identification of genes expressed in petals important not only to fragrance, but also for other floral traits. The future of this research will involve linking the genes to specific phenotypes through functional analysis. In addition, we will see more EST databases developed from various rose tissues under specific conditions, attempts to mine those databases for known genes, to identify genes specific to tissues and conditions, and to assign functions to ESTs via functional analysis. Unique ESTs and those that have identified functions will be mapped on the rose consensus genetic map as a part of the integration of structural and functional genomic information.

Functional Genomics

As more and more genes and EST sequences are being identified from rose, the next phase of rose genetics research will belong to understanding gene function. In contrast to past studies that examined expression of one gene at a time, new functional genomic tools allow simultaneous expression analysis of all genes and sequences identified from rose. This has been made possible by the microarray or gene-chip technology, in which all rose genes and expressed sequences can be bound to a glass slide and probed with a selected sequence. Such global gene expression analyses can identify all genes regulated by tissue or given conditions and help elucidate gene networks leading to the production of certain compounds.

Microarray-based gene expression analysis has other advantages as well. It is particularly suitable for isolating genes that cannot be cloned by traditional genomic methods. It has already been applied to discover genes related to floral scent production by analysing rose petal ESTs. In one study, gene-chips consisting of ESTs developed from petals of two contrasting floral phenotypes were tested for differential gene expression. ESTs selected by the functional analysis, combined with expression of some of the sequences in *Escherichia coli* and metabolic analysis, have yielded several candidate genes for floral-scent production. Future research will subject these candidate sequences to additional testing. In general, the trend in future rose research is to move towards the functional analysis of sequences generated from specific tissue under specific conditions. This will probably be applied to study the expression of genes involved in a wide range of situations such as extended shelf-life, high oil content, floral development, cold-hardiness and response to pathogen invasion.

Physical Mapping

A physical map consists of large overlapping DNA fragments (such as the BAC clones) ordered along the length of each one of the seven rose chromosomes. It is unlikely that the entire rose genome will be physically mapped soon. Rather, physical maps will be constructed for a number of localized regions in the chromosomes carrying important rose genes, such as the BAC contig assembled for the *Rdr1* locus. These local physical maps will be developed in the course of gene isolation projects to find out basepair distances between genes and markers, and to facilitate chromosome walking to the gene.

A physical map is now being assembled for peach, the species well poised to become a model for future genomics research in the Rosaceae. What is most likely to happen in the future is that genetic maps as well as local physical maps developed for other species in the family, including rose, will be anchored to the peach physical map. This will be accomplished by hybridizing peach BAC clones in the physical map with probes containing specific markers selected from the genetic maps.

Gene Tagging, Gene Introgression and Marker-Assisted Selection

One of the most practical applications of genetics research in producing new roses by breeding programmes is the ability to select for phenotypic traits using markers tightly linked to genes controlling the traits. A few important rose traits have already been tagged with molecular markers, but the markers currently available are not very effective for screening germplasm and progeny in breeding programmes. Most of the markers currently linked to rose traits are amplified fragment length polymorphism (AFLP) markers, which cannot be readily tested in other roses. In the future, these markers will be converted to simple PCR-based markers such as sequence-tagged sites (STS) or new SSR markers linked to the traits discovered. Consolidation of current genetic maps will facilitate this to some extent.

Another area in which the markers can be effectively used for crop improvement is in introgression of beneficial genes from wild germplasm into crop cultivars. Markers linked to the trait of interest in the wild (donor) parent, as well as markers distributed throughout the genome in the improved cultivar (recurrent parent), in the form of a genetic map, could be used in selection of breeding progeny. This would not only help in tracking the movement of beneficial alleles from wild species to modern rose cultivars, but it can also be used to reduce the genetic background of the donor parent in the progeny. Graphical displays of progeny genotypes along the length of chromosomes will be made to visualize the contribution of different chromosomal regions from the two parents, and to select those offspring with minimum contribution from the wild parent.

Improvements are needed in the future to the current DNA marker technology to lift constraints on their application by breeding programmes. Markers that can be effectively applied in selecting progeny should be technically simple methods to perform and applicable to all roses in general. For screening a large numbers of progeny for marker-assisted selection, simple PCR-based techniques are the most appropriate. Allele-specific markers that would allow PCR-based allele identification are also quite useful for the breeder. Further technological advancements are needed in marker analysis to realize fully the potential of molecular markers to breeding. For example, developing methods to perform PCR directly from crushed leaf discs would avoid lengthy DNA extraction and purification procedures. In addition, replacing currently used gel electrophoresis methods with nongel-based plus or minus assays would facilitate more widespread marker application. These nongel methods are routinely used in animal and human genetic diagnostic work, but not applied to plants well. One way of developing a simplified diagnostic method is to convert initial PCR-based allele specific markers into single nucleotide polymorphic (SNP) markers and use colorimetric assays to select alleles. Alternatively, SNP markers could be combined with DNA chips to test the presence or absence of specific alleles.

With future advancements in the marker technology, more 'breeder-friendly' markers will be available. It is hoped that the rose industry will reap the benefits of the current knowledge in rose genomics by utilizing the markers to make marker-assisted selection of rose breeding progeny a reality.

Genetic Engineering

Over the past years several regeneration and transformation protocols have been developed and it is now possible genetically to engineer roses by both *Agrobacterium*-mediated and particle bombardment methods. All regeneration protocols use somatic embryogenesis. Roses have been reported to be successfully transformed with genes for chalcone synthase, chitinase and other proteins. Future research in this area will examine the stability and inheritance of transgenes, number of copies inserted, expression of transgenes in target tissues and silencing of native genes. Some transgenic roses developed aiming at improving disease resistance have not shown promising results. It remains to be seen whether this is due to specific rose pathogen combinations or that some gene constructs simply do not function in the genetic background of cultivated roses. Instances where the phenotype of the original rose plant has been altered due to somaclonal variation have also been reported. The much-sought-after blue rose still remains elusive, but it is being pursued anew with novel genes derived from humans.

Emerging techniques for engineering the much smaller chloroplast genome appear to overcome many of the problems associated with the transformation of the nuclear genome. Low levels of transgene expression, transgene dispersal through pollen, silencing of native genes and some other disadvantages in nuclear genetic engineering can be overcome by engineering the maternally inherited chloroplast genome. The transgenic chloroplast approach has been successfully applied to confer disease resistance in plants and may be applied to roses.

Genotyping and Phylogenetic Analysis

Various types of DNA markers, especially the microsatellite markers, generated as parts of mapping experiments will find uses in other applications such as resolving questions in identity of rose bushes, parentage and errors in experimental sampling. Perhaps one day we may even have a rose cultivar identification database based on diagnostic molecular markers. Marker data generated from rose species, hybrids and cultivars will be used as tools to understand natural genetic variation, geographical variation, and to shed light on the rose domestication process. As effectively carried out in the past, markers will be applied to understand the unbalanced meiosis observed in dog roses.

Future basic research will also focus on investigating evolutionary relationships among roses. Furthermore, systematic studies with other related species in the Rosaceae will provide basic information on evolutionary relationships in the family. This will help identify chromosomal regions conserved between rose and other related species, identify possible translocations, duplication and other chromosomal rearrangements and other evolutionary changes that have taken place.

Conclusion

Of all ornamental species, roses are the most genomically studied plants to date. It is anticipated that roses will continue to receive heightened interest in the future as well. Because of this reason, and the known small genome size, roses will probably be used as a model species in the future to discover and understand genes important to ornamental and perfume industries. Floral fragrance, floral oil, floral quality and other detailed floral characteristics such as petal number, size, shape and colour and postharvest flower longevity are some of these traits. Model plant genomes such as *Arabidopsis* and rice are poorly suited to study these traits, though they can be useful in providing related genomic information.

The immediate future will bring rapid advances in the structural organization of the rose genome. As more EST data are generated, bioinformatics and gene mining will play a major role in gene identification. Functional analysis of genes and ESTs is expected to be a major research direction for the future. Comparative mapping methods also hold great potential for detailed rose genomics research. We will also see a shift from studying genes to their protein products. Depending on the advantages and ease of adaptation, new techniques that may be discovered in the future will also play a significant role in deciding future research directions.

A small number of research groups worldwide have contributed to most of the current body of knowledge on rose genetics and genomics. This number is expected to increase in the future. As with all other research, progress and the pace will be driven and determined by funding available for research from academic, government and industry sponsors and the cooperation between scientists in sharing of information gathered. While the former has not been very stimulating, the cooperation among the major rose research groups so far has been very encouraging.

See also: **Breeding:** Overview; Selection Strategies for Cut Roses; Selection Strategies for Disease and Pest Resistance. **Disease:** Black Spot; Powdery Mildew. **Fragrance and Pigments:** Functional Genomics. **Genetics:** Gene Mapping; Cloned Genes.

Further Reading

Bolar JP, Schneiter N, Smith FD *et al.* (2003) Genetic improvement of rose cultivars for disease resistance. *Acta Horticulturae* (in press).

Channelière S, Rivière S, Scalliet G *et al.* (2002) Analysis of gene expression in rose petals using expressed sequence tags. *Federation of European Biochemical Societies Letters* 515: 35–38.

Crespel L, Chirollet M, Durel CE *et al.* (2002) Mapping of qualitative and quantitative phenotypic traits in *Rosa* using AFLP markers. *Theoretical and Applied Genetics* 105: 1207–1214.

Debener T (2001) Molecular markers as a tool for analyses of genetic relatedness and selection in ornamentals. In: Vainstein A (ed.) *Breeding for Ornamentals: Classical and Molecular Approaches*, pp. 329–345. Dordrecht: Kluwer Academic Publishers.

Debener T and Mattiesch L (1999) Construction of a genetic linkage map for roses using RAPD and AFLP markers. *Theoretical and Applied Genetics* 99: 891–899.

Debener T, Mattiesch L and Vosman B (2001) A molecular marker map for roses. *Acta Horticulturae* 547: 283–287.

Debener T, Linde M and Dohm A (2003) The utilisation of molecular tools for rose breeding and genetics. *Acta Horticulturae* (in press).

Dohm A, Ludwig C, Schilling D and Debener T (2001) Transformation of roses with genes for antifungal proteins. *Acta Horticulturae* 547: 27–33.

Düring K (1996) Genetic engineering for resistance to bacteria in transgenic plants by introduction of foreign genes. *Molecular Breeding* 2: 297–305.

Guterman I, Shalit M, Menda N *et al.* (2002) Rose scent: genomics approach to discovering novel floral fragrance-related genes. *Plant Cell* 14: 2325–2338.

Malek BV, Weber WE and Debener T (2000) Identification of molecular markers linked to *Rdr1*, a gene conferring resistance to blackspot in roses. *Theoretical and Applied Genetics* 101: 977–983.

Marchant R, Davey MR, Lucas JA *et al.* (1998b) Expression of a chitinase transgene in rose (*Rosa hybrida* L.) reduces development of blackspot disease (*Diplocarpon rosae* Wolf). *Molecular Breeding* 4: 187–194.

Rajapakse S, Byrne DH, Zhang L *et al.* (2001) Two genetic linkage maps of tetraploid roses. *Theoretical and Applied Genetics* 103: 575–583.

Souq F, Coutos-Thevenot P, Yean H *et al.* (1996) Genetic transformation of roses, 2 examples: one on morphogenesis, the other on anthocyanin biosynthetic pathway. *Acta Horticulturae* 424: 381–388.

Genome Analysis *see* **Fragrance and Pigments**: Functional Genomics. **Genetics**: Gene Mapping.

GROWTH REGULATION

Contents

Hormones in Growth and Development

M Le Bris, Université Aix–Marseille III, Marseille, France

Introduction

Since the beginning of the twentieth century, the plant hormones, or phytohormones, have been the subject of more and more detailed scientific researches. Their list is constantly increasing and their mechanisms of action are beginning to be elucidated.

The definition of a hormone in animal physiology is unambiguous. It is an organic compound, active at very low concentrations, synthesized by endocrine glands and excreted in the internal milieu which transports it towards an organ or a tissue sensitive to its action, the function of which is controlled by this hormone. Hence, an animal hormone is considered as a chemical messenger that transmits an order originating from a regulating centre. It thus contributes to the function harmony of the whole organism.

In plants, the hormone concept is much less restrictive concerning the emission, transport and reception of the hormonal signal. There are no endocrine glands but zones of preferential synthesis that are not specifically differentiated. Transport from the synthesizing organ to the sensitive tissue does not always take place. The activation of the synthesis is generally directly determined by the environment. The specificity of cell response is weaker than in the animal environment. Moreover, a phytohormonal response does not necessarily result from the action of only one hormone; it results rather from interactions between several substances. The response is a function of the hormone concentration in the target cells. The hormone content of a cell results from different mechanisms such as *in situ* synthesis, import and export, catabolism, temporary inactivation and storage as conjugated forms. The response to hormones is also determined by the sensitivity of the target cell. Differences in sensitivity depend on genetic, anatomical, physiological and environmental factors as well as interactions with other hormones. Hence a phytohormone can be considered as an organic, endogenous, oligodynamic substance that carries an information to a sensitive target cell the functioning of which is influenced by this hormone.

In *Rosa* spp., as compared to some other plant species, specific knowledge concerning the phytohormones is still low. However, among higher plants, especially angiosperms, the involvement of the phytohormones in the physiological processes controlling growth and development are very similar. Roses are not known to behave fundamentally differently from the majority of angiosperms. Data concerning hormonal response of rose plants derive mainly from horticultural applications such as vegetative propagation, branching and the control of plant architecture, and postharvest flower care. In this article, the four classical plant hormones are examined, i.e. auxin, cytokinin, gibberellin and abscisic acid, plus ethylene which is often considered as a gaseous hormone (**Figure 1**).

Auxins

The primary auxin in plants is indole-3-acetic acid (IAA). Although some other natural compounds with auxin activity (e.g. auxesis) and close biochemical conformation are also present in plants such as indole-3-butyric acid (IBA), phenyl acetic acid and 4-chloro IAA, little is known about their specific physiological action.

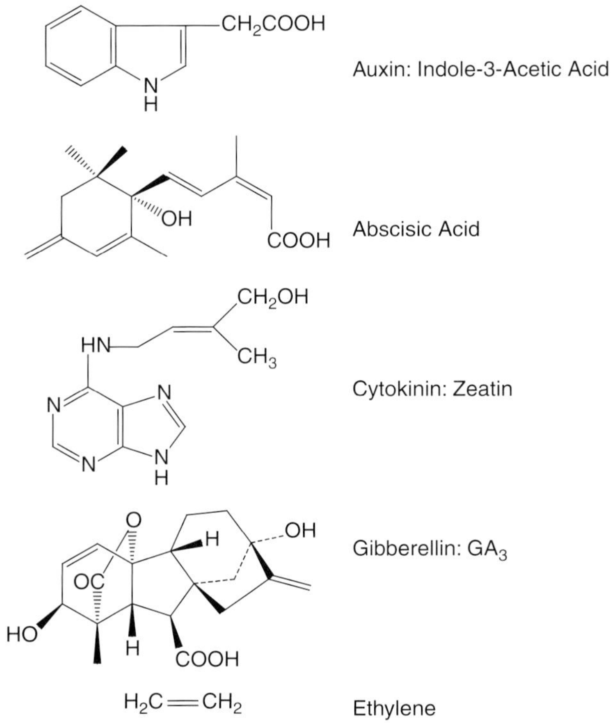

Figure 1 Chemical structure of representatives of different main groups of hormones in plants.

Auxin is mainly biosynthesized from the precursor tryptophan in the shoot apical parts, in meristems and young growing leaves in terminal buds. IAA can be catabolized both by photooxidation and by enzymatic degradation involving auxin oxidases. Forms conjugated with amino acids are known and they are considered as temporary inactive storage forms. IAA transport exhibits a polarization: it is mainly basipetal and follows the phloem sap flow. This transport can be inhibited by compounds such as 2,3,5-triiodobenzoic acid or *N*-(1-naphthyl)phthalamic acid.

The major role of auxin is to stimulate cell elongation. IAA loosens the cell wall and modifies its plasticity and elasticity characteristics. It thus contributes to the elongation of shoots and flower stems up to a threshold concentration over which it becomes phytotoxic. Leaf blade growth in monocotyledons is increased by auxin whereas, in some dicotyledons, it sometimes inhibits limb growth. Auxin also delays leaf abscission. As a cell growth stimulator, auxin is believed to mediate gravitropism and phototropism. Contrary to its effects on shoot growth, IAA inhibits root elongation.

At low concentrations and in the presence of cytokinins, auxin can stimulate cell proliferation as, for example, in the cambial zone. It contributes to the differentiation of the vascular system, i.e. xylem and phloem, and in this way influences stem enlargement. By stimulating cell proliferation, IAA also allows the neoformation and differentiation of buds. Moreover, at high concentrations, auxin actively stimulates root differentiation (rhizogenesis). In the practice of vegetative propagation, all these properties are used to generate new plants starting from isolated stem fragments: in a stem segment, polarized transport of auxin leads to a decreasing concentration gradient of IAA from the apical zone to the basal part. This difference in concentration along the stem segment promotes bud and root differentiation in apical and basal zones, respectively.

Furthermore, auxin is involved in apical dominance. Indeed, axillary bud breaking is often stopped because of correlative inhibitions, or paradormancies, that are attributed to auxin synthesized in the apical parts of the stem migrating downwards. Hence subjacent axillary buds are blocked in their growth and development. In all the different existing models explaining the phenomenon of apical dominance, auxin is considered to play a key role, although it may interact through a secondary inhibitor such as ethylene or abscisic acid.

Concerning sexual reproduction, mature seeds are known to release auxin which accumulates in the surrounding fruit pericarp. Auxin also delays fruit abscission.

In *Rosa*, among the numerous phenomena involving auxin, apical dominance has been widely studied because of its major influence on plant architectural development and resulting plant shape and potential for cut-flower production. Apical dominance is expressed in three main ways: inhibition of axillary bud break, inhibition of axillary shoot growth and control of the branch angle that the axillary shoots make with their bearing axis. Shoot decapitation, different stem orientation, stem bending or transverse incision above latent axillary buds promote axillary bud outgrowth. Their effect has been interpreted as corresponding to a limitation of phloem auxin flow originated from the apex. Likewise, during terminal flower initiation, the development of sylleptic branching in acropetal position is considered to result from a decrease in the auxin synthesis in the apical parts of the shoot. The rhizogenous properties of auxin are commonly used in rose vegetative multiplication from cuttings to induce root differentiation by exogenous applications of synthetic auxins such as IBA at the base of the cutting. In addition, transformation with a combination of *ROL A,B,C* genes of plants of the rootstock *Rosa hybrida* L. cv. Moneyway enhanced adventitious root formation on explants of stems and leaves. Such a phenotype was associated with an increase in auxin sensitivity.

Regarding postharvest physiology, auxins have been shown to delay peduncle abscission and to reduce peroxidase activity.

Cytokinins

Cytokinins are compounds with a structure close to adenine usually linked to an isopentenyl side chain. Kinetin was the first cytokinin to be discovered. It was isolated from herring sperm and was named after its ability to promote cytokinesis. The most common form of naturally occuring cytokinin in plants is zeatin. Many other natural cytokinins have already been isolated, including dihydrozeatin, isopentenyladenine and benzyladenine.

Cytokinins are present in most plant tissues. They are mainly biosynthesized *in situ*. Cytokinin concentrations are highest in the meristematic regions and in the areas of continuous growth, such as the root subapical zone, young leaves, developing fruits and seeds. Generally, cytokinin transport is acropetal via the xylem sap, but there is no real polarized transport. Cytokinins are irreversibly inactivated through two different reactions: the formation of *N*-conjugates with glucose, ribose or alanine, and the oxidative cleavage of the isopentenyl side chain by cytokinin oxidase.

Through their auxin antagonistic action, the cytokinins allow axillary bud release from apical dominance. They can stimulate the differentiation or neoformation and outgrowth of the axillary buds. Besides, they sometimes counteract seed or bud endodormancy, stimulate flower primordia differentiation and delay leaf and flower senescence.

In *Rosa*, several cytokinins, mainly belonging to zeatin and isopentenyladenine types, have been identified in different plant parts, including roots, axillary buds, xylem and phloem sap, leaves and petals.

Concerning their counteracting of apical dominance, it is accepted that most cytokinins are synthesized at the root level; they migrate to aerial parts via the xylem sap. The axillary bud outgrowth is intimately correlated with an increase in its cytokinin content. The hormone is then imported from roots and leaves. They promote bud outgrowth by counterbalancing the inhibitive action of the auxins that are synthesized in the shoot apex. Stimulating cytokinin production in root system alters plant architecture. For instance, *ROL C* gene encodes a cytokinin-β-glucosidase which hydrolyses cytokinin glucosides and thus liberates active free cytokinins; grafting of a scion onto a rootstock transformed with a combination of *ROL A, B, C* genes results in an increase in both axillary and basal bud outgrowth. Additionally, the direct application of exogenous cytokinins to rose plants has been found to stimulate axillary bud break and subsequent shoot growth as well as basal bud break and renewal shoot formation.

In horticultural practice, grafting on different rootstocks dramatically modifies branching and the resulting plant architecture. This influence is generally related to the differing capacity of the different rootstocks to synthesize cytokinins. Hence, analysis of the cytokinin content in xylem sap has been described as a potential criterion for rootstock selection.

They also influence source–sink relationships (metabolite translocation). Indeed, in the case of outgrowth of several lateral buds, often some of them give rise to shoots with aborted flowers due to trophic competitions. The direct application of cytokinins on buds limits flower atrophy by strengthening the sink ability (metabolite translocation) of treated buds.

Foliar application of cytokinins reduces the leaf abscission generally observed in winter and also the leaf abscission and flower bud drop that occurs during shipment of pot roses. Regarding postharvest physiology, cytokinins applied on the foliage, or sometimes in a holding solution, have been shown to delay petal senescence and thus to prolong average vase-life.

Gibberellins

All the gibberellins are acidic compounds. Many different gibberellins have been identified and according to the classical nomenclature they are assigned A-numbers in the order of their discovery (GAs). GAs are widespread and so far ubiquitous in angiosperms, gymnosperms and ferns. They have also been isolated from mosses, algae, fungal species and, more recently, from bacteria. All the gibberellins derive from the ent-gibberellane skeleton and correspond to diterpenes. As such, they are synthesized from acetyl CoA, via the mevalonic acid pathway. Gibberellin synthesis occurs in various plant parts characterized by active cell proliferation: immature tissues (e.g. maturing seeds and fruits), young shoot apical zones or the young leaf petiole. It can be blocked by numerous molecules such as cycocel or paclobutrazol. Temporary inactivation resulting from conjugation to glucosyl esters or glucosides is observed, predominantly in maturing seeds.

The gibberellins have many physiological effects, depending on the GA type and plant species. They are known to stimulate dramatically cell proliferation and elongation at the intercalary meristem level, thus leading to internodal growth. They also contribute to flower initiation stimulated by long day periods. Like the auxins, they interact with fruit pericarp development and they even lead to eventual parthenocarpic fruit formation. Considered as abscisic acid antagonists, the GAs can also break seed dormancy and promote germination of several species which require stratification or light postharvest treatments. Furthermore, they promote α-amylase synthesis in germinating cereal grains that are involved in the mobilization of seed reserves.

In *Rosa*, different gibberellins have been identified. The predominant gibberellins are probably GA_3 and GA_1, which have the main physiological roles in plant development. Most GA responses have been studied following exogenous applications. First, GAs have been shown to promote dormancy release and germination of rose achenes. GAs are also involved in the developmental processes of rose shoots and flowers. It was thus demonstrated that GA applications increase shoot length by stimulating internodal elongation and, when originating from sepals, contribute to peduncle and flower bud growth. Conversely, antigibberellins such as paclobutrazol are commonly used in order to block GA synthesis and subsequently reduce the plant dimensions, in the production of miniature pot plants. In addition, peduncle abscission and bud drop are reduced by exogenous GA applications.

Flowering of roses is also under control of gibberellins. In general, gibberellins, when exogenously applied, inhibit flowering. Indeed, applications of GA_3 reduced the number of flowers per plant and the proportion of plants in flower. However, flower abortion could be overcome by low-concentration spray applications of GA_3, but when the concentration was increased flowering was inhibited.

It is considered that floral initiation in short-season roses is permitted shortly after bud break in spring when concentrations of GAs are low and is inhibited by the subsequently higher levels of GAs. In contrast, in recurrent-flowering roses, low concentrations of GAs permit floral initiation throughout the growing season. It has been shown nevertheless that flowering could be suppressed in the recurrent-flowering roses by spray applications of GA_3 at a sufficiently high concentration.

Regarding postharvest physiology, it has been shown that the vase-life of rose flowers was significantly extended by gibberellin (GA_3) loading via the stem base. The extension of flower longevity was accompanied by an increase in dry and fresh weights of petals. The increase in weight, and consequently improved longevity, were attributed to the gibberellin-promoted transport of assimilates and the enhancement of the flower sink-strength by gibberellin.

Moreover, this inhibitory role of gibberellins on rose flower senescence may result from direct effects on cell membrane properties in the petal, for example in preserving membrane integrity and controlling the cell membrane permeability. Large variations are observed among cultivars, probably due to differences in their content of endogenous gibberellins.

Abscisic Acid

Unlike auxins, cytokinins and gibberellins, abscisic acid (ABA) is a single compound. It also plays an important role in plant growth and development. Its denomination derives from its first determined influence on the abscission of leaves and fruits. It corresponds to a sesquiterpenoid, which was originally isolated in abscising cotton fruits and maple dormant buds.

ABA is a natural plant product. Although it is a relatively simple molecule, some steps of its biosynthesis are still not certain. Although it is accepted that pathogenic fungi synthesize ABA directly from farnesyl pyrophosphate, it seems that ABA biosynthesis in higher plants is indirect and involves xanthophylls, i.e. oxidized forms of carotenoid, as the precursor. This biosynthesis pathway has been confirmed by mutant studies which carotenoid biosynthesis blockage was concomitant with ABA-deficiency. Other ABA-deficient mutants were impaired in zeaxanthin epoxidation into antheraxanthin and violaxanthin, which is considered to be the first ABA biosynthesis step. The last biosynthesis step corresponds to the oxidation of ABA aldehyde into ABA. Additionally, ABA biosynthesis is blocked by fluridone or norflurazon (known inhibitors of carotenoid synthesis) applications. The synthesis mainly occurs in the parenchyma of mature leaves and roots. It is strongly stimulated by environmental conditions including short day photoperiods or conditions that create hydric deficiency. Hence, ABA is often considered as a 'stress hormone'.

The inactivation of the molecule can result from its attachment to glucose, with formation of ABA glucose ester (ABA-GE) or from its oxidation with formation of phaseic and dihydrophaseic acids (PA and DPA respectively). ABA can be transported both up and down the stem. It is thus found in both xylem and phloem saps.

ABA is involved in different physiological processes. It is known to inhibit shoot growth and development. According to some available apical dominance models, it may be considered as a secondary inhibitor induced by the auxins (as ethylene). Indeed, exogenous applications of ABA on axillary buds cause growth inhibition and, following decapitation, ABA content usually decreases in the axillary buds. Besides, ABA contributes to stomatal regulation, with a closing effect. It is also responsible for induction and maintenance of endodormancy in seeds, buds and bulbs. It induces seeds to synthesize storage proteins and it inhibits the effect of gibberellins on α-amylase synthesis.

In *Rosa*, ABA, ABA-GE, PA and DPA have been identified. As well as apical dominance, the axillary buds are also subjected to short-distance correlative inhibitions originated from the stem and subtending leaf. This bud inhibition may be determined by a growth inhibitor such as ABA, that is synthesized or accumulated in stems or leaves. Axillary bud topophysis, which is observed along rose stems, may result from the increasing endodormancy in the latent state of the axillary buds encountered in the basipetal direction. Topophysis may be associated with an ABA sensitivity or *de novo* ABA synthesis in the axillary buds, which increases in the basipetal direction. The proximal buds exhibit an endodormancy that can be counteracted by fluridone (an ABA synthesis inhibitor). Likewise, ABA has been shown to be involved in the inhibition of seed germination.

As a response to environmental stress such as water stress, ABA induces a decrease of both the stomatal conductance and the net CO_2 assimilation in rose leaves which can subsequently affect photosynthesis. Moreover, it has been demonstrated that ABA accumulates in senescing petals during the postharvest phase and participates in the endogenous regulation of senescence. The higher the ABA level at harvest, the shorter the subsequent vase-life. Furthermore, exogenously applied ABA has a dual effect, according to the mode of application: when added to vase solution, ABA accelerates senescence processes and shortens the flower life duration, whereas ABA sprayed on the cut stem foliage reduces water loss and delays wilting and petal senescence.

Ethylene

Ethylene, unlike the other growth regulators, is a gaseous hormone. Among all these regulators, ethylene possesses the simplest structure. It is produced in all higher plants from methionine in almost all the tissues. The first step in ethylene biosynthesis is the conversion of *S*-adenosyl-methionine (SAM) into 5′-methylthioadenosine and ACC, catalysed by ACC synthase. The subsequent conversion of ACC into ethylene is catalysed by ACC oxidase. Ethylene production depends on respiration, environmental light quality and temperature, and it is stimulated by auxins. As a gas, ethylene can readily diffuse from all the plant tissues, except through cuticles.

Ethylene mainly influences maturation and senescence processes. It plays an important role in petal wilting, and in flower bud desiccation and abscission. It is also heavily produced during leaf and fruit senescence and abscission. It also contributes to fruit ripening. It is considered as a growth inhibitor because of its inhibitory effects on root elongation and axillary budbreak. Moreover, ethylene can stimulate the release of seed and bud dormancy, shoot and root growth differentiation, and the induction and opening of flowers.

In *Rosa*, it has been shown that spraying basal parts of rose plants with ethephon (a liquid ethylene-releasing compound) promotes the release of bottom-break dormancy. In addition, ethylene causes severe leaf abscission and flower bud atrophy. Indeed, it is known to induce defoliation in roses depending on the concentration and on the season of its application. These seasonal variations in response may be associated with modulations of receptor levels.

Moreover, the postharvest quality and longevity of flowers mainly depends on ethylene production and/or sensitivity to exogenous ethylene. It has generally been observed that high ethylene production levels in the senescing flowers and leaves result in reduced vase-life. But, in some cultivars, a low sensitivity to ethylene can counterbalance the potential effects of high ethylene production. Rose genes encoding ethylene biosynthetic pathway enzymes such as ACC oxidase and ACC synthase, and others belonging to different subgroups of ethylene receptor (*R. hybrida* Ethylene Receptor, RhETR) have already been isolated. It was shown that the expression of RhETR was all the higher than the cultivar exhibited a short flower life. Therefore, the different responses of rose cultivars could result from different expression levels of receptor genes during flower development.

Conclusion

Because of its numerous annual development cycles, the recurrent-flowering rose plant provides a remarkable model for the physiology of perennial plants. However, our knowledge of the implications of hormones in *Rosa* is still low. It is often restricted to the observation of plant reaction as a consequence of exogenous applications, or to occasional content measurements. As demonstrated on other plants, better understanding of the hormonal response requires a wide knowledge in hormonal contents as a result from dynamics like *de novo* synthesis, import and export, catabolism or storage, and in plant- or organ-specific sensitivity, generally associated with receptor and signal transduction pathways.

The development of molecular genetics on model plants such as *Arabidopsis*, *Antirrhinum* or *Lycopersicon* has led to major advances in the characterization of biosynthesis pathways, the identification of receptors or the understanding of hormone-dependent mechanisms. The understanding of rose plant development and growth physiology should benefit from these advances. Some results have already been directly transposed to the rose, such as the isolation of orthologous genes encoding ethylene receptors, previously identified in *Arabidopsis*. Moreover, recent progress in rose transformation may allow molecular studies of hormonal mechanisms. Transgenic plants with modified hormone levels or altered hormone responses may represent a valuable alternative to the experiments based on the spraying of plants with hormones to influence their growth and development.

See also: **Cell, Tissue and Organ Culture**: Micropropagation; Rooting of Micropropagated Shoots; *In Vitro* Regeneration Techniques; Cell Culture and Secondary Metabolites. **Growth Regulation**: Influences of Abiotic Factors in Growth and Development. **Postharvest Physiology**: Ethylene. **Propagation**: Seed Maturation and Germination. **Rootstocks**: Scion–Rootstock Relationships.

Further Reading

Agbaria H, Zamski E and Zieslin N (2001) Effects of gibberellin on senescence of rose flower petals. *Acta Horticulturae* 547: 269–279.

Arteca RN (1996) *Plant Growth Substances: Principles and Applications*. New York: Chapman & Hall.

Barthe P, Vaillant V and Gudin S (1991) Definition of indicators of senescence in the rose: effect of the application of plant hormones. *Acta Horticulturae* 298: 61–68.

Bianco J, Garello G and Le Page-Degivry M-T (1991) Gibberellins and abscisic acid in reproductive organs of *Rosa hybrida*. *Acta Horticulturae* 298: 75–82.

Buchanan BB, Gruissem W and Jones R (2000) *Biochemistry and Molecular Biology of Plants*. Rockville, MD: American Society of Plant Physiologists.

Davies PJ (1995) *Plant Hormones: Physiology, Biochemistry, and Molecular Biology*, 2nd edn. Dordrecht, The Netherlands: Kluwer Academic Publishers.

Klee H and Estelle M (1991) Molecular genetic approaches to plant hormone biology. *Annual Review of Plant Physiology and Plant Molecular Biology* 42: 529–551.

McCourt P (1999) Genetic analysis of hormone signaling. *Annual Review of Plant Physiology and Plant Molecular Biology* 50: 219–243.

Mor Y and Zieslin N (1987) Plant growth regulators in rose plants. *Horticultural Reviews* 9: 53–73.

Tucker GA and Roberts JA (2000) *Plant Hormone Protocols: Methods in Molecular Biology*. Totowa, NJ: Humana Press.

Influences of Abiotic Factors in Growth and Development

L Urban, INRA/CIRAD, Saint-Pierre de la Réunion, France

Introduction

Roses are grown commercially in many places in the world, from northern regions to the tropical highlands, suggesting a high capacity to adapt to a broad range of growing conditions. Roses appear to be resistant or tolerant to salinity stress, drought, low relative humidity of the air and high temperature and irradiance. Many outdoor cultivars thrive in polluted environments, suggesting that roses are also resistant to oxidative stress. But the highest crop performance can be achieved only if roses are provided with optimal growing conditions. Depending on local climatic constraints, research has focused on the effects on growth, development and resulting quality of cut flowers and pot plants, mainly of light and temperature in the north of Europe and Canada, and of relative humidity of the air and water availability in the Mediterranean area and the south of the United States. It has been recently suggested that greenhouses could be enriched in CO_2 all year round and all day long. This promising prospect has triggered new research about the effects of elevated CO_2 and its interactions with temperature and irradiance.

Salinity

Thresholds of salt injury have been reported to vary with EC (electrical conductivity) within the range 0.5 to $3\,\mathrm{mS\,cm^{-1}}$ in roses. Careful examination of the conditions of the trials from which these diverging figures were drawn, suggests that the response to salinity depends not only on cultivars, but also on evaporative demand and water supply. Under conditions of moderate evaporative demand or truly nonlimiting water supply, as can be achieved in soilless systems when nutrient solutions are provided at high concentration (with a lot of leaching), roses subjected to high salinity do not exhibit any substantial decrease in leaf stomatal conductance (g_s), net photosynthesis (A_{net}), cell elongation and apparent quality of flower stems. It is generally agreed that for greenhouse soilless conditions, EC of well-balanced nutrient solutions should be within the range of 1 to $1.7\,\mathrm{mS\,cm^{-1}}$, depending mainly on cultivars, evaporative demand and water supply. When considered separately, HCO_3^-, Na^+ and Cl^- have been reported to be harmful to rose production for single-element concentrations exceeding 2, 4 and 4 meq l^{-1}, respectively. Other elements, including SO_4^{2-}, may become harmful only through their contribution to the total salinity of the solution.

Relative Humidity of the Air

Different techniques, such as high pressure mist, canopy wetting and evaporative cooling have been used to increase relative humidity (RH) of the air and thus decrease the evaporative demand in rose greenhouses. Increasing RH results in an increase in leaf water potential (Ψ) and relative water content, but has no positive effect on g_s and A_{net} when water supply is truly nonlimiting (**Table 1**). If not, techniques aiming at increasing RH of the air appear to be effective in increasing g_s, A_{net} and consequently dry matter accumulation, bud break and commercial yield.

Increasing RH of the air often improves the quality of flower stems, leaves and size of buds, as indicated by the proportion of saleable first-grade flowers. Such positive effects may be attributed to an increase in leaf Ψ and the associated increase in turgor, which is the driving force behind cell elongation and expansion of tissues and organs. In addition, high RH has also been reported to increase the cuticular transpiration rate. Epicuticular wax and cells of the epidermis of roses cultivated *in vitro* have been observed to undergo morphological changes when RH is increased from 75% to 100%. An increase in cuticular transpiration rate may hasten postharvest damage of the water balance, which has been clearly demonstrated to be correlated with early appearance of senescence symptoms in cut flowers (*see* **Postharvest Physiology**: Xylem Structure and Function in Cut Roses). In other words, increasing RH of the air generally has a positive effect on apparent quality of cut roses, but may be detrimental to vase-life.

It has been observed that the high RH in tissue culture vessels has a morphogenetic effect on rose plantlets, by stimulating basal breaks (*see* **Cell, Tissue and Organ Culture**: Micropropagation; Rooting of Micropropagated Shoots). Similar effects have been observed on adult plants in greenhouse conditions. The physiological mechanisms involved are not clear.

Water Availability in the Root Zone

Roses subjected to water stress or to an osmotic shock have been demonstrated to resort to all three adaptative mechanisms known in plant cells: osmoregulation, turgor adjustment and modifications of water partitioning between the symplasmic and apoplastic fractions. Osmoregulation consists in an active accumulation of osmoticums in the symplasm which results in an increase in osmotic pressure and thus contributes to maintaining turgor at a high level. Increases in elastic modulus, i.e. in cell wall rigidity, have been observed in rose leaves, along with osmoregulation, as a consequence of water stress. Both osmoregulation and turgor adjustment have been reported to be favoured by mycorrhizal colonization and high photosynthetic photon flux density (PPFD). Increases in cell wall rigidity, because of their negative effect on turgor, are usually considered as a negative feature. But the decrease in Ψ associated with the decrease in turgor also favours uptake of water by cells from their dehydrating environment. The apoplastic water fraction (A_p) is astonishingly high in rose leaves. There are consistent reports of A_p values above 60%. It has been suggested that part of the water found in rose leaves may be associated to macromolecules such as starch. A_p fluctuates seasonally, reaching its maximum during the winter period in heated greenhouses. A_p increases as a consequence of water stress in rose leaves. Since increases in A_p occur at the expense of the symplasmic water fraction, they favour passive increases in concentration of osmoticums in the symplasm, i.e. decreases in both osmotic and water potentials. By combining all these three mechanisms – osmoregulation, turgor adjustment and increases in A_p – rose cells prove to be capable of maintaining a high water content in adverse conditions, without having to sacrifice turgor too much or too fast. In the

Table 1 Effect of relative humidity of the air (RH) and electrical conductivity of the nutrient solution (EC) on water potential, stomatal conductance (g_s) and net photosynthesis (A_{net}) of rose plants cv. 'Sonia' grown in the south of France in soilless conditions

	Low EC, low RH	*Low EC, high RH*	*High EC, low RH*	*High EC, high RH*
Leaf water potential (MPa)	−1.32 a	−0.84 b	−1.35 a	−0.78 b
g_s (mol H_2O m^{-2} s^{-1})	0.26 a	0.25 a	0.17 b	0.25 a
A_{net} (μmol CO_2 m^{-2} s^{-1})	12.6 a	11.8 a	9.4 b	12.0 a

Measurements were performed between 8.00 a.m. and 4.00 p.m. on one selected clear day during the summer period. RH was maintained above 80% in the high RH treatment with the help of high-pressure mist. Average RH was about 65% in the control (low RH treatment). EC of the nutrient solution was set at 1.8 mS cm^{-1}. EC increased to 2.0 mS cm^{-1} in the drainage solution when oversupply was more than 40% (low EC) and to 2.9 mS cm^{-1} in the drainage solution when oversupply was less than 10% (high EC). Different letters within the same row indicate significant differences ($P < 0.05$).

medium term, the survival strategy of roses outruns maintenance of turgor, negatively affecting cell elongation and the resulting flower size and weight.

Leaf stomata do not respond so much to water availability *per se* as to the build-up of ionic concentration in the root zone usually associated to the reduction of water supply. In soilless systems, under conditions of high evaporative demand, defects in water supply lead to an exponential build-up of ionic concentration near the roots, which may be at the origin of substantial decreases in g_s and A_{net}. Such observations indicate that roses are well protected against excessive losses of water by transpiration. Conversely, they also indicate that high crop performance can only be achieved provided that supply of water and associated minerals is carefully carried out. In soil conditions, mycorrhizae may enable roses to maintain g_s at low leaf and soil water potentials when compared to nonmycorrhizal plants.

Nitrogen

The saturating nitrogen concentration per unit leaf area has been estimated at $2.5\,g\,N\,m^{-2}$. Above this threshold, there is theoretically no gain for photosynthesis any more. Nitrogen is usually in excess in nutrient solutions (*see* **Production and Marketing**: Mineral Nutrition). But the possibility of decreasing nitrogen concentration should be considered with some caution. Deprivation of nitrogen may result in an accumulation of starch in all aerial plant parts, especially in flowers and leaves, because suppressing nitrogen suppresses the need to import carbohydrates from leaves for nitrogen assimilation in the roots. The maximal rate of net photosynthesis is reduced, while photosynthetic efficiency (α) and dark respiration are not affected, suggesting that A_{net} is downregulated by the accumulation of end-products of photosynthesis (*see* **Morphology and Anatomy**: Leaves).

Light

The light compensation point was found to be at 8 and $12\,\mu mol\,m^{-2}\,s^{-1}$ for well-lit 20-day-old leaves at 20 °C and 30 °C, respectively. In single-stem roses, compensation points of 65 and $30\,\mu mol\,m^{-2}\,s^{-1}$ were obtained in summer and winter, respectively. The light compensation point calculated from measurements performed on entire plants of miniature roses was found to be close to $70\,\mu mol\,m^{-2}\,s^{-1}$. All these values certainly deserve to be interpreted in the light of the corresponding concentrations of nitrogen per unit leaf area (*see* **Morphology and Anatomy**: Leaves).

Increasing PPFD has a strong positive effect on photosynthesis in roses as in other plant species. Results from leaf gas-exchange measurements performed on 20–30-day-old leaves (leaf temperature at 20 °C, ambient concentration of CO_2) and measurements of whole-plant net CO_2 exchange rate (NCER) performed at 22 °C showed that A_{net} becomes saturated when PPFD is around $800–1000\,\mu mol\,m^{-2}\,s^{-1}$ (maximum A_{net} ranging from 10 to 15 $\mu mol\,CO_2\,m^{-2}\,s^{-1}$). However, saturating PPFD may depend on cultivars and growing conditions. It has been observed that some cultivars are better adapted to low light conditions, while others take advantage of high PPFD. Leaves developed during winter, under low light conditions, accumulate less nitrogen and thus have a lower photosynthetic capacity than leaves developing in summer. The time of the day when measurements are performed may also play a role. Downregulation of A_{net} by the accumulation of photosynthetates in leaves during daytime is a process known to be more marked in the afternoon than in the morning. All these effects have not been clearly accounted for in the above-mentioned trials.

The relationship between photosynthesis and crop performance is a relatively straightforward one in roses, probably because reserves of carbohydrates in woody parts are not very important. This explains why supplementary lighting, from both high pressure sodium or metal halide lamps, thanks to its positive effect on A_{net}, has been generally observed to be beneficial to cut-flower yield. Bud break is increased by increasing PPFD before pruning. Moreover development of flower stems is accelerated in all three development phases (from excision/planting to the onset of bud growth, from the onset of bud growth to visible flower bud, and from visible flower bud to anthesis), when the daily light integral is increased. The duration of phase two is the most sensitive to photosynthetic photon flux and has been demonstrated to be a power function of PPFD. Increased bud break, combined with a shorter time period required for completion of flower development, accounts for the higher yields observed in conditions of high photosynthetic photon flux.

Increasing photosynthetic photon flux generally increases overall quality of flower stems (fresh and dry weights and size of all plant parts at flowering), because of it positive effect on A_{net}, osmotic pressure, turgor and cell elongation. There are some contradictory reports of a negative effect of supplementary lighting on apparent flower quality. Such negative effects may be attributed to a decrease in the net photosynthesis-to-number of breaks ratio, and the associated decrease in osmotic pressure, turgor and cell elongation. There are not many reports about the effect of light spectrum on growth and development in roses, but blue light has been observed to have a positive effect on shoot length.

The benefits of high photosynthesis and consequent carbohydrate accumulation extend to bud break and flower stem growth. It has been often reported that flower abortion rates and seasonal changes in irradiance are correlated, while shading results in an increase in flower bud abortion. Supplementary lighting enhances not only vegetative but also reproductive growth in northern regions where light may be a limiting factor. Because high PPFD increases sink strength of the apical bud, supplementary lighting has been consistently reported to reduce flower abortion. Light integrals play an essential role in flower bud formation. Minimal light integrals of 0.8 to 3.6 $mol\,m^{-2}\,day^{-1}$ are required, depending on cultivars. This effect probably reflects the high demand for photoassimilates of flower bud formation (*see* **Morphology and Anatomy**: Flower Development and the Abscission Zone).

There is some evidence that short-day conditions have a positive effect on the branching rate, but the genus *Rosa* is usually considered as insensitive to photoperiod *per se*. The sensitivity to light of lateral buds varies according to their position. Buds of the proximal part of decapitated rose stems are completely inhibited, regardless of light intensity. Buds of the distal part break even in the dark, while bud break in the intermediary position is promoted by high PPFD and red-to-far red ratio (*see* **Growth Regulation**: Axillary Bud Growth; **Morphology and Anatomy**: Flower Development and the Abscission Zone). Although there are some contradictory reports, a high red-to-far red ratio may also decrease the proportion of flower abortion.

Temperature

Frost sensitivity is highly cultivar-dependent in roses. Some cultivars have been reported to survive at −35 °C while others are killed at −5 °C. Measurements of chlorophyll fluorescence have been suggested as a tool to screen genotypes for chilling susceptibility. For PPFD above 600 $\mu mol\,m^{-2}\,s^{-1}$, A_{net} of 20-day-old leaves increases with leaf temperature up to 18 °C, remains on a plateau up to 35–40 °C, and then starts to decrease. Dark respiration rate increases also with temperature either linearly or exponentially (Arrhenius equation). Respiratory carbon losses at night have been observed to be proportional to daytime net photosynthesis, which is consistent with the coordination theory which states that all parameters related to the major biochemical limitations to photosynthesis vary in parallel. But other results indicate that dark respiration increases more than photosynthesis with temperatures from 15 to 35 °C. It would certainly be useful to assess the response of photosynthetic capacity to temperature more precisely, as has been done recently in other woody species (*see* **Morphology and Anatomy**: Leaves).

The effect of temperature on growth and development of roses has been rather extensively studied. High air temperature has a positive effect on bud break (number of breaks per unit area and time from pruning to bud break) (*see* **Growth Regulation**: Axillary Bud Growth). Duration of all the three phases of flower stem development is strongly temperature-dependent. Number of days from bud break to flowering may decrease threefold when temperature increases from 15 to 30 °C. By acting on both levels of bud break and duration of the time period required for completion of flower formation, high temperature has a strong positive effect on yield.

But increasing temperature also generally results in shorter flower stems as a consequence of a reduction in the number and length of internodes. When daytime temperature is above 17 °C and night temperature above 14 °C, lower fresh weight and dry weight of stems, leaves and flower buds are to be expected. This negative effect of high temperature on weight and size of flower stem parts, which is commonly observed in the summer period in southern regions, notwithstanding the conditions of high irradiance, may be attributed to a decrease in the net photosynthesis-to-number of breaks ratio, and the associated decrease in osmotic pressure, turgor and cell elongation. Moreover, high temperature may also result in an increase in the occurrence of malformation, as has been shown in 'Meidanclar' miniature roses submitted to a regime of day temperature of 29 °C and night temperature of 24 °C.

The effect of temperature on development has been well studied (*see* **Morphology and Anatomy**: Flower Development and the Abscission Zone; **Growth Regulation:** Axillary Bud Growth). Low leaf temperature may induce dormancy, depending on cultivars, while favouring the appearance of basal breaks. On the contrary, high leaf temperature favours bud breaks in the upper part of flowering stems. Day temperature lower than night temperature (so-called 'negative DIF') delays flowering of miniature roses by a few days at low CO_2, but advances it at high CO_2. Negative DIF increases total number of flowers and buds.

The effect of temperature in the root zone is not well understood, but it is generally observed that increasing the temperature in root zone has a positive effect on yield. An optimal temperature of 18 °C in the root zone has been recommended in soilless systems.

Table 2 Effect of high pressure mist and CO_2 enrichment on some growth and production parameters

	Control	*High RH*	*High RH and CO_2*
Harvested dry matter (kg m^{-2})	0.58 a	0.57 a	0.90 b
Number of breaks m^{-2}	74.9 a	76.6 a	91.8 b
Number of saleable flowers m^{-2}	89.0 a	87.1 a	104.7 b
Average stem length (cm)	56.0 a	53.7 a	63.4 b
Average stem diameter (mm)	5.6 a	5.6 a	6.0 b
Average bud diameter (mm)	15.0 a	15.7 b	16.3 c

Measurements were performed from 1 August to 30 October on rose plants cv. 'Sonia' grown in the South of France in soilless conditions with overabundant supplies of nutrient solution. Average relative humidity of the air (RH) was about 65% in the control. RH was maintained above 80% in the high RH treatment with the help of high-pressure mist. The concentration in CO_2 of the ambient air was about 350 μmol CO_2 mol^{-1} air in the control and about 900 μmol CO_2 mol^{-1} air from 9.00 a.m. to 4.00 p.m. in the high CO_2 treatment. Different letters within the same row indicate significant differences ($P < 0.05$). (Unpublished data and data adapted from Urban L (1994) *Acta Horticulturae* 361: 317–324.)

CO_2

The CO_2 compensation point has been evaluated at around 50 μmol CO_2 mol^{-1} air, notwithstanding leaf nitrogen concentration. Elevated CO_2 has a strong positive effect on A_{net}. A_{net} of 20-day-old leaves may double when the concentration of CO_2 of the air is doubled from 360 μmol CO_2 mol^{-2}. No acclimation has been observed to long-term enrichment in CO_2 during the cool season, probably because plants grown in heated greenhouses in soilless systems can take advantage of elevated CO_2 without experiencing any of the usual limiting factors that occur in open field conditions such as sink activity, and mineral and water availability. More recent observations, made during the summer season, in the south of France, in almost airtight greenhouses, with air temperatures currently exceeding 40 °C, showed that stomatal conductance was not reduced by CO_2 enrichment. No down-regulation of photosynthetic electron transport was observed as the consequence of the accumulation of nonstructural carbohydrates (glucose, fructose, sucrose and starch) associated with CO_2 enrichment. Moreover, CO_2 enrichment did not cause any increase in the risk of photodamage. Nonphotochemical quenching was even higher in conditions of high temperature and PPFD, suggesting that elevated CO_2 may have a protective effect on the photosynthetic machinery under such harsh conditions. Eventually, these observations confirm that photosynthesis of roses grown in conditions characterized by nonlimiting water and minerals supply benefits strongly from high concentrations of atmospheric CO_2 at both high and moderate temperatures and PPFD (*see* **Morphology and Anatomy:** Leaves).

Other observations suggest that CO_2 enrichment results in an increase in A_p, while osmoregulation and an increase in cell wall rigidity appear as additional adaptative mechanisms. All these mechanisms are known to favour both turgor and water uptake. The increase in turgor associated with CO_2 enrichment favours cell elongation and consequently flower stem and bud size (**Table 2**). CO_2 enrichment may also result in lower cuticular transpiration rate and higher leaf water potential at complete stomatal closure. These last features indicate a higher capacity of rose leaves to protect themselves against water losses by transpiration, and may explain why cut roses from greenhouses with CO_2 enrichment keep a positive water balance longer and have an extended vase-life when compared to standard cut roses (*see* **Production and Marketing:** Cut Rose Production).

Although these positive effects need to be reassessed for cultivars other than the ones used in the above-mentioned trials, CO_2 enrichment all day long and all year round appears to be a promising technique for greenhouse growers. If the availability of cheap CO_2 is confirmed, it will be interesting to test the effects of CO_2 enrichment at levels well above the usually recommended ones (700–1000 μmol CO_2 mol^{-1} air).

Conclusion

Although the agronomic response of roses to growing conditions has been extensively studied, there is still a lack of basic knowledge about the way roses respond to environmental factors. This may impair our capacity to improve crop performance further or to adapt rose production to new conditions. It is only very recently that the first biochemical model of rose-leaf photosynthesis has been published. The widespread use of such models, combined with measurements of chlorophyll fluorescence and analysis of leaf nitrogen and carbohydrate content, will help us to assess more precisely photosynthetic performance of roses in response to environmental conditions and to compare it among cultivars. Additionally, a fresh look at carbohydrate distribution in relationship with water status and temperature, could help to improve our understanding of the way development is affected by environmental factors. Incorporating such elements in existing models of production will greatly improve their predictive capacity and open new horizons for the improvement of crop performance.

See also: **Cell, Tissue and Organ Culture**: Micropropagation; Rooting of Micropropagated Shoots. **Growth Regulation**: Axillary Bud Growth. **Morphology and Anatomy**: Leaves; Flower Development and the Abscission Zone. **Postharvest Physiology**: Xylem Structure and Function in Cut Roses. **Production and Marketing**: Mineral Nutrition; Cut Rose Production.

Further Reading

Durkin DJ (1992) Roses. In: Larson R (ed.) *Introduction to Floriculture*, pp. 67–92. San Diego, CA: Academic Press.

Van den Berg GA (1987) *Bud Break: Influence of Temperature on Bud Break, Shoot Growth, Flower Bud Atrophy and Winter Production of Glasshouse Roses*. PhD thesis, Wageningen Agricultural University, Wageningen, The Netherlands.

Zieslin N and Agbaria H (eds.) (2001) *Proceedings of the 3rd International Symposium on Rose Research and Cultivation*, Herzliya, Israël, May 2000. *Acta Horticulturae* 547.

Zieslin N and Mor Y (1990) Light on roses: a review. *Scientia Horticulturae* 43: 1–14.

Axillary Bud Growth

N Bredmose, Danish Institute of Agricultural Sciences, Aarslev, Denmark

Introduction

The axillary buds develop in leaf axils and are of particular importance since aspects of growth readiness and plant architecture are integrated into their organogenic capacity (**Figure 1**). They provide propagation material and generate flowering shoots.

Axillary bud growth is under the influence of the shoot apex. It is subjected to apical dominance. The apical meristem within the axillary bud remains inhibited as long as the shoot apex is intact and active. However, leaf initiation occurs during growth inhibition, and leaf primordia slowly accumulate in the axillary bud. Axillary buds elongate visibly only if apical dominance is removed, the bud has matured and the environmental conditions for growth are suitable. At the moment of release from inhibition, or when a shoot is decapitated or an axillary bud is excised, most axillary bud meristems produce a specific number of leaves before initiating a terminal flower bud (*see* **Growth Regulation**: Floral Induction). Some basal axillary buds may remain inhibited even after the shoot has been decapitated. Abscisic acid is thought to inhibit their growth. In *Rosa canina*, for example, low temperature stimulates the onset of growth in these axillary buds, leading to the formation of basal shoots.

In rose, the apically positioned axillary buds are sylleptic, i.e. their growth continues, without interruption, immediately on formation. This is because the apical dominance ceases when the apical bud is converted into an inflorescence. In contrast, most medial and all basal axillary rose buds are proleptic, i.e. they undergo a period of inhibition prior to the onset of axillary bud growth. Sylleptic axillary buds are mainly formed in the axils of leaves with one to three leaflets and proleptic axillary buds in the axils of leaves with five leaflets. In some cut-rose cultivars, the borderline between them is only two to three nodes below the flower.

The development of an axillary bud may be divided into three phases: (1) from excision to onset of axillary bud elongation; (2) from onset of axillary bud growth to visible flower bud; and (3) from visible flower bud to anthesis (flowering, harvestable stage) (**Figure 2**).

The potential for growth of axillary buds can be studied by pruning the parent shoot of a growing rose plant above the axillary bud, by grafting the axillary bud on a growing plant, by growing cuttings (*see* **Propagation:** Budding in the Field; Own-Rooted Cuttings; Stenting; Seed Propagation). comprising a node with the axillary bud, a five-leaflet leaf and a piece of

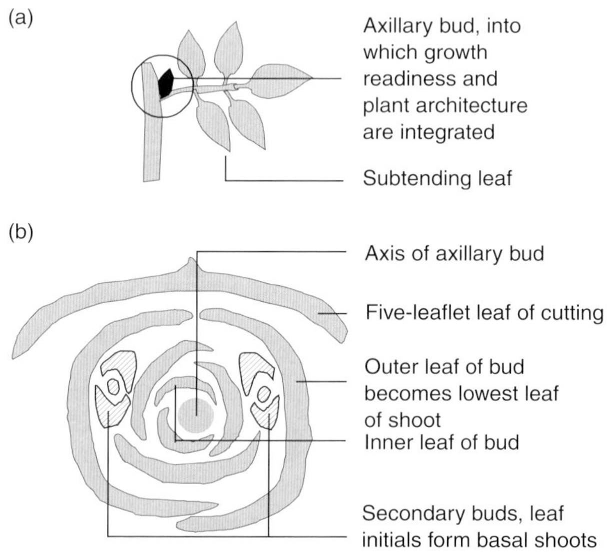

Figure 1 The axillary bud of a rose. (a) Single-node cutting with five-leaflet leaf. (b) Schematical cross-section of ontogenetically mature primary axillary rose bud showing seven leaves and leaf primordia, and two secondary axillary buds with two leaf initials.

internode (**Figures 1** and **2**), or by culturing the axillary bud (nodal sections) *in vitro* (*see* **Cell, Tissue and Organ Culture:** Micropropagation; Rooting of Micropropagated Shoots). Developmental processes are similar in each of these test conditions. Blind (flowerless) shoots as material for propagation by cuttings can perform as well as flowering shoots with respect to growth of the axillary bud and early development of the shoot (**Table 1**).

Control of axillary bud growth can be decisive in achieving the synchronous growth and development essential for automation of plant-handling processes.

Internal factors on which axillary bud growth depends include genotype, assimilate availability, nutritional status, age (chronological and ontogenetical), competition between plant parts (stem/leaf/petiole/axillary buds), hormone concentrations and topophysis. External factors include the quality and quantity of light, photoperiod, relative humidity, growing medium and temperature. In addition, gravitropism is believed to be due to a lateral redistribution of auxin with a resulting difference in rate of cell elongation on the two sides of the bud. Rose cultivars may respond differently to the same growing conditions (**Table 1**).

The control mechanism of axillary bud growth is thought to be mainly mediated by plant hormones (*see* **Growth Regulation:** Hormones in Growth and Development; Influences of Abiotic Factors in Growth and Development). Auxin has long been known to inhibit axillary bud growth, while cytokinins promote axillary bud growth. Cultivars (scions) on low-yielding rootstocks have lower cytokinin concentrations than scions on high-yielding rootstocks.

In this article the roles of assimilate supply, axillary bud age, topophysis, root formation, temperature, irradiance and cytokinins in axillary bud growth regulation are discussed, together with some future prospects for axillary bud growth regulation. The article mainly deals with recurrent-flowering roses.

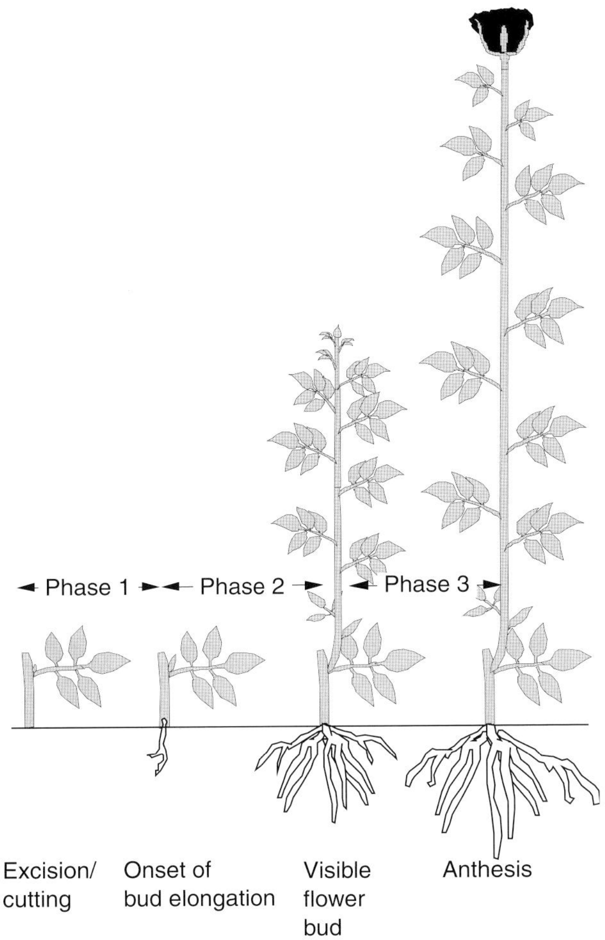

Figure 2 Stages of development divided schematically into three development phases: from excision/planting of cutting to onset of axillary bud growth, from onset of axillary bud growth to visibility of the flower bud, and from visibility of flower bud to anthesis.

Influence of the Mother Plant on the Growth of Excised Axillary Buds

The growth potential of excised axillary buds is influenced by the growth conditions to which the parent

Table 1 Effect of propagation material (blind and flowering shoots) in two rose cultivars, 'Frisco'® and 'Gabriella'®, on axillary bud growth and shoot length after 28 days of propagation of single-node cuttings. Average figures of two propagation methods are shown. Values followed by different letters are significantly different ($P < 0.05$) within rows

Response	*'Frisco'®*		*'Gabriella'®*	
	Blind shoot	*Flowering shoot*	*Blind shoot*	*Flowering shoot*
Axillary bud growth (%)	84.2^{c}	98.4^{a}	96.9^{ab}	90.0^{bc}
Initial shoot growth (cm)	3.0^{c}	7.6^{a}	4.3^{b}	3.9^{bc}

Interactive effect (cultivar × propagation material): $P = 0.001$ for both axillary bud and shoot growth

Based on Bredmose N and Hansen J (1995) Regeneration, growth and flowering of cut rose cultivars as affected by propagation material and method. *Scientia Horticulturae* 64: 103–111.

shoot on the stock plant was exposed. In this section effects of preexcision conditions like nutrition/assimilates, bud age and topophysis are discussed. Possible influences of preexcision temperature and irradiance will be mentioned in the following section.

Assimilate Supply

With an increase in assimilate supply (*see* **Production and Marketing:** Mineral Nutrition) the fresh and dry weight and size of the parent shoot and its axillary buds increase. However, the time taken for the axillary bud to elongate is not clearly affected by assimilate supply during axillary bud development. An increase in assimilate supply slightly increases the number of leaves and leaf primordia in the axillary bud. Consequently, the resulting number of leaves, preceding the flower on the shoot grown from the axillary bud, increases substantially. This increase is mainly in the number of five-leaflet leaves.

After release from inhibition, axillary bud elongation and subsequent shoot growth are, to a major extent, dependent on the assimilate supply available while growth is taking place. The axillary bud, with its growing leaf primordia, is a powerful nutrient sink (*see* **Morphology and Anatomy:** Leaves). Thus, young shoots depend primarily on old mature leaves for their supply of assimilates, and single-node cuttings grow better if the leaf remains attached.

Assimilate supply may affect cell division and cell enlargement. Cell number is only affected during development of the axillary bud since cell number across the stem diameter remains constant after the onset of axillary bud elongation. Cell enlargement is mainly affected after the onset of axillary bud elongation. Thus, pith diameter, which is primarily affected by cell enlargement, is largely influenced by growing conditions during shoot growth.

Generally, both cell division and cell enlargement are promoted by increased cytokinin content. It has been suggested that changes in the concentration of cytokinins modify the nutrient distribution pattern, which allows axillary buds to grow out.

Axillary Bud Age

The apical meristem of the axillary rose bud remains unfinished, for up to 1 year, while it is subject to apical dominance. With increasing chronological age of the axillary bud, its dry weight, dry-matter content, number of leaves and leaf primordia increase.

The axillary bud needs to reach a certain developmental stage before it can start to elongate. Upon release from inhibition, additional leaves and eventually a flower are formed. With increasing age of the axillary bud, the number of additional leaves formed decreases, and this number of leaves becomes more or less constant for axillary buds of shoots at the harvestable stage (**Figure 2**, right) when the axillary buds are mature. The total number of leaves preceding the flower increases with increasing axillary bud age, as do the number of scales, transitional leaves and non-elongated internodes of the subsequent shoot.

When an axillary bud is released from inhibition before the harvestable/mature stage, further leaf initiation occurs before the onset of axillary bud elongation. The time taken for the axillary bud to start elongating decreases slightly with increasing bud age, and the number of pith cells in the subsequent shoot increases.

Topophysis

The influence of axillary bud position along a shoot axis on subsequent growth and development is termed topophysis. Differences, related to position of origin, are expressed when cuttings are excised and grown as separate plants. In rose, these differences may be maintained after 2 years. Topophysis can determine the potential for axillary bud growth in rose, and this knowledge may be used in the industry to control growth.

When single-node cuttings with five-leaflet leaves are excised from different stem positions on the stock plant, and grown as rose plants, the onset of axillary bud elongation and initial shoot growth occurs faster in axillary buds originating from apical positions and slower in axillary buds from basal positions (**Table 2**). The vitality of cuttings originating from the two most basal positions is lowest. This growth effects of topophysis persist through to flowering. In 'Red Velvet'® the axillary buds originating from positions 4–9 possess the highest growth potential, and those from position 3–7 generally have the highest growth efficiency (**Figure 3**). Axillary bud growth potential and growth efficiency of roses can be improved by up to 49% and 34%, repectively, by selecting proper axillary bud positions.

Reduced vitality may be related to the assimilate supply and/or reflect responsiveness to hormonal signals. Inhibited axillary buds in rose plants can be released by exogenous application of cytokinin. Excision of stem cuttings removes the inhibition of the axillary rose bud and also promotes processes initiating root formation and axillary bud growth.

Auxin and cytokinins interact synergistically to stimulate cell division and antagonistically to control axillary bud growth. Most likely, root-derived cytokinins are transported via the xylem into the buds and are involved in the acceleration of axillary bud development. However, following rose shoot decapitation,

Table 2 Topophysic effects on time from excision and planting of single node cuttings until axillary bud growth and visible flower bud in the rose 'Red Velvet'. Axillary bud positions are numbered basipetally

Axillary bud position	*Time (days) from excision/planting until*		
	Axillary bud elongation (1)	*Visible flower bud (2)*	*(2) − (1)*
1 (most apical)	15.4	26.1	10.7
2	16.5	27.8	11.3
3	16.7	28.5	11.8
4	17.4	30.0	12.6
5	17.3	29.2	11.9
6	18.5	30.1	11.6
7	20.7	34.9	14.2
8	22.3	36.5	14.2
9 (most basal)	22.0	39.4	17.4
LSD	2.0	3.5	—

LSD, least significant differences at $P = 0.05$. Based on Bredmose N, Hansen J and Nielsen J (1999) Factors intrinsic to the axillary bud determine topophysic effects on bud and shoot growth and flower development in *Rosa hybrida*. *International Journal of Plant Sciences* 160: 819–825.

cytokinins are also transported from the subtending leaf to the axillary bud. Thus, upon excision, resumption of growth in an axillary rose bud may depend on its own cytokinin content and, following root formation, also on the supply from adventitious roots. This needs further investigation. However, results suggest that topophysis in *Rosa* is an independent phenomenon determined by factors intrinsic to the axillary bud, but the physiology of topophysis is still unclear.

The Growth of Axillary Buds after Excision

As outlined in the introduction, in the rose the shoot apex controls the onset of axillary bud growth, by apical dominance. Upon shoot decapitation or excision of the node the dominance is released and the axillary bud can grow out.

In this section possible effects of mainly postexcision conditions like root formation, temperature and irradiance are discussed.

Root Formation

The number of roots formed on cuttings depends on the original nodal position of the cutting. Generally, the best rooting response is obtained from medial to basal cutting positions. Relationships have been found between time taken for cuttings to form the first visible root and time to the onset of axillary bud growth (**Figure 4**). Similarly, root weight of cuttings is positively correlated to days to axillary bud elongation.

Significant correlations between shoot yield and length of longest root and number of primary roots are also reported in the rose. Such results may indicate that there is a positive relation between the number of roots formed by cuttings and the subsequent onset of axillary bud growth. This needs further investigation, however.

Differentiation of root initials is dependent upon either endogenous or applied auxins. Applied auxin increases root formation and inhibits axillary bud growth in cuttings. Endogenous auxin from the stem is thought to move down the stele to the root apex, where it reverses direction and moves upwardly through the root cortex. The intensity and polarity of auxin transport in stem cuttings are reported to be positively related to the formation and growth of adventitious roots. Grafting the rose cultivar 'Madelon'® on a *ROL A + B + C*-expressing transformant of the rootstock 'Moneyway' results in advanced axillary bud growth of the untransformed scion due to stimulated root formation of the transformed rootstock.

Thus, high auxin concentrations that promote root initiation and root elongation in cuttings may initially inhibit axillary bud growth. However, it is generally accepted that roots synthesize cytokinins, which are transported to the axillary buds and the promotive effect of cytokinins on axillary bud growth in intact rose plants has been documented. Thus auxin treatments may indirectly promote axillary bud development in the longer term.

Temperature

Air temperature Although the rate of leaf initiation increases with increasing temperature during axillary bud formation in the mother plant (preexcision), the number of leaves and leaf primordia in the axillary bud at the time of release from inhibition decreases slightly. Results indicate that raised temperature reduces the

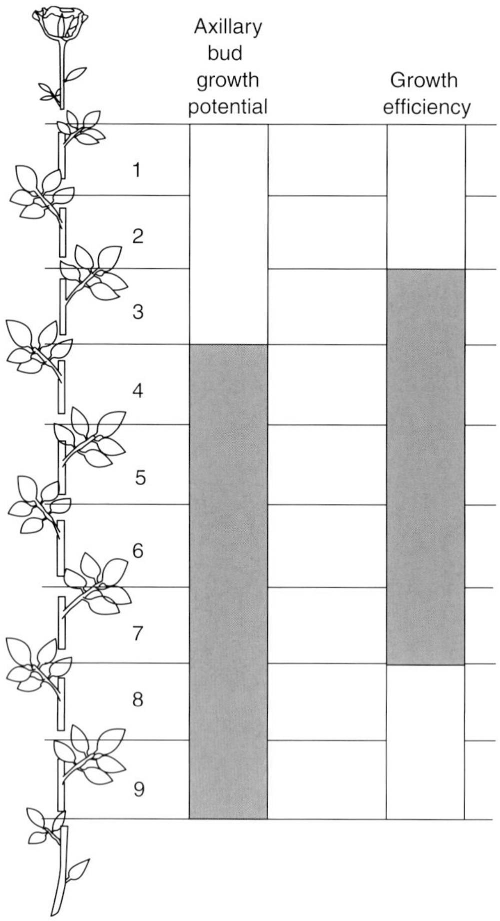

Figure 3 Topophysic influences on 'Red Velvet' rose axillary bud growth potential, expresssed by fresh biomass accumulation, and general plant growth efficiency, expressed by fresh-weight efficiency. Shaded area shows beneficial axillary bud positions. The nine axillary bud positions are shown schematically on the left. Adapted from Bredmose N, Hansen J and Nielsen J (1999) Factors intrinsic to the axillary bud determine topophysic effects on bud and shoot growth and flower development in *Rosa hybrida*. *International Journal of Plant Sciences* 160: 819–825, with permission.

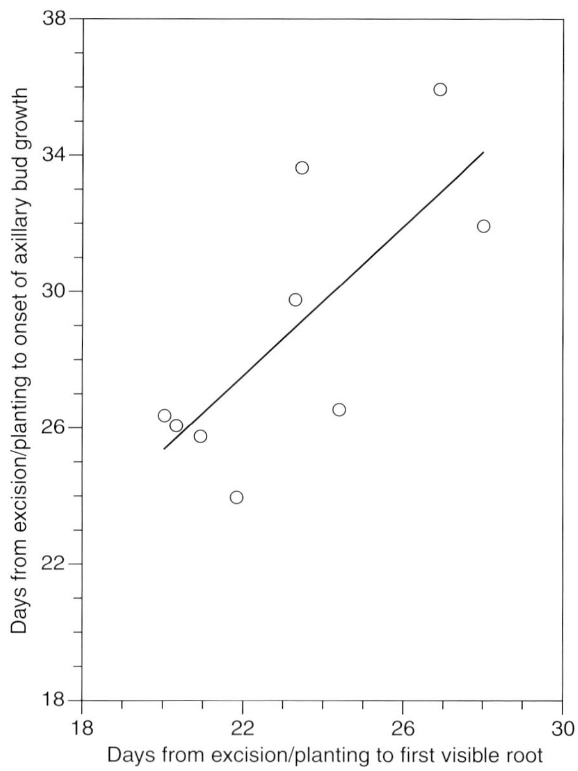

Figure 4 *Rosa* root formation–bud growth relation: relationship between time from excision/planting of cuttings to first visible root and time to onset of axillary bud growth in cuttings of 'Grand Gala' which received various treatments. The relation is described by the equation $y = 1.08x + 3.75$; $r = 0.56$.

degree of inhibition of the axillary bud and thus reduces the time until onset of bud growth after release from apical dominance. The total number of leaves preceding the flower is also reduced, mainly by the effect of temperature during axillary bud formation. The number of pith cells in cross-section decreases only slightly with increasing temperature. The pith contributes to a large extent to the rose shoot diameter.

Root temperature The onset of axillary bud growth in the rose is earlier at higher root temperatures, indicating mediation by cytokinins produced in the roots and/or translocated to the bud. At higher root temperatures roots are more branched, and the number of root tips is increased. Corresponding to earlier axillary bud growth, the cytokinin zeatin riboside concentration peaks earlier in bleeding sap at high root temperature than at low root temperature. However, the zeatin riboside concentration decreases slightly after onset of bud growth, but the concentration is not significantly affected by root temperature. The involved processes need clarification.

Irradiance

Inhibition of axillary bud growth is sensitive to changes in photosynthetic active radiation (PAR) and to changes in light quality. High PAR and light in the red spectrum both weaken inhibition and promote axillary bud growth.

Supplementary lighting stimulates the growth of the axillary buds of rose plants. Increasing the daily light quantum integral from 17.8 to 21.0 mol m^{-2} per day during three development phases – excision/planting to onset of bud growth, onset of bud growth to visible flower bud, and visibility of flower bud to anthesis (**Figure 2**) – significantly reduces the time needed for axillary bud growth and rose plant development (**Table 3**). The effects, however, are more apparent after the onset of bud growth, which include the

Table 3 Development and growth period of rose plants as affected by light integral. Effects of two levels of light quantum integral on time from cutting excision/planting until axillary bud growth, visible flower bud and anthesis. Average figures of four cultivars ('Lambada' 'Manhatten Blue', 'Red Velvet', and 'Sonia')

Light quantum integral (mol m^{-2} day^{-1})	*Time (days) from excision/planting until*		
	Axillary bud growth (first phase)	*Visible flower bud (second phase)*	*Anthesis (third phase)*
17.8	22.6	35.8	49.2
21.0	21.4	33.1	45.0
P (quantum integral) ≤	0.05	0.001	0.001
LSD	1.03	1.24	1.35

Least significant differences (LSD) are at $P = 0.05$. Based on Bredmose N (1997). Chronology of three physiological development phases of single-stemmed rose (*Rosa hybrida* L.) plants in response to increment in light quantum integral. *Scientia Horticulturae* 69: 107–115.

Table 4 Percentage of rose plants recorded with or without axillary bud growth or with flowerless axillary shoots (blind shoots) at two levels of light quantum integral in rose plants, average of four cultivars ('Lambada', 'Manhattan Blue', 'Red Velvet', and 'Sonia')

Light quantum integral (mol m^{-2} day^{-1})	*Axillary bud growth (%)*	*Blind shoots (%)*	*No bud growth (%)*	*Effect of treatment*
17.8	82.0	16.4	1.6	
21.0	89.1	8.6	2.3	
P (Fisher's exact test)				0.024

Based on Bredmose N (1997). Chronology of three physiological development phases of single-stemmed rose (*Rosa hybrida* L.) plants in response to increment in light quantum integral. *Scientia Horticulturae* 69: 107–115.

second (**Table 3**) and third growth phases. Increased irradiance also increases fresh biomass efficiency (fresh weight at anthesis divided by number of days to anthesis), stem diameter and specific fresh mass (fresh weight per unit length of stem) in the roses. The modest increment in irradiance results in a substantial reduction in the percentage of blind shoots formed (**Table 4**), indicating that the critical PAR level for completing rose flower bud development is relatively high.

These effects of irradiance are thought to be related to assimilate supply and translocation, and light perception by the roses. Light perception by axillary buds or shoot tips of roses has been documented. Effect of light on onset of bud growth is related mainly to red-to-far-red ratios, while the effect on axillary bud growth and further development is mainly related to PAR. PAR received by the rose shoot tip promotes shoot sink activity by increasing the sucrose-unloading process. Thus, by applying different light quantum integrals to rose, assimilate supply and translocation can be adjusted, thereby the rate of axillary bud growth can be influenced.

Light quantum integral is suggested as a means of synchronizing rose plant development. High light integral may promote the synthesis of cytokinins, which in turn may reverse the auxin effect.

Regulative Mechanism of Bud Break

In this section the possible mechanisms of axillary bud growth control and regulation are discussed, emphasizing cytokinins, which play a key role in the initiation of growth of axillary buds.

Influences of Cytokinins

Endogenous cytokinin levels are precisely regulated in plants with respect to developmental events like axillary bud growth, and also with respect to environmental factors such as light and nutrition. Evidence from hormonal studies suggests that apically produced auxin suppresses the onset of axillary bud growth which is promoted by cytokinins originated from roots/shoots. Generally, axillary bud growth promotion is thought to be a consequence of endogenous auxin lowering and cytokinin enhancement, raising the endogenous cytokinin-to-auxin ratio.

Concentrations of endogenous cytokinins in rose plant parts have been quantified and related to axillary bud growth. The rose stems contain high concentrations of cytokinins caused by lateral movement of zeatin riboside (zeatin precursor) (*see* **Growth Regulation:** Hormones in Growth and Development; Influences of Abiotic Factors in Growth and Development) from the xylem to adjacent stem tissue, and the ability of the stem to metabolize cytokinins. Young rose leaves are capable of *de novo* synthesis of cytokinins. In older leaves storage forms of cytokinins are found. In young axillary shoots the cytokinins are isopentenyl adenine compounds, suggesting import of cytokinins via the phloem or *de novo* synthesis of cytokinins. In the rose axillary buds forming basal shoots, zeatin riboside and zeatin riboside monophosphate are the main cytokinins, indicating that these buds receive their cytokinins from the roots.

The concentration of zeatin riboside in bleeding sap of rose plants is high at the onset of axillary bud growth, decreases when the axillary buds grow out, and increases thereafter till the onset of basal axillary bud growth appears, whereafter it decreases again.

The high zeatin riboside concentration in the bleeding sap of rose plants appears to coincide with the onset of axillary bud growth. Thus, in rose, axillary bud growth may be mediated by altered metabolism of cytokinins in the green rose plant parts, in addition to synthesis in the roots. After excision of a cutting, cytokinin may be part of an endogenous signal involved in axillary bud growth stimulation.

In addition, to cytokinins, a regulatory role of abscissic acid or an inhibitory complex has been reported for basal axillary bud growth. For axillary bud growth regulation a balance between cytokinins and abscisic acid/inhibitory complex is also possible, as is an interaction with other growth regulators such as ethylene. However, in rose the supply of cytokinins to culture medium is necessary for the axillary bud to complete its developmental programme *in vitro*. Furthermore, cytokinin application to the stem stump stimulates basal axillary bud growth. These relationships, especially the control mechanisms of axillary bud growth, need further clarification.

Future Prospects for Axillary Bud Growth Regulation

The introduction of new cultivars with more homogeneous axillary bud growth and new cultivation procedures, like the single-stemmed rose production method, may facilitate the regulation of axillary bud growth. The inhibitory effects of one dominant shoot on one or more subordinate shoots in a rose bush are absent in single-stemmed roses, thus potentially enabling improved control of growth of only one apical bud. Also the method of bending shoots reduces the competition of different plant parts for metabolites, improves the plant energy budget and better satisfies the need for assimilates associated with axillary bud growth.

It is suggested that the ability of an axillary bud to initiate growth may be the result of a combined effect of root-produced cytokinins transported to the axillary bud and the cytokinins synthesized in the stem, leaves and the axillary bud. Thus, lowering the endogenous auxin-to-cytokinin ratio favours axillary bud growth. Consequently, to promote axillary bud growth, treatments enhancing the natural endogenous content of cytokinins should be applied. However, this needs further investigation.

Molecular biology methods, now available for the transfer of genes to enrich the genome of rose (*see* **Biotechnologies for Breeding:** Genetic Transformation), could overcome some of the physiological problems which cannot be resolved by conventional methods. Manipulating the endogenous auxin-to-cytokinin ratio in rose plant tissue by gene insertion and modulation, and the use of hormone-insensitive/resistant mutants, together with other efforts discussed above, may provide more detailed information and allow the refinement of knowledge about the regulation of axillary bud growth in *Rosa*.

See also: **Biotechnologies for Breeding:** Genetic Transformation. **Cell, Tissue and Organ Culture**: Micropropagation; Rooting of Micropropagated Shoots. **Growth Regulation:** Hormones in Growth and Development; Influences of Abiotic Factors in Growth and Development. **Morphology and Anatomy**: Leaves. **Production and Marketing**: Mineral Nutrition. **Propagation**: Budding in the Field; Own-Rooted Cuttings; Stenting; Seed Propagation.

Further Reading

Brault M and Maldiney R (1999) Mechanisms of cytokinin action. *Plant Physiology and Biochemistry* 37(5): 403–412.

Bredmose NB (1998) Growth, flowering, and postharvest performance of single-stemmed rose (*Rosa hybrida* L.) plants in response to light quantum integral and plant population density. *Journal of the American Society of Horticultural Science* 123: 569–576.

Cleland RE (1999) Introduction: nature, occurrence and functioning of plant hormones. In: Hooykaas PJJ, Hall MA and Libbenga KR (eds.) *Biochemistry and Molecular Biology of Plant Hormones*, pp. 3–22. Amsterdam: Elsevier Science.

De Vries DP (1993) *The Vigour of Glasshouse Roses: Scion–Rootstock Relationships, Effects of Phenotypic and Genotypic Variation.* Thesis. Wageningen, The Netherlands: Wageningen Agricultural University.

De Vries DP and Dubois LAM (1997) Relations between early root growth and flower yield of clonal rose rootstocks. In: Altman A and Waisel Y (eds.) *Biology of Root Formation and Development*, pp. 283–288. New York: Plenum Press.

Dieleman JA (1998) *Cytokinins and Bud Break in Rose Combination Plants*. Thesis. Wageningen, The Netherlands: Wageningen Agricultural University.
Halevy AH (1996) Rose. In: Zamski E and Schaffer AA (eds.) *Photoassimilate Distribution in Plants and Crops Source–Sink Relationships*, pp. 883–892. New York: Marcel Dekker.
Marcelis-van Acker CAM (1994) *Axillary Bud Development in Rose*. Dissertation. Wageningen, The Netherlands: Wageningen Agricultural University.
Mor Y and Zieslin N (1987) Plant growth regulators in rose plants. In: Janick J (ed.) *Horticultural Reviews*, vol. 9, pp. 53–73. New York: AVI.
Van den Berg GA (1987) *Influence of Temperature on Bud Break, Shoot Growth, Flower Bud Atrophy and Winter Production of Glasshouse Roses*. Thesis. Wageningen, The Netherlands: Wageningen Agricultural University.
Zieslin N and Moe R (1985) Rosa. In: Halevy AH (ed.) *Handbook of Flowering Plants*, vol. 3, pp. 280–287. Boca Raton, FL: CRC Press.
Zieslin N and Mor Y (1990) Light on roses: a review. *Scientia Horticulturae* 43: 1–14.

Floral Induction

A V Roberts, University of East London, London, UK
P S Blake, Horticulture Research International, East Malling, Kent, UK

Introduction

Most wild roses are described as seasonal-flowering because they produce flowers only during a short flowering season. Recurrent-flowering cultivars possess a recessive mutant gene that permits flowering throughout the year. In this article, the morphological relationships between vegetative growth and flowering are compared in seasonal and recurrent-flowering roses. Genetic and hormonal factors that are known to affect floral induction are discussed.

Patterns of Vegetative Growth and Floral Initiation

A delay between seed germination and flowering is common in woody species. A plant is described as juvenile until it is competent to produce flowers and described as mature thereafter. The transition between the juvenile and mature phases is often accompanied by changes in the vegetative characters. In many woody plants, changes occur in growth rates, the shape of leaves and the ability of cuttings to form roots. Shoots that are formed at the base of mature woody plants are typically juvenile and it has been postulated that mature shoots are only formed when there has been a 'physiological distancing' of the shoot apical meristem from the root system. It has been suggested that diffusible substances produced by the roots, possibly gibberellins, retain juvenile characters in shoots within their sphere of influence.

In their first growing season, seedlings of seasonal-flowering roses produce no flowers and can be regarded as juvenile. In their second or, sometimes, third growing season they produce flowers and can be regarded as mature. Basal breaks–shoots that are formed at the base of a mature plant–do not produce flowers in their first year of growth and the leaves of these shoots have more leaflets than flowering shoots. Thus they exhibit juvenile characters.

Seedlings of recurrent-flowering roses produce flowers soon after germination. In hybrid tea roses and most recurrent-flowering bush roses, the flowers appear on shoots with only six to seven leaves. Seedlings of recurrent-flowering climbers form flowers on shoots with approximately 14 leaves. Basal breaks flower in the same year. The juvenile period of these roses can be regarded, therefore, as absent or of very short duration.

In temperate climates, most garden and field roses shed their leaves in late autumn and enter a dormant period. However, within the axillary and terminal buds on the dormant, leafless stems of mature shoots, cells of the terminal meristem continue to divide and form new leaf primordia. This slow increase in the number of leaf primordia is accompanied by a gradual increase in the length of the bud. At bud-break in spring, the bud elongates and the leaves are exposed (**Figure 1**). The first leaves to be exposed at the base of the new shoot are scale leaves that consist of fused stipules. These increase in number with the period of dormancy to which the buds were subjected. They are followed by compound leaves, the number of which varies with genotype and according to position on the stem from which the shoots originated. On average, the lower a shoot's point of origin, the more compound leaves it produces. Soon after bud-break, the shoot apical meristem undergoes a physiological change. It stops producing leaves and starts to initiate an inflorescence of one or more flowers. The switch from vegetative to reproductive growth is accompanied by a flattening of the meristematic dome. Several weeks later, in late spring or early summer, the flowers eventually emerge from the terminal bud. The growth of the primary axis is terminated by the inflorescence. Further growth occurs from secondary shoots, which originate from buds in the axils of leaves on the primary shoots. The number of secondary shoots that

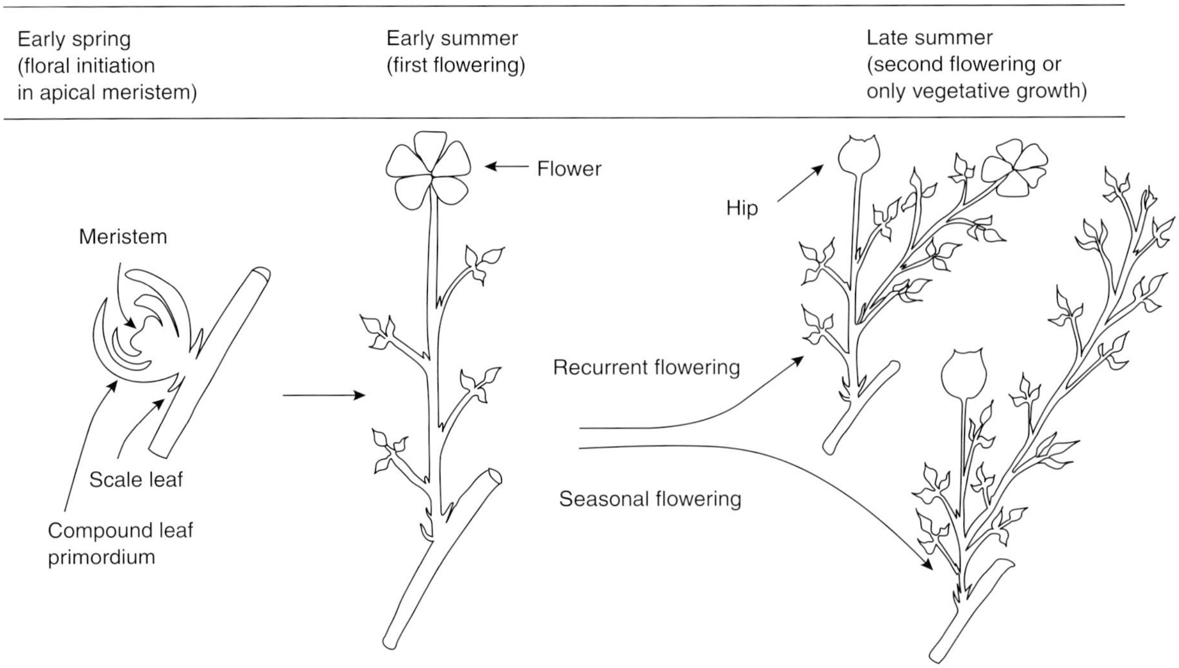

Figure 1 A scheme illustrating the development of primary and secondary shoots in seasonal- and recurrent-flowering roses. In both types of rose, floral initiation occurs at bud-break and terminates vegetative growth of the primary shoot. The development of secondary shoots in recurrent-flowering roses is terminated by an inflorescence but vegetative growth continues indefinitely in seasonal-flowering roses.

form on a primary shoot varies with genotype and environmental conditions, including the amount of light they receive.

In most seasonal-flowering roses, the secondary shoots form leaves continuously throughout the remainder of the growing season. They rarely form flowers. Thus, resources are not wasted on production of new flowers that would be unlikely to form ripe hips in the same growing season. Recurrent-flowering cultivars possess a mutant gene that overcomes this restraint and each secondary shoot, after producing a certain number of leaves, produces an inflorescence. The number of compound leaves is approximately the same for equivalent points of origin as those on the primary stems. If recurrent-flowering plants are maintained in a warm greenhouse, the development of axillary buds, each with a certain number of leaves then as inflorescence, occurs indefinitely in a repeating cycle of events. The determinate growth pattern of recurrent-flowering roses, in which the growth of each shoot is terminated by an inflorescence, contrasts with the indeterminate growth of the secondary shoots of seasonal roses (**Figure 1**).

Comparisons between recurrent-flowering bush roses and their climbing mutants shows that the climbing roses differ only in producing more compound leaves before the inflorescence is initiated. Their longer shoots result from the greater number of internodes, not the length of the internodes.

The number of flowering shoots that are produced on a rose bush depends on the number of axillary shoots that are stimulated into growth. The number of shoots varies directly with the strength of correlative inhibition (*see* **Growth Regulation:** Hormones in Growth and Development; Axillary Bud Growth). The strength of correlative inhibition can be reduced by increasing exposure to light and by training shoots to grow horizontally. Low temperatures and weak lighting can result in blind shoots that reduce the number of flowering shoots. Blind shoots that are occasionally formed by recurrent-flowering roses result from a failure in the growth of the inflorescence, not from a failure in floral initiation. Cultural practices, nutrition and environmental manipulations used to increase the productivity of a rose crop have no influence on whether or not flowers will be initiated and little influence on the number of compound leaves that will be initiated before flowering. However, floral initiation is markedly affected by genotype and can be changed by genetic mutation.

Mutations that Affect Floral Initiation

Sports result from gene mutation (*see* **Genetics:** Mutation). Mutation is of interest because it can sometimes improve the horticultural qualities of a plant. But it is also of interest because studies of the disrupted physiology of a sport may indicate how floral initiation works in a normal plant. Three categories of mutants that directly affect floral induction are described below.

The Gene for Recurrent Flowering

Recurrent-flowering sports of *Rosa chinensis* were found, cultivated and bred in China about 1000 years ago (*see* **History of Roses in Cultivation:** Ancient Chinese Roses). The gene responsible for this characteristic is a mutant gene that is silent in the presence of the wild-type allele and F_1 hybrids between recurrent-flowering cultivars and wild roses are seasonal-flowering. In other words, the mutant gene is recessive and the wild-type gene is dominant (*see* **Genetics:** Inheritance of Characteristics). If the F_1 hybrid is selfed or crossed with a recurrent-flowering rose, a proportion of the offspring will be homozygous for the mutant gene and express the recurrent-flowering character. Crosses between recurrent-flowering cultivars produce only recurrent-flowering offspring. It is generally the case that a recessive mutant gene is a damaged gene that is unable to function normally. At one level it can be said that the function that has been lost is the ability to restrain flowering, but it would be of interest to understand, at a molecular level, how this is achieved (*see* **Genetics:** Marker-Assisted Selection). The mutant gene that originated in *R. chinensis* is now widespread in cultivation but it is possible that the same gene may have mutated on more than one occasion. It is also possible that mutations of different genes may also result in recurrent-flowering. Differently mutated genes may confer slightly different characteristics. One example of a mutation leading to recurrent flowering is seen in the origin of the recurrent-flowering rose 'Little White Pet' as a sport of the seasonal rose 'Félicité et Perpétue'. 'Félicité et Perpétue' is a diploid rose that is thought to be an F_1 hybrid between *R. sempervirens* and a Noisette rose. From this pedigree it would be expected to be heterozygous for the wild-type and mutant gene and is, of course, seasonal-flowering. The mutational event that gave rise to 'Little White Pet' is likely to have involved mutation of the wild-type allele of the recurrent-flowering gene. *Rosa rugosa* is an unusual example of a wild rose that produces flowers throughout the growing season. It is possible that the genetic determination of the recurrent-flowering character of this species is different from that in the China roses.

Climbing Sports

Most climbing cultivars were not bred from seed. They arose as sports of bush roses. In *Modern Roses XI*, they are identified by the name of the original bush cultivar followed by the word 'Climbing'. The climbing sports differ from the original cultivar in having more nodes below the inflorescence, not by an increase in the length of internodes. Climbing sports have appeared frequently and, in those cultivars that have produced several sports, the first recorded sport has almost invariably been a climbing sport (*see* **Genetics:** Mutation). One can speculate, in the absence of firm evidence, that the high frequency of sporting may result from a single gene that is unstable or because the mutation of any one of several different genes can result in a climbing sport.

Polyploidy

Chromosome doubling may be carried out artificially with chemicals, such as colchicine, to obtain fertile plants from sterile hybrids (*see* **Biotechnologies for Breeding:** Manipulation of Ploidy Level). However, chromosome doubling has, on some occasions, resulted in plants in which the ability to form flowers is greatly reduced. A tetraploid form of the seasonal flowering diploid hybrid *R. persica* × *xanthina* failed to produce any flowers over a 10-year period. Also a hexaploid form of the recurrent-flowering triploid climber 'New Dawn' produced very few flowers over a 3-year period (most of the flowers that were formed appeared shortly after transfer from *in vitro* culture to soil). It is well known that chromosomally doubled plants often differ in having larger leaves, larger flowers and longer internodes than the original plant. These morphological changes reflect internal physiological changes which may interfere with floral induction.

Mechanisms of Floral Induction

In species which are sensitive to day length, a flowering stimulus can diffuse from the leaves of a florally induced plant to an uninduced plant through a graft union. In 1936, M.K. Chailakhyan proposed the name 'florigen' for this diffusible inducer of flowering. However, no substance uniquely responsible for floral induction has yet been identified. The existence of a flowering inhibitor, antiflorigen, has also been postulated but has not yet been identified. It seems quite likely that floral induction is not the responsibility of a unique molecule but rather of one or more hormones or metabolites that are already known to affect plant development in other ways.

Floral initiation cannot occur in buds that are subject to correlative inhibition. Axillary buds are not subject to correlative inhibition as the plant emerges from dormancy, after floral initiation on a main shoot terminates vegetative growth or if the apical meristem is excised (however, correlative inhibition may be reimposed on slowly developing buds by faster-growing shoots higher up the stem). In recurrent-flowering roses, the only other factor that limits the initiation of flowers on new shoots is the requirement that a certain number of leaves must first be initiated. As these roses have no specific environmental requirements for floral initiation (such as long or short days, or a cold spell), they are described as self-inductive. However, in seasonal-flowering roses, a period of dormancy is also required. Plants of *R. canina* that have overwintered in a warm greenhouse, for example, will not flower in spring. Thus any environmental factors involved in inducing dormancy may be considered to influence floral induction.

The class of plant hormones known as gibberellins are able to repress flowering in several woody Rosaceae, including apple, cherry and rose. The endogenous concentrations of gibberellins vary according to season and differ between seasonal-flowering roses in ways described below. Furthermore, applications of gibberellins have been shown to suppress flowering in seasonal but not recurrent-flowering roses. Some of the evidence in roses comes from experiments involving the seasonal-flowering climber 'Félicité et Perpétue' and its recurrent-flowering sport 'Little White Pet', that were referred to above. These roses are closely similar except for the seasonal vs recurrent flowering character and the consequential effect of these flowering patterns on their respective climbing and dwarf habits. Therefore, any physiological differences that are detected between them can be attributed to the effect of the recurrent-flowering mutation.

The gibberellins in roses have been identified and quantified by gas chromatography–mass spectrometry. They have been investigated at bud-break in March, in April just before the preformed flowers emerged and in September towards the end of the growing season (**Table 1**). The gibberellins that showed most variability were GA_1 and GA_3. In 'Félicité et Perpétue', their concentrations were low in March, increased substantially by April but by September had fallen again to similar levels to those found in March. In 'Little White Pet' these hormones were present in much lower concentrations than in 'Félicité et Perpétue', particularly in April, when they were, respectively, 17-fold and 12-fold lower than in 'Félicité et Perpétue'. To interpret floral initiation in these roses, a model has been proposed in which floral initiation proceeds if the combined concentrations of GA_1 and GA_3 stay below a certain threshold level but is prevented if they rise above that threshold. Thus at bud-break in 'Félicité et Perpétue', floral initiation proceeds because the concentrations of GA_1 and GA_3 are below the threshold. A second flowering is prevented because, by April, the threshold has been exceeded. In 'Little White Pet' the concentrations of GA_1 and GA_3 remain below the threshold throughout the growing season and flowering is unrestrained. Supporting evidence came from experiments in which solutions of GA_1 or GA_3 in 70% ethanol ($2\ mg\ ml^{-1}$) were repeatedly applied, in 2 μl quantities, to growing shoots at bud-break. In treated shoots of 'Félicité et Perpétue', the transition from vegetative to floral development was almost completely suppressed by both of these gibberellins and branches of the inflorescence developed as vegetative shoots, whereas in 'Little White Pet' flowering proceeded normally (**Table 2**, **Figure 2**).

The applications of gibberellins did not significantly affect the number of nodes formed before floral initiation but, in accordance with the known properties of these gibberellins, increased internode length. According to the proposed model, the combined concentrations of endogenous and exogenously applied gibberellins exceeded the threshold in 'Félicité et Perpétue' in March and prevented the first flowering. However, in 'Little White Pet', as the concentration of endogenous gibberellins was lower, the combined concentrations did not exceed the threshold and floral initiation was not inhibited. Further investigations are needed to clarify various issues. For example, are the

Table 1 Concentrations of gibberellins A_1 and A_3 at bud-break (March), prior to flowering (April) and late season (September)

		Concentration of gibberellins (*ng* g^{-1} *fresh weight* ± SE)		
		March	*April*	*September*
'Félicité et Perpétue'	GA_1	0.7 ± 0.30	36.8 ± 1.73	2.2 ± 0.07
	GA_3	13.7 ± 0.09	285.7 ± 27.00	12.5 ± 0.42
'Little White Pet'	GA_1	0.2 ± 0.05	2.2 ± 0.03	2.6 ± 0.02
	GA_3	7.3 ± 0.50	22.9 ± 0.62	12.5 ± 0.28

Based on the data of Roberts AV, Blake PS, Lewis R, Taylor JM and Dunstan DI (1999). The effect of gibberellins on flowering in roses. *Journal of Plant Growth Regulators* 18: 113–119.

Table 2 Response at flowering of 'Félicité et Perpétue' and 'Little White Pet' to applications of GA_3 at bud-break

		No. of flowers per inflorescence	*Mean internode length*	*No. of nodes below the inflorescence*
'Félicité et Perpétue'	Control	15.6 ± 2.83	31.6 ± 1.73	6.3 ± 0.37
	GA_3	0.3 ± 0.25	52.2 ± 2.21	6.9 ± 0.59
'Little White Pet'	Control	11.4 ± 1.94	34.4 ± 2.39	6.0 ± 0.38
	GA_3	12.0 ± 3.77	54.6 ± 3.10	6.0 ± 0.38

Based on the data of Roberts AV, Blake PS, Lewis R, Taylor JM and Dunstan DI (1999). The effect of gibberellins on flowering in roses. *Journal of Plant Growth Regulators* 18: 113–119.

Figure 2 Control shoot of 'Félicité et Perpétue' (left) and an equivalent shoot that was treated with GA_3 at bud-break (right). Note that, in the treated shoot, flowers have been replaced by vegetative shoots.

actions of GA_1 and GA_3 similar but independent and do they themselves directly repress floral induction or is their influence indirect, via a different type of molecule?

There is some evidence that implicates cytokinins as floral inducers. In one investigation, flowering shoots of the seasonal-flowering rose *R. damascena* were found to contain more cytokinin-like substances than nonflowering shoots. Furthermore, when shoots were treated with the cytokinin kinetin (sprays containing $10\,mg\,l^{-1}$) at approximately the time of bud-break, the treated plants produced more flowers than untreated shoots. It is not clear if the increase in the number of flowering shoots in response to kinetin was caused by an increase in the branching of shoots or a conversion of potentially vegetative shoots to flowering shoots.

The possibility of an interaction between cytokinins and gibberellins would be an interesting subject for further investigation. New protocols for metabolic profiling, based on gas chromatography–mass spectrometry, now allow a wide range of metabolites to be detected in a single chromatographic run. This approach would help to detect interrelated fluctuations between gibberellins and cytokinins during the course of the growing season.

Future Prospects

It would be interesting to know what range of genetic variants might be present amongst recurrent-flowering roses. In addition to the roses derived from the *R. chinensis* sport, there are at least three apparently independent sources of recurrent flowering. *Rosa rugosa* and *R. fedschenkoana* are recurrent-flowering. Also 'Mermaid', an F_1 hybrid between *R. bracteata* and a tea rose, is recurrent-flowering, which suggests that it received a recurrent-flowering allele of the *R. chinensis*-derived gene from *R. bracteata*. Questions that could be resolved by the appropriate crosses include whether all the recurrent-flowering genes are allelic and whether novel physiological and morphological characters might result from different combinations of these genes. Studies of endogenous hormones and the effects of exogenously applied hormones might also suggest differences in the mechanism of action of recurrent-flowering genes from different sources.

One possibility that has not been rigorously tested is that applications of abscissic acid or antigibberellins, such as paclobutrazol, may be able to induce abnormal flowering in seasonal-flowering hybrids. This might, for example, enable breeders to make crosses with first-year seedlings of F_1 hybrids.

It is possible that our knowledge of genetic and molecular control of flowering in roses will benefit from studies in other genera. The most detailed studies have been undertaken in herbaceous species, in particular, in *Arabidopsis*, in which 80 genes involved in the control of flowering have been identified. However, the transition from vegetative to flowering growth in shoot meristems of *Arabidopsis*, unlike *Rosa*, is responsive to day length and it is uncertain how similar the control mechanisms might be in these two genera. The control mechanisms in *Rosa* may be more similar to those of two other genera in the family Rosaceae, *Fragaria* (strawberry) and *Malus* (apple). In *Fragaria*, like *Rosa*, there is a dominant gene that causes seasonal instead of continuous flowering and in *Malus*, like *Rosa*, there is evidence that gibberellins suppress floral induction.

See also: **Biotechnologies for Breeding:** Manipulation of Ploidy Level. **Breeding:** Selection Strategies for Cut Roses. **Genetics:** Inheritance of Characteristics; Inheritance in the Dogrose; Mutation; Marker-Assisted Selection. **Growth Regulation:** Hormones in Growth and Development; Axillary Bud Growth. **History of Roses in Cultivation:** Ancient Chinese Roses. **Morphology and Anatomy:** Flower Development and the Abscission Zone. **Production and Marketing:** Overview of Roses and Culture.

Further Reading

Bernier G (1988) The control of floral evocation and morphogenesis. *Annual Reviews of Plant Physiology and Plant Molecular Biology* 39: 175–219.

Colasanti J and Sundaresan V (2000) 'Florigen' enters the molecular age: long-distance signals that cause plants to flower. *Trends in Biochemical Sciences* 25(5): 236–240.

Evans MMS and Barton MK (1997) Genetics of angiosperm shoot apical meristem development. *Annual Reviews of Plant Physiology and Plant Molecular Biology* 48: 673–710.

Hackett WP (1987) Juvenility, maturation and rejuvenation in woody plants. *Horticultural Reviews* 7: 109–155.

Zhao D, Yu Q, Chen C and Ma H (2002) Genetic control of reproductive meristems. In: McManus MT and Veit BE (eds.) *Meristematic Tissues in Plant Growth and Development*, pp. 89–141. Sheffield: Sheffield Academic Press.

Zimmerman RH, Hackett WP and Pharis RP (1985) Hormonal aspects of phase change and precocious flowering. In: Pharis RP and Reid DM (eds.) *Encyclopedia of Plant Physiology*, vol. II, pp. 79–115. Heidelberg: Springer-Verlag.

H

HISTORY OF ROSES IN CULTIVATION

Contents

Ancient Chinese Roses

W Guoliang, Jiangsu Forestry Academy, Dongshanqiao, Nanjing, China

Introduction

It is acknowledged that the rose is one of the four major cultivated flowers in the world. Fortunately, the Chinese rose has the same status among native Chinese flowers and its cultivation has a long history which can be traced back 2000 years, to the period of the Han Dynasty (141–87 BC), when wild roses were seen everywhere in the gardens of the imperial palace.

China is the distribution centre of the genus *Rosa*. Fossils of rose leaflets unearthed in Fushun, Liaoning province (northern China) have proved that the Chinese rose existed at least 40 million years ago (in the Eocene epoch). In 1940, fossils of two kinds of wild rose were discovered in the Shanwang palaeobotanic region of Shandong province, and they were then named *Rosa shanwangensis* Hu et Chaney (**Figure 1**), and dated 25 million years old. Since then many archaeological discoveries have been made. Pieces of clay pots engraved with designs of pot plants made 7000 years ago were excavated at the New Stone Age relic site of Yuyao county, Zhejiang province. Some coloured pots and flat basins painted with colourful five-petalled flowers dating from 3500 BC were unearthed at Dawenkou relics site in Taihu Lake region, Jiangsu Province (**Figure 2**). In ancient inscriptions on bones and tortoise shells, there are Chinese characters like *yuan* (garden), *pu* (nursery) and *you* (hunting park). In the Chinese language, *yuan* relates to planting and cultivating trees or fruit trees, *pu* to vegetables and *you* to early parks and woods. In addition, there

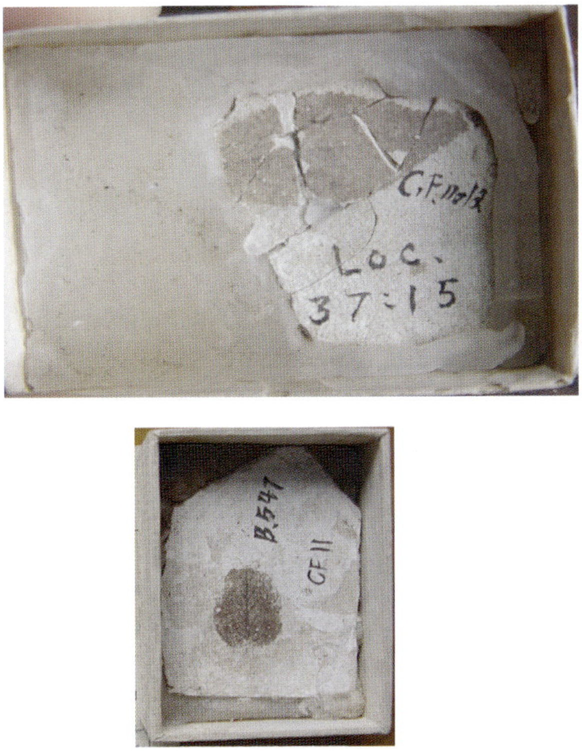

Figure 1 Wild rose leaflet fossils preserved in the Nanjing Institute of Geology and Palaeontology, Chinese Academy of Science, which were dug out from the Shanwang palaeobotanic region in Shandong province. Designated *Rosa shanwangensis* Hu et Chaney: 25 million years old.

Figure 2 Ancient painted pottery, preserved in Nanjing Museum, Jiangsu province, which was dug out from Dawendu culture ruins in Taihu, Jiangsju province.

were *hua* (flowers), *hui* (plants), *chong* (insects) and *cao* (weeds). However, the first appearance of the phrase *hua hui* (flowers and plants) was in *Liang History—Biography of He Dian*, written 1350 years ago. In the Tang Dynasty (618–907 AD), working staff specialized in planting flowers and other plants. An eminent poet, Bai Juyi, wrote then: 'I bought flower trees to plant them along the hill side of a mount to the east of the city. Any flowers will be greatly appreciated.'

The Origin and Evolution of China Roses

The origin and evolution of China roses can be traced from ancient books of Chinese paintings, classical poetry and agronomy so that their existence can be proved.

Roses in Ancient Flower and Bird Paintings

Old Chinese art can be classified into two periods: prehistoric and posthistoric. The prehistoric paintings are those before the Xia Dynasty (about 2000–1500 BC). The major artistic achievements of that period included coloured pottery characteristically showing bold design, although coarse rather than delicate. The definition of rock paintings, ground paintings and fresco paintings is so rough that plants and flowers cannot be definitely identified.

During the Suitang Dynasty (581–907 AD) the realistic painting genre was established, specializing in painting natural flowers and flying birds. During the Five-Dai period (907–960 AD), a group of realistic artists, represented by Huang Jian, recreated the natural beauty of flowers and birds. The period of the Northern to Southern Song Dynasty (960–1279 AD) saw the golden prime of the flower and bird paintings when many artists, among whom were Zhao Chang, Cui Bai, Ma Yuan and, even the Emperor, Zhao Ji (1082–1135 AD), recorded the panorama of real beauty of Chinese roses and plants, detailing large, upright, high-centred double flowers (**Table 1**), just as clear and precise as photographs of plants.

Literate painting sprang up during the Yuan Dynasty (1271–1368 AD), featuring elegance, refinement and beauty, reflecting the artists' personalities. At that time artists mainly painted plum blossoms, orchids, bamboos and chrysanthemums, while wild rugosa and China roses were occasionally featured. However, from the Ming Dynasty (1368–1644 AD), some famous Chinese artists, such as Qiu Ying and Lu Ji, began to paint more Chinese roses in the flower and bird paintings. So until the Qing Dynasty (1636 AD) there were two parallel styles: antique-imitating and creative. Langsning, an Italian artist who had been greatly appreciated by Emperor Kang Xi (1662–1720 AD), painted, using great technical skill, many large old multiple-petal roses cultivated in the palace gardens (**Table 1**).

As we can see from **Table 1**, 300 years ago there were red, double wild roses in the frescos of No. 3 Temple in Nuoxiang county, Xinjiang province. Also in the silk painting *Guiding Buddha* of the Later Tang Dynasty, there were clearly five-petalled wild roses. It is easy to see that large multiple-petalled Chinese roses in ancient Chinese paintings (Northen Song Dynasty, 960–1127 AD) are much the same as their modern cousins.

Chinese Roses in Ancient Books

China would still have numerous ancient gardening books if there had not been many wars and riots. Many books have been lost due to careless preservation, among which is the great work *Annals of Gardening by Empire Wei*, written during the South–North Dynasty (420 AD).

Table 1 Selected Chinese paintings dealing with roses in China

Name of painting	*Artist*	*Time (year)*	*Kind of painting*	*Size (cm)*	*Main features*
Portrait of the youth	Unknown	300 BC	Fresco		Deep crimson flower, 12 petals, double; no foliage and branches
Guiding Buddha	Unknown	AD 618–907	Silk painting	85 × 53.6	Six petals, five leaflets; red flower with vivid colour
Peacock and flowers	Cui Bai	AD 1061	Silk painting	183.1 × 109.8	Flowering shoots upright; reddish to red, high-centred; shoots, leaflets and flowers very similar to present rose
Happy new year	Zhao Chang	AD 960–1127	Silk painting	103.8 × 51.2	Branches upright, large flowers, double, mild-centred, like a photo
Rose and quail	Unknown	AD 1127–1279	Brocade	103 × 42.9	Bushy, upright; large flowers, double and high-centred; fine leaflets, shoots, and prickles
White roses	Ma Yuan	AD 1127–1280	Silk painting		Seven to nine leaflets; large flowers, white, double; features similar to *R. xanthina* L.
Eight flowers	Qian Xuan	AD 1239	Silk painting		Branch upright, large flowers, double, reddish; similar to China rose
One hundred flowers	Unknown	AD 1301	Paper painting	1679.5 × 31.5	Seven to nine leaflets; double flowers,easy to identify as *R. xanthina* L.
Flowers and birds	Xiao Ron	AD 1271–1368	Silk painting		Bushy, upright; large flowers, double or single; single similar to wild species of China rose.
Bamboo picture	Ke Jiusi	AD 1271–1369	Silk painting	126.3 × 75.2	Climbing stems with hooked prickles and fruits; similar to *R. cymosa* Tratt.
Orchid and bamboo	Xue Chuang	AD 1271–1370	Silk painting	106 × 46	Climbing stems; rare leaflets and fruits; vertical prickles to stems
Flowers and plants	Chen Chun	AD 1368–1644	Silk painting	118 × 96.9	Bushy, upright; large flowers, pinkish, double; fine buds and shoots
High-ranking imperial concubine	Qiu Ying	AD 1368–1645	Silk painting	41.1 × 33.8	Planted in parterre; pinkish, double, large flowers; a woman pouring water from kettle
Return to the ancients	Chen Hongsou	AD 1368–1646	Silk painting	24.6 × 27	Five to seven leaflets; large flowers, pinkish, double, high-centred
Flowers and birds	Han Ximen	AD 1368–1647	Brocade	25.2 × 23.7	One flower shoots upright; pinkish to red, prickles and shoot very similar to modern rose
China rose	Li Chan	AD 1738	Paper painting		Three shoots with five flowers; double, reddish; like a present-day rose

It is worth noting that Chinese Tang poetry is known all over the world. There are many versions which reveal people's interest and enthusiasm in planting Chinese roses, such as the famous poet Jia Dao's 'Every family plants roses instead of plum blossoms'; Xu Yin's 'It is easy to transplant roses from Hill Yuewangtai' and Shao Chuchang's 'We smell a beautiful spring banksia rose just as we have a pretty young lady approaching towards us'. All this explains that, even in the Tang Dynasty, differences between wild rose, rugosa and banksia roses were realized.

Tumi (tea rose) and *yue yue hong* (China rose) appeared in the Chinese poetic prose of the North-Song Dynasty (960–1279 AD) when the famous poet Yang Wanli highlighted the fact that the China rose blossoms all year round. *Tumi* is *Rosa odorata* (Andr.), sweet-smelling, distributed over Sichuan and Yuannan (south-west China) and with three colours of flower: white, yellow and red.

According to statistics there are over 30 gardening books of antiquity involving the wild rose and its classification. Among these are *Notes on Dwelling on Mount Ping Yuan Shan* by Li Deyu, the prime minister of the Middle Tang Dynasty, in which he mentioned *Rosa centifolia* Lindl, the cabbage rose; *Records of Flowers and Plants in Luyang* by Zhou Shihou (1082 AD), giving 37 varieties of Chinese roses, for instance, Baoxiang (*R. chinensis* var. baoxiang), which still exists, today; *Dreaming of Liang* by Wu Zimu (1127–1279 AD), describing the widespread occurrence of roses over Suzhou and Hangzhou (south-east China); and *Manual of Roses* by Yu Sou, listing 41 varieties, four of which are excellent ones. During the Ming Dynasty, there were more books on Chinese roses, such as *On Chinese Medical Herbs* (1596 AD) by Li Shizhen, a famous ancient Chinese doctor; *Qun Fang Pu*, *Lists of Famous Flowers* (1621 AD) by Wang Xiangjin, *Herbs for Starvation* (1640 AD) by Zhu Su, and *the Flower Mirror* (1688 AD) by Chen Min, which not only records the ordinary characteristics of different roses but also classifies *R. laevigata* Michx into three provenances, Shu Zhou, Yi Zhou and Quan Zhou. Up to the Qing Dynasty there had been more books on Chinese roses, for example, *Manual of Roses* and *Roses* (by Wang Zongxin, during the Guangxu period). All these books are of significance in the palaeobotanic study of old roses.

Moreover, the Japanese Fujiwara-no Teika in Kamakura (1213 AD) mentioned in his book *Meigetsuki,* written in Chinese, that the Japanese 'Koushin Rose' and 'Perpetual Flower' were transplanted from China to Japan. They certainly belong to the China rose group, judging by their features of blossoming in every season.

Contributions that Chinese Roses make to World Roses

Fossils of rose leaves have been unearthed at South Park, Colorado, USA and specimens of wild roses were found in Egyptian ancient tombs. Roses with identifiable features that first appeared in European paintings date back to the Renaissance, for instance, in the painting *Saints in the roses plantation* by Martin Schongauer in 1473, and in the painting *Venice* in 1864. They can also be found in *Fresco with blue bird* on Crete and in the Cuneiform Tablets of Mesopotamia. All these historical traces prove the long history of rose cultivation in Europe. However, there is no record of yellow roses or roses blooming in every season.

Chinese roses were not introduced to Europe until 1792. After importing different rose species, such as *R. odorata, R. rugosa, R. wichurana*, 'Slater's Crimson China', 'Parsons' Pink China', 'Hume's Blush Tea-scented China', and 'Parks' Yellow Tea-scented China', some new hybrid groups were raised. For example, the hybrid perpetual rose first appeared in 1837. It was hybridized again with a Chinese tea-scented rose. Finally, the first hybrid tea-scented rose named 'La France' was obtained in 1867. World roses have come into the times of modern roses since then. Mikinori Ogisu, a Japanese botanist, once wrote: 'Rose is a Chinese flower rather than a European flower'.

Mystery of the Original Species of 'Slater's Crimson China'

Is there an original species of 'Slater's Crimson China' (*yueyuehong* in Chinese)? It is an unsolved mystery. It is recorded that three westerners and one Japanese had seen the original species in China. A. Henry (1857–1930) found it for the first time at Yichang, Hubei province, in 1885. During his research, another English botanical collector, E.H. Wilson (1876–1930), came across the original species in the central and northern areas of Sichuan province in 1910. The botanical explorer Joseph Rock (1884–1962) also found it in Gansu province. Mikinori Ogisu also found the original species in 1983. All of the plants they found, however, belong to *R. chinensis* Jacq. var. *spontanea* (Rehd. et Wils.) Yu et Ku. Compared with 'Slater's Crimson China', they lack some important properties, such as multiple petals, short cane and repeated flowering.

From the information mentioned above, it can be concluded that evolution of a wild rose to China rose, when the Tang Dynasty was superseded by the Song Dynasty, is an important qualitative change. There are three probable reasons for this evolution. Firstly, when various kinds of wild roses were cultivated on a large scale, the probability of finding natural sports greatly increased. The mutant was propagated again, so the genetic pattern tended to be changed. After bud mutation, the new breed was shaped to bloom in every season and fruit infrequently. This is closer to our modern ornamental plants. Secondly, another probability is artificial pollination by ancient Chinese gardeners. Thirdly, they may have screened the new breed for particular properties after natural hybridization. All of these factors were likely to happen at that time. There is no denying that the evolution of Chinese roses results from ancient Chinese constant, long-term efforts. With great skill, they cultivated and improved breeds of rose plants. As a result, the mystery of whether or not there was an original species is merely an academic problem.

Figure 3 The old rose variety *R. multiflora* 'Nanjingfen', climbing, reddish to red, once-flowering, strong and vigorous, with good resistance to disease.

Present Situation of Chinese Old Roses

According to an investigation made by the Chinese Rose Association and my own observations, there still exist many precious Chinese old roses in some areas, such as in Yangzhou and Huiyin, Jiangsu province, Laizhou and Pingyin, Shandong province, Luoyang, Henan province, Kunming, Yunnan province and Kushui, Gansu province (**Figures 3–7**). From **Table 2**, it can be seen that Chinese old roses, with their property of disease resistance, are good ornamental plants. They are, therefore, precious breeding resources.

Distribution and Spread of *Rosa* Species Plant Resources

Throughout the world, there are about 150 naturally occurring species in the rose genus. In China, there are 82 (*see* **Distribution and Ecology**: Continental Asia and Japan), and still some new species are waiting to be identified. Plants of the China roses are mainly distributed in the central and south-western parts of China, especially in the roughly triangular area including Shanxi province, Tibet and south-western provinces, centred on the Sichuan basin. That area is the world centre of rose species distribution.

Since the nineteenth century, foreign botanical collectors and explorers have come to China and taken away botanical resources. More than 30 rose species have left their native sites and evolved in new habitats (**Table 3**).

Figure 4 A China rose, called *yueyuehong* (*R. chinensis* Jacq.), distributed in north and south parts of China. It has good resistance to cold and is recurrent, bushy and upright.

Figure 6 An old garden rose, called *yipengfen*. Charming, pale pink, upright bushy and recurrent-flowering.

Figure 5 An old rose called *fojianxiao* (*R. odorata* var. *gigantea* (Crép.) Rehd. et Wils.). Strongly fragrant, pinkish to red, double, high-centred, climbing, with very strong vigorous shoots and large leaflets.

Figure 7 A old rose of pure white colour, called *baiyutang* (*R. multiflora* Thunb. var. *alba-plena* Yu et Ku). Double-flowered, borne in clusters, with strong resistance to cold, it is mainly distributed in the northern part of China.

Conclusions and Prospects

The history of Chinese civilization extends back for 5000 years. Throughout this history, Chinese ancestors, with respect for nature and affection for plants, cultivated and bred new varieties of plants. Known as the 'mother of gardens' in the world, Chinese ancestors have introduced new varieties of roses, world-famous flowers, which are sometimes also named the queen of the flower kingdom. Their brilliant technique and knowledge of breeding new plant varieties can be seen from the following evidence. Seven thousand years ago, coloured pottery on which five-petalled flowers were painted was fashioned. Two thousand years ago, the Chinese Emperor Hanwudi had roses planted at the palace. More than 1000 years ago, there were roses that were bushy, big-flowered, double and recurrent-blooming. Over 200 years ago, Chinese roses were brought to Europe and began their important new phase of development.

At present, the traditional techniques of cultivating rugosa roses have been improved. There have been more than 50 varieties, and many of them have short

Table 2 A list of Chinese old roses existing in Jiangsu province

Chinese name	*Fragrant*	*Flower colour*	*Flower diameter (cm)*	*Petals*	*Other features*
Baiyutang	Slightly fragrant	Pure white	6	20	Climbing, strong resistance to cold; 7–10 flowers borne together on short stems; often in Northern China
Chilongtuzhu	Slightly fragrant	Deep red	11	35	Recurrent; strong, vigorous; resistant to disease
Dafugui	Fragrant	Pink to red	15	40	Foliage wrinkled; small prickles on main vein; peony-like flowers
Danyunxiyu	Slightly fragrant	Pink to red	10	40	Slim peduncle; many straight prickles on stems
Ezhangjinbo	Fragrant	Slightly yellow	5	50	Stems spreading; small foliage; bent shoot
Feigeliuyun	Fragrant	Yellow to reddish	10	40	Calyces like leather; half upright growth; foliage deep green
Fojianxiao	Very fragrant	Pinkish to red	9	13	Climbing, very vigorous; once-flowering, borne on short stems; large foliage
Jinfenlian	Very fragrant	Slightly pink	12	60	Long peduncle with prickles and bristles
Jinoufanlu	Fragrant	Slightly pink	12	6	Reddish spots on petals in autumn; resistant to disease
Mibohuang	Slightly fragrant	Yellow			Recurrent; strong, vigorous; resistant to disease
Nanjingfen	Slightly fragrant	Pinkish to red	10	15	Climbing, vigorous, prickles hooked, once-flowering
Ruanhongxiang	Very fragrant	Purple to reddish	10	35	Rare branching; borne in clusters; foliage wrinkled; strong resistance to disease
Shuangcuiniao	Fragrant	Green to white	15	45	Petals with waving margins
Shuimeiren	Fragrant	Pink	15	60	Larger foliage; obviously lateral veins
Simianjing	Fragrant	Deep red	8	50	Four-centred flowers
Yangchunbaixue	Fragrant	White	10	15	Flowers high-centred; hooked prickles and small bristles; upright
Yipengfen	Fragrant	Pinkish to reddish	12	60	Climbing; once-flowering; strong resistant to disease
Yuelinglong	Very fragrant	Pinkish to white	5	100	Bushy growth; branching in clusters
Yueyuehong	Slightly fragrant	Deep red	8	15	Foliage evergreen; borne on short stem; recurrent flowering
Yunzhengxiawei	Slightly fragrant	Pink to red	10	25	Branching upright; foliage wrinkled
Yushizhuang	Slightly fragrant	Orange to yellow	10	45	Long peduncle and easily bent; prickles hooked

Table 3 A list of Chinese wild roses spread abroad

Latin name	*Place of origin*	*Year*	*Introducer*
R. anemnifolia Fort.	Shanghai	1844	R. Fortune
R. banksiae Ait.	Guangdong	1807	William Kerr
R. banksiopsis Baker	Western China	1910	
R. bella Rehd. et Wils.	Southern China	1910	
R. bracteata Wendl.	Guangdong	1793	George Staunton
R. brunonii Lindl.	Tibet	1822	
R. chinensis Jacq.	Guangdong	1800	
R. chinensis var. *minima (Sims)*	Guangdong	1800	
R. corymbulosa Rolfe	Western China	1808	
R. davidii Crép.	Shichuang	1903	E.H. Wilson
R. farreri Stape	Ganshu	1915	Reginald Farrer
R. filipes Rehd. et Wils.	Western China	1908	E.H. Wilson
R. forrestiana Boulenger	Western China	1918	
R. gigantea Collett	Western China	1889	
R. giraldii Crép.	Southern China	1897	
R. helenae Rehd. et Wils.	Central China	1907	E.H. Wilson
R. hugonis Hemsl.	Central China	1899	
R. latibracteata Boulenger	Western China	1894	A.E. Pratt.
R. moyesii Hemsl. et Wils.	Western China	1936	
R. multibracteata Hemsl. et Wils.	Western China	1908	
R. odorata Sweet.	Ningbo	1845	Robert Fortune
R. omeiensis f. pteracantha Rehd. et Wils.	Shichuang	1901	
R. omeiensis Rolfe	Shichuang	1890	
R. persetosa Rolfe	Western China	1895	
R. prattii Hemsl.	Western China	1908	
R. primula Boulenger	Southern China	1910	
R. roxburghii Tratt.	Western China	1908	
R. rubus Thunb.	Western China	1907	
R. sericea Lindl.	Tibet	1822	
R. setipoda Hemsl. et Wils.	Western China	1895	
R. sinowilsonii Hemsl.	Western China	1904	E.H. Wilson
R. soulieana Crép.	Western China	1896	
R. sweginzowii Koehne	Eastern China	1909	
R. webbiana Royle	Tibet	1899	
R. willmottiae Hemsl.	Tibet	1904	E.H. Wilson
R. xanthina f. spontanea Rehd.	Northern China	1907	F.N. Meyer
R. xanthina Lindl.	Northern China	1907	

stems and bloom every season. Some have very few thorns. There are four main areas for growing rugosa roses in China: Pingyin (Shandong province) Miaofengshan (Beijing), Meishan (Sichuan province) and Kushui (Gansu province). In addition to their traditional ornamental function, roses have been employed in tea-making, wine-making, medicine, diet, drink and perfume (*see* **History of Roses in Cultivation:** History of the Perfume Industry. **Secondary Metabolites:** Culinary Uses and Nutritional Value; Medicinal and Pharmaceutical Uses). Land dedicated to the production of cut roses amounts to 820 hm^2, and products have been exported to Japan and South-Eastern Asia. Some research institutions and universities have carried out joint studies on how to protect and apply the resources of rose family plants and old Chinese roses. It can be predicted that the breeding of rose species will come to a new stage in the near future.

See also: **Classification**: Conventional Taxonomy (Wild Roses). **Distribution and Ecology**: Continental Asia and Japan. **History Roses in Cultivation**: European (Pre-1800); History of the Perfume Industry. **Secondary Metabolites**: Culinary Uses and Nutritional Value; Medicinal and Pharmaceutical Uses.

Further Reading

Len S (1993) *Appreciation of the Treasures Preserved in the Imperial Palace*, pp. 86–293. Shanghai Science and Technology Press.

Qinzhong C (2001) *The History of Chinese Painting*, pp. 1–242. Zhejiang Education Press.

Qixiang Z (ed.) (1999) *Famous Flowers in China*, pp. 89–104. Yunnan People's Press.

Yingkui S (1989) *Dictionary of Appreciation of Poetry in Ancient China*, pp. 407–558. Jiangsu Science and Technology.

Yinglan X (1993) *Ancient Flowers and Plants*, pp. 119–127. Agricultural Press.
Yuwen L and Wenqing H (eds.) (1999) *Masterpieces of Chinese Painting Overseas*, pp. 27–124. Shanghai Literature Press.
Zhongming S (1999) *Famous Flowers from Yunnan*, pp. 126–135. Yunnan Science and Technology Press.

European (Pre-1800)

F Joyaux, Roseraie de la Cour de Commer, Commer, France

Introduction

We are all aware of the difficulties of nomenclature presented by nineteenth-century botanical roses and even the horticultural varieties, despite the fact that there is a plethora of literature on the subject. So it is easy to imagine how difficult it would be to attempt to go back even further in time. Leaving to one side the roses of Ancient Greece and Rome or the medieval era, about which there is infinite uncertainty, the roses of the Renaissance and of the seventeenth and eighteenth centuries are not well known. The nomenclature varies widely from one author to the next and the morphological descriptions are often very basic. Moreover, the wide range of botanical and horticultural roses makes identification difficult. Furthermore, the few herbaria which have been conserved from the sixteenth century present only a limited number of species and varieties: before the nineteenth century the rose was not a plant that was judged worthy of scientific study. This article considers, with relevant evidence, the origins and history of cultivated roses found in Europe from the sixteenth to the eighteenth centuries.

The Contribution of Renaissance Herbaria

The oldest conserved herbaria provide us with the first accurate information on which to base our knowledge of roses of the modern era. Thanks to them, we have a fairly accurate idea of at least the species and varieties that were used for medicinal purposes in the sixteenth century; these herbaria were often established by doctors or apothecaries.

A very old herbarium, perhaps the oldest in Western Europe, is that of Gherardo Cibo, which is preserved in the Angelica Library in Rome, and which dates from 1532–53. In the past the German botanist Penzig studied it, as did Costantino Proietti more recently. **Table 1** lists the 17 roses classified by Cibo, as well as the designations proposed by Proietti. Some roses of this herbarium are worthy of comment.

First, *Rosa moschata* Herrm.: Cibo's herbarium seems to have the oldest conserved examples of this rose. We are aware of the great importance of this rose in the genealogy of a number of subsequent varieties (Damask and Noisette roses, for example) and this rose still causes debate today. Let us merely remember that this is a botanical rose, probably from the Middle East (although originating in more eastern regions, such as Persia and north India) and known in the time of the Roman Empire. It was probably lost from sight for centuries, at least in Western and Northern Europe. So its introduction at the very beginning of the sixteenth century, which is duly mentioned in the bibliography, was in all probability really a reintroduction.

Rosa lutea Mill., otherwise known as *R. foetida* (Herrm.) or Yellow Persian rose, is another remarkable rose which features in this herbarium. It is usually considered to have been introduced from Austria into Holland in 1583 by Charles de l'Ecluse, before crossing to England ('Austrian briar'). In fact, as shown by its presence in the herbarium, this rose had been introduced into Europe by the early sixteenth century. (It has even been hypothesized that it was introduced by the Moors to Spain in the thirteenth century.) At the

Table 1 The 17 roses classified by Cibo and designations proposed by C. Proietti

Cibo's designation	*Proietti's designation*
(Not designated)	*R. rubiginosa*
(Not designated)	*R. cinnamomea*
R. alba	*R. alba* var. *maxima*
R. purpurea	*R. gallica* var. *officinalis*
R. communis	*R. damascena* var. *bifera*
R. damascena	*R. moschata* var. *corneola*
R. damascena	*R. moschata* var. *corneola*
R. sine spinis	*R. moschata*
R. montana	*R. canina*
R. lutea	*R. foetida*
R. lutea	*R. foetida*
R. sine spinis	
R. sine spinis	
Rubus canis montana	*R. canina*
Rubus canis alius	*R. canina*
Rubus canis alpina	*R. spinosissima*
Rubus canis campestris	*R. canina*

Data from Proietti C (2001) Contributo allo studio delle rose presenti en Italia nel XVI secolo. In: *Annuario della Rosa 2001*, pp. 35–61. Monza: Associazione Italiana della Rosa. A partial translation into French can be found in *Rosa Gallica* 13: 28–33 and in *Rosa Gallica* 14: 37–41.

time it was the only known yellow rose, which is why it generated such great interest.

Note also the presence in this herbarium of the horticultural variety *R.* × *damascena* var. *bifera* (*R.* × *damascena* var. *semperflorens*) which at that time was the only known remontant rose. It was also an ancient rose (Virgil's '*Biferque rosaria Paestia*'), traditionally used in the Middle East in the manufacture of attar, an essential oil which is obtained by distillation. But, like *R. moschata* (which probably plays a part in its genealogy), this rose must have been forgotten in Western Europe for a very long time, before being reintroduced there at the beginning of the sixteenth century. Since the Cibo herbarium dates from the years 1532–53, these specimens of *R.* × *damascena* var. *bifera* must therefore be almost contemporary with this reintroduction.

Another important source, from around the same era, is the *Libri Picturari* of the library of the University of Jagiellon at Krakow: Claudia Swan recently published plates from these books. This painted herbarium, dating from the 1560s used to belong to an apothecary from Delft, called D.O. Cluyt (1546–98) (Theodorus Clutius), who taught at the University of Leyden and was in charge of its botanical garden after the famous botanist Charles de l'Ecluse (Carolus Clusius).

On two different pages of this publication are featured eight roses, called respectively *R. lutea*, *R. littoralis*, *R. graeca* and *R. canina* (on the first page), and on the second page, *R. alba*, *R. incarnata*, *R. rubra*, with a fourth rose which is not named. Most of these roses can be identified easily. On the first plate, which groups together the botanical roses, *R. lutea* corresponds well to the present-day *R. lutea* Mill., mentioned above; *R. littoralis* corresponds to *R. pimpinellifolia* L., or Great Burnet rose; *R. graeca* corresponds to *R. rubiginosa* L., or perfumed dog rose; while *R. canina*, whose name is unchanged, corresponds to *R. canina* L. or common dog rose.

The second plate groups together cultivated varieties. *Rosa alba*, which also has not changed name, is our current *R.* × *alba*, or white rose of York. The three following roses are different forms of gallica roses: *R. rubra* corresponds to Provins rose (*R. gallica* var. *officinalis*). (Our identifications in this second plate differ from those of Sam Segal, who identified all the plants in this herbarium.) Note that, whether botanical or cultivated, all of these examples are very classical roses, with the sole exception of *R. lutea* Mill. (see above for comments on the date of its reintroduction).

Of course we could quote many other herbaria or books of plates dating from the first two-thirds of the sixteenth century and which include a certain number of roses, whether botanical or, to a lesser degree, horticultural. Studying them would merely serve to confirm the conclusions that we can draw about the two examples given above.

Around the turn of the sixteenth century, it is mainly the botanical rose that is referred to. It is of interest for mainly medicinal reasons, as it was in the Middle Ages, and it was doctors who made the first illustrations of it. For example, this is the case with the rose (perhaps *R. gallica* var. *officinalis*) which is depicted in *Hortus Sanitatis Germanice* (Mainz, 1485) by the German doctor Johann von Cune, from the end of the fifteenth century.

But throughout the sixteenth century, the rose progressively becomes an object of scientific study. However, it has a much more modest place than other plants (and this is the case until the nineteenth century). On the other hand, these studies express wider preoccupations and not just medicine as the sole goal. Despite this, scholars note the differences between simple and double roses, but do not draw conclusions about their respective origins: they are otherwise mixed in with the herbaria and plates in books.

At the same time, as also in the Middle Ages, the rose continues to play an important role in religious symbolism. The red rose evokes the passion of Christ: the rose without thorns is the 'heavenly rose', which flowers unceasingly; the white rose brings to mind the Virgin Mary. All of these medieval images were still very much alive at the beginning of the sixteenth century (and still are today, to some extent). This explains why religious houses, monasteries and parishes grow some roses (and other equally symbolic flowers, like the iris and lily) to decorate places of worship.

But it would probably be unwise to deduce from its strong presence in Christian iconography that the rose was present to the same degree in private gardens (which, moreover, did not develop until the mid sixteenth century). How could it be otherwise? Leaving to one side the botanical roses of European origin that men of science cultivate for their studies or those roses which are still rare in Europe – *R. moschata* or *R. foetida*, for example – that humanists cultivate in an encyclopedic spirit, roses which can play a uniquely decorative role are few in number: some old Alba or rare remontant Damasks – not very many.

Great Novelties of the Turn of the Seventeenth Century

In the period 1580–1640, this situation would change noticeably as a result of the appearance or propagation of new varieties, whose decorative interest, and consequently horticultural interest, was considerable. To be convinced of this one needs only examine two important works, one from 1588 and the other from 1613. The first is the *Neuw Kräuterbuch* (Frankfurt, 1588) written by Jakob Theodor of Bergzabern (Tabernaemontanus), which features 11 engravings of roses. The list of roses included here is given in **Table 2**.

If the equivalents proposed by G. Krussmann are correct, this means that, from the end of the sixteenth century, the centifolia and even the Moss varieties were known at this time.

In fact, it has been quite clearly established that *R.* × *centifolia* dates from the end of the sixteenth century (and not from the seventeenth century, as too many works still state). Some years ago, Gisèle de La Roche had pieced together this story in a way which is as erudite as it is interesting. It has now been conceded that the 'rose with 100 leaves' is definitely not the one which Theophrastus and Pliny spoke about in their time, and as a result, it does not go back to antiquity as had long been repeated. Rather, sources seem to indicate that *R.* × *centifolia*, which counts *R.* × *damascena* in its genealogy, dates from the end of the sixteenth century and originates in The Netherlands. In fact, in 1581, a *R. damascena* var. *maxima* was already being cultivated at Anvers, and this is actually the 'hundred leaves' rose which Matthias de l'Obel cites in his *Kruydtboeck*. It is this same rose which is called '*Rosa provincialis* var. *major*' in the *Neuw Kräuterbuch* of Jakob Theodor of Bergzabern in 1588. And it is this same rose that John Gerard reproduced in his *Herball* of 1597 under the name of '*Rosa Hollandica sive Batava*', 'The Great Holland Rose or Great Province', specifying that the Dutch had obtained it from *R.* × *damascena*. So then we were in the presence of a new family of roses which would enjoy enormous success and fill pleasure gardens until the present day. The Dutch, at the same time as they were going on to become great producers of roses, would also become great painters: thus these roses featured in *Bouquet of Flowers in a Niche*, by Roeleandt Savery, in 1603, the oldest dated painting of flowers that has reached us from The Netherlands.

Table 2 Roses featured in the *Neuw Kräuterbuch* (Frankfurt, 1588) written by Jakob Theodor of Bergzabern (Tabernaemontanus)

Neuw Kräuterbuch	*Current designation according to Krussmann*
R. alba	*R. alba* var. *semiplena*
R. rubra	*R. gallica*
R. provincialis major	*R. centifolia*
R. provincialis minor	*R. centifolia*
R. sine spinis	*R. francofurtana*?
R. muscata alba	*R. moschata*?
R. muscata alba multiplex	*R. centifolia* var. *muscosa alba*
R. lutea	*R. foetida*
R. eglanteria	*R. pimpinellifolia*
R. sylvestris	*R. canina*
R. arvina	*R. arvensis*

Data from Krussmann G (1982) *Roses*. London: Batsford.

The other interesting comment on the *Neuw Kräuterbuch* of J.T. de Bergzabern is the '*Rosa centifolia muscosa* var. *alba*', at least according to Krussmann's interpretation. Others have also considered that the description of the 'Velvet Rose' of John Gerard (1597) may refer to a Moss rose. But the hypothesis that Moss roses have been in existence from the end of the sixteenth century has not yet found many followers. More serious sources focus instead on the beginning of the eighteenth century, and we will come back to this. Anyway, if some cases of mutations of Moss roses were already being observed in the sixteenth century, they remained very isolated.

The second work which we would like to take as an example is *Hortus Eystettensis*. Its history is known. In 1609 the Bishop–Prince of Eichstatt, J.K. von Gemmingen (1593 or 1595–1612), undertook to renovate his castle at Saint-Willebald, as well as the gardens surrounding it. Charles de l'Ecluse supplied some of the plants that were introduced to these gardens: in 1611 'in particular the beautiful roses' were remarked upon. At the same time as these works were going on, the Bishop–Prince commissioned an apothecary from Nuremberg, Basilius Besler (1561–1629), to publish a book that would reproduce some of the plants from these gardens: this book, *Hortus Eystettensis*, appeared in 1613 (**Table 3**).

The first question raised by this list concerns *R. gallica* var. *versicolor*, if our interpretation is correct. When did this mutation of *R. gallica* var. *officinalis* appear? The English love this rose, and certain English authorities have suggested the twelfth century – the era of 'Fayre Rosamonde', mistress of King Henry II – but this appears to be a complete fantasy. In contrast, others propose very late dates: in his *Garden Book* of 1659, Thomas Hanmer, who was the first to call it 'Rosa Mundi', indicates that it appeared several years earlier in Norfolk, while Thomas Fuller proposed 1661 at Norwich. In fact, it is possible that *R. gallica* var. *versicolor* comes to us from antiquity. According to Italian archaeologists, it can be seen on the (recently restored) frescos of the Casa del Bracciale d'Oro at Pompei: but on a fresco how could you distinguish between *R. gallica* var. *versicolor* and *R. damascena* var. *versicolor* ('York and Lancaster')? And in the period that interests us – the end of the sixteenth century – certain commentators have guessed that '*Rosa praenestina (vulgo provincialis) alba*', described by Charles de l'Ecluse

Table 3 Roses of *Hortus Eystettensis* (1613)

Hortus Eystettensis	*Probable identification*
Plate 94	
I *Rosa damascena flore pleno*	
Plate 95	
I *Rosa centifolia rubra*	*R.* × *centifolia*
II *Rosa praestina variegata*	*R. gallica* var. *versicolor*
III *Rosa provincialis flore incarnato pleno*	*R. gallica* var. *officinalis*
IV *Rosa lutea maxima flore pleno*	*R.* × *hemisphaerica*
Plate 96	
I *Rosa flore albo pleno*	*R.* × *alba*
II *Rosa flore albo simplici*	*R.* × *alba* var. *semiplena*
III *Rosa milesia flore rubro pleno*	*R. holosericea* (?)
IV *Rosa provincialis flore albo*	?
Plate 97	
I *Rosa ex rubro nigricans flore pleno*	*R.* × *francofurtana*
II *Rosa lacteola camerarii*	?
III *Rosa rubicunda saccharina dicta*	?
IV *Rosa damascena flore simplici*	?
Plate 98	
I *Rosa lutea flore simplici*	*R. foetida*
II *Rosa cinnamomea*	*R. majalis*
III *Rosa rubra praecox flore simplici*	*R. pendulina*
IV *Rosa praecox spinosa flore albo*	*R. spinosissima*
Plate 99	
I *Rosa sylvestris odorata incarnato flore*	*R. rubiginosa*
II *Rosa sylvestris flore rubro*	*R. canina*
III *Rosa milesia rubra flore simplici*	*R. gallica* (?)
IV *Rosa eglantaria*	*R. canina* (?)

Data from Aymonin G (1987) *L'Herbier des Quatre Saisons*. Paris: Mazenod; Aymonin G (2000) *Le Jardin d'Eichstätt*. Paris: Taschen.

in 1583, was none other than *R. gallica* var. *versicolor*. Gisèle de La Roche refutes this interpretation and even thinks that '*Rosa praenestina variegata*', depicted in *Hortus Eystettensis* (1613), is not *R. gallica* var. *versicolor* but rather *R. damascena* var. *versicolor*.

We do not have the means to resolve this controversy. Nevertheless, what is certain is that *R. gallica* var. *versicolor* precedes 1640 – which excludes Norfolk and Norwich – because in 1640, under the name of 'Versicolor', it is magnificiently reproduced by Nicolas Robert on a 'Vellum of the King'. This work is preserved at the National Museum of Natural History in Paris and was published in 1998. Now, above and beyond these questions of chronology, the important point is that, at this date, *R. gallica* var. *versicolor* was therefore greatly appreciated for its decorative value since it was the object of a precious painting which had a princely destination. The mutation from 'officinalis' to 'versicolor' illustrates admirably the transformation of a rose which was cultivated for centuries for its medicinal virtue to a rose which was henceforth admired for its decorative effect.

The other remarkable rose in the *Hortus Eytettensis* is '*Rosa lutea maxima flore pleno*' which, according to all the evidence, corresponds to our *R.* × *hemisphaerica* Herrm. It is a horticultural hybrid, originating from Turkey and Persia, which is a descendant of *R. rapinii* Boiss. et Bal. It is thought to have been introduced from Constantinople to the Viennese botanical garden by Charles de l'Ecluse – to whom the history of roses perhaps attributes slightly too much – at the end of the sixteenth century. In England, Gerard mentions it in his *Herball* of 1597, under the name of '*Rosa lutea multiplex*', indicating that 'it seldome fully shewes it self about London where it is kept in our chiefe gardens as a prime raritie'. At a time when the vast majority of double roses were white, pink or scarlet, the golden yellow colour of *R.* × *hemisphaerica* was evidently a great novelty. But it was also a great rarity (as it still is today) because it was very difficult to grow this flower in a cold and wet climate. It made such an impression on people of the time that seventeenth-century pictures of Dutch roses invariably include *R.* × *hemisphaerica*.

Incidentally, we should note that this rose was then called 'yellow Centifolia'. This parallel, which was introduced by popular language between *R.* × *hemisphaerica* and *R.* × *centifolia*, could lead to an interesting research hypothesis: would not the Centifolia, offspring of Damask, also be born in Asia Minor (and not The Netherlands), which also suggests their initial designation as *R. damascena* var. *maxima* by Mattias de L'Obel?

All things considered, these printed herbaria nevertheless give an impression of the species and varieties that were found in gardens in the mid seventeenth century. Botanical roses were numerous: they demonstrate the concern for collection which then motivated scholars, who looked on their gardens as a sort of living 'cabinet of curiosities'. The abundance of botanicals also resulted from the small number of horticultural varieties whose details were available.

In fact, leaving to one side rarities like *R.* × *hemisphaerica*, it was mostly a matter of gallica and white roses (Alba) and, increasingly, in the second half of the century, of Centifolias: in particular the 'great rose of Holland with 100 leaves' was hugely successful, perhaps because of the commercial dynamism of the Dutch horticulturalists.

The first available gardening works clearly convey the poor choice available to amateurs. For example, in 1669, Jan van der Groen's *Le Jardinier Hollandois* (*The Dutch Gardener*), published in Amsterdam in

French and Dutch, citing roses that could be found in The Netherlands, only listed about 20. Moreover their place in the work was much more modest than other flowers such as tulips or carnations. Often in the mid seventeenth century these gardening books did not even mention roses, with the sole exception of those reserved for gardens of apothecaries with some medicinal value, notably the 'rose of Provins'. In France you would have to wait for the first edition of *La Maison Rustique* (*The Rustic House*) in the 1680s to find even slightly more detailed practical advice on cultivation.

Introduction of Exotic Roses in the Eighteenth Century

In fact it was in the eighteenth century that the range of roses grown in Europe would suddenly increase. The development of maritime trade and European colonization as well as an increase in botanical journeys would, within a few years, find expression in the introduction into Europe of very many new species from North America and Eastern Asia.

However, one of the first novelties of the century has no connection with distant lands: more prosaically, it is the appearance of Moss roses. Their origin has been the object of a number of controversies: the paternity of this mutation of *R. centifolia* has been disputed by the French, and even more so by the Germans and the English. The Germans, for their part, proposed *Hortus Bosianus*, edited by Elias Peine in 1699. As described by Jäger this book mentions a '*Rosa centifolia, fructu muscoso*', which was found in the garden belonging to Caspar Bose, senior member of the Council of State in Leipzig. However we do not know when or where this mutation appeared.

But two authors have studied this question in a scholarly way and their conclusions concur. They are C.C. Hurst (1922) in England and G. de La Roche (1978) in Belgium. According to them, although they do not definitively reject the German theory, the first reliable mention was in 1720. At this time Dr Boerhaave, in a second edition of the list of plants grown in the botanical garden of Leyden, mentions a '*Rosa rubra plena, spinosissima pedunculo muscoso*', which is in fact *R.* × *centifolia* var. *muscosa*. Philip Miller, in his *Gardner's Dictionary* of 1760, moreover confirms that the first Moss rose he saw was that of Boerhaave, at Leyden, in 1727, and which may already have crossed to England in 1724, or at least did so through his care, in that same year, 1727. But even then, we do not know whether the mutation was in fact produced at Leyden or if it was an older rose which had been introduced to the town's botanical collection. (Incidentally, note that Miller, despite being very classical, is unreliable. For example, he writes that 'the *yellow rose* as well as the *Austrian rose* both originate in America: the French brought it from Canada' (!). Moreover, he classifies the Moss rose as a 'Moss Rose of Provins'.)

Whatever the case, this rose remained rare for several years: it is not until the second half of the eighteenth century that it can regularly be seen in nursery catalogues. And it is not until the end of the nineteenth century that the group of Moss roses increases in importance, mainly in France.

That being so, the real novelties of the century are to be found elsewhere, with the species from North America and above all China which were introduced to Europe at that time.

In fact, the first species were few in number and their introduction never had important horticultural consequences. *R. carolina* L. var. *plena*, the 'rosebush of Caroline with double flowers', was introduced before 1722; after 1726 came *R. palustris*, the 'Hudson rosebush'; and finally, half a century later, *R. blanda* Ait., the 'Labrador rose', was introduced. But all these roses held more interest for botanists than gardeners: none had any significant horticultural descendants. Nevertheless, they could be found in nursery catalogues: at the close of the eighteenth century, North America was both topical and fashionable.

The real innovation, in the eighteenth century, was therefore the introduction of roses from China (*see* **History of Roses in Cultivation**: Ancient Chinese Roses). This brought new nuances and forms, as well as cultivars more bold than the old Damasks. We are aware of the many novelties that emerge from the cultivation of roses in the nineteenth century.

The introduction of roses from China into Europe features among the most confused subjects of the history of roses. If the sources also seem contradictory, it is probably for two simple reasons (not to mention the fact that *Hibiscus sinensis* was at that time called 'China rose'!). On the one hand, Chinese horticulture, which is very ancient, had produced a considerable number of varieties, certain of which were probably very close to one another and which, in the eighteenth century and later, were quite possibly confused by Europeans. On the other hand, a large number of European countries were trading with China during this period and so it was possible that there were multiple introductions of the same variety, which would give rise to the apparent contradictions. So, for a rose which is as classic as *R. chinensis* var. *semperflorens*, the English themselves, who claim to have introduced it in 1792, think that perhaps it was cultivated in Italy from the sixteenth century, as seems to be proved by a painting by the Florentine Angelo Bronzino, which dates from 1529. (Note that in the past, 'China rose' was also called 'Siena rose': was this referring to the town or the colour – although Siena is brown?)

In any case, if we refer to the most current historical accounts, the first rose introduced from China would be a certain *Rosa chinensis*, so named in 1768 by Nikolaus Joseph von Jacquin, director of the gardens of Schönbrunn, according to a specimen conserved since 1733, under the name of '*Chineeshe Eglantier Roosen*' in the herbarium of the Dutch botanist Gronovius. According to this version, the first rose from China would not then have been introduced into Europe (by the Dutch?) until the beginning of the eighteenth century.

Then it was not until the end of this same century that there was the introduction of the rose that the Chinese call '*yue yue hong*', which is to say 'red (*hong*) of all the months (*yue yue*)' and which we in Europe have christened *R. chinensis* var. *semperflorens* (Koehne). The story of its introduction is well known. In 1789, this rose was noticed by an agent of the English Indies Company at Calcutta; he brought it back to England in 1792 and gave it to one of the directors of the company, Gilbert Slater, and from here it gets its English name of 'Slater's Crimson China'. The great novelty of this rose, in addition to its obvious remontance, was its colour, which is openly red, while our traditional European roses were purple and not red. (The first description was by William Curtis in the *Botanical Magazine* of December 1794.)

But the story of these introductions was probably much more complicated. Let us take the example of an obscure French nurseryman from Rouen, called Mustel. In 1772, he published a catalogue featuring three Chinese roses: the 'rosebush of China', the 'rosebush of China which is evergreen' and the 'rampant rosebush of China'. The 'rosebush of China' was probably *R. chinensis*. But what about the 'rosebush of China which is evergreen'? This could not be *R. chinensis* var. *semperflorens* since, at least according to English sources, it was not introduced until 20 years later. (Anyway, 'evergreen' would be the translation of *sempervirens*, not *semperflorens*.)

Another introduction from China in the eighteenth century was the ancient horticultural variety that the Chinese call '*yue yue fen*' and that the English rechristened 'Old Blush' or 'Parsons' Pink China'. As ever, the circumstances of its introduction are obscure. Perhaps it arrived in The Netherlands in the mid eighteenth century, or at least, from the Fa Ti nurseries, near Canton, to the botanic garden of Harlem in 1781. It would be from there that Joseph Banks, Director of Kew Gardens, would have introduced it into England in 1789. It has also been written that perhaps it would have been sent from Canton to England by George Staunton who accompanied the Englishman Macartney's mission to China. Its name derives from being grown in the garden of a certain Parsons, at Rickmansworth, in 1793, and it is from this date that the English nurseries of Colville began to multiply it. (This is according to the unlikely hypothesis proposed in an introduction by Staunton. Since the mission left England in 1792, it is implausible that this rose could have been cultivated and multiplied there from 1793.) It is a very fertile variety, easily reproduced by propagation, which in the nineteenth century played a prominent role in obtaining new roses, notably in France.

Among these Chinese introductions at the very end of the eighteenth century, we must also mention, around 1793–4, *R. bracteata* Wendl., the 'Macartney rose', introduced at the time of Macartney's mission to China. But this rose has few horticultural descendants.

A New Way of Seeing Roses

The enumeration of all these roses may give the impression that rose cultivation in the eighteenth century was radically modified by these different introductions. This would be a big mistake. In fact, these novelties remained quite rare and were mainly of interest to botanists for scientific reasons and to the aristocracy for reasons of prestige. On the other hand, it is true that, in comparison to the sixteenth and seventeenth centuries, the view of the rose was very different.

An examination of eighteenth-century gardening works clearly shows that the roses that were most grown at that time remained those of previous centuries and that, when all is said and done, the new introductions had a small place in ordinary gardens.

Let us take the French example of the classic *Maison Rustique*, 1762 edition, which may give an impression of the typical rosebushes found in gardens. Here is a list of those mentioned:

- Perfumed rose
- Unperfumed rose
- Pale-red rose
- Rose of Provins
- Siena or China rose
- Virginia rose
- Rose of all the months
- Musk or Damask rose
- Rose of Holland with 100 leaves
- Milk-white rose
- Flesh-coloured rose
- Variegated rose
- Simple dark-red rose
- Simple and double yellow rose

In addition to the classical European varieties, there are rarer roses: the Musk or Damask roses (*R. moschata*) or 'simple and double yellow rose' (*R. lutea* and *R. hemisphaerica*). As for the novelties of the eighteenth century, they were represented by the China rose (*R. chinensis*) – at least if we reject the hypothesis of its ancient introduction in Italy – and by the 'Virginia rose' (perhaps *R. virginiana* Mill. var. *plena*), which would perhaps not be very widespread in gardens.

It may be of interest here to report some advice on cultivation given by this same *Maison Rustique*:

> Rosebushes appear without a care everywhere, in the borders, along the walls, in the groves, at the feet of cradles, on fences and in pots; all we have to do is to prevent their roots from trailing, occasionally dig them over slightly, and rid them of too much wood or anything dead. We multiply them in October from rooted plants or runners, and from cuttings in spring. We can clip them and train them along a fence.
>
> The Siena rosebush is planted from cuttings in March.
>
> The rosebush with double yellow flowers requires very strong ground; it hates being pruned; you only have to remove the dead wood.
>
> Rosebushes flourish in light, sandy soil. Cultivation is general, with the exception of the 'rosebush of all the months' which flowers seven to eight times a year, if you take a small amount of care, without which it will only yield once like the others. Rosebushes are pruned back to the ground in September, to give early shoots in spring: they are pruned again at the end of March, approaching the new shoots right to the bud close to the stalk. They are pruned again after each sprouting, cutting the branches below the nodes where the flowers were after they have fallen: this is how they are forced to be in continual bloom; and to promote this frequent flowering, a foot of compost is spread at the base of the tree, so that the salts fall on to the roots.
>
> In the same way, in order to bring to the most common rosebushes a second flower at the end of summer, you only have to prune them under the first flower as soon as it has finished.
>
> It is also our experience that if you prune a rosebush very close to the ground a few days before Christmas, it will give flowers at Christmas the following year.
>
> Normally we put Dutch rosebushes at the feet of trees, so that they grow up, which gives a pretty effect: we prune these rosebushes in the spring, to one foot or one and a half foot, so that they bring forth in autumn.
>
> You can graft Dutch rosebushes, variegated, or rosebushes of all the months, on to common roses, as a shield graft or as an approach: this is quicker than raising one from seedlings.

In spite of the eminently traditional character of this advice on cultivation, the rose at the end of the eighteenth century was no longer perceived as it had been in previous centuries. The return to nature led to greater interest in 'wild' roses, notably those from distant countries. 'English' gardens multiplied: roses were much more at home there than in 'French' gardens, which were too regular and ordered to adapt to the natural bearing, and occasially the disordered bearing, of the great rosebushes. If *La Maison Rustique* could only report 15 species and varieties of rosebushes for gardens, Miller's *The Gardener's Dictionary*, published in England from 1731, would enumerate 46 and was of the opinion that the rose was 'the main ornament of a beautiful garden'. Infatuation with the rose was evident. The place of the rose in all the decorative arts from the end of the eighteenth century clearly expressed the growing place that it occupied in parks and gardens, at least those of the aristocracy.

This fashion was also a result of the extraordinary development of European freemasonry. In fact, for masons, the rose was a symbol of a better future for humanity, so it played an important role in their rituals. Because of this, in most parks adapted 'in an English style' in the eighteenth century by freemason owners, roses were increasingly apparent, which had not at all been the case before, in 'French' gardens.

In France, the list would be long, from the Parc Monceau (Paris), belonging to the duke of Chartres, Grand Master of the Great East, to that of Ermenonville, belonging to the marquis of Girardin, and so many others. Even nurserymen were sometimes linked to freemasonry; the most significant case was that of Descemet (1761–1839), a member of the Société Nationale des Neuf Soeurs (National Society of Nine Sisters), who, between 1804 and 1814, would become the first French breeder of roses.

The same applies to most European countries. The fine book recently published in Germany about the roses of Wilhelmshöhe, near Kassel, provides another example which is just as significant. Roses were introduced into this park from 1767, in the reign of the Landgrave Friedrich II (1760–85), thanks to its chief gardener Schwarzkopf (1738–1817). Now the creation of this rose garden was probably linked to the widespread existence of freemasonry in the region. This is what W. Grimm wrote about it:

> It is unlikely that the rosarium was created only for reasons of botanical interest, if we recall that at this time, Kassel was one of the centres of freemasonry... Some princes in Hesse were freemasons, and so were even some court employees ... The Weissenstein (i.e. Wilhelmshöhe) rose garden may have been a secret meeting place.

The case of Wilhelmshöhe is interesting from another point of view, as can be seen clearly here – it is one of the first examples of the role of the chief gardeners of the great princely estates in obtaining new varieties of roses. In addition to the botanical rose, the horticultural rose – the rose for pleasure – was beginning to develop. Just as the prince demands from his chapel master that he should write music for him, or of his chief cook that he should create new dishes, the prince demanded of his head gardener that he should obtain for him original varieties that cannot be found in other aristocratic parks. At Kassel, this gave rise to 'Hesse Centifolia Rose', 'scarlet Weissenstein', 'Little Hesse Rose' and many others: we are still growing the 'Perle de Weissenstein'.

The extraordinary development of the rose in the nineteenth century will merely be the continuation of this fashion – which at the beginning was very aristocratic – begun during the second half of the eighteenth century. The rose grown for medicinal purposes, such as the 'apothecary rose', began to lose importance; the same was true, under the influence of the 'philosophy of lights', for the rose of Christian symbols. On the other hand, the 'demedicalized' and 'dechristianized' rose, that is to say, the rose for beauty and pleasure, invaded gardens. The French Revolution of 1789 delayed this evolution for a short time, but once the storm had passed, this fashion of the end of the eighteenth century would very quickly be reborn: the nineteenth century will be truly that of the triumph of the rose.

See also: **Classification**: Conventional Taxonomy (Wild Roses). **Distribution and Ecology**: Europe. **History of Roses in Cultivation**: Ancient Chinese Roses; Modern (Post-1800). **Rose Collections and Trials**: Rose Collections and Herbaria.

Further Reading

Aymonin G (1987) *L'Herbier des Quatre Saisons*. Paris: Mazenod.
Aymonin G (1991) Les plantes du jardin épiscopal d'Eichstätt, vues par Basilius Besler (1613). *Bulletin de la Société Botanique* 1: 5–14.
Aymonin G (2000) *Le Jardin d'Eichstätt*. Paris: Taschen.
De La Roche G (1978) Commentaires: Etude sur les roses citées dans l'ouvrage. In: *Commentaires sur les Roses de P. J. Redouté: Apport à l'histoire du Genre Rosa*, vol. 4 *Redouté Les Roses*. Anvers: De Schutter.
Eeckhout P (1996) Petite histoire des tableaux de fleurs du XVIe au XVIIe siècle. In: *L'Empire de Flore*, pp. 269–270. Brussels: La Renaissance du Livre.
Hurst CC (1922) Notes on the history of the Moss Rose. *Rose Annual* 39–42.
Jäger A (1960) *Rosenlexicon*. Leipzig: Zentralantiquariat der Deutschen Demokratischen Republik.
Joyaux F (1998) *La Rose de France*. Paris: Imprimerie Nationale.
Joyaux F (2000–1) Jacques-Louis Descemet, premier grand obtenteur français. *Rosa Gallica* 6–8.
Joyaux F (2001) *La Rose, une Passion Française (1778–1914)*. Brussels: Complexe.
Krussman G (1982) *Roses*. London: Batsford.
Miller P (1785) *Dictionnaire des Jardiniers*. Paris: Guillot.
Opoix (2001) Sur les roses rouges de Provins. *Rosa Gallica* 10: 30–36.
Paterson A (1983) *The History of the Rose*. London: Collins.
Proietti C (2001) Contributo allo studio delle rose presenti en Italia nel XVI secolo. In: *Annuario della Rosa 2001*, pp. 35–61. Monza: Associazione Italiana delle Rosa.
Schnell and Steiner (eds.) (2001) *Rosen-Sammlung zu Wilhelmshöhe*, Regenburg.
Swan C (1998) *The Clutius Botanical Watercolors*. New York: Harry N. Abrams.
Wuyts L (1996) Des fleurs pour la foi, l'amour et la mort. In: *L'Empire de Flore*. Brussels: La Renaissance du Livre.

Modern (Post-1800)

M Marriott, David Austin Roses Ltd, Wolverhampton, UK

Introduction

The last two centuries have seen the most intense breeding programme that has ever been lavished on any horticultural plant. Since 1800, tens of thousands of varieties have been introduced and this represents only a minute fraction of the vast numbers of seedlings that were rejected by breeders as part of the selection process.

The period saw the development of the garden rose from a relatively homogeneous group to one varying hugely in size of flower and plant, in colour, in fragrance and in its ability to grow in a wide range of habitats.

During this time five large industries centred around roses have existed: (1) roses for the garden; (2) for cut flowers; (3) as house plants (pot roses); (4) for producing rose oil; and (5) for medicinal use. Varieties of the last two categories predate 1800 and, therefore, do not come under the remit of this article. Each industry is very different and each, often, has its own specific range of varieties, breeders and growers.

Garden Roses

The Old Roses

Once-flowering old roses The beginning of the nineteenth century saw the rose as one of the most popular plants in the garden. They were then almost wholly restricted to the Gallica, Damask, Alba, Centifolia and Moss roses, the origins and varieties of which are described elsewhere (*see* **History of Roses in Cultivation:** European (Pre-1800)). True red, yellow (with the exception of the *Rosa hemispherica* Herrm. and *R. foetida* Herrm.) and apricot-coloured flowers were unknown in gardens and the flowers variants included single, semidouble and the classic double old rose shape, often with over 100 petals.

Most of the once-flowering old roses that we see in nurserymen's catalogues today come from the nineteenth century and include many choice varieties (**Table 1**).

Huge numbers of varieties have been lost from this period but it is likely that the majority of varieties that are in cultivation today represent the toughest and the most beautiful. The Gallica, Damask, Alba, Centifolia and Moss roses were all known in gardens prior to 1800 but we see their greatest development after this year and then a rapid decline around 1875. At the time of the French Empress Josephine, at the beginning of the nineteenth century, the main breeders were Du Pont and Descemet, the latter probably being the first to sow rose seeds in large quantities.

Towards the middle of the nineteenth century the main breeders were Vibert, Laffay, Pernet Père, Parmentier and Mme Hebért (one of the few female rose hybridists). In France breeding was made particularly easy as the hips of the old roses, and the Gallica roses particularly, ripened out-of-doors with no difficulty. In terms of seedlings produced and varieties introduced, Vibert is one of the major breeders of all time and certainly was one of the most organized at that time. In his catalogue of 1820, he listed *c.* 700 varieties, with 264 Gallicas, of which 44 were of his own breeding. Between 1815 and 1851 he created an amazing 600 varieties, the majority being Gallica roses or varieties closely associated with them, although he also took great interest in other groups of roses – China, Tea and Noisette. However it is interesting to note few, if any, of his Gallica roses are in the front rank of varieties available today.

The Alba roses are, strangely, of rather mysterious origin, their breeders and dates of origin often being unknown although they are thought to be of *R. canina* origin.

The Damask roses reached their peak in the 1850s but then soon lost their popularity. The Moss roses were extremely popular in Victorian times; however, the great majority were most unsatisfactory plants and have subsequently been lost to cultivation. Indeed, many repeat-flowering mosses were introduced but are all but lost today.

Repeat-flowering old roses The reason for the rapid decline in the popularity of the once-flowering old roses was, of course, the introduction from China of roses with repeat flowering. With the exception of 'Quatre Saisons' (see also 'Autumn Damask'), repeat-flowering roses were unheard of.

The first introduction to Europe of the China roses was in 1751 in the form of 'Old Blush China' by the Swedish botanist Peter Osbeck and then a little later four more varieties – 'Slater's Crimson China' (1792), 'Parson's Pink China' (1793), 'Hume's Blush Tea-Scented China' (1809) and 'Parks' Yellow Tea-Scented China' (1824). They are all of unknown origin except for 'Parks' Yellow Tea-Scented China', which is a cross between *R. gigantea* Coll., a rose with very large waxy petals, and a China rose. At first, little use was made of them and the seedlings would have remained once-flowering since the genes controlling repeat flowering are recessive (*see* **Growth**

Table 1 A selection of seasonal flowering roses of the nineteenth century

Horticultural class	*Cultivar*	*Breeder*	*Date*
Gallica	'Tuscany Superb'	Rivers	Prior to 1837
	'Empress Josephine'	Hébert	Prior to 1854
	'Président de Sèze'	Hébert	1828
Damask	'La Ville de Bruxelles'	Vibert	1849
	'Mme Hardy'	Hardy	1832
	'Marie Louise'	Malmaison	1813
Alba	'Queen of Denmark'	—	1826
	'Mme Legras de St Germain'	—	Prior to 1842
	'Félicité Parmentier'	—	1831
Centifolia	'Chapeau de Napoléon'	—	1845
	'Cottage Maid' (see also 'Village Maid')	Vibert	1845
	'Ipsilanté'	—	1821
Moss	'Comtesse de Murinais'	Vibert	1843
	'Maréchal Davoust'	Robert	1853
	'William Lobb'	Laffay	1855

Regulation: Floral Induction) and breeders relied, at that time, on open pollination. These four roses were responsible for laying the foundations of our modern roses as well as further development of more China roses.

One of the first and most significant crosses between Chinese and European roses occurred purely by chance on the Ile de Bourbon in the Indian Ocean. Both 'Old Blush China' and Autumn Damask roses were planted here and, in 1817, a repeat-flowering old rose came to light and was brought back to Europe. This was the rose that subsequently gave rise to the Bourbon roses. These are still essentially old roses but do show some novel characteristics, for example, flower shape, colour (although no yellow) and slender growth. However some negative characteristics also became evident – increased susceptibility to black spot disease and balling of flowers, as in 'La Reine Victoria' (Schwartz 1872). More successful varieties are 'Louise Odier' (Margottin 1851) and 'Mme Isaac Perèire' (Bourbon Garcon 1881).

The Portland roses are of uncertain provenance but are likely to be also a Damask × China cross. They are truly old rose in character, are generally healthy and fragrant and have gained repeat flowering. It is unfortunate that they were not developed further. Prominent among them are 'Comte de Chambord' (Moreau-Robert 1860), 'Jacques Cartier' (Moreau-Robert 1868) and 'Rose de Rescht'. Although well over 100 were introduced, few are available today.

The other group of repeat-flowering old roses that were being developed in the nineteenth century are the hybrid perpetual roses. They were the result of various crosses between Portland and Bourbon roses with hybrid China roses. Most of the varieties extant today are very large, to the extent of being coarse. However, it may have been the fault of this group that roses started to gain a reputation for being delicate plants needing a great deal of attention, as many were very weak and rather unsatisfactory plants. Most of them have been lost to cultivation or are only grown successfully in warm climates. Most are extremely fragrant. The most successful varieties are 'Ferdinand Pichard' (Tanne 1921), 'Reine des Violettes' (Millet-Malet 1860), and 'Gloire de Ducher' (Ducher 1865) although, even in these, disease resistance, repeat-flowering and elegance of growth leave something to be desired.

While all these repeat-flowering, but essentially old roses were being developed, a radically different group were being bred – the tea roses. In them we see the first glimpse of the hybrid tea with their long petals, and new range of colours that included true reds and soft yellows. The tea roses are the result of crossing two of the original China roses – 'Hume's Blush Tea-Scented China' and 'Parks' Yellow Tea-Scented China' with various Bourbon and Noisette roses, the first being 'Adam' (Adam 1835). It also saw the beginning of growing roses in greenhouses for cut flowers – 'Catherine Mermet' (Guillot et fils 1869), with its more upward-pointing flowers, was the main variety of that time. In growth habit the teas are rather weak and sparse, especially in colder climates, and their flowers tend to hang down. However, in their time they were extremely popular as they were seen as a complete novelty. In warmer climates they are much more successful garden plants and some have quite extraordinary colours: 'Francis Dubreuil' (Dubreuil 1894) is a deep red, 'Lady Hillingdon' (Lowe and Sawyer 1910) is a rich apricot and 'Maman Cochet' (Cochet 1893) is an amazing mixture of pink, strawberry, lavender and soft yellow.

Modern Bush Roses

Hybrid teas The rose world was now poised for the development of the hybrid tea, the result of crosses between hybrid perpetual and tea roses. The first generally recognized is 'La France' (Guillot 1867), but in fact the first reliably documented hybrid tea was 'Victor Verdier', a cross between the hybrid perpetual 'Jules Margottin' (Margottin 1853) and the tea rose 'Safrano' (Beauregard 1839), and was bred by Lacharme of Lyons in 1859. There is some evidence that there were varieties that could have been called hybrid tea in the 1830s. Initially they were called *R.* × *odorata indica*; however, the arrival of 'La France' helped to confirm that a new class was in the making, although they were not officially recognized until some years later.

The development of hybrid tea roses and, indeed, roses generally, was greatly aided by Henry Bennet in England and Jean Sisley in France, who both made crosses between specific varieties with a definite aim in mind. Bennet also set up heated greenhouses so that he could control the environment and the hips would have a greater chance of ripening.

The early hybrid tea roses were not so different from the tea roses, although their high pointed centres were more pronounced. They had rather stronger, more reliable growth and the flowers were borne on stiffer pedicels so that they looked up rather than down, as in the teas. It was the perfection of the bud-shaped flowers that breeders and gardeners were obsessively seeking, often to the detriment of the rest of the plant in terms of disease resistance, plant shape and fragrance.

The hybrid tea was largely responsible for the huge popularity of the rose to the extent that in the 1950s and 1960s many people planted them almost exclusively in their gardens. In the 1960s it is estimated that 50 million garden roses were being sold annually in the United Kingdom alone and the likes of Harry Wheatcroft and Sam McGredy became household names. Unfortunately, as inevitably happens with any product that is extremely popular, a period of decline follows and certainly the 1980s and 1990s have seen a sharp fall in annual sales. Breeders, however, are still introducing many hybrid tea roses each year and roses generally are still bought in relatively large numbers – *c.* 15 million in 2000 in the UK – although the proportion of hybrid tea roses is much smaller than 40 years ago.

During its development the hybrid tea has seen many major breakthroughs. During the nineteenth century there were no varieties with flowers of a strong yellow. Therefore Joseph Pernet-Ducher worked with *R. foetida* in the hope of introducing this colour into the colour range. His first success was 'Soleil d'Or' in 1900 and subsequent varieties were known as Pernettia roses, although they gradually became absorbed into the hybrid tea group. They suffered badly from black spot.

Lack of vigour and susceptibility to disease were major issues with most of the hybrid teas in the first half of the twentieth century and, had it not been for the introduction of 'Peace' by Meilland in 1945, their further popularity and development may have been very different. Not only was 'Peace' much stronger and healthier than any other variety, it was also very fertile and soon became a pivotal parent in the history of the modern hybrid tea.

'Super Star' ('Tropicana') (Tantau 1960) was the first hybrid tea to contain pelargonidin (a bright vermilion pigment) in its flowers.

'Silver Jubilee'® (Cocker 1978) is outstanding for its density of foliage, quality of bloom and productivity. It is also exceptional in being excellent for both garden and exhibition use. Other hybrid tea roses which mark significant advances or are important in the history of their development include 'Lady Mary Fitzwilliam' (Bennet 1882), 'Mme Caroline Testout' (Pernet-Ducher 1890), 'Ophelia' (Paul 1912), 'Mme Edouard Herriot' (Pernet-Ducher 1913), 'Shot Silk' (Dickson 1924), 'Crimson Glory' (Kordes 1935), 'Elina'® (Dickson 1984) and 'Warm Wishes' (Fryer 1994). Mention should also be made of the hybrid tea roses with single, i.e. five-petalled blooms: Dickson introduced a number of varieties in the 'Irish' series, notably 'Irish Fireflame' (1912), and Krebs introduced 'White Wings' (1947).

Floribundas The hybrid tea roses led to the Floribunda and the Grandiflora roses. The name for the former (they were initially called hybrid polyantha) was first coined in 1930 by Dr J.H. Nicolas for crosses between hybrid tea and polyantha roses and so, in effect, the first true Floribunda was 'Grüss an Aachen' (1908 Gedulberg). It has flowers very much in the style of the old roses but it is obvious that the vegetative growth has China rose blood in it. The aim of Floribunda breeders was maximum flower production in a range of bright, unfading colours on a tough bushy plant. Poulsen of Denmark was instrumental in their early development. P.T. Poulsen introduced 'Red Riding Hood' in 1912 and Svend Poulsen introduced 'Kirsten Poulsen' and 'Else Poulsen' in 1924. These last two were great successes and sold in great numbers. Many of the early varieties were single or semidouble; however, the flowers have gradually taken on more of the character of the hybrid tea.

One cannot talk about the Floribunda roses without mentioning 'Iceberg' (Kordes 1958), certainly one of the most popular roses of all time and one that grows well in most rose-growing areas of the world. Using the strict definition of Floribunda, it should not be part of this group as it is a cross between the hybrid musk 'Robin Hood' (Pemberton 1927) and the hybrid tea 'Virgo' (Meilland-Richardier 1947), although the latter does have a polyantha as one of its parents. This is probably an early example of a variety being placed in a group for marketing purposes as opposed to its true parental origin.

Grandifloras The Grandiflora roses are, in appearance, a halfway house between hybrid tea and Floribunda roses and that is exactly how they are bred. 'Queen Elizabeth'®, the most famous, is a cross between the Hybrid Tea 'Charlotte Armstrong' (Lammerts 1940) and the Floribunda 'Floradora' (Tantau 1944) and was bred by Lammerts in 1954. Many of the modern hybrid tea roses introduced today could more accurately be called Grandiflora, with more than one flower per stem and rather smaller blooms.

Miniatures The smallest group of roses is the miniature, that arose from a miniature form of a China rose first introduced into Europe around 1800. It was used by breeders to develop a number of varieties that were particularly useful as pot plants for the house. Interest waned but in 1918 some plants were discovered in Switzerland and named 'Rouletii' (after the finder). This was used initially by Van Vink and Dot and so the miniatures were born. Undoubtedly the greatest influence on them has been by Ralph Moore of California introducing such varieties as 'Little Flirt' (1961) and 'Stars 'n Stripes' (1980).

Minifloras The miniature roses are very distinct in size from the Floribunda roses and, as a significant number of varieties were developed intermediate in size between the two, the Miniflora roses were identified as a group by the World Federation of Rose Societies, although the name has yet to be commonly accepted. They come from a mixed background and, as might be expected, vary greatly in shape and size of flower; however they are generally characterized by being very free-flowering and brightly coloured.

Rambling and Climbing Roses

The period of the late nineteenth century and early twentieth century saw the development of some interesting shrub and rambling roses that were, and are still, largely overshadowed by the better-known bush roses. However they contain some beautiful and worthwhile varieties. Most originate from crosses between various species roses and assorted hybrid roses such as hybrid tea, tea and hybrid perpetual roses.

Ramblers The ramblers are mostly hybrids from *R. arvensis* Huds. (Ayrshire roses), *R. wichurana* Crép., *R. multiflora* Thunb., *R. pendulina* L. Bours. and *R. sempervirens* L. with the odd few varieties coming from *R. soulieana* Crép., *R. filipes* Rehd. & Wils., *R. helenae* Rehd. & Wils., *R. laevigata* Mich., *R. canina* L. and *R. sinowilsonii* Hemsl. They are often beautiful, extremely floriferous and fragrant, although very few repeat-flower. It was from a sport of a *R. wichurana* hybrid – 'Dr W. Van Fleet' (Van Fleet 1910) – that one of the best-known repeat-flowering climbers arose – 'New Dawn' (Dreer 1930). This was the first variety to be protected by plant breeders' rights. Their further development has been hampered by being triploid (and hence sterile), by the requirement for smaller roses for the smaller gardens and a wish that all roses be repeat-flowering. Recent breeding by Austin, Hetzel and Warner has seen the development of excellent, smaller, repeat-flowering ramblers.

Climbers A number of climbers are the result of naturally occurring sports of hybrid tea and Floribunda roses (in one case only – 'Iceberg'). However, the Noisette climbers are thought to be the result of a species cross, *R. moschata* with a China, probably 'Parson's Pink China', during the nineteenth century. The first was 'Champneys' Pink Cluster' (Champneys 1811), soon followed by the free-flowering 'Blush Noisette' (Noisette before 1814). The latter was crossed with 'Parks' Yellow Tea-Scented China' to give the range of yellow Noisette roses that include 'Desprez à Fleur Jaunes' (Desprez 1826). They all prefer warmer climates where they flower most freely and produce flowers rather in the character of old roses. The Noisette roses were crossed with the tea roses and were important in the development of the hybrid tea class.

The Kordesii climbers were developed by Wilhelm Kordes from an open pollinated and normally infertile 'Max Graf' in the 1930s. One of the seedlings proved particularly hardy and was used in the breeding programme for climbers suitable for the cold German winter. The first two introductions were 'Hamburger Phoenix' and 'Leverkusen' in 1954 followed by 'Parkdirektor Riggers'® (1957), 'Dortmund'® (1955) and others. The Kordesii genes were also introduced into the hybrid tea roses: 'Silver Jubilee' (Cocker 1978) is an F_3 seedling from 'Parkdirektor Riggers'.

Shrub Roses

Many shrub roses have resulted from crosses being made with various species. For example, the Rugosa roses from *R. rugosa* Thunb. were developed by a number of different breeders around the world. With its complete disease resistance and natural repeat flowering, *R. rugosa* has attracted attention from breeders for over 100 years, with recent renewed interest. Most are fragrant.

The hybrid musk roses from *R. moschata* Herrm. were bred initially by the Rev. Pemberton and further developed by Kordes, Bentall and, more recently, Lens. They make large shrubs in cooler climates and climbers in warmer areas. Most are fragrant.

The polyantha roses originate from a dwarf *R. multiflora* Thunb. and are extremely tough and free-flowering but scentless. They were first bred by Sisley in the early 1870s; the first variety was introduced by Guillot in 1875. They were instrumental in the development of floribundas.

The Fruhling series was developed by Kordes from the 1930s to the 1950s using *R. pimpinellifolia* L. They are once-flowering shrubs with single or semi-double flowers. *Rosa pimpinellifolia* also gave rise to 'Golden Wings' (Shepherd 1956) and 'Jacqueline du Pré' (Harkness 1989).

Lord Penzance crossed *R. eglantaria* L. with various hybrid perpetual, Bourbon and species roses to produce the Penzance hybrids, some retaining the scented foliage.

Pedro Dot of Spain crossed *R. moyesii* Hemsl. & Wils. with a large hybrid tea called 'La Giralda' to produce 'Nevada' (1927) that then sported to 'Marguerite Hilling' (1959). Also from *R. moyesii* is 'Eddie's Jewel' (Eddie 1962).

'Cerise Bouquet' is the result of a cross between *R. multibracteata* Hemsl. & Wils. and 'Crimson Glory'. Tantau was responsible for the breeding, although it was introduced by Kordes in 1958.

Hulthemia persica Mich., a close relative of roses, was used by Harkness in his breeding programme, introducing 'Tigris' (1985) and 'Euphrates' (1986). Both, unfortunately, are very susceptible to disease. More recently, Warner, Ilsink and Moore have been working with them and promise worthwhile introductions in the near future.

Ground-cover roses The ground-cover roses are defined by being of rather spreading habit; the main breeders are Kordes and Poulsen. However, mention should also be made of the flower carpet series bred by Noack. Their parentage generally stems from *R. wichurana*.

Recent years have seen a renewed interest in using species specifically as a way of introducing greater disease resistance. Scrivens used *R. davidii elongata* Rehd. & Wils. in the development of 'Pretty Lady' (1996) and 'Baby Love'® (1995). Horner used *R. bella* Reh. & Wils. for his variety 'Golden Future' (1999).

English Roses The breeding of roses in many ways, came full circle in the latter part of the twentieth century with David Austin's English Roses. His aim was and is to 'combine the charm and fragrance of an old rose with the wide colour range and repeat-flowering of the modern roses'. The first 'Constance Spry'® (1961) was a cross between the Gallica 'Belle Isis' (Parmentier 1845) and the Floribunda 'Dainty Maid' (Le Grice 1938) but did not repeat-flower. The first repeat-flowering varieties came in 1969 and included 'Wife of Bath'. Major breakthroughs in their history were 'Graham Thomas'® (1983), 'Mary Rose'® (1983) and 'Heritage'® (1984), all being significant improvements on the previously introduced English Roses. The year 2001 saw the introduction of 'The Mayflower', an extremely healthy variety which exemplified the move towards a much greater level of disease resistance in the English Roses.

Although initially they were crosses between old roses and modern bush roses, their provenance in more recent years has been from a wide variety of sources. The chief unifying distinctions of the English Roses are an attractive and more or less informal growth habit, a flower more in the style of the old roses (itself very variable) and fragrance. Many make excellent short climbers, especially in warm climates.

Following from David Austin's success, nearly all professional breeders around the world have steered part of their breeding programme towards roses of a similar appearance, for example Guillot with the Generosas and Meilland with the Romanticas.

Future Developments in Garden Roses

The great diversity to be found in roses is a reflection of how easy they are to breed. They are extremely interfertile, their flowering parts are relatively large and the seedlings, if repeat-flowering, will bloom within a few months of germinating. The only major difficulty is ploidy level as most garden varieties are tetraploid and most species and their hybrids are diploid. However a few successful and highly significant crosses have been made with triploids and it is possible, although difficult, to double ploidy level artificially (*see* **Biotechnologies for Breeding:** Manipulation of Ploidy Level). Genetic manipulation remains, at the present time, an extremely difficult option for roses.

The future is likely to see varieties that are far more disease-resistant as gardeners are far less willing and able (because of the restricted availability of insecticides and fungicides) to spray. Indeed, good inroads have been made in this direction with the likes of Austin, Kordes, Noack and Warner; however, the challenge is to produce a variety that is both attractive and healthy.

Diversity is ever-increasing in terms of shape and colour of flower (although no true blue yet) and form of the plant. Fragrance, having been a largely forgotten attribute for many years, is again high on the list of desirable characteristics by breeders such as Austin and Delbard and is seen as increasingly desirable by gardeners. The end of the twentieth century and beginning of the twenty-first century has seen a move in favour of more informal, mixed gardens and therefore varieties that associate more easily in this style of garden are more likely to be popular. Refer to **Figure 1** which shows how introductions have changed since 1910.

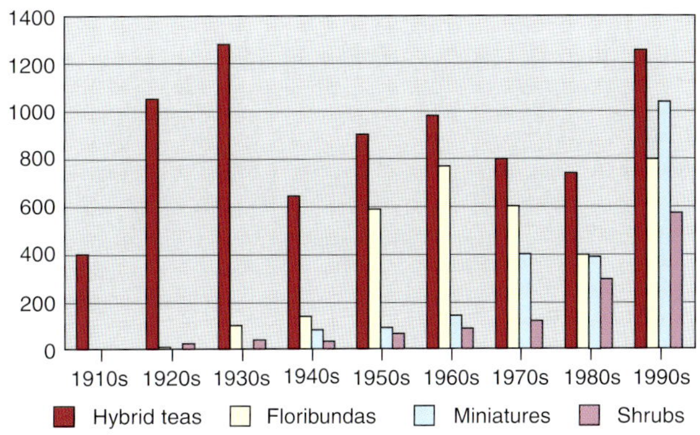

Figure 1 All official rose registrations received up to December 1999 plus roses of botanical importance. Courtesy of T. Cairns/ARS.

Pot Roses

Roses have been grown in pots for a long time but, as is to be expected, the cultural techniques and varieties used have changed hugely in the last 200 years. During the nineteenth century the varieties selected for pot culture were those deemed suitable according to various criteria and would not have been bred with this end-use in mind. In fact, one can list the criteria that Paul wrote in 1863 in *The Rose Garden* and still largely apply them today:

1. Elegance of growth
2. Contrast of colour
3. Abundance of bloom
4. Form of flower
5. Duration and constant succession of blooms
6. Sweetness.

Paul writes that the main advantages for pot culture are to be able to enjoy blooms for 12 months of the year and to grow those varieties that are more susceptible to damage from wind and rain. The popularity of pot culture was undoubtedly increased by the varieties often being – at that time and indeed for some decades afterwards – rather weak and susceptible to disease and the flower balling. He gives a long list of varieties that he thinks are suitable, including those shown in Table 2.

Gertrude Jekyll and Edward Mawley, in their book *Roses for English Gardens* (1901), also devoted some space to growing roses in pots and listed a number of hybrid perpetual, tea and – relatively new at that time – hybrid tea roses such as 'La France' (Guillot 1867), 'Mme Caroline Testout' (Pernet-Ducher 1890) and 'Liberty' (Dickson 1900).

During the twentieth century the Polyantha roses were further developed and, with their compact growth and tough constitution, their suitability for pot culture was recognized. Koster introduced a number of varieties that were used in large quantities, starting with 'Dick Koster' in 1929 and continuing with varieties such as 'Margo Koster' (1931) and 'White Koster' (1929). At that time 'Triomphe Orléanais' (Peauger 1912) was also very popular. The 1950s and 1960s saw the Floribunda roses coming to the fore, notable among them being 'Garnette' (Tantau 1951) and its many hybrids, including 'Carol Amling' (Amling Bros 1953) and 'Marimba'® (Dekkers 1965). These were particularly popular in the USA and were, until the start of the 1990s, the mainstay of the industry.

Research in the 1970s on growing roses in pots paved the way for the introduction of a whole range of new varieties that exhibit uniform growth, good flowering after severe pruning on a year-round basis, long shelf-life and reliable growth on their own roots. The breeding of these varieties has been dominated by Poulsen Roses from Denmark, with their Parade series.

In the United States, some pot roses are produced from bare-root-grafted plants; the Sunblaze series bred by Meilland of France has been particularly popular and is responsible for rejuvenating the traditional pot rose trade.

Cut Roses

The varieties used for cutting in the nineteenth century would have been much the same as those used for growing in containers. Because of low productivity and costs of production they would have been grown on a limited scale and limited to the wealthy. 'Catherine Mermet' (Guillot 1869) was prominent among varieties at that time.

In Europe the modern cut-flower industry could be said to have started in 1896 in Holland when the first purpose-built greenhouse was constructed. The varieties used at that time included 'Ulrich Brünner' (Levet 1881), 'Maréchal Niel' (Pradel 1864) and 'Kaiserin Auguste Viktoria' (Lambert 1891). Roses for cutting were also grown outdoors but only in areas with suitable climates, such as the southern parts of Germany, France and Italy. In both cases production was for local markets as vase-life would have been short and transportation slow.

Table 2 A selection of pot roses

Horticultural class	*Cultivar*	*Breeder*	*Date*
Hybrid perpetual	'Baroness Rothschild'	Pernet Père	1868
	'Général Jacqueminot'	Roussel	1853
Bourbon	'Souvenir de la Malmaison'	Beluze	1843
Tea	'Catherine Mermet'	Guillot	1869
	'Devoniensis'	Foster	1838
	'Maréchal Niel'	Pradel	1864
Polyantha	'Cécile Brünner'	Pernet-Ducher	1881
Noisette	'Rêve d'Or'	Ducher	1868

In the United States 'Mme Ferdinand Jamain' (Ledechaux 1875), with its long straight stems, was responsible for the rapid growth in the American cut-flower industry.

During the first half of the twentieth century production increased although suitable varieties were found by extensive testing rather than as part of a dedicated breeding programme. Important varieties at this time were 'Briarcliff' (Pierson 1926), 'Golden Rapture' Geheimr. Duisb. (Kordes 1934) and 'Better Times' (Hill 1934).

The implementation of plant breeders' rights in several European countries in 1948 was a big incentive to rose breeders to develop both garden and cut-flower roses. Among the first at that time were 'Spek's Yellow' (Spek 1950), 'Baccará'® (Meilland 1954), 'Super Star' (Tantau 1960) and 'Garnette' (Tantau 1951). The latter saw a great increase in both vase-life and productivity. It has been a crucial parent in the history of cut-flower varieties and has sported to several different colours. 'Motrea' (Mosselman and Terreehorst 1968) is thought to be related to 'Garnette' and was grown in large numbers. 'Garnette' and its derivatives were instrumental in the reduction in price of cut roses and hence a great increase in sales. It is interesting to note that 'Garnette' and other closely related varieties are all Floribunda roses and have been largely responsible for the classic hybrid tea flower that we see in the cut roses today.

During the 1960s Jackson and Perkins were successful in breeding outstanding varieties, including 'Bridal Pink'™ (1967) and 'Zorina' (Robinson, 1963). The latter was renowned for its productivity and was also a very important parent.

In the next two decades mention should be made of 'Mercedes'® (Kordes 1974) and 'Frisco'® (Kordes 1986); the latter, with a vase-life of 15 days, is still grown today.

Red roses have always been a prime goal in the breeding of cut roses but good varieties (as in garden roses) are elusive. However, 'Ilona' (Verbeek 1973), 'First Red'® (Pekmez 1983) and 'Mme Delbard' (Delbard 1982) have been extremely popular.

Breeding for cut-flower varieties continues on a large scale as it is more profitable than for garden forms. Selection is based mainly on vase-life, productivity and novel colours. Fragrance has not been high on the list of desirable characters and there are only a few varieties that are even lightly fragrant. However, David Austin has extended his breeding programme to include fragrant cut roses in the style of his English Roses. The first introductions are planned for early 2004.

Spray roses have been grown since the 1920s with varieties such as 'Gloria Mundi' (de Ruiter 1929) and 'Pink Delight' (Kordes 1936) but have remained a small part of the market. In recent years, Interplant from Holland have dominated their breeding with such varieties as 'Evelien' (1985) and 'Jay' (1985). Spray roses may well demand a bigger share of the market in the future. They are popular in Japan and the 'Naturals' bred by Schreurs have a vase-life of 21 days.

The cut-flower industry is no longer dominated by a few varieties around the world as it was 20 years ago. Both production and marketing have spread to many countries around the world and this has resulted in a great range of varieties being developed to satisfy each.

See also: **Biotechnologies for Breeding**: Manipulation of Ploidy Level. **Breeding**: Selection Strategies for Cut Roses; Selection Strategies for Pot Roses; Selection Strategies for Disease and Pest Resistance. **Distribution and Ecology**: Overview. **Growth Regulation**: Floral Induction. **History of Roses in Cultivation**: Ancient Chinese Roses; European (Pre-1800); History of the Perfume Industry. **Intellectual Property**: Plant Patents and Trademarks. **Production and Marketing:** Overview of Roses and Culture; Pot Rose Production; Cut Rose Production. **Rose Collections and Trials**: Rose Collections and Herbaria.

Further Reading

Austin D (1992) *Old Roses and English Roses.* Woodbridge, Suffolk: Antique Collectors Club.

Austin D (1992) *Shrub Roses and Climbing Roses.* Woodbridge, Suffolk: Antique Collectors Club.

Botanica's Roses (1998) Milson's Point. NSW, Australia: Random House.

Cairns T (ed.) (2000) *Modern Roses XI: The World Encyclopedia of Roses.* London: Academic Press.

Harkness J (1985) *The Makers of Heavenly Roses.* London: Souvenir Press.

Jekyll G and Mawley E (1901) *Roses for English Gardens. Country Life:* reprinted by Penguin books. Harmondsworth, Middlesex (1983).

Joyeaux F (1998) *La Rose de France.* Paris: Imprimerie National.

Krussman G (1981) *The Complete Book of Roses.* Portland, Oregon: Timber Press.

Paul W (1903) *The Rose Garden*, 10th edn. London: Simpkin, Marshall, Hamilton.

Pemberton B, Kelly W and Ferare J (1997) *Production of Pot Roses.* Portland, Oregon: Timber Press.

Phillips R and Rix M (1993) *The Quest for the Rose.* London: BBC Books.

Spek H (1987) *Geschichte und Entwicklung des Schnittrosenanbaus in den Niederlanden.* Pinneberg: Partner-Werbe-Druck und Verlagsgesellschaft.

Thomas G (1994) *The Graham Stuart Thomas Rose Book.* London: John Murray.

History of the Perfume Industry

P A van de Pol, Wageningen University, Wageningen, The Netherlands

Introduction

The Book of Perfumes, written by Eugene Rimmel in 1863, is still the standard work on the history of perfume:

> We may say that perfumery was studied and cherished by all the various nations which held in turn the sceptre of civilization. It was transmitted by the Egyptians to the Jews, then to the Assyrians, the Greeks, the Romans, the Asiatics, the Arabs, and at last to the European nations.

The scope of this article is to give a brief survey of the long history of the perfume industry including the role of the rose.

Use of Perfumes in Ancient Times

The first man-made perfumes were a byproduct of the discovery of fire. The scents that surrounded early man were natural ones – they were the smells of flowers, trees, animals and the earth itself. None of these was controllable or usable. The fresh, scented air after a rainstorm could not be bottled and released at will. Fire was the key that made perfume available. As soon as men noticed that burning woods and gums gave off odours, they had the knowledge to produce perfume when they wanted it. They had only to light a fire. The memory of these first deliberately created scents is preserved in our word 'perfume', which comes from the Latin *per* meaning 'through', and *fumus* meaning 'smoke'. The scented smoke rose into the air and became invisible as though it had become part of the sky itself. So early man used it as a vehicle or messenger, to carry his prayers to the heavens. The burning of incense, resinous gums, myrrh and many other fragrant herbs was employed in all forms of primitive worship. In the great temples dedicated to Isis, Osiris and the other divinities that watched over Egypt, the air was heavy with burning incense. The early Egyptians became familiar with the art of floral extraction. The reliefs on the walls of old temples show perfume being distilled from flowers such as lily and henna. The priests alone were in charge of perfumery and the greater part of all the aromatics brought into the country was stored in the temples. In the British Museum there is a scent bottle of Egyptian design, made of richly coloured glass and dating from the fifth century BC. The use of perfumes for sacred rites was gradually adapted to private use. Only the wealthy could afford such luxuries. Although most of the information that has come down to us is of ceremonial use of perfume, it is probably true that the private consumption of scent became greater than the religious. Both men and women of high rank considered perfume an essential accessory. The Egyptians were a hygienically minded people who believed that cleanliness was next to godliness. It was they who invented the elaborate system of baths that the Romans took over and became famous for. Abundant amounts of perfumes were used for scented ointments and baths. The Romans were obsessed by the rose. Rose-water perfumed their public baths. It flowed from fountains in the emperors' palaces, in the houses of the rich and powerful, and in the amphitheatres. At banquets and feasts roses were strewn everywhere. Even the cushions were stuffed with petals. The man who was perhaps the biggest spender at the perfumer's of all time was Antiochius IV, king of Syria from 175 to 164 BC. At the games he organized at Daphne, not far from Antioch, a procession of 200 glamorous women, each carrying a golden bowl of perfume, sprinkled the spectators with costly and exquisite scents. The thousands of special guests were anointed on their arrival with the perfume of their choice from a selection including saffron, cinnamon, spikenard and lilies. At one banquet, doves whose wings had been saturated with a variety of perfumes were released to fly around in the room. As a parting gift, when the long-drawn-out celebrations ended, each guest received a crown of myrrh and frankincense.

Religious Uses

In many religions scent plays an important role. In the book of Exodus of the Bible there are directions given to Moses for erecting an altar of incense and for making the holy anointing oil and scented offers. In the East, scented flowers and perfumes are used in Hindu worship, and incense is burned on all ceremonial occasions, whilst freshly gathered fragrant flowers adorn the temples. In Tibet, were fragrant gums are scarce, scented wood is burned in the temples, usually the wood of juniper. In China Confucius recorded that scented flowers like jasmine, peach and magnolia were scattered about on all festive occasions in temples and houses. The consumption of incense sticks burned here was enormous.

The use of perfume disappeared from Europe as the Romans withdrew. The Europe of the Dark Ages threw away everything of the Roman civilization that was thought to be decacadent and cause of the fall of the Roman empire. The Arabs kept the flag of perfume flying during Europe's Dark Ages. It was a rose-red flag. Rose-water was the Arabs' favourite scent, as it was to become the Europeans'. Legend has it that some of the caliphs had fountains of rose-water playing in their palaces. One of them, El-Mutawkkel, had the rugs in all his apartments continually sprinkled with rose-water and declared 'I am the king of all sultans, as the rose is the king of all fragrances'! Saladin, as soon as he entered Jerusalem in 1187, ordered the floors and walls of Omar's Mosque to be thoroughly washed down with rose-water. It were the Crusaders who really brought scent back to the men of Europe. They brought back exotic perfumes for their women and rose-water for themselves. They also brought back the idea of the bath and personal cleanliness.

The Perfume Industry

The historian of perfume has some difficulties. One is that we often know only the scented practices of religion and the great and famous people. Although we may with a fair degree of safety generalize from this knowledge – because the great and famous were usually the leaders of fashion – our generalizations may be off target. This is more particularly true because the public use of scent is often all we have to go on, when it is the intimate details of private lives that would be more to our purpose.

After the early history European perfume industry started in Genoa (Italy), one of the home harbours of the Crusaders. But the Crusaders were only one element in this resurgence; rapidly increasing wealth was a more important factor. Most of the perfumes were at first imported from the Middle East, but local business talent quickly got into the act. By the end of the twelfth century manufacturing perfumers were supplying their own products to the women of France. For some time these perfumers, and others in Europe, were mainly copyists. The first post-Roman European invention of a perfume seems to have been that of lavender water during the twelfth century. The water was made in England, France and Germany and has retained its popularity right up to our own day. The distillation of lavender, which began early in the seventeenth century, was the first attempt at any form of commercial scent production in the British Isles. Another European toilet water had a life almost equally long. 'Hungary water' first began to appear as part of the cosmetic armoury of European ladies in about 1370. Distilled from rosemary, it was the 'eau de Cologne' of the Middle Ages and it was still in use in Victorian times.

At the beginning of the fifteenth century Venice also became an important centre of the perfume industry. Some of the most expensive aromatics of the East reached the city through its trading connections with Constantinople. Also the city of Florence became famous for her perfumers and beauticians, and when Catherine de Medici became the wife of Henry II in 1533 she took with her a Florentine skilled in the art of concocting perfumes, a man named René who opened a shop in Paris which soon became the meeting place of the fashionable. In Rome, the most popular perfume was Neroli, a distillate of the orange flower, prepared from the fresh blossoms of the bitter orange, the bigarrade, one of the principal ingredients in the later developed Eau de Cologne.

The distillation of flowers and leaves was one of the chores in all large houses from the beginning of the sixteenth century. The household books (1502) of the 5th Earl of Northumberland contain the names of 'herbes to stylle', to make sweet waters to use for personal cleanliness, in cooking and for medicine. Balm and sage, marigold and tansy were amongst the most widely grown plants for distillation, and every garden had its rosemary and lavender bushes. All large houses possessed a room for the drying of herbs and another for their distillation where the ladies took lessons in the art of making sweet waters. These waters were given at Christmas or for birthdays. Silver bowls filled with rose-water were kept in the bedrooms to be used for washing the hands and face. The Eastern fashion of sprinkling rose-water over one's clothes was prevalent. Perfumes were also burnt in rooms. Perfumers travelled around, calling at the houses of the rich to fumigate musty apartments with fragrant woods of juniper or cypress. After that the room was made to smell as though it were filled with roses by burning hot embers soaked with rose-water. Clothes were kept in coffers made of fragrant wood such as juniper, cedar or sandalwood which impregnated the clothes.

Perfumes did not come into general use until the reign of Elizabeth I when Edward de Vere, Earl of Oxford, brought perfumed gloves from Italy in 1573. From that time onwards, the use of perfumes became fashionable, with civet and musk the basis of most costly preparations. The gloves presented to Elizabeth would be scented with frangipani, one of the most famous of all perfumes, which was distilled from the flowers of *Plumera alba*, which grows on several of the West Indian islands. Elizabeth's perfumers concocted a number of perfumes for the Queen's use, made to their own secret formula. One of Elizabeth's favourite perfumes was made of musk and roses. The perfumes were kept in small bottles of chastened silver. Towards the end of Elizabeth's reign, the first glass scent bottles were made in England by Huguenots who had fled from France.

It all sounds very idyllic – 'sweet' flower waters, odours 'most delicate'. We have to make an effort of imagination to appreciate that the scents of Elizabeth's court would have been far too pungent for our modern noses. They had to overcome the stench of unwashed bodies and clothes! The Tudors certainly had no fine feelings about cleanliness, but by their time scents had become very much part of the everyday life of the wealthy and middle classes. Rose-water was still the most popular scent. There is a recipe dating from 1602 for rose-water to which is added small quantities of spikenard, thyme, lemon, cloves and a touch of civet.

The idea of scented gloves was extended to other perfumed leather articles like jerkins and even shoes. Perfumed leather was the original product of the French centre of perfumery at Grasse, made to overcome the bad smell of new leather.

In the seventeenth century the French king Louis XIV was dubbed the 'sweetest smelling monarch that had yet been seen', and was so fond of perfumes and luxurious living that he would insist on his perfumer being in attendance while his scents were made up to his own requirements. And perfumes continued to increase in favour with the accession of Louis XV. Court etiquette demanded that, on each day, a different scent was to be worn, so that Versailles became known as 'la cour parfumée'. Madame de Pompadour, mistress of Louis XV, lavished perfume on herself on a scale that makes Queen Elizabeth seem a beginner. The French perfumers of the time sought to increase the sale of their products by making use of itinerant vendors who toured the country, visiting the large towns with a horse and cart stocked with favourite perfumes and aids to beauty. Others dispensed their wares in the streets of Paris. When, in 1770, Marie Antoinette became the wife of the future Louis XVI, she set a new fashion by her love of the more natural perfumes, distilled from the violet and rose, neglecting the heavier Eastern scents. Not surprisingly, at the time of the French Revolution little interest was taken in the use of perfumes but, when Bonaparte was crowned Emperor in 1804, his love of sweet-smelling substances gave perfumery a new popularity. Napoleon's wife Josephine followed Marie Antoinette's tastes and preferred simple, natural scents. Eau de Cologne was a new scent in Josephine's time. At the beginning of the eighteenth century, Paul Feminis, an Italian living in the German city of Cologne, had brought out a toilet water that at last pushed Hungary water into second place on the sales charts. It was based on citrus oils, neroli, lemon, bergamot and lavender. Under the name of 'l'eau admirable' it lasted for almost a hundred years. Then Jean Marie Farina, a descendant of Feminis, altered the formula slightly, incoporating rosemary into it, and marketed the new product under the name 'eau de cologne'. It was an immediate runaway success, was copied by many perfumers, and began a fashion for gentle toilet waters that has lasted until our own day.

Mixing Perfumes

An important question is: Why do perfumes have to be mixed at all? Why not be content with a single fragrance? There are several answers to these questions. One is that the perfumer is in business to make a profit; he has to have a product that is different from those of his rivals. Another is that there is no such thing as a standard consistent scent. Yet another is that a perfume has to have certain qualities, such as stability and staying power, that can be achieved only by mixing. So it is not just a matter of extracting the perfume of, say, the rose, bottling it, marketing it, and so on, all without the help of the creative perfumer. Apart from anything else, undiluted single scents are unlikely to be best-sellers. By themselves, and in bulk, they have a singularly overpowering and unpleasant smell.

The perfumer, then, is forced to be a mixer. At the time when Rimmel was writing in 1863, the perfumer had at his disposal some 500 fragrant ingredients, all of them natural products – essential oils, gums, and flower and fruit extracts. Today, he has about 4000 materials available. In practical business terms, this reduces to about 200 or 300 most-used substances. With these the perfumer concocts his new perfume. It may take years to perfect a new perfume. A fine perfume is rather like a three-course meal. It gives off first the hors d'oeuvre, known as its topnote. This is the first scent that hits the nose as soon as the stopper is removed from the bottle. Often it is a 'light' aroma, perhaps lemon or bergamot, which fades fairly quickly. Then comes the main course, the heart or the theme of the perfume. Finally, for desert, the perfume gives off its endnote, the fall that gives it its characteristic, individual flavour.

Synthetic Ingredients

Most perfumes, even the most expensive, contain a high proportion of synthetic materials. It has been calculated that two-thirds of the raw materials used in perfumery today are synthetic. This is partly because the cost of the natural products has become prohibitive. Another reason is that there can be a shortage of the natural product. Two tons of roses, or a hundred million petals, are needed to produce one pound of attar of roses. Yet not only the perfumers but also the soap-flake and detergent producers need thousands of tons of perfume ingredients a year.

There are two groups of synthetic aromatics: those that are copies of natural products and those that are not only man-made but man-invented. These are the scents that do not exist in nature but are now part of the perfumer's armoury. Some of the most sophisticated 'modern' perfumes could not exist without artificial chemicals, in particular the group of the aldehydes. Synthetics are not necessarily cheaper than natural products, but they can be produced in large quantities at consistent quality.

One of the first perfumes to use synthetic ingredients was *Chanel No. 5*, one of the all-time classics of perfumery, which was blended by Ernst Beaux in 1921. Coty's *L'Aimant* and Lanvin's *Arpège* followed within a few years. Both used synthetics, both became classics. Perfumers have always been understandably secretive about their formulae and have described their products only in the most general terms. Yardley's *Khadine* – 'sophisticated, soft and haunting, offering excitement and femininity in the same breath' – is a blend of spices of clove and pepper, attar of roses, jasmine, lily of the valley, sandalwood, vetivert and patchouli. Elizabeth Arden's *Blue Grass* – 'inspired by the freshness of a summer's day' – mingles 'jasmine, geranium, roses and lilies of the valley with other fragrant blossoms. Crisp touches of lavender and an underlying base of green moss add a cooling note.'

Rose Oil

The most valuable and useful of the natural perfume oils are what are known as 'absolutes'. These are the essential oils extracted from certain flowers – especially the rose, the jasmine and the violet. Every quality perfume contains a percentage of one or more of these flower oils, which impart what the trade calls 'smoothness' to the mix. The most precious of all these oils to the perfumer is that obtained from the rose.

When rose petals were first extracted is unknown, but the earliest written reference is in the ninth century when production was well established, for 30 000 bottles of attar of rose were delivered annually between AD 810 and 817 as tribute from the Iranian province of Faristan to Maman, the then Caliph of Baghdad. Rose-water was also traded along the Silk Road to China. Distillation probably began independently in Europe, the Middle East and the Far East. An Italian, Geronimo Rossi of Ravenna, is credited as the first European to do so, in 1574, describing his methods and oils produced in *De Distillatione Liber* (1582). The Chinese Li Shih-Chin in his work the *Pen T'sao* (1578) noted that essence of rose is made by distilling the flowers of *Rosa rugosa* and he also mentions essence of *R. indica*. Distillation began in Persia in 1612 and production of rose oil quickly became a major industry, with local cultivars and their processing spreading eastwards and westwards along the Silk Road. This distribution can be followed in the name for rose oil, attar of roses; *attar* from the Turkish is derived from the Farsi, *aettr*, which then meant a fat. In Hindi *itr* describes a liquid produced by absorbing in sandalwood oil a distillate of rose flowers. In modern Turkish the attar is known as *gulyagi* from the Farsi *gul* for rose flower.

Cultivation of *R. damascena* and production of rose oil began in what was then the Turkish province of Eastern Roumelia (now Kazanlik in Bulgaria) in the fifteenth century. Commercial rose growing to produce oil was well established in the Kazanlik region of Bulgaria by the end of the seventeenth century. So dominant did Bulgarian attar become on world markets that rose cultivars originating there became universally known as Kazanlik roses. On-going selection and breeding produced higher oil-yielding cultivars with more flowers blooming over a longer period and Bulgaria remains the leading producer of rose oil. It was not until 1894 that *R. damascena* was reintroduced to Turkey from Kazanlik, and large-scale production recommenced in the southwestern provinces of Burdur and Isparta. The first modern record of an export is 1110 kg attar from the then Turkish Empire to England and France in 1860. Turkey is now the world's second-largest producer of rose oil. Besides Bulgaria and Turkey other important countries producing rose oil are: China, Egypt, France, India and Morocco.

Rose Species Used for Rose Oil Production

The more important sources of commercial essential oil are *R. damascena*, *R. gallica*, *R. centifolia* and their cultivars, although other species may be grown locally to produce an oil or rose extract. *Rosa alba* L. (the white rose), whose yield was generally less than half that of *R. damascena*, is now little cultivated for its oil. Selection for high oil yield has been extensive, with crosses between *R. damascena*, *R. alba*, *R. gallica* and other oleiferous species and cultivars; the greatest increase in oil content of flowers in Bulgaria occurred with *R. damascena* as the pollen parent. Tea rose cultivars have been developed in China and Europe, whose oil has different odour characteristics.

The Art of Smelling

Of the five primary senses – touch, taste, smell, sight and sound – smell is unique in that there is no word in the English vocabulary to define it. There is no word to describe a particular odour as there is 'salt' or 'sweet' as applied to taste, or 'sharp' or 'soft' as applied to sound. Sight functions only in light; there are periods when the ear listens but hears nothing; the sluggish taste sleeps most of the time. But night or day there is no odourless place for the sensitive nose. However, scent is the most neglected of the senses and although it plays such an important role in daily life it is surprising that during education all senses are trained with exception of the nose.

See also: **Fragrance and Pigments**: Fragrance Profiles of Wild and Cultivated Roses; Functional Genomics. **History of Roses in Cultivation**: European (Pre-1800); Modern (Post-1800). **Secondary Metabolites**: Extraction and Industrial Processes.

Further Reading

Genders R (1972) *A History of Scent*. London: Hamish Hamilton.

Genders R (1977) *Scented Flora of the World*. London: Robert Hale.

Rimmel E (1863) *The Book of Perfumes*. London: Chapman & Hall.

Trueman J (1975) *The Romantic Story of Scent*. London: Aldus Books.

Weiss EA (1997) *Essential Oil Crops*. Wallingford, UK: CAB International.

Widrlechner MP (1981) History and utilization of *Rosa damascena*. *Economic Botany* 35: 42–58.

ISBN 0122276205

9 780122 276200

ENCYCLOPEDIA OF ROSE SCIENCE

ENCYCLOPEDIA OF ROSE SCIENCE

Editor-in-Chief
ANDREW V ROBERTS

Editors
THOMAS DEBENER
SERGE GUDIN

ELSEVIER
ACADEMIC
PRESS

Amsterdam Boston Heidelberg London New York Oxford
Paris San Diego San Francisco Singapore Sydney Tokyo

Elsevier Ltd., The Boulevard, Langford Lane, Kidlington, Oxford, OX5 1GB, UK

Elsevier Inc., 525 B Street, Suite 1900, San Diego, CA 92101-4495, USA

First edition 2003

Library of Congress Control Number: 2003112413

A catalogue record for this book is available from the British Library

ISBN: 0-12-227620-5 (set)

This book is printed on acid-free paper

Printed in Spain

EDITORIAL ADVISORY BOARD

FOREWORD

No other flower is as universally loved and grown or has a more illustrious history than the rose. Both reality and symbol, the rose has been a constant source of inspiration throughout history in such different fields as religion, architecture, literature, art, the perfume industry, fashion, gastronomy and, of course, horticulture and gardening. Throughout the centuries the rose has been regarded as the queen of flowers, and it is a fact that there is very little to surpass the sight of well-grown roses in full bloom. Whether grown by themselves or among other plants, roses bring to bear many years of development in an unparalleled array of functionality and beauty. However, the incredible history of roses starts in the hands of passionate, meticulous and attentive breeders, who imagine and develop colours, fragrances and shapes for the queen of the garden. What are the secrets of their alchemy? How is a rose born? The *Encyclopedia of Rose Science* attempts to answer these mysterious questions.

The research methods used by rose breeders to create and screen new varieties at this stage are mainly based on classical processes. Rose-breeding efficiency is still very dependent on the breeder's experience, despite recent scientific advances that have contributed to plant improvements. Moreover, science plays an increasingly important role in the production of this flower which is at the centre of one of the world's biggest floricultural industries. Rose breeders such as myself recognise this and have long backed scientific research, beginning in the 1970s and continuing to the present day. It has always seemed to me to be important to drive the organisation of international scientific symposia, to develop internal programs, and facilitate national and international cooperation in order to improve and speed up the whole creation and selection process.

The *Encyclopedia of Rose Science* draws together the whole range of rose techniques and scientific knowledge. For breeders, rose growers and avid amateurs, as well as researchers and students, this book will be an invaluable reference guide which will supply a treasury of information.

Alain Meilland
Le Cannet-des-Maures

September 2003

INTRODUCTION

Roses have been admired for their beauty and fragrance since the dawn of civilization. Over the centuries, poets, artists, historians, gardeners, apothecaries and scientists, each in their own way, have contributed to our appreciation, enjoyment and knowledge of roses. The *Encyclopedia of Rose Science* focuses on scientific contributions but also contains articles, including those on the history of cultivated roses, pruning and intellectual property rights, which provide background information and advice for an international readership of scientists, breeders, growers and knowledgeable rose enthusiasts.

Approximately 8 billion rose stems, 80 million potted plants and 220 garden roses are sold annually. This flourishing horticultural trade is served by many scientists in commercial organizations, research institutes and universities throughout the world, who have made substantial but sometimes under-appreciated contributions to knowledge of the rose. The *Encyclopedia of Rose Science* brings together in a single publication and on an unprecedented scale, overviews, explanations and interpretations that no recent publication has attempted. Whilst exploring their areas of special interest, readers will be able to follow cross-links (manually or electronically) to less familiar but related topics.

The editors have been helped in their tasks of identifying topics for inclusion in the *Encyclopedia*, choosing the best authors and reviewing the completed articles, by an Editorial Advisory Board that consists of 20 distinguished scientists and practitioners from many countries. Many topics, including those concerned with breeding, propagation, production, *in vitro* culture, physiology, fragrance, pests and pathogens, are well subscribed to in the scientific literature, but particular care was needed to provide essential information about some relatively neglected topics. For example, information about the distribution and ecology of the rose is difficult to obtain because few rose scientists presently specialize in these areas and literature sources are scattered in diverse, often inaccessible publications. To cover these areas, information was sought from representatives of the national rose societies affiliated to the World Federation of Rose Societies. Their knowledge of the local conditions, the distribution of wild species and the cultivars that grow best in their localities, provides unique information that will provide a valuable reference for formal ecological studies in the future.

The description of species of roses is an activity that was once at the very centre of scientific investigation and exploration. However, there is no recent work that is as comprehensive and detailed as the chapter on *Rosa*, in Rehder's *Manual of Cultivated Trees and Shrubs Hardy in North America* (2nd edition, 1960). As an important reference source, the chapter concerning *Rosa* has been reproduced in this *Encyclopedia*. For descriptions of species and classification, schemes are also given in some articles on Distribution and Ecology and Classification. Clearly, a monograph covering recent perspectives on the taxonomy, classification and description of the genus *Rosa* is urgently needed. Readers should note that there have been name changes, including the change from *R. wichuriana* Crép. to *R. wichurana* Crép. One name that appears in many scientific papers, but not in taxonomic works, is the name *R. hybrida* L. or (*R.* × *hybrida* L.). The origin of this name is mysterious! Searches by professional taxonomists have failed to identify any such species named by Linnaeus. It is, therefore, unclear as to what the name signifies. The origin of this name needs to be clarified or a new name introduced to describe an unambiguously defined category of hybrid roses.

It is likely that readers of the *Encyclopedia* will wish to refer repeatedly to descriptions of cultivated roses that are cited in the articles. To facilitate this, the reader can refer manually or electronically to *Modern Roses XI* which is reproduced in the *Encyclopedia*. This is the 11th edition of a periodically revised publication owned by the American Rose Society (the official registration body of cultivar names), edited by Tommy Cairns and published by Academic Press (an Elsevier imprint). It lists 24 000 cultivars. The reader will find that some cultivars referred to in the articles are not listed, often because they are old cultivars that are rarely seen, new cultivars that will appear in the next edition of *Modern Roses*, or rootstock varieties.

Most articles in the *Encyclopedia of Rose Science* include a statement of future developments anticipated by the authors. Perhaps an additional editorial comment is needed on one area of development that may have a significant impact. Biotechnology has arrived and is likely to be an area of much activity and discussion! It has been said, disparagingly, that modern cultivars are made by man but wild roses were made by God. However, most of us applaud the achievements of rose breeders. Over the centuries, the processes of genetic modification and selection that fashioned the wild roses have been extended by human intervention. Early introductions to rose gardens, including roses with multipetalled flowers and the recurrent-flowering habit, resulted from selections of naturally occurring mutants. Hybridization that enabled breeders to introduce new combinations of characters was a later development. Now modern rose breeders are able to plan breeding programmes, carry out difficult crosses and select seedlings with greater skill than ever before, to provide us with ever more attractive and healthy roses. However, the exploitation of new techniques in molecular biology may significantly challenge old practices. The possibilities are, of course, enormous and some of these are regarded with apprehension. The use of genetic probes that will enable selection to be carried out with infinitely greater efficiency than before will not change the essential genetic nature of rose crops. More controversially, genetic manipulation, already carried out in some laboratories, allows genes to be introduced that do not naturally occur in the genus *Rosa*. The verdict on genetic manipulation will be delivered by a discriminating public that is wary of innovation but also recognizes the environmental benefits, for example, of not spraying roses for infectious diseases. We hope that the articles on Genetics in this *Encyclopedia* will assist the reader in understanding the techniques involved and potential applications of the new approaches.

The editors gratefully acknowledge the expert contribution of the Major Reference Works unit at Elsevier, who have ensured that the *Encyclopedia of Rose Science* is presented to a high standard. We particularly appreciate the contribution of the Senior Developmental Editor, Mark Knowles: his patient advice, friendly support and, when necessary, tactful coercion of editors and authors alike have ensured that publication has kept to its schedule. The support of the Editorial Advisory Board has been invaluable and the distinguished contributions of the authors of the articles themselves will be apparent to all readers. We hope that the success of the *Encyclopedia* will be a fitting tribute to all who have made its production possible.

Andrew V Roberts
Thomas Debener
Serge Gudin

September 2003

GUIDE TO USE OF THE ENCYCLOPEDIA

Structure of the Encyclopedia

The material in the Encyclopedia is arranged as a series of entries in alphabetical order. Most entries consist of several articles that deal with various aspects of a topic and are arranged in a logical sequence within an entry. Some entries comprise a single article.

To help you realize the full potential of the material in the Encyclopedia we have provided three features to help you find the topic of your choice: a Contents List, Cross-References and an Index.

I. Contents List

Your first point of reference will probably be the contents list. The complete contents list, which appears at the front of volumes one and two will provide you with both the volume number and the page number of the entry. On the opening page of an entry a contents list is provided so that the full details of the articles within the entry are immediately available.

Alternatively you may choose to browse through a volume using the alphabetical order of the entries as your guide. To assist you in identifying your location within the Encyclopedia a running headline indicates the current entry and the current article within that entry.

You will find 'dummy entries' where obvious synonyms exist for entries or where we have grouped together related topics. Dummy entries appear in both the contents list and the body of the text.

Example

If you were attempting to locate material on the molecular biology of flavonoids via the contents list:

MOLECULAR BIOLOGY *see* BIOTECHNOLOGIES FOR BREEDING: Genetic Transformation. CLASSIFICATION: Chemotaxonomy and Molecular Taxonomy. FRAGRANCE AND PIGMENTS: Anthocyanins and other Flavonoid Pigments; Flavonoid Molecular Biology; Functional Genomics. GENETICS: Nuclear DNA Amounts; Inheritance in the Dogrose; DNA Fingerprinting; Gene Mapping; Molecular Markers; Cloned Genes; Marker-Assisted Selection; Research Trends. GROWTH REGULATION: Floral Induction. POSTHARVEST PHYSIOLOGY: Ethylene.

The dummy entry directs you to the Flavonoid Molecular Biology article, in the FRAGRANCE AND PIGMENTS entry. At the appropriate location in the contents list, the page numbers for articles under Fragrance and Pigments are given.

If you were trying to locate the material by browsing through the text and you looked up Molecular Biology then the following information would be provided in the dummy entry:

Molecular Biology *see* **Biotechnologies for Breeding**: Genetic Transformation. **Classification**: Chemotaxonomy and Molecular Taxonomy. **Fragrance and Pigments**: Anthocyanins and other Flavonoid Pigments; Flavonoid Molecular Biology; Functional Genomics. **Genetics**: Nuclear DNA Amounts; Inheritance in the Dogrose; DNA Fingerprinting; Gene Mapping; Molecular Markers; Cloned Genes; Marker-Assisted Selection; Research Trends. **Growth Regulation**: Floral Induction. **Postharvest Physiology**: Ethylene.

Alternatively, if you were looking up **Fragrance and Pigments** the following information would be provided:

FRAGRANCE AND PIGMENTS

Contents

2. Cross-References

All of the articles in the Encyclopedia have been extensively cross-referenced.

The cross-references, which appear at the end of an article, serve three different functions. For example, at the end of the GENETICS/DNA Fingerprinting article, cross-references are used:

i. To indicate if a topic is discussed in greater detail elsewhere.

> GENETICS/DNA Fingerprinting.
> *See also:* **Classification**: Conventional Taxonomy (Wild Roses); Horticultural Classification Schemes; Cultivar Identification by Image Analysis; Chemotaxonomy and Molecular Taxonomy. **Genetics**: Inheritance in the Dogrose; Species Crosses; Mutation; Gene Mapping; Molecular Markers; Marker-Assisted Selection. **Intellectual Property**: Plant Patents and Trademarks; Technical Examination and Conventional Characteristics. **Rose Collections and Trials**: Rose Collections and Herbaria.

ii. To draw the reader's attention to parallel discussions in other articles.

> GENETICS/DNA Fingerprinting.
> *See also:* **Classification**: Conventional Taxonomy (Wild Roses); Horticultural Classification Schemes; Cultivar Identification by Image Analysis; Chemotaxonomy and Molecular Taxonomy. **Genetics**: Inheritance in the Dogrose; Species Crosses; Mutation; Gene Mapping; Molecular Markers; Marker-Assisted Selection. **Intellectual Property**: Plant Patents and Trademarks; Technical Examination and Conventional Characteristics. **Rose Collections and Trials**: Rose Collections and Herbaria.

iii. To indicate material that broadens the discussion.

> GENETICS/DNA Fingerprinting.
> *See also:* **Classification**: Conventional Taxonomy (Wild Roses); Horticultural Classification Schemes; Cultivar Identification by Image Analysis; Chemotaxonomy and Molecular Taxonomy. **Genetics**: Inheritance in the Dogrose; Species Crosses; Mutation; Gene Mapping; Molecular Markers; Marker-Assisted Selection. **Intellectual Property**: Plant Patents and Trademarks; Technical Examination and Conventional Characteristics. **Rose Collections and Trials**: Rose Collections and Herbaria.

3. Index

The index will provide you with the page number where the material is located, and the index entries differentiate between material that is a whole article, is part of an article or is data presented in a figure or table. Detailed notes are provided on the opening page of the index.

4. Appendices

In addition to the articles that form the main body of the Encyclopedia, there are a number of appendices, which provide an historical list of Rose Awards, extensive lists of refereed papers relevant to articles, Rehder's classification scheme for *Rosa*, reproduced from the *Manual of Trees and Shrubs, Hardy in North America* (Second Edition), and a listing of rose names, reproduced from the American Rose Society's *Modern Roses XI*.

The appendices are located in volume 2, before the index, and in volume 3.

5. Contributors

A full list of contributors appears at the beginning of volumes one and two.

CONTRIBUTORS

A J J Akkerman
DLV Plant BV
Wageningen, The Netherlands

J P André
INRA
Antibes, France

J O Becker
University of California
Riverside, CA, USA

H Ben-Sade
The Hebrew University of Jerusalem
Rehovot, Israel

J-P Biolley
Université Claude Bernard
Lyon, France

J Blaas
Business Unit Biosciences
Plant Research International
Wageningen, The Netherlands

P S Blake
Horticulture Research International
East Malling,
Kent, UK

T J Blom
University of Guelph
Guelph, ON, Canada

W A Brandenburg
Business Unit Biosciences
Plant Research International
Wageningen, The Netherlands

N Bredmose
Danish Institute of Agricultural Sciences
Aarslev, Denmark

H Brichet
World Federation of Rose Societies
San Terenziano, Italy

D H Byrne
Texas A&M University
College Station, TX, USA

R I Cabrera
Texas A&M University
Dallas, TX, USA

T Cairns
American Rose Society
Studio City, CA, USA

C Casey
North Carolina State University
Raleigh, NC, USA

A Chaanin
W. Kordes Soehne
Klein Offenseth-Sparneshoop, Germany

D Chimonidou
Agricultural Research Institute
Nicosia, Cyprus

A C M Clerkx
Enschedé, The Netherlands

H A Collin
University of Liverpool
Liverpool, UK

J M Costa
Wageningen University
Wageningen, The Netherlands
Fundação para a Ciência e Tecnologia
Lisbon, Portugal

Y M Crane
Purdue University
West Lafayette, IN, USA

L Crespel
Meilland Star Rose
Le Luc en Provence, France

R R Cutler
University of East London
London, UK

D P de Vries
Plant Research International
Wageningen, The Netherlands

Present address: Foundation Sub Rosa
Ede, The Netherlands

T Debener
Federal Centre for Breeding Research on Cultivated Plants
Institute for Ornamental Plant Breeding
Ahrensburg, Germany

A Dohm
Federal Centre for Breeding Research on Cultivated Plants
Institute of Ornamental Plant Breeding
Ahrensburg, Germany

R Drewes-Alvarez
University of Applied Sciences Dresden
Dresden, Germany

L A M Dubois
Plant Research International
Wageningen, The Netherlands

D Esmenjaud
INRA
Antibes, France

J Ferare
The Conard–Pyle Co.
Hollister, CA, USA

F Ferrero
INRA
Saint Aygulf, France

B Feuerhahn
Berlin, Germany

J-L Fiasson
Université Claude Bernard
Lyon, France

K Fiasson
Université Claude Bernard
Lyon, France

R G Foottit
Agriculture and Agri-Food Canada
Ottawa, ON, Canada

G Forkmann
Technical University of Munich
Freising-Weihenstephan, Germany

H W M Fuchs
University of Delft
Delft, The Netherlands

M Fukuchi-Mizutani
Institute for Fundamental Research
Suntory Ltd
Osaka, Japan

M-H Gandelin
Geves Cavaillon
Les Vignères, France

N García Victoria
PPO Glasshouse Horticulture
Aalsmeer, The Netherlands

M L Gleason
Iowa State University
Ames, IA, USA

J-F Gonnet
Université Claude Bernard
Lyon, France

C Grossi
Université Claude Bernard
Lyon, France

S Gudin
Université d'Aix-Marseille III,
Marseille, France

W Guoliang
Jiangsu Forestry Academy
Dongshanqiao, Nanjing, China

S J Helland
Iowa State University
Ames, IA, USA

A P Hert
University of Florida
Gainesville, FL, USA

R K Horst
Cornell University
Ithaca, NY, USA

Y Jacob
INRA
Saint Aygulf, France

M Jay
Université Claude Bernard
Lyon, France

J B Jones
University of Florida
Gainesville, FL, USA

F Joyaux
Roseraie de la Cour de Commer
Commer, France

J F Karlik
University of California
Bakersfield, CA, USA

J W Kelly
Clemson University
Clemson, SC, USA

P G Kevan
University of Guelph
Guelph, ON, Canada

M Kilian
Bayer Environmental Science
Monheim, Germany

T Kwakkenbos
Community Plant Variety Office (CPVO)
Angers, France

M Le Bris
Université d'Aix-Marseille III
Marseille, France

R Lewis
University of East London
London, UK

M Linde
Federal Centre for Breeding Research on Cultivated Plants
Ahrensburg, Germany

P T Lynch
University of Derby
Derby, UK

N Marissen
PPO Glasshouse Horticulture
Aalsmeer, The Netherlands

M Marriott
David Austin Roses Ltd
Wolverhampton, UK

J G Mason
Florigene Ltd
Collingwood, Victoria, Australia

J Mattock
Oxford, UK

H E L Maw
Agriculture and Agri-Food Canada
Ottawa, ON, Canada

J Meynet
INRA
Saint Aygulf, France

J Mottley
University of East London
London, UK

J Mouchotte
Meilland Star Rose
Le Luc en Provence, France

R Müller
Royal Veterinary and Agricultural University
Frederiksberg, Denmark

B C Murphy
Cal Agri Products
Davis, CA, USA

T M Murphy
University of California
Davis, CA, USA

R Noack
Guetersloh, Germany

H Nybom
Swedish University of Agricultural Sciences
Kristianstad, Sweden

C O'Donnell
University of California
Davis, CA, USA

G Otto
Dresden, Germany

M P Parrella
University of California
Davis, CA, USA

H B Pemberton
Texas A&M University
Overton, TX, USA

T Pettitt
Horticulture Research International
Wellesbourne, Warwick, UK

F Piola
Université Claude Bernard
Lyon, France

M Podwyszynska
Research Institute of Pomology and Floriculture
Skierniewice, Poland

H M C Put
Cut Flower Hygiene
Deventer, The Netherlands

S Rajapakse
Clemson University
Clemson, SC, USA

O Raymond
Université Claude Bernard
Lyon, France

M Reid
University of California
Davis, CA, USA

A V Roberts
University of East London
London, UK

A Samach
The Hebrew University of Jerusalem
Rehovot, Israel

C Sanlaville-Boisson
Université Claude Bernard
Lyon, France

U K Schuch
University of Arizona
Tucson, AZ, USA

H Schulz
Federal Centre for Breeding Research on Cultivated Plants
Institute for Plant Analysis
Quedlinburg, Germany

A Schum
Federal Centre for Breeding Research on Cultivated Plants
Institute of Ornamental Plant Breeding
Ahrensburg, Germany

R C Shattock
University of Wales
Bangor, UK

N Shishkoff
Cornell University
Long Island Horticultural Research and Extension Center
Riverhead, NY, USA

J D Shorthouse
Laurentian University
Sudbury, ON, Canada

A Skytt Andersen
Royal Veterinary and Agricultural University
Frederiksberg, Denmark

W Spethmann
University of Hanover
Hanover, Germany

U Steiner
Rheinische Friedrich-Wilhelms-Universität Bonn
Bonn, Germany

B M Stummann
Royal Veterinary and Agricultural University
Frederiksberg, Denmark

M S Szyndel
Warsaw Agricultural University
Warsaw, Poland

Y Tanaka
Institute for Fundamental Research, Suntory Ltd
Osaka, Japan

S A Tjosvold
University of California Cooperative Extension
Watsonville, CA, USA

S Torre
Agricultural University of Norway
Aas, Norway

M J Tsujita
University of Guelph
Guelph, ON, Canada

Y Ueda
Chiba University
Matsudo, Japan

L Urban
INRA/CIRAD
Saint-Pierre de la Réunion, France

A Vainstein
The Hebrew University of Jerusalem
Rehovot, Israel

P A van de Pol
Wageningen University
Wageningen, The Netherlands

U van Meeteren
Wageningen University
Wageningen, The Netherlands

H A Verhoeven
Plant Research International
Wageningen, The Netherlands

M-R Viricel
Université Claude Bernard
Lyon, France

D Weiss
The Hebrew University of Jerusalem
Rehovot, Israel

G Werlemark
Swedish University of Agricultural Sciences
Kristianstad, Sweden

V Wissemann
Friedrich-Schiller-Universität Jena
Jena, Germany

X Xu
Horticulture Research International
East Malling, West Malling, UK

D Zamir
The Hebrew University of Jerusalem
Rehovot, Israel

D Zhang
Geves Le Magneraud F-17700
Surgères, France

CONTENTS

VOLUME 1

VOLUME 2

I

J

M

N

O

P

R

S

T

V

VOLUME 3

I

INSECTS AND OTHER ANIMALS

Contents

Overview of Insects

J D Shorthouse, Laurentian University, Sudbury, ON, Canada

Introduction

Most modern roses, and the gardens, greenhouses and propagation fields in which they grow, are biologically simplified and much effort is directed at avoiding our sharing them with other animals. Unfortunately, we are not always successful because garden roses, bred for large showy flowers, are sufficiently similar to their wild ancestors that most insects cannot distinguish them. To most rose fanciers, these insects are the enemy and all attempts are made to eradicate them.

There is no doubt that some rose-feeding insects cause intolerable damage and control or eradication is warranted (*see* **Insects and other Animals**: Insecticides). Some damaging insects suck the juices from the cells of roses while others cause unsightly bite marks on leaves and blossoms or consume entire leaves leading to loss of vigour, death of shoots or branches, or worse, death of the entire shrub. However, some predacious insects spend their lives on roses consuming pest species, whereas other insects visit rose flowers to feed on pollen. In the natural world, pollen feeders are not only harmless, but necessary for rose survival. Accidentally killing flower visitors, predators of rose pests, or those simply using roses as resting sites as they undertake their daily activities would be unfortunate.

Both garden and wild roses are an integral part of the food web of a diverse community of insects. Understanding the role of these insects and distinguishing harmful from beneficial ones is a challenge for rose growers. The purpose of this article is to survey some of the common insects that feed on garden roses and describe aspects of their feeding biology that will not only aid in identifying the insects in question, but also help in deciding whether or not control efforts are warranted. No attempt is made to recommend insecticides for controlling damaging species, as these chemicals vary from country to country and year to year. Current information on insecticidal control is readily available from local nursery owners, district agriculturalists, etc.

Basic Insect Biology

Insects undergo a series of changes or metamorphoses which may be either incomplete or complete. Those with incomplete metamorphosis include thrips, leafhoppers, aphids and scale insects. The eggs of these insects hatch into nymphs which resemble the parents except that they are smaller, wingless and sexually immature. Such insects continue to feed on the same foods from the time the eggs hatch into nymphs until they die as adults. Insects with incomplete metamorphosis do not have a pupal stage. Insects with complete metamorphosis include beetles, moths, flies, sawflies and gall wasps. Here eggs hatch into active, feeding and growing larvae which bear no resemblance to the adults. The feeding period is followed by the inactive pupal stage during which the insect changes completely into a sexually mature and winged stage. The stage at which injury is inflicted on roses varies with the species of insect. For example, thrips, aphids and leafhoppers are injurious in both the nymphal and adult stages, moths and sawflies are injurious only in the larval stage and some leaf-eating beetles and leaf-cutting bees are injurious as adults only.

All insects in northern environments must be able to tolerate cold temperatures and each species has a particular stage in which it overwinters. Aphids, for example, overwinter as eggs while many moths overwinter as pupae. Some insects overwinter on the roses while others overwinter in the soil. Some insects enter into an arrested state of development called diapause which prevents them from becoming active during warm spells in midwinter.

Feeding Biology of Rose Insects

The type and location of feeding damage are the most important clues in determining that damage is insect-caused and in identifying the insect responsible. Insects feed in one of two methods. Those with **mandibulate** mouthparts bite off and chew plant tissues whereas those with **haustellate** mouthparts imbibe liquid nourishment. Sucking insects are usually smaller than chewers. Mandibulate insects use mandibles with their toothlike ridges to cut and grind plant tissues, and other mouthparts such as the maxillae to manipulate and guide food towards the mouth. Insects with haustellate mouthparts have their mandibles and other mouthparts extended into thin, needle-like stylets. When the stylets are held together, they form two tubes, one for injecting saliva and the other for sucking up fluids. Plant bugs and aphids transmit viruses both directly and indirectly, resulting in damage to their hosts. Most of these insects feed on the complex sugar solutions of the phloem vessels or sap. Plant sap is a nutritionally dilute food, rich in sugars, but deficient in nitrogen and other mineral elements. Large quantities must be ingested to sustain growth.

The types of mouthparts may stay the same or change during the insect's life cycle. For example, caterpillars of moths have chewing mouthparts whereas the adults have tubular mouthparts for imbibing fluids. Conversely, other insects such as weevils have biting mouthparts in both immature and adult stages. Aphids and leafhoppers feed both in immature and adult stages. Suckers usually inflict less mechanical injury; however, they often cause serious deformation and stunting of shoots. Some insects such as spittle bugs feed singly whereas others, such as the Japanese beetle, aggregate when they feed.

Insects feeding on roses are specialists or generalists. Specialists either have a unique feeding manner with their hosts or feed exclusively on roses. Host-specific insects such as the rose aphid (*see* **Insects and other Animals**: Aphids) and rose gall wasps (*see* **Insects and other Animals**: Galls Induced by Cynipid Wasps) are called **monophagous**. Generalists such as the Japanese beetle and larvae of the gypsy moth feed on plants of many different families and are called **polyphagous**. Insects also show specialization with regard to the plant organ on which they feed. Some insects are specialist feeders on the tender tissues of leaf buds, the palisade or mesophyll cells within leaves, fluids within the vascular bundles or tissues inside the stems. Other insects are specialist feeders on flowers or tissues within the hips.

Insects on roses can also be categorized into different **guilds** which are groups of unrelated species that exploit the same resource in a similar manner. Examples of herbivorous guilds on roses include leaf-feeding, stem-boring, root-feeding, leaf-mining, gall-inducing and pollen-eating. Some of these guilds can be further divided. For example, leaf feeders can be divided into strip-feeders where insects remove large pieces of leaves, skeletonizers where all of the leaf except the vascular bundles are consumed or pit-feeders where small holes are made into the leaf surface.

Where Do Rose-Feeding Insects Come From?

All garden roses are derived from wild or species roses which are fed upon by many species of insects. Since garden roses are biologically similar to wild roses, it has been relatively easy for insects to move from one host to the other. Most rose insects are good fliers or small enough to be blown by the wind and can either fly directly to garden roses or land on them by chance. Some gardeners inadvertently introduce insects to their gardens when transplanting wild roses from their natural habitats. More serious is the accidental introduction of insects from one continent to another. This occurs when potted or bare-root stock roses are packaged on one continent along with insect eggs, immatures, or the adults attached, and shipped along with the plants to another continent.

Rose Insects with Haustellate Mouthparts

Spider Mites

Spider mites are not insects, but tiny animals of the Class Arachnida with simple, oval-shaped bodies and no wings or antennae (**Figure 1a**). Most species

Figure 1 (a) Two-spotted spider mite, *Tetranychus urticae* (reproduced with permission, Jack Kelly Clark, University of California). (b) Adult tarnished plant bug, *Lygus lineolaris* (reproduced with permission of Scott Bauer, ARS/USDA). (c) Nymph of spittlebug, *Philaenus spumarius*. (d) Rose leafhopper, *Edwardsiana rosae* (reproduced with permission, Jack Kelly Clark, University of California). (e) Whiteflies, *Trialeurodes vaporariorum*; two adult whiteflies are near the centre while the three spiny structures are cast skins of the last instar (reproduced with permission of the University of California). (f) Rose aphids, *Macrosiphum rosae*.

produce silk webbing on infested leaves. Spider mites pass through egg, six-legged larval, and two eight-legged nymphal stages before transforming into eight-legged adults with numerous bristles covering the legs and body (*see* **Insects and other Animals**: Mites). Adult females live for several weeks and lay numerous eggs during their lifetime. Under ideal conditions, they can complete their development from egg to adult in less than 2 weeks. Spider mites have needle-like mouth-parts and feed by piercing the leaves of host plants and sucking fluids from individual cells. The killing of individual or groups of cells results in transparent, yellow or tan patchworks. This causes leaves to have a stippled or flecked appearance, with pale dots where the cellular contents have been removed. Heavily infested plants may be discoloured, stunted or even killed. Females disperse to other plants by the wind. Most mites feed in colonies on the undersides of leaves, making them difficult to control with insecticides (*see* **Insects and other Animals**: Mites). Damage is most evident from the upper leaf surface. Early detection can be done by taking a piece of white cardboard and striking it with foliage. If present, the tiny, dot-sized mites will be seen (best seen with a 10× hand lens) walking slowly on the card. Spider mites have many natural enemies, including predacious mites, that reduce populations. Broad-spectrum insecticides often kill mite predators.

The most common spider mite on both field and greenhouse roses in North America is the two-spotted spider mite, *Tetranychus urticae* Koch. They are yellowish-green and usually have two large dark areas on either side of the body at about mid length (**Figure 1a**). They are easily seen wandering over the plants. Most spider mites spend the winter in the egg stage but the two-spotted spider mite overwinters as adult females resting in protected places. Females become active in April and May when they seek out the undersides of leaves. It is a global species and easily moved onto many kinds of plants.

Plant Bugs

The green capsid bug *Lygocoris pabulinus* (L.), common in Europe, is one of the few hemipteran pests of roses. Hemipterans, or the true bugs (only insects of the Order Hemiptera should be referred to as ‘bugs’), are distinguished from those in the Order Homoptera by the thickened and leathery-like basal portions of their front wings, whereas the apical portion is membranous (**Figure 1b**). The entire wings of homopterans are membranous or have a uniform texture throughout. Adults are fast-moving insects and fly when alarmed. Both adults and nymphs drop from the foliage when disturbed making them difficult to find. Nymphs and adults feed throughout the summer and overwinter in the egg stage. Eggs give rise to nymphs that feed by inserting their stylets, encased in a slender, segmented beak, into the leaf tissues and extracting the sap. Chlorophyll is removed and a toxic salivary secretion is injected which digests the glue that holds plant cells together. When feeding is concentrated, individual feeding spots may coalesce to form large brown blotches. The tarnished plant bug, *Lygus lineolaris* (Palisot de Beauvois) (**Figure 1b**), frequently feeds on roses, as does the four-lined plant bug *Poecilocapsus lineatus* (Fabricius). The four-lined plant bug is bright yellow with four distinct stripes and is widely distributed throughout Canada and the United States.

Spittlebugs

Spittlebugs, as they are called in the nymphal stage, or froghoppers in the adult stage, are small hopping insects rarely over 13 mm in length that vaguely resemble small frogs. They are found in the Order Homoptera. Adults are brown and like the nymphs, feed by sucking fluids. The presence of frothy, spittle-like masses on the shoots of roses denotes the presence of spittlebug nymphs (**Figure 1c**), which later develop into the froghopper. Each mass of spittle contains a yellowish-green, soft-bodied nymph whose feeding causes the wilting of the flower buds and bending of young shoots. The most common spittlebug in North America is the meadow spittlebug, *Philaenus spumarius* (L.) whereas a common species on roses in Britain is *P. leucophthalmus* L.

Rose Leafhopper

Leafhoppers are small insects of the Order Homoptera, about 10 mm in length, that hold their membranous and often colourful wings in a tent-like manner over their bodies. The most common species on North American roses is *Edwardsiana rosae* L. (**Figure 1d**), an introduced European native. It is an active, pale yellowish insect which takes flying leaps into the air when disturbed. Its feeding causes mottled areas on leaves during spring and summer. Severe damage can cause premature leaf fall. Roses serve as the food plant for only one of the several generations. The autumn brood of eggs is deposited within the bark of wild or cultivated roses. The young white nymphs, with their red eyes and small black dots on the thorax and wing pads, hatch in the spring. They establish themselves on the underside of a leaf and if the food quality is satisfactory, their entire development is spent at the same spot. Most first-brood adults vacate roses for other woody plants. They injure their hosts in three ways: by destruction of the chlorophyll, by removal of the plant fluids and by winter egg deposition. Fungal diseases often enter rose canes when tissues are cut by the egg-laying apparatus.

Whiteflies

Whiteflies are minute, snow-white insects of the Order Homoptera, rarely more than 2–3 mm in length, that resemble small moths (**Figure 1e**). They are not flies but are closely related to aphids and scale insects. Adults of both sexes are winged with the wings covered with a white dust or waxy powder. They are mostly found in groups on the underside of new leaves and readily fly when the plants are disturbed. The first instars are transparent to opaque, and range from light green to black. They move a short distance after hatching, settle and begin sucking juices. The first instars are active, but subsequent instars do not move and look like scales. The last instar looks like a pupa (**Figure 1e**) from which the adults appear. Infested plants become chlorotic. Whiteflies secrete honeydew which is a good medium for growing black sooty moulds. These moulds, which interfere with photosynthesis and lessen plant vigour, can be washed off with a strong spray of water. The common greenhouse whitefly, *Trialeurodes vaporariorum* (Westwood), is a worldwide pest of greenhouse-grown roses. Insecticide applications often fail to provide adequate control and an integrated pest management programme with cultural and biological methods is essential (*see* **Integrated Pest Management**). The greenhouse whitefly has developed resistance to some insecticides and few insecticides are registered for greenhouse use.

Aphids

Aphids are soft-bodied insects of the Order Homoptera that feed on the sap of cells on the underside of rose leaves and on young shoots and flower buds (**Figure 1f**). Aphids have complex life cycles which frequently occur on more than one species of plant (*see* **Insects and other Animals**: Aphids). Eggs are laid in the autumn and in the following spring give rise to females that produce living young. Their rate of reproduction is very high during the spring and summer phases as each aphid matures quickly and produces other individuals, all of which are females. Both male and female winged individuals are produced in the late summer. These individuals mate and lay eggs.

Aphids feed by inserting their stylets between the cells until they reach and puncture phloem cells of a vascular bundle. Saliva is secreted at the end of the stylets which breaks down the bonding between cells. Viruses can be transmitted by aphid saliva. Aphids must imbibe large amounts of fluid in order to acquire sufficient protein. The residual sugary fluid is ejected to the exterior in the form of honeydew, which can coat plants with a sticky residue.

The vigour of roses is reduced by aphid feeding as large amounts of sap is extracted. Although roses can tolerate many aphids, the quantity and quality of flowers is reduced by large populations and insecticides are sometimes needed. In many cases, leaves may become malformed and disfigured. Water jets are effective because, once removed, the wingless nymphs will not climb back onto the bush.

Many species of aphids attack cultivated roses (*see* **Insects and other Animals**: Aphids), with one of the most common being the large (2.5 mm) rose aphid, *Macrosiphum rosae* (L.). Originally described from Europe, this species feeds only on roses and is common throughout Canada and the United States. The entire life cycle can be spent on one host plant.

Aphids are commonly attacked by predators such as the larvae and adults of ladybird beetles and lacewings and the larvae of syrphid hoverflies. The larvae of small parasitic wasps develop within the body of an aphid; later the dead aphid's body, called a 'mummy', is stuck to the leaf.

Scale Insects

Scale insects, also of the Order Homoptera, produce waxy, round or oval scales over their bodies which protect the small, sessile insects beneath for most of their lives. Like gall-inducing insects (*see* **Insects and other Animals**: Galls Induced by Cynipid Wasps), scale insects are better known for the structures they produce than the insects themselves. Scales housing the females are larger than those of males. Winged males appear during early summer and mate with the wingless females. Eggs are laid beneath the scaly covering in July and August and the orange nymphs, called 'crawlers', wander over the plants. The nymphs settle into a sedentary existence and feed by inserting their stylets into the tissues and suck sap. Foliage wilts, turns yellow and drops from the plant. Growth is stunted and flowers are not produced. Severe infestations are unsightly owing to masses of the incrustations covering the leaves and stems. Adult scales are generally not affected by insecticides; however, the crawlers without their protective covers are vulnerable, including alternatives to conventional insecticides such as soaps and oils. It is best to cut out and burn the old and badly infested stems.

The rose scale, *Aulacaspis rosae* Bouché, is a common North American species, known as an armoured scale, which attacks canes and twigs. It overwinters in all stages from the egg to pregnant females, but mostly as young scales. Eggs are laid and protected by the scale of the female until they hatch and disperse to favourable spots on the canes. They insert their stylets, feed on cell sap and then moult and lose their legs and antennae. The scale, formed from wax secreted from the body wall, appears over their bodies (**Figure 2a**). Females remain under the scales for the duration of their lives. The males become small, two-winged, yellowish insects with legs, eyes, antennae and a prominent appendage extending from their abdomens. Males seek out and mate with the females and after mating, the females continue feeding and produce eggs.

The cottony maple scale *Pulvinaria innumerabilis* (Rathvon) is one of the largest and most conspicuous scales that attacks roses and other ornamental trees. Found throughout the United States and much of southern Canada, it overwinters as an immature, flat inconspicuous female on twigs of its host. It produces a white egg sac in the spring, and the crawlers appear in June and disperse to the leaves. Some scales found mainly in the southern United States, such as the California red scale, *Aonidiella aurantii* (Maskell), attack both citrus and ornamental shrubs. The cottony

(a) (b) (c) (d) (e) (f)

Figure 2 (a) Rose scale, *Aulacaspis rosae*. One of the scales has been lifted to show the crawlers beneath. (b) Western flower thrips, *Frankliniella occidentalis* (reproduced with permission, Jack Kelly Clark, University of California). (c) Larvae of scarabs feeding on rose roots. (d) Japanese beetle, *Popillia japonica*. (e) Rose chafer, *Macrodactylus subspinosus*. (f) Mature larva of rose-stem girdler, *Agrilus aurichalceus*.

cushion scale, *Icerya purchasi* Maskell, which was introduced into California from Australia in 1868, now occurs from California to North Carolina and feeds on many plants and trees besides roses. Other scales attacking roses are the peach scale *Eulecanium corni* Bouché and the nut scale *E. coryli* L.

Thrips

Thrips are small (1.5–3 mm in length), slender insects of the Order Thysanoptera, seldom exceeding 3 mm in length with two pairs of narrow wings (**Figure 2b**) that are fringed with long fine hairs. The life cycle includes the egg, two actively feeding nymphal stages, non-feeding prepupal and pupal stages, and the adult. Females of most species insert their tiny eggs into the leaves or buds of plants. The pale prepupae and pupae of most species drop to the soil or leaf litter or lodge within plant crevices.

Thrips have stylet-like rasping mouthparts used by feeding nymphs and adults to lacerate and break up surface tissues of soft petals and buds and suck the mashed food and juices with their imperfect suctorial mouthparts. Petals become blotched or streaked with discoloured areas, whereas buds become distorted and do not open properly. Areas where thrips have been feeding are covered with characteristic black specks, which are their faeces, and older leaves can develop a silver sheen.

Most thrips breed on various grasses and weeds then move to ornamental plants. White roses seem more attractive to thrips than other colours. Thrips are poor fliers but can spread long distances by floating with the wind or being transported on infested plants. To detect thrips, shake a blossom over a white sheet of paper and examine the insects with a hand lens. Thrips are difficult to control because plants are continually infested by dispersing adults and they tend to hide in flowers or crevices on the plant, making them difficult to reach with pesticides.

The most serious pest on roses in Europe is *Thrips fuscipennis* Haliday. Another species is *Frankliniella intonsa* Trybom which attacks flower buds and open blooms. The western flower thrips *F. occidentalis* (Pergande) is the most abundant and widely distributed species on roses in Canada and the United States and a target for integrated pest management (*see* **Insects and other Animals**: Thrips; **Integrated Pest Management**).

Rose Insects with Mandibulate Mouthparts

Chafers

Beetles, of the Order Coleoptera, are among the world's most diverse and ecologically successful animals and many phytophagous species feed on roses. Both larvae and adults have biting mouthparts; however, in most cases only the adults consume aboveground tissues. The most damaging and readily noticed are chafers or scarabs and weevils. Larvae of metallic wood-boring beetles and long-horned beetles feed within the stems of some roses. The most common chafers on roses in Europe are the cockchafers or May bug *Melolontha melolontha* L., the rose chafer *Cetonia aurata* L., the garden chafer or bracken clock *Phyllopertha horticola* L. and the summer chafer *Amphimallon solstitialis* L.

The larvae of scarabs are thick, fleshy, dirty-white, C-shaped grubs with the last segments of the abdomen swollen and black or purple. They are only observed when a rose patch is dug up (**Figure 2c**) or plants are transplanted. They have strong mandibles with which they sever roots of roses and other plants and girdle underground stems.

Melolontha melolontha are large beetles. They feed for 3 to 4 years and cut off food supplies to the aerial portions. Adults fly at dusk from mid May to early June. *Cetonia aurata* are smaller than the cockchafer and are distinguished by their metallic, golden–green elytra flecked with white lines. The larva is shaped like that of the cockchafer but is clothed with rows of reddish hairs. Adults feed on blooms devouring the petals and anthers and sometimes the leaves. *Phyllopertha horticola* is smallest of the four European chafers. The pronotum is metallic bluish-green and the elytra is reddish-brown. Adults fly in bright sunlight during June and often appear in swarms on flower buds and open blooms. Grubs live for only 1 year. Grubs of all species should be hand-picked when turning over the soil.

The most damaging chafer in North America is the Japanese beetle *Popillia japonica* Newman. This greenish-bronze beetle (**Figure 2d**), about 1 cm long, has white tufts of hair along sides of abdomen and wing covers. It was introduced into New Jersey in 1916 and has now spread throughout most of the eastern United States and southern Ontario and Nova Scotia. Larvae feed on the roots of grasses; however, they will also feed on rose roots. They overwinter as larvae in the soil below the frost line and the adults emerge in May and feed on numerous shrubs and trees; however, roses seem to be their favourite where they chew on flower buds and open flowers and skeletonize foliage. Adults are most active during the warmest part of the day and prefer roses exposed to the sun. They are controlled biologically with a disease-causing bacterium commonly called milky disease, caused by *Bacillus popilliae* (Doom).

Other scarabs, including *Macrodactylus subspinosus* (Fabricius) (**Figure 2e**), also called the rose chafer, are biologically similar to the Japanese beetle. Found mainly in eastern Canada and the northeastern United States, they aggressively feed on the foliage and flower buds. They are tan, long-legged, slender beetles about 8–12 mm long. Birds are killed by eating adult rose chafers. Light infestations can be hand-picked. Insecticidal control is difficult especially when adults keep flying in from surrounding untreated areas. *Trichiotinus texanus* (Horn), a pubescent black beetle with four white stripes on the elytra, is common in eastern Canada and adjoining parts of the United States whereas the western rose chafer, *Macrodactylus uniformis* Horn, and the green rose chafer, *Dichelonyx backi* Kirby, are common in the west.

The hoplia beetle, *Hoplia callipyge* LeConte, is reddish-brown, about 1 cm in length and is found from California to British Columbia. It is a day flier and prefers white flowers, on which it chews round holes in the petals. This beetle is best controlled by hand-picking. In Hawaii, the Chinese rose beetle, *Adoretus sinicus* Burmeister, is the most destructive chafer. Adults of this beetle come out at night and feed in groups on the undersides of leaves while the larvae feed on decaying organic matter.

Wood-Boring Beetles

Beetles with larvae that bore into the stems of rose include the flatheaded apple tree-borer, *Chrysobothris femorata* (Olivier) with adults known as metallic wood-borers. It feeds on a variety of trees and shrubs besides rose. Eggs are laid in bark crevices and the larvae bore into inner phloem making irregular tunnels which are filled with powder-like frass. It is found throughout southern Canada and the United States. The rose-stem girdler *Agrilus aurichalceus* (Redtenbacher) damages wild roses and *Rosa rugosa* Thunberg across Canada and the United States by causing the canes to desiccate and break off. The young larvae make spiral tunnels just under the bark and when they mature, they enter the stem and feed on pith. They overwinter as larvae inside the canes (**Figure 2f**) and pupate early in the spring, and the adults emerge to lay eggs in the bark of canes. One remedy is to clip off and dispose of infested canes in the spring before the beetles emerge. Another species, *Chrysobothris mali* Horn, causes the same kind of injury from British Columbia to California. Regular fertilization and watering promotes vigorous growth and reduces attacks.

Weevils

The most common weevil, beetles with distinctive long snouts and chewing mouthparts at their tips, is the rose curculio *Merhynchites bicolor* (Fabricius). This beetle is reddish to black (**Figure 3a**) and is destructive to roses throughout southern Canada and the United States. The greatest damage is caused by adults puncturing the floral parts inside the flower buds. When the buds open, the petals are riddled with holes resulting in ragged, unsightly blossoms. If flower buds are not common, the adults will feed on the tips of new rose shoots causing the death of the terminals. Eggs are laid in flower buds and the legless larvae feed within the blossoms. Full-grown larvae leave the bud and enter the soil where they overwinter. Pupation occurs in the spring. Frequent monitoring for the presence of rose weevils is important and regular hand-picking of the adults and removal of damaged buds and spent blossoms helps prevent future damage.

The Fuller rose beetle, *Asynonychus godmani* (Crotch), an ash-grey beetle introduced from Mexico, is a common pest of outdoor roses in the southern United States, and is also a greenhouse pest in other parts of the United States and Canada. Adults are flightless and all are females. They feed on the leaves and blossoms, leaving ragged or scalloped edges.

Diptera

The rose midge *Dasineura rhodophaga* (Coquillett) is one of the few flies, of the Order Diptera, that feed on roses. Adults of rose midges are mosquito-like and about 2 mm in length. The legless larvae tunnel through developing leaf and rose buds causing them to bend or become misshapen and then to blacken and die. They can devastate blooms in fields or greenhouses. They overwinter as pupae in the soil and in the spring eggs are laid inside the sepals of flower buds. The foliage and the stem surrounding affected buds may also blacken and die. Midge populations can be reduced by pruning infested buds. The pink larvae of another midge, the red bud borer *Thomasiniana oculiperda* Rübs, live between the bud grafts and the stock. Legless larvae of this species feed on sap in the layers between the bud-scion and shoot with the result that the buds wither and die. The best control is to prune and destroy infested plant parts. Systemic insecticides can be applied to the ground around the plants where the larvae pupate. A larger fly, the rose hip fly *Rhagoletis alternata* (Fallèn), has maggots that feed on the pulpy tissues of rose hips.

Larvae of Moths

All moths belong to the Order Lepidoptera and although several polyphagous species feed on roses, only a few are serious pests. Most larvae of moths feed on other species of ornamentals besides roses. Larvae of moths have five or fewer prolegs on the

Figure 3 (a) Rose curculio, *Merhynchites bicolor*. (b) Fuller rose beetle *Asynonychus godmani* (reproduced with permission, Jack Kelly Clark, University of California). (c) Redhumped caterpillar, *Schizura concinna* (reproduced with permission, Jack Kelly Clark, University of California). (d) Forest tent caterpillar, *Malacosoma disstria*. (e) Larva of sawfly, *Arge ochropa* (reproduced with permission of György Csóka). (f) Rose leaves damaged by leaf-cutter bee.

abdomen. All feed externally during their larval stages except the rose leaf-miner which feeds within the leaves.

Common species in Europe are the buff-tip, lackey, vapourer and winter moths, the leaf-rolling tortrix moths, the leaf skeletonizing moths and the rose leaf-miner.

In North America, the redhumped caterpillar, *Schizura concinna* (J. E. Smith) is common on roses and other ornamentals throughout southern Canada and the United States. Mature larvae are yellow with a red head and lined longitudinally with reddish, black and white stripes (**Figure 3c**). A number of prominent black tubercles arise from the back and there is a large reddish hump on the top of the first abdominal segment, giving this insect its name. Eggs are laid on the underside of leaves. Newly hatched larvae feed gregariously at first, but then disperse to skeletonize leaves. They pupate in the soil where they overwinter.

The ugly nest caterpillar, *Archips cerasivoranus* Fitch, is found throughout southern Canada and northern United States. It has a yellowish-green larva with a black head. The caterpillars are gregarious and spin a dense web around feeding sites that become filled with their excrement and pieces of leaves. They pupate in the mess they create. Adults emerge from July to September and lay eggs on branches for the winter. Damage is not severe but the ugly nests alter the aesthetic appearance of the plants.

The forest tent caterpillar, *Malacosoma disstria* Hübner (**Figure 3d**), is an irruptive species that feeds on roses, along with many other shrubs and trees, when populations are high. Roses are quickly defoliated during such times and neither hand-picking nor insecticides can control them. It is found across Canada and the northern United States and despite its name, it does not construct tents. Larvae are heavily clothed with hair and reach about 50 mm in length. Eggs are laid in small sleeves encircling branches. They overwinter in these masses and roses should be checked carefully in either the winter or early spring and the eggs removed by hand.

The white-marked tussock moth, *Hemerocampa leucostigma* (Smith), also occurs throughout southern Canada and northern United States and feeds on many hosts besides roses. It overwinters in the egg stage, eggs hatch in the spring and larvae disperse to skeletonize leaves. Older larvae chew the entire leaf except for the main veins and petiole. Full grown larvae are about 30 mm long with a red head and yellowish body with distinctive hair. Mature larvae pupate in a greyish cocoon which contains many body hairs and is constructed on twigs or branches.

Sawflies

All sawflies, bees and wasps belong to the Order Hymenoptera. Adult sawflies are primitive wasps with their abdomen bluntly attached to the thorax. They actively fly above the bushes during bright sunny days in May then deposit eggs into stems or leaves with their sawlike ovipositors. Larvae of sawflies are caterpillar-like (**Figure 3e**) except they have seven or eight pairs of prolegs compared to the five pairs of moths. Only the larvae feed on leaf tissues. Once mature, they drop to the ground and pupate in the soil. The larvae of about 20 species feed on roses in Europe. One species, the large rose sawfly, *Arge ochropa* (Gmelin), has bluish-green larvae with yellow–black along the back (**Figure 3e**) and six rows of black shining bristle-bearing tubercles. Larvae eat leaves from July to October and there are two generations per year. This species has been introduced into Canada and the United States and is becoming a serious pest of roses. The antler sawfly, *Cladius pectinicornis* Geoffroy, has flat, green larvae that taper towards the head and posterior. It occurs on the underside of leaves and eats irregular holes between the veins and along the margin. Larvae of the banded rose sawfly, *Emphytus cinctus* L., eat small circular areas in the leaves without breaking through to the upper epidermis. At rest they remain curled in a ball on the lower leaf surface. When full grown, the larvae tunnel into the stems to pupate.

The shiny, black adults of the rose slug sawfly, *Endelomyia aethiops* Fabricius, have yellowish larvae and the green contents of the food canal can be seen through the skin. Rose slugs are not slimy and do not have rasping mouthparts like true slugs. They feed on the upper side of the leaves and devour the entire leaf tissue, leaving only the lower epidermis unbroken. Attacked leaves are skeletonized and later turn brown. When mature, larvae drop to the ground and spin a cocoon. Some cocoons adhere to the roots making them easy to move from one site to another.

The shoot-borer sawfly, *Ardis brunniventris* Hartig, is a true stem borer in the larval stage. The leaf-rolling rose sawfly, *Blennocampa pusilla* Vitasaari & Vikberg, is most destructive causing a complete folding of the leaflets. Severe attacks result in loss of vigour due to reduced photosynthesis and premature leaf fall. Folding occurs before the eggs hatch and is caused by a chemical injected at oviposition.

There are three injurious rose slug sawflies on roses in North America: the bristly rose slug *Cladius isomerus* Norton, the introduced European rose slug *Endelomyia aethiops* Fabr. and the curled rose slug *Allantus cinctus* L. The bristly rose slug is found from Nova Scotia to British Columbia and down into southern California. Young larvae begin feeding as skeletonizers on the underside of the leaves and then chew large holes as they mature. Mature larvae are 16 mm in length, pale green and covered with hair-like bristles. There are as many as six generations each year. The European rose slug is found from the Rocky Mountains east to the Atlantic coast and is only found on roses. The curled rose sawfly, largest of the three species, begins by skeletonizing leaves and then devours the entire leaflet except for the main vein. When ready to pupate, the larva bores into the pith of twigs which kills a portion of the stem and opens it to fungus infections. Larvae are easily washed from roses with a strong jet of water from a garden hose. The stem-boring sawfly, *Hartigia trimaculata* Say, feeds inside current-year shoots. Adults appear in June and lay eggs in punctures in the shoots and the larvae girdle the twig causing the twig to wilt. As the larvae mature, they enter the pith and feeds downward. A related species, *Hartigia cressoni* (Kirby), is found in California and Nevada.

Leaf-Cutter Bees

Leaf-cutter bees are stout-bodied, solitary bees about the size of a honeybee, but darker and with light bands on the abdomen. They line their nest chambers in decaying logs or spaces between brick and stone walls with pieces cut from leaves. Roses are one of their favoured plants and it is common to find leaves with neatly cut circles or ovals (**Figure 3f**). Pieces are cut with mandibles of the adult bees and carried to the nest. A common species in Europe is *Megachile centuncularis* L. and in Canada *M. latimanus* Say. Adults are rarely observed cutting the leaves and even though the damage is sometimes severe, little can be done to control the bees.

Rose Seed Wasps

Minute wasps of the genus *Megastigmus* sp. lay their eggs in the seeds within rose hips. The larvae consume the soft internal tissues of seeds.

Cynipid Gall Wasps

Cynipids are small, sturdy reddish brown to black wasps about 6 mm in length. All are of the genus *Diplolepis* and the adults only live for about a week. They are rarely seen by gardeners; however, the complex galls they induce are most noticeable (*see* **Insects and other Animals**: Galls Induced by Cynipid Wasps). At least three species of cynipid wasps induce galls on roses in gardens of Europe and they have been accidentally established in North America. At least five species of North American *Diplolepis* have moved from wild roses in Canada to rugosa roses.

Beneficial Insects

Numerous species of bees, wasps, flies and beetles feed on the pollen of roses and are common visitors to wild and garden roses. Roses provide little or no nectar to flower-visiting insects, but they supply copious amounts of pollen. Most of the rose pollen produced is provided as a reward for insect visitors and in return, the plants are pollinated and gain from outcrossing (*see* **Insects and other Animals**: Pollination). The five petals of wild roses form a bowl-like landing pad for insects as do the multipetalled flowers of garden roses. Insects consume pollen to obtain high-quality food, rich in protein, fats, vitamins and trace elements. Common bees visiting rose flowers are honeybees, bumble bees, sweat bees and carpenter bees and wasps including metallic cuckoo wasps. Many species of syrphid flies and beetles such as long-horned wood borers, checkered beetles, scarabs and small carabid beetles feed on rose pollen. Common predators of rose-feeding insects include the larvae of syrphids, lacewings and ladybird beetles that feed on aphids, and vespid wasps and damsel bugs that feed on caterpillars.

Concluding Comments

It is unlikely that garden roses will ever be free of damaging insects and there is always the concern that new ones will move from wild roses or be introduced from other continents. Rose fanciers must remain vigilant over the possibility of damaging insects colonizing their shrubs, engage chemical controls only when necessary, and accept that in some cases when insect populations are low or the damage minimal, that the best tactic is to physically remove those unwanted. In an era of increasing public concern over misuse of insecticides and the need to conserve biodiversity, there is perhaps more tolerance towards sharing rose bushes with at least some species of insects that include roses in their life history strategies. In the words of American entomologist May Berenbaum:

> Like it or not, insects are part of where we have come from, what we are now, and what we will be. It seems to me that's a pretty good reason for getting better acquainted with them.

See also: **Insects and other Animals**: Aphids; Mites; Thrips; Galls Induced by Cynipid Wasps; Pollination; Insecticides. **Integrated Pest Management**.

Further Reading

Berenbaum MR (1995) *Bugs in the System: Insects and Their Impact on Human Affairs*. Reading, MA: Addison-Wesley.

Bernhardt P (1999) *The Roses's Kiss: A Natural History of Flowers*. Washington, DC: Island Press.

Borror DJ, Triplehorn CE and Johnson NF (1992) *An Introduction to the Study of Insects*, 6th edn. Fort Worth, TX: Harcourt College Publishers.

Flint ML and Karlik JF (2001) *Integrated Pest Management for Floriculture and Nurseries*. Oakland, CA: University of California.

Grissell E and Goodpasture C (2001) *Insects and Gardens: In Pursuit of a Garden Ecology*. Portland, OR: Timber Press.

Johnson WT and Lyon HH (1991) *Insects That Feed on Trees and Shrubs*. Ithaca, NY: Cornell University Press.

Milne L and Milne M (2001) *National Audubon Society Field Guide to North American Insects and Spiders*. New York: Alfred A. Knopf.

Potter DA and Held DW (2002) Biology and management of the Japanese beetle. *Annual Review of Entomology* 47: 175–205.

Price PW (1997) *Insect Ecology*, 3rd edn. New York: John Wiley.

Aphids

R G Foottit and H E L Maw, Agriculture and Agri-Food Canada, Ottawa, ON, Canada

Introduction

Aphids are small soft-bodied insects with sucking mouthparts, closely related to other plant-feeding sucking insects, such as plant bugs, leafhoppers, whiteflies, scale insects and psyllids. They are most easily recognized by the pair of dorsal openings on the posterior part of the abdomen, usually situated at the end of tubular appendages called siphunculi (**Figure 1**). Characteristics of their biology and the means by which they exploit their host plants have made them a very successful group, and many species have become economically significant plant pests. Populations can grow rapidly, resulting in the relatively sudden appearance of obvious dense colonies on new plant growth. Aphids occur throughout the world, but the greatest numbers of species are found in temperate regions. In this article, aspects of aphid biology, damage and control, with particular reference to species occurring on rose, are discussed, with a synopsis of relevant aphid species.

Feeding Damage

Mouthparts of aphids consist of slender stylets, enclosed within a sheath. The stylets are inserted between the cells of the surface tissues of the plant, and into the phloem cells. Because of the relatively low concentration of proteins and other essential nutrients in the phloem, large volumes are ingested, and the excess fluid and sugars excreted as sticky 'honeydew'. The amount of honeydew produced by a colony of aphids can be considerable, fouling the plant and providing a medium for the growth of sooty moulds, and attracting ants and wasps. Most species of aphids feed on new growth where nutrient flow is high, or on senescent leaves where tissue breakdown products are being mobilized and resorbed by the plant. Feeding on buds and unfurling leaves produces distortion, both as a result of physical damage and from reaction of the plant to components of the saliva. Plant vigour may be reduced, resulting in abortion of flower buds. In addition to the direct effects of feeding, plant pathogens may be transmitted on the mouthparts and in the saliva of aphids. However, aphids are apparently not significant vectors of viruses of rose.

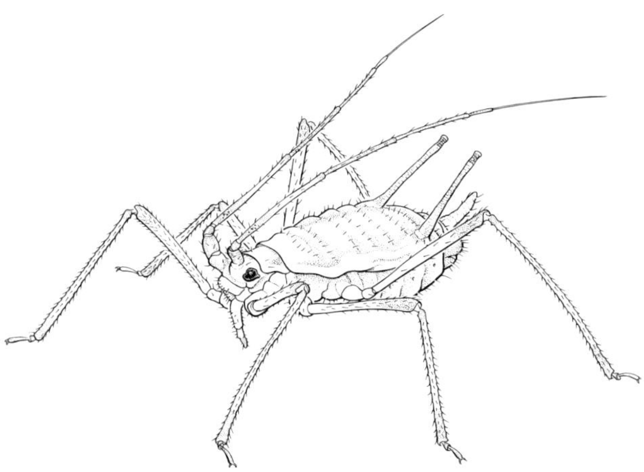

Figure 1 *Macrosiphum rosae* L. habitus. (Courtesy of Barry Flahey.)

Aphid Biology

Aphids are well known for the peculiarities of their life cycle. A distinguishing feature of aphids is the presence of both winged and wingless adults in the same colony. Also characteristic of aphids for at least part of their life cycle is continuous asexual production of larvae (parthenogenesis) by live birth (viviparity). In this mode of development, embryos arise from cells produced without chromosomal reduction and individuals are thus all female, genetically identical to the parent. Embryonic development of the next generation begins within the larva, sometimes even before birth, so that generations are telescoped. Newly deposited larvae are fully active and are quite similar to the adults in overall form. They undergo four moults in rapid succession, producing adults capable of bearing young in as little as 10 days. This overall strategy enables aphids to produce large numbers of offspring in a relatively short time, resulting in a large colony that can overcome the plant's defences and by sheer numbers ensure that at least a few individuals survive parasitoid and predator attack.

A few aphids restrict their life cycle to continuous parthenogenetic reproduction, particularly in warmer climates. However, most species also go through a sexual phase. Embryos produced in the autumn mature into sexual females and males, the males being produced by meiotic reduction of the X chromosome. The wingless female laying the overwintering egg is usually morphologically distinct from the usual parthenogenetic wingless female, having shorter appendages and swollen hind tibiae covered with scent glands. The wingless individual hatching from the egg in the spring (the stem mother) is usually larger, and has shorter appendages, than other wingless parthenogenetic females. Males may be either winged or wingless (sometimes both forms are found in the same species).

About 10% of aphid species exhibit host alternation (**Figure 2**). There are various patterns of alternation, but the typical pattern described here is followed by all alternating rose aphids. In the autumn, winged parthenogenetic females migrate to a winter host, usually a woody plant, and there deposit larvae which mature into sexual females. Winged males are produced on the summer host and migrate separately to the winter host, where they mate with the females. The eggs are laid on twigs, usually near the bases of the buds. The eggs hatch in the spring at about the time that the buds begin to expand. Development of the individuals hatching from the egg, as well as one or more following parthenogenetically produced generations, takes place on the new growth. In early summer, as the new growth matures, migration to a summer host, usually an herbaceous plant, takes place, and subsequent generations reproduce parthenogenetically until autumn, when remigration to the winter host takes place.

Most aphid species have a relatively narrow host range, usually being restricted to a single plant genus (referred to as monophagous species) or to a few, usually closely related, plant genera (oligophagous), but some have a very broad range of acceptable hosts (polyphagous). Host-alternating species are generally quite specific in their choice of winter host, but often less so in choice of summer hosts.

Rose Aphid Diversity

About 45 aphid species are associated with roses worldwide. They exhibit a wide range of life histories. While many of these species are locally distributed and are not frequently noted on ornamental roses, there are a number of species which have become increasingly cosmopolitan as world trade patterns have expedited movement. Life history characteristics and distribution for these species are summarized in **Tables 1** to **6**. The aphid species most frequently encountered on ornamental roses are discussed below.

The best known of these is the common, widely distributed rose aphid, *Macrosiphum rosae*. It is a relatively large green or reddish aphid with long black siphunculi and appendages with contrasting black and pale areas; the head is usually also quite dark. The winged forms have black sclerites along the sides of the abdomen. It feeds in groups on new growth on the soft stem tissue, young leaves and flower buds. This species usually alternates between roses and Dipsacaceae or Valerianaceae, but this alternation is not obligatory, and it may be found throughout the year on rose. *Macrosiphum mordvilkoi* is a virtually identical nonalternating species found on *Rosa rugosa* in the Russian Far East, Korea and Japan. In much of eastern Asia, *M. rosae* is largely replaced by the superficially similar *Sitobion ibarae*, but the general colour

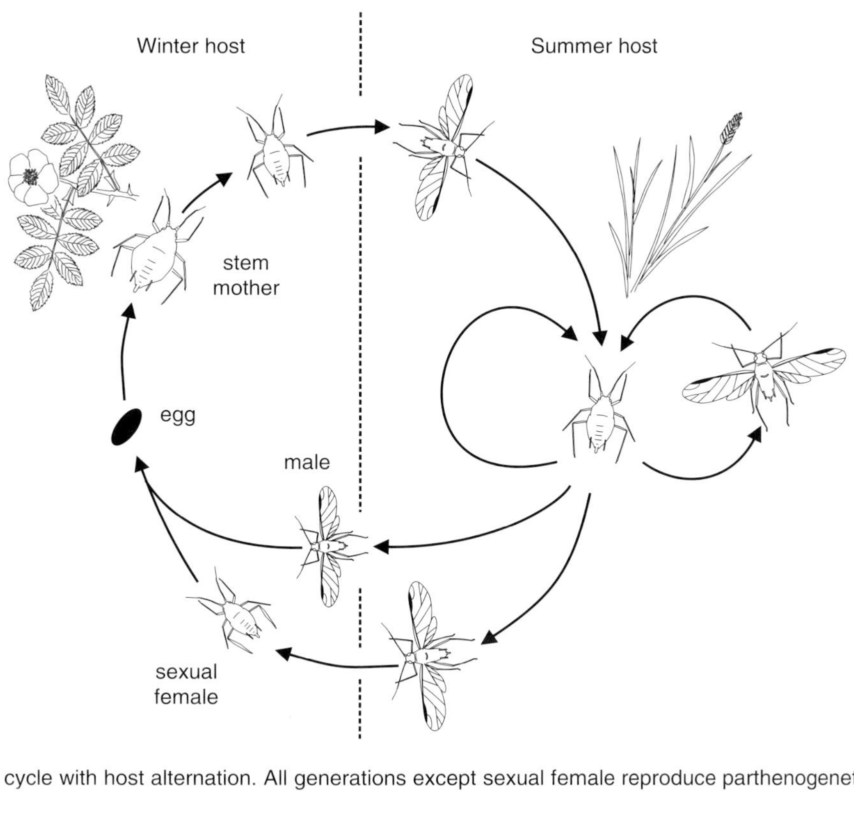

Figure 2 Aphid life cycle with host alternation. All generations except sexual female reproduce parthenogenetically.

Table 1 Host-alternating species using *Rosa* as sole winter host

Species	*Summer hosts*	*Distribution*
Macrosiphum euphorbiae (Thomas 1878)[a]	polyphagous	Cosmopolitan, (origin eastern North America)
impatientis (Williams 1911)	*Impatiens*	Eastern North America
rosae (Linnaeus 1758)[b]	Dipsacaceae, Valerianaceae (infrequently, other plants)	Cosmopolitan
Longicaudus trirhodus (Walker)	*Thalictrum, Aquilegia*	Holarctic (origin Eurasia)
Metopolophium alpinum Hille Ris (Lambers 1966)	Poaceae	Central Europe (Alps)
chandrani (David and Narayanan 1968)	Poaceae	India
dirhodum (Walker 1849)	Poaceae, occasionally other monocotyledons	Widespread in temperate regions (origin Europe)
montanum Hille Ris (Lambers 1966)	Poaceae	Southern Europe (Alps, Pyrenees)
Acyrthosiphon titigula (Hottes 1933)	*Corydalis*	Western North America (Colorado)
Wahlgreniella nervata (Gillette 1908)	Ericaceae	North America, introduced locally in western Europe and Brazil

[a] *Macrosiphum euphorbiae* does not alternate, and is sometimes entirely parthenogenetic outside of eastern North America.
[b] All or part of a particular *M. rosae* colony may remain on rose without alternation.

Table 2 Host-alternating species with several primary hosts, including *Rosa*

Species	*Other winter hosts*	*Summer hosts*	*Distribution*
Sitobion fragariae (Walker 1848)	*Fragaria, Rubus*	Poaceae	Cosmopolitan
kamtshaticum (Mordvilko 1919)	*Rubus*	Poaceae	East Asia

Table 3 Non-host-alternating species restricted to *Rosa*

Species	*Distribution*
Amphorophora catharinae (Nevsky 1928)[a]	Central Asia
Aphiduromyzus rosae (Uramov and Ibaranova 1967)	
Chaetosiphon chaetosiphon (Nevsky 1928)	Montane Europe to Kashmir
coreanum (Paik 1965)	Korea, Japan
glabrum (David, Rajasingh and Narayanan 1971)	India
gracilicorne (David, Rajasingh and Narayanan 1971)	India
tetrarhodum (Walker 1849)	Cosmopolitan, except southeast Asia
Eomacrosiphon nigromaculosum (MacDougal 1926)	Western North America
Ericaphis siphunculata (Richards 1961)	North America
wakibae (Hottes 1933)	North America
Macrosiphum floridae (Ashmead 1882)[a]	Southeastern USA
mordvilkoi (Miyazaki 1968)	Korea, Japan, Russian Far East
Maculolachnus sijpkensi (Hille Ris Lambers 1962)	North America, Mongolia
submacula (Walker 1848)	Europe to Ukraine, India
Myzaphis avariola (David, Rajasingh and Narayanan 1971)	India
bucktoni (Jacob 1946)	Europe to Mongolia, locally introduced in eastern North America
turanica (Nevsky 1929)	Mid-East through Asia to northwest India and Mongolia; also in Sweden
Myzus japonensis (Miyazaki 1968)	Japan
Pseudaphis abyssinica (Hille Ris Lambers 1954)	Northeast Africa
Pseudocercidis rosae (Richards 1961)	Western North America
Sitobion ibarae (Matsumura 1917)	Southeast Asia north to Korea
rosaeformis (Das 1918)	Pakistan, northern India to Malaysia
rosivorum (Zhang 1980)[a]	China

[a] Life histories of *Amphorophora catharinae*, *Macrosiphum floridae* and *Sitobion rosivorum* are not known but are likely not to be host-alternating.

Table 4 Oligophagous non-host-alternating species found on *Rosa*

Species	*Other hosts*	*Distribution*
Acyrthosiphon pseudodirhodum (Patch 1919)[a]	*Spiraea*	Eastern North America
Chaetosiphon thomasi (Hille Ris Lambers 1953)	*Fragaria, Potentilla*	North America
Ericaphis fimbriata (Richards 1959)[a]	*Fragaria*	North America
Macrosiphum pachysiphon (Hille Ris Lambers 1966)[a]	*Rubus, Potentilla*	India, Pakistan
pallidum (Oestlund 1887)[b]	Various plants, especially Rosaceae and Asteraceae	North America
Matsumurajana capitophoroides (Hille Ris Lambers 1966)[a]	*Rubus*	Pakistan, northern India
Myzaphis rosarum (Kaltenbach 1843)	*Potentilla* (*Dasiphora*), *Fragaria*	Throughout temperate area of world (origin Eurasia)
Rhodobium porosum (Sanderson 1900)	*Fragaria*	Temperate areas (origin North America)
Wahlgreniella arbuti (Davidson 1910)[a]	*Arbutus, Arctostaphylos*	Western North America, locally introduced in western Europe and Mid-East

[a] Species much less common on roses than on the other hosts listed.
[b] Life cycle of *Macrosiphum pallidum* is uncertain; it may exhibit facultative host alternation.

Table 5 Polyphagous host-alternating species occasionally using *Rosa* as secondary host. These species use a very wide range of plants as summer hosts. All may persist year-round as exclusively parthenogenetic populations in regions with mild winters

Species	*Winter host*	*Distribution*
Aphis fabae (Scopoli 1763)	*Euonymus, Viburnum, Philadelphus*	Cosmopolitan
spiraecola (Patch 1914)	*Spiraea, Malus*	Cosmopolitan
Myzus persicae (Sulzer 1776)	*Prunus*	Cosmopolitan

Table 6 Polyphagous non-host-alternating species

Species	*Distribution*
Aulacorthum solani (Kaltenbach 1843)	Cosmopolitan
Macrosiphum centranthi (Theobald 1915) (Indian populations only)	(India)
Myzus ornatus (Laing 1923)	Cosmopolitan
Toxoptera aurantii (Boyer de Fonscolombe 1841)	Tropical/subtropical
odinae (van der Goot 1917)	Tropical/subtropical
Aphis craccivora (Koch 1854)	Cosmopolitan
gossypii (Glover 1877)	Cosmopolitan

of the latter species is yellowish-green to brown. Also similar are *S. rosaeformis* of the Indian subcontinent and *S. kamtshaticum* of the Kamchatka Peninsula and Japan. Most records of *M. rosae* from east Asia refer to *M. mordvilkoi* or *S. ibarae*.

Macrosiphum euphorbiae, commonly known as the potato aphid, has a body form similar to *M. rosae*, but is on average somewhat smaller and has dusky to pale appendages not strongly contrasting with the general green (less often pinkish) body colour, and the winged form is without distinct dark sclerites on the abdomen. It is host-alternating in its apparent area of origin in eastern North America, but in the rest of the world it usually occurs as entirely asexual populations on a large number of plants, including roses. A nonalternating life cycle with sexual stages and overwintering egg occurs less frequently on various plants. In North America, a very similar aphid, *M. pallidum*, is commonly encountered on roses as well as various other unrelated plants.

The rose-grain aphid, *Metopolophium dirhodum*, is a green aphid, usually with a darker green longitudinal stripe, with long pale appendages. The siphunculi are relatively shorter than in the species discussed above. It is found clustered on new growth in the spring, and sexual forms on senescent leaves in the autumn. The summer is spent on various grasses, including grain crops, and it is thus especially common in grain-growing regions.

The yellow rose aphid, *Rhodobium porosum*, is green to yellow with a dusky head. The winged form has dusky lateral abdominal sclerites and fairly heavily pigmented wing veins. It feeds on the upper leaves and new shoots. It is fairly common throughout the tropics and subtropics, where it only reproduces parthenogenetically. In eastern North America, where it is a pest of strawberries as well as rose, it has a complete life cycle with sexual forms. It is uncommon in other temperate regions, except in greenhouses.

Chaetosiphon species are small, rather flattened, pale whitish green to green aphids (rarely pinkish). The wingless forms are easily recognized by their long knobbed hairs, and the winged forms have a dark patch on the dorsum of the abdomen. *Chaetosiphon tetrarhodum* is very common on roses throughout most of the world. Early in the year, this species is common on new shoots between and in the folds of the expanding leaves; later it is found scattered on the underside of older leaves. Because of their small size and scattered distribution on the plant, these aphids are often overlooked. Other *Chaetosiphon* species are similar in their appearance and habits, but more local in their geographic distribution (see **Table 3**).

Myzaphis rosarum is similar to *Chaetosiphon tetrarhodum* in size, shape and colour, but has inconspicuous hairs and the wingless form has the dorsal cuticle thick and pitted; the front of the head has a square projection. These aphids feed scattered over the underside of the leaves, on flowers and fruit, and on young shoots. *Myzaphis rosarum* is also very common on *Potentilla fruticosa*. Three other *Myzaphis* species (see **Table 3**) of more limited distribution are restricted to rose.

Maculolachnus species are quite different from other rose aphids both in morphology and feeding habits. The body is compact, yellow–brown to dark chestnut, and very hairy; the siphunculi are in the form of short broad cones. Both species feed on the root crown, stem base and lower woody branches and are attended by ants.

Several polyphagous aphids are sometimes found on rose, especially in glasshouse culture (**Tables 5** and **6**). *Myzus persicae* is very common in glasshouses throughout the world and is of particular concern as it is a known vector of many plant viruses.

Control

In general, aphid populations are effectively limited by numerous parasitoids (parasitic wasps), predators (ladybird beetles, lacewings, flower flies, predatory midges, predatory thrips) and pathogens (fungi, bacterial and viral diseases). Thrips, including normally herbivorous species, feed on aphid eggs in the spring.

The larger rose aphids (*Macrosiphum*, *Sitobion* and *Metopolophium* species) which cluster on the new growth are easily dislodged by shaking or by a jet of water from a hose, and, with vigilance and a little investment in labour, chemical control is usually not necessary. A water spray also serves to remove the water-soluble honeydew deposits. The smaller aphids, such as *Chaetosiphon* and *Myzaphis*, living in less exposed locations, are less susceptible to physical removal. However, because these species do not generally form dense aggregations, the effects on the host are not locally concentrated, and their feeding is usually easily tolerated unless population levels are high. If control is necessary, addition of a surfactant, such as insecticidal soap, to a water spray is often sufficient to reduce populations to a satisfactory level. Dormant oils are effective against overwintering eggs.

If chemical control is desired, there are a number of suitable compounds available. Systemic insecticides are particularly effective against aphids. Because regulations regarding pesticide use vary among jurisdictions, local extension specialists should be consulted, and no recommendations are given here.

Nomenclature Changes

The following indication of name changes in the major rose-inhabiting species is provided as an aid to tracing the literature. *Macrosiphum euphorbiae* was known in the older literature in North America as *M. solanifolii*. *Macrosiphum pallidum* is also called *M. pseudorosae* Patch. *Sitobion* was formerly often treated as a synonym or as a subgenus of *Macrosiphum*. *Metopolophium* has been placed as a subgenus or synonym of *Acyrthosiphon*. *Rhodobium porosum* has been placed in *Acyrthosiphon* and *Aulacorthum*, and was originally described in *Myzus*; *Acyrthosiphon rosaefoliae* Takahashi and *Rhodobium rosaefolium* (Theobald) are synonyms. *Chaetosiphon* species were formerly placed in *Capitophorus*; subgenus *Pentatrichopus* of *Chaetosiphon*, containing most of the rose-feeding species, has been treated by some workers as a full genus. *Maculolachnus rosae* (Cholodkovsky) is a synonym of *M. submacula*.

See also: **Insects and other Animals**: Overview of Insects; Thrips. **Integrated Pest Management**. **Production and Marketing**: Overview of Roses and Culture.

Further Reading

Blackman RL (1981) *Aphids*. Oxford, UK: Pergamon Press.

Blackman RL and Eastop VF (2000) *Aphids on the World's Crops: An Identification and Information Guide*, 2nd edn. Chichester, UK: John Wiley.

Borror DJ, Triplehorn CA and Johnston NF (1992) *An Introduction of the Study of Insects*, 6th edn. Philadelphia, PA: Saunders College Publishing.
Dixon AFG (2003) *Aphid Ecology*, 2nd edn. London: Kluwer Academic Press.
Minks AK and Harrewijn P (eds.) (1987) *Aphids, their Biology, Natural Enemies and Control*, vol. A, *World Crop Pests 2A*. Amsterdam, The Netherlands: Elsevier.
Minks AK and Harrewijn P (eds.) (1988) *Aphids, their Biology, Natural Enemies and Control*, vol. B, *World Crop Pests 2B*. Amsterdam, The Netherlands: Elsevier.
Minks AK and Harrewijn P (eds.) (1989) *Aphids, their Biology, Natural Enemies and Control*, vol. C, *World Crop Pests 2C*. Amsterdam, The Netherlands: Elsevier.

Mites

S A Tjosvold, University of California Cooperative Extension, Watsonville, CA, USA
J F Karlik, University of California Cooperative Extension, Bakersfield, CA, USA

Introduction

Mites are one of the most important pests of roses grown in glasshouses, nurseries and landscapes throughout the world. In one survey, over 40% of the total pest and disease management cost for glasshouse cut roses was attributed to the control of two-spotted spider mites (*Tetranychus urticae* Koch).

Spider mites are significant pests because they damage leaves and thereby cause loss of photosynthetic capacity. Spider mite infestations can lead to defoliation, sunburn of canes in outdoor roses and plant death in extreme cases. Spider mites have a wide host range, and can be introduced into a glasshouse from other crops or develop on weed species and subsequently move on to roses. Spider mites are tiny and often hide on the undersides of leaves, thus escaping early detection. They have tremendous reproductive capacity in warm and dry conditions, and can rapidly develop resistance to pesticides.

An integrated pest management (IPM) approach to spider mite control is essential, including consideration of sanitation, plant culture and the surrounding environment (*see* **Integrated Pest Management**). Monitoring the mite population is essential for sound decisions regarding timing and selection of control strategies. Chemical suppression can be effective, especially with pesticides that have specific activity against mites. Natural biological control can be nurtured in some rose production areas or biological control can be supplemented with insectary-raised predators. *Phytoseiulus persimilis* Athias-Henriot, a mite itself, is the primary predator used in glasshouse conditions where temperatures and humidity are moderate. These predators are more effective in newer bent-cane-trained rather than traditional hedgerow-trained cut roses. Natural biological control should be preserved in outdoor rose plantings.

Background

Taxonomy of Mites

Mites are in the same taxonomic class (Arachnida) as spiders, scorpions, harvestmen and ticks. Five families contain most of the plant-feeding mites. These include spider mites (Tetranychidae), false spider mites and flat mites (Tenuipalpidae), thread-footed mites (Tarsonemidae), bulb mites (Acaridae) and blister, bud, gall and rust mites (Eriophyidae). Many species of mites feed on roses (**Table 1**), the majority of which are spider mites. A notable exception is the eriophyid mite (*Phyllocoptes fructiphylus* Keifer) that transmits the virus or virus-like rose rosette pathogen to *Rosa multiflora* Thunb. and many other cultivated roses in the USA.

Of the spider mites, by far the most damaging mite in glasshouse-, nursery- and landscape-grown roses is the two-spotted spider mite (*Tetranychus urticae*). Sometimes this species is called the glasshouse spider mite, red spider mite or red spider. It has worldwide distribution. Its life cycle and management typify the biology and management of other spider mites, and this species will be used as a paradigm in the following discussion.

Two-Spotted Spider Mite

Body colour can be affected by the mite's stage of development, its health and the host plant material upon which it is feeding. Generally, there are two relatively large dark spots or irregular blotches, one on either side of the body. Adults vary in colour from light yellow to green, tan and various shades of orange. The adult female is oval-shaped with a rounded posterior, and is about 0.5 mm long. The adult male is smaller, narrower and has a distinctly pointed posterior. Females are normally about three times more abundant than males (**Figure 1**).

Females develop from fertilized eggs and have the normal two sets of chromosomes, while males develop from unfertilized eggs and have only one set of chromosomes. Therefore, new genetic features arising from mutation may be expressed quickly in males and could through selection be quickly added to the

Table 1 Mites that feed on rose

Mite	*Occurrence*	*Hosts other than rose*	*Mite family*
Bryobia lagodechiana (Reck)	New Zealand, Russia	Cucumber	Tetranychidae
Bryobia rubioculus (Scheuten)	Turkey, North and South America, Europe, South Africa, Australia and Asia	Apples, pears, peaches and other deciduous fruit	Tetranychidae
Cenopalpus spinosus (Donnadieu)	France, Germany and Monaco	Primrose, blackberries, dewberries, raspberries and dogwood	Tenuipalpidae
Eotetranychus frosti (McGregor)	Arizona, California, Louisiana, Missouri, North Dakota, Ohio and New York	Blackberry and raspberry	Tetranychidae
Eotetranychus smithi (Pritchard and Baker)	North-eastern USA and Japan	Berries	Tetranychidae
Eotetranychus tiliarum (Herman)	Turkey, Europe, Atlantic coast of USA	Linden, lime, sycamore, horse chestnut, hawthorn, hazel and willow	Tetranychidae
Lupotarsonemus randsi (Ewing)	Portugal, Florida and Virginia	European plum, sweet cherry, citrus and sugar cane	Tarsonemidae
Oligonychus (= Pritcharinychus) biharensis (Hirst)	Mauritius, India, Hawaii, Thailand, Malaya, Philippines, Antigua, Brazil and Mexico	Loquat, litchi, contoneaster, camphor, *Euphorbia longana*	Tetranychidae
Oligonychus gossypii (Zacher)	Central America, Brazil and equatorial Africa	Cotton, cassava, beans, citrus and peach	Tetranychidae
Oligonychus newcomeri (McGregor)	North-eastern USA, Washington State, North Carolina	Serviceberry, hawthorn, apple, pear, Bradford pear, *Prunus*, other Rosaceae	Tetranychidae
Oligonychus mangiferus (Rahman and Sapra)		Cotton, mango, loquat, peach, quince, pear, pomegranate, grapes and Eugenia	Tetranychidae
Panonychus citri (McGregor)	California, South Africa, Japan, Florida, China, South America, former USSR and India	Citrus	Tetranychidae
Panonychus ulmi (Koch)	USA, southern Canada, Europe, Georgia in the former USSR, China, Bermuda, Argentina, South Africa, India, Tasmania, New Zealand and Japan	Deciduous fruit orchards	Tetranychidae
Phyllocoptes fructiphilus (Koch)	North America		Eriophyoidae
Tetranychus canadensis (McGregor)	USA, Canada, the Middle East, Africa and Poland	Peach, plum, apple, crabapple, tomato, beans, sweet potato, okra, cotton, red clover, barley, rye, corn, wheat, ornamentals, elm, linden, horse chestnut, Osage orange, poplar and umbrella tree	Tetranychidae
Tetranychus cinnabarinus (Boisduval)	Hawaii, Bulgaria, Taiwan	Carnation, chrysanthemum, cymbidium, gladiolus, marigold, ficus, dieffenbachia, giant taro and gerbera	Tetranychidae
Tetranychus evansi (Baker and Pritchard)	Mauritius island, Brazil, Texas and California	Tomato, aubergine (eggplant), potato, peanut, rose, Plaine de Papayes, nightshade, Solanum, sweet potato, Asystasia, coromandeliana, red pepper, *Nicotianaglauca Phacelia*, lily of the valley, *Salpichora rhomboidea*	Tetranychidae
Tetranychus pacificus (McGregor)	Oregon and California	Mock orange, vetch, wild currant, china berry	Tetranychidae
Tetranychus urticae (Koch)	Worldwide	Many agricultural crops and ornamentals	Tetranychidae
Tetranychus turkestani (= T. atlanticus) (Ugarov and Nikolski)	USA, Europe, Russia, Japan, the Near East, the Middle East	Alfalfa, beans, castor bean, clover, cotton, carrot, cucumber, aubergine (eggplant), melons, parsley, soybean, squash, strawberry, sunflower	Tetranychidae
Tetranychus yusti (McGregor)	Delaware, Louisiana, Mexico, Central America and Ecuador	Soybean, Compositae, Leguminosae and Graminae. Cotton, okra, sweet potato, sunflower, white clover, marigold, peas, beans, cowpeas and peanuts	Tetranychidae
Tyrophagus perniciousus (Zachvatkin)	Turkey		Acaridae

Figure 1 Adult female two-spotted spider mite (*Tetranychus urticae*). To the left is a newly laid egg. (Reprinted with permission from UCCE.)

population. With a high reproductive rate and selection pressure brought about by pesticide use, pesticide resistance can develop in a short time.

Life Cycle

The life cycle of the two-spotted spider mite consists of five stages of development: the egg, larva, protonymph, deutonymph and adult. Newly laid eggs are round, about 0.14 mm in diameter, of translucent pale yellow colour, becoming opaque and straw-coloured with time. The egg hatches into a clear six-legged larva with noticeable crimson-coloured eye spots. The larva develops colour after feeding and the two characteristic dark spots are formed in the middle of the body. The larva becomes an eight-legged protonymph, slightly larger than the larva after a quiescent stage. The protonymph is transformed into the slightly larger deutonymph and, although this stage has similar coloration, males and females can usually be distinguished. The eight-legged adult emerges after feeding and a final quiescent stage. After a waiting period of up to 3 days, the adult female begins to lay eggs. Egg laying is rapid at first and then declines slowly. The period during which eggs are deposited can last from 10 days (34 °C) to 40 days (15 °C). Over 100 eggs can be laid by a single female.

The development time varies with temperature, humidity, host plant, leaf age and other factors, with temperature the most important factor. The lower threshold for development is about 12 °C and the upper limit for development is about 40 °C. The life cycle usually ranges from 10 days to 2 weeks. Developmental times for each life stage have been determined for constant temperatures (**Table 2**). Mites reared on detached rose leaves under two alternating night/day temperature regimes, 10/20 °C and 25/35 °C, took 8.3 and 28.2 days, respectively, to complete their life cycle.

Table 2 Developmental times (days) for various stages of two-spotted spider mite, *Tetranychus urticae*

Temperature (°C)	*Developmental stage*					
	Egg	*Larva*	*PN*	*DN*	*PO*	*Total*
15	14.3	6.7	5.3	6.6	3.5	36.3
20	6.7	2.8	2.3	3.1	1.7	16.6
30	2.8	1.3	1.2	1.4	0.6	7.3

PN, protonymph; DN, deuteronymph; PO, length of time before an adult female begins to oviposit (data from Sabelis, 1981).

There is a winter hibernation period or diapause in temperate or cooler climates, which is a result of the influence of temperature, photoperiod and mite nutrition. Adult females turn orange and hibernate under leaves, in cracks and crevices, or other protected places. These mites do not feed or reproduce until favourable conditions resume. Mites can flourish even in winter where the climate is warm or in glasshouse conditions where host plants are available.

Spider Mite Damage

All mobile life stages of spider mites feed. Feeding occurs mainly on the underside of rose leaves but damage usually becomes obvious on the upper leaf surfaces. Spider mites feed by inserting a pair of needle-like stylets into plant cells and then withdrawing cell sap. Feeding causes rose leaves to become stippled or flecked with pale dots. As the damaged areas coalesce, foliage appears bleached or bronzed (**Figure 2**). Leaves can dry and drop from the rose plant with large mite populations, and, left uncontrolled, the plant can die. When pest populations are high, female mites produce fine webbing that can cover leaves and flower buds, suggesting its common name of ‘spider’ mite.

Dispersal

Mites move to adjacent stems and leaves as their density increases and the leaf becomes inadequate to support the population nutritionally. When accompanied by high temperatures, mites sometimes quickly migrate upward to the plant’s terminal points. Female mites produce dense webbing, often covering several leaves and the terminal portion of the shoot. Many webbing-threads are formed, eventually creating dangling rope-like structures. Mites eventually free-fall, sometimes attached to threads of webbing (**Figure 3**). Wind can catch mites and blow them to new areas. Mites in a glasshouse may land on the ground and crawl considerable distances (30 m or more) to find

Figure 2 Rose leaf showing symptoms of mite damage from the feeding of two-spotted spider mite (*Tetranychus urticae*) on underside of leaf. (Reprinted with permission from UCCE.)

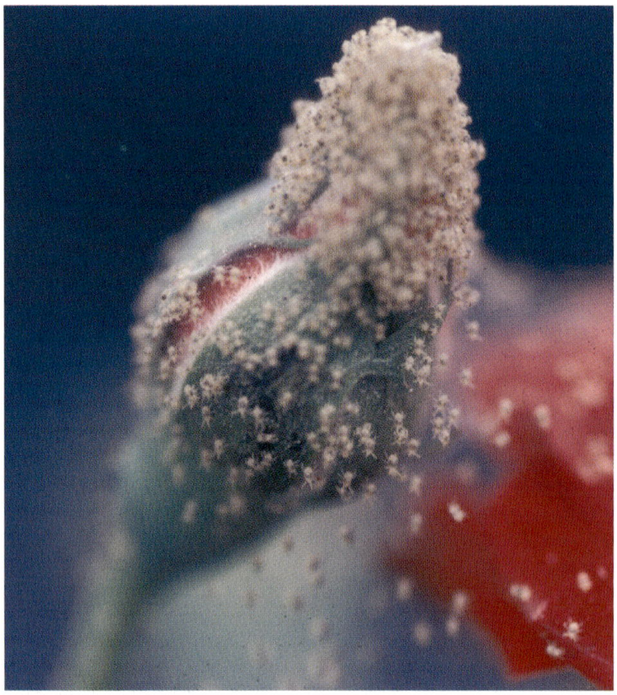

Figure 3 Two-spotted spider mite (*Tetranychus urticae*) migrating to terminal shoot and massing on rose flower bud. Mites are suspended by webbing.

new host plants. Mites can also be dispersed on farm workers' clothing and through movement of contaminated plant material.

Monitoring

Mite populations are often found on the older portions of the plant canopy, and as the population develops the mites are increasingly found higher, on the younger portions of the canopy. In traditional hedgerow cut roses, mites are typically found in decreasing densities as leaves are sampled from lower (older) to the upper (younger) portions of the canopy. In bent-cane-trained roses, mite populations are found at highest densities on the oldest leaves near the bend of the canes, where leaves have remained the longest without pruning or removal. In rose plant production fields, the middle-third stratum of the canopy should be sampled.

Mite populations are often found in patches or hot spots within the field or glasshouse. Once established on plants, the infestation usually expands slowly outward as new adjacent stems and leaves are colonized. Hot spots can be found where chemical sprays or predators were incompletely applied, near warm parts of the greenhouse (often near a sunny exposure), where particularly susceptible cultivars are found, or near an entrance or walkway where the mites were introduced. In the field, plants near roads may have elevated numbers of mites because dust and soil particles on leaves cause leaf temperatures to rise and impede predatory activity.

In the glasshouse, scouting of production plants should occur at least at weekly intervals in order to monitor the dynamics of a mite population and allow enough time to control the mites if deemed necessary. Scouts should always be wary of telltale signs of mites as they walk the production area. They should look very carefully in areas known to have hot spots, as described above. If stippled leaves are observed, the leaf is removed and examined on the underside with a 10× to 15× hand lens to determine if live mites are present. In addition to this active observation and sampling, a blind random sampling of leaves, with targeted leaf samples, can help find mites at very low levels and quantify the overall population density. The first leaf above or below the hook (in hedgerow systems) and the first leaf on a harvestable shoot at or near the bend of the cane (in bent-cane systems) are good standard areas from which to sample. Mites are often found in relatively high densities in these areas and an increasing mite population will move upward from these areas on to the harvested stem. Therefore the sampling of the population here is critical to when action must be taken to control mites.

Greenhouse cut roses on bent-cane-trained roses were sampled as described above in a California IPM demonstration. Four leaves per 100 m^2 were necessary to sample the average mite populations properly at normal statistical precision levels used in pest management. Plants were classified as infested if the scout found more than five mobile mites (eggs are not counted) on the sampled leaf or classified as not infested if there were five or fewer mobile mites. These samples were also used to determine co-occurrence of two-spotted spider mites with predator mites,

Phytoseiulus persimilis, and they were inspected for secondary pests and for diseases.

A rapid presence/absence (P/A) field sampling method and threshold levels were developed for field-grown rose plants in California. The P/A method is well suited for use in the field because it gives a rapid evaluation and is easy for field scouts to learn. Such P/A methods are widely used in Californian agriculture to evaluate pest presence. In the P/A method for roses, five-leaflet leaves in the middle third of the plant canopy were selected at random from 30 plants within a rose field. Rather than counting the mites on each leaf, leaves were considered infested if a single live spider mite was present, and the percentage of infested leaves noted. The percentage of infested leaves was found to be well-correlated to mite numbers obtained by brushing and counting leaves. The slope of the regression line may not be the same for all rose species or cultivars, nor for the first versus second year of a rose plant crop.

Action Threshold Levels

The pest density or plant damage at which control action should be taken is determined by many factors: the type of pest and associated damage, crop species and cultivar, stage of plant development, time until harvest or sale and market conditions. Thresholds can often be higher if infested plant parts are not marketed or if very effective methods are available for controlling the problem. Conversely, if available controls are slow-acting or only partially effective, as is often the case with biological control, thresholds may be relatively low for that pest. Sometimes regulations such as quarantines impose zero tolerance for certain pests.

A moderate mite population can be tolerated in landscape rose plantings and in rose plant production fields. In a field experiment in the California San Joaquin Valley using 'Peace' on 'Dr. Huey' rootstock, yield as measured by grade was similar at a 40% infestation threshold level to that of lower treatment threshold levels. Leaves were infested with *Tetranychus* species, and infestation measured by the P/A sampling scheme. These plants retained a full canopy, i.e. did not defoliate.

Glasshouse cut roses can tolerate low mite populations, especially if predators are active or can be applied. In a California IPM demonstration, mite treatments were initiated according to the percentage of infested leaves. If less than 10% of leaves were infested, no control action was taken. If greater than 10–25% of the leaves were infested, the mite predator *Phytoseiulus persimilis* was released. If greater than 25% of the leaves were found infested, a chemical control was applied. Release rates of predatory mites were based on co-occurrence of two-spotted spider mite and predator on the sampled leaf. The greater the occurrence of the predator, the fewer predators were applied.

Management

Cultural and Physical Methods

Use pest-free plants and cuttings. Train workers to inspect newly introduced plants or when plants are moved from one area to a new area. Follow strict sanitary rules for visitors and workers in the field or glasshouse. Mites on bent canes that lean into glasshouse walkways can easily be brushed on to clothing and carried by workers who pass through as they perform normal cultural practices. Workers could wear overalls that can be cleaned or removed before moving into noninfested production areas.

Destroy weeds that may harbour mites. Nightshade (*Solanum nigrum* L.), pigweed (*Amaranthus* spp.) and many other field weeds make very suitable hosts for spider mites. Creeping woodsorrel (*Oxalis corniculata* L.), lesser-seeded bittercress (*Cardamine oligosperma* Nutt.), common chickweed (*Stellaria media* L.) and henbit (*Lamium amplexicaule* L.) make very suitable hosts in the glasshouse. Once the mite population builds up on the weeds, they may move onto rose crops.

Spider mites develop best under hot, dry conditions. A summer ventilation temperature of 24 °C should be maintained if possible in glasshouses. This provides good rose quality and production, while higher temperatures would only cause quicker development of mites. Roses that are water-stressed are more likely to experience mite outbreak and resulting damage may be more significant. Regular overhead sprinkling, irrigation or fogging slows mite dispersal and reduces dust and low humidity that negatively impact predators. A forceful stream of water directed at leaves dislodges mites. Increasing humidity may, however, improve the conditions for many plant diseases (*see* **Disease**: Overview). Reduce dust around crops by imposing speed limits on adjacent dirt roads.

High foliar nitrogen levels also encourage mite development. Rose nutrition should be closely monitored to provide sufficient nitrogen levels, but no more.

Chemical Control

The introduction of organophosphate pesticides in 1947 resulted in the virtual elimination of spider mites on cut roses in glasshouses; however, by 1949, mite resistance became evident, and by 1950 mite resistance was prevalent in the eastern USA. New organophosphates and organochlorines became available in 1948–1958, but most lost activity in 1–4 years. Although modern insecticides such as pyrethroids may initially control spider mites, they too quickly lose their effectiveness with repeated applications.

Minimizing the use of a pesticide is the best way to delay resistance and lengthen the period it effectively controls the mite. There are various resistance management theories as to the best way to use chemicals in order to delay resistance, but all relate to avoiding overusing any one chemical class. If possible, chemical classes need to be rotated (*see* **Insects and other Animals**: Insecticides). Consult the pesticide label for information on resistance management specific to that product.

Mites are not insects, and therefore many of the insecticides used in insect management have limited or no effect on mites. Spider mites may even become a problem after the application of certain insecticides. Outbreaks commonly result from insecticides killing natural enemies of mites, but also occur when certain insecticides stimulate mite reproduction. For example, exposure to carbaryl, a carbamate insecticide, can cause mites to reproduce faster. Carbaryl and some organophosphate and pyrethroid insecticides also favour spider mites, apparently by increasing the level of nitrogen in leaves. Eriophyid mites, however, provide a notable exception, and chemical control for them is distinctly different than for spider mites. For example, the eriophyid mite *Phyllocoptes fructiphylus* is controlled by carbaryl and some pyrethroids.

Some pesticides have relatively specific activity for mites and are called acaricides or miticides. These include bifenazate, clofentazine, dicofol, dienochlor, hexythiazox, fenbutatin-oxide, fenpyroximate and pyridaben. Abamectin is often used as a miticide but has general insecticidal properties as well. Hexythiazox is active on eggs and hatching larvae but not the older life stages, while most of the other miticides have no, or sometimes very limited, activity on eggs.

Transitioning from a chemical into a biological control programme in a glasshouse is often difficult. Apparently, residual chemicals in the glasshouse itself or on plants can be toxic to predators. The associated insecticides and fungicides used for other pests can also preclude the use of biological control for mites. Insecticides with broad-spectrum activity and even some fungicides, such as thiophanate-methyl, should be avoided for several weeks before predators are first introduced. The use of directed sprays towards the target pest may be helpful to preserve predators. For example, a spray for western flower thrips can be directed towards the flower buds while avoiding coverage to the lower canopy where predatory mites could be feeding on spider mites. Mite densities may warrant immediate control before predators are applied. If so, a pesticide with relatively low impact on predators can be applied such as insecticidal soap or horticultural oil. Oils and soaps must contact mites to kill them so excellent coverage on the undersides of leaves is essential and multiple applications may be required. Soaps or oils should not be used on water-stressed plants or when temperatures exceed 30 °C.

Biological Control

Spider mites have many natural enemies that limit their numbers. Some of the most important are the predatory mites, including the western predatory mite, *Galendromus* (= *Metaseiulus* = *Typhlodromus*) *occidentalis* Nesbitt, *Phytoseiulus* species and *Neoseiulus* (= *Amblyseius*) species. Predatory mites can usually be distinguished from spider mites by their translucent colour, pear shape, longer legs and more active habits. Various insects are also important mite predators, such as the six-spotted thrips (*Scolothrips sexmaculatus* Pergande), the larvae and adults of the lady beetle (*Stethorus picipes* Casey), the larvae of certain flies, including the midges *Feltiella acarivora* Zehnter and *F. acarisuga* Vallot and various general predators such as minute pirate bugs, big-eyed bugs and lacewing larvae.

The major predacious mites commercially available for release are the western predatory mites *Neoseiulus californicus* McGregor and *Phytoseiulus persimilus*. *Neoseiulus californicus* is more effective under hot, dry conditions than is *P. persimilis*. It is a prey generalist so it can survive without spider mites present. It can be released at the start of the growing season in the field or at any time in a glasshouse before any spider mite infestation is obvious.

Phytoseiulus persimilis favours moderate temperatures (15–25 °C) and humidity (70%) and is the most important spider mite predator in glasshouses. It prefers feeding on eggs and young spider mites (**Figure 4**) This predator disperses more effectively with physical continuity between plants and between leaves within the plant canopy. It moves more effectively on bent-cane roses than on traditional hedgerow-trained roses. Better movement of the predator on potted rose plants has been accomplished by placing a large-meshed net over the crop, allowing dispersal from pot to pot.

Regular scouting is necessary to implement successful biological control of spider mites so that one can be confident in the population status of the spider mite and predator. At least one predator (*N. californicus* or *P. persimilis*) is usually needed for every 10 spider mites to provide control. Application rates of around

Figure 4 *Phytoseiulus persimilis*, a predatory mite, feeding on two-spotted spider mite egg. (Reprinted with permission from UCCE.)

two predatory mites per m^2 for a general application or over 200 per m^2 for hot spots are needed. More than one application of predatory mites is usually required, especially to distribute the predator to all the hot spots. Releases should be concentrated in these hot spots. Coloured flags can be placed at the hot spots when the scout is walking through the crop and later the applicator can appropriately apply the predator. Different flag colours could represent different scouting weeks.

In California, *P. persimilis* were successfully used in an IPM demonstration in cut-rose nurseries and almost eliminated the need for miticide applications. The cost of IPM during the first 8 weeks was higher than the cost of conventional control, primarily because of the higher number of predatory mites required during this trial period as growers learned proper scouting and release techniques. Some growers started predator releases before two-spotted spider mite densities were suitably low. When the predators failed to control the high density of spider mites, miticides were applied and predators were introduced again when the spider mites were at suitably low levels. After several releases had been made and predators became established, the release rate dropped and costs for the two control programmes were comparable.

The balance of prey and predator is often a juggling act and can be tipped to the favour of the spider mite with pesticide use, a change in environment, poor predator quality from the insectary or other unknown factors. In those situations where biological control has been lost, a fairly specific miticide known to be relatively compatible with the predator should be used to lower the mite density. Insecticidal soap and hexythiazox are examples of pesticides in which the residuals do not adversely affect *P. persimilis*, and could drop the mite density down to a level where predators could be reapplied.

The predatory midge *Feltiella acarisuga* is a widely distributed spider mite predator around the world and is available commercially for biological control. It will feed on all stages of spider mites, but generally prefers eggs or young mites. It attacks several species of spider mite, including the two-spotted spider mite and carmine spider mite (*Tetranychus cinnabarinus* Boisduval). The midge can provide good spider mite control when used in conjunction with *P. persimilis*. The predatory mite has relatively low dispersal ability, so it may fail to find patches of the pest, but *Feltiella*, having the ability to fly, is good at finding hot spots. The two predators can therefore be complementary.

See also: **Disease**: Overview. **Insects and other Animals**: Overview of Insects; Thrips; Insecticides. **Integrated Pest Management**.

Further Reading

Dreistadt SH (2001) *Integrated Pest Management for Floriculture and Nurseries*. Integrated Pest Management Project. University of California, Division of Agricultural and Natural Resources, Oakland, CA, USA. Publication 3402.

Helle W and Sabelis MW (1985) *Spider Mites: Their Biology, Natural Enemies and Control*, vols 1A and 1B. Amsterdam: Elsevier.

Jeppson LR, Keifer HH and Baker WE (1975) *Mites Injurious to Economic Plants*. Berkeley and Los Angeles, CA: University of California Press.

Osborne LS, Ehler LE and Nechols JR (1985) Biological control of the twospotted spider mite in greenhouses. IFAS. University of Florida, Florida Bulletin, pp. 853.

Sabelis MW (1981) *Biological Control of Twospotted Spider Mites using Phytoseiid Predators*. Part I. Agricultural Research Report 910. Wageningen, the Netherlands: Centre for Agricultural Publishing and Documentation.

Thrips

M P Parrella and C O'Donnell, University of California, Davis, CA, USA
B C Murphy, Cal Agri Products, Davis, CA, USA
C Casey, North Carolina State University, Raleigh, NC, USA

Introduction

Thrips have become major pests of roses during the past two decades. Their small size and propensity to develop resistance to insecticides and feed directly on developing rose buds and flowers have catapulted them to the forefront of arthropod pests attacking roses. The worldwide movement of thrips, especially the western flower thrips, *Frankliniella occidentalis* (Pergande), through the marketing of ornamental plants, has elevated thrips to major pest status on roses on a cosmopolitan scale. Cut flowers and propagation materials have long been implicated as posing the greatest problem in spreading thrips. This in turn puts greater pressure on rose growers to eradicate thrips from their greenhouses, especially if their flowers are destined for export. Research to develop a thrips management programme on roses in California was initiated in the mid-1990s. It quickly became apparent that information in the literature regarding thrips as pests of roses (and for that matter as a pest of all glasshouse crops) was inadequate on which to build a programme. Therefore a plan was developed to examine the following areas: accurate identification of thrips, ascertaining within and between plant distribution to develop rationale and accurate sampling plans, economic thresholds and biological control. Each of these areas will be expanded on in this article.

Early Literature

Reviews of ornamental pests prior to 1950 made reference to thrips as pests of roses, and most of these focused on outdoor roses. Usually these articles referred to thrips in a general way, without giving the names of species. It was not until the work of Compton, Dodge and Rickett, and Prichard that thrips were identified as major pests of floricultural crops, including roses. They listed several species attacking rose including the onion thrips (*Thrips tabaci* Lindeman), the flower thrips (*Frankliniella tritici* 'Fitch' and *F. moultoni* Hood), the western flower thrips (*F. occidentalis*), the greenhouse thrips (*Heliothrips haemorrhoidalis* 'Bouche'), and the banded greenhouse thrips (*Hercinothrips femoralis* 'Reuter'). Thrips were beginning to emerge as major pests of roses and some excellent control recommendations were made based on the biology of these species. For example, plants likely to be infested were recommended for spraying in May before the buds have developed and infested buds and open flowers should be picked off and burned. Also, thrips gain entrance to the greenhouse through open doors and ventilators, often swarming in on warm days in the autumn and to a lesser extent in the early spring. Growers routinely applied a stomach poison such as lead or copper arsenate (mixed with sugar) for control. The sugar was used to increase ingestion of the poison, thereby enhancing thrips mortality. It was reported that white sugar was safer to 'American Beauty' roses than brown sugar. Fumigation with products such as nicotine and cyanide were also commonly used prior to the advent of synthetic organic insecticides. In view of the fact that eggs are laid in plant tissues and the transformation stages (i.e. the development from larva to pupa) may occur in the ground, the most satisfactory control of thrips was by means of an insecticide which possessed residual action. In the late 1940s and the 1950s, the most satisfactory control of thrips was by means of insecticides such as DDT. During this same period, it was demonstrated that flower thrips had a propensity for more intricate blossoms (as opposed to *T. tabaci*). It was noted that rose petals would turn brown and affected buds would fail to open.

Many of the pest management guidelines developed for thrips in floriculture crops that followed these earlier reports were descriptive in nature and each seemed to repeat the same information. This was understandable because there was little research being done to develop new management strategies for thrips other than evaluating insecticides. This changed in the 1980s with the movement and establishment of the western flower thrips in roses and other crops around the world. More publications were written about *F. occidentalis* during the 1980s than during the entire time period from its identification in 1895 to 1980.

Development of an Integrated Pest Management (IPM) Programme

Proper Identification

When the California project on the development of integrated pest management (IPM) in roses was initiated, the inability of many individuals (growers and those who advised them) to distinguish various species of thrips caused concern. For many who believed that 'the only good thrips was a dead thrips', identifying thrips to the species level was not considered important. However, not all thrips pose a threat to the rose crop. Many nonpest species of thrips can move into a rose greenhouse by accident and do not warrant control action. Growers who fail to recognize this make unnecessary applications of pesticide that exacerbate problems with the development of insecticide resistance and add to the public perception that too many pesticide sprays are used on rose plants. The longest running set of data on thrips population fluctuations is that of Davidson and Andrewartha who sampled adults of *Thrips imaginis* Bagnall in 20 flowers from a rose hedge in south Australia each day for 14 years. Davidson and Andrewartha were simply using the rose hedge as a crop on which to base their ecological studies. *Thrips imaginis* Bagnall is a major problem on tree fruit in Australia, but it does not often warrant control action in roses.

Of the 5000 known species of thrips, 13 have been found attacking roses. These include *F. occidentalis*, *F. tritici* (Fitch), *F. bispinosa* (Morgan), *F. intonsa* (Trybom), *F. moultoni*, *T. tabaci*, *T. fucipennis* (Haliday), *Retithrips syriacus*, *Hercinothrips femoralis*, *Heliothrips haemorrhoidalis Rhipiphorothrips cruentatus* Hood, *Selenothrips rubrocinctus* Giard and *Anaphothrips obscurus* (Muller). Unfortunately, few individuals working in the field are capable of identifying thrips to species, and many often assume the western flower thrips is the culprit without engaging the services of a taxonomist. While this is important to fully comprehend all species of thrips attacking roses, it is not required at the field level where pest management decisions are made. Correct species identification is not necessary – being able to tell a pest from a nonpest thrips is all that is required. The first step in this process is the separation of thrips from rose flowers and foliage which involves breaking apart the roses in a Petri dish containing alcohol, decanting the alcohol over filter paper (often assisted by vacuum filtration), and examining the filter paper under a microscope. In addition, the examination of thrips trapped on sticky cards using a microscope or hand lens is a more convenient way to determine what thrips are in the glasshouse. However, a combination of both flower samples and sticky cards yields the best understanding of the species of thrips present in the glasshouse and the potential for them to cause damage. In an effort to advance the understanding of thrips identification for greenhouse managers, a grower-friendly key to separate the 21 species of thrips found attacking floriculture crops in the United States (including Hawaii) has been developed.

Thrips Monitoring and Economic Injury in Roses

A cornerstone of any IPM programme (*see* **Integrated Pest Management**) is the ability to detect accurately the presence of a pest and to monitor its population within the crop. Armed with this information, growers can time control measures only to pest densities that exceed the economic injury level for the crop. Development of quantitative sampling schemes to aid pest control decision-making can benefit growers by reducing costs of pest control, minimizing pesticide resistance and conserving natural enemies. The objectives of the research programme in California were to develop a rapid and easy-to-use sampling scheme for monitoring the density of thrips and to quantify the relationship between density and rose injury. The latter was used to estimate the economic injury level and set action thresholds for pest control measures.

Within-Plant Thrips Distribution

How does one know where to sample to estimate populations of thrips when confronted with a large glasshouse full of roses? The most cost-efficient plans use sampling units where pest densities tend to be highest and consistently high over time. The name 'flower' thrips suggests that these insects will be found most often in rose buds and open flowers; however, reliable data to reinforce this assumption are scarce. It has been shown that the western flower thrips may be abundant in the flowers of some plant species while they are unable to survive on the leaves of the same plant, but this will vary with the species of thrips and the host plant. To ensure that the most efficient sampling unit is selected for thrips monitoring, the within-plant distribution of thrips numbers along the length of flower canes was measured rather than for the entire plant. Since rose canes yielding flowers and buds reflect the marketed portion of the crop, the analysis was concentrated on these and ignored the sea of foliage at the base of the rose crop. Two rose greenhouses were selected with mature (blooming) rose plants and with an established thrips population. Terminals (flowering canes) were divided into a top, middle and bottom section and samples were taken weekly for 16 weeks. Several common varieties of roses were chosen and all leaves, buds, and flowers were removed from a single terminal, washed in alcohol, passed through vacuum filtration, and the various life stages of thrips were counted under a microscope. When data for all thrips were combined, there was a clear preference for the top strata of the rose terminal, i.e. the developing bud and flower (**Table 1**). However, no significant differences in larval densities were seen between the terminal and lower portions of the rose canes. It appears that rose leaves are acceptable to the western flower thrips for feeding and oviposition, but apparently there is a strong preference of adults for growth terminals, developing buds and flowers. This was most evident for female thrips and probably reflects their love of tight places and pollen.

As a direct result of this research, it is now recommended that pesticide sprays targeting thrips be directed more toward the developing buds and upper canes and need not be applied to the entire crop. By targeting only the developing buds for conventional pesticide application, it was demonstrated that a 65% reduction in the volume of water normally used in a rose crop spray adequately controlled thrips. Based on where the majority of thrips are found in a rose range, the developing bud was chosen as the sample unit. This sample unit is especially relevant since it is the marketed portion of the crop that growers must protect from damage.

Between-Plant Thrips Distribution

A complicating factor in developing a sampling plan for roses is the large number of cultivars present in a production range. Western flower thrips, like most herbivorous arthropods, show distinct preferences for some cultivars relative to others, which affects thrips distribution across a rose range. Other researchers found that when 16 rose cultivars were examined in a commercial operation, there was a high correlation between the number of *F. occidentalis* per flower and the brightness of floral reflectance at 475 nm. In addition it was demonstrated that selected rose cultivars exhibit different levels of resistance to thrips. In the present study, thrips also showed a marked preference for some rose cultivars over others; the largest differences in preference was also associated with flower colour (**Table 2**). Correlation was also high between adult and larval counts indicating that attractive rose cultivars were also preferred reproductive hosts.

Western flower thrips does exhibit significant preferences for some cultivars relative to others. The differences in susceptibility require that each cultivar be treated independently when monitoring for thrips and making pest control decisions. Therefore, decisions were based on distinct areas called pest management area units which are defined as the area planted (at approximately the same time) with the same or similar rose cultivar.

Once the sampling unit and pest management unit were defined, the between-plant distribution for thrips could be established. However, a sampling plan based on the developing rose bud or flower as the sampling unit would be unrealistic at a practical level because of the time required and the difficulty in extracting thrips. A more rapid method would be to use thrips captured on sticky traps as a predictor of thrips numbers in flowers and buds. This is particularly convenient because many growers currently rely on traps to detect pests and are familiar with their use. Therefore we concentrated on determining the between-trap distribution of thrips captured on blue and yellow traps within pest management units (defined above). Mean trap capture was then used as a predictor of numbers of thrips in developing flowers and buds. Trials established in commercial rose production facilities to compare yellow and blue sticky traps revealed that although both trap colours caught thrips throughout the year, blue traps tended to capture more. If a grower was interested in capturing primarily thrips, then blue cards should be used. If other pests are also a concern (such as whiteflies and leaf miners), then a yellow card would be more appropriate since it will capture a broad array of pests. It has been demonstated that either trap type can be used, but growers should stick to one trap colour and placement once monitoring begins.

The study to evaluate the predictive ability of sticky cards was conducted in 6 commercial rose facilities with infestations of western flower thrips in coastal greenhouses. One yellow and blue sticky card (Seabright Laboratories, 4026 Harlan Street, Emeryville, CA 94608) was placed just above the crop height in the middle of the rose bed. The plot size for each pair of cards varied within each greenhouse and ranged from 100 to 200 m^2 and there were at least four plots monitored in each pest management unit. Thrips were counted on these cards weekly for both the spring and summer crops. Ten randomly selected rose flowers were also taken each week from

Table 1 Within-plant distribution of western flower thrips on roses (cultivars Royalty, Lavon and Cara Mia) (mean number per flower ± 1 standard error)

Level	*Larvae*	*Males*	*Females*	*Total thrips*
Top	0.33 ± 0.11	0.11 ± 0.04	0.53 ± 0.09	0.87 ± 0.13
Middle	0.21 ± 0.12	0.03 ± 0.01	0.16 ± 0.04	0.46 ± 0.13
Bottom	0.20 ± 0.11	0.05 ± 0.03	0.15 ± 0.08	0.41 ± 0.16

Table 2 Distribution of western flower thrips on selected rose cultivars (mean number per flower ± 1 standard error)

Cultivar	*Larvae*	*Males*	*Females*	*Total thrips*
Royalty	0.13 ± 0.08	0.03 ± 0.02	0.19 ± 0.08	0.41 ± 0.12
Cara Mia	0.34 ± 0.11	0.09 ± 0.03	0.31 ± 0.07	0.71 ± 0.14
Lavande	0.17 ± 0.06	0.03 ± 0.01	0.31 ± 0.13	0.48 ± 0.13

each plot containing a sticky trap. These were brought back to the laboratory and all thrips were extracted. Significant correlations were found between thrips captured on cards and thrips per flower, indicating that traps could be used as a predictor of adult thrips in flowers. For example, a mean trap capture between 25 and 50 adult western flower thrips per card per week corresponded to 1 to 2 thrips per flower (**Figure 1**). Subsequent work has demonstrated that most rose cultivars can safely tolerate 1 to 2 thrips per flower without visible injury, so this action threshold was selected (**Figure 2**).

Thrips Sampling plan

A common tool in the development of sampling plans, Taylor's power law, was then used to describe the relationship between the variance and the mean of adult western flower thrips caught on individual cards within each pest management unit. Variables from Taylor's power law were then used to calculate fixed precision sampling plans. Such sampling plans calculate the number of traps needed per pest management unit to estimate populations of thrips at a predetermined fixed precision. Precision was calculated: $d=$ (standard error of the mean/mean)/(100). A precision of 0.25 is considered acceptable for pest management sampling although a more conservative precision level of 0.20 was used. Sampling decision lines are presented (**Figure 3**). Current recommendations are to use seven to eight blue or yellow cards per pest management unit and these are checked on a weekly basis.

Validation of the Sampling Program

Validation took place for 7 weeks and 20 weeks at commercial rose greenhouses. Yellow sticky cards were placed in each pest management unit within each glasshouse and each week the number of western flower thrips on cards was counted and each card was replaced. Ten harvestable flowers were also collected

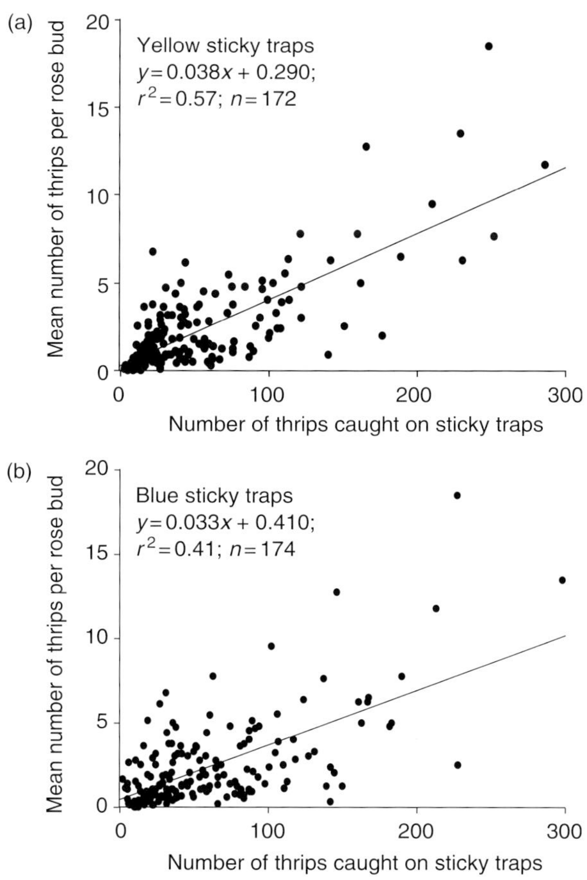

Figure 1 Relationship between single traps and mean western flower thrips numbers in rose buds: (a) yellow sticky traps, (b) blue sticky traps.

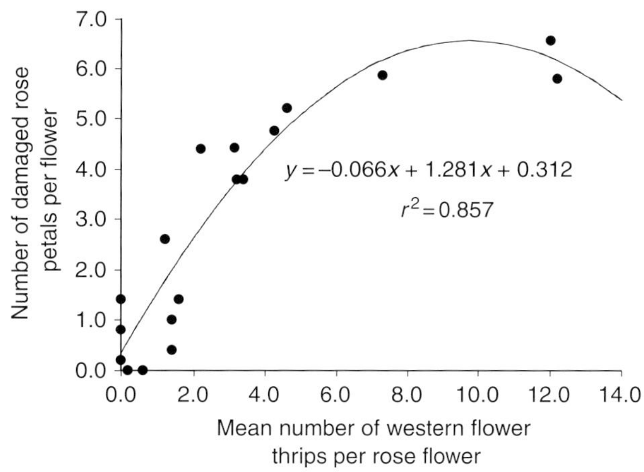

Figure 2 The relationship between density of western flower thrips and injury to rose petals.

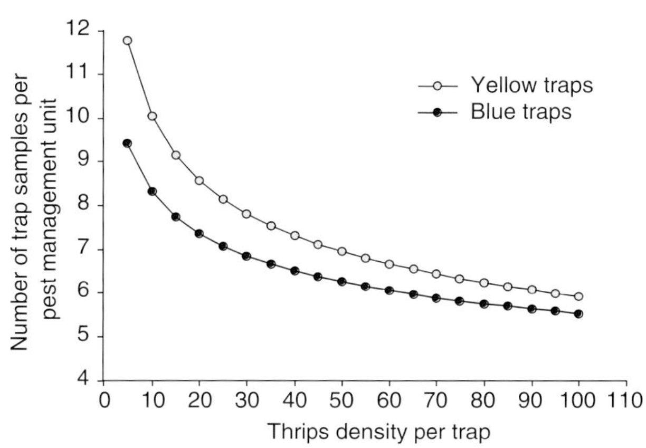

Figure 3 Estimated sticky trap sample size with a 20% level of precision.

Table 3 Predicted number of thrips per flower based on sticky trap counts

Flower colour	r^2	p	*Equation of the regression line*	*Predicted thrips per flower at 25 thrips per card*	*Predicted thrips per flower at 50 thrips per card*
All	0.36	< 0.0001	$y = 0.06119 + 0.02265x$	0.65	1.19
Yellow/orange	0.26	0.1053	$y = 0.60131 + 0.01962x$	1.09	1.58
White/pink	0.45	0.0002	$y = -0.0754 + 0.01936x$	0.41	0.89

weekly from four of the pest management units in each greenhouse and all thrips per flower were counted. Regression analysis was performed with thrips in flowers and thrips on yellow cards as the dependent and independent variables, respectively. In Monterey County, it was found that the regression equation generated from the data predicted that 25 to 50 thrips per yellow trap would correspond to 0.65 to 1.19 thrips per flower (**Table 3**). In Santa Barbara County, the regression equation predicted that 25 to 50 thrips per yellow trap would correspond to 0.28 to 0.75 thrips per flower. In both locations, there was considerable variation with flower colour. However, regardless of flower colour we were still able to predict the level of thrips in flowers using yellow sticky cards at levels below economic injury (i.e. 2 thrips per flower).

Sampling Strategy

Growers interested in using the information presented here to implement a sampling programme for western flower thrips in their glasshouses should consider the following steps:

1. Decide on the colour and size of the trap.
2. Find an appropriate supplier.
3. Determine how the cards will be hung just above the crop height. Keep in mind that the height of the cards may have to increase to accommodate the growing plants.

Once these decisions are made, the grower should not deviate from this protocal otherwise thrips numbers will be hard to interpret from week to week. At least a 10× hand lens will be necessary to view thrips on the cards. With low densities, grower scouts should remove the thrips from the card. At higher densities, this will not be feasible and the card may have to be replaced. In addition, it may be necessary to cover the card with some sort of clear plastic wrap which will facilitate removing the card so that is can be counted under a hand lens or microsope in the laboratory. The number of cards used will depend on two things: the size of pest management unit (defined earlier) and the number of thrips the grower believes can be tolerated per flower. The information presented here suggests that with a threshold of 1–2 thrips per flower, the traps must accurately capture 25–50 thrips per card. With these numbers, the growers will need seven or eight sticky cards per pest management unit. Research done to obtain these numbers was collected from glasshouses of various sizes and should only be used as a general guide. In other words, in a very small pest management unit, fewer cards may be adequate. In contrast, a pest management unit greater than 0.5 ha may require more cards. Keep in mind that the cards are an additional tool in the decision-making process; as thrips numbers begin to approach threshold values, sampling of flowers and developing buds should be done as well.

Biological Control

Successful biological control of any pest in floricultural crops is difficult (*see* **Integrated Pest Management**). It is even more difficult with thrips because only a few can be tolerated and maintaining these low levels with natural enemies is a major challenge. The predatory mites *Neoseiulus cucumeris* (Ouds.) and *Hypoaspis miles* (Berlese) and predatory bugs in the family Anthocoridae have been the focus of much study, but research on using these natural enemies on roses has been limited. It is difficult to pin down what works and what does not work. A study in Quebec, Canada showed that natural enemies of thrips on cut roses do not provide adequate control, but on pot roses in Denmark, many species of natural enemies are used successfully. Research on the utility of *N. cucumeris* as a biological control agent of thrips on roses in Finland has found that it is not effective and its use is not recommended. As noted by Finnish researchers, some biological control agent producers have recommendations on the use of *N. cucumeris* for thrips control, but detailed studies on the performance of this mite as a thrips biocontrol agent in roses are lacking.

At this time, the complex of natural enemies available for release onto roses for biological control of thrips may not be adequate to provide the level of control required. The search for new, more effective natural enemies continues; however, even if new natural enemies are identified, considerable hurdles have to be overcome to move these new biological control agents into commercial production and to obtain the necessary regulatory permits to ship them from country to country.

The entomopathogenic fungus *Beauveria bassiana* (Botanigard) has also been considered for biological control of western flower thrips. There are numerous fungi that attack thrips, but few are available commercially. There are reports of good success in keeping thrips populations below the economic threshold using regular applications of this fungus. It has also been found the fungal spores will penetrate deep into the petals of developing flowers and rapidly infect the concealed adult and immature stages. However, the fungus also operates relatively slowly and cannot adequately reduce large thrips outbreaks that threaten the rose crop. Instead, the fungus is best used in a maintenance programme when thrips populations are relatively stable.

Reduced Risk Pest Management Strategy for Rose Growers

The information presented in this article has been put into practice in California and combined with additional information obtained through a larger multi-investigator approach to managing thrips in floriculture crops. Roses are plagued by many other pests (*see* **Insects and other Animals**: Overview of Insects), and managing thrips cannot be done in a vacuum. Suggestions for the management of other pest problems on roses must be combined with strategies of thrips management to provide a comprehensive pest management programme for the rose crop.

Acknowledgments

Research reported in this chapter was supported, in part, by the American Floral Endowment and the USDA Floriculture and Nursery Research Initiative.

See also: **Breeding**: Selection Strategies for Disease and Pest Resistance. **Insects and other Animals**: Overview of Insects; Mites; Insecticides. **Integrated Pest Management**. **Production and Marketing**: Overview of Roses and Culture.

Further Reading

BioBest (2002) Rose. http://www.biobest.be

Buntin GD (1994) Developing a primary sampling program. In: Pedigo LP and Buntin GD (eds.) *Handbook of Sampling Methods for Arthropods in Agriculture*, pp. 99–115. Boca Raton, FL: CRC Press.

Butt TM and Brownbridge M (1997) Fungal pathogens of thrips. In: Lewis T (ed.) *Thrips as Crop Pests*, pp. 399–433. Wallingford, UK: CAB International.

Casey C and Parrella M (2002) Demonstration and implementation of a reduced risk management strategy in fresh cut roses. In: Enkegaard A (ed.) *Integrated Control in Protected Crops: Temperate Climate*, pp. 45–48. IOBC WPRS Bulletin 25.

Lambert L, Cecyre A, Chouffot T *et al.* (2002) An overview of biological control in ornamental greenhouses in Quebec, Canada. In: Enkegaard A (ed.) *Integrated Control in Protected Crops: Temperate Climate*, pp. 135–138. IOBC WPRS Bulletin 25.

Lewis T (1997) Field and laboratory techniques. In: Lewis T (ed.) *Thrips as Crop Pests*, pp. 435–475. Wallingford, UK: CAB International.

Loomans AJM, Murai T and Greene ID (1997) Interactions with hymenopterous parasitoids and entomopathogenic nematodes. In: Lewis T (ed.) *Thrips as Crop Pests*, pp. 355–397. Wallingford, UK: CAB International.

Mantel WP (1989) Bibliography of the western flower thrips, *Frankliniella occidentalis* (Pergande) (Thysanoptera: Thripidae). In: van Lenteren J and Wardlow LR (eds.) *Working Group: Integrated Control in Glasshouses*, pp. 29–66. IOBC WPRS Bulletin 12.

Parrella MP (2003) Development of IPM strategies in floriculture crops: roses as a model system. *Sting 22*. http://www.agrsci.dk/plb/iobc/sting/sting22.htm.

Parrella MP (2003) http://entomology.ucdavis.edu/faculty/parrella/index.cfm.

Parrella MP and Jones V (1987) Development of integrated pest management strategies in floriculture crops. *Bulletin of the Entomological Society of America* 33: 28–34.

Sabelis MW and Van Rijn PCJ (1997) Predation by insects and mites. In: Lewis T (ed.) *Thrips as Crop Pests*, pp. 259–354. Wallingford, UK: CAB International.

Steiner M and Goodwin S (2002) Development of a new thrips predator, *Typhlodromips montdorensis* (Schicha) (Acari: Phytoseiidae) indigenous to Australia. In: Enkegaard A (ed.) *Integrated Control in Protected Crops: Temperate Climate*, pp. 245–247. IOBC WPRS Bulletin 25.

Svendsen MS and Hansen EW (2002) Biological and integrated control in ornamentals: successes and challenges. In: Enkegaard A (ed.) *Integrated Control in Protected Crops: Temperate Climate*, pp. 261–264. IOBC WPRS Bulletin 25.

Terry LI (1997) Host selection, communication and reproductive behavior. In: Lewis T (ed.) *Thrips as Crop Pests*, pp. 65–118. Wallingford, UK: CAB International.

Vanninen I and Lannamaki M (2002) Performance of *Neoseiulus cucumeris* as a biocontrol agent of the western flower thrips in cut roses. In: Enkegaard A (ed.) *Integrated Control in Protected Crops: Temperate Climate*, pp. 289–292. IOBC WPRS Bulletin 25.

Vierbeigen G (1993) International movement, detection and quarantine of Thysanoptera pests. In: Parker BL, Skinner M and Lewis T (eds.) *Thrips Biology and Management*, pp. 119–132. New York: Plenum Press.

Nematodes

D Esmenjaud, Laboratoire de Biologie des Invertebres, INRA, Antibes, France

Introduction

Plant parasitic nematodes are microscopic worms which are generally filiform (at least at one of their stages), ranging from 0.5 mm (e.g. *Meloidogyne*, *Pratylenchus*) to 4–5 mm (e.g. *Xiphinema*). A wide variety of nematodes affect plants and some of them have a significant economic impact on commercial crops. On roses, a few species of nematodes are predominant, causing severe losses, and have a worldwide distribution. Because they are in the soil and they give rise to nonspecific symptoms (general decline and stunting of plants, poor plant growth and reduction of flowering), their presence is frequently ignored, and thus their role and importance as rose pests are underestimated. Nematodes may be a real concern for rose growers because they can be easily disseminated by vegetative rootstock material from nurseries and become one of the components of replant sickness. In this article, the symptoms, biology and lifecycle of the major nematode species parasitizing roses together with the control measures against these pests will be reported.

Major Nematode Pests

The major nematode pests of roses are listed in **Table 1**. The nematodes with the highest economic impact are the endoparasitic species of the genera *Meloidogyne* and *Pratylenchus*. Most life stages of *Meloidogyne* are sedentary within the root tissues whereas *Pratylenchus* are migratory at all developmental stages. Ectoparasitic nematodes such as the virus vector nematode *Xiphinema diversicaudatum* Thorne and other soil species may also limit rose growth and production.

Root-knot Nematodes of the Genus *Meloidogyne* (Figure 1a)

Meloidogyne hapla Chitwood is the most common nematode pest of roses. This typical species of temperate climates parasitizes roses from these climatic regions, but its distribution also extends to rose crops grown in Mediterranean areas and in protected conditions under glasshouses all over the world. This root-knot nematode (RKN) is frequently detected on *Rosa indica* and *R. multiflora* (**Figure 2a, b**). Other

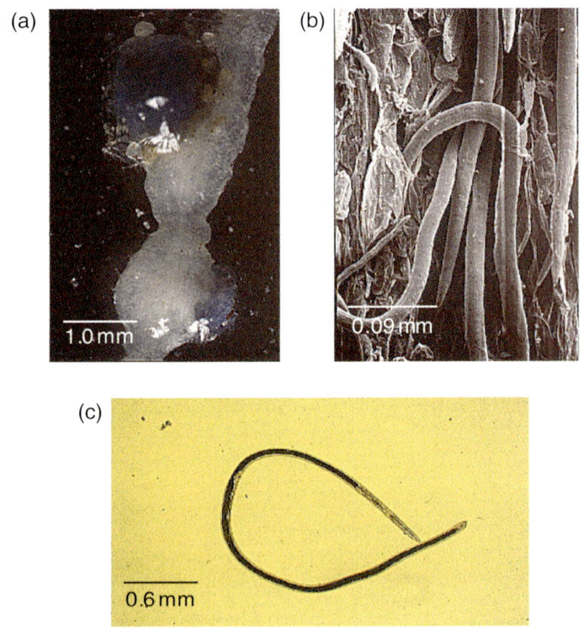

Figure 1 Important nematode pests of roses. (a) Galls of *Meloidogyne* on roots with external egg masses (stained in blue); (b) root-lesion nematode *Pratylenchus vulnus* in the cortical parenchyma of a root (scanning electron microscopy); (c) dagger nematode *Xiphinema diversicaudatum*. (a, c) © INRA; (b) courtesy of J. Pinochet.

Table 1 Major nematodes associated with roses

Species	*Common name*	*Type of parasitism*	*Symptoms*
Meloidogyne hapla	Root-knot	Endoparasitic; sedentary	Galls on smaller roots; root branching from galls
Pratylenchus vulnus and *P. penetrans*	Lesion	Endoparasitic; migratory	Root lesions (necrosis)
Xiphinema index and *Longidorus macrosoma*	Dagger	Ectoparasitic; migratory	Galls on root tips; swelling and curling tips
Paratylenchus	Pin		Root lesions (necrosis)
Criconematidae	Ring	Ectoparasitic;	
Rotylenchus	Spiral	migratory	Root browning or discoloration
Helicotylenchus	Spiral		

Meloidogyne species (*M. arenaria* Chitwood, *M. incognita* Chitwood and *M. javanica* Chitwood) are limited to Mediterranean and hot climates and have only been reported on *R. polyantha*.

All four RKN species are extremely polyphagous and also develop on other crops of the family Rosaceae as well as on several hundreds of botanical families. They form characteristic galls on the roots of their host plants which are the result of the compatible plant–nematode interaction. The reproductive cycle is described in **Figure 3**. Nematodes survive in the soil or plant residues as eggs, grouped in egg masses. The mobile second-stage juveniles hatch directly from the egg masses, move to the root tips, penetrate and migrate intercellularly into the root tissues. They become attached to the stele and develop into swollen third-

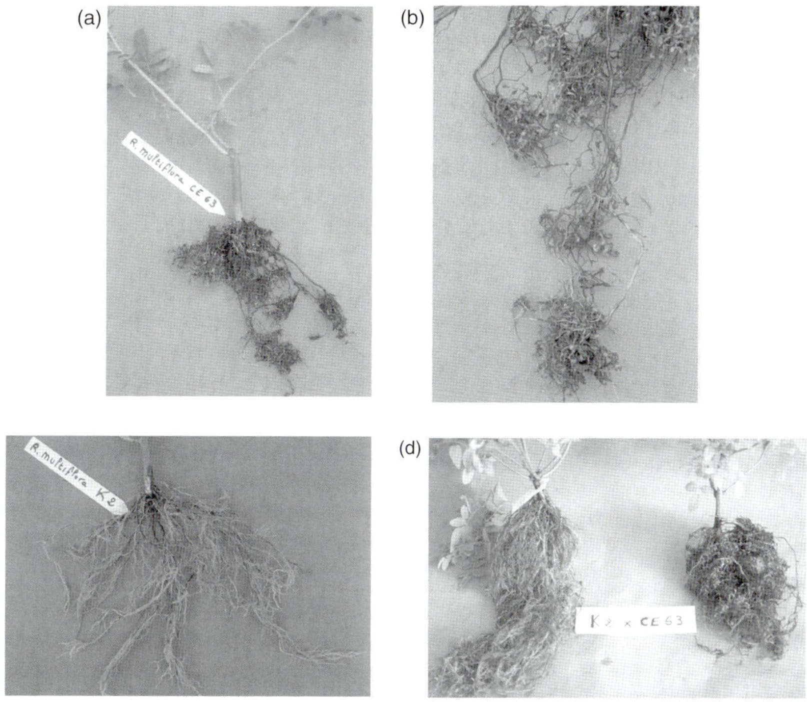

Figure 2 Roots of *Rosa multiflora* attacked by root-knot nematodes. (a and b) Susceptible accession 'CE63'; (c) resistant accession 'K2'; (d) two individuals from the cross K2 × CE63; resistant (left) and susceptible (right).

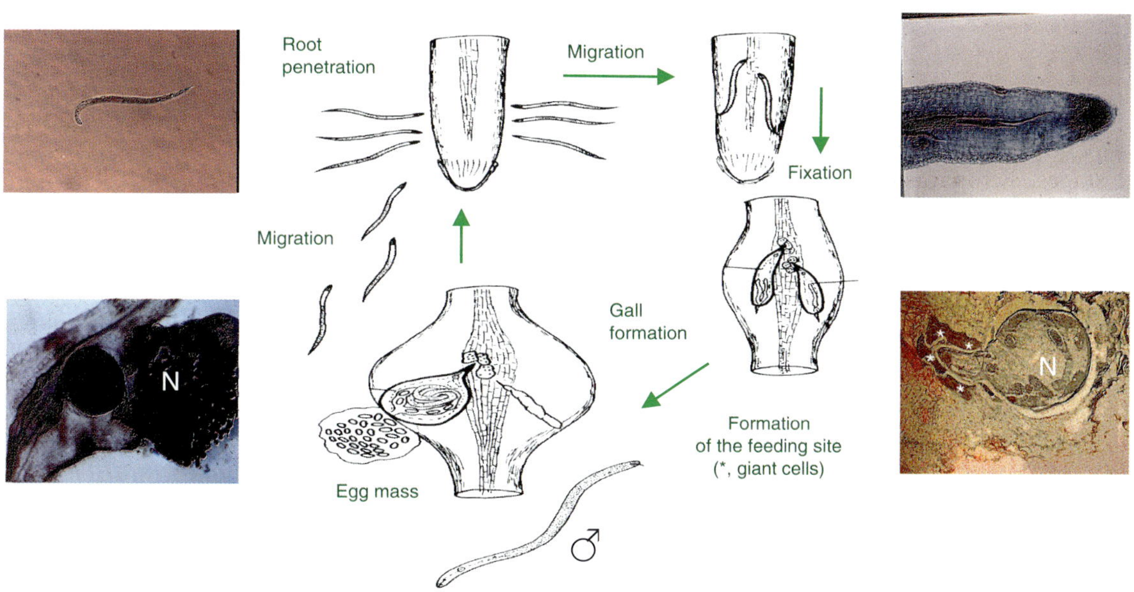

Figure 3 Reproductive cycle of *Meloidogyne hapla*. N, female nematode; *, giant cells. Photos © INRA.

and fourth-stage juveniles and then into sedentary females which each lay several hundred eggs gathered into a single egg mass. Feeding larvae and adults induce the formation of a few giant cells (feeding sites) around the nematode head, along with the multiplication of cells surrounding cortical and vascular parenchyma. Nutrient uptake by the RKN causes reduction and retardation of plant development. As shown in other plant species, it is suspected that nematodes facilitate penetration of the bacterium *Agrobacterium tumefaciens* into the root tissues and the subsequent development of symptoms of rose crown gall.

Meloidogyne hapla multiplies by both sexual (amphimixis) and asexual (parthenogenesis) reproduction, in contrast to the three other RKN species occasionally found on rose (*M. arenaria*, *M. incognita* and *M. javanica*), which are exclusively parthenogenetic. Numerous host tests performed in different countries show high variations in aggressivity among the populations grown on rose rootstocks and on other hosts. The results cast doubt on the existence of races or other infraspecific categories in this *Meloidogyne* species but confirm the existence of genetic polymorphism which is higher than in the three other species with parthenogenetic reproduction. Because of the increasing exchanges of rose plant material at national and international levels, populations originating from distant geographic origins are expected to be widely dispersed.

Root-lesion Nematodes *Pratylenchus vulnus* and *P. penetrans*

Root-lesion nematodes (RLN) are migratory species that are filiform at all stages. Two endoparasitic species are found in roses: *Pratylenchus vulnus* Allen & Jensen, and *P. penetrans* Chitwood & Oteifa. *Pratylenchus vulnus*, the most frequently recovered, has a worldwide distribution in protected rose crops and is also present in rose fields grown under Mediterranean conditions. *Pratylenchus penetrans* only develops in temperate climates (in both field and greenhouse conditions) and is mainly detected in oceanic areas. Both RLNs are polyphagous and can parasitize many fruit and nut crops as well as many annuals.

In constrast to RKN, they do not cause typical root symptoms but completely destroy the parenchyma of the cortical tissues (**Figure 1b**); they are thus responsible for extensive root necrosis and drastic reductions in plant growth. Filiform adults lay eggs in the necrotic tissues or in the soil. The four larval stages all feed directly on cortical root cells. Both species have male and female adult stages, which suggests a sexual mode of reproduction. Under favourable greenhouse conditions, reproduction can occur all year long, explaining the high nematode numbers that can be observed in some young rootlets (up to several thousand individuals per gram of roots). Important differences in the aggressiveness of the populations of *P. vulnus* to host plants such as *Prunus* and apple have been observed and a similar variability also probably occurs in roses, for which no specific data are available.

The Dagger Nematodes

Xiphinema diversicaudatum Thorne is often present in greenhouse rose crops. This ectoparasitic nematode, measuring 4–5 mm in length at the adult stage (**Figure 1c**), produces root galls at the tip of feeder rootlets. These galls are characterized by swollen and curled tips. In contrast, galls induced by *M. hapla* have a more symmetrical shape and are spread more regularly along the rootlet tips. Besides the impact of its direct feeding on the roots, this dagger nematode can transmit viruses such as arabis mosaic virus (ArMV) and strawberry latent ringspot virus (SLRV). As *X. diversicaudatum* and SLRV share many other hosts in the Rosaceae family, the risk of virus transmission by nematodes between different plant species is high. Consequently, growing roses and particularly propagating material must be avoided in fields where plant species belonging to the same botanical family have been previously cultivated. Another dagger nematode, *Longidorus macrosoma* Hooper, has been reported in Belgium and the UK as damaging greenhouse roses.

Other Ectoparasitic Nematodes

The most frequent genera are *Paratylenchus* (pin nematode), Criconematidae (ring nematodes), *Helicotylenchus* and *Rotylenchus* (spiral nematodes). These nematodes, although they may be frequently found in soils where roses are cultivated, do not justify specific control measures.

Detection and Identification

The presence of galls formed by RKN or the dagger nematode *X. diversicaudatum* allows direct diagnosis by a trained nematologist. Nevertheless, this primary diagnosis must be confirmed by reliable identification of the causal species in order to plan the appropriate control. This identification as well as that of other causal nematodes requires specific procedures for sampling by the grower and for subsequent sample processing and nematode extraction in the nematology laboratory. Identification relies on classical morphological and morphometrical data which can now be completed by enzymatic (RKN) or molecular tools (RKN and *X. diversicaudatum*). It may be performed from a single individual at any stage when polymerase chain reaction or another hybridization technique is run on abundant genomic sequences, such as ribosomal DNA or satellite DNA.

Control

No efficient control method can be applied on material already planted. As crop rotation usually employed for annuals is also not suitable, control relies mainly on chemical treatments, the use of nematode-free planting material and the use of resistant rootstocks.

Preplanting Soil Nematicides

Applying soil nematicides is justified when severe attacks are observed (high nematode densities in the previous crop) and in cases of replant sickness symptoms following repeated rose crops in the same field or greenhouse. In different countries, different ranges of fumigants (e.g. methyl bromide and dichloropropene), precursors of fumigants (e.g. dazomet) and nonvolatile nematicides (in granular form such as aldicarb) are still authorized for use but the list is becoming more and more restricted because of the negative environmental impact of the chemicals. Methylbromide, for example, which has a remarkably wide-spectrum activity (including weeds, soil fungi, nematodes and insects), will be banned in all countries of the European Union in 2005.

Nematicides protect young plantlets during the first stages of their development and allow their satisfactory installation and subsequent productivity. None the less, none of the nematicides can eradicate nematodes from the soil because individuals can usually survive in deeper levels where they can escape chemical diffusion; reinfestation of the plants will inevitably occur after a few months or years. It is absolutely necessary that preplanting chemical treatments, which act only against nematode stages surviving in the bare soil, are combined with the use of nematode-free nursery plants.

Postplanting Soil Nematicides

Nonvolatile nematicides, such as aldicarb, are not phytotoxic and thus can be used on growing plants at any time (but with strict respect for the conditions of use provided by the manufacturers, given their high acute toxicity). Aldicarb has a systemic action that reduces densities of nematodes and allows plants to recover from the attacks but, because of its nematostatic effect, this reduction has a short duration and repeated applications may be necessary.

Treatment of the Planting Material

Hot-water treatment effectively controls the nematodes *M. hapla* and *P. vulnus*, but it often damages the roots of roses. Combinations of temperature and duration of treatment may be specific for a particular rootstock (e.g. 44 °C for 12 min for the rootstock *R. chinensis* Jacq. cv 'Indica Major' in French tests). Pretreatment at a lower temperature to harden the roots is recommended. Immersion of roots for various durations in nematicide solutions such as oxamyl or phenamiphos has been performed but this treatment does not eliminate the nematodes completely and subsequent plant damage may occur.

Use of Nematode-free Plant Material

Because of the vegetative propagation of rose rootstocks, nematodes have been introduced in the fields or greenhouses almost exclusively by means of planting material from infected nurseries. Given that the long-distance exchanges of plant material between producing countries have progressively intensified, it is easy to understand why the major endoparasitic nematode pests of rose (RKN, RLN) are found all over the world. To try and restrict spread of this type, certification schemes have been developed with the aim of ensuring that planting material for export is free of pests and diseases. The European and Mediterranean Plant Protection Organization, for example, prepared a certification scheme for roses in 1997 which includes measures to ensure freedom from RKN, RLN and virus vector nematodes. The scheme is voluntary, but nurseries following its requirements are given a certificate of health by a national official plant protection service. Such certified planting material may be more easily and safely traded both nationally and internationally.

When certified healthy planting material is used in soil free of pestiferous species of nematode, problems caused by nematodes are unlikely to occur. Even when planted into soil where nematodes may be present in small numbers, the onset of nematode attack will be sufficiently delayed to enable the plants to establish properly.

Use of Resistant Rootstocks

Selection of nematode-resistant rootstocks is based on evaluation of the multiplication rate of a given nematode species on diverse plant material. To take into account the variability in nematode pathogenicity, plant cultivars or accessions have to be evaluated to a representative range of populations within the nematode species considered. As such tests on perennial plants are time- and space-consuming, data frequently refer to only one population per species (i.e. the local population where the problem occurs) and results obtained with the tested population cannot be extended systematically to other populations.

Resistance evaluations to *M. hapla* have been performed in several countries mainly on two diploid species of rose rootstocks, *R. multiflora* and *R. indica*. A wide range of clones of both species has been evaluated in France. In *R. multiflora*, major differences in host suitability between clones have been observed (**Figure 2a–c**). The clones 'K1' and 'K2' selected in Japan express a consistently good level of resistance to several geographic populations of *M. hapla*. These accessions perform better than the South African accession 'Floradale', and 'K2' in particular constitutes the best source for RKN resistance. Recent studies on inheritance of resistance suggest that the trait is polygenic (**Figure 2d**), which is crucial information for further breeding programmes. In *R. indica*, more reduced differences in host suitability are observed between the accessions evaluated, and resistance is much lower than in previous Japanese selections. Among the polyploid species *R. manetti* and *R. canina*, evaluations carried out in different countries agree to a large extent: *R. manetti* appears highly resistant to *M. hapla* and *R. canina* shows a good resistance level.

Several authors have mentioned differences in host suitabilities to *P. vulnus*. A Dutch *R. multiflora* Thunb. cv 'Ludiek' exhibits a lower host suitability than control rootstocks such as Floradale, *R. indica* 'Major' or one *R. canina* cultivar, and is also tolerant. Nevertheless, these data must be confirmed by using different geographic populations of this nematode.

While a majority of rose accessions allow the multiplication of *P. penetrans*, *R. multiflora* 'K1' and *R. laevigata* var. *anemoides* result less favourable to the reproduction of this nematode.

Conclusion and Prospects

Some nematode species have a real economic impact on rose production. As nematicides are progressively removed from the market, the development of more environmentally friendly control alternatives is stimulated. Breeding resistant rootstocks is a promising alternative, and the challenge will be to add resistance to both RKN and RLN in the same rootstock. When replant sickness due to nematode impact and to severe damage by soilborne diseases (*Verticillium*, crown gall) occurs, the development of new production systems based on soilless crops (e.g. on rockwool substrate) is justified. None the less, a few cases of attacks by *P. vulnus* introduced with the plants have been reported in such systems, thus illustrating the need for nematode-free planting material and the establishment of rose certification schemes.

Preventive rather than curative control is still recommended for rose nurseries, cut flowers or garden roses. Successful nematode management will continue to rely on the use of nematode-free planting material to prevent nematode introduction into new areas and the early detection of nematodes in the field or the greenhouse to limit the spread of these polyphagous pests. It is also important to eliminate weeds from rose fields or greenhouses as they represent potential alternate hosts for the nematodes.

See also: **Breeding**: Selection Strategies for Disease and Pest Resistance. **Disease**: Overview; Replant Problems and Soil Sickness; Viruses. **Integrated Pest Management**. **Rootstocks**: Clonal Rootstocks.

Further Reading

Coolen WA and Hendricks GJ (1972) *Monograph on the Nematological Situation in Belgian Rose Culture*. Merelbeke, Belgium: Rijksstation Voor Nematologie en Entomologie (in Dutch).

EPPO/OEPP (1998) *Certification Scheme: Pathogen-tested Material of Rose. EPPO Standard PM 4/21(1)*. Paris, France: European and Mediterranean Plant Protection Organization.

Horst RK (1983) *Compendium of Rose Diseases*. St Paul, MN: APS Press.

Ohkawa K (1986) Rootstock native to Japan. *Acta Horticulturae* 189: 61–63.

Peng YL, Chen WR and Moens M (2003) Resistance of *Rosa* species and cultivars to *Pratylenchus penetrans*. *HortScience* 38: 560–564.

Scotto La massese C (1986) Nematodes. In: Ferrer Marti F and Salvador Palomo PJ (eds.) *The Production of Roses under Protected Conditions*, pp. 258–266. Seville, Spain: Universal Plantas (in Spanish).

Voisin R, Minot JC, Esmenjaud D *et al.* (1995) Host suitability of rose rootstocks to the root-knot nematode *Meloidogyne hapla* using a high-inoculum-pressure test. *Acta Horticulturae* 424: 237–239.

Wang X, Jacob Y, Mastrantuono S *et al.* (1999) Preliminary study of the inheritance of resistance to the root-knot nematode *Meloidogyne hapla* in rose. *Proceedings of the 51st International Symposium on Crop Protection* (University of Gent) 64: 359–366.

Galls Induced by Cynipid Wasps

J D Shorthouse, Laurentian University, Sudbury, ON, Canada

Introduction

Gall-inducing insects, along with some mites and nematodes, are highly specialized herbivores with the ability to stimulate host plants into surrounding feeding sites of their immatures with thick layers of modified plant cells. The organ-like structures formed by these cells and tissues are called galls which are atypical plant growths where control of developmental events has passed from the plant to the insects. Some types of insect-controlled gall cells are found nowhere else in host plants and as these cells enlarge and proliferate, they encase the immature gall insects in individual chambers where they are provided with high-quality food and protection from the elements and predators. Gall-inducing insects are one of three common guilds of insects that live within plant organs. The other two are leaf miners and stem borers (*see* **Insects and other Animals**: Overview), both of which consume cells as they tunnel through the tissues of their host organs. Gall insects, in contrast, are sessile and spend their immature stages feeding on nutrient-rich cells that form around the inner surface of their chambers.

The relationship between gallers and their host plants is parasitic-like with the gall-inducers being the parasites and the host plants the provider of food. About 13 000 species of insects in seven orders have independently evolved the gall-inducing habit with the most complex galls being those induced by wasps of the Family Cynipidae in the Order Hymenoptera and midges in the Family Cecidomyiidae in the Order Diptera. There are about 2000 species of cynipids worldwide with the largest number inducing galls on oaks. About 40 species of cynipids of the genus *Diplolepis* induce galls on wild roses and some of these now attack garden roses. This article provides an overview of cynipid wasps and their galls found on roses and describes the biology of those found on garden roses of Europe and North America.

Attributes of Cynipid Galls

One of the most amazing aspects of cynipid gall biology is that each species of wasp induces structurally distinct galls. This is particularly evident when strikingly different galls of closely related species occur on the same host plant. How this happens is one of the key unsolved mysteries of gall biology, but it has been speculated that structurally unique galls result either from differences in the way larvae of each species of cynipid physically contact gall cells as they feed, or from salivary enzymes that vary in chemical composition and mimic plant hormones. Whatever the mechanism, galls of each species are so different that it is often easier to identify cynipid wasps from their galls than by examining the adult insects themselves.

Specificity

Cynipids are host-specific insects with each genus of wasp only able to induce galls on closely related host species. The genera of wasps inducing galls on oaks, for example, cannot induce galls on roses, or vice versa. Besides being host-specific, cynipids are also organ-specific and depending on the species of wasp, induce galls either on leaves, shoots, stems, acorns or roots. Cynipids that induce leaf galls deposit their eggs on the surface of immature leaflets still folded with leaf buds. Many of the stem gallers deposit their eggs within tissues at the base of leaf buds. Further, some cynipids induce single-chambered galls while others induce multichambered galls. Some multichambered galls house over 900 larvae.

Life Cycle of Cynipids

The life cycles of cynipids vary depending upon the genera of inducers. Those on oaks have complex life cycles with two morphologically distinct forms of adults each responsible for structurally distinct galls. This type of life cycle is called heterogeny, but it does not occur in *Diplolepis* and those on other hosts. *Diplolepis* have a simple life cycle with one form of adult and a single generation per year. Eggs of cynipids are deposited on immature or undifferentiated tissues and galls are initiated by freshly hatched larvae. Gall cells proliferate rapidly and as they grow, each larva is isolated in a separate chamber. All larval feeding and growth occurs within these chambers. In early autumn, the mature larvae enter a prepupal stage in which they overwinter. The prepupae pupate the following spring, turn into adults and the adults then chew an exit channel to the outside.

Structure of Galls

Galls of cynipids are composed of five types of plant cells arranged concentrically around the larval chamber. These include an innermost layer of cytoplasmically dense nutritive cells which line the larval chamber, an adjacent region of vacuolated cells referred to as parenchymatous nutritive cells or storage nutritive tissue which contain starch, a layer of sclerenchyma cells, gall cortical parenchyma cells and an outer layer of epidermal cells. The outside surface of galls, depending upon the species of inducer, may be smooth or covered with fibrous outgrowths or spines. Vascular tissue connected to that of the host organ extends through the gall to the nutritive cells and brings nutrients from other parts of the host plant. Most cells of cynipid galls are larger than cells in other parts of the host plant; however, it is the rapid proliferation of cells that contributes most to gall growth.

Cynipid galls, like those of gallers in other orders of insects, undergo three stages of development referred to as initiation, growth and maturation. Each stage is characterized by distinct periods of tissue activation, proliferation and differentiation in response to stimuli from the inducer. Initiation begins with oviposition and is the stage when the inducer isolates and withdraws a group of cells from the typical growth patterns of the host organ. Tissues chosen by females for oviposition are immature, developmentally flexible and most easily controlled by the young inducers. Cells near the freshly hatched larvae begin to enlarge and proliferate soon after feeding begins and the larvae are quickly isolated within their chambers. Cells lining the larval chambers differentiate into nutritive tissue soon after gall initiation. Part of the control exerted by inducers over the physiology of gall cells results in the nutritive cells becoming filled with high concentrations of proteins, enzymes, lipids and soluble carbohydrates. The growth stage is the period when masses of proliferating gall cells encircle each larva. Cynipid larvae allow the gall to become almost fully grown before extensive feeding begins. Vascular tissues differentiate within the mass of growing cells during the growth stage. Galls in the growth stage become physiological sinks for nutrients and assimilates produced elsewhere. These nutrients are pulled towards the galls and chemically altered as they accumulate in nutritive cells.

Proliferation of gall tissues slows near the end of the growth stage and the inducer begins to feed more actively on nutritive cells. As the nutritive cells are consumed, adjoining parenchymatous nutritive cells enlarge, become filled with nutrients, and in turn are consumed. This results in the rapid enlargement of both the larvae and the size of their chambers. Another important characteristic of the maturation stage is the differentiation of several layers of sclerenchyma cells between the parenchymatous nutritive cells and the gall cortex. The sclerenchyma layer provides galls with support, prevents some parasitoids from reaching the larvae with their ovipositors, helps prevent desiccation of the larvae and slows the winter formation of insect-damaging ice along the inside surface of the chambers. Mature larvae gradually consume all nutritive and parenchymatous nutritive cells down to the sclerenchyma such that by autumn, larval chambers are lined with a smooth layer of sclerenchyma. Actively feeding cynipid larvae must be present for all stages of cellular growth and differentiation in galls to occur. If the inducer is killed, nutritive cells lose their characteristics within 24 h and the galls stop growing.

Parasitoids and Inquilines

The presence of inducer larvae within clusters of single-chambered galls or large multichambered galls has not gone unnoticed by small insect-eating wasps, called parasitoids, equipped to locate their hosts within plant tissues. Adults of these minute chalcid and ichneumonid wasps oviposit when the galls are small enough to be penetrated by their ovipositors. Eggs are laid on the surface of immature inducer larvae which in many cases are too small to provide enough food for the parasitoid larvae to mature. The solution is for the immature parasitoids to feed on the inducer larvae without killing them. Once the inducer larva is nearly full grown, it is consumed by the parasitoid. Other types of cynipid wasps called inquilines are also attracted to galls of certain species. Inquilines are not able to induce galls of their own, but must develop inside the galls of others. Female inquilines deposit their eggs inside the chambers when galls are immature and in the process, stab and kill the inducer larvae. When the inquiline larvae hatch, they feed on gall cells initiated by the inducer, but soon they induce gall cells of their own making gall-like structures inside the galls of others. Inquilines commonly increase the size or alter the shape of galls they inhabit. The assemblage of all the inhabitants associated with a population of galls induced by the same species in a particular habitat form what are referred to as component communities. Many species of cynipids have a specific component community that is similar throughout the natural range of their galls.

Biology of *Diplolepis* Wasps and Their Galls

About 40 species of *Diplolepis* have been described in the world; about 30 are found in North America and the rest in Europe, Central Asia and Japan. All organs of roses except the roots are galled by *Diplolepis*, with most species being found on the leaves. Some leaf galls are single-chambered whereas others are coalesced or multichambered. The surface of some leaf galls is smooth whereas in galls of other species, it is clothed with sharp spines. Leaf galls vary in colour from yellowish-green to bright red. Stem galls of some species are single-chambered and only slight enlargements of the stem whereas others are multi-chambered and the size of golf balls. One species induces multichambered galls at the tips of sucker shoots just as they break the surface of the ground.

All adult *Diplolepis* are small, inconspicuous wasps (**Figure 1**), averaging 4–8 mm in length and are either black or reddish brown. The short, thick thorax is arched giving them a hump-backed appearance and the legs are usually amber. The wings are clear and venation is reduced. The ovipositor is curved and concealed within the abdomen. All species of female *Diplolepis* have a ploughshare-shaped plate at the ventral end of the abdomen called a hypopygium that is used to guide the ovipositor when depositing eggs.

Figure 1 Female *Diplolepis spinosa* resting on a rose leaf bud.

Figure 2 Mature larvae of *Diplolepis* within their chambers.

Diplolepis adults live for about a week, but are rarely seen by gardeners.

Diplolepis in northern regions survive temperatures as low as −38 °C by forming cryoprotectants (similar to antifreeze) in their blood that keeps them from freezing. Pupation occurs in the spring and the appearance of adults is synchronized with the presence of immature host tissues such as unforced leaf buds. *Diplolepis* are able to lay viable eggs without mating, a type of reproduction called parthenogenesis. Males are rare, usually comprising less than 5% of the population, or absent altogether.

Eggs are banana-shaped with a stalk at one end and are pinned to a single cell on the outside of a developing leaflet or within stem cells at the base of a bud. Eggs are glued to the plant with a translucent, yellowish drop of an unidentified substance. Larvae of all species are legless, have a smooth hairless surface and are cream-coloured (**Figure 2**). Larvae of stem gallers are cylindrical whereas those of leaf gallers are comma-shaped. Larvae slice into nutritive cells with their mandibles and the enriched fluid contents that spew over their heads are sucked into their mouths. The gut is blind throughout the larval stage to prevent fouling the confined space of their chambers. Larvae are full grown by the end of August and begin physiological preparations for overwintering well before cold temperatures are detected.

Diplolepis Galls Found in European Gardens

At least 20 species of wild roses are found in Europe (*see* **Distribution and Ecology**: Europe) and 6 endemic species of *Diplolepis* induce galls on several of these species. Of interest, roses such as *Rosa canina* L. (dog rose) and *R. eglanteria* L. (also known as *R. rubiginosa* L.) (sweetbrier or eglantine rose) appear more susceptible to galling than others. Dog and sweetbrier roses are commonly grown in gardens and this is where gardeners usually become familiar with *Diplolepis* galls. Galls here are undoubtedly the result of adult wasps flying from rural rose patches to urban habitats. Fortunately, cultivated roses including *R. rugosa*, are rarely attacked.

Figure 3 Galls of *Diplolepis centifoliae*.

Five of the six European species of *Diplolepis*, *D. centifoliae* (Hartig) *D. eglanteriae* (Hartig), *D. mayri* (Schlechtendal), *D. nervosa* (Curtis) and *D. rosae* (L.), induce galls on leaves, whereas *D. spinosissimae* (Giraud) induces galls on both stems and leaves. Galls of *D. centifoliae*, *D. eglanteriae* and *D. nervosa* induce single-chambered galls whereas *D. mayri* and *D. rosae* induce multichambered galls. Galls of *D. eglanteriae* and *D. centifoliae*, commonly known as pea galls (**Figure 3**), are spherical, light greenish or pink, and are found weakly attached on either the upper or lower surface of leaves. Galls of both species are uncommon and, along with the adults, are difficult to distinguish. Galls of *D. nervosa*,

Figure 4 Gall of *Diplolepis nervosa*. (Photograph courtesy of György Csóka.)

Figure 6 Gall of *Diplolepis mayri*.

Figure 5 Gall of *Diplolepis spinosissimae*. (Photograph courtesy of György Csóka.)

Figure 7 Gall of *Diplolepis rosae*.

commonly known as the spiked pea gall (**Figure 4**), are found on the upper surface of leaflets and are either rounded with a smooth surface or with blunt, horn-like papillae. Immature galls of *D. nervosa* are yellow–green or reddish, then turn brown when mature. Most stem galls of *D. spinosissimae* are single-chambered and the adults are the smallest *Diplolepis*. When on leaves, galls of this species are globular (**Figure 5**) and visible from both surfaces of the leaf.

Most galls of *D. mayri* (**Figure 6**), are multichambered and as large as 3 cm in diameter, while smaller galls appear as a series of coalesced single chambers. The surface of *D. mayri* galls are either smooth or clothed with fine, unbranched spines. Galls are soft and green or reddish when immature and turn brittle and brown to grey when mature. Some galls remain firmly attached to the host throughout the winter.

Galls of *D. rosae*, known as the bedeguar gall or robin's-pincushion gall (**Figure 7**), are the most commonly observed on roses in Europe. The name bedeguar comes from a French word meaning 'wind-brought'. They vary in size from small, bright red, single-chambered galls on the upper surface of leaves to large greenish-pink masses with up to 200 larvae that appear to encircle stems. The surface of this gall is covered with long, branched, wiry hairs which form a tangled mass of 6 cm or more across. Maturing galls are soft and succulent and are easily opened with a fingernail to reveal the larvae within (**Figure 8**). However, as the galls mature, walls of the chambers become so brittle they can only be opened with a firm, sharp knife.

Although galls of *D. rosae* are initiated on leaves, they commonly cause such an enlargement of the attacked leaf buds that masses of gall tissues surround the stem preventing the galls from abscising in the autumn. Such galls become unsightly masses over the winter and can only be removed by cutting the stems. Some galls are also induced on the hips. Most species of wild roses in Europe are hosts to *D. rosae*; however, *R. canina*, *R. eglanteria* and *R. dumetorum* Thuill. are the most commonly galled.

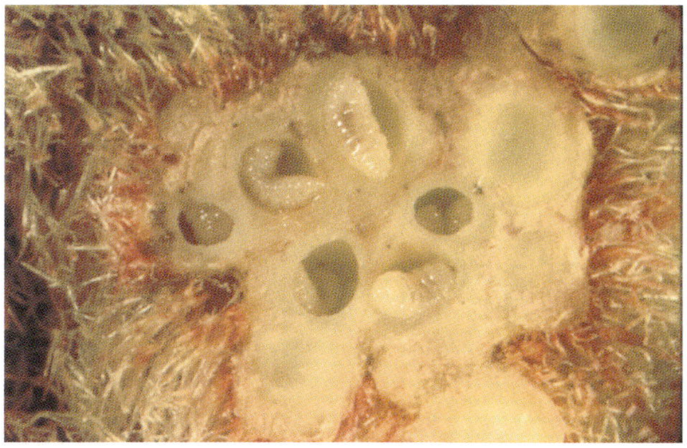

Figure 8 Gall of *Diplolepis rosae* opened to show larvae of inducers.

Figure 9 Galls of *Diplolepis spinosa*.

Most European populations of *D. rosae* are inhabited by the inquiline cynipid *Periclistus brandtii* (Ratzeburg) and up to 12 species of parasitoids. Larvae of *Periclistus* feed on tissues of the gall whereas larvae of the parasitoids feed on larvae of the inducers and the inquiline, or on each other. Galls of *D. rosae* are the best to use if one wants to examine gall inhabitants. Simply place a handful of galls in a glass jar in the spring, bring it indoors, and within a few weeks the sides will be crawling with reddish-brown inducers, black inquilines and parasitoids.

Galls on roses induced by species other than cynipids are the crown gall, induced by *Agrobacterium tumefaciens* (Smith & Townsend) Conn. (*see* **Disease**: Crown Gall), rolled leaflet galls induced by the sawfly *Blennocampa phyllocolpa* Vitasaari & Vikberg (also known as *B. pusilla* (Klug)) and the cecidomyiid *Wachtliella rosarum* (Hardy). The single larva of *B. phyllocolpa* rolls down both sides of the leaf along the midrib forming a tube-like gall which is often spirally bent. This gall is often common on both wild and cultivated roses. Gall formation by *W. rosarum* results from individual leaflets being folded upwards along the thickened midrib. Each gall contains several red cecidomyiid larvae. It is also found on both wild and cultivated roses.

Diplolepis Galls Found in North American Gardens

Wild roses are common shrubs over much of North America, with the densest patches found in southern Canada and the northern United States. The exact number of species is not known and the genus is in need of revision (*see* **Distribution and Ecology**: North America). It is likely that all North American roses are susceptible to galling by *Diplolepis*, but it appears that the more northern species such as *R. acicularis* Lindl., *R. blanda* Aiton, *R. woodsii* Lindl. and *R. arkansana* Porter have the most species.

Garden roses in North America become host to *Diplolepis* and their galls in two ways. First, at least six species have become established on the introduced Japanese shrub rose *R. rugosa* Thunb. *Rosa rugosa* is a popular garden shrub in southern Canada and the northern United States because of its vigorous growth and cold-hardiness. Unfortunately, it can also be heavily attacked by the stem galler *D. spinosa* (Ashmead) and the shoot galler *D. radicum* (Osten Sacken).

Galls of *D. spinosa* on both wild and domestic roses are spherical or irregularly rounded and clothed with stout, sharp spines (**Figure 9**), although sometimes the surface is smooth. Immature galls are yellowish-green and soft whereas mature galls are reddish-brown and so firm they are difficult to cut open. Removing mature galls with bare fingers usually results in numerous spines imbedded in your skin. *Diplolepis spinosa* galls on *R. rugosa* are often more successful than those on the hosts *R. blanda* and *R. woodsii*. For example, those on wild roses average 2.3 cm in diameter whereas those on *R. rugosa* average 3.0 cm in diameter. Galls on wild roses usually occur sparsely in patches of wild roses whereas dense clusters of 30 galls per square metre of ground surface are common on rugosa hedges (**Figure 10**). Hedges of *R. rugosa* about 15 m in length have been found with about 600 golf-ball-sized galls. One shrub found in a rural garden in northern Ontario had 226 galls, 50 of them current-year galls and 176 from the previous year. Further, galls on *R. blanda* commonly have an average of 16.5 larvae per gall whereas galls on *R. rugosa* have an average of 32.5 larvae per gall.

Galls of *D. radicum* (**Figure 11**) are found on several species of wild roses and frequently occur on garden *R. rugosa*. They are often overlooked because they form below the surface of the ground on the tips of

Figure 10 Hedge of *Rosa rugosa* heavily galled by *Diplolepis spinosa*.

Figure 12 Galls of *Diplolepis radicum* impeding the growth of sucker shoots.

Figure 11 Gall of *Diplolepis radicum*.

sucker shoots arising from rhizomes and only part of the galls protrude above the ground. Galls are variable in size and shape, but most are irregularly rounded with numerous bulges. The surface is smooth and tan to reddish-brown throughout all stages of development. Galls on wild roses range in size from 2 to 7 cm in diameter with the larger ones containing up to 200 larvae. Galls up to 10 cm in diameter and containing 350 larvae have been found on *R. rugosa* in botanical gardens. Galls sometimes become so abundant and large they terminate shoot growth (**Figure 12**) and impede the growth of hedges (**Figure 13**).

Other Nearctic species that only periodically gall *R. rugosa* are the leaf gallers *D. polita* (Ashmead) and *D. rosaefolii* (Cockerell) and the stem gallers *D. nodulosa* (Beutenmüller) and *D. triforma* Shorthouse & Ritchie. *Diplolepis polita* induces clusters of spherical, single-chambered, spiny galls (**Figure 14**) on the surface of leaves. *Diplolepis rosaefolii* induces lens-shaped, single-chambered galls (**Figure 15**) that often coalesce on the surface of leaves. As is the case in Europe, *Diplolepis* attacking shrub roses are undoubtedly flying, or being blown, into urban areas from galled wild roses in rural areas. Galls of Nearctic *Diplolepis* have not been found on tea roses and other species of garden roses.

The second way by which galls become established on North American garden roses is by accidental introductions from Europe. Galls of three European species of *Diplolepis, D. rosae, D. mayri* and *D. centifoliae*, have been introduced along with their host plants *R. canina* and *R. eglanteria*. By far the most common gall is that of *D. rosae* which is found in eastern Canada and the northeastern United States and in western Canada and northeastern United States. One botanical garden in Nova Scotia has records of *R. eglanteria* being imported from The Netherlands in 1981 and the galls have been abundant on resulting hedges ever since. The same probably occurred at all other sites where galls are abundant. Some gall populations have survived on ancestors of imported roses at the site of introduction for over 100 years. Further, both *R. canina* and *R. eglanteria* have become feral and widespread in many regions of North America and *D. rosae* has tracked them.

Galls of *D. mayri* are known from two sites in southern Ontario where large populations occur on feral *R. canina* growing along roadsides in agricultural areas. As with galls of *D. rosae*, *D. mayri* must have been introduced along with its host and persisted in gardens at nearby farms. The roses would then have been spread by birds eating the hips and wasps from the garden galls were able to locate them. Similarly, a botanical garden in Niagara Falls, Ontario introduced a few *R. canina* from The Netherlands in 1937 that must have been galled by *D. centifoliae*. Since then, the shrub has spread to the forested regions along the river below Niagara Falls, and *D. centifoliae* and their galls have spread with them. Galls here were abundant

Figure 13 Hedge of *Rosa rugosa* with growth of sucker shoots impeded by galls of *Diplolepis radicum*.

Figure 14 Galls of *Diplolepis polita*.

Figure 15 Galls of *Diplolepis rosaefolii*.

58 years later both on one shrub at the original site and on feral shrubs about 0.5 km away along the Niagara River. One isolated *R. canina* found in a deciduous forest near London, Ontario, Canada was host to all three species (**Figure 16**).

Control and Future Prospects

For most gardeners, the odd cynipid gall appearing on their roses is an annoyance rather than a host-damaging pest and no control measure need be taken. However, when the larger galls such as those of *D. rosae* and *D. spinosa* become abundant, they can weaken branches and sometimes the entire shrub. Attempting to control galls by killing the small adults with insecticides is usually futile because they are rarely seen as they fly from plant to plant. Further, most adults rest beneath the foliage when not searching for oviposition sites and there is always the potential of new wasps flying in from elsewhere. The best approach is to check roses for stem and nonabscising leaf galls in the autumn and either remove them with snips or twist them off with gloved hands. When leaf galls such as *D. nervosa* or *D. rosaefolii* are present, disposing abscised leaves in the autumn helps to reduce populations of the wasps. There is some evidence that galls are more successful on water-stressed plants, so adequate watering may help reduce the risk of galls being established.

The greatest concern caused by *Diplolepis* galls is probably the threat of non-native species becoming established in new lands when host plants are moved

Figure 16 Three galls of European *Diplolepis* found on feral shrub in a forest in southern Ontario, Canada. Note the gall of *D. centifoliae* to the left of centre, the gall of *D. rosae* towards the top and galls of *D. mayri* to the bottom and right.

from one continent to another. It is unlikely that any remaining species of European *Diplolepis* will become establish in North America, but the threat of North American species becoming established, for example in Europe, Australia or Japan, is real. Species such as *D. spinosa* and *D. radicum* could easily become established wherever *R. rugosa* is grown and there is always the threat that some species of *Diplolepis*, once established on one host, will attain the ability to expand onto another. Further, there is the possibility that other species of *Diplolepis* currently found only on wild roses, will become established on garden roses, and then spread to other continents when their host plants are exported.

Some gardeners interested in learning about complexities of the natural world may not find a few *Diplolepis* galls a threat to the health of their roses. Instead they may enjoy watching a few galls grow to maturity, then removing them in the autumn to prevent further spread. Further, if *D. rosae* are present, gardeners can gain some satisfaction by knowing that Shakespeare watched and wrote about the same galls. Thus roses with a few endemic *Diplolepis* galls can provide knowledgeable gardeners and their visitors with a subject for intriguing conversation.

See also: **Disease**: Crown Gall. **Distribution and Ecology**: Overview; Europe; North America. **Insects and other Animals**: Overview of Insects.

Further Reading

Bagatto G and Shorthouse JD (1994) Mineral nutrition of galls induced by *Diplolepis spinosa* (Hymenoptera: Cynipidae) on wild and domestic roses in central Canada. In: Williams MAJ (ed.) *Plant Galls: Organisms, Interactions, Populations*, pp. 405–428. Systematics Association Special Volume no. 49. Oxford: Clarendon Press.

Csóka G (1997) *Plant Galls*. Budapest, Hungary: Agroinform. (Volume in both Hungarian and English.)

Meyer J (1987) *Plant Galls and Gall Inducers*. Berlin: Gebrüder Borntraeger.

Redfern M and Askew RR (1992) *Plant Galls*. Slough, UK: Richmond Publishing.

Redfern M, Shirley P and Bloxham M (2002) British plant galls: identification of galls on plants and fungi. *Field Studies* 10: 207–531.

Shorthouse JD (1988) Occurrence of two gall wasps of the genus *Diplolepis* (Hymenoptera: Cynipidae) on the domestic shrub rose, *Rosa rugosa* Thunb. (Rosaceae). *Canadian Entomologist* 120: 727–737.

Shorthouse JD (1993) Adaptations of gall wasps of the genus *Diplolepis* (Hymenoptera: Cynipidae) and the role of gall anatomy in cynipid systematics. *Memoires of the Entomological Society of Canada* 165: 139–163.

Shorthouse JD (2001) Galls induced by cynipid wasps of the genus *Diplolepis* (Cynipidae: Hymenoptera) on cultivated shrub roses in Canada. *Acta Horticulturae* 547: 83–92.

Shorthouse JD and Brooks SE (1998) Biology of the galler *Diplolepis rosaefolii* (Hymenoptera: Cynipidae), its associated component community, and host shift to the shrub rose Thérèse Bugnet. *Canadian Entomologist* 130: 357–366.

Shorthouse JD and Rohfristch O (1992) *Biology of Insect-Induced Galls*. New York: Oxford University Press.

Williams MAJ (ed.) (1994) *Plant Galls: Organisms, Interactions, Populations*. Systematics Association Special Volume no. 49. Oxford: Clarendon Press.

Pollination

Kevan P G, University of Guelph, Guelph, ON, Canada

Introduction

Roses attract many insects. Why is this, and what do the insects gain? What do the plants gain? *The Rose's Kiss* by Peter Bernhardt (1999) provides an entertaining and informative look into the natural history and sex life of flowers, yet information about roses is scarce. Just what is known about roses and their pollination?

First, it is necessary to define pollination. Pollination is the transfer of pollen from the anthers of a flower to the stigmas of a different flower on a different plant or on the same plant, or within the same flower. Pollination does not necessarily result in fertilization of the ovules within the flower's ovary. Self-compatibility is required if self-pollination is to be fruitful.

A pollinator is the agent that causes that transfer. Pollinators range from physical agents, especially the wind (wind pollination is called anemophily), or biotic agents such as insects, birds, bats and other animals (pollination by insects is called entomophily, by birds ornithophily, by bats chiropterophily). There is a large technical vocabulary associated with pollination. Roses are entomophilous. Similarly, plant-breeding systems are highly varied and complex.

When one thinks about insects as visitors to flowers, one thinks of pollinators, nectar and pollen feeders, and sometimes about feeding damage to flowers. All are part of the anthecology (floral ecology) of roses. The form of the flowers is crucial to understanding how pollination takes place.

Figure 1 Rosa rugosa.

Floral Biology

Flowers of rose species are mostly hermaphroditic and homogamous, meaning that they open with both male (stamens numbering from as few as 50 to as many as 250, depending on species) and female (pistils) maturing simultaneously. *Rosa setigera* Michx., in which functional male and female flowers are present on different plants, is the only exception. In most species of roses, the flowers are large with rose-coloured, white or, less frequently, yellow-coloured petals that overlay the much smaller green sepals that form the sheath of the floral bud. The flowers are referred to as perigynous because the petals and sepals are borne on a raised, fleshy receptacle, the hip, which surrounds but is not fused with the numerous ovaries that it encloses (**Figure 1**). Most roses have a delightful and characteristic scent. The flowers are devoid of nectar (although some species may secrete minute amounts on the edge of the receptacle at the base of the flower between the filaments and the ovary) but produce large amounts of pollen, which is the main reward sought by insect visitors.

That was the state of knowledge when, at the turn of the twentieth century, Paul Knuth reviewed what was known about floral biology in the genus *Rosa*. He also synthesized the information available on insect visitors to the flowers, noting mostly a wide range of beetles (Coleoptera), flies (Diptera) and bees (Hymenoptera). As is typical of research in pollination ecology of that day, little information is given on the role of pollinators in the setting of hips and production of seeds.

Fertilization of the ovules within the hip ensues following pollination; however, the pollen first germinates on the stigma, grows through the style and liberates its gametes (sperm cells) into the waiting ovule. It may come about through automatic or insect-mediated selfing within one flower (autogamy), or through pollen transfer from flower to flower on the same plant (geitenogamy). Cross-pollination (xenogamy) results when pollen is transferred from the anthers of one plant to the stigmas of the flowers of another.

Figure 2 Pollination bag on rose excludes insect- and wind-borne pollen, allows controlled cross- and self-pollinations, and is essential to understanding the nature of sexual reproductive and floral biology.

Self- and Cross-Pollination (Figure 2)

One might expect that autogamy in roses, with their mainly homogamous flowers, would be general.

Although there are only a few studies on pollination and breeding systems in roses, it appears that the situation is more complex than originally thought and that xenogamy is important in several species. How do the flowers of roses function to favour the transfer of pollen between flowers and between plants?

The petals of many roses, such as *R. canina* L. and *R. arvensis* Huds., are more or less erect when the flowers first open. The stamens then spread outwards, exposing the stigmas in the centre of the flower. Insect visitors to these flowers tend to land and perch on the stigmas while feeding at the anthers. As the flowers age over a period of a day, the petals increasingly open, as do the stamens, and visiting insects land on them and become liberally dusted with pollen. Thus, an insect that has visited an older flower readily transfers pollen to a younger one. The mechanism to favour cross-pollination in *R. rubiginosa* L. is further enhanced by the flowers being slightly (by a matter of an hour or so) protogynous (i.e. with the stigmas receptive to pollen germination before the anthers of the flower have dehisced and started to shed pollen). These mechanisms do not prevent self-pollination, the likelihood of which would seem to be enhanced in species in which the stamens are tall and curved in over the stigmas, but self-fertilization is another matter.

The sperm cells in the pollen and egg cells in the ovules are normally produced in plants through a series of cell divisions involving meiosis (*see* **Genetics**: Meiosis) and some additional mitotic divisions. As in humans, and most animals and plants, meiotic divisions produce gametes (sex cells: sperms and eggs) that have half the number of chromosomes (haploid) as the parents (who are diploid). When a sperm and egg unite during fertilization, the original (diploid) number of chromosomes is restored. During the processes of meiosis and fertilization, the chromosomes become shuffled and recombined so that variability results. This variability can be used by plant breeders for making selections, and is used by nature in natural selection for the fittest offspring (i.e. those that are healthiest in turn produce the most and healthiest offspring through Darwinian selection). Self-fertilization severely constrains the amount of variability of the offspring. The Darwin–Knight law notes that nature disfavours self-pollination and abhors perpetual self-fertilization.

There are species of roses that are not self-pollinating, some that are weakly self-pollinating, and others that are easily self-pollinated, but the published data of Eileen Macfarlane are difficult to decipher in regard to the important difference between self-pollinating and self-fertility, although she wrote: 'My experiments showed that the roses are self-fertile.' It is known that two species she records as highly self-pollinating, *R. setigera* and *R. rugosa* Thunb., both require cross-pollination by insects to set fruit. Recently published data indicate that most diploid species, but only a minority of polyploid species, are self-incompatible. More information is needed about the breeding systems and pollination requirements of the whole genus.

Roses of the section *Caninae* are polyploids with $2n = 28$. In each of these species the pollen contains seven chromosomes whereas the egg cells contain $2n - 7$ chromosomes. This situation arises from unusual meiotic divisions which differ on the male and female side (*see* **Genetics**: Inheritance in the Dogrose). This strange mechanism may have advantages in providing a way to remove deleterious genes from the genome (gene complement) that is kept in the seeds. Deleterious genes may be segregated during meiosis to be aborted. Deleterious genes that are retained during microsporogenesis (pollen production) may be selected against by differential capacity for pollen of higher genetic quality to be favoured during fertilization. This sort of mate selection is common in plants. It is important for plants to ensure that seeds in fruits are of the highest quality because investment in the seeds far outweighs that in pollen. Such a strategy may be especially valuable in maintaining genetic variability and fitness in highly self-fertilizing plants.

Insect Visits, Pollination and Fertilization

Insect visits to roses may be variously important to the plants for sexual reproduction, fruit and seed production, and hence dispersal. Some roses are self-pollinating and self-fertilizing and can set hips without insect pollination, whereas others cannot. However, little is known about most species in the genus *Rosa*. Many of the horticultural varieties do not produce flowers that can be used by pollinators. Some lack pollen, do not set fruit and must be propagated vegetatively. Some have been so selected for floral beauty in horticulture that even sexual organs and the characteristic scent are missing.

Floral Advertisements and Rewards for Pollinators

To understand how insects visit the flowers of roses and bring about pollination, one must turn to studies on species, or wild roses. It is well known that the flowers of wild roses attract a great diversity of insect visitors and almost all feed on pollen (*see* **Insects and other Animals**: Overview of Insects). The wide variety of bees, of which the most conspicuous are

Figure 3 *Bombus terrestris* visits a flower of *Rosa rugosa*. Photograph courtesy of Heid Dobson.

bumblebees (*Bombus* spp.) (**Figure 3**), carpenter bees (*Xylocopa* spp.) and the familiar honeybees (*Apis mellifera*), all forage at rose flowers to collect pollen. Bumblebees are often heard making high-pitched buzzing sounds as they harvest pollen. This sound and vibration serve to release pollen from the anthers. The bees take the pollen back to their nests where it is used for feeding the larvae. Rose pollen, like pollen of many other flowers, is highly nutritious, being rich in proteins, amino acids, carbohydrates, lipids and vitamins. Other insects, including some small bees, eat the pollen while on the flowers. The nutrients in the pollen are converted into yolk in developing eggs of flies, beetles and other insects. Particularly conspicuous on roses are hover flies (Syrphidae), many of which are beneficial as their larvae consume aphids.

A wide variety of other insects can be found on the flowers of roses. Most are considered destructive, such as thrips (Thysanoptera), chafers (Scarabeidae) and earwigs (Dermaptera) (*see* **Insects and other Animals**: Overview of Insects). Population studies of thrips on rose flowers allowed the great Australian ecologists Andrewartha and Birch to model insect population growth over time and under different weather conditions.

Leaf-cutting bees (Megachilidae) sometimes cut circular holes from the petals, but their damage is more commonly seen on the leaves. These bees are beneficial in nature because of their importance in pollination. The leaf and petal segments they take are used in the construction of their nests. Crab spiders (Thomisidae) often sit in rose flowers and ambush flower-visiting insects.

The colours of the petals, and the colours of the stamens and pistils, provide a visual target for foraging pollinators. Against the background of vegetation, the arrangement of colours, mostly pink petals with a centre of yellow, is conspicuous to the human eye. Insects see differently from human beings, having the capacity to see ultraviolet reflections, but mostly not red ones. Insects also have compound eyes, made of many minute hexagonal facets, that allow them to see shapes in a sort of mosaic. Other differences also render floral colours difficult to comprehend in the insect world. Green leaves of most plants are uniformly reflecting across the insect visual spectrum and can be thought of as neutrally coloured to pollinators (i.e. the equivalent of dull grey). As far as we are aware, roses do not reflect ultraviolet patterns and the large amount of red reflection is probably not useful to most insects. Thus, most pink rose petals are usefully reflective to insects in the blue part of the spectrum. The yellow stamens and pistils strongly absorb blue light. Thus, the petals are coloured distinctively from the vegetation, as are the stamens and pistils coloured differently from the petals.

Rose flowers tend to be large and it is assumed that they would be visible to pollinators from great distances. This may not be the case as individual roses are probably invisible to approaching pollinators until they are within about 30–40 cm, perhaps even much less, as recent research in bee vision physiology suggests.

The odours of roses are produced on various floral parts. The petals are well known for their delightful aroma and are used extensively in the perfume industry. The best known species for the chemistry of scents in relation to their importance in pollination is *R. rugosa*. The main chemical is 2-phenyl ethanol with lesser amounts of monoterpenoid alcohols. The anthers and pollen also emit different scents, eugenol and methyl eugenol, that smell somewhat spicy, and the pollen itself is characteristically scented with long-chain ketones, aldehydes and esters.

How pollinators orient to and choose flowers at which to forage is complex, involving several sensory systems (vision, scent, touch and time). Honeybees can distinguish between the male and female flowers of *R. setigera*, and between newly opened (brightly coloured, pink with yellow anthers) and older flowers (with paler petals and browning anthers). Interchanging fresh petals on to older flowers did not increase the older flowers' attractiveness, but placing older petals on to fresh flowers did not detract greatly from floral attractiveness by comparison with that of unmanipulated fresh flowers. Scent differences were not measured. Bumblebees (*Bombus*) on *R. rugosa* (**Figure 3**) primarily use the pollen scent to assess the availability of forage in the flowers, but visual stimuli from the anthers are also important. Petal odours and colours are of secondary use in the bees' selection of rewarding flowers.

Little is known about the role of other floral signals from roses, such as microtextural features that bees can distinguish or timing of opening. It has been suggested that dish-shaped flowers, such as those of roses, could act as parabolic sonic reflectors that could amplify the sound of the bees buzzing in the flowers to release pollen and that could be attractive to other bees.

See also: **Genetics**: Meiosis; Inheritance in the Dogrose. **Insects and other Animals**: Overview of Insects. **Morphology and Anatomy**: Pollen Grains and Tubes. **Propagation**: Seed Maturation and Germination.

Further Reading

Bernhardt P (1999) *The Rose's Kiss: A Natural History of Flowers*. Washington, DC: Island Press.

Faegri K and van der Pijl L (1979) *The Principles of Pollination Ecology*, 3rd edn. Oxford: Pergamon Press.

Flament I, Debonneville C and Furrer A (1993) Volatile constituents of roses. In: Teranishi R, Buttery RG and Sugsawa H (eds.) *Bioactive Volatile Chemicals from Plants*. American Chemical Society Symposium Series no. 525, pp. 269–281.

Kemp JR (1994) *Floral Morphology and Pollination Biology of* Rosa setigera *Michaux*. PhD dissertation. Guelph, ON, Canada: University of Guelph.

Knuth P (1908) *Handbook of Flower Pollination*, vol. II. Translated by JR Ainsworth Davis. Oxford: Clarendon Press.

Macfarlane EWE (1963) A self-pollination mechanism and other items in rose species. *American Rose Annual* 48: 188–193.

Proctor M, Yeo P and Lack A (1996) *The Natural History of Pollination*. London: Harper-Collins.

Richards AJ (1986) *Plant Breeding Systems*, 2nd edn. London: George Allen and Unwin.

Willson MF (1983) *Plant Reproductive Ecology*. New York: John Wiley.

Insecticides

J F Karlik, University of California Cooperative Extension, Bakersfield, CA, USA

Introduction

The production of rose flowers and plants free of damage caused by insects and related arthropods (e.g. spider mites) is required for product quality, aesthetic enjoyment and at times by government regulation. Genera and species of aphids, thrips, scales, caterpillars and spider mites are important rose pests (*see* **Insects and other Animals:** Overview of Insects; Aphids; Mites; Thrips). Strictly speaking, insecticides are agrochemicals which kill insects, whereas acaricides are used to control arachnid pests such as spider mites. The information presented in this article pertains to both insecticides and acaricides, and where specific chemicals are acaricides they will be identified as such.

In a rose-growing system using integrated pest management (IPM) methods (*see* **Integrated Pest Management**), plant species and varieties are selected, where possible, to be resistant to insect pests. The environment is managed so as to favour the desired plants and accordingly to suppress pest problems. However, an IPM system does not preclude the use of insecticides to suppress pests injurious to rose flowers, leaves, stems or roots. Systematic monitoring of crops and plants for pests is utilized to identify and quantify populations of damaging and beneficial insects. Data are used to make judicious decisions about whether or not an insecticide application may be needed, which insecticide should be used, and when single or repeated applications should be made. Insecticide treatments are selected to be effective and yet, in so far as possible, the least disruptive to pest predators and parasites and the most environmentally benign.

The location and purpose for roses may influence the number of insecticide applications required for their maintenance. In landscape situations, low levels of damage to rose plants can usually be tolerated without significant aesthetic loss, and natural biological controls may be abundant. These may include predatory and parasitic insects that can suppress populations of rose pests, and allow for infrequent insecticide applications. In rose plant production fields, insects attacking only rose flowers can often be tolerated, but those infesting wood often cannot, so insecticide application for them may be required. In commercial cut-rose production, little damage is permissible on the harvested crop. The confined nature of glasshouses along with warm temperatures and abundant host plants can allow rapid reproduction of insect pests. Rose crops may require multiple applications of insecticides for pest suppression, but insecticide selection may be constrained by pest resistance and consideration of biological control introduced into the glasshouse.

Insecticides differ among themselves in biological and environmental characteristics, such as selectivity for pests, effects on nontarget organisms, persistence and degradation products. Most insecticides in contemporary use are organic chemicals, i.e. are carbon-based, and owe their effects to biochemical reactions with organs or systems within the target insect. Insecticides may be natural products extracted from plants (e.g. pyrethrum), microbial products (e.g. *Bacillus thuringiensis*) or may be synthesized (e.g. malathion). Some synthetic insecticides (e.g. pyrethroids) have a structure resembling naturally occurring compounds. Some insecticides also have effects on non-insect rose pests (e.g. azidirachtin against powdery mildew) or are used in medicine (e.g. avermectins). All insecticides possess unique physicochemical properties, such as aqueous solubility and volatility, which critically affect the environmental fate of the specific chemical. Insecticides both dissipate and degrade in the environment, although the rates due to various processes vary enormously among them.

The insecticides available to producers of rose plants and flowers vary from place to place, and over time new products enter the market and others are removed. Government regulations, research and development results and decisions of pesticide manufacturers affect product availability. Therefore, this article presents the topic of insecticides primarily from a conceptual perspective and gives principles for selecting and using these materials.

Background

Several terms are important for a discussion of this subject. 'Selective' refers to insecticides which affect certain species of insects or spider mites, while having little effect on others. Nonselective insecticides, also called broad-spectrum insecticides, are indiscriminate toward the insect species affected. Systemic insecticides are mobile to some degree within plants to which they are applied, whereas insecticides without systemic activity remain on plant surfaces. Reduced-risk insecticides are those which have low mammalian acute toxicity, have short persistence in the environment, and are generally benign toward beneficial insects and mites.

Insecticides have three names: a trade or brand name,a common name and a chemical name. For example, the insecticide with the chemical name 1-napthyl methylcarbamate has the common name carbaryl, with a tradename Sevin. Chemical names are invariant throughout the world, common names are often consistent and tradenames vary more widely. Sometimes the trade and common names are the same, with diazinon as an example.

Insecticides may be categorized according to several schemes, including origin (natural versus synthetic, for example), by toxicity or by other characteristics. The approach taken in this article is to group them by chemical structure. Compounds with similar structures usually share the same mode of action, and may have similar physicochemical properties and resulting environmental behaviours.

Considerations for Insecticide and Acaricide Selection for Roses

Fit within an IPM Programme

Many insecticides are available for use on roses, but if good IPM practices are followed the need for them will be lessened. If insecticide application is required, the most effective for the pest problem is favoured. Secondary considerations include selectivity for the pest problem and minimal ancillary environmental impacts. For example, abamectin has been very effective for controlling spider mites on outdoor rose plants in the south-west USA. It is very selective toward leaf-feeding mites, thus preserving biological control, and its use has resulted in a sharp reduction in number of acaricide applications made per growing season. Although broad-spectrum insecticides can be effective in the short term, their use can result in loss of predatory and parasitic insects important for biological control, allowing a strong rebound of the pest population. This occurrence is well-known in field rose plant production in California where the number of spider mites may 'explode' following application of a broad-spectrum insecticide. Rose plant growers there have also found that use of broad-spectrum compounds such as methomyl resulted in a need for repeated applications during the remainder of the growing season, since biological control became impaired.

Regulatory Considerations

When confronting an insect problem, the first question in selecting an insecticide may be that of current availability due to government regulation. Insecticides available commercially are subject to regulatory review by national environmental protection agencies, although the scope of review and admissible products vary widely among countries. To be allowed for sale as an insecticide, a chemical typically must show effectiveness (efficacy) against specific insects as measured in research trials. Environmental behaviour and safety to agricultural workers are also considered to be of primary importance. In the USA, California has its own environmental protection agency and is more restrictive toward pesticides than the corresponding national agency.

Regulatory considerations also include the means of permissible application, worker safety, re-entry interval for a treated area and reporting requirements. The pesticide label lists application rates and specifies protective clothing for workers. A re-entry interval may also be given. Additional rules for usage and specific interpretations of the label may be developed and enforced by the local regulatory authority. Pesticide sales and use are little regulated in some countries, more so in others. The most regulated locale is probably California, where all pesticide applications to agricultural crops, including roses in plant production fields, glasshouses and commercial landscapes, must be made to a state agency which compiles these data.

Reading the Label

The label of an insecticide is developed through extensive and expensive research, and is an authoritative guide to the use of the product. The label gives the tradename, the common name, the chemical name, the percentage of active ingredient and the formulation, and gives detailed directions for use, storage and disposal. Insecticide labels also contain a signal word placing the chemical into one of three toxicity categories (**Table 1**).

Active Ingredients and Formulations

Insecticides differ in their active ingredients and formulations. The active ingredient is responsible for the insecticide mode of action, and may be composed of identical molecules, of molecules differing stereochemically or may be a mixture of several chemical compounds. The active ingredient also determines whether the insecticide will move systemically through the plant. The formulation affects how the chemical is applied, its uptake and whether adjuvants (additional chemicals which enhance activity) are needed. For example, insecticides are often applied after mixing with water, so the formulation may be that of an emulsifiable concentrate where the insecticide is dissolved in an organic solvent, or a wettable powder or flowable material in which the active ingredient forms a suspension in water. Alternatively, some insecticides can be applied as granules to the soil, where water will cause them to be released. Some insecticides designed for use in glasshouses are delivered as an aerosol (particles suspended in air) from a smoke generator.

Although most insecticides have activity only against arthropods and often affect only a single insect order, a few have activity against both insects and diseases of roses. For example, commercial formulations of the botanical derivatives neem oil and jojoba wax are also able to control powdery mildew. These latter compounds are effective in controlling spider mites as well, and may be a good choice when both mites and powdery mildew are present. A difficulty with these compounds is that they must be applied at relatively high concentrations to be sufficiently active and therefore pose an increased risk of phytotoxicity. Glasshouse roses grown in moderate climates are more susceptible to pesticide phytotoxicity than those grown in hotter, drier areas and are generally more susceptible to phytotoxicity than field-grown roses.

Toxicity

Insecticides by their nature and purpose are toxic, but the degree of toxicity varies enormously among them. A full discussion of this subject is beyond the scope of this chapter, but mention of acute toxicity is prudent. The toxicity to an organism of an insecticide, like any other substance, depends on the dose an organism receives and the potency of the material. A common measure of a compound's potency is acute toxicity normalized to body weight, expressed as LD_{50}. This number has units of mg toxicant per kg body weight which results in a lethal dose to 50% of the experimental population. LD_{50} values are obtained through animal exposure studies. Formulated insecticide products contain some fraction of active ingredient, so the corresponding toxicity of the formulation is lower than for the active ingredient alone.

Treatment Thresholds

Treatment thresholds, often referred to as economic injury levels, are used to provide guidance on decisions for insecticide application. They are used to determine at what point the benefits of treatment outweigh the costs of application. Thresholds are determined through replicated experiments that correlate plant damage to a given density of an insect pest. In lieu of experiments, semiquantitative thresholds can be

Table 1 Insecticide signal words and respective acute toxicities

Category	*Relative toxicity*	*Signal word*	*LD_{50} oral (mg kg^{-1})*	*LD_{50} dermal (mg kg^{-1})*	*Probable oral lethal dose, 70 kg human*
I	Highly toxic	Danger[a] Poison[b]	< 50	< 200	Few drops to 5 ml
II	Moderately toxic	Warning	50–500	200–2000	5–30 ml
III	Slightly toxic to relatively nontoxic	Caution	> 500	> 2000	30–500 ml or more

[a] Pesticides may have the Danger label and be in Category I because of specific hazards, such as eye injury.
[b] Pesticides with the additional signal word 'Poison' also have the skull and crossbones displayed on the label, indicating high acute toxicity.

developed over the long term by growers who regularly monitor crops, make, retain, organize and evaluate good records. A record of insecticide application to a rose crop, with its success or failure, is important for a grower or manager.

Resistance Management

An insect population usually has members with a range of genotypes, some of whom may be resistant to the insecticide. From these surviving individuals a resistant population can arise. Frequent application of the same insecticide coupled with short generation times leads to rapid development of resistance. If resistance is conferred by a single gene, a population unaffected by the insecticide may arise within a few generations. A short-term approach to overcome resistance is to increase the insecticide dose, but this usually results in a pest population derived from the strongest individuals despite the use of the insecticide more often and at higher rates. A better approach is to use a given insecticide as infrequently as possible to preserve sensitivity in the population, and to rotate with other insecticides having different modes of action. Indeed, some insecticide labels specify that only one or two applications can be made during the cropping season. Resistance management is a major consideration in insecticide usage for glasshouse roses, since the insect population is confined and cannot outcross with susceptible individuals from outside.

Mode of Action

The mode of action of an insecticide refers to how it interferes with insect metabolism. The mode of action within a chemical class is usually the same. Most insecticides act to interfere with insects at a biochemical level, although a few, such as the oils, act through physical means.

The nervous system of insects is the site of activity for several classes of insecticides. Nerve impulses are transmitted by a change in electrical potential along the nerve fibres. This potential is achieved by differing concentrations of potassium and sodium ions inside and outside the nerve fibre. A narrow ($\sim$30 nm) gap, called a synapse, exists at the end of a nerve cell. To trigger the next nerve in sequence, a chemical neurotransmitter is secreted and diffuses across the synapse and binds to receptor sites on the next nerve, causing that nerve to fire. An enzyme is secreted within a fraction of a second after binding, which catalyses destruction of the neurotransmitter remaining in solution in the synapse. Without removal of the neurotransmitter, the stimulated nerve would continue to fire, causing paralysis.

The principal neurotransmitter in insects and spider mites is acetylcholine. The enzyme causing destruction of acetylcholine is acetylcholinesterase, often abbreviated as cholinesterase. Organophosphate and carbamate insecticides react with cholinesterase, binding to it so as to prevent its repeated reaction. Acetylcholine therefore remains undestroyed, causing repeated nerve firing and paralysis. Nicotine and the neonicotinoid insecticides are acetylcholine mimics, causing continued firing of nerves and resulting in respiratory paralysis. The acetylcholine–cholinesterase reaction scheme is found in the nervous system of mammals including humans, so insecticides interfering with transmission of nerve impulses are toxic to them as well. However, due to their much smaller size the effective dose for insects is a tiny fraction of that deleterious to humans.

Insecticide Classes

The chemical classes as presented here are based on similarities in chemical structures. Analogues refer to chemicals with the same basic structure but with differences in attached atoms or groups of atoms. Some naturally occurring chemicals with insecticidal properties have been used as models for analogue synthetic insecticides.

Oils are a reduced-risk insecticide class composed of carbon chains to which hydrogen atoms are attached. Oils owe their insecticidal properties to their low viscosity and low surface tension, and are therefore able to flow into insect breathing tubes and clog them, causing suffocation. As chain length increases, the molecular weight of the oil increases, the melting and boiling points of the oil decrease and the oil becomes more viscous – and more phytotoxic. 'Heavy' oils refer to longer chain lengths while 'light' or 'summer' oils refer to smaller molecules with shorter chain lengths. Horticultural oils applied to rose plants outdoors for insect control during winter dormancy, sometimes called dormant oils, can have longer chain lengths than those applied during the summer.

Soaps are another reduced-risk insecticide class, and have a water-soluble head and an oil tail in which other oil-like substances can dissolve. Soaps act to dislodge insects from feeding and to disrupt surface membranes. Their persistence in the environment and toxicity to humans are low.

Rotenone is a natural product derived from the cube plant. It is a broad-spectrum insecticide, very toxic to fish, and like most natural products has a very short life in the environment. Rotenone is not used in commercial production but may find a use in rose gardens.

Pyrethrum is extracted from flowers of the tropical flower *Chrysanthemum cinerariaefolium*. Pyrethrum is a mixture of four compounds of similar structure called pyrethrins. They have broad-spectrum activity against insects and are especially noted for their rapid knockdown of insects and short residual life in the environment. Pyrethrins are used as household insecticides and sold in spray dispensers mixed with a synergist, which increases their activity. Pyrethrum is seldom used in commercial rose production because of its short life and broad-spectrum activity, but may find occasional use in landscapes or rose gardens.

Pyrethroids are synthetic analogues of pyrethrins, and the numbers and uses of pyrethroid insecticides are increasing as they are promoted as organophosphate replacements. Pyrethroids have broad-spectrum activity which may interfere with organisms used in biological control. They are rapidly degraded in the environment. Their mode of action is similar to the organochlorines, and cross-resistance between these two groups of insecticides has occurred.

Insect pathogens, including species of fungi, bacteria, protozoa, nematodes and viruses, can cause a significant decline in a pest population in the field or glasshouse. These pathogens can occur naturally and may be introduced or augmented through application of reduced-risk microbial insecticides. Perhaps the most familiar products of this type are *Bacillus thuringiensis* powders, which contain protein crystals containing a toxin produced when bacteria sporulate. The toxin is released when the crystals are ingested by susceptible insects. *Bacillus thuringiensis* is not able to infect humans or higher animals, and so is quite safe for workers. The *B. thuringiensis* var. *kurstaki* strain produces toxins active against Lepidoptera (butterflies and moths) larvae, several of which feed on roses. This product should be targeted at the younger stages of the caterpillar. Other specific subspecies of *B. thuringiensis* produce toxins active against larvae of leaf beetles, mosquitoes, black flies and fungus gnats. Other commercially available insect pathogens are available, such as the fungus *Beauveria bassiana* (Balsamo) Vuillimen, and the nematodes of the genera *Heterorhabditis* and *Steinernema*.

The avermectins are a group of eight large cyclic compounds produced by the actinomycete *Streptomyces avermitilis* (Kim and Goodfellow). Avermectins are also used in veterinary and human medicine as antihelminthic drugs. The most active member of the group has the common name abamectin, and is used as an acaricide in the production of rose plants and flowers. It is very selective toward plant-feeding spider mites, has low toxicity toward predatory mites, beneficial insects and humans, and is readily biodegraded. It is a reduced-risk compound.

Organochlorines were the first major group of synthetic insecticides, and became available in the 1940s. Dichlorodiphenyltrichloroethane (DDT) is the most famous member of this group. Organochlorines vary in toxicity but are generally much more toxic to insects than to humans, due to their ease of penetration of the insect cuticle compared to human skin. Organochlorines as a group have long (months to decades) persistence in the environment and high lipid-solubility, resulting in bioaccumulation and biomagnification in the food chain. Most organochlorine compounds have been removed from the market in developed countries, but are still used in other parts of the world. Their activity results from keeping sodium channels open on nerve fibres, causing inability of the nerve to re-establish an electical potential after an impulse has passed. The organochlorine acaricide dicofol is useful in rose production and is selective toward spider mites, although both leaf-feeding and predatory mites are killed.

Organophosphates share a central phosphorus atom with a double bond to oxygen or sulphur, and three side chains. For insecticidal activity, one of those side chains called the 'leaving group' has to be able to break off in a reaction known in organic chemistry as an SN2 nucleophilic reaction. The more readily the bond breaks between the phosphorus atom and the leaving group, the more active (and toxic) the insecticide. Organophosphates vary over four orders of magnitude in toxicity, but all owe their activity to nucleophilic reaction with cholinesterase and inactivation of that enzyme. They have moderate persistence in the environment and are broad-spectrum. Use of organophosphates on outdoor roses in arid climates can result in flare-ups of spider mites.

Carbamates are derivatives of carbamic acid, and have a central carbon atom with a double bond to oxygen and another bond to nitrogen. Their mode of action is similar to organophosphates, causing depression of cholinesterase activity. Like organophosphates, these compounds have broad-spectrum activity against insects and have moderate persistence in the environment. The highly toxic aldicarb is a soil-applied systemic sometimes used in glasshouse production and occasionally in outdoor rose plant production because of its nematicidal activity. Its use has been circumscribed in some parts of the world because of its leaching potential in soil.

Nicotine is a natural product which acts as an acetylcholine mimic. It is broad-spectrum and is very toxic to humans. Its environmental persistence is short. The neonicotinoids are analogues of nicotine. Some, such as imidacloprid, are systemic and may be applied to soil to give season-long control of insects such as aphids. However, these compounds have

Table 2 Some insecticides and acaricides listed by common name and grouped by chemical class, with category of acute toxicity and comments

Common name	Category	Comments
Organochlorines		
Dicofol	II	Selective for spider mites
Dienoclor	II	Selective for spider mites
Endosulfan	I	Broad-spectrum for insects and spider mites
Organophosphates		
Acephate	III	Foliar – applied with limited systemic movement
Diazinon	II	Can be used as a soil insecticide or applied to foliage
Dichlorvos (DDVP)	I	Glasshouse use, much more volatile than most organophosphates
Disulfoton	I	Systemic, applied to soil
Malathion	III	Widely used contact insecticide
Methidathion	I	Scale insects on outdoor roses
Naled	I	Glasshouse use, applied as smoke, converts to dichlorvos
Sulfotep	I	Glasshouse use, applied as smoke
Carbamates		
Aldicarb	I	Soil-applied systemic, mostly glasshouse use, has nematicidal activity
Carbaryl	II	Chewing insects on outdoor roses
Methomyl	I	Broad-spectrum, for outdoor roses
Nicotine and neonicotinoids		
Nicotine sulphate	I	Glasshouse use, broad-spectrum
Imidacloprid	III	Systemic from soil or foliage application, 2–4 months' residual activity in the plant
Pyrethroids		
Bifenthrin	II	Broad-spectrum, may be phytotoxic to rose flowers
Cyfluthrin	II	Broad-spectrum, may be phytotoxic to rose flowers
Fluvalinate	I	Broad-spectrum, may be phytotoxic to rose flowers
Permethrin	II	Broad-spectrum, may be phytotoxic to rose flowers
Resmethrin	III	Broad-spectrum, may be phytotoxic to rose flowers
Insect growth regulators		
Kinoprene	II	Glasshouse use, needs 2 weeks for activity
Fenoxycarb	III	Glasshouse use, needs 2 weeks for activity
Miscellaneous compounds		
Fenbutatin (Hexakis)	I	Highly selective for plant-feeding spider mites, outdoor use
Propargite	I	Selective for spider mites in outdoor roses
Reduced-risk compounds of various classes		
Azidirachtin	III	Some fungicidal activity, low level of insect control
Bacillus thuringiensis	III	Several strains, excellent for Lepidoptera larvae outdoors
Abamectin	III	Highly selective for plant-feeding spider mites
Soaps	III	Short residual activity and low toxicity
Oils	III	Dormant season use on outdoor roses for scale insects May be phytotoxic if applied during warm weather

broad-spectrum activity and can interfere with biological control. Increasing scale-insect populations seem to be associated with their use in California.

Characteristics of Specific Insecticides for Roses

Many insecticides and acaricides have been used for rose plant and rose flower production. The chemical classes, relative toxicity and uses of several materials may be found in **Table 2**. Trade names and availability of these compounds will vary.

See also: **Breeding**: Selection Strategies for Disease and Pest Resistance. **Insects and other Animals**: Overview of Insects; Aphids; Mites; Thrips; Galls Induced By Cynipid Wasps. **Integrated Pest Management**.

Further Reading

Cremlyn RJ (1991) *Agrochemicals: Preparation and Mode of Action*. West Sussex, England: Wiley.

Dreistadt SH (2001) *Integrated Pest Management for Floriculture and Nurseries*. Statewide Integrated Pest

Management Project. University of California Division of Agriculture and Natural Resources publication 3402. Oakland, CA: University of California.

Farm Chemicals Handbook (1995). Willoughby, OH: Meister Publishing.

Flint ML and Karlik J (2000) *Healthy Roses*. Statewide Integrated Pest Management Project. University of California Division of Agriculture and Natural Resources publication 2158. Oakland, CA: University of California.

INTEGRATED PEST MANAGEMENT

J F Karlik, University of California Cooperative Extension, Bakersfield, CA, USA
S A Tjosvold, University of California Cooperative Extension, Watsonville, CA, USA

Introduction

The production of insect-, disease- and nematode-free rose plants is required both for product quality and by government regulation. For the glasshouse manager, landscape manager, commercial nurseryman or gardener, suppression of disease, insect and weed problems is necessary to realize the full growth potential, economic return and aesthetic benefit of roses.

Integrated pest management (IPM) considers plants, the environment and pests to be part of an interrelated system. In IPM, plant species and varieties are selected and the environment is managed so as to favour the desired plants, and accordingly to suppress pest problems. Monitoring – the regular, systematic inspection of crops and plants for pests and diseases – is necessary to identify and quantify pest and disease incidence and make judicious decisions on when and how to initiate controls. Where pest problems occur, they may be managed with techniques specific for them. IPM favours nonchemical approaches, including physical, cultural and biological control methods, but does not exclude chemical treatment. If pesticides are needed, treatments are selected to be the most environmentally benign and the least disruptive to the predators and parasites of the pests. In landscape situations, low levels of damage to rose plants can usually be tolerated without significant aesthetic loss, yet in commercial cut- or potted-plant rose production little damage is permissible on the harvested crop. In outdoor roses, the greater tolerance for damage along with a larger range in environmental conditions may afford additional flexibility in fashioning IPM strategies.

IPM Approach

IPM implies a multifaceted approach toward mitigating pest problems in managed biological systems. The plants under cultivation, the surrounding environment and the pests may be considered as the three principal components of this interrelated system. Through the careful selection of plant varieties and rootstocks and environmental management, pest problems on rose plants may be reduced, possibly obviating the need for pesticide applications. In addition to the pest problems posed by pathogens, insects and weeds, roses may display symptoms that are the result of chemical toxicities, nutrient deficiencies or excesses, or environmental problems. Such abiotic disorders may lead to pest outbreak, but can often be remedied by changing environmental conditions or correcting the deficient or excessive factor causing the disorder.

IPM stresses the early detection and characterization of pests. Monitoring, also called scouting, is the regular, systematic inspection of crops and growing areas for pests. Inspections should target plant parts and growing areas that historically harbour insects, diseases and weeds. Ideally, rose plants should be systematically observed and a quantitative measurement of the pest population made and recorded. Such data supply valuable information to determine timing and method of control and to help evaluate control practices.

When pests are present, an IPM approach emphasizes nonchemical control through measures such as cultural control (e.g. removal of diseased plant parts to reduce inoculum production), biological control (e.g. introduction of predatory insects) and mechanical control (e.g. cultivation to reduce a weed population). An IPM approach might include treatment with pesticides, but with consideration given to the level of pest pressure, the level of biological control in the field, the season and the availability of nonchemical control methods. Pesticides differ in biological and environmental characteristics, such as selectivity for pests, effects on nontarget organisms, persistence and degradation products. If pesticide applications are required, the most specific for the pest problem with least persistence and ancillary environmental impacts are favoured.

Ideally, plant damage or economic thresholds for pests are determined through replicated experiments that provide guidance for pesticide application. In lieu of experiments, semiquantitative thresholds can be developed over the long term by growers who regularly monitor crops, make, retain, organize and evaluate good records.

In the remainder of this article we address the three components of IPM for a rose-growing system: plant selection, environmental management and the pests themselves.

Plant Selection

Rose species, scion varieties and rootstocks differ in susceptibility to specific pest problems. For example, the *Rosa* sp. 'Dr. Huey' rootstock is much more susceptible to powdery mildew (*Podosphaera pannosa* (Wallr. ex Fr.) Lev. var. *rosae* Wor.) than is *R. noisettiana* Theory 'Manetti' (Crivelli). As a group, landscape (shrub) rose varieties are among the most resistant to powdery mildew, and glossy-foliaged varieties of hybrid teas and grandifloras can have good resistance to mildew as well. Black spot, caused by the fungus *Diplocarpon rosae* Wolf, causes black spots to develop on the upper surface of leaves and stems, and miniature roses are more susceptible than other types. Anthracnose, caused by the fungus *Sphaceloma rosarum* (Pass.) Jenkins, causes spots on leaves, affecting hybrid teas and old-fashioned climbing and rambler roses most often. Lists of varieties resistant or susceptible to diseases may be obtained as outcomes of experimental trials or experience. Variety selection for glasshouses may be constrained by marketing considerations, such as bud shape, stem length and vase-life, whereas it may be possible to have greater flexibility in choosing varieties for pest resistance for the landscape.

Selecting disease-free planting stock is the primary management strategy for certain diseases. Viruses, including rose mosaic virus and others, may infect rose plants, although damage may be mostly cosmetic with little reduction in plant vigour. There is no known treatment for eradicating viruses in established plants. Infected plants should be destroyed and virus-free stock obtained for future planting. Crown gall, caused by the bacteria *Agrobacterium tumefaciens* (Smith et Townsend) Conn., affects roses along with many woody plants as well as some herbaceous plants, including chrysanthemums and daisies. Crown-gall bacteria invade tissue following wounding, resulting in distorted tissue growth at the base of the plant, on roots or on stems. Roses should not be planted in infested soil or near infested plants, and it is best to decline the offer of 'gift' roses having questionable appearance of wood.

Species and varietal differences may be less pronounced for insect problems. For thrips, fragrant, light-coloured, yellow or white roses are most often attacked and can be severely damaged. Cultivars with sepals that remain tightly wrapped around the bud until blooms open have fewer problems. The hoplia beetle (*Hoplia callipyge* LeConte) prefers feeding on light-coloured roses (white, pink, apricot and yellow) rather than darker-coloured varieties.

When rose crops or plants are monitored, attention to the most sensitive varieties should be given because they can often be the first indicator of disease or insect presence. They therefore alert the manager of a possible developing epidemic. Not only may these varieties become infested or diseased first among other varieties, they may require the most intensive pest management once established.

Environmental Management

Site Selection and Modification

Site selection is a consideration mostly for outdoor roses; however, glasshouses should be oriented and constructed to allow for maximum light transmission to the rose crop. This not only helps roses maintain health for resisting pests and diseases, but can also help maximize crop production and quality, especially in low-light winter conditions.

Planting sites for outdoor roses may have full sun to partial shade, but roses do best with 6 h or more of direct sun. Site selection is particularly important for minimizing foliar disease problems, so sunny locations with good air circulation should be favoured. Downy mildew, caused by the fungus *Peronospora sparsa* Berk., requires moist, humid conditions and is difficult to control with fungicides. Downy mildew incidence can be reduced by increasing air circulation through pruning and avoiding frequent overhead irrigation. Botrytis blight, caused by the fungus *Botrytis cinerea* Pers. ex Fr., is favoured by high humidity. To help control botrytis blight, humidity can be reduced around plants by modifying irrigation, pruning and reducing ground cover. Rust and black spot, caused by the fungi *Phragmidium* spp. and *Diplocarpon rosae*, respectively, are favoured by cool, moist conditions, so providing good air circulation around plants helps minimize these diseases.

Winter injury from cold temperature results in dead or dying flowers, twigs and stems, and may be followed by stem canker diseases caused by weak pathogens that move into injured tissue. Roses may be protected over the winter with a thick layer (0.5–1 m) of leaf mulch in areas with winter minima below about −10 °C without snow cover or about −20 °C with snow cover. Winter injury may also occur in mountain areas where late frosts affect developing tissue.

Sunburn is caused by excessive temperatures on rose canes, often as an indirect result of defoliation caused by drought stress or pressure from spider mites. Damage appears as blackened areas on the sunny side of canes. Sunburn is more likely to occur in the summer season and in climates with daytime temperatures greater than 30 °C. Reflected heat from masonry, vinyl siding or rock mulch may also cause canes to become sunburned. Injury to rose canes from sunburn may lead to stem canker diseases or infestation of wood borers.

Glasshouse Humidity Management

The ability to manipulate the glasshouse environment, especially the relative humidity, can be a very powerful management tool to help prevent disease. The majority of fungi, aerial nematodes and bacteria that cause plant diseases require either very high humidity adjacent to, or free water on, the plant surfaces they infect. Many fungi need high relative humidity to produce spores. Moreover, free water in the form of rain, sprinkler irrigation water, syringing water or pesticide spray can move bacteria and many fungi from one site to another. Often, the longer free water exists on plant surfaces, the greater is the disease severity.

Condensation of water vapour can occur as glasshouse temperatures decline at night. Once the glasshouse air reaches saturation, water will condense on the coldest surfaces, which may be plant parts, leading to the undesirable outcome of free water on rose tissues. Cold plant surfaces are associated with stratified, stagnant air. By mixing air through the leaf canopy with circulation fans, the leaf temperature stays close to air temperature, and air temperature throughout the house is uniform. Condensation may instead first occur on structures in the glasshouse or glazing. In cold winter climates and with single-layer glazing the condensate that forms on the glasshouse cover can have a desirable effect of keeping the relative humidity of the glasshouse lower so plants stay drier, as long as they do not get dripped upon.

The condensation point can be altered by the amount of water added, or not added, to the glasshouse air and by manipulating air temperature. A typical pattern of the relationship of greenhouse air temperature, relative humidity and leaf wetness is

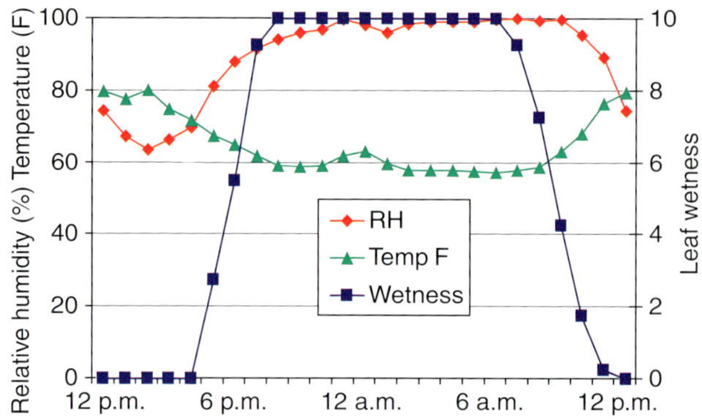

Figure 1 Typical relationship of air temperature, relative humidity and leaf wetness in a glasshouse. Leaf wetness is measured by a sensor that determines electrical resistance.

shown in **Figure 1**. At sundown, and during the night, the grower can manually or through environmental-control software increase air circulation with top-vents and heat. The rising column of heated air carries away excess moisture and is replaced by outside air which is normally cooler, thus holding less water. The dehumidification process does not work as well in warm humid climates or rainy periods because the air outside may hold more water than air within the glasshouse. It is desirable to keep the humidity at least 10–20% below the condensation point by balancing heat and ventilation. This allows a buffer for sudden temperature drops and temperature differences between glasshouse air and leaf and flower surfaces.

Soil Management

Certain abiotic problems may be directly attributed to soil management. For example, nutrient deficiencies or excesses may cause specific foliar symptoms. Excessive salt levels favour the development of root disease (*see* **Production and Marketing:** Field Rose Production).

Irrigation

Drought stress leads to necrosis of leaf margins, defoliation and sunburn of canes, and may contribute to problems from spider mites. However, overwatering of poorly drained soils may lead to root disease and nutritional deficiencies. Washing of leaf surfaces may be especially important for reducing spider mite populations, even where other irrigation methods are practised.

Since most fungi need free water for spore germination, the placement, frequency and duration of rainfall and irrigation may affect disease pressure. To minimize foliar diseases such as rust and black spot, irrigation should not wet rose leaves for extended periods of time. Irrigation with over-the-top delivery should take place in the morning so foliage dries during the day. Drip or trickle irrigation systems allow irrigation of rose plants without wetting leaves. Conversely, in warm arid climates, overhead irrigation for outdoor roses can suppress both powdery mildew and spider mites. Powdery mildew is recognized by white to grey powdery growth on leaves, shoots, sepals, buds and occasionally on petals. Powdery mildew does not require free water on the plant surfaces for spore germination and is active during warm, dry summer conditions. Overhead irrigation or washing at midday may limit the disease by disrupting the daily spore-release cycle; there should be time for the foliage to dry.

Pest Management

Natural Biological Control of Insects on Outdoor Roses

Some invertebrates ought to be tolerated, and a number of antagonists to rose pests are often present around rose plants. For example, spiders such as crab spiders, jumping spiders, cobweb spiders and orb-weavers may be found outdoors in landscapes and gardens. All spiders are predators and many contribute significantly to biological control. Parasitic wasps are important for the control of aphids, scale insects, caterpillars and many other insect pests. Certain predaceous beetles and other generalized and specific predators may also be found. Insecticide sprays may kill predators, resulting in a resurgence of pest problems greater than those observed before spraying. Tolerating low levels of infestation by pests and using nonchemical control methods will help preserve natural biological control in outdoor roses.

Aphids are probably the most common insect pest of roses (*see* **Insects**: Aphids), but are held in check by a number of antagonists. For example, tiny parasitic wasps of the genera *Praon* and *Aphidius* lay their eggs within aphids, rapidly immobilizing them. Larvae of the braconids devour the aphids from the inside, turning them into bronze or black crusty, bloated 'mummies'. The parasite pupates within the mummy and then cuts a neat round hole and emerges as a fully grown wasp. Green lacewings in the genera *Chrysopa* and *Chrysoperla* are also common natural enemies of aphids and other soft-bodied insects. Perhaps the most recognizable aphid predators are the many red and black species of lady beetles, including the convergent lady beetle (*Hippodamia convergens* Guérin-Méneville) and the multicoloured Asian lady beetle (*Harmonia axyridis* Pallas). These lady beetles are predators in both the adult and larval stages and may be recognized by the black, alligator-shaped larva with orange dots and the oblong, yellow eggs that are laid on end in groups.

Syrphid flies, sometimes called flower flies or hover flies, are also important predators of aphids and very common on roses. Adults, which superficially resemble wasps, feed on nectar and pollen before reproducing and are often seen hovering above flowers. The legless and maggot-shaped larvae are often found within aphid colonies. Species vary in coloration from dull-brown or yellow to bright green, but most have a yellow longitudinal stripe on the back.

Other rose pest predators include minute pirate bugs (*Orius tristicolor* White), which are tiny true bugs with black and white markings as adults. They are often among the first predators to appear in spring, and feed on mites, insect and mite eggs, immature scales and thrips.

Leatherwings or soldier beetles are moderate to large-sized beetles having leatherlike dark wings, and orange or red heads and thoraxes. They feed on aphids and are very common on roses. Many people mistake them for pests, but they are insect predators both as adults and larvae (in the soil).

In rose plant production or in landscapes, control of flower-feeding thrips is not imperative because flowers are not harvested, and the western flower thrips (*Frankliniella occidentalis* Pergande) is therefore desirable as a predator of spider mites (*see* **Insects and other Animals**: Overview of Insects; Thrips).

Insect Releases

Releases of insect predators outdoors have been successful in some situations, less so in others. Releases of commercially available convergent lady beetles can reduce an aphid population, but large numbers must be released on each individual rose plant. For best results, lady beetles should be misted with a water spray before release and releases should be made in the evening at dusk by placing beetles on canes at the base of plants. However, expect 90% of the lady beetles to fly away in the first 24 h. In any event, all released lady beetles are unlikely to lay eggs and will fly away once aphid populations have been substantially reduced. Lady beetles naturally present on-site often increase in number when aphid populations are high. Releasing green lacewings against the rose aphid has not been shown to offer significant control in research trials. Releases of predator mites have been used in glasshouses for spider mite suppression (*see* **Insects and other Animals**: Mites).

Cultural Control of Pests

Cultural control of insects and spider mites Ants should be kept out of rose plants with sticky barriers or traps to improve biological control. In many landscape situations, knocking aphids off with a forceful spray of water early in the day is all that is needed to supplement natural control. Insecticidal soaps or neem oil can also be used to increase mortality of aphids, with only moderate impact on natural enemies.

Limited pruning may also reduce the incidence of some insect problems. For example, canes harbouring scale insects can be removed by pruning during the dormant season. Frequent clipping and disposal of spent blooms may reduce thrips problems. Some caterpillars, like the tobacco budworm, may occasionally bore into flower buds. Buds showing telltale frass can be removed and destroyed. Pruning off occasional terminals of canes infested with larvae of the raspberry horntail (*Hartigia cressoni* Kirby) may be sufficient for control. Eggs of flat-headed borers tend to be laid on stressed rose plants, especially in bark wounds caused by sunburn or disease. Infested wood should be removed and plants given sufficient irrigation to minimize drought stress. Excessive pruning in summer can lead to sunburn and borer invasion, so moderation in pruning is best at that time of year. Damage from leaf-feeding caterpillars is usually not severe and treatment not usually necessary; rolled leaves may be hand-picked or clipped.

Disease management through cultural methods IPM techniques include removal of diseased or dead tissue by making cuts in healthy tissue at an angle and just above a node. For field-grown roses, severely affected canes can be pruned back during the dormant season to limit black spot, rust and powdery mildew. Botrytis often begins on dead or dying herbaceous tissue, such as senescing flower petals and leaves, but may then move into living tissue, including twigs and canes. Limiting the sources of inoculum can be helpful, such as removing and disposing of fallen leaves and petals. Also, all nonharvested flower buds should be removed from a cut-flower crop since an old bud can become an excellent source of *Botrytis* conidia once it is infected. Pruning and carefully disposing of leaves can limit disease infestations but may not entirely eradicate them, since airborne spores from other locations can provide fresh inoculum.

Stem cankers and dieback can be caused by a number of different fungi, including *Coniothyrium* spp. and *Nectria* spp. Cankers are often brown to deep purple or black, often with gray centres or small, black, spore-producing structures on dead tissue. Cankers often develop after injury due to cold weather, so additional late-spring pruning may be necessary to eliminate them if late frosts occur. Keeping plants vigorous through sufficient irrigation and fertilization will minimize the incidence of stem cankers. Making cuts just above nodes helps prevent canker diseases in production of cut flowers.

Nematode management Nematodes have been suppressed in rose plant production fields in some locations with nematicides or fumigants (*see* **Insects and other Animals**: Nematodes). Nonchemical approaches may be effective for nematode suppression, including steam treatment of soils and development of biological antagonists.

Weed management through cultural practices Mulching with 5–10 cm of organic material such as wood chips will help reduce annual weeds and make hand-weeding easier. Woven landscape fabrics placed under organic mulch will provide weed control for several years. In most home gardens, mulches supplemented with regular hand-weeding or rogueing (digging out the entire plant, roots and all) should provide satisfactory weed control. Mechanical cultivation devices such as hoes must be used with care because roses are shallow-rooted.

Considerations in Pesticide Selection in an IPM Programme

Many pesticides are available for use on roses, but if good IPM practices are followed the need for them will be lessened. In general, compounds with short persistence in the environment, specificity to the target pest and reduced human health risk are preferred. An example of one of these reduced-risk pesticides is horticultural oil. It can be applied in winter or early spring and can effectively control scale insects on outdoor roses. Horticultural oil is not inherently selective toward pests; however, it becomes selective when it is applied in the dormant season when the target pests are present but desirable parasites and predators are not. Conversely, broad-spectrum insecticides should be avoided, since they tend to eliminate biological control and may promote flare-ups of spider mites or other arthropods. In rose plant production fields an application of the organophosphate insecticide methomyl often results in loss of biological control, and frequent follow-up pesticide applications are then necessary (*see* **Insects and other Animals**: Insecticides).

Insect pathogens, including species of fungi, bacteria, protozoa, nematodes and viruses, can cause diseases of insects and result in a significant decline in a pest population in the field or glasshouse. These pathogens can occur naturally or may be introduced through application of microbial insecticides. The most familiar product of this type is *Bacillus thuringiensis*, known as *Bt*, a bacterium which attacks butterfly and moth larvae, yet has almost no toxicity to mammals. This product should be targeted at the younger stages of the caterpillar. Specific subspecies of *B. thuringiensis* are active against larvae of leaf beetles, mosquitoes, black flies and fungus gnats. Other commercially available insect pathogens are available, such as the fungus *Beauveria bassiana* (Bals.) Vuill. and the nematodes *Heterorhabditis* and *Steinernema* species.

Many fungicides are available for suppression of foliar diseases but most must be applied to prevent rather than to eradicate infections. Therefore, timing is critical and repeat applications may be necessary, which may not be justified in garden or landscape situations. The reduced-risk fungicide potassium bicarbonate is effective against powdery mildew. Commercial formulations of the botanical derivatives neem oil and jojoba wax are also able to control powdery mildew. These latter compounds are effective in controlling spider mites as well, and may be a good choice when both mites and powdery mildew are present. A difficulty with these compounds is that they must be applied at relatively high concentrations to be sufficiently active, and therefore pose an increased risk of phytotoxicity. Glasshouse roses grown in moderate climates are more susceptible to pesticide phytotoxicity than those grown in hotter, drier areas, and are generally more susceptible to phytotoxicity than roses grown outdoors.

In extensive outdoor plantings or professionally managed public or commercial landscapes, mulches and hand-weeding may be supplemented with herbicides. The herbicides oryzalin and pendimethalin can be used around roses to prevent germination of weed seedlings, but will not control established weeds. These herbicides primarily control grasses but also interfere with germination of certain broad-leaf plants. In field experiments, the herbicides fluazifop-*p*-butyl and sethoxydim were not phytotoxic to rose plants when applied over-the-top to remove surrounding grasses.

Roses are sensitive to herbicides used for control of broad-leaf weeds, such as 2,4-dichlorophenoxyacetic acid (2,4-D), triclopyr and dicamba. Roses are also very sensitive to glyphosate (e.g. Roundup™), which can be absorbed through the green stems in addition to rose leaves. Glyphosate damage may appear at budbreak the spring after inadvertent application to leaves or stems the previous autumn. Symptoms include a proliferation of small, narrow shoots and leaves.

Chemical weed control in glasshouses is limited to nonselective postemergent herbicides. Perennial weed problems can only be controlled by the tedious application of these herbicides, such as glyphosate, to the individual weed plants, carefully avoiding rose plants (leaves and green canes). Prevention of weed problems by not introducing weed seed or vegetative parts should therefore be a priority. Annual weeds can be controlled by carefully directed application of contact herbicides such as diquat or herbicidal soap.

IPM Management for Specific Insects

Aphids are probably the most common insect pest on outdoor roses (*see* **Insects and other Animals**: Aphids). The actual species involved depends on where the roses are grown. Aphids favour rapidly growing tissue such as buds and shoots. Low to moderate levels of aphids do little damage to roses, although many gardeners are concerned with their very presence. Moderate to high aphid populations can produce copious amounts of honeydew, resulting in the growth of sooty mould, which blackens leaves. Very high numbers may kill buds or reduce the size of flowers. As noted above, aphids have many natural enemies, including lady beetles, that may rapidly reduce increasing populations. In many locales, aphids are normally a problem for only about 4–6 weeks in spring and early summer before high summer temperatures reduce their number. In glasshouses, aphids may occur in any season, but they are found most commonly in spring. Their damage can be significant because many species attack developing rose buds.

Thrips are serious pests of glasshouse roses, but can often be tolerated in the outdoor environment. Western flower thrips cause injury primarily to rose flowers rather than other plant parts, causing blossom petals to streak with brown or become distorted. The torpedo-shaped 1–1.5-mm yellow or black thrips insects are found within the flowers. To monitor the population in glasshouses, yellow sticky cards can be placed vertically just above the top of the plant canopy at a density of at least one trap for every 900 m^2 (*see* **Insects and other Animals**: Thrips). Such placement will capture adults of whiteflies as well as thrips. Control of thrips with insecticides is difficult because materials are mostly effective on early developmental stages, which are commonly found within buds or flowers where most pesticide applications cannot reach. Thrips problems are more likely to be severe where many rose plants located close together provide a continuously blooming habitat. In glasshouses, mature flowers left inadvertently unharvested provide a very favourable location for thrips' reproduction and development. The additional time that the thrips are able to remain on the flower may be sufficient to complete their life cycle, and the pollen provided by open flowers contributes greatly to the nutrition of the adults. Thorough cleaning and disposal of old roses are necessary.

Certain beetles, such as Fuller rose beetle (*Asynonychus godmani* Boheman) chew flowers and foliage, leaving notched or ragged edges (*see* **Insects and other Animals**: Overview of Insects). Adult beetles are pale brown weevils about 1 cm long. They are flightless and hide during the day, often on the undersides of leaves; feeding takes place at night. The larvae are root-feeders but do not seriously damage roses. Low numbers can be ignored, but when numbers are high, beetles can be controlled by hand-picking, using sticky material on stems, or trimming branches that create bridges to walls and other plants. Japanese beetle (*Popillia japonica* Newman) is a leaf-feeding beetle which can cause extensive leaf and flower damage to roses, although the larval stage is more damaging to the herbaceous plants upon which it feeds. Beetles are difficult to control with insecticides because they have a long emergence period that extends from late spring to autumn. Parasitic nematodes may be helpful if applied to the soil in early to mid-summer.

Rose curculio (*Merhynchites* spp.) is a red-to-black snout weevil about 0.5 cm long that prefers yellow and white roses (*see* **Insects and other Animals**: Overview of Insects). It punches holes in flowers and buds and may create ragged holes in blossoms or kill the developing bud. Terminal shoots may be killed as well, if rose curculio weevils are numerous. Larvae feed within buds, often killing them before they open. Adults should be hand-picked off plants and infested buds destroyed. A broad-spectrum insecticide can be applied to kill adults if the infestation is severe.

Caterpillars such as orange tortrix, tussock moth, fruittree leafroller, tent caterpillar and omnivorous looper may feed on rose leaves; some of these caterpillars may also tie leaves with silk. Pheromone-baited traps are commercially available that capture adults of many of these caterpillars (**Figure 2**). These traps consist of a sticky surface and a dispenser containing a pheromone to lure the adults (usually males). Counts of trapped insects indicate when adults are active and therefore when more intensive monitoring may be warranted to look for egg masses and young larvae.

Some insects may be difficult to control, especially if natural biological control is to be maintained, and damage should be tolerated. Rose leafhopper (*Edwardsiana rosae* L.) causes stippling larger than mite stippling but tends to be a problem only in certain localities. Cast skins and the absence of webbing on the underside of leaves are good indications that these pests are present. Plants can tolerate moderate stippling. Leafcutter bees (*Megachile* spp.) cause limited damage to rose leaves by cutting semicircular holes in the margins of leaves and carry leaf material back to use in lining their nests. Bees are important pollinators and should not be killed.

Figure 2 An insect species-specific pheromone lure is placed in a sticky trap such as this one to attract moths and other flying adult insects.

Future Prospects for IPM in Rose Production and Management

Rose breeding, at least for outdoor plants, is leading to varieties requiring less maintenance and having inherent pest resistance. Where pest populations need rapid suppression, the trend toward use of reduced-risk pesticides steadily continues. Many of the newer compounds are botanical or microbial derivatives or analogues, offering selectivity toward pests along with low toxicity to humans and desirable environmental properties. The trend in IPM for roses also includes increasing use of active biological control, including introduction of organisms antagonistic to pests. Research and development of fungal, microbial and invertebrate antagonists of rose pests are leading to additional situations where nonchemical intervention is feasible and effective.

See also: **Insects and other Animals**: Overview of Insects; Aphids; Insecticides; Thrips; Mites. **Production and Marketing**: Field Rose Production.

Further Reading

Dreistadt SH (2001) *Integrated Pest Management for Floriculture and Nurseries*. Publication 3402. Oakland, CA, USA: University of California Agriculture and Natural Resources.

Flint ML and Dreistadt SH (1998) *Natural Enemies Handbook: The Illustrated Guide to Biological Pest Control*. Publication 3386. Oakland, CA, USA: University of California Agriculture and Natural Resources.

Flint ML and Karlik JF (2000) *Healthy Roses*. Publication 21589. Oakland, CA, USA: University of California Agriculture and Natural Resources.
Horst RK (1983) *Compendium of Rose Diseases*. St Paul, MN, USA: American Phytopathological Society.
Karlik J, Becker JO and Schuch UK (2001) IPM for field-grown rose plants in California. *Acta Horticulturae* 547: 97–102.
Strider DL (1985) Roses. In: *Diseases of Floral Crops* vol. 2, Chapter 30. New York: Praeger Publishing.

INTELLECTUAL PROPERTY

Contents

Plant Patents and Trademarks

L A M Dubois, Plant Research International, Wageningen, The Netherlands

Introduction

Since the beginning of history, human beings have been selecting in the vegetable kingdom for the necessities of daily life. Thanks to the discovery of systematic inheritance of plant characters by the monk Gregor Mendel in 1865, plant breeding was greatly accelerated, resulting in a range of new plant varieties. As a consequence, plant breeders demanded, as in industrial inventions, a system for granting exclusive rights for their breeding products. Such a right would mean that breeders of new varieties would have the exclusive right to exploit their new creations. Consequently, breeders would have both the opportunity to get returns on their investments in the new products, and be stimulated to create subsequent varieties.

The first country that recognized intellectual property rights on plants was the USA. Exclusively directed at vegetatively propagated crops, the so-called Plant Patent Act of 1930 was established. Better still, the first plant patent ever was granted to Henry F. Bosenberg, of New Brunswick, NJ, for a climbing rose on 18 August 1931. The Netherlands was the second country to proclaim a similar act: Het Kwekersbesluit, in 1941, was the first independent intellectual property law for both generatively and vegetatively propagated crops. The first rose that obtained Plant Breeder's Right was 'DeRuiter's Herald' of deRuiter, Hazerswoude, The Netherlands on 27 July 1948. Owing to different laws in different countries, the same variety may be protected either by Plant Patent or by Plant Breeder's Right.

Before World War II, breeders from different countries had united in Association Internationale des Sélectionneurs Professionnels pour la Protection des Obtentions Végétales (ASSINSEL), an association for the protection of plant varieties. A diplomatic conference on the subject of plant variety protection, initiated by ASSINSEL in 1957, resulted in the Union pour la Protection des Obtentions Végétales (UPOV) convention in 1961. That convention formulated a number of formal and material criteria for the protection of new plant varieties. For several decades, the UPOV convention of 1961 operated satisfactorily. However, owing to further developments in genetics, particularly the rise of biotechnology, the danger of a conflict between two different intellectual property rights increased – between Plant Breeder's Right and Patent Right. For that reason, the UPOV convention of 1961 was revised in 1972, 1978 and 1991.

Organizations

ASSINSEL, the oldest organization promoting the intellectual property rights of plants, was founded in 1938. This worldwide operating organization of public and private plant breeders has its registered office at Nyon, Switzerland. ASSINSEL was the leader in the creation of the UPOV convention of 1961. It is committed to the legal protection of new plant varieties, particularly the creation of industrial protection rights for new plant varieties, respectively new genetic basic material for plants. The organization represents breeders of all important plant species of agriculture, horticulture, industrial crops and ornamentals.

UPOV, an intergovernmental organization based on 'The International Convention for the Protection of New Varieties of Plants', was founded 2 December 1961 in Paris. The main aims of UPOV are: (1) to promote the protection of the rights of plant breeders in favour of the development of agriculture, and to safeguard the interest of breeders; and (2) to obtain a uniform system for the protection of the rights of plant breeders in all member states. UPOV consists of two permanent organs, the Council and the Office. The Council has established a number of subgroups with different tasks, e.g. preparing recommendations and guidelines. Important guidelines as the basis of granting breeders' rights for a new variety are the Guidelines for the Conduct of Tests for Distinctness, Homogeneity, and Stability, the so-called DUS test.

In January 2003, there were 52 UPOV member states. UPOV has been revised in 1972, 1978 and 1991. The most important revision took place in 1991. Compared to UPOV 1961, the most conspicuous modifications were the introduction of (1) the system of essentially derived varieties (EDVs) and (2) the limitation of farmers privilege.

Communauté Internationale des Obtenteurs de Plantes Ornementales et Fruitières de Reproduction Asexuée (CIOPORA) is an international organization, founded in 1961 by breeders of vegetatively propagated fruit and ornamental crops. Its registered office is in Geneva, Switzerland. CIOPORA represents large numbers of breeders and growers of ornamentals (e.g. roses, carnations, chrysanthemums, freesias or gerberas) and fruit (e.g. apples, pears, apricots or peaches). National branches of CIOPORA are active in various countries.

Intellectual Property Rights on Plants

In most branches of law concerning intellectual property rights, one or more treaties have been 'signed'. The most important and overall governing treaty is the Treaty of Paris of 1883, which protects industrial property. Currently, 107 countries are treaty members, which together form a union that is supervised by World Intellectual Property Organization (WIPO), based in Geneva, Switzerland. WIPO was founded in 1967. From that time, industrial property right was renamed in intellectual property right. By granting such rights, the products of the human spirit are protected against unauthorized exploitation by third parties. This means that if intellectual property rights are infringed, owners have the right to take legal action. In addition, owners of intellectual property rights may exploit their rights by licensing. As to plant varieties, various possibilities to protect the results of a successful breeding programme are available. Depending on the country, the plant breeders may choose between (Plant) Patent Right, Plant Breeder's Right or trademark.

Patent Rights

In general, 'patents shall be granted for any inventions which are susceptible of industrial application, which are new, and which involve an inventive step' (article 52 § 1 European Patent Convention (EPC)). The term of a patent shall be 20 years from the date of filing the application. Exceptions to patentability in Europe are 'plant- and animal **varieties** or essentially biological processes for the production of plants or animals' (article 53a EPC). Specific national and international law systems are available to protect new plant varieties.

Plant Patent

Granting an industrial patent on plants has raised serious problems in the past because of the impossibility of reproducibility, that is, in vegetatively propagated crops, breeding exactly the same variety again. Already in 1930, the USA had recognized this problem. Hence, a specific law to protect plant varieties, the Plant Patent Act was proclaimed in that year. The Plant Patent Act is exclusively applicable to vegetatively propagated crops, e.g. roses, fruit trees or chrysanthemums. The Plant Patent Act is composed of a mixture of elements of the traditional (Industrial) Patent and the protection of plant varieties. The act can be seen as a *lex speciales* of the industrial patent law system. The applicant of a Plant Patent is referred to as an inventor. The application is a legal case. For a Plant Patent to be granted the inventor must demonstrate that the new variety is really new and distinctly different in at least one distinguishing characteristic from its antecedents and known varieties. Extra criteria for a new plant variety are distinctness and novelty. In practice, new plant varieties can be distinguished by flower colour, fragrance, flavour, habit, disease resistance or vase-life. To that end, specific characteristics have been established for every crop. The Plant Patent system is applied in the USA, Japan and several South American countries.

Plant Breeder's Right

In Europe, Plant Breeder's Right is the most common form to protect the intellectual property rights of the breeder. All EU member states have implemented the 1961, 1978 or 1991 UPOV convention. In the Netherlands, a protection system for breeding activities was established in Kwekersbesluit in 1941. In 1967, following implementation of the UPOV convention of 1961, a specific law was established – the Zaaizaad-en Plantgoedwet (ZPW). Breeder's Right (UPOV 1991) are granted when the variety is (1) new; (2) distinct; (3) uniform; and (4) stable – the so-called DUS test. The DUS test is solely based on phenotypic characteristics. Besides these conditions, the variety must be denominated. Regarding the propagating material of the protected variety, authorization of the breeder is required for:

- production or reproduction (multiplication)
- conditioning for the purpose of propagation
- offering for sale, selling or other marketing
- importing, exporting and stocking.

Breeders may make their authorization subject to conditions and limitations. Generally, the breeder's Right is granted for a period of 25 years, but for trees and vines it is 30 years. On a European Community (EC) level, a harmonization of regulations on intellectual property rights for plant varieties has resulted in a Community Plant Variety Right, which is mainly based on UPOV 1991. The EC Council Regulation on Community Plant Variety Right of 27 July 1994 (no. 2100/94) became operative on 27 April 1995. As in national breeder's right, the duration of a Community Plant Variety Right is 25 years, while for trees and vines it is 30 years (article 19).

Trademark

Trademark is based on a registration system, of which four different forms are available: (1) registration in the Benelux (Eenvormige Beneluxwet op de merken, Trb. 1996, 225); (2) registration in other countries based on national law in each country; (3) international registration based on the Madrid Agreement (1891) and The Protocol to the Madrid Agreement; (4) registration in the EU, based on the EC Council Regulation on the Community Trade Mark (1994). Individual marks include denominations, drawings, prints, stamps and letters, all serving to distinguish the products and services of a company. The term of a Trademark registration is 10 years, which may be repeatedly extended for 10-year periods thereafter. According to breeder's right, a new plant variety should be denominated. Although the use of a Trademark as the name of a variety is forbidden, UPOV Convention 1991 (article 20) has determined that, besides the obligation to bring into the market propagation material under the registered variety name, the addition of a Trademark or Trade name is allowed. In practice, many new varieties are protected by both Breeder's Right and Trademark. The former name is usually a more or less artificial name, often partly designating the breeder's company, while the Trademark is often a well-sounding or romantic name. In commerce, however, the Trademark should always go together with the registered variety name. In spite of the foregoing, some (rose) breeders only protect their new products by Trademark because it is much cheaper than Plant Breeder's Right or Plant Patent.

In applying for Breeder's Right, using an old variety name for a new variety is not allowed. It is often seen, however, that in the course of time breeders use the same Trademark for successive new varieties. In practice, variety denominations and trademarks are often confused, which gives rise to a great deal of irritation by third parties. Most irritations are due to: (1) the contractual obligation to use both names; (2) lack of clarity as to the correct denomination of the variety; (3) the contractual obligation to use the Trademark after untimely withdrawal or termination of the Breeder's Right; and (4) prolonged charging of royalties over Trademark only.

Phenotypic vs Genotypic Characteristics

As mentioned above, Plant Patent or Plant Breeder's Right being granted in fruit and ornamentals is based on the description of a number of externally visible (phenotypic) characteristics, which are specific to each crop (*see* **Intellectual Property**: Technical Examination and Conventional Characteristics). In contrast to these vegetatively propagated crops, in a number of sexually propagated crops (vegetables, maize, etc.) Plant Patent or Breeder's Right is granted on the basis of a partial analysis of the genotype of a variety. Several genotypic characteristics in these crops can be detected by DNA fingerprinting according to randomly amplified polymorphic DNA (RAPD) or amplified fragment length polymorphism (AFLP), that make visible molecular markers. Particularly in the case of essential derivation or genetically very close relationships, fingerprinting appears to be an accurate method to distinguish between varieties. However, varieties may be so close that genetic distances have been agreed upon for several crops. For example, in maize, distances are indicated by coloured zones of similarity: red zone represents 90–100% similarity, orange zone 80–90% similarity and green zone $<80\%$ similarity.

In contrast to sexually propagated crops, in the new regulations for the application for Community Plant Variety Right, a phenotypic description has been maintained for vegetatively propagated crops. Not unjustly, it has been questioned by research institutions, biotech companies and even by breeders why this 'old-fashioned' method is preferred to the much faster and cheaper method of genotypic characterization by so-called molecular markers. There are two main answers to this question: (1) in negotiating UPOV 1991, the breeders of fruits and ornamentals pressed for phenotypic characterization because they 'want to see the plant'; and (2) despite new DNA techniques, flower colour mutants, which so often occur in roses, cannot be distinguished from the original variety. In addition, aversion to new techniques may be partly inspired by the large differences between breeders of sexually propagated and those of vegetatively propagated crops. It cannot be denied that breeding methods of sexually propagated crops have had more support from research, that the methods used are more refined than those in vegetative crops, and that their breeders are better scientifically trained. The current situation in rose breeding does not differ from this rule.

Essentially Derived Varieties

Prior to UPOV 1991, increasing problems as to the ownership of mutants in ornamentals, as well as novel developments in biotechnology, raised the question whether or not the existing system would meet the future demands of breeders as to adequate protection of their intellectual property rights. Particularly the introduction of new techniques in breeding and propagation, and the fear of 'double protection' have urged the revision of the UPOV convention 1961/1978 in 1991. The discussions and preparations for the revision of UPOV 1961/1978 ran parallel to the design of the Directive on legal protection of Biotech Material (98/44/EC) and the introduction of the Community Plant Variety Right (no. 2100/94). In UPOV 1991, this resulted in the introduction of a new concept, the so-called 'system of essentially derived varieties', sometimes indicated as a 'dependency system'. The system of EDVs applies when a new variety, with the exception of the differences which result from the act of derivation, essentially conforms to another variety, the initial variety, in the expression of characteristics that result from the genotype(s) of that initial variety. Such a new variety is called an 'essentially derived variety' or EDV. It should be remembered that, particularly in ornamentals, Breeder's Right are solely granted on phenotypic characteristics.

EDVs occur in both seed- and vegetatively propagated crops. In seed-propagated crops EDVs are e.g. hybrid varieties, varieties from self-pollination or from back-crosses. In vegetatively propagated crops EDVs may consist of mutants or of genetically modified organisms (GMOs). In seed crops, which will not be further considered here, problems have already arisen when new varieties could not be phenotypically distinguished from progenitors or existing varieties, or when the repeated use of protected varieties was needed for the breeding of a new variety. It is notable that one original variety may produce more than one EDV, and that one EDV may give rise to a complete family of EDVs. All such EDVs, however, are essentially derived from the original or mother variety. Even when the Breeder's Right of the mother variety has expired or been untimely terminated, its breeder may claim property rights on the EDV. Designating a new variety as an EDV, however, remains the responsibility of the breeder of the original variety. Jurisprudence on the question for how many generations new varieties should be considered EDVs is not yet available.

Mutants

The uncontrolled occurrence of mutants in protected varieties of vegetatively propagated crops, either natural (sports) or induced, has always been a thorn in the flesh of the breeder of the original variety. This was particularly true in ornamentals, where one single flower colour mutant in a large monoclonal growing area can be easily perceived. In UPOV 1961/1978, mutants could still be protected as original varieties. However, breeders considered this compensation for the finder was too good for so little breeding effort. This art of breeding was sometimes even considered as 'imitation' or 'cosmetic breeding'. On the other hand, it was argued that isolation, propagation and selecting for a 'true-to-type' mutant is an effort that is comparable to selection in a seedling population. To prevent protection of mutants by third parties, rose breeders had fixed in contracts with growers that rights of mutants would be awarded to both parties, a clause that appeared to be unlawful as it conflicted with Article 85 of the EU Treaty. The above problems have now been resolved by the newly introduced system of EDVs of UPOV 1991.

Genetically Modified Organisms

Modern biotechnology enables the introduction of 'foreign genes' into the genome of a certain plant. Foreign genes may originate from other plant species, insects, bacteria or otherwise. Because the location of the insertion of the new gene in the original genome cannot be directed, its phenotypic expression should be awaited. Modification of living material with foreign genes is called genetic modification and the result is a GMO. Commonly, foreign genes are the result of clever genetic engineering, in which various pieces of DNA fragments are cemented to form a new construct. Fragments or completely new constructs may be protected by various industrial patents, owned by one or more, usually multinationally operating biotechnological companies or research institutes. Because genetic engineering is very expensive, biotechnological companies will only invest when high returns are expected. In practice this means that the development of constructs for and the genetic modification of large food crops have priority over ornamentals. In crops, GMOs may either be a derivation of protected varieties, or just from 'a plant'. In the former case they are considered as EDVs, in the latter case as a 'patented plant'. Both types of plants may complicate the right of breeder's exemption.

It is notable that most genes or gene constructs can be detected in plants by means of molecular marker techniques. In spite of high expectations about new types of resistance to pest and diseases, new plant habits and flower colours of ornamentals, and promises to solve the world food problem, the genetic modification of plants has currently been subjected to strict regulations in many countries (e.g. in the EU regulations), which hamper the development and cultivation of GMOs.

Practical Consequences of a Dependency System for Roses

In ornamentals, particularly roses, the occurrence of mutants is an important source of new varieties. Commonly, a mutant arises from an accidental deviation in the reproduction of a somatic cell. In principle two kinds of mutants may occur: (1) natural mutants, generally called sports; and (2) induced mutants, which result from treatment of plant parts with ionizing radiation or mutagenic chemicals. The latter acts are called mutation breeding.

More than in any crop, plants of outstanding chrysanthemum selections resulting from cross-breeding are perpetually treated with ionizing radiation before being introduced as a variety. This treatment quickly results in a family of colour mutants, of which the order of occurrence is predictable. The advantage for the breeder is the control over a range of varieties with more or less the same plant habit, but with different colours, thus preventing the discovery of mutants by a third party. In this way breeders breed their own EDVs. However, not all crops mutate so easily as chrysanthemum.

In rose varieties the occurrence of mutants is relatively rare, but mutant varieties may be of great importance. As an example of mutants in the rose, **Table 1** presents the number of registered varieties in which 1–16 mutants occurred and that were protected by Plant Breeder's Right until the year 1999.

Table 1 shows that, until 1999, 399 mutants originating from 151 rose varieties were protected by Plant Breeder's Right. Like the number of unprotected rose mutants, the number of mutants that arose from other mutants is unknown.

Table 1 The number of registered rose varieties with 1–16 mutants

Number of varieties	*Number of mutants per variety*	*Total number of mutants*
76	1	76
29	2	58
14	3	42
7	4	28
4	5	20
6	6	36
4	7	28
2	8	16
4	9	36
1	10	10
3	11	33
1	16	16

Table 2 Mutants derived from 'Frisco' (Korflapei)

Variety denomination	*Trade name*	*Mutant of*
Kobra	—	'Korflapei'
Korbus	—	'Korflapei'
Korfleur	'Black Beauty'	'Korflapei'
Korhuis	'Fresh'	'Korflapei'
Korhulpei	'Sweet Lady'	'Korflapei'
Korkameel	—	'Korflapei'
Korkouw	'Edon'	'Korflapei'
Kormolen	'Aura'	'Korflapei'
Kornyfla	'Amore'	'Korflapei'
Kororco	—	'Korflapei'
Korrufri	—	'Korflapei'
Korsalton	'Safari'	'Korflapei'
Korsaphil	'Christine'	'Korflapei'
Korstapei	—	'Korflapei'
Naomi	'Naomi'	'Korflapei'
Spekra	'Rodeo'	'Korflapei'
Korpocus	'Hocus Pocus'	'Korfleur'
Spehocsel	'Abracadabra'	'Korpocus'
Spesimsala	'Simsalabim'	'Korpocus'
Korbushoc	'Fidibus'	'Korpocus'

Table 2 shows an important supplier of successful colour mutants, the rose cultivar 'Frisco' (Korflapei). For 16 mutants and four mutants-of-mutants, Plant Breeder's Right have been applied for.

An important criterion for granting breeder's right is the specific description of a variety. For roses this description contains 46 phenotypic characters. The complete description of a new variety determines whether it differs sufficiently from already existing varieties. Commonly, this description is supported by the judgement of a committee of specialists. Since the introduction of the dependency system, the characteristic description for roses has not been changed.

If a difference is based on one characteristic only, as often occurs in EDVs, breeder's right granted, provided that the other criteria are fulfilled.

Breeder's Exemption vs Research Exemption

UPOV 1991 has maintained breeder's exemption (article 15). This means that any breeder may use, without any restriction, any protected variety as a progenitor in breeding programmes. The results of those breeding programmes may apply for Breeder's Right or Plant Patent. Breeder's exemption is not without reason, because in vegetatively propagated crops free use of all varieties is almost the only possible way to keep abreast of the requirements of modern cultivation, as well as to prevent inbreeding by using restricted germplasm. In Industrial Patent, research exemption offers the possibility to industrial inventors to use, for free, the patented invention but only for further research and development purposes in their branch of industry. It is thus seen that, on the one hand there is a large similarity between the two exemptions during the research phase, but that on the other hand the exclusive ownership of the Industrial Patent is continued, which is not the case in Breeder's Right. That Plant Breeder's Right and industrial patent may meet is illustrated by four examples:

1. In the first example a breeder/owner of a protected variety uses his or her variety to introduce a patented gene owned by a third party. As long as the new variety is not commercialized, this action is legal, according to the research exemption in patent law. But it may be clear that commercialization of the new variety, which is a GMO/EDV, needs authorization of the patent owner.
2. A second, slightly different case occurs when the breeder uses 'a plant' which is not yet protected by Breeder's Right, for the introduction of a patented gene. In this case too, commercialization of that plant needs authorization of the patent holder.
3. A third situation occurs when the owner of a patented gene uses a protected variety owned by a third party to introduce that gene. For commercialization of that new variety, which again is a GMO/EDV, authorization of the owner of Breeder's Right is required.
4. A fourth possibility is when the owner of a patented gene, uses 'a plant' to introduce that gene. For the protection and commercialization of that plant, there seem to be two possibilities: (1) apply for Breeder's Right as a new variety; or (2) apply for industrial patent.

It is thus seen that when a commercially interesting plant variety contains patented genes or traits, it is important to the breeder that, in accordance with breeder's exemption, it should remain freely available for further breeding.

For the patent-holder, however, it is important that each time such a variety is used, either in commerce or in breeding, royalties for the patent should be charged.

Consequences of Intellectual Property Rights

This article may have given the impression that the ownerships of plant varieties and industrial patents are well-regulated by various laws, and that difficulties with EDVs have been nicely cleared away by the new system regulations of UPOV 1991. However, there are still some cloudlets in the sky. Since the introduction of UPOV 1991, multinational biotech companies have continuously tried to obtain industrial patent on plants or plant varieties. Two case laws, concerning the patenting of GMOs have offered significant jurisprudence. In the first case, the so-called Plant Cells/Plant Genetic Systems (PGS) case, it was attempted to obtain an industrial patent on a plant that had been provided with a patented gene which codes for herbicide resistance (European patent no. 0 242 236, granted 10 October 1990, 'Plant cells resistant to glutamine synthetase inhibitors, made by genetic engineering'). Initially, the application was held up, but in a final process, the Technical Board of Appeal at the European Patent Office (EPO) permitted the patenting of all plant material (including plant groupings which encompass plant varieties), simultaneously maintaining the ban on patentability of a plant variety as such (T0356/93). This decision was, in fact, a compromise between the two protection systems, and in agreement with the Directive on legal protection of Biotech Material (98/44/EC).

In the second case, however, the so-called Transgenic Plant/Novartis case, where plants were provided with genes coding for controlling plant pathogens, the compromise reached in the PGS case has been confused by the decision of the Enlarged Boards of Appeal. It concerned the European patent no. EP546.090, 'Anti-pathogenically effective compositions comprising lytic peptides and hydrolytic enzymes' (G0001/98). In this case the decision read that the ban on patentability of plant varieties can be bypassed.

The state of the art in the year 2003 thus seems to be that the system of EDVs of UPOV 1991 has been surpassed. Plants and plant varieties can now be protected in three ways – by Plant Breeder's Right,

Plant Patent and by Industrial Patent – developments that were already forecast in legal studies.

See also: **Classification:** Cultivar Identification by Image Analysis. **Genetics:** Mutation. **Intellectual Property:** Technical Examination and Conventional Characteristics.

Further Reading

Anonymous (1975) Information from UPOV. *Newsletter* 1: 2–5.

Anonymous (1985) EU-Beschikking IV 30.017 13 December 1985.

Anonymous (1996) CIOPORA-International. *CIOPORA-Info*: 30–31.

Anonymous (2001) *Intellectual Property and Biotechnology: A Training Handbook*. A training resource for developing countries in the APEC region funded under the APEC Support Program of the Australian Agency for International Developments (AID).

Bostyn SJR (2000) Witte rook boven het Europees Octrooibureau. *Bijblad Industriële Eigendom*: 201–207.

Broertjes C and Van Harten AM (1988) *Applied Mutation Breeding for Vegetatively Propagated Crops*. Amsterdam: Elsevier.

Debener T, Janakiram T and Mattiesch L (2000) Sports and seedlings of rose varieties analysed with molecular markers. *Plant Breeding* 119: 71–74.

De Riek J (2001) Are molecular markers strengthening plant variety registration and protection? *Acta Horticulturae* 552: 215–223.

Dorhout Mees TJ (1989) Nederlands handels – en faillissementsrecht. II. Industriële eigendom en mededingingsrecht: pp. 77 e.v. Arnhem: Gouda Quint B.V.

Dubois LAM (1998) Thesis. Juridische en praktische aspecten van de invoering van afhankelijk kwekersrecht bij vegetatief vermeerderde gewassen. Afstudeerscriptie opleiding Nederlands Recht Rijks Universiteit Utrecht: 46 pp.

Dubois LAM (2001) Intellectual property rights on plants, with special reference to vegetatively propagated ornamentals. *Acta Horticulturae* 552: 207–214.

Espenhain F (1989) Developments in plant breeders rights in the view of UPOV. *Vortr. Pflanzenzüchtg.* 16: 283–292.

Heitz A (1998) The concept of essentially derived varieties. Annex to the UPOV Model Law on the Protection of New Varieties of Plants.

Lange P (1993) Derived varieties and dependency under the revised UPOV convention. *Plant Variety Protection* 72: 51–63.

Reid BC (1999) Intellectual property guides. A practical guide to patent law. *Plants* 2.35–2.41: 26–28.

Royon R (1998) Effective protection is long overdue. *Prophyta Annual* 32–35.

Thumm N (2000) Intellectual property rights. National systems and harmonisation in Europe. *Patentability* 111–115.

Van der Kooij PACE (1997) Thesis. *Introduction to the EC Regulation on Plant Variety Protection.*

Van der Walt W (1998) Identification technologies. EDVs urge for reliable testing methods. *Prophyta* April 55–57.

Van Nispen CJJC (1990) Thesis. Octrooirecht en biotechnologie. Lezing gehouden op 13 februari te Bleiswijk voor de Vereniging voor Agrarisch Recht.

Van Overwalle G (1996) Octrooieerbaarheid van plantenbiotechnologische uitvindingen. Een rechtsvergelijkend onderzoek naar een rechtvaardiging van een uitbreiding van het octrooirecht: 745 pp.

Van Overwalle G (1997) Patents for plants and plant varieties. Towards the abolition of article 53[b] EPC. *Prophyta Annual* 32–35.

Van Overwalle G (1999) Patent protection for plants: a comparison for American and European approaches. *IDEA – The Journal of Law and Technology* 39(2): 143–194.

Whealy CA (2002) United States. Plant patents and legal issues. *FloraCulture International* 28–32.

Wichers Hoeth L (2000) Kort begrip van het intellectuele eigendomsrecht: 501 pp.

Wolf AHJ (1987) 'Mutanten-problematiek' in verband met de door de Commissie van de Europese Gemeenschap gewezen 'Rozenbeschikking'. *Prophyta* 2: 48–49.

Wolf AHJ (1988) Kwekersrecht-merkenrecht. *Prophyta* 7: 190–191.

Technical Examination and Conventional Characteristics

T Kwakkenbos, Community Plant Variety Office (CPVO), Angers, France

Introduction

This article aims to give an overview of the very broad area of plant breeders' rights, technical examination and the conventional characteristics used for the preparation of a variety description. Useful information is collected to give guidance on where to find information about protected varieties and for further indepth study of the subject. Other forms of variety registration are also briefly mentioned, for instance the International Registration Authority for Roses, and the Vaste Keurings Commissie (VKC) for registration of varieties for the Dutch flower auctions.

Plant Breeders' Rights

The Union pour la Protection des Obtentions Végétales (UPOV) system for protection of plant breeders' rights of new plant varieties dates from the early 1960s. In November 1961 in Paris, the inaugural UPOV convention took place. Since that convention, UPOV has grown exponentially to almost 10 times its original size and many improvements have been implemented worldwide. In order to learn about UPOV and the systems in the different countries this first article will provide a guide to the basics of plant breeders' rights.

National Systems inside the European Union

Inside the European Union (EU), a national system of plant breeders' rights exists in most member states, which gives protection for plant varieties within the borders of that state. Legislation within different countries differs, depending on national legislation and the UPOV treaty that was signed by that country. Depending on this legislation it is possible either to protect plant varieties of all species of the plant kingdom (legislation based on UPOV 1991) or a limited number of species (legislation based on UPOV 1978 or earlier).

Community Plant Variety Rights

In 1994 the legislation for unique protection for plant varieties came into force for all plant species in the EU. The main regulation (based on UPOV 1991), Council Regulation EU 2100/94, came into force on 1 September 1994. The system is managed by an autonomous body of the EU, the Community Plant Variety Office (CPVO). This office opened during summer 1995 in Brussels, and since 1997 has had its permanent seat in Angers, France. Plant varieties of all species that meet the criteria of novelty, distinctness, uniformity and stability (DUS) can be protected. This protection covers all member states of the EU. When the enlargement of the EU takes place in 2004, the protected territory will be larger, including the new member states. When a plant variety is protected in one or more member states of the EU and receives Community plant variety rights (European plant breeders' rights), these rights will overrule national rights. Some countries, like the Netherlands, offer the possibility of so-called 'sleeping rights'. It is possible to keep your national rights during the Community protection in a nonactive (sleeping) way for a reduced annual fee. Once Community rights have ended, national rights can be reactivated. Other countries, like Germany, offer the possibility to file an application for national rights, during a limited period of time after the community plant variety rights have ended. For more details, contact the relevant national authority. A community plant variety right for a rose variety has a maximum duration of 25 years.

National Systems outside the European Union

The number of countries that are members of UPOV is growing very fast. At the end of 2001 the Republic of Korea became the 50th member of UPOV. The legislation for plant breeders' rights in different UPOV member states differs. A good overview of the status of ratification of different member states can be found on the UPOV website.

UPOV

The UPOV was established by the International Convention for the Protection of New Varieties of Plants, which was signed in 1961 in Paris. The convention entered into force in 1968. It was revised in Geneva in 1972, 1978 and 1991. UPOV's seat is in Geneva at the premises of the World Intellectual Property Organization (WIPO). UPOV has more than 50 member states that have ratified one of its conventions.

Technical Examination (DUS Test)

A variety that has plant breeders' rights will have to meet the following criteria:

- Distinctness: needs to be distinct, which means sufficiently different from all existing varieties known at the time of the application. The existing varieties need to be in so-called 'common knowledge'. This common-knowledge principle has been designed from the angle of agriculture and vegetables. In these sectors each variety that enters the market is entered on an official list. If not listed, these varieties cannot be commercialized. This process does not exist for ornamentals. In the ornamental sector many varieties are not listed or protected and it is often difficult to establish whether a variety is in common knowledge or not. For example, when an (amateur) breeder makes a crossing and selects a superb variety out of the seedlings, and neither puts this variety on the market nor shows the variety in an exhibition, it is probably not in common knowledge. If the breeder gives away plants to friends and one of the cuttings from cuttings shows up as a variety that is applied for plant breeders' rights, it is difficult to prove the ownership of the original breeder.
- Uniformity means all plants have to show the same UPOV characteristics (**Table 1**). Some minor differences can be tolerated, but if you want to buy

Table 1 Characteristics of the Union pour la Protection des Obtentions Végétales (UPOV) guideline for roses TG/11/7

Number	*Characteristics*	*Description*	*Score*
1	Plant: growth habit (excluding climbing varieties)	Narrow bushy	1
		Bushy	3
		Broad bushy	5
		Flat bushy	7
		Creeping	9
2	Plant: height (as for 1)	Short	3
		Medium	5
		Tall	7
3	Plant: width (excluding creeping varieties)	Narrow	3
		Medium	5
		Broad	7
4	Young shoot: anthocyanin coloration (shoot about 20 cm long)	Absent or very weak	1
		Weak	3
		Medium	5
		Strong	7
		Very strong	9
5	Young shoot: hue of anthocyanin coloration (as for 4)	Bronze	1
		Bronze to reddish brown	2
		Reddish brown	3
		Reddish brown to purple	4
		Purple	5
6	Prickles	Absent	1
		Present	9
7	Prickle: shape of lower side	Deep concave	1
		Concave	3
		Flat	5
		Convex	7
		High convex	9
8	Short prickles: number	Absent or very few	1
		Few	3
		Medium	5
		Many	7
		Very many	9
9	Long prickles: number	Absent or very few	1
		Few	3
		Medium	5
		Many	7
		Very many	9
10	Leaf: size	Very small	1
		Small	3
		Medium	5
		Large	7
		Very large	9
11	Leaf: green colour (at time of first flowering)	Very light	1
		Light	3
		Medium	5
		Dark	7
		Very dark	9
12	Leaf: glossiness of upper side	Absent or very weak	1
		Weak	3
		Medium	5
		Strong	7
		Very strong	9

Table 1 Continued

Number	*Characteristics*	*Description*	*Score*
13	Leaflet: cross-section	Concave	3
		Slight Concave	4
		Flat	5
		Slight convex	6
		Convex	7
14	Leaflet: undulation of margin	Absent or very weak	1
		Weak	3
		Medium	5
		Strong	7
		Very strong	9
15	Terminal leaflet: length of blade	Short	3
		Medium	5
		Long	7
16	Terminal leaflet: width of blade	Narrow	3
		Medium	5
		Broad	7
17	Terminal leaflet: shape of base	Wedge-shaped	1
		Obtuse	2
		Rounded	3
		Cordate	4
18	Flowering shoot: number of flowers	Very few	1
		Few	3
		Medium	5
		Many	7
		Very many	9
19	Flower pedicel: number of hairs or prickles	Few	3
		Medium	5
		Many	7
20	Flower bud: shape of longitudinal section (just before separation of sepals)	Round	3
		Broad-ovate	5
		Ovate	7
21	Flower: type	Single	1
		Semidouble	2
		Double	3
22	Semidouble or double varieties only Flower: number of petals	Very few	1
		Few	3
		Medium	5
		Many	7
		Very many	9
23	Flower: diameter	Very small	1
		Small	3
		Medium	5
		Large	7
		Very large	9
24	Flower: view from above	Round	1
		Irregularly rounded	2
		Star-shaped	3
25	Flower: side view of upper part (fully opened flower)	Flat	1
		Flattened convex	2
		Convex	3
26	Flower: side view of lower part (as for 25)	Concave	1
		Flat	2
		Flattened convex	3
		Convex	4

Table 1 Continued

Number	*Characteristics*	*Description*	*Score*
27	Flower: fragrance	Absent or very weak	1
		Weak	3
		Medium	5
		Strong	7
		Very strong	9
28	Sepal: extensions	Absent or very weak	1
		Weak	3
		Medium	5
		Strong	7
		Very strong	9
29	Petal: size	Very small	1
		Small	3
		Medium	5
		Large	7
		Very large	9
30	Petal: colour of middle zone of inner side	RHS Colour Chart (indicate reference number)	
31	Petal: colour of marginal zone of inner side	RHS Colour Chart (indicate reference number)	
32	Petal: spot at base of inner side	Absent	1
		Present	9
33	Petal: size of spot at base of inner side	Very small	1
		Small	3
		Medium	5
		Large	7
		Very large	9
34	Petal: colour of spot at base of inner side	RHS Colour Chart (indicate reference number)	
35	Petal: colour of middle zone of outer side	RHS Colour Chart (indicate reference number)	
36	Petal: colour of marginal zone of outer side	RHS Colour Chart (indicate reference number)	
37	Petal: spot at base of outer side	Absent	1
		Present	9
38	Petal: size of spot at base of outer side	Very small	1
		Small	3
		Medium	5
		Large	7
		Very large	9
39	Petal: colour of spot at base of outer side	RHS Colour Chart (indicate reference number)	
40	Petal: reflexing of margin	Absent or very weak	1
		Weak	3
		Medium	5
		Strong	7
		Very strong	9
41	Petal: undulation of margin	Absent or very weak	1
		Weak	3
		Medium	5
		Strong	7
		Very strong	9
42	Outer stamen: predominant colour of filament	White	1
		Yellow	2
		Orange	3
		Orange red	4
		Pink	5
		Red	6
		Purple	7
		Green	8

Table 1 Continued

Number	*Characteristics*	*Description*	*Score*
43	Seed vessel: size (at petal fall)	Very small	1
		Small	3
		Medium	5
		Large	7
		Very large	9
44	Hip: shape of longitudinal section	Funnel-shaped	1
		Pitcher-shaped	2
		Pear-shaped	3
45	Time of beginning of flowering	Very early	1
		Early	3
		Medium	5
		Late	7
		Very late	9
46	Flowering: habit	Once flowering	1
		Twice flowering	2
		Almost continuous flowering	3

RHS, Royal Horticultural Society.

Figure 1 Example of nonuniformity. This pink-coloured variety shows light-yellow sectors in the flower. Courtesy of Dr B. Spellerberg, Bundessortenamt, Germany.

a plant of a protected variety, you need to be sure that you get what you pay for (**Figure 1**).

- Stability means that all plants show the same characteristics after propagation. After a variety has been propagated for several years it should look the same.
- A variety also has to meet the novelty criteria. This means that, within certain limits, a variety should be new, and not previously commercialized.

In order to see whether a variety meets these criteria a so-called DUS test is performed. The different criteria are examined as follows:

1. Distinctness: there are no geographical borders to common knowledge: this means in practice world-wide. Examples of varieties of common knowledge are varieties that are protected for plant breeders' right in a UPOV country, or that are sold on the market. For ornamental plants the borders of common knowledge are not always 100% clearly defined. Since it cannot be expected that the expert who performs the trial has knowledge of all varieties of common knowledge worldwide, a number of tools are available. In many countries, the technical expert makes use of a panel of experts, a group selected for its knowledge of roses and appointed by the national board of plant breeders' rights, to advise on distinctness of the candidate variety.
2. Uniformity: in the case of a UPOV DUS test, some off-types (plants that show a different expression of characteristics) are allowed. For vegetatively propagated varieties and in a sample size between six and 35 plants, one off-type is allowed.
3. Stability: it is difficult to observe stability, since the candidate variety needs to be propagated one or more times to see whether the expression of its characteristics remains stable. A variety is considered to be stable when it is uniform.
4. Novelty: the question whether a variety is novel at the application date depends on the legislation of the country of application. When legislation is based on UPOV 1991, such as the European plant breeders' rights system, a variety is novel when it has either not been sold at all, or has been sold up to a maximum of 1 year before application within the territory of protection (country, or in the case of European protection, within the borders of the EU) or a maximum of 4 years outside this territory. In other cases, usually based on older UPOV treaties,

the variety is not allowed to have been sold before the application, nor even been in a catalogue or shown in an exhibition. When considering applying for plant breeders' rights, it is wise to ask the relevant authority for information on this subject.

Providing Samples

In order to perform a technical examination, the competent authority will ask the applicant to deliver a sample of plants for technical examination. As an example of such a technical examination, the requirements for different types of roses are given for examination for Community plant variety rights in the EU (**Table 2**).

Additional Requirements

- Plants must be vegetatively propagated, visually healthy and not treated in any way that will affect subsequent development.
- The plant material must not have undergone any treatment that influences the growth and development of the plant, unless the examination office allows or requests such treatment. If it has been treated, full details of the treatment must be given.

Test Guidelines and UPOV Guidelines

In order to obtain a harmonized approach, test guidelines have been developed to establish a uniform way of describing varieties worldwide. For plant breeders' rights the UPOV guidelines have been developed by the national experts of the member states. Since UPOV is expanding very quickly, general guidance is also being developed. The main document for technical examination is UPOV's General Introduction (document TG/1/3, adopted on 19 April 2002) that gives guidance on all aspects of technical examination. Associated with this General Introduction are the so-called TGP (Test Guidelines' Procedures) documents, which are currently being written. These TGP documents give detailed information on all kinds of technical matters related to the technical examination of plant varieties for protection. In the case of roses the current UPOV Guideline is TG/11/7, dating from 12 October 1990. This guideline is available in a trilingual version (English, German and French). A list of characteristics can be found in **Table 1**. For the UPOV guidelines, the expression of the characteristics is given in scores. For example, for characteristic plant height the scores 3–5–7 can be given, representing short (3)–medium (5)–tall (7). This will make these guidelines internationally usable. In order to calibrate the expression of characteristics, reference varieties are taken into account in these test guidelines.

Since the UPOV guideline dates back to 1990, a revision can be expected soon. The guideline can be used for all types of roses: garden roses, pot roses, rootstocks and greenhouse roses for cut-flower production.

The characteristics are observed or measured. When a characteristic is measured, such as characteristic 29: Petal: size, these characteristics are calibrated with

Table 2 Examples of sample requirements for technical examination of roses for community plant variety rights, according to Community Plant Variety Office (CPVO) technical protocol CPVO-TP/11/1

Examination office	*Closing date for applications*	*Plant material delivery*	*Sample*
United Kingdom Garden roses	28/02	Week 45-13	Six plants suitable for winter planting. Plants should be of sufficient size and maturity to flower and/or show their other representative characteristics in the first year
Germany Garden roses	15/02	Week 45-13	Six 1-year-old plants with at least three shoots on a frost-hardy rootstock or on own roots
Germany Pot roses	15/02	Not later than 31/03	Six plants on their own roots
The Netherlands Cut flowers[a]	15/11	01/02–15/02	10 rooted cuttings in a 5-cm cocos plug
The Netherlands Cut flowers[b]	15/11	01/02–15/02	20 rooted cuttings in a 5-cm cocos plug
Germany Rootstocks	15/02	Week 45-13	Six 1-year-old plants with at least three shoots on their own roots

[a] Seedling variety.
[b] Mutation variety.

reference varieties that are used every year to ensure continuity. Values of measured characteristics are never given in a variety description. The value of the expression of the measured characteristic is always given in the form of a score with the help of the calibation of the reference varieties. For the observation of colours, the Royal Horticultural Society (RHS) (UK) colour chart is used. This colour chart is built up from fans of colours, based on the observation of a human eye. To describe these RHS colours, UPOV has developed a list of 50 colour groups.

CPVO Test Guidelines

The CPVO has the obligation to develop test protocols for technical examination. These test protocols are prepared by the technical experts of the EU member states involved in technical examination, in cooperation with the experts of the CPVO. For roses, the technical protocol is CPVO-TP/11/1 dating from 31 October 2002.

Reference Collections

In order to prepare the technical examination and to be able to judge distinctness of the candidate variety, examination offices can make use of the following tools:

- Plant collections: collections of protected varieties and other varieties that are in common knowledge. These plants are maintained in the field (for garden roses and rootstock varieties) or in containers in the greenhouse (as with pot roses). When a candidate needs a direct comparison with a reference variety, usually plants of the reference variety are propagated in a similar way as the candidate variety and planted side by side in a trial.
- Nonliving reference collections: in some cases it is not possible to have a living collection, or it may not be complete. Alternatives or supplementary methods have been developed, such as:
  (i) collection of catalogues of rose breeders and propagators
  (ii) collection of (digital) pictures or slides
  (iii) collection or a database of variety descriptions
  (iv) information obtained from the internet.

For future collections one could also think of DNA banks or collections of DNA fingerprints of varieties. Another technique that could be developed in the future is the use of image analysis and matching images to compare candidate varieties with reference collections.

Variety Descriptions

On the basis of the guidelines mentioned, a variety description of the candidate variety is prepared. The candidate variety should differ from existing varieties in at least one important characteristic. In a modern plant breeders' rights system, where plant varieties may be very similar, developments are ongoing towards a system in which an accumulation of small differences between the candidate and a reference variety gives a sufficient base for a positive decision on distinctness. Within UPOV there are discussions to make the variety descriptions commonly available. Some countries, like Australia, publish variety descriptions of all varieties that are granted plant breeders' rights in their national gazette.

Variety Denominations

A variety should have a name. There is an obligation for the holder of a plant variety right to give his or her variety a denomination. This denomination is a unique name under which this particular variety is known. The principle of the acceptability of a variety denomination is simple:

- It should not cause confusion with other variety names.
- It should be correct and it should not cause offence in any language.
- It should be pronounceable.

Since the number of rose varieties is increasing rapidly every year, it is becoming increasingly difficult to find a denomination that is acceptable. Breeders found the solution of combining a fantasy name with an abbreviation of the breeder (usually three letters, like MEI-, KOR-, RUI-, JAC- and NOA-). Since a few years ago, a code (a combination of figures and letters) is also allowed as a variety denomination. Within the EU a unique system of variety denominations has been introduced that is used both for plant breeders' rights and variety listing for the Common Catalogue (for fruits, vegetables and agricultural crops). In addition to the variety denomination, a trade name or trademark is allowed under certain conditions.

Registers of Varieties

Registers for Plant Breeders' Rights

Each country or organization like the CPVO has a register where all applications and granted rights for plant variety protection are registered. Depending on the legislation in force, the registers are, in general, public. Most countries publish their applications and decisions in an official gazette. A list of addresses of all UPOV member states can be found on the UPOV website (**Table 3**). For further information, contact the national plant breeders' rights authorities (**Table 4**).

Where to find Information Concerning Protected Varieties

Information on applications, decisions, protected varieties and variety denominations is published in the gazettes of the plant breeders' rights authorities. Usually there is also an annual report with an overview of all protected varieties of that country. In many countries plant breeders' rights authorities also publish their protected varieties on their website (**Table 4**).

UPOV Information

Information on protected varieties in most UPOV countries can be obtained on the UPOV CD-ROM. It is possible to subscribe to receive an update of this important information resource every 2 months.

Other Registrations

In addition to official registrations in the different UPOV member states (**Table 3**), other important registration authorities have been created. Although the basis of these registrations has not been formalized by a legal framework, these registrations are generally accepted by the rose world. A nonexhaustive list will be discussed here, starting with the most important registration authority for roses – the International Registration Authority for Roses (IRAR).

American Rose Society: International Registration Authority for Roses In 1955, the American Rose Society was designated as the IRAR by the International Society for Horticultural Science. In 1981 the IRAR was designated to serve as the Registrar of Communauté Internationale de Plantes Ornementales et Fruitières de Reproduction Asexuée (CIOPORA). It keeps track of all those registrations in a large database. A list of varieties is published regularly in the form of a book. The latest version of this book, *Modern Roses XI: World Encyclopedia of Roses*, was published in 2000. This book contains over 25 000 varieties of roses (see Appendix X). Registration of a new rose variety is simple – by an electronic registration system via the internet.

VKC/Plantscope NL VKC takes care of the registration of ornamental varieties for the Dutch Flower Auction system (Vereniging voor Bloemenveilingen Nederland, VBN). In order to obtain a registration number, which is necessary to market a variety under its (commercial) name, some plants or flowers have to be delivered to the VKC office in Aalsmeer, Netherlands. In this office, all the administrative data are registered, a description is made and a photo is taken. All the data are included in a database called Plantscope (**Table 5**). This database can be consulted through the internet (although it is restricted to professional use at the time of writing).

Table 3 Websites of registration organizations

UPOV	www.upov.int
CPVO	www.cpvo.eu.int
WIPO	www.wipo.int
ARS	www.ars.org

Table 4 Websites of plant variety rights offices

Australia	www.affa.gov.au/pbr
Bolivia	www.semillas.org
Canada	www.cipo.gc.ca
Denmark	www.agrsci.dk/afs
Estonia	www.plant.agri.ee
France	www.geves.fr
Germany	www.bundessortenamt.de
Hungary	www.hpo.hu
Kyrgyzstan	www.patent.kg
Mexico	www.sagar.gob.mx/Snics
Netherlands	www.kwekersrecht.nl
New Zealand	www.pvr.govt.nz
Republic of Korea	www.seed.go.kr
Republic of Moldovia	www.agepi.md
Russian Federation	fax.aris.ru/ARIS/desel/
UK	www.defra.gov.uk/plant/plantfrm.htm
USA	www.ams.usda.gov/science/PVPO/pvp.htm

Table 5 Other relevant websites

VKC (see also Plantscope for flower auction registration)	www.vkc.nl
Plantscope	www.plantscope.nl
International Society for Horticultural Science (ISHS)	www.ishs.org
Allgemeine Deutsche Rosenneuheitenprüfung (ADR)	www.ADR-Rose.de

See also: **Breeding**: Selection Strategies for Cut Roses; Selection Strategies for Pot Roses. **Classification**: Cultivar Identification by Image Analysis. **Intellectual Property**: Plant Patents and Trademarks.

Furthing Reading

Anonymous (1978) *International Convention for the Protection of New Varieties of Plants*. Text of 1961,

1972 and 1978. UPOV publication 293. Geneva, Switzerland: UPOV.

Anonymous (1990) *Guidelines for the Conduct of Tests for Distinctness, Homogeneity and Stability for* Rosa *L.* G/11/7. Geneva, Switzerland: UPOV.

Anonymous (1991) *International Convention for the Protection of New Varieties of Plants.* Text of 1991 only. UPOV publication 221. Geneva, Switzerland: UPOV.

Anonymous (1994) Council Regulation (EC) no. 2100/94 of 27 July 1994 on Community Plant Variety Rights. *Official Journal of the European Communities,* L 227: 0001–0030. Luxembourg: OPOCE.

Anonymous (1994) *Grouping of Colours of the RHS Colour Chart.* UPOV document TWO/27/3 (will be included in future UPOV TGP/11: Observation of colours).

Anonymous (2000) Guidelines of the administrative council on variety denominations. *Official Gazette of the CPVO* 15 June: 78.

Anonymous (2001) *RHS Colour Chart.* London, UK: Royal Horticultural Society.

Anonymous (2002) *General Introduction to the Examination of Distinctness, Uniformity and Stability as the Development of Harmonised Descriptions of New Varieties of Plants.* TG/1/3. Geneva, Switzerland: UPOV.

Anonymous (2002) *Protocol for Distinctness, Uniformity Stability Tests* Rosa *L.* CPVO-TP/11/1. Angers, France: CPVO.

Anonymous (2002) *International Union for the Protection of New Varieties of Plants. What it is, What it Does.* UPOV publication no. 437 (E). Geneva, Switzerland: UPOV.

Anonymous Bimonthly Update. *Database on Plant Varieties on CD-ROM.* UPOV-ROM. Geneva, Switzerland: UPOV.

Cairns T (ed.) (2000) *Modern Roses XI. World Encyclopedia of Roses.* London: Academic Press.

Dubois LAM (1998) *Juridische en Praktische Aspecten van de Invoering van Afhankelijk Kwekersrecht bij Vegetatief Vermeerderde Gewassen.* Utrecht: Afstudeerscriptie opleiding Nederlands Recht RijksUniversiteit Utrecht.

Dubois LAM (2001) Intellectual property rights on plants, with special reference to vegetatively propagated ornamentals. *Acta Horticulturae* 552: 207–214.

Van der Kooij PACE (1997) *Introduction to the EC Regulation on Plant Variety Protection.* The Hague, The Netherlands: Kluwer Academic.

Royon R and CIOPORA (2002) *CIOPORA Green Paper on Plant Variety Protection.* Policy statement. Geneva, Switzerland: CIOPORA.

J

Judging *see* **Rose Collections and Trials**: Awards.

M

Molecular Biology *see* **Biotechnologies for Breeding**: Genetic Transformation. **Classification**: Chemotaxonomy and Molecular Taxonomy. **Fragrance and Pigments**: Anthocyanins and other Flavonoid Pigments; Flavonoid Molecular Biology; Functional Genomics. **Genetics**: Nuclear DNA Amounts; Inheritance in the Dogrose; DNA Fingerprinting; Gene Mapping; Molecular Markers; Cloned Genes; Marker-Assisted Selection; Research Trends. **Growth Regulation**: Floral Induction. **Postharvest Physiology**: Ethylene.

MORPHOLOGY AND ANATOMY

Contents

Shoots and Stems

J P André, INRA, Antibes, France

Introduction

There are enough morphological and anatomical similarities between the numerous rose varieties to justify a generalized description of their shoots. The rose bush architecture results from the development of a small number of main branches that issue from a grafted bud or cutting and which, in turn, bear a succession of axillary shoots. When the rose plants are managed for intensive production, only the growing pre-flowering shoots remain intact; all the others are cut at harvest or occasionally pruned (**Figure 1**).

The growth habit of the stem, its dimensions, colour, epidermis and bark, and the disposition of adjacent leaves and flowers represent its main morphological characters that constitute the visible expression of morphogenesis and organ maturation. The same processes are at the origin of its anatomical characters; the nature of its tissues, their cellular structure, their relative arrangements and the connections they establish between the stem, its leaves, the axis from which it originates, and the flower peduncle it forms.

Shoot and Stem Morphology

After bud-break, the very young shoots make an angle of approximately 45° with the vertical. This incline is often the same for the short sylleptic shoots, axillary to the uppermost leaves and growing together with them, whereas the long proleptic shoots, that issue from lower buds released from apical dominance, straighten up as soon as they are a

Figure 1 Rose bush with some leaves of a long proleptic shoot. 1, 2, 3, connection between the shoot and the flower-bearing stem, the foliar petiole, and the flower pedicel; s, location of the transverse section shown in **Figure 3 (b)**.

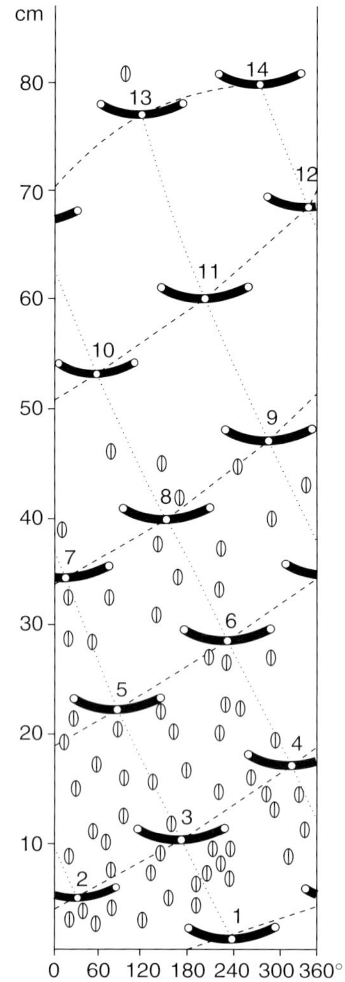

Figure 2 Surface of the proleptic shoot spread on one plane, like that of a cylinder. The leaf scars numbered from 1 to 14 are regularly arranged on virtual helices: the main helix (broken line) and the even and uneven parastichies (dotted lines). The prickle distribution is random (crossed ovals), decreasing in number from the bottom to the top.

few centimetres long and hence have a more or less vertical bearing.

In cut-flower production, the flowering stems are harvested when they are around 60–80 cm long for the proleptics, and around 30 cm long for the sylleptics. For a given variety, the length L and the base diameter D of the proleptic stems, their dry weight M (without leaves and flower) and volume V are strongly correlated parameters. Thus, in 'Diplomat', when L and D are expressed in mm, M in mg and V in mm^3:

$$M \cong 0.13.L^{1.55};\ D \cong 1.15.L^{0.28};\ M \cong 0.2V$$

Anthocyanins often colour the subepidermis, stem and leaves of the young shoot. The reddish colour disappears during further growth and the chlorophyllous pigmentation appears. The stem parenchyma which contains chloroplasts communicates with the atmosphere through the stomata that are positioned in short longitudinal files along the epidermis. The density of stomata in the stem is around 10–15 mm^{-2}, which is four to eight times less than that of the leaves. With time, several new cellular layers are formed under the epidermis, which is fragmented under their pressure and which they progressively replace to form a brown bark, the rhytidome, visible at the base of the flower-bearing branches.

The presence and abundance of prickles are varietal characters. They are randomly distributed on the stem. Their density generally decreases from the proximal to the distal ends of the stem (**Figure 2**). They originate from the epidermis and subepidermis, unlike the leaves that are initiated in the shoot apical meristem.

The phyllotaxic arrangement of the leaves on the stem is alternate. Their insertion can thus be connected by virtual helicoidal lines. The direction of the helix that links them together, often wrongly referred to as generative spiral, is either right- or left-oriented (**Figure 2**), in equal proportions, according to a great number of observations.

Shoot and Stem Anatomy

The anatomical characters of typical dicotyledon stems are observed in roses and correspond to a simple and regular arrangement of the tissues. The pith stands in the centre, surrounded by successive concentric layers of both kinds of sap-conducting tissues (xylem and phloem), with the cambium in between the cortical tissues and the epidermis. In the growing shoot, tissue formation (histogenesis) continues and accelerates the combination of three processes – cell division, growth and differentiation – that are already active, although to a lesser extent, during bud formation. In plants, the ability to divide is a property of the meristematic cells and is transmitted, without interruption, from one stem to another, and from one plant to another. Three different kinds of meristem function in the shoot: an apical meristem, which is constituted by a minute group of cells located at the very extremity of the shoot, and two lateral meristems, both longitudinally arranged. The lateral meristems are the cambial meristem, located between the two conductive tissues, and the phellogen, under the epidermis.

The apical divisions continuously add new cells to the files of cells already present at bud stage and generate new organs at the same time: leaves and axillary buds in a first vegetative phase, then floral parts in the reproductive phase. The cell dimensions increase, especially in length (elongation process), as newly formed cells acquire, through the differentiation process, different morphologies and functions that characterize the tissues they progressively constitute.

The tissues issuing from the apical divisions are called primary tissues. The pith is an example. It is the first-formed supporting tissue in the young shoot: its cell files are grouped in a rigid and turgid central column; its large polyedric cells are like braces and surround smaller cylindrical cells (**Figure 3**).

> In the primary bud the cells of the pith were isodiametric and equal in size. The cells were vital... and contained starch and sugars... During pith development, two types of cells were differentiated: small cells (20–80 μm in diameter), having thick walls with many pits and large cells (at least 80 μm in diameter), characterised by thin walls and few pits. The small cells appeared to form a network throughout the pith that was connected with the rays. They were vital and often accumulated starch and sugars. However, the large cells died and became filled with air... The fraction of the pith in the total diameter of a harvestable flower shoot decreased during shoot development (Marcelis van Acker 1994).

Simultaneously, at the sides of the pith column, the cell files of the cambial meristem develop and elongate. In this actively growing part, they constitute the procambium. Although they form a narrow structure at the apex, the procambium files expand below it by divisions and growth, first producing new procambium files on the sides of the previous ones, then files of the first conductive tissues, primary xylem, on their inner sides towards the pith, and primary phloem, on their outer sides towards the peripheral cortical tissues that are also forming. Along any of these files of primary tissues newborn cells can be observed at the apex, elongating ones below it and full-grown ones at the base. The upper limit of this basal part rises as the shoot elongates, until it catches up at the apex when stem growth is complete.

Together, the cell files of the procambium and of both of the primary conductive tissues it generates form the vascular bundles; a part of these bundles has a major role in the junction zones of stem and leaves. The bundles are separated from each other in the shoot-growing part. They join at their enlarged base, thus forming a continuous cylinder in the basal part, where elongation is already completed. This part of the meristem is the cambium *sensu stricto*. The cambial cells correspond to the mature state of the procambial cells and together form continuous files. The cambium generates so-called secondary vascular tissues (their constituent cells do not elongate any more). Hence, secondary xylem and phloem are formed on both sides of the meristem, between previously formed primary xylem and phloem (**Figure 3**). When observed in transverse sections of a flower stem, these four conductive tissues form cell clumps with trapezium shapes separated by wide bands of radial parenchyma that arise from the divisions in another part of the cambium located between the bundles. This radial parenchyma extends on both sides of the cambium cylinder and belongs to both secondary phloem and xylem (**Figure 3**).

At this stage, the rigidity of the flower stem is ensured by its lignified secondary xylem, while the pith cells are dead and filled with air. In comparison, the unlignified phloem forms groups of relatively fragile cells of small dimension: they are located between the rigid but brittle xylem, and the wide fibrous bands of the pericyclic sclerenchyma which seem to act as mechanical shrouds in the stem.

At the periphery of the stem, the radial parenchyma enlarges at the point of contact with the wide cells of a chlorophyllous parenchyma with variable width, itself surrounded by four to six cell layers of a non-chlorophyllous collenchyma. The smooth epidermis, covered by a wax cuticle, is punctuated with short longitudinal files of stomata located close to the most pigmented and thicker parts of the chlorophyllous parenchyma. The photosynthetic activity of this

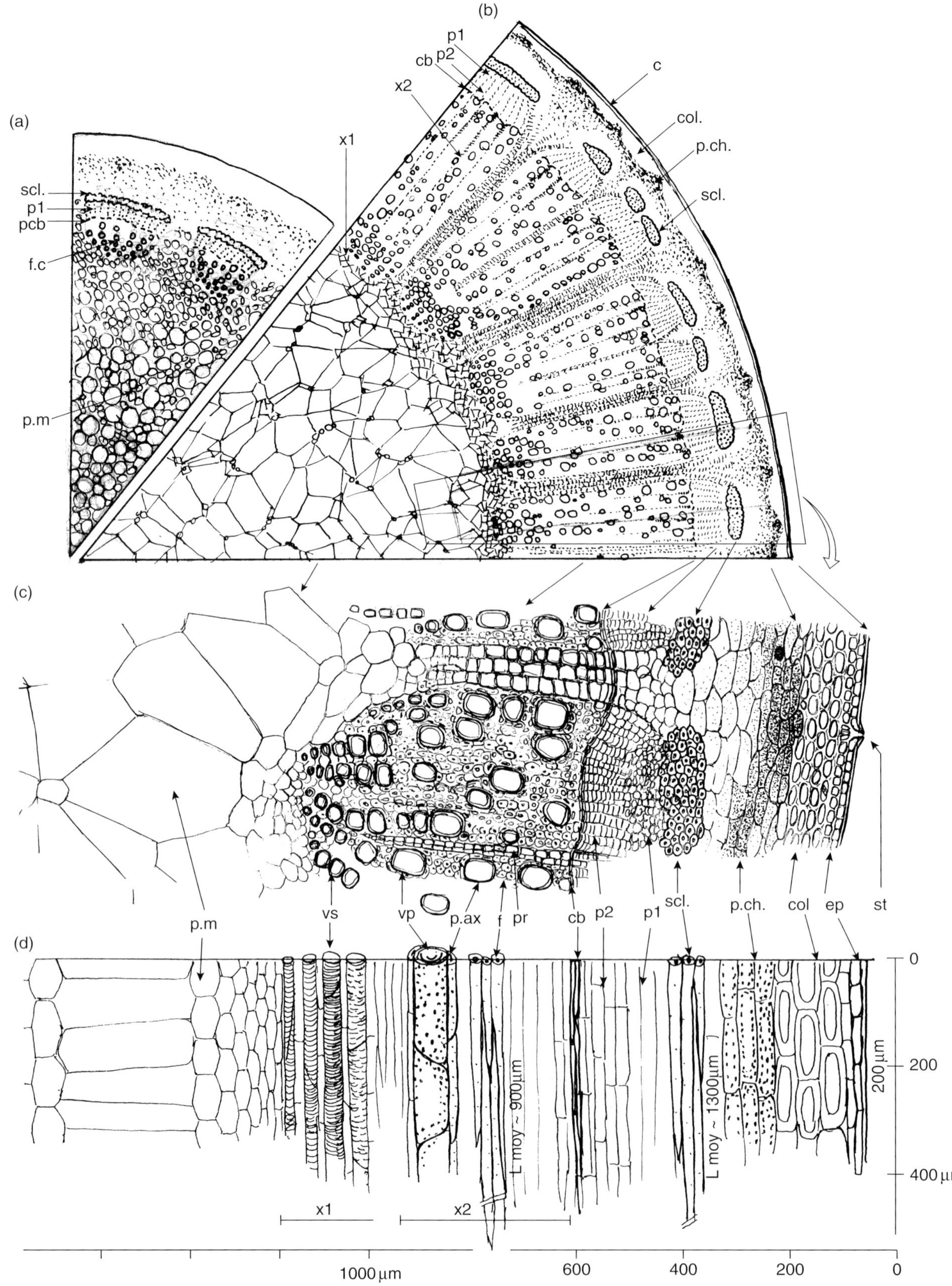

Figure 3 Arrangement of the tissues in a growing shoot (a) and in a flowering stem (b–d). (a) Partial transverse section of a shoot (diameter 4 mm). The conducting bundles are located at the periphery of the pith. (b) Partial transverse section of a stem (diameter 7 mm). (c) enlargement of the framed part of (b). (d) Schematic radial section showing the longitudinal cell files of the tissues (radial parenchyma not shown). f.c; primary conducting bundle; pcb, procambium; cb, cambium; c, cuticule; st, stomata; ep, epidermis; col, collenchyma; p.ch., chlorophyllous parenchyma; scl., pericyclic sclerenchyma; p 1, 2, primary, secondary phloem; x1, 2, primary, secondary xylem; vs, vp helical, pitted vessels; f, fibres; p.r, radial parenchyma; p.ax, axial parenchyma; p.m, medullary parenchyma.

parenchyma ends when the third stem meristem, the phellogen, begins to develop under the epidermis, forming new cell layers that will further constitute the suberized bark of the mature stem.

Tissue Connections Between the Stem and Adjacent Organs

Primary tissues each extend as a continuum between developing organs. The extension of secondary tissues is similar, but it mainly takes place in perennial organs and scarcely, or not at all, in leaves and inflorescences.

At the level of transverse connecting areas between two perennial organs or between a perennial and a deciduous organ, there is a functional continuity of tissue. However, local modifications in the assembly and morphology of connecting cells are often observed.

Connection Between the Stem and Leaves

The sap flow between the stem and its associated leaves gives special importance to the xylem and phloem connections.

The xylem vessels consist of files of dead cylindrical cells. Only the lateral cell walls remain, they are sealed and arranged lengthwise in the axis (**Figure 3**). There are more than 1000 of these vessels in a flower stem, and several hundreds, with a slightly different morphology, in the bundles of its primary xylem. As primary xylem is the only xylem formed in the young absorbing roots and in the leaves, and as it is no longer functional in the mature roots and the flowering branches, the water flows from the bottom to the top first in the primary xylem vessels, then in the secondary xylem, then again in primary xylem, through lateral transfers. As soon as the growing shoot apex is initiated, every young leaf diverts to its benefit a group of three developing vascular bundles, named 'foliar traces'. The three foliar traces develop and mature at the same time as the leaf. They appear at the level of each leaf insertion, as revealed by the superficial dissection of a mature stem. These three bundles group the primary phloem that emerges from its sclerenchyma sheath, the primary xylem that emerges from an opening in the secondary xylem, and, located between these two tissues, a meristematic layer that remains at the procambial stage and continues the cambium (**Figure 4**). In the petiole, the foliar traces are grouped together and divide again into several small bundles in the direction of the leaflets.

The phloem sap first flows from the adult leaves to the stem and from the stem towards the growing organs, leaves, roots and flowers (**Figure 5**). While entering the stem, it often releases some of its sugar content to the cells of the rays adjacent to the foliar traces, where important starch reserves accumulate.

The contact areas between the perennial stem and the deciduous petioles correspond to future abscission zones. At this level, the cell walls are barely lignified and the sclerenchyma that is present on both sides is interrupted. The same thing occurs between the stem and the peduncle.

Connection Between the Stem and Flower Peduncle

The flower peduncle is located between the uppermost leaves of the stem and the receptacle. When rose flowers are withered, the peduncle dries up during the next season and blackens down to its abscission zone, at the junction with the stem that remains alive and green.

The transverse plane of the future abscission zone represents an important anatomical transition between both organs: it corresponds to the upper limit of the secondary conductive tissues (**Figure 4d–g**).

By using the new technique of vascular microcasting, it appears that a part of the primary xylem vessels changes morphology while going through this zone, as is also the case at the level of the emerging point of the foliar traces and, more generally, when any abscission zone has to be crossed: these vessels are formed by elements with pitted cell walls in the stem and spiral thickening of cell walls in the peduncle.

Connection Between Two Axes

The disposition of the tissues is identical in all axes. Between the main axis and its axillary axis, the only observable difference lies in the development stage of the secondary tissues, which is more advanced in the bearing axis from between 10 days and several months, depending on whether the rose bush is continuously grown under glass or outdoors and then submitted to the seasons. In both cases, the flower-bearing axis is cut above the axillary, at harvest. Hence, at the connection level, a physiological weakening of the remaining section of the bearing axis and of its cambium occurs between this point and the cut surface. In reaction, a healing tissue is produced by the axillary axis cambium that develops on both sides of the connection over the secondary xylem of the main axis (**Figure 4h–l**). Between the bud-break and the full development of the axillary shoot, its primary, then secondary tissues develop in continuity with those of the main axis; the primary xylem of the young axillary axis extends under the secondary xylem of the main axis, maintaining continuity with its primary xylem (**Figure 4j**).

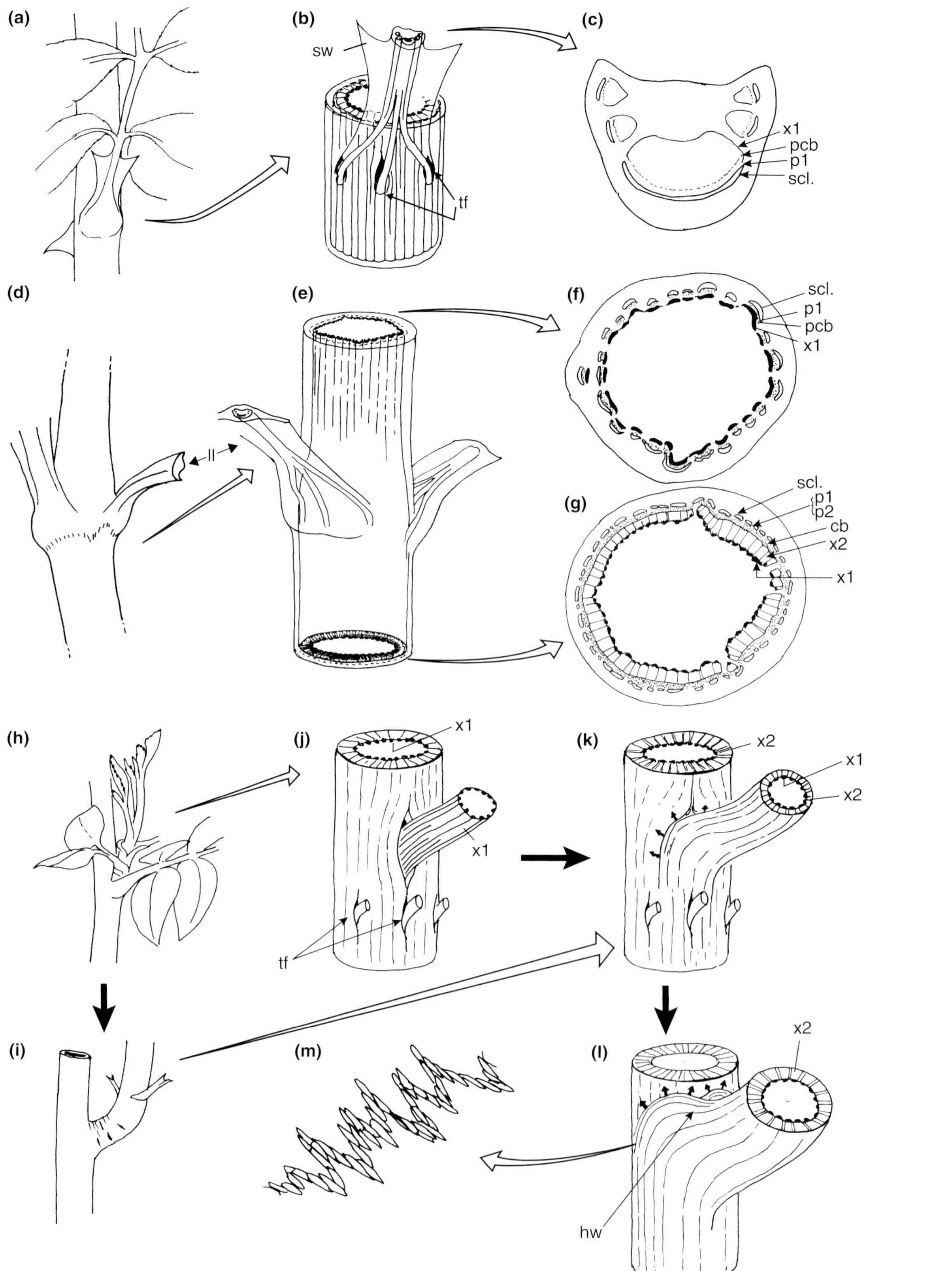

Figure 4 Connections between the stem and the leaf (**a, b**), the floral pedicel (**d, e**) and the supporting axis (**h**–**m**); transverse sections of the petiole over the stipular wings (**c**), of the stem under the last leaves (**g**) and of the pedicel (**f**). In the stem section, the secondary tissues form a continuous ring (**g**), whereas the primary conducting bundles of the pedicel are separated, but close to one another, and form a wavy line (**f**). As it grows (**h**–**i**), the axillary shoot forms its primary xylem (**j**) and phloem, then its secondary xylem (**k**) and phloem, connected to the similar tissues of its supporting axis. The secondary xylem of the shoot extends over that of the cut-bearing axis (**l**), forming healing wood. The vessels in the healing wood commonly show a zigzag shape (**m**). sw, stipular wing; ll, last leaves; tf, foliar traces; hw, healing wood: for other abbreviations, see **Figure 3**.

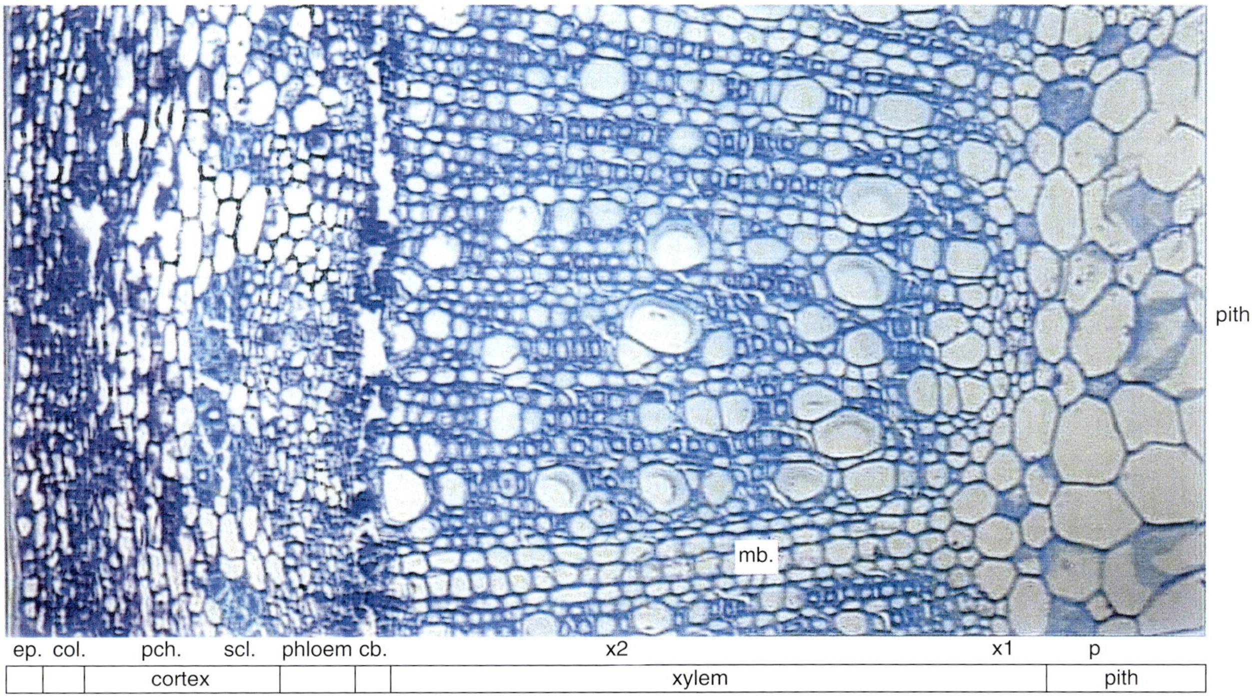

Figure 5 Light micrograph of a thin transverse section of the stem of *Rosa* 'Madelon' shows: epidermis (ep.), cortex, collenchyma (col.), chlorophyllous parenchyma (pch.), sclerenchyma (scl.), phloem, cambium (cb.), xylem secondary (x2) and primary xylem (x1) medullary rays (mb.) and pith structure (p). Reproduced from Da Cunha Costa JMR (2002) *The Role of the Leaf on the Dynamics of Growth and Rooting of Leafy Stem Cuttings of Rose*. PhD Thesis. Wageningen, The Netherlands: Wageningen University, with permission.

See also: **Growth Regulation**: Hormones in Growth and Development; Influences of Abiotic Factors in Growth and Development. **Morphology and Anatomy**: Leaves; Flower Development and the Abscission Zone; Roots; Pollen Grains and Tubes. **Postharvest Physiology**: Xylem Structure and Function in Cut Roses.

Further Reading

Darlington AB and Dixon MA (1991) The hydraulic architecture of roses (*Rosa hybrida*). *Canadian Journal of Botany* 69(4): 702–710.

Esau K (ed.) (1965) *Plant Anatomy*. Chichester: John Wiley.

Marcelis Van Acker CAM (1993) Morphological study of the formation and development of basal shoots in roses. *Scientiae Horticulturae – Amsterdam* 54(2): 143–152.

Marcelis Van Acker CAM (1994) Ontogeny of axillary buds and shoots in roses: leaf initiation and pith development. *Scientiae Horticulturae* 57: 111–122.

Marcelis Van Acker CAM and Scholten HJ (1995) Development of axillary buds of rose in-vitro. *Scientiae Horticulturae-Amsterdam* 63(1–2): 47–55.

Rouffa AS and Gunkel JE (1951) A comparative study of vegatative shoot apices in the Rosaceae. *American Journal of Botany* 38: 290–300.

Rouffa AS and Gunkel JE (1951) Leaf initiation, origin and pattern of pith development in the Rosaceae. *American Journal of Botany* 38: 290–300.

Vandoorn WG and Reid MS (1995) Vascular occlusion in stems of cut rose flowers exposed to air – role of xylem anatomy and rates of transpiration. *Physiologia Plantarum* 93(4): 624–629.

Leaves

S Torre, Agricultural University of Norway, Aas, Norway

Introduction

Leaves are perhaps the most obvious and well-recognized parts of higher plants. Their morphological characters have been much exploited as a taxonomic identifier for a large proportion of plants, including for the genus *Rosa*. The leaves have a variety of functions. Most important, they are the primary photosynthetic structures. They control the entry of CO_2 and the exit of water, and are thus critical for growth and survival. Rose leaf morphology and structure has been studied in relation to postharvest physiology, micropropagation and rooting of leafy stem cuttings, because leaves control the loss of water, and are the primary source of carbohydrates, mineral nutrients and hormones.

In roses, the leaves also have an ornamental value. Due to the rose's popularity as an ornamental plant, morphological characters like leaf shape, size and colour are important qualities. The genus *Rosa* is broad indeed, and leaves of a rose can show great structural and morphological diversity depending on cultivars and their locations on the shoot. Leaves can also be conditioned and modified by the environment in which they develop. This article will cover general aspects of leaf morphology and cellular structure of modern roses, primary roses for greenhouse production.

General Rose Leaf Morphology

Roses have compound leaves. This means they have individual laminar units borne on a rachis (**Figure 1**). Thorns may be borne on the rachis. A compound leaf may act to improve heat transfer efficiency in a plant by allowing better air exchange over the leaf surface and it may reduce herbivore damage. In roses, the compound leaves are pinnate and multiple orders of the structure occur. With only one exception (*Rosa persica*), the leaves of the roses are always odd-pinnate. Hybrid Teas usually have five leaflets and Floribundas mostly five to seven. The wild species normally have seven to nine leaflets. Many Asiatic species often have 13–15 (up to 19) leaflets while a few American species (*R. stellata*, *R. minutifolia*) have only three small leaflets. Where the leaf stalk meets the stem, it has a pair of rudimentary green growths known as stipules (**Figure 1**).

Some species also have bracts, which are modified and reduced leaves. They occur in the inflorescence, on the pedicle or on the peduncle, and are simple, never pinnate. Leaf morphology varies within the same shoot, depending on the position along the stem axis. In greenhouse-grown cut roses three groups of leaves can in general be distinguished: (1) uppermost leaves beneath the flower with one or three leaflets, (2) main group of six to eight middle leaves with five to seven leaflets, and (3) the lower leaves with one or three leaflets or bractlike leaves without leaf blades of the shoot base (**Figure 2**).

The leaflets have a typical shape. They are more or less ovate or elliptical, with a rounded, wedgeform or slightly heart-shaped base. The blade of the leaflets is mostly tough and leathery. The size and shape of a leaflet is a product of the number of cells and the expansion of the cells, and as already mentioned, dependent on environmental conditions. Rose leaf area is found to be affected by temperature, light level, plant water status, water availability, salinity, CO_2 supplementation and air humidity, depending on the scale of

Figure 1 An assortment of rose leaves taken from the middle part of the stems, demonstrating the morphological diversity of the genus *Rosa*. Note the stipules, which are small leaflike appendages adnate to both sides of the base of the leaf stalk. (a) *R.* × *hybrida* 'Syr'; (b) *R. micrugosa*; (c) *R. palustris*; (d) *R. rugosa*; (e) *R. forrestiana*; (f) *R. pimpinellifolia*; (g) *R. helenae* 'Hybrida'; (h) *R. alba*; (i) *R. multibracteata*.

the survey. The thickness of a rose leaf is also affected by environmental conditions during growth, especially temperature, light and air humidity (**Table 1**).

Leaf area is important regarding photosynthesis and growth of a rose plant. Further, leaf area influences root formation and growth of leafy stem rose cuttings. It may also directly influence growth of the primary shoot of a cutting because the initial growth of shoots in rose plants depends on assimilates supplied by mature leaves. It has recently been found that the original leaf area of a cutting is a good indicator for growth of roots from cuttings. The presence of the leaf during the first week of propagation was critical for survival and its removal caused stem rot. This was caused by low carbohydrate concentrations. The leaf area may also influence postharvest life of a cut rose and the survival rate of *in vitro* cultured rose plantlets after transfer to greenhouse conditions. This is, however, mainly caused by water stress (high area leads to high water loss). More important are however the presence of functional or nonfunctional stomata and the reaction of stomata to a developing water stress.

Figure 2 Types of leaves usually found in *Rosa* × *hybrida* 'Baccará'®. (a) Bract; (b) one-leaflet leaf; (c) two-leaflet leaf; (d) three-leaflet leaf; (e) five-leaflet leaf; (f) seven-leaflet leaf. (Reproduced with permission from Moe R (1971) Factors affecting flower abortion and malformation in roses. *Physiologia Plantarum* 24: 291–300.)

The Cellular Basis of Rose Leaf Structure

The cellular structure of a rose leaf is faintly typical of a dicotyledonous leaf. It comprises basically three types of tissue systems: epidermis, mesophyll and vascular tissue (**Figures 3** and **4**). First, cells of an adaxial epidermis form the upper boundary of the leaf. Immediately below are layers of palisade mesophyll cells with relatively small air spaces in between. Below this is a layer of spongy mesophyll cells, which contains large air spaces that are next to the stomata in the lowermost layer of the leaf, the abaxial epidermis. Within these layers, other specialized cell types are present, the most important of which is the vascular system of the leaf. This develops as a layer approximately midway between the two layers of the epidermis, and forms a reticulate network throughout the lamina. The vascular strand consists of a xylem and a phloem element, with the phloem on the abaxial side. Below, the different cell types will be described in more detail.

Epidermis

Epidermis of a rose leaf is composed of various types of cells: epidermal cells, composing the main mass of the epidermal tissue, the guard cell of the stomata and epidermal appendages termed trichomes and emergences. The size and shape of the cells, the thickness of the walls and the distribution and number of specialized cells per area may all vary (**Table 2**). The epidermal cells usually have wavy anticlinal walls (**Figure 3a**). The waviness of the walls is variable. It is not known what induces this shape during development, but it is known to be dependent on environmental condition and location. Often the undulations occur only on the lower side of the leaf or are more pronounced here than on the upper side. Epidermal cells are also found to be more undulated in greenhouse-grown plants than in micropropagated plants.

The epidermal cell wall is thick and lignified, and measures some 1–2 μm. Outside the outer cellulose wall there is a cuticle layer, which also varies in thickness. In plants cultured *in vitro* it is 0.04 μm thick compared to about 0.3 μm in greenhouse-grown plants. The cuticle surface is covered with wax

Table 1 Leaf characteristics of different rose species and cultivars

Cultivar/species	*Leaf thickness (μm)*	*Palisade cell size (μm^2)*	*Number of palisade cells per mm^2*	*Number of mesophyll cells per mm^2*	*Reference*
R. × *hybrida* 'Mme Isaac Pereire'	80–160[a]	144–500[a]	—	—	Johansson *et al.* (1992)
R. × *hybrida* 'Baroness'	140–167[a]		911–1003[a]	628–797[a]	Torre (1999)
R. × *hybrida* 'Aruba'	145–151	—	—	—	Torre (unpublished data)
R. × *hybrida* 'Souvenir'	152–166	—	—	—	Torre (unpublished data)
R. bracteata	158.3	—	—	—	Ueda *et al.* (2000)
R. rugosa	186.7	—	—	—	Ueda *et al.* (2000)

[a] The intervals respresenting different air humidities during growth, where the lowest value is found on leaves developed at the highest air humidity and the highest value found on leaves developed at the lowest air humidity.

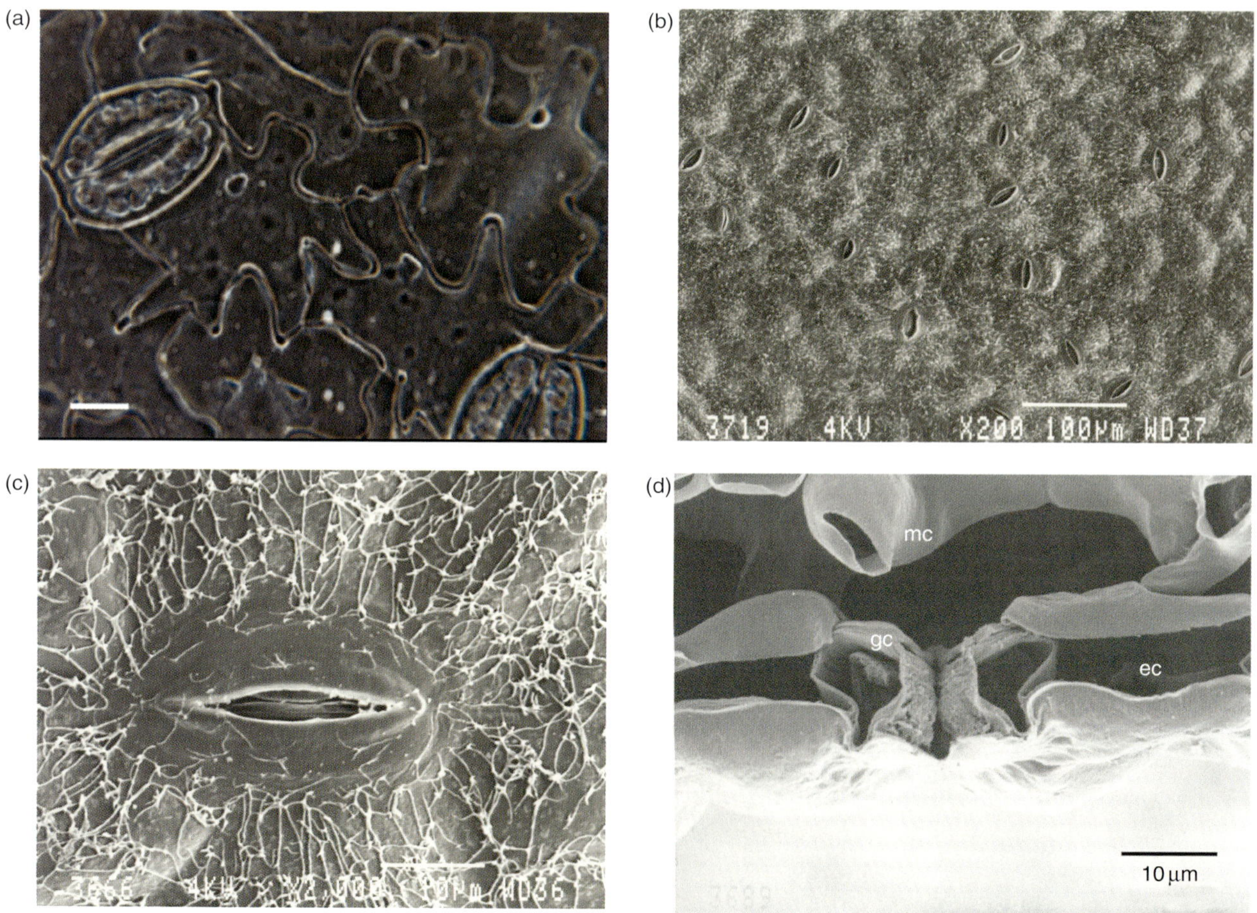

Figure 3 Epidermal cells of a rose leaf. (a) Abaxial epidermis peel from *R.* × *hybrida* 'Baroness' shows highly irregularly shaped epidermal pavement cells and stomata each consisting of a pair of guard cells. Note the starch grains in the guard cells. Scale bar = 10 μm. (b) Scanning electron micrograph of the abaxial epidermal surface of a 'Baroness' leaf showing wax cover and numerous stomata. (c) Scanning electron micrograph of the abaxial epidermal surface of a 'Baroness' leaf showing a closed stoma and wax flakes with bare areas in between. (d) Scanning electron micrograph of a cross-section of two guard cells and epidermal cells of 'Baroness' stomata. Note the ledges of wall material on the upper side. ec, epidermal cell; gc, guard cell; mc, mesophyll cell.

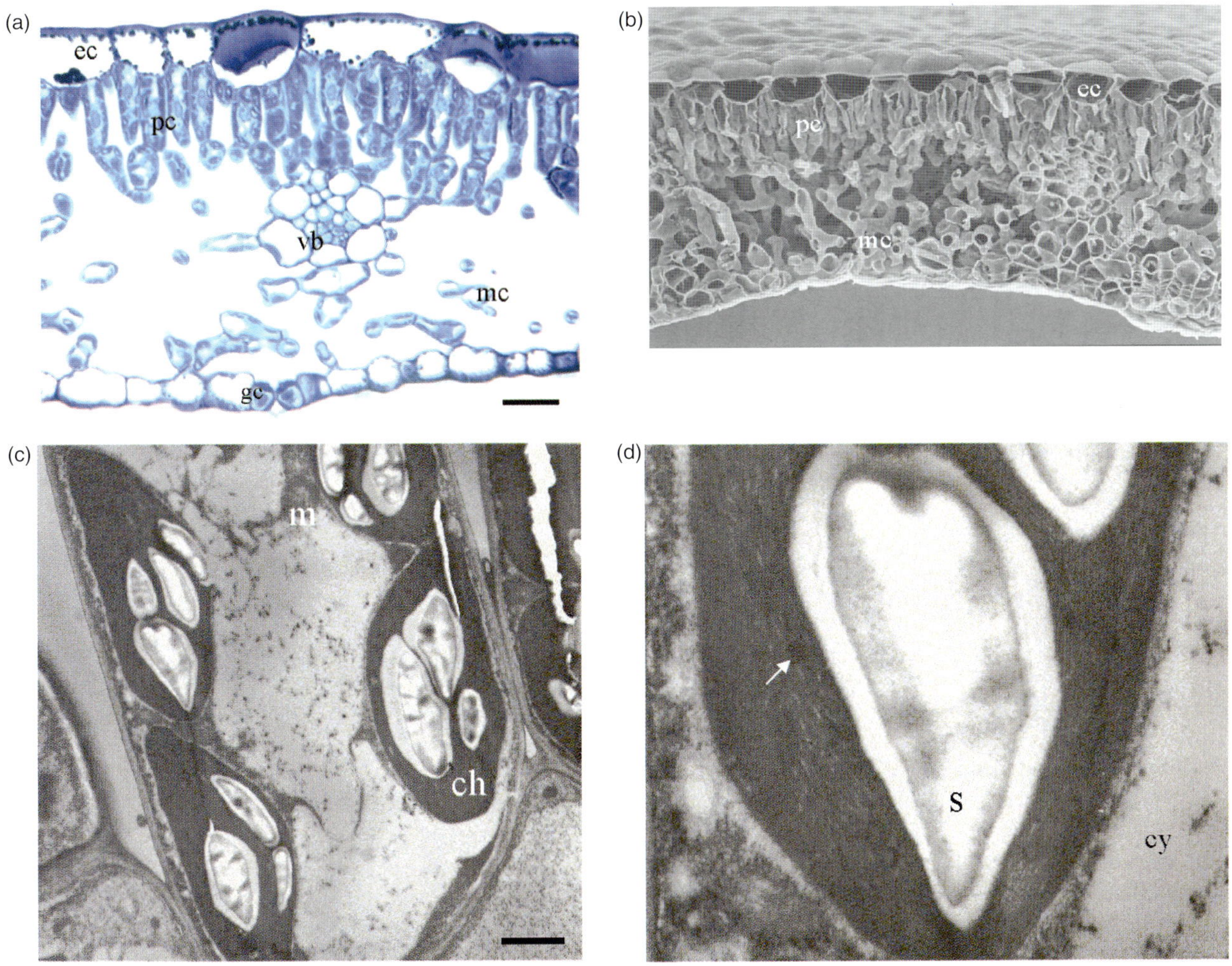

Figure 4 Rose leaf stucture. (a) Light photomicrograph of a longitudal cross-section through a *R.* × *hybrida* 'Baroness' leaf showing the various cell layers with the adaxial epidermis uppermost and the abaxial epidermis on the lower surface. Scale bar = 25 μm. (b) Scanning electron micrograph of a longitudal section of a 'Baroness' leaf showing the various cell types. Note the large air gaps between both the palisade cells and the spongy mesophyll cells. (c) Transmission electron micrographs of a palisade parenchyma cell from a 'Baroness' leaf. Scale bar = 2 μm. (d) Chloroplasts with lamellar structure from the same 'Baroness' leaf. Scale bar = 400 nm. ch, chloroplast; cy, cytoplasm; ec, epidermal cell; gc, guard cell; m, mitochondria; mc, mesophyll cell; pc, palisade cell; vb, vascular bundle; s, starch; ⟶, plastoglobuli.

(epicuticular wax) on both the adaxial and the abaxial side. Usually the highest amount of wax is found on the abaxial side (**Figure 3b**). This has two effects: diminution of water loss and enhancement of reflection of the sunlight, thus protecting the plant from too much heat. In roses, the epicuticular wax can vary both in amount and in structure, depending on the stage, species and environment in which it has developed. In greenhouse-grown roses, the deposition of wax may appear in the forms of platelets or scales with bare areas in between (**Figure 3c**). In some species of *Rosa*, wax structures may be used as a taxonomic identifier. In the dog roses (*Rosa*, section *Caninae*), microscopy studies revealed that all taxa of the subsection *Rubiginosae* are characterized by a granulare type of epicuticular waxes, whereas members from the other subsection form triangular rodlets (i.e. *R. canina*, *R. glauca*). Environmental factors can influence the development of the cuticle and wax of rose leaves. In micropropagated material the covering is usually less pronounced than in greenhouse-grown plants.

Stomata occur on the abaxial surface of rose leaves and belong to the type anomocytic (also known as the ranunculaceous or agenous type). In this group the guard cells are surrounded by cells indistinguishable from other epidermal cells, i.e. no subsidiary cells are present (**Figure 3a**). The guard cells of roses are generally kidney-shaped, forming elliptical stomata and have ledges of wall material on the upper side (**Figure 3d**). Guard cells may occur at the same level as the adjacent epidermal cells, or they may protrude

Table 2 Epidermal characteristics of different rose cultivars

Cultivar	*Epidermal cell length (μm)*	*Epidermal thickness (μm)*	*Stomata frequency (mm²)*	*Stomata length (μm)*	*Hydathode number*	*Reference*
R. × *hybrida* 'Mme G. Delbard'®	—	—	5.0–9.7[a]	—	—	Ghashghaie *et al.* (1992)
R. multiflora 'Montse'	26–72[a]	—	95–193[a]	—	0–175[a] per mm²	Capellades *et al.* (1990)
R. × *hybrida* 'Baroness'®	—	13–19[a]	39–52[a]	26–36[a]	0	Torre (1999)
R. × *hybrida* 'Souvenir'	—	—	32–45[a]	—	0	Torre (unpublished data)
R. × *hybrida* 'Aruba'	—	—	45–56[a]	—	0	Torre (unpublished data)
R. × *hybrida* 'Mme Isaac Pereire'	18–80[a]	16–60[a]			—	Johansson *et al.* (1992)
R. × *hybrida* 'Queen Elizabeth'®					20–35[a] per leaf apices	Donnelly and Skelton (1989)
R. × *hybrida* 'Grand Prix'			96–116			Sevelius and Hyttinen (2001)
R. × *hybrida* 'Tineke'			94–109			Sevelius and Hyttinen (2001)

[a] The intervals respresenting different air humidities during growth, where the lowest value is found on leaves developed at the highest air humidity and the highest value found on leaves developed at the lowest air humidity.

above or be sunken below the surface of the epidermis, dependent on the environment in which they develop.

The size and the number of stomata are dependent on stage, cultivar and growing conditions (**Table 2**). Experiments with rose leaves grown under different humidities have shown that stomatal frequency decreases with an increase in relative air humidity. In contrast to the basic rose epidermal cells, the guard cells contain chloroplasts and starch. In addition, the cytoplasm of guard cells contains a full assortment of organelles and cellular components. Roses may also form hydathodes, which are structures through which water is secreted in liquid form (guttation) (**Table 2**). They are formed during the early stages of leaf development and can often be found at the end of vascular bundles. Their guard cells still look like those of normal stoma complexes, but they are very small and spherical and they cannot be closed. The presence of hydathodes is very much dependent on the environment, especially relative humidity. In the cultivar 'Montse' it was found that the relative air humidity had to be close to 100% before any hydathodes developed. However, studies on other cultivars showed hydathodes in both tissue-cultured plantlets and greenhouse-grown roses. In the latter case, the observation was dependent on age. Probably, the development of hydathodes is dependent on climate, cultivar and age of the leaf.

Epidermal appendages on the leaves are common in roses. Common types of appendages are called trichomes. In roses, glandular trichomes, often called glands, are common and usually found at the edges of leaves and in proximity to the vascular strand. Glands may be sessile or stalked. They are involved in the secretion of sticky substances, frequently a mixture of terpenes and mucilage. The use of trichomes in taxonomy of roses is well known. *Rosa rugosa* exudes a large amount of syruplike droplets from mushroom-shaped glands located on the lower surface of the running leaf veins. Many of the hybrid rugosas have glandular trichomes, while others possess only sparse or some completely lack glandular trichomes. Also more massive structures termed emergences (thorns) are common. Thorns are common on the leaf margins and consist of epidermal as well as subepidermal tissue.

Mesophyll Parenchyma

Mesophyll is usually specialized as a photosynthetic tissue. In roses, as in many plants, particularly in dicotyledons, the mesophyll is differentiated into palisade parenchyma and spongy parenchyma (**Figure 4**). The palisade cells are always found on the adaxial surface of the leaf. They have an elongated shape, and in cross-section of the leaf they are rod-shaped and appear to be arranged in rows, while in a section parallel to the leaf surface these cells are seen to be rounded and separated or only slightly attached to one another. The cells of the palisade parenchyma may be arranged in one or more layers. In the latter case the length of the cells in the different rows may be equal, or they may become shorter towards the centre of the mesophyll. The number of palisade layers, the size of the palisade cells and the density of the cell structure depend largely upon light intensity, CO_2 content of the atmosphere and air humidity. Leaves that have been exposed to low air humidity and large quantities of light during ontogenesis will usually results in a multilayered palisade parenchyma. In contrast, shaded leaves and high air humidity usually give rise to a single-layered palisade parenchyma. The age of the leaves is also of importance. Old leaves usually shows a reduction in palisade parenchyma thickness.

The spongy parenchyma appears less regular. Spongy parenchyma shows many forms of cells, either nearly isodiametric, or elongated in the same direction as the palisade cells and connected with each other by lateral extensions of various lengths. Since the mesophyll tissue is the major photosynthetic tissue in roses a large majority of the chloroplasts are found here, both in the palisade cells and mesophyll cells (**Figure 4**). However, the palisade cells usually contain more chloroplasts than the spongy cells. The ultrastructure of a palisade cell is shown in **Figure 4c** and **d**. From other plants it is well known that shade induces smaller and thicker leaves with bigger cells than are found in leaves in the sun. Also the number and morphology of chloroplast will be affected. Clear anatomical or morphological differences are not found in rose leaves grown under low and high light levels. However, this is of course dependent on the scale of the survey. Probably, an extreme difference in light level during the development of the leaf is needed before any anatomical differences can be recognized. A similar situation is found for relative air humidity: the leaf has to develop under high humidity before differences can be seen in mesophyll structure. Very high air humidity (>90%) throughout the development period tends to give fewer mesophyll cells and higher intercellular air space compared to moderate air humidity (70%).

Vascular Tissue in the Leaf

The arrangement of the vascular bundles of a rose leaf is called reticulate, because the main branches of the vascular system form a network (**Figure 5a**). The veins are of different sizes depending on the degree of branching. A cross-section through the central vein (midrib) is shown in **Figure 5b**. The bundle is crescent-shaped, with xylem on the upper part and phloem on the lower part supported by ground tissue. From the central vein smaller vein branch and show similar anatomic characters (**Figure 4a**). The density of veins in leaves is found to be different depending on environment. In leaves developed at high relative air humidity (>90%), the distance between the veins becomes larger compared to the situation at low relative air humidity (<70%), where the distance between the veins became smaller.

See also: **Cell Tissue and Organ Culture**: Micropropagation; Rooting of Micropropagated Shoots. **Morphology and Anatomy**: Shoots and Stems; Flower Development and the Abscission Zone; Roots; Pollen Grains and Tubes. **Postharvest Physiology**: Xylem Structure and Function in Cut Roses; Packing Systems and Precooling. **Production and Marketing**: Cut Rose Production.

(a)

(b)

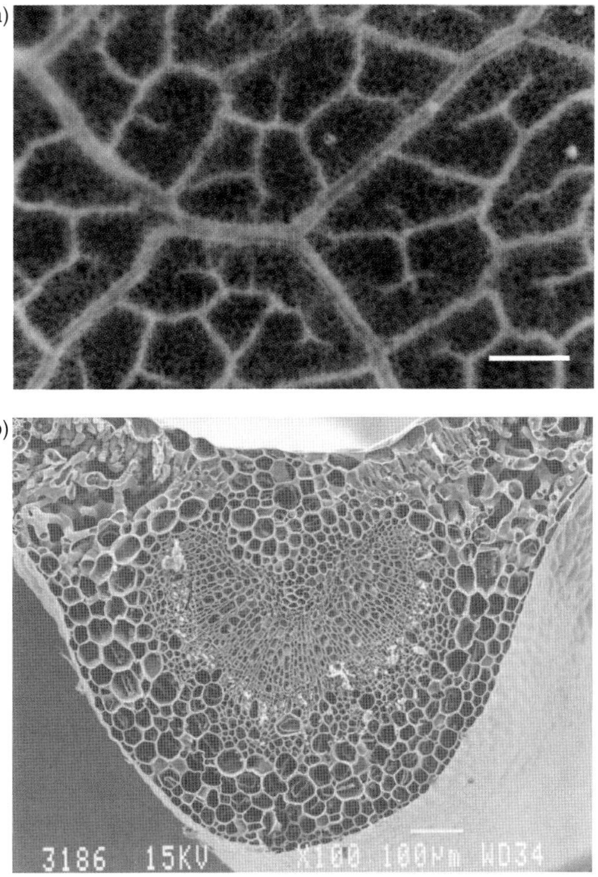

Figure 5 Vascular tissue. (a) Portions of a cleared *R.* × *hybrida* 'Baroness' leaf seen in surface view. The photograph was taken interveinal close to the central vein; (b) Scanning electron micrographs of a cross-section of the central vein from a 'Baroness' leaf. Scale bar = X µm.

Further Reading

Björkman O (1981) Responses to different quantum flux densities. In: Lange OL, Nobel PS, Osmond CB and Ziegel H (eds.) Physiological plant ecology. I. Responses to the physical environment. *Encyclopedia of Plant Physiology*, vol. 12A, pp. 57–107. Berlin: Springer-Verlag.

Capellades M, Fontarnau R, Carulla C and Debergh P (1990) Environment influences anatomy of stomata and epidermal cells in tissue-cultured *Rosa multiflora*. *Journal of the American Society for Horticultural Science* 115: 141–145.

Costa J-MR-C (2002) *The Role of the Leaf in Growth Dynamics and Rooting of Leafy Stem Cuttings of Rose*. PhD Thesis, University of Wageningen, Wageningen, The Netherlands.

Donnelly DJ and Skelton FE (1989) Comparison of hydathode structure in micropropagated plantlets and greenhouse-grown 'Queen Elizabeth' rose plants. *Journal of the American Society for Horticultural Science* 114: 841–846.

Fahn A (1990) The leaf. *Plant Anatomy*, 4th edn, pp. 187–221. Oxford: Pergamon Press.
Ghashghaie J, Brenckmann F and Saugier B (1992) Water relations and growth of rose plants cultured *in vitro* under various relative humidities. *Plant Cell Tissue and Organ Culture* 30: 51–57.
Johansson M, Kronestedt-Robards EC and Robards AW (1992) Rose leaf structure in relation to different stages of micropropagation. *Protoplasma* 166: 165–176.
Krüssmann G (1981) The leaf. In: Parey P (ed.) *Rosen, Rosen, Rosen* (English translation: *The Complete Book of Roses*, transl. Krüssmann G and Raban N.) pp. 238–239. Portland, OR: Timber Press.
Levitt J (1980) *Responses of Plants to Environmental Stresses*, 2nd edn, vol. 2, *Water, Radiation, Salt and Other Stresses*. New York: Academic Press.
Moe R (1971) Factors affecting flower abortion and malformation in Roses. *Physiologia Plantarum* 24: 291–300.
Sevelius N and Hyttinen T (2001) Selection of rose cultivars for low light greenhouse production by photosynthetic features. *Acta Horticulturae* 547: 159–166.
Short KC, Price L and Roberts AV (1981) Micropropagation of roses. In: Harkness J (ed.) *The Rose Annual*, St Albans, UK: Royal National Rose Society.
Sinha N (1997) Simple and compound leaves: reduction or multiplication? *Trends in Plant Science* 2: 365–403.
Torre S (1999) *The Role of Air Humidity and Calcium in Determining Postharvest Quality of Roses*. PhD Thesis, Agricultural University of Norway, Aas, Norway.
Willmer C and Fricker M (1996a) The distribution of stomata. In: Black M and Charlwood B (eds.) *Stomata*, pp. 12–30. London: Chapman & Hall.
Willmer C and Fricker M (1996b) The structure and development of stomata. In: Black M and Charlwood B (eds.) *Stomata*, pp. 36–82. London: Chapman & Hall.

Flower Development and the Abscission Zone

D Chimonidou, Agricultural Research Institute, Nicosia, Cyprus

Introduction

Recurrent-flowering roses are perennial woody shrubs, which continuously form new shoots (*see* **Growth Regulation:** Floral Induction). Each shoot has the potential to form a terminal flower. After harvesting a flowering shoot, the most distal axillary buds will sprout and develop into flowering shoots that produce the next flush of blooms.

Much research on the rose has been focused on the effects of various scion–rootstock combinations, harvesting procedures, growth regulations and environmental factors on flower yield. However, relatively little attention has been paid to axillary buds, although these buds are important in four aspects of the development of a rose plant: (1) when propagated vegetatively, an axillary bud gives rise to the above-ground part of a new rose plant; (2) axillary buds are assumed to give rise to basal shoots, which determine the potential flower production; (3) the degree of branching of shoots depends on the growth of axillary buds, which is also an important determinant of potential flower production; (4) axillary buds give rise to the flowering shoots, which determine the actual flower production.

Since axillary buds form the basis for the potential flower production of a rose plant, knowledge of the development and internal structure of an axillary bud will lead to a better understanding of the growth of a plant. Furthermore, it enables us to define cultural practices.

In this article particular emphasis will be given to the study of the axillary buds of greenhouse roses, from their initiation up to the stage when the flower has its full complement of floral parts. In order to do so, it is necessary to define the stages of rose development, categorize them into a scale and correlate the external and internal faces of the bud.

Stages of Development of a Shoot

Developmental stages, starting with new axillary buds after marketable flowering stems had been removed, were defined and described (**Table 1** and **Figure 1**) in 'Madelon'. Buds were chosen on the basis of the diameter of the subtending shoot, so that the shoots of the selected buds would have approximately the same diameter – 1 cm. Daily measurements on flower bud development were recorded. Macroscopic records of the different stages of development were based on shoot elongation and the appearance of the leaves (three-leaflet and five-leaflet leaves), as well as on the flower bud. The different stages of rose bud development were recorded externally, and correlated with the records of the internal face viewed by scanning electron microscopy (SEM) (**Figure 1**). The SEM techniques used are described in Appendix 1.

The aim was to record the development of the apex from the vegetative condition to the stage when the flower had its full complement of floral parts. Representative stages were then selected and allocated the scale shown in **Table 1**.

Table 1 Stages of development from bud break to the development of a flower bud (see **Figure 1** for illustration)

Stage	*External indicators*	*Internal indicators*
1	Bud expanded before elongation (A_1)	First three-leaflet leaf starts to form (A_2)
2	Bud elongated more than 20 mm but first three-leaflet leaf not yet visible (B_1)	All leaf primordia well formed; apical dome is flattened with surface undulating (B_2)
3	First three-leaflet leaf expanded but second leaf not yet visible (C_1)	Floral differentiation started: sepals well formed and first petal primordial initiated (C_2)
4	Second three-leaflet leaf expanded; first five-leaflet leaf not expanded (D_1)	Second cycle of petal primordial initiated (D_2)
5	First five-leaflet leaf expanded; second leaf not clearly visible (E_1)	All petals initiated; stamen primordial initiated in centre of receptacle (E_2)
6	The second five-leaflet leaf expanded (F_1)	Stamens well developed; carpel primordial initiated (F_2)
7	All leaves expanded; flower bud visible but not yet separated (G_1)	Carpel primordial initiated in centre of receptacle (G_2)
8	Flower bud 5–7 mm diameter (H_1)	Carpel primordia well developed (H_2)

The stages of development are adapted from those used by Moe R and Kristofferson T (1969) The effect of temperature and light on growth and flowering of *Rosa* 'Baccara' in greenhouses. *Acta Horticulturae* 14: 157–163 and Horridge JS and Cockshull KE (1974) Flower initiation and development in the glasshouse rose. *Scientia Horticulturae* 2: 273–284.

Flower Bud Development

Flower Initiation

Flower initiation in recurrent-flowering roses is considered to be self-inductive, since this process is not obviously regulated by the environment. This implies that if an endogenous flower-inducing agent is required, it is synthesized with no external stimulus and is, therefore, present in all plants at all times. It was demonstrated in 1974 that both stem extension and flower initiation occurred in axillary buds soon after the subtending shoot had been removed. The buds are not competent to initiate a flower while their extension growth is inhibited by the upper parts of the shoot (apical dominance) (*see* **Growth Regulation**: Floral Induction).

Flower initiation takes place shortly after the start of axillary bud growth. When the flowering shoot reaches a length of approximately 20 mm (stage 2), the apical dome is flattened with its surface undulating. This stage can be considered as the transition stage from vegetative to reproductive development. Stage 3 (sepals well formed and initiation of first cycle of petal primordial), which is more readily identified than stage 2, is the first indication that an apex is commencing reproductive development.

In 'Madelon', vegetative apices were never observed on shoots longer than 35 mm. Thus a rose apex is ready to initiate a flower soon after rapid cell division commences and the endogenous flowering stimulus is apparently present throughout the plant. This is also supported by reports on four rose cultivars – 'Baccará'®, 'Pink Sensation', 'Super Star' and 'Sonia' – under different temperature regimes. Except after prolonged low-temperature treatment, vegetative apices are never observed on shoots longer than 50 mm. Moreover, it has been reported that, 10 days after the release of the primary bud of 'Sweet Promise' and at a stem length of approximately 50 mm, all leaves and a terminal flower bud are formed.

Stages of Development Critically Affected by Water Supply

Water is essential for plant growth and optimum supply of water is vital in species of commercial interest, such as roses. A reduction in water availability has detrimental effects on the quantity and quality of the production. It is important to identify the critical periods during which water stress will have the most pronounced detrimental effect on growth and yield of a crop. In order to investigate this experimentally, it is essential to observe the changes in growth stages, so that water stress can be applied at exactly the right time to the investigated plants at different stages of development.

In experiments carried out with 'Madelon', plants were pruned to leave two five-leaflet leaves on each cane. Buds subtended by the uppermost leaf on canes of similar-diameter bud are observed as they sprout and develop to commercial flowers. Six treatments were applied at defined stages, as described in **Table 2**. When the meristems were at the required stages of development, as determined by the external stage of the shoot, water stress was induced by suspending irrigation for 5–6 days, normally resulting in a tensiometer reading of −0.09 to −0.10 MPa. At this soil moisture, leaf stomatal resistance is 40–45 s cm^{-1} (**Figure 2**). After the drought treatment, the plants were watered again to restore the soil tension to −0.03 to −0.04 MPa, and thereafter a daily irrigation was applied, so as to maintain this soil moisture level.

Figure 1 Stages 1–8 (external and internal faces) of development from the axillary bud to the flower bud (see **Table 1** for definitions of the stages). Internal faces are scanning electron microscopy photographs. B, earliest bud break; 3L, three-leaflet leaf; Sh, shoot elongation; H, hairs; 5L, five-leaflet leaf; Fb, flower bud; C, carpel; P, petal; R, receptacle; S, sepal; St, stamen. (Reproduced with permission from ISHS.)

Figure 1 Continued.

In treatment 1 (**Table 2**), water stress applied at the stage of leaf primordium formation (stage 1) resulted in a delay in the rose production cycle but did not have any negative effects on the quality of the flower shoot or the flower bud (**Table 3**). Treatment 2, prior to petal initiation, is more sensitive. Water stress at this stage affected the quality of flower buds (reduction of the number of well-formed petals and of the height of the flower bud). Water stress imposed in treatment 3, prior to stamen initiation, was the most damaging to rose development, affecting the quantity (up to 70% reduction of production, **Table 4**) and the quality of the flowering shoots and buds (reduction in stem length and fresh weight of the flowering shoot, diameter and height of bud and the number of well-formed petals of the flower bud; **Table 3**, **Figure 3**).

Water stress applied in treatments 4 and 5, after stamen formation and carpel formation, did not have any negative effect, either on the number of flowering shoots that reached the marketable stage, or on the quality of the flowers produced. The results were similar to those obtained in treatment 6, where irrigation was given throughout all stages of development.

Table 2 Water stress treatments defined in terms of stages of shoot development (**Table 1**) at which the stress was applied. The water stress was applied by suspending irrigation for 5–6 days, generating the soil tension up to −0.09 to −0.10 MPa. After the drought treatment, the plants were watered again to restore the soil tension to −0.03 to −0.04 MPa, and thereafter a daily irrigation was applied, so as to maintain this soil moisture level

Treatments	*Stages of shoot development at which stress was applied*
1	1
2	2 and 3
3	4 and 5
4	6 and 7
5	8
6 (control)	(no stress applied)

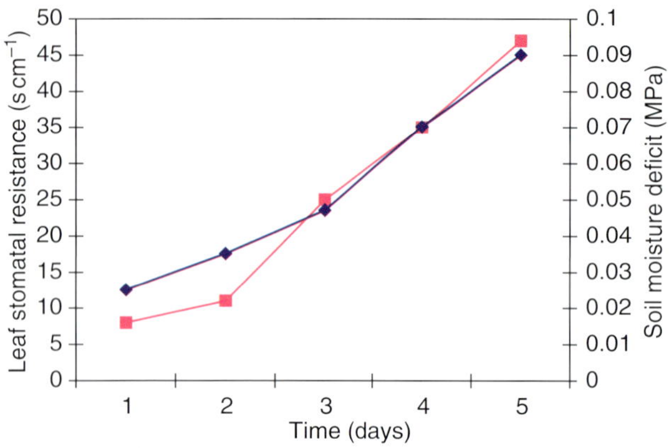

Figure 2 Drought stress applied in all treatments. Data points represent the means of 32 replications for soil moisture deficit (diamonds) and leaf stomatal resistance (squares) in relation to the days required to reach the level of water stress in summer.

The Effect of Irrigation and Shading on Flower Bud Development

Macroscopic and microscopic (SEM) observations were recorded in order to study any differences in bud development (at the first stages of rose development, from bud break to the stage of floral parts complement) under the different irrigation and shading treatments.

Macroscopic observations show that the stem length as well as the time taken to reach the different stages of rose development is affected by irrigation (as defined in **Table 5**) and shading.

Shading was introduced during April to July, when outside radiation was too high. In experiments starting in April, three levels of shading were used. The shading level 1 (S1) was a 39% reduction of global radiation outside the greenhouse; shading level 2 (S2) was a reduction of 62% and shading level 3 (S3) was a reduction of 79%. The global solar energy ($MJ\,m^{-2}\,s^{-1}$) outside during the 4 months (April to July) was 20.29, 19.83, 26.34 and 26.67 $MJ\,m^{-2}\,s^{-1}$ respectively.

Although there was no effect due to irrigation or shading on the first six stages of development

Table 3 Effect of water stress treatments on the characteristics of flowers that developed to the marketable stage

Variable	*Treatments*						
	1	*2*	*3*	*4*	*5*	*6*	*LSD*
Number of well-formed petals per flower	25.0	21.6	16.9	25.7	26.2	25.4	1.47
Number of distorted (small) petals per flower	1.8	2.6	2.6	1.7	1.3	1.1	0.78
Bud diameter (cm)	1.9	1.8	1.6	1.9	1.9	1.9	0.08
Bud length (cm)	2.7	2.5	1.9	2.9	2.7	2.8	0.14

LSD, least significant differences.

Table 4 Effect of water stress at different stages of bud development on the number of normal, malformed and blind shoots. Each of six irrigation treatments was applied to six plants in each of four replicates. Values represent the proportion of the theoretical maximum of 24 potential flowering shoots for each treatment during the three seasons

Treatments	*Number of flowers*		
	Normal	*Malformed*	*Blind shoots*
Winter			
1	0.67	0.08	0.25
2	0.42	0.25	0.33
3	0.29	0.33	0.33
4	0.58	0.08	0.33
5	0.67	0.08	0.25
6	0.67	0.08	0.25
LSD	0.14	0.13	0.11
Spring			
1	0.79	0.04	0.17
2	0.83	0.08	0.08
3	0.46	0.29	0.21
4	0.92	0.00	0.08
5	0.96	0.00	0.04
6	0.92	0.04	0.08
LSD	0.13	0.11	0.12
Summer			
1	0.92	0.00	0.08
2	0.83	0.00	0.12
3	0.67	0.25	0.08
4	0.83	0.12	0.08
5	1.00	0.00	0.00
6	0.92	0.00	0.08
LSD	0.15	0.07	0.13

LSD, least significant differences.

(**Figures** 4 and 5), acceleration of flower bud appearance was observed on the flowering shoots under the most severe irrigation treatment (T4) (**Figure 4**). Flowering shoots under this treatment, irrespective of shading, reached the stage of flower bud appearance earlier, resulting in a shorter stem length compared to flowering shoots under the less severe treatment. Shading had an effect at the stage of flower bud appearance with longer flowering shoots under the heavier shadings S2 and S3, compared to those under S1, and a trend for acceleration of flower bud appearance under the least shaded treatment S1 (**Figure 5**). However, observations under SEM of samples originating from different irrigation and shading treatments showed that the subsequent development is the same under all irrigation and shading treatments. It appears that imposed water stress affects growth rates but not development.

Abscission Zone and Peduncle

The presence of an abscission zone at the junction of the stem and peduncle that is apparent and potentially active throughout floral developments is verified in greenhouse roses. The different components of different regions of a rose offer varying resistances to water flow. The region of least conductance is the distal 15–20 mm of the peduncle, at the base of the receptacle (conductance is approximately 20% of that in the stem). The second major region of low conductance is the abscission zone, which may function as a safety zone (*see* **Postharvest Physiology:** Xylem Structure and Function in Cut Roses).

Figure 3 (a) Normal flower bud at the marketable stage. (b) Malformed flower bud at the marketable stage (height of the bud reduced and equal to diameter), observed under treatment 2 (water stress prior to petal initiation) and, more frequently, under treatment 3 (water stress prior to stamen initiation).

Table 5 The monthly pan evaporation (*Et*), water requirements (WR) and four different irrigation treatments applied (T1–T4). The rate of irrigation water was the same for all treatments and was applied in different irrigation intervals by self-compensating drippers with a discharge of 3.75 l h^{-1}

Month	Et (*mm day*$^{-1}$)	*WR* (*l day*$^{-1}$ *pot*$^{-1}$)	*T1* (*min*)	*T2* (*min*)	*T3* (*min*)	*T4* (*min*)
January	1.45	0.1024	2	4	6	12
February	2.00	0.1413	3	6	9	18
March	2.97	0.2098	3	6	9	18
April	4.76	0.3363	5	10	15	30
May	6.91	0.4882	8	16	24	48
June	9.30	0.6570	11	22	33	66
July	10.00	0.7065	11	22	33	66
August	9.06	0.6401	10	20	30	60
September	7.08	0.5002	8	16	24	48
October	4.51	0.3186	5	10	15	30
November	2.48	0.1752	3	6	9	18
December	1.53	0.1081	2	4	6	12

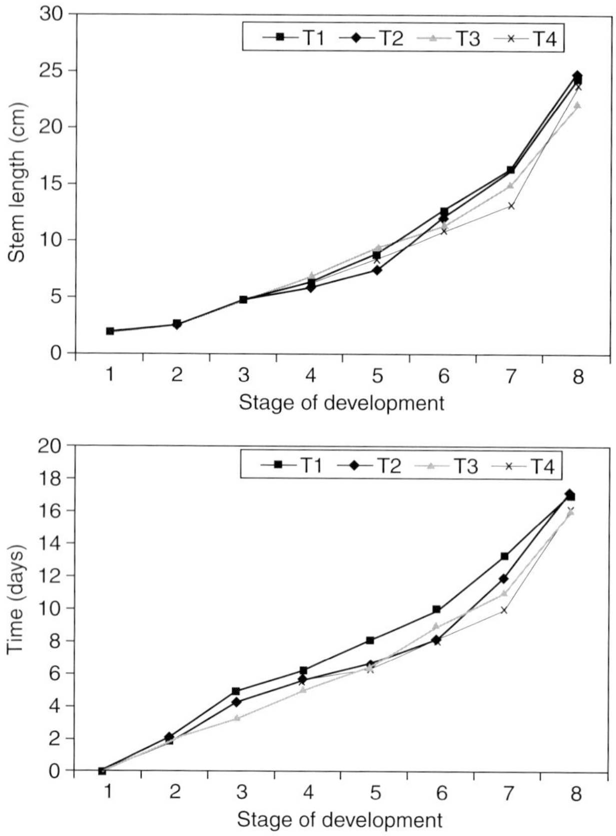

Figure 4 Progress of the stages of development under four irrigation treatments (T1–T4) in relation to stem length and the time from bud break. (Reproduced with permission from ISHS.)

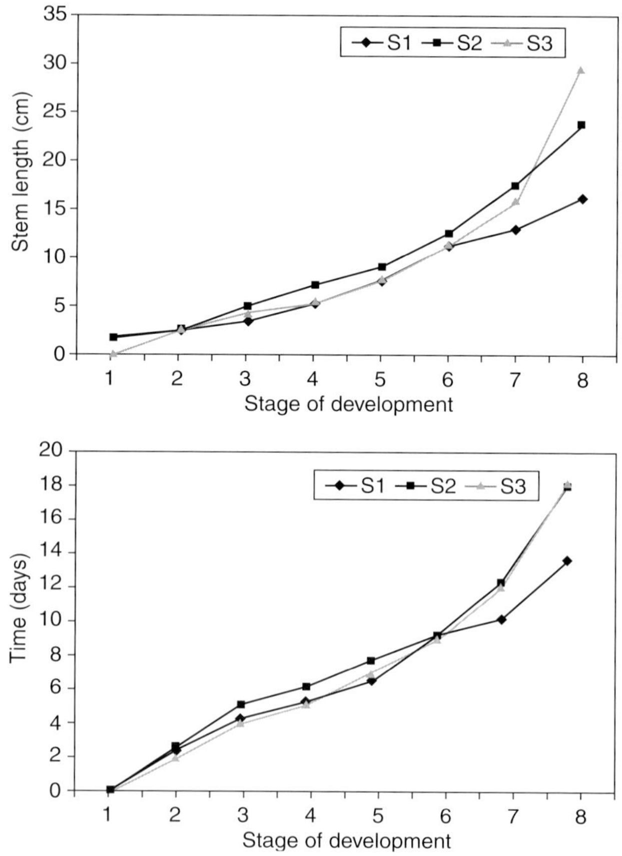

Figure 5 Progress of the stages of development under three shading treatments (S1–S3) in relation to stem length and the time from bud break. (Reproduced with permission from ISHS.)

The abscission zone and the region of low conductance in the peduncle may work in conjunction in the control of growth and development of the rose plant. During times of stress, many species of plants, including roses, tend to sacrifice peripheral organs to protect the main axis of the plant (**Figure 6**).

The flower is a principal site of water loss from the rose plant (*see* **Postharvest Physiology:** Xylem Structure and Function in Cut Roses). If water loss from the flower becomes excessive, cavitation-induced embolisms in the distal region of the peduncle will limit water flow to the flower, thereby minimizing water

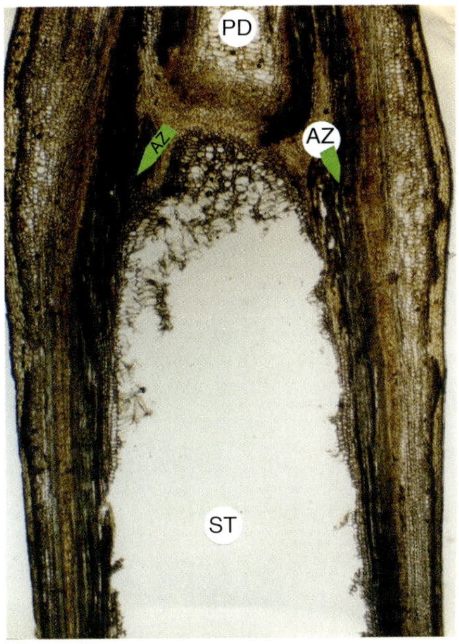

Figure 6 Abscission zone (AZ) between stem (ST) and peduncle (PD).

loss and reserving the water for the main axis of the plant.

The safety zone at the base of the peduncle, as well as limiting the influence of water stress associated with the abscission process, may limit the influence of drought stress. Such a process protects the stem at the expense of the flower. The accelerated appearance of the flower bud under water stress conditions, mentioned above, at stage 7 (**Figure 5**) is related to the protection mechanism of the rose plant to accelerate the development of the safety zone at the base of the peduncle to limit the influence of drought stress. As this zone is located at the base of the receptacle, the appearance of the flower bud is accelerated.

Malformations

The transition to flowering is the consequence of a progressive series of changes at the apex in which the indeterminate pattern of leaf and internode production is modified to produce a determinate floral structure.

In roses the apical meristem of the uppermost lateral buds usually contains primordial for the floral organs, at stage 2. Anatomical studies of the rose flower-bud development shows that, following the flattening of the dome-shaped apex at this stage, sepal primordia and an axillary bud are visible on the flanks of the apical meristem. The sepal primordia and axillary bud are followed sequentially by petal, stamen and pistil primordia. The formation of the latter two occurs very close in time. The orderly sequence of development of the floral organs has led to a model in which

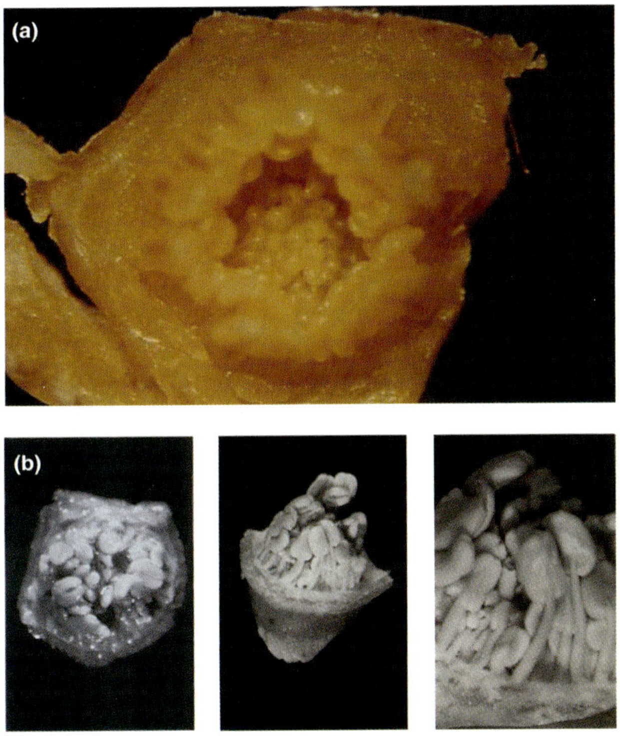

Figure 7 Flower buds at stage 8. (a) Normal flower bud: stamens arranged at the periphery and carpels well formed at the centre of the receptacle (stereomicrograph). (b) Three views of a malformed flower bud observed on flowering shoots under treatment 2 (water stress prior to petal initiation) and treatment 3 (water stress prior to stamen initiation). Stamens tightly packed at the periphery and the centre of the receptacle (stereomicrograph).

flower development has been likened to a relay race; the formation of one organ type depends on the formation of a previous organ type. However, numerous exceptions are known, including flowers in which some organ types are missing, the formation of stamens directly on the apical meristem and reversion types in which stamens are followed by petals.

For roses it is reported that the application of water stress during stamen initiation (stage 5), or during petal initiation (stage 3) in winter, causes abortions and malformations of the flower bud once the stem has elongated to approximately 15 cm. Such malformed buds have no carpels at stage 8, and stamens are tightly packed on the centre of the receptacle (**Figure 7**, compare (a) and (b)).

Flowerless shoots During winter, a high percentage of flowerless shoots are observed. The production of flowerless shoots is caused by events such as low temperature and low light intensity when they occur after flower initiation. Under these conditions, long vegetative shoots are never observed. 'Blind wood' is formed when the floral axis fails to develop a complete set of floral organs. However abortion of the developing flower is reported to occur before carpel primordia are initiated. The partially formed flower then abscises, leaving a blind shoot. Various environmental factors, such as low temperature and low light, increase the incidence of 'blind wood' (*see* **Morphology and Anatomy:** Shoots and Stems).

See also: **Morphology and Anatomy**: Shoots and Stems. **Postharvest Physiology**: Xylem Structure and Function in Cut Roses. **Rootstocks**: Scion–Rootstock Relationships.

Appendix

Preparation of Specimens for Scanning Electron Microscopy

Dissected apices were prepared for scanning electron microscopy as follows:

1. *Fixation.* Dissected specimens were immediately submerged in 'Kew Coctail' (i.e. 11 parts industrial methylated spirits; 1 part 40% formalin; 8 parts water; 1 part glycerine) for a minimum of 12 hours.
2. *Dehydration.* Specimens were sequentially submerged for 30 minutes in the following increasingly dry acetone solutions:
   30% acetone
   50% acetone
   70% acetone
   90% acetone
   95% acetone
   100% acetone
   100% absolutely dry acetone
   100% absolutely dry acetone
   100% absolutely dry acetone
3. *Critical point drying.* Dehydrated apices were critically point dried using a Tousimis Samdri-780 as per manufacturer's instructions.
4. *Mounting.* Critically point dried apices were carefully mounted on 'stubs' using double-sided tape. Forceps and the dissecting microscope were used.
5. *Gold coating.* The specimens were coated with gold (30 nm thick) using a Polaron SEM Coating Unit-E5150.
6. The specimens were then examined and photographed under a Joel JSM-35R scanning electron microscope.

Further Reading

Battey NH and Lyndon RF (1984) Changes in apical growth and phyllotaxis on flowering and reversion in *Impatiens balsamina* L. *Annals of Botany* 54: 553–567.

De Vries DP (1993) *The Vigour of Glasshouse Roses.* Dissertation. Wageningen, The Netherlands: Agricultural University of Wageningen.

De Vries DP and Dubois LAM (1983) Relations between basal bottom-breaks and harvested shoots in own-rooted hybrid tea-rose seedlings and their clones. *Gartenbauwiss* 48: 189–192.

Fuchs HWM (1994) Scion–Rootstock Relationships and Root Behaviours in Glasshouse Roses. Dissertation. Wageningen, The Netherlands: Agricultural University of Wageningen.

Halevy AH (1972) Phytohormones in flowering regulation of self-inductive plants. *Proceedings of the 18th International Horticultural Congress* 5: 187–198.

Marcelis-van Acker CAM (1994) Axillary Bud Development in Rose. Dissertation. Wageningen, The Netherlands: Wageningen Agricultural University.

Milburn JA (1979) *Water Flow in Plants.* London: Longman.

Mor Y and Zieslin N (1987) Plant growth regulators in rose plants. *Horticultural Review* 9: 53–73.

Pruitt RE, Chang C, Pang PPY and Meyerowitz EM (1987) Molecular genetics and development of *Arabidopsis*. In: Loomis WR (ed.) *Genetic Regulation of Development,* pp. 327–338. New York: Liss.

Steeves TA and Sussex IM (1989) *Patterns in Plant Development.* Cambridge: Cambridge University Press.

Van den Berg GA (1987) *Influence of Temperature on Bud-break, Shoot Growth, Flower Bud Atrophy and Winter Production of Glasshouse Roses.* Dissertation. Wageningen, The Netherlands: Agricultural University of Wageningen.

Zieslin N and Moe R (1985) Rosa. In: Halevy AH (ed.) *Handbook of Flowering,* vol. 4, pp. 214–225. Boca Raton, FL: CRC Press.

Roots

A Skytt Andersen, Royal Veterinary and Agricultural University, Frederiksberg, Denmark
H W M Fuchs, University of Delft, Delft, The Netherlands

Introduction

THE ROSE aloft in sunny air,
Beloved alike by bird and bee,
Takes for the dark root little care
That toils below it ceaselessly.
John James Piatt

Roots of rose plants serve the same purposes as plant roots in general: keep the plant steady in the ground, provide uptake surfaces for water and nutrients and sites for synthesis and storage of various substances of importance for the growth and function of the whole plant. These functions are reflected in the morphology and anatomy of rose roots.

The morphology and to some extent the anatomy of roots are highly dependent on several factors:

1. Genetic: Species, cultivar.
2. Environmental: The soil in which the roots grow, the climatic conditions to which the above ground parts are subjected and treatments such as pruning or removal of flowering branches.

On the other hand roots have a profound influence on the growth of tops and on the production of shoots and flowers. Thus the poem above does not reflect reality very strictly. If one searches the Internet for items on 'roses and their roots' an astounding response of 619 000 hits appears; if we add the modifiers 'anatomy and morphology' it is narrowed down to 892, still a wealth of information. A number of very thorough investigations of root growth and development in different types of roses have been carried out; some of these are cited below. However, in most of the general books and other descriptions of roses there has been very little emphasis on the below-ground parts of the rose plant.

Morphology

The roots of rose plants differ according to type of plant: whether it is a seed-propagated plant or root stock on which a scion is grafted or budded, or the plant is on 'own' roots, i.e. a vegetatively propagated plant with adventitious roots.

The root systems of seed-propagated rose plants, whether of plants propagated for the purpose of producing rootstocks for grafting or of seedlings produced in the course of a breeding programme, originate from the single radicle emerging from the seed upon germination (**Figures 1** and **2**). From this singular structure lateral roots develop in profusion giving rise to a fibrous root system (**Figure 2**) the form and architecture of which is strongly related to the species. Some species develop a very shallow and widespread root system while others grow with a dominant taproot and few lateral roots. Cultivar differences are seen in several ways; even in cultivars as closely related as two potted rose cultivars in the same breeder's series have conspicuously differing root structures (**Figure 3**). Several researchers who have found that this process is also highly dependent on soil type and water and nutrient supply have also studied the development of the architecture of a rose seedling root. This may influence the performance of the above-ground plant parts in several aspects, such as resistance to wilting and the longevity of flowers. In particular, lack of certain major nutrients such as nitrogen and phosphorous restricts the root size (**Figure 4**). On the other hand low P supply has been used successfully in pot rose production as a means for restricting top growth, which can be interpreted as a result of changed distribution of photosynthates between root and top. Plants with minimal P available have a much higher root/top

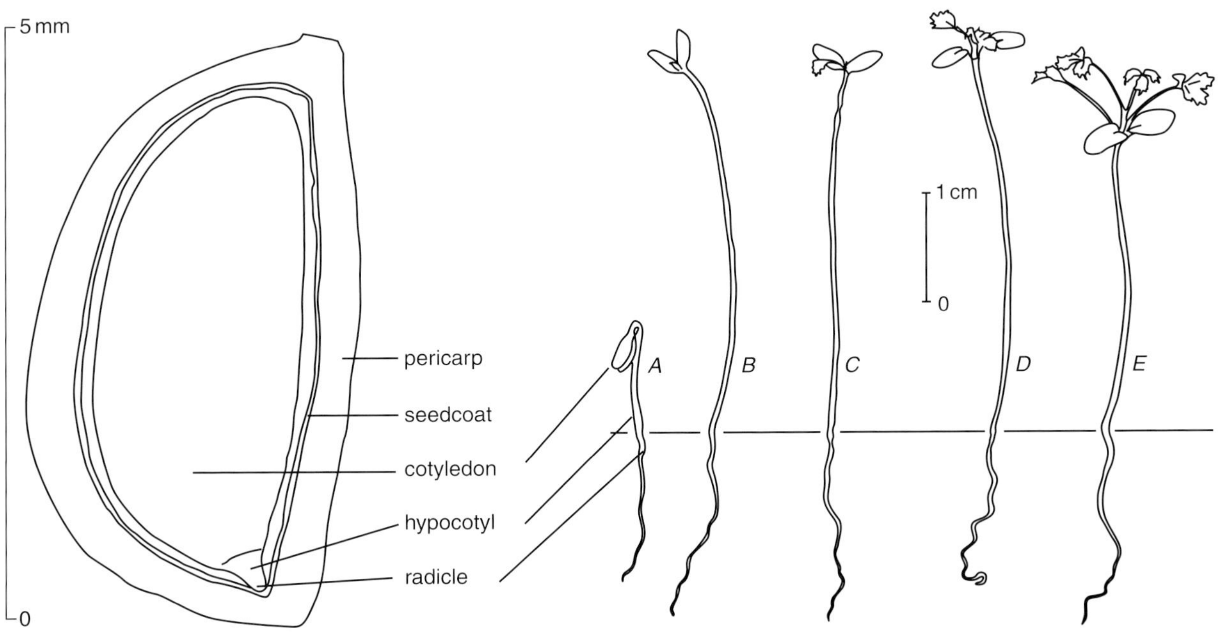

Figure 1 Primary seedling roots. (Reproduced from Schopmeyer CS (1974) *Seeds of Woody Plants in the United States*, USDA Forest Service Agriculture Handbook no. 450. Washington, DC: US Government Printing Office.)

Figure 2 Root systems of four different species of 3-month-old rose seedlings. From left: *R. canina 'Inermis'*, *R. canina 'Penta'*, *R. canina 'Polemiana'* and *R. multiflora*.

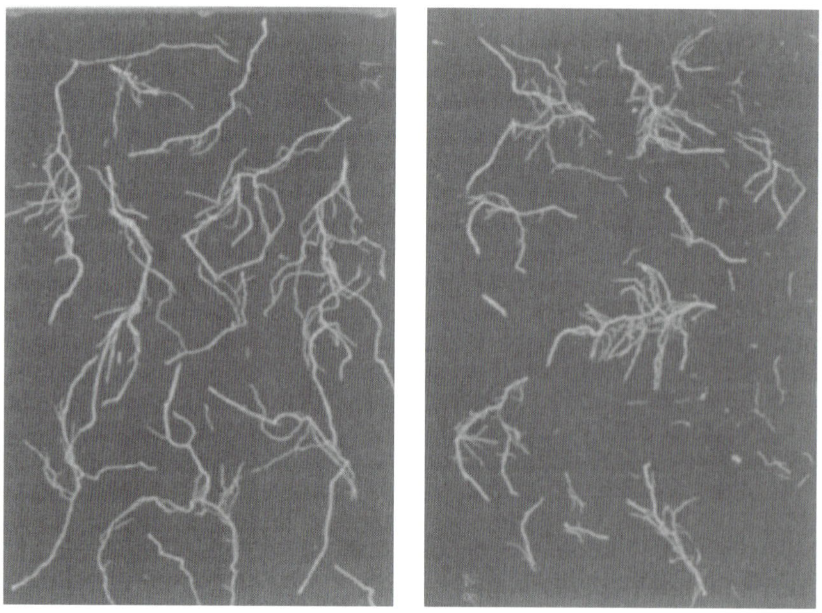

Figure 3 Differing root architecture of two pot rose cultivars. 'Charming' Poulhappy^(N), left and 'Bianca' Poulbian^(N), right. Roots were cut from the cuttings two weeks after rooting in an inert medium, note how 'Bianca' roots are highly dense while the roots of 'Charming' are longer and with less laterals. The density influence the water uptake properties of the cultivars rendering 'Bianca' a shorter display life than 'Charming'. (Photographs courtesy of Kristian Borch.)

ratio than plants with an adequate or luxurious P supply (**Figure 5**). They also keep their flowers considerably longer. During the years following the transplanting of a garden or greenhouse rose plant to its final growing place, roots grow and develop continuously so that a 5-year-old rose root (**Figure 4**) consists of a crown several centimetres thick where the vascular connections to the trunk above ground takes place through an intricate anatomical pattern.

Rose roots, like those of most other plants, live in close association (symbiosis) with mycorrhizal fungi. These enlarge the surface of the active absorption

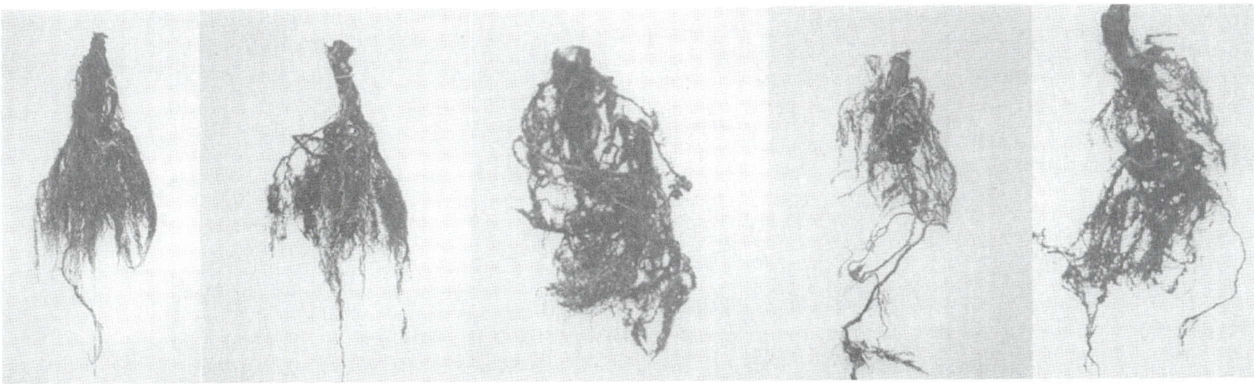

Figure 4 Five-year-old rose roots (cv. Parel van Aalsmeer grafted on *R. canina* 'Bröggs Stachellose' grown in nutrient solutions with different major ions missing. (Reproduced with premission from Johansson J (1977) *The Nutrient Requirement of Greenhouse Roses as Determined by Leaf and Soil Analysis*. PhD Thesis, Swedish University of Agricultural Sciences, Stockholm.)

Figure 5 Roots of 'Bianca' Poulbian$^{\circledR}$(Poulsen Parade) after 10 weeks with different phosphate concentrations in the root zone. Left to right: 0%, 50 μM and 150 μM phosphate. (Photograph courtesy of C.W. Hansen.)

zones of roots (**Figure 6**) aiding the plant, especially under suboptimal conditions regarding nutrient and water uptake. Mycorrhizal fungi (species of *Glomus* and *Gigaspora*) that are commonly present in many soils have been shown to increase root mass by 25–30% especially when the soil offers low phosphate availability.

Morphology of Cutting Roots

More and more roses are propagated by cuttings. At the present time (2002) potted rose plant production worldwide numbers around 150 million pots per year with several cuttings in each. The larger roses are also increasingly propagated this way.

The roots on a cutting differ from those of a seedling root in several aspects. First, there are from five to ten 'primary' roots on each cutting. The formation of these is treated elsewhere (*see* **Propagation: Own-Rooted Cuttings**), but after the initial formation in the stem tissues adventitious roots develop anatomically in a similar way to seedling roots. The morphology, however, remains different (compare **Figure 2**, with **Figures 3** and 7). Of the multitude of original adventitious roots one or a few take the lead in a form of apical dominance. It has been claimed that the roots of cuttings are more vulnerable than seedling roots, a postulate that has not been verified by experiments in which cutting-propagated plants thrive just as well as grafted or budded plants. The discrepancy may be found in the early weakness of the roots of cuttings which when handled roughly may break from the stem.

Root growth is affected by treatments of the aboveground parts in many ways. Pruning or cutting flower stems from growing rose plants will cause the death of roots and reduce the new growth of roots (**Table 1**). This is due to diminished availability of carbohydrates produced by the leaves. Roots function as storage organs for carbohydrates and the stored quantity of carbohydrates, mostly in the form of starch, varies throughout the year for plants growing under garden conditions.

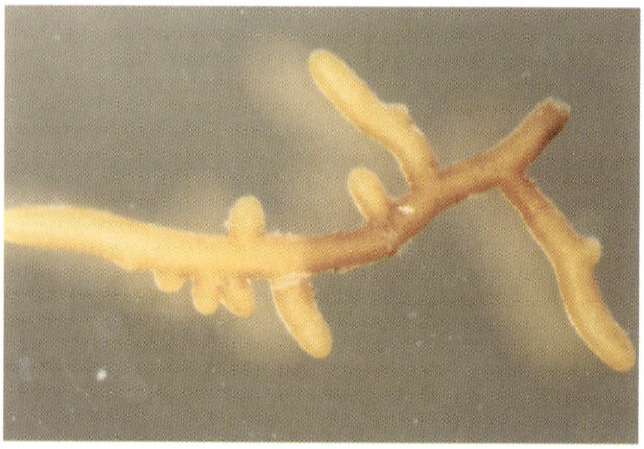

Figure 6 Root with mycorrhiza. (Photograph courtesy of B. Kieliszewska.)

For greenhouse-grown roses the term root regenerating potential (RRP) is often used. It indicates the ability of the root system to regenerate after a hard cutback when many roots die back due to lack of nutrient support from above ground. The RRP is strongly affected by the number of buds on the remaining stems since growing buds act as competitive sinks to root meristems for carbohydrates and hormones (**Table 1**).

Anatomy

Root anatomy is distinct from that of the shoot in several ways. And *Rosa* roots, especially the older ones, are distinguishable from those of many other woody plant roots.

The primary root apex is covered with a root cap consisting of gelatinous cells that can be sloughed off during the roots penetration of the soil through growth. The root meristem produces new cells by division where these cells elongate in a zone immediately behind the meristem; they also differentiate into the various cell types like cortex and conductive tissue: phloem and xylem, fibres and rays. Roots have a specialized tissue surrounding the central core of the root (stele); this is called the pericycle and the endodermis. The function of the endodermis is to control the uptake of ions and water into the conductive tissues. Young roots have extensions of the epidermal cells called root hairs; these are in direct contact with soil particles and

Figure 7 Rooted rose cuttings: five different basis internode lengths, 0–4 cm, cv. Mistral – Poulsen Parade.

Table 1 The effect of different pruning strategies at harvest of cut stems on fresh weight changes (grams per root) of greenhouse rose roots. Rose plants of the cultivar 'Motrea' were grown on after the pruning treatments and dug up at weekly intervals

Treatment	*Day 7*	*Day 14*	*Day 21*	*Day 28*	*Day 38*
Normal harvest	+1.3	−0.2	−1.5	−1.0	+1.8
plus removal of laterals	+1.7	+1.6	+2.1	+2.1	+2.0
plus removal of leaves	−0.3	−2.8	−2.8	−2.7	−2.5

Modified after Fuchs, HWM (1994) *Scion–Rootstock Relationships and Root Behaviour in Glasshouse Roses*. PhD Thesis, Wageningen University, Wageningen, The Netherlands.

water-filled pores of the ground, and a large part of solution uptake goes through them.

Growth in length of the root is accomplished through cell extensions in the region behind the meristem (the elongation zone). Roots normally grow downwards; this is called positive gravitropism. A complicated series of events of hormonal character is initiated through small starch granules (statholites) in cells of the root cap; these can move around in the cell in response to gravity thus giving the root cells a sense of direction in relation to the earth's gravity. Roots, on the other hand, react negatively towards light; i.e. they are negatively phototropic. Young roots can grow around a stone or other hard object in the soil by a mechanism called differential growth: when a root touches the hard particle it is stressed and produces stress hormones which restrict growth on the side not directly in touch with the particle.

Lateral roots are not initiated in the meristem-like leaves on the shoot meristem. Instead they develop further away as new centres of meristimatic activity in the pericycle, from where they grow out through and penetrate the outer layers of cells in cortex and epidermis. Connections to the main root conducting tissues are established early in this process.

Older roots (**Figure 8**) develop protective tissues such as cork and layers of cortex consisting of unspecialized cells replacing the epidermis of younger roots. The cork layer (phellem) may be several cells thick and is formed from a cork cambium (phellogen); the outer layers of cork consist of dead, thin-walled, air-filled cells. Some of these cells are dark-coloured. In the cortex below are found fibre cells. The phloem here is present in triangular or quadrangular groups of cells as seen in cross-section. In many of the rose root phloem cells are deposited prominent crystals of calcium oxalate. All these tissues – cork, phellogen, cortex, and phloem – are collectively called the bark of the root.

Secondary thickness growth of the root is accomplished through meristimatic activity, cell divisions in the areas between primary xylem and phloem called the cambium. Cambium may form a continuous cylindrical layer producing xylem elements towards the centre of the root and phloem cells towards the perimeter. The xylem vessels of older rose roots are found solitary or paired without any clearly marked annual rings present. Among the xylem cells are rays which can be from one to 25 cell layers thick; these are living cells providing a symplastic (continuous cytoplasmic) connection through the root in radial direction from the pith in the centre to the phloem in the bark and vice versa.

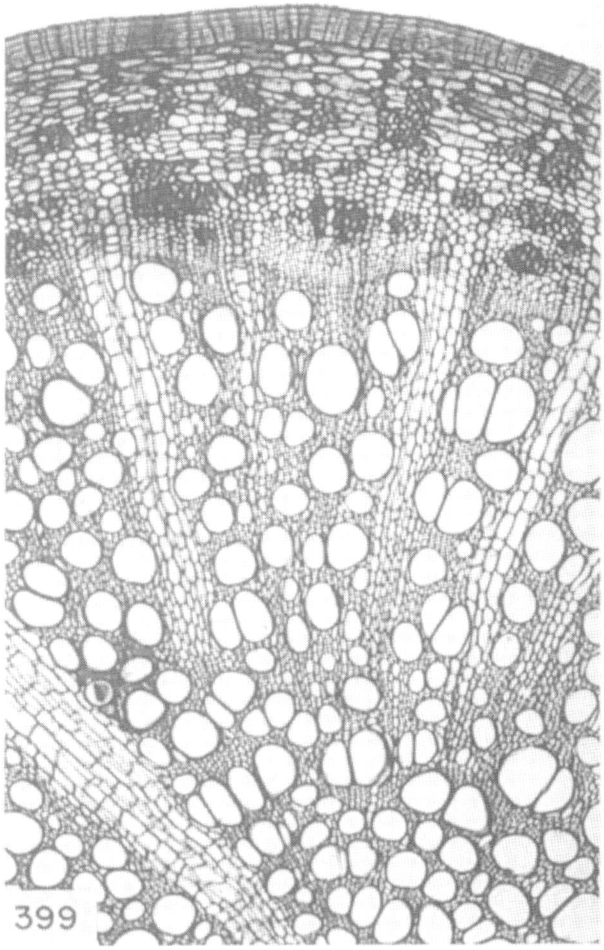

Figure 8 Cross-section of a rose root. Enlarged ×50. (Reproduced with permission from Cutler DF *et al.* (1987) *Root Identification Manual of Trees and Shrubs*. London: Chapman Hall.)

It is possible for the trained observer to distinguish even small parts of rose root tissues from those of other woody plant roots by concentrating on the above-mentioned features of missing annual rings and the angular complexes of cells in the phloem. Also the presence of crystals in some of the phloem cells is rather unusal, and characteristic of roses.

See also: **Morphology and Anatomy**: Shoots and Stems; Leaves; Flower Development and the Abscission Zone; Pollen Grains and Tubes. **Postharvest Physiology**: Xylem Structure and Function in Cut Roses. **Propagation**: Own-Rooted Cuttings.

Further Reading

Atkinson D (ed.) (1991) *Plant Root Growth: An Ecological Perspective*. Oxford: Blackwell Scientific Publications.

Borch K (1998) *Plant Stress and Root Growth Response to Low Phosphorous Nutrition and the Implication of Ethylene*. PhD Thesis, Odense University, Denmark.

Costa JM-RC (2002) *The Role of the Leaf in Growth Dynamics and Rooting of Leafy Stem Cuttings of Rose*. PhD Thesis, University of Wageningen, The Netherlands.

Cutler DF, Rudall PJ, Gasson PE and Gale RMO (1987) *Root Identification Manual of Trees and Shrubs*. London: Chapman & Hall.

Davies FT Jr (1986) Effects of mycorrhizal fungi on *Rosa* multiflora 'Brooks 56' understock. *Acta Horticulturae* 189: 117–121.

Fuchs HWM (1986a) Root regeneration of rose plants as influenced by applied auxins. *Acta Horticulturae* 189: 101–107.

Fuchs HWM (1986b) Harvesting, pruning and root reactions of roses. *Acta Horticulturae* 189: 109–115.

Fuchs HWM (1994) *Scion–Rootstock Relationships and Root Behaviour in Glasshouse Roses*. PhD Thesis, Wageningen University, Wageningen, The Netherlands.

Johansson J (1977) *The Nutrient Requirement of Greenhouse Roses as Determined by Leaf and Soil Analysis*. PhD Thesis, Swedish University of Agricultural Sciences, Stockholm.

Waisel Y, Eshel A and Kafaki U (eds.) (1996) *Plant Roots: The Hidden Half*. New York: Marcel Dekker.

Pollen Grains and Tubes

Y Jacob and F Ferrero, INRA, Saint Aygulf, France

Introduction

Fundamental biological functions of pollen in the sexual reproduction of plants are now well known. Firstly, the haploid microgametophyte is the widespread vector of recombinant characters after male meiosis. Secondly, the pollen grain, deposited on the female stigma, germinates by forming a pollen tube from an aperture of pollen wall, in order to ensure the double fertilization. In genus *Rosa*, as in many other genera, pollen, as a reproductive unit, plays a major role in various aspects of the life cycle of the plant. Its formation is one of the highest energy-consuming processes of the plant. As a male sexual component, pollen gives theoretically contributes half of the characters of the father plant (except for the *Caninae* section, roses with asymmetric male/female meiosis (*see* **Genetics**: Meiosis; Inheritance in the Dogrose). For the wild roses, the pollen contributes significantly to spatial and temporal gene flows. For the 'manmade' roses, pollen is manipulated and managed principally in order to give oriented progenies with valuable characteristics. A tentative review of current knowledge of rose pollen in relation to these different aspects is made here.

Pollen Morphology

At macroscopic level, rose pollen, like the pollen of many rosaceous species, looks like a very fine dust with a more or less yellow colour, constituted by numerous elongated pollen grains. This fine powder is released when the pollen sacs open at the stage of anther dehiscence (**Figure 1**). Under the light microscope, fresh rose pollen has an elongated form (**Figure 2**) with a constant ratio of length to width equal to 2.

Rose pollen structure is similar to pollen structure of other rosaceous species. The visible part of the pollen grain is the exine which is the outer layer of a living pollen grain. This layer is mainly constituted of a

Figure 1 Fresh pollen released from isolated stamens in a plastic Petri dish, 24 h after collection from fresh flowers. Pollen 'dust' is visible on the surface of the dish.

polymer called sporopollenin, and small quantities of polysaccharides. Sporopollenin, formed from the polymerization of carotenoids, is chemically very stable and is resistant to almost all kinds of environmental damage. In roses, exine is equipped with three apertures (three pores: pollen is thus described as 'tricolpate' or 'triporate') from which the pollen tube emerges upon germination and fertilization process. When observed under the scanning electron microscope, the exine of rose pollen grains shows some structural and anatomical patterns that are correlated with phylogenetic background of the species that produced them. The exine is divided into two sublayers: the outermost sexine and the unsculptured underlying nexine. In roses species or cultivars, the sexine patterns differ mainly in their aperture membranes, length and density of ridges and furrows and in the density and spatial patterns of small perforations. These give to the exine a striated or reticulated appearance with several levels of combinations of one or both characteristics. The inner layer, called the intine, is mainly composed of cellulose and is very similar in construction to ordinary plant cell walls (**Figure 3**).

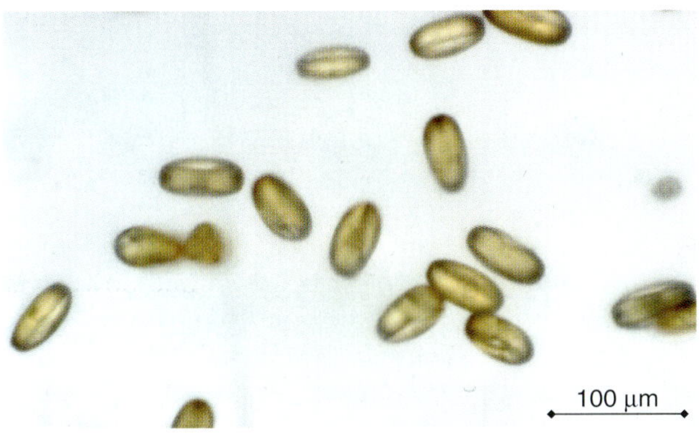

Figure 2 Fresh pollen of a diploid rose (*Rosa rugosa* var. *rubra*) observed under optical microscope without any staining. The elongated form of pollen and some apertures are visible.

Pollen Formation

Pollen is the result of two sequential stages. The first stage is the microsporogenesis (*see* **Genetics:** Meiosis) followed by the microgametogenesis, the process of transformation of a microspore in a pollen grain, which is basically the mature microgametophyte. Microspore development consists of an expansion of its volume, which is commonly associated with the formation of a vacuole. This is also the stage of sporopollenin polymerization and the formation of the final pollen wall structure as previously described. During this stage, microspore nucleus moves from a central position to an eccentric one, close to the wall. In this position the first pollen mitosis (pollen mitosis I) occurs and gives two unequal cells. The larger one is the vegetative cell which contains a large and diffuse haploid nucleus. The other cell is the generative cell with a small and condensed nucleus. This cell is contained within the cytoplasm of the vegetative cell. In roses, the generative nuclei in mature pollen contains a $2n$ DNA amount whereas the vegetative cell contains only a $1n$ DNA amount. This means that, even if the second pollen mitosis (pollen mitosis II) occurs in the pollen tube (**Figure 4**), DNA replication

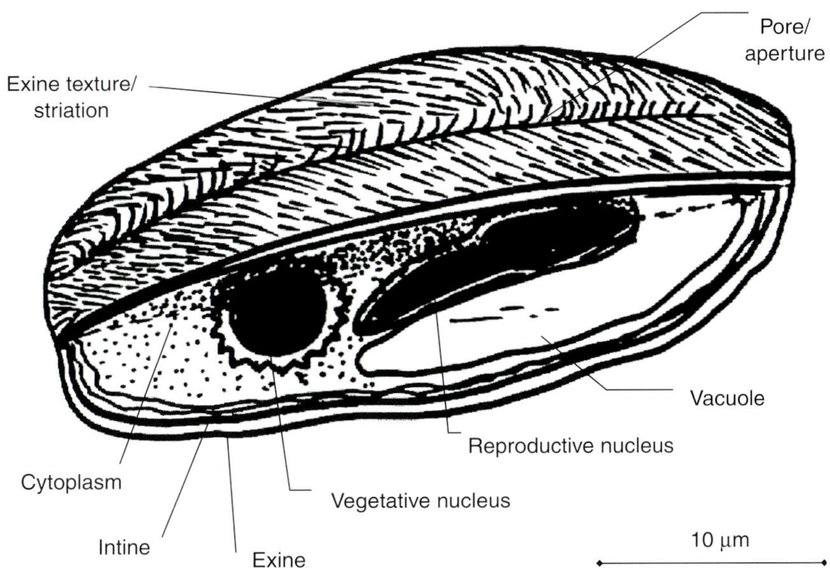

Figure 3 Schematic representation of the morphology of a rose pollen grain.

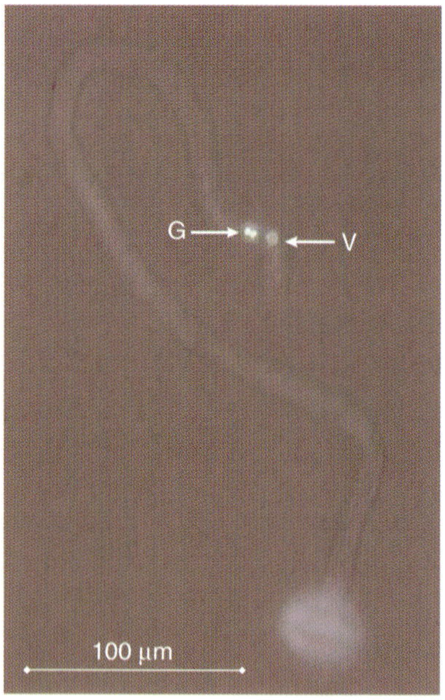

Figure 4 Pollinic mitosis II of the generative cell inside a rose pollen tube after 24 h of *in vitro* germination. Observation under epifluorescent light microscope after DNA staining with DAPI (4′,6-diamidino-2-phenylindole, which is a specific-DNA fluorescent stain). Two sperm nuclei (G) are clearly visible whereas the vegetative nucleus (V) is in proximal position into the pollen tube extremity. (Reproduced from Jacob Y *et al.* (2001) *Acta Horticulturae* 547: 383–385 with permission of ISHS.)

is performed into the generative cell, inside the pollen grain. This pollinic mitosis occurs without any cytokinesis (cell division).

Pollen Release

In rose flowers, stamens (when they are present) are constituted by the filament, which is the stalk connected with the floral receptacle, and the anthers, where the pollen is present, which are attached to the filament by the connective. Each anther is constituted with two symmetric pollen bags, (thecae), fixed to the connective. At flower maturity, pollen is generally released by conjunction of external and internal factors but the critical stage of dehiscence is basically the initial stage of pollen spreading. For example, on *Rosa rugosa* Thunb. ex Murray, pollen is generally released from the stamens when the flower bud is still closed. In contrast, on rose cultivars with a high level of duplication, pollen may be released later after flower bud opening. External factors such as ambient air humidity, temperature and solar radiation can also influence rose pollen release. Natural vectors of pollen are insects (*see* **Insects and other Animals:** Overview of Insects). Rose breeders currently use some techniques to accelerate pollen release in order to manage the hybridization process. Anthers can be collected after removal of petals on defined flower bud stages. They are placed in glass Petri dishes at temperatures above 20 °C and hygrometry equal to 60%, in order to provoke anther opening. When pollen is produced in great quantity, anther dehiscence can also occur after 12 h in a growth chamber at 30 °C (*see* **Breeding:** Methods of Cross Breeding).

Pollen Germination and Fertilization

Pollen tube growth which has been studied in roses displays similarities to that in other Rosaceae and, by extension, other genera. When a rose pollen grain lands on a stigma, it adsorbs water and a rehydration process occurs. Various signals, especially those of pollen–stigma interaction, induce an intense metabolic activity in the pollen grain which start to grow and forms a pollen tube. The pollen tube invades the pistil, growing between the walls of the stigmatic cells, then travelling through an extracellular matrix within the transmitting tissue of the style. The pollen tube finally arrives at the ovary, where it targets an ovule that contains an egg. This phase of pollen development, called the progamic phase, corresponds to the phase occurring between pollination and fertilization. In roses, the pollen tube reaches the ovaria between 12 and 24 h after pollination. Fertilization is believed to occur within this interval of time. The growth of the pollen tube has been recently studied at the molecular level on model plants such as *Arabidopsis thaliana*, and several genes involved in pollen hydration, germination and tube guidance have been identified. Pollen tubes are themselves considered to be a model for cellular growth of plant cells.

Some external factors are well known to influence rose pollen germination strongly. Important variation in stigma pH can be found between genotypes. This factor, known to be involved in the pollen germination response, depends on the genotype, and plays a role in the fertilization response and consequently in seed set. Ionic elements like boron and calcium which are involved in the metabolism on pollen tube growth, and at different concentrations can stimulate or inhibit it. Naturally occurring polyamines such as spermine also stimulate pollen tube germination and elongation when they are sprayed on stigma.

Self-Incompatibility

In roses, as in other Rosaceae, reproductive barriers like mechanisms of self-incompatibility (inability of pollen to fertilize the ovule of the same plant) occur. It has been determined that the type of self-incompatibility found in *R. rugosa* is a gametophytic mechanism. This means that all compatible or incompatible

Table 1 Cytological staining procedures

Procedure	*Materials*		*Method*
Acetocarmine staining	Glacial acetic acid	45 ml	Heat the 45% acetic acid solution to boiling. Add the 0.5 g carmine and continue heating for several minutes with stirring. Filter and cool. Pollen is put on slides, stained with a few drops of acetocarmine solution and cover-slipped. Under light microscope, the pollen cytoplasm of viable pollen grains appears in red.
	Water	55 ml	
	Carmine	0.5 gm	
Alexander staining	Ethanol 95%	10 ml	Pollen is put on slides, stained with a few drops of Alexander solution and cover-slipped. After 15–20 min, the pollen is observed under light microscope, its wall being coloured in green and the cytoplasm of viable pollen grains in red–purple.
	Malachite green (1% in 95% ethanol)	1 ml	
	Fuchsin acid (1% in water)	5 ml	
	Orange G (1% in water)	0.5 ml	
	Phenol	5 g	
	Chloral hydrate	5 g	
	Glacial acetic acid	2 ml	
	Glycerol	25 ml	
	Distilled water	50 ml	
Aniline blue staining	Aniline blue (1% in water)	1 part	Several hours after pollination, flowers are collected and finely sliced from stigma to the floral peduncle. They are placed in the mix and boiled for 10 min. Before observation, they were placed in the following mix to improve fluorescence: samples are put on slides, mounted with a drop of aniline blue (1%) in K_3PO_4 0.1M, cover-slipped and observed under fluorescence microscope equipped with subsequent filters (340–380 nm excitation filter).
	Tween 20	2 parts	
	K_3PO_4 (0.2 mol l^{-1})	7 parts	
	Sodium hydroxide (1 mol l^{-1})	1 part	

pollen starts to germinate, but growth of incompatible pollen tubes stops in the style while compatible tubes go on to fertilize the egg in the ovary. By extension, this gametophytic self-incompatibility (GSI) probably occurs in the majority of self-incompatible roses. The GSI system, which is genetically controlled by a single-locus, multiallelic family of genes, has also recently been dissected at molecular level in other Rosaceae and also in other model plants.

To assess self-incompatibility or compatibility in wide crosses, pollen germination and pollen tube growth can be observed by fluorescence microscopy with classical protocols developed on other species such as the aniline blue staining procedure (**Table 1**). In roses, fertilization is difficult to observed with these techniques, due to the presence of natural fluorescence of stylar tissues that mask fluorescence in the pollen tube wall. (**Figure 5**).

Pollen Management

Rose pollen has largely been studied in order to optimize its management, mainly for the benefit of rose breeders. To facilitate crosses in breeding programmes, efficient protocols to assess rose pollen vitality have been developed, and protocols for long-term conservation of pollen have been studied. The first characteristic of rose pollen that can be underlined is its great variability in terms of differences of pollen characteristics between different genotypes. The large genetic diversity in the genus, the various interspecific combination abilities, the range of ploidy level and the intense effort to produce horticultural roses are each a part of the reasons that can explain this variability. A wild rose will generally produce a homogeneous and functional pollen population with null or a very low proportion of dead, empty or abnormal pollen grains. In contrast, many horticultural hybrids from various origins produce pollen populations with a low to very high percentage of inefficient pollen grains (unable to fertilize). Various male meiotic or postmeiotic disturbances can generally explain the presence of dead, sterile or abnormal pollen. Microscopic observations under visible or ultraviolet light, after specific staining, can be performed to detect easily a part of these abnormalities. For example, staining with acetocarmine stain, or Alexander stain is currently used for this purpose (**Table 1**). Among these pollen characteristics, the presence of unreduced male gametes (pollen with chromosome number higher than half of the chromosome number of the father plant) can easily be detected due to the relation between pollen size and ploidy level of the pollen (**Figures 6** and **7**). The ploidy level of rose pollen can also be estimated with flow cytometry techniques, which measure the amount of DNA in individual pollen grains.

The occurrence of unreduced pollen can significantly disturb heredity studies. Indeed, a mix of normal and unreduced pollen is able to induce progenies with different ploidy levels. However, presence of

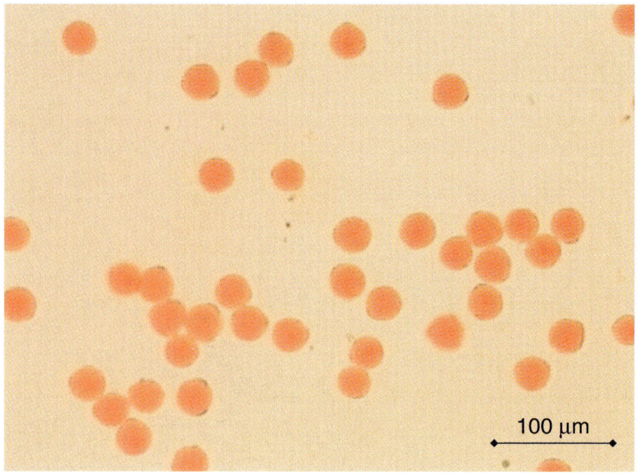

Figure 6 Pollen of a diploid species (*Rosa rugosa* var. *rubra*) stained with acetocarmine (only viable pollen is stained in red) and observed under light microscope. Pollen size is very regular. Viability level is very high (close to 100%).

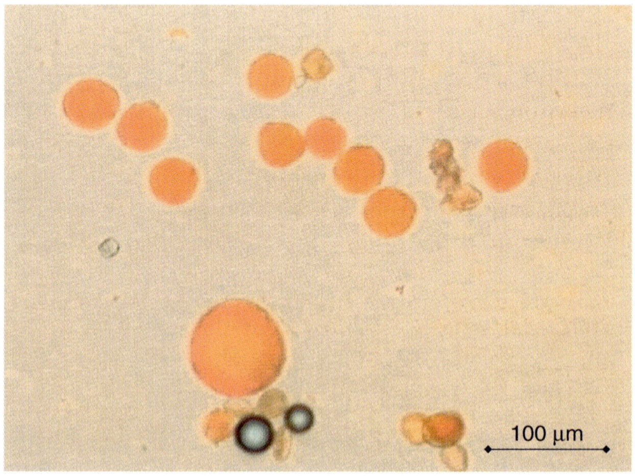

Figure 7 Pollen of a tetraploid variety (JV6 cv. Saint Trop) stained with acetocarmine and observed under light microscope. Pollen size is larger than the pollen of diploid roses and some 'jumbo pollen' attached to unreduced pollen occurs. Viability level is highly variable depending on the particular cultivar.

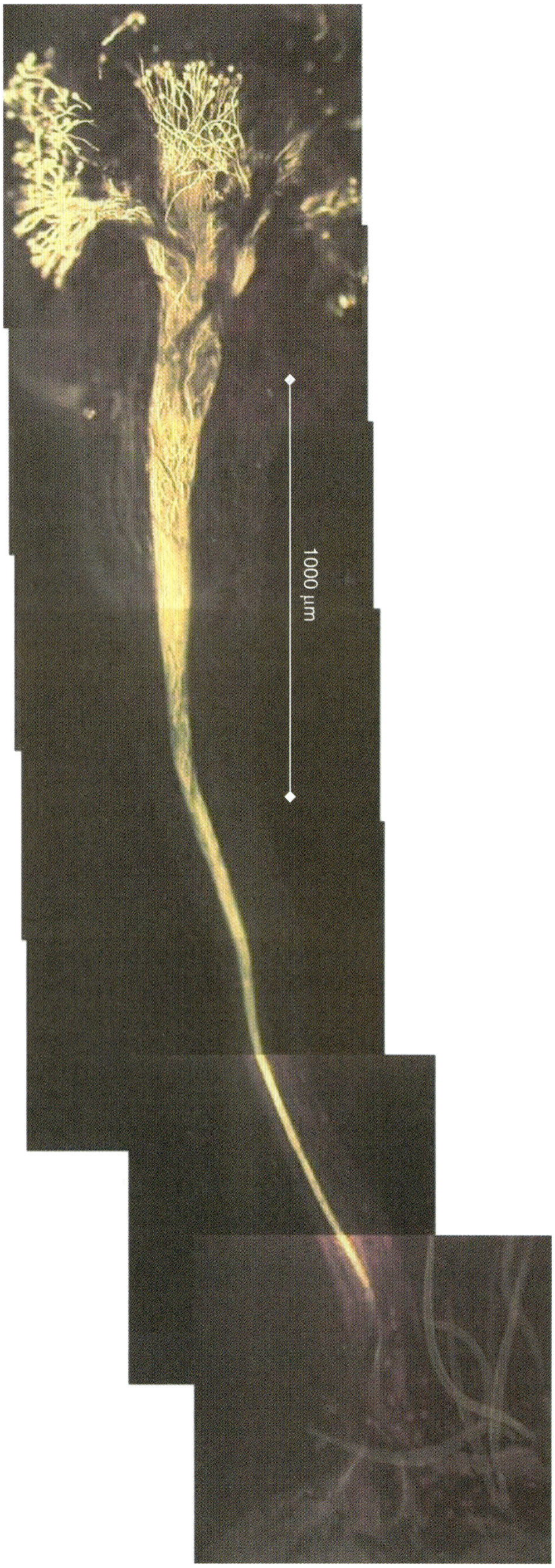

Figure 5 *In situ* germination of tetraploid pollen observed under epifluorescent light microscope, 24 h after pollination. Protocol used is a classical protocol of clearing the pistil followed with staining with aniline blue. The tubes are visible in the highest part of the style. The extracellular matrix is also fluorescent and makes observation in the bottom part of the style difficult. (Reproduced from Jacob Y *et al.* (2001) *Acta Horticulturae* 547: 383–385 with permission of ISHS.)

unreduced pollen must be turned as an advantage for the breeder, particularly when adjustment at a superior ploidy level is found. For example, a diploid rose, able to produce unreduced pollen and crossed with a tetraploid plant, can give theoretically tetraploid plants in the progeny. More generally, unreduced pollen grains can be a way to sexually polyploidize roses. This mechanism is believed to have played a significant role in the origin of rose polyploidy (*see* **Genetics**: Karyology).

Pollen Storage and Viability

Rose pollen is functional when it is freshly collected and keeps a certain viability during the following days of pollen release. When the pollen is stored at ambient temperature during several weeks its efficiency decreases rapidly. Different freezing conditions have shown that normal freezing can increase the period of usability of pollen up to 2 to 3 months. For long-term storage, cryopreservation of rose pollen is an efficient method. Pollen can apparently retain a high level of viability after 1 year in liquid nitrogen. Various investigations have been carried out to estimate the pollen viability and male fertility of rose cultivars. Viability and fertility can be estimated by *in vitro* pollen germination tests. These tests, which are done in a simple medium containing sucrose, boric acid and deionized water, allow the observation of pollen tube growth. However, these tests often overestimate pollen ability to grow and to fertilize.

Conclusion and Prospects

The pollen of roses shows morphological, anatomical and genetic characteristics which find applications in several domains. Pollen studies contribute to a better knowledge of phylogenetic relationships in the genus *Rosa* and constitute an aid for the complex botanical and horticultural classifications of roses. In various species, recent developments in the knowledge of pollen biology have been achieved particularly at molecular level (i.e. fine mechanisms of pollen stigma interaction, pollen tube guidance, etc.). These developments applied to rose pollen should allow progress towards a better management of hybridizations and consequently a means to optimize rose breeding programmes. At both professional and amateur levels, rose breeding activities should be stimulated by transfer of simple methods (i.e. microscopy control of pollen quality) to ensure the best possible efficacy of rose pollen manipulation. Finally, in terms of pollen storage, modern protocols could significantly contribute to long-term management of rose germplasm.

See also: **Biotechnologies for Breeding**: Manipulation of Ploidy Level. **Breeding**: Methods of Cross Breeding. **Genetics**: Karyology; Meiosis; Inheritance in the Dog-rose. **Insects and other Animals**: Overview of Insects; Pollination.

Further Reading

Cheung AY and Wu H (2001) Pollen tube guidance: right on target. *Science* 293: 1441–1442.

De Nettancourt D (1997) Incompatibility in angiosperms. *Sexual Plant Reproduction* 10(4): 185–199.

Jacob Y and Pierret V (2000) Pollen size and ploidy level in the genus *Rosa*. *Acta Horticulturae* 508: 289–292.

Jacob Y, Priol V, Ferrero F, Coudret A and Sallanon H (2001) Fluorescent staining of roses pollen tubes and nuclei by microscopy and flow cytometry analysis. *Acta Horticulturae* 547: 383–385.

McCormick S (1998) Self-incompatibility and other pollen–pistil interactions. *Current Opinion in Plant Biology* 1: 18–25.

Raudaskoski M, Ström H and Laitiainen L (2001) Pollen tube cytoskeleton: structure and function. *Journal of Plant Growth Regulation* 20(2): 113–130.

Shivanna KR and Sawhney VK (1997) *Pollen Biotechnology for Crop Production and Improvement*. Cambridge, UK: Cambridge University Press.

Taylor LP and Hepler PK (1997) Pollen germination and tube growth. *Annual Review of Plant Physiology and Plant Molecular Biology* 48: 461–491.

Ueda Y, Takeshita D and Ando T (1996) Pollination in *Rosa rugosa* Thunb. ex Murray. *Acta Horticulturae* 424: 309–310.

N

Nutrition *see* **Production and Marketing**: Mineral Nutrition.

O

Organ Culture *see* **Cell, Tissue and Organ Culture**: Micropropagation; Rooting of Micropropagated Shoots; *In Vitro* Regeneration Techniques.

P

Perfume Industry *see* **Fragrance and Pigments**: Odoriferous Substances and Pigments; Fragrance Profiles of Wild and Cultivated Roses. **History of Roses in Cultivation**: History of the Perfume Industry. **Secondary Metabolites**: Extraction and Industrial Processes.

Pest Resistance *see* **Breeding**: Selection Strategies for Disease and Pest Resistance.

Pigments *see* **Fragrance and Pigments**: Odoriferous Substances and Pigments; Anthocyanins and other Flavonoid Pigments; Flavonoid Molecular Biology.

Ploidy *see* **Biotechnologies for Breeding**: Manipulation of Ploidy Level; Amphidiploidy. **Genetics**: Nuclear DNA Amounts.

POSTHARVEST PHYSIOLOGY

Contents

Xylem Structure and Function in Cut Roses

H M C Put, Cut Flower Hygiene, Deventer, The Netherlands
A C M Clerkx, Enschedé, The Netherlands

Introduction

As cut flowers lack roots, root-supplied hormones and dissolved materials are no longer available. As transpiration continues unabated, water moves upwards from the cut end and allows air to enter the vascular system. This makes it difficult to reestablish the supply of water to leaves and flowers. Additionally, when water uptake is established, dissolved materials, microbes and particulate matter carried into the vascular system restrict the flow of water and shorten postharvest life.

The vase-life of cut greenhouse roses is, above all, dependent on water balance, which is affected by the water uptake by the flower, water transport through the stem and transpiration. Although the maintenance of an optimal water status is the most important factor in cut-flower longevity, many of the underlying mechanisms leading to disturbed water balance are still unresolved.

This article summarizes methods of assessing the water status of the cut-rose flower, analysis of factors leading to a disturbed water status, the relation of a disturbed water status to the xylem structure and the consequences of disturbed water status for vase-life.

Methods

Methods to Assess the Water Status of the Cut-Rose Flower

Preparation of stems for vase-life tests The plants are harvested from the greenhouse at one of the standard stages of bud development, mainly at stage 2 (**Figure 1**). The stems are cut with a sterile razor blade to a length of 50 cm, followed immediately by 30 min in running tap or deionized water to remove air bubbles. The tap water (see **Table 1** for ionic composition) or deionized water is decontaminated by filtration through a membrane with a pore diameter of 0.47 μm. Vase-life is then evaluated.

Buffer solutions for evaluation of pH influence Sterilized distilled water is used to prepare stock solutions of citric acid (0.1 mol l^{-1}) and sodium phosphate (0.2 mol l^{-1}, McIlvain buffer). Dilutions of the stock solutions are added to the vase-water: 2 mmol l^{-1} citric acid and 16 mmol l^{-1} sodium phosphate (pH 7), 5 mmol l^{-1} citric acid and 10 mmol l^{-1} sodium

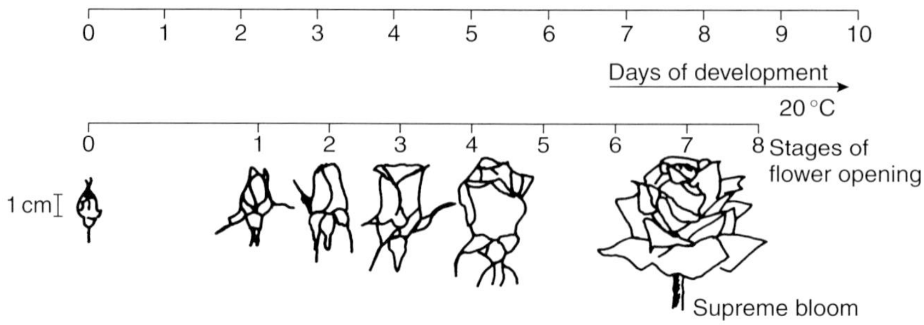

Figure 1 Stages in the flower opening and development of *Rosa* cultivars. Reproduced from Berkholst CEM (1980) De waterhuishouding van afgesneden rozen. *Bedrijfsontwikkeling* 11: 332–336, with permission.

Table 1 Composition of tap water[a] at Wageningen, Netherlands

Compound	*mg l^{-1}*	*mmol*
Chloride (Cl^-)	7	0.197
Nitrite (NO_2^-)	< 0.02	$< 4.35 \times 10^{-4}$
Nitrate (NO_3^-)	< 1	< 0.0016
Sulphate (SO_4^{3-})	6	0.00625
Hydrocarbonate (HCO_3^-)	91	1.492
Carbon dioxide (CO_2)	2	0.045
Carbonate (CO_3^{2-})	< 1	< 0.0167
Phosphate (PO_4^{3-})	0.08	8.42×10^{-4}
Ammonium (NH^{4+})	< 0.05	$< 2.78 \times 10^{-3}$
Iron (Fe)	0.02	3.58×10^{-4}
Manganese (Mn)	< 0.01	$< 1.82 \times 10^{-4}$
Oxygen (O_2)	7.3	0.228
Calcium (Ca^{2+})	27	0.675
Magnesium (Mg^{2+})	2.2	0.091
Sodium (Na^+)	6	0.261
Potassium (K^+)	0.5	0.013
Copper-dissolving ability	0.6	
pH 7.9		
Electrical conductivity	0.017 S m^{-1}	

[a] The tap water in the Netherlands does not contain particulate matter not passing a micropore filter of 0.45 and 0.22 μm diameter. This was the water used in experiments on *Rosa* 'Sonia'.

The ionic composition of tap water at Rutgers, USA is similar to that detailed in **Table 1** but contains a higher concentration of Fe (0.03–0.5 mg Fe l^{-1}) and some Fe-containing microparticles which do not pass through a 0.47 μm micropore filter. Rutgers water was used in experiments involving *Rosa* 'Kardinal'® described in this article.

Reproduced from Van Meeteren U, Van Gelder H, and Van Ieperen W (2000) Reconsideration of the use of deionized water as vase water in postharvest experiments on cut flowers. *Postharvest Biology and Technology* 18: 169–181, with permission.

phosphate (pH 5), 8 mmol l^{-1} citric acid and 4 mmol l^{-1} sodium phosphate (pH 3). The pH of the vase-water is measured when cut flowers are introduced and at intervals over the following 1–6 days.

Vase-water inoculation Suspensions of viable or heat-killed bacterial or fungal cells are washed and homogenized and added to sterile tap or deionized water at concentrations ranging from 10^8 to 10^4 ml^{-1}.

Preparation of inocula The bacteria *Bacillus subtilis*, *B. licheniformis*, *Enterobacter agglomerans*, *Pseudomonas aeruginosa* and *P. putida* are inoculated on nutrient agar (NA; Beckton Dickinson Bioquest) and incubated for 72 h at 30 °C. The fungi *Fusarium oxysporum* and *Kluyveromyces marxianus* are inoculated on malt extract agar (MA; Difco) and incubated for 2–4 days at 25 °C. The microbial growth on the surface of the plates is washed off into sterile deionized water of 10 °C, vortex-mixed and centrifuged at 10 °C and 10 000 **g**. The supernatant is decanted and the microbial cells suspended in sterile deionized water, mixed and centrifuged a second time. The washed cells are then diluted in sterile deionized water to a cell concentration of approximately 5×10^8 ml^{-1}. Cell concentrations are assessed microscopically, and by a colony-forming unit (cfu) most probable number (MPN) method on PA or MA agar. The suspensions are used within 24 h and stored at 5 °C.

Microorganisms: preparation of microbial metabolites Microbial metabolites are prepared by inoculating 5 ml of a microbial cell suspension into a growth medium of sucrose (5% w/v) and incubated at ambient temperature (25–30 °C) and time (5–8 days). The growth culture is centrifuged at 10 °C. The supernatant is decanted, and microbial cells and cell particles removed by pressure filtration through a 0.22 μm sterile disposable filter unit (Millipack-60, MPGL-06-Hz). For molecular sieving at a nominal cut-off level of molecular weight 10 kDa, a Millipore PT packet 10 K NMWL, retention separator 10/PK and pump motor 23 OV 50/60 HZ at 10 PSI is used. Filtrate and retained materials are made up to equal volumes with sterilized tap water, and then stored at 2–5 °C for up to 60 h.

Vases Cylindrical vases, calibrated to 200 ml, are presterilized. Each vase is filled with 200 ml sterilized vase-water with a composition as specified in the test protocol and one rose. The vases are covered with aluminium foil to prevent airborne contamination and maintained at 28 ± 1 °C. Light is delivered for 24 h each day or a 12-h light/dark period is used at 65 μmol m^{-2} s^{-1} PAR (Photosynthetic Active Radiation).

Vase-life observations Observations include water uptake by the rose shoot, stage of bud development and ornamental value. Vase-life is measured from placing in the vase solution to the loss of flower turgor. After 4 days' vase-life under normal conditions, bud development is mainly up to stage 7, a fully blooming rose (**Figure 1**), and may remain as such for another 4 days. The end of vase-life is marked by a visibly decreased ornamental value in approximately 50% of flower buds.

At time zero and 4 days of vase-life, the microbiological state of the vase fluids is assessed by determining bacterial cfu ml^{-1} vase fluid after inoculating decimal dilutions on plate count agar, and fungal cfu on potato dextrose agar (Becton Dickinson Bioquest) and incubating for 2–4 days at 25–30 °C. In addition, microbial cell numbers are assessed microscopically by Gram staining and a modified Breed smear technique showing the total number of viable and nonviable bacterial cells per ml of vase fluid.

Xylem hydraulic conductivity Water conductivity is expressed per transverse sectional area in a stem segment of 5 cm length. The stem segments are from identical plant material with regard to bud stage, vessel thickness and vessel height. Water conductivity is the reciprocal value of resistivity to water flow, thus the higher the water conductivity, the lower the water-flow resistivity of the xylem vessels. Three consecutive stem sections 5 cm each are cut off from the basal end upwards and inserted on to the manifold of the apparatus (**Figure 2**) in an inverted position.

The apparatus consists of an overhead 10-l bottle filled with glass-distilled water and is connected by medical plastic tubing to a glass manifold of 24 pieces. A Mariotte tube is inserted through a rubber stopper into the 10-l bottle to maintain constant water pressure. The distance or height of the water column measured from the base of the Mariotte tube to the upper end of the stem segment is 60 cm. A piece of soft plastic tubing is attached to each T-piece for connection to a single stem segment. Triplicate samples of roses are taken from the vase and three consecutive stem sections of 5 cm each are cut, under water, from the basal end upwards and, in an inverted position, inserted on to the manifold. The water solution is then allowed to flow through the segments for 60 min to purge air in the lines and to obtain an equilibrated water flow (**Figure 3**). Thereafter, each stem segment is connected to a plastic tube of known weight. The water flow through the segments is assessed after 30,

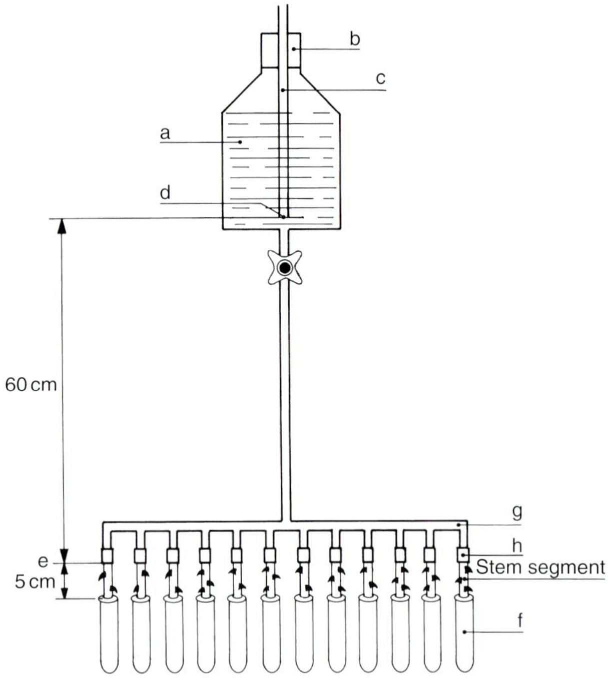

Figure 2 Apparatus designed to measure xylem conductivity. a, bottle containing 10 l sterile glass-distilled water; b, rubber stopper; c, Mariotte tube; d, base of the Mariotte tube; e, lower end of the inserted stem segment of 5 cm length; f, plastic tube of known weight to collect water flowing through the stem segment in 30 min; g, glass manifold with 12 connectors; h, connector of soft plastic tubing. Reproduced from Put HMC and Rombouts FM (1989) The influence of purified microbial pectic enzymes on the xylem anatomy, water uptake and vase life of *Rosa* cultivar 'Sonia'. *Scientia Horticulturae* 38: 147–160, with permission.

60 and 90 min flow and the average conductivity is calculated in ml per 30 min per stem segment of 5 cm.

Note: Unlike many chrysanthemum and gerbera cultivars, greenhouse roses do not contain a hollow cavity in the centre of the stem, giving two possible pathways for water uptake.

Measurement of water potential At day 4 (96 h), two rose stems from each factorial combination are removed at the last internodes, trimmed to a flower with 5-cm peduncle, and inverted into a Scholander pressure bomb for water potential measurements.

Methods for Microscopic Observations of the Xylem Structure

Xylem conduit length distribution Stem segments of 10–30 cm length are connected to a pumping system with silicone rubber tubing. With a pulling pressure difference of 50 kPa applied to the upper end of the segments, the lower end of the segments is placed in an aqueous 1% (w/w) suspension of red latex particles. Latex particles of $<2\,\mu m$ diameter are used; they can easily enter an open conduit, but are too large to pass bordered pit pairs. The particles are carried upwards into the open conduits by the mass of water flow. Once the water flow through the stem segments ceases, it is assumed that all cut open conduits are stained.

The segments are removed, placed in plastic bags and stored in a refrigerator at 4 °C for, at most, 2 weeks

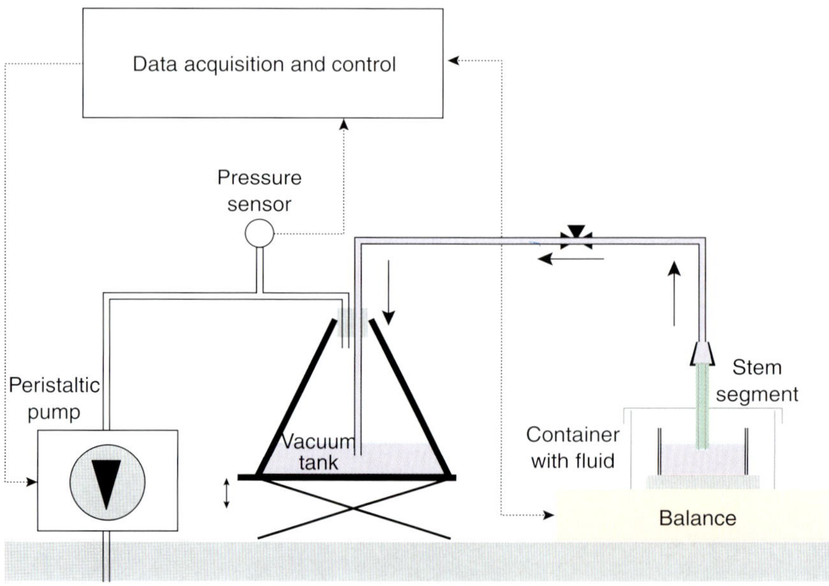

Figure 3 Layout of a recently developed apparatus for a rapid sensitive registration of hydraulic conductance of stem segments influenced by small variations of the ionic composition of the vase fluid. The arrows indicate the flow direction. Reproduced from Van Ieperen W, Van Meeteren U and Gelder H (2000) Fluid ionic composition influences hydraulic conductance of xylem conduits. *Journal of Experimental Botany* 51: 769–776, with permission.

before being subjected to further processing to determine xylem conduit length distribution. This is done by light microscope analysis on cross-sections of 2-mm-thick transverse sections (made with a razor-blade) of a series of samples from the upper and lower stem ends and of two or more sections of all internodes, to have both nodal and internodal intervals. Digital images of the sections are made with a colour camera attached to a binocular microscope.

Cross-section area analysis Portions of stem segments are embedded in Kayser's glycerol gelatine (Merck, 9242) on microscope slides and digital images are made. More than 30 images are needed to acquire a detailed record of the entire xylem area of one transverse section. The number of conduits (average), conduit cross-sectional area and the longest and shortest axis of an ellipse superimposed on the conduits are measured using the image analysis system.

Conduits are distinguished from fibres by eye and marked by mouse clicks using the presence of bordered pits as a criterion. Hydraulic conductivity of stem segments from different heights of the stem is calculated, assuming that conduits are infinite, uniform elliptic, parallel-arranged pipes.

The latex red test (*accumulation of micromaterials on cross-section xylem*) After 24, 48 and 72 h of vase-life, individual flowers are transferred to a shallow beaker containing a 0.3% w/v red acrylic latex dye in deionized water. The stem base is immersed up to at most 10 mm in the paint solution for 4 min. Flower stems are then held in air laterally for 30 min, allowing the paint to dry, and transverse slices of 1–2 mm thickness are cut from the stem base for examination under a 45 × binocular stereo objective showing red-stained microparticles or microcompounds accumulated on the xylem cross-wall (**Figure 4**).

The acid fuchsin test *(plugging of xylem vessels)* After 48 and 96 h flower life, roses are removed and placed for 30 min in an acid fuchsin solution of 0.5% w/v in 50% v/v ethanol. The epidermis and cortex of the stems are then peeled off and the extent and height of the uptake of the red-coloured solution into the vessels can be seen by the naked eye. This indicates blockage of the water transport into the vessels, physically by microorganisms and particulate matter, and physiologically by loss of functional vessels.

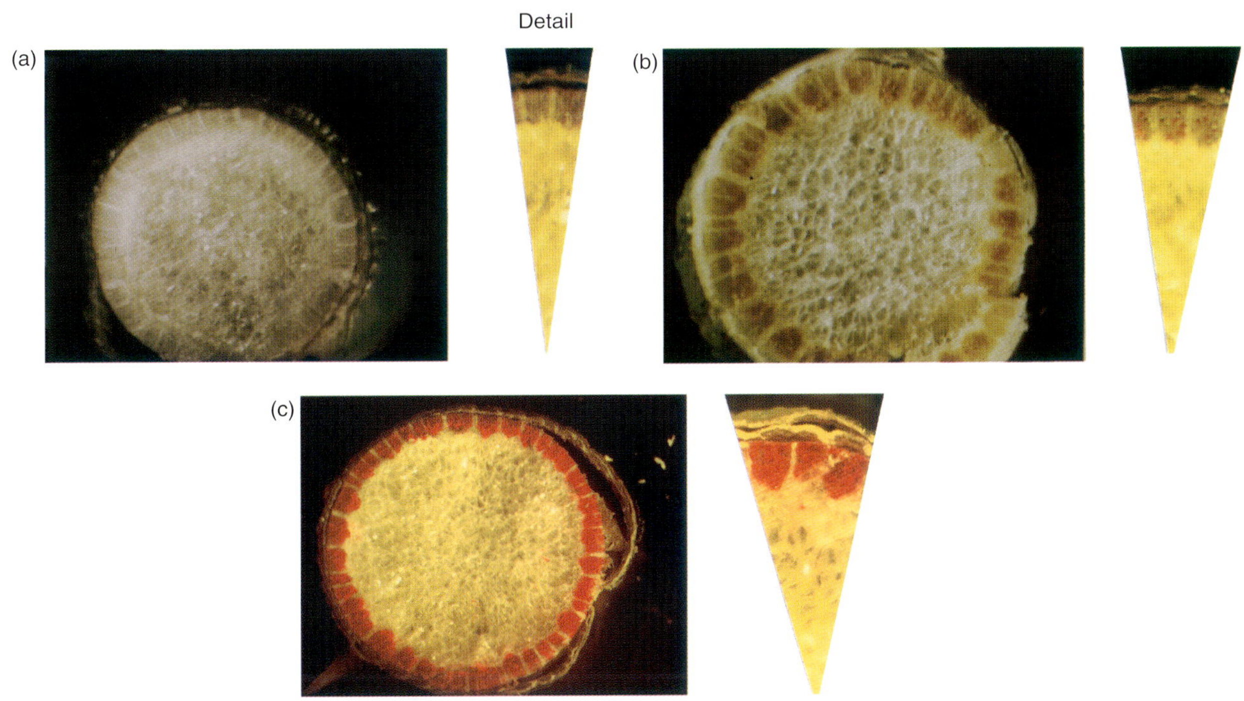

Figure 4 *Rosa* 'Kardinal' cut surface cross-sections examined under 45 × binocular stereo objective after 2 days vase-life in: (a) tap water; (b) tap water given a latex red test; (c) deionized water inoculated with *Bacillus subtilis* (10^6 ml^{-1}). Reproduced from Durkin DJ, Put HMC and Clerkx ACM (1998) SEM and quality control of cut roses: the latex red test. *Microscopy and Analysis* July: 9–12, with permission.

Microscopic observations *(vascular sections)* (a) Fresh thin stem sections of 10 μm thickness made by a swinging arm microtome are stained with either toluidine blue O (TBO) or safranin-alcian green, or unstained, viewed by a microscope. (b) Fixation of vascular sections of 10 μm is overnight in formaldehyde : aceticacid : ethanol 5 : 5 : 90 after 60 min vacuum conditions and the pith is excised for faster air removal. Dehydration is followed by a graded series of ethanol/water. The dehydrated specimens are embedded in 7100 Technovit (hydroxy ethyl metacrylate) and stained or not. (c) To determine the cellular composition of the xylem, macerates of shoot tissue are made, using peroxide, glacial acetic acid and water 1 : 1 : 1 at 50 °C for 24 h and stained with TBO or acid fuchsin.

Scanning electron microscope (SEM) specimen preparations After 0, 24, 48 or 72 h vase-life, and/or after execution of a latex red test, *c.* 10-mm pieces are cut from the individual flower stem end using a sterile razor blade. Pieces of the stem ends are fixed in 2% w/v gluteraldehyde in cacodylate buffer 0.1 mol l^{-1}, pH 7.2. The fixed specimens are transported and stored at 5 °C until the dehydration process, which is executed within 4 weeks of sample preparation. The specimens are then washed in deionized water and dehydrated through a graded series of ethanol/water of 10% v/v to 100%, leaving the specimens in each solution for 10 min. The specimens are freeze-fractured in the 96% solution, critical-point-dried and mounted with Leit-C conductive carbon paste on to aluminium stubs. Finally, the specimens are coated with gold-palladium (80–20) and examined in an SEM at 15 kV accelerating voltage.

Cryo-SEM specimen preparation Pieces of a rose stem are taken at the first centimetre, and at 5, 10 and 20 cm from the cutting end. The pieces are mounted on brass stubs and subsequently frozen in nitrogen slush. They are directly sputtered with gold and examined for air emboli, or they are transferred to the preparation chamber of a Hexland–Oxford CT 1000/CP2000 system and freeze-fractured. The specimens are etched for *c.* 30 min at 190 K (temperature anticontaminator is 90 K). Finally the specimens are sputtered with gold and examined in an SEM at *c.* 110 K at 15 kV accelerating voltage. In the cryo-SEM images empty (embolized), water (solutions) and air in vessels can easily be distinguished.

X-ray microanalysis The specimens prepared for SEM are coated with carbon. Each stem piece is measured five times at different points by EDX (Energy Dispersive X-ray) analysis. Conditions: 100 live counting time and 15 kV accelerating voltage, spot 500 nm.

Analysis of Factors Leading to a Disturbed Water Status of the Cut Rose

Microbial Particles in Vase-Water

The influence of washed viable and washed heat-killed bacteria on the water conductivity (**Figure 5**), the vase-life (**Figure 6**), the bud development (**Figure 7**) and daily water uptake (**Figure 8**) of 'Sonia' shows that vessels are entirely blocked (acid fuchsin test) when 10^8 ml^{-1} bacteria are added to the vase fluid. When 10^4 ml^{-1} bacteria are added, only some of the smaller vessels are blocked, slightly reducing water conductivity and vase-life.

Small differences occur between equal initial numbers of viable and heat-killed cells and between washed suspensions of equivalent concentrations of *Bacillus*, *Pseudomonas* and *Enterobacter*. These bacterial species are furthermore unable to pass the intervessel pit membranes and only a minority of the bacteria initially suspended in the vase fluid are able to infiltrate, with the sap stream, into open vessels. Their number decreases with increasing distance from cutting point to sampling point, contact time and increasing concentrations of initially suspended bacteria (**Figure 9**).

Microbial Metabolites in Vase-Water

Roses cut at bud stages 0–1 (**Figure 1**) need metabolic sugars for respiratory energy. The natural sugar source of the young bud and leaves is insufficient for complete flower development (**Figure 10**). Sugars added to the vase-water as so-called 'cut-flower food' reduce the sugar deficit and contribute towards osmotic adjustment and protection of membrane integrity. However, sugars promote multiplication of microorganisms that depend on carbohydrates for their growth, leading to formation of capsular slime and bacterial secretions of high-molecular materials like exopolysaccharides.

Microbial exopolysaccharides of high molecular weight (>10 kDa) in vase-water show xylem vessel blockage at the cut end. Injury and disorganization to the structure of the lower parts of the vessels result in a decrease in water conductivity, water uptake, vase-life and flower bud development. Exopolysaccharides do not pass through the xylem pit membranes (**Figures 11** and **12**). The higher the conversion of sucrose into polysaccharides, the more disturbed are the water relations of the flower.

Microbial metabolites of low molecular weight (<10 kDa) pass the pit membranes without causing xylem blockage, but disturb bud development even when diluted 100-fold (**Figures 13** and **14**).

If purified microbial pectic enzymes, pectate lyase and polygalacturonase are added at concentrations of

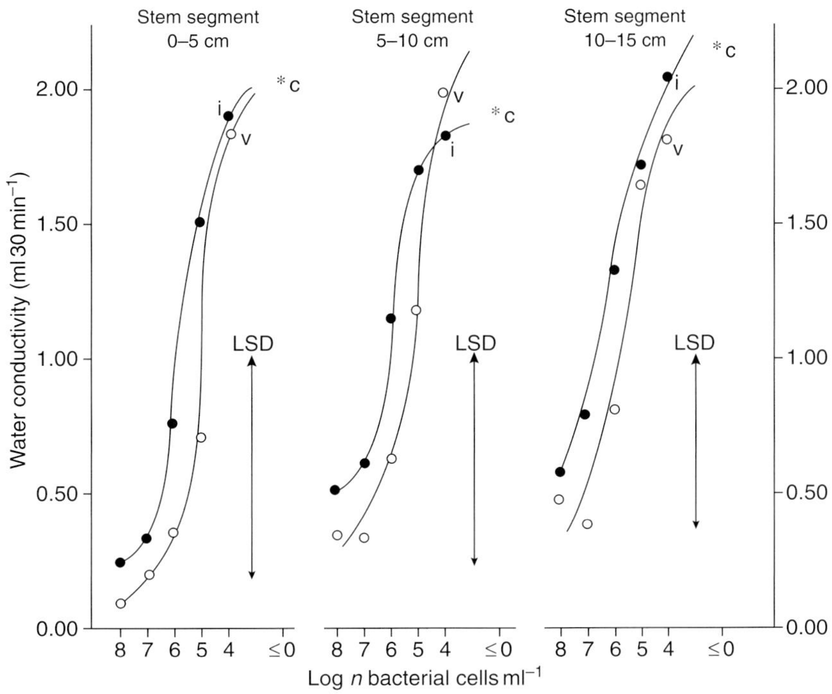

Figure 5 The influence of the number of viable or heat-inactivated bacteria, added per ml vase water, on the hydraulic conductivity of stem segments of *Rosa* 'Sonia' after 2 days vase-life. (○), viable bacteria (v); (•), inactivated bacteria (i); (*), sterilized tap water (c). Values are means of three triplicate tests of five different bacterial strains. Reproduced from Put HMC and Jansen L (1989) Effects on the vase-life of cut *Rosa* cv. 'Sonia' of bacteria added to the vase-water. *Scientia Horticulturae* 39: 167–179, with permission.

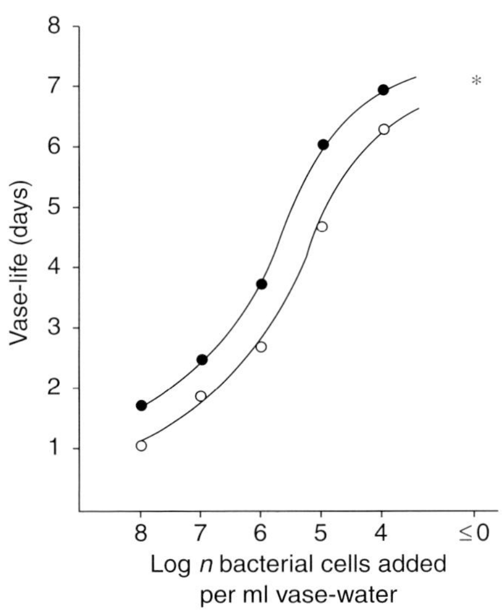

Figure 6 The influence of the number of bacteria, added per ml vase-water, on the vase-life of *Rosa* 'Sonia'. (○), viable bacteria; (•), inactivated bacteria; (*), sterilized tap water. Values are means of three triplicate tests of five different bacterial strains. Reproduced from Put HMC and Jansen L (1989) Effects on the vase life of cut *Rosa* cv. 'Sonia' of bacteria added to the vase water. *Scientia Horticulturae* 39: 167–179, with permission.

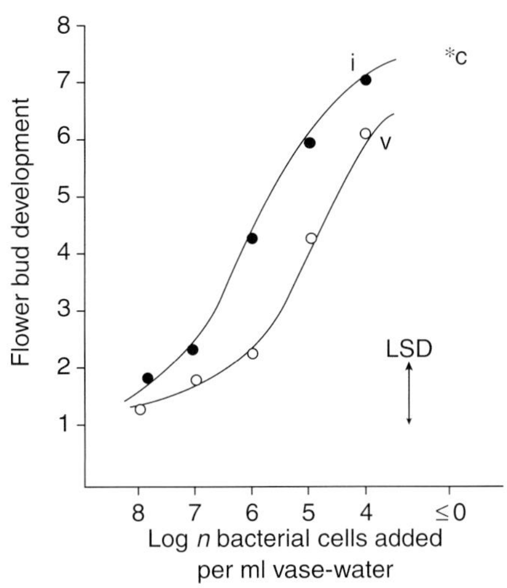

Figure 7 The influence of the number of bacteria, added per ml vase-water, on the bud development of *Rosa* 'Sonia'. (○), viable bacteria (v); (•), inactivated bacteria (i); (*), sterilized tap water. Values are means of three triplicate tests of five different bacterial strains. Reproduced from Put HMC and Jansen L (1989) Effects on the vase-life of cut *Rosa* cv. 'Sonia' of bacteria added to the vase-water. *Scientia Horticulturae* 39: 167–179, with permission.

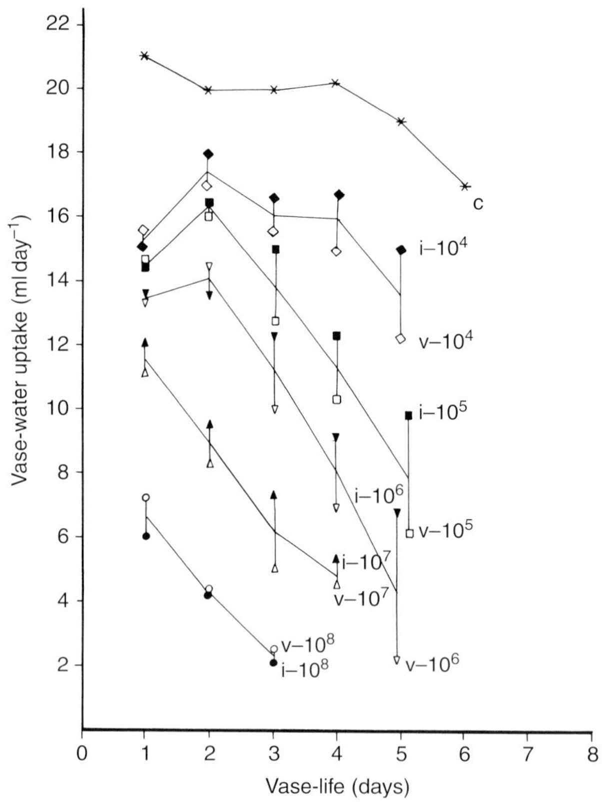

Figure 8 The influence of the number of bacteria, added per ml vase-water, on the daily water uptake (ml) of *Rosa* 'Sonia' in relation to time in vase. Open symbols indicate viable bacteria and solid symbols indicate inactivated bacteria: (○/●), 10^8 ml^{-1}; (▲/△), 10^{-7} ml^{-1}; (▼/▽), (v) 10^6 ml^{-1}; (■/□), (v) 10^5 ml^{-1}; (◆/◇), 10^5 ml^{-1}; (*), sterilized tap water. Values are means of three triplicate tests of five different bacterial strains. Reproduced from Put HMC and Jansen L (1989) Effects on the vase life of cut *Rosa* cv. 'Sonia' of bacteria added to the vase water. *Scientia Horticulturae* 39: 167–179, with permission.

0.1–1 units ml^{-1}, disturbance of the water status leads to injury of the vessel pit structure and desiccation of flower petals, probably by attacking the pectic substances of the middle lamella. A slightly disturbed bud development is also observed at a low enzyme concentration (0.1 units ml^{-1}) in the vase fluid (**Figures 15** and **16**).

Phytoalexin and their Elicitors

Small particles in vase-water, even at very low concentrations, and xylem cell wall fragments may markedly decrease water uptake and disturb the water status of the flower, although the vessel structure remains normal and no vessel occlusions are observed. This phenomenon reveals that these micromaterials act as biotic or abiotic phytoalexin elicitors, leading to strong physiological reactions, a decreased water conductance and decreased vase-life.

Air in Xylem Vessels of Cut Roses

When a rose stem is cut, air is trapped in the open xylem vessels, disturbing the water transport and the flow of water to the flower. After a period of dry transport, the flower needs to be rehydrated or deaerated. Rehydration ability is cultivar-dependent, and is related to unknown factors of the vessel structure (pit border) and/or hormone function. The rose 'Sonia' does easily rehydrate, even after a long dry period of transport and storage. 'Cara Mia' and 'Kardinal'®, however, need a sensitive rehydration even shortly after cutting. Air in xylem vessels can be demonstrated by cryo-SEM.

Relation of a Disturbed Water Status to the Xylem Structure of Cut Roses

In cross-sections of a stem, the xylem can be seen to contain numerous (about 1500) vessels, 2.5–50 μm wide and 2.0–250 mm long. They contain many intervessel pit membranes with capillary holes of about 50 nm which are too small for bacteria and macromolecular materials to pass through (**Figure 17**). The smaller vessels show capillary activity (*see* **Morphology and Anatomy**: Shoots and Stems).

The xylem vessel wall, in cross-section, 1.5–3.5 μm wide, acts as a sieve or filter on which macromolecular material, particulate matter and microbial cells present in vase-water may accumulate (**Figure 18**).

The cross-section xylem vessel filter phenomenon indicates that only a small fraction of microbial cells, affirmed by bacterial counts (**Figure 9**), particulate matter and macromolecular materials, is able to infiltrate the sap stream and enter the vascular system, obstructing the smaller vessels, while the remainder accumulates at the cut surface on the xylem cross-wall. Simultaneously and reciprocally there is particle repulsion into the surrounding vase solution.

Microbial cells may loosely (reversibly) attach on to the xylem cross-walls, the pores of which are generally wider than the microbial cell, microparticle or particulate matter suspended in the surrounding vase fluid. They then act as a superlayered microfilter obstructing in part the water uptake.

Formation of a microbial biofilm and carbohydrates leaking from the phloem may locally initiate chemostatic responses to nutrient gradients around the cut surface, leading to movement towards, accumulation, and possible adhesion of viable microbial cells on to the cut surface.

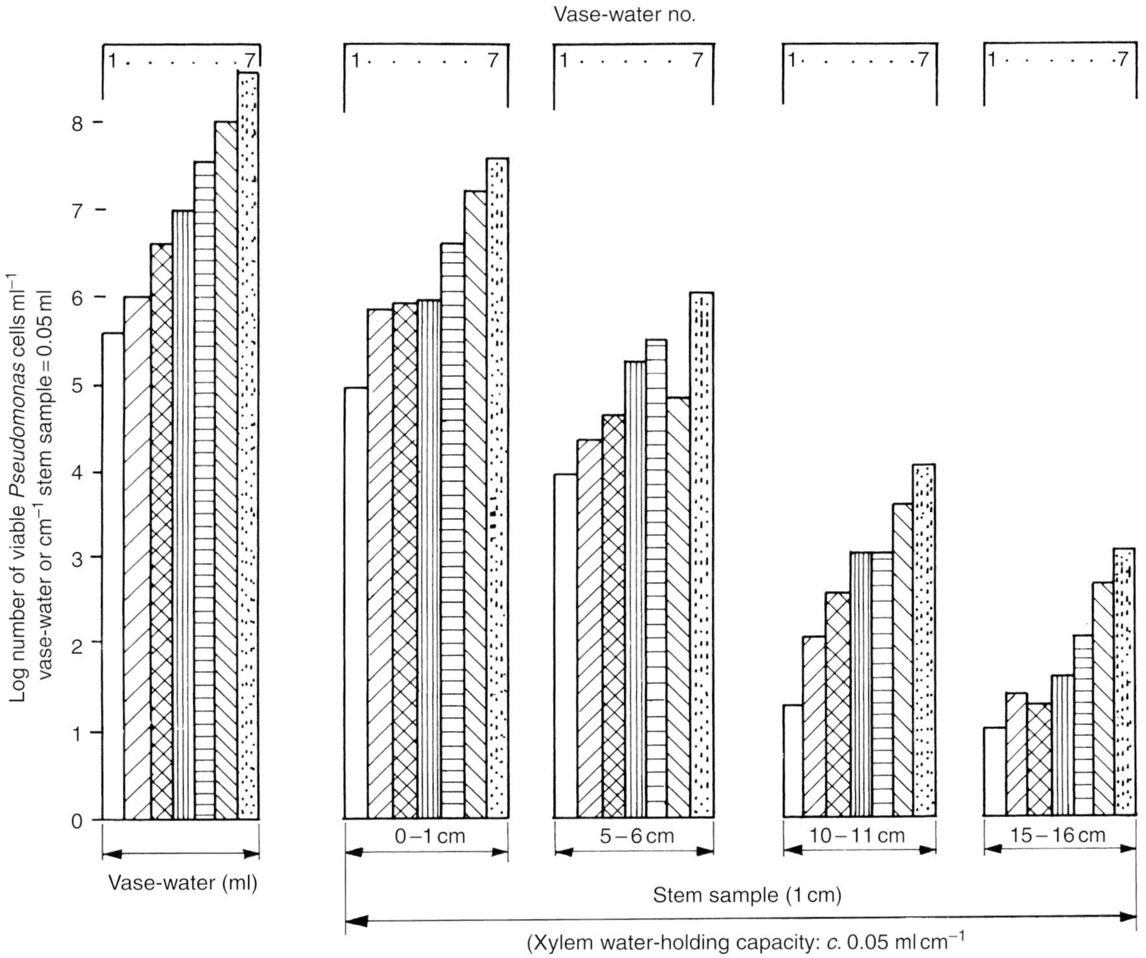

Figure 9 The number of infiltrated *Pseudomonas putida* in 1-cm vessel segment at four different levels of the xylem vessel of cut *Rosa* 'Sonia' after 24 h vase-life in vase water inoculated with seven subsequent dilutions (vase water no.) of *P. putida*. Maximum 10^8, 5×10^7, 10^7, 5×10^6, 10^6, 5×10^5 minimum $10^5\,ml^{-1}$ *P. putida*. The average water capacity of 1-cm vessel segment $= 0.05 \pm 0.003$ ml. Reproduced from Put HMC and Van der Meyden T (1988) Infiltration of *Pseudomonas putida* cells, strain 48, into xylem vessels of cut *Rosa* cv. 'Sonia'. *Journal of Applied Bacteriology* 64: 197–208, with permission.

Fluid dynamic forces serve to disperse microorganisms and microparticles evenly in proximity to the flower stem. Attachment phenomena – chemotaxic, physicochemical or physical – tend to direct the microorganisms towards the stem base where they accumulate, as shown by SEM. Properties of the interface are determined by both the rose stem cut surface and the aqueous environment (the vase solution), in which the microorganisms are suspended and the rose stem is immersed.

When a rose stem is immersed in an aqueous ionic environment, it usually acquires surface charge, either by adsorption of ions or by ionization of surface groups. The surface charge depends, to a great extent, upon the composition of the aqueous phase. In an aqueous environment, these surface charges are counterbalanced by oppositely charged ions; a part is bound to the surface and the remainder is distributed in the diffuse stem surface layers. This attraction–repulsion phenomenon is related to a Derjaguin, Landau, Verwey and Overbeek (DLVO)-type Van der Waals attraction and electrostatic repulsion in aqueous vase-water with a low ionic strength ($< 0.001\,mol\,l^{-1}$), such as tap water. Bacteria and xylem surface are both negatively charged. At increasing ionic strength of the vase fluid (up to $0.1\,mol\,l^{-1}$) deposition occurs, determined by the extension of the macromolecules into the surrounding medium. It either promotes deposition by bridging or inhibits it by steric repulsion. In media like tap water, most bacteria carry a surface charge of negative polarity due to the ionization of surface groups. The isoelectric point of these bacteria is *c.* pH 3.0. Near this pH bacteria tend to agglutinate spontaneously by

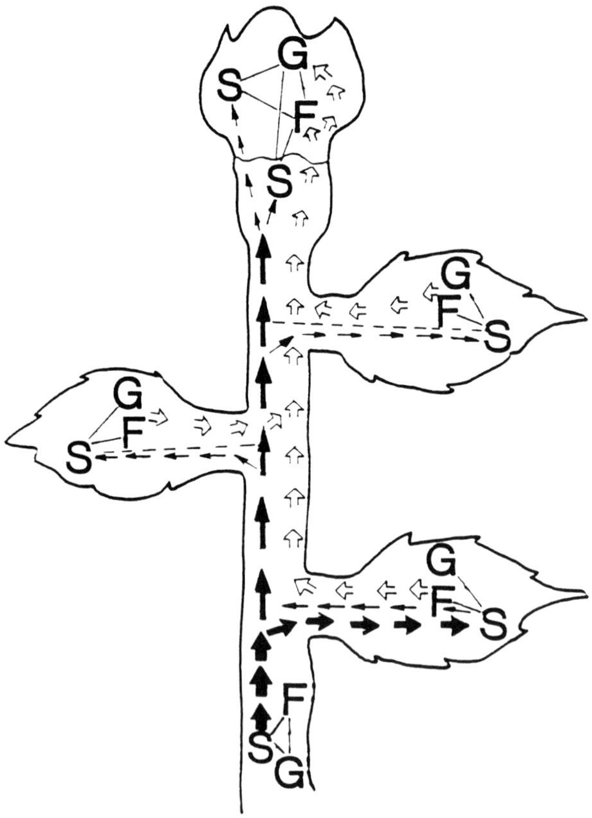

Figure 10 Diagram representing (semiquantitatively by means of arrow thickness) carbohydrate transport in the floral branch of 'Carina'® rose supplied with a solution of glucose or sucrose (G, glucose; F, fructose; S, saccharose). When the floral branch of a rose is supplied with a solution of 0.16 mol l^{-1} glucose or 0.08 mol l^{-1} sucrose, the level of carbon rose at the same rate in both cases. Glucose is immediately transformed into sucrose in the stem axis. Sucrose migrates to the leaves and flower and is equally distributed between the ovary and the petals. In the flower, sucrose is immediately hydrolysed; in the leaves its hydrolysis occurs more slowly. A strong isomerase activity occurs in the petals. Reproduced from Paulin A, Souter D and Bureau JM (1981) Evolution comparée des glucides dans les diverse organes de la rose coupée alimentée temporairement avec une solution glucose ou de saccharose. *Physiologie Végétale* 19: 59–76, with permission.

clumping and precipitation into the surrounding medium; this explains the repulsion phenomenon of bacteria on the xylem cross-wall and lower parts of the xylem vessels at pH 3.

Vase-water Pretreatment

In vase fluids containing a certain number of microorganisms, macromolecular material and/or particulate matter, lowering the pH to 3 results in a decreased accumulation of these micromaterials on the cut surface of the xylem and the lower parts of the xylem vessel wall. This improves the water status and increases the vase-life of the flower (**Tables 2** and **3**; **Figures 19** and **20**).

Particulate matter, as observed in US tap water, shows a comparable phenomenon by deposition of a thick layer of amorphic materials on the xylem cross-wall and lower parts of the xylem vessels, which increases when bacteria (up to 10^6 ml^{-1}) are added to the US tap vase-water. They show particle repulsion to a certain extent at pH 3 (**Figures 21** and **22**).

In addition, monovalent or divalent cations normally present in sterile Dutch tap water (**Table 1**), even at concentrations of less than 0.1 mol l^{-1}, may increase hydraulic conductance in xylem conduits by modifying the resistance of the xylem walls due to the presence or absence of unmasked negatively charged groups along and across the xylem cell walls (**Figures 3** and **23**).

Reduced infiltration of bacteria and other anionic inorganic and organic substances on to and into xylem vessels is also obtained by adding 1.5 mmol l^{-1} Al^{3+} in vase-water of pH 3–4. This leads to the formation of flocky deposits or coprecipitation of $Al(OH)_3$ around bacterial cells and microparticles. These act as a prefilter, leaving only a relatively coarse and water-porous plaque on the stem base. They do not obstruct water uptake but interfere with the accumulation of capsular slime, preventing microbial and microparticle accumulation and infiltration into the vessels (**Figures 24** and **25**).

The xylem conduits are not straight cylinders from the cut surface to the top; however, only a minor increase in the diameter of individual vessels is observed by SEM, showing relatively larger conduits in the lower parts of the stem. Additional data concerning the influence of the vessel shape on the hydraulic conductivity are obtained by measures of hydraulic conductivity at short 30-s intervals and along >15 cm stem (**Figure 3**).

The Peduncle

The zone between the flower and the stem, the peduncle, represents, by its hydraulically complex and mainly high-resistance structure, an important source of water deficit, just as the young bud needs water and energy for floral development (*see* **Morphology and Anatomy**: Flower Development and the Abscission Zone). This phenomenon almost certainly results from the transfer of water from peduncle to flower which, in practical terms, deprives the peduncle of turgor. It leads to some kinking of the weak wall of the tracheids and this translates to a further loss of turgor and ultimately 'bent neck'.

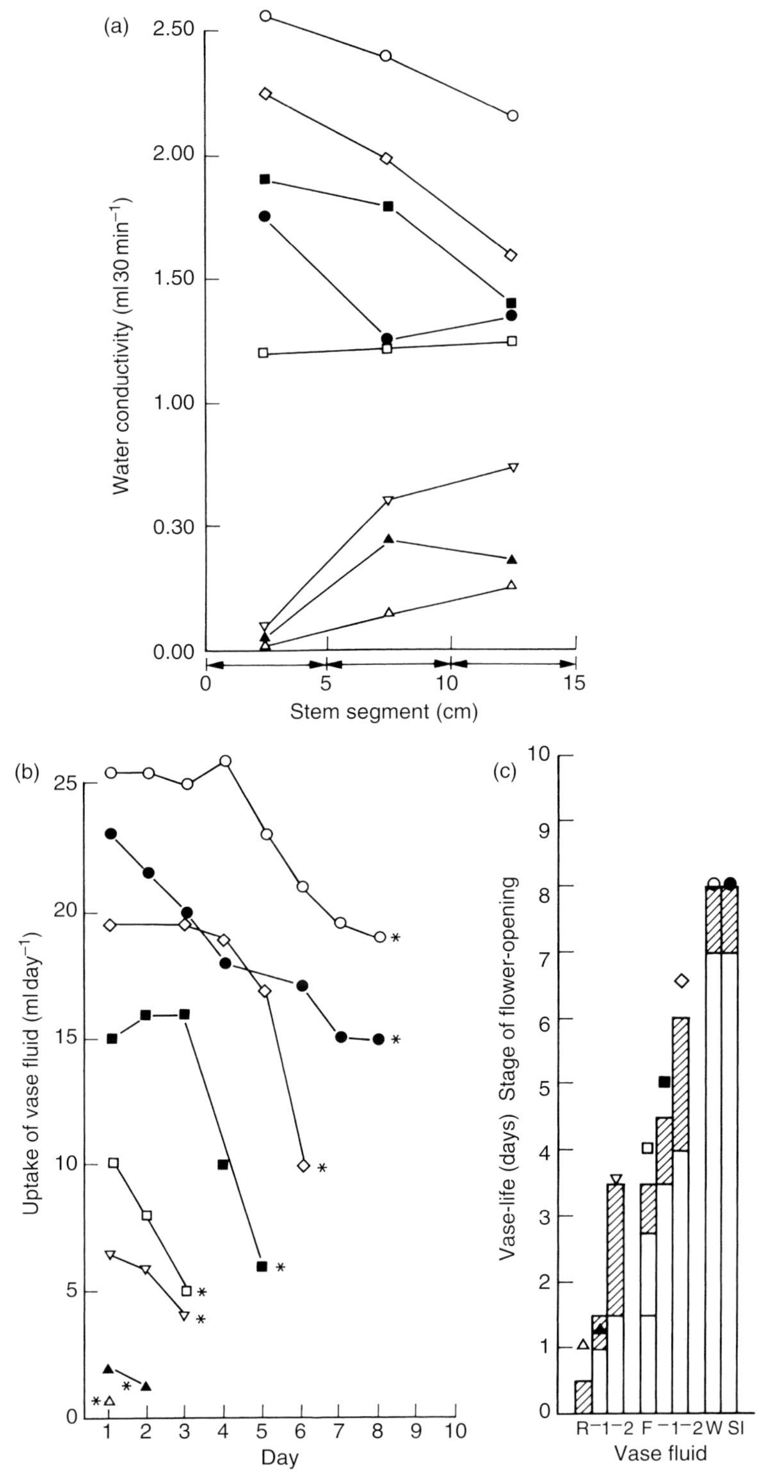

Figure 11 Water relations and vase-life of *Rosa* 'Sonia' in vase-water containing molecular filtrate and retained culture fluids of *Bacillus subtilis*. (a) Hydraulic conductivity (ml 30 min^{-1}) of three consecutive *Rosa* stem segments of 5 cm each after 1 day vase-life. Vase water additives are prepared from supernatant fluid of a shaken culture of *B. subtilis*, separated by a molecular separator at 10 kDa. Dilutions of retained culture and molecular filtrate are prepared in sterilized tap water. (△), retained culture, R; (▲), R, 1 : 10; (▽), R, 1 : 100; (□), molecular filtrate, F; (■), F, 1 : 10; (◇), F, filtrate 1 : 100; (○), sterilized tap water, W; (●), sucrose 1% in sterilized tap water, S. (b) Uptake of vase fluid (ml day^{-1}); for symbols see (a); (*), end-point of the vase-life of rose flowers; (c) vase-life (days). The shaded part of the vase-life bar indicates the number of days during which the ornamental value of the roses is mediocre. Reproduced from Put HMC and Klop W (1990) The effects of microbial exopolysaccharides (EPS) in vase water on the water relations and vase life of *Rosa* cv. 'Sonia'. *Journal of Applied Bacteriology* 68: 367–384, with permission.

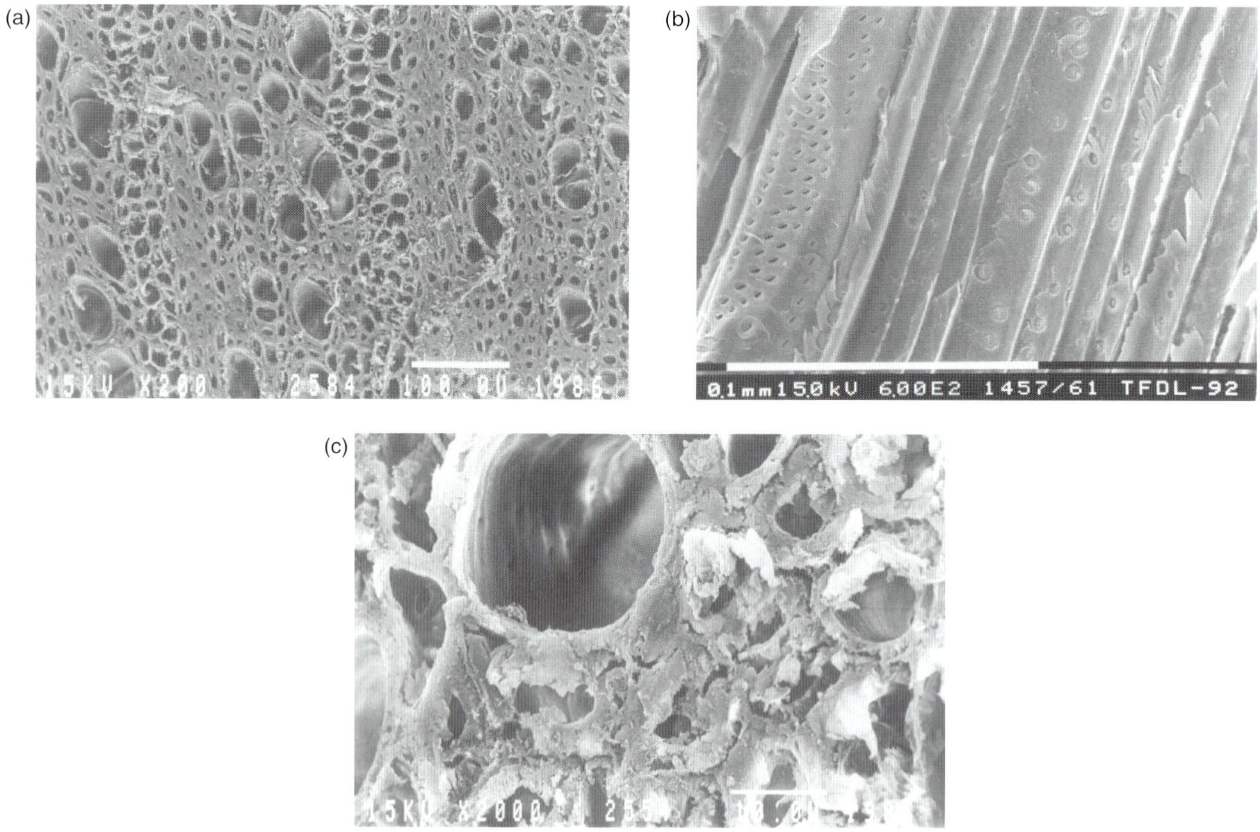

Figure 12 Scanning electron microscopy. (a) Transverse section near the cut surface of *Rosa* 'Sonia' after 1 day vase-life in sterilized tap water. Bar = 100 μm. (b) Longitudinal section near the cut surface of the same rose. Bar = 0.1 mm. (c) Cut surface transverse section of the stem of *Rosa* 'Sonia' after 1 day vase-life in exopolysaccharides (EPS) containing retained culture of *Bacillus licheniformis*. It shows amorphic material on the vessel wall and occlusions of the smaller vessels. Bar = 10 μm. Reproduced from Put HMC and Clerkx ACM (1988) The infiltration ability of microorganisms: *Bacillus, Fusarium, Kluyveromyces* and *Pseudomonas* spp. into xylem vessels of *Gerbera* cv. 'Fleur' and *Rosa* cv. 'Sonia' cut flowers: a scanning electron microscope study (SEM). *Journal of Applied Bacteriology* 64: 515–530, with permission.

Figure 13 Leaves of *Rosa* 'Sonia' held 2 days in: culture filtrate (FO) of *Fusarium oxysporum*; RO, retained culture; W, sterilized tap water. The RO and FO leaves show chlorotic spots. Reproduced from Put and Klop (1990) *Journal of Applied Bacteriology* 68: 367–394, with permission.

Figure 14 The ornamental value of *Rosa* 'Sonia' after 2 days vase-life in: F–1, vase-water prepared from a 1/10 dilution of a molecular filtrate $M < 10\,000$ of *Bacillus subtilis*, shows delay in bud development; R–1 vase-water prepared from a 1/10 dilution of retained culture of *B. subtilis* $M > 10\,000$ shows bent neck; W, sterilized tap vase-water, showing normal bud development. Reproduced from Put HMC and Klop W (1990). The effects of microbial exopolysaccharides (EPS) in vase water on the water relations and vase life of *Rosa* cv. 'Sonia'. *Journal of Applied Bacteriology* 68: 367–394, with permission.

Figure 15 Purified pectate lyase: Pect 1E; 0.5E; 0.1E, pectate lyase units ml^{-1} added to 1/80 mol l^{-1} Tris buffered vase-water pH 7.5 of *Rosa* 'Sonia' after 3 days vase-life shows curling dark-red coloured petal edges and dried necrotic leaves compared to normal bud development of roses in pH 7.5 buffered vase water (far right). Reproduced from Put HMC and Rombouts FM (1989) The influence of purified microbial pectic enzymes on the xylem anatomy, water uptake and vase life of *Rosa* cultivar 'Sonia'. *Scientia Horticulturae* 38: 147–160, with permission.

The variable conductance across the abscission zone and peduncle regions in response to water stress is related to differences in the internal anatomical features of the xylem conducting system (**Figure 26**). The peduncle has a highly developed xylem with relatively large-diameter elements but the distal segment of the peduncle is poorly vascularized and the remainder of the peduncle, including the abscission zone, is a transition between the two extremes. The abscission zone and peduncle regions are a site of reduced hydraulic conductance due not to a reduction in the number or size of xylem conduits but to changes in the alignment of the elements (**Figure 27**).

The length of peduncle is typically 105 ± 15 mm, depending on the cultivar. The separation between the peduncle and the stem is frequently indicated by differences in colour. The distal 15–20 mm segment of the peduncle, measured from the ovary, shows a much weaker vascular development than the stem. Vessel lengths in the mid-region of the peduncle are shorter than in the stem (**Figure 28**). However, neither a reduction in the number of elements nor changes in their diameter are apparent when the region is examined in cross-section.

Regardless of where the stem is cut below the abscission zone, microparticles of dye infused from the stem end do not penetrate the abscission zone, and no vessels are continuous through the zone. Thus the hydraulic systems of the two organs are semi-isolated from each other by the abscission zone (**Figure 29**).

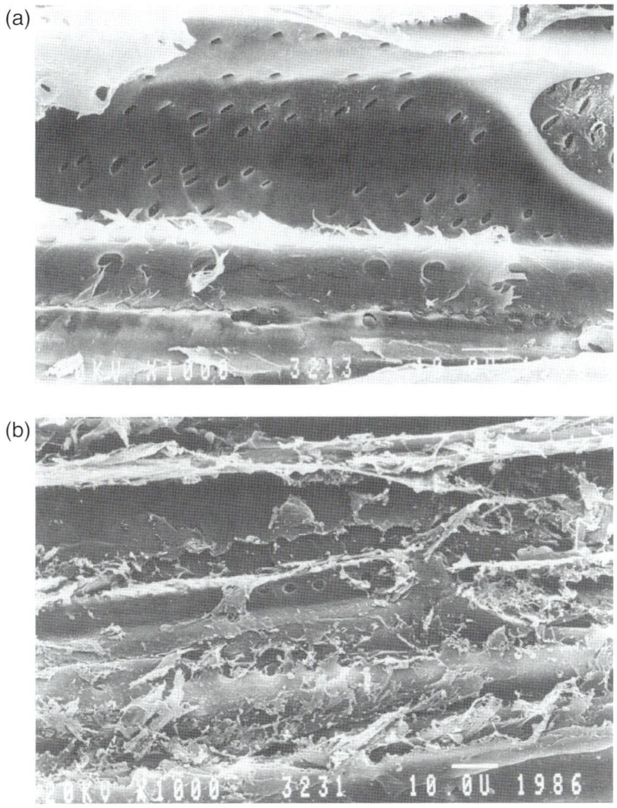

Figure 16 Scanning electron microscopy. (a) Longitudinal section of a xylem vessel of *Rosa* 'Sonia' near the base of the stem after 2 days vase-life in buffered vase water pH 7.5 showing a normal vessel structure. Bar = 10 μm. (b) Longitudinal section of a xylem vessel of *Rosa* 'Sonia' near the base of the stem after 2 days vase-life in buffered vase water pH 7.5 and addition of 5 units purified pectate lyase ml^{-1} showing degradation of vessel structure. Bar = 10 μm. (b) Reproduced from Put HMC and Rombouts FM (1989) The influence of purified microbial pectic enzymes on the xylem anatomy, water uptake and vase life of *Rosa* cultivar 'Sonia'. *Scientia Horticulturae* 38: 147–160, with permission.

The hydraulic architecture of the peduncle and abscission zone represents the most sensitive part for water stress of the xylem vessels system of cut roses.

The Consequences

Genetic studies are needed to assist breeders to obtain a reinforced hydraulic architecture of the peduncle and abscission zone. This would lead to a decrease in sensitivity to water stress and an increase in the stability of the water status of the cut flower.

Improved protocols are needed for aseptic handling and transport of cut roses.

Improved protocols are needed for additives that will give a standardized and controlled composition of vase-water.

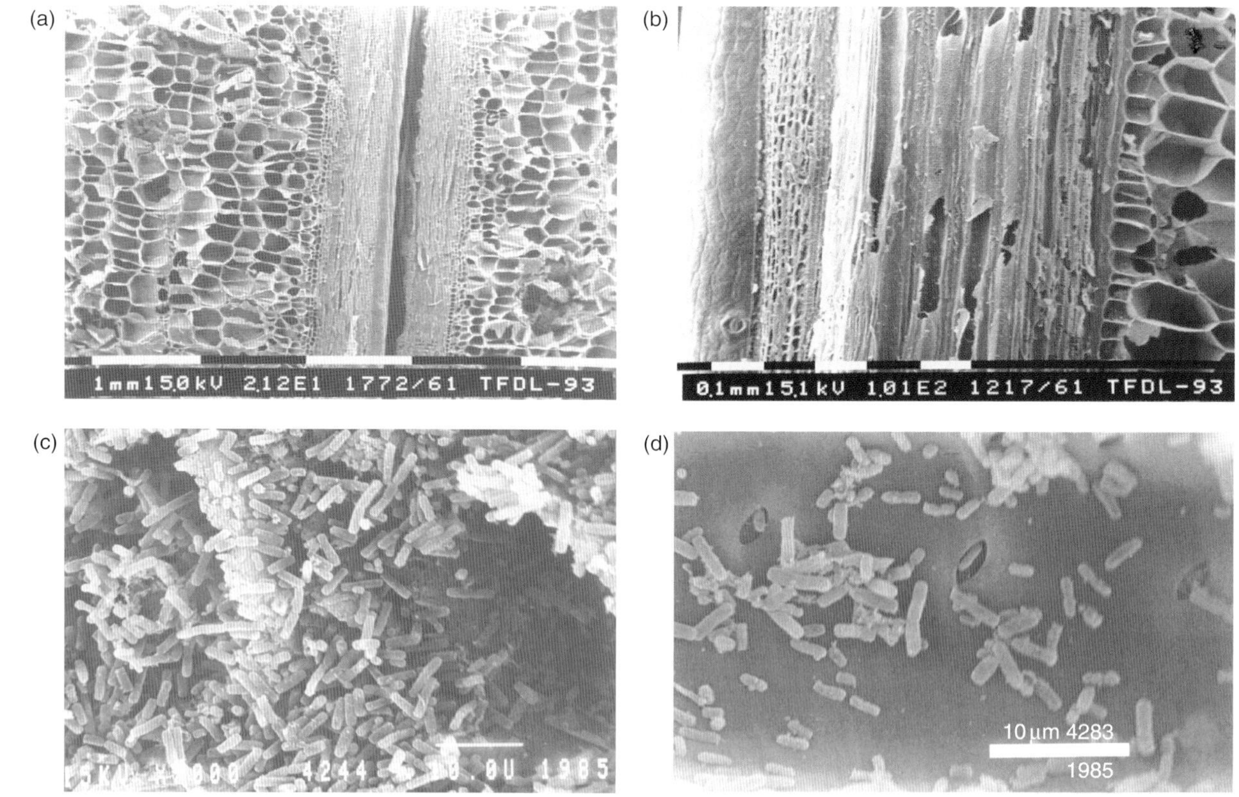

Figure 17 Scanning electron microscopy. (a) Overview of a longitudinal section of two parts of a *Rosa* 'Kardinal' vessel. Bar = 1 mm. (b) Overview of a longitudinal section at 2.5 mm from the cutting point of a hydrated *Rosa* 'Kardinal' shows cortex, phloem, cambium, xylem and pith structure. Bar = 0.1 mm. (c) Transverse section of the cut surface of *Rosa* 'Sonia' after 4 h vase-life in vase-water inoculated with 10^7 ml^{-1} *Bacillus polymyxa* shows a thick layer of *B. polymyxa* bacteria accumulating on the cut surface. Bar = 10 μm. (d) Vessel wall 10 mm from the cut surface of a *Rosa* 'Sonia' after 1 day vase-life in vase-water composed as in (c) shows a relatively low number of infiltrated bacteria. Bar = 10 μm. (b) Reproduced from Put HMC and Clerkx ACM (2001) Anatomy of cut *Rosa* xylem observed by Scanning Electron Microscope. *Acta Horticulturae* 547: 331–339, with permission. (c–d) Reproduced from Put HMC and Clerkx ACM (1988) The infiltration ability of microorganisms: *Bacillus, Fusarium, Kluyveromyces* and *Pseudomonas* spp. into xylem vessels of *Gerbera* cv. 'Fleur' and *Rosa* cv. 'Sonia' cut flowers: a scanning electron microscope study (SEM). *Journal of Applied Bacteriology* 65: 515–530, with permission.

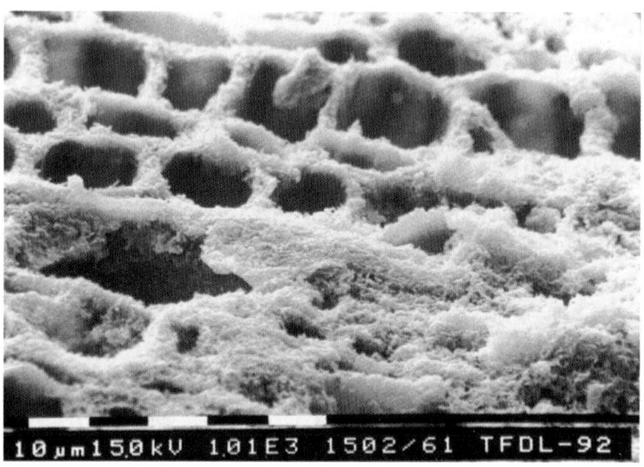

Figure 18 Scanning electron microscopy. Xylem vessel cross-wall filter phenomenon. Transverse section of the cut surface of *Rosa* 'Kardinal' after 2 days' vase-life in sterilized deionized tap water inoculated with 5×10^6 ml^{-1} *Pseudomonas aeruginosa* showing bacterial cells accumulating on the xylem cross-wall. Bar = 10 μm.

Table 2 Water potential measurements (MPa)[a] of *Rosa* 'Kardinal'® peduncles at 96 h of vase-life

pH	*Micropore-filtered tap water*	*Micropore-filtered deionized water*
Zero samples: vase water not inoculated		
N	−1.93	−0.88
3	−0.29	−0.46
5	−0.83	−0.36
Vase water inoculated with *c.* 10^6 ml^{-1} *B. subtilis*		
N	−2.24	−1.47
3	−0.59	−0.29
5	−1.40	−1.07
Vase water inoculated with *c.* 10^6 ml^{-1} *Ps. aeruginosa*		
N	−1.88	−1.46
3	−0.66	−0.29
5	−2.30	−1.27

[a] Mean values of duplicate measurements.
N, no pH regulation = pH 6.5–7.
Reproduced from Durkin DJ, Put HMC and Clerkx ACM (2001) The influence of vase water pretreatment on the accumulation of micro-particles, micro-compounds and bacterial cells on the cut surface xylem of *Rosa* 'Kardinal' observed by SEM. *Gartenbauwissenschaft* 66: 93–101, with permission.

Table 3 Water potential measurements (MPa)[a] of *Rosa* 'Kardinal' peduncles at 96 h of vase-life

pH	*Micropore-filtered tap water*	*Micropore-filtered deionized water*
Zero samples: vase water not inoculated		
N	−1.62	−1.09
3	−0.15	−0.13
5	−0.64	−0.80
Vase water inoculated with *c.* 5 × 10⁵ ml⁻¹ *Fusarium oxysporum*		
N	−2.33	−2.16
3	−2.06	−1.09
5	−2.62	−2.38
Vase water inoculated with *c.* 5 × 10⁵ ml⁻¹ *Kluyveromyces marxinus*		
N	−1.59	−1.07
3	−0.30	−0.32
5	−1.31	−1.28

[a] Mean values of duplicate measurements.
N, no pH regulation = pH 6.5–7.

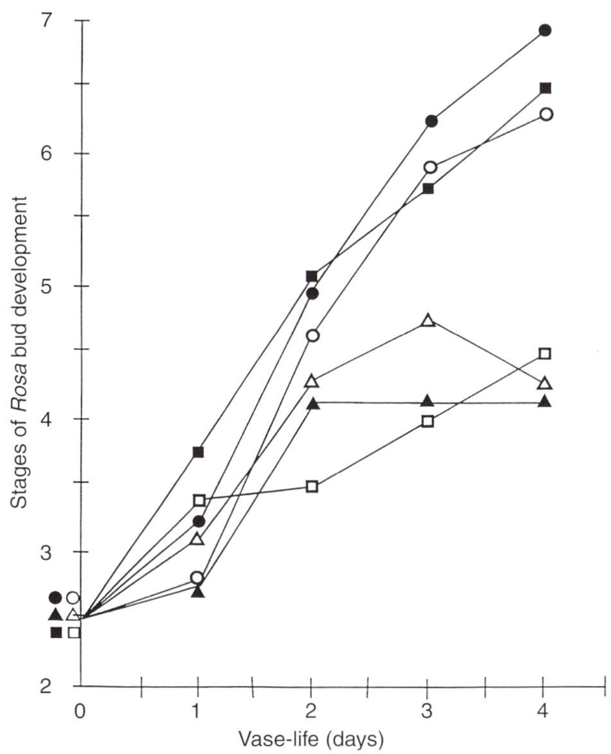

Figure 19 Stages of bud development of *Rosa* 'Kardinal' during 4 days' vase-life in micropore-filtered tap water. (○), no pH regulation = pH 6.5; (●), buffer pH 3; (□), inoculation of *Bacillus subtilis* 10^6 ml^{-1}, pH 6.5; (■), buffer pH 3; (△), inoculation of *Fusarium oxysprorum* spores 10^6 ml^{-1}, pH 6.5; (▲), buffer pH 3. Reproduced from Durkin DJ, Put HMC and Clerkx HCM (1995) Scanning Electron Microscope observations of the cut surface of *Rosa* 'Kardinal' influenced by vase water composition. *Acta Horticulturae* 405: 97–100, with permission.

Figure 20 *Rosa* 'Kardinal' after 2 days' vase-life in: (a) micropore filtered tap water: weak delay in bud development; (b) tap water: delay in bud development; (c) deionized water pH 3 inoculated with *Bacillus subtilis* 10^6 ml^{-1}: almost normal bud development; (d) deionized water pH 3 inoculated with *B. subtilis* 10^6 ml^{-1}, normal pH 6.5: bud development delay.

Figure 21 Scanning electron microscopy. pH 3 phenomenon. (a) Transverse section of the cut surface of *Rosa* 'Kardinal' after 2 days' vase-life in micropore-filtered tap water shows granular material that passes the 0.45 µm pore filter and accumulates on the vessel wall, covering the smaller fibre lumen. Bar = 10 µm. (b) Another rose after 2 days' vase-life in micropore-filtered tap water pH 3 shows clean vessel walls. Bar = 10 µm. (c) Longitudinal section at 1 mm from the cut surface of a rose in pH 3 vase-water shows clean vessel walls. Bar = 0.1 mm. Reproduced from Put HMC and Clerkx ACM (2001) Anatomy of cut *Rosa* xylem observed by Scanning Electron Microscope. *Acta Horiculturae* 547: 331–339, with permission.

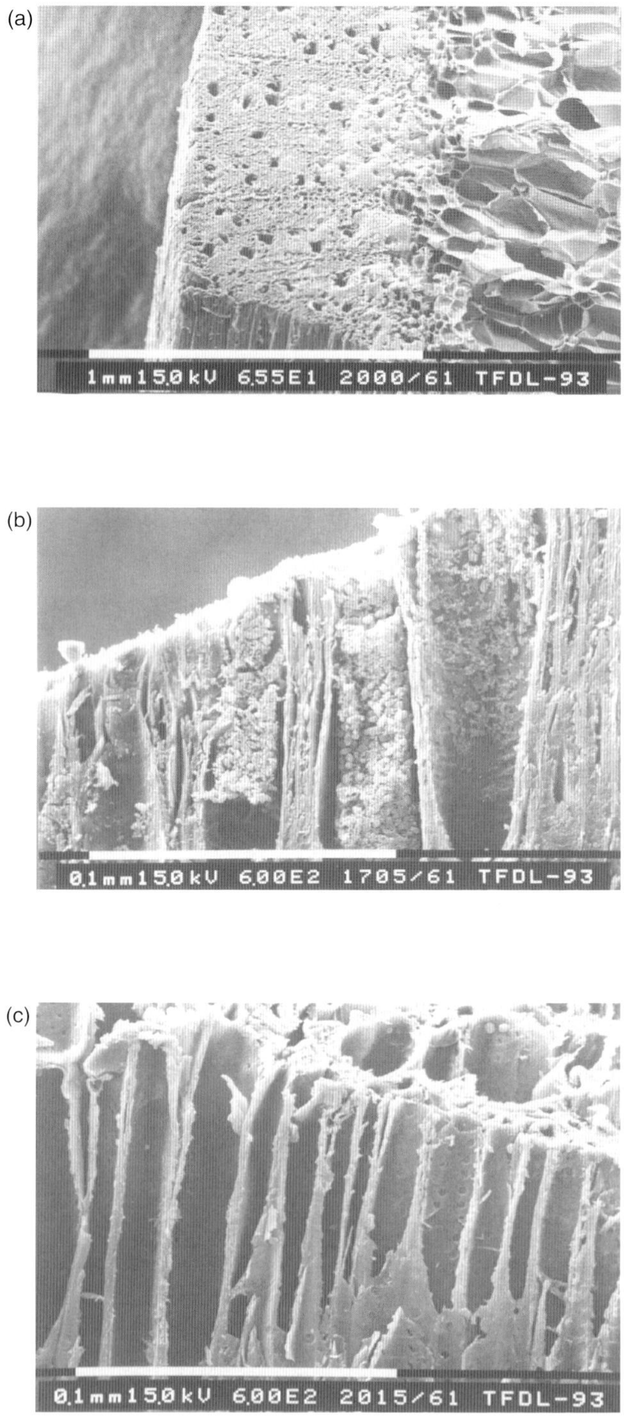

Figure 22 Scanning electron microscopy. pH 3 phenomenon. (a) Overview of a transverse section of the cut surface of *Rosa* 'Kardinal' after 2 days' vase-life in micropore-filtered tap water inoculated with $5 \times 10^5\,ml^{-1}$ *Kluyveromyces bulgaricus* cells shows thick clusters of microbial materials on the xylem cut surface. Bar = 1 mm. (b) Longitudinal section near the cut surface shows infiltration of a mass of *Kluyveromyces* cells into the lower parts of xylem vessels. Bar = 0.1 mm. (c) When vase water is buffered to pH 3, a longitudinal section of the flower stem shows nearly no *Kluyveromyces* infiltration and a few cells on the cut surface part of the section. Bar = 0.1 mm.

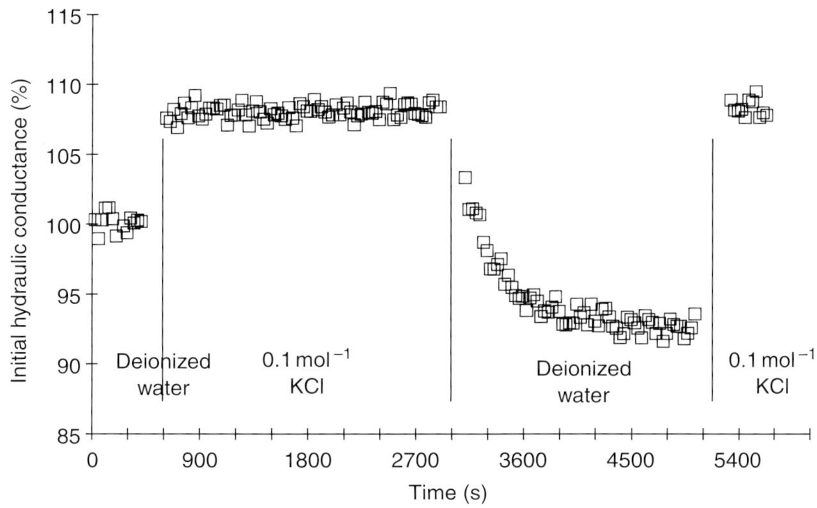

Figure 23 Changes in hydraulic conductance, measured in a 20-cm-long stem segment of chrysanthemum, induced by changing the flowing solution from deionized water to 0.1 mol l^{-1} KCl, to deionized water and again to 0.1 mol l^{-1} KCl at a constant difference of 40 kPa over the stem segment. Reproduced from Van Ieperen W, Van Meeteren U and Gelder H (2000) Fluid ionic composition influences hydraulic conductance of xylem conduits. *Journal of Experimental Botany* 51: 769–776, with permission.

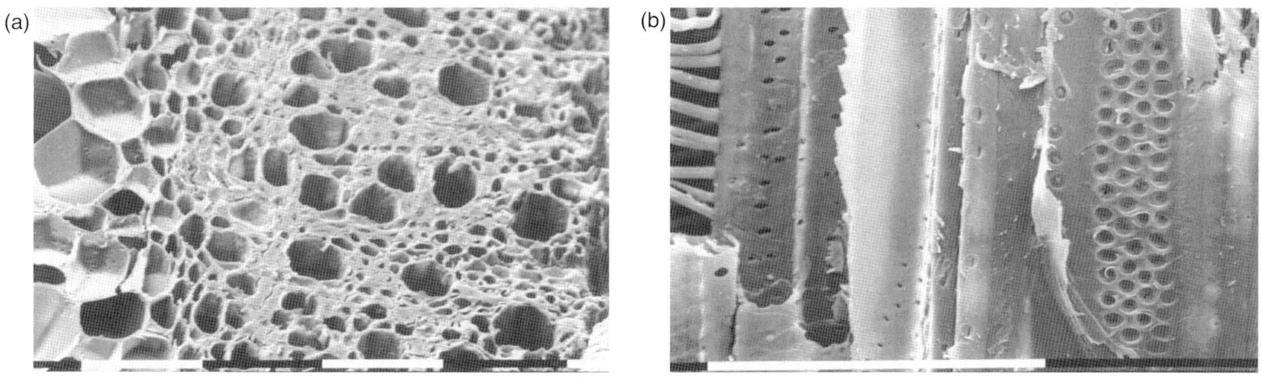

Figure 24 Cryo-scanning electron microscope. Aluminium sulphate phenomenon. (a) Overview of the cut surface cross-section of *Rosa* 'Sonia' after 2 days' vase-life in sterilized deionized vase-water. Bar = 0.1 mm. (b) Longitudinal section of xylem vessels of the same rose near the cuttting point. Bar = 0.1 mm. Reproduced from Put, Klop and Clerkx (1991) *Acta Horticulturae* 298: 303–312, reproduced with permission of the ISHS.

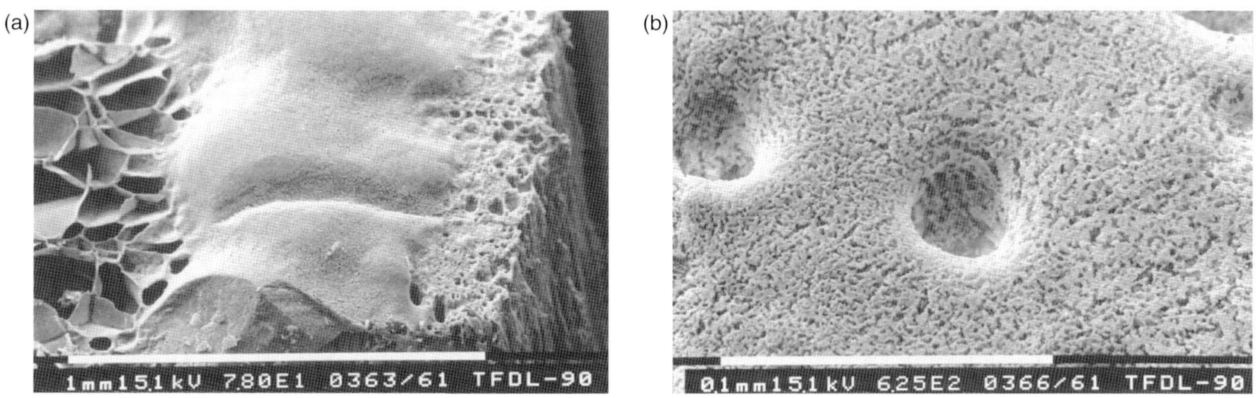

Figure 25 Cryo-scanning electron microscope. Aluminium sulphate phenomenon. (a) Overview of the cut surface cross-section of *Rosa* 'Sonia' after 2 days' vase-life in a solution of $Al_2(SO_4)_3 \cdot 18H_2O$ (500 μg ml^{-1} deionized water pH 4) inoculated with 10^6 ml^{-1} *Bacillus subtilis*. Bar = 1 mm. (b) Detail of (a). Bar = 0.1 mm. *Bacillus* cells are surrounded by a thin granular layer of $Al(OH)_3$. The Al-complex forms a protective layer near the cut surface, and prevents *Bacillus* accumulation and infiltration and blockage of water uptake. Reproduced from Put HMC, Klop W, Clerkx ACM and Boekestein A (1992) Aluminium sulphate restricts migration of *Bacillus subtilis* in xylem of cut roses: a scanning electron microscope study. *Scientia Horticulturae* 51: 261–274, with permission.

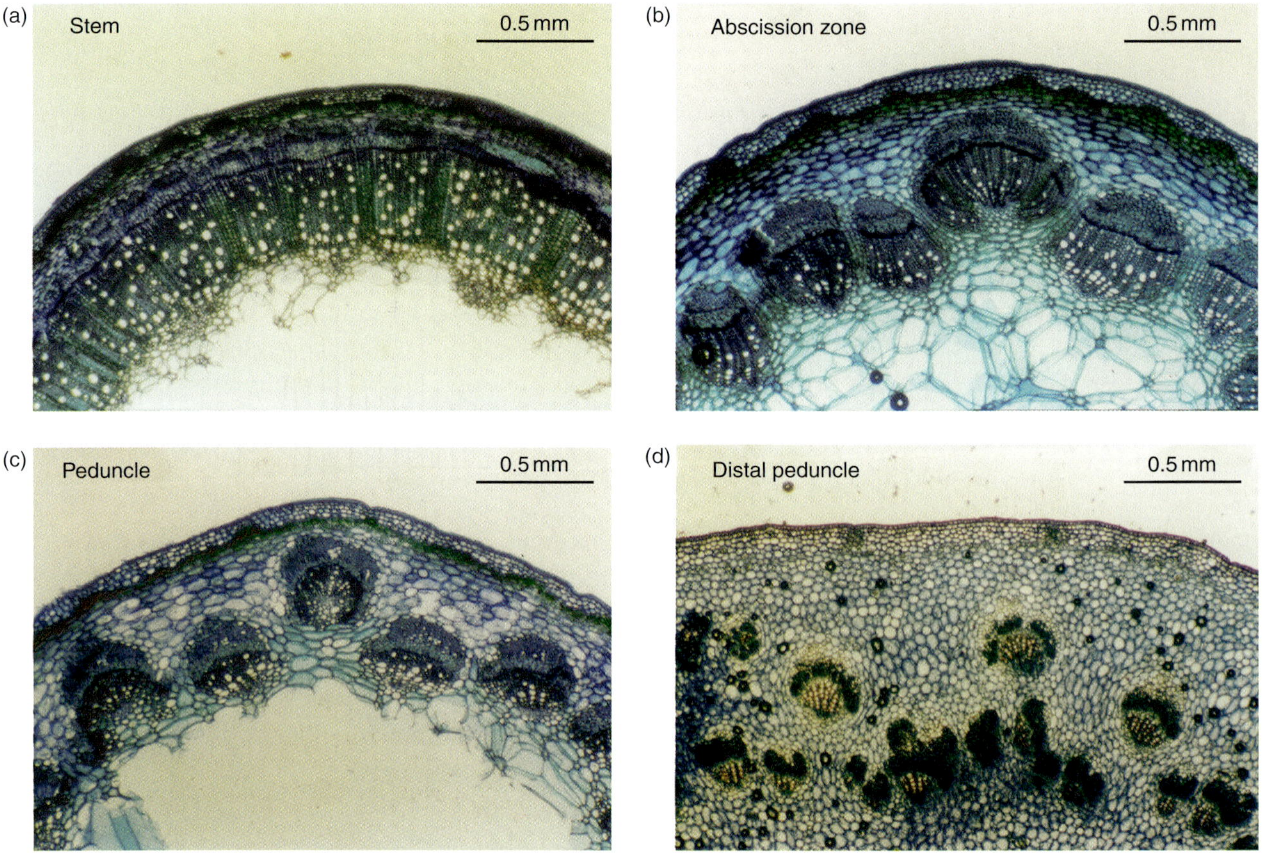

Figure 26 Fresh transverse sections showing general features of the xylem of various components of the cut rose. Sections are stained with toluidine blue O. (a) stem; (b) abscission zone; (c) mid-region of peduncle; (d) distal end of peduncle. Bar = 0.5 mm. Reproduced from Darlington AB and Dixon MA (1991) The hydraulic architecture of roses (*Rosa hybrida*). *Canadian Journal of Botany* 69: 702–710, with permission.

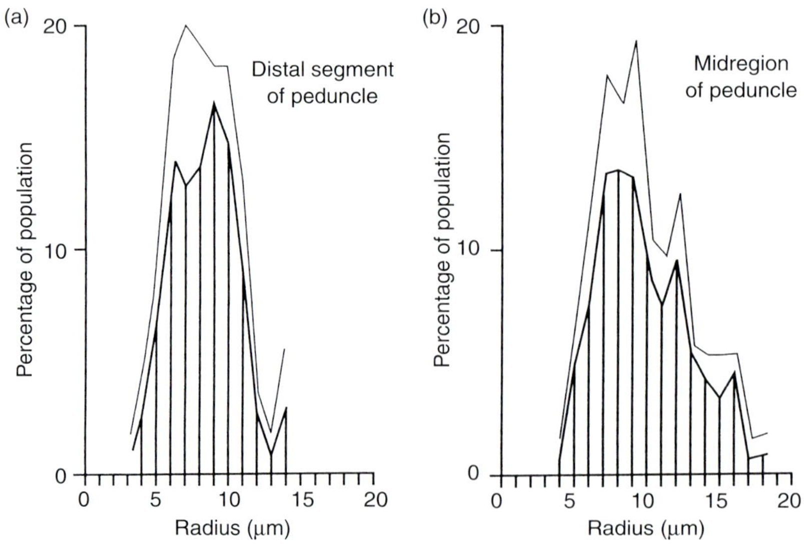

Figure 27 Distribution of sizes: radii (a–d) and relative calculated conductance (e–h) of xylem conduits in the various components of the cut rose. The relative conductance of the size class is a function of its mean radius to the fourth power and its frequency in the population. Lines above the bars of a–d indicate standard errors ($n = 8$). Modified from Darlington AB and Dixon MA (1991) The hydraulic architecture of roses (*Rosa hybrida*). *Canadian Journal of Botany* 69: 702–710, with permission.

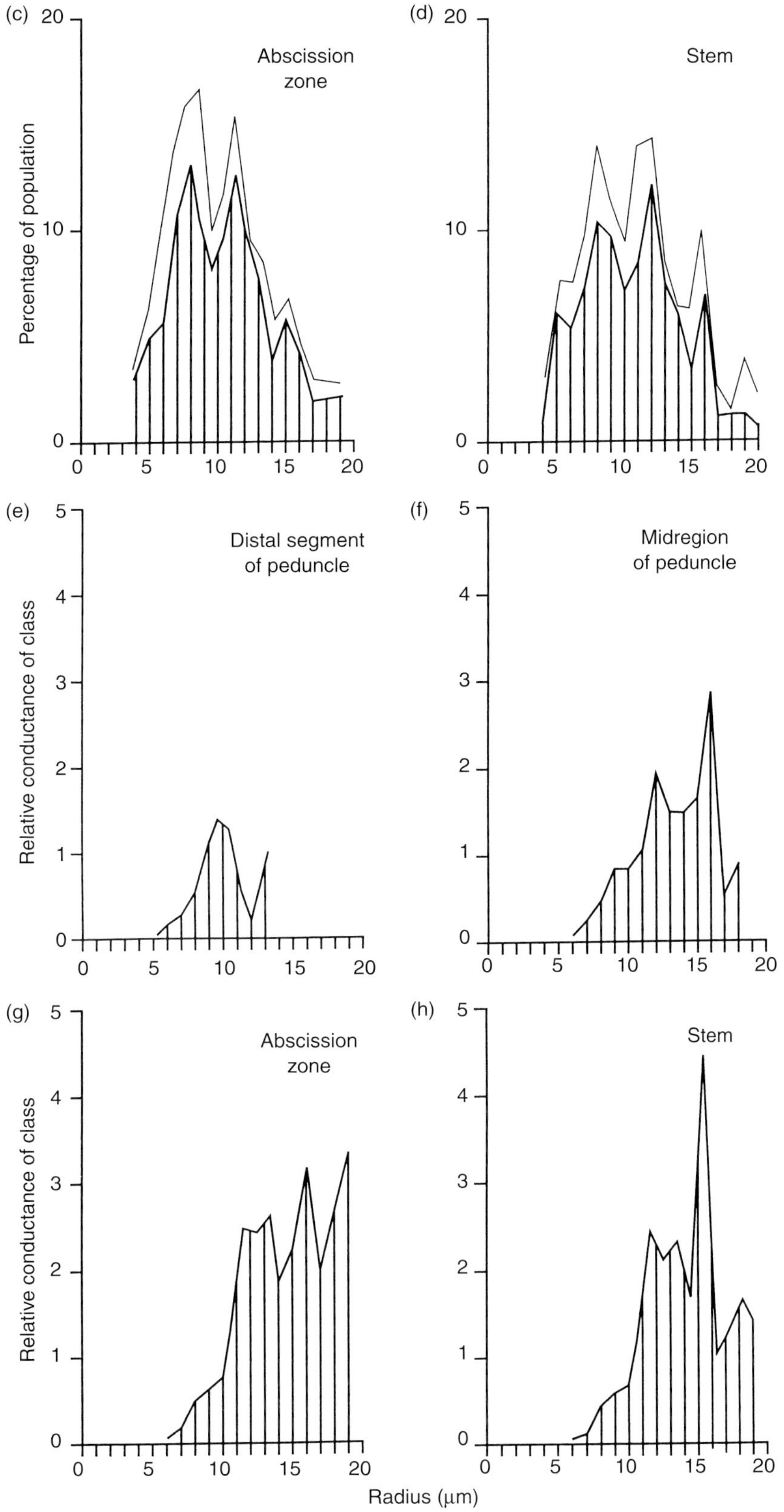

Figure 27 Continued.

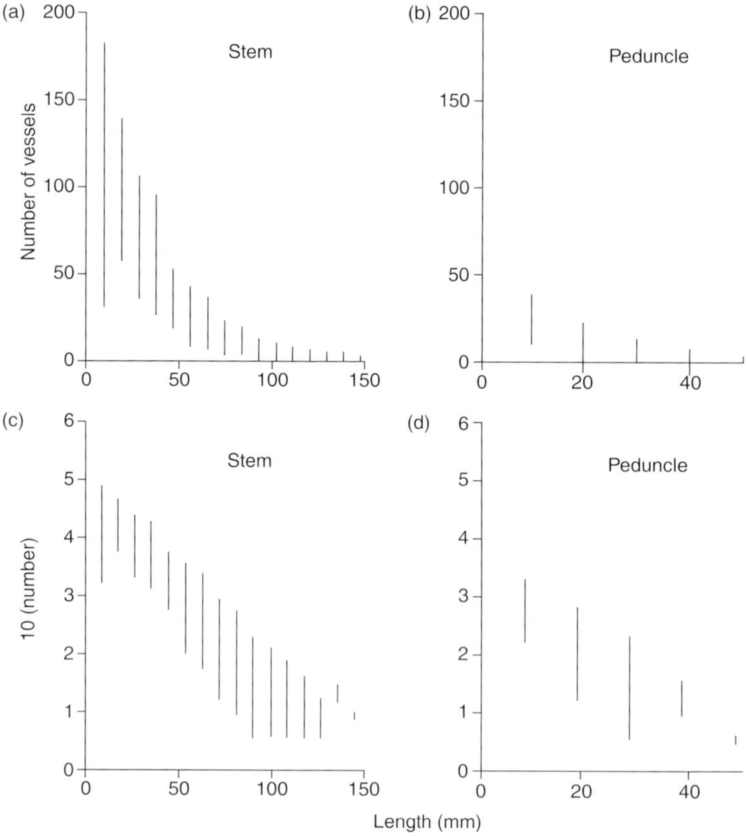

Figure 28 Vessel length distribution of the stem and peduncle of *Rosa hybrida* L. is determined by the paint-infiltration method. The counts of the number of paint-filled (acid fuchsin) vessels present in transverse view are presented as a function of distance from (a) cut end and (b) peduncle; (c) and (d) are natural logarithm transformations of (a) and (b), respectively. Modified from Darlington AB and Dixon MA (1991) The hydraulic architecture of roses (*Rosa hybrida*). *Canadian Journal of Botany* 69: 702–710, with permission.

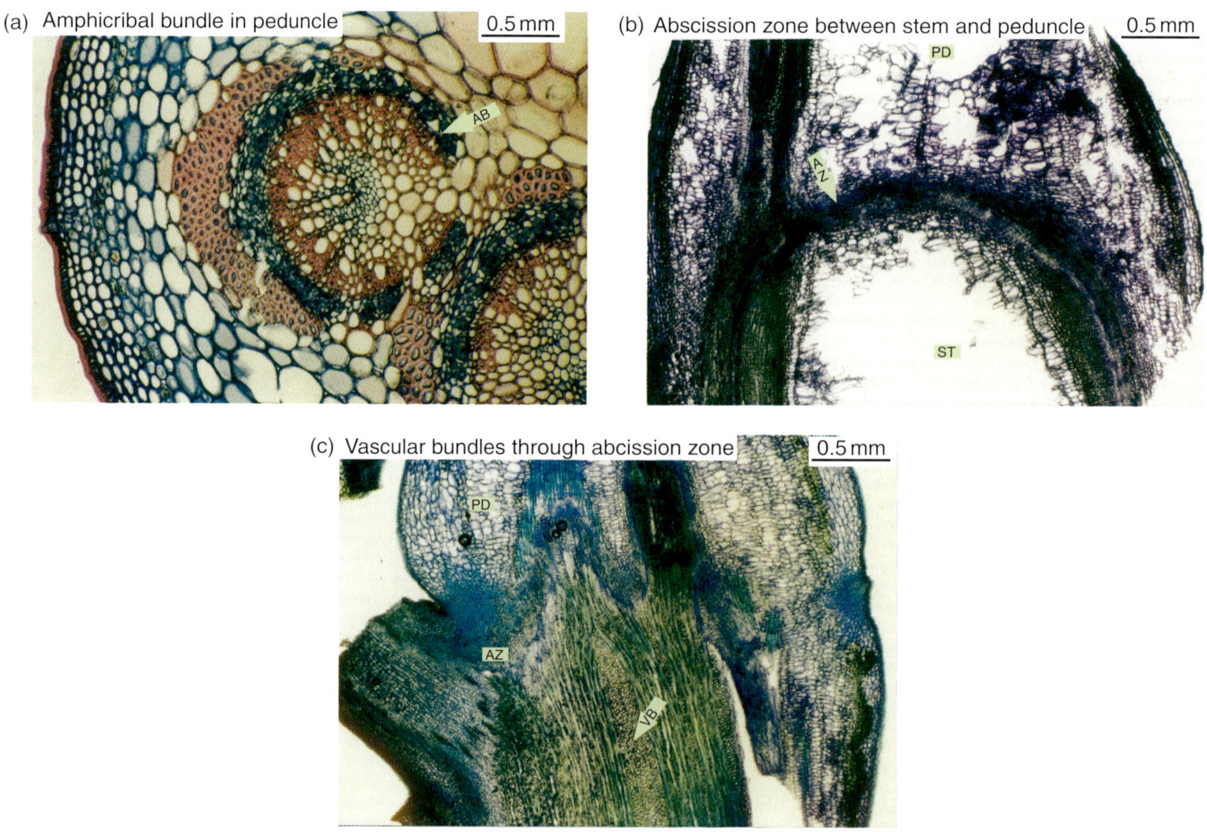

Figure 29 Micrographs of vessel sections of a *Rosa hybrida* L. stained with toluidine blue O. (a) Transverse view of amphicribal bundle (AB) in the peduncle. (b) Longitudinal view of abscission zone (AZ) between stem (ST) and peduncle (PD). (c) Tangential view of vascular bundles (VB) through the abscission zone. Photographs courtesy of Mike Dixon.

See also: **Growth Regulation**: Floral Induction. **Morphology and Anatomy**: Shoots and Stems; Flower Development and the Abscission Zone. **Production and Marketing**: Cut Rose Production.

Further Reading

Aarts JFTh (1957) *On the Keepability of Cut Flowers*. Thesis. Wageningen, The Netherlands: Mededelingen van de Landbouwhogeschool.

Durkin DJ (1979) Effect of Millipore filtration, citric acid, and sucrose on peduncle water potential of cut rose flower. *Journal of the American Society for Horticultural Science* 104: 60–863.

Fletcher M (1996) Bacterial attachment in aquatic environments: a diversity of surfaces and adhesion strategies. In: Fletcher M (ed.) *Bacterial Adhesion, Molecular and Ecological Diversity*, pp. 1–24. New York: Wiley-Liss.

Halevy AH and Mayak S (1979, 1981) Senescence and postharvest physiology of cut flowers I and II. *Horticulture Reviews* 1, 3: 204–236, 59–143.

Halverson LJ and Stacey G (1986) Signal exchange in plant–microbe interactions. *Microbiological Reviews* 50: 193–225.

Harris JR (ed.) (1991) *Electron Microscopy in Biology: A Practical Approach*. Oxford: IRL Press/Oxford University Press.

Lovitt RW and Wright CJ (1999) Instruments for studying morphology and anatomy. In: *Encyclopedia of Food Microbiology*, vol. 2, pp. 1379–1425. London: Academic Press.

Marcelis-Van Acker CAM (1994) *Axillary Bud Development in Rose*. PhD thesis. Agriculture University, Wageningen, The Netherlands.

Marshall KC (1996) Adhesion as a strategy for access to nutrients. In: Fletcher M (ed.) *Bacterial Adhesion: Molecular and Ecological Diversity*, pp. 59–88. New York: Wiley-Liss.

Nijsse J, Van Meeteren U and Keijzer CJ (2001) Quantification of emboli by visualisation of air filled xylem vessels. *Acta Horticulturae* 543: 245–250.

Poxton IR (1993) Prokaryote envelope diversity. *Journal of Applied Bacteriology, Symposium Supplement* (22) 74: 1S–11S.

Put HMC (1991) *Factors Affecting the Vase Life of* Rosa *Cultivar 'Sonia': Microbiological and Scanning Electron Microscope Investigations*. PhD Thesis. Agriculture University, Wageningen, The Netherlands.

Rijnaarts HHM, Norde W, Lykema JJ and Zehnder AJB (1999) DLVO and steric contributions to bacterial deposition in media of different ionic strengths. *Colloids and Surfaces B: Biointerfaces* 14: 179–195.

Salisbury FB and Ross CW (eds.) (1985) *Plant Physiology*, pp. 1–540. Belmont, CA: Wadsward.

Skillman IC, Sutherland IW and Jones MV (1999) The role of exopolysaccharides in dual species biofilm development. *Journal of Applied Microbiology Symposium Supplement* 83: 13S–18S.

Van der Kooij D (1999) Potential for biofilm development on drinking water distribution systems. *Journal of Applied Microbiology Symposium Series* 28: 39S–44S.

Zimmermann MH (1938) Xylem structure and the accent of sap. Berlin: Springer Verlag.

Effects of Supplemental Carbohydrates

N García Victoria and N Marissen, PPO Glasshouse Horticulture, Aalsmeer, The Netherlands

U van Meeteren, Horticultural Production Chains Group, Plant Sciences, Wageningen University, The Netherlands

Introduction

Adding sugars to the vase solution of cut roses has long been known to delay senescence and increase vase-life and is conventional wisdom in many cultures. Consequently, most commercial preservative formulations contain sugars as the dominant ingredient.

Extensive research has been conducted to determine the physiological effect of the mechanism by which these added carbohydrates promote vase-life. Many authors have studied the carbohydrate contents of intact roses allowed to bloom and to wilt attached to the plant, and have compared them to those of flowers harvested in different development stages, with or without the supply of carbohydrates through the stem. Others have tried to increase the carbohydrate reserves of leaves and buds by means of light and temperature during growth and development as a way of improving vase-life without depending on exogenous sugar sources. Much work has been dedicated to following, using radioactive carbon isotopes, the pathways of the added carbohydrates, and has thus contributed to our understanding of the role of the stem and thus the leaves on internal sugar transport.

This article briefly reviews all this gathered knowledge on how the endogenous and exogenously applied sugars are transported, where they accumulate, how they are converted into metabolites which are directly useful for the cut flower, and what senescence processes they delay.

Why do Added Carbohydrates Prolong the Vase-Life of Cut Roses?

Many beneficial effects have been attributed to applied carbohydrates, although some of these effects are not well documented with roses:

- They are used as substrate for dry-matter production and for respiration (energy)
- They help to maintain the osmotic potential of the petal cells
- They delay the loss of membrane integrity of the petal cells
- They help maintain flower colour
- They promote partial stomatal closure of the leaves, contributing to a better water balance of the cut flower
- They reduce ethylene sensitivity
- They regulate senescence at a genetic level.

Flower Bud Opening and Vase-Life in Cut Roses

Commercial cut roses are harvested while the flower is still at the bud stage. Their ornamental life consists of a gradual petal expansion (= bud opening), followed by a short period of maximal flower diameter (= full bloom). After full bloom, symptoms of senescence like turgor loss (due to the leakage of electrolytes as a consequence of reduced membrane integrity), colour changes and petal wilting or petal abscission indicate the end of the vase-life or ornamental life.

Failure of buds to open is one important cause of reduced vase-life in cut roses, since it often results in wilting before opening takes place. We owe a great deal of our knowledge about the effects of endogenous and exogenous sugars and their physiological pathways to the variety ‘Madelon’, which opens badly. For a long time this red variety was one of the best-sellers in the Netherlands among the red varieties, but failed to open in conditions where other roses would normally open. For years it was used as a model for understanding bud opening. The variety disappeared from the Dutch Auctions top 10 in 1997.

Effect of Internal and External Carbohydrates on Flower Bud Opening

The opening of a rose bud is a process of irreversible petal growth. The growth is not a consequence of cell divisions, but of expansion of the existing cells with an increase of fresh (water uptake) and dry weight. The enlargement process of the cells starts in the petals of the outer whorls and requires assimilates for energetic, biosynthetic and osmotic purposes.

The growing corolla of intact flowers (attached to the plant) continues to import dry matter throughout its development, until the stage of fully expanded petals. Unlike intact flowers, cut roses must rely on their stored sugars in the petals, which decline very fast after harvest, and on supply from other parts of the cut rose. Indeed, while a continuous accumulation of starch is measured on the opening intact bud, the amount of starch in the cut flower declines until it reaches a level roughly 20% of that of the intact flower. Under normal growth conditions, the corolla sugar demand is fulfilled by the vegetative parts of the plant. Attached flowers reach a bigger corolla diameter than cut roses without exogenous carbohydrate supply, as well as a higher fresh weight. The rate of diameter increase of the flower bud on attached roses is however slower than that of successfully developing (sucrose-fed) cut roses. Exogenous sugar supply has proven to improve bud opening, as has been reported by many authors. Moreover, there is a correlation between the rate of flower opening and the sucrose concentration of the solution (**Figure 1**).

In roses that do not open well, complete flower opening can also be achieved experimentally, without adding any sugar, by reducing the number of petals of the corolla. A reduction in the number of remaining petals (always leaving at least the inner 10 petals) improved flower opening. The petal area, fresh and dry weight, thickness, water content and dry-matter content of the 10 innermost petals were all increased by increasing the number of petals removed. This confirms once again the sink activity of the flower, and that there is competition between petals for the carbohydrates stored and supplied by other parts of the flower.

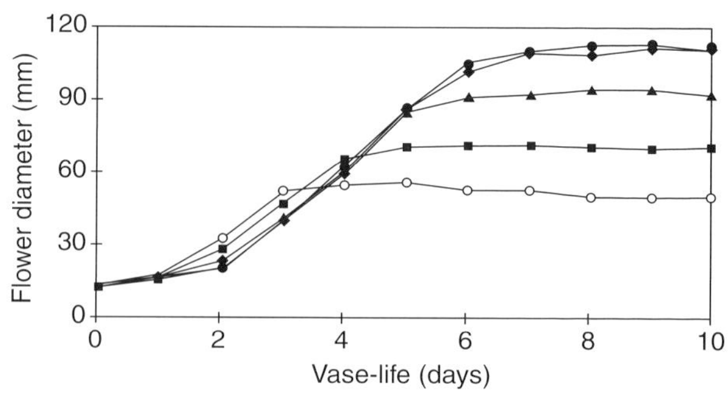

Figure 1 Effect of the sucrose concentration in the 1 mg l^{-1} $AgNo_3$ solution on the final flower diameter (mm) of cut ‘Madelon’ roses under postharvest conditions. Sucrose concentrations: 0 m mol l^{-1} (○), 15 m mol l^{-1} (■), 30 m mol l^{-1} (▲), 45 m mol l^{-1} (•), 60 m mol l^{-1} (◆). Standard error (SE) < 4%. (Reproduced with permission from Kuiper D *et al.* (1995) *Scientia Horticulturae* 60: 325–336.)

Premature Ageing

Flower tissues use carbohydrates to maintain normal functioning by respiration. The energy is needed to counteract the continuous degradation of proteins and ion concentrations. This maintenance respiration is assumed to be related to the amount of tissue present, especially amounts of proteins and minerals. As a consequence, larger or more developed flower buds have a higher maintenance respiration compared to small or less developed flower buds. When the available carbohydrates in the cut flower become exhausted, the normal functioning of the tissue cannot be maintained and the result is premature ageing, finally leading to death of the flower bud. However, small buds use more carbohydrates for growth, due to synthesis of new structural biomass and to achieve the osmotically derived turgor pressure essential for cell expansion. As a result, there is an optimal combination of development stage at harvest and carbohydrate content.

Senescence-Related Genes

Although in many flower species exogenous application of sugars increases the endogenous amount of soluble sugars two to four fold, delaying the visible and physiological changes associated with senescence, carbohydrates are not always exhausted when visible senescence begins. Evidence of the metabolic regulation of plant genes is emerging in relation to genes involved in senescence regulation. A number of flower senescence-related genes have been isolated. Therefore the role of sugars in regulating senescence of flowers at a genetic level was proposed by Eason.

The Role of the Leaves in Carbohydrate Transport

The carbohydrates are transported to the flower through the phloem, while uptake of water (and supplementary sugars dissolved in the vase water) mainly takes place through the xylem. Supplied glucose is immediately converted into sucrose by the stem tissues before being transported to the leaves. Besides being a source of carbohydrates to the petals, the leaves are the place where the sucrose from the vase is hydrolysed, and the phloem is loaded with the resulting glucose and fructose to be transported to the flower.

Experiments conducted at the Aalsmeer Research Station in 1995 by Marissen and colleagues proved that either there is also loading capacity in the stem, or the xylem can also transport sucrose directly to the corolla. In these experiments the carbohydrate contents of 'Madelon' corollas from stems with partially or totally removed leaves were determined. When no sugar is supplied, decreasing the leaf area on the stems does indeed cause a lower sugar concentration in the corollas, but when external sugar is supplied, more sugar is found in the corollas of leafless stems (**Figure 2**).

Effects of Preharvest Conditions on Carbohydrate Content and Vase-Life

The stage of maturity of the corolla at the moment of harvest affects the carbohydrate reserves of the corolla. Consequently, flowers harvested at a younger development stage, with a lower carbohydrate content, are less likely to open without external carbohydrate supply than flowers harvested at a more mature stage of development. Van Doorn reported varietal differences in the required development stage for successful opening: 'Madelon' only opened as attached flowers when cut at development stage 3 (open sepals, loosened petals) without exogenous sugar supply. 'Sonia', on the other hand, did not open without sugar supply if harvested at stage 0 (closed bud), but will readily open when harvested at stage 1–2 (pointed to cylindrical bud).

From this and other observations, a high concentration of carbohydrate in the harvested corolla is assumed to be a prerequisite for a long vase-life. With

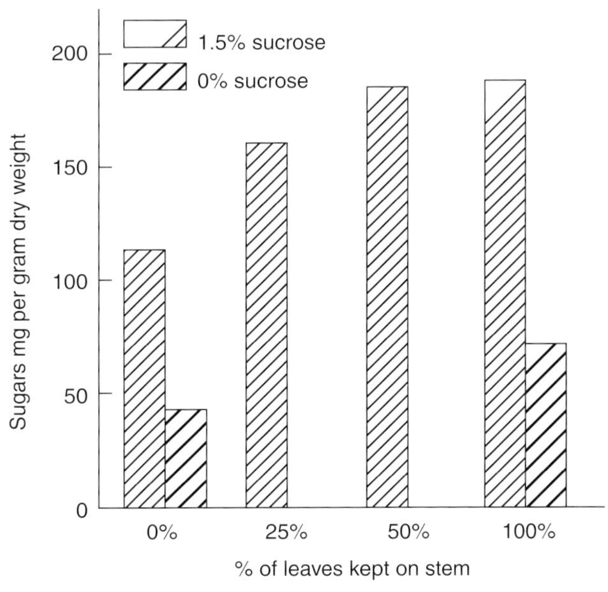

Figure 2 Total sugar content (glucose + fructose + sucrose) in corollas of 'Madelon' after 7 days vase-life in a solution containing 1 mg l^{-1} $AgNO_3$ with or without 15 g l^{-1} sucrose. (Reproduced with permission from Marissen N and La Brijn L (1995) *Acta Horticulturae* 405: 81–87.)

certain varieties (like the rose 'Madelon') this appears to be true and a good flower opening correlates with high carbohydrate level in the corolla. This is not seen in 'Sonia', where differences in carbohydrate content in the corollas (provided they are not harvested in cutting stage 0), do not seem to affect opening. Moreover, Kuiper calculated that the total amount of stored carbohydrates in petals at bud stage is only sufficient to supply respiratory substrate during the first 2 days after harvest. These calculations, together with experiments on flowers with detached leaves, led to the conclusion that not only the carbohydrate contents of the corolla, but also the carbohydrate pool of the leaves and the stem are important. This was clearly demonstrated for freesia flowers as well.

Increasing the internal carbohydrate reserve of the flower bud, the leaves and the stem by means of manipulating growth conditions (light and temperature) might be expected to lead to a better flower bud opening and vase-life. However, differences in carbohydrates caused by growing plants at different light intensities did not correlate with the vase-life, while differences in growing temperature affected vase-life without affecting the carbohydrate contents of leaves and buds. Moreover, the four varieties involved in this research appeared to produce contradictory results (**Figure 3**). An exogenous supply of carbohydrate seems, at present, to be the only way to meet the demands of cut roses to ensure a long vase-life and a proper opening flower bud.

The Osmotic Role of Carbohydrates During Bud Opening

Some authors postulate the osmotic activity of the reducing sugars in the petals. Ho and Nichols, for instance, reported an increase in soluble sugar content of rose petals during flower blooming due mainly to starch hydrolysis. Its function is that of lowering the osmotic potential in petal cells, thereby promoting an influx of water to drive cell expansion. But studies by Evans and Reid on the rhythmic expansion of petals showed that during petal expansion the osmotic potential of the cells increased, rather than decreased. Besides, the dry-weight gains during expansion could not be explained by net total carbohydrate changes

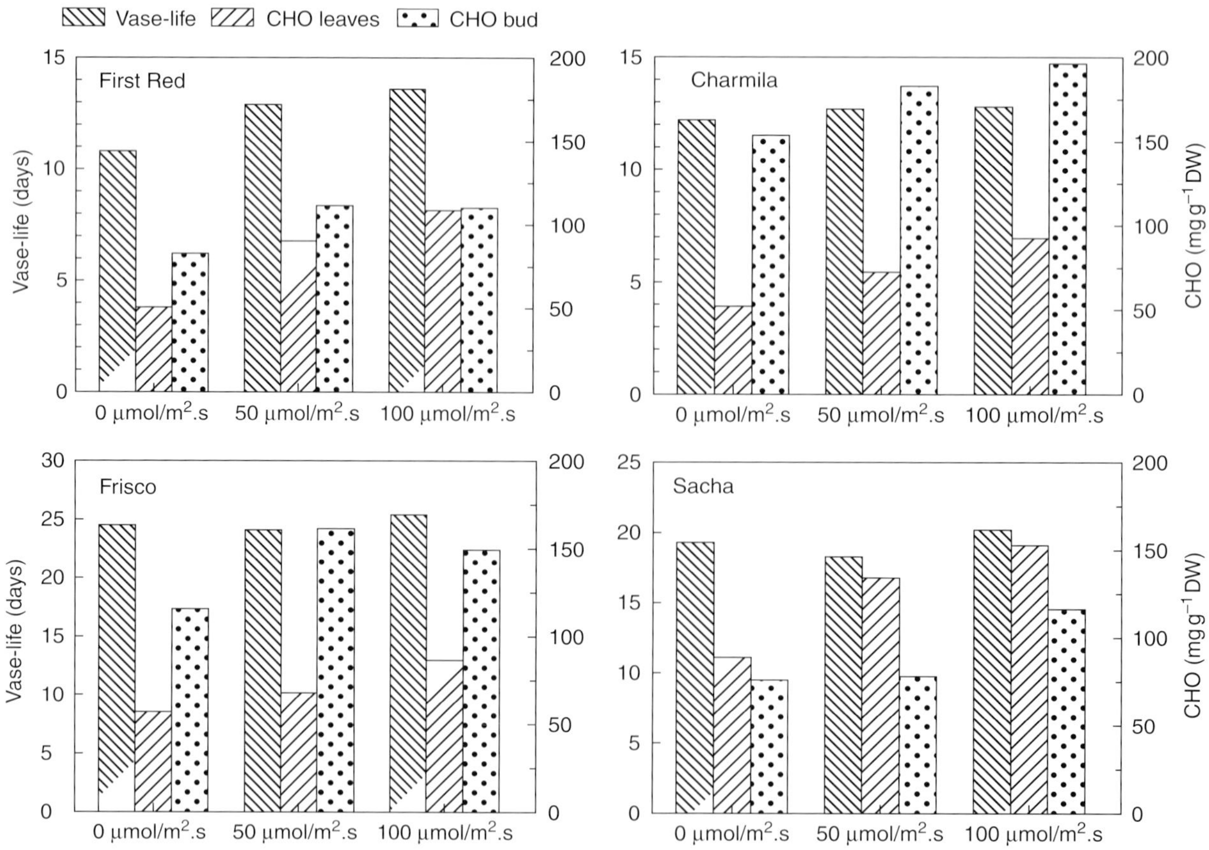

Figure 3 Vase-life and carbohydrate (CHO) levels of leaves and buds of roses grown under different levels of supplemental lighting. (Reproduced with permission from Marissen N (2001) Effects of pre-harvest light intensity and temperature on carbohydrate levels and vase life of cut roses. *Acta Horticulturae* 543: 331–336.)

(starch + soluble sugars). This suggests the dry weight might have increased due to the translocation of other solutes into the petals. Solutes other than sugars apparently account for a large portion of the osmotic potential of petal cells. The authors calculated the osmotic potential due to other solutes from the difference between the measured osmotic potential of the petal sap and the calculated osmotic potential due to soluble sugars.

If the effect of sucrose in the holding solution is (partly) due to its osmotic properties, it might be possible to replace it by another osmoticum. Experiments conducted by Kuiper demonstrated that replacement of sucrose with KNO_3 is possible without affecting flower opening up to $15\,m\,mol\,l^{-1}$. Below this level, flower opening is reduced as compared to a $45\,m\,mol\,l^{-1}$ sucrose solution; also, the colour changes. The role of sucrose in osmoregulation seems to be taken over by the KNO_3, while the cellular processes not related to osmoregulation can be influenced by the smaller amount of sucrose supplied.

In the experiment described above, water uptake for the KNO_3 and sucrose solutions was also measured. Potassium nitrate is, unlike sucrose, readily transported to the flower. The flowers held in $21.7\,m\,mol\,l^{-1}$ KNO_3 solutions consumed much more water than those in the solutions containing sucrose, but only 1.5% of the absorbed water was directed to the flowers. The authors attributed this to the inhibitory effect of sugar on stomatal transpiration. In the longer term, sucrose attracted a higher percentage of the absorbed water to the flower.

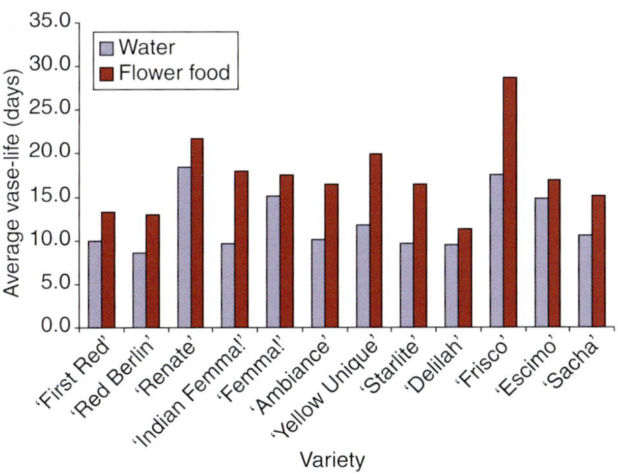

Figure 4 Average vase-life of some of the most important varieties in The Netherlands (in numbers of stems sold), with and without addition of flower food (Chrysal) to the vase-water. Values are means of 20 replicates per testing period. Each variety was tested in three consecutive periods (autumn, spring and summer). Courtesy of A. van Nieuwenhoven, Test Centre Aalsmeer Flower Auction.

The Role of Bactericides in Sucrose-Based Postharvest Formulations

An undisturbed water flow is of basic importance for positive effects on bud opening by the supplied sugars. It is well known (*see* **Postharvest Physiology:** Xylem Structure and Function in Cut Roses) that bacteria added to the vase-water in concentrations above 10^6 cfu ml^{-1} reduce the water potential of the flower heads, the flower diameter and consequently vase-life of 'Sonia' in direct proportion to the bacteria counts. According to Van Doorn's research, the mode of action of bacteria in vascular blockage appears to be a physical plugging of the xylem vessels. Therefore, most experiments (including many of those mentioned above) have been conducted with the addition of a bactericide to the sucrose solution. The choice of bactericide is rather wide. A frequently used one is silver nitrate in deionized water in concentrations around $1\,mg\,l^{-1}$. Also chlorine formulations, from household bleach (sodium hypochlorite) to slow-release compounds (DICA, Chloramine-T) have often been combined with the sucrose addition. Other commonly used bactericides have been 8-hydroxyquinolinecitrate (8-HQC) or 8-hydroxyquinolinesulfate (8-HQS) at rates of about $300\,mg\,l^{-1}$.

In experiments with the addition of several bactericides to the vase-water, aluminium sulphate appeared to have a better effect on flower opening than 8-HQC. When the inclusion of sucrose in the vase-water was combined with aluminium sulphate, flower opening of 'Madelon' roses was similar to uncut flowers, and much better than when either of the compounds was supplied alone. The positive effect of sugar was independent, but somehow additive, to those of the antimicrobial compound.

Together with aluminium sulphate, citric acid is the most extensively used bactericide, next to sucrose or glucose, in the commercial formulations packed in sachets to be used by flower consumers. Sometimes other osmotica (like potassium salts) are included in these formulations, known by the general public as 'flower food'. The vase-life of commercial roses is spectacularly improved in many cases by the use of flower food as compared to plain water (**Figure 4**).

See also: **Breeding**: Selection Strategies of Cut Roses. **Morphology and Anatomy**: Flower Development and the Abscission Zone. **Postharvest Physiology**: Effects of Temperature; Packing Systems and Precooling; Xylem Structure and Function in Cut Roses; Ethylene. **Production and Marketing**: Cut Rose Production.

Further Reading

Halevy AH and Mayak S (1979) Senescence and postharvest physiology of cut flowers. In: Janick J (ed.) *Horticultural Reviews*, vol. 1, 204–236 and vol. 3, 59–143. West Point, CT: ACI.

Marissen N (2001) Effects of pre-harvest light intensity and temperature on carbohydrate levels and vase life of cut roses. *Acta Horticulturae* 543: 331–336.

Paulin A (1986) Influence of exogenous sugars on the evolution of some senescence parameters of petals. *Acta Horticulturae* 181: 183–193.

Van Doorn WG (1993) *Vascular Occlusion in Stems of Cut Rose Flowers*. Thesis. Wageningen, The Netherlands: Agricultural University.

Effects of Temperature

M Reid, University of California, Davis, CA, USA

The most important factor affecting the postharvest life of cut roses is the temperature at which they are held. Although refrigerated coolers, forced air precoolers and refrigerated transportation vehicles are frequently employed during the distribution chain, they are often held at temperatures well above the optimal temperature for storing cut roses, which is close to freezing point (0 °C). For example, the temperature of roses transported from Quito, Ecuador, to Davis, California (**Figure 1**) varied from a starting temperature of 7 °C to as high as 16 °C during air transport to Miami, and even after complete precooling in Miami, returned to warm temperatures during subsequent transport to the West Coast. Although the transit times are relatively short, exposure to elevated temperatures even for these brief periods can have a very detrimental effect on the quality and subsequent vase-life of the flowers.

The reason for the requirement for low temperature during transportation and storage of cut roses is the dramatic effect of temperature on the respiration of the flowers. Respiration, the basis for all life processes, is required not only for growth and development, but also for ageing and senescence. All parts of a cut rose have extremely high rates of respiration (**Figure 2**), and the rate and pattern of change in respiration after harvest are strongly influenced by temperature (**Figure 3**). Like many other perishable commodities, the respiration of rose flowers increases exponentially with temperature (**Figure 4**). This dramatic increase in respiration with temperature is normally described using Q_{10} values (the ratio of respiration at one temperature to that at a temperature 10 °C lower). For cut-rose cultivars that have been measured (**Table 1**) these values range from 1.5 to as high as 6 depending on cultivar and the temperature range over which the Q_{10} is measured. Because the rate of flower development and ageing is closely correlated to their rate of respiration, the values in **Table 1** can be interpreted

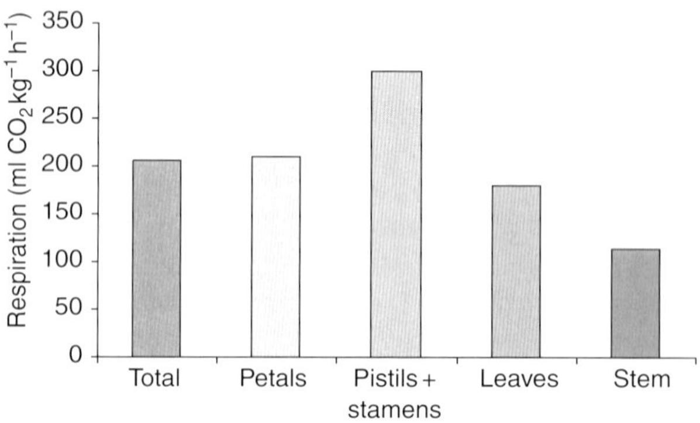

Figure 2 Respiration (at 20 °C) of different portions of a cut rose.

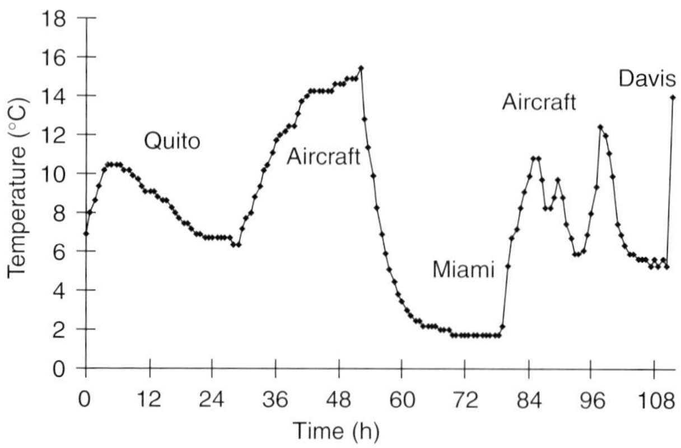

Figure 1 Temperature profile during transportation of packed roses from Quito, Ecuador, to Davis, California, via Miami, Florida.

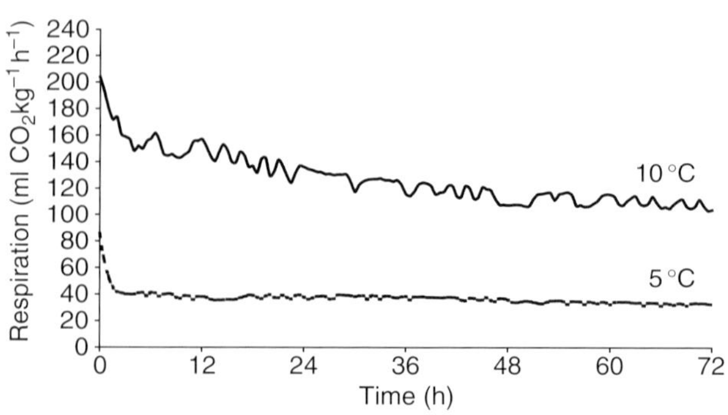

Figure 3 Changing respiration after harvest of cut roses ('Cara Mia') held at two different temperatures.

to predict the effect of temperature on the rate of ageing of the different cultivars. Flowers of 'Preference' (Deforge 1955), for example, would be expected to age more than 22 times faster at 20 °C than at 0 °C. In contrast, the same temperature change would result in only a fourfold increase in ageing rate for 'Raphaela' flowers.

The vase-life of roses that are stored even for a few days is closely correlated with their respiration during storage. As an example, flowers of the cultivar 'Raphaela' were stored for 5 days at different temperatures then held at 20 °C to permit estimation of their vase-life. Despite the relatively low Q_{10} values for this cultivar, there was a dramatic effect of higher storage temperatures in reducing the subsequent vase-life of the flowers (**Figure 5**). Plotting the vase-life of the flowers as a function of their respiration rate during the storage period generates a highly significant correlation (**Figure 6**), indicating the importance of flower respiration on overall vase-life.

A photograph of flowers of 'First Red'® roses (**Figure 7**) taken after 8 days in a standard interior environment following storage for 5 days at different temperatures further illustrates the importance of maintaining low temperatures during the postharvest chain. Flowers held at temperatures above 0 °C not only had reduced vase-life, but were of poorer quality and failed to open as fully as freshly cut flowers (not shown) or those held at the proper storage temperature (0 °C).

There have been claims that storage in water or commercial floral preservative solutions might reduce the importance of low temperatures during the postharvest handling of cut roses, and this has stimulated the use of Procona and Aquapak systems for handling the flowers. As can be seen in **Figure 8** roses benefit little from being stored in water except at higher storage temperatures, when the slight advantage achieved by storage in water is offset by the substantial reduction in vase-life caused by the elevated storage temperature.

In addition to increasing the rate of respiration, and therefore ageing of the blooms, temperature also affects other processes related to ageing of rose

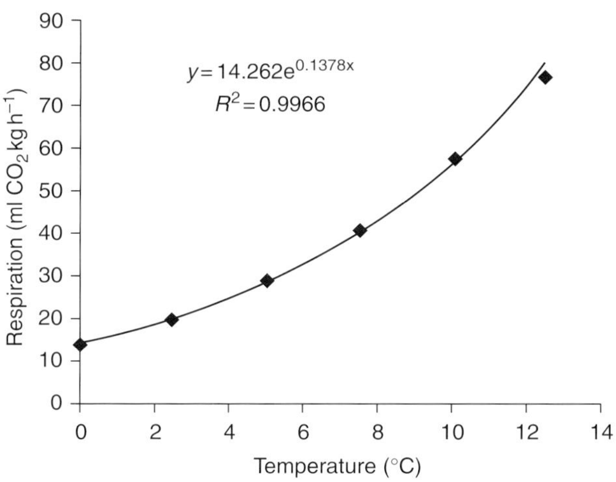

Figure 4 Effect of temperature on the respiration of flowers of 'Ambiance' roses.

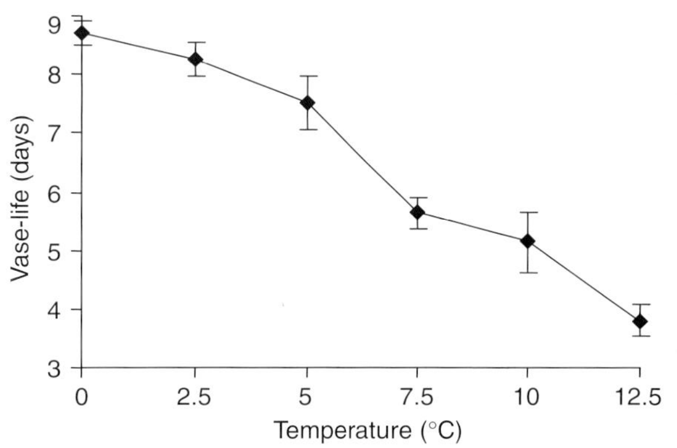

Figure 5 Effect of temperature during 5 days' storage on the subsequent vase-life at 20 °C of 'Raphaela' roses.

Table 1 Q_{10} values, respiration rates (ml CO_2 kg^{-1} h^{-1}) and equations relating respiration to storage temperature of selected cut-rose cultivars (r = respiration, t = storage temperature)

Cultivar	Q_{10} *0–10 °C*	Q_{10} *10–20 °C*	*Respiration 10 °C*	*Respiration/temperature equation*
'Ambiance'	4.13	3.85	57.84	$r = 14.262e^{0.138t}$
'Cara Mia'	5.74	3.17	70.52	$r = 12.912e^{0.149t}$
'Fire 'n' Ice'™	5.35	3.05	81.95	$r = 17.154e^{0.141t}$
'First Red'	5.67	3.00	56.84	$r = 9.3104e^{0.16t}$
'Kardinal'	4.11	2.71	78.88	$r = 21.115e^{0.121t}$
'Preference'	4.82	4.76	31.11	$r = 6.0271e^{0.161t}$
'Raphaela'	2.40	1.54	67.51	$r = 27.253e^{0.089t}$
'Tineke'	2.39	2.78	38.42	$r = 14.726e^{0.097t}$

To calculate heat production, multiply ml CO_2/kg hr by 440 to get BTU per ton per day or by 122 to get kcal per metric tonne per day.

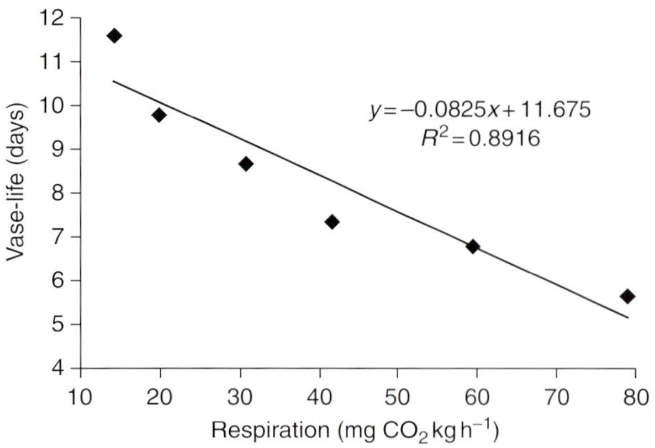

Figure 6 Relationship between respiration rate during storage and vase-life at 20 °C after storage for flowers of 'Ambiance'.

Figure 7 Effect of storage for 5 days at different temperatures (as indicated) on quality of 'First Red' roses after 8 days in the vase at 20 °C.

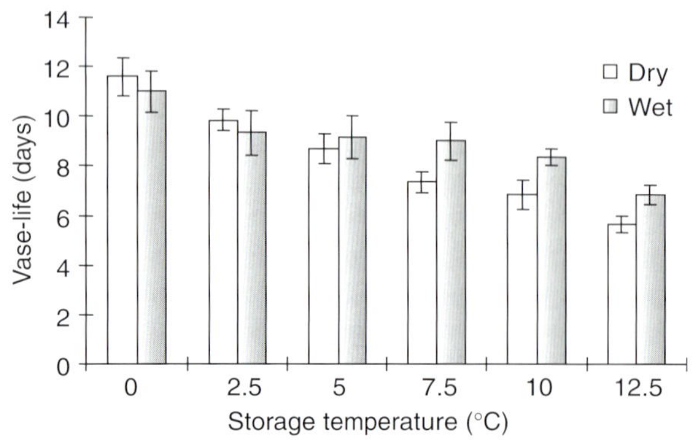

Figure 8 Effect of storage temperature on the subsequent vase-life of 'Ambiance' roses. The flowers were held for 5 days at different temperatures, either dry (wrapped in polyethylene to prevent water loss) or in water containing a biocide.

flowers. In particular, flowers respond more rapidly to ethylene at warmer temperatures and lose water much more rapidly (and therefore are likely to be more water-stressed). One of the common sources of postharvest loss in roses is attack by grey mould (*Botrytis cinerea*). Although this fungus grows more rapidly at warm temperatures, it can also grow (albeit more slowly) at proper storage temperatures. Temperature management plays an important role in the aetiology of this disease because the spores of *Botrytis* can germinate wherever free moisture is present. Poor temperature management, particularly transfer of flowers from cold storage to warmer handling areas, often results in condensation of water on the harvested flowers, leading to accelerated growth of *Botrytis* (Figure 9).

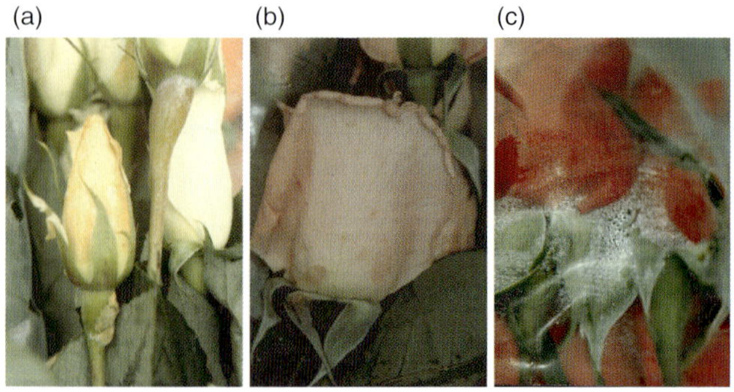

Figure 9 Slow precooling, or alternating temperatures results in condensation on sleeves and petals (a) that can lead to germination and growth of *Botrytis cinerea* on rose petals (b), and finally total loss of the flowers (c).

See also: **Disease**: *Botrytis*. **Postharvest Physiology**: Xylem Structure and Function in Cut Roses; Effects of Supplemental Carbohydrates; Packing Systems and Precooling; Ethylene. **Production and Marketing**: Cut Rose Production.

Further Reading

Anonymous (1988) *The Flower Care Booklet*. New York: Flower Council of Holland.

Cevallos JC and Reid MS (2001) Effect of dry and wet storage at different temperatures on the vase life of cut flowers. *Horttechnology* 11: 199–202.

Faragher J, Slater T, Joyce D and Williamson V (2002) *Postharvest Handling of Australian Flowers*. Rural Industries Research and Development Corporation 02/021:217PP.

Nell TA and Reid MS (2001) *Flower and Plant Care: The 21st Century Approach*. Alexandria, VA: Society of American Florists.

Nowak J, Rudnicki RM and Duncan AA (1990) *Postharvest Handling and Storage of Cut Flowers, Florist Greens, and Potted Plants*. Portland, OR: Timber Press.

Worrall R, Gollnow B and Wade N (2000) *Rose Flower Care for Professionals*. Australia: NSW Agriculture.

Ethylene

R Müller and B M Stummann, Royal Veterinary and Agricultural University, Frederiksberg, Denmark

Introduction

The gaseous plant hormone ethylene plays a central role in physiological and developmental processes, such as germination, growth, flower initiation and opening, senescence of leaves and flowers, organ abscission and fruit ripening. Ethylene is sometimes called the senescence hormone. Certain plants are ethylene-sensitive and flowers of these plants deteriorate rapidly in the presence of the gas. Ethylene can be produced by almost all plant parts, although production depends on the type of tissue and stage of development. Ethylene production increases during leaf abscission, fruit ripening and flower senescence. Ripening fruits and vegetables produce large amounts of ethylene and should therefore not be stored or transported with flowers. Wounding, physical and chemical injury (e.g. pathogen attack) and physiological stresses (e.g. disease and temperature stress) increase ethylene production. Other plant hormones such as auxins and abscisic acid (ABA) can stimulate ethylene biosynthesis. Ethylene is biologically active at very low concentrations. Ethylene-mediated flower senescence is a significant problem in horticulture, most obviously in the floriculture industry.

Ethylene During Postharvest Handling of Roses

Ethylene has a negative effect on postharvest quality and may either be present in the environment (exogenous ethylene) or may be produced by senescing plant organs and in response to stress (endogenous ethylene) (**Figure 1**). Sensitivity to ethylene has important implications during the transport and handling of potted roses and cut roses in supermarkets and other areas where the air is commonly contaminated with ethylene. Moreover, unfavourable transport conditions such as high temperature, darkness and shaking in the trucks can reduce the display life of roses as a result of endogenous ethylene produced in response to stresses in the postharvest environment. Ethylene production in rose flowers clearly increases in some cultivars during transportation, e.g. in response to shaking in the trucks. In potted roses, the main causes of reduced postharvest quality are accelerated flower wilting, leaf yellowing, bud and leaf abscission and infection by grey mould (*Botrytis cinerea*). While ethylene is not assumed to be the primary agent of leaf yellowing in miniature potted roses, it is known to cause accelerated flower senescence as well as bud, flower and leaf drop.

Extensive work has been done in order to define the optimal transport conditions to reduce ethylene-induced postharvest losses in miniature roses. Temperature is one of the most important factors affecting flower longevity and quality during transport and postharvest handling. Experiments with miniature potted roses have shown that plants stored at 4 °C

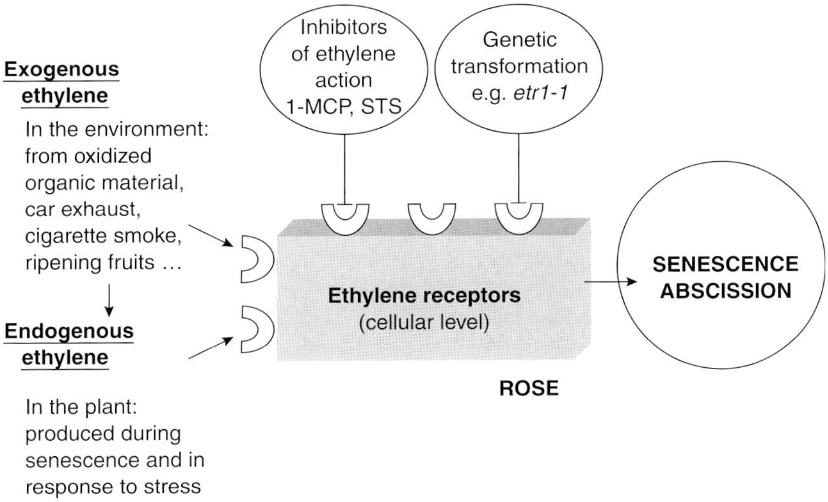

Figure 1 Ethylene has an effect on postharvest quality (accelerated senescence, abscission of plant organs) of rose either exogenous from contamination or endogenous, produced from senescing plant organs or in response to stress. 1-MCP, 1-methylcyclopropene; STS, silver thiosulphate.

have the longest poststorage floral longevity, the best flower quality and the least abscission, regardless of cultivar, storage duration or stage of flower at the beginning of storage. Low temperatures are known to inhibit ethylene effects during the cold storage period, and should therefore be aimed during postharvest handling of both potted miniature roses and cut-rose cultivars.

Postharvest quality of commercially important miniature cultivars varies significantly, even in an ethylene-free environment. In some short-lived cultivars 100% of the flowers had faded after 12 days; in long-lasting cultivars less than 10% loss in display quality took place after the same time period. The postharvest quality of potted roses differs among growers with different growing conditions. Temperature and day length, light intensity, composition of the nutrient solution and water availability during the production period affect the longevity of flowers.

Effects of Exogenous Ethylene

For many flowers the importance of ethylene in senescence was examined by testing the effects of exogenous ethylene. These effects were first noted as a practical problem resulting from environmental pollution with ethylene. As early as in the 1930s, researchers treated different plant tissues, including flowers, with ethylene gas and recorded the responses. The reactions varied, including abscission of petals, failure to open and senescence of the open flowers. In the 1980s Rosaceae were classified as a family where petal abscission occurs as an initial symptom of flower senescence. At that time it was assumed that ethylene is not a natural regulator of rose flower senescence, and that it only affects the postharvest life of roses when present as an atmospheric pollutant or when its synthesis is stimulated by stress.

Since the late 1990s it has been known that senescence in rose flowers is regulated by ethylene. Miniature potted roses mostly show petal wilting as flowers senesce under greenhouse and postharvest conditions; less frequently, flower or petal abscission or corolla colour change is observed. The ageing process of the flowers is clearly accelerated by exogenous ethylene. Increasing concentrations of exogenous ethylene result in a significant reduction of longevity in many commercially produced cultivars, expressed as a percentage of healthy open flowers per plant (**Figure 2**). Increasing amounts of exogenous ethylene in the postharvest environment accelerate leaf, bud abscission and flower senescence, and there are marked differences in sensitivity and response to ethylene among miniature rose cultivars. For example, the long-lived cultivar 'Charming Parade' is very sensitive to exogenous ethylene, suggesting that natural longevity (i.e. flower life in an ethylene-free environment) is not necessarily correlated with low sensitivity to exogenous ethylene. 'Vanilla Kordana', with excellent longevity in air, is almost insensitive to exogenous ethylene, indicating the potential for a breeding approach to increasing postharvest life of miniature roses.

Figure 2 Symptoms of *Rosa hybrida* after ethylene exposure for 6 days in the postharvest environment. Leaf, bud and flower abscission of *Rosa hybrida* 'Charming Parade' after exposure to (from left) 0, 0.1, 0.5, 1.0 $\mu l\, l^{-1}$ ethylene.

Both miniature roses and cut-rose cultivars are sensitive to exogenous ethylene, but there are great differences in sensitivity and symptoms among cultivars. In cut-rose cultivars, the application of ethylene has dramatically variable effects on opening and postharvest life of flowers. Depending on the cultivar, flower maturity and ethylene concentration, opening and senescence are unaffected, accelerated, completely inhibited or dramatically altered. In miniature rose cultivars, a variety of symptoms can be observed in response to ethylene, depending on cultivar. In addition to the mostly observed increase in flower wilting, some cultivars abscise petals after exposure to high ethylene concentrations (approx. 1 $\mu l\, l^{-1}$ for 4–6 days). Abscission in response to ethylene can even be observed in cultivars which senesce by petal wilting in an ethylene-free environment. Other common symptoms are abscission of buds, flowers, leaves, bud yellowing, 'bent neck', change in flower colour or shape. Some of the typical symptoms of a range of cultivars observed in ethylene environment are shown in **Table 1**. Furthermore, some cultivars react with abscission of stipules, even before other symptoms are visible. Another ethylene symptom which is often noticed is yellowing of petioles and sepals.

Ethylene Production

Ethylene is synthesized from methionine through the intermediates *S*-adenosyl methionine (SAM, also abbreviated as AdoMet) and the cyclic amino acid 1-aminocyclopropane-1-carboxylic acid (ACC). The

Table 1 Symptoms of *Rosa hybrida* after treatment with 0.5 µl l^{-1} ethylene for 4 days in the postharvest environment

	Abscission				*Yellowing*		*Changed flower shape or colour*	*'Bent neck'*
	Bud	*Flower*	*Petals*	*Leaves*	*Buds*	*Leaves*		
'Charming'	x	(x)		x				
'Sunset'	x					x	x	
'Orange'	(x)							
'Bianca'	x			x				x
'Bronze'	x		x		x		x	
'Vanilla'®								
'Pink Marina'				x			x	x
'Foxy'				x				x

enzyme converting methionine to SAM is SAM synthetase. ACC synthase converts SAM to ACC and *S*-adenosyl methanethiol. While ACC is oxidized to ethylene, HCN and CO_2 by ACC oxidase, methanethiol is reincorporated into methionine via the Yang cycle (**Figure 3**).

In many plant species the plant hormone ethylene regulates flower senescence. These so-called climacteric flowers exhibit a rise in ethylene production during senescence processes. Most of the present knowledge concerning the ethylene-mediated flower senescence is derived from studies on carnation flowers, used as a model for climacteric flowers. The senescence of carnation flowers is associated with a climacteric rise in ethylene production and an increase in respiration. Similar relationships have been observed for other flowers and this pattern of CO_2 and ethylene production resembles that observed in climacteric fruits such as bananas, apples, peaches, avocados and tomatoes. Similar to many fruits, the onset of senescence is associated with a coordinated increase in the activities of ACC synthase and ACC oxidase. The ethylene production seen in climacteric flowers is autocatalytic, which means that exposure to ethylene stimulates ethylene biosynthesis.

In several studies it has been shown that rose flower senescence is also regulated by ethylene. Many miniature rose cultivars show a climacteric rise in ethylene production during flower senescence, even though huge differences are found between cultivars (**Figure 4**). In some miniature rose cultivars flower senescence is accompanied by a clear climacteric rise in ethylene production, and in others there is a moderate or very low ethylene production. Longevity in the absence of exogenous ethylene is presumably a function of endogenous ethylene production. For example, the short life of individual flowers in 'Bronze' is associated with a clear climacteric peak in ethylene

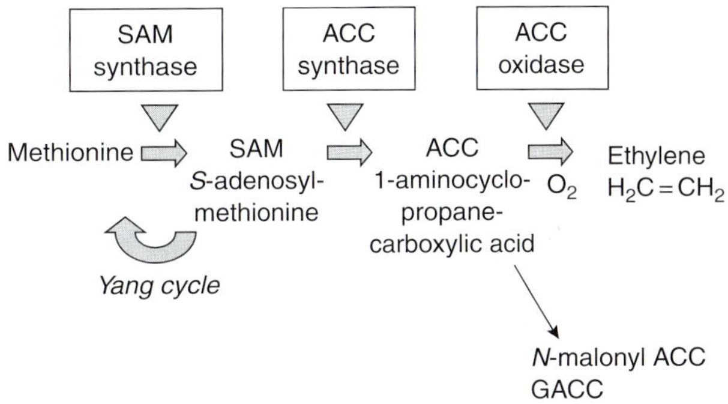

Figure 3 Schematic overview of ethylene biosynthesis. Ethylene is synthesized from methionine through the intermediates *S*-adenosyl methionine (SAM) and the cyclic amino acid 1-aminocyclopropane-1-carboxylic acid (ACC). The enzyme converting methionine to SAM is SAM synthetase, ACC synthase converts SAM to ACC and ACC is oxidized to ethylene by ACC oxidase. Reproduced from Muller R and Stummann BM (2003) Genetical control of ethylene perception and signal transduction related to flower senescence. *International Journal of Food, Agriculture and Environment* 1: 87–94, with permission.

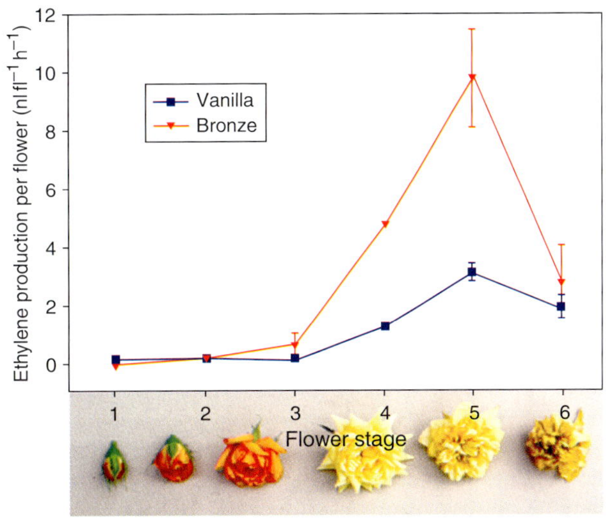

Figure 4 Ethylene production during flower senescence in flowers of *Rosa hybrida* 'Bronze Parade' and 'Vanilla Kordana'.

production, similar to that found in typical climacteric flowers. The excellent longevity of 'Charming Parade' in an ethylene-free environment is associated with a very low ethylene production of flowers. However, we have to keep in mind that there is no correlation between longevity and sensitivity to ethylene. The cultivar 'Charming Parade' has an excellent longevity in an ethylene-free environment and at the same time this cultivar is highly sensitive to exogenous ethylene (**Figure 2**).

On the other hand, differences in flower longevity cannot always be explained as a result of ethylene production during flower senescence. The cultivar 'Vanilla Kordana' exhibited a climacteric rise in ethylene production during flower senescence and excellent longevity, which made this cultivar interesting for basic studies on postharvest characteristics.

In many rose cultivars, ethylene exposure results in an increased ethylene production of flowers, presumably as a result of an upregulation of genes for ACC oxidase and ACC synthase. This phenomenon that ethylene causes an increase of its own production is termed autocatalytic ethylene production. Interestingly, differences in autocatalytic ethylene production seem to be a major reason for the short display life of several cultivars. Diffencences in autocatalytic ethylene production exist for both flowers and leaves of miniature rose cultivars. Apparently, not only initial ethylene production but also degree of autocatalytic stimulation is a major factor determining the display life of roses.

Inhibition of Ethylene Biosynthesis

The effects of ethylene can be inhibited at the level of ethylene biosynthesis or ethylene action. If ethylene biosynthesis is blocked, no protection against exogenous ethylene takes place. Aminoetoxyvinylglycine (AVG) and aminoxyacetic acid (AOA) block the conversion of SAM to ACC by ACC synthase, which requires pyridoxal phosphate as cofactor. ACC oxidase activity can be blocked by Co^{2+}, Triton X-100, α-aminoisobutyric acid (AIB) and the allocoronamic acid derivates.

Furthermore, salicylic acid was shown to inhibit ethylene production, apparently by blocking the conversion of ACC to ethylene. Inhibitors of ACC synthase and ACC oxidase only prevent ethylene biosynthesis and not ethylene action, for example in ethylene-polluted environments. Since these inhibitors do not offer a protection against exogenous ethylene, their use is valuable for studies of ethylene biosynthesis, but they do not play an important role in horticultural practice.

Inhibitors of Ethylene Action to Improve Rose Display Life

Many organic compounds interact with the ethylene receptor: some are agonists and mimic ethylene, while others are antagonists, preventing ethylene action by blocking the receptor. Some cyclic olefins counteract ethylene in a competitive manner. Of these compounds, 2,5-norbornadiene (2,5-NBD) blocks at the lowest concentration, but all of these cyclic olefins require continuous exposure to be effective. The relative blocking ability of these compounds also parallels their ability to bind to silver and silver ions negates the effects of ethylene in plants. Silver, applied in the form of thiosulphate (STS), is an effective inhibitor of ethylene responses, and its use is commercial practice for a number of horticultural commodities. It has been used with success on cut flowers and potted plants, e.g. miniature roses. Bud abscission and flower senescence can be decreased and the longevity of flowers and whole plants can be improved by applying STS. In several countries, the use of STS has become common practice for miniature roses in order to reduce ethylene-induced postharvest losses. However, the use of silver raises environmental concerns and it should be replaced by less harmful chemicals. In several countries the use of STS is restricted by law. Therefore intensive research is going on in this area aiming at commercialization of a nontoxic inhibitor of ethylene action.

Some cyclopropenes have been shown to counteract ethylene effects in a number of plants. Cyclopropene (CP), 1-methylcyclopropene (1-MCP) and 3,3-dimethylcyclopropene (3,3-DMCP) prevent ethylene effects for 10–12 days in tissues given a single exposure at very low concentration. 1-MCP, CP and 3,3-DMCP presumably bind to a Cu ion in the ethylene binding site in the ethylene receptor, preventing ethylene from binding to treated tissues. While ethylene diffuses rapidly from the receptor, 1-MCP and other inhibitors remain bound for long periods, preventing the binding of ethylene.

1-MCP in particular appears to be suitable for many commercial applications and can improve postharvest characteristics of ornamentals by competing with ethylene for binding sites at the receptor level. As low a concentration as $0.5\ nl\,l^{-1}$ 1-MCP is sufficient to protect some flowers for several days against ethylene. Some plant species and different plant organs require higher concentrations of the inhibitor. The concentration required to protect plants depends on the time of exposure. The longer the exposure time, the lower the required concentration. For miniature roses, a pretreatment with concentrations in the range from 5 to $100\ nl\,l^{-1}$ 1-MCP for 6 h has been reported.

However, the optimal concentration and the length of the protection period have not been accurately determined for roses. Some flowers may deteriorate for other reasons, e.g. fungal infection, before they become sensitive to ethylene after treatment with inhibitors, thus preventing an accurate determination of protection period. Carnations treated with 1-MCP appear to remain insensitive for 12–15 days at 24 °C. At lower temperatures higher concentrations of the inhibitors are required for complete protection of carnations against ethylene, suggesting that binding to the receptor is less than at higher temperatures. Experiments indicated that higher concentrations of 1-MCP may be required at lower temperatures, and no inhibition of ethylene responses is found where ethylene production is already high in 1-MCP-treated tissues. Cultivar differences in the efficiency of the inhibitor can be observed. In miniature potted roses, pretreatment with 1-MCP clearly reduces the rise in autocatalytic ethylene production in flowers of some investigated cultivars, while no effect can be registered in flowers of another cultivar with low endogenous ethylene level.

1-MCP and its use are patented and the licence is held by an American company. Until now, its use as postharvest treatment for cut flowers and potted plants is registered in North America, but registration for Europe is expected. The commercialized product is a powder, which after addition of water releases the gaseous 1-MCP. In potted miniature rose, 1-MCP protects against exogenous ethylene, increasing display life and reducing abscission of buds, flowers and leaves. The ethylene production in response to exogenous ethylene is totally inhibited by treatment with 1-MCP, while the effectiveness of 1-MCP is less efficient in senescing flowers. Additionally, it is assumed that during further plant development new binding sites are synthesized and such receptor proteins are not protected by 1-MCP and thus senesce in response to ethylene. While STS and 1-MCP prolong the postharvest performance of miniature roses in ethylene, this effect is less pronounced in an ethylene-free environment.

The Ethylene Signal Transduction Pathway in Plants

In plants not only hormone concentration but also sensitivity of the tissue determines ethylene responses, and the development of sensitivity to growth substances such as ethylene appears to precede the developmental process it induces. The primary steps in ethylene action are assumed to be similar in all plant cells, in spite of the following diverse effects on plant development. Ethylene binds to a receptor, followed by activation of one or more signal transduction pathways, leading to a cellular response. Finally, ethylene exerts its effect by altering the expression of various target genes. The similarity of the first steps in ethylene signal transduction in different plants makes it possible for rose science to benefit from the progress achieved by molecular studies using the model plant *Arabidopsis thaliana*. In this model plant, several genes encoding components involved in the ethylene signal transduction pathway have been studied, mainly by screening for alterations in the so-called triple response, which refers to morphological changes in dark-grown seedlings treated with ethylene. A number of ethylene-insensitive or constitutive-response mutants have been isolated in *Arabidopsis* and several genes have been cloned. The essential features of the current model of ethylene signal transduction in *Arabidopsis*, and other plants affected by ethylene, are summarized in **Figure 5**. The gene *ETR1* (ethylene-resistant 1) encodes a protein (ETR1) with amino acid similarity to certain bacterial receptor proteins. Certain dominant mutant versions of this gene (e.g. *etr1-1*) confer ethylene-insensitivity because they encode a receptor that cannot bind ethylene and at the same time constitutively suppresses ethylene responses in the flowers. The suppressed ethylene response, e.g. partial or full ethylene insensitivity, of *etr1-1* plants is due to the fact that the mutant receptor activates constitutive triple response 1 (CTR1) even in the presence of ethylene (cf. **Figure 5**). Thus, the effect of such mutations mimics the effect of STS, which probably eliminates ethylene binding because Ag ions substitute the Cu ions in the binding sites, and of 1-MCP, which binds competitively to the ethylene binding sites. In *Arabidopsis*, five members of the ethylene receptor gene family have been cloned. Some of the other components shown in **Figure 5** also exist in families of two or more closely related versions.

Ethylene Receptor Genes are Differentially Expressed During Rose Flower Senescence

In minature roses, fragments of four rose ethylene receptor genes, termed *RhETR1-4*, have been cloned. Although only parts of the sequences have been determined, RhETR1, RhETR2 and RhETR4 exhibit high amino acid identity with *Arabidopsis* and other receptors with three membrane-spanning domains (cf. **Figure 5**), while RhETR3 has high identity to receptors with four membrane-spanning domains.

The relative transcript abundance for the genes *RhETR1-3* varies significantly during flower

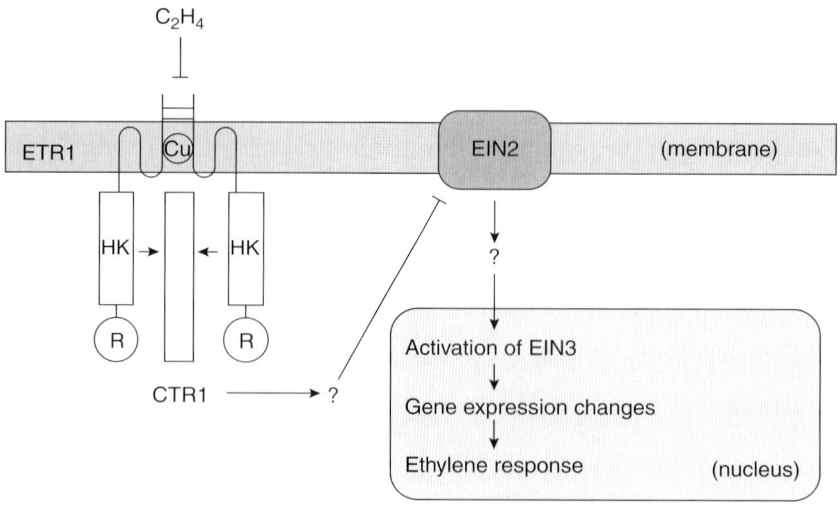

Figure 5 The model for ethylene signal transduction in plants. The receptor shown is the *Arabidopsis* ethylene-resistant 1 (ETR1) receptor but other receptors function in a similar way. All receptors have a histidine kinase (HK) domain but some lack the receiver (R) domain. Binding of ethylene requires a Cu ion. The two monomers of the receptor are joined by two disulphide bonds. In the absence of ethylene, ETR1 keeps constitutive triple response 1 (CTR1) active. Active CTR1 keeps the membrane protein ethylene-insensitive 2 (EIN2) inactive, presumably via intermediates. EIN2 may be an ion channel placed in the plasma or ER membrane and may open to allow ion influx into the cytosol when the inhibiting effect of active CTR1 stops. Active EIN2 leads, probably via intermediate steps, to activation of the transcription factor EIN3. The resulting changes in gene expression cause the ethylene response.

development and after hormone treatment. *RhETR2* expression barely changes during flower development and in response to ethylene and ABA treatment, but the transcript levels for *RhETR2* are different for cultivars with different longevity. In contrast, *RhETR1* and *RhETR3* exhibit differential expression within each cultivar. Exposure to low ethylene concentrations results in an upregulation of *RhETR1* and *RhETR3* expression in rose flowers. These two receptor genes are also regulated during flower development and may be rate-limiting for ethylene perception and a determinant of flower longevity. The expression of *RhETR1* is distinctly higher in a cultivar with short flower life than in another long-lasting rose cultivar. While expression of *RhETR1* precedes ethylene production from the flowers, the abundance of *RhETR3* transcript increased during flower senescence, indicating that the ethylene response system in rose flowers comprises several receptor types with overlapping expression. In a cultivar with excellent flower longevity despite moderate ethylene production, the expression of *RhETR1* and *RhETR3* is reduced. These results indicate that differences in flower life among rose cultivars – in an ethylene-free environment and in response to exogenous ethylene – may be due to differences in receptor levels. These observations are thus in apparent contrast to the standard model of ethylene signal transduction that predicts that a reduction in the level of receptors results in increased ethylene sensitivity, whereas an increase in receptor number results in decreased sensitivity.

Strategies for Improvement of Postharvest Quality in Roses: Inhibition of Ethylene Biosynthesis or Transformation with *etr1-1*

Various biotechnological modifications may be suggested to improve the postharvest quality of roses. Several studies on the effects of decreased ACC oxidase activity in ornamentals have been performed. Using an ACC oxidase cDNA from carnation, transgenic carnation plants constitutively expressing antisense ACC oxidase RNA were produced. Flowers of one transformant exhibited lowered climacteric ethylene production and distinctly delayed flower senescence. Similar strategies may be applicable for roses to improve postharvest quality.

Plants in which ethylene biosynthesis is inhibited, either by chemicals or by genetic modification, are still sensitive to exogenous ethylene. A strategy aimed at blocking ethylene perception can be expected to be more effective at increasing flower life in roses, since it should protect against exogenous as well as endogenous ethylene.

Recently, several examples of genetic transformation of horticultural crops with the mutant *Arabidopsis* receptor gene *etr1-1* have been reported. Expression of the *etr1-1* gene in transgenic carnations and *Petunia* delayed flower senescence, resulting in a significant increase in flower life. In *etr1-1*-transgenic carnation flowers, expression of the *ACO1* (ACC Oxidase 1) gene was downregulated, indicating that the normally observed autocatalytic induction of ethylene biosynthesis by ethylene is absent due to the inability of ethylene to bind to the etr1-1 protein. The delay in senescence observed in *etr1-1* transgenic flowers was longer than in flowers pretreated with inhibitors of ethylene biosynthesis (AOA) or ethylene receptors (STS).

The Model of Ethylene Signal Transduction and Future Prospects

Ethylene-sensitive flowers offer unique model systems for studying biological responses mediated by the plant hormone ethylene. Research on rose and several other flowering plants has indicated that flower senescence is the result of a complex regulation of both ethylene biosynthesis and perception. The standard model of ethylene signal transduction (**Figure 5**) indicates that an increase in the level of receptors or CTR1, or a decrease in the level of ethylene-insensitive 3 (EIN3) should result in decreased ethylene sensitivity, whereas a decrease in the level of receptors or CTR1, or an increase in the level of EIN3, should result in increased ethylene sensitivity. Under the physiologically reasonable assumption that ethylene sensitivity increases during rose flower senescence and in response to ethylene, a decrease in the level of receptors or CTR1, or an increase in the level of EIN3, may therefore be expected.

Reports of floral *ETR* homologue regulation have sometimes been conflicting. Different results have been obtained in ethylene receptor expression studies of *Dianthus* and rose flowers during development and senescence. The results in *Dianthus* indicate that there is an inverse relationship between the level of ethylene receptors and the sensitivity to ethylene and that a reduction in the amount of ethylene receptor proteins increases ethylene sensitivity in carnations. This is what the standard model predicts. In contrast, in roses a parallel relationship between the level of ethylene receptor transcript and flower longevity was found. The expression of *RhETR3*, one of the four ethylene receptor genes in rose, increases in senescing flowers of a cultivar with short floral life, while it remains at low levels in another cultivar with a long flower life. Furthermore, the expression of *RhETR3* is increased by ABA and ethylene treatment, which stimulated flower senescence in rose.

The biological explanation for the multiplicity of receptor genes in rose and other plants is currently unknown, but it may be suggested that individual hormone receptor types maintain distinct functional identities via the capacity to respond differentially to developmental hormonal cues. Differential regulation of expression of the receptor gene family by ethylene may provide a mechanism to achieve differential sensitivities even in the same response under different conditions. Increased receptor transcript levels after ABA treatment suggests an involvement of ABA in the senescence process. It is not known whether ABA promoted flower senescence is due to increased ethylene production or to a change in tissue sensitivity to ethylene.

The transcript levels of CTR1, the similar protein CTR2 and EIN3 have also been investigated for senescing rose flowers. Transcription of the two *CTR* genes is increased in response to exogenous ethylene. In contrast, the transcription factor RhEIN3 is constitutively expressed in senescing flowers and in response to ethylene and ABA.

Other options may be suggested for decreasing ethylene sensitivity than transformation with a gene for a receptor, such as *etr1-1*, that does not bind ethylene but still activates CTR1 (cf. **Figure 5**). If the standard model is correct, which would imply a need for explanation of the observations that do not seem to be in accordance with it, then an upregulation of receptor level, upregulation of CTR1 level or downregulation of EIN3 could conceivably lead to resistant plants or resistant selected plant organs. Presumably further basic research on the mechanism of ethylene signal transduction will reveal further ways in which that goal may be reached.

See also: **Postharvest Physiology**: Xylem Structure and Function in Cut Roses; Effects of Supplemental Carbohydrates; Effects of Temperature; Packing Systems and Precooling. **Production and Marketing**: Pot Rose Production; Cut Rose Production.

Further Reading

Abeles FB, Morgan PW and Saltveit ME Jr (1992) *Ethylene in Plant Biology*, 2nd edn, p. 414. New York: Academic Press.

Chang C and Stadler R (2001) Ethylene hormone receptor action in *Arabidopsis*. *BioEssays* 23: 619–627.

Müller R and Stummann BM (2003) Genetic control of ethylene perception and signal transduction related to flower senescence. *International Journal of Food, Agriculture and Environment* 1: 87–94.

Müller R, Andersen AS and Serek M (1998) Differences in display life of miniature potted roses (*Rosa hybrida* L.). *Scientia Horticulturae* 76: 59–71.

Müller R, Sisler EC and Serek M (2000) Stress induced ethylene production, ethylene binding, and the response to the ethylene action inhibitor 1-MCP in miniature roses. *Scientia Horticulturae* 83: 51–59.

Müller R, Lind-Iversen S, Stummann BM and Serek M (2000) Expression of genes for ethylene biosynthetic

enzymes and an ethylene receptor in senescing flowers of miniature roses. *Journal of Horticultural Science and Biotechnology* 75: 12–18.
Müller R, Stummann BM and Serek M (2000) Characterization of an ethylene receptor family with differential expression in rose (*Rosa hybrida* L.) flowers. *Plant Cell Report* 19: 1232–1239.
Serek M, Sisler EC and Reid MS (1994) Novel gaseous ethylene inhibitor prevents effects in potted flowering plants. *Journal of the American Society of Horticultural Science* 119: 1230–1233.
Serek M, Sisler EC and Reid MS (1996) Ethylene and the postharvest performance of miniature roses. *Acta Horticulturae* 424: 145–149.
Sisler EC and Serek M (1997) Inhibitors of ethylene responses in plants at the receptor level. *Physiologia Plantarum* 100: 577–582.
Taiz L and Zeiger E (2002) *Plant Physiology*, 3rd edn, pp. 519–538. Sunderland, MA: Sinauer.

Packing Systems and Precooling

M Reid, University of California, Davis, CA, USA

Introduction

For cut roses, the first steps in the postharvest chain, namely rehydration, grading, packing, and precooling, are critical to the eventual quality and vase-life of the flowers displayed in the consumer's home. This article provides an overview of the systems that are presently employed in the major growing regions, recognizing that the sequence of the steps and the details differ considerably from grower to grower and country to country. Whatever the system that is employed, the objectives of these first manipulations are to deliver a product that can be transported (or stored) under cool temperatures and will arrive at destination with as much potential vase-life as it had at the time of harvest.

Packing Systems

Harvest to Packing Shed

At harvest where blooms of different sizes, maturities, and qualities are collected, the flowers are typically placed in a plastic sheet or net and placed in a bucket containing water (**Figure 1**). This 'first drink' is a critical step in the life of the flower, and the water in the bucket should be fresh, of good quality, and should contain a bactericide (50 ppm hypochlorite is a frequent choice). Frequently flowers are allowed to stand under the warm, high light conditions of the greenhouse for some time after harvest, and this is undesirable, even if the stems are in water, because of the potential for excessive water loss and later poor water relations. Although there is a tendency in many cut flowers and most perishable crops for the grading and packing operations to be moved to the field (thereby reducing repeated handling and reducing the time until the produce is cooled), this trend has not been seen with cut roses, which are always taken to a packing shed for grading, packing and cooling.

Rehydration

Roses harvested during the day are inevitably somewhat dehydrated, and a universal part of the postharvest handling system is rehydration. The water in the buckets into which the flowers are placed in the greenhouse will start this rehydration process, and that is why its freshness, quality and sterility are so important. A very common strategy for handling roses is to transport them from the field, in water, directly into a cooler, where rehydration is accomplished rapidly. At cool temperatures the flowers lose very little water, so the water taken up by the flowers is employed solely for restoring the water deficit. There are several disadvantages to this common strategy. If warm flowers are wrapped in plastic sheet, the cooling process will be relatively slow, and accompanied by condensation, which can lead to the germination and growth of grey mould (*Botrytis cinerea*). Even when the flowers are in nets, cooling will be relatively slow. Cooling flowers prior to grading and packing has the additional disadvantage that the flowers will warm up when they are brought to the packing shed. In addition to the dangers of the condensation that will ensue, growers need to be aware of the costs of cooling the product twice – after harvest, and after packing. Alternative strategies that have been demonstrated in research studies include:

- harvesting at night (when flowers are fully hydrated, do not need to be placed in water, and typically have longer vase-life)
- rehydration in cool but not cold conditions
- harvesting, grading and bunching *prior* to placing in rehydration solution.

Grading

Although there are no formal international grade standards for cut roses, agreements between growers and wholesalers, and national standards (whether mandatory or voluntary) typically include stem length, stem

Figure 1 After harvest, roses are typically wrapped in plastic or net (a) and placed in water. An alternative system is to lay the harvested flowers into a plastic container (b) that can be placed in rehydration solution (c) when the flowers arrive at the packing shed or cooler.

straightness, bud maturity, bud size, leaf quality and freedom from pests, diseases and damage. Modern equipment that utilizes machine vision technology for size and quality grading, and robotic systems for bunching and tying are available for automatic grading of roses (**Figure 2**). However, the high cost and throughput of these machines make them appropriate only for the largest production operations. Most rose grading is still carried out by hand, sometimes utilizing graders that separate flowers by stem length, more often entirely by hand (**Figure 2**).

Bunching

Cut roses are normally bunched after grading. The number of flowers in the bunch varies according to growing area, type, quality and market needs. Groups of 10, 12 and 25 are common for single-stemmed roses. Many different simple systems have been devised to assist in the construction of uniform bunches (**Figure 3**). Central American roses are often bunched by hand on plastic sheets that are then rolled up to provide a spiral pack. The disadvantage of this system is the likelihood of condensation that will accompany cooling the flowers. Spray-type roses may be bunched by the number of open flowers, by weight or by bunch size.

Lower leaves and thorns are removed from the stems, sometimes by hand, using a variety of hand tools (a potato masher works quite well!), most often using a modified chicken plucker (**Figure 4**), whose vibrating rubber fingers are relatively gentle. Bunches are held together using string, paper-covered wire or elastic band, are trimmed to length (**Figure 5**) and are frequently sleeved soon after harvest to separate them, protect the flower heads, prevent tangling and identify the grower or shipper. Materials used for sleeving include paper (waxed or un-waxed), corrugated card (smooth side towards the flowers) and polyethylene (perforated, un-perforated and blister) (**Figure 6**). Sleeves can be preformed (although variable bunch size can be a problem), or they can be formed around each bunch using tape, heat sealing (polyethylene) or staples.

There are many shapes of packing containers for cut rose, but most are long and flat (**Figure 7**). This design restricts the depth of the flowers in the box, which may in turn reduce physical damage. In addition, flower heads can be placed at both ends of the container for better use of space. With this kind of flower placement, whole layers of newspaper have often been used to prevent the layers of flowers from injuring each other. The use of small pieces of newspaper to protect only the flower heads, however, is probably the better practice, since it allows for more efficient cooling of flowers after packing. It is critically important that containers be packed in such a way that transport damage is minimized. Some packers anchor the product by using enough flowers and foliage in the box so that the package, after banding, holds itself firmly. To avoid longitudinal slip, packers in many flower-producing countries use one or more 'cleats'. These are normally foam- or newspaper-covered wood pieces that are placed over the product, pushed down and stapled into each side of the box. Padded metal straps, elastic bands, high-density polyethylene blocks or

Figure 2 Roses are most often graded by hand (a) but simple machiners for length grading (b) and robotic sorter/bunchers that use machine vision (c) are used by large-volume producers.

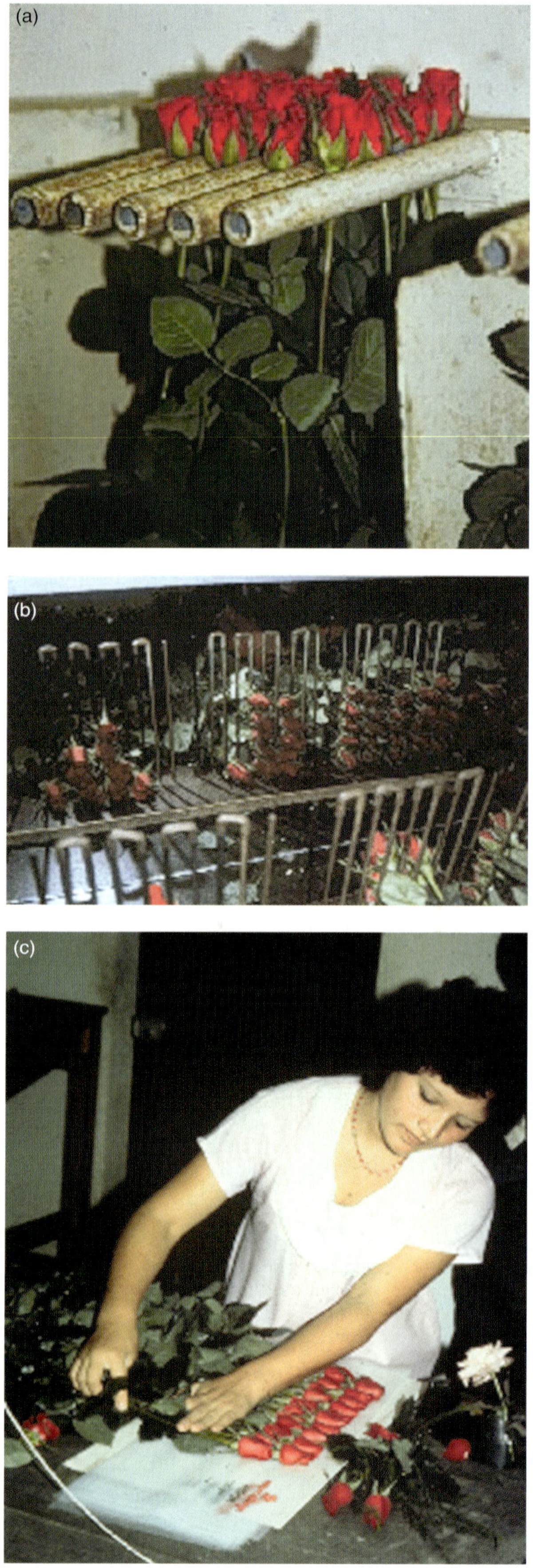

Figure 3 A range of simple devices is used to assist in preparation of rose bunches. Vertical pipe or wire racks (a) assist in preparation of uniform bunches. Horizontal systems can use mirrors (b) for the same purpose. Central American producers commonly lay the flowers on plastic, then roll them into a spiral (c).

(a) (b)

Figure 4 Thorns and leaves are removed (a) from bunched roses using a range of devices; commonly a rubber-fingered chicken plucker (b).

Figure 5 After grading, roses are typically bunched in groups of 10 to 25 (sometimes as many as 70), and then trimmed to uniform length.

cardboard tubes can also be used as cleats. The heads of the flowers should be placed 6–10 cm from the end of the box to allow effective precooling and to eliminate the danger of petal bruising should the contents of the box shift. Cleats are really only effective if one is placed in the bottom of the box to prevent 'bowing' of the base of the box, and slippage of the flowers under the top cleats.

Upright packing, in systems like the 'Procona' package, is increasingly popular in the commercial handling of cut roses. Procona packs comprise plastic bases and a cardboard sleeve to allow transport of flowers upright in water. This system is more expensive than traditional boxes, and a smaller volume of product can be packed in it, but the presence of water may improve outturn when roses are not transported under proper temperature conditions.

Precooling

By far the most important part of maintaining the quality of cut roses is ensuring that they are cooled as soon as possible after harvest and that optimum temperatures are maintained during distribution. Roses should ideally be transported and stored at 0–1 °C. Once packed, cut roses are difficult to cool. Their high rate of respiration and the high temperatures of most greenhouses and packing areas result in heat build-up in packed rose boxes unless measures are taken to ensure temperature reduction. It is therefore necessary to cool them as soon as possible after packing. Individually, roses cool (and warm) rather rapidly (half-cooling times of a few minutes). So, while individual roses can be cooled quickly, it is also true that individual flowers brought out of cool storage

Figure 6 Exported roses are usually sleeved in absorbent material such as blister polyethylene (a) or soft corrugated fibreboard (b).

into a warmer packing area will warm quickly and develop condensation prior to packing. The simplest method of ensuring that packed roses are adequately cooled and dry is, therefore, to pack them in the cool room. Although this method is not always popular with packers, will probably increase labour cost, and may slow down packing somewhat, it ensures a cooled, dry product.

Forced-air cooling of boxes with end holes or closeable flaps is the most common and effective method for precooling cut flowers (**Figure 8**). Cool air is sucked or blown through the boxes. Care must be taken to pack in such a way that air can flow through the box and not be blocked by the packing material (**Figure 9**). In general, packers use less paper when packing roses for precooling. The half-cooling time for roses (the time taken for the flower temperature to fall from the initial temperature half-way to the temperature of the cool room) ranges from 20 to 40 min, depending on the packing system. Roses should be cooled for three half-cooling times (by which time they are 7/8 cool).

To ensure adequate cooling, cool-room operators use a needle-probe digital thermometer to measure the

Figure 7 Horizontal boxes used in California include (from left) plain fibreboard, polyurethane-sprayed fibreboard, polystyrene caskets and fibreboard boxes with sprayed-in-place polyurethane.

Figure 8 In a typical forced-air precooler for roses, cool air is sucked through the packed flowers.

temperature of the flowers, inserting the needle through the side of the package into the middle of the flower bunches (**Figure 10**). If the packages are to remain in a cool environment after precooling, vents may be left open to assist removal of the heat of respiration. Flowers that are to be transported at ambient temperatures can be packed in polyethylene caskets, foam-sprayed boxes or boxes with the vents resealed. Ice that is used after precooling is only effective if it is

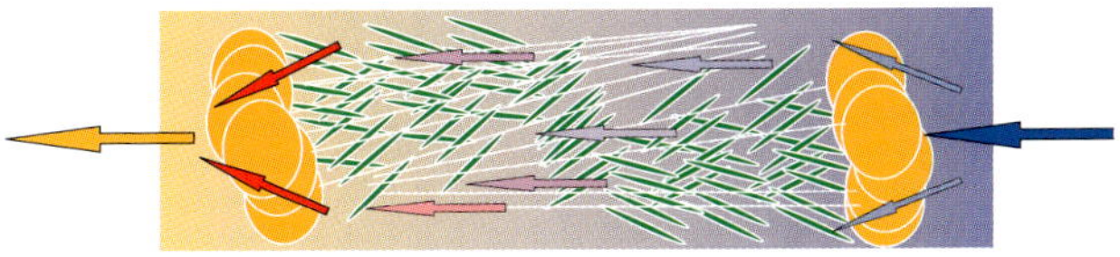

Figure 9 In forced air cooling, cool air is drawn or blown through the roses to remove heat.

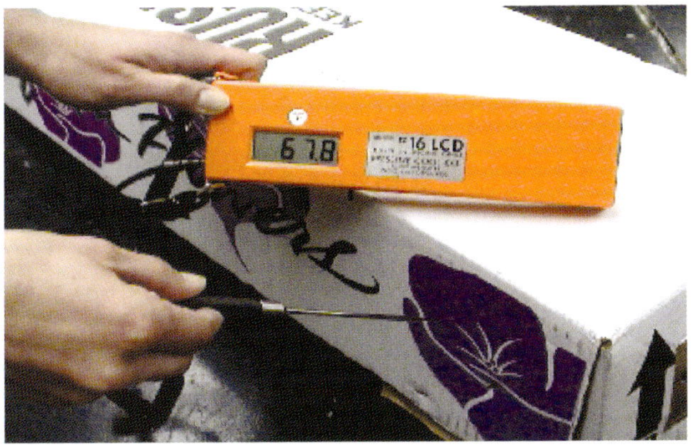

Figure 10 Temperatures of packed roses are measured through the side of the carton, using a needle-probe digital thermometer.

placed so that it intercepts heat entering the carton (i.e. it must surround the product), and care must be taken to ensure that the ice does not melt onto the flowers or cause freezing damage.

Precooling of vertical hampers or Proconas presents a particular challenge, but can be achieved using a down-draft forced-air cooling system that adapts the normal forced air precooling system and uses the pallet as a plenum for air sucked down through the vertical boxes as shown in **Figure 11**.

Figure 11 In the 'Downdraft' system for cooling 'Procona' flower boxes, air is sucked through the flowers and into a conventional forced air-cooling wall through the pallet on which the boxes are standing. A fibreboard or canvas shroud around the base of the boxes ensures that the air is drawn through the flowers.

See also: **Breeding**: Selection Strategies for Cut Roses. **Postharvest Physiology**: Xylem Structure and Function in Cut Roses; Effects of Supplemental Carbohydrates; Effects of Temperature; Ethylene. **Production and Marketing**: Cut Rose Production.

Further Reading

Anonymous (1988) *The Flower Care Booklet*. New York: Flower Council of Holland.

Faragher J, Slater T, Joyce D and Williamson V (2002) *Postharvest Handling of Australian Flowers*. Melbourne, Australia: Rural Industries Research and Development Corporation, 217 p.

Nell TA and Reid MS (2001) *Flower and Plant Care: The 21st Century Approach*. Alexandria, Virginia: Society of American Florists.

Nowak J, Rudnicki RM and Duncan AA (1990) *Postharvest Handling and Storage of Cut Flowers, Florist Greens, and Potted Plants*. Portland OR: Timber Press.

Worral RT, Gollnow B and Wade N (2000) *Rose Flower Care for Professionals*. Sydney, Australia: NSW Agriculture.

PRODUCTION AND MARKETING

Contents

Overview of Roses and Culture

H B Pemberton, Texas A&M University, Texas, USA

The rose is one of the most commonly cultivated plants in the world. When considering only cut flower imports into main European markets and the USA, the wholesale value approached 1 billion euros in 1999. Despite being endemic to the northern hemisphere, production centres can be found in all sections of the globe. Equal to the diversity found in production location are the many different forms of production that have evolved over the centuries. These different forms of production follow closely the development of cultivars for specific uses. The production of garden roses, cut roses and miniature potted roses is all part of production schemes that reflect restrictions of local climate, the feasibility of moving material through marketing channels, and consumption patterns of all those who buy roses for personal or public use (**Table 1**).

Garden Roses

Garden rose plants are produced on all continents except Antarctica (*see* **Production and Marketing:** Field Rose Production). One reason for this is the high level of government restriction on the movement of intact plants from one country to another to prevent the spread of pests and diseases. For this reason, garden rose plant production can be very regional, with large centres of production in Canada, the USA, Mexico, Brazil, the UK, France, Germany, Spain, Italy, some Eastern European countries, Morocco, Israel, Japan, South Africa, Australia and India.

Most garden roses are produced by grafting using the T-bud technique. The most popular varieties for this type of production are from the hybrid tea and floribunda hybrid groups, though some miniature and shrub types are grown this way as well. Though field production schemes are very similar worldwide, regional differences exist. These differences are usually based on the use of different rootstock species and/or the use of seedling in comparison with cutting rootstocks. For example, *Rosa multiflora* is used in the south-central USA, Canada and Japan, whereas *R.* 'Dr Huey' is used in the western USA and *R. Corymbifera* 'Laxa' is used in northern Europe. Also, cutting-grown rootstocks are used in the USA whereas seedling rootstocks are used in Canada, Japan and northern Europe (*see* **Rootstocks:** Usage of Rootstocks). Production schemes utilizing cutting-grown rootstock require a 2-year period, whereas the use of seedling-grown rootstock requires a 3-year production scheme. Another regional difference is the technique of bud preparation for T-budding. For example, in the USA, the bud is cut with a sliver of wood still attached, whereas in the UK the wood is removed before being inserted into the T-shaped cut on the stock.

Field rose plant production tends to be concentrated in local areas. This can be due to local environmental factors such as soil type and water availability, but just as important a reason is the availability of knowledgeable growers and trained workers. This type of rose production requires a high degree of skill and knowledge that in many cases has been passed down for many generations. In addition, field rose plant production requires a large capital investment such that new producers do not readily enter the business without a serious commitment, knowledge gained from experience about the complex process of growing a grafted rose plant and financial support.

A production method that is still growing in popularity uses cutting-grown plants for containerized nursery production. This method has always been

Table 1 Popularity of different cut flowering types by producing/consuming country

Flower size	*Country with significant production and/or consumption*													
	Columbia	*Ecuador*	*France*	*Germany*	*Israel*	*Italy*	*Japan*	*Kenya*	*Mexico*	*Morocco*	*The Netherlands*	*South Africa*	*USA*	*Zimbabwe*
Large	X	X	X			X	X		X	X			X	X
Medium	X			X	X	X	X	X	X		X			X
Small				X	X			X			X		X	
Spray[a]							X							
Miniature[b]						X	X					X	X	

[a]Spray types occasionally seen in markets in Europe and the USA.
[b]There is no large-scale production of cut miniature types. They are occasionally seen in the markets indicated.
Adapted from Pertwee J (2000) *Production and Marketing of Roses.* The Netherlands: Elsevier International Business Information.

the method of choice for miniature roses as grafting is not required. As more and more cultivars classified as shrubs enter the market, own-root plants are becoming the production method of choice as these plants also perform well without grafting. In addition, most antique roses, many of which have survived for decades in abandoned landscapes as own-root plants, are produced this way. This type of production is not as concentrated in local areas as field rose production as it is not tied to specific soil types or other aspects of climate. This is because the plants are grown in an artificial medium in containers under many different types of nursery structures that can protect against temperature extremes. Also, the expertise for grafting is not required (*see* **Propagation:** Own-Rooted Cuttings).

Traditionally, field-grown roses were dug when dormant and sold in nurseries as bare-root plants during late winter or early spring. With the advent of cold-storage facilities, it became common for these plants to be packaged for the mass market. In recent years, containerization of dormant bare-root plants has become more and more popular. Plants are then sold in bud and bloom in the mass market as well as garden centres.

Cut Roses

As with garden roses, cut-rose production is usually concentrated in local areas due to climate and availability of skilled labour, but the distance from production centre to market is usually much greater than that for garden roses as planes can be used for shipping. This is due to relatively low shipping costs and a general lack of phytosanitary restrictions for importation. In contrast to the shipping of intact garden rose plants, international shipments of cut flower stems is generally allowed. Since the 1980s, cut-rose production has shifted from many developed countries where consumption is high to lesser developed countries where costs of facilities and labour are lower. Also, light availability in the lesser developed countries is greater than it is in the more developed countries during the winter and spring season when consumption is higher than other times of the year. Thus, we now see the US market being supplied with cut roses from Colombia, Ecuador and Mexico. In Europe, the Dutch have traditionally been the main producers and the Germans the main consumers, but many roses are now imported from Israel and Africa. The largest exporters of cut roses are The Netherlands, Colombia, Ecuador, Kenya, Zimbabwe and Israel. Significant increases in production are occurring in Ecuador, Colombia, Kenya, Zimbabwe, South Africa, Tanzania and Zambia. The largest importers are Germany, the

USA, The Netherlands (some are also re-exported), France, Switzerland and the UK. This is not to say that domestic production in countries where consumption is high is totally stagnant. In southern Europe, for example, demand is increasing to the point that domestic production is increasing. Also, there is always a market for high-quality roses such as those with longer stems and this provides a niche for domestic producers. Consumption is also increasing in many countries where production is increasing, such as in Italy, Mexico, Brazil, South Africa, India, China, Japan and Taiwan.

Cut-rose producers can be divided into two categories. The larger growers usually realize the economies of scale that are needed to sell to the mass market. Smaller growers usually emphasize quality and service domestic speciality markets with lower shipping costs. Many larger growers are located in exporting countries and use production methods such as in-ground beds and structures that require less financial input. Many producers in developed countries emphasize quality and utilize a high-technology approach to growing, including supplemental lighting during the winter and substrate culture. New harvesting techniques such as the arching or bending method are also frequently used. Production of a high-quality product year-round is necessary to justify the increased cost. Another type of production can be found in countries such as Israel where the climate favours winter production, but summer conditions result in lower-quality flowers.

The cultivars used for cut-flower production are predominantly from the Hybrid Tea and Floribunda (also known as Sweetheart) groups. Red is the predominant colour followed by yellow and pink. In northern Europe, shorter-stemmed roses outsell the large-flowered types, but in southern Europe and the USA the larger types are preferred. Larger flowering types can be more profitable than other types and are generally sold through speciality markets. Medium-flowered roses are relatively new, but are popular with the mass markets and growers in Kenya and Zimbabwe. Small-flowered roses are popular in northern Europe, but are less profitable due to high harvesting costs. Spray roses are popular in Japan, but consumption is low in Europe and the USA. Miniatures are occasionally seen in Japan, the USA, Italy and South Africa, but production is limited (*see* **Production and Marketing:** Cut Rose Production. **Breeding:** Selection Strategies for Cut Roses).

Miniature Potted Roses

Production of miniature potted roses is the most recent type of rose culture to become popular in many parts of the world. Traditionally, bare-root grafted plants of the floribunda and polyantha classes were forced as potted plants for winter and spring sales. However, breeding efforts led to the introduction of a whole new group of miniature cultivars starting in the 1980s that were bred specifically for pot plant production. These cultivars are characterized by ease of cutting propagation, strong uniform growth following a pinch, which is advantageous for greenhouse production, and good postharvest longevity, which is necessary for consumer acceptance and the commercial success of any potted plant crop. Since most of these cultivars have been recently introduced, plant patents are still in force. This has limited pot rose production to those countries where strong laws protecting breeders' rights are in force. This also limits the use of new cultivars of garden and cut roses in many parts of the world, but there are many cultivars that are either unpatented or have expired patents that can be used for those types of culture. This is not the case with miniature potted roses. Major centres of production include Denmark, The Netherlands, the USA, Canada and Japan, with production also in France, Germany and Italy.

Miniature potted roses are produced from cuttings by either direct sticking into the container used for finishing or by using liners. Direct sticking is used by larger growers who produce crops year-round, whereas liners are used by smaller growers who grow seasonally or for holidays. The larger growers tend to have more specialized growing facilities that utilize technology to a high degree. Liners tend to be more economical for producers who have less sophisticated growing facilities and a more heterogeneous product mix.

There are two main groups of potted miniature cultivars – the true miniatures and the large growing types. The true miniatures are suitable for smaller pots and tend to be more popular in Europe, whereas the larger growing types need larger pots for production and tend to be more popular in North America. Also, Europeans (especially in northern areas) seem to prefer pastel colours, whereas North Americans seem to prefer stronger colours. Other types of miniature pot rose production that are also popular are products such as patio rose trees and pot sizes suitable for patio display (*see* **Production and Marketing:** Pot Rose Production).

Conclusion

Worldwide consumption of rose plants and flowers remains strong and many segments of the market are still increasing in popularity. This demand continues to be fuelled by the introduction of new cultivars.

Breeding goals differ somewhat depending on the final use. For example, breeders of garden roses are emphasizing disease resistance and low-maintenance landscape use as well as colour, flower form and fragrance. Breeders of cut roses are emphasizing colour, flower size and fragrance for specific markets. Breeders of miniature potted roses are emphasizing postharvest longevity, colour, flower size and garden performance. Many of these efforts include gene manipulation techniques and all future breeding will be impacted by biotechnological advances. One fact remains – the rose is one of the most popular flowers in the world.

See also: **Breeding**: Selection Strategies for Pot Roses. **Intellectual Property**: Plant Patents and Trademarks; Technical Examination and Conventional Characteristics. **Production and Marketing**: Field Rose Production; Pot Rose Production; Cut Rose Production. **Propagation**: Own-Rooted Cuttings. **Rootstocks**: Usage of Rootstocks.

Further Reading

Pemberton HB, Kelly JW and Ferare J (1997) *Production of Pot Roses*. Timber Press Growers Handbook Series, vol. 7. Portland, OR: Timber Press.

Pertwee J (2000) *Production and Marketing of Roses*. The Netherlands: Elsevier International Business Information.

Zieslin N (1995) Influence of climatic and socio economical factors on mode of cultivation and research of rose plants. *Acta Horticulturae* 424: 21–22.

Mineral Nutrition

R I Cabrera, Texas A&M University, Dallas, TX, USA

Introduction

Compared to other ornamental plants or crops, roses grown for cut-flower production (*see* **Production and Marketing:** Cut Rose Production) are known for their high water and mineral nutrient (or fertilizer) requirements. High-maintenance requirements can also be found in garden roses possessing the perpetual or recurrent-blooming trait (*see* **Growth Regulation:** Floral Induction).

Fertilization practices in cut-rose production have evolved significantly in the last century, going from a mostly organic and rudimentary methodology based on mineral soils to highly technical, computerized nutrient delivery systems based on soilless media or hydroponics. The mineral nutrition and fertilization of cut roses have historically received the most research attention and documentation due to their economic interest, and thus they will be the main focus in this chapter, although most of the information covered here is also applicable to garden roses.

Early Fertilization Practices

While it is acknowledged that cut-rose production is ancient in its origin (*see* **History of Roses in Cultivation:** Ancient Chinese Roses; European (Pre-1800); Modern (Post-1800), it was not until the late 1800s and early 1900s, with the discovery of the role and function of essential mineral elements, that soil fertility practices took a more rational meaning and management.

Animal manures and meat industry byproducts, along with residues from agricultural crops, were the main constituents of early fertilization practices. Their incorporation and use in rose production were rustic, based largely on frustrating trial-and-error applications. Use of fairly raw or fresh animal manures was, for example, the cause of many crop failures due to ammonia toxicity and the addition of high levels of soluble salts. In other instances, when using organic materials that had been adequately composted (or aged) and applied at rates considered adequate, temporary nutrient deficiencies were observed. This result was because the release and availability of nutrients from organic materials are highly dependent on the frequency of application, and, more importantly, on soil-related factors like temperature, moisture and microorganism population. The source (origin) and composition of these organic fertilizers were also fairly variable and usually ignored, further contributing to relatively unpredictable nutrient releases and crop responses.

The nutrient composition of fresh animal manures (i.e. cow, horse, pig, sheep) is relatively small compared to other materials (**Table 1**) and perhaps their major beneficial effect was to supply organic matter to the soil. These products continue to be used today in soil-based (ground beds) greenhouse rose production operations found in South American and African countries. Furthermore, garden or landscape roses also benefit from the use of these materials. It should be noted that the most important consideration for the use of manures and nutrient-rich organic materials is that they need to have undergone a proper composting process, including a monitoring of quality parameters such as particle size distribution and a chemical characterization that includes total soluble salts content

Table 1 Organic fertilizers used in the production of greenhouse roses

Material	*Nutrient analysis (%)*			*Application rate (kg ha^{-1})*	*Comments*
	N	*P*	*K*		
Fresh ('wet') manures (cow, horse, pig, sheep)	0.5–0.8	0.1–0.3	0.4–0.7	Up to 40 000 per year	Contain high levels of N as NH_3, causing toxicity when released. May also lead to salt build-up. Sterilization prior to use is recommended to minimize weed and pathogen problems
Dried cow manure	1.8–2.0	0.7–0.8	1.7–1.9	—	More expensive than inorganic fertilizers
Dried chicken manure	2.5–2.7	1.3–1.4	2.0–2.1	—	More expensive than inorganic fertilizers
Dried blood	9–14	0	0	1000–1500	N is readily available
Animal tankage	5–10	0.9–2.2	2–3	1500–2000	Material made of slaughterhouse residues that have been pressure-cooked
Raw bone meal	3–4	8–10	0	2000–2500	Slowly available N and P
Steamed bone meal	1.5–2.5	3–5	0	1500–2000	Steaming enhances nutrient availability

Note: The listed application rates for these organic materials, with the exception of manures, were often repeated four to eight times per year. For example, recommendations for animal tankage applications were 1500 kg ha^{-1} every 6 weeks during the growing season.
Adapted from data from White JW (1987).

(salinity). Manures often contain significant amounts of weed seeds, which need appropriate control measures.

Modern Fertilization Practices

Solid Inorganic Fertilization

The development of commercial inorganic fertilizers during the early 1900s drastically changed fertilization practices for greenhouse roses. The advent of granulated and chemically uniform inorganic fertilizer salts with high nutrient content, like urea, ammonium nitrate and ammonium sulphate, reduced labour during application (there was no need to incorporate by tilling) as compared to organic fertilizers. Perhaps more importantly, these new fertilizers allowed for a more direct control and precision in the delivery and availability of nutrients that would maintain soil fertility levels that supported higher flower yields. Given their high and fast nutrient availability, and thus an enhanced propensity to leaching (losses below the root zone), the frequency of use of inorganic fertilizers became more intensive, ranging from monthly to biweekly to weekly applications. These early inorganic fertilizers often contained only one or two nutrient elements of interest, and thus various fertilizer materials were required to provide the crops with a balanced nutrition.

Liquid Feeding

During the mid-1900s a new line of synthetic inorganic fertilizers would further change rose fertility practices. These were complex (multinutrient) proprietary fertilizers with high solubility rates, which, along with the use of proportioners or fertilizer injectors, led to the development of continuous applications of nutrients in the irrigation water, a practice known as fertigation. This fertigation, or the fertilization of crops with small quantities of water-soluble fertilizers at each watering, provided several advantages:

- It reduced labour and time by combining watering and fertilization in one operation
- It reduced the potential for undesirable concentrations of soluble salts in the root zone
- There was better control of crop growth and higher quality due to a steady supply of nutrients.

Complete water-soluble fertilizers were originally developed to include the major nutrients nitrogen (N), phosphorus (P) and potassium (K), dictated mostly by their high plant demand relative to other elements. Examples of these fertilizers include formulas such as 20-20-20, 25-10-10 and 25-0-25. The numbers in these fertilizer formulas represent, in order, the weight percentage of N, P (as P_2O_5) and K (as K_2O) contained in the fertilizer product.

The first water-soluble fertilizers often had some levels or traces (impurities) of other essential elements required by the rose plants, but these were often not enough to meet the demands of the crop. This meant that other elements besides NPK had to be supplemented separately, often at lower application frequencies, and applied directly to the soil or growing medium in formulations that were soluble, quickly available or slow-release. For example, calcium (Ca) and magnesium (Mg) are often incorporated to the soil once a year as dolomitic limestone, a product that is also used to adjust soil pH. Individual trace (or minor) elements and mixtures were used once or twice a year, also incorporated into the soil.

Recent water-soluble fertilizer formulations are more complex, often including all the essential mineral nutrients (macro- and micronutrients) required by roses. Whilst their advantages are obvious, these fertilizers are expensive and often lack flexibility in their nutrient ratios or proportions due to solubility and chemical reactivity constraints. Thanks to the development of new and sophisticated fertilizer injection systems, nowadays controlled by computers, more growers are preparing their own fertigation formulations using base fertilizers salts such as KNO_3, NH_4NO_3, $(NH_4)_2SO_4$, $Ca(NO_3)_2$, K_2HPO_4, $MgSO_4$, K_2SO_4 and others. Depending on the specific solubility and reactivity of each salt, a concentrated solution of these salts, either alone or in combination, is placed in individual containers or tanks. These concentrated solutions are then injected in the main irrigation lines by means of proportioners that are either mechanically or electronically adjusted by the grower. The rate of release or injection from these proportioners is calibrated or adjusted so that the ratio (or proportion) of concentrated nutrient solution to irrigation water volume produces the nutrient concentrations desired for application to the crop. **Table 2** presents the composition and range of nutrient concentrations in solutions used to irrigate greenhouse roses.

Essential Nutrient Elements

Besides the elements carbon, hydrogen and oxygen, which are provided by air and water, most cultivated plants, including roses, need 14 other essential mineral elements (**Table 3**). The term 'essential' basically denotes that the plant cannot complete its life cycle in the absence of any of these elements. These elements have been divided into two categories, macro- and micronutrients, mostly to denote their relative requirement or accumulation in plant tissues. Nevertheless, this classification does not detract any of these elements of their essential role and function in the life of the plant, and all are equally considered when planning and managing the fertilization of a rose plant or crop. By convention, macronutrient concentrations in plant tissue are commonly expressed in terms of percentage of dry weight, whereas micronutrients are expressed in milligram per kilogram of dry weight.

Table 2 Chemical composition and concentration of a nutrient solution used to irrigate a greenhouse rose crop. The listed range of nutrient concentrations is generic in nature, representing those solutions used to irrigate roses growing in mineral soil beds and organic-based (peat) growing media. Formulations for roses growing in hydroponic systems are in general less concentrated

Nutrient	*Concentration (mg l^{-1})*
Nitrogen (N)	100–250
Phosphorus (P)	15–30
Potassium (K)	150–250
Calcium (Ca)	250–350
Magnesium (Mg)	70–120
Iron (Fe)	1.0–2.0
Boron (B)	0.2–0.5
Manganese (Mn)	0.3–0.6
Zinc (Zn)	0.02–0.05
Copper (Cu)	0.01–0.02

Tissue and Soil Analyses

Even though some of the essential nutrients can be absorbed by leaves, when naturally or artificially deposited on their surfaces, they are mostly absorbed from the soil or substrate by the roots. The absorbed nutrients are subsequently distributed and accumulated throughout the plant, supporting all growth and maintenance processes. Given that leaf tissues play the most important role in the physiology of the plant, namely the process of photosynthesis (i.e. carbohydrate production), it is also the tissue that often and most readily responds to external and internal changes in nutrient supply. It is thus a common and widely accepted practice to report the nutritional status of a plant or a crop using the nutrient concentrations found in the leaves. As leaf nutrient concentration is affected by leaf age, studies have concluded that recently matured leaves in a ready-to-harvest rose flower shoot are the most representative to sample for tissue analyses. For practical purposes, these leaves are defined as the first two five-leaflet leaves counting down from the flower bud when its petals are beginning to show colour.

The 'ideal or normal' range of leaf nutrient concentrations for roses is presented in **Table 3**. It should be noted that, while these guidelines are based on a review of the literature on the mineral nutrition and fertilization of roses, they are still generic in nature and should not constitute the principal basis to dictate major decisions on the fertility management of roses. Such a simplistic approach is not recommended as it ignores the soil, environmental and plant factors that could be affecting nutrient uptake. The determination of this 'normal' range of tissue nutrient concentrations has mostly been done empirically, based primarily on subjective visual ratings of what constitutes a desirable

Table 3 Essential mineral nutrients and the leaf tissue concentration ranges accepted as normal or adequate for roses

Element	*Normal concentration in leaves*[a]	*Comments*
Macronutrients	**(%)**	
Nitrogen	3.0–4.0	Main element of interest in rose fertilization and management. Deficiency symptoms expressed as reduced plant growth and general chlorosis in older foliage
Phosphorus	0.2–0.4	Deficiency often expressed in reduced leaf size and dull grey-green leaf colour; buds slow to develop
Potassium	1.5–2.5	Along with N, this element takes a central place in fertilization management. Deficiency commonly expressed with scorching and necrosis of older leaves, often confounded with chlorine toxicity or water stress symptoms
Calcium	1.0–2.0	Deficiency causes dieback and browning of growing points (both shoot and root tips), similar to B-deficiency symptoms
Magnesium	0.2–0.4	Deficiency symptoms shown as interveinal chlorosis and dark or necrotic blotches in older foliage
Sulphur	0.15–0.25	Rarely involved in nutrient disorders in roses. Its ionic form (sulphate) is commonly found in many fertilizer salts
Micronutrients	**($mg\ kg^{-1}$)**	
Chlorine	Up to 4000	Typically avoided in production due to salinity concerns, but recently shown to benefit rose growth in moderate levels. Toxicity shown as necrosis of older foliage, beginning at leaf borders
Boron	30–80	Deficiency causes death of shoot and root tips. Toxicity symptoms include intraveinal necrotic spots with burning of leaf borders, affecting oldest leaves first
Iron	50–150	Perhaps the most critical of the micronutrients due to its widespread deficiency problems. Deficiency shown as interveinal chlorosis in young leaves. Veins become chlorotic under advanced deficiency
Manganese	50–200	Similar to iron, deficiency symptoms shown as interveinal chlorosis in new foliage. Toxicity causes small black spots in older leaves
Zinc	20–50	Deficiency severely stunts growth of new leaves (dwarfing or rosetting) and causes death of growing points. Chlorosis in tips of new leaves is common
Copper	3–15	Deficiency conditions very similar to Zn
Molybdenum	0.5–2.0	Rarely a problem; deficiency causes poor growth and drying of tips and edges of leaves
Nickel	—	Not yet determined in roses

[a]Concentrations on a dry-tissue basis.
Information compiled from data from Cabrera RI (1992) *MS Thesis*. University of California-Davis; Cabrera RI (1994) PhD dissertation, University of California-Davis; Carlson WH and Bergman EL (1966) Tissue analysis of greenhouse roses (*Rosa hybrida*) and correlation with flower yield. *Proceedings of the American Society for Horticultural Science* 88: 671–677; Johansson J (1978) Effects of nutrient levels on growth, flowering and leaf nutrient content of greenhouse roses. *Acta Agriculturae Scandinavica* 28: 363–386; Sadasivaiah SP and Holley WD (1973) Ion balance in nutrition of greenhouse roses. *Roses Inc. Bulletin* (November): 1–27.

foliage greenness or colour and with relatively less consideration to yield parameters. There are few studies that have attempted to correlate nutrient concentration values with actual rose flower and dry-matter yield responses. A study on the N nutrition of roses recently revealed that, while leaf N concentrations between 3.0 and 3.5% produced the greenest foliage, lower N concentrations (down to 2.5%) did not affect flower production or dry-matter yields.

Soil and growing media analyses are also conducted in roses with the goals of helping maintain adequate and balanced nutrient supplies in the root zone and aiding in the identification and confirmation of potential or incipient nutritional disorders (i.e. deficiencies or toxicities). As in most other agronomic and horticultural crops, complete soil tests in roses are highly recommended in preplant, helping to determine the supply, availability and balance of essential mineral nutrients (**Table 4**). Of particular interest are those elements whose chemical reactivity in soils is relatively slow or complex, and whose correction or adjustment is greatly limited once the crop is established. Examples of elements that are critically important in preplant include P, Ca, Mg and Fe. In addition to preplant, systematic and frequent soil and growing media testing are also recommended and practised in greenhouse roses due to their high productivity and dynamic growth and development phases, conditions that place a great demand for most nutrients, but particularly N and K.

A soil test should also include results on the soil chemical properties of pH and total soluble salts. Soil pH, which is basically a logarithmic measure of the concentration of hydrogen ions, is of paramount importance as it directly and significantly affects the chemical reactivity of most elements in the soil. The

Table 4 Interpretative nutrient concentrations and chemical properties used in the diagnosis of soil and growing media analyses for roses

Parameter	*Optimum concentration or value*	
	Soil[a]	*Growing media*[b]
Nitrogen (as NO_3-N)	25–100 mg kg^{-1}	100–200 mg l^{-1}
Nitrogen (as NH_4-N)	5–25 mg kg^{-1}	—
Phosphorus	25–100 mg kg^{-1}	5–10 mg l^{-1}
Potassium	100–300 mg kg^{-1}	100–200 mg l^{-1}
Calcium	1000–6000 mg kg^{-1}	150–250 mg l^{-1}
Magnesium	200–800 mg kg^{-1}	40–80 mg l^{-1}
pH[c]	6.0–6.8	5.5–6.5
EC[c]	1.0–3.0 dS m^{-1}	2.0–3.0 dS m^{-1}[d]

[a] Dry soil basis, except pH and electrical conductivity.
[b] On saturated media extract.
[c] Always measured in a saturated soil or media extract.
[d] Electrical conductivity in deciSiemens per metre, equivalent to mmhos cm^{-1}.
Information compiled from data from Agricultural Analytical Services Laboratory (Pennsylvania State University, College Park, PA); Cabrera RI and Evans RY (2001) Inorganic nitrogen loading and distribution in soil profiles beneath rose greenhouses. *Acta Horticulturae* 547: 227–233; Perry Laboratory (Watsonville, CA); Peters RB and Knauss JF (1984) Making use of soil and leaf analysis of roses. *American Rose Annual*, pp. 124–130. American Rose Society, Shreveport, Louisiana, USA; Soil and Plant Laboratory (Orange, CA); Warncke DD (1986) Analyzing greenhouse growth media by the saturation extraction method. *HortScience* 21: 223–225; White JW (1987) Fertilization. In: Langhans RW (ed.) *Roses: A Manual of Greenhouse Rose Production*, pp. 87–135. Haslett, MI: Roses Incorporated.

pH scale of 0–14 is comprised of an acidic range (values <7) and a basic range (values >7). The ideal soil and growing media pH range for roses is slightly acidic, ranging between 5.5 and 6.5. Values above neutrality (pH 7) commonly lead to deficiencies of most metallic micronutrients (Fe, Mn, Cu, Zn) and should be properly monitored and adjusted during production. Preplant adjustments of soil pH are, fortunately, relatively easy to accomplish, often involving the application and incorporation of liming chemicals (like dolomitic limestone) to increase soil pH, and sulphur or sulphate-based salts (like gypsum) to lower it. Treatment and/or adjustment of the irrigation water may also be required under high-pH conditions.

Like pH, the concentration of soluble salts is an important component of tests on a soil or a growing medium. Salt concentrations in the soil water (technically known as soil solution) and inside the root cells to a large extent determine the flow of water between the two. Water flows in the direction of the higher salt concentration, which generally occurs inside the roots. This water movement or plant water uptake is negatively affected when the concentration of salts in the soil solution build up beyond certain threshold levels, leading to plant wilting and death even under adequate soil moisture conditions. Soluble salts come from various sources, but the major contributors are the irrigation water and the applied fertilizer salts. While a certain level of soluble fertiliser salts is needed to supply the nutritional needs of a rose crop, that level must not be exceeded or it will start affecting plant water uptake. Levels of total soluble salts were originally reported in the concentration unit of parts per million (same as mg l^{-1}), but nowadays they are measured indirectly by reading the electrical conductivity (EC) of the soil solution, and reported in concentration units of deciSiemens per metre (dS m^{-1}) or its older equivalent millimhos per centimetre (mmhos cm^{-1}). Soluble salts monitoring in roses is extremely important, as these plants are considered a salt-sensitive species. It is recommended that EC values for soils or growing media used in rose production should not exceed 3 dS m^{-1} (**Table 4**). Fortunately, soluble salt levels can be easily diluted and leached from soils by making adjustments to irrigation water volume and the incoming nutrient (fertilizer) concentrations, along with changes in the frequency of their applications.

Compared to soil-based systems, monitoring and adjustment of nutrient concentrations, pH and soluble salts in modern soilless (inert growing media and hydroponics) rose production systems are extremely dynamic and require a fundamental and thorough knowledge of plant biology, mineral nutrition and soil chemistry. Nutrition management of soilless-grown roses is literally an hourly to daily endeavour that does not allow much room for mistakes. The consequences of such mistakes or disregards could be experienced within matter of hours. This is primarily due to the lack of a large and stable matrix, like a mineral or field soil that can buffer or withstand major chemical changes while still providing a relatively steady and balanced supply of nutrients. Conversely, this same large and chemically stable nature of mineral soils could also be considered their major hindrance to ground rose production, as corrections or changes to fertilization programmes and problems often require weeks to months to produce a desirable crop response.

Regardless of the type of production system (soil, organic growing media, hydroponics), a systematic and frequent monitoring of plant and root substrate nutritional conditions or status is recommended to maintain adequate and balanced nutrient supplies that sustain rose growth, productivity and quality.

Overfertilization and Nutrient Leaching

Compared to any other agronomic and horticultural crop, greenhouse roses have no equal when it comes to fertiliser applications. For instance, until recent times, a typical greenhouse rose crop could receive up to 7000 kg of N ha^{-1} $year^{-1}$, compared to the

100–200 kg applied to an agronomic (field) crop. The maintenance of a vigorous and sustained year-round productivity (growth) of roses led to the continuous application of water-soluble fertilizers at every irrigation event, adding up to the staggering figure just reported. Surveys of soil profiles beneath commercial greenhouse rose operations in Europe and North America have, not surprisingly, revealed that nutrient loading below the root zone (upper 30 cm) is extreme. These surveys reported total soil N loading on an area basis as high as 2200 kg ha^{-1} in the upper 1.5 m profile of soils under a rose greenhouse. A recent controlled fertilization study supports these results, pointing to maximum N leaching losses of 50–60% of the applied fertilizer, which corresponded to yearly N leaching losses of 2000–3000 kg ha^{-1}. These reports immediately suggested that greenhouse roses were receiving fertilizer applications in excess of the actual nutrient demand by the crop.

Such extreme nutrient leaching losses have been pointed to as sources of surface and groundwater pollution, and are the focus of concern and scrutiny by governments and the general public in several developing countries. For instance, legislative measures in northern European countries have forced the greenhouse industry to convert all of its growing operations to closed production systems that eliminate nutrient leaching and runoff. This has resulted in the development of nutrient-recirculating technologies and management practices that result in fertilizer use efficiencies approaching 100%. Rose growers in the Netherlands, Israel and some parts of the United States and Canada who have converted and retrofitted their operations to drainage water recirculation and hydroponic systems significantly reduce their nutrient leaching losses, all without affecting flower production and quality.

Estimating Nutrient Requirements

Traditional and conventional fertilization monitoring and management by soil and tissue analyses, while being effective in sustaining flower yield and quality, are primitive in nature by ignoring the actual nutrient requirements of a rose crop over a production cycle or season.

A relatively simplistic way of projecting or calculating the nutrient requirements of a rose crop over a given period of time involves the use of harvested dry-matter yields and average tissue nutrient concentrations. In other words, the simple multiplication of dry-matter yields per unit area per unit time by the average nutrient concentration found in that harvested dry matter gives a good estimate of how much nutrient is being harvested and will need to be absorbed by the rose crop to sustain a subsequent and similar production cycle. This assumes that the nonharvested rose plant components remain constant (no changes in dry matter or tissue nutrient concentrations). As fertiliser use efficiency in most crops is fairly poor, rarely exceeding 50%, the projected nutrient requirements by a rose crop growing in soil or in an organic growing medium will need to be adjusted accordingly when calculating how much fertilizer will be applied to the crop.

A recent research study measured the N requirements of a rose crop by collecting and analysing all the harvested flower shoot tissues over the course of 1 year. The measured annual crop N requirement ranged from 1200 to 1700 kg ha^{-1}, levels which were corroborated by a later hydroponic study that directly measured N uptake over the course of one full year using the same rose cultivar and rootstock. These figures on actual rose crop nutrient requirements support the above-mentioned concerns about overfertilization and undesirable nutrient losses to leaching.

Nutrient Uptake by Roses

While fairly simple and relatively effective, the estimation of nutrient requirements by a rose crop using projections of dry-matter productivity and average tissue nutrient concentrations ignores the dynamics of nutrient uptake as affected by environmental and plant factors. Temperature, light, relative humidity, and soil water and nutrient supply and availability are some of the environmental factors involved in crop nutrient uptake and utilization. Critical plant factors include plant age, internal nutrient concentration and the different stages or phases of crop growth and development.

Under conditions of steady and nonlimiting supply of nutrients (i.e. fertigation, hydroponics), prevalent in the controlled environment of modern rose greenhouses, the plants have a tight control and regulation of nutrient uptake. Under these nonlimiting conditions root cells selectively absorb nutrients from the soil solution, in quantities dictated by plant demand and not in relation to what is being supplied externally. **Figure 1** shows the absorption of N, P, K, Ca and Mg in mature greenhouse rose plants over the course of one cycle of growth and flowering. The plants were growing in a recirculating hydroponic system, where the roots were constantly bathed with a complete and balanced nutrient solution. Day 0 in the graphic represents the day when the flowers from the previous flowering flush were harvested. As expected, reductions in nutrient uptake were visible as soon as the flowers from the previous flush were harvested, and this reduced uptake continued even as the new flower shoots began to develop. The lowest nutrient uptake was observed when the shoots were rapidly elongating, in fact when they were achieving their fastest elongation rate. It was not until shoot elongation rates began to decrease that nutrient uptake increased and reached its maximum when the flower shoots were ready to be harvested. This pattern of nutrient uptake clashes with the intuitive assumption or notion that nutrient demand by the plants will be directly related to their growth rate, that is, high nutrient uptake during periods of fast or vigorous growth.

Often forgotten is the fact that plant nutrient uptake is an active physiological process that requires energy. This energy is provided in the form of photoassimilates (carbohydrates) produced by the leaves during photosynthesis, and, as such, roots have to compete for them with the rest of the plant. The pattern of nutrient uptake described here (**Figure 1**) is explained by competition within the rose plant for photoassimilates and nutrients. According to this proposition, nutrient uptake is regulated by the balance between the demand for photoassimilates and the demand for nutrients within the various plant organs. During periods when flower shoots are rapidly elongating, shoots and leaves become major sinks (importers) for assimilates. This greatly diminishes their supply to the roots, which in turn reduces root growth and nutrient uptake activities. A diminished nutrient uptake eventually cannot meet the nutrient demands of the developing shoots, slowing and eventually leading to a cessation of shoot growth. When shoot growth slows down or ceases, carbohydrates become available again for translocation to the roots, permitting an increase in the rate of nutrient absorption until the shoots are ready to be harvested. Various research studies have verified the timing of these oscillations in nutrient uptake with actual patterns of photoassimilate distribution along the various growth phases of rose flower shoots.

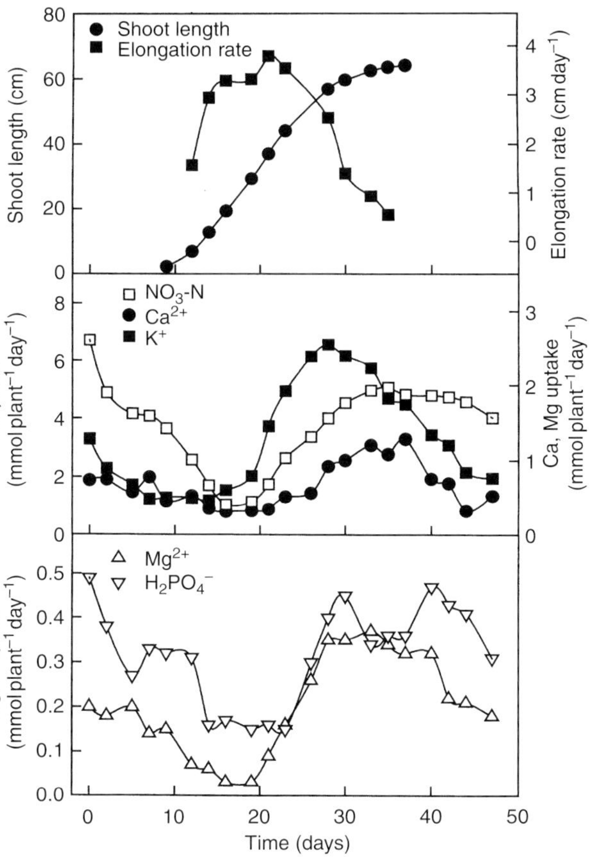

Figure 1 Patterns of nutrient uptake in greenhouse roses over the course of one flush of growth and flowering. Each data point is the average of 20 plants. Reproduced from Cabrera RI, Evans RY and Paul JL (1993) Cyclic nitrogen uptake by greenhouse roses. *Scientia Horticulturae* 63: 57–66, with permission from Elsevier Science.

This cyclic or rhythmic pattern of nutrient uptake has significant practical implications for the greenhouse rose industry. First, these observations bring into question the use of a continuous supply of high nutrient concentrations in the irrigation water, which, by disregarding the actual patterns of low and high plant nutrient uptake, is contributing to the high leaching losses reported earlier. The close and predictable relationship between nutrient uptake and flower shoot development reported here could be the basis for developing nutrition programmes tailored to respond to changes in rose plant nutrient demand over time, enhancing fertilizer use efficiency and reducing fertilizer losses to leaching. Since flowering flushes in greenhouse roses can be and are manipulated by pruning practices (*see* **Growth Regulation**: Axillary Bud Growth. **Production and Marketing:** Cut Rose Production), modulation of nutrient applications in relation to shoot elongation could potentially be practised in commercial production systems. Hydroponic production systems, such as one using rockwool, may be best suited to this practice. However, the suitability of supplying nutrients according to plant demand in roses growing in soil systems needs to be ascertained, as the dynamics of nutrient mobility, availability and reactivity in mineral soils are significantly different to those found in soilless substrates and hydroponic systems.

Prospects for Nutrition in Roses

A worldwide awareness of critical environmental and water-related issues is having a major impact on agricultural activities, and perhaps more significantly, on intensive cropping systems such as greenhouse crops. At present there are very few crops that come close to the high, and sometimes exorbitant, water and fertilizer inputs used in rose cut-flower production. Dwindling fresh-water resources, along with poor use efficiencies and their pollution with urban and agricultural chemicals are radically and rapidly changing the cultural practices of this crop.

Various countries and regions, particularly in Europe and North America, have already developed legislative measures that order minimum or zero discharge of nutrients from greenhouses to the environment and also limit the use of high-quality irrigation water. Rose growers and researchers in these regions are already adapting and generating technologies and crop management practices that comply with these environmental regulations, and at the same time have been able to sustain flower production and quality. Technologies such as recirculating nutrient solution systems, while still under development and refining, have been shown to be most viable alternatives to these challenges. As it turns out, the basic concepts and tenets of plant mineral nutrition are being revisited and applied to roses, with recent research advances detailing peculiarities on the nutrition of this crop. Current research efforts on this crop are being focused on salt tolerance and nutritional requirements as affected by the genetic makeup of rose cultivars, rootstocks, management and manipulation of the greenhouse environment, quality of irrigation water, and irrigation scheduling and management under effluent or drainage recirculation conditions.

With regard to roses used in gardens and landscapes, there is also a strong interest, as well as research efforts, aimed to identify and select cultivars that are resource-efficient (i.e. requiring minimum fertilization and irrigation). Rose plants that are able to thrive under minimum management conditions or under marginal garden or landscape conditions are actively being sought and tested.

See also: **Growth Regulation**: Axillary Bud Growth. **History of Roses in Cultivation**: Ancient Chinese Roses; European (Pre-1800); Modern (Post-1800). **Production and Marketing**: Overview of Roses and Culture; Cut Rose Production.

Further Reading

Boodley JW and White JW (1969) Fertilization. In: Mastalerz JW and Langhans RW (eds.) *Roses: A Manual on the Culture, Management, Diseases, Insects, Economics and Breeding of Greenhouse Roses* pp. 78–92. Haslett, MI: PA Flower Growers Association, NY State Flower Growers Association, Inc. and Roses Incorporated.

Cabrera RI (1994) *Characterization of Nitrogen Uptake by Greenhouse Roses*. PhD dissertation. Davis, CA: University of California at Davis.

Joiner JN, Poole RT and Conover CA (1983) Nutrition and fertilization of ornamental greenhouse crops. *Horticultural Reviews* 5: 317–403.

Nelson PV (1991) *Greenhouse Operation and Management*, 4th edn. Englewood Cliffs, NJ: Prentice Hall.

Oertli JJ (1968) Nutrient deficiencies in rose plants. *Roses Inc. Bulletin* (June): 20–26.

Peters RB and Knauss JF (1984) Making use of soil and leaf analysis of roses. *American Rose Annual* pp. 124–130. Shreveport, LA: American Rose Society.

Reed DW (1996) *Water, Media, and Nutrition for Greenhouse Crops*. Batavia, IL: Ball.

White JW (1987) Fertilization. In: Langhans RW (ed.) *Roses: A Manual of Greenhouse Rose Production*, pp. 87–135. Haslett, MI: Roses Incorporated.

Field Rose Production

J F Karlik, University of California Cooperative Extension, Bakersfield, CA, USA
J O Becker, University of California, Riverside, CA, USA
H B Pemberton, Texas A&M University, Overton, TX, USA
U K Schuch, University of Arizona, Tucson, AZ, USA

Introduction

Most roses destined for outdoor planting, including hybrid teas, grandifloras and floribundas, require 2 years for field production from rootstock planting to harvest of the finished plants. If the rootstock plants are grown from seed, an additional year is needed, and tree roses may also require 3 years in the field. In contrast, varieties grown on their own roots (including shrub or landscape roses) and roses grown for greenhouse planting ('started eyes') are planted and harvested within 1 year. Outdoor rose plant production requires a long growing season, well-drained soils, preferably of sandy texture, and sufficient water. Rose plants are well-suited to irrigated production, but may also be grown in areas with well-distributed annual rainfall. Cultural practices may include field preparation (which may include soil fumigation), growth of the rootstock either from seed or by planting of rootstock cuttings in autumn or winter, budding of scion varieties and pruning to force the scion to grow and develop additional canes. Sufficient water must be provided and plant nutrients added during the growing seasons. Pests such as weeds, insects and diseases must be managed. The crop is harvested during the autumn and winter seasons, graded and placed in cold storage to await shipment. Most US rose plant production is located in the south-western part of the country (California and Arizona), with additional production in the south-central region (Texas). Other major field production centres include England, Germany, Spain, Italy, France, some eastern European countries, Morocco, Israel, Australia and Japan.

The following scheme is used in the south-west USA, where approximately 50 million rose plants are produced annually, and is a plan close to what is used in areas where roses are grown using rootstock propagated as cuttings (such as southern Europe and north Africa). Many variations of the general production scheme are possible which take into account local conditions and the varieties being grown, such as omission of the budding step for varieties grown on their own roots (*see* **Propagation:** Own-Rooted Cuttings).

General Production Scheme for Rose Plants

Rose crops usually follow rotational crops, such as lucerne followed by grain, since continuous production of rose plants results in soil compaction and degradation of soil structure. After spring or early-summer harvest of the final rotational crop, the field is cultivated to incorporate crop residue and then laser-levelled. A fumigant may be applied in August or September prior to planting, and if so, the soil is preirrigated. Soil fumigation is primarily for nematode control but also has benefit for disease and weed suppression. After fumigation, the field is furrowed for preplant irrigation. Prior to planting, a soil residual herbicide, such as oryzalin or trifluralin, may be applied and incorporated to suppress germinating weeds.

Rootstock cuttings treated with auxin are placed in the field in November or early December, and begin to form callus after planting. In early spring (mid-February in the south-west USA), small roots extending 1 cm from the cutting can be found about the same time bud-break occurs. Rose rootstocks used for grafting scion varieties destined for outdoor planting include 'Dr Huey', *Rosa multiflora* Thunb. and *R. odorata* Andr. (Sweet), with 'Dr Huey' the most commonly used in the south-western USA and *R. multiflora* most commonly used in south-central USA. The rootstock 'Manetti' is often used for greenhouse varieties (*see* **Rootstocks:** Usage of Rootstocks; Clonal Rootstocks).

The scion variety is T-budded on to the rootstock during the April 1–June 15 period in the US south-west. After 21 days the rootstock may be cut back to force growth of the scion, or rootstock top removal can be done at a later time, depending on the exact production scheme used. Application of fungicides, insecticides and acaricides may be necessary during the growing season, although chemical applications may be minimized through an integrated pest management approach (*see* **Integrated Pest Management**). Fertilizer can be added via foliar sprays, injection into soil, mixed with irrigation water or as a granular side-dress application. Weeds are removed through tractor cultivation and hand-hoeing. Although rose plants may be grown where rainfall supplies sufficient water, planned irrigation helps assure steady growth, and is required in arid climates and in areas with prolonged dry periods. Delivery of water to rose plants may be accomplished through furrows, overhead sprinklers or buried or above-ground drip tubing.

If the rootstock top was not removed in the first year, it is removed during the next winter to force growth of the scion. In spring, when the scion shoots reach about 20 cm, shoots are pruned to increase the number of basal breaks and resulting canes. The growing period during the second year allows the plant to develop canes with increased diameter and a more extensive root system. Additional fertilizer, water and pesticides are applied, depending on field conditions. Plants are defoliated mechanically just prior to harvest, which usually begins in early November. Plants are gathered by field workers after being lifted by mechanical diggers, rinsed to remove soil and transported to indoor processing areas. Plants are placed on conveyor belts and individually graded. They may be treated with an antitranspirant on their aerial part to retard moisture loss before being placed in cold storage for later shipping.

Climate and Soil Requirements for Rose Plant Production

Rose production is favoured by an extended growing season with a majority of sunny days. In North America the primary production centres have more than 300 frost-free days, and less than 10% cloudy days in summer. Coarse-textured soils with low content of silt and clay are favoured because plants must be lifted from the field at harvest and soil removed from the roots. Rose plants can be produced in unirrigated fields, but the limited water-holding capacity of coarse-textured soils can lead to stress if rainfall is infrequent. Irrigation provides much more certainty of the crop outcome, and for irrigated rose production, soils of sandy loam texture are preferred for their rapid infiltration rates. Such soils in current production areas are often uniform to more than 2 m in depth.

Soil Fumigation

Sanitation and good field preparation are key elements for any successful crop, and even more important for plant production systems with a long growing season. Prevention of nematode-infested rose rootstocks is a desirable quality goal and can be mandated by regulation. For example, the California Department of Food and Agriculture specifies and monitors, via its local offices, the use of cultural, physical and chemical soil treatments for nematode control before utilizing a property for nursery purposes. Certain pests and soilborne pathogens, such as endo- or semiendoparasitic nematodes (*Meloidogyne hapla* Chitwood, *Pratylenchus vulnus* Allen & Jensen) and the wilt fungus (*Verticillium dahliae* Kleb.) can remain dormant and infective for extended time periods in old plant roots. In recent decades, management of these problems has relied almost completely on preplant soil fumigation that delivered biocidal compounds to the target organism. For the past 40 years, methyl bromide (MeBr) or MeBr–chloropicrin mixtures have been the principal compounds chosen for soil fumigation because of their physicochemical properties, which resulted in extraordinary biocidal activity, vertical distribution in soil and diffusion into plant tissues. In rose production, the benefits of such treatment have been seen not only in the wide-spectrum control of organisms that are of known economic impact, such as plant-parasitic nematodes, pathogens and weeds, but also in counteracting a soil condition that is responsible for 'specific replant disease' or a 'replant problem' (*see* **Disease:** Replant Problems and Soil Sickness). Such situations become apparent as difficulties in establishing certain plant species into nursery sites where the same or closely related plant species were previously grown. The stunting and limited root growth are generally attributed to microbial activity, but the complexity of the rhizosphere interactions has often obscured the identification of the causal agents.

The identification of MeBr as a potentially significant contributor to stratospheric ozone depletion has resulted in its phase-out for the industrialized and developing nations by 2005 and 2015, respectively. The intensive search for a suitable replacement has so far not resulted in many innovations, although iodomethane appears promising. Some of the other currently registered soil fumigants, such as 1,3-dichloropropene, metam sodium and chloropicrin, have good activity individually against plant-parasitic nematodes or soilborne fungi. But, even in combination, they do not have the spectrum of activity growers have been accustomed to with MeBr. Moreover, replacing MeBr by pesticide 'cocktails' to cover a similar spectrum of target organisms is likely to increase the pesticide load per hectare. Further regulatory restrictions for all soil fumigation treatments are to be expected and signal the end of an era when past agronomic practices (or mistakes) could be mitigated by preseason soil fumigation.

Soilborne Disease and Plant-Parasitic Nematode Management

In rose production in the US south-west, biotic soilborne problems are limited to a few disease-causing organisms, although each has potentially serious economic consequences.

The only important bacterial disease on roses is bacterial crown gall, caused by *Agrobacterium tumefaciens* (Smith & Townsend) Conn. (*see* **Disease:** Crown Gall). The soilborne bacteria enter into the rose tissues through natural or inflicted fresh wounds and transform the host cells into tumour cells which produce convoluted woody growth. Galls occur most frequently near the soil surface at the crown, less frequently on roots, the bud union or on other parts. The bacteria can survive in soil for many years and have a wide host range, although in the field 'Dr Huey' is more resistant than 'Manetti'. Soil fumigation with broad-spectrum biocides reduces the population of *A. tumefaciens*, but only heat is effective in eliminating the bacteria. Sanitation is the most important disease-control strategy. Diseased plants should be removed. Before planting, suspending rose cuttings in a 0.5% sodium hypochlorite solution to wet the wood thoroughly helps to disinfest the surface areas. Exposure time of the cuttings to the sodium hypochlorite solution is 1 min or less, followed by washing. Cutting tools are also frequently dipped into a disinfectant solution. Good results have been achieved with the commercially available biocontrol agent *A. radiobacter* K84 that inhibits the pathogen if K84 is sprayed on to injured tissues soon after wounding and is present in a sufficient population density. A preplant dip of rootstock cuttings into a suspension of K84 has been effective in preventing crown gall in rose fields.

The fungus *Verticillium dahliae* affects a wide range of plant species and is one of the most destructive soilborne plant diseases. The fungus penetrates the host roots, invades the xylem and its conidia disseminate upward in the plant. External symptoms such as wilting and yellowing of foliage are followed by dieback, starting from the base to the tip of a shoot.

Streaking of purple-black colour may be seen on the outside of the stem, although vascular discolouration does not occur. (Purple colour on stems in blotches, sections or along one side may result from other causes.) The presence of all three symptoms – wilting, yellowing of foliage and streaking – strongly indicates the presence of *Verticillium*. The fungus survives as mycelia in plant debris or as tiny, black microsclerotia that stay infective in soil up to 15 years. *Verticillium dahliae* is extremely difficult to control due to its wide host range and survival abilities. Soil steaming and fumigation with chloropicrin or in combination with MeBr are the only effective methods to suppress the fungus. The rootstock 'Manetti' is considered resistant to most strains of *V. dahliae*.

Although numerous plant-parasitic nematodes are able to feed and reproduce on roses, the most important species in the United States are the northern root-knot nematode (*M. hapla*) and lesion nematode (*P. vulnus*) (*see* **Insects and other Animals:** Nematodes). Above-ground symptoms of nematode damage are often nonspecific, such as wilting, chlorosis, and reduced vigor and growth. Small galls on roots and excessive root branching are symptomatic of *M. hapla* while much-reduced feeder root systems with root lesions may indicate the presence of *P. vulnus*.

Management of plant-parasitic nematodes should be based on avoidance of infestation, population reduction and plant tolerance/resistance. Exclusion measures are the most economic antinematode strategies and include sanitation, nematode-free planting material, nematode-free irrigation water, avoidance of potentially contaminated run-off water and plant-parasitic nematode-free soil or other growth media. In field production, the management of the land between production cycles with rotation crops that do not support the reproduction of *M. hapla* and *P. vulnus* will be of special importance if effective chemical treatments, i.e. soil fumigants, are no longer available. The beneficial effect of rotation crops in the reduction of plant-parasitic nematode populations is based on the ability of non-host plants to stimulate nematode hatch. Newly hatched nematodes die without reproducing, resulting in a reduction in the overall population. Furthermore, an active and diverse rhizosphere microfauna and flora contains numerous nematophagous organisms that reduce plant-parasitic nematode populations. Unfortunately, nonhost rotation is a difficult task, as both *M. hapla* and *P. vulnus* have a wide range of hosts, including many weeds. However, cereals have not been reported as hosts of either species.

Considerable variation has been reported in the host rootstock susceptibility to the two nematode species. 'Manetti', 'Dr Huey' and *R. odorata* and, to a lesser degree, *R. multiflora* rose rootstocks were good hosts for *P. vulnus*. *Meloidogyne hapla* reproduced poorly on 'Manetti' and no root-galling was observed. In contrast, the root-knot nematodes reproduced well on 'Dr Huey', *R. odorata* and *R. multiflora*. Roots of both 'Dr Huey' and *R. odorata* had more and larger galls than those of *R. multiflora*. The development of nematode-resistant rootstocks may be accelerated in the future by new discoveries about the mechanisms of plant resistance and nematode virulence.

Nutritional Requirements for Rose Plants and Fertilizer Addition

Roses prefer well-drained soil with a pH near the neutral value of 7.0 (*see* **Production and Marketing:** Mineral Nutrition). Many soils are still suitable despite a pH above 7.0, but the likelihood of micronutrient deficiencies becomes greater as pH increases, especially for pH values above 7.5. Roses are also tolerant of a pH below 7.0 as long as it is not below 5.5, at which point micronutrient imbalances can also occur. Roses are not salt-tolerant, so electrical conductivity values (EC), which indicate the concentration of salt ions in the soil, should be less than $2.0\,dS\,m^{-1}$ as measured by a saturated paste extract.

Nitrogen is the nutrient typically in shortest supply, and deficiency causes chlorosis and drop of older leaves. Most sandy loam soils have low concentrations of available and reserve nitrogen, and although nitrate levels may be lower than the suggested value as measured by soil tests, this nutrient is easy to add so that a low supply of nitrogen does not usually pose a difficulty. Nutrient concentrations in soil suitable for roses are given in **Table 1**. In field production, nitrogen is usually added in four to six increments during the growing season. Nonpressurized solutions of fertilizer nutrients rather than granular materials are typically applied to fields. Some solutions often used are UN32, a solution of urea and ammonium nitrate containing 32% nitrogen, and an 8-8-8 solution containing nitrogen, phosphorus and potassium. These are applied to fields by injection at a 10–15 cm depth into soil through shanks placing the solution 15 cm from plants, or may be added to irrigation water. The amount of nitrogen added annually is typically $225–335\,kg\,ha^{-1}$.

Phosphorus and potassium are less mobile in soil than is nitrogen. The amounts needed for rose crops are thought to be $90–135\,kg\,ha^{-1}$ of each, although some growers prefer much greater additions of potassium. Phosphorus and potassium are added before planting, or during the growing season (e.g. January and April in California).

Table 1 Suggested ranges for soil pH, electrical conductivity values (EC) and nutrient levels for rose plants. Units are dS m^{-1} for EC and p.p.m. for nutrients

Soil characteristic	*Range*
pH	6.0–7.5
EC	0.5–2.0
NO_3-N	35–180
NH_4-N	0–20
Phosphorus	5–50
Potassium	50–300
Calcium	40–200
Magnesium	20–100
Boron	0.05–0.5
Iron	0.3–3.0
Manganese	0.2–3.0
Zinc	0.03–3.0
Copper	0.001–0.5
Molybdenum	0.01–0.10

Table 2 Suggested ranges for nutrient levels in rose tissue

Nutrient	*Range*	*Unit*
Nitrogen	3.0–5.0	%
Phosphorus	0.2–0.3	%
Potassium	2.0–3.0	%
Calcium	1.0–1.5	%
Magnesium	0.25–0.35	%
Boron	30–60	p.p.m.
Iron	50–150	p.p.m.
Manganese	30–250	p.p.m.
Zinc	15–50	p.p.m.
Copper	5–15	p.p.m.

Micronutrient deficiencies, especially iron and zinc, appear as interveinal chlorosis of new leaves. These elements may be deficient because soils are too wet, are too alkaline, or correspond to types such as sandy loam which have a low content of micronutrients. Tissue tests may be used to provide information on the current nutritional status of rose plants, and suggested ranges for nutrient concentrations in leaves are presented in **Table 2**. Because inorganic forms of iron and zinc form insoluble precipitates in alkaline soils, iron and zinc may be applied directly to foliage. Iron and zinc in a chelated form may be applied to either soil or foliage. Where it is deficient in soil due to supply or pH, zinc is added two or three times per season as a foliar spray of 4.5 l ha^{-1} of 6.5–9% zinc chelate.

Nutrient excesses may limit rose growth if the total salt level becomes too high, with resulting EC above 2 dS m^{-1}, which causes lack of vigour and short shoots. Salt levels above this value may not necessarily result in chlorotic or necrotic leaf symptoms, even though rose growth is affected. A few nutrients cause specific toxicities. Boron is found in excess in some soils, causing stunting of plants, chlorosis and marginal necrosis of newest leaves. Boron should be below 1 p.p.m. in soil. Manganese toxicity can be a problem in extremely acid soils, causing black speckling of the leaves. Liming of soils, especially when the pH is below 5.0, is the best prevention for this problem. Dolomitic lime at a general rate of 2240 kg ha^{-1} can be incorporated before planting to raise soil pH; however, it is best to test the soil prior to lime application to determine the appropriate amount.

Plant Spacing and Field Topography

In virtually all production fields where winter precipitation is limiting and rootstock cuttings are rooted in the field, furrow irrigation is used initially because of the need to seal soil around the cuttings and keep the soil moist until rooting has occurred. Unlike most crops, roses grown under these conditions are planted in the bottom of the furrows to ensure water availability to the cutting as it roots. This early cultural requirement often leads to continuation of furrow irrigation with fields arranged to provide delivery of irrigation water at the higher end, and a collection ditch at the lower end of the field. Placing the rose plants in the furrow implies that tractor drivers pay heed to the location of tyres, since it is natural for the tyre to find the lowest terrain and slide into the furrow. In field production with furrow irrigation, roses are usually grown with a 15 cm in-row spacing and 107 cm between rows, resulting in 62 000 plants ha^{-1}. Some growers prefer a between-row spacing of 122 cm, resulting in fewer but larger plants ha^{-1}.

In areas of high winter rainfall, rootstock cuttings are planted in raised beds. Deep planting in the bed helps maintain moisture around the cutting as rooting occurs during the winter. The beds are then levelled in the spring to expose the shank of the cutting in preparation for budding. An in-row spacing of 15 cm is normally used with a 120 cm between-row spacing. This results in an essentially flat-ground system that can be used in areas where irrigation is not normally required. However, buried or above-ground drip irrigation can be invaluable for periods of drought due to poor distribution of yearly rainfall.

In arid climates it is possible to establish roses on raised beds with irrigation provided via buried drip tubing. Roses grown on beds grow more quickly in the early part of the season, probably due to increased soil temperature, and produce more biomass than roses grown in the conventional furrow system. Weed management is easier since roses occupy the higher ground. However, rose cuttings on raised beds need overhead irrigation during the rooting period to seal the soil and keep the cuttings sufficiently moist. To prevent root disease, it is important not to apply water via both drip and overhead delivery systems during the rooting period for cuttings.

In arid climates it is also possible to grow roses on flat ground irrigated with buried drip tubing, which allows more flexibility for in-row and between-row spacings, including increasing in-row distance and reducing between-row distance, so as to make both distances more nearly equal, reducing interplant competition. However, as with raised beds, keeping cuttings sufficiently moist during the rooting period is critical, and supplemental overhead irrigation is necessary during establishment to ensure sealing of soil and preservation of hydration of wood.

Preparation and Planting of the Rootstock Cuttings and Seedlings

Wood for rootstock cuttings is harvested in November–December from virus-free mother blocks or first-year fields. This wood is usually prepared for immediate planting, but may be placed in cold storage. Harvested wood is current season growth of about pencil diameter, or 6–7 mm. The wood is dipped in a mild bleach solution, and cut into 23-cm lengths without regard to location of nodes, since rooting occurs without regard to node position. Thorns and all but the uppermost two buds are removed in a process called de-eyeing. The cuttings are bundled into groups of 25 or 50, and the basal ends dipped into a synthetic auxin powder which promotes rooting and also makes it easy for field workers to identify the proper end for planting. The bundles are placed in groups of four into plastic bags and either sent to the field for immediate planting or held in cold storage until planting. If the cuttings don't desiccate, the resulting stand should contain approximately 98% live plants.

In a furrow-irrigated system, furrows are made in the field and depressions are marked at 15-cm intervals by measuring wheels pulled by a tractor. Cuttings are planted by hand at a depth of 7–8 cm in the bottom of the furrows. Irrigation water is then introduced to seal the soil around the cuttings and to keep the base of the cuttings wet. Furrow irrigation provides excellent conditions for cutting establishment.

In a raised-bed system, beds are first raised and shaped. The beds are then narrowly opened down the centre with a blade and the planting intervals are marked by measuring wheels, as in the furrow system. Cuttings are planted 18 cm deep because the shank of the cutting will be exposed upon levelling the field the following spring. The opening in the bed is then packed around the cuttings. The deep planting maintains moisture and provides protection from temperature extremes. Moisture is usually maintained by winter rainfall in this system, but irrigation can insure against drying out when precipitation is inadequate.

In many locations, rootstocks are grown from seed instead of cuttings (*see* **Propagation**: Seed Propagation). For rootstock seedling production, seeds are sown directly into nursery seedling beds. After one season of growth, the seedlings are dug and graded. Uniform seedlings (usually 5–7 mm diameter for *R. multiflora*) are selected and placed in the field using a transplanter during late winter or early spring for budding later in the same growing season. They are then grown using the same practices as for a budded cutting.

Budding of Scion Varieties

Most roses in the USA are budded in the field through a process called T-budding or shield budding, although chip budding and bench grafting may be used for some varieties (*see* **Propagation:** Budding in the Field). Budwood is harvested the previous season, or purchased from virus-free foundation sources, and stored under temperature and relative humidity conditions of $-1.5\,^{\circ}\mathrm{C}$ and $\geq 90\%$, respectively, until needed in the field. For T-budding in the United States, teams of two workers travel along each row, with each worker horizontally suspended in a push-cart. The first worker cuts the bud from the budstick about 1 cm below the bud and finishing about 1.5 cm above the bud. This tissue contains the bud with attached phloem, cambium and xylem tissue 1–2 mm in thickness. The first worker then makes a T-shaped cut in the rootstock, and inserts the bud. The second worker follows, tying the bud in place with a 10-cm length of rubber wrapped below and above the bud. This tie provides mechanical support and reduces desiccation of the bud and surrounding wood. Normal exposure to sun will cause the rubber to break in a few weeks, so the stem is not girdled. Other materials, such as plastic film, have been tested as substitutes for rubber, with less satisfactory results. A typical field team requires less than 15 s to bud a rose plant. Per day, a good team can bud 3500 or more plants, and daily counts above 4500 have been reported. After 30 days, the number of live buds is counted in sample sections of the field with $\geq 90\%$ expected for a proficient team.

Forcing the Scion

The scion may be forced in either the first or second year depending on grower preference. In recent years, practice in California has shifted to favour forcing of the scion the first year, 21 days after budding, through partial severing of the rootstock above the bud, followed by removal either in late summer or during the next winter. (If the rootstock is completely removed 21 days after budding, the resulting plant has too many canes, and canes are curved upward. Growers prefer three straight canes.) Forcing the scion the first year seems to prevent a problem of weak shoot attachment during spring of the second year, and increases the yield of number 1-grade plants by 3–5%. Of course, if a grower does not have a mother block, the rootstock is left intact during the first year so wood can be harvested for winter planting for the next crop.

Irrigation

Drought stress leads to defoliation and sunburn of canes and may contribute to spider mite problems (*see* **Integrated Pest Management. Insects and other Animals:** Mites). However, overwatering or poorly drained soils may lead to root disease and nutritional deficiencies. Water may be supplied via overhead sprays, furrow irrigation or drip tubing. Frequency and duration of irrigation will depend on weather conditions and soil texture, respectively. It is best to fill the entire root zone during an irrigation, and for a given depth of rooting the amount of water held will depend on the soil texture. Therefore, it is frequency of irrigation rather than duration which changes with season and weather conditions.

An estimate can be made of irrigation need based on pan evaporation or reference evapotranspiration. For example, a standard US class A pan is surrounded by irrigated cool-season turfgrass and the evaporation is measured daily. Crop water use can be estimated as a fraction of pan evaporation. However, the more common reference value for water loss is reference evapotranspiration (ETo), calculated from data of wind speed, solar radiation, temperature and humidity, and is the water used by a cool-season turfgrass at 100% cover under no moisture stress (Penman reference crop). Crop water use is then obtained from:

$$\mathrm{ETo} \times Kc = \mathrm{CWR}$$

where Kc is the crop coefficient and CWR is crop water use. Both ETo and Kc will change during the growing season, although average values for Kc are often used. The Penman ETo is a fraction of about 0.8 of pan evaporation. The crop coefficient for roses has not been reported, but is thought to be about 80–90% of ETo. For the southern San Joaquin Valley of California, with mean annual pan evaporation of 162 cm and ETo of 125 cm, about 110 cm of water are usually applied annually to a rose crop – indicating that additional water conservation may be possible.

Irrigations can be scheduled through a water budget method, in which crop water use is estimated based on daily ETo. However, feedback obtained through field monitoring of soil moisture content by looking, feeling soil or using moisture-measuring devices (e.g. tensiometres placed at different depths) is essential for proper scheduling. Comparison of soil moisture-measuring devices is beyond the scope of this article, but it is essential for a grower to calibrate and compare new devices to familiar ones. Roses grow best when about 50% of available water is depleted between irrigations, and even during the warmest summer weather of daily maxima >40 °C in the San Joaquin Valley, irrigation is at weekly intervals for first-year plants and every 10 days for second-year plants.

Drip irrigation via buried tubing or tape may confer benefits in field rose plant production, including limiting weed growth, conserving water and as a delivery method for fertilizers or pesticides. However, the subsurface drip tape can interfere with harvest and has to be removed and collected – often a messy field operation – before other crops can be planted.

Buried drip irrigation tape has been used commercially on a field-scale for water delivery during the irrigation season in both Arizona and California. Field experience coupled with limited experiments suggests that a depth of 20–30 cm was optimal. Establishment of rose cuttings is difficult with drip irrigation, unless rain occurs frequently or overhead irrigation is available to wet surface soil, keep cuttings moist and seal the soil surface. This need alone may prevent widespread adoption of drip irrigation in the south-western USA, unless water costs become very high.

In the south-central USA, drip irrigation is a preferred method for water delivery as topography does not permit the use of furrow irrigation. Tubing can be located above or below ground. Winter rainfall is usually sufficient for rooting of cuttings.

Pruning

Pruning provides an opportunity to increase the number of stems and therefore to improve the grade of rose plants (*see* **Pruning of Garden Roses, Practical Guidelines**). Pruning requirements vary among types and varieties of rose plants, and growers keep records of varieties which will probably need pruning to produce three canes, i.e. those plants which usually produce only one or two canes after budding. Pruning is done to force an additional cane at the bud union, and only one more cane can be expected after each pruning operation. Plants should have sufficient carbohydrate reserves and enough time to grow after the cut is made, so it is usually possible to prune only once per growing season. A heading cut is made in the spring to a shoot about 15–18 cm in length, leaving about 3 cm. Plants grown during 1-year production cycles, such as shrub roses and ‘started eyes’, are usually not pruned.

Harvesting

Prior to harvest of the rose plants, budwood may be gathered from mid-November to December and placed in cold storage. Shoots from the middle of the plant that are the thickness of a pencil or slightly thicker (~7 mm) are cut and leaves removed. Shoots are cut to about 22 cm so as to contain three eyes, then bundled, wrapped in wet newspaper and placed in plastic bags. These 'budsticks' are then placed in cold storage. The storage temperature is critical and should be −2 °C to −1 °C, uniform throughout the facility. If the temperature is too high the wood will not remain completely dormant and damage will occur during dethorning. If too cold, the wood will be injured or killed.

The harvest process begins with defoliation of rose plants. Chemical desiccants have been evaluated, but results suggest that desiccants damage canes and reduce bud-break of the plant during the next season. Therefore, mechanical defoliation is the standard practice, accomplished by a machine pulled by a tractor. This machine has rubber 'fingers' attached to spinning vertical cylinders which jostle the leaves from rose plants.

For harvest, soils must be dry enough to support equipment, but contain sufficient moisture to facilitate digging and soil removal from plants. Water may also be withheld to harden plants. Digging machines consist of tracked vehicles with raised centres. A U-shaped blade is inserted at the beginning of the row, and the forward motion of the unit and reciprocating action of the blade shakes and lifts roses from the soil. Rose plants are gathered by field workers, tied and loaded on to flatbed trailers for transport to processing areas.

Preventing plant desiccation is the primary consideration during the harvest operation. Under ideal harvest conditions ambient temperature is 4 °C with enough humidity to produce ground fog, but warm dry weather often occurs instead. Roses on the trailers are rinsed with water to remove clinging soil and to keep the plants wet. Tarps can be used to prevent plant drying during transport, and growers attempt to have plants at processing sheds within 30 min of digging. Plants are placed in a shaded environment under overhead sprinklers which provide a steady mist to ensure hydration. Plants are then graded individually, trimmed and placed into cold storage. They also may be treated with waxes or other antitranspirant compounds prior to storage. Like budwood or rootstock cuttings, storage temperature for plants is critical and should be −2 °C to −1 °C, uniform throughout the facility, with high relative humidity.

See also: **Disease**: Crown Gall; Replant Problems and Soil Sickness. **Insects and other Animals**: Mites; Nematodes. **Integrated Pest Management**. **Production and Marketing**: Mineral Nutrition. **Propagation**: Budding in the Field; Own-Rooted Cuttings; Seed Propagation. **Pruning of Garden Roses, Practical Guidelines**. **Rootstocks**: Usage of Rootstocks; Clonal Rootstocks.

Further Reading

Davies FT Jr, Hambrick CE III and Pemberton HB (1987) Grafting and adventitious root formation of Texas field rose bushes. *Acta Horticulturae* 189: 89–100.

Flint ML and Karlik JF (2000) *Healthy Roses*. Publication no. 21589. Oakland, CA: University of California.

Krüssmann G (1981) *The Complete Book of Roses*. Portland, OR: Timber Press.

Pemberton HB (1992) The Texas rose industry. *Combined Proceedings of the International Plant Propagators' Society* 42: 391–393.

Pemberton HB, Haby VA, Roberson WE and Davis JV (1986) Root and shoot growth of *Rosa multiflora* cuttings taken from fertilized stock plants. *Acta Horticulturae* 189: 123–126.

Pryor M, Harwood CA and Wessig H (1987) Plant production. In: Langhans RW (ed.) *Roses: A Manual of Greenhouse Rose Production*. Haslett, MI: Roses Incorporated.

Pot Rose Production

H B Pemberton, Texas A&M University, Overton, TX, USA
J W Kelly, Clemson University, Clemson, SC, USA
J Ferare, The Conard–Pyle Co., Hollister, CA, USA

Introduction and World Market

The potted miniature rose is a popular greenhouse crop in many parts of the world. Traditional spring pot forcing of field-grown, bare-root grafted plants of the Floribunda and Polyantha types reached its peak in the 1950s and 1960s, but this market was in decline by the 1970s. The introduction of new varieties in the 1980s and the development of an economical system of year-round pot production using cutting propagation and supplemental light to produce plants in small pots for mass-market sales resulted in a significant rejuvenation and expansion of the pot rose market (*see* **History of Roses in Cultivation:** Modern (Post-1800)). The impact of the new cutting-grown products on the European and North American markets has been dramatic. Current world production is estimated at 60–80 million pots yearly. Major centres of production include Denmark, The Netherlands, the United States, Canada and Japan with production also in France, Germany and Italy. Most production is in 10-cm to 12.5-cm pots, but growers have a tendency to use larger, 12.5-cm to 15-cm pots in North America and smaller, 6-cm to 7.5-cm pots in Europe.

Market expansion has been driven by the introduction of miniature cultivars that exhibit greenhouse and postproduction performance superior to those traditionally used. Recently introduced cultivars have also been used for spring forcing of bare-root grafted plants, resulting in a rejuvenation of the traditional market. Most modern pot rose production is and will continue to be with miniature-type cultivars.

Miniature Pot Rose Types

Two main groups of miniature cultivars all classed as miniatures; (*see* **Classification:** Horticultural Classification Schemes) suitable for pot forcing are the true miniatures (small growing types) and the large growing types (see **Figures 1** and **2**). The true miniatures are most suitable for a 10-cm pot programme because they are compact, well branched and heavy flowering, and may be forced under low light intensity. Compared to the larger miniatures, the true miniatures are usually easier to grow and produce more but smaller flowers. True miniatures are well adapted for a St Valentine's Day crop, but are also suitable as an impulse purchase year-round. They bloom continuously throughout the growing season in the garden, although they may not be winter hardy in the northern United States and Canada.

The large growing types are increasing in popularity. The vividly coloured flowers, which are smaller than those of the Floribundas but, in most cases, larger than the true Miniatures, are highly resistant to shattering. Plants are best adapted to an 11-cm to 15-cm pot programme. They are also available as budded, field-grown plants which are forced in 18-cm to 20-cm pots. Most of these cultivars exhibit good winter hardiness and disease tolerance.

Propagation

The most common method for producing pot roses is by cuttings. Field-grown bare-root plants (*see* **Production and Marketing:** Field Rose Production) are forced in containers by some growers for specific markets. Cutting propagation (*see* **Propagation:** Own-Rooted Cuttings) is generally used to produce plants of all cultivar types from the Miniature class. The key to efficient production of asexually propagated cuttings is the use of 'clean' (i.e. free of pests and diseases) stock plant material. Cuttings can be obtained from either blocks of stock plants maintained specifically for cutting production or from the shoots obtained from a crop in progress that is being pinched. Commercial propagators use the first option, while large growers involved in a continuous, year-round growing cycle tend to favour the second method, especially in

Figure 1 An example of a small growing miniature pot rose type is 'Charming'^(TM) 'Parade'® 'Poulhappy'^(N). Reproduced with permission from Poulsen Roser ApS.

Figure 2 An example of a larger growing miniature pot rose type is Lady Sunblaze® (*Rosa* 'Meilarco').

modern, fully controlled greenhouse environments. Many smaller and/or seasonal producers purchase rooted cuttings from a specialist propagator. In this case, obtaining the highest quality product possible on a reliable basis becomes the main concern. Another consideration when deciding between cutting propagation and purchase of rooted cuttings is the fact that a licence is required to propagate patented cultivars (*see* **Intellectual Property**: Plant Patents and Trademarks). Propagation of patented cultivars without a licence is illegal.

Production Methods

After the propagation phase, rooted cuttings are exposed to various types of production environments depending on the type of plant to be finished. To produce liners, cuttings are first rooted in cell trays. The cell size used depends upon the cultivar, the number of cuttings rooted in each cell, and the size liner desired. After the propagation phase, the newly rooted cuttings are moved into an optimum growing environment and grown to the size needed for forcing.

In addition to or instead of liners, some growers use a direct stick programme in which they propagate the cuttings in the actual pot that is used for forcing. This technique is mostly used by large, specialized growers who produce year-round crops. This system is designed for 6-cm to 11-cm pot production and requires specific equipment and growing techniques not suitable for many growers. Direct sticking can also be used for producing a prefinished plant for year-round finishing or for outdoor summer propagation.

Potted roses grown from cuttings are produced using one of two different methods, the short cycle or the long cycle. Short-cycle plants are typically miniature cultivars that remain in active growth from propagation to finishing (*see* **Table 1**). The forcing phase begins with either rooted cuttings propagated in the final forcing container or liner. Depending on plant size, liners can be potted into pots 10–15 cm in diameter. Plants prefinished in 10-cm pots can be forced in the same pot or finished in 15-cm pots.

As a general rule, direct sticking is used by large, specialist growers. For producers with less sophisticated growing facilities or needing a more heterogeneous product mix, liners are probably more economical and allow more flexibility in the type of product finished. Liners usually consist of one to four cuttings rooted in each cell of a large-cell tray and are shipped in active growth. They should be potted as soon as possible after receipt, but can be stored for up to 5 days at 1.5 °C. These plants should never be exposed to freezing temperatures, because plants in active growth cannot withstand being frozen.

One liner (three to four rooted cuttings) is usually potted into a 10-cm to 15-cm pot. Small liners (one to two rooted cuttings) are available for potting into 7.5-cm pots. Larger-size pots would require two or more liners or more than one pinch before finishing.

The final forcing stage begins by pinching the plants to 4–5 cm above the pot rim. However, doing this

Table 1 Commonly used production schemes for cutting-grown pot rose crops

Short-cycle production using liners		*Time (weeks)*	*Short-cycle production using direct sticking*	
Insert cuttings in liner cells for rooting		0		Insert cuttings in pots to be used for final forcing
	Rooting period	0–3	Rooting period	
	Acclimation of rooted cuttings to greenhouse conditions and first growth flush	3–6	Acclimation of rooted cuttings to greenhouse conditions, first growth flush, and root development	
Pot liners into pot size desired for finishing		6		Pinch growth to start the final forcing period
	Establishment of roots to the bottom of the pot	6–9	Plants exposed to optimum greenhouse conditions for flowering	
Pinch growth to start the final forcing period		9		
	Plants exposed to optimum greenhouse conditions for flowering	9–14	Plants exposed to optimum greenhouse conditions for flowering	
		14		Flowering
	Plants exposed to optimum greenhouse conditions for flowering	14–17		
Flowering		17		

without a well-established root system results in poor and uneven bud break leading to a finished plant with nonuniform flowering and poor balance. After direct-stuck cuttings have rooted, they must be allowed to develop until the root system fills the pot prior to pinching. Root systems must be similarly developed on freshly potted liners prior to pinching for final forcing. It takes 2 to 3 weeks to establish plants in 10-cm pots. A slightly longer time is needed for liners being potted into 12.5-cm or 15-cm pots, but the finished plants will be larger. Bottom heat is beneficial during this stage when greenhouse conditions are cool. Finishing time for a short-cycle plant is 5–8 weeks from the final pinch, depending on the time of year and local conditions. This is the most commonly used forcing method for finishing plants from March to November in North America. In northern Europe, supplemental high intensity lighting and CO_2 are added to produce plants year-round. Winter forcing with supplemental lighting is also common in the United States and Canada.

With the long-cycle method, better results are obtained when forcing for winter and St Valentine's Day flowering. Long-cycle plants are grown under outdoor conditions during the summer and autumn by potting liners into or direct stick propagating in the desired final pot size to be used for winter forcing. Shade at 25–35% can be used if needed for moderating the temperature, but plants should be exposed to as much light as possible. Plants should be sheared periodically to encourage branching from the crown and in the basal part of the plant; they should also be watered and fertilized like a crop being forced. Pesticides should be applied to prevent black spot and powdery mildew (*see* **Disease**: Black Spot; Powdery Mildew). After the plants have gone dormant in autumn, they must be protected from severe winter cold by being covered or placed into cold storage. We have observed these plants to be difficult to force successfully in October, but they are usually ready to force by early December in most areas of North America.

Long-cycle plants should be unpacked and placed into the forcing greenhouse immediately upon arrival. As the plants are unpacked, the shoots should be cut back to 5–7 cm above the pot and cleaned of all dead branches and debris. If potting into larger pots, repotting should be done as soon after cutback as possible.

Long-cycle plants are typically grown in 10-cm or 11-cm pots. Forcing can be done in the same container or plants can be potted into a larger size prior to forcing. Plants grown in 10-cm pots and potted into 15-cm pots at the start of forcing finish about the same time as those forced directly in the original pot, but are larger because of the increased rooting volume. This allows a larger plant to be finished for Easter or Mother's Day than can be produced using the short-cycle method with only one pinch. However, plants potted from 10-cm pots to pots larger than 15 cm will probably need a pinch to produce a large enough plant. In this case, the use of a dormant bare-root plant could be a better choice. Extremely vigorous, succulent, sucker-type growth will appear occasionally and should be completely removed. If allowed to continue growing, these sucker-type shoots will outgrow the rest of the plant and result in nonuniform flowering. Pinching this type of growth is ineffective.

Long-cycle plants are characterized by well-developed buds and root systems full of carbohydrate reserves; therefore they can be forced under the lower light intensity of the winter months much better than a short-cycle plant. In addition, long-cycle plants can be used to force for Easter or Mother's Day in less expensive growing structures such as cold frames. For even growth, a good balance between light and temperature is needed. It takes 7–8 weeks to finish long-cycle plants for St Valentine's Day, or 6 weeks if using high-intensity supplemental lighting. They can also be used for Easter or Mother's Day crops, but cold storage could be necessary for a late spring crop in mild climates to prevent bud break prior to forcing. Long-cycle production is more common in large nurseries where the plants are used internally rather than sold to finishers.

Greenhouse Environment

Light intensity is the most important factor affecting rose plant growth and flowering. As a general recommendation, 16 h day^{-1} of 5000 lx of supplemental lighting should be regarded as a minimum amount of light needed for successful winter production of pot roses in North America, particularly for St Valentine's Day crops. For long-cycle or bare-root plants, 3500 lx may be adequate. A longer photoperiod of 18–20 h is recommended at higher latitudes (approximately 60 °N, i.e. Alaska). The type of light source used is also critical. High-pressure sodium vapour lamps are the most commonly used source of supplemental lighting because of their high efficiency for converting electrical power to radiant energy. Incandescent lighting, which has a low red to far-red light ratio, should never be used for pot rose production.

Pot roses are commonly grown with a 17–18 °C night and 20–21 °C day temperature. A constant temperature between 18 and 20 °C can be used in greenhouses with a high degree of climate control. However, night temperatures should not be higher than day temperatures. Drastic changes in temperature over a short period of time should be avoided, because malformed flowers could result. For example, if the night temperature has to be raised by a few degrees to speed cropping time, it should be done by 0.5–1 °C per night instead of all at once.

Not surprisingly, time to flowering is directly related to forcing temperature, but extremely high temperatures should be avoided. Cultivar selections should be made if attempting summer production in hot summer areas. Shade and fan and pad cooling systems are commonly used to control greenhouse day temperature in the summer. Shade at 20–35% can be used, with 30% being most common. Shade greater than 35% should not be used even under high light and temperature conditions. Many cultivars will grow acceptably as long as day temperatures are kept under 32 °C.

CO_2 enrichment is most commonly practised during the winter months at latitudes greater than 40 °N, or anywhere infrequent ventilation allows economical maintenance of elevated CO_2 levels. A level of 1000 ppm is generally recommended, especially for winter production of short-cycle grown plants. However, the effect of CO_2 enrichment on flowering and final plant form should be assessed under local conditions to determine whether this practice is needed for enhancing plant quality.

Media, Water and Nutrition

Pot roses should be grown in a well-drained medium with good water-holding capacity. Any peat-lite mix such as 1 peat : 1 perlite : 1 vermiculite by volume is desirable. Optimum pH for pot rose production is 5.5 to 6.0, with readings above 7.0 resulting in chlorosis. Incorporating superphosphate and dolomitic limestone in the mix helps balance the pH and provides calcium and magnesium.

Either too much or too little moisture can result in leaf yellowing, abscission and root loss on rose plants. Overwatering is especially harmful when starting to force dormant long-cycle plants into growth, on newly potted liners (long- or short-cycle plants), and on plants in all types of production schedules that have just been pinched or cut back while in active growth. On the other hand, allowing plants to dry out even on one occasion can also ruin a crop. Roses do not recover well from a missed watering, so careful monitoring is very critical. Excessive drying of the medium can also result in soluble salt injury if fertility levels are high. Drought stress becomes especially critical as flowers reach pea-bud size. At this stage, overwatering is hard to do at normal forcing temperatures.

Water quality should also be considered for successful pot rose production. Water supply, pH and soluble salt content can vary widely across production areas. Total soluble salts above 910 ppm and a bicarbonate ion (HCO_3^-) level of over 90 ppm should be avoided for pot rose production. Sodium and chloride ions can also be harmful and should be maintained below 46 ppm and 70 ppm, respectively.

A commonly used fertilization programme consists of 180 to 200 ppm nitrogen with a two-feed, one-leach schedule and a reduction of nitrogen upon finishing. Any good, commercially available fertilizer such as 20-20-20 or 15-10-12 can be used. Supplementing a constant feed schedule with a topical application of a balanced, slow-release fertilizer can be beneficial for maintaining nutrient levels. A weekly liquid application of iron chelate and magnesium at 5 and 10 ppm, respectively, can be used to ensure good foliage colour. Micronutrients should be incorporated into the feeding programme to ensure proper growth (*see* **Production and Marketing**: Mineral Nutrition).

Nutritional demand can vary depending on climatic, environmental and geographical conditions. For example, plants grown under high levels of supplementary light or in a geographic location with a warm, high-light climate may need more fertilization than plants under ambient light conditions or a cool, cloudy climate. Also, plants suddenly exposed to sunny, warm conditions after being held in shady, cool areas can exhibit symptoms of deficiency of micronutrients (e.g. copper) in the young foliage. This is usually an uptake problem which can occur when the warm conditions result in a high shoot demand for nutrients combined with the lack of root activity in the still cold media that has not had enough time to warm because of the suddenness of the environmental change. These conditions occur most frequently during late winter and spring when forcing in cold frames or structures without cooling systems.

Fertilization programmes should also be tailored to the type of plant being used for forcing. Fertilization of short-cycle plants should be maintained continuously so that no checks in growth occur during production. Long-cycle plants usually exhibit a loss in root activity during winter storage. This can manifest itself as an interveinal chlorosis which is visible on new foliage at the start of the forcing period. This is usually a transient problem, but care should be taken not to overwater or fertilize excessively until root activity is restored. If the chlorosis persists, other problems such as a root rot could be present.

Spacing and Plant Growth Regulators

Spacing can be pot to pot until 2 weeks after the final pinch. Final spacing should be 32 to 43 pots m^{-2} for 10-cm pots, 27 to 32 pots m^{-2} for 11-cm pots, or 11 to 16 pots m^{-2} for 12.5-cm to 16-cm pots. Larger pot sizes should be spaced proportionally farther apart.

Many of the newer miniature cultivars grow more in proportion to finished pot size than the older Polyantha and Floribunda types without a growth regulator. Although few studies have been published concerning efficacy, paclobutrazol (Bonzi™) has been observed to be effective on pot roses and is the most commonly used product. Two sprays of 50 ppm or three sprays of 33 ppm applied 7 to 10 days apart have been found to reduce plant height. For maximum effectiveness, sprays should be applied before buds are visible. Sumagic™ has also been found to be effective. But, as with all chemicals, the label should be consulted and applications on small groups of plants under local conditions should be tested before large-scale applications are made.

Pests and Pathogens

Forced pot roses are usually sold after 6–17 weeks in the greenhouse, and are not subject to as many insect and mite problems as longer-grown pot crops. However, some pests can be bothersome for the greenhouse grower. The most common insect and mite pests are aphids, thrips and spider mites (*see* **Insects and other Animals**: Aphids; Mites; Thrips). Systemic insecticides, which are carried through the intercellular fluids, are most effective on aphids and spider mites. Contact poisons are also effective, since the residue is located where the insect or mite will make contact with it while feeding.

Two-spotted spider mite is a serious pest associated with rose production (*see* **Insects and other Animals**: Mites). Spider mites on roses can be difficult to combat because of rather dense foliage. Therefore, it is critical to get good spray penetration into the foliage and to contact the underside of the leaf with the spray because this is where spider mites tend to reside. Mite problems increase substantially as greenhouse temperatures increase and during times of lower humidity.

Thrips are the most common insect pest found in pot rose production (*see* **Insects and other Animals**: Thrips). They feed primarily on the rose flower, but will also cause foliar damage. Systemic insecticides and contact poisons are effective on this pest. Screening the greenhouse with fine mesh materials to reduce the entry of thrips and other insects will reduce the number of spray applications required.

There are several different types of diseases that affect pot roses. *Cylindrocladium scoparium* causes black, water-soaked cracking of the bark near the soil line in pot roses. This fungal stem rot has become a significant problem as the production of pot roses has increased. It is most damaging during propagation, where its action is cyclical. The fungus is always present, but seems to be more active under warm, humid, low-light conditions and when plants are under stress. Very little is known about the life cycle of *Cylindrocladium* and about the conditions that affect its activity on roses; disease control is difficult. The best control of *Cylindrocladium* is the implementation of good sanitary practices including the prompt removal and destruction of diseased plants.

Powdery mildew is one of the most common fungal disease problems encountered in pot rose production (*see* **Disease**: Powdery Mildew). Powdery mildew symptoms include a white, powdery substance which may develop on all plant parts including upper and lower surfaces of leaves. There are effective chemicals on the market for control of powdery mildew. The critical factor for chemical control is to begin fungicide applications as soon as weather conditions are conducive and symptoms occur. Environmental controls such as reducing night greenhouse humidity through adequate ventilation is one sound approach to discourage powdery mildew development. Continuous air movement from horizontal airflow fans will also help prevent powdery mildew.

Botrytis blight is a common pot rose fungal disease problem (*see* **Disease**: *Botrytis*). It is frequently a problem during shipping and storage of plant materials, but it also occurs during production. Controlling relative humidity levels and dampness in the greenhouse helps reduce the incidence of *Botrytis*.

Another fungal disease that can cause severe losses in pot rose production is downy mildew (*see* **Disease**: Downy Mildew). Symptoms include purplish red to dark brown, irregular to angular spots, and severe leaf abscission. Under cool, humid conditions, this disease can strike suddenly and spread very quickly. However, maintaining humidity below 85% with ventilation can prevent the disease's development.

Guidelines for Postproduction Handling and Marketing

While pot roses have great market potential, there must be better postproduction handling of the crop for it to reach its full potential. The following recommendations should serve as guidelines to help ensure consumer satisfaction:

- Keep plants well irrigated during production and in all phases of postproduction handling, including retail and consumer display areas.
- If using a constant liquid feed system, 200 to 300 ppm nitrogen from a balanced fertilizer is

recommended with a clear water leaching every two to three waterings. Prior to sleeving and shipping, media should be leached with clear water. Adequate calcium and low ammonia levels in the flowers can prolong postharvest life of the plants and help prevent grey mould (*Botrytis cinerea*) development.

- If CO_2 enrichment is used during production, plants must be shipped at low temperatures and for short durations to prevent etiolated shoot growth.
- Avoid producing roses during mid-winter without adequate light or during the hottest summer periods without adequate cooling of growing facilities. Moderate day/night temperatures (22/18 °C) during forcing are recommended for longest postproduction performance.
- Spring and early summer production is usually optimal for postproduction longevity. However, this recommendation is highly dependent on location.
- Keep crops free from pest problems.
- Select cultivars carefully according to growing region and postproduction characteristics.
- Harvest pot roses when most of the flowers are showing colour.
- Supply a care tag with each pot rose.
- Sleeve roses before shipping, and box if necessary to prevent physical damage during shipping.
- Roses should be shipped and stored for as short a time as possible. Ideal storage temperature is 1.5–4 °C.
- Avoid exposing plants to any source of ethylene gas including combustion engines, fruits, vegetables and cut flowers (*see* **Postharvest Physiology**: Ethylene).
- Benzyladenine (BA) (25 ppm) or 6-(benzylamino)-9-(2-tetrahydropyranyl)-9H-purine (PBA) (50 ppm) applied as a preshipping spray reduces leaf chlorosis if warm temperatures or long storage durations are anticipated; they will also increase flower longevity for some cultivars. Follow label instructions closely.
- Silver thiosulphate (STS) (adjust concentration depending on cultivar) applied as a preshipping spray will substantially reduce flower abscission that occurs during shipping and interior use. Follow label instructions closely.
- Providing minimum light levels needed for plant quality maintenance is critical for successful marketing and consumer use.

See also: **Classification**: Horticultural Classification Schemes. **History of Roses in Cultivation**: Modern (Post-1800). **Intellectual Property**: Plant Patents and Trademarks. **Postharvest Physiology**: Ethylene. **Production and Marketing**: Mineral Nutrition; Field Rose Production. **Propagation**: Own-Rooted Cuttings.

Further Reading

Borch K, Williams MH and Hoyer L (1995) Influence of simulated transport on postharvest longevity of three cultivars of miniature rose. *Acta Horticulturae* 424: 175–180.

Dole JM and Wilkins HF (1999) *Rosa*. In: *Floriculture Principles and Species*, pp. 495–508. Upper Saddle River, NJ: Prentice-Hall.

Hammer PA (1992) Other flowering pot plants. In: Larson RA (ed.) *Introduction to Floriculture*, pp. 477–509. San Diego, CA: Academic Press.

Heins RD (1981) Forcing pot roses for Valentine's Day. *Florists' Review* 169: 14–15.

Jiao J, Wang X and Tsujita MJ (1990) Whole plant net photosynthesis of miniature roses influenced by light, CO_2, and temperature. *Acta Horticulturae* 272: 261–265.

Jørgensen E (1992) *Growing Parade-Roses*®. Hillrød, Denmark: PejseTrykcenter A/S.

Mackay I (1985) Roses: the spring pot crop. In: Ball V (ed.) *Ball Red Book*, 14th edn, pp. 677–685. Reston, VA: Reston.

Maxie EC, Hasek RF and Sciaroni RH (1974) Keep potted roses cool. *California Flower and Nursery Reports* March: 9–10.

McCann KR (1991) Mini potted roses hit the big time. *Greenhouse Grower* 9(13): 20–23.

Miller R (1987) Tips for the mini pot rose grower. *GrowerTalks* 51: 80–82, 85.

Moe R (1973) Propagation, growth and flowering of potted roses. *Acta Horticulturae* 31: 35–50.

Mor Y, Halevy AH, Kofranek AM and Kubota J (1986) Forcing pot roses from own-root cuttings: effect of growth retardants and light. *Acta Horticulturae* 189: 201–208.

Pellett G, Ferguson R and Zary K (1998) *Rosa* (Rose). In: Ball V (ed.) *Ball Red Book*, 16th edn, pp. 705–726. Batavia, IL: Ball Publishing.

Pemberton HB, Kelly JW and Ferare J (1997a) *Production of Pot Roses*. Portland, OR: Timber Press.

Pemberton HB, Kelly JW and Ferare J (1997b) *Rose*. In: Gaston ML, Carver SA, Irwin CA and Larson RA (eds) *Tips on Growing Specialty Potted Crops*, pp. 112–121. Columbus, OH: Ohio Florists' Association.

Pobudkiewicz A and Goldsberry KL (1989) Controlling the growth habit of dwarf pot roses with uniconazole (Sumagic™). *Colorado Greenhouse Growers' Association Research Bulletin* 471: 1–2.

Serek M, Sisler EC and Reid MS (1995) Ethylene and the postharvest performance of miniature roses. *Acta Horticulturae* 424: 145–149.

Smith RB and Skog LJ (1992) *Cold and Controlled Atmosphere Storage of Miniature Potted Roses*. Ontario, Canada: Horticultural Research Institute of Ontario.

Cut Rose Production

T J Blom and M J Tsujita, University of Guelph, Guelph, ON, Canada

Introduction

Cut-rose production worldwide under protected cover is estimated to be about 8500 ha, with an annual production of about 15–18 billion stems worldwide. The greatest production areas are in Colombia (1000 ha), Ecuador (2000 ha) and Kenya (1000 ha). Some European countries, such as The Netherlands (900 ha), Italy (900 ha) and France (400 ha) are intermediate producers, while Japan may have about 400 ha in production. Rose prices are determined by bud size (hybrid tea, sweethearts or miniatures) and within each group by stem length. Grading on stem length generally occurs on 10-cm intervals, e.g. 30–40 cm; 40–50 cm up to 1.0 m in length. As a general rule, the taller the stems, the higher the price. Hybrid tea roses demand a higher price than sweetheart roses of similar length. Rose prices worldwide are somewhat dependent on the auction price at the Dutch Auction system (The Netherlands).

Cut roses can be shipped worldwide by plane due to the relatively low shipping costs and generally a lack of phytosanitary restrictions for importing cut flowers. Rose production has shifted during the period of 1985–2000 from developed countries (western Europe, North America, Japan) to lesser developed countries (Ecuador, Colombia and Kenya) to take advantage of the ideal climate, lower labour and production costs during periods when there is lower production and high energy costs in developed countries (winter season). In North America, producers sell generally through standing orders with a specific price for given qualities and volumes, while exporters generally depend more on world market price, which may change from day to day.

Production facilities have become larger and operations of over 10 ha (not uncommon in underdeveloped countries) allow for more effective marketing and shipping procedures and labour efficiency. Greenhouse ranges at higher latitudes often use single-layered glass, double-layered acrylic or air-inflated double polyethylene, while at lower latitudes a single-layered polyethylene film is the most common glazing material. The production areas dedicated to hybrid tea and sweetheart roses are about equal, with a small percentage dedicated to sprays. Red has been the predominant colour, followed by yellow, pink and white. Colour demand depends on time of year, trend and specific holidays.

Production Systems

Rose production in protected environments started at the end of the nineteenth century in order to extend the marketing period for this crop. During the winter period, the rose plants became dormant and were forced again in early spring. When heating systems were installed, and using suitable cultivars, production became year-round. The introduction of climatic computers (analog followed by digital) allowed optimization of environmental conditions such as temperature, carbon dioxide, light and humidity, and later irrigation and fertilizer delivery as well.

Up to 1990, roses were primarily grown in the soil of the greenhouse. In order to accommodate the various soil and moisture conditions and optimize production, roses were grafted or budded on rootstock (synonym: understock). Budded or grafted plants were normally planted into the soil beds at 10 plants m^{-2} of bed (or 6–7 plants m^{-2} of greenhouse) during the winter/spring season and allowed to grow and be harvested. Plants were allowed to sprout and growers either harvested or initially pinched the shoots. The top axillary bud or buds at the node below the cut or pinch would then develop into shoots. Depending on the shoot-to-shoot cycle (6–10 weeks), the flowers were harvested on a daily basis, usually above a five-leaflet leaf. The result was that the crop grew taller over time, and created a tall hedge (**Figures 1** and **2**). After each year, the plants were pruned back with or without a dormant period. In North America, this usually occurred after the spring season (May/June), as rose growers are dependent on special holidays such as Valentine's and Mother's Day for a large percentage of their revenue. To facilitate crop timing for special holidays, rose growers usually applied a soft or hard pinch; then, following bud-break, shoot development was controlled by temperature, and sometimes supplemental lighting.

Plants were usually watered with sprinkler lines which were placed about 30 cm above the soil level. Soluble fertilizer was added to the water on a regular basis or was added to the soil in a dry form a number of times per year.

Around 1985–1990, the production system started to change quite drastically. First, substrate culture was introduced. This technique involved the use of a relatively small volume (3–4 l per plant) of substrate (rockwool, perlite or coco coir) and plants were watered including nutrients a number of times per day through a low-volume drip system (**Figure 3**). This more precise technique of feeding and watering based on crop demand led to increased production per m^2.

Another development which affected rose production is the arching or bending technique which

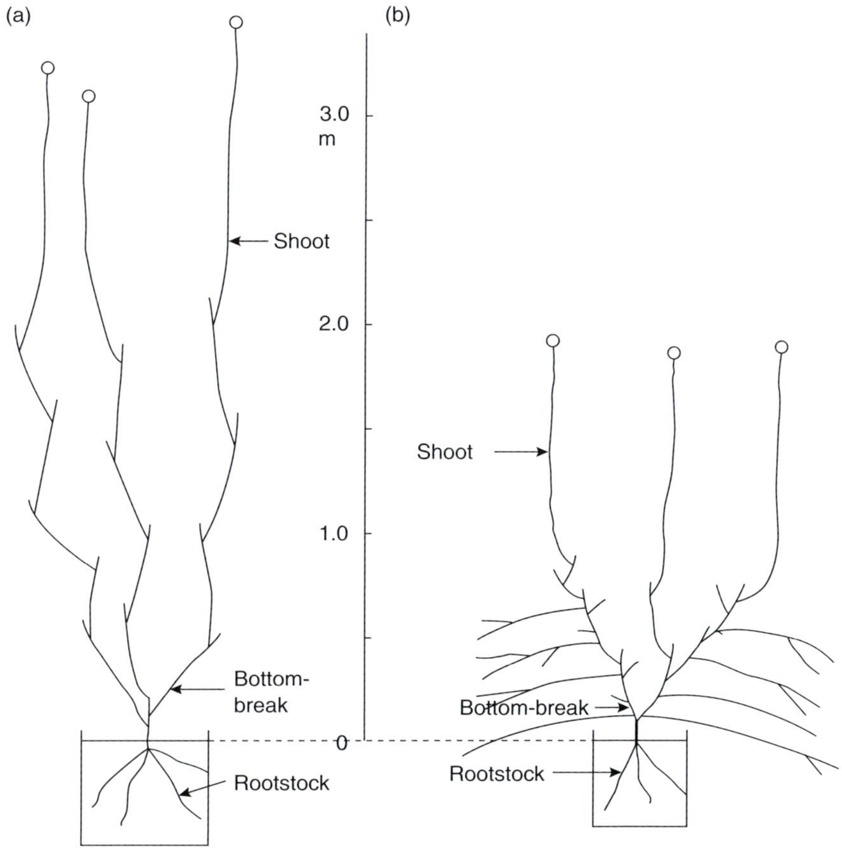

Figure 1 A schematic drawing of (a) a hedge and (b) an arching system for the production of cut roses.

Figure 2 A view of a hedge system (background) and a new planting (foreground) with sprinkler irrigation lines and heating pipes.

originated in Japan. This technique allows harvesters to cut longer stems by cutting back to the knuckle rather than a five-leaflet leaf. All weak shoots are bent downward to fill any area void of foliage (mostly the aisle) and thus attain a desirable leaf area index to optimize photosynthetic potential and facilitate the transport of sugars to the developing shoots (**Figure 1**). The shoots which arise from dormant buds following knuckle cuts are generally more vigorous and produce high-quality and superior grades. This vigour may be attributed to the fact that the cane diameter is larger at the knuckle and, since the weak shoots are bent, all remaining shoots are exposed to better light and less competition. Thus, the exposure of the knuckle cane to a higher red to far-red ratio may facilitate the movement of cytokinin to the dormant buds, allowing them to break more freely. Two practical benefits of the arching technique is that the annual cutback is eliminated and that roses are grown at working-level height (**Figure 3**). The advantage of the arching technique is reduced labour, longer stem roses with high quality and a more manageable crop compared to the upright hedge system. The negative side is that the production is lower.

Propagation

Cut-rose cultivars are vegetatively propagated so that all roses are genetically similar. However, there are a number of techniques used to propagate:

- cuttings
- grafting/budding
- miniplants

In the past 20 years, a small number of growers have been producing plants grown on their own roots. Plants are produced from a single- or double-node cutting, rooted in a misting bed and subsequently transplanted in a production greenhouse. However, the majority of growers use rootstock (*see* **Rootstocks**: Usage of Rootstocks) for added vigour, better colour and resistance to disease. The rootstock is either seed-propagated (e.g. *Rosa canina* 'Inermis') or vegetatively propagated (e.g. *R. indica* 'Manetti') (*see* **Rootstocks**: Clonal Rootstocks). The plants are budded in the field followed by a 4–6-month growth period (North America) or grafted in the greenhouse (Europe). The rootstock used depends on the type of rose (*see* **Rootstocks**: Scion–Rootstock Relationships), the production system, and expected duration of the crop (3–7 years).

For substrate production (rockwool, coco coir, etc.) many propagation operations produce miniplants. These plants are produced by grafting the scion on the rootstock while they are in the rooting process (stenting). Basically the scion and the rootstock cutting are held together by means of a small clothes-pin during rooting in a propagation bed. The main advantage for the propagator is the short turnaround time of the plant material (4–6 weeks); fewer plants have to be propagated on speculation, and there is less chance of infection with disease organisms and/or nematodes.

Figure 3 Rose plants planted in elevated troughs and irrigated with a low-volume drip system. The troughs are sloped and have an internal drain system, so that excessive irrigation water can be collected (black bucket across the ends of the troughs) for recirculation.

Basic Anatomy

A rose shoot is composed of a succession of 8–15 reiterated units: each unit consists of a leaf (or scale), an axillary bud, a node and an internode. The lower units are preformed in an axillary bud and form the lower part of the shoot, whereas the upper units are neoformed, i.e. formed after release from inhibition and making up the upper part of the shoot. Axillary buds are partly developed shoots arising from an axil and often protected by leaves or leaf-like scales (**Figure 4**). The formation and subsequent development of axillary buds into flowering shoots are vital for optimal flower production. These axillary buds determine the flower stem production but also build the plant structure so that sufficient growing points are available to achieve optimal production.

Axillary buds are usually formed slightly later than the subtending leaf primordium. These axillary buds may develop into a sprout and/or flower, but a number of processes may prevent this from occurring. The growth of an axillary bud at a node continues over time but does not necessarily develop into a shoot as this is controlled by apical dominance and correlative inhibition. Apical dominance is the feature whereby the apex is the dominant sink for assimilates, thereby inhibiting the development of axillary buds. Apical dominance is believed to be controlled by the production of auxin in the shoot

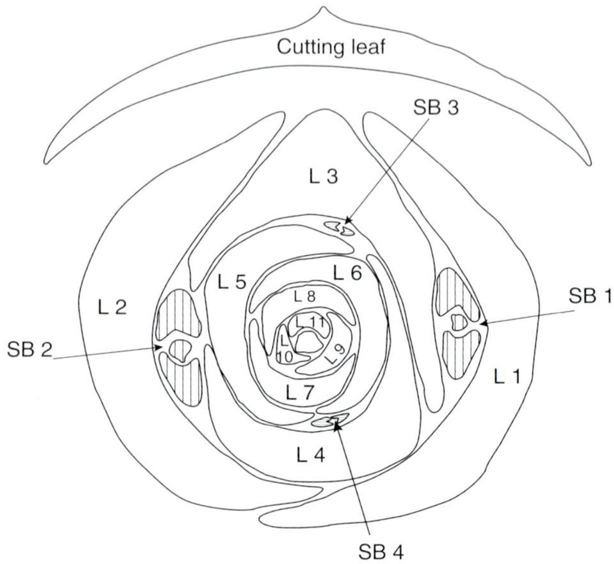

Figure 4 Transverse section of an axillary rose bud 'Sweet Promise'. The bud is composed of a total of 11 leaves and leaf primordia (L). The outermost leaves of the bud will become the lowermost leaves of the shoot. Secondary buds (SB) are present in the axils of the outermost four leaves. Reproduced from PhD thesis by Marcelis-van Acker C. *Axillary Bud Development in Rose*. The Netherlands: Wageningen University, with permission.

apex and then migrates down the stem into axillary buds. Apical dominance is only present when shoots are growing. Cytokinins, which are produced in the terminal points of the roots, counteract the effect of the auxins and promote axillary bud growth. Upper buds are less inhibited than lower-formed buds. This process is called correlative inhibition. In correlative inhibited buds, the capacity of auxin and gibberellin synthesis is absent and this synthesis of the latter two growth regulators does not start until release from inhibition. This process of apical dominance prevents roses from freely branching. Correlative inhibition can be removed when one removes the flower from a flowering shoot. Upon cutting the terminal shoot, the most distal (upper) bud(s), stimulated by cytokinins from the roots and the lack of production of auxins in the terminal bud which was removed, will grow into a shoot.

When a shoot is bent (arching), the uppermost buds near the bent area will become the apically dominant buds. There is a shorter response time for the uppermost bud when compared to bending. When the terminal section of a shoot is cut, the uppermost (remaining) bud will break within 3–5 days while the uppermost bud from a bent shoot may take longer and is also less predictable (5–15 days).

An axillary bud is comprised of the primary meristem, several leaves and a number of secondary buds (**Figure 4**). The primary meristem is in the centre while the secondary buds are present in the outer (often scale-like and also oldest) leaves. No flower is formed as yet. Axillary buds need to reach a certain developmental stage before they are able to break. After release of apical dominance, the apex rapidly initiates a number of leaves and switches from a vegetative to reproductive stage. The primary bud contains only leaves while the flower primordia are initiated shortly after the termination of the apical dominance. The secondary buds are forming leaves during the unfolding of the primary bud. Although bud age increases towards the base of the shoot, permitting an enhanced development, the rate of development of lower buds is less compared to that of higher-positioned buds. This correlative inhibition of the bud by the upper parts is typical for roses.

There are two distinct stages during the development of axillary buds: (1) a stage of inhibition on the outgrowth of the meristem; and (2) a stage of growth into a shoot after release of apical dominance. During inhibition, the apical meristem of a primary axillary bud remains vegetative and continues to form leaves, secondary axillary buds and stem tissue. Therefore, the older the axillary bud is, the more leaves are formed and the longer it takes to break. After a shoot has been cut, the remaining upper axillary bud on that stem will develop and becomes the apical dominant shoot which will form a number of additional leaves before the flower is induced (generally 8–12 days after the cut). Flower induction is independent of photoperiod and/or light sum. The size of the emerging shoot depends on assimilate supply (size of cane from which it originates), the age, location of the bud on the stem and temperature.

Up to the point of flower initiation, the shoot is a carbon sink and relies on subtending leaves for growth and development. After the flower is initiated, and about the time that the leaves change colour from red to green, the shoot becomes self-sufficient in the carbohydrate supply and subsequently becomes a carbohydrate exporter. If the latter is not the case (e.g. under low light conditions), then the flower may abort, resulting in a blind shoot. The breaks below the blind shoot looks similar to witches' broom. The process from cut to cut takes about 6–10 weeks, depending on type of flower (sweetheart vs hybrid tea), temperature, carbon dioxide, total radiation and number of leaves on the shoot.

Light

Roses are considered to be a high-light-requiring crop. The light factor is one of the important reasons why much rose production has moved to a lower latitude with more light and less variation between summer and winter. It has been shown that both photosynthetic and carbon export rates of rose leaves increase with increased light intensity.

Environmental Influence on Photosynthesis and Carbon Export in Greenhouse Roses during Development of the Flowering Shoot

Optimal rose production is achieved with about 12–15 mol m^{-2} day^{-1} of visible light at plant level. Productivity of rose stems generally follows changes in light sum. It has been shown that at a given temperature the time from bud-break or flower cut increases by 2–5 days as light integral decreases from 11 to 6 mol m^{-2} day^{-1}.

Optimizing Greenhouse Temperature for Rose Flower Head Size and Productivity

In many areas of the world, the light sum during the winter period is well below this level (1–3 mol m^{-2} day^{-1}) and supplemental lighting is often used by many growers during this period. Light intensities of 40 μmol m^{-2} s^{-1} and higher are achieved with the use of high-pressure sodium (SON-T) lamps. These lamps are very efficient in converting electrical power to photosynthetically active radiation (PAR) and also have a high-red to far-red ratio, which may enhance bud-breaks. These lights are usually turned on during cloudy days for a maximum duration of 18–20 h, while allowing a minimum dark period of 4 h. Even low irradiation (30–60 μmol m^{-2} s^{-1}) of the plants into the extended night period will compensate for the carbon loss from respiration which would have occurred during this period, thus maintaining a higher accumulative daily carbon gain during the cropping period. During low-light days, it is not unusual for roses at the pea bud stage to lose more than 50% of the carbon gained in photosynthesis through respiratory process.

More than 75% of the biomass production for a rose crop can be explained by light interception. A leaf area index (leaf surface area/area of greenhouse) of 4–5 is considered optimal as it intercepts about 80–85% of the available light. Therefore, the management of a rose crop is primarily concerned with this factor. With the traditional hedge system, roses are cut at the first or second five-leaf above the place of origin (knuckle), which means that there are always some leaves remaining on the shoot at harvest. With the arching technique, roses are usually cut at or close above the knuckle (e.g. 3–4 cm above the origin of the shoot) and thus few leaves are left behind. In order to maintain sufficient leaf mass, shoots of low quality are bent over and thus create the leaf mass. In the latter case, there is a very shallow plant canopy (**Figure 1**). This arching is to be done on a regular basis as leaves lose their capacity to contribute to the total biomass of the plant over time. It is estimated that leaves are only contributing to the carbohydrate gain (net carbon exporter) between 3 and 10 weeks after being developed. Excessive numbers of leaves may hinder the number of developing shoots due to the reduced ratio of the red/far-red light within the canopy.

Temperature

The daily 24-h average temperature controls the development rate of a rose shoot. Generally the development rate increases linearly between 10 and 25 °C. However, the optimal temperature depends on the amount of photosynthates available and thus, its relationship with CO_2 is very critical. For example, at ambient CO_2 (360 μl l^{-1}) and light saturation for whole-plant photosynthesis (1000 μmol m^{-2} s^{-1}), the optimum temperature is near 22 °C whereas at 200 μmol m^{-2} s^{-1} it is 16 °C. If CO_2 is maximized at 1000 μl l^{-1} the optimal temperature will increase by 1–2 °C. Above 25 °C photosynthetic and carbon export rates of rose leaves will decrease while respiration rate will significantly increase. The daily net carbon gain is a function of total carbon gain in photosynthesis minus carbon loss through respiration. Under low-light conditions with fewer carbohydrates produced in photosynthesis, the optimal temperature is generally lower (16–19 °C) than when there is ample light available (18–20 °C). Temperature has a significant effect on quality aspects such as stem length, bud and leaf size and stem diameter. Under constant day/night temperatures, stem length, bud size and stem diameter increase with decreasing temperatures from 25 to 16°C (24-h average temperatures). Also the number of blind shoots increases with temperature. Rose shoots grown under a higher day (T_d) than night (T_n) temperature are generally longer than those grown under $T_d < T_n$. Temperature management is aimed at balancing productivity with quality of the roses under a given set of light conditions.

Carbon Dioxide

In enclosed environments such as greenhouses with a growing crop, CO_2 levels will usually drop below ambient levels (360 μl l^{-1}) during the daytime due to the CO_2 assimilation process. During the night, carbon dioxide levels will rise above ambient. Increasing CO_2 levels during the daytime to about 800–1000 μl l^{-1} are beneficial to rose production due to the increased dry-matter production and consequently a quicker growth cycle time. The source of CO_2 supplementation is usually from the combustion of fuels such as natural gas or propane (burners or boilers), but sometimes liquid CO_2 is considered to prevent the chance of incomplete combustion. During times of limited ventilation, the level of supplementation is usually decreased to ambient levels. During the summer CO_2 supplementation may be limited, due to high levels of ventilation.

Humidity

Humidity affects rose production for several reasons:

1. Leaf size, and thus the capacity to photosynthesize, increases with humidity. Humidification of the air is a suitable technique to prevent low humidity which may occur in cold climates and greenhouses covered with a single layer of glass (due to condensation on glazing).
2. Nutrient uptake is governed to a large degree by the water (+nutrient) uptake of the plants, which in turn is controlled by the transpiration rate.
3. When air and plant temperatures differ, there is a chance that water condensation will take place on the leaves, leading to the occurrence of botrytis on flowers and canes or diseases such as downy or

powdery mildew (*see* **Disease:** *Botrytis*; Downy Mildew; Powdery Mildew).

Generally, high humidity is controlled by venting inside air out of a greenhouse and allowing drier air to enter in combination with heating. Low humidity is controlled by high-pressure misting systems or by the use of fan and pad cooling.

Substrates

One of the biggest changes which has occurred in protected rose production at the end of the twentieth century is the change from soil to substrate. In soil, roses were planted in beds (30 × 30 cm centre), generally three or four rows wide. Water and nutrients were provided by means of a sprinkler system, which was placed above the soil surface. Each bed was usually underdrained to direct excess irrigation water away from the root zone. In the 1980s, production in the western world shifted to substrate culture. This technique uses small volumes of growing substrate which is rockwool, peat or coco coir-based. Individual plants are planted on the rockwool slabs, in troughs or pots filled with organic substrate. Water and nutrients are provided on a per-plant basis through microirrigation systems ($1-2\,l\,h^{-1}$). Irrigation delivery is generally computer-controlled and driven by amount of incoming radiation, stage of crop (leaf area index), heat demand and/or time. The volume of water applied is usually $2-7\,l\,m^{-2}\,day^{-1}$ and the majority is given during the daytime in gifts of about 75–125 ml per plant per irrigation. The advantages of the substrate are the more precise delivery system based on crop needs, uniformity of substrate, lack of need of steam sterilization, water management and capability to recirculate the excess nutrient solution. Also, substrate in troughs could be placed at a more convenient working height, which became more important with the arching system. Recirculation water which can be more easily collected with the substrate system can be heat-sterilized, ultraviolet-treated or passed through a slow sand filter. After sterilization, the recirculation water is mixed with fresh water and additional nutrients are added to control the electrical conductivity and pH of the solution.

Nutrition

With the substrate systems, all macro- and micronutrients are delivered as soluble fertilizers. The recommended target values are based on the substrate and whether the irrigation solution is used as a once-through or a recirculating system. Target values for an inert medium are given in **Table 1**.

Table 1 Target values of nutrient solution for roses grown in an inert medium and using a once-through system

Macroelement	*mmol l^{-1}*	*Microelement*	*μmol l^{-1}*
NO_3^-	11.0	Fe	25.0
NH_4^+	1.5	Mn	5.0
$H_2PO_4^-$	1.25	Zn	3.5
K^+	4.5	B	20.0
Ca^{2+}	3.25	Cu	0.75
Mg^{2+}	1.125	Mo	0.5
SO_4^{2-}	1.25		

To deliver the solution, soluble fertilizers and micronutrients are usually dissolved in two concentrate tanks, to avoid precipitation of incompatible ions, and diluted to obtain a target electrical conductivity (EC) of $1.6-1.8\,mS\,cm^{-1}$. Calculation schedules for fertilizer amounts generally incorporate the presence of nutrients in the water source if it deviates from rainwater. Recirculated water generally has an EC which is usually higher than that of the irrigation solution. The difference between EC_{rec} and EC_{sol} provides an estimate of nutrient accumulation in the substrate and whether or not to increase irrigation frequency. Water with high levels of bicarbonates is generally treated with an acid so that the final pH of the nutrient solution is between 5.5 and 6.0. Sophisticated nutrient injectors can control both EC and pH at the desired level. Further adjustments to the target values are generally made depending on specific cultivar demands, rootstock, season and stage of crop.

Harvesting and Postharvest Handling Practices

The stage of flower bud development determines whether or not a rose has to be cut. Usually this is done when the outermost petal begins to unfurl but some cultivars are cut before this stage. Due to the rapid opening of rose buds, roses are generally harvested once or twice a day using hand secateurs. After roses are cut, the stems are generally placed immediately in a hydrating solution to maintain turgidity and reduce the chances of air embolism at the bottom of the stem. After roses are cut, they are either cooled or graded to length. Containers used for storage of cut roses are disinfected prior to use to reduce the incidence of microbial growth in the storage solution, which may lead to stem blockage. Roses are stored at 2–3 °C at a relative humidity of 90–95% in the cooler. Dry storage of cut roses in a cooler is not commonly done, except for special high-demand days.

See also: **Breeding**: Selection Strategies for Cut Roses. **Growth Regulation**: Axillary Bud Growth. **Postharvest Physiology**: Xylem Structure and Function in Cut Roses; Effects of Temperature; Ethylene; Packing Systems and Precooling. **Production and Marketing**: Overview of Roses and Culture. **Rootstocks**: Scion–Rootstock Relationships; Usage of Rootstocks; Clonal Rootstocks.

Further Reading

Jiao J and Grodzinski B (1998) Environmental influences on photosynthesis and carbon export in greenhouse roses during development of the flowering shoot. *Journal of the American Society of Horticultural Science* 123: 1081–1088.

Jiao J, Tsujita MJ and Grodzinski B (1991) Influence of temperature on net CO_2 exchange in roses. *Canadian Journal of Plant Science* 71: 235–243.

Jiao J, Tsujita MJ and Grodzinski B (1991) Influence of radiation and CO_2 exchange in roses. *Canadian Journal of Plant Science* 71: 245–252.

Lieth JH and Pasian CC (1990) A model for net photosynthesis of rose leaves as a function of photosynthetically active radiation, leaf temperature, and leaf age. *Journal of the American Society of Horticultural Science* 115: 486–491.

Marcelis-van Acker CAM (1994) Development and growth potential of axillary buds in roses as affected by bud age. *Annales Botanicae* 74: 437–443.

Mortensen LM, Gislerød HR and Mikkellsen H (1992) Maximizing the yield of greenhouse roses with respect to artificial lighting. *Norwegian Journal of Agricultural Science* 6: 27–34.

Vanden Berg GA (1987) *Influence of Temperature on Bud Break, Shoot Growth, Flower Bud Atrophy and Winter Production of Glasshouse Roses*. PhD thesis. Wageningen, The Netherlands: Agriculture University.

PROPAGATION

Contents

Budding in the Field

J Mattock, Oxford, UK

Introduction

The propagation of rose cultivars is a precise operation involved with the multiplication, worldwide, of many millions of plants both as a product to grow ornamentally or as an economic crop for the propagation of cut-flower cultivars. In simple terms rose plants can be propagated either by seed or vegetatively. Because the modern rose is a product of many years (or even centuries) of selection and latterly calculated and carefully controlled hybridizing it is impractical to propagate true to type from seed. This can only be achieved by using the fabric of the plant to reproduce pure replicas, i.e. clones.

Methods

There are several techniques to achieve this and they are collectively encompassed under the title of vegetative propagation. This can be further interpreted by method. Principally this is achieved by using the whole fabric of the plant, i.e. rooting cuttings or else by transferring the bud in the axil of a leaf stalk onto a rootstock either by budding or grafting.

Cuttings

Historically multiplying roses by using hardwood cuttings has been the universal method used, and although some skill was involved (particularly concerning the age of the wood and the seasons) it was the general practice until 1800 in Europe, the United States and the Far East. In many Third World countries this method is still used to produce stocks for budding. Other forms of producing plants from cuttings include micropropagation and softwood cuttings, which are becoming economically viable in Europe and the United States.

Budding and Grafting Defined

Budding is the method used to transfer the eye of the cultivar from a stock plant onto a root system. The eye is defined as the bud lying inside the leaf axil and is not to be confused with the term flower bud. Grafting is the method used to transfer a length of cultivar onto a rootstock and is primarily used in the production of miniature cultivars.

Why Bud rather than Propagate by Using Cuttings?

Since about 1800 with the introduction of *Rosa chinensis* into breeding programmes there has been a marked reduction in the vigour of rose plants grown from cuttings, particularly Hybrid Teas, because of their inability to produce a satisfactory root system to sustain a reasonable rate of growth and (probably of greater importance) a good quality of flower. The fact that the growth of a cultivar could be improved by using a vigorous rootstock had been common knowledge since biblical times particularly in the production of fruit and other economically important plants. It was no great surprise therefore that a rose cultivar should be budded onto a compatible vigorous rootstock, e.g. *R. canina*. In the last 200 years this method of propagation has become commonplace and worldwide some 240 million plants (commonly called maidens) are produced annually through the process of budding. It is important to remember that whilst the majority of stocks give greater vigour to a variety than can be obtained by propagating by cuttings they cannot determine any other form of growth as for instance a dwarfing stock on fruit trees.

Stocks

For approximately 130 years *R. canina* grown from seed was the common stock used in most temperate regions of the world where rose plants are grown and sold. Since that time many different species, hybrid species and cultivated briars have been trialled and there are now some 18 varieties which are in common usage each with its own advantages and disadvantages (**Table 1**). The stock most extensively used in Europe is *R. corymbifera* 'Laxa'.

Table 1 Rose varieties used as stocks for propagation by budding

Standard stems	*Seedling stocks*
Rosa rugosa	*Rosa canina* (Dog rose)
Rosa canina	'Brögs Thornless'
'Pfänder'	'Inermis'
'Chessums choice' (*Chesstock*)	'Pfänder'
'Rocket' (*Harwhippet*)	'Pollmer'
	'Schmid's Ideal'
	'Superb' ('Superbe')
	Rosa corymbifera 'Laxa'
	Rosa chinensis 'Major'
	Rosa multiflora
	Rosa manetti[a]
	Rosa fortuniana[a]
	Rosa 'Dr Huey'[a]

[a] These varieties are particularly suitable for semitropical environments.

Except for standard stems all the seedling stocks are marketed as 1-year-old plants and are priced relative to the diameter of the barrel when purchased and are graded 3–5 mm, 4–6 mm, 5.8 mm and 8–12 mm. Standard stems are produced and calibrated by height in centimetres, thus a first-quality stem is 175–200 cm in height.

Raising seedling stocks for the rose trade is a specialized operation that is carried out principally in The Netherlands but also in Denmark, Germany, Poland and various of the United States of America. Stocks are also produced by taking cuttings of various forms of *R. canina* and the last four varieties in the above list. Each stock has its foibles, for instance 'Laxa' must be budded early, usually in June/July in the United Kingdom as they dry up quite quickly whereas *R. canina* will run until the end of September but is very prone to develop suckers.

Season for Budding

The operation of budding can only be successful if the cambium of both the stock and the scion material and the maturity of the bark are compatible and, in layman's terms, 'the sap is running freely'. In Europe this is a progressive trek for contract budders from April–May in Spain, May in Italy, June in France, June–July in Britain and The Netherlands and July in Germany, Denmark and Poland. A similar progression takes place in the southern hemisphere in December and January.

Planting Stocks

Good briar ground must be clean of perennial weeds with a pH of 6.5. It must be well drained and free of growing a previous crop of roses for at least 5 years. Preferably it should be a light loam wheat stubble sprayed in early September (in Britain) with 'Round-Up' (glyphosate) followed by a heavy dressing of farmyard manure after about 8 weeks. Most good soils will benefit with a deep ploughing and in slightly heavier soils a subsoiling. After a few hard frosts, cultivation with discs will produce a good tilth for planting. The stocks are planted at approximately 20 000 to 22 000 per acre usually in late winter or early spring depending on climatic conditions and the state of the soil (**Figure 1**). An extremely sophisticated planting machine, developed from a cabbage planter, is the universal implement used to plant stocks. They are generally planted at about some 17–20 cm (7–7½ inches) apart in rows and 90 cm (3 ft) row to row. Since the development of modern lifting machinery it is not uncommon to plant in rows 60 cm (2 ft) and 120 cm (4 ft) apart. This enables two rows to be lifted at once and thus reduces both time and the consolidation of the soil with heavy machinery. The stock size is controlled by cost and adaptability to the nursery soil. Generally light soils produce an acceptably buddable plant much earlier in the season than heavy tilths. Alternatively the nurseryman may want to plant a bigger size if he is producing budwood inside and fortuitously able to commence budding earlier. Stocks must be planted very precisely. The barrel of the stock, that is the operative part, should be at ground level or a fraction deeper. When planted too high the resultant plant will be subject to drying out and wind damage, if too deep the removal of excess soil to gain access to the barrel before budding is a labour-intensive and a laborious exercise.

An important cultural factor is keeping the stocks clean of weeds which can be very intrusive and if allowed to smother the stocks makes the development of the plants difficult and preparation for budding impossible. The most efficient method is control with a pre-emergent weedkiller.

Figure 1 Field of stocks preparatory to budding: early summer.

Figure 2 Potential budwood, untrimmed: midsummer.

Scion Production and Collection

Scion wood (budwood) is taken from stock plants or maiden plants propagated the previous year or budwood specifically produced in a protected environment, usually a greenhouse or tunnels. This budwood (**Figure 2**) is usually collected as close to the time of budding as is practical although invariably some compromise is made as in many instances the wood has to travel some distance and in many instances across national boundaries when introducing new varieties to other countries.

The ripeness of the budwood is generally determined by the maturity of the bark. An accepted criterion is that a stem has just borne a flower but the petals have not dropped. Wood taken from younger material is called softwood and can be difficult to handle; older wood can be what is called 'dry wood' and causes problems extracting the bud without damaging it or is usually so mature that the cambium is dry. Budwood when harvested must never be allowed to dry out, and it requires careful handling. The stems must be trimmed of any top growth, that is the flowering bract and usually two buds immediately below that junction which are generally considered blind (**Figure 3**). The leaves are also removed to leave a leafstock which makes for easy handling when budding. The resultant stem is called a stick of buds (**Figure 4**). In commercial terms when cutting budwood from a typical Hybrid Tea 1000 buds is approximately 100–150 sticks which are usually bundled in 100s and labelled. Because of the sheer numbers involved and the speed at which budding is done these days, care has to be taken storing budwood. There are many theories about removing spines (thorns). They vary from 'spinning' at the stage of leaf-trimming to spinning immediately before budding. Most budwood sent some distance either domestically or exported by air does not have the thorns removed as this operation is deemed to dry out stems; care therefore has to be taken when packing for dispatch.

Storing and Despatching Budwood

Common practice is to store budding material in cool rooms at about 2–4 °C with a high humidity. This is achieved by placing budwood between sheets of newspaper soaked in a weak solution of a common plant fungicide. Properly maintained, the budwood will be viable for up to 8 weeks, although this practice has been used to keep budwood alive indefinitely. It was common practice to keep budwood fresh plunged in buckets of shallow water but this has now been discontinued. This propagating material is regularly dispatched from nursery to nursery and country to country. Care must be taken to maintain the viability of the scions, and careful packing is paramount. Damp moss is used and/or newspaper soaked in a 2% solution of any common rose fungicide in water, wrapped in polythene and bubblewrap. The destination must be clearly marked as time is of the essence swiftly activated by overnight express. When the material is to cross national boundaries many importing countries demand the inclusion of a phytosanitary certificate which is normally obtainable from state or national agricultural department in the country of origin and in certain countries by local agricultural departments in the importing country. Wherever the scion material is eventually being used it is important to obey strict rules of hygiene and collect budwood from virus-free material and store it in a sterile environment.

Figure 3 Trimmed budwood and budding knife.

Tools

The importance of high-quality, very sharp tools and implements cannot be overemphasized. Bruised and torn wood is a haven for many problems.

Secateurs

The finest implements are side by side, that is the blade slices or produces a cut comparable with that made by a pruning knife. The very best are produced from high-quality steel. They must be maintained in a clean and sharp condition and wiped with a surgical spirit or methylated spirit between handling varieties. Some manufacturers run a reconditioning service for secateurs.

Figure 4 Stick of buds, with a bud taken, and a bud trimmed.

Budding Knives

The shape and size of this essential piece of equipment has varied little over the years. The important criterion is the quality of the blade of surgical steel but probably of equal importance is the design of the handle which must be comfortable in the hands of the operator. The handles are commonly made from bone or plastic and the majority are designed with a chisel edge in the reverse of the handle to assist with opening the cut once the blade has functioned. If a high-quality blade is used and kept sharp, the other important point is to clean the blade thoroughly at the end of a working day and wipe it with surgical spirit. Small carborundum stones are an invaluable asset to maintain a keen edge.

Tying In

When the budding operation is completed, the bud should in theory be secured by the natural elasticity of the bark; however in reality some form of mechanical method must be employed.

Raffia

Raffia is commercially obtained in skeins by weight and qualified by its quality. As its origin is the leaves of a palm of the genus *Raffia* its harvesting can be casual resulting in a very coarse fibre; therefore an examination and knowledge of the source is essential. High-quality raffia is soft and friable making it easy to use, especially when it is given a dunking in water and flagellated, then subsequently dried out to give extra softness. The skeins are divided into controllable bunches and cut up into lengths of about 40–45 cm (16 in). This is an approximate length to tie in the buds. And because of its quality raffia will quickly break down as the stock continues to grow and the bud remains dormant in the T-cut.

Rubber Ties

These are a German invention. They were introduced into the rose trade some 50 years ago and are commonly sold under the generic name of Fleishauer ties. This is a clever invention consisting of very friable polypropylene patch which degenerates quickly and is wrapped round the budded barrel and secured at the reverse with a staple. This was a remarkable invention and reduced the skill of tying in to a very simple operation.

Latex Strips

These are strips of latex which are used very much in the manner of raffia.

The important point is that all three of these materials perish after a short period allowing the budded stock to develop. If a stronger material is used it could strangulate the plant.

Preparation for Budding

When the stocks have developed they are deemed to be buddable when the barrel is a minimum of 8–10 mm in diameter. Conversely the stock with a barrel of 15–16 mm in diameter is too big and the T-cut fails to heal and splits open. To gain access to the barrel the soil may have to be removed to a depth of about 7 cm (2–3 in). Removing the topsoil around the stock to reveal the barrel can be a time-consuming, labour-intensive exercise which was originally done manually with a narrow hoe. Machines have now been developed to blow off this excess soil, leaving a clean barrel.

Method of Budding

The bud or eye is removed with sufficient bark surrounding it to ensure its safe transfer (**Figure 5**). The knife is inserted 1 cm (½ in) above the bud to make a shallow cut downwards when the cut bud is then torn off the stem and reversed to remove the thin sliver of wood which is still attached to the bud. Finally the bud is trimmed leaving about 2–3 cm (1 in) of bark carrying the bud. The stock is then prepared by making a T-cut in the barrel (**Figure 6**), a simple operation which commences with a horizontal cut some 10–15 mm below the top growth followed by a vertical cut into the stock across the first cut. There is considerable skill employed with this operation as it must not be too

Figure 5 Prepared bud.

heavy and cut into the wood of the plant but at the same time with a practised 'flick' open up the cut to receive the prepared bud, in short lifting the bark. Once this operation is complete the bud is inserted (**Figure** 7) and if necessary trimmed to be a comfortable fit. The bud is then immediately secured with a raffia or rubber tie (**Figure** 8). Thus the operation is completed and quickly heals up (becomes callused), usually within 2 weeks.

Contract Budders

The operation of budding has now become a thoroughly skilled operation and is now a time-and-motion directed exercise. Whereas at one time the staff of a rose nursery consisted of experienced budders demanding an expensive wage structure it has become common practice to employ itinerant contract budders who work at amazing piece/time rates and have been known to bud at the speed of 4000 per day, always accepting that the budwood material has been previously thorned and that a 'patcher', follows on.

The Following Stages

When budded very early these budded stocks can go into growth by the end of the summer; 'shot out' is the term used. To experienced nursery managers this is a mixed blessing: if there is much autumn growth there is the temptation to lift these plants and pot on for instant sale, which has become common practice particularly when the stocks were budded early and headed back very shortly afterward (**Figure** 9). Alternatively they can be very soft and prone to frost damage during the winter.

In the normal course of events the budded stocks lie dormant and it is the practice in some countries to earth them up during very hard weather but the success of this operation is a debatable point. In late winter the stock is headed back to just above the bud leaving above ground the stump of the stock (**Figure** 10). The budded stock will develop with considerable

Figure 6 Stock with T-cut opened ready to receive bud.

Figure 8 Budded stock with Fleishauer tie.

Figure 7 Bud *in situ*.

Figure 9 Heading back rose stock: late winter.

acceleration as the season progresses (**Figure 11**) and mature into a good flowering saleable plant by early autumn (**Figure 12**). As the budded stock develops in late spring/early summer it is prone to two impediments. The most damaging is 'blowouts': as the buds develop they are prone to wind damage, that is the union of the stock and cultivar breaks down and is destroyed irrevocably. To obviate this it is common practice to top the developing shoots down to about 10 cm (4 in) with a very sharp knife. This procedure is usually completed by early summer and obviously reduces the risk of wind damage but in addition probably produces an improved mature plant with several breaks. The other problem is that some geographical areas are subject to late spring frosts which can devastate a field of plants. Although very rarely will this kill a young plant, it can delay the maturing of a rose crop and many early summer flower shows have experienced postponement because of this unfortunate climatic catastrophy. The remedy is to allow the young plants to recover naturally; never remove frosted growth.

Figure 10 Headed back rose stock.

Figure 11 Buds beginning to grow: late spring.

Diseases

Whilst freshly planted stocks have a remarkable resilience to most of the normal diseases of garden roses they can be prone to two enemies. Powdery mildew (*Podosphaera pannosa* var. *rosae*) can sometimes have a devastating effect on *R. canina* particularly from midsummer. The remedy is to spray with accepted systemic rose fungicides. Rose rust (*Phragmidium mucronatum*) is the great enemy of *R. corymbifera* 'Laxa' and can debilitate seedling stocks from midsummer. Systhane has recently proved a deterrent if applied thoroughly. As a systemic fungicide it is remarkably efficacious provided the plants are completely wetted. When budding has been completed a form of *Botrytis* can sometimes attack the open cut and the bud is killed before it is callused and healed. Known in the United Kingdom as 'black bud' this disease is encouraged by lack of attention to plant hygiene. Apart from systematically sterilizing all equipment, e.g. knives, secateurs, etc., any water used in storing budwood must have a 2% solution of a rose fungicide.

Predators

Where rabbits, hares and deer are indigenous they can do tremendous damage and treat a rose crop as a staple diet. Many sophisticated methods of excluding them have been used but a strong wire fence is the only 100% practical solution. For rabbits and hares a small-mesh wire fence 130 cm high and very strong with at least 25 cm buried in the ground is the only remedy and deer can only be dissuaded by the conventional forestry design of fence some 2 m high. Constant vigilance must be maintained particularly where badgers can demolish lengths of fence overnight.

In very cold winters fieldmice have been known to eat the dormant eyes of a budded stock. The only

Figure 12 Field of maiden roses: midsummer.

solution to this problem is to earth up the stock at the beginning of the winter and keep the stocks completely free of weeds thus preventing access to the barrel of the stock.

See also: **Propagation**: Own-Rooted Cuttings; Stenting. **Rootstocks**: Scion–Rootstock Relationships; Rootstock Breeding; Usage of Rootstocks; Clonal Rootstocks; Seedling Stocks.

Further Reading

Garner RJ (1988) *The Grafter's Handbook*, 5th edn. London: Cassell.

Hartmann HT, Kester DE, Davis FT and Geneve RL (1997) *Plant Propagation: Principles and Practices*, 6th edn. Englewood Cliffs, NJ: Prentice-Hall.

Own-Rooted Cuttings

J M Costa, Horticultural Production Chains Group, Wageningen University, Wageningen, The Netherlands and Fundação para a Ciência e Tecnologia, Lisbon, Portugal

P A Van de Pol, Wageningen University, Wageningen, The Netherlands

Introduction

Asexual reproduction of roses by stem cuttings is probably the simplest method of propagating roses. Most old roses, English roses, miniatures and some rootstocks are commonly propagated by cuttings because they can easily form roots and, further, they grow vigorously on their own roots. Some hybrid teas (cut roses) and floribundas can also grow vigorously on their own roots, although they are often propagated by grafting on to different rootstocks. Every year, millions of roses are propagated by cuttings to support the needs of different sectors, like the cut-rose flower sector (in The Netherlands, Israel, Ecuador and Kenya), garden roses and rootstock production in open fields (e.g. in the USA, Germany, France, Italy, Spain) or pot roses (e.g. Denmark, The Netherlands, USA) (**Figure 1**). In this article, some of the main anatomical, physiological and morphological aspects related to adventitious root formation in rose cuttings and the main environmental factors influencing this process are described and discussed.

Types of Stem Cuttings

Three main types of cuttings can be used in rose propagation: hardwood, semihardwood and softwood.

Hardwood Cuttings

These are cuttings made from mature, quiescent, lignified shoots after the leaves have abscised. Hardwood cuttings are normally used for field propagation of rootstocks (e.g. *Rosa multiflora, R. chinensis* 'Indica Major') or other vigorous types, like climbers and old English varieties. These cuttings often contain a great amount of starch. Their length can vary between 20 cm and a maximum of 100 cm for propagation of tree roses (standards). Prior to the insertion of these cuttings into the ground, all the lower basal axillary buds should be removed to prevent formation of suckers (shoots growing out from the rootstock part of the plant below the graft). In colder climates the hardwood material is harvested in late autumn or early winter and stored for planting in spring, whereas in mild climates (e.g. south Europe, California) cuttings are harvested and planted in autumn (**Figure 1**).

Semihardwood Cuttings

These are cuttings with a partly lignified stem, taken from shoots after full growth (including flowering) of a new shoot. This material is commonly used for the propagation of garden roses (summer cuttings) and cut roses (cuttings taken all year round when mother plants are grown in a greenhouse). Single or multinode stem cuttings with one or more leaves can be used but, for reasons of economy of plant material, single-node stem cuttings, 4–6 cm in length with a single leaf, are often used in cut-rose propagation (**Figure 1**).

Softwood Cuttings

These are cuttings propagated from nonlignified material, with a very low dry-matter content, taken after cessation of the new spring growth, in the case of garden plants, or from newly developed shoots before flowering, in the case of pot roses. Softwood cuttings, in particular shoot tips, require more careful environmental control during propagation because they are very soft, have scarce reserves and their young leaves are very sensitive to wilting. Miniature pot roses are usually propagated by softwood stem cuttings (single node, with a stem length of about 3–4 cm (**Figure 1**), or multinode with a maximum length of about 10 cm when availability of mother plant material is not a limiting factor). In some cases, cut roses and garden roses may also be propagated by softwood material provided that the proper environmental conditions are guaranteed.

Figure 1 (a) Single-node leafy softwood cutting of cut roses. Propagation of cut roses in greenhouse in (b, c) the Netherlands and (d) Ecuador (courtesy of Walter Barentzen). (e, f) Propagation of cut roses in a climate chamber in the Netherlands. (g) Recently rooted cuttings before hardening. (h) Greenhouse facilities for hardening rooted cuttings of cut rose. (i) Planting material of cut roses in rockwool cubes (courtesy of Berry Verlaan) and (j) coco fibre (courtesy of Walter Barentzen). (k) Young rooted plant 1 week after plantation in greenhouse. (l) Propagation of leafless stem hardwood cuttings of the rootstock 'Dr Huey', Wasco, TX (courtesy of John Karlick) and of (m) *Rosa multiflora* in south Europe. Stem segment (3–4 cm) of single node cuttings of pot rose (n) 5 days and (o) 11 days after severance (courtesy of Bert Rombouts).

Advantages and Disadvantages of Rooted Cuttings Compared to Grafted Plants

Advantages

Roses on their own roots often have an indefinite lifespan as there is no problem with the formation of suckers and they are also cheaper. For cut-rose propagation, cuttings are cheaper than grafted plants because they require less labour and there are no costs for production of the rootstock.

Disadvantages

Some rose cultivars are not adapted to all kinds of soil or are not resistant to diseases or nematodes (e.g. *Meloidogyne* spp.). This can be very important for garden roses and in particular for cut roses when they are not cultivated in artificial substrate. Plants from cuttings take longer to reach a saleable size (mostly for garden roses) and not all the cultivars/varieties can be grown vigorously without a rootstock. When less vigorous, they often produce fewer and smaller blooms than grafted plants, although this greatly depends on the genetical vigour characteristics of the cultivar. Propagation by cuttings also has the inconvenience that it requires a larger number of stock plants and can be more expensive than propagation by budding in the field because it requires investment in expensive facilities.

Basic Organogenetic Processes During Propagation

Adventitious Root Formation

Roots arising on stems, leaves or explants from these tissues are termed adventitious. In rose, adventitious roots are formed in response to wounding because root primordia are not preexisting. The formation of adventitious roots is a process characterized by different developmental phases, involving specific anatomical developments. In a simplified but generally accepted way, adventitious root formation can be divided into two main phases: initiation and elongation (**Figure 2**). The initiation phase includes the dedifferentiation of specific cells that become meristematic, the formation of root initials (groups of meristematic cells) and the differentiation of organized root primordia and xylem vessels. The elongation phase is characterized by growth of the root primordia along the cortex until emergence, enlargement of vessels and acquisition of functionality of the new roots by connection of

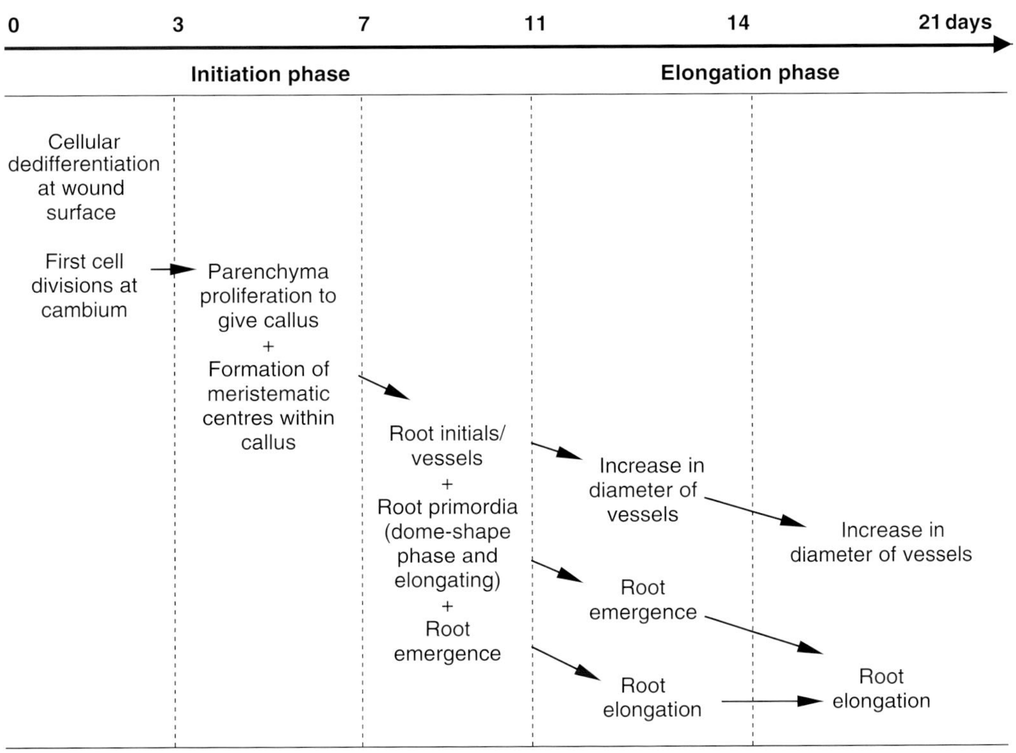

Figure 2 A schematic illustration of the phases and related anatomical events during the rooting process for single-node leafy stem cuttings of *R.* 'Madelon' not treated with auxins and propagated at a temperature of 25 °C, under a light intensity of $16\ W\ m^{-2}$ and photoperiod of $16\ h\ day^{-1}$. Reproduced from Costa JM (2002) *The role of the leaf in growth dynamics and rooting of leaf stem cuttings of rose.* PhD thesis, Wageningen University, the Netherlands with permission.

root xylem vessels to the stem vessels of the cutting (**Figure 2**).

A relevant anatomical development following severance and preceding root initiation on rose cuttings is the proliferation of undifferentiated parenchyma tissue (callus) at the basal end of the stem segment by activation of cambium activity. Callus formation increases the size of the cortex and is externally visible by the swelling at the base of cuttings. Callus proliferation, when not abnormally excessive, may be considered as an early indicator of a cutting's viability and a prerequisite for optimal rooting because the new vessels and root initials are formed within the new parenchyma tissue (**Figure 2**).

Axillary Bud Outgrowth

Bud outgrowth in rose plants is a typical hormonal process, promoted by the supply of cytokinins formed in the roots (root tips). Thus, rooting normally has a positive influence on bud outgrowth. Growth of the axillary bud is also positively influenced by the supply of carbohydrates because it requires energy. Therefore, under source-limited conditions, bud outgrowth should occur preferentially after root formation because roots have less capacity to compete for assimilates than growing buds. This is supported by findings showing that rooting is negatively correlated with axillary bud-break during the first days of propagation although, at a later stage, root elongation is accompanied by larger growth of the axillary shoot.

Some of the Main Factors and Processes Influencing Propagation by Cuttings

Internal Factors

Carbohydrates and photosynthesis Carbohydrate relations during propagation of stem cuttings are characterized by four main aspects: (1) the existing carbon (reserves) at severance; (2) the carbon fixed by leaf photosynthesis; (3) the carbon lost due to maintenance respiration (of leaf or stem) or growth respiration (formation or growth of the callus, the roots and the axillary bud); and (4) carbon partitioned between preexisting sinks (stem and leaf) and new sinks (roots and the axillary bud/primary shoot). The ability of cuttings to sustain a positive carbohydrate balance determines their survival and further growth, especially when the initial reserves are limiting, as happens with semihardwood or softwood material. Depletion of carbohydrates at the rooting zone of cuttings, by leaf removal, very low light intensity (1.5 W m^{-2} photosynthetic active radiation) or by low CO_2 (80–100 p.p.m.) has been shown to increase the incidence of stem rot and reduce callus and root formation in single-node stem cuttings of *R.* 'Madelon' and 'Sonia' see 'MEIHELVET'. The ability to photosynthesize in the first days of propagation is not only a prerequisite for survival but also a requisite for optimal rooting and growth of softwood or semi-hardwood cuttings. Furthermore, growth (dry-weight accumulation) and rooting (number and dry weight of roots) were linearly related to the integral of leaf photosynthesis of cuttings during propagation.

Leafless hardwood cuttings have a different logic of carbohydrate relations because maintenance and initial growth of this type of cuttings are totally supported by the reserves existing at the moment of severance. Therefore, starch content at severance can be a good predictor of growth of cuttings, as has been found for hardwood cuttings of *R. multiflora*.

Water status Severance from the mother plant represents a disruption of the normal water relations in cuttings. Leafy cuttings are particularly sensitive to this process because they can lose water via leaf transpiration, which does not happen with leafless hardwood cuttings. Negative water balances in leafy cuttings may also result from the combination of excessive leaf transpiration with difficult water uptake due to poor hydraulic conductance of the stem or due to very low substrate water content. Leaves are the most sensitive structure of cuttings to water stress, especially when very young, as happens with leaves of softwood material. A quite common visible consequence of water stress is leaf yellowing, which can be followed by leaf abscission and death of cuttings in more extreme cases. Even when water stress does not cause death, it decreases the photosynthetic activity of cuttings and promotes synthesis of hormones, like abscisic acid (ABA), which decrease rooting. At a more cellular level, it is known that water stress diminishes turgidity of cells, limiting their capacity to divide and, consequently, to initiate roots.

Hormones and nutrients Auxins are the most important group of hormones involved in adventitious root formation. They are synthesized in leaves, buds and growing shoots. Auxins promote cell division and may influence carbohydrate metabolism by increasing starch hydrolysis. However, root formation is mainly the result of a balance between auxins and other hormones (cytokinins, gibberellins) (*see* **Cell, Tissue and Organ Culture**: Micropropagation; Rooting of Micropropagated Shoots). This balance influences not only root initiation but also further elongation of roots and growth of the axillary shoot and consequently, the future shoot-to-root ratio of the new rooted plant. Furthermore, the effectiveness of auxins may be conditioned by the physiological status of cuttings in terms of carbohydrates and/or other nutrients.

External Factors

Mother plant management The yield of cuttings in rose depends on the number of buds breaking from the stock plants. Thus, practices that increase mother plants' vigour (e.g. adequate substrate and fertilization, increased light integrals, CO_2 air enrichment and adequate temperature regimes) (*see* **Production and Marketing:** Overview of Roses and Culture) should increase the yield of good-quality cuttings. Cuttings taken from mother plants grown under different conditions (e.g. by different growers or in different countries) showed different rooting and growth. In practice, this is supported by findings showing that cuttings harvested from mother plants grown in a greenhouse rooted better than cuttings taken from mother plants grown outdoors. Sanitation should be a priority of the mother plant management to keep stock plant material free from pests or diseases (e.g. *Botrytis, Podosphaera pannosa*). However, propagators have to be aware that the application of certain biocides to mother plants may promote leaf abscission during propagation.

Harvesting of mother plant material In general, better rooting and higher uniformity have been achieved in cut-rose propagation by harvesting cuttings from the intermediate part of shoots, at the time the colour of the flower buds starts to be visible and sepals are reflexing. However, the effect of the position of the cutting within the flower shoot also depends on the cultivar. For example, the cut-rose cultivar 'First Red' rooted better when cuttings were taken from the most apical part of the shoot. On the other hand, hardwood cuttings of *R. multiflora* and semihardwood cuttings of *R. centifolia* rooted better when harvested from both the intermediate and the basal parts of the shoots. The possibility of using less ripe material should still be considered because it allows faster rooting (e.g. *R. canina* 'Sturcinq', a selected seedling from *R. canina* 'Inermis') given that the proper propagation environment (light, humidity) is guaranteed. The use of blind shoots has also been shown not to have a negative effect on rooting of cuttings of certain cultivars when compared to normal flower shoots.

Another important aspect concerns the process of severance itself. Since cambium activity is responsible for proliferation of the new parenchyma tissue (callus) where roots are initiated (**Figure 2**), the damage caused to the stem structure and cambium layer should be reduced to a minimum during severance. Thus, wounding should be sharp, especially for softwood material, which is more prone to damage. However, wounding has been shown to promote rooting of hardwood cuttings, as reported for *R. damascena*, *R. centifolia* or *R. multiflora*.

Effect of leaf and stem morphology Leaves are very important for the propagation of softwood or semihardwood leafy stem cuttings of rose because they guarantee survival, rooting and growth of the axillary bud via the supply of carbohydrates, auxins, nutrients or other rooting factors. Leaf area is probably the most important morphological parameter of leafy stem cuttings influencing their rooting performance. This has been shown by experiments with different cultivars of *R. polyantha* and also with *R.* 'Sonia'. More recently, experiments involving the total or partial removal of the original leaf of cuttings of *R.* 'Madelon' at different moments of propagation showed that rooting and dry-weight accumulation were linearly related to leaf area duration (leaf area in cm^2 multiplied by the number of days the leaf is attached to the cutting). Likewise, applications of sodium silicate ($50-100\ mg\ l^{-1}$) to increase leaf resistance against black spot have been shown to enhance leaf retention time and improve rooting of rose cuttings. Thus, keeping the original leaf intact and healthy during propagation is a prerequisite for optimal rooting and growth of cuttings, and, therefore, for good-quality planting material.

Stem morphology is also important, in particular for leafless hardwood cuttings because regrowth depends entirely on the reserves existing in the stem at severance. Thinner cuttings have smaller storage capacity than the thicker ones, which can cause death before cuttings form roots. Also double or multinode leafy cuttings of *R.* 'Motrea' have been shown to produce larger plants and better root systems compared to single-node cuttings due to the beneficial combination of larger photosynthetic and storage capacities.

Hormone treatments To guarantee faster and more uniform rooting, rooting hormones are often used in commercial propagation. Faster rooting reduces exposure to stress and diseases and improves uniformity. Indolebutyric acid (IBA) and naphthaleneacetic acid (NAA) are the most commonly used synthetic auxins (*see* **Cell, Tissue and Organ Culture**: Micropropagation; Rooting of Micropropagated Shoots). More lignified material usually requires higher auxin concentrations than softer material, but high concentrations may also be phytotoxic and decrease rooting (**Table 1**). Leaves have been shown to have a synergetic effect on the activity of exogenous auxin. In fact, the original leaf needs to be present and photosynthetically active to promote rooting of auxin-treated leafy stem cuttings of *R.* 'Madelon' and *R.* 'Sonia' because

Table 1 Effect of different concentrations (p.p.m.) of indolebutyric acid (IBA) on rooting and bud outgrowth in cuttings of *Rosa* 'Sonia' propagated under a photoperiod of 17 h (Philips 160 W) and a temperature of 28 °C, substrate sand/peat (1 : 1) (v/v) measured after 15 days of propagation

	IBA concentration (p.p.m.)						
	0	*250*	*500*	*1000*	*2500*	*5000*	*10000*
Stem rot (%)	0	13	28	25	40	62	
Bud-break (%)	90	54	61	46	21	21	11
Longest root (mm)	29	57	49	37	30	24	22
Root number	3.6	17.2	23.2	28.2	17.3	21.8	15.1

Adapted from Abbing R (1987) *De invloed van een auxine-behandeling op stekken van de roos 'Sonia'* (in Dutch). Thesis, Department of Horticulture, Wageningen University, the Netherlands, with permission.

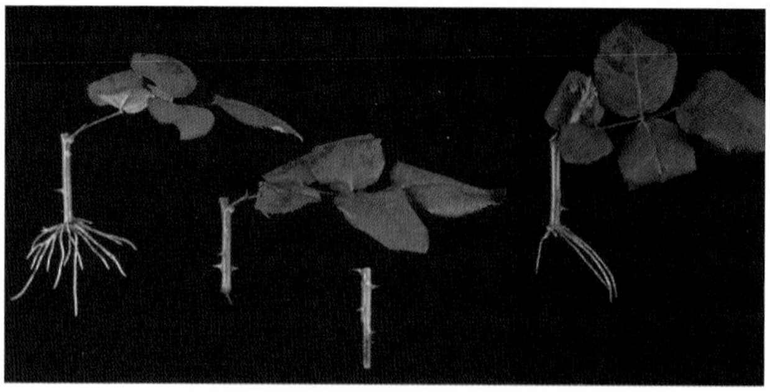

Figure 3 The combined effect of leaves and auxin treatments (indolebutyric acid (IBA) 0.4%) on the rooting of single-node leafy stem cuttings of *Rosa* 'Madelon' measured after 14 days of propagation. The cuttings were propagated at 25 °C under a light intensity of 16 W m^{-2} for 16 h day^{-1}. Left to right: IBA-treated cutting propagated under standard light; IBA-treated cutting with the original leaf covered with aluminium foil; IBA-treated cutting without the original leaf, and nonauxin-treated cutting under standard light conditions.

leafless or covered leaf cuttings or cuttings propagated in darkness failed to root and/or were affected by stem rot (**Figure 3**).

Another important effect of auxins is the inhibition of the axillary bud outgrowth (**Table 1**) which seems to be mediated by ethylene synthesis. This effect is very useful in practice, especially when propagating cultivars with an early bud-break. IBA and NAA are available commercially in talc or liquid formulations. A list of several possible treatments with auxins is presented in **Table 2**.

Environmental factors (light, temperature, CO_2, watering, substrate and nutrients) Optimal light levels (intensity and photoperiod) for propagation of leafy cuttings are difficult to determine because they depend on several factors, like the growing conditions of the mother plants, the cultivar and the levels of other factors such as temperature and relative humidity. Callus and root formation in leafy stem cuttings of *R.* 'Madelon' occurred at very low light integrals (3.4 MJ m^{-2}) but responded positively to increased light integrals within the range of 0–27.4 MJ m^{-2}. Air temperatures of 20–30 °C have been shown to be adequate for the propagation of most roses by cuttings (**Table 3**), although they are dependent on the levels of other environmental factors such as light and relative humidity. Excessively high air temperature (35 °C) has been shown to reduce severely rooting of cuttings *R.* 'Motrea' when applied for a 2-day period during the first 9 days of propagation. Bottom heating in the propagation beds hastens and improves rooting of roses because it accelerates and increases meristematic activity and speeds up root initiation. Substrate temperatures ranging between 22 and 27 °C are suggested in diverse literature on rose propagation (**Table 3**). Temperature also influences axillary bud-break. As an example, it has been shown that combining low air temperature (15 °C) with high substrate temperature (27 °C) reduced bud growth in *R. polyantha* (miniature rose) cuttings.

CO_2 enrichment is known to promote rooting of leafy stem cuttings, mostly by promoting the photosynthesis activity of cuttings. Levels between 500 and 1200 p.p.m. are reported in the literature although values between 700 and 1000 p.p.m. seem to be the most used either in the propagation of cut or pot roses.

Water stress should be minimized during the propagation of leafy stem cuttings by reducing water losses via transpiration to a minimum and optimizing water uptake. Mist or fog systems in combination or not with shadowing screens are common ways to ensure a relative humidity close to saturation levels, decrease leaf temperature and maintain the turgidity of leaves. The rooting substrate should guarantee, in turn, good aeration and drainage as rooting involves respiratory activity, which requires oxygen. In general, 25–40% volume of air is the ideal proportion of air in the substrate. In practice, the combination of organic (peat, sphagnum moss, coco fibre) and nonorganic (sand, vermiculite, perlite) is often used (**Table 4**). The mix of perlite with peat is the most common substrate for propagation of pot roses, although rockwool or foams can also be used. In the propagation of cut roses by leafy stem cuttings, coco fibre and rockwool are the main substrates. Furthermore, the salinity levels of the

Table 2 Auxin treatments proposed in the literature to promote rooting of rose cuttings

Rose species/cultivars	*Treatment*	*References*
R. damascena	20 s dip in solution 2000 p.p.m. IBA/NAA, with a 10 s dip in H_2SO_4 (2 mol l^{-1})	Khosh-khui and Tafazoli (1979) *Sci. Hortic.* 10: 395–399
R. centifolia	3500 p.p.m. IBA in a 0.03% solution of potassium hydroxide	Al-Saqri and Alderson (1996) *J. Hortic. Sci.* 71: 729–737
R. 'Frisco'	24 h dip in a solution of NAA 10^{-5} mol l^{-1}	Baas and Gislerød (1996) Beworteling van rozenstek is steenwol kan beter. *Vakblad voor de Bloemisterij*, 36: 36–37
R. 'Madelon'	0.4% IBA talc preparation	Costa *et al.* (2001) *Acta Hortic.* 547: 167–174
R. polyantha	5 s dip in a solution of 1800 p.p.m. IBA	Moe (1973) *Acta Hortic.* 31: 35–51
R. 'Frisco', *R.* 'Gabriella'	0.15% Captan + 0.3% IBA	Bredmose and Hansen (1997) *Acta Hortic.* 424: 23–28

IBA, indolebutyric acid; NAA, naphthaleneacetic acid.

Table 3 Substrate and air temperature set points, light intensity (photosynthetic active radiation: PAR) and photoperiod proposed in the literature for the propagation of cut and pot roses by cuttings

Rose species/ cultivars	*Substrate temp. (°C)*	*Air temp. (°C)*	*Light intensity (PAR) W m^{-2}*	*Photo period (h)*	*References*
Rosa spp.	22–25	25–30	15–20 (SON-T/HPI-T)	16–20	Dubois *et al.* (1990) *Vakblad Bloemisterij* 45: 26–31
R. 'First Red', *R.* 'Frisco'	23	20	4.5 (SON-T/HPI-T)	18	Hazendonk *et al.* (2000) *Proefstation voor bloemen en glassgroenten Report* n. 271
R. 'Madelon'	23–25	25	16 (SON-T/HPI-T)	16	Costa and Challa (2002) *Sci. Hortic.* 95: 111–121
R. polyantha (miniature)	21–22	21–27	4	24	Moe (1973) *Acta Hortic.* 31: 35–51; Pemberton *et al.* (1997) *Production of Pot Roses*. Timber Press Growers Handbook Series, Vol. 7. Timber Press, Portland, Oregon
R. polyantha (miniature)	24–27	18–21			Borowski *et al.* (1987) *Acta Agrobot.* 40: 27–39
R. 'Sonia'		22	10–12 (SON-T)	20	Berninger *et al.* (1990) *Agronomie*, 10: 787–796

SON-T, high pressure sodium lamps, Philips SON-T Agro 400 W; HPI-T, metal halide vapour are lamps, Philips HPI-T 4000 W.

Table 4 Different types of substrates used in the propagation of rose by cuttings

Rose species/cultivars	*Substrate*	*References*
Rosa centifolia	Peat : perlite : vermiculite (1 : 1 : 1) (v/v/v)	Al-saqri and Alderson (1996) *J. Hortic. Sci.* 71: 729–737
R. 'Royalty'	Perlite : sphagnum moss peat (3 : 1)	Sun and Bassuk (1993) *J. Am. Soc. Hort. Sci.* 118: 638–643
R. 'Madelon'	Sand: peat (1 : 1)	Costa *et al.* (2001) *Acta Hortic.* 547: 167–174
R. 'Red Success'	Vermiculite or sand	Pivetta *et al.* (1999) *Acta Hortic.* 482: 333–338
R. 'First Red', *R.* 'Frisco'	Coco fibre plugs	Hazendonk *et al.* (2000) *Proefstation voor bloemen en glassgroenten Report* 271
R. 'Frisco', *R.* 'Gabriella'	Rockwool cubes (7 × 7 × 6.5 cm)(l,w,h)	Bredmose and Hansen (1997) *Acta Hortic.* 424: 23–28

substrate should not be excessively high nor should it contain unwanted seeds.

Fertilization during propagation improves the growth of cuttings and would be more efficient after roots are formed. The use of slow-release fertilizers and the addition of vesicular–arbuscular mycorrhizal fungi (VAMF) inoculum to the substrate are possible ways of increasing efficiency of the applied fertilizers. Consequently, the use of mycorrhizal fungi has been shown to increase the percentage of the better-rooted cuttings in pot-rose propagation and to improve the growth of rooted cuttings of the rootstock *R. multiflora*.

Sanitation before and during propagation *Botrytis cinerea* and *Cylindrocladium scoparium* are probably the two most serious fungi affecting cuttings during propagation. Other important fungi causing losses during propagation are *Podosphaera pannosa, Verticillium, Fusarium, Phytophthora* and *Pythium* spp. The crown-gall disease caused by *Agrobacterium tumefaciens* and the rose mosaic virus are also easily spread by propagation since they are difficult to detect (e.g. rootstocks of *R. chinensis* 'Indica Major' for the crown gall). Therefore, the use of certified clean mother plant material is the prime way of avoiding losses during propagation and the spread of diseases to the field or greenhouse after transplantation. To prevent incidence of diseases during propagation the propagation area (greenhouse) and equipment (knives, used flats and trays) should be periodically disinfected (e.g. with chlorine solutions). The cuttings may also be disinfected with fungicide solutions before being inserted in the substrate. The water should be clean and the substrates disinfected. Moreover, optimal environment control during propagation is another way of controlling diseases. Healthy plant tissue is less susceptible to infection by fungi like *Phytophthora* or *Fusarium*. Therefore, proper environment control should avoid any physiological predisposition of the material for disease. Low content of carbohydrates in stem tissues of rose cuttings has been suggested to increase susceptibility to *Fusarium* whereas the supply of sugar solutions (glucose and sucrose) to cuttings delayed the appearance of stem black rot in leafless cuttings of *R.* 'Madelon'.

Trends in Propagation by Stem Cuttings

At present, one of the relevant current trends is the use of cuttings as the preferred method to propagate either garden or cut roses, instead of grafted plants. However this may change, especially in the case of cut roses, as soon as a new vigorous rootstock appears in the market (*see* **Propagation**: Budding in the Field).

Moreover, the increasing demand for more environmentally friendly horticultural practices requires that the future development of rose propagation be focused in two main topics: the use of more environmentally friendly practices and greater quality control in the propagation chain. The first aspect will force propagators to use more 'natural' methods (e.g. no use of hormones, integrated pest and disease management, the use of mycorrhizal fungi). Recent research into environmentally friendly practices in cut-rose propagation has shown that, without application of auxin, rooting is excessively delayed and this is not acceptable in practice. However, the inability of rose cuttings easily to form roots will probably be solved in the near future. Breeding companies, for example, are selecting for easy-to-root cut-rose cultivars and genetic studies have shown that the introduction of ROL genes (genes of *Agrobacterium rhizogenes* responsible for the phenotypic effects in the hairy-root disease) enhances the sensitivity of stem tissues to auxin and improves rooting either *in vivo* or *in vitro*. The use of genetically modified material may, however, be limited by the negative perception of genetically modified organisms by the general public. The use of auxin-conjugates, which permit the slow release of auxin and improve rooting, may also increase in the future. Micropropagation has also been shown to solve the difficulties of rooting *in vivo* (the clone of *R.* 'Sturcinq' is an example) and may thus be more widely used in the future as a tool for rose propagation (*see* **Cell, Tissue and Organ Culture**: Micropropagation).

It is also our opinion that higher quality in the propagation chain (at least for leafy stem cuttings) will be achieved in future by guaranteeing a stricter control of pests, diseases of the mother plants and by optimizing the control of the environment (light, CO_2, temperature and humidity) during the propagation. This last aspect would permit the optimization of the photosynthetic activity of cuttings, which has been considered a major determinant of the rooting and growth of leafy stem cuttings.

Another trend is the automatization of the propagation process to reduce labour costs. This should be mostly a trend for countries where labour is expensive. In The Netherlands, for example, robots are already being used in the propagation of pot roses. Finally, the search for methods to assess the internal quality (potential for growth) of cuttings, such as chlorophyll fluorescence imaging, may also be a tool to increase the efficiency and quality of the propagation process.

Acknowledgements

This work has been financed by a scholarship of the Program Praxis XXI from the Fundação para a Ciência e Tecnologia (FCT), Lisbon, Portugal.

See also: **Cell, Tissue and Organ Culture:** Micropropagation; Rooting of Micropropagated Shoots. **Morphology and Anatomy:** Roots. **Production and Marketing:** Overview of Roses and Culture.

Further Reading

Baas R, Gislerød HR, Van den Berg D and Roeber RU (1997) Do roots of rose cuttings suffer from oxygen deficiency during propagation in rockwool? *Acta Horticulturae* 450: 123–131.

Bredmose N and Hansen J (1997) Influence of propagation material and method on regeneration, growth and flowering of cut rose cultivars Frisco and Gabriella. *Acta Horticulturae* 424: 23–28.

Costa JM (1994) *Ethanol and Adventitious Root Formation in Cuttings of* Rosa *cv Motrea as Influenced by High Temperature Treatments and Substrate Water Contents.* Thesis, Wageningen University, the Netherlands.

Costa JMRC (2002) *The Role of the Leaf in Growth Dynamics and Rooting of Leafy Stem Cuttings of Rose* PhD thesis, Wageningen University, the Netherlands.

Costa JM, Challa H, Van Meeteren U and Van de Pol PA (2001) Photosynthates: mainly stored and yet limiting in propagation of rose cuttings. *Acta Horticulturae* 547: 167–174.

Davies FT, Hembrik CE, Pemberton HB and Faan Y (1987) Grafting and adventitious root formation of Texas field rose bushes. *Acta Horticulturae* 189: 89–100.

De Hoog J (1998) *Handbook for Modern Greenhouse Cultivation.* Aalsmeer, The Netherlands: PPO.

Harrison-Murray RS and Howard BH (1998) Environmental requirements as determined by rooting potential in leafy cuttings. In: Cockshull KE, Gray D, Seymour GB and Thomas B (eds.) *Genetical and Environmental Manipulation of Horticultural Crops*, pp. 75–94. Wallingford: CAB International.

Hartmann HT, Kester DE, Davies FT and Geneve RL (1997) *Plant Propagation, Principles and Practices*, 6th edn. Upper Saddle River, NJ: Prentice Hall.

Loach K (1988) Water relations and adventitious rooting. In: Davis TD, Haissig BE and Shankla N (eds.) *Adventitious Root Formation in Cuttings*, pp. 102–116. Portland, OR: Discorides Press.

Moe R (1973) Propagation, growth and flowering of potted roses. *Acta Horticulturae* 31: 35–51.

Pemberton HB, Kelly JW and Ferare J (1997) *Production of Pot Roses.* Timber Press Growers Handbook Series, vol. 7. Portland, OR: Timber Press.

Pivetta KFL, Pereira FM, Banzatto DA and Graziano P (1999) Effect of type of cuttings and indolbutyric acid on the rooting of rose (Rose sp. 'Red Success') leafy cuttings during two seasons. *Acta Horticulturae* 482: 333–338.

Stoltz LP and Anderson RG (1988) Rooting of single node cuttings of roses. *Acta Horticulturae* 227: 230–235.

Van Labeke MC, Dambre P, Bodson M and Pien H (2001) Developmental changes in carbohydrate content in young rose shoots (*Rosa hybrida* 'Frisco'). *Acta Horticultarae* 547: 193–201.

Stenting

P A van de Pol, Wageningen University, Wageningen, The Netherlands

Introduction

Stenting is a method of quick propagation of roses based on grafting a cultivar on an unrooted cutting of a rootstock, resulting in a complete plant in 3 weeks. Stenting (to stent) is a contraction of the Dutch words 'stekken' (to strike a cutting) and 'enten' (to graft). The stented plant is called a 'stentling'. Other authors call this technique cutting–grafting. The technique of combining cutting and grafting is not new. It was described in 1896 for the combination of *Morus alba* and *M. rubra*. Stenting of roses is applicable all year-round and therefore is ideal for flexible propagation, selection of new rootstocks and studying scion–rootstock interrelations and development of the graft union.

From a physiological point of view a stentling is a complicated plant: wound healing, root formation and bud development occur simultaneously and interact, competing as well as cooperating. Meanwhile physiological processes like photosynthesis, respiration and hormone synthesis play an important role.

In this article the factors influencing vascular development at the graft union and starch accumulation and depletion around the graft union during the early development of a rose stentling are presented and discussed.

Grafting Methods

The technique used is whip grafting. A piece of stem with a good five-leaflet leaf and a single dormant axillary bud is used as a scion. A piece of stem consisting of a single internode without buds or leaves is used as a rootstock. The rootstock type must be easy to root. The general stage of ripeness of the material must be flowers just opening for the scions and for the rootstocks well-developed leaves and thorns which can be broken off easily.

The top of the rootstock internodes and the basal part of the scions are cut at an angle of 30°. Scion and internode are combined and held together with a clothes-peg (**Figure 1**). For a good development of the graft union it is necessary that at least at the side where the leaf is located the cambia of the graft partners are in close contact. To prevent wilting the leaves are moistened. If auxin is applied, the basal wound of the rootstock internode is dipped in talcum powder with indolebutyric acid (IBA) 0.4%, to

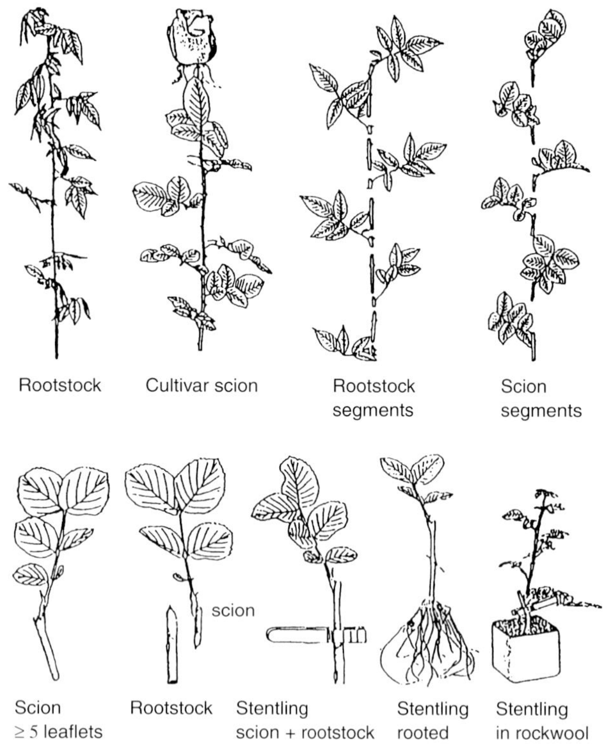

Figure 1 Schematic representation of stenting, a grafting technique based on grafting a scion on a piece of stem of a rootstock which has to be rooted simultaneously.

accelerate root formation. The plants are placed in a propagation bench, covered with double-layered polyacrylic. The rooting substrate is a mixture of sand and peat at a ratio of 1 : 1 (v/v) at pH 5.5, cocos fibres or rockwool. The relative humidity of the atmosphere must be kept at almost 100% by misting a few times a day and the temperature of the medium and atmosphere should be kept at about 25 °C.

Starch Depletion and Accumulation

In preliminary tests it was found that the process of starch depletion and accumulation involves various tissues in a more or less sequenced pattern. Thus, via staining and microscopical examination, operating an arbitrary scale became feasible, marking the inner cortex and medullary rays as the two tissues of major importance. To examine the changes of starch in a stentling, *Rosa hybrida* 'Varlon' with Ilona® was grafted on unrooted internodes of the same cultivar. The changes in starch level in the stem of the scion and rootstock are presented in **Figure 2**.

A marked increase in starch level of the scion is shown, particularly during the first week after grafting. Starch in the rootstock shows a sharp decrease, directly after grafting, reaching level zero on day 3 and subsequently increasing (**Figure 2**).

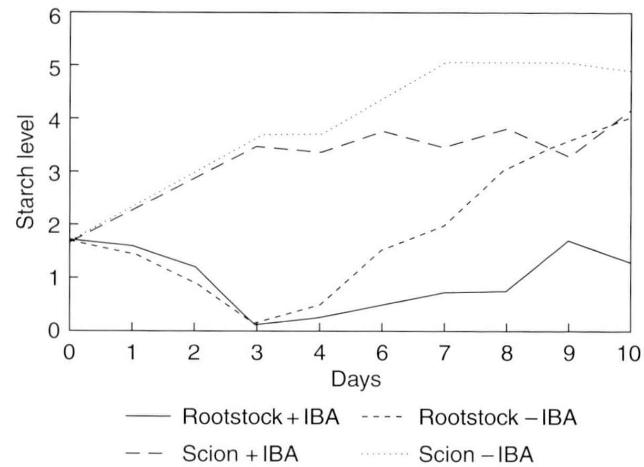

Figure 2 Changes in starch level of rootstock and scion stems in the 10 days after stenting *R. hybrida* Ilona® on Ilona®. The influence of basal application of indolebutyric acid (IBA: 0.4% in talcum powder) on the starch level is indicated. The starch level was made feasible operating an arbitrary scale after staining and microscopic examination.

Effect of Auxin on Starch Patterns

It is remarkable that basal application of IBA, to promote rooting, delays the starch increase from day 3 to 10 in the scion as well as in the rootstock (**Figure 2**). This can be explained by energy consumption due to stimulated cell activity.

Vessel Connection of the Graft Union

The development of the phloem and xylem connection of the graft union after stenting was investigated. Labelled CO_2 was applied to the leaflets of the grafted scion and, by measuring radioactivity below the graft union, the development of the phloem could be recorded (**Figure 3**). The xylem connection was investigated by colouring the connected vessels with acid fuchsine, applied to a freshly cut new basal wound of the rootstock (**Figure 3**). The results, presented as the percentage of plants with vessel connection, show that phloem connection begins at day 5 and xylem connection at day 6, while after 1 week all the stentlings show phloem as well as xylem connection.

Effect of Auxin on the Xylem Connection

The effect of basal application of IBA on the development of the percentage of stentlings with xylem connection was studied. The results show that the development of the xylem graft union is delayed when root formation is promoted by IBA (**Figure 4**). This delay points to competition between the development of the two wound regions.

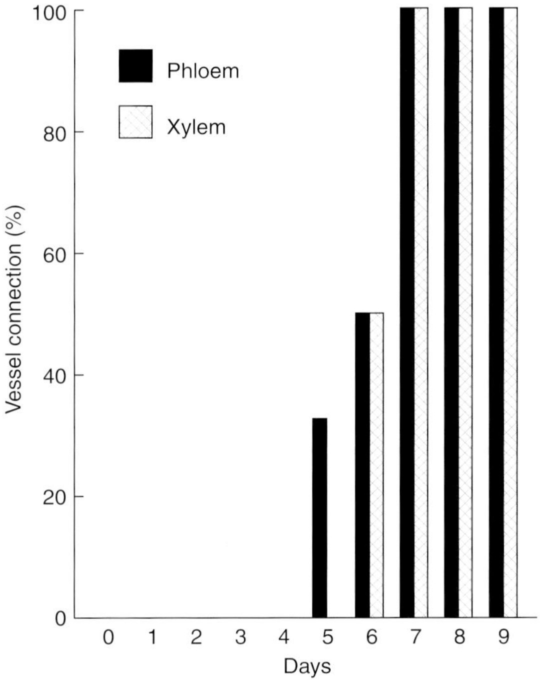

Figure 3 Development of phloem and xylem connection in the 9 days after stenting *R. hybrida* 'Ilona' on 'Ilona'. Phloem connection was observed by labelled CO_2 application; xylem connection was made visible by acid fuchsin colouring.

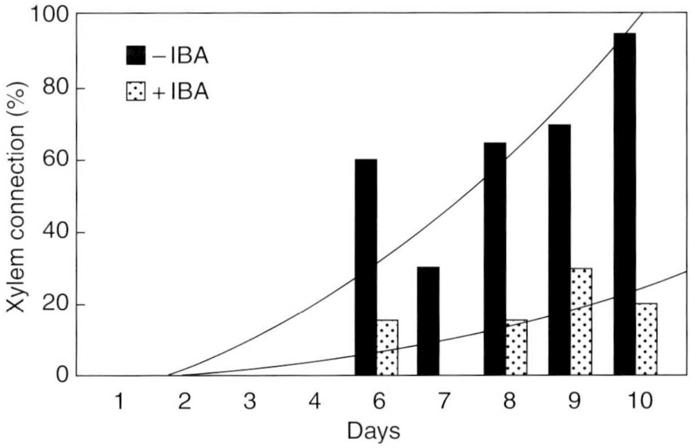

Figure 4 The effect of basal application of indolebutyric acid (IBA: 0.4% in talcum powder) on the percentage of plants with xylem connection during 10 days after stenting *R. hybrida* Ilona® on Ilona®.

Effect of Environmental Conditions on Development of the Graft Union

A large-scale facility was built in the Netherlands to propagate roses under completely controlled environmental conditions of light, temperature, CO_2 as well as humidity of air and substrate. The facility contains four units of six layers in which around 100 000 roses can be stented every 3 weeks. The important question is, what are the optimal conditions per period of stage of development for the total duration of 500 h (= 3 weeks) of the stenting process?

Table 1 Influence of 0, 1, 2, 3 or 4 days at 32 °C on the xylem connection, 9 days after stenting of *Rosa hybrida* 'Ruimeva' on 'Ludiek' ($n = 14$)

Days at 32 °C	*Xylem development*		*% of plants with damage*[a]
	% of plants with connection	*% of connected vessels*	
0	100	25	0
1	100	40	0
2	100	67	29
3	65	10	64
4	0	0	42

[a] Leaf damage or brown colouring of the cambium of the scion.

Effect of Temperature on the Xylem Connection

Hypothetically it was supposed that the first phases of wound development with cell division could be affected by high temperatures. The temperature was set at 32 °C in the first 0–4 days after stenting. It can be concluded that 1 day at 32 °C promotes xylem vessel connection without negative effects, while 2 days at 32 °C promotes development of the graft union but can cause damage to the scion leaf (**Table 1**).

Effect of Humidity on the Xylem Connection

Because of the sensitivity of rose leaves for water loss the humidity of the air during propagation (cuttings or stentlings) was normally kept at almost 100% relative humidity. Hypothetically, a strictly controlled mild daily water loss would cause some foliar transpiration with some gradient of humidity inside the stentling, which might promote the vascular development of the graft union. Continuous 100% air humidity during early development was compared with daily water losses of the stentlings. The water loss was measured by weighing the plants during dry conditions varying from 0 to 2.5 h day^{-1}. Subsequently the plants were replaced in wet conditions. Daily water loss up to 9% proved to promote the percentage of connected xylem vessels after 10 days, while 13% of water loss had a clearly negative effect (**Table 2**).

Pretreatment of Rootstock Internodes

In some cases, such as immature rootstock material or scion–rootstock combinations with difficulties in wound healing or rooting, pretreatment of the rootstock before stenting can be useful. Rootstock internodes with leaves were prerooted as cuttings until the stage of formation of root primordia, indicated by a swollen base (**Figure 5**). Depending on the conditions and rootstock type, this pretreatment will last around 7–10 days. When this material is grafted, its leaf is removed, together with the corresponding axillary bud. The advantages of this method are that immature rootstock material can be used. Normally immature rootstock material will be very sensitive to black rot, caused by fungi. This effect proved to be stronger when IBA was applied. A low energy level due to immaturity and IBA effect may explain this effect. Another advantage of pretreating the rootstock is that the vulnerable period between grafting and hardening-off can be shortened to 2 weeks. Moreover, the pretreated rootstock material can be easily stored in a plastic bag at around 5 °C (after removing the leaf) for a couple of weeks, which makes planning easier and makes long-distance transport possible.

Table 2 Influence of daily water loss on xylem connection, 10 days after stenting *R. hybrida* 'Sonia'® on 'Indica major'. The percentage of connected xylem vessels is presented as a score: 1 = 1–25%; 2 = 26–50%.

Daily water loss (%)	*Xylem development*	
	% of plants with connection	*Score*
0	60	0.95
3	75	1.05
5	70	1.50
7	75	1.30
9	75	1.90
11	50	0.90
13	35	0.55

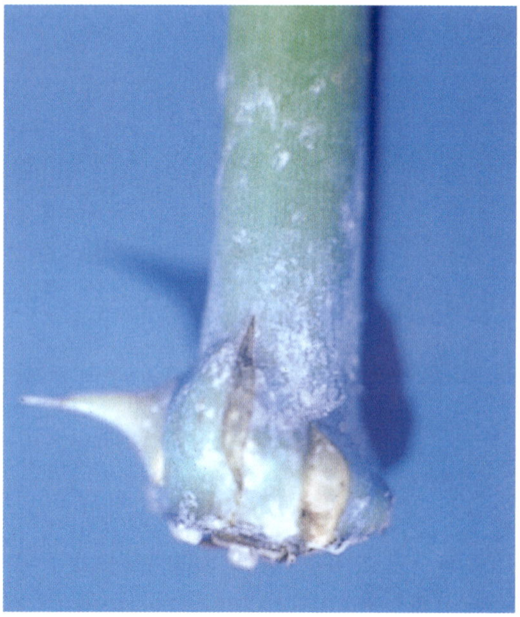

Figure 5 Basal wound of a rootstock internode after 9 days of prerooting treatment. At the base root initials are just appearing.

Effect of Rootstock Pretreatment on the Xylem Connection

The effect of a pretreated rootstock on the xylem connection was investigated (**Table 3**). It has been shown that pretreatment up to 10 days promotes xylem development of the graft union. An explanation is that when a rootstock internode is pretreated, IBA treatment is given around 1 week before grafting and so it is not a cause of competition for the graft union. The percentage of variance is the relative standard deviation of the percentage of connected xylem vessels. It can be seen that xylem development is more homogeneous after pretreatment of the rootstock.

Xylem Development of a Stentling with Two Graft Unions

To examine the wound development of two graft unions in one plant, double stentlings were made of the combination *R. hybrida* 'Sonia/Sonia/Sonia'. In this study a leafless and budless internode of Sonia® was used as an interstock (**Figure 6**).

The rate of healing of the xylem of the two graft unions was examinated after 12 days by the acid fuchsin test (**Table 4**). It can be concluded that

Table 3 Effect on the xylem connection when the rootstock was pretreated as a cutting and observed 11 days after stenting of the combination of *Rosa hybrida* 'Madelon' on 'Madelon' (*n* = 16)

Period of pretreatment of rootstock as cutting (days)	*Xylem development*		*% of variance (= relative standard deviation of % of connected vessels)*
	% of plants with connection	*% of connected vessels*	
0	94	33	61
4	100	45	45
8	100	46	43
10	100	53	42
12	100	49	32

simultaneous development of two graft unions in one stentling is possible and that the upper graft union is connected much faster than the lower one. The explanation is that the upper wound is closer to the leaf, and is a source of carbohydrates and auxin. Moreover the upper wound will, have less competition with the basal wound of the rootstock as a result of longer distance. The technique of double stenting offers possibilities for studying various graft combinations which differ in compatibility. In this case the scion and rootstock can be kept constant, to avoid foliage or rooting effects, while the interstock can be varied.

Influence of Roots on the Early Development of Rose Plants

The effect of three root genotypes on plant parameters and early production 6 months after cutting or stenting is presented in **Table 5**. Plant development as well as flower production are accelerated by rootstocks, compared to own-rooted cuttings.

Figure 6 Stenting with an interstock. Combination: *Rosa hybrida* 'Sonia/Sonia/Sonia'.

Table 4 Xylem development of a stentling with two graft unions, 12 days after stenting of the combination *Rosa hybrida* Sonia on Sonia

Position	*Xylem development*	
	% of plants with connection	*% of connected vessels*
Upper position	100	34
Lower position	17	3

Development of the Basal Wound of a Rose Stem

It is remarkable that the basal wound of a rose stem can have three destinations. In a vase as a cut flower in most cases almost nothing will happen to the wound in 2 weeks. However, the same wound inserted in rooting substrate as a cutting will form roots, while when placed on another wound it will form vessels. An important question is: what are the driving forces behind this behaviour? Preliminary experiments showed that a grafted basal wound of a scion has preference for vessel formation over rooting, even when the conditions for rooting from the graft union are made favourable, by bringing the graft union into contact with the rooting substrate. In most cases only rooting from the grafted scion occurs when the vessel connection fails. A hypothesis to explain this could be that auxin is the driving force for phloem connection and rooting. The downward movement of auxin is interrupted at the graft union: the concentration will rise until the threshold value for phloem vascularization is attained. Then the phloem vessel connection is made and the auxin level will decrease because of continued downward movement. At the basal wound of the rootstock the movement of auxin will be interrupted again; the concentration will rise and induce rooting when the threshold value for rooting is attained. In a preliminary experiment the auxin content of a stentling over 16 days was analysed: indeed, after 2 days a peak at the basal wound of the scion and after 6 days a peak at the basal wound of the rootstock could be found.

Application of Stenting and Future Prospects

Stenting is applied nowadays by millions, especially in countries like Ecuador and Kenya, where the use of rootstocks is hardly a point of discussion. In northern countries the additional cost of using rootstocks is the main reason for giving preference to own-rooted cuttings. However, examples such as are presented in **Table 5** show that additional costs can be rapidly repaid by an accelerated start of a new crop and higher productivity. At this time Natal Briar is the most popular rootstock. This rootstock is already more than 50 years old. Therefore the future prospects of the application of stenting depend on the development of a new generation of rootstocks as a result of breeding programmes.

Table 5 Influence of roots on the early development of *Rosa hybrida* 'First Red'® 6 months after cutting or stenting ($n = 18$)

Stock name or code	*Diameter of root collar (mm)*	*Number of bottom-breaks*	*Diameter of bottom-breaks (mm)*	*Number of harvested shoots*
Own roots	12.5	1.6	8.6	3.6
'Natal Briar'	16.2	1.9	10.2	5.3
18–31	18.9	2.1	10.3	5.6

See also: **Propagation**: Own-Rooted Cuttings. **Rootstocks**: Scion–Rootstock Relationships; Rootstock Breeding; Usage of Rootstocks; Clonal Rootstocks; Seedling Stocks.

Further Reading

Bailey LH (1896) *The Nursery-Book*. London: McMillan.

Costa JM, Challa H, Van Meeteren U and Van de Pol PA (2001) Photosynthates: mainly stored and yet limiting in propagation of rose cuttings. *Acta Horticulturae* 547: 167–174.

Garner RJ (1979) *The Grafter's Handbook*, 4th edn. London: Faber and Faber.

Grueber KL and Hanan JJ (1980) Rose rootstocks and miniplant propagation: preliminary results and observations. *Bulletin Colorado Growers Association* 364: 1–2.

Hartmann HT, Kester DE and Davies FT (1990) *Plant Propagation: Principles and Practices*, 5th edn. New Jersey: Prentice Hall.

Marcelis-van Acker CAM (1994) *Axillary Bud Development in Rose*. PhD thesis. The Netherlands: Wageningen University.

McFadden SE (1963) Grafting leafy stem cuttings, a technique for propagation roses. *Proceedings of the Florida State Horticulture* 76: 412–416.

Ohkawa K (1980) Cutting-grafts as a means to propagate greenhouse roses. *Scientia Horticulturae* 13: 191–199.

Ohkawa K (1984) Cutting grafts as a technique for producing started-eye rose bushes outdoors. *HortScience* 19: 527–528.

Shininger TL (1979) The control of vascular development. *Annual Review of Plant Physiology* 30: 313–337.

Skene DS, Shephard HR and Howard BH (1983) Characteristic anatomy of union formation in T- and chip-budded fruit and ornamental trees. *Journal of Horticultural Science* 58: 295–299.

Van de Pol PA and Breukelaar A (1982) Stenting of roses: a method for quick propagation by simultaneously cutting and grafting. *Scientia Horticulturae* 17: 187–196.

Van de Pol PA and Van der Vliet G (1979) Rozen stekken en enten in één handeling (in Dutch). *Vakblad Bloemisterij* 26: 40–41.

Van de Pol PA, Joosten MHAJ and Keizer H (1986) Stenting of roses, starch depletion and accumulation during the early development. *Acta Horticulturae* 189: 51–59.

Van de Pol PA, Kleemans AAM and Oudenes JH (1993) The use of genetical variation in roots to improve productivity. In: Struik PC, Vredenberg WJ and Renkema JA (eds.) *Plant Production on the Threshold of a New Century*, pp. 331–337. Dordrecht: The Netherlands: Kluwer Academic.

Wareing PF, Hanney CEA and Digby J (1964) The role of endogenous hormones in cambial activity and xylem differentiation. In: Zimmerman H (ed.) *The Formation of Wood in Forest Trees*, pp. 323–324. New York: Academic Press.

Ypema HL, Van de Pol PA and Bollen GJ (1987) Black rot of stentlings of roses: a disease caused by various soil fungi. *Scientia Horticulturae* 33: 269–280.

Seed Propagation

S Gudin, University d'Aix-Marseille III, Marseille, France

Introduction

Roses are characterized by high levels of heterozygosity, resulting from their compulsory or preferential (according to the species) outcrossing. Therefore, true-to-type multiplication can only be obtained through clonal propagation by grafting, cutting or microcutting. Moreover, the proportion of rose seed that is produced and germinates is often lower than many other crops: hip sets and seed germination in modern roses are both often lower than 50%, with hip contents usually ranging between one and 30 seeds (according to the genotype). Hence, seed propagation is limited in horticultural practice to the production of a few species used as understocks and for landscaping, some pot plants, such as *Rosa multiflora* 'Blue Angel', and to varietal creation which mainly results from cross-breeding. Some roses used in forestry or landscaping, essentially in the USA or Europe, are also seed-propagated.

Natural Propagation

Although poorly documented, the age of most wild rose species at first flowering is probably 2–4 years, with 1–2 years seed crop intervals, as reported for five very different species. Seed dispersal corresponds to a case of endozoochoria, as it seems to be mostly carried out by several birds and mammals, included in the at least 24 species listed as rose hip eaters and belonging to these groups. As the rose seed is in fact an achene, i.e. a fruitlet containing the true seed surrounded by a thin testa and a hard bony pericarp, increased germination after rose achene digestion by pheasants or sharp-tailed grouse has been reported, as a probable result of the partial destructuring of the pericarp, which is a well-known physical obstacle to rose germination. Although it is not documented in *Rosa*, it can be hypothesized that, as for other hard-coated seeds such as *Symphoricarpos racemosa* and *S. orbiculatus* (also belonging to the Rosaceae) or *Larrea tridentata*, microorganisms in the vicinity of the fallen seed improve its germination by permeabilizing its pericarp. Pericarp permeabilization has indeed been shown to be a crucial step in rose achene imbibition and subsequent germination, and unpublished results comparing the imbibition and germinability of surface-sterilized and untreated rose seeds placed on sterilized and unsterilized substrates, respectively, have shown the positive action of the seed surrounding microorganisms. Moreover, as humidity and cold, in the form of 1–4 °C stratifications, has frequently been shown to overcome the dormancy of rose achenes, it is logical to assume that naturally overwintering rose seeds possess an enhanced germinability. This need for cold in order to germinate better might be linked to the global distribution of some species, for example *R. gallica* L., which is especially frequent at altitudes around 1000 m in the Mediterranean part of the southern French Alps. However, the germination of a unique species, *R. persica* Michx., has been shown not to be improved by a cold pretreatment. In fact, this lack of reaction to cold is linked to the global absence of seed dormancy in this species and to the unequalled ability of freshly produced seed to achieve complete imbibition. *R. persica* is also characterized by a thick exocarp (or epicarp), the most external layer of the pericarp, known to be the most absorbent, and an exceptionally thick endocarp, the most internal layer of the pericarp, known to be the least permeable to water. These different specificities have been interpreted as putative adaptative features for this Far East desert region species corresponding to the promotion of rapid swelling by water (at the exocarp level) during the rare rainfalls, and to the protection of the embryo by the endocarp from drying until germination in the long term.

Artificial Propagation

In horticulture, rose seed propagation is mainly concerned with understock multiplication. Although some species are propagated from cuttings, for example, *R. indica* 'Major', which is essentially used in cut-rose greenhouses of the Mediterranean, most species, especially used for garden rose multiplication, are seed-propagated. These include *R. canina* L., *R. rubiginosa* L., 'Pollmeriana' (a hybrid resulting from a cross between *R. canina* and *R. setigera* Michx.), *R. coriifolia* var. *froebelii* (Lamb.) Rehd. (syn. *R. laxa* Hort. non Retz.), and *R. multiflora* Thunb. ex J. Murr. These garden rose stocks all belong to the *Caninae* section except for *R. multiflora* (Synstylae) and 'Pollmeriana' (an intersection hybrid, as *R. setigera* belongs to the *Synstylae*) which, however, can be grouped with the *Caninae* because it inherited the hemisexual habit of the *Caninae*, discussed below, from *R. rubiginosa*). This is certainly not by chance. Indeed, when an understock is used, it is expected that its propagation results in a sufficiently homogenous population, in order that the grafted plants behave similarly. In that sense, the *Caninae* stock species, as seed-propagated, possess an advantage. They are all polyploid, $2n = 28$, 35 or 42 ($x = 7$), with 35 as the most common, and, unlike the other *Rosa* sections, they are characterized by a very peculiar meiosis (*see* **Genetics**: Inheritance in the Dogrose) with unequal distribution of chromosomes to the daughter cells, which results in a pentaploid such as *R. canina*, in female gametes with 28 chromosomes and male gametes with only seven. Fertilization, also when resulting from a cross with a diploid species (such as *R. setigera*) as the pollen parent, restores the original set of chromosomes. The seed parent in such an example thus contributes four-fifths of the chromosomes. Thus the seedling offspring are fairly homogeneous and closely resemble the seed parent; furthermore it has been shown that self-pollination and apomixis occur in *Caninae*. Besides their use as seed-propagated understocks, this also explains why it was proposed, some years ago, to propagate the *Caninae* species used to produce the hips necessary to make the popular rose hip dessert soup in Sweden (*R. canina*, *R. rubiginosa*, *R. dumalis* Bechst., *R. sherardii* Davies and *R. villosa* var. *mollisima* Rau) by seed.

Varietal creation through cross-breeding also uses seed propagation. Following the emasculation of the bisexual flowers of the varieties chosen as seed parents, they are pollinated with pollen previously collected from the varieties chosen as pollen parents. This leads, in large rose-breeding companies, to the collection of hundreds of thousands of achenes annually. The know-how concerning pollen handling and storage, pollination conditions, seed maturation conditions as mainly determined by the seed parent plant nutrition and surrounding greenhouse climate, seed selection, postmaturation and germination varies from one company to another and is often a jealously kept secret. However, there is published evidence that the following treatments have a positive effect on seed set: (1) a 1-month cold treatment at 4 °C of the future seed parent plants just after cut-back; (2) the right choice of the pollination season, characterized by day temperatures of 23–30 °C and a relative humidity of 60–65%; (3) the choice of female parental cultivars with stigmatic exudates of low pH; and (4) a putrescine treatment of the stigmata prior to pollination. These positive effects are apparently determined by increased pollen germination and female fertility. Possibly, some of these findings are used by the breeders.

Rose seeds are also collected on different wild or naturalized species for forestry use. Hence plantlets of *R. blanda* Ait., *R. canina*, *R. rubiginosa*, *R. gymnocarpa* Nutt, *R. multiflora*, *R. nutkana* Presl., *R. rugosa* Thunb., *R. setigera*, *R. wichurana* Crép. and *R. woodsii* Lindl., germinated in US Forestry Department nurseries, are utilized in environmental forestry, as habitat or food for wildlife, shelterbelts or watersheds in North America.

Whereas production of seeds for varietal creation often involves controlled pollinations under glass, the seeds for understocks, production of hip-bearing plants and forestry use are collected on outdoor plants and result from open pollination. Colours of the ripe hips range from orange-red to scarlet and purple, according to the genotypes. For the outdoor plants, flowering and hip ripening occur at different dates according to the species and locations. For example, in Europe and North America, *R. canina* blooms from May to June and produces ripe hips in September to October, and *R. multiflora* blooms from May to July and produces ripe hips in August. The hips of all species remain on the plants long after ripening. The hips can be hand-picked soon after their original green colour turns to reddish, or at any time thereafter. It has been reported that the achenes collected shortly after hip ripening germinate more readily than those extracted from dried hips. The achenes can be extracted either manually after cutting the hips in half, or after macerating the hips in water and allowing the pulp and empty seeds to float away. They are then immediately sown, stored or stratified.

Average numbers of seeds per kilogram range from 68 000 (for *R. canina*) to 100 000 (for *R. setigera*). When stored dry in sealed containers kept at 1–4 °C, they can keep a fair germinability for at least 4 years. Except for *R. persica,* rose seeds have been shown to exhibit only embryo dormancy (as suspected in *R. nutkana* Presl.) or, more commonly, both a slight embryo and a deeper seed coat dormancy. Thus, in most species, excised embryos have germinated much more readily than entire achenes. However, since embryo extraction is difficult, because the achene's pericarp has first to be softened, for example, by acid treatment, germinating isolated embryos is not practised in horticultural propagation. As immature embryos are easier to isolate from their surrounding soft pericarp, early *in vitro* embryo rescue may be practised, but as it is very time-consuming and delicate, this is limited, to our knowledge, to some 'difficult' crosses in varietal creation. Various chemical scarification treatments have been said to improve germination slightly, including sulphuric acid application (very common for seeds not belonging to *Rosa*). However, opposing views to the acid treatment have been expressed and its practicability has been judged doubtful and unnecessary by many.

As the rose achene pericarp is very hard and lignified, with a composition close to that of hardwoods such as oak or chestnut tree, softening treatments with bases, such as soda, which is for instance used to treat olive seeds, would probably deserve some attention, as bases are classically utilized in cellulose preparation for their strong delignifying effect. Recently, an impressive improvement in *R. multiflora* germination has been demonstrated after a seed treatment with macerating enzymes, but this will probably have to be more widely experimented on different species before it can be largely used. In fact, in order to break dormancy (at least partially), nurserymen carry out different types of stratification, in moist sand or vermiculite. Most commonly, the seeds are cold-stratified, at 0–5 °C, for varying lengths of time. Sometimes, warm stratification at 20–26 °C for 1–2 months preceding the cold treatment is used. A positive effect of a controlled desiccation for 20 days at 2 °C following warm and cold stratification has even been reported. In the nurseries, the freshly collected seeds can either be sown outdoors immediately, warm-stratified for sowing later in autumn, stratified over autumn and winter for sowing in the spring, or stored dry for delayed sowing, within 1–4 years. The seeds in cold stratification are often examined in the early spring and subsequently sown if they show signs of germination; if they do not, they are held in cold stratification until the following spring. Indeed, it has often been reported that, for steady germination, the longer the cold stratification lasts, the better. When germination is carried out for varietal creation, as the breeders want to see the result of their crosses as soon as possible, the time available for stratification is limited to a period of a few months between the end of summer or beginning of autumn (seed collection) and mid-winter, if sowing is carried out under glass, or spring if it takes place outdoors. When this period of time, for instance, corresponds to 3 months, stratification for 1 month at 23 °C followed by 2 months at 4 °C has been stated to be optimal.

The seeds, according to their usage, can either be broadcast or sown in drills. The recommended planting depth depend on size of seed and is usually 1 cm. Oat straw and sawdust have been reported as successful mulches, but sand and vermiculite are also commonly used. Unpublished results have shown that the germination substrate has a strong impact on the final result, most probably linked with the possibility it provides for the development of pericarp-softening microorganisms. Finally, although isolated rose embryos have been shown to germinate steadily at 23 °C, it is very difficult to recommend an average temperature for achene germination, as it is quite dependent on the residual level of dormancy following postmaturation treatment. In practice, breeders often just maintain their germination greenhouses at nonfreezing temperatures after sowing, which usually takes place in mid-winter.

See also: **Biotechnologies for Breeding:** Early Embryo Rescue. **Breeding**: Overview; Methods of Cross Breeding. **Genetics**: Inheritance in the Dogrose. **Production and Marketing**: Overview of Roses and Culture; Field Rose Production. **Propagation**: Seed Maturation and Germination. **Rootstocks**: Seedling Stocks.

Further Reading

Bailey LH (1914) *The Standard Encyclopedia of Horticulture*. New York: Macmillan.

Crocker W and Barton LV (1931) After-ripening, germination and storage of certain roseaceous seeds. *Contributions to the Boyce Thompson Institute* 3: 385–404.

Dubois LAM and De Vries DP (1988) Factors affecting fruit and seed set in the hybrid tea-rose 'Sonia'. *Acta Horticulturae* 226: 223–230.

Gill JD and Pogge FL (1974) Seeds of woody plants in the United States. In: *Agriculture Handbook*, pp. 732–737. Washington, DC: US Department of Agriculture, Forest Services.

Gordon AG and Rowe DCF (1982) Seed manual for ornamental trees and shrubs. *Forestry Commission Bulletin (London)* 59: 35–47.

Gudin S (2000) Rose: genetics and breeding. *Plant Breeding Review* 17: 159–189.

Heit CE (1967) Propagation from seed. Part 6. Hardseededness – a critical factor. *American Nurseryman* 125: 10–12.

Heit CE (1967) Propagation from seed. Part 11. Storage of deciduous tree and shrub seeds. *American Nuseryman* 126: 12–13.

Jackson GAD and Blundell JB (1963) Germination in *Rosa*. *Journal of Horticultural Science* 38: 10–320.

Mayhall MP (1956) Roses from seed. *American Rose Magazine* 13: 8–9, 18–19.

Rowley GD (1956) Germination in *Rosa canina*. *American Rose Annual* 41: 70–73.

Tinker MAH and Wisley MA (1935) Rose seeds: their after-ripening and germination. *Journal of the Royal Horticultural Society* 60: 399–417.

Ueda Y, He H, Kurosawa T *et al.* (2001) Morphological character and germination in achenes of *Rosa persica* Michx. *Acta Horticulturae* 547: 129–140.

Seed Maturation and Germination

Y Ueda, Chiba University, Matsudo, Japan

Introduction

Although rose breeding is now at an advanced stage, the low rate of success in germinating roses, needed for their long-term propagation, has become a serious problem. Another problem is the different germination rates amongst species and cultivars. All breeders have to deal with these problems, drawing on their own knowledge and experience. Both the morphological characteristics and the natural dormancy of seeds of *Rosa* may inhibit germination. The genus *Rosa* is so widespread, ranging from subtropical to temperate regions in the northern hemisphere, that special germination mechanisms appropriate for each climate have been suggested. There may also be different reactions to germination in different cultivars, depending on their origins.

This article will explain those morphological and physiological features of rose seeds which inhibit germination, and discuss how the problem may be overcome.

Obstacles for Germination

Morphological Parameters

What is commonly called a rose seed is technically an achene (**Figures 1** and **2**), a one-seeded fruit, the wall arising from the integuments of the seed fused to the ovary wall. The achenes are contained in a fruit-like hip which is vase shaped and consists of an enlarged receptacle characteristically seen in the Rosaceae. Three to 4 weeks after fertilization, achenes rapidly expand in size, and then their growth slows down until they become morphologically complete in 8 weeks' time. Seeds are surrounded by a hard-coated pericarp, comprising three layers – the outer exocarp, middle mesocarp and inner endocarp (**Figure 3**). An exocarp is normally formed from a single layer of cells and can absorb a large amount of water. Because the exocarp of *Rosa persica* Michx. forces cells to elongate lengthwise, it is characteristically thicker than that of other species. Mesocarp is the thickest layer, and it is ligneous and hard. Cells of the mesocarp of *R. rugosa* Thunb. are large and empty, so that they float easily on water, increasing the floating pressure. An endocarp has a different cell structure from mesocarp and is fibrous. The testa consists of two layers and is thin (**Figure 3**).

As a result, the advantage of the thick, hard pericarp of an achene is that it protects the embryo it contains, but its disadvantage is that it prevents water absorption and air diffusion. It seems that achenes which do not germinate well have an especially thick pericarp, and that germination depends on the thickness of the endocarp, which does not allow water absorption. In an achene, water is taken in for germination and gases are exchanged through a micropyle on the narrow suture (**Figure 2**).

Figure 1 Achenes of *Rosa persica* Michx.

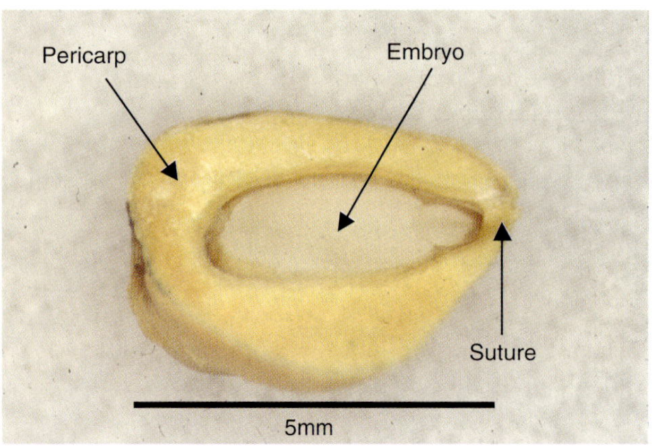

Figure 2 A longitudinal section of achene in *Rosa platyacantha*.

In the following section, the term 'achene' is interchangeable with seed.

Physiological Parameters

Roses fruit in autumn, when the temperature drops and the days become shorter. These conditions are not suitable for seed germination or for seedlings to grow. Once they are mature, seeds need to halt their germination until the weather is better suited to the growth of young seedlings.

The inhibition of germination, when seeds are dormant, occurs in the afterripening stage of an embryo, and is controlled by the balance between abscisic acid (ABA) and gibberellic acid or cytokinin. ABA is a major germination inhibitor and it is contained in the pericarp of the achene. The amount of ABA in a rose achene is 10–1000 times more than that in other plants, and it differs amongst species and cultivars. For example, there is 2.7 μg (g fresh weight)$^{-1}$ ABA in *R. rugosa*, while 0.4 μg (g fresh weight)$^{-1}$ is found in *R. persica*, which is thought not to have a period of seed dormancy. On the other hand, the amount of gibberellic acid needed to interrupt dormancy is higher in immature seeds and decreases as the seeds mature. Different types of seed dormancy induced by the embryo, or by the achene pericarp or testa, have been studied. As the amount of abscisic acid differs, so does the variation in depth of dormancy among species and cultivars. Roses from the *Synstylae*, such as *R. multiflora* Thunb. var. *adenochaeta* (Koidz.) Makino and *R. wichurana* Crép., have a relatively shallow level of dormancy, while that of

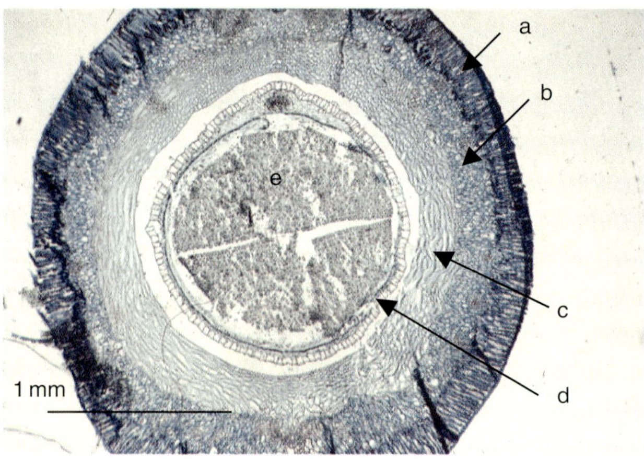

Figure 3 Microscopic picture of cross-section of achene in *Rosa persica* Michx. × 40. a, exocarp; b, mesocarp; c, endocarp; d, testa; e, embryo.

R. canina L. belonging to section *Caninae* is comparatively deep. This finding may be a result of the place of origin of the species, and the depth of dormancy may be a result of adapting to the environmental conditions of the natural habitat.

Postharvest Treatments to Improve Germination

Various methods have been developed to encourage the germination of rose seeds. Efforts have been made to remove both physical and physiological disturbances of germination in rose seeds. The following section discusses those methods.

Temperature

The most common treatment is stratification, which interrupts dormancy by reducing the period of afterripening of seeds stored in cold and humid conditions (**Figure 4**). This is because ABA, which is believed to inhibit germination, leaches out, so that there is less ABA in the achene. In *R. rugosa* at 2 weeks after treatment, there is one-third of total ABA, while 4 weeks later the level of ABA has again reduced by half. Treatment with alternating temperatures is the most effective, and a programme such as 20 °C for 1–2 months followed by 2–5 °C for 2–4 months is used. The period of treatment varies amongst species and cultivars, as well as the condition of the seeds. The temperature is so low because the character of protein and other chemicals, which encourage dormancy, changes at those degrees: treatment at lower or higher temperatures is less effective. On the other hand, germination inhibitors also leach out when treated at intermediate temperatures. The effect of low-temperature treatment to encourage germination is that not only do inhibitors leach out, but also other physiological reactions are initiated. Indole-3-acetic acid (IAA) also decreased during the first 2 weeks of stratification at both 4 °C and at 17 °C. The results suggest that IAA is without a regulating role in dormancy release. It is suggested that there are cytokinin-like substances such as hormones other than IAA that regulate the dormancy-breaking process. In the process of stratification, there may be progress of afterripening, breakdown of secondary dormancy, metabolism of germination inhibitors and others.

Figure 4 Stratification of rose achenes using moist peatmoss. Dormant breaking achenes are germinating.

In terms of sowing temperature, a germination test on *R. multiflora* var. *adenochaeta* suggests that the optimal temperature for germination of rose seeds ranges between 10 and 20 °C; at lower and higher temperatures germination is slower.

Chemicals

As seed dormancy is controlled by the balance between ABA and gibberellic acid or cytokinin, gibberellic acid and cytokinin have been used to improve germination rates. The achenes are soaked for 6 h in a solution of 6-benzylaminopurine at a concentration of 10 μg ml^{-1} or in gibberellic acid at 150 μg ml^{-1}. Activated charcoal improves seed germination by leaching ABA from the pericarp.

Differing from the hormone treatments, acid chemicals are used to remove physical obstacles from the pericarp. In an acid treatment, achenes are treated with concentrated sulphuric acid or Cross and Bevan's reagent (CB), made by dissolving zinc chloride in concentrated hydrochloric acid. In addition, macerating enzymes such as driselase, cellulose, pectinase and pectolyase have been tested. Driselase probably acts to loosen the bond between cells along the suture of the pericarp and forces the pericarp to split. As a result, the germination inhibitor leaches from inside the split pericarp and germination is improved.

Microorganisms

Under natural conditions, the physical problems of the achene pericarp may be overcome by decay caused by microorganisms in organic accumulations, such as dead leaves. Following this theory, rose germination has been tested in an experiment aiming to accelerate the decay of the achene pericarp by introducing into the sowing compost commercial compost activator, which would produce more microorganisms. Using seeds of *R. corymbifera* 'Laxa', whose germination rate is low when seeds are treated normally, a germination test was conducted by sowing them on vermiculite mixed in with compost activator. The achene pericarp was broken and the effect of encouraging germination was observed. When the seed

Figure 5 Seedlings of *Rosa persica* Michx.

surface was sterilized, a similar effect was not seen, demonstrating the importance of the microorganisms living originally in the achene pericarp.

Future Prospects

As explained, various methods have been tried for those germinating rose seeds which are difficult to germinate (**Figure 5**). Currently, stratification using low temperatures or alternating temperatures is most commonly practised. However, in the field of rose breeding, there are still problems with some hybrids whose germination rates are extremely low, even with this kind of method. It is anticipated that more effective methods integrating these treatments will be developed. A new method aiming to activate internal microorganisms using compost activator could be of great benefit to breeders.

See also: **Insects and other Animals**: Pollination. **Morphology and Anatomy**: Pollen Tubes. **Propagation**: Seed Propagation. **Rootstocks**: Seedling Stocks.

Further Reading

Bewley JD and Black M (1982) *Physiology and Biochemistry of Seeds in Relation to Germination*, 60–125. Berlin: Springer-Verlag.

Blundel JB and Jackson GAD (1971) Rose seed germination in relation to stock production. In: L Hollis (ed.) *The Rose Annual 1971*. Royal National Rose Society, St Albans, UK, pp. 129–135.

Cullum FJ, Bradley SJ and Williams ME (1990) Improved germination of *Rosa corymbifera* 'Laxa' seed using a compost activator. *Combined Proceeding-International Plant Propagators' Society* 40: 244–250.

Dubois LAM and De Vries DP (1988) Factors affecting fruit and seed set in the hybrid tea-rose 'Sonia'. *Acta Horticulturae* 226: 223–230.

Finch-Savage WE (1998) *Farm Woodland Tree Seed*. Wellesbourne: Horticultural Research International.

He H, Ueda Y, Kurosawa T *et al.* (2001) Morphological character and germination in achenes of *Rosa persica* Michx. *Acta Horticulturae* 547: 129–140.

Hudson T, Dale EK, Fred TD and Robert LG (1997) *Plant Propagation: Principles and Practices*. Upper Saddle River: Prentice Hall.

Jackson GAD and Blundel JB (1963) Germination in *Rosa*. *Journal of Horticultural Science* 38: 310–320.

PRUNING OF GARDEN ROSES, PRACTICAL GUIDELINES

T Cairns, American Rose Society, Studio City, CA, USA

Introduction

Like all fruit trees of the *Rosaceae* family, roses require an annual pruning to reinvigorate the plant before it produces a new harvest of blooms in the spring. The job is best performed after the chance of frost has disappeared from the calendar. To ignore the pruning process will allow the plant to enlarge in size but some new stems, being so thin and weak, will be unable to support the flowers. The result is a bigger bush that spreads along the ground. Moreover, the previous year's foliage can be a source of fungal infections from spores that survived the winter. Hence, roses require an annual cut-back to ensure attractive and productive plants year after year.

This article is primarily concerned with pruning outdoor roses. Management practices of greenhouse roses and nursery stocks are described elsewhere (*see* **Growth Regulation**: Axillary Bud Growth).

The Process

Of all gardening practices, pruning seems to create a sense of unease because of a lack of appropriate logic and understanding. The art and technique of rose pruning are not a mystery. The secret is to remember to apply a few basic principles derived from practical experiences gathered over hundreds of years. Pruning know-how provides an annual opportunity for the gardener to correct, adjust and modify the growth patterns to increase flower production, and influence the plant architecture. Historically, emphasis has been misplaced so that it is believed that not performing the right steps will result in failure. Failure to follow the directions carefully will only translate into an awkward overall bush shape. However, the rose bush will still perform in the spring. Following the basic logic of why certain functions are performed can be a mechanism to help master the technique. The advice offered here has been directed towards pruning rose bushes that are least 1 year old in temperate climates.

While some measure of education can be gained from reading books and attending public pruning demonstrations, it is the practical aspects that will ultimately teach the finer points of rose pruning. Experimentation will offer guidance in which varieties can be pruned hard (canes of 60–90 cm or less) and which ones lightly (canes of 120 cm). This basic understanding of rose varieties and their characteristics will enhance the enjoyment of rose growing, bringing a degree of satisfaction and pride. If one approaches rose pruning like an artist, optimum bush shape can be created.

In warm-climate zones, roses rarely experience the extended freezing conditions of winter that occur in temperate or cold climates. Rose growers in these warm climate zones are more preoccupied with the principles of correct pruning and dormancy. The practice of regulating the shape can enhance the landscape ambience of the rose garden, ensuring a vigorous first bloom that makes all the effort worthwhile. Pruning, if approached logically, can encourage new basal growth from the bud union – usually regarded as strong evidence of good health. Removal of old wood and damaged or diseased parts can allow a recuperative process to take place for increased growth. This removal of branches and sometimes old canes serves as a form of dormancy in warm climates since it cannot be achieved naturally. It is a healthy habit to allow rose bushes a period of nonproduction. Pruning permits a time period when plant processes are slowed down and then redirected to produce that first magnificent spring bloom.

Equipment

Above all else, investment in a pair of high-quality pruning shears is mandatory. One manufacturer even has a special hand grip designed for left-handed people, swivel handles and a model with blade removal for maintenance. For miniature roses, there are smaller versions of these pruning shears which rely on a smaller, straight-edged blade surface. For removal of large woody canes at the bud union, a pruning saw will allow access for flush removal. Attempts to use pruning shears for these jobs usually result in damage to the bud union. It is best to approach cane removal with a proper saw designed specifically for the job. For cutting large-diameter canes a pair of lopping shears with 30- or 45-cm handles can facilitate the cutting without placing too much pressure on the hands. Again, attempts to cut large-diameter canes with pruning shears will require a lot of extra strength. Lopping shears with long handles solve the strength problem and make the cut clean and sharp. Invest in a small wire brush (about 5 cm wide by 75 cm deep) to help remove loose bark from the bud union. Such treatments can often encourage basal breaks and stimulate new growth since growth often finds it impossible to break through the heavy tree-like bark encountered on older bushes. Finally, save on profanities while pruning by buying a good strong pair of leather gauntlet gloves or hand gloves that are puncture-proof. There is nothing as irritating as a thorn under the nail to cause a string of words rarely heard in a rose garden!

General Guidelines (Figure 1)

1. Always prune to good healthy tissue, which is easily recognized by the green bark on the outside of the cane and the white pith core revealed after the cut is made. If the central pith is not pure white, this is an indication of necrosis (dead or dying tissue).
2. After the cut is made, a drop of wood glue can ensure a quicker recovery as well as provide an instant protective barrier against cane borers. Normally, the sap will rise, containing a natural polysaccharide, pectin, which hardens to form a protective barrier.
3. Prune the centre of the bush to open for maximum air circulation and light penetration. Creating an environment for good air circulation is crucial to

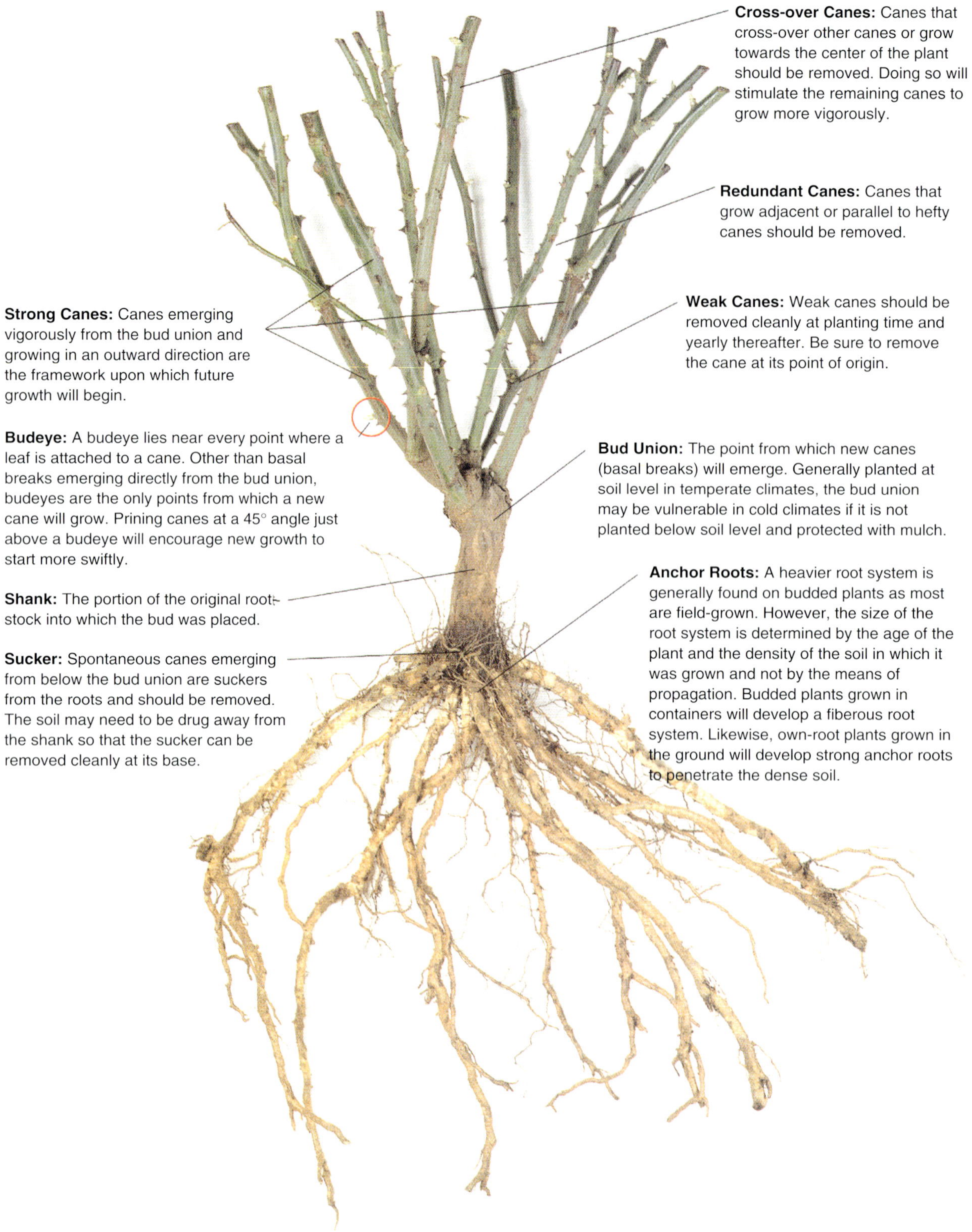

Figure 1 Illustrated guidelines for pruning. Adapted with permission from Arena Roses, Paso Robles, California (1999).

avoid microclimatic conditions suitable for fungi growth.

4. Plan to remove all growth on main canes that is not capable of sustaining a reasonable thickness of stem. Removing stems no thicker than a pencil will ensure stronger straighter stems to support large blooms.
5. If suckers are present (i.e. there is growth from the root structure), remove them from as close to the main root cane below the bud union as possible. Failure to remove suckers from the shank of the understock may result in their recurrence.
6. Remove old canes that appear woody by sawing them off as close to the bud union as possible.
7. After pruning has been completed, remove any remaining foliage from the canes and clean up around the bush, discarding all foliage as trash. Last year's foliage may contain dormant spores of various fungi and their retention can lead to infections.

Cutting Technique

By far the most important technique to master in pruning roses is the correct angle and direction of the primary cut. The final pruning cut should be made at approximately a 45° angle, about 0.5 cm above a leaf axle where there is a dormant eye. If the bush has foliage present, the cut location is easy to find. However, cuts are often made further down canes where there is no foliage to guide to the appropriate cut. Under these circumstances look for the dormant eye on such canes by locating where foliage was once connected. The eye is normally visible as a slight swelling above the surface of the cane. Making these primary cuts correctly is the real key to success in pruning. **Figure 2** illustrates the location of the bud eye, the correct way to make the cut, and the resultant growth.

The cut should be made in such a way that the new growth eventually emerges from eyes that naturally point away from the centre of the bush. This deliberate planning of new spring growth patterns gives the rose bush a pleasing overall circular shape that spreads from the centre outwards. Such cuts should avoid the new growth being directed inward and colliding with other stems. The logic behind making the cut slope down and away from the eye is that the excess natural sap that will rise from the inside of the cane to seal the cut naturally can pour down the opposing side of the cane and not interfere with the developing eye. **Figure 3** illustrates the consequence of improper cutting around the bud eye.

The essential principles of pruning are to cut just above an eye at the prescribed angle, directing the new growth away from the centre of the bush. These principles also apply when cutting stems for display for the home and removing spent blooms from the bush. Furthermore, keeping the centre of the bush open and not congested will help combat fungal diseases such as mildew due to increased air circulation.

Pruning of Different Rose Types

Hybrid Teas and Grandifloras

The growth habit of hybrid teas and grandifloras is such that by winter they are generally 2.5–3 m tall and looking very lanky. The height of the canes remaining after pruning on an established bush can be as tall as 60–120 cm in some instances. In general, about four to five major canes can remain, at an average height of 90 cm. Removing older canes is a trigger mechanism to the rose bush to attempt to promote basal breaks or new canes in the spring. This regenerative process is fundamental to the health of the bush and, unless performed carefully, the rose bush will gradually become nonproductive. **Figure 4** illustrates the pruning technique as applied to hybrid teas.

Floribundas

Since floribundas are mainly for garden display, many older canes are allowed to remain to permit a dense development of inflorescence. Therefore, the number of canes remaining after pruning is much greater than that recommended for hybrid teas. Two main reasons drive this strategy. First, there are considerably fewer potential dormant eyes for first growth because of the length of the canes. Second, the flowering habit emphasis of floribundas is to produce large numbers of flower clusters but not always necessarily on long stems. Based on these logical parameters, the pruning of floribundas does not necessitate the removal of all growth to leave only three or four main canes. Leaving a greater number is advised. **Figure 5** illustrates the pruning technique as applied to floribundas.

Miniature Roses

First of all, most miniature roses are grown on their own roots, i.e. there is no bud union and therefore no suckers. Because precise pruning when applied to miniature roses is very labour-intensive, most rosarians prefer to take garden clippers and trim off the tops at no higher than 30 cm above soil level (height varies according to growth habit of the variety). However, after such treatment it is wise to remove, on an individual plant basis, any twiggy growth and open up

(a) (b) (c) (d)

Figure 2 (a) Locating the dormant bud eye by looking for a small circular swelling pimple at the junction where a leaf once resided. (b) Locating the dormant eye (stem rotated 90 degrees to accentuate the spot). (c) Making that all important cut to promote the new growth in the correct direction away from the centre of the bush. (d) Emerging new growth after 3–4 weeks providing stems bearing flowers.

the centre of the plant to air circulation and light penetration.

Old Garden Roses

When attempting to prune old garden roses, avoid treating them as modern hybrid teas and floribundas. For maximum bloom production, pruning should be more of a light grooming than a severe pruning. Prune only last year's growth. After a few years, however, this practice makes for a very lanky bush. Therefore, each year prune back some of the oldest canes to promote basal and postbasal breaks. Keeping a proper balance between new growth and continuing old growth patterns is the secret to growing old garden roses.

Climbers and Shrubs

Climbers will generally not flower profusely unless the canes are trained into the horizontal position.

(a) (b) (c) (d)

Figure 3 Misplaced pruning cuts. (a) Pruning cut made too close to developing bud eye. (b) Pruning cut made too far above eye causing partial die-back. (c) Pruning cut made too far above the bud eye causing severe die-back almost to the junction. (d) Pruning cut made causing damage to outer epidermis and therefore permitting potential infection.

(a) (b)

Figure 4 Hybrid tea bush (a) before pruning and (b) after pruning.

(a) (b)

Figure 5 Floribunda bush (a) before pruning and (b) after pruning.

Cut the ends of these long-established canes to about the place where the canes are slightly larger than pencil thickness. Then, rather than remove each side stem that has flowered, cut them to the lowest possible five-leaflet, i.e. about 2–4 cm from the main cane. This process will cause the cane to flower along its complete length for a terrific spring display.

Pruning in Temperate and Cold Climates

Depending on the severity of the temperatures experienced, precise pruning cannot be exercised. Pruning simply involves cutting back the canes until the central pith is pure white with no decaying or dying tissue. In many cases the canes are often cut to as low as 10 cm above the surface. Indeed, test trials at the gardens of the Royal National Rose Society, UK, concluded that superior results were obtained by ignoring all the rules and merely cutting the canes with a hedge trimmer above 1.5 m above the ground, leaving all the twiggy bits. Some gardeners, however, prefer to winter-protect to ensure survival of more cane structure. In these cases some of the principles of precision pruning can be applied with success.

After Pruning

The following recommendations are offered to help reduce the potential for disease and encourage vigorous growth:

1. To ensure the destruction of all insects and fungi, apply a dormant spray immediately after pruning. Do not wait 1–2 weeks to perform this spraying since in that space of time some eyes may have developed and such a spraying could damage them. Spray immediately after pruning when there is no development of the eyes. A return to the old-fashioned oil and sulphur spray technique may be a wise decision after pruning. A combination of horticultural oil and a lime-sulphur preparation should aid in the destruction of the spores (including both powdery and downy mildew) residing in the soil and on the immediate surface and subsurface of the canes.
2. After brushing the bud union with a wire brush to remove the old bark, cover the bud union with about 15–25 cm of the surrounding mulch. This protective mound of mulch keeps the bud union moist and therefore receptive to basal breaks and hence new canes. Additionally, this mound can protect the bud union from mild frost conditions and wind-chill factors.
3. Avoid fertilization until about 3–4 weeks after pruning. Then apply one to two cups of solid rose food formulation around the base of the mound covering the bud union and then uncover the bud union. The mulch then covers the rose food and provides a clean landscaping surface to start the year off.

See also: **Growth Regulation**: Axillary Bud Growth. **Morphology and Anatomy**: Flower Development and the Abscission Zone. **Production and Marketing**: Overview of Roses and Culture.

Further Reading

Cairns T (ed.) (1999) All About Roses: The Easiest Roses to Grow. Des Moines, IA, USA: Ortho Books.

Beales P, Cairns T, Duncan W *et al.* (2002) *Botanica's Roses: The Enyclopedia of Roses*. Australia: Random House.

R

Resistance *see* **Breeding**: Selection Strategies for Disease and Pest Resistance.

ROOTSTOCKS

Contents

Scion–Rootstock Relationships

D P de Vries, Plant Research International (PRI), Wageningen, The Netherlands
Present address: Foundation Sub Rosa, Ede, The Netherlands

Introduction

The propagation of (woody) scion varieties by grafting onto rootstocks is a very old horticultural practice. Scientific research on scions, rootstocks and the performance of combination plants, however, is as recent as the end of the nineteenth century. Initially, research focused on difficulties in propagation, the poor growth of certain combinations and even the total incompatibility of plant parts. After a range of apple rootstocks of varying vigour was introduced by East Malling Research Station, UK, research to explain differences in vigour began.

In combination plants of various kinds, the term 'scion–rootstock relationship' has been interpreted in several ways. Depending on the crop and the plant parts under investigation, it relates to: (in)-compatibility between scion and stock, the manner or degree in which the stock affects the scion variety, the manner or degree in which the scion variety affects the stock and the manner or degree in which scion and stock contribute to the growth and development of the combination plant. Sometimes, instead of 'relationship' the term 'interaction' is used, in which case it is often presented as a statistical parameter.

Scion–rootstock relationships have been studied in fruit more than 80 years before the subject became of interest in the rose. Currently, scion–rootstock relationships in the rose indicate the variation in the vigour of rose combination plants, resulting from the variation in the individual vigour of scion varieties and rootstocks. The vigour of garden rose varieties on various stocks, which has been poorly investigated, is commonly expressed as the number or length of shoots of combination plants in the nursery. Data about the growth and development of combination plants when they are planted in the garden for a number of years are rare. For various reasons the vigour of cut-rose combination plants is commonly expressed as 'shoot yield in a certain period of time', which is close to that for garden roses. It is notable that the above parameters of vigour are a practical approach to a much more involved matter. Basically, vigour should be determined as 'dry matter production in time', in which the scientific parameters absolute growth rate (AGR) or relative growth rate (RGR) should be considered. This kind of research is highly unpopular among researchers in woody crops, particularly rosarians. In the new research, which started around 1980, scion–rootstock relationships in the rose have been studied at Plant Research International (PRI) and at Wageningen Agricultural University, The Netherlands, and at the Hebrew University of Jerusalem, Israel. From about the year 2000 onwards, research on scion–rootstock relationships in the rose has slowly wound down.

The Vigour of Combination Plants

In a number of crops other than roses, where varieties are commonly grown on clonal rootstocks, research workers have been reluctant to draw conclusions about scion–rootstock relationships. It may be that this is because too large an effect has always been alotted to stock influences. A clarifying example, however, is one of the last investigations in apple rootstocks, carried out at East Malling Research Station in the UK in the late 1950s. In a cleverly designed experiment, a series of clonal rootstocks of different vigour were alternately used as a scion variety and as a stock. The important conclusion of this long-term experiment was that the vigour of the scion variety and that of the stock probably exerted additive effects on the vigour of combination plants, when vigour was expressed as the dry weight of complete plants (roots + shoots + pruning material + fruits). Only when these original apple data were recalculated, however, did it become clear that variety–stock interaction as to dry weight was indeed absent, which in fact gives the same effect as additive vigour of stock and scion. An additional but disturbing effect in apple stocks is that, besides vegetative growth, weak (= dwarfing) apple stocks have a specific effect on 'earliness', viz. on the onset of flowering and fruiting of combination trees. Earliness, by the way, does not seem to play a role in the rose. Also in experiments with *Picea* or *Cucumber* varieties, grafted on a series of stocks, similar conclusions on the additive action of varieties and stocks for the length of the trees – respectively the overall mean daily length increase – were revealed. It was likely therefore that there would be additive effects of varieties and stocks in most crops, including roses. The absence of scion–rootstock interaction for vigour indicates that, given a series of rootstocks arranged for induced vigour in one scion variety, other varieties of different vigour grafted on these stocks will perform similarly.

When during the twentieth century fast global expansion of the production areas of glasshouse cut roses on clonal stocks began, demands for high-yielding scion varieties and stocks which induced still higher yields of excellent quality, increased at the same time. Due to the fast reaction of several European institutes and universities, new research began. So great was the demand for information that sometimes research was even funded by private enterprises. All these experiments concentrated on the factors that would affect the shoot yield of combination plants. However, because few similar investigations were available in other crops, most projects had to start from the beginning. In fact, the new research had to answer three basic questions:

1. Which part of the combination plant determines its overall vigour: the rootstock, the scion variety or both?
2. Will there be scion–rootstock interaction which has an effect on vigour?
3. What is the external or internal basis of vigour of combination plants?

Scion – Rootstock Relationships in Cut Roses

In the rose, experiments with several stocks and varieties were few and far between before new research began in 1980. However, an experiment with the cut-rose varieties 'Sonia' and 'Super Star', each grafted onto the same eight stocks, is a good introduction to probable rootstock action in this crop. In **Figure 1** the individual shoot yields of the two scion varieties have been plotted against the overall mean yield of these varieties on each stock (arrows on abscissa). It can be seen that, despite the individual differences between 'Sonia' and 'Super Star', stocks exerted pronounced differences in shoot yield. On average, 'Sonia' yielded about 7.5 shoots more (98%) than 'Super Star', but differences between weak stocks ('Laxa') and vigorous stocks ('Multiflora') were larger in the lower-yielding 'Super Star' (83%) than in the high-yielding 'Sonia' (28%). Because these varieties were chosen on purpose for their large difference in yield, a conclusion as to which plant part had the most effect on additive vigour cannot be drawn. From other experiments in the rose, indications are that stock effects are more pronounced in low- than in high-yielding varieties. Most important, however, was that in this early rose experiment an additive effect on vigour was demonstrated. In other words, in this experiment the effect of scion–rootstock interaction on vigour was probably absent.

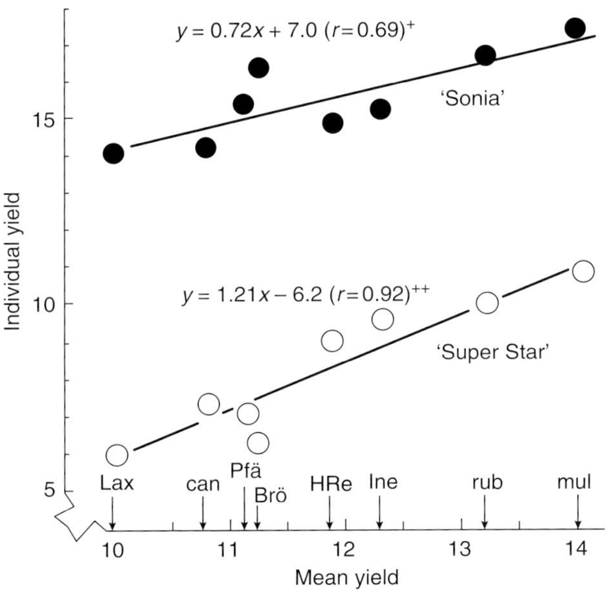

Figure 1 The relationship between the overall mean shoot yield in the second year of the cut rose scion varieties 'Sonia' and 'Super Star' grafted on to eight different rootstock varieties (indicated by arrows on the abscissa), and the individual yield of the two varieties on these stocks. Correlations (r) significant at $P = 0.05$ (+) or 0.01 (+ +). Lax: *R. dumetorum* 'Laxa'; can: *R. canina* L.; Pfä: *R. canina* 'Pfänder'; Brö: *R. canina* 'Brögs Stachellose'; HRe: *R. canina* 'Heinsohns Record'; Ine: *R. canina* 'Inermis'; rub: *R. rubiginosa* L.; mul: *R. multiflora* Thunb. *Note*: See Leemans, J. A. (1964) Rootstocks for roses, Plant Publicity Holland, 72 p. © 2001, D. P. de Vries and L. A. M. Dubois.

Inherent Vigour of Genotypes

In fruit culture the term 'induced vigour' is commonly used to indicate the level of vigour which a range of rootstocks communicates to or induces in a scion variety. Because scion–rootstock relationships had not been studied at all in roses, the term was unknown before about 1980. At PRI, Wageningen, a range of experiments on induced vigour in the rose was carried out in soil in heated glasshouses. In previous experiments at PRI it had already been ascertained that when they were planted after first flowering in a heated glasshouse, after a short period own-rooted hybrid tea seedlings grew and developed exactly like combination plants. In the new research at PRI which started in 1990, two simultaneously running experiments were carried out with Hybrid Tea seedlings. In three successive years, batches of 60–80 Hybrid Tea seedlings, each year consisting of different genotypes, were planted in the glasshouse and followed for their shoot yield for 8 months. In the first experiment the genotypes were dug up, graded for largely differing classes of shoot yield, and subsequently bench-grafted onto one stock: *Rosa canina* 'Inermis', 8–12 mm seedlings. Combination plants were planted in a heated glasshouse. The shoot yield of each batch of these combination plants was recorded for one full year. In all 3 years highly significant correlations occurred between the yields of own-rooted genotypes and yields of the same genotypes on 'Inermis' stocks. This clearly indicated that the vigour of a rose genotype is a factor that is present at least in the aerial part of the plant.

In the second experiment, the root systems of the same 60–80 Hybrid Tea genotypes used in the first experiment were used as a rootstock for just one variety: 'Sonia'. Each year, highly significant correlations occurred between the shoot yields of the original seedlings on their own roots, and the shoot yields of 'Sonia' when grafted on the root systems of these seedlings (**Figure 2**). These correlations indicated that the vigour of the original seedlings returned as 'induced vigour' when they were used as a rootstock. By digging them up and weighing them, it was shown that the root systems of high-yielding rootstocks were also larger and heavier than the low-yielding ones.

From these two experiments it became clear that vigour, expressed as shoot yield, is an inherent factor that is present in both the shoots and the roots of the plants of a genotype. It was likewise clear that the additive vigour of stock and scion determine the vigour of rose combination plants. The significance for future rootstock breeding is that selection for highest vigour in seedling populations may yield vigorous stocks. In successive experiments, however, closer study of the growth and development of combination plants had to be carried out.

Growth and Development of Rose Combination Plants

The shoot yield of cut-rose varieties, as the main practical parameter of vigour, has been the subject of many studies. From the middle of the twentieth century, the annual reports of several experimental stations regularly published differences in yield between cut-rose varieties grafted on the seedling rootstock 'Inermis' only. Only in the last decades have yields on various stocks been compared in scientific publications. Close study has revealed that, independent of the rootstock or the culture system applied, there is a fundamental similarity in the growth and development of cut-rose and garden-rose plants.

Differences between the two categories are seem to be limited to harvesting of shoots and seasonal influences, owing to different pruning methods. Growth and development are illustrated in **Figure 3** for a bench-grafted plant, from the time of grafting until first flower harvest in the glasshouse. Ready-to-plant

Figure 2 The vigour of 'Sonia' combination plants as induced by a range of stocks. From left: a very weak stock (one bottom-break) to far right: vigorous stock (four bottom-breaks).

combination plants consist of the scion of the variety (1) with one shortened leaf and one quiescent axillary bud (4), and the rootstock (2). In most varieties the quiescent axillary bud will break 10–20 days after grafting and grow out as the primary shoot (3 in **Figure 3b**). When the flower bud is visible in 8–12 weeks, several quiescent buds below the flower bud will break and form shoots that are longer than the primary shoot. A well-known cultural measure is to pinch the primary shoots' flower bud and successively bend this shoot downward over 90°. The consequences of bending are twofold: lateral buds of the primary shoot will stop sprouting, and the lowest two or more quiescent buds of the primary shoot will break after several weeks and develop into unusually vigorous shoots, named 'bottom-breaks' or 'renewal canes' (6 in **Figure 3c**). In the traditionally applied plant architecture, bottom-breaks form the frame of the rose bush. When bottom-breaks are pinched or pruned, their laterals (8) are the first marketable product of the combination plant.

Harvesting of cut-rose flowers is a continuous process of cutting new flowering shoots from previously cut laterals. Because a piece of shoot of 10–15 cm, containing several quiescent buds, remains on the plant after each harvest, plants gradually grow taller. To keep plants under control, from time to time shoots are cut 'beneath', i.e. below the

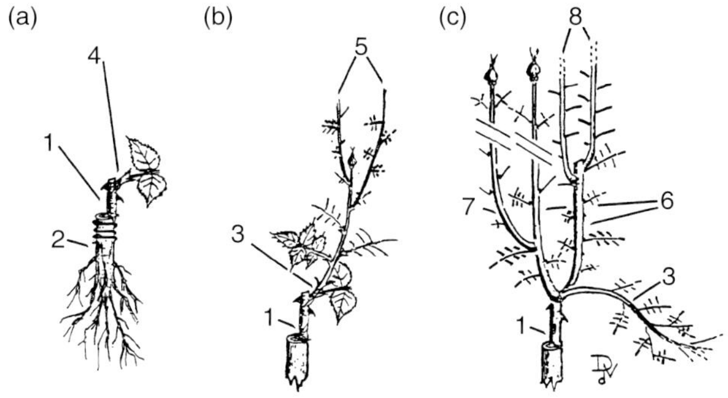

Figure 3 Stages of initial growth and development of a bench-grafted rose combination plant. (a) Day 0: scion (1) grafted on to a seedling rootstock (2); (b) combination plant 10 weeks old: primary shoot (3) grown from the quiescent bud (4); two upper axillary buds grown out into laterals (5); (c) 18 weeks old: primary shoot bent-down, basal bottom-breaks (6) grown from lowest axillary buds; axillary bottom-break (7); basal bottom-break hard-pinched, two marketable laterals (8). © 1993, D.P. de Vries.

latest or last but one harvest site. The foregoing indicates that the number of flowering shoots harvested from one plant in 1 year is determined by the combined effects of the number of bottom-breaks per plant, the number of axillary buds that break and grow into new shoots, the time of bud-break after each harvest and the speed of growth.

Apical Dominance

Apical dominance means that the growing apex of a plant controls the quiescence of buds sitting in lower positions along a shoot. This kind of bud inhibition is termed 'correlative inhibition'. Correlative inhibition usually has a gradient: the lower on the shoot, the stronger the inhibition. In plants with strong apical dominance, the growth is very upright, with one dominant central axis, as for example in firs. Such trees are called 'excurrent'. Low apical dominance is seen in the bushy growth of many shrubs, which are indicated as 'decurrent'. Correlative inhibition is explained by the action of auxin, mainly indole-3-acetic acid, which is synthesized by the growing shoot apex. Auxin, which migrates slowly via the phloem in a basipetal direction, inhibits the sprouting of axillary buds sitting lower on the shoot. The correlative inhibition of trees is usually lifted after taking away a shoot's apex, or after bending the shoot into a horizontal position.

In roses too, the habitus of plants is controlled by correlative inhibition. An important difference from trees is that correlative inhibition is only active as long as the apices of growing shoots produce auxin. When left unpruned (unharvested), auxin synthesis in a rose shoot terminates whenever the shoot apex transcends from the vegetative to the generative phase, which occurs when the apically placed flower bud is initiated.

Correlative inhibition of a rose shoot is also lifted after decapitation. It is likely that, in contrast to the more bushy garden-rose varieties, apical dominance in the upright-growing cut roses is strongest. It has been shown in various ways that shoot yield in cut roses is determined by degree of bud-break. When grafted on the same rootstock, scion varieties show large genotypic variation for the number of bottom-breaks as well as for the number of axillary buds that break after each harvest. This means that genotypic variation for bud-break may be explained by a genotypic variation for apical dominance. However, there are differences in induced shoot yield of various rootstocks. Because rootstocks clearly affect bud-break of the scion variety, the question is: how do they manage?

A Model of Vigour

In explaining the control of bud-break of the scion variety by the rootstock, the internal balance of two major endogenous phytohormones plays an important role. These hormones are indole-3-acetic acid (auxin) and one or more cytokinins, probably zeatin or zeatin riboside. The role of auxin in apical dominance and indications of an internal hormone balance of plants came from clever experiments as long ago as the 1930s. At that time it was demonstrated that, apart from favourable synergy as regards the sprouting of buds, auxins might antagonize cytokinins and vice versa. Following indications from physiological research in the apple, and from *in vitro* and *in planta* investigations in the rose, it was hypothesized that apically produced auxin in scion varieties of the rose is antagonized by root-produced cytokinins. Bud-break would be controlled by the ratio of the two endogenous phytohormones, rather than their actual levels. The growth of combination plants would be increasingly vigorous as the ratio of auxins to cytokinins shifts towards cytokinins, and the other way round. This hypothesis has been put into a model (**Figure 4**) for an established cut-rose combination plant in a closed plantation. For simplicity, the effects of other endogenous hormones, which play their roles, have been omitted. In spite of regular shoot harvests, always followed by temporarily arrested root growth, the shoot-to-root (S/R) ratio in established plants is more or less in equilibrium. Another effect of regular harvests in a closed crop is that plants cannot grow wider any more. Apically synthesized auxin not only controls axillary bud-break. When auxin migrates basipetally in the phloem and into the roots, it stimulates

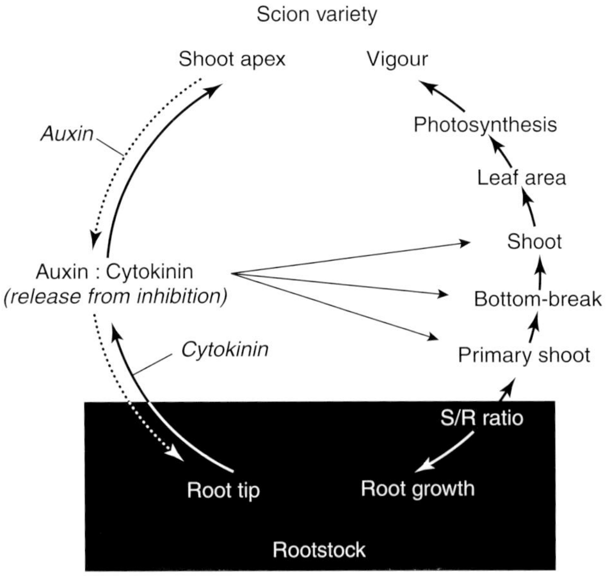

Figure 4 A model of the vigour control of (cut) rose combination plants. S/R, shoot-to-root. © 1993, D. P. de Vries.

root growth and consequently the formation of new root tips, which are a rich source of cytokinins. The more root tips available per plant, the higher the level of transportable cytokinins. These water-soluble cytokinins are acropetally transported into the shoot and its organs via wood vessels. The antagonistic actions of auxin and cytokinin on axillary bud inhibition, respectively bud-break, are now displayed. When the auxin production of the shoot apex has sufficiently decreased, and cytokinin level is sufficiently high, buds sitting beneath the apex are released from apical dominance and will sprout. Of course, new shoots will build their own systems of apical dominance in several weeks. As more axillary buds sprout, the plant will bear more shoots, thus inducing a larger, photosynthetically active leaf area, which produces more carbohydrates to be used for higher growth rates. Unbridled growth is kept in check by regular harvests of flowering shoots. Objections to the model are: (1) the role of neither abscisic acid (ABA) in bud inhibition, nor that of gibberellins in flower induction is explained; (2) cytokinins are not exclusively formed in the root tips, but in young leaves as well; (3) when supraoptimal auxin concentration occurs in the roots, root growth may be inhibited rather than promoted. In one of the last research projects in rootstocks at PRI, cytokinins were determined in the bleeding sap of decapitated combination plants grown in nutrient solution. It concerned three scion varieties to graft onto a small selection of rootstocks of varying vigour. The research confirmed that the endogenous cytokinins mainly consisted of zeatin riboside and its metabolites. The most important conclusion, however, was that the levels of rootstock-produced cytokinins were positively correlated with the induced vigour of these stocks, thus confirming the supposition that the cytokinin levels produced by stocks are the basis of their induced vigour. Unfortunately, financial funding did not permit continuation of this research on the simultaneous determination of auxin levels.

The current state of the art as to knowledge of scion–rootstock relationships in the rose may be summarized as follows:

- The vigour of combination plants is determined by the additive vigour of the scion variety and the rootstock, hence interaction as regards vigour is absent.
- The basis of vigour of combination plants is probably determined by the ratio of the endogenous hormones auxin and cytokinin.
- In further research, involving several scion varieties and rootstocks, it would be highly recommended to determine both auxin and cytokinin levels.

See also: **Growth Regulation:** Hormones in Growth and Development. **Propagation:** Budding in the Field; Own-Rooted Cuttings. **Rootstocks:** Rootstock Breeding; Usage of Rootstocks; Seedling Stocks.

Further Reading

DeVries DP (1993) *The Vigour of Glasshouse Roses. Scion–Rootstock Relationships. Effects of Phenotypic and Genotypic Variation*. Thesis, Agricultural University of Wageningen, The Netherlands.

De Vries DP and Dubois LAM (1997) Relations between early root growth and flower yield of clonal rose rootstocks. In: Altman A and Waisel Y (eds.) *Biology of Root Formation and Developoment*, pp. 283–288. New York: Plenum Press.

Dieleman JA (1998) *Cytokinins and Bud Break in Rose Combination Plants*. Thesis, Agricultural University of Wageningen, The Netherlands.

Edwards RE (1955) *A Survey of Rootstocks for Roses*. Dissertation, University of Nottingham, UK.

Fuchs HWM (1994) *Scion–Rootstock Relationships and Root Behaviour in Glasshouse Roses*. Thesis, Agricultural University of Wageningen, The Netherlands.

Kool MTN (1996) *System Development of Glasshouse Roses*. Thesis, Agricultural University of Wageningen, The Netherlands.

Kramer PJ and Kozlowski TT (1979) *Physiology of Woody Plants*. New York: Academic Press.

Krüssmann G (1986) *Rosen, Rosen, Rosen*. Berlin: Parey.

Shepherd RE (1954) *History of the Rose*. New York: Macmillan.

Ximing Hu (2002) *Growth and Productivity of Cut Rose as Related to the Rootstock*. Thesis, Agricultural University of Wageningen, The Netherlands.

Rootstock Breeding

D P de Vries, Plant Research International (PRI), Wageningen, The Netherlands
Present address: Foundation Sub Rosa, Ede, The Netherlands

Introduction and History of Breeding

Plant breeding may be simply defined as: 'the improvement of crops by humans, for human benefit'. In conventional rootstock breeding three disciplines were available: (1) selection and propagation of desirable genotypes from the wild; (2) selection of deviating genotypes within existing varieties; and (3) cross-breeding of suitable progenitors. In 1996 biotechnological research succeeded in genetically transforming the clonal rootstock 'Moneyway'. The earliest rootstock breeding was recorded in the seventeenth century when British nurserymen established the first mother gardens for seed production. At that time, important rootstocks were Damask (*R. damascena* Mill.), 'The White' (*R. alba* L.), 'Francford' (*R. francofurtana* Muench) and the 'Wild Eglantine' (Sweetbrier or *R. rubiginosa* L.). Seedlings of these genotypes satisfied the needs of horticulturists for more than two centuries until, in the nineteenth century, the dog rose (*R. canina* L.) came into vogue. Dog rose seedlings were much more uniform than the stocks hitherto known. Their discovery was the beginning of a success story of the seedling canina stocks. Even today, *R. canina* varieties, known as the refined and rather uniform Edelcaninas, are the most important and best-known stocks for garden roses all over the world. Owing to their system of reproduction, named heterogamy, in which more chromosomes are contributed by the female than male parent (*see* **Genetics**: Inheritance in the Dogrose), dog roses are ideal objects to breed new varieties within that species. Breeding new canina varieties has been a hobby, giving status to the breeder, rather than a scientifically based project. Edelcanina varieties like 'Schmid's Ideal', 'Heinsohn's Record' or 'Brög's Stachellose' still remind us of the breeder's name. Caninas were still so important in the 1950s that experiments were initiated to breed for improved resistance to powdery mildew and a reduced stratification period.

Dog roses have served their purpose for the multiplication of both garden and glasshouse roses for more than 100 years, but ultimately even varieties like 'Inermis', 'Brög's Stachellose' and 'Pollmers' could no longer meet the demands of modern glasshouse cultivation. As a consequence, both the number of dog rose varieties and their area in north-western European nurseries have gradually decreased. For the glasshouse industry in artificial substrates, caninas have completely given way to clonal stocks. It is only for garden roses, and in countries where cut-rose cultivation is practised in soil, that caninas are still being used. As to rootstocks for the cut-rose industry, current emphasis is on vigour control of scion varieties rather than just on their multiplication. Greatest induced vigour in combination with shoot quality has become a breeding aim. A number of recently developed clonal stocks have initially met this requirement. 'Moneyway', 'Multic', 'Indica Major' (virus-free) and 'Popeye' induce early and high flower yields, are easy to propagate by cutting–grafting and cause few problems in artificial substrates. These stocks, however, have become less significant with the introduction of one simple clonal stock: 'Natal Briar'. It did not result from a modern breeding programme, was not selected from a population, but is just a genotype that became highly successful once it was accidentally tried out as a stock. Possibly only for financial reasons, 'Natal Briar' has displaced most other clonal stocks. However, the use of combination plants in the cut-rose industry is decreasing anyway. The present trend is to grow cut-rose varieties on their own roots, simultaneously applying new plant architectures. This means that the success of future stock must depend on properties that greatly surpass those of the current ones. Without the ability to induce significantly greater vigour than of own-rooted varieties, while maintaining high shoot quality, preferably in combination with resistance to substrate-borne diseases, new stocks are doomed to failure.

From a scientific point of view, present possibilities of breeding new stocks have never been so good because the questions about induced vigour have been answered, for the most part. However, the circumstances under which breeding was undertaken in the twentieth century have radically changed. Currently, the breeding of stocks, either seed- or clonal propagated, has become expensive, to the extent that just one company can hardly fund it. Even when scientifically trained breeders, with perseverance and a good knowledge of the crop, are involved, rootstock breeding remains laborious, time-consuming and needs great creativity, as similar research is scarce. For the current breeder, the first choice is between seedling and clonal stocks. Since seedling stocks should be uniform, two obvious choices are cross-breeding using heterogamy, or breeding for homozygosity using diploid progenitors. In the category of clonal stocks, only one genotype needs to be selected from many segregating seedling populations. A similar situation occurs for the introduction of one useful gene by biotechnology.

Induced Vigour

For several decades, the approach to breeding new rose stocks has been comparable to that in the apple. In apple, rootstock breeding was initiated at East Malling Research Station, UK in 1900. In roses, breeding did not start before 1980. The main reasons for lagging behind by 80 years were contentment with seedling stocks, lack of interest in clonal stocks and general ignorance of the concept of (induced) vigour among rosarians. Induced vigour means that, within a certain range, the vigour of the stock modifies the genotypically determined vigour of the scion. Hence, in apple, weak stocks have a dwarfing effect on the combination tree, while vigorous stocks promote its growth. In general, vigour is related to dry-matter production in time, in which endogenous hormones play an important role in the distribution of leaf-synthesized carbohydrates over plant parts. Physiological parameters of vigour include absolute or relative growth rate. In cut roses, vigour is expressed by shoot yield in a certain period of time, either in number or in weight. However, whereas in apple small, easy-to-handle, precocious (early onset of fruiting) trees are currently desired, hence weak stocks, in the rose vigorously growing plants with many long shoots are aimed at, hence the highest stock vigour possible is desired. Precocity does not play a role in the rose (*see* **Rootstocks**: Scion–Rootstock Relationships).

Phenotypic Expression of Vigour

With regard to selection for highest vigour in rootstock populations, an important question is how the individual vigour of seedlings can be phenotypically recognized without grafting complete populations. Apple research contributes to that quest, since vigorous apple rootstock genotypes could be empirically recognized after sawn-off plants had been earthed and allowed to produce shoots. Genotypes which produced (many) long shoots in one season were termed 'vigorous', while those with (few) short shoots were 'weak' rootstocks. The long-term programme of multiplication, grafting and recording fruit yield and growth of combination trees ultimately proved the selectionists to be scientifically right.

In the rose, research on the selection of vigorous individuals in segregating populations, is based on a series of experiments with Hybrid Tea (HT) seedlings, of which the results can be transmitted to rootstock populations. An important parameter of vigour of HT seedlings is their shoot yield in a period of time. At Plant Research International (PRI), Wageningen, The Netherlands, several investigations into vigour in the rose were carried out that are described more fully elsewhere (*see* **Rootstocks**: Scion–Rootstock Relationships). These experiments, which involved various comparisons between own-rooted and grafted plants, demonstrated that vigour, expressed as shoot yield, is an inherent factor that is present in both the shoots and the roots of the plants of a scion genotype. For rootstock breeding the significance is that, in segregating seedling populations, direct selection for vigour can be made via the character shoot production. However, in contrast to HT populations, rootstock populations probably consist of nonrecurrent flowering individuals of which the one-season-old shoots do not bear flowers. Hence, following seed germination in spring, rootstock seedling plants may be grown in the field for one season.

In nondestructive measurements, the parameter shoot yield of the HT stocks is substituted by 'number of shoots' and 'sum of shoot length per plant' at the end of season. The advantage of growing populations in the nursery is that several other characters – resistance to powdery mildew or black spot, leaf size, thorniness, internode length and straightness of the shoots – may be evaluated as well. Particularly in cutting – grafting (*see* **Propagation:** Stenting. **Rootstocks:** Usage of Rootstocks) the latter two characteristics are highly valued. Moreover, there is ample plant material available for the tests of rootability. It is self-evident that selected individuals, or complete populations, should always be tested in combination with one or more scion varieties.

Indirect Testing of Vigour

Indirect testing of vigour means that instead of grafting followed by long-term observation of combination plants (in the glasshouse), the vigour of genotypes may be derived via correlations with plant or root characteristics other than yield. Indirect testing of vigour has made considerable progress, but not all methods have been checked in rose stocks. It has been demonstrated, however, that the induced vigour of (clonal) rose stocks can be reliably forecast by growing young combination plants in the phytotron in nutrient solution under controlled light and temperature conditions. In an experiment at PRI, the scion variety 'Sonia' was cutting-grafted on to eight clonal stocks, of which the induced vigour (shoot yield) of that scion variety had previously been ascertained in long-term glasshouse experiments. 'Sonia' scions, of which the axillary bud was removed, were selected for identical leaf area, while the rootstocks were selected for diameter and length. Combination plants were grown in aerated nutrient solution under controlled conditions in a phytotron chamber. Owing to abundant root growth, observations could not be prolonged beyond 15 days.

Daily recorded parameters were: the number of first- and second-order roots, number of root tips, total root length, root weight, specific root length, absolute and relative root growth, root porosity and shoot-to-root ratio. Unexpectedly, highest correlations (*r*) between the known shoot yield of 'Sonia' occurred with: specific root length (+ 0.77), number of root tips (+ 0.83) number of primary roots (+ 0.83), total root length (+ 0.83) and length of longest root (+ 0.89). It thus appeared that the relatively simple-to-measure parameter 'length of the longest root' in combination with 'number of root tips' may indicate the most vigorous stocks. The expenses of experiments in the phytotron should be weighed against another very highly priced, but probably even more accurate method, viz. analysis of root-produced cytokinins. These analyses, which were likewise carried out at PRI, demonstrated the level of endogenous cytokinins in the bleeding sap of various rose stocks (mainly zeatin riboside and its metabolites), to be positively correlated with their induced vigour. This indicates that, in a series of stocks, those stocks that produce the highest level of cytokinins are the most vigorous (**Figures 1** and **2**).

Currently, molecular markers for several characters of rose, including resistance to disease, are being developed or are already available. The greatest effectiveness of markers is commonly for monogenic characters, but those for so-called quantitative trait loci (QTLs) are presently being investigated. It is possible that markers for cytokinin loci may be discovered in the course of research. The same may apply to the quantitatively determined characteristic vigour. Another method of determining vigour of rootstocks has been developed at East Malling Research Station for apple and other rootstocks as well. It has been shown that there is an interesting inverse relation between vigour and the ratio of bark to wood in roots of about 5-mm diameter. Accordingly, stocks with a relatively thick bark are weak, while those with a thin bark are vigorous. This ratio has not yet been investigated in the rose.

Figure 1 Root system of 16-day-old combination plant of 'Sonia' on 'Moneyway', grown in nutrient solution in the PRI phytotron, Wageningen, The Netherlands.

Figure 2 Root system of 12-day-old combination plant of 'Sonia' on 'Marleen', grown in nutrient solution in the PRI phytotron, Wageningen, The Netherlands.

Breeding Heterogamous Stocks

The phenotypic uniformity of seedlings of Edelcanina varieties struck both horticulturists and scientists in the nineteenth century. Like *R. corymbifera* 'Laxa', and *R. rubiginosa*, Edelcanina varieties are pentaploids ($2n = 5x = 35$). Initially, botanists reasoned that Edelcaninas were homozygous, which was not the case. Later it was supposed that seeds were formed apomictically. Although apomixis does play a minor role, it has been ascertained that Edelcaninas are heterogamous – the ovules are tetraploid ($4x$), while the pollen is haploid (x) (**Figure 3**). As a consequence of heterogamy, there is a pronounced maternal influence on seed formation, which is phenotypically expressed in the seedlings.

As long as the ovules are fertilized by pollen of the same variety, seedlings resemble the mother and each other. Even when different Edelcanina varieties act as pollinators, seedlings may still show great uniformity, but when such 'hybrid' seedlings are repeatedly used as mother plants in seed gardens, the percentage of deviating genotypes ('rogues') increases with each successive generation. Since rogues may be the basis of a new stock variety, their selection in the field was a hobby of many Dutch and German nurserymen in

Figure 3 *Rosa canina* 'Inermis', a major progenitor to breed for heterogamy.

Figure 4 Ripe hips of *Rosa canina* 'Inermis' after cross-breeding.

the nineteenth and twentieth centuries. However, today, heterogamy is still a powerful system for breeding new seedling varieties, but it should be noted that the great advantage of heterogamy for uniformity simultaneously contains its greatest disadvantage in breeding: F_1 populations will hardly show selectable segregation. This means that segregating F_2 generations should be awaited, and only the F_3 may present populations of uniform stock with desired plant characters. Other drawbacks are the 2-year stratification period of seeds and the long (2–3 years) juvenile period of seedlings. Because each generation may take 6–7 years from the moment of pollination, breeders should be very patient and persevering because the complete programme may take over 20 years. The probable result of this long-term programme may be a rootstock for garden roses with improved properties of vigour, and environment-friendly fungus resistance.

The basis of cross-breeding should be one of the better *R. canina* varieties, e.g. 'Inermis', that has proved its merits in the twentieth century (*see* **Rootstocks**: Seedling Stocks) (**Figure** 4). Using 'Inermis' to raise F_1 populations means that its ($4x$) ovules are pollinated with diploid ($2n = 2x = 14$) fathers, producing (x) pollen. Diploid fathers occur among various wild roses or in nonperpetual flowering ancient ramblers. Although pollen incompatibility does not occur often, the use of different species in crosses challenges this phenomenon. It is not a luxury, therefore, to involve 10 or more different fathers. Obvious requirements of fathers are nonrecurrent-flowering, cross-compatibility with caninas, high vegetative vigour, well-developing root systems, resistance to the fungus diseases powdery mildew (*Podosphaera pannosa* (Wallr.: Fr.) de Bary var. *rosae* Lév.) downy mildew (*Peronospora sparsa* Berk.) and black spot (*Diplocarpon rosae* Wolf), winter-hardiness, absence of thorns and nematode resistance(?)no. These characters should also be selected for in successive F_1 to F_3 generations.

Although modern rose breeding is carried out in the glasshouse, breeding with the large canina bushes may be realized in the field, where the greatest dangers are late frosts and unwanted insect pollination. Father plants, however, may be kept in the glasshouse to ensure timely, sufficient quantities of pollen. The number of seedlings per population depends on fruit and seed set, but should not be fewer than 100 plants. Planting rows of various seedlings side by side in the field will facilitate overall evaluation of F_1 to F_3 populations. Spraying with fungicides should be omitted to test natural fungus resistance. While for the F_2s there is still the choice whether F_1 plants should be individually selfed or poly-crossed (pollen mixture), in all successive populations plants must be selfed. In principle, plants of each former generation may be discarded when they have yielded good numbers of seedlings of the next one. Ploidy levels of the retained F_2 and F_3 plants should be validated. It is likely that, after self-pollination of each F_2 plant, several F_3 populations will show sufficient uniformity, but comparison with seedlings of the original mother variety, i.e. 'Inermis', is inevitable. The result of these actions may be one or more offspring that are 'better' than the original mother varieties. The ultimate proof is, of course, grafting with a suitable scion variety. Final steps are: observation on bud take with several scion varieties and performance as to vigour of combination plants in the nursery and in the glasshouse. It is only when the qualities of the new creations have been validated that an application for breeder's right can be made and the new strain can be introduced to commerce. Not without reason have horticulturists shunned the breeding of seedling stocks. A very practical, but much less scientifically based approach is the approved method of selecting rogues from canina seedling fields. Rogues, which probably originate from accidental cross-pollination with scion varieties or other stock varieties, are easily recognized in the field in autumn. It cannot be excluded that by retaining such plants, followed by isolated planting, selfing and merely observing their offspring, a lucky hit may be scored.

Breeding Homozygous Stocks

Although it can be expected, theoretically, that inbreeding leads to genetic homozygosity and hence to some loss in vigour, it is needed to obtain a uniform crop of seedlings.

Crops are generally divided into self-pollinated or inbreeding crops, and cross-pollinated or outbreeding crops. Few crops fall strictly in one category only. Many species of roses are capable of self-pollination and are described as self-compatible if their pollen will fertilize ovules from the same plant, producing normal seeds and seedlings. Many rose species are diploids ($2n = 2x = 14$), that are self-compatible, although there are also reports of self-incompatibility in some species. Interspecific crosses between diploid species have not always been successful. Most current rose scion varieties are tetraploids ($2n = 4x = 28$), and may set seeds on selfing, but rarely produce normal seedlings. For that reason rose varieties are classed as a mainly cross-pollinated crop that needs pollen from a different genotype to reproduce in a normal way. A general rule is that after crossing two different, uniform, probably (partly) homozygous partners of the same crop, the F_1 generation is phenotypically uniform but highly heterozygous. Desired segregation for characters only occurs in the F_2, where favourable genotypes are retained and others are discarded. In successive steps, F_3 and probably F_4 generations are raised by selfing the retained genotypes until sufficiently uniform, probably homozygous individuals are obtained. In a theoretical example of breeding homozygous stocks, the diploid ($2n = 2x = 14$) *R. multiflora* Thun. variety 'Smit's Thornless' might be used as a female parent (**Figure 5**). 'Smit's Thornless' produces uniform seedlings and thus must be at least partially homozygous. It is nonrecurrent-flowering, easy to cross-pollinate, bears large sprays of many flowers, has few seeds per fruit, a favourable short stratification period and a juvenile period of 1 year.

Figure 5 Thornless *Rosa multiflora*, an anticipated progenitor in breeding homozygous seedling stocks.

As in the *R. canina* programme, suitable fathers should be vigorous diploids with few thorns and should be resistant to fungal diseases. They may be either species roses or healthy diploid varieties. It should be realized that such crosses are interspecific. Interspecific crosses may result in pollen incompatibility or sterility of F_1 plants. To improve fruit and seed set, as well as to speed up the programme, breeding should preferably be carried out in the glasshouse. Even in the F_1 generation, some populations may show segregation for characters. Weak or unsuitable F_1 individuals should be immediately discarded, and vigorous or otherwise good-looking ones should be retained. When F_1 plants are self-compatible, segregation for characters is expected in the F_2. Depending on the degree of segregation of important characters, selfing of suitable plants may be continued for several generations. As in the *R. canina* example, the value of new varieties is only validated after the performance of combination plants has been tested. Amongst *R. multiflora* seedlings, rogues occur, probably in an even higher percentage than in *R. canina*. Also in this case selection for rogues may be a cheap basis of a new seedling rootstock variety or, when exceptionally vigorous, of a clonal stock.

Breeding Clonal Rootstocks

Unlike seedling stocks, of which the plants are uniform by heterogamy or by homozygosity, plants in a clone are uniform owing to vegetative reproduction (*see* **Rootstocks**: Usage of Rootstocks). For the breeding of clonal stocks four disciplines may be practised: (1) selection and propagation of desirable genotypes from the wild; (2) selection of deviating genotypes within existing varieties; (3) cross-breeding of suitable progenitors; (4) introduction of foreign genes via biotechnology. As mentioned in the foregoing paragraphs, there are various ways of screening populations and selection for desirable characteristics, particularly vigour. In contrast to finding clonal stocks in the wild, selection in existing populations has been quite successful. Hence, *R. indica* 'Major' and 'Manetti', the oldest clonal stocks known, were selected from populations that aimed at scion varieties. Also the stock 'Dr Huey' (*R.* 'Shafter'), raised by Thomas from 'Ethel' × 'Grüss an Teplitz' in 1914, was selected from a population aiming at a new scion variety. Even the recently introduced clones 'Multic' and 'Ludick' have resulted from selection in a natural *R. multiflora* var. *cathayensis* Rhed. & Wils. population, at Wageningen University, The Netherlands, around 1990.

Two other Dutch projects have aimed at selection of clonal *R. canina* stocks from seedling populations. The first one, a funded programme, aimed at a clonal 'Inermis' in the last decade of the twentieth century (*see* **Rootstocks**: Seedling Stocks). Here, the experience in apple was applied in a reversed way, viz. a large number of good-looking individual 'Inermis' seedlings were cloned by root grafting them with 'Sonia'. After 2 years of recording yield data, four clones, STUR 1, 2, 3 and 5, were selected. Unfortunately, the retained mother plants of these clones had meanwhile become adult and would no longer root as softwood cuttings. In another (private) programme, 'Spektaculair', a root clone of 'Inermis' was selected. Hence, each spring *ex vitro* plants of the stock had to be planted in the field and uprooted in autumn. Combination plants were composed via root grafting. Owing to undesirable *Phytophtora* infection, this clone has never become successful.

The category 'populations from cross-breeding' has been better documented than the former ones. Probably the first project was initiated in the 1930s, at Iowa State University, USA. The use of a *R. multiflora* variety and other progenitors has been described, but the results are as yet unclear. It is not improbable that the stocks IT 9 and 18, which are seemingly lost, have originated from that programme. Around 1980, modern year-round cut-rose culture in north-western Europe gave a new impetus to research in adapted stocks. Within a larger programme, the breeding and selection of new clonal stocks for the glasshouse were initiated. Using the dominant inheritance of the gene for nonrecurrent flowering (R), crosses between recurrent flowering HT varieties and nonrecurrent species roses have yielded vigorous nonrecurrent clonal stocks. Bred at PRI, Wageningen, the stocks 'Moneyway', a triploid ($2n=3x=21$) from the cross 'White Weekend' (HT, $2n=4x=28$) × *R. multiflora* var. *adenochaeta* ($2n=2x=14$) and 'Marleen', likewise a triploid from the cross 'Sonia' (HT, $2n=4x=28$) × *R. multiflora* var. *adenochaeta* ($2n = 2x=14$), resulted from that programme. The underlying ideas of using HT varieties as progenitors were: better fruit and seed set, probable avoidance of scion–graft incompatibility and the likely occurrence of hybrid vigour (heterosis). Another clonal stock, 'Popeye', which was developed by the private Dutch breeder and plant propagator Olij, was introduced in 1998. It resulted from the cross *R. multiflora* var. *adenochaeta* (tetraploid form) × 'Kardinal' (HT). Besides the fact that 'Popeye' induces shoot yields that equal those of own-rooted varieties, it inherited the character resistance to the nematode *Meloidogyne hapla*.

As to actual breeding of clonal stocks, the views on tactics and selection methods to be followed have changed drastically since the 1980s. Selection criteria for vigour are now available, construction of combination plants has improved and problems of winterhardiness are avoided by planting rootstock fields in the tropics. In spite of that, questions as to which progenitors to use for even more vigorous populations and the introduction of resistance genes against 'new' diseases like *Cylindrocladium* or *Conyothrium* keep the breeder busy. With regard to breeding for improved vigour, it should be noted that, as in many crops, vigour in the rose commonly increases with ploidy level, until an optimum is reached. In the rose, the nonrecurrent-flowering triploids ($2n=3x=21$) seem to be the most vigorous ones. For that reason, triploid rootstocks seem a good option. Aiming at nonrecurrent-flowering triploids has the additional advantage that easy-to-pollinate tetraploid HT varieties may be used as mothers. In combination with a multitude of diploid species roses available as fathers, vigorous, triploid offspring can be raised. This philosophy was successfully applied in the creation of 'Moneyway' and 'Marleen'.

A question yet to be answered is, which fathers should be used? Obviously, species with pronounced winter dormancy should be avoided because light quantity or quality in the temperate zones may affect the growth of their offspring. Hence, the Asiatic species roses might be considered. As to species roses in general, it should be admitted that available knowledge is about botanical rather than horticultural properties. Some authors, however, have listed vigorous but little-used species. Thus *R. multiflora* selections (*R. multiflora* var. *adenochaeta*, *R. multiflora* var. *cathayensis* Rehd. et Wils. and *R. multiflora* 'Inermis'), *R. bella* Rehd. et Wills. or *R. davidii* Crép. have been proposed as a good choice in breeding. Alternatively, healthy, vigorous, ancient ramblers like 'Albéric Barbier' (*R wichurana* hybrid), or early varieties of *R. odorata* (Andr.) Sweet and *R. gigantea* Coll. ex Crép. might be interesting partners.

Another approach may be direct cross-breeding, viz. without the use of hybrid roses, between suitable rose species. In addition to cross-breeding with inherently vigorous genotypes, science offers the possibility to breed highly vigorous plants by the inbreeding of various lines, followed by crossing between these lines. Application of that method has effected heterosis or hybrid vigour in many crops. In the rose, inbreeding by selfing is known to lead to inbreeding depression at such a fast rate that experiments were not continued. On the other hand, the few varieties that have escaped that phenomenon may be a valuable basis of vigour.

In addition to traditional cross-breeding and selection, biotechnology promises fantastic opportunities to improve existing (stock) varieties by the insertion of one or more highly effective foreign genes. Unlike several other crops, however, the genetic manipulation of roses is a difficult affair. Despite this, biotechnological research at PRI has been successful in breeding three genetically modified organisms (T1, T2, T3) derived from the clonal stock 'Moneyway', containing *rolABC* or *rolB* genes, in 1992. When used as stocks for 'Madelon', 'Vivaldi' and 'First Red'®, the shoot yield induced did increase about 20% compared to that on untransformed 'Moneyway', but unfortunately the shoot quality was highly inferior. Because *rol* genes interfere with the endogenous hormone balance, this result is enough to make one think about how little extra cytokinin is needed, for example, to disturb the performance of combination plants. Another foreign gene, the *IPT* gene, seems useful for enhancing the endogenous cytokinin production as well, thus contributing to the distribution of carbohydrates over more shoots than normal. Again the question arises about how tender the endogenous auxin-to-cytokinin ratio is. As long as the practical use of genetically modified organisms is prohibited by European Union legislation, further development of stocks by genetic modification is seriously hampered.

See also: **Propagation**: Stenting. **Rootstocks**: Scion–Rootstock Relationships; Usage of Rootstocks; Seedling Stocks. **Breeding**: Selection Strategies for Disease and Pest Resistance.

Further Reading

Allard RW (1999) *Principles of Plant Breeding* (2nd edn). New York: John Wiley.

DeVries DP (1993) *The Vigour of Glasshouse Roses. Scion–Rootstock Relationships. Effects of Phenotypic and Genotypic Variation.* Thesis. Wageningen, The Netherlands: Agricultural University of Wageningen.

De Vries DP and Dubois LAM (1997) Relations between early root growth and flower yield of clonal rose rootstocks. In: Altman A and Waisel Y (eds.) *Biology of Root Formation and Development*, pp. 283–288. New York: Plenum Press.

Dieleman JA (1998) *Cytokinins and Bud Break in Rose Combination Plants.* Thesis. Wageningen, The Netherlands: Agricultural University of Wageningen.

Ferguson JF (1933) A botanical study of rose stocks. *Journal of the Royal Horticultural Society* 62: 344–371.

Garner RJ (1979) *The Grafter's Handbook* (4th edn). London: Faber and Faber.

Hartmann HT, Kesler DE, Davies IT and Geneve RL (1997) *Plant Propagation: Principles and Practices.* NJ: Prentice-Hall.

Hurst CC (1941) Notes on the origin and evolution of our garden roses. *Journal of the Royal Horticultural Society* 66: 73–82, 242–250, 282–289.

Kramer PJ and Kozlowski TT (1979) *Physiology of Woody Plants.* New York: Academic Press.

Krüssmann G (1986) *Rosen, Rosen, Rosen*. Berlin: Paul Parey Verlag.

Werlemark G (2000) *Genetic Variability and Reproductive Strategies in Nordic Dogroses, Rosa Section Caninae.* PhD Thesis. Alnarp: Swedish University of Agricultural Science.

Usage of Rootstocks

D P de Vries, Plant Research International (PRI), Wageningen, The Netherlands
Present address: Foundation Sub Rosa, Ede, The Netherlands

Introduction

A rootstock is a plant or plant part onto which the scion of a variety is grafted. Because rose varieties are highly heterozygous for all characters, they must be propagated vegetatively to conserve the genotype. In addition to cuttings, layering or root sowing, grafting varieties onto rootstocks serves that purpose. The act of grafting creates a so-called 'combination plant'. Combination plants commonly consist of two different genotypes: the stock that is subsoil and the scion variety that forms the aerial part. In principle, two kinds of rootstocks are being used for grafting: seedling or generatively propagated stocks (*see* **Root Stocks:** Seedling Stocks) and clonal or vegetatively propagated stocks (*see* **Rootstocks:** Clonal Rootstocks). It is notable, by the way, that any rose may be used as a stock, but not any stock will prosper in any root substrate. Although rose varieties may prosper for some time on stocks that do not belong to the genus *Rosa*, in practice rose varieties are always grafted onto another rose.

Current garden rose bushes are mainly combination plants, of which the subsoil part is a rose seedling of some named stock variety. Although in glasshouse culture in Europe rootstocks are still being used, for example in The Netherlands, current cut-rose varieties are for the greater part grown on their own roots. This is not only because in modern glasshouse cultivation the yield of own-rooted plants is almost equal to those on a rootstock, but also a matter of cost of plant material for the production of young plants. For example, in western Europe in the year 2002, one-season-old seedling stocks of *Edelcanina* varieties cost approximately €0.20 apiece, and the internodes of clonal stocks cost approximately €0.05. Depending on the variety and time of year, one internode of a scion variety to be used for grafting or for softwood cuttings costs about €0.30 (excluding royalties). Hence, the price of material for a plant on *Edelcanina* stock would be €0.50, on clonal stock €0.35, and on its own roots €0.30. At a normal plant density of 80 000 plants per hectare, differences between the cheapest and most expensive plants would be about €16 000. Outside Europe, however, seedling and clonal rootstocks are still extensively used for both outdoor and glasshouse cultivation.

Rootstocks enable the relatively fast and cheap multiplication of large numbers of scion varieties; they give combination plants a good hold in the soil, control their vigour, guarantee a good uptake of water and nutrients and bring about improved winter-hardiness. In particular clonal stocks can be used to control the growth and development of scion varieties. Stocks do not directly influence the disease resistance of varieties, but may do so indirectly as a consequence of rankly or rather of inhibited growth. Some varieties of stock affect the flower colour of scion varieties unfavourably, which particularly occurs in combination with pink or light yellow varieties.

History of the Usage of Rootstocks

The cultivation of roses has been mentioned within living memory, although it is unclear whether or not the plants of earliest scion varieties were obtained by grafting onto rootstocks, by cuttings or by just sowing their seeds. The application of rootstocks in southern and western Europe is not mentioned in the literature before 1551. Over 100 years later, around 1665, it appeared to be customary to dig up suitable seedlings of various species of roses in the wild in autumn, and subsequently graft them on to a valuable scion variety. In contrast to current views on the function of stocks, originally varieties were grafted as an intermediate step for the purpose of getting them on their own roots, with the underlying idea that self-rooted plants were more true to type than those on the roots of a strange kind. To get varieties on their own roots, one-season-old combination plants were dug up, the variety parts were lightly scarified and the plants were replanted deep into the soil. Next autumn, when the scion variety had formed roots, the stocks were removed. In the eighteenth century, however, owing to a considerable increase in the number of newly bred scion varieties, the demand for rootstocks could no longer be met by haphazardly collecting stocks from the wild. Henceforth, seedlings were obtained from seeds of deliberately selected mother plants of various descent, thus establishing the first seed gardens. Gradually, it became clear that, when grown on a seedling stock, varieties were as beautiful as on their own roots. The best seedling stocks in the UK were 'Damask' (*Rosa damascena* Mill.), 'The White' (*R. alba* L.), 'Francford' (*R. francofurtana* Muench) or the 'Wilde Eglantine' (Sweet Briar or *R. rubiginosa* L.).

The dog rose (*R. canina* L.) was not mentioned as a useful stock before the beginning of the nineteenth century. The discovery of dog roses as a stock was the beginning of the success story of the seedling canina stocks, which, as the refined and rather uniform *Edelcaninas*, even today are the most important and best-known stocks for garden roses all over the world. Next to the caninas, however, *R. corymbifera* 'Laxa', and the uniform *R. multiflora* Thunb. seedlings were used in both Europe and the USA, from the nineteenth century onwards. In addition to these stocks, seedlings of *R. acicularis* Lindl., *R.* × *fortuneana* Lindl., *R. laevigata* Michaux, *R. rugosa* Thunb., *R. setigera* Michaux and *R. villosa* L. have been used but have never become of general importance. Although the gene centres of the rose occur throughout the northern hemisphere, most of the species used for stocks were endemic in either Europe or the USA. It seems justified to conclude that horticulturists favoured those stocks because of climatic adaptation to their homeland. For that reason several stocks that were successful in southern USA did not survive in Europe. It should be mentioned that all stocks are of the 'nonrecurrent' type – they do not flower in the year of germination or cutting, but only after 1 year or more, and when they flower, it is in spring only.

Parallel to the development of seedling stocks in the nineteenth century, breeders who meanwhile had imported recurrent flowering China roses and used them in crosses selected two clonal stocks of lasting importance – 'Manetti' and *R. indica* 'Major'. Both clones are still being used for garden roses or glasshouse cut roses in the Mediterranean area and the USA. Another old clonal stock, Dr Huey (*R.* 'Shafter'), has been in vogue in Australia for some time. In the late twentieth century, several newly selected clonal stocks for the glasshouse were introduced: 'Moneyway', 'Marleen', 'Multic', 'Popeye', 'STUR' selections and 'Spektakulair' in The Netherlands, as well as a number of K-stocks in Japan. Despite the favourable characteristics of many new stocks they were replaced by 'Natal Briar', which has almost completely dominated the western European cut-rose industry from about 1990 onwards. In the course of the twentieth century, the initially large assortment of seedling stocks has gradually narrowed. Decreasing prices of garden rose plants and heavy competition between propagators of cut-rose plants in Europe have accelerated this process. This situation is not a stimulus for the development of new stocks. As a consequence, prospects of future breeding of new stocks appear to be not very good; the private sector fears high investments for little gain, while the public sector awaits the opportunity of funded projects for research.

Production of Combination Plants

For the multiplication of scion varieties by grafting, either complete rootstock plants or their shoots or roots are commonly used. Grafting means to prepare and place together plant parts so that they may grow together. Of the many methods of grafting, those usually applied to produce rose combination plants are: (1) budding; (2) bench grafting; (3) root grafting; or (4) cutting–grafting ('stenting'), each with several variants. When complete one-season-old seedling stocks are involved, axillary buds or scions of the variety are grafted onto the root collar of these stocks. In the case of clonal stocks, the axillary buds are either budded in the nursery onto the aerial part of rooted cuttings, or the scion is placed onto yet unrooted internodes of the stock in the glasshouse.

Budding

Budding of garden roses in northwest Europe is usually carried out in the outdoor nursery at the height of the growing season, which is in June–July. Stocks are the one-season-old plants (6–8 mm grading) of seedling rootstock varieties that had been planted in rows in the nursery in late winter or early spring. In the southern hemisphere, budding is likewise carried out during the height of the season. Because stocks have many shoots at the time of budding, these are pushed to one side of a row of plants by ingeniously designed heavy bars, so that there is easy access to the root collar for the budder. Invariably, the method known as 'shield budding' or 'T-budding' is used for roses (**Figure 1**). To that aim the axillary buds are taken from shoots of the current season's growth of the scion variety – the so-called 'bud sticks'. Buds exist in the leaf axils of bud sticks.

In shield budding the axillary bud is removed from the bud stick by a shallow slicing cut which begins about 1 cm below the bud, passes under the bud and comes out well above it. In advance, the root collars of the stocks are wiped free of soil and prepared by making a T-shaped incision through the rind down to the wood. Then the rind is raised and the shield slid downwards between the edges of the rind and well below the horizontal incision. The surplus material of the shield is cut off exactly at the horizontal incision. The shield is held in position and the wound covered by a Fleishauer. Fleishauers, which have replaced tying raffia, consist of thin 3 × 3-cm slabs of pararubber, on one side provided with a long metal staple. The rubber slab, covering the T-incision with the shield, is stretched around the inserted bud in the root collar and pierced on the opposite side by the staple. Lightly earthed in autumn, the budded stocks hibernate with a quiescent bud that will sprout next spring after the

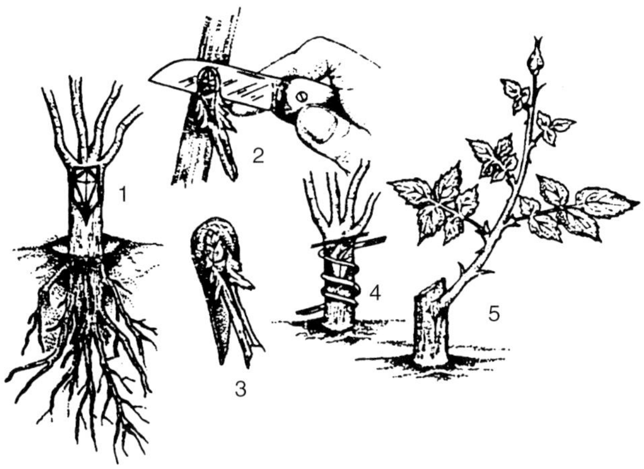

Figure 1 Shield budding. (1) Seedling rootstock with T-incision in root collar; (2) cutting the shield of the scion variety; (3) shield with axillary bud and leaf petiole; (4) shield inserted in T-incision and tied with raffia or Fleishauer; (5) combination plant with primary shoot. © 1986, D. P. de Vries and L. A. M. Dubois.

(a)

(b)

Figure 2 (a) Successfully rooted and sprouted 25-day-old cutting-graft on Natal Briar rootstock, Olij Nursery, De Kwakel, the Netherlands. (b) Clonal interstock 'Push Up' with two inserted buds of a scion variety.

rootstocks' shoots have been removed. This method will yield vigorous bushes with several branches in autumn. The quality of uprooted combination plants of these '1-year bushes' is determined by their number of branches, hence AA = three branches or more, A = at least two branches, or B = one branch. For the production of standard roses (90–180 cm tall) the long and relatively upright stems of 3-year-old *R. canina* 'Pfänder' seedlings are budded with three buds of a scion variety.

In the early twentieth century, the sturdy, vigorous stems of *R. rugosa* 'Hollandica', multiplied by cuttings, were also in vogue. A seeming novelty in the production of standard roses is the current use of an intermediate or interstock (**Figure 2**). To that aim vigorous clonal stocks, upright-growing, with few spines, are budded on 1-year-old *R. corymbifera* 'Laxa' seedlings. In the next year the interstock is budded in July, so that the final combination plant consists of three instead of two different genotypes. Favourable interstocks are 'Push-Up', 'Moneyway', 'Chessum's Choice' or 'Rocket'. So-called 'half-year' bushes of cut roses, to be exclusively planted in the glasshouse, are budded in May or early June, likewise on 6–8-mm or 8–12-mm stocks that had been planted in early winter in Dutch nurseries. When the bud has taken after 4–6 weeks, the shoots of the rootstock are removed, thus inducing the quiescent bud to sprout and yield a light bush that can be directly planted in the glasshouse in October. For these half-year bushes the same quality standards are applied as for the 1-year bushes. In the Mediterranean area, mainly to produce cut-rose bushes, budding is carried out in the field onto half-year-old rooted hardwood cuttings of the clonal stock *R. indica* 'Major' in April. Combination plants are ready for planting in the glasshouse in October. In the USA, 'Manetti' is being treated in a similar way, where this stock is also used for garden roses. In Japan and Korea budding is sometimes carried out in the glasshouse on hardwood cuttings of clonal stocks. Another form of budding is currently applied in Kenya, where branches of clonal rootstocks are shield-budded in the field with a series of buds of cut-rose scion varieties. For budding the axillary buds with subtending leaves of the rootstock are removed and substituted by buds of the scion variety. Three weeks later the branch is harvested, cut into pieces that contain the variety's bud, and subsequently rooted under mist.

Bench Grafting

Bench grafting of cut roses is carried out under glass in specialized nurseries, mainly in winter. Stocks are the one-season-old seedlings, of mainly the *Edelcaninas* 'Inermis', 'Brögs Stachellose' or 'Pollmers', when these are available from November onwards. Stocks of the 8–12-mm grading are prepared for grafting by removing the shoots just at the top of the root collar (**Figure 3**). Because the stock is always thicker than the scion, the method of grafting is so-called 'rind grafting'. Hence, in the rind of the root collar a basipetally directed incision of 3–5 cm is made, and the rind is raised at both sides of the incision to insert the scion. Scions for bench grafting which originate from actively growing varieties in heated glasshouses are named 'warm' scions; those from varieties in winter rest are 'cold' scions. The latter are commonly without leaves. Currently, in year-round culture, only warm scions are being used; these consist of one internode of a flowering shoot with one leaf that is shortened to two leaflets, and its quiescent axillary bud. These scions, prepared as 'splice grafts', are placed between the rind

Figure 3 Bench grafting. (1) Scion of the variety; (2) rootstock with T-incision ready for the scion; (3) tied rind graft; (4) young combination plant about 6 weeks old. © 1986, D. P. de Vries and L. A. M. Dubois.

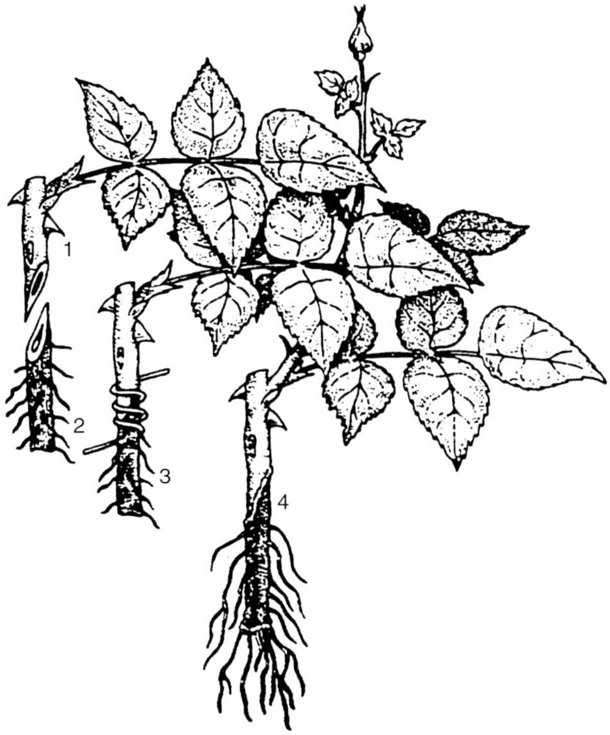

Figure 4 Root grafting. (1) Scion of the variety; (2) piece of root of the stock; (3) scion and stock united with tying raffia or washing peg; (4) young combination plant 6 weeks old. © 1986, D. P. de Vries and L. A. M. Dubois.

and wood of the stock's root collars, with the scions' leaf always pointing 'outside'. In rind grafting the contact between the cambiums of the scion and stock is optimal. To keep the scion in position, it is tied to the root collar by adhesive tape, which has been substituted for raffia or cotton twine for several years. When placed under 'double glass' (windows covering tables in a glasshouse) at high temperatures (23–25 °C), high air humidity and under 24-h supplementary lighting, young plants are ready for planting in 4–6 weeks. In some nurseries small-scale bench grafting is used to supplement summer budding.

Root Grafting

For root grafting, pieces of roots of *Edelcanina* seedlings about 5–7 cm long and 2–5 mm thick are splice-grafted with a full-leafed scion of the variety, likewise under glasshouse conditions (**Figure 4**). In this action it is important that, at least on one side of the constructed plantlet, the cambiums of scion and stock are in touch. Scions are held in position by washing pegs. The grafted young plants are stuck either in rockwool blocks or in coco plugs and, like the young cutting grafts, placed in a conditioned glasshouse. Root grafting is a relatively cheap method, since several pieces of root may be cut from one 8–12-mm rootstock plant.

Cutting–Grafting

Cutting–grafting, which originated in the USA, has been further developed in The Netherlands for the propagation of cut roses in the last decade of the twentieth century. The method was renamed there as 'stenting'. Currently, cutting–grafting is the main way to produce cut-rose combination plants in western Europe. As in root grafting, in cutting–grafting the scion consists of one internode of a flowering shoot of the variety, with one full leaf and its quiescent axillary bud (**Figure 5**). This scion is splice-grafted with a 5–7-cm-long piece of stem of a clonal stock, preferably of the same diameter as the scion. As in root grafting, scion and stock parts are kept in place by washing pegs. In this case too, the emphasis is on a good connection between cambiums. The basal part of the stock internode is treated with a rooting-promoting auxin, commonly indole-3-butyric acid (IBA), in concentrations of about 0.5% in talcum powder. Growing together of the scion and the stock, the root formation of the stock and sprouting of the scion occur almost simultaneously under water mist, high temperature and 24-h supplementary lighting.

A slightly different form of cutting–grafting is applied in the south of France, where softwood pieces of

Figure 5 Cutting–grafting. (1) Scion of the variety; (2) piece of internodium of the clonal stock; (3) scion and stock united with cotton twine or washing peg; (4) young combination plant 6 weeks old. © 1986, D. P. de Vries and L. A. M. Dubois.

clonal stocks are shield-budded under glasshouse conditions. When miniature rose varieties came into vogue in the 1930s, yet another form of cutting–grafting was applied in Dutch nurseries. 'Foot stems' (30 cm) or 'standards' (90–120 cm) cut in autumn from outdoor-grown thornless *R. multiflora* shoots are cutting-grafted with dormant wood of miniature rose varieties. These kinds of plants are exclusively produced in late winter to be ready for auction at Mother's Day.

For seedling stocks it has been demonstrated that the diameter of the root collar has a pronounced effect on the vigour of combination plants in the glasshouse during the first year. Hence, when the diameter of 'Inermis', 'Brögs Stachellose' or 'Pollmers' seedling stocks used for bench grafting increased from 3 to 14 mm in diameter, the first year's flower shoot production of the scion variety 'Sonia' increased by about 25%. For that reason nursery practice for glasshouse roses has always favoured the 8–12-mm grading of stocks.

In the practice of grafting, it is generally accepted that different plant parts will unite and continue growing as one combination plant. This is not the case when so-called incompatibility occurs. Scion–rootstock incompatibility means that the scion and stock completely fail to unite, but commonly the term indicates that the union between scion and rootstock is incomplete so that it may snap in the course of time. The phenomenon is well-known to occur in apple, pear, plum or apricot, when varieties are grafted on stocks of a different species. Frequently, incompatibility is accompanied by sometimes large swellings at the graft union, by unhealthy growth or by 'suckering' – wildly growing shoots of the stock. Although incompatibility in roses is rare, it has been reported that when grafted onto the seedling stock 'Heinsohns Rekord', several varieties show poor growth and true signs of incompatibility after 1 year or more. The same may apply to late varieties grafted onto *R. rubiginosa*. In addition, scions of *R. banksiae* cultivars that are compatible with *R. indica* 'Major' do not take at all on *Edelcanina* stocks.

Fading of flower colours has been reported in mainly light yellow or pink-coloured scion varieties when these are grafted on the seedling stocks 'Laxa' or 'Multiflora'. Also the clonal stock 'Multic' induces paler flower colours in most scion varieties than when these are grafted on canina. Particularly for glasshouse-grown varieties, paling detracts from the quality of the auctioned product.

Despite the fact that most rose varieties are easily propagated by softwood cuttings, own-rooted garden rose plants commonly lack drought resistance, and particularly winter-hardiness, compared to the same varieties on an *Edelcanina* stock. These effects may be explained from: (1) the rather flat, superficial and therefore drought-sensitive root system of cuttings, compared to the more vertical-growing and deep-rooting seedling stocks; and (2) later entering winter dormancy and earlier bud-break in spring of own-rooted varieties. Similar plant performance has been noticed in likewise own-rooted *ex vitro* garden roses. For those reasons (seedling) stocks should be preferred to own-rooted plants. Only a number of garden varieties of winterhardy species like *R. rugosa*, which tend to form root suckers, prosper on their own roots.

The current certificates of 'true to type and freedom of viruses' for both seedling and clonal stocks, as issued by Dutch Naktuinbouw, largely contribute to the quality of combination plants. Particularly for clonal rootstocks, of which the mother plants are cultivated for a length of time in the same location, often in the open, freedom from virus has become an important factor to prevent spreading of production-inhibiting viruses. In Kenya, for example, new fields are often laid out with locally produced softwood cuttings, so the chances of virus infection increase with each vegetative generation. For that reason perpetually founding new fields with plant material from a previous vegetative generation is against the strict rules for maintaining the virus-free status of stocks. This means that when laying out new fields, plants grown from virus-free stock plants should be used.

An intriguing, but poorly investigated factor is the longevity of combination plants. Empirical knowledge is, however, that as long as good care is taken, garden roses may live for 25 years and longer. For glasshouse-grown cut roses the question is more or less a theoretical one, because growers tend to switch to new varieties every 4–5 years.

See also: **Propagation:** Budding in the Field; Stenting; Own-Rooted Cuttings. **Rootstocks:** Scion–Rootstock Relationships; Rootstock Breeding; Clonal Rootstocks; Seedling Stocks.

Further Reading

Allard RW (1999) *Principles of Plantbreeding* (2nd edn). New York: John Wiley.

Buck GJ (1951) Varieties of rose understocks. *American Rose Annual* 101–116.

De Vries DP (1993) *The Vigour of Glasshouse Roses. Scion–rootstock Relationships. Effects of Phenotypic and Genotypic Variation.* Thesis, Agricultural University of Wageningen, The Netherlands.

De Vries DP and Dubois LAM (1997) Relations between early root growth and flower yield of clonal rose rootstocks. In: Altman A and Waisel Y (eds.) *Biology of Root Formation and Development*, pp. 283–288. New York: Plenum Press.

Dieleman JA (1998) *Cytokinins and Bud Break in Rose Combination Plants.* Thesis, Agricultural University of Wageningen, The Netherlands.

Edwards RE (1955) *A Survey of Rootstocks for Roses.* Dissertation, University of Nottingham.

Ferguson JF (1933) A botanical study of rose stocks. *Journal of the Royal Horticultural Society* 58: 344–371.

Garner RJ (1979) *The Grafter's Handbook* (4th edn). London: Faber & Faber.

Hartmann HT, Kester DE and Geneve RL (1997) *Plant Propagation: Principles and Practices.* Upper Saddle River, NJ: Prentice-Hall.

Kramer PJ and Kozlowski TT (1979) *Physiology of Woody Plants.* New York: Academic Press.

Krüssmann G (1954) Die Vermehrung der Laubgehölze. In: Krüssmann G (ed.) *Die Baumschule*, pp. 422–439. Berlin: Parey.

Krüssmann G (1986) *Rosen, Rosen, Rosen.* Berlin: Parey.

Leemans JA (1964) Rootstocks for roses. *Plant Propaganda Holland*, 72.

Ohkawa K (1986) Rootstock native to Japan. *Acta Horticulturae* 189: 61–66.

Shepherd RE (1954) *History of the Rose.* New York: Macmillan.

Van De Pol PA (1986) Root grafting and screening super canina rootstocks. *Acta Horticulturae* 189: 81–87.

Werlemark G (2000) *Genetic Variability and Reproductive Strategies in Nordic Dogroses, Rosa Section Caninae.* Doctoral Thesis, Swedish University of Agricultural Science, Alnarp.

Clonal Rootstocks

D P de Vries, Plant Research International (PRI), Wageningen, The Netherlands
Present address: Foundation Sub Rosa, Ede, The Netherlands

Introduction

Among the rootstocks which are being used for the propagation of rose varieties, two categories basically exist: seedling rootstocks, also called generatively propagated stocks (*see* **Rootstocks:** Seedling Stocks), and clonal rootstocks, also named vegetatively propagated stocks. The use of rootstocks was already known as early as the year 1550, when good-looking seedlings were dug up in the wild, to be grafted with precious scion varieties. It was not until the end of the nineteenth century that varieties of seedling stocks came into vogue. These same rootstock varieties are still being used for garden roses and for a restricted number of glasshouse cut roses. Despite the fact that the first clonal stocks were only used for garden roses, their current use is mainly in the cut-rose industry. Recently, a new market for clonal rootstocks has been discovered, in which a special kind of vigorous clonal stock is used as an interstock for standard garden rose trees.

Today, in the application of stocks, the emphasis is on the control of the vigour of combination plants, rather than on just the multiplication of varieties. Only new clonal stocks meet that requirement. Rooting hormones and the production of combination plants under controlled environments have greatly contributed to the renewed popularity of clonal stocks. Because clonal stocks literally form the basis of the production of many generations of scion varieties, the mother plants must be treated with great care. Owing to their long life and way of cultivation, the chances of infection with *Agrobacterium* and various viruses that inhibit the growth and development of the combination plant are obvious. Despite the availability of a number of highly vigorous clonal stocks at the turn of the century, there is a tendency to limit the assortment to just one or two varieties. Owing to a newly developed system of cultivation for cut roses in western Europe, in which most varieties are grown on their own roots, the general use of (clonal) stocks has been forced back. However, it is possible that new clonal stocks may be developed which improve the vigour of cut-rose combination plants without the loss of shoot quality.

History of Clonal Rootstocks

In contrast to the historical development of seedling stocks from wild rose species, the first clonal stocks did not spring from endemic European or North American roses. They were either direct descendants of *Rosa chinensis* Jacq. and *R. odorata* (Andr.) Sweet, or hybrids with these species. This indicates that clonal stocks appeared later in history than seedling ones. It should also be considered that, despite the fact that bench grafting had already been mastered before the year 1500, owing to a lack of technical apparatus the operation of grafting must have been a difficult one for early horticulturists. In addition, it should be realized that the glasshouse-grown 'softwood' material of roses propagates faster and easier than the dormant 'hardwood' material used in those times. For these reasons, the development of clonal stocks initially lagged behind that of seedling stocks. The discovery of root-promoting auxins, and the application of the method of cutting-grafting under a controlled environment, have been highly conducive to the use of clonal stocks in the modern cut-rose industry.

Already in 1740, *R. indica* 'Major' (syn. *R. odorata* (falso)) was mentioned as a rootstock. In the Mediterranean area *R. indica* 'Major', which is called 'Niszr' in Egypt, is still a much-esteemed stock for cut roses. Another clonal stock, R. 'Manetti' (syn. *R. noisettiana* 'Manetti', *R. chinensis* 'Manetti' hort.), was named after its probable breeder Dr Manetti, director of the botanical gardens in Monza, Italy (**Figure 1**). It arrived in England in 1835 and was spread by the famous breeder Thomas Rivers. 'Manetti' is also written as 'Manettii'. In contrast to *R. indica* 'Major', 'Manettii' is winter-hardy. Owing to the discovery of the technique of propagation via hardwood cuttings, *R. indica* 'Major' and 'Manettii' became very popular in southern Europe in a short time. Because in north-west Europe *R. indica* 'Major' was too soft to be used in propagation and could not be handled by propagators before 1980, growers had to depend on imports of combination plants from the Mediterranean area.

Figure 1 Well-rooted, 5-month-old hardwood cuttings of Inermis # 29, Brögs, 'Floradale' and 'Manettii', grown in the Plant Research International nursery, Wageningen, the Netherlands.

Another old clonal stock, 'Dr Huey' (syn. R. 'Shafter'), raised from 'Ethel' × 'Gruss an Teplitz' by Thomas in 1914, was once in vogue in Australia but is now little used. For the above-mentioned reasons, the breeding and selection of new clonal stocks have been less popular than those of seedling stocks. However, in the 1930s, at Iowa State University, USA, the first scientific breeding programme for clonal stocks was initiated, using a *R. multiflora* variety as a progenitor. Unfortunately, the stocks IT-9 and IT-18, which originated from that programme, are probably now lost.

Around 1980, the demands of modern year-round cut-rose culture in north-west Europe gave a new impulse to research into adapted stocks by the Dutch public sector. Besides fundamental research on scion–rootstock interactions, endogenous hormonal balance in combination plants, propagation methods or root growth in artificial substrates, part of the programme aimed at the breeding and selection of new clonal stocks for the glasshouse. Using the knowledge about dominant inheritance of the gene for nonrecurrent flowering (R), crosses between recurrent-flowering hybrid tea varieties and nonrecurrent species roses have yielded several new vigorous clonal stocks. These were: 'Moneyway', a triploid ($2n = 3x = 21$) from the cross White 'Weekend' × *R. multiflora adenochaeta*; 'Marleen', a triploid from the cross 'Sonia' × *R. multiflora adenochaeta*, both bred at Plant Research International (PRI), Wageningen, the Netherlands (**Figure 2**), 'Multic' and 'Ludiek' selections from *R. multiflora cathayensis* Rhed. & Wils., respectively *R. multiflora* Thunb. populations made at Wageningen University; four clonal *R. canina* 'Inermis', STUR 1, 2, 3 and 5, and another 'Inermis' clone 'Spektaculair', to be used for root grafting only. 'Popeye', which was developed by the Dutch breeder Olij, was introduced in 1998. With the probable exception of the STUR clones, the winter-hardiness of the new stocks is questionable. This means that in northwest Europe these stocks should either be cultivated under glass or imported from countries with more favourable climates.

Figure 2 *R. multiflora* var. *adenochaeta* has been used at Plant Research International, Wageningen, the Netherlands, as a progenitor to breed the clonal rootstocks 'Moneyway' and 'Marleen'.

Figure 3 One-season-old hedge of 'Moneyway' in the Dutch province of Groningen.

Despite the highly favourable characteristics of these new stocks as to rootability or induced vigour, they were pushed aside by the newcomer 'Natal Briar'. This stock, which originated from South Africa, was introduced into The Netherlands via Australia in about 1990. Presently, 'Natal Briar' determines over 80% of the market of cut-rose combination plants in western Europe. In a different programme, aiming at adaptation to low-energy environments, genetic transformation of the clonal stock 'Moneyway' (**Figure 3**) was carried out at PRI. This research has yielded three new genetically modified organisms (GMOs: T1, T2, T3) containing *rolABC* or *rolB* genes, which would improve the adaptation to low temperature of combination plants. When actually used as stocks at normal temperatures, the induced shoot yield of the cut-rose varieties 'Madelon', 'Vivaldi' and 'First Red' increased about 20% compared to that of untransformed 'Moneyway'. Unfortunately, the increase in yield was combined with the inferior quality of the harvested shoots, at least compared to those on untransformed 'Moneyway'. Because the practical use of GMOs is still prohibited by European Union legislation, the development of new GMO stocks is presently in cold storage.

Properties of Clonal Stocks

A clone is defined as 'the total vegetative offspring of one genotype', which applies to both clonal rootstocks and scion varieties. This means that all plants of a clone are phenotypically identical when grown under the same environmental conditions. In principle, a clone indefinitely remains the same as long as the original genotype is conserved in a safe manner. If not, various small additive factors may modify a clone in the course of time, in which respect the rather short-lived scion varieties have an advantage over rootstock clones. Probably the most important factor of the modification of clones is the elusive factor of ageing, which seems irreversible in animal and plant life. Owing to both internal and external factors, cells grow older in the course of time. Where in the ontogeny of woody crops the juvenile and adult phases have been well studied, senescence is much less so. It should be accepted that over decades senescence will affect rootstock performance in a negative way. In this connection it is remarkable, however, that rejuvenation or reinvigoration of stocks by prolonged *in vitro* culture has been successfully applied in improving the vigour and rootability of old and even relatively new stock clones.

Another factor in the modification of rootstocks is the likelihood of successive small mutations. Those that do not visibly affect the vegetative phenotype, such as the frequently occurring flower colour mutants, remain unnoticed in clonal rootstocks. Also mutations of physiological properties, such as changes in endogenous hormone levels or in the distribution of carbohydrates, will hardly show phenotypically. Even changes in the ability of root formation or root branching of stocks are not easily detected (**Figure 4**).

Figure 4 Comparison of the root systems of about 6-month-old rootstock plants. (a) Seedling *R. canina* 'Inermis' (8–12 mm) #1623, and softwood cuttings of (b) *R. indica* 'Major' #1624, (c) 'Moneyway' #1625 and (d) 'Multic' #1629, Plant Research International nursery, Wageningen, the Netherlands.

Infection with viruses, bacteria or other pathogens contribute to the slow degradation of clones. Because rootstocks are mainly grown in the open, it can hardly be prevented that the plants are visited by sucking insects or nematodes that may spread visible, or, even worse, latent viruses that affect the vigour or habitus of combination plants. Besides viruses, clonal stocks may spread the soilborne *Agrobacterium tumefaciens* (root-knot disease) or *A. rhizogenes* (hairy root disease). The former induces large tumours at the graft union, roots and even leaves, causing growth reduction of the combination plant; the latter generates an abundant, typical thin type of root, causing drought damage. Both bacteria, of which several biovars are known, have been widely used in the genetic transformation of plants. Indications are that in the field *A. rhizogenes* may transfer foreign DNA from one plant species to another, thus inducing 'natural' genetic transformation. It is unclear whether *A. tumefaciens* is spread by the rootstock, which is likely, or by the scion variety. In order to avoid financial claims from rose growers, rootstock growers and producers of young combination plants take meticulous care to prevent infection by disinfecting pruning shears and grafting knives.

Another newly described growth-inhibiting disease was found to be caused by *Phytoplasma* in Poland. To make matters worse, it has been confirmed that accidentally flowering rootstocks produced seeds which germinated in the rootstock field. The resulting vigorously growing hybrid seedlings, looking like the mother plant, have been propagated by inattentive nurserymen, thus introducing a new genotype within an old clone.

Despite the above disadvantages, most of which can be taken care of by reselection and health measures, clonal stocks have great advantages over seedling stocks, particularly for the cut-rose industry. Experiences with modern clonal stocks have fundamentally changed views about the use of stocks. Where initially stocks were used for the easy multiplication of scion varieties, present emphasis is on the control of the vigour of those varieties. Current clonal stocks enhance the vigour of combination plants, and are compatible with all cut-rose varieties tested so far. The performance of scion varieties on clonal stocks is more predictable and reliably uniform over the seasons. Clonal stocks are less sensitive to changes in light quantity in the course of the year, and likewise to continuous flower harvests, because of fast regrowth of the roots. In addition, clonal stocks are more reliable for research purposes than the ever-varying Edelcaninas. For these reasons, clonal stocks should be generally preferred to seedling stocks for cut roses.

Protection of clonal stocks by plant patent or plant breeder's rights is on one hand an advantage to breeders, because next to the scion variety royalties may be imposed on the stock. On the other hand, royalties on both the scion variety and the stock are not beneficial for propagators and rose growers, because of extra investments per hectare. In this connection it is notable that the net price of one young cut-rose plant comprises the following factors: price of the scion, price of the rootstock, royalties on the scion, royalties on the stock, price of the rockwool block or cocos plug, cost of labour and cost of conditioned glasshouse equipment.

Various tests have demonstrated that particularly the old clonal stocks have been seriously infected with a complex of viruses, usually indicated as rose mosaic virus. It has been demonstrated that this complex probably consists of a mixture, of which apple mosaic virus (ApMV), *Prunus* necrotic ringspot virus (PNRSV) and *Arabidopsis* mosaic virus (ArMV) are the main ones. Owing to recent private endeavour, a number of virus-free rootstock clones now sustain the quality of combination plants. This includes clones of *R. indica* 'Major' and of 'Natal Briar', produced by NakTuinbouw in the Netherlands, and of 'Manettii' in the USA.

Production of Clonal Stocks

The basis of propagation of scion varieties on clonal stocks is shoots about 1 m long of a clonal stock variety. Unlike cut-rose varieties, shoots of clonal stocks do not flower, at least not when the plants are regularly harvested. Because the major use of clonal stocks is in cutting-grafting, the shoots are defoliated after harvest and subsequently cut into 8–10-cm-long pieces. Both the bases and the apexes of these shoots are discarded as being either too thick or too thin. Preparing the stocks for cutting-grafting consists of (partly) removal of the thorns and cutting away axillary buds. To obtain sufficient numbers of shoots, large quantities of stock plants must be grown. In mild climates the plants are commonly grown in the field, but in temperate zones, where winter frosts occur, stock plants are grown in heated glasshouses. Since glasshouse cultivation is expensive, propagators of cut roses have sought and found cheaper ways. Hence, in The Netherlands shoots of *R. indica* 'Major' were initially imported from the South of France or Israel. However, when *R. indica* 'Major' became less popular in northwest Europe and gave way to 'Natal Briar', the latter stock too was initially grown in the glasshouse. In the middle of the 1990s, Dutch propagators founded fields of 'Natal Briar' in Kenya and Uganda. In addition to cheaper production in the tropics, the advantage is that the wood of these plants is better ripened than that from the glasshouse, which induces greater success in cutting and grafting. When defoliated after harvest, shoots are bundled and shipped to Europe, where they may be kept in cool storage for several weeks before being used in cutting or grafting. The high costs of air freight are easily counterbalanced by the advantages of cheap production and of well-ripened shoots.

Multiplication of Clonal Stock Plants

Multiplication of clonal stocks is traditionally by hardwood cuttings, and only recently by softwood cuttings. Traditionally, in the Mediterranean area, new rootstock fields are laid out in autumn. In these fields, rootstock plants are usually planted in rows at least 2 m apart, and spaced about 50 cm within rows. A field may contain several thousands of plants which, owing to regular shoot harvests, are in various stage of development. Owing to the fact that clonal stocks are nonrecurrent-flowering, the plants form long vegetative shoots in a relatively short time. In the mild climates of the South of France, Spain, Morocco or Israel, *R. indica* 'Major' is grown in hedges, and is only harvested in autumn for the production of hardwood cuttings. Besides for direct use as a stock, such cuttings may be used to found new fields. With the introduction of the method of cutting-grafting, in these countries too the demand for year-round production of combination plants has increased. As in The Netherlands, this has compelled the cultivation of rootstocks towards protected environments.

See also: **Intellectual Property:** Plant Patents and Trademarks. **Propagation:** Budding in the Field. **Rootstocks:** Rootstock Breeding; Seedling Stocks.

Further Reading

De Vries DP (1993) *The Vigour of Glasshouse Roses. Scion–Rootstock Relationships. Effects of Phenotypic and Genotypic Variation.* Thesis. Wageningen, The Netherlands: Agricultural University.

Dieleman JA (1998) *Cytokinins and Bud Break in Rose Combination Plants.* Thesis. Wageningen, The Netherlands: Agricultural University.

Edwards RE (1955) *A Survey of Rootstocks for Roses.* Dissertation. Nottingham: University of Nottingham.

Ferguson JF (1933) A botanical study of rose stocks. *Journal of the Royal Horticultural Society* 58: 344–371.

Garner RJ (1979) *The Grafter's Handbook* (4th edn). London: Faber & Faber.

Hartmann HT, Kester DE and Geneve RL (1997) *Plant Propagation: Principles and Practices.* New Jersey: Prentice-Hall.

Jackson MB (1986) *New Root Formation in Plants and Cuttings.* Dordrecht: Martinus Nijhoff.

Krüssmann G (1986) *Rosen, Rosen, Rosen.* Berlin: Parey.

Shepherd RE (1954) *History of the Rose.* New York: Macmillan.

Seedling Stocks

A J J Akkerman, DLV Plant BV, Wageningen, The Netherlands

Introduction

A seedling stock is a 1-year-old rootstock grown from seed that has been produced from seedbushes especially grown for that purpose. There are about 10 main types of rootstock (including seven types of *Rosa canina* – the so-called Edelcaninas) that are still in use these days, although the importance of some of them has changed. The stock *R. canina* 'Inermis' was the main stock for a long time but since rootstocks from cuttings have been used for glasshouse roses, the amount of *R. canina* 'Inermis' has decreased dramatically. The main stocks these days are *R. corymbifera* 'Laxa' (**Figure 1**) and *R. rubiginosa*. Unsold rootstocks can be used for forestry or landscaping purposes; unused hips can be used for making jam, though *R. pendulina* is more suitable for this purpose. In this article rose rootstocks grown from seed, from the Dutch perspective, are dealt with in a practical way, from the planting of shrubs for a seed plantation to the grading of 1-year-old stocks.

Propagation

The Seed Plantation

Usually a rootstock-nursery grows its own seed in vast planted areas where the rose shrubs are planted in long rows (**Figures 2** and **3**). The main reasons why nurseries grow their own seed are: (1) growers can be sure that the seed is of the right species; (2) they do not have to depend on the seed trade; and (3) home-grown seed is usually of better quality than bought seed. The plantations can be in the immediate neighbourhood, several kilometres away or even in other countries with warmer summers. The germination of seed grown in a warm climate is often of better quality.

Figure 1 *Rosa corymbifera* 'Laxa'. Courtesy of PPO, Noordbroek, the Netherlands.

Figure 2 Sowing of rose rootstock.

It is best to plant the shrubs in long rows, with grass between the rows. In this way the growers do not lose time by turning their machinery. The distance between the rows is 5 m and between plants in the row is 2 m; so on one hectare you can find approximately 1000 plants. A plantation is started with 1-year-old plants. In the first 2 years there is no harvest; harvest starts in the third year and in the fourth year a full harvest can be expected.

Every Edelcanina plant produces about 2–3 kg hips annually and those hips can give around 0.5 kg of clean seed. This indicates that, when in full production, a plantation can give approximately 500 kg clean seed every year. The seed/hip proportion is different with *R. multiflora*; this species gives 1 kg clean seed for every 11 kg hips.

When the mother-plants have grown too heavy and tall, they can be pruned to a height of about 20 cm by a machine. After 3 years one can expect full harvest again. Diseases do not do much harm; the only serious pest is the strawberry blossom beetle (*Anthonomus rubi*) which makes the flowerbuds fall off at the end of May.

Figure 3 Lifting of rose rootstocks.

Harvest takes place at the end of August and throughout September. The hips are then colouring from orange to red and the seed is physiologically ripe. Harvest starts with *R. corymbifera* 'Laxa'; the Edelcanina selections 'Pollmer' and 'Inermis' are the last to ripen. Harvesting is mostly still done by hand, although there has been some testing of blackcurrant harvesters. Recently a blueberry harvester has been put into use in the Netherlands. This machine has a capacity of 50 kg hips per hour. The rates of harvesting by hand amount to about 6 kg hips per hour. If the hips are picked when they turn dark red, they are difficult to handle (the fruit starts to rot) and, worse, the seed falls into deep dormancy.

The selections of the Edelcaninas are uniform and, therefore, it was thought that propagation by seed was apomictic. This is not so; heterogamy occurs. *Rosa canina* has 35 chromosomes, of which 28 come from the egg cells and seven originate from the pollen (*see* **Genetics**: Inheritance in the Dogrose). The influence of the mother plant is much greater than the pollen parent and so the plants look uniform. This leads to the conclusion that it is preferable to plant different selections of *R. canina* several kilometres apart, although there is no concerted opinion on this distance.

The Seed Treatment

After the harvest, the seed is ground and sieved. Some selections (*R. canina* 'Inermis', *R. canina* 'Heinsohn's Record' and *R. canina* 'Pfänder') need an extra long stratification period and should be put in the stratification bin in October. *R. multiflora* germinates after a shorter period and can be sown in the next spring when stratified in October. The seed of most selections is dried after sieving. This can be done mechanically or outside, on a cloth in the sun. After that, seed can be stored dry (35% relative humidity) and cool (10 °C) until March–April of the next year. Dried seed should then be put into water for 2 days and then stratification starts; the purpose is to break the seed dormancy. Dormancy involves two factors: the mechanical resistance of the woody pericarp of the seed and the physiological dormancy of the seed. The woody pericarp is best broken at temperatures of *c.* 15–20 °C and the dormancy at temperatures of approximately 5 °C. Warm stratification thus lasts for about 7 months and cold stratification lasts for about 3–4 months. The seed starts to germinate by then and has to be stored at −4 °C until the moment of sowing. There are two serious problems: the whole treatment lasts a long time and the germination rate is often disappointing. It is preferable to stratify in a climate cell under conditions that control both the temperature and humidity. In general, climate control gives better germination but has not yet found acceptance in the nurseries.

There are some opportunities to shorten the process. The pericarp can be attacked using chemical scarification (concentrated sulphuric acid (H_2SO_4) for at least 30 min). After that, the warm stratification can be shortened to 2–3 weeks. There is a risk in this treatment because the pericarp is permeable to water and the acid can damage the embryo. Acid treatment is common in England. It is less risky to use compost makers (e.g. Garotta). This encourages the activity of microorganisms, so the pericarp can be degraded quicker. Tests with *R. canina* 'Inermis' have shown that with 50 g of compost maker per kg seed and a warm treatment of 8 weeks, followed by 28 weeks of cold stratification, there was 80% germination. In spite of these results, stratification outside, in the old-fashioned way, is still very common in the Netherlands and Germany.

The stratification bin should have a floor at ground level which is permeable to water so that the seed is never too wet. The best location is in shadow, for instance on the north side of a barn. One should place the bins in a cage to protect them against birds, which have a preference for *R. canina* 'Pfänder'. The seed should be mixed with sand, or sand and peat in equal parts. The quantity is 1–2 kg sand to 1 kg seed. Care should be taken to prevent drying out and damage by severe frost. The layer of seed and sand should not exceed 20–25 cm in depth (**Figure 4**).

Germination starts in winter, so seed bins must be regularly checked in December and January. When about 2–3% of the seed has germinated, the seed should be put in a freezer to hold it back until the right sowing-time. Sowing of germinating seeds results in bent root-necks. Openness of the seed can be checked by pushing the flat of a knife-blade on the seed while it is lying on a table. Several small plastic bags with 10 or 50 g of seed should be stored separately in the freezer to make checks on germination in February–April. The cases should be covered with sand or plastic to prevent drying out. The seed stays in the freezer until a day or two before sowing.

Figure 4 Stratification bin with several compartments for different kinds of seed. In this stratification bin, overhead sprinkling of the seed is possible to be sure of the right percentage of moisture.

Germination Analysis

The number of plants on the seed bed is very important. The financial results are related to this figure. While germination can be low or seed can revert to dormancy, it is very important to know how many seeds one has to sow to achieve a reasonable number of plants per metre of seed bed. The way to do this is as follows:

- plastic bags with the measured weights of seed plus sand should be defrosted
- the seed should be sown in rows in a case with common soil
- the cases should be placed in a room that is not too cold
- the seedlings should be counted after 2 weeks, so that the germination rate in the measured bag is known.

When the number of germinable seeds to be sown per hectare is known, one can easily calculate the required weight of seeds. For instance, if there are 175 germinable seeds in 10 g of seed and you want to sow 2 000 000 per ha, you need to sow (2 000 000/175) × 10 g of seed per ha. As a general rule, sowing two germinable seeds leaves one plant to lift in autumn. It is common to sow 2 million seeds of *R. corymbifera* 'Laxa' per ha, 1.25 million seeds of *R. canina* 'Inermis' per ha, 2.25 million seeds of *R. multiflora* per ha and 1.5 million seeds per ha of the other rootstocks.

The Sowing

One should choose a soil that supplies the following:

- no nematodes (check on *Pratylenchus penetrans* and *Meloidogyne hapla*)
- a fresh soil, where roses have not been grown before
- good porosity (not easily silted up), to avoid bent necks
- no wind-erosive soil; the soil can bury or expose the seedlings and grains of sand can damage the seedlings by abrasion.

Seeds should be sown around 15–30 April in the Netherlands. Seedlings that have been sown before

Figure 5 Field of rose rootstocks at the beginning of June.

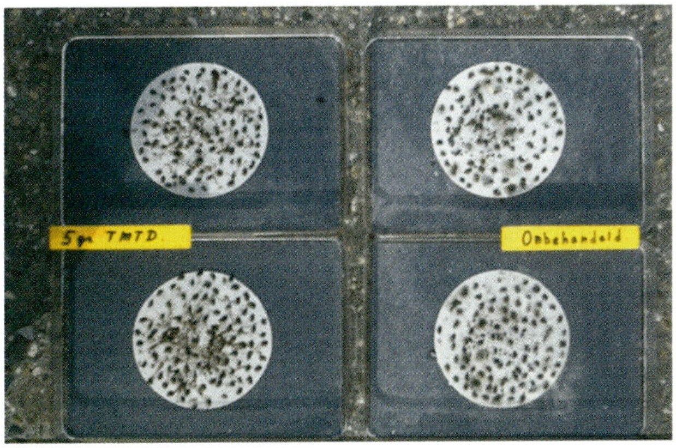

Figure 6 Test of the germinal force of rose seed. Treatment with thiram to prevent damping off (note that fungicides like Previcur, Topsin M, Rovral are superior to Thiram).

that date can be damaged by late frosts and seedlings that are sown later may not grow thick enough. The seed is sown with a sowing machine in four rows on a bed (**Figure 5**). The depth should be around 3 cm for the Edelcaninas and a little less, about 1.5–2 cm, for *R. multiflora*. The depth is necessary to get long necks (2 cm) for budding. The outer two rows are often sown slightly thicker because these rows have more space. Before sowing, the soil is prepared with a harrow and, after sowing, with a Cambridgeroll. It is important that all soil clods are crushed because a seedling that meets a clod on its way up grows bent. When there are many clods, the soil is raked after sowing. The adjustment of the sowing machine is mostly done by trial and error. First, the weight of seed that a case contains is known. Also the area one wishes to sow with that weight of seed is known. One should just sow the case and check where it ends; then adjust the sowing machine for a better result with the next case. Also one should note the adjustment of the machine and how many metres and kilograms have been sown with that adjustment. This is important for good sowing results in the future. One should empty the case slowly and little by little (**Figure 2**).

To prevent the symptoms of a complex of soil fungi (e.g. *Phytophthora cactorum*, *Pythium*, *Fusarium*, *Rhizoctonia*) the seeds can be disinfected with fungicides before sowing (**Figure 6**). These fungi constrict the necks of the seedlings and the effect is that they fall over ('damping-off') (**Figure 7**). These diseases are likely to occur when temperatures are changeable and there is a lot of rain during the germination period. *R. corymbifera* 'Laxa' is the most susceptible of the common rootstocks to these soil fungi.

In the following period, the main priority is to control the weeds. This is done by spraying chemicals and by mechanical hoes. Since the usage of chemicals is restricted, investigation into mechanical weeding is an important topic in research.

Figure 7 Damping off of young rose rootstocks, at the beginning of May.

Disease Organisms

Diseases that causes damping-off of the newly germinated seedlings are mentioned above. Other important diseases are as follows:

- Nematodes: one should sow in a soil that is free of nematodes like *Pratylenchus penetrans* and *Meloidogyne hapla*. So, one should take samples of the soil and have them analysed. *R. corymbifera* 'Laxa', one of the most widely used rootstocks, is the most susceptible rootstock for *P. penetrans*. *R. canina* 'Inermis' has a high resistance to *P. penetrans* in comparison with *R. corymbifera* 'Laxa'. *P. penetrans* multiplies more rapidly on *R. multiflora* than on the Edelcaninas. *R. multiflora* is the most susceptible rootstock to *M. hapla*. The least susceptible are *R. rubiginosa* and *R. canina* 'Pfänder'. The other Edelcaninas and *R. corymbifera* 'Laxa' are intermediate; their resistance varies.

  Susceptibility of the main rootstocks to the nematodes *M. hapla* and *P. penetrans* is shown in **Table 1**.

Table 1 The susceptibility of the main rootstocks to the nematodes *Meloidogyne hapla* and *Pratylenchus penetrans*

Rootstock	*Susceptibility to*	
	M. hapla	P. penetrans
R. corymbifera 'Laxa'	Moderate	High
R. multiflora	General high, but variable	General high, but variable
R. canina 'Pollmer'	Moderate	High
R. canina 'Schmid's Ideal'	Moderate	Moderate
R. canina 'Inermis'	Moderate	Moderate
R. canina 'Brög's Stachellose'	Low–moderate	High
R. canina 'Pfänder'	Low	Moderate
R. canina 'Superbe'	Moderate	Moderate
R. canina 'Heinsohn's Record'	Low–moderate	Moderate
R. rubiginosa	Low	Low

The usage of soil sterilizers (like metam-natrium or dichloropropeen) has been restricted in recent years, so it is very likely that the spread of nematodes will be one of the main concerns for the future. The green manure *Tagetes patula* (especially 'Single Gold') has been commonly used since *c.* 1995 to combat successfully the nematode *P. penetrans* but its efficacy against *M. hapla* is doubtful.

- Powdery mildew (*Podosphaera pannosa*) is a very common disease, especially in hot, dry summers. A severe attack of mildew causes much loss of growth and, therefore, stem thickness. When the disease spreads on the branches, there will be problems with storage in the cool store; those branches easily turn black. The susceptibility to mildew can be compared between the common rootstocks as follows.

  The most susceptible rootstock is doubtless *R. canina* 'Pfänder' but the following rootstocks are moderately susceptible: *R. canina* 'Inermis', *R. canina* 'Heinsohn's Record', *R. canina* 'Schmid's Ideal', *R. canina* 'Superbe' and *R. rubiginosa*. There are a number of rootstocks that are fairly resistant: *R. multiflora*, *R. canina* 'Brög's Stachellose' and *R. canina* 'Pollmer'. Finally, *R. corymbifera* 'Laxa' is more or less resistant. In the Netherlands, since 1994, a computerized system has been used; data from a local weather station are processed in a software programme that predicts the right time of spraying. Growers who use this system spray about 33% less than before but produce the same quality in autumn.
- Downy mildew (*Peronospora sparsa*) is a disease that is common in wet, cold conditions in spring and summer. It causes fall of leaves until only the top leaves are left. When the disease attacks early (in spring), many seedlings die. Like powdery mildew, it is a disease that can be very harmful. The most susceptible rootstock is *R. corymbifera* 'Laxa', but all rootstocks are more or less susceptible.

Figure 8 Nitrogen administered to *R. corymbifera* 'Laxa'. Courtesy of PPO, Noordbroek, the Netherlands.

Diseases like black spot or rust seldom occur when seedlings are regularly sprayed against powdery mildew and downy mildew. Of course, pests like aphids or caterpillars can also occur. In general, inspection service control is important so that buyers can rely on quality.

Fertilizers

Rootstocks need a slow, sturdy growth. So nitrogen should be given carefully, in small portions (**Figure 8**). It is important to start with soil analysis to learn what the soil contains and, after that, to decide what kind of fertilizers are necessary to fill the gaps. If manure has been added, one should calculate the appropriate NPK (nitrogen, phosphorus, potassium) supplement. When no manure has been added and no soil analysis data are available, then growers give a NPK fertilizer before sowing in the proportions 7 + 14 + 28 and approximate quantity of 400 kg ha^{-1}. At the end of June around 300 kg ha^{-1} ammonium nitrate limestone is given. It is even better to add 200 kg ha^{-1} limestone nitrate twice at intervals of *c.* 3 weeks. Magnesium deficiency is common in roses and should be carefully checked in the soil analysis.

Rootcutting

Root cutting should be carried out with a sharp, flat cutting knife of bed-width at a depth of about 17–20 cm between the end of July and the month of August. This stimulates the production of hairy side-roots. Irrigation before and afterwards is necessary. Cutting the tap roots causes less growth of the root neck-diameter, but in the end a rootstock is produced that has a better quality because of a well-branched root system.

Harvest and Grading

Harvesting of rootstocks (shown in June in **Figure 5**) starts around mid-October with *R. corymbifera* 'Laxa', *R. multiflora* and *R. canina* 'Pfänder', then in November with *R. canina* 'Heinsohn's Record', *R. canina* 'Superbe', *R. canina* 'Schmid's Ideal' and *R. rubiginosa*. Lastly, in December, *R. canina* 'Brög's', *R. canina* 'Pollmer' and *R. canina* 'Inermis' are lifted. A lifting machine with a capacity of approximately 0.5 ha day^{-1} is used (**Figure 3**) and superfluous leaves and twigs are cut off with a knife on the lifting machine. The rootstocks leave the machine in bundles and are transported to the barn in pallets (one should be careful to avoid doing this when there are frosts). There they can be graded according to the diameter of the root necks. The most common grades are 3–5, 5–8 and 8–12 mm, but 3–4, 4–6, 6–8 and 8–12 grades are also used. Grading is time-consuming (about 1000–1100 plants can be graded in 1 h – equivalent to 600–700 h ha^{-1}), so there is a tendency to do the grading in countries like Hungary or Poland where wages are lower. The main criteria in grading are: diameter of the root neck, long straight root neck, no tap roots, no disease or damage. Trueness to type is checked in the summer and cannot be assessed on the grading table. Daylight is unsuitable, so one should not grade in front of windows. Good artificial light should be used right above the grading table.

After grading, plants can be cold-stored (0–0.5 °C), or sometimes stored in the ground. Care should be taken to prevent drying out, which is the main cause of loss of quality.

Types and Usage

Types

In the Netherlands, Germany and Denmark, *c.* 10 types of rootstock are grown. The most popular rootstocks are *R. corymbifera* 'Laxa' and *R. canina* 'Inermis'. During the last few years *R. canina* 'Inermis' has lost much of its importance. The main reason is the usage of clonal rootstocks propagated by cuttings (e.g. 'Natal Briar'). Rootstocks like *R. canina* 'Inermis', *R. canina* 'Brög's' and *R. canina* 'Pollmer', which were used a lot as rootstock for glasshouse roses, have lost their value for that purpose. Seedling rootstocks are mainly used nowadays for budding to make shrubs like hybrid tea, Floribunda, English and climbing roses. Park shrub roses and patio roses are mostly made from cuttings.

The 10 most important rootstocks are shown below, with their degree of suckering in a scale from 1 to 10 where 1 = no suckering at all and 10 = a lot of suckering:

- *R. corymbifera* 'Laxa' 3
- *R. multiflora* 3
- *R. canina* 'Schmid's Ideal' 4
- *R. canina* 'Heinsohn's Record' 5
- *R. canina* 'Pfänder' 6
- *R. canina* 'Pollmer' 6
- *R. canina* 'Brög's Stachellose' 8
- *R. canina* 'Superbe' 8
- *R. canina* 'Inermis' 8
- *R. rubiginosa* 9

Usage

The main rootstock for budding of shrub roses in the Netherlands is still *R. rubiginosa*. Other rootstocks that are being used in considerable amounts are *R. multiflora*, *R. corymbifera* 'Laxa', *R. canina* 'Heinsohn's Record', *R. canina* 'Inermis', *R. canina* 'Schmid's Ideal' and *R. canina* 'Pfänder'. The rootstocks *R. canina* 'Superbe', *R. canina* 'Pollmer' and *R. canina* 'Brög's Stachellose' are seldom used . The advantages, disadvantages and various characteristics of these rootstocks are shown in **Table 2**, which should be read in conjunction with the statements on suckering and disease-susceptibility (above).

Some rootstocks that may have significance in the future are *R. canina* 'Meyer's Zukunft', also known as *R. canina* 'Rosso'. It is in use in Italy as rootstock for standards, has a well-branched root system and is thus suitable to grow in containers. Another rootstock for standards is *R. canina* 'Veendam', which might be better than *R. canina* 'Pfänder', especially when it has fewer spots on the stem.

Table 2 Important characteristics of the most important rootstocks

Species/cv.	*Advantages*	*Disadvantages*	*Suitable scions*	*Unsuitable scions*	*Recognition*	*Remarks*
R. rubiginosa	Long budding period; good flower colour; long-lasting	Sometimes root neck produces suckers; sometimes hard to transplant	Suitable with most scions	'New Dawn' 'Paul's Scarlet Climber' (curly heads)	Round leaves with apple fragrance; many prickles	Good all-round rootstock
R. multiflora	Strong growth; dense root system, therefore suitable for containers	Necks form hair-roots if not freed of soil in time; susceptible to *Verticillium*; short lifetime; weak flower colour	Climbers and 'Koster' roses	'Nina Weibull' 'Iceberg' 'Paul's Scarlet Climber'	Often nine leaflets; stipules pectinate	Use only at buyer's request due to variable budding results; less hardy – unsuitable for Scandinavia
R. corymbifera 'Laxa'	Can be budded early; buds easily	Tap roots, therefore unsuitable for containers; unsuitable for soil long used for roses; needs lime and potassium: shortage of these causes rust	Smaller roses ('Koster' roses) for vigorous growth; Floribundas with dark red colour; hybrid teas	'Rumba'® 'Summer Holiday'	Leaf colour greyish-green	In demand in the UK; unsuitable for Scandinavia; a good rootstock if grown on an adequate soil
R. canina 'Heinsohn's Record'	Can be budded early; gives a good yellow colour	Short budding period– unsuitable for Floribundas generally	Hybrid teas; species roses; climbers with pink colour	Incompatible with many varieties listed below[a]	Shoot tips reddish	Good rootstock for specific purposes
R. canina 'Inermis'	Gives strong growth, one of the very best for growing A-quality		Suitable with most scions	'Europeana'® 'Duftwolke' (incompatibility) 'Morsdag' 'Orange Morsdag' 'Tom-tom' (chance of curly heads)	Colour light green; leaves narrow and shoot tips pinched	Until recently, much in use for glasshouse roses
R. canina 'Pfänder'	120-cm-long cuttings, in cold storage and on heated bin gives 80% rooting. Cuttings can be budded in July, so culture reduced to 2 years	Gives fewer branches	Very much in use for standards (4-year culture), when 'Pfänder' is budded on 'Laxa', a 3-year culture	'Duftwolke' (incompatible)	Leaf colour dull greyish-green; Prone to mildew	New English selections of 'Pfänder' include 'Rocket'

R. canina 'Schmid's Ideal'	Suitable for severe winters (northern and eastern Europe)	Roses grown on this rootstock give shoots that ripen 2 weeks later, thus lengthening the budding programme	Hybrid teas; species roses; red-flowered Floribundas	'Landora'	Shoots with zig-zag growth	
R. canina 'Superbe'	Gives strong growth, thus a good rootstock for slow-growing roses		Climbers; Floribundas; 'Koster' roses		Large leaves with reddish stem	Good all-round rootstock; necks tend to get corky, so must be stored under ground
R. canina 'Pollmer'	Suitable for early budding (short growth period)	Short growing period	Suitable for roses in a warm climate (France)		Large top leaf	Has lost a lot of its popularity
R. canina 'Brög's Stachellose'					Top leaves brownish-green	Has lost a lot of its popularity

[a] Known incompatibility of *R. canina* 'Heinsohn's Record' with: 'Alain', 'Allotria'®, 'Amsterdam'®, 'Betty Prior', 'Europeana'®, 'Friedrich Heyer', 'Highlight', 'Masquerade', 'Morsdag', 'Orange Morsdag', 'Orange Triumph'®, 'Schweizer Grüss', 'Tom-tom', 'Centurion', 'Duftwolke', 'Eagle', 'Marlena', 'Nina Weibull'®, 'Prins Claus', 'Iceberg', 'Feuerwerk'®, 'F.J. Grootendort', 'Mozart', 'Persian Yellow' and 'Pink Grootendorst'.

See also: **Classification**: Conventional Taxonomy (Wild Roses); Horticultural Classification Schemes. **Genetics**: Inheritence in the Dogrose. **Rootstocks**: Scion–Rootstock Relationships; Rootstock Breeding; Usage of Rootstocks.

Further Reading

De Bresser WAA (1984) *De Teelt van Rozenonderstammen*. Tilburg: Leaflet Consulentschap Akkerbouw en Tuinbouw.

Derkx MPM (1997) *Geconditioneerde Opheffing van de Kiemrust in Boom – en Heesterzaden*. Boskoop: Boomteeltpraktijkonderzoek.

de Vries DP (1993) *The Vigour of Glasshouse Roses: Scion–Rootstock Relationships*. PhD thesis, University of Wageningen, Wageningen, The Netherlands.

Krüssmann G (1986) *Rosen, Rosen, Rosen*, pp. 67–99. Berlin: Parey.

Leemans JA (1966) *Rootstocks for Roses – Characteristics and Cultural Value*. Boskoop: PPH.

Lösing H (1995) *Bedeutung und integrierter Bekämpfung wandernder Nematoden* (Pratylenchus *sp.) als eine Ursache von Nachbauschäden bei Rosen*. Dissertation. Berlin: Humboldt-Universität zu Berlin.

van Hees-Boukema EM (1993) *Het Zaaien van houtige Boomkwekerijgewassen*. Boskoop: IKC Boomteelt.

ROSE COLLECTIONS AND TRIALS

Contents

Rose Collections and Herbaria

T Cairns, American Rose Society, Studio City, CA, USA

Introduction

Rose gardens, while often perceived as extravagant, are ubiquitous throughout the civilized world. There are thousands of rose gardens worthy of attention. In creating gardens, humanity has taken the beauty and entropy of nature and brought high degrees of architectural order to otherwise disarranged botanical assets. The intervention of mankind to organize and complement plants of varying habit, shape, colour and fragrance is evidence of the required partnership of man's imagination working synergistically with nature's disorder to create a garden not only pleasing to the eye but also to the various senses.

The collaboration of man with nature usually originates with the vision of humanity creating the basic concept based on botanical intelligence. But the changing will and divergence of nature constantly alters the equation of geometry to produce a changing effect that constantly affects the overall architecture of a rose garden. Unlike great buildings, rose gardens are ephemeral, giving rise to a multitude of transient displays, each different and passing. This advanced knowledge permits garden designers to have premature visions of how their creations will look in the future. While rose gardens may appear natural, they are indeed inventions of mankind introducing design where it had not existed beforehand.

The great rose gardens of the world have been expanding since the inception of serious garden design in Italy and France during the seventeenth and eighteenth centuries where astonishment was the goal more than order and beauty for the eye. These rose gardens ranged from massive to large to sometimes simple, each dependent on the vision of the creator. Each country has developed major rose collections based on the availability of plants and experience learned from other designs throughout the world. In the twenty-first century the emphasis in maintaining large rose collections is not just for beauty and order, but also for historical relevance. Additionally, many of these fine gardens have now embraced the concept of the incorporation of international rose trials initiated by Parc du Bagatelle in Paris, France.

World Federation of Rose Societies Gardens of Merit

The World Federation of Rose Societies (WFRS) makes awards from time to time to 'Gardens of Merit', for outstanding rose gardens. The following list gives the dates and the recipients of this award.

1995: La Roseraie de l'Hay du Val-de-Marne, France

8, rue Albert Watel, 94200 L'Hay-les-Roses, France.

The world's oldest existing garden devoted exclusively to roses celebrated its centenary in 1994. It was created by architect Edouard André and rosarian Jules Graveraux, who assembled a collection of all species and varieties known at the time. Together they created a rose garden of historical value and breathtaking beauty. In 1968 ownership of the garden passed to the department of Val-de-Marne, which has conscientiously continued the maintenance and renewal of the collection while encouraging public participation in this setting of unrivalled charm.

1996: Il Roseto Botanico 'Carla Fineschi', Italy

52022 Cavriglia (Arezzo), Italy.

Professor Gianfranco Fineschi started collecting roses 30 years ago and today he has one of the world's largest private collections – more than 6500 varieties, each represented by a single plant. In this living museum the genus *Rosa* is presented from a historical and a botanical point of view. Species, subspecies and later hybrids are systematically presented, while climbers and ramblers form walls between the scientific divisions. Many of the beds of modem varieties are grouped according to their hybridizers. In this garden overlooking the hills of Tuscany one may trace the history of the rose from antiquity to the present day while marvelling at the beauty of the garden itself.

1998: The Garden of Roses at Hex Castle, Belgium

Kasteel Hex, B3870 Heers, Belgium.

This historical site was constructed by Prince Bishop Velbruck in 1770. The gardens are of a formal Renaissance style while the surrounding parkland reflects the natural English style. About 250 varieties of roses are found throughout the property. Since 1970 Countess Michel d'Ursel has renewed and increased the collection of roses. Some bushes are originals which date back to the construction of the castle. Others are plants which the Countess d'Ursel brought back from her travels to replace those that had died. She also added many modern varieties, but always those that were in harmony with these gracious surroundings.

2000: City of Belfast International Rose Garden, Northern Ireland

Upper Malone Road, Belfast, Northern Ireland

The City of Belfast International Rose Garden is situated in one of the city's most beautiful areas in an outstanding landscape in the form of Sir Thomas and Lady Dixon Park. The park extends to 50 ha on the southern outskirts of Belfast, beside the Upper Malone Road and bounded to the west by the meandering River Lagan. Up to 35 000 rose bushes, each with dozens of blooms, are seen by many thousands of residents and visitors alike during the summer making the garden a 'must see' attraction in Northern Ireland. Sir Thomas and Lady Dixon Park was chosen as an ideal setting for a large rose display garden, which would incorporate an area for trial purposes. The first roses were planted in the winter of 1964/65, including entries for Trial I, 1965/66. In the late 1980s the gardens were completely redesigned, offering a series of circular rose beds skilfully integrated into the hillside landscaping.

2003: The Europa-Rosarium at Sangerhausen, Germany

Steinberger Weg 3, D-06526 Sangerhausen

This world-famous garden was officially opened in 1903 in the presence of its Royal Patron, the Empress Augusta. In 1989, after the Reunification of Germany, the Rosarium rapidly gained its well-deserved international reputation as the comprehensive repository of many old garden roses in danger of complete extinction. The collection of old garden roses is the most extensive in the world today comprising 50 000 plants (6500 varieties). In 2003 this garden celebrated its 100th birthday in conjunction with the German Rose Society.

Rose Gardens of the World

The information that follows is a selected alphabetical listing (derived from WFRS publications) by country of the various notable rose gardens available for public viewing with an extended narrative of national gardens and others most worthy of special note.

Argentina

'El Rosedal' Municipal Rose Garden Palermo Park, Buenos Aires

Best blooms to be seen in October–November. This garden was created in 1914 by inspiration drawn from Parc du Bagatelle in the Bois de Boulogne near Paris. In 1994 a massive renovation programme was undertaken. By 1996 the garden boasted nearly 14 000 plants on display, and was home to an international trial ground.

Australia

Royal Tasmanian Botanic Garden Queens Domain 7000, Tasmania

National Rose Trial Garden of Australia Adelaide Botanic Garden, Hackney Road, Adelaide 5001, South Australia

This garden is situated within the Botanic Garden of Adelaide, capital city of South Australia. It is a joint venture between the National Rose Society of Australia, the Rose Introducers of Australia, and the Botanic Garden of Adelaide, and is administered by a Management Council of seven. Each partner is represented by two members, together with a further representative from the Nursery and Landscape Industry Association. First trials were initiated in 1997.

Adelaide International Rose Garden Adelaide Botanic Garden, Hackney Road, Adelaide 5001, South Australia

5000 roses in a landscaped garden.

Rymill Park Rose Garden Bartels Road and Dequetteville Terrace, Adelaide 5001, South Australia

Main rose garden of the City of Adelaide.

Carrick Hill Fullarton Road, Springfield 5062, South Australia

Tea roses, bedding roses and the Alister Clark Rose Garden.

Ruston's Rose Garden Moorna Street, Renmark 5341, South Australia

Largest collection of roses in Australia. 50 000 roses, old and new.

Mount Lofty Botanic Garden Piccadilly, South Australia

Australian species collection.

Chateau Barrosa Rose Garden Barossa Valley, South Australia

5000 roses in a landscaped setting.

Quatre Saisons Heritage Rose Garden Lot 1, Mills Road, Glen Forrest 6009, Western Australia

Large collection of heritage and modern roses.

The Rose Garden Royal Botanic Gardens, Sydney, New South Wales

Sunraysia Oasis Botanic Garden River Road, Buronga 2739, New South Wales

Good use of climbing roses on pergolas and portholes.

Riversdale Maud Street, Goulburn 2580, New South Wales

A National Trust property with the garden recently restored.

Victoria State Rose Garden K Road, Werribee 3030, Victoria

Established in 1986 with over 4000 plants with 300 species on approx. 5 ha designed in the shape of a Tudor rose with five petals. Awarded WFRS Plaque of Merit in 1999.

Royal Botanic Gardens South Yarra 3141, Victoria

Flemington Racecourse Epsom Road, Flemington 3031, Victoria

18 000 rose plants in massive displays.

Morwell Centenary Rose Garden Commercial Road, Vicroads 340.C6, Morwell, Victoria

Latrobe City Council rose garden. Imaginatively landscaped gardens of 1.6 ha with 100 beds planted with 3000 roses. Professional selection of recent hybrids and traditional favourites make up 250 varieties.

Alister Clark Memorial Rose Garden Corner Somerton and Bulla Roads, Bulla, Victoria

Collection of Alister Clark roses.

Austria

Donaupark Wien Donauturmstrasse 2, 1220 Wien

Best viewed in June–July.

Burggarten Burgring, 1010 Wien

Standard roses.

Rosarium Baden bei Wien Doblhoffpark, 2500 Baden

Inaugurated in 1969 with approx. 9 ha with 25 000 rose bushes. Home to Austrian International Rose Trials.

Belgium

Bellevaux 4960 Malmedy
Approx. 1100 bushes on 1.2 ha. Best viewed at the end of June–July and in September–October.

Coloma Domein Coloma, 1610 St Pieters Leeuw
Flemish rose garden on a world scale with 5000 plants on 5 ha. Best viewed in June–July.

Hex Castle Kasteel Hex, B-3870 Heers
This historical site was constructed by Prince Bishop Velbruck in 1770. The gardens are of a formal Renaissance style while the surrounding parkland reflects the natural English style. About 250 varieties of roses are found throughout the property. Since 1970 Countess Michel d'Ursel has renewed and increased the collection of roses. Some bushes are originals which date back to the construction of the castle. Others are plants which the Countess d'Ursel brought back from her travels to replace those that had died. She also added many modern varieties, but always those that were in harmony with these gracious surroundings.
Castle grounds are about 60 ha with a formal rose garden of 3 ha containing 450 species of roses as well as other historic plants and climbers. The garden is open, on weekdays only, to private groups upon written request. Garden awarded WFRS Plaque of Merit in 1997.

Huise Château de l'Ast Korte Aststraat 12, 9750 Huise Zingem
This chateau and garden in West Flanders was the property of the parents of founding WFRS President and President Emeritus, Baroness Lily de Gerlache de Gomery who recently retired as President of the Belgian National Rose Society after 40 years of long and distinguished service. Her father started the garden around 1925 and was the first President of the Belgian National Rose Society. There are approx. 1000 bushes on 0.5 ha. Roses were selected by family members because they loved them or because they grow well in the garden. These include Hybrid Teas, climbers, old polyanthas and grandifloras, modern roses, landscape roses and a collection of *R. wichurana*s. Visits must be arranged by special request or made in connection with special events.

Provinciale Rozentuin Rivierenhof Turnhouttsebaan 232, 2100 Deurne
7300 plants on 8 ha, 250 vareities of historic roses, a collection of species and a collection of *R. moschata* hybrids.

Le Roeulx Home St Jacques
Approx. 1500 rose bushes on 1.5 ha. Home to an Annual International Rose Trial.

Rosarium West Vlaanderen Kortrijk Kasteel t'Hooge, Doorniksesteenweg 221, 8500 Kortrijk
Nearly 5000 plants located on 1.5 ha, mostly historical and modern. Home to an Annual International Rose Trial. Best viewed in June–July.

Select Delforge–Reynaert Rozentiun Rozenlaan 12, 9111 Belsele (St Niklaas)
Small garden on 0.3 ha but with 2500 bushes. Worth a visit during end of June–July.

Den Blakken Wegvoeringstraat, 9230 Wetteren
Modern roses, landscape roses, groundcovers, climbing roses; approx. 1500 roses on 0.5 ha.

Vrijbroekpark Provinciaal Domein Ridder Dessainlaan, 2800 Mechelen
This garden has 11 000 roses on 2 ha, species and modern.

Roseraie de Terre Franche Rue Collean 2, 1325 Chaumont-Gistoux
Approx. 1000 species roses, old roses and trials entries on 0.75 ha. Best viewed in the first part of June.

Den Blakken Wegvoeringstraat, 9230 Wetteren
Modern roses, landscape roses, groundcovers, climbing roses (about 1500 rose bushes) on 0.5 ha. Best viewed at the end of June–July.

Bermuda

Millenium Rose Garden at Waterville Waterville Paget parish by The National Trust
Established in 1987 it is a repository garden (0.5 ha) for roses growing in Bermuda. Open to visitors during daylight hours. Best viewed in April.

Camden Paget by the Botanical Gardens
A collection of 40 historic roses growing on 0.25 ha, best viewed in April.

Canada

Memorial University Garden Sir Wilfred Grenfell Campus, Cornerbrook, Newfoundland

Memorial University Campus Garden Elizabeth Avenue, St John's, Newfoundland

Le Jardin Botanique de New Brunswick Saint-Jacques, New Brunswick

Le Jardin Botanique de Montréal 4101 Sherbrooke Street E., Montreal, Quebec
There are 10 000 roses.

Connaught Rose Garden Mount Royal, Quebec, Gordon Park, Saint-Lambert, Quebec

Royal Botanical Gardens 680 Plains Road, West Burlington, Ontario

University of Guelph Arboretum Highway 6, Guelph, Ontario

Agriculture Canada Central Experimental Farm Prince of Wales Drive, Ottawa, Ontario

Rideau Hall Historic Rose Garden 1 Sussex Drive, Ottawa, Ontario

Canadian National Exhibition Park Lakeshore Avenue West, Toronto, Ontario

Queen Elizabeth Gardens Jackson Park, Tecumseh Road, Windsor, Ontario

Agriculture Canada Research Station Morden, Manitoba

Calgary Zoo Rose Garden Calgary, Alberta

Olds College Rose Garden Olds, Alberta
Species and shrubs, many early Canadian.

Burnaby Centennial Rose Garden Burnaby Mountain Park, Burnaby, British Columbia

Centennial Rose Garden Dogwood Pavilion, 621 Poirier Street, Coquitlam, British Columbia

Butchart Gardens Saanich, British Columbia
World renowned for the splendour of the rose gardens.

Agriculture Canada Station Summerland, British Columbia

Stanley Park Rose Garden Vancouver, British Columbia

Van Dusen Botanical Garden 37th and Oak Street, Vancouver, British Columbia

Horticulture Centre of the Pacific Vancouver Island, British Columbia
The Peninsular Rose Club's test garden.

Victorian Garden Government House, Victoria, British Columbia

Czech Republic

Rosarium, Exhibition Grounds FLORA Olomouc Olomouc, Wolkerova
Founded in 1972 with the assistance of the Rosa Club, a botanical garden setting with about 300 cultivars (plus a strong representation of roses from Czech breeders) on 3.4 ha. Best viewed in the middle of June.

Rosarium, Institute of Botany, Czech Academy of Science Pruhonice, near Prague
A 250-ha natural park dating back to 1885 with a rose garden begun in 1964 by Prof. Svoboda. Managed by the Institute of Botany, it has approx. 600 modern roses and 210 historical roses, co-located with the Research Institute of Decorative Horticulture. A showcase setting for roses from Czech hybridizers.

Rosarium, International Contest of Rose Novelties Hradec Králové-Kukleny Pardubicka
Established in 1974, it is home to an outstanding Annual International Rose Trials with an attractive rose garden featuring about 100 varieties. Best viewed in the middle of June.

Rosarium and Rose Garden of Prague Hill Petrin, Prague
Founded in 1932, both these gardens are situated on the remains of the medieval fortifications of Prague. There are approx. 2000 rose bushes. Best viewed in the middle of June.

Denmark

Rosenhaven i Valbyparlken Hammelstrupvej, DK-2500 Valby, Copenhagen
Located on 1.5 ha, this is frequently voted for as the most beautiful rose garden by the people of Copenhagen. It has about 12 000 rose bushes featuring species roses, historic roses and climbers. Best viewed in July.

Gerlevparken Gerlev, DK-3630 Jægerspris
Located on approx. 1.2 ha with 6000 plants representing predominantly old garden roses and a wonderful showcase for varieties bred by Poulsen. Best viewed in June–July.

Kellerishaven Kellerisvej 64, DK-3490 Kvistgård
Species roses and old garden roses (about 4000 plants) on 0.9 ha, best viewed in July. Also voted the most beautiful rose garden in mid-July by the local residents of Kvistgård.

Rosenhaven i Østre Anlæg Stockholmsgade, DK-2100 Copenhagen

This garden has 1000 plants, mainly species roses. Best viewed in July.

Rosenhaven i Enghaven Enghavevej, DK-1674 Copenhagen

There are about 1000 plants, mainly species roses and climbers. Best viewed in July.

Rosenhaven i Kongens Have Georg Brandes Plads, DK-1307 Copenhagen

There are approx. 1700 bushes (mainly species) on 0.2 ha.

Hellerup Strandpark DK-2900 Hellerup, Gentofte

About 3500 rose bushes, mainly species and climbers, in 0.4 ha.

Roserne i Botanisk Have Øster Farimagsgade 2, University of Copenhagen, DK-1353 Copenhagen

Mainly species roses and old garden roses. Best seen in July.

Gråsten Slotspark DK-6300 Gråsten

Open daily except when the Royal Family is in residence. Best viewed in July.

Geografisk Have Chr.d.4.Vej, DK-6000 Kolding

About 10 000 rose bushes on 11 ha, mainly species and old garden roses, voted most beautiful rose garden in mid-July by the local residents.

Dronningens Rosenhave Kongevejen 1, DK-8000 Aarhus C, Park og Kirkegårdsforvaltningen

Less than 5000 rose bushes on about 2 ha, including species roses, historic roses and climbers. When the Royal Family are in residence at the castle, entrance is by special permission from the Royal Court.

Den Jyske Rosenpark Nygade 18c, DK-9620 Aalestrup

About 10 000 rose plants on display.

Finland

Helsingin Yliopiston Viikin Laitokset/Viik Garden of Helsinki University Kasvibiologian Laitos, F-00014 Helsinki

Oulun Yliopiston Kasvitieteellinen Puutarha/The Garden of Oulu University Linnanmaa, FIN-90100 Oulu

Tampereen Kaupungin Hatanpään Arboretumin Ruusutarha Hatampäänkatu 20, FIN-33100 Tampere

Turun Yliopiston Kasvitieteellinen Puutarha/The Garden of Turku University Ruissalo 25, FIN-20100 Turku

Kärsälä Rose Farm Vahliokulmantie 155, FIN-31400 Somero

This baroque garden, privately administered, covers 1 ha and contains 250 different taxa. It is the largest collection of pimpinellifolias in Finland, as well as many other bush roses. Best viewed June 15–August 15.

Oulujoen Taimisto (The Nursery of Oulu River) Taimikuja 10, FIN-91519 Rova

The show garden contains some 100 different hardy bush roses suitable for the climate (65° N). Best seen in July.

France

Roseraie du Château Château Arnay-le-Vieil, F-18200 Saint-Amand Montrond

Roseraie du Parc Floral de la Source Parc Floral d'Orléans la Source, Avenue du Parc Floral, F-45100 Orléans

Jardin des Roses Comité des Fils de la Rose, Les Arènes, F-49700 Doué la Fontaine

Jardins de la Roseraie Service des Espaces Verts, Avenue Maurice Dardat, F-49000 Angers

Roseraie du Jardin des Plantes Rue d'Oran, F-59800 Lille

Roseraie de Saverne Route de Paris, F-67700 Saverne

Annual International Rose Trials.

Roseraie de Schiltigheim Schiltigheim (Mairie de Strasbourg)

Roseraie du Parc de la Tête d'Or Parc de la Tête d'Or, F-69006 Lyon

Viewed with great pride by the citizens of Lyon, this formal rose garden, founded in 1961, is part of a 105 ha park established in 1855. With almost 60 000 rose bushes mostly planted in mass displays in 5 ha, this garden has held rose trials since 1930.

Roseraie du Parc de Loisirs Saint Nicolas Parc de Loisirs, F-71380 Chatenoy en Bresse (Mairie de Châlons-sur-Saône)

Roseraie du Parc du Buisson Rond Parc du Buisson Rond, Rue Sainte Rose, F-73000 Chambéry

Roseraie du Jardin des Plantes Muséum d'Histoire Naturelle, Jardin des Plantes, 2 rue Cuvier, F-75005 Paris

Roseraie du Parc de Bagatelle Parc de Bagatelle, Route de Sèvres Neuilly, F-75016 Paris

Close to the centre of Paris, Bagatelle offers visitors a delightful space with a charming castle. The most prestigious efforts of French and foreign rose growers are gathered here. From mid-May to mid-July 10 000 rose bushes of 1200 varieties are in bloom. Then in September, remontant varieties grace the garden once more. Here, one can discover old roses no longer available commercially, winning roses from the International Rose Competition, and classic varieties which have delighted the public for over a century. The International New Rose Competition has been held annually since 1907, at Bagatelle.

Roseraie du Parc de Chambord Direction des Espaces Verts, 9 rue de l'Hôtel de Ville, F-82000 Montauban

Roseraie Municipale de Limoges Rue de la Roseraie, F-87000 Limoges

Roseraie des Jardins Albert Kahn 10 Quai du 4 Septembre, F-92100 Boulogne-sur-Seine

Roseraie du Château de la Malmaison 15 Avenue du Château, F-92500 Rueil Malmaison

Formal rose garden, once the home of Empress Josephine who collected old garden roses.

Roseraie Municipale Honoré Daumier Avenue du Maréchal Juin, F-93600 Aulnay-sous-Bois

La Roseraie de l'Hay du Val de Marne 8 rue Albert Watel, F-94200 L'Hay-les-Roses

The world's oldest existing garden devoted exclusively to roses celebrated its centenary in 1994. It was created by architect Eduard André and rosarian Jules Graveraux, who assembled a collection of all species and varieties known at the time. Together they created a rose garden of historical value and breathtaking beauty. In 1968 ownership of the garden passed to the department of Val-de-Marne, which has conscientiously continued the maintenance and renewal of the collection while encouraging public participation in this setting of unrivalled charm. Awarded WFRS garden Plaque of Merit in 1995.

Roseraie du Parc de la Beaujoie Rue des Pays de la Loire, F-44300 Nantes

Site of Annual International Rose Trials.

Germany

Rosengärtnerei Kalbus Altdorf Hagenhausen, D-90518 Altdorf

Mainly old garden roses and 'English Roses'; 3000 plants (400 varieties), established in 1984.

Gönneranlage Lichtentaler Allee, D-76530 Baden-Baden

The popularity of Baden-Baden was due in part to Konrad Adenauer who was a keen rosarian. During a conference of foreign ministers in Baden-Baden, he was persuaded to become 'Patron of the Rose'. So in 1952 the first rose trials took place in the Gomier garden. This garden celebrated its Golden Jubilee in 2002, a proud 50 years of accomplishment. There are 25 000 plants (350 varieties).

Rosenneuheitengarten Auf dem Beutig, Moltkestrasse, D-76530 Baden-Baden

Established in 1981, this garden has 5000 plants (200 varieties).

Bad Langensalza Rosengarten Vor dem Klagetor, D-99947 Bad Langensalza

Established 1989, this garden has 10 100 plants (400 varieties) in 1.7 ha.

Schau- und Prüfungsgarten der Rosen-Union Steinfurther Haupstrasse 25, D-61231 Bad Nauheim-Steinfurth

This garden has 8000 plants (250 varieties).

Schaugarten Gönewein Steinfurther Haupstrasse 1-5, D-61231 Bad Nauheim-Steinfurth

There are 4000 plants (150 varieties) in this garden.

Rosengarten im Kurpark Bad Sassendorf Im Kurpark, D-59505 Bad Sassendorf

Rosengarten der neuen Residenz Domplatz 8, D-96049 Bamberg

There are 4500 plants (50 varieties).

Britzer Garden Sangerhauser Weg 1, D-12349 Berlin

Approximately 18 000 rose bushes, modern and historical, in 155 ha.

Rosengarten auf der Pfaueninsel Am Ende des Nikolskoer Weges, Fähre zur Insel, Berlin

About 1000 plants, old to modern, 300 varieties.

Rosengarten im Grossen Tiergarten Strasse des 17. Juni 31, D-10785 Berlin

This garden has 5300 plants (110 varieties).

Bürgerpark Pankow Schönholzer Strasse, Berlin-Pankow

Mass displays of 40 varieties, total plants 6000.

Rosengarten Bremen Rhododendronpark, Marcus-allee, Bremen

About 10 000 roses (representing 200 varieties) are displayed in this garden, which was renovated in 1997/8.

Rosengarten Coburg Am Anger, D-96450 Coburg

Originally established in 1929, the modernized garden now contains 7000 plants (72 varieties).

Rosenhöhe Mathildenhöhe, Darmstadt

This garden was established in 1982; it has 10 000 plants (290 varieties).

Deutsches Rosarium VDR Dortmund-Westfalen-park, An der Buschmühle 3, D-44139 Dortmund

Established in 1972, this garden has 50 000 plants (3500 varieties).

Lehr- und Sichtungsgarten Dresden-Pillnitz Lohmener Strasse 12, D-01326 Dresden-Pillnitz

Founded in 1993, this garden has 2000 plants (270 varieties).

Rosengarten im Burggraben Kurfürstliche Burg, Burgstrasse, D-65343 Eltville

This garden contains an impressive 10 000 plants representing 350 varieties.

Rosengarten und Roseneneuheiten-Garten in Palmengarten Siesmayerstrasse 61, D-60323 Frankfurt-am-Main

There are about 10 300 plants in mass plantings of 900 varieties.

Rosengarten im Planten und Blomen Tiergarten-strasse, Hamburg Mitte

Established in 1992, this garden has 3500 plants (300 varieties).

Rosengarten im Stadtpark Hannover Theodor-Heuss-Platz, Hannover

About 5300 plants (140 varieties) are grown in this garden.

Italienischer Rosengarten Strasse der Wild-und Strauchrosen, Roseninformationsgarten, D-78465 Insel Mainau

Almost 30 000 plants (1300 varieties) are displayed here.

Herzogenriedpark Hochuferstrasse 27, D-68167 Mannheim

Established in 1975, this garden has 6000 plants (240 varieties).

Rosensichtungsgarten Sachsenstrasse 8, München

There are almost 5000 plants (220 varieties).

Rosengarten im Westpark Westendstrasse, München

The garden is 200 ha with 25 000 plants (420 varieties).

Rosengarten Pinneberg Fahltsweide, D-25421 Pinneberg

Founded in 1932, the garden has 8300 plants (80 varieties).

Europa-Rosarium Sangerhausen Steinberger Weg 3, D-06526 Sangerhausen

This world famous garden, situated in the centre of Germany, was officially opened in 1903 in the presence of its Royal Patron, the Empress Augusta. The Europa-Rosarium at Sangerhausen must rate as the most important rose garden in the world. It has an inventory of almost 7000 cultivars spanning the complete history of the evolution of the rose. Thanks to the initiative of rose breeder Peter Lambert, many old garden roses and species in danger of extinction were gathered together to form a valuable collection and the City of Sangarhausen had just the right plot of land to accommodate this unique collection. In 1903 Sangerhausen opened its gates for the first time during the German Rose Society Congress to display over 2000 cultivars on about 1.5 ha. Today the land devoted to roses is now 12.5 ha. Thanks to many benefactors throughout the years, this garden has received patronage and protection from the ravages of two world wars. In 1989, after the Reunification of Germany, the Rosarium rapidly gained its well-deserved international reputation as the comprehensive repository of many old garden roses in danger of extinction. The collection of old garden roses is the most extensive in the world today, consisting of 50 000 plants (6800

Table 1 Inventory of cultivars at the Europa-Rosarium

Classification	*Number of cultivars*
Alba	25
Bengal	55
Bourbon	65
Centifolia	40
Climbers	850
Damask	30
Floribunda	780
Gallica	160
Groundcovers	80
Hybrid Polyantha	225
Hybrid Tea	2380
Modern Shrubs	310
Moss	130
Noisette	55
Pernetiana	170
Polyantha	280
Ramblers	465
Rugosa	80
Tea	115
Wild species	500

varieties) (**Table 1**). In 2003 this garden celebrated its 100th birthday in conjunction with the German Rose Society.

Rosengarten von W. Kordes' Söhne Rosenstrasse 54, D-25365 Sparrieshoop

A collection of 5000 plants (250 varieties).

Tal der Rosen, Killesberg Am Kocherhof 16, D-70192 Stuttgart

Approx. 18 000 plants (400 varieties) are displayed.

Rosarium Uetersen D-25436 Uetersen

Established in 1934, this garden has 35 000 plants (900 varieties).

Rosenhöhe im Vogelpark Walsrode Am Rieselbach, D-29664 Walsrode

There are 13 000 plants (25 varieties).

Europas Rosengarten Zweibrücken Rosengartenstrasse, Zweibrücken

A huge collection of 60 000 plants (2000 varieties).

Wildrosengarten Fasaneriestrasse, Zweibrücken

A unique collection of species roses, with 2000 plants (700 varieties).

Greece

Historic Rose Garden of Schinos Centre for Hellenism, Damianos Foundation at Schinos, Corinth

This rose garden is part of a 4-ha idyllic setting inspired by the Royal National Rose Society's 'Gardens of the Rose' at St Albans, UK. Planting started in winter 1995 using a spiral design with a pavilion at the centre. Roses featured are hybrid teas, shrubs and groundcovers as well as many historical roses.

India

Mughal Gardens Adjacent to the Presidential Mansion, Delhi

As the name would imply, this 6-ha rose garden established in 1922 was designed to amplify the Mughal influence using both elegant Persian and British landscaping trends expressed in herculean size and dimensions. In particular the 'Butterfly Garden' contains only fragrant roses. There are over 250 varieties, some domestic and some imported.

Rose Garden at New Delhi Airport This municipal garden established in 1993 has more than 24 000 rose bushes planted in 6.5 ha in mass displays in 300 separate beds, and maintained in peak condition. It is now the largest rose garden in the city featuring hybrid teas, floribundas and polyanthas.

Israel

Wohl Rose Park of Jerusalem 91101 Jerusalem, Municipality of Jerusalem and the Jerusalem Foundation

In this garden species roses, historic roses, climbers and modern varieties (15 000 plants, 400 varieties in 6 ha) are imaginatively displayed. They are best viewed in May. The park is laid out in a natural way, with hills, grottos, ponds, a waterfall and natural rock faces. It includes a 'Garden of Nations' where each section has been donated by a different country, province or city. Each section has a special feature and a representative collection of roses from the respective country.

Ramat Hanadiv Rothschild Foundation, 30900 Zichron Yaakov

Ramat Hanadiv spreads out to the south of picturesque Zichron Yaakov. At the heart of the park are the Memorial Gardens, the burial site of Baron Edmond Rothschild and his wife, Baroness Adelaide. The area contains several individual sections: the Cascade, the Rose, the Fragrance and Palm Gardens and an Entrance Court, each with its own distinctive vegetation and design character. Together they create a rich floral kaleidoscope. A visit to the Memorial Gardens is only part of the year-round attractions of Ramat Hanadiv. Nature trails and archaeological sites are located in the nature reserve. There are 1000 plants in 0.1 ha.

Italy

Roseto Comunale Marcia 7-8, Via di Valle, Rome

The garden is about 1.4 ha containing a wonderful mixture of species roses, historic roses, climbers and modern varieties; there are 1200 plants plus 350 varieties on rose trials. Best viewing times are May–June and September–October.

Il Roseto Niso Fumagalli Villa Reale, I-20052 Monza

This wonderful garden, founded by Associazione Italiana della Rosa and established in 1964, has 3000 plants (mainly modern with some species and old garden roses) in 1 ha. It is home to now famous annual International Rose Trials. Best viewed in late May and June. Closed July and August.

Il Roseto di Nervi Villa Grimaldi Villa Grimaldi, Via Capolungo 9, I-16167 Genoa

Principally recurrent blooming varieties, 7000 rose bushes are displayed on 1 ha, best viewed from the end of April to the middle of June.

Il Roseto di Cavriglia I-52022 Cavriglia, Arezzo

The garden is privately administered by La Fondazione Carla Fineschi. Professor Gianfranco Fineschi started collecting roses 30 years ago and today he has one of the world's largest private collections: more than 7000 varieties, each represented by a single plant. Originally planted between 1850 and 1910 the garden featured many cultivars that were no longer in commerce or were in danger of being lost. Professor Fineschi established and revived the garden, beginning in 1967. In this living museum the genus *Rosa* is presented from an historical and a botanical point of view. Species, subspecies and later hybrids are systematically presented, while climbers and ramblers form walls between the scientific divisions. Many of the beds of modern varieties are grouped according to their hybridizers. In this garden overlooking the hills of Tuscany one may trace the history of the rose from antiquity to the present day while marvelling at the beauty of the garden itself. Prime viewing is May–June and September.

Il Roseto di Valleranello Via Darwin 300, I-00134 Roma

Private garden containing 1300 species roses, historic roses, climbers and modern roses on 1.3 ha. Wonderful landscaping of roses with trees. By appointment; best viewed in May–June.

Le Rose della Ginestrella 'La Ginestrella', I-06070 San Egidio, Perugia

Privately administered. Species roses, historic roses, climbers and modern roses are displayed: there are 3000 plants on 1.5 ha. By appointment; best viewed May–June and September.

Santa Maria I-06058 San Terenziano

Privately administered by Mr and Mrs André Brichet. Species roses, historic roses, climbers (about 600), old roses only. By appointment only.

Il Roseto dell'Orto Botanico Largo Cristina di Svezia 24, I-00165 Roma

A historical rose garden, belonging to L'Università 'La Sapienza', Dipartimento di Biologia Vegetale, on 0.8 ha. Best in May, June and September; open all year except August, Sundays and public holidays.

Japan

Yubari City Rose Garden Takamatsu, Yubari City, Hokkaido 068-04

Well over 20 000 rose bushes planted on 2 ha. Best viewed in July and September.

Rose Garden Chizaki 22-35 Fushimi, Chuo-ku, Sapporo City, Hokkaido 064

This privately administered rose garden located on a hillside is about 1.8 ha. Best viewed June–July and September–October.

Nakajima-Park, Hyakka-en Nakajima Koen, Chuo-ku Chuo-ku, Sapporo City, Hokkaido 064

This city-administered 1 ha rose garden has about 2000 rose bushes. Best viewed in July and September.

Hakodate Rose Hill Rose Garden 144 Higashiyama-Cho, Hakodate City, Hokkaido

Privately administered rose garden with 6000 bushes, mostly modern, on 0.5 ha with a focal centroid of a hugh floral clock.

Kacho Keikoku Rose Garden Utarubenosato, Towadako-Machi, Aomori

Privately administered garden on 33 ha featuring over 20 000 roses (species and old garden roses). This is the largest rose garden in Japan. Best viewed in July and September.

Hanamaki-Onsen Rose Garden 1-58-1 Yumotodai, Hanamaki City, Iwate 025-03

Privately administered garden, with 4000 roses on 1.7 ha. Best viewed June–July and September–October.

Yokote Park Rose Garden Matsuzaki, Aza Shirotsuki, Yokote City, Akita 013

This city-administered rose garden has about 3000 roses planted on 0.3 ha. Best viewed in June and September.

Higashizawa Park Rose Garden 1-3-6 Chuo, Murayama, Murayama City, Yamagata 995

This city garden has 15 000 rose bushes on 1.5 ha. Best viewed in June and September.

Ibaragi Prefectural Flower Park 200 Shimoaoyanagi, Yasato-Machi, Ibaragi 315-01

Town-administered garden boasts of 30 000 rose bushes on 2 ha, specializing in climbers. Best viewed in June and October.

Igashira Park 99 Shimokagoya, Mooka City, Tochigi 321-44

Administered by the Prefecture, this garden has 1500 roses on 0.5 ha. Best viewed May–June and October.

Shikishima Park Rose Garden Shikishima-Machi, Maebashi City, Gunma 371

City-administered garden with 2000 roses on 4.5 ha. Best viewed in June and October.

Keisei Rose Nurseries Inc. 755 Owada-Shimden, Yachio City, Chiba 755

Private rose garden with 5500 plants on 10 ha. This is biggest rose nursery in Japan.

Hibiya Park Rose Garden Hibiya-Koen, Chiyoda-ku, Tokyo 100

This small but delightful oasis in the heart of Tokyo is maintained by the Prefecture. There are 500 well-maintained plants on just 0.05 ha.

Tokyo Metropolitan Jindai Botanical Park 5-31-10 Jindaiji, Chofu City, Tokyo 182

The Japan Rose Concours is held in this 1.2 ha rose garden with 5000 plants. Best viewed in May and October.

Mukogaoka-Yuen Rose Garden Nagao, Tama-ku, Kawasaki City, Kanagawa 044

This privately administered 1 ha garden with 20 000 bushes has 1000 varieties including species and climbers. Best viewed May–June and October–November.

Kanagawa Prefectural Ofuna Botanical Garden 1018 Okamoto, Kamakura City, Kanagawa 247

Wonderful collection of 60 species (1000 plants) on 0.1 ha. Best viewed in May and October.

Shimada Rose-Hill Park 1652-1 Noda, Shimada City, Shizuoka

This city-maintained garden has 8000 plants on 0.4 ha; roses can be enjoyed throughout the year in the the greenhouses. Best viewed in May and October.

Hamamatsu City Flower Park 195 Kanzanji-Machi, Hamamatsu City, Shizuoka 432-12

This city garden showcases almost 2000 bushes on 0.3 ha. Best viewed May–June and October–November.

Higashiyama Zoo and Botanical Garden 3-70 Higashiyama-Motomachi, Chigusa-ku, Nagoya City, Aochi 464

City-maintained rose garden featuring mostly floribundas; about 1000 plants on 0.2 ha.

Tsuruma Park Tsuruma 1-Chome, Showa-ku, Nagoya City, Aichi 466

This city rose garden has 6000 roses planted on 0.6 ha. Best viewed May–June and October.

Flower Festival Commemorative Park 1584-1 Seta, Kani-city, Gifu-Prefecture 509-0213

This magnificent rose garden under the patronage of the Governor of Gifu has 41 000 plants (1250 varieties) in 80.7 ha. It is home to annual International Rose Trials.

Kyoto Botanical Garden Hanmoku-cho, Sakyo-ku, Kyoto City, Kyoto 606

This Prefecture-sponsored rose garden is 2.5 ha in extent with 2000 roses.

Peace Rose Garden 1-1 Senri-Banpaku-Koen, Suita City, Osaka 565

Built as the rose garden of the Osaka Exposition of 1970, the garden is now maintained by the Japan Rose Society and has 5400 roses on 0.6 ha.

Hirakata Park Rose Garden 1-1 Hirakatakoen-Machi, Hirakata City, Osaka 573

This privately administered rose garden is about 1 ha with 5000 bushes representing 1000 varieties, mainly species and old garden roses. Best viewed May–June and October–November.

Itami City Aramaki Rose Park Aza Matsubama, Aramaki, Itami City, Osaka 664
The city-maintaned rose garden has 10 000 roses on 1.7 ha.

Hyogo Prefecture Flower Centre 128201 Iimori, Toyokura-Machi, Kasai City, Osaka 679-01
Society-maintained garden of 0.3 ha with 4700 plants.

Himeji Rose Garden 2222 Toyotomi, Toyotomi-Machi, Himeji City, Hyogo 679-21
Privately administered garden with 4000 plants (1200 varieties) on 0.2 ha.

R.S.K. Rose Garden 1592-1 Natsukawa, Okayama City, Okayama 701-01
This privately administered garden has a circular design featuring 25 000 rose bushes on 3 ha.

Hiroshima Botanical Garden 495 Kurashige 3-Chome, Saeki-ku, Hiroshima City, Hiroshima 731-51
City garden features 1000 species and old garden roses on 0.1 ha.

Rose Park Hanazono-Machi 1-Chome, Fukuyama City, Hiroshima 720
This city rose garden has 4500 rose plants on 1.5 ha.

Kashii Flower Garden 7-2-1 Kasumigaoka, Higashi-ku, Fukuoka City, Fukuoka 813
A privately administered garden with 3000 bushes on 0.3 ha. Its main feature is a large pergola of roses.

Miyazaki Kodomono–Kuni Rose Garden 7275-1 Aza Matsuzoe, Oaza Kaeda, Miyazaki City, Miyazaki Prefecture
A privately administered garden of 5000 roses on 0.2 ha.

Luxembourg

Merl Park City of Luxembourg
This city rose garden has about 3000 species and old garden roses. Best viewed in June.

Mondorf Rosegarden Mondorf City and Domaine Thermal
Administered by Domaine Thermal and Mondorf City, this rose garden was viewed by the delegates to the WFRS World Rose Conference in 1997 during Opening Ceremonies.

The Netherlands

Westbroekpark The Hague, Municipality of The Hague
This world-famous rose trial ground boasts 20 000 rose bushes in mass display beds (300 varieties). The 5-ha site with its layout and charm is a marvellous tribute to the rose.

New Zealand

Gardens abound in New Zealand and almost every city, town and hamlet has a small rose garden that the local rose society actively maintains. Due to space restrictions many gardens have been omitted from the following list, but they remain a testimony to the love of roses by the citizens.

Takapuna Rose Garden Anzac Street, Takapuna
Garden features 630 rose bushes.

Belmont Rose Garden Lake Road, Belmont
There are 600 rose bushes.

Botanic Gardens and Auckland Regional Trial Grounds Hill Road, Manurewa
There are 200 roses in the Botanic Gardens and 4500 in the Trial Grounds.

Morrinsville Rose Gardens Morehouse Street, Morrinsville
Gardens with almost 2000 rose bushes in mass displays.

Rogers Rose Garden Cobham Drive, Hamilton
Established in 1969 to celebrate the first World Rose Convention in the Hamilton Botanic Gardens (a 58-ha site), this garden has approximately 4500 roses planted in 2 ha, mainly in large informal beds for mass effect. In 2000 Hamilton City Council in agreement with Rose Introducers of New Zealand named this garden the permanent home of the 'Rose of The Year' trial held for many years previously in Auckland Regional Botanical Gardens. Best viewed in October.

Te Awamutu Rose Garden Gorst Avenue off Highway, Te Awamutu
Well maintained by the local rose society, the 2500 bushes are examples of roses growing exceptionally well.

King Edward Rose Garden High Street, Hawera
About 300 rose bushes are on display.

Dugald McKenzie Rose Garden and International Trial Grounds Park Road, Palmerston North
Established in 1969 this garden is home to the world-renowned International Trial Grounds

administered by Palmerston North under the auspices of the National Rose Society of New Zealand to test roses from abroad as well as of domestic origin. The trial garden is situated in the Dugald McKenzie Rose Garden, part of the city's 25-ha esplanade of garden, trees, shrubs and parkland.

Lady Norwood Rose Garden Glenmore Street, Wellington

Situated within Wellington Botanic Gardens, this rose garden opened in 1953 covering an area of 4500 m^2 in a spectacular circular formal design featuring about 2000 modern roses as well as 1200 miniatures with climbers elegantly draped over the outer perimeter strucure.

Bolton Street Cemetery Kinross Street, Wellington

Old garden roses and Noisettes, are the speciality here.

Queen Elizabeth Park Rose Garden Dixon Street, Masterton

This garden features about 800 modern roses.

Frimley Rose Garden Frimley Avenue, Hastings

There are almost 4000 roses in this fine garden.

Kennedy Park Rose Garden Kennedy Park, Napier

Another proud rose garden, this has 4000 bushes supported by the local society.

Murray Linton Rose Garden Fenton Park, Trigg Avenue, Rotorua

About 1500 rose bushes are on display.

Whangarei Rose Garden Water Street, Whangarei

A small but attractive rose garden with 700 plants.

Centennial Rose Garden Pollard Park, Parker Street, Blenheim

There are mass plantings of 1500 rose bushes.

Mona Vale Fendalton Road, Christchurch

In this small but beautiful garden planted in 1970 there are approximately 300 bushes, standards and climbers adorning walls, niches and beds. In 1991 the gardens were redesigned in preparation for the 1994 World Rose Convention. A total of 17 beds, arches, pillars, raised miniature beds, a computerized watering system and new soil all attributed to the specular condition of the garden maintained by the Canterbury Rose Society.

Trevor Griffiths Historical Garden of Heritage Roses Main Highway, Temuka

An oasis for lovers of old garden roses and species.

Public Gardens Severn Street, Oamaru

Attractive garden featuring about 700 modern roses and 200 Miniatures.

Botanic Garden Great King Street, Dunedin

A collection of 1500 roses.

Queen's Park Gala Street, Invercargill

A beautiful collection of 3000 rose plants.

Samuels Rose Garden Broadgreen Park, Nayland Road, Stoke

An attractive collection of 2800 bushes.

Norway

Tomb Rose Garden Tomb, N-1640 Raade

Designed in 1990 to be a living rose catalogue of old garden roses for Nordic climates, administered by the Norwegian Rose Society, this garden has 2000 rose bushes, modern roses, historic roses, and climbers. This is a well-maintained garden in which all the roses are clearly labelled and a host to rose trials. Best viewed in July.

Rosariet NLH Postbox 5022, N-1430 Aas

Administered by the Agricultural College at Aas, this garden has a collection of 2000 species. Best viewed in July.

Frogner Parken Rosegarden Kingos Gate 17, N-0457 Oslo

This well-maintained garden is supervised by the Oslo Park Institution, and contains about 800 modern rose bushes. It is also home to a rose trial ground. Best viewed in July.

Pakistan

Gulab Bagh Islamabad

Established in 1992 in the very heart of the city and administered by the Pakistan National Rose Society, this garden has 12 000 rose bushes planted for mass display (170 varieties) using a pleasing architectural design on 2 ha. The fine collection of modern roses is a fitting tribute to roses and the city. Best viewed end of March to mid May.

Rose and Jasmine Garden Islamabad

About 8000 rose bushes are on display in a garden supervised by the Capital Development Authority. Best viewed early March to mid-May.

Gulistan-e-Fatima Jinnah Lahore

This garden is administered by Lahore Municipal Corporation. Best viewed from early March to the end of April.

Cantonment Garden Multan

This garden, supported by Multan Cantonment Board, has 4500 roses planted in neat beds for mass display. Best viewed early March to the end of April.

Poland

Stacja Doswiadczalna Ochrony Roslin Plant Protection Station, Srem Polska, 63-100 Srem Wójtostwo

Administered by COBO – Research Centre for Cultivars, this garden is the repository of a reference rose collection for plant breeders' rights and registrations of cultivars in Poland. Best viewed June 15–30 and August–September.

Research Institute of Pomology and Floriculture Ul. Warynskiego 14, 96-100 Skierniewice.

Administered by the Research Institute of Pomology and Floriculture in Skierniewice, this garden has 2500 rose bushes (250 varieties) planted in 0.3 ha. It is noted for its work in the evaluation of rose cultivars for fungus disease susceptibility. Best viewed in June.

Didactic Collection, Agricultural University in Poznan Ul. Dabrowskiego 159, 60-594 Poznan

This large collection of 6000 roses (700 varieties) on 0.8 ha is administered by the Department of Ornamental Plants. Best viewed June 15–30 and August–September.

Didactic Collection, Agricultural University in Lublin Ul. Leszczynskiego 58, 20-032 Lublin

Administered by the Department of Ornamental Plants, this garden has approximately 2500 plants (250 varieties) on 6 ha. Best viewed June 15–30 and August–September.

Rosarium Wojewódzki Park Kultury im gen. Jerzego Zietka Ul. Katowicka 2, 41-501 Chorzów

This rose garden was initiated by the Polish Rose Society in 1964 and opened in 1968. It now contains about 40 000 rose bushes, mostly modern hybrids, on 1 ha of land in a 17-ha rosarium site. Best viewed June 15–30 and August–September.

Rosarium Cytadela Ul. Za Cytadela 1a, 61-663 Poznan

There are almost 23 000 roses (150 varieties) in a 5-ha rosarium with approx. 0.5 ha of rose beds. Best viewed June 15–30 and August–September.

Romania

Rosariul Gradinii Botanice Gradina Botanica, Dumbrava R. 9, RO-6600 Iasi

The largest rose garden in Romania, this is administered by the University of Iasi. The garden has 10 000 rose bushes on 1.55 ha, mostly species, old garden roses and climbers. Best viewed in June and July.

Rosariul Gradinii Botanice Gradina Botanica, RO-3400 Cluj

Approx. 2000 species, historic roses and climbers on 0.2 hectares is garden is supervised by the University of Sc., best viewed June and July.

Rosariul Statiunea Pomicola C.P. 1-86, RO-3400 Cluj

This garden has approx. 3500 roses, mostly modern Hybrid Teas, Floribundas, groundcovers and some climbers; it is administered by Statiunea Pomicola, Cluj. There are trial grounds for domestic rose breeding programmes. Best viewed in June–July.

Rosariul Institutul de Cercetare pt Legumicultura si Floricultura RO-8268 Vidra/Sai

This garden has 5500 plants representing 170 varieties, mainly new Hybrid Teas and Floribundas. It is administered by Institutul de Cercetare pt Legumicultura si Floricultura. Best viewed May–June and August–September.

Rosariul Complexul Muzeal de Stiintele Naturii Domneasca 91, RO-6200 Galati

Mainly new varieties, Hybrid Teas, Floribundas, Polyanthas, Shrubs and Climbers, are represented in the 2500 plants of 150 varieties on 0.4 ha, administered by Complexul Muzeal de Stiintele Naturii. Best viewed from the end of May–July.

Slovakia

Arborétum Borová Hora The Technical University in Zvolen, 960 53 Zvolen

Administered by the Technical University in Zvolen, the arboretum is 48 ha in extent with a rose garden of 700 bushes mainly representing species and hardy old garden roses. Best viewed in mid-June.

South Africa

Travelling the Wine Route in the Western Cape, one finds roses planted at every vineyard. There are well-kept gardens at Vergelegen, Somerset West, Stellenberg, Claremont and Rustenberg, and Stellenbosch. The most popularly grown variety is the floribunda 'Iceberg' as bushes, climbers and standards.

Ludwig's Rose Farm Ludwig Taschner, P.O. Box 28165, Sunnyside 0132

A wonderful collection of 1000 modern roses and old garden roses featuring many of the newer varieties from German, British and American breeders. Best viewed October–June.

Johannesburg Botanical Garden Thomas Bowler Street, Emmarentia, Johannesburg

Large rose beds planted in terraces, with 100 varieties in mass displays. Best viewed from October to the end of May.

Robert Jameson Park Montpelier Street, Durban

An interesting collection of about 150 varieties of Hybrid Teas in a tropical landscape setting. Best viewed mid-September to June.

Alexandra Park Commercial Road, Pietermaritzburg

A rose park developed in the city centre. All the varieties are named after a place or a city.

Durbanville Rose Garden N1 motorway (25 km from Cape Town)

An outstanding rose garden with about 7000 rose bushes featuring trial grounds for varieties yet to be released in South Africa. Special attention has been paid to miniature roses by devoting a sunken garden in a circular design to showcasing many of the world's top miniature varieties. Best viewed from mid-October to May.

Duncan's Roses at Arieskraal Farm Grabouw, Cape Town 7160

A beautifully landscaped show garden of several hundred varieties, set amid huge apple orchards. Best viewed from the end of October to May.

Freshwoods Elgin, Cape

Privately administered by Peter and Barbara Knox-Shaw, this rose garden with about 500 varieties an abundance of trees and other plants in a natural setting and is an oasis for any rose travellers to visit. Best viewed in early November. By appointment only.

Sonskyn Rose Nursery De Doorns, Boland

Set between vineyards with majestic mountains as a backdrop, this garden displays 300 varieties.

City of Bloemfontein Bloemfontein, Orange Free State

Roses are planted in most major traffic islands, in every park and in the gardens of museums, schools and government institutions. Best viewed from the second half of October to early May.

Sweden

Göteborgs Rosarium Trädgårdsföreningen, Box 225, 401 23 Göteborg

Established in 1987 and administered by the Göteborgs Fritidsförvaltning (Göteborg Community), this rose garden occupies 1 ha featuring about 7000 plants (2700 varieties) with an emphasis on old garden roses and species. Also includes a substantial rose library. Best viewed June–July.

Fredriksdal, Helsingborgs Museum Box 1283, 251 12 Helsingborg

Administered by the Helsingborgs Community, this garden has approx. 1800 rose bushes (450 varieties) on 0.5 ha, with a speciality in old garden roses. Best viewed June–July.

Rosarium Norrköping Utställningsvägen 28, Himmelstalundsskdan, 602 36 Norrköping

This rose garden contains 350 plants (300 varieties) on 1 ha supervised by the Himmelstalund School of Horticulture. Best viewed in June–July.

Switzerland

Parc de la Grange Genève Quai de Colony, CH-1211 Genève 2

This world-famous rose garden opened in 1947 by the City of Genève, situated on the lakeside, has about 12 000 plants (modern and old garden roses) on 1 ha. International Rose Trials have been held there since 1948. Best viewed in June and September.

Blind Fragrance Rose Garden Schanze, CH-8640 Rapperswil

Supported by Verkehrs- und Verschönerungsverein Rapperswil, the rose garden is about 1600 m^2 and contains about 1500 fragrant roses; rose markers are also in Braille. Best viewed in June.

Rosengarten beim Einsiedlerhaus Endingerstrasse, CH-8640 Rapperswil

In this old historic town, administered by Verkehrsund Verschönerungsverein Rapperswil, this small rose garden has 1200 plants on 200 m^2, featuring species and old garden roses. Best viewed in June.

Schlossgarten Heidegg Schloss Heidegg, CH-6284 Gelfingen

The Baroque-style garden, administered by Kanton Luzern, contains 1500 plants representing about 500 varieties on 1000 m^2.

Rose Gardens Kartause Ittingen Stiftung Kartause Ittingen, CH-8532 Warth/TG

On the site of a Carthusian monastery founded in 1461 (now a Cultural Centre), this rose garden is on 2.3 ha with 750 plants, modern and old garden roses.

Rosengarten Feldbachareal Gemeindeverwaltung, CH-8266 Steckborn/TG

An ancient refectory is surrounded by the rose garden, on the Untersee. There are 1500 plants on 4200 m^2.

Arboretum du Vallon de l'Aubonne–Wildroses Association de l'Arboretum, En Plan, CH-1170 Aubonne

A unique collection of wild roses from Europe, Asia and America (100 plants).

Roseraie du Jardin Botanique Chambésy 1 chemin de l'Impératrice, CH-1292 Chambésy

Unique display of old garden roses and species sponsored by the City of Geneva.

Roseraie du Jardin Botanique Porrentruy 22 route de Fontenais, CH-2900 Porrentruy

Formal monastery garden maintained by the Musée Jurassien des Sciences Naturelles.

United Kingdom

Rose gardens abound in England where the rose is the most loved flower in the land.

England

The Gardens of the Rose The Royal National Rose Society, Chiswell Green, St Albans, Herts AL2 3NR

This major rose garden has about 30 000 plants on 11 ha featuring roses of all types including emphasis on rose trails, a special tribute to 'Peace' and its descendants, miniatures, shrubs, climbers and old garden roses, all in massive display beds. Best viewed from mid-June to mid-October.

Mottisfont Abbey 2 Oakley Road, Mottisfont, Romsey, Hants SO5 1LQ

Operated by The National Trust, the rose garden is a landscaping marvel featuring 1500 old garden roses in 0.6 ha surrounding a magnificient old abbey. Best viewed in June and July.

Queen Mary Rose Garden Regents Park, London

Established as a tribute to Queen Mary, the rose garden is impeccably maintained by The Royal Parks, London. It features 15 000 modern rose bushes in wonderful mass display beds in 0.93 ha. Best viewed June to October.

Royal Horticultural Society Wisley, Ripley, Woking, Surrey GU23 6QB

Within the vast gardens maintained by the Royal Horticultural Society at Wisley, the rose garden features mostly modern roses in classical beds with small box hedges. There are about 5000 roses neatly displayed in several gardens.

Woburn Abbey Woburn, Bedfordshire MK17 9WA

The Duchess's rose garden includes mass displays of 'Iceberg'.

Manor Farm Little Horwood, Winslow, Bucks (north east of Winslow off A413)

As well as wide variety of roses, this garden has a hosta border and wild-flower meadow.

Elton Hall Elton, Peterborough, Cambs PE8 6SH

A fine collection of old garden roses is displayed here.

Arley Hall Great Budworth, Northwich, Cheshire CWg 6NA

This garden has Hybrid Teas, shrubs and species roses.

David Austin Roses Bowling Green Lane, Albrighton, Staffs, WV7 3HB

Over 900 varieties of hybrid teas, shrubs, climbers, species and old garden roses are grown at this specialist nursery.

Haddon Hall Bakewell, Derbyshire (2 miles south of Bakewell on the A6)

The walls of the sixteenth-century hall are covered with an extensive collection of climbers and ramblers.

Royal Horticultural Society Garden at Hyde Hall Rettendon, Chelmsford, Essex CM3 8ET

A wide assortment of roses of all kinds is on display.

Kiftsgate Court Gardens Chipping Campden, (Gloucestershire), GL55 6LW

This garden is famous for its roses, especially *filipes* 'Kiftsgate'.

Hidcote Manor Garden Hidcote, Chipping Campden, Glos

A National Trust property, famous for old garden roses. Best viewed in June.

Hatfield House Hatfield, Herts

Seventeenth century gardens restored to their former beauty.

NORTHERN IRELAND

City of Belfast International Rose Garden Upper Malone Road, Belfast

Situated in a park setting this 4-ha garden has 20 000 rose bushes comprising species roses, historic roses and moderns. Trials grounds were established in 1964. In the late 1980s the gardens were completely redesigned, offering a series of circular rose beds skilfully integrated into the hillside landscaping. Formally opened for viewing in 1990, this garden has quickly gained a high reputation and deserved the award by WFRS in 2000 of 'Garden Plaque of Merit' for a spectacular and dramatic tribute to roses. Best viewed mid-July to mid-August.

SCOTLAND

Tollcross Rose Garden Tollcross, City of Glasgow

Established in 1989, this city rose garden occupies about 1 ha on a sloping hillside with the design for the rose beds representing rose petals unfurling to provide an absolutely beautiful display. International rose trials are held in this parkland setting every August. The garden has about 10 000 plants.

United States of America

There are so many public rose gardens located within the 50 states that the selection below is only a very small representation. A note [a] after a garden name denotes an All-America Rose Selections (AARS) display.

Bellingrath Gardens[a] 12401 Bellingrath Gardens Road, Theodore, AL 36582

3500 plants, 50 varieties.

Descanso Gardens Rose Garden[a] 1418 Descanso Drive, La Canada, CA 91011

600 plants.

Exposition Park Rose Garden[a] 701 State Drive, Los Angeles, CA 90037

15 000 plants, 145 varieties.

Tournament of Roses Wrigley Garden[a] 391 South Orange Grove Boulevard, Pasadena, CA 91184

Headquarters for the Pasadena Tournament of Roses Association which holds the world-famous 'Rose Parade' each New Year's Day. 1500 plants, 100 varieties.

Golden Gate Park Rose Garden[a] Golden Gate Park, Section 7, San Francisco, CA 94117

1650 plants, 130 varieties.

San Jose Municipal Rose Garden[a] Naglee and Dana Avenues, San Jose, CA 95112

5000 plants, 184 varieties.

Huntington Library, Art Collections and Botanical Gardens 1151 Oxford Road, San Marino, CA 91108

4000 plants, 2000 varieties.

A.C. Postal Memorial Rose Garden[a] 555 Plaza Rubio, Santa Barbara, CA 93103

1200 plants, 150 varieties.

Rose Hills Pageant of Roses Garden[a] 3900 South Workman Mill Road, Whittier, CA 90601

Home to AARS Trails, ARS Award of Excellence Miniature Rose Trails, and most recently International Rose Trials initiated in 2000. The 'Elegy Garden of Remembrance' is mass planted in red and white roses, bushes and standards. 7000 plants, 600 varieties.

Longmount Memorial Rose Garden[a] Roosevelt Park, 700 Longs Peak, Longmount, CO 80501

Reported to be the oldest public rose garden in the United States. 1300 plants, 90 varieties.

Elizabeth Park Rose Garden[a] Prospect Avenue and Asylum Avenue, Hartford, CT 06119

15 000 plants, 800 varieties.

Atlanta Botanical Rose Garden[a] Piedmont Park at the Prado, Atlanta, GA 30309

335 plants, 55 varieties.

Kentucky Memorial Rose Garden[a] 937 Phillips Lane, Louisville, KY 40209

1600 plants, 120 varieties.

American Rose Center[a] 8877 Jefferson-Paige Road, Shreveport, LA 71119

Home of the American Rose Society, with 20 000 roses planted in 60 beds within a wooded landscape complete with fountains, sculpture and a church for weddings.

Stanley Park[a] 400 Western Avenue, Westfield, MA 01085
2500 plants, 50 varieties.

Lyndale Park Municipal Rose Garden[a] 4125 East Lake Harriet Parkway, Minneapolis, MN 55409
3000 plants, over 150 varieties.

Hattiesburg Area Rose Society Garden[a] University of Southern Mississippi, Hattiesburg, MS
1600 plants, 40 varieties.

Missoula Memorial Rose Garden[a] Blaine and Brooks Streets, Missoula, MT 59807
1180 plants, 75 varieties.

Prospect Park Rose Garden[a] 8205 Apache Avenue Northeast, Albuquerque, NM 87110
1100 plants, 225 varieties.

Queens Botanical Garden[a] 43-50 Main Street, Flushing, NY 11355
7000 plants, 85 varieties.

The Biltmore Estates[a] North Park Square, Asheville, NC 28801
2000 plants, 100 varieties.

Columbus Park of Roses[a] 3923 North High Street, Columbus, OH 43214
10 669 plants, 558 varieties.

Tulsa Municipal Rose Garden[a] Woodward Park, 21st and Peoria, Tulsa, OK 74114
9000 plants, 291 varieties.

Corvallis Rose Garden[a] Avery Park, Corvallis, OR 97330
1400 plants, 350 varieties.

Portland International Rose Test Garden[a] 400 Southwest Kingston Avenue, Portland, OR 97201
The City of Portland is world famous for its rose gardens. Washington Park is the largest of three rose gardens (1.8 ha with some 10 000 roses, 450 varieties) within the site and hosts annual international rose trials. Established in 1917, it is one of the oldest rose gardens in the United States. The Peninsula Park Garden, initiated in 1909, is a sunken design featuring about 9000 roses. Finally Ladd's Addition is near the downtown area and has about 3000 roses (50 varieties).

Hershey Gardens[a] 170 Hotel Road, Box 416, Hershey, PA 17033
10 000 plants, 260 varieties.

Longwood Gardens[a] Department of Horticulture, Box 501, Kennett Square, PA 19348
950 plants, 95 varieties.

Tyler Municipal Rose Garden[a] 420 South Rose Park Drive, Tyler, TX 75702
Home to an annual rose parade with a great garden displaying 30 000 plants, 400 varieties.

All-America Rose Selections Garden[a] 7931 East Boulevard Drive, Alexandria, VA
850 plants, over 50 varieties.

Norfolk Botanical Gardens Azalea Garden Road, Norfolk, VA 23518
4000 plants, 300 varieties.

Woodland Park Rose Garden[a] 5500 Phinney Aveune North, Seattle, WA 98103
4850 plants, 264 varieties.

Boerner Botanical Garden[a] 5879 South, 92nd Street, Hales Corner, WI 53130
Over 4000 plants, over 500 varieties.

Uruguay

Rosaleda del Prado (Prado Municipal Rose Garden) Avenida Buschental, Montevideo
Operated by the municipal authorities of Montevideo, this garden was founded in 1912 and has beautiful Art Nouveau metallic structures and rose plants imported from France. It is about 1 ha in size with 1000 rose plants including some old garden roses.

See also: **History of Roses in Cultivation**: Ancient Chinese Roses; European (Pre-1800); Modern (Post-1800). **Rose Collections and Trials**: Rose Societies; Awards.

Further Reading

Beales P (1985) *Classic Roses*. London: Harvill Press.
Cairns T (ed.) (2000) *Modern Roses: The World Encyclopedia of Roses*, 11th edn. San Diego, CA: Academic Press.
Krüssman G (1986) *Rosen, Rosen, Rosen*, 2nd edn. Berlin: Verlag Paul Parey.

Rose Societies

T Cairns, American Rose Society, Studio City, CA, USA

Introduction

Over the last 100 years national rose societies have been established in most countries to provide, in the most fundamental way, an opportunity of sharing their participants' hobby, expertise and camaraderie. This propagation has been a gradual and natural development, with many local rose societies within each country being formed to support national organizations with their mission statements. There are over 40 national rose societies in the world. Such a development has led to the need for an international organization to blend and bring together different cultures for a common good.

World Federation of Rose Societies

(Address: 46 Alexandra Road, St. Albans, Hertfordshire AL1 3AZ, UK; www.worldrose.org; fax (44) 1727 833 648.)

The foundation for the World Federation of Rose Societies (WFRS) was laid down during the Fourth International Rose Conference in London in 1968. Baroness Lily de Gerlache de Gomery of Belgium was selected as its Founding President. Since the National Rose Society of New Zealand had already initiated arrangements for an international rose conference in 1971, it was mutually agreed to select Hamilton, New Zealand as the venue for the First World Rose Convention, appropriately named Roseworld '71. It was at this conference that the first formal steps were taken to becoming an organization. Representatives of nine founding member countries were:

1. Australia: Mr L. Lawrence, Dr A.S. Thomas
2. Canada: Mr R.H. Keith, Mrs W. Lyzaniwsky
3. India: Mr S.M. Dahanukar, Dr S.Y. Sirdar
4. Japan: Mr Y. Kameoka
5. New Zealand: Mr D. Butcher, Mr F.W. Smith, Mr H. Taylor
6. South Africa: Mr H. Buss, Mr and Mrs J. Wise
7. Switzerland: Mr W. Gunthart
8. UK Mr R.C. Balfour, Mr F.M. Bowen
9. USA Dr R.C. Allen, Dr E. Lyle, Mr L. Satterlee, Mr H. Goldstein.

The WFRS was registered in the UK as a company limited by guarantee and as a charity under the number 1063582. The objectives of the society, as stated in the constitution, are:

- to encourage and facilitate the interchange of information and knowledge of the rose between national rose societies
- to coordinate the holding of international conventions and exhibitions
- to encourage, and where appropriate, sponsor research into problems concerning the rose
- to establish common standards for judging new rose seedlings
- to establish a uniform system of rose classification
- to grant international honours and/or awards
- to encourage and advance international cooperation on all other matters concerning the rose.

Membership is open to all national rose societies throughout the world. To date, 37 countries representing over 100 000 rose growers throughout the world have become members.

National Rose Societies

Argentina

Asociacion Argentina de Rosicultura (AAR), Av. del Libertador 408 (C1001 ABR), Buenos Aires, Argentina; www.ars@rosicultura.com.ar; tel. (54) 114328-7695; fax (54) 11 4394-2054.

Located in Buenos Aires, members meet every month for an educational presentation. Society produces a rose annual as well as quarterly bulletins. Rose shows are held in the autumn in March and in the spring in October. Society has four affiliate societies in Bariloche, Cordoba, Mendoza and San Juan. International rose trials are conducted at the Municipal Rose Garden in Buenos Aires in November of each year.

Australia

National Rose Society of Australia, 29 Columbia Crescent, Modbury North, South Australia 5092, Australia; tel. (61) (8) 8264 0084.

The society is composed of five State Rose Societies (Queensland, New South Wales, Victoria, South Australia and Western Australia) providing the President in rotation on an annual basis. Rose shows are held in the spring (October–November) and in the autumn (April–May).

Austria

Osterreichische Rosenfreunde in der Osterreichischen Gardenbau-Gesellschaft; Parkring 12, A-1010 Vienna, Austria; tel. (43) 1 512 84 16; fax (43) 1 512 84 16/17.

The Austrian Rose Society publishes a magazine, *Garten*, for all its members. In has an extensive library of over 7000 volumes available to members on a lending basis.

Bangladesh

Bangladesh National Rose Society; c/o Prof. Z.N. Tahmida, Department of Botany, University of Dhaka, Dhaka-1000, Bangladesh.

Belgium

Société Royale Nationale 'Les Amis de la Rose' Koninklijke Nationale Maatschappij 'De Vrienden van de Roos'; Korte Aststraat 12, B-9750 Huise Zingem, Belgium; tel. (32) (0) 9 384 83 39.

Society, founded in 1926 by M. Georges van Oost, is one of the founding members of the WFRS. Members receive three newsletters and an annual, *Rosa Belgica*.

Bermuda

Bermuda Rose Society; PO Box PG 162, Paget PG BX, Bermuda; tel. (1) (441) 236-2970.

Located some 1100 km from the eastern seaboard of the USA, Bermuda is a treasure trove of old garden roses and is often referred to as 'a living museum of roses'. The society has about 100 active members and holds regular meetings from October to May.

Canada

Canadian Rose Society; 17 Kintyre Avenue, Toronto, Ontario, Canada M4M 1M2; www.mirror.org/group/crs/; tel. (1) (416) 466-1879.

Founded in 1913 in Toronto, the society quickly became the national society in 1954. Members receive a *Rose Annual* and a magazine, *The Rosarian*, three times a year. The Society sponsors two national rose shows, the National Rose Show in June and the National All Miniature Rose Show in July.

Chile

Associación Chilena de la Rosa; Avda Pres. Kennedy 5334, Depto 131, Vitacura-Santiago, Chile; tel. (56) 2-218 51 28.

Established in 1998, the society has a membership of 50 (growing rapidly), holding regular monthly meetings from March to December. The annual Spring Rose Show is held at the end of October.

China

China Rose Society; c/o Beijing Botanic Gardens, Wofusi Road, Xiangshan, Beijing 10093, People's Republic of China; tel. (86) 19062591561 ext. 2236; fax (86) 10-62591283.

Established in 1986, the society is based in Laizhou City in the province of Shandong where *Rosa chinensis* flourishes, as it does all over the nation. The society holds an annual rose festival in Laizhou every May.

Czech Republic

Czech Rosa Club; Livornska 440, 109 00 Prague 10, Czech Republic; www.zahradkari.cz/czs/szo/rosa-klub; tel. (420) 2 2162 4508; fax (420) 2 2162 4710.

The Czech Rosa Club is an associated organization of the Czech Amateur Gardener Society. Founded in 1968, the membership is 500–600. Local groups are located in Prague, Brno, Hradec Kralove, Kralupy n. V., Louny, Olomonouc and Kromeriz. The society publishes a *Bulletin for Growers and Lovers of Roses* on an irregular basis and has held annual international rose trials at Hradec Kralove since 1982.

Denmark

Danish Rose Society; Kjirkedalsvej 63, Raarup, DK-7130 Juelsminde, Denmark; www.rosenselskabet.dk; tel. (45) 7568 5232; fax (45) 7568 5746.

Currently the Society has 1770 members and issues a regular bulletin, *RosenNyt*, in February, May, September and December.

Finland

Suomen Ruususeura r.y. – Finska Rosensallskapet r.f; Isolammenkatu 8 B 66, FIN-33 710 Tampere, Finland; www.puutarha.net/ruususeura; tel. (358) 3 356 2919.

The society was established in 1989, has approximately 700 members and publishes a quarterly journal, *Ruusunlehti – RosenVladet*.

France

Société Franaise des Roses 'Les Amis des Roses'; Roseraie du Parc de la Tête d 'Or, F-69459 Lyon Cedex 06, France; tel. (33) (0) 474 94 04 36; fax (33) (0) 474 95 54 16.

Germany

Verein Deutscher Rosenfreunde; Waldseestr. 14, D-76530 Baden-Baden, Germany; www.rosenfreunde.de; tel. (49) 7221/31302; fax (49) 7221/38337.

Society issues the *Rosenbogen* four times a year and the *Rosen-Jahrbuch* annually.

Greece

Hellenic Rose Society; Akti Miaouliu 93, Piraeus 185-38, Greece; tel. (30) 1 429 0620; Fax (30) 1 429 0375.

Founded in 1996 by Damianos Conatantinou, the gardens of the society are located in the Centre for Hellenism Damianos Foundation, located 75 km south-west of Athens. The society publishes a biannual report.

India

Indian Rose Federation; c/o Hyderabad Decan Cigarette Factory Pvt. Ltd, Musheerabad, Hyderabad 500-048, India; tel. (91) 40 7616 167; fax (91) 40 7612 760.

Israel

Israel Rose Society; Wohl Rose Park, PO Box 10185, 91101 Jerusalem, Israel; http://roses.jerusalem.muni.il; tel. (972) 2 5637233.

Italy

Associazione Italiana della Rosa; Roseto Niso Fumagalli, Villa Reale, 20052 Monza. Italy; www.augustea.it/airosa; tel. (39) 039-320 994; fax (39) 039-2086 237.

Founded in 1964 by Niso Fumagalli, the society garden is located in front of the Villa Reale in Monza where the international rose trials are held every year. The society publishes *Notiziario* three times a year and *Annuario* immediately after the trial winners are announced in May.

Japan

Japan Rose Society; 101 Tama-Haimu, 5-9-1 Todoroki, Setagaya-ku, Tokyo, Japan; www.hanaippai.com/japan-rose-society; tel. (81) 3-3702-9413.

Founded in 1948, the Ministry of Agriculture in 1962 approved its juridical status. The society conducts international rose trials in cooperation with the city of Tokyo, publishes a regular monthly *Rose Letter* as well as an *Annual Report*. Its annual rose show is held in Tokyo for six consecutive competitive days in the middle of April.

Luxembourg

Letzeurger Rousefrenn 'Les Amis de la Rose'; 361, rue de Rollingergrund, L-2441 Luxembourg, Grand Duchy of Luxembourg; tel. (352) 44 54 92; fax (352) 44 54 11.

Netherlands

Nederlandse Rozenvereniging; Hoekenburgplein 4, 4023 AS Rijswijk (Gld), The Netherlands; tel. (31) 0345-558465.

Society has 1800 members who receive a quarterly bulletin. The Westbroekpark in the Hague is home to the famous international rose trials, held annually since 1962.

New Zealand

National Rose Society of New Zealand; PO Box 66, Bunnythorpe, Palmerston North, New Zealand; www.nzroses,org.nz; tel. (64) 6 3292 700.

The society has 43 district rose societies located throughout the country and holds two annual rose conventions. Members receive a cultural handbook, a quarterly publication *The New Zealand Rosarian*, and a *Rose Annual*. International rose trials are held in Hamilton and Palmerston North.

Northern Ireland

Rose Society of Northern Ireland; 15 Carnduff Road, Larne, Co. Antrim BT40 3NJ, Northern Ireland; www.rosemerald.co.uk; tel. (44) (0) 28 2827 2658.

Activities of the society include lectures held in Belfast during the winter months and two rose shows, during the summer together with two newsletters. Belfast is the home of the international rose trials held in cooperation with the city of Belfast at the Sir Thomas and Lady Dixon Park.

Norway

Norwegian Rose Society; Delebukta 71, 1390 Vollen, Norway; http://home.enitel.no/norskrose/; tel. (47) 66 90 11 56; fax (47) 66 98 89 41.

Founded in 1982, the Society has over 1200 members. The national rose show takes place in Oslo in early July. The society has established a demonstration garden at Tomb Agricultural School near Moss, south of Oslo, to illustrate to the public old garden roses as well as modern roses under trial. Members receive *Rosebladet*, the official quarterly publication of the society.

Pakistan

Pakistan National Rose Society; House 38, Street 5 Sector F8/3, Islamabad, Pakistan; tel. (92) 228 2060.

Members of the society enjoy the privileges of free advice on growing roses, inspection of their gardens by experts to diagnose problems, the publications of the society and assistance on obtaining plants. Two national rose shows are held each year – in Islamabad in April and elswhere in the country in the autumn.

Poland

Polish Society of Rose Fanciers; ul. Broniewskiego 19/7, 01-780 Warszawa, Poland; tel. (48) 22 621 19 64.

The society is particularly noted for holding an annual rose show to benefit cancer hospices in Cracow and Warsaw.

Romania

Asociatia Amicii Rozelor din Romania; RO-3400 Cluj, CP 1-86, Romania; tel. (40) 64-433264; fax (40) 64-1545989.

The society has about 1600 members located in nine district rose societies primarily in the western part of the country in Cluj, Arad, Oradea, Satu-Mare, Zalau, Baia Mare, Tg. Mures, Sibiu and Barlad. The official publication, *Rosarium*, appears twice a year with a short summary in English. The society is active in holding demonstrations, rose shows and lectures throughout the country.

Slovakia

Rosa Klub; Sturova 40, 960 01 Zvolen, Slovakia; www.tuzvo.sk/rosaklub; tel. (421) (0) 45 5206 565; fax (421) (0) 45 532 00 15.

Established in 1973 as a special club of the Slovak Union Gardeners by an interested group of rosarians at the Borova Hora Arboretum of the Technical University in Zvolen. Each June the club holds conferences entitled Beauty of Roses and Shapes.

Slovenia

Sekeija Ljubiteljev Vrtnic; Celovska 43, 1000 Lujubljana, Slovenia; tel. (386) 611326170.

The society has about 150 members and is growing rapidly, publishing a magazine entitled *My Little World*.

South Africa

Federation of Rose Societies of South Africa (ROSA); PO Box 28188, 0132 Sunnyside, Pretoria, Republic of South Africa; tel. (27) 12 544 0144; fax (27) 12 544 0813.

Rose shows are held throughout the country during the bloom cycle of October to June. An international trial garden is located just outside Cape Town at the Durbanville Rose Garden.

Spain

Asociación Española de la Rose; Cortijo Castellanos, Apdo de Correos 17, 41300 S. José de la Rincoñado (Sevilla), Spain; tel. (34) 95 479 57 10; fax (34) 95 479 57 11.

The administration of the association is sponsored by a private institution. There are about 1400 members in the association. Publications include *The Bulletin Rosa-Rosae*, *Separata* and *The First Catalog of Spanish Varieties (1999)*.

Sweden

Svenska Rosensallskapet; Alvsbacka – Grasas 209, S 660 60 Molkom, Sweden; www.welcome.to/swerose; tel. (46) 0553 30 100.

The society has approximately 1300 members who receive the magazine *RosenBladet* on a quarterly basis.

Switzerland

Gesellschaft Schweizerischer Rosenfreunde; Bahnhofstr. 11, 8640 Rapperswil, Switzerland; www.Rosenfreunde.ch; tel. (41) 055/210 24 09.

Founded in 1959, the society has about 3500 members. They receive the monthly magazine *Rosenblatt* and the annual *Rosa Helvetica*.

United Kingdom

Royal National Rose Society (RNRS); The Gardens of The Rose, Chisell Green, St Albans, Herts AL2 3NR, UK; www.roses.co.uk; tel. (44) 1727 850461; fax (44) 1727 850360.

The RNRS is the world's oldest rose society, maintaining the world-famous Gardens of the Rose at its headquarters in St Albans just north of London. Perhaps best known for the wonderful rose shows held throughout the UK, the RNRS also holds an annual international rose trial grounds with emphasis on vigour and disease resistance.

Uruguay

Asociación Uruguaya de la Rosa; Casilla de Correo 18914 Dto 9, CP 11500 Montevideo, Uruguay; tel. (598) 575 36 84.

The society holds monthly meetings and special lectures throughout the year as well as a spring rose show. The *Annual Report* is published in both Spanish and English and is distributed throughout the world.

United States of America

American Rose Society (ARS); PO Box 30 000, Shreveport, Louisiana, USA; www.wrs.org; tel. 1 (318) 938-5402; fax 1-318-938-5405.

Organized in 1892, the ARS is the largest rose society in the world, with almost 22 000 members and about 400 affiliated local societies and chapters distributed throughout the 50 states. The society publishes, as membership benefits, a monthly full colour magazine, annual and a *Handbook for Selecting Roses*.

In 2000 the WFRS extended its family by offering associate membership to other horticulture organizations. To date six organizations have joined:

1. Australia: Heritage Roses Australia Inc.
2. France: Société Nationale d'Horticulture de France

3. France: Rosaraie de la Cour de Commer
4. Italy: Roseto Botanica 'Carla Fineschi' di Cavriglia
5. Japan: Hana no Miyako Gifu for The Promotion of Flowers and Greenery
6. Scotland: Scottish Sweet Pea, Rose and Carnation Society.

World Rose Conventions

Every 3 years the WFRS holds a world rose convention under the sponsorship of the host nation selected for the event. A primary function of the convention is to concentrate on fulfilling its mission, as stated above. To carry out the declared duties the WFRS has established a number of standing committees: awards, technical (classification and registration), conservation, convention liaison, fund-raising, publications and show standardization. In the years between these world conventions, regional rose conferences take place in various member countries. WFRS communicates with its member countries via regular publication of *World Rose News*.

Any WFRS member society may bid to host a convention. World rose conventions have been held as follows:

1968	Founding meeting in London, UK
1971	Hamilton, New Zealand (Roseworld '71)
1974	Chicago, USA (Roseworld '74)
1976	Oxford, UK
1979	Pretoria, South Africa (Rosafari)
1981	Jerusalem, Israel (Rose Pilgrimage)
1983	Baden Baden, Germany (Roseromantic)
1985	Toronto, Canada (Rosecapades)
1988	Sydney, Australia (*Rosa australis*)
1991	Belfast, Northern Ireland (Rose Emerald)
1994	Christchurch, New Zealand (Roseworld '94)
1997	The Benelux (Belgium, Netherlands and Luxembourg jointly)
2000	Houston, USA (Texas 2000)
2003	Glasgow, Scotland (Glasgow, City of the Rose 2003)
2006	Osaka, Japan (awarded to the Japan Rose Society).

Rose Hall of Fame

Every 3 years the WFRS elects a rose to The World's Favourite Rose by a global voting procedure. Roses that have been elected to this prestigious title have been:

- 'Peace' in 1976 – hybrid tea, yellow blend, introduced in 1945, hybridized by Francis Meilland, France
- 'Queen Elizabeth'® in 1979 – Grandiflora, medium pink, introduced in 1954, hybridized by W.E. Lammerts, USA
- 'Fragrant Cloud' in 1981 – hybrid tea, orange, introduced in 1963, hybridized by M. Tantau, Germany
- 'Iceberg' in 1983 – Floribunda, white, introduced in 1958, hybridized by R. Kordes, Germany
- 'Double Delight'™ in 1985 – hybrid tea, red blend, introduced in 1977, hybridized by H. Swim, USA
- 'Papa Meilland'® in 1988 – hybrid tea, dark red, introduced in 1963, hybridized by A. Meilland, France
- 'Pascali'® in 1991 – hybrid tea, white, introduced in 1963, hybridized by L. Lens, Belgium
- 'Just Joey' in 1994 – hybrid tea, orange blend, introduced in 1972, hybridized by R. Pawsey, UK
- 'New Dawn' in 1997 – large flowered climber, light pink, introduced in 1930, sport of 'Dr. W. Van Fleet', discovered by Dreer, USA
- 'Ingrid Bergman'® in 2000 – hybrid tea, dark red, introduced in 1984, hybridized by Olesen, Denmark.

In 1988 at the World Rose Convention in Sydney, Australia it was decided to establish an Old Rose Hall of Fame to recognize roses of historical or genealogical importance and those roses which have enjoyed continued popularity over a great many years. To date the following roses have been so recognized:

- 'Cécile Brünner'
- 'Gloire de Dijon'
- 'Old Blush'
- 'Souvenir de la Malmaison'
- 'Grüss an Teplitz'

Gardens of Merit

Every 3 years the WFRS also awards outstanding rose gardens with the WFRS Plaque of Merit by a global voting procedure. Rose gardens that have been elected to this prestigious title are as follows.

1995 La Roseraie de l'Hay du Val de Marne (France)

The world's oldest existing garden devoted exclusively to roses celebrated its centenary in 1994. Situated at L'Hay-les-Roses, it was created by architect Eduard André and rosarian Jules Graveraux, who assembled a collection of all species and varieties known at the time. Together they created a rose garden of historical value and breathtaking beauty. In 1968 ownership of the garden passed to the municipality of Val de Marne, which has conscientiously continued the maintenance and renewal of the collection while encouraging public participation in this setting of unrivalled charm.

1996 Il Roseto Botanico 'Carla Fineschi' (Italy)

Professor Gianfranco Fineschi started collecting roses 30 years ago and today he has one of the world's largest private collections – more than 6500 varieties, each represented by a single plant. In this living museum the genus *Rosa* is presented from an historical and botanical point of view. Species, subspecies and later hybrids are systematically presented, while climbers and ramblers form walls between the scientific divisions. Many of the beds of modern varieties are grouped according to their hybridizers. In this garden overlooking the hills of Tuscany one may trace the history of the rose from antiquity to the present day while marvelling at the beauty of the garden itself.

1998 The Garden of Roses at Hex Castle (Belgium)

This historical site was constructed by Prince Bishop Velbruck in 1770. The gardens are of a formal Renaissance style while the surrounding parkland reflects the natural English style. About 250 varieties of roses are found throughout the property. From 1970 Countess Michel d'Ursel renewed and increased the collection of roses. Some bushes are originals which date back to the construction of the castle. Others are plants which the Countess d'Ursel brought back from her travels to replace those that had died. She also added many modern varieties, but always those that were in harmony with these gracious surroundings.

2000 Sir Thomas and Lady Dixon Park Rose Garden (Northern Ireland)

Situated in one of Belfast's most beautiful areas, the Sir Thomas and Lady Dixon Park has about 128 acres. The Dixon estate was formally presented to Belfast Corporation by the late Lady Edith Stewart Dixon in December 1959, in memory of her husband. Throughout the ensuing years, the park became a favourite place for visitors to enjoy the fine expanses of rolling lawns and meadows, mature trees, copse and woodland, rhododendrons and other horticultural features as well as the roses in season. The gardens are now home to an international rose trial ground. The gardens were completely redesigned in the late 1980s. The unique design took on the architectural symmetry of a series of circular rose gardens skilfully integrated into the rolling hills and informal landscape. Many of the older and poorer varieties of roses were discarded and the best retained for planting in the newly designed display beds. Today the rose garden is world renowned, receiving visitors from all over the world.

See also: **Distribution and Ecology**: South Africa; Australia and New Zealand; Continental Asia and Japan; Europe; Israel; North America; South America. **Rose Collections and Trials**: Rose Collections and Herbaria; Awards.

Further Reading

World Rose News published by the World Federation of Rose Societies: two 24-page issues per year; www.worldrose.org.

Awards

T Cairns, American Rose Society, Studio City, CA, USA

Introduction

Competitions or evaluations of actual rose plants *in situ* in different countries and climates are generally referred to as 'International Rose Trials' and have taken place at the various international rose gardens throughout the world. For instance, Bagatelle in France has been conducting rose trials since 1907 and Lyon in France since 1931. The main purpose of such trials is to plant and display modern rose varieties for critical evaluation in a parklike setting for the benefit of parks departments, nurserymen and the general public as to the intrinsic qualities of those varieties in that particular microclimate. To date there are over 30 international trials conducted throughout the world.

Varieties submitted by breeders, professional and amateur, for test generally cannot have been introduced more than 5 years ago. While named varieties are preferred, varieties not yet introduced are permitted under their 'variety denomination'. In some gardens there are limits on the number of entries permitted per breeder.

Breeders/introducers are often required to submit a letter of intention to provide entries to rose trials indicating the variety name, colour, growing habit and classification. Entries in the following classifications are generally accepted:

- Hybrid Tea and Grandiflora (large-flowered)
- Floribunda (cluster flowered)
- Mini-Flora (dwarf-flowered clustered)
- Miniature roses
- Shrubs (shrubs, groundcovers, etc.)
- Climbers and Wichuranas.

Varieties are generally judged throughout the 2-year period by a permanent jury panel followed by an international jury. The composition of the jury panels normally reflects diversity as well as an international flavour, e.g. growers, breeders, parks departments, landscape architects and members of rose organizations at home and abroad.

Guidelines are formulated by the curator of the gardens using similar methods and techniques as employed in other international trials throughout the world. The permanent judging panel, during each scheduled site inspection, scores based on the following criteria: novelty, bud and flower form, colour opening, color finishing, substance, habit, quantity of flowers, vigor/renewal, foliage, disease/insect resistance. Fragrance may also be judged by the permanent judging panel on a system graded from 1 (no detectable scent) up to 10 (very strong scent).

The international jury, an invited and distinguished group of judges, usually follows a simpler point-scoring system, except that sometimes they are not required to judge fragrance. With these generalizations discussed above, it should be stressed that there are no regulatory guidelines in place and many gardens adopt their own unique scoring systems. The following gardens have conducted international trials.

Belgium

Rosarium West Flanders

The trials are conducted in the rose garden of the Castle of t'Hooghe, Kortrijk, established in 1959 by the Provincial Centrum in Bertem-Rumbeke, in partnership with the city of Kortrijk. In the first year of planting, roses are judged three times by the permanent jury. In their second year the roses are judged twice by the permanent and once by an international jury. Winning varieties receive a golden or silver award (**Table 1**).

Rose Garden of Le Roeulx

This rose trial garden is in the small city of Le Roeulx in the Hainaut province. Unlike many other European trials this competition takes place in the first weekend of September each year (i.e. in the second bloom cycle) (**Table 2**). Five plants of each entry are grown for 2 years under the watchful eyes of the members of the permanent jury. The final winners are judged in various criteria, namely, resistance to diseases, blooming, shape, number of petals and fragrance. A Gold Medal, a Silver Medal or a Certificate are granted to the top scoring varieties.

Czech Republic

Hradec Kralove

Although a recent edition to the European international rose trial scene, this garden has established itself as a premier competition (**Table 3**) attracting a wide diversity of breeders from all over the world.

France

Parc du Bagatelle, Paris

International Rose Trials have been held annually at the Parc du Bagatelle since 1907 (**Table 4**). This competition was organized by Jean Claude Nicolas Forestier assisted by Jules Gravereaux, a businessman (owner of the department store 'Le Bon Marché') and a rose enthusiast (he created his own rose garden in L'Hay les Roses, a southern Paris suburb).

This competition, being the first of its kind, established the model that eventually has been emulated all over the world. Traditionally held on the third Thursday of June, the international jury scores the roses based on two criteria, overall presentation and fragrance. Theses scores are then combined with the permanent jury scores collected throughout the trial period, namely number of blooms, plant habit, resistance to diseases, flower form, inflorescence and fragrance.

Since 1986, the Parc du Bagatelle has also been host to a competition for roses suitable for landscaping purposes (**Table 5**).

Table 1 Awards by Rosarium West Flanders (1991–1999)

Year	*Cultivar*	*Class*	*Breeder*
1999	'Dawn Sunsation'	Floribunda	Kordes
1998	'Jardins de France'	Floribunda	Meilland
1997	'Betty Harkness'	Floribunda	Harkness
1996	'Bordure Rose'®	Floribunda	Delbard
1995	'Mondiale'®	Hybrid Tea	Kordes
1994	'NDR 1 Radio Niedersachsen'	Floribunda	Kordes
1993	'Sommermorgen'	Floribunda	Kordes
1992	'Clos Fleuri Rose No. 2'	Floribunda	Delbard
1991	'Professor Boesman'	Floribunda	DVP Melle

Table 2 Awards by Rose Garden of Le Roeulx (1991–2000)

Year	*Award*	*Cultivar*	*Class*	*Breeder*
2000	Gold Medal	'Andre Brichet' (DVP 94-73)	Floribunda	Rijksstat.Voor Sierplantent
1999	Silver Medal	'RVS93104'	Floribunda	DVP Melle
		'ADAlexcor'	Hybrid Tea	Adam
	Most Fragrant	'MEIgripine'	Hybrid Tea	Meilland
1998	Gold Medal	'ORYbee'	Floribunda	Orye
		'Acapella'®	Hybrid Tea	Tantau
	Most Fragrant	'Acapella'®	Hybrid Tea	Tantau
1997	Gold Medal	'91/1193-4'	Floribunda	Poulsen
	Most Fragrant	'MEIsionver'	Hybrid Tea	Meilland
1996	Gold Medal	'Bolcholi'	Hybrid Tea	Meilland
	Most Fragrant	'Bolcholi'	Hybrid Tea	Meilland
1995	Gold Medal	'MEItekit'	Hybrid Tea	Meilland
	Most Fragrant	'MEItekit'	Hybrid Tea	Meilland
1994	Gold Medal	'Esme Euvrard'	Floribunda	Kordes
	Most Fragrant	'Frederic Mistral'®	Hybrid Tea	Meilland
1993	Silver Medal	'Schneekonigin'®	Shrub	Tantau
		'KORlamaus'	Floribunda	Kordes
	Most Fragrant	'Auguste Renoir'	Hybrid Tea	Meilland
1992	Silver Medal	'Warm Wishes'	Hybrid Tea	Fryer
	Most Fragrant	'Warm Wishes'	Hybrid Tea	Fryer
1991	Silver Medal	'Pink Bassino'	Shrub	Kordes
		LAPdev	Hybrid Tea	Laperrière
	Most Fragrant	'Plein Soleil'®	Hybrid Tea	Laperrière

Table 3 Awards at the International Rose Trials of Hradec Kralove (2000)

Year	*Award*	*Cultivar*	*Class*	*Breeder*
2000	Golden Rose	'RvS-93-102'	Floribunda	RVS
	First Prizes	'Sebastian Kneipp' (KORpastato)	Hybrid Tea	Kordes
		'Rosenprofessor Sieber' (KORparesni)	Floribunda	Kordes
		'LLX 9001'	Shrub	Lens
		'Manita' (KORberhig)	Climber	Kordes
		'Yellow Palette' (POUlsendo)	Miniature	Poulsen
		'Tumbling Waters' (POUltumb)	Groundcover	Poulsen

Table 4 Awards by Parc du Bagatelle, Paris (1991–2002)

Year	*Category of award*	*Cultivar*	*Class*	*Breeder*
2002	First Prize	'Amber Flush' (INT B39796)	Hybrid Tea	Interplant
2001	First Prize	'Amber Cover' (POUlbambe)	Shrub	Poulsen
2000	First Prize	'Viborgh' (POUllug)	Hybrid Tea	Poulsen
1999	First Prize	DICyeti	Hybrid Tea	Dickson
1998	First Prize	KORdatura	Shrub	Kordes
1997	First Prize	'Riberhus'	Hybrid Tea	Poulsen
1996	First Prize	'Manureva'	Hybrid Tea	Sauvageot
1995	First Prize	No award		
1994	First Prize	'Golden Nugget'®	Shrub	Interplant
1993	First Prize	'Bluhwunder'®	Floribunda	Kordes
		'Rose de Rennes'®	Hybrid Tea	Adam
1992	First Prize	'Zambra 93'®	Floribunda	Meilland
		'Gypsy Dancer'	Polyantha	Dickson
1991	First Prize	'Les Amoureux de Peynet'®	Floribunda	Meilland

Table 5 Landscape awards by Parc du Bagatelle, Paris (1991–2002)

Year	*Award*	*Cultivar*	*Class*	*Breeder*
2002	First Prize	'Gartnerfreude'	Groundcover Shrub	Kordes
2001	First Prize	'Easy Cover'	Groundcover Shrub	Poulsen
2000	First Prize	'Sweet Cover'	Groundcover Shrub	Poulsen
		'Bordure Blanche'	Shrub	Delbard
1999	First Prize	'Diamond Border'	Shrub	Poulsen
1998	First Prize	'Bayernland Cover'	Floribunda	Poulsen
1996	First Prize	'Relax Meillandecor'	Groundcover Shrub	Meilland
1995	First Prize	'Douceur Normande'®	Shrub	Meilland
1994	First Prize	'Emera'	Floribunda	Noack
1993	First Prize	'Carmagnole'®	Floribunda	Delbard
1992	First Prize	'Denise Grey'®	Shrub	Meilland
1991	First Prize	'Comtesse Jeanne de Flandres'	Shrub	Meilland

Parc de la Téte d'Or, Lyon

This garden, with a rich historical legacy and theme, has a diversity of plants representing species, old garden roses and moderns. Located just to the south of the city of Lyon where the Rhône and Saône rivers meet, the garden boasts an area of 17 000 m^2 (57 000 plants) in a parkland setting. Established in 1930, this garden evaluates the rose creations of French hybridizers, awarding annually a special prize entitled '*La Plus Belle Rose de France*' (The Most Beautiful Rose in France) (**Table 6**). In 1977 the awards programme was expanded to include an international prize entitled '*Grandes Roses du Siècle*' (Great Roses of the Century). Yet another award, '*Prestige de la Rose*', is presented by the international jury to the breeder of the finest Hybrid Tea. There is also an award for the best perfumed rose (**Table 7**).

La Roseraie de Saverne

Located about 400 km from Paris, Saverne is a delightful town in the Alsace region of France. La Roseraie de Saverne has a rich history dating back to 1900 when 940 roses were planted in an attractive garden setting. In 1925 the first rose trials were held emulating the Bagatelle system but were discontinued. Trials resumed in 1983 (**Table 8**). In 2000 there were about 7000 roses in the gardens.

Germany

International Rose Trials of Baden-Baden

In 1952 the first rose trials took place in the charming city of Baden-Baden within the Black Forest region. This competetion was due in part to the patronage of Konrad Adenauer who was himself was a keen rose grower. Trials are held annually in the last week of June (**Table 9**). The scoring system adopted is based on a maximum of 100 points derived as follows: general impression, 15 points; petals, 10 points; disease resistance, 15 points; blooms and buds, 15 points; wilting, 5 points; scent, 15 points; novelty, 15 points; charm, 10 points.

German Rose Trials

The 'Anerkannte Deutsche Rose' (ADR) programme is parallel to the 'All America Rose Selections' (AARS) in the United States in that trials are carried out in nine official test gardens throughout Germany. Sites are deliberately chosen to include varying climatic and soil conditions. Hybrid teas, floribundas and miniatures are tested for a 3-year period while climbers, shrubs and groundcover roses are under a 4-year test period. About 40 cultivars are judged by permanent panels in 11 cities (Kiel, Osnabruck, Marquardt, Rethmar, Dortmund, Sangerhausen, Wiesbaden, Zwelbrucken, Vetschochheim and Weihenstephan). Test criteria included are form, colour, fragrance, repeat, hardiness, vigour, disease resistance, etc. However the main emphasis is placed upon disease resistance. Test roses are not given any protection from fungus disease during the test period.

The ADR judges consists of the Bund Deutscher Baumschulen (German Nurseries Association), rose breeders and independent trial gardens. On the basis of test scores derived from the various test sites, the ADR trial results by the Bundessortenamt (Federal Office of Plant Varieties), the awards are made annually (**Table 10**).

Ireland

St Anne's Park, Dublin

Situated in the historic St Anne's Park, Raheny, close to the north side of Dublin City and officially opened in July 1975, this garden has over 25 000 rose plants, grouped according to type, with a special section reserved for the prestigious International Rose Trials

Table 6 Awards of the title 'Medaille d'Or/La Plus Belle Rose de France' by Parc de la Tête d'Or, Lyon (1991–2000)

Year	*Cultivar*	*Class*	*Breeder*
2000	'Privilège de St-Verny'		Orard
1999	'Harmonia Sub Rosa'	Hybrid Tea	Dorieux
1998	'ADAbuco'	Floribunda	Adam
1997	'Novaia'	Floribunda	Orard
1996	'Rose d'Or de Montreux'	Hybrid Tea	Adam
1995	'Aramis'	Floribunda	Laperrière
1994	'Abbaye de Cluny'™	Hybrid Tea	Meilland
1993	'No Name'	Floribunda	Dorieux
1992	'MEIkame'	Floribunda	Meilland
1991	'Le Grand Huit'	Hybrid Tea	Adam

Table 7 Additional Awards by Parc de la Tête d'Or, Lyon (1938–1984)

Year	*Award*	*Cultivar*	*Class*	*Breeder*
1984	Prix Prestige de la Rose	'Ferline'®	Hybrid Tea	Laperrière
1979	Grandes Roses du Siècle	'Lady Rose'®	Hybrid Tea	Kordes
		'Pénélope'®	Hybrid Tea	Gaujard
1973	Medaille d'Or du Parfum	'Idao'	Hybrid Tea	Gaujard
1972	Medaille d'Or du Parfum	'Gamma'	Hybrid Tea	Gaujard
1939	Prix de la Rose Parfumée	'Neige Parfum'	Hybrid Tea	Mallerin
1938	Prix de la Rose Parfumée	'Guinée'	Hybrid Tea, Climbing	Mallerin

Table 8 Awards by La Roseraie de Saverne (1999–2000)

Year	*Prize*	*Cultivar*	*Class*	*Breeder*
2000	Gold Medal	'Neige Parfum' (DORblan)	Hybrid Tea	Dorieux
	Fragrance	'Caruso' (MEIcarouze)	Hybrid Tea	Meilland
1999	Gold Medal	ORAcol		Orard
	Fragrance	'Wischoten' (MEIfani)		Meilland

(**Table 11**). Dublin is one of the few cities that holds trials with the newest unnamed varieties of roses judged by a jury of local and world rose experts each year.

Italy

'Roseto Niso Fumagalli' of Monza

The building of the Roseto began in 1964. The plot of land donated by the town for the site was a wild and ignored piece of land just inside the court of the Villa Reale di Monza (the summer Royal Palace in the eighteenth century). In September 1965 the first International Rose Trials were held in Monza. On that occasion the fragrance award was given to the Dutch hybridizer Jan Lendeers, who obligingly named his rose 'Monza'. This recognition of the importance of scent over other criteria set the Monza Trials apart from other gardens. The fragrance prize was a small-size replica, in gold and precious stones, of the Crown of Queen Teodelinda dating from the sixth century and currently preserved in Monza cathedral. Since 1966, rose trials under the auspices of the Associazione Italiana della Rosa (AIR) have been held every year at the end of May (**Table 12**).

Il Roseto de Roma

This delightful rose garden and its trials date back to 1933. The city of Rome emulated the Bagatelle, Paris trials by devoting a civic rose garden as the site for rose competitions, establishing itself as the second city in Europe to do so. Trials are normally held on the third Saturday in May each year (**Table 13**).

The Netherlands

Rosarium Westbroek in The Hague, The Netherlands, is perhaps one of the most beautiful rose gardens in the world. Each year distinguished judges and breeders

Table 9 Awards at International Rose Trials of Baden-Baden (1991–2000)

Year	*Award*	*Cultivar*	*Class*	*Breeder*
2000	Gold Medals	'Black Magic'	Hybrid Tea	Tantau
		'Andre Brichet' (DvP 94-73)	Floribunda	RVS
		(DICyeti)	Floribunda	Dickson
		'Dietrich Woessner' (HUBar)	Floribunda	Huber
		'Teasing Georgia'	Shrub	Austin
1999	Golden Rose of Baden-Baden	'Malaga'	Hybrid Tea	Interplant
	Gold Medals	'Malaga'	Hybrid Tea	Interplant
		'93/3237-1'	Floribunda	Poulsen
		'Gärtnerfreude'	Groundcover	Kordes
		'CHEwsally'	Climber	Warner
1998	Casino of Baden-Baden Award	No Name	Hybrid Tea	Bear Creek
	Gold Medals	'JACbrant'	Hybrid Tea	Bear Creek
		'Pink Hedge'	Floribunda	Interplant
		'CF 93.7651.1'	Floribunda	Delbard
		'4437 FRYwhoppa'	Floribunda	Fryers
		'Sugar Baby'	Miniature	Tantau
		'Diamond Border'	Climber	Poulsen
1997	Casino of Baden-Baden Award	'Tintinara'™	Floribunda	Dickson
	Fragrance Award	'Rêve d'un Soir'	Hybrid Tea	Croix
	Gold Medal	'Goldelse'	Floribunda	Tantau
1996	Casino of Baden-Baden Award	'Rouge Adam'	Hybrid Tea	Michel Adam
	Fragrance Award	'Chartreuse de Parme'	Hybrid Tea	Delbard
	Gold Medals	'Pimpernelle'	Floribunda	Delbard
		'KO 87/15833-01'	Groundcover	Kordes
		'Ranger'	Climber	Barni
1995	Casino of Baden-Baden Award	'Old Port'	Hybrid Tea	McGredy
	Fragrance Award	'Melody Perfumée'™	Hybrid Tea	Dorieux
	Gold Medals	'Rose de Rennes'®	Hybrid Tea	Michel Adam
		'90-9066-8'	Grandiflora	Poulsen
		'CG 88.375.306'	Floribunda	Delbard
1994	Casino of Baden-Baden Award	'Concerto'™	Floribunda	Meilland
	Fragrance Award	'Happy Day'	Climber	Warner
	Gold Medals	'POUlsol'	Hybrid Tea	Poulsen
		'POUlket'	Floribunda	Poulsen
1993	Casino of Baden-Baden Award	'3430/JC 27693A'	Hybrid Tea	Cocker
	Fragrance Award	'Frederic Mistral'®	Hybrid Tea	Meilland
	Gold Medals	'Zambra'®	Floribunda	Meilland
		'1100-87'	Floribunda	Noack
		'POUlter'	Groundcover	Poulsen
		'Boreal Adaribu'		Michel Adam
		'Benny Goodman'	Miniature	McGredy
1992	Casino of Baden-Baden Award	'Imperatrice Farah'	Hybrid Tea	Delbard
	Fragrance Award	'Fragrant Lady'	Hybrid Tea	Meilland
	Gold Medals	'Papi Delbard'	Hybrid Tea	Delbard
		'Fragrant Lady'	Hybrid Tea	Meilland
		'JACbeau'	Floribunda	Bear Creek
		'Sommermarchen'	Groundcover	Kordes
		'F 153-88'	Climber	Interplant
1991	Casino of Baden-Baden Award	'MT 82381'	Hybrid Tea	Tantau
	Fragrance Award	'MEInuiton'	Hybrid Tea	Meilland
	Gold Medals	'Golden Medaillon 91'	Hybrid Tea	Kordes
		'G106-84'	Hybrid Tea	Interplant
		'Fairysnow'	Floribunda	Fryers
		'Matchball'®	Climber	Lens
		'Climbing Rimosa'	Climber	Meilland

from around the world are invited to participate in awarding Gold Medals to various classes and the top prize 'The Golden Rose of The Hague'. Roses are planted in massive display beds. Judges award gold medals to new varieties after a 2-year period (**Table 8**) and the prestigious 'Golden Rose of The Hague' after a 5-year trial period (**Tables 14** and **15**).

Table 10 Awards in the ADR German Rose Trials (1991–2000)

Year	*Cultivar*	*Class*	*Breeder*
2000	'Nemo'	Shrub	Noack
	'Vinesse'	Floribunda	Noack
	'Bad Birnbach'	Floribunda	Kordes
	'Maxi Vita'	Floribunda	Kordes
1999	'Celina'	Floribunda	Noack
	'Ravenna'	Shrub	Noack
	'Rotfassade'	Climber	Noack
	'Neon'	Hybrid Tea	Kordes
	'Kronjuwel'	Floribunda	Noack
	'Brautzauber'	Floribunda	Noack
1998	'Rote Apart'	Shrub	Uhl
	'Estima'	Shrub	Noack
	'Saremo'	Shrub	Noack
1997	'Manita'	Climber	Kordes
	'Danica'	Shrub	Noack
	'Arcadia'	Shrub	Noack
1996	'Postillion'	Shrub	Kordes
	'Felicitas'	Shrub	Kordes
	'Sommerabend'®	Shrub	Kordes
	'Queen Mother'	Floribunda	Kordes
	'Rosenprofessor Sieber'	Floribunda	Kordes
	'Crimson Meidiland'	Shrub	Meilland
1995	'Schneekonigin'	Shrub	Tantau
	'Magic Meidiland'™	Shrub	Meilland
	'Medusa'	Shrub	Noack
	'Hannovers Weisse'	Shrub	Noack
	'Windrose'™	Shrub	Noack
	'Northern Lights'	Shrub	Noack
	'Melissa'	Floribunda	Noack
	'Aspirin'	Floribunda	Tantau
	'Bayernland'	Floribunda	Rosenunion
1994	'Bingo Meidiland'®	Shrub	Meilland
	'Dortmunder Kaiserhain'	Shrub	Noack
	'Rotes Meer'	Shrub	Baum
	'Baum's Rokkoko'	Shrub	Baum
	'Bluhwunder'®	Floribunda	Kordes
1993	'Pink Bassino'	Shrub	Kordes
	'Hagenbecks Tierpark'	Shrub	Meilland
	'Schneesturm'®	Shrub	Tantau
	'Mirato'	Shrub	Tantau
	'Foxi'	Shrub	Tantau/Uhl
	'Armada'®	Shrub	Harkness
	'Mazurka'	Floribunda	Meilland
	'Vicky'	Floribunda	Noack
1992	'Palmengarten Frankfurt'®	Shrub	Kordes
	'Pierette'	Shrub	Tantau
	'Rugelda'®	Shrub	Kordes
	'Rote Woge'	Shrub	Meilland
	'Schone Dortmunderin'®	Floribunda	Noack
1991	'Apfelblute'	Shrub	Noack
	'Wildfang'®	Shrub	Noack
	'Richard Strauss'®	Shrub	Noack
	'Super Excelsa'®	Climber	Hetzel
	'Schneeflocke'®	Floribunda	Noack

New Zealand

Palmerston North

Established at Palmerston North, New Zealand in 1969, these trial grounds are comanaged by the National Rose Society of New Zealand and the Palmerston North City Council. They were selected as they were ideally situated, both climatically and geographically, near the centre of New Zealand. For the first year of trials an assessment is made

Table 11 Awards at the International Rose Trials, Dublin (1991–2000)

Year	*Cultivar*	*Class*	*Breeder*
2000	'Josie Whitney' (HARfancy)	Hybrid Tea	Harkness
	'HORcohabit'	Miniature	Horner
1999	'RSV93-104'	Floribunda	Melle Experimental Station
	'Lucky Duck' (DICzest)	Patio	Dickson
	'K091/2544-94'	Groundcover	Kordes
1998	'Jardins de France'	Floribunda	Meilland
	'Danny Boy' (DICxcon)	Patio	Dickson
	'93/3030-2'	Groundcover	Poulsen
1997	'Jill's Rose'(GANjil)	Floribunda	Gandy
	'DICxotic'	Patio	Dickson
	'Majorie Marshall'	Shrub	Harkness
1996	'Easy Going'™	Floribunda	Harkness
	'Riverdance'	Patio	Kenny
	'DICvivid'	Shrub	Dickson
1995	'89/8137-11'	Floribunda	Poulsen
	'Crème de la Crème'	Climber	Gandy
	'INT B35590'	Patio	Interplant
1994	'Isobel Derby'	Hybrid Tea	Horner
	'Bewitched'	Floribunda	Poulsen
	'Our Molly'	Shrub	Dickson
1993	'G240-88'	Hybrid Tea	Interplant
	'Dormstadt Fulda'	Floribunda	Kordes
	'Playtime'	Shrub	Kordes
1992	'Benita'®	Hybrid Tea	Dickson
	'In The Pink'	Floribunda	Limes
	'Top Marks'	Patio	Fryers
1991	'Dawn Chorus'	Hybrid Tea	Dickson
	'Golden Holstein'®	Floribunda	Kordes
	'Colleen'	Shrub	Interplant

Table 12 Awards at International Rose Trials of Monza (1991–2000)

Year	*Award*[a]	*Cultivar*	*Class*	*Breeder*
2000	Gold Medals	(JACpetor)	Hybrid Tea	Jackson & Perkins
		'Amber Cover'	Groundcover	Poulsen
	Rosa Italiana	BAR 5069	Hybrid Tea	Barni
	Rosa profumata	MEIceppus	Hybrid Tea	Meilland
1999	Arredo Urbano	'Michelle d'Hoop' (RVS 93-101)	Floribunda	RVS
1998	Gold Medals	'Cumbaya'	Floribunda	Meilland/Richardier
		'MEIcurzis'	Hybrid Tea	Meilland
	Rosa Italiana	'Fatima' (VHKO)	Hybrid Tea	Cazzaniga
	Rosa Profumata	'Hacienda' (ORAdal)	Hybrid Tea	Orard
1997	Gold Medals	'Diamond Border'	Shrub	Poulsen Roser
		'Easy Going'®	Floribunda	Harkness
		'Michelangelo'(MEItelov)	Hybrid Tea	Meilland
	Rosa Italiana	'Avance'	Miniature	Barni
	Rosa Profumata	'Naema'	Climber	Delbard
1996	Gold Medal	'PANborclass BP01'	Floribunda	Panozzo
	Rosa Italiana	'Tempi Moderni'	Floribunda	Barni
	Rosa Profumata	'Jude the Obscure' (N22B89)	Shrub	Austin
1995	Gold Medals	'Sommermorgen'	Shrub	Kordes
		'Jardins de Villandry'	Hybrid Tea	Delbard
	Rosa Italiana	'Sans Souci'	Floribunda	Barni
	Rosa Profumata	'James Perière'	Hybrid Tea	Meilland
1994	Gold Medals	'Fredensborg'®	Floribunda	Poulsen Roser
		'LAPvi'	Hybrid Tea	Laperrière
		'Gold Symphonie'	Miniature	Meilland
	Rosa Italiana	'Sandra Mondaini' (FLI)	Floribunda	Cazzaniga
	Rosa Profumata	'Frédéric Mistral'®	Hybrid Tea	Meilland

Table 12 Continued

Year	*Award*[a]	*Cultivar*	*Class*	*Breeder*
1993	Gold Medals	'Leonardo da Vinci'®	Floribunda	Meilland
		'Abbaye de Cluny'®	Hybrid Tea	Meilland
	Rosa Italiana	'Suni'®	Floribunda	Barni
	Rosa Profumata	'Toulouse Lautrec'®	Hybrid Tea	Meilland
1992	Gold Medal	'Clos Fleuri Rose No. 2'	Floribunda	Delbard
	Rosa Italiana	'Luciana Savignano' (VQEI)	Hybrid Tea	Cazzaniga
	Rosa Profumata	'Souvenir de Marcel Proust'	Hybrid Tea	Delbard
1991	Gold Medal	'Colleen'®	Shrub	Interplant
	Rosa Italiana	'Frine'®	Shrub	Barni
	Rosa Profumata and Gold Medal	'MEIyoma'	Hybrid Tea	Meilland

[a] Rosa dell'Anno First prize, Gold Medal; Rosa Profumata: Fragrance Award, Crown of Queen Theodelinde; Rosa Italiana: Italian Rose Prize, Gold Medal; Rose Arredo Urbano: Landscape Rose Prize, Gold Medal.

Table 13 Gold Medal awards at the International Rose Trials of Rose, Italy (1991–2000)

Year	*Cultivar*	*Class*	*Breeder*
2000	'Neige Parfum'		Dorieux
1999	'Bella di Todi'		Barni
	'DELstroja'		Delbard
1998	'Hacienda'		Orard
	'Cotillion'®	Floribunda	Jackson & Perkins
1997	'Caprice de Meilland'	Hybrid Tea	Meilland
	'Dorlobla'		Dorieux
1996	Unnamed seedling		Meilland
	Unnamed seedling		Meilland
1995	'Nette Ingeborg'		Havel
	'Wonderland'®	Shrub	Dickson
1994	'Solo Mio'	Hybrid Tea	Poulsen
	'Berleburg'	Floribunda	Poulsen
1993	'JAChov'		Jackson & Perkins
	'Lady Penelope'	Climber	Warner
1992	'Imperatrice Farah'		Delbard
	'Mr JCB'	Floribunda	Dickson
1991	'Karen Blixen'	Hybrid Tea	Poulsen
	'Giant Pink'®	Shrub	Lens

Table 14 Gold Medals awarded at the Hague (1982–1996)

Year	*Cultivar*	*Class*	*Breeder*
1996	'Marie-Louise Velge'	Floribunda	RvS-Melle
1990	'Flower Carpet'™	Shrub	Noack
1988	'Graceland'	Hybrid Tea	Warriner.
	'Rainbow Yellow Parade'™	Miniature	Olesen
1985	'Kiboh'	Floribunda	Suzuki
1984	'Goldstar'	Hybrid Tea	Cants
1982	'Disco Dancer'®	Floribunda	Dickson

Table 15 'Golden Rose of The Hague' awards (1981–1997)

Year	*Cultivar*	*Class*	*Breeder*
1997	'Sunset Celebration'™	Hybrid Tea	Fryer
1990	'Mariandel'®	Floribunda	Kordes
1987	'Ingrid Bergman'®	Hybrid Tea	Olesen
1986	'Mountbatten'®	Floribunda	Harkness
1981	'Maria Mathilda'®	Miniature	Lens

Table 16 Award Winners at the New Zealand International Rose Trial Ground (1991–2000)

Year	*Award*	*Cultivar*	*Class*	*Breeder*
2000	Gold Star of the South Pacific	'Racy Lady'(TM)	Hybrid Tea	Dickson
	Novelty	'Fourth of July'(TM)	Climber	Carruth
	Fragance	'CHEwadmore'	Climber	Warner
1999	Gold Star of the South Pacific	'Serendipity' (HARamity)	Floribunda	Harkness
	Silver Star of the City of Palmerston North	'Mikayla Paige'	Floribunda	Flinn
	Fragrance	'Emperor' (MACgeorgi)	Hybrid Tea	McGredy
1998	Gold Star of the South Pacific	'Taffeta'	Climbing Mini-Flora	Warner
	Silver Star of the City of Palmerston North	'Blackberry Nip'(SOMnip)	Hybrid Tea	Somerfield
	Fragrance	'Taffeta' (CHEwcrest)	Climbing Mini-Flora	Warner
	Novelty and Certificate of Merit	'Scentimental'(TM)	Floribunda	Carruth
1997	Gold Star of the South Pacific	'Anniversary'	Floribunda	Matthews
	Silver Star of the City of Palmerston North	'Millennium'	Floribunda	Grant
1996	Gold Star of the South Pacific	'Good as Gold'	Climbing Mini-Flora	Warner
	Fragrance and Certificate of Merit	'The McCartney Rose'(TM)	Hybrid Tea	Meilland
1995	Gold Star of the South Pacific and Novelty	'Patio Honey'	Climbing Mini-Flora	Warner
	Silver Star of the City of Palmerston North	'Aorangi'	Hybrid Tea	Sandbrook
	Fragrance and Certificate of Merit	'Carolyn'	Hybrid Tea	McGredy
1993	Gold Star of the South Pacific	'Sheila's Perfume'	Hybrid Tea	Sheridan
	Novelty and Certificate of Merit	'Little Bo-Peep' (POUllen)	Groundcover	Olesen
	Fragrance	'Sheila's Perfume'	Hybrid Tea	Sheridan 1994
	Gold Star of the South Pacific	'Spirit of Hope'	Hybrid Tea	McGredy
	Novelty and Certificate of Merit	'Warm Welcome'	Climbing Mini-Flora	Warner
1992	Gold Star of the South Pacific	'Enchantment'	Floribunda	Olesen
	Fragrance and Certificate of Merit	'City of London'®	Climber	Harkness
	Novelty and Certificate of Merit	'Phantom' (MACcatsan)	Shrub	McGredy
1991	Gold Star of the South Pacific	'Foster's Wellington Cup'	Floribunda	McGredy

with only three main criteria: bush, flower and fragrance. In the second year of trial a more in-depth evaluation is conducted using the following criteria: habit, growth, disease resistance, freedom of flower, flower form, flower substance, flower colour, novelty. Each category carries a 10% weighting with the exception of freedom of flowering and fragrance, which carry a 15% and 5% weighting respectively.

A Certificate of Merit is awarded to varieties receiving a minimum of 70 points (**Table 16**). The rose with the highest points receives the Gold Star of the South Pacific. Amateur breeders are acknowledged with the Silver Star of the City of Palmerston North given to the highest pointed amateur-bred rose that reaches award status.

Auckland Regional Botanic Gardens, Manurewa

Developed from a 65 ha farm site south of the city at Manurewa in 1973, this trial site was officially opened in 1982. The garden boasts over 6000 roses planted in the gardens making it one of the largest rose gardens in the country. Rose varieties for the trials require 25 plants of each cultivar and only cultivars already in commerce or those about to be introduced are eligible. Awards are given for the best in each of the categories hybrid tea, floribunda, miniature/patio, ground cover/shrub and fragrance (**Table 17**). The best overall cultivar is named 'Auckland's Rose of the Year'. Winners from previous trials are planted out in permanent beds in the gardens. As of November 2000, these trials were moved to Hamilton Gardens and the contest renamed 'New Zealand Rose of The Year'.

South Africa

Durbanville Rose Trials, Cape Town

Located in the residential area of Durbanville on the eastern slopes of Table Mountain, these trial gardens have a terrific panoramic view across the Cape. The first plantings took place in 1979, with the official opening of the garden in 1983. Today the rose gardens boast over 6000 plants.

For the trials four plants of each entry are required and are judged over a 3-year period using the following criteria: growth habit and vigour, quality and quantity, disease resistance, floriferousness, shape of bud and flower, colour, weather resistance/self-cleaning, novelty value and overall impression (**Table 18**).

Spain

Spanish International Rose Trials, Madrid

Created in 1955 by Ramon Ortiz, the Madrid Rose Garden is located in the Parque del Oeste within the

Table 17 Award winners at the Rose Introducers of New Zealand (RINZ) International Rose Trial Ground (1990–2000)

Year	*Award*	*Cultivar*	*Class*	*Breeder*
2000	Hybrid Tea	'Tintinara'	Hybrid tea	Dickson
	Climber	'Little Rambler'	Climbing Miniature	Warner
	Shrub	'Country Touch'		Kordes
	Miniature	'Cinderella'		Schuurman
	Most Fragrant Rose	'Blackberry Nip'		Somerfield
	Auckland Rose of the Year	'Heartache'		Kordes
1999	Hybrid Tea	'Tintinara'™	Hybrid Tea	Dickson
	Floribunda	'Heartache'	Floribunda	Kordes
	Miniature/Patio	'Cinderella'(SUNmapa)		Schuurman
	Climber	'Little Rambler'	Climbing Miniature	Warner
	Groundcover/Shrub	'Country Touch'	Shrub	Kordes
	Most Fragrant	'Blackberry Nip'(SOMnip)	Hybrid Tea	Sommerfield
	Auckland Rose of The Year	'Heartache'	Floribunda	Kordes
1998	Hybrid Tea	'Spirit of Hope'(MAChahei)	Grandiflora	McGredy
	Floribunda	'Summer Dream'	Floribunda	Fryer
	Miniature/Patio	'Hauraki'	Patio	McGredy
	Climber	'Patio Honey'	Climbing Miniature	Warner
	Groundcover/Shrub	'Country Prince' (KORfungo)		Kordes
	Most Fragrant	'Celebration' (COCgrand)		Cocker
	Auckland Rose of The Year	'Hauraki'	Patio	McGredy
1997	Hybrid Tea	'Valencia'®	Hybrid Tea	Kordes
	Floribunda	'Enchantment'	Shrub	Poulsen
	Miniature/Patio	'Manapouri'	Patio	McGredy
	Climber	'Patio Queen'	Climbing Miniature	Warner
	Groundcover/Shrub	'Cottage Maid'	Shrub	Poulsen
	Most Fragrant	'Mamy Blue'®	Hybrid Tea	Delbard
	Auckland Rose of The Year	'Cottage Maid'	Shrub	Poulsen
1996	Hybrid Tea	'Warm Wishes'	Hybrid Tea	Fryer
	Floribunda	'Triple Treat' (DELroblu)		Delbard
	Miniature/Patio	'Scentasia'	Miniature	Schuurman
	Climber	'High Hopes'	Climber	Harkness
	Groundcover/Shrub	'Glowing Cushion'	Shrub	Ilsink
	Most Fragrant	'Super Bowl'	Grandiflora	McGredy
	Auckland Rose of The Year	'Sunnyside'	Shrub	Poulsen
1995	Hybrid Tea	'Candella'®	Hybrid Tea	McGredy
	Floribunda	'Rosemary Harkness'	Hybrid Tea	Harkness
	Miniature/Patio	'Sequoia Gold'™	Miniature	Moore
	Climber	'Morning Light'	Climber	Harkness
	Groundcover/Shrub	'Sunnyside'	Shrub	Poulsen
	Most Fragrant	'Rosemary Harkness'	Hybrid Tea	Harkness
	Auckland Rose of The Year	'Sequoia Gold'™	Miniature	Moore
1994	Hybrid Tea	'Gold Medal'®	Hybrid Tea	Christensen
	Floribunda	'Raspberry Ice'	Floribunda	Kordes
	Miniature/Patio	'Rosabell'	Floribunda	Cocker
	Climber	'Warm Welcome'	Climbing Miniature	Warner
	Groundcover/Shrub	'Robin Redbreast'	Mini-Flora	Ilsink
	Most Fragrant	'Dioressence'®	Grandiflora	Delbard
	Auckland Rose of The Year	'Eldorado'	Floribunda	Christensen
1993	Hybrid Tea	'Player'	Hybrid Tea	Schuurman
	Floribunda	'Happy Go Lucky'	Grandiflora	Schuurman
	Miniature/Patio	'Pretty Polly'	Miniature	Meilland
	Climber	'City of London'®	Floribunda	Harkness
	Most Fragrant	'Big Purple'	Hybrid Tea	Stephens
	Auckland Rose of The Year	'Dublin Bay'	Climber	McGredy
1992	Hybrid Tea	'Paddy Stephens'	Hybrid Tea	McGredy
	Floribunda	'Eldorado'	Floribunda	Christensen
	Miniature/Patio	'Little Opal'	Patio	Schuurman
	Climber	'Dublin Bay'	Climber	McGredy
	Most Fragrant	'Margaret Merril'®	Floribunda	Harkness
	Auckland Rose of The Year	'Little Opal'	Patio	Schuurman

Table 17 Continued

Year	*Award*	*Cultivar*	*Class*	*Breeder*
1991	Hybrid Tea	'Amberlight'	Hybrid Tea	Cocker
	Floribunda	'Margaret Merril'®	Floribunda	Harkness
	Miniature/Patio	'Golden Rosamini'	Miniature	Ilsink
	Most Fragrant	'Heart Throb'	Hybrid Tea	Harkness
	Auckland Rose of The Year	'Sexy Rexy'	Floribunda	McGredy
1990	Hybrid Tea	'Aotearoa–New Zealand'	Hybrid Tea	McGredy
	Floribunda	'Sexy Rexy'	Floribunda	McGredy
	Miniature	'Whiteout'	Miniature	McGredy
	Most Fragrant	'Aotearoa–New Zealand'	Hybrid Tea	McGredy
	Auckland Rose of The Year	'Aotearoa–New Zealand'	Hybrid Tea	McGredy

Table 18 Gold Medal Winners at Durbanville, South Africa (1991–1999)

Year	*Cultivar*	*Class*	*Breeder*
1999	TAN 92114		Tantau
1995	'TAN 91172'		Tantau
1994	'Melrose'		Melle Research Station
1993	'DELserex'		Delbard
	'Molineux'	Shrub	Austin
1992	'Ficksburg'	Floribunda	Delbard
1991	'Efekto 21' (MEItolba)		Meilland

Table 19 Gold Medal Awards at the International Rose Trials of Madrid (1991–1999)

Year	*Cultivar*	*Class*	*Breeder*
1999	'Unnamed seedling'		Harkness
1998	'Heavenly Pink'	Hybrid Musk	Lens
1997	No award		
1996	No name		Kordes
1995	No name		Meilland
1994	'Guimalda de Amor'		Lens
1993	No name		Kordes
1992	No name		Lens
1991	'Frisson Frais'®	Hybrid Musk	Lens

city of Madrid and covers an area of approximately 32 000 m^2. The rose garden has been divided into three distinct areas. The area devoted to roses is about 14 000 m^2 divided into 356 growing plots. Rose trials have been conducted since 1956 without interruption (**Table 19**) and take place towards the middle of the month of May to coincide with celebrations of San Isidoro Labrador, Patron Saint of Madrid.

Switzerland

Geneva

The 'La Grange' Rose Garden was established and opened to public viewing in 1947. Consequently the City of Geneva sponsored the first international rose trials with the help and assistance of the Horticultural Society of Geneva.

The title 'Gold Rose of Geneva' is given to the variety with the highest score (**Table 20**). The breeder is given a golden rose created by the famous goldsmith Yves-G. Piaget. Perfume experts also select the most fragrant variety which receives the 'Perfume Cup'. In 2002, the City of Geneva celebrated the 55th anniversary of their rose trails.

United Kingdom

Royal National Rose Society International Trials

The Royal National Rose Society purchased its present home in St Albans Hertfordshire, England, in 1957. By 1963 the plans for an international trial ground were implemented. The Trial Grounds Committee assessed the seedlings with recommendations for awards at the

Table 20 Awards at International Rose Trials of Geneva (1991–2000)

Year	*Award*	*Cultivar*	*Class*	*Breeder*
2000	Golden Rose	(LAPomira)	Hybrid Tea	Laperrière
	Gold Medals	'Neige Parfum'	Hybrid Tea	Dorieux
		'Raymond Nazereau' (ADAquethom)	Floribunda	Adam
		'Amber Cover' (94/4197-6)	Miniature	Poulsen
		(SB-93201)	Shrub	Sauvageot
		(92/2234-1)	Climber	Poulsen
	Most Fragrant Rose	(MEIceppus) (AM 988)		Meilland
1999	Golden Rose	DORcafe	Hybrid Tea	Dorieux
	Gold Medals	'AM 979'	Hybrid Tea	Meilland
		'RT 9268'	Floribunda	Tantau
		CHEwcreepy'	Miniature	Warner
		'LLX 9001'	Shrub	Lens
		'WHOL'	Climber	Wuest
	Most Fragrant Rose	'MR 977	Hybrid Tea	Richardier
1998	Golden Rose	'AM 963' (MEInnzincaro)	Hybrid Tea	Meilland
	Gold Medals	'LAPasu'	Hybrid Tea	Laperrière
		'Jardins de France'	Floribunda	Richardier
		'CHEwayaye'	Miniature	Warner
		'BAR 4475'	Shrub	Barni
		'CROpal'	Climber	Croix
	Most Fragrant Rose	'ORAdal'	Hybrid Tea	Orard
1997	Golden Rose	'Marie-Louise Velge'	Floribunda	RVS
	Gold Medals	'Michelangelo'	Hybrid Tea	Meilland
		'RvS 90-94'	Floribunda	RVS
		'CHEwgreat'	Miniature	Warner
		'Linderof'	Shrub	Kordes
	Most Fragrant Rose	'Caprice de Meilland'	Hybrid Tea	Meilland
1996	Golden Rose	'Eclipse'	Floribunda	Sauvageot
	Gold Medals	'Poker'	Hybrid Tea	Meilland
		'LAPema'	Floribunda	Laperrière
		'CHEwdarkeye'	Miniature	Warner
		'Hugues Aufray'	Shrub	Dorieux
		'Soirée de Bonheur' (CROpal)	Climber	Croix
	Most Fragrant Rose	'Chartreuse de Parme'	Hybrid Tea	Delbard
1995	Golden Rose	'Rose d'Annecy'®	Shrub	Adam
	Gold Medals	'Comtesse Brigitte Chandon-Moet'	Hybrid Tea	Dorieux
		'United Nations Rose'	Floribunda	Harkness
		'Red Paillette'	Miniature	Poulsen
		'MT 86360'	Shrub	Tantau
		'KORlehnig'	Climber	Kordes
	Most Fragrant Rose	'DELrousi'	Hybrid Tea	Delbard
1994	Golden Rose	'Cherry Meillandecor'®	Shrub	Meilland
	Gold Medals	'HARwippee'	Hybrid Tea	Harkness
		'NDR 1 Radio Niedersachsen'	Floribunda	Kordes
		'AM 923'	Miniature	Meilland
		'INT B82688'	Shrub	Interplant
		'KORwerti'	Climber	Kordes
1993	Golden Rose	'Suni'®	Floribunda	Barni
	Gold Medals	'ADAsilli'	Hybrid Tea	Adam
		'KORgezak'	Floribunda	Kordes
		'Shine On'™	Miniature	Dickson
		'Devon'®	Shrub	Poulsen
		'INt B12185'	Climber	Interplant
	Most Fragrant Rose	'Selbar 3694'	Hybrid Tea	Barni
1992	Golden Rose	'Helga Piaget'	Hybrid Tea	Meilland
	Gold Medals	'Imperatrice Farah'	Hybrid Tea	Delbard
		'Sommermarchen'	Shrub	Kordes
		'INT B20586'	Miniature	Interplant
		'KORmalkus'	Climber	Kordes
	Most Fragrant Rose	No name	Hybrid Tea	Ferrer

Table 20 Continued

Year	*Award*	*Cultivar*	*Class*	*Breeder*
1991	Golden Rose	'Selbar 3016'	Floribunda	Barni
	Gold Medals	'Signature'®	Hybrid Tea	Jackson & Perkins
		'Selbar 3016'	Floribunda	Barni
		'INT B30885'	Miniature	Interplant
		'Explorer'	Climber	Dickson
	Most Fragrant Rose	'Selbar 3050'	Hybrid Tea	Barni

Table 21 Awards by Royal National Rose Society, St Albans (1991–2000)

Year	*Award*	*Cultivar*	*Class*	*Breeder*
2000	President' International Trophy	'Rotfassade' (TG2468)	Climber	Noack
1999	Gold Medal	'Rhapsody in Blue'	Shrub	Cowlishaw
1998	President's International Trophy	'CHEwreepy'		Warner
	Gold Medals	'JP90-2668'		Bear Creek
		'RVS 90-94'		RVS
1997	President's International Trophy	'Greetings'		Bear Creek
1996	President's International Trophy	'Molineux'™	Shrub	Austin
	Gold Medal	'Starry Eyed'	Shrub	Horner
1995	President's International Trophy	'Broadlands'	Shrub	Tantau
1994	President's International Trophy	'Twilight'	Shrub	Noack
1993	Gold Medals	'Baby Love'™	Miniature	Scrivens
		'Oxfordshire'	Kordes	Shrub
1992	President's International Trophy	'Blenheim'	Shrub	Tantau
1991	President's International Trophy	'Little Bo-Peep'	Miniature	Poulsen
	Gold Medals	'Flower Carpet White'	Shrub	Noack
		'Singin' in the Rain'	Floribunda	McGredy

end of the third year. Four major award levels were established: President's International Trophy, Gold Medals, Certificates of Merit and Trial Grounds Certificates. With some minor modifications these procedures are still practiced today (**Table 21**).

Varieties are evaluated during each growing season by a panel of 16 permanent judges using a points system determined by the Society's Trials Committee and approved by the Board. Currently these are: vigour and habit, 20 points; health, 20 points; beauty of form or cluster, 15 points; abundance of flowering, 10 points; general effect (colour and impact), 15 points; novelty, 10 points; fragrance, 10 points. The standards for awards are normally: Trial Ground Certificate requires 70 points minimum; Certificate of Merit requires 72.5 points minimum; and Gold Medal requires 75 points minimum. The Harry Edland Memorial Medal is awarded to the most fragrant award-winning rose and the Torridge Award for the best award-winning rose raised by an amateur breeder.

Sir Thomas and Lady Dixon Park, Belfast, Northern Ireland

Sam McGredy of the famous Portadown rose firm of Samuel McGredy & Son was the guiding force that culminated in the autumn of 1963 in the selection of Sir Thomas and Lady Dixon Park as the location for the City of Belfast International Rose Garden. This lovely park was presented to Belfast by the late Lady Edith Stewart Dixon in December 1959, in memory of her husband. The first roses were planted in the winter of 1964/5, including entries for Trial I, 1965/66. In March 1964 the Rose Society of Northern Ireland was formed to facilitate the organization and judging of the trials (**Table 22**). Today up to 35 000 rose bushes are planted and during the summer the garden is an important tourist attraction.

Tollcross Park International Rose Trials, Glasgow, Scotland

Conceived by the Glasgow City Council in consultation with the British Rose Growers Association in 1986, this wonderful rose garden has been designed like the petals of a rose on a gently sloping mound. Situated in Tollcross Park in the East End of Glasgow, the Rose Trial Garden currently covers an area of approximately 1 hectare. The top award, The Golden Prize of the City of Glasgow, is awarded to the rose variety receiving the highest score. Other awards include the Tollcross Fragrance Prize and an award for the Best Established Rose, given to the best variety that has been growing in the trial beds for 10 years. The public are also encouraged to take part in the judging and for a 6-week period each summer, visitors to the gardens are invited to make their choice from the varieties in their final year of assessment (**Table 23**). The variety receiving the most votes is recognized as 'The People's Choice'.

United States of America

All-America Rose Selections

The All-America Rose Selections (AARS) is a non-profit organization of US rose producers and introducers. As their prime objective, they arrange to test new varieties and through extensive evaluation recommend to the general public only those limited varieties deemed exceptional. The few top varieties each year bear the prestigious title of being an 'AARS Winner' and this coveted award is a hallmark of excellence indicating that the variety can be grown in almost any climate. The testing protocol undergone by all the varieties submitted by the various producers and introducers is rigorous. Varieties are specially numbered and are closely evaluated for a period of 2 years in 25 specially designated AARS test garden sites throughout the various climatic zones of the United States. In this way, the varieties are given a true test of varying climates from heat, drought, cold winds, insects, diseases and even to hard-to-please judges. Within the United States there exists a large number of different climatic zones. The

Table 22 Awards at International Rose Trials of Belfast, Northern Ireland (1991–2000)

Year	*Award*[a]	*Cultivar*	*Class*	*Breeder*
1999/2000	Gold Medal and Prize of the City of Belfast	'Pride of England'	Hybrid Tea	Harkness
	DARD Award	'White Gold'	Floribunda	Cocker
	Sainsbury Award	'Charming Cover'	Shrub	Poulsen
	Fragrance	'White Gold'		Cocker
1998/99	Gold Medal and Prize of the City of Belfast	'Poetry in Motion'	Hybrid Tea	Harkness
	Dani Award and Certificate of Merit	'Schloss Balthasar'	Floribunda	Kordes
	R. J. Frizzell Memorial Award	'Constance Finn'	Floribunda	Harkness
	The Sainsbury Award	'Wine and Dine'®	Shrub	Dickson
1997/98	Gold Medal and Prize of the City of Belfast	'St Christopher'	Hybrid Tea	Harkness
	Dani Award and Certificate of Merit	'Pretty Lady'	Floribunda	Scrivens
	R. J. Frizzell Memorial Award	'Helena Renaissance'	Shrub	Poulsen
	The Sainsbury Award	'Sommerabend'®	Shrub	Kordes
1996/97	Gold Medal and Prize of the City of Belfast	'Pure Bliss'™	Hybrid Tea	Dickson
	Dani Award and Certificate of Merit	'Georgette'	Floribunda	Kordes
	R. J. Frizzell Memorial Award	'Benelux Star'	Hybrid Tea	Landbow
	City of Belfast Award	'Our Molly'	Shrub	Dickson
1995/96	Gold Medal and Prize of the City of Belfast	'Warm Wishes'	Hybrid Tea	Fryer
	Dani Award and Certificate of Merit	'Domstadt Fulda'®	Floribunda	Kordes
	R. J. Frizzell Memorial Award	'Frederic Mistral'®	Hybrid Tea	Meilland
	Tyrone Crystal Prize	'Turbo'™	Hybrid Rugosa	Meilland
1994/95	Gold Medal and Prize of the City of Belfast	'Abbaye de Cluny'™	Hybrid Tea	Meilland
	Dani Award and Certificate of Merit	'Ruby Anniversary'	Floribunda	Harkness
	R. J. Frizzell Memorial Award	'Renaissance'	Hybrid Tea	Harkness
	Tyrone Crystal Prize	'Roselina'™	Shrub	Kordes
1993/94	Gold Medal and Prize of the City of Belfast	'Memoire'®	Hybrid Tea	Kordes
	Dani Award and Certificate of Merit	'Sunseeker'	Floribunda	Dickson
	R. J. Frizzell Memorial Award	'Marco Polo'	Hybrid Tea	Meilland
	Tyrone Crystal Prize	'Yellow Fleurette'®	Shrub	Ilsink
1992/93	Gold Medal and Prize of the City of Belfast	'New Horizon'	Floribunda	Dickson
	Dani Award and Certificate of Merit	'La Vie en Rose'	Floribunda	Kameyama
	R. J. Frizzell Memorial Award	'The McCartney Rose'™	Hybrid Tea	Meilland
	Tyrone Crystal Prize	'Eye Appeal'®	Shrub	Ilsink
1991/92	Gold Medal and Prize of the City of Belfast	'Painted Moon'	Hybrid Tea	Dickson
	Dani Award and Certificate of Merit	'Day Light'®	Floribunda	Ilsink
	R. J. Frizzell Memorial Award	'Sweet Juliet'	Shrub	Austin
	Tyrone Crystal Prize	'Colleen'		Ilsink

[a] Gold Medal and Prize of the City of Belfast: presented by Belfast City Council for the best Hybrid Tea rose; Dani Award and Certificate of Merit: presented by the Department of Agriculture for Northern Ireland for the best Floribunda, Miniature, Climber and including those described as Patio-type roses; R. J. Frizzell Memorial Award: presented by the Northern Ireland Tourist Board for the most fragrant rose of the meritorious varieties; Tyrone Crystal Prize: presented Sainsbury Award (1987–1996): presented by Sainsbury's for the best shrub rose; City of Belfast Award (1996–97): presented by the Belfast City Council for the best shrub rose.

Table 23 Awards at the International Rose Trials, Glasgow (1991–2000)

Year	*Award*	*Cultivar*	*Class*	*Breeder*
2000	Golden Prize of the City of Glasgow	'Amber Nectar'	Floribunda	Eurosa
	Gold Medal	'Hearbeat 97'	Floribunda	Cocker
	The People's Choice	'Perception'	Hybrid Tea	Harkness
	Best Established Rose	'Fellowship'	Floribunda	Harkness
1999	Golden Prize of the City of Glasgow	'Scarlet Patio'	Miniature	Kordes
	Gold Medal	'Molineux'	Shrub	Austin
	Tollcross Fragrance Prize	'Noble Anthony'	Shrub	Austin
	The People's Choice	'ARC Angel'	Hybrid Tea	Fryer
	Best Established Rose	'Mary Hayley Bell'	Shrub	Kordes
1998	Golden Prize of the City of Glasgow	'Sunset Boulevard'	Floribunda	Harkness
	Gold Medal	'Scotland Wink' (Adavapma)		Adam
	Tollcross Fragrance Prize	'L'Aimant'	Floribunda	Harkness
	The People's Choice	'Aberdeen Celebration'	Floribunda	Cocker
	Best Established Rose	'Anna Livia'	Floribunda	Kordes
1997	Golden Prize of the City of Glasgow	'Friend for Life'	Floribunda	Cocker
	Gold Medal	'The Holt'	Shrub	Mehring
	Tollcross Fragrance Prize	'Sally's Rose'	Hybrid Tea	Cant
	The People's Choice	'Julie Y'	Hybrid Tea	Harkness
	Best Established Rose	'Abbeyfield Rose'	Hybrid Tea	Cocker
1996	Golden Prize of the City of Glasgow	'Berkshire'		Kordes
	Gold Medal	'Our Molly'	Shrub	Dickson
	Tollcross Fragrance Prize	'Renaissance'	Hybrid Tea	Harkness
	The People's Choice	'Our Molly'		Dickson
	Best Established Rose	'Rosy Cushion'®	Shrub	Ilsink
1995	Golden Prize of the City of Glasgow	'Remembrance'	Floribunda	Harkness
	Gold Medal	'Flair'	Floribunda	Dickson
	Tollcross Fragrance Prize	'Thomas Barton'		Meilland
	The People's Choice	'Gordon's College'	Floribunda	Cocker
1994	Golden Prize of the City of Glasgow	'Mandarin'	Miniature	Kordes
	Gold Medal	'Ena Baxter'	Hybrid Tea	Cocker
	Tollcross Fragrance Prize	'Marty'	Floribunda	Pearce
	The People's Choice	'Valencia'®	Hybrid Tea	Kordes
1993	Golden Prize of the City of Glasgow	'Hello'	Miniature	Cocker
	Gold Medal	'Flower Carpet'™	Shrub	Noack
	Tollcross Fragrance Prize	'Indian Summer'	Hybrid Tea	Pearce
1992	Golden Prize of the City of Glasgow	'Fulton MacKay'	Hybrid Tea	Cocker
	Gold Medal	'Pretty Polly'	Miniature	Meilland
1991	Golden Prize of the City of Glasgow	'Anna Livia'	Floribunda	Kordes
	Gold Medal	'Biddulph Grange'	Shrub	Fryers
	Tollcross Fragrance Prize	'Rosemary Harkness'	Hybrid Tea	Harkness

varieties being tested should be capable of surviving anywhere from the cold climates in Illinois to the tropical climates of Florida and California. To survive and win under such diverse climatic conditions is a mark of excellence for the variety finally chosen as a winner. Not all varieties selected for this honour are exhibition roses. In many cases good garden display varieties are finally chosen for their abundance, bloom production, or speed of bloom cycle.

Judges are assigned by AARS to the various test gardens. Most AARS judges are either professors in college departments of horticulture or perhaps key supervisors of large public gardens. During the 2-year evaluation period, roses are graded by the judges by a predetermined scoring system based on vigour, growth habit, disease resistance, foliage, flower production, bud and flower form, opening and finishing colour, fragrance, stem and overall value. Scores are compiled and those varieties considered exceptional by the judges are declared the winners for that particular year. By this rigorous process AARS has promoted and encouraged the rose industry to improve the vitality, strength and beauty of roses. Since its inception in 1940, this hallmark of excellence has been bestowed upon 171 varieties (**Table 24**).

Criteria for AARS Judging

Novelty Novelty may be regarded as the first impression of an entry. Novelty is that quality which makes a new rose unique, exciting and memorable.

Table 24 Awards by All-America Rose Selections (1991–2002)

Year	*Cultivar*	*Class*	*Breeder*
2000	'Crimson Bouquet'(TM)	Grandiflora	Kordes
	'Gemini'(TM)	Hybrid Tea	Zary
	'Knock Out'(TM)	Shrub	Radler
1999	'Betty Boop'(TM)	Floribunda	Carruth
	'Fourth of July'(TM)	Large Flowered Climber	Carruth
	'Kaleidoscope'(TM)	Shrub	Walden
	'Candelabra'(TM)	Grandiflora	Zary
1998	'Fame!'(TM)	Grandiflora	Zary
	'Opening Night'(TM)	Hybrid Tea	Zary
	'First Light'(TM)	Shrub	Marciel
	'Sunset Celebration'(TM)	Hybrid Tea	Fryer
1997	'Artistry'(TM)	Hybrid Tea	Zary
	'Timeless'(TM)	Hybrid Tea	Zary
	'Scentimental'(TM)	Floribunda	Carruth
1996	'St Patrick'(TM)	Hybrid Tea	Strickland
	'Livin' Easy'(TM)	Floribunda	Harkness
	'Mount Hood'	Hybrid Tea	McGredy
	'Carefree Delight'(TM)	Shrub	Meilland
1995	'Brass Band'(TM)	Floribunda	Christensen
	'Singin' in the Rain'	Floribunda	McGredy
1994	'Secret'(TM)	Hybrid Tea	Tracy
	'Caribbean'(TM)	Grandiflora	Kordes
	'Midas Touch'(TM)	Hybrid Tea	Christensen
1993	'Solitude'(TM)	Grandiflora	Olesen
	'Rio Samba'(TM)	Hybrid Tea	Warriner
	'Sweet Inspiration'	Floribunda	Warriner
	'Child's Play'(TM)	Miniature	Saville
1992	'Brigadoon'(TM)	Hybrid Tea	Warriner
	'Pride 'n' Joy'(TM)	Miniature	Warriner
	'All That Jazz'(TM)	Shrub	Twomey
1991	'Sheer Elegance'(TM)	Hybrid Tea	Twomey
	'Perfect Moment'(TM)	Hybrid Tea	Kordes
	'Shining Hour'(TM)	Grandiflora	Warriner
	'Carefree Wonder'(TM)	Shrub	Meilland

Form, buds A pleasing assemblage of petals, considered from calyx break to the half-open stage.

Form, flowers A measure of the balance and appeal of petal arrangement from half-open to full blown, as influenced by the number of petals, their shape, configuration and overall size.

Colour, buds The overall attractiveness of petal colour from calyx break to the half-open stage of the bloom; colours are considered for their own value (irrespective of personal colour preferences), the cleanness of uniform hues and the harmony and graceful progression of combined colours.

Colour, open blooms Continuation of the previous category, considered from the half-open stage to the freshly full-blown flower.

Ageing quality The measure of how well open blooms hold up as they pass their peak and then finish; whether the colour and form remain pleasing to the end.

Flowering effect The impact of the whole rose plant as it comes into and passes each full bloom cycle; the number of flowers, their size, proportions and distribution, colour and finishing quality will all play a part in the score.

Fragrance Since there are different types and degrees of perfume in different roses, and these are influenced by temperature and other factors, scoring this category is recommended several times on different days and at different times of the day.

Stem/cluster Stems should display their blooms to advantage, holding them upright and in harmonious balance between flower and plant.

Plant habit The growth structure of the entire plant as influenced by its height, breadth, general shape and balance.

Vigour The continuing growth action of the plant, as typified by rapid growth response after blooming or recovery from climatic extremes like cold weather; a measure of the plant's vitality and willingness to keep displaying its talents.

Foliage The amount, size, shape, quality and distribution of greenery has a great deal to do with the overall decorativeness of a rose bush and can significantly enhance or detract from the appearance of a cut bloom.

Disease resistance The entry's ability to withstand the various rose diseases, specifically black spot, mildew and/or rust is considered.

Repeat bloom The main duty of the rose is to produce flowers, and this category gives the judge a chance to consider how well the plant does that job throughout the growing season; whether it repeats at all, and how rapidly it repeats.

AARS Test Gardens in the United States

- Garden Valley Ranch, Petaluma, California
- International Rose Test Gardens, Portland, Oregon
- Disney Institute, Gardening and The Great Outdoors, Florida
- Tyler Rose Garden, Tyler, Texas
- Woodland Park Rose Garden, Seattle, Washington
- Edisto Memorial Gardens, Orangeburg, South Carolina
- Woodward Park, Tulsa, Okalhoma
- Fernbank Science Center, Atlanta, Georgia
- E.F.A. Reinisch Rose Test Garden, Topeka, Kansas
- Iowa State University, Ames, Iowa
- Old Westbury Gardens, Old Westbury, New York
- The Stanley Park of Westfield, Inc., Westfield, Massachusetts
- John E. Voight Trial Garden, Hales Corners, Wisconsin
- Chicago Botanic Gardens, Glencoe, Illinois
- Missouri Botanical Gardens, St Louis, Missouri
- Brigham Young University, Provo, Utah
- City of Buffalo/Joan Fuzak Memorial Garden, Buffalo, New York
- Doane College, Crete, Nebraska
- Elizabeth Park Rose Garden, West Hartford, Connecticut
- Michigan State University, East Lansing, Michigan
- Lyndale Park Gardens, Minneapolis, Minnesota
- Descanso Gardens, La Canada/Flintridge, California
- Norfolk Botanical Garden, Norfolk, Virginia
- Mesa Community College, Sun Lakes, Arizona (new garden as of 1998)
- Rose Hills Memorial Park, Whittier, California (new garden as of 1999)

American Rose Society Award of Excellence for Miniature Roses

The prospectus presented to the Board of Directors of the American Rose Society in 1973 in Columbus, Ohio, outlined the reasons for organizing the American Rose Society (ARS) Miniature Rose Test Garden Program. Miniature roses have gained increasing popularity in recent years becoming an integral part of all national, regional and local shows. Therefore, it was resolved that the ARS undertake an active, impartial testing, scoring and rating programme exclusively for miniature roses. Starting in 1975 each entry has to be tested for a 2-year period. No more than five Awards of Excellence (AOE) may be given in any year (**Table 25**) and all awards may be withheld if no entry warrants it. (More awards were made in 1975, the inaugural year; see **Table 25**.)

The By-laws of the All-America Rose Selections list purposes which are parallel to the American Rose Society test purposes:

1. To foster the development of new and better roses in and for the United States.
2. To foster the establishment of rose test gardens in the United States for the purpose of testing, judging and rating new kinds, species and varieties and acquainting people with the most worthy roses.
3. To make awards to new rose varieties of superior quality and marked distinction as determined by their performance in the official test gardens and to publicize and recommend those varieties which receive such awards.

Miniature roses are evaluated in six test gardens by teams of ARS members:

- American Rose Society, Shreveport, Louisiana
- Fernbank Science Center, Atlanta, Georgia
- Hershey Gardens, Hershey, Pennsylvania
- Toledo Botanical Gardens, Toledo, Ohio
- Rose Hills Memorial Park, Whittier, California
- International Rose Test Garden, Portland, Oregon.

The elements used in the 2-year evaluation include: novelty, bud form, flower form, colour opening, colour finishing, substance, habit, quantity of flowers, vigour/renewal, foliage and disease/insect resistance.

Recently Established Trial Grounds

A number of trial grounds have recently been established in various countries including Buenos Aires,

Table 25 Awards of Excellence for Miniature Roses by the American Rose Society (1991–2000)

Year	*Cultivar*	*Breeder*
2000	'Applause'	Saville
	'Lemon Gems'™	Walden
	'Ralph Moore'	Saville
1999	'Little Flame'™	Walden
	'Autumn Splendor'	Williams
	'Little Tommy Tucker'™	Tucker
	'Sweet Caroline'	Williams
1998	'Playgold'	Moore
1997	'Tropical Twist'™	Walden
1996	'Angelica Renae'™	Wells
1995	'Jingle Bells'™	Zary
1994	'Hot Tamale'™	Zary
1993	'Billie Teas'	Hooper
	'Boomerang'	Sponner
	'Child's Play'™	Saville
	'Kristin'™	Benardella
	'Palmetto Sunrise'	Williams
1992	'Cal Poly'	Moore
	'Debidue'™	Williams
	'Figurine'™	Benardella
	'Sincerely Yours'	Moore
1991	'Golden Halo'™	Saville
	'Good Morning America'™	Saville
	'Just For You'	Moore
	'Suzy'	Bridges

Argentina; Adelaide, Australia; Gifu, Japan; and most recently Rose Hills, USA, all designed on the European models.

Note for Tables 1 to 25

Most of the tables 1 to 25 have been reduced in length in this article. Extended tables can be found in Appendix 1 at the back of Volume 2.

See also: **Distribution and Ecology**: South Africa; Australia and New Zealand; Continental Asia and Japan; Europe; North America; South America. **Rose Collections and Trials**: Rose Collections and Herbaria; Rose Societies.

Further Reading

Beales P, Cairns T, Duncan W *et al.* (1998) *Botanica's Roses*. Australia: Random House.

WFRS Triennial Report on Roses 2000. St. Albans, UK: American Rose Society for the World Federation of Rose Societies.

World Rose News features annual reports of trial winners from around the world. Published biannually. St. Albans, UK: World Federation of Rose Societies.

S

SECONDARY METABOLITES

Contents

Culinary Uses and Nutritional Value

R R Cutler, University of East London, London, UK

Introduction

Of what use will be a dish of roses to you?

Sheikh Muslih-uddin Sa'di Shirazi, thirteenth-century Persian poet

This quotation actually referred to a display of roses but this article considers a 'dish of roses' in a more literal sense. Most people may associate roses or indeed any flower recipes with ancient books of only historic interest but their use is reflected in modern recipes where the use of rose flavours in sweets and cakes is common. Sugar flavoured with rose essence was very popular in olden times and recipes containing this mixture were common in sixteenth- and seventeenth-century cookbooks. Although many of the recipes that will be given in this article originate in ancient times, when roses were more valued for their scent, sweet flavour and medicinal uses (*see* **Secondary Metabolites**: Medicinal and Pharmaceutical Uses), it will be demonstrated that some of these ancient recipes can be used as a basis for modern 'culinary delights'.

It should be appreciated that in many of the recipes used in this article the hips of the rose are used in similar ways to other fruits; many of the recipes relate to sweets, jellies and jams. Either rose petals or rose hips may be used in the recipes. Rose petals offer a wide variety of colours and scents and rose hips are similarly varied in size, shape and sweetness. Variations of the recipes described below can be made simply by changing the colour and variety of roses used in the preparation.

Wide selections of recipes are used to illustrate the culinary uses of roses. Where possible, each recipe will list the ingredients needed (and quantities) followed by the method used for preparation. The original weights and measures used have been converted to SI units where possible.

Nutritional Value and Flavour

A comparison was made of reported values of some of the chemicals found in rose hips with the same chemicals in other Rosaceae fruits and in oranges. This compared some of the potential nutritional values in these fruits (**Table 1**). Overall, hips were a good source of vitamin C, boron and pectins.

Vitamin C was one of the main chemicals found in hips. The main nutritional value from roses is thought to come from the vitamin C present in hips, in which it occurs at a high concentration compared to other fruits. Apart from variations in vitamin C concentrations due to the methods used, differences have been shown to be related to where the roses are grown and the species tested (*see* **Secondary Metabolites**: Medicinal and Pharmaceutical Uses). The amount of vitamin C affects pH. Usually the more bitter (acid) hips contain more of vitamin C (ascorbic acid).

The nutritional value of vitamin C in the diet is to promote the synthesis of collagen, a major structural protein in the body. Scurvy is the disease caused by

Table 1 Comparison of some of the nutritional chemicals[a] found in rose hips compared with other selected Rosaceae fruits and with oranges (Rutaceae)

	Rose hips (Rosa *subsp., Rosaceae*)	*Apple* (Malus domestica, *Rosaceae*)	*Pear* (Pyrus communis, *Rosaceae*)	*Strawberries* (Fragaria *subsp., Rosaceae*)	*Orange* (Citrus sinensis, *Rutaceae*)
Ascorbic acid (vitamin C)	1000–12 500	20–400	40– 250	400–6900	500–4100
Pectins	34 000–46 000	1400–66 500	40 000	54 000	1300–5900
Glucose	Trace	17 200–18 200	NA	NA	23 600
Carotenoids	100–500	120	NA	<10	10–40
Citric acid	Trace	Trace	Trace	40–80	5600–9800
Folic acid	100–180	NA	NA	70–97	NA
Boron	20	1–110	1–80	1–160	1–30

[a] Data presented in mg kg^{-1}, to nearest 10 mg kg^{-1} (except for boron).
NA, not available.

lack of vitamin C in the diet; the symptoms (loose teeth, bleeding gums, haemorrhaging and poor wound healing) are related to the poor synthesis of collagen. One needs 10 mg day^{-1} of vitamin C to prevent scurvy; the recommended daily intake is 60 mg day^{-1} which could easily and cheaply be supplied from fresh hips of *Rosa canina*. At higher concentrations than the recommended daily dose, vitamin C acts as an antioxidant; it also acts as a cofactor in the synthesis of nerve transmitters and is involved in the action of hormones. Small amounts of the vitamins A, B_3, D, E and K have also been identified in hip extracts. These could also have some nutritional value if not destroyed through processing.

The importance of boron has only recently been identified. Boron is thought important to maintain bone density, and it is thought it may have particular relevance to women (e.g. in relation to osteoporosis). Hips may provide some sugars for energy but the other fruits were a better source of these.

Pectins may not be nutritional but are well known for their gelling, thickening and stabilizing properties and therefore of potential use in food preparation.

Flavours

Volatile and semivolatile organic compounds and acidity have a great influence on the flavour and fragrance (*see* **Fragrance and Pigments**: Fragrance Profiles of Wild and Cultivated Roses) of foods. Flavours are complex blends of chemicals. Humans can identify five basic tastes: salty, sweet, sour, bitter and unami (monosodium glutamate). A sense of smell is however necessary to identify certain flavours such as those of the rose. The odours from the volatile oils are released in the saliva and molecules travel up the nasal passage from the throat. People who have a problem with tasting often have an underlying olfactory (smell) disorder. Acidity and volatile oils are important factors in our flavour perception.

Short-chain fatty acid esters of citronellol are important flavour and fragrance compounds widely used in the food, cosmetic and pharmaceutical industries. For over a century the main constituent of rose oil was identified as citronellol. It was not until the 1960s that the trace constituents essential to a rose's fragrance and flavour were reported. Then came the discovery of the rose ketones, damascenone and damascone, two important fragrance and flavour chemicals used in industry. Rose ketones are derived from the degradation of carotenoid.

In particular β-damascenone has a very low odour threshold of 0.002 parts per billion (p.p.b.), compared with 40 p.p.b. for citronellol and up to 100 p.p.b. for damascone, and a flavour threshold of only 0.009 p.p.b. It is considered to be one of the most important flavour compounds in the essential oil of roses even though it is normally only present in concentrations of 100 p.p.m.

The terpenoid ether (rose oxide) also contributes to rose aroma and flavour. The astringent taste found in rose petals is mainly due to gallic acid. The acidity produced by compounds such as ascorbic acid in hips (Vitamin C) affects their bitterness of hips and their overall flavour.

Use of Rose Hips for Cooking

For cooking purposes, hips are best when slightly underripe, firm and bright red. After picking, wash the hips and remove the tassels (remains of the petals, anthers and style and the tails, i.e. pedicels). Prepare them as soon as possible after picking, to preserve as much vitamin C as possible.

When Is a Hip Ripe?

Roses and apples are related species and the hip is like a little 'rose apple'. Rose hips differ in size and shape; they also differ in sweetness, colour and the time it takes to ripen, so choose your hips carefully. As with all fruit, you will know when the hip is ripe because the sides will 'give' slightly when gently squeezed. Make sure the hips are ripe but not soft before you use them.

Chemical Contamination

As you would with any fruit, be careful the roses have not been sprayed with insecticide or dusted with sulphur. This is very important: you need clean, but pure material.

Recipes

Rose Hip Juice and Dried Petals

In its most basic form, juice can be extracted from ripe hips and used as a source of flavour and vitamins in a variety of recipes. Dried rose petals can be also used for decoration or added to other dried fruit or cereal mixtures.

Rose hip juice (United States) To extract the juice, partly cover with water; bring quickly to a boil and cook slowly for about 15 min. The juice can then be extracted by draining through a jelly bag or cheesecloth. Let drip into large container. Mixed with other fruits, hip juice makes conserves, jellies or syrups with an excellent flavour.

The pulp sieved to remove the seeds and the skins can be used to make jam, marmalades and ketchups. The flavour can be enhanced by adding some fruit juice such as cranberry or rhubarb.

To dry hips (United States) To dry hips cut them in two and remove the seeds with the point of a knife. Dry quickly in an oven at 140–150 °C. Dried rose hips can be added to breakfast cereal, cooked with fruit sauce or pulverized and put in baked products.

Teas

There are many recipes for tea made by combining rose petals with other flavours such as spearmint or lemon. In teas the white, bitter part of the petal does not have to be removed before use. The fragrance of the petals used in making the tea is related to personal taste. Many tea manufacturers produce a variety of 'rose-flavoured' teas that can be purchased commercially.

When selecting a rose petal for tea infusions, the scent of the petals is more important than the colour. In an experiment, infusions were made from two types of petals; one with petals of 'Ulrich Brünner' with a strong scent and colour and the other using petals with an equally strong colour but little scent, 'George Dickson'. **Figure 1** shows the 'Ulrich Brünner' petals before and after infusion with the lightly coloured infusion liquid. **Figure 2** shows the infusion with the similarly coloured but unscented rose petals from 'George Dickson'. The infusion was darker (right glass in **Figure 3**) but the liquid had a very bitter flavour compared with the sweeter, scented flavour from the lighter-coloured infusion of 'Ulrich Brünner' (left glass in **Figure 3**).

Rose petal tea (Central Europe)
Ingredients 8 g packed rose petals; 1 fresh sprig spearmint; 1–2 l boiling water (depending on taste).

Method Rinse rose petals well. Place in a coffee press with fresh spearmint sprig. Add boiling water, steep for 3–5 min, press, drain and serve. Adding honey or sugar may bring out the flavour; one tablespoon (15 ml) per cup is suggested.

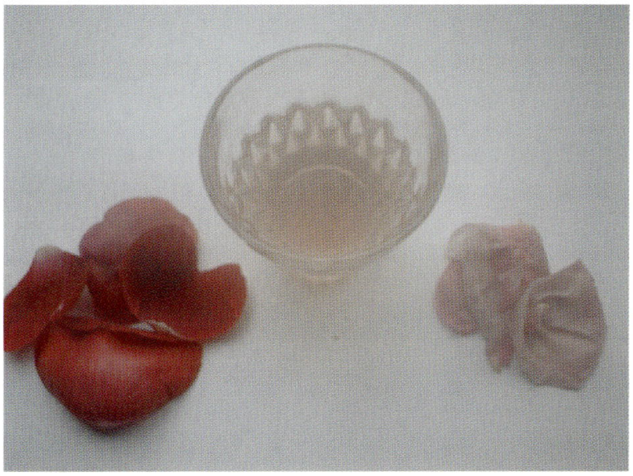

Figure 1 In the centre, rose tea made from the scented rose petals of 'Ulrich Brünner'. Petals shown before being made into tea (left) and after use (right).

Figure 2 In the centre, rose tea made from the unscented rose petals of 'George Dickson'. Petals shown before being made into tea (left) and after use (right).

Figure 3 Direct comparison of teas made from scented petals ('Ulrich Brünner') producing a sweet tea and a bitter darker tea ('George Dickson').

Rose-Water

Within the recipes cited in this text various uses of rose-waters, oils and scents are described. There are many ancient recipes to make 'rose-waters' varying from letting scented rose petals soak for days in spring water to boiling rose petals in an equal volume of water. One of the more common methods to produce 'an excellent rose-water' is described below; distillation apparatus is a necessity to make quality rose-water. Rose water can also be purchased commercially.

Rose water (Britain)

Ingredients Rose petals (Damask roses).

Method Beat the petals into a paste using a mortar and leave them soaking in their juice for up to 6 h. Strain the juice from this mixture and mix the juice with fresh rose petals adding 400 g of petals to each litre of juice. Let the infusion stand for 24 h. Distil the liquid under a gentle heat to produce the rose-water.

Cakes and Biscuits

In bakery the role of the rose is usually to provide rose-water to flavour the product. Below is one example of a variety of cakes that use rose-water as an ingredient in the cake mixture; the cake is decorated variously with rose petals (either candied or not) or rose hips.

Rose-water flavoured cake (United States and Britain)

Ingredients 200 ml egg whites, at room temperature; 2.5 ml cream of tartar; 150 ml sugar; 5 egg yolks; 150 ml cooking (salad) oil; 220 ml double-strength rose water; 700 ml sifted cake flour; 300 ml sugar; 10 ml baking powder; 4 ml salt.

Method Use a large flat nongreased pan. Make the egg-yolk mixture by mixing the oil, 5 egg yolks, and rose water together with the flour. Beat this mixture until smooth. Set aside.

Beat the egg whites until foamy. Add the cream of tartar. Beat the mixture then gradually mix the sugar into the mixture. Continue to beat until the mixture is stiff and peaks are formed. Spread the egg-yolk mixture over the egg white mixture. Fold together until just blended. Do not stir. Pour mixture into a cake pan.

Bake at 160 °C for 55 min, then at 180 °C for a further 10–15 min, or until cake surface seems baked. Let the cake cool completely before taking it out of the pan. Decorate the surface with candied rose petals and serve with cream and soft fruits.

Cookies (United States)

Ingredients 100 ml butter or margarine; 50 ml of light brown sugar; 1 egg, separated; 2.5 ml vanilla extract; 2.5 ml rose-water; 15 g finely chopped nuts; rose petal jam or rose hip jelly; powdered sugar or shredded rose petals for decoration.

Method Preheat the oven to 150 °C. Mix the sugar, softened butter and egg yolk. Add vanilla, rose-water and salt, mixing well. Slowly add the flour. Mix very well. Shape the dough into small balls. Whisk the egg white. Roll the dough balls first in the egg white, and then roll them again in the nuts. Place the dough balls on a sheet of greased paper about 5 cm apart. Bake for just 5 min.

Remove the greased paper sheet from the oven and use your thumb to make a dent in the top of each cookie. Put jam or jelly in each indent. Return the sheet to the oven and bake for another 8 min.

Remove finished cookies from the oven. Remove gently from cookie sheet and allow to cool. Cookies may be dusted with powdered sugar or shredded rose petals or left plain. Makes about 25 cookies.

Candied Rose Hips

Tests comparing different species of hips for production of candied hips A number of different hips (**Figure 4**) were tested to determine whether there was any significant difference in taste of candied hips from different varieties. In a small study to examine if there was a variation in basic flavour and shape after cooking, the hips from *R. canina*, *R. rugosa* var. *alba* (**Figure 5**), *R. andersonii*, 'Chessum's Choice' (**Figure 6**), *R. gallica* var. *officinalis* and *R. rubiginosa* were all processed. The best methods for removing the seeds, pulp and fine hairs from the hips was to chop off the end (with threadlike tassels) and scoop

Figure 4 Hips from a variety of roses: (a) *Rosa canina*; (b) *R. rubiginosa*; (c) 'Chessum's Choice'; (d) *R. rugosa* var. *alba*; (e) *R. andersonii*; (f) *R. gallica*.

Figure 5 *Rosa rugosa* var. *alba*: hips and white petals.

out the pulp with a large darning needle (**Figure 7**). To prepare the candy the sugar and water were boiled together and then the hips were added and cooked for a further 1–2 min in the mix (**Figure 8**). The hips from *R. canina*, 'Chessum's Choice' and *R. rugosa* var. *alba* gave the best results for flavour; the others were found to be either too bitter or virtually tasteless. *Rosa canina* and 'Chessum's Choice' produced a delicious bitter-sweet taste; *R. rugosa* var. *alba* was more fleshy and not as flavoursome. As for shape, the hips from *R. canina* lost their shape compared to those of 'Chessum's Choice' and *R. rugosa* var. *alba* (**Figure 9**).

The conclusion drawn from this small survey was that the hips you select to candy or make into jam may have a wide range of flavours and that it is up to individuals to select which is best for them. However, some hips may not be suitable to candy because they lose structure on cooking. These varieties can, however, still produce an excellent flavour and may be more suitable for jams, jellies and marmalades.

Figure 6 'Chessum's Choice': rose hips.

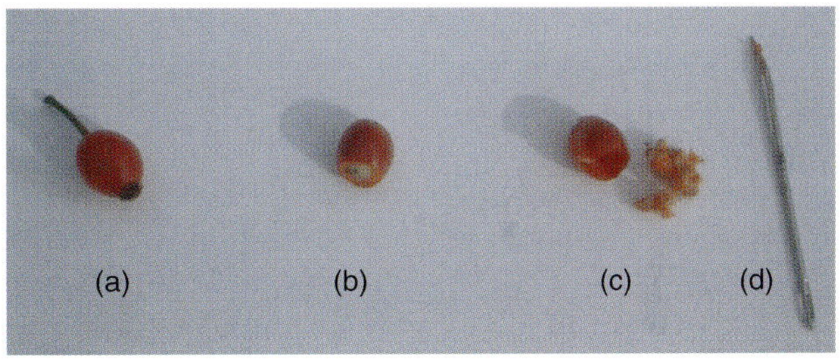

Figure 7 Preparation stages for candied rose hips: (a) hips; (b) hips topped and tailed; (c) pulp; (d) flat needle used for removing pulp from hips.

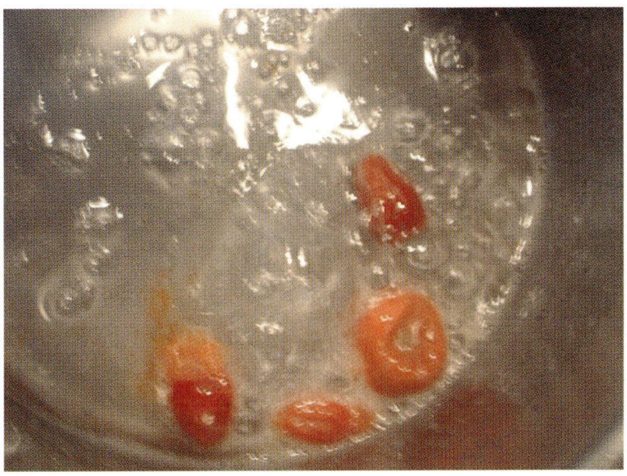

Figure 8 Rose hips boiling in sugar to coat with candy.

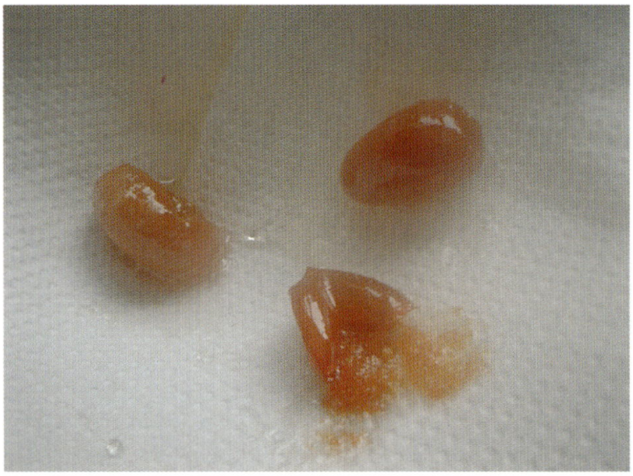

Figure 9 Candied rose hips (from left to right): *Rosa rugosa alba*, *R. canina* and 'Chessum's Choice'. Note the comparative loss of shape with *R. canina*.

Candied hips (United States, Britain)
Ingredients 300 g rose hips; 100 g sugar; 50 ml water.

Method Remove seeds and pulp from the ripe hips using a knife. Make syrup by combining the sugar and water, heating the mixture until the sugar is dissolved. Add the rose hips and boil the mixture for 10 min. Make sure the mixture penetrates to the heart of the hips. Remove the fruit from the syrup with a slotted spoon and drain on waxed paper. Sprinkle with sugar and leave to dry. Store in a closed container between sheets of waxed paper until used.

Candied Rose Petals

In the sixteenth and seventeenth centuries people could be environmentally friendly and use sunshine to help 'candy' their products.

Candied petals (France)
Ingredients Red rose petals; rose-water; castor sugar.

Method On a sunny day lay the rose petals on paper and the sprinkle them with double-refined sugar. Leave to bake in the sun until candied, turn them over and sprinkle the other side and leave to bake in the sun again. Sometimes sprinkle rose water on the mix to add flavour and scent. This process could be repeated to taste.

Using egg whites in the production of candied petals
Ingredients As above plus egg whites.

Method Beat the egg whites to a smooth but not stiff mixture (like paint). Using a paintbrush coat rose petals on both sides. Spread the petals on a dish and coat both sides with castor sugar. The recommendation is to bake in a hot sun as above but a very gentle bake in an oven can be used as a modern equivalent.

Conserves (Jellies), Jams and Marmalade

One of the simplest recipes is to make a conserve or jam from petals or hips. The following definitions cover these types of products:

- Conserves are made from mixtures of fruits. They usually contain citrus fruit, nuts and raisins and, of course, rose petals or hips. Conserves generally do not set as firmly as jams and their consistency is more syrupy. Modern jellies are of a similar makeup and consistency and are linked to this group in the text.
- Jams are made from crushed or ground fruits. They usually have a thicker consistency than conserves due to the high pectin content. To make jams the fruit is normally first simmered in water to release the pectin. Jams set when the pectin gels at about 105 °C.
- Marmalade is a jelly with pieces of fruit suspended in it. Citrus peel and juice are frequently the basis of marmalade.

Rose petal conserves (Britain)
Ingredients 450 g red rose petals, of any deep red, scented roses (e.g. 'Etoile de Hollande' or 'General MacArthur'); 450 g loaf sugar; 30 ml orange flower water.

Method Any petals should be gathered when fully open but before they fade. Remove the white parts which are bitter. Dry the roses out of direct sunlight. Put petals in a muslin bag and place in boiling water for 1 min; drain well. Make syrup by gently simmering the sugar with a little water and add the orange flower water. Add the petals and cook until the conserve thickens. Keep the petals under the syrup by pressing them down. Store the conserve in jam jars.

Rose hip conserve
Ingredients Rosehips (ripe) and sugar; use at 1 : 1 or 1 : 3 ratios.

Methods In the well-respected book '*The Complete Housewife*' published around 1736, it was recommended that the hips were collected before they were soft, slit in half and the seeds removed. They were then placed in an 'earthen pot' and left until they were soft enough to pulp. This mixture was mixed with equal amounts of sugar, heated without boiling and stored in sealed jars.

In an earlier, 'sweeter', method published by John Partridge in 1586, one should take the buds of roses, remove the white parts, beat and grind them adding one weight of hips with three equal weights of sugar until they are 'perfectly incorporated'.

***Rosa canina* petals conserve** Princess Alexandre Gazarine described an alternative method for making a conserve from the petals of *R. canina* in the *Russian Cookbook* of 1924. Rose oil is usually, though not exclusively, prepared by distillation from the fresh petals of *R. damascena*. It is a colourless or yellow liquid which smells of roses. It can be purchased commercially. Rose oil made from any rose species can be used for this recipe.

Ingredients 230 g *R. canina* petals; 920 g sifted sugar; juice of half a lemon; 100 ml water; one drop of rose oil.

Method Cut off all the damaged parts of the petals. Put the petals into boiling water and boil well. Pour the extract through a sieve to drain. Dry the petals and powder them with 230 g of sifted sugar. Make syrup by heating the remaining sugar with the water. When the syrup thickens add the lemon juice and the sugared rose petals. Boil the mixture once. Pour the mixture into a bowl and when cool add one drop of 'attar of roses' (rose oil), mix well and store in sealed jars.

Rose hip jelly (United States) This modern 'jelly' recipe bears a similarity to the conserve from '*The Complete Housewife*'; it is made from a mixture of hips and apples.

Ingredients 1 kg dried apple slices; 1 kg rosehips; 3 kg sugar; water.

Method Place the apple slices into a deep cooking pot, cover with warm water and stand overnight (at least 8 h). Next add the fresh rose hips and again cover with warm water. Bring the mixture to a boil and cook until soft. Drain the liquid through a jelly bag into a new pot to remove the plant material. Add sugar and boil for another 20 min or until mixture gels into a thick mass when dropped from a spoon into cold water. Pour into sterilized jars and seal.

One alternative method to prepare the hips (which may also be suitable for other recipes) is to first put them through a dry mincing machine and cut them up as fine as possible, so that when hot (not boiling) water is poured over them and they are left to steep (completely covered) the best extraction is obtained from them.

Rose hip jam The *Russian Cookbook* also describes how to make a jam out of *R. canina* hips. As you would expect with a jam the hips remain a part of the jam and are not sieved out as is the case with the conserves and jellies above. This preparation will produce a very coarse jam similar in many ways to the rose hip marmalade described below. If however like Sir Kenelm Digby (*c.* 1669) you found the presence of whole hips in the jam to be 'clogging as doth the ordinary rough conserve made of raw roses', then a simple alternative to this recipe is to puree the fruits (e.g. using a food processor) before use.

Ingredients 1 kg *R. canina* hips; 3 kg sugar; 600 ml water.

Method Choose ripe hard hips. Scrape each berry, cut off the tip and remove all of the pulp with a small spoon or some other tool. Be careful not to break the rose hip. Drop these hips into cold water, agitating several times to remove the fine hairs that cover the surface. Put the hips into a saucepan and pour boiling water over them. Boil the mixture further and when the water boils, stop, drain and allow it to cool.

To prepare the syrup add the water and the sugar together and bring to the boil, twice. Add the prepared hips and cook until they are tender. Remove the scum from the jam, pour the mixture into a bowl and let it set for several days. Gently shake the mixture so that the hollowed-out rose hips remain filled with the syrup. Store the jam in sealed containers.

Rose petal jam (United States) Although traditionally jams set more effectively than conserves due to their content of pectin from the plant material, today we can now add manufactured pectin (e.g. Suregel) directly to the mixture. This allows 'jams' to be made from rose petals. A typical example is shown below.

Ingredients 25 g fragrant rose petals, tightly packed (pesticide-free); 600 ml water; 1 packet of gelling/setting gel; 30 ml lemon juice; red food colouring (optional); 800 g sugar.

Method Boil the petals then steep them for 20 min, pushing petals into the liquid occasionally. Strain the petals mixture. Retain the liquid and the petals. Measure the liquid and make it up to 600 ml with water. Mix liquid with pectin mix, lemon juice, and a few drops of red food colouring.

Bring the liquid to a boil, add the sugar and boil for another minute. Immediately remove from the heat, and stir in the rose petals. Stir for a few minutes more to prevent the petals from floating. Store the jam in sealed sterilized jars.

Marmalade

Historically some of the recipes called rosehip marmalades were misnamed according to today's terminology. All of them were actually jams or even conserves. The one actual marmalade recipe, given below, is an example of how to include crudely chopped material into the mixture to produce what could be termed marmalade. It does however produce a very similar product to some of the cruder jams described above that use whole hips in their preparation.

Rose hip marmalade (United States) In this method crudely sliced hips are added to syrup and mixed.

Ingredients 1 kg hips; 1 kg sugar; 1 l water.

Method Soak washed rose hips in the water for 2 h to soften the skins. After soaking, bring to a boil in the same water and cook for 15 min. Strain the liquid into a smaller pot and add the sugar. Stir the mixture well then boil it carefully until it reaches a thick syrupy consistency.

Add the previously cooked rose hips to the syrup (you can chop them into chunks or slices). Continue to boil the syrup mixture until the hips are very tender. Pour the mixture into sterilized jars and seal them tight.

Honey

In the seventeenth century Thomas Tryon published the following recipe for a rose honey.

Rose honey (United States, Britain)
Ingredients 230 g rose petals; 2.3 kg honey; 1.7 l water.

Method Cut the white heels from red rose petals. Add boiling water, stir and let the petals soak for at least 12 h. Drain well then add the liquor to the honey. Boil the mixture until it once again has a thick syrupy consistency.

Ice Cream and Sorbet

Rose-flavoured ice cream (New Zealand, United States)
Ingredients 12 g packed rose petals prepared by cutting the white base from each petal; 15 ml lemon juice; 230 ml water; 550 ml sugar; 1 l heavy cream; 1 l whole fresh milk.

Method Blend the rose petals, the lemon juice, water and sugar until liquefied. Add the heavy cream and the milk to the liquefied mixture. Mix well and freeze the mixture until light and fluffy. Remove from freezer, pack in a container with a tight-fitting lid and store in freezer.

Raffles rose ice cream This recipe is based on the 'rose ice cream' served at the Raffles Hotel in Singapore.

Ingredients 1 l vanilla ice cream; 15 ml dried ground-up red rose petals; 15 ml rose syrup; 15 ml rose-water; 10 ml ground cardamom; 45 ml chopped pistachios.

Method Soften ice cream in a large bowl. Stir in the ingredients until well mixed. Re-freeze.

Rose-flavoured sorbet (United States) As an alternative to ice cream on a hot day a flavoursome sorbet can be made from mixture of rose and lemon flavours.

Ingredients 4 scented rose petals; 2 tablespoons rose-water; 3 tablespoons lemon juice; 450 g 5× powdered sugar; 1/2 tablespoons salt; 2 egg whites; 1 l water.

Method Wash the rose petals and boil in water, then simmer for 10 min. Strain the water, save the remains of the petals.

Put the sugar and water in a pan, sprinkle in the gelatine and leave to soak for 10 min. Gently heat the water and stir until the gelatine and sugar have dissolved to a clear syrup. Remove from the heat and stir in the lemon juice, rose-water and soggy petals. Set aside to cool. Freeze the rose mixture uncovered for 1.5 h or until the edges are slushy. Remove from freezer and beat until smooth. Fold the egg whites into the mixture and freeze for at least 3 h.

Omelette

Rose recipes tend to concentrate on the use of roses to flavour sweets; here is a modern savoury use for the rose.

Rose-flavoured omelette (United States)
Ingredients 3 eggs; 45 ml water; 2 ml rose-water; 30 ml chopped chives; 30 ml feta cheese (or other mild cheese); 4 rose petals; salt; sprig of spearmint; butter.

Method Rinse and dry rose petals, cut them (remove white) into slivers with sharp scissors. Blend eggs, water, rose-water, half the rose slivers and chives. Heat butter in thick pan. Once bubbling add the egg mixture. Cook on one side and add the cheese and the remaining petals to the centre of the omelette. Fold in half and serve.

Pickled Rose Hips

This recipe is even further from the use of roses in sweets. Not a popular dish today, it seems to have no modern equivalent. In the seventeenth century, however, it was a popular snack.

Pickled hips (Britain)
Ingredients Rose hips; white wine vinegar; sugar (to taste).

Method Hips were put into an earthenware jar, mixed with white wine vinegar and sugar, stored until soft and ‘pickled rosebuds’ were eaten, possibly as a source of vitamins, during the winter months.

Turkish Delight

Turkish delight flavoured with rose (New Zealand)
A delicious sweet enhanced with the flavour and scent of roses.

Ingredients 350 ml sugar; 180 ml triple-strength rose-water (or to flavour); 30 g plain gelatine powder; 180 ml icing sugar; 120 ml corn starch.

Method Gently heat sugar and rose-water in a deep, heavy pan until sugar is dissolved and bring to the boil for approximately 3 min. Dissolve the gelatine in 50 ml of water and bring to the boil. Add two drops of red food colouring to the sugar. Slowly add the boiling gelatine solution to the boiling sugar and whisk together. Boil for 2 min and set aside to cool. Butter a shallow pan, pour in the mixture and leave to set. Mix the icing sugar and corn starch and spread the mixture on a sheet of waxed paper. Put the Turkish delight onto the paper and cover the block with the sugar mixture. Cut the block into squares and recoat with sugar as required. Store for 2 days before eating.

Salads and Sandwiches

Fresh, fragrant and brightly coloured rose petals can be mixed into party salads by adding the petals from blooms directly (after washing) to any salad. The common salad dressing used for rose salads is a mixture of raspberry-flavoured vinegar and light olive oil, although the rose-flavoured vinegar described below may make an interesting alternative. The petals must be clean and pesticide-free. Rose petals bruise easily, so you might want to use them whole.

To follow on from the use of rose petals in salads they may equally be used in sandwiches. Below is an example but you may wish to experiment further.

Rose-flavoured sandwich (United States)
Ingredients 1 small baguette; 1 large leaf of curly lettuce; 10 rose petals fresh scented to taste (well washed); goat cheese; walnuts or pine nuts; red onion slices; rose and berry dressing (1/2 cup raspberry yogurt and 1/2 teaspoon rose-water).

Method Cut baguette in half and toast lightly. Rinse romaine and rose petals well and pat dry. Place one romaine leaf on 1/2 toasted baguette. Layer rose petals, goat cheese and walnuts. Top with rose and berry dressing. Place the second romaine leaf on the other flute; add onion slices, serve open-faced.

Sugar

A recipe for rose sugar was published in the sixteenth century by Sir Hugh Platt. This recipe required the rose petals to be powdered and mixed with sugar.

Rose-flavoured sugar (Britain)
Ingredients Scented red rose petals (as many as you want); sugar; water.

Method Dry the petals in an oven and beat them into a powder. Boil sugar with a small amount of water, mix in the powdered petals and let the mixture ‘boyle till it be sugar again’. The modern equivalent is to mix the sugar and roses in a food processor, one volume of roses to two volumes of sugar, and let the mixture stand for 3 weeks.

Wine and Vinegar

In the early nineteenth century, a crude but simple recipe for rose wine was described by John Farley. This basically mixed rose water with sugar and let the mixture ferment.

Rose wine (Britain)
Ingredients Rose petals; 1.5 kg sugar; yeast; 4.5 l water.

Method Add boiling water to an equal volume of petals and let them steep for 1 h. Continue to add more petals and steep them in this liquid to ‘take out the whole strength and tincture of the roses’. Mix this liquid with sugar, add the yeast and ferment the mixture for 30 days. Once settled, separate off the clear wine.

Rose vinegar (Britain) In the sixteenth century, a simple method for preparing rose vinegar was to mix vinegar with large amounts of rose petals and leave the mixture to stand in a sunny place over the summer, occasionally adding fresh roses. Below is a more modern version of the method.

Ingredients 250 ml rose petals; 2 sprigs of fresh tarragon; 1 l distilled white vinegar.

Method Place rose petals and fresh tarragon in the vinegar. Place in a sealed container for 3 weeks. Strain the petals from the vinegar. This mixture stores for up to 6 months in a refrigerator.

Conclusions

In the past the petals and fruits of roses have been used in a wide variety of recipes, mainly sweets. These rose-flavoured recipes seemed to be particularly popular in the fifteenth and sixteenth centuries. Perhaps surprisingly, there is still a vast range of modern recipes (not all sweets) in which roses play a major part. Today we recognize that the vitamin C content may be important for nutrition, while the flavour may relate to the volatile molecules responsible for the unique scent of the rose such as the rose ketone β-damascenone. So while older recipes tended to concentrate on the direct use of petals or hips, modern recipes often enhance the rose flavour using rose-water or oils in the recipe and the petals are included as a decorative element. Nevertheless it is obvious that the scent, colours and hips produced by roses remain an important ingredient to a wide variety of recipes from many different countries. So to answer the original question 'Of what use will be a dish of roses to you?', a dish of roses can provide a source of flavour, nutrition and culinary inspiration, enjoyed by many throughout the world.

See also: **Fragrance and Pigments**: Fragrance Profiles of Wild and Cultivated Roses. **History of Roses in Cultivation**: European (pre-1800); Modern (post-1800). **Secondary Metabolites**: Medicinal and Pharmaceutical Uses.

Further Reading

Ager A (1986) *Creative Cooking and Entertaining*. London: Readers Digest.

Cai JT and Ding ZH (1995) Nutrients composition of *Rosa laevigata* fruits. *Science and Technology of the Food Industry* 3: 26–29.

Duke JA (1992) *Handbook of Phytochemical Constituents of GRAS Herbs and Other Economic Plants*. Boca Raton, FL: CRC Press.

Dunnage-Roy A (1997) *Cooking with Roses*. Wellington, New Zealand: Daphne Brasell Associates Limited.

Demir F and Ozcan M (2001) Chemical and technological properties of rose (*Rosa canina* L.) fruits grown wild in Turkey. *Journal of Food Engineering* 47: 333–336.

Hocking GM (1997) *A Dictionary of Natural Products*. Medford, NJ: Plexus Publishing.

Law D (1970) *Herbs for Cooking and Healing*. Stroud, UK: W Foulsham.

Law D (1973) *The Concise Herbal Encyclopedia*. Edinburgh, UK: J Bartholomew and Son.

Riddle JM (1987) 'Folk tradition and folk medicine: recognition of drugs in classical antiquity'. In: Scarborough J (ed.) *Folklore and Folk Medicines*, pp. 33–61. Madison, Wisconsin: American Institute of the History of Pharmacy.

Rohde ES (1973) *Rose Recipes from Olden Times*. New York: Dover.

Webb GP (2002) *Nutrition: A Health Promotion Approach*, 2nd edn. London: Edward Arnold.

Zhao GY, Hou AW and Gao FT (1988) Study on the change of vitamin C content of rose fruit during its storage and processing. *Journal of the North-East Forestry University* 16: 102–105.

Medicinal and Pharmaceutical Uses

R R Cutler, University of East London, London, UK

In the long and colourful history of roses they have been recognized as symbols of love, beauty, war, religion and politics. Less well-known, however, is that they were also known as symbols of medicine.

This article will briefly cover details of the historic use of roses in medicine, the components present in rose petals and rose hips that have been identified as having biological activity and the pharmaceutical uses of extracts from petals and hips.

A Brief History of Medicine and Roses

The rose, according to fossil evidence found in the Colorado Rockies, is at least 35 million years old. In nature, the genus *Rosa* has around 150 species that originally extended throughout the northern hemisphere. The use of plants and plant extracts as medicines can be traced to 50 000–60 000 years ago. So why should ancient humans select roses to prepare medicines? In those times, plants were often selected for their colour or the shape of their leaves, so it would be surprising if roses were not an early candidate. Red plant parts and flowers, for example, were used to treat bleeding disorders (rose extracts are recognized astringents), heart-shaped leaves were used to treat heart problems (rose extracts are used as tonics). This approach was called the doctrine of signatures, where the discovery of medicinal activity was based on trial and error.

Two of the most ancient civilizations with surviving medical texts were those of Egypt and China. The cultivation of roses is thought to have begun over 5000 years ago, possibly in ancient China, although little is known of the history of roses in that country prior to the tenth century (*see* **History of Roses in Cultivation**: Chinese Roses; European (pre-1800)). What we do know is that roses were cultivated in China for many centuries prior to this; however, since the Chinese did not prize the rose as they did the chrysanthemum, there is little documented evidence of their existence. Ancient Chinese texts on medicinal plants state that rose extracts are 'effective for dropsy and constipation' and today *Rosa laevigata* (Cherokee rose) is used in China for treating diarrhoea (although too much can actually cause diarrhoea).

The most complete medical document still in existence is the *Papyrus Ebers* (1550 BC). This work lists a wide range of herbal treatments dating back beyond 2000 BC. Plants with strong aromas were important to Egyptian practitioners who sometimes made little distinction between medicines and perfumes and used aromatic blends for both.

The ancient Greeks were famous as medical practitioners, but much of early Greek medical knowledge was actually imported from Egypt. Famous physicians throughout history have used roses in medicinal mixtures. The Greek Hippocrates (400 BC), one of the first to reject the belief that illness was caused by the supernatural (gods) and to accept the idea that illness is caused by natural factors, an imbalance of 'bodily humours', used rose petals mixed with oil for diseases of the uterus.

During the period of the Roman Empire, roses were grown primarily in the Middle East and used extensively throughout the empire (*see* **History of Roses in Cultivation**: European (pre-1800)). Around AD 55 the Roman physician Dioscorides wrote *De Materia Medica*, an account of over 600 species of plants with medicinal value. This remained a major medical textbook for the next 1600 years. Interestingly, a number of the recipes in *De Materia Medica* are the same as those listed in *Papyrus Ebers*, and prescribed for the same ailments. Many Roman healers used rose extracts and the famous Roman scholar Pliny the Elder recorded numerous ways to extract medicines from roses.

By medieval times roses were used to make all sorts of salves, lotions and medicines to treat anything from hangovers to the plague! For a sore throat, apothecaries would prescribe a hot mixture of crushed rose petals and peppercorns. Sufferers of other ailments might put a pellet made of rose petals on their tongues. The most celebrated medicinal rose of the time was 'the apothecary's rose' (*R. gallica officinalis*, **Figure 1**). The apothecary's rose may have originated in Persia, and was brought to Europe in the twelfth or thirteenth century by knights and the monks who travelled with them, returning from the Crusades. Crusaders used a salve made of red roses to treat wounds.

During these unsettled times, Benedictine monks grew roses in their monastery gardens and used them for medicines. The rosary may have derived its name from the rose hips that were originally strung into prayer beads by monks. Later, rosary beads were made from a fragrant paste of crushed rose petals dried, rolled into beads, and strung into 'rosaries'. This second form may have developed from the fact that rose petals were believed to have a purifying effect against disease and were used as a preventive measure. People of that time often carried nosegays of roses to freshen foul air.

Around the same period, the noted twelfth-century German abbess and herbalist Hildegard von Bingen cited, amongst other herbal treatments, a prescription for rose hip tea. She recommended this tea for a wide range of complaints, including headaches, cramps and dizziness. Treatments based on rose extracts are also mentioned in another major medieval source of pharmaceutical and medical information, the *Compendium of Medicine* (*c.* 1250) by Gilbertus

Figure 1 The apothecary's rose, *Rosa gallica* var. *officinalis*. (Photograph courtesy of Prof. A. V. Roberts.)

Anglicus (Gilbert the Englishman). For a sore throat he recommended a tonic called 'honey rosette'. This was made from mixing a pound of minced roses with an equal amount of honey and heating the mixture over a fire. He also suggested that pale cheeks could be reddened with a lotion of 'brazil-wood chips' soaked in rose water.

Although medieval times were not generally noted as a period when medical knowledge made significant advances, there were selected areas in Europe that became known as centres of importance. The Provence region in France became a major centre for the apothecary's rose in the thirteenth century, so much so that the rose was locally renamed the 'rose de provence'. It was mainly cultivated by monks who were interested in the rose's medicinal properties, but later it was also grown for its perfume and dried for potpourri. In these medieval times, recipes from more ancient times were commonly used. One sixteenth-century recipe that originated from Roman times was to take dried petals, steep them in wine and drink the mixture as a cure for a hangover.

Herbalists of the time considered the red rose, *R. gallica*, to be more binding and more astringent than any of the other species. The famous seventeenth-century physician Nicholas Culpeper (1616–1654; **Figure 2**) had much to say on roses. In his book, *The English Physitian: or an Astrologo-Physical Discourse of the Vulgar Herbs of this Nation*, he

Figure 2 Nicholas Culpeper (1616–1654). (Courtesy of Yale University, Cushing/Whitney Medical Library.)

gives many medicinal uses for the rose – white, red and damask. In one discourse he states:

> Of the Red Roses are usually made many compositions, all serving to sundry good uses, viz. electuary of roses, conserve both moist and dry, which is usually called sugar of roses, syrup of dry roses and honey of roses; the cordial powder called aromatic rosarum, the distilled water of roses, vinegar of roses, ointment and oil of roses and the rose leaves dried are of very great use and effect.

A selection of Culpeper's rose-based medicines and their prescribed uses are given in **Table 1**. Some of the uses suggested by Culpeper are similar to those suggested by Hildegard von Bingen and by Gilbertus Anglicus.

By the nineteenth century, apothecaries predominated throughout Provence. Each shop had the Rose de Provence planted outside as a symbol of the profession.

Remedies containing the apothecary's rose were used to cure indigestion, sore throats, skin rashes and eye maladies. Women of the time believed that the petals could eliminate wrinkles and preserve their youth if rubbed on the skin. An example of a recipe for rose tea from that time was to take five teaspoons of diced rose petals, steep them in four cups of boiling water for 5–10 min then sweeten the drink with honey.

These and many other medicinal uses have been described over the ages for rose extracts. They have been used for treating fatigue, eye diseases, heart diseases, respiratory diseases, insomnia (sleeplessness), stomach troubles (e.g. indigestion) and women's diseases (e.g. period pains).

In the beginning of the nineteenth century apothecaries and scientists started to investigate which components of plant extracts contained bioactive materials and by the end of that century they had discovered that roses contained potassium and iron (which could provide nutritional benefit) and essential oils.

Although the apothecary's rose and other cultivated roses were commonly used in Europe, in North America, Native Americans had used wild roses in herbal medicines for many years. An example of a rose used by Native Indians was the prickly rose or *R. acicularis* (**Figure 3**).

The Blackfoot brewed the roots of roses into a bitter tea that was used in the treatment of diarrhoea. The Blood Indians made poultices out of rose hips which were used to treat boils. They also made rose hips into a decoction (boil in water for 15–20 min) for patients with diarrhoea. The leaves and fruits were also boiled to produce a cure for stomach aches. Possibly the most dramatic cure was to stick rose thorns into painful skin and set them on fire. The thorn then burned down and,

Table 1 Some of the uses of rose extracts prescribed by Culpeper (1616–1654)

Medicines	*Prescribed use in Culpeper's text*	*Summary*
Moist conserve	'is of much use both binding and cordial, the old conserve mixed with *aromaticum rosarum* is a very good preservative in the time of infection'	Astringent (local contraction of blood vessels, reduces secretion), anti-infective
Dry conserve	'called the sugar of roses is a very good cordial against faintings, swoonings, weakness and trembling of the heart, strengthens a weak stomach, promotes digestion and is a very good preservative in the time of infection. The dry conserve called the sugar of roses is a very good cordial to strengthen the heart and spirit'	General tonic, aids digestion, anti-infective
The syrup of roses	'cooleth an over-heated liver and the blood in agues, comforteth the heart and resisteth putrefaction and infection'. Culpeper's syrup was made of the pale red petals of the Damask rose by infusing them 24 h in boiling water, then straining off the liquor and adding twice the weight of refined sugar to it. This syrup was also stated as an 'excellent purge for children and adults of a costive habit, a small quantity to be taken every night'. A conserve of the buds was quoted as having the same properties as the syrup	General tonic, anti-infective
Honey of roses	'is used in gargles and lotions to wash sores, either in the mouth, throat or other parts, both to cleanse and heal them'	Antiseptic, mouthwash
Red rose water	'is well known, it is cooling, cordial, refreshing, quickening the weak and faint spirits, used either in meats or broths to smell at the nose, or to smell the sweet vapours out of a perfume pot, or cast into a hot fire-shovel. It is of much use against the redness and inflammation of the eyes to bathe therewith and the temples of the head'	General tonic, anti-inflammatory eyewash
Rose ointment	'is much used against heat and inflammation of the head, to anoint the forehead and temples and to cool and heal red pimples'	Antipyretic, antiseptic
Oil of roses	'is used to cool hot inflammation or swellings and to bind and stay fluxes of humours to sores and is also put into ointments and plasters that are cooling and binding'	Antipyretic, astringent
The decoction of red roses	'is made with white wine, and when used is very good for headache and pains in the eyes, ears, throat and gums' (decoction – boil in liquid for 10–20 min)	Analgesic (pain relief)

Extracts from *The English Physitian: or an Astrologo-Physical Discourse of the Vulgar Herbs of this Nation* by Nicholas Culpeper.

not surprisingly, a blister was left. This treatment was meant to 'cure' pain. Less dramatically, the Cherokee used an infusion (a tea) made from the bark and/or roots from the swamp rose (*R. palustris*) to treat intestinal infections by worms, diarrhoea and dysentery.

Current Perspectives on the Medicinal Use of Roses

Throughout history herbal practitioners have used various plant tissues for different uses and the same has been true for roses. The petals and bark were used in native medicines, some of which have been mentioned above.

Native Americans used the dried petals of red roses both topically and orally. Rose leaves and mint, heated and applied outwardly to the stomach, were thought to help a weak stomach and, when applied as a paste to the regions of the liver and heart, cooled and soothed them. The historic use of bark by Cherokee is mentioned above.

Today the medicinal use of roses is widespread in herbal medicines from many countries (many of which can be found on the internet) and in complementary therapies such as aromatherapy. These plant extracts can be variously made into teas, syrups, tinctures, tablets and capsules. In allopathic medicine the use of petals is restricted as a colourant, although the use of hip extracts has been used in some countries to treat scars and wounds.

The following sections on the biochemistry and pharmaceutical uses of roses will concentrate on extracts from petals and from hips alone. These are the areas which attract the most attention and, in the case of hips, attract current research into the formulation of new pharmaceutical products.

Figure 3 Rose hips of *Rosa acicularis* (prickly rose).

Biochemistry and Pharmaceutical Uses of Rose Petals

Red rose petals have been cited in nearly all pharmacopoeias. They were originally prescribed for their mild astringency and tonic value, but are used today mainly to give odour and colour to some pharmaceutical preparations.

Potential Bioactive/Medically Useful Agents Found in Red Rose Petals

Red colourant An important constituent of red rose petals is the acidic red colouring matter.

Quercitrin Quercitrin is an edible bioflavonoid also found in many other plants. It is a bitter pale-yellow crystalline glycoside (quercetin-3-L-rhamnoside) which produces quercetin and rhamnose on hydrolysis. In 2002 at the second International Conference of the Nitric Oxide Society, it was reported that quercitrin restores flavin-containing monooxygenase activity, which is normally suppressed by nitric oxide in rats. This research suggests that quercitrin-rich foods such as onions and tomatoes may be beneficial for individuals who suffer from hepatitis, cirrhosis or hepatic cancers with increased production of nitric oxide.

Gallic acid (3,4,5 trihydroxybenzoic acid) The astringency in rose petals is due to a little gallic acid, found in tannins. It has not yet been proved whether quercitannic acid (normally found in the tannin of oak bark) is also a constituent.

Essential oil (*see* **Fragrance and Pigments:** Fragrance Profiles of Wild and Cultivated Roses). The odour in rose petals is due to a very small amount of complex volatile oils, not identical to the official *Ol. rosae*. The principal constituents of the oil are citronellol (34–55%), geraniol, (around 14%) and nerol (around 7%). Other main constituents are paraffin (around 16%), polyethyl alcohol (around 2.8%), farnesol (around 1.2%), rose oxide (a terpenoid ether, around 0.5%), eugenol (also around 1.2%).

Essential oils are produced by steam distillation from fresh petals. In aromatherapy, rose oil is used in the treatment of a wide range of stress-related conditions and can also be used for hyperactivity in children.

Rose oil is recommended by aromatherapists, to be helpful for menopausal women or during heavy menstrual periods. The antiseptic and anti-inflammatory properties of rose oil can be used in the treatment of skin problems, relieving dryness, inflammation, heat and itching of the skin. Rose oils are also used in the treatment of digestive disorders such as gastroenteritis, gastric ulcers, to relieve feelings of nausea and to help regenerate damaged intestinal walls.

There are other claims by manufacturers (mainly unsubstantiated) for the use of rose petal oil, many of which can be traced to historical uses mentioned above. It has been variously described as antiseptic, antispasmodic, antiviral, astringent, bactericidal (kills bacteria), cicitrizant (helps against scarring), haemostatic (astringent), a regulator of appetite and a sedative. It is also a good tonic of the heart, liver, uterus and stomach. However, the oil extract from hips has been shown to have interesting biological properties, so there may be some links between the active agents in rose petal and rose hip oils.

Other components There is a reasonable amount of sugar, gum, fat, iron and potassium in rose petals which could provide some nutritional benefit.

Pharmaceutical Uses of Rose Petals

According to the *British Pharmacopoeia*, red rose petals should be obtained from *R. gallica*. However, there are many cultivars of this species; in fact, few plants of *R. gallica* grow in the wild or in cultivation, only hybrids, so that the exact requirements of the *British Pharmacopoeia* may be difficult to follow.

Plants used in preparations are generally scented roses of a deep red colour, or, when dried produce a deep red tint. Petals suitable for medicinal purposes must yield a deep rose-colour, an astringent and fragrant infusion when boiling water is poured upon them. The most suitable are the so-called hybrid perpetuals, flowering from June to October. Recommended varieties are shown in **Table 2** (**Figure 4**).

Table 2 Cultivars of *Rosa gallica* recommended for medicinal use

Hybrid	*Details of colour and smell*
'Eugène Fürst'	Coloured a deep dark red with a sweet scent
'Général Jacqueminot'	A fine, rich crimson colour with a good scent
'Hugh Dickson'	A large-petalled rose, but of a fine, deep red colour and sweet-scented
'Ulrich Brünner'	Bright red in colour
'Richmond'	Deep crimson red colour
'Liberty'	Scarlet red in colour

Figure 4 Cultivars of *Rosa gallica* recommended for pharmaceutical use, 'Hugh Dickson' and 'Ulrich Brünner'.

In the *British Pharmacopoeia* three main preparations have been described using rose petals: a confection (a sweetened medication), an acid infusion and a fluid extract (**Table 3**). Other formulations included rose water, dried rose petals and rose petal tincture.

Rose water The *British Pharmacopoeia* stated that rose water 'shall be prepared by mixing the distilled rose-water of commerce, obtained mostly from *R. damascena* (but also from *R. centifolia*, the hundred-leaved rose or cabbage rose, and other species), with twice its volume of distilled water immediately before use'. Rose water has been used as a vehicle for other medicines and as an eye lotion.

Triple or stronger rose water is water saturated with volatile oil of rose petals. It is a byproduct in the distillation of oil of roses. The finest rose water is obtained by 'distillation of the fresh petals. It should be clear and colourless, not mucilaginous, and to be of value medicinally must be free from all metallic impurities'. The excess volatile oil is separated from the clear aqueous portion of the distillate and stored in air-tight containers protected from sunlight. Stronger rose water can be diluted with water when rose water is needed.

Rose water ointment, or cold cream, was used as a soothing, cooling application for superficial lesions of the skin. For its preparation, the *British Pharmacopoeia* directed that '0.5 oz. (14.1 g) each of spermaceti and white wax be melted with 9 oz (255 g) of Almond oil, the mixture poured into a warmed mortar and 7 fluid ounces (198 mls) of rose-water and 8 minims (0.000000048 mls) of oil of Rose then incorporated with it'. An alternative ointment from the US

Table 3 British phamacopoeial preparations for rose extracts

Preparation	*Method*
Confection	The confection is made by 'beating 1 lb [453 g] of fresh red rose petals in a stone mortar with 3 lb (1360 g) of sugar'. It was used in pill-making and formerly prescribed for haemorrhage of the lungs and for coughs
	The alternative US pharmaceutical confection can be made by 'rubbing red rose petals, powdered, with heated rose-water, adding gradually fine, white sugar and heating the whole together till thoroughly mixed'
Acid infusion	The acid infusion is made 'from dried, broken-up, red rose petals, diluted with sulphuric acid, sugar and boiling water, infused in a covered vessel for 15 minutes and strained'
	The mixture has a red colour and an agreeable flavour. It was 'employed for its astringent effects in the treatment of stomatitis and pharyngitis'. It was thought to work because it contained aromatic sulphuric acid. This also made it a useful preparation in the treatment of 'nightsweats resulting from depression'
	A nonacid infusion is also made for use as a flavouring for other medicines and as a lotion for ophthalmia, etc.
Fluid extract	The fluid extract was made from 'powdered red rose petals with glycerine and dilute alcohol'. It had a deep red colour and pleasant odour and a 'mildly astringent taste'

Table 4 American pharmacopoeial formulations

Formulations	*Details*
Syrup of red rose	Used to impart a flavour and odour to other syrups and mixtures The syrup is red in colour and has an acid, astringent taste
Honey of roses	Mixture of clarified honey and fluid extract of roses Tastes better than ordinary honey and is an astringent. In olden days (cf. Culpeper), honey of roses was used for sore throats and mouth ulcers It was made by pounding fresh petals in a small quantity of boiling water, filtering and boiling the liquid with honey
Rose vinegar	Used in Europe for headaches caused by hot sun (sunstroke). Rose vinegar is made by steeping dried rose petals in best distilled vinegar (not boiled) Cloths or linen rags are soaked in the liquid and then applied to the head
Rose oil	Also known as otto or attar of rose, oleum rosae or esencia de Rosa Rose oil is made from the flowers of *R. damascena*, *R. centifolia* or other species. It is made by distilling fresh flowers. It is a colourless or yellow liquid with the odour and taste of roses As well as being used in perfumery, it can be used as a flavouring agent An acceptable daily dose is only 0.5 mg day^{-1} kg^{-1} of body weight

Pharmacopoeia contains 12.5 g of cetyl esters wax, 12 g of bees wax, 56 g of almond oil, 500 mg of borax, 2.5 ml of stronger rose water, 16.5 ml of water and 0.2 ml rose oil.

In the various US *Pharmacopoeias* a number of different rose formulations have been documented, including syrup and a honey of roses (**Table 4**) (*see* **Secondary Metabolites:** Culinary Uses and Nutritional Value).

Dried rose petals In modern herbal medicine the flowers of the common red rose, dried, can be given in infusions (teas) (*see* **Secondary Metabolites:** Culinary Uses and Nutritional Value) as a tonic and sometimes in powder for haemorrhage as an astringent.

Rose petal tincture A tincture can be made from dried rose petals by pouring 0.56 l of boiling water on 28 g of the dried petals, adding 15 drops of oil of vitriol and 180–240 g of white sugar, depending on taste. The tincture is red-coloured when strained. In Culpeper's time a rose tincture was recommended for daily use: 'Three or four spoons-full of the tincture taken two or three times a day are considered good for strengthening the stomach and a pleasant remedy in all haemorrhages'.

Potential Bioactive/Medically Useful Agents Found in Rose Hips

Rose hips are the fruit of a rose that develop after the petals have fallen. In Britain during World War II they were used as a source of vitamin C to prevent scurvy because of the shortage of citrus fruit. Rose hips are high in vitamin C.

Potential Bioactive Agents Found in Rose Hips

There are a wide range of potentially bioactive materials that have been identified from rose extracts.

Vitamins Essential to body function, some vitamins cannot be synthesized by the body in sufficient amounts and have to come from the diet. Generally, relatively small amounts of vitamins are needed by the body. Vitamins A, B_3, C (1000–12 500 mg kg^{-1} in hips), D, E and K have all been found in rose hip extracts. Vitamins A, D, E and K are fat-soluble vitamins; vitamins B and C are water soluble. A summary of the use and recommended doses for each of the vitamins are given in **Table 5**.

Pectin (34 000–46 000 mg kg^{-1}) Pectin is a natural fibre that is in many fruits and vegetables. A carbohydrate, a polymer of galacturonic acids and sometimes partially esterified, it is used primarily to treat stomach complaints. Its activity is based on its ability to form particulate solutions. Many current formulations are available in combination with kaolin to treat diarrhoea.

Bioflavonoids (100–3500 mg kg^{-1}) Quercitrin, an edible bioflavonoid found in roses, has been mentioned above. Bioflavonoids are complex compounds found primarily in citrus fruits. It is believed by some that they aid in the absorption of vitamin C (though vitamin C is normally rapidly absorbed by the body). If they do in some manner help the activity of vitamin C, bioflavonoids could assist in boosting the immune system to fight infection and help cellular repair.

One study conducted at the Ministry of Health in Finland hypothesized that bioflavonoids could reduce

Table 5 Vitamins found in roses, their potential uses and recommended daily dose

Vitamin	*Common name*	*Potential use and daily requirement*
A	*Trans*-retinol	Required to maintain epithelial tissues in various parts of the body, especially in the eye. Recommended intake 0.6–0.7 mg kg^{-1} day^{-1} (adult)
B_3	Niacin	A component of nicotinamide adenine dinucleotide (NAD) and NAD phosphate – important body coenzymes. Recommended intake 13–17 mg kg^{-1} day^{-1} (adult)
C	Ascorbic acid	May aid the function of the immune system and is therefore important to help the body fight disease. It also aids in collagen synthesis and aids the absorption of iron. Scurvy is a disease caused by vitamin C deficiency. This is characterized by bleeding gums and poor wound healing. Recommended intake 40 mg kg^{-1} day^{-1} (adult)
D	Calciferol	Important to the development and maintenance of the skeletal system. Essential for hormone synthesis. Lack of vitamin D can result in skeletal abnormalities such as rickets. It can also increase the body's susceptibility to infection. However, if the body is exposed to sufficient sunlight there is no need for this vitamin to be supplemented (0.005 mg kg^{-1} day^{-1})
E	Alpha-tocopherol	In normal people, deficiencies are almost never seen. Its main function is to act as an antioxidant to prevent oxidative damage by free radicals. Safe intake >4 mg kg^{-1} day^{-1}
K	Phylloquinone	Lack of this vitamin can cause bleeding in newborn babies. Human gut bacteria are the human body's main source of this vitamin. Safe intake >0.8 mg kg^{-1} day^{-1}

the risk of heart disease by preventing the build-up of arterial plaque. The sources cited for these substances included citrus fruits and rose hips. Quercitrin may also regulate cell growth and may therefore be partially responsible for the regulation of new cell growth in the body.

Leucoanthocyanins, another bioflavonoid found in hips, are potent antioxidants *in vitro* and may be more effective than vitamin E. It seems that leucoanthocyanins increase the tone and elasticity of capillary walls, making them more resistant to haemorrhages, ruptures and infection. Capillary strength also results from the antioxidant/free radical scavenging of the leucoanthocyanins. In a review in 1994 Leibovitz added to this list of bioactive properties, reporting that 'Anthocyanidins, like catechins, possess anti-inflammatory, antihepatotoxic, cardioprotective, antimicrobial, and immunostimulatory effects'.

Tannins (20 000–27 000 mg kg^{-1}) Rose hips contain about 2% tannin. Condensed tannins are polyphenols, widespread in plants. They are part of the plant's defences against environmental stress. Gallic acid, a main component in hydrolysable tannins, was mentioned above. Tannin formulations can contain glucose in addition to gallic acid. The high concentration of tannin-binding proteins in human saliva may be related to the diet of human ancestors. Long-term ingestion of large quantities of certain tannin-containing plant materials may be carcinogenic and these tannin-binding proteins could be a defence against this.

Tannic acid is a conjugate of glucose and gallic acid residues. It is an astringent and was used in the treatment of burns. Topical use is now restricted to bed sores and minor ulcers. Tannins precipitate alkaloids and proteins and have been used in cases of alkaloid poisoning. Tannins are also antioxidants and, as such, are active free radical scavengers.

Citric acid (<100 mg kg^{-1}) Citric acid is an antioxidant, and a major component of the citric acid cycle, which is a source of biosynthetic precursors and essential for the oxidative metabolism of glucose and other simple sugars.

Malic acid (<100 mg kg^{-1}) Malic acid functions in cosmetic formulations as a pH adjuster; its salt sodium malate functions as a skin-conditioning agent. Malic acid has been reported to be used in more than 50 cosmetic formulations across a range of product types. Malic acid was shown to be relatively nontoxic in acute toxicity studies in animals.

Essential oils These are discussed in the previous section.

Other components Sugars (fructose and sucrose), iron, zinc, calcium and phosphorus may produce some nutritional benefit (*see* **Secondary Metabolites:** Culinary Uses and Nutritional Value).

Pharmaceutical Uses of Rose Hips

Researchers have found that rose hips may contain 400% more vitamin C than oranges. In 1941, more than 200 tons of rose hips from Britain's *R. canina* (dog rose) (**Figure 5**) were collected and made into syrup for civilian and military consumption to prevent scurvy. Different cultivars of rose hips contain different amounts of vitamin C. *Rosa villosa* and *R. canina* supposedly contain the highest amounts, but this may vary according to soil conditions. The importance of vitamin C in the diet is given in **Table 5**.

Pharmaceutically, rose hip preparations come as capsules, tablets, syrup, tincture, teas and cream. Specific indications mentioned in the *British Herbal Pharmacopoeia* are as a dietary supplement, a natural source of vitamin C, with small amounts of A and B. Professor Fluck, in his beautifully illustrated book on *Medicinal Plants*, recommended the use of hips as a tisane (tea infusion) for treating fatigue (use as a tonic). Add two tablespoonfuls of chopped dried seeds to 1 l of cold water and bring to the boil.

Brier hips have also been recommended to produce a diuretic effect without irritating the kidneys and to prevent kidney stones. The method for a typical tea infusion would be: use about 1 g of hips (without seeds) with one cup boiling water.

Recent studies using the hips from two forms of wild roses, *R. canina* and *R. rubiginosa*, have produced some interesting new possibilities for the medicinal use of roses.

Figure 5 *Rosa canina* (dog rose). (Photograph courtesy of Prof. A. V. Roberts.)

Rosa canina (Dog Rose)

As mentioned above, *R. canina* hips are well known to contain a large amount of vitamin C; however, they also contain tannins. A French study in 2002 focused on the polyphenolics (tannins) contained in hips to evaluate their antioxidative properties. Their extracts were found to contain mainly phenolics such as proanthocyanidins and flavonoids. The authors concluded that the antioxidative effects of *R. canina* were due to the presence of polyphenolics and vitamin C.

In clinical studies carried out in Germany in 2001, serum creatinine levels declined in healthy subjects and in test patients with osteoarthritis under rose hip treatment. Additionally, patients suffering from osteoarthritis reported that their physical symptoms improved. Data from a randomized controlled trial indicated not only that pain significantly declined but also that 'flexion of the hip' improved in the active treatment group.

Rose hips could be used as a natural treatment in patients with osteoarthritis. It is still unproven which components are responsible for the anti-inflammatory effects; however, there is some evidence that it is due to the high content of vitamins and bioflavonoids in the extract.

Rosa rubiginosa (also known as *Rosa mosqueta* or Sweet Briar)

Some people in Chile may be surprised to learn that a plant they considered indigenous to the Andes is in fact an ancient import from Europe. Experts have agreed that the rose known locally as *R. mosqueta* is in fact *R. rubiginosa* (**Figure 6**). Oil extracts from this migrant may possess useful medical properties.

A 2-year study in Chile focused on the cicatrizing (scar-reducing) properties of an oil extracted from the

Figure 6 *Rosa rubiginosa*, sometimes called *R. mosqueta*: petals and hips.

Table 6 Suggested uses of rose extracts and possible agents involved

Suggested medical use	*Part of plant*	*Potential active agent found in roses*
Astringent	Petals	Gallic acid
Sweetener/colour	Petals	Sugars and colorant
Antiseptic/antibacterial agent	Petals	Essential oils
Help to prevent heart disease	Hips	Bioflavonoids
Help in skin repair	Hips	Vitamin A
Tonic	Hips	Vitamin C
Antioxidants	Hips	Vitamin C and polyphenolics
Boost for immune system	Hips	Vitamin C and possibly quercitrin

rose hips of *R. rubiginosa*. The oil preparation (now called *Rosa mosqueta* rose hip oil), was applied to 180 patients with surgical, traumatic and postburning scars, as well as to a group suffering from premature ageing. It was reported that the oil produced an effective regenerative action on the skin. Continuous application of the oil helped to reduce scars and wrinkles; to prevent advancement of premature ageing; and to improve natural skin colour and tone.

Since then, numerous studies have been completed at various universities in Chile and in other countries. The positive action of *Rosa mosqueta* rose hip oil in helping to regenerate damaged skin tissues was originally attributed to its high content of unsaturated essential fatty acids, namely oleic (15–20%), linoleic (44–50%) and linolenic (30–35%). These acids were originally thought important because they are components of the cell membrane and precursors of prostaglandins and leucotrienes that provide tissue-regenerating and rejuvenating effects. However, there was not enough evidence to support this theory. More recently, Chilean scientists have identified vitamin A as the component responsible for the pharmacological properties of the oil. It is well recognized that vitamin A (**Table 6**) is important in maintaining epithelial cells. There are known epithelial changes that take place in cases of vitamin A deficiency. In the west, natural and synthetic retinoids have been successfully used in the treatment of skin diseases, including acne and psoriasis. This could be information to back the use of rose hip extracts for skin repair. They have been used topically and orally, but, like vitamin A, oral retinoids can be toxic at high concentrations.

So why is there not a greater use of extracts from European *R. rubiginosa* for skin repair, given the wide-spread use of herbal medicines throughout Europe? It is recognized that roses grown in different soils and conditions, even if they are the same species, can produce hips with different chemical contents. Small amounts of vitamin A are present in most rose hips; however, it is recognized that the vitamin yield from rose buds of the same species can alter dramatically according to their origins and species. Norwegian (*R. canina*) hips have been reported to contain 60 000 mg kg^{-1} of vitamin C. Equivalent English hips contain only 2000 mg kg^{-1}. In addition, some rose hips have been reported to contain up to 5000 units of vitamin A per kg. Possibly, the hips from *R. rubiginosa* grown on the slopes of the Andes have a higher content of useful vitamins than the sweet briar grown in Europe.

Conclusions

These and future studies may show that the medicinal effects of roses may still have some benefit, even to modern medicine. Only time will tell if Herrick knew something that modern science is still learning: 'Gather ye rosebuds while ye may'. But, perhaps scientists would change these immortal words to: 'Gather the rosebuds *where* ye may'. There are many biologically active agents that can be extracted from roses (**Table 6**). Further investigations of the contents of extracts from wild roses growing far from their original habitat, with changes in soil and conditions and potential minor genetic modifications, may still produce some surprises. Only time will tell.

See also: **Classification**: Chemotaxonomy and Molecular Taxonomy. **Fragrance and Pigments**: Fragrance Profiles of Wild and Cultivated Roses. **History of Roses in Cultivation**: European (pre-1800); Modern (post-1800). **Secondary Metabolites**: Culinary Uses and Nutritional Value.

Further Reading

British Herbal Pharmacopoeia (1983) Bournemouth, UK: British Herbal Medicine Association.

British Pharmacopoeia (2000) London: Pharmaceutical Press.

Culpeper N (1992) *Culpeper's Complete Herbal*. London: Bloomsbury Books.

Fluck H (1976) *Medicinal Plants*. Slough, UK: W Foulsham.

Hocking GM (1997) *A Dictionary of Natural Products*. Medford, NJ: Plexus.

Law D (1973) *The Concise Herbal Encyclopedia*. Edinburgh: J Bartholomew.
Leibovitz BE (1994) Polyphenols and Bioflavonoids: the medicines of tomorrow. *Tawnsend Letter for Doctors*, Part 1, p. 310, April, 1994.
Reynolds JEF (1989) *Martindale – The Extra Pharmacopoeia*, 29th edn. London: Pharmaceutical Press.
Riddle JM (1987) Folk tradition and folk medicine: recognition of drugs in classical antiquity. In: Scarborough J (ed.) *Folklore and Folk Medicines*. Madison, Wisconsin: American Institute of the History of Pharmacy.
Robbers JE, Speedie MK and Tyler VE (1996) *Pharmacognosy and Pharmacobiotechnology*. Baltimore, MD: Williams & Wilkins.
Rossnagel K and Willich SN (2001) Value of complementary medicine exemplified by rose hips. *Gesundheitswesen* 6: 412–464.
Rubin S (1974) *Medieval English Medicine*. New York: Barnes & Noble.
Webb GP (2002) *Nutrition: A Health Promotion Approach*, 2nd edn. London: E Arnold.

Extraction and Industrial Processes

H A Collin, University of Liverpool, Liverpool, UK

Introduction

The rose flower is a source of the most expensive perfumes. In addition rose extracts flavour food and drinks (*see* **Secondary Metabolites**: Culinary Uses and Nutritional Value) and the fruit is a rich source of vitamin C (*see* **Secondary Metabolites**: Medicinal and Pharmaceutical Uses). Roses used for extraction of perfume components are grown in a small number of sites about the Mediterranean. Since it is a labour-intensive industry the original production was based on a large number of smallholdings with each person extracting the perfume in mobile field-based stills. These have been gradually replaced by larger centrally placed stills and then by solvent extraction using either petroleum ether or benzene. There is still a very traditional element to the industry and some of the earlier methods of maceration and enfleurage (the process of exposing odourless oils to the scent of fresh flowers) which are thought to retain more accurately the exact perfume from the petals are still in use. The industry has been given a large boost by the expansion of interest in alternative medicine which makes extensive use of rose extracts particularly those produced traditionally. Now in an age of global competition new methods of extraction are being examined to simplify and automate the procedures without losing the quality of the queen of perfumes. In this article the rose varieties used for perfume extraction, the different sites at which the roses are grown and the varied methods of extraction and purification of the extracts will be examined.

Areas of Production of Rose Perfume

The expansion of the Ottoman Empire brought the methods of rose growing and distillation to the Balkans. Here as a result of a fortunate combination of climate and soil, rose growing became concentrated in a town in Bulgaria, Kazanlik, in the Balkan Mountains. On the southern aspects of these mountains the roses are protected from the cold north winds and at an altitude of 300–800 m the summer temperatures are not too high. The soil of well-drained limestone, sand and acid-free loam meant that it is easy to work to the depth required to propagate and grow the plants. This area rapidly surpassed the earlier sites of cultivation in Persia, but the production industry was still characterized by many smallholdings with primitive field-based stills used to extract the rose oil. After the turn of the twentieth century the industry changed to larger more modern distilleries, which then processed the petals supplied by the peasant farmers, but hydrodistillation was still the principle on which the methods of extraction were based. Since growing and collection of rose plants and petals is a very labour-intensive business, the smallholdings continue to produce the petals and supply the distilleries.

Another important centre of the industry is Turkey, which provides rose oil for the perfume industry as well as to the food and pharmaceutical industries. A larger industry became established after the major oil-producing rose species were smuggled out of Bulgaria and were planted in an area of Anatolia. A moist sandy loam at an altitude of 800–1100 m on protected southern facing slopes provided comparable production conditions to those in Bulgaria. Initially perfume production was on mostly small sites and the distillation was field-based. Production was subsequently modernized to compete more effectively with the Bulgarian industry and now the perfumes from Turkey are considered to be comparable to those from Bulgaria.

The French perfume industry is based in southern France. The producers are once again small and sites restricted to the regions of Grasse, Pegomas, Mougin and Montauroux, with the highest yields from altitudes of 1100–1200 m. In Morocco the rose-producing regions are between the Atlas Mountains in the north and the Djebel Sarro in the south. Unlike the other areas the roses in Morocco are grown between trees or as dense hedges surrounding gardens and arable fields. The unopened floral buds are removed from the parent plants, the buds dried then extracted by hydrodistillation or by solvent extraction for the volatiles. In another area near Meknès at an altitude of 800 m the hybrid *Rosa centifolia* has been planted in a loamy sandy soil as dispersed plants, rather similar to the way in which grapevines are planted, which allowed for much easier access to the plants. Petals were collected and extracted by modern centrally placed distilleries. Although initially the product was considered inferior to that of Bulgaria, the area is now renowned for the production of high-quality rose oil.

Varieties Used for Perfume Extraction

Of the many varieties of roses known, only relatively few exhibit the marked fragrance required for commercial extraction. Most of the fragrance is found almost exclusively in the so-called Centifolia group of cultivars of the genus *Rosa* and of these only three are exploited commercially for the extraction of their perfume. The most popular variety for extraction of rose oil is *R. damascena* which according to the rose taxonomists resides in the section *Gallicanae*. This section contains the parent species *R. gallica* which is found in both wild and cultivated forms and hybridizes readily with other species in the section. *Rosa damascena* differs from *R. gallica* by being taller and more robust, the prickles are all the same size, the inflorescence is many-flowered and the fruit oblong not globose. The *R. damascena* also shows recurrent flowering. It is supposed that the *R. damascena* may have arisen through a chance hybridization between *R. gallica* and a local species such as *R. phoenicia* and the hybrid was then propagated vegetatively and selected over a period of time. Another theory proposes that *R. damascena* was a hybrid between *R. gallica* and *R. canina*. The hybrid has now escaped cultivation and plants can be found growing wild in the Caucasus, Syria, Morocco and Andalusia. *Rosa damascena* as the most important of all the perfumery roses; it is planted extensively in Bulgaria and to a smaller extent in Anatolia in Turkey.

Of lesser importance is the *R. alba* L. or *R. damascena* Mill. var. *alba*, the so-called 'white cottage rose'; this rose contains less volatile oil than *R. damascena* and the oil is also of lower quality. Being hardier and more resistant to unfavourable climatic factors than the pink damask rose, the white cottage rose is grown in Bulgaria as a hedge around the pink roses and in the higher altitude where the *R. damascena* does not grow well. The rose *R. centifolia*, the so-called 'light pink cabbage Rose', is grown extensively in the Grasse region of southern France and in Morocco. The flower is referred as the 'Rose de Mai' which, because of its smaller amount of oil, is not extracted for oil but is used primarily as a source of rose-water.

Planting and Harvesting of Roses

The planting pattern of perfume roses does show some variation between sites, but there is a certain similarity in the way they are cultivated. A description of the procedures in Bulgaria would be sufficiently representative. The ground used for cultivation of the rose is deep ploughed then parallel ditches 1 m deep and 40–50 cm wide and 2.5 m apart are dug in the autumn. The exposed soil is aerated and broken down by the winter frosts, then part of the fine tilth is placed back in the ditch. Cuttings from an older plant are placed horizontally in four rows followed by 5 cm of fine soil then 5 cm of mature stable manure. In May the shoots appear and more soil is added into the ditches. During the first year weed removal and addition of soil are repeated at intervals. In the second and third year the land is kept weed free and ploughed. The first harvest is gathered in the third year when it can produce about 1500 kg ha^{-1} of flowers. From then on the harvest will increase to 5000 kg ha^{-1}. After 10 years the plantation may show signs of decay when the bushes are cut down to the ground to rejuvenate. Given this treatment a field may remain productive for 30 to 40 years before the bushes are replaced.

The flowers begin to bloom in the second half of May with the exact timing being determined by the temperature and rainfall. Harvesting is begun when the flowers begin to open and continues until all the flowers are removed. Picking starts early in the morning and continues until 8 or 9 a.m. Flowers are collected by hand in the early morning when they are clipped off just below the calyx and dropped into bags which will hold about 25–30 kg. These are transported to the distillery. If in order to remove all the flowers, harvesting has to continue after 9 a.m. when the dew has left the plant the yield and quality of the oil is poor.

Mild and humid weather prolongs the flowering and the flowers produce more oil of better quality. The more prolonged harvesting period also makes it easier to manage the large quantities of petals. In hot dry weather the harvesting period is much shorter and the oil tends to evaporate more readily from the petals. The shorter harvesting period also means that large quantities of petals are delivered to the distillery when the amounts may be too large for the distillery to cope with rapidly. The delay leads to fermentation and loss of oil from the petals.

Extraction by Distillation

Field Stills

Field stills are small portable stills simple in design that can be set up in the fields near a source of running water. These stills which can accommodate 120–150 l of water and 15–20 kg of roses are made of copper, tinned on the inside and heated by wood-burning open fires. The retort, which is wider at the base, has a removable head with a goose neck and this leads to a pipe which passes through a barrel of warm water to condense the distillate. The condenser tube ends above a glass bottle in which the condensate is collected. During the distillation 20–30 l of water is distilled over initially and no oil is separated. The distillate from several stills is combined to give about 100 l then this volume is redistilled (cohobated) slowly until about 8 l is collected in a long slender bottle then the next 4 l in another similarly shaped bottle. The first and second collections are allowed to stand for a period to separate the oil, then the oil decanted from the top. The remaining water from the distillate referred to as the first and second waters are combined, added to another still and redistilled. The oil from the mixed distillate is referred to as 'peasant oil'. The residual water remaining after the first extraction of the petals is separated from the petals and added to a new batch of roses. Using this technique of cohobation there never is any loss of oil. Although a higher yield of oil is achieved in these stills compared with the more modern ones, it is at the expense of quality since there is a carry-over of decomposition products and higher boiling point compounds with the repetitive redistillation.

Direct Fire Stills

The same principle of hydrodistillation and cohobation applies to the larger more modern stills. Rose petals cannot be distilled by injecting live steam directly onto the petals, since they would agglutinate and form an impenetrable mass. The petals have to float freely on the surface of boiling water and although they still tend to form a mass the petals are processed effectively by the boiling water. A more efficient system would be to have stirrers or baffles to break up the mass. Since the rose oil is relatively soluble in water the distilled water must again be redistilled or cohobated in order to obtain the oil.

Most of the industrial stills are of the same basic design as the field stills, but larger, i.e. 1000–2000 l capacity, and they are heated by an open fire. Because they do not make use of steam they are of simple and inexpensive construction. The stills are made of copper heavily tinned on the inside and insulated with bricks to permit an even heat distribution and retention of the temperature. The still top has a large hole through which the flower petals and water are added. A grid 10 cm above the base prevents the petals from coming into contact with the hot lower surface. At the level of the grid is another covered hole through which the extracted flowers and residual water can be discharged rapidly after completion of the distillation process. The still also has a pipe and valve to allow the residual waters to be drawn off when the still is used for cohobation. Retorts of 1200–1500 l capacity are filled with 150–300 kg of roses then covered with 1200 l, i.e. 4–8 times their weight in water. Once the still is charged with flowers, the mass is stirred thoroughly then the top closed and the fire started. After about 1.5 h distillation starts and continues for another 1.5 h. It is important to keep the heat low initially so as to avoid a rapid expansion of the air in the retort, which would take the most volatile and valuable components away too fast and they would escape before condensing. The water inlet for the condenser is kept at a temperature of 35–40 °C. At lower temperatures stearoptene in the oil will solidify and block the condenser. Distillation of the flowers is complete under these conditions when approximately 140 l per 250 kg has distilled over. The small quantity of greenish oil collected in the receiver flask is called 'direct oil'. This is drawn off with a pipette and poured into a glass bottle; the remaining 140 l of water, i.e. 'first waters' are pumped into a storage tank and bulked with other first waters. The exhausted flowers and the residual water with them are thrown away. To isolate the remaining oil in the distilled water, the combined first waters are redistilled. A total of 1200 l of the combined first waters is redistilled in a retort of 1800 l and 140 l slowly distilled off. The temperature of the water is again kept low. The process takes 1.5 h to heat the water and 1 h for redistillation. The condensate of 140 l of 'second water' and a small volume of oil, 'water oil', is pipetted off as before and added to the direct oil. These two combined oils provide the commercial rose oil. The combined oil in glass bottles is exposed to the sunlight for several days, which cause the impurities and colloidal matter to sink and water to separate. The clear oil is decanted, filtered and stored in tinned copper containers. In general this oil consists of 25% direct oil and 75% water oil. The 140 ml of redistilled water, left after removing the water oil, is pumped into the storage tank holding the first waters. The residual water left in the retort after the second waters has been distilled off is kept in the retort and used for the distillation of a new batch of roses. The only part of the process that is removed from this cycle is the exhausted flowers, the residual water left after extracting them and the essential oil.

Steam Stills

These are similar to the direct-fire stills except that a steamjacket is used to heat the retort and a steam coil is used for the injection of live steam into the boiling still (**Figure 1**). The flowers must still be covered with boiling water and be free to move within it. The use of steam allows for quicker heating of the contents of the retort and for a more efficient control of the distillation (**Figure 2**). The live steam also stirs the bulk of the petals. A modification of this is to have a number of baskets on a horizontal axis so that the roses are inundated by the water alternating with the steam. The flowers are kept moving and thus reduce the possibility of agglutination. This permits the use of less water and reduces the time taken for the distillation. The quality of the oil is excellent with considerably less stearoptene.

Maceration and Enfleurage

An earlier process for abstracting volatiles from rose petals which is little used today is maceration. The method relies on the ability of boiling fat to absorb the volatiles. Alternatively the volatiles are absorbed into cold fat when the process is referred to as enfleurage and this does seem to still have a place in perfume production. A purified fat base suitable for both maceration and enfleurage is prepared by mixing one part of highly purified tallow and two parts of lard. The crude fat is cleaned by hand then crushed mechanically and beaten in a current of cold water. After removal of all impurities the fat is slowly melted over a steam bath. Small quantities of benzoin (0.6%) and alum (0.15–0.30%) are then added. The benzoin acts as a preservative while the alum causes the impurities to coagulate so that they can be skimmed off. The warm fat is filtered through cloth then left to cool so that water separates. The fat (corps) has a smooth white uniform consistency free of water and is odourless.

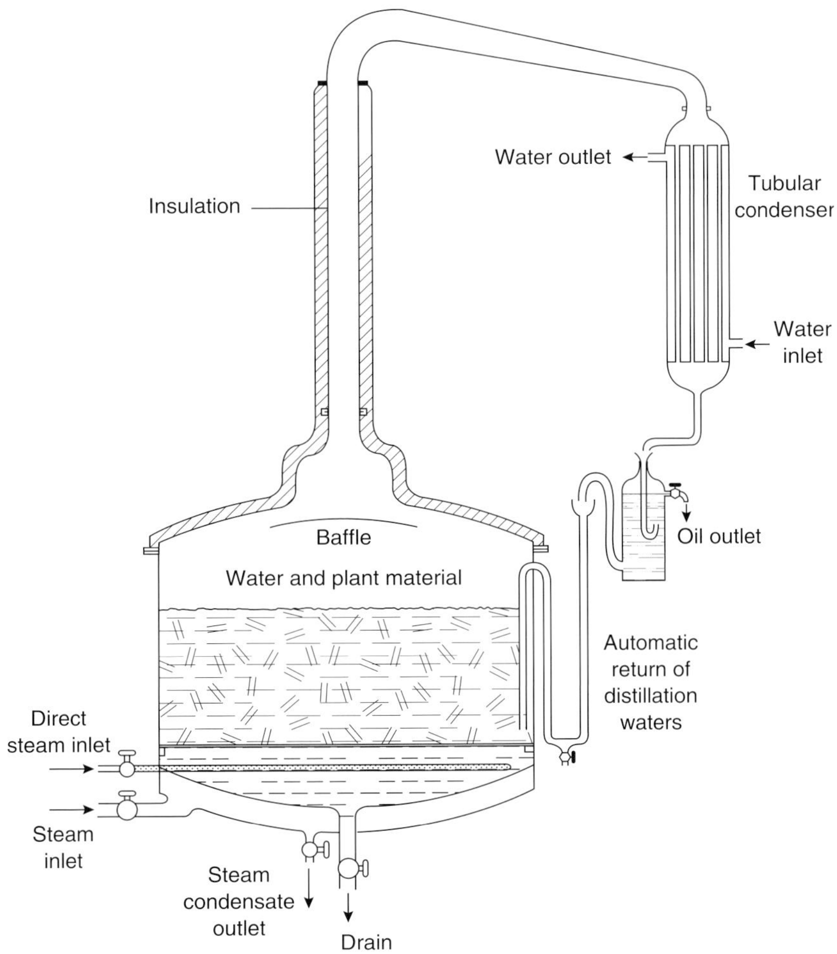

Figure 1 Diagram of steam-heated still for distillation of rose oil. Reproduced from Guenther *The Essential Oils* Volume I with permission.

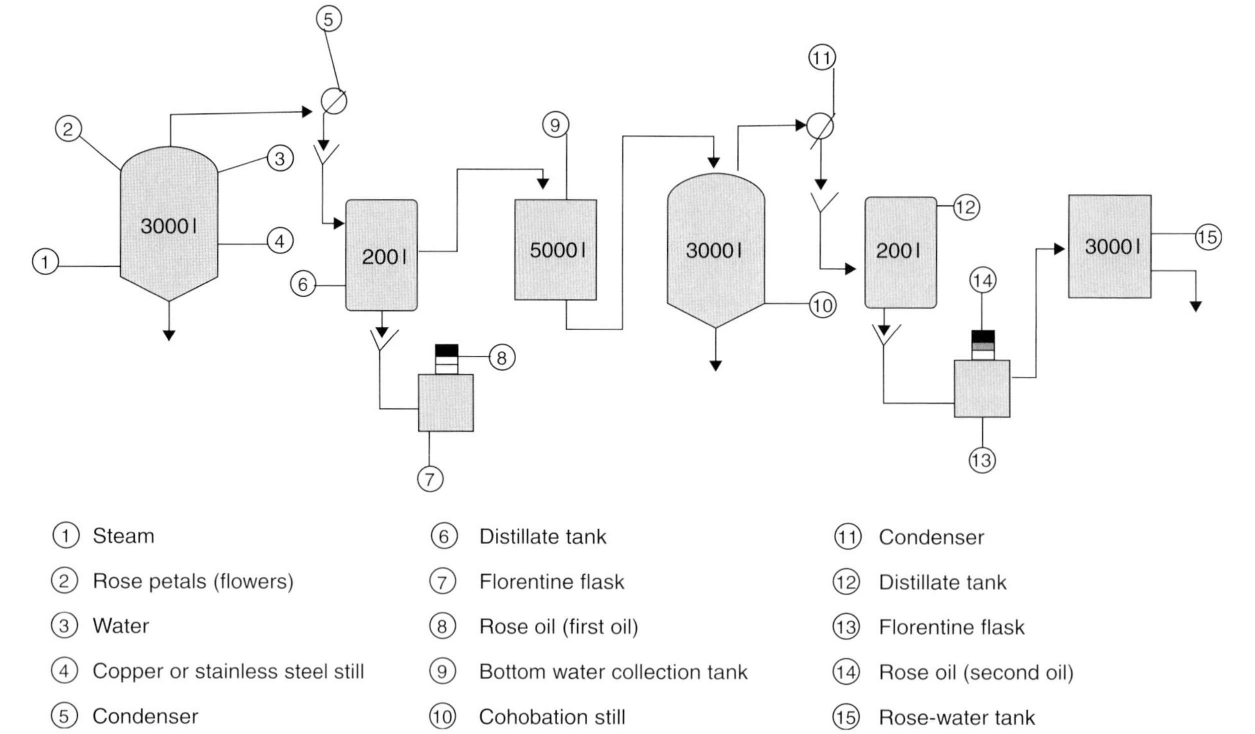

Figure 2 Flow diagram for rose oil extraction by distillation. Reproduced from Guenther *The Essential Oils* Volume I with permission.

In the hot extraction repeated batches of fresh flowers are immersed in the same hot fat until the fat becomes saturated with flower perfume. In a typical extraction a batch of 80 kg of corps is heated to 80 °C and at that temperature is charged with 20 kg flowers and left at that temperature for 0.5 h. This is repeated until 2–2.5 kg flowers have been extracted for every 1 kg of corps. The mass is left for 1 h then reheated. The melted mass is strained through metal sieves to remove the debris. Residual fat on the sieves is removed by boiling water. The extracted flowers are bagged and placed in a press and after subjecting the bags to high pressure the squeezed out fat is dissolved in boiling water. The cooled and separated fat is called 'pomade de rose', and is sold as such. The pomade is cold-extracted by stirring with alcohol under cool conditions in closed copper tin-lined vessels called batteuses. The pomade is repeatedly extracted with alcohol until the alcohol is saturated with flower oil and the corps odourless. The latter is normally used for making soap. The saturated alcohol is reduced *in vacuo* at a temperature of −15 °C and filtered to remove the fat when it becomes 'extract de rose' or the alcohol is reduced in volume *in vacuo* giving the 'absolute'.

In the much more laborious enfleurage, the petals are placed by hand on glass plates whose both surfaces have been smeared with a layer of the corps. The individual plates are contained in a wooden frame referred to as a chassis and these can be stacked in piles. The petals stuck to the upper surface of the glass are left for 24 h, when the volatiles released by the petals are absorbed by the fat layer above them, the petals are removed, the plate is turned over and the new upper surface charged with fresh petals. The plate is turned over daily so that each layer of corps becomes alternatively an absorbing and an immobilizing layer. Since no heat is applied during the extraction the fat contains the flower oils as emitted by the living flowers. At the end of the harvest the pomade is cold-extracted by stirring with alcohol as in the hot fat extraction process to give the concentrate of pomade and absolute.

Extraction with Volatile Solvents

The volatile compounds in rose flowers can be removed more efficiently by extraction with solvents. Now the quantities of perfume prepared in this way exceed the amount produced by hydrodistillation. Major areas of production are from *R. centifolia* in the Grasse region of France and to a smaller extent Morocco. Distillation of this variety only yields a small amount of oil whereas petals from the much more odour-rich *R. damascena* in Bulgaria are both hydrodistilled and solvent extracted. In the solvent extraction fresh flowers are added to specially constructed extractors which are run at room temperature with a solvent, either petroleum ether or benzene.

Before use both solvents are purified to remove sulphur and nitrogenous compounds by repeated washing in turn with strong sulphuric acid, water, hot sodium hydroxide, water then drying. The benzene is further purified by repeated crystallization. The criterium for purity is that if 50 ml of the solvent is evaporated at 40 °C it should leave no residue. Although petroleum ether is the best solvent since it is chemically inert, removes the perfume volatiles very readily and leaves a minimum residue, it has the disadvantage of low boiling point and ready flammability. Benzene is ranked second. Its disadvantage is that its high boiling point make it difficult to remove and it also extracts large quantities of wax and colouring matter. Ether tends to be used for the higher-priced perfumes and benzene for the lower.

During the extraction the solvent penetrates the flowers and dissolves the natural flower perfume together with some waxes and colouring matter. The extract is then concentrated at low temperature when the solvent is driven off leaving the extract. The extraction temperature is kept low at 60–70 °C to avoid breakdown of the components so that the extracted flower oils represent the natural perfume as closely as possible. The disadvantages of the solvent extraction method is that the use of ether requires much more sophisticated machinery and trained personnel to run it, whereas hydrodistillation is relatively simple. The other problem is that all extracted oils are darker in colour and require a much higher proof alcohol to dissolve them. The higher costs of the solvent extraction is worthwhile since the rose is a high-value product.

Stationary Solvent Extractor

Solvent extraction is undertaken in either a stationary or a rotating extractor. The stationary extractors have a capacity of 1200 l and the ability to accommodate 180 kg flowers with a ratio of 400–450 l of solvent per 100 kg flowers. The stationary extractors are vertical with several perforated metal grids arranged horizontally around a vertical support shaft. The flowers are spread over the grids to avoid lumping and so that the solvent can penetrate the mass. After closing the extractor the flowers are washed three times with solvent. The third washing is used as the second washing for the next flower batch then as the first washing and is finally pumped into the evaporator. The solvent is distilled off and used as fresh solvent for a third washing which serves as second of the next flower batch, etc. A fourth extraction yields only small quantities of wax and other inert substances. The actual flower oil is in the first and second washes while the third wash serves as a wash for the remains of the second wash. After the third washing the flowers are exhausted and can be discarded (**Figure 3**). The solvent on them is regained by steam distillation, i.e. by blowing live steam through the mass then separating water and solvent in specially constructed Florentine flasks. The first washing takes 45 min, the second 35 min and the third 25 min. Including the period for transfer and steam distillation the complete batch extraction takes 4.5–5 h.

Rotary Solvent Extractor

The rotating extractor is made up of a heavily tinned iron drum which rotates around a horizontal axle. Perforated baffles divide the interior into four parts which are filled by flowers via separate manholes. The whole system then rotates and dips and drains the flowers alternatively in the solvent at the bottom of the vessel. Three extractions are made in the same way as for the stationary extractor. The first and the most concentrated is pumped into the evaporator and concentrated by distilling off the solvent. After the third washing has been drawn off the exhausted flower material has steam blown over it to recover the ether. The advantages of the rotary system is that the action of the solvent is more penetrating and more effective, resulting in a higher yield. Much less solvent is in circulation, only 160–170 l is required for each 100 kg flowers and evaporation losses are reduced. One rotary extractor does the work of three or four stationary extractors in a row. The major disadvantage is that it cannot accommodate such large quantities of material at any one time.

Production of Concrete and Absolute

The solvent extract from the petals is evaporated in a still heated by a steam jacket. The recovered solvent serves as a fresh solvent for the third washing of the original solvent extraction, while the concentrated solution is transferred to a smaller still and concentrated completely *in vacuo*. The concentration is done at as low a temperature as possible and all traces of the solvent are eliminated. The remaining solid is the floral 'concrete' which still contains the natural pigments plus all the waxes and colourings. The concrete is extracted with alcohol to remove the waxes and to concentrate it. After dissolving the concrete in warm alcohol, eight to ten times the volume of alcohol is added and the amount stirred in a batteuse. Usually the concrete has five to six washings, i.e. a third washing serves as a second one for a following batch of concrete, the first washing subsequently representing the most concentrated solution. The alcohol of the first washing is drawn off from the wax, cooled to

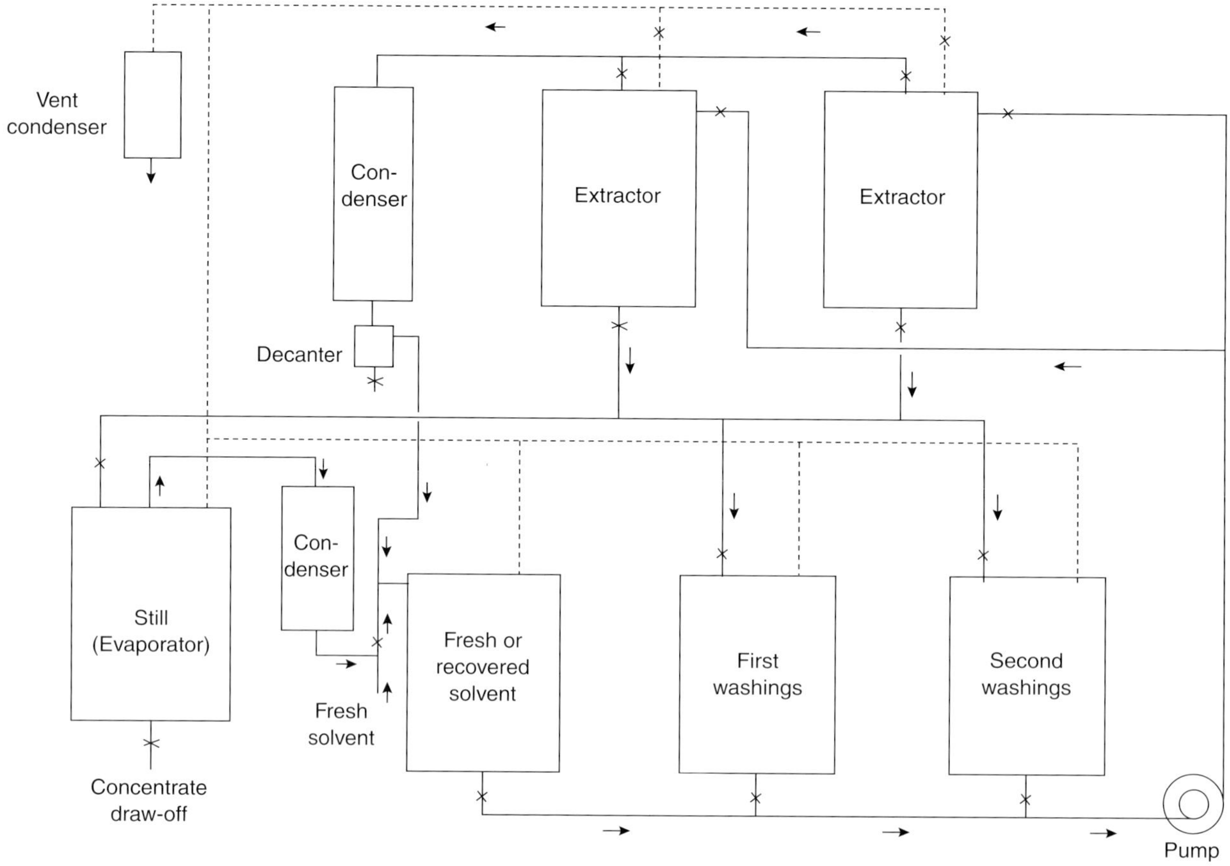

Figure 3 Flow diagram for rose oil extraction by solvent showing flow path (solid lines) and safety vents (dotted lines) Reproduced with permission from Agaoglu Y S (2000).

20–25 °C when more wax precipitates and is filtered to give the absolute. The clear solution can be used directly in alcoholic perfumes. The liquid absolutes represent a further concentration of this alcoholic solution. The alcohol is distilled off which causes some loss of the odorous principles but it provides a more convenient source of perfume than the concrete. The absolutes are viscous oils with a pronounced colour, are soluble in high proof alcohol and represent the most concentrated oil used in commerce. The absolutes still contain about 50–80% soluble wax and 20–25% volatile oils, which can be isolated by steam distillation. Upon ageing concretes and absolutes usually acquire a reddish colour. There is an improvement upon ageing then the product will deteriorate acquiring a rancid acid note which is due to the formation of acetic acid and ethyl acetate.

Yields of Perfume

The highest yields were from the old-fashioned field stills; only 2500–3200 kg of flowers were required to produce 1 kg of oil. This high yield is due to the fact that in the field stills the water remaining after the initial distillation was often reused. The peasant oils derived from this distillation had a weaker honeylike and less pungent odour than the oils from the modern direct-heated stills. The steam-heated stills produced a yield of oil from *R. damascena* of 1 kg of oil per 2600–8000 kg flowers, with 4000 kg required under average growing conditions. Directly fired stills produce a slightly higher yield than the steam stills but with a less powerful aroma. The oil from *R. damascena* is a light yellow occasionally slightly greenish mass, semisolid at room temperature. The odour of the oil is powerful and characteristic of fresh roses; the flavour is strong and sweet and somewhat honeylike.

Roses grown in the Grasse region of southern France are *R. centifolia*, a light-coloured pink rose. A small proportion of the harvested roses are hydrodistilled to give the fragrant rose-water of the region. To produce the rose-water, 1000 kg flowers are placed in a still and boiled until 1000 l of water has been distilled over. The water is not redistilled and represents the commercial rose-water. In the process of distillation about 100 g of direct oil can be separated off. This represents a yield of 1 kg of rose oil per 10 000 kg of *R. centifolia* compared with 1 kg of rose oil from 4000 kg of *R. damascena*. The higher yields of the latter represent the contribution that redistillation makes to the yield. Because rose-water is a solution it has to be stored in clean containers in a dark cool place.

An amount of 400–500 kg *R. centifolia* is required to produce 1 kg of concrete, from which 500–600 g of absolute is obtained. Concrete of rose obtained from *R. centifolia* by extraction with petroleum ether is a waxy light to dark brown mass with a characteristic and strong rose odour. Compared with the Bulgarian product, the French oil is softer and more spicy. The corresponding absolute is a yellowish to light brown viscous liquid with a strong and lasting rose odour. About 400–450 kg *R. damascena* is required to yield 1 kg of rose concrete which in turn gives 520 g of alcohol-soluble absolute. Concrete of *R. damascena* is a light to dark brown waxy mass and is only partially soluble in alcohol. The absolute is a viscous greenish-brown liquid soluble in alcohol. Its odour is strong and characteristic of rose but softer, more mellow and much more lasting than the distilled oil. The odour of extracted rose oil differs from that of distilled products since the extract still contains the natural waxes phenyl alcohol and eugenol. During hydrodistillation these compounds remain largely in the residual petals or water and are thrown away.

While the composition of the oil may vary from place to place and with rose variety the major compounds in the oil, in order of importance are constant. These are: L-citronellol, geraniol, L-linanool and phenylethyl alcohol (the amounts in distilled oil are small because of its solubility in water, whereas the extracted concretes and absolutes contain large quantities of phenylethyl alcohol); farnesol, esters of phenyl ethylalcohol, nonyl aldehyde and other higher aldehydes; citral, eugenol, eugenol methyl ether, carvone and sesquiterpenes such as stearoptene. There are also a large number of minor compounds that contribute to the odour of the oil. The principle components of the absolute are phenylethyl alcohol, geraniol, citronellol, nerol, farnesol and small quantities of eugenol, and eugenol acetate.

A mixture of the major compounds in the same proportion as rose oil would have some similarity to the original rose oil, but it is the trace compounds that provide the characteristic and powerful original odour. These are numerous and as yet not able to be reproduced in a synthetic perfume.

Modern Developments in Extraction Methods

Steam distillation and solvent extraction are the two most commonly used techniques. The disadvantage of the former is that the high temperatures of the boiling water may causes losses in the thermolabile compounds in the flowers, whereas solvent extraction can leave solvent residues in the extract with an associated toxicological risk. Solvent extraction shows low selectivity, which means that apart from the desirable volatiles there are also high molecular weight non-volatile components such as fatty oils, waxes and colouring matters. Both techniques have a long extraction time with little possibility of the processes being able to be automated. Modern methods have sought to overcome these disadvantages. The two most appropriate are continuous subcritical water extraction (CSWE) and supercritical-CO_2 extraction (SC-CO_2).

Continuous Subcritical Water Extraction (CSWE)

In essence the effectiveness of CSWE is based on the use of water as an extractant when used at temperatures of 100–374 °C and high pressures to maintain the liquid state. The dielectric constant of water as an effective leaching agent can be manipulated by temperature and pressure. At ambient temperatures and pressures water has a dielectric constant of about 80 making it an extremely polar solvent. At a temperature of 250 °C and a pressure of 40 bars the dielectric constant is 27, which is similar to ethanol and allows for leaching of low polarity compounds. When used for the extraction of essential oils at a temperature of 125–175 °C it can extract fragrance and flavour compounds from rosemary for example and very little of the undesirable waxes and lipids. The ratio of water to plant material and the temperature of the water were found to be the key variables in determining the composition and quality of the essential oil. The extraction apparatus consists of a water reservoir, a high-pressure pump, an extractor with a stainless steel cylinder extraction chamber, which is closed with screws at either end that permits the circulation of leaching fluid through them. This chamber with a stainless steel preheater is located in an oven designed to work at temperatures up to 300 °C. A cooler system is used to cool the fluid from the oven. A variable restrictor allows the desired pressure to be maintained in the system.

There are two major drawbacks with the CSWE technique. These are the reactive nature of water under the extraction conditions of CSWE and the high temperature of the process both of which could lead to decomposition of the rose oil. The water has to be ultra pure and degased before use which will reduce its reactivity. To assess the possibility of temperature induced loss, an analysis of the extract composition after CSWE extraction needs to be compared with the composition of oil from the standard hydrodistillation or solvent extraction.

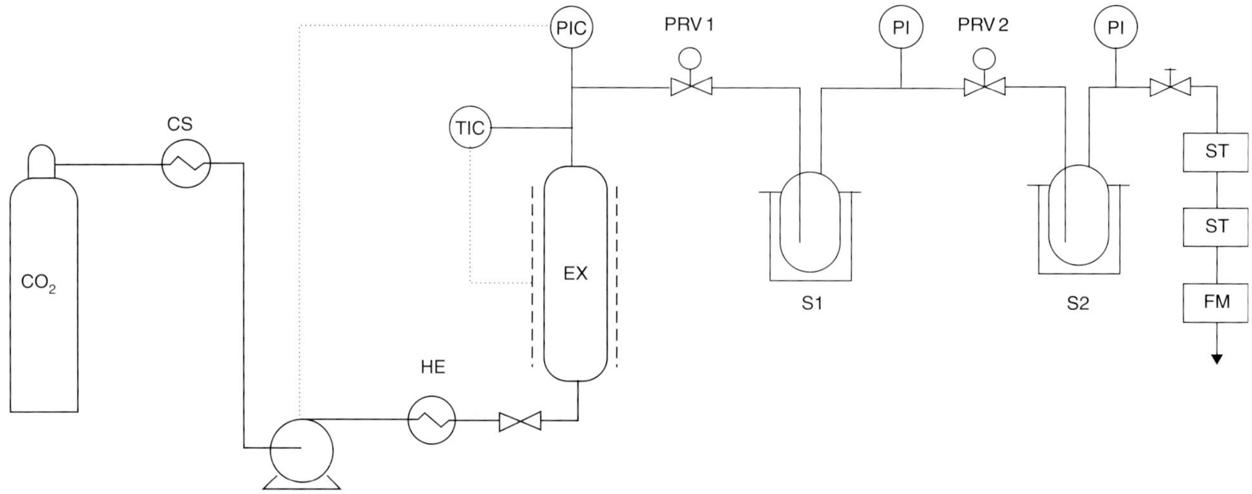

Figure 4 Flow diagram of a supercritical-CO_2 system suitable for extraction of rose oil. CO_2, CO_2 cylinder; CS, cooling system; HE, heat exchanger; EX, extractor; PRV, pressure regulating valve; FM, flow meter; S, separator; ST, sorbent trap. Reproduced from Rincón J, De Lucas A and Gracia I (2000) Isolation of rock rose essential oil using supercritical CO_2 extraction. *Separation Science and Technology* 35: 2745–2763.

Supercritical-CO_2 Extraction (SC-CO_2)

The technique of SC-CO_2 is a modification of supercritical fluid extraction, which is based on the enhanced solvating power of fluids above their critical point. Most studies have been based on the use of CO_2 because of its critical powers, low toxicity and chemical inertness. This technique overcomes the possible loss of thermolabile compounds that might occur during the established hydrodistillation process. In the extraction process (**Figure 4**) liquid CO_2 is provided from a steel cylinder and after being cooled in a cold bath (about 0 °C), it is compressed by a membrane dosage pump. The CO_2 passes through a tubular heat exchanger then to a 2-l stainless steel cylinder containing the material to be extracted. After leaving the extractor the CO_2 loaded with extracted compounds flows through separator 1 where waxes are precipitated then through extractor 2 where the oil is collected. Finally the depressured CO_2 is vented through sorbent traps. At the low temperatures (30–60 °C) and pressures (8–10 MPa) used it is unlikely that decomposition would occur.

Selective extraction to reduce the inclusion of waxes can be obtained by varying the pressure or temperature although this selectivity is only able to exclude the fatty acids, colouring matter and resins. Because the cuticular waxes are soluble in CO_2 these will be included in the extractant and are removed by fractional separation in a separate stage. Another more serious limitation is the need for the material to be dry and finely divided. The drying process would lead to loss of volatiles from the petals.

The advantage of the CSWE over the SC-CO_2 method is that there is no extraction of the undesirable cuticular waxes, and it is unnecessary to dry the material before use. In addition to this cost saving, the equipment in the CSWE method is cheaper, as the pressures are much lower, and there is no outlay on very expensive supercritical grade CO_2. The CSWE would appear to be the better technique for the extraction of rose oil. It has the added advantage of short extraction time, good reproduction of the aromas of the volatile oils examined so far and the potential for automation. However as yet hydrodistillation and solvent extraction are still the preferred industrial methods of extraction.

See also: **Fragrance and Pigments**: Fragrance Profiles of Wild and Cultivated Roses. **History of Roses in Cultivation**: European (pre-1800); Modern (post-1800); History of the Perfume Industry. **Secondary Metabolites**: Culinary Uses and Nutritional Value; Medicinal and Pharmaceutical Uses.

Further Reading

Agaoglu YS (2000) Rose oil industry and the production of oil rose (*Rosa damascena* Mill.) in Turkey. *Biotechnology and Biotechnology Equipment* 14: 8–15.

Grieve M (1998) Roses. In: Leyel CF (ed.) *A Modern Herbal*, pp. 683–694. London: Tiger Books.

Guenther E (1948) The production of essential oils. In: Guenther E (ed.) *The Essential Oils*, pp. 87–226. Princeton, NJ: Van Nostrand.

Guenther E (1952) Essential oils of the plant family Rosaceae. In: Guenther E (ed.) *The Essentials Oils*, 2nd edn, pp. 3–48. Princeton, NJ: Van Nostrand.

Jarvis AP and Morgan D (1997) Isolation of plant products by supercritical-fluid extraction. *Phytochemical Analysis* 8: 217–222.

Luque de Castro MD, Jiménez-Carmona MM and Fernández-Perez V (1999) Towards more rational techniques for the isolation of valuable essential oils from the plants. *Trends in Analytical Chemistry* 18: 708–716.

Rincón J, De Lucas A and Gracia I (2000) Isolation of rock rose essential oil using supercritical CO_2 extraction. *Separation Science and Technology* 35: 2745–2763.

Widrlechner MP (1981) History and utilization of *Rosa damascena*. *Economic Botany* 35: 42–58.

Selection Strategies *see* **Breeding**: Selection Strategies for Cut Roses; Selection Strategies for Pot Roses; Selection Strategies for Disease and Pest Resistance.

Soil Sickness *see* **Disease**: Replant Problems and Soil Sickness.

Sports *see* **Genetics**: Mutation.

T

Taxonomy *see* **Classification**: Conventional Taxonomy (Wild Roses); Chemotaxonomy and Molecular Taxonomy.

Tissue Culture *see* **Cell, Tissue and Organ Culture**: *In Vitro* Regeneration Techniques; Cell Culture and Secondary Metabolites.

V

Viruses *see* **Disease**: Overview; Viruses.

APPENDIX 1

ROSE COLLECTIONS AND TRIALS/Awards

This Appendix contains extended tables 1 to 25 of the above article, which mainly showed data for the years 1991–2000.

Belgium

Table 1 Awards by Rosarium West Flanders (1985–1999)

Year	*Cultivar*	*Class*	*Breeder*
1999	'Dawn Sunsation'	Floribunda	Kordes
1998	'Jardins de France'	Floribunda	Meilland
1997	'Betty Harkness'	Floribunda	Harkness
1996	'Bordure Rose'®	Floribunda	Delbard
1995	'Mondiale'®	Hybrid Tea	Kordes
1994	'NDR 1 Radio Niedersachsen'	Floribunda	Kordes
1993	'Sommermorgen'	Floribunda	Kordes
1992	'Clos Fleuri Rose No. 2'	Floribunda	Delbard
1991	'Professor Boesman'	Floribunda	DVP Melle
1990	'Floranije'	Floribunda	DVP Melle
1989	'Paul Shirville'	Hybrid Tea	Harkness
1988	'MEIlustrel'	Floribunda	Meilland
1987	'Matilda'	Floribunda	Meilland
1986	'L'Oréal Trophy'	Hybrid Tea	Harkness
1985	'Puccini'	Shrub	Lens

Table 2 Awards by Rose Garden of Le Roeulx (1991–2000)

Year	*Award*	*Cultivar*	*Class*	*Breeder*
2000	Gold Medal	'Andre Brichet'(DVP 94-73)	Floribunda	Rijksstat.Voor Sierplantent
1999	Silver Medal	'RVS93104'	Floribunda	DVP Melle
		'ADAlexcor'	Hybrid Tea	Adam
	Most Fragrant	'MEIgripine'	Hybrid Tea	Meilland
1998	Gold Medal	'ORYbee'	Floribunda	Orye
		'Acapella'®	Hybrid Tea	Tantau
	Most Fragrant	'Acapella'®	Hybrid Tea	Tantau
1997	Gold Medal	'91/1193-4'	Floribunda	Poulsen
	Most Fragrant	'MEIsionver'	Hybrid Tea	Meilland
1996	Gold Medal	'Bolcholi'	Hybrid Tea	Meilland
	Most Fragrant	'Bolcholi'	Hybrid Tea	Meilland
1995	Gold Medal	'MEItekit'	Hybrid Tea	Meilland
	Most Fragrant	'MEItekit'	Hybrid Tea	Meilland
1994	Gold Medal	'Esme Euvrard'	Floribunda	Kordes
	Most Fragrant	'Frederic Mistral'®	Hybrid Tea	Meilland
1993	Silver Medal	'Schneekonigin'®	Shrub	Tantau
		'KORlamaus'	Floribunda	Kordes
	Most Fragrant	'Auguste Renoir'	Hybrid Tea	Meilland
1992	Silver Medal	'Warm Wishes'	Hybrid Tea	Fryer
	Most Fragrant	'Warm Wishes'	Hybrid Tea	Fryer
1991	Silver Medal	'Pink Bassino'	Shrub	Kordes
		LAPdev	Hybrid Tea	Laperrière
	Most Fragrant	'Plein Soleil'®	Hybrid Tea	Laperrière

Czech Republic

Table 3 Awards at the International Rose Trials of Hradec Kralove (2000)

Year	*Award*	*Cultivar*	*Class*	*Breeder*
2000	Golden Rose	'RvS-93-102'	Floribunda	RVS
	First Prizes	'Sebastian Kneipp' (KORpastato)	Hybrid Tea	Kordes
		'Rosenprofessor Sieber' (KORparesni)	Floribunda	Kordes
		'LLX 9001'	Shrub	Lens
		'Manita' (KORberhig)	Climber	Kordes
		'Yellow Palette' (POUlsendo)	Miniature	Poulsen
		'Tumbling Waters' (POUltumb)	Groundcover	Poulsen

France

Table 4 Awards by Parc du Bagatelle, Paris (1907–2002)

Year	*Category of award*	*Cultivar*	*Class*	*Breeder*
2002	First Prize	'Amber Flush' (INT B39796)	Hybrid Tea	Interplant
2001	First Prize	'Amber Cover' (POUlbambe)	Shrub	Poulsen
2000	First Prize	'Viborgh' (POUllug)	Hybrid Tea	Poulsen
1999	First Prize	DICyeti	Hybrid Tea	Dickson
1998	First Prize	KORdatura	Shrub	Kordes
1997	First Prize	'Riberhus'	Hybrid Tea	Poulsen
1996	First Prize	'Manureva'	Hybrid Tea	Sauvageot
1995	First Prize	No award		
1994	First Prize	'Golden Nugget'®	Shrub	Interplant
1993	First Prize	'Bluhwunder'®	Floribunda	Kordes
		'Rose de Rennes'®	Hybrid Tea	Adam
1992	First Prize	'Zambra 93'®	Floribunda	Meilland
		'Gypsy Dancer'	Polyantha	Dickson
1991	First Prize	'Les Amoureux de Peynet'®	Floribunda	Meilland
1990	First Prize	'Ville du Roeulx'	Floribunda	Van Landdoliw
1989	First Prize	'Cocorico'®	Floribunda	Meilland
1988	Gold Medal	'Mme Fernandel'®	Hybrid Tea	Meilland
1987	First Prize	'Charles Aznavour'	Floribunda	Meilland
1986	First Prize	'Summer Sérénade'®	Floribunda	Smith
1985	First Prize	'Pareo'®	Floribunda	Meilland
1984	Gold Medal	'Le Pink'	Miniature	Jackson & Perkins
1983	Gold Medal	'Morgenrot'®	Shrub	Kordes
1982	Gold Medal	'L'Oréal Trophy'	Hybrid Tea	Harkness
1981	Gold Medal	'New Face'®	Shrub	Interplant
	Gold Medal	'Sun Flare'	Floribunda	Jackson & Perkins
1976	Gold Medal	'Grand Siècle'®	Hybrid Tea	Delbard
1975	Gold Medal	'Nil Bleu'	Hybrid Tea	Delbard-Chabert
1974	Gold Medal	'Matangi'®	Floribunda	McGredy
1973	Gold Medal	'Grand Nord'®	Floribunda	Delbard-Chabert
1972	Gold Medal	'Bon-Bon'	Floribunda	Warriner
1971	Gold Medal	'Satura'	Hybrid Tea	Meilland
1970	Gold Medal	'Diablotin, Climbing'®	Climbing Floribunda	Delbart-Chabert
1969	Gold Medal	'Iskra'	Climber	Meilland
1968	Gold Medal	'Parure d'Or'®	Climber	Delbard
1967	Gold Medal	'Antigone'®	Hybrid Tea	Gaujard
1966	Gold Medal	'Majorette'®	Hybrid Tea	Meilland
1965	Gold Medal	'Arianna'®	Hybrid Tea	Meilland
1964	Gold Medal	'Joseph's Coat'®	Climber	Armstrong
1963	Gold Medal	'Grande Coquette'	Hybrid Tea	Delbard-Chabert
1962	Gold Medal	'Mme René Cassin'	Hybrid Tea	Delbard
1961	Gold Medal	'Zambra'®	Floribunda	Meilland
1960	Gold Medal	'Clair Matin'®	Climber	Meilland
1959	Gold Medal	'Garden Party'®	Hybrid Tea	Swim

Table 4 Continued

Year	*Category of award*	*Cultivar*	*Class*	*Breeder*
1958	Gold Medal	'Miracle'	Floribunda	Verbeek
1957	Gold Medal	'Sarabande'®	Floribunda	Meilland
1956	Gold Medal	'Paris-Match'	Hybrid Tea	Meilland
1955	Gold Medal	'Golden Fleece'	Floribunda	Boerner
1954	Gold Medal	'Mme René Coty	Hybrid Tea	Meilland
1953	Gold Medal	'Mojave'	Hybrid Tea	Swim
1952	Gold Medal	'Flambée'	Hybrid Tea	Mallerin
1951	Gold Medal	'Confidence'	Floribunda	Meilland
1950	Gold Medal: French Breeder	'Mme Louis Laperrière'	Hybrid Tea	Laperrière
1949	Gold Medal: Foreign Breeder	'Fashion'	Floribunda	Boerner
1948	Gold Medal: French Breeder	'Mme Vincent Auriol'	Hybrid Tea	Caron
	Gold Medal: Foreign Breeder	'Sutter's Gold'	Hybrid Tea	Swim
1947	Gold Medal: Foreign Breeder	'Applause'	Hybrid Tea	Swim
1946	Gold Medal: French Breeder	'André le Troquer'	Hybrid Tea	Meilland
1945	Gold Medal: French Breeder	'Gloire de France'	Hybrid Pernetiana	Gaujard
1944	Gold Medal: French Breeder	'Mme Kriloff'	Hybrid Tea	Meilland
1943	Gold Medal: French Breeder	'Vive la France'	Hybrid Pernetiana	Gaujard
	Gold Medal: Foreign Breeder	'Sondermeldung'	Floribunda	Kordes
1941	Gold Medal: French Breeder	'Rose Impériale'	Hybrid Pernetiana	Gaujard
	Gold Medal: Foreign Breeder	'Lorenzo Pahissa'	Climber	Pahissa
1940	Gold Medal: French Breeder	'Dunkerque'	Hybrid Tea	Laperrière
	Gold Medal: Foreign Breeder	'Aurora'	Hybrid Polyantha	Leenders
1939	Gold Medal: French Breeder	'Mme Auguste Chatain'	Hcap	Mallerin
	Gold Medal: Foreign Breeder	'Mme Virgilio Pirola'	Hybrid Tea	Lens
1938	Gold Medal: French Breeder	'Mme Edmond Labbé'	Hybrid Tea	Mallerin
	Gold Medal: Foreign Breeder	'Marquesa de Urquijo'	Hybrid Tea	Van Rossem
1937	Gold Medal: French Breeder	'Golden State'	Hybrid Tea	Mallerin
	Gold Medal: Foreign Breeder	'Prince Bernhard'	Hybrid Tea	Van Rossem
1936	Gold Medal: French Breeder	'Mme Henri Guillot'	Hybrid Tea	Mallerin
	Gold Medal: Foreign Breeder	'Eclipse'	Hybrid Pernetiana	Nicolas
1935	Gold Medal: French Breeder	'Princesse Amédée de Broglie'	Hybrid Tea	Mallerin
	Gold Medal: Foreign Breeder	'Mrs Oswald Lewis'	Hybrid Tea	Cants
1934	Gold Medal: French Breeder	'Mme Joseph Perraud'	Hybrid Tea	Gaujard
	Gold Medal: Foreign Breeder	'Angels Mateu'	Pernetiana	Dot
1933	Gold Medal: Foreign Breeder	'Mevrouw van Straaten van Nes'	Polyantha	Leenders
1932	Gold Medal: French Breeder	'Mme Cochet-Cochet'	Hybrid Tea	Mallerin
	Gold Medal: Foreign Breeder	'Luis Brinas'	Pernetiana	Dot
1931	Gold Medal: French Breeder	'Dance of Joy'	Hybrid Wichurana	Sauvageot
	Gold Medal: Foreign Breeder	'Comtesse Vandal'	Hybrid Tea	Leenders
1930	Gold Medal: French Breeder	'Mme Raymond Gaujard'	Pernetiana	Pernet-Ducher
	Gold Medal: Foreign Breeder	'Cayetana Stuart'	Pernetiana	Dot
1929	Gold Medal: French Breeder	'Mrs Pierre S. DuPont'	Hybrid Tea	Mallerin
	Gold Medal: Foreign Breeder	'Président Deville'	Hybrid Tea	Leenders
1928	Gold Medal: Foreign Breeder	'Lady Margaret Stewart'	Hybrid Tea	Dickson
1927	Gold Medal: Foreign Breeder	'Mme Grégoire Staechelin'	Climber	Dot
1926	Gold Medal: French Breeder	'Helen Fox'	Hybrid Tea	Buatois
	Gold Medal: Foreign Breeder	'Kardinal Piffl'	Pernetiana	Leenders
1925	Gold Medal: French Breeder	'Ville de Paris'	Pernetiana	Pernet-Ducher
	Gold Medal: Foreign Breeder	'Gooiland Beauty'	Hybrid Tea	Van Rossem
1924	Gold Medal: French Breeder	'Angèle Pernet'	Pernetiana	Pernet-Ducher
	Gold Medal: Foreign Breeder	'John Russell'	Hybrid Tea	Dobbie
1923	Gold Medal: French Breeder	'Président Chérioux'	Pernetiana	Pernet-Ducher
	Gold Medal: Foreign Breeder	'Sunstar'	Hybrid Tea	Dickson
1922	Gold Medal: French Breeder	'Toison d'Or'	Hybrid Tea	Pernet-Ducher
	Gold Medal: Foreign Breeder	'Elvira Aramayo'	Hybrid Tea	Looymans & Zonen
1921	Gold Medal: French Breeder	'Souvenir de Georges Pernet'	Pernetiana	Pernet-Ducher
	Gold Medal: Foreign Breeder	'Miss Lolita Armour'	Pernetiana	Howard & Smith
1920	Gold Medal: French Breeder	'Souvenir de Claudius Pernet'	Pernetiana	Pernet-Ducher
	Gold Medal: Foreign Breeder	'Frances Gaunt'	Hybrid Tea	Dickson
1919	Gold Medal	'Jean C. N. Forestier'	Pernetiana	Pernet-Ducher

Table 4 Continued

Year	*Category of award*	*Cultivar*	*Class*	*Breeder*
1918	Gold Medals	'Los Angeles'	Pernetiana	Howard & Smith
		'Paul's Scarlet Climber'	Hybrid Wichurana	Paul & Son
1917	Gold Medal: French Breeder	'Mme Caristie Martel'	Hybrid Tea	Pernet-Ducher
	Gold Medal: Foreign Breeder	'Margaret Dickson Hamill'	Hybrid Tea	Dickson
1916	Gold Medals	'Constance'	Pernetiana	Pernet-Ducher
		'Mrs Wemyss Quin'	Pernetiana	Dickson
1915	Gold Medals	'Mme Raymond Poincaré'	Hybrid Tea	Gravereaux
		'Mme Marcel Delauney'	Hybrid Tea	Leenders
1914	Gold Medal: French Breeder	'Souvenir de E. Guillard'	Hybrid Tea	Chambard
	Gold Medal: Foreign Breeder	'Mevrouw Dora Van Tets'	Hybrid Tea	Leenders
1913	Gold Medal: French Breeder	'Mme Charles Lutaud'	Hybrid Tea	Pernet-Ducher
	Gold Medal: Foreign Breeder	'Mabel Drew'	Hybrid Tea	Leenders
1911	Gold Medal: French Breeder	'Beauté de Lyon'	Pernetiana	Pernet-Ducher
	Gold Medal: Foreign Breeder	'Jonkheer J.L. Mock'	Hybrid Tea	Leenders
1909	Gold Medal: French Breeder	'Lyon Rose'	Pernetiana	Pernet-Ducher
	Gold Medal: Foreign Breeder	'Mme Segond Weber'	Tea	Soupert & Notting
1908	Gold Medal	'Rhea Reid'	Hybrid Tea	Richmond
1907	Gold Medal	'Marquise de Sinéty'	Hybrid Tea	Pernet-Ducher

Table 5 Landscape awards by Parc du Bagatelle, Paris (1986–2002)

Year	*Award*	*Cultivar*	*Class*	*Breeder*
2002	First Prize	'Gartnerfreude'	Groundcover Shrub	Kordes
2001	First Prize	'Easy Cover'	Groundcover Shrub	Poulsen
2000	First Prize	'Sweet Cover'	Groundcover Shrub	Poulsen
		'Bordure Blanche'	Shrub	Delbard
1999	First Prize	'Diamond Border'	Shrub	Poulsen
1998	First Prize	'Bayernland Cover'	Floribunda	Poulsen
1996	First Prize	'Relax Meillandecor'	Groundcover Shrub	Meilland
1995	First Prize	'Douceur Normande'®	Shrub	Meilland
1994	First Prize	'Emera'	Floribunda	Noack
1993	First Prize	'Carmagnole'®	Floribunda	Delbard
1992	First Prize	'Denise Grey'®	Shrub	Meilland
1991	First Prize	'Comtesse Jeanne de Flandres'	Shrub	Meilland
1989	First Prize	'Vent d'Été'	Groundcover Shrub	Kordes
1988	First Prize	'Marjorie Fair'®	Shrub	Harkness
1987	First Prize	'Rosy La Sevilliana'	Floribunda/groundcover	Meilland
1986	First Prize	'Rush'®	Shrub	Lens

Table 6 Awards of the title 'Medaille d'Or/La Plus Belle Rose de France' by Parc de la Tête d'Or, Lyon (1931–2000)

Year	*Cultivar*	*Class*	*Breeder*
2000	'Privilège de St-Verny'		Orard
1999	'Harmonia Sub Rosa'	Hybrid Tea	Dorieux
1998	'ADAbuco'	Floribunda	Adam
1997	'Novaia'	Floribunda	Orard
1996	'Rose d'Or de Montreux'	Hybrid Tea	Adam
1995	'Aramis'	Floribunda	Laperrière
1994	'Abbaye de Cluny'™	Hybrid Tea	Meilland
1993	'No Name'	Floribunda	Dorieux
1992	'MEIkame'	Floribunda	Meilland
1991	'Le Grand Huit'	Hybrid Tea	Adam
1990	'MEIrolange'	Floribunda	Meilland
1989	'No Name'	Hybrid Tea	Meilland

Table 6 Continued

Year	*Cultivar*	*Class*	*Breeder*
1988	'MEIgilan'	Hybrid Tea	Meilland
1987	'Starion'®	Hybrid Tea	Dorieux
1986	'Prince Igor, Climbing'	Climber	Meilland
1985	'MEIcapula'	Hybrid Tea	Dorieux
1984	No Awards		
1982	'Satellite'	Hybrid Tea	Delbard-Chabert
1981	'Epidor'®	Hybrid Tea	Delbard
1980	'Florella'	Hybrid Tea	Meilland
1979	'Gilbert Bécaud'	Hybrid Tea	Meilland
1978	'Mme Bruno Coquatrix'	Hybrid Tea	Pineau
1977	'Président Léopold Senghor'®	Hybrid Tea	Meilland
1976	'La Marseillaise'®	Hybrid Tea	Delbard-Chabert
1975	'Krioga'	Floribunda	Kriloff
1974	'Bordure Rose'®	Polyantha	Delbard
1973	'Scherzo'®	Hybrid Tea	Meilland
1972	No awards		
1971	'Atoll'®	Hybrid Tea	Meilland
1970	'Illisca'®	Hybrid Tea	Laperrière
1969	'Imperator'	Hybrid Tea	Meilland
1968	'Baronne Edmond de Rothschild'	Hybrid Tea	Meilland
1967	'Pharaon'®	Hybrid Tea	Meilland
1966	'Apogée'®	Hybrid Tea	Delbard
1965	'Tourmaline'®	Hybrid Tea	Delbard
1964	'No Name'	Hybrid Tea	Meilland
1963	'No Name'	Hybrid Tea	Meilland
1962	'Mme René Cassin'	Hybrid Tea	Delbard
1961	'No Name'	Hybrid Tea	Meilland
1960	'Saint-Exupéry'	Hybrid Tea	Delbard
1959	'Semis inedit'	Hybrid Tea	Meilland
1958	'Christian Dior'	Hybrid Tea	Meilland
1957	'Rose Gaujard'®	Hybrid Tea	Gaujard
1956	'Astrée'	Hybrid Tea	Croix
1955	'Soraya'®	Hybrid Tea	Meilland/Richardier
1954	'Mme Plumecocq'	Hybrid Tea	Gaujard
1953	'Radar'	Hybrid Tea	Meilland
1951	'Confidence'	Hybrid Tea	Meilland
1950	'Altesse'	Hybrid Tea	Meilland
1948	'Opéra'	Hybrid Tea	Gaujard
1947	'Charles Mallerin'	Hybrid Tea	Meilland
1946	'Alain'	Polyantha	Meilland
1945	'Michèle Meilland'	Hybrid Tea	Meilland
1944	'Profusion'	Hybrid Tea	Meilland
1943	'Profusion'	Hybrid Tea	Meilland
1942	'Mme A. Meilland'	Hybrid Tea	Meilland
1941	'Mme Guimet'	Hybrid Tea	Colombier
1940	'Ville de Prague'	Hybrid Tea	Chambard
1939	'Mme Noël'	Hybrid Tea	Chambard
1938	'Marie-Rose Besson'	Hybrid Tea	Mallerin
1937	'Golden State'	Hybrid Tea	Meilland
1936	'La Parisienne'	Hybrid Tea	Mallerin
1935	'Princesse Amédée de Broglie'	Hybrid Tea	Mallerin
1934	'Mme Joseph Perraud	Hybrid Tea	Gaujard
1931	'Mme Nicolas Aussel'	Hybrid Tea	Pernet Ducher

Table 7 Additional Awards by Parc de la Tête d'Or, Lyon (1938–1984)

Year	Award	Cultivar	Class	Breeder
1984	Prix Prestige de la Rose	'Ferline'®	Hybrid Tea	Laperrière
1979	Grandes Roses du Siècle	'Lady Rose'®	Hybrid Tea	Kordes
		'Pénélope'®	Hybrid Tea	Gaujard
1973	Medaille d'Or du Parfum	'Idao'	Hybrid Tea	Gaujard
1972	Medaille d'Or du Parfum	'Gamma'	Hybrid Tea	Gaujard
1939	Prix de la Rose Parfumée	'Neige Parfum'	Hybrid Tea	Mallerin
1938	Prix de la Rose Parfumée	'Guinée'	Hybrid Tea, Climbing	Mallerin

Table 8 Awards by La Roseraie de Saverne (1999–2000)

Year	Prize	Cultivar	Class	Breeder
2000	Gold Medal	'Neige Parfum' (DORblan)	Hybrid Tea	Dorieux
	Fragrance	'Caruso' (MEIcarouze)	Hybrid Tea	Meilland
1999	Gold Medal	ORAcol		Orard
	Fragrance	'Wischoten' (MEIfani)		Meilland

Germany

Table 9 Awards at International Rose Trials of Baden-Baden (1989–2000)

Year	Award	Cultivar	Class	Breeder
2000	Gold Medals	'Black Magic'	Hybrid Tea	Tantau
		'Andre Brichet' (DvP 94-73)	Floribunda	RVS
		(DICyeti)	Floribunda	Dickson
		'Dietrich Woessner' (HUBar)	Floribunda	Huber
		'Teasing Georgia'	Shrub	Austin
1999	Golden Rose of Baden-Baden	'Malaga'	Hybrid Tea	Interplant
	Gold Medals	'Malaga'	Hybrid Tea	Interplant
		'93/3237-1'	Floribunda	Poulsen
		'Gärtnerfreude'	Groundcover	Kordes
		'CHEwsally'	Climber	Warner
1998	Casino of Baden-Baden Award	No Name	Hybrid Tea	Bear Creek
	Gold Medals	'JACbrant'	Hybrid Tea	Bear Creek
		'Pink Hedge'	Floribunda	Interplant
		'CF 93.7651.1'	Floribunda	Delbard
		'4437 FRYwhoppa'	Floribunda	Fryers
		'Sugar Baby'	Miniature	Tantau
		'Diamond Border'	Climber	Poulsen
1997	Casino of Baden-Baden Award	'Tintinara'™	Floribunda	Dickson
	Fragrance Award	'Rêve d'un Soir'	Hybrid Tea	Croix
	Gold Medal	'Goldelse'	Floribunda	Tantau
1996	Casino of Baden-Baden Award	'Rouge Adam'	Hybrid Tea	Michel Adam
	Fragrance Award	'Chartreuse de Parme'	Hybrid Tea	Delbard
	Gold Medals	'Pimpernelle'	Floribunda	Delbard
		'KO 87/15833-01'	Groundcover	Kordes
		'Ranger'	Climber	Barni
1995	Casino of Baden-Baden Award	'Old Port'	Hybrid Tea	McGredy
	Fragrance Award	'Melody Perfumée'™	Hybrid Tea	Dorieux
	Gold Medals	'Rose de Rennes'®	Hybrid Tea	Michel Adam
		'90-9066-8'	Grandiflora	Poulsen
		'CG 88.375.306'	Floribunda	Delbard
1994	Casino of Baden-Baden Award	'Concerto'™	Floribunda	Meilland
	Fragrance Award	'Happy Day'	Climber	Warner
	Gold Medals	'POUlsol'	Hybrid Tea	Poulsen
		'POUlket'	Floribunda	Poulsen

Table 9 Continued

Year	*Award*	*Cultivar*	*Class*	*Breeder*
1993	Casino of Baden-Baden Award	'3430/JC 27693A'	Hybrid Tea	Cocker
	Fragrance Award	'Frederic Mistral'®	Hybrid Tea	Meilland
	Gold Medals	'Zambra'®	Floribunda	Meilland
		'1100-87'	Floribunda	Noack
		'POUlter'	Groundcover	Poulsen
		'Boreal Adaribu'		Michel Adam
		'Benny Goodman'	Miniature	McGredy
1992	Casino of Baden-Baden Award	'Imperatrice Farah'	Hybrid Tea	Delbard
	Fragrance Award	'Fragrant Lady'	Hybrid Tea	Meilland
	Gold Medals	'Papi Delbard'	Hybrid Tea	Delbard
		'Fragrant Lady'	Hybrid Tea	Meilland
		'JACbeau'	Floribunda	Bear Creek
		'Sommermarchen'	Groundcover	Kordes
		'F 153-88'	Climber	Interplant
1991	Casino of Baden-Baden Award	'MT 82381'	Hybrid Tea	Tantau
	Fragrance Award	'MEInuiton'	Hybrid Tea	Meilland
	Gold Medals	'Golden Medaillon 91'	Hybrid Tea	Kordes
		'G106-84'	Hybrid Tea	Interplant
		'Fairysnow'	Floribunda	Fryers
		'Matchball'®	Climber	Lens
		'Climbing Rimosa'	Climber	Meilland
1990	Casino of Baden-Baden Award	'Karen Blixen'	Hybrid Tea	Poulsen
	Fragrance Award	'MEIfonze'	Hybrid Tea	Meilland
	Gold Medals	'Karen Blixen'	Hybrid Tea	Poulsen
		'RVS 86-46'	Floribunda	Rijksstation
		'Adrien Mercier'®	Floribunda	Dorieux
		'Mainaufeuer'®	Groundcover	Kordes
		'Lenidora'	Climber	Lens
1989	Casino of Baden-Baden Award	'CROpal'	Hybrid Tea	Croix
	Fragrance Award	'Velvet Fragrance'	Hybrid Tea	Fryers
	Gold Medals	'Milky Way'	Hybrid Tea	Interplant
		'Cocorico'®	Floribunda	Meilland
		'POUlcov'	Groundcover	Poulsen

Table 10 Awards in the ADR German Rose Trials (1952–2000)

Year	*Cultivar*	*Class*	*Breeder*
2000	'Nemo'	Shrub	Noack
	'Vinesse'	Floribunda	Noack
	'Bad Birnbach'	Floribunda	Kordes
	'Maxi Vita'	Floribunda	Kordes
1999	'Celina'	Floribunda	Noack
	'Ravenna'	Shrub	Noack
	'Rotfassade'	Climber	Noack
	'Neon'	Hybrid Tea	Kordes
	'Kronjuwel'	Floribunda	Noack
	'Brautzauber'	Floribunda	Noack
1998	'Rote Apart'	Shrub	Uhl
	'Estima'	Shrub	Noack
	'Saremo'	Shrub	Noack
1997	'Manita'	Climber	Kordes
	'Danica'	Shrub	Noack
	'Arcadia'	Shrub	Noack
1996	'Postillion'	Shrub	Kordes
	'Felicitas'	Shrub	Kordes
	'Sommerabend'®	Shrub	Kordes
	'Queen Mother'	Floribunda	Kordes
	'Rosenprofessor Sieber'	Floribunda	Kordes
	'Crimson Meidiland'	Shrub	Meilland

Table 10 Continued

Year	*Cultivar*	*Class*	*Breeder*
1995	'Schneekonigin'	Shrub	Tantau
	'Magic Meidiland'™	Shrub	Meilland
	'Medusa'	Shrub	Noack
	'Hannovers Weisse'	Shrub	Noack
	'Windrose'™	Shrub	Noack
	'Northern Lights'	Shrub	Noack
	'Melissa'	Floribunda	Noack
	'Aspirin'	Floribunda	Tantau
	'Bayernland'	Floribunda	Rosenunion
1994	'Bingo Meidiland'®	Shrub	Meilland
	'Dortmunder Kaiserhain'	Shrub	Noack
	'Rotes Meer'	Shrub	Baum
	'Baum's Rokkoko'	Shrub	Baum
	'Bluhwunder'®	Floribunda	Kordes
1993	'Pink Bassino'	Shrub	Kordes
	'Hagenbecks Tierpark'	Shrub	Meilland
	'Schneesturm'®	Shrub	Tantau
	'Mirato'	Shrub	Tantau
	'Foxi'	Shrub	Tantau/Uhl
	'Armada'®	Shrub	Harkness
	'Mazurka'	Floribunda	Meilland
	'Vicky'	Floribunda	Noack
1992	'Palmengarten Frankfurt'®	Shrub	Kordes
	'Pierette'	Shrub	Tantau
	'Rugelda'®	Shrub	Kordes
	'Rote Woge'	Shrub	Meilland
	'Schone Dortmunderin'®	Floribunda	Noack
1991	'Apfelblute'	Shrub	Noack
	'Wildfang'®	Shrub	Noack
	'Richard Strauss'®	Shrub	Noack
	'Super Excelsa'®	Climber	Hetzel
	'Schneeflocke'®	Floribunda	Noack
1990	'Heidetraum'™	Shrub	Noack
1989	'Burghausen'®	Shrub	Kordes
	'Vogelpark Walsrode'	Shrub	Kordes
	'Ricarda'®	Floribunda	Noack
	'Play Rose'	Floribunda	Meilland
1988	'Rodinghausen'®	Shrub	Noack
1987	'Pink Meidiland'®	Shrub	Meilland
	'Sommerwind'	Shrub	Kordes
	'Lavender Dream'®	Shrub	Ilsink
	'Elina'®	Hybrid Tea	Dickson
	'Dolly'	Floribunda	Poulsen
1986	'Pink La Sevillana'®	Floribunda	Meilland
	'Berolina'	Hybrid Tea	Kordes
	'Ravensberg'®	Shrub	Noack
	'Romanze'	Shrub	Tantau
1985	'Banzai 83'®	Hybrid Tea	Meilland
1984	'Rosenresli'®	Climber	Kordes
	'Bonanza'®	Shrub	Kordes
1982	'Angela'®	Shrub	Kordes
	'Bonica 82'	Floribunda	Meilland
	'Iga 83 München'	Floribunda	Meilland
	'Aachener Dom'	Hybrid Tea	Meilland
1980	'Robusta'®	Shrub	Kordes
	'Red Yesterday'	Shrub	Harkness
1979	'La Sevillana'®	Floribunda	Meilland
1977	'Chorus'®	Floribunda	Meilland
	'Grandhotel'®	Shrub	McGredy
1976	'Andalusien'®	Floribunda	Kordes
	'Compassion'®	Climber	Harkness

Table 10 Continued

Year	*Cultivar*	*Class*	*Breeder*
1975	'Morning Jewel'®	Climber	Cocker
1974	'Westerland'®	Shrub	Kordes
	'Montana'®	Floribunda	Tantau
1973	'Escapade'®	Floribunda	Harkness
	'Friesia'	Floribunda	Kordes
1972	'Tornado'®	Floribunda	Kordes
1970	'Edelweiss'	Floribunda	Poulsen
1968	'Lichtkönigin Lucia'®	Shrub	Kordes
	'Bischofsstadt Paderborn'®	Shrub	Kordes
1966	'Sympathie'®	Climber	Kordes
1960	'Schneewittchen'	Shrub	Kordes
1958	'Dirigent'®	Shrub	Tantau
1954	'Dortmund'®	Climber	Kordes
1952	'Flammentanz'®	Climber	Kordes

Ireland

Table 11 Awards at the International Rose Trials, Dublin (1981–2000)

Year	*Cultivar*	*Class*	*Breeder*
2000	'Josie Whitney' (HARfancy)	Hybrid Tea	Harkness
	'HORcohabit'	Miniature	Horner
1999	'RSV93-104'	Floribunda	Melle Experimental Station
	'Lucky Duck' (DICzest)	Patio	Dickson
	'K091/2544-94'	Groundcover	Kordes
1998	'Jardins de France'	Floribunda	Meilland
	'Danny Boy' (DICxcon)	Patio	Dickson
	'93/3030-2'	Groundcover	Poulsen
1997	'Jill's Rose'(GANjil)	Floribunda	Gandy
	'DICxotic'	Patio	Dickson
	'Majorie Marshall'	Shrub	Harkness
1996	'Easy Going'™	Floribunda	Harkness
	'Riverdance'	Patio	Kenny
	'DICvivid'	Shrub	Dickson
1995	'89/8137-11'	Floribunda	Poulsen
	'Crème de la Crème'	Climber	Gandy
	'INT B35590'	Patio	Interplant
1994	'Isobel Derby'	Hybrid Tea	Horner
	'Bewitched'	Floribunda	Poulsen
	'Our Molly'	Shrub	Dickson
1993	'G240-88'	Hybrid Tea	Interplant
	'Dormstadt Fulda'	Floribunda	Kordes
	'Playtime'	Shrub	Kordes
1992	'Benita'®	Hybrid Tea	Dickson
	'In The Pink'	Floribunda	Limes
	'Top Marks'	Patio	Fryers
1991	'Dawn Chorus'	Hybrid Tea	Dickson
	'Golden Holstein'®	Floribunda	Kordes
	'Colleen'	Shrub	Interplant
1990	'DICosis'	Hybrid Tea	Dickson
	'RSV 85-66'	Floribunda	Melle Experimental Station
	'Orange Jewel'®	Patio	Kordes
1989	'High Sheriff'	Hybrid Tea	Harkness
	'Birthday Girl'(MEIlasso)	Floribunda	Meilland
	'Ramira'®	Climber	Kordes
1988	'Savoy Hotel'	Hybrid Tea	Harkness
	'Jacqueline Nebout'	Floribunda	Meilland
	'Red Dot'	Shrub	Interplant

Table 11 Continued

Year	*Cultivar*	*Class*	*Breeder*
1987	'Tequila Sunrise'	Hybrid Tea	Dickson
	'K0802949'	Floribunda	Kordes
	'Sweet Symphony'	Patio	Meilland
1986	'DICmuffinman'	Floribunda	Dickson
	'Conservation'	Patio	Cocker
	'Essex'	Shrub	Poulsen
1985	'Selfridges'	Hybrid Tea	Kordes
	'Holstein'	Floribunda	Kordes
	'Star Child'®	Patio	Dickson
1984	'Spirit of Youth'	Hybrid Tea	Meilland
	'Princess Alice'	Floribunda	Harkness
	'87/6002-12'	Miniature	Poulsen
1983	'Woods of Windsor'(KORprill)	Hybrid Tea	Kordes
	'Hedgefire'	Floribunda	Kordes
	'Snowball'	Miniature	McGredy
1982	'Pinky'	Hybrid Tea	McGredy
	'326-77'	Floribunda	Interplant
	'MEIdisentis'	Shrub	Meilland
1981	'Loving Memory'	Hybrid Tea	Kordes
	'New Face'®	Floribunda	Interplant
	'MACsupsid'	Patio	McGredy

Italy

Table 12 Awards at International Rose Trials of Monza (1965–2000)

Year	*Award*[a]	*Cultivar*	*Class*	*Breeder*
2000	Gold Medals	(JACpetor)	Hybrid Tea	Jackson & Perkins
		'Amber Cover'	Groundcover	Poulsen
	Rosa Italiana	BAR 5069	Hybrid Tea	Barni
	Rosa profumata	MEIceppus	Hybrid Tea	Meilland
1999	Arredo Urbano	'Michelle d'Hoop' (RVS 93-101)	Floribunda	RVS
1998	Gold Medals	'Cumbaya'	Floribunda	Meilland/Richardier
		'MEIcurzis'	Hybrid Tea	Meilland
	Rosa Italiana	'Fatima' (VHKO)	Hybrid Tea	Cazzaniga
	Rosa Profumata	'Hacienda' (ORAdal)	Hybrid Tea	Orard
1997	Gold Medals	'Diamond Border'	Shrub	Poulsen Roser
		'Easy Going'®	Floribunda	Harkness
		'Michelangelo' (MEItelov)	Hybrid Tea	Meilland
	Rosa Italiana	'Avance'	Miniature	Barni
	Rosa Profumata	'Naema'	Climber	Delbard
1996	Gold Medal	'PANborclass BP01'	Floribunda	Panozzo
	Rosa Italiana	'Tempi Moderni'	Floribunda	Barni
	Rosa Profumata	'Jude the Obscure' (N22B89)	Shrub	Austin
1995	Gold Medals	'Sommermorgen'	Shrub	Kordes
		'Jardins de Villandry'	Hybrid Tea	Delbard
	Rosa Italiana	'Sans Souci'	Floribunda	Barni
	Rosa Profumata	'James Perière'	Hybrid Tea	Meilland
1994	Gold Medals	'Fredensborg'®	Floribunda	Poulsen Roser
		'LAPvi'	Hybrid Tea	Laperrière
		'Gold Symphonie'	Miniature	Meilland
	Rosa Italiana	'Sandra Mondaini' (FLI)	Floribunda	Cazzaniga
	Rosa Profumata	'Frédéric Mistral'®	Hybrid Tea	Meilland
1993	Gold Medals	'Leonardo da Vinci'®	Floribunda	Meilland
		'Abbaye de Cluny'®	Hybrid Tea	Meilland
	Rosa Italiana	'Suni'®	Floribunda	Barni
	Rosa Profumata	'Toulouse Lautrec'®	Hybrid Tea	Meilland

Table 12 Continued

Year	*Award*[a]	*Cultivar*	*Class*	*Breeder*
1992	Gold Medal	'Clos Fleuri Rose No. 2'	Floribunda	Delbard
	Rosa Italiana	'Luciana Savignano' (VQEI)	Hybrid Tea	Cazzaniga
	Rosa Profumata	'Souvenir de Marcel Proust'	Hybrid Tea	Delbard
1991	Gold Medal	'Colleen'®	Shrub	Interplant
	Rosa Italiana	'Frine'®	Shrub	Barni
	Rosa Profumata and Gold Medal	'MEIyoma'	Hybrid Tea	Meilland
1990	Gold Medal and Rosa Italiana	'Rinascimento'®	Shrub	Barni
	Gold Medal	'Holfairywhite' (FRYhfwhite)	Miniature	Fryers/Holmes
	Rosa Profumata	'Biapro'	Hybrid Tea	CEA
1989	Gold Medals	'Cocorico'®	Floribunda	Meilland
		'Bellissima'®	Hybrid Tea	Laperrière
	Rosa Italiana	'Ornella Muti'®	Shrub	Barni
	Rosa Profumata	'KORlorca'	Hybrid Tea	Kordes
1988	Gold Medal	'Orange Meillandina, Climbing'	Climber	Meilland
	Rosa Italiana	'Sorriso di Pace'	Floribunda	Cazzaniga
	Rosa Profumata	'The McCartney Rose' (MEIzeli)	Hybrid Tea	Meilland
1987	Gold Medals	'Wapiti'®	Floribunda	Meilland
		'Charlotte Rampling'®	Hybrid Tea	Meilland
	Rosa Italiana	'Dalila di Lazzaro' (IGMI)	Floribunda	Cazzaniga
	Rosa Profumata	'Déclic'	Climber	Croix
1986	Gold Medal	'Kordes' Rose Wurzburg'	Hybrid Tea	Kordes
	Arredo Urbano	'Lavender Dream'®	Shrub	Interplant
1985	Gold Medals	'Lovita'	Floribunda	Meilland-Keisei
		'Révolution Française'	Hybrid Tea	Meilland
	Rosa Italiana	'CATE'	Hybrid Tea	Cazzaniga
	Rosa Profumata	'Duftzauber 84' (KORzaun)	Hybrid Tea	Kordes
1984	Gold Medals	'International Herald Tribune'®	Floribunda	Harkness
		'Solidor'®	Hybrid Tea	Meilland
	Rosa Italiana	'Profumo d'Italia' (Eleonora)	Hybrid Tea	Motta
	Rosa Profumata	'Dioressence'®	Grandiflora	Delbard
	Arredo Urbano	'Celeste' (AFB174)	Miniature	Motta
1983	Gold Medals	'Céline Delbard'	Floribunda	Delbard
	Rosa Italiana	'Loretta Goggi'	Hybrid Tea	Cazzaniga
	Rosa Profumata	'Themis'	Hybrid Tea	Gaujard
1982	Gold Medals	'Rush'®	Shrub	Lens
		'Axelle'	Hybrid Tea	Gaujard
	Rosa Profumata and Rosa Italiana	'Gilles Villeneuve'	Climber	Cazzaniga
1981	Gold Medals	'Mariandel'	Floribunda	Kordes
		'Miriana'	Hybrid Tea	Meilland
	Rosa Italiana	'Gocce d'Oro'	Floribunda	Cazzaniga
	Rosa Profumata	'Fragrant Mist'	Floribunda	Smith
1980	Gold Medals	'Mitzi'	Floribunda	Meilland
		'Waverland'®	Hybrid Tea	Lens
	Rosa Profumata and Rosa Italiana	'Oriella Dorella'	Hybrid Tea	Cazzaniga
1979	Gold Medals	'Sally Holmes'®	Floribunda	Fryer/Holmes
		'Blanche Comète'	Hybrid Tea	Delbard
	Rosa Profumata	'Sérénité'®	Hybrid Tea	Gaujard
1978	Gold Medals	'Letchworth Garden City'®	Floribunda	Harkness
	Rosa Italiana	'Sylva Koscina'	Hybrid Tea	Cazzaniga
	Rosa Profumata	'Margaret Merril'	Floribunda	Harkness
1977	Gold Medals	'Candeur'®	Floribunda	Delbard
		'Love'	Grandiflora	Jackson & Perkins
	Rosa Italiana	'Sondrio'	Hybrid Tea	Cazzaniga
	Rosa Profumata	'Cucci'	Hybrid Tea	Cazzaniga
1976	Gold Medals	'Simplicity'®	Floribunda	Jackson & Perkins
	Rosa Italiana	'Valentina Cortese'	Floribunda	Cazzaniga
		'Louis de Funès'	Hybrid Tea	Meilland
1975	Gold Medals	'RHJ93'	Floribunda	Harkness
		'Plicata'	Hybrid Tea	Jackson & Perkins
	Rosa Italiana	'Ester Fumagalli' (VWS110)	Hybrid Tea	Cazzaniga
	Rosa Profumata	'Nil Bleu'®	Hybrid Tea	Delbard

Table 12 Continued

Year	*Award*[a]	*Cultivar*	*Class*	*Breeder*
1974	Gold Medals	'Yesterday'®	Polyantha	Harkness
		'Vivre'	Hybrid Tea	Delbard
	Rosa Italiana	'Pironti-Critow'	Climbing Hybrid Tea	Pironti
	Rosa Profumata	'Charles de Gaulle'®	Hybrid Tea	Meilland
1973	Gold Medal & Rosa Italiana	'IMS106'	Floribunda	Cazzaniga
	Rosa Italiana	'Zandrina' (BZ-CA)	Hybrid Tea	Bortolotti
	Rosa Profumata	'Ennio Morlotti'	Floribunda	Cazzaniga
1972	Gold Medals	(MEImaine)	Climber	Meilland
		'MEIscova'	Hybrid Tea	Meilland
	Rosa Italiana:	'Romina'	Hybrid Tea	Cazzaniga
	Rosa Profumata	'Jadis'	Hybrid Tea	Jackson & Perkins
1971	Gold Medals	'SMG 65449'	Floribunda	McGredy
		'Satura' (MEIlouk)	Hybrid Tea	Meilland
	Rosa Italiana	'M5814''	Landscape	Mansuino
	Rosa Profumata	'Mambo' (MT6494)	Floribunda	Tantau
1970	Gold Medal	'Prince Igor'®	Floribunda	Meilland
	Rosa Italiana	'Lyra'	Floribunda	Pironti
	Rosa Profumata	'Coppa'	Hybrid Tea	Mansuino
1969	Rosa Profumata	'Baronne Edmond de Rothschild'®	Hybrid Tea	Meilland
	Gold Medal	'Herzog von Windsor'	Hybrid Tea	Tantau
		'Sarabande, Climbing'®	Climbing Floribunda	Meilland
1968	Gold Medal	'Sundra'®	Floribunda	Gaujard
	Rosa Italiana	'Saragat'	Hybrid Tea	Malandrone
	Rosa Profumata	'Anne Marie Treschlin'	Hybrid Tea	Meilland
1967	Gold Medal	'Escurial'®	Hybrid Tea	Delbard
	Rosa Profumata	'Fontainebleau'	Hybrid Tea	Delbard-Chabert
1966	Gold Medals	'Ice White'	Floribunda	McGredy
		'Versailles'®	Hybrid Tea	Delbard
	Rosa Italiana	'Tiziana'	Floribunda	Cazzaniga
	Rosa Profumata	'Red Devil'	Hybrid Tea	Dickson
1965	Gold Medals	'Mimi'	Miniature	Meilland
		'Mme Neige'	Hybrid Tea	Meilland
	Gold Medal and Rosa Italiana	'Amica' (PAP06)	Hybrid Tea	Cazzaniga
	Rosa Profumata	'Monza' (JC GQ 70)	Hybrid Tea	Leenders
		'Fascination'	Hybrid Tea	Jackson & Perkins

[a] Rosa dell'Anno First prize, Gold Medal; Rosa Profumata: Fragrance Award, Crown of Queen Theodelinde; Rosa Italiana: Italian Rose Prize, Gold Medal; Rose Arredo Urbano: Landscape Rose Prize, Gold Medal.

Table 13 Gold Medal awards at the International Rose Trials of Rose, Italy (1933–2000)

Year	*Cultivar*	*Class*	*Breeder*
2000	'Neige Parfum'		Dorieux
1999	'Bella di Todi'		Barni
	'DELstroja'		Delbard
1998	'Hacienda'		Orard
	'Cotillion'®	Floribunda	Jackson & Perkins
1997	'Caprice de Meilland'	Hybrid Tea	Meilland
	'Dorlobla'		Dorieux
1996	Unnamed seedling		Meilland
	Unnamed seedling		Meilland
1995	'Nette Ingeborg'		Havel
	'Wonderland'®	Shrub	Dickson
1994	'Solo Mio'	Hybrid Tea	Poulsen
	'Berleburg'	Floribunda	Poulsen
1993	'JAChov'		Jackson & Perkins
	'Lady Penelope'	Climber	Warner
1992	'Imperatrice Farah'		Delbard
	'Mr JCB'	Floribunda	Dickson

Table 13 Continued

Year	*Cultivar*	*Class*	*Breeder*
1991	'Karen Blixen'	Hybrid Tea	Poulsen
	'Giant Pink'®	Shrub	Lens
1990	'Paul Ricard'®	Hybrid Tea	Meilland
	Unnnamed seedling		Fryers
1989	Unnamed seedling		Dorieux
	'Cocorico'®	Floribunda	Meilland
1988	'Ecole d'Ecuilly'	Hybrid Tea	Laperrière
	'Orange Meillandina, Climbing'	Climber	Meilland
1986	'Cary Grant'™	Hybrid Tea	Meilland
	'Pupa'		Vivai Aurelia
1985	Unamed seedling		Fryer
1984	'Modern Art'®	Hybrid Tea	Poulsen
	'Lavender Dream'®	Shrub	Interplant
1983	'Miss Eleganza'		Vivai Aurelia
	'Gianlauro'		Vivai Aurelia
1982	'Kan-pai'	Hybrid Tea	Keisei
	'Rush'®	Shrub	Lens
1981	'Epidor'®	Hybrid Tea	Delbard
	'Snow Carpet'®	Miniature	McGredy
1980	'Pristine'®	Hybrid Tea	Jackson & Perkins
	Unnamed seedling		Meilland
1979	'Catherine Deneuve'®	Hybrid Tea	Meilland
	'Sally Holmes'®	Shrub	Fryer
1978	Unnamed seedling		McGredy
	'Margaret Merril'®	Floribunda	Harkness
1977	Unnamed seedling		Howard
	'Marjorie Fair'®	Shrub	Harkness
1976	'Double Delight'™	Hybrid Tea	Armstrong
	'Baby Bio'	Floribunda	Smith
1975	Unnamed seedling		Jackson & Perkins
	'Anne de Bretagne'®	Shrub	Meilland
1974	'Matangi'®	Floribunda	McGredy
	Unnamed seedling		Meilland
1973	'Fernand Majorel'	Floribunda	Gaujard
	'Grand Nord'®	Hybrid Tea	Delbard
1972	'Rusticana'		Meilland
	Unnamed seedling		Meilland
1971	'Florian'	Floribunda	Meilland
	'Pirotiti Xoia'		Pironti
1970	'Diablotin, Climbing'®	Climber	Delbard
	'Vol de Nuit'	Hybrid Tea	Delbard
1969	'Atlantis'	Floribunda	Harkness
	'El Cid'	Hybrid Tea	Armstrong
1968	Unnamed seedling		Meilland
	'Baronne Edmond de Rothschild'®	Hybrid Tea	Meilland
1967	'Madame Dimitriu'®	Polyantha	Delbard
	Unnamed seedling		Meilland
1966	'Sahara'		Tantau
	Unnamed seedling		Meilland
1965	Unnamed seedling		Meilland
	'Arianna'®	Hybrid Tea	Meilland
1964	'Zorina'	Floribunda	Jackson & Perkins
	'Blue Moon'®	Hybrid Tea	Tantau
1963	'Sea Foam'®	Shrub	Schwartz
	'Eiffel Tower'	Hybrid Tea	Armstrong
1962	Unnamed seedling		Jackson & Perkins
	'Allegro'®	Hybrid Tea	Meilland
1961	'Zambra'®	Floribunda	Meilland
	'La Scala'	Hybrid Tea	Howard
1960	'Purezza'	Climber	Mansuino
	'Piccadilly'®	Hybrid Tea	McGredy

Table 13 Continued

Year	*Cultivar*	*Class*	*Breeder*
1959	'Farandole'	Floribunda	Meilland
	'Pink Peace'®	Hybrid Tea	Meilland
1958	No award		
1957	No award		
1956	No award		
1955	'Fanfare'	Floribunda	Armstrong
	'Grand'mère Jenny, Climbing'	Climber	Meilland
1954	'Rosemary Rose'	Floribunda	DeRuiter
1953	'Pastorella'	Hybrid Tea	Meilland
1952	'Favorita'™	Hybrid Tea	Jackson & Perkins
1951	'Helen Traubel'	Hybrid Tea	Armstrong
1940	No award		
1939	'Flash'	Climber	Conard Pyle
1938	'Grande Duchesse Charlotte'	Hybrid Tea	Ketten Frees
1937	'Cardinal de la Puma'	Hybrid Tea	Leenders
1936	'Madame Jean Croibier'		Gaujard
1935	'Eclipse'	Hybrid Tea	Jackson & Perkins
1934	No award		
1933	'Condesa de Sástago'	Hybrid Tea	Dot

The Netherlands

Table 14 Gold Medals awarded at the Hague (1962–1996)

Year	*Cultivar*	*Class*	*Breeder*
1996	'Marie-Louise Velge'	Floribunda	RvS-Melle
1990	'Flower Carpet'™	Shrub	Noack
1988	'Graceland'	Hybrid Tea	Warriner.
	'Rainbow Yellow Parade'™	Miniature	Olesen
1985	'Kiboh'	Floribunda	Suzuki
1984	'Goldstar'	Hybrid Tea	Cants
1982	'Disco Dancer'®	Floribunda	Dickson
1978	'Marijke Koopman'	Hybrid Tea	Fryer
	'View'®	Polyantha	Lens
1976	'City of Belfast'®	Floribunda	McGredy
1973	'Scarlet Queen Elizabeth'®	Floribunda	Dickson
1972	'Amsterdam'®	Floribunda	Verschuren
1970	'Jan Spek'®	Floribunda	McGredy
	'Satchmo'®	Floribunda	McGredy
	'Electron'®	Hybrid Tea	McGredy
1968	'Espéranza'	Floribunda	Delforge
1967	'Pharaoh'	Hybrid Tea	Meilland
1966	'Sangria'®	Floribunda	Meilland
	'Irish Gold'	Hybrid Tea	Dickson
1965	'Scented Air'	Floribunda	Dickson
	'Arianna'®	Hybrid Tea	Meilland
1964	'Ronde Endiablée'	Floribunda	Combe
	'Silva'®	Hybrid Tea	Meilland
1963	'My Girl'	Floribunda	DeRuiter
	'Pascali'®	Hybrid Tea	Lens
1962	'Allegro'®	Hybrid Tea	Meilland
	'Europeana'®	Floribunda	DeRuiter

Table 15 'Golden Rose of The Hague' awards (1961–1997)

Year	*Cultivar*	*Class*	*Breeder*
1997	'Sunset Celebration'™	Hybrid Tea	Fryer
1990	'Mariandel'®	Floribunda	Kordes
1987	'Ingrid Bergman'®	Hybrid Tea	Olesen
1986	'Mountbatten'®	Floribunda	Harkness
1981	'Maria Mathilda'®	Miniature	Lens
1976	'Muchacha'®	Floribunda	Meilland
1973	'Scarlet Queen Elizabeth'®	Floribunda	Dickson
1968	'Orange Sensation'®	Floribunda	deRuiter
1966	'Lilli Marleen'®	Floribunda	Kordes
1965	'Peace'	Hybrid Tea	Meilland
1964	'Wendy Cussons'	Hybrid Tea	Gregory
1962	'Colibri'	Miniature	Meilland
1961	'Paprika'®	Floribunda	Tantau

New Zealand

Table 16 Award Winners at the New Zealand International Rose Trial Ground (1971–2000)

Year	*Award*	*Cultivar*	*Class*	*Breeder*
2000	Gold Star of the South Pacific	'Racy Lady'™	Hybrid Tea	Dickson
	Novelty	'Fourth of July'™	Climber	Carruth
	Fragance	'CHEwadmore'	Climber	Warner
1999	Gold Star of the South Pacific	'Serendipity' (HARamity)	Floribunda	Harkness
	Silver Star of the City of Palmerston North	'Mikayla Paige'	Floribunda	Flinn
	Fragrance	'Emperor' (MACgeorgi)	Hybrid Tea	McGredy
1998	Gold Star of the South Pacific	'Taffeta'	Climbing Mini-Flora	Warner
	Silver Star of the City of Palmerston North	'Blackberry Nip'(SOMnip)	Hybrid Tea	Somerfield
	Fragrance	'Taffeta' (CHEwcrest)	Climbing Mini-Flora	Warner
	Novelty and Certificate of Merit	'Scentimental'™	Floribunda	Carruth
1997	Gold Star of the South Pacific	'Anniversary'	Floribunda	Matthews
	Silver Star of the City of Palmerston North	'Millennium'	Floribunda	Grant
1996	Gold Star of the South Pacific	'Good as Gold'	Climbing Mini-Flora	Warner
	Fragrance and Certificate of Merit	'The McCartney Rose'™	Hybrid Tea	Meilland
1995	Gold Star of the South Pacific and Novelty	'Patio Honey'	Climbing Mini-Flora	Warner
	Silver Star of the City of Palmerston North	'Aorangi'	Hybrid Tea	Sandbrook
	Fragrance and Certificate of Merit	'Carolyn'	Hybrid Tea	McGredy
1993	Gold Star of the South Pacific	'Sheila's Perfume'	Hybrid Tea	Sheridan
	Novelty and Certificate of Merit	'Little Bo-Peep' (POUllen)	Groundcover	Olesen
	Fragrance	'Sheila's Perfume'	Hybrid Tea	Sheridan 1994
	Gold Star of the South Pacific	'Spirit of Hope'	Hybrid Tea	McGredy
	Novelty and Certificate of Merit	'Warm Welcome'	Climbing Mini-Flora	Warner
1992	Gold Star of the South Pacific	'Enchantment'	Floribunda	Olesen
	Fragrance and Certificate of Merit	'City of London'®	Climber	Harkness
	Novelty and Certificate of Merit	'Phantom' (MACcatsan)	Shrub	McGredy
1991	Gold Star of the South Pacific	'Foster's Wellington Cup'	Floribunda	McGredy
1990	Gold Star of the South Pacific	'Class Act'	Floribunda	Warriner
	Fragrance and Certificate of Merit	'Aotearoa–New Zealand'	Hybrid Tea	McGredy
1989	Gold Star of the South Pacific	'Remuera'	Floribunda	McGredy
1988	Gold Star of the South Pacific	'Amber Queen'®	Floribunda	Harkness
	Silver Star of the City of Palmerston North	'Iris Gee'	Floribunda	Gee
	Fragrance and Certificate of Merit	'Embruixada'	Hybrid Tea	Dot
1987	Gold Star of the South Pacific	'Elina'®	Hybrid Tea	Dickson
1986	Gold Star of the South Pacific	'Hot Chocolate'	Floribunda	Simpson
1985	Gold Star of the South Pacific	'Fyvie Castle'	Hybrid Tea	Cocker
		'Beachcomber' (MAClapaz)	Floribunda	McGredy
	Silver Star of the City of Palmerston North	'Offbeat'	Floribunda	Simpson

Table 16 Continued

Year	*Award*	*Cultivar*	*Class*	*Breeder*
1984	Gold Star of the South Pacific	'Golden Queen'	Hybrid Tea	Kordes
		'Sexy Rexy'®	Floribunda	McGredy
		'Sierra Sunrise'	Miniature	Moore
	Fragrance and Certificate of Merit	'Heart Throb'	Floribunda	Harkness
1983	Gold Star of the South Pacific	'Gold Medal'®	Grandiflora	Christensen
		'MACchrijo'	Floribunda	McGredy
		'Angelita'®	Miniature	McGredy
	Fragrance and Certificate of Merit	'English Miss'	Floribunda	Cants
1982	Gold Star of the South Pacific	'Lady Meilland'	Hybrid Tea	Meilland
		'Margaret Merril'®	Floribunda	Harkness
		'Clarissa'®	Miniature	Harkness
1981	Gold Star of the South Pacific	'Te Moana'	Hybrid Tea	Simpson
		'Sue Lawley'	Floribunda	McGredy
	Fragrance and Certificate of Merit	'Esmeralda'®	Hybrid Tea	Kordes
1980	Gold Star of the South Pacific	'Manou Meilland'®	Hybrid Tea	Meilland
		'Uetersen'®	Climbing Floribunda	Kordes
1979	Gold Star of the South Pacific	'True Love'	Hybrid Tea	DeRuiter
		'Brown Velvet'	Floribunda	McGredy
	Silver Star of the City of Palmerston North	'Anne Kercher'	Hybrid Tea	Simpson
	Fragrance and Certificate of Merit	'Deep Secret'	Hybrid Tea	Tantau
1978	Gold Star of the South Pacific	'Benson & Hedges Gold'®	Hybrid Tea	McGredy
		'Strawberry Ice'	Floribunda	Delbard
	Silver Star of the City of Palmerston North	'Velvet Lustre'	Hybrid Tea	Simpson
1977	No award			
1976	Gold Star of the South Pacific	'John Simpson'	Hybrid Tea	McGredy
		'Julischka'®	Floribunda	Tantau
	Silver Star of the City of Palmerston North	'Sport of Eutin'	Floribunda	Burnett
1975	Gold Star of the South Pacific	'Vienna Woods'	Hybrid Tea	Kordes
		'Friesia'	Floribunda	Kordes
	Silver Star of the City of Palmerston North	'Simpson's Red'	Hybrid Tea	Simpson
1974	Gold Star of the South Pacific	'Pink Silk'	Hybrid Tea	Gregory
		'Coventry Cathedral'	Floribunda	McGredy
	Silver Star of the City of Palmerston North	'Avril Sherwood'	Floribunda	Sherwood
	Fragrance	'Pink Silk'	Hybrid Tea	Gregory
1973	Gold Star of the South Pacific	'Landora'	Hybrid Tea	Tantau
		'Picasso'	Floribunda	McGredy
	Fragrance	'Westerland'®	Floribunda	Kordes
1972	Gold Star of the South Pacific	'Olympic Torch'	Hybrid Tea	Keisei
		'Elka Gaarlandt'	Floribunda	Buisman
	Fragrance	'Courvoisier'®	Floribunda	McGredy
	Silver Star of the City of Palmerston North	'601'	Floribunda	Sherwood
1971	Gold Star of the South Pacific	'Pania'	Hybrid Tea	McGredy
		'City of Belfast'®	Floribunda	McGredy
	Fragrance	'Happy Anniversary'	Floribunda	Delbard

Table 17 Award winners at the Rose Introducers of New Zealand (RINZ) International Rose Trial Ground (1990–2000)

Year	*Award*	*Cultivar*	*Class*	*Breeder*
2000	Hybrid Tea	'Tintinara'	Hybrid tea	Dickson
	Climber	'Little Rambler'	Climbing Miniature	Warner
	Shrub	'Country Touch'		Kordes
	Miniature	'Cinderella'		Schuurman
	Most Fragrant Rose	'Blackberry Nip'		Somerfield
	Auckland Rose of the Year	'Heartache'		Kordes
1999	Hybrid Tea	'Tintinara'™	Hybrid Tea	Dickson
	Floribunda	'Heartache'	Floribunda	Kordes
	Miniature/Patio	'Cinderella'(SUNmapa)		Schuurman
	Climber	'Little Rambler'	Climbing Miniature	Warner

Table 17 Continued

Year	*Award*	*Cultivar*	*Class*	*Breeder*
	Groundcover/Shrub	'Country Touch'	Shrub	Kordes
	Most Fragrant	'Blackberry Nip' (SOMnip)	Hybrid Tea	Sommerfield
	Auckland Rose of The Year	'Heartache'	Floribunda	Kordes
1998	Hybrid Tea	'Spirit of Hope' (MAChahei)	Grandiflora	McGredy
	Floribunda	'Summer Dream'	Floribunda	Fryer
	Miniature/Patio	'Hauraki'	Patio	McGredy
	Climber	'Patio Honey'	Climbing Miniature	Warner
	Groundcover/Shrub	'Country Prince' (KORfungo)		Kordes
	Most Fragrant	'Celebration' (COCgrand)		Cocker
	Auckland Rose of The Year	'Hauraki'	Patio	McGredy
1997	Hybrid Tea	'Valencia'®	Hybrid Tea	Kordes
	Floribunda	'Enchantment'	Shrub	Poulsen
	Miniature/Patio	'Manapouri'	Patio	McGredy
	Climber	'Patio Queen'	Climbing Miniature	Warner
	Groundcover/Shrub	'Cottage Maid'	Shrub	Poulsen
	Most Fragrant	'Mamy Blue'®	Hybrid Tea	Delbard
	Auckland Rose of The Year	'Cottage Maid'	Shrub	Poulsen
1996	Hybrid Tea	'Warm Wishes'	Hybrid Tea	Fryer
	Floribunda	'Triple Treat' (DELroblu)		Delbard
	Miniature/Patio	'Scentasia'	Miniature	Schuurman
	Climber	'High Hopes'	Climber	Harkness
	Groundcover/Shrub	'Glowing Cushion'	Shrub	Ilsink
	Most Fragrant	'Super Bowl'	Grandiflora	McGredy
	Auckland Rose of The Year	'Sunnyside'	Shrub	Poulsen
1995	Hybrid Tea	'Candella'®	Hybrid Tea	McGredy
	Floribunda	'Rosemary Harkness'	Hybrid Tea	Harkness
	Miniature/Patio	'Sequoia Gold'™	Miniature	Moore
	Climber	'Morning Light'	Climber	Harkness
	Groundcover/Shrub	'Sunnyside'	Shrub	Poulsen
	Most Fragrant	'Rosemary Harkness'	Hybrid Tea	Harkness
	Auckland Rose of The Year	'Sequoia Gold'™	Miniature	Moore
1994	Hybrid Tea	'Gold Medal'®	Hybrid Tea	Christensen
	Floribunda	'Raspberry Ice'	Floribunda	Kordes
	Miniature/Patio	'Rosabell'	Floribunda	Cocker
	Climber	'Warm Welcome'	Climbing Miniature	Warner
	Groundcover/Shrub	'Robin Redbreast'	Mini-Flora	Ilsink
	Most Fragrant	'Dioressence'®	Grandiflora	Delbard
	Auckland Rose of The Year	'Eldorado'	Floribunda	Christensen
1993	Hybrid Tea	'Player'	Hybrid Tea	Schuurman
	Floribunda	'Happy Go Lucky'	Grandiflora	Schuurman
	Miniature/Patio	'Pretty Polly'	Miniature	Meilland
	Climber	'City of London'®	Floribunda	Harkness
	Most Fragrant	'Big Purple'	Hybrid Tea	Stephens
	Auckland Rose of The Year	'Dublin Bay'	Climber	McGredy
1992	Hybrid Tea	'Paddy Stephens'	Hybrid Tea	McGredy
	Floribunda	'Eldorado'	Floribunda	Christensen
	Miniature/Patio	'Little Opal'	Patio	Schuurman
	Climber	'Dublin Bay'	Climber	McGredy
	Most Fragrant	'Margaret Merril'®	Floribunda	Harkness
	Auckland Rose of The Year	'Little Opal'	Patio	Schuurman
1991	Hybrid Tea	'Amberlight'	Hybrid Tea	Cocker
	Floribunda	'Margaret Merril'®	Floribunda	Harkness
	Miniature/Patio	'Golden Rosamini'	Miniature	Ilsink
	Most Fragrant	'Heart Throb'	Hybrid Tea	Harkness
	Auckland Rose of The Year	'Sexy Rexy'	Floribunda	McGredy
1990	Hybrid Tea	'Aotearoa–New Zealand'	Hybrid Tea	McGredy
	Floribunda	'Sexy Rexy'	Floribunda	McGredy
	Miniature	'Whiteout'	Miniature	McGredy
	Most Fragrant	'Aotearoa–New Zealand'	Hybrid Tea	McGredy
	Auckland Rose of The Year	'Aotearoa–New Zealand'	Hybrid Tea	McGredy

South Africa

Table 18 Gold Medal Winners at Durbanville, South Africa (1981–1999)

Year	*Cultivar*	*Class*	*Breeder*
1999	TAN 92114		Tantau
1995	'TAN 91172'		Tantau
1994	'Melrose'		Melle Research Station
1993	'DELserex'		Delbard
	'Molineux'	Shrub	Austin
1992	'Ficksburg'	Floribunda	Delbard
1991	'Efekto 21' (MEItolba)		Meilland
1990	'Christopher Columbus'	Hybrid Tea	Meilland
1989	'Glow Achievement'	Floribunda	Tantau
1988	'Glad Tidings'	Floribunda	Tantau
1987	'Diadem'	Floribunda	Tantau
1986	'Pink La Sevillana'®	Floribunda	Meilland
1985	'Satellite'	Hybrid Tea	Delbard
1984	'Royal Show' (MEIrasimac)	Shrub	Meilland
1983	'Little Red Hedge' (KORtaba)	Floribunda	Kordes
1982	'Lady Meilland'	Hybrid Tea	Meilland
1981	'Bella Rosa'®	Floribunda	Kordes

Spain

Table 19 Gold Medal Awards at the International Rose Trials of Madrid (1956–1999)

Year	*Cultivar*	*Class*	*Breeder*
1999	'Unnamed seedling'		Harkness
1998	'Heavenly Pink'	Hybrid Musk	Lens
1997	No award		
1996	No name		Kordes
1995	No name		Meilland
1994	'Guimalda de Amor'		Lens
1993	No name		Kordes
1992	No name		Lens
1991	'Frisson Frais'®	Hybrid Musk	Lens
1990	'Pleins Feux'	Polyantha	Dorieux
1989	'Karine Sauvageot'®	Floribunda	Sauvageot
1988	'Yellow Fairy'®	Shrub	Poulsen
1987	'Horsholm City-Rose'		Poulsen
1986	'Ingrid Bergman'®	Hybrid Tea	Poulsen
1985	'JACara'		Jackson & Perkins
1984	'Bordure Vive'®	Hybrid Tea	Delbard
1983	'Encarnita de Tiemo'		Muftoz
1982	'Satellite'	Hybrid Tea	Delbard
1981	'Snow Carpet'®	Miniature	McGredy
1980	'Soleil de Feu'		Delbard
1979	'Pénélope'®	Hybrid Tea	Gaujard
1978	'Priscilla Burton'®	Floribunda	McGredy
1977	'Manou Meilland'®	Hybrid Tea	Meilland
1976	'Julie Delbard'®	Floribunda	Delbard
1975	'Fragola'	Hybrid Tea	Croix
1974	'Bordure Rose'®	Floribunda	Delbard
1973	'The Sun'	Floribunda	McGredy
1972	'Tony Jacklin'	Floribunda	McGredy
1971	'Argental'	Hybrid Tea	Croix
1970	'Crêpe de Chine'®	Hybrid Tea	Delbard
1969	'Kalinka'	Floribunda	Meilland
1968	'Prince Tango'®	Polyantha	Delbard

Table 19 Continued

Year	*Cultivar*	*Class*	*Breeder*
1967	'Pharaon'®	Hybrid Tea	Meilland
1966	'Samourai'	Groundcover	Meilland
1965	'Tourmaline'®	Hybrid Tea	Delbard Chabert
1964	No name		Meilland
1963	'Roi Soleil'	Hybrid Tea	Dorieux
1962	'Royal Highness'	Hybrid Tea	Swim & Weeks
1961	'Isabel de Ortiz'	Hybrid Tea	Kordes
1960	'Piccadilly'®	Hybrid Tea	McGredy
1959	'Santa Tereza d'Avila'	Hybrid Tea	Da Silva
1958	'Fugue'®	Climber	Meilland
1957	'Champs-Elysées'®	Hybrid Tea	Meilland
1956	'Condesa de Mayalde'	Hybrid Tea	Dot

Switzerland

Table 20 Awards at International Rose Trials of Geneva (1948–2000)

Year	*Award*	*Cultivar*	*Class*	*Breeder*
2000	Golden Rose	(LAPomira)	Hybrid Tea	Laperrière
	Gold Medals	'Neige Parfum'	Hybrid Tea	Dorieux
		'Raymond Nazereau' (ADAquethom)	Floribunda	Adam
		'Amber Cover' (94/4197-6)	Miniature	Poulsen
		(SB-93201)	Shrub	Sauvageot
		(92/2234-1)	Climber	Poulsen
	Most Fragrant Rose	(MEIceppus) (AM 988)		Meilland
1999	Golden Rose	DORcafe	Hybrid Tea	Dorieux
	Gold Medals	'AM 979'	Hybrid Tea	Meilland
		'RT 9268'	Floribunda	Tantau
		CHEwcreepy'	Miniature	Warner
		'LLX 9001'	Shrub	Lens
		'WHOL'	Climber	Wuest
	Most Fragrant Rose	'MR 977	Hybrid Tea	Richardier
1998	Golden Rose	'AM 963' (MEInnzincaro)	Hybrid Tea	Meilland
	Gold Medals	'LAPasu'	Hybrid Tea	Laperrière
		'Jardins de France'	Floribunda	Richardier
		'CHEwayaye'	Miniature	Warner
		'BAR 4475'	Shrub	Barni
		'CROpal'	Climber	Croix
	Most Fragrant Rose	'ORAdal'	Hybrid Tea	Orard
1997	Golden Rose	'Marie-Louise Velge'	Floribunda	RVS
	Gold Medals	'Michelangelo'	Hybrid Tea	Meilland
		'RvS 90-94'	Floribunda	RVS
		'CHEwgreat'	Miniature	Warner
		'Linderof'	Shrub	Kordes
	Most Fragrant Rose	'Caprice de Meilland'	Hybrid Tea	Meilland
1996	Golden Rose	'Eclipse'	Floribunda	Sauvageot
	Gold Medals	'Poker'	Hybrid Tea	Meilland
		'LAPema'	Floribunda	Laperrière
		'CHEwdarkeye'	Miniature	Warner
		'Hugues Aufray'	Shrub	Dorieux
		'Soirée de Bonheur' (CROpal)	Climber	Croix
	Most Fragrant Rose	'Chartreuse de Parme'	Hybrid Tea	Delbard
1995	Golden Rose	'Rose d'Annecy'®	Shrub	Adam
	Gold Medals	'Comtesse Brigitte Chandon-Moet'	Hybrid Tea	Dorieux
		'United Nations Rose'	Floribunda	Harkness
		'Red Paillette'	Miniature	Poulsen

Table 20 Continued

Year	*Award*	*Cultivar*	*Class*	*Breeder*
		'MT 86360'	Shrub	Tantau
		'KORlehnig'	Climber	Kordes
	Most Fragrant Rose	'DELrousi'	Hybrid Tea	Delbard
1994	Golden Rose	'Cherry Meillandecor'®	Shrub	Meilland
	Gold Medals	'HARwippee'	Hybrid Tea	Harkness
		'NDR 1 Radio Niedersachsen'	Floribunda	Kordes
		'AM 923'	Miniature	Meilland
		'INT B82688'	Shrub	Interplant
		'KORwerti'	Climber	Kordes
1993	Golden Rose	'Suni'®	Floribunda	Barni
	Gold Medals	'ADAsilli'	Hybrid Tea	Adam
		'KORgezak'	Floribunda	Kordes
		'Shine On'™	Miniature	Dickson
		'Devon'®	Shrub	Poulsen
		'INt B12185'	Climber	Interplant
	Most Fragrant Rose	'Selbar 3694'	Hybrid Tea	Barni
1992	Golden Rose	'Helga Piaget'	Hybrid Tea	Meilland
	Gold Medals	'Imperatrice Farah'	Hybrid Tea	Delbard
		'Sommermarchen'	Shrub	Kordes
		'INT B20586'	Miniature	Interplant
		'KORmalkus'	Climber	Kordes
	Most Fragrant Rose	No name	Hybrid Tea	Ferrer
1991	Golden Rose	'Selbar 3016'	Floribunda	Barni
	Gold Medals	'Signature'®	Hybrid Tea	Jackson & Perkins
		'Selbar 3016'	Floribunda	Barni
		'INT B30885'	Miniature	Interplant
		'Explorer'	Climber	Dickson
	Most Fragrant Rose	'Selbar 3050'	Hybrid Tea	Barni
1990	Golden Rose	'Roxane'®	Hybrid Tea	Laperrière
	Gold Medals	'Roxane'®	Hybrid Tea	Laperrière
		'RVS 85-66'	Floribunda	RVS
		'INT B24384'	Miniature	Interplant
		'KO 813363'	Climber	Kordes
	Most Fragrant Rose	'Detty'	Hybrid Tea	Adam
1989	Golden Rose	'AM 873'	Floribunda	Meilland
	Gold Medals	'Miriam Tudor' ('FRYmartini')	Hybrid Tea	Fryer
		'AM 873'	Floribunda	Meilland
		'83/2395'	Miniature	Poulsen
		'Ramira'®	Climber	Kordes
	Most Fragrant Rose	'Nella Martinetti'®	Hybrid Tea	Meilland
1988	Golden Rose	'AM 860'	Hybrid Tea	Meilland
	Gold Medals	'AM 860'	Hybrid Tea	Meilland
		'DIC17108'	Floribunda	Dickson
		'INT B13882'	Miniature	Interplant
	Most Fragrant Rose	'Zepigpu'	Hybrid Tea	McGredy
1987	Golden Rose	'LM 854'	Floribunda	Meilland
	Gold Medals	'Don Bosco' (LAPour)	Hybrid Tea	Laperrière
		'LM 854'	Floribunda	Meilland
		'Prima'	Miniature	Harkness
		'Pink Robusta'®	Climber	Kordes
	Most Fragrant Rose	'LAPab'	Hybrid Tea	Laperrière
1986	Golden Rose	'KORwelle'	Hybrid Tea	Kordes
	Gold Medals	'KORwelle'	Hybrid Tea	Kordes
		'ES.9/1/81'	Floribunda	Smith
		'INT B10181'	Miniature	Interplant
	Most Fragrant Rose	'AM 841'	Hybrid Tea	Meilland
		'ES 47/1/81'	Floribunda	Smith
		'78-6093'	Climber	Poulsen
1985	Golden Rose	'Sourire d'Orchidée'	Climber	Croix
	Gold Medals	'AM 832'	Hybrid Tea	Meilland

Table 20 Continued

Year	*Award*	*Cultivar*	*Class*	*Breeder*
		'Fragrant Delight'®	Floribunda	Wisbech
		'Sourire d'Orchidée'	Climber	Croix
	Most Fragrant Rose	'MT 77137'	Hybrid Tea	Tantau
1984	Golden Rose	'Jardins de Bagatelle'®	Hybrid Tea	Meilland
	Gold Medals	'Jardins de Bagatelle'®	Hybrid Tea	Meilland
		'Artiste'®	Floribunda	Dorieux
	Most Fragrant Rose	'JL 80/4'	Hybrid Tea	Leenders
1983	Golden Rose	'International Herald Tribune'®	Floribunda	Harkness
	Gold Medals	'Louis de Funès'	Hybrid Tea	Meilland
		'International Herald Tribune'®	Floribunda	Harkness
	Most Fragrant Rose	'MT 77247'	Hybrid Tea	Tantau
1982	Golden Rose	'MEIdauvival'	Hybrid Tea	Meilland
	Gold Medals	'MEIdauvival'	Hybrid Tea	Meilland
		'JL 1977/6'	Floribunda	Leenders
	Most Fragrant Rose	'MEIdauvival'	Hybrid Tea	Meilland
1981	Golden Rose	'Tobago'®	Floribunda	Delbard
	Gold Medals	'Black Night'®	Hybrid Tea	Huber
		'Tobago'®	Floribunda	Delbard
	Most Fragrant Rose	'MEIcerdonite'	Hybrid Tea	Meilland
1980	Golden Rose	'Neville Gibson'	Hybrid Tea	Harkness
		'Soleil de Feu'	Floribunda	Delbard
	Gold Medal	'Neville Gibson'	Hybrid Tea	Harkness
	Most Fragrant Rose	'Radox Bouqet'	Floribunda	Harkness
1979	Golden Rose	'Farouche'	Floribunda	Meilland
	Gold Medals	'Helmut Schmidt'®	Hybrid Tea	Kordes
		'Farouche'	Floribunda	Meilland
	Most Fragrant Rose	'GAUdabon'	Hybrid Tea	Gaujard
1978	Gold Medals	'Lord Gold'®	Hybrid Tea	Delbard
		'Margaret Merril'®	Floribunda	Harkness
	Most Fragrant Rose	'JL 3325-306'	Hybrid Tea	Leenders
1977	Gold Medals	No name		DeRuiter
		'Candeur'®	Floribunda	Delbard
	Most Fragrant Rose	No Name		Herhold
1976	Gold Medals	'Estelle'	Hybrid Tea	Croix
		'Parador'®	Hybrid Tea	Meilland
	Most Fragrant Rose	'Double Delight'®	Hybrid Tea	Armstrong
1975	Gold Medals	'MEIvilese'	Hybrid Tea	Meilland
		'Nirvana'®	Floribunda	Meilland
		'Compassion'®	Climber	Harkness
	Most Fragrant Rose	'Destin'	Hybrid Tea	Croix
1974	Gold Medals	'MEIzera'	Hybrid Tea	Meilland
		'Légion d'Honneur'®	Floribunda	Delbard
		'Pironti Critow'	Climber	Pironti
	Most Fragrant Rose	'JP69 461'		Jackson & Perkins
1973	Gold Medals	'KO 66267'	Hybrid Tea	Kordes
		'MEInoemi'	Floribunda	Meilland
	Most Fragrant Rose	'MEInausic'	Hybrid Tea	Meilland
1972	Gold Medals	'JPC 63-3930'	Hybrid Tea	Jackson & Perkins
		'Rusticana'	Floribunda	Meilland
	Most Fragrant Rose	'Gamma'	Hybrid Tea	Gaujard
1971	Gold Medals	'Atoll'	Hybrid Tea	Meilland
		'MT 6614'	Floribunda	Tantau
	Most Fragrant Rose	'Argental'	Hybrid Tea	Croix
1970	Gold Medals	'Aquarius'	Grandiflora	Armstrong
		'Sweet Song'	Floribunda	Meilland
	Most Fragrant Rose	'Duftzauber'	Hybrid Tea	Kordes
1969	Gold Medals	'Silent Night'	Hybrid Tea	McGredy
		'MEIsibar'	Floribunda	Meilland
	Most Fragrant Rose	'Merci'	Hybrid Tea	Gaujard
1968	Gold Medals	'Star Burst' (MEIgasco)	Hybrid Tea	Meilland
		'City of Belfast'®	Floribunda	McGredy

Table 20 Continued

Year	*Award*	*Cultivar*	*Class*	*Breeder*
	Most Fragrant Rose	'HD 1.880.1'		Delbard
1967	Gold Medals	'Pharaon'®	Hybrid Tea	Meilland
		'Mme Dimitriu'®	Polyantha	Delbard
		'Obélisque'®	Climber	Delbard
	Most Fragrant Rose	'Red Chief'	Hybrid Tea	Armstrong
1966	Gold Medals	'Versailles'®	Hybrid Tea	Delbard
		'Sangria'®	Floribunda	Meilland
	Most Fragrant Rose	'26 AN17'	Hybrid Tea	Armstrong
1965	Gold Medals	'Pink Supreme'	Hybrid Tea	DeRuiter
		'MEldiam'	Floribunda	Meilland
	Most Fragrant Rose	'Youri San'	Hybrid Tea	Meilland
1964	Gold Medals	'Divine'	Hybrid Tea	Delbard
		'56.6021'	Floribunda	Jackson & Perkins
	Most Fragrant Rose	'Sweet Afton'	Hybrid Tea	Armstrong
1963	Gold Medals	'Eiffel Tower'	Hybrid Tea	Armstrong
		'56-17211'	Floribunda	Jackson & Perkins
	Most Fragrant Rose	'Papa Meilland'®	Hybrid Tea	Meilland
1962	Gold Medals	'Maman'	Hybrid Tea	Delbard
		'56-2342'	Floribunda	Jackson & Perkins
	Most Fragrant Rose	'56-20722'		Jackson & Perkins
1961	Gold Medals	'Columbus Queen'	Hybrid Tea	Armstrong
		'1622-57-3'	Floribunda	Kordes
	Most Fragrant Rose	'5096-56-3'		Kordes
1960	Gold Medals	'Super Star'	Hybrid Tea	Tantau
		'MElmen'	Floribunda	Meilland
	Most Fragrant Rose	No name		Tantau
1959	Gold Medals	'Pink Peace'®	Hybrid Tea	Meilland
		'Tantau's Favorit'	Floribunda	Tantau
		'Bettina, Climbing'®	Climber	Meilland
1958	Gold Medals	'53-17342'	Hybrid Tea	Jackson & Perkins
		'Edith de Martinelli'®	Floribunda	Arles
1957	Gold Medals	'MEltrans'	Hybrid Tea	Meilland
		'Sarabande'®	Floribunda	Meilland
1956	Gold Medals	'Liliana'	Hybrid Tea	Nadal
		'Paprika'®	Floribunda	Tantau
1955	Gold Medals	'Montezuma'	Hybrid Tea	Armstrong
		'Circus'	Floribunda	Armstrong
1954	Gold Medals	'Roundelay'	Hybrid Tea	Armstrong
		'Java'	Floribunda	Mallerin
1953	Gold Medals	'Mojave'	Hybrid Tea	Armstrong
		'Feuerbrand'	Floribunda	Tantau
1952	Gold Medals	'Buccaneer'	Hybrid Tea	Armstrong
		'Moulin Rouge'®	Floribunda	Meilland
1951	Gold Medals	'Mme Roger Verlomme'	Hybrid Tea	Mallerin
		'Cocorico'®	Floribunda	Meilland
1950	Gold Medals	'Ville de Gand'	Hybrid Tea	Gaujard
		'Vogue'	Floribunda	Jackson & Perkins
1949	Gold Medals	'Sutter's Gold'	Hybrid Tea	Armstrong
1948	Gold Medals	'Marcelle Gret'	Hybrid Tea	Meilland
		'Alain'	Floribunda	Meilland

United Kingdom

Table 21 Awards by Royal National Rose Society, St Albans (1963–2000)

Year	*Award*	*Cultivar*	*Class*	*Breeder*
2000	President' International Trophy	'Rotfassade' (TG2468)	Climber	Noack
1999	Gold Medal	'Rhapsody in Blue'	Shrub	Cowlishaw
1998	President's International Trophy	'CHEwreepy'		Warner
	Gold Medals	'JP90-2668'		Bear Creek
		'RVS 90-94'		RVS
1997	President's International Trophy	'Greetings'		Bear Creek
1996	President's International Trophy	'Molineux'™	Shrub	Austin
	Gold Medal	'Starry Eyed'	Shrub	Horner
1995	President's International Trophy	'Broadlands'	Shrub	Tantau
1994	President's International Trophy	'Twilight'	Shrub	Noack
1993	Gold Medals	'Baby Love'™	Miniature	Scrivens
		'Oxfordshire'	Kordes	Shrub
1992	President's International Trophy	'Blenheim'	Shrub	Tantau
1991	President's International Trophy	'Little Bo-Peep'	Miniature	Poulsen
	Gold Medals	'Flower Carpet White'	Shrub	Noack
		'Singin' in the Rain'	Floribunda	McGredy
1990	President's International Trophy	'Kent'®	Shrub	Poulsen
	Gold Medal	'Top Marks'	Miniature	Fryer
		'Fellowship'	Floribunda	Harkness
1989	President's International Trophy	'Pretty Polly'	Miniature	Meilland
1988	President's International Trophy	'Warm Welcome'	Climbing Miniature	Warner
	Gold Medal	'Tango'	Shrub	McGredy
1987	President's International Trophy	'Mary Hayley Bell'	Shrub	Kordes
	Gold Medal	'Surrey'	Shrub	Kordes
1986	No award			
1985	President's International Trophy	'Solitaire'	Hybrida Tea	McGredy
	Gold Medal	'Lincoln Cathedral'	Hybrid Tea	Langdale
1984	President's International Trophy	'Anisley Dickson'	Floribunda	Dickson
	Gold Medals	'Grouse'	Shrub	Kordes
		'Laughter Lines'	Floribunda	Dickson
1983	President's International Trophy	'Ards Beauty'	Floribunda	Dickson
	Gold Medal	'Freedom'®	Hybrid Tea	Dickson
1982	President's International Trophy	'The Times'		Kordes
	Gold Medal	'POUlbright'	Floribunda	Poulsen
1981	President's International Trophy	'Anna Ford'®	Miniature	Harkness
1980	President's International Trophy	'Rebecca Claire'	Hybrid Tea	Law
1979	No award			
1978	No award			
1977	President's International Trophy	'Silver Jubilee'®	Hybrid Tea	Cocker
1976	President's International Trophy	'Priscilla Burton'®	Floribunda	McGredy
1975	No award			
1974	President's International Trophy	'Matangi'®	Floribunda	McGredy
1973	No award			
1972	President's International Trophy	'Topsi'®	Floribunda	Tantau
1971	President's International Trophy	'Fountain'	Hybrid Tea	Tantau
1970	President's International Trophy	'Alec's Red'®	Hybrid Tea	Cocker
	Gold Medal	'News'®	Floribunda	LeGrice
1969	President's International Trophy	'Red Planet'	Hybrid Tea	Dickson
	Gold Medal	'Mullard Jubilee'	Hybrid Tea	McGredy
1968	President's International Trophy	'Molly McGredy'	Floribunda	McGredy
1967	President's International Trophy	'City of Belfast'®	Floribunda	McGredy
	Gold Medal	'Fred Loads'	Shrub	Holmes
1966	No awards			
1965	President's International Trophy	'Grandpa Dickson'	Hybrid Tea	Dickson
	Gold Medals	'City of Leeds'	Floribunda	McGredy
		'Ernest H. Morse'	Hybrid Tea	Kordes
1964	President's International Trophy	'Fragrant Cloud'	Hybrid Tea	Tantau
1963	President's International Trophy	'Elizabeth of Glamis'®	Floribunda	McGredy
	Gold Medal	'Evelyn Fison'	Floribunda	McGredy
		'Fragrant Cloud'	Hybrid Tea	Tantau

Table 22 Awards at International Rose Trials of Belfast, Northern Ireland (1965–2000)

Year	*Award*[a]	*Cultivar*	*Class*	*Breeder*
1999/2000	Gold Medal and Prize of the City of Belfast	'Pride of England'	Hybrid Tea	Harkness
	DARD Award	'White Gold'	Floribunda	Cocker
	Sainsbury Award	'Charming Cover'	Shrub	Poulsen
	Fragrance	'White Gold'		Cocker
1998/99	Gold Medal and Prize of the City of Belfast	'Poetry in Motion'	Hybrid Tea	Harkness
	Dani Award and Certificate of Merit	'Schloss Balthasar'	Floribunda	Kordes
	R. J. Frizzell Memorial Award	'Constance Finn'	Floribunda	Harkness
	The Sainsbury Award	'Wine and Dine'®	Shrub	Dickson
1997/98	Gold Medal and Prize of the City of Belfast	'St Christopher'	Hybrid Tea	Harkness
	Dani Award and Certificate of Merit	'Pretty Lady'	Floribunda	Scrivens
	R. J. Frizzell Memorial Award	'Helena Renaissance'	Shrub	Poulsen
	The Sainsbury Award	'Sommerabend'®	Shrub	Kordes
1996/97	Gold Medal and Prize of the City of Belfast	'Pure Bliss'™	Hybrid Tea	Dickson
	Dani Award and Certificate of Merit	'Georgette'	Floribunda	Kordes
	R. J. Frizzell Memorial Award	'Benelux Star'	Hybrid Tea	Landbow
	City of Belfast Award	'Our Molly'	Shrub	Dickson
1995/96	Gold Medal and Prize of the City of Belfast	'Warm Wishes'	Hybrid Tea	Fryer
	Dani Award and Certificate of Merit	'Domstadt Fulda'®	Floribunda	Kordes
	R. J. Frizzell Memorial Award	'Frederic Mistral'®	Hybrid Tea	Meilland
	Tyrone Crystal Prize	'Turbo'™	Hybrid Rugosa	Meilland
1994/95	Gold Medal and Prize of the City of Belfast	'Abbaye de Cluny'™	Hybrid Tea	Meilland
	Dani Award and Certificate of Merit	'Ruby Anniversary'	Floribunda	Harkness
	R. J. Frizzell Memorial Award	'Renaissance'	Hybrid Tea	Harkness
	Tyrone Crystal Prize	'Roselina'™	Shrub	Kordes
1993/94	Gold Medal and Prize of the City of Belfast	'Memoire'®	Hybrid Tea	Kordes
	Dani Award and Certificate of Merit	'Sunseeker'	Floribunda	Dickson
	R. J. Frizzell Memorial Award	'Marco Polo'	Hybrid Tea	Meilland
	Tyrone Crystal Prize	'Yellow Fleurette'®	Shrub	Ilsink
1992/93	Gold Medal and Prize of the City of Belfast	'New Horizon'	Floribunda	Dickson
	Dani Award and Certificate of Merit	'La Vie en Rose'	Floribunda	Kameyama
	R. J. Frizzell Memorial Award	'The McCartney Rose'™	Hybrid Tea	Meilland
	Tyrone Crystal Prize	'Eye Appeal'®	Shrub	Ilsink
1991/92	Gold Medal and Prize of the City of Belfast	'Painted Moon'	Hybrid Tea	Dickson
	Dani Award and Certificate of Merit	'Day Light'®	Floribunda	Ilsink
	R. J. Frizzell Memorial Award	'Sweet Juliet'	Shrub	Austin
	Tyrone Crystal Prize	'Colleen'		Ilsink
1990/91	Gold Medal and Prize of the City of Belfast	'Tequila Sunrise'	Hybrid Tea	Dickson
	Dani Award and Certificate of Merit	'Cocorico'®	Floribunda	Meilland
	R. J. Frizzell Memorial Award	'Fragrant Dream'	Hybrid Tea	Dickson
	Tyrone Crystal Prize	'Biddulph Grange'	Shrub	Fryer
1989/90	Gold Medal and Prize of the City of Belfast	'Malcolm Sargent'	Hybrid Tea	Harkness
	Dani Award and Certificate of Merit	'City of London'®	Floribunda	Harkness
	R. J. Frizzell Memorial Award	'Rendez-vous'	Hybrid Tea	Meilland
	Tyrone Crystal Prize	'Dortmunder'		Ilsink
1988/89	Gold Medal and Prize of the City of Belfast	'Cherry Brandy '85'	Hybrid Tea	Tantau
	Dani Award and Certificate of Merit	'Orange Fire'®	Floribunda	Ilsink
	Tyrone Crystal Prize	'Eye Opener'®	Shrub	Ilsink
1987/88	Gold Medal and Prize of the City of Belfast	'Lovely Lady'	Hybrid Tea	Dickson
	Dani Award and Certificate of Merit	'Conqueror's Gold'	Floribunda	Harkness
	Tyrone Crystal Prize	'Pink Torch'®	Shrub	Ilsink
1986/87	Gold Medal and Prize of the City of Belfast	'Rosemary Harkness'	Hybrid Tea	Harkness
	Dani Award and Certificate of Merit	'Fruité'	Floribunda	Meilland
	R. J. Frizzell Memorial Award	'Rosemary Harkness'		Harkness
1985/86	Gold Medal and Prize of the City of Belfast	'Remember Me'®	Hybrid Tea	Cocker
	Dani Award and Certificate of Merit	'Amber Queen'®	Floribunda	Harkness
	R. J. Frizzell Memorial Award	'Yves Piaget'®	Hybrid Tea	Meilland
1984/85	Gold Medal and Prize of the City of Belfast	'Ingrid Bergman'®	Hybrid Tea	Poulsen
	Dani Award and Certificate of Merit	'Lutin'	Shrub	Meilland
1983/84	Gold Medal and Prize of the City of Belfast	'L'Oréal Trophy'	Hybrid Tea	Harkness
	Dani Award and Certificate of Merit	'Mountbatten'®	Floribunda	Harkness
	Dani Award and Certificate of Merit	'Fragrant Gold'	Hybrid Tea	Tantau

Table 22 Continued

Year	*Award*	*Cultivar*	*Class*	*Breeder*
1982/83	Gold Medal and Prize of the City of Belfast	'Pacemaker'	Hybrid Tea	Harkness
	Dani Award and Certificate of Merit	'Pimlico'	Floribunda	Meilland
	R. J. Frizzell Memorial Award	'Radox Bouquet'	Floribunda	Harkness
1981/82	Gold Medal and Prize of the City of Belfast	'Basildon Bond'	Hybrid Tea	Harkness
	Dani Award and Certificate of Merit	'Bright Smile'®	Floribunda	Dickson
	R. J. Frizzell Memorial Award	'Sweetheart'	Hybrid Tea	Cocker
1980/81	Gold Medal and Prize of the City of Belfast	'Lady Rose'®	Hybrid Tea	Kordes
	Dani Award and Certificate of Merit	'Colibri 79'®	Miniature	Meilland
	R. J. Frizzell Memorial Award	'Young Venturer'	Floribunda	Mattock
1979/80	Gold Medal and Prize of the City of Belfast	'Silver Jubilee'®	Hybrid Tea	Cocker
	Dani Award and Certificate of Merit	'Memento'®	Floribunda	Dickson
	R. J. Frizzell Memorial Award	'Double Delight'™	Hybrid Tea	Armstrong
1978/79	Gold Medal and Prize of the City of Belfast	'Mascotte '77'®	Hybrid Tea	Meilland
	Dani Award and Certificate of Merit	'Tempo'	Floribunda	Meilland
	R. J. Frizzell Memorial Award	'Burgund'®	Hybrid Tea	Kordes
1977/78	Gold Medal and Prize of the City of Belfast	'Young Quinn'®	Hybrid Tea	McGredy
	Dani Award and Certificate of Merit	'Eyepaint'®	Floribunda	McGredy
	R. J. Frizzell Memorial Award	'Charles de Gaulle'®	Hybrid Tea	Meilland
1976/77	Gold Medal and Prize of the City of Belfast	'Evening Star'®	Floribunda	Jackson & Perkins
	Dani Award and Certificate of Merit	'Dreaming Spires'	Climber	Mattock
	R. J. Frizzell Memorial Award	'Evening Star'		Jackson & Perkins
1975/76	Gold Medal and Prize of the City of Belfast	'Sunsilk'	Floribunda	Fryer
	Dani Award and Certificate of Merit	'Matangi'®	Floribunda	McGredy
	R. J. Frizzell Memorial Award	'Scented Star'	Hybrid Tea	Lowe
1974/75	Gold Medal and Prize of the City of Belfast	'Fragrant Hour'	Hybrid Tea	McGredy
	Dani Award and Certificate of Merit	'Scherzo'®	Floribunda	Meilland
	R. J. Frizzell Memorial Award	'Fragrant Hour'		McGredy
1973/74	Gold Medal and Prize of the City of Belfast	'Alexander'®		Harkness
	Dani Award and Certificate of Merit	'Southampton'	Floribunda	Harkness
	R. J. Frizzell Memorial Award	'Harriny'	Hybrid Tea	LeGrice
1972/73	Gold Medal and Prize of the City of Belfast	'Mala Rubinstein'	Hybrid Tea	Dickson
	Dani Award and Certificate of Merit	'Picasso'	Floribunda	McGredy
	R. J. Frizzell Memorial Award	'Mala Rubinstein'		Dickson
1971/72	Gold Medal and Prize of the City of Belfast	'Mullard Jubilee'	Hybrid Tea	McGredy
	Dani Award and Certificate of Merit	'Pink Wonder'	Floribunda	Meilland
	R. J. Frizzell Memorial Award	'Alec's Red'®	Hybrid Tea	Cocker
1970/71	Gold Medal and Prize of the City of Belfast	'Adair Roche'	Hybrid Tea	McGredy
	Dani Award and Certificate of Merit	'Molly McGredy'	Floribunda	McGredy
	R. J. Frizzell Memorial Award	'Baronne Edmond de Rothschild'®	Hybrid Tea	Meilland
1969/70	Gold Medal and Prize of the City of Belfast	'Peer Gynt'®	Hybrid Tea	Kordes
	Dani Award and Certificate of Merit	'City of Belfast'®	Floribunda	McGredy
	R. J. Frizzell Memorial Award	'Grande Amore'		Kordes
1968/69	Gold Medal and Prize of the City of Belfast	'Red Devil'	Hybrid Tea	Dickson
	Dani Award and Certificate of Merit	'Escapade'®	Floribunda	Harkness
	R. J. Frizzell Memorial Award	'City of Hereford'	Hybrid Tea	LeGrice
1967/68	Gold Medal and Prize of the City of Belfast	'Grandpa Dickson'	Hybrid Tea	Dickson
	Dani Award and Certificate of Merit	'Jan Spek'®	Floribunda	McGredy
	R. J. Frizzell Memorial Award	'Silver Star'®	Hybrid Tea	Kordes
1966/67	Gold Medal and Prize of the City of Belfast	'Pink Supreme'	Hybrid Tea	DeRuiter
	Dani Award and Certificate of Merit	'Scented Air'	Floribunda	Dickson
	R. J. Frizzell Memorial Award	'Arthur Bell'	Floribunda	McGredy
1965/66	Gold Medal and Prize of the City of Belfast	'Colour Wonder'	Hybrid Tea	Kordes
	Dani Award and Certificate of Merit	'Marlena'	Floribunda	Kordes
	R. J. Frizzell Memorial Award	'Sea Pearl'	Floribunda	Dickson
	R. J. Frizzell Memorial Award	'Fragrant Air'	Floribunda	Pearce

[a] Gold Medal and Prize of the City of Belfast: presented by Belfast City Council for the best Hybrid Tea rose; Dani Award and Certificate of Merit: presented by the Department of Agriculture for Northern Ireland for the best Floribunda, Miniature, Climber and including those described as Patio-type roses; R. J. Frizzell Memorial Award: presented by the Northern Ireland Tourist Board for the most fragrant rose of the meritorious varieties; Tyrone Crystal Prize: presented Sainsbury Award (1987–1996): presented by Sainsbury's for the best shrub rose; City of Belfast Award (1996–97): presented by the Belfast City Council for the best shrub rose.

Table 23 Awards at the International Rose Trials, Glasgow (1989–2000)

Year	*Award*	*Cultivar*	*Class*	*Breeder*
2000	Golden Prize of the City of Glasgow	'Amber Nectar'	Floribunda	Eurosa
	Gold Medal	'Hearbeat 97'	Floribunda	Cocker
	The People's Choice	'Perception'	Hybrid Tea	Harkness
	Best Established Rose	'Fellowship'	Floribunda	Harkness
1999	Golden Prize of the City of Glasgow	'Scarlet Patio'	Miniature	Kordes
	Gold Medal	'Molineux'	Shrub	Austin
	Tollcross Fragrance Prize	'Noble Anthony'	Shrub	Austin
	The People's Choice	'ARC Angel'	Hybrid Tea	Fryer
	Best Established Rose	'Mary Hayley Bell'	Shrub	Kordes
1998	Golden Prize of the City of Glasgow	'Sunset Boulevard'	Floribunda	Harkness
	Gold Medal	'Scotland Wink' (Adavapma)		Adam
	Tollcross Fragrance Prize	'L'Aimant'	Floribunda	Harkness
	The People's Choice	'Aberdeen Celebration'	Floribunda	Cocker
	Best Established Rose	'Anna Livia'	Floribunda	Kordes
1997	Golden Prize of the City of Glasgow	'Friend for Life'	Floribunda	Cocker
	Gold Medal	'The Holt'	Shrub	Mehring
	Tollcross Fragrance Prize	'Sally's Rose'	Hybrid Tea	Cant
	The People's Choice	'Julie Y'	Hybrid Tea	Harkness
	Best Established Rose	'Abbeyfield Rose'	Hybrid Tea	Cocker
1996	Golden Prize of the City of Glasgow	'Berkshire'		Kordes
	Gold Medal	'Our Molly'	Shrub	Dickson
	Tollcross Fragrance Prize	'Renaissance'	Hybrid Tea	Harkness
	The People's Choice	'Our Molly'		Dickson
	Best Established Rose	'Rosy Cushion'®	Shrub	Ilsink
1995	Golden Prize of the City of Glasgow	'Remembrance'	Floribunda	Harkness
	Gold Medal	'Flair'	Floribunda	Dickson
	Tollcross Fragrance Prize	'Thomas Barton'		Meilland
	The People's Choice	'Gordon's College'	Floribunda	Cocker
1994	Golden Prize of the City of Glasgow	'Mandarin'	Miniature	Kordes
	Gold Medal	'Ena Baxter'	Hybrid Tea	Cocker
	Tollcross Fragrance Prize	'Marty'	Floribunda	Pearce
	The People's Choice	'Valencia'®	Hybrid Tea	Kordes
1993	Golden Prize of the City of Glasgow	'Hello'	Miniature	Cocker
	Gold Medal	'Flower Carpet'™	Shrub	Noack
	Tollcross Fragrance Prize	'Indian Summer'	Hybrid Tea	Pearce
1992	Golden Prize of the City of Glasgow	'Fulton MacKay'	Hybrid Tea	Cocker
	Gold Medal	'Pretty Polly'	Miniature	Meilland
1991	Golden Prize of the City of Glasgow	'Anna Livia'	Floribunda	Kordes
	Gold Medal	'Biddulph Grange'	Shrub	Fryers
	Tollcross Fragrance Prize	'Rosemary Harkness'	Hybrid Tea	Harkness
1990	Golden Prize of the City of Glasgow	'Abbeyfield Rose'	Hybrid Tea	Cocker
	Gold Medal	'Champagne Cocktail'	Floribunda	Horner
1989	Golden Prize of the City of Glasgow	'Sexy Rexy'®	Floribunda	McGredy
	Gold Medal	'Anna Ford'®	Miniature	Harkness
	Tollcross Fragrance Prize	'Sheila's Perfume'	Floribunda	Sheridan

United States of America

Table 24 Awards by All-America Rose Selections (1940–2002)

Year	*Cultivar*	*Class*	*Breeder*
2000	'Crimson Bouquet'™	Grandiflora	Kordes
	'Gemini'™	Hybrid Tea	Zary
	'Knock Out'™	Shrub	Radler
1999	'Betty Boop'™	Floribunda	Carruth
	'Fourth of July'™	Large Flowered Climber	Carruth
	'Kaleidoscope'™	Shrub	Walden
	'Candelabra'™	Grandiflora	Zary

Table 24 Continued

Year	*Cultivar*	*Class*	*Breeder*
1998	'Fame!'™	Grandiflora	Zary
	'Opening Night'™	Hybrid Tea	Zary
	'First Light'™	Shrub	Marciel
	'Sunset Celebration'™	Hybrid Tea	Fryer
1997	'Artistry'™	Hybrid Tea	Zary
	'Timeless'™	Hybrid Tea	Zary
	'Scentimental'™	Floribunda	Carruth
1996	'St Patrick'™	Hybrid Tea	Strickland
	'Livin' Easy'™	Floribunda	Harkness
	'Mount Hood'	Hybrid Tea	McGredy
	'Carefree Delight'™	Shrub	Meilland
1995	'Brass Band'™	Floribunda	Christensen
	'Singin' in the Rain'	Floribunda	McGredy
1994	'Secret'™	Hybrid Tea	Tracy
	'Caribbean'™	Grandiflora	Kordes
	'Midas Touch'™	Hybrid Tea	Christensen
1993	'Solitude'™	Grandiflora	Olesen
	'Rio Samba'™	Hybrid Tea	Warriner
	'Sweet Inspiration'	Floribunda	Warriner
	'Child's Play'™	Miniature	Saville
1992	'Brigadoon'™	Hybrid Tea	Warriner
	'Pride 'n' Joy'™	Miniature	Warriner
	'All That Jazz'™	Shrub	Twomey
1991	'Sheer Elegance'™	Hybrid Tea	Twomey
	'Perfect Moment'™	Hybrid Tea	Kordes
	'Shining Hour'™	Grandiflora	Warriner
	'Carefree Wonder'™	Shrub	Meilland
1990	'Pleasure'	Floribunda	Warriner
1989	'Class Act'	Floribunda	Warriner
	'Debut'™	Miniature	Meilland
	'New Beginning'™	Miniature	Saville
	'Tournament of Roses'	Grandiflora	Warriner
1988	'Amber Queen'®	Floribunda	Harkness
	'Mikado'™	Hybrid Tea	Suzuki
	'Prima Donna'™	Grandiflora	Shirakawa
1987	'Bonica'™	Shrub	Meilland
	'New Year'®	Grandiflora	McGredy
	'Sheer Bliss'	Hybrid Tea	Warriner
1986	'Broadway'™	Hybrid Tea	Perry
	'Touch of Class'™	Hybrid Tea	Kriloff
	'Voodoo'™	Hybrid Tea	Christensen
1985	'Showbiz'	Floribunda	Tantau
1984	'Impatient'	Floribunda	Warriner
	'Intrigue'	Floribunda	Warriner
	'Olympiad'™	Hybrid Tea	McGredy
1983	'Sun Flare'	Floribunda	Warriner
	'Sweet Surrender'	Hybrid Tea	Weeks
1982	'Brandy'™	Hybrid Tea	Swim & Christensen
	'French Lace'	Floribunda	Warriner
	'Mon Cheri'™	Hybrid Tea	Christensen
	'Shreveport'™	Grandiflora	Kordes
1981	'Bing Crosby'	Hybrid Tea	Weeks
	'Marina'®	Floribunda	Kordes
	'White Lightnin'™	Grandiflora	Swim & Christensen
1980	'Love'	Grandiflora	Warriner
	'Honor'™	Hybrid Tea	Warriner
	'Cherish'	Floribunda	Warriner
1979	'Friendship'™	Hybrid Tea	Lindquist
	'Paradise'™	Hybrid Tea	Weeks
	'Sundowner'	Grandiflora	McGredy

Table 24 Continued

Year	*Cultivar*	*Class*	*Breeder*
1978	'Charisma'	Floribunda	Hill
	'Color Magic'	Hybrid Tea	Warriner
1977	'Double Delight'™	Hybrid Tea	Swim & Ellis
	'First Edition'	Floribunda	Delbard-Chabert
	'Prominent'®	Grandiflora	Kordes
1976	'America'™	Large Flowered Climber	Warriner
	'Cathedral'	Floribunda	McGredy
	'Seashell'	Hybrid Tea	Kordes
	'Yankee Doodle'®	Hybrid Tea	Kordes
1975	'Arizona'	Grandiflora	Weeks
	'Oregold'	Hybrid Tea	Tantau
	'Rose Parade'	Floribunda	Williams
1974	'Bahia'	Floribunda	Lammerts
	'Bon-Bon'	Floribunda	Warriner
	'Perfume Delight'	Hybrid Tea	Weeks
1973	'Electron'®	Hybrid Tea	McGredy
	'Gypsy'	Hybrid Tea	Swim & Weeks
	'Medallion'®	Hybrid Tea	Warriner
1972	'Apollo'®	Hybrid Tea	Armstrong
	'Portrait'	Hybrid Tea	Meyer
1971	'Aquarius'	Grandiflora	Armstrong
	'Command Performance'	Hybrid Tea	Lindquist
	'Redgold'	Floribunda	Dickson
1970	'First Prize'	Hybrid Tea	Boerner
1969	'Angel Face'	Floribunda	Swim & Weeks
	'Comanche'	Grandiflora	Swim & Weeks
	'Gene Boerner'	Floribunda	Boerner
	'Pascali'®	Hybrid Tea	Lens
1968	'Europeana'®	Floribunda	DeRuiter
	'Miss All-American Beauty'	Hybrid Tea	Meilland
	'Scarlet Knight'	Grandiflora	Meilland
1967	'Bewitched'	Hybrid Tea	Lammerts
	'Gay Princess'	Floribunda	Boerner
	'Lucky Lady'	Grandiflora	Armstrong & Swim
	'Roman Holiday'	Floribunda	Lindquist
1966	'American Heritage'®	Hybrid Tea	Lammerts
	'Apricot Nectar'	Floribunda	Boerner
	'Matterhorn'®	Hybrid Tea	Armstrong & Swim
1965	'Camelot'	Grandiflora	Swim & Weeks
	'Mister Lincoln'®	Hybrid Tea	Swim & Weeks
1964	'Granada'	Hybrid Tea	Lindquist
	'Saratoga'	Floribunda	Boerner
1963	'Royal Highness'	Hybrid Tea	Swim & Weeks
	'Tropicana'	Hybrid Tea	Tantau
1962	'Christian Dior'	Hybrid Tea	Meilland
	'Golden Slippers'	Floribunda	Abrams
	'John S. Armstrong'	Grandiflora	Swim
	'King's Ransom'®	Hybrid Tea	Morey
1961	'Duet'	Hybrid Tea	Swim
	'Pink Parfait'	Grandiflora	Swim
1960	'Fire King'	Floribunda	Meilland
	'Garden Party'®	Hybrid Tea	Swim
	'Sarabande'®	Floribunda	Meilland
1959	'Ivory Fashion'	Floribunda	Boerner
	'Starfire'	Grandiflora	Lammerts
1958	'Fusilier'	Floribunda	Morey
	'Gold Cup'	Floribunda	Boerner
	'White Knight'	Hybrid Tea	Meilland
1957	'Golden Showers'®	Large Flowered Climber	Lammerts
	'White Bouquet'	Floribunda	Boerner
1956	'Circus'	Floribunda	Swim

Table 24 Continued

Year	Cultivar	Class	Breeder
1955	'Jiminy Cricket'	Floribunda	Boerner
	'Queen Elizabeth'®	Grandiflora	Lammerts
	'Tiffany'	Hybrid Tea	Lindquist
1954	'Lilibet'	Floribunda	Lindquist
	'Mojave'	Hybrid Tea	Swim
1953	'Chrysler Imperial'	Hybrid Tea	Lammerts
	'Ma Perkins'	Floribunda	Boerner
1952	'Fred Howard'	Hybrid Tea	Howard
	'Helen Traubel'	Hybrid Tea	Swim
	'Vogue'	Floribunda	Boerner
1951	No selection		
1950	'Capistrano'	Hybrid Tea	Morris
	'Fashion'	Floribunda	Boerner
	'Sutter's Gold'	Hybrid Tea	Swim
	'Mission Bells'	Hybrid Tea	Morris
1949	'Forty-niner'	Hybrid Tea	Swim
	'Tallyho'	Hybrid Tea	Swim
1948	'Diamond Jubilee'	Hybrid Tea	Boerner
	'High Noon'	Climbing Hybrid Tea	Lammerts
	'Nocturne'	Hybrid Tea	Swim
	'Pinkie'	Polyantha	Swim
	'San Fernando'	Hybrid Tea	Morris
	'Taffeta'	Hybrid Tea	Lammerts
1947	'Rubaiyat'	Hybrid Tea	McGredy
1946	'Peace'	Hybrid Tea	Meilland
1945	'Floradora'	Floribunda	Tantau
	'Horace McFarland'	Hybrid Tea	Mallerin
	'Mirandy'	Hybrid Tea	Lammerts
1944	'Fred Edmunds'	Hybrid Tea	Meilland
	'Katherine T. Marshall'	Hybrid Tea	Boerner
	'Lowell Thomas'	Hybrid Tea	Mallerin
	'Mme Chiang Kai-shek'	Hybrid Tea	Duehrsen
	'Mme Marie Curie'	Hybrid Tea	Gaujard
1943	'Grande Duchesse Charlotte'	Hybrid Tea	Ketten
	'Mary Margaret McBride'	Hybrid Tea	Pernet-Ducher
1942	'Heart's Desire'	Hybrid Tea	Howard
1941	'Apricot Queen'	Hybrid Tea	Howard
	'California'	Hybrid Tea	Howard
	'Charlotte Armstrong'	Hybrid Tea	Lammerts
1940	'Dickson's Red'	Hybrid Tea	Dickson
	'Flash'	Large-Flowered Climber	Hatton
	'The Chief'	Hybrid Tea	Lammerts
	'World's Fair'	Floribunda	Kordes

Table 25 Awards of Excellence for Miniature Roses by the American Rose Society (1975–2000)

Year	*Cultivar*	*Breeder*
2000	'Applause'	Saville
	'Lemon Gems'™	Walden
	'Ralph Moore'	Saville
1999	'Little Flame'™	Walden
	'Autumn Splendor'	Williams
	'Little Tommy Tucker'™	Tucker
	'Sweet Caroline'	Williams
1998	'Playgold'	Moore
1997	'Tropical Twist'™	Walden
1996	'Angelica Renae'™	Wells
1995	'Jingle Bells'™	Zary
1994	'Hot Tamale'™	Zary
1993	'Billie Teas'	Hooper
	'Boomerang'	Sponner
	'Child's Play'™	Saville
	'Kristin'™	Benardella
	'Palmetto Sunrise'	Williams
1992	'Cal Poly'	Moore
	'Debidue'™	Williams
	'Figurine'™	Benardella
	'Sincerely Yours'	Moore
1991	'Golden Halo'™	Saville
	'Good Morning America'™	Saville
	'Just For You'	Moore
	'Suzy'	Bridges
1990	'Regine'™	Hefner
1989	'Dee Bennett'™	Saville
	'Jim Dandy'™	Benardella
	'Nighthawk'	Hardgrove
	'Tipper'™	Jolly
1988	'Old Glory'™	Benardella
	'Heavenly Days'™	Saville
1987	'Ring of Fire'™	Moore
	'Sequoia Gold'™	Moore
1986	'Jean Kenneally'™	Bennett
	'Rainbow's End'™	Saville
1985	'Black Jade'™	Benardella
	'Centerpiece'™	Saville
	'Jennifer'™	Benardella

Table 25 Continued

Year	*Cultivar*	*Breeder*
	'Loving Touch'	Jolly
	'Winsome'™	Saville
1984	'Baby Eclipse'™	Moore
	'Hot Shot'	Bennett
	'Julie Ann'™	Saville
	'Little Jackie'™	Saville
1983	'Cornsilk'™	Saville
	'Cupcake'™	Spies
	'Hombre'	Jolly
	'Snow Bride'	Jolly
	'Valerie Jeanne'	Saville
1982	'Center Gold'™	Saville
1981	'Pacesetter'	Schwartz
	'Party Girl'™	Saville
1980	'Holy Toledo'™	Christensen
	'Pink Petticoat'	Strawn
1979	'Cuddles'	Schwartz
	'Puppy Love'	Schwartz
	'Red Flush'	Schwartz
	'Zinger'	Schwartz
1978	'Avandel'	Moore
	'Gloriglo'	Williams
	'Humdinger'	Schwartz
	'Rise 'n' Shine'	Moore
1977	'Jeanne Lajoie'	Sima
	'Peaches 'n' Cream'	Woolcock
1976	'Hula Girl'	Williams
	'Peachy White'	Moore
	'Red Cascade'	Moore
1975	'Beauty Secret'	Moore
	'Judy Fischer'	Moore
	'Lavender Lace'	Moore
	'Magic Carrousel'®	Moore
	'Mary Marshall'	Moore
	'Over the Rainbow'	Moore
	'Sheri Anne'	Moore
	'Starglo'	Williams
	'Toy Clown'	Moore
	'White Angel'	Moore

APPENDIX 2

SELECTION OF REFEREED PAPERS

Biotechnologies for Breeding: Manipulation of Ploidy Level

Basye R (1990) An amphidiploid of *Rosa banksiae* and *Rosa laevigata* induced by colchicines. *American Rose Annual* 75: 82–88.

Crespel L, Meynet J, Gudin S and Zhang D (2002) AFLP-based estimation of 2*n* gametophytic heterozygosity in two parthenogenetically derived dihaploids of *Rosa hybrida* L. *Theoretical and Applied Genetics* 104: 451–456.

El Mokadem H, Crespel L, Meynet J and Gudin S (2002) The occurrence of 2*n*-pollen and the origin of sexual polyploids in dihaploid roses (*Rosa hybrida* L.). *Euphytica* 125: 169–177.

El Mokadem H, Meynet J and Crespel L (2002) The occurrence of 2*n* eggs in dihaploids derived from *Rosa hybrida* L. *Euphytica* 124: 327–332.

Ma Y, Byrne DH and Chen J (1997) Amphidiploid induction from diploid rose interspecific hybrids. *HortScience* 32: 292–295.

Marchant R, Power JB, Davey MR and Chartier-Hollis J (1994) Embryo rescue, for the production of F_1 hybrids, in English rose. *Euphytica* 74: 187–193.

Meynet J, Barrade R, Duclos A and Siadous R (1994) Dihaploid plants of roses (*Rosa hybrida*, cv 'Sonia') obtained by parthenogenesis induced using irradiated pollen and *in vitro* culture of immature seeds. *Agronomie* 2: 169–175.

Roberts AV, Lloyd D and Short C (1990) *In vitro* procedures for the induction of tetraploidy in a diploid rose. *Euphytica* 49: 33–38.

Semeniuk P and Arisumi T (1968) Colchicine-induced tetraploid and cytochimeral roses. *Botanical Gazette* 129: 190–193.

Tabaeezadeh Z and Khosh-Khui M (1981) Anther culture of Rosa. *Scientia Horticulturae* 15: 61–66.

Wissemann V, Mollers C and Hellwig FH (1996) Microspore and anther culture in the genus Rosa, investigations and current status. *Journal of Applied Botany* 70: 218–220.

Wissemann V, Mollers C and Hellwig FH (1998) Microspore and anther culture in the genus Rosa, further investigations. *Journal of Applied Botany* 72: 7–9.

Biotechnologies for Breeding: Amphidiploidy

Basye RE (1987) A probable amphidiploid of *Rosa abyssinica* and *Rosa rugosa*. *American Rose Annual* 72: 19–24.

Basye RE (1990) An amphidiploid of *Rosa banksiae* and *Rosa laevigata* induced by colchicine. *American Rose Annual* 75: 83–88.

Ma Y, Byrne DH and Chen J (1997) Amphidiploid induction from diploid rose interspecific hybrids. *HortScience* 32: 292–295.

Ma Y, Crane CF and Byrne DH (2000) Meiotic behavior in a tetraploid rose and its hybrid progeny. *HortScience* 35: 1127–1131.

Roberts AV, Lloyd D and Short KC (1990) Procedures for the induction of tetraploidy in diploid rose. *Euphytica* 49: 33–38.

Semeniuk P and Arisuma T (1968) Colchicine-induced tetraploid and cytochimeral roses. *Botanical Gazette* 129: 190–193.

Svejda FJ (1988) Rose germplasm L83. *HortScience* 23: 415–416.

Breeding: Overview

Crespel L, Chirollet M, Durel CE *et al.* (2002) Mapping of qualitative and quantitative phenotypic traits in *Rosa* using AFLP markers. *Theoretical and Applied Genetics* 105: 1207–1214.

Crespel L, Gudin S, Meynet J and Zhang D (2002) AFLP-based estimation of 2n gametophytic heterozygosity in two parthenogenetically derived dihaploids of *Rosa hybrida* L. *Theoretical and Applied Genetics* 104: 451–456.

Debener T and Mattiesch L (1999) Construction of a genetic linkage map for roses using RAPD and AFLP markers. *Theoretical and Applied Genetics* 99: 891–899.

Debener T, Bartels C and Spethmann W (1997) Parentage analysis in interspecific crosses between rose species with RAPD markers. *Gartenbauwissenschaft* 62: 180–184.

El Mokadem H, Crespel L, Meynet J and Gudin S (2002) The occurrence of 2n-pollen and the origin of sexual polyploids in dihaploid roses. *Euphytica* 125: 169–177.

Gudin S (2000) Rose: genetics and breeding. *Plant Breeding Review* 17: 159–189.

Li X, Krasnyanski SF and Korban SS (2002) Optimization of the *uidA* gene transfer into somatic embryos of rose via *Agrobacterium tumefaciens*. *Plant Physiology and Biochemistry* 40: 453–459.

Rajapakse S, Byrne DH, Zhang L *et al.* (2001) Two genetic linkage maps of tetraploid roses. *Theoretical and Applied Genetics* 103: 575–583.

Schum A, Hofmann K, Ghalib N and Tawfik A (2001) Factors affecting protoplast isolation and plant

regeneration in *Rosa* spp. *Gartenbauwissenschaft* 66: 115–122.
von Malek B and Debener T (1998) Genetic analysis of resistance to blackspot (*Diplocarpon rosae*) in tetraploid roses. *Theoretical and Applied Genetics* 96: 228–231.
Werlemark G, Uggla M and Nybom H (1999) Morphological and RAPD markers show a highly skewed distribution in a pair of reciprocal crosses between hemisexual dogrose species. *Theoretical and Applied Genetics* 98: 557–563.

Breeding: Methods of Cross-Breeding

Cole P and Melton B (1986) Self- and cross-compatibility relationships among genotypes and between ploidy of the rose. *Journal of the American Society of Horticultural Science* 111: 122–125.
Gudin S (1992) Influence of bud chilling on subsequent reproductive fertility in roses. *Scientia Horticulturae* 51: 139–144.
Gudin S (2000) Rose: genetics and breeding. *Plant Breeding Review* 17: 159–189.
Gudin S and Arene L (1991) Influence of the pH of the stigmatic exudate on male–female interaction in *Rosa hybrida* L. *Sexual Plant Reproduction* 4: 110–112.
Gudin S and Arene L (1992) Putrescine increases effective pollination period in roses. *HortTechnology* 2: 211–213.
Gudin S, Arene L and Bulard C (1991) Influence of season on rose pollen quality. *Sexual Plant Reproduction* 4: 113–117.
Gudin S, Arene L and Pellegrino C (1991) Influence of temperature and hygrometry on rose pollen germination. *Advances in Horticultural Science* 5: 96–98.
Shepherd RE (1958) The ABCs of Hybridizing. *American Rose Annual* 43: 78.
Vries de DP and Dubois LAM (1983) Pollen and pollination experiments. X. The effect of repeated pollination on fruit- and seed set in crosses between the Hybrid Tea-rose cvs. Sonia and Ilona. *Euphytica* 32: 685–689.

Breeding: Selection Strategies for Pot Roses

Chu-CY, Knight SL and Smith MAL (1993) Effect of liquid culture growth and development of miniature roses (*Rosa chinensis* Jacq. 'Mimima'). *Plant Cell, Tissue and Organ Culture* 32: 329–334.
Cushman LC, Pemberton HB and Kelly JW (1994) Cultivar, flower stage, silverthiosulphate and BA interactions affect performance of potted miniature roses. *HortScience* 29: 805–808.
Cushman LC, Pemberton HB, Miller JC and Kelly JW (1998) Interactions of flowerstage, cultivar, and shipping temperature and duration affect pot rose performance. *HortScience* 33: 736–740.
De Vries DP and Dubois LAM (1984) Inheritance of the recurrent flowering and moss characters in F_1 and F_2 Hybrid Tea × *R. centifolia muscosa* (Aiton) Seringe populations. *Gartenbauwissenschaft* 49: 97–100.
De Vries DP and Dubois LAM (1987) The effects of BAP and IBA on the sprouting and adventitious root formation of 'Amanda' rose softwood cuttings. *Scientia Horticulturae* 34: 115–121.
Dot P (1951) Meine Miniatur-Rosen. *Deutsche Baumschule* 3: 264.
Dubois LAM (1983a) Moederdag-potrozendag. Omschakeling bloemisterij beïnvloedt sortiment. *Vakbl. Bloemisterij* 17: 40–43.
Dubois LAM (1983b) *Study of Descent, Origin and Breeding of Small Roses* (in Dutch). Report IVT #187, 21p.
Dubois LAM and De Vries DP (1984) Züchtung bietet Perspektieven für Erweiterung des niederländisches Topfrosensortiments. *Zierpflanzenbau* 24: 1068–1069.
Dubois LAM and De Vries DP (1987) On the inheritance of the dwarf character in Polyantha × *Rosa chinensis minima* (Sims) Voss F_1-populations. *Euphytica* 36: 535–539.
Dubois LAM and De Vries DP (1991) Variation in adventitious root formation of softwood cuttings of *R. chinensis minima* (Sims) Voss cultivars. *Scientia Horticulturae* 47: 345–349.
Dubois LAM and De Vries DP (1994) Comparison of the growth and development of unpinched and pinched pot roses, raised from single-node softwood cuttings. *Gartenbauwissenschaft* 59: 18–20.
Dubois LAM and De Vries DP (1997) Comparison of *R. chinensis minima* (Voss) Sims miniature seedlings and their clonal plants, with reference to selection for pot rose cultivars. *Journal of Genetics and Breeding* 51: 181–184.
Dubois LAM, Roggemans J, Soyeurt G and De Vries DP (1988) Comparison of the growth and development of dwarf rose cultivars propagated *in vitro* and *in vivo* by softwood cuttings. *Scientia Horticulturae* 35: 293–299.
Gregory CW (1974) What's new in miniature roses? *Rose Annual* 55–58.
Gregory CW and Moore RS (1977) Future of Miniature bush and Miniature climbing roses. *Rose Annual* 130–134.
Haakana K, Sarkka L and Somersalo S (2001) Gaseous ethanol penetration of plant tissues positively affects the growth and commercial quality of miniature roses and dill. *Scientia Horticulturae* 88: 71–84.
Hurst CC (1941) Notes on the origin and evolution of our garden roses. *Journal of the Royal Horticultural Society* 66: 73–82; 242–250; 282–289.
Krüssmann G (1951) Unbekanntes über der Miniatur-rose 'Pompon de Paris' (*R. rouletii*). *Deutsche Baumschule* 3: 259–260.
Lata P (1983) Inheritance in triploid hybrids of *Rosa*. *Environmental and Experimental Botany* 23: 85–92.
Monteiro JA, Nell TA and Barrett JE (2001) High production temperature increases postproduction longeveity and reduces bud drop of potted miniature roses 'Meirutral' and 'Meidanclar'. *HortScience* 36: 953–954.
Moore RS (1968) A study of moss and miniature roses. *American Rose Annual* 53: 49–60.

Mortensen LM (1991) Effects of temperature, light and CO_2 level on growth and flowering of Miniature roses. *Norwegian Journal of Agricultural Sciences* 5: 295–300.
Mortensen LM, Ottosen CO and Gislerod HR (2001) Effects of air humidity and K : Ca ratio on growth, morphology, flowering and keeping quality of pot rose. *Scientia Horticulturae* 90: 1–2.
Muller R, Stumman BM, Anderson AS and Serek M (1999) Involvement of ABA in postharvest life of miniature potted roses. *Plant Growth Regulation* 29(3): 143–150.
Muller R, Stumman BM, Sisler EC and Serek M (2001) Cultivar differences in regulation of ethylene production in miniature rose flowers (*Rosa hybrida* L.) *Gartenbauwissenschaft* 66: 34–38.
Rasmussen K, Pedersen AR and Ottosen CO (1998) Increased levels of calcium in the nutrient solution improves post harvest life of potted rose. *Acta Horticulturae* 464: 540.
Serek M and Anderson AS (1993) AOA and BA influence floral development and longevity of potted 'Victory Parade' miniature roses. *HortScience* 28: 1039–1040.
Tjosvold SA, Wu MJ and Reid MS (1995) Reduction of post-production loss in potted miniature roses. *Acta Horticulturae* 405: 408–414.
Wilson CT (1974) Some thoughts on Miniatures. *American Rose Annual* 59: 28–34.
Wulff HD (1955) Cytologische Untersuchungen an Zwergwüchsigen Gartenrosen. *Der Züchter* 25: 122–128.
Wylie AP (1954) The history of garden roses. Part I. Masters Memorial Lecture. *Journal of the Royal Horticultural Society* 79: 8–24.

Breeding: Selection Strategies for Disease and Pest Resistance

Debener T (1999) Genetic analysis of important morphological and physiological characters in diploid roses. *Gartenbauwissenschaft* 64: 14–20.
Debener T, Janakiram T and Mattiesch L (2000) Sports and seedlings of rose varieties analysed with molecular markers. *Plant Breeding* 119: 71–74.
Malek Bv and Debener T (2000) Identification of molecular markers linked to Rdr 1, a gene conferring resistance to blackspot in roses. *Theoretical and Applied Genetics* 101: 977–983.
Schum A, Hofmann K, Ghalib N and Tawfik A (2001) Factors affecting protoplast isolation and plant regeneration in *Rosa* spp. *Gartenbauwissenschaft* 66: 115–122.

Cell, Tissue and Organ Culture: Micropropagation

Carelli BP and Echeverrigaray S (2002) An improved system for the *in vitro* propagation of rose cultivars. *Scientia Horticulturae* 92: 69–74.
DeRiek J, Piqueras A and Debergh PC (1997) Sucrose uptake and metabolism in a double layer system for micropropagation of *Rosa multiflora. Plant Cell, Tissue and Organ Culture* 47: 269–278.
De Vries DP and Dubois LAM (1994) Re-invigoration of clonal rose rootstocks by sustained micropropagation. *Gartenbauwissenschaft* 59: 81–85.
Dubois LAM, Roggemans J, Soyeurt G and De Vries DP (1988) Comparison of the growth and development of dwarf rose cultivars propagated *in vitro* and *in vivo* by softwood cuttings. *Scientia Horticulturae* 35: 293–299.
Douglas GC, Rutledge CB, Casey AD and Richardson DHS (1989) Micropropagation of floribunda, ground cover and miniature roses. *Plant Cell, Tissue and Organ Culture* 19: 55–64.
Ghashghaie J, Brenckmann F and Saugier B (1992) Water relations and growth of rose plants cultured *in vitro* under various relative humidities. *Plant Cell, Tissue and Organ Culture* 30: 51–57.
Horan I, Walker S, Roberts AV *et al.* (1995) Micropropagation of roses: the benefits of pruned mother-plantlets at stage II and a greenhouse environment at stage III. *Journal of Horticultural Science* 70: 799–806.
Hsia C-N and Korban SS (1996) Factors affecting *in vitro* establishment and shoot proliferation of *Rosa hybrida* L. and *Rosa chinensis minima. In Vitro Cellular Development and Biology – Plant* 32: 217–222.
Johansson M, Kronestedt-Robards EC and Robards AW (1992) Rose leaf structure in relation to different stages of micropropagation. *Protoplasma* 166: 165–176.
Martin C, Carré M and Vernoy R (1981) La multiplication végétative *in vitro*. Les végétaux ligneux cultivés: cas des rosiers. *Comptes Rendus de l' Academie de Science de Paris* 293: 175–177.
Onesto JP, Poupet R, Julien P and Jacquemont R (1985) Production de potées fleuries de rosier à partir de plantules obtenues par multiplication *in vitro* conforme automne 1983–printemps 1984. *Horticulture Française* 176: 3–10.
Podwyszynska M and Goszczynska DM (1998) Effect of inhibitors of ethylene biosynthesis and action, as well as calcium and magnesium on rose shoot rooting, shoot-tip necrosis and leaf senescence *in vitro*. *Acta Physiologiae Plantarum* 20: 91–98.
Reist A (1985a) Culture *in vitro* (civ) en pepinière de rosiers: Premiers résultats d'exploitation de plantes civ pour la coupe. *Revue Suisse de Viticulture, d'Arboriculture et d'Horticulture* 17: 173–176.
Reist A (1985b) Culture *in vitro* (civ) en pepinière de rosiers: une alternative au bouturage ou au greffage des variétés? *Revue Suisse de Viticulture, d'Arboriculture et d'Horticulture* 17: 361–364.
Reist A and Lê LC (1989) Comportement au champ de rosiers propagés *in vitro*: bilan d'un essai comparatif de 5 ans. *Revue Suisse de Viticulture, d'Arboriculture et d'Horticulture* 21: 19–23.
Rogers RB and Smith MAL (1992) Consequences of *in vitro* and *ex vitro* root initiation for miniature rose production. *Journal of Horticultural Science* 67: 535–540.

Rout GR, Samantaray S, Mottley J and Das P (1999) Biotechnology of the rose: a review of recent progress. *Scientia Horticulturae* 81: 201–228.

Sallanon H, Berger M, Genoud C and Coudret A (1998) Water stress and photoinhibition in acclimatization of *Rosa hybrida* plantlets. *In Vitro, Cellular and Developmental Biology – Plant* 34: 169–172.

Sing SK and Syamal MM (2001) A short pre-culture soak in thidiazuron or forchlorfenuron improves axillary shoot proliferation in rose micropropagation. *Scientia Horticulturae* 91: 169–177.

Van der Salm TPM, Van der Toorn CJG, ten Cate CHH *et al.* (1994) Importance of the iron chelate formula for micropropagation of *Rosa hybrida* L. 'Moneyway'. *Plant Cell, Tissue and Organ Culture* 37: 73–77.

Yan M, Byrne DH and Jing C (1996) Propagation of rose species *in vitro*. *In Vitro, Cellular and Developmental Biology – Plant* 32: 103–108.

Cell, Tissue and Organ Culture: Rooting of Micropropagated Shoots

Arnold NP, Binns MR, Cloutier DC, Barthakur NN and Pellerin R (1995) Auxin, salt concentration, and their interaction during *in vitro* rooting of winter-hardy and hybrid tea roses. *HortScience* 30: 1436–1440.

Bertazza G, Baraldi R and Predieri S (1995) Light effects on *in vitro* rooting of pear cultivars of different rhizogenetic ability. *Plant Cell, Tissue and Organ Culture* 41: 139–143.

Bressan PH, Kim YJ, Hyndman SE, Hasegawa PM and Bressan RA (1982) Factors affecting *in vitro* propagation of rose. *Journal of the American Society of Horticultural Science* 107: 979–990.

Campos PS and Pais MSS (1990) Mass propagation of the dwarf rose cultivar 'Rosamini'. *Scientia Horticulturae* 43: 321–330.

Debergh PC, De Meester J, De Riek J, Gillis S and Van Huylenbroeck J (1992) Ecological and physiological aspects of tissue-cultured plants. *Acta Botanica Neerlandica* 41: 417–423.

Douglas GC, Rutledge CB, Casey AD and Richardson DHS (1989) Micropropagation of floribunda, ground cover and miniature roses. *Plant Cell, Tissue and Organ Culture* 19: 55–64.

Drew RA, McComb JA and Considine JA (1993) Rhizogenesis and root growth of *Carica papaya* L. in vitro in relation to auxin sensitive phases and use of riboflavin. *Plant Cell, Tissue and Organ Culture* 33: 1–7.

Genoud-Gourichon C, Sallanon H and Coudret A (1996) Effect of sucrose, agar, irradiance and CO_2 concentration during the rooting phase on the acclimatization of *Rosa hybrida* to *ex vitro* conditions. *Photosynthetica* 32: 263–270.

Horan I, Walker S, Roberts AV, Mottely J and Simpkins I (1995) Micropropagation of roses: the benefits of pruned mother–plantlets at stage II and a greenhouse environment at stage III. *Journal of Horticultural Science* 70: 799–806.

Hyndman SE, Hasegawa PM and Bressan RA (1982a) Stimulation of root initiation from cultured rose shoots through the use of reduced concentrations of mineral salts. *HortScience* 17: 82–83.

Hyndman SE, Hasegawa PM and Bressan RA (1982b) The role of sucrose and nitrogen in adventitious root formation on cultured rose shoots. *Plant Cell, Tissue and Organ Culture* 1: 229–238.

Khosh-Khui M and Sink KC (1982) Rooting-enhancement of *Rosa hybrida* for tissue culture propagation. *Scientia Horticulturae* 17: 371–376.

Khunachak A, Chin CK, Le T and Gianfagna T (1987) Promotion of asparagus shoot and root growth by growth retardants. *Plant Cell, Tissue and Organ Culture* 11: 97–110.

Kornova KM and Michailova J (1994) Study of the in vitro rooting of Kazanlak oil-bearing rose (*Rosa damascena* Mill.). *Journal of Essential Oil Research* 6: 485–492.

Kumar A, Sood A, Palni UT, Gupta AK and Palni LMS (2001) Micropropagation of *Rosa damascena* Mill. from mature bushes using thidiazuron. *Journal of Horticultural Science and Biotechnology* 76: 30–34.

Nag S, Saha K and Choudhuri A (2001) Role of auxin and polyamines in adventitious root formation in relation to changes in compounds involved in rooting. *Journal of Plant Growth Regulation* 20: 182–194.

Orlikowska T (1988) Propagation of quince S1 (*Cydonia oblonga* Mill.) in vitro. *Fruit Science Report* 15: 157–165.

Orlikowska T (1992) Influence of arginine on in vitro rooting of dwarf apple rootstock. *Plant Cell, Tissue and Organ Culture* 31: 9–14.

Podwyszynska M and Goszczynska D (1998) Effect of inhibitors of ethylene biosynthesis and action, as well as calcium and magnesium on shoot rooting, shoot-tip necrosis and leaf senescence *in vitro*. *Acta Physiologiae Plantarum* 20: 91–98.

Podwyszynska M and Olszewski T (1995) Influence of gelling agents on shoot multiplication and uptake of macroelements by in vitro culture of rose, cordyline and homalomena. *Scientia Horticulturae* 64: 77–84.

Podwyszynska M, Wojtania A and Gabryszewska E (2000) Application of m-topolin for plant micropropagation. *Zeszyty Naukowe Instytutu Sadownictwa i Kwiaciarstwa* 7: 173–180 (in Polish with English abstract and table descriptions).

Rahman SM, Hossain M, Rafiul Islam AKM and Joarder OI (1992) Effects of media composition and culture conditions on in vitro rooting of rose. *Scientia Horticulturae* 52: 163–169.

Rogers RB and Smith MAL (1992) Consequences of *in vitro* and *ex vitro* root initiation for miniature rose production. *Journal of Horticultural Science* 67: 535–540.

Rogers RB, Knight SL and Smith MAL (1991) Effects of growing media and aerial environments on acclimatization of in vitro grown miniature rose plantlets. *Journal of Environmental Horticulture* 9: 181–184.

Sallanon H, Berger M, Genoud C and Coudret A (1998) Water stress and photoinhibition in acclimatization of

Rosa hybrida plantlets. *In Vitro Cellular and Developmental Biology – Plant* 34: 169–172.

Scagel CF (2001) Cultivar specific effect of mycorrhizal fungi on the rooting of miniature rose cuttings. *Journal of Environmental Horticulture* 19: 15–20.

Taslim A and Patel BA (1995) Micropropagation of rose – a review. *Agricultural Reviews (Karnal)* 16: 211–218.

Van der Salm TPM, Van der Toorn CJG, Hanischten Cate CH *et al.* (1994) Importance of the iron chelate formula for micropropagation of *Rosa hybrida* L. 'Moneyway'. *Plant Cell, Tissue and Organ Culture* 37: 73–77.

Wardrop J, Lowe KC, Davey MR, Marchant R and Power JB (1997) Carbon dioxide-gassed fluorocarbon enhances micropropagation of rose (*Rosa chinensis* Jacq.). *Plant Cell Reports* 17: 17–21.

Williamson B, Cooke DEL, Duncan JM *et al.* (1998) Fungal infection of micropropagated plant at weaning: a problem exemplified by downy mildew in *Rubus* and *Rosa*. *Plant Cell, Tissue and Organ Culture* 52: 89–96.

Wilson D and Nayar NK (1995) Effect of activated charcoal on in vitro rooting of cultured rose shoots. *South Indian Horticulture* 43: 32–34.

Cell, Tissue and Organ Culture: *In Vitro* Regeneration Techniques

Arene L, Pellegrino C and Gudin S (1993) A comparison of the somaclonal variation level of *Rosa hybrida* L. cv. Meirutral plants regenerated from callus or direct induction from different vegetative and embryogenic tissues. *Euphytica* 71: 83–90.

Arif MB and Khatamian H (1989) *In vitro* morphogenesis from leaf callus of *Rosa* species. *HortScience* 23: 102.

Berardi G (1989) Indagine preliminare sulla rigenerazione di plantule *in vitro* da petali di rosa. *Colture Protette* 18: 111–113.

Burger DW, Liu L, Zary KW and Lee CI (1990) Organogenesis and plant regeneration from immature embryos of *Rosa hybrida* L. *Plant Cell, Tissue and Organ Culture* 21: 147–152.

De Wit JC, Esendam HF, Honkanen JJ and Tuominen U (1990) Somatic embryogenesis and regeneration of flowering plants in rose. *Plant Cell Reports* 9: 456–458.

Dubois LAM and de Vries DP (1995) Preliminary report on the direct regeneration of adventitious buds on leaf explants of *in vivo* grown glasshouse rose cultivars. *Gartenbauwissenschaft* 60: 249–253.

Hill GP (1967) Morphogenesis of shoot primordia in cultured stem tissue of a garden rose. *Nature* 216: 596–597.

Hsia C and Korban SS (1996) Organogenesis and somatic embryogenesis in callus cultures of *Rosa hybrida* and *Rosa chinensis minima*. *Plant Cell, Tissue and Organ Culture* 44: 1–6.

Ishioka N and Tanimoto S (1990) Plant regeneration from Bulgarian rose callus. *Plant Cell, Tissue and Organ Culture* 22: 197–199.

Jacobs G, Allan P and Bornman CH (1969) Tissue culture studies on rose: use of shoot tip explants. I. Auxin: cytokinin effects. *Agroplantae* 1: 179–188.

Jacobs G, Allan P and Bornman CH (1970) Tissue culture studies on rose: use of shoot tip explants. II. Cytokinin: gibberellin effects. *Agroplantae* 2: 25–28.

Jacobs G, Allan P and Bornman CH (1970) Tissue culture studies on rose: use of shoot tip explants. Auxin: gibberellin effects. *Agroplantae* 2: 45–50.

Jacobs G, Bornman CH and Allan P (1968) Tissue culture studies on rose: use of pith explants. III. *South African Journal of Agricultural Science* 11: 673–678.

Kintzios S, Manos C and Makri O (1999) Somatic embryogenesis from mature leaves of rose (*Rosa* sp.). *Plant Cell Reports* 18: 467–472.

Kintzios S, Drossopoulos JB and Lymperopoulos C (2000) Effect of vitamins and inorganic micronutrients on callus growth and somatic embryogenesis from young mature leaves of rose. *Journal of Plant Nutrition* 23: 1407–1420.

Khosh-Khui M and Sink KC (1982) Callus induction and culture of *Rosa*. *Scientia Horticulturae* 17: 361–370.

Kunitake H, Imamizo H and Mii M (1993) Somatic embryogenesis and plant regeneration from immature seed-derived calli of rugosa rose (*Rosa rugosa* Thunb.). *Plant Science* 90: 187–194.

Leffring L and Kok E (1990) Regeneratie via bladstukjes van roos. *Prophyta* 8: 244–245.

Li X, Krasnyanski SF and Korban SS (2002) Somatic embryogenesis, secondary somatic embryogenesis, and shoot organogenesis in Rosa. *Journal of Plant Physiology* 313–319.

Lloyd D, Roberts AV and Short KC (1988) The induction *in vitro* of adventitious shoots in *Rosa*. *Euphytica* 37: 31–36.

Lu CY (1993) The use of thidiazuron in tissue culture. In Vitro *Cellular and Developmental Biology* 29: 92–96.

Marchant R, Davey MR, Lucas JA and Power JB (1996) Somatic embryogenesis and plant regeneration in Floribunda rose (*Rosa hybrida* L.) cvs. Trumpeter and Glad Tidings. *Plant Science* 120: 95–105.

Matthews D, Mottley J, Horan I and Roberts AV (1991) A protoplast to plant system in roses. *Plant Cell, Tissue and Organ Culture* 24: 173–180.

Murali S, Sreedhar D and Lokeswari TS (1996) Regeneration through somatic embryogenesis from petal-derived calli of *Rosa hybrida* L. cv. Arizona (hybrid tea). *Euphytica* 91: 271–275.

Noriega C and Söndahl MR (1991) Somatic embryogenesis in hybrid tea roses. *BioTechnology* 9: 991–993.

Rosu A, Skirvin RM, Bein A, Norton MA, Kushad M and Otterbacher AG (1995) The development of putative adventitious shoots from a chimeral thornless rose (*Rosa multiflora* Thunb. ex J. Murr.) *in vitro*. *Journal of Horticultural Science* 70: 901–907.

Rout GR, Debata BK and Das P (1991) Somatic embryogenesis in callus cultures of *Rosa hybrida* L. cv. Landora. *Plant Cell, Tissue and Organ Culture* 27: 65–69.

Rout GR, Debata BK and Das P (1992) *In vitro* regeneration of shoots from callus cultures of *Rosa hybrida* L. cv. Landora. *Indian Journal of Experimental Biology* 30: 15–18.

Sarasan V, Roberts AV and Rout GR (2001) Methyl laurate and 6-benzyladenine promote the germination of somatic embryos of a hybrid rose. *Plant Cell Reports* 20: 183–186.

Van der Salm TPM, van der Toorn CJG, Hänisch ten Cate CH and Dons HJM (1996) Somatic embryogenesis and shoot regeneration from excised adventitious roots of the rootstock *Rosa hybrida* L. 'Moneyway'. *Plant Cell Reports* 15: 522–526.

Vissesuwan R, Kawai T and Mill M (1997) Plant regeneration systems from leaf segment culture through embryogenic callus formation of *Rosa hybrida* and *R. canina*. *Breeding Science* 47: 217–222.

Cell, Tissue and Organ Culture: Protoplast Culture

Bottcher UF, Aviv D and Galun E (1989) Complementation between protoplasts treated with either of two metabolic inhibitors results in somatic-hybrid plants. *Plant Science* 63: 67–77.

Kim SW, Oh SC, In DS and Liu JR (2003) Plant regeneration of rose (*Rosa hybrida*) from embryogenic cell-derived protoplasts. *Plant Cell, Tissue and Organ Culture* 73: 15–19.

Marchant R, Davey MR and Power JB (1997) Isolation and culture of mesophyll protoplasts from *Rosa hybrida*. *Plant Cell, Tissue and Organ Culture* 50: 131–134.

Matthews D, Mottley J, Horan I and Roberts AV (1991) A protoplast to plant system in roses. *Plant Cell, Tissue and Organ Culture* 24: 173–180.

Schum A, Hofmann K, Ghalib N and Tawfik A (2001) Factors affecting protoplast isolation and plant regeneration in *Rosa* spp. *Gartenbauwissenschaft* 66: 115–122.

Cell, Tissue and Organ Culture: Cell Culture and Secondary Metabolites

Amorim HV, Dougall DK and Sharp WR (1977) The effect of carbohydrate and nitrogen concentration on phenol synthesis in Paul's Scarlet rose cells grown in tissue culture. *Physiologia Plantarum* 39: 91–95.

Auh C-K and Murphy TM (1995) Plasma membrane redox enzyme is involved in the synthesis of O_2^- and H_2O_2 by *Phytophthora* elicitor-stimulated rose cells. *Plant Physiology* 107: 1241–1427.

Baydoun EA-H and Fry SC (1988) [2-^{3}H]Mannose incorporation in cultured plant cells: investigation of L-galactose residues in the primary cell wall. *Journal of Plant Physiology* 132: 484–490.

Berger RG, Akkan Z and Drawert F (1990) Catabolism of geraniol by cell suspension cultures of *Citrus limon*. *Biochimica Biophysica Acta* 1055: 234–239.

Bolwell GP, Davies DR, Gerrish C *et al.* (1998) Comparative biochemistry of the oxidative burst produced by rose and French bean cells reveals two distinct mechanisms. *Plant Physiology* 116: 1379–1385.

Davies ME (1972) Effects of auxin on polyphenol accumulation and the development of phenylalanine ammon-lyase activity in dark-grown suspension cultures of Paul's Scarlet rose. *Planta* 104: 66–77.

Edelmann HG and Fry SC (1992) Kinetics of integration of xyloglucan into the walls of suspension-cultured rose cells. *Journal of Experimental Botany* 43: 463–470.

Edelmann HG and Fry SC (1992) Factors that affect the extraction of xyloglucan from the primary cell walls of suspension-cultured rose cells. *Carbohydrate Research* 228: 423–431.

Eliot RF (1970) Axenic culture of meristem tips of *Rosa multiflora*. *Planta* 95: 183–186.

Fletcher JS (1974) Metabolism of glutamate in suspension cultures of Paul's Scarlet rose cells. *Plant Physiology* 54: 503–505.

Fletcher JS (1974) Control of amino acid synthesis in tissue culture cells. *Plant Physiology* 56: 450–451.

Fry SC and Street HE (1980) Gibberellin-sensitive suspension cultures. *Plant Physiology* 65: 472–477.

García-Romera I and Fry SC (1995) The longevity of biologically-active oligogalacturonides in rose cell cultures: degradation by exo-polygalacuronase. *Journal of Experimental Botany* 46: 1853–1857.

García-Romera I and Fry SC (1997) Upper limits for endogenous oligogalacturonides and free galacturonic acid in rose cell-suspension cultures: implications for the action of exo- and endo-polygalacturonases *in vivo*. *Journal of Plant Physiology* 150: 241–246.

Huerta AJ and Murphy TM (1989) Effects of extracellular pH on UV-induced K^+ efflux from cultured rose cells. *Plant Physiology* 90: 749–753.

Huerta AJ and Murphy TM (1989) Control of intracellular glutathione and its effect on ultraviolet radiation-induced K^+ efflux in cultured rose cells. *Plant, Cell and Environment* 12: 825–830.

Imbrie CW and Murphy TM (1982) UV-action spectrum (254–405 nm) for inhibition of a K^+-stimulated adenosine triphosphatase from the plasma membrane of *Rosa damascena*. *Photochemistry and Photobiology* 36: 537–542.

Imbrie CW and Murphy TM (1984) Solubilization and partial purification of two adenosine triphosphatases from a rose cell plasma membrane fraction. *Plant Physiology* 74: 611–616.

Imbrie CW and Murphy TM (1984) Photoinactivation of two detergent solubilized plasma membrane ATPase from *Rosa damascena*: action spectra. *Plant Physiology* 74: 617–621.

Imbrie CW and Murphy TM (1984) Proton pumping by vesicles reconstituted from two fractions of solubilized rose-cell plasma membrane ATPase. *Biochimica et Biophysica Acta* 778: 229–232.

Komalavilas P, Zhu JK and Nothnagel EA (1991) Arabinogalactan-proteins from the suspension culture medium and plasma membrane of rose cells. *Journal of Biological Chemistry* 266: 15956–15965.

Lam TH and Street HE (1977) Effects of selected aryloxyalkanecarboxylic acids on growth and levels of soluble phenols in cultured cells of *Rosa damascena*. *Zeitschrift für Pflanzenphysiologie* 84: 121.

Langan KJ and Nothnagel EA (1997) Cell surface arabinogalactan-proteins and their relation to cell proliferation and viability. *Protoplasma* 196: 87–98.

Lorences EP and Fry SC (1990) Absolute measurement of cell expansion in plant cell suspension cultures. *Plant Cell, Tissue and Organ Culture* 24: 211–215.

Murashige T (1974) Plant propagation through tissue culture. *Annual Review of Plant Physiology* 25: 135–166.

Murphy TM (1984) The effect of sulfhydryl reagents on K^+ efflux from rose cells: relationship to UV-stimulated efflux. *Plant Physiology* 75: 138–141.

Murphy TM (1988) Induced K^+ efflux from cultured rose cells: effects of protein-synthesis inhibitors. *Plant Physiology* 86: 830–835.

Murphy TM (1990) Effect of broad-band ultraviolet and visible radiation on hydrogen peroxide formation by cultured rose cells. *Physiologia Plantarum* 80: 63–68.

Murphy TM and Auh CK (1992) Phytophthora elicitor inhibits ferricyanide reduction by cultured rose cells. *Environmental and Experimental Botany* 32: 487–496.

Murphy TM and Vu HD (1996) Photoinactivation of superoxide synthases of the plasma membrane from rose (*Rosa damascena* Mill.) cells. *Photochemistry and Photobiology* 64: 106–109.

Murphy TM and Wilson C (1982) The UV-stimulated K^+ efflux from rose cells. *Plant Physiology* 70: 709–713.

Murphy TM, Hamilton CM and Street HE (1979) A strain of *Rosa damascena* cultured cells resistant to ultraviolet light. *Plant Physiology* 64: 936–941.

Murphy TM, Matson GB and Morrison SL (1983) Ultraviolet-stimulated $KHCO_3$ efflux from rose cells: regulation of cytoplasmic pH. *Plant Physiology* 73: 20–24.

Murphy TM, Hurrell HC and Sasaki TL (1985) Wavelength dependence of ultraviolet radiation-induced mortality and K^+ efflux in cultured cells of *Rosa damascena*. *Photochemistry and Photobiology* 42: 281–286.

Murphy TM, Vu HD and Nguyen T (1998) The superoxide synthases of rose cells: comparison of assays. *Plant Physiology* 117: 1301–1305.

Qian YC, Nguyen T and Murphy TM (1993) Effects of washing on the plasma membrane and on stress reactions of cultured rose cells. *Plant Cell, Tissue and Organ Culture* 35: 245–252.

Serpe MD and Nothnagel EA (1994) Effects of Yariv phenylglycosides on *Rosa* cell suspensions: evidence for the involvement of arabinogalactan-proteins in cell proliferation. *Planta* 193: 542–550.

Serpe MD and Nothnagel EA (1995) Fractionation and structural characterization of arabinogalactan-proteins from the cell wall of rose cells. *Plant Physiology* 109: 1007–1016.

Serpe MD and Nothnagel EA (1996) Heterogeneity of arabinogalactan-proteins on the plasma membrane of rose cells. *Plant Physiology* 112: 1261–1271.

Skirvin RM and Chu MC (1979) *In vitro* propagation of 'Forever Yours' rose. *HortScience* 14: 608–610.

Smith CK and Fry SC (1999) Biosynthetic origin and longevity in vivo of α-D-mannopyranosyl-(1→4)-α-D-glucuronopyranosyl-(1→2)-myo-inositol, an unusual extracellular oligosaccharide produced by cultured rose cells. *Planta* 210: 150–156.

Smith CK, Hewage CM, Fry SC and Sadler IH (1999) α-D-Mannopyranosyl-(1→4)-α-D-glucuronopyranosyl-(1→2)-myo-inositol, a new and unusual oligosaccharide from cultured rose cells. *Phytochemistry* 52: 387–396.

Smith RC and Fry SC (1991) Endotransglycosylation of xyloglucans in plant cell-suspension cultures. *Biochemical Journal* 279: 529–535.

Smith TA, Best GR, Abbott AJ and Clements ED (1978) Polyamines in Paul's Scarlet rose suspension cultures. *Planta* 144: 63–68.

Svetek J, Yadav MP and Nothnagel EA (1999) Presence of a glycosylphosphatidylinositol lipid anchor on rose arabinogalactan proteins. *Journal of Biological Chemistry* 274: 14724–14733.

Thompson JE and Fry SC (1997) Trimming and solubilization of xyloglucan after deposition in the walls of cultured rose cells. *Journal of Experimental Botany* 48: 297–305.

Thompson JE and Fry SC (2000) Evidence for covalent linkage between xyloglucan and acidic pectins in suspension-cultured rose cells. *Planta* 211: 275–286.

Thompson JE and Fry SC (2001) Density-labelling of cell wall polysaccharides in cultured rose cells: comparison of incorporation of ^2H and ^{13}C from exogenous glucose. *Carbohydrate Research* 332: 175–182.

Walko RM and Nothnagel EA (1989) Lateral diffusion of proteins and lipids in the plasma membrane of rose protoplast. *Protoplasma* 152: 46–56.

Weinstein LH, Tulecke W, Nickell LG and Laurencot HJ (1962) Biochemical and physiological studies of tissue cultures and plant parts from which they are derived. III. Paul's Scarlet rose. *Contributions from Boyce Thompson Institute* 21: 371–386.

Cell, Tissue and Organ Culture:
Storage and Cryopreservation

Arene L, Pellegrino C and Gudin S (1993) A comparison of the somaclonal variation level of *Rosa hybrida* L. cv. Meirutral plants regenerated from callus or direct induction from different vegetative and embryogenic tissues. *Euphytica* 71: 83–90.

Benson EE, Lynch PT and Stacey GN (1998) Advances in plant cryopreservation technology: current applications in crop plant biotechnology. *AgBiotech News Info* 10: 133N–141N.

Blakesley D and Kiernan RJ (2001) Cryopreservation of axillary buds of a *Eucalyptus grandis* × *Eucalyptus camaldulensis* hybrid. *Cryo-Letters* 22: 13–18.

Channuntapipat C, Collins G, Bertozzi T and Sedgley M (2000) Cryopreservation of *in vitro* almond shoot tips by vitrification. *Journal of Horticultural Science and Biotechnology* 75: 228–232.

Harding K, Benson EE and Clacher K (1997) Plant conservation biotechnology: an overview. *Agro Food Industry Hi-Tech* 8: 24–29.

Lauzer D and Laberge C (1996) Establishment of a collection of *Rosa* species through *in vitro* embryo culture. *HortScience* 31: 458–459.

Lynch PT, Harris W and Chartier-Hollis JM (1996) Cryopreservation of shoot tips of *Rosa multiflora*. *Plant Growth Regulation* 20: 43–45.

Martin C, Iriondo JM, Gonzalez-Benito E and Perez C (1998) The use of tissue culture techniques in the conservation of plant biodiversity. *Agro Food Industry Hi-Tech* 9: 37–40.

Reed BM (2001) Implementing cryogenic storage of clonally propagated plants. *Cryo-Letters* 22: 97–104.

Reed BM, Dumet D, Denoma JM and Benson EE (2001) Validation of cryopreservation protocols for plant germplasm conservation: a pilot study using *Ribes*. *Biodiversity and Conservation* 10: 939–949.

Rosu A, Skirvin RM, Bein A *et al.* (1995) The development of putative adventitious shoots from a chimeral thornless rose (*Rosa Multiflora* Thunb Ex J Murr) *in-vitro*. *Journal of Horticultural Science* 70: 901–907.

Sakai A, Matsumoto T, Hirai D and Niino T (2000) Newly developed encapsulation-dehydration protocol for plant cryopreservation. *Cryo-Letters* 21: 53–62.

Turner SR, Touchell DH, Senaratna T *et al.* (2001) Effects of plant growth regulators on survival and recovery growth following cryopreservation. *Cryo-Letters* 22: 163–174.

Yan M, Byrne DH and Jing C (1996) Propagation of rose species *in vitro*. In Vitro *Cellular Developmental Biology: Plant* 32: 103–108.

Zhao YH, Wu YJ, Engelmann F *et al.* (1999) Cryopreservation of apple shoot tips by encapsulation–dehydration: effect of preculture, dehydration and freezing procedure on shoot regeneration. *Cryo-Letters* 20: 103–108.

Classification: Conventional Taxonomy (Wild Roses)

Reichert H (1996) Die Legitimität der von Linné publizierten Namen mitteleuropäischer Wildrosen. *Gleditschia* 24: 13–23.

Wissemann V (2000) Molekulargenetische und morphologisch-anatomische Untersuchungen zur Evolution und Genomzusammensetzung von Wildrosen der Sektion Caninae (DC.) Ser. *Botanische Jahrbucher fur Systematik, Pflanzengeschichte und Pflanzen Geographie* 122: 357–429.

Classification: Chemotaxonomy and Molecular Taxonomy

Antonelli A, Fabbri C, Giorgioni ME and Bazzocchi R (1997) Characterization of 24 old garden roses from their volatile compositions. *Journal of Agriculture and Food Chemistry* 45: 4435–4439.

Ben-Meir H and Vainstein A (1994) Assessment of genetic relatedness in roses by DNA fingerprint analysis. *Scientia Horticulturae* 58: 115–121.

Biolley JP, Jay M and Barbe JP (1992) Chemometric approach (flavonoids) in an automatic recognition of modern rose cultivars. *Biochemistry and System Ecology* 20: 697–705.

Biolley JP, Jay M and Viricel MR (1994) Flavonoid diversity and metabolism in 100 *Rosa* × *hybrida* cultivars. *Phytochemistry* 35: 413–419.

Brubaker CL, Paterson AH and Wendel JF (1999) Comparative genetic mapping of allotetraploid cotton and its diploid progenitor. *Genome* 42: 184–203.

Crespel L, Chirollet M, Durel CE *et al.* (2002) Mapping of qualitative and quantitative phenotypic traits in *Rosa* using AFLP markers. *Theoretical and Applied Genetics* 105: 1207–1214.

Cronn RC, Small R and Wendel JF (1999) Duplicated genes evolve independently after polyploid formation in cotton. *Proceedings of the National Academy of Science of the USA* 96: 14406–14411.

Debener T and Mattiesch L (1998) Effective pairwise comparison of long primers for RAPD analyses in roses. *Plant Breeding* 117: 147–151.

Debener T and Mattiesch L (1999) Construction of a genetic linkage map for roses using RAPD and AFLP markers. *Theoretical Applied Genetics* 99: 891–899.

Debener T, Bartels C and Mattiesch L (1996) RAPD analysis of genetic variation between a group of rose cultivars and selected wild rose species. *Molecular Breeding* 2: 321–327.

Eugster CH and Märki-Fischer E (1991) The chemistry of rose pigments. *Angewandte Chemie* 30: 654–672.

Ferguson D and Sang T (2001) Speciation through homoploid hybridization between allotetraploids in peonies (*Paeonia*). *Proceedings of the National Academy of Science of the USA* 98: 3915–3919.

Fernandez-Romero MD, Torres AM, Millan T, Cunero JI and Cabrera A (2001) Physical mapping of ribosomal DNA on several species of the subgenus *Rosa*. *Theoretical and Applied Genetics* 103: 835–838.

Grossi C, Raymond O and Jay M (1997) Isozyme polymorphism of *Rosa* spp and cultivar identification. *Euphytica* 98: 11–19.

Grossi C, Raymond O and Jay M (1998) Flavonoid and enzyme polymorphisms and taxonomic organisation of *Rosa* sections: *Carolinae*, *Cinnamomeae*, *Pimpinnellifoliae* and *Synstylae*. *Biochemistry and System Ecology* 26: 857–871.

Grossi C, Raymond O, Sanlaville-Boisson C and Jay M (1999) *Rosa* taxonomy and hierarchy of markers defined by Act Statis. *Zeitschrift für Naturforschung* 54c: 25–34.

Iwata H, Kato T and Ohno S (2000) Triparental origin of damask roses. *Gene* 259: 53–59.

Jan CH and Byrne DH (1999) Rose germplasm analysis with RAPD markers. *HortScience* 34(2): 341–345.

Jiang CX, Wright RJ, El-Zik K and Paterson AH (1998) Polyploid formation created unique avenues for response to selection in *Gossypium* (cotton). *Proceedings of the National Academy of Science of the USA* 95: 4419–4424.

Kovats E (1987) Composition of essential oils. Part 7. Bulgarian oil of rose (*Rosa damascena* Mill.). *Journal of Chromatography* 406: 185–222.

Lata P (1982) Cytological studies in the genus *Rosa*: meiotic analysis of 10 cultivars. *Cytologia* 47: 639–648.

Ma Y, Islam-Faridi MN, Crane CF *et al.* (1996) A new procedure to prepare slides of metaphase chromosomes of roses. *HortScience* 31: 855–857.

Ma Y, Crane CF and Byrne DH (1997) Karyotypic relationships among some *Rosa* species. *Caryologia* 50: 317–326.

Ma Y, Islam-Faridi MN, Crane CF, Ji Y, Stelly DM, Price HJ and Byrne DH (1997) *In situ* hybridization of ribosomal DNA to rose chromosomes. *Journal of Heredity* 88: 158–161.

Martin M, Piola F, Chessel D, Jay M and Heizmann P (2001) The domestication process of the modern rose: genetic structure and allelic composition of the rose complex. *Theoretical and Applied Genetics* 102: 398–404.

Matsumoto S, Wakita H and Fukui H (1997) Molecular classification of wild roses using organelle DNA probes. *Scientia Horticulturae* 68: 191–196.

Matsumoto S, Kouchi M, Yabuki J *et al.* (1998) Phylogenetic analyses of the genus *Rosa* using the *matK* sequence: molecular evidence for the narrow genetic background of modern roses. *Scientia Horticulturae* 77: 73–82.

Mikanagi Y, Yokoi M, Ueda Y and Saito N (1995) Flower flavonol and anthocyanin distribution in subgenus *Rosa*. *Biochemistry and System Ecology* 23: 183–200.

Mikanagi Y, Saito N, Yokoi M and Tatsuzawa F (2000) Anthocyanins in flowers of genus *Rosa*, sections *Cinnamomeae* (= *Rosa*), *Chinenses*, *Gallicanae* and some modern garden roses. *Biochemistry and System Ecology* 28: 887–902.

Millan T, Osuna F, Cobos S, Torres AM and Cubero JI (1996) Using RAPDs to study phylogentic relationships in *Rosa*. *Theoretical and Applied Genetics* 96: 621–627.

Nakamura S (1987) Scent and component analysis of the hybrid tea rose. *Perfumer and Flavorist* 12: 43–45.

Nybom H, Olsson A and Werlemark G (1996) Morphometric variation in nordic dogroses (*Rosa* sect. *Caninae*, Rosaceae). *Acta Universitatis Upsaliensis, Symbolae Botanicae Upsaliensis* 31: 59–68.

Nybom H, Carlson-Nilsson U, Werlemark G and Uggla M (1997) Different levels of morphometric variation in three heterogamous dogrose species (*Rosa sect. Caninae*). *Plant System Evolution* 204: 207–224.

Olsson A, Nybom H and Prentice HC (2000) Relationships between Nordic dogroses (*Rosa* L. sect. *Caninae*, Rosaceae) assessed by RAPDs and elliptic Fourier analysis of leaflet shape. *Systematic Botany* 25(3): 511–521.

Piola F, Raymond O, Fiasson JL *et al.* (2001) Modern tools in rose taxonomy: contribution of chemical, biochemical and molecular approaches to the study of the domestication of the modern rose. *Recent Research Developments in Plant Biology* 2: 77–88.

Raamsdonk van LWD (1995) The effect of domestication on plant evolution. *Acta Botanica Neerlandica* 44: 421–438.

Rajapakse S, Byrne DH, Zhang L *et al.* (2001) Two genetic linkage maps of tetraploid roses. *Theoretics and Applied Genetics* 103(4): 575–583.

Ramsey J and Schemske DW (1998) Pathways, mechanisms, and rates of polyploid formation in flowering plants. *Annual Review of Ecology Systems* 29: 467–501.

Raymond O, Biolley JP and Jay M (1995) Fingerprinting the selection process of old roses by means of floral phenolic metabolism. *Biochemistry and System Ecology* 23: 555–565.

Raymond O, Fiasson JL and Jay M (2000) Synthetic taxonomy of *Rosa* races using Act Statis. *Zeitschrift für Naturforschung* 55: 399–409.

Raymond O, Fiasson JL and Jay M (2001) The domestication syndrome within hybrid perpetuals roses: the effect of unconscious selection on flavonoids. *Biochemistry and System Ecology* 29: 513–529.

Raymond O, Piola F and Sanlaville-Boisson C (2002) Inference of reticulation in outcrossing allopolyploid taxa: caveats, likelihood and perspectives. *Trends in Ecology and Evolution* 17(1): 3–6.

Sang T, Crawford DJ and Stuessy TF (1995) Documentation of reticulate evolution in peonies (*Paeonia*) using internal transcribed spacer sequences of nuclear ribosomal DNA: implications for biogeography and concerted evolution. *Proceedings of the National Academy of Science of the USA* 92: 6813–6817.

Small RL, Ryburn JA, Cronn RC, Selanan T and Wendel JF (1998) The tortoise and the hare: choosing between non-coding plastome and nuclear *Adh* sequences for phylogeny reconstruction in recently diverged plant group. *American Journal of Botany* 85: 1301–1315.

Tackholm G (1920) On the cytology of the genus *Rosa*. *Svensk Botanisk Tidskrift* 14: 300–311.

Trapp SC and Croteau RB (2001) Genomic organization of plant terpene synthase and molecular evolutionary implications. *Genetics* 158: 811–832.

Wendel JF (1989) New world tetraploid cottons contain old world cytoplasm. *Proceedings of the National Academy of Science of the USA* 86: 4132–4136.

Wendel JF (2000) Genome evolution in polyploids. *Plant Molecular Biology* 42: 225–249.

Wendel JF, Schnabel A and Seelanan T (1995) Bidirectional interlocus concerted evolution following allopolyploid speciation in cotton (*Gossypium*). *Proceedings of the National Academy of Science of the USA* 92: 280–284.

Werlemark G, Uggla M and Nybom H (1999) Morphological and RAPD markers show a highly skewed distribution in a pair of reciprocal crosses between hemisexual dogrose species, *Rosa* sect. *Caninae*. *Theoretical and Applied Genetics* 98: 557–563.

Wissemann V (1999) Genetic constitution of *Rosa* sect. *Caninae* (*R. canina*, *R. jundzilli*) and Sect. *Gallicanae* (*R. gallica*). *Journal of Applied Botany* 73: 191–196.

Wolfe KH (2001) Yesterday's polyploids and the mystery of diploidization. *Nature Review* 2: 334–341.

Disease: Crown Gall

Bouzar H (1994) Request for a judicial opinion concerning the species of *Agrobacterium*. *International Journal of Systematic Bacteriology* 44: 373–374.

Jones JB (1988) Systemic movement of *Agrobacterium tumefaciens* in symptomless stem tissue of *Chrysanthemum morifolum*. *Plant Disease* 72: 51–54.

Nautiyal CS, Dion P and Chilton WS (1991) Mannopine and mannopinic acid as substrates for *Arthrobacter* sp. Strain MBA209 and *Pseudomonas putida* NA513. *Journal of Bacteriology* 9: 2833–2841.

Nester EW, Gordon MP, Amasino RM and Yanofsky MF (1984) Crown gall: a molecular and physiological analysis. *Annual Review of Plant Physiology* 35: 387–413.

Pionnat S, Keller H, Héricher D, Bettachini A, Dessaux Y, Nesme X and Poncet C (1999) Ti plasmids from *Agrobacterium* characterize rootstock clones that initiated a spread of crown gall disease in Mediterranean countries. *Applied and Environmental Microbiology* 65: 4197–4206.

Reynders-Aloisi S, Pelloli G, Bettachini A and Poncet C (1998) Tolerance to crown gall differs among genotypes of rose rootstocks. *HortScience* 33: 296–297.

Disease: *Botrytis*

Chang AY, Gladon RJ, Gleason ML *et al.* (1997) Postharvest quality of cut roses following electron-beam irradiation. *HortScience* 32: 698–701.

Elad Y and Volpin H (1988) The involvement of ethylene and calcium in gray mold of *Pelargonium*, *Ruscus*, and rose plants. *Phytoparasitica Israel Journal of Plant Protection Science* 16: 119–131.

Elad Y, Ayish N, Ziv O and Katan J (1990) Control of grey mould (*Botrytis cinerea*) with film-forming polymers. *Plant Pathology* 39: 249–254.

Elad Y, Kirshner B and Gotlib Y (1993) Attempts to control *Botrytis cinerea* on roses by pre- and postharvest treatments with biological and chemical agents. *Crop Protection* 12: 69–73.

Gupta GK (1979) Bud and twig blight of rose as caused by *Botrytis cinerea* Pers. ex. Fr. *Indian Journal of Horticulture* 36: 473.

Hammer PE and Marois JJ (1989) Nonchemical methods for postharvest control of *Botrytis cinerea* on cut roses. *Journal of the American Society of Horticultural Science* 114: 100–106.

Hammer PE, Evenson KB and Janisiewicz WJ (1993) Postharvest control of *Botrytis cinerea* on cut rose flowers with pyrrolnitrin. *Plant Disease* 77: 283–286.

Jarvis WR and Slingsby K (1975) Tolerance of *Botrytis cinerea* and rose powdery mildew [*Sphaerotheca pannosa*]. *Canadian Plant Disease Survey* 55: 44.

Karlik JF and Flint ML (2002) Diseases and abiotic disorders of outdoor roses. APSnet Feature Story. http://www.apsnet.org/online/feature/rose/.

Kerssies A and Frinking HD (1996) Relations between glasshouse climate and dry weight of petals, epicuticular wax, cuticle, pre-harvest flowering period and susceptibility to *Botrytis cinerea* of gerbera and rose flowers. *European Journal of Plant Pathology* 102: 257–263.

Kerssies A, Bosker-van Zessen AI and Frinking HD (1995) Influence of environmental conditions in a glass on conidia of *Botrytis cinerea* on post-harvest infection of rose flowers. *European Journal of Plant Pathology* 101: 201–216.

Kerssies A, Bosker-van Zessen AI, Wagemakers CAM and van Kan JAL (1995) Variation in pathogenicity and DNA polymorphism among *Botrytis cinerea* isolates sampled inside and outside a glasshouse. *Plant Disease* 81: 781–786.

Magie RO (1969) Control of *Botrytis cinerea* diseases on chrysanthemum, carnation, rose, snapdragon, petunia and *Phalaenopsis* flowers. *Florida State Horticultural Society Proceedings* 82: 373–378.

Marois JJ (1992) Biological control of *Botrytis cinerea*. *NATO ASI Series A Life Science* 230: 109–111.

McCain A and Welch KE (1982) Controlling gray mold of rose flowers. University of California Extension. Davis, CA.

Morandi MAB, Sutton JC and Maffia LA (2000) Effects of host and microbial factors on development of *Clonostachys rosea* and control of *Botrytis cinerea* in rose. *European Journal of Plant Pathology* 106: 439–448.

Shaul O, Elad Y and Zieslin N (1996) Suppression of *Botrytis* blight in cut rose flowers with gibberellic acid. Effects of exogenous application of abscisic acid and paclobutrazol. *Postharvest Biology and Technology* 7: 145–150.

Shaul O, Elad Y and Zieslin N (1995) Suppression of botrytis blight in cut rose flowers with gibberelic acid: effect of concentration and mode of application. *Postharvest Biology and Technology* 6: 321–330.

Sman RGM, Evelo RG, Wilkinson EC and van Doorn WG (1996) Quality loss in packed rose flowers due to *Botrytis cinerea* infection as related to temperature regimes and packaging design. *Postharvest Biology and Technology* 7: 341–350.

University of Illinois Extension (1997) *Botrytis Blight or Gray Mold of Ornamental Plants*. RPD no. 623.

Williamson B, Duncan GH, Harrison JG *et al.* (1995) Effect of humidity on infection of rose petals by dry-inoculated conidia of *Botrytis cinerea*. *Mycology Research* 99: 1303–1310.

Yunis H and Elad Y (1989) Survival of dicarboximide-resistant strains of *Botrytis cinerea* in plant debris during summer in Israel. *Phytoparasitica Israel Journal of Plant Protection Science* 17: 13–21.

Disease: Downy Mildew

Achar P (1997) First report of downy mildew disease of rose caused by *Peronospora sparsa* in KwaZulu Natal, Southern Africa. *Plant Disease* 81: 695.

Breese WA, Shattock RC, Williamson B and Hackett C (1994) *In vitro* spore germination and infection of cultivars of *Rubus* and *Rosa* by downy mildew from both hosts. *Annals of Applied Biology* 125: 73–85.

Gill DL (1977) Downy mildew of roses in Georgia. *Plant Disease Reporter* 61: 230–231.

Hall H and Shaw C (1987) Brambles: downy mildew on wild and cultivated *Rubus* species in New Zealand. *New Zealand Journal of Experimental Agriculture* 15: 57–60.

Harwood C (1995) Adventures of a rose pathologist. *Advances in Plant Pathology* 11: 303–316.

Koponen H, Hellqvist S, Lindqvist-Kreuze H, Bang U and Valkonen J (2000) Occurrence of *Peronospora sparsa* (*P. rubi*) on cultivated and wild *Rubus* species in Finland and Sweden. *Annals of Applied Biology* 137: 107–112.

Lindqvist H, Koponen H and Valkonen JPT (1998) *Peronospora sparsa* on cultivated *Rubus arcticus* and its detection by PCR based on ITS sequences. *Plant Disease* 82: 1304–1311.

O'Neill TM, Pye D and Locke T (2002) The effect of fungicides, irrigation and plot density on the development of *Peronospora sparsa*, the cause of downy mildew in rose and black berry. *Annals of Applied Biology* 140: 207–214.

Tate K (1981) Aetiology of dryberry disease of boysenberry in New Zealand. *New Zealand Journal of Experimental Agriculture* 9: 371–376.

Tate K and Mespel G (1983) Control of dryberry disease (*Peronospora sparsa*) in boysenberry with fungicides. *New Zealand Journal of Experimental Agriculture* 11: 141–146.

Tian X, Lu G and Bai J (1998) A new record of *Peronospora* in China. *Mycosystema* 17: 287–288.

Williamson B, Breese WA and Shattock RC (1995) A histological study of downy mildew (*Peronospora rubi*) infection of leaves, flowers and developing fruits of Tummelberry and other *Rubus* spp. *Mycological Research* 99: 1311–1316.

Williamson B, Cooker DEL, Duncan JM *et al.* (1998) Fungal infections of micropropagated plants at weaning: a problem exemplified by downy mildew in *Rubus* and *Rosa*. *Plant Cell, Tissue and Organ Culture* 52: 89–96.

Disease: Powdery Mildew

Conti GG, Bassi M, Maffi D and Bocci AM (1986) Host–parasite relationship in a susceptible and a resistant rose cultivar inoculated with *Sphaerotheca pannosa*: II. Deposition rates of callose, lignin and phenolics in infected or wounded cells and their possible role in resistance. *Phythopathology* 117: 312–320.

Perera RG and Wheeler BEJ (1975) Effect of water droplets on the development of *Sphaerotheca pannosa* on rose leaves. *Transactions of the British Mycology Society* 64: 313–319.

Sivapalan A (1993) Effects of impacting rain drops on the growth and development of powdery mildew fungi. *Plant Pathology* 42: 256–263.

Vankuik AJ (1995) A weather-based forecasting model of powdery mildew in rose seedlings *Mededelingen – Faculteit Landbouwkundige en Toegepaste Biologische Wetenschappen*. Gent: Universiteit Gent. 60, pp. 439–445.

Wisniewska-Grzeszkiewicz H and Wojdyla AT (1996) Evaluation of rose cultivars to fungal diseases susceptibility. In: Morisot A and Ricci P (eds.) *2nd International Symposium on Roses*. Leuven, Belgium: International Society for Horticultural Science (ISHS).

Xu XM (1999) Effects of temperature on the latent period of the rose powdery mildew pathogen, *Sphaerotheca pannosa*. *Plant Pathology* 48: 662–667.

Yamaoka N and Takeuchi Y (1999) Morphogenesis of the powdery mildew fungus in water (4). The significance of conidium adhesion to the substratum for normal appresorium development in water. *Physiological and Molecular Plant Pathology* 54: 145–154.

Yarwood CE (1957) Powdery mildews. *Botanical Review* 23: 235–300.

Disease: Rust

Anthony VM, Shattock RC and Williamson B (1985) Interaction of red raspberry cultivars with isolates of *Phragmidium rubi-idaei*. *Plant Pathology* 34: 521–527.

Vries DP de and Dubois LAM (2001) Developments in breeding for horizontal and vertical fungus resistance in roses. *Acta Horticulturae* 552: 103–112.

Disease: Replant Problems and Soil Sickness

Braun PG (1995) Effects of *Cylindrocarpon* and *Pythium* species on apple seedlings and potential role in apple replant disease. *Canadian Journal of Plant Pathology* 17(4): 336–341.

Coolen WA and Hendrickx GJ (1972) Investigations of the resistance of rose root stocks to *Meloidogyne hapla* and *Pratylenchus penetrans*. *Nematologica* 18: 155–158.

Isutsa DK and Merwin IA (2000) *Malus* germplasm varies in resistance or tolerance to apple replant disease in a mixture of New York orchard soils. *HortScience* 35(2): 262–268.

Klaus H (1939) Das Problem der Bodenmüdigkeit unter Berücksichtigung des Obstbaus. *Landwirtschaftliche Jahrbücher* 89: 413–459.

Kloczko M and Spethmann W (1994) Untersuchungen über Schäden durch Nematoden an Rosen in Baumschulen. *Taspo Gartenbaumagazin* 3(5): 10–11.

Mazzola M (1998) Elucidation of the microbial complex having a causal role in the development of apple replant disease in Washington. *Phytopathology* 88(9): 930–938.

Morgan GT and Maclean AA (1968) Influence of soil pH on an introduced population of *Pratylenchus penetrans*. *Nematologica* 14: 311–312.

Otto G (1972a) Untersuchungen über die Ursache der Bodenmüdigkeit bei Obstgehölzen. I. Versuche zur Übertragung der Bodenmüdigkeit durch Wurzeln. *Zentralblatt für Bakteriologie, Parasitenkunde, Infektionskronkheiten und Hygiene, II Abt.* 127: 279–289.

Otto G (1972b) Untersuchungen über die Ursache der Bodenmüdigkeit bei Obstgehölzen. II. Versuche zur Übertragung der Bodenmüdigkeit durch wurzelfreien müden Boden. *Zentralblatt für Bakteriologie, Parasitenkunde, Infektionskronkheiten und Hygiene, II Abt.* 127: 601–611.

Otto G (1972c) Untersuchungen über die Ursache der Bodenmüdigkeit bei Obstgehölzen. III. Versuche zur Beseitigung der Bodenmüdigkeit durch Dämpfung bei verschiedenen Temperaturen. *Zentralblatt für Bakteriologie, Parasitenkunde, Infektionskronkheiten und Hygiene, II Abt.* 127: 777–782.

Otto G (1973) Untersuchungen über die Ursache der Bodenmüdigkeit bei Obstgehölzen. V. Einfluß verschiedener Dämpftemperaturen auf die Mikroflora eines müden Bodens. *Zentralblatt für Bakteriologie, Parasitenkunde, Infektionskronkheiten und Hygiene, II Abt.* 128: 377–382.

Otto G and Winkler H (1976) Nachweis der Bodenmüdigkeit in wachsenden Apfelanlagen. *Zentralblatt für Bakteriologie, Parasitenkunde, Infektionskronkheiten und Hygiene, II. Abt.* 131: 730–735.

Otto G and Winkler H (1977) Untersuchungen über die Ursache der Bodenmüdigkeit bei Obstgehölzen. VI. Nachweis von Aktinomyzeten in Faserwurzeln von Apfelsämlingen in Böden mit verschiedenen Müdigkeitsgraden. *Zentralblatt für Bakteriologie, Parasitenkunde, Infektionskronkheiten und Hygiene, II. Abt.* 132: 593–606.

Otto G and Winkler H (1989) Stand und Anwendungsmethoden bei der Bekämpfung von Nachbauschäden in Baumschulen. *Gartenbau* 36: Supplements 5–6.

Otto G, Winkler H and Szabó K (1993b) Investigations about the course of infestation of rootlets of apple seedlings by root pathogenic actinomycetes in soils with specific apple replant disease. *Zentralblatt für Mikrobiologie* 148: 467–476.

Santo GS and Lear B (1976) Influence of *Pratylenchus vulnus* and *Meloidogyne hapla* on the growth of rootstocks of roses. *Journal of Nematology* 8: 18–22.

Spethmann W (1993) Neue Methoden der Vermehrung und Züchtung von Rosen. *Rosenjahrbuch Verein Deutscher Rosenfreunde* 1993: 67–82.

Szabó K (1999a) Bodenmüdigkeit bei Rosengewächsen. *Gartenpraxis* 8: 43–48.

Szabó K (1999b) Ist Bodenmüdigkeit bekämpfbar? *Deutsche Baumschule* 51(1): 14–15.

Thomae K (1953) Bodenmüdigkeit im öffentlichen Grün. *Gartenamt* 9: 176–180.

Westcott III SW, Beer SV and Stiles WC (1986) Infection of apple roots by *Actinomycetes* associated with soils conductive to apple replant disease. *Plant Diseases* 70: 1125–1128.

Winkler H and Otto G (1972) Untersuchungen über die Ursache der Bodenmüdigkeit bei Obstgehölzen. IV. Einfluß verschiedener Dämpftemperaturen auf freilebende Nematoden in müden Böden. *Zentralblatt für Bakteriologie, Parasitenkunde, Infektionskronkheiten und Hygiene* 127: 783–788.

Wunderlich B and Wolf A (1993) Biotest zum Nachweis von Bodenmüdigkeit bei Rosen. *Gartenbau-Magazin* 2(1/2): 57–59.

Disease: Viruses

Amrine Jr JW and Hindal DF (1988) Rose rosette: a fatal disease of multiflora rose. *West Virginia University Agricultural and Forestry Experimental Station Morgantown*, Circular 147, August.

Basit AA and Francki RJB (1970) Some properties of rose mosaic virus from South Australia. *Australian Journal of Biological Science* 23: 1197–1206.

Bos L (1976) Symptom expression and variation of rose mosaic. *Netherlands Journal of Plant Pathology* 82: 239–249.

Bos L and Perguin FW (1975) Rose bud proliferation, a disorder of still unknown aetiology. *Netherlands Journal of Plant Pathology* 81: 187–198.

Brierley P (1935a) Streak, a virus disease of roses. *Phytopathology* 25: 7.

Brierley P (1935b) Symptoms of rose mosaic. *Phytopathology* 25: 8.

Brierley P and Smith FF (1940) Mosaic and streak disease of rose. *Journal of Agricultural Research* 61: 625–660.

Cammack RH (1966) Soil-borne viruses in rose. *Plant Pathology* 15: 47–48.

Cardin L and Devergne JC (1971) The isolation of prunus necrotic ringspot and tobacco necrosis virus from roses. *Annales de Phytopathologie* 3: 539–540.

Cheo P (1970) Rose wilt or dieback. *Lasca Leaves* 20: 88–89.

Cochran LC (1950) Infection of apple and rose with the ringspot virus. *Phytopathology* 40: 964.

Converse RH and Bartlett AB (1979) Occurrence of viruses in some wild *Rubus* and *Rosa* species in Oregon USA. *Plant Disease Reporter* 63: 441–444.

Crowe FJ (1984) Witches' broom: a new outbreak in several central states. *American Rose Annals* 69: 83–87.

Devergne JC and Goujon C (1975) Studies on growth abnormalities on roses grown in glasshouses. *Annales de Phytopathologie* 7: 71–79.

Epstein AH and Hill JH (1995) The biology of rose rosette disease: a mite-associated disease of uncertain aetiology. *Journal of Phytopathology* 143: 353–360.

Farrar EH and Frost RR (1972) A new disease of roses in Britain. *Plant Pathology* 21: 97.

Fletcher JT and Kingham HG (1962) Rose line pattern virus. *Plant Pathology* 11: 92.

Fry PR and Hammett KRW (1971) Rose wilt virus in New Zealand. *New Zealand Journal of Agricultural Research* 14: 735–743.

Fry PR and Hunter JA (1956) Rose mosaic virus in New Zealand. *New Zealand Journal of Science and Technology Section A* 37A: 478–482.

Fulton RO (1952) Mechanical transmission and properties of rose mosaic virus. *Phytopathology* 42: 413–416.

Fulton RO (1967) Purification and serology of rose mosaic virus. *Phytopathology* 57: 1197–1201.

Fulton RO (1968) Serology of some viruses isolated from cherry necrotic ringspot, plum line pattern, rose mosaic and apple mosaic. *Phytopathology* 58: 635–638.

Fulton RO (1970) A disease of rose caused by tobacco streak virus. *Plant Disease Reporter* 54: 949–951.

Fulton RO (1972) Apple mosaic virus. In: Harrison BD and Murant AF (eds.) *Commonwealth Mycological Institute and Association of Applied Biologists; Description of Plant Viruses*, Sheet No. 83, pp. 4.

Grieve BJ (1931) 'Rose wilt' and 'dieback'. A virus disease of roses occurring in Australia. *Australian Journal of Experimental Biology and Medical Science* 8: 107–121.

Gualaccini F (1963) Singolare vinescenza, prolungamento dell'asse fiorale e anomalie forliari su *Rosa* ottenuta da seme di pianta affetta da'mosdaico giallo ('rose yellow mosaic'). *Bolletino della R Stazione Patologie Vegetale, Roma* 21: 45–46.

Gumpf DJ and Weathers LG (1974) Detection of a new infectious disease of roses in California. *Acta Horticulturae* 36: 53–57.

Halliwell RS and Milbrath JA (1962) Isolation and identification of tomato ringspot virus associated with rose plants and rose mosaic virus. *Plant Disease Reporter* 46: 555–557.

Hammett KRW (1971) Symptom differences between rose wilt virus and *Verticillium* wilt of rose. *Plant Disease Reporter* 55: 916–920.

Handley MK (1986) *Identification and Partial Characterization of Two Cryptic Viruses in Rose by Analysis of Double-stranded RNA Content*. PhD thesis, University of California, Davis.

Hicks RGT and Frost RR (1984) Isolation of tobamovirus from rose. *Plant Pathology* 33: 581–587.

Hollings M, Stone OM, Brunt AA, Atkey PT and Fry PR (1973) Rose *Rosa* spp.; rose wilt. *Glasshouse Crops Research Institute Annual Report for 1972*, 105.

Hollings M, Stone OM, Atkey PT and Pawley RR (1974) Rose *Rosa* spp.; rose dieback (rose wilt). *Glasshouse Crops Research Institute Annual Report for 1973*, 118.

Hunter JA (1965) Further studies of rose mosaic virus in New Zealand. *New Zealand Journal of Agricultural Research* 8: 327–335.

Hunter JA (1966) Rose colour break: a virus disease of roses. *New Zealand Journal of Agricultural Research* 9: 1070–1072.

Ikin R and Frost RR (1974) Virus disease of roses. Part I. Their occurrence in the United Kingdom. *Phytopathologische Zeitschrift* 79: 160–168.

Ikin R and Frost RR (1976) Virus diseases of roses. Part II. Strawberry latent ringspot virus R/1 : 2.6 or 2 × 1.6/38 : S/S : S/Ne. *Phytopathologische Zeitschrift* 87: 205–223.

Kirkpatrick HC, Lindner RC, Cheney PW and Graham SO (1962) Rose as a source of prunus ringspot virus. *Plant Disease Reporter* 46: 722–723.

Kirkpatrick HC, Lindner RC and Cheney PW (1968) Symptomless infection of stone fruits with mechanically transmissible virus from rose. *Plant Disease Reporter* 52: 449–502.

Klesser PJ (1967) Occurrence of rose streak virus in South Africa. *South African Journal of Agricultural Science* 10: 849–856.

McDaniel GL, Buck GJ and Ford RE (1971) Isolation of tobacco ringspot virus from rose. *Phytopathology* 61: 45–49.

McWhorter FP (1931) Further report on rose mosaic in Oregon. *Plant Disease Reporter* 15: 1–3.

Meyer MA (1960) Virus diseases of roses in South Africa. *South African Journal of Agricultural Science* 3: 467–471.

Moury B, Cardin L, Onesto JP, Candresse T and Poupet A (2001) Survey of prunus necrotic ringspot virus in rose and the variability in rose and *Prunus* species. *Phytopathology* 91: 84–91.

Sastry KSM (1966) Yellow vein mosaic of *Rosa bourbiana*. *Indian Phytopathology* 19: 316–317.

Schmelzer K (1966) Zur Aetiologie des Rosengelbmosaiks. *Zentralblatt für Bakteriologie, Parasitenkunde, Infektionskrankheiten und Hygiene* 120: 233–236.

Schmelzer K (1967) Die Strichelkrankheit der Rose (rose streak) in Europa. *Phytopathologische Zeitschrift* 58: 92–95.

Secor GH and Nyland G (1978) Rose ring pattern. A component of the rose mosaic complex. *Phytopathology* 68: 1005–1010.

Secor GA, Kong M and Nyland G (1977) Rose virus and virus-like diseases. *California Agriculture* 31(3): 4–7.

Slack SA, Traylor JA, Williams HE and Nyland G (1976a) Rose leaf curl, a distinct component of a disease complex which resembles rose wilt. *Plant Disease Reporter* 60: 178–182.

Slack SA, Traylor JA, Nyland G and Williams HE (1976b) Symptoms, indexing and transmission of rose spring dwarf disease. *Plant Disease Reporter* 60: 183–187.

Smith KM (1937) Rosa virus 1–4. In: *A Texbook of Plant Virus Diseases*, pp. 152–156. Philadelphia: P Blakiston's.

Sweet JB (1975a) Strawberry latent ringspot virus in glasshouse roses. *Plant Pathology* 24: 93–96.

Sweet JB (1975b) Soil-borne viruses occurring in nursery soils and infecting some ornamental species of *Rosaceae*. *Annals of Applied Biology* 79: 49–54.

Thomas BJ (1975) Rose *Rosa* spp.; rose dieback (rose wilt). *Glasshouse Crops Research Institute Annual Report for 1974*, 118–119.

Thomas BJ (1976) Rose *Rosa* spp.; rose dieback. *Glasshouse Crops Research Institute Annual Report for 1975*, 123.

Thomas BJ (1980) Some degeneration and dieback diseases of the rose *Glasshouse Crops Research Institute Annual Report for 1979*, 178–190.

Thomas BJ (1981) Studies on rose mosaic disease in field grown roses produced in United Kingdom. *Annals of Applied Biology* 98: 419–429.

Thomas BJ (1982a) The effects of prunus necrotic ringspot virus on field grown roses. *Annals of Applied Biology* 100: 129–134.

Thomas BJ (1982b) Rose – *Rosa* spp. The effect of arabis mosaic and strawberry latent ringspot viruses on growth and flower quality. Seed transmission in *Rosa* spp. Detection viruses in new varieties. *Glasshouse Crops Research Institute Annual Report for 1981*, 143–145.

Thomas BJ (1984a) Epidemiology of three viruses infecting the rose in the United Kingdom. *Annals of Applied Biology* 105: 213–222.

Thomas BJ (1984b) Rose mosaic disease: symptoms induced in roses by graft inoculation with both prunus necrotic ringspot virus and apple mosaic virus. *Plant Pathology* 33: 155–160.

Thomas HE and Scott CE (1953) Rosette of rose. *Phytopathology* 43: 218–219.

Traylor JA, Williams HE and Nyland G (1966) Symptoms caused by strains of prunus necrotic ringspot virus in rose resemble typical rose mosaic. *Phytopathology* 56: 152.

Traylor JA, Wagnon HK and Williams E (1971) Rose spring dwarf, a graft-transmissible disease. *Plant Disease Reporter* 55: 294.

Viehmeyer G (1961) A 'new' disease of roses. *American Rose Annals* 46: 98–101.

Wagnon HK (1966) Report of a new or unusual plant pathogen: witches' broom (rosette of rose). *Californian Department of Agriculture Plant Pathology* A-70-7.

Weiss F and McWhorter FP (1930) Pacific coast survey for rose mosaic. *Plant Disease Reporter* 14: 203–205.

White RP (1928) An infectious chlorosis of roses. *Plant Disease Reporter* 12: 33–34.

White RP (1930) An infectious chlorosis of rose. *Phytopathology* 20: 130.

White RP (1932) Chloroses of the rose. *Phytopathology* 22: 53–69.

Wong SM and Horst RK (1993) Purification and characterization of an isolate of apple mosaic virus from rose in the USA. *Journal of Phytopathology* 139: 33–47.

Wong SM, Horst RK and Langhans RW (1988) Symptomatology and occurrence of apple mosaic and prunus necrotic ringspot on rose in New York. *Acta Horticulturae* 234: 437–447.

Fragrance and Pigments: Fragrance Profiles of Wild and Cultivated Roses

Aharoni A, de Vos CH, Verhoeven HA *et al.* (2002) Nontargeted metabolome analysis by use of Fourier transform ion cyclotron mass spectroscopy. *Omics* 6: 217–234.

Babu KGD, Singh B, Joshi VP and Singh V (2002) Essential oil composition of Damask rose (*Rosa damascena* Mill.) distilled under different pressures and temperatures. *Flavour and Fragrance Journal* 17: 136–140.

Gora J and Kalemba D (1995) Chemical composition of the essential oil of *Rosa centifolia* L. petals. *Journal of Essential Oil Research* 7: 89–90.

Guterman I, Shalit M, Menda N *et al.* (2002) Rose scent: genomics approach to discovering novel floral fragrance-related genes. *The Plant Cell* 14: 2325–2338.

Helsper JPFG, Davies JA, Bouwmeester HJ *et al.* (1998) Circadian rhythmicity in emission of volatile compounds by flowers of *Rosa hybrida* L. cv. Honesty. *Planta* 207: 88–95.

Kaul VK, Gurjal RV and Singh B (1999) Volatile constituents of oil of flowers of *Rosa brunonii* Lindl. *Flavour and Fragrance Journal* 14: 9–11.

Kim H-J, Kim K, Kim N-S and Lee D-S (2000) Determination of floral fragrances of *Rosa hybrida* using solid-phase trapping solvent extraction and gas chromatography–mass spectrometry. *Journal of Chromatography* A 902: 389–404.

Knudsen JT, Tollstent L and Bergstrom LG (1993) Floral scents: a checklist of volatile compounds isolated by head-space techniques. *Phytochemistry* 33: 253–280.

Fragrance and Pigments: Flavonoid Molecular Biology

Biolley JP and Jay M (1993) Anthocyanins in modern roses: chemical and colorimetric features in relation to the colour range. *Journal of Experimental Botany* 44: 1725–1734.

Biolley JP, Jay M and Forkmann G (1993) Pigmentation patterns of modern rose mutants throw light on the flavonoid pathway in *Rosa* × *hybrida*. *Phytochemistry* 36: 1189–1196.

Channeliere S, Riviere S, Scalliet G *et al.* (2002) Analysis of gene expression in rose petals using expressed sequence tags. *FEBS Letters* 515: 35–38.

Holton T and Tanaka Y (1994) Blue roses – a pigment of our imagination? *Trends in Biotechnology* 121: 40–42.

Martens S, Teeri T and Forkmann G (2002) Heterologous expression of dihydroflavonol 4-reductases from various plants. *FEBS Letters* 531: 453–458.

Schlangen K, Gosch C, Roemmelt S, Knott J, Fischer TC, Treutter D, Forkmann G, Stich K and Halbwirth H (2003) Can prohexadione-Ca induce antimicrobial flavonoids in rose? *European Journal of Horticultural Science* 68(3): in press.

Suzuki K, Tsuda S, Fukui Y, Fukuchi-Mizutani M, Yonekura-Sakakikara K, Tanaka Y and Kusumi T (2000) Molecular characterization of rose flavanoid biosynthesis genes and their application in petunia. *Biotechnology and Biotechnological Equipment* 14: 56–62.

Tanaka Y, Fukui Y, Fukuchimizutani M *et al.* (1995) Molecular cloning and characterisation of *Rosa hybrida* dihydroflavonol 4-reductase gene. *Plant Cell Physiology* 36: 1023–1031.

Fragrance and Pigments: Functional Genomics

Channeliere S, Riviere S, Scalliet G *et al.* (2002) Analysis of gene expression in rose petals using expressed sequence tags. *FEBS Letters* 515: 35–38.

Chen F, Tholl D, D'Auria JC *et al.* (2003) Biosynthesis and emission of terpenoid volatiles from arabidopsis flowers. *Plant Cell* 15: 481–494.

Dudareva N, Murfitt LM, Mann CJ *et al.* (2000) Developmental regulation of methyl benzoate biosynthesis and emission in snapdragon flowers. *Plant Cell* 12: 949–961.

Guterman I, Shalit M, Menda N *et al.* (2002) Rose scent: genomic approach to discover novel floral fragrance-related genes. *Plant Cell* 14: 2325–2338.

Lavid N, Wang J, Shalit M *et al.* (2002) O-Methyltransferases involved in the biosynthesis of volatile phenolic derivatives in rose petals. *Plant Physiology* 129: 1899–1907.

Lavy M, Zuker A, Lewinsohn E *et al.* (2002) Linalool and linalool oxide production in transgenic carnation flowers expressing the *Clarkia breweri* linalool synthase gene. *Molecular Breeding* 9: 103–111.

Shalit M, Guterman I, Volpin H *et al.* (2003) Volatile ester formation in roses: identification of an acetyl-CoA: geraniol acyltransferase in developing rose petals. *Plant Physiology* 131: 1866–1876.

Zuker A, Tzfira T, Ben-Meir H *et al.* (2002) Modification of flower color and fragrance by antisense suppression of the flavanone 3-hydroxylase gene. *Molecular Breeding* 9: 33–41.

Genetics: Karyology

Hanson RE, Zwick MS, Choi S *et al.* (1995) Fluorescent *in situ* hybridization of a bacterial artificial chromosome. *Genome* 38: 645–651.

Ma Y and Chen JY (1992) Chromosome studies of 6 species of *Rosa* in China. *Guihaia* 12(4): 333–336.

Ma Y, Islam-Faridi MN, Crane CF *et al.* (1996) A new procedure for making metaphase chromosome slides in roses. *HortScience* 31(5): 855–857.

Ma Y, Crane CF and Byrne DH (1997) Karyotypic relationships among some *Rosa* species. *Caryologia* 50: 317–326.

Ma Y, Islam-Faridi MN, Crane CF *et al.* (1997) *In situ* hybridization of ribosomal DNA to rose chromosomes. *Journal of Heredity* 88(2): 158–161.

Romero Zarco C (1986) A new method for estimating karyotype asymmetry. *Taxon* 35: 526–530.

Subramanian D (1987) Cytotaxonomic studies of south Indian Rosaceae. *Cytologia* 52: 395–403.

Yokoya K, Roberts AV, Mottley J, Lewis R and Brandham PE (2000) Nuclear DNA amounts in roses. *Annals of Botany* 85: 557–561.

Genetics: Meiosis

Berninger E (1992) Le rosier de serre pour fleurs à couper. In: Gallais A and Bannerot H (eds.). *Amélioration des Espèces Végétales Cultivées*, pp. 490–504. Paris: INRA Editions.

Lata P (1982) Cytological studies in the genus *Rosa* II. Meiotic analysis of ten species. *Cytologia* 47: 631–637.

Lata P (1982) Cytological studies in the genus *Rosa* III. Meiotic analysis of ten cultivars. *Cytologia* 47: 639–647.

Ma Y, Crane CF and Byrne DH (1996) Meiotic behavior of hybrids involving *Rosa laevigata*. *Cytologia* 61: 457–463.

Ma Y, Byrne DH and Chen J (1997) Colchicine-induced amphidiploids of rose interspecific hybrids. *HortScience* 32: 292–295.

Ma Y, Crane CF and Byrne DH (2000) Meiotic behavior in a tetraploid rose and its hybrid progeny. *HortScience* 35: 1127–1131.

McConnell J and Kamemoto H (1993) Morphology and meiotic behavior of three *Dendrobium* amphidiploids and their diploid counterparts. *HortScience* 28: 935–937.

Teoh SB and Ong EC (1983) Differential meiotic behaviour in hybrid clones of *Aranda* 'Christine' (Orchidaceae). *Euphytica* 32: 799–806.

Genetics: Inheritance of Characteristics

Biolley JP and Jay M (1993) Anthocyanins in modern roses: chemical and colorimetric features in relation to the colour range. *Journal of Experimental Botany* 44: 1725–1734.

De Vries DP and Dubois LAM (1977) Early selection in hybrid tea-rose seedlings for cut stem length. *Euphytica* 26: 761–764.

De Vries DP and Dubois LAM (1978) On the transmission of the yellow flower colour from *Rosa foetida* to recurrent flowering hybrid tea-roses. *Euphytica* 27: 205–210.

De Vries DP and Dubois LAM (1984) Inheritance of the recurrent flowering and moss characters in F_1 and F_2 hybrid tea × *Rosa centifolia muscosa* (Aiton) Seringe populations. *Gartenbauwissenschaft* 49: 97–100.

Dubois LAM and De Vries DP (1987) On the inheritance of the dwarf character in Polyantha × *Rosa chinensis minima* (Sims) Voss F_1 populations. *Euphytica* 36: 535–539.

Debener T, Bartels C and Mattiesch L (1996) RAPD analysis of genetic variation between a group of rose cultivars and selected wild species. *Molecular Breeding* 2: 321–327.

Debener T, Bartels C and Spethmann W (1997) Parentage analysis in interspecific crosses between rose species with RAPD markers. *Gartenbauwissenschaft* 62: 180–184.

Forkmann G (1993) Genetics of flavonoids. In: Harborne J (ed.) *The Flavonoids: Advances in Research since 1986*, pp. 537–564. London: Chapman & Hall.

Jicínská D (1976) Morphological features of F_1-generation in *Rosa* hybrids 1. Hybrids of some species of the Sect. Caninae with *Rosa rugosa*. *Folia Geobotanica et Phytotaxonomica Praha* 11: 301–311.

Lewis WH and Bayse RE (1961) Analysis of nine crosses between diploid *Rosa* species. *Proceedings of the American Society of Horticultural Science* 78: 572–579.

Malek B v and Debener T (1998) Genetic analysis of resistance to blackspot (*Diplocarpon rosae*) in tetraploid roses. *Theoretical and Applied Genetics* 96: 228–231.

Malek B v and Debener T (2000) Identification of molecular markers linked to *Rdr1*, a gene conferring resistance to blackspot in roses. *Theoretical and Applied Genetics* 101: 977–983.

Marshall HH, Campbell CG and Collicut LM (1983) Breeding for anthocyanin colors in *Rosa*. *Euphytica* 32: 205–216.

Rahtlef H v (1928) Vererbungsstudien an der Edelrose. *Zeitschrift für Pflanzenzüchtung* 13: 9–47.

Rajapakse S, Byrne D, Zhang L *et al.* (2001) Two genetic linkage maps of tetraploid roses. *Theoretical and Applied Genetics* 103: 575–583.

Semeniuk P (1971) Inheritance of recurrent blooming in *R. wichuraiana*. *Journal of Heredity* 62: 203–204.

Semeniuk P (1971) Inheritance of recurrent and non-recurrent blooming in 'Goldilocks' × *Rosa wichuraiana* progeny. *Journal of Heredity* 62: 319–320.

Stougaard B (1984) Analysis of variation in progeny from *Rosa multiflora* crosses. *Tidsskrift for Pnateavl* 88: 287–297.

Svejda F (1977) Breeding for improvement of flowering attributes of winterhardy *Rosa rugosa* hybrids. *Euphytica* 26: 697–701.

Svejda F (1977) Breeding for improvement of flowering attributes of winterhardy *Rosa kordesii* Wulf hybrids. *Euphytica* 26: 703–708.

Svejda F (1979) Inheritance of winterhardiness in roses. *Euphytica* 28: 309–314.

Genetics: Inheritance in the Dogrose

Almquist S (1919) *Sveriges Rosae*. Stockholm: PA Norstedt.

Blackhurst HT (1948) Cytogenetic studies on *Rosa rubiginosa* and its hybrids. *Proceedings of the American Society of Horticultural Science* 52: 510–516.

Fagerlind F (1940) Sind die *Canina*-Rosen agamospermische Bastarde? *Svensk Botanisk Tidskrift* 34: 334–354.

Gatt M, Hammett K and Murrey B (1999) Confirmation of ancient polyploidy in *Dahlia* (Asteraceae) species using genomic *in situ* hybridization. *Annals of Botany* 84: 39–48.

Gustafsson Å (1944) The constitution of the *Rosa canina* complex. *Hereditas* 30: 405–428.

Gustafsson Å and Håkansson (1942) Meiosis in some *Rosa* hybrids. *Botaniska Notiser* 331–343.

Hatton TJ (1989) Spatial patterning of sweet briar (*Rosa rubiginosa*) by two vertebrate species. *Australian Journal of Botany* 14: 199–205.

Jicinska D (1976) Morphological features of F_1 generation in *Rosa* hybrids. I. Hybrids of some species of the sect. *Caninae* with *Rosa rugosa*. *Folia Geobot Phytotax Praha* 11: 301–311.

Kroon GH and Zeilinga AE (1974) Apomixis and heterogamy in rose rootstocks (*Rosa canina* L.). *Euphytica* 23: 345–352.

Ma Y, Islam-Faridi MN, Crane CF *et al.* (1997) *In situ* hybridization of ribosomal DNA to rose chromosomes. *Journal of Heredity* 88: 158–161.

Nybom H, Olsson Å and Werlemark G (1996) Morphometric variation in Nordic dogroses (*Rosa* sect. *Caninae*, Rosaceae). *Acta Universitatis Upsaliensis, Symbolae Botanicae Upsaliensis* 31: 59–68.

Nybom H, Carlson-Nilsson U, Werlemark G and Uggla M (1997) Different levels of morphometric variation in three heterogamous dogrose species (*Rosa* sect. *Caninae*, Rosaceae). *Plant Systematics and Evolution* 204: 207–224.

Nybom H, Esselink D, Werlemark G and Vosman B (2002) Microsatellite DNA marker inheritance indicates preferential pairing between two highly homologous genomes in polyploid and hemisexual dogroses, *Rosa* L. sect. *Caninae*. Accepted in *Heredity* and in press.

Olsson P, Nybom H and Prentice H (2000) Relationships between Nordic dogroses (*Rosa* L., Sect. *Caninae*, Rosaceae) assessed by RAPDs and Fourier analysis of leaflet shape. *Systematic Botany* 25: 511–521.

Roberts AV (1975) The nature and taxonomic significance of the system of inheritance in *Rosa nanothamnus* (Rosaceae). *Botanical Journal of the Linnean Society* 71: 59–66.

Täckholm G (1920) On the cytology of the genus *Rosa*. *Svensk Botanisk Tidskrift* 14: 300–311.

Vega JM and Feldman M (1998) Effect of the pairing gene Ph1 and premeiotic colchicine treatment on intra- and interchromosome pairing of isochromosomes in common wheat. *Genetics* 150: 1199–1208.

Werlemark G (2000) Evidence of apomixis in hemisexual dogroses, *Rosa* section *Caninae*. *Sexual Plant Reproduction* 12: 353–359.

Werlemark G and Nybom H (2001) Skewed distribution of morphological character scores and molecular markers in three interspecific crosses in *Rosa* section *Caninae*. *Hereditas* 134: 1–13.

Werlemark G, Uggla M and Nybom H (1999) Morphological and RAPD markers show a highly skewed distribution in a pair of reciprocal crosses between hemisexual dogroses species, *Rosa* sect. *Caninae*. *Theoretical and Applied Genetics* 98: 557–563.

Winther K, Rein E and Kharazmi A (1999) The anti-inflammatory properties of rose hips. *Inflammopharmacology* 7: 63–68.

Wissemann V (1999) Genetic constitution of *Rosa* section *Caninae* (*R. canina*, *R. jundzillii*) and sect. *Gallicanae* (*R. gallica*). *Journal of Applied Botanics* 73: 191–196.

Zielinski J (1985) Studia nad radzajem *Rosa* L. Systematyka sekcji *Caninae* DC. em. *Christ. Arboretum Körnickie Rocznik* 30: 3–109.

Genetics: Species Crosses

Basye RE (1987) A probable amphidiploid of *Rosa abyssinica* and *Rosa rugosa*. *American Rose Annual* 72: 19–24.

Blackhurst HF (1948) Cytological studies on *R. rubiginosa* and its hybrids. *Proceedings of the American Society of Horticultural Science* 52: 510.

Cole RD (1917) Imperfection of pollen and mutability in genus *Rosa*. *Botanical Gazette* 63: 110–123.

Cullum FJ, Bradley SJ and Williams ME (1990) Improved germination of *Rosa corymbifera* 'Laxa' seed using a compost activator. *International Plant Propagator's Society* 40: 244–250.

Erlanson EW (1929) Cytological conditions and evidence for hybridity in North American wild roses. *Botanical Gazette* 87: 443–506.

Erlanson-MacFarlane EW (1963) A self-pollination mechanism and other intims in rose species. *American Rose Annual* 48: 188–193.

Fagerlind F (1940) Sind die *Canina*-Rosen agamospermische Bastarde? *Svensk Botanisk Tidsskrift* 34(4): 334–354.

Fagerlind F (1944) Kompatibilität und Inkompatibilität in der Gattung *Rosa*. *Acta Horti Bergiani* 13: 247–302.

Fagerlind F (1945) Die Bastarde der *Canina*-Rosen, ihre Syndese- und Formbildungsverhältnisse. *Acta Horti Bergiani* 14: 3–37.

Feuerhahn B and Spethmann W (1995) Kompoststarter verbessert die Keimung von Rosensaatgut. *TASPO-Gartenbaumagazin* 4(8): 25–26.

Gudin S, Areen L and Bullard C (1991) Influence of season on rose pollen quality. *Sexual Plant Reproduction* 4: 113–117.

Gustaffson A (1944) The constitution of the *Rosa canina* complex. *Hereditas* 30: 405–428.

Hanelt P (1982) Fortpflanzungssysteme und kritischer Status von Sippen der Flora der DDR. *Biologische Rundschrift* 20: 62–75.

Hurst CC (1928) Differential polyploidy in the genus *Rosa*. Verhandlungen des V. internationalen Kongresses für Vererbungswissenschaften in Berlin 1927. Supplementband II: 866–906.

Jicinska D (1975) Diversity of pollination in some *Rosa* species. *Preslia* 47: 267–274.

Khosh-Khui M, Bassiri A and Niknejad M (1976) Effects of temperature and humidity of pollen viability of six rose species. *Canadian Journal of Plant Science* 56: 517–523.

Klasterska AJ and Natarajan AT (1974) Cytological studies of the genus *Rosa* with special reference to the section *Caninae*. *Hereditas* 76(1): 97–108.

Malmgren U (1986) Wild-growing species of *Rosa* in Sweden. *Svensk Botanisk Tidsskrift* 80(4): 209–227.

Mayer A (1904) Flora von Tübingen und Umgebung, Schwäbische Alb vom Plettenberg bis zur Teck, Balingen, Hechingen, Reutlingen, Urach, Rottenburg, Herrenberg, Böblingen. Tübingen.

Meenakshi B (1977) Sterility, its incidence and causes with special reference to roses. *Indian Journal of Horticulture* 34(3): 437–445.

Müller K (1957) Ulmer Flora. *Mitteilungen des Vereins für Naturwissenschaft und Mathematik Ulm* 25: 105–109.

Paris CD and Maney TJ (1939) *Rosa multiflora* and its progeny. *Journal of the Iowa Academy of Science* 46: 149–160.

Pearson HM and Harney PM (1984) Pollen viability in *Rosa*. *HortScience* 19(5): 710–711.

Peimbert P, Krewinkle J and Morey D (1963) Factors influencing seed set in roses, vol 2. A preliminary bioassay of stored and fresh pollen. *American Rose Annual* 48: 195–198.

Persiel F (1974) Erste Versuche und Beobachtungen für die Resistenzzüchtung bei Rosenunterlagen im Rahmen des Arbeitsprogrammes der Bundesanstalt für gartenbauliche Pflanzenzüchtung. *Acta Pruhoniciana* 32: 58–66.

Persiel F (1977) Resistenzuntersuchungen gegen Mehltau und Sternrußtau bei den als Unterlagen genutzten Rosenarten. *Deutsche Baumschule* 29(7): 302–304.

Popova M and Kozhukharova K (1983) Studies on the hybridization of the species *Rosa canina* L. und *Rosa rugosa* Thunb. *Nauch Tr Vis Selskostopanske Institut Vasil Kalarvov* 28(4): 59–68.

Preil W (1980) Pollen lagern – aber wie? *Gartenbaupraxis* 6(5): 210–211.

Ratzek JC, Yarnell SH and Flory WS (1939) Crossing relation of some diploid species of roses. *Proceedings of the American Society of Horticultural Science* 37: 983–992.

Roberts AV (1977) Relationship between species in the genus *Rosa*, section *Pimpinellfoliae*. *Botanical Journal of the Linnean Society* 74: 309–328.

Schnetz J (1910) Einige Beobachtungen über individuelle Variation und temporäre Merkmalsschwankungen bei wilden Rosen. *Mitteilungen der Bayrischen Botanischen Gesellschaft zur Erforschung der heimischen Flora* 2(14): 229–237.

Schnetz J (1911) Die Geschichte eines Rosenbastards. *Mitteilungen der Bayrischen Botanischen Gesellschaft* 20(2): 219–222.

Schwertschlager J (1915) Beobachtungen und Versuche zur Biologie der Rosenblüte und Rosenbefruchtung. *Berichte der Bayrischen Botanischen Gesellschaft München* 15: 1–16.

Schwertschlager J (1926) Die Rosen Bayerns. *Berichte der Bayrischen Botanischen Gesellschaft München* 18: 1–126.

Stougaard B (1983) Pollination in *Rosa multiflora*. *Tidsskrift Plantearl* 87: 633–642.

Svedja F (1974) Reproductive capacity of F_1-hybrids from *Rosa rugosa* and *Rosa chinensis* cultivars. *Euphytica* 23(3): 665–669.

Täckholm G (1922) Zytologische Studien über die Gattung *Rosa*. *Acta Horti Bergiani* 7: 97–381.

Uilenburg NG (1985) De geschiedenis van de roos. 4a: Botanische rozen en parkrozen. *Groen* 41(3): 26–29.

Vogel M (1937) Beobachtungen an Wildrosensämlingen im Rosarium des VDR Sangerhausen. *Rosenjahrbuch (Verein Deutscher Rosenfreunde)* 1937: 23–25.

Vries de DP and Dubois LAM (1983) Pollen and pollination experiments, vol. X. The effect of repeated pollination on fruit and seed set in crosses between the hybrid tea-rose cvs. 'Sonia' and 'Ilona'. *Euphytica* 32: 685–689.

Wulff HD (1951) *Rosa* × *kordesii*, eine neue amphidiploide Rose. *Der Züchter* 21: 123–132.

Wulff HD (1952) Zytologische Betrachtungen an Rosenbastarden. *Der Züchter* 22: 233–244.

Genetics: Mutation

Arnold NP, Barthakur NN and Tanguay M (1998) Mutagenic effects of acute gamma irradiation on miniature roses: target theory approach. *HortScience* 33: 127–129.

Datta SK (1985) Radiosensitivity of garden roses. *Journal of Nuclear Agricultural Biology* 14: 133–135.

Datta SK (1986) Effects of recurrent gamma irradiation on rose cv. 'Contempo'. *Journal of Nuclear Agricultural Biology* 15: 125–127.

Datta SK and Gupta MN (1982) Gamma-ray induced white flower mutant in rose cv. 'Junior Miss'. *Journal of Nuclear Agricultural Biology* 11: 32–33.

Debener T, Janakiram T and Mattiesch L (2000) Sports and seedlings of rose varieties analysed with molecular markers. *Plant Breeding* 119: 71–74.

Dommergues P (1976) Mutagénèse induite chez les rosiers. *Horticulture Française* 74: 1–3.

Ibrahim R, Mondelaers W and Debergh PC (1998) Effects of X-irradiation on adventitious bud regeneration from in vitro leaf explants of *Rosa hybrida*. *Plant Cell, Tissue and Organ Culture* 54: 37–44.

Lata P (1975) Effects of ionising radiation on roses: meiotic studies on control and gamma ray induced mutants of rose cultivar 'Monteszuma'. *Cytologia* 40: 213–219.

Lata P (1975) Effects of ionising radiation on roses II. Meiotic studies on rose cultivar 'Pink Parfait' and its gamma ray induced mutants. *Cytologia* 40: 633–638.

Lata P (1980) Effect of ionizing radiation on roses: induction of somatic mutations. *Environmental and Experimental Botany* 20: 325–333.

Smilansky Z, Umiel N and Zieslin N (1986) Mutagenesis in roses (cv Mercedes). *Environmental and Experimental Botany* 26: 279–283.

Zimmerman PW and Hitchcock AE (1951) Rose 'sports' from adventitious buds. *Contributions to the Boyce Thompson Institute* 16: 221–224.

Genetics: DNA Fingerprinting

BenMeir H and Vainstein A (1994) Assessment of genetic relatedness in roses by DNA fingerprint analysis. *Scientia Horticulturae* 58: 115–121.

Crespel L, Gudin S, Meynet J and Zhang D (2002) AFLP-based estimation of 2*n* gametophytic heterozygosity in two parthenogenetically derived dihaploids in *Rosa hybrida* L. *Theoretical Applied Genetics* 104: 451–456.

Debener T and Mattiesch L (1998) Effective pairwise combination of long primers for RAPD analyses in roses. *Plant Breeding* 117: 147–151.

Debener T, Bartels C and Mattiesch L (1996a) RAPD analysis of genetic variation between a group of rose cultivars and selected wild rose species. *Molecular Breeding* 2: 321–327.

Debener T, Bartels C and Spehtmann W (1996b) Parentage analysis in interspecific crosses between rose species with RAPD markers. *Gartenbauwissenschaft* 62: 180–184.

Debener T, Janakiram T and Mattiesch L (2000) Sports and seedlings of rose varieties analysed with molecular markers. *Plant Breeding* 119: 71–74.

Gallego FJ and Martinez I (1996) Molecular typing of rose cultivars using RAPDs. *Journal of Horticultural Science* 71: 901–908.

Holm S (1995) Unexpected high levels of genetic variation in *Potentilla argentea* L. (s.l.) in southern Sweden. *Hereditas* 123: 127–139.

Hubbard M, Kelly J, Rajapakse S, Abbott AG and Ballard RE (1992) Restriction fragment length polymorphisms in rose and their use for cultivar identification. *Horticultural Science* 27: 172–173.

Iwata H, Kato T and Ohno S (2000) Triparental origin of Damask roses. *Gene* 259: 53–59.

Jan CH, Byrne DH, Manhart J and Wilson H (1999) Rose germplasm analysis with RAPD markers. *Horticultural Science* 34: 341–345.

Kim Y-J and Byrne DH (1996) Interspecific hybrid verification of *Rosa* with isozymes. *Horticultural Science* 31: 1207–1209.

Martin M, Piola F, Chessel D, Jay M and Heizmann P (2001) The domestication process of the modern rose: genetic structure and allelic composition of the rose complex. *Theoretical and Applied Genetics* 102: 398–404.

Matsumoto S and Fukui H (1996) Identification of rose cultivars and clonal plants by random amplified polymorphic DNA. *Scientia Horticulturae* 67: 49–54.

Matsumoto S, Wakita H and Fukui H (1997) Molecular classification of wild roses using organelle DNA probes. *Scientia Horticulturae* 68: 191–196.

Millán T, Osuna F, Cobos S, Torres AM and Cubero JI (1996) Using RAPDs to study phylogenetic relationships in *Rosa*. *Theoretical and Applied Genetics* 92: 273–277.

Olsson ÅME (1999) *Morphometric and Molecular Variation in the Nordic Dog Roses (*Rosa *L. sect.* Caninae*, Rosaceae)*. PhD dissertation, Department of Systematic Botany, Lund University, Sweden.

Olsson ÅME, Nybom H and Prentice HC (2000) Relationships between Nordic dog roses (*Rosa* L. sect. *Caninae*, Rosaceae) assessed by RAPDs and elliptic Fourier analysis of leaflet shape. *Systematic Botany* 25: 511–521.

Rajapakse S, Hubbard M, Kelly JW, Abbott AG and Ballard RE (1992) Identification of rose cultivars by restriction fragment length polymorphism. *Scientia Horticulturae* 52: 237–245.

Torres A, Millán T and Cubero JI (1993) Identifying rose cultivars using random amplified polymorphic DNA. *Horticultural Science* 28: 333–334.

Vainstein A and Ben-Meir H (1994) DNA fingerprint analysis of roses. *Journal of the American Society for Horticultural Science* 119: 1099–1103.

Walker CA and Werner DJ (1997) Isozyme and randomly amplified polymorphic DNA (RAPD) analyses of Cherokee rose and its putative hybrids 'Silver Moon' and 'Anemone'. *Journal of the American Society for Horticultural Science* 122: 659–664.

Werlemark G (2000) Evidence of apomixis in hemisexual dogrose species, *Rosa* sect. *Caninae*. *Sexual Plant Reproduction* 12: 353–359.

Werlemark G (2001) *Genetic Variability and Reproductive Strategies in Nordic Dog Roses*, Rosa *section* Caninae. PhD dissertation, Department of Horticultural Plant Breeding, Swedish University of Agricultural Sciences, Sweden.

Werlemark G and Nybom H (2001) Skewed distribution of morphological character scores and molecular markers in three interspecific crosses in *Rosa* section *Caninae*. *Hereditas* 134: 1–13.

Werlemark G, Uggla M and Nybom H (1999) Highly skewed distribution of morphological and RAPD markers in a pair of reciprocal crosses between hemisexual dog rose species, *Rosa* sect. *Caninae*. *Theoretical and Applied Genetics* 98: 557–563.

Zhang D, Germain E, Aloisi S and Gandelin MH (1999) Development of amplified fragment length polymorphism markers for variety identification in rose. *Acta Horticulturae* 508: 113–120.

Genetics: Gene Mapping

Gillham NW, Boynton JE and Harris EH (1991) Transmission of plastid genes. In: Bogorad L and Vasil IK (eds.) *The Molecular Biology of Plastids*, pp. 55–92. Orlando, FL: Academic Press.

Lander ES and Botstein D (1989) Mapping Mendelian factors underlying quantitative traits using RFLP linkage maps. *Genetics* 121: 185–199.

Matsumoto S, Wakita H and Fukui H (1997) Molecular classification of wild roses using organelle DNA probes. *Scientia Horticulturae* 68: 191–196.

Matsumoto S, Kouchi M, Yabuki J *et al.* (1998) Phylogenetic analyses of the genus *Rosa* using the *mat*K sequence: molecular evidence for the narrow genetic background of modern roses. *Scientia Horticulturae* 77: 73–82.

Michelmore RW, Paran I and Kesseli RV (1991) Identification of DNA markers linked to disease resistance genes by bulked segregant analysis: a rapid method to detect markers in specific genomic regions using segregation populations. *Proceedings of the National Academy of Sciences USA* 88: 9828–9832.

Moore G, Devos KM, Wang Z and Gale MD (1995) Gramineouses, line up and form a circle. *Current Biology* 5: 737–739.

Newton KJ (1988) Plant mitochondrial genomes: organization, expression and variation. *Annual Review of Plant Physiology and Plant Molecular Biology* 39: 503–532.

Stam P (1993) Construction of integrated genetic linkage maps by means of a new computer package JoinMap. *Plant Journal* 3: 739–744.

Zhang H-B and Wu C (2001) BAC as tools for genomic sequencing. *Plant Physiology and Biochemistry* 39: 195–209.

Genetics: Molecular Markers

Ben-Meir H and Vainstain A (1994) Assessment of genetic relatedness in roses by DNA fingerprint analysis. *Scientia Horticulturae* 58: 158–164.

Crespel L, Chirollet M, Durel CE *et al.* (2002) Mapping of qualitative and quantitative phenotypic traits in *Rosa* using AFLP markers. *Theoretical and Applied Genetics* 105: 1207–1214.

Debener T and Mattiesch L (1998) Effective pairwise combination of long primers for RAPD analyses in roses. *Plant Breeding* 117: 147–151.

Debener T and Mattiesch L (1999) Construction of a genetic linkage map for roses using RAPD and AFLP markers. *Theoretical and Applied Genetics* 99: 891–899.

Debener T, Bartels C and Mattiesch L (1996) RAPD analysis of genetic variation between a group of rose cultivars and selected wild rose species. *Molecular Breeding* 2: 321–327.

Hubbard M, Kelly J, Rajapakse S, Abbott A and Ballard R (1992) Restriction fragment length polymorphisms in rose and their use for cultivar identification and patent protection. *HortScience* 27: 172–173.

Iwata H, Kato T and Ohno S (2000) Triparental origin of Damask roses. *Gene* 259: 53–59.

Jan CH, Byrne DH, Manhart J and Wilson H (1997) Rose germplasm analysis with RAPD markers. *HortScience* 34: 341–345.

Malek BV, Weber WE and Debener T (2000) Identification of molecular markers linked to *Rdr1*, a gene conferring resistance to blackspot in roses. *Theoretical and Applied Genetics* 101: 977–983.

Matsumoto S, Wakita H and Fukui H (1997) Molecular classification of wild roses using organelle DNA probes. *Scientia Horticulturae* 68: 191–196.

Millan F, Osuma F, Cobos S, Torres A and Cubero JI (1996) Using RAPDs to study phylogenetic relationships in *Rosa*. *Theoretical and Applied Genetics* 92: 273–277.

Rajapakse S, Hubbard M, Kelly JW, Abbott AG and Ballard RE (1992) DNA 'fingerprinting' of rose cultivars by restriction fragment length polymorphism. *Scientia Horticulturae* 52: 237–245.

Rajapakse S, Byrne DH, Zhang L *et al.* (2001) Two genetic linkage maps of tetraploid roses. *Theoretical and Applied Genetics* 103: 575–583.

Torres AM, Millan T and Cubero JT (1993) Identifying rose cultivars using random amplified polymorphic DNA markers. *HortScience* 28: 333–334.

Werlemark G, Uggla M and Nybom H (1999) Morphological and RAPD markers show a highly skewed distribution in a pair of reciprocal crosses between hemisexual dogrose species, *Rosa* sect. *Caninae*. *Theoretical and Applied Genetics* 98: 557–563.

Genetics: Cloned Genes

Channeliere S, Riviere S, Scalliet G *et al.* (2002) Analysis of gene expression in rose petals using expressed sequenced tags. *FEBS Letters* 515: 35–38.

Chmelnitzky I, Zieslin N, Azizbekova N, Lifschitz E and Khayat E (1997) Abnormal organogenesis in roses (*Rosa hybrida* cv Montrea). *Plant Molecular Biology Reporter* Supplement 15: 413.

Evans RC, Alice LA, Campbell CS, Kellogg EA and Dickinson TA (2000) The granule-bound starch synthase (GBSSI) gene in the Rosaceae: multiple loci and phylogenetic utility. *Molecular and Phylogenetic Evolution* 17: 388–400.

Fukuchi-Mizutani M, Savin K, Cornish E *et al.* (1995) Senescence-induced expression of a homologue of Δ9 desaturase in rose petals. *Plant Molecular Biology* 29: 627–635.

Fukuchi-Mizutani M, Ishiguro K, Nakayama T *et al.* (2000) Molecular and functional characterization of a rose lipoxygenase cDNA related to flower senescence. *Plant Science* 160: 129–137.

Gutterman H, Shalit M, Menda N *et al.* (2002) Rose scent: genomics approach to discovering novel floral fragrance-related genes. *Plant Cell* 14: 2325–2338.

Iwata H, Kato T and Ohno S (2000) Triparental origin of Damask roses. *Gene* 259: 53–59.

Kitahara K and Matsumoto S (2000) Rose MADS-box genes 'MASAKO C1 and D1' homologous to class C floral identity genes. *Plant Science* 151: 121–134.

Lavid N, Wang J, Shalit M *et al.* (2002) O-Methyltransferase involved in the biosynthesis of volatile phenotic derivatives in rose petals. *Plant Physiology* 129: 1899–1907.

Matsumoto S, Kouchi M, Yabuki J *et al.* (1998) Phylogenetic analyses of the genus *Rosa* using the *mat*K sequence: molecular evidence for the narrow genetic background of modern roses. *Scientia Horticulturae* 77: 73–82.

Matsumoto S, Kouchi M, Ueda Y and Fukui H (2000) Phylogenetic analyses of the subgenus *Eurosa* using the ITS nrDNA sequence. *Acta Horticulturae* 521: 193–202.

Matsumoto S, Nishio H, Fukui H and Ueda Y (2001) Phylogenetic analyses of genus *Rosa*: polyphyly of section Pimpinellifoliae and origin of *Rosa* × *fortuniana* Lindl. *Acta Horticulturae* 547: 357–363.

Morgan DR, Soltis DE and Robertson KR (1994) Systemic and evolutionary implications of *rbc*L sequence variation in Rosaceae. *American Journal of Botany* 81: 890–903.

Savolainen V, Manen JF, Douzeri E and Spichiger R (1994) Molecular phylogeny of families related to celastrales based on rbcL 5′ flanking sequences. *Molecular and Phylogenetics Evolution* 3: 27–37.

Scalliet G, Journot N, Jullien F *et al.* (2002) Biosynthesis of the major scent components 3,5-dimethoxytoluene and 1,3,5-trimethoxybenzene by novel rose O-methyltransferases. *FEBS Letters* 523: 113–118.

Tanaka Y, Fukui Y, Fukuchi-Mizutani M *et al.* (1995) Molecular cloning and characterization of *Rosa hybrida* dihydroflavonols 4-reductase gene. *Plant Cell Physiology* 36: 1023–1031.

Wu S, Nishihara S and Ueda Y (2001) Phylogenetic analysis of section *Synstylae* in the genus *Rosa* based on rapid markers. *Acta Horticulturae* 547: 391–402.

Wu S, Ueda Y, Nishihara S and Matsumoto S (2001) Phylogenetic analysis of Japanese *Rosa* species using DNA sequence of nuclear ribosomal internal transcribed spacers (ITS). *Journal of Horticultural Science and Biotechnology* 76: 127–132.

Genetics: Research Trends

Daniell H, Muhammad D, Khan S and Allison L (2002) Milestones in chloroplast genetic engineering: an environmentally friendly era in biotechnology. *Trends in Plant Science* 7(2): 84–91.

Delseny M, Salses J, Cooke R *et al.* (2001) Rice genomics: present and future. *Plant Physiology and Biochemistry* 39: 323–334.

Gale MD and Devos KM (1998) Comparative genetics in the grasses. *Proceedings of the National Academy of Sciences, USA* 95: 1971–1974.

Lee PS and Lee KH (2000) Genomic analysis. *Current Opinion in Biotechnology* 11: 171–175.

Moore G, Devos KM, Wang Z and Gale MD (1995) Gramineouses, line up and form a circle. *Current Biology* 5: 737–739.

The Arabidopsis Genome Initiative (2000) Analysis of the genome sequence of the flowering plant *Arabidopsis thaliana*. *Nature* 408: 796–813.

Growth Regulation: Hormones in Growth and Development

Clark DG, Kelly JW and Pemberton HB (1991) Postharvest quality characteristics of cultivars of potted rose in response to holding conditions and cytokinins. *HortScience* 26: 1195–1197.

Cline MG (1994) The role of hormones in apical dominance: new approaches to an old problem in plant development. *Physiologia Plantarum* 90: 230–237.

Dieleman JA, Verstappen FWA and Kuiper D (1998) Bud break and cytokinin concentration in bleeding sap of *Rosa hybrida* as affected by the genotype of rootstock. *Journal of Plant Physiology* 152: 468–472.

Dieleman JA, Verstappen FWA, Nicander B *et al.* (1997) Cytokinins in *Rosa hybrida* in relation to bud break. *Physiologia Plantarum* 99: 456–464.

Dieleman JA, Verstappen FWA, Perik RRJ and Kuiper D (1997) Quantification of the export of cytokinins from roots to shoots of *Rosa hybrida* and their degradation rate in the shoot. *Physiologia Plantarum* 101: 347–352.

Ganevelin R and Zieslin N (2002) Contribution of sepals and gibberellin treatments to growth and development of rose (*Rosa hybrida*) flowers. *Plant Growth Regulation* 37: 255–261.

Le Bris M, Michaux-Ferrière N, Jacob Y *et al.* (1999) Regulation of bud dormancy by manipulation of ABA in isolated buds of *Rosa hybrida* cultured *in vitro*. *Australian Journal of Plant Physiology* 26: 273–281.

Lukaszewska AJ, Bianco J, Barthe P and Le Page-Degivry MT (1994) Endogenous cytokinins in rose petals and the effect of exogenously applied cytokinins on flower senescence. *Plant Growth Regulation* 14: 119–126.

Meyer S and Genty B (1998) Mapping intercellular CO_2 mole fraction (C_i) in *Rosa rubiginosa* leaves fed with abscisic acid by using chlorophyll fluorescence imaging. *Plant Physiology* 116: 947–957.

Meyer S and Genty B (1999) Heterogeneous inhibition of photosynthesis over the leaf surface of *Rosa rubiginosa* L. during water stress and abscisic acid treatment: induction of a metabolic component by limitation of CO_2 diffusion. *Planta* 210: 126–131.

Müller R, Lind-Iversen S, Stummann BM and Serek M (2000) Expression of genes for ethylene biosynthetic enzymes and an ethylene receptor in senescing flowers of miniature potted roses. *Journal of Horticultural Science and Biotechnology* 75: 12–18.

Müller R, Owen CA, Xue ZT, Welander M and Stummann BM (2002) Characterization of two CTR-like protein kinases in *Rosa hybrida* and their expression during flower senescence and in response to ethylene. *Journal of Experimental Botany* 53: 1223–1225.

Müller R, Sisler EC and Serek M (2000) Stress induced ethylene production, ethylene binding, and the response to the ethylene action inhibitor 1-MCP in miniature roses. *Scientia Horticulturae* 83: 51–59.

Müller R, Stummann BM and Serek M (2000) Characterization of an ethylene receptor family with differential expression in rose (*Rosa hybrida* L.) flowers. *Plant Cell Reports* 19: 1232–1239.

Müller R, Stummann BM, Andersen AS and Serek M (1999) Involvement of ABA in postharvest life of miniature potted roses. *Plant Growth Regulation* 29: 143–150.

Roberts AV, Blake PS, Lewis R, Taylor JM and Dunstan DI (1999) The effect of gibberellins on flowering in roses. *Journal of Plant Growth Regulation* 18: 113–119.

Sabehat A and Zieslin N (1994) GA_3 effects on postharvest alterations in cell membranes of rose (*Rosa* × *hybrida*) petals. *Journal of Plant Physiology* 144: 513–517.

Sabehat A and Zieslin N (1995) Promotion of postharvest increase in weight of rose (*Rosa* × *hybrida* cv. Mercedes) petals by gibberellin. *Journal of Plant Physiology* 145: 296–298.

Serek M, Prabucki A, Sisler EC and Andersen AS (1998) Inhibitors of ethylene action affect final quality and rooting of cuttings before and after storage. *HortScience* 33: 153–155.

Van der Salm TPM, Bouwer R, van Keizer LCP *et al.* (1998) Stimulation of scion bud release by *rol* gene transformed rootstocks of *Rosa hybrida* L. *Journal of Experimental Botany* 49: 847–852.

Van der Salm TPM, Van der Toorn CJG, Bouwer R *et al.* (1997) Production of *ROL* gene transformed plants of *Rosa hybrida* L. and characterization of their rooting ability. *Molecular Breeding* 3: 39–47.

Visser PB, Keizer PLCP, Van der Salm TPM and De Jong J (2000) Rootstocks transformed with *rol* A, B, C genes enhance rose flower production. *Biotechnology and Biotechnological Equipment* 14: 63–70.

Growth Regulation: Influences of Abiotic Factors in Growth and Development

Augé RM, Schekel KA and Wample RL (1986) Osmotic adjustment in leaves of VA mycorrhizal and nonmycorrhizal rose plants in response to drought stress. *Plant Physiology* 82: 765–770.

Augé RM, Schekel KA and Wample RL (1987) Rose leaf elasticity changes in response to mycorrhizal colonization and drought acclimation. *Physiologia Plantarum* 70: 175–182.

Augé RM, Stodola AJW and Pennell BD (1990) Osmotic and turgor adjustment in *Rosa* foliage drought-stressed under varying irradiance. *Journal of the American Society of Horticultural Science* 115: 661–667.

Beeson RC and Graham MED (1991) CO_2 enrichment of greenhouse roses affects neither Rubisco nor carbonic anhydrase activity. *Journal of the American Society of Horticultural Science* 116: 1040–1045.

Berninger A (1994) Development rate of young greenhouse rose plants (*Rosa hybrida*) rooted from cuttings in relation to temperature and irradiance. *Scientia Horticulturae* 58: 235–251.

Hopper DA, Hammer PA and Wilson JR (1994) A simulation model of *Rosa hybrida* growth response to constant irradiance and day and night temperatures. *Journal of the American Society of Horticultural Science* 119: 903–914.

Jiao J, Tsujita MJ and Grodzinski B (1991) Influence of radiation and CO_2 enrichment on whole plant net CO_2 exchange rate in roses. *Canadian Journal of Plant Science* 71: 245–252.

Maas FM and Bakx EJ (1995) Effects of light on growth and flowering of *Rosa hybrida* 'Mercedes'. *Journal of the American Society of Horticultural Science* 120: 571–576.

Marcelis-van Acker CAM (1995) Effect of temperature on development and growth potential of axillary buds in roses. *Scientia Horticulturae* 63: 241–250.

Menard C and Dansereau B (1995) Differential response of rose cultivars to light source and nitrogen fertilization. *Scientia Horticulturae* 64: 117–132.

Mor Y and Halevy AH (1984) Dual effect of light on flowering and sprouting of rose shoots. *Physiologia Plantarum* 61: 119–124.
Mortensen LM and Fjeld T (1998) Effects of air humidity, lighting period and lamp type on growth and vase life of roses. *Scientia Horticulturae* 73: 229–237.
Pasian CC and Lieth JH (1994) Prediction of flowering rose shoot development based on air temperature and thermal units. *Scientia Horticulturae* 59: 131–145.
Pasian CC and Lieth JH (1996) Prediction of rose shoot development: model validation for the cultivar 'Cara Mia' and extension to cultivars 'Royalty' and 'Sonia'. *Scientia Horticulturae* 66: 117–124.
Urban L and Langelez I (1992) Effect of high pressure mist on leaf water potential, leaf diffusive conductance, CO_2 fixation and production of 'Sonia' rose plants grown in rockwool. *Scientia Horticulturae* 50: 229–244.
Urban L, Jaffrin A and Chraibi A (1993) Analysis of pressure volume curves of leaves of *Rosa hybrida* cv. Sonia. *Journal of Experimental Botany* 260: 605–613.
Urban L, Fabret C and Barthelemy L (1994) Interpreting changes in stem diameter in rose plants. *Physiologia Plantarum* 92: 668–674.
Urban L, Barthélémy L, Bearez P and Pyrrha P (2001) Effect of elevated CO_2 on photosynthesis and chlorophyll fluorescence of rose plants grown at high temperature and high photosynthetic photon flux density. *Photosynthetica* 39: 275–281.
Urban L, Six S, Barthélémy L and Bearez P (2002) Effect of elevated CO_2 on water relations, water balance and senescence of cut roses. *Journal of Plant Physiology* 159: 717–723.
Zieslin N and Tsujita MJ (1990) Response of miniature rose to supplementary illumination. I. Effect of light intensity. *Scientia Horticulturae* 42: 113–121.

Growth Regulation: Axillary Bud Growth

Al-Saqri F and Alderson PG (1996) Effects of IBA, cutting type and rooting media on rooting of *Rosa centifolia*. *Journal of Horticultural Science* 71: 729–737.
Bredmose N (1995) High utilization ratio of supplementary light by cut roses at all times of the year. *Gartenbauwissenschaft* 60: 125–131.
Bredmose N (1997) Chronology of three physiological development phases of single-stemmed rose (*Rosa hybrida* L.) plants in response to increment in light quantum integral. *Scientia Horticulturae* 69: 107–115.
Bredmose NB (1998) Growth, flowering, and postharvest performance of single-stemmed rose (*Rosa hybrida* L.) plants in response to light quantum integral and plant population density. *Journal of the American Society of Horticultural Science* 123: 569–576.
Bredmose N and Hansen J (1995) Regeneration, growth and flowering of cut rose cultivars as affected by propagation material and method. *Scientia Horticulturae* 64: 103–111.
Bredmose N and Hansen J (1996) Topophysis affects the potential of axillary bud growth, fresh biomass accumulation and specific fresh weight in single-stem roses (*Rosa hybrida* L.). *Annals of Botany* 78: 215–222.
Bredmose N, Hansen J and Nielsen J (1999) Factors intrinsic to the axillary bud determine topophysic effects on bud and shoot growth and flower development in *Rosa hybrida*. *International Journal of Plant Sciences* 160(5): 819–825.
Cockshull KE and Horridge JS (1977) Apical dominance and flower initiation in the rose. *Journal of Horticultural Science* 52: 421–427.
Dieleman JA, Verstappen FWA, Perik RRJ and Kuiper D (1997) Quantification of the export of cytokinins from roots to shoots of *Rosa hybrida* and their degradation rate in the shoot. *Physiologia Plantarum* 101: 347–352.
Dieleman JA, Verstappen FWA, Nicander B *et al.* (1997) Cytokinins in *Rosa hybrida* in relation to bud break. *Physiologia Plantarum* 99: 456–464.
Dieleman JA, Verstappen FWA and Kuiper D (1998) Bud break and cytokinin concentration in bleeding sap of *Rosa hybrida* as affected by the genotype of the rootstock. *Journal of Plant Physiology* 152: 468–472.
Dieleman JA, Verstappen FWA and Kuiper D (1998) Root temperature effects on growth and bud break of *Rosa hybrida* in relation to cytokinin concentration in xylem sap. *Scientia Horticulturae* 76: 183–192.
Kapchina-Toteva V, van Telgen H-J and Yakimova E (2000) Role of phenylurea cytokinin CPPU in apical dominance release in *in vitro* cultured *Rosa hybrida* L. *Journal of Plant Growth Regulation* 19: 232–237.
Le Bris M, Champeroux A, Bearez P and Le Page-Degivry M-T (1998) Basipetal gradient of axillary bud inhibition along a rose (*Rosa hybrida* L.) stem: growth potential of primary buds and their two most basal secondary buds as affected by position and age. *Annals of Botany* 81: 301–309.
Le Bris M, Michaux-Ferriere N, Jacob Y *et al.* (1999) Regulation of bud dormancy by manipulation of ABA in isolated buds of *Rosa hybrida* cultured *in vitro*. *Australian Journal of Plant Physiology* 26: 273–281.
Marcelis-van Acker CAM (1994) Ontogeny of axillary buds and shoots in roses: leaf initiation and pith development. *Scientia Horticulturae* 57: 111–122.
Marcelis-van Acker CAM (1994) Effect of assimilate supply on development and growth potential of axillary buds in roses. *Annals of Botany* 73: 415–420.
Marcelis-van Acker CAM (1994) Development and growth potential of axillary buds in roses as affected by bud age. *Annals of Botany* 74: 437–443.
Marcelis-van Acker CAM (1995) Effects of temperature on development and growth potential of axillary buds in roses. *Scientia Horticulturae* 63: 241–250.
Marcelis-van Acker CAM and Scholten HJ (1995) Development of axillary buds of rose *in vitro*. *Scientia Horticulturae* 63: 47–55.
Moe R (1972) Effect of daylength, light intensity and temperature on growth and flowering in roses. *Journal of the American Society of Horticultural Science* 97: 796–800.
Mor Y and Halevy AH (1984) Dual effect of light on flowering and sprouting of rose shoots. *Physiologia Plantarum* 61: 92–97.

Mor Y, Halevy AH and Porath D (1980) Characterization of the light reaction in promoting the mobilizing ability of rose shoot tips. *Plant Physiology* 66: 996–1000.

Ohkawa K (1984) Effects of benzyladenine on bud break of roses. *Scientia Horticulturae* 24: 379–383.

Särkkä LE and Rita HJ (1999) Yield and quality of cut roses produced by pruning or by bending down shoots. *Gartenbauwissenschaft* 64: 173–176.

Sun W-Q and Bassuk NL (1993) Auxin-induced ethylene synthesis during rooting and inhibition of budbreak of 'Royalty' rose cuttings. *Journal of the American Society of Horticultural Science* 118: 638–643.

Van der Salm TPM, Bouwer R, van Dijk AJ *et al.* (1998) Stimulation of scion bud release by *rol* gene transformed rootstocks of *Rosa hybrida* L. *Journal of Experimental Botany* 49: 847–852.

Van Staden J (1982) Transport of (8 ^{14}C)-zeatin from mature rose leaves after shoot decapitation. *Botanical Gazette* 143: 201–205.

Zamski E, Oshri S and Zieslin N (1985) Comparative morphology and anatomy of axillary buds along a rose shoot. *Botanical Gazette* 146: 208–212.

Zieslin N and Halevy AH (1976) Components of axillary bud inhibition in rose plants. I. Effects of different plant parts (correllative inhibition). *Botanical Gazette* 137: 291–296.

Zieslin N and Khayat E (1983) Involvement of cytokinin, ABA and endogenous inhibitors in sprouting of basal buds in rose plants. *Plant Growth Regulation* 1: 279–288.

Zieslin N, Haase N and Halevy AH (1976) Components of axillary bud inhibition in rose plants. II. The effect of bud position on degree of inhibition. *Botanical Gazette* 137: 297–300.

Growth Regulation: Floral Induction

Abad Farooqi AH, Shukla YN, Sharma S and Bansal RP (1994) Relationship between gibberellin and cytokinin activity and flowering in *Rosa damascena* Mill. *Plant Growth Regulation* 14: 109–113.

Araki T (2001) Transition from vegetative to reproductive phase. *Current Opinions in Plant Biology* 4: 63–68.

Battey NH and Tooke F (2002) Molecular control and variation in the floral transition. *Current Opinions in Plant Biology* 5: 62–68.

Cockshull KE and Horridge JS (1977) Apical dominance and flower initiation in the rose. *Journal of Horticultural Science* 42: 421–427.

de Vries DP (1976) Juvenility in hybrid tea roses. *Euphytica* 25: 321–328.

Horridge JS and Cockshull KE (1974) Flower initiation and development in the glasshouse rose. *Scientia Horticulturae* 2: 273–284.

Maas FM, Hofman-Eijer LB and Hulstein K (1995) Flower morphogenesis in *Rosa hybrida* 'Mercedes' as studied by electron and light microscopy. Effects of light and shoot position on a branch. *Annals of Botany* 75: 199–205.

Marcelis van Acker CAM (1994) Development and growth-potential of axillary buds in roses as affected by bud age. *Annal of Botany* 74: 437–443.

Marcelis van Acker CAM (1994) Effect of assimilate supply on development and growth-potential of axillary buds in roses. *Annals of Botany* 73: 415–420.

Roberts AV, Blake PS, Lewis R, Taylor JM and Dunstan DI (1999) The effect of gibberellins on flowering in roses. *Journal of Plant Growth Regulation* 18: 113–120.

Zieslin N and Halevy AH (1976) Flower bud atrophy in Baccara roses. (IV) The effect activity of various growth substances in leaves of flowering and non-flowering shoots. *Physiologia Plantarum* 37: 317–325.

Zieslin N and Halevy AH (1976) Flower bud atrophy in Baccara roses. (V) The effects of different growth substances on flowering. *Physiologia Plantarum* 37: 326–330.

History of Roses in Cultivation: Ancient Chinese Roses

Guoliang W (1998) Principle and technology of micropropagation in modern roses. *Journal of Jiangsu Forestry Science and Technology* 24: 52–56.

Guoliang W and Youliang W (1997) Quantitative study on rose grafting in pots year round. *Journal of Jiangsu Forestry Science and Technology* 24: 26–30.

Junyu C (1986) Some thoughts on roses. *World Agriculture* 8: 51–53.

Ogisu M (1996) Some thoughts on the history of China roses. *The New Plantsman* 3: 152–157.

Yanshi Z (1994) Introduction of old roses existing in Huiying. *Chinese Gardening* 2: 51–52.

Ziming L (19xx) Roses of China and roses of Europe. *Acta Horticulturae Sinica* 3: 387–394.

Insects and other Animals: Aphids

Chakrabarti S and Ghosh AK (1971) On the rose infesting aphids (Insecta: Homoptera) in India. *Indian Journal of Horticulture* 27: 226–232.

David SK (1976) A taxonomic review of *Macrosiphum* (Homoptera: Aphididae) in India. *Oriental Insects* 9: 461–493.

David SK and Narayanan K (1968) Three new species of aphids from south western Himalayas in India. *Bulletin of Entomology* 9: 99–103.

David SK, Rajasingh SG and Narayanan K (1970) The Myzaphidines (Homoptera) of India with descriptions of three new species. *Oriental Insects* 4: 395–406.

Hand SC and Williams CT (1981) The overwintering of the rose-grain aphid *Metopolophium dirhodum* on wild roses. In: Thresh JM (ed.) *Pests, Pathogens and Vegation*, pp. 285–298. London: Pitman.

Heie OE (1992) The Aphidoidea (Hemiptera) of Fennoscandia and Denmark. IV. Family Aphididae: Part 1 of tribe Macrosiphini of subfamily Aphidinae. *Fauna Entomologica Scandinavica* 25: 1–188.

Heie OE (1994) The Aphidoidea (Hemiptera) of Fennoscandia and Denmark. V. Family Aphididae: Part 2 of tribe Macrosiphini of subfamily Aphidinae. *Fauna Entomologica Scandinavica* 28: 1–239.

Hille Ris Lambers D (1939) Contributions to a monograph of the Aphididae of Europe. II. The genera *Dactynotus* Rafinesque, 1818; *Staticobium* Mordvilko, 1914; *Macrosiphum* Passerini, 1860; *Masonaphis* nov. gen.; *Phalaris* Leach, 1826. *Temminckia* 4: 1–134, plates i–vi. Leiden, Netherlands: E. J. Brill.

Hille Ris Lambers D (1947) Contributions to a monograph of the Aphididae of Europe. III. The genera *Phalaris* Leach, 1826; *Microsiphum* Chol., 1902; *Anthracosiphon* nov. gen.; *Delphiniobium* Mordv., 1914, *Subacyrthosiphon* nov. gen.; *Silenobium* Börner, 1939; *Titanosphon* Nevsky, 1928; *Metopolophium* Mordv., 1914; *Cryptaphis* nov. gen.; *Rhodobium* nov. gen.; *Impatientinum* Mordv., 1914; *Aulacorthum* Mordv., 1914. *Temminckia* 7: 179–319, plates xii–xviii. Leiden, Netherlands: E. J. Brill.

Hille Ris Lambers D (1953) Contributions to a monograph of the Aphididae of Europe. V. The genera *Rhopalosiphoninus* Baker, 1920; *Eucarazzia* Del Guercio, 1921; *Rhopalomyzus* Mordv., 1921; *Chaetosiphon* Mordv., 1914; *Cryptomyzus* Oestl., 1922; *Pleotrichophorus* Börner, 1930; *Capitophorus* v.d. Goot, 1913. *Temminckia* 9: 1–176, plates i–vi. Leiden, Netherlands: E. J. Brill.

Jensen AS (1998) Redescription of *Macrosiphum impatientis* (Williams), another rose aphid from eastern North America, with a key to related species (Hemiptera: Aphididae). *Proceedings of the Entomological Society of Washington* 100: 32–42.

MacGillivray ME (1963) The yellow rose aphid, *Rhodobium porosum* (Sanderson) (Homoptera: Aphididae), on strawberry. *Canadian Entomologist* 95: 892–896.

Miyazaki M (1968) On three aphids of *Rosa rugosa* occurring in Japan with description of a new species (Homoptera: Aphididae). *Kontyû* 36: 274–284.

Miyazaki M (1971) A revision of the tribe Macrosiphini of Japan (Homoptera: Aphididae, Aphidinae). *Insecta Matsumurana* 34: 1–247.

Nevsky VP (1928) The plant-lice of Middle-Asia. *Entomologische Mitteilungen* 17: 182–199.

Paik WH (1965) *Aphids of Korea*. Seoul, South Korea: Seoul National University.

Paik WH (1972) *Illustrated Encyclopedia of Fauna and Flora of Korea*, vol. 13, *Insecta (V)*. Seoul, South Korea: Samwha Publishing Co. (in Korean).

Richards WR (1961) New genera and species of rose-infesting aphids. *Canadian Entomologist* 93: 622–625.

Richards WR (1963) The Myzaphidines of Canada (Homoptera: Aphididae). *Canadian Entomologist* 95: 680–704.

Shands WA, Simpson GW and Wave HE (1972) Seasonal population trends and productiveness of the potato aphid on swamp rose in north-eastern Maine. *Technical Bulletin, University of Maine* 52: 1–35.

Tomiuk J and Wöhrmann K (1980) Population growth and population structure of natural populations of Macrosiphum rosae L. *Zeitschrift für angewandte Entomologie* 90: 464–473.

Zhang G and Zhong T (1980) New species of Chinese Macrosiphinae (II) (Homoptera, Aphididae). *Entomotaxonomia* 2: 215–225 (in Chinese, with English summary).

Insects and other Animals: Mites

Karlik J and Harwood CA (1994) Spider mites in rose plant fields, Wasco, CA, 1992. *Arthropod Management Tests* 19: 336.

Karlik J and Wollesen P (1994) Spider mites in rose plant fields, Wasco, CA, 1993. *Arthropod Management Tests* 19: 334–335.

Karlik J, Ali AD and Harwood CA (1994) Spider mites in rose plant fields, Wasco, CA, 1988. *Arthropod Management Tests* 19: 335.

Karlik J, Goodell PB and Osteen GW (1995) Improved mite sampling may reduce acaricide use in roses. *California Agriculture* 49: 38–40.

Karlik J, Goodell PB and Osteen GW (1995) Sampling and treatment thresholds for spider mite management in field-grown rose plants. *HortScience* 30: 1268–1270.

Tjosvold SA, Greene ID and Colbert D (1996a) Effect of spray methods and miticides on control of twospotted spider mites infesting commercial greenhouse roses, 1994. In: *Arthropod Management Tests*. Lanham, MD: Entomological Society of America, 21: 361.

Tjosvold SA, Greene ID and Colbert D (1996b) Effect of spray methods and miticides on control of twospotted spider mites infesting commercial greenhouse roses, 1995. In: *Arthropod Management Tests* 21: 362.

Tjosvold SA and Chaney WE (2001) Evaluation of reduced risk and other biorational miticides on the control of spider mites (*Tetranychus urticae*). *Acta Horticulturae* 547: 93–96.

Insects and other Animals: Thrips

Bryan DE and Smith RF (1956) The *Frankliniella occidentalis* (Pergande) complex in California (Thysanoptera: Thripidae). *University of California Publications in Entomology* 10: 359–410.

Compton CC (1930) *Greenhouse Pests*. Entomological Series Circular no. 12. Urbana, IL: State of Illinois, Department of Registration and Education, State Natural History Survey Division.

Davidson J and Andrewartha HG (1948) Annual trends in a natural population of *Thrips imaginis* (Thysanoptera). *Journal of Animal Ecology* 17: 193–199.

Dodge BO and Rickett HW (1943) *Diseases and Pests of Ornamental Plants*. Lancaster, PA: Jacques Cattell Press.

Guam WG, Giliomee JH and Pringle KL (1994) Resistance of some rose cultivars to the western flower thrips, *Frankliniella occidentalis* (Thysanoptera: Thripidae). *Bulletin of Entomological Research* 84: 487–492.

Henneberry TJ, Taylor EA and Smith FF (1961) Foliage and soil treatments for the control of thrips on outdoor roses. *Journal of Economic Entomology* 54: 233–235.

Immaraju J, Paine TD, Bethke JA *et al.* (1992) Western flower thrips (Thysanoptera: Thripidae) resistance to insecticides in coastal California Greenhouses. *Journal of Economic Entomology* 85: 9–14.

Murphy BC, Morisawa TA, Newman JP *et al.* (1998) Fungal pathogen controls thrips in greenhouse flowers. *California Agriculture* 52(3): 32–36.

Parrella MP (1995) IPM: approaches and prospects. In: Parker BL, Skinner M and Lewis T (eds.) *Thrips Biology and Management*, pp. 357–363. New York: Plenum Press.

Parrella MP (2000) Glasshouses: Integrating natural enemy release and biological pesticides for insect management in roses. In: Hoddle M (ed.) *2nd California Conference on Biological Control 2000, July 11–12*, pp. 57–58.

Prithchard AE (1949) *California Greenhouse Pests and Their Control*. Berkeley: California Agricultural Experimental Station.

Southwood TRE (1978) *Ecological Methods*, 2nd edn. New York: Chapman & Hall.

Tenenbaum D (2002) Would a rose not smell as sweet? *Environmental Health Perspectives* 110(5): A240–A247.

Taylor LR (1961) Aggregation, variance and the mean. *Nature* 189: 732–735.

Warrick J (2000) Fresh cut flowers: fragrant, beautiful – and often dosed with pesticides. *National Wildlife Magazine* June/July: 12–13.

Weiss HB (1916) *The More Important Greenhouse Insects*. New Brunswick: New Jersey Agricultural Experiment Stations.

Insects and other Animals: Nematodes

Coolen WA and Hendricks GJ (1972) Investigation on the resistance of rose rootstocks to *M. hapla* and *P. penetrans*. *Nematologica* 18: 155–158.

Daiber KC (1985) Resistance of the rose rootstock selection 'Floradale' to *M. hapla* Chitwood (Tylenchida: *Meloidogynidae*). *Phytophylactica* 17: 167.

Dalmasso A and Bergé JB (1975) Genetic variability in *Meloidogyne* specially in *M. hapla*. *Cahiers ORSTOM, Série Biologie, Nématologie* 10: 233–238 (in French).

Ohkawa K and Saigusa T (1981) Resistance of rose rootstocks to *Meloidogyne hapla, Pratylenchus penetrans* and *Pratylenchus vulnus*. *HortScience* 16: 559–560.

Santo GS and Lear B (1976) Influence of *Pratylenchus vulnus* and *Meloidogyne hapla* on the growth of rootstocks of rose. *Journal of Nematology* 8: 18–23.

Schneider JHM, Jacob JJs' and Pol PA Van de (1995) *Rosa multiflora*, 'Ludiek', a rootstock with resistant features to the root-lesion nematode *Pratylenchus vulnus*. *Scientia Horticulturae* 63: 37–45.

Winfield AL (1974) Observations on the occurrence, pathogenicity and control of *Pratylenchus vulnus*, *P. thornei* and *Xiphinema diversicaudatum* associated with glasshouse roses. *Annals of Applied Biology* 77: 297–307.

Winfield AL (1974) Damage to roses by *Longidorus macrosoma*. *Plant Disease Reporter* 58: 913–914.

Insects and other Animals: Pollination

Dobson HEM, Bergström G and Groth I (1990) Differences in fragrance chemistry between flower parts of *Rosa rugosa*. *Israel Journal of Botany* 39: 143–156.

Dobson HEM, Danielson EM and van Wesep I (1999) Pollen odor chemicals as modulators of bumble bee foraging on *Rosa rugosa* Thunb. (Rosaceae). *Plant Species Biology* 14: 153–166.

Kevan PG, Eisikowitch D, Ambrose JD and Kemp JR (1990) Cryptic dioecy and insect pollination in *Rosa setigera* Michx. (Rosaceae), a rare plant of Carolinian Canada. *Biological Journal of the Linnean Society* 40: 229–243.

Ueda Y and Akimoto S (2001) Cross- and self-compatibility in various species of the genus *Rosa*. *Journal of Horticultural Science and Biotechnology* 76: 392–395.

Insects and other Animals: Insecticides

Karlik J and Harwood CA (1994) Spider mites in rose plant fields, Wasco, CA, 1992. *Arthropod Management Tests* 19: 336.

Karlik J and Wollesen P (1994) Spider mites in rose plant fields, Wasco, CA, 1993. *Arthropod Management Tests* 19: 334–335.

Karlik J, Ali AD and Harwood CA (1994) Spider mites in rose plant fields, Wasco, CA, 1988. *Arthropod Management Tests* 19: 335.

Karlik J, Goodell PB and Osteen GW (1995) Improved mite sampling may reduce acaricide use in roses. *California Agriculture* 49: 38–40.

Karlik J, Goodell PB and Osteen GW (1995) Sampling and treatment thresholds for spider mite management in field-grown rose plants. *HortScience* 30: 1268–1270.

Integrated Pest Management

Karlik J, Goodell PB and Osteen GW (1995) Improved mite sampling may reduce acaricide use in roses. *California Agriculture* 49: 38–40.

Karlik J, Goodell PB and Osteen GW (1995) Sampling and treatment thresholds for spider mite management in field-grown rose plants. *HortScience* 30: 1268–1270.

Van Lenteren C and Woets J (1988) Biological and integrated control in greenhouses. *Entomology* 33: 239–269.

Morphology and Anatomy: Leaves

Brown VK and Lawton JH (1991) Herbivory and the evolution of leaf size and shape. *Philosophical*

Transactions of the Royal Society of London B 333: 265–272.

Gurevitch J and Schuepp PH (1990) Boundary layer properties of highly dissected leaves: an investigation using electochemical fluid tunnel. *Plant Cell Environment* 13: 783–792.

Sallanon H, Laffray D and Coudret A (1993) Ultrastructure and functioning of guard cells of *in vitro* cultured rose plants. *Plant Physiology and Biochemistry* 29: 333–339.

Sinha N (1997) Simple and compound leaves: reduction or multiplication? *Trends in Plant Science* 2: 365–403.

Ueda Y, Nishihara S, Tomita H and Oda Y (2000) Photosynthetic response of Japanese rose species *Rosa bracteata* and *Rosa rugosa* to temperature and light. *Scientia Horticulturae* 84: 365–371.

Williams M, Rosenqvist E and Buchhave M (1999) Response of potted miniature roses (*Rosa* × *hybrida*) to reduced water availability during production. *Journal of Horticultural Science and Biotechnology* 74: 301–308.

Wissemann V (2000) Epicuticular wax morphology and the taxonomy of *Rosa* (section *Caninae*, subsection Rubiginosae). *Plant Systematics and Evolution* 221: 107–112.

Yaron B, Zieslin N and Halevy AH (1979) Response of Baccara roses to saline irrigation. *Journal of the American Society for Horticultural Science* 94: 481–484.

Yasuyuki H, Endoh K, Kudo T and Tahara S (2001) Capability of wild *Rosa rugosa* and its varieties and hybrids to produce sesquiterpene components in leaf glandular trichomes. *Bioscience, Biotechnology, Biochemistry* 65(9): 2037–2043.

Zeroni M and Gale J (1988) Responses of Sonia roses to continuous daytime CO_2 supplementation under controlled environment conditions. *Australian Journal of Agricultural Research* 39: 863–870.

Zieslin N, Kohl HC, Kofranek AM and Halevy AH (1978) Changes in the water status of cut roses and its relationship to bent-neck phenomenon. *Journal of the American Society for Horticultural Science* 103: 176–179.

Zieslin N, Plaut Z, Marek J and Halevy AH (1979) The responses of rose plants to evaporative cooling: plant water status and CO_2 fixation. *Scientia Horticulturae* 11: 191–196.

Morphology and Anatomy: Flower Development and the Abscission Zone

Chimonidou-Pavlidou D (1996) Effect of water stress at different stages of rose development. *Acta Horticulturae* 424: 45–51.

Chimonidou-Pavlidou D (1999) Irrigation and sensitive stages of rose development. *Acta Horticulturae* 481: 393–401.

Chimonidou-Pavlidou D (2000) Correlation between internal and external stages of rose development. *Acta Horticulturae* 515: 233–244.

Chimonidou-Pavlidou D (2001) Effect of irrigation and shading at the stage of flower bud appearance. *Acta Horticulturae* 547: 245–251.

Cockshull KE and Horridge JS (1977) Apical dominance and flower initiation in the rose. *Journal of Horticultural Science* 52: 421–427.

Darlington AB and Dixon MA (1991) The hydraulic architecture of roses (*Rosa hybrida*). *Canadian Journal of Botany* 69: 702–710.

Horridge JS and Cockshull KE (1974) Flower initiation and development in the glasshouse rose. *Scientia Horticulturae* 2: 273–284.

Khayat E and Zieslin N (1982) Environmental factors involved in the regulation of sprouting of basal buds in rose plants. *Journal of Experimental Botany* 33: 1286–1292.

Kool MTN and Van de Pol PA (1993) Controlling the plant development of *R. hybrida* 'Motrea'. *Scientia Horticulturae* 53: 239–248.

Lindenbaum S, Ginzburg C and Halevy AH (1975) A morphological study of the 'Bulhead' malformation in 'Baccara' rose. *Annals of Botany* 39: 219–223.

Lyndon RF (1979) Aberrations in flower development in Silene. *Canadian Journal of Botany* 57: 233–235.

Marcelis-van Acker CAM (1994) Ontogeny of axillary buds and shoots in roses: leaf initiation and pith development. *Scientia Horticulturae* 57: 111–122.

Mayak S, Halevy AH, Sagie S, Bar-Yoseph A and Bravdo B (1974) The water balance of cut rose flowers. *Physiologia Plantarum* 31: 15–22.

Moe R (1971) Factors affecting flower abortion and malformation in roses. *Physiologia Plantarum* 24: 291–300.

Moe R (1972) Effect of daylength, light intensity, and temperature on growth and flowering in roses. *Journal of the American Society of Horticultural Science* 97: 796–800.

Moe R and Kristoffersen T (1969) The effect of temperature and light on growth and flowering of *Rosa* 'Baccara' in greenhouses. *Acta Horticulturae* 14: 157–163.

Roberts GL, Tsujita MJ and Dansereau B (1993) Supplemental light quality affects budbreak, yield and vase life of cut roses. *HortScience* 28: 621–622.

Wilkings HF (1988) Techniques to maximize cutting production. *Acta Horticulturae* 226: 137–143.

Zamski E, Oshri S and Zieslin N (1985) Comparative morphology and anatomy of axillary buds along a rose shoot. *Botanical Gazette* 146: 208–212.

Zieslin N (1981) Plant management of greenhouse roses. Flower cutting procedure. *Scientia Horticulturae* 15: 179–186.

Zieslin N (1992) Regulation of flower formation in rose plants: a reappraisal. *Scientia Horticulturae* 49: 305–310.

Zieslin N and Halevy AH (1976) Components of axillary bud inhibition in rose plants. I. Effect of different plant parts (correlative inhibition). *Botanical Gazette* 137: 291–296.

Zieslin N and Halevy AH (1978) Components of axillary bud inhibition in rose plants. III. Effect of stem orientation and changes of bud position on the stem by budding. *Botanical Gazette* 139: 60–63.

Zimmermann MH (1978) Hydraulic architecture of some diffuse-porous trees. *Canadian Journal of Botany* 56: 2286–2295.

Morphology and Anatomy: Roots

Auge RM and Stodola AJW (1990) Apparent increase in symplastic water contributes to greater turgor in mycorrhizal roots of droughted *Rosa* plants. *New Phytologist* 115: 285–295.

Hambrick III CE, Davies Jr FT and Pemberton HB (1984) Correlation between carbohydrate/nitrogen ratio and shooting ability in *Rosa multiflora* stem cuttings. *HortScience* 19: 566.

Pirch H (2000) Wie gut sind wurzelechte Rosen? *Gartenpraxis* 8: 23–29.

Paterson DR, Taber RA, Earhart DR *et al.* (1983) Influence of methyl bromide soil fumigation on VA mycorrhizal infection and growth of *Rosa multiflora* understock. *HortScience* 18: 607.

Taber RA, Paterson DR, Pemberton HB *et al.* (1983) Influence of indigenous rose root mycorrhiza and nutrient solution concentration on growth of *Rosa multiflora*. *HortScience* 18: 571.

Morphology and Anatomy: Pollen Grains and Tubes

El Mokadem H, Crespel L, Meynet J and Gudin S (2002) The occurrence of 2*n* pollen and the origin of sexual polyploids in dihaploid roses (*Rosa hybrida* L.). *Euphytica* 125: 169–177.

Gudin S (1992) Influence of bud chilling on subsequent reproductive fertility in roses. *Scientia Horticulturae* 51: 139–144.

Gudin S and Arene L (1991) Influence of the pH of the stigmatic exudate on male–female interaction in *Rosa hybrida* L. *Sexual Plant Reproduction* 4: 110–112.

Gudin S, Arene L and Bulard C (1991) Influence of season on rose pollen quality. *Sexual Plant Reproduction* 4: 113–117.

Luzny J and Burivalova J (1989) A contribution to the determination of pollen viability in selected garden rose and rootstock collections. *Acta Universitatis Agriculturae, Facultas Horticulturae* 4: 39–48.

Marchant R, Power JB, Davey MR, Chartier HJM and Lynch PT (1993) Cryopreservation of pollen from two rose cultivars. *Euphytica* 66: 235–241.

Meynet J, Barrade R, Duclos A and Siadous R (1994) Dihaploid plants of roses (*Rosa* × *hybrida*, cv. Sonia) obtained by parthenogenesis induced using irradiated pollen and *in vitro* culture of immature seeds. *Agronomie Paris* 14: 169–175.

Rajasekharan PE and Ganeshan S (1994) Freeze preservation of rose pollen in liquid nitrogen: feasibility, viability and fertility status after long-term storage. *Journal of Horticultural Science* 69: 565–569.

Ueda Y (1994) Systematic studies in the genus *Rosa*. *Technical Bulletin of Faculty of Horticulture Chiba University* 70: 241–328.

Ueda Y and Okada Y (1994) Discrimination of rose cultivar groups by pollen surface structure. *Journal of Horticultural Science* 69: 601–607.

Visser T, Vries DPd, Welles GWH and Scheurink JAM (1977) Hybrid tea-rose pollen. I. Germination and storage. *Euphytica* 26: 721–728.

Visser T, Vries DPd, Scheurink JAM and Welles GWH (1977) Hybrid tea-rose pollen. II. Inheritance of pollen viability. *Euphytica* 26: 729–732.

Voyiatzi CI (1995) An assessment of the *in vitro* germination capacity of pollen grains of five tea hybrid rose cultivars. *Euphytica* 83: 199–204.

Postharvest Physiology: Xylem Structure and Function in Cut Roses

Camper AA, Burr M, Ellis B *et al.* (1999) Development and structure of drinking water biofilms and techniques for their study. *Journal of Applied Microbiology Symposium Supplement* 85: 1S–12S.

Canny JC (1997) Vessel contents of leaves after excision. A test of Scholander's assumption. *American Journal of Botany* 84: 1217–1222.

Canny MJ (1995) A new theory for the ascent of sap – cohesion supported by tissue pressure. *Annals of Botany* 75: 343–357.

Darlington AB and Dixon MA (1991) The hydraulic architecture of roses (*Rosa hybrida*). *Canadian Journal of Botany* 69: 702–710.

Darvill AG and Albersheim P (1984) Phytoalexins and their elicitors – a defence against microbial infection in plants. *Annual Reviews of Plant Physiology* 35: 243–275.

Dixon MA and Peterson CA (1989) A re-examination of stem blockage in cut roses. *Scientia Horticulturae* 38: 277–285.

Dixon MA, Butt JA, Murr DP and Tsujita MJ (1988) Water relations of cut greenhouse roses: the relationships between stem water potential, hydraulic conductance and cavitation. *Scientia Horticulturae* 36: 109–118.

Durkin DJ, Put HMC and Clerkx ACM (1998) SEM and quality control of cut roses: the latex red test. *Microscopy and Analysis* July: 9–12.

Durkin DJ, Put HMC and Clerkx ACM (2001) The influence of vase water pretreatment on the accumulation of microparticles, microcompounds and bacterial cells on the cut surface xylem of *Rosa cv.* '*Kardinal*' obseved by SEM. *Gartenbauwissenschaft* 66: 93–101.

Nijsse J, Van der Heijden GWAM, Van Ieperen W *et al.* (2001) Xylem hydrolic conductivity related to conduit dimensions along chrysanthemum stems. *Journal of Experimental Botany* 52: 319–327.

Put HMC (1990) Microorganisms from freshly harvested cut flower stems and developing during the vase life of chrysanthemum, gerbera and rose cultivars. *Scientia Horticulturae* 43: 129–144.

Put HMC and Clerkx ACM (1988) The infiltration ability of microorganisms: *Bacillus, Fusarium, Kluyveromyces* and *Pseudomonas* spp. into xylem vessels of *Gerbera* cv. 'Fleur' and *Rosa* cv. 'Sonia' cut flowers: a scanning electron microscope study (SEM). *Journal of Applied Bacteriology* 65: 515–530.

Put HMC and Jansen L (1989) Effects on the vase life of cut *Rosa* cv. 'Sonia' of bacteria added to the vase water. *Scientia Horticulturae* 39: 167–179.

Rijnaarts HHM, Norde W, Lykema J and Zehnder JH (1999) DLVO and steric contributions to bacterial deposition in media of different ionic strengths. *Colloids and Surfaces B: Biointerfaces* 14: 179–195.

Sperry JS and Tyree MT (1988) Mechanism of water stress-induced xylem embolism. *Plant Physiology* 88: 581–587.

Tyree MV, Salleo S, Nardini A, Lo Gullo MA and Mosca R (1999) Refilling of embolized vessels in young stems of Laurel. Do we need a new paradigm? *Plant Physiology* 120: 11–21.

Van Ieperen W, Van Meeteren U and Gelder H (2000) Fluid ionic composition influences hydraulic conductance of xylem conduits. *Journal of Experimental Botany* 51: 769–776.

Van Meeteren H, Van Gelder H and Van Ieperen W (2000) Reconsideration of the use of deionized water as vase water in postharvest experiments on cut flowers. *Postharvest Biology and Technology* 18: 169–181.

Zimmermann MH and Jeje AA (1981) Vessel length distribution in stems of some American woody plants. *Canadian Journal of Botany* 59: 1828–1829.

Zwieniecki MA and Holbrook NM (2000) Bordered pit stucture and vessel wall surface properties. Implications for embolism repair. *Plant Physiology* 123: 1015–1020.

Postharvest Physiology: Effects of Supplemental Carbohydrates

Dai N, Schaffer A, Petreikov M *et al.* (1999) Overexpression of arabidopsis hexokinase in tomato plants inhibits growth, reduces photosynthesis, and induces rapid senescence. *Plant Cell* 11: 1253–1266.

Eason JR, De Vré LA, Somerfield SD and Heyes JA (1997) Physiological changes associated with *Sandersonia aurantiaca* flower senescence in response to sugar. *Postharvest Biology and Technology* 12: 43–50.

Evans RY and Reid MS (1988) Changes in carbohydrates and osmotic potential during rhythmic expansion of rose petals. *Journal of the American Society of Horticultural Science* 113: 884–888.

Fujiki Y, Yoshikawa Y, Sato T *et al.* (2001) Dark-inducible genes from *Arabidopsis thaliana* are associated with leaf senescence and repressed by sugars. *Physiologia Plantarum* 111: 345–352.

Ho LC and Nichols R (1977) Translocation of ^{14}C-sucrose in relation to changes in carbohydrate content in rose corollas cut at different stages of development. *Annales Botanicae* 41: 227–242.

Kuiper D, Ribot S, Van Reenen HS and Marissen N (1995) The effect of sucrose on the flower bud opening of 'Madelon' cut roses. *Scientia Horticulturae* 60: 325–336.

Kuiper D, Ribot S and Van Reenen HS (1996) Characterisation of flower bud opening in roses: a comparison of Madelon and Sonia roses. *Postharvest Biology and Technology* 9: 75–86.

Ono K and Watanabe A (1997) Levels of endogenous sugars, transcripts of rbcS and rbcL, and of RuBisCo protein in senescing sunflower leaves. *Plant Cell Physiology* 38: 1032–1038.

Paulin A (1980) Influence d'une alimentation continue avec une solution glucosée sur l'évolution des glucides et des protéines dans les divers organes de la rose coupée. *Physiologia Plantarum* 49: 55–61.

Van Doorn WG, Groenewegen G, Van de Pol PA and Berkholst CEM (1991) Effects of carbohydrate and water status on flower opening of cut Madelon roses. *Postharvest Biology and Technology* 1: 45–57.

Van Meeteren U, Van Gelder H and Van de Peppel AC (1995) Aspects of carbohydrate balance during floret opening in freesia. *Acta Horticulturae* 405: 117–122.

Postharvest Physiology: Ethylene

Müller R, Stummann BM, Andersen AS and Serek M (1999) Involvement of ABA in postharvest life of miniature potted roses. *Plant Growth Regulation* 29: 145–150.

Müller R, Andersen AS and Stummann BM (2001) Comparison of postharvest properties of closely related miniature rose cultivars (*Rosa hybrida* L.). *Scientiae Horticulturae* 91: 325–337.

Müller R, Stummann BM, Sisler EC and Serek M (2001) Cultivar differences in regulation of ethylene production in miniature rose flowers (*Rosa hybrida* L.). *Gartenbauwissenschaft* 1: 34–38.

Müller R, Owen CA, Xu Z-T, Welander M and Stummann BM (2002) Characterization of two CTR-like protein kinases in *Rosa hybrida* and their expression during flower senescence and in response to ethylene. *Journal of Experimental Botany* 53: 1223–1225.

Müller R, Owen CA, Xu Z-T, Welander M and Stummann BM (2003) The transcription factor EIN3 is constitutively expressed in miniature rose flowers with differences in postharvest life. *Journal of Horticultural Science and Biotechnology* 78: 10–14.

Serek M (1993) Ethephon and silver thiosulfate affect postharvest characteristics of *Rosa hybrida* 'Victory Parade'. *HortScience* 28: 199–200.

Serek M and Andersen AS (1993) AOA and BA influence on floral development and longevity of potted 'Victory Parade' miniature rose. *HortScience* 28: 1039–1040.

Serek M, Sisler EC and Reid MS (1994) Novel gaseous ethylene binding inhibitor prevents ethylene effects in potted flowering plants. *Journal of the American Society of Horticultural Science* 119: 1230–1233.

Tjosvold SA, Wu M-J and Reid MS (1994) Reduction of postproduction quality loss in potted miniature roses. *HortScience* 29: 293–294.

Production and Marketing: Mineral Nutrition

Cabrere RI (2002) Rose yield, dry matter partitioning and nutrient status responses to rootstock selection. *Scientia Horticulturae* 95: 75–83.

Cabrera RI, Evans RY and Paul JL (1993) Leaching losses of nitrogen from container-grown roses. *Scientia Horticulturae* 53: 333–345.

Cabrera RI, Evans RY and Paul JL (1995) Cyclic nitrogen uptake by greenhouse roses. *Scientia Horticulturae* 63: 57–66.

Cabrera RI, Evans RY and Paul JL (1996) Enhancement of short-term nitrogen uptake by greenhouse roses under intermittent N-deprivation. *Plant and Soil* 179: 73–79.

Gabriels R and Meneve R (1973) Nutritional requirements of roses grown in peat. *Scientia Horticulturae* 1: 341–349.

Johansson J (1978) Effects of nutrient levels on growth, flowering and leaf nutrient content of greenhouse roses. *Acta Agriculturae Scandinavica* 28: 363–386.

Lorenzo H, Cid MC, Siverio JM and Caballero JM (2000) Influence of additional ammonium supply on some nutritional aspects in hydroponic rose plants. *Journal of Agricultural Science* 134: 421–425.

Rey FA (1987) Copper nutrition of greenhouse roses relative to supplementary irradiation and growing medium. *Journal of Plant Nutrition* 10: 47–65.

Sadasivaiah SP and Holley WD (1973) Ion balance in nutrition of greenhouse roses. *Roses Inc. Bulletin* (November) (suppl.): 1–27.

St. John MG and Shorthouse JD (2000) Allocation patterns of organic nitrogen and mineral nutrients within stem galls of *Diplolepis spinosa* and *Diplolepis triforma* (Hymenoptera: Cynipidae) on wild roses (Rosaceae). *Canadian Entomology* 132: 635–648.

Tamimi YN, Matsuyama DT, Ison-Takata KD and Nakano RT (1999) Distribution of nutrients in cut-flower roses and the quantities of biomass and nutrients removed during harvest. *HortScience* 34: 251–253.

Woodson WR and Boodley JW (1982) Effects of nitrogen form and potassium concentration on growth, flowering, and nitrogen utilization of greenhouse roses. *Journal of the American Society for Horticultural Sciences* 107: 275–278.

Production and Marketing: Field Rose Production

Hambrick CE III, Davies FT Jr and Pemberton HB (1991) Seasonal changes in carbohydrate/nitrogen levels during field rooting of *Rosa multiflora* 'Brooks 56' hardwood cuttings. *Scientia Horticulturae* 46: 137–146.

Henderson JC, Davies FT Jr and Pemberton HB (1991) Landscape rose response to low moisture levels and a hydrophilic gel. *Scientia Horticulturae* 46: 129–135.

Paterson DR, Taber RA, Pemberton HB and Earhart DR (1986) Interaction between an indigenous endomycorrhizal fungus and mineral nutrition of *Rosa multiflora* understock. *HortScience* 21: 312–313.

Production and Marketing: Pot Rose Production

Andersson NE (1991) The influence of constant and diurnally changing CO_2 concentrations on plant growth and development. *Journal of Horticultural Science* 66: 569–574.

Asaoka M and Heins RD (1982) Influence of supplemental light and preforcing storage treatments on the forcing of 'Red Garnette' rose as a pot plant. *Journal of the American Society for Horticultural Science* 107: 548–552.

Clark DG, Kelly JW and Decoteau DR (1991a) Influence of end-of-day red and far-red light on potted roses. *Journal of Environmental Horticulture* 9(3): 123–127.

Clark DG, Kelly JW and Pemberton HB (1991b) Post-harvest quality characteristics of cultivars of potted rose in response to holding conditions and cytokinins. *HortScience* 26(9): 1195–1197.

Clark DG, Kelly JW and Rajapakse NC (1993) The effects of carbon dioxide enrichment on production and postharvest characteristics of *Rosa hybrida* L. 'Meijikatar'. *Journal of the American Society for Horticultural Science* 118: 613–617.

Cushman LC, Pemberton HB and Kelly JW (1994) Cultivar, flower stage, silver thiosulfate, and BA interactions affect performance of potted miniature roses. *HortScience* 29: 805–808.

Cushman LC, Pemberton HB, Miller Jr JC and Kelly JW (1998) Interactions of flower stage, cultivar, and shipping temperature and duration affect pot rose performance. *HortScience* 33: 736–740.

Halevy AH and Kofranek AM (1976) The prevention of flower bud and leaf abscission in pot roses during simulated transport. *Journal of the American Society for Horticultural Science* 101: 658–660.

Jenks MA, Andersen L, Teusink RS and Williams MH (2001) Leaf cuticular waxes of potted rose cultivars as affected by plant development, drought and paclobutrazol treatments. *Physiologia Plantarum* 112: 62–70.

Kyalo TM, Pemberton HB and Zajicek JM (1996) Seasonal growing environmental affects quality characteristics and post-production longevity of potted miniature roses. *HortScience* 31: 120–122.

Moe R (1970) Growth and flowering of potted roses as affected by temperature and growth retardants. *Meldinger Fra Norges Landbrukshoegskole* 49: 1–16.

Monteiro JA, Nell TA and Barrett JE (2001) Postproduction of potted miniature rose: flower respiration and

single flower longevity. *Journal of the American Society for Horticultural Science* 126: 134–139.
Mortensen LM (1991) Effects of temperature, light and CO_2 level on growth and flowering of miniature roses. *Norwegian Journal of Agricultural Science* 5: 295–300.
Müller R, Andersen AS and Serek M (1998) Differences in display life of miniature potted roses (*Rosa hybrida* L.). *Scientia Horticulturae* 76: 59–71.
Müller R, Sisler EC and Serek M (1999) Stress induced ethylene production, ethylene binding, and the response to the ethylene action inhibitor 1-MCP in miniature roses. *Scientia Horticulturae* 83: 51–59.
Müller R, Stummann BM, Andersen AS and Serek M (1999) Involvement of ABA in postharvest life of miniature potted roses. *Plant Growth Regulation* 29: 143–150.
Müller RS, Lind-Iversen S, Stummann BM and Serek M (2000) Expression of genes for ethylene biosynthetic enzymes and an ethylene receptor in senescing flowers of miniature potted roses. *Journal of Horticultural Science and Biotechnology* 75: 12–18.
Nell TA and Noordegraaf CV (1991) Simulated transport, postproduction irradiance influence postproduction performance of potted roses. *HortScience* 26: 1401–1404.
Nell TA and Noordegraaf CV (1992) Postproduction performance of potted roses under simulated transport and low irradiance levels. *HortScience* 27: 239–241.
Rajapakse NC and Kelly JW (1994) Influence of spectral filters on growth and postharvest quality of potted miniature roses. *Scientia Horticulturae* 56: 245–256.
Rajapakse NC, Clark DG, Kelly JW and Miller WB (1994) Carbohydrate status and postharvest leaf chlorosis of miniature roses as influenced by carbon dioxide enrichment. *Postharvest Biology and Technology* 4: 271–279.
Risemen A, Jensen C and Williams M (2001) Stomatal conductivity and osmotic adjustment during acclimation to multiple cycles of drought stress in potted miniature roses (*Rosa* × *hybrida*). *Journal of Horticultural Science and Biotechnology* 76: 138–144.
Serek M (1993) Ethephon and silver thiosulfate affect postharvest characteristics of *Rosa* hybrida 'Victory Parade'. *HortScience* 28: 199–200.
Serek M and Anderson AS (1993) AOA and BA influence on floral development and longevity of potted 'Victory Parade' miniature rose. *HortScience* 28: 1039–1040.
Starkey KR and Pedersen AR (1997) Increased levels of calcium in the nutrient solution improves the postharvest life of potted roses. *Journal of the American Society for Horticultural Science* 122(6): 863–868.
Tjosvold SA, Wu M and Reid MS (1994) Reduction of postproduction quality loss in potted miniature roses. *HortScience* 29: 293–294.
Williams MH, Rosenqvist E and Buchhave M (1999) Response of potted miniature roses (*Rosa* × *hybrida*) to reduced water availability during production. *Journal of Horticultural Science and Biotechnology* 74: 301–308.
Williams MH, Rosenqvist E and Buchhave M (2000) The effect of reducing production water availability on the post-production quality of potted miniature roses (*Rosa* × *hybrida*). *Journal of Horticultural Science and Biotechnology* 18: 143–150.
Zieslin N and Tsujita MJ (1990a) Response of miniature rose to supplementary illumination. 1. Light intensity. *Scientia Horticulturae* 42: 113–121.
Zieslin N and Tsujita MJ (1990b) Response of miniature rose to supplementary illumination. 2. Effect of stage of plant development and cold storage. *Scientia Horticulturae* 42: 123–131.

Production and Marketing: Cut Rose Production

Brednose NB (1998) Growth, flowering and postharvest performance of single-stemmed rose (*Rosa hybrida* L.) plants in response to light quantum integral and plant population density. *Journal of the American Society for Horticultural Science* 123: 569–576.
Maas FM and Bakx EJ (1995) Effects of light on growth and flowering of *Rosa hybrida* 'Mercedes'. *Journal of the American Society for Horticultural Science* 20: 571–576.
Roberts AV, Blake PS, Lewis R, Taylor JM and Dunstan DI (1999) The effect of gibberellins on flowering in roses. *Journal of Plant Growth Regulation* 18: 113–119.
van der Salm TPM, Bouwer R, van Dijk AJ, Keizer LCP, Hanisch ten Cate CH, van der Plas LHW and Dons JJM (1998) Stimulation of scion bud release by rol gene transformed rootstock of *Rosa hybrida* L. *Journal of Experimental Botany* 49: 847–852.
Xu XM (1999) Effects of temperature on the length of the incubation period of rose powdery mildew (*Sphaerotheca pannosa* var. *rosae*). *European Journal of Plant Pathology* 105: 3–21.

Propagation: Own-Rooted Cuttings

Al Saqri F and Alderson PG (1996) Effects of IBA, cutting type and rooting media on rooting of *Rosa centifolia*. *Journal of Horticultural Science* 71: 729–737.
Berninger E, Barrade R and Pudelska K (1990) Étude du comportement de différents génotypes de rosiers de serre (*Rosa hybrida*) propagés par bouturage. I. Bouturage et culture de 'miniplants' en serre. *Agronomie* 10: 787–796 (in French).
Borowski E, Hagen P and Moe R (1987) Air and root medium temperature and the rooting of chrysanthemum and rose cuttings. *Acta Agrobotica* 40: 1–2, 27–39.
Costa JM and Challa H (2002) The effect of the original leaf of rooting and initial growth of planting material of rose. *Scientia Horticulturae* 95: 111–121.
Davies FT (1987) Effects of VA-mycorrhizal fungi on growth and nutrient uptake of cuttings of *Rosa multiflora* in two container media with three levels of fertiliser application. *Plant and Soil* 104: 31–35.

Dubois LAM and De Vries DP (1985) Directions for satisfactory production of glass cut rose cuttings; experiments with 'Sonia' softwood cuttings. *Vakblad voor de Bloemisterij* 45: 26–31 (in Dutch).

Dubois LAM and De Vries DP (1991) Variation in adventitious root formation of softwood cuttings of *Rosa chinensis* minima (Sims) Voss cultivars. *Scientia Horticulturae* 47: 345–349.

Dubois LAM, Kool M, Van de Pol PA and De Vries DP (1990) Rozen steken, stenten of wortelenten. *Vakblad voor de Bloemisterij* 45: 26–31 (in Dutch).

Gillman JH and Zlesak DC (2000) Mist applications of sodium silicate to Rose (*Rosa* L. × 'Nearly Wild') cuttings decrease leaflet drop and increase rooting. *HortScience* 35: 773.

Gislerød HR (1983) Physiological conditions of propagation media and their influence on the rooting of cuttings. III. The effect of air content and temperature in different propagation media on rooting of cuttings. *Plant and Soil* 75: 1–14.

Khosh-Khui M and Tafazoli E (1979) Effect of acid or base treatment on auxin response of Damask rose cuttings. *Scientia Horticulturae* 10: 395–399.

Marcelis-van Acker CAM (1994) Effect of assimilate supply on development and growth potential of axillary buds in roses. *Annals of Botany* 73: 415–420.

Marcelis-van Acker CAM and Leutscher KJ (1993) Effect of the type of cutting on heterogeneity and growth of *Rosa hybrida* cv. 'Motrea' and *Schefflera arboricola* cv. 'Compacta'. *Scientia Horticulturae* 54: 59–67.

Pierik RLM and Van de Pol PA (1995) Factors controlling adventitious root formation on stem explants of rose (*Rosa hybrida* 'Motrea') in vitro. *Revista Chapingo Horticultura* 3: 23–29.

Scagel CF (2001) Cultivar specific effects of mycorrhizal fungi on the rooting of miniature rose cuttings. *Journal of Environmental Horticulture* 19: 15–20.

Sun WQ and Bassuk NL (1993) Auxin-induced ethylene synthesis during rooting and inhibition of bud break of 'Royalty' rose cuttings. *Journal of the American Society of Horticultural Science* 118: 638–643.

Van der Salm TPM, Van der Toorn CJG, Bouwer R, Ten Cate CHH and Dons HJM (1997) Production of ROL gene transformed plants of *Rosa hybrida* L. and characterisation of their rooting ability. *Molecular Breeding* 3: 39–47.

Ypema HL, van de Pol PA and Bollen GJ (1987) Black rot of stentlings of roses: a disease caused by various soil fungi. *Scientia Horticulturae* 33: 269–280.

Propagation: Stenting

Dubois LAM, Kool M, Van de Pol PA and De Vries DP (1990) Rozen stekken, stenten of wortelenten (in Dutch). *Vakblad Bloemisterij* 45: 26–31.

Kool MTN and Van de Pol PA (1993) Controlling the plant development of *Rosa hybrida* 'Motrea'. *Scientia Horticulturae* 53: 239–248.

Zieslin N and Halevy HA (1976) Components of axillary bud inhibition in rose plants. Effect of different plant parts (correlative inhibition). *Botanical Gazette* 137: 291–296.

Propagation: Seed Propagation

De Vries DP and Dubois LAM (1987) The effect of temperature on fruit set, seed set and germination in 'Sonia' × 'Hadley' hybrid tea-rose crosses. *Euphytica* 36: 117–120.

Dubois LAM and De Vries DP (1983) Pollen and pollination experiments X. The effect of repeated pollination on fruit- and seed set in crosses between the hybrid tea-rose cvs. 'Sonia' and 'Ilona'. *Euphytica* 32: 685–689.

Erlanson EW (1963) A self-pollination mechanism and other items in rose species. *American Rose Annual* 188–193.

Gudin S (1992) Influence of bud chilling on subsequent reproductive fertility in roses. *Scientia Horticulturae* 51: 139–144.

Gudin S (1994) Embryo rescue in *Rosa hybrida* L. *Euphytica* 72: 205–212.

Gudin S and Arene L (1991) Influence of the pH of the stigmatic exudate on male–female interaction in *Rosa hybrida* L. *Sexual and Plant Reproduction* 4: 110–112.

Gudin S and Arene L (1992) Putrescine increases effective pollination period in roses. *Horticultural Technology* 2: 211–213.

Gudin S, Arene L, Chavagnat A and Bulard C (1990) Influence of endocarp thickness on rose achene germination: genetic and environmental factors. *HortScience* 25: 786–788.

Gudin S, Arene L and Bulard C (1991) Influence of season on rose pollen quality. *Sexual and Plant Reproduction* 4: 113–117.

Gudin S, Arene L, Chavagnat A and Bulard C (1991) Structure and water absorption of different parts of the rose achene pericarp during imbibition. *Biologia Plantarum (Praha)* 33: 262–267.

Gudin S, Arene L and Chavagnat A (1992) Relation entre imbibition, densité, taux de remplissage et faculté germinative chez l'akène de *Rosa hybrida* L. *Agronomie* 12: 123–126.

Kroon GH and Zeilinga AE (1974) Apomixis and heterogamy in rose rootstocks (*Rosa canina* L.). *Euphytica* 23: 345–352.

Peimbert R, Krewinkle J and Morey D (1963) Factors influencing seed production in hybrid roses. II. A preliminary bio-assay of stored and fresh pollen. *American Rose Annual* 194–198.

Roberts L (1979) Practical aspects of the acid treatment of rose seed. *Plant Propagator* 25: 13–14.

Suszka B and Bujarska-Borkowska B (1987) Seed after-ripening, germination and seedling ermergence of *Rosa canina* L. and some of its rootstock selections. *Arboretum Kornickie* 32: 231–296.

Von Abrams GJ and Hand ME (1956) Seed dormancy in *Rosa* as a function of climate. *American Journal of Botany* 43: 7–12.

Werlemark G, Carlson-Nilsson U, Uggla M and Nybom H (1995) Effects of temperature treatments on seedling emergence in dogroses, *Rosa* sect. *Caninae* (L). *Acta Agricultura Scandinavica Section B, Soil and Plant Science* 45: 278–282.

Wohlers MA and Morey D (1963) Factors influencing seed production in hybrid roses. III. Determining the actual germinability of rose pollen. *American Rose Annual* 199–204.

Wohlers MA, Peimbert R and Morey D (1962) Factors influencing seed production in hybrid roses. 1. Pollen color changes as related to deterioration during storage. *American Rose Annual* 103–109.

Yambe Y and Takeno K (1992) Improvement of rose achene germination by treatment with macerating enzymes. *HortScience* 27: 1018–1020.

Propagation: Seed Maturation and Germination

Dubois LAM and De Vries DP (1983) Pollen and pollination experiments. X. The effect of repeated pollination on fruit-and seed set in crosses between the hybrid tea-rose cvs. 'Sonia' and 'Ilona'. *Euphytica* 32: 685–689.

Gudin S, Arene L, Chavagnat A and Bulard C (1990) Influence of endocarp thickness on rose achene germination: genetic and environmental factors. *HortScience* 25: 786–788.

Gudin S, Arene L, Chavagnat A and Bulard C (1991) Structure and water absorption of different parts of the achene pericarp during imbibition. *Biologia Plantarum (Praha)* 33: 262–267.

Jackson GAD (1968) Hormonal control of fruit development, seed dormancy and germination with particular reference to *Rosa*. *Society of Chemical Industry Monograph* 31: 127–156.

Jackson GAD and Blundell JB (1965) Germination of *Rosa arvensis*. *Nature* 205: 518–519.

Jin B, Dong H and Yang X (1995) Shortening hybridization breeding cycle of rose – a study on mechanisms controlling achene dormancy. *Acta Horticulturae* 404: 40–47.

Morey D (1956) The use of chemicals in breaking seed dormancy in hybrid tea roses. *American Rose Annual* 41: 64–68.

Morpeth DR and Hall AM (2000) Microbial enhancement of seed germination in *Rosa corymbifera* 'Laxa'. *Seed Science Research* 10: 489–494.

Tillberg E (1983) Levels of endogenous abscisic acid in achenes of *Rosa rugosa* during dormancy release and germination. *Physiologia Plantarum* 58: 243–248.

Von Abrams GJ and Hand ME (1956) Seed dormancy in *Rosa* as a function of climate. *American Journal of Botany* 43: 7–12.

Yambe Y and Takeno K (1992) Improvement of rose achene germination by treatment with macerating enzymes. *HortScience* 27: 1018–1020.

Yambe Y, Hori Y and Takeno K (1992) Levels of endogenous abscisic acid in rose achenes and leaching with activated charcoal to improve seed germination. *Journal of the Japanese Society for Horticultural Science* 61: 383–387.

Yambe Y, Takeno K and Saito T (1995) Light and phytochrome involvement in *Rosa multiflora* seed germination. *Journal of the American Society for Horticultural Science* 120: 953–955.

Rootstocks: Scion–Rootstock Relationships

DeVries DP and Dubois LAM (1989) Variation for the shoot production of *Rosa* hybrida 'Sonia', as induced by different Edelcanina rootstock clones. *Gartenbauwissenschaft* 54(5): 211–215.

DeVries DP and Dubois LAM (1990) Shoot production of 'Sonia' on hybrid tea rootstock clones of different vigour. *Gartenbauwissenschaft* 55(6): 268–271.

Kool MTN, Van de Pol PA and Berentzen WTJ (1991) Formation and early development of bottom breaks in 'Motrea' roses. *Scientia Horticulturae* 48: 293–298.

Tubbs FR (1980) Growth relations of rootstock and scion in apples. *Journal of Horticultural Science* 55(2): 181–189.

Visser PB, Keizer PLCP, Van der Salm TPM and De Jong J (2000) Rootstocks transformed with *rol* A, B, C genes enhance rose flower production. *Journal of Biotechnology and Biotechnological Equipment* 14: 63–74.

Vyvian MC (1955) Interrelation of scion and rootstock in fruit trees. *Annals of Botany* 75: 401–423.

Zieslin N and Mor Y (1981) Plant management of greenhouse roses. Formation of renewal canes. *Scientia Horticulturae* 15: 67–75.

Rootstocks: Rootstock Breeding

Buck GJ (1978) I.T.-9 and I.T.-18 rose rootstocks. *HortScience* 13: 601–602.

DeVries DP and Dubois LAM (1989) Variation for the shoot production of *Rosa* hybrida 'Sonia', as induced by different Edelcanina rootstock clones. *Gartenbauwissenschaft* 54: 211–215.

DeVries DP and Dubois LAM (1990) Shoot production of 'Sonia' on hybrid tea rootstock clones of different vigour. *Gartenbauwissenschaft* 55: 268–271.

Gallais A (1984) Use of indirect selection in plant breeding. In: Lange W, Zeven AC and Hogenboom NG (eds.) *Efficiency in Plant Breeding*, pp. 45–60. Wageningen, The Netherlands: PUDOC.

Krickl M (1958) Züchtungsversuche über die Möglichkeit einer unbedingten Keimfähigkeit der Samen von *Rosa canina* im ersten Frühjahr. *Mitteilungen Klosterneuburg* VIII B: 198–207.

Persiel F (1977) Resistenzzüchtungen gegen Mehltau und Sternrusstau bei den als Rosenunterlagen genutzten Rosenarten. *Deutsche Baumschule* 29: 214–215.

Stougaard B (1984) Analysis of variation in progeny from *Rosa multiflora* crosses. *Tidsskrift for Planteavl* 88: 287–297.

Visser PB, Keizer PLCP, Van der Salm TPM and De Jong J (2000) Rootstocks transformed with *rol* A, B, C genes enhance rose flower production. *Journal of Biotechnology and Biotechnological Equipment* 14: 63–74.

Rootstocks: Usage of Rootstocks

Carlson M (1929) Microchemical studies of rooting and non-rooting rose cuttings. *Botanical Gazette* 87: 64–80.

Chu C-Y (1990) Budded cuttings for propagating roses. *Scientia Horticulturae* 43: 163–168.

De Vries DP and Dubois LAM (1988) Shoot and root characters of one-season-old *Rosa canina* 'Inermis' rootstock seedlings in relation to the root collar diameter. *Gartenbauwissenschaft* 53(3): 30–33.

De Vries DP and Dubois LAM (1988) Effects of plant density on root and shoot characters of one-season-old *Rosa canina* 'Inermis' seedlings. *Gartenbauwissenschaft* 53(3): 117–121.

De Vries DP and Dubois LAM (1989) Variation for the shoot production of *Rosa* hybrida 'Sonia', as induced by different Edelcanina rootstock clones. *Gartenbauwissenschaft* 54(5): 211–215.

De Vries DP and Dubois LAM (1990) Shoot production of 'Sonia' on hybrid tea rootstock clones of different vigour. *Gartenbauwissenschaft* 55(6): 268–271.

Glenny F (1927) Rose stocks. *Rose Annual*: 169–186.

Haenchen E (1982) Zum Verhalten autovegetativ vermehter Rosen im Freiland. *Archiv für Gartenbau* 30(1): 1–14.

Hanan JJ and Grueber KL (1984) Rose rooting and grafting. Colorado greenhouse growers. *Research Bulletin* 410: 1–6.

Lothian L (1953) Rose rootstocks. *Horticultural Advertiser*: 6–7.

Ohkawa K (1984) Cutting grafts as a technique for producing started-eye rose bushes outdoors. *HortScience* 19: 527–528.

Rowley GD (1961) Rose rootstocks: a first field trial. *Journal of Horticultural Science* 36: 160–167.

Saul J (1850) Rose stocks. *Journal of the Royal Horticultural Society* 5: 70–73.

Schupper F (1945) De Boskoopse rugosa-roos (in Dutch). *De Boomkwekerij*, 145.

Stangler BB (1956) *Origin and Development of Adventitious Roots in Stem Cuttings of Chrysanthemum, Carnation and Rose.* Cornell University, Agricultural Experimental Station. Ithaca, New York, Memoir 342.

Van De Pol PA and Breukelaar A (1982) Stenting of roses: a method for quick propagation by simultaneously cutting and grafting. *Scientia Horticulturae* 17: 187–196.

Visser PB, Keizer PLCP, Van der Salm TPM and De Jong J (2000) Rootstocks transformed with *rol* A, B, C genes enhance rose flower production. *Journal of Biotechnology and Biotechnological Equipment* 14: 63–74.

Warne LGG and Fothergill L (1942) The growth of rose stocks. *Journal of the Royal Horticultural Society* 67: 359–361.

Rootstocks: Clonal Rootstocks

Buck GJ (1978) I.T.-9 and I.T.-18 rose rootstocks. *HortScience* 13(5): 601–602.

De Vries DP and Dubois LAM (1988) Shoot and root characters of one-season-old *Rosa canina* 'Inermis' rootstock seedlings in relation to the root collar diameter. *Gartenbauwissenschaft* 53(3): 30–33.

De Vries DP and Dubois LAM (1988) Effects of plant density on root and shoot characters of one-season-old *Rosa canina* 'Inermis' seedlings. *Gartenbauwissenschaft* 53(3): 117–121.

De Vries DP and Dubois LAM (1989) Variation for the shoot production of *Rosa hybrida* 'Sonia', as induced by different Edelcanina rootstock clones. *Gartenbauwissenschaft* 54(5): 211–215.

De Vries DP and Dubois LAM (1990) Shoot production of 'Sonia' on hybrid tea rootstock clones of different vigour. *Gartenbauwissenschaft* 55(6): 268–271.

De Vries DP and Dubois LAM (1994) Re-invigoration of clonal rose rootstocks by sustained micropropagation. *Gartenbauwissenschaft* 59(2): 81–85.

McFadden SE (1962) *Rosa fortuneana* compared with other rose rootstocks grown in Florida. *Proceedings of the Florida State Horticultural Society* 75: 497–500.

Van Der Linde PCG (2003) Ageing (in Dutch). Submitted.

Visser PB, Keizer PLCP, Van der Salm TPM and De Jong J (2000) Rootstocks transformed with *rol* A, B, C genes enhance rose flower production. *Journal of Biotechnology and Biotechnological Equipment* 14: 63–74.

Tukey HB and Brase KD (1931) The propagation of multiflora rootstocks for roses by softwood cuttings. Bulletin no. 598. Cornell: Cornell University.

Rootstocks: Seedling Stocks

Akkerman AJJ (1978) Rozenonderstammen. *Bedrijfsontwikkeling* 9: 573–576.

Akkerman AJJ (1978) Rozezaadstruiken. *Groen* 11: 441–442.

Akkerman AJJ (1987) Dunner zaaien levert dikkere maten op. *Vakblad voor de Bloemisterij* 28: 30–31.

Derkx MPM (2001) Compostversneller verbetert kieming rozenzaad. *de Boomkwekerij* 31/32: 20–21.

De Vries DP and Dubois LAM (1988) Uniformiteit rozenonderstammen kan worden verbeterd. Verschillen tussen boven- en ondergrondse eigenschappen. *de Boomkwekerij* 2: 12–13.

De Vries DP and Dubois LAM (1988) Rose rootstocks: causes of variation and the clonal propagation of rootstocks, with reference to *Rosa canina* 'Inermis'. *Professional Horticulture* 2: 62–66.

De Vries DP and Dubois LAM (1988) Effects of plant density on root and shoot characters of one-season-old

Rosa canina 'Inermis' seedlings. *Gartenbauwissenschaft* 53: 117–121.
De Vries DP and Dubois LAM (1988) Shoot and root characters of one-season-old *Rosa canina* 'Inermis' rootstock seedlings in relation to the root collar diameter. *Gartenbauwissenschaft* 53: 30–33.
De Vries DP and Dubois LAM (1989) Variatie in Edelcanina's groter dan aangenomen. *de Boomkwekerij* 48: 20–21.
Dubois LAM and De Vries DP (1990) Genetische variatie in planteigenschappen, stekbaarheid en bloemproduktie van Edelcaninaonderstammen. *Prophyta* 2: 43–45.
Dubois LAM, De Vries DP and Jansen H (1990) Rozenonderstammen in beweging. *Prophyta* 5: 117–119.
Kool MTN and Van De Pol PA (1992) Aspects of growth analysed for *Rosa hybrida* 'Motrea' as affected by six rootstocks. *Gartenbauwissenschaft* 57: 120–125.
Kroon GH (1975) Uniformiteit en bastaardering bij rozenonderstammen. *Groen* 31: 27.
Kroon GH and Zeilinga AE (1974) Apomixis and heterogamy in rose rootstocks. *Euphytica* 23: 345–352.
Naus JW (1989) Kennis rozenonderstammen levert geld op. *de Boomkwekerij* 35: 8–11.
Schmid R (1965) Meine Canina-Züchtungen. *Deutsche Baumschule* 8: 194–195.
Smit W (1974) De geschiedenis van de teelt van rozenonderstammen in de provincie Groningen. *De Plantenbeurs* 86: 2.
van der Meer M (1994) Moderne teelt kasrozen vereist modern uitgangsmateriaal. *de Boomkwekerij* 4: 10–13.
van Rijsewijk V (1997) Waarschuwingssysteem voor echte meeldauw bespaart rozenkweker geld. *de Boomkwekerij* 27/28: 22–23.
Versteegen FHMM and Verstraelen MWA (1993) Afpennen rozenonderstammen goed voor kwaliteit en opbrengst. *de Boomkwekerij* 24/25: 18–19.
Wylie A (1976) Why are the caninae rose different? *New Zealand Rose Annual* 67–79.

Secondary Metabolites: Culinary Uses and Nutritional Value

Albert-Matesz R (2000) Hip, hip. *The Herb Companion* 12(3): 36.
Anonymous (2001) Herb basics. *Herbs for Health* 6: 28.
Flores A (2002) Rose hips: the harvest festival. *American Nurseryman* 195: 24–27.
Healing Garden (2000) Rose hips. *Organic Gardening* 47(5): 20.
http://www.kattitudes.com/rosemenu.html.
Ozcan M and Akbulut M (1998) Some physical and chemical properties of myrtle (*Mytus communis* L.) fruits. *Food* 23: 121–123.

Secondary Metabolites: Medicinal and Pharmaceutical Uses

Albert-Matesz R (2000) Hip, hip. *Herb Companion* 12(3): 36.
Anon. (2000) Healing Garden: Rose hips. *Organic Gardening* 47(5): 20.
Anon. (2001) Herb basics. *Herbs for Health* 6: 28.
Anon. (2001) *Nursing Herbal Medicine Handbook*. Pennsylvania: Springhouse.
British Herbal Pharmacopoeia (1996). Bournemouth, UK: British Herbal Medicine Association.
Ebbell B (transl.) (1937) *The Papyrus Ebers*. Copenhagen: Levin & Munksgaard.
Facmo R (1994) Free radicals scavenging action and anti-enzyme activities of procyanidines from *Vitis vinifera*. *Arzneimittel-Forschung/Drug Research* 44(1): 592–601.
Flores A (2002) Rose hips: the harvest festival. *American Nurseryman* 195: 24–27.
Haslam E (1996) Natural polyphenols (vegetable tannins) as drugs: possible modes of action. *Journal of Natural Produce* 59(2): 205–215.
Law D (1970) *Herbs for Cooking and Healing*. London, UK: W Foulsham.
Mills S and Bone K (2000) *Principles and Practice of Phytotherapy (Modern Herbal Medicine)*. London: Churchill Livingstone.
Okuda T, Yoshida T and Hatano T (1989) Ellagitannins as active constituents of medicinal plants. *Planta Medicina* 5(2): 117–122.
Okuda T, Yoshida T and Hatano T (1992) Pharmacologically active tannins isolated from medicinal plants. *Basic Life Science* 59: 539–569.
Rossnagel K and Willich SN (2001) Value of complementary medicine exemplified by rose-hips. *Gesundheitswesen* 63(6): 412–416.

Secondary Metabolites: Extraction and Industrial Processes

Babu KGD, Bikram S, Joshi VP and Singh V (2002) Essential oil composition of Damask rose (*Rosa damascena* Mill.) distilled under different pressures and temperatures. *Flavour and Fragrance Journal* 17: 136–140.
Basile MM, Jimenez C and Clifford AA (1998) Extraction of rosemary with superheated water. *Journal of Food and Agricultural Chemistry* 46: 5205–5209.
Boelens MH (1997) Differences in chemical and sensory properties of orange flower and rose oils obtained from hydrodistillation and from supercritical CO_2 extraction. *Perfumer and Flavorist* 22: 31–35.
Illés V and Ottó A (1997) Extraction of hiprose fruit by supercritical CO_2 and propane. *Journal of Supercritical Fluids* 10: 209–218.
Lawrence BM (1991) Progress in essential oils: rose oil and extract. *Perfumer and Flavorist* 16: 43–77.
Reverchon E (1997) Supercritical fluid extraction and fractionation of essential oils and related products. *Journal of Supercritical Fluids* 10: 1–37.
Reverchon E and DellaPorta G (1996) Rose concrete fractionation by supercritical CO_2. *Journal of Supercritical Fluids* 9: 199–204.

APPENDIX 3

REHDER'S CLASSIFICATION OF ROSES

Introduction

The entry on the genus *Rosa* in Alfred Rehder's *Manual of Trees and Shrubs Hardy in North America*, first published by MacMillan, New York, has been reproduced in this Appendix. It was first published in 1927. The second edition was published in 1940 and reissued in 1960. It starts with a key to subgenera, sections and to the species that Rehder considered of primary importance. Detailed descriptions of species identified in the key then follow, accompanied by listings of varieties and forms, hybrids and related species. In his descriptions of species, Rehder used abbreviated references to illustrations, as shown below. The additional abbreviation 't' indicates a plate (tabula) and 'c' indicates that an illustration is in colour.

As the scheme for *Rosa* is reproduced without the original introduction to the book, some comments are needed for guidance to its usage. In his description of species, Rehder refers to flowering times and fruiting times using the abbreviations fl. and fr., respectively, followed by Roman numerals indicating the months by number. He described seven zones in North America, in terms of minimum temperatures (originally °F, converted here to °C) as follows: Zone I, lower than −46 °C, (the treeless zone in extreme northern Canada), Zone II −46 to −37 °C, Zone III −37 to −29 °C, Zone IV −29 to −23 °C, Zone V −23 to −21 °C, Zone VI −21 to −15 °C, Zone VII −15 to −12 °C (see **Figure 1**). Many of the botanical names appear as a trinomial, followed by the names of the authorities and var. (variety), or f. (forma) indicating the taxonomic status of the third name. For example, '*R. spinosissima hispida* (Sims) Koehne, var.' indicates that *hispida* is a variety of *R. spinosissima* Sims, and that Koehne published the varietal name.

Rehder's classification of roses has been cited in many publications and acclaimed for its excellence. However, additional species have been described subsequently and fresh evidence of relationships, some based on recent genomic investigations, has been published. Undoubtedly there is a need for a new monograph on the genus.

Abbreviations of References to Illustrations

A.F. *The American Florist*, vol. 1. Chicago, 1885
A.G. *American Gardening*, vols 1–25. New York, 1880–1904
B.B. Britton NL and Addison Brown (1913) *An Illustrated Flora of the Northern United States, Canada, etc.*, vols 1–3, 2nd edn. New York
B.C. Bailey LH (1914–1917) *Standard Cyclopedia of Horticulture*, vols 1–6. New York
B.M. Curtis W *et al.* (eds) (1787) *The Botanical Magazine*. London
B.R. *The Botanical Register*, edited by Edwards S (vols 1–14) and Lindley J (vols 15–33). London, 1815–1847
B.S. Bean WJ (1914–1933) *Trees and Shrubs Hardy in the British Isles*, vols 1–3. London
C.L. *Country Life* (USA), vol. 1. New York, 1901
D.L. Dippel L (1899–1893) *Handbuch der Laubholzkunde*, vols 1–3. Berlin
Em. Emerson GB (1875) *A Report on the Trees and Shrubs Growing Naturally in the Forests of Masssachusetts*, 4th edn. Boston
F.D. von Schlechtendal DFL, Langethal and Schenk (eds) *Flora von Deutschland* (1880–1887), vols 1–30, 5th edn. Gera-Untermhaus
F.S. *Flore des Serres et des Jardins de l'Europe*, vols 1–23. Gand, 1845–1880
G. *Gardening Illustrated*, vols 1–33. London, 1880–1925
G.C. *Gardeners' Chronicle* (1841–1873) London. New series, vols 1–26, London, 1874–1886. Series III, vol. 1, London, 1887
G.F. Sargent CS (ed.) (1888–1897) *Garden and Forest*, vols 1–10. New York
Gg. *Gardening*, vol. 1. Chicago, 1892
G.H. Guimpel F, Willdenow and Hayne (1815–1820) *Abbildungen der deutschen Holzarten*. Berlin
Gn. *The Garden*, founded by Robinson W, vols 1–91. London, 1871–1927
Gn.M. *The Garden Magazine*, vols 1–39. New York, 1905–1924. Continued as *Garden Magazine and Home Builder*, vols 40–47. Boston, 1924–1928. Continued as *American Home*, vol. 48
Gn.W. *Gardening World Illustrated*, vols 1–26. London, 1884–1909
Gs. Förster K, Schneider C and Kühl O (eds) (1920) *Gartenschönheit*, vol. 1. Berlin-Westend
Gt. Regel E *et al.* (ed.) (1852–1938) *Gartenflora*, vols 1–87. Erlangen
G.W. Hesdörffer M *et al.* (1897–1933) *Gartenwelt*, vols 1–87. (In 1937 united with *Blumen- und Planzenbau*)
H.I. *Hookes' Icones Plantarum*, vol. 1. London, 1837
I.S. Hu HH and Chun WY (1927) *Icones Plantarum Sinicarum*. Fase
I.T. van den Bossche L (ed.) (1899–1909) *Icones Selectae Horti Thenensis*, vols 1–6. Brussels
J.L. *Journal of the Royal Horticultural Society of London*, vols 1–9. New series, vol. 1. London, 1846
K.S. Keeler HI (1903) *Our Northern Shrubs*. New York
L.A. Loudon JC (1833) *Arboretum et Fruticetum Britannicum*, vols 1–8. London
M.D. *Mitteilungen der deutschen Dendrologischen Gesellschaft*. Berlin, 1893
M.G. *Möller's deutsche Gärtner-Zeitung*, vol. 1. Erfurt, 1886
M.N. Mechan T (1878–1879) *The Native Flowers and Ferns of the United States*, vols 1–2. Boston
N.D. Duhamel du Monceau HL (1804–1819) *Traité des Arbres et Arbustes*, vols 1–7. Paris
N.F. Cox EHM (ed.) (1928) *New Flora and Silva*, vol. 1. London
N.K. Nakai T (1915) *Flora Sylvatica Koreana*, vol. 1. Tokyo
N.S. Nuttall T (1865) *The North American Sylva*, vols 4–6. Philadelphia
R.B. *Revue de l'Horticulture Belge et Etrangère*, vol. 1. Gand, 1875
R.H. *Revue Horticole*. Paris, 1829
R.R. Redouté PJ (1835) *Les Roses*, vols 1–3, 3rd edn. Paris, 1835
S.E. Sowerby J (1863–1902) *English Botany or Colored Figures of British Plants*, vols 1–13, 3rd edn. London
S.H. Schneider C (1906–1912) *Illustriertes Handbuch der Laubholzkunde*, vols 1–2. Jena
W.A. Wilson EH (1925) *America's Greatest Garden; the Arnold Arboretum*. Boston
W.R. Willmott E (1910–1914) *The Genus* Rosa, vols 1–2. London

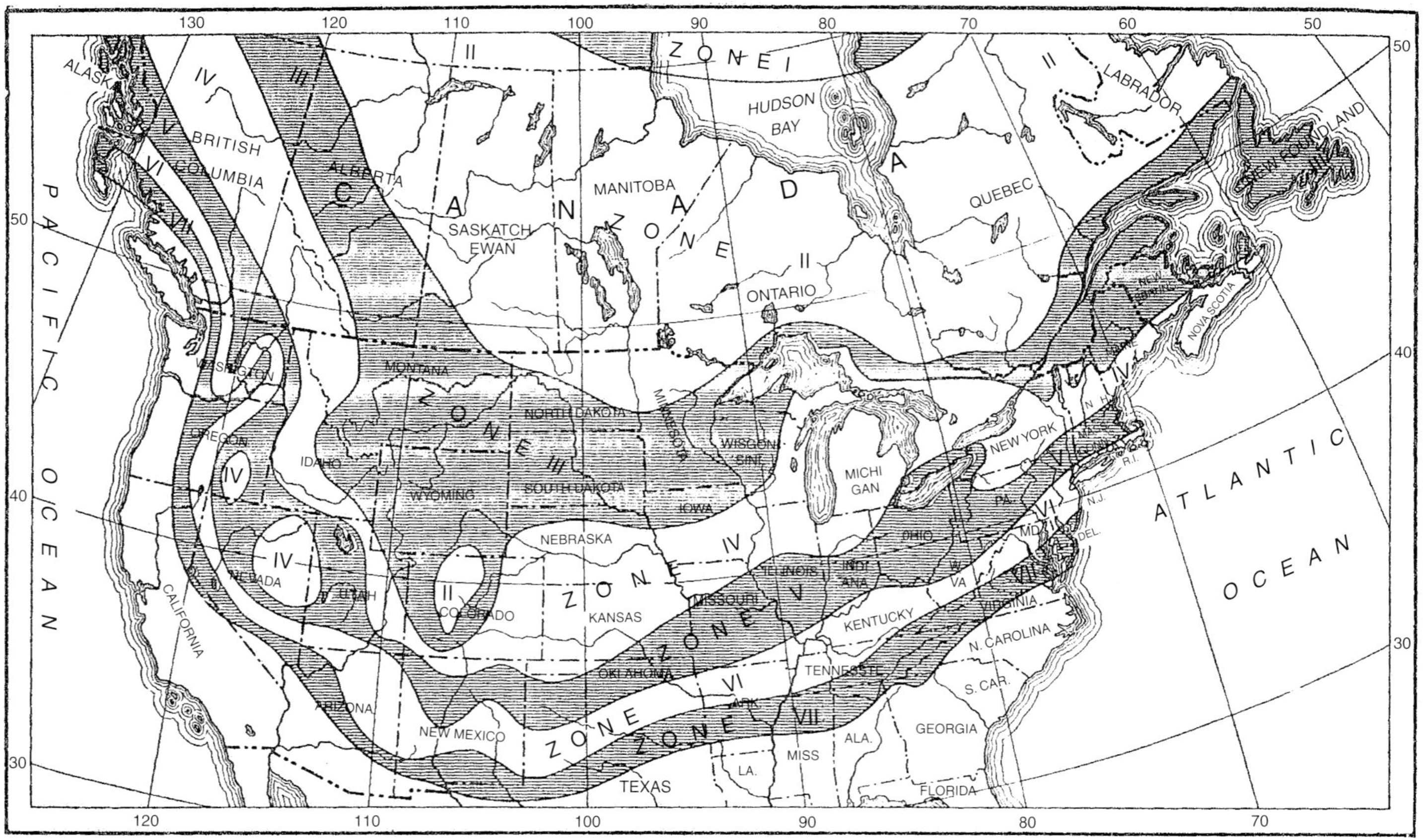

Figure 1 Map of Climatic Zones. Reproduced from Rehder A (1960) *Manual of Cultivated Trees and Shrubs, Hardy in North America*. 2nd Edn. New York: Macmillan 426–451, 979–981 © The Blackbum Press: Caldwell, New Jersey. Reproduced with permission.

ROSA L. Rose. Deciduous or sometimes evergreen shrubs, with upright, climbing or trailing stems, usually prickly, rarely unarmed; leaves alternate, odd-pinnate, rarely simple, stipulate: flowers solitary or corymbose, at the end of short bracts; sepals and petals five, rarely four; stamens numerous; pistils numerous, enclosed in a usually urn-shaped receptacle (hip) which becomes fleshy and berrylike at maturity, enclosing several or many bony achenes. (Rosa, its ancient Latin name.) One hundred to 200 species throughout the temperate and subtropical regions of the northern hemisphere; in North America to north Mexico, in Africa to Abyssinia, in Asia to the Himalayas and the Philippine Islands.

The species of this genus are very variable and hybridize easily and the conception of the species varies greatly according to the views of different botanists. While Bentham & Hooker recognized only about 30 species, Gandoger enumerates from Europe and West Asia alone 4266 species. There is no recent monograph; Crépin's *Primitiae Monographiae Rosarum* (1869–82) is the most important general treatment; for American roses see Rydberg (1918) in *North American Florist* 22: 483–533; for Asiatic roses see Boulenger (1933–36) in *Bulletin du Jardin Botanique de Bruxelles* 9: 203–279; 13: 167–266; and 14: 115–221; for European roses see Boulenger, *Roses d'Europe*, 2 vols. (1924–31) and Keller in Ascherson & Graebner, *Synthese Mitteleuropaische Flora*, vol. 6, pt. 1, 32–384 (1900–2). For cultivated roses see Ellen Willmott's *The Genus* Rosa (1910–14). Within the limited space of this book it has not been possible to describe all the spontaneous species which could be grown within our area nor to delve deeply into the numerous horticultural forms.

The rose has been one of the most favorite flowers, if not the most favorite of all ages, at least in the occidental world. The economic products of roses are few, attar of rose being perhaps the most important; the hips of some large-fruited species are sometimes used for culinary purposes.

Leaves simple, without stipules: flowers yellow.......... Subgenus I. *Hulthemia*
Leaves pinnate, stipulate.
A Receptacle smooth or bristly; urn-shaped; achenes on wall and bottom (or only on the bottom in Sect. 4).......... Subgenus II. *Eurosa*
  B Stipules adnate about 1/2, persistent.
    C Styles little or not exserted forming a head closing the mouth of the receptacle (more or less exserted in nos. 7, 8, 17, 37, 38, 47).
      D Flowers solitary, without bracts, rarely several-flowered.
        E Leaflets 5–9 or more, 1–4 cm long; flowers usually white or yellow.......... Sect. 1. *Pimpinellifoliae*
        EE Leaflets 3–5, 1–6 cm long: flowers usually pink or crimson.......... Sect. 2. *Gallicanae*
      DD Flowers corymbose, with bracts if solitary: leaflets usually 5–11.
        E Stems with usually stout hooked prickles, sometimes mixed with glandular bristles: outer sepals usually pinnate.......... Sect. 3. *Caninae*
        EE Stems at least at base with usually straight prickles and bristles: sepals usually entire.
          F Sepals after flowering spreading, deciduous: achenes inserted only at the bottom of receptacle.......... Sect. 4. *Carolinae*
          FF Sepals after flowering upright, usually persistent: achenes inserted on wall and bottom.......... Sect. 5. *Cinnamomeae*
    CC Styles exserted.
      D Styles united into a column, usually about as long as stamens.......... Sect. 6. *Synstylae*
      DD Styles free, about half as long as stamens: leaflets usually 3–5.......... Sect. 7. *Indicae*
  BB Stipules free or nearly so, caducous.
    C Bracts glabrous: leaflets 3–5.
      D Pedicels and receptacle smooth: flowers small, umbellate, yellow or white: stipules subulate.......... Sect. 8. *Banksianae*
      DD Pedicels and receptacle bristly; flowers large, solitary, white: stipules denticulate.......... Sect. 9. *Laevigatae*
    CC Bracts tomentose or pubescent: leaflets 7–9; stipules pectinate: flowers one or few with large bracts at base.......... Sect. 10. *Bracteatae*
AA Receptacle prickly, cup-shaped; achenes only at bottom; leaflets not exceeding 2.5 cm.
  B Leaflets 7–15 on flowering bracts: achenes on a slightly raised torus.......... Subgenus III. *Platyrhodon*
  BB Leaflets 3–7 on flowering bracts: flowers without disk: achenes on a conical torus.......... Subgenus IV. *Hesperhodos*

Subgenus I. *Hulthemia* — 1. *R. persica*

Subgenus II. *Eurosa*

Sect. 1. *Pimpinellifoliae*

Flowering bracts usually prickly and bristly, rarely unarmed; flowers white to pink, sometimes yellow:
fruit black .... 2. *R. spinosissima*
Flowering bracts only prickly: fruit red.
A Flowers with 5 petals, yellow, sometimes orange or red.
B Prickles straight.
C Leaflets simply serrate, eglandular.
D Shoots at least at base bristly: leaflets glabrous, rather sharply serrate .... 3. *R. hugonis*
DD Shoots without bristles: leaflets usually villous when young, crenate–serrate .... 4. *R. xanthina*
CC Leaflets doubly glandular–serrate, glandular beneath.
D Leaflets glabrous and glandular beneath: sepals not leafy at tip .... 5. *R. primula*
DD Leaflets pubescent and glandular beneath: sepals leafy at tip .... 6. *R. foetida*
BB Prickles hooked: leaflets simply serrate, bluish green .... 7. *R. hemisphaerica*
AA Flowers with 4 petals, white .... 8. *R. omeiensis*

Sect. 2. *Gallicanae*

Prickles very unequal: leaflets usually doubly glandular–serrate.
A Leaflets firm, leathery: pedicels upright .... 9. *R. gallica*
AA Leaflets thin, sometimes simply serrate: pedicels nodding .... 10. *R. centifolia*

Prickles uniform: leaflets simply, serrate, not glandular.
A Receptacle glandular–hispid: leaflets ovate–oblong, often pubescent beneath .... 11. *R. damascena*
AA Receptacles usually glabrous: leaflets broad–ovate or broad–elliptic, pubescent beneath .... 12. *R. alba*

Sect. 3. *Caninae*

Leaflets more or less glandular beneath or often doubly serrate with gland-tipped teeth.
A Prickles slender, straight or slightly curved: leaflets tomentose.
B Sepals upright, persistent: fruit large, about 2 cm across: auricles incurved, falcate .... 13. *R. pomifera*
BB Sepals finally deciduous: fruit smaller: auricles short, divergent .... 14. *R. tomentosa*
AA Prickles hooked: leaflets glabrous or pubescent beneath.
B Leaflets suborbicular or broad–oval, usually obtuse, glandular on both sides: styles pubescent: sepals long-persistent .... 15. *R. eglanteria*
BB Leaflets ovate or oval, acute or short–acuminate, eglandular above; styles glabrous or nearly so; sepals soon deciduous .... 16. *R. micrantha*

Leaflets not glandular except sometimes on midrib; teeth usually simple.
A Style exserted: sepals reflexed, deciduous .... 17. *R. stylosa*
AA Styles not or little exserted.
B Leaflets pubescent, at least beneath .... 18. *R. corymbifera*
BB Leaflets glabrous.
C Leaflets bright or dark green, lustrous above .... 19. *R. canina*
CC Leaflets bluish-green, tinged with red, dull above, simply serrate, 7–9: bracts bloomy, with few prickles .... 20. *R. rubrifolia*

Sect. 4. *Carolinae*

A Stipules more or less convolute, at least on the shoots; prickles hooked: leaflets serrulate .... 21. *R. palustris*
AA Stipules flat: leaflets rather coarsely serrate.
B Leaflets 5–7: prickles slender, straight .... 22. *R. carolina*
BB Leaflets 7–11.
C Infrastipular prickles usually curved: leaflets 7–9, usually elliptic .... 23. *R. virginiana*
CC Infrastipular prickles straight, slender: leaflets 7–11, elliptic–oblong to narrow–oblong.
D Bracts densely bristly: pedicels 1–3 cm long .... 24. *R. nitida*
DD Bracts not bristly: pedicels 5–10 mm long .... 25. *R. foliolosa*

Sect. 5. *Cinnamomeae*

Infrastipular prickles not present.

A Bracts and prickles tomentose: leaflets thick, rugose above .... 26. *R. rugosa*
AA Bracts and prickles glabrous.
B Bracts and stems bristly.
C Flowers solitary or few: leaflets 3–7, rarely 9 .... 27. *R. acicularis*
CC Flowers corymbose: leaflets 7–11: low plants.
D Leaflets glabrous, acute .... 28. *R. arkansana*
DD Leaflets pubescent at least beneath, usually obtuse .... 29. *R. suffulta*
BB Bracts and stems unarmed or the latter with weak bristles when young.
C Leaflets simply serrate, 5–7: fruit subglobose .... 30. *R. blanda*
CC Leaflets doubly serrate, 7–9: fruit usually oblong to ovoid .... 31. *R. pendulina*

Infrastipular prickles present.

A Stipules at least on shoots more or less convolute; prickles curved .... 32. *R. cinnamomea*
AA Stipules flat.
B Sepals persistent.
C Leaflets 3–7: bracts prickly, not bristly.
D Styles not exserted.
E Flowers usually several, less than 5 cm across: leaves usually simply serrate.
F Prickles straight or nearly so.
G Leaflets puberulous beneath, bright green; stipules broad .... 33. *R. pisocarpa*
GG Leaflets glabrous, bluish-green; stipules narrow .... 34. *R. woodsii*
FF Prickles curved: leaflets villous beneath .... 35. *R. californica*
EE flowers solitary, 5–6 cm across: leaflets doubly glandular–serrate .... 36. *R. nutkana*
DD Styles exserted; inflorescence subumbellate: leaflets doubly serrate .... 37. *R. corymbulosa*
CC Leaflets 7–15 (5–9 in *R. persetosa* under no. 40, and *R. forrestiana* under no. 46, and in no. 41).
D Flowers corymbose: leaflets 2–6 cm long.
E Sepals entire.
F Styles exserted: leaflets acute, 3–6 cm long .... 38. *R. davidi*
FF Styles not exserted.
G Petals glabrous outside; inflorescence exceeded by the leaves; leaflets simply serrate .... 39. *R. caudata*
GG Petals slightly pubescent outside; inflorescence not exceeded by the leaves; leaflets usually doubly serrate .... 40. *R. setipoda*
EE Sepals pinnate .... 41. *R. hemsleyana*
DD Flowers solitary, rarely several, on short lateral bracts; if corymbose, leaflets less than 2 cm long.
E Style not exserted.
F Leaflets pubescent beneath, at least on the midrib (see also *R. fedtschenkoana* under no. 45).
G Bracts with prickles and bristles, leaflets usually doubly serrate, 3–5 cm long. .... 42. *R. sweginzowii*
GG Bracts only with prickles: leaflets usually simply serrate, 1–3 cm long .... 43. *R. moyesii*
FF Leaflets glabrous, 8–20 mm long.
G Leaflets acutish, elliptic or ovate, 1.5–2.5 cm long .... 44. *R. bella*
GG Leaflets obtuse, orbicular to oblong, 6–20 mm long.
H Leaflets serrate above the middle: pedicels short .... 45. *R. webbiana*
HH Leaflets sharply serrate from below the middle: pedicels 1.5–3 cm long .... 46. *R. sertata*
EE Styles exserted, nearly as long as stamens: leaflets obtuse, 8–15 mm long .... 47. *R. multibracteata*
BB Sepals deciduous, falling off as a whole, together with their base: flowers solitary: leaflets 5–9, 8–25 mm long.
C Stems without bristles: leaflets simply serrate .... 48. *R. willmottiae*
CC Stems bristly and prickly.
D Prickles straight: leaflets glabrous, doubly serrate .... 49. *R. gymnocarpa*
DD Prickles hooked: leaflets pubescent beneath, simply serrate .... 50. *R. beggeriana*

Sect. 6. *Synstylae*

Stipules pectinate; prickles usually infrastipular; styles glabrous. 51. *R. multiflora*

Stipules entire or denticulate: prickles scattered.

A Habit upright, usually with arching branches.
B Leaflets 3–5 on flowering bracts, pubescent at least on veins beneath.
C Flowers small, about 1 cm across: leaflets linear–lanceolate .... 52. *R. watsoniana*
CC Flowers about 5 cm across: leaflets usually ovate–oblong .... 53. *R. setigera*
BB Leaflets 5–9 on flowering bracts.
C Leaflets acute or acuminate, 3–10 cm long; stems often sarmentose.

D Leaflets pubescent, at least beneath.
E Fruit ovoid, 1.5 cm long; corymb umbel-like: leaflets usually 7–9, pubescent chiefly on the veins .... 54. *R. helenae*
EE Fruit about 1 cm long: corymb not umbel-like: leaflets 5–7, pubescent beneath all over.
F Leaflets usually 5, glabrous above, sharply and coarsely serrate: bracts glabrous: fruit subglobose .... 55. *R. rùbus*
FF Leaflets usually 7, pubescent above: bracts often puberulous: fruit ovoid. .... 56. *R. brunonii*
DD Leaflets glabrous or nearly so.
E Leaflets 5–7, green beneath.
F Styles pubescent: flower-buds gradually acuminate .... 57. *R. moschata*
FF Styles glabrous; flower-buds abruptly acuminate .... 58. *R. maximowicziana*
EE Leaflets usually 5, glaucescent beneath .... 59. *R. henryi*
CC Leaflets obtuse or acutish, 1–2.5 cm long, glaucescent: upright shrub .... 60. *R. soulieàna*
AA Habit trailing or prostrate.
B Leaves evergreen or half-evergreen, lustrous on both sides: flowers usually several.
C Leaflets usually 9, obtuse, 8–20 mm long; stipules dentate .... 61. *R. wichuraiana*
CC Leaflets usually 5, acuminate, 2–5 cm long; stipules entire .... 62. *R. sempérvirens*
BB Leaves deciduous, glaucescent beneath; leaflets usually 7: flowers often solitary .... 63. *R. arvensis*

Sect. 7. *Indicae*

Stems with uniform prickles: leaves glabrous.
A leaves persistent or semipersistent.
B Flowers white, yellowish or light pink, very fragrant; sepals usually entire: fruit globose or depressed–globose; stipules without or with few marginal glands at the auricles .... 64. *R. odorata*
BB Flowers red or pink, rarely whitish, not or slightly fragrant; sepals usually pinnate: fruit ovoid or pyriform: stipules glandular–ciliate .... 65. *R. chinensis*
AA Leaves deciduous: flowers many in a corymb .... 66. *R. noisettiana*

Stems with prickles and bristles: leaves slightly pubescent beneath: flowers solitary or few .... 67. *R. borboniana*

Sect. 8. *Banksianae* — 68. *R. banksiae*

Sect. 9. *Laevigatae* — 69. *R. laevigata*

Sect. 10. *Bracteatae* — 70. *R. bracteata*

Subgenus III. *Platyrhodon* — 71. *R. roxburghii*

Subgenus IV. *Hesperhodos* — 72. *R. stellata*

Subgenus I. *Hulthemia* (Dumort.) Focke.

Leaves simple, estipulate: flowers solitary; receptacle urnshaped, constricted at the mouth by a disk, prickly. (*Hulthemia* Dumort., *Lowea* Lindl., Sect. *Simplicifoliae* Lindl.)

1. R. persica *Michx.*

Low shrub, to 0.5 m, with slender yellowish-brown prickly branches: leaves short-petioled, elliptic to oblong, acute at ends, 1.5–3 cm long, serrate, bluish-green, finely pubescent: flowers solitary, yellow with purple eye, about 2.5 cm across: fruit globose, prickly. Fl. V–VIII. B.M.7096(c). B.R.1261(c). W.R.1(c). G.C.III.6:8,9,78. (*R. simplicifolia* Salisb., *R. berberifolia* Pall., *Lowea b.* Lindl.) Persia, Afghanistan to Song. Introduced 1790. Zone VII? A very distinct and remarkable species, but difficult to grow.

Hybrid

R. persica × *clinophylla* = *R. hardii* Paxt. Leaflets 1–7, oblong–obovate, glabrous, stipulate: flowers 5 cm across, pale yellow, with orange eye: calyx tube pubescent, with few prickles. W.R.7,t(c). P.M.10:t.195(c). G.C.II.24:469. Gn.19:473. Origin before 1836.

Subgenus II. *Eurosa* Focke.

Leaves pinnate, stipulate: flowers usually corymbose, sometimes solitary; receptacle urn-shaped, constricted at apex by a disk; achenes basal and parietal: fruit smooth, glandular or setose.

Sect. 1. Pimpinellifoliae *DC.*

Usually low shrubs; stems usually with straight prickles and bristles: leaflets small, 7–9; stipules narrow, with divergent and dilated auricles: flowers solitary, without bracts; sepals entire, erect and persistent. Europe and Asia. (Includ. Sect. *Luteae* Crép. and *Sericeae* Crép.)

2. R. spinosissima *L.*

Burnett Rose (Scotch Rose) Low shrub, rarely exceeding 1 m, with spreading or recurving branches; bracts usually densely prickly and bristly: leaflets 5–11, usually 7–9, orbicular to oblong–ovate, 1–2 cm long, simply serrate, or doubly glandular–serrate, glabrous, sometimes glandular beneath; stipules entire, rarely glandular–dentate: flowers solitary, but usually very numerous on short bracts along the stems, pink, white or yellow, 2–5 cm across; pedicels usually glandular-hispid: fruit subglobose, black or dark brown, 1–1.5 cm across. Fl. V–VI; fr. IX. W.R.247,t(c). B.R.431(c). F.D.25:t. 2623(c). Gn.60:24;74:598;86:306. G.C.70:218. (*R. pimpinellifolia* L., *R. illinoensis* Bak.) Europe, West Asia; occasionally naturalized in North America. Cultivated before 1600. Zone IV.

Varieties and forms
A very variable species.

R. s. pimpinellifolia (L.) Hook., var. Pedicels smooth.

R. s. altaica (Willd.) Rehd., var. More vigorous, less bristly: flowers large, white; pedicels glabrous. W.R.257,t(c). B.R.888(c). Gn.53:t.170(c). Gng.5: 307(h). W.A.104,t(h). (*R. altaica* Willd., *R. grandiflora* Lindl.)

R. s. myriacantha (DC.) Koehne, var. Bracts very prickly: leaflets doubly glandular–serrate, very small: flowers small, white, blushed. W.R. 261,t(c). R.R.1:gr.6,7.

R. s. inermis (DC.) Rehd., var. Bracts almost unarmed: flowers pink. R.R.1:gr.6,6. (*R. spinosissima* var. *mitissima* Koehne, *R. pimpinellifolia* var. *inermis* DC.)

R. s. andrewsii Willm., var. With double red flowers W.R.263, t(c).

R. s. nana Andrews. With double white flowers Andrews, Roses, 122(c).

R. s. hispida (Sims) Koehne, var. Upright shrub to 2 m: leaflets usually simply serrate; pedicels smooth: flowers sulfur-yellow, 6–7 cm across. B.M.1570(c). W.R.259,t(c). Gn.56,t.398(c);62:17. J.L.27:508. (*R. hispida* Sims, *R. lutescens* Pursh.)

R. s. luteola Andr. Similar to the preceding variety: leaflets 7: flowers pale yellow, about 5 cm across. W.R.255,t(c). G.28:281. (*R. ochroleuca* Swartz.)

R. s. lutea Bean, var. About 1 m tall: leaflets broad–oval, to 2.5 cm long, pubescent beneath: flowers bright yellow. Origin unknown, possibly a hybrid.

Hybrids
R. spinosissima × *canina* = *R. hibernica* Smith. Shrub to 3 m, with upright or arching stems, bristly and prickly: leaflets 5–9, bluish-green, pubescent beneath: flowers usually 3, pink, about 3 cm across; sepals lobed: fruit globose, red. W.R.289,t(c). Introduced 1802.

R. spinosissima × *pendulina* = *R. reversa* Waldst. & Kit. Shrub to 1.5 m; stems upright, with slender prickles and bristles: leaflets 5–9, 6–20 mm long, dark green: flowers solitary, red, 4–5 cm across; pedicels glandular–bristly: fruit ovoid, 2 cm long, pendulous, red. F.D.25:t.2624(c). (*R. rubella* Sm.) Europe. Cultivated 1820. Possibly of similar parentage is *R. malyi* Kern. Shrub to 2 m: leaflets oval or suborbicular, 1.5–3 cm long, usually doubly serrate, glabrous: flowers usually solitary, deep red, 3–4 cm across: fruit oblong–ovoid. W.R.297,t(c). (*R. pendulina* var. *m.* Kern.) Dalmatia. Cultivated 1902.

R. spinosissima × ? *tomentosa* = *R. involuta* Smith. Shrub to 3 m; stems with prickles and bristles: leaflets 5–9, elliptic, simply or doubly glandular–serrate, pubescent beneath: flowers pink or whitish, solitary: fruit globose, red. W.R.281,t(c). Origin in Scotland.

R. spinosissima × *carolina* = *R. kochiana* Koehne. Bracts densely bristly: leaflets 9–11, small: flowers 1–3, deep rose. (*R. oxyacantha* K. Koch, not Bieb.) Origin before 1870.

R. spinosissima × *foetida* see under no. 6.

Related species
R. koreana Komar. Dense shrub to 1 m: bracts densely bristly, dark red: leaflets 7–11, elliptic or obovate–elliptic, 1–2 cm long, obtusish, sharply glandular–serrate, glabrous or slightly pubescent beneath: flowers 2.5–3 cm across, white, flushed pink; pedicels stipitate–glandular; fruit ovoid, 1–1.5 cm long, orange–red. N.K.7:t.14. Act. Hort. Petrop. 22:t.11. Korea. Introduced 1917.

3. R. hugonis *Hemsl.*

Shrub to 2.5 m, with arching branches; prickles straight, compressed, mixed on the shoots with bristles: leaflets 5–13, oval to obovate or elliptic, 8–20 mm long, obtuse, sometimes acutish, finely serrate, glabrous or slightly villous on the veins when unfolding: flowers solitary on short bracts, light yellow, 5 cm across; pedicels 1.5–2 cm long, like the receptacle glabrous: fruit depressed–globose, about 1.5 cm across, dark scarlet to blackish red. Fl. V–VI; fr. VII–VIII. B.M.8004(c). W.R.279,t(c). B.S.2:429. Gn.87:589. S.L. 324(h). Gn.M.31:373(h);38:37(h). Central China. Introduced 1899. Zone V. Free-flowering and early rose, one of the best single yellows.

4. R. xanthina *Lindl.*

Shrub to 3 m; bracts brown with stout and straight prickles, shoots without bristles: leaflets 7–13, broad–oval or nearly suborbicular, rarely elliptic, 8–20 mm long, obtuse, serrate, glabrous above, slightly villous beneath, at least when young: flowers solitary, yellow, double or semidouble, 4 cm across. Fl. V–VI; fr. VII–VIII. N.K.7:t.6. (*R. xanthinoides* Nakai.) Cultivated in North China and Korea. Introduced 1906. Zone V.

Varieties and forms

R. x. spontanea Rehd. f. Flowers single. The wild form. Franchet, Pl. David 1:t.15. North China and Korea. Introduced 1906. Zone V.

Related species

R. graciliflora Rehd. & Wils. Shrub to 4 m; branches with few prickles: leaflets 7–9, oval or elliptic, 1–2 cm long, rounded or acutish at apex, doubly glandular–serrate, glabrous beneath or pubescent on midrib; stipules broad, glandular–ciliate: flowers pale pink, 3–4 cm across; receptacle ovoid; sepals enlarged at apex and sometimes toothed: fruit red. West China. Introduced 1932. Zone VI?

R. farreri Stapf. Stem densely bristly below: leaflets 7–9, ovate to elliptic, 5–18 mm long, acutish, glabrous; stipules narrow, ending in a short tooth: flowers pale pink to white, 2–2.5 cm across; pedicels 1.2–2.5 cm long, glabrous; sepals scarcely leafy at tip: fruit ovoid–oblong, coral-red. Northwest China. Introduced 1915. Zone V.

R. farreri persetosa Stapf, f. Stems densely bristly throughout: flowers salmon-pink. B.M.8877(c). G.C.94:239.

5. R. primula *Boulenger*

Upright shrub to 2 m; bracts slender with numerous stout wide-based straight prickles: leaflets 9–15, rarely 7, elliptic or elliptic–obvate to elliptic–oblong, 6–20 mm long, acute or obtuse, minutely and doubly glandular-serrulate, glabrous, glandular beneath: flowers solitary, 3–4 cm across, pale yellow or yellowish white; pedicels short, smooth: fruit globose or obovoid, red like its stalk, about 1 cm long, with reflexed sepals. Fl. V–VI; fr. VII, soon dropping. N.F.8:f.83;11:f.95(h). (*R. xanthina* auth., not Lindl., *R. ecae* Rehd., not Aitch.) Turkestan to North China. Introduced 1910. Zone V. Valued for its early flowers and aromatic young foliage.

Closely related species

R. ecae Aitch. Shrub to 1 m, densely prickly: leaflets 5–9, mostly obovate, 4–8 mm long, rounded or truncate at apex: flowers 1.5–2 cm across, yellow, short-stalked, often without leaves at base. H.I.1329. B.M.7666(c). Afghanistan. Introduced 1880. Zone VI?

6. R. foetida *Herrm.*

Austrian Brier. Shrub to 3 m; stems slender, often sarmentose, brown, with straight prickles: leaflets 5–9, broad–oval or obovate, 1.5–4 cm long, usually rounded at base, doubly glandular–serrate, dark green above, glandular and slightly pubescent beneath; stipules glandular–serrate: flowers solitary or sometimes several, deep yellow, 5–7 cm across, of unpleasant odor; pedicels and receptacle smooth; sepals leafy at tip: fruit globose, red. Fl. VI. B.M.363(c). W.R.267,t(c). F.D.25:t.2622(c). Gn.53:22,t(c). (*R. lutea* Mill., *R. eglanteria* Mill., not L.) West Asia. Introduced before 1600. Zone IV.

Varieties and forms

R. foetida bicolor (Jacq.) Willm., var. Austrian Copper Brier. Flowers orange–scarlet or coppery-red inside. B.M.1077(c). W.R.269,t(c). Gn.53:23; 55:425. (*R. lutea* prol. *punicea* (Mill.) Aschers. & Graebn., *R. punicea* Mill., *R. b.* Jacq.) Introduced before 1590.

R. foetida persiana (Lem.) Rehd., var. Persian Yellow. With double flowers F.S.4:t.374(c). More floriferous than the double Sulfur Rose and flowers smaller, but more double. Introduced 1838.

Hybrids

R. foetida × *spinosissima* = *R. harisonii* Rivers, Harison's Yellow Rose. Lower, with shorter branches: flowers paler and less double than Persian Yellow; pedicels and receptacle bristly: fruit nearly black. S.B.II.4:410(c). F.E.18:6. H.B.5:t.99(c). (*R. foetida* var. *H.* Rehd., *R. lutea* var. *hoggii* Sweet, *R. vorbergii* Graebn.) Origin about 1830. Hardier than Persian Yellow.

R. foetida × *eglanteria* = *R. penzanceana* Rehd. Lady Penzance Rose. Low shrub with arching branches and hooked prickles: leaflets 7, fragrant: flowers pink with yellow eye. Origin about 1890.

7. R. hemisphaerica *Herrm.*

Sulfur Rose. Shrub to 3 or 4 m; stems slender, often sarmentose, with hooked prickles: leaflets 5–9, obovate, 1–3.5 cm long, rounded at apex, cuneate, coarsely toothed, bluish green, glaucescent beneath and pubescent; rachis pubescent, sparingly glandular: flowers double, scentless, light yellow, 5 cm across; pedicels glabrous or glandular–hispid; sepals with leafy toothed apex. Fl. VI. W.R.273,t(c). B.R.46(c). (*R. sulphurea* Ait., *R. glaucophylla* Ehrh.) West Asia. Introduced before 1625. Zone VI? The wild form (*R. rapinii* Boiss.) with single yellow flowers and globose fruit is not in cultivation.

8. R. omeiensis *Rolfe*

Shrub to 3 or 4 m; stems upright, with flattened wide-based prickles, young shoots often bristly: leaflets 9–17, oblong or elliptic–oblong, 8–30 mm long, acutish, cuneate, serrate, glabrous or puberulous on midrib beneath; petioles puberulous and prickly: flowers white, 2.5–3.5 cm across; pedicels and receptacle smooth: fruit pear-shaped, 8–15 mm long, bright red, on a thickened yellow stalk of about equal length. Fl. V–VI; fr. VII. B.M. 8471(c). I.S.2:t.81. B.C.5:2996. West China. Introduced 1901. Zone VI. Distinct shrub with finely divided fernlike foliage; coloring of fruit unique; the stems of f. *pteracantha* conspicuous in winter.

Varieties and forms

R. omeiensis chrysocarpa Rehd., f. Fruit yellow.

R. omeiensis pteracantha (Franch.) Rehd. & Wils., f. Prickles much enlarged at base, decurrent and often confluent, forming wide wings, red and translucent when young. B.M.8218(c). Gn.69:300,t(c). G.C.38:260. J.L.27:491. V.F.99(h). (*R. sericea* var. *p.* Fr.) Introduced 1890.

Closely related species

R. sericea Lindl. Leaflets 7–11, oval or obovate, 1–2 cm long, serrate only near apex, silky-pubescent beneath: fruit red, on a slender, not thickened stalk. B.M.5200(c). W.R.163,t(c). R.H.1897:444,445. Himalayas. Introduced 1822. Zone VII?

Sect. 2.* Gallicanae *DC.

Upright usually low shrubs: stems usually with hooked prickles mixed with bristles; stipules narrow, adnate: leaflets 3–5, usually firm: flowers few, often with narrow bracts or solitary on a usually bractless pedicel; sepals often lobed, reflexed after flowering, deciduous; styles not exserted. Three or possibly one species in Europe and West Asia.

9. R. gallica *L.*

French Rose. Upright shrub, with creeping rootstock, rarely 1.5 m tall; stems usually densely covered with prickles and bristles: leaflets 3–5, broad-elliptic or ovate, 2–6 cm long, obtusish or abruptly acuminate, rounded or subcordate at base, doubly or sometimes simply glandular–serrate, dark green and rugose above, paler and pubescent beneath, firm, deflexed; petiole and rachis bristly and glandular–hispid; stipules glandular–ciliate: flowers solitary, pink or crimson, 4–7 cm across, on rather long stout pedicels glandular–hispid and bristly; receptacle glandular–hispid; sepals glandular outside: fruit subglobose or turbinate, brick-red. Fl. VI. W.R.325,t(c). F.D.25:t.2634(c). B.M.1377(c). G.H.1:t.89(c). Central and South Europe, West Asia; occasionally naturalized in North America. Long cultivated. Zone V.

Varieties and forms

R. gallica pumila (Jacq.) Reg., var. Dwarf form, with creeping rootstock: flowers red, single. R.R.2:gr. 16,12(c).

R. gallica officinalis Thory. Apothecary Rose (Double French Rose.) With double flowers, otherwise like the typical form. W.R.359,t(c). (*R. gallica* var. *plena* Reg., *R. provincialis* Mill.)

R. gallica Agatha Thory. With rather small double pink flowers, the outer petals spreading, the inner ones concave. R.R.2: gr.17,17–21.

R. gallica versicolor L., var. Flowers semidouble; petals striped, white and red. R.R.2:gr.16,12(c). (*R. gallica* var. *rosa-mundi* West.) It has been confused with a similar form of *R. damascena*.

Hybrids

R. gallica × *canina* = *R. Waitziana* Tratt. With scattered, usually slightly curved and unequal prickles: leaflets rather firm, usually simply serrate, nearly glabrous or pubescent on the midrib beneath; pedicels stipitate–glandular; sepals glandular on back: flowers large, usually deep rose-pink, 6–8 cm across, often solitary. Europe. Cultivated 1874.

R. waitziana macrantha (Desportes) Rehd., var. Prickles nearly equal: leaflets nearly glabrous beneath: flowers several; pedicels slightly glandular; flowers flushed rose at first, becoming nearly white. Gn.52: 465.t.f.1(c). R.H.1901:541.t.f.2(c). W.R.2:403,t(c). Introduced about 1825.

R. gallica × *tomentosa* = *R. marcyana* Boullu. To 1 m tall; stems prickly and bristly: leaflets 5–7, broad–oval, 2–5 cm long, pubescent on both sides: flowers pink to light purple, 6–7 cm across, long-stalked. W.R.335,t(c). (? *R. therebinthinacea* Bess.) Spontaneous in Southern France.

R. gallica × *cinnamomea* = *R. francofurtana* Muenchh. To 2 m; stems with straight or hooked prickles, flowering bracts nearly unarmed: leaflets 5–7, elliptic, serrate. pubescent beneath; upper stipules much dilated: flowers 5–7 cm across, single or double, purple, slightly fragrant, receptacle at base and pedicels glandular–hispid; sepals erect after flowering, entire or nearly so: fruit turbinate. Fl. VI. F.D.25:t.2628(c). R.R.3:gr.23.1(c). (*R. turbinata* Ait.) Origin before 1770.

R. gallica × *moschata*: see under no. 57.

R. gallica × *arvensis*: see under no. 63.

R. gallica × *corymbifera:* see no. 12.

10. R. centifolia *L.*

Cabbage Rose. Closely related to the preceding; taller, to 2 m; rootstock less creeping; prickles more unequal, partly stouter: leaflets usually 5, pubescent on both sides or only beneath, often simple serrate, larger and thinner; rachis not prickly: flowers usually pink, very double, nodding, on long and slender, very glandular pedicels, fragrant; petals inflexed; sepals persistent: fruit ellipsoid to subglobose. Fl. VI–VII. W.R.341,t(c). R.R.2:gr.14, 1–12. G.1:340. (*R. gallica* var. *centifolia* (L.) Reg.) Eastern Caucasus. Zone V. Only the double form in cultivation; known to the ancients.

Varieties and forms

R. centifolia muscosa (Mill.) Ser., var. Moss Rose. Flowers rose, with pedicels and calyx glandular–mossy. B.M.69(c). Gn.18:84(c). G.W.7:125.

R. centifolia albo-muscosa Willm. Like the preceding, but flowers white. B.R.102(c). W.R.349,t(c).

R. centifolia cristata Prevost, var. Similar to var. *muscosa*, but the mossy excrescences confined to the edges of the sepals. B.M.3475(c). W.R.351,t(c). G.W.7:125.

R. centifolia pomponia (DC.) Lindl., var. Pompon Rose. Dwarf: leaflets elliptic, about 2.5 cm long, glabrous above: flowers bright red, double, 3–3.5 cm across; pedicels densely setose. W.R.353,t(c). (*R. pulchella* Willd., *R. dijonensis* Roess.)

R. centifolia parvifolia (Ehrh.) Rehd., var. Burgundian Rose. Similar to the preceding: leaflets 1–2 cm long: flowers about 2.5 cm across; pedicels slightly setose. B.R.452(c). W.R.355,t(c). (*R. pulchella* Ehrh., *R. burgundensis* West., *R. burgundiaca* Roess., *R. gallica* var. *pulchella* (Ehrh.) Ser.)

Related species

R. richardii Rehd. Low shrub with unequal scattered hooked prickles: leaflets 3–5, elliptic to ovate–oblong, rugose above, pubescent beneath: flowers few, pale rose, 5–7 cm across; sepals large, pinnately lobed, leafy at apex. W.R.337,t(c). (*R. sancta* Rich., not Andrews.) Ethiopia. Cultivated 1902. Zone VII?

11. R. damascena *Mill.*

Damask Rose. Shrub to 2 m; stems usually with numerous stout hooked prickles, sometimes mixed with glandular bristles: leaflets usually 5, sometimes 7, ovate to ovate–oblong, 2–6 cm long, simply serrate, glabrous above, more or less pubescent beneath; petioles prickly; stipules sometimes pectinate: flowers blush to red, double, in corymbs, on slender glandular–hispid and prickly pedicels; sepals reflexed during anthesis, deciduous, glandular–hispid on back like receptacle: fruit obovoid, 2.5 cm long, bristly, red, Fl. VI–VII. W.R.369,t(c). R.R.2:gr.16,6(c). (*R. belgica* Mill., *R. polyanthos* Roess., *R. gallica* var. *damascena* Voss.) Introduced from Asia Minor in the sixteenth century. Zone IV.

Varieties and forms

R. damascena versicolor West., var. York and Lancaster Rose. Flowers semidouble, white and striped and blotched pink, or some flowers white and some pink. Lawrance, Coll. Ros. t.10(c). (*R. damascena variegata* Thory.)

R. damascena trigintapetala Dieck, var. With semi-double red flowers Gt.38:129. G.C.III.7:45. Cultivated in Southern Europe for the manufacture of attar of roses.

Hybrid

R. damascena × *chinensis*: see no. 67.

12. × R. alba *L. (?* R. corymbifera × gallica*)*

Upright shrub to 2 m; stem with unequal hooked prickles, sometimes mixed with bristles: leaflets usually 5, broad–elliptic or ovate to oblong–ovate, 2–6 cm long, simply serrate, pubescent beneath; upper stipules dilated: flowers more or less double, 6–8 cm across, white or blush, fragrant, corymbose; pedicels glandular–hispid; receptacle usually smooth above; sepals glandular–bristly, deciduous: fruit oblong–ovoid, 2 cm long, scarlet. Fl. VI. Andrews, Ros. t.10(c). Origin unknown. Introduced 1597. Zone IV.

Varieties and forms

R. alba suaveolens Dieck, var. Cultivated in Southeast Europe for attar of roses.

R. alba incarnata (Mill.) West., var. Flowers double, white, tinged with pink. W.R.413.t(c). (*R. incarnata* Mill., *R. alba* var. *rubicunda* Roess.) The closely related *R. collina* Jacq. with rose-colored flowers and shorter sepals is probably of the same origin and referable to *R. alba*.

Sect. 3. Caninae

Stems upright or arching, usually with numerous hooked or rarely straight prickles; upper stipules dilated: corymbs usually many-flowered, with dilated bracts; outer sepals lobed, reflexed after anthesis and deciduous or erect and persistent.

13. R. pomifera *Herrm.*

Densely branched shrub to 2 m, with almost straight slender spines: leaflets usually 5–7, on shoots sometimes 9, elliptic to elliptic–oblong, acute or obtuse, 3–5 cm long, doubly glandular–serrate, grayish green and pubescent above, tomentose beneath and glandular, rarely glabrescent; auricles of stipules spreading or converging: flowers 1–3, pink, 3–5 cm across; pedicels bristly and stipitate–glandular: fruit usually subglobose, to 2.5 cm across, hispid, with persistent upright sepals. Fl. VI–VII. B.M.7241(c). W.R.435,t(c). G.H.t.88(c). F.D.25:t.2632(c). (*R. villosa* L., in part.) Europe, West Asia. Cultivated 1771. Zone V.

Varieties and forms

R. pomifera duplex (West.) Rehd., f. With semidouble flowers. W.R.436,t(c).

Closely related species

R. mollis Smith. Low shrub; bracts purplish, bloomy: leaflets silky–tomentose beneath, less glandular, smaller: pedicels sparingly glandular: fruit smaller, not hispid, sometimes stipitate–glandular; sepals short, nearly entire. W.R.417,t(c). (*R. villosa* var. *mollissima* Rau.) Europe, West Asia. Introduced 1818. Zone V.

R. sherardi Davies. Dense shrub to 2 m; bracts often zigzag; spines often curved: leaflets broad–oval to elliptic, pubescent above, tomentose beneath, bluish green: flowers deep pink, usually several, exceeded by the basal leaves; sepals appendaged: fruit ovoid or pyriform, 1.2–2 cm across, with persistent sepals. (*R. omissa* Desegl.) North and Central Europe. Cultivated 1993. Zone IV.

R. hawrana Kmet. Flowering bracts usually unarmed; leaflets elliptic, glabrous above, pubescent beneath: flowers pink, 5 cm across: fruit globose, densely bristly. W.R.431,t(c). Hungary. Cultivated 1914.

Other related species

R. orientalis Dup. Dwarf: prickles subulate; young bracts densely pubescent: leaflets 1–2 cm long, simply serrate, pubescent on both sides: flowers solitary, pink, short-stalked; pedicels villous, glandular and bristly: fruit ellipsoid, sepals upright. S.H.1:f.321k-n, 322g. Southeast Europe, West Asia. Introduced ?

R. heckeliana Tratt. Very similar, but prickles stouter, hooked, and flowers smaller. South Europe Introduced ?

R. elymaitica Boiss. & Hausskn. Shrub to 1 m, with zigzag pruinose bracts; prickles stout, curved: leaflets usually 5, 5–15 mm long, tomentose below: flowers pink, small; styles exserted, tomentose; pedicels and receptacle stipitate–glandular: fruit globose, about 1 cm across; sepals reflexed. S.H.1:f.322g′(1). North Persia. Introduced 1900. Zone VII.

14. R. tomentosa *Smith.*

Shrub to 2 m; bracts often zigzag; young bracts often bloomy; prickles stout, curved or straight: leaflets 5–7, elliptic or ovate, acute or short–acuminate, 2–4 cm long, doubly serrate, finely pubescent above, tomentose and glandular beneath; stipules with short triangular spreading auricles: flowers pale pink or nearly white, about 4 cm across; pedicels often glandular–hispid; sepals lobed: fruit subglobose, 1–2 cm across, stipitate–glandular; sepals usually deciduous before maturity. Fl. VI. W.R.421,t(c). S.E.3:t.467(c). F.D.25:t.2631(c). Gn.77:511. Europe. Cultivated 1820; occasionally escaped in the eastern states of the USA. Zone V.

Hybrids

R. tomentosa × *gallica*: see under no. 9.

R. tomentosa × *spinosissima*: see under no. 2.

15. R. eglanteria *L.*

Sweetbrier. (Eglantine.) Much branched shrub to 2 m; prickles strong, hooked, sometimes mixed with bristles: leaflets 5–7, orbicular to oval, 1–3 cm long, obtuse or acutish, dark green and glabrous above, pubescent beneath, glandular on both sides, fragrant: flowers 1–3, bright pink, 3–5 cm across, on short glandular–hispid pedicels; receptacle glandular–hispid: fruit subglobose or ovoid, orange to scarlet, with more or less spreading, tardily deciduous sepals. S.E.3:468(c). D.F.25:t.2630(c). N.D.7:t.7(c). (*R. rubiginosa* L.) Europe; often naturalized in North America. Zone IV. Planted for its aromatic foliage.

Varieties and forms

R. eglanteria duplex West., var. With double flowers. W.R.449,t(c).

Hybrid

R. eglanteria × *foetida*: see under no. 6.

Closely related species:

R. inodora Fries. Dense shrub to 2 m; leaflets elliptic to obovate–oblong, 1.5–3 cm long, cuneate, pubescent beneath: flowers pinkish or white; pedicels short, usually eglandular. S.H.1:f.323h-i. W.R.469,t(c). (*R. elliptica* Tausch, *R. caryophyllacea* Bess.) Europe. Cultivated 1875. Zone V.

R. sicula Tratt. Low shrub with slender, nearly straight, subequal prickles: leaflets 5–9, orbicular to elliptic, glabrous and glandular, 6–12 mm long: flowers usually solitary, 2.5–3 cm across: fruit small, globose. B.M.7761(c) and G.W.8:17(h), as *R. seraphinii*. South Europe, Africa. Introduced before 1894. Zone VII?

R. glutinosa Sibth. & Sm. Dwarf; prickles unequal, partly scattered, straight, to 1 cm long, and partly bristly, gland-tipped and very numerous: leaflets elliptic to obovate, 1–3 cm long, densely glandular: flowers small, pink: fruit globose, small. S.F.5:t.482(c). W.R.467,t(c). (? *R. tuschetica* Boiss.) Southeast Europe, West Asia. Introduced 1821.

R. glutinosa dalmatica (Kern.) Borb., var. Fruit ellipsoid, 2–2.8 cm long. B.M.8826(c) (*R. d.* Kern.) Cultivated 1882.

16. R. micrantha *Sm.*

Much-branched shrub, to 2 m, with arching branches armed with uniform curved prickles, without bristles: leaflets 5–7, broad–ovate, 1.5–3 cm long, short–acuminate, doubly glandular–serrate, glabrous or pubescent above, pubescent and glandular beneath: flowers 1–4, pink to white, about 3 cm across; styles glabrous, somewhat exserted; pedicels glandular–hispid: fruit ovoid or subglobose, with spreading or reflexed deciduous sepals. Fl. VI. W.R.461,t(c). S.E.3:t.469(c). D.F.25:t.2630,IIB(c). B.B.2:f.2316(as *R. rubiginosa*). Europe, Mediterranean; often escaped throughout North America. Zone V.

Related species

R. horrida Fisch. Dwarf, with unequal prickles: leaflets broad–ovate to elliptic, 1–1.5 cm long, glandular beneath and often above: flowers white, usually 3 cm across; pedicels short; styles short. W.R.477,t(c). S.H.1:f.$323_{11\text{-}0}$.324e. (*R. ferox* Bieb., not Lawrance.) Southeast Europe, Caucasus, Asia Minor. Introduced 1796.

R. agrestis Savi. Shrub to 3 m with elongated branches; prickles subequal: leaflets oblong to oblong–obovate, 1.5–5 cm long, usually cuneate, pubescent, or nearly glabrous: flowers whitish or pinkish; pedicels eglandular; style somewhat exserted. W.R.457,t(c). S.E.3:t,470(c). (*R. sepium* Thuill.) Europe, North Africa. Cultivated 1878.

R. serafinii Viv. Low shrub with copious, unequal, partly hooked prickles: leaflets orbicular-ovate, 5–12 mm long, glandular beneath, glabrous: flowers solitary, pink, on very short eglandular pedicels; sepals lobed, reflexed after anthesis, finally deciduous. W.R.475,t(c). (*R. seraphinii* Viv.) Mediterranean. Cultivated 1914.

17. R. stylosa *Desv.*

Shrub to 3 m, with arching branches; prickles stout, hooked: leaflets 5–7, elliptic to elliptic–oblong, 1.5–5 cm long, acute, serrate, lustrous above, pubescent beneath, rarely glabrous, eglandular; rachis pubescent, usually prickly: flowers 1–18, white or light pink, 3–5 cm across; pedicels glandular; styles usually shorter than stamens: fruit ovoid, outer sepals with many lobes. W.R.47,t(c). S.E.3:t.475(c). S.H.1:f.323s-s^2,-324i-k. (*R. systyla* Bast.) Europe. Cultivated 1838. Zone V.

18. R. corymbifera *Borkh.*

Upright shrub to 3 m, with spreading or arching branches; prickles stout, hooked: leaflets, 5–7, closely set, orbicular–oval to elliptic, 2.5–4 cm long, rounded or subcordate, usually simply serrate, pubescent on both sides or only below on the veins: flowers one to many, 4–5 cm across; pedicels usually long, glabrous or glandular; sepals soon reflexed, deciduous: fruit ovoid or subglobose, 1.5–2 cm long, orange–red, usually smooth. Fl. vi. W.R.397,t(c). S.E.3:t.491(c). Gs.3:134(h). (*R. dumetorum* Thuill.) Europe, West Asia, Africa. Cultivated 1838; rarely escaped. Zone V.

Hybrid

R. corymbifera × *gallica*: see no. 12.

Closely related species

R. coriifolia Fries. Lower: leaflets usually cuneate at base: flowers short-stalked; bracts large; sepals upright, long-persistent, W.R.391, t(c). S.E.3:t. 472(c). Europe, West Asia. Cultivated 1878.

R. corymbifera froebelii (Christ) Rehd., var. Of more vigorous habit: leaflets simply of doubly serrate, bluish green: flowers small, white. (*R. canina* var. *F.* Christ, *R. laxa* Froeb., not Retz.) Cultivated 1890.

19. R. canina *L.*

Dog Rose. Shrub to 3 m, with often arching branches; prickles stout, hooked: leaflets 5–7, oval or elliptic, 2–4 cm long, doubly or simply serrate, glabrous or slightly pubescent beneath; upper stipules dilated: flowers 1–3, light pink or white, 4–5 cm across; pedicels usually glabrous, 1–3 cm long; receptacle ellipsoid, usually smooth; sepals usually lobed, later reflexed and finally deciduous: fruit scarlet, ellipsoid, 1.5–2 cm long. Fl. vi. W.R.379,t(c). S.E.3:t.474(c). D.F.25:2629(c). N.D.7:t.11(c). B.B.2:f.2312. Europe; occasionally naturalized in North America. Zone III. Much used as stock for grafting.

Varieties and forms

R. canina exilis (Crép.) Keller, var. Low form: leaflets small, 6–10 mm long: flowers pink, 2.5 cm across. W.R.385,t(c). (*R. exilis* Crép.)

Hybrids

R. canina × *gallica*: see under no. 9.

R. canina × *spinosissima*: see under no. 2.

R. canina × *rubrifolia*: see under no. 20.

Related species

R. dumalis Bechst. Bracts often pruinose, very prickly: leaflets close set, ovate oblong–ovate, sharply and doubly serrate, bluish-green, glaucous beneath: flowers rose-red: pedicels short; sepals lobed, upright, long-persistent. S.H.1:f.324p,325a. (*R. glauca* Vill., not Pourr., *R. reuteri* God.) Europe, West Asia. Cultivated 1872.

R. britzensis Koehne. Prickles slender: leaflets 7–11, elliptic, 2.5–3.5 cm long, usually simply serrate: flowers 1–2, pale pink, changing to white. 3–4 cm across: fruit brown, with upright sepals. M.D.1910:94. Gs.4:102. S.H. 2:f.593a-e. Kurdistan. Introduced 1901.

R. montana Chaix. Prickles nearly straight: leaflets broad–ovate, cuneate, obtuse, often tinged bluish-violet: flowers pink; pedicels with purplish black stalked glands and bristles: fruit large, oblong–ellipsoid, with persistent lobed sepals. S.H.1:f.325c-c^1,326b. South Europe, North Africa. Cultivated 1872.

R. marginata Wallr. Prickles straight or nearly straight, rather slender: leaflets 3–7, ovate to elliptic, doubly glandular–serrate, nearly glabrous, usually glandular beneath: flowers pink, 5–7 cm across; pedicels long, glandular and bristly: fruit subglobose to ellipsoid, smooth; sepals deciduous. W.R.463,t(c). S.H.1:f.321e-f,322f. (*R. jundzillii* Bess.) Europe, West Asia. Cultivated 1870. Has some resemblance to *R. gallica*.

20. R. rubrifolia *Vill.*

Upright shrub to 2 m, with slender purplish bloomy bracts; prickles few, hooked or straight, rather small: leaflets 7–9, elliptic to ovate–lanceolate, 2–3.5 cm long, simply serrate, bluish-green and more or less tinged with purplish-red, glabrous: flowers few, deep red, 3–3.5 cm across; pedicels glabrous or with few bristles; sepals usually entire, longer than petals, deciduous; fruit subglobose, about 1.5 cm across, bright red, smooth. Fl. vi–vii. W.R.399,t(c). B.R.430(c). G.W.7:139(h). F.D.25:2627(c). (*R. ferruginea* Déségl., not Vill., *R. glauca* Pourr.) Mountains of central and southern Europe. Introduced 1814. Zone II. Distinct with its purplish leaves and purplish stems; flowers rather inconspicuous.

Hybrids

R. rubrifolia × *canina* = *R. pokornyana* Borbas. Stems with scattered prickles, purplish when young: leaflets usually 7, elliptic or oval, slightly doubly serrate, bluish-green, tinged purple when young: flowers 1–3, deep red, 3.5–4 cm across; sepals slender, with a few narrow appendages. Hungary. Cultivated 1916.

R. rubrifolia × *rugosa*: see under no. 26.

Sect. 4. Carolinae *Crép.*

Upright, mostly low shrubs; stems slender, with usually straight paired prickles and often with bristles; upper stipules usually narrow: corymbs usually few-flowered; sepals spreading after flowering, soon deciduous, the outer entire or with few erect lobes; pedicels and receptacle glandular–hispid, rarely smooth; achenes inserted only at the bottom of the usually depressed–globose receptacle.

21. R. palustris *Marsh.*

Upright shrub to 2 m, with slender glabrous sometimes reddish stems: leaflets 7, rarely 9, elliptic to narrow–oblong, acute at ends, 2–5 cm long, finely serrate, usually pubescent beneath at least on the veins; stipules narrow, involute: flowers usually corymbose, pink, about 5 cm across; bracts leafy; pedicels short, usually glandular–hispid: fruit about 8 mm high. Fl. VI–VIII. W.R.211,t(c). Em.2:488,t(c). M.N.1:t.43(c). B.B.2:285. (*R. carolina* L. 1762 partly, not 1753, *R. pennsylvanica* Michx.) Northern States, Minnesota, Mississippi, and Florida. Introduced 1726. Zone IV.

Hybrids

R. palustris × *virginiana* = *R. mariae-graebnerae* Aschers. Leaflets more shining, broader, more coarsely toothed. Fl. VI–IX. Gt.51:t.1504,t.4(c). Origin about 1880. Also found wild with the parents.

R. palustris × *rugosa*: see under no. 26.

22. R. carolina *L.*

Shrub to 1 m, suckering; stems with slender straight prickles, usually very bristly when young; bracts sometimes unarmed: leaflets usually 5, rarely 7, elliptic to lance–elliptic, rarely oblanceolate, 1–3 cm long, acute or obtuse, sharply serrate with ascending teeth, glabrous, not or slightly lustrous above, paler beneath, pubescent on veins or nearly glabrous; stipules narrow: flowers usually solitary, rose, about 5 cm across; pedicels and receptacle glandular–hispid; outer sepals lobed: fruit about 8 mm across. Fl. VI–VII. W.R.201,t(c). M.N.2:t.9(c, as *R. lucida*). B.B.2:f.2314. (*R. humilis* Marsh., *R. parviflora* Ehrh., *R. virginiana* var. *humilis* Schneid.) Maine to Wisconsin, Kansas, Texas, and Florida. Introduced 1826. Zone IV. Handsome rose well suited for borders of shrubberies.

Varieties and forms

R. carolina grandiflora (Bak.) Rehd., var. Leaflets usually 7, obovate or oval; flowers about 6 cm across. W.R.207,t(c). Lindl., Ros. Mon.t.3(c). (*R. obovata* Raf., *R. humilis* var. *g.* Bak., *R. virginiana* var. *g.* Bean.) Maine to Michigan, Delaware and Montana. *R. carolina glandulosa* (Crép.) Rehd., var. Leaflets doubly glandular–serrate and rachis glandular–hispid or stipitate–glandular. Ontario and Massachusetts to Florida and northern Mexico. K.S.174 (as *R. humilis*). (*R. serrulata* Raf., *R. parviflora* var. *g.* Crép.) Cultivated 1902.

R. carolina villosa (Best) Rehd., var. Leaflets 5–7, pubescent beneath: flowers 1–4. House, Wild flower N. Y. t.102(c). (*R. humilis* var. *villosa* Best, *R. lyonii* Pursh.) Massachusetts to Minnesota, south to Georgia and Kansas. Cultivated 1887.

Varieties and forms

R. carolina alba (Rehd.) Rehd., f. A form of the preceding with white flowers. W.R.198,t(c). A.F.12:1098. Gg.5:306. Introduced about 1870.

Hybrid

R. carolina × *spinosissima*: see under no. 2.

23. R. virginiana *Mill.*

Shrub to 2 m, with few or no suckers; prickles often hooked, only young shoots bristly: leaflets 7–9, elliptic to obovate–elliptic, 2–6 cm long, usually acute at ends, serrate with ascending teeth, dark green and lustrous above, glabrous or pubescent on veins beneath; upper stipules dilated: flowers usually few, sometimes solitary, 5–6 cm across; pedicels and receptacle glandular–hispid: fruit 1–1.5 cm across. Fl. VI–VII. W.R.197,t(c),199, t.308,t(c). N.D.7:t.7, f.2(c). Gt.56:t.1564,f.2(c). Gn.55:428;71:493. B.B.2:285. K.S.175. (*R. lucida* Ehrh., *R. humilis* var. *lucida*, (Ehrh.) Best, *R. blanda* var. *willmottiana* Bak.) Newfoundland to Virginia, Alabama, and Montana. Introduced before 1807. Zone III. Handsome rose with lustrous leaves and bright pink flowers.

Varieties and forms

R. virginiana plena Rehd., var. Flowers double. Andrews, Ros. t.102(c). R.R.2:t,opp.p.7.(c). (*R. rapa* Bosc.) Introduced before 1820.

R. virginiana lamprophylla Rehd., var. Of very dense habit, usually lower than 1 m; leaves of shoots usually 9-foliolate; leaflets elliptic to obovate, acute or sometimes obtuse, cuneate, very lustrous above; petioles usually reddish. Maine to Connecticut. Introduced 1881.

Hybrid

R. virginiana palustris = *R. mariae-graebnerae*: see under no. 21.

24. R. nitida *Willd.*

Shrub to 0.5 m; stems densely bristly and with slender prickles 3–5 mm long: leaflets 7–9, close set, elliptic to oblong, 1–3 cm long, finely serrate, lustrous above, slightly pubescent or glabrous beneath; stipules, particularly the upper ones, dilated: flowers solitary or few, rose, 4–5 cm across; pedicels and receptacle glandular–hispid; sepals usually entire: fruit about 1 cm across. Fl. VI–VII. W.R.215,t(c), 217,t. B.B.2:285. Newfoundland to Connecticut. Cultivated 1807. Zone III. Handsome dwarf rose.

Hybrid

R. nitida × *californica* = *R. scharnkeana* Graebn. Low shrub; stem, with slender prickles, often bristly: leaflets 7–9, oblong, cuneate: flowers 1–5, rose–purple. Gt.51:t.1504,f,2(c). Origin before 1900.

Related species

R. rudiuscula Greene. Stems to 1 m, densely bristly: leaflets 5–9, subcoriaceous, elliptic, acute at ends, densely pubescent and pale beneath: flowers few; pedicels usually glandular–hispid. Iowa and Montana to Oklahoma. Introduced 1917. Zone V.

25. R. foliolosa *Nutt.*

Low shrub to 0.5 m; stem with short, straight, slightly reflexed prickles or nearly unarmed, rarely bristly; leaflets 7–9, rarely 11, narrow- to linear-oblong, 1–3 cm long, acute, finely serrate, glabrous and lustrous above, glabrous or pubescent on the veins beneath; stipules usually narrow: flowers solitary or few, rose, sometimes white, about 4 cm across; pedicels and receptacle sparingly glandular–hispid or the latter sometimes smooth: fruit subglobose, about 8 mm across. Fl. VI–VIII. B.M.8513(c). W.R.219,t(c). G.F.3:101. B.C.5:f.3445. Arkansas and Oklahoma to Texas. Cultivated 1880. Zone V.

Varieties and forms

R. foliolosa alba Bridwell, var. Flowers white. Cultivated 1919.

Sect. 5. Cinnamomeae DC.

Upright shrubs; prickles usually straight, in infrastipular pairs or scattered; stems often bristly: leaflets 5–11: inflorescence usually many-flowered with dilated bracts; sepals usually entire, upright after flowering and persistent, rarely deciduous; receptacle usually smooth.

26. R. rugosa *Thunb.*

Upright shrub to 2 m; stems stout, tomentose, densely bristly and prickly: leaflets 5–9, elliptic to elliptic–obovate, 2–5 cm long, acute or obtusish, serrate, lustrous and dark green, rugose and glabrous above, glaucescent, reticulate and pubescent beneath, thick and firm; petioles tomentose and bristly: flowers solitary or few, 6–8 cm across, purple to white; pedicels short, bristly; receptacle smooth: fruit depressed–globose, smooth, brick-red, 2–2.5 cm across. Fl. VI(–VIII). W.R.181,t(c). S.Z.1:t.28(c). B.R.420(c). Gt.30:t.1049;42:537. Gn.52:384,t(c). G.C.II.14:372. C.L.A.2:76;7:624. (*R. ferox* Ait., *R. regeliana* André & Lind.) North China, Korea, Japan; escaped in the northeastern states of the USA. Introduced about 1845. Zone II.

Forms of typical *R. rugosa* (*R. rugosa* var. *thunbergiana* C. A. Mey.) are: *R. rugosa typica* Reg., var. With purple flowers; the typical form (*R. rugosa rubra* Hort.). *R. rugosa rosea* Rehd., var. With rose-colored flowers. *R. rugosa alba* W. Robins. With white flowers Gn.9:452,t(c). G.8:261. (*R. r.* var. *albiflora* Koidz.). *R. rugosa alboplena* Rehd., var. With double white flowers.

Geographical varieties are: *R. rugosa chamissoniàna* C. A. Mey., var. Bristles almost absent from the branches: leaflets narrower and smaller, less rugose. (*R. pubescens* Bak., not Roxb.) A form with double purple flowers is *R. rugosa plena* Reg., var. Gt.24:t.846. (*R. rugosa f. rubroplena* Rehd.) *R. rugosa kamtchatica* (Vent.) Reg., var. With slenderer, less densely armed branches, thinner and less rugose leaflets and smaller flowers and fruit B.M.3149(c). B.R.419(c). N.K.7:t.8. (*R. kamtchatica* Vent.) Introduced about 1770.

R. rugosa is one of the most ornamental and at the same time most vigorous and hardy of the wild roses, much planted for its dark green lustrous foliage turning orange and scarlet in autumn, for its large flowers appearing during the whole summer and its large red fruits. Many hybrids have been raised.

Hybrids

R. rugosa × *arvensis* = *R. paulii* Rehd. Procumbent shrub with large white flowers in corymbs. Gn.86:286(h). (*R. rugosa repens alba* Paul.) Origin before 1903.

R. rugosa × *chinensis* = *R. calocarpa* Willm. With single rose-colored flowers, and handsome abundantly produced fruit W.R.189,t(c),199,t. R.H.1895:446,447. Gn.46:548;52:384. Origin before 1891.

R. rugosa × *odorata* (or *dilecta*) = *R. bruantii* Rehd. Here belongs 'Madame Georges Bruant' (B.C.5:2992) and 'Blanc Double de Coubert'. Origin before 1887.

R. rugosa × *borboniana* = *R. arnoldiana* Sarg. Large semidouble deep crimson flowers. Origin before 1914. The typical form is a cross between *R. rugosa* and 'Général Jacqueminot'. The same cross is represented by 'Mrs Anthony Waterer'. Origin before 1900.

R. rugosa × *palustris* = *R. spaethiana* Graebn. Flowers corymbose, large, purple. Gt.51:t.1504,f.3(c) Cultivated 1902.

R. rugosa × *blanda* = *R. warlesyensis* Willm. Leaflets 5–7, small: flowers pink. W.R.185,t(c). Origin before 1910.

R. rugosa × *carolina*. Procumbent shrub with purple flowers. W.R.203.t(c). (*R. humilis* × *rugosa* Koehne.) Origin before 1893.

R. rugosa × *roxburghii* = *R. micrugosa* Henkel. Upright, very prickly shrub: leaflets small, pubescent. GT.59:t.1581(c). R.H.1905:144. (*R. vilmorinii* Bean.)

R. rugosa × *multiflora*: see under no. 51.

R. rugosa × *wichuraiana*: see under no. 61.

R. rugosa × *rubrifolia* = *R. rubrosa* Preston. Bracts less densely spiny and pubescent: leaflets leathery, glabrous and glaucescent beneath; rachis glabrous: flowers large pink. G.C.80:329(h). Cultivated 1903; also raised in Ottawa 1920.

27. R. acicularis *Lindl.*

Low shrub, to 1 m; stems densely bristly with weak prickles; bracts sometimes unarmed: leaflets 3–7, broad–elliptic to oblong, 1.5–5 cm long, usually acute at ends, simply serrate, dull and glabrous above, pubescent beneath; stipules usually broad: flowers solitary, deep rose, 4–5 cm across, fragrant; pedicels glabrous, rarely slightly glandular–hispid: fruit usually pear-shaped with distinct neck, 1.5–2 cm long. Fl. V–VI. W.R.147,149,235,t(c, as *R. woodsii*). D.L.3:f.244. N.K.7:t.10. (*R. sayi* Schwein.) Alaska to Wyoming, Michigan and New York; Northeast Asia. Introduced 1805. Zone IV.

Varieties and forms

R. acicularis engelmannii (S. Wats.) Crép., var. Flowering bracts usually bristly; leaflets doubly glandular–serrate, glabrous and glandular beneath; fruit ellipsoid or pear-shaped, smaller. G.F.2:377. B.C.5:2995. D.L.3:f.245. (*R.E.*S. Wats.) North Dakota and Montana to Colorado. Introduced 1891.

R. acicularis bourgeauiana (Crép.) Crép., var. Leaflets broad–elliptic, usually obtuse, more or less doubly glandular–serrate, slightly pubescent and glandular beneath: fruit subglobose with very short neck. B.C.5:2994. B.B.2:f.2309 (as *R. acicularis*). (*R. B.* Crép.) Ontario to Colorado, Montana and Mackenzie.

R. acicularis nipponensis (Crép.) Koehne, var. Leaflets 7–9, 1–3 cm long, simply serrate, glabrous; petioles bristly: pedicels glandular–bristly: fruit, ovoid, 1.5–2 cm long. B.M.7646(c). W.R.151,t(c). (*R.n.* Crép.) East Asia. Introduced 1894.

28. R. arkansana *Porter.*

Shrub to 0.5 m; stems very prickly and bristly: leaflets 9–11, elliptic 2–6 cm long acute, coarsely and sharply serrate, glabrous and lustrous above, glabrous or sparingly pilose on the veins beneath; stipules dilated, glandular–dentate: flowers corymbose, pink, about 4 cm across; pedicels glabrous: fruit subglobose, 1–1.5 cm across. Fl. VI-VII. Wisconsin and Minnesota to Kansas and Colorado. Cultivated 1917. Zone IV.

29. R. suffulta *Greene.*

Low shrub to 0.5 m; stems densely prickly and bristly, usually green; leaflets 7–11, broadly elliptic to obovate–oblong, 1.5–4 cm long, usually obtuse, cuneate, simply serrate, finely pubescent on both sides or finally glabrous above; stipules dilated, glandular–dentate; petiole and rachis finely pubescent: flowers corymbose, pink, about 3 cm across; pedicels and receptacle glabrous; sepals sometimes lobed: fruit globose, about 1 cm across. Fl. VI. B.B.2:284. (*R. pratincola* Greene, not H. Br., *R. heliophila* Greene.) Alberta and Manitoba to Illinois and west Texas. Cultivated 1880. Zone III.

Varieties and forms

R. suffulta alba Rehd., f. Flowers white. Introduced 1901.

30. R. blanda *Ait.*

Shrub to 2 m with slender stems, unarmed or with scattered bristles when young, brown: leaflets 5–7, rarely 9, elliptic to obovate oblong, 2–6 cm long, usually acute, coarsely serrate, dull and glabrous above, paler and finely pubescent or glabrous beneath: stipules dilated: flowers solitary or few, pink, 5–6 cm across; pedicels and receptacle smooth: fruit subglobose, about 1.2 cm across, sometimes ellipsoid. Fl. V–VI. W.R.307,t(c). Miller & Whiting, Wild Fl. 152. B.B.2:283. (*R. solandri* Tratt., *R. subblanda* Rydb.)

Newfoundland to Pennsylvania, Missouri, North Dakota, and Manitoba. Introduced 1773. Zone II?

Hybrids

R. blanda × *chinesis* = *R. aschersoniana* Graebn. Stems with hooked prickles: flowers very numerous, small, bright light purple; styles usually exserted. Gt.51:t.1504,f1(c). Origin about 1880.

R. blanda × *rugosa*: see under no. 26.

31. R. pendulina *L.*

Stems slender, to 1 m, usually nearly unarmed, rarely bristly and prickly: leaflets 5–13, oblong–ovate or oblong–elliptic, 2–6 cm long, obtuse, doubly glandular–serrate, usually pubescent, rarely pubescent on both sides, more densely so beneath; stipules dilated: flowers usually solitary or 2–5, about 4 cm across, rose or purple; pedicels and receptacle smooth or glandular–bristly: fruit oblong, with neck, usually nodding, 2–2.5 cm long, bright red. Fl. V–VI. B.R.424(c). W.R.2:293,t(c). F.D.25:t.2625(c). (*R. alpina* L., *R. cinnamomea* L. 1753, not 1759.) Mountains of southern and central Europe. Cultivated 1789. Zone V. Handsome rose with conspicuous large fruit.

Varieties and forms

R. pendulina pyrenaica (Gouan) R. Keller, var. Dwarf with glaucous bracts and glandular–bristly pedicels and receptacle. B.M.6724(c). G.H.t.93(c). Gn.27:t.544. (*R. pyrenaica* Gouan) Cultivated 1815.

R. pendulina oxyodon (Boiss.) Rehd., var. Leaflets 5–9, usually glabrous, teeth acuminate, sometimes simple: flowers 2–7; receptacle smooth; pedicels glandular–bristly. (*R. oxyodon* Boiss., *R. alpina* var. *oxyodon* Boulenger.) Caucasus. Cultivated 1896. Zone V.

Hybrids

R. pendulina × *chinensis* = *R. lheritierana* Thory. Boursault Rose. Climbing to 4 m, sparingly prickly: leaflets 3–7, oblong–ovate, simply serrate, glabrous: flowers in corymbs, light or dark purple, double or semidouble: fruit subglobose, smooth. R.R.3: gr.26,3. W.R.301,t(c). (*R. reclinata* Thory, *R. boursaultii* Hort.) Origin before 1820.

R. pendulina × *spinosissima*: see under no. 2.

32. R. cinnamomea *L.*

Cinnamon Rose. Shrub to 2 m; stems slender with short curved prickles, often unarmed: leaflets 5–7, rarely 3, elliptic to oblong, obtuse or short–acuminate, 1.5–4 cm long, simply serrate, dull green and pubescent above, densely pubescent beneath: stipules dilated, more or less convolute: flowers solitary or few, purple, about 5 cm across; pedicels 2–3 cm long, like the receptacle smooth: fruit depressed–globose, 1.2–1.5 cm thick, scarlet. Fl. V–VI. W.R.141,t(c). D.F.25:t.2626(c). A.G.13:343. (*R. spinosissima* L. partly.) Europe, North and West Asia, occasionally escaped in North America. Cultivated 1596. Zone IV.

Varieties and forms

R. cinnamomea plena West., var. With double flowers W.R.143,t(c). R.R.1:gr.9,1(c). B.C.5:2993. (*R. majalis* Herrm., *R. foecundissima* Muenchh.) Cultivated before 1770. Sometimes escaped in the eastern states of the USA.

Hybrid

R. cinnamomea × *gallica*: see under no. 9.

Closely related species

R. davurica Pall. Prickles straight and slender: leaflets doubly serrate, glandular; stipules narrow: pedicles longer, glandular: fruit ovoid N.K.7:t.9. S.H.l:f.327c,328b-b[1]. (*R. cinnamomea* var. *davurica* C. A. Mey.) North China, northeast Asia. Introduced 1910. Zone V.

R. laxa Retz. Prickles hooked or straight: leaflets 7–9, glabrous or pubescent beneath: flowers several, white: pedicels glandular: fruit ovoid. S.H.l:f.325l-l[2]. Turkestan. Cultivated about 1800.

R. amblyotis C. A. Mey. Prickles slender, straight, pointing upward; leaflets rarely 9, elliptic–oblong or oblong–ovate, 3–5 cm long, usually acute, sharply serrate, puberulous beneath at least on midrib: pedicels slender, to 2 cm long, glabrous; flowers 5 cm across, red; sepals caudate: fruit depressed–globose or pyriform. Kamchatka. Cultivated 1917. Zone II?

R. marretii Lévl. Bracts dark purple; prickles slender, straight: leaflets usually 7, oblong, 2–4 cm long, obtusish to acute, thinly pubescent beneath: flowers 1–3, pink; pedicels and calyx-tube smooth; sepals caudate, entire, stipitate-glandular on back: fruit subglobose. W.R.2:495,t. Saghalin. Cultivated 1908. Zone II?

33. R. pisocarpa *A. Gr.*

Shrub to 2 m; stems slender, with weak slender, usually ascending prickles 2–5 mm long; leaflets 5–7, elliptic–oblong to oblong–obovate, 1–4 cm long, acutish or obtuse, coarsely serrate, green and finely pubescent beneath, short-petioluled; upper stipules dilated, slightly glandular-dentate; flowers corymbose with leafy bracts, pink, about 3 cm across; sepals glandular–hispid on back; fruit globose, sometimes with short neck, 8 mm across. Fl. VI–VIII. B.M.6857(c). W.R.227,t(c). British Columbia to California and Idaho. Introduced about 1882. Zone V.

34. R. woodsii *Lindl.*

Shrub to 2 m; stem with numerous straight or slightly curved prickles: leaflets 5–7, obovate to oblong–obovate, petioluled, 1–3 cm long, simply serrate, glaucous and glabrous or nearly glabrous beneath; stipules entire or slightly toothed, like the petiole usually glandless, rather narrow; flowers 1–3, pink, rarely white, 3–4.5 cm across; pedicels and receptacle glabrous; fruit globose, rarely ellipsoid and with distinct neck, 8–10 mm thick. Fl. VI–VIII. W.R.236,t(c). (*R. maximiliani* Nees, *R. sandbergii* Greene.) Saskatchewan and British Columbia to Kansas and Utah. Cultivated 1880. Zone III.

Varieties and forms

R. woodsii fendleri (Crép.) Rydb. Lower, with straight slender prickles; stipules and petioles glandular; leaflets usually doubly glandular–serrate. W.R.175.t(c).177,t. B.B.2:f.2311 (as *R. woodsii*). (*R. fendleri* Crép.) British Columbia and Minnesota to Arizona and Chihuahua. Cultivated 1888.

Closely related species

R. macounii Greene. Stems with straight prickles, bristly when young: leaflets obovate, pale or glaucous and finely puberulent beneath. B.R.926 (c, as *R. woodsii*). (*R. woodsii* Lindl. 1826, not 1820.) British Columbia to Saskatchewan, Oregon, Nebraska, west Texas. Introduced before 1826.

35. R. californica *Cham & Schlecht.*

Shrub to 3 m, with stout flattened, recurved prickles 5–8 mm long; young shoots sometimes bristly; flowering bracts usually prickly: leaflets 5–7, broad–elliptic to oblong–obovate, 1–3 cm long, usually obtuse, usually simply serrate, not, glandular, dull and appressed–pubescent above, villous beneath: flowers in leafy-bracted corymbs, pink, about 4 cm across; sepals villous outside; pedicels glabrous or slightly villous: fruit globose, with a distinct neck, 1–1.5 cm across. Fl. VI–VIII. Armstrong, Field Book Wild Fl. 221,f. Oregon to lower California. Cultivated 1878. Zone V.

Varieties and forms

R. californica plena Rehd., f. With double or semi-double flowers. W.R.223,t(c).

R. californica nana Bean., var. Dwarf form.

Hybrid

R. californica × *nitida*: see under no. 24.

36. R. nutkana *Presl.*

Shrub to 1.5 m; stems usually dark brown, upright, with rather large straight prickles; young stems sometimes bristly; leaflets 5–9, broad–elliptic or oblong–ovate, 1.5–5 cm long, acute or obtuse, doubly glandular–serrate, dark green and glabrous above, somewhat gladular–puberulous beneath, rarely slightly pubescent on veins; upper stipules dilated, glandular–dentate, glabrous; flowers usually solitary, 5–6 cm across, rose, rarely white; pedicels 2–3 cm long, usually slightly glandular–hispid; receptacle glabrous: fruit globose, without neck, 1.5–1.8 cm across. Fl. VI–VII. W.R.231,t(c). G.F. 1:449. B.C.5: 2993. Alaska to Wyoming and California. Introduced about 1876. Zone V.

Varieties and forms

R. nutkana hispida Fern., var. Leaflets coarsely serrate; receptacle glandular–hispid. (*R. macdougalii* Holz.) British Columbia to Utah.

Related species

R. spaldingii Crép. Leaflets 5–7 oval, usually obtuse, with simple scarcely glandular teeth, puberulent beneath: pedicels glabrous. British Columbia to Oregon and Utah. Cultivated 1915. Zone V.

R. oreophila Rydb. Stems usually with weak curved prickles: leaflets 5–7, oval or ovate, 2–3.5 cm long, simply serrate, finely pubescent beneath; pedicels glabrous. Colorado. Introduced 1898. Zone V.

37. R. corymbulosa *Rolfe*

Shrub to 2 m; stems upright or sarmentose: branches with few slender straight paired prickles or unarmed: leaflets 3–5, ovate–oblong, 2–4 cm long, acute, finely doubly serrate, dull green and slightly pubescent above at first, glaucescent and pubescent beneath; rachis puberulous and sparingly glandular; stipules narrow and glandular–ciliate: flowers red with white eye, 2–2.5 cm across, in umbel-like many-flowered corymbs; pedicels slender, about 2 cm long, like the receptacle glandular–bristly; styles pubescent, exserted; fruit subglobose or ovoid 1–1.3 cm long, scarlet. Fl. VI–VII. B.M.8566(c). G.C.79:23. West

China. Introduced 1908. Zone V. Handsome rose with numerous small flowers.

38. R. davidi *Crép.*

Shrub to 3 m; stems with straight stout scattered prickles 4–6 mm long, much enlarged at base: leaflets 7–9, rarely 11, elliptic to ovate–oblong, 2–4, rarely 6 cm long, simply serrate, glabrous above, glaucescent and puberulous beneath; rachis puberulous, sparingly prickly: flowers pink, 3.5–5 cm across, corymbose; pedicels like the oblong receptacle glandular–bristly or sometimes only puberulous; styles exserted about 3 mm: fruit ovoid or oblong–ovoid, 1.5–2 cm long, with long neck, scarlet. Fl. VI–VII. B.M.8679(c). G.C.78:423. West China. Introduced 1908. Zone V.

Varieties and forms

R. davidi elongata Rehd. & Wils., var. Leaflets 5–7 cm long: flowers fewer: fruit 2–2.5 cm long. Gs.16:1(c). Introduced 1908. Handsome in fruit.

39. R. caudata *Bak.*

Shrub to 4 m; stems with scattered straight stout prickles to 8 mm long, broad at base: leaflets 7–9, elliptic–ovate or ovate to oblong–ovate, 2.5–5 cm long, acute, simply serrate, glabrous and glaucescent beneath: rachis smooth or sparingly prickly and glandular–bristly; stipules rather broad, glandular–ciliate; flowers red, 3.5–5 cm across, in few-flowered corymbs; pedicels and receptacle glandular–bristly, rarely nearly smooth; sepals entire, long–caudate, with leafy apex: fruit oblong–ovoid, with long neck, 2–2.5 cm long, orange or coral–red; sepals upright. Fl. VI. West China. Introduced about 1896. Zone V.

Related species

R. banksiopsis Bak. Prickles few or stems unarmed above: leaflets 7–9, oblong, usually pubescent beneath: flowers 2–3 cm across; pedicels and receptacle glabrous. W.R.505(fr). West China. Introduced 1907 or 1909.

R. macrophylla Lindl. Large shrub, with few prickles or unarmed: leaflets 9–11, elliptic–ovate to elliptic–oblong, 2.5–6 cm long, acute, pubescent beneath: flowers 1–3, red, about 5 cm across; pedicels and receptacle glandular–setose or smooth: fruit oblong–ovoid, 2.5–3 cm long. W.R.157,t(c).159,t. Himalayas. Introduced 1818. Zone VI.

40. R. setipoda *Hemsl. & Wils.*

Shrub to 3 m; stem with infrastipular, straight wide-based prickles to 8 mm long: leaflets 7–9 elliptic or ovate, 3–6 cm long, obtuse or acutish, usually doubly serrate, dark green and glabrous above, glaucous and usually glandular beneath and puberulous on the veins or sometimes glabrous; stipules broad, densely glandular–ciliate; rachis sparingly prickly and stipitate–glandular: flowers pale pink to rose–purple, about 5 cm across, in loose corymbs; pedicels and receptacle copiously glandular–setose; sepals long–caudate, with foliaceous serrate tips; petals slightly pubescent outside: fruit oblong–ovoid, with narrow neck, about 2.5 cm long, deep red. Fl. VI. J.L.27:486. (*R. macrophylla* var. *crasse-aculeata* Vilm.) Central China. Introduced 1895. Zone (V). Conspicuous in June with its large pink flowers and in fall with its nodding clusters of red hips.

Hybrid

R. setipoda × *moyesii*: see under no. 43.

Related species

R. persetosa Rolfe. Stems densely bristly; leaflets 5–9, 2–5 cm long, simply serrate, puberulous beneath; rachis puberulous: flowers deep pink, 2–3 cm across, in large corymbs; pedicels and receptacle glabrous; sepals entire. J.L.27:487,488. (*R. davidi* var. *persetosa* Boulenger.) West China. Introduced 1895. Zone V.

41. R. hemsleyana *Täckholm.*

Shrub to 2 m; stem with few mostly infrastipular prickles: leaflets 7–9, rarely 5, oval or elliptic, 2–5 cm long, acute or obtuse, slightly doubly glandular–serrate, glabrous beneath or pubescent on the veins; stipules broad, glandular–ciliate; rachis stipitate–glandular and with few small prickles: flowers 3–11, pink, 3–5 cm across; pedicels 1–3 cm long, densely glandular–setose; sepals caudate with broad serrate tip and lateral appendages: fruit ovoid–oblong, with long neck, about 2.5 cm long. Fl. VI; fr. IX–XI. B.M.8569(c). Central China. Introduced 1904. Zone V?

42. R. sweginzowii *Koehne.*

Shrub to 5 m; stems with large compressed prickles, sometimes mixed with smaller prickles; leaflets 7–11, elliptic to ovate–oblong, acute, 2–5 cm long, doubly serrate, glabrous above, pubescent beneath, more densely on the veins; rachis prickly and glandular; stipules wide, glandular–ciliate: flowers 1–3, rose, about 4 cm across; pedicels and receptacle glandular–hispid; sepals sparsely lobed and serrate at apex: fruit oblong, about 2.5 cm long, bright red. Fl. VI. I.S.2:t.80. M.D.1910:96. G.C.72:135. Gn.87:588. Northwest China. Cultivated 1910. Zone V.

Related species

R. murielae Rehd. & Wils. Shrub to 2.5 m; stems with slender prickles and bristles: leaflets 9–15, elliptic to elliptic–oblong, 1–4 cm long, acutish, pubescent on midrib beneath: flowers white, in 3–7-flowered corymbs; sepals entire: fruit ellipsoid, 1–2 cm long. West China. Introduced 1904. Zone (V).

43. R. moyesii *Hemsl. & Wils.*

Shrub to 3 m; stems with scattered and paired short straight prickles: leaflets 7–13, ovate or elliptic to oblong–ovate, 1–4 cm long, acute, closely serrulate, glabrous except pubescent on midrib beneath; rachis puberulous, glandular and bristly; stipules wide, glandular–ciliate: flowers 1–2, deep blood-red, 4–6 cm across; pedicels short, stipitate–glandular; receptacle stipitate–glandular or smooth; sepals ovate, long–caudate: fruit oblong–ovoid, narrowed into a neck, 5–6 cm long, deep orang–red. Fl. VI. B.M.8338(c). W.R.229,t(c),230,t. J.L.27:489. B.S.2:435. G.C.70:219. Gn.72:313. V.F.95(h). (*R. macrophylla* var. *rubrostaminea* Vilm.) West China. Introduced 1894 and 1903. Zone (V). Handsome rose, particularly the type with dark red flowers.

Varieties and forms

R. moyesii fargesii Rolfe, var. Leaflets broad–oval to suborbicular, 1.2–2 cm long, obtuse. G.C.65:19. Gn.86:543;87:590. Cultivated 1900?

R. moyesii rosea Rehd. & Wils., f. Leaflets more coarsely serrate; flowers light pink. B.M.9248(c). (*R. holodonta* Stapf.) Introduced 1908.

Hybrids

R. moyesii × *willmottiae*: see under no. 48.

R. wintoniensis Hillier (*R. moyesii* × *setipoda*) and *R. highdownensis* Hillier (*R. moyesii* × ?) Gs.18:166.

Related species

R. saturata Bak. Stems unarmed or with few small prickles: leaflets usually 7, ovate to ovate–lanceolate, 2.5–6 cm long, acute or short–acuminate, sharply serrate, pubescent on the veins beneath or nearly glabrous: flowers solitary, dark red, 5 cm across; anthers purple; pedicels glabrous: fruit globose–ovoid, 1.5–2 cm long, coral-red. Central China. Introduced 1907. Zone (V).

44. R. bella *Rehd. & Wils.*

Shrub to 3 m; stems with slender straight prickles, bristly near base: leaflets 7–9, elliptic or ovate, 1–2.5 cm long, acutish, rarely obtusish, simply serrate, glaucescent beneath, glabrous except stipitate–glandular on midrib; upper stipules broad, glandular–ciliate: flowers 1–3, bright rose, 4–5 cm across; pedicels 5–10 mm long, like the receptacle stipitate–glandular; sepals caudate, leafy and entire at apex: fruit ellipsoid, attenuate into the neck, 1.5–2 cm long, scarlet. Fl. VI. I.S.2:t.79. North China. Introduced 1910. Zone V.

Varieties and forms

R. bella pallens Rehd. & Wils., f. Flowers light rose-colored. Introduced 1910.

45. R. webbiana *Royle.*

Slender shrub to 2 m; branches slender, often pruinose when young, with straight yellow prickles about 1 cm long: leaflets 5–9, orbicular to obovate or broad–elliptic, 6–20 mm long, rounded at apex, rarely acutish, serrate, entire toward the base, glabrous or sometimes slightly pubescent beneath; petiole prickly: flowers pale pink, 4–5 cm across; pedicels short, glabrous or slightly glandular; sepals lanceolate, entire, glandular outside: fruit ovoid, 1.5–2 cm long, bright red. Fl. VI. W.R.233,t(c). Himalayas to Afghanistan and Turkestan. Introduced 1879. Zone (V).

Related species

R. fedtschenkoana Reg. Stems with prickles and bristles: leaflets 7–9, oblong, 2–3 cm long, pubescent beneath: flowers 1–4, white; pedicels and receptacle hispid. B.M.7770(c). W.R.155,t(c). Turkestan. Introduced 1876?

46. R. sertata *Rolfe.*

Slender shrub to 2 m; branches glaucous, with straight slender prickles: leaflets 7–11, oval to narrow–elliptic, 6–20 mm long, obtuse or acutish, sharply serrate, glabrous, glaucescent beneath; stipules glandular–ciliate: flowers one or few, rose or rose–purple, 5–6 cm across; pedicels 1.5–3 cm long, glandular or glabrous; sepals caudate, entire: fruit ovoid, 1.5–2 cm long, deep red. Fl. VI. B.M.8473(c). W.R.153,t(c). G.C.54:166. Central and West China. Introduced 1904. Zone V. Handsome graceful rose.

Related species

R. giraldii Crép. Bracts with slender prickles: leaflets usually 7, suborbicular to ovate or elliptic, serrate, pubescent: flowers pink, 2.5 cm across; pedicels short, like the receptacle glandular–setose: fruit ovoid, about 1 cm long, scarlet. Central China. Introduced 1897?

R. giraldi venulosa Rehd. & Wils., var. Leaflets firmer, reticulate beneath, often glabrous or nearly glabrous above. Introduced 1907. Zone V?

R. prattii Hemsl. Stems prickly and bristly or without bristles; leaflets 7–15, elliptic to oblong, 6–15 mm long, obtusish, serrate, pubescent on the veins beneath; rachis puberulous: flowers one or several, 2 cm across, pink, sepals ovate, abruptly acuminate; pedicels and receptacle glandular–bristly: fruit subglobose to ovoid, 6–10 mm long; sepals finally deciduous. West China. Introduced 1903 or 1908. Zone V.

R. elegantula Rolfe. Shrub to 2 m; bracts usually densely bristly: leaflets 7–9, elliptic or elliptic–ovate, 1–2 cm long, obtuse, usually simply serrate, glaucous: flowers deep pink, 2.5–3 cm across; pedicels and the ovoid–oblong receptacle smooth. West China. Cultivated 1900. Zone V.

R. forrestiana Boulenger. Shrub to 2 m; leaflets 5–7, rarely 9, oval to obovate, 1–2 cm long, rounded at apex, glabrous or pubescent on midrib beneath, teeth simple or nearly so: flowers one or few, with large bracts at base, rose, 2–3.5 cm across; styles exserted; pedicels 8–15 mm long, stipitate-glandular. Bull. Jard. Bot. Bruxelles, 14:127. West China. Cultivated 1922. Zone VII?

47. R. multibracteata *Hemsl. & Wils.*

Shrub to 2 m; bracts slender, glabrous, with slender straight paired prickles: leaflets 7–9, broad–oval to suborbicular, 6–15 mm long, obtuse, doubly serrulate, glabrous or pubescent on midrib beneath; rachis glandular and often with few small prickles: flowers pink, 3 cm across, paniculate or corymbose, with many crowded bracts; pedicels and receptacle glandular–bristly; sepals ovate, abruptly acuminate, entire; styles villous, exserted about 4 mm: fruit ovoid, 1–1.5 cm long, orange–red. Fl. VI. G.C.100:23. Gt.81:253. West China. Introduced 1910. Zone VI.

48. R. willmottiae *Hemsl.*

Densely branched shrub to 3 m; stems with straight paired prickles, glaucous when young: leaflets 7–9, elliptic to obovate or nearly orbicular, 6–15 mm long, usually rounded at apex, closely and often doubly serrate, glabrous: flowers usually solitary, rose–purple, about 3 cm across; pedicels about 1 cm long, glabrous; sepals lanceolate, caudate, smooth: fruit subglobose, about 1 cm long, orange–red. Fl. VI. B.M.8186(c). Gn.87:588. S.L.323(h). West China. Introduced 1904. Zone V?

Hybrid

R. willmottiae × *moyesii* = *R. pruhoniciana* Zeman. Dense shrub with bright pink flowers (*R. Hillieri* Hillier.) Origin about 1920.

49. R. gymnocarpa *Nutt.*

Shrub to 3 m; stems with paired very slender prickles and bristles: leaflets 5–9, broad–elliptic elliptic–oblong, 1–3.5 cm long, doubly glandular–serrate, glabrous; stipules glandular–ciliate and dentate: flowers usually solitary, rose, about 3 cm across; pedicels glabrous or glandular–hispid: fruit ellipsoid or subglobose, 6–8 mm across. Fl. VI–VII. W.R.221,t(c). British Columbia to California and Montana. Introduced 1893. Zone V.

Related species

R. alberti Reg. Shrub to 1 m; stems densely prickly and bristly: leaflets orbicular to obovate or ovate, 6–30 mm long, often doubly serrate, appressed–pubescent beneath: flowers white, about 3 cm across; pedicels glandular; receptacle smooth: fruit pear-shaped, 1.5–2 cm long. Turkestan. Introduced 1877. Zone V.

50. R. beggeriana *Schrenk.*

Much-branched shrub to 2.5 m; stems with hooked paired and scattered prickles: leaflets 5–9, oval or elliptic–obovate, 8–25 mm long, simply serrate, pubescent beneath or glabrous, bluish or grayish green; flowers in several- or many-flowered corymbs, rarely solitary, white, rarely pink, 2–3 cm across; pedicels and receptacle glandular or smooth; fruit subglobose, 6–8 mm across, red, finally dark purple. Fl. VI; fr. VIII–IX. W.R.171,t(c). I.T.5:t.122. S.H.1:f.325 i-k,[1] 326k-l. North Iran to Altai and Song. Introduced 1881. Zone III.

Sect. 6.* Synstylae *DC.

Stems climbing, trailing or sarmentose, rarely upright, with hooked prickles: flowers in corymbs, outer sepals pinnate, rarely entire, caducous; style connate into a slender exserted column.

51. R. multiflora. *Thunb.*

Vigorous shrub with long reclining or climbing branches: leaflets usually 9, obovate to oblong, 1.5–3 cm long, acute or obtuse, serrate, pubescent: flowers in many-flowered pyramidal corymbs, 1.5–2 cm across, usually white; sepals ovate, abruptly acuminate; styles glabrous: fruit small, globular. Fl. VI–VII. B.M.7119(c). G.F.3:405,4:535(h). Gt.67: 185(h). Gg.5:120(h),121. (*R. polyantha* Sieb. &

Zucc., *R. intermedia* Carr., *R. Wichurae* Koch.) Japan, Korea. Introduced before 1868. Zone V.

Varieties and forms

R. multiflora thunbergiana Thory is the typical form described above, with small white flowers and pubescent pedicels.

R. multiflora calva Franch. & Sav., var. Leaflets pubescent beneath only on midrib: flowers small, white or pinkish; pedicels glabrous. (*R. calva* Boulenger.) Japan, North and Central China.

R. multiflora cathayensis Rehd. and Wils. Flowers pink, 2–4 cm across, in few- to many-flowered rather flat corymbs; pedicels glabrous, sometimes glandular. Gn.89:35(h). (*R. calva* var. *cathayensis* Boulenger, *R. cathayensis* Bailey, *R. gentiliana* Lévl. & Vant.) China. Introduced 1907.

R. multiflora carnea Thory. A form of the preceding with double light pink flowers B.M.1059(c). B.R.425(c). (*R. multiflora* var. *plena* Reg., *R. florida* Poir.) Introduced 1804.

R. multiflora platyphylla Thory. Seven Sisters Rose. Similar to the preceding, but leaves larger and flowers deep pink. B.R.1372(c). Introduced 1817. Closely related to this form is the well known 'Crimson Rambler.' A.G.16:233.

Hybrids

R. multiflora × *rugosa* = *R. iwara* Sieb. With single, rather small white flowers. W.R.193,t(c). (*R. yedoensis* Mak.)

R. multiflora × *chinensis* Here belongs the Polyantha roses (*R. polyantha* Hort., not Sieb. & Zucc. nor Roessig.) Upright low shrubs with numerous, rather small double flowers. Gn.29:118,t. R.H.1884:501(h). Also 'Baby Rambler' (*R. multiflora* × *chinensis minima*) belongs here.

R. multiflora × *moschata* Here probably belongs the rose known as *R. polyantha grandiflora*. W.R.34,t(c).

R. multiflora × *borboniana* 'Général Jacqueminot.' Here belongs the Dawson Rose (*R. dawsoniana* Hort., not Ellw. & Barry which is typical *R. multiflora*.) G.F.6:316(h),317. G.W.7:125(h).

R. multiflora × *wichuraiana*: see under no. 61.

52. R. watsoniana *Crép.*

Shrub with sarmentose or recurving branches: leaflets 3–5, linear–lanceolate, with entire wavy margin, 2.5–6 cm long, pubescent beneath: flowers about 1 cm across, white or pink in many-flowered pyramidal corymbs; style glabrous. Fl. VI. W.R.53,t(c); G.F.3:477. R.B.14:183. (*R. multiflora* var. *watsoniana* Matsum.) Cultivated in Japan. Introduced 1870. Zone VI. Possibly a mutation of *R. multiflora*; peculiar rather than ornamental.

53. R. setigera *Michx.*

Prairie Rose. Shrub with glabrous climbing or recurving stems to 5 m long: leaflets 3, rarely 5, ovate to oblong–ovate, 3–9 cm long, short–acuminate, serrate, pubescent on the veins beneath: flowers about 5 cm across, rose, fading to whitish, nearly scentless, in rather few-flowered corymbs; pedicels and receptacle glandular–hispid; styles glabrous: fruit globose, about 8 mm across. Fl. VI–VIII. W.R.71,t(c). Mn.8:t.5(c). G.F.10:323. Gng.1:325. C.L.A.4:339; 7:473. S.L.320(h). Ontario to Nebraska, Texas and Florida. Introduced 1810. Zone IV.

Varieties and forms

R. setigera inermis Palm. & Steyerm., f. Stems unarmed or with few weak spines: leaves glabrous. Introduced 1923.

R. setigera serena Palm. & Steyerm., var. Stems unarmed: leaves pubescent beneath. Introduced 1924.

R. setigera tomentosa Torr. & Gr., var. Leaves tomentose beneath, dull above: flowers smaller, more numerous. R.R.1:gr.5,3(c). L.A.2:f.516. (*R. rubifolia* Ait.) Introduced 1800.

Hybrids

There are several double-flowered forms as 'Baltimore Belle' (B.C.5:f.3439) which are probably hybrids with *R. noisettiana*, *R. gallica*, *R. multiflora* and others.

Related species

R. triphylla Roxb. Climbing; prickles few, small, recurved: leaflets 3, rarely 5, ovate–lanceolate, acuminate, glabrous, glaucous beneath; stipules narrow, dentate: flowers corymbose, small, white, double, the inner petals shorter and narrower than the outer ones; pedicels and calyx-tube glabrous. R.H.1842:281,t(c). W.R.67:t(c). (*R. anemoneflora* Fort.) East China. Introduced 1844. Zone VII.

54. R. helenae *Rehd. & Wils.*

Sarmentose shrub to 5 m, with stout hooked prickles: leaflets 7–9, rarely 5, oblong–ovate to ovate–lanceolate, 2.5–5 cm long, short–acuminate, rounded or broad–cuneate at base, sharply serrate, grayish-green beneath and pubescent on the veins: flowers white, fragrant, 3 cm across, in many-flowered umbel-like corymbs 7–15 cm across; pedicels slender, 2–3 cm long, stipitate–glandular like the receptacle; sepals lanceolate, sparingly pinnate; styles pubescent: fruit ovoid or oblong–ovoid, 1–1.5 cm long, scarlet. Fl. VI. I.S.2:t.77. J.L.47:195. Gn.89:35(h). Gs.6:108(h). (*R. floribunda* Bak., not Stev.) Central China. Introduced 1907. Zone (V).

55. R. rubus *Lévl. & Vant.*

Sarmentose shrub to 6 m, with small hooked prickles: leaflets usually 5, ovate–elliptic to obovate, 3–8 cm long, abruptly acuminate, sharply or coarsely serrate pubescent beneath, rarely nearly glabrous: flowers white, fragrant, 2.5–3 cm across, in dense corymbs; pedicels 1–2 cm long, tomentose, glandular: fruit subglobose, 8 mm across, scarlet. Fl. VI. W.R.2:507,t. (*R. ernesti* Bean.) Central and West China. Introduced 1907. Zone VI?

Related species

R. mulligani Boulenger. Prickles strongly dilated at base: leaflets 5–7, elliptic, to 6 cm long, acute or acuminate, pubescent beneath on the veins: flowers 4.5–5.5 cm across; pedicels 2.5–3.5 cm long; sepals 12–15 mm long; long–pinnate: fruit 10–13 mm high. West China. Introduced 1917–19. Zone VI?

R. glomerata Rehd. & Wils. Leaflets usually 5–7, oblong to oblong–obovate, 3–10 cm long, serrulate, rugose above, reticulate beneath and pubescent on the veins: fruit orange–red. West China. Introduced 1908. Zone VI?

56. R. brunonii *Lindl.*

Himalayan Musk Rose. Tall shrub with arching or sarmentose branches, glabrous or thinly villous while young; prickles hooked, short and stout: leaflets 5–7, elliptic–oblong to oblong–lanceolate, 3–6 cm long, acute or acuminate, serrulate, slightly pubescent or glabrous above, soft-pubescent beneath; petioles and rachis pubescent, usually with scattered prickles: flowers white, 3–5 cm across, fragrant, in many-flowered corymbs; sepals lanceolate, acuminate, much longer than the receptacle; pedicels slender, glandular–pubescent; styles pubescent: fruit ovoid, about 1 cm long, glabrous. Fl. VI–VII. B.M.4030(c). B.R.829(c). W.R.37,t(c). G.C.36:152,153. Gn.77:511;79:113(h), 519(h);86:363(h; as *R. moschata*). (*R. brownii* Tratt.) Himalayas. Introduced 1822. Zone VII? Usually cultivated as *R. moschata*.

57. R. moschata *Herrm.*

Musk Rose. Shrub with arching or sarmentose glabrous branches; prickles straight or slightly curved; leaflets 5–7, elliptic–ovate to oblong–ovate; 2.5–5 cm long, acute or acuminate, serrulate, glabrous or nearly so beneath, except pubescent on the midrib; petioles and rachis nearly glabrous, usually prickly: flowers white, 3–5 cm across, in usually 7-flowered corymbs, of musky fragrance; pedicels slender, slightly pubescent and glandular; styles pubescent: fruit ovoid, small. Fl. VI. W.R.33.t(c). (*R. ruscinonensis* Déségl.) South Europe, North Africa, West Asia. Long cultivated. Zone VII?

Varieties and forms

R. moschata plena West., var. With double flowers. R.R.1:99,t(c).

R. moschata nastarana Christ, var. More robust: leaflets nearly glabrous beneath: flowers over 5 cm across, tinged pinkish, more numerous. W.R.39,t.(c). R.H.1880:314,315. (*R. Pissardi* Carr.) Iran. Introduced 1879. Zone VI?

Hybrids

R. moschata × *gallica* = *R. dupontii* Déségl. Stems prickly and bristly: leaflets fewer and broader; flowers larger; styles shorter; pedicels glandular. B R.861(c). W.R.43,t(c). Gn.67:254;77:510. Origin before 1817. Here belongs also *R. freundiana* Graebn. Gt.57:471. G.25:311.

Related species

R. longicuspis Bertol. Sarmentose and climbing: leaflets coriaceous, lustrous, ovate or elliptic to ovate–oblong, 5–10 cm long, acuminate or long–acuminate, usually rounded at base, reticulate beneath and usually quite glabrous; stipules finely denticulate: flowers in many-flowered corymbs; petals silky outside; pedicels pubescent. Bertoloni, Misc. Bot. t.3. (*R. lucens* Rolfe.) Himalayas, West China. Cultivated 1915. Zone VII?

R. sinowilsoni Hemsl. Similar to the preceding: leaflets glabrous or slightly pubescent beneath; stipules very narrow, glandular–denticulate: flowers subumbellate or paniculate, in bud obtuse or abruptly pointed: pedicels glabrous, not or slightly glandular. West China. Introduced 1904. Zone VI?

R. cerasocarpa Rolfe. Leaflets subcoriaceous, 5, sometimes 3, ovate or elliptic–ovate, acute or acuminate, glabrous; stipules narrow, entire: flowers 2.5–3 cm across, paniculate, pedicels and receptacle glabrous, glandular; bud abruptly pointed: fruit globose, 1 cm diam. B.M.8688(c). Central China. Cultivated 1914. Zone VI?

58. R. maximowicziana *Regel.*

Shrub with sarmentose or arching branches; young stems with scattered hooked prickles and bristles: leaflets 7–9, elliptic–ovate to oblong, 3–6 cm long, acute or acuminate, finely serrate, paler beneath, glabrous; stipules glandular–serrulate: flowers white, 3–3.5 cm across, in many-flowered corymbs with persistent bracts; sepals ovate, abruptly caudate–acuminate; pedicels stipitate–glandular or glabrous; styles glabrous: fruit subglobose or broad–ovoid, about 1 cm across. Fl. VI–VII. N.K.7:t.1. (*R. fauriei* Lèv.) Manchuria, Korea. Introduced ?

R. maximowicziana jackii (Rehd.) Rehd., var. Stems without bristles. N.K.7:t.3. (*R. jackii* Rehd., *R. kelleri* Bak., not Dalla Torre & Sarnth.) Korea. Introduced 1905. Zone V.

59. R. henryi *Boulenger.*

Sarmentose shrub; branches with hooked prickles, often purplish; floral bracts unarmed: leaflets usually 5, elliptic to elliptic–ovate, 4.5–8 cm long, acuminate, rounded or broad–cuneate at base, simply serrate, glabrous and glaucescent beneath; stipules narrow, entire: flowers white, fragrant, 3–3.5 cm across, in umbel-like corymbs; bud gradually acuminate; pedicels 1.2–2 cm long, pubescent and stipitate–glandular; receptacle pubescent; sepals pubescent and slightly glandular, with a few appendages; styles pubescent: fruit globose, 8–10 mm across, dark red. Fl. VI–VII. J.L.27:483. (*R. gentiliana* Rehd. & Wils., not Lévl. & Vant., *R. moschata* var. *densa* Vilm.) Central and East China. Introduced 1907. Zone VI.

Closely related species

R. filipes Rehd. & Wils. Leaflets usually 5–7, oblong–ovate to lanceolate, 3.5–7 cm long, acuminate, pale beneath and glabrous, but glandular: pedicels glabrous and stipitate–glandular, 2–3.5 cm long; flowers 2–2.5 cm across: fruit globose, about 1 cm across, scarlet. West China. Introduced 1908. Zone (V).

R. crocacantha Boulenger. Shrub to 2.5 m; prickles orange–yellow: leaflets 5–7, elliptic, to 5 cm long, doubly serrate, glabrous or pubescent beneath: flowers very numerous, 1–1.5 cm across; pedicels 1–1.8 cm long; styles glabrous. West China. Introduced 1917–19. Zone VI?

R. phoenicea Boiss. Leaflets usually 5, acute or obtuse; corymbs pyramidal, usually many-flowered; styles glabrous. Asia Minor. Introduced about 1885. Zone VI?

60. R. souleiana *Crép.*

Upright shrub to 4 m with spreading branches; prickles stout, hooked: leaflets usually 7, elliptic or obovate to ovate–oblong, 1–3 cm long, acute or obtuse, serrulate, glabrous beneath: flowers white, 3–3.5 cm across, in many-flowered corymbs 10–15 cm across: pedicels and receptacle glandular; sepals ovate, acuminate: fruit ovoid, about 1 cm long, orange–red. Fl. VI. B.M.8158(c). W.R.57,t(c). Gn.89:35(h). West China. Introduced 1896. Zone VI.

61. R. wichurana *Crép.*

Memorial Rose. Half-evergreen, with long prostrate or trailing branches; prickles strong, curved: leaflets 7–9, suborbicular to broad–ovate or obovate, 1–2.5 cm long, usually obtuse, coarsely serrate, dark green and lustrous above, lighter green and lustrous beneath, glabrous; stipules dentate: flowers 4–5 cm across, white, fragrant, in few- to many-flowered pyramidal corymbs; styles pubescent; pedicels slightly glandular–hispid: fruit ovoid, about 1 cm long. Fl. VII–IX. B.M.7421(c). W.R. 59,t(c). R.H.1898:105,t(c),106. G.F.4:569;6:337. G.C.II.22:99. M.G.13:580–5(h). (*R. luciae* var. *wichurana* Koidz., *R. bracteata* Hort., not Wendl.) Japan, Korea, Formosa, East China. Introduced 1891. Zone V. Adapted for covering banks and rocky slopes. Many hybrids have been raised.

Hybrids

R. wichurana × *rugosa* = *R. jacksonii* Willm. Leaflets larger, rugose above; flowers bright crimson. R.W.63,t(c). Here belongs 'Lady Duncan.'

R. wichurana × *multiflora* 'Crimson Rambler' = *R. barbierana* Rehd. Leaflets 5–7, larger: flowers single, carmine, 4–5 cm across; petals whitish at the base. R.H.1901:20,t(c). (*R. wichurana rubra* Andre.) Origin before 1900.

There are also numerous hybrids of *R. wichurana* with Hybrid Teas and Hybrid Perpetuals (G.F.6:337. Mn.8:27,156).

Closely related species
R. luciae Franch. & Rochebr. Of more upright habit: leaflets usually 7, narrower and thinner: flowers smaller. East China, Korea, Japan. Introduced ?

62. R. sempervirens *L.*

Evergreen or half-evergreen shrub with trailing or climbing stems: leaflets 5–7, ovate–lanceolate, 2–5 cm long, acuminate, serrulate, lustrous above and beneath, glabrous: flowers 3–5 cm across, in few-, rarely many-flowered corymbs, slightly fragrant; pedicels glandular–hispid; styles usually pubescent: fruit subglobose or ovoid, about 1 cm long, orange–red. Fl. VI–VII. B.R.465(c). W.R.19,t(c). South Europe, North Africa. Zone VI?

Varieties and forms
R. sempervirens prostrata (DC.) Desv. Leaflets elliptic, acute, smaller: styles glabrous; fruit ovoid. (*R. prostrata* DC.)

R. sempervirens scandens (Mill.) Nichols. Leaflets elliptic to oblong, obtuse: fruit subglobose. (*R. scandens* Mill.) Some double-flowered forms like 'Felicité perpetuelle' are probably hybrids with *R. chinensis*.

63. R. arvensis *Huds.*

Deciduous shrub with trailing stems: leaflets usually 7, ovate to elliptic–ovate, 1–3.5 cm long, acute, serrate, dull above, glabrous or slightly pubescent beneath, rather thin: flowers 3–5 cm across, scentless, in few-flowered corymbs or solitary; styles glabrous: fruit ovoid. Fl. VI–VII. B.M.2054(c). W.R.11,t(c). S.E.3,t.476(c). F.D.25:t.2633(c). Gn.60:223;77:510. (*R. repens* Scop., *R. silvestris* Herrm.) Europe. Cultivated 1750. Zone V.

Varieties and forms
R. arvensis ayreshirea Ser., var. Ayrshire Rose. More vigorous: leaves green on both sides, remaining longer on the branches B.M.2054(c). (*R. capreolata* Neill, *R. arvensis* var.*scandens* Sweet, *R. arvensis* var. *capreolata* Bean.) The forms with double flowers often called Ayrshire Roses are probably hybrids with other roses.

Hybrids
R. arvensis × *chinensis* = *R. ruga* Lindl. Trailing: flowers large, pink, double, fragrant, in several-flowered corymbs. B.R.1389(c). W.R.55,t(c). Origin before 1830.

R. arvensis × *gallica* = *R. polliniana* Spreng. Upright or sarmentose: leaflets usually 5, of firmer texture, slightly pubescent beneath: flowers long-stalked, large, white to pink. W.R.333,t(c). Europe. Cultivated 1820.

R. arvensis × *rugosa*: see under no. 26.

Sect. 7. Indicae *Thory.*

Stems upright or procumbent, with sparse, hooked prickles: leaflets 3–5, rarely 7, glabrous; stipules and bracts narrow: inflorescence one-to many-flowered; sepals entire or sparingly lobed, reflexed after flowering: styles exserted, free. Two Chinese species.

64. R. odorata *Sweet.*

Tea Rose. Evergreen or half-evergreen, with long, sarmentose, often climbing branches; prickles scattered, hooked: leaflets 5–7, elliptic or ovate to oblong–ovate, 2–7 cm long, acute or acuminate, sharply serrate, lustrous above, glabrous: flowers solitary or 2–3, pink, on rather short, often glandular stalks, 5–8 cm across. Fl. VI–IX B.R.804(c). R.R.3:gr.25,19(c). (*R. thea* Savi, *R. chinensis* var. *fragrans* (Thory) Rehd., *R. indica fragrans* Thory, *R. indica* var. *odoratissima* Lindl.) China. First introduced in the double blush form in 1810. Zone VII.

Varieties and forms
R. odorata ochroleuca (Lindl.) Rehd., f. With pale yellow double flowers R.R.3:gr.25,20(c). Andrews, Ros.2:t,77(c). Introduced 1824.

R. odorata pseudindica (Lindl.) Rehd., var. 'Fortune's Double Yellow'. ('Beauty of Glazenwood', 'Gold of Ophir'.) Flowers salmon–yellow, outside tinged with red, double, 7–10 cm across. B.M.4679(c). W.R.85,t(c). F.S.8:t.769(c).

R. odorata gigantea (Coll.) Rehd. & Wils., var. Very vigorous, climbing to 10 m: flowers creamy white, single, 10–15 cm across; pedicels and receptacle smooth. B.M.7972(c). W.R.34(c). G.C.III.6:13;37:136;51:314. Gn.67:179;71:67. (*R. g.* Collett.) Southwest China, Burma. Introduced 1889. Zone VII.

R. odorata erubescens Rehd. & Wils., f. A form of the preceding var. with blush or pale pink flowers, often smaller. West China. Introduced ? This form or the preceding var. is probably the phylogenetic type of the Tea Rose.

Hybrids
R. odorata gigantea × *brunonii*. Here belong 'Madelaine', 'Lemoine', 'Montanosa', and 'Montecito' with paniculate flowers.

R. odorata × *borboniana* = *R. dilecta* Rehd. Here belong the hybrid Tea Roses which are hybrids between the Tea Rose and the Hybrid Perpetuals and of which 'La France' may be considered a typical representative.

R. odorata (or *dilecta*) × *rugosa*: see under no. 26.

R. odorata × *laevigata*: see under no. 69.

65. R. chinensis *Jacq.*

China Rose. Low upright shrub with stout more or less hooked prickles, sometimes nearly unarmed: leaflets 3–5, broad–ovate to ovate–oblong, 2.5–6 cm long, acuminate, serrate, lustrous and dark green above, pale beneath, glabrous: flowers usually several, rarely solitary, on long, usually glandular stalks, rarely short-stalked, about 5 cm across, crimson to pink or nearly white: fruit 1.5–2 cm long. Fl. VI–IX. W.R.79,t(c). Jacquin, Obs. Bot. 3:t.55(c). (*R. indica* Lindl., not L., *R. sinica* L.—Bengal Rose.) China. Cultivated 1768. Zone VII?

Varieties and forms

R. chinensis semperflorens (Curtis) Koehne., var. Crimson China Rose. (Chinese Monthly Rose). Stems slender, prickly or nearly unarmed; leaflets rather thin, mostly stained with purple: flowers usually solitary on long slender pedicels, crimson or deep pink. B.M.284(c). W.R.89,t(c).

R. chinensis minima (Sims.) Voss, var. Fairy Rose. Dwarf shrub, rarely exceeding 25 cm, with small rose-red flowers, about 3 cm across, single or double. B.M.1762(c). B.R.538(c). R.R.3:gr.25,6,7(c). Gs.14:138. H.B.11:117. (*R. lawranciana* Sweet, *R. roulettii* Corr.) Hybrids of this var. with *R. multiflora* forms are the 'Baby Ramblers.'

R. chinensis viridiflora (Lav.) Dipp., var. With monstrous green flowers; the petals transformed into small green leaves. F.S.11:1136(c). Origin about 1855.

R. chinensis mutabilis (Corr.) Rehd., f. Flowers simple, first sulfur-yellow, changing to orange, red and finally crimson, 4.5–6 cm across. R.H.1934:60,t(c). (*R. m.* Corr.) Cultivated 1934.

Hybrids

R. chinensis × *arvensis*: see under no. 63.

R. chinensis × *moschata*: see no. 66.

R. chinensis × *rugosa*: see under no. 26.

R. chinensis × *blanda*: see under no. 30.

R. chinensis × *pendulina*: see under no. 31.

R. chinensis × *damascena*: see no. 67.

66. × R. noisettiana *Thory (*R. chinensis × moschata*)*

Noisette Rose (Champney Rose). Shrub to 3 m with upright or spreading stems; prickles reddish: leaflets 5–7, usually oblong–ovate to oblong–lanceolate, glabrous: flowers usually in many-flowered corymbs, white to pink or red, sometimes yellow; styles exserted, pubescent, loosely cohering. W.R.32(c). Gn.71:335. B.S.2:437. Origin about 1816. Here belongs the Manetti Rose (*R. manetti* Crivelli, *R. chinensis* var. *manetti* Dipp.), a vigorous shrub with semidouble or single flowers (B.C.5:2988), often used as stock for Tea and China Roses.

67. × R. borboniana *Desp. (*R. chinensis × damascena*)*

Bourbon Rose. Upright shrub with prickly and often glandular–hispid stems: leaflets usually 7, ovate to lance–ovate, acute, lustrous above, slightly pubescent beneath: flowers solitary or few in a corymb, on glandular pedicels, double or semidouble, usually purple. W.R.339,t(c). (*R. borbonica* Morr.) Fl. VII–IX. Origin before 1819. By crossing this hybrid with the different roses of the *Gallicae* group, the Hybrid Bourbon Roses have originated and these crossed with *R. chinensis* and its varieties have produced the Hybrid Perpetuals or Remontant Roses. All these hybrids may be included in *R. borboniana* in an enlarged sense.

Hybrids

R. borboniana × *odorata*: see under no. 64.

R. borboniana × *rugosa*: see under no. 26.

Sect. 8. **Banksianae *Lindl.***

Evergreen climbing shrubs with slender glabrous branches armed with hooked prickles or unarmed: leaflets 3–7; stipules free, subulate, caducous: flowers in corymbs, white or yellow; sepals entire, reflexed after flowering, deciduous. Two species in China.

68. R. banksiae *Ait.*

Banks's Rose. Climbing to 6 m, with sparingly prickly or unarmed branches: leaflets 3–5, rarely 7, elliptic–ovate to oblong–lanceolate, 2–6 cm long, acute or obtusish, serrulate, glabrous except at base of midrib beneath; rachis pubescent: flowers white or yellow, 2.5 cm across, slightly fragrant, on slender smooth stalks in many-flowered umbels: fruit small globose. Fl. VI–VIII. Gn.83:107(h). China.

Varieties and forms
R. banksiae normalis Reg., var. Flowers single, white. G.C. 31:539. J. L.27:501. Cultivated 1877.

R. banksiae alboplena Rehd., var. Flowers double, white, fragrant. B.M.1954(c). Introduced 1807. The typical form.

R. banksiae lutea Lindl., var. Flowers double, yellow, scentless. B.R.1105(c). (*R. banksiae luteoplena* Rehd.) Introduced 1824.

R. banksiae lutescens Voss., f. Flowers single, yellow. B.M.7171(c). Introduced about 1870. Zone VII.

Hybrids
R. banksiae × *laevigata?* = *R. fortuneana* Lindl. Climbing shrub with sparingly prickly stems: leaflets 3–5, ovate–lanceolate, lustrous: flowers solitary, large, double, white; pedicels hispid. W.R.36,t(c). F.S.7:256(c). J.F.2:27. Introduced 1850. Zone VII?

Related species
R. cymosa Tratt. Bracts slender, with hooked prickles: leaflets. 3–7, elliptic to ovate–lanceolate, glabrous: flowers small, white, in many-flowered corymbs; styles exserted, pubescent: fruit small, globose. G.C.37:227. (*R. indica* L., in part, *R. microcarpa* Lindl., not Bess., nor Retz., *R. sorbiflora* Focke.) China. Introduced 1904? Zone VII?

Sect. 9. Laevigatae *Thory.*

Evergreen shrubs with scattered hooked prickles: leaflets. usually 3; stipules free or adnate at base, caducous: flowers solitary, large, without bracts; sepals erect, entire, persistent; pedicels and receptacles densely bristly.

69. R. laevigata *Michx.*

Cherokee Rose. To 5 m high: leaflets 3, rarely 5, elliptic–ovate to lance–ovate, 3–6 cm long, acute or acuminate, sharply and finely serrate, glabrous and lustrous, reticulate beneath: flowers white, rarely rose-colored, 6–8 cm across, fragrant: fruit pyriform, bristly, 3.5–4 cm long. Fl. v. B.M.2847(c)., B.R.1922(c). W.R.117,t(c). G.C.III.6:497. Gn.68: 206(h);71:225(h). M.G.21:397. Gn.M.38:174(h). (*R. cherokeensis* Donn, *R. sinica* Murr., *R. ternata* Poir., *R. camellia* Hort.) China; naturalized from Georgia to Florida and Texas. Introduced before 1780 to America. Zone VII.

Hybrids
R. laevigata × ? *odorata* = *R. anemonoides* Rehd. ('Anemone Rose'). With large single light pink flowers R.H.1901:548,t(c). W.R.41,t(c). M.G.11: 345. Gn.62:413;77:340;80:309. Origin before 1896.

R. laevigata × *dilecta*. Here belongs 'Silver Moon'.

R. laevigata × *banksiae*: see under no. 68.

R. laevigata × *bracteata*: see under no. 70.

Sect. 10. Bracteatae *Thory.*

Evergreen shrubs with erect or sarmentose, tomentose or pubescent branches; prickles in pairs; stipules slightly adnate and pectinate: inflorescence with large bracts; sepals entire, reflexed after flowering; receptacle tomentose. Two Asiatic species.

70. R. bracteata *Wendl.*

Macartney Rose. Stems sarmentose or procumbent, tomentose, with stout hooked prickles: leaflets 5–9, elliptic to obovate,1.5–5 cm long, obtuse or acutish, crenate-serrulate, lustrous above, glabrous or pubescent on midrib beneath; rachis glabrous: flowers usually solitary, white, 5–7 cm across, short-stalked, with large laciniated pubescent bracts at base: fruit globose, 3–3.5 cm across, orange–red, woolly. Fl. VIII. B.M.1377(c). B.B.2:268. Gn.70:192. W.R.125,t(c). (*R. macartnea* Dum.-Cours.) China; naturalized from Virginia to Florida and Texas. Introduced 1793. Zone VII.

Hybrids
R. bracteata × *laevigata* Here belongs 'Marie Leonida' (*R. alba odorata*) with creamy white double flowers. W.R.1:127,t(c). Cultivated 1832.

Closely related species
R. clinophylla Thory, with straight prickles and acute leaflets often pubescent beneath; rachis pubescent. B.R.739(c). R.R.3,gr.24.2. (*R. involucrata* Roxb., *R.lyellii* Lindl.) India. Introduced before 1817; probably not hardy. *R. clinophylla* × *persica*: see under no. 1.

Subgenus III. *Platyrhodon* (Hurst) Rehd.

Leaves pinnate: stipules adnate, very narrow with subulate divergent auricles: flowers 1 or 2, with small caducous bracts: receptacle prickly, cup-shaped, achenes on a slightly elevated torus: stamens 150–175; sepals broad, erect, persistent, pinnate. (Sect. *Microphyllae* Crép., *Saintpierrea* Germ, de St. Pierre, *Platyrhodon* Dcne, ex Hurst).

71. R. roxburghii *Tratt.*

Spreading shrub to 2.5 m; bark gray, peeling: leaflets elliptic to elliptic–oblong, 1–2 cm long, acute, sharply serrate, glabrous beneath: flowers pale pink, 5–6 cm across; pedicels and receptacle prickly: fruit depressed–globose, 3–4 cm across, very prickly. (*R. microphylla* Roxb., not Desf.) China, Japan.

Varieties and forms

R. roxburghii plena Rehd., var. With double flowers Fl. VI. B.R.919(c). B.M.3490(c). W.R.135,t(c). G.C.103:87. Gn.81:484,548. S.L.325(h). Introduced 1828. Zone VI? The type of the species.

R. roxburghii normàlis Rehd. & Wils., f. The wild form with single flowers. China. Introduced 1908.

R. roxburghii hirtula (Reg.) Rehd. & Wils., var. Leaflets elliptic to ovate–oblong, 1–2.5 cm long, pubescent beneath. B.M.6548(c). Japan. Introduced before 1880.

Hybrid

R. roxburghii × *rugosa* = *R. micrugosa*: see under no. 26.

Subgenus IV. *Hesperhodos* Cockerell.

Leaves pinnate; stipules adnate, with divergent dilated auricles: flowers solitary, without bracts; receptacle prickly, cup-shaped, without disk; achenes on a conical elongated torus; sepals pinnate, erect, persistent: fruit not fleshy. (Sect. *Minutifoliae* Crèp., *Hesperhodos* Boulenger.)

72. R. stellata *Woot.*

Shrub to 0.6 m; branches with numerous slender yellowish white prickles and closely stellate–pubescent when young: leaflets 3, sometimes 5, broad–cuneate–obovate, 5–10 mm long, incisely dentate, pubescent on both sides: flowers solitary, deep rose–purple, 3.5–5.5 cm across: fruit broad–turbinate, 1–15 cm across, dull reddish. Fl. VI–VIII. W.R.305,t(c). J.L.27:456. Bull. Torr. Bot. Club 25:t.335,f.1–5,9. (*Hesperhodos stellatus* Boulenger.) West Texas to Arizona. Introduced 1902. Zone (V).

Varieties and forms

R. stellata mirifica (Greene) Cockerell, var. More vigorous, to 1.2 m high; bracts without stellate pubescence: leaflets usually 5, sometimes nearly glabrous: flowers 3.5–6 cm across. B.M.9070(c). Gs.20:179. (*R. mirifica* Greene.) North Mexico. Introduced 1916. Zone (V).

Related species

R. minutifolia Engelm. Bracts pubescent, with slender brown prickles: leaflets 5–7 ovate to elliptic, 3–10 mm long, pubescent: flowers pink or white, 2.5 cm across. G.F.1:102. B.C.5:2996. J.L.27:456. Lower California. Cultivated 1910. Zone VII?

INDEX OF REHDER'S CLASSIFICATION OF ROSES

INDEX

NOTE

This index is in letter-by-letter order, whereby hyphens and spaces within index headings are ignored in the alphabetization. Prefixes and terms in parentheses are excluded from the initial alphabetization.

The index is arranged in set-out style with a maximum of three levels of heading. Major discussion of a subject is indicated by bold page numbers. Page numbers suffixed by T and F refer to Tables and Figures respectively.

Cross-reference terms in italics are general cross-references, or refer to subentry terms within the main entry (the main entry is not repeated to save space).

Readers are also advised to refer to the end of each article for additional cross-references – not all of these cross-references have been included in the index.

To save space in the index certain abbreviations have been used:

AARS - All-American Rose Selections
ABA - abscissic acid
AFLP - amplified fragment length polymorphism
ARS- American Rose Society
CIOPORA - Communauté Internationale des Obtenteurs de Plantes Ornementales et Fruitières de Reproduction Asexuée
IAA - indole-3-acetic acid
IPM - integrated pest management
RAPD - random amplified polymorphic DNA
RFLPs - restriction fragment length polymorphisms
SSR - simple sequence repeat
STS - sequence tagged sites
UPOV - Union pour la Protection des Obtentions Végétales
vs. indicates a comparison.

A

B

C

D

F

G

H

I

J

K

L

M

O

P

Q

R

T

U

V

W

X

Y

Z

ISBN 0122276205

9 780122 276200

Encyclopedia of ROSE SCIENCE

ENCYCLOPEDIA OF
ROSE SCIENCE

Editor-in-Chief
ANDREW V ROBERTS

Editors
THOMAS DEBENER
SERGE GUDIN

ELSEVIER
ACADEMIC
PRESS

Amsterdam Boston Heidelberg London New York Oxford
Paris San Diego San Francisco Singapore Sydney Tokyo

Elsevier Ltd., The Boulevard, Langford Lane, Kidlington, Oxford, OX5 1GB, UK

Elsevier Inc., 525 B Street, Suite 1900, San Diego, CA 92101-4495, USA

First edition 2003

Library of Congress Control Number: 2003112413

A catalogue record for this book is available from the British Library

ISBN: 0-12-227620-5 (set)

This book is printed on acid-free paper

Printed in Spain

APPENDIX 4

LISTING OF ROSE NAMES

MODERN ROSES XI reproduced by permission of American Rose Society as International Cultivar Registration Authority for Roses, by Tommy Cairns, PhD, DSc Vice President, American Rose Society

Explanation of a Typical Entry

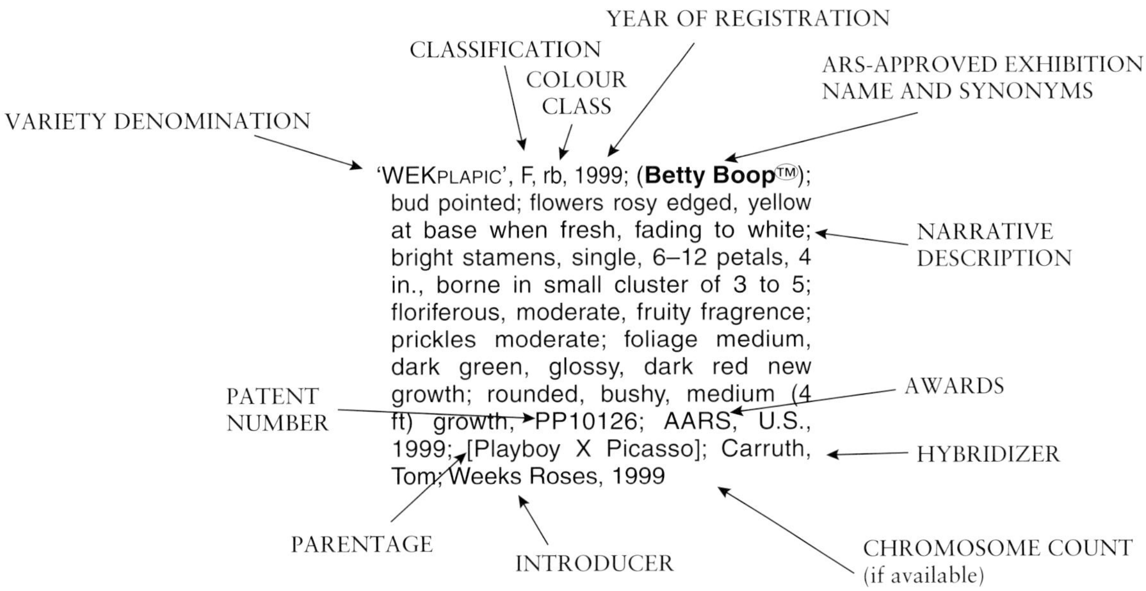

VARIETY DENOMINATION appears in single quotes, usually a code name with the first three letters in capitals to denote a particular hybridizer or introducer. Formal registration by IRAR is further denoted by the remainder of the name appearing in small capital letters.

CLASSIFICATION is according to the ARS-approved list.

COLOUR CLASS is according to the ARS-approved classifications.

YEAR OF REGISTRATION. In the case of some roses, the year of introduction, or when first cultivated has been used in the absence of the year of registration.

ARS-APPROVED EXHIBITION NAME appears in bold face. It is usually a fancy name but can also be the approved variety denomination.

SYNONYMS follow the code name and represent alternative names for the rose assigned in other countries. Such names are not approved for use in ARS rose shows.

NARRATIVE DESCRIPTION is generally an edited version of the official registration information on file with IRAR/ARS.

PARENTAGE appears with the seed parent listed first followed by the pollen parent seperated by the × symbol. With complex parentage descriptions, parentheses are used.

HYBRIDIZER. This information has been taken directly from official registration files maintained by ARS.

INTRODUCER. Where possible the introducer or nursery marketing the rose has been listed.

AWARDS. Where a rose has been given a special award, this information is listed at the end of the narrative statement. A list of abbreviations describing these awards appears elsewhere.

Awards

Generally only Gold Medals (GM) or their equivalent are listed for the various awarding agencies throughout the world.

All-American Rose Selection (AARS)
American Rose Society Awards:
- Award of Excellence (AOE) for Miniature Roses
- American Rose Center Trial Ground Gold, Silver and Bronze Certificates
- Daved Fuerstenberg Prize
- Dr W Van Fleet Medal
- Gertrude M. Hubbard GM
- James Alexander Gamble Rose Fragrance Award
- John Cook Medal
- Miniature Rose Hall of Fame

ADR Anerkannte Deutsche Rose (Germany)
Baden-Baden GM
Bagatelle (Paris) GM
Belfast GM
Copenhagen GM
Dublin GM
Geneva GM
Genova (Genoa, Italy) GM
Glasgow (Tollcross, Scotland) GM and Lord Provost Prize
Le Roeulx (Belgium) GM
Lyon GM
Madrid GM
Monza GM
New Zealand GM
NZ Gold Star of the South Pacific
NZ Silver Star of the South Pacific
Orleans GM
Portland (Oregon, USA) GM
RINZ Auckland Rose of the Year
Rome GM
Rose of The Year, Great Britain (ROTY)
RNRS President's International Trophy (PIT)
RNRS GM
RNRS James Mason Medal
The Hauge GM and the Golden Rose
Tokyo GM
World Federation of Rose Societies (WFRS) Hall of Fame

Abbreviations

AARS	All-American Rose Selections
ADR	Anerkannte Deutsche Rose (Germany)
AOE	Award of Excellence for Miniature Roses
ARS	American Rose Society
ARF	American Rose Foundation
B&A	Bobbink & Aitkins
B.C.	British Columbia
ca.	circa
CIOPORA	Communauté Internationale des Obtenteurs de Plantes Ornamentales et Fruitières de Reproduction Asexuée
cms	centimetres
C-P	The Conard-Pyle Co.
dbl.	double
Dr	Doctor, Docteur
EFR	Edition Français des Roses
ft	foot, feet
GM	Gold Medal (or equivalent) awards
H&S	Howard & Smith
in.	inch, inches
IRAR	International Registration Authority for Roses
J&P	Jackson & Perkins
Mlle	Mademoiselle
Mme	Madame
Mr	Mister
Mons	Monsieur
Mt	Mount
NRS (RNRS)	National Rose Society of Great Britain, (now Royal National Rose Society)
OGR	Old Garden Rose
Prof.	Professor, Professeur
ROTY	Rose of the Year
Souv	Souvenir
URS	Universal Rose Selection
WFRS	World Federation of Rose Societies

ARS-Approved Horticultural Classifications

Species Roses

Sp	Species

Old Garden Roses

A	Alba
Ayr	Ayrshire
B	Bourbon & Climbing Bourbon
Bslt	Boursalt
C	Centifolia
D	Damask
HBc	Hybrid Bracteata
HCh	Hybrid China & Climbing Hybrid China
HEg	Hybrid Eglanteria
HFt	Hybrid Foetida
HGal	Hybrid Gallica
HMult	Hybrid Multiflora
HP	Hybrid Perpetual & Climbing Hybrid Perpetual
HSem	Hybrid Sempervirens
HSet	Hybrid Setigera
HSpn	Hybrid Spinosissima
Misc. OGR	Miscellaneous OGRs
M	Moss & Climbing Moss
N	Noisette
P	Portland
T & Cl T	Tea & Climbing Tea

Modern Roses

F & Cl F	Floribunda & Climbing Floribunda
Gr & Cl Gr	Grandiflora & Climbing Grandiflora
HKor	Hybrid Kordesii
HMoy	Hybrid Moyesii
HMsk	Hybrid Musk
HRg	Hybrid Rugosa
HWich	Hybrid Wichurana
HT & Cl HT	Hybrid Tea & Climbing Hybrid Tea
LCl	Large-Flowered Climber
Min & Cl Min	Miniature & Climbing Miniature
MinFl	Mini-Flora
Pol & Cl Pol	Polyantha & Climbing Polyantha
S	Shrub

ARS Colour Classifications

w	White, Near White & White Blend
ly	Light Yellow
my	Medium Yellow
dy	Deep Yellow
yb	Yellow Blend
ab	Apricot & Apricot Blend
ob	Orange & Orange Blend
op	Orange-Pink & Orange-Pink Blend
or	Orange-Red & Orange-Red Blend
lp	Light Pink
mp	Medium Pink
dp	Deep Pink
pb	Pink Blend
mr	Medium Red
dr	Dark Red
rb	Red Blend
m	Mauve & Mauve Blend
r	Russet

List of Rose Breeder Code Designations

ADA	Adams (Washington, USA)
ADA	Adams (Liffre, France)
AND	Anderson (Aberdeen, Scotland, UK)
ARD	Burden (Canada)
ARM	Armstrong (California, USA)
ANT	Perry (California, USA)
ARO	Armstrong (California, USA)
AUS	Austin (Wolverhampton, England, UK)
BAI	Bailey (Minnesota, USA)
BAR	Barrett (Oxford, England, UK)
BAR	Barni (Psitola, Italy)
BAT	Battersby (Cleveland, England, UK)
BEA	Beales (Norfolk, England, UK)
BED	Chesum (Bedford, England, UK)
BEE	Bees (Chester, England, UK)
BEL	Bell (USA)
BEN	Benardella (New Jersey, USA)
BER	Berg (Connecticut, USA)
BER	Berggren (Germany)
BOS	Bossom (Middlesex, England, UK)
BRA	Bracegirdle (Manchester, England, UK)
BRI	Bridges (North Carolina, USA)
BRI	Briant (Anjou, France)
BRO	Brown (British Columbia, USA)
BUC	Buck (Iowa, USA)
BUL	Bull (UK)
BUR	Burks (Tyler, Texas, USA)
BUR	Burrows (Derby, England, UK)
BUR	Burston (Hertfordshire, England, UK)
BUS	Buster (Iowa, USA)
CAN	Cants (Essex, England, UK)
CAP	Pallek (Ontario, Canada)
CAR	Carlton (Oregon, USA)
CHE	Chessum (Bedford, England, UK)
CHE	Warner (Devon, England, UK)
CLE	Clements (Oregon, USA)
COC	Cocker (Aberdeen, Scotland, UK)
COM	Combe (Grenoble, France)
COO	Cooper (USA)
CRO	Croix (Loire, France)
DAV	Davis (Powys, Wales, UK)
DEL	Delbard (Paris, France)
DES	Desmet (Michigan, USA)
DES	Desamero (California, USA)
DEV	DeVor (California, USA)
DIC	Dickson (Portadown, Northern Ireland, UK)
DOH	Hardgrave (California, USA)
DOR	Dorieux (Montagny, France)
DOT	Dot (Barcelona, Spain)
DRI	Driscoll (Walsalt, England, UK)
DUC	Ducher (Saint-Priest, France)
DUG	Douglas (New Zealand)
DYK	Dykstra (Missouri, USA)
EVE	Everitt (UK)
FER	Fergusson (Devon, England, UK)
FOX	Fox (USA)
FRA	Cowlishaw (Derby, England, UK)
FRY	Fryer (Cheshire, England, UK)
GAU	Gaujar (France)
GAN	Gandy (Leicester, England, UK)
GEE	van Geest (Lincolnshire, England, UK)
GEL	Giles (Florida, USA)
GEN	Genesis (England, UK)
GIL	Gilmore
GLA	Langdale (Lincoln, England, UK)
GRA	Grant (New Zealand)
GRE	Gregory (Lincolnshire, England, UK)
GRI	Griffiths (New Zealand)
GROOT	Grootendorst (Boskoop, The Netherlands)
GUE	Guest (Worcester, England, UK)
HAD	Davidson (California, USA)
HAR	Harkness (Hertfordshire, England, UK)
HAV	Verschuren (Haps, The Netherlands)
HEL	Hetzel (Germany)
HEN	Henson (England, UK)
HER	Herholdt (Eikenhof, South Africa)
HIL	Hillier (England, UK)
HIL	E.G. Hill (Indiana, USA)
HIN	Hintlian (Washington, USA)
HOL	Holmes (Cheshire, England, UK)
HOO	Hooper (Tennessee, USA)
HOR	Horner (Essex, England, UK)
HUB	Huber (Dottikon, Switzerland)
HUX	Huxley (Australia)
INT	Interplant (Leersum, The Netherlands)
JAC	Jackson & Perkins (California, USA)
JAL	Jalbert (British Columbia, Canada)
JAY	Jellyman (Gloucester, England, UK)
JEL	Jelly (Indiana, USA)
JEL	Jellyman (Gloucester, England, UK)
JOH	J.H. Hill (Indiana, USA)
JOHN	Johnson (Illinois, USA)
JOL	Jolly (Maryland, USA)
JON	Jones (California, USA)
JOL	Julius Muha (Canada)
JUS	Justice (Oregon, USA)
KAM	Kammeraad (Aalsmeer, The Netherlands)
KEI	Keisei (Tokyo, Japan)
KEL	Kelly (New York, USA)
KEN	Kenny (Ireland)
KIN	King (Lousiana, USA)
KOR	Kordes (Sparrieshoop, Germany)
KRI	Kriloff (Antibes, France)
LAM	Lammerts (California, USA)
LAN	Sealand (Chester, England, UK)
LAP	Laperrière (France)
LAV	Laver (Ontario, Canada)
LEC	LeCroy (Alabama, USA)
LEE	Lee (Washington, USA)
LEG	LeGrice (Norfolk, England, UK)
LEN	Lens (Oudenbury, Belgium)
LET	Lettunich (California, USA)
LIN	Lindquist (California, USA)
LIN	Linder (South Africa)
LOU	Lougheed (Canada)
LUD	Ludwig Taschner (Pretoria, South Africa)
MAC	McGredy (Auckland, New Zealand)
MAL	Mallerin (Isère, France)
MAL	Maltagliati (California, USA)
MAL	Mander (British Columbia, Canada)
MAR	Martin (California, USA)
MAS	Guillot-Massad (Cremieu, France)
MAT	Mattock (Oxford, England, UK)
MAT	Mather (Australia)
MAT	Matthews (New Zealand)
MEI	Meilland (Antibes, France)
MEH	Mehring (Berkshire, England, UK)
MEJ	Mekdecci (Canada)
MEK	Mekdecci (Canada)
MEY	Meyer (Ohio, USA)
MIC	Michael Williams (California, USA)
MIN	Mini Rose (Texas, USA)
MOR	Moore (California, USA)
NIRP	Nirp (France)
NOA	Noack (Germany)
NOB	Nobbs (New Zealand)
NOL	Nolan (Australia)
NOS	Nostell Priory (West Yorkshire, England, UK)
ORA	Orard (Feyzin, France)
ORT	Ortega (California, USA)
PAL	Peter Long (Australia)
PEA	Pearce (Devon, England, UK)
PEK	Pekmez (France)
PER	Astor Perry (North Carolina, USA)
PIN	Pineau (France)
PIX	Pixie Treasures (California, USA)
POU	Poulsen (Denmark)
PRA	Pratt (New Zealand)
PRO	Prophyl (Tasmania, Australia)
REA	Rearsby (Leicester, England, UK)
REN	Rennie (Ontario, Canada)
RES	Sheldon (Illinois, USA)
REX	Rex (California, USA)
REY	Reynolds (New Zealand)
ROG	Rogan (Croatia)
RUI	deRuiter (The Netherlands)
RUP	Rupert (California, USA)
SAB	Burrows (Derby, England, UK)
SAN	Sanday (Bristol, England, UK)
SAN	Sandbrook (New Zealand)
SAV	Sauvageot (Dijon, France)
SAV	Saville (Massachusetts, USA)
SCH	Schluetar (Texas, USA)
SCH	Schwartz (Maryland, USA)
SCRIV	Scrivens (Worcester, England, UK)
SEA	Sean McCann (Dublin, Ireland)
SHE	Sheridan (London, England, UK)
SHE	Sherwood (New Zealand)
SIE	Sievers (Germany)
SIM	Simpson (Hamilton, New Zealand)
SMI	Smith (Nottingham, England, UK)
SOC	Southern Cross (Christchurch, New Zealand)
SOM	Somerfield (New Zealand)
SPE	Spek (Boskoop, The Netherlands)
SPI	Spies (Virginia, USA)
SPO	Spooner (Oregon, USA)
SPR	Sproul (California, USA)
STRO	Strobel (Germany)
STA	Starkl (Hungary)
STA	Strickland (California, USA)
STE	Stephens (New Zealand)
STI	Stibbard (Australia)
STR	Strickland (California, USA)
SUN	Schuurman (Henderson, New Zealand)
TAL	Taylor (Alabama, USA)
TAN	Tantau (Uetersen, Germany)
TAY	Taylor (Michigan, USA)
THO	Thompson (Wales, UK)
TIN	Tiny Petals (California, USA)
TOM	Thompson (Australia)
TRO	Thomas Robinson (Channel Islands, England, UK)
TSCH	Tschanz (Switzerland)
TUC	Tucker (Tennessee, USA)
TWO	Twomey (California, USA)
UHL	Uhl (Germany)

A

1997 Traditional Home™, *see* 'WAMhome'

21 Again!, *see* 'MEInimo'

A Caen la Paix®, *see* ORAkah

'**A Capella**', HT, mr, 1984; flowers large blooms, dbl., 35 petals, borne singly, slight fragrance; foliage large, medium green, matt; upright growth; Bronze, ARC TG, 1984; [Command Performance X Tiffany]; Stoddard, Louis

'**A Feuilles de Chanvre**', A, lp, 1800; flowers medium, dbl.; Eurosa, Gallicanae; Braun, H.; (Sangerhausen)

A Fleurs de Rose Tremier de la Chine, C; Unknown; (Hay)

A Fleurs Doubles Violettes, C, m; (Cavriglia)

'**A Fleurs Gigantesques**', HGal, mp, 1813; flowers rich deep pink, large, very dbl., intense fragrance; bushy growth

A Fleurs Roses de Laffay, HSem; Eurosa, Synstylae; Laffay, M.

'**A Longs Pédoncules**', M, lp, 1854; bud pale green, mossy, long peduncles; flowers pink, flushed lilac, many petalled, small blooms, nodding in clusters; foliage small, round, soft green; vigorous growth; Eurosa, Gallicanae; Robert

'**A Night in June**', HT, or, 1935; flowers coral-red, very large, dbl., slight fragrance; foliage leathery; vigorous, bushy growth; [Evening Star sport]; Elmer's Nursery

A Shropshire Lad, *see* 'AUSled'

'**A. Denis**', HT, my, 1935; flowers lemon-yellow, streaked carmine, large, dbl., intense fragrance; [Gorgeous X Marion Cran]; Böhm, J.

'**A. Drawiel**', HP, dr, 1887; flowers large, dbl., moderate fragrance; Lévêque; (Sangerhausen)

'**A. Dvorak**', HT, op, 1933; flowers light pinkish orange, open, large; foliage glossy, dark; bushy growth; [Mme Butterfly X Gorgeous]; Böhm, J.

'**A. G. A. Rappard**', HT, op, 1934; (A.G.A. Ridder van Rappard); flowers salmon, verging on neyron pink, well-shaped, dbl., slight fragrance; foliage glossy, bright green; vigorous, bushy growth; Buisman, G. A. H.

'**A. G. Furness**', HT, mr, 1941; flowers rich red, semi-climbing growth; [Sensation seedling]; Clark, A.

'**A. Geoffroy de st Hilaire**', HP, mr, 1878; flowers medium, dbl., intense fragrance; tall growth; Verdier; (Sangerhausen)

A. H. Kirk, HT, op; flowers salmon pink with carmine, large, dbl., moderate fragrance; (Sangerhausen)

A. J. Herwig, F, 1965; Buisman, G. A. H.; (Cavriglia)

'**A. MacKenzie**', S, rb, 1985; (Alexander MacKenzie); bud ovoid; flowers medium red, reverse lighter, dbl., 45 petals, 3 in., cupped, blooms in clusters of 6–12, repeat bloom, moderate fragrance; purple prickles; foliage yellow-green, glossy, leathery; upright growth; [Queen Elizabeth X (Red Dawn X Suzanne)]; Svedja, Felicitas; Agriculture Canada, 1994

'**A. N. W. B. Rose**', HT, w, 1933; flowers white, tinted yellow, open, large, dbl., slight fragrance; very vigorous growth; [Frau Karl Druschki X Souv. de Claudius Pernet]; Buisman, G. A. H.

'**A. W. Jessep**', HT, dp, 1952; flowers rich cerise-pink, large, dbl., 38 petals, moderate fragrance; vigorous growth; Clark, A.

A.G.A. Ridder van Rappard, *see* '**A. G. A. Rappard**'

A.R.S. Centennial, *see* 'SAVars'

Aachener Dom, *see* 'MEIcapinal'

'**Aafje Heynis**', HT, mr, 1964; flowers bright red, well-shaped, large; foliage glossy, light green; vigorous, upright growth; [Prima Ballerina X Salvo]; Buisman, G. A. H.

Aaland, S, dp

'**Aalsmeer Gold**'®, HT, dy, 1978; (Bekola); bud long, pointed; flowers deep yellow, dbl., 23 petals, exhibition form, borne singly, slight fragrance; foliage glossy; vigorous, upright, bushy growth; [Berolina X Seedling]; Kordes

Aasmeer, HT, lp; flowers peachish-orange with yellow tones

Aba Saheb, HT, ob, 1993; Chiplunkar

'**Abaillard**', HGal, pb; (Abalard); flowers rose marbled, dbl.; Eurosa, Gallicanae; Vibert, 1845

Abalard, *see* '**Abaillard**'

'**Abasanta**', HT, dp, 1956; bud pointed, cerise; flowers carmine, occasionally streaked white, semi-dbl., 15–25 petals, 3.5 in., moderate fragrance; foliage dark; compact growth; [Red Columbia X (Red Columbia X Tausendschön sport)]; Motose

'**Abba Dabba**', Min, mr, 1979; bud long, pointed; flowers, dbl., 38 petals, borne 1–5 per cluster, slight fragrance; tiny, curved prickles; foliage small, medium green; compact, bushy growth; [Red Can Can X Seedling]; Lyon

Abbaye de Cluny™, *see* 'MEIbrinpay'

Abbé Berlèze, HP, dp; Eurosa, Chinensis; Guillot et Fils, 1864; (Roseraie de l'Hay)

'**Abbé Bramerel**', HP, dr, 1871; flowers large, dbl., slight fragrance; Guillot et Fils; (Sangerhausen)

'**Abbé Girardin**', B, mp, 1881; flowers carmine-pink, center darker, large, dbl., moderate fragrance; [Louise Odier X Hermosa]; Bernaix, A.

'**Abbé Giraudier**', HP, mr, 1869; flowers cherry red, very large, dbl.; Eurosa, Chinensis; Levet, 1869; (Sangerhausen)

Abbe Lemire, *see* ORAnat

'**Abbé Millot**', HT, lp, 1900; flowers large, dbl.; Corboeuf; (Sangerhausen)

Abbeyfield Rose, *see* 'COCbrose'

'**Abbotswood**', S, mp, 1954; dbl.; habit similar to R. canina; [Chance hybrid of R. canina X Unknown garden variety]; Hilling

Abdul Hamid, Misc OGR, mr

'**Abel Carrière**', HP, dr; flowers velvety crimson, brighter center, large, dbl., 45 petals; Eurosa, Chinensis; [Baron de Bonstetten X Seedling]; Verdier, E., 1875

'**Abel Grant**', HP, lp, 1866; flowers large, dbl., intense fragrance; medium growth; Damaizin; (Sangerhausen)

Abelzieds, HRg, mp, 1957; flowers large, semi-dbl., slight fragrance; medium height; Rieksta; (Sangerhausen)

'**Abendröte**', HP, op, 1919; flowers light coral-red; compact growth; [Frau Karl Druschki X Juliet]; Ebeling; Teschendorff

Aberdeen Celebration, *see* COCmystery

'**Aberdonian**', F, rb, 1974; flowers golden bronze and scarlet, large, dbl., 20 petals, slight fragrance; foliage glossy; [(Evelyn Fison X Manx Queen) X (Sabine X Circus)]; Cocker

'Abhaya', HT, pb, 1997; (**Magic Medley**); flowers pink with dark red markings, white reverse, dbl., 26–40 petals, 6 in., borne mostly singly, slight fragrance; foliage large, medium green, dull; bushy, medium (3 ft) growth; [Pristine X Priyatama]; Viraraghavan, M.S.

'**Abhisarika**', HT, rb, 1977; bud long, pointed; flowers red blend, striped, small, dbl., 48 petals, 2.5–3 in., exhibition form, slight fragrance; foliage glossy, yellowish; vigorous growth; [Induced mutant X Kiss of Fire]; IARI

'**Abiding Faith**', HT, dp, 1954; bud semi-ovoid; flowers deep rose-pink, dbl., 20–35 petals, 4–5 in., intense fragrance; vigorous, upright growth; [La France sport X (Senator X Florex)]; Motose

'**Abiding Faith, Climbing**', Cl HT, dp, 1957; Motose

Abigaile®, *see* 'TANELAIGIB'

Abington Park Northampton, HT, op, 1997; flowers very full, large, very dbl., 41 petals, 7+ in., borne in small clusters, intense fragrance; some prickles; foliage medium, medium green; bushy, medium (3 ft) growth; [Cynthia Brooke X Little Darling]; Jones, L.J.

'ABOL', HT, w, 1927; flowers ivory-white tinted blush, large, dbl.; intense fragrance; Evans; Beckwith

'ABONDANT', Pol, mp, 1914; flowers medium-large, dbl.; Turbat; (Sangerhausen)

Abracadabra, *see* 'JACBUTE'

Abraham Darby™, *see* 'AUSCOT'

'ABRAHAM ZIMMERMAN', HP, mr, 1879; flowers large, dbl.; moderate fragrance; Lévêque; (Sangerhausen)

Abraxas, S, dr, 1973; flowers medium-large, dbl.; VEG; (Sangerhausen)

'ABRICOT', HT, ab, 1929; flowers apricot and coral-salmon, reverse coral-red; [Mrs Aaron Ward X Jean C.N. Forestier]; Barbier

'ABRICOTÉ', T, ab; flowers apricot, margins flesh, dbl., cupped; Dupuis, Prior to 1848

Absolute Hit, *see* POUlrougel

Absolutely™, *see* 'SAVALUTE'

Abu, HT, dp; flowers large, dbl., intense fragrance; VEG; (Sangerhausen)

'ABUNDANCE', F, mp, 1974; dbl., 30 petals, 4.5 in., slight fragrance; foliage dark; low, bushy growth; [Seedling X Firecracker]; Gandy, Douglas L.

'AC DE MONTARVILLE', HKor, mp, 1999; (**De Montarville**); flowers medium pink, reverse light pink, dbl. (17–25 petals), medium (7.8 cms) blooms bourne mostly single; slight fragrance; few prickles; foliage small, blue-green, dull; compact, medium growth; Agriculture et Agroalimentaire Canada

'AC MARIE VICTORIN', HKor, pb, 1999; (**Marie-Victorin**); flowers medium pink-yellow, reverse pink-yellow, full (26 petals), medium (7.6 cms) blooms borne in small clusters; slight fragrance; few prickles; foliage medium. Light green, semi-glossy; arched, medium (1.5 m) growth; [Arthur Bell X (R. kordesii X Max Graf O.P.)]; Agriculture et Agroalimentaire Canada

'AC WILLIAM-BOOTH', HKor, mr, 1999; (**William-Booth**); flowers medium red, reverse medium pink, single (4–11 petals), small (5 cm) blooms borne in large clusters; fragrant; moderate prickles; foliage medium, dark green, semi-glossy; arching, tall (2 cm) growth; [(R. kordesii X Max Graf O.P.) X (Arthur Bell X Apple Jack)]; Agriculture et Agroalimentaire Canada

Academy®, *see* 'MACGUTSY'

'ACADIAN', HCh, dp, 1986; flowers deep, bright pink, medium, single, 5 petals, intense fragrance; fine prickles; foliage small, medium green, matt; upright, bushy, hardy growth; [R. nitida X R. chinensis semperflorens]; James, John

AcapellaR, *see* TANallepal

'ACAPULCO', F, yb, 1962; bud pointed; flowers yellow and light orange, often flushed pink, medium, dbl., 30–40 petals, cupped, clusters, slight fragrance; foliage dark, glossy; bushy, compact growth; [Seedling X Masquerade]; Von Abrams; Peterson & Dering

Acapulco, *see* DICblender

Acaritha, HRg; Eurosa, Cinnamomeae; Wartz; (Hay)

'ACCENT', F, mr, 1977; bud ovoid; flowers cardinal-red, dbl., 25 petals, 2–2.5 in., flat, slight fragrance; foliage small, dark, leathery; bushy, compact growth; [Marlena X Seedling]; Warriner, William A.; J&P

Acclaim, *see* 'JACAIM'

'ACCOLADE', HT, rb, 1979; bud ovoid; flowers bright red, shaded darker, dbl., 48 petals, exhibition form, borne mostly singly, slight fragrance; hooked, brown prickles; foliage dark, matt; vigorous growth; [(Daily Sketch X Charles Mallerin) X Peter Frankenfeld]; Dawson, George; Rainbow Roses

'ACCORD', F, ab, 1965; flowers apricot-peach, reverse carmine, well formed, exhibition form, clusters, moderate fragrance; foliage dark; bushy growth; [Circus X Seedling]; Mason, P.G.

Ace of Diamonds, *see* 'BRIACE'

Ace of Hearts, *see* 'KORRED'

'ACERVATE', Pol, mp, 1934; Miers

Acey Deucy™, *see* 'SAVATHREE'

Achantha, Gr, dr, 1987; Viraraghavan, M.S.

'ACHIEVEMENT', HWich, dp, 1925; foliage variegated; [Dorcas sport]; English

'ACHILLE', Pol, mp, 1936; flowers medium-large, semi-dbl.; Guinoisseau; (Sangerhausen)

'ACHILLE CESBRON', HP, mr, 1894; flowers very large, dbl., moderate fragrance; Rousset; (Sangerhausen)

Achille Gonod, HP; Eurosa, Chinensis; Gonod, 1864; (Roseraie de l'Hay)

'ACHILLE GONOD', HP, mr, 1864; flowers large, dbl., intense fragrance; Gonod; (Sangerhausen)

'ACIDALIE', B, w; flowers white with blush center, large, dbl., globular, recurrent bloom; vigorous growth; Eurosa, Chinensis; Rousseau, 1838

'ACQUA CHETA', HT, rb, 1962; flowers magenta-red, reverse silvery, dbl., 50 petals; foliage dark; compact growth; [Crimson Glory X Peace]; Giacomasso

'ACTRICE', HT, rb, 1966; bud ovoid; flowers red and light pink, large, dbl.; foliage dark; [Tzigane X Kordes' Perfecta]; Verschuren, A.; Van Engelen

'ADA PERRY', Min, op, 1978; bud ovoid; flowers soft coral-orange, medium, dbl., 40 petals, exhibition form, slight fragrance; foliage dark; vigorous, upright growth; [Little Darling X Coral Treasure seedling]; Bennett, Cecilia 'Dee'; Tiny Petals Nursery

'ADABEL', S, rb, 1987; (**Belles and Beaus**); flowers red-white bicolor, floribunda type, medium, dbl., 15–25 petals, exhibition form, no fragrance; foliage medium, dark green, glossy, disease resistant; upright, moderately hardy growth; [Little Darling X Seedling]; Adams, Dr Neil D.

ADAburi, F, w, 1994; (**Dame de L'Etoile**®); Adam, M.

'ADADISRES', S, mr, 1987; (**Disraeli**); flowers medium, dbl., 26–40 petals, no fragrance; foliage large, dark green, glossy; upright, bushy, tall, broad growth; [Hamburger Phoenix X Seedling]; Adams, Dr Neil D.

ADAfetap, *see* **Rose d'Annecy**®

'ADAGIO', HT, dr, 1971; bud very long, pointed; flowers blood-red, dbl., 28–35 petals, 3.5–4 in., globular; foliage dark; vigorous, bushy growth; [Seedling X Uncle Walter]; Lens

ADAharos, HT, lp, 1995; (**Rose de Rennes**®); Adam, M.

'ADAIR ROCHE', HT, pb, 1968; flowers deep pink, reverse silver, well-formed, large, dbl., 30 petals, slight fragrance; foliage glossy; GM, Belfast, 1971; [Paddy McGredy X Femina seedling]; McGredy, Sam IV; McGredy

'ADAJAN', S, yb, 1993; (**Jan's Wedding**); dbl., 15–25 petals, 2–2.5 in., borne in large and medium clusters; some prickles; foliage medium, dark green, semi-glossy; bushy growth; tall (6 ft) winter hardy; [Dornroschen X Lichtkonigin Lucia]; Adams, Dr Neil D.; Rosehaven Nursery, 1993

'ADAM', T, mp; (President); flowers rich rosy-salmon, very large, dbl., globular; Adam, 1833

'ADAM MESSERICH', B, mr; flowers rose-red, semi-dbl., cupped, recurrent bloom; foliage glossy, light; vigorous, bushy growth; Eurosa, Chinensis; [Frau Oberhofgärtner Singer X (Louise Odier seedling X Louis Philippe)]; Lambert, P., 1920

'ADAM RACKLES', T, pb; flowers marbled and spackled pink; Rommel, 1905

ADAmona, F, mr; (**Santiago**®); Adam

ADAmonbu, F, w; (**Montauban de Bretagne**®); Adam

Adam's Smile™, *see* 'SAVASMILE'

ADAnuamn, HT, mp, 1992; (**Parfum d'Armor**®); Adam, M.

'ADASNOW', S, w, 1993; (**Snow Bear**); dbl., 15–25 petals, 3–3.25 in., borne in clusters of 3–5; some prickles; foliage medium, dark green, semi-glossy; upright (4 ft) growth; [Dornroschen X Rosanna]; Adams, Dr Neil D.; Rosehaven Nursery, 1993

'ADASUN', LCl, my, 1993; (**Summer Sun**); dbl., 15–25 petals, 1.5–2.75 in., borne in large clusters, repeat bloom; some prickles; foliage medium, dark green, semi-glossy; tall (6 ft), spreading growth; hardy; [Prairie Princess X Lichtkonigin Lucia]; Adams, Dr Neil D.; Rosehaven Nursery, 1993

ADAtapora, S, mr, 1996; (**Bourgogne No. 2**); Adam, M.

ADAtonysil, F, dy, 1995; (**Rose d'Or de Montreux**); Adam, M.

ADAzombar, F, lp; (**Venise**); Adam, 1997

Added Touch, *see* 'LYOAD'

Addo Heritage, *see* POUIdotage

Adela®, *see* FEgama

'**ADÉLAIDE D'ORLÉANS**', HSem, w, 1826; bud very small, well-formed; flowers pale rose, yellow stamens, semi-dbl., borne in clusters; vigorous (15 ft) growth; Jacques

'**ADELAIDE HOODLESS**', S, dp, 1975; bud ovoid; flowers light red, medium, semi-dbl., repeat bloom, slight fragrance; foliage glossy; vigorous, bushy growth; [Fire King X (J.W. Fargo X Assiniboine)]; Marshall, H.H.; Canadian Ornamental Plant Foundation, 1972

'**ADELAIDE LEE**', Min, pb, 1986; flowers white with red petal edges, urn-shaped, dbl., 25 petals, 1 in., borne singly, slight fragrance; few, green, straight prickles; foliage medium green, matt; upright growth; [Gene Boerner X Magic Carrousel]; Stoddard, Louis

'**ADÉLAÏDE MOULLÉ**', HWich, pb, 1902; flowers lilac-pink, center carmine, clusters, midseason bloom; [R. wichurana X Souv. de Catherine Guillot]; Barbier

Adelaide Tonight, *see* 'TOMDEL'

'**ADÈLE**', HT, dy, 1935; flowers clear gold, base deeper; [Roselandia X Clarice Goodacre]; Lens

'**ADÈLE**', HGal, pb, 1815; Vibert

'**ADÈLE COURTOISÉ**', HGal, dp; flowers rosy red, small, very dbl.

'**ADÈLE CROFTON**', HT, yb, 1928; flowers yellow overlaid scarlet-orange, dbl., moderate fragrance; Dickson, A.

'**ADÈLE DESCEMET**', HGal; Descemet, M., prior to 1814

'**ADÈLE FREY**', Cl HT, dp, 1911; flowers large, very dbl., slight fragrance; Walter; (Sangerhausen)

'**ADÈLE HEU**', HGal, 1816; flowers purple-pink with white shadings, medium large, dbl., moderate fragrance; Vibert, 1816; (Sangerhausen)

'**ADÈLE PAVIC**', lp, 1850; flowers medium large, dbl., moderate fragrance; Vibert; (Sangerhausen)

Adèle Pradel, *see* '**MME BRAVY**'

'**ADÈLE PRÉVOST**', HGal, lp; flowers blush, center pink, large, dbl.; vigorous, upright growth; Eurosa, Gallicanae

'**ADELINE**', C, pb, 1830; flowers vivid rose, paler toward edge, dbl.; foliage dark; compact, branching growth; Vibert

'**ADELINE**', M, dp; bud well mossed; flowers lilac-rose, dbl., 2 in.; foliage light green; compact, well-branched growth

'**ADELINE GENÉE**', F, my, 1967; bud plump; dbl., 50 petals, 4 in., blooms in clusters, slight fragrance; foliage glossy; low, bushy growth; [Paddy McGredy X Seedling]; Harkness

Adesmano®, HT, pb; Adam

ADharman, HT, dr, 1993; (**Commandant Cousteau**); Adam, M.

'**ADIANTIFOLIA**', HRg, mp, 1907; Eurosa, Cinnamomeae; Cochet-Cochet; (Hay)

'**ADIEU DE BORDIER**', HGal, mr; flowers vivid red, very dbl.

Aditya, HT, dy, 1990; K&S

Admirable, *see* 'SEARODNEY'

Admiral, *see* '**waves**'

'**ADMIRAL DEWEY**', HT, mr, 1899; flowers large, dbl.; Taylor; (Sangerhausen)

'**ADMIRAL RODNEY**', HT, pb, 1973; flowers pale rose-pink, reverse deeper, dbl., 45 petals, 4–4.5 in., intense fragrance; foliage large, glossy, dark; vigorous growth; Trew, C.; Warley Rose Gardens

'**ADMIRAL SCHLEY**', HT, r, 1901; flowers red; [Col. Joffé X Général Jacqueminot]; Cook, J.W.

'**ADMIRAL WARD**', HT, rb, 1915; flowers crimson-red, shaded fiery red and velvety purple, large, dbl., globular, moderate fragrance; [Seedling X Château de Clos Vougeot]; Pernet-Ducher

'**ADMIRATION**', HT, rb, 1922; flowers cream, shaded vermilion, pointed, large, dbl., intense fragrance; foliage light green; moderately bushy growth; McGredy

Admired Miranda®, *see* 'AUSMIR'

'**ADOLF DEEGEN**', HT, pb, 1935; flowers rosy pink with fiery streaks, large, moderate fragrance; [Ophelia X Wilhelm Kordes]; Böhm, J.

'**ADOLF GRILLE**', F, dp, 1940; bud pointed, ovoid; flowers scarlet-carmine, dbl., 25 petals, 4–5 in., cupped, in clusters, slight fragrance; foliage leathery, dark, wrinkled; vigorous, bushy, compact growth; [Dance of Joy X (Cathrine Kordes X E.G. Hill)]; Kordes; J&P

'**ADOLF HORSTMANN**'®, HT, yb, 1971; (Adolph Horstmann); flowers yellow-orange, dbl., 5 in., exhibition form, slight fragrance; foliage glossy; vigorous, upright growth; [Colour Wonder X Dr A.J. Verhage]; Kordes, R.; Kordes

'**ADOLF KÄRGER**', HT, my, 1918; flowers golden yellow, fading, slight fragrance; [Cissie Easlea X Sunburst]; Kordes, H.

'**ADOLF KOSCHEL**', HT, ab, 1918; flowers orange-yellow, moderate fragrance; [Harry Kirk X Louise Catherine Breslau]; Kordes, H.

Adolf Papeleu, F, w, 1998; RvS-Melle

'**ADOLPH GUDE**', HT, pb, 1941; flowers rose-pink, reverse darker, dbl., 30 petals, 5.5–6 in., exhibition form, moderate fragrance; very vigorous growth; [Red Radiance sport]; Gude

Adolph Horstmann, *see* '**ADOLF HORSTMANN**'®

'**ADONIS**', HT, ly, 1921; flowers ivory-yellow, shaded lemon, dbl., moderate fragrance; GM, NRS, 1920; [Sunburst X American Beauty]; Bees

'**ADORA**', Pol, mp, 1936; bud deep flame-pink; flowers rose-pink, dbl., intense fragrance; vigorous, dwarf growth; Beckwith

Adora, HT, op, 1986; Kasturi

'**ADORABLE**', HT, lp, 1930; flowers flesh-pink; [Columbia sport]; Eichholz

'**ADORATION**', HT, op, 1940; (Ile de France); bud long, pointed; flowers bright salmon, dbl., 22 petals, 5 in., moderate fragrance; foliage leathery; vigorous, bushy growth; [(Mme Joseph Perraud X Seedling) X Seedling]; Gaujard; J&P

'**ADORN**', Min, pb, 1986; flowers medium pink blending lighter, reverse medium pink, urn-shape, dbl., 30 petals, exhibition form, borne singly, no fragrance; slender, light prickles; foliage medium, dark, semi-glossy; medium, upright, bushy growth; [Seedling X Seedling]; McDaniel, Earl; McDaniel's Min. Roses

Adrian, F; Camprubi, C.; (Cavriglia)

'**ADRIAN BAILEY**', F, op, 1972; bud globular; flowers orange-scarlet, dbl., 28 petals, 3.5 in.; foliage glossy; bushy, upright growth; [Fragrant Cloud X Evelyn Fison]; Bailey

'**ADRIE STOKMAN**', HT, mp, 1948; flowers darker pink than briarcliff; Stokman

Adrien Mercier®, F, rb

'**ADRIENNE BERMAN**', HT, mr, 1996; flowers very large blooms spiraled, thick-textured, dbl., 26–40 petals, borne mostly singly, moderate fragrance; some prickles; foliage large, dark green, semi-glossy; upright, medium growth; [(Precious Platinum X Silver Jubilee) X Cardiff Bay]; Poole, Lionel; F. Haynes & Partners, 1997

'**ADRIENNE DE CARDOVILLE**', B, lp, 1859; Guillot Père

'Adrienne Leal', F, dp, 1965; bud ovoid; flowers deep pink to soft mauve, medium, dbl., slight fragrance; foliage dark, leathery; vigorous, upright growth; [Roundelay sport]; Leal

'Adrienne Martin', HT, dp, 1930; flowers carmine, base yellow, very dbl., cupped, moderate fragrance; foliage glossy, dark; vigorous, bushy, low growth; [Recuerdo de Angel Peluffo X The Queen Alexandra Rose]; Buatois

'Advance', HT, rb, 1940; flowers orange-flame, shaded cerise, reverse cerise, large, dbl., slight fragrance; foliage leathery, dark; long, strong stems; vigorous, bushy growth; [Comtesse Vandal X Mrs Sam McGredy]; LeGrice; C-P

Advance Guard, *see* **'Wilhelm Teetzmann'**

Adventure, *see* **'Aventure'**

'Advocate', HT, mr, 1928; flowers crimson-scarlet, dbl., moderate fragrance; Dickson, A.

Aebleblomst, HMsk, lp, 1955; Petersen

'Aëlita', S, w, 1952; flowers white tinted green, large, dbl., 50 petals, intense fragrance; foliage glossy; vigorous growth; [New Dawn seedling]; Shtanko, E.E.

Ænnchen Müller, *see* **'Annchen Müller'**

'Aenne Burda', HT, mr, 1973; flowers blood-red, large, dbl., exhibition form, slight fragrance; foliage large, glossy; vigorous, upright, bushy growth; Kordes

'Aenne Kreis', HT, ab, 1930; flowers orange-yellow, reverse lighter, well-formed, dbl.; vigorous, branching growth; [Wilhelm Kordes sport]; Kreis

Affirm, HT, mp, 1998; (McMillan's Pink); flowers medium pink full (26–40 petals), high pointed centre, medium (4 in.) Blooms borne singly or small clusters; slight frangance; numerous prickles; foilage medium, medium green, dull; spreading, low (36 in.) Growth; [Seedling X Seedling]; McMillan, Thomas; Hortico, Inc., 1998

'Aflame', F, op, 1954; flowers orange-strawberry, semi-dbl., 4 in.; blooms in clusters, intense fragrance; foliage dark, glossy; vigorous, low, spreading growth; [(Poulsen's Pink X Ellinor LeGrice) X Mrs Pierre S. duPont]; LeGrice

'Africa Star', HT, m, 1965; dbl., 65 petals, 3.5–4 in.; foliage coppery; bushy growth; West; Harkness

African Queen, *see* TANeliet

'African Sunset', HT, ab, 1966; flowers apricot-orange, pointed, dbl., 4.5 in., moderate fragrance; free growth; [Sutter's Gold X Chantré]; Herholdt, J.A.

African Sunset, *see* JACpik

After Glow, *see* JOHilmar

After Midnight™, *see* 'KINnight'

'Afterglow', HT, ab, 1930; flowers apricot-yellow, dbl., intense fragrance; [Seedling X Souv. de Claudius Pernet]; Hill, Joseph H., Co.

'Afterglow', HT, ob, 1938; (Sam Buff); bud long, pointed, orange; flowers golden yellow, reverse golden orange, large, dbl., slight fragrance; foliage bluish green, glossy; long stems; vigorous, bushy growth; [Mrs Sam McGredy sport]; LeGrice

Afterglow, F, dp

Afternoon Delight, *see* 'JOLaft'

'Agar', HGal, pb, 1843; flowers dark rose, spotted, medium, dbl.; Vibert, 1843

'Agate Pourpre', HT, mr, 1965; flowers currant-red, dbl., 5–6. in., cupped, intermittent bloom; foliage dark, serrated; [Impeccable X Rome Glory]; Delbard-Chabert; Cuthbert

Agatha, *see* **R. gallica agatha**

'Agatha Christie', F, dp, 1966; bud ovoid; flowers pink-red, blooms in clusters, dbl.; foliage dark; compact growth; Buisman, G. A. H.

'Agathe Fatime', HGal, mp; Vibert, before 1815

'Agathe Incarnata', HGal, mp; flowers pink, quartered; habit intermediate between the Gallicas and the Damasks

Agathe Nabonnand, *see* **'Mme Agathe Nabonnand'**

'Age Tendre', HT, dp, 1966; bud long, pointed; flowers rose, large, dbl., exhibition form; strong stems; vigorous growth; [Queen Elizabeth X Spartan]; Croix, P.

Agemy, HT, ob

Agemy Sport, HT, dy

Agéna®, *see* 'DELcus'

'Agénor', HGal, m; flowers reddish purple, medium, dbl.; Vibert, ca 1832

Agkon, HT, dp; Agel

'Aglaia', HMult, ly, 1896; (Yellow Rambler); flowers straw-yellow to white, dbl., in clusters, moderate fragrance; foliage glossy; vigorous growth; [R. multiflora X Rêve d'Or]; Schmitt; Lambert, P.

'Agnes', HRg, my, 1900; flowers pale amber, center deeper, open, dbl., profuse, non-recurrent bloom, moderate; foliage light green, glossy, rugose; short stems; vigorous (6 ft), bushy growth; very hardy; (21); [R. rugosa X R. foetida persiana]; Saunders, 1900; Central Exp. Farm, 1922

'Agnès Ageron', HT, mr, 1958; flowers cerise-red, reverse tinted currant-red, well formed; very vigorous growth; [Mme Méha Sabatier X Léonce Colombier]; Arles; Roses-France

'Agnes Barclay', HT, yb, 1927; flowers yellow and reddish salmon, moderate fragrance; Clark, A.; NRS Victoria

Agnes Bernauer®, *see* KORnauer

'Agnes De Puy', HT, dp, 1930; flowers geranium-red veined gold, large blooms, dbl., 22 petals, exhibition form, moderate fragrance; long stems; very vigorous growth; hardy.; [Lady Battersea X (Honeymoon X Mme Butterfly)]; De Puy

'Agnes Emily Carman', HRg, mr, 1898; flowers bright crimson, large, dbl., some repeat bloom; foliage large, rugose; vigorous (5 ft) growth; Carman

'Agnes Glover', HT, dr, 1924; flowers deep velvety crimson, moderate fragrance; [Admiral Ward X George Dickson]; Chaplin Bros.

'Agnes Kruse', F, 1936; flowers velvety red, fiery, large; vigorous growth; [Mme Edouard Herriot X Eblouissant]; Tantau

'Agnes Laura Blackburn', F, yb, 1989; bud pointed; flowers bicolor yellow, medium, single, 5 petals, borne in sprays; [City of Portsmouth X Seedling]; Cants of Colchester, Ltd.

'Agnes Lucké', HT, pb, 1959; bud long, pointed; flowers cerise-pink, reverse whitish pink, large to medium size, dbl., 50 petals, exhibition form, moderate fragrance; foliage leathery; vigorous, upright growth; [Happiness X Peace, Climbing]; Armbrust; Langbecker

'Agnes Marguerite', HT, or, 1953; flowers reddish orange, shaded apricot; foliage glossy; Bide

'Agnes Roggen', HT, pb, 1926; flowers pale pink, reverse carmine, dbl., moderate fragrance; Leenders, M.

'Agnes und Bertha', HMult, lp, 1926; flowers small, single; Bruder Alfons; (Sangerhausen)

'Agnes Winchel', HT, pb, 1989; bud pointed; flowers light pink with deep pink border, medium, dbl., 28 petals, exhibition form, borne singly, slight, fruity fragrance; globular, medium, orange fruit; prickles slightly hooked, medium, green; foliage medium, dark green, semi-glossy; upright, medium growth; PP007826; [Dorothy Anne X Seedling]; Winchel, Joseph F.; Coiner Nursery, 1990

Agni, HT, or, 1985; G&L

Agnihotri, HT, dp, 1981; Kasturi

'Agreement', F, pb, 1971; flowers deep glowing pink, base golden, dbl., 30 petals, 3 in., exhibition form; foliage glossy, bright green; tall, very free growth; LeGrice

'Agrippina', HCh, lp; flowers crimson with typical china white line down center of petals

Agrippina, Climbing, *see* **'Cramoisi Supérieur, Climbing'**

Ah Mow, T, lp; Hay

Ahalya, F, lp, 1969; Kasturi

'AHIMSA', HT, my, 1996; (Orient Silk); dbl., 26–40 petals, 5 in., borne mostly singly, moderate fragrance; no prickles; foliage large, medium green, semi-glossy; upright, medium (3 ft) growth; [Mme Charles Sauvage X Seedling]; Viraraghavan, M.S.; Hortico Roses, 1996

'AHLIN', F, ob, 1979; bud ovoid; flowers orange, 15 petals, exhibition form, borne 4-6 per cluster, slight fragrance; hooked prickles; foliage large, leathery; upright growth; [Seedling X Little Darling]; Fong

'AHOI', F, or, 1964; bud urn shaped; flowers bright orange-red, dbl., blooms in clusters; low, bushy growth; Tantau, Math.

'AÏCHA', HSpn, dy, 1966; bud long, pointed; flowers deep yellow, large, semi-dbl., intense fragrance; foliage light green; vigorous, bushy growth; [Souv. de Jacques Verschuren X Guldtop]; Petersen

'AÏDA', HT, mr, 1956; bud urn shaped; flowers clear rose-red, dbl., 25 petals, 5 in., cupped, intense fragrance; foliage leathery; vigorous, upright, symmetrical growth; [Crimson Glory X Signora]; Mansuino, Dr Andrea; J&P

'AIGLON', HT, rb, 1961; bud long, pointed; flowers coppery red, reverse yellow, large, dbl., 38 petals, moderate fragrance; foliage glossy, light green; vigorous, upright growth; [Viola X Opera seedling]; Gaujard

'AILEEN', F, ab, 1975; flowers light apricot, peach blended, dbl., 3.5 in., moderate fragrance; upright, free growth; [Elizabeth of Glamis sport]; Wood

'AIMABLE AMIE', HGal, dp; flowers medium, dbl.; Pronville, before 1900

'AIMABLE ROUGE', G, prior to 1845; Godefroy

'AIMABLE ROUGE', HGal; (Amiable Rouge); Vibert, 1819 or 1820

'AIMÉE VIBERT', N, w, 1828; (Bouquet de la Mariée, Nivea); flowers pure white, dbl., blooms in small clusters, recurrent bloom; vigorous growth; [Champneys' Pink Cluster X R. sempervirens hybrid]; Vibert

'AIMÉE VIBERT SCANDENS', N, w, 1841; Curtis

Ain't Misbehavin', *see* 'SEABLA'

Ain't She Sweet™, *see* 'WEKFRAG'

Air France, *see* 'MEIFINARO'

'AIR FRANCE', HT, dr, 1958; flowers red to cherry-red, medium, dbl., cupped, moderate fragrance; vigorous, bushy growth; RULED EXTINCT 9/82 ARM; [Seedling X Poinsettia]; Asseretto, A.; Pin-Blanchon

Air France Meillandina, *see* 'MEIFINARO'

Airain, *see* **'CHÂTELET'**

'AIRBORNE', F, dp, 1949; flowers deep pink, large, dbl., moderate fragrance; vigorous growth; [Donald Prior X Rosamunde]; Leenders, M.

'AISLING', F, w, 1966; flowers ivory-cream, dbl., 4 in., slight fragrance; foliage small, glossy; very vigorous growth; [Queen Elizabeth X Allgold]; Slattery

Ajaccio, HT, 1968; Croix, P.; (Cavriglia)

'AJANTA', HT, m, 1978; bud tapered; dbl., 35–40 petals, 5–5.5 in., exhibition form, moderate, spicy fragrance; foliage large, glossy; bushy growth; [Lady X seedling X Memoriam]; Thakur; Doon Valley Roses

Aka Tombo, Min, rb

Akashdeep, F, or, 1993; Chiplunkar

Akashi, S, lp

Akashsundari, HT, m, 1982; Pal, Dr B.P.

'AKEBONO', HT, yb, 1964; flowers light yellow flushed carmine, high pointed, large, dbl., 56 petals; foliage dark, glossy; vigorous, upright growth; [Ethel Sanday X Narzisse]; Kawai

'AKEMI', F, or, 1977; bud ovoid; dbl., 33 petals, 2–2.5 in., cupped, slight fragrance; foliage dark; vigorous, upright growth; [(Sarabande X Hawaii) X (Sarabande X Ruby Lips)]; Keisei Rose Nurseries, Inc.

Akito, *see* 'TANITO'

Al Moise, N, lp

'ALABAMA', HT, pb, 1976; bud long, pointed; flowers deep pink, reverse near white, dbl., 25 petals, 3.5–4 in., exhibition form, moderate, tea fragrance; foliage dark, leathery; upright growth; [Mexicana X Tiffany]; Weeks, 1977

'ALABASTER', HT, w, 1961; dbl., 70–80 petals, 5 in., exhibition form, moderate fragrance; moderate, upright growth; [Blanche Mallerin X McGredy's Ivory]; Wyant

'ALADDIN', F, mr, 1965; bud ovoid; flowers red, very dbl., blooms in clusters; many prickles; foliage dark; [Miracle X Edith Piaf]; Verbeek

Aladdin, *see* 'SUNALAD'

Aladdins Dream, *see* TWOlad

Aladin®, Min, ob

'ALAIN', F, mr, 1948; bud ovoid; flowers bright carmine-red, semi-dbl., blooms in clusters, slight fragrance; foliage glossy, dark; upright, bushy growth; (28); GM, Geneva, 1948; [(Guineé X Skyrocket) X Orange Triumph]; Meilland, F.; C-P

'ALAIN BLANCHARD', HGal, m, 1839; flowers purplish-crimson, becoming mottled giving a spotted look, semi-dbl., cupped, moderate fragrance; foliage medium green; vigorous, medium growth; [Probably R. centifolia X R. gallica]; Vibert

'ALAIN, CLIMBING', Cl F, mr, 1957; (Grimpant Alain); Delforge

'ALAIN, CLIMBING', Cl F, mr, 1957; Roth

'ALAMEIN', F, mr, 1963; flowers scarlet, semi-dbl., 10 petals, 3 in., flat, blooms in clusters, slight fragrance; foliage dark; vigorous, bushy growth; [Spartan X Queen Elizabeth]; McGredy, Sam IV; McGredy

Alamode, *see* 'BROALA'

'ALANNA HOLLOWAY', HT, dr, 1974; flowers ovoid, slight fragrance; foliage glossy, leathery; very vigorous growth; [Uncle Walter sport]; Holloway

'ALASKA', HT, w, 1949; flowers ivory-white, well-formed, large, dbl., 50 petals, moderate fragrance; vigorous, upright growth; [Peace X Blanche Mallerin]; Meilland, F.

'ALASKA CENTENNIAL', Gr, dr, 1967; bud long, pointed; flowers dark blood-red-scarlet, dbl., 3.5–5 in., exhibition form, moderate fragrance; foliage dark, leathery, glossy; vigorous, upright growth; [Siren X Avon]; Morey, Dr Dennison; Country Garden Nursery

Alba, 1965; Mansuino, Q.; (Cavriglia)

'ALBA FOLIACEA', A, w, 1824

Alba Garcia®, *see* FEstro

'ALBA MAXIMA', A, w; (Great Double White, Maxima, R. alba maxima, The Jacobite Rose); flowers similar to Maiden's Blush, but center creamy; height 6–8 ft

Alba Meidiland™, *see* 'MEIFLOPAN'

Alba Meidiland™, *see* 'MEICOUBLAN'

Alba Meillandécor, *see* 'MEIFLOPAN'

Alba Meillandina®, *see* 'MEINABRON'

Alba Minima, A, mp, 1997; Scarman

'ALBA ODORATA', HBc, w, 1834; (R. bracteata alba odorata, R. microphylla alba odorata); flowers anthers yellow, large, moderate fragrance; straight prickles; foliage 5–7 oval leaflets; wood greenish-purple; Mariani

Alba Rosea, *see* **'MME BRAVY'**

Alba Semi-plena, A, w, prior to 1867; flowers white, golden stamens, semi-double (8–12 petals); large crop of red hips in fall; a very old form observed as a sport from Alba Maxima

Alba Suaveolens, A, w, cultivated prior to 1970; (R. alba suaveolens Dieck); flowers white, semi-double, 2 in. Diameter; source of attar of roses

Albania, Min, 1991; Dot, Simon; (Cavriglia)

'ALBAST', HT, op, 1928; bud ovoid; flowers salmon-pink, open, large, dbl., slight fragrance; foliage bronze, glossy; vigorous growth; [Morgenglans X Mrs Wemyss Quin]; Van Rossem

Albatross, HWich, w, 1988

'ALBÉRIC BARBIER', HWich, ly, 1900; bud small, yellow; flowers creamy white, center yellow, dbl., borne in clusters, non-recurrent, moderate; foliage glossy, dark; vigorous growth; (14); [R. wichurana X Shirley Hibberd]; Barbier

Alberich®, *see* **'HAPPY'**

'ALBERT', F, op, 1962; flowers orange salmon, dbl., 50 petals, exhibition form, blooms in clusters; foliage glossy, bronze; vigorous, bushy, compact growth; Jones; Hennessey

'ALBERT EDWARDS', S, ly, 1961; flowers creamy yellow, single, 5 petals, 1.5–2 in., moderate fragrance; vigorous growth; [R. spinosissima altaica X R. hugonis]; Hillier

'ALBERT GILLES', HT, op, 1943; flowers pink tinted coral; [Julien Potin X Mme Joseph Perraud]; Mallerin, C.; A. Meilland

'ALBERT HOFFMAN', HT, ly, 1904; flowers very large, dbl., intense fragrance; Welter; (Sangerhausen)

'ALBERT LA BLOTAIS', HP, dr, 1881; Moreau et Robert

'ALBERT LA BLOTAIS, CLIMBING', Cl HP, mr, 1888; flowers large, very dbl.; Pernet Père; (Sangerhausen)

'ALBERT MAUMENÉ', S, or, 1934; bud large, pointed, ovoid, copper-red; flowers carrot red shaded copper, semi-dbl., cupped, profuse, repeated bloom; foliage glossy, dark; very vigorous, bushy, open habit growth; [Mme Edouard Herriot X R. hugonis]; Sauvageot, H.

'ALBERT PAYÉ', HP, lp, 1873; flowers flesh-pink, large; vigorous growth; Touvais

'ALBERT PIKE', HT, dp, 1926; flowers glowing cerise, flushed peach at times; [Columbia sport]; Vestal

Albert Poyet, LCl, pb, 1979; Eve, A.

Albert Weedall, *see* 'SCRIVLEO'

'Alberta', F, dp; flat, 10 petals, slight; medium matt green; [Carefree Beauty X Red Hot]; Fleming, Joyce L., 1995

'ALBERTA HUNTER', F, ly, 1985; flowers red stamens, large, dbl., 25 petals, cupped, borne singly, moderate fragrance; small, globular, dull orange fruit; small, straight, red prickles; foliage medium, dark, semi-glossy; medium, upright, bushy growth; [Eleanor Perenyi X Lillian Gish]; French, Richard

'ALBERTAN', S, mp, 1962; flowers bright pink, semi-dbl., 15–18 petals; (14); [Athabasca seedling]; Erskine

Albert-Georg Pluta Rose, F, mr, 1986; flowers luminous red, large, dbl.; Tantau, Math.; (Sangerhausen)

'ALBERTINE', LCl, op, 1921; bud ovoid; flowers light salmon pink, large, dbl., cupped, blooms in clusters, non-recurrent, moderate fragrance; vigorous growth; [R. wichurana X Mrs Arthur Robert Waddell]; Barbier

Alberton Amor, *see* LUDsportiana

Albinia, S, op, 1996; Poulsen

'ALBION', HT, mp, 1969; flowers urn-shaped, large, dbl., borne singly, slight fragrance; foliage semi-glossy; bushy growth; [Frileuse X Seedling]; Poulsen, Niels D.; Vilmorin-Andrieux

'ALBION', S, w; bud cream; dbl., recurrent bloom; foliage leaflets small, firm, dark; height 4 ft; hardy; Skinner

Albuquerque Enchantment, *see* 'MORALBUQUE'

'ALCAZAR', F, or, 1961; bud pointed; flowers coppery red, semi-dbl., 18 petals, 3 in., slight fragrance; foliage glossy, bronze; bushy growth; [Jolie Princesse X (Opera X Miss France)]; Gaujard

Alcha, S, mp

Alchemist, *see* **'ALCHYMIST'**

'ALCHYMIST', S, ab, 1956; (Alchemist, Alchymiste); bud ovoid; flowers yellow shaded orange, pink and red, large, very dbl., heavy, non-recurrent bloom, moderate fragrance; foliage glossy, bronze; vigorous, upright (6 ft) growth; [Golden Glow X R. eglanteria hybrid]; Kordes

Alchymiste, *see* **'ALCHYMIST'**

'ALCIME', HGal, dp; (Alcine); flowers deep rosy pink, spotted white, edged lilac-blush, large, dbl., cupped; vigorous, upright growth; Vibert, 1834; (Sangerhausen)

Alcine, *see* **'ALCIME'**

Alcmaria, F, mr, 1971; flowers blood red, large, dbl.; Verbeek; (Sangerhausen)

Alconbury, *see* 'TALALCONBURY'

Alden Biesen, HMsk, 1996; Lens, Louis; (Cavriglia)

'ALEC ROSE', F, or, 1969; flowers scarlet, well-formed, blooms in trusses; free growth; [Hassan X John Church]; McGredy, Sam IV; McGredy

Alec's Red®, *see* 'CORED'

'ALEC'S RED, CLIMBING'®, Cl HT, mr, 1975; [Alec's Red sport]; Harkness

'ALEGRIAS', F, rb, 1978; bud globular; flowers medium yellow, reverse orange-red, blooms, 23 petals, cupped, borne 3-6 per cluster, no fragrance; straight, reddish-green prickles; foliage medium, glossy; upright, bushy growth; [Seedling X Charleston seedling]; Rose Barni-Pistoia

Alejandra Conde®, *see* FEgosa

'ALENA', Gr, ob, 1970; flowers orange, large, dbl., cupped, moderate fragrance; foliage large, glossy, dark; vigorous, upright, bushy growth; [Seedling X Tropicana]; Raffel; Port Stockton Nursery

Alessa, LCl, mr, 1988; Jensen

Alex C. Collie, *see* COCocrust

Alexander®, *see* 'HARLEX'

'ALEXANDER EMSLIE', HT, dr, 1918; flowers deep crimson, base slightly white, moderate fragrance; Dickson, A.

'ALEXANDER HILL GRAY', T, dy, 1911; (Yellow Cochet, Yellow Maman Cochet); flowers deep lemon-yellow, aging deeper, large, intense fragrance; vigorous growth; Dickson, A.

Alexander Laquemont, *see* **'ALEXANDRE LAQUEMENT'**

Alexander MacKenzie, *see* **'A. MACKENZIE'**

'ALEXANDER MARGHILOMAN', HT, w, 1928; flowers cream-white, center salmon; [Harry Kirk seedling]; Mühle

Alexander Milne, *see* 'LAVMILNE'

'ALEXANDER VON HUMBOLDT', HKor, mr, 1960; flowers scarlet, borne in large clusters; foliage glossy; vigorous (9–12 ft) growth; [R. kordesii X Cleopatra]; Kordes, R.; Kordes

Alexandra, *see* 'HARLEX'

'ALEXANDRA'®, T, yb, 1900; flowers buff and yellow; Paul, W.

Alexandra, *see* KORbaxand

ALEXANDRA LEEK, F, ob, 1995; flowers medium, soft orange, single (5 petals), medium blooms borne in small clusters; fragrant; some prickles; foilage medium, medium green, semi-glossy; upright, bushy, tall growth; [Masquerade X Mrs John Lang]; Fleming, Joyce; Hortico Roses, 1995

Alexandra Rose, *see* 'AUSDAY'

'ALEXANDRE CHOMER', HP, m, 1875; Liabaud

'ALEXANDRE DUMAS', D, lp, 1969; very dbl., cupped, blooms in clusters, abundant, non-recurrent bloom, intense, damask fragrance; foliage small, light green, glossy, leathery; vigorous, upright (6–8 ft) growth; [Ma Perkins X R. damascena versicolor]; Fankhauser

'ALEXANDRE DUPONT', HP, dr, 1892; flowers dark, velvety red, very large, dbl., moderate fragrance; Liabaud; (Sangerhausen)

'ALEXANDRE GIRAULT', LCl, pb, 1909; flowers carmine-red, base salmon, large, dbl., non-recurrent, moderate fragrance; vigorous growth; [R. wichurana X Papa Gontier]; Barbier

'ALEXANDRE LAQUEMENT', HGal, m; (Alexander Laquemont); flowers violet spotted with red

Alexandre Tremouillet, HWich, dp, 1988

Alexandria Rose, *see* 'WILALEX'

'ALEXANDRINE CHAPUIS', HT, my, 1935; bud long, pointed, yellow, shaded carmine; flowers well formed, large, dbl.; foliage bright green; [Feu Joseph Looymans X Seedling]; Vially

Alexia, *see* 'CANLOT'

'ALEXIA WILSON', F, ob, 1972; flowers cream veined orange; [Elizabeth of Glamis sport]; Horner

Alexis, *see* 'HARLEXIS'

'ALEXIS LEPÈRE', HP, mr, 1875; flowers very large, dbl.; Vigneron; (Sangerhausen)

'ALEZANE', HT, ab, 1935; bud urn shaped, reddish brown; flowers deep apricot, large, dbl., cupped, moderate, fruity fragrance; foliage glossy, bronze, dark; very vigorous growth; [Angèle Pernet X Comtesse de Castilleja]; Pahissa; J&P

Alfi(TM), *see* 'POULFI'

Alfie(TM), *see* 'POULFI'

Alfie Luv, *see* 'TALALF'

'ALFIERI', HGal, m, 1833; flowers medium large, dbl.; Vibert; (Sangerhausen)

'ALFRED A. BUCKWELL', HT, mr, 1952; flowers red, dbl., 30 petals, 3 in., exhibition form, intense fragrance; foliage dark, leathery; vigorous growth; [Hector Deane X Betty Uprichard]; Buckwell

'ALFRED COLOMB', HP, dp, 1865; flowers strawberry-red, reflexes crimson-carmine, large, dbl., 45 petals, exhibition form, recurrent bloom, intense fragrance; growth dense; [Général Jacqueminot X Seedling]; Lacharme, F.

'ALFRED DE DALMAS', M, lp, 1855; (Mousseline); bud rose-colored; flowers light pink with blush edges, of poor quality, small, dbl., in corymbs, some recurrent bloom, moderate fragrance; very prickly; vigorous, straggling growth; Laffay, M.

'ALFRED DE ROUGEMONT', HP, dr, 1863; flowers crimson-magenta, well formed, very large, dbl.; vigorous, upright growth; [Général Jacqueminot seedling]; Lacharme, F.

Alfred Dietrich, LCl, rb, 1980

'ALFRED K. WILLIAMS', HP, mr, 1877; flowers carmine-red, changing to magenta; [Général Jacqueminot sport]; Schwartz, J.

'ALFRED LEVEAU', HP, 1860; Vigneron; (Cavriglia)

'ALFRED NEWTON', HT, yb, 1959; flowers pale yellow edged crimson, dbl., 35–40 petals, 5 in., exhibition form, intense fragrance; foliage dark; vigorous growth; [Moonbeam X Karl Herbst]; Kemp, M.L.

'ALFRED PÉTOT', HT, dr, 1935; bud elongated; flowers crimson, well formed, large blooms, dbl., intense fragrance; foliage dark; strong stems; very vigorous growth; [Jeanne Excoffier X Yves Druhen]; Buatois

'ALFRED W. MELLERSH', HT, op, 1918; flowers salmon-yellow, shaded rose, center amber; Paul, W.

'ALFREDO MOREIRA DA SILVA', HT, yb, 1946; bud long, pointed; flowers golden yellow tinted coral, large, semi-dbl.; foliage dark; very vigorous, upright growth; [Dr Kirk X Peace]; Mallerin, C.; A. Meilland

'ALGER', HT, op, 1943; bud pointed; flowers clear pink, reverse salmon-pink, very large, dbl.; foliage leathery; vigorous growth; Gaujard

'ALGONQUIN', S, pb, 1928; flowers purplish rose, center white, large, single, flat, nonrecurrent; large, bottle shaped, red fruit; foliage dull, yellow-green; very vigorous (10 ft) growth; hardy; [R. rubrifolia X R. rugosa hybrid seedling]; Central Exp. Farm

'ALI-BABA', F, or, 1963; flowers large, semi-dbl., 15–20 petals, blooms in clusters; [(Corail X Baccará) X Seedling]; Croix, P.

'ALIBI', F, w, 1960; flowers well formed, dbl., 30 petals, 3–4 in.; vigorous growth; [Kaiserin Auguste Viktoria X Pink Fragrance]; deRuiter

'ALICE', HT, lp, 1978; bud long; flowers light pink, darker center, dbl., 45 petals, exhibition form, borne singly, shy bloom, no fragrance; long, red prickles; foliage large; spreading (to about 4 ft) growth; [Royal Highness X Christian Dior]; Allender, Robert William

'ALICE', Pol, 1925; flowers pink, fringed, borne in large clusters; RULED EXTINCT 11/80 ARM; [Echo X Orléans Rose]; Spek

'ALICE', 1935; dbl.; RULED EXTINCT 11/80 ARM; Wright, Percy H.

'ALICE ALDRICH', HRg, lp, 1901; flowers clear, bright pink, large, dbl., repeat bloom; [R. rugosa X Caroline de Sansal]; Lovett, J.T.; Conard & Jones

'ALICE AMOS', F, pb, 1922; bud long, pointed; flowers cerise, white eye, single, blooms in clusters on strong stem; very vigorous growth; [Tip-Top X Seedling]; Spek; Prior

Alice Faye, *see* 'SEAODD'

'ALICE GRAHAM', HT, w, 1903; flowers large, very dbl., moderate fragrance; Dickson, A.; (Sangerhausen)

'ALICE GRAY', Ayr, w; flowers creamy white, semi-dbl.; moderately vigorous growth; P.T., 1848

'ALICE HAMILTON', Ch, dr, 1903; Nabonnand

'ALICE HARDING', HT, my, 1937; bud ovoid; flowers golden yellow, large, dbl., moderate fragrance; foliage glossy, dark; vigorous growth; [Souv. de Claudius Pernet X Mrs Pierre S. duPont]; Mallerin, C.; J&P

'ALICE HOFFMANN', Ch, rb, 1897; Hoffmann

Alice Jarrett, *see* 'EVEJAUNTY'

'ALICE KAEMPFF', HT, pb, 1921; flowers silvery rose-pink, center coppery yellow, dbl., moderate fragrance; [Gen. MacArthur X Radiance]; Felberg-Leclerc

'ALICE KING', Cl F, mr, 1988; flowers medium, luminous red, aging lighter, small, very dbl., 40–50 petals, borne in sprays, no fragrance; prickles small, medium red; foliage light green, glossy; vigorous, slight perpetual growth; [Dublin Bay X Seedling]; Harrison, G.

Alice Lee, *see* 'TALALI'

'ALICE LEROI', M, pb, 1842; bud well-mossed; flowers lilac-blush shaded rose, center deep rose, very large, dbl.; vigorous growth; Vibert

'ALICE LINDSELL', HT, lp, 1902; flowers large, dbl.; Dickson, A.; (Sangerhausen)

'ALICE MANLEY', HT, my, 1958; bud long, pointed; flowers mimosa-yellow, dbl., 45–50 petals, 3.5–4 in., exhibition form, slight fragrance; foliage leathery; vigorous, upright, bushy growth; [Seedling X Golden Rapture]; Hill, Joseph H., Co.

'ALICE MARION WHYTE', HT, lp, 1932; flowers soft pink, fading pure white, moderate fragrance; Evans

Alice Mavis, S, lp; Peden, G.H., 1994; (Weatherly, L.)

'ALICE PAT', F, rb, 1981; bud ovoid; flowers red shading to white, reverse white shaded pink, dbl., 38 petals, flat, borne 1-6 per cluster, slight fragrance; slightly hooked, reddish-tan prickles; foliage dark, glossy; upright, short growth; Bronze, ARC TG, 1983; [Seedling X Seedling]; Jerabek, Paul E.

Alice Springs, LCl, lp; cupped, moderate; Peden; (Weatherly, L.)

'ALICE STERN', HT, w, 1926; bud long, pointed; flowers white, center cream, sometimes salmon, large, dbl., 30–40 petals, moderate fragrance; foliage dark, bronze; vigorous, bushy growth; [Grange Colombe X Sunburst]; Gillot, F.

'ALICE VENA', HGal, m; flowers plum-purple, large, blooms in clusters

'ALICE WIEMAN', Pol, dp, 1965; flowers pink, open, small, dbl., continuous bloom, moderate fragrance; growth moderate, bushy; [Seedling X Rita Sammons]; Bodley

'ALICIA', HT, pb, 1972; flowers rose-pink, edged cream, full, dbl., 36 petals, 3.5 in., moderate fragrance; foliage dark; vigorous growth; [Golden Scepter X Ena Harkness]; Lees

Alicja, LCl, lp

'ALIDA', HT, mr, 1938; bud long, pointed; flowers well formed, very dbl.; foliage bright; vigorous, bushy growth; [Charles P. Kilham X E.G. Hill]; Lens

'ALIDA LOVETT', LCl, lp, 1905; bud long, pointed; flowers shell-pink, base shaded sulfur, large, dbl., blooms in clusters, seasonal bloom, slight fragrance; foliage glossy; vigorous, climbing growth; [Souv. du Prés. Carnot X R. wichurana]; Van Fleet, 1905; J.T. Lovett

'ALIETTE', HT, 1958; flowers salmon-pink tinted orange, well formed, slight fragrance; vigorous, bushy growth; [Pres. Herbert Hoover X Signora]; Arles; Roses-France

'ALIJ PACHA CHERIFF', HP, 1886; Lévêque, P.; (Cavriglia)

'ALIKA', HGal, mr, 1906; (Gallica Grandiflora, R. gallica grandiflora); flowers brilliant red with no purple, many stamens, petalage variable, large, semi-dbl., moderate fragrance; vigorous growth; Hansen, N.E., 1906; In commerce, 1930

'ALINE', HT, or, 1950; flowers peach, compact, dbl., 22 petals, 4 in., moderate fragrance; foliage dark; vigorous growth; [Picture sport]; Astolat Nursery

Aline, Min, or, 1994; Laperrière

Alinka, HT, ob, 1985; Kordes

Alisha, *see* 'SPODAY'

Alison, *see* 'COCLIBEE'

'ALISON BROWN', HT, dp, 1993; flowers deep pink, dbl., 26–40 petals, 3–3.5 in., borne mostly singly, some small clusters, intense fragrance; some prickles; foliage medium, dark green, semi-glossy; medium (125 cms), bushy growth; [Ena Harkness sport]; Brown, Mrs. Ann; Brown, 1993

'ALISON WHEATCROFT', F, ab, 1959; flowers apricot flushed crimson; [Circus sport]; Wheatcroft Bros.

Alistair Sheridan, *see* 'PEARICH'

'ALISTER CLARK', F, lp, 1999; dbl., 26–40 petals, 3–3.5 in., borne in small clusters, intense fragrance; prickles moderate; foliage small, dark green, glossy; spreading, medium (3 ft) growth; [Marjory Palmer sport]; Newman, Laurie, 1990; Reliable Roses, 1990

Alister Clark's Pink, HT, lp; Clark, A.; (Weatherly, L.)

'ALISTER STELLA GRAY', N, ly, 1894; (Golden Rambler); bud long, pointed; flowers pale yellow, center orange, fading to white, dbl., blooms in blusters, recurrent bloom, moderate fragrance; vigorous, climbing growth; (14); Gray, A.H.

'ALIX ROUSSEL', T, yb, 1908; Gamon

All Ablaze™, *see* 'WEKSAMSOU'

All Gold, *see* **'ALLGOLD'**®

All In One, *see* 'MEILIDER'

All That Jazz™, *see* 'TWOADVANCE'

All-American Bride, *see* 'WILBRID'

'ALLAMAND-HO', S, pb, 1984; bud ovoid; flowers pink and yellow blend, petals edged ruby red, color intensifying with age, dbl., 38 petals, cupped, blooms in clusters of 1–6, repeat bloom, moderate, sweet fragrance; awl-like, red-brown prickles; foliage dark, leathery, semi-glossy; erect, bushy growth; hardy; [(Hawkeye Belle X Prairie Star) X Iobelle]; Buck, Dr Griffith J.; Iowa State University

Allambie, HT, 1973; Dovanan, R.; (Weatherly, L.)

'ALLARD', (form or hybrid of R. xanthina), dy; (R. xanthina Allard); flowers chrome-yellow, dbl.

Allegeo '80, *see* 'MEIFIKALIF'

'ALLÉGRESSE', HT, mr, 1952; flowers red, becoming lighter, well formed, very dbl.; strong stems; vigorous, upright growth; [Fantaisie X Sensation]; Robichon; Vilmorin-Andrieux

'ALLEGRETTO', HT, or, 1975; bud long, pointed; semi-dbl., 14–17 petals, 4 in., slight fragrance; foliage leathery; upright growth; [Fragrant Cloud X Sutter's Gold]; Huber

Allegro®, *see* 'MEIARLO'

Alleluia, Cl Pol, 1960; Cazzaniga, F. G.; (Cavriglia)

Allelulia®, *see* 'DELATUR'

Allen Box, Pol, lp; Box, S., 1928; (Weatherly, L.)

Allen Brundrett, HT, dr, 1994; Brundrett

'ALLEN CHANDLER', Cl HT, mr, 1923; bud long, pointed; flowers brilliant crimson, large blooms, semi-dbl., borne 3-4 per cluster, recurrent bloom, slight fragrance; foliage dark, leathery, glossy; vigorous, pillar growth; (28); GM, NRS, 1923; [Hugh Dickson X Seedling]; Chandler; Prince

'ALLEN'S FRAGRANT PILLAR', Cl HT, rb, 1931; bud long, pointed; flowers cerise, base flushed yellow, open, large blooms, dbl., recurrent bloom; foliage glossy, bronze; long, strong stems; moderate climbing growth; [Paul's Lemon Pillar X Souv. de Claudius Denoyel]; Allen

'ALLEN'S GOLDEN CLIMBER', LCl, ob, 1933; flowers orange, large, dbl., non-recurrent; foliage glossy; Allen

Allen's Jubilee, *see* **'JUBILEE'**

'ALLGOLD'®, F, my, 1958; (All Gold); flowers bright buttercup-yellow, 15–22 petals, 3 in., borne singly and in large trusses, slight fragrance; foliage small, glossy, dark; vigorous growth; GM, NRS, 1956; [Goldilocks X Ellinor LeGrice]; LeGrice, 1956

'ALLGOLD, CLIMBING', Cl F, my, 1961; (Grimpant All Gold, Grimpant Allgold); [Allgold sport]; Gandy, Douglas L.

'ALLIANCE', F, dp, 1966; bud ovoid; flowers light red and deep pink, large blooms, dbl., moderate fragrance; foliage soft; strong stems; very vigorous, bushy growth; [Rosita X Queen Elizabeth]; Delforge

Alliance, *see* 'MEIBLERI'

'ALLIANCE FRANCO-RUSSE', T, my, 1899; Goinard

'ALLISON SWEETIE', Gr, mr, 1975; bud long, pointed, oval; dbl., 21–25 petals, 3–3.5 in., exhibition form, moderate fragrance; foliage large, glossy, dark; tall, vigorous growth; [Tropicana X Mister Lincoln]; Miller, F.

'ALLISWELL', HT, w, 1956; bud ovoid; flowers white overcast pink, center cameo-pink, peony shaped, dbl., 50–60 petals, 5–6 in., moderate fragrance; vigorous, bushy growth; [Neige Parfum X (Charlotte Armstrong X Blanche Mallerin)]; Motose

Allotria®, *see* 'TANAL'

Allspice, *see* 'AROALL'

Allumbie

'ALLURE', HT, pb, 1950; bud long, pointed, carmine-rose; flowers neyron rose, base yellow, dbl., 28–30 petals, 5–6 in., exhibition form, slight fragrance; foliage leathery, glossy, light green; very vigorous, upright growth; [Mrs Pierre S. duPont X Charlotte Armstrong]; Swim, H.C.; Inter-State Nursery

Allux Symphony, *see* 'AUSLETT'

'ALMA', F, dr; flowers dark crimson, dbl., blooms in large clusters; dwarf growth; [Orange Triumph X Eutin]; Riethmuller

Alma Bierbauer, S, op, 1997; flowers salmon pink with yellow eye and pink stamens, single; Williams, J. Benjamin

'ALMA MATER', HT, lp, 1929; flowers lighter in color; [Columbia sport]; Good & Reese

Alma-atinskaj Aromatnaja, HT, m, 1959; flowers large, very dbl., intense fragrance; Besschetnowa; (Sangerhausen)

Almandet, HT, mp, 1966; Croix

'ALMIRANTE AMÉRICO TOMÁS', HT, rb, 1955; (Miramar); flowers geranium-red, reverse old-gold, large, dbl.; very vigorous growth; [Sultane X Peace]; da Silva, Moreira

Almond Glory, *see* 'ZIPALM'

Almondeen, *see* 'AROFRAP'

'ALOHA', Cl HT, mp, 1949; bud ovoid; flowers rose-pink, reverse deeper, dbl., 58 petals, 3.5 in., cupped, recurrent bloom, moderate fragrance; foliage leathery, dark; vigorous, pillar (8–10 ft) growth; [Mercedes Gallart X New Dawn]; Boerner; J&P

'ALOIS JIRÁSEK', HT, ob, 1931; bud long, pointed; flowers dark orange, tinted brownish yellow, large, dbl., moderate fragrance; foliage glossy; vigorous growth; [Mme Butterfly X Mrs George Shawyer]; Böhm, J.

'ALOUETTE', Pol, op, 1971; bud long, pointed; flowers salmon-orange, medium, semi-dbl., cupped, profuse, continuous bloom, slight fragrance; foliage leathery; vigorous, bushy growth; [Ambassadeur Baert X Seedling]; Delforge

'ALOYSIA KAISER', S, my, 1937; bud ochre-yellow; flowers reddish bright yellow, round, large, dbl., blooms in clusters, seasonal bloom; vigorous growth; [Miss G. Mesman X Belle Doria]; Lambert, P.

'ALPAIDE DE ROTALIER', HP, lp, 1863; Campy

'ALPENFEE', HSet, m, 1890; flowers medium large, dbl.; Geschwind, R.; (Sangerhausen)

Alpenglühen, *see* **'ALPINE GLOW'**

Alpengrüss, F, 1974; Noack, Werner; (Cavriglia)

Alpenkonigin®, F, mr

Alpha, *see* 'MEInastur'

Alpha Moe, *see* 'MOEalpha'

Alphee, HT, 1972; Croix, P.; (Cavriglia)

Alphonse Daudet, *see* MEIrouve

'Alphonse de Lamartine', HP, lp, 1853; flowers rosy blush, medium, dbl.; moderate growth; Ducher

'Alphonse Karr', T, m, 1878; flowers crimson-purple, center lighter, large, dbl.; [Duchess of Edinburgh seedling]; Nabonnand, G.

'Alphonse Soupert', HP, mp, 1883; flowers large, dbl., moderate fragrance; Lacharme, F.; (Sangerhausen)

'Alphonse Troussard', HT, 1951; Buatois; (Cavriglia)

'Alpin', LCl, or, 1960; flowers bright orange-red, large, dbl., recurrent bloom, slight fragrance; foliage dark, glossy; vigorous growth; [Spectacular X Seedling]; Combe; Minier

'Alpine', HT, my, 1954; bud long, pointed; flowers clear yellow, open, dbl., 30 petals, 5.5 in., moderate fragrance; foliage leathery; vigorous, upright growth; [Sunnymount sport]; Grillo

'Alpine Glow', F, or, 1954; (Alpenglühen); bud deep orange-red; flowers vermilion-red, dbl., 28 petals, 3–4 in., cupped, blooms in clusters of 4-10, moderate fragrance; foliage glossy; vigorous, bushy growth; [Cinnabar X (Cinnabar X Käthe Duvigneau)]; Tantau, Math.; J&P

Alpine Rock, HT, w

Alpine Rose, *see* **R. pendulina**

'Alpine Sunset'®, HT, ab, 1973; flowers peach-pink, apricot reverse, dbl., 30 petals, 7–8 in., moderate fragrance; foliage glossy, medium green; vigorous, upright growth; [Dr A.J. Verhage X Irish Gold]; Cants of Colchester, Ltd., 1974

'Alsace', HT, op, 1946; flowers salmon-pink, base gold, dbl., 6 in., moderate fragrance; foliage dark; free growth; [Peace X Mme Joseph Perraud]; Meilland, F.

Alsace, *see* PEKtarampe

'Alsace-Lorraine', HP, dr, 1879; flowers dark velvety red, large; Duval

'Alsterufer', HT, dr, 1909; flowers medium large, dbl., intense fragrance; Lambert, P.; (Sangerhausen)

'Alt Wien', HT, pb, 1965; flowers carmine-rose, dbl., 4 in.; foliage dark, glossy; compact growth; [Queen of Bermuda sport]; Prinz; Wohlt

Altaica, *see* **R. spinosissima altaica**

'Altalaris', S, w, 1941; bud large, pointed; flowers white, sometimes flushed pink, open, single, abundant, non-recurrent bloom; apple shaped, bright red fruit; foliage leathery; vigorous, bushy, prickly growth; [R. spinosissima altaica X R. acicularis]; Skinner

Alte Liebe, HT, or, 1974; flowers large, dbl., slight fragrance; Berger, A.; (Sangerhausen)

Alte Liebe, F, op, 1983; flowers salmon-orange and pink, medium large, dbl.; Kordes, W. Söhne; (Sangerhausen)

Altenburg, F, or, 1977; flowers medium large, dbl.; Berger, A.; (Sangerhausen)

'Altesse', HT, dp, 1950; bud ovoid; flowers strawberry-red, large, dbl., 35–40 petals, globular; vigorous, upright growth; [Vercors X Léonce Colombier]; Meilland, F.; URS

Altesse, *see* 'MEIdinro'

Altesse 75, HT, 1975; Meilland, L.; (Cavriglia)

'Althea', HT, mr, 1930; flowers glowing crimson, moderate, damask fragrance; Pemberton; Bentall

Altissimo®, *see* 'DELmur'

Alt-marburg, S, lp, 1979; flowers large, dbl., moderate fragrance; Weihrauch; (Sangerhausen)

'Altmärker', HT, yb, 1908; flowers ochre-yellow tinted garnet, large, dbl., moderate fragrance; vigorous growth; [Kaiserin Auguste Viktoria X Luciole]; Türke

'Altonia', (natural variation of R. setigera), mp; flowers brilliant pink, single or semi-double, semi-dbl.; (found in Illinois, 1835)

'Alt-Rothenburg', HT, dp, 1939; flowers large, dbl., moderate fragrance; Burkhardt; (Sangerhausen)

Altus, *see* 'DELmur'

'Alupka', N, w, 1800; flowers medium large, dbl., moderate fragrance; (Sangerhausen)

'Alvares Cabral', F, mr; flowers bright red; [Pinocchio X Alain]; da Silva, Moreira

'Always', HT, dr, 1959; bud long, pointed; flowers dark cardinal-red, dbl., 28 petals, 6 in., moderate fragrance; foliage leathery; vigorous growth; [(Charlotte Armstrong X Applause) X Ena Harkness]; Leon, Charles F., Sr.

Always A Lady, *see* 'TINlady'

Always Mine™, *see* 'DEVsiem'

'Ama', F, or, 1955; bud ovoid; flowers deep orange-scarlet, large, dbl., exhibition form, blooms in clusters (up to 20); foliage dark, glossy, leathery; vigorous, bushy growth; [Obergärtner Wiebicke X Independence]; Kordes

Amadeus®, *see* 'BARamad'

Amadeus®, HT, ob

Amadine, P, mp

'Amadis', Bslt, dr, 1829; (Crimson Boursault); flowers deep crimson-purple, large, semi-dbl., cupped; no prickles; young wood whitish green, old wood red-brown; vigorous, upright growth; excellent pillar; [R. chinensis X R. pendulina]; Laffay, M.

'AMAevelina', Min, lp, 1992; (**Evelina**); dbl., 26–40 petals, 1.5–2.75 in., borne mostly singly, no fragrance; foliage medium, medium green, semi-glossy; medium (100–150 cms), bushy growth; [Rosa Maria X Seedling (Pink 078)]; Mansuino, Dr Andrea, 1990

Amaglia, LCl, w, 1980

'Amalfi', Pol, dy, 1971; bud long, pointed; flowers deep yellow to salmon-pink, small, dbl., cupped, abundant, continuous bloom, slight fragrance; foliage small, dark; moderate, bushy growth; Delforge

Amalia™, *see* 'MEIcauf'

'Amalia Jung', HT, mr, 1934; flowers crimson-red, large, dbl., moderate fragrance; foliage glossy, cedar-green; vigorous growth; [Mrs Henry Winnett X Lady Helen Maglona]; Leenders, M.

'Amalie de Greiff', HT, pb, 1912; flowers brick-rose, center salmon and orange-yellow, dbl.; [Herrin von Lieser X Mme Mélanie Soupert]; Lambert, P.

Amalinda, HT, mr

'Amami', HT, op, 1927; flowers peach-pink, overlarge, semi-dbl., 12–15 petals; foliage light; very vigorous growth; Easlea

Amanda, *see* 'BEEsian'

Amanda™, *see* 'DEVnina'

Amanda, *see* **'Amruda'**

Amanda, F, ob, 1979; Noack, Werner

Amanda Kay, *see* 'JUSamanda'

Amanda Marciel™, *see* 'DEVnina'

'Amanda Patenaude', P, mp

Amanecer, HT; Dot, Pedro; (Cavriglia)

Amara, *see* GAUhari

Amarancha, F, ab

'Amarante', Pol, mr, 1916; flowers dark crimson, sometimes striped white, blooms in clusters of 25–70, Barbier

'Amarillo', HT, dy, 1961; bud pointed; flowers deep yellow, dbl., 30 petals, 5 in., moderate fragrance; foliage leathery, light green; vigorous, upright growth; [Buccaneer X Lowell Thomas]; Von Abrams; Peterson & Dering

Amarillo, F, dy, 1994; Select Roses, B.V.

'Amateur E. Biron', HT, op, 1928; flowers shrimp-pink, center tinted copper, reverse old-rose, camellia-shaped; Biron

'Amatsu-Otome', HT, yb, 1960; flowers golden yellow edged deep orange, dbl., 48 petals, 4.5–5 in., exhibition form, slight fragrance; foliage semi-glossy; vigorous, compact growth; [Chrysler Imperial X Doreen]; Teranishi, K.; Itami Rose Nursery

'AMAURY FONSECA', Pol, w, 1914; flowers white, suffused light pink in fall, well formed, Soupert & Notting

'AMAZING GRACE', HT, mp, 1973; flowers rich pink, high pointed, dbl., 43 petals, 5 in., slight fragrance; [Carina X Mischief]; Anderson's Rose Nurseries

Amazon, *see* JACap

'AMAZONE', LCl, mr, 1961; flowers bright red; [Spectacular sport]; Delforge

'AMAZONE', T, my, 1872; flowers yellow, reverse veined rose, well formed; [Safrano X ?]; Ducher

Amazone(TM), F, mr; de Ruiter, G.

Ambasciatore Marco Fracisci, T, lp

'AMBASSADEUR BAERT', F, op, 1964; flowers salmon; foliage bronze; low, compact growth; [Sumatra X Seedling]; Delforge

'AMBASSADEUR NEMRY', F, dp, 1949; flowers deep rose-pink, reverse salmon-carmine, large, dbl., moderate fragrance; vigorous growth; Leenders, M.

'AMBASSADOR', HT, ob, 1930; bud long, pointed; flowers bronze-salmon, large, dbl., moderate fragrance; foliage leathery, dark; vigorous, bushy growth; [Mme Butterfly X Souv. de Claudius Pernet]; Premier Rose Gardens

Ambassador®, *see* 'MEInuzeten'

Ambassador, *see* MEIbigoud

'AMBER', Cl Pol, 1908; flowers pale amber, single, early; low growth; RULED EXTINCT; [Jersey Beauty seedling]; Paul, W.

'AMBER', HT, 1930; flowers amber; RULED EXTINCT; [Ophelia sport]; Jordan, B.L.; Beckwith

'AMBER BEAUTY', F, pb, 1962; flowers pink tinted brown, dbl., intense fragrance; [Goldilocks X Lavender Pinocchio]; Leenders, J.

Amber Flash(TM), *see* 'WILdak'

'AMBER GOLD', HT, dy, 1962; bud ovoid; flowers deep golden yellow, dbl., 40–45 petals, 5–5.5 in., cupped, moderate fragrance; foliage leathery; vigorous, upright growth; [Golden Rapture X Golden Scepter]; Moro; J&P

Amber Light, *see* **'AMBERLIGHT'**

Amber Nectar, *see* 'MEHamber'

Amber Queen®, *see* 'HARroony'

Amber Ribbon, *see* 'ZIPamb'

'AMBER SANDS', Min, ab, 1984; flowers large, dbl., 35 petals, exhibition form, slight fragrance; foliage medium, medium green, semi-glossy; upright, bushy growth; [Fragrant Cloud X Poker Chip]; Hardgrove, Donald & Mary; Rose World Originals

Amber Spire, S, ab, 1995; Kordes

Amber Star, HT, ab

Amber Star, *see* 'MANstar'

Amber Sunset, *see* 'MANamsun'

Amber Waves(TM), *see* 'JACamque'

Amberglo(TM), *see* 'MINapco'

'AMBERLIGHT', F, yb, 1961; (Amber Light); flowers clear amber, dbl., 3.5–4 in., blooms in open clusters, intense, fruity fragrance; vigorous, upright, bushy growth; [(Seedling X Lavender Pinocchio) X Marcel Bourgouin]; LeGrice, 1962

Amberlight, *see* 'COCbamber'

'AMBIANCE', HT, lp, 1955; bud long, pink, passing to cream; [Comtesse Vandal X Pres. Macia]; Delforge

Ambiance, *see* NIRPnufdeu

Ambiance, HMsk, pb, 1994; Lens

'AMBOSSFUNKEN', HT, rb, 1961; (Anvil Sparks); flowers coral-red streaked golden yellow, well-formed, large, dbl., 33 petals, moderate fragrance; bushy growth; Meyer; Kordes

'AMBRA'®, F, ob, 1982; flowers deep, large, dbl., 20 petals, cupped, blooms in clusters of 3–5, slight fragrance; reddish prickles; foliage medium, brownish-green, glossy; upright, bushy growth; [Seedling X Seedling]; Rose Barni-Pistoia

'AMBRE', HT, ob, 1950; bud ovoid; flowers brilliant orange and yellow, very large, dbl., moderate fragrance; foliage glossy, dark; very vigorous, bushy growth; [Peace X Seedling]; Gaujard

'AMBRE SOLAIRE', Pol, pb, 1966; bud ovoid; flowers red, pink and yellow, dbl., blooms in clusters, slight fragrance; foliage dark; [Masquerade X Seedling]; Ebben

Ambridge Rose, *see* 'AUSwonder'

'AMBROGIO MAGGI', HP, mr, 1879; flowers very large, dbl.; Pernet Père; (Sangerhausen)

'AMBROISE PARÉ', HGal, m, 1846; Vibert

'AMBROSIA', F, ob, 1962; flowers amber, single, 7–10 petals, 2.5–3 in., flat, blooms in large clusters; foliage dark; vigorous, bushy growth; [Seedling X Shepherd's Delight]; Dickson, Patrick; A. Dickson

Ambrosia, *see* 'SUNambro'

'AMDO', HRg, mp, 1927; semi-dbl., 16 petals, in clusters of 7–10, non-recurrent; [Tetonkaha X La Mélusine]; Hansen, N.E.

'AMEDÉE DE LANGLOIS', B, 1872; Vigneron; (Cavriglia)

'AMÉDÉE PHILIBERT', HP, m; flowers purple and red; Lévêque, 1879

Amélia, A, mp; flowers bright pink, anthers deep yellow, large, dbl., non-recurrent bloom, moderate fragrance; Vibert, J. P., 1823

Amelia Anderson, T, pb

'AMELIA BARTER', HT, mr, 1963; flowers bright scarlet tinted silvery, dbl., 5.5 in.; foliage dark, glossy; [Queen Elizabeth X Claude]; Barter

'AMELIA EARHART', HT, yb, 1932; (Président Charles Hain); flowers golden yellow, center flushed pink, large, dbl., intense fragrance; vigorous growth; [Souv. de Claudius Pernet X (Louise Catherine Breslau X Paul Neyron)]; Reymond; J&P

'AMELIA FLEMING', F, mp, 1995; flowers medium pink with pink tones, prominent stamens, single, 5 petals, 1.5–2.75 in., borne in clusters, moderate fragrance; foliage medium, medium green, matt; upright (120–140 cms) growth; [Marchenland X Bambula]; Fleming, Joyce L.; Hortico Roses, 1994

'AMELIA LOUISE', Min, mp, 1994; flowers small, dbl., 26–40 petals, 1.5–3 in., borne mostly singly, slight fragrance; few prickles; foliage small, medium green, semi-glossy; low, bushy growth; Haynes, F., 1993; F. Haynes & Partners, 1993

'Amelia Louise', HP, mp, 1844; flowers purple-violet, dbl., cupped; [Seedling X Seedling]

Amelia Rey Colaco, F, 1960; Moreira da Silva, A.; (Cavriglia)

'AMÉLIE DE BETHUNE', HT, rb, 1923; flowers coral-red shaded carmine, dbl.; Pernet-Ducher

'AMÉLIE DE MANSFIELD', HGal, mp; flowers vivid pink, medium, dbl.

'AMÉLIE GRAVEREAUX', HRg, mr, 1903; bud ovoid; flowers medium red, fading, dbl., recurrent bloom, foliage dark, rugose; vigorous, spiny growth; [(Général Jacqueminot X Maréchal Niel) X Conrad Ferdinand Meyer]; Gravereaux

'AMERICA', HMult, mp, 1915; (Walsh's Rambler); flowers pink, center white, single, in large clusters (often to 75); [R. wichurana X R. multiflora]; Walsh

America(TM), *see* 'JACclam'

'AMERICAN BANNER', N, pb, 1879; flowers carmine striped white, small, semi-dbl.; [Bon Silène sport]; Cartwright; P. Henderson

'AMERICAN BEAUTY', HP, dp; bud globular; flowers deep pink, large, dbl., 50 petals, cupped, sometimes recurrent bloom, intense fragrance; vigorous growth; Lédéchaux, 1885; Bancroft and Field Bros., 1886

'AMERICAN BEAUTY, CLIMBING', LCl, dp, 1909; flowers deep rose-pink, large, cupped, non-recurrent, moderate fragrance; height 12–15 ft; [(R. wichurana X Marion Dingee) X American Beauty]; Hoopes, Bro. & Thomas

'AMERICAN CLASSIC', HT, dr, 1994; (America's Classic); flowers moderately full, large blooms, dbl., 15–25 petals, 3–3.5 in., borne mostly singly, slight fragrance; some prickles; foliage large, medium green, matt; tall, upright growth; [Seedling X Seedling]; Winchel, Joseph F., 1997; Certified Roses, Inc., 1997

'AMERICAN DAWN', HT, rb, 1976; bud ovoid; flowers rose-red, base white, dbl., 30 petals, 4–5 in., exhibition form, slight fragrance; foliage glossy, dark; upright growth; [Personality X Seedling]; Warriner, William A.; J&P

American Dream™, *see* 'WINBUR'

American Fantasy™, *see* 'TWOFAN'

'AMERICAN FLAGSHIP', HT, mr, 1946; bud urn-shaped; flowers bright scarlet, semi-dbl., 15 petals, 3.5–4.5 in., slight, spicy fragrance, foliage leathery; vigorous, upright, bushy growth; [Crimson Glory X Self]; Lammerts, Dr Walter; C.R. Burr

'AMERICAN GIRL', HT, mr, 1929; flowers large, dbl., exhibition form, moderate fragrance; foliage leathery, dark; long stems; vigorous growth; [Hollywood sport]; Maton

American Glory, *see* 'TWOADMIRE'

American Heritage®, *see* 'LAMLAM'

'AMERICAN HERITAGE, CLIMBING', Cl HT, yb, 1971; Arora, Bal Raj; The Rosery, India

'AMERICAN HOME', HT, dr, 1960; bud ovoid; dbl., 30 petals, 4.5 in., cupped, intense fragrance; foliage leathery; vigorous, upright growth; [Chrysler Imperial X New Yorker]; Morey, Dr Dennison; J&P

American Honor, *see* 'TWOHONOR'

American Independence, *see* 'MEIFINARO'

American Legion, *see* **'LEGION'**

'AMERICAN PILLAR', HWich, pb, 1902; flowers carmine-pink, white eye, golden stamens, single, blooms in large clusters, non-recurrent; red fruit; foliage leathery, glossy; vigorous (15–20 ft) growth; (21); [(R. wichurana X R. setigera) X Red Hybrid Perpetual]; Van Fleet, 1902; Conard & Jones, 1908

American Pride, *see* 'JACARED'

'AMERICAN PRIDE', HT, 1928; bud long, pointed; flowers pure white, outside petals occasionally tinted pink, dbl., 35 petals, 4.5 in., moderate fragrance; foliage leathery; very vigorous growth; RULED EXTINCT 2/79 ARM; [Grillodale sport]; Grillo

American Rose Centennial™, *see* 'SAVARS'

'AMERICAN ROSEATE', HEg, dp; flowers bright rose, semi-dbl.; vigorous growth; Prince Nursery, prior to 1846

American Spirit, *see* 'JACTRED'

'AMERICAN WHITE', HEg, w; flowers creamy white, semi-dbl.; vigorous growth; Prince Nursery, about 1840

'AMERICANA', HT, mr, 1961; bud ovoid; flowers bright red, dbl., 28 petals, 5.5 in., exhibition form, moderate fragrance; foliage leathery; vigorous, upright growth; [Poinsettia seedling X New Yorker]; Boerner; J&P

America's Classic, *see* **'AMERICAN CLASSIC'**

America's Choice, *see* 'POULANDER'

'AMERICA'S JUNIOR MISS', F, lp, 1964; (Junior Miss); bud ovoid; flowers soft coral-pink, medium, dbl., moderate fragrance; foliage glossy; vigorous, bushy growth; [Seventeen X Demure seedling]; Boerner; J&P

Amerlock®, S, op

'AMES 5', S, mp, 1932; (Ames Climber); flowers good size, in clusters, non-recurrent; thornless stems; red in winter; very vigorous growth; very hardy; [R. multiflora X R. blanda]; Maney; Iowa State College

'AMES 6', S, mp; (14)

Ames Climber, *see* **'AMES 5'**

Amethyst, HT, m, 1978; flowers violet-pink, large, dbl., moderate fragrance; Urban, J.; (Sangerhausen)

'AMÉTHYSTE', HWich, m, 1911; flowers violet-crimson; [Non Plus Ultra sport]; Nonin

Ametista®, HT, m, 1985; Barni, V.

'AMI CHARMET', HP, dp, 1900; flowers very large, dbl., moderate fragrance; Dubreuil; (Sangerhausen)

Ami Chenault, HT; Moreira da Silva, A.; (Cavriglia)

Ami Clement®, F, dp

Ami des Jardins, *see* 'KORAMI'

'AMI DESVIGNES', HT, mr, 1954; flowers base of petals veined coral, petals waved, large, semi-dbl.; foliage glossy; vigorous growth; Privat

Ami Dietrich, HT; Moreira da Silva, A.; (Cavriglia)

'AMI F. MAYERY', HT, mr, 1938; bud long, pointed, vermilion-red; flowers poppy-red, very large blooms, dbl., cupped; foliage dark; stiff stems; vigorous growth; [Huguette Vincent X Seedling]; Denoyel, Mme.; C. Chambard

'AMI L. CRETTÉ', HT, rb, 1931; bud long, pointed, coppery oriental red; flowers crimson-red, reverse light coral-rose and yellow, medium to, semi-dbl., cupped, moderate fragrance; foliage dark; very vigorous, bushy growth; Chambard, C.; C-P

'AMI LÉON CHENAULT', F, mr, 1929; flowers dark garnet, slightly striated white, blooms in clusters; [Lafayette sport]; Nonin

'AMI LÉON PIN', HT, ly, 1947; bud long, pointed; flowers pale yellow, reverse pink, very large, dbl., moderate fragrance; foliage dark; very vigorous growth; Gaujard

'AMI MARTIN', HP, or, 1906; flowers vermilion orange-red, very large, very dbl., intense fragrance; Chédane-Guinoisseau; (Sangerhausen)

'AMI QUINARD', HT, dr, 1927; bud long, pointed; flowers blackish garnet and coppery scarlet, medium, semi-dbl., 17 petals, cupped, moderate fragrance; foliage leathery; vigorous growth; [Mme Méha Sabatier X (Mrs Edward Powell X R. foetida bicolor]; Gaujard, 1927; C-P, 1930

'AMI RENÉ BADEL', F, mr, 1961; bud pointed; flowers carthamus-red, dbl., 36 petals, exhibition form, intense fragrance; foliage bronze; [Belle Créole X (Gloire du Midi X Paul Crampel)]; Arles; Roses-France

Amiable Rouge

'AMICA', HT, 1966; flowers cinnabar-red, large, moderate fragrance; foliage light green; [Coup de Foudre X Lampo]; Cazzaniga, F. G.; F. Giuseppe

'AMIGA MIA', S, mp, 1978; bud ovoid, pointed; flowers empire-rose, dbl., 30 petals, 4–5 in., exhibition form, moderate fragrance; foliage large, dark, leathery; vigorous, upright, bushy growth; [Queen Elizabeth X Prairie Princess]; Buck, Dr Griffith J.; Iowa State University

'AMIGO', F, mr, 1951; bud short, pointed; flowers currant-red, semi-dbl., 17–20 petals, 3–3.5 in., cupped, moderate, spicy fragrance; foliage leathery; upright, bushy growth; [World's Fair X Adolf Grille]; Whisler; Germain's

Amigo Roger Sucret, F, 1970; Dot, Simon; (Cavriglia)

'AMIRAL GRAVINA', HP, m, 1860; flowers blackish purple, shaded with scarlet; Moreau et Robert

'AMISTAD SINCERA', HT, w, 1963; (Sincera); bud ovoid; flowers pure white, dbl., 55 petals, 4–4.5 in., exhibition form; foliage leathery; vigorous, well-branched growth; [Alaska X Virgo]; Camprubi, C.; C-P

'AMITIÉ', HT, ob, 1951; flowers coppery orange-yellow, well shaped, large, dbl.; vigorous, branching growth; [Mandalay X Schéhérazade]; Mallerin, C.; EFR

Amitié, HGal, pb

'AMLEGER', HT, mr, 1972; flowers blood-red to cardinal-red, imbricated, dbl., 40 petals, 4.5 in., cupped, slight, tea fragrance; foliage dark, leathery; vigorous growth; [Baccará sport]; Molina; URS

Ammerland®, S, my, 1986; Noack, Werner

'AMOR', F, mp, 1957; bud pointed; flowers clear pink, medium, dbl., 22 petals; vigorous, bushy growth; [(Golden Rapture X Floribunda seedling) X Self]; deRuiter; Blaby Rose Gardens

'AMORE', F, mp, 1957; flowers rosy pink, reverse lighter, medium, dbl., moderate fragrance; foliage semi-glossy; moderate growth; [Orange Triumph X Spring Song (F)]; Riethmuller

Amoretta®, *see* 'AmoRU'

Amorette, *see* 'AmoRU'

Amorosa, F; Moreira da Silva, A.; (Cavriglia)

Amorosa, *see* KORignale

Amorous, *see* 'JACARINA'

'AmoRU', Min, w, 1980; (Amoretta®, **Amorette**, Snowdrop); bud pointed; flowers white, ivory center, dbl., borne 10–15 per cluster, slight fragrance; long, narrow, red prickles; foliage light, mid-green; short, dense, bushy growth patio; [Rosy Jewel X Zorina]; deRuiter; Fryer's Nursery, Ltd., 1979

Amour Ardent, *see* **'BURNING LOVE'**

'AMOURETTE', HCh, mp; flowers rose-pink, edged lighter, large, pointed, petals recurving; foliage dark; stems red

'AMOUREUSE', HT, ab, 1966; flowers coppery yellow, large, dbl., 34 petals, borne singly; foliage large, medium green; [Peace X Rose Gaujard seedling]; Gaujard

Amours de Savernem, *see* SAUdime

'AMPÈRE', HT, or, 1937; flowers nasturtium-red, edges lighter, reverse orange-yellow, very large, dbl., intense fragrance; foliage bright green; [Charles P. Kilham X Condesa de Sástago]; Meilland, F.

'AMRUDA', Min, dr, 1979; (Amanda, Red Ace); flowers medium, dbl., 20 petals, slight fragrance; foliage small, medium green, semi-glossy; bushy growth; [Scarletta X Seedling]; deRuiter, 1977; Fryer's Nursery, Ltd., 1982

Amstelveen, S, op, 1993; Meilland

Amsterdam®, *see* 'HAVAM'

'AMULETT', *see* TANaluma

Amulett®, Min, pb, 1991; dbl.; foliage medium green; wide, bushy (20 in.). growth; Tantau

'AMURENSIS', (form of R. blanda carpohispida), mp; flowers large, semi-dbl.; sparsely armed; wood red; tall growth

'AMY', F, mp, 1954; bud ovoid; flowers carmine-rose, dbl., 38 petals, 2.5–3 in., exhibition form, blooms in loose clusters, moderate fragrance; foliage leathery; compact, dwarf growth; [Show Girl X Fashion]; Von Abrams; Peterson & Dering

Amy Brown®, *see* 'HARKUSHI'

Amy Grant™, *see* 'TUCKAMY'

'AMY JOHNSON', LCl, mp, 1931; bud ovoid; flowers large, dbl., cupped, moderate fragrance; foliage wrinkled; vigorous (12–15 ft) growth; [Souv. de Gustave Prat X ?]; Clark, A.; NRS Victoria

Amy Powell, F, op, 1996

'AMY REBECCA', Min, dy, 1986; flowers deep yellow, small, dbl., 38 petals, exhibition form, borne usually singly, slight fragrance; small, brown prickles; foliage small, medium green, semi-glossy; medium, upright growth; [Rise 'n' Shine X Summer Butter]; Jolly, Marie; Rosehill Farm, 1987

'AMY ROBSART', HEg, dp, 1894; flowers deep rose, large, semi-dbl., summer bloom, moderate fragrance; foliage fragrant; vigorous growth; [R. eglanteria X HP or B]; Penzance; Keynes, Williams & Co.

Amy Sis, Min, 1999

'AMY VANDERBILT', F, m, 1956; bud globular; flowers lavender-lilac, dbl., 70 petals, 3 in., cupped, blooms in pyramidal clusters, moderate fragrance; foliage dark, glossy; upright, bushy growth; [Lavender Pinocchio seedling X Lavender Pinocchio]; Boerner; J&P

'AMY'S DELIGHT', Min, mp, 1980; bud ovoid; flowers clear medium pink, opening imbricated, dbl., 60 petals, borne usually singly, no fragrance; straight, tan prickles; foliage small, medium green, very glossy; compact, bushy growth; [Little Darling X Little Chief]; Williams, Ernest D.

Ana de Cuevas, *see* FEmiento

Anabell®, *see* 'KORBELL'

'ANACRÉON', HGal, mp, 1836; flowers carmine-pink, medium large, very dbl.; Vibert; (Sangerhausen)

'ANAÏS', B, mp

'ANAÏS SÉGALAS', HGal, pb, 1837; flowers rosy crimson, edged rosy lilac, expanded, large, dbl.; branching growth; Vibert

Anaise, B, mp; flowers medium large, dbl.; (Sangerhausen)

Anant, HT, pb, 1991; Chiplunkar

'ANASTASIA', HT, w, 1980; bud ovoid, pointed; flowers, dbl., 30 petals, exhibition form, borne usually singly, no fragrance; bronze prickles, turning brown with age; foliage large, dark; vigorous, dense growth; [John F. Kennedy X Pascali]; Greff, N.P.

'ANATOLE', HGal, mr, 1827; Noisette

'ANATOLE DE MONTESQUIEU', HSem, w, 1860; Van Houtte

Ancestry™, *see* JAYANC

Anchieta, HT; (Brazil); (Cavriglia)

Anci Böhm, *see* **'ANCI BÖHMOVA'**

'ANCI BÖHMOVA', HMult, mp, 1929; (Anci Böhm); [Marietta Silva Tarouca sport]; Böhm, J.

'ANCIENT ART', Min, op, 1985; flowers orange-pink blend, yellow reverse, medium, dbl., slight fragrance; foliage medium, medium green, matt; upright, bushy growth; [Rise 'n' Shine X Picnic]; Hardgrove, Donald L.; Rose World Originals

Anda, *see* 'LENDA'

Andalusien®, *see* 'KORDALU'

'ANDANTE', S, op, 1962; bud ovoid, pointed; flowers light salmon-pink, medium, dbl., cupped, repeat bloom, slight fragrance; foliage dark, bronze, leathery; vigorous (5–6 ft), upright, arching growth; [Sea of Fire X (Josef Rothmund X R. laxa)]; Buck, Dr Griffith J.; Iowa State University

'ANDANTE', HT, mr, 1962; flowers large, dbl., 40 petals; foliage bronze; vigorous, bushy growth; Laperrière; EFR

'ANDELI', HT, rb, 1977; (**Double Delight**™); bud long, pointed to urn-shaped; flowers creamy white becoming strawberry-red, dbl., 30–35 petals, 5.5 in., exhibition form, borne mostly singly, moderate, spicy fragrance; foliage large, deep green; upright, spreading, bushy growth; AARS, US, 1997 GM, Baden-Baden, 1976 GM, Rome, 1976 Hall of Fame, WFRS, 1985 James Alexander Gamble Fragrance Medal, ARS, 1986; [Granada X Garden Party]; Swim, H.C. & Ellis, A.E.; Armstrong Nursery

'ANDENKEN AM J. DIERING', Cl HT, 1902; Hinner, W.; (Cavriglia)

'ANDENKEN AN ALMA DE L'AIGLE', HMsk, lp, 1948; (Andenken an Alma de l'Aigle Ilsabella); Kordes

Andenken an Alma de l'Aigle Ilsabella, *see* **'ANDENKEN AN ALMA DE L'AIGLE'**

'ANDENKEN AN BRESLAU', HWich, mr, 1913; flowers carmine-red, medium large, dbl.; Kiese; (Sangerhausen)

'ANDENKEN AN FRANZ HEINSOHN', F, dr, 1938; flowers urn-shaped, medium, dbl., in clusters, slight fragrance; upright, bushy growth; [D.T. Poulsen X Seedling]; Poulsen, S.

'ANDENKEN AN GARTENDIREKTOR SIEBERT', HMult, pb, 1923; flowers carmine-rose, with yellow, blooms in clusters; [Eisenach X Polyantha seedling]; Kiese

Andenken an Gustav Frahm, Pol, mr

Andenken an Hermann Thiess, F, or, 1959; flowers medium large, semi-dbl.; Krause; (Sangerhausen)

'ANDENKEN AND JOHANNES GEHLHAAR', HT, w, 1925; flowers large, dbl., moderate fragrance; Gehlhaar; (Sangerhausen)

'ANDENKEN AND MORITZ VON FRÖHLICH', HT, mr, 1905; flowers carmine red, large, dbl., slight fragrance; Hinner, W.; (Sangerhausen)

'ANDERSEN'S YELLOW', HT, my, 1984; flowers large, dbl., slight fragrance; foliage medium, dark, semi-glossy; [Queen Elizabeth X Seedling]; Walter, J.C.; Kimbrew-Walter Roses

'ANDERSONII', S, mp, 1935; flowers pink, medium, single, 2–2.5 in.; spreading habit growth; [Chance hybrid of R. canina X possibly R. arvensis]

'ANDGEO', F, ob, 1982; (**Georgie Anderson**); flowers shades of orange, medium, dbl., slight fragrance; foliage medium, dark, semi-glossy; upright growth; [Elizabeth of Glamis X Unnamed seedling]; Anderson's Rose Nurseries

'ANDGLO', F, yb, 1985; (**Glowing Embers**); flowers yellow, red reverse, medium, dbl., 35 petals, slight fragrance; foliage medium, medium green, glossy; bushy growth; [Manx Queen X Daily Sketch]; Anderson's Rose Nurseries, 1982

'ANDMAC', Cl HT, yb, 1985; (**Whisky Mac, Climbing**, Whisky, Climbing); Anderson's Rose Nurseries

'ANDORRA', HT, ab, 1973; bud long, pointed; flowers large, dbl., 24 petals, exhibition form, intense fragrance; foliage glossy, dark; vigorous, upright, bushy growth; [Dr A. J. Verhage X Seedling]; Kordes; Fey

'ANDOUR', HT, yb, 1984; (**Our Love**); flowers yellow-orange; [Doris Tysterman sport]; Anderson's Rose Nurseries

'ANDPAI', F, dr, 1987; (**Paisley Anniversary**); flowers medium, dbl., 15–25 petals, slight fragrance; foliage medium, dark green, glossy; upright growth; [Michele X Smiling Through]; Anderson's Rose Nurseries

'ANDPIN', F, w, 1984; (**Pink Ice**); flowers white with deep pink petal edges, medium, dbl., 20 petals, no fragrance; foliage medium, light green, semi-glossy; bushy growth; [Unnamed seedling X Iceberg]; Anderson's Rose Nurseries

'ANDRÉ GAMON', HT, mp, 1908; flowers carmine; Pernet-Ducher

'ANDRÉ LE TROQUER', HT, ob, 1946; flowers orange shading to apricot, dbl., 30 petals, 5 in., cupped, intense fragrance; foliage very dark; very vigorous, upright growth; GM, Bagatelle, 1946 GM, NRS, 1951; Mallerin, C.; A. Meilland

'ANDRÉ LEROY D'ANGERS', HP, mr, 1862; bud dark violet; flowers crimson, shaded with violet, often ill-formed, large, dbl.; vigorous growth; Trouillard; Standish

'ANDRÉ LOUIS', HMult, w, 1920; flowers white, center flesh-pink; Tanne; Turbat

'ANDRÉ PERNET', HT, mr, 1956; bud long; flowers red, becoming lighter at center and purplish on outer petals, dbl.; foliage dark; very vigorous growth; [Peace X Mme Elie Dupraz seedling]; Gaujard

'ANDRÉ SCHWARTZ', T, mr; flowers crimson, sometimes striped white; Schwartz, J., 1884

'ANDREA', Min, pb, 1971; bud pointed; flowers deep pink, silver reverse, small, dbl., 20 petals, 1.5 in., exhibition form; foliage dark; vigorous, bushy, spreading growth; [Little Darling X ?]; Moore, Ralph S.; Sequoia Nursery, 1978

Andrea, HT, ob, 1968

Andrea Stelzer, *see* KORfachrit

'ANDRÉE JOUBERT', HT, op, 1952; bud very long, dark orange-coral; flowers pastel salmon, large; [Soeur Thérèse X Duquesa de Peñaranda]; Mallerin, C.; EFR

'ANDRÉE LENOBLE', Pol, dp, 1915; flowers unfading rose or red; Turbat

'ANDRÉE PALTHEY', HT, mr, 1946; bud long, pointed; flowers bright red, large, very dbl., moderate fragrance; foliage bronze; vigorous growth; [Mme Joseph Perraud X Seedling]; Gaujard

'ANDRÉE PERRIER', HT, ob, 1932; bud long; flowers orange-yellow, shaded carmine, very large, dbl., cupped, [Souv. de F. Bohé X Seedling]; Chambard, C.

'ANDRÉE ROUX', HT, rb, 1927; flowers coral-red, tipped carmine, reverse yellow; Pernet-Ducher; Gaujard

'ANDRÈE SAUVAGER', HT, yb, 1935; flowers light orange-yellow, large, dbl., moderate fragrance; Mallerin, C.; (Sangerhausen)

'ANDRÉE VANDERSCHRICK', HWich, w, 1935; bud greenish; flowers opening well, small, very dbl., blooms in clusters, profuse seasonal bloom, moderate fragrance; foliage dark; very vigorous, climbing growth; Buatois

'ANDRÉE-SOPHIE GIRARD', F, mr, 1958; flowers currant-red, edged silvery, dbl.; [Alain X Independence]; Arles; Roses-France

Andreo Vettori, T, mp

'ANDRES BATLLE', HT, dr, 1951; bud long, pointed; flowers crimson, large blooms, dbl., exhibition form, slight fragrance; strong stems; vigorous growth; [Comtesse Vandal X Sensation]; Camprubi, C.

Andrew Barton Patterson, HT, lp; Allender, Robert William, 1999; (Weatherly, L.)

Andrew's Comfort, S, ly, 1996

'ANDREWSII', M, mp, 1807

'ANDROI', F, rb, 1986; (**Royal Mail**); flowers red with yellow petal edges, small, dbl., 20 petals, no fragrance; foliage small, medium green, glossy; upright, bushy growth; [Seedling X Manx Queen]; Anderson's Rose Nurseries

'ANDROS', Cl HT, rb, 1985; (**Rosy Cheeks, Climbing**); Anderson's Rose Nurseries

'ANDROY', F, rb, 1983; (**Royal Touch**); flowers red, reverse silver, medium, dbl., 20 petals, slight fragrance; foliage medium, medium green, semi-glossy; upright growth; [Orange Sensation X Elizabeth of Glamis]; Anderson's Rose Nurseries

'ANDSHY', F, rb, 1984; (**Shy Maiden**); flowers white with red petal edges, large, dbl., 35 petals, slight fragrance; foliage medium, light green, glossy; bushy growth; [Iceberg seedling X Iceberg]; Anderson's Rose Nurseries

'ANDSUN', F, yb, 1983; (**Sunshine Princess**); flowers medium, dbl., 20 petals, slight fragrance; foliage medium, medium green, semi-glossy; bushy growth; Anderson's Rose Nurseries

'ANDULKA', HT, op, 1935; bud long, pointed; flowers pink to salmon-pink, large, dbl.; vigorous, bushy growth; Brada, Dr.; Böhm

'ANDWEE', Min, mr, 1988; (**Wee Topper**); flowers small, dbl., 15–25 petals, slight fragrance; foliage small, light green, semi-glossy; bushy growth; [Starina sport]; Anderson's Rose Nurseries

'ANDWEL', F, mr, 1984; (**Welcome Home**); dbl., 20 petals; foliage medium, light green, semi-glossy; upright growth; [Orange Sensation X Michelle]; Anderson's Rose Nurseries

'ANDWIT', HT, yb, 1983; (**With Love**); flowers yellow, petal edges pink, moderate fragrance; foliage medium, medium green, semi-glossy; upright growth; [Irish Gold X Daily Sketch]; Anderson's Rose Nurseries

'ANEMONE', S, lp, 1896; (Anemone Rose, Pink Cherokee, R. X anemonoides); flowers silver pink, single, 5 petals, 4 in., spring bloom, then scattered bloom, moderate fragrance; foliage glossy; vigorous, bushy growth; (14); [Possibly R. laevigata X Tea]; Schmidt, J.C.; J.C. Schmidt

Anemone, *see* R. X anemonoides

Anemone Rose, *see* **'ANEMONE'**

Anemone Rose, *see* R. X anemonoides

'ANEMONOIDES', C, mr, 1814; Poilpre

Anette, F, 1978; Noack, Werner; (Cavriglia)

'ANGARA', HT, rb, 1983; flowers dark red with hues of orange; [Montezuma sport]; Gupta, Dr M. N. & Shukla, R.; National Botanical Research Institute

Angel™, *see* 'DEVITE'

Angel Bells®, *see* 'HERMELA'

'ANGEL CREAM', F, w, 1989; bud broadly ovate; flowers creamy white, turning pure white; [Angel Face sport]; Ravi, Professor N., 1988

'ANGEL DARLING', Min, m, 1976; flowers lavender, small, 10 petals, 1.5 in., slight fragrance; foliage leathery; vigorous growth; [Little Chief X Angel Face]; Moore, Ralph S.; Sequoia Nursery

'ANGEL DELIGHT', HT, ab, 1976; flowers peach shaded salmon; [Femina sport]; Fryers Nursery, Ltd.

'ANGEL DUST', Min, w, 1978; bud ovoid; flowers small, 18–20 petals, exhibition form; foliage dark; vigorous, upright, spreading growth; [Magic Carrousel X Magic Carrousel]; Bennett, Cecilia 'Dee'; Tiny Petals Nursery

'ANGEL EYES', Min, dp, 1978; bud ovoid; flowers spinel-red, semi-dbl., 10 petals, 1 in., slight fragrance; foliage small; very compact, bushy growth; Lyon

'ANGEL FACE', F, m, 1968; bud pointed; flowers deep mauve-lavender, petal edges blushed ruby, dbl., 25–30 petals, 3.5–4 in., exhibition form, borne in large clusters, intense, strongly citrus fragrance; foliage dark, leathery, glossy; vigorous, upright, bushy growth; AARS, 1969 John Cook Medal, ARS, 1971;

[(Circus X Lavender Pinocchio) X Sterling Silver]; Swim & Weeks; C-P

Angel Face, Climbing, Cl F, 1981; Haight; (Cavriglia)

Angel Girl, *see* 'WELAN'

'**Angel Guimera**', HT, my, 1926; flowers amber-yellow, dbl.; [Frau Karl Druschki X Souv. de Claudius Pernet]; Dot, Pedro; S. Dot

Angel Pink, *see* 'MORGEL'

Angel Rose, *see* **R. chinensis minima**

'**Angel Wings**', HT, yb, 1958; bud ovoid; flowers yellow, shading to white, edged pink, dbl., 23 petals, 3.5–4 in., cupped, moderate fragrance; foliage leathery; upright growth; GM, Portland, 1959; [Golden Rapture X Girona]; Lindquist; Howard Rose Co.

'**Angela**', F, yb, 1957; flowers golden yellow shaded crimson, dbl., 28 petals, 2.5 in., trusses of 15–20, very free bloom, slight fragrance; foliage glossy, dark; vigorous, upright growth; [Masquerade X Golden Scepter]; Kordes; Morse

Angela®, *see* 'KORDAY'

Angela, *see* JACcream

Angela Daffey, Min, lp; Hannemann, F., 1989; (Weatherly, L.)

Angela Lansbury, *see* TWOangel

Angela Merici, HT, w, 1987; Lens

Angela Rippon®, *see* 'OCARU'

'**Angela's Choice**', F, pb, 1973; flowers light pink, reverse deep pink, semi-dbl., 15 petals, 3 in.; foliage matt; vigorous, upright growth; [Dainty Maid X Anna Wheatcroft]; Gobbee, W.D.

'**Angèle**', HT, w, 1933; bud ovoid; flowers creamy white, large, dbl., blooms in clusters, moderate fragrance; foliage leathery; long stems; vigorous, bushy growth; [Seedling X Kaiserin Auguste Viktoria]; Vestal

'**Angèle Pernet**', HT, ob, 1924; bud ovoid; flowers reddish-orange shaded yellow, large, dbl., globular, moderate, fruity fragrance; foliage dark, bronze, leathery; vigorous, bushy growth; GM, Bagatelle, 1924 GM, NRS, 1925; [Bénédicte Seguin X HT]; Pernet-Ducher

Angelglo™, *see* 'MINACO'

Angelica, *see* 'KORDAY'

Angelica Renae™, *see* 'WELANGEL'

Angelika, F, dp

'**Angelina**', S, pb, 1976; flowers rose-pink, white eye and reverse, 11 petals, 3 in., blooms in clusters, moderate fragrance; foliage matt, light green; bushy, spreading growth; [(Tropicana X Carine) X (Cläre Grammerstorf X Frühlingsmorgen)]; Cocker, 1975

Angelina Lauro, *see* '**Angeline Lauro**'

'**Angeline Lauro**', HT, or, 1967; (Angelina Lauro); flowers medium, dbl.; foliage dark; Lens, 1968; Spek

'**Angelique**', F, op, 1961; bud ovoid; flowers coral-pink to salmon-pink, open, dbl., 20–25 petals, 2.5–3 in., in clusters, slight fragrance; vigorous, spreading growth; [World's Fair X Pinocchio]; Swim, H.C.; C.R. Burr

Angelique®, *see* 'KORANGELI'

'**Angélique Quétier**', m, 1839; flowers violet-pink, large, very dbl., slight fragrance; Quétier; (Sangerhausen)

'**Angelis**', HT, w, 1960; bud long; single, 8–10 petals, intense fragrance; upright, bushy growth; [Virgo X Ibiza]; Dot, Pedro

Angelita®, *see* 'MACANGELI'

'**Angelita Ruaix**', HT, ob, 1940; flowers orange-yellow, large, dbl., exhibition form; foliage glossy, dark; very vigorous growth; [Duquesa de Peñaranda X Pres. Herbert Hoover]; Dot, Pedro

Angelo Sgaravatti, HT, 1981; (Cavriglia)

Angel's Blush, *see* 'MICANGEL'

Angel's Blush

Angel's Camp White, T, w

'**Angels Mateu**', HT, ab, 1934; bud ovoid; flowers salmon, overlaid gold, large, dbl., 40 petals, globular, moderate, blackberry fragrance; foliage dark; vigorous, bushy growth; GM, Bagatelle, 1934 GM, Rome, 1934; [Magdalena de Nubiola X Duquesa de Peñaranda]; Dot, Pedro; C-P

'**Angelus**', HT, w, 1921; flowers white, center cream, large, dbl., 40–45 petals, moderate fragrance; foliage leathery, dark; vigorous growth; [Columbia X Ophelia seedling]; Lemon

Angelus, HT, ob, 1980; Gaujard

'**Angelus, Climbing**', Cl HT, w, 1933; Dixie Rose Nursery

'**Angie Heatwole**', F, mp, 1996; flowers, dbl., 15–25 petals, 1.5–2.75 in., borne in small clusters, slight fragrance; few prickles; foliage medium, medium green, semi-glossy; medium, upright growth; [Seedling X Dreamer]; Heatwole, Robert E.

Angkor®, *see* 'DELTRAC'

Angle, Ayr, lp; ('Jessica' , Rose Angle); flowers flesh tinged with rose, large, semi-dbl.

Anglica Minor, A, w; flowers center muddled, dbl.; foliage dark gray-green; dwarf growth

'**Angola**', F, dr; [Seedling X Alain]; da Silva, Moreira

Angora, HT; (Cavriglia)

'**Angus MacNeil**', F, op, 1967; flowers salmon-pink and cream, well formed, in clusters; foliage dark; free growth; [The Optimist X Ma Perkins]; Vincent

'**Anibal David**', HT, my, 1961; da Silva, Moreira

'**Anicka**', Pol, dr, 1940; flowers small, dbl.; Valàsek; (Sangerhausen)

Anika®, HT, dy

Anikall, HT, dy, 1987; Haschke

'**Animo**', F, yb, 1962; flowers yellow, becoming copper-red, open, semi-dbl., 12–16 petals, 2.5 in., blooms in clusters; bushy growth; [Masquerade X Beauté]; deRuiter

Anisley Dickson, *see* 'DICKIMONO'

'**Anita**', F, pb, 1982; bud ovoid, pointed; flowers large, dbl., 43 petals, blooms in clusters of 3–7, slight, tea fragrance; large prickles; foliage large, glossy; medium growth; [Rumba X Marmalade]; Swim, H.C. & Christensen, J.E.; Armstrong Nursery

Anita Charles™, *see* 'MORNITA'

Anita Pereire, *see* DRAdiwu

'**Anita Stahmer**', F, dp, 1973; bud long, pointed; flowers deep pink, dbl., 27 petals, 2.5 in., exhibition form, slight fragrance; foliage soft; vigorous, upright, bushy growth; [Zorina sport]; Kordes

Anita's Apricot Fantasy, Min, ab

Anita's Briar Rose, S, lp

Anita's Pink Splash, Min, pb

Anita's Quartered Tropicana, HT, mr, 1998

Anja, *see* KORkompo

Anjani, Pol, rb, 1970; Kasturi

Anjou®, *see* KORNICKEN

Anjou Festival®, F, mr

'**Ankara**', HT, op, 1940; flowers salmon-orange, center coppery, very large, dbl., cupped, slight fragrance; foliage leathery; vigorous, upright growth; [Joanna Hill X Mme Joseph Perraud]; Meilland, F.; A. Meilland

Ankori, *see* 'KORANGELI'

Ann, *see* 'AUSFETE'

'**Ann Aberconway**', F, ab, 1976; flowers apricot-orange, dbl., 20 petals, 3 in., moderate fragrance; foliage dark, leathery; [Arthur Bell X Seedling]; Mattock

'**Ann Barter**', HT, mr, 1962; flowers cerise, semi-dbl., 18 petals, 5 in., moderate fragrance; foliage dark; vigorous growth; [Peace X Ena Harkness]; Barter

Ann Delforge, HT, ob

'**Ann Elizabeth**', F, mp, 1962; flowers clear rose-pink, large, semi-dbl., 15 petals, open clusters, slight fragrance; foliage glossy; vigorous, quite tall growth; Norman; Harkness

'**Ann Endt**', HRg, dr, 1978; bud long-sepaled; flowers medium, single, 5 petals, moderate, cinnamon fragrance; foliage small, soft; [R. rugosa X R. foliolosa]; Nobbs

'ANN FACTOR', HT, ab, 1975; bud ovoid; flowers pastel apricot, large, very dbl., exhibition form, intense fragrance; foliage large, glossy, bronze, leathery; vigorous, bushy growth; [Duet X Jack O'Lantern]; Ellis & Swim; Armstrong Nursery

'ANN HOLBROOK', Min, yb, 1981; bud globular; flowers yellow-pink blend, small, 15–25 petals, slight fragrance; foliage small, medium green, glossy; spreading growth; [Patricia Scranton X Little Darling]; Dobbs, Annette E.

Ann Kercher, HT, lp; flowers variable form, small

Ann Moore, *see* 'MORBERG'

Anna®, *see* PEKCOUGEL

Anna Aguilera, F, 1972; Dot, Simon; (Cavriglia)

'ANNA ALEXIEFF', HP, mr, 1858; flowers rose pink, large, dbl., cupped, borne in clusters of 3, moderate fragrance; medium, upright growth; Margottin, 1858

'ANNA CHARTRON', T, w; flowers white tinted pink; [Kaiserin Auguste Viktoria X Luciole]; Schwartz, Vve., 1897

'ANNA DE DIESBACH', HP, dp; (Anna Von Diesbach, Gloire de Paris, Glory of Paris); bud long, pointed; flowers deep pink, center darker, dbl., 40 petals, 5–6 in., cupped, intense fragrance; vigorous, tall growth; [La Reine X Seedling]; Lacharme, F., 1869

'ANNA DE NOAILLES', HT, dr, 1941; flowers crimson-red, medium, semi-dbl., exhibition form, intense fragrance; very vigorous, bushy growth; [Étoile de Hollande X Seedling]; Gaujard

Anna Ford®, *see* 'HARPICCOLO'

'ANNA HARTMANNOVÁ', HT, w, 1933; flowers cream-white, very dbl., moderate fragrance; [Frau Luise Kiese sport]; Brada, Dr.; Böhm

Anna Katherine, *see* 'ORTANA'

Anna Livia, *see* 'KORMETTER'

'ANNA LOUISA', F, lp, 1967; flowers soft pink, dbl., 2.5 in., blooms in large clusters; vigorous, low, bushy growth; [Highlight X Valeta]; deRuiter

'ANNA MARIA', HT, dp, 1948; (Anna Marie); bud ovoid; flowers rosy pink, dbl., 40–70 petals, 5 in., exhibition form, intense fragrance; foliage leathery, dark; vigorous, upright growth; [Soeur Thérèse X (Duquesa de Peñaranda X Mrs Pierre S. duPont)]; Ohlhus; C-P

'ANNA MARIA', HSet, dp, 1843; Feast

Anna Marie, *see* **'ANNA MARIA'**

'ANNA MÜLLER-IDSERDA', F, mp, 1966; bud ovoid; flowers pink, medium, dbl., intense fragrance; foliage dark; [Duet X Juliette E. van Beuningen]; Buisman, G. A. H.

'ANNA NEAGLE', HT, rb, 1937; flowers bright currant-red, base sunflower-yellow, dbl.; foliage dark; free, branching growth; McGredy

'ANNA OLIVIER', T, pb, 1872; flowers yellowish flesh, shaded salmon, reverse rose, well-formed, large, dbl.; vigorous growth; (14); Ducher

Anna Pavlova, HT, lp, 1981; Beales, Peter

'ANNA RÜBSAMEN', HWich, mp, 1904; flowers clear pink, large, dbl., blooms in compact clusters; vigorous growth; Weigand, C.

Anna Saheb, HT, pb, 1993; Chiplunkar

'ANNA SCHARSACH', HP, mp, 1890; flowers large, dbl., intense fragrance; Geschwind, R.; (Sangerhausen)

'ANNA SOUPERT', HT, yb, 1934; flowers yellow, center orange, very large, cactus-dahlia form, dbl., slight fragrance; foliage bronze, dark; dwarf growth; [Sunburst X Prince de Bulgarie]; Soupert, G.; C. Soupert

'ANNA STAVE', HT, w, 1973; bud ovoid; flowers white, tipped pink, medium, dbl., exhibition form, intense fragrance; foliage dark, soft; moderate, upright growth; [Pink Parfait X Kordes' Perfecta]; Curtis, E.C.; Kimbrew

Anna Vena, Pol, dr, 1972; flowers medium, dbl.; Zyla; (Sangerhausen)

Anna Von Diesbach, *see* **'ANNA DE DIESBACH'**

'ANNA WHEATCROFT', F, or, 1958; flowers light vermilion, gold stamens, single, 4 in., blooms in clusters, slight fragrance; foliage dark, glossy; vigorous growth; [Cinnabar seedling X Seedling]; Tantau, Math.; Wheatcroft Bros.

Anna Zinkeisen, *see* 'HARQUHLING'

'ANNABELLA'®, HT, my, 1981; bud ovoid; flowers deeply, medium, dbl., 35 petals, cupped, borne singly, no fragrance; straight, light yellow prickles; foliage medium, dark, matt; upright growth; [Ambassador X Seedling]; Rose Barni-Pistoia

'ANNABELLA', HT, 1940; flowers light buff-gold, dbl., 55 petals, 5 in., moderate fragrance; RULED EXTINCT 5/83 ARM; [Joanna Hill sport]; Grillo

Annabelle, *see* 'KORBELL'

Annabelle Kolle, F, dp, 1993; Hetzel

Anna-Maria de Montravel, *see* **'ANNE-MARIE DE MONTRAVEL'**

Annan's Orchard, M, dp

'ANNAROY', HT, pb, 1951; bud ovoid; flowers pink with slight salmon undertone, imbricated, very dbl., 110 petals; foliage glossy; [Pink Princess X Los Angeles]; Shepherd; Bosley Nursery

Anna's Red Glory, HT, mr, 1994

Anna's Yellow Glory, HT, my

Ænnchen Müller, *see* **'ANNCHEN MÜLLER'**

'ANNCHEN MÜLLER', Pol, dp, 1907; (Ænnchen Müller, Annie Mueller); flowers warm rose, fading, large, dbl., cupped, blooms in clusters, slight fragrance; foliage glossy; vigorous, bushy growth; [Crimson Rambler X Georges Pernet]; Schmidt, J.C.

'ANNCHEN VON THARAU', HMult, w, 1866; flowers medium large, dbl.; Geschwind, R.; (Sangerhausen)

'ANNE', HT, mr, 1925; flowers cherry-red, large, dbl., globular, moderate, damask fragrance; foliage leathery; vigorous, bushy, compact growth; Pemberton

'ANNE COCKER'®, F, op, 1970; flowers vermilion, medium, dbl., 36 petals, 2.5 in.; foliage glossy, light to medium green; vigorous, upright growth; [Highlight X Colour Wonder]; Cocker, A.; Cocker, 1971

Anne Dakin, LCl, pb, 1974; Holmes

Anne de Bretagne®, *see* 'MEITURAPHAR'

Anne Diamond, *see* **'LANDIA'**

'ANNE D'ORNANO', HT, m, 1967; bud long, pointed; flowers bright purple-crimson, very large, dbl., moderate fragrance; foliage dark, leathery; vigorous, bushy growth; [John S. Armstrong X Rose Gaujard]; Gaujard

Anne Elizabeth, HT, lp; Thomas, Dr A.S., 1979; (Weatherly, L.)

'ANNE FARNWORTH', HT, rb, 1964; flowers like parent, in sunset shades; [Tzigane sport]; Court

Anne Ford, *see* 'HARPICCOLO'

'ANNE GREGG', HT, ab, 1994; flowers moderately full, large blooms, dbl., 15–25 petals, 3–3.5 in., slight fragrance; some prickles; foliage large, dark green, matt; medium (80 cms), upright growth; [Diorama sport]; Rearsby Roses, Ltd., 1994; Rearsby Roses, Ltd., 1994

Anne Harkness®, *see* 'HARKARAMEL'

'ANNE JACKSON', LCl, dp, 1973; flowers cerise, medium, dbl., 30–35 petals, cupped, moderate fragrance; foliage glossy, bronze; vigorous growth; [Spectacular X LCl]; Jackson, J.R.

Anne Kercher, *see* 'SIMJEZBEL'

'ANNE LAFERRÈRE', HP, dr, 1916; flowers deep velvety blood-red; vigorous growth; Nabonnand, C.

Anne Laure, *see* MASlako

'ANNE LETTS', HT, pb, 1954; flowers rose-pink, reverse silvery, pointed, dbl., 28 petals, 4.5 in., moderate fragrance; foliage glossy; bushy growth; [Peace X Charles Gregory]; Letts

'ANNE LEYGUES', T, lp, 1905; flowers flesh-pink; [Gén. Schablikine X Comtesse Bardi]; Nabonnand, P.&C.

'ANNE MARIA', HSet, dp, 1843; flowers pale rose, large, dbl., blooms in clusters; Feast

Anne Marie, *see* 'MEIFOUR'

Anne Marie Laing, *see* 'JOSPINK'

Anne Marie Trechslin®, *see* 'MEIFOUR'

'ANNE MCDONALD', F, pb, 1992; flowers rose pink and creamy yellow, classical, 30 petals, 3.25 in., exhibition form, borne in clusters of 10–30, slight fragrance; foliage medium green, glossy; medium to tall (1–3 m), upright growth; [Granada X Kordes Perfecta]; Spriggs, Ian Raymond, 1984; Treloar Roses Pty. Ltd., 1991

Anne Morrow Lindbergh, *see* 'JACYAP'

'ANNE OF GEIERSTEIN', HEg, dr, 1894; flowers deep crimson, single, summer bloom, moderate fragrance; bright scarlet fruit; foliage fragrant; vigorous growth; [R. eglanteria X HP or B]; Penzance; Keynes, Williams & Co.

'ANNE POULSEN', F, mr, 1935; (Anne-Mette Poulsen); bud long, pointed; flowers bright crimson-red, darkening, large, dbl., blooms in clusters, moderate fragrance; vigorous growth; [Ingar Olsson X Seedling]; Poulsen, S.; J&P, 1935; McGredy, 1935

'ANNE SCRANTON', F, lp, 1971; bud ovoid; flowers light pink, center flesh-white, medium, dbl.; foliage leathery; vigorous, upright growth; [Queen Elizabeth X Katherine T. Marshall]; Dobbs, Annette E.

'ANNE VANDERBILT', HT, or, 1941; bud pointed; flowers reddish orange, open, dbl., 28 petals, 4–5 in., intense fragrance; foliage leathery, glossy; very vigorous, bushy growth; [Seedling X Stargold]; Brownell, H.C.

'ANNE WATKINS', HT, ab, 1962; flowers deep cream shaded yellow, reverse apricot, well-shaped, dbl., 30 petals, 5 in.; foliage dark, glossy; vigorous, upright growth; [Ena Harkness X Grand'mère Jenny]; Watkins Roses

Anne-Aymone Giscard d'Estaing®, *see* DORalp

Anneka, *see* 'HARRONVER'

'ANNEKE DOORENBOS', F, pb, 1956; flowers silver-pink, reverse darker; [Buisman's Triumph sport]; Doorenbos; Boerma

'ANNEKE KOSTER', Pol, dr, 1927; flowers deep red; [Greta Kluis sport]; Koster, D.A.

Anneliese, HT, dr, 1971; Miessler, Herbert; (Sangerhausen)

Anneliese, HT, lp, 1990; flowers light pink to white, large, semi-dbl.; Wänninger, Franz; (Sangerhausen)

Anneliesse Rothenberger, *see* 'TANOLG'

'ANNE-MARIE COTTE', N, w, 1875; flowers medium, dbl.; Guillot; (Sangerhausen)

'ANNE-MARIE DE MONTRAVEL', Pol, w, 1879; (Anna-Maria de Montravel); flowers pure white, of irregular form when fully open, sometimes showing stamens, dbl., 1.5 in., moderate, lily-of-the-valley fragrance; very few prickles; foliage dark green above, grayish beneath, glossy, 3–5 leafl; dwarf, compact growth; [Dbl. flowered Multiflora X Mme de Tartas]; Rambaux

'ANNE-MARIE MILLIAT', HT, w, 1939; bud very long; flowers large, very dbl.; very vigorous growth; Gaujard

'ANNEMARIE VAN ONSEM', F, mr, 1971; bud ovoid; flowers vivid red, open, medium, semi-dbl.; foliage large, glossy, dark; very vigorous, upright, bushy growth; [Circus X Korona]; Institute of Ornamental Plant Growing

Anne-Mette Poulsen, *see* **'ANNE POULSEN'**

Annerose, F, or, 1976; Croix

'ANNE'S DELIGHT', Min, dp, 1981; bud pointed; flowers, dbl., 40 petals, exhibition form, borne usually singly, slight fragrance; long, thin prickles; foliage small, dark, glossy; upright, bushy growth; [Little Darling X Over the Rainbow]; Williams, Ernest D.; Mini-Roses

'ANNETTE', HT, op, 1952; bud long, pointed; flowers salmon-pink, dbl., 20–25 petals, 4–4.5 in., exhibition form, moderate fragrance; foliage glossy, leathery, dark; vigorous, compact growth; [Charlotte Armstrong X Contrast]; Swim, H.C.; Earl May Seed Co.

Annette Dobbs, *see* 'MORNET'

Annette Elizabeth, LCl, lp; [Clair Matin X Wedding Day]; Stewart, L., 1992; (Weatherly, L.)

'ANNETTE GATWARD', HT, ab, 1954; flowers peach, large, dbl., moderate fragrance; foliage light green; upright growth; [Mrs Charles Lamplough X Barbara Richards]; Gatward

'ANNETTE GRAVEREAUX', HT, my, 1929; flowers lemon-yellow, shaded orange, dbl.; [Mev. C. van Marwijk Kooy X Golden Emblem]; Leenders, M.

'ANNI JEBENS', HT, dy, 1932; bud large, long, pointed, golden yellow; flowers blood-red, reverse golden yellow, dbl., exhibition form, moderate fragrance; foliage leathery, glossy, bronze; bushy, dwarf growth; [Charles P. Kilham X Mev. G.A. van Rossem]; Kordes

'ANNI WELTER', M, mp, 1906; flowers large, moderate fragrance; [Crested Moss X La France]; Welter

Annie Beaufais, F, or, 1962; deRuiter

'ANNIE BRANDT', HT, op, 1932; bud long, pointed; flowers pink tinted coral, open, large, semi-dbl., moderate fragrance; foliage leathery, glossy; very vigorous, bushy growth; [Mrs Pierre S. duPont X Colette Clément]; Mallerin, C.; C-P

'ANNIE BURGESS', HP, lp, 1926; flowers pale pink, in clusters, early; [Lyon Rose X Frau Karl Druschki]; Burgess, S.W.

'ANNIE COOK', T, rb, 1888; flowers blush-tinted; [Bon Silène sport]; Cook, J.W.

'ANNIE CRAWFORD', HP, mp, 1915; (Miss Annie Crawford); flowers bright pink, very large, dbl., 30–35 petals, exhibition form, recurrent bloom; vigorous growth; GM, NRS, 1914; Hall

'ANNIE DE METZ', HT, or, 1932; bud large; dbl., exhibition form; [Golden Emblem X R. foetida bicolor seedling]; Mallerin, C.; C-P

'ANNIE DREVET', HT, rb, 1939; bud long, yellow; flowers fiery red, reverse yellow, large, semi-dbl., cupped, slight fragrance; foliage leathery, glossy, dark; vigorous growth; [Charles P. Kilham X (K. of K. X Mari Dot)]; Caron, B.; A. Meilland, 1939; Port Stockton Nursery, 1941

'ANNIE DUPEYRAT', HT, op, 1935; bud ovoid; flowers very large, dbl., moderate fragrance; foliage leathery; bushy growth; [Mrs T. Hillas X Elvira Aramayo]; Mallerin, C.; C-P

Annie East, *see* 'BOSREXEVER'

Annie Girardot, HT, 1979; Kriloff, Michel; (Cavriglia)

'ANNIE LAURIE', HT, pb, 1918; (Double Mme Butterfly); bud long, pointed; flowers flesh-pink, base yellow, large, dbl., cupped, moderate fragrance; foliage glossy; very vigorous growth; [Ophelia sport]; Stuppy Floral Co.

Annie Mueller, *see* **'ANNCHEN MÜLLER'**

Annie R. Mitchell, *see* 'MORMITCHELL'

'ANNIE VIBERT', N, w, 1871; Vibert

Annie Wood, *see* **'MLLE ANNIE WOOD'**

Annie Wood Mille, HP, mr

Annie's Red, F, dr

Annie's Song, F, lp, 1990; Spriggs, Ian Raymond

'ANNIVERSARY', HT, dr, 1981; bud long, pointed; flowers, dbl., 23 petals, exhibition form, borne 1–3 per cluster, moderate fragrance; prickles hooked down; foliage medium to large; upright growth; [Love Affair X Seedling]; Hoy, Lowel L.; Joseph H. Hill, Co.

'ANNIVERSARY', HT, 1961; bud ovoid; flowers yellow, dbl., 55–60 petals, 4–5 in., exhibition form, moderate fragrance; foliage leathery, dark; strong stems; vigorous, upright growth; RULED EXTINCT 5/81 ARM; [Mary Jo X Lamplighter]; Hill, Joseph H., Co.

Anniversary, *see* MATtlace

Ann's Rose, *see* 'JUDANN'

ANN'S WEDDING', HT, my, 1975; flowers very full, dbl., 40 petals, 4 in., moderate fragrance; foliage glossy, dark; free growth; [Whisky Mac sport]; Rosemount Nursery

Annulet, Pol, lp; Miers, A., 1935; (Weatherly, L.)

'ANNY', Min, w, 1949; flowers pale pink fading white, micro-mini, dbl., 30 petals,

.5 in.; height 6 in.; [Rouletii X Perla de Montserrat]; Dot, Pedro

'ANNY BRANDT', HT, yb, 1951; bud pointed; flowers creamy yellow, edge and reverse tinted lilac, petals waved, dbl.; vigorous growth; Mallerin, C.; EFR

'ANOTHER CHANCE', HT, w, 1994; flowers, dbl., 26–40 petals, 1.5–2.75 in., borne mostly singly, slight fragrance; some prickles; foliage medium to large, medium green, matt; medium-tall, bushy growth; [Mount Shasta X Seedling]; Heyes, Alex, 1996; Rainbow Roses, 1996

Anri, HT, yb, 1997; flowers full, large, very dbl., 26–40 petals, 4.5 in., slight fragrance; few prickles; foliage large, dark green, glossy; bushy, upright, medium (180 cms) growth; [Mme. Sachi X Yarna]; Ohtsuki, Hironaka

Ans, *see* 'BENBURGUN'

Antares, *see* BARtares

'ANTHEA', HT, yb, 1949; flowers pale yellow flushed rose, compact, dbl., 20–25 petals, 4–5 in., slight fragrance; foliage dark; vigorous growth; [McGredy's Yellow X Phyllis Gold]; Bees

Anthea Fortescue, *see* 'PEASHINE'

Anthea Turner, HT, yb, 1997; flowers very full, large, very dbl., 41 petals, 4 in., slight fragrance; few prickles; foliage medium, dark green, semi-glossy; compact, medium growth; [Alec's Red X Harry Wheatcroft]; Thomas, D.

'ANTHÉOR', HT, ab, 1948; bud long, furled; flowers reddish apricot, dbl., intense fragrance; [(Joanna Hill X Duquesa de Peñaranda) X (Charles P. Kilham X Mme Joseph Perraud)]; Meilland, F.

Anthony Meilland, *see* 'MEITALBAZ'

Anticipation, *see* 'MEIDINRO'

Antico Amore®, *see* 'BARANAM'

Antigone®, *see* 'GAUHTI'

Antigua, *see* 'JACTIG'

Antike, F, yb, 1966; flowers reddish-yellow, medium, dbl.; Kordes, W. Söhne; (Sangerhausen)

Antike 89®, *see* KORdalen

Antique Gold, *see* 'LAVTIQUE'

'ANTINEA', HT, op, 1934; flowers salmon-orange, base yellow, very large, dbl., moderate fragrance; foliage glossy; very vigorous, bushy growth; [Julien Potin X Seedling]; Gaujard; Dreer; H & S

'ANTIQUE', F, rb, 1967; flowers crimson and gold, blooms in clusters; bushy growth; [Honeymoon X Circus]; Kordes; A. Dickson

Antique 89, LCl, rb; Kordes, 1988

'ANTIQUE LACE', F, mp, 1991; bud ovoid; flowers medium pink, light pink center, urn-shaped, medium, semi-dbl., 25–35 petals; foliage medium, dark green, semi-glossy; bushy, medium growth; [Seedling X Little Cameo]; Strahle, B. Glen, 1985; Coyier's Roses, 1990

Antique Rose™, *see* 'MORCARA'

Antique Silk, *see* 'KORAMPA'

Antique Tapestry, *see* 'CLETAPE'

'ANTIQUE VELVET', Min, dr, 1993; flowers velvet dark red, moderately full (15–25 petals), medium (4–7 cms) blooms borne in large clusters; patio; fragrant; few prickles; foliage large, medium green, semi-glossy; tall, upright, bushy growth; [Valerie Jeanne X (Party Girl X Pillow Talk)]; Jobson, Dan; Jobson. 1993

'ANTOINE DUCHER', HP, rb, 1866; flowers violet-red, large; (28); Ducher

Antoine Mouton, HP, mp; Levet, 1874

'ANTOINE NOAILLY', HT, mr, 1958; bud long; flowers clear red, scalloped petals; [Seedling X Mme G. Forest-Colcombet]; Croix, P.

'ANTOINE RIVOIRE', HT, lp, 1895; bud ovoid; flowers light pink shaded darker, imbricated, dbl., moderate fragrance; foliage dark; vigorous growth; [Dr Grill X Lady Mary Fitzwilliam]; Pernet-Ducher

'ANTOINE SCHURZ', HP, 1890; Geschwind, R.; (Cavriglia)

'ANTOINE VERDIER', HP, pb, 1871; flowers pink, shaded muddy lilac, dbl.; Jamain, H.

'ANTOINETTE', HT, ab, 1968; bud long, pointed; flowers open, medium size, dbl., moderate fragrance; foliage large, glossy, leathery; vigorous, upright growth; [Queen Elizabeth X Peace]; Patterson; Patterson Roses

Antoinette, *see* KORiganta

'ANTONELLA', HT, mp, 1964; bud globular; flowers camellia-pink, dbl., 5 in., exhibition form, moderate fragrance; strong stems; vigorous growth; Mondial Roses

Antonella Fineschi, F, 1985; Fineschi, G.; (Cavriglia)

'ANTONELLIANA', HT, ob, 1952; flowers orange and deep yellow tipped, well-formed, dbl.; foliage glossy; vigorous growth; [Gaiezza X Margaret McGredy]; Giacomasso

'ANTONIA', F, w, 1979; bud long, pointed; flowers blushed white, palest pink flush, well-formed, dbl., blooms in large clusters, slight fragrance; vigorous, bushy, low growth; [Tantau's Tip Top sport]; Bazeley, B.L.; Highfield Nursery

'ANTONIA D'ORMOIS', HGal, lp; flowers blush, fading at edge, medium, dbl., cupped; Vibert, 1835

'ANTONIA PAHISSA', HT, ob, 1935; bud long, pointed; flowers rich orange, large blooms, dbl., cupped, moderate fragrance; foliage glossy, dark; long stems; very vigorous, bushy growth; Pahissa

Antonia Ridge®, *see* 'MEIPARADON'

'ANTONIA ROLLERI DE PELUFFO', HT, dr, 1926; flowers brilliant red, center darker, dbl., moderate fragrance; [Gen. MacArthur X Mme Edouard Herriot]; Soupert & Notting

'ANTONIETTA INGEGNOLI', Pol, pb, 1923; flowers golden pink, opening in two distinct tones on same plant, dbl., moderate fragrance; [R. wichurana X R. chinensis]; Ingegnoli

'ANTONIN DVORÁK', HT, op, 1933; flowers light salmon-pink, very large, very dbl., intense fragrance; Böhm, J.; (Sangerhausen)

Antony Meilland, *see* 'MEITALBAZ'

'ANTOPP', Gr, dy, 1996; (**Golden Opportunity™**); bud urn-shaped; flowers golden yellow, light golden yellow, blooms borne in small clusters, dbl., 15–25 petals, slight fragrance; some prickles; foliage medium, medium green, semi-glossy; medium growth; [Broadway X Delta Gold]; Qerry, Anthony; Certified Roses, Inc., 1995

'ANURAG', HT, pb, 1980; bud long, pointed; flowers tyrian rose, dbl., 54 petals, exhibition form, borne singly, intense fragrance; hooked, brown prickles; foliage large, smooth, light green; upright, bushy growth; [Sweet Afton X Gulzar]; Division of Vegetable Crops and Floriculture

Anurupa, HT, mp, 1997; Friends Rosery

Anuschka®, *see* 'TANKANUSCH'

Anusheh, *see* 'PAYABLE'

Anvil Sparks, *see* **'AMBOSSFUNKEN'**

'ANYTIME', Min, op, 1973; flowers salmon-orange, purplish eye, semi-dbl., 12 petals, .5–1 in., moderate fragrance; foliage dark; [New Penny X Elizabeth of Glamis]; McGredy, Sam IV; McGredy

'ANZAC', HT, op, 1943; flowers azalea-pink with coppery scarlet sheen, camellia form, dbl., 42–50 petals, 5 in., moderate, fruity fragrance; foliage leathery; long stems; very vigorous, upright, compact growth; [Miss Rowena Thom X Seedling]; Howard, F.H.; H&S

'AORANGI', F, w, 1979; bud pointed; flowers cream, dbl., 41 petals, 3 in., exhibition form, slight fragrance; foliage large; upright growth; [Arthur Bell X Red Devil]; Murray, Nola

Aorangi, *see* SANaran

Aotearoa-New Zealand, *see* 'MACGENEV'

'AOZORA', HT, m, 1973; (Blue Sky); flowers deep lilac-blue, large, dbl., exhibition form, moderate fragrance; foliage large, leathery; vigorous, upright growth; [Sterling Silver seedling X Seedling]; Suzuki, Seizo; Keisei Rose Nursery, 1972

'APACHE', S, yb, 1961; bud ovoid, flushed red; flowers medium to buff-yellow, dbl., 60 petals, 5–6. in., intense fragrance; foliage leathery; vigorous (5–6 ft), spreading, open growth; [Fred Howard X

Buccaneer]; Von Abrams; Peterson & Dering

'APACHE BELLE', HT, rb, 1969; bud ovoid; flowers orange-red, very dbl., 5 in., slight fragrance; foliage glossy; vigorous, upright, compact growth; [The Alamo sport]; Sitton; Co-Operative Rose Growers

Apache Princess, *see* 'TWOMIN'

'APACHE TEARS', F, rb, 1971; flowers cream to creamy pink, petals edged red, medium, dbl., exhibition form, slight fragrance; foliage large, light; vigorous, bushy growth; [Karl Herbst X China Doll]; Pikiewicz; Edmunds Roses

'APACHE WELLS', F, yb, 1971; bud ovoid; flowers canary-yellow, washed pink, medium-small, dbl., exhibition form, slight fragrance; foliage leathery; vigorous, bushy growth; [Circus X The Optimist]; Williams, J. Benjamin

Apachi, *see* **'TAI-GONG'**

Apart, HRg, m, 1981; flowers mauve blend, dbl.; prolific fruit; 2.5 ft growth; Uhl, J.

Aparte, *see* **'SPARTAN'**

'APELES MESTRES', LCl, dy, 1931; flowers sunflower-yellow, large, dbl., globular, moderate fragrance; foliage dark, glossy; vigorous, climbing growth; [Frau Karl Druschki X Souv. de Claudius Pernet]; Dot, Pedro; C-P

Aperitif®, HT, lp, 1998; bud pointed, yellow; dbl., slight fragrance; foliage medium, dark green, semi-glossy, disease-resistant; vigorous, upright (3 ft) growth; McGredy, Sam IV

'APÉRITIF', F, pb, 1972; bud ovoid; flowers ivory, petals edged rose-pink, medium, dbl., exhibition form, slight fragrance; foliage leathery; vigorous, upright, bushy growth; [Seedling X Starbright]; Boerner; J&P

Apertif, *see* 'MACWAIRAR'

Apfelblute, S, w, 1991; flowers light pink-white, medium, semi-dbl., no fragrance; Noack, Werner; (Sangerhausen)

'APFELBLÜTE', Pol, w, 1907; flowers light pink-white, small, dbl., moderate fragrance; Wirtz & Eicke; (Sangerhausen)

'APHRODITE', HT, or, 1928; bud long, pointed; flowers coral-red, shaded gold, large, semi-dbl., intense fragrance; foliage dark, glossy; vigorous, bushy growth; [Hortulanus Budde X Toison d'Or]; Easlea

'APHRODITE, CLIMBING', Cl HT, or, 1933; Hillock

Apogée®, *see* 'DELBAT'

'APOLLINE', B, mp, 1848; flowers bright rose-pink, cupped; [Pierre de St. Cyr seedling]; Verdier, V.

Apollo®, *see* 'ARMOLO'

'APOLLO', HT, dy, 1941; flowers golden yellow; [Mme Joseph Perraud sport]; Armstrong, J.A.; Armstrong Nursery

Apollo, HT, pb, 1998

Apollo Tribute, HT, ab

Apoman®, HT, yb; Adam

Apothecary's Rose, *see* **R. GALLICA OFFICINALIS**

(The) Apothecary's Rose of Provins, *see* **R. GALLICA OFFICINALIS**

'APOTHEKER FRANZ HAHNE', S, op, 1919; flowers salmon-rose on orange-yellow ground; Müller, Dr F.

'APOTHEKER GEORG HÖFER', HT, or, 1900; flowers coppery-red, very large, very dbl., moderate fragrance; Welter; (Sangerhausen)

'APOTHEKER GEORGE HÖFER, CLIMBING', Cl HT, or, 1941; flowers very large, dbl., intense fragrance; Vogel, M.; (Sangerhausen)

'APOTHEOSE', F, yb, 1963; flowers indian yellow edged red, becoming garnet-red, vigorous growth; [Arc-en-Ciel X Seedling]; Delforge

'APPEAL', HT, mp, 1957; flowers clean pink, dbl., intense fragrance; long stems; vigorous, bushy growth; [Ena Harkness X Treasure]; Fletcher; Tucker

'APPLAUSE', HT, dp, 1949; bud long, pointed; flowers light red, dbl., 50 petals, 4–4.5 in., exhibition form, slight fragrance; foliage leathery, dark; vigorous, upright, bushy growth; GM, Bagatelle, 1947; [Contrast X Charlotte Armstrong]; Swim, H.C.; Armstrong Nursery

'APPLAUSE', *see* 'SAVAPPLE'

'APPLE BLOSSOM', HMult, lp, 1932; flowers light pink, center lighter, petals crinkled, dbl., borne in huge clusters; vigorous growth; [Dawson X R. multiflora]; Burbank; Stark Bros.

Apple rose, *see* **R. POMIFERA**

'APPLEBLOSSOM', S, lp, 1963; flowers apple blossom-pink, dbl., cupped, recurrent bloom; bushy (2 ft) growth; Skinner

Appleblossom, S, mp, 1997; Noack, Werner

Appleblossom Festival, S, pb, 1999; Williams, J. Benjamin

'APPLEDORE', F, lp, 1962; flowers medium, dbl., 24 petals, moderate fragrance; foliage glossy; vigorous growth; [Karl Herbst X Pink Charming]; Allen, E.M.

'APPLEJACK', S, pb, 1973; bud small, long, pointed, ovoid; flowers neyron rose, stippled crimson, large, semi-dbl., repeat bloom, intense fragrance; foliage leathery; vigorous, upright, bushy growth; [Goldbusch X (Josef Rothmund X R. laxa Retzius)]; Buck, Dr Griffith J.; Iowa State University

'APPLETON'S LIMELIGHT', HT, dy, 1934; bud long, pointed; flowers deep golden yellow, open, large, semi-dbl.; foliage leathery, glossy; vigorous, bushy growth; [Lady Forteviot sport]; Appleton

'APPRECIATION', HT, mr, 1971; flowers light red shading crimson, pointed, dbl., 27 petals, 4 in., slight fragrance; foliage glossy; vigorous growth; [Queen Elizabeth X Seedling]; Gregory

Apps Rose, *see* KORalogen

Apricot Angel, HT, ab, 1999

Apricot Beauty, S, ab, 1980; Rendu

'APRICOT BRANDY', F, ab, 1972; flowers apricot, base yellow, dbl., 22 petals, 4 in., slight fragrance; foliage bronze-green; Fryers Nursery, Ltd.

Apricot Charm™, *see* 'MINAECO'

Apricot China, F, mr, 1996

'APRICOT CRÈME', Min, ab, 1988; bud pointed; flowers light apricot, edges cream, reverse apricot to cream; [Yellow Doll, Climbing sport]; Bell, Douglas & Judy; Michigan Mini Roses, 1989

'APRICOT DAWN', HT, ab, 1938; flowers apricot, base yellow, moderate fragrance; [Golden Dawn sport]; Wyant

Apricot Delicious, Min, ab, 1996; Welsh

Apricot Delight, *see* ATTlight

Apricot Doll, *see* 'LAVDOLL'

Apricot Gem, F, ab, 1978; Delbard-Chabert

'APRICOT GLOW', LCl, ab, 1936; flowers apricot, turning apricot-pink, dbl., large trusses, intense fragrance; foliage very glossy; long stems; very vigorous (20 ft) growth; [(Emily Gray X Dr W. Van Fleet) X Jacotte]; Brownell, H.C.; B&A, 1936; C-P, 1937; Dreer, 1937

Apricot Hit®, Min, ab

Apricot Kisses, *see* 'GEACOT'

Apricot Medinette, *see* 'POULCOT'

Apricot Midinette, *see* 'POULCOT'

Apricot Mist™, *see* 'SAVAMIST'

Apricot Mist, *see* 'SOCSPO'

Apricot Moon, F, ab, 1995; Lowery/Eisen

'APRICOT NECTAR', F, ab, 1965; bud ovoid; flowers pink-apricot, base golden, dbl., 4–4.5 in., cupped, intense, fruity fragrance; foliage glossy, dark; vigorous, bushy growth; AARS, 1966; [Seedling X Spartan]; Boerner; J&P

'APRICOT PARFAIT', HT, ab, 1978; bud ovoid; flowers apricot-pink blend, dbl., 53 petals, 4 in., exhibition form, slight fragrance; foliage large, dark; upright growth; [Seedling X South Seas]; Warriner, William A.; J&P

Apricot Perfection, *see* 'CLEPERF'

Apricot Prince, *see* 'AROSNAP'

'APRICOT QUEEN', HT, ab, 1940; bud pointed; flowers salmon-pink, base apricot-orange, large, dbl., 45 petals; foliage leathery;

very vigorous, bushy growth; AARS, 1941; [Mrs J.D. Eisele X Glowing Sunset]; Howard, F.H.; H&S

Apricot Queen Elizabeth, Gr, ab, 1980; Verschuren

'Apricot Queen, Climbing', Cl HT, ab, 1950; Maranda

'Apricot Silk', HT, ab, 1965; flowers apricot, large, dbl., exhibition form, moderate fragrance; foliage dark, glossy; vigorous, upright growth; [Souv. de Jacques Verschuren X ?]; Gregory

Apricot Spice, *see* 'SANspic'

Apricot Summer, *see* KORpapiro

Apricot Sunblaze, *see* 'SAVamark'

Apricot Sunblaze®, *see* 'MEIfruije'

Apricot Surprise, S, ab

Apricot Twist, *see* 'MORbrown'

'Apricot Wine', F, ab, 1978; flowers burnt apricot, semi-dbl., 12 petals; foliage dark, glossy; low, compact growth; [Allgold X Seedling]; Slack

'Apriheart', Min, ab, 1983; bud small; flowers light apricot, center deeper, small, dbl., borne singly, intense fragrance; very few prickles; foliage small, medium green, semi-glossy; bushy growth; [Picnic X Rise 'n' Shine]; Hardgrove, Donald L.

April Fools Day, *see* 'GREgsil'

'April Hamer', HT, pb, 1983; flowers shell pink with bright pink edges, large, dbl., 40 petals, exhibition form, moderate fragrance; foliage dark; vigorous, upright growth; [Mount Shasta X Prima Ballerina]; Bell, Ronald J.; Treloar Roses Pty. Ltd.; Wisconsin Roses, 1998

'April Moon', S, my, 1984; bud small; flowers lemon yellow, dbl., 28 petals, cupped, borne 5–10 per cluster, repeat bloom, moderate, sweet fragrance; awl-like, tan prickles; foliage dark, leathery; erect, short, bushy growth; hardy; [Serendipity X (Tickled Pink X Maytime)]; Buck, Dr Griffith J.; Iowa State University

'Aprilia', HT, dp, 1937; bud ovoid; flowers old-rose, open, very large, dbl., slight fragrance; foliage leathery; vigorous, upright growth; GM, Rome, 1937; Cazzaniga, F. G.

Aprutina, F, 1969; Borgatti, G.; (Cavriglia)

'Apsara', HT, op, 1983; flowers light pink shaded salmon, medium, dbl., 75 petals, flat, intense fragrance; hooked, brown prickles; foliage medium, roundish, leathery; upright growth; [Sonia X Sabine]; Pal, Dr B.P.; K.S.G. Son's Roses, 1982

'Apsara', F, 1966; bud ovoid; flowers salmon-pink, open, medium, semi-dbl.; foliage glossy; vigorous, upright, open growth; RULED EXTINCT 3/84 ARM; Pal, Dr B.P.; Indian Agric. Research Inst.

'Aquarelle', F, dy, 1969; bud ovoid; flowers open, dbl., 18–25 petals, 2 in., flat, moderate, fruity fragrance; foliage dark; very vigorous growth; [Gold Strike X Golden Garnette]; Lens

Aquarelle, HT, r; Croix

Aquarius, *see* 'ARMaq'

'Aquilla Bright', HT, ob, 1988; flowers medium, dbl., 26–40 petals, slight fragrance; foliage medium, medium green, semi-glossy; upright growth; [Sunblest X Matador]; Lea, R. F. G.

'Ara Pacis', HT, w, 1955; bud tubular, well formed; flowers ivory-white edged reddish purple, very large, dbl., 50 petals, foliage glossy, bright green; long stems; vigorous growth; [Peace X Marguerite Chambard]; Giacomasso

'Arabella', HT, dp, 1918; flowers crimson-pink, pointed, large, moderate fragrance; vigorous growth; [Mme Caroline Testout sport]; Tantau

'Arabesque', F, lp, 1978; bud pointed; flowers soft pink, 10 petals, 3 in., slight fragrance; vigorous growth; [(Gavotte X Tropicana) X Tropicana]; Sanday, John

Arabesque, *see* INTerstreep

Arabia®, *see* 'TANibara'

'Arabian Nights', F, op, 1963; flowers light salmon-orange, well-formed, dbl., 25 petals, 4.5 in., moderate fragrance; vigorous growth; [Spartan X Beauté]; McGredy, Sam IV; McGredy

'Araby', Gr, lp, 1973; bud long, pointed; flowers light orchid-pink, center white, open, large, dbl., intense fragrance; foliage large, glossy, dark, leathery; vigorous, upright growth; [Honey Chile X Rose Merk]; Thomson

'Araceli Leyva', HT, op, 1940; bud long, pointed; flowers rose-salmon, large blooms, dbl., cupped, moderate fragrance; foliage leathery; strong stems; vigorous, upright growth; [Mme Butterfly X Comtesse Vandal]; Dot, Pedro

'Arakan', F, lp, 1968; flowers, dbl., blooms in trusses, slight fragrance; [Pink Parfait X Ivory Fashion]; Harkness

'Aramis', F, dr, 1964; flowers bright scarlet, semi-dbl., 3 in., blooms in clusters of 7–8; foliage dark; very bushy, compact growth; [Bel Ami X (Java X Alain)]; Laperrière; EFR

'Arashiyama', HT, w, 1997; (Kyoto 1200; YKH 501); flowers ivory white with pink shade, very full (41+ petals), large blooms, fragrant; moderate prickles; foliage medium, medium green, semi-glossy; medium (1–2 m) growth; [Garden Party X Pristine]; Kameyama, Yasushi

'Aratama', HT, yb, 1976; bud pointed; flowers yellow and red, dbl., 25 petals, 6 in., exhibition form, slight fragrance; foliage glossy, light green; vigorous growth; [Kordes' Perfecta X (Garden Party X Christian Dior)]; Takahashi, Takeshi

Aravali Princess, HT, op, 1987; Pal, Dr B.P.

Arbelle, *see* GAUrama

ARC Angel, *see* FRYorst

'Arc de Triomphe', HT, yb, 1955; bud globular; flowers yellow-copper, large, dbl., moderate fragrance; foliage glossy, olive-green; upright growth; RULED EXTINCT 1/86; Frères, Buyl

Arc de Triomphe, *see* 'JACale'

Arcade, Min, mr, 1996; Spooner, Raymond A.

'Arcadia', HWich, dr, 1913; flowers crimson-scarlet, dbl., rosette, blooms in clusters, very vigorous growth; Walsh

'Arcadia', HT, rb, 1938; bud ovoid; flowers reddish copper, very large, dbl.; foliage glossy; long stems; vigorous growth; Gaujard

Arcadia Louisiana Tea, T, pb

Arcadian, *see* 'MACnewye'

Arcanto, *see* BARarc

Arcata Light Yellow Wichurana, HWich, ly

Arcata Pink Globs, N, lp

'Arc-en-Ciel', F, pb, 1961; flowers rich yellow to salmon-pink, carmine and crimson, dbl., 65–70 petals, borne in tight clusters; foliage dark; growth moderate; [Masquerade X Maria Delforge]; Delforge

'Arch. Reventós', HT, ab, 1935; bud ovoid, apricot; flowers cream-yellow, large, dbl.; foliage glossy, dark; vigorous, bushy growth; Leenders, M.

Arch. Reventós, Climbing, Cl HT; Leenders, M.; (Cavriglia)

'Archangel', S, lp, 1979; flowers delicate pink, frilled, single, 9–10 petals, 3.5 in., slight fragrance; foliage light green; tall growth; [Little Darling X Gypsy Moth]; Hawker, U.

'Archduchess Charlotte', HCh, dp, 1975; bud pointed; flowers intense deep solid pink, very dbl., 76 petals, 2.5–3 in., cupped, profuse bloom early summer, moderate fragrance; foliage glossy, smooth; climbing growth; [Archduke Charles sport]; Earing, F.E.; Kern Rose Nursery

'Archduke Charles', HCh, rb; flowers rose with paler edges, aging to rich crimson; moderate growth; Laffay, M., prior to 1837

'Archiduc Joseph', T, pb, 1892; flowers purplish pink, center flesh-pink; vigorous growth; (14); [Mme Lombard seedling]; Nabonnand, G.

'Archiduchesse Elisabeth d'Autriche', HP, mp, 1881; flowers rose-pink, dbl.; Moreau et Robert

'Archiduchesse Elisabeth-Marie', Pol, ly, 1898; flowers canary-yellow fading white, imbricated, medium, dbl., intense fragrance; vigorous growth; [Mignonette X Luciole]; Soupert & Notting

'Archie Gray', HT, dr, 1920; flowers deep crimson, shaded scarlet, dbl.; Dickson, H.

'Arctic Emerald', S, w, 1982; bud globular, pointed; flowers white, with yellow-green center, small to medium, semi-dbl., 12 petals, borne 1–5 per cluster, repeat bloom, slight fragrance; foliage small, light green; low, compact growth; [Thérèse Bugnet X Europeana]; James, John

'Arctic Flame', HT, mr, 1955; bud pointed; flowers bright red, dbl., 55 petals, 5 in., moderate fragrance; vigorous, bushy growth; [(Queen o' the Lakes X Pink Princess) X Mirandy]; Brownell, H.C.; Stern's Nursery

'Arctic Glow', S, rb, 1982; bud globular, pointed; flowers scarlet shading to white center, large, dbl., 28 petals, borne singly, repeat bloom, slight fragrance; foliage dark, rough; compact growth; [Pike's Peak X Show Girl]; James, John

'Arctic Pink', F, lp, 1966; flowers pink fading lighter, 3 in., cupped, blooms in trusses, intense fragrance; vigorous growth; [Dearest sport]; Smith, E.

Arctic Rose, *see* **R. acicularis**

Arctic Snow™, *see* 'MINlco'

Arctic Sunrise, *see* 'BARarcsun'

'Ardelle', HT, w, 1957; bud long, pointed; flowers creamy white, dbl., 72 petals, 5 in., exhibition form, moderate fragrance; foliage glossy; very vigorous, compact growth; [Mrs Charles Lamplough X Peace]; Eddie; Harkness; Wyant

Ardennes, *see* 'INTerpeel'

'Ardente', Cl F, op; flowers salmon and orange; [Seedling X Alain]; da Silva, Moreira

'Ardoisée de Lyon', HP, m, 1858; flowers violet-rose, vigorous growth; Damaizin

'Ardon', HWich, mp, 1925; flowers bright neyron rose, stained white, borne in pyramidal clusters of 30–40; Turbat

'Ardore', HT, mr, 1973; bud ovoid, globular; flowers orient red, large, dbl., cupped, moderate fragrance; foliage large, dark, leathery; very vigorous, upright, bushy growth; [Seedling X Ninfa]; Calvino

'ARDour', S, ly, 1999; (**Joyce Barden**); dbl., 26–40 petals, 4 in., borne in small clusters, intense fragrance; prickles moderate, outer shoots mostly thornless; foliage medium, dark green, dull, light green when new; upright, spreading, medium (5–6 ft) growth; [Sweet Juliet X Souv. de la Malmaison]; Barden, Paul

Ards Beauty, *see* 'DICjoy'

'Ards Pillar', HT, mr, 1903; flowers large, dbl.; Dickson, A.; (Sangerhausen)

'Ards Rambler', Cl HT, or, 1908; flowers large, dbl., intense fragrance; Dickson, A.; (Sangerhausen)

'Ards Rover', Cl HP, dr, 1898; flowers crimson, shaded maroon, large, dbl., sometimes recurrent bloom, moderate fragrance; pillar growth; Dickson, A.

Arejay, *see* 'KIRrad'

Arena's Dream, *see* KORdreweer

'Arend Herwig', F, ob, 1966; bud ovoid; flowers orange-red, medium, dbl.; foliage dark; [Korona X Heureux Anniversaire]; Buisman, G. A. H.

'Arethusa', HCh, yb, 1903; flowers yellow, tinted apricot; vigorous growth; Paul, W.

Argental, HT, w; Croix

'Argentina', F, ab, 1941; flowers reddish apricot, semi-dbl., moderate fragrance; [Mev. Nathalie Nypels X Orange Glory]; Leenders, M.

'Argentine Cramon', HT, w, 1915; flowers white, center tinted salmon-rose; Chambard, C.

'Argosy', HT, op, 1938; flowers salmon, flushed pink, dbl.; long stems; [Souv. de Gustave Prat X Seedling]; Clark, A.; NRS New South Wales

Argovia, HT, pb, 1998; Huber

'Argyle', HT, w, 1921; flowers pure white; vigorous growth; [Mme Caroline Testout X Marquise de Sinéty]; Dobbie

'Aria', F, op, 1957; flowers salmon shaded pink, large, dbl.; bushy growth; [Duchess of Rutland X Fashion]; deRuiter

'Ariadne', HCh, rb, 1918; flowers bright crimson, center shaded yellow; Paul, W.

'Ariake', HT, w, 1976; bud globular; flowers ivory, dbl., 47–50 petals, 5 in., cupped, moderate fragrance; foliage light green; vigorous, upright growth; [(Lady X (Rose = 'Lady X') X Garden Party) X Seedling]; Teranishi, K.; Itami Rose Nursery

Arianna®, *see* 'MEIdali'

Arianna, *see* MEIkruza

'Aribau', HT, mr, 1936; bud long, pointed; flowers brilliant red, large blooms, dbl.; foliage glossy; long stems; very vigorous growth; [K. of K. X Director Rubió]; Dot, Pedro; H. Guillot

'Ariel', HT, my, 1921; flowers golden yellow, streaked crimson, large blooms, dbl., globular, moderate fragrance; foliage dark; long stems; vigorous, bushy growth; GM, NRS, 1920; [Mme Edouard Herriot X Natalie Boettner]; Bees

Arielle Dombasie®, *see* 'MEIhourag'

Arioso, HT, or, 1970; flowers light vermilion, full, dbl., 25 petals, 5 in., slight, fruity fragrance; foliage glossy, dark; very vigorous, upright growth; [(Paris-Match X Baccará) X Marella]; Paolino; URS

Arioso, *see* MEImucas

'Ariste', HT, my, 1960; bud very pointed, deep yellow; flowers light yellow, medium, single, 5 petals, moderate fragrance; foliage dark, leathery; vigorous, bushy growth; [Joanna Hill X Souv. de Mme Boullet]; Jones; Hennessey

'Aristide Briand', HWich, m, 1928; flowers mauve-pink, borne in clusters of 10–29, moderate fragrance; [Yseult Guillot X Seedling]; Penny

'Aristide Dupuy', HP, 1866; Trouillard; (Cavriglia)

'Aristobule', M, dp, 1849; flowers dark rose with touches of clear rose, dbl.; growth good; Foulard

'Aristocrat', HT, pb, 1949; bud long, pointed; flowers clear light pink, reverse darker, dbl., 28–35 petals, 4.5–5 in., exhibition form, moderate fragrance; foliage leathery, dark; very vigorous, upright growth; [Pink Delight sport]; Holmes, M.A.; Avansino; Mortensen

Arizona, *see* 'WErina'

'Arizona Sunset', Min, yb, 1985; flowers light yellow, flushed orange-red, medium, cupped, 20 petals, cupped, slight; prickles slanted downward; foliage small, medium green, semi-glossy; bushy, spreading growth; PP006559; [(Orange Sweetheart X Zinger) X Party Girl]; Jolly, Nelson F.; Rosehill Farm

'Arjun', HT, or, 1980; bud long, pointed; flowers, dbl., 35 petals, cupped, borne singly or 8 per cluster, slight fragrance; hooked prickles; foliage large, smooth; tall, upright growth; [Blithe Spirit X Montezuma]; Division of Vegetable Crops and Floriculture

'Arkansas', HT, or, 1980; bud ovoid, pointed; flowers urn-shaped, dbl., 48 petals, borne singly or 2–4 per cluster, slight, spicy fragrance; long, oval-based prickles, hooked downward; foliage leathery; upright, vigorous growth; [Seedling X Seedling]; Weeks

Arkansas Rose, *see* **R. arkansana**

'Arkansas Sunshine', F, my, 1962; flowers golden yellow, dbl., 30 petals, cupped, blooms in clusters; foliage dark, leathery; vigorous, bushy growth; [Goldilocks X Seedling]; Jones; Hennessey

Arkavathi, F, dp, 1971; Kasturi

'Arkle', HT, or, 1977; bud cupped; flowers dark tangerine, dbl., 45–50 petals, 4 in., cupped, intense fragrance; foliage glossy; vigorous growth; [Whisky Mac sport]; Hughes Roses

'Arlene Francis', HT, my, 1957; bud long, pointed; flowers golden buttery yellow, dbl., 25–30 petals, 5 in., exhibition form, borne mostly singly, intense, sweet licorice fragrance; foliage dark, glossy; vigorous growth, upright, medium sized; [Eclipse seedling X Golden Scepter]; Boerner; J&P

'ARLEQUIN', HT, ob, 1945; flowers orange-yellow and coppery red, very large, dbl., globular, moderate fragrance; foliage dark, glossy; bushy growth; Gaujard

'ARLEQUIN', HGal, m, 1837; flowers purple-violet, dbl.; medium growth; Paillard

Arles, *see* **'FERNAND ARLES'**

Arlette, *see* POUlske

Armada®, *see* 'HARUSEFUL'

'ARMAGH', HT, ab, 1950; flowers creamy pink, apricot and buff, pointed, large, dbl., 49 petals, slight fragrance; foliage dark; free growth; [Sam McGredy X Admiration]; McGredy, Sam IV

'ARMAQ', Gr, pb, 1971; (**Aquarius**); bud ovoid; flowers medium pink blend, medium, dbl., exhibition form, slight fragrance; foliage large, leathery; vigorous, upright, bushy growth; AARS, 1971 GM, Geneva, 1970; [(Charlotte Armstrong X Contrast) X (Fandango X (World's Fair X Floradora))]; Armstrong, D.L.; Armstrong Nursery

'ARMÉNIE', HT, ab, 1936; bud purple-garnet; flowers blood-red, shaded, large, very dbl., moderate fragrance; very vigorous, bushy growth; [Rhea Reid X Yves Druhen]; Buatois

'ARMIDE', A, w, 1817; flowers medium, moderate fragrance; Vibert

'ARMILLA', Gr, pb, 1973; (**Cherry-Vanilla**); bud pointed; flowers pink, center creamy yellow, medium, dbl., cupped, moderate fragrance; foliage dark, leathery; vigorous, upright, bushy growth; [Buccaneer X El Capitan]; Armstrong, D.L.; Armstrong Nursery

'ARMINDA', HT, op, 1956; bud ovoid; flowers bright pink tinted coral, large blooms, very dbl., globular, moderate fragrance; foliage glossy; strong stems; vigorous growth; [Peace X Symphonie]; Camprubi, C.

'ARMLU', Gr, lp, 1966; (**Lucky Lady**); bud long, pointed; flowers light pink, reverse darker, large, dbl., 28 petals, exhibition form, slight fragrance; foliage dark, glossy; vigorous, upright growth; AARS, 1967; [Charlotte Armstrong X Cherry Glow]; Armstrong, D.L. & Swim; Armstrong Nursery

'ARMMA', HT, w, 1965; (**Matterhorn**®); flowers medium to large, dbl., exhibition form; foliage leathery; very tall, upright growth; AARS, 1966 GM, Portland, 1964; [Buccaneer X Cherry Glow]; Armstrong, D.L. & Swim; Armstrong Nursery

'ARMOLO', HT, my, 1971; (**Apollo**®); bud long, pointed; flowers large, dbl., moderate fragrance; foliage large, glossy, dark, leathery; vigorous, upright, bushy growth; AARS, 1972; [High Time X Imperial Gold]; Armstrong, D.L.; Armstrong Nursery

'ARMONIA', HT, mr, 1951; bud long, pointed; flowers bright red, blooms, dbl., exhibition form, moderate fragrance; foliage dark, leathery; long stems; very vigorous, upright growth; Cazzaniga, F. G.

Armorique Nirpaysage, *see* 'POULAPS'

Armosa, *see* **'HERMOSA'**

'ARM-ROY BEAUTY', HT, mr, 1945; very vigorous, tall growth; [Better Times sport]; Armacost & Royston

Arnaud Delbard®, *see* 'DELTEP'

'ARNDT', LCl, lp, 1913; bud yellowish red; flowers pale pink, dbl., in large clusters, recurrent bloom; foliage dark; half climbing growth; [Hélène X Gustav Grünerwald]; Lambert, P.

'ARNELDA MAE', Min, lp, 1983; bud small; flowers small blooms, dbl., 30 petals, exhibition form, borne singly and in clusters up to 7, slight fragrance; foliage small, light green, matt; bushy growth; [Sheri Anne X Seedling]; Pencil, Paul S.

'ARNHEM GLORY', HT, dr, 1959; flowers deep velvety red, large, intense fragrance; vigorous growth; Verschuren

'ARNOLD', HRg, mr, 1893; (Arnoldiana); flowers scarlet, single, some recurrent bloom; vigorous growth; [R. rugosa X Génèral Jacqueminot]; Dawson; Eastern Nursery

Arnold Greensitt, *see* 'NOSTARN'

Arnoldiana, *see* **'ARNOLD'**

'AROALL', HT, my, 1977; (**Allspice**); bud ovoid, pointed, deep yellow; dbl., 35 petals, 4.5 in., moderate, honey and tea rose fragrance; foliage large, olive-green; vigorous, upright, bushy growth; [Buccaneer X Peace]; Armstrong, D.L.; Armstrong Nursery

'AROART', HT, dr, 1981; (**Jennifer Hart**); bud ovoid, pointed; flowers, dbl., 45 petals, exhibition form, borne singly, slight, tea fragrance; foliage medium, medium green, semi-glossy; medium, upright, bushy growth; PP005219; [Pink Parfait X Yuletide]; Swim, H.C. & Christensen, J.E.; Armstrong Nursery, 1982

'AROBIPY', HT, w, 1987; (**Crystalline**™, Valerie Swane); flowers large, pure white, dbl., 30–35 petals, exhibition form, borne usually singly, moderate, spicy or sweet tea fragrance; globose, large, orange fruit; normal, light green-tan prickles; foliage medium, medium green, semi-glossy; upright, bushy, tall growth; PP006714; [Bridal Pink X (Blue Nile X (Ivory Tower X Angel Face))]; Christensen, Jack & Carruth, Tom; Armstrong Nursery, 1986

'AROBLAVEET', HT, dr, 1982; (**Centenary College**); flowers deep red, large, dbl., 35 petals, intense fragrance; foliage large, medium green, matt; vigorous, upright, tall growth; [Angel Face X Typhoo Tea]; Christensen, Jack E.; Armstrong Nursery

'AROBRI', Min, ab, 1978; (**Holy Toledo**™); bud ovoid, pointed; flowers brilliant apricot-orange, reverse yellow-orange, imbricated, dbl., 28 petals, 1.5–2 in.; foliage small, glossy, dark; vigorous, bushy growth; AOE, 1980; [Gingersnap X Magic Carrousel]; Christensen, Jack E.; Armstrong Nursery

'AROBRISP', HT, dr, 1988; (**Tuxedo**); bud ovoid; flowers urn-shaped, large, borne singly, dbl., 45 petals; prickles broad, hooked downward, red-brown; foliage medium, medium green, semi-glossy; upright growth; [Portland Trailblazer X Olympiad]; Christensen, Jack E.; Bear Creek Gardens, 1988

'AROCAD', HT, ab, 1981; (**Brandy**™); bud long, pointed; flowers deep apricot, dbl., 25–30 petals, 5 in., exhibition form, borne mostly singly, slight, tea fragrance; straight prickles; foliage large, dark green; vigorous, medium growth; PP005168; AARS, 1982; [First Prize X Dr A.J. Verhage]; Swim, H.C. & Christensen, J.E.; Armstrong Nursery

'AROCANT', HT, ob, 1985; (**Spring Fever**); flowers apricot, pink and orange blend, large, dbl., 35 petals, slight fragrance; foliage large, dark, semi-glossy; [Gingersnap X Brandy]; Christensen, Jack E.; Armstrong Nursery

'AROCHARM', F, pb, 1987; (**Origami**™); flowers clear, soft pink, outstanding form, medium, borne usually singly, dbl., 25 petals, exhibition form, moderate, spicy fragrance; normal, light green-tan prickles; foliage medium, medium green, semi-glossy; upright, bushy, medium growth; [Coquette X Zorina]; Christensen, Jack & Carruth, Tom; Armstrong Nursery, 1991

'AROCHER', HT, rb, 1981; (**Mon Cheri**™); bud ovoid, pointed; flowers medium pink, suffusing to near yellow at base, aging to dark, dbl., 38 petals, slight, spicy fragrance; short prickles; foliage semi-glossy, medium green; upright, medium growth; PP005156; AARS, 1982; [(White Satin X Bewitched) X Double Delight]; Christensen, Jack E.; Armstrong Nursery

'AROCLIDD', Cl HT, rb, 1982; (**Double Delight, Climbing**, Grimpant Double Delight); PP005155; [Double Delight sport]; Christensen, Jack E.; Armstrong Nursery, 1985

'AROCOMU', HT, m, 1983; ('AROCUMU' **Patsy Cline**™); flowers light lavender, petals edged ruby lavender, large, dbl., 35 petals, exhibition form, intense fragrance; foliage medium, dark, matt; upright, bushy growth; PP005556; [Angel Face X Double Delight]; Christensen, Jack E.; Armstrong Nursery

'AROCORE', HT, mr, 1986; (**Rodeo Drive**™, Sunset Strip); flowers bright deep red, large, dbl., 32 petals, exhibition form, borne usually singly, slight fragrance; fruit not observed; many attenuated, medium, reddish prickles aging light brown; foliage medium to large, medium green, semi-glossy; bushy, medium growth; PP006813;

[Merci X Pharaoh]; Christensen, Jack E.; Armstrong Nursery

AROcruby, F, rb, 1987; (**Park Place**); Christensen, Jack E.

'ARODER', HT, dr, 1972; (**Kentucky Derby**); flowers large, dbl., exhibition form, slight fragrance; foliage large, glossy, leathery; vigorous, upright, bushy growth; [John S. Armstrong X Grand Slam]; Armstrong, D.L.; Armstrong Nursery

'ARODI', Min, mp, 1978; (**Heidi**™); bud mossy; flowers clear medium pink, mini-moss, small, dbl., 35 petals, 1.5 in., intense fragrance; foliage glossy; very vigorous, bushy growth; [Fairy Moss X Iceberg]; Christensen, Jack E.; Armstrong Nursery

'ARODOUSNA', HT, rb, 1986; (**Givenchy**™, Paris Pink); flowers pink, blushed red, reverse pink, yellow base, blooms in sprays, dbl., 30 petals, exhibition form, intense, spicy fragrance; no fruit; medium, brown, hooked prickles; foliage medium, dark; medium, upright, bushy growth; [Gingersnap X Double Delight]; Christensen, Jack E.; Armstrong Nursery, 1985

'AROFEIGEL', HT, lp, 1982; (**Peggy Lee**); flowers pale pink; PP005467; [Century Two sport]; Feigel, John R.; Armstrong Nursery

'AROFIRIC', F, rb, 1987; (**Fire 'n' Ice**™); flowers red, reverse white, fading purplish-red, medium, borne usually singly, dbl., 40 petals, exhibition form, slight fragrance; no fruit; foliage medium, dark green, glossy, very attractive, pointed; upright, bushy, tall growth; [Bluhwunder X Love]; Christensen, Jack & Carruth, Tom, 1981; Armstrong Nursery, 1985

'AROFRAP', HT, pb, 1982; (**Almondeen**); flowers almond and pink blend, good form, large, dbl., slight fragrance; foliage large, medium green, semi-glossy; upright, bushy growth; PP005704; [Angel Face X First Prize]; Christensen, Jack E.; Armstrong Nursery, 1984

'AROFRICHEE', HT, yb, 1986; (**Pink Lemonade**); flowers yellow turning bright pink, well-formed, large, dbl., 35 petals, moderate fragrance; foliage large, medium green, semi-glossy; upright, bushy growth; [Friendship X Rosy Cheeks]; Christensen, Jack E.; Armstrong Nursery

'AROFUTO', HT, mr, 1989; (City of Warwick, **Grand Marshall**™); bud ovoid, pointed; flowers large, borne usually singly, dbl., 35 petals, exhibition form, slight fragrance; prickles hooked, medium, red to brown; foliage medium, medium green, semi-glossy; upright, bushy, medium growth; [Futura X Olympiad]; Christensen, Jack E.; Michigan Bulb Co., 1989

'AROGLOFY', F, dy, 1988; (**Bugle Boy**™); bud ovoid, pointed; flowers deep yellow, good substance, medium, borne in sprays of 5–6, dbl., 35 petals, cupped, slight, tea fragrance; fruit unknown; prickles hooked slightly downward, medium, red to tan; foliage medium, medium green, very glossy; bushy, medium growth; [Sunsprite X (Katherine Loker seedling X Gingersnap)]; Christensen, Jack E.; Bear Creek Gardens, 1988

'AROGLOR', HT, m, 1983; (**Governor's Lady Gloria**); flowers pastel mauve, well-formed, large, dbl., 35 petals, intense fragrance; foliage large, medium green, matt; upright, bushy growth; [Sweet Afton X Blue Nile]; Christensen, Jack E.; Armstrong Nursery

'AROGOBI', Min, ob, 1984; (**Fool's Gold**); flowers gold, reverse bronze, well-formed, small, dbl., 20 petals, slight fragrance; foliage medium, dark, semi-glossy; upright growth; [Cricket X Dr A.J. Verhage]; Christensen, Jack E.; Armstrong Nursery

'AROGRAJU', HT, m, 1985; (**Plum Crazy**); flowers deep lavender, well-formed, large, dbl., 35 petals, moderate fragrance; foliage medium, dark, matt; bushy growth; [(Ivory Tower X Angel Face) X Blue Nile]; Christensen, Jack E.; Armstrong Nursery

'AROGRESH', HT, w, 1983; (**Mint Julep**); flowers pale green and pink blend, large, dbl., 35 petals, slight fragrance; foliage medium, medium green, semi-glossy; upright growth; [White Masterpiece X Queen Elizabeth]; Christensen, Jack E.; Armstrong Nursery

'AROGREWOD', HT, m, 1987; (**Silverado**); flowers soft silver blushed ruby, reverse white, large blooms, dbl., 28 petals, exhibition form, borne singly, slight fragrance; large prickles, hooked slightly downward; foliage medium, dark, matt; medium, bushy growth; PP006861; [(Ivory Tower X Angel Face) X Paradise]; Christensen, Jack E.; Armstrong Nursery

Aroha, *see* 'MURHA'

'AROHAICLO', Cl Gr, or, 1982; (**Olé, Climbing**™); Haight, George S. & Swim, H.C.; Armstrong Nursery

'AROJECHS', F, rb, 1980; (**Confetti**); bud ovoid; flowers deep yellow, aging orange-red, 18–25 petals, exhibition form, borne 3–7 per cluster, slight, tea fragrance; prickles hooked downward; foliage medium green; upright growth; PP005399; [Jack O'Lantern X Zorina]; Swim, H.C. & Christensen, J.E.; Armstrong Nursery, 1983

'AROKET', Min, ob, 1978; (**Cricket**™); bud ovoid; flowers light orange to yellow, dbl., 25 petals, 1–1.5 in., globular, slight fragrance; foliage dark; upright, bushy growth; [Anytime X Katherine Loker]; Christensen, Jack E.; Armstrong Nursery

'AROKR', F, my, 1978; (**Katherine Loker**); bud pointed; flowers medium golden yellow, classic form, imbricated, dbl., 28 petals, 3.5 in., slight fragrance; upright, spreading growth; [Zorina X Dr A.J. Verhage]; Swim, H.C. & Christensen, J.E.; Armstrong Nursery, 1979

AROkris, F, dy, 1990; (**Golden Wedding**); Bear Creek Gardens

'AROKUNCE', HT, ob, 1982; (**Inferno**); flowers well-formed, large, slight fragrance; foliage medium, dark, semi-glossy; upright, bushy growth; PP005558; [Zorina X Yankee Doodle]; Christensen, Jack E.; Armstrong Nursery

'AROLALA', HT, m, 1983; (**Azure Sea**); flowers silvery lavender, petals edged ruby, deeper reverse, well-formed, large, dbl., 30 petals, slight fragrance; foliage large, dark, matt; upright, bushy growth; PP005693; [(Angel Face X First Prize) X Lady X]; Christensen, Jack E.; Armstrong Nursery

'AROLAQUELI', Gr, m, 1986; (**Lagerfeld**™, Starlight); flowers silvery lavender, blooms in sprays of 5–15, dbl., 30 petals, 4–5 in., exhibition form, intense fragrance; large, globular, orange fruit; medium, light brown prickles, hooked downward; foliage medium, medium green, matt; tall, upright, bushy growth; [Blue Nile X (Ivory Tower X Angel Face)]; Christensen, Jack E.; Armstrong Nursery, 1985

'AROLEMO', HT, ly, 1986; (**Lanvin**™); dbl., 30 petals, exhibition form, blooms in sprays of 3–5, moderate fragrance; no fruit; medium, straight, light brown to red prickles; foliage medium, dark green tinted red, semi-glossy; medium, upright bushy growth; [Unnamed seedling X Katherine Loker]; Christensen, Jack E.; Armstrong Nursery, 1985

'AROLICAL', HT, m, 1984; (**Blue Ribbon**); bud ovoid; flowers lilac-blue, well-formed, dbl., 35–40 petals, 5–6 in., borne mostly singly, intense sweet rose fragrance; foliage very large, dark, semi-glossy; upright growth; PP006000; [(Angel Face X First Prize) X Blue Nile]; Christensen, Jack E.; Armstrong Nursery, 1986

'AROLYME', HT, my, 1987; (**Sundial**™); flowers medium, bright yellow, medium, borne usually singly, dbl., 30 petals, exhibition form, slight, spicy fragrance; no fruit; many, normal, small and large, light green-tan prickles; foliage medium, medium green, glossy; upright, bushy, medium growth; [(Golden Wave X (American Heritage X First Prize)) X ((Camelot X First Prize) X Yankee Doodle)]; Christensen, Jack E.; Armstrong Nursery, 1986

'**AROMA**', HT, mr, 1931; bud ovoid; flowers crimson, large, dbl., vigorous, bushy growth; Cant, B. R.; J&P

Aromatic, HT, pb, 1970

'AROMICLEA', HT, ob, 1984; (**Voodoo**™); flowers salmon, yellow, orange and pink blend, large, dbl., 30–35 petals, 5–6 in., exhibition form, intense, rich, sweet fragrance; foliage medium, dark, very glossy; upright, bushy, tall growth; PP006121; AARS, 1986; [((Camelot X First Prize) X Typhoo Tea) X Lolita]; Christensen, Jack E.; Armstrong Nursery, 1986

'AROMIKEH', Min, mr, 1981; (**Hotline**™); bud ovoid, pointed, lightly mossed; flowers bright medium red, mini-moss, dbl., 22 petals, exhibition form, borne usually singly, moderate, moss fragrance; straight, thin prickles; foliage medium green; compact (near 12 in) growth; PP005672; [Honest Abe X Trumpeter]; Christensen, Jack E.; Armstrong Nursery

'AROMONTELIB', Cl HT, yb, 1980; (Caribia, Climbing, **Harry Wheatcroft, Climbing**); Mungia, Fred A., Sr; Montebello Rose Co., Inc.

'ARON', Min, dr, 1978; (**Honest Abe**™); bud mossy; flowers deep velvety crimson-red, mini-moss, small, dbl., 33 petals, 1.5 in., slight, tea fragrance; foliage glossy; vigorous, bushy growth; [Fairy Moss X Rubinette]; Christensen, Jack E.; Armstrong Nursery

'ARONEMO', Min, ly, 1977; (**Honey Moss**); bud mossy; flowers near white, toward honey, mini-moss, dbl., 52 petals, 1 in., flat, moderate fragrance; foliage dark, leathery; spreading growth; [Fairy Moss X ?]; Sudol, Julia

'ARONESUF', F, dy, 1988; (**South Pacific**™); flowers color holds well, medium, borne in sprays of 4–6, dbl., 26 petals, cupped, slight fragrance; no fruit; prickles hooked slightly downward, medium, few, green to tan; foliage medium, medium green, glossy; bushy, medium growth; [Sunsprite X Unnamed seedling]; Christensen, Jack E.; Bear Creek Gardens, 1988

'ARONEWP', HT, yb, 1988; (**New Peace**™); flowers yellow-cream with bright red margins aging larger red margin, dbl., 48–52 petals, exhibition form, slight fragrance; prickles pointed, small, dark tan; foliage medium, medium green, matt; upright, medium growth; [Gingersnap X Young Quinn]; Christensen, Jack E.; Michigan Bulb Co., 1988

'AROPICLU', S, pb, 1984; (**Dream Cloud**); flowers light to dark salmon-pink, dbl., 20 petals, blooms in large pyramidal clusters, slight fragrance; foliage medium, long, narrow, medium green, matt; spreading, bushy growth semi-pendulous habit; PP005998; [Zorina X Gartendirektor Otto Linne]; Christensen, Jack E.; Armstrong Nursery, 1985

'AROPLUMI', HT, m, 1990; (**Fragrant Plum**); bud long, ovoid; flowers light lavender blushing purple, smoky edges, dbl., 20–25 petals, 4–4.5 in., exhibition form, borne singly and in large clusters, intense, fruity fragrance; tall, upright growth; [Shocking Blue X (Blue Nile X (Ivory Tower X Angel Face))]; Christensen, Jack E.; Armstrong Nurseries, 1990

'AROPRAWN', Min, lp, 1982; (**Helen Boehm**); flowers soft pink, small, dbl., 20 petals, exhibition form, slight fragrance; foliage small, medium green, semi-glossy; upright, bushy growth; PP005397; [Foxy Lady X Deep Purple]; Christensen, Jack E.; Armstrong Nursery, 1983

'ARORAJU', F, ob, 1986; (Lady Glencora, **Orange Juice**); flowers clear orange, blooms in sprays of 3–5, dbl., 33 petals, exhibition form, slight fragrance; ovoid, medium, orange-red fruit; long, red prickles; medium, upright, bushy growth; [Katherine Loker X Gingersnap]; Christensen, Jack E.; Michigan Bulb Co.

'ARORASP', Min, dy, 1981; (**Sunspray**™); bud ovoid, long, pointed; flowers bright deep yellow, semi-dbl., 16 petals, borne singly or several per cluster, slight, tea fragrance; straight prickles; foliage semi-glossy, dark; vigorous, upright growth; PP005035; [Gingersnap X Magic Carrousel]; Christensen, Jack E.; Armstrong Nursery

'AROREROY', HT, mr, 1987; (Royal Delight, **Royal Success**™); flowers large, dbl., 30 petals, exhibition form, borne usually singly; fruit not observed; normal, light green to tan prickles; foliage medium, medium green, semi-glossy; upright, bushy, tall growth; PP006910; [Red Success X Royalty]; Christensen, Jack & Carruth, Tom; Armstrong Nursery, 1986

'ARORESAS', Gr, mr, 1987; (**Scarlett O'Hara**™); flowers medium, dbl., 35 petals, exhibition form, borne usually singly, slight fragrance; nearly thornless; foliage medium, medium green, semi-glossy; upright, tall growth; PP006670; [Red Success X Mary DeVor]; Christensen, Jack & Carruth, Tom; Armstrong Nursery, 1986

AROshrel, HT, op, 1991; (**Michelle Joy**); Bear Creek Gardens

'AROSHRIM', Min, op, 1980; (**Foxy Lady**™); bud ovoid, pointed; flowers salmon and creamy blend, imbricated, dbl., 25 petals, 1.5 in.; foliage small; tall, vigorous, bushy growth; [Gingersnap X Magic Carrousel]; Christensen, Jack E.; Armstrong Nursery

Arosia, HT, mp, 1998; Noack, Werner

'AROSILHA', HT, m, 1986; ('AROSILMA', **Silk Hat**™); flowers red purple, cream reverse, large blooms borne usually singly, dbl., 45 petals, exhibition form, moderate fragrance; no fruit; foliage large, medium green, matt; medium, upright, bushy growth; [Ivory Tower X (Night 'n' Day X Plain Talk)]; Christensen, Jack E.; Armstrong Nursery, 1985

'AROSNAP', F, ob, 1978; (Apricot Prince, **Gingersnap**, Prince Abricot); bud long, pointed; flowers pure orange, imbricated to ruffled, dbl., 30–35 petals, 4 in., borne in small clusters, slight fragrance; foliage dark green; vigorous, upright, bushy growth; PP4330; [(Zambra X (Orange Triumph X Floradora)) X (Jean de la Lune X (Spartan X Mandrina))]; Delbard-Chabert; Armstrong Nursery

'AROSTAL', HT, m, 1985; (**Quicksilver**); flowers pale lavender gray, large blooms borne singly, dbl., 25 petals, exhibition form, moderate fragrance; medium, yellow-gray prickles; foliage large, dark, matt; tall, upright, bushy growth; [Blue Nile X Brandy]; Christensen, Jack E.; Michigan Bulb Co.

AROsumo, HT, my, 1983; (**Morning Sun**); Christensen, Jack E.

'AROTIGY', HT, yb, 1986; (**Polo Club**™); flowers yellow bordered red, fading cream with pink edges, dbl., slight fragrance; fruit not observed; straight, small, few, greenish to light brown prickles; foliage medium, dark green, semi-glossy; upright, bushy, tall growth; PP006758; [Gingersnap X Young Quinn]; Christensen, Jack E.; Armstrong Nursery, 1986

'AROTRUSIM', S, mr, 1988; (**Bloomin' Easy**™, Blooming Easy); flowers clear, bright red, urn-shaped, medium, semi-dbl., 22–25 petals, borne singly, repeat bloom, moderate, tea fragrance; round, medium, bright red-orange fruit; prickles hooked slightly downward, medium, red; foliage large, medium green, glossy; upright, bushy, medium growth; PP007157; [Trumpeter X Simplicity]; Christensen, Jack E.; Armstrong Nursery, 1987

'AROVIDIL', Min, mr, 1980; (**Little Red Devil**); bud ovoid, pointed; flowers imbricated blooms borne 1–8 per cluster, dbl., 44 petals, 1.5 in., slight fragrance; small, narrow prickles; foliage small, semi-glossy, irregularly serrated; vigorous, bushy, fairly tall growth; [Gingersnap X Magic Carrousel]; Christensen, Jack E.; Armstrong Nursery

'AROVULC', HT, ob, 1984; ('AROVULE', Harlequin, **Miss Liberté**, Miss Liberty); flowers coral orange to dusty deep red, well-formed, large, dbl., 35 petals, slight fragrance; foliage large, dark, semi-glossy; upright, bushy growth; [(Camelot X First Prize) X Gingersnap]; Christensen, Jack E.; Armstrong Nursery

'AROVULE', HT, ob, 1984; ('AROvulc', Harlequin, **Miss Liberté**, Miss Liberty); flowers coral orange to dusty deep red, well-formed, large, dbl., 35 petals, slight fragrance; foliage large, dark, semi-glossy; upright, bushy growth; [(Camelot X First Prize) X Gingersnap]; Christensen, Jack E.; Armstrong Nursery

'AROWAGO', HT, rb, 1986; (Show Off, **Show-off**); flowers velvety brilliant red, reverse silvery blend, large, dbl., 35 petals, slight fragrance; foliage large, dark, semi-glossy; upright, bushy growth; [Typhoo Tea X Snowfire]; Christensen, Jack E.; Armstrong Nursery

'AROWEDYE', HT, dy, 1982; (**Candlelight**™); bud ovoid, pointed; flowers deep yellow, spiraled, dbl., 30 petals, borne mostly singly, slight fragrance; large-based prickles; foliage large, semi-glossy; medium-tall, upright, branching growth; PP005398; [Shirley Laugharn X (Bewitched X King's Ransom)]; Christensen, J. E. & Swim, H. C.; Armstrong Nursery

'AROWHIF', Gr, w, 1980; (**White Lightnin'**™); bud ovoid, pointed; flowers clear white, ruffled, dbl., 30 petals, 3.5–4 in., exhibition form, borne in small clusters, intense, citrus fragrance; foliage glossy, medium green; upright, bushy growth; PP4670; AARS, 1981; [Angel Face X Misty]; Swim, H.C. & Christensen, J.E.; Armstrong Nursery

AROwillip, MinFl, lp; (**Strawberry Fayre**); Bear Creek Gardens, 1991

AROyefel

'AROYEFELT', F, my, 1987; (AROyefel, **Haiku**™); flowers medium, borne usually singly or in sprays of 2–3, dbl., 38 petals, exhibition form, slight fragrance; no fruit; normal, light green to tan prickles; foliage medium, dark green, glossy; upright, bushy, medium growth; [Bridal Pink X Sunspray]; Christensen, Jack & Carruth, Tom; Armstrong Nursery, 1986

'AROYOL', Min, my, 1979; (**Hopscotch**); bud ovoid, pointed; flowers golden yellow, imbricated, small, dbl., 28 petals, 1.5 in.; foliage small; vigorous, bushy growth; [Gingersnap X Magic Carrousel]; Christensen, Jack E.; Armstrong Nursery

'AROYQUELI', Gr, my, 1982; (**Gold Medal**®); bud ovoid, long, pointed; flowers deep golden yellow sometimes flushed orange, classic shape, dbl., 30–35 petals, 4.5–5 in., exhibition form, mostly borne singly, slight, fruity fragrance; foliage large, dark; tall, upright, bushy growth; PP005177; Gold Star of the South Pacific, Palmerston North, NZ, 1983; [Yellow Pages X (Granada X Garden Party)]; Christensen, Jack E.; Armstrong Nursery

'AROYUMI', Min, yb, 1984; (**Ferris Wheel**); flowers yellow, turning pink, orange and red (striped), medium, dbl., 20 petals, slight fragrance; foliage small, dark, semi-glossy; bushy growth; PP005703; [Golden Angel X Cricket]; Christensen, Jack E.; Armstrong Nursery

Arpége, HT, 1978; Dorieux, Francois; (Cavriglia)

'ARPEGGIO', F, dp, 1961; bud pointed; flowers light red, semi-dbl., 12–18 petals, 3 in., blooms in clusters, slight fragrance; foliage dark, glossy; vigorous, compact growth; Von Abrams; Peterson & Dering

'ARRAS', Pol, mr, 1924; flowers crimson-red, [Triomphe Orléanais sport]; Turbat

'ARRILLAGA', HP, lp, 1929; flowers light pink, base golden, large, dbl., 50 petals, moderate fragrance; vigorous growth; [(R. centifolia X Mrs John Laing) X Frau Karl Druschki]; Schoener; B&A

'ARROGANCE', HT, pb, 1970; flowers coral-pink, dbl., 25 petals, 4 in., slight fragrance; foliage dark, leathery; compact growth; [Mischief X John S. Armstrong]; Poulsen, Niels D.; Poulsen

Arromanches, F, op, 1975; Eve, A.

Arrowtown, LCl, dy, 1982; Martin

Art Deco, HMsk, mr, 1993; Lens

'ARTAMA', HT, yb, 1978; bud pointed; 25 petals, exhibition form, slight fragrance; few prickles; foliage light green, pointed, glossy; sturdy growth; [Kordes' Perfecta X (Garden Party X Christian Dior)]; Takahashi, Takeshi

'ARTEK', HT, dr, 1939; flowers dark, velvety red, medium, dbl.; Kosteckij; (Sangerhausen)

Artful Dodger, *see* 'SABBELIEF'

'ARTHUR BELL', F, my, 1965; flowers yellow to creamy yellow, large, semi-dbl., 15 petals, intense fragrance; foliage heavily veined; vigorous growth; [Cläre Grammerstorf X Piccadilly]; McGredy, Sam IV; McGredy

'ARTHUR BELL, CLIMBING', Cl F, my, 1979; Pearce, C.A.; Limes Rose Nursery

'ARTHUR COOK', HT, dr, 1924; bud long, pointed; flowers deep crimson, large blooms, dbl., moderate fragrance; foliage light, glossy; long, strong stems; vigorous, bushy, compact growth; GM, NRS, 1925; McGredy

Arthur Cox, *see* JAYART

'ARTHUR DE SANSAL', P, m, 1855; flowers rich crimson-purple, fully damask-like, dbl.; Cochet, Sc.

'ARTHUR HILLIER', HMoy, dp, 1961; flowers rose-crimson, single, 5 petals, 2.5–3 in., repeat bloom, slight fragrance; vigorous growth; [R. macrophylla X R. moyesii]; Hillier

'ARTHUR J. TAYLOR', HT, dr, 1947; flowers large, dbl.; Wheatcroft Bros.; (Sangerhausen)

'ARTHUR OGER', HP, m, 1875; flowers purple/pink, very large, dbl.; Oger; (Sangerhausen)

'ARTHUR R. GOODWIN', HT, or, 1909; flowers coppery orange-red, passing to salmon-pink; [Seedling X Soleil d'Or]; Pernet-Ducher

'ARTHUR R. GOODWIN, CLIMBING', Cl HT, or

'ARTHUR SCARGILL', Min, mr, 1985; [Amruda sport]; Thompson, M.L.

'ARTHUR SCHULTE', HT, rb, 1986; flowers cherry to blood red with ivory white at base of petals, large, dbl., 43 petals, exhibition form, borne singly and in sprays of 1–3, moderate, damask fragrance; rounded, medium, pumpkin-orange fruit; medium-large, light green-bronze prickles; foliage large, dark green, semi-glossy, disease resistant; upright, bushy, medium, very vigorous, some interbranching growth; [Colorama seedling X Chrysler Imperial]; Williams, J. Benjamin; Krider Nursery, 1989

'ARTHUR YOUNG', M, m, 1863; flowers dark purple, large, cupped, recurrent bloom, vigorous growth; Portemer fils

Artiste®, *see* 'DORISTEL'

'ARTISTIC', F, ob, 1971; flowers orange, fading red, pointed, semi-dbl., 10–15 petals, 2 in., moderate fragrance; foliage small; LeGrice

Artistry™, *see* 'JACIRST'

Arturo Toscanini, *see* 'MEILUCRE'

Aruba, *see* SPEcawijk

'ARUBA-CARIBE', HT, pb, 1967; bud ovoid; flowers rose-pink and ivory, dbl., 38 petals, 5 in., exhibition form, intense, fruity fragrance; foliage leathery; vigorous, upright growth; [Diamond Jubilee seedling X Fashion seedling]; Boerner; J&P

'ARUNA', HT, or, 1968; flowers bright orange-scarlet, medium, dbl., cupped; foliage glossy; bushy, compact growth; [Independence X Seedling]; Pal, Dr B.P.; Indian Agric. Research Inst.

'ARUNIMA', F, dp, 1975; bud ovoid; flowers deep pink, dbl., 50 petals, 2 in.; foliage glossy; vigorous, bushy, compact growth; [Frolic X Seedling]; IARI

As de Coeur, *see* **'HERZ AS'**®

'ASABORAKE', HT, dy, 1979; flowers deep yellow, sometimes tipped pink, large, dbl., 30 petals, exhibition form, slight fragrance; foliage medium green, semi-glossy; bushy, upright growth; [(Golden Scepter X Narzisse) X Kordes' Perfecta]; Ota, Kaichiro

'ASAGUMO', HT, yb, 1973; (Oriental Dawn); bud ovoid; flowers large, dbl., cupped, moderate fragrance; foliage glossy, dark; vigorous, upright growth; [Peace seedling X Charleston seedling]; Suzuki, Seizo; Keisei Rose Nursery

ASAlav, Min, m, 1985; (**Lavender Crystal**); (ASAlav)

'ASBACH', F, or, 1960; [Ambassadeur Nemry X Cinnabar]; Leenders, M.

'ASCHENBRÖDEL', Pol, 1903; flowers rose with salmon, small, dbl., moderate fragrance; bushy, dwarf growth; [Petite Léonie X R. foetida bicolor]; Lambert, P.

'ASCHERMITTWOCH', LCl, w, 1955; (Ash Wednesday); bud silvery gray; flowers snow-white, large, dbl., blooms in large trusses; vigorous growth; Kordes

'ASCHERSONIANA', S, m; (R. X aschersoniana); flowers bright, light purple, very numerous, small; growth to 6 ft; (14); [R. blanda X R. chinensis]

'ASCOT', F, ab, 1962; flowers salmon-coral, semi-dbl., 18 petals, 4 in., blooms in clusters; low growth; [Brownie X Seedling]; Dickson, Patrick; Dickson, A.

Ascot Jubilee, HT, lp; Welsh, Eric; (Weatherly, L.)

'ASEPALA', M, w; flowers white, shaded flesh, sometimes edged rose, petal edges curled, small, dbl.; compact, erect growth

Ash Wednesday, *see* **'ASCHERMITTWOCH'**

Asha, HT, pb, 1968; Agarwal

Ashgrove Jubilee, *see* 'WELCO'

Ashley Marie, *see* 'JUDMARIE'

Ashley's Surprise, Min, m, 1997; King

Ashwini '89, HT, dr, 1989; K&S

Asja, HT, or, 1998; McGredy, Sam IV

Askari, *see* JACkari

'ASMODÉE', HGal, mr, 1849; flowers rosy crimson, large, dbl.; Vibert

Aspen, *see* 'POULURT'

'ASPIRANT MARCEL ROUYER', HT, ab, 1919; bud long, pointed; flowers apricot, tinted salmon-flesh, veined yellow, large blooms, dbl., moderate fragrance; foliage glossy, bronze; long, strong stems; very vigorous growth; [Sunburst X Seedling]; Pernet-Ducher

'ASPIRANT MARCEL ROUYER, CLIMBING', Cl HT, ab, 1934; Brenier, E.C.

Aspirin, *see* TANiripsa

'ASSEMBLAGE DES BEAUTÉS', HGal, dr, 1823; flowers brilliant crimson, medium, dbl.; erect growth; Delaage

'ASSINIBOINE', S, dp, 1962; flowers purplish red, large to medium blooms, semi-dbl., intermittent bloom, slight fragrance; foliage glossy; weak stems; vigorous growth; [Donald Prior X R. arkansana]; Marshall, H.H.; Canada Dept. of Agric.

Asso di Cuori®, *see* 'KORRED'

Asso di Cuori, Climbing, Cl HT, dr

'ASSO FRANCESCO BARACCA', HT, pb, 1936; bud long; flowers golden salmon, center deeper, well formed, dbl., intense fragrance; long stems; vigorous growth; [Julien Potin X Seedling]; Giacomasso

'ASTA VON PARPAT', HMult, m, 1909; flowers medium, dbl.; Geschwind, R.; (Sangerhausen)

Asterix®, *see* 'LENPON'

'ASTOLAT CHARM', HT, pb, 1951; flowers flesh-pink, base apricot, dbl., 30 petals, 4 in., intense fragrance; foliage dark; vigorous growth; [McGredy's Salmon sport]; Astolat Nursery

'ASTORIA', HT, ob, 1963; bud pointed; flowers bright orange shaded coral, moderate fragrance; foliage bronze, glossy; [Opera X Demoiselle]; Delforge

Astra™, *see* 'WILSMA'

Astra, *see* 'BENSTAR'

Astra, *see* RUIzesac

Astra Desmond, LCl, yb

'ASTRAL', HT, dp, 1976; flowers deep rose-pink, low-centered, dbl., 24 petals, 4 in., intense fragrance; foliage glossy, dark; vigorous growth; [Tropicana X Pink Favorite]; Bees

Astral, LCl, dr; Croix

Astrea®, *see* 'BARorp'

'ASTRÉE', HT, pb, 1956; flowers medium pink, reverse shaded orange, large, dbl., intense fragrance; [Peace X Blanche Mallerin]; Croix, P.

Astreé, Climbing, Cl HT, op; Croix

Astrée, Climbing, Cl HT, 1969; Croix, P.; (Cavriglia)

Astrid Lindgren®, S, lp, 1991; Poulsen

Astrid Späth, *see* **'FRAU ASTRID SPÄTH'**

'ASTRID SPATH STRIPED', F, rb, 1933

'ASTROROSE', HT, mr, 1971; bud long, pointed; flowers cardinal-red to crimson, medium, dbl., moderate fragrance; foliage large, dark, bronze, leathery; vigorous growth; [Helene Schoen X Chrysler Imperial]; Porter; General Bionomics

'ASTURIAS', F, rb, 1956; flowers red, reverse carmine, large, dbl., 40 petals, borne in corymbs; foliage glossy; strong stems; very vigorous growth; [Méphisto X Coralín]; Dot, M.

'ASUN GALINDEZ DE CHAPA', HT, pb, 1923; flowers salmon-pink, reverse darker, base yellow; [Mons. Paul Lédé X Jacques Vincent]; Ketten Bros.

Atago, HT, or; Keisei Rose Nurseries, Inc.

'ATALANTA', LCl, 1927; bud coppery pink; flowers flesh-pink, dbl., moderate fragrance; [Paul Ploton X William Allen Richardson]; Williams, A.

'ATALANTE', F, mr, 1958; flowers geranium-red, large; very vigorous growth; Frères, Buyl

'ATARA', HT, rb, 1975; flowers medium red, flecked and striped near white, reverse near white with red; [Suspense sport]; Nevo, Motke; Maoz Haim Rose Nursery, 1974

Atco Royal, *see* FRYwinner

Atena®, HT, dy

'ATHABASCA', (R. macounii variant), dp, 1930; flowers deep pink, semi-dbl., non-recurrent; very vigorous growth; hardy; (found growing wild at Vilna, Alberta, Canada, about 1930)

'ATHALIN', B, mr, 1830; flowers cherry-red; (28); Jacques

Athena, *see* **'RÜHKOR'**

'ATHENE', HT, lp, 1975; bud green; flowers porcelain-pink, base yolk-yellow, dbl., 37 petals, 5.5 in., intense, apple fragrance; foliage large; moderate, bushy growth; [Peace X Diamond Jubilee]; Murray & Hawken; Rasmussen's

Athene, *see* BENcreberg

'ATHERTON', HT, my, 1981; bud long, pointed; flowers urn-shaped, 35 petals, borne 1–3 per cluster, moderate, fruity fragrance; small, recurved prickles; foliage medium, matt; medium growth; [Seedling X Sunblest]; Perry, Astor

'ATHLETE', HT, rb, 1964; flowers nasturtium-red shaded gold, medium, semi-dbl., 16 petals, slight fragrance; foliage light green; bushy growth; [Tzigane X Claude]; Barter

'ATHLONE', F, yb, 1965; flowers cream edged orange-scarlet, open, blooms in clusters, slight fragrance; foliage small, dark; free growth; [Circus X Cinnabar]; McGredy, Sam IV

Athos®, *see* 'LAPWON'

'ATIDA', HT, dr, 1941; flowers velvety dark scarlet, well formed, quite large, semi-dbl., moderate fragrance; very vigorous growth; [Mme Van de Voorde X Dance of Joy]; Sauvageot, H.; Sauvageot

Atida, *see* SAUblim

Atida 93, HT, 1993; Sauvageot, H.; (Cavriglia)

Atkins Beauty, *see* 'SAVACENT'

Atlanta, *see* 'WEKDOCLEM'

Atlanta, Min, mp, 1992; deRuiter

Atlantic®, *see* 'GAVAL'

Atlantic City, *see* 'JOLLMEA'

Atlantic Star, *see* FRYworld

'ATLANTIDA', HT, op, 1939; flowers coppery salmon, shaded peach; vigorous growth; Pahissa

'ATLANTIS', F, m, 1970; flowers deep mauve, single, 5 petals, 3 in., moderate fragrance; foliage glossy, purple tinted; GM, Rome, 1969; [Orangeade X Lilac Charm]; Harkness

Atlantis Palace, *see* POUIsiana

Atlas®, *see* 'DELKORT'

Atoll, *see* 'MEIBYSTER'

Atom Bomb, *see* **'ATOMBOMBE'**

'ATOMBOMBE', F, mr, 1953; (Atom Bomb, Atomflash, Velvet Robe); bud pointed; flowers deep scarlet-orange, dbl., 28 petals, 2.5 in., clusters, very vigorous growth; [Obergärtner Wiebicke X Independence]; Kordes

Atomflash, *see* **'ATOMBOMBE'**

'ATOMIC WHITE', HT, ly, 1948; bud long, pointed; flowers white, center tinted yellow, large, dbl., exhibition form, moderate fragrance; vigorous, bushy growth; [Pink Princess X Shades of Autumn]; Brownell, H.C.

'ATROPURPUREA', HRg, mr, 1899; flowers carmine-crimson, single, blooms in clusters; vigorous (3–5 ft) growth; [R. rugosa X R. damascena]; Paul

'ATROPURPUREA', Pol, m, 1910; flowers purple-red; [Mme Norbert Levavasseur X Perle des Rouges]; Levavasseur

'ATTAR OF ROSES', HT, w, 1936; bud globular; flowers creamy white, edged pink, dbl., cupped, moderate fragrance; foliage glossy, bronze; long, strong stems; vigorous, bushy, compact growth; Cant, B. R.

Attila, A, dp; flowers deep pink, large, semi-dbl., cupped

ATTlight, HT, ab, 1999; (**Apricot Delight**); Attfield

'**ATTRACTION**', HT, yb, 1931; flowers yellow and orange, dbl., globular, slight fragrance; foliage glossy, bronze; dwarf, bushy growth; Dickson, A.

'**ATTRAKTION**'®, F, op, 1963; bud long, pointed; flowers salmon-copper, reverse golden yellow, dbl., borne in large clusters, intense fragrance; foliage glossy, dark; vigorous, bushy growth; Tantau, Math.

'**AUBADE**', HT, ob, 1963; flowers yellow-orange, medium, dbl.; foliage dark; [Docteur Valois X Seedling]; Verbeek

'**AUBREY COBDEN**', HT, my, 1948; flowers clear yellow, medium, dbl., 30 petals, intense fragrance; foliage glossy, stems red; free growth; [Oswald Sieper X Seedling]; Mee; Fryer's Nursery, Ltd.

Auckland Metro, *see* 'MACBUCPAL'

Audace®, *see* 'BUCBI'

'**AUDIE MURPHY**', HT, mr, 1957; bud long, pointed, crimson; dbl., 20 petals, 4.5–5.5 in., exhibition form, moderate, spicy fragrance; foliage semi-glossy, dark, bronze; vigorous, upright, bushy growth; GM, Portland, 1957; [Charlotte Armstrong X Grande Duchesse Charlotte]; Lammerts, Dr Walter; Roseway Nursery

'**AUDINE**', HT, lp, 1950; flowers shell-pink, reflexed, dbl., 22 petals, exhibition form, moderate fragrance; foliage dark; vigorous, upright growth; [Percy Izzard X William Moore]; Gatward

'**AUDREY**', HT, dr, 1922; flowers deep crimson, well formed, exhibition form, moderate fragrance; Paul, W.

Audrey Gardner, *see* 'PEASPECIAL'

'**AUDREY HARRISON**', Cl HT, ab, 1969; flowers pale apricot, semi-dbl., 4 in., free bloom, early, intense fragrance; foliage glossy, dark; [Shot Silk sport]; Harrison; Harkness

Audrey Hepburn, *see* 'TWOADORE'

'**AUDREY MARIE**', F, mr, 1994; flowers medium red, dbl., 26–40 petals, 1.5–2.75 in., borne mostly singly, slight fragrance; some prickles; foliage medium, medium green, semi-glossy; low (2 1/2 ft), bushy growth; [Tamanga X Anne Scranton]; Dobbs, Annette E., 1994; Michael's Roses, 1994

Audrey McCormack, *see* 'KENOBSESS'

Audrey Mieklejohn, HT, my

'**AUDREY STELL**', HT, pb, 1937; flowers soft strawberry-pink, reverse sulfur-yellow, semi-dbl.; foliage light; long, strong stems; vigorous, bushy growth; [Soeur Thérèse sport]; Stell; Stell Rose Nursery

Audrey Wilcox®, *see* 'FRYWILREY'

Augie Boy, *see* 'RENBOY'

August Kordes, *see* '**LAFAYETTE**'

'**AUGUST NOACK**', HT, dr, 1928; bud long, pointed; flowers deep scarlet, open, large, dbl., exhibition form, moderate fragrance; foliage bronze, dark; vigorous, bushy growth; [Columbia sport]; Kordes

'**AUGUST NOACK, CLIMBING**', Cl HT, dr, 1935; Lens

'**AUGUST SEEBAUER**', F, dp, 1944; (The Queen Mother); bud long, pointed; flowers deep rose-pink, large, dbl., exhibition form, blooms in clusters; foliage glossy; vigorous growth; (28); [Break o' Day (HT) X Else Poulsen]; Kordes

Augusta, *see* 'POULDAVA'

'Augusta', HRg, lp; flat, intense; Wright, Percy H., 1953

'**AUGUSTE CHAPLAIN**', HP, mr, 1921; flowers large, dbl.; Tanne; (Sangerhausen)

'**AUGUSTE COMTE**', T, mp, 1896; flowers large, dbl.; Soupert & Notting; (Sangerhausen)

'**AUGUSTE DELOBEL**', LCl, mr, 1924; flowers carmine, large white eye, yellow stamens, borne in clusters; Turbat

'**AUGUSTE FINON**', HMult, yb, 1923; flowers golden yellow passing to coppery and salmon, borne in clusters, moderate fragrance; [Goldfinch X Seedling]; Turbat

'**AUGUSTE GERVAIS**', LCl, ab, 1918; flowers coppery yellow and salmon, fading white, semi-dbl., 4–5 in., blooms in clusters, seasonal bloom, moderate fragrance; very vigorous, climbing growth; [R. wichurana X Le Progrès]; Barbier

'**AUGUSTE HALEM**', HT, 1891; Guillot, J.B.; (Cavriglia)

'**AUGUSTE KORDES**', Cl F, mr, 1928; (Lafayette, Climbing); flowers light scarlet, borne in clusters, fine seasonal bloom, later intermittent; vigorous, climbing growth; [Lafayette sport]; Kordes

Auguste Renoir®, *see* 'MEITOIFAR'

'**AUGUSTE ROUSSEL**', LCl, pb, 1913; flowers salmon-pink to flesh-pink, well-formed, petals undulated, large, semi-dbl., seasonal bloom; very vigorous, climbing growth; [R. macrophylla X Papa Gontier]; Barbier

'**AUGUSTE VERMARE**', HT, or, 1958; bud pointed; flowers coral-red; vigorous, upright growth; [Comtesse Vandal X Seedling]; Arles; Roses-France

'**AUGUSTINE GUINOISEAU**', HT, w, 1889; (White La France); flowers white, tinted light flesh; [La France sport]; Guinoiseau, B.

Augustine Halem, HT, op

'**AUGUSTUS HARTMANN**', HT, dp, 1914; flowers tyrian rose, large, dbl., slight fragrance; foliage rich green, leathery; vigorous growth; GM, NRS, 1914; Cant, B.R.

Augustus Hartmann, Climbing, Cl HT, 1980; Fineschi, G.; (Cavriglia)

'**AUGUSTUS STONE**', HT, ab, 1999; flowers light apricot/amber, resist bad weather, dbl., 26–40 petals, 4–4.5 in., borne in small clusters, moderate fragrance; some prickles; foliage medium, dark green, dull; bushy, medium (3 ft) growth; [Gavotte X Pot of Gold]; Poole, Lionel; David Lister, Ltd., 2000

'**AULD LANG SYNE**', HT, ob, 1956; bud ovoid; flowers orange edged pale orange-yellow, dbl., 35–45 petals, 4 in., moderate, fruity fragrance; bushy growth; [Tawny Gold X (Talisman X Tawny Gold)]; Motose

Aunt Belle's Tea, T, pb

'**AUNT GERRY**', HT, ly, 1992; flowers golden yellow, more golden yellow than Lanvin, slight petal, dbl., 35 petals, 3–3.5 in., slight, spicy fragrance; prickles prickly peduncle; foliage medium, medium green, semi-glossy; medium (90–120 cms), upright, very vigorous growth; [Lanvin sport]; Sheldon, John & Jennifer; Trophy Roses, Ltd., 1992

'**AUNT HARRIET**', HMult, rb, 1918; flowers scarlet-crimson, white eye, yellow stamens; vigorous growth; [Apolline X R. wichurana]; Van Fleet; Phila. Farm Journal

'**AUNT HONEY**', S, mp, 1984; bud large, ovoid; flowers, dbl., 38 petals, exhibition form, borne 5–10 per cluster, repeat bloom, moderate, damask fragrance; awl-like, tan prickles; foliage medium, dark olive-green; erect, short, bushy growth; hardy; [Music Maker X Habanera]; Buck, Dr Griffith J.; Iowa State University

'**AUNT RUTH**', LCl, pb; 1995; Flowers streaked medium pink, white eye, reverse white streaked pink, semi-dbl, (6–14 petals), medium (3 in.) blooms borne in small and large clusters, some singly; slight fragrance; few prickles; foliage medium, medium green, semi-glossy, medium, spreading, bushy growth; [Unknown X Unknown]; Jerabek, Paul E.

Auntie's Favorite, HT, dr

'**AUNTY DORA**', F, m, 1970; flowers magenta, semi-dbl., blooms in trusses, slight fragrance; low growth; [Dearest X Lilac Charm]; Deamer; Warley Rose Gardens

Aunty Lil, *see* 'KENMASNIA'

'**AUREA**', HT, yb, 1948; flowers yellow, reverse red, well formed, moderate, fruity fragrance; foliage glossy; vigorous growth; Dot

Aurea, *see* MEIbosat

'**AUREATE**', HT, ob, 1932; flowers orange and scarlet to yellow, large, dbl., moderate fragrance; foliage bronze; vigorous, bushy growth; GM, NRS, 1929; Dickson, A.

'AURELIA CAPDEVILA', HT, pb, 1933; flowers pink, base salmon, moderate fragrance; foliage dark; vigorous growth; Dot, Pedro

'AURELIA LIFFA', HSet, mr, 1896; Geschwind, R.

Aurelia Weddle, HT, or, 1995; Weddle, Von C.

Aurelian Gold, LCl, my

'AURÉLIEN IGOULT', HMult, m, 1924; flowers purple; vigorous, climbing growth; [Veilchenblau X Seedling]; Igoult; Bruant

'AUREOLA', HT, my, 1934; flowers golden yellow; [Mev. G.A. van Rossem sport]; Böhm, J.

'AURÉOLE', HT, ob, 1951; bud large; flowers orange-yellow, semi-dbl., moderate fragrance; foliage glossy, dark; very vigorous growth; Gaujard

'AURIOL', Pol, rb; Alderton, 1930

Auriu de Cluj, F, lp; flat, 15 petals, slight; medium, light green, glossy; [Orangeade X Western Sun]; St. Wagner, 1984; Res. Stn. f. Horticulture, Cluj, 1991, Romania

'AURORA', LCl, ob, 1956; flowers orange-yellow, large, dbl.; vigorous growth; Kordes

'AURORA', HMsk, ly, 1928; flowers yellow, passing to creamy white, single, moderate fragrance; [Danaë X Miriam]; Pemberton

'AURORA', F, op, 1941; flowers salmon-pink tinted golden yellow and orange, dbl., moderate fragrance; GM, Bagatelle, 1940; [Mev. Nathalie Nypels X Seedling]; Leenders, M.

'AURORA', (variety of R. acicularis), dp; flowers have wide, thick petals, single, 5 petals; foliage red and purple in fall; ((found by Erskine in Alberta, Canada, about 1950)

Aurora, F, my

Aurora, HT, lp; Strnad; (Czech Rosa Club)

'AURORA BOREAL', HT, rb, 1935; flowers geranium, shaded fiery red, semi-dbl.; foliage dark; very vigorous growth; [Étoile de Hollande X (Ville de Paris X Sensation)]; Munné, B.

'AURORE', HCh, yb; flowers yellow passing to salmon-pink, loose form, recurrent bloom, (14); Touvais

'AURORE', HT, pb, 1936; flowers pink, base yellow, dbl., moderate fragrance; vigorous growth; Capiago

Aurore, F, or

'AURORE', HP, mp, 1861; Laffay, M.

'AURORE D'ESPAGNE', HT, pb, 1966; bud long, pointed; flowers salmon-pink to nankeen yellow, medium, semi-dbl., cupped, slight fragrance; vigorous growth; [Zambra X Queen Elizabeth]; Dot; Minier

'AURORE DU MATIN', HP, mr, 1857; flowers light red, large, dbl., intense fragrance; Rolland; (Sangerhausen)

'AURORE SAND', LCl, pb, 1964; flowers two-tone soft pink and copper, well formed, recurrent bloom, moderate fragrance; vigorous growth; GM, Bagatelle, 1963; [Mme Moisans X Odette Joyeux]; Robichon

AUSap, S, ly, 1988; (**Queen Nefertiti®**); Austin, David

AUSapple, S, lp, 1983; (**Dapple Dawn**); Austin, David

'AUSBAKER', S, yb, 1998; (**Teasing Georgia**); very dbl., 110 petals, 3.5 in., borne mostly singly, moderate fragrance; few prickles; foliage medium, semi-glossy, disease-resistant; branching, strong, graceful, medium (3.5 ft) growth; [Charles Austin X seedling]; Austin, David; David Austin Roses, Ltd., 1998

'AUSBAND', S, dy, 1998; (**Buttercup™**); flowers loosely cupped, large heads of semi-double golden yellow florets, 21 petals; bright green semi-glossy foliage; almost thornless; light airy growth 3–4 ft; excellent repeat, disease resistant; Austin, David; David Austin Roses, Inc., 1998

'AUSBEAM', S, ab, 1992; (**Moonbeam**); flowers apricot yellow, blooms borne in small clusters, semi-dbl., 6–14 petals, 1.5–2.75 in., intense fragrance; some prickles; foliage medium, light green, semi-glossy; medium (100 cms), bushy growth; Austin, David; David Austin Roses, Ltd., 1983

'AUSBELLS', S, dp, 1994; (**Bow Bells™**); flowers deep pink, dbl., 15–25 petals, 1.5–2.75 in., borne in large clusters, moderate fragrance; some prickles; foliage medium, medium green, semi-glossy; upright, bushy (120 cms) growth; [Seedling X Graham Thomas]; Austin, David, 1991; David Austin Roses, Ltd., 1991

'AUSBLOOM', S, dr, 1994; (**The Dark Lady**); flowers dark red, very full, dbl., 140 petals, 2.75–3 in., borne in small clusters, intense fragrance; some prickles; foliage medium, dark green, semi-glossy; medium (3 ft), upright, bushy growth; [Mary Rose X Prospero]; PP8677 Austin, David, 1991

'AUSBLOSSOM', S, lp, 1994; (**Peach Blossom**); flowers blush pink, semi-dbl., 6–14 petals, 1.5–2.75 in., borne in large clusters, slight fragrance; some prickles; foliage medium, medium green, semi-glossy; bushy, spreading (120 cms) growth; [The Prioress X Mary Rose]; Austin, David, 1990; David Austin Roses, Ltd., 1990

'AUSBLUSH', S, lp, 1985; (**Heritage®**); flowers medium, dbl., cupped, recurrent bloom, intense fragrance; foliage small, dark, semi-glossy; upright, bushy growth; [Unnamed seedling X Iceberg seedling]; Austin, David; David Austin Roses, Ltd., 1984

AUSboard, S, mp, 1986; (**Gertrude Jekyll®**); PP7220 Austin, David

'AUSBREAK', S, my, 1993; (**Jayne Austin**); flowers old fashioned, blooms borne in small clusters, very dbl., 110–130 petals, 1.5–2.75 in., intense fragrance; few prickles; foliage medium, dark green, semi-glossy; medium (110 cms), upright growth; [Graham Thomas X Tamora]; Austin, David; David Austin Roses, Ltd., 1990

'AUSBRED', S, ab, 1985; (**Bredon®**); flowers medium, dbl., repeat bloom, moderate fragrance; foliage small, light green, matt; upright growth; [Wife of Bath X Lilian Austin]; Austin, David; David Austin Roses, Ltd., 1984

AUSbrid, S, lp, 1997; (**Mayor of Casterbridge**); flowers very full, old rose form, medium, very dbl., 91 petals, 5.5–6.5 in., borne in small clusters, moderate fragrance; some prickles; foliage medium, light green, leathery; upright, medium (3.5 × 2.5 ft) growth; Austin, David

'AUSBUFF', S, ab, 1990; (**English Garden™**); flowers soft apricot yellow, very large, very dbl., cupped, intense fragrance; foliage clear green; bushy, vigorous growth; PP007214; [Lilian Austin X (Seedling X Iceberg)]; Austin, David; Delbard Roses, 1988

'AUSBURN', S, lp, 1986; (**Robbie Burns**); flowers light pink, white center, small, single, 5 petals, moderate fragrance; foliage small, medium green, matt; bushy, strong growth; [Wife of Bath X R. pimpinellifolia]; Austin, David, 1985

'AUSBURTON', S, lp, 1994; (**Emily**); flowers pale pink, very dbl., 3–3.5 in., borne in small clusters, moderate fragrance; many prickles; foliage medium, medium green, semi-glossy; upright, bushy (75 cms) growth; [The Prioress X Mary Rose]; PP88388 Austin, David; David Austin Roses, Ltd., 1992

'AUSCA', S, w, 1983; (**Fair Bianca®**); flowers light yellow to white, very dbl., quartered, moderate fragrance; foliage medium, light green, semi-glossy; upright growth; hardy, medium; Austin, David; David Austin Roses, Ltd., 1982

AUScam, S, mp, 1995; (**Marinette**); flowers double, large, semi-dbl., 15–25 petals, borne singly or in small clusters, slight fragrance; prickles few to some; foliage medium, medium green, semi-glossy; bushy, medium growth; [Lucetta X Red Coat]; Austin, David

'AUSCANTERBURY', S, mp, 1969; (**Canterbury**); flowers rose-pink, medium, semi-dbl., 12 petals, repeat bloom, intense fragrance; [(Monique X Constance Spry) X Seedling]; Austin, David

'AUSCAT', S, w, 1995; (**Winchester Cathedral**); flowers white, with a tendency to revert to pink, very dbl., 2–2.75 in., cupped, borne in small clusters, moderate fragrance; some prickles; foliage medium, medium green, semi-glossy; medium (120 cms), upright, bushy growth; PP008141; [Mary Rose sport]; Austin, David; David Austin Roses, Ltd., 1988

'AUSCENT', S, dp, 1994; (**John Clare**); flowers very full, medium, very dbl., 110–130 petals, borne in small clusters, slight fragrance; few prickles; foliage medium, dark green, semi-glossy; upright, medium (90 cms) growth; [Wife of Bath X Seedling]; Austin, David

'AUSCHAR', S, yb, 1994; (**Charity**™); flowers full, 90 + petals, soft apricot yellow; strong myrrh fragrance; medium upright growth, few thorns; Austin, David; David Austin Roses, Inc., 1994

'AUSCHILD', S, dr, 1987; (**Fisherman's Friend**®); flowers deep crimson, reverse lighter, fading crimson-purple, large, very dbl., cupped, repeat bloom, intense, damask fragrance; no fruit; broad, straight, large, red-brown prickles; foliage medium, dark green, semi-glossy; bushy growth; [Lilian Austin X The Squire]; Austin, David

AUSclough, S, dp, 1987; (**Sir Clough**); Austin, David

'AUSCLUB', S, lp, 1995; (**Kathryn Morley**); flowers pale pink, very dbl., 3–3.5 in., cupped, borne in small clusters, moderate fragrance; many prickles; foliage medium, dark green, glossy; tall (5–6 ft), bushy growth; [Mary Rose X Chaucer]; Austin, David; David Austin Roses, Ltd., 1990

'AUSCOMP', S, my, 1994; (**Happy Child**); flowers bright yellow, very dbl., 3–3.5 in., borne in small clusters, intense fragrance; some prickles; foliage medium, medium green, glossy; medium (100 cms), bushy growth; [Seedling X Hero]; PP9007; Austin, David; David Austin Roses, Ltd., 1993

'AUSCON', S, mp, 1981; (**Chaucer**®); bud globular; cupped, borne 1–5 per cluster, intense fragrance; slightly hooked, red prickles; foliage medium green; vigorous, upright, bushy growth; [Seedling X Constance Spry]; PP10618; Austin, David; David Austin Roses, Ltd., 1970

'AUSCOOK', S, rb, 1997; (**Heather Austin**); flowers very full, medium, very dbl., 50 petals, 1.5–2.5 in., borne in small clusters, moderate fragrance; some prickles; foliage medium, medium green, semi-glossy; upright to bushy, medium (3 × 2.5 ft) growth; Austin, David

'AUSCOT', S, op, 1990; (**Abraham Darby**™); bud rounded, dark pink base with yellow; flowers pink peach-apricot, cup form, very large, very dbl., 3–4 in., quartered, intense fragrance; foliage dark green, shiny; vigorous, bushy, angular growth; PP007215; [Yellow Cushion X Aloha]; Austin, David; 1985

'AUSCOUNTRY', S, lp, 1994; (**Country Living**); flowers pale pink, very dbl., 3–3.5 in., borne in small clusters, moderate fragrance; some prickles; foliage medium, medium green, semi-glossy; upright, bushy (105 cms) growth; [Wife of Bath X Graham Thomas]; Austin, David; David Austin Roses, Ltd., 1991

'AUSCRESS', S, ab, 1992; (**Cressida**); flowers apricot peach, blooms borne in small clusters, very dbl., 3–3.5 in., intense fragrance; some prickles; foliage small, light green, semi-glossy; tall (180 cms), upright growth; Austin, David; David Austin Roses, Ltd., 1983

'AUSCRIM', S, dr, 1993; (L. D. Braithwaite, **Leonard Dudley Braithwaite**, Braithwaite); flowers bright red, informal, large, very dbl., 3–3.5 in., borne in small clusters, intense fragrance; some prickles; foliage medium, dark green, semi-glossy; medium (110 cms), bushy growth; [Mary Rose X The Squire]; Austin, David; David Austin Roses, Ltd., 1983

'AUSCROSS', S, lp, 1997; (**Windflower**); flowers light pink, full (26–40 petals), medium blooms; slight fragrance; foliage medium, medium green, semi-glossy, very disease resistant; bushy, medium (120 cms) growth; [Shropshire Lass X Seedling]; Austin, David; David Austin Roses, Ltd.

'AUSCUP', S, ab, 1985; (**Ellen**®); flowers old rose form, large, dbl., intense fragrance; foliage large, medium green, semi-glossy; bushy growth; Austin, David, 1984

'AUSDAY', S, pb, 1994; (**The Alexandra Rose**); flowers coppery pink, single, 5 petals, 1.5–2.75 in., borne in large clusters, slight fragrance; few prickles; foliage medium, medium green, semi-glossy; bushy, compact (130 cms) growth; [Shropshire Lass X Heritage]; Austin, David; David Austin Roses, Ltd., 1992

'AUSDIMINDO', S, dp, 1995 (**Bibi Maizoon**; Bibi Mezoon); flowers rich pink, very full (41+ petals), large (7+ cms) blooms borne in small clusters; very fragrant; some prickles; foliage medium, dark green, semi-glossy; medium (100 cms), arching, bushy growth; [The Reeve X Chaucer]; Austin, David; David Austin Roses, Ltd. 1989

'AUSDIR', S, dr, 1994; (**Tradescant**); flowers very dark red, tends to burn in hot weather, very dbl., 1.5–2.75 in., rosette, borne in small clusters, moderate fragrance; some prickles; foliage medium, dark green, semi-glossy; bushy, spreading (75 cms) growth; PP9009; [Prospero X Seedling]; Austin, David; David Austin Roses, Ltd., 1993

'AUSDOCTOR', S, mr, 1992; (**Doctor Jackson**); flowers scarlet, golden stamens, blooms borne mostly singly, single, 5 petals, 1.5–2.75 in., no fragrance; few prickles; foliage medium, medium green, semi-glossy; medium (120 cms), spreading growth; Austin, David; David Austin Roses, Ltd., 1987

'AUSDOR', HRg, mp, 1994; (**Mrs Doreen Pike**); flowers medium, very dbl., 1.5–2.75 in., borne in large clusters, moderate fragrance; many prickles; foliage small, light green, semi-glossy; medium (90 cms), bushy, spreading, compact growth; [Martin Frobisher X Roseraie de l'Hay]; Austin, David; David Austin Roses, Ltd., 1993

'AUSDOVE', S, lp, 1986; (**Dove**); flowers medium, dbl., slight fragrance; foliage medium, dark, semi-glossy; spreading growth; [Wife of Bath X Iceberg seedling]; Austin, David, 1984

AUSecret, S, lp; flowers soft warm pink, very dbl., 110 petals, 3 in., rosette, light fragrance; almost thornless; foliage shiny, disease-resistant; arching growth; Austin, David, 1999

'AUSELLE', S, lp, 1985; (**Belle Story**®); flowers light pink, yellow stamens, large, dbl., 35 petals, cupped, intense fragrance; foliage medium, medium green, semi-glossy; bushy growth; PP007213; [(Chaucer X Parade) X (The Prioress X Iceberg)]; Austin, David; David Austin Roses, Ltd., 1984

'AUSEMI', S, ab, 1992; (**Lucetta**®); flowers pale peach, blooms borne in small clusters, semi-dbl., 6–14 petals, 5.25 in., moderate fragrance; some prickles; foliage medium, medium green, semi-glossy; medium (120 cms), spreading growth; Austin, David; David Austin Roses, Ltd., 1983

'AUSFAR', S, lp, 1999; (**Dr Herbert Gray**); flowers clear light pink, reverse medium pink, very dbl., 41 petals, 3.5 in., cupped, borne in small clusters, moderate fragrance; few prickles; foliage small, dark green, glossy; spreading, medium (3 ft) growth; [Heritage X seedling]; Austin, David; David Austin Roses, Ltd., 1998

'AUSFATHER', S, ab, 1981; (**Charles Austin**®); bud globular; flowers apricot tinged pink, fading to light pink, rosette, borne 1–3 per cluster, dbl., 70 petals, moderate fragrance; hooked, red prickles; foliage medium green, dense; vigorous, upright, bushy growth; [Chaucer X Aloha]; Austin, David; David Austin Roses, Ltd., 1973

'AUSFETE', S, mp, 1999; (**Ann**); single, 5–11 petals, 2.5 in., borne in clusters, moderate fragrance; prickles moderate; foliage medium, medium green, semi-glossy; bushy, medium (3 ft) growth; Austin, David; David Austin Roses, Ltd., 1997

'AUSFIN', S, dp, 1994; (**Financial Times Centenary**); flowers deep pink, very dbl., 3–3.5 in., borne in small clusters, intense fragrance; some prickles; foliage medium, dark green, semi-glossy; medium (110 cms), upright, bushy growth; PP8142; Austin, David; David Austin Roses, Ltd., 1988

AUSfire, S, pb, 1997; (**Morning Mist**); flowers single, medium, single, 5 petals,

4 in., borne in small clusters, slight fragrance; some prickles; foliage large, light green, leathery; attractive, shrubby, spreading, tall (4.5 × 4 ft) growth; Austin, David; 1996

'AUSFIRST', S, lp, 1961; (**Constance Spry**®); flowers rose pink, blooms in clusters, dbl., 5 in., cupped, moderate, myrrh fragrance; foliage dark; vigorous (5–6 ft) growth; [Belle Isis X Dainty Maid]; Austin, David

'AUSGLISTEN', S, mp, 1994; (**Cottage Rose**, The Cottage Rose); flowers warm pink, very dbl., 1.5–2.75 in., borne in large clusters, moderate fragrance; many prickles; foliage medium, medium green, semi-glossy; medium (90–105 cms), upright, bushy growth; [Mary Rose X Wife of Bath]; Austin, David; David Austin Roses, Ltd., 1991

'AUSGLOBE', S, mp, 1994; (**Brother Cadfael**™); flowers rich pink, dbl., 45 petals, cupped, intense, old rose fragrance; few thorns; foliage dark green; straight stems; strong, bushy growth; PP008681; Austin, David, 1990

'AUSGOLD', S, dy, 1993; (**Golden Celebration**); flowers old fashioned, blooms borne in small clusters), very dbl., 55–75 petals, 3–3.5 in., borne in small clusters, intense fragrance; some prickles; foliage large, dark green, semi-glossy; medium (120 cms), bushy growth; [Charles Austin X Abraham Darby]; Austin, David; David Austin Roses, Ltd., 1992

AUSguard, S, lp; rich deep pink, very dbl., 110 petals, 3 in., shallow-cupped, intense, old rose fragrance; foliage dark, slightly glossy; broad, bushy growth; Austin, David, 1999

'AUSHAM', S, mp, 1999; (**Geoff Hamilton**); flowers rounded, very dbl., 108 petals, 2.5 in., borne in small clusters, moderate fragrance; prickles moderate; foliage medium, medium green, semi-glossy, good disease resi; sturdy, compact, upright, medium (4 ft) growth; [Heritage X seedling]; Austin, David; David Austin Roses, Ltd., 1997

'AUSHERO', S, mp, 1983; (**Hero**®); flowers glistening medium pink, deeply, large, dbl., 20 petals, cupped, intense fragrance; foliage medium, medium green, semi-glossy; spreading growth; [The Prioress X Unnamed seedling]; Austin, David; David Austin Roses, Ltd., 1982

'AUSIRE', S, dr, 1976; (**The Squire**®); bud globular; flowers very dark red, rosette-shaped blooms borne 1–3 per cluster, very dbl., 120 petals, intense fragrance; straight prickles; foliage dark; open, bushy growth; [The Knight X Château de Clos Vougeot]; Austin, David, 1977

'AUSJESS', S, dp, 1992; (**Pretty Jessica**); flowers deep pink, blooms borne in small clusters, very dbl., 1.5–2.75 in., intense fragrance; some prickles; foliage medium, dark green, semi-glossy; low (65 cms), bushy growth; [Wife of Bath X Unnamed seedling]; Austin, David; David Austin Roses, Ltd., 1983

AUSjo, S, my, 1997; (**Jude the Obscure**); flowers very full, large, globular, very dbl., 55–70 petals, globular, borne singly or in small clusters, intense fragrance; some prickles; foliage medium, medium green, semi-glossy; bushy, medium growth; [Abraham Darby X Windrush]; Austin, David

'AUSJOLLY', S, pb, 1999; (**Mary Magdalene**); very dbl., 41 petals, 4 in., borne in small clusters, intense, myrrh fragrance; prickles moderate; foliage medium, medium green, dull; spreading, medium (3 ft) growth; [seedling X seedling]; Austin, David, 1998

'AUSJUNO', S, dp, 1992; (**Immortal Juno**); flowers deep pink, blooms borne in small clusters, very dbl., 3–3.5 in., intense fragrance; some prickles; foliage medium, medium green, semi-glossy; tall (150 cms), upright growth; Austin, David; David Austin Roses, Ltd., 1983

AUSky, S, mp, 1997; (**Mistress Quickly**); flowers very full, informally arranged, small, very dbl., 41 petals, borne in very large clusters, slight fragrance; few prickles; foliage small to medium, medium green, semi-glossy; bushy, medium growth; disease resistant; [Blush Noisette X Martin Frobisher]; PP10617; Austin, David, 1998

'AUSLAND', S, lp, 1997; (**Scepter'd Isle**); flowers double, cupped, medium, dbl., 45 petals, 4–5 in., borne in small clusters, intense fragrance; some prickles; foliage medium, dark green, semi-glossy; narrow, bushy, low (2.5 x 2 ft) growth; PP10969; Austin, David

'AUSLEA', S, ab, 1983; (**Leander**®); flowers small blooms in clusters, very dbl., flat, moderate fragrance; foliage medium, medium green, semi-glossy; spreading growth; [Charles Austin X Unnamed seedling]; Austin, David; David Austin Roses, Ltd., 1982

'AUSLEAF', S, ob, 1985; (Autumn Leaves, **English Elegance**®); flowers large, dbl., moderate fragrance; foliage medium, medium green, semi-glossy; upright growth; PP007557; Austin, David

'AUSLEAN', S, lp, 1983; (**Cymbeline**); flowers medium, dbl., intense fragrance; foliage medium, medium green, semi-glossy; spreading growth; [Seedling X Lilian Austin]; Austin, David; David Austin Roses, Ltd., 1982

'AUSLEAP', S, ab, 1994; (**Sweet Juliet**); very dbl., 1.5–2.75 in., cupped, borne in small clusters, moderate fragrance; some prickles; foliage medium, medium green, semi-glossy; upright, bushy (100 cms) growth; GM, PP8153; Belfast, 1992; [Graham Thomas X Admired Miranda]; Austin, David; David Austin Roses, Ltd., 1989

'AUSLED', S, yb, 1997; (**A Shropshire Lad**); flowers very full, rosette form, large, very dbl., 100 petals, 3 in., rosette, borne in small clusters, intense fragrance; prickles moderate; foliage medium, medium green, semi-glossy; bushy, medium (4 × 3 ft) growth; Austin, David, 1996

'AUSLETT', S, ly, 1994; (Allux Symphony, **Symphony** flowers soft yellow, very dbl., 3–3.5 in., borne in small clusters, slight fragrance; foliage medium, light green, glossy; medium (120 cms), upright, bushy growth; [The Friar X Yellow Cushion]; Austin, David, 1986; David Austin Roses, Ltd., 1986

'AUSLEVEL', S, w, 1994; (**Glamis Castle**); very dbl., 1.5–2.75 in., cupped, borne in small clusters, intense, myrrh fragrance; many prickles; foliage medium, medium green, semi-glossy; medium (100 cms), bushy growth; [Graham Thomas X Mary Rose]; PP8765; Austin, David; David Austin Roses, Ltd., 1992

'AUSLIAN', S, dp, 1992; (**Warwick Castle**®); very dbl., 3–3.5 in., borne in small clusters, moderate fragrance; some prickles; foliage small, medium green, matt; low (75 cms), spreading growth; [The Reeve X Lilian Austin]; Austin, David; David Austin Roses, Ltd., 1986

'AUSLIGHT', S, mp, 1990; (**Claire Rose**); flowers very large, cupped, intense fragrance; foliage clear green; bushy, vigorous growth; [Charles Austin X (Seedling X Iceberg)]; Austin, David; 1988

'AUSLILAC', S, pb, 1994; (**Lilac Rose**); flowers lilac pink, very dbl., 3–3.5 in., borne in small clusters, intense fragrance; some prickles; foliage medium, medium green, semi-glossy; upright (90 cms) growth; PPAF; [Seedling X Hero]; PP8837; Austin, David; David Austin Roses, Ltd., 1990

'AUSLO', S, mr, 1990; (**Othello**); flowers clear red, fading to purple-red, very large, very dbl., 4–5 in., cupped, intense fragrance; many large thorns; foliage dark green; vigorous, bushy growth; PP007212; [Lilian Austin X The Squire]; Austin, David; 1989

'AUSLOT', S, rb, 1999; (**Sophy's Rose**); flowers red-purple, dome-shaped, very dbl., 82 petals, 3.5 in., borne in small clusters, moderate fragrance; prickles moderate; foliage medium, medium green, semi-glossy; strong, bushy, vigorous, medium (3.5 ft) growth; [Prospero X seedling]; Austin, David; David Austin Roses, Ltd., 1997

'AUSMAK', S, lp, 1994; (**Eglantyne**, Eglantyne Jebb); flowers light, delicate pink, very dbl., 3–3.5 in., cupped, borne in small clusters, intense fragrance; some prickles; foliage medium, medium green, matt; medium (115–130 cms), bushy growth; [Seedling X Mary Rose];

PP9526; Austin, David; David Austin Roses, Ltd., 1994

'AUSMAN', S, mp, 1987; (**The Countryman**®); flowers rosette, medium, borne usually singly and in sprays of 3–5, dbl., 40 petals, repeat bloom, intense, damask fragrance; ovoid, medium, red fruit; hooked, small, pale prickles; foliage medium, medium green, matt; spreading, medium growth; PP007556; [Unnamed seedling X Comte de Chambord]; Austin, David

'AUSMARY', S, mp, 1983; (**Mary Rose**®); flowers large, very dbl., cupped, recurrent bloom, intense fragrance; foliage medium, medium-green, matt; upright, bushy growth; [Unnamed seedling X The Friar]; Austin, David; David Austin Roses, Ltd.

'AUSMAS', S, dy, 1983; (**Graham Thomas**®); flowers rich deep yellow, medium, dbl., 35 petals, cupped, recurrent bloom, intense fragrance; foliage small, dark, glossy; bushy growth; [Unnamed seedling X (Charles Austin X Iceberg seedling)]; Austin, David; David Austin Roses, Ltd.

'AUSMASH', S, pb, 1997; (**Heavenly Rosalind**); flowers soft pink center deepening to medium pink on outer edges, wild-rose effect, single (5 petals), medium blooms borne in small clusters; slight fragrance; some prickles; foliage medium, dark green, dull, leathery; upright, bushy, medium growth; Austin, David

'AUSMIAN', S, mp, 1983; (**Charmian**®); flowers large, dbl., intense fragrance; foliage medium, medium green, semi-glossy; spreading growth; [Seedling X Lilian Austin]; Austin, David; David Austin Roses, Ltd., 1982

'AUSMIR', S, lp, 1983; (**Admired Miranda**®); flowers opening flat, then reflexing, large, dbl., flat, intense fragrance; foliage medium, medium green, semi-glossy; upright growth; [The Friar X The Friar]; Austin, David, 1982

'AUSMIT', S, my, 1987; (**St Cecilia**); flowers pale buff yellow, fading cream, medium, borne usually singly, dbl., 40 petals, cupped, repeat bloom, moderate, myrrh fragrance; no fruit; small, brown prickles; foliage small, medium green, matt; bushy, low, medium growth; [Unnamed seedling X Unnamed seedling]; PP8157; Austin, David

'AUSMOL', S, dy, 1994; (**Molineux**™); very full rosettes of flowers; strong tea fragrance; 110–120 petals borne in small-medium clusters; bushy upright growth; few thorns; 3–4 ft tall; very good repeat flowering; excellent disease resistance; foliage dark-green, semi-glossy; PP9624; Austin, David; David Austin Roses, Inc., 1994

'AUSMOON', S, 1997; (**Pegasus**); flowers very full, large, camelia-like, very dbl., 110 petals, borne in small clusters, intense, rich tea-rose fragrance; few prickles; foliage large, dark green, semi-glossy, leathery; bushy, branching, medium growth; Austin, David

'AUSMOUND', S, op, 1981; (**Lilian Austin**®); bud globular; dbl., 33 petals, flat, borne 1–5 per cluster, moderate fragrance; hooked, brown prickles; foliage glossy, dark; spreading growth; [Aloha X The Yeoman]; Austin, David, 1973

'AUSMOVE', S, dr, 1999; (**Tess of the D'Urbervilles**); flowers dark, crimson red, very dbl., 41 petals, 3.5 in., cupped, borne in small clusters, moderate fragrance; many prickles; foliage large, dark green, semi-glossy; branching, medium (3 ft) growth; [The Squire X seedling]; Austin, David; David Austin Roses, Ltd., 1998

'AUSMUM', S, 1997; (**Pat Austin**™); bud pointed, ovoid; flowers very full, large, bright copper inside, paler outside, very dbl., 50 petals, 1–2 in., borne in small clusters, some prickles; foliage large, dark green, glossy; branching, rounded, medium growth; PP9527; Austin, David

'AUSMURR', S, mp, 1985; (**Hilda Murrell**®); flowers old rose form, large, dbl., intense fragrance; foliage large, medium green, matt; bushy growth; [Unnamed seedling X (Parade X Chaucer)]; Austin, David, 1984

'AUSNUN', S, w; (**The Nun**); Austin, David, 1987

'AUSOIL', S, ab, 1992; (**Troilus**); flowers creamy apricot, blooms borne in large clusters, very dbl., 3–3.5 in., intense fragrance; some prickles; foliage large, dark green, semi-glossy; medium (110 cms), upright growth; [(Duchesse de Montebello X Chaucer) X Charles Austin]; Austin, David; David Austin Roses, Ltd., 1983

'AUSOLD', S, dp, 1997; (**Trevor Griffiths**™); flowers very full, old rose form, large, very dbl., 41 petals, borne in small clusters, intense fragrance; some prickles; foliage medium, dark green, dull, rough texture; upright, medium (105 cms) growth; Austin, David

'AUSONIUS', HMsk, pb, 1932; bud oval, yellow-red; flowers yellowish pink, center white, semi-dbl., borne in pyramidal trusses of 20–50, free, recurrent bloom, moderate fragrance; foliage leathery; semi-climbing, bushy growth; [(Chamisso X Léonie Lamesch) X (Geheimrat Dr Mittweg X Tip-Top)]; Lambert, P.

'AUSPALE', S, lp, 1994; (**Redoute**); pale pink sport of Mary Rose; more petals, otherwise identical; Austin, David; David Austin Roses, Inc., 1994

'AUSPERD', S, ab, 1992; (**Perdita**®); flowers blush apricot, blooms borne in small clusters, very dbl., 3–3.5 in., intense fragrance; some prickles; foliage medium, medium green, semi-glossy; medium (100 cms), bushy growth; Edland Fragrance Medal, ARS, 1984; [The Friar X (Seedling X Iceberg)]; Austin, David; David Austin Roses, Ltd., 1983

'AUSPERO', S, dr, 1983; (**Prospero**®); flowers large, dbl., flat, intense fragrance; foliage medium, dark, matt; weak, upright growth; [The Knight X Unnamed seedling]; PP9008; Austin, David, 1982

'AUSPOLY', S, ly, 1994; (**Charlotte**, Elgin Festival); flowers soft yellow, very dbl., 3–3.5 in., borne in small clusters, moderate fragrance; few prickles; foliage medium, medium green, semi-glossy; medium (100 cms), upright, bushy growth; [Seedling X Graham Thomas]; Austin, David; David Austin Roses, Ltd., 1993

'AUSPOM', S, w, 1997; (**Snow Goose**); flowers full, small, very dbl., 26–40 petals, 1–2 in., borne in small clusters, slight fragrance; few prickles; foliage small, dark green, smooth; bushy, (8 ft); Austin, David

'AUSPORT', S, m, 1983; (**Wise Portia**); flowers large, dbl., intense fragrance; foliage medium, dark, semi-glossy; bushy growth; [The Knight X Seedling]; Austin, David, 1982

AUSpot, S, mp, 1988; (**Potter & Moore**); Austin, David

'AUSPRIMA', S, mr, 1995; (**Sir Edward Elgar**); flowers cerise-crimson, very dbl., 1.5–2.75 in., borne mostly singly, intense fragrance; many prickles; foliage medium, medium green, semi-glossy; medium (75 cms), upright, bushy, compact growth; [Mary Rose X The Squire]; Austin, David; David Austin Roses, Ltd., 1992

AUSrace, S, lp; flowers pale apricot in center, whole bloom fading to cream, rounded, very dbl., 118 petals, 3 in.; rounded, moderate, tea fragrance; foliage slightly glossy, dark green; bushy growth; Austin, David, 1999

'AUSRAM', S, w, 1994; (**Francine Austin**); very dbl., 1.5 in., borne in large clusters, slight fragrance; some prickles; foliage small, medium green, semi-glossy; groundcover; bushy, spreading (90 cms) growth; PPAF; [Alister Stella Gray X Ballerina]; PP8156a; Austin, David; David Austin Roses, Ltd., 1988

'AUSREEF', S, lp, 1995; (**Sharifa Asthma**, Sharifa); flowers pale pink, fading to pinkish white, very full (41+ petals), large (7+ cms) blooms borne in small clusters; very fragrant; many prickles; foliage medium, dark green, semi-glossy; medium (90 cms), bushy strong growth; [Mary Rose X Admired Miranda]; Austin, David; David Austin Roses Ltd., 1989

'AUSREEVE', S, dp, 1981; (**The Reeve**®); bud globular; flowers deep pink, blooms borne 1–5 per cluster, dbl., 58 petals, cupped, repeat bloom, intense fragrance; hooked, red prickles; foliage red, turning to green; spreading, shrubby growth; [Lilian Austin X Chaucer]; Austin, David, 1979

'AUSREN', S, pb, 1994; (**Charles Rennie Macintosh**); flowers lilac pink, very dbl., 1.5–2.75 in., borne in small clusters; many prickles; foliage medium, dark green, matt; medium (110 cms), bushy growth; PP8155; [Seedling X Mary Rose]; Austin, David; David Austin Roses, Ltd., 1988

'AUSRON', S, lp, 1983; (**Lordly Oberon**); flowers large, very dbl., cupped, intense fragrance; foliage large, medium green, matt; upright growth; [Chaucer X Seedling]; Austin, David; David Austin Roses, Ltd., 1982

'AUSROYAL', S, dr, 1987; (**William Shakespeare**®); flowers deep crimson-purple, fading rich purple, rosette, small, bor, dbl., repeat bloom, intense, damask fragrance; no fruit; broad based, straight, medium, red prickles; foliage large, dark green, semi-glossy; upright, tall growth; [The Squire X Mary Rose]; Austin, David

'AUSRUSH', S, ly, 1985; (**Windrush**®); flowers large, semi-dbl., intense fragrance; foliage medium, light green, matt; vigorous, branching growth; [Unnamed seedling X (Canterbury X Golden Wings)]; Austin, David, 1984

'AUSSAL', S, mp, 1997; (**Radio Times**); flowers very full, old fashioned form, medium, very dbl., 95–110 petals, borne in small clusters, intense fragrance; many prickles; foliage medium, medium green, semi-glossy; bushy, low (75 cms) growth; PP9525; Austin, David

'AUSSAUCER', S, ab, 1992; (**Evelyn**); flowers old fashioned rosette form, blooms borne in small clusters, very dbl., 4–5 in., intense fragrance; some prickles; foliage medium, medium green, semi-glossy; medium (110 cms), upright, bushy growth; [Graham Thomas X AUStamora]; PP8680; Austin, David; David Austin Roses, Ltd., 1991

AUSschool, S, lp; flowers soft yellow, dbl., 51 petals, 2.25 in., loose, produced in large sprays, moderate, musk fragrance; foliage bright green, matte; bushy growth; Austin, David, 1999

'AUSSEMI', S, dp, 1994; (Herbalist, **The Herbalist**™); flowers deep pink, semi-dbl., 6–14 petals, 1.5–2.75 in., borne in small clusters, slight fragrance; some prickles; foliage medium, medium green, semi-glossy; medium (90 cms), bushy, spreading growth; [Seedling X Louise Odier]; Austin, David; David Austin Roses, Ltd., 1991

'AUSSNOW', HWich, w, 1985; (**Mountain Snow**); flowers small blooms in large sprays, dbl., 20 petals, slight fragrance; foliage large, dark, semi-glossy; vigorous (to 20 ft) growth; Austin, David

'AUSSPRY', S, mp, 1994; (**Sir Walter Raleigh**™); flowers warm medium pink, very dbl., 3–3.5 in., borne mostly singly, intense fragrance; few prickles; foliage medium, medium green, semi-glossy; medium (120 cms), upright, bushy growth; PP7213; [Lilian Austin X Chaucer]; Austin, David; David Austin Roses, Ltd., 1985

'AUSTAMORA', S, ab, 1992; (**Tamora**); flowers apricot yellow, deeply, old rose form, blooms borne in small, very dbl., 3–4 in., cupped, intense, unusual myrrh fragrance; some prickles; foliage small, dark green, semi-glossy; medium (90 cms), bushy growth; [Chaucer X Conrad Ferdinand Meyer]; Austin, David; David Austin Roses, Ltd., 1983

'AUSTANIA', S, w, 1983; (**Proud Titania**); dbl., 35 petals, flat, intense fragrance; foliage small, medium green, semi-glossy; upright growth; [Unnamed seedling X Unnamed seedling]; Austin, David, 1982

'AUSTOP', S, lp, 1999; (**Barbara Austin**); very dbl., 73 petals, 3 in., borne mostly singly, moderate fragrance; prickles moderate; foliage medium, medium green, semi-glossy, good disease resi; upright, medium (4 ft) growth; [Fair Bianca X seedling]; Austin, David, 1997; David Austin Roses Ltd., 1997

Austragold®, *see* 'MACKUNG'

Austral, HT, 1976; Bees; (Cavriglia)

'AUSTRALIA FELIX', HT, pb, 1919; bud small, globular; flowers pink and silver shaded lavender, semi-dbl., cupped, intense fragrance; foliage dark, glossy; vigorous, bushy growth; [Jersey Beauty X La France]; Clark, A.

'AUSTRALIAN BEAUTY', Cl HT, dr, 1912; Kerslake

'AUSTRALIAN BEAUTY', Cl HP, lp; [President X Lord Macaulay]; Kerslake, 1907; (Weatherly, L.)

Australian Bicentennial, HT, lp; [Daily Sketch X Red Planet]; Bell, Ronald J., 1987; (Weatherly, L.)

Australian Centre Gold, Min, my

Australian Gold®, *see* 'KORMAT'

'AUSTRALIE', HT, dp, 1907; flowers large, dbl.; Kerslake; (Sangerhausen)

'AUSTREAM', S, w, 1997; (**Rushing Stream**); flowers single, small, single, 4–7 petals, 1 in., borne in large clusters, clove fragrance; few prickles; foliage large, light green, glossy; broad, bushy, medium (1.5 ft) growth; Austin, David

Austrian Briar, *see* **R. FOETIDA**

Austrian Brier Rose, *see* **R. FOETIDA**

Austrian Copper Rose, *see* **R. FOETIDA BICOLOR**

Austrian Yellow Rose, *see* **R. FOETIDA**

Austriana, *see* TANanairan

'AUSUEL', S, ab, 1992; (**Emanuel**®, Emmanuelle); flowers apricot pink, opening rosette, blooms borne in small clusters, dbl., 3–3.5 in., flat, intense fragrance; some prickles; foliage small, medium green, semi-glossy; medium (110 cms), bushy growth; [(Chaucer X Parade) X (Seedling X Iceberg)]; Austin, David; David Austin Roses, Ltd., 1985

'AUSVELVET', S, dr, 1993; (**The Prince**™); flowers large, very dbl., 125 petals, deeply cupped, intense fragrance, strong fragrance; many thorns; foliage dark green; short, upright, bushy growth; PP008813; Austin, David, 1990

AUSverse, S, lp; flowers dark crimson turning to purple, very dbl., 105 petals, 3.25 in., cupped; dark, thorny stems; foliage tough, dark green, slightly glossy; vigorous upright bush growth; Austin, David, 1999

'AUSWALKER', S, my, 1993; (**The Pilgrim**™); very dbl., 170 petals, cupped, intense, strong fragrance; few large thorns; foliage bright green leaves; strong, upright growth; PP008678; Austin, David, 1991

AUSway, S, mr, 1997; (**Noble Antony**); flowers very full, deeply domed, outer petals recurve, large, 85–90 petals, borne singly or in small clusters, moderate fragrance; some prickles; foliage medium, dark green, semi-glossy; bushy, medium, growth; [Seedling X Seedling]; PP010779; Austin, David

'AUSWEBB', S, ab, 1985; (**Mary Webb**®); flowers very large, dbl., cupped, intense fragrance; foliage large, light green, matt; bushy growth; [Unnamed seedling X Chinatown]; Austin, David, 1984

'AUSWEN', S, mr, 1985; (**Wenlock**®); flowers large, dbl., intense fragrance; foliage large, dark, semi-glossy; vigorous growth; [The Knight X Glastonbury]; Austin, David, 1984

'AUSWHITE', S, w, 1987; (**Swan**); flowers white, tinged buff, reverse white, rosette, large, very dbl., borne usually singly, repeat bloom, moderate, fruity fragrance; no fruit; hooked, medium, red prickles; foliage large, light green, semi-glossy; upright, tall growth; PP007564; [Charles Austin X (Seedling X Iceberg)]; Austin, David

'AUSWIFE', S, pb, 1969; (The Wife of Bath, **Wife of Bath**); flowers deep rose-pink, reverse blush, medium, semi-dbl., cupped, repeat bloom, intense fragrance; foliage small; [Mme Caroline Testout X (Ma Perkins X Constance Spry)]; Austin, David

'AUSWILL', S, ab, 1999; (**William Morris**); flowers apricot blend, reverse light pink, very dbl., 120 petals, 3.5 in., borne in small clusters, intense fragrance; prickles moderate; foliage medium, dark green, glossy; branching, medium (4 ft) growth; [Abraham Darby X seedling]; Austin, David; David Austin Roses, Ltd, 1998

'AUSWINE', S, m, 1967; (**Chianti**); flowers purplish maroon, borne in small clusters, semi-dbl., repeat bloom, slight fragrance; foliage dark, glossy; vigorous growth; [Dusky Maiden X Tuscany]; Austin, David; Sunningdale Nursery, 1965

'AUSWING', S, ly, 1994; (**Wildflower**); flowers pale yellow, single, 5 petals, 1.5 in., borne mostly singly, slight fragrance; some prickles; foliage small, light green, semi-glossy; low (60 cms), bushy, spreading growth; [Canterbury X Seedling]; Austin, David; David Austin Roses, Ltd., 1986

AUSwinter, S, lp; flowers bright apricot orange, very dbl., 120 petals, 3 in., cupped, intense, fruity fragrance; almost thornless; dark glossy foliage; strong, arching growth; Austin, David, 1999

'AUSWITH', S, lp, 1994; (**St Swithun**™); flowers pale pink, very dbl., 3–3.5 in., borne in small clusters, moderate fragrance; some prickles; foliage medium, medium green, semi-glossy; medium (90 cms), bushy growth; [Mary Rose X Seedling]; PP9010; Austin, David; David Austin Roses, Ltd., 1993

'AUSWONDER', S, ab, 1994; (**Ambridge Rose**); flowers apricot pink, very dbl., 1.5–2.75 in., cupped, borne in small clusters, intense fragrance; some prickles; foliage medium, dark green, semi-glossy; small (2.5 ft), bushy growth; PPAF; [Charles Austin X Seedling]; PP8679; Austin, David; David Austin Roses, Ltd., 1990

'AUTOCRAT', HT, pb, 1925; flowers ochre to flesh-pink, reverse prawn-red, large; Beckwith

'AUTUMN', HT, ob, 1928; flowers burnt-orange, streaked red, dbl., 70 petals, cupped, moderate fragrance; foliage dark, glossy; [Sensation X Souv. de Claudius Pernet]; Coddington

'AUTUMN BOUQUET', S, mp, 1948; bud long, pointed; flowers carmine-rose-pink, large, dbl., recurrent bloom, intense fragrance; foliage leathery; vigorous, upright, compact growth; [New Dawn X Crimson Glory]; Jacobus; B&A

'AUTUMN DAMASK', D, mp; (Castilian, Four Seasons, Quatre Saisons, Rose of Castille, Old Castilian, R. bifera, R. bifera semperflorens, R. damascena bifera, R. damascena semperflorens, R. semperflorens, Rose des Quatre Saisons); tending to bloom in autumn as well as early summer; (28)

Autumn Dawn, *see* 'BRIAUTUMN'

'AUTUMN DELIGHT', HMsk, w, 1933; flowers stamens red, single, borne in large clusters; Bentall

'AUTUMN DUSK', Gr, pb, 1976; bud ovoid, pointed; flowers pale tyrian rose and white, dbl., 33 petals, 3.5–4.5 in., exhibition form, moderate fragrance; foliage leathery; upright, bushy growth; [Music Maker X (Dornröschen X Peace)]; Buck, Dr Griffith J.; Iowa State University

Autumn Fire, *see* 'MORANIUM'

Autumn Fire, *see* **'HERBSTFEUER'**

'AUTUMN FLAME', HT, ob, 1953; bud long, pointed; flowers orange-yellow tinted red, dbl., 35–40 petals, 4 in., cupped; foliage dark, leathery; upright, bushy, compact growth; [Ednah Thomas X Autumn]; Thomson

Autumn Frost, *see* 'LITLIN'

'AUTUMN GLOW', HT, ob, 1978; bud pear shaped; flowers yellow-orange, dbl., 55 petals, 5 in., exhibition form, slight fragrance; foliage small; bushy growth; [Pascali X Bayadère]; Anderson's Rose Nurseries

'AUTUMN GOLD', HT, yb, 1969; bud pointed; flowers brown-butterscotch-yellow, dbl., 42 petals, 3.5–4 in., globular, moderate fragrance; foliage glossy, dark, leathery; tall, upright growth; [Seedling X Seedling]; Weeks

'AUTUMN HUES', F, ob, 1962; bud pointed; flowers orange, yellow and scarlet, dbl., 35 petals, 3–3.5 in., exhibition form, blooms in clusters, slight fragrance; foliage glossy; upright growth; [Pinocchio X Fred Edmunds]; Von Abrams; Peterson & Dering

Autumn Kiss, *see* 'WALKISS'

Autumn Leaves, *see* 'AUSLEAF'

Autumn Magic, *see* 'FOUTUM'

'AUTUMN QUEEN', HT, ob, 1933; (Vice-President Curtis); bud long, pointed; flowers burnt-orange, pink, gold, open, semi-dbl., slight fragrance; foliage leathery, bronze; vigorous, bushy growth; Vestal

Autumn Shades, Min, lp; [Silver Jubilee X Oz Gold]; Hannemann, F., 1997; The Rose Paradise; (Weatherly, L.)

Autumn Song, F, pb

Autumn Splendor, *see* 'MICAUTUMN'

'AUTUMN SPRAY', F, yb, 1964; flowers gold edged red, dbl., 40 petals, 3 in., flat; foliage glossy; [Masquerade X Isobel Harkness]; Norman; Harkness

Autumn Sunblaze™, *see* 'MEIFERJAC'

'AUTUMN SUNLIGHT', LCl, or, 1965; flowers orange-vermilion, semi-globular, dbl., 30 petals, blooms in clusters, moderate fragrance; foliage glossy, bright green; [Spectacular X Goldilocks, Climbing]; Gregory

'AUTUMN SUNSET'™, S, ab, 1986; flowers medium apricot, with touches of orange and golden yellow, fading lighter, cupped, 20 petals, cupped, repeat bloom, intense, fruity fragrance; round, medium, orange fruit; curved, medium, red prickles; foliage medium, medium green, glossy, disease resistant; bushy, tall, climbing growth; [Westerland sport]; Lowe, Malcolm; Lowe*s Own Root Roses, 1988, U.S.A.

'AUTUMN TINTS', yb, 1914; flowers coppery-pink with yellow, medium, dbl.; Cant, B. R.; (Sangerhausen)

'AUTUMN, CLIMBING', Cl HT, ob, 1951; deVor, W.L.; Amling-DeVor Nursery

Avalanch II, S, w, 1998; Williams, J. Benjamin

'AVALANCHE', HT, w, 1922; flowers creamy white, center deeper, dbl.; Lippiatt

'AVALANCHE', HT, w, 1936; bud long; flowers large; foliage slightly bronze; vigorous, bushy, compact growth; Chambard, C.

Avalanche, *see* 'JACAY'

Avalanche Rose®, *see* 'DELAVAL'

'AVALON', HT, ab, 1935; bud ovoid; flowers apricot-yellow, center deeper, very large, dbl., globular, slight fragrance; foliage glossy, bronze; vigorous, bushy growth; [Duchess of Atholl sport]; Western Rose Co.; Germain's

Avance, *see* BARvan

Avandel, *see* 'MORVANDEL'

Avanti, *see* 'JACSAY'

'AVE MARIA', HT, w, 1957; flowers pure white, well-formed, dbl., 4–6 in.; vigorous, bushy growth; RULED EXTINCT 12/85; [Seedling X Break o' Day, Climbing]; Brownell, H.C.; Stern's Nursery

Ave Maria®, *see* 'KORAV'

Avebury, LCl, lp; McLeod, J., 1995; Honeysuckle Cottage; (Weatherly, L.)

'AVENANT', HGal, lp; flowers deep flesh-pink, fading, large, very dbl.; erect growth

'AVENIR', HWich, lp, 1910; flowers small, semi-dbl.; Corboeuf; (Sangerhausen)

'AVENTURE', HT, or, 1964; (Adventure); flowers large, dbl., 55 petals, exhibition form, slight fragrance; foliage leathery, glossy; vigorous, upright, bushy growth; [(Corail X Baccará) X Seedling]; Croix, P., 1964; C-P, 1965

Aventure No. 2, *see* CROpal

Avenue's Red, *see* 'KORJET'

Aveu, HT, dp; Croix

'AVIATEUR BLÉRIOT', HWich, yb, 1910; flowers pale orange-yellow, fading white, dbl., 34 petals, borne in clusters, non-recurrent, moderate, magnolia fragrance; foliage glossy, dark; vigorous, climbing growth; [R. wichurana X William Allen Richardson]; Fauque

'AVIATEUR MICHEL MAHIEU', HT, mp, 1912; flowers large, dbl., intense fragrance; Soupert & Notting; (Sangerhausen)

'AVIATOR PARMENTIER', HT, op, 1941; bud long, pointed; flowers peach-pink to orange, dbl., 30–35 petals, 4–4.5 in., exhibition form; foliage glossy; very vigorous growth; [Seedling X Briarcliff]; Verschuren-Pechtold; J. Parmentier

'AVIGNON', F, my, 1974; flowers medium, dbl., 23 petals, 2 in., slight fragrance; foliage light, glossy; vigorous growth; [Zambra X Allgold]; Cants of Colchester, Ltd.

'AVIORA', HT, mr, 1960; flowers clear red, dbl., 40 petals; foliage glossy, dark; very large growth; [Happiness X Seedling]; Verbeek

'AVÔ ALBINA', HT, rb, 1956; flowers crimson, reverse silvery, large, exhibition form; very vigorous growth; [Peace X Crimson Glory]; da Silva, Moreira

'AVÔ ALFREDO', HT, rb, 1956; flowers spectrum-red, reverse carmine, [Seedling X Independence]; da Silva, Moreira

'AVOCA', HT, mr, 1907; flowers crimson-scarlet, large, dbl., exhibition form, moderate fragrance; very vigorous growth; Dickson, A.

'AVOCAT DUVIVIER', HP, m, 1875; dbl., moderate fragrance; Lévêque; (Sangerhausen)

Avocet, *see* 'HARPLUTO'

'AVON'®, HT, dr, 1961; bud ovoid; dbl., 23 petals, 4.5–5.5 in., exhibition form, intense fragrance; foliage leathery; vigorous, upright growth; [Nocturne X Chrysler Imperial]; Morey, Dr Dennison; J&P

Avon, *see* POUlmulti

'AVON, CLIMBING', Cl HT, dr, 1975; Kumar

'AVRIL SHERWOOD', F, my, 1975; bud ovoid; flowers golden yellow fading to buttercup-yellow, semi-dbl., 18 petals, 3–3.5 in., slight fragrance; foliage glossy, dark, leathery; upright growth; [Pink Parfait X Allgold]; Sherwood; F. Mason

'AWAKENING', LCl, lp, 1992; dbl., 26–40 petals, 1.5–2.75 in., quartered, moderate fragrance; foliage light green, glossy; 10′ x 8′ growth; [New Dawn sport]; Hortico, Inc., 1992; Peter Beales Roses, 1992

'AWARD', HT, mr, 1953; flowers velvety red, intense fragrance; foliage very glossy, dark; short stems; vigorous growth; [Will Rogers X Mme Henri Guillot]; Taylor, C.A.

Awareness, *see* FRYbingo

Awayuki, S, w

Axeline, LCl, mp

Axelle, HT, 1964; Gaujard; (Cavriglia)

Ayako™, *see* 'MALMIYA'

Ayaori, F, rb

'AYLSHAM', S, dp, 1948; bud ovoid; flowers deep pink, large, dbl., non-recurrent, slight fragrance; foliage light green, glossy; vigorous (to 5 ft) growth; [Hansa X R. nitida]; Wright, Percy H.

'AYRSHIRE QUEEN', Ayr, dr, 1835; (Ayrshire Splendens); flowers purplish crimson, semi-dbl.; [Blush Ayrshire X Tuscany]; Rivers

Ayrshire Rose,; (**R. ARVENSIS AYRESHIREA**, R. arvensis capreolata, R. capreolata)

Ayrshire Splendens, *see* **'AYRSHIRE QUEEN'**

'AZALEA ROSE', Pol, mp, 1940; flowers bright rose-pink, resemble azaleas; [Ellen Poulsen sport]; Griffing Nursery

'AZEEZ', F, op, 1965; bud pointed; flowers coral-pink, reverse lighter, medium, dbl., exhibition form; foliage leathery; upright, compact growth; Pal, Dr B.P.; Indian Agric. Research Inst.

'AZELDA', HT, op, 1963; bud long, pointed; flowers coral-rose, dbl., cupped, intense fragrance; foliage glossy, light green; long stems; vigorous, upright growth; [Queen Elizabeth X Rosenelfe]; Matthews

'AZTEC', HT, or, 1957; bud long, pointed; flowers scarlet-orange, dbl., 25 petals, 4–5 in., exhibition form, moderate fragrance; foliage glossy, leathery; vigorous, spreading growth; [Charlotte Armstrong X Seedling]; Swim, H.C.; Armstrong Nursery

Azulabria, *see* 'RUIBLUN'

'AZUR', F, m, 1967; bud ovoid; flowers deep lavender-mauve, pointed, semi-dbl., 15 petals, 2.5–3 in., intense fragrance; foliage dark reddish-green; vigorous, compact, bushy growth; [Sterling Silver X (Gold Strike X Golden Garnette)]; Lens

Azure Sea, *see* 'AROLALA'

B

B. C.™, *see* 'KINbee'

'B. S. Bhatcharji', HT, my, 1933; flowers buttercup-yellow, well formed, large; foliage glossy, dark; vigorous growth; Dickson, A.

'B. W. Price', HT, mp, 1943; bud long, pointed; flowers cerise-pink, open, large, single, 6–8 petals, foliage soft; vigorous, upright, bushy growth; [Night X Mme Butterfly]; McGredy; J&P

Babe, Pol, pb, 1958; Hazelwood

'Babe Ruth', HT, op, 1950; bud ovoid; flowers coral, reverse rose-coral, dbl., 35–40 petals, 4–4.5 in., cupped, moderate fragrance; foliage leathery, glossy, bronze; vigorous, upright growth; [Los Angeles X Seedling]; Howard, F. H.; H&S

'Babette', F, m, 1969; flowers lavender-pink, small, single, cupped, slight fragrance; foliage leathery, small; vigorous, bushy growth; [Seedling X Eminence]; Gaujard

'Babette Rambler', HWich, mr, 1906; flowers carmine-red, medium, dbl.; Walsh; (Sangerhausen)

'BABs', HT, dy, 1984; (**Coton Gold**); flowers glowing yellow; [Whisky Mac sport]; Babb, J. T.

'Babt Deitz', Pol, or, 1924; flowers oriental red, edged salmon, dbl.; Opdebeeck

'Baby Alan', Pol, mp, 1930; flowers shining pink, resembling English daisy in form, very dbl., borne in clusters; Kessler; J. T. Lovett

'Baby Alberic', Pol, ly, 1932; bud yellow; flowers creamy white, small, dbl., recurrent bloom; vigorous growth; Chaplin Bros.

Baby Ashley™, *see* 'KINash'

Baby Baccará®, *see* 'MEIbyba'

Baby Ballerina, Min, pb, 1997; Villegas

Baby Bath, Pol, mp, 1989; Robinson

'Baby Betsy McCall', Min, lp, 1960; flowers micro-mini, dbl., 20 petals, 1 in., cupped, moderate fragrance; foliage leathery, light green; vigorous, dwarf, compact (8 in.) growth; [Cécile Brünner X Rosy Jewel]; Morey, Dr Dennison; J&P

Baby Bettina, *see* 'MEIdacinu'

'Baby Betty', Pol, mp, 1929; bud ovoid, yellow, tinged red; flowers pink, base lighter, small, dbl., cupped, abundant, recurrent bloom, moderate fragrance; foliage leathery, dark, bronze; vigorous, compact, bushy growth; [Eblouissant X Comtesse du Cayla]; Burbage Nursery

'Baby Bio', F, dy, 1977; flowers deep golden yellow, patio, dbl., 28 petals, 3 in., slight fragrance; foliage glossy; GM, Rome, 1976; [Golden Treasure X Seedling]; Smith, E.; Rosemont Nursery Co.

Baby Blanket, *see* 'KORfullwind'

'Baby Blaze', F, mr, 1954; (Lund's Jubiläum); bud ovoid; flowers cherry-red, white eye, dbl., 33 petals, 3 in., cupped, blooms in clusters of 10–25, moderate fragrance; foliage light, glossy; vigorous, bushy, compact growth; [World's Fair X Hamburg]; Kordes, W.; J&P

'Baby Bunting', Min, dp, 1953; flowers deep pink, small, dbl., 20 petals, 1.5 in., moderate fragrance; [Ellen Poulsen X Tom Thumb]; deVink, J.; T. Robinson, Porchester Nursery

Baby Cakes, *see* 'TALbaby'

Baby Carnaval, *see* 'TANbakede'

Baby Carnival, *see* 'TANbakede'

'Baby Caroline Louise', Min, w, 1998; bud tall; flowers white, slight lemon center, white reverse, very dbl., 26–40 petals, 2 in., exhibition form, borne in small, well-formed clusters, opens flat, slight fragrance; prickles moderate; foliage medium, medium green, semi-glossy; tall, bushy growth; [Sexy Rexy X Laura Ford]; Barker, S.J.L.

Baby Cécile Brunner, *see* 'MORcebru'

'Baby Château', F, mr, 1936; (Baby Vougeot, Château); bud ovoid, crimson; flowers red shaded garnet, large, dbl., moderate fragrance; foliage glossy, bronze; very vigorous, bushy growth; (28); [Aroma X (Eva X Ami Quinard)]; Kordes

'Baby Cheryl', Min, lp, 1965; bud pointed; flowers light pink, reverse lighter, micro-mini, small, dbl., moderate, spicy fragrance; foliage leathery; vigorous, dwarf growth; [Spring Song X Seedling]; Williams, Ernest D.; Mini-Roses

Baby Chicks, Min, my

Baby Crimson, *see* **'Perla de Alcañada'**

Baby Crimson, Min, mr

'Baby Darling', Min, ab, 1964; bud pointed; flowers apricot-orange, small, dbl., 20 petals; dwarf, bushy (12 in.) growth; [Little Darling X Magic Wand]; Moore, Ralph S.; Sequoia Nursery

'Baby Darling, Climbing', Cl Min, ab, 1972; Trauger, F.; Sequoia Nursery

Baby Diana™, *see* 'SAVadi'

Baby Doll, *see* **'Tip-Top'**®

Baby Dominic, *see* 'JUDnic'

Baby Donnie, Min, mr, 1972

Baby Dorothy, *see* **'Maman Levavasseur'**

Baby Eclipse™, *see* 'MORecli'

'Baby Elegance', Pol, ab, 1912; flowers pale yellow-orange, single; Hobbies

Baby Face, *see* 'LAVaby'

'Baby Faurax', Pol, m, 1924; flowers violet, small, dbl., blooms in large clusters, moderate fragrance; dwarf growth; Lille, L.

'Baby Garnette', Min, mr, 1962; flowers blood-red, small, dbl.; foliage dark; vigorous (10–12 in.), compact growth; [Red Imp X Sparkler]; Morey, Dr Dennison; J&P

'Baby Gloria', Pol, or, 1936; flowers salmon-red, larger than parent; very dwarf (6–8 in.) growth; [Gloria Mundi sport]; Böhm, J.

'Baby Gold Star', Min, dy, 1940; (Estrellita de Oro); bud pointed; flowers golden yellow, semi-dbl., 14 petals, slight fragrance; foliage small, soft; [Eduardo Toda X Rouletii]; Dot, Pedro; C-P

'Baby Gold Star, Climbing', Cl Min, dy, 1964; Williams, Ernest D.; Mini-Roses

Baby Grand™, *see* 'POUlit'

Baby Herriot, *see* **'Étoile Luisante'**

'Baby Jayne', Cl Min, mp, 1957; (Fairy Hedge, Pixie Hedge); flowers soft pink, very small, dbl., 45 petals, borne in clusters; foliage small, glossy; height 3–4 ft; [Violette X Zee]; Moore, Ralph S.; Germain's

'Baby Katie'™, Min, pb, 1978; bud ovoid, pointed; flowers cream and pink blend, small, dbl., 28 petals, exhibition form, slight fragrance; foliage matt, green; vigorous, compact, bushy growth; PP4471; [Sheri Anne X Watercolor]; Saville, F. Harmon; Nor'East Min. Roses

'Baby Lilian', Min; flowers begonia-rose, base tinted orange-yellow, rather large for the class, dbl., slight fragrance; foliage small, light, glossy; vigorous (12–15 in.) growth

Baby Love™, *see* 'SCRivluv'

Baby Love®, Min, lp

'Baby Lyon Rose', Pol, rb, 1916; flowers coral-rose, shaded chrome-yellow or shrimp-red, dbl.; Turbat

Baby Maria, Min, dp

Baby Mascarade, *see* 'TANbakede'

Baby Maskarade, *see* 'TANbakede'

Baby Maskerade, *see* 'TANbakede'

Baby Masquerade®, *see* 'TANbakede'

'Baby Masquerade, Climbing', Cl Min, rb, 1974; Sykes, R. O.; Mini-Roses

Baby Michael, *see* 'JUSmichael'

'BABY MINE', Pol, my, 1929; flowers sulfur to butter-yellow; [Cécile Brunner, Climbing seedling]; Moore, Ralph S.

'BABY OPHELIA', Min, lp, 1961; bud pointed; flowers soft pink, dbl., 33 petals, 1 in., moderate fragrance; foliage glossy; vigorous, bushy (8–12 in.) growth; [(R. wichurana X Floradora) X Little Buckaroo]; Moore, Ralph S.; Sequoia Nursery

Baby Orange Triumph, Min, or; flowers small, dbl.; (Sangerhausen)

'BABY PEACE', Min, yb, 1962; bud urn shaped; flowers ivory-yellow tipped pink, dbl., 50–55 petals, .5–1 in., slight fragrance; very vigorous growth; [Peace sport]; De Mott & Johnson, G.E.

'BABY PINOCCHIO', Min, pb, 1967; bud ovoid; flowers salmon-pink blend, small, dbl., moderate fragrance; foliage glossy, leathery; vigorous, bushy growth; [Golden Glow X Little Buckaroo]; Moore, Ralph S.; Mini-Roses;, Sequoia Nursery

Baby Rambler, *see* 'CHEWRAMB'

Baby Rambler, Climbing, *see* **'MISS G. MESMAN'**

'BABY ROSAMUNDE', Pol, mp, 1930; flowers rose-pink, semi-dbl.; Kessler; J.T. Lovett

Baby Secret, *see* BRIbaby

'BABY SUNBEAM', LCl, ab, 1934; flowers light apricot, passing to cream-yellow, large yellow center, borne in large clusters; foliage bronze; very vigorous growth; Burbank; Stark Bros.

Baby Sunrise, *see* 'MACPARLEZ'

'BABY SYLVIA', F, op, 1959; flowers flesh-salmon-pink, dbl., 25–30 petals, 3 in.; very vigorous growth; [Lady Sylvia X Seedling]; Fryers Nursery, Ltd.

Baby Talisman, *see* **'PRESUMIDA'**

'BABY TALK', F, m, 1980; bud small, ovoid, pointed; flowers dusty mauve-pink, patio, dbl., 26 petals, borne mostly singly, moderate, tea fragrance; foliage small to medium, moderately thin; low to medium, compact, dense growth; [Plain Talk X Angel Face]; Weeks

Baby Tausendschön, *see* **'ECHO'**

Baby Tom, *see* **BOSEYEBALL**

'BABY TYPHOON', Min, ab, 1988; flowers bright, clear apricot, center golden-yellow, medium, dbl., 35–40 petals, exhibition form, borne singly or in sprays of 3–5, moderate, damask fragrance; prickles pointed, large, straw; foliage medium, dark green, semi-glossy; bushy, spreading, tall growth; [Gold Coin sport]; Halevi, A. M., 1987

Baby Vougeot, *see* **'BABY CHÂTEAU'**

Babyflor®, *see* TANrolfy

'BABYLON', HT, op, 1976; flowers deep coral-pink, low-centered, dbl., 36 petals, 5 in., intense fragrance; foliage glossy, dark; moderately vigorous growth; [Tropicana X Pink Favorite]; Bees; Sealand Nursery

Baby's Blush, HCh, mp, 1995; Scarman

'BACARDI', HT, ob, 1987; flowers uniform iridescent coral-orange, fading lighter, small, dbl., 30 petals, exhibition form, borne singly, slight, fruity fragrance; straight, small, reddish prickles; foliage medium, medium green, semi-glossy; upright, medium growth; PP006862; [Seedling self pollination sport]; Weeks, O.L.; Weeks Wholesale Rose Growers, 1988

Baccará®, *see* 'MEIGER'

Baccará, Climbing®, *see* 'MEGERSAR'

'BACCHANTE', HGal, dr, 1848

'BACCHUS', HT, dp, 1951; flowers cherry-pink, medium, dbl., 25 petals, moderate fragrance; vigorous growth; GM, NRS, 1952; Dickson, A.

Back Home, *see* 'GELBACK'

Bad Bergzabern, HT, mr, 1993; Hetzel

Bad Birnbach, F, dp, 1999; Kordes

Bad Ems, S, mp, 1990; Schultheis

Bad Füssing®, *see* 'KORBAD'

Bad Homburg, HT, ab, 1995; Eichelmann

'BAD LANGENSALZA', F, pb, 1954; flowers carmine-pink, large, dbl., moderate fragrance; Berger, W.; (Sangerhausen)

Bad Nauheim, *see* **'NATIONAL TRUST'**

Bad Naukeim, *see* **'NATIONAL TRUST'**

'BAD NEUENAHR', HKor, mr, 1958; bud ovoid; flowers scarlet, dbl., 50 petals, 4 in., cupped, blooms in clusters (to 15), moderate fragrance; foliage dark, leathery; vigorous (6 ft) growth; Kordes, W.

'BAD PYRMONT', F, or, 1976; bud pointed; dbl., 40 petals, 4 in., exhibition form, abundant bloom, slight fragrance; foliage dark, soft; vigorous, upright, bushy growth; [Duftwolke X Seedling]; Kordes; Horstmann

Bad Salzuflen®, F, mp

'BAD WÖRISHOFEN'®, F, mr, 1972; (Gruss an Worishofen); bud ovoid; semi-dbl., 18 petals, 2.5 in., cupped, slight fragrance; foliage dark, soft; vigorous, bushy growth; [Sarabande X Marlena]; Kordes, W.

'BADEN-BADEN', HT, dr, 1952; flowers deep crimson, large, dbl., intense fragrance; foliage dark, leathery; vigorous, upright growth; [Poinsettia X Crimson Glory]; Kordes

Badener Gold®, F, ob, 1974; McGredy, Sam IV

Badia, HT; Mansuino, Dr Andrea; (Cavriglia)

'BADINAGE', HT, op, 1951; flowers salmon-pink flushed coppery, dbl., 30 petals, 4–4.5 in., moderate fragrance; foliage leathery, dark; vigorous growth; [Peace X Seedling]; Gaujard

Badner Traberchampion, HT, lp

'BAGATELLE', HT, mr, 1943; bud large, oval; flowers bright red, very dbl., moderate fragrance; foliage leathery; very vigorous, bushy growth; Gaujard

'BAGDAD', HT, or, 1953; bud ovoid to urn shape; flowers nasturtium-red to orange, becoming cupped, dbl., 48–55 petals, 4.5 in., exhibition form, intense fragrance; foliage glossy; very vigorous, bushy growth; [Charlotte Armstrong X Signora]; Swim, H. C.

Bagheera, *see* 'KORGERA'

'BAGLIORE', HT, mr, 1955; flowers crimson-red, intense fragrance; stiff stems; very vigorous growth; Aicardi, D.

'BAHAMA', F, dr, 1968; semi-dbl., intense fragrance; vigorous, low growth; [Fidélio X Hanne]; Soenderhousen

'BAHIA', F, ob, 1974; bud ovoid; flowers orange, medium, dbl., cupped, moderate, spicy fragrance; foliage glossy, dark, leathery; vigorous, upright, bushy growth; AARS, 1974; [Rumba X Tropicana]; Lammerts, Dr Walter; Armstrong Nursery

'BAHRS LIEVELING', F, mr, 1950; flowers velvety crimson-red; [Donald Prior X Orange Triumph]; Leenders, M.

Bailey Red, F, mr

'BAISER', HT, w, 1953; flowers pearly white, edges tinted pink, dbl., 40 petals, 3 in., exhibition form, moderate fragrance; vigorous growth; [Mme Joseph Perraud X Independence]; Mallerin, C.; EFR

'BAJAZZO', HT, rb, 1961; flowers velvety blood-red, reverse white, well-formed, large, intense fragrance; vigorous, upright growth; Kordes, R.; Kordes

Bakels, F, 1980; Gaujard; (Cavriglia)

Bakewell Scots Briar, HSpn, w

Bakker's Newcomer®, HT, m

Bakker's Orange, HT, ob

'BAKTAF', Min, yb, 1988; **(Taffy)**; flowers soft, pastel butter yellow, tipped light pink, reverse pure, dbl., 33 petals, exhibition form, slight fragrance; prickles nearly straight, small, light brown; foliage small, medium green, matt; bushy, low, compact, full growth; [Unnamed seedling X Unnamed seedling]; Baker, Larry

Baladin, F; Combe, M.; (Cavriglia)

Balaji, HT, ob, 1998; Patil, B. K.

'BALALAIKA', F, dr, 1978; bud ovoid; flowers blood-red, dbl., moderate fragrance; foliage glossy; low, bushy growth; Hubner; O. Baum

'BALCON', LCl, or, 1960; bud ovoid; flowers geranium-red, open, medium, dbl., slight fragrance; foliage dark, glossy; very vigorous growth; [Spectacular X Seedling]; Combe; Minier

'BALDUIN', HT, pb, 1896; (Crimson Maman Cochet, Helen Gould, Red Maman Cochet); flowers pink, edged darker, large, dbl., moderate fragrance; vigorous growth; [Charles Darwin X Marie van Houtte]; Lambert, P., 1896; Dingee & Conard, 1901

Baldwin, *see* **'SLEEPY'**

'BALÉARES', HT, w, 1957; bud pointed; flowers base and reverse of petals white suffused carmine, deeper at edges, dbl., 35 petals, intense fragrance; foliage dark, glossy; stiff stems; upright growth; [Peace X Flambee]; Dot, Simon

'BALI', Pol, 1960; flowers orange-yellow, open, large, single, 6 petals; foliage glossy, light green; bushy growth; [Masquerade X Golden Rain]; Leenders, J.

'Bali-Hi', HT, mp, 1959; flowers peach-pink to shell-pink, medium, exhibition form, moderate fragrance; foliage glossy; vigorous, upright growth; Lowe

'BALINESE', HT, m, 1963; bud ovoid; flowers brown and lavender tones, dbl., 35–40 petals, 4.5–5 in., cupped, intense, fruity fragrance; foliage glossy; vigorous, upright growth; [Grey Pearl X Brownie]; Boerner; J&P

'BALJIT', HT, dp, 1979; bud pointed; flowers spirea-red, open but full, dbl., 30–40 petals, 5 in., long-lasting, profuse bloom; foliage dark, leathery; free growth; [Velsheda sport]; Lucknow

Balkan Star, HT, m, 1997; Tapanchev

'BALL OF SNOW', N, w, 1887; Henderson; (Sangerhausen)

'BALLADE', F, op, 1960; flowers deep orange-salmon, well formed, dbl., 25 petals, borne in clusters; bushy growth; [Signal Red X Polyantha seedling]; deRuiter

Ballade, *see* TANedallab

'BALLADY', C, lp, 1934; flowers large, dbl.; Perrot; (Sangerhausen)

'BALLERINA', HMsk, mp, 1937; flowers bright soft pink, white eye, small, single, blooms in very large clusters; vigorous (3 ft) growth; Bentall

'BALLERINA', F, my, 1941; flowers naples yellow, large, dbl., moderate fragrance; Leenders, M.

'BALLERINE', HT, w, 1955; bud long; flowers satiny snow-white, open, medium, dbl., intense fragrance; foliage glossy; vigorous, bushy growth; Buyl Frères; Delforge

Ballet®, *see* 'KORFLOT'

'BALLET, CLIMBING', Cl HT, w, 1962; Kordes

BALlgance, HT, dr, 1996; (**Red Elegance™**); Twomey, Jerry

'BALLILA', HT, mr, 1935; flowers medium, dbl.; Bräuer; (Sangerhausen)

Ballindalloch Castle, *see* 'COCNEEL'

Balmain Climber, LCl, lp

'BALPARTY', HT, w, 1999; (**Rose Odyssey 2000**); flowers cream with pink edge, dbl., 26–40 petals, 4–5 in., exhibition form, borne mostly singly, moderate fragrance; prickles moderate; foliage medium to large, medium to dark green, semi-glossy; upright, bushy, medium (3.5 to 4.5 ft) growth; [Garden Party sport]; Ballin, Don & Paula, 2000; Arena Rose Co., 2000

Baltik, F, m; flowers violet, large, dbl., slight fragrance; VEG; (Sangerhausen)

'BALTIMORE', HT, lp, 1898; flowers blush tinted; [Mme Antoine Rivoire X Lady Mary Fitzwilliam]; Cook, J.W.

'BALTIMORE BEAUTY', LCl, ly, 1927; flowers buff-yellow, fading white, semi-dbl., borne in clusters, moderate fragrance; long, strong stems; Schluter

'BALTIMORE BELLE', HSet, lp, 1843; (Belle de Baltimore); flowers pale blush to rose-white, very dbl., borne in clusters, nonrecurrent, moderate fragrance; vigorous growth; [Probably R. setigera X Noisette]; Feast

Balwant, HT, ab; [Christian Dior sport]; Phadtare

'BAMBEY', HT, mr, 1979; bud long, pointed; dbl., 60 petals, exhibition form, intense, fruity fragrance; small, recurved prickles; foliage matt; medium growth; [(Fragrant Cloud X Peace) X Alec's Red]; Perry, Astor, 1981

'BAMBI', F, mp, 1962; flowers bright pink, dbl., 24 petals, 2 in., blooms in clusters, slight fragrance; foliage dark; vigorous, bushy, low growth; [The Optimist X Korona]; Watkins Roses

'BAMBI', F, mp, 1962; bud pointed; flowers light apricot-pink, semi-dbl., 20 petals, 3 in., cupped, blooms in clusters, slight fragrance; foliage glossy; vigorous, bushy, compact growth; Von Abrams; Peterson & Dering

Bambi, *see* DELmistri

'BAMBINA', F, w, 1962; bud yellow; flowers large, semi-dbl., borne in clusters; [Seedling X Virgo]; da Silva, Moreira

'BAMBINO', Min, mp, 1953; [Perla de Alcañada pink sport]; Dot, Pedro

Bambino, *see* 'SAVABINO'

Bambolina®, Min, pb

Bamboo Rose, *see* **R. MULTIFLORA WATSONIANA**

'BAMBULA'®, F, op, 1970; dbl., 28 petals, 5 in., slight fragrance; foliage glossy, dark; Tantau, Math.; Wheatcroft & Sons, 1969

Banana Split, *see* 'ZIPBAN'

'BANARAS DAWN', HT, ab, 1977; bud tapered; flowers apricot-buff, dbl., 30 petals, 4.5 in., exhibition form, intense, fruity fragrance; foliage glossy, light green; vigorous growth; Saxena; Doon Valley Roses

'BANATER ROSE', HT, w, 1927; flowers cream-white, center orange-yellow, dbl., slight fragrance; [Harry Kirk seedling]; Mühle

'BANBRIDGE', F, pb, 1967; flowers rose-red and yellow, well shaped, 3 in., blooms in clusters, slight fragrance; [Mme Léon Cuny X Cläre Grammerstorf]; McGredy, Sam IV

'BANCO', HT, op, 1956; flowers salmon-pink becoming gold tinted, well-formed, large, dbl., 50 petals; foliage bright green; vigorous, upright growth; [Peace X Seedling]; Laperrière; EFR

Banco 86®, *see* 'LAPDEF'

'BANGOR', F, or, 1972; flowers geranium-lake, ovate, dbl., 24 petals, 3.5 in., slight fragrance; foliage leathery; free growth; [Jubilant X Marlena]; Dickson, A.

Bangor Cathedral, *see* KIRmelody

Bangsbo, *see* POUlrine

'BANJARAN', F, yb, 1969; flowers gold and orange-red, small, dbl., cupped, blooms in clusters, slight fragrance; foliage leathery; vigorous, upright, compact growth; Pal, Dr B. P.; K. S. G. Son's Roses

Banks' Rose, *see* **R. BANKSIAE**

Banksiae Alba, *see* **R. BANKSIAE BANKSIAE**

'BANKSIAEFLORA', HSem, w; flowers white, center cream, small, very dbl.; [Possibly R. arvensis X Noisette hybrid]

Banksian Rose, *see* **R. BANKSIAE**

'BANNER', HT, pb, 1951; flowers deep pink, striped white, [Charlotte Armstrong sport]; Raffel; Port Stockton Nursery

Banquise®, HT, w

Banquise, *see* LAPruni

'BANSHEE', S, mp; flowers pink, troubled by balling, of poor texture, very dbl., nonrecurrent, moderate fragrance; Origin unknown, 1928

'Bantry Bay'®, LCl, mp, 1967; flowers soft pink, reverse bright pink, 4 in., blooms in clusters, slight fragrance; [New Dawn X Korona]; McGredy, Sam IV; McGredy

Banzai, *see* 'MEILIMONA'

Banzai, T, or

Banzai '76, *see* 'MEILIMONA'

Banzai 83®, HT, or

'BAPTISTE LAFAYE', Pol, dp, 1910; flowers carmine-pink, medium, dbl.; Puyravaud; (Sangerhausen)

BARafne, S, pb, 1995; (**Dafne**); Barni, V.

BARafne, S, op, 1998; (**Meriggio**); Barni, V.

Barakura, S, lp, 1999

'BARAMAD', S, rb, 1991; (**Amadeus®**); Barni, V.

'BARANAM', HT, mp, 1988; (**Antico Amore®**); Barni, V.

BARarc, S, mp, 1995; (**Arcanto**); Barni, V.

'BARARCSUN', Min, w, 1989; (**Arctic Sunrise**); bud pointed; flowers small, dbl., 30 petals, flat, borne in sprays of 40–60, no fragrance; no fruit; prickles long, thin, small, pale tan; foliage small, medium green, glossy; spreading, low growth; [Snow Carpet X Tranquillity]; Barrett, F. H., 1991

Barba Blue, HT, m

'BARBARA', HT, rb, 1923; flowers bright red, base yellow, reverse pale yellow, Paul, W.

'BARBARA', HT, yb, 1962; flowers amber-yellow lightly flushed pink, semi-dbl., 20 petals, 5 in., exhibition form; foliage dark; very vigorous growth; Gaujard; Gandy Roses, Ltd.

Barbara®, *see* GAUbor

Barbara Allen, S, w, 1999; flowers ivory with pink washing on petal edges, semi-dbl.; spreading growth; Williams, J. Benjamin

Barbara Austin, *see* 'AUStop'

Barbara Bush™, *see* 'JACBUSH'

Barbara Carrera, F, op, 1994; Beales, Peter

Barbara Dawson, HT, lp, 1987; [Mount Shasta X Saffron]; Dawson

'BARBARA FRIETCHIE', HT, dp, 1959; bud long, pointed; flowers rose-red, open, large, dbl.; vigorous, upright growth; [Heart's Desire X The Chief]; Silva; Plant Hybridizers of Calif.

Barbara Hauenstein, HT, mp, 1957; flowers large, dbl., moderate fragrance; Poulsen; (Sangerhausen)

Barbara Hendricks, *see* DORgran

'BARBARA JOYCE', Min, mr, 1999; very dbl., 41 petals, 1.5 in., borne in small clusters, no fragrance; few prickles; foliage medium, dark green, resistant, semi-glossy; upright, medium (16 in.) growth; [Miss Dovey X Kristin]; Jolly, Betty J.

Barbara Mandrell™, *see* 'KINBARB'

'BARBARA MASON', HT, my, 1947; bud long, pointed; flowers medium, semi-dbl., exhibition form, moderate fragrance; foliage glossy; vigorous, upright growth; [Eclipse X Luis Brinas]; Moss; F. Mason

'BARBARA MEYER', Gr, or, 1971; bud ovoid; flowers medium, very dbl., exhibition form, slight fragrance; foliage glossy, dark; vigorous, upright growth; [Queen Elizabeth X Tropicana]; Meyer, H.M.; Aloe Vera Nursery

'BARBARA RICHARDS', HT, yb, 1930; flowers yellow, reverse flushed pink, very large, dbl., moderate fragrance; weak necks; vigorous, bushy growth; Dickson, A.

'BARBARA ROBINSON', HT, w, 1925; bud long, pointed; flowers creamy white, large, dbl., exhibition form, slight fragrance; Dickson, A.

'BARBARA STRAUS', HT, dr, 1978; bud long, pointed; flowers, dbl., 22 petals, borne singly, slight fragrance; foliage glossy; spreading, upright growth; [Mister Lincoln X Unknown]; Schwartz, Ernest W.; Flora World, Inc.

'BARBARA WARD', HT, mr, 1931; flowers crimson-scarlet, large, dbl., 42 petals, moderate fragrance; vigorous growth; [Royal Red X Columbia]; Ward, F.B.

Barbara Worl, B, mp, 1991

Barbara's Rose, HBc, yb

'BARBARELLA', Min, pb, 1981; flowers deep pink, reverse creamy yellow, small, dbl., 20 petals, slight fragrance; foliage small, dark, matt; bushy growth; [Seedling X Seedling]; Barni-Pistoia, Rose

Barbarina, F, yb, 1965; flowers large, dbl., moderate fragrance; Berger, A.; (Sangerhausen)

Barbarosa, S, lp, 1978

'BARBAROSSA', HP, m, 1906; flowers carmine-purple, large, dbl., 55 petals, moderate fragrance; vigorous growth; [(Frau Karl Druschki X Captain Hayward) X Princesse de Béarn]; Welter

BARbea, HT, w, 1995; (**Beatrice**); Barni, V.

'BARBECUE', F, mr, 1961; flowers rich red, dbl., 30 petals, 4 in., flat, blooms in large clusters, moderate fragrance; foliage dark; vigorous, upright, bushy growth; [Seedling X Lilli Marleen]; Dickson, Patrick; A. Dickson

Barbeuce, F, mr, 1961; flowers large, dbl., moderate fragrance; Dickson, Patrick; (Sangerhausen)

'BARBIE', F, mp, 1977; bud pointed, ovoid; flowers sweetheart, dbl., 35–40 petals, 2.5 in., slight fragrance; foliage small, glossy; upright growth; [Escort X Jazz Fest]; Swim, H.C. & Ellis, A.E.; Armstrong Nursery

Barbie®, *see* 'JACMOBLI'

Barbra Streisand, *see* 'WEKQUANEZE'

BARbrio, F, yb; (**Briosa**); Barni, V.

Barby, *see* KORaby

'BARCAROLLE', HT, mr, 1959; flowers crimson tinted geranium-red, large, dbl., 50 petals; vigorous, bushy growth; [Paulette X Tonnerre]; Laperrière; EFR

BARcast, S, lp, 1987; (**Castore®**); Barni, V.

'BARCELONA', HT, dr, 1932; bud long, pointed; flowers crimson, large, dbl., 75 petals, moderate, spicy fragrance; foliage dark; vigorous growth; [(Sensation X Templar) X Lord Charlemont]; Kordes

Barcelona 95®, *see* FEcasa

BARciak, LCl, rb, 1994; (**Ciak®**); Barni, V.

BARcora, MinFl, ob, 1993; (**Corallina**); Barni, V.

BARdem, F, lp; (**Tempi Moderni**); Barni, V., 1996

BARdon, HT, lp, 1988; (**Donna Marella Agnelli®**); Barni, V.

BARdord, HT, ab, 1998; (**Dorada**); Barni, V.

'BARDORG', Min, yb, 1993; (**Newbury Angel**); flowers open yellow turning to pale orange, blooms borne in large clusters, dbl., 26–40 petals, 1.5 in., no fragrance; some prickles; foliage small, medium green, matt; low (40 cms), bushy growth; [Freegold X Orange Honey]; Barrett, F.H.; Hills Nurseries, 1993

'BARDOU JOB', B, dr, 1887; flowers crimson, shaded blackish, semi-dbl.; vigorous, semi-climbing growth; [Gloire des Rosomanes X Général Jacqueminot]; Nabonnand, G.

BARemoz, HT, ab, 1999; (**Emozione**); Barni, V.

BARenic, S, lp, 1993; (**Berenice**); Barni, V.

BARfeel, F, lp, 1994; (**Feeling®**); Barni, V.

BARfiob, F, w, 1988; (**Fiocco Bianco**); Barni, V.

BARflam, S, dr, 1993; (**Flaming Star**); Barni, V.

BARflor, MinFl, dp, 1994; (**Florita®**); Barni, V.

BARfort, Min, op, 1999; (**Fortunella**); Barni, V.

BARfri, S, mp, 1992; (**Frine®**); Barni, V.

BARfunn, F, ob, 1998; (**Funny Girl**); Barni, V.

BARgai, MinFl, pb, 1993; (**Gaia®**); Barni, V.

BARger, S, mr, 1996; (**Ranger**); Barni, V.

BARgold, F, my, 1991; (**Golden Dance®**); Barni, V.

BARhop, HT, mr, 1999; (**Hot Point**); Barni, V.

'Barillet', M, mr, 1850; flowers dark carmine, large, dbl., cupped; Verdier, V.

Barillet, HP, lp; flowers brilliant rose; Deschamps, 1867

BARIMP, HT, mr, 1990; (**Impulse®**); Barni, V.

BARinas, S, pb, 1989; (**Rinascimento®**); Barni, V.

BARitzia, HT, w, 1996; (**Letizia**); Barni, V.

BARjou, S, dy, 1999; (**Bijou**); Barni, V.

Barkarole, *see* 'TANELORAK'

BARkel, HT, lb; (**Helen Keller**); Barni, V., 1991

'BARKHATNAIA KRASAVITSA', HT, dr, 1938; (Velvet Beauty); flowers velvety red, medium, semi-dbl., cupped, moderate fragrance; foliage dark; low growth; Gubonen

BARlar, F, mr, 1999; (**La Rossa**); Barni, V.

BARleo, LCl, dy; (**Eleonora**®); Barni, V.

BARlev, F, ab, 1991; (**Rita Levi Montalcini**®); Barni, V.

Barley Sugar, Cl Min, lp; Peden, R., 1997; (Weatherly, L.)

BARliz, MinFl, mr, 1993; (**Liza**®); Barni, V.

BARluc, Min, ob, 1992; (**Lucy**); Barni, V.

BARmar, HT, dr, 1992; (**Marina Marini**®); Barni, V.

BARmeri, HT, op, 1994; (**Meridiana**®); Barni, V.

BARmiss, HT, yb, 1999; (**Rosita Missoni**); Barni, V.

BARmiz, S, dy, 1999; (**Mizar**); Barni, V.

'**Barn Dance**', S, op, 1975; bud ovoid, long, pointed; flowers light salmon-pink, dbl., 23 petals, 2.5–3.5 in., cupped, slight fragrance; foliage light to dark, leathery; [Tickled Pink X Prairie Princess]; Buck, Dr Griffith J.; Iowa State University

BARnec, HT, pb, 1986; (**Mirella**®); Barni, V.

Barni, HT, ob

Barock, *see* TANkorab

BARoly, S, rb, 1998; (**Olympus**); Barni, V.

'**Baron de Bonstetten**', HP, dr, 1871; flowers dark velvety crimson, large, dbl., 80 petals, sometimes recurrent bloom, moderate fragrance; vigorous, compact growth; [Général Jacqueminot X Géant des Batailles]; Liabaud

'**Baron de Wassenaer**', M, dp, 1854; flowers light crimson, cupped, blooms in clusters, vigorous growth; Verdier, V.

'**Baron de Wolseley**', HP, 1882; Verdier, E.; (Cavriglia)

'**Baron Girod de l'Ain**', HP, rb, 1897; flowers bright red, petals edged white; [Eugène Furst sport]; Reverchon

'**Baron Haussmann**', HP, mr, 1867; flowers light red, large, very dbl.; Lévêque; (Sangerhausen)

'**Baron J.B. Gonella**', B, pb, 1859; Guillot Père

'**Baron Jacques Riston**', HT, op, 1936; bud long, pointed; flowers salmon-pink, well shaped, large, dbl., 45–50 petals, moderate fragrance; very vigorous growth; [Mme Butterfly X (Rev. David R. Williamson X Gorgeous)]; Ketten Bros.

Baron Meillandina®, *see* 'MEItifran'

'**Baron Palm**', HT, mr, 1913; flowers large, dbl., moderate fragrance; Lambert, P.; (Sangerhausen)

Baron Sunblaze, *see* 'MEItifran'

'**Baron Taylor**', HP, mr, 1880; flowers light red, large, dbl.; Dugat; (Sangerhausen)

'**Baron T'Kind de Roodenbecke**', HP, m, 1897; flowers shaded purple; Lévêque

'**Baronesa de Ovilar**', HT, 1935; flowers carmine tinted yellow, very large, dbl., intense fragrance; foliage dark; strong stems; very vigorous growth; [Sensation X Souv. de Claudius Pernet]; Munné, B.

Baroness Henrietta Snoy, *see* '**Baronne Henriette de Snoy**'

'**Baroness Krayenhoff**', HT, pb, 1931; flowers peach-pink, center orange, very large, dbl., moderate fragrance; foliage light green; long stems; vigorous, compact growth; [Mrs Henry Bowles X Lady Roundway]; Buisman, G. A. H.

'**Baroness Rothschild**', HP, lp, 1868; (Baronne Adolphe de Rothschild); flowers very soft rose, tinted white, large, dbl., 40 petals, cupped, some recurrent bloom, vigorous, erect growth; [Souv. de la Reine d'Angleterre sport]; Pernet Père

Baronesse®, *see* 'TANessenor'

'**Baronesse A. van Hövell tôt Westerflier**', HT, m, 1933; flowers carmine-purple, base yellow, large, dbl., moderate fragrance; foliage bronze; very vigorous, bushy growth; Leenders, M.

'**Baronesse H. von Geyr**', HT, pb, 1928; flowers pale flesh and vermilion-red, dbl., moderate fragrance; [(Farbenkonigin X Juliet) X Sunburst]; Leenders, M.

'**Baronesse M. van Tuyll van Serooskerken**', HT, pb, 1922; flowers rose and lilac-white, base apricot, semi-dbl.; [Jonkheer J.L. Mock X Mme Mélanie Soupert]; Leenders, M.

'**Baronesse Manon**', F, mp, 1938; flowers clear pink, firm petals, 3 in., moderate fragrance; foliage glossy, holly-like; vigorous, bushy growth; [Else Poulsen X Dame Edith Helen]; Poulsen, S.; C-P, 1952

'**Baronesse S. H. W. van Dedem**', HT, yb, 1923; flowers yellow and coppery, open, large, semi-dbl., moderate fragrance; foliage dark, glossy; vigorous, bushy growth; Leenders, M.

'**Baronesse van Ittersum**', HMult, rb, 1910; flowers light crimson, shaded deeper, open, semi-dbl., borne in clusters, free, non-recurrent bloom, slight fragrance; foliage dark, glossy; vigorous, climbing (15 ft) growth; [Crimson Rambler X Mme Laurette Messimy]; Leenders, M.

'**Baronin Anna von Lüttwitz**', HMult, lp, 1909; flowers cream-pink, medium, dbl.; Walter; (Sangerhausen)

'**Baronin von Adelebsen**', S, mr, 1938; flowers medium, single; Vogel, M.; (Sangerhausen)

Baronne Adolphe de Rothschild, *see* '**Baroness Rothschild**'

'**Baronne Charles d'Huart**', HT, pb, 1910; flowers lilac-rose suffused white, reflexed, dbl.; [Pharisaer X Seedling]; Ketten Bros.

'**Baronne de Maynard**', N, w, 1865; flowers white, edges tinged pink, cupped; moderately vigorous growth; [Blanche Lafitte X Sappho (an unrecorded variety)]; Lacharme, F.

'**Baronne de Medem**', HP, 1876; Verdier, E.; (Cavriglia)

Baronne de Nervo, S, 1969; Delbard-Chabert; (Cavriglia)

Baronne de Noirmont, B, mp; Granger

'**Baronne de Prailly**', HP, 1871; Liabaud, J.; (Cavriglia)

Baronne de Rothschild, *see* 'MEIgriso'

Baronne de Rothschild, Climbing, *see* 'MEIgrisosar'

'**Baronne de Saint Didier**', HP, 1886; Leveque, P.; (Cavriglia)

Baronne de Schorlemmer, HT, mp, 1985; flowers large, dbl.; Lens, Louis; (Sangerhausen)

'**Baronne de Stael**', HGal, ob, 1820; flowers salmon-pink, open, large, dbl.; vigorous, branching growth; Vibert

'**Baronne de Vivario**', Pol, w, 1925; flowers, dbl., in clusters, [Orléans Rose X Jeanny Soupert]; Soupert & Notting

Baronne Edmond de Rothschild®, *see* 'MEIgriso'

Baronne Edmond de Rothschild, Climbing®, *see* 'MEIgrisosar'

'**Baronne Finaz**', HT, mp, 1961; flowers bright pink, large blooms, dbl., 35 petals, moderate fragrance; foliage glossy, dark; long stems; very vigorous, upright growth; [Peace seedling X Opera]; Gaujard

'**Baronne G. de st Paul**', HP, lp, 1894; flowers large, dbl.; Glatenet; (Sangerhausen)

'**Baronne Henriette de Loew**', T, w, 1888; flowers pinkish white, center yellow, dbl., moderate fragrance; Nabonnand, G.

'**Baronne Henriette de Snoy**', T, pb, 1897; (Baroness Henrietta Snoy, Baroness Henriette Snoy); flowers flesh, reverse carmine-pink, well-formed, large, dbl.; vigorous growth; [Gloire de Dijon X Mme Lombard]; Bernaix, A.

'**Baronne Maurice de Graviers**', HP, 1866; Verdier, E.; (Cavriglia)

'**Baronne Nathalie de Rothschild**', HP, lp, 1885; flowers very large, very dbl., moderate fragrance; Pernet Père; (Sangerhausen)

Baronne Presval, HP, dp

'**Baronne Prévost**', HP, mp, 1842; flowers rose-pink, shading lighter, large, dbl., flat, recurrent bloom, moderate fragrance; vigorous, erect growth; Desprez

Baronne Prévost, Climbing, Cl HP, mp

Baronne Surcouf, *see* 'BRIancouf'

Baroque, *see* HARbaroque

'BARORP', Min, m, 1985; (**Astrea®**); Barni, V.

Barossa Dream, *see* 'SUNAUCK'

'BAROUT', HT, ab, 1982; (Handout, Tranquility, **Tranquility**); flowers pale apricot, peach and yellow blend, blooms borne singly, dbl., 35 petals, slight fragrance; foliage medium, dark, glossy; upright growth; [Whisky Mac X Pink Favorite]; Barrett, F.H.; John Mattock, Ltd.

BARpipp, MinFl, pb, 1995; (**Pippy**); Barni, V.

BARpoll, S, mr, 1987; (**Polluce®**); Barni, V.

BARprett, LCl, dp, 1992; (**Pretty Pink**); Barni, V.

BARprimo, S, ly, 1995; (**Primo Passo**); Barni, V.

BARpris, HT, my, 1987; (**Primo Sole®**); Barni, V.

BARred, F, mr, 1990; (**Red Sea®**); Barni

BARrez, HT, op, 1996; (**Carezza**); Barni, V.

BARrian, MinFl, mr, 1995; (**Brian**); Barni, V.

Barricade, LCl, 1971; Combe, C.; (Cavriglia)

'**BARRIE**', F, pb, 1973; bud ovoid; flowers salmon-pink, medium, dbl., cupped, free bloom, slight fragrance; foliage glossy, light; [Sumatra X Fashion]; Schloen, J.

BARros, F, mp, 1990; (**Rosellana®**); Barni, V.

Barry Fearn, *see* KORschwama

Barry Stephens, *see* 'HORCABELLERO'

BARsan, F, w, 1995; (**Sans Souci**); Barni, V.

BARsav, F, op, 1999; (**Mafalda di Savoia**); Barni, V.

BARser, HT, op, 1998; (**Serenata**); Barni, V.

'BARSHIFLO', F, w, 1986; (**Snowgoose**); flowers small blooms in clusters, dbl., 20 petals, no fragrance; foliage medium, medium green, semi-glossy; upright growth; [Seaspray X Iceberg]; Barrett, F. H.

BARsib, HT, dr, 1987; (**Sibilla®**); Barni, V.

BARsor, HT, ob, 1990; (**Eros**); Barni, V.

BARstel, F, lp; (**Stelvio Coggiatti**); Barni, V., 1999

BARsus, F, pb; (**Suni®**); Golden Rose, Geneva, 1993; Barni, V., 1993

BARtares, S, pb, 1996; (**Antares**); Barni, V.

'**BARTER'S PINK**', HT, ob, 1963; flowers bright coral-pink, deeply veined, dbl., 4 in., slight fragrance; foliage light green; vigorous, bushy growth; [Queen Elizabeth X Claude]; Barter

'**BARTHELEMY JOUBERT**', HP, 1877; Moreau et Robert; (Cavriglia)

BARtusc, HT, mr; (**Tuscia®**); Barni, V., 1993

Barunka, F, op, 1983; flowers salmon-pink, large, very dbl., slight fragrance; Urban, J.; (Sangerhausen)

BARvan, S, mp, 1999; (**Avance**); Barni, V.

BARvelv, LCl, lp; (**Velvet Star**); Barni, V., 1998

BARwast, HT, ob; (**Wall Street®**); Barni, V., 1988

BARwrock, LCl, lp; (**White Rock**); Barni, V., 1996

'**BASHFUL**', Pol, pb, 1955; (Giesebrecht); flowers reddish pink, white eye, small, single, blooms in trusses, bushy, compact growth; deRuiter; Gregory & Willicher Baumschulen

'**BASILDON BELLE**', F, or, 1964; flowers vermilion, [Anna Wheatcroft sport]; Maarse, G.; Basildon Rose Gardens

Basildon Bond, *see* 'HARJOSINE'

Basilika, HT, op; flowers salmon-pink, large, dbl., slight fragrance; VEG; (Sangerhausen)

Bassino®, *see* 'KORMIXEL'

Bastei, S, or, 1973; Schmadlak, Dr

'**BASTOGNE**', HT, or; flowers vermilion, shaded darker, well formed, moderate fragrance; foliage dark; very vigorous growth; Grandes Roseraies

Basye's Blueberry, S, mp, 1982; Basye, Robert

Basye's Myrhh-Scented, S; Basye, Robert

Basye's Purple Rose, S, m, 1968; Basye, Robert

'BATMERCURY', HT, my, 1994; (**Freddie Mercury**); dbl., 26–40 petals, 1.5–2.75 in., borne mostly singly, moderate fragrance; some prickles; foliage large, dark green, glossy; medium, bushy growth; [Tina Turner sport]; Stainthorpe, Eric; Battersby Roses, 1994

'BATSYBIL', HT, dr, 1993; (**Dr Sybil Johnson**); dbl., 26–40 petals, 1.5–2.75 in., borne mostly singly, moderate fragrance; some prickles; foliage medium, medium green, semi-glossy; medium (85 cms), upright growth; [Tropicana X Prima Ballerina]; Stainthorpe, Eric; Battersby Roses, 1994

Battersby Beauty, *see* **'HOBATBEAUT'**

'**BATTLE OF BRITAIN**', HT, yb, 1970; flowers yellow-orange, medium, dbl., 30 petals, moderate fragrance; [Miss Ireland X Summer Sunshine]; Gandy, Douglas L.; Wheatcroft & Sons

'BATTOO', HT, pb, 1992; (**Myra**); dbl., 15–25 petals, 1.5–2.75 in., borne in small clusters, slight fragrance; many prickles; foliage medium, dark green, glossy; medium (75 cms), bushy growth; [Matangi X Mood Music]; Stainthorpe, Avril E.; Battersby Roses, 1990

Baty's Pink Pillar, LCl, mp; (found rose)

Bavarian Girl, *see* 'KORLEEN'

Baveria, *see* 'KORMUN'

Baxter Beauty, HT, ab, 1938; light yellow to sulphur overlaid outside with light salmon pink; [Lorraine Lee sport]; Clark, A.

Bay Glow, Min, yb

'**BAYADÈRE**', HT, ab, 1954; flowers salmon-pink to canary yellow, tinted pink, large, dbl., 52 petals, exhibition form, slight fragrance; foliage dark, bronze; vigorous, bushy growth; GM, NRS, 1954; [R.M.S. Queen Mary X Seedling]; Mallerin, C.; EFR

Bayerngold®, *see* 'TANYAB'

Bayreuth, S, rb, 1965; flowers medium, dbl., moderate fragrance; Kordes, W. Söhne; (Sangerhausen)

'**BE GLAD**', HT, rb, 1988; flowers white changing to red, reverse white, aging deep pink, medium, dbl., 25–30 petals, exhibition form, moderate, fruity fragrance; no fruit; prickles normal; foliage medium, dark green, glossy; upright, medium growth; Silver, ARC TG, 1988; [Paradise X Color Magic]; McMillan, Thomas G.

'BEAcath', HT, my, 1997; (**Norwich Cathedral**); flowers full, medium, very dbl., 26–40 petals, 4 in., borne mostly singly, intense fragrance; some prickles; foliage medium, medium green, leathery, semi-glossy; upright, low (3 ft)growth; [Diamond Jubilee sport]; Beales, Peter

'**BEACH BOY**', Min, my, 1981; bud pointed; flowers golden buff, blooms borne usually singly, dbl., 48 petals, exhibition form, slight fragrance; short, tan prickles; foliage small, dark, slightly glossy; upright, bushy growth; [Tom Brown X Seedling]; Williams, Ernest D.; Mini-Roses

Beachcomber, *see* 'MACLAPAZ'

'**BEACON**', HT, 1952; Swim, H.C.; (Cavriglia)

'**BEACON BELLE**', Ayr, lp, 1919; flowers flesh passing to white, borne in clusters, dbl., moderate fragrance; [Orléans Rose X (Katharina Zeimet X R. arvensis hybrid)]; Farquhar

'**BEACON LODGE**', HT, dr, 1976; flowers deep red, full, dbl., 40–45 petals, 4–5 in., moderate fragrance; foliage glossy, dark; very vigorous growth; [Chopin X Ena Harkness]; Ellick

'BEAdix', Cl HT, ab, 1997; (**Dixieland Linda**); flowers very full, large, very dbl., 41 petals, 4.5 in., borne in small clusters, intense fragrance; foliage medium, dark green, glossy; upright, medium (8–10 ft.)growth; [Aloha sport]; Beales, Peter

'BEAHOR', S, mp, 1998; (**Horatio Nelson**); flowers clear pink, very full, opens flat, very dbl., 41–50 petals, rosette, borne in small clusters; prickles moderate; foliage medium, medium green, glossy; compact (3 ft) growth; [Centenaire de Lourdes X

Aloha]; Beales, Peter; Peter Beales Roses, 1997

Beales' Mons Tillier, T, mr

Beales' The Bride, T, w

'BEAMAC', S, w, 1998; (**Macmillan Nurse**); flowers white, very full, very dbl., 41–50 petals, borne in large clusters, moderate fragrance; prickles moderate; foliage medium, dark green, glossy; compact, 3 ft growth; [Bonica X Maigold]; Beales, Peter; Peter Beales Roses, 1998

'BEAPAW', S, mp, 1999; (**Paws**); very dbl., 41 petals, 4 in., borne mostly singly, intense fragrance; many prickles; foliage medium, medium green, dull; bushy, medium (2.5 × 2 ft) growth; [Silver Jubilee X Constance Spry]; Beales, Peter; Peter Beales Roses, 1999

'BEATRICE', F, dp, 1968; flowers deep rose pink, in large sprays, dbl., slight fragrance; [Paddy McGredy X (Kordes' Perfecta X Montezuma)]; McGredy, Sam IV

Beatrice, *see* BARbea

Beatrice Berkeley, HT, lp; Fitzhardinge, before 1932; (Weatherly, L.)

'BEATRICE BOEKE', HT, dp, 1966; bud ovoid; flowers pink-red, large, dbl.; [Montezuma X Detroiter]; Buisman, G. A. H.

'BEATRICE MCGREGOR', HT, dr, 1938; flowers large, dbl., intense fragrance; [Sensation X Seedling]; Clark, A.; NRS Victoria

Beatrice Samzin, F, mp, 1970; flowers large, very dbl., slight fragrance; Scholle, E.; (Sangerhausen)

'BEATRIX', HP, 1855; Cherpin; (Cavriglia)

'BEATWE', F, dr, 1997; (**Twenty-Fifth**); flowers semi double, medium, semi-dbl., 8–14 petals, 2.5 in., borne in large clusters, slight fragrance; some prickles; foliage medium, dark green, semi-glossy; compact, low (2 ft) growth; [Redbreast X Pearl Drift]; Beales, Peter

'BEAU NARCISSE', HGal, m; flowers purple, striped, Miellez, before 1828

Beau Rose, Misc OGR, mp

Beau Rose Primaplant, Pol, dp, 1963; flowers large, dbl.; Vlaeminck; (Sangerhausen)

'BEAUCAIRE', Pol, rb, 1937; flowers edges chamois-pink, center coppery yellow, reverse coppery-red, dbl.; vigorous, dwarf growth; Grandes Roseraies

'BEAUJOLAIS', HT, mr, 1932; flowers crimson-carmine, very large, dbl., globular, intense fragrance; very vigorous growth; [Hadley X Laurent Carle]; Croibier

Beaulieu, *see* 'DICOBEY'

'BEAULIEU', HP, 1883; Moreau et Robert; (Cavriglia)

Beaulieu, *see* 'TANZAHDE'

'BEAULIEU ABBEY', F, ab, 1964; flowers creamy yellow suffused pink, well formed, dbl., 4 in., moderate fragrance; foliage dark, glossy, leathery; vigorous growth; [Masquerade X Docteur Valois]; Cobley; Blaby Rose Gardens

Beauregard, LCl, 1972; Croix, P.; (Cavriglia)

'BEAUTÉ', HT, ab, 1953; bud long; flowers light apricot, well-formed, large, dbl., moderate fragrance; vigorous growth; [Mme Joseph Perraud X Seedling]; Mallerin, C.; EFR

'BEAUTÉ D'AUTOMNE', Pol, dp, 1918; flowers bright rose-pink, borne in clusters of 50–70, dbl.; [Phyllis X Seedling]; Turbat

'BEAUTÉ DE FRANCE', HT, w, 1920; flowers creamy white to pure white, inside yellow, dbl., moderate fragrance; [Mme Mélanie Soupert X Kaiserin Auguste Viktoria]; Toussaint Mille Fils

'BEAUTÉ DE FRANCE', HT, 1952; bud long, pointed; flowers brick flushed coppery, medium, dbl., moderate fragrance; foliage leathery, dark; upright growth; [(Comtesse Vandal X Seedling) X Seedling]; Gaujard

'BEAUTÉ DE L'EUROPE', T, op, 1881; flowers light orange-pink, yellow base, large, dbl., intense fragrance; Gonod; (Sangerhausen)

'BEAUTÉ DE LYON', HP, or, 1910; flowers coral-red, tinted yellow, dbl., moderate fragrance; GM, Bagatelle, 1911; [Seedling X Soliel d'Or]; Pernet-Ducher

'BEAUTE FRANCAISE', HP, mr, 1862; Lartay

'BEAUTÉ INCONSTANTE', T, ob, 1892; flowers orange to coppery red, shaded carmine and yellow, moderate fragrance; vigorous growth; Pernet-Ducher

'BEAUTÉ LYONNAISE', HT, w, 1895; flowers white tinted pale yellow, large; Pernet-Ducher

'BEAUTE ORLÉANAISE', HWich, w, 1919; flowers white to flesh-pink, borne in clusters of 20–25, dbl., slight fragrance; Turbat

Beauté Spatiale®, *see* 'DELDROP'

Beauté Virginale, A, w

Beautiful Black, HP, dr

Beautiful Britain, *see* 'DICFIRE'

'BEAUTIFUL DOLL', Min, mp, 1982; dbl., 25 petals, exhibition form, borne singly and in small clusters, moderate fragrance; foliage small, dark, semi-glossy; bushy, spreading growth; [Seedling X Zinger]; Jolly, Betty J.; Rosehill Farm

'BEAUTIFUL DREAMER', F, ob, 1977; flowers orange to sunset-gold, full, pointed, dbl., 35 petals, 2–2.5 in., slight fragrance; foliage glossy; [Seedling X Seedling]; Herholdt, J. A.

'BEAUTIFUL SUNDAY', Min, mr, 1977; bud ovoid; flowers bright geranium-red, small, very dbl., 60 petals, cupped, moderate fragrance; foliage glossy, dark, leathery; compact, bushy growth; [Camelot X Camelot seedling]; Takatori, Yoshiho; Japan Rose Nursery

'BEAUTY', HT, mr, 1931; flowers American Beauty red, but darker, very large, dbl., moderate fragrance; long stems; vigorous growth; [(Crusader X Premier) X American Beauty]; Ward, F. B.

'BEAUTY CREAM', F, w, 1956; flowers cream, large, moderate fragrance; foliage dark; vigorous, upright growth; Verschuren; Gandy Roses, Ltd.

Beauty of Badgen, HT, French, A., 1924; (Weatherly, L.)

'BEAUTY OF BEESTON', HP, 1882; Frettingham; (Cavriglia)

Beauty of Brisbane, Pol, lp; [Goldlachs sport]; Perrot, 1932; (Weatherly, L.)

'BEAUTY OF DROPMORE', S, w, 1956; dbl., non-recurrent; bushy, erect growth; [R. spinosissima altaica X R. spinosissima cultivar]; Skinner

Beauty of England, HP, dp

Beauty of Festival, HT, 1955; Klimenko, V. N.; (Cavriglia)

BEAUTY OF GLAZENWOOD, *see* **'FORTUNE'S DOUBLE YELLOW'**

Beauty of Glenhurst, HCh, dp, 1983; Morley, Dr B.

'BEAUTY OF HURST', LCl, w, 1926; flowers creamy buff, dbl., moderate fragrance; foliage dark; very vigorous growth; Hicks

Beauty of Leafland, HSpn, pb; dbl., intense; [Butterball X Haidie]; Erskine

'BEAUTY OF NEW SOUTH WALES', Pol, rb, 1931; flowers bright crimson, center white, small blooms in clusters of 20, single, moderate fragrance; bushy, dwarf growth; [Orléans Rose X Alice Amos]; Knight, G.

'BEAUTY OF ROSEMAWR', T, pb, 1903; flowers carmine rose veined vermilion and white, medium, dbl.; Conard & Jones

'BEAUTY OF STAPLEFORD', HT, rb, 1879; flowers red and violet, well formed, large; moderate growth; [Mme Bravy X Comtesse d'Oxford]; Bennett

Beauty of the Prairies, *see* **'QUEEN OF THE PRAIRIES'**

'BEAUTY OF WALTHAM', HP, mr, 1862; flowers rosy crimson, large, dbl., recurrent bloom, moderate fragrance; vigorous growth; Paul, W.

Beauty Queen, *see* 'CANMISS'

'BEAUTY SECRET', Min, mr, 1965; bud pointed; flowers cardinal-red, small, dbl., 1.5 in., exhibition form, intense fragrance; foliage small, glossy, leathery; vigorous, bushy growth; AOE, ARS, 1975; Miniature Rose Hall of Fame, ARS, 1999; [Little Darling X Magic Wand]; Moore, Ralph S.; Sequoia Nursery

Beauty Star, *see* 'FRYSTAR'

Beautyglo™, *see* 'MINWCO'

'BEAUTY'S BLUSH', HRg, dp, 1955; flowers deep pink, becoming lighter, dbl., non-recurrent, moderate fragrance; vigorous (6 ft) growth; hardy on the Canadian prairies.; [Tetonkaha X Pink Pearl]; Univ. of Saskatchewan

'BEAYAR', HT, ly, 1999; (**Yardley Baroque**); dbl., 26–40 petals, 4.5 in., borne mostly singly, moderate fragrance; many prickles; foliage medium, dark green, semi-glossy; upright, medium (2 ft) growth; [Alpine Sunset sport]; Beales, Peter; Peter Beales Roses, 1997

'BÉBÉ BLANC', Pol, w, 1922; dbl., borne in large clusters; dwarf growth; Turbat

'BÉBÉ FLEURI', HCh, mp, 1906; flowers small, semi-dbl.; Dubreuil; (Sangerhausen)

'BÉBÉ LEROUX', Pol, w, 1901; flowers medium, in trusses of 20–40, moderate fragrance; compact growth; [Mignonette X Archiduchesse Elisabeth-Marie]; Soupert

Bebe Lune®, *see* 'DELFIB'

Becca Godman, F, pb, 1998; Wells

Becker Street, HT, op

Beckett's Single, HGal, mp; Quest-Ritson, before 1995

'BECKY', HT, dp, 1925; flowers glowing rose-pink, single, vigorous growth; Beckwith

Bedazzled™, *see* 'JACHOTTA'

BEDCHILD, HT, dy, 1996; (**Pudsey Bear**); Chessum, Paul

'BEDFORD CRIMSON', HT, dr, 1926; flowers velvety crimson, well formed, dbl., 40 petals, moderate fragrance; [Richmond X Château de Clos Vougeot]; Laxton Bros.

'BEDFORDIA', HT, pb, 1931; flowers pink, outer petals lighter, center salmon, dbl., 52 petals, moderate fragrance; vigorous growth; Laxton Bros.

'BEDRICH SMETANA', HT, w, 1933; flowers pearly white, open, very large, semi-dbl.; vigorous, bushy growth; [Modesty X Ophelia]; Böhm, J.

BEDswap, MinFl, lp; (**Rest in Peace**); Chessum, Paul

'BEEBOP', HT, lp, 1981; (**Cleo**); flowers soft light pink, blooms borne 1–2 per cluster, dbl., 37 petals, exhibition form, slight fragrance; red prickles; foliage light green, semi-matt; strong, bushy growth; [Kordes' Perfecta X Prima Ballerina]; Bees

'BEEJES', HT, yb, 1982; (**Jessie Mathews**); flowers light yellow, petals edged pink, medium, dbl., 35 petals, slight fragrance; foliage medium, light green, semi-glossy; bushy growth; [Ernest H. Morse X Rosenella]; Bees

'BEELAH', HT, yb, 1983; (**Tallulah**); flowers gold with deep pink petal edges, large, dbl., 20 petals, slight fragrance; foliage medium, dark, semi-glossy; bushy growth; [Astral X Piccadilly]; Bees

'BEERIL', HT, w, 1983; (**Cyril Fletcher**); flowers creamy white, cabbage rose form, large, dbl., 35 petals, intense fragrance; foliage medium, dark, semi-glossy; upright, bushy growth; [Fragrant Cloud X Whisky Mac]; Bees

'BEE'S FROLIC', S, lp, 1976; bud small, pointed; semi-dbl., 14 petals, borne 2–3 per cluster, repeat bloom, intense fragrance; small prickles; foliage medium, matt; upright, strong growth; [Schneezwerg X Clair Matin]; Dawnay, Mrs. E.

Bees Knees™, *see* 'JACKEE'

'BEESIAN', F, my, 1979; (**Amanda**); bud globular; dbl., 25 petals, 3.5 in., exhibition form, slight fragrance; foliage small, light green; upright growth; [Arthur Bell X Zambra]; Bees

'BEEVAL', F, pb, 1983; (**Vale of Clwyd**); flowers yellow, pink petal edges, medium, dbl., no fragrance; foliage large, medium green, glossy; upright growth; [Handel X Arthur Bell]; Bees

Begonia, HT, op

Behold™, *see* 'SAVAHOLD'

Bekola, *see* **'AALSMEER GOLD'**

Bel Ami®, *see* 'LAPCI'

Bel Ami, *see* **'PHILIPPE'**

'BEL ANGE', HT, mp, 1962; (Bella Epoca, Belle Ange, Belle Epoque); flowers soft pink, reverse darker, dbl., 35 petals, 4.5 in., moderate fragrance; foliage dark; vigorous growth; GM, Kortrijk, 1965; [(Independence X Papillon Rose) X (Charlotte Armstrong X Floradora)]; Lens; J&P, 1965

Bel Ange, Climbing, ClHT, lp; [Bel Ange sport]; Ruston, D., 1970; (Weatherly, L.)

'BEL CANTO', HT, or, 1964; flowers bright geranium-red, well formed; foliage coppery; upright growth; Mondial Roses

Bel Esprit, HMsk, pb, 1992; Lens

Bel-Air, HT, dr, 1986; Swane

Belami, *see* 'KORPRILL'

'BELCHARM', Min, ob, 1992; (**Bell Charmer**); flowers distinctive soft orange with light yellow petal base and reverse, dbl., 26–40 petals, 1.5–2.75 in., slight fragrance; some prickles; foliage medium, medium green, semi-glossy; medium (45–55 cms), upright, bushy growth; [(Cherish X Avandel) X Seedling]; Bell, Charles E., Jr.; Kimbrew Walter Roses, 1993

Belfast Belle, *see* 'DICROBOT'

Belfield, HCh, mr

Belgian Lace®, S, lp, 1994; Lens, Louis; (Sangerhausen)

'BELGICA', HT, dr, 1929; flowers crimson-red, shaded garnet, large, very dbl.; vigorous growth; Buyl Frères

'BELINDA', HMsk, mp, 1936; semi-dbl., borne in very large, erect trusses, moderate fragrance; vigorous (4–6 ft) growth; a good hedge or pillar rose; Bentall

Belinda, *see* 'TANBEEDEE'

'BELINDA'S DREAM', S, mp, 1992; (Belinda's Rose); very dbl., 3.25–4.25 in., borne in small clusters, repeat bloom, moderate, fruity, raspberry fragrance; foliage medium, medium green, matt; medium (5 x 4'), bushy growth; [Jersey Beauty X Tiffany]; Basye, Robert, 1988

Belinda's Rose, *see* **'BELINDA'S DREAM'**

Bell Charmer, *see* 'BELCHARM'

'BELL RINGER', Min, dp, 1989; (Bellringer); flowers deep pink, silver at base, small, dbl., 34 petals, slight, fruity fragrance; foliage medium, medium green, semi-glossy; upright, bushy growth; [(Fragrant Cloud X Avandel) X Bonny]; Bell, Charles E., Jr.; Kimbrew Walter Roses, 1989

Bella, *see* DEVsmooth

Bella, *see* POUlijill

'BELLA DONNA', D, lp; flowers soft lilac-pink, large, dbl.

'BELLA DONNA', F, mr, 1964; bud ovoid; flowers bright red medium, in clusters, dbl.; foliage dark; [Baccará seedling X Miracle]; Verbeek

Bella Epoca, *see* **'BEL ANGE'**

Bella Multiflora, HMult, mp, 1994; flowers large, borne in sprays; few prickles; arching growth; Uhl, J., 1994

Bella Nitida, S, mp, 1994; flowers large; medium growth; very hardy; Uhl, J.

Bella Renaissance, *see* POUljill

Bella Rosa®, *see* 'KORwondis'

Bella Via, *see* 'ZIPVIA'

Bella Weiss®, *see* 'KORTUEL'

'BELLARD', HGal; Roseraie de l'Hay, Prior to 1857

Belle, *see* 'MARBEL'

'BELLE ADÉLAIDE', HGal, dp; flowers cerise-red, very dbl., flat; Miellez

'BELLE AMOUR', A, lp; flowers soft pink with salmon tones, prominent yellow stamens, semi-dbl., moderate, myrrh fragrance; height 5–6 ft; Found at a convent at Elboeuf in the 1940's

Belle Ange, *see* **'BEL ANGE'**

Belle Anglaise, HT, dy, 1995

Belle au Bois Dormant, S, pb, 1960; Kordes

'BELLE AURORE', A, lp; flowers flesh shaded purple, large, very dbl.; Descemet, about 1815

'BELLE BIBLIS', HGal, m, 1815; flowers medium, dbl., moderate fragrance; Descemet, M.; (Sangerhausen)

'BELLE BLANCA', LCl, w; [Probably a Belle Portugaise white sport]

Belle Blonde, *see* 'MENAP'

Belle Champenoise, HT, my, 1997; Orard

'BELLE CLEMENTINE', A, lp; flowers mottled flesh-color

Belle Courtisanne, *see* **'KÖNIGIN VON DÄNEMARK'**

'BELLE CRÉOLE', F, rb, 1958; flowers variegated red and brick-red, borne in clusters; foliage dark; vigorous, upright growth; [(Gruss an Teplitz X Independence) X (Floradora X Independence)]; Arles; Roses-France

'BELLE CUIVRÉE', HT, or, 1924; flowers coral-red, shaded coppery yellow, semi-dbl.; Pernet-Ducher

Belle d'Automne, HT, 1969; Ducher, Ch.; (Cavriglia)

Belle de Baltimore, *see* **'BALTIMORE BELLE'**

'BELLE DE BORDEAUX', T, pb, 1861; (Gloire de Bordeau, Gloire de Bordeaux); flowers pink with crimson center, large, dbl.; vigorous growth; Lartay

'BELLE DE CRÉCY', HGal, m; flowers cerise and purple, becoming lavender-gray, center green, dbl., flat, moderate fragrance; lax growth; prior to 1829

Belle de Dom, *see* MASdomo

Belle de Juin, F, 1974; Croix, P.; (Cavriglia)

Belle de Londres, *see* **'COMPASSION'**®

'BELLE DE MARLY', HGal, dp; flowers bright rose, shaded violet, large, dbl.

'BELLE DE PARME', F, mp, 1962; flowers lilac-mauve; low, spreading growth; [(Lafayette X (Gruss an Teplitz X Independence)) X R. rugosa rubra]; Arles; Roses-France

'BELLE DE PROVINS', HT, dr, 1954; flowers velvety dark red, well formed, large, dbl., intense fragrance; vigorous growth; [Crimson Glory X E. G. Hill]; Robichon

Belle de Remalard, HMult, pb, 1998; d'Andlau

'BELLE DE SÉGUR', A, lp; (Joséphine Beauharnais, Joséphine de Beauharnais); flowers soft rosy flesh, edges blush, dbl., cupped; foliage dark; vigorous, upright growth; Vibert, prior to 1848

'BELLE DE YÈBLES', HGal, mr; flowers bright red; Desprez, before 1835

'BELLE DES JARDINS', HGal, m, 1872; flowers purplish violet-red, variegated carmine, striped white, dbl.; vigorous growth; [Village Maid X Seedling]; Guillot et Fils

Belle des Jardins, *see* **'VILLAGE MAID'**

'BELLE DORIA', HGal, pb; flowers pink with red flecks and stripes, cupped; Parmentier, before 1847

'BELLE D'ORLÉANS', LCl, or, 1958; flowers large blooms in clusters, dbl., recurrent bloom; foliage glossy; vigorous growth; [Seedling X Independence]; Robichon

Belle Epoque, *see* **'BEL ANGE'**

Belle Epoque, *see* FRYyaboo

'BELLE ÉPOQUE', HT, pb, 1962; (Royale); flowers fuchsia-pink, reverse creamy, dbl., 40 petals, 6 in., moderate fragrance; foliage glossy; vigorous, upright growth; [Peace X Independence]; Kriloff, Michel; Cramphorn's Nursery

'BELLE ÉTOILE', Gr, my, 1961; bud long, pointed; flowers golden yellow, well formed, blooms in clusters of 5–7, dbl., 25 petals; long stems; vigorous, upright growth; [Joanna Hill X Tawny Gold]; Lens

'BELLE HÉLÈNE', HGal, m; Vibert, before 1815

'BELLE HERMINIE', HGal, m; Coquerel (?), before 1838

'BELLE ISIS', HGal, lp, 1845; flowers pale flesh-pink, dbl.; Parmentier

'BELLE IVRYENNE', HP, dp, 1891; flowers large, dbl.; Lévêque; (Sangerhausen)

'BELLE LILETTE', HT, mr, 1926; flowers carmine-red, very large, dbl., moderate fragrance; Gemen & Bourg; (Sangerhausen)

Belle Loire, S; Croix, P.; (Cavriglia)

'BELLE LYONNAISE', Cl T, ly, 1870; flowers canary yellow, fading white, large, dbl., moderate fragrance; vigorous, climbing growth; [Gloire de Dijon seedling]; Levet, F.

Belle Meillandina®, *see* 'MEIDANEGO'

'BELLE NANON', B, dp, 1872; Lartay

'BELLE NORMANDE', HP, w; flowers silvery rose; [La Reine sport]; Oger, 1864

Belle of Berlin, *see* TANireb

Belle of Portugal, *see* **'BELLE PORTUGAISE'**

'BELLE OF PUNJAB', F, mp, 1965; bud ovoid; flowers pink, dbl., exhibition form, slight fragrance; foliage dark, glossy; very vigorous, compact growth; [Montezuma X Flamenco]; Pal, Dr B.P.; Indian Agric. Research Inst.

'BELLE OF TASMANIA', F, or, 1968; bud pointed; flowers velvety scarlet, center darker, medium, dbl., exhibition form; foliage glossy; vigorous, tall, compact growth; [Korona X Étoile de Hollande]; Holloway

'BELLE POITEVINE', HRg, mp, 1894; bud long, pointed; flowers rose-pink to magenta-pink, large, semi-dbl., recurrent bloom; foliage dark, rugose; vigorous (3 1/2–4 ft), bushy growth; (14); Bruant

'BELLE PORTUGAISE', LCl, lp, 1903; (Belle of Portugal); bud very long, pointed (to 4 in.); flowers light flesh-pink, semi-dbl., 4–6 in., long spring bloom; foliage glossy; very vigorous (20 ft) growth; not hardy north; (21); [R. gigantea X Reine Marie Henriette]; Cayeux, H.

'BELLE ROSINE', HGal, dp; flowers deep pink, edged lighter, open, large, dbl.; erect growth; Vibert, 1829 or 1830

'BELLE ROUGE'™, HT, dr, 1956; flowers dark velvety red becoming carmine-purple, large; [Happiness seedling X Impeccable]; Delbard-Chabert

Belle Rubine, *see* **'VILLAGE MAID'**

'BELLE SANS FLATTERIE', HGal, m, 1820; Hardy, before 1806

Belle Siebrecht, *see* **'MRS W.J. GRANT'**

Belle Siebrecht, Climbing, *see* **'MRS W.J. GRANT, CLIMBING'**

Belle Story®, *see* 'AUSelle'

'BELLE SUISSE', HT, mr, 1936; flowers large, dbl.; Heizmann, E.; (Sangerhausen)

'BELLE SULTANE', m, 1800; flowers purple-red, medium, dbl.; (Sangerhausen)

Belle Sunblaze, *see* 'MEIDANEGO'

Belle Symphonie, *see* MEIrivoui

'BELLE VICHYSOISE', N, lp, 1897; flowers pink or pinkish white, small blooms in clusters of 20–50; very vigorous growth; Lévêque; Lévêque

Belle Villageoise, *see* **'VILLAGE MAID'**

Belle Yvrienne, HP, rb

Belles and Beaus, *see* 'ADABEL'

Bellevue®, *see* 'POULENA'

'BELLINA', F, op, 1959; flowers shrimp-pink, blooms in large clusters, dbl., 40 petals, 2–2.5 in., exhibition form, moderate fragrance; foliage glossy; low, compact growth; [Pinocchio X (Fashion X Orange Triumph)]; Von Abrams; Peterson & Dering

Bellisima, *see* 'SUNLAMPO'

Bellissima®, *see* 'ZIPBELL'

Bellissima®, *see* 'LAPBAU'

Bellona, *see* 'KORILONA'

Bellringer, *see* **'BELL RINGER'**

Belmont, HCh, lp

'BELOVED', HT, w, 1978; bud tinted pink; flowers reflexing, dbl., 33–35 petals, 6–7 in., moderate fragrance; foliage glossy; [Memoriam sport]; Hill, E. H.; Shropshire Roses

'BELVÉDÈRE', F, dr, 1928; flowers velvety dark red, large, dbl.; vigorous, dwarf growth; [Eblouissant X Château de Clos Vougeot]; Kiese

'BELVÉDÈRE', HT, dr, 1955; [Reine Elisabeth X Christopher Stone]; Delforge

Belvedere Park, *see* 'LECBELPA'

'BEN ARTHUR DAVIS', HT, yb, 1935; flowers yellow, reverse pinkish gold, dbl., cupped;

foliage leathery, glossy, dark; bushy growth; [Edith Nellie Perkins sport]; Bostick

'BEN CANT', HP, dr; flowers crimson, center darker, very large, dbl., 25 petals, exhibition form, intense fragrance; vigorous growth; [Suzanne-Marie Rodocanachi X Victor Hugo]; Cant, B. R., 1901

'BEN STAD', LCl, pb, 1925; (Rev. Floris Ferwerda); flowers pink, center yellow, edged white, reverse flesh-pink; [Silver Moon X Mme Jules Grolez]; Undritz; B&A

'BENALAV', Min, m, 1987; (**Lavender Jade**(TM)); flowers lavender-white bicolor-mauve blend, large, borne usually singly, dbl., 32–35 petals, exhibition form, intense, damask fragrance; fruit not observed; short, straight prickles; foliage medium, dark green, semi-glossy; upright, tall growth for a miniature.; [Rise 'n' Shine X Laguna]; Benardella, Frank A.; Nor'East Min. Roses, 1987

'BENBLACK', Min, dr, 1985; (**Black Jade**(TM)); bud near black; flowers deep red, medium, dbl., 35 petals, 1–1.5 in., exhibition form, borne singly, no fragrance; foliage medium, dark, semi-glossy; upright growth; PP005925; AOE, 1985; [Sheri Anne X Laguna]; Benardella, Frank A.; Nor'East Min. Roses

'BENBURGUN', Min, dr, 1990; (**Ans**); Benardella, Frank A.

BENcamelia, HT, pb, 1993; (**Roedean**); Benardella, Frank A.

BENcreberg, HT, w, 1996; (**Athene**); Benardella, Frank A.

'BENDAY', Min, mr, 1988; (**Old Glory**(TM)); flowers bright post office red, aging blood-red to crimson, large, dbl., 23–25 petals, 1.5–2 in., exhibition form, borne singly and in small clusters, no fragrance; prickles long, thin, curved downward, gray-red; foliage medium, medium green, semi-glossy; mini-flora upright, tall, vigorous growth; PP005658; AOE, 1988; [Rise 'n' Shine X Harmonie]; Benardella, Frank A.; Nor'East Min. Roses, 1988

'BENDIGOLD', F, or, 1979; bud globular; dbl., 3 in., moderate fragrance; foliage glossy, bronze; vigorous, upright growth; [Rumba X Redgold]; Murley, J.J.; Brundrett

'BÉNÉDICTE SEGUIN', HT, ob, 1918; flowers ochre, shaded coppery orange, dbl.; Pernet-Ducher

'BENEDICTUS XV', HT, w, 1917; flowers rosy white, dbl., moderate fragrance; [Jonkheer J.L. Mock X Marquise de Sinéty]; Leenders, M.

'BENEDIKT ROEZL', HRg, lp, 1925; flowers light carmine-rose, large, very dbl., moderate fragrance; foliage rugosa-like; vigorous, very bushy growth; [R. rugosa seedling X La France]; Berger, V.; Faist

'BENELUX', F, dr, 1949; flowers crimson-red, semi-dbl., moderate fragrance; [Donald Prior X Rosamunde]; Leenders, M.

Benelux Star, HT, 1997; 37 petals, 4 in., cupped, moderate fragrance; foliage matt; strong grower growth; RvS-Melle; (Cavriglia)

Benevolence, *see* 'SANOLENCE'

'BENFIG', Min, w, 1991; (**Figurine**(TM)); bud delicate coloration, well-formed; flowers ivory white tinged pink, moderately full, semi-dbl., slight fragrance; few prickles; foliage medium, dark green, matt; very long stems suitable for cutting; medium (40–50 cms), upright, bushy growth; AOE, 1992; [Rise 'n' Shine X Laguna]; Benardella, Frank A.; Weeks Roses, 1992

'BENGAL CENTIFOLIA', HCh, 1804; Noisette

Bengal Cramoisi Double, *see* **'SANGUINEA'**

Bengal Ordinaire, *see* **'PALLIDA'**

Bengal Rose, *see* **R. CHINENSIS**

'BENGALE ANIMÉE', HCh, 1832; Laffay, M.; (Cavriglia)

Bengale Cerise, HCh, mr

'BENGALE D'AUTOMNE', HCh, dp, 1825; Laffay, M.

'BENGALE ROUGE', HCh, mr, 1955; bud ovoid; flowers bright carmine-red, open, very large, recurrent bloom; foliage abundant; very vigorous growth; [Gruss an Teplitz X Seedling]; Gaujard

Bengali(R), *see* 'KORAL'

'BENGEE', Min, ly, 1987; (**Gee Gee**(TM)); flowers medium yellow, fading lighter, loose, small, dbl., 20–25 petals, cupped, borne usually singly, slight, fruity fragrance; no fruit; pointed, beige prickles; foliage medium, light green, matt, edges toothed; upright, bushy, medium growth; PP006783; [Rise 'n' Shine X Patricia]; Benardella, Frank A., 1981; Kimbrew Walter Roses, 1987

'BENGT M. SCHALIN', S, dp, 1956; flowers rose-red, blooms in clusters (up to 10), semi-dbl., non-recurrent, slight fragrance; foliage light green, leathery, glossy; very vigorous growth; [R. kordesii X Eos]; Kordes

'BENHILE', Min, rb, 1999; (**Hilde**); flowers white with red washing, reverse ivory, dbl., 17–25 petals, 1.5–2 in., borne mostly singly, moderate fragrance; few prickles; foliage medium, dark green, glossy; bushy, medium (12–18 in.) growth; [Figurine X Kristin]; Benardella, Frank A.

'BEN-HUR', Gr, mr, 1960; bud long, pointed; flowers crimson, dbl., 23 petals, 4–5 in., exhibition form, moderate fragrance; foliage leathery, glossy; vigorous growth; [Charlotte Armstrong X (Charlotte Armstrong X Floradora)]; Lammerts, Dr Walter; Germain's

Benihime, Min, mr

BENimbro, Min, yb, 1990; (**Imbroglio**); Benardella, Frank A.

Benita, *see* 'DICQUARREL'

'BENJAMIN DROUET', HP, 1878; Verdier, E.; (Cavriglia)

'BENJAMIN FRANKLIN', HT, lp, 1969; bud ovoid; flowers dawn-pink, large, dbl., 55 petals, exhibition form, slight fragrance; foliage dark, leathery; upright growth; Von Abrams

'BENJEN', Min, pb, 1985; (**Jennifer**(TM)); flowers light pink, white reverse, small, dbl., 35 petals, exhibition form, intense fragrance; foliage medium, dark, semi-glossy; bushy, spreading growth; PP005857; AOE, 1985; [Party Girl X Laguna]; Benardella, Frank A.; Nor'East Min. Roses

'BENJIM', Min, rb, 1988; (**Jim Dandy**(TM)); bud pointed; flowers medium red, reverse yellow flushed red, aging lighter, medium, exhibition form, slight, spicy fragrance; no fruit; no prickles; foliage medium, medium green, semi-glossy; upright, bushy, medium growth; PP007166; AOE, 1989; [Rise 'n' Shine X Marina]; Benardella, Frank A.; Nor'East Min. Roses, 1989

Benkey, HT, 1969; Itami, B.; (Cavriglia)

BENlavscent, Min, m, 1997; (**Moon River**); Benardella, Frank A.

'BENMAGIC', Min, rb, 1992; (**Kristin**(TM)); flowers white/red bicolor, do not open beyond 1/2 open stage, dbl., 27–30 petals, 1.5 in., exhibition form, borne singly, no fragrance; foliage large, dark green, semi-glossy; upright, bushy, medium growth; PP8603; AOE, 1993; [DICmickey X Tinseltown]; Benardella, Frank A.; Nor'East Min. Roses, 1993

BENmfig, Min, w, 1996; (**Jilly Jewel**); Benardella, Frank A.

'BENNETT'S SEEDLING', Ayr, w, 1840; (Thoresbyana); flowers pure white, expanded form, medium, dbl., moderate fragrance; Lord Manners at Thoresby, England

'BENOIST PERNIN', HP, 1889; Kiard; (Cavriglia)

'BENOIT FRIART', HT, m, 1978; flowers rosy lilac, medium blooms borne 1–7 per cluster, dbl., 47 petals, flat, intense fragrance; foliage matt; upright growth; [Fragrant Cloud X Astée]; Rijksstation Voor Sierplantenteelt

Benoni '75, *see* 'POULONI'

BENorchid, Min, m, 1995; (**Orchid Lace**); Benardella, Frank A.

BENraar, HT, rb; (**Zebra**); Benardella, Frank A., 1995

'BENRAD', Min, or, 1987; (**Radiant**(TM)); flowers brilliant orange-red, urn-shaped, medium, dbl., 23–27 petals, exhibition form, borne singly, moderate, spicy fragrance; no fruit; prickles long, straight, pointed slightly downward, gray-red; foliage dark green, semi-glossy; mini-flora long, straight stems, upright, tall growth; PP006569; [Sheri Anne X Sheri Anne]; Benardella, Frank A.; Nor'East Min. Roses, 1988

'BENROS', Min, pb, 1987; (**Rosie**™); flowers cream with pink edges, excellent form, medium, borne in sprays, dbl., 30–33 petals, exhibition form, slight fragrance; fruit not observed; long, thin, straight prickles; foliage medium, medium green, exceptional substance; upright, bushy, medium growth; [Rise 'n' Shine X (Sheri Anne X Laguna)]; Benardella, Frank A.; Nor'East Min. Roses, 1987

Ben's Gold, HT, dy, 1996; flowers star-shaped, semi-dbl.; Williams, J. Benjamin

Benson & Hedges Gold®, *see* 'MACGEM'

Benson & Hedges Special, *see* 'MACSHANA'

'BENSTAR', Min, pb, 1995; (Astra, **Soroptimist International**); flowers shrimp pink and ivory, opening to star shape, very dbl., 1.5 in., slight fragrance; no prickles; foliage large, dark green, glossy; tall (24in), upright, bushy growth; [Party Girl X Rosie]; Benardella, Frank A.; Bell Roses, Ltd., 1994

'BENTALL'S SCARLET', HT, or, 1935; flowers bright scarlet, slight fragrance; very vigorous growth; Bentall

'BENTEM', Min, dr, 1989; (**Rosetime**); flowers medium red, brushed with dark red, upright, medium, dbl., 26–40 petals, slight fragrance; foliage medium, medium green, matt; upright growth; [Rise 'n' Shine X Black Jade]; Benardella, Frank A.; Kimbrew Walter Roses, 1989

BENtintot, Min, lp; (**Tiny Tot**); Benardella, Frank A., 1990

'BENTVELD', HT, op, 1932; flowers carmine-orange, edges lighter than parent; vigorous growth; [Charles P. Kilham sport]; Posthuma; Low

Benvenuto®, *see* 'MEIELPA'

'BÉRANGÈRE', M, lp, 1849; flowers delicate pink, large, dbl.; Vibert

'BERbow'™, Min, or, 1997; (**Oxbow**); flowers double, medium, dbl., 15–25 petals, borne mostly singly, no fragrance; foliage medium, medium green, dull; compact, medium (12–15 in.) growth; [Luis Desamero X Rainbow's End]; Berg, David H.

'BERBUT', Min, w, 1994; (**Butter 'n' Sugar**); flowers vary from white to yellow shading to white, very dbl., 1.5–2.75 in., borne mostly singly, slight fragrance; few prickles; foliage medium, medium green, semi-glossy; low (15–18 in.), spreading growth; [Klima X (Intrigue X Poker Chip)]; Berg, David H.

'BERCEUSE', HT, my, 1950; bud globular, yellow; flowers chamois, very large, very dbl., intense fragrance; foliage leathery; very vigorous, upright growth; [Signora X Mrs Pierre S. duPont]; Robichon

Berenice, F, 1979; Croix, P.; (Cavriglia)

Berenice, *see* BARenic

'BÉRÉNICE', HGal, rb, 1818; flowers rose and crimson, shaded with slate, large, dbl., globular; pendulous growth habit; Vibert

Berenice Neville, *see* 'KIRBELL'

'BERGERS ERFOLG', HRg, mr, 1925; flowers fire-red, stamens yellow, blooms in clusters, single, 3–4 in., occasionally recurrent bloom; foliage dark; very vigorous, bushy growth; [R. rugosa seedling X Richmond]; Berger, V.; Pfitzer

Bergers Koralle, F, op, 1956; flowers dark salmon-pink, large, dbl.; Berger, W.; (Sangerhausen)

Bergers Morgenröte, S, lp, 1959; flowers creamy pink, medium, dbl.; Berger, W.; (Sangerhausen)

Bergers Roma, HT, rb, 1965; flowers large, dbl.; Berger, W.; (Sangerhausen)

'BERGESLOH', Pol, mr, 1929; flowers carmine-red, small, dbl.; Vogel, M.; (Sangerhausen)

Bergfeuer, F, 1959; Leenders, J.; (Cavriglia)

Bergfeuer Superior, F, 1968; Leenders, J.; (Cavriglia)

'BERGME', F, mr, 1977; (**Gabriella**®, Gabrielle); bud ovoid; dbl., 33 petals, 3 in., cupped, slight fragrance; foliage glossy; vigorous, bushy growth; PP004452; Berggren; W. Kordes Sons

'BERGRAT OTTO BERGER', HT, w, 1924; flowers creamy white to sulfur, center deeper, dbl., moderate fragrance; [Pharisaer X Prince de Bulgarie]; Berger, V.; Faist

Bering Renaissance, *see* POUlberin

Berkeley, *see* 'JACIENT'

Berkeley Beauty™, *see* 'MORBERK'

Berleburg, *see* POUlbella

'BERLEI', Min, ob, 1995; (**Leila**); flowers light orange with yellow reverse, blooms borne mostly single, dbl., 26–40 petals, 1.5 in., slight fragrance; some prickles; foliage medium, medium green, semi-glossy; medium (15 in.), upright growth; [Arizona Sunset X June Laver]; Berg, David H.

Berlengas, HT; Moreira da Silva, A.; (Cavriglia)

'BERLIN', S, ob, 1949; bud long, pointed; flowers orange-scarlet, center golden, large, single, borne in large clusters, repeat bloom, moderate fragrance; prickles large; foliage leathery, dark; very vigorous, upright growth; (28); [Eva X Peace]; Kordes

Berlin Beauty, HT, lp

Berliner Luft®, F, ob, 1985; (Jura); Hauser

'BERMUDA PINK', HT, lp, 1974; bud ovoid; flowers flesh-pink, dbl., 35–40 petals, 4–5 in., globular, slight, rose fragrance; foliage glossy, light; compact, moderate growth; [Queen of Bermuda X Montezuma]; Golik; Dynarose

Bermuda's Catherine Mermet, T, w

Bermuda's Emmie Gray, Ch, rb

Bermuda's Kathleen, S, mr

'BERMUDIANA', HT, mp, 1966; bud ovoid; dbl., 35–60 petals, 5–6 in., exhibition form, moderate fragrance; foliage leathery; vigorous, upright growth; [Golden Masterpiece seedling X Golden Masterpiece seedling]; Boerner; J&P

'BERN'®, HT, dp, 1975; bud long, pointed; flowers deep pink, shallow, dbl., 24 petals, 4 in., intense, spicy fragrance; foliage dark, leathery; spreading growth; [Crimson Glory X Lilac Charm]; Huber

'BERNADETTE', HT, lp, 1964; flowers large, dbl., exhibition form, slight fragrance; foliage dark, leathery; vigorous, bushy growth; [Peace sport]; Kelly

Bernadette, HT, w, 1976; Dorieux

Bernadette Chirac®, *see* 'DELBÉCHIR'

'BERNAIX, CLIMBING', Cl HT, dr, 1935; [Souv. d'Alexandre Bernaix sport]; Shamburger, C. S.

Bernalene, *see* KORcountry

'BERNALIA', S, mp, 1961; flowers pink, center white, small blooms in clusters, semi-dbl., recurrent bloom, moderate fragrance; foliage bright green; vigorous growth; [Mosqueta X Cecilia]; Bernal

'BERNARD', P, op; (Pompon Perpetual); flowers salmon-pink; [Rose du Roi sport]; (France), 1846

'BERNARD VERLOT', HP, rb; flowers red-scarlet, center violet; Baron-Veillard, 1874

'BERND CLÜVER', F, mp, 1974; bud pointed; 2.5–3 in., moderate, fruity fragrance; upright growth; [Nordia X Sans Souci]; Reinold

Bernensis, S, mr, 1994; Meilland

Bernhard Daneke Rose, *see* 'TANWEIEKE'

'BERNICE', HT, mp, 1927; flowers carmine-pink on yellow base, semi-dbl., moderate fragrance; Pemberton

'BERNICE', Pol, mp, 1937; bud ovoid; flowers brilliant cerise-pink, small blooms in clusters, dbl., globular, recurrent bloom, intense fragrance; foliage glossy, light; dwarf growth; [Baby Tausendschon (probably Echo) X Gloria Mundi]; Nicolas; J&P

Bernina®, F, w, 1979; deRuiter

'BERnora'™, Min, yb, 1997 (**Nora**); flowers double, medium, dbl., 15–25 petals, borne mostly singly, slight fragrance; foliage medium, medium green, dull; upright, tall (15–18 in.) growth; [Rainbow's End X Leila]; Berg, David H.

Bernstein, F, dy, 1987; flowers golden yellow with copper tints, large, very dbl.; Tantau, Math.; (Sangerhausen)

Bernstein, Pol, dy; flowers dark yellow and copper, medium, dbl.; VEG; (Sangerhausen)

Bernstein-Rose®, *see* TANeitber

'**Berolina**', F, my, 1977; bud long, pointed; dbl., 24 petals, 3.5 in., exhibition form, slight fragrance; foliage dark, soft; vigorous, upright growth; [Mabella X Seedling]; Kordes

Berolina, *see* 'KORpriwa'

'BERpar', Min, w, 1988; (**Wintonbury Parish**); flowers white with top third of petal light red, medium, borne usually singly, semi-dbl., 14 petals, exhibition form, slight fragrance; prickles curved, large, light greenish-white; foliage medium, medium green, semi-glossy; upright, medium growth; PP006843; [Poker Chip X Lady X]; Berg, David H.

Berries 'n' Cream™, *see* 'POUlclimb'

Berry Berry Grape, Min, m, 1997; King

Berry Berry Red, *see* 'TALber'

'**Bersagliera**', HT, dp, 1960; bud globular; flowers crimson, phlox-pink and fuchsia-pink, medium to large blooms, dbl.; strong stems; vigorous growth; Luigi

Bert Hinkler, Pol, lp; Harrison, A., 1928; (Weatherly, L.)

'**Bertha**', S, lp, 1946; delicate pink, very large, borne like hollyhock flowers, single, non-recurrent; erect stems; height 8 ft; [(R. rugosa X Hybrid Perpetual) X (R. multiflora X R. blanda)]; Wright, Percy H.

'**Bertha Aikman**', HT, op, 1977; bud high pointed; flowers two-toned salmon-pink, dbl., 45 petals, 4 in., moderate fragrance; foliage matt; moderate, slightly spreading growth; [Gypsy Moth X Percy Thrower]; Simpson, J.W.; Manawatu Rose Soc.

'**Bertha Gorst**', HT, mr, 1933; flowers crimson-cerise, base gold, veined bronze, very large, dbl., intense fragrance; foliage bronze; [Autumn sport]; Beckwith

'**Bertha Kiese**', HT, my, 1913; flowers medium, semi-dbl., slight fragrance; Jacobs; (Sangerhausen)

'**Bertha Turner**', HT, op, 1925; flowers salmon-peach, moderate fragrance; Pemberton

'**Bertha von Suttner**', HT, ly, 1918; flowers light yellow with copper tints, medium, dbl.; Verschuren; (Sangerhausen)

'**Berthe Baron**', HP, lp, 1869; flowers delicate rose shaded with white, large, dbl.; [Jules Margottin X Seedling]; Baron-Veillard

'**Berthe Dumesnil de Montchauvau**', HP, lp, 1876; flowers medium, dbl.; Jamain, H.; (Sangerhausen)

'**Berthe Lévêque**', HP, 1866; Cochet, A.; (Cavriglia)

'**Berthe Mallerin**', HT, or, 1960; flowers red tinted orange, dbl.; strong stems; vigorous growth; Mallerin, C.

'**Berti Gimpel**', HP, mp, 1913; flowers large, semi-dbl.; Altmüller; (Sangerhausen)

Bertram, *see* '**Sneezy**'

'**Bertram Park**', HT, mr, 1929; (Coquette); flowers rosy crimson, base yellow, single, slight fragrance; [Eblouissant X Mme Edouard Herriot]; Burbage Nursery

'**Berwick**', HSpn, pb; flowers rose shading to white at edges, large, semi-dbl.; dwarf growth

Beryl, HT, ly

'**Beryl Ainger**', HT, w, 1955; flowers cream, base golden yellow, moderate fragrance; [The Doctor sport]; F. Cant

Beryl Bach, *see* 'HARtesia'

'**Beryl Formby**', HT, yb, 1948; flowers golden yellow shaded crimson, dbl., 36–40 petals, intense fragrance; foliage glossy; bushy growth; [McGredy's Sunset sport]; Fryers Nursery, Ltd.

'**Beryl Formby, Climbing**', Cl HT, yb, 1956; Letts

'**Beryl Wearmouth**', F, mp, 1973; semi-dbl., 19 petals, 4.5 in.; foliage light green, matt; [(Ann Elizabeth X Orange Sensation) X Sea Pearl]; Harkness

Besançon, HT, 1968; Sauvageot; (Cavriglia)

'**Bess Lovett**', LCl, dp, 1915; flowers light red, well-formed blooms in clusters, dbl., cupped, moderate fragrance; foliage glossy, dark; vigorous, climbing growth; Van Fleet; J.T. Lovett

'**Bessie Brown**', HT, ly, 1899; flowers yellowish white, large, very dbl., moderate fragrance; foliage light, leathery, glossy; Dickson, A.

'**Bessie Chaplin**', HT, mp, 1921; flowers bright pink, center deeper, very dbl., slight fragrance; GM, NRS, 1923; [Lady Pirrie X Gorgeous]; Chaplin Bros.

'**Bessie Johnson**', HP, 1873; Curtis; (Cavriglia)

Bessie Lee, Pol; (Cavriglia)

Bessy, S, ob, 1998; Interplant

'**Best Regards**', HT, pb, 1944; flowers pink bicolor, dahlia form, dbl., 50–60 petals, 6–7 in., moderate fragrance; foliage leathery, dark; very vigorous, compact growth; [Soeur Thérèse X Signora]; Morris; Germain's

'**Best Regards, Climbing**', Cl HT, pb, 1940; Elmer, C. A.; Germain's

'**Best Wishes**', HT, dp, 1959; bud long, pointed; flowers currant-red, dbl., 25–35 petals, 5.5–6 in., exhibition form, moderate fragrance; foliage leathery; strong stems; very vigorous growth; Fisher, G.; Arnold-Fisher Co.

Best Wishes, *see* CHESnut

'BETahat', HT, rb, 1987; (**Malahat**); flowers scarlet, reverse white, fading slightly, medium, borne usually singly, dbl., 40 petals, exhibition form, intense fragrance; round, medium, brown fruit; foliage medium, red-brown to dark green, glossy; bushy, vigorous growth; [Pristine X Shocking Blue]; Betts, John; Wisbech Plant Co., 1988

'**Betano Beach**', F, rb, 1966; bud long, pointed; flowers light salmon-pink, reverse scarlet and burnt crimson, medium, dbl., exhibition form; foliage glossy, dark, leathery; vigorous, upright growth; [Ma Perkins X Detroiter]; Fankhauser; A. Ross & Son

'**Beth**', F, ly, 1966; flowers buff-yellow; [Elizabeth of Glamis sport]; O'Connell; Rumsey

Beth, HSpn, lp, 1973; Mertens

Bethany Helena, HT, pb, 1997; flowers full, large, very dbl., 26–40 petals, borne mostly singly, moderate fragrance; foliage large, dark green, semi-glossy; upright, medium growth; [Tom Foster X Gavotte]; Poole, Lionel

'**Betinho**', HT, rb, 1958; flowers velvety red and brown; [Charles Mallerin X Monte Carlo]; da Silva, Moreira

'BETsat', HT, dr, 1987; (**Black Satin**); dbl., 30 petals, exhibition form, borne singly, slight fragrance; rounded, medium, red fruit; prickles few, pointed slightly down, small, light brown; foliage medium, dark green, semi-glossy; upright, medium growth; [Folklore X Loving Memory]; Betts, John; Wisbech Plant Co., 1987

'**Betsie Jane**', HT, ab, 1976; bud long, pointed; flowers soft apricot-pink, full, large, dbl., moderate fragrance; foliage light green; tall, vigorous growth; [Bewitched sport]; Tresise

'BETsue', F, or, 1987; (**Sue Betts**); flowers long lasting color, large blooms borne in sprays of 65–120, dbl., cupped, moderate fragrance; round, red fruit; prickles sharp, thin, brown; foliage medium, medium green, semi-glossy; bushy, low growth; [Europeana sport]; Betts, John; Wisbech Plant Co., 1987

'**Betsy McCall**', F, op, 1956; bud ovoid; flowers shrimp-pink, open, blooms in large clusters, dbl., 25–30 petals, 3–3.5 in., moderate fragrance; foliage glossy; vigorous, bushy growth; [Seedling X Fashion]; Boerner; J&P

Betsy Murchison, *see* DESbet

Betsy Ross, *see* 'DELup'

'**Betsy Ross**', HT, ob, 1931; flowers like parent but marked russet-orange; [Talisman sport]; Samtmann Bros.

Betsy Taaffe, S, lp; [Abraham Darby sport]; Taaffe, 1996; (Weatherly, L.)

'BETSY VAN NES', Pol, mr, 1914; flowers small, semi-dbl.; van Ryn; (Sangerhausen)

'BETTE IRENE', S, ab, 1986; flowers apricot, reverse deeper, fading light pink, imbricated, medium, semi-dbl., 9 petals, repeat bloom, moderate fragrance; rounded, medium, ornamental, bright red fruit; awl-like, medium, brown prickles; foliage medium, dark green, matt; upright, medium-tall growth; [Dairy Maid X Seedling]; Schneider, Peter

'BETTER HOMES & GARDENS', HT, pb, 1976; bud ovoid; flowers rose, ivory reverse, dbl., 38 petals, 3–3.5 in., exhibition form, slight fragrance; foliage glossy, dark; medium-tall, upright growth; [Tropicana X Peace]; Warriner, William A.; J&P

Better Homes & Gardens Diamond Jubilee™, *see* 'WINBILEE'

'BETTER TIMES', HT, mr, 1934; flowers cerise, large, dbl., exhibition form, slight fragrance; foliage dark, leathery; very vigorous, compact growth; [Briarcliff sport]; Hill, Joseph H., Co.

'BETTER TIMES, CLIMBING', Cl HT, mr, 1937; Parmentier, J.

Bettie Herholdt, *see* 'HERTIE'

Bettina®, *see* MEpal

Bettina '78, *see* 'MEIBRICO'

Bettina, Climbing®, *see* 'MEPALSAR'

'BETT'S CARDINAL SPIRIT', F, dr, 1999; flowers cardinal red, dbl., 15–25 petals, 4.5 in., borne in small clusters, slight fragrance; few prickles; foliage medium, medium green, semi-glossy; upright, medium (3 ft) growth; [High Spirit X (Moody Blues X Melina)]; Walters, Betty & Richard

'BETT'S LITTLE GEM', Min, mp, 1996; flowers medium pink with white ring around stamens, open, cup-shaped, dbl., 26–40 petals, slight fragrance; few prickles; foliage medium, light green, semi-glossy; upright, medium growth; [Pink Petticoat X Unknown]; Walters, Betty & Richard

Bett's Little Rhapsody, *see* 'WALTRAP'

'BETT'S PINK LACE', Min, pb, 1996; flowers white with pink edge, changing to salmon pink, blooms borne, dbl., 26–40 petals, intense fragrance; few prickles; foliage medium, dark green, glossy; compact, low growth; [Pink Petticoat X Seedling]; Walters, Betty & Richard

'BETT'S SNOW DANCER', F, w, 1999; bud hint of green, opening to white; flowers outer edge white with cream center, numerous golden stamens, dbl., 26–40 petals, 4.5 in., borne in large clusters, slight fragrance; some prickles; foliage medium, medium green, semi-glossy; compact, medium (4 ft) growth; [Mt Hood X Aorangi]; Walters, Betty & Richard

'BETT'S WHITE DELIGHT', Min, w, 1992; flowers creamy white opening to pure white, rosette, very dbl., 58–86 petals, 2.5 in., borne in clusters of 5, slight fragrance; foliage medium to dark green, matt; tall, compact growth; [Pink Petticoat X Seedling]; Walters, Betty & Richard, 1987

'BETTY', HT, pb, 1905; flowers coppery rose, shaded yellow, large, dbl., moderate fragrance; very vigorous growth; Dickson, A.

'BETTY ALDEN', Cl Pol, lp, 1919; flowers appleblossom-pink passing to white, single; [Orléans Rose X (Katharina Zeimet X R. arvensis hybrid)]; Farquhar

'BETTY BAUM', HT, lp, 1927; flowers delicate pink, base yellow, dbl., moderate fragrance; [Premier sport]; Baum

Betty Bee™, *see* 'BLABEE'

'BETTY BERKELEY', T, mr, 1904; flowers bright red; Bernaix, A.

'BETTY BLAND', S, dp, 1925; flowers deep rose, fading pink, center deeper, dbl., non-recurrent bloom, moderate fragrance; foliage rich green, soft; twigs ruby-red; vigorous (6 ft), bushy growth; very hardy; (14); [R. blanda X HP]; Skinner

'BETTY BLOSSOM', HWich, mp, 1923; flowers clear rose-pink, loose cluster, semi-dbl.; vigorous pillar or bush growth; [R. wichurana X Mrs W.J. Grant]; Dawson; Eastern Nursery

Betty Boop™, *see* 'WEKPLAPIC'

'BETTY CUTHBERT', HT, or, 1964; bud long, pointed; flowers medium, very dbl., intense fragrance; foliage soft, glossy; very vigorous, upright growth; [Roundelay sport]; Palmer; Palmer & Engall

Betty Driver, *see* 'GANDRI'

'BETTY FREE', F, mp, 1950; bud pointed; flowers neyron rose, borne in clusters, dbl., 25 petals, slight fragrance; vigorous growth; [Fortschritt sport]; LeGrice

'BETTY GRACE CLARK', HT, ob, 1933; flowers orange-yellow, reverse streaked red, exhibition form, moderate fragrance; vigorous, bushy growth; [Marie Adélaide sport]; Clarke Bros.

Betty Harkness, *see* HARette

Betty Herholdt, *see* 'HERTIE'

'BETTY HULTON', HT, dy, 1923; flowers deep saffron-yellow, dbl., moderate fragrance; GM, NRS, 1923; Dickson, A.

'BETTY MAY WOOD', HT, ab, 1968; flowers apricot to buff, reverse coral-salmon, well formed, dbl., 30 petals; foliage light green; free growth; [Mischief sport]; Wood

'BETTY MORSE', HT, mr, 1950; bud long, pointed; dbl., 25 petals, 4 in., slight fragrance; foliage olive-green; vigorous growth; [Crimson Glory X (Crimson Glory X Cathrine Kordes)]; Kordes; Morse

'BETTY NEUSS', HT, mp, 1973; bud small, long, pointed; flowers pure pink, medium, dbl., slight fragrance; foliage small; very vigorous, upright growth; Dawson, George; C. Brundrett

'BETTY PAUL', HT, mp, 1988; flowers medium pink blend, medium, borne usually singly or in sprays, dbl., 20–25 petals, exhibition form, slight fragrance; ovoid, small, yellow-orange fruit; prickles straight, medium, light brown; foliage medium, medium green, semi-glossy, disease resistant; upright, medium, tall growth; Bronze, ARC TG, 1987; [Queen Elizabeth X (Tiffany X Tropicana)]; Warner, A.J.

'BETTY PEARSON', HT, w, 1929; flowers cream, center apricot, petals shell shaped, large; Burbage Nursery

'BETTY PRIOR', F, mp, 1935; bud ovoid, dark carmine; flowers carmine-pink, borne in clusters, single, 5 petals, cupped, moderate fragrance; vigorous, bushy growth; GM, NRS, 1933; [Kirsten Poulsen X Seedling]; Prior; J&P, 1938

Betty Prior, Climbing, *see* 'COOPRIOR'

Betty Sheriff, LCl, w; Sheriff

'BETTY STIELOW', HT, dp, 1928; flowers dark pink, almost red at times, dbl., moderate fragrance; [Premier sport]; Stielow Bros.

'BETTY SUTOR', HT, pb, 1929; flowers pale pink, veined rose, reverse rosy, large, dbl., moderate fragrance; foliage light, glossy; vigorous, bushy growth; McGredy

'BETTY UPRICHARD', HT, ab, 1922; flowers delicate salmon-pink, reverse carmine with coppery sheen, large, dbl., 20 petals, intense fragrance; foliage light, leathery, glossy; very vigorous, tall growth; GM, NRS, 1921; Dickson, A.

'BETTY UPRICHARD, CLIMBING', Cl HT, ab, 1936; Krause

'BETTY WILL', S, pb, 1963; flowers bright pink, lighter reverse, dbl., 35 petals; some large prickles; foliage dark, leathery; red canes; tall growth; [George Will X Betty Bland]; Erskine

Betty Wilson, *see* 'SHERISCENT'

Betty Wright™, *see* 'BURBET'

'BETTY, CLIMBING', Cl HT, pb, 1926; Hohman

Betty's Baby, *see* 'JUDBABY'

Betty's Pride®, Min, rb

'BETZEL'S PINK', HT, op, 1955; bud long; flowers coral-pink, blooms, dbl., 55–60 petals, 5 in., exhibition form, moderate fragrance; foliage leathery, glossy; strong stems; vigorous, bushy growth; [Pres. Herbert Hoover sport]; Betzel; Edmunds Roses

'BEVERLEY ANNE', HCh, mr, 1987; flowers red-purple, open, borne in sprays of 3–7, dbl., 21 petals, flat, slight fragrance; no prickles; foliage pivoted, serrated; semi-dwarf growth; [Seedling X Seedling]; Nobbs, Kenneth J., 1986

Beverley Stoop, *see* 'JAYDON'

Beverley Watson, *see* KORdreischi

Beverly Hills, *see* 'DELMATOR'

'BEVERLY JAYNE', S, lp, 1999; flowers center deep pink, outer light pink, very dbl., 41 petals, 1.3 in., borne in large clusters, slight fragrance; few prickles; foliage small, light green, dull; upright, bushy, medium (3 ft) growth; [Angela Rippon X New Dawn]; Jones, L.J.

'BEVERLY NICOLS', HT, op, 1939; flowers cream, reverse salmon, well formed, large, exhibition form, moderate fragrance; vigorous growth; Burbage Nursery

'BEWITCHED', HT, mp, 1967; bud urn shaped; flowers cotton candy pink, large, dbl., 27–30 petals, 5 in., exhibition form, borne singly, moderate, damask fragrance; foliage large, apple green, glossy; new stems and foliage are red; vigorous, medium, rounded growth; AARS, 1967 GM, Portland, 1967; [Queen Elizabeth X Tawny Gold]; Lammerts, Dr Walter; Germain's

Bewitched, *see* POUlbella

Bewitched, Climbing, Cl HT, mp, 1989

Beyreuth, S, rb

'BEZRUC', HT, mr, 1938; flowers large, dbl.; Böhm, J.; (Sangerhausen)

'BHAGMATI', F, mr, 1977; bud ovoid; 15–20 petals, 2.5–3 in., cupped, slight fragrance; foliage glossy, light green; dwarf, vigorous, bushy growth; [Charleston X ((Roman Holiday X Flamenco) X Goldgleam)]; Viraraghavan, M.S.; Gopalsinamiengar

Bhanu, HT, yb, 1990; K&S

Bharami, *see* **'BHARANI'**

'BHARANI', Min, m, 1979; (Bharami); flowers small, semi-dbl., slight fragrance; foliage small, light green, matt; upright growth; [Seedling X Seedling]; Goralaswamiengar, K.S.; Gandy Roses, Ltd., 1973

Bhargav, HT, op, 1993; K&S

Bhavani, HT, op, 1986; Kasturi

'BHIM', HT, dr, 1970; bud long, pointed; flowers scarlet-red, open, large, very dbl., abundant, intermittent bloom, moderate fragrance; vigorous, upright growth; [Charles Mallerin X Delhi Princess]; IARI; Div. of Vegetable Crops & Flori.

'BIANCA', HT, 1913; bud long, pointed; flowers pale peach tinted pink and violet, center shaded rose, dbl.; RULED EXTINCT 3/83 ARM; Paul, W.

'BIANCA', HT, 1927; flowers white, sometimes lightly flushed cream or pink, well-formed, dbl., moderate fragrance; RULED EXTINCT 3/83 ARM; Pemberton

Bianca®, HT, w, 1987; Kuhn

'BIANCA CAMELIA', HT, w, 1933; bud ovoid, pointed; flowers snow-white, center light blush-yellow, very large, dbl., 23–25 petals, intense fragrance; foliage light green; very vigorous, upright growth; [Nuntius Pacelli X Sachsengruss]; San Remo Exp. Sta.

Bianco, *see* 'COCBLANCO'

Bibi Maizoon™, *see* AUSdimindo

Bibiché®, *see* DORflo

Bibiché, Climbing, Cl F, 1980; Dorieux, Francois; (Cavriglia)

'BICENTENNIAL', F, rb, 1975; flowers deep pink and red blend, medium, dbl., exhibition form, slight fragrance; foliage leathery; vigorous, upright, bushy growth; Meyer, C.; C-P

'BICHETTE', HT, mp, 1968; flowers persian rose, large, dbl., 20–25 petals; foliage dark, leathery; vigorous, upright growth; [Diamond Jubilee X Seedling]; Verschuren, A.; Stassen

Bicolette, HT, rb, 1980; Tschanz, E.

'BICOLORE', HP, 1877; Oger; (Cavriglia)

'BICOLORE INCOMPARABLE', HGal, 1861; Touvais; (Cavriglia)

Biddulph Grange, *see* FRYdarkeye

Biddy, Min, w

Bienkie, *see* KORsisten

'BIENVENU', Gr, ob, 1969; bud long, pointed; flowers reddish orange, large, dbl., 70 petals, exhibition form, intense fragrance; foliage leathery, matt; vigorous, upright growth; [Camelot X (Montezuma X War Dance)]; Swim & Weeks; Weeks Wholesale Rose Growers

Bifera Italica, HGal, lp; flowers medium, dbl.; (Sangerhausen)

Big and Beautiful, *see* 'TINBAB'

'BIG APPLE', HT, mr, 1983; flowers large, dbl., moderate fragrance; foliage large, medium green, matt to semi-glossy; upright, spreading growth; [Mister Lincoln X (Suspense X King's Ransom)]; Weeks, O.L.; Weeks Wholesale Rose Growers, 1984

Big Apricot, HT, ab; Wells

'BIG BANG'®, F, or, 1980; bud pointed; flowers deep orange-red, shallow-blooms borne 5–10 per cluster, semi-dbl., 13 petals, cupped, no fragrance; reddish-green prickles; foliage matt, light green; bushy growth; [Sarabande X Sarabande seedling]; Barni-Pistoia, Rose

'BIG BEN', HT, dr, 1964; flowers well-formed, 5–6 in., intense fragrance; foliage dark; tall growth; [Ena Harkness X Charles Mallerin]; Gandy, Douglas L.

Big Cabbage of Holland, *see* **'GROS CHOUX D'HOLLANDE'**

Big Chief, *see* **'PORTLAND TRAILBLAZER'**

Big Daddy®, *see* 'MACCHOME'

'BIG DUKE', HT, dp, 1991; (Duke Wayne); bud pointed; flowers deep pink, silvery pink reverse, large blooms borne usually, dbl., exhibition form, moderate, fruity fragrance; foliage large, medium green, semi-glossy; upright, tall growth; [The Duke X Seedling]; Weddle, Von C., 1984

'BIG JACK CHARLTON', HT, w, 1994; flowers ivory, pale pink edge, blooms borne mostly single, dbl., 26–40 petals, 3–3.5 in., slight fragrance; some prickles; foliage large, dark green, glossy; medium (85 cms), upright growth; [Gavotte X Queen Esther]; Poole, Lionel; Battersby Roses, 1995

Big Jim Larkin, *see* 'SEALARK'

'BIG JOHN', Min, mr, 1979; bud pointed; flowers deep medium red, base yellow, dbl., 42 petals, 1–1.5 in., exhibition form; foliage small, glossy, bronze; upright, bushy growth; [Starburst X Over the Rainbow]; Williams, Ernest D.; Mini-Roses

Big John, *see* **'GALLERIA'**

Big John, LCl

Big Purple, *see* 'STEBIGPU'

'BIG RED', HT, dr, 1967; bud pointed; dbl., 52 petals, 4.5–6 in., exhibition form, slight fragrance; foliage leathery; vigorous, upright, bushy growth; [Chrysler Imperial X Seedling]; Meilland, Mrs. Marie-Louise; C-P

'BIG SPLASH', Cl HT, rb, 1969; bud pointed; flowers flame-red, reverse lighter, base yellow, large, dbl., exhibition form, moderate fragrance; foliage glossy, leathery; vigorous, climbing (8–10 ft) growth; [Buccaneer X Bravo]; Armstrong, D. L.; Armstrong Nursery

'BIGGI', HT, yb, 1975; bud ovoid; dbl., 27 petals, 4.5 in., exhibition form, intermittent bloom, moderate fragrance; foliage glossy, dark; vigorous, upright growth; [Dr. A.J. Verhage X Seedling]; Kordes; Fey

'BIJOU', Pol, lp, 1932; flowers old rose; deRuiter; Sliedrecht & Co.

Bijou, *see* BARjou

Bijou, MinFl, pb, 1991; Lens

'BIJOU DE ROYAT-LES-BAINS', HCh, mp, 1891; flowers medium, dbl.; Veysset; (Sangerhausen)

'BIJOU DES AMATEURS', HGal, dr, 1848

Bijou des Prairies, *see* **'GEM OF THE PRAIRIES'**

Bijou d'Or, *see* TANledolg

'BIJOU SUPERIOR', Pol, lp; flowers have more lasting color; [Bijou sport]; deRuiter

'BIKINI RED', HT, mr, 1974; bud ovoid; flowers rose-red, tinged white, ruffled, dbl., 40 petals, 5 in., moderate, fruity fragrance; foliage glossy, dark; [Queen of Bermuda X Peace]; Golik; Dynarose

Bila Junior Miss, HT, lp; Strnad; (Czech Rosa Club)

'BILFAN', Min, yb, 1986; (**Jack's Fantasy**); flowers yellow blushed with orange-red from edge, reverse medium yellow, dbl., 21 petals, exhibition form, slight fragrance; no fruit; few, beige prickles slightly sloped downward; foliage medium, medium green, semi-glossy; upright, bushy, medium growth; [Little Darling X Over the Rainbow]; Bilson, Jack M., Jr. & Bilson, Jack M. III

'BILICE', Min, pb, 1987; (**Phoebe's Choice**™); flowers pink, edges yellow, reverse same, pink softens, white edges, dbl., exhibition form, no fragrance; round, light orange & green fruit; reddish-green prickles, sloped downwards; foliage medium, medium green, glossy; upright, bushy growth; [Little Darling X Over the Rainbow]; Bilson, Jack M., Jr. & Bilson, Jack M. III

'BILITIS', F, or, 1969; flowers vermilion-red, center yellow; [Tabarin X Golden Slippers]; Gaujard

'BILL BEAUMONT', F, mr, 1982; flowers crimson, dbl., 20 petals, slight fragrance; foliage medium, medium green, matt; bushy growth; [Evelyn Fison X Redgold]; Fryer, Gareth; Fryer's Nursery, Ltd.

Bill Cone, *see* 'MICONE'

'BILL HUNT', HT, op, 1976; flowers deep coral, full, medium, dbl., 30 petals, moderate fragrance; foliage glossy; fairly vigorous growth; [Mischief X Serenade]; Blakemore

Bill Slim, *see* 'HARQUITO'

'BILL TEMPLE', HT, w, 1975; flowers cream, dbl., 30 petals, 5–6 in., slight fragrance; foliage glossy, dark; [Crimson Halo X Piccadilly]; Harkness

Bill Warriner™, *see* 'JACSUR'

'BILLARD ET BARRÉ', Cl T, my, 1898; flowers golden yellow, dbl., intense fragrance; [Mlle Alice Furon X Duchesse d'Auerstadt]; Pernet-Ducher

Billie and Lew, HWich, m, 1995; Nobbs

Billie Teas, *see* 'HOOTEA'

'BILLIONAIRE', HT, mr, 1972; bud ovoid, long, pointed; flowers large, dbl., exhibition form, slight fragrance; foliage leathery; vigorous, upright growth; [Fragrant Cloud X Proud Land]; Warriner, William A.; J&P

Billy, S, rb, 1993; Hauser

'BILLY BOILER', Cl HT, mr, 1927; vigorous, tall growth; Clark, A.; NRS Victoria

'BILLY BOY', HT, my, 1926; flowers sunflower-yellow, semi-dbl.; [Golden Emblem X Christine]; McGredy; Beckwith

Billy Boy, *see* 'MORBOY'

Billy Graham™, *see* 'JACGRAY'

'BILPAT', Min, pb, 1989; (**Naughty Patricia**); flowers medium pink, edges blush, outer petals quill, large, borne u, dbl., 29 petals, exhibition form, slight, fruity fragrance; globular, medium, green with orange-red blotch fruit; prickles slight downward slope, reddish-tan; foliage large, medium green, matt; upright, tall growth; [Rise 'n' Shine X Redgold]; Bilson, Jack M., Jr. & Bilson, Jack M. III

'BIMBORO'®, HT, dr, 1978; bud globular; dbl., 46 petals, 3–3.5 in., exhibition form, moderate fragrance; foliage glossy; vigorous, bushy, upright growth; [Seedling X Kardinal]; Kordes, W. Söhne

'BING CROSBY', HT, ob, 1980; bud ovoid; flowers strong clear orange, very dbl., 40–45 petals, 5 in., cupped, borne mostly singly, slight, light spice fragrance; long prickles hooked downward; foliage medium, heavy, leathery, wrinkled, dark; long stems; vigorous, upright, bushy growth; PP4695; AARS, 1981; [Seedling X First Prize]; Weeks

'BINGO', HT, dr, 1955; (Dyna); bud pointed; flowers cardinal-red, dbl., 55 petals, 4–5 in., exhibition form, intense fragrance; foliage glossy; upright growth; [(Hadley seedling X Ami Quinard) X Crimson Glory]; Robichon; Ilgenfritz Nursery

Bingo Meidiland®, *see* 'MEIPOTAL'

Bingo Meillandecor®, *see* 'MEIPOTAL'

Bingo Queen, *see* 'TINQUEEN'

Bipontina, S, mr

Bird of Fire, *see* 'TAYBIRD'

'BIRDIE BLYE', S, mp, 1904; bud long, pointed, carmine; flowers rose-pink, large blooms in clusters, dbl., cupped, recurrent bloom, slight fragrance; foliage light; vigorous (4–5 ft) growth; [Helene X Bon Silene]; Van Fleet; Conard & Jones

'BIRDSONG', LCl, mp, 1991; semi-dbl., 6–14 petals, 1.5 in., borne in very large clusters, moderate fragrance; few prickles; foliage small to medium, medium green, semi-glossy; tall (300 + cms), spreading, large, rambling growth; [Seedling X Seedling]; Seward, Grace

'BIRGITTA', HT, dr, 1961; bud ovoid; flowers large, dbl.; de Boer

Birgitte de Villenfagne, HMsk, pb

'BIR-HACKEIM', HT, mr, 1946; bud long, pointed; flowers fiery red, open, semi-dbl., cupped; very vigorous, upright, bushy growth; Mallerin, C.; A. Meilland

Birichina, F; (Italy); (Cavriglia)

'BIRMINGHAM BOERNER', F, w, 1975; flowers light flesh-pink turning white; [Gene Boerner sport]; Schoepfle

'BIRMINGHAM POST', F, dp, 1968; flowers deep pink, large, dbl., intense fragrance; foliage leathery; vigorous growth; [Queen Elizabeth X Wendy Cussons]; Watkins Roses

'BIRTHDAY PARTY', Min, mp, 1979; bud ovoid; flowers pink, medium, dbl., 28 petals, 1.5–2 in., exhibition form, moderate fragrance; foliage dark; upright, spreading growth; [Attraktion X Sheri Anne]; Strawn; Pixie Treasures Min. Roses

'BIRTHDAY PRESENT', Cl HT, dr, 1950; bud ovoid; dbl., 20 petals, exhibition form, non-recurrent, intense fragrance; foliage dark, leathery; vigorous, climbing growth; [Guinee X Rouge Mallerin]; Toogood

Birthday Wishes, *see* GUESDELAY

'BISCAY', Min, mp, 1988; bud ovoid; flowers medium pink, reverse slightly darker, medium, borne usually, dbl., 26 petals, exhibition form, slight fragrance; prickles slightly downward pointed, medium, deep pink; foliage medium, medium green, semi-glossy; upright, medium growth; [Summer Spice X Seedling]; Bridges, Dennis A.; Bridges Roses, 1989

'BISCHOF DR KORUM', HP, pb, 1921; flowers yellowish rose, dbl., moderate fragrance; [Frau Karl Druschki X Laurent Carle]; Lambert, P.

'BISCHOFSSTADT PADERBORN'®, S, or, 1964; (Fire Pillar); flowers cinnabar-scarlet, saucer shaped, semi-dbl.; vigorous (3–4 ft), bushy growth; ADR, 1968; Kordes, R.

'BISCOF', Min, op, 1983; (**Sadler**); flowers orange pink, small, dbl., 43 petals, exhibition form, no fragrance; foliage medium green, matt; upright growth; [Faberge X Darling Flame]; Bischoff, Francis J.; Kimbrew-Walter Roses

'BISFRA', Min, lp, 1983; (**Penny Annie**); flowers small, dbl., 35 petals, exhibition form, no fragrance; foliage medium, medium green, matt; bushy growth; [Little Darling X Unnamed seedling]; Bischoff, Francis J.

'BISHOP DARLINGTON', HMsk, ab, 1926; bud ovoid; flowers cream to flesh-pink, with yellow glow, large, semi-dbl., 17 petals, cupped, recurrent bloom, moderate, fruity fragrance; foliage bronze, soft; semi-climbing growth; [Aviateur Bleriot X Moonlight]; Thomas; Dreer, 1928;, H&S

Bishop Elphinstone, *see* 'COCJOLLY'

'BISHOP OF SHERWOOD', HT, op, 1976; flowers pale salmon flushed pink, full, dbl., 34 petals, 5 in., intense fragrance; vigorous, upright growth; [Mischief X (Wendy Cussons X Peace)]; Bracegirdle

Bishop's Rambler, HWich, dr, 1996; Scarman

'BISHOP'S ROSE', C, pb; vigorous (to 5 ft) growth; [Apparently R. gallica X R. centifolia]

'BISJEN', Min, rb, 1981; (**Ginny**); bud ovoid; flowers white edged medium red, yellow at hinge, yellow stamens, small, dbl., 45 petals, exhibition form, no fragrance; straight red prickles; foliage dark, leathery, reddish tinge on new growth; upright, compact growth; PP005275; [Little Darling X Toy Clown]; Bischoff, Francis J.; Kimbrew-Walter Roses

'BISMAR', Min, op, 1984; (**Marty's Triumph**™); flowers bright coral pink, white reverse, small, dbl., 28 petals, exhibition form, borne singly, slight fragrance; pale green, straight prickles; foliage medium, dark, semi-glossy; upright, bushy growth; PP006099; [Little Darling X Unnamed seedling]; Bischoff, Francis J., 1985

'BISRED', Min, rb, 1996; (**JuJu**™); flowers dark red with a little white at base, good color, very good, dbl., 26–40 petals, no fragrance; few prickles; foliage medium, dark green, glossy; upright, bushy, medium growth; [Little Darling X Black Jade]; Bischoff, Francis J.; Kimbrew Walter Roses, 1997

'BISyel', Min, my, 1997; (**Lida O**) flowers full, small, very dbl., 26–40 petals, 1.5 in., borne in small clusters, no fragrance; foliage medium, medium green, semi-glossy; medium (23 cms) compact growth; [Party Girl X Miss Dovie]; Bischoff, Francis J.

'BIT O' GOLD', Min, dy, 1981; bud ovoid; flowers deep yellow, imbricated form, micro-mini, dbl., 40 petals, moderate fragrance; foliage small, dark, semi-glossy; compact, bushy growth; PP005306; [Seedling X Golden Angel]; Williams, Ernest D.; Mini-Roses

'BIT O' MAGIC', Min, pb, 1979; bud pointed; flowers deep pink, reverse nearly white, micro-mini, dbl., 50 petals, 1 in., exhibition form; foliage small, dark, glossy; compact, spreading growth; [Over the Rainbow X Over the Rainbow]; Williams, Ernest D.; Mini-Roses

'BIT O' SPRING', Min, pb, 1980; bud long, pointed; flowers medium buffy pink, reverse lighter yellow-pink, borne usually singly, dbl., 45 petals, exhibition form, moderate fragrance; thin, tan prickles curved down; foliage deep green, matt; upright, bushy growth; [Tom Brown X Golden Angel]; Williams, Ernest D.

'BIT O' SUNSHINE', Min, dy, 1956; (Little Bit o' Sunshine); flowers bright buttercup-yellow, small, 18–20 petals, 1.5 in., moderate fragrance; bushy (12–14 in.), compact growth; [Copper Glow X Zee]; Moore, Ralph S.; Sequoia Nursery

'BIT OF HONEY', Min, dy, 1980; dbl., 48 petals; [Sunnydew sport]; Vastine, Gilbert; Gulf Stream Nursery

Bizarre Triomphante, *see* **'CHARLES DE MILLS'**

'BLABEE', Min, pb, 1983; (**Betty Bee**™); flowers pink, white reverse, small, dbl., exhibition form, no fragrance; foliage small, medium green, semi-glossy; compact, bushy growth; PP005448; [Little Darling X Toy Clown]; Blazey, Daniel; Nor'East Min. Roses

'BLABY COURIER', HT, mr, 1956; bud deep crimson; flowers vivid scarlet, moderate fragrance; strong stems; Verschuren; Blaby Rose Gardens

Blaby Jubilee, *see* **'DRIES VERSCHUREN'**

'BLABY MONARCH', HT, lp, 1960; flowers rose-pink, dbl., 40 petals, 5–6 in., exhibition form, intense fragrance; foliage light green; vigorous growth; [Briarcliff X Seedling]; Verschuren; Blaby Rose Gardens

'BLACK BEAUTY', HT, dr, 1973; flowers garnet-red, large, dbl., slight fragrance; bushy growth; [(Gloire de Rome X Impeccable) X Papa Meilland]; Delbard; Bees

'BLACK BESS', F, dr, 1939; bud long, pointed; flowers blackish crimson, borne in clusters, semi-dbl., exhibition form, slight fragrance; foliage dark, bronze; vigorous, bushy growth; [Dance of Joy X Crimson Glory]; Kordes; Morse

Black Bourbon, B, dr

'BLACK BOY', LCl, dr, 1919; (Blackboy); flowers very dark red, large, semi-dbl., moderate fragrance; foliage sparse, wrinkled, light; vigorous growth; [Étoile de France X Bardou Job]; Clark, A.; NRS South Australia

'BLACKBOY', M, dr, 1958; bud ovoid, lightly mossed; flowers deep crimson, large, very dbl., non-recurrent, intense fragrance; foliage light green, leathery; vigorous, upright, bushy growth; [World's Fair X Nuits de Young]; Kordes

Black Delight, HT, dr, 1985; G&L

'BLACK FIRE', Pol, dr, 1969; bud ovoid; flowers open, medium, dbl., slight fragrance; foliage dark, soft; vigorous, bushy growth; [Red Favorite X Seedling]; Delforge

'BLACK GARNET', HT, dr, 1980; bud ovoid, pointed; flowers dark, black-red, velvety, dbl., 50–55 petals, 5.5 in., exhibition form, borne singly or 2–3 per cluster, slight, tea fragrance; short to medium prickles, hooked downward; foliage medium to large, moderately leathery, dark grayish-g; bushy, upright, branching growth; PP4738; [Mister Lincoln X Mexicana]; Weeks

Black Gold, *see* CLEblack

'BLACK ICE', F, dr, 1971; dbl., 24 petals, 4 in., slight fragrance; foliage glossy, dark; [(Iceberg X Europeana) X Megiddo]; Gandy, Douglas L.

Black Jack™, *see* 'MINKCO'

Black Jade™, *see* 'BENBLACK'

'BLACK KNIGHT', HT, dr, 1934; flowers crimson shaded blackish, dbl., 30–35 petals, moderate fragrance; foliage glossy, dark; vigorous growth; [Ami Quinard X Château de Clos Vougeot]; Hillock

Black Lady®, *see* 'TANBLADY'

'BLACK MAGIC', LCl, dr, 1953; flowers blackish crimson, medium, occasional repeat bloom, intense fragrance; [Guinee seedling]; Hamilton

Black Magic, *see* TANkalgic

Black Night®, *see* 'HUBAR'

'BLACK OPAL', HT, dr, 1957; flowers dark velvety red; [Mirandy X Tassin]; Ulrick, L.W.

Black Pearl, *see* 'DELURT'

'BLACK PRINCE', HP, dr, 1866; flowers dark crimson shaded black, large, dbl., cupped, recurrent bloom, intense fragrance; vigorous growth; Paul, W.

'BLACK RUBY', HT, dr, 1965; flowers center crimson, large, dbl., 40 petals, cupped, tall growth; [Rome Glory X Impeccable]; Delbard-Chabert; Cuthbert

Black Ruby, LCl, mr

Black Sapphire, *see* 'LAVSAPH'

Black Satin, *see* 'BETSAT'

'BLACK TAQUIN', LCl, dr, 1955; flowers dark crimson-maroon; vigorous growth; [Honour Bright X Guinee]; Eacott

'BLACK TEA', HT, r, 1973; flowers brown, urn-shaped, blooms borne usually singly, dbl., 32 petals, slight fragrance; deep brown, hooked prickles; foliage medium, dark, semi-glossy; medium, bushy growth; [Hawaii X (Aztec X (Goldilocks X Fashion))]; Okamoto, K.; K. Hirakata Nursery

'BLACK VELVET', HT, dr, 1960; bud ovoid; dbl., 28 petals, 5–5.5 in., exhibition form, intense fragrance; foliage leathery, dark; vigorous, upright growth; [New Yorker X Happiness]; Morey, Dr Dennison; J&P

Blackberry Blossom, Pol, w, 1980; Semple

Blackberry Nip, *see* 'SOMNIP'

Blackberry Rose, *see* **R. RUBUS**

Blackboy, *see* **'BLACK BOY'**

'BLAIRII No. 1', HCh, mp; flowers bright rose, sometimes tinged red, large, semi-dbl., cupped, intense fragrance; branching growth; liable to injury from severe cold (less hardy than Blairii No. 2); [Parks' Yellow Tea-scented China X Hardy rose]; Blair, 1845

'BLAIRII No. 2', HCh, lp, 1845; flowers rosy blush, large, dbl., moderate fragrance; vigorous (up to 15 ft hardy; [Parks' Yellow Tea-scented China X Hardy rose]; Blair

'BLAKENEY'S RED', HT, mr, 1962; bud long, pointed; flowers currant-red, base yellow, dbl., 45 petals, 5 in., exhibition form, moderate fragrance; foliage leathery; moderate, bushy growth; [Karl Herbst X Peace]; Blakeney; Eddie

'BLANC DE VIBERT', P, w, 1847; dbl., sometimes recurrent bloom; foliage light green; Vibert

Blanc Dot, *see* **'BLANCHE DOT'**

'BLANC DOUBLE DE COUBERT', HRg, w, 1892; bud pleasing; flowers half-open white, fairly large, dbl., repeat bloom, intense fragrance, even at night; foliage very rugose; vigorous (5–7 ft) growth; [R. rugosa X Sombreuil]; Cochet-Cochet

Blanc Lafayette, *see* **'DAGMAR SPÄTH'**

Blanc Meillandécor, *see* 'MEIcoublan'

Blanc Parfait, *see* **'POMPON BLANC PARFAIT'**

'BLANC PUR', Misc OGR, w, 1827; Mauget

Blanc Queen Elizabeth, *see* **'WHITE QUEEN ELIZABETH'**

'BLANCA', F, w, 1965; flowers pure white, large blooms in clusters, dbl., slight fragrance; vigorous growth; [Purpurine X (Papillon Rose X Sterling Silver)]; Lens

'BLANCHE AMIET', HT, op, 1921; flowers coppery salmon, passing to clear rose; Turbat

Blanche Colombell, *see* DELgribla

Blanche Comète, *see* 'DELoblan'

'BLANCHE DE BELGIQUE', A, w; (Blanche Superbe); flowers pure white, very large, dbl.

'BLANCHE DE BELGIQUE', A, lp; (Blush Belgique); flowers blush, large, semi-dbl.; [R. alba X R. canina]; Vibert, 1817

'BLANCHE DOT', HT, w, 1962; (Blanc Dot); flowers snow-white, well-formed, large, dbl., 33 petals, slight fragrance; [White Knight X Virgo]; Dot, Pedro

'BLANCHE DURANTHON', N, w, 1800; flowers medium, dbl.; Nabonnand; (Sangerhausen)

'BLANCHE FROWEIN', LCl, yb, 1916; flowers coppery yellow, moderate fragrance; Leenders, M.

'BLANCHE MALLERIN', HT, w, 1941; bud long, pointed; flowers pure white, dbl., 33 petals, 4 in., exhibition form, foliage leathery, glossy; vigorous growth; [Edith Krause X White Briarcliff]; Mallerin, C.; A. Meilland; C-P

'BLANCHE MESSIGNY', HT, ly, 1923; flowers creamy yellow, large, dbl., 45 petals, moderate fragrance; vigorous, bushy growth; Gillot, F.

'BLANCHE MOREAU', M, w, 1880; bud well mossed; flowers pure white, large blooms in clusters, dbl., repeat bloom, moderate fragrance; lax growth; [Comtesse de Murinais X Perpetual White Moss]; Moreau et Robert

Blanche Mousseuse, *see* **'SHAILER'S WHITE MOSS'**

'BLANCHE NABONNAND', T, w, 1883; flowers creamy white tinged lemon or flesh, dbl., globular, moderate fragrance; Nabonnand, G.

Blanche Neige, *see* 'MACcarpe'

'BLANCHE NEIGE', Pol, w, 1929; Koster

Blanche Nouvelle, M, w

'BLANCHE ODORANTE', HT, w, 1952; bud very long, pointed; flowers purest white, large, very dbl., intense fragrance; vigorous growth; [Pole Nord X Neige Parfum]; Caron, B.; EFR

Blanche Pasca®, *see* 'LENip'

'BLANCHE REBATEL', Pol, pb, 1889; flowers carmine and white, Bernaix, A.

'BLANCHE SIMON', w, 1862; flowers large, dbl.; Moreau et Robert; (Sangerhausen)

Blanche Simple, HGal, w

Blanche Superbe, *see* **'BLANCHE DE BELGIQUE'**

Blanche Unique, *see* **'UNIQUE BLANCHE'**

'BLANCHE WIMER', Min, pb, 1986; flowers light yellow and pink blend, blooms borne usually singly, dbl., 66 petals, cupped, slight, fruity fragrance; no fruit; slightly hooked, reddish-green prickles; foliage medium, medium green, semi-glossy; tall, upright growth; [Pink Petticoat X Seedling]; Shaw, Dr John, 1985

'BLANCHEFLEUR', C, w, 1835; flowers white tinted blush, dbl.; vigorous growth; Vibert

'BLANDA EGRETA', HMult, mp, 1926; flowers small, dbl.; Bruder Alfons; (Sangerhausen)

Blanik, F, dr, 1964; flowers medium, dbl.; Vecera, L.; (Sangerhausen)

Blank, HP, mp

'BLARNEY', Cl HT, mp, 1934; flowers pink, base apricot, dbl.; recurrent bloom, moderate fragrance; foliage leathery, dark; vigorous, climbing growth; [Irish Charm sport]; Howard Rose Co.

Blastoff™, *see* 'MORflash'

'BLATENSKÀ KRÀLOVNA', HWich, mp, 1937; flowers medium, dbl.; Böhm, J.; (Sangerhausen)

'BLATNÁ', HT, dr, 1927; flowers velvety dark red, dbl., slight fragrance; dwarf growth; [Lieut. Chaure X Oskar Cordel]; Böhm, J.

'BLAUWE DONAU', F, m, 1971; (Blue Danube); bud ovoid; flowers full, open, semi-dbl., 10–12 petals, 3 in., intense fragrance; foliage glossy; bushy growth; [Orangeade X Sterling Silver]; Verschuren, Ted; Verschuren

'BLAYDON RACES', F, yb, 1976; flowers scarlet shading to yellow, 2.5 in., slight fragrance; foliage glossy, dark, leathery; vigorous, tall, upright growth; [Bobby Shafto X (Arthur Bell X Piccadilly)]; Wood

'BLAZE', LCl, mr, 1932; flowers bright scarlet, dbl., 20–25 petals, 2–3 in., cupped, blooms in large clusters, recurrent bloom, slight fragrance; foliage leathery, dark; very vigorous, climbing growth; [Paul's Scarlet Climber X Gruss an Teplitz]; Kallay; J&P

'BLAZE AWAY', F, or, 1979; bud pointed; flowers scarlet-vermilion, single, 7 petals, 3.5 in., slight fragrance; vigorous, bushy growth; [(Karl Herbst X Crimson Glory) X Sarabande]; Sanday, John

Blaze Superier, *see* **'DEMOKRACIE'**

Blazing Lights, HT, ob, 1995; K&S

Bleak House, HRg, lp; Bleak House, 1995; (Weatherly, L.)

'BLEBAR', F, ob, 1981; (**Wagbi**); bud pointed; flowers orange-pink, blooms borne 3–5 per cluster, dbl., 20 petals, flat, moderate fragrance; large, red prickles; foliage large, dark; vigorous, upright growth; Barrett, F. H.; John Mattock, Ltd.

Blenheim, *see* TANmurse

Bles Bridges, *see* KORampli

'BLESMA SOUL', HT, lp, 1982; bud long, pointed; dbl., 36 petals, borne singly or in small clusters, intense fragrance; large based, reddish-brown prickles; foliage light green; upright growth; [Pascali X Fragrant Cloud]; Anderson's Rose Nurseries

'BLESS MY TIME', Gr, lp, 1983; flowers clear light pink, veins sometimes darker; [Queen Elizabeth sport]; Orr, Rudolph F.

Blessed Event, *see* 'LAVfun'

'BLESSINGS'®, HT, op, 1967; flowers medium coral-salmon, large, dbl., 30 petals, exhibition form, moderate fragrance; [Queen Elizabeth X Seedling]; Gregory, 1968

Blessings, Climbing, ClHT, 1975; (Cavriglia)

'BLEstogil', S, pb, 1982; (**Festival Fanfare**); flowers dark pink with paler pink stripes, which become nearly white; [Fred Loads sport]; Ogilvie, W. D.

'BLEU MAGENTA', HMult, m; flowers dark crimson-purple, borne in clusters, dbl., slight fragrance; foliage dark

Blickfang, F, dr; flowers reddish-violet, large, very dbl., no fragrance; VEG; (Sangerhausen)

'BLITHE SPIRIT', HT, lp, 1964; flowers medium, dbl., slight fragrance; foliage leathery; vigorous, spreading, upright growth; [Fandango X Seedling]; Armstrong, D.L. & Swim; Armstrong Nursery

Blizzard, *see* 'JACdrift'

Bloemfontein, *see* KORtrolle

'BLÖHM & VOSS', F, or; flowers bright orange-scarlet good trusses

'BLONDE BOMBSHELL', F, my, 1995; flowers medium yellow fading to pale yellow with pink edging, full (26–40 petals), medium blooms borne in small clusters; fragrant, numerous prickles; foliage medium, dark green, glossy, upright, medium growth; [Seedling X Seedling]; Bees of Chester; L. W. van Geest Farms Ltd., 1995

Blondie, *see* 'JACliy'

Blondie, *see* LAPdil

'BLONDINE', HT, lp, 1954; bud pointed, globular; flowers blush-pink, dbl., 60 petals, 4 in., moderate fragrance; foliage leathery; upright growth; [Catalina sport]; Grillo

'**BLONDINE**', Cl HT, w; flowers pearl-white, well formed, large, intense fragrance; foliage clear green; vigorous growth; [Comtesse Vandal, Climbing X Michèle Meilland, Climbing]; Arles; Roses-France

Blood-red ChinaRose, *see* '**SANGUINEA**'

'**BLOODSTONE**', HT, or, 1951; flowers large, dbl., 24 petals, exhibition form, slight fragrance; foliage dark, coppery green; vigorous growth; [The Queen Alexandra Rose X Lord Charlemont]; McGredy, Sam IV

Bloomer Girl, *see* 'TINGIRL'

Bloomfest™, *see* 'JUDFEST'

'**BLOOMFIELD ABUNDANCE**', F, lp, 1920; flowers light salmon-pink, dbl.; foliage glossy, dark; bushy growth; GM, Portland, 1919; [Sylvia (probably R.) X Dorothy Page-Roberts]; Thomas; B&A

'**BLOOMFIELD BEVERLY**', HT, or, 1924; flowers orange-crimson, dbl., moderate fragrance; vigorous growth; [Mary, Countess of Ilchester X Mme Edouard Herriot]; Thomas

'**BLOOMFIELD BRILLIANT**', LCl, op, 1931; flowers light salmon with orange glow, large, semi-dbl., 18 petals, moderate fragrance; vigorous growth; [Mme Abel Chatenay X Kitty Kininmonth]; Thomas; H&S

Bloomfield Cemetary, T, ab

'**BLOOMFIELD COMET**', HMsk, op, 1924; bud long, pointed, reddish orange; flowers orange, base yellow, single, 5 petals, 3.5 in., moderate fragrance; foliage sparse, light bronze, soft; [Duchess of Wellington X Danaë]; Thomas; B&A

'**BLOOMFIELD COMPLETENESS**', HMsk, ab, 1931; flowers deep orange-yellow, dbl., moderate fragrance; [Bloomfield Perfection X Mme Butterfly]; Thomas

'**BLOOMFIELD COURAGE**', HWich, rb, 1925; flowers dark velvety red, center white, prominent yellow stamens, small, single, non-recurrent; foliage dark; vigorous, climbing or pillar (20 ft) growth; Thomas; B&A; H&S

'**BLOOMFIELD CULMINATION**', HMsk, pb, 1924; bud long, pointed; flowers rose-pink, center white, single, 3 in., recurrent bloom, slight fragrance; foliage leathery; moderately vigorous growth; [Sheila Wilson X Danaë]; Thomas; B&A

'**BLOOMFIELD DAINTY**', HMsk, my, 1924; bud long, pointed, deep orange; flowers clear canary-yellow, single, 2 in., moderate fragrance; foliage glossy; moderately vigorous growth; [Danaë X Mme Edouard Herriot]; Thomas; B&A

'**BLOOMFIELD DAWN**', HMsk, pb, 1931; bud long, slender, rose-pink; flowers light pink, base yellow, reverse deep pink, large blooms, semi-dbl., moderate fragrance; long, strong stems; [Seedling, Climbing X Bloomfield Progress]; Thomas; Armstrong Nursery

'**BLOOMFIELD DECORATION**', HMsk, pb, 1925; flowers cerise-pink, center white, prominent golden stamens, open, single, moderate fragrance; foliage glossy; [Sylvia X Arndt]; Thomas; B&A, 1927; H&S, 1927

'**BLOOMFIELD DISCOVERY**', HMsk, pb, 1925; flowers pink, reverse darker, single; foliage dark; moderately vigorous growth; [Danaë X (Frau Karl Druschki X Mme Caroline Testout)]; Thomas; B&A

Bloomfield Endurance, *see* '**W. FREELAND KENDRICK**'

'**BLOOMFIELD EXQUISITE**', Cl HT, mp, 1924; flowers clear pink, dbl., recurrent bloom, moderate fragrance; vigorous growth; [Gloire de Dijon X Gruss an Teplitz]; Thomas; H&S

'**BLOOMFIELD FASCINATION**', HMsk, ly, 1924; flowers light canary yellow, small, dbl., slight fragrance; foliage rich bronze green, soft; [Danaë X Mme Lautette Messimy]; Thomas; B&A

'**BLOOMFIELD FAVORITE**', HMsk, w, 1924; flowers pinkish cream, dbl., slight fragrance; [Debutante X Moonlight]; Thomas

'**BLOOMFIELD FLAME**', HT, rb, 1930; bud long, pointed, flame-red; flowers crimson-flame, center orange-yellow, large, dbl., 22 petals, moderate, spicy fragrance; foliage leathery, glossy, dark bronze; vigorous, bushy growth; [Louise Crette X Mme Charles Lutaud]; Thomas; H&S

Bloomfield Improvement, *see* '**EDNAH THOMAS**'

Bloomfield Loveliness, *see* '**SOPHIE THOMAS**'

'**BLOOMFIELD LUSTRE**', LCl, op, 1931; flowers salmon-pink, base yellow, dbl., moderate fragrance; [Hortulanus Budde X Souv. de Mme Léonie Viennot]; Thomas; H&S

'**BLOOMFIELD MAGIC**', LCl, op, 1924; flowers light salmon to cream, flat, moderate fragrance; moderately vigorous growth; [Gloire de Dijon X Frau Berta Gurtler]; Thomas

'**BLOOMFIELD MYSTERY**', LCl, lp, 1924; flowers silver-pink, tinged yellow, 2 in., flat, moderate fragrance; moderately vigorous growth; [Blanche Frowein X Bloomfield Abundance]; Thomas; B&A

'**BLOOMFIELD PERFECTION**', HMsk, w, 1925; flowers cream-yellow suffused lilac, dbl., recurrent bloom, slight fragrance; [Danaë X Bloomfield Abundance]; Thomas; B&A, 1927

'**BLOOMFIELD PERPETUAL**', HP, w, 1920; flowers resembling the Cherokee Rose, single; [Iceberg X Frau Karl Druschki]; Thomas; B&A

'**BLOOMFIELD PROGRESS**', HT, mr, 1920; flowers glowing red, dbl., intense fragrance; vigorous growth; [Mary, Countess of Ilchester X Gen. MacArthur]; Thomas; B&A

'**BLOOMFIELD QUAKERESS**', Cl T, ly, 1931; flowers small blooms, semi-dbl., free recurrent bloom, foliage glossy, light; long stems; vigorous growth; [Chance Safrano seedling]; Thomas; Armstrong Nursery

'**BLOOMFIELD ROCKET**', LCl, dp, 1925; flowers dark pink, center lighter, very large, single, slight fragrance; very vigorous (6–8 ft) growth; [Mme Caroline Testout X Ulrich Brunner Fils]; Thomas; B&A

Bloomin' Easy™, *see* 'AROTRUSIM'

Bloomin' Pretty, LCl, w, 1999; flowers white with pale pink paint, semi-dbl.; Williams, J. Benjamin

Blooming Easy, *see* 'AROTRUSIM'

'**BLOOMSDAY**', F, ob, 1981; flowers orange, marked brown, reverse deep gold, loose, 4–5 per cluster, dbl., 28 petals, moderate fragrance; straight, red-brown prickles; foliage matt, green; vigorous, upright growth; [Belinda X (Maxi X Joyfullness)]; McCann, Sean; Hughes Roses

Bloomtown, *see* 'TALBLO'

'**BLOSSOM**', HT, pb, 1925; flowers red over peach-pink, reverse pinkish yellow; Beckwith

'**BLOSSOM HILL**', F, ob, 1957; flowers orange, medium in clusters, single, 10 petals, 2.5 in., cupped, slight fragrance; foliage dark, glossy; vigorous, upright, bushy growth; Kordes; Morse

Blossom Magic, *see* MEImanoir

'**BLOSSOMTIME**', LCl, mp, 1951; bud pointed; flowers pink, reverse deeper, blooms in clusters of 3–8, dbl., 38 petals, 4 in., exhibition form, repeat bloom, intense fragrance; tall shrub or moderate climbing (6–7 ft) growth; [New Dawn X Seedling]; O'Neal; Bosley Nursery

Blue Angel, *see* 'RENANGEL'

Blue Bajou®, *see* KORkultop

Blue Bell, *see* MEInalpir

Blue Bird, HT; Poulsen; (Cavriglia)

'**BLUE BOY**', M, m, 1958; bud ovoid; flowers deep reddish violet, large, dbl., exhibition form, non-recurrent, intense fragrance; foliage light green, glossy; vigorous (3 ft), upright, bushy growth; [Louis Grimmard X Independence]; Kordes

Blue Boy, HT, m

Blue Carpet, *see* 'HARQUILLYPOND'

'**BLUE CHATEAU**', HT, m, 1999; flowers pale violet, very dbl., 50 petals, exhibition form, intense fragrance; to 4 ft growth; [Madame Violet X seedling]; Teranishi, K.; Itami Rose Nursery, 1994, Japan

Blue Chip, *see* 'JACAVE'

Blue Danube, *see* '**BLAUWE DONAU**'

Blue Delight, HT, m, 1980; Kasturi

'**BLUE DIAMOND**', HT, m, 1963; flowers lavender, dbl., 35 petals, 4 in., moderate fragrance; foliage dark, coppery; vigorous, compact, bushy growth; [Purpurine X (Purpurine X Royal Tan)]; Lens

Blue Friendship, S, m, 1984; Verschuren

Blue Girl, *see* 'KORgi'

'Blue Glow', HT, m, 1982; bud pointed; flowers pinkish-mauve, large blooms borne singly and 3–5 per cluster, dbl., 45 petals, exhibition form, slight fragrance; light brown, triangular prickles; foliage light green; upright, branching growth; [Silent Night X Blue Moon]; Cattermole, R. F.

'Blue Heaven', HT, m, 1971; bud ovoid; flowers large, dbl., exhibition form, intense fragrance; foliage large, glossy; vigorous, upright growth; [(Sterling Silver X Simone) X Song of Paris]; Whisler, D.; Gro-Plant Industries

Blue Ice, *see* 'LAVchip'

Blue Jay, F, m

Blue Magic, HT; Dot, Pedro; (Cavriglia)

Blue Magic, Min, m, 1986

Blue Mikey, HP, m

'Blue Mist', Min, m, 1970; bud short, rounded; flowers soft pink to lavender, micro-mini, small, dbl., 23 petals, intense fragrance; foliage soft; vigorous, bushy, rounded growth; [Seedling X Seedling]; Moore, Ralph S.; Sequoia Nursery

Blue Monday, *see* 'TANsi'

Blue Monday, Climbing, *see* **'Blue Moon, Climbing'**

Blue Moon®, *see* 'TANsi'

'Blue Moon, Climbing', Cl HT, m, 1981; (Blue Monday, Climbing, Mainzer Fastnacht, Climbing, Sissi, Climbing); [Blue Moon sport]; Mungia, Fred A., Sr.; Montebello Rose Co., Inc.

Blue Nile®, *see* 'DELnible'

Blue Ocean, HT, m, 1983; Hande

Blue Parfum®, *see* 'TANtifum'

Blue Perfume, *see* 'TANtifum'

Blue Peter, *see* 'RUIblun'

Blue Rambler, *see* **'Veilchenblau'**

Blue Ribbon, *see* 'AROlical'

Blue River®, *see* 'KORsicht'

'Blue River', HT, m, 1973; bud long, pointed; flowers mauve-magenta, large, dbl., cupped, intense fragrance; foliage glossy; vigorous, upright growth; RULED EXTINCT 6/84 ARM; [Mainzer Fastnacht X Silver Star]; Kordes; Horstmann

Blue Rosalie, *see* **'Veilchenblau'**

Blue Skies, *see* 'BUCblu'

Blue Sky, *see* **'Aozora'**

Blue Sky, HT, m, 1982; Dot

'Blue Star, Climbing', Cl HT, m, 1963; flowers lavender-blue, small blooms in cluster, dbl., flat, profuse, non-recurrent bloom, slight fragrance; foliage glossy; vigorous (8 ft) growth; Thompson, M.L.

Blue Violet, HT, m, 1981; Perry

Blue Work, HT, m

Blueberry Hill™, *see* 'WEKcryplag'

Blueblood™, *see* 'LAVblu'

Bluenette, *see* 'RUIblun'

Bluesette®, *see* 'LENmau'

Bluewunder, *see* KORedan

Bluhendes Barock, *see* NOAbell

Bluhwunder®, *see* KORedan

'Blumen Dankert', HMult, mp, 1904; flowers carmine-pink, medium, semi-dbl.; Kiese; (Sangerhausen)

'Blumenschmidt', T, yb, 1906; flowers primrose-yellow, outer petals rose-pink, [Mlle Franziska Kruger sport]; Schmidt, J.C.

'Blumenschmidt's Elfenkönigin', HT, w, 1939; flowers ivory-white, center orange, very large, dbl., exhibition form, moderate, lily-of-the-valley fragrance; long, strong stems; bushy, dwarf growth; [Ophelia X Julien Potin]; Weigand, C.; Schmidt, J.C.

'Blumenschmidts Sonntagskind', F, mp, 1945; flowers medium, dbl.; Vonholdt; (Sangerhausen)

Blush Baby, Min, 1986; Pearce, C. A.; (Cavriglia)

Blush Belgique, *see* **'Blanche de Belgique'**

'Blush Boursault', Bslt, pb; (Calypso); flowers blush, center deep flesh, very large blooms, tending to ball, very dbl., globular; foliage remains longer than other boursaults; pendulous growth habit

'Blush China', HCh, lp, 1903; flowers medium, semi-dbl., slight fragrance; Cant, B. R.; (Sangerhausen)

'Blush Damask', D, lp, 1759; flowers center rose, shading to pale blush on outside petals, small, dbl.

'Blush Hip', A, lp; flowers soft pink, green eye, dbl.; vigorous growth

Blush Maman Cochet, *see* **'William R. Smith'**

'Blush Moss', M, lp; flowers blush, center pinkish when first open, well-mossed, large, dbl., cupped; moderate branching growth

Blush Musk, *see* **'Fraser's Pink Musk'**

'Blush Noisette', N, w; flowers pinkish white; (14); [Champneys' Pink Cluster seedling]; Noisette, before 1817

'Blush Queen', HT, lp, 1924; flowers blush-pink, dbl., moderate fragrance; Cant, F.

'Blush Rambler', HMult, lp, 1903; flowers blush-pink, blooms in clusters, semi-dbl., cupped, non-recurrent bloom; vigorous, climbing (10–12 ft) growth; [Crimson Rambler X The Garland]; Cant, B. R.

'Blushing Beauty', LCl, lp, 1934; flowers three-toned shell-pink, very large; very vigorous growth; Burbank; Stark Bros.

Blushing Blue, *see* 'RENblue'

'Blushing Bride', HT, w, 1918; flowers white, center blush, dbl.; RULED EXTINCT 5/90; Dickson, H.

'Blushing Bride', HT, W, 1930; flowers white tinged pink, large, dbl., intense fragrance; vigorous growth; RULED EXTINCT 5/90; [Mme Butterfly X Premier]; Hill, Joseph H., Co.

Blushing Bride, *see* 'GREblub'

Blushing Bride, *see* HARfling

Blushing Dawn, *see* 'WEEblush'

Blushing Groom, *see* 'SEAgru'

'Blushing Jewel', Min, lp, 1958; bud ovoid; flowers blush-pink, overcast rose-pink, open, dbl., 45–50 petals, .5–.75 in., moderate fragrance; low, compact growth; [Dick Koster sport X Tom Thumb]; Morey, Dr Dennison; J&P

Blushing June, *see* 'LECblu'

'Blushing Lucy', LCl, lp, 1938; flowers pale pink, white eye, large cluster, semi-dbl., profuse, late bloom, intense fragrance; foliage glossy; vigorous growth; Williams, A.

Blushing Maid, *see* 'MACkepa'

'Blushing Queen', Gr, w, 1976; bud pointed; flowers near white, center blush-pink, dbl., 38–40 petals, exhibition form, moderate fragrance; vigorous, upright growth; [Queen Elizabeth sport]; Baker, Larry

Blushing Rose, *see* **'Coy Colleen'**

Bo, *see* PEKoubo

Bo, HT, lp; flowers small

'Bob Collard', F, mr, 1986; flowers very luminous, brilliant red, reverse deeper red, medium, single, 11 petals, cupped, borne singly or in sprays of 5–10, slight, fruity fragrance; round, cupped, average, red fruit; foliage medium, medium green, glossy; upright, bushy, medium growth; [Happiness X Copenhagen]; Turley, V.G., 1987

Bob Greaves, *see* FRYzippy

'Bob Hope', HT, mr, 1966; bud urn-shaped; flowers scarlet red, dbl., 35–40 petals, 6 in., exhibition form, borne mostly singly, intense, damask fragrance; foliage dark, leathery; vigorous, tall, upright growth; [Friedrich Schwartz X Kordes' Perfecta]; Kordes, R.; J&P

Bob Kennedy, HT; Ingegnoli; (Cavriglia)

'Bob Woolley', HT, ab, 1970; flowers peach-pink, reverse lemon, dbl., 60 petals, 5 in., slight fragrance; foliage matt, green; [Gavotte X Golden Scepter]; Sanday, John

'Bobbie James', HWich, w, 1961; flowers creamy white, blooms in trusses, single,

7–9 petals, 2 in., cupped, intense fragrance; foliage glossy; vigorous growth; Sunningdale Nursery

'BOBBIE LUCAS', F, op, 1967; flowers deep salmon-orange, well-formed, blooms in clusters, 3.5 in., slight fragrance; foliage dark; [Margot Fonteyn X Elizabeth of Glamis]; McGredy, Sam IV

Bobbie Robbie, *see* **'SOUR. DE J. CHABERT'**

Bobbie Vesely, *see* 'RESBOBIE'

'BOBBINK WHITE CLIMBER', LCl, w, 1951; flowers creamy yellow to pure white, dbl., 45–50 petals, 3 in., recurrent bloom, moderate fragrance; foliage dark, glossy; vigorous, climbing growth; [Dream Girl X Seedling]; Jacobus; B&A

'BOBBY CHARLTON', HT, pb, 1974; flowers deep pink, reverse silver, well-formed, dbl., 38 petals, 6 in., moderate, spicy fragrance; foliage dark, leathery; GM, Baden-Baden, 1976 GM, Portland, 1980; [Royal Highness X Prima Ballerina]; Fryer, Gareth; Fryer's Nursery, Ltd.

Bobby Dazzler, *see* 'ROSELLA'

'BOBBY SHAFTO', HT, my, 1967; [Piccadilly sport]; Wood

Bobino, F, mr

'BOBO', HT, mr, 1979; bud long; flowers urn-shaped blooms borne singly, dbl., 30 petals, intense, fruity fragrance; Perry, Astor, 1981

'BOBOLINK', Min, dp, 1959; flowers rose-pink, base near white, dbl., 50 petals, 1–1.5 in., slight, fruity fragrance; foliage leathery, glossy; vigorous, bushy (18 in.) growth; [(R. wichurana X Floradora) X (Oakington Ruby X Floradora)]; Moore, Ralph S.; Sequoia Nursery

Bobravka, S, op, 1975; flowers dark orange-pink with red tones, medium, dbl., moderate fragrance; Urban, J.; (Sangerhausen)

Bob's Koster, Pol, dp

Bob's Peach, *see* 'WEBBREAD'

'BOCCA NEGRA', HMult, dr, 1910; flowers purple-crimson, center white, borne in clusters of 15–20, cupped; vigorous growth; Dubreuil

'BOCCACCIO', F, or, 1963; flowers bright scarlet-red, large blooms in clusters, dbl., 30–40 petals; foliage glossy; vigorous, upright growth; [Atombombe X Seedling]; Verschuren, A.; Stassen

'BOCCACE', HP, 1859; Moreau et Robert; (Cavriglia)

'BOCHER', HT, pb, 1990; very dbl., slight fragrance; foliage medium, medium green, matt; upright growth; [Prima Ballerina X Alec's Red]; Rodgers, Shafner R.

Bodhisatva, *see* 'VIRSPLASH'

'BOHEMIA', HT, mp, 1928; flowers pure rose-pink, semi-dbl., moderate fragrance; [Mrs Franklin Dennison X Mrs Henry Morse seedling]; Böhm, J.

'BOHÉMIENNE', HT, or, 1954; dbl., 45 petals; bushy growth; Mallerin, C.; EFR

'BÖHM JUNIOR', HT, dp, 1935; flowers carmine-red, dbl., intense fragrance; foliage glossy; bushy growth; [Laurent Carle X (Paul's Scarlet Climber X Ethel Somerset)]; Böhm, J.

'BÖHM SENIOR', HT, mr, 1938; flowers large, dbl.; Böhm, J.; (Sangerhausen)

'BÖHMOROSE', HT, mp, 1935; flowers rosy carmine to light rose-pink, very large, dbl.; foliage glossy, dark; very vigorous growth; [Gen. MacArthur X Laurent Carle]; Böhm, J.

'BÖHMOVA AZUROVÀ', HP, m, 1934; flowers dark violet-pink, large, dbl.; Böhm, J.; (Sangerhausen)

'BÖHMOVA POPELKA', Pol, dr, 1934; flowers dark blood-red; foliage curiously variegated; Böhm, J.

'BÖHM'S CLIMBER', Cl HT, mr, 1935; flowers medium, dbl., moderate fragrance; Böhm, J.; (Sangerhausen)

'BÖHM'S TRIUMPH', HT, dr, 1934; flowers very large, semi-dbl., exhibition form; foliage dark; vigorous, bushy growth; [Vaterland X Lord Charlemont]; Böhm, J.

Bojangles, *see* 'JACSUN'

Boksburg Fantasia, *see* 'KORNITZEL'

Bolchoi, *see* MEIruchka

Bolcholi, *see* MEIzuzes

Bolero, *see* POUlbo

Bolero, HT, 1963; Pironti, N.; (Cavriglia)

'BOLÉRO', HT, rb, 1937; flowers nasturtium-red and gold, dbl.; foliage leathery; long stems; very vigorous, bushy growth; Gaujard

'BOLÉRO', F, ob, 1958; bud pointed; flowers orange, large, dbl.; very vigorous growth; Buyl Frères

Bombon, Min, 1967; Dot, Simon; (Cavriglia)

'BON ACCORD', HT, pb, 1967; flowers pink shaded silver, high-pointed, 4.5 in., moderate fragrance; foliage glossy; [Prima Ballerina X Percy Thrower]; Anderson's Rose Nurseries

'BON SILÈNE', T, dp; bud well-formed; flowers deep rose, large, dbl., recurrent bloom, moderate fragrance; vigorous growth; Hardy, before 1837

'BON SILÈNE BLANC', T, ly, 1885; (White Bon Silène); flowers pale yellow to creamy white, large; Morat

'BON VOYAGE', HT, 1960; (Voeux de Bonheur); Delbard-Chabert; Stark Bros., 1969

'BONA WEILSCHOTT', HT, mp, 1889; flowers large, dbl., moderate fragrance; Soupert & Notting; (Sangerhausen)

Bonanza®, *see* 'KORMARIE'

'BONANZA', HT, 1957; flowers yellow shaded red, slight fragrance; foliage dark; vigorous growth; RULED EXTINCT 5/82 ARM; Verbeek

'BONAVISTA', HRg, lp, 1977; bud ovoid; flowers open, dbl., 20 petals, 2 in., abundant, repeat bloom, intense fragrance; foliage yellow-green; upright, bushy growth; [Schneezwerg X Nemesis]; Svedja, Felicitas; Canada Dept. of Agric.

'BON-BON', F, pb, 1974; bud ovoid; flowers deep rose-pink, reverse white, large, semi-dbl., slight fragrance; foliage dark; vigorous, bushy growth; AARS, 1974; [Bridal Pink X Seedling]; Warriner, William A.; J&P

Bonbon Hit®, *see* POUlbon

'BOND STREET', HT, op, 1965; flowers deep salmon-pink, dbl., 75 petals, 4.5 in., intense fragrance; [Radar X Queen Elizabeth]; McGredy, Sam IV; McGredy

Boneyard Yellow, F, my, 1996

'BONFIRE', HMult, mr, 1928; flowers scarlet, in clusters of 20–25, dbl.; foliage light; very vigorous growth; [Crimson Rambler X R. wichurana]; Turbat

'BONFIRE NIGHT'®, F, rb, 1971; flowers red, shaded yellow-orange, 19 petals, 3.5 in., globular, slight fragrance; foliage matt; [Tiki X Variety Club]; McGredy, Sam IV; McGredy

Bonfire of Artec, HT; (Cavriglia)

Bonhomme, *see* 'LAVHOMME'

'BONICA'®, F, or, 1958; flowers scarlet, dbl., 70 petals, exhibition form, blooms in clusters, slight fragrance; foliage dark, leathery; vigorous, bushy growth; [(Alain X Independence) X Moulin Rouge]; Meilland, F.; C-P;, URS

Bonica™, *see* 'MEIDOMONAC'

Bonica '82, *see* 'MEIDOMONAC'

'BONITA', LCl, dr, 1958; flowers crimson, medium, in clusters of 3–8, semi-dbl., 15 petals, moderate fragrance; foliage dark, glossy; vigorous (10–15 ft) growth; Knight, A.T.

'BONJOUR', Gr, mr, 1965; flowers bright red, medium, dbl., slight fragrance; foliage leathery; very vigorous, bushy growth; [Mignonne X Miss Universe]; Gaujard

'BONN', HMsk, or, 1950; flowers orange-scarlet, blooms in trusses of 10, dbl., 25 petals, 4 in., moderate, musk fragrance; foliage glossy; upright, bushy (4–5 ft) growth; [Hamburg X Independence]; Kordes

'BONNE CHERE', S, rb, 1985; flowers bright velvet red, creamy eye, blooms in large clusters, single, 5 petals, 3.5 in., repeat blooming, slight fragrance; foliage medium, dark, leathery; vigorous, arching (to 4 ft) growth; [(R. nutkana X Baronne Prevost) X Alika]; James, John

'BONNE FÊTE', HT, or, 1960; bud long, pointed; flowers orange-coral, large blooms, dbl., 25–30 petals, slight fragrance; foliage leathery, glossy; strong stems; vigorous, bushy growth; [Independence X Barcelona]; Delbard-Chabert

Bonne Nouvelle, *see* **'GOOD NEWS'**

'BONNE NUIT', HT, dr, 1966; bud pointed; flowers blackish red, medium, dbl., exhibition form, moderate fragrance; foliage sparse, glossy; bushy growth; Combe; Wyant, 1956

'BONNE NUIT, CLIMBING', Cl HT, dr, 1964; Kashimoto; Itami Rose Nursery

'BONNIE', S, dp, 1956; similar to Aylsham but taller and more rapid and vigorous in growth; [Hansa X R. nitida]; Wright, Percy H.

'BONNIE ANNE', HT, pb, 1975; flowers pink, reverse yellow and pink, dbl., 30 petals, 3–3.5 in., moderate fragrance; foliage large, dark, glossy; [Prima Ballerina X Wendy Cussons]; MacLeod

'BONNIE BELLE', HWich, mp, 1911; flowers rose, large, single, slight fragrance; height 8–10 ft; Walsh

'BONNIE BESS', HT, op, 1929; flowers deep coral-pink, suffused copper, dbl., slight fragrance; [Wilhelm Kordes X (Crusader X Sunburst)]; Dale

'BONNIE DOONE', Cl HT, lp; [City of Little Rock seedling X Seedling]; Clark, A., 1941

'BONNIE HAMILTON', F, or, 1976; flowers vermilion-red, well-formed, dbl., 26 petals, 2.5 in., slight fragrance; foliage dark; [Anne Cocker X Allgold]; Cocker

'BONNIE JEAN', HT, rb, 1933; flowers carmine-cerise, base white, large, single; vigorous growth; Archer

'BONNIE MAID', F, pb, 1951; flowers silvery pink reverse deep pink, blooms in clusters of 5–8, semi-dbl., 17 petals, 3 in.; foliage leathery, dark; vigorous, bushy growth; GM, NRS, 1955; LeGrice

'BONNIE PINK', F, lp, 1964; bud ovoid; flowers geranium-pink overcast begonia-rose, with upright center petals, dbl., 35–40 petals, 4.5 in., exhibition form, moderate fragrance; foliage glossy; vigorous, upright growth; [White Garnette seedling X Hawaii]; Boerner; J&P

'BONNIE PRINCE', HMult, w, 1916; flowers white, center tinged yellow, dbl., moderate fragrance; excellent pillar growth; [Tausendschön X Seedling]; Cook, T. N.; Portland Rose Soc., 1924

'BONNIE PRINCE CHARLIE', HT, mr, 1960; bud pointed; flowers bright red, well formed, dbl., moderate fragrance; Cuthbert

Bonnie Prince Charlie's Rose, *see* White Rose of York

'BONNIE SCOTLAND', HT, dp, 1976; flowers light red, dbl., 43 petals, 5 in., exhibition form, intense, damask fragrance; foliage glossy; [Wendy Cussons X Percy Thrower]; Anderson's Rose Nurseries

'BONNY', Min, mp, 1974; flowers deep pink, lighter reverse, dbl., globular, slight fragrance; foliage small, light, wrinkled; dwarf growth; [Zorina X Seedling]; Kordes

Bonny, *see* KORniebon

Bonsoir, *see* 'DICBO'

'BOOKER T. WASHINGTON', HMult, dr, 1930; flowers deep maroon, borne in clusters, Turbat; C-P

Boomerang, *see* 'SPOBOOM'

'BOOYOL', S, mp, 1977; (**Rachel**); flowers medium, dbl., 35 petals, slight, delicate fragrance; foliage medium, medium green, matt; upright growth; Booth, Mrs. Rachel Y.

'BO-PEEP', Min, mp, 1950; bud ovoid, pointed; flowers rose-pink, micro-mini, very small, dbl., 28 petals, cupped, slight fragrance; foliage small, glossy; bushy, dwarf (5–8 in.) growth; [Cécile Brunner X Tom Thumb]; deVink; C-P

'BORDEAUX', HMult, mr, 1908; flowers wine-red, small blooms in large cluster, dbl.; very vigorous growth; [Crimson Rambler X Blanche Rebatel]; Soupert & Notting

'BORDER BEAUTY', F, mr, 1957; flowers bright scarlet, medium in very large trusses (10–12 in.), semi-dbl.; 2–2.5 in., foliage dark, glossy; vigorous, upright growth; [Floribunda seedling X Signal Red]; deRuiter; Gregory

'BORDER CORAL', F, op, 1957; flowers coral-salmon, blooms in trusses, semi-dbl., 2–2.5 in.; foliage dark, glossy; vigorous, spreading growth; [Signal Red X Fashion]; deRuiter; Gregory

'BORDER CORAL, CLIMBING', Cl F, op, 1966; Sanday; Gregory

'BORDER GEM', F, pb, 1961; bud ovoid; flowers geranium-pink, center light salmon-orange, blooms in clusters, dbl., 20 petals, 2.5–3 in., moderate fragrance; foliage leathery; compact, low growth; [(Navajo X Golden Dawn) X Pinocchio]; Morey, Dr Dennison; J&P

'BORDER GOLD', F, dy, 1966; bud ovoid; flowers medium to deep yellow, very dbl., slight fragrance; foliage dark, leathery; vigorous, low, bushy growth; [Allgold X Pigmy Gold]; Morey, Dr Dennison; Country Garden Nursery

'BORDER KING', Pol, mr, 1952; (Roi des Bordures); flowers bright strawberry-red, small large truss, semi-dbl., 16 petals; foliage dark, glossy; very vigorous growth; deRuiter; Gregory

'BORDER KING, CLIMBING', Cl Pol, mr, 1960; Gregory

'BORDER PRINCESS', F, op, 1951; flowers coral-pink shaded orange, reflexed, dbl., 22 petals, 3 in., exhibition form; foliage light, glossy; vigorous growth; Verschuren-Pechtold; Bentley

'BORDER QUEEN', F, op, 1951; (Reine des Bordures); flowers salmon-pink, center paler, reverse darker, large truss, single, 9 petals, 2.5 in.; foliage leathery, olive-green; vigorous, compact growth; (28); GM, NRS, 1950; deRuiter; Gregory; Harkness

'BORDERER', Pol, pb, 1918; flowers salmon, fawn and pink, semi-dbl., slight fragrance; dwarf, spreading growth; [Jersey Beauty seedling]; Clark, A.; NRS Victoria

Bordura de Nea, F, lp; short; flowers small to medium sized, white-pink, flat, 8 petals, borne in clusters, slight; small, semi-glossy, medium green; [Bonica '82 X Incandescent]; St. Wagner, 1994; Res. Stn. f. Fruit Growing, Cluj, 1995, Romania

'BORDURE', Pol, mp, 1911; flowers carmine-pink, small, dbl.; Barbier; (Sangerhausen)

Bordure Blanche, *see* DELbobla

Bordure de Nacrée, *see* 'DELCROUF'

Bordure d'Or®, *see* DELbojaune

Bordure Magenta, F, 1994; Delbard, Georges; (Cavriglia)

Bordure Nacrée®, *see* 'DELCROUF'

Bordure Rose®, *see* 'DELBARA'

Bordure Rose #2, *see* DELcoussi

Bordure Vermillon®, *see* DELbover

Bordure Vive®, *see* DELboviv

Bordurella®, F, pb

'BOREALIS', HKor, w, 1980; bud globular, pointed; dbl., 27 petals, exhibition form, borne singly, repeat bloom, moderate, sweet fragrance; gray prickles; foliage shiny; bushy growth; [Blanche Mallerin X Leverkusen]; James, John

'BORN FREE', Min, or, 1978; bud long, pointed; flowers brilliant orange-red, small, dbl., 20 petals, 1.5 in.; foliage dark; bushy, upright growth; [Red Pinocchio X Little Chief]; Moore, Ralph S.; Sequoia Nursery

Borsalino®, *see* INTerhyro

Borussia®, S, ly, 1992; Poulsen

'BORYANA', HT, dr, 1977; flowers large, dbl., 55 petals; foliage dark, glossy; vigorous, upright growth; [Tallyho X Spartan]; Staikov, Prof. Dr V.; Kalaydjiev and Chorbadjiiski

'BOSADSON', LCl, dr, 1997; (**Louisa Jane Morris**); flowers double, frilly, medium, 15–25 petals, slight fragrance; some prickles; foliage medium, medium green, semi-glossy; climbing (15 ft) growth; [Admiral Rodney X Evelyn Fison]; Bossom, W.E.

'BOSANNCOAT', Cl F, dp, 1996; (**Country Garden**); flowers bright pink, white center, blooms borne in large clusters, semi-dbl., 8–14 petals, 4 in., slight fragrance; few prickles; foliage medium, light green, glossy; spreading, climbing, tall (15 ft) growth; [Anne Harkness X Seedling]; Bossom, W.E.

'BOSANNE', F, op, 1990; (**Middlesex County**); bud pointed; flowers golden peach, medium, borne in sprays of 20–36, semi-dbl., 10 petals, flat, slight fragrance; foliage medium, medium green, glossy; tall, upright, very vigorous growth; [Anne Harkness X Greensleeves]; Bossom, W.E., 1985

'BOSANNEVES', F, my, 1996; (**Enfield in Bloom**); flowers deep yellow, reddish tinge to petal edge, reverse lighter yellow, dbl., 26–40 petals, 3.5 in., slight fragrance; moderate prickles; foliage large, medium green, semi-glossy; upright, tall (115 cms) growth; [Anne Harkness X Greensleeves]; Bossom, W.E.

'BOSARTHRIC', Cl F, pb, 1997; (**Caring For You**); flowers very full, large, very dbl., 41 petals, borne in small clusters, moderate fragrance; some prickles; foliage medium, medium green, semi-glossy; climbing (15 ft) growth; [English Miss X Summer Wine]; Bossom, W.E.

'BOSBIGSOUTH', F, lp, 1999; (**Lady Yvonne**); dbl., 17–25 petals, 3.5 in., borne in small clusters, moderate fragrance; prickles moderate; foliage medium, medium green, semi-glossy; medium (3 ft) growth; [Southampton X seedling]; Bossom, W.E.

'BOSCHERRY', F, pb, 1995; (**Darling Diane**); flowers lavender pink blend, blooms, dbl., 26–40 petals, 3–3.5 in., slight fragrance; foliage large, dark green, semi-glossy; tall, upright growth; [Champagne Cocktail X Seedling]; Bossom, W.E.

'BOSCHERRYDRIFT', F, pb, 1997; (**Deborah Devonshire**); flowers old fashioned form, very full, medium, very dbl., 41 petals, 2.75 in., borne in small clusters, moderate fragrance; few prickles; foliage medium, dark green, glossy; bushy, medium (90 cms) growth; [Pearl Drift X CherryAde]; Bossom, W.E.

Boscobel(TM), *see* 'KINBOSCO'

'BOSCONPEA', F, pb, 1997; (Fab, **Lady of Hertford**); flowers semi double, medium, semi-dbl., 8–14 petals, 2.5 in., borne in small clusters, slight; few prickles; foliage small, medium green, glossy; upright, tall (120 cms.) growth; [Conversation X Pearl Drift]; Bossom, W.E.

'BOSELADNEE', F, dr, 1996; (**Linda Mary**); flowers dark red, silver reverse, blooms borne in small clusters, dbl., 15–25 petals, 3.25 in., slight fragrance; few prickles; foliage medium, dark green, semi-glossy; bushy, medium (70 cms) growth; [Guinee X Glad Tidings]; Bossom, W.E.

'BOSELFTAY', F, mr, 1997; (**Connie Crook**); flowers full, large, very dbl., 26–40 petals, 3.5 in., borne in large clusters, slight fragrance; few prickles; foliage large, dark green, glossy; upright, tall (120 cms)growth; [Seedling X Selfridges]; Bossom, W.E.

'BOSEMWINE', Cl F, dp, 1995; (**Mrs Cecily McMullen**); flowers salmon pink, blooms borne in small clusters, very dbl., 2.75 in., moderate fragrance; few prickles; foliage large, medium green, semi-glossy; tall, spreading (10 ft × 10 ft), climbing growth; [English Miss X Summer Wine]; Bossom, W.E.

'BOSEYEBALL', S, mr, 1997; (**Baby Tom**); flowers single, single, 4–7c petals, borne in small clusters, slight fragrance; few prickles; foliage small, light green, glossy; spreading, medium (60 cms) growth; [Eyepaint X Ballerina]; Bossom, W.E.

'BOSGREEN', F, w, 1992; (**Peppermint Ice**); flowers creamy green, blooms borne singly or in sprays of 3–5, semi-dbl., 10 petals, 3.25 in., cupped, slight fragrance; foliage medium, medium green, semi-glossy; upright, medium growth; [Anne Harkness X Greensleeves]; Bossom, W.E.; E.B. LeGrice Roses, Ltd., 1991

'BOSHIPEACON', F, op, 1994; (**Emily Victoria**); flowers salmon pink, blooms borne in small clusters, very dbl., 1.5–2.75 in., moderate fragrance; some prickles; foliage medium, medium green, semi-glossy; low to medium (60 cms), bushy, compact growth; [Conservation X (Pearl Drift X Highfield)]; Bossom, W.E.

'BOSHOTEARL', F, lp, 1997; (**Lady Barbara Blossom**); flowers full, frilly, large, very dbl., 26–40 petals, 4 in., borne in small clusters, slight fragrance; some prickles; foliage medium, light green, glossy; upright, tall (90 cms) growth; [Savoy Hotel X Pearl Drift]; Bossom, W.E.

'BOSILJURIKA', F, mr, 1995; (**Preservation**); flowers bright red, blooms borne in small clusters, very dbl., 2.75 in., slight fragrance; some prickles; foliage medium, medium green, glossy; medium, upright growth; [Silver Jubilee X Paprika]; Bossom, W.E.

'BOSLAURAMBER', Min, dy, 1998; (**Good Morning Sunshine**); flowers small, golden yellow, very dbl., 26–40 petals, 2.5 in., borne mostly singly, moderate fragrance; prickles numerous; foliage small, dark green, semi-glossy; low, compact growth; [Laura Ford X Forever Amber]; Bossom, W.E.

'BOSLORVET', F, ob, 1993; (**Sharon Anne**); flowers orange salmon, blooms borne in large clusters, dbl., 15–25 petals, 1.5–2.75 in., slight fragrance; few prickles; foliage large, medium green, semi-glossy; tall to medium (120 cms), upright growth; [Sharon Lorraine X Brown Velvet]; Bossom, W.E.; Bossom, 1993

'BOSNEECH', S, dr, 1996; (**William Stubbs**); dbl., 15–25 petals, 3.25 in., borne in small clusters, early flowering (May), slight fragrance; moderate prickles; foliage small, medium green, glossy; spreading (10 × 12 ft) growth; [Seedling X Guinee]; Bossom, W.E.

Bosom Buddy, HT, or, 1992; Wells

'BOSPARDON', F, m, 1998; (**Nancy Bennett**); flowers lavender pink, dbl., 15–25 petals, 2.75 in., borne in small clusters, slight fragrance; prickles moderate; foliage large, dark green, semi-glossy; tall, spreading growth; [City of London X Paprika]; Bossom, W.E.

'BOSPEABAY', F, mp, 1993; (**Tender Loving Care**); flowers medium pink, moderately full, dbl., 20 petals, 4 in., borne in large clusters, moderate fragrance; some prickles; foliage medium, dark green, semi-glossy; medium to tall (90 cms), upright growth; [Pearl Drift X (Dublin Bay X Seedling)]; Bossom, W.E.; F. Haynes & Partners, 1995

'BOSREXCITY', F, w, 1995; (**Frances Perry**); dbl., 26–40 petals, 3–3.5 in., borne in small clusters, moderate fragrance; some prickles; foliage medium, medium green, semi-glossy; low, bushy growth; [Sexy Rexy X City of London]; Bossom, W.E.

'BOSREXEVER', S, ly, 1998; (**Annie East**); flowers cream yellow, semi-dbl., 8–14 petals, 3 in., large clusters, slight fragrance; prickles moderate; foliage large, dark green, semi-glossy; spreading, medium growth; [Sexy Rexy X Forever Amber]; Bossom, W.E.

'BOSREXEYE', F, or, 1998; (**Irene Smith**); flowers coral with lighter reverse, very dbl., 26–40 petals, 2.75 in., borne in large clusters, slight fragrance; prickles moderate; foliage medium, dark green, glossy; bushy, medium growth; [Sexy Rexy X Eyepaint]; Bossom, W.E.

'BOSSA NOVA', HT, dy, 1964; flowers deep golden yellow, dbl., 28 petals, 4 in., exhibition form; foliage dark; [Leverkusen X Buccaneer]; McGredy, Sam IV; McGredy

Bossa Nova, *see* POUloma

'BOSSEXEYE', F, mr, 1999; (**Olive Elsie**); flowers medium red, lighter reverse, 3 in., borne in small clusters, slight fragrance; prickles moderate; foliage medium, light green, semi-glossy; medium (2.5 ft) growth; [Sexy Rexy X Eyepaint]; Bossom, W.E.

'BOSSUET', HGal, mr; flowers scarlet, edges darker

'BOSTON', HT, mp, 1917; dbl., intense fragrance; [Mrs George Shawyer X Seedling]; Montgomery Co.

'BOSTON BEAUTY', Pol, mp, 1919; flowers clear pink, dbl., moderate fragrance; [Orléans Rose X (Katharina Zeimet X Old Ayrshire rose)]; Farquhar

Botanica, *see* 'TOMBOT'

Botaniste Abrial, *see* **'LOWELL THOMAS'**

'BOTANISTE HENRI GRIMM', F, my, 1958; flowers golden to straw-yellow, becoming pink tinted, large, blooms in clusters; vigorous growth; [Goldilocks X Fashion]; Gaujard, R.; G. Truffaut

Botany Bay, HT, rb

'Botzaris', D, w, 1856; flowers creamy white, dbl., flat; foliage light green

'Boudoir', HT, pb, 1942; (Paul Fromont); flowers tyrian rose, reverse white, large, dbl., 50 petals, exhibition form; foliage leathery; vigorous, upright, bushy growth; [Ampere X (Charles P. Kilham X Margaret McGredy)]; Meilland, F.; C-P

Boudoir, *see* CLEbou

'Bougainville', N, pb, 1822; bud red; flowers pink in center, becoming paler and tinged with lilac at the base, very dbl., cupped; branches very prickly; foliage narrow, glossy; Vibert

Boule de Nanteuil, *see* **'Comte de Nanteuil'**

'Boule de Neige', B, w, 1867; flowers pure white, compact, dbl., occasional recurrent bloom, moderate fragrance; foliage dark; [Blanche Lafitte X Sappho (Vibert's)]; Lacharme, F.

Boulie's Dream™, *see* 'WILfolk'

'Bountiful', F, pb, 1972; flowers strawberry-salmon, reverse deeper, dbl., 33 petals, 3 in., exhibition form, moderate fragrance; foliage small; tall, erect growth; [Vesper X Seedling]; LeGrice

'Bouquet', F, dp, 1940; (Lied, Siegeslied); flowers deep pink, dbl., 33 petals, cupped, slight fragrance; foliage dark, leathery; vigorous, bushy, compact growth; [Ingar Olsson X Heidekind]; Tantau; C-P

'Bouquet Blanc', Pol, w, 1914; flowers small, dbl.; Conard; (Sangerhausen)

'Bouquet Blanc', HP, w, 1856; flowers white to light pink, medium, dbl.; Robert; (Sangerhausen)

'Bouquet Charmant', HGal, mp; flowers rosy purple, large, dbl.; Hardy, before 1811

Bouquet de la Mariée, *see* **'Aimée Vibert'**

Bouquet de Marie, HP, w; Damaizin, 1858

'Bouquet de Neige', Pol, w, 1900; flowers medium, dbl.; Vilin; (Sangerhausen)

Bouquet de Venus, HGal, lp; Lerouge

'Bouquet d'Or', N, yb, 1872; flowers yellow, center coppery salmon, large, dbl.; [Gloire de Dijon X Seedling]; Ducher

'Bouquet d'Or', HT, ly, 1922; flowers light golden yellow, well formed; vigorous growth; Lippiatt

Bouquet Fait, *see* 'LEN 1'

Bouquet Parfait®, HMsk, pb, 1989; Lens

'Bouquet Rose', Pol, lp, 1928; flowers flesh-pink and peach-blossom, clusters of 30–40; Granger-Gaucher; Turbat

'Bouquet Rouge', F, mr, 1963; flowers large blooms in clusters of 14–16, dbl., 45 petals; vigorous growth; [(Gruss an Teplitz X Independence) X (Independence X Floradora)]; Arles; Roses-France

'Bouquet Tout Fait', N, w; Laffay, M., before 1836

Bouquet Vanille®, *see* DELblatine

Bouquetterie, Min, my

Bourbon Queen, *see* **'Queen of Bourbons'**

'Bourbon Rose', 1817; (R. canina borboniana, R. X borboniana); flowers pink, red or purple solitary or in few-flowered corymbs, semi-dbl., 3 in., some are recurrent

Bourgogne®, S, mr, 1983; Interplant

Bourgogne No. 2, *see* ADAtapora

'Boursault Rose', (R. boursaultii, R. l'heritierana, R. reclinata); flowers pink to purple, nodding, in corymbs, semi-dbl., non-recurrent; subglobose, smooth fruit; climbing to 12 ft growth; (14 and probably 21)

Bouton Red Bourbon, B, dr

'Boutonniere', HT, op, 1940; flowers salmon-pink, dbl., 40–50 petals, 3 in.; foliage dark, glossy, bronze; vigorous, bushy, compact growth; [Lulu X Mrs Sam McGredy]; Lammerts, Dr Walter; Armstrong Nursery

'Bouzloudja', HT, mr, 1974; flowers large, dbl., 40 petals, slight fragrance; foliage dark, glossy; vigorous, bushy growth; [Sarah Arnot X Rina Herholdt]; Staikov, Prof. Dr V.; Kalaydjiev and Chorbadjiiski

Bow Bells™, *see* 'AUSbells'

Bowie Pink, S, mp

Bowie Pink Lady, HT, lp, 1995; bud dark pink; flowers soft pink with hint of lemon at heart, exhibition form; Williams, J. Benjamin

Bowie White Patio, Min, w; Williams, J. Benjamin

Bowie Yellow Patio, MinFl, my, 1995; exhibition form; Williams, J. Benjamin

Boy Crazy, *see* 'DICrevival'

Boy O Boy™, *see* 'DICuniform'

'Boy Scout', HT, or, 1946; (Scout); bud long, pointed; flowers flame, large, dbl., exhibition form, moderate fragrance; foliage dark, glossy; very vigorous, upright, bushy growth; [Joanna Hill X Olympiad]; Duehrsen; California Roses

Boys' Brigade®, *see* 'COCdinkum'

'Bozena Nemcová', HT, dp, 1931; flowers pure dark pink, very large, dbl., cupped, intense fragrance; foliage bronze, thick; strong stems; vigorous growth; [Sylvia X Priscilla]; Böhm, J.

'Brad', HT, ab, 1999; dbl., 26–40 petals, 5.5 in., borne mostly singly, moderate fragrance; few prickles; foliage medium, medium green, dull; bushy, medium (3 ft) growth; [Gavotte X Ravenswood]; Poole, Lionel

'Bradgate', HT, yb, 1970; flowers yellow and deep red bicolor, pointed, dbl., 27 petals, 4–5 in., slight fragrance; foliage dark; free growth; [Piccadilly sport]; Worth; Lowe

Bradley Craig®, *see* 'MACstewar'

Bradley Graig, *see* 'MACstewar'

'Bradova Germania', HFt, op, 1932; flowers coppery-pink, large, semi-dbl.; Brada, Dr; (Sangerhausen)

'Bradova Lososova Druschki', HP, op, 1937; flowers salmon-pink, large; Brada, Dr; (Sangerhausen)

Braine l'Alleud, HT, 1984; Delforge, H.; (Cavriglia)

'Braiswick Charm', HWich, ob, 1914; flowers orange-yellow, edges almost white clusters, moderate fragrance; foliage dark, glossy, leathery; long, strong stems; very vigorous growth; Cant, F.

Braithwaite, *see* 'AUScrim'

'Brandenburg', HT, or, 1965; flowers deep salmon, reverse darker, dbl., 40 petals, 5 in., exhibition form; vigorous, upright growth; [(Spartan X Prima Ballerina) X Karl Herbst]; Kordes, R.; McGredy

Brandenburg Gate, *see* 'JACgate'

'Brandon', LCl, mr, 1964; bud long, pointed; flowers dark carmine-red, large, dbl., very free, recurrent bloom; vigorous growth; Combe

Brandy™, *see* 'AROcad'

'Brandy Butter', HT, ob, 1982; bud long, pointed; flowers pale gold, blooms borne 4–5 per cluster, dbl., 28 petals, exhibition form, moderate fragrance; thick, triangular-shaped prickles; foliage mid-green; tall, upright growth; [Fred Gibson X Royal Gold]; Northfield, G.

Brandy Snap™, *see* 'DICqueue'

Brandy, Climbing, Cl HT, lp; [Brandy sport]; Swane, 1994; (Weatherly, L.)

'BRAndyglow', F, yb, 1993; (**Vanessa Belinda**); flowers yellow blend, dbl., 15–25 petals, 3–3.5 in., borne in small clusters, moderate fragrance; some prickles; foliage medium, dark green, glossy; medium (100 cms), bushy growth; [Champagne Cocktail X Seedling (Glenfiddich X Priscilla Burton)]; Bracegirdle, A.J.; Bracegirdle, 1993

'BRAndypink', F, mp, 1992; (**Ronald George Kent**); dbl., 15–25 petals, 3–3.5 in., borne in small clusters, slight fragrance; few prickles; foliage large, dark green, semi-glossy; medium (86 cms), bushy growth; [Pink Favorite X Piccasso]; Bracegirdle, A.J.

Brandysnap, F, ab, 1999

Brandysnap, *see* 'DICqueue'

'BRANDYWINE', HT, ly, 1941; bud long, pointed; flowers buff-yellow, dbl., 25–30 petals, 4–5 in., slight fragrance; foliage olive-green, leathery; very vigorous, upright growth; John Cook Medal, ARS, 1945; [Seedling X Souvenir]; Thompson's, J.H., Sons; C-P; J&P

Brasero, F, mr

Brasier, *see* **'BRAZIER'**

'BRASILIA', HT, rb, 1968; flowers scarlet, reverse gold, dbl., 4 in., slight fragrance; [Kordes' Perfecta X Piccadilly]; McGredy, Sam IV; McGredy

Brass Band™, *see* 'JACCOFL'

Brass Monkey, *see* HARxaglen

Brass Ring, *see* 'DICGROW'

'BRATOWIN', HT, dp, 1993; (**George Armer**); flowers deep pink, blooms borne mostly single, dbl., 15–25 petals, 1.5–2.75 in., exhibition form, slight fragrance; some prickles; foliage medium, bronze-red, matt; medium (12 cms), upright growth; [Gavotte X First Prize]; Bracegirdle, A.J.; Bracegirdle, 1993

Braunwald®, S, w

Bravado, *see* 'JACRO'

Brave Heart, *see* 'HORBONDSMILE'

Braveheart, *see* CLEbravo

'BRAVO', HT, mr, 1951; bud ovoid; flowers cardinal-red, borne in clusters, dbl., 35 petals, 4–5 in., exhibition form, slight fragrance; foliage leathery; vigorous, upright, moderately bushy growth; [World's Fair X Mirandy]; Swim, H.C.; Armstrong Nursery

Braz Ornelas, HT; Moreira da Silva, A.; (Cavriglia)

'BRAZIER', HT, mr, 1937; (Brasier); flowers flame-scarlet, large, dbl., slight fragrance; foliage glossy; very vigorous, bushy growth; [Charles P. Kilham X Seedling]; Mallerin, C.; A. Meilland; C-P

'BRAZIL', HT, rb, 1947; bud long, pointed; flowers saturn-red, reverse saffron-yellow, well shaped, dbl.; foliage light green; Caron, B.; URS

Bread 'n' Butter, *see* 'TINBUTT'

Break o' Dawn, *see* 'MINFCO'

'BREAK O' DAY', Pol, pb, 1937; flowers copper-pink, center yellow, fading to shell-pink, blooms in, semi-dbl., cupped, recurrent bloom; vigorous growth; Archer

'BREAK O' DAY', HT, ob, 1939; (Delta); flowers orange shades, large, dbl., 50 petals, intense fragrance; vigorous growth; [Seedling X Glenn Dale]; Brownell, H.C.

'BREAK O'DAY, CLIMBING', Cl HT, ob, 1944; Brownell, H.C.

'BREAKAWAY', Min, mr, 1980; bud ovoid, pointed; dbl., 23 petals, borne singly, slight fragrance; straight, tiny, brownish-green prickles; foliage small, medium green; compact, upright, bushy growth; [Dandy Lyon X Seedling]; Lyon, Lyndon

Breath of Life, *see* 'HARQUANNE'

Breathless, *see* 'JACCHRY'

Breathtaking, *see* 'LEOBRETAK'

Bredon®, *see* 'AUSbred'

'BREEZE HILL', LCl, ab, 1926; flowers flesh tinted apricot, center rose, paling, blooms in clusters, dbl., 55 petals, 3 in., cupped, non-recurrent, moderate fragrance; bushy growth, heavy canes; [R. wichurana X Beauté de Lyon]; Van Fleet; American Rose Society

Breezy™, *see* 'SAVABREZ'

'BREGINA', F, or, 1965; flowers vermilion-red; [Mandrina sport]; deRuiter

'BRENDA', HEg, lp, 1894; flowers peach-blossom-pink; foliage fragrant; very vigorous growth; Penzance; Keynes, Williams & Co.

'BRENDA ANN', HT, dy, 1974; flowers amber-yellow, full, dbl., 37 petals, 5 in., slight fragrance; foliage glossy, tinted bronze; vigorous, tall, upright growth; [Piccadilly sport]; Watts

Brenda Burg, *see* CLEswan

'BRENDA COLVIN', LCl, lp, 1970; (R. filipes 'Brenda Colvin'); single, 5 petals, 1 in., blooms in trusses, intense fragrance; foliage dark, glossy; very vigorous growth; Colvin; Sunningdale Nursery

Brenda Lee, *see* 'MICLEE'

'BRENDA OF TASMANIA', F, w, 1971; bud ovoid; flowers white, center pink, large, dbl., cupped, moderate fragrance; foliage large, glossy; upright, bushy growth; [Queen Elizabeth sport]; Holloway

'BRENDA'S FRAGRANCE', HT, pb, 1967; flowers cerise edged white, exhibition form, intense fragrance; foliage leathery; vigorous growth; [Lieut. Chaure X Hector Deane]; Smith, W.H.

Brennende Liebe, *see* **'BURNING LOVE'**

Brennpunkt, F, 1972; Haenchen, E.; (Cavriglia)

'BRENNUS', HCh, dr, 1830; flowers deep red, shaded with violet; Laffay, M.

'BRESILIENNE', Pol, dr, 1971; bud long, pointed; flowers deep red, medium, dbl., cupped, slight fragrance; foliage bronze, leathery; vigorous, dwarf, bushy growth; [Red Favorite X Ena Harkness]; Delforge

Brewood Belle, LCl, mr, 1996; Scarman

'BRIACE', Min, mr, 1998; (**Ace of Diamonds**); flowers bright medium red, good substance, very dbl., 26–40 petals, 1–1.5 in., exhibition form, borne mostly singly and small clusters, intense fragrance; few prickles; foliage small, dark green, semi-glossy; strong; bushy, spreading, medium growth; Disease resistant; Bridges, Dennis A.; Bridges Roses, 1998

Brian, *see* BARrian

Brian Donn, *see* 'TINDONN'

Brian Lee, *see* 'TINLEE'

'BRIANCOUF', S, mp, 1989; (**Baronne Surcouf**); Briant

'BRIAND-PANEUROPA', HT, mp, 1931; flowers carmine-rose, base yellow, very large, dbl., slight fragrance; vigorous growth; [Franklin X Willowmere]; Böhm, J.

Brianna, *see* 'DEVCARLOS'

'BRIAN'S SONG', S, op, 1977; dbl., 35 petals, 4 in., exhibition form, intermittent bloom, moderate fragrance; foliage large, glossy, dark; [Independence X Pike's Peak]; Smith, R.L.; Smith's Greenhouse & Nursery

Briant Hill®, Min, ob, 1994; Poulsen

'BRIARCLIFF', HT, pb, 1926; flowers center deep rose-pink, outer petals lighter; [Columbia sport]; Pierson, P. M.

'BRIARCLIFF BRILLIANCE', HT, dp, 1932; flowers rose to rose-red, large, dbl.; vigorous growth; [Briarcliff sport]; Pierson, P.M.

'BRIARCLIFF SUPREME', HT, pb, 1947; bud long, pointed; flowers briarcliff pink, very large, dbl., intense fragrance; foliage dark; very vigorous, upright growth; [Briarcliff sport]; Hinner, P.; Bauské Bros. & Hinner

'BRIARCLIFF, CLIMBING', Cl HT, pb, 1929; Parmentier, J.

'BRIAUTUMN', Min, op, 1999; (**Autumn Dawn**); flowers orange-pink, reverse slightly lighter, dbl., 17–25 petals, 1.3 in., slight fragrance; prickles moderate; foliage medium, dark green, semi-glossy; spreading, medium (2 ft) growth; [Carrot Top X Unknown]; Bridges, Dennis A.

BRIbaby, Min, w, 1997; (**Baby Secret**); Bridges, Dennis A.

'BRIBEN', HT, pb, 1993; (**Dr Jack Bender**); flowers pink and white, blooms borne mostly single, dbl., 26–40 petals, 3–3.5 in., slight fragrance; many prickles; foliage medium, dark green, semi-glossy; bushy, spreading (100 cms) growth; [Lady X X Flaming Beauty]; Bridges, Dennis A.; Bridges Roses, 1993

'BRIBOLT', HT, pb, 1990; (**Thunderbolt**); bud ovoid; flowers pink blend, white reverse, color intensifies with age, dbl., 45 petals, slight fragrance; foliage medium, dark green, matt; bushy, medium growth; [Lady X X Flaming Beauty]; Bridges, Dennis A., 1980; Bridges Roses, 1991

'BRICHARM', Min, yb, 1992; (**Southern Charm**); flowers shades of light yellow and pink, pink intensifying with sun, dbl., 25–40 petals, 1.5–2.75 in., slight fragrance; no prickles; foliage large, medium green, semi-glossy; medium (40–45 cms), upright, bushy, slightly spreading growth;

[Baby Katie X Unknown (select pollen)]; Bridges, Dennis A.; Bridges Roses

'BRIchat', Min, my, 1999; (**Chit Chat**); dbl., 26–40 petals, .75 in., borne mostly singly, no fragrance; few prickles; foliage medium, dark green, semi-glossy; compact, bushy, medium (18–20 in.) growth; [Cal Poly X Unknown]; Bridges, Dennis A.

'BRIclass', HT, pb, 1992; (**Carolina Classic**); flowers medium pink edged deeper pink, blooms borne mostly singly, very dbl., 3–3.5 in., slight fragrance; few prickles; foliage medium, medium green, matt; medium (90–100 cms), upright growth; [Just Lucky X Flaming Beauty]; Bridges, Dennis A.; Bridges Roses

'BRIcos', Min, ob, 1997; (**Cosmic**); flowers light to medium orange with light yellow reverse; full (26–40 petals), medium (4–7 cms) blooms borne in small clusters; slight fragrance; foliage large, medium green, semi-glossy; upright (24–26 in.), bushy growth; [Suzy X Seedling]; Bridges, Dennis A.; Bridges Roses

Bridal Blush™, HT, w; DeVor Nurseries, Inc, 1989

'Bridal Bouquet', HT, w, 1976; dbl., 23 petals, 3–4 in., slight fragrance; foliage matt, green; vigorous, compact growth; [Lady Sylvia sport]; Ormerod

Bridal Delight, HT, lp; [Bridal White sport]; Day, 1992; (Weatherly, L.)

Bridal Pink™, *see* 'JACbri'

'Bridal Robe', HT, w, 1955; flowers ivory-white, high-pointed,, dbl., 54 petals, 4 in., moderate fragrance; foliage glossy, olive-green; vigorous growth; GM, NRS, 1953; [McGredy's Pink X Mrs Charles Lamplough]; McGredy, Sam IV

Bridal Shower™, *see* 'JACtafl'

Bridal Shower, *see* DEVblush

Bridal Sonia, Gr, mp

Bridal Sunblaze®, *see* 'MEIlmera'

'Bridal Veil', F, ly, 1954; bud ovoid, cream; flowers white overcast seafoam-yellow, open, medium in clusters, dbl., 75–80 petals, 2.5 in., intense fragrance; vigorous, bushy growth; [Pinocchio seedling X Pigmy Gold]; Boerner; Stark Bros.

Bridal White™, *see* 'JACwhy'

'BRIdawn', Min, m, 1991; (**Purple Dawn**); flowers velvety mauve, long-lasting, urn-shaped, blooms borne usuall, dbl., 25 petals, 1.5 in., exhibition form, slight fragrance; foliage medium, dark green, semi-glossy; upright, tall growth; [Party Girl X Unknown]; Bridges, Dennis A.; Bridges Roses

Bride, *see* FRYyearn

'Bride's Blush', HT, w, 1923; flowers creamy white, at times blush-pink, single, 6 petals, moderate fragrance; [Columbia sport]; Amling Co.

Bride's Dream, *see* 'KORoyness'

'Bride's White', F, w, 1968; flowers pure white, small, dbl., cupped; vigorous, upright, bushy growth; Mansuino, Q.; Carlton Rose Nurseries

'Bridesmaid', T, lp, 1893; [Catherine Mermet sport]; Moore

'Bridesmaid, Climbing', Cl T, lp, 1893; Moore

'Bridget', HT, op, 1947; bud long, pointed; flowers brilliant orange-scarlet, base bright golden yellow, dbl., 35–40 petals, 4–5 in., flat; foliage glossy, bright green; [Mrs Henry Bowles X Phyllis Gold]; Fletcher; Tucker

Bridget, *see* CLEbridge

Bridget, *see* KOReibei

Bridgets Joy, HT, lp, 1997; flowers full, large, very dbl., 26–40 petals, 4 in., borne mostly singly, slight fragrance; foliage medium, medium green, semi-glossy; compact, medium (140 cms) growth; [Coral Fiesta sport]; Johnstone, Leonard E.

'Bridgwater Pride', F, op, 1982; flowers rich salmon, medium, dbl., 20 petals, moderate fragrance; foliage medium, dark, semi-glossy; bushy, compact growth; [Vera Dalton X Allgold]; Sanday, John; Sanday Roses, Ltd.

'BRIdixie', Min, rb, 1992; (**Dixieland**); flowers in shades of vibrant pink, red and white intensifying with sun, dbl., 15–25 petals, 1.5–2.75 in., exhibition form, slight fragrance; no prickles; foliage medium, medium green, semi-glossy; tall (50–60 cms), upright growth; [Fancy Pants X Seedling]; Bridges, Dennis A.; Bridges Roses

Brier Bush, *see* **R. canina**

'BRIfire', Min, rb, 1991; (**Gabriel's Fire**); bud pointed; flowers creamy light yellow turning red, urn-shaped, blooms borne usually singly, semi-dbl., 20–22 petals, 2 in., exhibition form, intense fragrance; foliage medium, medium green, semi-glossy; bushy, spreading, medium growth; [Sachet X Unknown]; Bridges, Dennis A., 1988; Bridges Roses, 1992

'BRIfree', HT, w, 1994; (**Freemont**); flowers near white blushed with salmon, blooms borne mostly singly, very dbl., 3–3.5 in., moderate fragrance; some prickles; foliage large, dark green, semi-glossy; tall, upright growth; [Thriller X Tiki]; Bridges, Dennis A.; Bridges Roses, 1994

'Brigadeiro França Borges', F, or, 1960; flowers bright orange-red, [Independence X Seedling]; da Silva, Moreira

Brigadoon™, *see* 'JACpal'

BRIgarold, HT, mr, 1989; (**Roland Garros**®); Briant

'BRIgaunt', Min, dr, 1991; (**Gauntlet**); bud ovoid; flowers dark red, lighter reverse, lightens slightly with age, medium, semi-dbl., 20–22 petals, exhibition form, slight fragrance; foliage large, medium green, semi-glossy; spreading, medium growth; [Kitty Hawk X Unnamed seedling]; Bridges, Dennis A., 1986; Bridges Roses, 1990

'BRIgentle', Min, mr, 1998; (**Gentleman's Agreement**); bud long, slender; flowers medium red, excellent substance, very dbl., 26–40 petals, 1–1.5 in., exhibition form, borne mostly singly, no fragrance; few prickles; foliage large, medium green, semi-glossy; tall, upright, bushy growth; Disease resistant; [Purple Dawn X Select pollen]; Bridges, Dennis A.

'Bright Angel', S, yb, 1977; bud ovoid; flowers light yellow, edged pink, dbl., 48 petals, 4 in., exhibition form, intermittent bloom, intense fragrance; foliage large, glossy, dark; vigorous, bushy growth; [Dornroschen X Golden Wings]; Smith, R.L.; Smith's Greenhouse & Nursery

'Bright Beam', Gr, pb, 1972; bud ovoid; flowers cream, edged pink, large, dbl., cupped, profuse, intermittent bloom, moderate fragrance; foliage glossy, dark; very vigorous, upright growth; [Peace X Little Darling]; Fuller; Wyant

Bright Beauty, HT, or

'Bright Boy', HT, lp; brilliant red, Clark, A., 1948

Bright Eyes, *see* 'SANmar'

'Bright Eyes', F, 1948; bud ovoid; flowers light yellow to primrose, medium, large truss, dbl., 25 petals, slight fragrance; foliage leathery, glossy, dark; dwarf growth; RULED EXTINCT 6/83 ARM; [Joanna Hill X (Heidekind X Betty Uprichard)]; Duehrsen; H&S

'Bright Fire', *see* PEAxi

'Bright Garbs', F, or, 1977; bud pointed; semi-dbl., 12 petals, 2.5 in., moderate fragrance; foliage large; dwarf growth; [Orangeade X Seedling]; Hardikar, Dr M.N.

'Bright Jewel', Min, pb; bud pointed; flowers rose-pink, white center, very small blooms in clusters, semi-dbl.; low, compact growth

Bright Lights, *see* 'KSEelite'

Bright Meadow, *see* **'Yasnaya Poliana'**

'Bright Melody', S, mr, 1984; bud ovoid, pointed; flowers large, dbl., 30 petals, cupped, borne 1–10 per cluster, repeat bloom, slight fragrance; awl-like, tan prickles; foliage medium-large, dark olive green, leathery; erect, bushy growth; [Carefree Beauty X (Herz As X Cuthbert Grant)]; Buck, Dr Griffith J.; Iowa State University

'Bright Morning', HT, my, 1958; flowers golden yellow, moderate fragrance; Ratcliffe

'BRIGHT RED', Pol, dr, 1938; flowers velvety dark red, large cluster, dbl.; bushy growth; deRuiter

Bright Sight, *see* 'JUDBRIGHT'

Bright Smile®, *see* 'DICDANCE'

Bright Spark, *see* RUbrispa

Bright Star, MinFl, 1999; single; Williams, J. Benjamin

'BRIGHT WINGS', HT, op, 1942; (Gitane); bud long, pointed, rosy orange; flowers centers orange shading to pink, large, dbl., 22 petals, cupped, moderate, fruity fragrance; foliage bronze; strong stems; vigorous, bushy growth; [Mme Arthaud X Annie Drevet]; Mallerin, C.; A. Meilland;, C-P

'BRIGHTNESS', HT, rb, 1958; bud long, pointed; flowers scarlet, reverse golden yellow, merging to orange, dbl., 28–35 petals, 4.5–5.5 in., exhibition form, intense fragrance; foliage leathery, glossy; bushy, moderately vigorous growth; [Doreen sport]; Fryers Nursery, Ltd.

'BRIGHTSIDE', Min, or, 1974; dbl., 25 petals, 1 in., exhibition form, moderate fragrance; foliage small, matt; upright, bushy growth; [Persian Princess X Persian Princess]; Moore, Ralph S.; Park Seed Co.

'BRIGHTSIDE CREAM', N, w; flowers creamy white, yellow stamens blooms in clusters, semi-dbl., 18 petals, 3 in., foliage dark; vigorous growth, long canes

Brigitte de Landsvreugd, LCl, w

Brigitte de Villenfagne, HMsk, lp, 1993; flowers small to medium, single, moderate fragrance; Lens, Louis; (Sangerhausen)

'BRIGITTE JOURDAN', F, or, 1960; bud ovoid; flowers pomegranate-red, large, dbl., globular, slight fragrance; foliage dark, glossy; vigorous, bushy growth; [Belle Créole X Independence]; Arles; Roses-France

'BRIGRAN', HT, mp, 1990; (**Grand Romance**); bud ovoid; flowers reverse slightly lighter, aging slightly lighter, urn-shaped, very dbl., 50 petals, moderate fragrance; foliage medium, dark green, semi-glossy; upright, medium growth; [Lady X X Wini Edmunds]; Bridges, Dennis A., 1987; Bridges Roses

'BRIHEART', Min, dr, 1996; (**Hearts A'Fire**); flowers velvety, blooms borne mostly singly, single, 15–25 petals, 1.25 in., slight fragrance; moderate prickles; foliage medium, dark green, glossy, disease-resistant; strong, upright growth; [Merrimac X Seedling]; Bridges, Dennis A.; Bridges Roses, 1996

'BRIICE', HT, lp, 1994; (**Ice Breaker**); dbl., 15–25 petals, 1.5–2.75 in., borne mostly singly, intense fragrance; some prickles; foliage medium, medium green, semi-glossy; tall, upright growth; [Thriller X Just Lucky]; Bridges, Dennis A.; Bridges Roses, 1994

'BRIINCOG', Min, m, 1995; (**Incognito**™); flowers mauve blend with yellow reverse, dbl., 15–25 petals, 1.5 in., borne mostly singly, slight fragrance; few prickles; foliage medium, dark green, semi-glossy; tall (30–34 in.), upright, bushy growth; [Jean Kenneally X Twilight Trail]; Bridges, Dennis A.; Bridges Roses, 1995

BRIletjui, S, rb, 1989; (**Juillet®**); Briant

'BRILLANT', HT, mr, 1952; flowers somewhat large, semi-dbl., moderate fragrance; Kordes, W. Söhne; (Sangerhausen)

Brillant, S, or, 1983; flowers medium, dbl., slight fragrance; Kordes, W. Söhne; (Sangerhausen)

'BRILLIANCE', F, or, 1958; bud ovoid; flowers coral, dbl., 50 petals, 2 in., cupped, moderate fragrance; foliage glossy; compact growth; [Seedling X Independence]; Boerner; J&P

'BRILLIANCY', HT, or, 1936; flowers brilliant scarlet, dbl., intense fragrance; foliage leathery; short stems; bushy growth; [Étoile de Hollande X Daily Mail Scented Rose]; LeGrice

Brilliant, *see* **'DETROITER'**

Brilliant Cover, *see* POUlterp

'BRILLIANT ECHO', Pol, mp, 1927; flowers rosy pink; [Echo sport]; Western Rose Co.

'BRILLIANT KING', F, mr, 1961; flowers bright red, shallow, dbl., 32 petals, 4 in., cupped, slight fragrance; [Cocorico X Orange Delight]; Leenders, J.

Brilliant Light, *see* **'KAGAYAKI'**

Brilliant Meillandina, *see* 'MEIRANOGA'

Brilliant Pink Iceberg™, *see* 'PROBRIL'

'BRILLIANT RED', HT, mr, 1938; bud very long, pointed; flowers brilliant red, well formed, very large, dbl.; foliage bright green; vigorous, bushy growth; [Charles P. Kilham X Étoile de Hollande]; Lens

'BRILLIANT STAR', F, mr, 1965; flowers bright red, center shaded yellow, semi-dbl., 12–20 petals, 2.5–3 in., borne in clusters; foliage glossy; vigorous growth; [Masquerade X Dicksons Flame]; Watkins Roses

'BRIMAN', Min, mp, 1993; (**Manteo**); dbl., 15–25 petals, 1.5 in., borne mostly singly, slight fragrance; some prickles; foliage small, medium green, semi-glossy; tall (55–60 cms), upright, bushy growth; [Rise 'n' Shine X Seedling]; Bridges, Dennis A.; Bridges Roses, 1993

'BRINDIS', LCl, rb, 1962; flowers geranium-red, center yellow, single; foliage dark; vigorous growth; [Orange Triumph, Climbing X (Phyllis Bide X Baccará)]; Dot, Simon

Brinessa®, HT, pb, 1985; Delforge

'BRINEY', Min, w, 1994; (**Whitney**); flowers near white with a hint of pink, blooms borne mostly single or, dbl., 15–25 petals, 1.5 in., moderate fragrance; few prickles; foliage small, medium green, semi-glossy; medium (38 cms), upright, spreading growth; [Jennifer X Unknown]; Bridges, Dennis A.; Bridges Roses, 1994

Briosa, *see* BARbrio

'BRIROSE', Min, m, 1998; (**Moonlight and Roses**); flowers light lavender with rosy edges, with darker lavender and lavender overlay, very dbl., 26–40 petals, 2–2.5 in., exhibition form, borne mostly singly, slight fragrance; some prickles; foliage medium, dark green, glossy; bushy, upright growth; [Seedling X Select pollen]; Bridges, Dennis A.; Bridges Roses, 1998

Brisbane Blush, *see* 'PALOK'

'BRISE PARFUMÉE', F, mr, 1950; bud pointed; dbl., borne in clusters, intense fragrance; many spines; foliage glossy, bronze; bark and twigs reddish brown; very vigorous growth; [Böhm's Triumph X Baby Chateau]; Truffaut, G.

BRIseflu, S, w; (**White Fleurette®**); Briant, 1989

BRIsiness, Min, w, 1990; (**Riviere de Diamant**); Briant

'BRISIS', Min, w, 1998; (**Sis**); flowers delicate white with a hint of pink, firm substance, dbl., 15–25 petals, 1.5 in., exhibition form, borne mostly singly, slight fragrance; few prickles; foliage small, medium green, semi-glossy; strong, bushy, slightly spreading growth; Disease resistant; [Party Girl X Cape Hatteras]; Bridges, Dennis A.; Bridges Roses, 1998

'BRISTEP', HT, mp, 1995; (**Steppin' Out**); flowers medium pink, full (26–40 petals), large (7 cms) blooms borne mostly single; fragrant; few prickles; foliage medium, dark green, semi-glossy; medium (4 1/2 ft), bushy growth; [Kardinal X Thriller]; Bridges, Dennis; Bridges Roses, 1995

'BRISTOL', HT, rb, 1968; flowers bright crimson, reverse lighter, large, dbl., moderate fragrance; foliage dark; compact growth; [Gavotte X Tropicana]; Sanday, John

'BRISTOL POST', HT, op, 1972; flowers pale salmon-pink, base orange, pointed, dbl., 29 petals, 4.5 in., slight fragrance; foliage slightly glossy, dark; upright growth; [Vera Dalton X Parasol]; Sanday, John

'BRISUN', Min, yb, 1993; (**Summer Sunset**); flowers bright yellow with salmon pink shading, blooms borne mostly, dbl., 15–25 petals, 1.5 in., slight fragrance; some prickles; foliage small, medium green, semi-glossy; medium (45 cms), bushy, compact growth; [Fancy Pants X Seedling]; Bridges, Dennis A.; Bridges Roses, 1993

'BRISUZY', Min, mp, 1990; (**Suzy**); bud pointed; flowers medium pink, near white at base blending to light pink, light, dbl., 38–40 petals, exhibition form, slight fragrance; foliage medium, dark green, semi-glossy; bushy, medium growth; PP007738; AOE, 1991; [Party Girl X Unknown seedling]; Bridges, Dennis A., 1986; Bridges Roses, 1991

'BRITANNIA', Pol, rb, 1929; flowers crimson, center white, small, borne in clusters of 30–40, single, recurrent bloom; foliage small, leathery, light; compact, bushy growth; [Coral Cluster X Eblouissant]; Burbage Nursery

Britannia, *see* FRYcalm

Britannica, Pol, rb

Brite Blue, HT, m, 1984

Brite Lites®, *see* 'HARTANNA'

Britestripe, *see* 'CLEBRITE'

Britestripes, *see* 'CLEBRITE'

'BRITIDE', Min, w, 1991; (**Tidewater**); bud pointed; flowers white with slight pink tints, blooms borne usually singly, dbl., 30–32 petals, 2.5 in., exhibition form, moderate fragrance; foliage medium, medium green, matt; long stems for cutting; bushy, spreading, medium growth; [Jennifer X Unknown]; Bridges, Dennis A., 1988; Bridges Roses

'BRITISH COLUMBIA CENTENNIAL', HT, pb, 1971; bud ovoid; flowers light rose-pink, reverse white, large, dbl., exhibition form, slight fragrance; foliage leathery; vigorous growth; [Pink Masterpiece X Seedling]; Boerner; Pan American Bulb Co.

'BRITISH QUEEN', HT, w, 1912; flowers creamy white, center flushed, open, dbl., slight fragrance; foliage light green, soft; long, weak stems; bushy growth; GM, NRS, 1912; McGredy

'BRITRICK', Min, rb, 1995; (**Trickster**); flowers red with white reverse, blooms borne mostly singly, dbl., 28–30 petals, 1.5 in., intense fragrance; foliage small, medium green, semi-glossy; medium (45 cms) growth; [Jennifer X Red Beauty]; Bridges, Dennis A.; Bridges Roses, 1995

Brittany's Glowing Star, *see* 'MANSTAR'

'BRITY', HT, mr, 1997; (**Tycoon**); flowers very full, large, very dbl., 41 petals, 2.75 in., borne mostly singly, slight fragrance; foliage medium, dark green, semi-glossy; medium (4 ft) upright, bushy growth; [Kardinal X Thriller]; Bridges, Dennis A.

'BRIYOUNG', Min, pb, 1997; (**Forever Young**); flowers full, small, very dbl., 26–40 petals, 1.5 in., borne mostly singly, slight fragrance; foliage medium, dark green, semi-glossy; low (12–14 in.) compact growth; [Trickster X Select Pollen]; Bridges, Dennis A.

'BRNO', HT, ob, 1933; flowers maroon, orange, gold, open, large, semi-dbl.; foliage glossy; vigorous, bushy growth; [Souv. de George Beckwith X Rosemary]; Böhm, J.

'BROADCASTER', Gr, or, 1969; bud ovoid; flowers medium, dbl., slight fragrance; foliage dark; vigorous growth; [Queen Elizabeth X Circus]; Perry; Conklin

Broadlands, *see* TANmirson

Broadway™, *see* 'BURWAY'

Broadway, LCl, lp; large single apricot flowers in spring, flat; healthy evergreen foliage; Clark, A.

Broadway (F), F, dp

'BROADWAY (S)', S, yb, 1933; Clark

'BROALA', Min, w, 1997; (**Alamode**); flowers single, medium, single, 4–7 petals, borne in small clusters, slight fragrance; foliage medium, dark green, glossy; bushy, medium (2.5 ft) growth; [Esprit X Party Girl]; Brown, Ted

'BROCADE', HT, pb, 1960; (Jeune Fille); bud ovoid; flowers soft rose, base cream-white, dbl., 54 petals, 4–5 in., cupped, moderate fragrance; foliage leathery; vigorous, upright growth; [Charlotte Armstrong X Baiser]; Combe; Hémeray-Aubert

Brocade, S, lp, 1998; flowers ivory with mauve tint, dbl.; Williams, J. Benjamin

'BROCAKE', Min, w, 1997; (**Cheese Cake**); flowers double, medium, dbl., 15–25 petals, borne mostly singly, slight fragrance; foliage medium, dark green, glossy; bushy, medium (5 ft) growth; [Esprit X Seedling]; Brown, Ted

'BROCARP', Min, dp, 1997; (**Fragrant Carpet**); flowers double, medium, dbl., 15–25 petals, borne in small clusters, intense fragrance; foliage medium, dark green, semi-glossy; spreading (2′ × 5′) growth; [Seedling X Seedling]; Brown, Ted

'BROCOR', F, ab, 1997; (**Corinne**); flowers double, medium, dbl., 15–25 petals, borne in small clusters, moderate fragrance; foliage medium, dark green, glossy; bushy, medium (30 in.) growth; [Esprit X Indian Summer]; Brown, Ted

'BRODAN', Min, mp, 1997; (**Danielle**); flowers medium pink fading to light pink as bloom opens, reverse light pink, full (26–40 petals), medium ($1\frac{1}{2}$ in.) blooms borne in small clusters; very fragrant, foliage medium, medium green, semi-glossy; medium (18 in.) growth; [Esprit X Seedling]; Brown, Ted.

'BRODREAM', F, m, 1997; (**Canadian Dream**); flowers single, medium, single, 4–7 petals, 3 in., borne in small clusters, moderate fragrance; foliage medium, dark green, glossy; upright, tall (4.5 x 6 ft) growth; [Nymphenburg X Seedling]; Brown, Ted

'BROEMP', F, dr, 1997; (**Red Emperor**); flowers semi double, large, semi-dbl., 15–25 petals, borne in small clusters, moderate fragrance; foliage medium, dark green, glossy; spreading, low (2 ft) growth; [Esprit X Seedling]; Brown, Ted

'BROFOX', F, ob, 1997; (**Foxtrot**); flowers single, medium, single, 4–7 petals, borne mostly singly, slight fragrance; foliage medium, burgundy turning dark green, glossy; spreading, low (1 ft) growth; [Esprit X Stretch Johnson]; Brown, Ted

'BROGAR', F, rb, 1997; (**Red Garter**); flowers single, medium, single, 4–7 petals, borne in small clusters, slight fragrance; foliage medium, dark red turning dark green, glossy; upright, medium (4 ft) growth; [Esprit X Stretch Johnson]; Brown, Ted

'BRÖG'S CANINA', (strain of R. canina), lp; almost thornless; vigorous growth; in use as an understock; Brög

'BROHEART', Min, mr, 1997; (**Heartstrings**); flowers very full, large, opening to OGR-type bloom, very dbl., 41 petals, borne in small clusters; foliage medium, dark green, dull; bushy, medium (16 in.) growth; [Esprit X Seedling]; Brown, Ted

'BROMAN', Min, ob, 1997; (**Music Man**); flowers double, medium, dbl., 15–25 petals, borne mostly singly, slight fragrance; foliage medium, medium green, semi-glossy; upright, tall (20–24 in.) growth; [Esprit X Seedling]; Brown, Ted

Bronce Masterpiece, *see* **'BRONZE MASTERPIECE'**

'BRONEW', Min, lp, 1997; (**New Antique**); flowers full, large, very dbl., 26–40 petals, borne in small clusters, no fragrance; foliage medium, dark green, semi-glossy; bushy, tall (30 in.) growth; Brown, Ted

Bronte, LCl, dy

Bronze Baby, *see* PEAxanthous

'BRONZE BEAUTY', F, ob, 1974; bud ovoid-pointed; flowers golden yellow to orange-yellow, open, dbl., 27 petals, 4 in., slight fragrance; foliage large, leathery; vigorous growth; [Electra X Woburn Abbey]; Warriner, William A.; J&P

'BRONZE BEDDER', HT, yb, 1920; flowers bronzy yellow, large, single; Paul, W.

'BRONZE MASTERPIECE', HT, ab, 1960; (Bronce Masterpiece); bud long; flowers bronze-apricot, becoming orange-yellow, dbl., 48 petals, 5.5–6 in., exhibition form, moderate fragrance; foliage leathery, glossy; vigorous, upright growth; GM, Geneva, 1958; [Golden Masterpiece X Kate Smith]; Boerner; J&P

'BROOK SONG', S, my, 1985; flowers imbricated form, large blooms borne 1–8 per cluster, dbl., 40 petals, repeat bloom, moderate fragrance; awl-like, tan prickles; foliage leathery, dark; erect growth; hardy; [Prairie Star X Tom Brown]; Buck, Dr Griffith J.; Iowa State University, 1984

Brookdale Giant White, *see* **'JACKMAN'S WHITE'**

Brooks Red, HT, lp

'BROOKVILLE', HT, ly, 1942; flowers cream-yellow; [Leonard Barron sport]; Brookville Nursery

Broomfield Novelty, *see* **'MARGARET ANDERSON'**

'BROROY', Gr, dp, 1997; (**Viceroy**); flowers double, medium, dbl., 15–25 petals, borne in large clusters, intense fragrance; foliage medium, dark green, glossy; upright, medium (4.5 ft) growth; [Esprit X Harmonie]; Brown, Ted

'BROSUN', F, ob, 1997; (**Harvest Sun**); flowers semi double, large, semi-dbl., 8–14 petals, borne in large clusters (8–10), moderate fragrance; foliage medium, dark green, glossy; upright, tall (4.5 ft) growth; [Esprit X Mountbatten]; Brown, Ted

Brother Cadfael™, *see* **'AUSGLOBE'**

'BROTHER SUN', Gr, my, 1998; flowers medium yellow, deeper color at base, flattened top, long lasting, very dbl., 41–50 petals, 2.5–3 in., borne mostly singly, moderate fragrance; prickles moderate; foliage medium, medium green, dull; compact, upright growth; [Goldener Olymp X Australian Gold]; Fleming, Joyce L.; Hortico, Inc., 1997

'BROTHER WILFRID', HT, mp, 1976; [Alec's Red sport]; Wood

Brown County Splendor™, *see* 'WILKBSP'

'BROWN STUDY', F, r, 1999; flowers russet, turning brown, darker at edges, dbl., 15–25 petals, 2.75 in., borne singly and in small clusters, slight fragrance; prickles moderate; foliage medium, dark green, semi-glossy; spreading, medium (3 ft) growth; Jerabek, Paul E.; Freedom Gardens, 1998, U.S.

Brown Sugar, *see* 'TALBRO'

Brown Velvet, *see* 'MACULTRA'

'BROWNELL YELLOW RAMBLER', HMult, my, 1942; (Yellow Rambler); flowers petals recurved, dbl., borne in more open clusters than Dorothy Perkins, slight fragrance; [(Emily Gray X Ghislaine de Feligonde) X Golden Glow]; Brownell, H.C.

'BROWNIE', F, r, 1959; bud ovoid, tan shades, edged pinkish; flowers brownish tan, reverse yellow, blooms in small clusters, dbl., 38 petals, 3.5–4 in., cupped, moderate fragrance; foliage leathery; vigorous, upright, bushy growth; [Lavender Pinocchio seedling X Grey Pearl]; Boerner; J&P

'BROWSHOLME ROSE', Ayr, w, 1900

Brundrette Centenary, S, dp

Brunella®, Min, mr

'BRUNETTE', F, yb, 1970; bud ovoid; flowers amber-yellow to orange, dbl., 25 petals, 1.5–2.5 in., cupped; foliage glossy; [(Purpurine X Lavender Pinocchio X Fillette) X (Gold Strike X Golden Garnette)]; Lens; Spek

Bruocsella®, *see* 'LENBRU'

Brushstrokes, *see* 'GUESCOLOUR'

Bruun, F, 1995; Delforge, H.; (Cavriglia)

Bryan, LCl, my, 1993; Mekdeci

Bryan Freidel Pink Tea, T, w

Bryte White, HT, w

Bubble Bath, HMsk, lp, 1980; Matson

Bubble Gum, *see* 'BUSBUGUM'

Bubbles, *see* 'ZIPBUB'

Bubbles, *see* FRYbubbly

Bubikopf®, *see* TANkobi

'BUCBI', S, mp, 1977; (Audace®, **Carefree Beauty**™); bud ovoid, long, pointed; flowers light rose, 15–20 petals, 4.5 in., repeat bloom, moderate fragrance; foliage olive-green, smooth; vigorous, upright, spreading growth; PP004225; [Seedling X Prairie Princess]; Buck, Dr Griffith J.; C-P

'BUCBLU', HT, m, 1983; (**Blue Skies**); flowers large, dbl., 35 petals, moderate fragrance; foliage large, semi-glossy; upright, bushy growth; PP005756; [((Sterling Silver X Intermezzo) X (Sterling Silver X Simone)) X (Music Maker X (Blue Moon X Tom Brown))]; Buck, Dr Griffith J.; J.B. Roses, Inc., 1988

'BUCCANEER', Gr, my, 1952; bud urn-shaped; flowers buttercup-yellow, dbl., 30 petals, 3–3.5 in., cupped, moderate fragrance; foliage dark, leathery; vigorous, upright, tall growth; GM, Geneva, 1952; [Golden Rapture X (Max Krause X Capt. Thomas)]; Swim, H.C.; Armstrong Nursery

'BUCKEYE BELLE', S, pb, 1956; bud globular; flowers pale to deep pink, open, small, borne in compact clusters, semi-dbl., 15 petals, 1.5 in., abundant, recurrent bloom, moderate fragrance; foliage dark; vigorous, upright, bushy growth; [R. hugonis X Seedling]; Garwood

Bucks Fizz, *see* 'POULGAV'

Budapest, HT, ab

'BUENOS AIRES', F, r, 1957; flowers burnt brick to dark red-ochre, dbl.; low growth; [Mme Henri Guillot X Pinocchio]; Silva

'BUFF BEAUTY', HMsk, ab, 1939; flowers apricot-yellow blooms in clusters of 12, dbl., 50 petals, 4 in., moderate fragrance; foliage large, medium green, semi-glossy; vigorous (to 6 ft (21)) growth; [William Allen Richardson X Seedling]; Bentall, Ann

'BUFF KING', LCl, ab, 1939; bud ovoid, deep amber; flowers amber and buff, large, cupped, foliage glaucous green; long, strong stems; very vigorous (10–12 ft) growth; Horvath; Wayside Gardens Co.

Buffalo, HRg, dp, 1989; Uhl, J.

Buffalo Bill, *see* 'MACYOUMIS'

Buffalo Gal, *see* 'UHLATER'

Buffy™, *see* 'KINBUFF'

Buffy Sainte-Marie, *see* 'MANCLASSIC'

Bugle Boy™, *see* 'AROglofy'

'BUISMAN'S GLORY', F, mr, 1952; flowers currant-red, open, medium, single; foliage light green; [Karen Poulsen X Sangerhausen]; Buisman, G.A.H.

Buisman's Gold, F, dy

'BUISMAN'S TRIUMPH', F, mp, 1952; flowers bright pink, becoming lighter, large, semi-dbl., 13 petals; foliage dark; vigorous growth; [Käthe Duvigneau X Cinnabar]; Buisman, G.A.H.

'BUISSON ARDENT', Gr, mr, 1956; flowers bright red, medium, moderate fragrance; foliage dark; [Peace X Seedling]; Gaujard

'BUISSON D'OR', HFt, my, 1928; flowers canary-yellow, dbl., good seasonal bloom, moderate fragrance; height 3–5 ft; [Mme Edouard Herriot X R. X Harisonii]; Barbier

Bukala, LCl, ob

Bukavu, F, rb, 1998; Lens

'BULLATA', C, mp; (R. X centifolia bullata, Rose à Feuilles de Laitue); flowers overlapping petals, very dbl., 3 in., globular, solitary or in clusters on long, slender peduncles, summer bloom, intense fragrance; foliage very large and crinkled like lettuce; (28)

Bulls Red, *see* 'MACRERO'

'BULLMEG', F, mr, 1995; (**Meghan's Arrival**); flowers medium red, full (26–40 petals), small blooms borne in large clusters; slight fragrance; some prickles; foliage medium, light green, glossy; upright, bushy, medium growth; [Doris Tysterman X Seedling]; Bull, Derek Gordon; Bull, 1986

'BUNDESRAT HÄBERLIN', HFt, yb, 1940; flowers medium, dbl.; Soupert & Notting; (Sangerhausen)

'BUNKER HILL', HT, dp, 1949; bud pointed; flowers rose-red, large, dbl., 25–40 petals, exhibition form, intense fragrance; foliage leathery, dark; very vigorous, bushy growth; [Rome Glory X Better Times]; Fisher, G.; Arnold-Fisher Co.

Bunny Hop, *see* 'JACCLIP'

'BUNTE FRAU ASTRID SPÄTH', Pol, lp, 1940; flowers light pink with red, medium, semi-dbl.; Vogel, M.; (Sangerhausen)

Bunter Kobold, MinFl, yb, 1995; Dickson, Patrick

'BURALP', HT, dr, 1988; (**National Velvet**™); flowers dark, deep velvet red, urn-shaped, medium, semi-dbl., 35 petals, exhibition form, borne usually singly, slight fragrance; globular, average, orange fruit; prickles recurved, average, dark; foliage large, medium green, semi-glossy; upright, tall growth; PP007236; [Poinsettia X National Beauty]; Burks, Larry; Co-Operative Rose Growers, 1990

'BURAPWARD', HT, yb, 1996; (**Dreamward**); flowers light yellow with pink blush, blooms borne mostly singly, dbl., 15–25 petals, 4.5 in., slight fragrance; moderate prickles; [Peace X Granada]; Perry, Astor; Certified Roses, Inc., 1997

'BURBANK', T, mp, 1900; (Santa Rosa); flowers bright rose-pink shading lighter, dbl., moderate fragrance; [Hermosa X Bon Silene]; Burbank; Burpee

Burbank White Rambler, HWich, w; Burbank

'BURBET', Gr, m, 1988; (**Betty Wright**™); flowers lilac-mauve, darker at tips, aging lighter lilac, urn-shaped, semi-dbl., cupped, slight, fruity fragrance; globular, average, orange-yellow fruit; prickles slight recurve, average, brown; foliage medium, medium green, matt; bushy, medium growth; [Angel Face X Seedling]; Burks, Larry; Co-Operative Rose Growers

'BURBRAHMVIDYA', HT, w, 1996; (**Simply Irresistible**, gust Simon); flowers white with pink tones, reverse cream with pink tones, blooms, dbl., 15–25 petals, moderate fragrance; few prickles; foliage medium, medium green, semi-glossy; upright, medium (4 1/2 ft) growth; [First Prize X Garden Party]; Datt, Braham; Certified Roses, Inc.

BURbrindley, F, or, 1995; (**Katie Crocker**); Burrows, Steven

'BURCEN', HT, or, 1988; (**Redlands Century**™); flowers medium orange-red, aging slightly lighter, urn-shaped, medium, semi-dbl., slight fragrance; globular, small, orange fruit; prickles average, yellow-green; foliage medium, medium green, semi-glossy; upright, medium growth; [World Peace X Command Performance]; Perry, Anthony; Co-Operative Rose Growers, 1988

'BURDANCER', Gr, w, 1995; (**Cloud Dancer**™); flowers white blend, double, blooms borne in small clusters, 15–25 petals, slight fragrance; few prickles; foliage medium, medium green, semi-glossy; upright, medium growth; [White Queen Elizabeth X Unknown]; Burks, Larry; Certified Roses, Inc., 1995

'BURDEL', HT, yb, 1988; (**Delta Gold**™); flowers red and yellow blend, reverse red-yellow, urn-shaped, medium, semi-dbl., 35 petals, slight fragrance; round, average, orange fruit; prickles slight recurve, average, dark; foliage medium, dark green, glossy; strong stems; bushy, medium growth; [Arizona X World Peace]; Perry, Anthony; Co-Operative Rose Growers, 1989

'BUREURO', Cl F, dr, 1987; (**Europeana, Climbing**); [Europeana sport]; Burks, Joe J.; Cooperative Rose Growers

Burg Baden, LCl, mp, 1955; Kordes

'BURGEMEESTER BERGER', HT, lp, 1934; flowers soft pink, marked white; vigorous growth; [Dame Edith Helen sport]; Leenders Bros.

'BURGEMEESTER SANDBERG', HT, op, 1920; flowers silvery pink, shaded coral-rose-pink, dbl., moderate fragrance; [Pharisaer X Lady Alice Stanley]; Van Rossem

'BURGEMEESTER VAN OPPEN', HT, my, 1939; flowers golden yellow, moderate fragrance; [Golden Ophelia X Pardinas Bonet]; Leenders, M.

'BÜRGERMEISTER CHRISTEN', HT, dp, 1911; flowers large, dbl., intense fragrance; Bernaix, A.; (Sangerhausen)

Burghausen®, *see* KORonto

Burgund®, *see* 'KORGUND'

Burgund '81' *see* 'KORGUND '81'

'BURGUNDIAN ROSE', C, pb, 1664; (Burgundy Rose, Parvifolia, Pompon de Bourgogne, Pompon de Burgogne, R. burgundensis, R. burgundica, R. centifolia parvifolia, R. ehrrhartiana, R. gallicaremensis, R. parvifolia, R. pomponia, R. remensis); flowers deep pink suffused purple, center paler, rosette form, dbl., 1 in.; few prickles; foliage dark gray-green; 3–5 ft growth; (14)

'BURGUNDY', HT, mr, 1939; flowers wine-red broad, evenly arranged petals, dbl.; very vigorous growth; [Vaterland X Seedling]; H&S

Burgundy Queen, *see* 'WILJAME'

Burgundy Rose, *see* **'BURGUNDIAN ROSE'**

'BURJES', HT, mp, 1996; (**Jesse's Jewels**); semi-dbl., 6–14 petals, borne mostly singly, intense fragrance; few prickles; foliage medium, medium green, dull; upright, medium growth; [(Unknown X Pink Apache Belle) X Unknown]; Burks, Larry; Certified Roses, Inc.

Burkhard, *see* **'GRUMPY'**

Burkhardt, *see* **'GRUMPY'**

'BURLOV', HT, w, 1988; (**Pure Love**™); flowers urn-shaped, large, semi-dbl., borne singly, slight fragrance; globular, average, gray-red fruit; prickles slight recurve, average, brown-green; foliage medium, dark green, semi-glossy; upright, medium growth; PP006781; [Queen Elizabeth X World Peace]; Perry, Anthony; Co-Operative Rose Growers, 1987

'BURMA STAR', F, ab, 1974; flowers light apricot yellow, dbl., 22 petals, 3.5 in., moderate fragrance; foliage large, glossy; [Arthur Bell X Manx Queen]; Cocker

'BURNABY', HT, w, 1954; (Gold Heart, Golden Heart); flowers creamy white, dbl., 56 petals, 4–6 in., exhibition form, slight fragrance; foliage dark, glossy; vigorous, bushy growth; GM, NRS, 1954 GM, Portland, 1957; [Phyllis Gold X Pres. Herbert Hoover]; Eddie; Peterson & Dering

Burnet Rose, *see* R. spinosissima

Burning Gold, F, ob, 1990; Poulsen

'BURNING LOVE', Gr, mr, 1956; (Amour Ardent, Brennende Liebe); flowers scarlet, blooms in trusses of 3–5, dbl., 22 petals, 4 in., moderate fragrance; foliage dark, glossy; vigorous, bushy growth; GM, Baden-Baden, 1954; [Fanal X Crimson Glory]; Tantau, Math.

Burning Sky, *see* 'WEZIP'

'BURNT ORANGE', F, ob, 1973; flowers deep orange; [Woburn Abbey sport]; Hamilton

Burr Rose, *see* **R. ROXBURGHII**

'BURRED', HT, dr, 1987; (**Theodore Roosevelt**); flowers deep red, fading slightly lighter, medium, borne singly, semi-dbl., 6–14 petals, exhibition form, slight fragrance; globular, large, dark red fruit; average, light yellow-green prickles; foliage large, medium green, semi-glossy; upright, tall growth; [Alamo sport]; Burks, Larry; Co-Operative Rose Growers, 1986

'BURR'S MULTIFLORA', (clone of R. multiflora); vigorous; used as understock

BURspec, F, w, 1996; (**Good Luck**); Burston

'BURSRTPARA', HT, ob, 1996; (**Tropical Paradise**); flowers orange blend, reverse orange blend with yellow gold, blooms, dbl., 15–25 petals, 4.5 in., moderate fragrance; few prickles; foliage large, dark green, glossy; upright, tall (5 ft) growth; [South Seas X First Prize]; Rodgers, Shafner R.; Certified Roses, Inc.

'BURSTEIN', HT, op, 1982; (**Easter Bonnet**, Super Derby); [Queen Elizabeth sport]; Burks, Joe J.; Co-Operative Rose Growers

'BURWAH', Cl HT, dp, 1953; flowers deep rose-pink, large, dbl., cupped; very vigorous climbing growth; [Editor McFarland X Black Boy (Cl HT)]; Ulrick, L.W.

'BURWAY', HT, yb, 1985; (**Broadway**™); bud pointed; flowers golden yellow, blended orange-pink, well-formed, medium, dbl., 30–35 petals, 4–4.5 in., exhibition form, borne singly, moderate, spice and damask fragrance; foliage medium to large, dark, semi-glossy; upright growth, moderately tall; PP5827; AARS, 1986; [(First Prize X Gold Glow) X Sutter's Gold]; Perry, Anthony; Cooperative Rose Growers, 1986

'BURWIN', HT, rb, 1988; (**Double Perfection**™); flowers red, reverse white, aging darker red, reverse cream, medium, semi-dbl., exhibition form, slight fragrance; round, average, orange fruit; prickles slight recurve, average, brown; foliage medium, dark green, semi-glossy; bushy, medium growth; [(My Dream X First Prize) X Seedling]; Winchel, Joseph F.; Co-Operative Rose Growers, 1987

'BURWINLADHOM', HT, m, 1998; (**Ladies Home Journal**®); flowers pinkish mauve, medium, semi-double, semi-dbl., 8–14 petals, borne mostly singly, slight fragrance; few prickles; foliage medium, medium green, semi-glossy; upright, medium, 4 1/2–5 ft growth; [Unknown X

Unknown]; Winchel, Joseph F.; Certified Roses, 1998

'BURwintradhom', HT, m, 1998; (**Traditional Home**®); flowers pinkish, double, exhibition, dbl., 15–25 petals, 4 in., exhibition form, borne mostly singly, slight fragrance; prickles moderate; foliage medium, dark green, semi-glossy; upright, medium, 5 ft growth; [Unknown X Unknown]; Winchel, Joseph F.; Certified Roses Inc., 1998

'BURworpe', HT, pb, 1987; (**World Peace**™); flowers pink blend suffused with cream, tips of petals tinged dark pink, dbl., 30–35 petals, exhibition form, moderate, fruity fragrance; globular, orange fruit; average prickles; foliage large, medium green, semi-glossy, disease resistant; upright, tall growth; [First Prize X Gold Glow]; Perry, Anthony, 1972; Co-Operative Rose Growers, 1989

'BURyear', Cl, Gr, ob, 1995; (**New Year, Climbing**); flowers spanish orange, semi-dbl. (6–14 petals), medium blooms borne in small clusters; slight fragrance; some prickles; foliage medium, medium green, glossy; spreading, tall growth; [New Year sport]; Burks, Joe J.; Certified Roses, Inc., 1995

'BURyellow', Cl F, my, 1987; (**Sun Flare, Climbing**™, Yellow Blaze); flowers medium yellow, fading lighter, medium, semi-dbl., 15–25 petals, 3.5 in., cupped, borne in sprays of 3–5, slight, licorice fragrance; globular, medium, orange-red fruit; short, yellow-green prickles, slightly curved; foliage medium, medium green, semi-glossy; spreading, tall growth, 14 ft; PP006509; [Sun Flare sport]; Burks, Joe J.; Co-Operative Rose Growers, 1987

'BUSbugum', Min, pb, 1999; (**Bubble Gum**); flowers pink, reverse ivory, dbl., 17–25 petals, 1.2 in., borne mostly singly, slight fragrance; prickles moderate; foliage medium, medium green, semi-glossy; upright, tall (2 ft) growth; [Garden State X seedling]; Buster, Larry S.; Kimbrew-Walter Roses, 1998

Bush Baby, *see* 'PEAnob'

Bush Garden Climber, LCl, op; Thomas

'Bushfire', HWich, mr, 1917; flowers bright crimson, yellow zone around center, small, very large, dbl.; Clark, A.

Bushfire, *see* POUlbufi

'Bushu', HT, ob, 1984; bud ovoid; flowers orange-red, reverse lighter, large, dbl., 35 petals, exhibition form, no fragrance; prickles slanted downward; foliage medium, dark, semi-glossy; tall, vigorous, upright growth; [Dolce Vita X Roklea]; Yasuda, Yuji

Bushveld Dawn, *see* 'MEImainger'

'BUSimel', Min, m, 1993; (**Simply Elegant**); flowers light lavender inside with darker reverse, blooms borne most, dbl., 26–40 petals, 1.5–2.75 in., slight fragrance; some prickles; foliage medium, medium green, matt; upright, compact (50 cms) growth; [Lady X X Winsome]; Buster, Larry S.; Kimbrew Walter Roses, 1993

'BUSmalou', MinFl, m, 1996; (**Mary Louise**); flowers red lavender, lighter lavender reverse, blooms borne mostly, dbl., 26–40 petals, moderate fragrance; moderate prickles; foliage medium green, dull; upright, tall growth; RR; [Lady X X Winsome]; Buster, Larry S.; Kimbrew Walter Roses, 1996

'BUStev', Min, w, 1992; (**Stevie**); flowers well formed, blooms borne mostly singly opens to 3+ cms, very dbl., 53 petals, 1.5 in., slight fragrance; some prickles; foliage small, medium green, matt; low (42 cms), upright, compact growth; [Frau Karl Druschki X Miniature seedling]; Buster, Larry S.

Busy, Min, dp

Busy Lizzie, *see* 'HARbusy'

'Busybody', HT, dy, 1929; flowers rich chrome-yellow, small; [Georges Schwartz X Lena]; Clark, A.; Hazlewood Bros.

'Buta', HT, w; flowers pure white, well formed, intense fragrance; Cant, B. R.

Butter 'n' Sugar, *see* 'BERbut'

'Butterball', S, ly, 1950; flowers creamy yellow, single, non-recurrent, large, roundish, reddish fruit; prickles prickly, arching branches; foliage small; spinosissima type; height 6 ft; Skinner

'Buttercup', HT, my, 1930; flowers buttercup-yellow, well-formed, semi-dbl., moderate fragrance; vigorous growth; Dobbie

'Buttercup', HT, ab, 1929; flowers apricot-yellow, dbl., cupped; Towill

Buttercup™, *see* 'AUSband'

'Buttercup', HWich, my, 1909; Paul

Butterflies, LCl, rb

Butterflies, *see* 'MEKkada'

Butterflies, *see* INTermug

Butterflies, *see* POUlbut

'Butterflies of Gold', HT, my, 1939; [Mrs Arthur Curtiss James seedling]; Brownell, H.C.

Butterfly®, HT, rb; Herholdt

'Butterfly Glow', HT, or, 1969; flowers vermilion, full, dbl., 20 petals, 3.5 in., slight fragrance; foliage dark; free growth; [Centre Court X Carla]; Barter

Butterfly Kisses, *see* 'GELfly'

Butterfly Papilio, *see* **'Motylek'**

'Butterfly Wings', F, pb, 1976; flowers ivory, petals edged pink, blooms in clusters, semi-dbl., 12 petals, 4–4.5 in., flat, moderate fragrance; foliage large; [Dainty Maid X Peace]; Gobbee, W.D.; Harkness

'Buttermere', HWich, yb, 1932; flowers creamy yellow flushed pink, large truss; foliage glossy; vigorous, erect growth; Chaplin Bros.

Buttermint, *see* 'MORsnop'

'Butterscotch', HT, yb, 1942; bud long, pointed; flowers lemon-chrome, reverse pale orange-yellow, dbl., 28 petals, 4.5–5.5 in., slight fragrance; foliage leathery, glossy, dark; upright, compact growth; [Souv. de Claudius Pernet X R.M.S. Queen Mary]; Hill, Joseph H., Co.; H&S, 1946; Wayside Gardens Co., 1946

Butterscotch, *see* 'JACtan'

Buttons, *see* 'LEMbut'

Buttons, *see* 'DICmickey'

'Buttons 'n' Bows', Min, dp, 1981; (Felicity II, Teeny-Weeny); bud small; flowers deep pink, and, reflexing at maturity, borne singly or in sprays, dbl., 28 petals, exhibition form, moderate, fruity fragrance; straight prickles; compact, upright growth; [Mini-Poul X Harriet Poulsen]; Poulsen Roser APS; Windy Hill Nursery, 1982

'Buzby', F, or, 1976; flowers light vermilion, conical, small, semi-dbl., 11 petals, 1.5 in., slight fragrance; dwarf, compact, upright growth; [Irish Mist X Topsi]; Plumpton, E.

Buzzy's White Seedling, N, w

By Appointment, *see* 'HARvolute'

By Design, *see* 'LAVsign'

'By Joe', Min, w, 1990; bud pointed; flowers ivory white, blooms borne singly, dbl., 25 petals, exhibition form, no fragrance; foliage medium, dark green, matt; upright, bushy, tall growth; [Pink Petticoat X Pink Petticoat]; Gatty, Joseph, 1990; Keith Keppel, 1990

'Byala Valentina', Gr, w, 1974; flowers creamy white, large, dbl., 25 petals, slight, delicate fragrance; foliage dark, glossy; vigorous, upright growth; [Queen Elizabeth X Seedling]; Staikov, Prof. Dr V.; Kalaydjiev and Chorbadjiiski

'BYRbradby', HT, op, 1999; (**Mary Bradby**™); flowers unusual orange and russet, reverse salmon, ruffled petals, dbl., 17–25 petals, 3 in., borne mostly singly, slight fragrance; few prickles; foliage medium, dark green, semi-glossy, disease-resistant; upright, medium (5 ft) growth; hardy; [Queen Elizabeth X Taboo]; Byrnes, Robert; Overbrooke Gardens, 1999

'BYRlynne', S, pb, 1999; (**Lynne Elizabeth**™); flowers light pink, reverse medium pink, dbl., 15–25 petals, 4 in., borne mostly singly, slight fragrance; few prickles; foliage medium, medium green, semi-glossy, disease-resistant; spreading, bushy, medium (4.5 ft) growth; winter hardy; [Carefree Beauty X Carefree Beauty]; Byrnes, Robert; Overbrooke Gardens, 1999

'BYRSHERRILL', S, pb, 1999; (**Sherrill Anne**™); flowers blush pink, reverse medium pink, very dbl., 41 petals, 4 in., slight fragrance; few prickles; foliage medium, medium green, semi-glossy; upright, medium (4–5 ft) growth; [Country Dancer X Country Dancer]; Byrnes, Robert; Overbrooke Gardens, 1999

C

'C. A. FLETCHER', HT, dp, 1947; flowers clear rose-crimson, well formed, dbl., 35–40 petals, 5–6 in., slight fragrance; vigorous growth; [May Wettern X Mrs Henry Bowles]; Fletcher; Tucker

'C. CHAMBARD', HT, 1934; flowers deep yellow, reverse tinted red, dbl., cupped; foliage dark; vigorous growth; Bel

'C. H. MIDDLETON', HT, dr, 1939; flowers dark crimson, large, very dbl., exhibition form, intense fragrance; foliage glossy; long stems; vigorous, bushy growth; Cant, B. R.

'C. V. HAWORTH', HT, dr, 1917; flowers intense black-scarlet with rich crimson bloom, massive shell-shaped petals, moderate fragrance; vigorous growth; GM, NRS, 1919; Dickson, A.

'C. V. HAWORTH, CLIMBING', Cl HT, dr, 1932; Cant, F.

'C. W. COWAN', HT, mr, 1912; flowers warm carmine-cerise, dbl., moderate fragrance; Dickson, A.

C.F. Worth, *see* **'MME CHARLES FREDERIC WORTH'**

'C.S.R.', HP, 1934; flowers white with pink striping, large, dbl.; Böhm, J.; (Sangerhausen)

C.W.S., *see* **'CANADIAN WHITE STAR'**®

Cabaret, *see* 'JACARET'

'CABARET', F, or, 1963; flowers vermilion-salmon, camellia shaped, dbl., 35–40 petals, 2.5 in., borne in clusters, vigorous, upright growth; RULED EXTINCT 6/83 ARM; [Dacapo X Floribunda seedling]; deRuiter; Blaby Rose Gardens

Cabaret®, *see* LAPed

'CABBAGE ROSE', C, mp; (Provence Rose, R. centifolia, R. gallica centifolia); flowers overlapping petals, blooms borne singly or in clusters, very dbl., 3 in., summer blooming, moderate fragrance; (28, 21)

'CACAPHONY', HT, op, 1974; flowers pink tinged orange, dbl., 28 petals, 4 in., moderate, fruity fragrance; foliage glossy, dark; [Baccará X Golden Showers]; Golik; Dynarose

Cachet™, *see* 'TUCKACH'

Cacilda Backer, HT; (Brazil)

Cactus Blanc®, *see* 'DELTROB'

'CADDY', HT, my, 1943; bud large, well formed, yellow; vigorous growth; [Soeur Thérèse X Prof. Deaux]; Meilland, F.

'CADENZA', LCl, dr, 1967; bud ovoid; dbl., 2.5–3 in., blooms in clusters, recurrent bloom, slight fragrance; foliage glossy, dark, leathery; compact, moderate growth; [New Dawn X Embers, Climbing]; Armstrong, D.L.; Armstrong Nursery

'CADETTE', F, ab, 1971; bud pointed; flowers pastel pink-apricot, dbl., 35 petals, 2 in., cupped, moderate, fruity fragrance; foliage glossy, dark; vigorous, upright growth; [Poupee X Fillette]; Lens

Cadillac, *see* KORveril

Cadillac DeVille, *see* 'WEKCRYLAND'

'CAECILIE SCHARSACH', HP, lp, 1887; flowers large, very dbl., intense fragrance; Geschwind, R.; (Sangerhausen)

Caesar, *see* 'VARBOLE'

Caesar's Rose, *see* 'JUDSAR'

'CAFÉ', F, r, 1956; flowers coffee-with-cream color, blooms in clusters, very dbl., flat, moderate fragrance; foliage olive-green; vigorous growth; [(Golden Glow X R. kordesii) X Lavender Pinocchio]; Kordes; McGredy

Café Olé, *see* 'MOROLÉ'

'CAFOUGNETTE', HT, op, 1956; flowers soft orange-salmon, reverse carmine-red, open, well shaped,; foliage dark; strong stems; [Happiness X Peace]; Dorieux; Pin

Cagul, HT, op, 1971; flowers salmon-orange and pink, large, very dbl., moderate fragrance; Moreira da Silva, A.; (Sangerhausen)

Cahto Maid, S, lp

Caid, *see* 'DELSIRP'

'CAIRNGORM', F, ob, 1973; flowers tangerine and gold, dbl., 25 petals, 2.5 in., slight fragrance; foliage glossy, dark; upright growth; [Anne Cocker X Arthur Bell]; Cocker

Caitlin, *see* 'TRACAIT'

Cajun Dancer, *see* 'TALCAJ'

Cajun Spice, *see* 'TALSPI'

Cajun Spice, HT, or, 1996; McMillan, Thomas G.

Cal Poly, *see* 'MORPOLY'

Calay, *see* 'TROBELLE'

Calcutta 300, HT, pb, 1994; V&B

Caldwell Pink, Pol, mp

Caldwill Pink Cl., Cl Pol, mp

'CALEDONIA', HT, w, 1928; bud long, pointed; flowers large, dbl., 25 petals, exhibition form, slight fragrance; foliage leathery, dark; vigorous growth; RULED EXTINCT 7/83 ARM; Dobbie

'CALEDONIA, CLIMBING', Cl HT, w, 1936; Bel

'CALEDONIAN', HT, ly, 1983; flowers creamy yellow, large, dbl., slight fragrance; foliage medium, medium green, semi-glossy; bushy growth; [Kordes' Perfecta X Irish Gold]; Mayle, W.J.

Calgary, HT, mr, 1997; Twomey, Jerry

'CALGOLD', Min, dy, 1977; bud pointed; flowers deep clear yellow, small, dbl., 23 petals, 1.5 in., slight fragrance; foliage small to medium, glossy; bushy growth; [Golden Glow (Brownell) X Peachy White]; Moore, Ralph S.; Sequoia Nursery

Calibra, *see* KORcrisett

'CALICO', HT, pb, 1976; flowers pink, yellow reverse, dbl., 3.5–4 in., globular, slight, tea fragrance; foliage dark; vigorous, upright to spreading growth; [Seedling X Granada]; Weeks; Weeks Wholesale Rose Growers

Calico Doll, *see* 'SAVADOLL'

'CALICO STAR', F, yb, 1977; bud ovoid; flowers golden yellow, edged red, dbl., 25–30 petals, 4 in., exhibition form, slight fragrance; upright growth; [Circus X Lavender Girl]; Fong; United Rose Growers

'CALIENTE', F, dr, 1974; bud ovoid-pointed; flowers deep pure red, medium, dbl., exhibition form, slight fragrance; foliage large, leathery; very vigorous, bushy growth; [Seedling X Seedling]; Warriner, William A.; J&P

'CALIFORNIA', HT, ob, 1940; bud long, pointed; flowers ruddy orange, reverse overlaid pink, dbl., 30 petals, 5–6 in., moderate, fruity fragrance; foliage leathery, glossy; vigorous, bushy, spreading growth; AARS, 1941; [Miss Rowena Thom X Lady Forteviot]; Howard, F.H.; H&S

'CALIFORNIA', HT, ob, 1916; flowers deep orange, base golden yellow; H&S

'CALIFORNIA BEAUTY', HT, my, 1926; Pacific Rose Co.

'CALIFORNIA BEAUTY', HT, dp, 1935; bud long, pointed to ovoid; flowers deep bright pink, very large, dbl., foliage leathery, dark; vigorous growth; [Dame Edith Helen X Hollywood]; Proietti

California Blonde, *see* 'RENBLONDE'

'CALIFORNIA CENTENNIAL', HT, dr, 1949; bud long, pointed; dbl., 28 petals, 3.5–4 in., exhibition form, intense fragrance; foliage leathery, bronze; vigorous, upright growth; [Tango X Mauna Loa]; Howard, F.H.; H&S

California Dreaming, *see* 'RENOVE'

California Girl, *see* 'RENIRL'

California Glory, *see* 'ORTCAL'

'CALIFORNIA GOLD', Pol, ab, 1934; flowers orange-yellow, clusters, dbl., globular, slight fragrance; foliage leathery, light;

long stems; very vigorous, bushy growth; [Gloria Mundi sport]; Smith, J.

California Ground Rose, *see* **R. SPITHAMEA**

California Sun, *see* 'RENSUN'

California Surf, *see* 'RENURF'

California Wild Rose, *see* **R. CALIFORNICA**

'CALIFORNIA, CLIMBING', Cl HT, ob, 1953; Howard, A.P.; H&S

'CALIFORNIA'S FAVORITE', HT, op, 1949; bud long, pointed; flowers light salmon-pink, base yellow, large, moderate fragrance; foliage soft; vigorous, bushy growth; [Stockton Beauty sport]; Raffel; Port Stockton Nursery

'CALINE', HT, op, 1958; flowers soft geranium-red, medium, dbl., 33 petals; foliage clear green; vigorous, upright growth; Ducher, Ch.; EFR

Calliope®, S, mp, 1985; Barni, V.

Calliope, *see* HARfracas

'CALLISTO', HMsk, my, 1920; flowers golden yellow, rosette form, borne in clusters, recurrent bloom; height 3–4 ft; [William Allen Richardson X Self]; Pemberton

'CALOCARPA', HRg, mp; (R. rugosa calocarpa, R. X calocarpa); flowers rose-colored, single; fruit handsome, in abundance; (14); [R. rugosa X Form of R. chinensis]

Calumet, *see* 'SOCAPAN'

Calypso, *see* **'BLUSH BOURSAULT'**

'CALYPSO', D, lp; flowers rosy blush with paler edges, large, dbl., cupped; Vibert, prior to 1848

'CALYPSO', F, or, 1957; bud globular; flowers orange-red, reverse red, loose, blooms in large pyramidal, semi-dbl., 18 petals, 3–3.5 in., cupped, intense, damask fragrance; foliage dark, leathery; vigorous, bushy growth; [Geranium Red X Fashion]; Boerner; Stuart

Calypso®, *see* 'POULCLIMB'

'CALYPSO', N, lp

'CAMAIEUX', HGal, m; flowers white and pale rosy purple, striped, well-shaped, dbl., intense fragrance; vigorous, rather dwarf growth; 1830

Camaieux Fimbriata, HGal, m, 1980; Bell

Camaieux Reversion, HGal, m, 1985; Robinson

Camara®, *see* 'DELCAMA'

'CAMAY', HT, dp, 1959; flowers deep rose-pink to carmine, dbl., 40 petals, 5 in., moderate fragrance; foliage light green; vigorous growth; [Ena Harkness X C.A. Fletcher]; Fletcher; Tucker

'CAMBRAI', HFt, yb, 1920; flowers light orange-yellow, medium, dbl., Smith; (Sangerhausen)

Cambria, S, lp

Cambridgeshire, *see* KORhungen

'CAMELEON', F, yb, 1961; flowers yellow, pink and orange, large clusters, dbl., 22–27 petals, intense fragrance; foliage dark, glossy; upright, compact, bushy, symmetrical growth; [Masquerade X Seedling]; Verschuren, A.; van Engelen

'CAMELIA', F, dp, 1953; flowers cerise-pink, medium, clusters; strong stems; moderate growth; [Pinocchio X Seedling]; Klyn

'CAMÉLIA', HT, or, 1948; bud pointed; flowers fiery vermilion-red, medium, semi-dbl., cupped, moderate fragrance; foliage glossy, dark; vigorous, bushy growth; GM, Geneva, 1948; [Vainqueur X (Charles P. Kilham X Katharine Pechtold)]; Heizmann & Co.

'CAMELIA ROSE', HCh, 1830; Prévost, prior to 1830

'CAMELIAROSE', F, lp, 1960; flowers soft pink, camellia form, very free bloom, especially in autumn; upright growth; [Mme Joseph Perraud X Incendie]; Croix, P.; Minier

'CAMELOT', Gr, op, 1964; bud ovoid; flowers shrimp-pink, blooms in clusters, dbl., 48 petals, 3.5–4 in., cupped, moderate, spicy fragrance; foliage leathery, glossy, dark; vigorous, tall growth; AARS, 1965; [Circus X Queen Elizabeth]; Swim & Weeks; C-P

'CAMEO', Pol, op, 1932; flowers salmon-pink, turning soft orange-pink; [Orléans Rose sport]; deRuiter; J&P

Cameo Cream, *see* 'MEIPIERAR'

Cameo Cream, HP, lp; flowers purple-violet, dbl., cupped

'CAMEO QUEEN', Min, pb, 1986; flowers light pink, reverse blends of pink, mini-flora, dbl., 34 petals, exhibition form, borne usually singly, slight fragrance; medium, long, light prickles; foliage large, medium green, semi-glossy; medium, bushy growth; [Heartland X Seedling]; Bridges, Dennis A.

'CAMEO SUPERIOR', Pol, mp; flowers have more lasting color; [Cameo sport]; deRuiter

Cameo, Climbing, *see* **'PINK CAMEO'**

'CAMILLA', HT, dp, 1954; bud ovoid; flowers strawberry-red, loosely formed, large, dbl., 30–36 petals, moderate fragrance; very vigorous growth; [Fiamma X Talisman]; Aicardi, D.; Olivieri

Camilla Sunsation, *see* KORfibi

Camille, *see* 'JACFRUIT'

'CAMILLE BERNARDIN', HP, rb, 1866; flowers red with white, very large, dbl., intense fragrance; Gautreau; (Sangerhausen)

Camille Pisarro, *see* DELstricol

'CAMILLO SCHNEIDER', HT, mr, 1922; flowers clear blood-red, dbl., slight fragrance; [Lieut Chaure X Comte G. de Rochemur]; Kordes

'CAMOËNS', HT, mp, 1881; flowers bright rose, center shaded yellow, moderate fragrance; moderately vigorous growth; [Antoine Verdier X Seedling]; Schwartz, J.

Camp David, HT, dr, 1984; Tantau

Campanela, *see* 'DELTRUT'

Campanile®, *see* 'DELTRUT'

'CAMPFIRE', Cl Pol, or, 1956; flowers orange-scarlet, rosette shape, small, borne in clusters; vigorous (6–8 ft) growth; [Cameo sport]; Fryers Nursery, Ltd.

Campfire, *see* 'WILCAMP'

Campfire Arteka, *see* **'KOSTIOR ARTEKA'**

'CAMPFIRE GIRL', HT, op, 1946; bud long, pointed; flowers deep salmon, large, dbl., exhibition form, moderate fragrance; foliage dark, leathery; vigorous, upright, bushy growth; [Joanna Hill X Gruss an Aachen]; Duehrsen; California Roses

Camphill Glory, *see* 'HARKREME'

'CAMPINA', HT, lp, 1937; bud long; flowers flesh-pink; very vigorous growth; [Comtesse Vandal X White Briarcliff]; Lens

'CAMPING', Pol, pb, 1967; flowers deep lavender-pink with white eye, small, borne in large clusters, single, globular; foliage small, olive-green; low, bushy growth; [Paul Crampel sport]; Grabczewski

Camrose, S, mp, 1998; Twomey, Jerry

Can Can, *see* 'LEGGLOW'

'CANADIAN CENTENNIAL', F, or, 1965; bud ovoid; flowers coral-red, medium, dbl., cupped, moderate fragrance; foliage glossy; vigorous, upright, compact growth; [Pinocchio seedling X Spartan]; Boerner; J&P

Canadian Dream, *see* 'BRODREAM'

'CANADIAN JUBILEE', HT, op, 1927; flowers indian red to pink, base orange, dbl., slight fragrance; [Priscilla X Commonwealth]; Dunlop

'CANADIAN NORTHLIGHT', HT, dp, 1984; flowers deep pink, large blooms in clusters of 6–9, dbl., 28 petals, exhibition form, slight fragrance; dark red prickles; foliage medium green, leathery; spreading growth; [(Fragrant Cloud X Diamond Jubilee) X Super Sun]; Mander, George

Canadian Sunset, HT, rb, 1996; flowers red with yellow reverse; foliage disease-resistant; Williams, J. Benjamin

'CANADIAN WHITE STAR'®, HT, w, 1980; (C.W.S., Dr Wolfgang Pöschl); flowers opening to multi-pointed star, dbl., 43 petals, exhibition form, borne singly, slight fragrance; slightly hooked prickles; foliage dark, leathery, glossy; vigorous, upright growth; PP005852; [Blanche Mallerin X Pascali]; Mander, George, 1980; Hortico Roses, 1985

Canadiana, *see* **'IMPERIAL GOLD'**

Canadiana, HT, 1962; Swim, H.C.; (Cavriglia)

'CANana', HT, mp, 1983; (**Mary Donaldson**); flowers medium salmon-pink, medium, dbl., exhibition form, intense fragrance; foliage large, dark, glossy; upright growth; [Kathleen O'Rourke X Seedling]; Cants of Colchester, Ltd., 1984

'Canarias', HT, mr, 1964; flowers bright red, large; Dot, Pedro; Minier

'Canarienvogel', Pol, 1903; (Sunset Glow); flowers saffron-yellow and amber-yellow stained pink and purple, semi-dbl.; strong stems; vigorous growth; [Étoile de Mai X Souv. de Catherine Guillot]; Welter

Canarina, 1965; Mansuino, Q.; (Cavriglia)

'Canary', T, my, 1852; bud small, well formed; flowers canary-yellow; growth rather weak; Guillot Père

'Canary', HT, ly, 1929; bud golden yellow, edges flushed; flowers light yellow, deepening, spiral, exhibition form, moderate fragrance; branching growth; Dickson, A.

'Canary', Min, my; flowers primuline-yellow, star shaped, semi-dbl.

Canary, *see* 'TANcary'

Canary, HSpn, ly

'Canary Bird', S, dy, 1907; (R. xanthina 'Canary Bird'); flowers yellow stamens, single, 5 petals, blackish purple fruit; [Probably R. hugonis X R. xanthina]

'Canary Charm', HT, my, 1969; flowers very large, dbl., exhibition form; foliage dark, leathery; free growth; [Wiener Charme sport]; Knight, G.

Canary Island, HGal, m

'Canasta', HT, mr, 1966; bud long, pointed; flowers bright red, large, dbl., exhibition form, moderate fragrance; very vigorous, upright growth; [Karl Herbst X Miss Universe]; Gaujard

'Canberra', Pol, op, 1935; flowers salmon-coral-pink; [Gloria Mundi sport]; Knight, G.

'Canberra', HT, op, 1927; flowers carmine, reverse buff at base, shading to salmon-pink; [Donald MacDonald X The Queen Alexandra Rose]; Burbage Nursery

'Canberra', HT, mp, 1928; very dbl.; Harrison

Canberra Rose, *see* 'TOMfed'

'Cancan', F, ob, 1969; (Diorette); bud short, pointed; flowers mandarin-red, small, dbl.; moderate, spicy fragrance; foliage leathery; Jelly; E.G. Hill Co.

Can-Can, *see* 'LEGglow'

Cancun™, *see* 'WEKibertaz'

Candelabra™, *see* 'JACcinqo'

Candella®, *see* 'MACspeego'

Candeur®, *see* 'DELcande'

'Candeur Lyonnaise', HP, w, 1914; bud long, pointed; flowers white, sometimes tinted pale yellow, large, dbl.; very vigorous growth; [Frau Karl Druschki seedling]; Croibier

Candia™, *see* 'MEIbiranda'

Candice, *see* 'ZIPcan'

Candid, HT, w; Delbard

'Candida', HT, yb, 1964; flowers creamy yellow, center salmon, well formed; [Tawny Gold X Golden Scepter]; Leenders, J.

'CANdide', HT, dy, 1983; (Gold Star, Goldina, **Goldstar**, Point du Jour®); flowers medium, dbl., 35 petals, slight fragrance; foliage small, light green, matt; upright growth; GM, The Hague, 1984; [Yellow Pages X Dr A.J. Verhage]; Cants of Colchester, Ltd., 1984

Candide, HT, 1970; Gaujard; (Cavriglia)

Candle Light, Cl Min, yb, 1995; Warner, Chris

'Candleflame', Min, yb, 1956; bud slender; flowers red, yellow and orange, single, 5 petals; foliage leathery; vigorous (10 in.), bushy growth; [(Soeur Thérèse X Julien Potin) X (Eblouissant X Zee)]; Moore, Ralph S.; Sequoia Nursery

'Candleglow', HT, yb, 1951; bud long, pointed; flowers yellow washed shrimp-pink, dbl., 38 petals, 5–6 in., cupped, slight fragrance; foliage glossy, dark; vigorous, upright growth; [Golden Rapture X (Seedling X Joanna Hill)]; Whisler; Germain's

Candlelight™, *see* 'AROwedye'

'Candlelight', HT, 1932; flowers yellow, deeper in hot weather, large, dbl., exhibition form, moderate fragrance; foliage glossy; bushy growth; RULED EXTINCT 9/81 ARM; [Souv. de Claudius Pernet X Mme Butterfly]; Horvath; Bosley Nursery

CANdoodle, F, w, 1990; (**Lady Rachel**); Cants of Colchester, Ltd.

'Candy', HT, ob, 1950; bud long, pointed; flowers apricot-orange, medium, dbl., exhibition form, moderate fragrance; vigorous, upright, compact, bushy growth; [Pink Princess X Shades of Autumn]; Brownell, H.C.

'Candy Apple', Gr, mr, 1966; bud ovoid; flowers bright apple red, very full, very dbl., 40–45 petals, 5 in., cupped, blooms in small clusters, slight, tea fragrance; foliage matt, olive-green; upright, medium, bushy growth; [Jack O'Lantern X (Seedling X El Capitan)]; Weeks, O.L.; Weeks Wholesale Rose Growers

'Candy Cane', Cl Min, pb, 1958; flowers deep pink, striped white, blooms in loose clusters, semi-dbl., 13 petals, 1.5 in., vigorous, upright (to 4 ft) growth; [Seedling X Zee]; Moore, Ralph S.; Sequoia Nursery

Candy Corn, *see* 'TALcan'

'Candy Favorite', HT, pb, 1971; flowers carmine, striped pale rose, full, dbl., 25 petals, 4.5–5 in., slight fragrance; foliage very glossy; vigorous growth; [Pink Favorite sport]; Heath, W.L.

Candy Flo, HT, rb

'Candy Floss', F, lp, 1959; flowers bright pink, rosette shape, borne in large clusters, dbl., 40 petals, 3 in., slight fragrance; very vigorous growth; [Lilibet X Seedling]; Fryers Nursery, Ltd.

Candy Mountain™, *see* 'JACchari'

'Candy Pink', Min, lp, 1969; bud ovoid; flowers small, dbl.; foliage small, leathery; vigorous, dwarf, bushy growth; [(R. wichurana X Floradora) X (Oakington Ruby X Floradora)]; Moore, Ralph S.; Sequoia Nursery

Candy Rose®, *see* 'MEIranovi'

Candy Stick, *see* **'Candystick'**

'Candy Stripe', HT, pb, 1963; flowers dusty pink streaked (striped) lighter; [Pink Peace sport]; McCummings; C-P

Candy Sunblaze™, *see* 'MEIdanclar'

'Candystick', HT, pb, 1978; (Candy Stick, Red 'n' White Glory); flowers deep pink striped white; [Better Times sport]; Williams, J. Benjamin; Lakeland Nursery Sales

'Canigó', HT, w, 1927; dbl., slight fragrance; [Antoine Rivoire X Mme Ravary]; Dot, Pedro

'CANlish', F, mp, 1983; (**St Helena**, Union-Rose st Helena®); flowers medium lilac pink, medium, dbl., 20 petals, slight fragrance; foliage medium, medium green, semi-glossy; upright growth; [Jubilant X Prima Ballerina]; Cants of Colchester, Ltd.

CANlloyd, F, ob, 1991; (**Lloyds of London**); Cants of Colchester, Ltd.

'CANlot', F, pb, 1984; (**Alexia**); flowers cream blended pink, medium, dbl., 20 petals, moderate fragrance; foliage medium, medium green, semi-glossy; upright growth; [Jubilant X Seedling]; Cants of Colchester, Ltd.

'CANmiss', F, mp, 1984; (**Beauty Queen**); flowers medium, dbl., intense fragrance; foliage large, dark, glossy; [English Miss X Seedling]; Cants of Colchester, Ltd.

'Cannes Festival', HT, yb, 1951; flowers indian yellow veined amber, pointed, dbl., 35 petals, 4 in., moderate fragrance; foliage dark; vigorous, upright, branching growth; [Peace X Prinses Beatrix]; Meilland, F.; URS

Cannes Festival 83, *see* MEIlicafal

'CANrem', HT, pb, 1994; (**Sally's Rose**); dbl., 15–25 petals, 1.5–2.75 in., many prickles; foliage dark green, glossy; medium (60 cms), bushy growth; [Amber

Queen X Remember Me]; Cants of Colchester, Ltd.; Cants of Colchester Ltd., 1994

'CANsend', F, dp, 1992; (**Colchester Beauty**); flowers candy pink, blooms borne mostly singly, semi-dbl., 6–14 petals, 1.5–2.75 in., intense fragrance; some prickles; foliage medium, dark green, semi-glossy; medium (60–80 cms), bushy growth; [English Miss X Seedling]; Pawsey, P.R.; Cants of Colchester Ltd., 1989

CANsit, F, 1996; (**Jenny's Rose**); Cants of Colchester, Ltd.; (Cavriglia)

CANson, F, mp, 1991; (**Dame Wendy**); Cants of Colchester, Ltd.

'Cantab', S, dp, 1927; bud long, pointed; flowers deep pink, base white, stamens yellow, saucer-shaped, single, 3.5 in., non-recurrent; foliage dark, 7–9 leaflets; height 6–8 ft; [R. nutkana X Red-Letter Day]; Hurst

'Cantabile', S, lp, 1962; bud ovoid; flowers light camellia-rose shaded darker, medium, dbl., 25 petals, repeat bloom, moderate fragrance; foliage leathery, bronze; vigorous, upright (5 ft) growth; [Harmonie X (Josef Rothmund X R. laxa)]; Buck, Dr Griffith J.; Iowa State University

'Cantabrigiensis', S, ly; (R. X cantabrigiensis, R. X pteragonis cantabrigiensis, The Cambridge Rose); flowers pale yellow, single, 2.25 in., (14); [R. hugonis X R. sericea hookeri]

'Cantate', F, ob, 1959; flowers orange; [Red Favorite sport]; van de Water; Spek

Canterbury, *see* 'AUScanterbury'

Canterbury, *see* **'Scoop Jackson'**

Canterbury Pride, *see* 'REYcantpri'

Cantilena Bohemica, HT, dp, 1977; flowers carmine-pink, large, very dbl.; Havel; (Sangerhausen)

Cantilena Moravica, HT, yb, 1981; Havel

'Canzonetta', F, or, 1953; bud ovoid, cardinal-red; flowers orange-red, reverse yellow suffused red, open, large, borne, semi-dbl., 10–12 petals, moderate fragrance; foliage dark, glossy, leathery; vigorous, bushy growth; [Lawrence Johnston X Fashion]; San Remo Exp. Sta.; Sgaravatti

Cape Cod, *see* 'POUlfan'

'Cape Coral', HT, ab, 1964; bud ovoid; flowers orange-coral, dbl., 50–55 wavy petals, 5 in., cupped, moderate fragrance; foliage leathery, glossy, veined red; long, strong stems; vigorous, bushy growth; [Spartan X Golden Masterpiece]; Boerner; J&P

'Cape Hatteras', Min, w, 1988; flowers medium, borne singly, dbl., 40 petals, exhibition form, slight, spicy fragrance; prickles long, pointed, medium pink-tan; foliage medium, dark green, glossy; upright, medium, vigorous growth; [Rise 'n' Shine X Seedling]; Bridges, Dennis A.; Bridges Roses

'Capeline', LCl, mp, 1965; flowers tyrian rose, semi-dbl., exhibition form, recurrent bloom; foliage leathery; vigorous growth; [Etendard X Diane d'Urfe]; Hémeray-Aubert

Capella, *see* 'MEIrilocra'

'Caper', F, dr, 1976; bud long, pointed; flowers open, dbl., 25 petals, 2–2.5 in., slight fragrance; foliage dark; upright growth; [Seedling X Mary DeVor]; Warriner, William A.; J&P

'CAPflare', HT, or, 1996; (**Solarflash**, Solar Flare, Solarflare); flowers bright vermilion, vibrant dark orange, with dark petal edge, dbl., 26–40 petals, 3.5 in., some to numerous prickles; foliage medium, dark green, semi-glossy; medium to low ($2\frac{1}{2}$–3 ft), compact growth; [Unknown Seedling X Hot Pewter]; Pallek, Otto; Pallek, 1995

'Capistrano', HT, mp, 1949; bud ovoid; flowers bright pink, dbl., 36 petals, 6 in., globular, moderate fragrance; foliage leathery; vigorous, upright growth; AARS, 1950; Morris; Germain's

'Capistrano, Climbing', Cl HT, mp, 1952; Germain's

'Capitaine Basroger', M, rb, 1890; flowers bright carmine-red, shaded purple, large, dbl.; very vigorous, almost climbing growth; Moreau et Robert

'Capitaine Dyel de Graville', B, pb, 1843; Boutigny

'Capitaine Georges Dessirier', HT, dr, 1919; flowers dark velvety red, shaded crimson, dbl., moderate fragrance; [Seedling X Château de Clos Vougeot]; Pernet-Ducher

'Capitaine John Ingram', M, m; bud well mossed; flowers variously described as dark purple, velvety crimson and reddish purple, dbl.; vigorous growth; Laffay, M., 1855

'Capitaine Jouen', HP, mr, 1900; flowers very large, dbl.; Boutigny; (Sangerhausen)

'Capitaine Millet', T, mr, 1901; flowers bright red, reverse purplish, base golden, large, dbl., moderate fragrance; [Gén. Schablikine X Mme Etienne]; Ketten Bros.

'Capitaine Peillon', HP, m, 1893; flowers purple-pink, large, dbl.; Liabaud; (Sangerhausen)

'Capitaine Sisolet', HP, mp; flowers rose, large, very dbl., cupped; vigorous, branching growth; (France), Prior to 1848

'Capitaine Soupa', HT, dp, 1902; flowers carmine-pink, very large, dbl.; Laperrière; (Sangerhausen)

'Capitaine Soupa, Climbing', Cl HT, dp, 1938; flowers carmine-pink, large, dbl.; Vogel, M.; (Sangerhausen)

'Capitaine Williams', HGal, dr; flowers medium, very dbl.

Capitalia, HT; Ansaloni; (Cavriglia)

Capitole®, *see* LAPovi

'CAPmina', F, mp, 1996; (**Conservency Rose**); flowers vibrant pink, some variegation, not solid pink, long lasting, dbl., 30 petals, 1.5 in., slight fragrance; some prickles; foliage medium, medium green, glossy, disease resistant; medium (20 in.), bushy growth; Pallek, Ruth; Carl Pallek & Son Nurseries, 1995

Caporosso, HRg, mr, 1975; Mansuino, Q.

Cappa Magna®, *see* 'DELsap'

'Capreolata Ruga', Misc OGR, mp, 1820

'Capri', F, op, 1956; bud conical; flowers bright coral, reverse lighter, to, blooms in small clusters, dbl., 38 petals, 3–3.5 in., exhibition form, slight fragrance; foliage bright green, glossy, leathery; vigorous growth; [Fashion X Floradora]; Fisher, G.; Arnold-Fisher Co.

Capri, HT, ab

Capri Sun, *see* TANnus

'Caprice', HT, pb, 1948; (Lady Eve Price); bud ovoid; flowers deep pink, reverse cream, large, dbl., 24 petals, slight fragrance; foliage dark, leathery; vigorous, upright, bushy growth; [Peace X Fantastique]; Meilland, F.; C-P

'Caprice', HT, or, 1934; bud long, pointed; flowers orient red and peach-red, large, dbl., slight fragrance; foliage glossy, light; vigorous growth; [Seedling X Gwyneth Jones]; Leenders, M.

Caprice 86, S, 1986; Meilland, L.; (Cavriglia)

Caprice de Mielland, *see* MEIsionver

'Caprice, Climbing', Cl HT, or, 1951; Lens

'Capricorn', F, rb, 1961; flowers scarlet, reverse silver, borne in large clusters, semi-dbl., 16 petals, 2 in., slight fragrance; foliage dark, dull, leathery; very vigorous growth; Verschuren; Blaby Rose Gardens

Capriolata, Misc OGR, pb, 1970

'Capriole', F, mp, 1956; bud cherry-red; flowers vivid pink, stamens prominent, medium, in large clusters, semi-dbl.; foliage dark, semi-glossy; vigorous, dwarf, bushy growth; [Red Favorite X Fanal]; Tantau, Math.

'Capt. F. S. Harvey-Cant', HT, pb, 1923; bud long, pointed; flowers peach-pink, reverse deep pink, very large, dbl., moderate fragrance; foliage dark, leathery; vigorous, bushy growth; GM, NRS, 1922; Cant, F.

'Capt. Robinson', Sp, dp; semi-dbl., 13 petals, non-recurrent, moderate fragrance;

few prickles; foliage small, dark; height 1 ft; ((discovered in the sand dunes of Porter Co., In 1924))

'CAPT. RONALD CLERK', HT, or, 1923; flowers vermilion-scarlet, semi-dbl., slight fragrance; McGredy

'CAPT. RONALD CLERK, CLIMBING', Cl HT, or, 1935; Austin & McAslan

'CAPTAIN BLIGH', HT, mp, 1939; bud long, pointed; flowers silvery rose, large, very dbl.; foliage leathery, dark; very vigorous growth; [Gustav Grünerwald X Betty Uprichard]; Fitzhardinge; Hazlewood Bros.

'CAPTAIN BLOOD', HT, dr, 1938; bud ovoid; flowers scarlet and crimson, dbl., cupped, intense fragrance; foliage leathery; vigorous, bushy growth; [Gen. MacArthur X E.G. Hill]; Melville Bros.; R. Murrell

'CAPTAIN CHRISTY', HT, lp, 1873; flowers soft flesh-pink, center darker, large, dbl., 40 petals, globular, slight fragrance; wide, compact growth; [Victor Verdier X Safrano]; Lacharme, F.

'CAPTAIN CHRISTY, CLIMBING', Cl HT, lp, 1881; Ducher

Captain Cook, *see* 'MACAL'

'CAPTAIN F. BALD', HT, dr, 1919; flowers scarlet-crimson, velvety black sheen, dbl., slight fragrance; Dickson, A.

'CAPTAIN GLISSON', HT, dy, 1935; bud long, pointed; flowers dark yellow, edged lighter, large, dbl., 28–30 petals; very vigorous, compact growth; [Joanna Hill X Sweet Adeline]; Hill, Joseph H., Co.

'CAPTAIN HARRY STEBBINGS', HT, dp, 1980; bud long, pointed; flowers deep pink, dbl., 43 petals, 5–6.5 in., exhibition form, intense, fruity fragrance; foliage large, leathery; upright, bushy growth; [Seedling sport]; Stebbings; Country Garden Nursery

'CAPTAIN HAYWARD', HP, dp, 1893; flowers light crimson, edged lighter, large, dbl., 25 petals, exhibition form, sparsely recurrent, moderate fragrance; large, orange fruit; vigorous growth; (28); [Triomphe de l'Exposition seedling]; Bennett

'CAPTAIN HAYWARD, CLIMBING', Cl HP, dp, 1906; Paul

'CAPTAIN KIDD', HSet, mr, 1934; flowers blood-red, open, large, borne, dbl., cupped, moderate fragrance; foliage leathery, dark; long, strong stems; very vigorous, climbing or tall pillar growth; [R. setigera seedling X Hoosier Beauty]; Horvath

'CAPTAIN KILBEE STUART', HT, dr, 1922; bud long, pointed; flowers scarlet-crimson, very large, dbl., intense fragrance; moderate growth; GM, NRS, 1922; Dickson, A.

'CAPTAIN KILBY', F, mr, 1955; flowers blood-red, medium, large truss, semi-dbl., moderate fragrance; foliage glossy, parsley-green; very vigorous growth; deRuiter; Gandy Roses, Ltd.

'CAPTAIN SAMUEL HOLLAND', S, mr, 1991; flowers moderately medium, dbl., 15–25 petals, slight fragrance; foliage medium, medium green, glossy; spreading, medium growth; Ogilvie, Ian S.; Agriculture Canada, 1992

'CAPTAIN SASSOON', HT, dr, 1938; flowers dark crimson, moderate fragrance; vigorous growth; Gaujard

'CAPTAIN THOMAS', Cl HT, w, 1935; bud long, pointed; flowers lemon to cream, single blooms in clusters, recurrent bloom, moderate fragrance; foliage glossy, light; climbing or pillar (10 ft) growth; [Bloomfield Completeness X Attraction]; Thomas; Armstrong Nursery, 1938

Captain Watkins, Pol, mp; Heyde, C.W., before 1960

'CAPTAIN WOODWARD', HP, dp; flowers light red

Captivation(TM), *see* 'KINCAP'

'CAPTIVATOR', HT, mp, 1942; [Better Times sport]; Hill, Joseph H., Co.

'CAPUCINE CHAMBARD', HFt

'CARA MIA', HT, mr, 1969; (Danina, Dearest One, Maja Mauser, Natacha); bud ovoid; flowers large, dbl., moderate fragrance; foliage dark; vigorous, upright growth; McDaniel, G.K.; Carlton Rose Nurseries

'CARABELLA', F, yb, 1960; bud pointed, apricot; flowers cream edged pink, open, borne in clusters, single, moderate fragrance; foliage glossy, light green; moderate, bushy growth; [Gartendirektor Otto Linne X Seedling]; Riethmuller

Caraibes, HT, 1970; Combe, M.; (Cavriglia)

'CARALE', F, dy, 1993; (**Cisco**(TM)); dbl., 26–40 petals, 1.5–2.75 in., borne in small clusters, slight fragrance; foliage medium, medium green, glossy; medium (60 cms), upright growth; [Escort X Golden Nugget]; Strahle, Robert; Carlton Rose Nurseries, 1989

Caramba®, *see* 'TANCA'

'CARAMEL CREME', HT, my, 1980; bud ovoid; flowers caramel yellow, blooms borne singly, dbl., 30 petals, cupped, slight, spicy fragrance; long prickles, hooked downward; foliage finely serrated; bushy, moderate growth; [(Sunbonnet X Mister Lincoln) X Oldtimer]; Weeks, O.L.

Caramel Sunset, HT, lp; flowers butterscotch, good form

Caramella®, *see* MEInitper

CARange, HT, lp; (**Tropical Passion**); Strahle, B. Glen, 1999

'CARAVELLE', HT, mr, 1964; flowers bright cherry-red, large, dbl., exhibition form; strong stems; vigorous, upright growth; [Better Times X Seedling]; Mondial Roses

'CARBANE', HT, dp, 1993; (**Hot Spot**(TM)); flowers red pink, blooms borne mostly singly, dbl., 26–40 petals, 3–3.5 in., intense fragrance; some prickles; foliage large, dark green, semi-glossy; upright (90–100 cms) growth; [Duchess sport]; Strahle, B. Glen; Carlton Rose Nurseries, 1990

'CARCADE', HT, w, 1993; (**White Cascade**(TM)); flowers large, white, full, very dbl., 26–40 petals, 3–3.5 in., borne mostly singly, slight fragrance; some prickles; foliage medium, medium green, semi-glossy; medium, upright growth; [Coquette X Jack Frost]; Strahle, B. Glen; Carlton Rose Nurseries, 1984

'CARDEAL DE ROHAN', F, dr, 1957; flowers deep red, semi-dbl.; da Silva, Moreira

'CARDIFF BAY', HT, mr, 1993; dbl., 26–40 petals, 3–3.5 in., borne mostly singly, intense fragrance; some prickles; foliage medium, red changing to dark green, matt; medium (105 cms), upright growth; [Seedling X Loving Memory]; Poole, Lionel; F. Haynes & Partners, 1994

'CARDINAL', HT, 1904; flowers red; RULED EXTINCT 12/85 ARM; [Liberty X Seedling]; Cook, J.W.

'CARDINAL DE RICHELIEU', HGal, m; (Cardinal Richelieu); flowers dark purple, large, dbl.; bushy growth; Laffay, M., before 1847

Cardinal Hume®, *see* 'HARREGALE'

'CARDINAL MERCIER', HT, op, 1930; flowers salmon-pink tinted orange, overlarge, globular; foliage bronze; vigorous growth; Lens

Cardinal Richelieu, *see* **'CARDINAL DE RICHELIEU'**

Cardinal Song(TM), *see* 'MEIMOUSLIN'

'CARDINALE DE LA PUMA', HT, dp, 1938; flowers carmine-red, large, semi-dbl., slight fragrance; foliage glossy; vigorous, bushy growth; GM, Rome, 1937; [Mgr. Lemmens X Lord Charlemont]; Leenders, M.

'CARDINALS HAT', HT, mr, 1960; [Souv. de Jacques Verschuren X Charles Mallerin]; Leenders, J.

Cardinal's Robe, *see* 'MACRENO'

Cardinal's Rose, *see* 'MACRENO'

Care Deeply, *see* 'LYOCA'

'CAREFREE', F, dp, 1959; flowers rose bengal, camellia form, medium, clusters, dbl., 75 petals; foliage dark; free growth; [Alain X Pinocchio]; Fletcher

Carefree Beauty(TM), *see* 'BUCBI'

Carefree Delight, *see* 'MEIPOTAL'

Carefree Wonder(TM), *see* 'MEIPITAC'

'CARELESS LOVE', HT, pb, 1955; flowers deep pink, streaked (striped) and splashed white; [Red Radiance sport]; Conklin; Golden State Nursery

'CARELESS MOMENT', Min, pb, 1977; bud long, pointed; flowers white, lightly edged pink, dbl., 45 petals, 1–1.5 in., exhibition form, moderate fragrance; foliage small; bushy, spreading growth; [Little Darling X Over the Rainbow]; Williams, Ernest D.; Mini-Roses

'CARESS', HT, yb, 1935; flowers pale buttercup yellow, tinted rose, edged salmon-carmine, dbl., moderate fragrance; foliage bright green, leathery; vigorous growth; Dickson, A.

Carezza, *see* BARrez

'CARFAIT', HT, op, 1993; (**Peach Parfait**™); flowers pink coral, blooms borne mostly singly, dbl., 26–40 petals, 1.5–2.75 in., few prickles; foliage medium, medium green, semi-glossy; medium (70–100 cms), upright growth; [Indian Pink X Melissa]; Strahle, B. Glen; Carlton Rose Nurseries, 1992

Caribbean™, *see* 'KORBIRAC'

Caribbean Queen, *see* 'RENEEN'

Caribe, *see* 'JACIBE'

Caribia, *see* **'HARRY WHEATCROFT'**

Caribia, Climbing, *see* 'AROMONTELIB'

'CARIBOU', HRg, w, 1946; bud pointed; flowers large, single, 5 petals, flat, non-recurrent bloom, slight fragrance; ornamental fruit; foliage glossy, leathery, dark, rugose, scented like sweetbr; bushy, vigorous growth; very hardy.; [Ross Rambler X (R. rugosa X R. eglanteria) X Seedling)]; Preston; Central Exp. Farm

'CARILLON', HT, or, 1935; bud long, pointed, scarlet-orange; flowers brilliant flame, paling, large, semi-dbl.; foliage glossy, light; bushy growth; [Charles P. Kilham X Mrs Pierre S. duPont]; Nicolas; J&P

'CARILLON', HT, dy, 1953; bud long, pointed, deep yellow spotted carmine; flowers medium, dbl., 30–35 petals; vigorous growth; [(Soeur Thérèse X Orange Nassau) X Orange Nassau]; Moulin-Epinay; Vilmorin-Andrieux

Carina®, *see* 'MEICHIM'

'CARINA SUPERIOR', HT, w, 1978; bud pointed; flowers white, reverse light pink, dbl., 35–40 petals, 5 in., exhibition form, moderate fragrance; foliage leathery; upright, bushy growth; [Carina sport]; Takatori, Yoshiho; Japan Rose Nursery

Carina, Climbing®, *see* 'MEICHIMSAR'

'CARINE', HT, op, 1911; bud long, pointed; flowers orange-carmine tinted buff, well formed, intense fragrance; strong stems; vigorous growth; Dickson, A.

Caring For You, *see* 'BOSARTHIC'

Carinella, HT, w

Caring For You, *see* 'BOSARTH'

Carinita, *see* MEIvostro

'CARIOCA', HT, ab, 1942; bud long; dbl., 25–38 petals, 4.5 in., exhibition form, slight fragrance; foliage dark, glossy, leathery; vigorous, upright growth; [Talisman sport]; Chase; J&P

'CARIOCA', HT, ob, 1951; bud semi-ovoid; flowers tangerine-orange, large, dbl., globular, moderate fragrance; foliage bright bronze green; long, strong stem; very vigorous growth; [Rubin X Mme Henri Guillot]; Lens

'CARISSIMA', HWich, lp, 1905; flowers delicate flesh, carnation-like, dbl., moderate fragrance; Walsh

Caritas, HT, mp, 1970; flowers large, dbl., slight fragrance; Urban, J.; (Sangerhausen)

'CARITO MACMAHON', HT, my, 1934; bud large, ovoid; dbl., cupped, moderate fragrance; foliage glossy, dark; strong stem; vigorous growth; [Mrs Pierre S. duPont X Cayetana Stuart]; Dot, Pedro

'CARL KEMPKES', F, dr, 1937; bud long, pointed; flowers crimson, open, large, borne in clusters, semi-dbl., slight fragrance; foliage glossy, dark; strong stem; vigorous, bushy growth; [Dance of Joy X Mary Hart]; Kordes; Späth

Carl Luhn's Autumn Taffeta, HT, ob

Carl Philip Kristian IV, The Time Rose, *see* 'KORPEAHN'

Carl Red, HT, mr

'CARLA', HT, op, 1963; flowers soft salmon-pink, dbl., 26 petals, 3.5–5 in., moderate fragrance; foliage dark; vigorous growth; [Queen Elizabeth X The Optimist]; deRuiter; Ball Seed Co., 1968

'CARLA', S, mr, 1963; (14); [Will Alderman X Hansa]; Erskine

Carla, Cl HT, lp; Ross, A., & Son, 1969; (Weatherly, L.)

'CARLA, CLIMBING', Cl HT, mr, 1969; Ross, A., & Son

'CARLEA', S, mp, 1965; flowers similar to Victory Year but darker and more free blooming, non-recurrent, [Betty Bland seedling]; Wright, Percy H.

Carlita, HT, mp, 1975; flowers large, dbl., slight fragrance; Kamrad; (Sangerhausen)

'CARLOS BEAUTY', (variety of R. acicularis), mp; flowers bright pink, semi-dbl., 12–15 petals, intense fragrance; ((found in Alberta, Canada))

'CARLOS REIS', Cl HT, mr; flowers purplish red; [Étoile de Hollande, Climbing X Pres. Herbert Hoover]; da Silva, Moreira

Carl's Rose, *see* 'LEMCAR'

Carl's Rose, Min, m

'CARLSHAM', S, mp, 1964; flowers rose-pink, large, dbl., 25 petals, recurrent bloom; foliage glossy; (14); [Hansa X R. nitida]; Erskine

Carmagnole®, *see* 'DELROBLA'

Carmargue, *see* 'CHEWOZ'

'CARMEL BICE', F, pb, 1959; bud ovoid; flowers pink, reverse lighter, blooms in clusters, semi-dbl., intense fragrance; foliage leathery, glossy; vigorous, upright growth; [Gartendirektor Otto Linne X Seedling]; Riethmuller

Carmel Sunset, *see* 'MACCLOSEUP'

'CARMELA', Min, ob, 1980; bud lightly mossed; flowers orange, yellow center, small, semi-dbl., 15 petals, borne in sprays of 3–5, slight fragrance; small, ovoid to globular, orange fruit; small, brownish prickles; foliage small, light green, matt; medium, bushy, spreading growth; [Fairy Moss X Yellow Jewel]; Moore, Ralph S.; Moore Min. Roses

Carmela®, *see* 'INTERGEORGE'

'CARMELITA', HT, mr, 1933; bud ovoid; flowers vivid red, large, dbl., exhibition form; foliage dark; very vigorous, bushy growth; [Matchless X Milady]; Spanbauer

Carmelita, F, dy

'CARMEN', HRg, mr, 1907; flowers crimson, stamens yellow, large blooms in clusters, single; foliage dark; vigorous growth; [R. rugosa rosea X Princesse de Bearn]; Lambert, P.

'CARMEN', HT, dr, 1956; flowers deep red, well-formed, large, dbl., intense fragrance; foliage dark; moderate growth; [Crimson Glory X Seedling]; Delforge

Carmen, *see* 'JACLAM'

Carmen, *see* TANemrac

Carmen de Bencomo, HT; Camprubi, C.; (Cavriglia)

Carmen Miranda, HT; (Brazil); (Cavriglia)

'CARMEN PAPANDREA', Min, dp, 1989; flowers deep pink; [Magic Carrousel sport]; Papandrea, John T.

'CARMEN SISTACHS', Pol, 1936; flowers rose-pink, base old-gold, cluster, dbl.; foliage sparse, small, soft, light; small vigorous, bushy growth; Dot, Pedro

'CARMEN SYLVA', HT, yb, 1891; flowers yellowish white with pink, large, dbl.; Heydecker; (Sangerhausen)

'CARMEN TALÓN', HT, dr, 1953; bud oval; flowers velvety dark red, large, dbl., 35 petals, moderate fragrance; strong stems; vigorous growth; [Charles Mallerin X Satan]; Dot, Pedro

'CARMEN TESSIER', HT, dr, 1964; bud long, pointed; flowers crimson-red, open, dbl., slight fragrance; foliage leathery; vigorous, bushy growth; [Seedling X Independence]; Mondial Roses

'CARMENCITA', Min, w, 1954; bud ovoid; flowers pure white, dbl., 55 petals; foliage clear green; vigorous growth; [Lady Sylvia X Perla de Alcanada]; Camprubi, C.

'CARMENETTA', S, lp, 1923; flowers pale pink, borne in clusters, single, slight fragrance; foliage leathery, reddish; vigorous (7 ft), spreading (11 ft very hardy; (28); [R. rubrifolia X R. rugosa]; Central Exp. Farm

Carmine Button, Min, dp

Carmine Pillar, *see* **'PAUL'S CARMINE PILLAR'**

Carmosine, *see* LAPmiravi

Carnation, HCh, mp

Carnaval®, *see* 'KORFRILLA'

Carnaval de Rio®, *see* 'DELORFEU'

'CARNE', M, lp; flowers flesh-pink, large, dbl.; Robert

Carnet de Bal, HT, 1972; Dorieux, Francois; (Cavriglia)

Carnival, *see* 'KORFRILLA'

'CARNIVAL', HT, op, 1939; flowers glowing orange to soft cerise, dbl.; foliage glossy, light; vigorous growth; RULED EXTINCT, 1/88; Archer

'CARNIVAL GLASS', Min, ob, 1979; bud pointed; flowers yellow-orange blend,, dbl., 38 petals, 1–1.5 in., slight fragrance; foliage small, glossy, bronze; bushy, spreading growth; [Seedling X Over the Rainbow]; Williams, Ernest D.; Mini-Roses

'CARNIVAL PARADE', Min, yb, 1978; bud long, pointed; flowers golden yellow, edged red, dbl., 45 petals, 1 in., exhibition form, slight fragrance; foliage small, dark, glossy; upright, bushy growth; [Starburst X Over the Rainbow]; Williams, Ernest D.; Mini-Roses

'CARNIVAL QUEEN', HT, mr, 1965; bud long, pointed; flowers luminous red, large, semi-dbl., moderate fragrance; vigorous growth; [(Tassin X Priscilla) X Charlotte Armstrong]; Armbrust; Langbecker

'CAROL', HT, pb, 1964; flowers cyclamen-rose, becoming orchid at edge, center apricot, well-formed, dbl., 45 petals, slight fragrance; no prickles; moderate growth; [Queen Elizabeth X Confidence]; Herholdt, J.A.; Herholdt's Nursery

Carol, *see* **'CAROL AMLING'**

'CAROL AMLING', F, mp, 1953; (Carol, Garnette Carol, Garnette Pink); flowers deep rose-pink, edged lighter; [Garnette sport]; Amling, C.M. & Beltran; Amling Bros.

'CAROL ANN', Pol, op, 1940; flowers orange-salmon, small, tight clusters, dbl., 35–45 petals, 1–1.5 in., cupped, recurrent bloom; dwarf (12 in. or less) growth; [Marianne Kluis Superior sport]; Kluis; Klyn

Carol Ann, *see* 'PEAPOST'

Carol Ann, Min, lp; Welsh; (Weatherly, L.)

'CAROL HOWARD', HT, mp, 1935; bud long, pointed; flowers rose-pink, peony shape, very large, moderate fragrance; foliage dark; vigorous growth; Pfitzer; P.J. Howard

Carola®, S, dp, 1988; Noack, Werner

Carole Anne, HT, my

Carolin, *see* **'CORALIN'**

Carolin Reiberl, *see* HELreib

Carolina Classic, *see* 'BRICLASS'

Carolina Daza®, *see* FEpuma

'CAROLINA MOON', HT, my, 1989; bud pointed; flowers medium yellow, fading slightly, borne singly, dbl., 28 petals, exhibition form, intense, fruity fragrance; prickles pointed slightly downward, medium, deep pink; foliage medium, medium green, semi-glossy; upright, tall, vigorous growth; [Just Lucky X Thriller]; Bridges, Dennis A.; Bridges Roses, 1990

Carolina Morning, *see* 'MICAR'

Carolina Rose, *see* **R. CAROLINA**

'CAROLINA SUNSET', LCl, rb, 1990; bud ovoid; flowers crimson, gold reverse, blooms borne in sprays of 3–7, dbl., cupped, slight fragrance; foliage average, dark green, matt; upright, tall growth with 10–12 ft; [Seedling X Seedling]; Jeremias, Lephon L.

'CAROLINE', HT, op, 1955; flowers cinnabar-salmon, very large, dbl., 70 petals, moderate fragrance; foliage dark, glossy, leathery; very vigorous, bushy growth; [Peace X Seedling]; Gaujard

Caroline Anne, HT, my, 1986; Thomas

'CAROLINE BANK', HMult, lp, 1890; flowers pink with white, medium, dbl.; Geschwind, R.; (Sangerhausen)

Caroline Brian, MinFl, ob, 1997

'CAROLINE BUDDE', HMult, dr, 1913; (Lien Budde); flowers crimson-red, large, dbl.; foliage dark; vigorous growth; [Crimson Rambler X Léonie Lamesch]; Leenders, M.

Caroline Clarke, *see* 'HORBONDARC'

'CAROLINE D'ARDEN', HP, mp, 1888; flowers very large, dbl., intense fragrance; Dickson, A.; (Sangerhausen)

Caroline Davison, *see* 'HARHESTER'

Caroline de Berry, *see* **'FOLIACÉE'**

Caroline de Monaco, *see* 'MEIPIERAR'

'CAROLINE DE SANSAL', HP, mp, 1849; flowers pink, center darker, large, dbl., flat, recurrent bloom; vigorous growth; Desprez

'CAROLINE EMMONS', F, rb, 1962; bud ovoid; flowers scarlet-red, dbl., 55–60 petals, 3.5 in., cupped, intense, geranium fragrance; foliage leathery; vigorous, upright, bushy growth; [(Geranium Red X Fashion) X (Diamond Jubilee X Fashion)]; Boerner; Home Nursery Products Corp.

'CAROLINE ESBERG', LCl, mp, 1926; flowers dull rose, cluster, dbl.; Diener

'CAROLINE KAART', HT, op, 1964; flowers salmon-pink veined dark red, large; foliage glossy; upright growth; [Bayadère X Ballet]; Buisman, G. A. H.

'CAROLINE KUSTER', N, ly; Pernet, 1872

'CAROLINE MARNIESSE', N, w, 1848; flowers creamy white, dbl.; Roeser

'CAROLINE OLDREY', HT, lp, 1998; flowers pale pink, very full, exhibition, very dbl., 41–45 petals, 2.5–3 in., borne mostly singly, moderate fragrance; prickles moderate; foliage medium, medium green, semi-glossy; medium, compact, upright growth; [Solitaire X Prima Ballerina]; Jones, L.J.; Haynes Roses, 1997

'CAROLINE PLUMPTON', F, dp, 1975; flowers deep neyron rose, edged lighter, moderately full, dbl., 20–25 petals, 2.5 in., slight fragrance; foliage matt, green; very free growth; [Red Lion X Seedling]; Plumpton, E.

Caroline Schmitt, *see* **'MME CAROLINE SCHMITT'**

'CAROL-JEAN', Min, dp, 1977; (Indian Meillandina, Indian Sunblaze); bud pointed; flowers deep pink, dbl., 22 petals, 1 in., slight fragrance; foliage small to medium, dark; upright, very bushy growth; [Pinocchio X Little Chief]; Moore, Ralph S.; Sequoia Nursery

Carol-Joy, HT, mr, 1991; Allender, Robert William

Carolyn, *see* **'CORALIN'**

Carolyn, *see* 'MACCOLUMB'

'CAROLYN ANN', Min, mp, 1985; flowers small, dbl., 20 petals, moderate fragrance; foliage medium, medium green, semi-glossy; bushy growth; [Gene Boerner X Baby Katie]; Hooper, Clint; Kimbrew Walter Roses

'CAROLYN DEAN', LCl, mp, 1941; bud long, pointed; flowers bright pink, small, borne in clusters, single, 1.75 in., recurrent bloom; foliage glossy; height 5 ft; [Étoile Luisante X Sierra Snowstorm]; Moore, Ralph S.; Sequoia Nursery

'CAROLYN DIANNE', F, lp, 1964; flowers medium, dbl., moderate fragrance; foliage glossy; vigorous, compact growth; [Ma Perkins X Pinocchio]; Patterson; Patterson Roses

'CARON', F, pb, 1972; flowers white and pink, dbl., 38 petals, 3 in., slight fragrance; foliage semi-glossy; upright growth; [Kordes' Perfecta X Saratoga]; Langdale, G.W.T.

Carosi, HT, 1969; Wituszynski, B.; (Cavriglia)

'CAROUBIER', HWich, mr, 1912; flowers very bright light crimson-scarlet, single, early bloom, Nonin

'CAROUGE', HT, dr, 1976; flowers crimson-red, dbl.; fall growth; [Marylene X Credo]; Gaujard

Carpet of Color, *see* KORhaugen

'**Carpet of Gold**', LCl, my, 1939; flowers, dbl., 2–3 in., trailing growth; [(Emily Gray X Yellow Rambler) X Golden Glow]; Brownell, H.C.

'**Carrie Corl**', Gr, mr, 1969; bud ovoid; flowers large, dbl., moderate fragrance; foliage dark, leathery; vigorous growth; [Queen Elizabeth X (Queen Elizabeth seedling X Happiness)]; Germain's; Flower of the Month

'**Carrie Jacobs Bond**', HT, mp, 1935; bud ovoid; flowers rose-pink, center flushed crimson, large, very dbl., intense fragrance; foliage leathery, dark; long stems; vigorous growth; [Premier Supreme X Lady Leslie]; Howard, F.H.; Dreer

'**Carrie Jacobs Bond, Climbing**', Cl HT, mp, 1940; H&S

Carrot Top, *see* 'POUltop'

'**Carrousel**', Gr, mr, 1950; flowers medium, dbl., 20 petals, moderate fragrance; foliage leathery, dark, glossy; vigorous, upright, bushy growth; GM, ARS, 1956 GM, Portland, 1955; [Seedling X Margy]; Duehrsen; Elmer Roses Co.

'**Carrousel, Climbing**', Cl Gr, mr, 1958; Weeks; Elmer Roses Co.

'**Carry Nation**', F, w, 1959; bud ovoid; flowers white, center cream, small, borne in clusters, semi-dbl., globular, slight fragrance; foliage soft, glossy; vigorous, low growth; [Pinocchio X Katharina Zeimet]; Silva; Plant Hybridizers of Calif.

Carte Blanche™, *see* 'MEIringa'

Carte d'Or, *see* 'MEIrobidor'

Carte Rose, F, lp

Cartwheel, *see* 'JACcart'

Cary Grant™, *see* 'MEImainger'

'**Caryatide**', HT, dr, 1955; bud long, pointed; semi-dbl.; strong stems; upright growth; [Hens Verschuren X Poinsettia]; Buyl Frères

'**Casa Blanca**', LCl, w, 1968; bud tinged carmine-pink; semi-dbl., blooms in clusters, intermittent bloom, slight fragrance; foliage dark, glossy; vigorous, climbing growth; [New Dawn X Fashion]; Sima

'**Casanova**', HT, ly, 1964; flowers straw-yellow, dbl., 38 petals, 6 in., exhibition form, moderate fragrance; [Queen Elizabeth X Kordes' Perfecta]; McGredy, Sam IV; Fisons Horticulture

'**Cascabel**', F, rb, 1957; flowers red, reverse pearly, passing to carmine, borne in clusters, dbl., 45–50 petals, globular, foliage glossy; very vigorous, upright, compact growth; [Méphisto X Perla de Alcañada]; Dot, Pedro

'**Cascade**', LCl, mr, 1951; flowers bright crimson, almost single, large, in clusters, very dbl.; very vigorous growth; [Holstein X American Pillar]; Mallerin, C.; EFR

Cascade, HT, w, 1990; Strahle, B. Glen

'**Cascadia**', HMsk, lp, 1927; flowers blush-pink paling to white, borne in clusters, semi-dbl., 15 petals, 1 in., recurrent bloom, slight fragrance; foliage glossy, dark; tall growth; GM, Portland, 1922; [Mme d'Arblay X Bloomfield Abundance]; Thomas; B&A

'**Cashmere**', HT, ab, 1980; bud pointed; flowers soft apricot, borne singly and 3–5 per cluster, dbl., 30 petals, globular, slight, spicy fragrance; long, hooked prickles; foliage leathery, dark; vigorous, bushy growth; [Tanya X Jack O'Lantern]; Weeks, O.L.

'**Casimir Moullé**', HWich, pb, 1910; flowers purplish pink, reverse silvery pink; [R. wichurana X Mme Norbert Levavasseur]; Barbier

Casino®, *see* 'MACca'

'CASlon', HT, rb, 1996; (**Long Island**®); flowers creamy white with red edge, blooms borne mostly single, dbl., 26–40 petals, 4 in., moderate fragrance; foliage medium, light green, semi-glossy; upright, tall (3–4 ft) growth; [Tiki X Color Magic]; Castillo, Angel; Z&B Own Root Roses, 1998

Casper, *see* '**Magna Charta**'

Casque d'Or®, *see* 'DELcascor'

'**Cassandra**', HT, mr, 1966; flowers cherry, 4.5 in., exhibition form, slight fragrance; foliage dull, serrated; free growth; [(Karl Herbst X Ena Harkness) X (Christian Dior X Peace)]; Dorieux; Bees

Cassandra, F, or, 1998; Noack, Werner

Cassandre®, *see* MEIdenji

'**Casta Diva**'®, HT, w, 1982; flowers large, dbl., 35 petals, cupped, no fragrance; flat, yellow prickles; foliage large, dark, glossy; upright growth; [Pascali X Seedling]; Barni-Pistoia, Rose

'**Castanet**', F, op, 1960; bud ovoid; flowers orange-pink, reverse lighter, dbl., 45 petals, 3.5 in., moderate fragrance; upright, bushy growth; [Chic X Garnette seedling]; Boerner; J&P

Castel, *see* 'DELset'

Castella®, *see* 'TANallet'

Casteliana, F; (Italy); (Cavriglia)

Castilian, *see* '**Autumn Damask**'

Castilian, HT, ob, 1996; Williams, J. Benjamin

Castle Hill, *see* 'ZIPhill'

Castle Howard Tercentenary, *see* 'TANtasch'

Castle of Mey, *see* 'COClucid'

Castore®, *see* BARcast

'**Catalina**', HT, pb, 1939; bud long, pointed, carmine, reverse salmon; flowers carmine suffused yellow and rose, very large, dbl., 23 petals, intense fragrance; foliage dark, glossy; very vigorous growth; [Pres. Herbert Hoover X Katharine Pechtold]; San Remo Exp. Sta.

'**Catalina**', HT, op, 1940; bud very long, coral-pink edged old-rose; flowers salmon to shrimp-pink, dbl., 25–30 petals, 4.5 in., exhibition form, moderate fragrance; very vigorous, upright growth; [Joanna Hill sport]; Grillo

Catalina Frau®, *see* FErama

'**Catalonia**', HT, or, 1933; flowers bright orange-crimson, shaded gold, large, very dbl., globular; foliage dark; bushy growth; [(Shot Silk X Mari Dot) X Jean C.N. Forestier]; Dot, Pedro; C-P

'**Catalunya**', Cl HT, mr, 1918; [Gruss an Teplitz sport]; Nonin

Caterpillar®, *see* 'POUlcat'

'**Catharina Klein**', HT, pb, 1930; bud ovoid; flowers bright pink, yellow background, large, very dbl., moderate fragrance; foliage rich green; very vigorous growth; [Mrs Franklin Dennison X Hadley]; Berger, V.; Münch & Haufe

'**Cathay**', F, ly, 1957; dbl., 35–40 petals, 3.5–4.5 in., exhibition form, intense fragrance; foliage glossy, leathery; vigorous, compact growth; [Fandango X Pinocchio]; Swim, H.C.

'**Cathcart Bedder**', HT, op, 1939; flowers salmon-shrimp-pink; foliage bronze; vigorous, bushy growth; Austin & McAslan

'**Cathedral**', F, ab, 1975; (Coventry Cathedral, Houston); flowers apricot shading salmon, dbl., 22–25 petals, 4–5 in., borne in small clusters, slight, anise fragrance; foliage glossy, olive-green; bushy growth, medium, compact; AARS, 1976 GM, Portland, 1974 Gold Star of the South Pacific, Palmerston North, NZ, 1974; [Little Darling X (Goldilocks X Irish Mist)]; McGredy, Sam IV

Cathedral Peak, *see* JACwiht

Cathedral Splendour, *see* HARbell

'**Catherine Anne**', Min, w, 1999; flowers medium, dbl., 17–25 petals, borne in small clusters, slight fragrance; many prickles; foliage medium, medium green, semi-glossy; bushy, tall (3 ft) growth; [Rise 'n' Shine X City of London]; Goodall, G. T.

'**Catherine Blackburn**', Pol, lp; Matthews, W.J., 1921; (Weatherly, L.)

Catherine Bonnard, B, dp; Verdier, E., 1871

Catherine Cookson, *see* 'NOScook'

'**Catherine de Würtemberg**', M, lp, 1843; bud well mossed; flowers soft pink, large, very dbl., globular; vigorous growth; Robert

Catherine Deneuve®, *see* MEIpraserpi

'CATHERINE GHISLAINE', D, m, 1885; flowers violet-pink, fading to white, small, semi-dbl.; Roseraie de l'Hay; (Sangerhausen)

'CATHERINE GUILLOT', B, dp, 1860; flowers carmine-rose, large, dbl.; vigorous growth; [Louise Odier seedling]; Guillot Père

'CATHERINE II', HCh, lp, 1832; Laffay, M.

'CATHERINE LANGEAIS', HT, mr, 1965; flowers carmine-red, well formed, dbl., 46 petals, exhibition form; vigorous, upright growth; [Michele Meilland X Berthe Mallerin]; Hémeray-Aubert

'CATHERINE MARIE', HT, mp, 1991; bud pointed; flowers medium to shell pink, dbl., 26 petals, exhibition form, intense fragrance; foliage medium, medium green, semi-glossy; [Pristine X Captain Harry Stebbings]; Wambach, Alex A., 1990

Catherine McAuley, *see* JACibras

'CATHERINE MERMET', T, lp, 1869; bud well-shaped; flowers flesh-pink, edges tinted lilac-pink, large, dbl., moderate fragrance; vigorous growth; Guillot et Fils

Catherine Pericard, F, 1981; Pineau; (Cavriglia)

'CATHERINE SEYTON', HEg, lp, 1894; flowers soft pink, yellow stamens, single, non-recurrent, moderate fragrance; foliage fragrant (apple); vigorous growth; (35); Penzance; Keynes, Williams & Co.

'CATHERINE SOUPERT', HP, 1879; Lacharme, F.; (Cavriglia)

'CATHIE', HT, m, 1999; flowers mauve, outer petals edges shaded light red, dbl., 26–40 petals, 3.5 in., slight fragrance; foliage medium, dark green, dull; upright, tall (4 ft) growth; [Paradise X Unknown]; Priestly, James L.

Cathie Irwin, *see* KORlowi

'CATHRINE KORDES', HT, dr, 1930; bud long, pointed, blood-red, shaded black; flowers dark scarlet, large, dbl., moderate fragrance; foliage dark (blood-red when young), leathery; very vigorous growth; [(Mme Caroline Testout X Willowmere) X Sensation]; Kordes; Dreer; H&S

'CATHRINE KORDES, CLIMBING', Cl HT, dr, 1938; Krohn; Kordes

'CATHY ANNE', HT, r, 1990; bud pointed; semi-dbl., 25 petals, exhibition form, borne singly, no fragrance; prickles long, slender, red; foliage small, dark green; long, straight stems; medium growth; [Judith Morton X Sylvia]; Wilson, George D., 1986

'CATINAT', HGal, m; flowers violet, spotted; Robert, ca. 1850

'CATINAT', HP, mr, 1874; Oger

'CATO', HMult, lp, 1904; flowers medium, dbl.; Gratama; (Sangerhausen)

'CATORCE DE ABRIL', HT, yb, 1932; flowers yellow streaked red, Padrosa

Cato's Pink Cluster, N, lp

Cat's Meow™, *see* 'DESLING'

'CAULDRON', S, mp, 1984; flowers small, single, 5 petals, slight fragrance; foliage medium, mid-green, matt; upright growth; [R. rubrifolia X R. nutkana]; Holliger, Franc

'CAULDRON', HT, pb, 1971; flowers rose-pink to yellow, dbl., 23 petals, 4.5 in., moderate fragrance; foliage glossy, dark; RULED EXTINCT 6/83 ARM; Waterhouse Nursery

Cauvery, HT, dr, 1973; Kasturi

'CAVALCADE', F, rb, 1950; bud ovoid; flowers oxblood-red and yellow, changing daily to crimson, carmine and silvery pink, dbl., 32 petals, intense, fruity fragrance; foliage glossy, dark; vigorous, bushy growth; Verschuren-Pechtold; Stuart

'CAVALCADE, CLIMBING', Cl F, rb, 1957; Gandy, Douglas L.

'CAVALIER', HT, ob, 1939; flowers burnt-orange to cream-buff, dbl., moderate fragrance; [Mrs Franklin D. Roosevelt sport]; Samtmann Bros.

Cavriglia, HT, 1989; Fineschi, G.; (Cavriglia)

'CAYENNE', HT, ob, 1966; bud short, pointed; flowers deep orange, dbl., 38 petals, 3–4 in., slight fragrance; upright growth; [South Seas X Seedling]; Warriner, William A.; J&P

'CAYETANA STUART', HT, my, 1931; bud long, pointed; flowers very large, dbl., cupped, moderate fragrance; foliage dark, glossy; very vigorous, bushy growth; GM, Bagatelle, 1930; [Isabel Llorach X (Constance X Sunburst)]; Dot, Pedro; C-P

'CECHOSLAVIA', HT, w, 1921; flowers milky white suffused salmon-carmine, center golden yellow, dbl., moderate fragrance; [Pharisaer X Mme Antoine Mari]; Berger, V.; A. Berger

'CECIL', HT, my, 1926; flowers golden yellow, blooms in large clusters, single, 5 petals, 4 in., bushy growth; Cant, B. R.

'CECIL, CLIMBING', Cl HT, my, 1940; Chaffin; Armstrong Nursery

Cécile Brünner, *see* **'MLLE CÉCILE BRÜNNER'**

Cécile Brünner, Climbing, *see* **'MLLE CÉCILE BRÜNNER, CLIMBING'**

'CECILE CUSTERS', HT, pb, 1914; flowers lilac-rose, reverse deep rose-pink, dbl.; [Mme Abel Chatenay X Violet Liddell]; Leenders, M.

Cecile Lens, Min, mp

'CECILE MANN', HT, mr, 1939; vigorous growth; [Mrs Albert Nash X Seedling]; Clark, A.; Brundrett

'CÉCILE RATINCKX', HT, my, 1924; flowers coppery yellow; [Louise Catherine Breslau X Mme Edouard Herriot]; Vandevelde

'CÉCILE VERLET', HT, yb, 1926; flowers yellow, passing to rose-pink; [Marianna Rolfs X Nordlicht (HT)]; Walter, L.

'CÉCILE WALTER', HT, op, 1926; bud long, pointed; flowers coral-pink to coppery pink, base gold, very large, dbl., 28 petals, cupped, slight fragrance; foliage rich green, leathery; vigorous growth; [(Mme Mélanie Soupert X Mme Edouard Herriot) X Seedling]; Mallerin, C.; C-P

'CECILIA', S, pb, 1980; bud globular, pointed; flowers light pink blended darker pink, carnation-like form, dbl., 25 petals, borne singly, moderate, spicy fragrance; hooked prickles; foliage small, pointed, glossy; vigorous, compact, bushy growth; [(R. wichurana X Baronne Prevost) X (R. X odorata)]; James, John

'CECILIO RODRIGUEZ', HT, dr, 1950; bud long, pointed; flowers dark velvety red, large, dbl., exhibition form, intense fragrance; foliage dark; strong stems; very vigorous growth; [Tassin X Eugenio d'Ors]; Camprubi, C.

Cecilitas, Min, lp

Cecil's Bright Apricot, HT, ab, 1996; Godman

Cecily Gibson, *see* 'EVERBRIGHT'

Cedar Crest College, *see* 'WILCREST'

'CEDRIC ADAMS', HT, dr, 1949; bud ovoid; flowers scarlet to carmine, large, dbl., exhibition form, moderate fragrance; foliage dark, bronze; vigorous growth; [Pink Princess X Crimson Glory]; Brownell, H.C.

Cee Dee Moss, *see* 'MORCEEDEE'

Cearcee, HT, ob, 1995; Gokhale

Celebrate America™, *see* 'TANCRESSOR'

'CELEBRATION', F, op, 1961; flowers salmon-pink, reverse lighter, blooms in clusters, dbl., 30 petals, 3 in., cupped, slight fragrance; foliage light green; vigorous, bushy growth; [Dickson's Flame X Circus]; Dickson, Patrick; A. Dickson

Celebration 2000, *see* 'HORCOFFITUP'

Celebration Day, HT, mp, 1987; Thames Valley Rose Growers

'CELEBRITY', HT, pb, 1945; (Lustrous); bud pointed; flowers pink, base shaded yellow, dbl., 30–35 petals, 6–7 in., exhibition form, moderate, spicy fragrance; foliage leathery, dark; vigorous, bushy growth; RULED EXTINCT 1/88; [Golden Rapture X Carmelita]; Hill, E.G., Co.

'CELEBRITY', HT, dy, 1988; bud pointed, large; flowers deep yellow, aging clear yellow, large, dbl., 30–35 petals, 4.5–5 in., exhibition form, borne usually singly, moderate, fruity fragrance; prickles pointed slightly downward, small, yellow-brown; foliage large, dark green, glossy; upright, bushy, medium growth; PP7264;

[(Sunbonnet X Mister Lincoln) X Yello Yo Yo]; Weeks, O.L.; Weeks Roses, 1989

Céleste, *see* **'Celestial'**

'Céleste', A, lp; flowers medium to large, moderate fragrance

'Celeste Mahley', F, pb, 1972; bud long, pointed; flowers medium, dbl., slight fragrance; foliage leathery; vigorous, upright, bushy growth; [Seventeen X Gemini]; Byrum; J.H. Hill Co.

'Celestial', A, lp; (Céleste); flowers light blush, golden stamens, large, dbl., non-recurrent, intense fragrance; foliage bluish; vigorous (to 6 ft) growth; Kew, Prior to 1797

'Celestial', HSpn, lp; flowers pale flesh, small, dbl., intense fragrance; low growth; [R. eglanteria X R. spinosissima]

'Celestial', HT, lp, 1924; flowers light pink, edged paler, dbl., moderate fragrance; [Premier sport]; Myers & Samtmann

'Celestial Star', F, or, 1965; bud ovoid; dbl., borne in clusters, foliage glossy; vigorous, compact growth; Pal, Dr B.P.; Indian Agric. Research Inst.

'Celia Walker', F, rb, 1962; flowers cherry-red, reverse silver, clusters, semi-dbl., 15–20 petals, 3 in., slight fragrance; foliage dark, glossy; vigorous, upright growth; [Alain X Golden Scepter]; Fletcher; Tucker

Celica, *see* 'MEIfota'

Celientje, HT, ly, 1997

Celie's Slidell Pink, Cl T, lp

Celina, *see* NOAsun

'Célina', M, m, 1855; bud heavily mossed; flowers reddish-purple, center occasionally streaked white, large, dbl.; Hardy

'Celina Dubos', D, lp, 1849; flowers pale pink, sometimes recurrent bloom; Dubos

'Céline', B, mp, 1824; flowers pale rose, large large clusters, dbl., cupped, does not repeat, very vigorous growth; Laffay, M.; Laffay, about 1825

Céline Delbard®, *see* 'DELcet'

'Céline Forestier', N, ly, 1858; flowers pale yellow, large blooms in clusters of 3–4, dbl., repeat blooming, moderate fragrance; foliage dark, glossy; vigorous (to 6 ft) growth; (14); Leroy, A.

'Cels Multiflora', HMult, lp; Hardy/Cels, 1836

'Celsiana', D, lp; flowers pale pink, petals crinkled, blooms in cluster of 3–4, semi-dbl., 4 in., moderate fragrance; foliage smooth, grayish, fragrant; vigorous, upright (4–5 ft) growth

'Cendrillon', HT, mp, 1951; bud very large; flowers salmon, dbl., intense fragrance; foliage leathery; very vigorous, upright growth; Gaujard

Centenaire de Lourdes®, *see* 'DELge'

Centenaire de Lourdes Rouge®, *see* DELfloro

Centenaire du Vesinet, *see* 'PIRonia'

Centenary, *see* KOReledas

Centenary College, *see* 'AROblaveet'

Centennaire de Lourdes, *see* 'DELge'

'Centennial', HT, ob, 1953; bud ovoid; flowers peach-red, reverse orange-buff, dbl., 40–45 petals, 4–4.5 in., exhibition form, moderate fragrance; foliage rich green; vigorous, bushy growth; [Seedling X Orange Nassau]; Mallerin, C.; J&P

Centennial Gold, HT, my

'Centennial Miss', Min, dr, 1952; flowers deep wine-red, base tinged white, dbl., 60 petals, 1 in., moderate fragrance; no prickles; foliage small, dark, leathery, glossy; dwarf (10–12 in.), bushy growth; [Oakington Ruby X Self]; Moore, Ralph S.; Sequoia Nursery

Centennial Star(TM), *see* 'MEInereau'

'Centennial Sweetheart', Cl F, mr, 1959; bud globular; flowers bright red, borne in large clusters, dbl., 25 petals, 3.5 in., intense fragrance; foliage glossy, light green; very vigorous growth; [Alain sport]; Greene

Center Gold(TM), *see* 'SAVacent'

Centerpiece(TM), *see* 'SAVapiece'

Centifolia a Fleurs Doubles Violettes, C, m

'Centifolia Foliacea', C, mp, 1810

Centifolia Major, *see* **'Rose des Peintres'**

'Centifolia Minima', C, lp; flowers soft pink, very dbl., 80–100 petals, non-recurrent, moderate fragrance; height 8–15 in., branching

Centifolia Minima, *see* **'Rouletii'**

'Centifolia Muscosa', M, mp; (Common Moss, Communis, Moss Rose, Mousseux Ancien, R. centifolia muscosa, R. muscosa); flowers large, peduncles and calyx glandular, mossy, dbl., (28)

Centifolia Muscosa, *see* **Communis**

'Centifolia Rosea', HP, mp, 1863; flowers bright rose, large, cupped; foliage crinkled, light green wood with many red prickles; vigorous growth; Touvais

Centifolia Variegata, *see* **'Village Maid'**

Central Park, *see* 'POUlpyg'

'Centre Court', HT, w, 1964; flowers white, center tan, well shaped,, dbl., 38 petals, 4 in., moderate fragrance; foliage dark; vigorous growth; [Eden Rose X Ena Harkness]; Barter

Centre Piece, *see* 'SAVapiece'

'Centrex Gold', F, my, 1975; dbl., 25 petals, 3 in., slight fragrance; foliage matt, green; [Alison Wheatcroft X Chinatown]; Smith, E.; Wheatcroft

'Centro de Lectura', HT, mr, 1959; flowers crimson, dbl., 40 petals, moderate fragrance; foliage glossy, bright green; long strong stems; vigorous growth; [Texas Centennial X Peace]; Dot, M.

Centurio®, *see* 'HAVop'

'Centurion', F, dr, 1975; flowers blood-red, shaded crimson, dbl., 30 petals, 3 in., slight fragrance; foliage glossy, dark; [Evelyn Fison X Seedling]; Mattock

'Century 21', HT, lp, 1962; bud ovoid; flowers shell-pink, reverse slightly darker, large, dbl., exhibition form, moderate fragrance; foliage glossy; vigorous growth; [Condessa de Sástago X Soeur Therese]; Morey, Dr Dennison; J&P

Century Sunset, *see* 'TANsaras'

'Century Two', HT, mp, 1971; bud long, pointed; flowers large, hot pink, dbl., 30–35 petals, 5 in., exhibition form, floriferous, borne singly, moderate, damask fragrance; foliage leathery; vigorous, upright, bushy growth; [Charlotte Armstrong X Duet]; Armstrong, D.L.; Armstrong Nursery

'Ceremony', F, w, 1970; bud cupped; flowers pure white, dbl., 40–45 petals, 2.5 in.; foliage leathery; very bushy growth; [Tiara X Pascali]; Lens

'Ceres', HMsk, lp, 1914; flowers pale blush, tinted light yellow, stamens bright yellow, semi-dbl., profuse seasonal bloom; shrub growth; Pemberton

'Ceres', HT, ob, 1922; flowers deep orange, center salmon; [Sunburst X Mme Edmond Rostand]; Spek

Ceres, *see* TANegnaro

'Cerise', HT, dp, 1945; flowers deep pink, large, dbl., 25 petals, intense fragrance; foliage dark, leathery; vigorous, upright growth; [Crimson Glory X Sterling]; Tantau

Cerise, S, dp; Delbard

'Cerise Bouquet', S, dp, 1958; flowers cerise-crimson, semi-dbl., flat, moderate fragrance; foliage small, grayish; open, arching growth; [R. multibracteata X Crimson Glory]; Kordes

Cerise Dawn(TM), *see* 'DEVrise'

'Cerise d'Orlin', HGal, dp; flowers deep pink, reverse silver, loose, semi-dbl.

Cerise Margo, Pol, dp

Cerise Nights, S, dp

Cerise Rouge, S, mr

'Cerise Talisman', HT, dp, 1933; flowers cerise; [Talisman sport]; Clarke Bros.

Cérisette la Jolie, *see* **'Surpasse Tout'**

Cernousek, HT, lp; Strnad; (Czech Rosa Club)

Cervanek, HT, lp; Strnad; (Czech Rosa Club)

'CERVANKY', Cl HT, dy, 1935; flowers large, dbl.; Böhm, J.; (Sangerhausen)

Cervena Gloria Dei, HT, lp; Strnad; (Czech Rosa Club)

Cervena Super Star, HT, 1976; Strnad; (Cavriglia)

Cerys Ann, *see* 'GUEscan'

Cesar™, *see* MEIsardan

'CESARE BECCARIA', HGal, 1870; flowers large, dbl.; Moreau et Robert; (Sangerhausen)

'CESKÁ POHADKA', Pol, rb, 1933; flowers red, salmon, rose and white, borne in clusters; very vigorous, dwarf growth; [Golden Salmon sport]; Böhm, J.

Ceské 'Praci cest', LCl, 1970; flowers small, dbl.; Vecera, L.; (Sangerhausen)

'CÉSONIE', D, dp, 1859; flowers deep rose, large, dbl.; compact growth; Moreau et Robert, Prior to 1848

'CESTIFLORA', HSpn, ly; flowers sulfur-yellow, early bloom; glossy, black fruit; foliage finely divided; height 3–4 ft

Cevennes, *see* 'KORlirus'

Cha Cha®, *see* 'COCarum'

'CHABLIS', HT, w, 1983; flowers creamy white, medium-large, dbl., exhibition form, slight fragrance; foliage large, medium green, matt; upright growth; [Seedling X Louisiana]; Weeks, O.L.; Weeks Wholesale Rose Growers, 1984

'CHACITA', F, dr, 1947; bud ovoid, globular; flowers deep red, medium, borne in clusters, dbl., cupped, moderate fragrance; foliage dark, leathery; vigorous, bushy growth; [Pinocchio X Crimson Glory]; Boerner

Chacok, *see* 'MEIcloux'

'CHAGIP', F, mr, 1956; (**Embrasement**); flowers fiery red, in clusters of 8–12, dbl.; foliage bronze; vigorous growth; Delbard-Chabert

'CHALICE', HT, ab, 1959; bud ovoid, pointed; flowers apricot-yellow, dbl., 55–60 petals, 5.5–6 in., exhibition form, intense fragrance; foliage leathery, glossy; vigorous, upright growth; [Orange Delight X Golden Rapture]; Verschuren-Pechtold; J&P

Chalice Well, HMult, lp; [R. multiflora X Unknown]; McLeod, J., 1995; Honeysuckle Cottage; (Weatherly, L.)

'CHALLENGE', F, dr, 1962; flowers deep blood-red, camellia shape, borne in clusters, semi-dbl., 20 petals, 2 in.; foliage dark; bushy, low growth; [Alain X Pinocchio]; Fletcher; Tucker

'CHALLENGER', HT, dr, 1938; flowers dark crimson, very large, dbl., exhibition form, intense fragrance; foliage leathery, dark; long stems; vigorous, bushy growth; RULED EXTINCT, 6/89; Cant, B. R.

Challenger™, *see* 'HILred'

Challis Gold, *see* 'NOSchal'

Chalom, HT; Moreira da Silva, A.; (Cavriglia)

'CHAMBA PRINCESS', F, op, 1969; flowers salmon-pink, open, medium, semi-dbl., intermittent bloom, slight fragrance; foliage leathery; moderate, bushy growth; IARI

'CHAMBE DI KALI', HT, mp, 1983; flowers large blooms borne singly, dbl., 22 petals, exhibition form, slight fragrance; green to brown prickles; foliage large, dark, glossy; medium, compact growth; [Bewitched X Seedling]; Pal, Dr B.P.; K.S.G. Son's Roses

'CHAMBORD', F, mp, 1960; bud pointed; flowers pink, becoming darker, open, large, borne in clusters, dbl., slight fragrance; foliage bronze; bushy growth; [Roquebrune X Queen Elizabeth]; Delforge

'CHAMELEON', HT, or, 1918; flowers flame, edged cerise; [Lyon Rose seedling]; Dickson, A.

Chameleon, lp; [Wee Beth X (Starina X Unknown)]; Welsh, Eric, 1995; Biotech Plants Pty, Ltd.; (Weatherly, L.)

'CHAMI', HMsk, mp, 1929; flowers bright rose-pink, stamens yellow, single, recurrent bloom, intense fragrance; bushy growth; Pemberton

'CHAMISSO', HMult, pb, 1922; flowers flesh-pink, center yellowish white, borne in clusters, semi-dbl., profuse, recurrent bloom., moderate fragrance; foliage bronze; long stems; vigorous, trailing growth; [Geheimrat Dr Mittweg X Tip-Top]; Lambert, P.

Chamois Dore®, F, ab, 1990; Lens

'CHAMP WEILAND', HT, mp, 1915; flowers clear pink; foliage glowing, reddish; [Killarney sport]; Weiland & Risch

'CHAMPAGNE', HT, yb, 1961; bud pointed, ovoid; flowers buff shaded apricot, dbl., 28 petals, 4–5 in., exhibition form, moderate fragrance; foliage leathery, dark; vigorous, upright, bushy growth; [Charlotte Armstrong X Duquesa de Peñaranda]; Lindquist; Howard Rose Co.

Champagne Arches, LCl, lp

Champagne Cocktail, *see* 'HORflash'

Champagne Pearl, F, ab, 1983; (Champagnerperle); Kordes (Sangerhausen)

Champagner®, *see* 'KORampa'

Champagnerperle, *see* **Champagne Pearl**

'CHAMPION', HT, yb, 1976; flowers yellow-cream, flushed red and pink, dbl., 50–55 petals, 7–8 in., moderate fragrance; foliage large, light; [Irish Gold X Whisky Mac]; Fryer, Gareth; Fryer's Nursery, Ltd.

Champion®, F, or, 1988; Noack, Werner

'CHAMPION OF THE WORLD', HP, mp, 1894; (Mrs de Graw, Mrs DeGraw); flowers rose-pink, large, dbl., seasonal bloom, moderate fragrance; vigorous growth; [Hermosa X Magna Charta]; Woodhouse

'CHAMPION, CLIMBING', Cl HT, yb, 1999; flowers yellow-cream, flushed red and pink, very dbl., 50–55 petals, 3–3.5 in., exhibition form, moderate fragrance; foliage large, medium green; climbing (12 ft) growth; GM, Japan (JRC), 1997; [Champion sport]; Kameyama, Yasushi, 1990; La Vie en Rose Co., Ltd., 1998

'CHAMPLAIN', HKor, dr, 1982; flowers large, bright medium red with darker petal tips, dbl., 30 petals, 2.5 in., borne in small clusters, repeat bloom, slight fragrance; straight, yellow-green prickles; foliage small, dark yellow-green; bushy growth, 3 ft; hardy; [(R. kordesii X Seedling) X (Red Dawn X Suzanne)]; Svedja, Felicitas; Agriculture Canada

'CHAMPNEYS' PINK CLUSTER', N, lp; (Champneys' Rose); dbl., borne in large clusters, recurrent bloom; moderately vigorous growth; moderately hardy; (14); [R. chinensis X R. moschata]; Champneys, 1811

Champneys' Rose, *see* **'CHAMPNEYS' PINK CLUSTER'**

Champs-Elysées®, *see* 'MEIcarl'

Champs-Elysées, Climbing®, *see* 'MEIcarlsar'

'CHANDELLE', F, lp, 1958; flowers light pink, reverse darker, well formed, medium, borne in clusters, dbl., intense fragrance; foliage bronze; vigorous, compact growth; [Gretel Greul X (Lady Sylvia X Fashion)]; Lens

'CHANDERI', HT, lp, 1968; bud long, pointed; flowers light pink, edged deeper, medium, dbl., exhibition form, slight fragrance; foliage soft; moderate, upright, compact growth; [Peace X Seedling]; Singh

'CHANDRAMA', F, w, 1980; bud pointed; dbl., 25 petals, borne in clusters of 3–6, slight fragrance; straight, brown prickles; foliage dark; spreading growth; [White Bouquet X Virgo]; Division of Vegetable Crops and Floriculture

Chandrika, Min, w, 1978; Kasturi

Chanel, HT, dp, 1995; Juneida

'CHANELLE', F, op, 1959; flowers peach-pink shaded rose-pink, well-formed, blooms in clusters, dbl., 20 petals, 3 in., moderate fragrance; foliage dark, glossy, pointed; vigorous, bushy growth; GM, Madrid, 1959; [Ma Perkins X (Fashion X Mrs William Sprott)]; McGredy, Sam IV; McGredy

'CHANOINE TUAILLON', HT, dp, 1931; bud persian red; flowers carmine, very large, dbl., globular, moderate fragrance; very vigorous growth; [Betty Uprichard X Lucie Nicolas Meyer]; Gillot, F.

Chanson d'Été, *see* **'SUMMER SONG'**

'CHANTAL', HT, w, 1958; bud long, ovoid; flowers ivory, center yellow, edged carmine, very large, dbl., 60 petals, moderate fragrance; vigorous, bushy, upright growth; [Mme Charles Sauvage X Carillon]; Moulin-Epinay

Chantebrise, LCl, mr, 1969; Croix, P.; (Cavriglia)

'**CHANTECLERC**', F, mr, 1956; flowers bright red, large, foliage bright green; bushy growth; [Peace X Seedling]; Gaujard

Chantefleur, *see* LAParan

Chanterelle, *see* 'WARDROSA'

'**CHANTILLY**', HT, or, 1964; flowers orange to red-lead, dbl., 40 petals; foliage glossy, dark; strong stems; vigorous growth; [Baccará X Seedling]; Verschuren, A.; van Engelen

'**CHANTILLY LACE**', HT, m, 1978; bud long; flowers red-purple, dbl., 35 petals, 4–5 in., exhibition form, intense fragrance; foliage glossy; vigorous growth; PP004665; [Blue Moon X Angel Face]; deVor, Paul F.; DeVor Nurseries, Inc.

Chantoli, *see* 'FRYXOTIC'

'**CHANTRÉ**', HT, ob, 1958; bud long, pointed; flowers orange and golden yellow, dbl., 20–25 petals, 5 in., exhibition form, moderate fragrance; foliage dark, leathery; very vigorous, upright, bushy growth; [Fred Streeter X Antheor]; Kordes, R.

Chapeau de Napoléon, *see* '**CRESTED MOSS**'

'**CHAPELAIN D'ARENBERG**', HGal, mp; flowers bright pink

'**CHAPERON ROUGE**', F, mr, 1951; bud globular; flowers velvety crimson, center darker, open, medium, very dbl., moderate fragrance; foliage glossy, bronze, dark; vigorous, upright growth; [Crimson Glory X (Baby Chateau X Seedling)]; Vilmorin-Andrieux

'**CHAPLIN'S CRIMSON GLOW**', LCl, rb, 1930; flowers deep crimson, base white, large, dbl.; Chaplin Bros.

'**CHAPLIN'S PINK CLIMBER**', LCl, mp, 1928; flowers bright pink, stamens golden yellow, large blooms in large clusters, semi-dbl., flat, non-recurrent, very vigorous growth; (28); GM, NRS, 1928; [Paul's Scarlet Climber X American Pillar]; Chaplin Bros.

'**CHAPLIN'S PINK COMPANION**', LCl, lp, 1961; flowers silvery pink, blooms in clusters of up to 30, dbl., 22 petals, 2 in., moderate fragrance; foliage glossy; vigorous growth; [Chaplin's Pink Climber X Opera]; Chaplin, H.J.; Chaplin & Sons

'**CHAPLIN'S TRIUMPH**', HT, dr, 1936; bud long, pointed; flowers deep velvety crimson, intense fragrance; foliage dark; vigorous growth; Chaplin Bros.

'**CHARADE**', F, op, 1965; flowers coral-cerise tinted salmon, pointed, borne in clusters, dbl., 3 in.; foliage glossy; free growth; [Queen Elizabeth X Seedling]; Herholdt, J.A.

'**CHARADE**'™, F, dr, 1988; flowers cardinal-red, small, dbl., 20–25 petals, exhibition form, borne in sprays of 2–5, slight fragrance; prickles straight, medium, lilac; foliage medium, dark green, semi-glossy; mini flora bushy, profuse growth; PP006669; [Cindy X Sassy]; Hill, Joseph H., Co.; DeVor Nurseries, Inc., 1988

Chardonnay, *see* 'MACRELEASE'

Chardony, *see* 'MACRELEASE'

'**CHARIOT OF ROSES**', Min, dr, 1986; dbl., 25 petals, borne singly, no fragrance; small, globular fruit; small, nearly straight, reddish prickles; foliage small, dark, matt; long, pendulous growth; [Seedling (S) X Fairy Moss]; Fischer, C.&H.; Alpenflora Gardens

Chariots of Fire, LCl, dr, 1996; free-standing climber growth; Williams, J. Benjamin

Charisma, *see* 'JELROGANOR'

'**CHARISMA**', HT, or, 1973; RULED EXTINCT 3/77 ARM; Meilland

Charisma, *see* PEAtrophy

'**CHARITY**', HT, mr, 1953; flowers bright velvety red, dbl., 40–50 petals, 5 in., exhibition form, moderate fragrance; foliage glossy; vigorous, upright, bushy growth; [Will Rogers X Mme Henri Guillot]; Taylor, C.A.

Charity™, *see* 'AUSchar'

Charivari, *see* 'KORUB'

'**CHARLEMAGNE**', HGal, dp

Charlemagne, HP, lp

'**CHARLES ALBANEL**', HRg, mr, 1982; flowers medium, dbl., 20 petals, repeat bloom, moderate fragrance; straight, gray-green prickles; foliage yellow-green, rugose; groundcover; low, spreading growth; [Souv. de Philemon Cochet X Seedling]; Svedja, Felicitas; Agriculture Canada

Charles Austin®, *see* 'AUSfather'

Charles Aznavour, *see* 'MEIBEAUSAI'

'**CHARLES BONNET**', HP, dp, 1884; flowers dark rose, medium, dbl., repeat bloom; Bonnet

'**CHARLES CRETTÉ**', HT, mp, 1917; flowers velvety rose, large; Chambard, C.

'**CHARLES DARWIN**', HP, mp, 1879; flowers brownish crimson, large; [Mme Julian Daran seedling]; Laxton Bros.; W. Paul

Charles de Gaulle®, *see* 'MEILANEIN'

'**CHARLES DE LAPISSE**', HT, lp, 1910; flowers pale blush-pink; [Mme Caroline Testout sport]; Laroulandie

'**CHARLES DE LEGRADY**', T, 1884; Pernet-Ducher; (Cavriglia)

'**CHARLES DE MILLS**', HGal, m; (Bizarre Triomphante); flowers dark crimson and purple, many petaled, quartered, 4.5 in., quartered, very few prickles; height to 5 ft; Hardy, before 1746

'**CHARLES DICKENS**', F, op, 1970; flowers rosy salmon, semi-dbl., 16 petals, 3 in., slight fragrance; [Paddy McGredy X Elizabeth of Glamis]; McGredy, Sam IV; McGredy

'**CHARLES DILLON**', F, mp, 1970; flowers soft pink, large, dbl., 24 petals, slight fragrance; foliage dark; low, bushy growth; [Orangeade X Piccadilly]; Wood; Homedale Nursery

Charles Dingee, *see* '**WILLIAM R. SMITH**'

'**CHARLES DUVAL**', HP, mr, 1847; flowers scarlet, dbl., cupped; vigorous growth; Laffay, M.

'**CHARLES E. SHEA**', HT, mp, 1917; flowers rich pink; [Mrs George Shawyer sport]; Hicks

Charles Eyck, F, yb

'**CHARLES F. WARREN**', HT, mp, 1957; flowers rose-pink, well-formed, large, dbl., 40 petals, moderate fragrance; vigorous growth; [Wilfred Pickles X Karl Herbst]; Mee

'**CHARLES FARGAS**', HT, pb, 1935; flowers large, dbl.; Dot, Pedro; (Sangerhausen)

'**CHARLES GATER**', HP, mr, 1893; flowers red, dbl., 40 petals, globular, moderate fragrance; vigorous growth; Paul

'**CHARLES GREGORY**', HT, ob, 1947; flowers vermilion, shaded gold, well-formed, dbl., 22 petals, moderate fragrance; foliage dark, glossy; vigorous growth; Verschuren; Gregory

'**CHARLES GREGORY, CLIMBING**', Cl HT, ob, 1960; Gregory

'**CHARLES H. RIGG**', HT, mr, 1931; flowers bright red fading to pink, dbl., moderate fragrance; large stout, erect stems; vigorous growth; Chaplin Bros.

'**CHARLES HENRY**', HT, mr, 1967; flowers crimson, moderate fragrance; foliage dark; vigorous, upright growth; [Ena Harkness X Lady Sylvia]; Hooney

'**CHARLES J. GRAHAME**', HT, mr, 1905; flowers dazzling scarlet, well formed, large, intense fragrance; vigorous growth; Dickson, A.

'**CHARLES K. DOUGLAS**', HT, mr, 1919; bud long, pointed; flowers large, dbl., 28 petals, moderate fragrance; foliage dark; vigorous growth; Dickson, H.

'**CHARLES K. DOUGLAS, CLIMBING**', Cl HT, mr, 1934; Leenders Bros.

Charles Kuralt, *see* CLEtraveler

'**CHARLES LAMB**', HP, 1884; Paul, W.; (Cavriglia)

'**CHARLES LAWSON**', B, dp, 1853; flowers vivid rose, vigorous, compact growth; Lawson

'**CHARLES LEFÈBVRE**', HP, dr, 1861; flowers reddish crimson shaded purple, large, dbl., 70 petals, cupped, bloom often recurrent, moderate fragrance; vigorous, tall growth; [Général Jacqueminot X Victor Verdier]; Lacharme, F.

'CHARLES LOUIS', HP, mr, 1800; flowers large, very dbl., moderate fragrance; Guinoiseau, B.; (Sangerhausen)

'CHARLES MALLERIN', HT, dr, 1951; flowers blackish crimson, dbl., 38 petals, 6 in., flat, intense fragrance; foliage leathery, dark; vigorous, irregular growth; [(Rome Glory X Congo) X Tassin]; Meilland, F.; C-P

'CHARLES MALLERIN, CLIMBING', Cl HT, dr, 1960; Balducci & Figli

'CHARLES MARGOTTIN', HP, mr, 1864; flowers very large, dbl., moderate fragrance; Margottin; (Sangerhausen)

Charles Metroz, Pol, dp

Charles Notcutt, *see* KORhassi

'CHARLES P. KILHAM', HT, or, 1926; flowers red-orange, fading to lincoln red, well formed, large, dbl., 32 petals, slight fragrance; vigorous, bushy growth; GM, NRS, 1927; McGredy; Beckwith

'CHARLES P. KILHAM, CLIMBING', Cl HT, or, 1931; Howard Rose Co.

'CHARLES P. KILHAM, CLIMBING', Cl HT, or, 1934; Morse

'CHARLES QUINT', HGal, m; flowers lilac-rose and white; Robert, 1856

'CHARLES RAVOLLI', T, dp; flowers carmine-rose; moderate growth; Pernet Père, 1875

Charles Rennie Macintosh, *see* 'AUSREN'

Charles Turner, HP, mr, 1968; flowers large, dbl.; Margottin; (Sangerhausen)

Charles Walker's Mignonette, Pol, w

'CHARLES WILLIAM', HT, pb, 1989; bud urn-shaped; flowers carmine rose, reverse lighter, yellow at base of petals, pointed, reflexing, very dbl., 56 petals, no fragrance; prickles light brown; foliage matt, medium green, large; upright, tall growth; [Bradenburg X Command Performance]; Cattermole, R.F.

Charles Wood, HP, rb; Laffay, M., 1864

Charleston, *see* 'MEIRIDGE'

Charleston, HP, mp

Charleston 88, *see* 'MEIRESTIF'

Charleston Graveyard, HCh, rb

'CHARLESTON, CLIMBING', Cl F, yb, 1966; Rumsey, R.H.

Charlie(TM), *see* 'KINCHA'

Charlie Brown, *see* 'MORCHARLIE'

Charlie Chaplin, *see* TSCHaka

'CHARLIE MCCARTHY', F, w, 1955; bud creamy white; flowers pure white, blooms in clusters, dbl., 28 petals, 1.5–2 in., moderate fragrance; foliage glossy, leathery, dark; dwarf, compact growth; [Mrs Dudley Fulton X Mermaid]; Wiseman; H&S

'CHARLIE PERKINS', HT, dr, 1970; bud long, pointed; flowers overlaid black, full, large, semi-dbl., moderate fragrance; foliage large, glossy, dark, bronze, leathery; very vigorous, upright, bushy growth; [Carrousel X Circus]; Zombory; General Bionomics

'CHARLIE'S AUNT', HT, pb, 1965; flowers cream, heavily suffused rose, high-pointed, dbl., 65 petals, 5 in., moderate fragrance; foliage dark; [Golden Masterpiece X Karl Herbst]; McGredy, Sam IV; Geest Industries

'CHARLIE'S UNCLE', HT, rb, 1975; flowers cream, suffused carmine, full, dbl., 40 petals, 5–6 in., exhibition form, slight fragrance; foliage large, leathery; [Charlie's Aunt sport]; Haynes

Charlotte, *see* 'AUSpoly'

'CHARLOTTE', HT, op, 1941; bud long, pointed; flowers salmon-pink and coral, base gold, dbl., exhibition form, intense fragrance; foliage glossy; vigorous, bushy growth; [Joanna Hill X Golden Dawn]; Duehrsen; California Roses

Charlotte, *see* TANettola

'CHARLOTTE ANNE', F, m, 1993; flowers lavender, single, 5 petals, 1.5–2.75 in., borne in small clusters, slight fragrance; some prickles; foliage medium, dark green, glossy; medium, spreading growth; [Playgirl sport]; Blankenship, Paul; Roses Unlimited, 1994

'CHARLOTTE ARMSTRONG', HT, dp, 1940; bud long, pointed, blood-red; flowers deep pink, dbl., 35 petals, 3–4 in., moderate fragrance; foliage dark, leathery; vigorous, compact growth; AARS, 1941 David Fuerstenberg Prize, ARS, 1941 Gertrude M. Hubbard, ARS, 1945 GM, Portland, 1941 John Cook Medal, ARS, 1941 NRS, NRS, 1950; [Soeur Thérèse X Crimson Glory]; Lammerts, Dr Walter; Armstrong Nursery

'CHARLOTTE ARMSTRONG, CLIMBING', Cl HT, dp, 1942; Morris; Armstrong Nursery, 1942;, Mon Reve Nursery, 1950

Charlotte Brownell, HT, yb; Brownell, H.C.

'CHARLOTTE CHEVALIER', HT, dy, 1916; flowers dark canary-yellow; [Arthur R. Goodwin sport]; Chambard, C.

'CHARLOTTE E. VAN DEDEM', HT, my, 1937; bud long, pointed; flowers large, semi-dbl.; foliage glossy; vigorous growth; GM, Portland, 1938; [Roselandia X Ville de Paris]; Buisman, G.A.H.

'CHARLOTTE ELIZABETH', Gr, dp, 1965; flowers deep rose-pink, blooms in clusters, dbl., 26 petals, 3–4 in., exhibition form; foliage glossy; Norman; Harkness

'CHARLOTTE IVES', F, pb, 1965; bud globular; flowers rose-pink, center light yellow, open, medium, borne in clusters, single, slight fragrance; compact, upright, bushy growth; [Ma Perkins X Rose Gaujard]; Warren

Charlotte Kemp, HWich, m, 1997; Nobbs

'CHARLOTTE KLEMM', HCh, or, 1905; flowers red shaded orange, medium, semi-dbl.; Türke

'CHARLOTTE MACKENSEN', HMult, mr, 1938; flowers carmine red with silvery reverse, medium, dbl.; Vogel, M.; (Sangerhausen)

'CHARLOTTE MARIE', S, pb, 1998; flowers pale pink, light at the base, pale reverse, single, 4–7 petals, 3 in., borne in small clusters, slight fragrance; prickles moderate; foliage medium, narrow, dark green, semi-glossy; low, compact, upright growth; [Marchenland X Rosa virginiana]; Fleming, Joyce L.; Hortico, Inc., 1997

'CHARLOTTE PATE', HT, yb, 1971; bud globular; flowers pink, yellow reverse, dbl., 25 petals, 3.5 in., intense fragrance; foliage medium, medium green, glossy; [Wendy Cussons X Golden Sun]; MacLeod

Charlotte Rampling®, *see* MEIhirvin

Charlotte Searle, *see* KORnends

'CHARLOTTE VON RATHLEF', HWich, lp, 1936; very dbl., abundant seasonal bloom; foliage leathery, dark; vigorous, climbing growth; [Fragezeichen X American Pillar]; Vogel, M.; Heinemann

'CHARLOTTE WHEATCROFT', F, mr, 1957; flowers bright scarlet, large, large truss, single; foliage dark, glossy; vigorous, tall growth; Wheatcroft Bros.

'CHARLOTTE WIEREL', Pol, w, 1926; flowers cream-white, center bright rose-pink; vigorous growth; [Bebe Leroux X Helene Videnz]; Walter, L.

'CHARLOTTENHOF', S, mr, 1938; flowers medium, dbl.; Vogel, M.; (Sangerhausen)

Charlye Rivel®, *see* FEkusa

'CHARM', HT, op, 1920; bud reddish orange, shaded pink and copper; flowers coppery yellow; Paul, W.

'CHARM', *see* **'CHARME'**

Charm Bracelet, *see* 'JACFOG'

'CHARM OF PARIS', HT, mp, 1965; flowers pink, medium, dbl., 48 petals, 3.5 in., moderate fragrance; vigorous growth; Edland Fragrance Medal, ARS, 1966; [Prima Ballerina X Montezuma]; Tantau, Math.

'CHARMAINE', Pol, mp, 1929; flowers pink tinged salmon, open, sprays, dbl., recurrent bloom, moderate fragrance; foliage bright, glossy; long stems; very vigorous, bushy growth; [Evelyn Thornton seedling]; Burbage Nursery

Charmant®, *see* 'KORZIMKO'

Charmante, *see* KORzimko

'CHARME', HT, dp, 1930; ('CHARM', 'GERMANIA'); bud long, pointed; flowers cherry-red, large, dbl., moderate fragrance; foliage glossy; vigorous growth; Rice Bros. Co.

Charme d'Amour, *see* **'LIEBESZAUBER'**

Charme de Paris, F, 1979; Delbard, Georges; (Cavriglia)

Charme de Vienne, *see* 'KORSCHAPRAT'

'**CHARMENTE**', HT, lp, 1975; bud globular; dbl., 42 petals, 4 in.; foliage bright green, leathery; upright, spreading growth; [Fragrant Cloud X Ena Harkness]; Huber

'**CHARMER**', HT, lp, 1923; flowers silvery pink; [Pharisaer seedling X Joseph Hill]; Schoener; Doyle

'**CHARMER**', HT, lp, 1934; flowers light pink, center shaded salmon, large, dbl., exhibition form, moderate fragrance; foliage leathery; vigorous, free branching growth; GM, NRS, 1932; Dickson, A.

'**CHARMGLO**', Min, pb, 1980; bud long, pointed; flowers creamy white painted deep pink, reverse lighter, borne usually singly, dbl., 35 petals, exhibition form, slight fragrance; long, thin, brown prickles, curved down; foliage small, medium to dark green, slightly matt; bushy, compact growth; [Seedling X Over the Rainbow]; Williams, Ernest D.; Mini-Roses

'**CHARMI**', HMsk, mp, 1929; flowers light to dark pink, medium, semi-dbl., intense fragrance; Pemberton; (Sangerhausen)

Charmian®, *see* 'AUSmian'

'**CHARMING**', HT, mp, 1922; flowers salmon-pink, reverse coral-pink, semi-dbl., slight fragrance; [Alexander Hill Gray X Mme Edouard Herriot]; Van Rossem

Charming Cover, *see* POUlharm

Charming Diana™, *see* 'TWODI'

'**CHARMING MAID**', F, ob, 1953; bud orange-salmon; flowers salmon, base golden, blooms in trusses, single, 5–6 petals, 4–4.5 in., moderate fragrance; foliage dark, glossy; GM, NRS, 1953; [Dainty Maid X Mrs Sam McGredy]; LeGrice

'**CHARMING PRINCESS**', HT, yb, 1926; flowers deep yellow, edged vermilion large, dbl.; foliage dark; vigorous growth; [The Queen Alexandra Rose sport]; Hancock

Charming Rosamini, *see* RUIcharo

Charming Rose, HT, lp

Charming Vienne, *see* 'KORSCHAPRAT'

Charter 700, F, my, 1993; Fryer, Gareth

'**CHARTREUSE**', HT, my, 1940; bud long; flowers canary-yellow, dbl., moderate fragrance; foliage glossy; bushy growth; [Soeur Thérèse X Angels Mateu]; Mallerin, C.; A. Meilland

Chartreuse, *see* 'MACYEFRE'

Chartreuse de Parme, *see* DELviola

'**CHASE BEAUTY**', HT, dr, 1947; bud long; flowers rich dark red, dbl., 35–45 petals, 5.5 in., exhibition form, slight fragrance; foliage dark; very vigorous, tall growth; [Better Times sport]; Chase; Chase Gardens

Chasin' Rainbows™, *see* 'SAVACHASE'

Chasing Rainbows, *see* 'SAVACHASE'

'**CHASTITY**', Cl HT, w, 1924; flowers pure white, base lemon, dbl., exhibition form, non-recurrent bloom; foliage light, glossy; vigorous growth; Cant, F.

'**CHASTLETON**', HP, lp, 1800; flowers large, very dbl., slight fragrance; (Sangerhausen)

Château, *see* '**BABY CHÂTEAU**'

Chateau Canon, *see* MACghovie

Château d'Amboise®, *see* 'DELROUVEL'

Chàteau de Bagnols, *see* ORAreg

'**CHÂTEAU DE CHENONCEAUX**', HT, mp, 1973; flowers brilliant pink, large, dbl., 45 petals; tall growth; [Americana X Queen Elizabeth]; Gaujard

'**CHÂTEAU DE CLOS VOUGEOT**', HT, dr, 1908; flowers deep velvety red, dbl., 75 petals, intense, damask fragrance; foliage dark, leathery; sprawling growth; Pernet-Ducher

'**CHÂTEAU DE CLOS VOUGEOT, CLIMBING**', Cl HT, dr, 1920; Morse

Chateau de Filain, *see* SAUbima

'**CHATEAU DE LA JUVENIE**', S, lp, 1901

'**CHÂTEAU DE NAMUR**', HGal, pb; Quétier, before 1842

'**CHÂTEAU DE VAIRE**', S, dr, 1934; (Vaire); flowers deep red, dbl., cupped, non-recurrent; foliage bronze, dark; height $3\frac{1}{2}$–$6\frac{1}{2}$ ft bushy; [Charles K. Douglas X R. macrophylla]; Sauvageot, H.

Chateau de Versailles®, *see* 'DELRICOS'

Chateau Frontenac, S, dp

Chateau la Croix, *see* DORcroix

'**CHÂTEAU LA SALLE**', HT, yb, 1966; bud long, pointed; flowers buff-yellow, large, dbl., exhibition form, intense fragrance; foliage dark, leathery; vigorous, bushy, compact growth; [Joanna Hill X Ellinor LeGrice]; Morey, Dr Dennison; Country Garden Nursery

'**CHÂTEAU PELLES**', HT, lp, 1927; bud cream-white; flowers soft pink, shaded salmon, dbl., slight fragrance; [Harry Kirk seedling]; Mühle

Chateauroux, F; Croix, P.; (Cavriglia)

'**CHÂTELAINE**', F, op, 1957; bud pointed; flowers coral overcast salmon, well formed, borne in small clusters, dbl., 32 petals, 3 in., moderate fragrance; foliage glossy, coppery; vigorous growth; [(Peace X Seedling) X Fashion]; Lens

Châtelaine de Lullier, HT, mr, 1987; flowers carmine-red, large, dbl., moderate fragrance; Meilland; (Sangerhausen)

'**CHÂTELET**', HT, mp, 1952; (Airain); bud globular, coral; flowers pink heavily tinted salmon, medium, very dbl.; [Yvonne Plassat X Seedling]; Moulin-Epinay; Vilmorin-Andrieux

'**CHATILLON RAMBLER**', HWich, mp, 1913; flowers salmon-pink, small, semi-dbl., slight fragrance; height 15–20 ft; [Dorothy Perkins X Crimson Rambler]; Nonin

'**CHATILLON ROSE**', Pol, mp, 1923; flowers bright pink, blooms in large clusters, semi-dbl., cupped, moderate fragrance; foliage glossy; bushy (1–2 ft growth; [Orléans Rose X Seedling]; Nonin

Chatillon White, *see* '**WHITE Chatillon**'

Chatsworth, *see* TAnotax

'**CHATTEM CENTENNIAL**'™, Min, or, 1979; bud ovoid; flowers medium, dbl., 38 petals, cupped, slight, fruity fragrance; upright, bushy growth; [Orange Sensation X Zinger]; Jolly, Betty J.; Rosehill Farm

'**CHATTEM CENTENNIAL, CLIMBING**', Cl Min, or, 1990; bud ovoid; flowers orange-red, orange-red reverse, aging light orange, loose, medium, dbl., 38 petals, cupped, borne usually singly or in sprays of 3–5, slight, fruity fragrance; foliage medium, light green, matt; tall (4–6 ft) growth; [Chattem Centennial sport]; Jolly, Marie; Rosehill Farm, 1991

'**CHATTER**', F, mr, 1947; flowers velvety bright crimson, blooms in large clusters, semi-dbl., 14 petals, cupped, moderate fragrance; bushy, compact growth; (28); [World's Fair X Betty Prior]; Boerner; J&P

'**CHATTER, CLIMBING**', Cl F, mr, 1960; Schmidt, K.

'**CHATTERBOX**', F, ob, 1973; flowers bright orange-vermilion, rosette, semi-dbl., 16 petals, 2 in.; foliage glossy; dwarf growth; [Sarabande X Circus]; Sanday, John

Chaucer®, *see* 'AUScon'

Chaumant, *see* 'MACLOCKER'

Chaumont, *see* 'MACLOCKER'

Chaumont™, F, ob

Checkers, *see* 'PIXCHEK'

'**CHEER**', F, mp, 1941; flowers deep rose-pink, open, clusters, semi-dbl., 4 in., moderate fragrance; foliage leathery; vigorous, upright growth; RULED EXTINCT 1/84 ARM; [Dance of Joy X Golden Rapture]; Kordes; J&P

Cheer Up, *see* 'TINCHEER'

'**CHEERFUL**', HT, ob, 1915; flowers orange-flame, base yellow, very large, dbl., moderate fragrance; foliage rich green, glossy; McGredy

'**CHEERFULNESS**', F, op, 1981; flowers orange, pink and yellow blend, borne 10 per cluster, 8 petals, slight fragrance; small, brown prickles; foliage small; compact growth; [Seedling X Seedling]; Everitt, Derrick; Gandy Roses, Ltd.

Cheerie, *see* '**CHÉRIE**'

Cheerio, *see* '**CHÉRIE**'

Cheerio, *see* '**PLAYBOY**'®

Cheerleader™, *see* 'MORCHEER'

Cheers®, *see* 'SAVALOT'

Cheery Chatter, *see* 'LYOTER'

Chelsea, *see* 'MORSEA'

Cheese Cake, *see* 'BROCAKE'

'**CHELSEA**', HT, mr, 1950; flowers carmine shaded orient red, dbl., 4 in., moderate fragrance; vigorous, compact growth; RULED EXTINCT 6/86; LeGrice

Chelsea®, *see* 'ORTSEA'

Chelsea Belle, *see* 'TALCHELSEA'

Chelsea Gold, *see* 'LANWOOL'

Chelsea Pensioner, *see* 'MATTCHE'

'**CHÉNÉDOLÉ**', HCh, or; flowers crimson, small, dbl., cupped, moderate fragrance; prickles shoot very prickly; vigorous, upright growth; Thierry, prior to 1848

'**CHERÉ MICHELLE**', Min, op, 1986; (Cherie Michelle); flowers white with coral pink petal edges, reverse same, small, dbl., 30 petals, exhibition form, slight fragrance; medium, globular, orange-green fruit; small, cream to light brown prickles, hooked downward; foliage medium, medium green, semi-glossy; medium, upright growth; [Sheri Anne X Anita Charles]; Jolly, Marie; Rosehill Farm, 1987

'**CHÉRIE**', F, mp, 1931; (Cheerie, Cheerio); flowers bright rose-pink, small, borne in clusters, dbl., cupped, moderate fragrance; foliage leathery; vigorous, bushy growth; [Else Poulsen sport]; Morse

'**CHÉRIE**', F, ob, 1964; bud long; flowers bright orange, reverse coppery, well formed, dbl.; Gaujard

Cherie Michelle, *see* '**CHERÉ MICHELLE**'

Cherish, *see* 'JACSAL'

Cherokee Fire, *see* 'LYOCH'

Cherokee Rose, *see* **R. LAEVIGATA**

Cherries Jubilee, *see* CLEcherry

Cherries 'n' Cream, Min, rb, 1997; King

'**CHERRIO**', F, dp, 1937; flowers carmine-cerise, large, borne in clusters, semi-dbl., cupped, slight fragrance; vigorous, compact growth; Archer

'**CHERRIO**', F, mp, 1948; (Planten un Blomen); flowers light pink, reverse darker, semi-dbl.; vigorous, bushy growth; [Holstein X Sapho]; Kordes

'**CHERRY**', HT, pb, 1928; flowers brilliant carmine-pink flushed yellow, lower half yellow, large, dbl., exhibition form, moderate fragrance; vigorous, bushy growth; McGredy

'**CHERRY BLOSSOM**', F, mp, 1964; bud orient red; flowers rose-pink to camellia-pink, borne in clusters, dbl., 26 petals; foliage glossy, dark; compact growth; [Fashion X Seedling]; Verschuren, A.; van Engelen

Cherry Bomb, *see* 'JOLCHER'

'**CHERRY BRANDY**'®, HT, ob, 1965; flowers orange, dbl., 30 petals, 5 in., moderate fragrance; foliage dark, glossy, leathery; very vigorous, upright growth; GM, Belfast, 1989; Tantau, Math.

Cherry Brandy '85, *see* 'TANRYRANDY'

'**CHERRY CHARM**', LCl, pb, 1976; flowers deep pink, reverse silver pink, dbl., 25 petals, 5 in., non-recurrent, slight fragrance; foliage dark; [Norwich Salmon X (Seedling sport X Peeping Tom)]; MacLeod

CHERRY CHEERFUL, Pol, rb, 1995; flowers cherry red with white shading, semi-dbl. (6–14 petals), small blooms borne in large clusters; slight fragrance; few prickles; foliage medium, medium green, glossy; compact, spreading, bushy, medium (3 ft × 5 ft) growth; [Anytime X Mountain Mist]; Jobson, Daniel J.; Jobson, 1995

Cherry Cola, HT, dr

Cherry Cordial, *see* 'WELCORD'

Cherry Cream, Gr, w, 1995

'**CHERRY GLOW**', Gr, mr, 1959; flowers cherry-red, dbl., 23 petals, 3–4 in., cupped, moderate, spicy fragrance; foliage leathery, glossy; vigorous, upright growth; [Floradora X First Love]; Swim, H.C.; C.R. Burr

Cherry Glow, *see* '**SWEET CHERRY**'

Cherry Hi, *see* 'MORSHODOT'

Cherry Jubilee, *see* 'JACSOS'

Cherry Lips, F, dp, 1997; Williams, J. Benjamin

Cherry Magic®, *see* 'MORCHERMAG'

Cherry Meidiland, *see* 'MEIRUMOR'

Cherry Meillandecor®, *see*, 'MEIRUMOR'

'**CHERRY PAGE**', HT, pb, 1914; flowers carmine-pink, base yellow; [Duchess of Bedford X Le Progres]; Easlea

Cherry Pastel®, *see* 'WILCHER'

'**CHERRY PIE**', HT, dp, 1965; flowers deep rose-pink, dbl., 4.5 in., exhibition form, slight fragrance; Gaujard; Gandy Roses, Ltd.

'**CHERRY RIPE**', F, mr, 1949; flowers scarlet, small, dbl., 70 petals, intermittent bloom; vigorous growth; [Orange Triumph sport]; Heers; Pacific Nursery

Cherry Sunblaze®, *see* 'MEIBEKARB'

Cherry Sunblaze®, *see* 'MEIBOKARB'

Cherry Velvet®, *see* 'POULTRESS'

Cherry Wine, *see* 'JALWINE'

'**CHERRY, CLIMBING**', Cl HT, pb, 1934; Savage Nursery

'**CHERRYADE**', S, dp, 1961; flowers deep pink, well-formed, dbl., 40 petals, 4 in., moderate fragrance; foliage dark; vigorous, tall growth; [New Dawn X Red Wonder]; deRuiter

'**CHERRY-ROSE**', HT, mp, 1946; bud long, pointed; flowers cherry-rose, very large, dbl., exhibition form, moderate fragrance; foliage glossy; very vigorous, upright, compact growth; [Pink Princess X Crimson Glory]; Brownell, H.C.

Cherry-Vanilla, *see* 'ARMilla'

Cherry-Vanilla, Climbing, Cl Gr; (Cavriglia)

'**CHERUB**', HMult, pb, 1923; flowers pink and salmon, small, semi-dbl., cupped, profuse, non-recurrent bloom; foliage rich green, glossy, wrinkled; very vigorous, climbing growth; [Claire Jacquier seedling]; Clark, A.; Brundrett

'**CHERVENA GHITA**', HT, mr, 1974; flowers bright cerise, large blooms in clusters, dbl., 75 petals; foliage dark, glossy; vigorous, upright growth; [General Stefanik X Peace]; Staikov, Prof. Dr V.; Kalaydjiev and Chorbadjiiski

Cheryl's Delight®, *see* 'MINRCO'

'**CHESAPEAKE**', Min, lp, 1984; flowers small, dbl., 50 petals, slight fragrance; foliage medium, medium green, semi-glossy; bushy growth; [Rise 'n' Shine X (Helen Traubel X First Prize)]; Jolly, Nelson F.; Rosehill Farm

Chesapeake Sunset, F, ob, 1999; dbl.; Williams, J. Benjamin

CHEsdeep, MinFl, lp; (**Thank You**); Chessum, Paul

Cheshire, *see* 'FRYELISE'

'**CHESHIRE CREAM**', F, w, 1975; flowers soft buff, becoming cream, dbl., 50 petals, 2.5 in., moderate, spicy fragrance; foliage small, glossy; low growth; [(Anna Wheatcroft X Ivory Fashion) X (Buff Beauty X Masquerade)]; Holmes, R.; Fryer's Nursery, Ltd.

'**CHESHIRE LADY**', HT, pb, 1968; flowers bright pink to scarlet, large, dbl., 30–40 petals; free growth; [Fragrant Cloud X Gavotte]; Dale, F.

'**CHESHIRE LIFE**', HT, or, 1972; flowers vermilion, spiral, dbl., 36 petals, 5 in., slight fragrance; foliage dark, leathery; [Prima Ballerina X Princess Michiko]; Fryer, Gareth; Fryer's Nursery, Ltd.

Cheshire Regiment, *see* FRYzebedee

'**CHESHUNT HYBRID**', HT, mr, 1872; flowers red shaded violet, large, dbl.; vigorous growth; [Believed to be Mme de Tartas X Prince Camille de Rohan]; Paul

CHESnut, Cl HT, rb, 1996; (**Best Wishes**); Chessum, Paul

Chess®, *see* 'INTERTRA'

Chess Man, *see* 'GELMAN'

Chessum's Choice, *see* 'CHESTOCK'

CHEssupremo, MinFl, my, 1995; (**Golden Hands**); Chessum, Paul

'CHESTER', F, my, 1976; flowers golden yellow, semi-dbl., 15 petals, 3 in., slight fragrance; foliage glossy, dark; vigorous growth; [Arthur Bell X Zambra]; Bees

Chester Cathedral, *see* 'FRANSHINE'

Chestnut rose, *see* **R. ROXBURGHII**

'CHESTOCK', (strain of R. canina), w, 1988; (**Chessum's Choice**); used as a stem of standard (tree) roses; [Pfander's Canina sport]; Chessum, Paul; Alan Thompson

'CHEVREUL', M, mp, 1887; flowers salmon-pink, well mossed; large fruit, colorful in fall; Moreau et Robert

Chevreuse, *see* 'KORPLAVI'

'CHEVY CHASE', HMult, dr, 1939; flowers dark crimson, small blooms in clusters of 10–20, dbl., 65 petals, non-recurrent, moderate fragrance; foliage soft, light green, wrinkled; vigorous, climbing (to 15 ft) growth; Dr W. Van Fleet Medal, ARS, 1941; [R. soulieana X Eblouissant]; Hansen, N.J.; B&A

'CHEWALLOP', Cl Min, ob, 1992; (Rocketeer, **Rosalie Coral**); flowers bright coral-orange, opens flat, dbl., 15–25 petals, 1.5 in., flat, borne in small clusters, slight fragrance; few prickles; foliage small, medium green, glossy; medium (175–190 cms), bushy growth; PP9013; [[Elizabeth of Glamis X (Galway Bay X Sutter's Gold)] X Anna Ford]; Warner, Chris, 1991

CHEwapri, Cl Min, ab, 1994; (**Patio Charm**); Warner, Chris

'CHEWARVEL', Cl Min, my, 1989; (King Tut, **Laura Ford**®, Normandie); bud pointed; flowers medium yellow, reverse lighter yellow, aging pink flushes, dbl., 22 petals, 2 in., exhibition form, borne mostly singly, slight, fruity fragrance; round, large, average fruit; prickles straight, small, infrequent, light brown; foliage small, light green, glossy; upright, bushy, tall growth, 7 ft; PP9012; [Anna Ford X [Elizabeth of Glamis X (Galway Bay X Sutter's Gold)]]; Warner, Chris, 1990

'CHEWAZE', F, or, 1986; (**Pillar Box**, Pillar BoxWardlip); flowers vivid vermillion, moderately medium, dbl., 15–25 petals, slight fragrance; foliage medium, medium green, semi-glossy; upright growth; [Alexander X (Galway Bay X Elizabeth of Glamis)]; Warner, Chris

'CHEWBEAUT', F, or, 1992; (**Donald Davies**); flowers vermilion, blooms borne in small clusters, dbl., 15–25 petals, 1.5–2.75 in., slight fragrance; few prickles; foliage medium, medium green, semi-glossy; medium (90 cms), upright growth; [Anne Harkness X Beautiful Britain]; Warner, Chris

CHEwdainty, Cl Min, yb, 1997; (**Gilt Edged**); Warner, Chris

'CHEWDOR', LCl, op, 1998; (**Lady Penelope**); flowers salmon pink, pink reverse, looks like a pom-pom dahlia, repeats well, very dbl., 26–40 petals, 2.5–3 in., borne in small clusters, slight fragrance; few prickles; foliage medium, medium green, glossy; spreading, medium (8 ft) growth; [Laura Ford X Royal Baby]; Warner, Chris; Bransford Garden Plants, 1998

CHEWECREST, Cl Min, lp; (**Taffeta**); Warner, Chris, 1997

'CHEWELL', F, r, 1988; (**Iris Webb**); flowers tan, fading to slate gray, medium, dbl., 15–25 petals, moderate fragrance; foliage medium, dark green, semi-glossy; bushy growth; [Southampton X (Belinda X (Elizabeth of Glamis X (Galway Bay x Sutters Gold)))]; Warner, Chris; LeGrice Roses, 1990

'CHEWGLORIOUS', Cl Min, mr, 1999; (**Love Knot**); flowers crimson, reverse light red, dbl., 17–25 petals, 2.5 in., borne in small clusters, slight fragrance; few prickles; foliage small, medium green, glossy; upright, tall (6 ft) growth; [Laura Ford X Ingrid Bergman]; Warner, Chris; Warner's Roses, 2000

'CHEWHARLA', Cl Min, m, 1989; (**Laura Ashley**); bud pointed; flowers lilac-mauve pink, reverse pink, aging same, loose, small, single, moderate, fruity fragrance; oval, small, red fruit; prickles hooked, small, brown; foliage small, medium green, semi-glossy; spreading, low growth; [Marjorie Fair X Nozomi]; Warner, Chris; R. Harkness & Co., Ltd., 1991

'CHEWILY', S, mp, 1999; (**Mix 'n' Match**™); bud short, furled; flowers pastel pearl pink with golden boss of stamens, semi-dbl., 10–15 petals, 1.5–2 in., borne in large clusters, slight fragrance; foliage medium, dark green, quilted; medium, rounded, slight spreading growth; [Seaspray X (Rosa sinowilsonii X Marjorie Fair)]; Warner, Chris

CHEwily, S, lp, 1998; (**Pink Cottage**); Warner, Chris

'CHEWIZZ', Cl Min, or, 1992; (**Warm Welcome**); flowers orange vermilion, blooms borne in small clusters, semi-dbl., 6–14 petals, 1.5 in., moderate fragrance; few prickles; foliage small, dark green, semi-glossy; tall (200+ cms), upright growth; [[Elizabeth of Glamis X (Galway Bay X Sutter's Gold)] X Anna Ford]; Warner, Chris, 1991

'CHEWLEGACY', F, r, 1988; (**Edith Holden**, Edwardian Lady, The Edwardian Lady); flowers russet-brown with yellow center, reverse slightly paler, aging slate grey, semi-dbl., 15 petals, slight fragrance; rounded, small, orange fruit; prickles very few; foliage medium, medium green, glossy; upright, tall, robust growth; [Belinda X (Elizabeth of Glamis X (Galway Bay X Sutters Gold))]; Warner, Chris; E.B. LeGrice Roses, Ltd., 1988

'CHEWOZ', S, ob, 1992; (Carmargue, **Telford's Promise**); flowers coppery salmon, blooms borne in large clusters, semi-dbl., 6–14 petals, 1.5 in., slight fragrance; few prickles; foliage small, medium green, glossy; spreading (60 cms) growth; [Mary Sumner X Nozomi]; Warner, Chris, 1990

'CHEWPAN', Min, mr, 1997; (**Peter Pan**); foliage small, medium green, glossy; bushy, low (10–12 cms) growth; [Eyeopener X Seedling]; Warner, Chris

'CHEWPEARL', Cl Min, mp, 1999; (**Star Performer**); flowers satin pink, reverse medium pink, small, dbl., 17–25 petals, borne in small clusters, slight fragrance; few prickles; foliage small, medium green, glossy; upright, tall (7 ft) growth; for walls, fences, pillars; [Laura Ford X Congratulations]; Warner, Chris; Warner's Roses, 1999

CHEwpiwitel, Cl Min, ly, 1995; (**Patio Honey**); Gold Star of the South Pacific, Palmerston North, NZ, 1995; Warner, Chris

'CHEWPIXCEL'R, lp, 1995; (**Open Arms**); flowers shell pink, semi-dbl. (6–14 petals), small blooms borne in large clusters; fragrant; few prickles; foliage small, dark green, glossy; spreading, medium growth; [Mary Sumner X Laura Ashley]; Warner, Chris; Warner's Roses

'CHEWPOBEY', S, or, 1995; (**Pathfinder**); flowers vermillion, yellow eye, small blooms borne in large clusters, semi-dbl., 6–14 petals, slight fragrance; few prickles; foliage small, medium green, glossy; patio/groundcover; spreading, low growth; Breeders Choice, 1996 TGC, St. Albans, 1994; [(Little Darling X Anna Ford) X Eyeopener]; Warner, Chris & Barbara; Warner's Roses, 1996

'CHEWPOPE', Cl Min, m, 1997; (**Gloriana 97**); flowers double, medium, dbl., 15–25 petals, borne in small clusters, slight fragrance; foliage medium, medium green, semi-glossy; upright, tall (7 ft) growth; [Laura Ford X Big Purple]; Warner, Chris

'CHEWRAMB', Cl Min, lp, 1994; (**Little Rambler**); flowers pale pink, blooms borne in large clusters, dbl., 26–40 petals, 1.5 in., intense fragrance; few prickles; foliage small, dark green, semi-glossy; tall (7 ft × 7 ft), bushy, spreading growth; [(Cecile Brunner X Baby Faurax) X (Marjorie Fair X Nozomi)]; Warner, Chris; Warner's Roses, 1995

'CHEWSEA', Cl Min, op, 1992; (**Nice Day**, Patio Queen); flowers salmon pink, blooms borne in large clusters, dbl., 15–25 petals, 1.5 in., moderate fragrance; few prickles; foliage small, bronze turning medium green, glossy; upright (200 cms), bushy, climbing growth; [Seaspray X Warm Welcome]; Warner, Chris; Warner's Roses, 1993

'CHEWSOS', Min, ob, 1996; (**Society Special**); flowers yellow with red reverse, small blooms borne in small clusters, dbl., 26–40 petals, slight fragrance; few prickles; foliage small, medium green, semi-glossy; upright, bushy, tall growth; [Laura Ford X Anne Harkness]; Warner, Chris; Warner's Roses, 1997

'CHEWSUNBEAM', Cl Min, dy, 1994; (**Good as Gold**); dbl., 25–26 petals, 1.5 in., borne in small clusters, moderate fragrance; some prickles; foliage small, light green, semi-glossy; tall (7 ft), upright growth; Gold Star of the South Pacific, Palmerston North, NZ, 1996; [Anne Harkness X Laura Ford]; Warner, Chris; Warner's Roses, 1995

'CHEWSUNFORD', Min, dy, 1996; (**Golden Handshake**); flowers clear bright yellow, small blooms borne in small clusters, dbl., 15–25 petals, slight fragrance; few prickles; foliage small, light green, glossy; upright, bushy, tall growth; TGC, St. Albans, 1994; [Pam Ayres X Laura Ford]; Warner, Chris; Warner's Roses, 1997

CHEwwily, S, lp, 1995; (**Cottage Garden**); Warner, Chris

Cheyenne, *see* 'SPOCHEY'

'CHEYENNE', Cl HT, lp, 1962; bud long, pointed; flowers base coral pink, dbl., 30–40 petals, 4–5 in., exhibition form, slight fragrance; foliage leathery; vigorous (6–7 ft) growth; RULED EXTINCT 1/85 ARM; [Queen Elizabeth X Seedling]; Von Abrams; Peterson & Dering

'CHEYENNE FRONTIER', HT, w, 1971; bud large; flowers white, slowly changing to red, medium, dbl., exhibition form, slight fragrance; foliage large, glossy, dark; moderate, upright growth; [(Charlotte Armstrong X Vogue) X Peace]; Adams, M.R.

'CHEZ VITO', HT, mr, 1972; bud ovoid; flowers medium red, reverse lighter, medium, dbl., exhibition form, moderate fragrance; foliage large, leathery; vigorous, upright, bushy growth; [Paris-Match X (Baccará X Happiness)]; Meilland

'CHI LO SÀ?', HT, yb, 1965; flowers yellow suffused red; [(Peace X Fiaba) X Seedling]; Giacomasso, Fratelli

Chianti, *see* 'AUSwine'

'CHIARASTELLA', HT, rb, 1948; flowers rose-red and yellow bicolor, well formed, large, dbl.; foliage glossy; strong stems; vigorous growth; [Julien Potin X Mme G. Forest-Colcombet]; Giacomasso, Fratelli

'CHIC', F, pb, 1953; bud ovoid; flowers geranium-pink, blooms in clusters, dbl., 68 petals, 2.5 in., cupped, moderate fragrance; vigorous, branching growth; [Pinocchio seedling X Fashion]; Boerner; J&P

'CHIC PARISIEN', F, op, 1956; flowers coral-pink, center darker, well-formed blooms in clusters of 4–8, dbl., slight fragrance; foliage dark; vigorous growth; Delbard-Chabert

Chica, HT, lp, 1998; Kordes

'CHICAGO', HT, m, 1928; flowers soft mauve-pink, dbl., intense fragrance; [Premier sport]; Aldous

Chicago Peace®, *see* 'JOHNAGO'

Chicago Peace, Climbing, Cl HT, lp; [Chicago Peace sport]; Brundrett, 1978; (Weatherly, L.)

Chick-a-dee, *see* 'MORCHICK'

Chickasaw Rose, *see* **R. BRACTEATA**

Chidori, *see* **'RYOKKOH'**

Chief, *see* **'THE CHIEF'**

'CHIEF JUSTICE HOLMES', Cl HT, dr, 1935; flowers very dark red; GM, Portland, 1936; [Jules Margottin X Château de Clos Vougeot]; Schoener

'CHIEF SEATTLE', HT, yb, 1951; bud conical; flowers buff and old-gold, center shrimp-red, dbl., 55 petals, 4–5 in., exhibition form, moderate fragrance; foliage glossy; tall growth; [Charlotte Armstrong X Signora]; Swim, H.C.; Armstrong Nursery

'CHIEFTAIN', HT, rb, 1936; bud ovoid; flowers brilliant red, base yellow, large, dbl., exhibition form; foliage leathery, dark; vigorous growth; [Hadley X Talisman]; Montgomery Co.

'CHIFFON', HT, lp, 1940; flowers blush-pink, tinted light lavender, dbl., 30 petals, 5.5 in., moderate fragrance; [Regina Elena sport]; Grillo

Childhood Memories, *see* FERho

Child's Play™, *see* 'SAVACHILD'

Chili Pepper, *see* 'GELPEP'

Chill Out, *see* 'GELOUT'

Chilterns, *see* 'KORTEMMA'

Chimene, *see* HARazz

Chimo®, *see* 'INTERCHER'

Chin Chin, *see* **'Promise'**

'CHINA BELLE', HCh, pb, 1980; bud ovoid; flowers light and medium pink blend with yellow, blooms borne 5–7 per cluster, dbl., cupped, slight, fruity fragrance; curved, red prickles; foliage glossy; short stems; compact, bushy, upright growth; [(Doubloons X Holiday) X Slater's Crimson China]; James, John

'CHINA DOLL', Pol, mp, 1946; bud pointed; flowers china-rose, base mimosa-yellow, blooms in large trusses, dbl., 24 petals, 1–2 in., cupped, slight, tea fragrance; foliage leathery, with mostly 5 leaflets (similar to Pinkie); dwarf (18 in.), bushy growth; [Mrs Dudley Fulton X Tom Thumb]; Lammerts, Dr Walter; Armstrong Nursery

'CHINA DOLL, CLIMBING', Cl Pol, mp, 1977; (Weeping China Doll); Weeks; Weeks Wholesale Rose Growers

China Rose, *see* **R. CHINENSIS**

'CHINATOWN'®, F, dy, 1963; (Ville de Chine); flowers yellow, sometimes edged pink, blooms in clusters, dbl., 4 in., intense fragrance; foliage dark; vigorous, tall, bushy growth; GM, NRS, 1962; [Columbine X Clare Grammerstorf]; Poulsen, Niels D.; A. Dickson;, McGredy

Chinatown Moss, M, rb

Chinese Lantern, *see* 'FOUCHIN'

Chinese Monthly Rose, *see* **'SLATER'S CRIMSON CHINA'**

Chinese Puzzle, *see* 'GELPUZZLE'

'CHINGARI', F, yb, 1976; bud pointed; flowers aureolin to currant-red, open, semi-dbl., 17 petals, 3 in., slight fragrance; foliage glossy; vigorous, bushy, compact growth; [Charleston X Seedling]; Pal, Dr B.P.; Laveena Roses

Chinquapin Rose, *see* **R. ROXBURGHII**

Chipie, *see* 'POUBICARBE'

Chipmonk, *see* 'PIXICHIP'

Chipmunk, *see* 'PIXICHIP'

'CHIPPER', Min, ab, 1966; bud ovoid; flowers salmon-pink, small, dbl., slight fragrance; foliage glossy, leathery; vigorous, dwarf growth; [(Dany Robin seedling X Fire King) X Perla de Montserrat]; Meilland, Alain A.; C-P

'CHIPPEWA', S, mp; flowers rose-pink, borne in clusters, semi-dbl.; foliage leathery, bronze; Central Exp. Farm

'CHIQUITA', LCl, op, 1938; flowers orange-yellow to coppery orange and salmon-pink, base yellow, intense fragrance; RULED EXTINCT 1/88; [Sierra Snowstorm X Étoile Luisante]; Moore, Ralph S.; Brooks & Son

Chiquita™, *see* 'MORKITA'

'CHIRAZ'®, HT, pb, 1986; flowers creamy white, flushed pink, petals edged carmine red, large; foliage dark; dense growth; [Kordes' Perfecta X Peace]; Kriloff, Michel

'CHIRIPA', F, mr, 1957; bud pointed; flowers red, reverse carmine, borne in clusters of 3–5, dbl., 26 petals; foliage bright green; upright, compact growth; [Radar X (Rosalia Riviera X Independence)]; de Dot, G.F.

Chit Chat, *see* 'BRICHAT'

Chitchor, F, pb, 1996; Pal, Dr B.P.

Chitra, HT, ob, 1995; IARI

Chitrarajini, HT, pb, 1985; Kasturi

Chivalry®, *see* 'MACPOW'

'CHIYO', HT, dp, 1975; bud long, pointed; flowers deep pink, dbl., 25 petals, 4–4.5 in., exhibition form, slight, fruity fragrance; foliage glossy, medium to dark green; vigorous, upright growth; [Karl Herbst X Chrysler Imperial]; Ota, Kaichiro; Eastern Roses, 1970

'CHLORIS', A, lp; (Rosée du Matin); flowers soft pink, many petaled, reflexing with a button eye; few prickles; foliage dark, leathery; vigorous (to 4–5 ft) growth; (France), Prior to 1835

'CHOBEE', F, mr, 1993; dbl., 26–40 petals, 1.5–2.75 in., borne mostly singly, slight fragrance; few prickles; foliage medium, medium green, matt; medium, bushy growth; [Seedling X Congratulations]; Giles, Diann; Giles Rose Nursery, 1993

'CHOO-CHOO CENTENNIAL', Min, lp, 1980; bud ovoid; flowers light pink, edged darker, reverse white, small blooms borne, dbl., 68 petals, exhibition form, slight fragrance; straight prickles; foliage matt, light green; compact, bushy growth; [Rise 'n' Shine X Grand Opera]; Jolly, Betty J.; Rosehill Farm

'CHOO-CHOO'S BABY', Min, rb, 1980; bud urn-shaped; flowers red, shaded to yellow at base, borne usually 1–2 per cluster, dbl., 26 petals, flat, slight fragrance; no prickles; foliage tiny, light green; low, branching, dense growth; [Watercolor X Watercolor seedling]; Jolly, Betty J.; Rosehill Farm

'CHOPIN', HT, mr, 1968; dbl., 38 petals, 4–6 in., moderate fragrance; foliage medium to light green; vigorous growth; [Montezuma X Christian Dior]; Ellick

'CHORALE', S, lp, 1978; bud ovoid, pointed; flowers pale pink, dbl., 48 petals, 3.5 in., exhibition form, moderate fragrance; foliage dark, leathery; vigorous, upright, bushy growth; [(Ruth Hewitt X Queen Elizabeth) X (Morning Stars X Suzanne)]; Buck, Dr Griffith J.; Iowa State University

Chorus®, *see* 'MEIMORE'

'CHORUS GIRL', F, or, 1970; flowers vermilion, semi-dbl., 16 petals, 3 in., exhibition form, moderate fragrance; foliage dark, coppery; [Highlight X Seedling]; Robinson, H.; Victoria Nursery

Chorus Line, *see* 'JACDAZ'

Chorus, Climbing, *see* MEIjulitasar

'CHOT PESTITELE', HP, mp, 1932; bud oblong; flowers rose-pink, dbl., 22 petals, 6 in., flat, moderate fragrance; foliage glossy; [Frau Karl Druschki sport]; Böhm, J.

'CHOTA', F, or, 1975; flowers light vermilion, very full, dbl., 50 petals, 3.5 in., slight fragrance; foliage small, semi-glossy; moderately low growth; [Violet Carson X Korona]; Sheen

'CHOUETTE', Pol, dr, 1969; bud ovoid; flowers dark fire-red, full, large, dbl.; foliage dark, soft; vigorous, bushy growth; [Atlantic X Seedling]; Delforge

'CHOWAN', HT, ob, 1995; dbl., 26–40 petals, 4.75 in., borne mostly singly, moderate fragrance; some prickles; foliage large, medium green, dull; tall (160 cms), upright growth; [Folklore X Hot Pewter]; Perry, Astor; Hortico Roses, 1995

Chris, *see* KIRsan

Chris Evert™, *see* 'WEKJUVOO'

'CHRIS JOLLY', MinFl, or, 1985; flowers miniflora, medium, dbl., 40 petals, exhibition form, slight fragrance; foliage medium, medium green, semi-glossy; upright, bushy growth; [(Orange Sweetheart X Zinger) X Rise 'n' Shine]; Jolly, Nelson F.; Rosehill Farm

'CHRISGOBRO', HT, yb, 1998; (**Joshua Bradley**); flowers deep gold, bronze reverse, strong substance, dbl., 17–25 petals, 3–3.5 in., borne mostly singly, slight fragrance; prickles moderate, medium, hooked; foliage medium, dark green, dull; upright, medium growth; [(Gingersnap X Brandy) X Caramel Creme]; Christensen, Jack E.; Armstrong Garden Centers, 1999

CHRisgood, F, yb; (**Rainbow Warrior**); (CHRisgood), 1996

'CHRISJEVANS', HT, or, 1989; (**Vivacious Dianne**); flowers medium, dbl., 33 petals, slight fragrance; foliage medium, medium green, semi-glossy; upright, bushy growth; [Voodoo X Hello Dolly]; Christensen, Jack E.

'CHRISS AND DIANNI', (clone of R. multiflora), w; used as understock

Chris's Sunshine, HT, my

'CHRISSIE MACKELLAR', HT, op, 1913; bud crimson-carmine on deep madder; flowers orange-pink, reverse deeply zoned orange, semi-dbl., moderate fragrance; Dickson, A.

Christa, *see* 'PIXSTA'

Christel von der Post®, *see* KORpora

'CHRISTIAN CURLE', HWich, lp, 1910; flowers flesh-pink; [Dorothy Perkins sport]; Cocker

Christian Dior, *see* 'MEILIE'

'CHRISTIAN DIOR, CLIMBING', Cl HT, mr, 1966; Chang, Chi-Shiang

'CHRISTIANA WOOD', HT, mr, 1975; bud ovoid; dbl., 31 petals, 3.5 in., moderate fragrance; upright, bushy growth; [Tropicana sport]; Wood, J.

Christiane Horbiger, F, w

'CHRISTINA', HT, pb, 1977; flowers deep pink, reverse light pink, large, dbl., 45 petals; foliage leathery, glossy; vigorous, upright growth; [Rina Herholdt X Seedling]; Staikov, Prof. Dr V., Kalaydjieve & Chorbadjiiski

'CHRISTINA', HT, pb, 1959; flowers pink, reverse currant red, dbl.; very vigorous growth; RULED EXTINCT 12/85 ARM; [Granat X Radiance]; Crouch; Roseglen Nursery

Christina, *see* 'SELSTAR'

Christina Atherton, *see* 'MACSEV'

'CHRISTINA NILSSON', HP, 1867; Lévêque, P.; (Cavriglia)

'CHRISTINE', HT, dy, 1918; flowers deep golden yellow, well-shaped, small, moderate fragrance; foliage dark, glossy; GM, NRS, 1916; McGredy

Christine, *see* CLEalta

'CHRISTINE CL.', Cl HT, dy, 1936; Willink

'CHRISTINE GANDY', F, dp, 1958; flowers deep pink, blooms in small clusters, semi-dbl., 3 in., moderate fragrance; foliage dark; vigorous growth; [Polyantha seedling X Fashion]; deRuiter; Gandy Roses, Ltd.

Christine Horbiger, *see* 'MACMOUHOO'

'CHRISTINE HORBIGER', *see* 'MACMOUHOO'

Christine Lanson, HT, ob

'CHRISTINE PRIOR', HT, rb, 1924; flowers deep rosy red, flushed yellow and peach, base yellow, semi-dbl., moderate fragrance; McGredy

'CHRISTINE WEINERT', Min, or, 1976; flowers brilliant scarlet, shaded deeper, flat to rounded, dbl., 25 petals, 1 in., moderate fragrance; foliage small, leathery; upright, bushy growth; [(Little Darling X Eleanor) X (Little Darling X Eleanor)]; Moore, Ralph S.; Sequoia Nursery

'CHRISTINE WRIGHT', LCl, mp, 1909; flowers wild-rose-pink, very large, semi-dbl., cupped, moderate fragrance; foliage glossy; long stems; height 12–15 ft; [Seedling X Mme Caroline Testout]; Hoopes, Bro. & Thomas

'CHRISTINE WUNDERLICH', HT, op, 1934; flowers yellowish orange-pink, large, moderate fragrance; very vigorous growth; [Golden Ophelia sport]; Wunderlich

'CHRISTINE, CLIMBING', Cl HT, dy, 1936; Willink

Christingle, *see* 'HARVALEX'

'CHRISTMAS BEAUTY', HT, dr, 1942; flowers darker red than parent, dbl., 25–30 petals, 4.5–5 in., moderate fragrance; foliage blue-green, leathery; vigorous growth; [Better Times sport]; Krowka

Christmas Card, *see* 'RESCARD'

'CHRISTMAS CHEER', HT, mr, 1957; bud long, pointed; flowers cherry-red, dbl., 45–50 petals, 4.5–5.5 in., exhibition form, slight fragrance; foliage dark, leathery; vigorous, upright growth; [Sister Kenny X Happiness]; Hill, Joseph H., Co.

'CHRISTMAS RED', HT, mr, 1948; bud large, long, pointed; flowers spectrum-red, open, dbl., moderate fragrance; foliage glossy; bushy, dwarf growth; [Pink Princess X Crimson Glory]; Brownell, H.C.

'CHRISTOBEL', HT, ab, 1937; flowers apricot-yellow shaded salmon, very large, dbl.; foliage glossy; very vigorous, bushy growth; [Frau Karl Druschki X Mme Butterfly]; Croibier

Christoph Colombus, *see* 'MEIRONSSE'

'CHRISTOPH WEIGAND', HT, lp, 1928; flowers very large, dbl., exhibition form, slight fragrance; foliage rich green, wrinkled;

vigorous growth; [Frau Karl Druschki X Souv. de Claudius Pernet]; Weigand, C.

Christophe Colomb, *see* 'MEIRONSSE'

Christopher, *see* 'TINCHRIS'

Christopher, *see* COCopher

Christopher Columbus, *see* 'MEIRONSSE'

Christopher Columbus, *see* POUlstripe

'CHRISTOPHER MILTON', HT, dp, 1965; bud long, pointed, light red; flowers medium pink, edged lighter, large, dbl., exhibition form; foliage dark, glossy; very vigorous, upright growth; [Christian Dior sport]; Martin, W.A.

'CHRISTOPHER STONE', HT, mr, 1935; bud long, pointed; flowers large, dbl., 30 petals, moderate, damask fragrance; foliage bright green; vigorous growth; GM, NRS, 1934 GM, Portland, 1937; Robinson, H.; C-P, 1936;, Wheatcroft Bros., 1935

'CHRISTOPHER STONE, CLIMBING', Cl HT, mr, 1942; Marsh's Nursery

'CHROMATELLA', N, ly, 1843; (Cloth of Gold); flowers creamy white, center yellow, large, very dbl., globular, moderate fragrance; vigorous, climbing growth; [Lamarque seedling]; Coquereau

Chrysandra, S, dp, 1983; flowers carmine-pink, large, semi-dbl., intense fragrance; Weihrauch; (Sangerhausen)

'CHRYSLER IMPERIAL', HT, dr, 1952; bud long, pointed; flowers deep red, velvety, very dbl., 45–50 petals, 4.5–5 in., exhibition form, borne singly, intense fragrance; foliage dark, semi-glossy; vigorous, compact growth; AARS, 1953 GM, ARS, 1956 GM, Portland, 1951 James Alexander Gamble Fragrance Medal, ARS, 1965 John Cook Medal, ARS, 1964; [Charlotte Armstrong X Mirandy]; Lammerts, Dr Walter; Germain's

'CHRYSLER IMPERIAL, CLIMBING', Cl HT, dr, 1957; (Grimpant Chrysler Imperial); Begonia, P.B.; Germain's

Chrystelle, *see* 'GODIALING'

'CHRYZIA', HT, dp, 1970; bud long, pointed; flowers light red, large, dbl., moderate fragrance; vigorous, upright, bushy growth; [Chrysler Imperial X Lady Zia]; Wyant

'CHUCKLES', F, dp, 1958; bud long, pointed; flowers deep pink, white eye, blooms in large clusters, semi-dbl., 11 petals, 3.5 in., moderate fragrance; foliage dark, leathery; vigorous, bushy growth; [(Jean Lafitte X New Dawn) X Orange Triumph]; Shepherd; Bosley Nursery

Chula Vista, *see* 'TINCHULA'

Church Mouse, *see* 'FOUMOUSE'

Ciak®, *see* BARciak

'CIBLES', HRg, mr, 1893; flowers bright red, base yellow; vigorous, upright growth; [R. rugosa rubra X Perle de Lyon]; Kaufmann

Cicely Lascelles, Cl HT, ob; flowers pink shaded salmon, semi-dbl.; very vigorous pillar type; [Frau Oberhofgartner Singer X Scorcher]; Clark, A., 1937

'CICELY O'RORKE', Cl HT, mp, 1937; flowers pink, shaded salmon, large, semi-dbl., cupped, recurrent bloom; long stems; very vigorous, climbing or pillar growth; [Souv. de Gustave Prat X Seedling]; Clark, A.; NRS Victoria

'CIDADE DE LISBOA', LCl, pb, 1939; bud long, pointed; flowers salmon-pink, edged yellow, large, semi-dbl., exhibition form, slight fragrance; long stems; vigorous growth; [Belle Portugaise X Mme Edouard Herriot]; da Silva, Moreira

Cider Cup, *see* 'DICLADIDA'

Ciel Bleu®, HT, m, 1982; Dot, Simon

'CILLY MICHEL', HT, or, 1928; flowers nasturtium-red, large, dbl., moderate fragrance; [Mme Mélanie Soupert X Felbergs Rosa Druschki]; Felberg-Leclerc

Cimarosa®, F, ob, 1989; Lens

'CIMARRON', HT, pb, 1938; bud spiraled, almost red; flowers ruffled, salmon, reverse deep pink; vigorous, compact growth; [Nellie E. Hillock X Golden Dawn]; Hillock

'CINA', HT, ob, 1973; bud full; flowers brilliant coral, flushed salmon; foliage large; [Premiere Ballerine X Femina]; Gaujard

'CINDERELLA', Min, w, 1953; flowers satiny white tinged pale flesh, micro-mini, dbl., 55 petals, 1 in., moderate, spicy fragrance; no prickles; upright growth; [Cécile Brunner X Tom Thumb]; deVink; C-P

'CINDERELLA', HWich, dp, 1909; flowers deep pink, petal tips quilled, small, borne in large clusters, dbl., free, non-recurrent bloom; vigorous growth; Walsh

Cinderella Gold, *see* 'MORCINGOLD'

'CINDERELLA, CLIMBING', Cl Min, w, 1975; Sequoia Nursery

'CINDERELLA'S MIDNIGHT ROSE', HT, mp, 1976; flowers rich pink, 4–5 in., slight fragrance; foliage glossy; vigorous growth; England

Cindy™, *see* 'MINAACO'

Cindy, *see* MEItinor

'CINERAIRE', HCh, rb; flowers rich red, center white, very small blooms in trusses, single, 5 petals, cupped; E. Murrell, prior to 1964

'CINERAMA', HT, pb, 1966; flowers salmon, reverse buff-yellow, pointed, 4.5–5 in., slight fragrance; vigorous growth; [Seedling X Tzigane]; Herholdt, J.A.

'CINERARIA', Pol, m, 1934; flowers carmine-purple, center white, open, large, semi-dbl., slight fragrance; foliage soft; vigorous, bushy growth; [Miss Edith Cavell X Tip-Top]; Leenders, M.

'CINGALLEGRA', HT, yb, 1958; bud well shaped; flowers lemon-yellow, edged pinkish, dbl.; long, strong stems; [Golden Scepter X Crimson Glory]; Cazzaniga, F. G.

'CINNABAR', F, or, 1945; (Tantau's Triumph); bud small, globular; flowers scarlet, semi-dbl., cupped, blooms in clusters, slight fragrance; foliage leathery; bushy, upright growth; [Baby Chateau X R. roxburghii]; Tantau

'CINNABAR IMPROVED', F, or, 1951; (Verbesserte Tantau's Triumph); flowers orange-scarlet, borne in trusses, semi-dbl.; (28); [(Cinnabar X Kathe Duvigneau) X Cinnabar]; Tantau

Cinnamon Delight™, *see* 'MINICIN'

Cinnamon Rose, *see* **R. CINNAMOMEA**

Cinnamon Toast™, *see* 'SAVACIN'

Cinzia, 1996; Embriaco, B.; (Cavriglia)

'CIRCE', HT, lp, 1916; flowers whitish-pink, large, dbl.; Paul, W.; (Sangerhausen)

Circé, *see* 'GAUMOVA'

'CIRCUS', F, yb, 1956; bud urn-shaped; flowers yellow marked pink, salmon and scarlet, very dbl., 45–58 petals, 2.5–3 in., exhibition form, blooms in large, ruffled clusters, moderate, tea to spicy fragrance; foliage semi-glossy, leathery; bushy growth; AARS, 1956 GM, Geneva, 1955 GM, NRS, 1955; [Fandango X Pinocchio]; Swim, H.C.; Armstrong Nursery

Circus Clown, *see* 'MORPICO'

'CIRCUS KNIE'®, HT, yb, 1975; bud long, pointed; dbl., 32–38 petals, 3.5–4 in., cupped, slight fragrance; foliage dark, leathery; [Moulin Rouge X Peace]; Huber

'CIRCUS PARADE', F, yb, 1963; flowers multicolor, redder than Circus; [Circus sport]; Begonia, F.B. & DeVor, P.; Armstrong Nursery

'CIRCUS, CLIMBING', Cl F, yb, 1961; House; Armstrong Nursery

Cisco™, *see* 'CARALE'

'CISSIE', Min, mp, 1979; bud long, pointed; 18–20 petals, 1 in., exhibition form, borne singly, slight fragrance; small, thin prickles; foliage dark green with touches of red and bronze on edges; bushy, upright growth; [Gene Boerner X Elfinesque]; Bennett, Cecilia 'Dee'; Tiny Petals Nursery

'CISSIE EASLEA', HT, my, 1913; [Melanie Soupert X Rayon d'Or]; Pernet-Ducher

'CITATION', Gr, dr, 1982; flowers dark red, reverse lighter, blooms in sprays of 2–3, dbl., 35 petals, exhibition form; globular, light orange fruit; short, lilac prickles, hooked downward; foliage large, dark, semi-glossy; medium, bushy growth; [Seedling X Seedling]; Hoy, Lowel L.; Joseph H. Hill, Co.

Citrina, HT, dy

'CITRON', HT, ly, 1942; flowers buff, shaded copper, dbl., 28 petals, cupped, moderate fragrance; foliage reddish; vigorous growth; [Julien Potin X Seedling]; Gaujard; J&P

'CITRONELLA', F, ly, 1946; flowers lemon-yellow, borne in clusters, semi-dbl., 15 petals, globular, moderate fragrance; foliage glossy; branching growth; [Mev. Nathalie Nypels X Donald Prior]; Leenders, M.; Longley

Citronella®, *see* 'SUNSIEN'

Citron-Fraise, *see* DELcifra

City Girl, *see* 'HARZORBA'

City Lights, *see* POUlgan

City of Alexandria, *see* 'WILALEX'

City of Auckland, *see* 'MACTANE'

'CITY OF BATH', HT, pb, 1969; flowers deep candy-pink, reverse lighter, dbl., 55 petals, 4 in., moderate fragrance; foliage matt, green; [Gavotte X Buccaneer]; Sanday, John

City of Belfast®, *see* 'MACCI'

'CITY OF BENALLA', HT, op, 1983; bud globular, pointed; flowers carmine, opening with outer petals paling, inner petals coral, dbl., exhibition form, slight fragrance; brown prickles, hooked down; foliage dark, dense, glossy; vigorous, tall growth; [My Choice X Extravaganza]; Dawson, George; Rainbow Roses

City of Birmingham, *see* 'KORHOLST'

City of Bradford, *see* 'HARROTANG'

'CITY OF CARDIFF', HT, rb, 1992; very dbl., 3–3.5 in., borne mostly singly, moderate fragrance; some prickles; foliage medium, medium green, matt; medium, upright growth; [Lady Sylvia X Chicago Peace]; Poole, Lionel, 1993

City of Christchurch, *see* MACoffer

City of Dunedin, *see* 'REYCIDUN'

'CITY OF GISBORNE', HT, op, 1968; bud long, pointed; flowers pink shaded orange-yellow, large, semi-dbl., exhibition form, intense fragrance; foliage glossy; vigorous, upright growth; [Prima Ballerina X Prima Ballerina]; Appleyard

'CITY OF GLASGOW', HT, ab, 1970; flowers apricot, suffused pink; [Femina sport]; Haynes

'CITY OF GLOUCESTER', HT, dy, 1969; flowers saffron-yellow shaded gold, large, dbl., exhibition form; vigorous growth; [Gavotte X Buccaneer]; Sanday, John, 1970

'CITY OF HAMILTON', F, ob, 1972; flowers orange and gold, dbl., 3.5 in., slight fragrance; foliage dull; free growth; [Innisfree X Elizabeth of Glamis]; Dickson, A.

'CITY OF HARVEY', HWich, op, 1944; flowers pink tinted orange, small, borne in clusters, semi-dbl., cupped, profuse bloom, not repeated, slight fragrance; foliage dark, glossy; height 12 ft; [R. wichurana X Orléans Rose]; Wiseman

'CITY OF HEREFORD', HT, mp, 1967; flowers carmine-pink, pointed, 6 in., intense fragrance; foliage dark; [Wellworth X Spartan]; LeGrice

City of Ichalkaranji, HT, pb, 1989; Chiplunkar

City of Invercargill, *see* 'REYINVER'

'CITY OF KINGSTON', F, or, 1973; bud ovoid; flowers medium, very dbl., globular, slight fragrance; foliage dark, leathery; moderate, bushy growth; [Malibu X Independence]; Schloen, J.

'CITY OF LEEDS', F, op, 1966; flowers salmon, blooms in clusters, semi-dbl., 19 petals, 4.5 in., slight fragrance; foliage dark; GM, RNRS, 1965; [Evelyn Fison X (Spartan X Red Favorite)]; McGredy, Sam IV; McGredy

'CITY OF LITTLE ROCK', HT, mp, 1924; flowers hydrangea-pink, open, semi-dbl., moderate fragrance; vigorous growth; Hill, E.G., Co.; Vestal

City of London®, *see* 'HARUKFORE'

City of Manchester, *see* 'NOSMAN'

'CITY OF NEWCASTLE', HT, yb, 1976; flowers yellow, tinged salmon-orange, dbl., 35 petals, 5 in., intense fragrance; foliage semi-glossy; vigorous, tall, upright growth; [Arthur Bell X Mischief]; Wood

City of Newcastle Bicentennary, *see* 'JACOPPER'

'CITY OF NORWICH', HT, mr, 1949; bud ovoid; flowers scarlet-crimson, well formed, dbl., 35 petals, 6 in., moderate fragrance; foliage leathery; [Crimson Glory X (Crimson Glory X Cathrine Kordes)]; Kordes; Morse

'CITY OF NOTTINGHAM', F, or, 1962; flowers orange-scarlet, rosette shape, borne in clusters, dbl., 30–40 petals, 2.5 in., slight fragrance; foliage dark; vigorous, bushy, compact growth; [Seedling (F) X Moulin Rouge]; deRuiter

City of Panjim, HT, pb, 1972; Kasturi

City of Pilsen, *see* **'PLZEN'**

'CITY OF PORTLAND', F, mr, 1977; bud pointed; flowers geranium-red, base primrose-yellow, single, 5 petals, 2.5–3.5 in., slight fragrance; foliage glossy; very free growth; [Seedling X Cocktail seedling]; Takatori, Yoshiho; Japan Rose Nursery

'CITY OF PORTSMOUTH', F, ob, 1975; flowers copper, dbl., 25 petals, 3–4 in., moderate fragrance; foliage bronze; tall growth; Cants of Colchester, Ltd.

City of Pretoria, *see* KORseubel

City of Sheffield, *see* 'NOSSHEF'

'CITY OF SPRINGFIELD', F, rb, 1988; flowers non-fading, medium, borne in sprays of 6–10, dbl., 34 petals; bushy, hardy growth; [Pink Parfait X Roman Holiday]; Pencil, Paul S., 1989

City of Wangaratta, HT, mr, 1985; Dawson

City of Warwick, *see* 'AROfuto'

City of Welland, HT, yb, 1996; Kordes

City of Windsor, HT, ob

'CITY OF WORCESTER', HT, mr, 1983; flowers large, dbl., 35 petals, exhibition form, moderate fragrance; foliage medium, medium green, matt; [Red Planet X (Ena Harkness X Fragrant Cloud)]; Crivens, L.

'CITY OF YORK', LCl, w, 1945; (Direktör Benschop); flowers creamy white, large, in clusters of 7–15, semi-dbl., 15 petals, cupped, moderate fragrance; foliage glossy, leathery; vigorous, climbing growth; GM, ARS, 1950; [Prof. Gnau X Dorothy Perkins]; Tantau; C-P

'CIUDAD DE OVIEDO', C, mp; (R. centifolia simplex); single; (28)

Cl. All Ablaze™, *see* 'WEKSAMSOU'

Cl. Berries 'n' Cream, *see* 'POULCLIMB'

Cl. Shadow Dancer, *see* 'MORSTRORT'

'CLAIR DE LUNE', HT, m, 1967; bud pointed; flowers large, dbl., moderate fragrance; foliage leathery; vigorous, upright growth; [Eminence X Viola]; Gaujard

Clair Matin®, *see* 'MEIMONT'

Clair Renaissance, *see* POUlsyng

Claire, HT, pb, 1987; Dawson

'CLAIRE DESMET', HT, my, 1932; flowers golden yellow, very dbl., cupped, moderate fragrance; foliage leathery, bronze; long, strong stems; vigorous, bushy growth; [Margaret Dickson Hamill X Souv. de Claudius Pernet]; Buatois

'CLAIRE JACQUIER', N, ly, 1888; flowers yellow, fading to creamy yellow; very vigorous growth; [Possibly R. multiflora X Tea rose]; Bernaix, A.

Claire Jolly, HMsk, 1991; Lens, Louis; (Cavriglia)

Claire Rayner, *see* 'MACPANDEM'

Claire Rose, *see* 'AUSlight'

Claire Scotland, *see* 'COCDIMITY'

'CLAIRE-FRANCE', HT, mp, 1964; bud pointed; flowers clear pink, large, dbl., 40 petals, exhibition form, slight fragrance; foliage glossy; vigorous, upright growth; Mondial Roses

Clairette, HT, 1974; Croix, P.; (Cavriglia)

Clanwilliam, HT, op

Clara, *see* 'RESCLARA'

Clara, HT, rb

Clara®, *see* FEsato

'CLARA BOW', ClHT, yb, 1927; (Clara Bow, Climbing); flowers yellow stained crimson, dbl., slight fragrance; height 12–15 ft;

[Golden Emblem sport]; Padella Rose Co.; Germain's

Clara Bow, Climbing, *see* **'Clara Bow'**

'Clara Cochet', HP, lp, 1886; flowers very large, dbl.; Lacharme, F.; (Sangerhausen)

'Clara Curtis', HT, my, 1922; flowers rich golden yellow, very dbl., intense fragrance; GM, NRS, 1919; Dickson, A.

Clara d'Arcis, *see* **'Mme Clara d'Arcis'**

Clara Granato, F, mr

'Clara Munger', LCl, m; flowers lavender; Munger

'Clara Watson', HT, w, 1894; flowers mother-of-pearl white, center peach, dbl.; moderate fragrance; Prince

Clara's Surprise, *see* 'KIRwim'

'Clare', HT, pb, 1972; flowers cream, edged rose-pink, dbl., 35 petals, 5.5 in., exhibition form, slight fragrance; foliage large, medium green, semi-glossy; vigorous, tall growth; [Ethel Sanday X Rose Gaujard]; MacLeod

'Clare de Escofet', HT, w, 1920; flowers delicate flesh-white, dbl.; Easlea

'Cläre Grammserstorf', F, dy, 1957; bud ovoid; flowers large, dbl., exhibition form; foliage leathery, glossy; vigorous, bushy growth; [Harmonie X R. eglanteria seedling]; Kordes

'Clare Helen', HT, lp, 1985; [Gail Borden sport]; Owen, Fred

'Claret', Min, m, 1977; bud short, pointed; dbl., 48 petals, 1 in., cupped; very compact, spreading. growth; [Little Chief X Little Chief]; Saville, F. Harmon; Nor'East Min. Roses

'Claret Cup', Min, rb, 1962; bud globular; flowers dark red, white eye, small blooms in clusters, dbl., moderate fragrance; foliage leathery, dark; vigorous, bushy, compact growth; [Spring Song (Min) X Eutin]; Riethmuller; Hazlewood Bros.

'Clarice Goodacre', HT, w, 1916; bud pointed; flowers ivory-white, shaded chrome, dbl., exhibition form, moderate fragrance; foliage dark, soft; vigorous, bushy growth; Dickson, A.

Claridge, *see* 'PEKlipink'

'Clarion Call', F, op, 1995; flowers orange pink, red salmon hue, blooms borne 1–5 per cluster, semi-dbl., 6–14 petals, 1.5–2.75 in., slight fragrance; foliage medium, medium green, matt; bushy (up to 75 cms) growth; 1st prize, Unnamed Seedling, Norfolk, VA, National Show, ARS, 1991; [Canadian Centennial X Traumerei]; Fleming, Joyce L.; Hortico Roses, 1994

Clarissa®, *see* 'HARprocrustes'

'Clarissa Dana', HT, pb, 1933; flowers brilliant pink with amber glow, large, dbl., exhibition form, moderate fragrance; vigorous growth; [(HT X La France) X Marechal Niel]; Nicolas; J&P

Clarita, *see* 'MEIbyster'

'Clarity', F, w, 1993; flowers white with cream center, dbl., 15–25 petals, 1.5–2.75 in., borne in small clusters, slight fragrance; some large prickles; foliage large, dark green, glossy; medium, upright, bushy growth; [Pristine X (Party Girl X Laureate)]; Jobson, Daniel J.; Jobson, 1993

Clark Ochre Seedling, Cl HT, dy; Clark

Clark Pink, Cl HT, mp; Clark

'Clarke's Multiflora', (clone of R. multiflora), w; thornless; used as understock

Class Act, *see* 'JACare'

Classic, HT, dp, 1997; K&S

'Classic Beauty', HT, pb, 1994; dbl., 15–25 petals, 3–3.5 in., borne mostly singly, slight fragrance; some prickles; foliage medium, medium green, matt; medium, bushy growth; Winchel, Joseph F.; Certified Roses, Inc., 1997

'Classic Chick', Min, ob, 1986; flowers orange, blooms borne usually singly, dbl., 35 petals, exhibition form, no fragrance; few, slender prickles; foliage medium, dark, semi-glossy; medium, upright growth; [Seedling X Seedling]; McDaniel, Earl; McDaniel's Min. Roses

Classic Love, *see* 'LYOcl'

Classic Orange, Min, ob, 1997

Classic Sunblaze®, *see* 'MEIpinjid'

'Classic Touch', HT, lp, 1991; flowers large blooms borne mostly singly, dbl., 26–40 petals, slight fragrance; foliage large, medium green, semi-glossy; tall, upright growth; [Touch of Class sport]; Hefner, John; Co-Operative Rose Growers, 1993

Classical Velvet, *see* 'JOHillstar'

'Classie Lassie'™, Gr, pb, 1990; bud pointed; flowers ivory pink, with salmon pink edges, aging to salmon, oval-shaped, semi-dbl., 25–30 petals, exhibition form, moderate, fruity fragrance; prickles average, brown green, slight recurve; foliage medium, medium green, glossy, disease resistant; bushy, medium growth; [Touch of Class X Seedling]; Winchel, Joseph F.; Co-Operative Rose Growers, 1991;, Speciality Roses, 1991

Classy, *see* 'TRAcla'

Classy™, *see* HILrap

'Claude', HT, mr, 1950; flowers bright orient red, dbl., 35 petals, 6–7 in., moderate fragrance; foliage glossy, dark; vigorous, upright growth; [(Comtesse Vandal X Brazier) X Seedling]; Mallerin, C.; EFR;, Wheatcroft Bros.

'Claude Jacquet', HP, dr, 1892; flowers dark red/purple, very large, dbl., moderate fragrance; Liabaud; (Sangerhausen)

Claude Monet®, *see* JACdesa

'Claude Petit', HT, mp, 1936; flowers soft salmon-pink, stamens yellow, very dbl., exhibition form, intense fragrance; vigorous growth; [Mlle Marie Mascuraud X Beauté de Lyon]; Buatois

'Claude Rabbe', LCl, dp, 1941; flowers carmine-pink, medium, borne in clusters of 6–10; Buatois

'Claudia', F, pb, 1959; flowers deep cherry-coral, dbl.; vigorous growth; [Ma Perkins X Geranium Red]; Broadley; Roseglen Nursery

Claudia, *see* MEIyacom

'Claudia Augusta', N, pb; Damaizin, 1858

Claudia Cardinale, *see* MAScatna

'Claudy Chapel', HT, yb, 1930; flowers deep yellow and coppery, to salmon-pink, large, dbl., moderate fragrance; vigorous growth; Beaumez; Delhaye

Claus Groth, *see* **'Klaus Groth'**

Claysnow, *see* **'Snow Ballet'**®

CLEalta, Cl HT, ob, 1998; (**Christine**); Clements, John K.

CLEamour, MinFl, rb, 1992; (**Glamour Girl**); Clements, John K.

'CLEbeau', Min, yb, 1992; (**Golden Beauty**, Golden Girls, (CLEgirl); flowers gold, edged copper, blooms borne mostly singly, dbl., 26–40 petals, 1.5 in., exhibition form, slight fragrance; few prickles; foliage small, dark green, glossy; medium (30 cms), bushy, spreading growth; [Seedling X Seedling]; Clements, John K.; Heirloom Old Garden Roses, 1990

CLEblack, Min, dr, 1996; (**Black Gold**); Clements, John K.

CLEbou, S, pb, 1995; (**Boudoir**); Clements, John K.

CLEbravo, S, dr, 1998; (**Braveheart**); Clements, John K.

CLEbridge, HMsk, ab, 1996; (**Bridget**); Clements, John K.

'CLEbrite', Min, rb, 1991; (**Britestripe**, Britestripes); flowers soft pink, striped bright red, blooms borne in small clusters, dbl., 26–40 petals, 1.5 in., no fragrance; some prickles; foliage small, medium green, semi-glossy; medium (30 cms), bushy, compact growth; [Pinstripe X Seedling]; Clements, John K.; Heirloom Old Garden Roses, 1990

CLEcham, S, mp, 1995; (**Pink Champagne**); Clements, John K.

CLEcherry, HWich, mr, 1999; (**Cherries Jubilee**); Clements, John K.

CLEconcert, S, lp, 1999; (**Regina Louise**); Clements, John K.

CLEdan, HMsk, pb, 1995; (**Dancing in the Wind**); Clements, John K.

CLEdare, MinFl, lp; (**Virginia Dare**); Clements, John K., 1998

'CLEdrag', HCh, dr, 1992; (**Dragon's Eye**, Eye of the Dragon, The Dragon's Eye); very dbl., 1.5–2.75 in., exhibition form, borne mostly singly, moderate fragrance; some prickles; foliage small, dark green, semi-glossy; medium (70 cms), bushy, compact growth; [Seedling X Seedling]; Clements, John K.; Heirloom Old Garden Roses, 1991

'CLEfire', Min, or, 1991; (**Firestorm**); flowers fiery oriental lacquer orange-red, blooms borne mostly singly, very dbl., 1.5 in., exhibition form, no fragrance; few prickles; foliage small, medium green, semi-glossy; medium (35 cms), bushy, spreading growth; [Seedling X Seedling]; Clements, John K.; Heirloom Old Garden Roses

'CLEgran', S, lp, 1994; (**Grandma's Lace**); flowers pale pink, blooms borne in clusters, dbl., 26–40 petals, 1.5–2.75 in., moderate fragrance; foliage medium, medium green, semi-glossy; medium, spreading growth; [Sexy Rexy X Trier]; Clements, John K.; Heirloom Old Garden Roses

CLEheat, S, ly, 1997; (**Good Old Summertime**); Clements, John K.

CLEhelp, S, mr, 1997; (**Helping Hands**); Clements, John K.

'CLEhon', Min, w, 1991; (**Honey Mini-Delite**); flowers honey cream, blooms borne in small clusters, dbl., 26–40 petals, 1.5 in., no fragrance; few prickles; foliage small, medium green, matt; low (25 cms), bushy, compact growth; [Seedling X Seedling]; Clements, John K.; Heirloom Old Garden Roses

CLEhonor, S, dr, 1999; (**Ora Kingsley**); Clements, John K.

'CLEice', Min, w, 1991; (**Ice Crystal**); flowers crystal white, blooms borne in small clusters, very dbl., 1.5 in., exhibition form, no fragrance; some prickles; foliage small, medium green, semi-glossy; micro-mini; low (20 cms), bushy, compact growth; [Seedling X Baby Betsy McCall]; Clements, John K.; Heirloom Old Garden Roses, 1989

CLEjoan, S, w, 1996; (**Joan Fontaine**™); Clements, John K.

CLEjoy, S, mp, 1997; (**Memories**); Clements, John K.

CLEkate, S, mp, 1996; (**Kateryna**); Clements, John K.

'CLEkiss', Min, rb, 1991; (**Strawberry Kiss**); flowers white edged red, blooms borne mostly singly, dbl., 15–25 petals, 1.5 in., exhibition form, no fragrance; some prickles; foliage small, light green, semi-glossy; medium (35 cms), upright, bushy growth; [Seedling X Seedling]; Clements, John K.; Heirloom Old Garden Roses, 1990

'CLElady', Min, pb, 1991; (**Fancy Lady**); flowers medium pink, white reverse, high pointed, blooms borne mostly singly, very dbl., 1.5 in., exhibition form, no fragrance; few prickles; foliage small, dark green, glossy; tall (50 cms), upright growth; [Seedling X Seedling]; Clements, John K.; Heirloom Old Garden Roses, 1990

CLEland, HWich, w, 1996; (**Landscape Splendor**); rambler growth; Clements, John K.

'CLElight', Min, rb, 1991; (**Little Lighthouse**); flowers bright red with yellow to white eye, blooms borne in small clusters, single, 5 petals, 1.5 in., exhibition form, no fragrance; some prickles; foliage small, medium green, semi-glossy; low (18 cms), bushy, spreading, compact, growth; groundcover; [Robin Red Breast X Little Artist]; Clements, John K.; Heirloom Old Garden Roses, 1992

'CLElips', Min, rb, 1991; (**Lipstick 'n' Lace**); flowers cream, shaded lipstick red, blooms borne in small clusters, dbl., 26–40 petals, 1.5–2.75 in., exhibition form, moderate fragrance; no prickles; foliage medium, medium green, matt; medium (30 cms), upright, compact growth; [Seedling X Seedling]; Clements, John K.; Heirloom Old Garden Roses, 1992

'CLElock', Min, my, 1991; (**Turlock High**); flowers bright yellow, blooms borne mostly singly, dbl., 26–40 petals, 1.5 in., intense fragrance; some prickles; foliage small, dark green, semi-glossy; medium (40 cms), upright, bushy growth; [Rise 'n' Shine X Seedling]; Clements, John K.; Heirloom Old Garden Roses, 1992

CLElou, S, ob, 1996; (**Louise Clements**); Clements, John K.

'**Clematis**', HWich, rb, 1924; flowers dark red, prominent white eye, small, borne in clusters, single; vigorous, climbing growth; Turbat

Clémence Beaugrand, mp; flowers medium, dbl.; Roseraie de l'Hay; (Sangerhausen)

'**Clémence Raoux**', HP, mp; Granger, 1869

'**Clemence Robert**', M, mp, 1863; flowers heavily mossed, sometimes recurrent bloom; Robert et Moreau

CLEment, LCl, dr, 1995; (**Commitment**); Clements, John K.

'**Clément Pacaud**', HT, dp, 1916; flowers brilliant carmine; Chambard, C.

'**Clementina**', HT, pb, 1961; flowers carmine and rosy white bicolor; [Grand Gala X Vicky Marfá]; Dot, Pedro

'**Clementina Carbonieri**', T, yb, 1913; Bonfiglio, A.

'**Clementine**', HEg; flowers rosy blush

Clementine, *see* '**Janet's Pride**'

Clementine, *see* TANogrew

'**Clementine Duval**', B, mp, 1847; flowers bright rose, cupped; Laffay, M.

'**Clémentine Séringe**', HP, mp, 1840; flowers rose-pink, dbl.; Wood

CLEmist, MinFl, lp; (**Will-o'-the-Wisp**); Clements, John K., 1998

Cleo, *see* 'BEEbop'

'**Cleopatra**', *see* KORverpea

'**Cleora**', Min, mp, 1977; bud globular; flowers medium pink, reverse darker, small, dbl., 50 petals, 1.5 in., flat; foliage small, light, leathery; vigorous growth; [Fairy Moss X Fairy Moss]; Dobbs; Port Stockton Nursery

'CLEpeach', Min, ab, 1991; (**Peach Silks**); flowers rich peach, blooms borne in small clusters, dbl., 15–25 petals, 1.5 in., no fragrance; few prickles; foliage small, dark green, glossy; tall (50 cms), upright growth; [Seedling X Seedling]; Clements, John K.; Heirloom Old Garden Roses

'CLEperf', Min, ab, 1992; (**Apricot Perfection**); flowers soft apricot, moderately, dbl., 15–25 petals, 1.5 in., exhibition form, borne mostly singly, slight fragrance; some prickles; foliage small, medium green, matt; tall (50 cms), upright, bushy growth; [My Louisa X Seedling]; Clements, John K.; Heirloom Old Garden Roses, 1990

'CLEpurp', Min, m, 1991; (**Color Purple**); flowers rich deep purple, small, blooms borne mostly singly, very dbl., 1.5 in., moderate fragrance; few prickles; foliage medium, dark green, holly-like, glossy; low (25 cms), spreading, compact growth; [Angel Face X Seedling]; Clements, John K.; Heirloom Old Garden Roses, 1990

'CLErain', Min, yb, 1991; (**Oregon Rainbow**); bud high-pointed, exhibition; flowers golden yellow, shaded and edged red, blooms, dbl., 26–40 petals, 1.5 in., no fragrance; foliage small, dark green, semi-glossy; medium (40 cms), bushy growth; [Seedling X Seedling]; Clements, John K.; Heirloom Old Garden Roses

'CLEruff', S, rb, 1994; (**Ruffles 'n' Flourishes**); flowers violet red and gold, blooms borne in small clusters, dbl., 15–25 petals, 1.5–2.75 in.; few prickles; foliage medium, reddish green, glossy; low (18 in.), upright, bushy growth; [Sexy Rexy X Whistle Stop]; Clements, John K.; Heirloom Old Garden Roses, 1994

'CLEscrub', Min, or, 1991; (Patriot Flame, **Scudbuster**); flowers intense rocket flame orange, blooms borne mostly singly, very dbl., 1.5 in., exhibition form, slight fragrance; some prickles; foliage small, dark green, semi-glossy; medium (35 cms), upright growth; [Seedling X Seedling]; Clements, John K.; Heirloom Old Garden Roses, 1992

'CLEshir', Min, mp, 1991; (**Sweet Shirley**); flowers medium pink, lighter reverse, blooms borne mostly singly, dbl., 15–25 petals, 1.5 in., exhibition form, slight fragrance; some prickles; foliage small, medium green, semi-glossy; bushy (40 cms), spreading growth; [Tweedle Dee X Seedling]; Clements, John K.; Heirloom Old Garden Roses

'CLEsnob', Min, rb, 1991; ('CLEsnow', **Snowblush**); flowers purest white, edges red, blooms borne mostly singly, very dbl., 1.5–2.75 in., exhibition form, no fragrance; some prickles; foliage medium green, matt; tall (50 cms),upright growth; [Minuette X Seedling]; Clements, John K.; Heirloom Old Garden Roses, 1990

'CLEsnow', Min, rb, 1991; ('CLEsnob', **Snowblush**); flowers purest white, edges red, blooms borne mostly singly, very dbl., 1.5–2.75 in., exhibition form, no fragrance; some prickles; foliage medium green, matt; tall (50 cms), upright growth; [Minuette X Seedling]; Clements, John K.; Heirloom Old Garden Roses, 1990

CLEspirit, S, dp, 1999; (**Laura Clements**); Clements, John K.

CLEswan, S, ob, 1999; (**Brenda Burg**); Clements, John K.

'CLEtape', Min, rb, 1991; (**Antique Tapestry**); flowers burgundy and gold, dbl., 15–25 petals, 1.5–2.75 in., exhibition form, borne mostly singly, slight fragrance; few prickles; foliage medium, dark green, semi-glossy; medium (40 cms), upright growth; [Redgold X Seedling]; Clements, John K.; Heirloom Old Garden Roses, 1990

'CLEtess', Min, lp, 1991; ('CLEtest', **Tess**); flowers flesh pink, blooms borne mostly singly, semi-dbl., 6–14 petals, 1.5 in., exhibition form, no fragrance; few prickles; foliage small, medium green, semi-glossy; medium (30 cms), bushy growth; [Cupcake X Seedling]; Clements, John K.; Heirloom Old Garden Roses, 1992

'CLEtig', Min, yb, 1992; (**Tiger Stripes**); flowers unusual striped color combination, yellow, striped orange, b, dbl., 26–40 petals, 1.5 in., no fragrance; few prickles; foliage small, medium green, semi-glossy; low (25 cms), bushy, compact growth; [Seedling X Seedling]; Clements, John K.; Heirloom Old Garden Roses, 1991

CLEtraveler, S, mp, 1999; (**Charles Kuralt**); Clements, John K.

'CLEtwin', Min, pb, 1991; (**Twin Pinks**); flowers two-tone pink, blooms borne in small clusters, dbl., 15–25 petals, 1.5 in., exhibition form, slight fragrance; some prickles; foliage small, medium green, semi-glossy; low (30 cms), bushy, compact growth; [Seedling X Seedling]; Clements, John K.; Heirloom Old Garden Roses

'Cleveland', HT, pb, 1916; (H.P. Pinkerton); flowers reddish copper on old-rose, base coppery yellow, dbl., moderate fragrance; Dickson, H.

'Cleveland Bouquet', HT, op, 1940; flowers shrimp-pink with salmon-pink undertone, open, semi-dbl.; Horvath

Cleveland II, *see* **'Mrs Dunlop Best'**

CLEwedding, S, my, 1996; (**Morning Has Broken**); Clements, John K.

CLEwonder, S, lp; (**Wonderstripe**); Clements, John K., 1997

Cliff Richard, F, pb, 1993; Van Geest

Cliffs of Dover, *see* 'POUlemb'

Clifton Moss, *see* **'Shailer's White Moss'**

Climbing Bride, *see* **'Ruth Vestal'**

Climbing Geheimrat Duisberg, *see* **'Golden Rapture, Climbing'**

Climbing Rainbow's End™, *see* 'SAVAclend'

'Climbing Roses'

Climbing Spek's Yellow, *see* **'Golden Scepter, Climbing'**

'Climentina', HT, mp, 1955; flowers rosy pink, large, slight fragrance; [Independence X Peace]; Klimenko, V. N.

'Clinora', Cl HT, dy, 1978; (**Landora, Climbing**); bud pointed; flowers well shaped, 3.5 in., intense fragrance; [Sunblest sport]; Orard, Joseph; Pekmez

'Clio', HP, lp, 1894; flowers flesh, large blooms in clusters, very dbl., globular, seasonal bloom, moderate fragrance; foliage rich green; vigorous growth; Paul, W.

Clio, HT, or, 1972; Gaujard

Clio®, S, mr

Clive Lloyd, HT, dp

Clivia®, *see* 'KORtag'

Clochermerle®, *see* 'DELpétri'

Clo-Clo, *see* 'PEKatan'

Cloris Adriana, HT, dp, 1997; flowers very full, high centered, large, very dbl., 41 petals, borne mostly singly, intense fragrance; some prickles; foliage large, dark green, dull; spreading, bushy, medium growth; [Blue Moon X Gavotte]; Poole, Lionel

Clos de la Pellerie, *see* 'DELclopel'

Clos Fleuri, S, 1970; Delbard-Chabert; (Cavriglia)

Clos Fleuri Bicolore, F, 1992; Delbard, Georges; (Cavriglia)

Clos Fleuri Blanc®, *see* 'DELblan'

Clos Fleuri Champagne, F, w, 1990; Delbard

Clos Fleuri d'Or, F, dy, 1990; Delbard

Clos Fleuri Jaune®, *see* 'DELjaune'

Clos Fleuri Rose®, *see* 'DELpomp'

Clos Fleuri Rose No. 2, *see* DELodive

Clos Fleuri Rouge®, *see* 'DELecla'

Clos Vougeot®, *see* 'DELific'

'Clotaria', HT, mr, 1936; (Red Gruss an Coburg); flowers bright fuchsia-red, well formed, medium, dbl., 26–28 petals, intense fragrance; foliage dark, glossy; vigorous, upright, bushy growth; [Gruss an Coburg X J.C. Thornton]; San Remo Exp. Sta.

Cloth of Gold, *see* **'Chromatella'**

'Clotilde Soupert', Pol, w, 1890; flowers pearly white, center soft rose-pink, large blooms in clusters, very dbl., moderate fragrance; foliage rich green, soft; bushy (10–20 in.) growth; [Said to be Mignonette X Mme Damaizin]; Soupert & Notting

'Clotilde Soupert, Climbing', Cl Pol, w, 1902; Dingee & Conard

'Clotilde Soupert, Climbing', Cl Pol, w, 1896; Berckmans, P.J., Co.

Cloud Dancer™, *see* 'BURdancer'

Cloud Nine, *see* 'JAClite'

Cloud Nine, S, w, 1999; flowers white touched with apricot; Williams, J. Benjamin

'Cloudland', HT, m, 1994; dbl., 26–40 petals, 3–3.5 in., moderate fragrance; foliage large, medium green, matt; upright (150 cms) growth; [Blue Moon X Paradise]; Perry, Astor; Perry Roses, 1995

Clouseau, *see* 'TALclouseau'

Clove Scented Musk, Misc OGR, w

'Clovelly', HT, dp, 1924; flowers carmine-pink, well formed, moderate fragrance; vigorous growth; GM, NRS, 1924; Hicks

Cloverdene, *see* HERclov

'Club', F, mr, 1957; flowers bright red, open, medium, single, slight fragrance; foliage glossy, bronze; very vigorous, bushy growth; [Peace X Opera seedling]; Gaujard

Clubrose Lydia, *see* **'Lydia'**

'Clubrose Scala', F, rb, 1973; (Scala); flowers blood-red and orange, medium, dbl., globular, slight fragrance; foliage glossy, dark, leathery, bronze; vigorous, upright, bushy growth; [Marlena X Seedling]; Kordes

Clydebank Centenary, *see* 'COCdazzle'

'Clytemnestra', HMsk, op, 1915; bud copper; flowers salmon-chamois, ruffled, small blooms in clusters, recurrent bloom, moderate fragrance; foliage leathery, dark; bushy (3–4 ft), spreading growth; GM, NRS, 1914; [Trier X Liberty]; Pemberton

'Cnos X', HT, my, 1957; flowers golden yellow tinged copper, moderate fragrance; vigorous growth; [Bettina seedling]; Bronisze (Poland) State Nursery

'Coalite Flame', HT, or, 1974; very dbl., 60 petals, 5 in., moderate fragrance; foliage large, matt; [Fragrant Cloud X Red Planet]; Dickson, A.

'COBY FANKHAUSER', HT, my, 1971; flowers buttercup-yellow, large, dbl., exhibition form, moderate fragrance; foliage leathery; vigorous, upright, bushy growth; [John S. Bloomfield X Elizabeth Fankhauser]; Fankhauser

'COCABEL', Min, dp, 1980; (**Little Jewel**); flowers deep pink, patio, blooms borne 6–12 per cluster, dbl., 34 petals, slight fragrance; straight prickles; foliage small, glossy, dark; low, compact growth; [Wee Man X Belinda]; Cocker, James

'COCABEST', Min, mr, 1980; (**Wee Jock®**); bud pointed; flowers patio, blooms borne 9–15 per cluster, dbl., 50 petals, slight fragrance; red prickles; foliage small, fairly glossy, medium green; low, compact growth; [National Trust X Wee Man]; Cocker, James

'COCACERT', F, ab, 1982; (**Vital Spark**); flowers gold, flushed coral, medium, dbl., 35 petals, slight fragrance; foliage medium, medium green, semi-glossy; bushy growth; [(Anne Cocker X (Sabine X Circus)) X Yellow Pages]; Cocker, James; Cocker & Sons

'COCADILLY', HT, pb, 1985; (**The Coxswain**); flowers cream and pink blend, large, dbl., 35 petals, intense fragrance; foliage medium, medium green, semi-glossy; bushy growth; [(Tropicana X Ballet) X Silver Jubilee]; Cocker, Alexander M.; Cocker & Sons

'COCAGOLD', HT, my, 1981; (**Golden Jubilee**); dbl., 29 petals, exhibition form, moderate, tea fragrance; narrow, red-brown prickles; foliage large, matt green, glossy; [Peer Gynt X Gay Gordons]; Cocker, James

'COCAMOND', Min, op, 1981; (**Dainty Dinah**); flowers medium salmon-pink, patio, small, semi-dbl., slight fragrance; foliage small, medium green, semi-glossy; bushy growth; [Anne Cocker X Wee Man]; Cocker, James; Cocker & Sons

'COCANEILA', Min, m, 1982; ('COCANELIA', **Pink Posy**); flowers lilac, small, dbl., moderate fragrance; foliage dark, matt; bushy growth; [Trier X New Penny]; Cocker, James; Cocker & Sons

'COCANELIA', Min, m, 1982; ('COCANELIA', **Pink Posy**); flowers lilac, small, dbl., moderate fragrance; foliage dark, matt; bushy growth; [Trier X New Penny]; Cocker, James; Cocker & Sons

'COCAPEER', HT, mp, 1980; (**Sweetheart**); bud ovoid; flowers medium pink, yellow base, blooms borne singly, dbl., 52 petals, exhibition form, moderate fragrance; foliage large, medium green; vigorous, upright growth; Fragrance Award, Belfast, 1982; [Peer Gynt X (Fragrant Cloud X Gay Gordons)]; Cocker, James

COCaquil, F, dr, 1984; (**Cockadoo**); Cocker

'COCARDE JAUNE', HT, pb, 1933; bud large, long, pointed, reddish salmon; flowers yellowish salmon, base yellow, reverse coral-red, dbl., cupped, moderate fragrance; foliage dark; vigorous, bushy growth; [Diana X Marie Adélaide]; Ketten Bros.

'COCARDE PALE', HGal, 1820; Pradel, H.; (Cavriglia)

'COCARED', HT, ab, 1982; (**Roddy MacMillan**); flowers large, dbl., 35 petals, moderate fragrance; foliage medium, medium green, semi-glossy; bushy growth; [(Fragrant Cloud X Postillion) X Wisbech Gold]; Cocker, James; Cocker & Sons

'COCARUM', Min, ob, 1982; (**Cha Cha®**); flowers orange-red with yellow eye, dbl., slight fragrance; foliage small, medium green, semi-glossy; bushy growth; [(Wee Man X Manx Queen) X Darling Flame]; Cocker, James; Cocker & Sons, 1983

'COCASUN', HT, ab, 1981; (**Sunset Song**); bud pointed; flowers golden amber, borne 1–5 per cluster, dbl., 46 petals, exhibition form, slight fragrance; beak-shaped, red-brown prickles; foliage large, glossy, light olive green; upright growth; [(Sabine X Circus) X Sunblest]; Cocker, James

'COCATHES', F, mp, 1980; (**Crathes Castle**); 18 petals, borne 12–15 per cluster, moderate fragrance; triangular prickles; foliage large, dark, glossy; rounded, bushy growth; [Dreamland X Topsi]; Cocker, James

'COCAVOTER', F, ob, 1983; (**Sweet Nell**); flowers orange, large, dbl., 35 petals, slight fragrance; foliage medium, medium green, semi-glossy; upright growth; [Anne Cocker X ((Mischief X ((Sabine X Circus) X (Tropicana X Circus)))]; Cocker, Alexander M.; Cocker & Sons, 1984

'COCBADEN', HT, op, 1985; (**Doctor Dick**, Dr Dick); flowers orange-coral, large, dbl., exhibition form, slight fragrance; foliage large, medium green, matt; upright growth; [Fragrant Cloud X Corso]; Cocker, James & Sons, 1986

'COCBAMBER', HT, pb, 1985; (Amberlight, **Fyvie Castle**); flowers light apricot, amber and pink blend, well-formed, large, dbl., 35 petals, moderate fragrance; foliage large, medium green, semi-glossy; upright growth; Gold Star of the South Pacific, Palmerston North, NZ, 1985; [(Sunblest X (Sabine X Dr A.J. Verhage)) X Silver Jubilee]; Cocker, Alexander M.; Cocker & Sons

'COCBAY', HT, mp, 1982; (**May Lyon**); flowers large, dbl., 35 petals, slight fragrance; foliage large, medium green, glossy; bushy growth; [((Anne Cocker X Arthur Bell) X National Trust) X Silver Jubilee]; Cocker, James; Cocker & Sons, 1983

'COCBLANCO', Min, w, 1983; (**Bianco**); flowers patio, small, dbl., 35 petals, no fragrance; foliage medium, medium green, semi-glossy; bushy growth; [Darling Flame X Jack Frost]; Cocker, James; Cocker & Sons

'COCBONNE', HT, or, 1989; (**Ena Baxter**); bud pointed; flowers salmon pink, reverse salmon red, medium, borne in sprays of 5, dbl., 26 petals, exhibition form, slight fragrance; urn-shaped, large, brown fruit; prickles triangular, average, green; foliage large, medium green, glossy; bushy, medium growth; GM, Glasgow, 1994; [HARkrispin X Silver Jubilee]; Cocker, James & Sons

'COCBROSE', HT, dp, 1983; (**Abbeyfield Rose**); flowers rose red, large, dbl., 35 petals, slight fragrance; foliage medium, medium green, semi-glossy; bushy growth; Golden Prize, Glasgow, 1990; [National Trust X Silver Jubilee]; Cocker, A.; Cocker & Sons, 1985

'COCCAGES', HT, ob, 1988; (Country Heritage, **Our Jubilee**); flowers medium, dbl., 20 petals, slight fragrance; foliage medium, medium green, semi-glossy; upright growth; [Yellow Pages X Silver Jubilee]; Cocker, James & Sons, 1986

'COCCELESTE', F, mp, 1988; (**Rosabell**); flowers medium, dbl., slight fragrance; foliage medium, medium green, semi-glossy; patio; bushy growth; [(National Trust X Wee Man) X Darling Flame]; Cocker, James & Sons, 1986

'COCCINELLE', LCl, or, 1956; flowers bright geranium-red, Buyl Frères

'COCCLARE', S, my, 1989; (**Ray of Sunshine**); bud pointed; flowers small, borne in sprays of 3–9, semi-dbl., 15 petals, cupped, slight, spicy fragrance; round, small, green fruit; prickles small, green; foliage small, dark green, glossy; bushy, low growth; [Sunsprite X (Clare Grammerstorf X Frühlingsmorgen)]; Cocker, James & Sons, 1988

'COCCLARION', HT, op, 1992; (**Laura Anne**); flowers pink flushed orange, blooms borne in large clusters, dbl., 26–40 petals, 3–3.5 in., moderate fragrance; some prickles; foliage large, medium green, glossy; bushy (76 cms) growth; [((Sabine X Circus) X Maxi) X Harrinyl]; Cocker, James; James Cocker & Sons, 1990

COCcopna, F, op, 1982; (**Coppa Nob**); Cocker

'COCCORD', Min, or, 1982; (**Little Prince**); flowers orange-red, yellow eye, patio, small, semi-dbl., slight fragrance; foliage medium, medium green, semi-glossy; upright, bushy growth; [Darling Flame X (National Trust X Wee Man)]; Cocker, James; Cocker & Sons, 1983

'COCCRAZY', F, r, 1988; (**Gingernut**); flowers medium, dbl., 43 petals, moderate fragrance; foliage small, medium green, semi-glossy; patio; bushy growth; [(Sabine X Circus) X Darling Flame]; Cocker, James & Sons, 1989

'COCdana', HT, yb, 1988; (**Fulton MacKay**, Maribel); flowers large, dbl., 20 petals, moderate fragrance; foliage large, medium green, glossy; bushy growth; Golden Prize, Glasgow, 1992; [Silver Jubilee X Jana]; Cocker, James, 1989

'COCdandy', HT, ob, 1988; (**Royal Volunteer**); flowers orange-red and yellow blended, medium, dbl., 26–40 petals, slight fragrance; foliage medium, light green, semi-glossy; upright growth; [Yellow Pages X Alexander]; Cocker, James & Sons

'COCdapple', Min, lp, 1985; (**Scottish Special**); flowers patio, large, semi-dbl., slight fragrance; foliage small, medium green, semi-glossy; bushy growth; [Wee Man X Darling Flame]; Cocker, Ann G., 1987

'COCdarlee', Min, op, 1986; (**Coral Reef**); flowers orange, patio, medium, semi-dbl., slight fragrance; foliage small, medium green, glossy; bushy growth; [(Darling Flame X St. Albans) X Silver Jubilee]; Cocker, Ann G.; Cocker & Sons

'COCdazzle', F, or, 1988; (**Clydebank Centenary**); flowers orange-vermilion red, medium, dbl., 15–25 fimbriated petals, slight fragrance; foliage medium, medium green, matt; upright growth; [((Highlight X Colour Wonder) X (Parkdirector Riggers X Piccadilly)) X Darling Flame]; Cocker, James & Sons, 1987

'COCdestin', HT, ob, 1984; (**Remember Me**®); flowers orange and yellow blend, large, dbl., 20 petals, slight fragrance; foliage small, dark, glossy; bushy, spreading growth; [Ann Letts X (Dainty Maid X Pink Favorite)]; Cocker

'COCdimity', Min, ab, 1992; (**Claire Scotland**); flowers light apricot pink, blooms borne in small clusters, dbl., 15–25 petals, 1.5–2.75 in., slight fragrance; some prickles; foliage medium, medium green, semi-glossy; patio; low (50–80 cms), bushy growth; [(National Trust X Wee Man) X Darling Flame]; Cocker, James; James Cocker & Sons, 1990

'COCdimple', Min, pb, 1986; (**Conservation**); flowers patio, medium, semi-dbl., slight fragrance; foliage small, light green, semi-glossy; bushy growth; GM, Dublin, 1986; [((Sabine X Circus) X Maxi) X Darling Flame]; Cocker, James & Sons, 1988

'COCdinkum', Min, mr, 1983; (**Boys' Brigade**®); flowers medium, patio, blooms in clusters, single, 5 petals, no fragrance; foliage small, medium green, semi-glossy; bushy growth; [(Darling Flame X Saint Alban) X (Little Flirt X Marlena)]; Cocker, Ann G.; Cocker & Sons, 1984

COCember, HT, or, 1991; (**Roxburghe Rose**); Cocker

'COCflag', Min, mr, 1989; (**Highland Laddie**); bud pointed; flowers scarlet red, medium, borne in sprays of 5–11, dbl., 19 petals, cupped, slight fragrance; urn-shaped, medium, brown fruit; prickles medium, green; foliage medium, medium green, glossy; upright growth; [National Trust X Dainty Dinah]; Cocker, James & Sons

COCfoster, S, mr, 1988; (**Regal Red**); Cocker

'COCglen', F, yb, 1989; (**Honey Bunch**®, Honeybunch); bud ovoid; flowers yellow with salmon-red, reverse yellow, aging honey-yellow, dbl., 45 petals, cupped, moderate fragrance; round, small, green fruit; prickles small, green; foliage small, dark green, glossy; patio; bushy, low growth; [[(Sabine X Circus) X Maxi] X Bright Smile]; Cocker, James & Sons

'COCgold', F, dy, 1992; (Dania, **Toprose**); flowers bright yellow, blooms borne in large clusters, dbl., 26–40 petals, 3–3.5 in., slight fragrance; some prickles; foliage large, medium green, glossy; medium (75 cms), upright growth; GM, Baden-Baden, 1987; [((Chinatown X Golden Masterpiece) X Adolph Horstmann) X Yellow Pages]; Cocker, James; James Cocker & Sons, 1988

'COCgrand', HT, lp, 1992; (**Myriam**®); very dbl., 3–3.5 in., borne mostly singly, intense fragrance; some prickles; foliage large, medium green, semi-glossy; medium (75 cms), upright growth; [Typhoo Tea X Grandpa Dickson]; Cocker, James; James Cocker & Sons, 1991

'COCharod', F, mr, 1993; (**Shirley Spain**); flowers glowing russet red, blooms borne in small clusters, dbl., 15–25 petals, 1.5–2.75 in., slight fragrance; some prickles; foliage medium, dark green, glossy; medium, compact growth; [Seedling X Roddy McMillan]; Cocker; James Cocker & Sons, 1992

'COChello', Min, mr, 1992; (**Hello**); flowers crimson with white eye, blooms borne in large clusters, semi-dbl., 6–14 petals, 1.5–2.75 in., slight fragrance; some prickles; foliage medium, medium green, semi-glossy; patio; medium (50–80 cms), bushy growth; Golden Prize, Glasgow, 1993; [Darling Flame X Seedling]; Cocker, James; James Cocker & Sons, 1991

'Cochineal Glory', HT, mr, 1937; bud pointed; flowers open, large, semi-dbl.; vigorous, bushy growth; Leenders, M.

'Cochineal Glory, Climbing', Cl HT, mr, 1945; flowers large, semi-dbl.; Vogel, M.; (Sangerhausen)

'COChunter', HT, or, 1992; ('COChuster', **Ohshima Rose**); dbl., 26–40 petals, 1.5–2.75 in., borne mostly singly, moderate fragrance; some prickles; foliage medium, medium green, matt; tall (91.44 cms), upright growth; [(National Trust X Alexander) X Red Planet]; Cocker, James; James Cocker & Sons, 1991

'COChuster', HT, or, 1992; ('COChunter', **Ohshima Rose**); dbl., 26–40 petals, 1.5–2.75 in., borne mostly singly, moderate fragrance; some prickles; foliage medium, medium green, matt; tall (91.44 cms), upright growth; [(National Trust X Alexander) X Red Planet]; Cocker, James; James Cocker & Sons, 1991

'COCjabby', F, op, 1992; (**Gordon's College**); flowers coral salmon, blooms borne in large clusters, dbl., 26–40 petals, 3–3.5 in., moderate fragrance; some prickles; foliage large, dark green, glossy, purplish when young; medium, upright growth; [Abbeyfield Rose X Roddy McMillan]; James Cocker & Sons, Ltd.

COCjojo, F, ob; (**UNICEF**); Cocker, 1993

'COCjolly', F, dr, 1993; (**Bishop Elphinstone**); flowers crimson, blooms borne in large clusters, dbl., 26–40 petals, 1.5–2.75 in., slight fragrance; some prickles; foliage large, dark green, semi-glossy; tall (3 ft), upright growth; [Seedling X (Tropicana X Baccara)]; Cocker, James; James Cocker & Sons, 1994

Cockadoo, *see* COCaquil

Cockle Shells, *see* 'LEOcok'

Cocktail®, *see* 'MEImick'

Cocktail '80, *see* 'MEItabifob'

'COClager', F, ab, 1997; (**William Quarrier**); flowers full, medium, very dbl., 26–40 petals, borne in large clusters, slight fragrance; some prickles; foliage medium, light green, dull; compact, tall (2.5–3 ft) growth; [Silver Jubilee X Geraldine]; Cocker, Ann G.

'COClament', F, mr, 1998; (**Hope 98**); flowers sparkling geranium red, lighter reverse, frilly petals, dbl., 15–25 petals, 1–1.5 in., borne in large clusters, slight fragrance; prickles moderate, small, slightly hooked; foliage small, medium, glossy; compact, low, bushy growth; [Memento X Evelyn Fison]; Cocker, Ann G.; James Cocker & Sons, 1998

'COClands', HT, lp, 1992; (**Scotland's Trust**); flowers light pink, silver reverse, moderately well shaped lg. blooms, dbl., 15–25 petals, 1.5–2.75 in., intense fragrance; some prickles; foliage medium, medium green, matt; medium (16–20 cms), upright growth; [Sunblest X Prima Ballerina]; Cocker, James & Sons

'COClent', F, ab, 1993; (**Lady MacRobert**); flowers light apricot, blooms borne in large clusters, dbl., 15–25 petals, 1.5–2.75 in., slight fragrance; some prickles; foliage medium, light green, semi-glossy; medium ($2\frac{1}{2}$ ft), upright, compact growth; [Clydebank Centenary X Seedling]; Cocker; James Cocker & Sons, 1993

'COClibee', F, op, 1996; (**Alison**); flowers peach salmon, 3–3.5 in., borne in large clusters, moderate fragrance; some prickles; foliage medium, medium green, glossy; medium, upright, bushy growth; [Silver Jubilee X (Sabine X Circus)]; Cocker, James & Sons; James Cocker & Sons, 1995

COClion, HT, op, 1997; (**Press and Journal**); Cocker

'COCLUCID', F, ob, 1992; (**Castle of Mey**); flowers orange gold, blooms borne in small clusters, dbl., 15–25 petals, 1.5–2.75 in., moderate fragrance; some prickles; foliage medium, dark green, semi-glossy; medium, bushy growth; [Anne Cocker X (Yellow Pages X Silver Jubilee)]; Cocker, James; James Cocker & Sons

'COCMARRIS', F, or, 1996; (**Wee Cracker**); flowers orange vermilion, blooms borne in large clusters, dbl., 15–25 petals, 1.5–2.75 in., slight fragrance; some prickles; foliage small, medium green, semi-glossy; low (46–60 cms), bushy, compact growth; [Len Turner X Jean Thomson Harris]; Cocker, James & Sons; James Cocker & Sons, 1995

COCmystery, F, ob, 1994; (**Aberdeen Celebration**); Cocker

'COCNANNE', F, pb, 1993; (**Friend for Life**); flowers pink, semi-double, blooms borne in large clusters, 6–14 petals, 1.5–2.75 in., slight fragrance; some prickles; foliage medium, dark green, glossy; medium ($2\frac{1}{2}$ ft), bushy, compact growth; [Seedling X (Anne Cocker X Silver Jubilee)]; Cocker; James Cocker & Sons, 1994

'COCNEEL', F, dp, 1997; (**Ballindalloch Castle**); flowers full, medium, very dbl., 26–40 petals, borne in large clusters, slight fragrance; some prickles; foliage medium, medium green, glossy; bushy, medium (2.5 ft) growth; [(Anne Cocker X Maxi) X Silver Jubilee]; Cocker, Ann G.

'COCNEST', F, ab, 1993; (**Constance Fettes**); dbl., 26–40 petals, 1.5–2.75 in., borne in small clusters, moderate fragrance; some prickles; foliage medium, medium green, glossy; medium ($2\frac{1}{2}$ ft), upright, bushy growth; [(Fragrant Cloud X Alexander) X Sunblest]; Cocker; James Cocker & Sons, 1993

COCnilly, F, mr, 1996; (**Ray of Hope**); Cocker

'Coco', F, yb, 1975; flowers deep golden yellow to orange-pink, dbl., 20–25 petals, 3 in., moderate fragrance; foliage glossy, bright; vigorous, free growth; [Pernille Poulsen X Redgold]; Fryer, Gareth; Fryer's Nursery, Ltd.

Coco, *see* KORferse

COCocrust, F, mr, 1996; (**Alex C. Collie**); Cocker

'COCONUT ICE', HT, pb, 1991; bud large, pointed; flowers pink blend with mauve shadings, blooms borne 1–3 per stem, dbl., exhibition form, slight fragrance; foliage medium green, semi-glossy; tall, vigorous growth; [Alexander X Vol de Nuit]; Walker, D.R., 1981

COCopher, HT, mr, 1993; (**Christopher**); Cocker

'COCORAY', F, ab, 1997; (**Innocence 96**); flowers semi double, small, semi-dbl., 8–14 petals, borne in small clusters, slight fragrance; few prickles; foliage small, medium green, glossy; bushy, low (2.5 ft) growth; [Clydebank Centenary X Ray of Sunshine]; Cocker, Ann G.

'COCORICO', F, or, 1951; bud pointed; flowers geranium-red, blooms in clusters, single, 8 petals, 3 in.; moderate, spicy fragrance; foliage glossy, bright; vigorous, upright, bushy growth; (28); GM, Geneva, 1951; GM, NRB, 1951; [Alain X Orange Triumph]; Meilland, F.; C-P, 1953;, URS, 1951

Cocorico®, *see* MEIlasso

'COCORICO, CLIMBING', Cl F, or, 1964; Ruston, D.

'COCORONA', F, ab, 1997; (**Heartbeat 96**); flowers double, medium, dbl., 15–25 petals, slight fragrance; some prickles; foliage medium, dark green, glossy; upright, medium (2.5 ft) growth; Cocker, Ann G.

COCORUST, F, lp; (**Alex C Collie**); Cocker

'COCOSIMBER', HT, rb, 1998; (**President Heldar Aliyev**); flowers vibrant salmon red, lighter reverse, very full, very dbl., 41 petals, 2–2.5 in., borne singly or in large clusters, moderate fragrance; prickles moderate, medium, slightly hooked; foliage large, dark green, glossy; upright, medium growth; [Silver Jubilee X Remember Me]; Cocker, Ann G.; James Cocker & Sons, 1999

'COCOTTE', HT, mp, 1958; flowers bright salmon, large, dbl., moderate fragrance; foliage bronze; upright growth; [Peace X (Fashion X Vogue)]; Gaujard

COCover, HT, ab, 1997; (**Gwen Major**); flowers full, large, very dbl., 26–40 petals, borne mostly singly or in large clusters, moderate fragrance; some prickles; foliage medium, dark green, glossy; bushy, upright, medium (2.5 ft) growth; [Silver Jubilee X Remember Me]; Cocker, Ann G.

'COCQUAMBER', HT, r, 1998; (**Home of Time**); flowers cinnamon red with bronze shading, lighter reverse, very dbl., 26–40 petals, 2 in., mostly borne singly, slight fragrance; prickles moderate, medium, slightly hooked; foliage medium, dark green, glossy; upright, medium ($2\frac{1}{2}$ ft) growth; [(Sabine X Circus) X Amber Queen]; Cocker, Ann G.; James Cocker & Sons, 1998

'COCQUATION', F, lp, 1999; (**Hi Society**); flowers lilac pink, reverse lighter, dbl., 17–25 petals, 1 in., borne in large clusters, slight fragrance; prickles moderate; foliage small, light green, glossy; upright, bushy, medium (2–2.5 ft) growth; [Conservation X (Chanelle X (Golden Masterpiece X Adolf Horstmann))]; Cocker, A.; James Cocker & Sons, 1999

'COCQUETRUM', F, op, 1998; (**Rose 2000**, Rose 2000, Rose Two Thousand); flowers coral vermilion, lighter reverse, semi-dbl., 8–14 petals, .5–1.5 in., borne in large clusters, slight fragrance; prickles moderate amount, small, slightly hooked; foliage medium, dark green, glossy; bushy, low, rounded growth; [Trumpeter X Clydebank Centenary]; Cocker, Ann G.; James Cocker & Sons, 1998

'COCQUIRIAM', F, w, 1998; (**White Gold**, White Gold 98, Whitegold); flowers white with golden yellow center, reverse same, opens wide and flat, very dbl., 26–40 petals, 3 in.; borne in large clusters, moderate fragrance; prickles moderate, medium, slightly hooked; foliage large, medium green, semi-glossy; tall, bushy growth; [Morning Jewel X Myriam]; Cocker, Ann G.; James Cocker & Sons, 1998

'COCREDWARD', F, lp, 1997; (**Marguerite Anne**); flowers full, medium, very dbl., 26–40 petals, borne in large clusters, moderate fragrance; foliage medium, dark green, glossy; bushy, broad, medium (2.5 ft) growth; [Anisley Dickson X Roddy McMillan]; Cocker, Ann G.

'COCTY', HT, rb, 1971; (**Curiosity**); flowers scarlet, reverse gold, dbl., 35 petals, 4 in., cupped, slight fragrance; foliage variegated (only rose to date with this property); [Cleopatra sport]; Cocker

Coed™, *see* 'JELLO'

Coeur d'Alene, HRg, mp, 1996

Coeur d'Amour, *see* 'DICAM'

Coeur Farouche®, *see* DELfrei

Coffee Country, *see* 'VIRBROWN'

'COGAMO', LCl, pb, 1996; (Mon Amour, **My Love**™); flowers phlox pink, reverse coral pink, cream pink at base, blooms b, dbl., 26–40 petals, 4.5 in., moderate fragrance; few prickles; foliage medium, dark green, semi-glossy to glossy; spreading, tall (10 ft) growth; [Caprice X Peace, Cl.]; Coggiatti, Stelvio

'COGNAC', F, ab, 1956; flowers apricot, reverse darker, stamens dark amber, large, dbl., borne in small clusters, foliage glossy, dark olive-green; moderate growth; [Alpine Glow X Mrs Pierre S. duPont]; Tantau, Math.

'COGNAC, CLIMBING', Cl F, ab, 1962; Kordes

'COIMBRA', HT, dp, 1953; flowers cerise-pink, large, dbl., exhibition form, intense fragrance; [Heinrich Wendland X Crimson Glory]; da Silva, Moreira

'COISTOR', S, w, 1999; (**Hailstorm**™); semi-dbl., 12–16 petals, 1 in., borne in large clusters, slight fragrance; prickles moderate; foliage small, medium green, dull; upright, tall (3 ft) growth; [seedling X seedling]; Coiner, Jim; Coiner Nursery, 2000

Colbert®, *see* 'DELCOLB'

'COLCESTRIA', Cl HT, pb, 1916; flowers rose to silver-pink, petals reflexed, large, dbl., intense fragrance; foliage light; Cant, B. R.

Colchester Beauty, *see* 'CANSEND'

'COLCHESTER GAZETTE', F, mr, 1972; flowers bright red, rosette form, dbl., 40–50 petals, 2 in., slight fragrance; foliage light; very free growth; [Evelyn Fison X Etendard]; Cants of Colchester, Ltd.

'COLERAINE', HT, lp, 1970; flowers pale pink, classic form,, dbl., 49 petals, 3.5 in., slight fragrance; foliage light; free growth; [Paddy McGredy X (Mme Léon Cuny X Columbine)]; McGredy, Sam IV; McGredy

'COLE'S PINK LAFAYETTE', F, mp, 1930; flowers rose-pink, semi-dbl.; [Lafayette sport]; Cole Nursery Co.

'COLETTA MONTANELLI', HT, dy, 1975; bud globular; flowers deep yellow, dbl., 32 petals, 4 in., exhibition form, slight fragrance; foliage dark, soft; vigorous, upright, bushy growth; [Seedling X Peer Gynt]; Kordes; Barni

'COLETTE', N, pb, 1932; flowers center nankeen yellow, shaded salmon-pink, large; foliage glossy, light; very vigorous growth; RULED EXTINCT 7/90; [William Allen Richardson X Mme Laurette Messimy]; Schwartz, A.

Colette, *see* 'JACCOL'

Colette(TM), *see* MEIroupis

'COLETTE BERGES', HT, mr, 1940; flowers crimson, medium, dbl., 45 petals, exhibition form, moderate, damask fragrance; vigorous growth; [Red Columbia X Ami Quinard]; Dot, Pedro

'COLETTE CLÉMENTE', HT, ob, 1932; flowers reddish orange, medium, semi-dbl., slight fragrance; foliage glossy, dark; vigorous growth; [(Mme Mélanie Soupert X Mme Edouard Herriot) X (Mrs Edward Powell X R. foetida bicolor)]; Mallerin, C.; C-P

'COLETTE JELOT', F, yb, 1942; flowers amber-yellow shaded currant-red, open, medium, semi-dbl., borne in clusters, slight fragrance; foliage soft; dwarf growth; [Ampere X ((Charles P. Kilham X (Charles P. Kilham X Capucine Chambard))]; Meilland, F.; A. Meilland

Colette John Keats(TM), *see* 'MEIROUPIS'

'COLFRAGRASAR', Cl HT, or, 1973; (**Fragrant Cloud, Climbing**, Nuage Parfume, Climbing); Collin, W.C.; W.H. Collin & Sons

Colibre, *see* 'MEIMAL'

Colibre 79®, *see* 'MEIDANOVER'

Colibri, *see* 'MEIMAL'

'COLIBRI', Pol, ly, 1898; flowers soft yellow, fading, Lille

Colibri 80, Min, 1980; Meilland, Alain A.; (Cavriglia)

'COLIN KELLY', HT, dp, 1945; bud pointed; flowers cerise-red, very large, dbl., intense fragrance; foliage leathery; upright, bushy growth; [E.G. Hill X The Queen Alexandra Rose]; Krebs; Marsh's Nursery

'COLIN KELLY, CLIMBING', Cl HT, dp; Marsh's Nursery

Colinda, *see* JACredi

Colin's Salmon, HT, op; [Bel Ange sport]; Bell, Ronald J., 1970; (Weatherly, L.)

'COLISÉE', F, mr, 1965; (Colysée); bud pointed; flowers coppery pink, open, medium, semi-dbl.; foliage dark, glossy; very vigorous, bushy growth; [Atlantic X Circus]; Gaujard; Ilgenfritz Nursery

'COLLEEN', HT, mp, 1914; bud high pointed; flowers bright rose shaded rose-pink, large, dbl., moderate fragrance; vigorous growth; GM, NRS, 1913; McGredy

Colleen Little, *see* 'LEECAL'

'COLLEEN MOORE', HT, dr, 1944; flowers velvety carmine, open, semi-dbl., 15–25 petals, 4–4.5 in., intense fragrance; foliage leathery, dark; strong stems; vigorous, upright growth; [((De Luxe X Senior) X Premier) X Chieftain]; Hill, Joseph H., Co.; Bosley Nursery

College Avenue #10, HT, yb

College Avenue Columbia, HT, dp

Collegiate 110, *see* DICknowall

Collegiate Pride, HT, mr, 1989; Bell

Cologne, *see* **'KÖLN AM RHEIN'**

Cologne, *see* 'MACSUPBOW'

Cologne Carnival, *see* 'KORGI'

Coloma's Gold, Min, yb, 1998; [Work of Art sport]; Fischer

Colombe, HT, 1976; Croix, P.; (Cavriglia)

Colombina®, HT, pb, 1985; Barni, V.

Colombine, *see* **'COLUMBINE'**

'COLONEL CAMPBELL WATSON', HT, mp, 1936; bud pointed; flowers salmon-pink, dbl., exhibition form; foliage soft; long stem; vigorous growth; GM, NRS, 1935; [Joan Howarth X Portadown]; Bees

'COLONEL DAZIER', HT, w, 1927; flowers rosy white, reverse bright rose, base golden yellow, dbl., intense fragrance; [Le Progrés X Jonkheer J.L. Mock]; Ketten Bros.

'COLONEL DE SANSAL', HP, 1875; Jamain, H.; (Cavriglia)

'COLONEL FOISSY', HP, mp, 1849; flowers medium, dbl.; Margottin; (Sangerhausen)

'COLONEL GRAVEREAUX', HT, op, 1940; flowers salmon-coral, reverse yellow; strong stem; vigorous, upright growth; Mallerin, C.; A. Meilland

'COLONEL JOFFÉ', T, mr, 1893; flowers purplish red, petals wrinkled, flat; Liabaud

'COLONEL LECLERC', HT, mp, 1909; flowers tyrian pink, dbl., moderate fragrance; [Mme Caroline Testout X Horace Vernet]; Pernet-Ducher

'COLONEL LINDBERGH', Pol, op, 1928; flowers salmon-orange, [Juliana Rose sport]; Ouden, Den

'COLONEL NICOLAS MEYER', HT, mr, 1934; bud pointed; flowers brilliant velvety red, open, dbl., moderate fragrance; foliage leathery; strong stems; vigorous, bushy growth; [La Maréchale Petain X Edouard Mignot]; Sauvageot, H.

'COLONEL OSWALD FITZGERALD', HT, dr, 1917; flowers dark velvety crimson, well formed, dbl.; vigorous, branching growth; Dickson, A.

'COLONEL R. S. WILLIAMSON', HT, lp, 1907; flowers white, center deep blush, large, dbl., exhibition form, moderate fragrance; foliage glossy, dark; vigorous, open growth; Dickson, A.

'COLONEL ROBERT LEFORT', M, mr; flowers purple-red; Verdier, E., 1884

'COLONEL SHARMAN-CRAWFORD', HT, mr, 1933; flowers rich velvety crimson, large, dbl., exhibition form, intense fragrance; foliage leathery; long, strong stems; vigorous, bushy growth; GM, NRS, 1931; Dickson, A.

Colonel Svec, *see* **'PLUKOVNIK SVEC'**

Colonia, LCl, mr, 1988; flowers large, semi-dbl.; Meilland; (Sangerhausen)

'COLONIAL WHITE', LCl, w, 1959; bud ovoid; dbl., 3 in., flat, recurrent bloom, moderate fragrance; foliage light green; vigorous growth; [New Dawn X Mme Hardy]; Wyant

'COLONIAL WHITE', T, lp

'COLOR GIRL', F, pb, 1966; bud ovoid; flowers whitish, edged red, medium, dbl., exhibition form, intense fragrance; foliage dark, leathery; bushy, low growth; [Little Darling X Cocorico]; Fuller; Wyant

Color Guard, *see* 'JOLCOL'

Color Magic, *see* 'JACMAG'

Color Purple, *see* 'CLEPURP'

Colorado, HT, 1970; Combe, M.; (Cavriglia)

Colorado, F, mp; Select Roses, B.V.

Colorama, *see* 'MEIRIGALU'

'COLORANJA', HT, or, 1963; (Mustang); bud ovoid; flowers large to medium, dbl., exhibition form; foliage dark, leathery; vigorous, bushy growth; [Fandango X (Independence X Papillon Rose)]; Lens

Colorbreak, *see* 'MACULTRA'

Colorburst, *see* TWOcherish

'COLOSO', HT, mp, 1962; dbl., 50 petals, moderate fragrance; very vigorous growth; [Chrysler Imperial X (Peace X Queen Elizabeth)]; Dot, Simon

Colossus(TM), *see* 'PERSUS'

'COLOUR CARNIVAL', F, yb, 1962; flowers primrose-yellow edged pink, well formed, borne in clusters, dbl., 50 petals, 3 in., slight fragrance; vigorous, low, bushy growth; LeGrice; Wayside Gardens Co.

'COLOUR GLOW', F, ob, 1969; flowers orange, reddening to flame,, dbl., 36 petals, 3.5 in., globular, slight fragrance; vigorous growth; [Tropicana X Masquerade]; Butter; Wood End Gardens

Colour Magic, MinFl, pb, 1999

Colour Parade, *see* 'LAVCOAT'

'COLOUR SERGEANT', F, or, 1972; dbl., 20 petals, 4 in., slight fragrance; foliage glossy; [Queen Elizabeth X (Ann Elizabeth X Circus)]; Harkness

Colour Wonder, *see* 'KORBICO'

Colourama, *see* 'MEIRIGALU'

'COLUMBIA', HT, mp, 1916; bud long, pointed; flowers glistening rose-pink, large, dbl., 65 petals, intense fragrance; foliage dark; vigorous growth; Gertrude M. Hubbard, ARS, 1919 GM, Portland, 1919; [Ophelia X Mrs George Shawyer]; Hill, E.G., Co.

'COLUMBIA', HWich, mp, 1903;, non-recurrent; moderately vigorous growth; [Seedling X Mme Caroline Testout]; Hooper, Bro. & Thomas

'COLUMBIA, CLIMBING', Cl HT, mp, 1923; Vestal, 1923 & Lens, 1929

'COLUMBINE', F, yb, 1956; (Colombine); flowers creamy yellow tinged pink, well-formed blooms in open clusters, intense fragrance; foliage glossy; vigorous growth; [Danish Gold X Frensham]; Poulsen, S.; McGredy

Columbus™, *see* 'WEKUZ'

'COLUMBUS QUEEN', HT, pb, 1962; bud ovoid, pointed; flowers light pink, reverse darker, dbl., 24–30 petals, 4 in., exhibition form, slight fragrance; foliage leathery, dark; vigorous, upright growth; GM, Geneva, 1961; [La Jolla X Seedling]; Armstrong, D.L.; Armstrong Nursery

'COLUMELLE', HGal, 1860; Moreau et Robert; (Cavriglia)

Colysée, *see* **'COLISÉE'**

'COMANCHE', Gr, or, 1968; bud pointed; flowers medium, dbl., exhibition form, slight fragrance; foliage leathery; vigorous, upright, bushy growth; AARS, 1969; [Spartan X (Carrousel X Happiness)]; Swim & Weeks; C-P

Come Lei, HT, 1968; Giacomasso; (Cavriglia)

Comedie, F, 1977; Combe, M.; (Cavriglia)

'COMENDADOR NOGUEIRA DA SILVA', HT, rb, 1961; flowers dark red, reverse silvery; [Confidence X Independence]; da Silva, Moreira

'COMET', Cl F, pb, 1934; flowers flesh pink, shades salmon-yellow, large, very dbl., slight fragrance; foliage leathery; short strong stem; [Gruss an Aachen sport]; Mesman; Bosley Nursery

Comice de Tarn-et-Garonne, B, mr; flowers carmine red, intense fragrance; Pradel, 1852

'COMMAND PERFORMANCE', HT, or, 1970; bud ovoid; flowers medium, dbl., exhibition form, intense fragrance; foliage leathery; vigorous, tall, bushy growth; AARS, 1971; [Tropicana X Hawaii]; Lindquist; Howard Rose Co.

'COMMANDANT BEAUREPAIRE', B, pb, 1874; flowers bright rose-pink streaked (striped) purple-violet and marbled white, dbl., occassional repeats, moderate fragrance; foliage light green; vigorous growth; Moreau et Robert

Commandant Cousteau, *see* ADharman

'COMMANDANT FÉLIX FAURE', HP, mr, 1901; flowers light crimson-red, tinted vermilion, dbl., 25 petals, cupped, moderate fragrance; vigorous, upright growth; Boutigny

'COMMANDANT L. BARTRE', HT, rb, 1920; flowers dark carmine-red, tinted brilliant pink, dbl.; vigorous growth; [Lady Ashtown X Louis van Houtte]; Schwartz, A.

'COMMANDANT LETOURNEUX', HT, dp, 1903; flowers carmine-pink, very large, very dbl., intense fragrance; Bahaud; (Sangerhausen)

'COMMANDATORE FRANCESCO INGEGNOLI', Cl HT, or, 1923; flowers geranium-red, Ingegnoli

'COMMANDEUR JULES GRAVEREAUX', HP, mr, 1908; bud pointed; flowers dazzling red, center shaded maroon, large, peony-like, dbl., intense fragrance; vigorous growth; [Frau Karl Druschki X Liberty]; Croibier

'COMMANDEUR JULES GRAVEREAUX, CLIMBING', Cl HP, mr, 1925; Belouet; Grandes Roseraies

'COMMANDO', HT, op, 1945; bud long, pointed; flowers orange-buff, suffused pink, camellia form, dbl., 30–35 petals, 5 in., moderate fragrance; foliage leathery, glossy; upright, bushy growth; [Mrs J.D. Eisele X Glowing Sunset]; Howard, F.H.; H&S

Commitment, *see* CLEment

Common Blush China, *see* **'OLD BLUSH'**

Common Monthly, *see* **'OLD BLUSH'**

Common Moss, *see* **'CENTIFOLIA MUSCOSA'**

Common Moss, *see* **'COMMUNIS'**

'COMMONWEALTH', HT, dp, 1923; bud pointed; flowers deep pink, large, dbl., moderate fragrance; foliage leathery, rich green; [Ophelia X Seedling]; Montgomery Co.; A.N. Pierson

'COMMONWEALTH', F, pb, 1948; (Herzblut); flowers crimson with white eye, borne in clusters, semi-dbl., 10 petals, slight fragrance; foliage leathery; (28); [Col. Nicolas Meyer X Holstein]; Kordes; Morse

Commonwealth Glory, *see* HARclue

COMMUNIS, *see* Centifolia Muscosa

'COMMUNIS', M, mp; (Centifolia Muscosa, Common Moss, Mousseau Ancien, Old Pink Moss, Pink Moss, R. centifolia muscosa); bud mossed; flowers pale rose, very dbl., globular, vigorous growth

Compactilla, HSpn, ly

'COMPASSION'®, LCl, op, 1972; (Belle de Londres); flowers salmon pink shaded apricot, large blooms borne singly or in, dbl., 36 petals, moderate, sweet fragrance; large, reddish prickles; foliage large, dark; medium, bushy growth; Geneva, 1979, Orleans, 1979 ADR, 1976 Edland Fragrance Medal, ARS, 1973 GM, Baden-Baden, 1973; [White Cockade X Prima Ballerina]; Harkness, 1973

Compassionate Friend, *see* 'HARZODIAC'

'COMPLICATA', HGal, pb; flowers deep pink, white eye, yellow stamens, large, single, rampant (to 6 ft) growth

Comptoire des Bourbons, B, mp

'COMRADE', HT, mr, 1948; flowers scarlet-crimson, well formed, medium, moderate fragrance; Dickson, A.

COMsan, HT, w, 1993; (**Cosmos**®); Combe

'COMSOR', HT, dr, 1971; (**Norita**®, Norita-Schwarze Rose®); flowers very deep red, large, dbl., exhibition form, moderate fragrance; foliage dark, leathery; vigorous, bushy growth; [Charles Mallerin X Seedling]; Combe; Kern Rose Nursery, 1966

'COMTE ADRIEN DE GERMINY', HP, mp, 1881; flowers large, dbl.; Lévêque; (Sangerhausen)

Comte Boula de Nanteuil, *see* **'COMTE DE NANTEUIL'**

'COMTE DE BOBRINSKY', HCh, dp, 1849; flowers carmine-pink, medium, dbl.; Marest; (Sangerhausen)

'COMTE DE CHAMBORD', P, pb, 1860; flowers pink tinted lilac, very dbl., flat, intense fragrance; vigorous, erect growth; Robert et Moreau

'COMTE DE FALLOUX', HP, dp; Trouillard, 1860

'COMTE DE FLANDRES', HP, 1881; Lévêque, P.; (Cavriglia)

'COMTE DE MORTEMART', HP, mp, 1880; (Comtesse de Mortemart); flowers clear pink, intense fragrance; Margottin

'COMTE DE NANTEUIL', HGal, m; (Boule de Nanteuil, Comte Boula de Nanteuil); flowers crimson purple, center sometimes fiery crimson, compact, large, dbl., branching growth; Roeser, ca. 1834

'COMTE DE NANTEUIL', HP, pb, 1852; flowers light rose with darker edges, sometimes with a green center, dbl., cupped; vigorous growth; Quétier

Comte de Paris, HP, m; Laffay, M., 1839

Comte de Paris, HP, lp; Verdier, E., 1864

'COMTE DE TORRES', HT, op, 1906; flowers coppery salmon-pink, large, dbl.; Schwartz, A.; (Sangerhausen)

'COMTE F. DE CHAVANAC', HT, mp; flowers peach-blossom-pink, center rosy carmine; [Antoine Rivoire X Zephirine Drouhin]

'COMTE F. DE CHAVANAC, CLIMBING', Cl HT, mp, 1929; Siret-Pernet

Comte Florimond de Bergeyck, HP, pb

'COMTE FOY', HGal, lp; (Comte Foy de Rouen); flowers pale rose, very large, dbl., cupped; Savoureux, 1827

Comte Foy de Rouen, *see* **'COMTE FOY'**

'COMTE FRÈDÈRIC DE THUN-HOHENST', HP, dr, 1880; flowers large, dbl., moderate fragrance; Lévêque; (Sangerhausen)

'COMTE G. DE ROCHEMUR', HT, 1911; flowers bright scarlet, large, dbl., moderate fragrance; [Xavier Olibo X Gruss an Teplitz]; Schwartz, A.

'COMTE RAIMBAUD', HP, mr, 1867; flowers crimson, large; Rolland

COMteri, LCl, 1976; (**Cortege**); Combe, M.; (Cavriglia)

'COMTESSE ANNE DE BRUCE', HT, pb, 1937; bud pointed; flowers coppery pink to nasturtium-red, very large, semi-dbl., cupped, moderate fragrance; foliage glossy; very vigorous growth; [Charles P. Kilham X (Mrs Pierre S. duPont X R. foetida bicolor seedling)]; Mallerin, C.; H. Guillot

'COMTESSE BARDI', T, pb, 1896; flowers reddish fawn; Soupert & Notting

'COMTESSE BEATRIX DE BUISSERET', HT, 1900; flowers very large, dbl., moderate fragrance; Soupert & Notting; (Sangerhausen)

'COMTESSE BRANICKA', HP, lp, 1888; flowers large, dbl.; Lévêque; (Sangerhausen)

Comtesse Brigitte Chandon-Moet, *see* DORfaut

'COMTESSE CAHEN D'ANVERS', HP, mp, 1885; flowers large, dbl., moderate fragrance; Lédéchaux; (Sangerhausen)

'COMTESSE CÉCILE DE CHABRILLANT', HP, pb, 1858; flowers satiny pink, silvery reverse, medium, dbl., globular, moderate fragrance; numerous small, dark prickles; foliage dark, leathery; Marest

'COMTESSE CÉCILE DE FORTON', Cl T, pb, 1916; flowers rose-peach, very large, dbl., moderate fragrance; Nabonnand, G.

Comtesse d'Alcantara, *see* 'KRIBATIS'

'COMTESSE D'ANSEMBOURG', HT, ly, 1918; flowers yellowish white, dbl., intense fragrance; [Étoile de France X Marquise de Sinéty]; Leenders, M.

'COMTESSE DE BARBANTANE', B, lp, 1858; Guillot Père, 1858

Comtesse de Barry, F, 1994; Meilland, Alain A.; (Cavriglia)

Comtesse de Bouchard, *see* **'COMTESSE DE BOUCHAUD'**

'COMTESSE DE BOUCHAUD', N, my, 1890; (Comtesse de Bouchard); flowers saffron yellow, very large; Guillot et Fils

'COMTESSE DE BRESSON', HP, lp, 1873; flowers large, dbl.; Guinoiseau, B.; (Sangerhausen)

'COMTESSE DE CASERTA', T, mr, 1877; flowers coppery-red, large; Nabonnand, G.

'COMTESSE DE CASSAGNE', HT, mp, 1919; flowers coppery rose, shaded bright rose, sometimes entirely yellow, dbl., moderate fragrance; Guillot, M.

'COMTESSE DE CASTILLEJA', HT, op, 1926; bud orange; flowers coral, cupped, intense fragrance; foliage dark; strong stems; very vigorous growth; [(Mme Edouard Herriot X Juliet) X Seedling]; Chambard, M.

'COMTESSE DE CHAPONAY', LCl, lp, 1924; flowers cream-rose to salmon, dbl., intense fragrance; very vigorous growth; [R. gigantea X Mme Hoste]; Nabonnand, G.

'COMTESSE DE FLANDRE', HP, lp, 1878; flowers large, dbl.; Verdier, E.; (Sangerhausen)

'COMTESSE DE FRIGNEUSE', T, my, 1885; flowers large, dbl., moderate fragrance; [Mme Damaizin X Seedling]; Guillot et Fils

'COMTESSE DE GALARD-BÈARN', N, ly, 1894; flowers light yellow with pink tones, large, dbl.; Bernaix, A.; (Sangerhausen)

'COMTESSE DE LA MORANDIÈRE', HT, op, 1929; flowers shrimp-pink, reverse coral-red, dbl., moderate fragrance; Chambard, C.

Comtesse de Labarathe, *see* **'DUCHESSE DE BRABANT'**

'COMTESSE DE LACÉPÈDE', HGal, lp; flowers silvery blush, center sometimes rosy, large, dbl.; moderate growth

'COMTESSE DE LEUSSE', HT, dy, 1878; Nabonnand

'COMTESSE DE MARTEL', HT, mp, 1939; flowers carnation-pink, center coppery, very large, moderate fragrance; very vigorous growth; [Charles P. Kilham X Margaret McGredy]; Meilland, F.

Comtesse de Mortemart, *see* **'COMTE DE MORTEMART'**

'COMTESSE DE MURINAIS', M, w, 1843; (White Moss); flowers flesh, opening white, large, dbl., non-recurrent, moderate fragrance; height 4–5 ft; Vibert

'COMTESSE DE NOGHERA', T, op, 1902; flowers light salmon-pink, very large, very dbl., moderate fragrance; Nabonnand; (Sangerhausen)

Comtesse de Paris, HT, dy, 1993; Briant

'COMTESSE DE POLINGAC', HP, dr, 1862; flowers medium, dbl.; Granger; (Sangerhausen)

'COMTESSE DE ROCQUIGNY', B, w, 1874; flowers salmon-red, Vaurin

Comtesse de Segur, *see* DELtendre

'COMTESSE DE SERENYE', HP, lp, 1874; flowers soft pink, Lacharme, F.

Comtesse de Vezins, Misc OGR, mp

Comtesse Diana, *see* HELlux

'COMTESSE DORIA', M, m, 1854; flowers purple-pink, shaded salmon, heavily mossed; Portemer fils

'COMTESSE D'OXFORD', HP, dp, 1869; (Countess of Oxford); flowers deep pink/red with violet stripes, dbl., globular, recurrent bloom, moderate fragrance; Guillot Père

Comtesse d'Oxford, HP, dp

'COMTESSE DU CAYLA', HCh, ob, 1902; flowers nasturtium-red, tinted orange, semi-dbl., flat, recurrent bloom, moderate fragrance; foliage dark, glossy; vigorous growth; Guillot, P.

'COMTESSE EMMELINE DE GUIGNE', T, pb, 1903; Nabonnand

'COMTESSE FESTECTICS HAMILTON', T, pb, 1897; Nabonnand

'COMTESSE FRESSINET DE BELANGER', HP, mp, 1886; flowers rose; Lévêque

'COMTESSE G. DE ROQUETTE-BUISSON', HP, mp, 1899; flowers medium, dbl.; Lévêque; (Sangerhausen)

'COMTESSE HENRIETTA COMBES', HP, pb, 1881; flowers satin-rose; Schwartz, J.

'COMTESSE ICY HARDEGG', HT, dp, 1911; flowers carmine, large, dbl.; [Mrs W.J. Grant X Liberty]; Soupert & Notting

'COMTESSE MÉLANIE DE POURTALES', HT, w, 1914; flowers creamy white, outer petals shaded red; [Frau Karl Druschki X Mme Ravary]; Walter

'COMTESSE MOENS DE FERNIG', HT, dr, 1961; flowers wine-red, dbl., 25 petals, moderate fragrance; vigorous growth; [Poinsettia X Seedling]; Verbeek

'COMTESSE O'GORMAN', HP, rb, 1888; flowers red and violet; Lévêque

'COMTESSE O'GORMAN', T, lp; Nabonnand, 1892

Comtesse Ouwaroff, *see* **'DUCHESSE DE BRABANT'**

'COMTESSE PROZOR', LCl, pb, 1922; flowers salmon-rose, reverse coral red; [R. gigantea X Comtesse de Bouchaud]; Nabonnand, P.

'COMTESSE RENÉ DE BEARN', HP, 1896; Lévêque, P.; (Cavriglia)

Comtesse Risa du Parc, *see* **'COMTESSE RIZA DU PARC'**

'COMTESSE RIZA DU PARC', T, mp, 1876; (Comtesse Risa du Parc); flowers rose to carmine, large, dbl., globular, moderate fragrance; [Duchesse de Brabant seedling]; Schwartz, J.

Comtesse Vally de Serenye, Hp, lp, 1875; Fontaine

'COMTESSE VANDAL', HT, pb, 1932; (Comtesse Vandale, Countess Vandal); flowers salmon-pink, reverse coppery pink, large, dbl., 30 petals, exhibition form, moderate fragrance; foliage leathery; bushy growth; (28); GM, Bagatelle, 1931; [(Ophelia X Mrs Aaron Ward) X Souv. de Claudius Pernet]; Leenders, M.; J&P

'COMTESSE VANDAL, CLIMBING', Cl HT, op, 1936; (Comtesse Vandale, Climbing, Countess Vandal, Climbing, Grimpant Comtesse Vandal); J&P

Comtesse Vandale, *see* **'COMTESSE VANDAL'**

Comtesse Vandale, Climbing, *see* **'COMTESSE VANDAL, CLIMBING'**

Concertino®, *see* 'MEIBINOSOR'

Concertino™, *see* 'MEIGLUSOR'

'CONCERTO', F, mr, 1953; bud ovoid, pointed; semi-dbl., 12–15 petals, 2–2.5 in., cupped, slight fragrance; foliage dark; upright, bushy growth; GM, NRS, 1953; PIT, RNRS, 1953; [Alain X Floradora]; Meilland, F.; URS

Concerto™, *see* 'MEIROLOUR'

Concerto, *see* MEIhailoil

Concerto 94, S, 1994; Meilland, Alain A.; (Cavriglia)

'CONCERTO, CLIMBING', Cl F, mr, 1968; Truffant, G.; URS

'CONCHITA', Pol, mp, 1935; flowers clear salmon, borne in clusters, dbl., cupped, slight fragrance; foliage glossy; vigorous growth; Jordan, H.; Low

Concorde, *see* **'FOREVER YOURS'**

Concorde, *see* MEIreibat

Concorde, *see* MEIdorsun

Concorde, HT, lp

'CONCORDIA', HT, mp, 1924; flowers glowing pink, edged silver-pink, semi-dbl., moderate fragrance; Brix; Teschendorff

'CONCORDIA', HT, rb, 1946; flowers red, reverse deep yellow, well formed, dbl., moderate fragrance; vigorous growth; [Charles P. Kilham X Crimson Glory]; Giacomasso

Condesa de Barcelona®, *see* FEodase

'CONDESA DE BENAHAVIS', HT, mp, 1949; flowers salmon-pink, well formed, intense fragrance; [Étoile de Hollande X Sensation]; La Florida

'CONDESA DE GLIMES', F, pb; flowers begonia-pink, center ochre-yellow, single; Bofill; Torre Blanca

'CONDESA DE MAYALDE', HT, rb, 1956; bud pointed; flowers white edged carmine, dbl., 30 petals, exhibition form; foliage glossy; compact, upright growth; [Peace X Flambee]; Dot, Pedro

'CONDESA DE MAYALDE, CLIMBING', Cl HT, w, 1964; Samuels

'CONDESA DE MUNTER', HT, or, 1932; flowers geranium-red, tinted orange-yellow, large, semi-dbl., cupped; long, strong stems; vigorous growth; [Souv. de Josefina Plà X Souv. de Claudius Pernet]; Munné, B.

'CONDESA DE SALDANHA', HT, ob, 1961; flowers reddish orange, mottled yellow, large, dbl., 45 petals, moderate fragrance; vigorous growth; Munné, M.

'CONDESA DE SÁSTAGO', HT, pb, 1932; bud ovoid; flowers deep pink, reverse yellow, large, dbl., 55 petals, cupped, moderate fragrance; foliage glossy, dark; vigorous, tall growth; (28); GM, Rome, 1933; [(Souv. de Claudius Pernet X Maréchal Foch) X Margaret McGredy]; Dot, Pedro; C-P

'CONDESA DE SÁSTAGO, CLIMBING', Cl HT, pb, 1936; Vestal

'CONDESA DE VILLARREA', HT, mr, 1960; flowers crimson, reverse cardinal-red, large, dbl., 35 petals, long, stiff stem; vigorous growth; [Chrysler Imperial X Texas Centennial]; Dot, M.

Conditorum, *see* **R. GALLICA CONDITORUM**

'CONESTOGA', S, w, 1946; bud ovoid; flowers open, borne in clusters, dbl., 30 petals, 2 in., non-recurrent, slight fragrance; foliage soft, sparse, small; upright, vigorous growth; hardy.; [Betty Bland X Seedling]; Preston; Central Exp. Farm

Confection, *see* 'JACUTE'

'CONFEDERATION', HT, mp, 1964; flowers compact, large, dbl., 70 petals, intense fragrance; foliage dark, glossy; moderate, vigorous growth; [Queen o' the Lakes X Serenade]; Golik; Ellesmere Nursery

'CONFERENCE 63', HT, dr, 1965; flowers deep crimson, dbl., 4.5 in., slight fragrance; foliage light green; tall growth; Quentin

Confetti, *see* 'AROjechs'

'CONFIDENCE', HT, pb, 1951; bud ovoid; flowers pearly light pink to yellow blend, large, dbl., 28–38 petals, exhibition form, moderate fragrance; foliage dark, leathery; vigorous, upright, bushy growth; GM, Bagatelle, 1951; [Peace X Michèle Meilland]; Meilland, F.; C-P, 1953;, URS, 1951

Confidence™, *see* 'MEIGAFOR'

'CONFIDENCE, CLIMBING', Cl HT, pb, 1961; Hendrickx; URS

'CONGO', HT, dr, 1943; bud long, pointed; flowers velvety maroon, open, medium, dbl., intense fragrance; foliage leathery, bronze; bushy, dwarf growth; [Admiral Ward X Lemania]; Meilland, F.; A. Meilland

'CONGOLAISE', HT, dr; flowers velvety dark red, well formed, large, dbl., moderate fragrance; vigorous growth; Tantau

Congratulations, *see* 'KORLIFT'

Connie, *see* 'BOSELFTAY'

'CONNIE', Min, dy, 1991; bud pointed; flowers dark yellow fading to light yellow, semi-dbl., 11 petals, flat, borne singly or in small clusters, slight fragrance; foliage medium, dark green, glossy; bushy, medium (36 cms), very dense growth; Jerabek, Paul E.

Connie Crook, *see* 'BOSELFTAY'

'CONNIE MACK', F, dr, 1952; flowers dark velvety crimson, medium, borne in clusters, dbl., 25 petals, slight fragrance; foliage glossy, dark; vigorous growth; [Seedling X Margy]; Duehrsen; H&S

'CONQUEROR', HT, ly, 1929; flowers saffron-yellow, fading pale yellow, semi-dbl., 15 petals, moderate fragrance; vigorous, bushy growth; Chaplin Bros.

Conqueror's Gold, *see* 'HARTWIZ'

Conquest, *see* 'HARBRILL'

Conquete, HT, 1967; Combe, M.; (Cavriglia)

Conquistador, LCl, 1983; Pineau; (Cavriglia)

'CONRAD FERDINAND MEYER', HRg, lp, 1899; flowers silver pink, large blooms in clusters, dbl., cupped, repeat bloom, intense fragrance; foliage leathery; vigorous (8–10 ft), good pillar rose, bushy growth; (28); [R. rugosa hybrid X Gloire de Dijon]; Müller, Dr F.

'CONRAD HILTON', F, my, 1962; bud ovoid; flowers golden yellow, outer petals sometimes white, dbl., 30–45 petals, 2.5 in., flat, moderate fragrance; foliage leathery, dark, glossy, crinkled; very vigorous, upright, bushy growth; [(R. X dupontii X Pinocchio) X (Goldilocks X Feu Pernet-Ducher)]; Shepherd; Bosley Nursery

'CONRAD O'NEAL', S, dp, 1966; bud ovoid; flowers deep pink, medium, very dbl., intense fragrance; foliage dark, glossy; vigorous, upright growth; [Blossomtime seedling X Don Juan]; O'Neal; Wyant

'CONRAD'S CRIMSON', S, rb, 1972; flowers crimson, shaded purple, dbl., 30 petals, 3 in., flat, early bloom, moderate fragrance; foliage light green to bronze; [Sweet Sultan X Conrad F. Meyer]; Eacott

Conservation, *see* 'COCDIMPLE'

Conservency Rose, *see* 'CAPMINA'

'CONSOLATA', HT, yb, 1936; bud pointed; flowers yellow and coppery nasturtium-red; foliage bronze; long, strong stems; Capiago

'CONSPICUOUS', HT, mr, 1930; flowers glowing scarlet, very large, dbl.; vigorous growth; Dickson, A.; B&A, 1932

'CONSTANCE', HT, my, 1915; flowers yellow to golden yellow, dbl., exhibition form, slight fragrance; foliage rich green, glossy; bushy growth; GM, Bagatelle, 1916; [Rayon d'Or X Seedling]; Pernet-Ducher

'CONSTANCE CASSON', HT, pb, 1920; flowers carmine, flushed apricot, dbl., moderate fragrance; [Queen Mary X Gorgeous]; Cant, B. R.

Constance Fettes, *see* 'COCNEST'

Constance Finn, *see* HARedcn

'CONSTANCE MORLEY', HT, ob, 1981; flowers orange-gold tinged red, diffused center, borne singly, dbl., 39 petals, slight fragrance; elongated, orange prickles; foliage dark, glossy; spreading, bushy growth; [Piccadilly X Seedling]; Gregory, C. & Sons, Ltd.

Constance Spry®, *see* 'AUSfirst'

'CONSTANCE, CLIMBING', Cl HT, my, 1927; Pacific Rose Co.

'CONSTANTIA', HT, mp, 1960; bud pointed; flowers neyron rose, well formed, dbl., 40 petals, 4–4.5 in., moderate fragrance; moderate growth; [Baccará X Grace de Monaco]; Herholdt, J.A.; Herholdt's Nursery

'CONSTANZE', HT, ob, 1966; bud pointed; flowers orange, well-formed, large, dbl., 25–30 petals; foliage dark, glossy; upright, bushy growth; Tantau, Math.

Constanze, *see* TANzecon

'CONSTELLATION', HT, ob, 1949; bud long, pointed; flowers coppery orange, well formed, very large; very vigorous, erect growth; [Peace X Seedling]; Gaujard

Constellation, LCl, ab, 1999

Constellation, Min, lp

Consuelo, HT, mr

Consul M. Mezin, HT; Verbeek; (Cavriglia)

'CONTEMPO', F, ob, 1971; bud ovoid; flowers orange blending to gold, medium, dbl., exhibition form, moderate fragrance; foliage light, leathery; vigorous, bushy growth; [Spartan X (Goldilocks X (Fandango X Pinocchio))]; Armstrong, D.L.; Armstrong Nursery

'CONTENTMENT', HT, pb, 1956; bud globular; flowers soft pink suffused yellow, dbl., 65–70 petals, 5.5 in., exhibition form, intense fragrance; foliage glossy, leathery; vigorous, upright growth; [(Seedling X Lilette Mallerin) X Orange Delight]; Boerner; J&P

Contessa, *see* 'JACRIS'

Contessa Mona Bismark, Gr; (Cavriglia)

'CONTINENTAL', HT, mr, 1966; bud ovoid; flowers cardinal-red, large, dbl., moderate, fruity fragrance; foliage leathery; vigorous, upright growth; [Baccará X Yuletide]; Lammerts, Dr Walter; Amling-DeVor Nursery

'CONTRAST', HT, pb, 1937; flowers china-pink and bronze, reverse white and bronze, dbl., exhibition form, moderate fragrance; foliage leathery, glossy; very vigorous, bushy, compact growth; [Seedling X Talisman]; H&S

Contribute, HT, my, 1977; Dawson

Cool Dude, *see* 'RENDUDE'

Cool Wave, *see* 'ZIPCOOL'

'COOLNESS', F, w, 1958; bud ovoid, cream; dbl., 55–60 petals, 2.5–3 in., moderate fragrance; foliage leathery, glossy; vigorous, bushy, compact growth; [Glacier seedling X Starlite seedling]; Boerner; J&P

Coon Carnival®, *see* 'KORCOON'

'COOPERI', S, 1927; (Cavriglia)

'COOPEROO EMBLEM', Pol, lp; Harrison, A., 1927; (Weatherly, L.)

'COOPER'S BURMESE', S, w, 1927; (Gigantea Cooperi, R. X Cooperi); flowers near white, large, single; foliage glossy; vigorous (to 20 ft) growth; (14); [Possibly a natural hybrid of R. gigantea X R. laevigata]

'COOPRIOR', Cl F, mp, 1995; (**Betty Prior, Climbing**); flowers medium carmine pink, medium, single, 5 petals, borne in small clusters, slight fragrance; some prickles; foliage large, medium green, dull; upright, spreading, medium growth; Betty Prior sport; Cooper, Donell; Certified Roses, 1997

'COORAN', F, dp, 1953; bud long, pointed; flowers deep rose-pink, very large, borne in clusters, dbl., intense fragrance; very vigorous growth; [Mrs Tom Henderson X Ming Toy]; Ulrick, L.W.

Coorg, *see* 'VIRBROWN'

'COOROY', F, mp, 1953; flowers rose-pink, borne in clusters, very dbl., exhibition form; [Mrs Tom Henderson X Self]; Ulrick, L.W.

'COOSYL', Min, rb, 1977; (**My Own**); flowers red, yellow center, small, single, 5 petals, moderate fragrance; foliage medium, medium green, semi-glossy; [(Scarlet Knight X Soeur Therese) X Willie Winkie]; Cook, Sylven S.

'COOSYN', LCl, yb, 1981; (**Golden Penny**); flowers gold, reverse creamy, dbl., intense, lemon fragrance; foliage medium, medium green, semi-glossy; upright growth (7–8 ft); [Queen Elizabeth X Scarlet Knight]; Cook, Sylven S.

'COPACABANA', LCl, or, 1966; dbl., 40 petals, 3.5 in., globular, slight fragrance; foliage dark; [Coup de Foudre X Seedling]; Dorieux; Bees

Copacabana, *see* DORneye

'COPENHAGEN', Cl HT, mr, 1964; flowers scarlet, dbl., 5 in., moderate fragrance; foliage coppery; vigorous growth; [Seedling X Ena Harkness]; Poulsen, Niels D.; McGredy

'COPIA', F, mr, 1964; flowers bright cardinal-red, large, semi-dbl.; vigorous, bushy growth; [(Independence X Seedling) X Tour de France]; Mondial Roses

Coppa Nob, *see* COCcopna

'COPPÉLIA', HT, ob, 1952; flowers rosy shades, deepening to orange, medium, dbl., 28 petals, cupped, slight fragrance; foliage leathery; vigorous, upright growth; [Peace X Europa]; Meilland, F.

Coppélia 76, *see* MEIgurami

Copper Arch, *see* KORhurtlen

'COPPER CLIMBER', LCl, op, 1938; bud pointed, coppery; flowers glowing coppery salmon, edged pink, large, Burbank; Stark Bros.

'COPPER CORONET', HT, ab, 1986; flowers coppery amber, reverse blush pink, fading amber and cream, large, very dbl., 60 petals, exhibition form, slight fragrance; foliage medium, medium green, semi-glossy; upright, medium growth; [Ginger Rogers X Royal Highness]; Strange, J.F.

Copper Crown, *see* 'WILCROWN'

'COPPER DELIGHT', F, ob, 1956; flowers clear orange, large blooms in large clusters, semi-dbl., 14 petals, moderate fragrance; foliage olive-green; vigorous, upright, bushy growth; [Goldilocks X Ellinor LeGrice]; LeGrice

Copper Gem, HT, ab, 1984; Cocker

'COPPER GLOW', LCl, ob, 1940; flowers copper, dbl., 27 petals, 4 in., seasonal bloom, intense fragrance; foliage glossy; vigorous, climbing (20 ft), open habit growth; [Golden Glow X Break o' Day]; Brownell, H.C.

'COPPER KETTLE', F, ob, 1978; bud elongated; flowers brilliant copper-orange and yellow, dbl., 28 petals, 2.5–3.5 in., exhibition form, moderate fragrance; foliage glossy, bronze, dark; upright growth; [Queen Elizabeth X Golden Slippers]; Williams, J. Benjamin; J.B. Williams & Associates

Copper Kettle, S, ob, 1997; bud tangerine; flowers brilliant copper orange with yellow, moderate fragrance; foliage disease-resistant; hardy to zone 5; Williams, J. Benjamin

Copper King, *see* 'HERCOP'

'COPPER LUSTER', HT, op, 1945; flowers coppery pink, loosely, large, dbl., 23 petals, cupped, slight fragrance; foliage glossy, bronze; vigorous, upright growth; [Better Times X Orange Nassau]; Roberts; Totty

'COPPER NUGGET', HT, op, 1942; bud ovoid to urn shaped; flowers orange-salmon, small, dbl., 50–60 petals, exhibition form, slight fragrance; foliage leathery, glossy, dark; strong stems; dwarf, bushy growth; [Charles P. Kilham X Capt. Thomas]; Lammerts, Dr Walter; Armstrong Nursery

Copper Pot, *see* 'DICPE'

'COPPER RUFFLES', Min, or, 1982; flowers small, dbl., exhibition form, slight fragrance; foliage small, medium-green, semi-glossy; upright growth; [Anytime X Sheri Anne]; Dobbs, Annette E.

Copper Sunset™, *see* 'SAVacop'

'**Copperkins**', HT, ob, 1957; bud long, pointed, dark orange-flame; flowers orange, well formed, dbl., 25 petals, intense, fruity fragrance; foliage glossy; vigorous growth; [Mme Henri Guillot X Golden Scepter]; Ratcliffe

Coppers Grand, S, mr, 1997

Coppertone, *see* 'KORol'

'**Coppery Heart**', S, yb, 1958; flowers coppery yellow shaded red, large, dbl., repeat bloom, moderate fragrance; foliage dark, glossy; long stem; very vigorous growth; [Peace X Conrad Ferdinand Meyer]; Gaujard

Copy Cat, *see* 'MORcat'

'**Coq de Roche**', HT, dr, 1945; flowers blood-red, large, very dbl.; very vigorous growth; [Duquesa de Peñaranda X J. B. Meilland]; Meilland, F.

'**Coquellicot**', HT, or, 1942; flowers orange-red with gold, medium, semi-dbl., slight fragrance; Meilland; (Sangerhausen)

Coquette, *see* '**Bertram Park**'

Coquette, *see* 'JACco'

'**Coquette**', HT, lp, 1929; flowers pale flesh-pink, well formed, slight fragrance; vigorous growth; Dobbie

'**Coquette Bordelaise**', HP, pb, 1896; flowers dark pink with white stripes, very large, dbl., moderate fragrance; Duprat; (Sangerhausen)

'**Coquette de Lyon**', HP, lp, 1859; flowers flesh-pink, Lacharme, F.

'**Coquette de Lyon**', T, ly, 1872; (La Coquette de Lyon); flowers canary-yellow, medium size, Ducher

'**Coquette de Lyon**', T, lp, 1870; Ducher

'**Coquette des Alpes**', B, w, 1867; flowers white tinged blush, semi-, medium to large, cupped; vigorous growth; [Blanche Lafitte X Sappho (Vibert's)]; Lacharme, F.

'**Coquette des Blanches**', B, w, 1871; flowers white, lightly washed pink, dbl., cupped, moderate fragrance; vigorous growth; [Blanche Lafitte X Sappho (Vibert's)]; Lacharme, F.

'**Coquina**', HWich, pb, 1909; flowers rose-pink fading lighter, base creamy white, borne in large, single, cupped, moderate fragrance; foliage dark, almost evergreen; long strong stems; very vigorous (20–24 ft) growth; Walsh

'**Cora**', HGal, m, 1885; flowers small, dbl.; Savoureux; (Sangerhausen)

Cora Marie, *see* 'KORlimit'

'**Corail**', LCl, pb, 1931; flowers light peach-blossom-pink, reverse coral-pink and carmine, opening well, dbl., recurrent bloom; foliage bright, glossy; very vigorous growth; [William Allen Richardson X Orléans Rose]; Schwartz, A.

'**Coral**', LCl, mp; foliage massive; vigorous, climbing growth; half-hardy.; [R. sinowilsonii X Seedling]; F.C. Stern

'**Coral**', HT, pb, 1931; flowers bright coral, base buttercup-yellow, dbl., globular, moderate fragrance; very erect stems; vigorous growth; Dickson, A.

Coral®, *see* FEloma

'**Coral Anne Griffiths**', HT, mp, 1978; bud very tight; very dbl., 72 petals, moderate fragrance; foliage glossy, dark; very large (6 in.) very vigorous growth; [Red Devil sport]; Henson

'**Coral Bay**', HT, op, 1971; bud ovoid; flowers silvery coral-orange, medium, dbl., cupped, moderate fragrance; foliage glossy, leathery; vigorous growth; [Seedling X Seedling]; Swim & Weeks; Weeks Wholesale Rose Growers

'**Coral Beauty**', Pol, mp, 1941; flowers spinel-pink, small borne in clusters, dbl., 1.5 in., flat; vigorous, branching growth; [Sport of Orléans Rose sport]; deRuiter; J&P

'**Coral Belle**', F, ob, 1962; bud pointed, ovoid; flowers vermilion, medium, dbl., 45 petals, 1.5–2.5 in., exhibition form, moderate fragrance; foliage leathery; strong stems; vigorous, upright growth; [Stoplite X Orange Sweetheart]; Jelly; E.G. Hill Co.

Coral Bells, Pol, op, 1962; flowers coral pink, medium, dbl.; J&P; (Sangerhausen)

Coral Border, *see* POUlalo

Coral Button, Min, op

Coral Cameo, *see* 'MORcalyn'

Coral Carpet, MinFl, mp, 1997; flowers coral pink to light pink; groundcover growth; Williams, J. Benjamin

'**Coral Cascade**', F, or, 1980; flowers coral-red, blooms borne singly or 3–5 per cluster, dbl., 50–75 petals, globular, slight, carnation fragrance; reddish-gray prickles; foliage opens russet, turning dark green, glossy; vigorous, compact, bushy growth; [Van Bergen X Pink Hat sport]; James, John

'**Coral Cluster**', Pol, op, 1920; flowers coral-pink; GM, NRS, 1921; [Orléans Rose sport]; Murrell, R.

'**Coral Creeper**', LCl, dp, 1938; bud deep red; flowers coral to light pink, semi-dbl., 4 in., intense fragrance; foliage leathery; upright stems; very vigorous growth; [(Dr W. Van Fleet X Emily Gray) X Jacotte]; Brownell, H.C.

'**Coral Crown**', F, or, 1960; flowers coral-red, blooms in clusters, dbl., 35 petals, 3 in., exhibition form, moderate fragrance; foliage glossy; low, compact growth; [Else Poulsen X (Fashion X Orange Triumph)]; Von Abrams; Peterson & Dering

'**Coral Cup**', Pol, op, 1936; flowers soft coral, borne in clusters, very dbl., cupped, profuse, repeated bloom, slight fragrance; very vigorous, bushy growth; [Gloria Mundi sport]; B&A

'**Coral Dawn**'®, LCl, mp, 1952; bud ovoid; flowers rose-pink, blooms in clusters, dbl., 30–35 petals, 5 in., cupped, moderate fragrance; foliage leathery; vigorous (8–12 ft) growth; [(New Dawn seedling X Seedling) X Seedling]; Boerner; J&P

'**Coral Destiny**', HT, op, 1984; flowers medium coral pink, large, dbl., 35 petals, slight fragrance; foliage medium green, semi-glossy; upright growth; [Joanna Hill X Queen Elizabeth]; Perry, Anthony; Ball Seed Co.

'**Coral Drops**', HMoy, op; flowers pale coral-pink, single, non-recurrent

Coral Fairy, *see* 'MORcofair'

Coral Fantasy, *see* 'LYOco'

Coral Fiesta®, *see* 'DOTrames'

Coral Floc, F, or; Dot

'**Coral Gem**', F, pb, 1958; bud ovoid; flowers light coral-pink, open, borne in clusters, dbl., 40–45 petals, 2.5 in., cupped, moderate, fruity fragrance; foliage leathery, dark; vigorous, bushy growth; [(Pinocchio X Mrs Sam McGredy, Climbing) X (Pinocchio X Mrs Sam McGredy, Climbing)) X Fashion]; Boerner; J&P

'**Coral Glow**', LCl, op, 1964; flowers salmon-pink blooms in clusters; [Spectacular X Seedling]; Croix, P.; Minier

Coral Ice, *see* INTerice

Coral Island, Min, op

Coral Meidiland™, *see* 'MEIpopal'

Coral Midinette, Min, op, 1991; Moore

'**Coral Mist**', HT, op, 1966; bud ovoid; flowers coral-pink, dbl., exhibition form, moderate fragrance; foliage glossy, light green; vigorous, bushy, open growth; [Spartan X Good News]; Patterson; Patterson Roses

Coral 'n' Gold, *see* 'MANcoral'

Coral Palace, *see* 'POUldron'

'**Coral Pillar**', Cl HT, mr, 1945; flowers geranium-pink, large, dbl., exhibition form, intense fragrance; foliage glossy, dark; very vigorous, upright growth; [Crimson Glory X Capt. Thomas]; Lammerts, Dr Walter; Univ. of Calif.

'**Coral Princess**', F, ob, 1966; bud ovoid; flowers coral-orange, large, dbl., 25 petals, cupped, moderate fragrance; foliage leathery; vigorous, bushy growth; [(Fashion seedling X Garnette seedling) X Spartan]; Boerner; J&P

'**Coral Queen**', HT, or, 1928; flowers coral-red, large, dbl., cupped, slight fragrance; foliage dark, glossy; strong stems vigorous, bushy growth; [The Queen Alexandra Rose sport]; Reeves

'Coral Queen Elizabeth', F, op, 1966; flowers coral-salmon, borne in clusters, dbl., 3 in., moderate fragrance; foliage glossy; very vigorous growth; [Queen Elizabeth X Seedling]; Gregory

Coral Reef, *see* 'COCdarlee'

'Coral Reef', HT, mp, 1948; bud long, pointed; flowers pink, medium, dbl., 25–30 petals, moderate fragrance; foliage leathery, wrinkled, dark; vigorous, upright, bushy growth; RULED EXTINCT 4/86; [Joanna Hill X R.M.S. Queen Mary]; Hill, Joseph H., Co.

Coral Rosamini, Min, op

Coral Sand, *see* 'RENsand'

'Coral Satin', LCl, op, 1960; bud ovoid; flowers coral, dbl., 25 petals, 3.5–4 in., exhibition form, moderate fragrance; foliage leathery, glossy; vigorous (6–8 ft) growth; [New Dawn X Fashion]; Zombory; J&P

'Coral Sea', HT, dp, 1942; bud globular, old-rose; flowers light red, large, dbl., 40–45 petals, exhibition form, slight fragrance; foliage leathery, dark; long, strong stems; vigorous, upright, much branched growth; [Katharine Pechtold X R.M.S. Queen Mary]; Hill, Joseph H., Co.

Coral Sea, HT, op, 1999

'Coral Silk', F, op, 1972; flowers coral and peach, semi-dbl., 18 petals, 3 in., flat, moderate fragrance; foliage glossy, dark; very free growth; Gregory

Coral Spire, *see* KORkragor

Coral Sprite™, *see* 'JACoral'

'Coral Star', HT, op, 1967; flowers coral-pink, well-formed, medium, intense fragrance; vigorous, upright growth; [Tropicana X Stella]; Robinson, H.

'Coral Sunset', HT, or, 1966; (Coucher de Soleil); bud ovoid; flowers coral-red, large, dbl., moderate fragrance; foliage dark, leathery; vigorous growth; [Garnette seedling X Hawaii]; Boerner; J&P

'Coral Treasure', Min, ob, 1971; bud ovoid; flowers coral-orange, medium, dbl.; foliage glossy, leathery; dwarf, bushy growth; [Seedling X Little Buckaroo]; Moore, Ralph S.; Sequoia Nursery

'CoralGlo', F, op, 1960; bud ovoid; flowers orange-rose, large, dbl., 42 petals, cupped, slight fragrance; foliage bronze; bushy growth; [Independence X Fashion]; Boerner; Stark Bros.

'Corali', LCl, rb, 1931; flowers coral-red to rosy salmon, profuse seasonal bloom; foliage bright green; vigorous growth; Dot, Pedro

'Coralie', M, lp; flowers flesh, well-formed, medium, dbl.; Miellez, ca 1860

'Coralie', D, lp; flowers soft pink, cupped; foliage grayish; vigorous growth

'Coralie', LCl, pb, 1919; flowers coral-red to deep pink, dbl., non-recurrent; foliage glossy; vigorous growth; [Hiawatha X Lyon Rose]; Paul, W.

'Coralin', Min, or, 1955; (Carolin, Carolyn, Karolyn); flowers coral-red, dbl., 40 petals, low, compact growth; [Méphisto X Perla de Alcañada]; Dot, M.

'Coralín Superb', Min, dr, 1958; [Coralín sport]; Will; Kordes

'Coralita', LCl, or, 1964; bud ovoid, deep red; flowers orange-coral, dbl., 40–45 petals, 4 in., moderate fragrance; foliage dark, leathery; vigorous (6–8 ft) growth; David Fuerstenberg Prize, ARS, 1968; [(New Dawn X Geranium Red) X Fashion]; Zombory; J&P

Coralitos, *see* 'SUNcor'

'Corallina', T, rb, 1900; flowers coppery red, large, dbl.; Paul, W.

Corallina, *see* BARcora

'Coralline', Pol, or, 1938; flowers very large clusters, profuse, repeated bloom; very rampant growth; [Gloria Mundi X Golden Salmon Superieur]; Smith, J.; Eddie

Corallovy Surprise, HT, 1982; Klimenko, V. N.; (Cavriglia)

'Corbeille Royale', HT, mp, 1956; bud globular; flowers salmon-pink, dbl., moderate fragrance; vigorous, upright growth; Buyl Frères

'Cordelia', HT, ab, 1975; flowers peach shaded deeper, pointed, dbl., 20 petals, 4–4.5 in., moderate fragrance; foliage dark; tall growth; LeGrice

'Cordelia de Grey', S, dp, 1995; flowers deep, bright pink, small white eye, golden stamens, single, 5 petals, 2 in., borne in small clusters, slight fragrance; numerous prickles; foliage small, dark green, dull; medium, upright growth, arching canes; de Grey, Cordelia

'Cordial', HT, mp, 1965; bud ovoid; flowers pink, medium, in clusters, dbl.; foliage dark; [Satisfaction X Seedling]; Verbeek

Cordon Bleu, *see* 'HARubasil'

Cordon Rouge, LCl, or, 1970; flowers large, dbl.; Combe; (Sangerhausen)

Cordula®, *see* 'KORtri'

'COred', HT, mr, 1973; (**Alec's Red**®); dbl., 45 petals, 6 in., moderate fragrance; foliage matt, green; vigorous, upright growth; ADR, 1973 Edland Fragrance Medal, ARS, 1969 GM, RNRS, 1970 PIT, RNRS, 1970; [Fragrant Cloud X Dame de Coeur]; Cocker, 1970

'Corehead', HT, mr, 1960; flowers bright crimson, intense fragrance; [Ena Harkness X Hazel Alexander]; Dicksons of Hawlmark

Corina, *see* 'JACvep'

'Corinium', F, m, 1970; flowers mauvish pink, full rosette form, dbl., 48–50 petals, 3 in., intense fragrance; foliage semi-glossy; [Alamein sport]; Cooper

Corinne, *see* 'BROcor'

Corky, *see* 'ZIPcork'

'Corlia', HT, or, 1921; flowers terra-cotta, moderate fragrance; moderate growth; Bees

Corne Herholdt, F, pb

'Cornelia', HMsk, pb, 1925; flowers strawberry flushed yellow, rosette form, small, dbl., borne in flattish sprays, recurrent bloom, moderate fragrance; foliage dark bronze, leathery, glossy; very vigorous growth; (14); Pemberton

'Cornelia', HT, pb, 1919; flowers light pink, base orange, dbl., moderate fragrance; [Ophelia X Mrs Aaron Ward]; Scott, R.

Cornelia Cook, *see* **'Cornélie Koch'**

'Cornélie Koch', T, w, 1855; (Cornelia Cook); flowers creamy white, tinged lemon-yellow and flesh, well formed, dbl., moderate fragrance; vigorous growth; [Devoniensis seedling]; Koch

'Cornelis Timmermans', HT, pb, 1919; flowers clear pink, edged deep yellow, dbl., moderate fragrance; [Pharisaer X Le Progres]; Timmermans

'Corner', HT, pb, 1945; bud pointed; flowers cerise, base yellow, reverse yellow washed pink, semi-dbl., 20 petals, 4.5 in., moderate fragrance; foliage leathery, glossy, dark; vigorous, bushy growth; [Soeur Thérèse X Seedling]; Wyant

'Cornet', HP, m, 1845; flowers rose tinted with purple, very large, very dbl., cupped, moderate, centifolia fragrance; vigorous, branching growth; Lacharme, F.

'Cornice de Tarn-et-Garonne', B, mr, 1852; Pradel

Cornsilk™, *see* 'SAVasilk'

'COrob', F, dr, 1970; (**Rob Roy**®); dbl., 30 petals, 4.5 in., exhibition form, slight fragrance; foliage glossy; [Evelyn Fison X Wendy Cussons]; Cocker, 1971

'Corolle', LCl, mr, 1962; flowers medium, semi-dbl., 12 petals; vigorous growth; [Spectacular X Cocktail]; Dot, Simon

Corona de Oro, *see* **'Gold Crown'**

'Coronado', HT, rb, 1961; bud long, pointed; flowers red, reverse yellow, dbl., 40 petals, 5–6 in., exhibition form, moderate fragrance; foliage glossy, dark; vigorous, upright growth; [(Multnomah X Peace) X (Multnomah X Peace)]; Abrams, Von; Peterson & Dering

Coronado, Climbing, Cl HT; Ansaloni; (Cavriglia)

'Coronation', HWich, rb, 1911; flowers red, lightly striped white, small, large clusters; Turner

'Coronation', HP, mp, 1913; flowers flesh, shaded bright shrimp-pink, well formed, large, dbl., 50 petals, recurrent bloom, moderate fragrance; smooth wood; vigorous growth; GM, NRS, 1912; Dickson, H.

'CORONATION GOLD', F, ab, 1978; (Maja Oetker); bud globular; flowers golden yellow to apricot, dbl., 27 petals, 4 in., slight fragrance; foliage glossy; vigorous, upright growth; [(Sabine X Circus) X (Anne Cocker X Arthur Bell)]; Cocker, A.; Cocker

'CORONATION GOLD', HT, yb, 1954; flowers golden yellow flushed crimson, large, intense, damask fragrance; foliage glossy; vigorous growth; RULED EXTINCT 1980; [Signora X Peace]; Cox

'CORONET', F, dr, 1957; flowers deep crimson, blooms in clusters, semi-dbl., 17 petals, 3 in.; foliage dark, glossy; vigorous, upright growth; [Independence X Red Wonder]; deRuiter; Blaby Rose Gardens

'CORONET SUPREME', HT, mp, 1955; bud short, pointed; flowers rose-pink, dbl., 45–65 petals, 4–5 in., moderate fragrance; foliage dark, glossy; very vigorous, upright growth; [Seedling X Golden Rapture]; Jelly; E.G. Hill Co.

'CORPORAL JOHANN NAGY', HSet, dp, 1890; flowers carmine pink, medium, dbl.; Geschwind, R.; (Sangerhausen)

'CORPUS CHRISTI', HT, dr, 1985; flowers large, dbl., 35 petals, moderate fragrance; foliage large, medium green, semi-glossy; upright, bushy, spreading growth; PP005854; [Seedling X Night Time]; Weeks, O.L.; Weeks Wholesale Rose Growers

Corrida, Gr, 1969; Delbard-Chabert; (Cavriglia)

'CORRIE KOSTER', Pol, dp, 1923; flowers light coral-red, opening to deep pink, borne in clusters, semi-dbl., slight fragrance; [Juliana Rose sport]; Koster, M.; Royer

'CORROBOREE', HT, m, 1962; bud long, pointed; flowers lilac-mauve, open, very large, dbl., intense, damask fragrance; foliage glossy; vigorous, upright growth; [Baccará X My Choice]; Fankhauser

Corry, HT, or, 1978; Kordes

'CORSAGE', F, w, 1965; flowers small blooms in clusters, semi-dbl., globular, slight fragrance; foliage soft; vigorous, bushy growth; [Blanche Mallerin X White Swan]; Belden; Wyant

Corsair, *see* 'SANROC'

Corsaire, HT, 1974; Combe, M.; (Cavriglia)

'CORSICA', F, pb, 1964; flowers salmon-pink edged dark pink to red, semi-dbl., 20 petals, intense fragrance; foliage dark, glossy; vigorous, bushy growth; Verschuren, A.; van Engelen

'CORSO'®, HT, ob, 1976; flowers coppery orange, dbl., 33 petals, 4.5 in., slight fragrance; foliage glossy, dark; [Anne Cocker X Dr A.J. Verhage]; Cocker

'CORSO FLEURI', F, dr, 1956; flowers scarlet-red, petals waved, semi-dbl.; dwarf, compact growth; [Red Favorite sport]; Mondial Roses

Cortege, *see* COMteri

Corvette, *see* KORveco

'CORYANA', S, mp; (R. X coryana); flowers rich pink, single, 2.5 in.; sparingly prickly; tall, shrub growth; [R. roxburghii X R. macrophylla]; Hurst, C.C., 1926

'CORYLUS', S, mp, 1988; flowers open, medium, borne usually singly or in sprays of 1–4, single, slight fragrance; round, medium, scarlet fruit; prickles small, light brown; foliage medium, dark green, deep veined; bushy growth; [R. nitida X R. rugosa rubra]; LeRougetel, Hazel; Peter Beales, 1988

'COSETTA'®, F, ob, 1984; flowers light orange, reverse deeper with reddish shadings, medium, dbl., 30 petals, no fragrance; foliage medium, dark, matt; [Zorina X Sole di San Remo]; Bartolomeo, Embriaco

Cosette, *see* 'HARQUILLYPOND'

Cosima®, *see* 'TANCOFEUMA'

'COSIMO RIDOLFI', HGal, m, 1842; flowers old-rose to lilac, spotted crimson, medium, cupped; foliage soft green; compact growth; Vibert

Cosmic, *see* 'BRIcos'

Cosmopoliet, *see* **'COSMOPOLITAN'**

Cosmopolit, *see* **'COSMOPOLITAN'**

'COSMOPOLITAN', F, mp, 1955; (Cosmopoliet, Cosmopolit);, large clusters, bushy growth; [Silberlachs X Seedling]; Buisman, G. A. H.

Cosmos®, *see* COMsan

Costa Dorada, HT; Dot; (Cavriglia)

Cote d'Azur, HT, m, 1998

Cote Rotie, *see* **'GRAND AMOUR'**

'COTILLION', F, yb, 1967; bud ovoid; flowers yellow, edged pink, medium, exhibition form, slight fragrance; foliage leathery; vigorous, upright growth; [Rumba X Golden Garnette]; Hill, Joseph H., Co.

Cotillion™, *see* 'JACSHOK'

Coton Gold, *see* 'BABs'

'COTORRITA REAL', Pol, pb, 1931; flowers white, rose and yellow, small, borne in clusters, very dbl., globular, profuse, repeated bloom, slight fragrance; vigorous growth; Padrosa

'COTSWOLD CHARM', HT, dp, 1967; flowers magenta, exhibition form; foliage dark, leathery; moderate growth; [Ophelia seedling X William Moore]; Bennett, V.G.T.

Cotswold Gold, *see* 'JAYCOT'

Cotswold Sunset, *see* 'JAYSUN'

Cottage Dream, S, mp

Cottage Garden, *see* 'HARYAMBER'

Cottage Garden, *see* CHEWWILY

Cottage Maid, *see* **'VILLAGE MAID'**

Cottage Maid, *see* POUlspan

Cottage Pink, HMult, mp

Cottage Rose, *see* 'AUSGLISTEN'

Cottage White, HMult, w

'COTTON CANDY', HWich, mp, 1952; bud well-formed; very dbl., 2.5 in., blooms in clusters; foliage very glossy, turning to autumn colors; vigorous (10–15 ft) growth; [R. wichurana X Seedling (F with R. multibracteata ancestry)]; Moore, Ralph S.; Sequoia Nursery

'COTTON TOP', F, w, 1962; bud short, pointed; dbl., 45–50 petals, 2–3 in., flat, slight fragrance; foliage leathery; vigorous, upright, well-branched growth; [Seedling X White Butterfly]; Hill, Joseph H., Co.

Cottontail, *see* 'TINTAIL'

Coucher de Soleil, *see* **'CORAL SUNSET'**

Coucou, *see* DORgold

'COULEUR DE BRENNUS', HGal, mr; flowers medium, dbl.; slender shrub growth

'COUNTESS CADOGAN', HT, lp, 1978; bud long, pointed; dbl., 40–45 petals, 4.5 in., exhibition form, slight fragrance; foliage light green, leathery; very vigorous, upright growth; [Carlita sport]; Buss; H. Buss Nursery

Countess Celeste™, *see* 'POULDRON'

'COUNTESS CLANWILLIAM', HT, mr, 1915; flowers pinkish cherry-red, dbl., exhibition form, moderate fragrance; foliage rich green; bushy growth; GM, NRS, 1913; Dickson, H.

'COUNTESS MARY', Cl HT, mp, 1933; [Mary, Countess of Ilchester Climbing sport]; Dixie Rose Nursery

Countess Mary of Ilchester, *see* **'MARY, COUNTESS OF ILCHESTER'**

'COUNTESS OF DALKEITH', F, rb, 1957; flowers vermilion flushed orange, very dbl., intense fragrance; vigorous growth; [Fashion sport]; Dobbie

'COUNTESS OF DERBY', HT, w, 1906; flowers creamy white with pink tints, large, dbl., intense fragrance; Dickson, A.; (Sangerhausen)

'COUNTESS OF ELGIN', HT, op, 1925; flowers salmon-pink, reverse deep rose-pink; vigorous growth; [Mme Edouard Herriot sport]; Ferguson

'COUNTESS OF GOSFORD', HT, op, 1906; flowers salmon-pink, long pointed, large, dbl.; GM, NRS, 1905; McGredy

Countess of Ilchester, *see* **'MARY, COUNTESS OF ILCHESTER'**

'COUNTESS OF LIEVEN', Ayr, w; flowers creamy white, medium, semi-dbl., cupped

'COUNTESS OF LONSDALE', HT, dy, 1919; dbl., moderate fragrance; Dickson, H.

Countess of Oxford, *see* **'COMTESSE D'OXFORD'**

'COUNTESS OF PEMBROKE', HT, pb; flowers pink shaded darker, large, very dbl.; vigorous growth

'COUNTESS OF STRADBROKE', Cl HT, dr, 1928; bud ovoid; flowers dark glowing crimson, well shaped, very large, dbl., globular, free, recurrent bloom, intense fragrance; foliage rich green, wrinkled; vigorous growth; [Walter C. Clark X Seedling]; Clark, A.; Hazlewood Bros.

'COUNTESS OF WARWICK', HT, my, 1919; flowers lemon-yellow, edged pink, large; Easlea

Countess Vandal, *see* **'COMTESSE VANDAL'**

Countess Vandal, Climbing, *see* **'COMTESSE VANDAL, CLIMBING'**

'COUNTRY DANCER', S, dp, 1973; bud ovoid; flowers rose-red, large, dbl., repeat bloom; moderate fragrance; foliage large, glossy, dark, leathery; vigorous, dwarf, upright, bushy growth; [Prairie Princess X Johannes Boettner]; Buck, Dr Griffith J.; Iowa State University

'COUNTRY DOCTOR', HT, lp, 1952; bud long, pointed to ovoid; flowers silvery pink, large, dbl., exhibition form, moderate fragrance; foliage glossy; vigorous, bushy growth; [Pink Princess X Crimson Glory]; Brownell, H.C.

Country Fair, *see* HARbanner

Country Garden, *see* 'BOSANNCOAT'

'COUNTRY GIRL', F, mr, 1958; bud ovoid; flowers geranium-red, medium, dbl.; vigorous growth; [Independence X Salmon Perfection]; Temmerman; Schraven

Country Girl, *see* 'HORALCAMSTRIP'

Country Girl, *see* 'WILCGIR'

'COUNTRY GIRL, CLIMBING', Cl F, mr, 1962; Buyl Frères

Country Heritage, *see* 'COCCAGES'

Country Joy, *see* 'MORCOJO'

Country Lady, *see* 'HARTSAM'

Country Life, *see* HARzap

Country Living, *see* 'AUScountry'

'COUNTRY MAID', F, lp, 1971; slight fragrance; low, compact growth; [Tip Top sport]; Whartons Roses; Deamer

Country Morning, *see* 'FLOMOR'

'COUNTRY MUSIC', S, dp, 1973; bud ovoid; flowers neyron rose, large, dbl., intermittent bloom, moderate fragrance; foliage large, leathery; vigorous, dwarf, upright, bushy growth; [Paddy McGredy X ((World's Fair X Floradora) X Applejack)]; Buck, Dr Griffith J.; Iowa State University

Country Music, *see* HARcheer

'COUNTRY SONG', S, lp, 1984; bud ovoid; flowers large, dbl., 28 petals, cupped, borne 1–5 per cluster, repeat bloom, moderate, myrrh fragrance; awl-like, brown prickles; foliage leathery, dark; erect, bushy growth; hardy.; [Carefree Beauty X The Yeoman]; Buck, Dr Griffith J.; Iowa State University

'COUNTRYMAN', S, pb, 1978; bud ovoid, pointed; flowers light rose-bengal, dbl., 25–30 petals, 4–5 in., cupped, moderate fragrance; foliage large, dark, leathery; vigorous, upright, spreading, bushy growth; [(Improved Lafayette X Independence) X Maytime]; Buck, Dr Griffith J.; Iowa State University

'COUNTRYWOMAN', HT, my, 1978; bud globular; flowers lemon-yellow, medium, dbl., moderate fragrance; foliage leathery; bushy growth; [Seedling X Peace]; Dawson, George; Australian Roses

'COUNTY FAIR', F, mp, 1960; bud ovoid, pointed; flowers medium to dark pink, fading much lighter, blooms in clusters, 8–10 petals, 2.5–3 in., flat, slight fragrance; foliage leathery, dark, semi-glossy; vigorous, bushy growth; [Frolic X Pink Bountiful]; Swim, H.C.; Armstrong Nursery

County Girl, *see* 'HORALCAMSTRIP'

Coup de Coeur, *see* MALloeur

'COUP DE FOUDRE', F, or, 1956; bud well-formed; flowers fiery red, cupped; foliage glossy, bronze; vigorous growth; [(Peace X Independence) X Oiseau de Feu]; Hémeray-Aubert

Coup de Foudre, Climbing, Cl F, 1976; Fineschi, G.; (Cavriglia)

'COUPE D'HÉBÉ', B, dp, 1840; flowers deep pink, waxy texture, large, very dbl., cupped, intense fragrance; foliage glossy; vigorous, erect growth; [Bourbon X R. chinensis hybrid]; Laffay, M.

'COUPE D'OR', LCl, my, 1930; flowers canary-yellow, open, dbl., cupped, moderate fragrance; foliage rich green, leathery, glossy; vigorous, climbing or trailing growth; [Jacotte seedling]; Barbier

Coupe d'Or, *see* **'GOLD CUP'**

'COURAGE', HT, dr, 1923; flowers deep brilliant maroon-crimson, very large, dbl., exhibition form, intense fragrance; foliage rich green, leathery; bushy, dwarf growth; McGredy

'COURAGE', HT, rb, 1941; flowers red tinted yellow, open, very large, dbl., slight fragrance; vigorous, bushy growth; [Seedling X Brazier]; Mallerin, C.; A. Meilland

Courage, *see* POUlduff

Courageous Indira, HT, ob, 1991; Chandrakant

'COURIER', LCl, pb, 1930; flowers pink on white ground, borne in clusters; vigorous, climbing growth; Clark, A.; Brundrett

Couronne d'Or, *see* **'GOLD CROWN'**

'COURT JESTER', F, ob, 1980; flowers orange, reverse yellow, borne 5–7 per cluster, exhibition form, slight fragrance; large hooked, red-brown prickles; foliage mid-green, glossy; tall, upright growth; Cants of Colchester, Ltd.

Courtisane, HT, or, 1965; flowers large, dbl.; Gaujard; (Sangerhausen)

Courtney, *see* 'GELCOURT'

'COURTNEY PAGE', HT, dr, 1922; flowers velvety dark scarlet-crimson, dbl., intense fragrance; GM, NRS, 1920; McGredy

Courtoisie®, *see* 'DELCOURT'

'COURTSHIP', HT, mp, 1955; bud conical; flowers cerise-pink, reverse lighter, dbl., 28 petals, 4–5 in., exhibition form, foliage dark; vigorous, bushy growth; [Mme Henri Guillot X Peace]; Shepherd; Bosley Nursery

Courvoisier®, *see* 'MACSEE'

'COUSIN ESSIE', S, w, 1996; semi-dbl., 8–14 petals, 1.5 in., moderate fragrance; few prickles; foliage medium, medium green, semi-glossy; spreading, busy, tall, (6 ft) growth; [Seedling of Honeyflow sport]; Robertson, Myrtle; Honeysuckle Cottage Nursery, 1988

'COVA DA IRIA', HT, rb, 1963; flowers light red, reverse gold; [Seedling X Crimson Glory]; da Silva, Moreira

'COVENT GARDEN', HT, rb, 1919; flowers rich deep crimson flushed plum-black on reverse, intense fragrance; foliage leathery, glossy; vigorous growth; GM, NRS, 1918; Cant, B. R.

'COVENTRIAN', F, dp, 1962; flowers ruby-cerise, well formed, borne in large clusters, dbl., 30 petals, 3–3.5 in.; foliage dark, glossy; vigorous, bushy growth; [Highlight X Seedling]; Robinson, H.

Coventry Cathedral, *see* **'CATHEDRAL'**

'COVER GIRL', HT, ob, 1960; bud long, pointed; flowers orange, copper and gold, dbl., 28–35 petals, 5 in., exhibition form, slight fragrance; foliage glossy, dark; upright, bushy growth; [Sutter's Gold X (Mme Henri Guillot X Seedling)]; Von Abrams; Peterson & Dering

'COWICHAN SUPER', Cl HT, op, 1975; flowers orange-salmon, dbl., 30–35 petals, 5 in., bloom repeats, intense fragrance; foliage glossy; tall, very vigorous growth; Tamarack Roses

Cox's Pink Polyantha, Pol, mp

Cox's Red Polyantha, Pol, mr

'COY COLLEEN', HT, w, 1953; (Blushing Rose); bud pointed, rosy white; flowers milky white, well-formed, borne in clusters, slight fragrance; foliage glossy; vigorous growth; [(Modesty X Portadown Glory) X Phyllis Gold]; McGredy, Sam IV

'CRACKER', HT, lp; flowers bright red with prominent stamens; Clark, A., 1920

'CRACKER, CLIMBING', S, mr, 1927; flowers red with white, large, single, intense fragrance; Clark, A.; (Sangerhausen)

'CRACKERJACK', F, rb, 1959; flowers scarlet flushed yellow, borne in clusters, semi-dbl., 20 petals, moderate fragrance; free growth; [Fashion X Masquerade]; Fryers Nursery, Ltd.

Crackerjack, *see* POUlcrack

Crackling Fire(TM), *see* 'JACORG'

'CRAIGWEIL', HT, pb, 1929; flowers silvery cerise-pink, reverse deeper, dbl., moderate fragrance; strong stems; vigorous growth; [Mme Abel Chatenay seedling]; Hicks

'CRAMOISI DES ALPES', HGal, mr; flowers bright purplish red, dbl., heavy, non-recurrent bloom; Hardy (?), before 1829

'CRAMOISI ÉBLOUISSANT', HCh, dr, 1839; flowers medium, dbl.; (Sangerhausen)

'CRAMOISI FONCÉ VELOUTÉ', M, dr; flowers deep velvety crimson

'CRAMOISI PICOTÉ', HGal, rb, 1834; flowers crimson, streaked and mottled (striped & spotted) darker, medium, very dbl., 2 in.; Vibert

'CRAMOISI SUPÉRIEUR', HCh, mr, 1832; (Lady Brisbane); flowers crimson-red, small blooms in large clusters, dbl., cupped, recurrent bloom; vigorous growth; (21); Coquereau

'CRAMOISI SUPÉRIEUR, CLIMBING', Cl HCh, mr, 1885; (Agrippina, Climbing, Lady Brisbane, Climbing); (21); Couturier

'CRANBROOK', Pol, lp; Matthews, W.J., 1921; (Weatherly, L.)

'CRARAE', HT, yb, 1981; flowers deep yellow marked with scarlet, large, dbl., 35 petals, slight fragrance; foliage medium, dark, glossy; bushy growth; [Piccadilly X Fred Gibson]; McKirdy, J.M.; John Sanday Roses

Crathes Castle, *see* 'COCATHES'

Crazy Dottie, *see* 'SEADOT'

Crazy For You, *see* 'WEKROALT'

Crazy Horse, F; de Ruiter, G.; (Cavriglia)

Crazy Quilt, *see* 'MORTRIP'

Crazy Spire, HT, ob, 1995

'CREAM CRACKER', HT, w, 1979; bud pointed; flowers cream, classical pointed form, blooms borne singly, dbl., 38 petals, 4.5 in., intense, apple fragrance; often without prickles; foliage large, matt; bushy growth; [Columbine X Iceberg]; Murray, Nola; Rasmussen's

'CREAM CRACKER', HT, lp, 1933; flowers creamy buff, reverse shaded salmon, large, dbl.; vigorous growth; RULED EXTINCT 10/78 ARM; Dickson, A.

Cream Delight, *see* 'SUNCREDEL'

Cream Dream, *see* KORomtar

'CREAM GOLD', Min, my, 1978; bud long, pointed; flowers small, dbl., 38 petals, 1.5 in., exhibition form, moderate fragrance; compact, spreading growth; [Golden Glow X Seedling]; Moore, Ralph S.; Sequoia Nursery

'CREAM PEACH', F, w, 1975; flowers cream, edged pink, full, semi-dbl., 15–20 petals, 3–4 in., slight fragrance; foliage large, glossy; very free growth; [Paddy McGredy X Seedling]; Sheridan

'CREAM PUFF', MinFl, pb, 1981; bud ovoid; flowers cream blushed pink, borne singly or in clusters of 3, semi-dbl., 18 petals, moderate fragrance; long prickles; foliage dark, semi-glossy; spreading, bushy growth; [Little Darling X Elfinesque]; Bennett, Cecilia 'Dee'; Tiny Petals Nursery

Cream Puff, Climbing, Cl Min, lp; [Cream Puff sport]; Trimper, K., 1996; (Weatherly, L.)

Cream Sunsation, *see* KORgatine

'CREDO', HT, mr, 1965; flowers purplish red, medium, dbl., exhibition form, moderate fragrance; foliage leathery; vigorous, upright growth; [Eminence X John S. Armstrong]; Gaujard

'CREE', S, lp; flowers pale pink fading to white, large, single, early bloom, non-recurrent; foliage glossy, bright green; vigorous growth; [R. rugosa albo-plena X R. spinosissima hispida]; Central Exp. Farm

'CREEPING EVERBLOOM', LCl, mr, 1939; dbl., 30 petals, 4 in., borne in clusters, recurrent bloom, moderate fragrance; canes 3 ft long growth; [Frederick S. Peck X (Général Jacqueminot X Dr W. Van Fleet)]; Brownell, H.C.

Creepy, *see* 'MORPAPPLAY'

'CREINA MURLAND', HT, dy, 1934; flowers sunflower-yellow, deepening, dbl., slight fragrance; foliage glossy; very vigorous growth; Dickson, A.

'CREME', HWich, lp, 1895; flowers small, semi-dbl., intense fragrance; Geschwind, R.; (Sangerhausen)

Crème de la Crème, *see* GANcre

Crème Glacée, *see* 'LAVCREME'

'CRÉOLE', HT, dr, 1962; bud long, pointed; flowers purplish, red base coppery, large, dbl., slight fragrance; foliage dark, glossy; very vigorous, bushy growth; [Peace X Josephine Bruce]; Gaujard

Crêpe de Chine®, *see* 'DELTOP'

'CREPE MYRTLE', F, mr, 1937; (Harold Ickes); height 5 ft; [Permanent Wave sport]; Dixie Rose Nursery

Crepe Suzette, *see* 'TRASUZ'

'CRÉPUSCOLO', HT, ob, 1955; flowers copper, pointed, dbl.; strong stems; upright growth; [Julien Potin X Sensation]; Aicardi, D.; Giacomasso

'CRÉPUSCULE', N, ab, 1904; flowers orange, fading to apricot-yellow; makes a tall hedge growth; Dubreuil

'CRESCENDO', HT, rb; flowers bright red, base touched with white, well formed, large, dbl., moderate fragrance; tall growth

'CRESSET', Gr, mr, 1961; bud long, pointed; flowers scarlet, semi-dbl., exhibition form, blooms in clusters, moderate fragrance; foliage leathery, glossy, dark; vigorous, bushy growth; [Queen Elizabeth X Cocorico]; Francis; F. Mason

Cressida, *see* 'AUScress'

'CRESTED JEWEL', M, mp, 1971; bud long, pointed, 'mossed' similar to crested moss; flowers bright rose-pink, medium, semi-dbl., exhibition form; foliage leathery; vigorous growth; [Little Darling X Crested Moss]; Moore, Ralph S.; Sequoia Nursery

'CRESTED MOSS', C, mp, 1827; (Chapeau de Napoléon, Crested Provence Rose, Cristata, R. centifolia cristata, R. centifolia muscosa cristata); flowers similar to centifolia muscosa, but the mossy excrescences confined to sepal edges; Vibert

Crested Provence Rose, *see* **'CRESTED MOSS'**

Crested Sweetheart, *see* 'MORSWEET'

Cricket(TM), *see* 'AROKET'

Cricket Cl., Cl Min, ob

Cricri®, *see* 'MEICRI'

Cri-Cri, *see* 'MEICRI'

Crimean Night, HT, dr; USSR

'CRIMSON BEAUTY', HT, dr, 1930; flowers crimson, dbl.; very vigorous growth; [Hoosier Beauty X Crimson Champion]; Dingee & Conard

'CRIMSON BEAUTY', HT, dr, 1935; flowers red, shaded scarlet and maroon, large, dbl., intense fragrance; foliage leathery; vigorous growth; [Daily Mail Scented Rose X Étoile de Hollande]; LeGrice

'CRIMSON BEDDER', HP, mr, 1874; flowers medium, dbl.; Cranston; (Sangerhausen)

Crimson Blush, *see* SIEson

Crimson Bouquet(TM), *see* 'KORBETEILICH'

Crimson Boursault, *see* **'AMADIS'**

'CRIMSON BROCADE', HT, dr, 1962; flowers bright scarlet-crimson, large, exhibition form, moderate fragrance; strong stems; vigorous growth; Robinson, H.

Crimson Cascade, *see* FRYclimbdown

'CRIMSON CHAMPION', HT, dr, 1916; flowers velvety crimson-red, dbl., moderate fragrance; dwarf growth; [Étoile de France X Seedling]; Cook, J.W.

'CRIMSON CHATENAY', HT, dr, 1915;, moderate fragrance; [Mme Abel Chatenay X Leuchtfeuer (HCh)]; Merryweather

Crimson China Rose, *see* **'SLATER'S CRIMSON CHINA'**

'CRIMSON CONQUEST', Cl HT, dr, 1931; flowers scarlet crimson, small, semi-dbl., 2.5 in.; foliage light green; vigorous growth; [Red-Letter Day sport]; Chaplin Bros.

'CRIMSON CROWN', HT, mr, 1905; flowers medium, semi-dbl.; Dickson, A.; (Sangerhausen)

'CRIMSON DAMASK', D, dr, 1901; Turner

'CRIMSON DAWN', F, rb, 1970; flowers crimson to carmine, full, dbl., 20 petals, 3.5–4.5 in., moderate fragrance; foliage dark; vigorous growth; [(Anne Poulsen X Dainty Maid) X (Bonn X Opera)]; Ellick

Crimson Delight, *see* 'JOHILLGOLF'

'CRIMSON DESCANT', LCl, mr, 1972; flowers crimson, dbl., 30 petals, 5 in., slight fragrance; [Dortmund X Etendard]; Cants of Colchester, Ltd.

'CRIMSON DIAMOND', HT, dr, 1947; bud long, pointed; flowers crimson-red, open, large, dbl., 35–40 petals, moderate fragrance; bushy growth; [Crimson Glory X Charlotte Armstrong]; Lammerts, Dr Walter; L.C. Lovett

'CRIMSON DUKE', HT, mr, 1963; bud ovoid; flowers crimson, dbl., 45–55 petals, 4–5 in., exhibition form, moderate fragrance; foliage leathery, dark; vigorous, upright, bushy growth; [(Happiness X Independence) X Peace]; Meilland, Alain A.; C-P

Crimson Elegance, *see* 'LEOCREL'

'CRIMSON EMBLEM', HT, dr, 1916; flowers brilliant crimson-scarlet, large, dbl., cupped; vigorous growth; McGredy

'CRIMSON ERECTA', HT, 1954; Cazzaniga, F. G.; (Cavriglia)

'CRIMSON FRAGRANCE', HT, dr, 1979; bud long, slender; dbl., 38 petals, borne singly, intense fragrance; reddish-bronze prickles; foliage slender, mid-green; branching, upright growth; [Fragrant Cloud X Seedling]; Wright, R. & Sons

Crimson Gallica, HGal, mr

'CRIMSON GEM', Min, dr, 1974; (Flammette); bud ovoid; flowers deep red, medium, very dbl., cupped, slight fragrance; foliage bronze, soft; vigorous, bushy growth; [Lillan X Polyantha seedling]; deRuiter; C-P

'CRIMSON GLOBE', M, dr, 1890; flowers deep crimson, large, dbl., globular; vigorous growth; Paul, W.

'CRIMSON GLORY', HT, dr, 1935; bud long, pointed; flowers deep velvety crimson, large, dbl., 30–35 petals, 5 in., cupped, borne mostly singly, intense, damask fragrance; foliage leathery; vigorous, bushy, spreading growth; (28); GM, NRS, 1936 James Alexander Gamble Fragrance Medal, ARS, 1961; [Cathrine Kordes seedling X W.E. Chaplin]; Kordes; Dreer; J&P

'CRIMSON GLORY, CLIMBING', Cl HT, dr, 1946; (Grimpant Crimson Glory); J&P

'CRIMSON GLOW', Pol, dr, 1945; flowers oxblood-red, semi-dbl., cupped, intense fragrance; foliage glossy, dark; vigorous, upright growth; [Night X Mrs Dudley Fulton]; Lammetts; Univ. of Calif.

Crimson Glow, *see* **'OUR PRINCESS'**

'CRIMSON HALO', HT, dr, 1964; flowers deep rose-red, dbl., 30 petals, 5 in., globular, intense fragrance; vigorous growth; [Karl Herbst X Crimson Glory]; Park; Harkness

'CRIMSON KING', HT, dr, 1943; (Liebesglut); bud ovoid, long, pointed; flowers deep velvety crimson, large, dbl., exhibition form, intense, damask fragrance; foliage leathery; vigorous, bushy growth; [Crimson Glory X Kardinal]; Kordes; C-P

Crimson Lace™, *see* 'JACEZZ'

Crimson Maman Cochet, *see* **'BALDUIN'**

Crimson Masse, *see* **'LIBERTÉ'**

Crimson Medinette, *see* 'POULCRIM'

Crimson Meillandecor, *see* MElouscki

Crimson Midinette, *see* 'POULCRIM'

'CRIMSON MME DESPREZ', B, dr; flowers crimson, large, cupped

'CRIMSON MOSS', M, dr; flowers crimson, dbl.; vigorous growth; Lee, before 1846; Lee, prior to 1846

'CRIMSON ORLÉANS', Pol, dr, 1922; [Orléans Rose sport]; Koster, M.

Crimson Promise, *see* 'SPRINGPROMISE'

'CRIMSON QUEEN', HT, dr, 1912; flowers rich crimson, turning blue with age, very large, dbl., globular, intense fragrance; vigorous growth; [(Liberty X Richmond) X Gen. MacArthur]; Montgomery, A.

'CRIMSON QUEEN', HP, dr, 1890; Paul

Crimson Rambler, *see* **'TURNER'S CRIMSON RAMBLER'**

Crimson Rosamini, Min, mr

'CRIMSON ROSETTE', F, dr, 1948; bud small, ovoid; flowers dark crimson, rosette, blooms in clusters, dbl., 30 petals, 1–1.5 in., slight fragrance; foliage leathery, dark; vigorous, bushy, dwarf growth; Krebs; H&S

'CRIMSON SHOWER', HWich, mr, 1951; flowers clear crimson, pompon, dbl., 20 petals, 1.25 in., slight fragrance; foliage glossy, light; vigorous (10 ft) growth; [Excelsa seedling]; Norman; Harkness

Crimson Tide, *see* 'MACMOTA'

Crimson Velvet, F, dr; flowers medium, single, slight fragrance; (Sangerhausen)

Crimson Wave, *see* 'MEIPERATOR'

Crinkles, *see* 'FROCRIN'

'CRINOLINE', LCl, or, 1964; semi-dbl., cupped, recurrent bloom; foliage bronze; vigorous, climbing growth; [Diane d'Urfe X Étendard]; Hémeray-Aubert

Crispin-Morwenna, *see* 'HARKITTEN'

'CRISSY', Min, ob, 1979; bud pointed; flowers deep bright coral, dbl., 20–25 petals, 1.5–2 in., slight fragrance; foliage dark; [Liverpool Echo X Sheri Anne]; Strawn; Pixie Treasures Min. Roses

Crissy, HT, ob

'CRISTAL', HT, ob, 1937; bud long, pointed; flowers orange-yellow, large, dbl.; foliage leathery, light; vigorous, bushy growth; [Julien Potin X Seedling]; Gaujard

Cristata, *see* **'CRESTED MOSS'**

Cristian, *see* METset

'CRISTIN CIRA', Gr, op, 1996; flowers orange, yellow, salmon blend, blooms borne in small clusters, dbl., 15–25 petals, slight fragrance; some prickles; foliage large, medium green, semi-glossy; upright, medium growth; Vanderkruk, William (Discoverer); Hortico Roses, 1995

Cristina, HT, 1972; Zandri, R.; (Cavriglia)

Cristobal Colon, *see* 'MEIRONSSE'

Cristoforo Colombo, *see* 'MEIRONSSE'

'CRISTOFORO COLOMBO', HT, yb, 1953; (Christophe Colomb); bud long; flowers reddish yellow, reverse tinged pink, large, dbl., 25–35 petals, slight fragrance; foliage glossy; long stems; very vigorous growth; RULED EXTINCT 4/92; [Julien Potin X Frau Karl Druschki]; Aicardi, D.; V. Asseretto

'CRITERION', HT, dp, 1966; flowers rose-red, dbl., 5 in., moderate fragrance; foliage dark; vigorous, tall growth; [(Independence X Signal Red) X Peace]; deRuiter

Criterion Cl., Cl HT, mr

'CROCK O' GOLD', HT, my, 1970; flowers clear golden yellow, classic form, dbl., 30–35 petals, 4.5 in.; [Beauté sport]; Anderson's Rose Nurseries

'CROFT ORIGINAL', HT, RB, 1971; flowers old-gold and red, full, dbl., 30 petals, 5 in., moderate fragrance; foliage light; Cocker; Wheatcroft Bros.

Croix Blanc, HT, w, 1982; Croix, P.

Croix d'Or, Cl HT; Croix, P.; (Cavriglia)

Cromwell School, Misc OGR, rb

CROpal, HT, ob; (**Aventure No. 2**); Croix

'CROWN JEWEL', F, or, 1964; flowers bright orange-red, medium, in clusters, dbl., moderate fragrance; foliage dark, leathery; moderate growth; [Pink Bountiful X Spartan]; Boerner; J&P

'CROWN OF GOLD', HT, yb, 1937; bud pointed; flowers deep gold, edged lemon-yellow, large, dbl., exhibition form; foliage light green, leathery, vigorous growth; [Seedling X Joanna Hill]; Duehrsen; H&S

Crown of Jewels, *see* **'Little Beauty'**

'Crown Prince', HP, rb, 1880; flowers reddish crimson tinged with purple; Paul & Son

Crucencia, *see* **'Hot Pewter'**

Crucenia, *see* **'Hot Pewter'**

Crumble Bar, F, dy, 1982; LeGrice

'Crusader', HT, dr, 1920; flowers crimson-red, center brighter, large, dbl., 65 petals, slight fragrance; foliage leathery, rich green; vigorous growth; Montgomery Co.; A.N. Pierson

Cruzenia, *see* **'Hot Pewter'**

'Crystal', Min, pb, 1985; flowers light pink, deep pink reverse, small, dbl., 35 petals, exhibition form, moderate fragrance; foliage large, dark, glossy; bushy growth; [Zinger X Seedling]; Bridges, Dennis A.; Bridges Roses

Crystal Brook, HWich, pb, 1994; Sutherland

Crystal Palace, *see* POUlrek

'Crystal Star', Min, w, 1994; dbl., 26–40 petals, 1.5 in., borne mostly single and in small clusters, slight fragrance; some prickles; foliage small, medium green, semi-glossy; medium growth; [Pink Petticoat X Tooth of Time]; Muha, Julius; Mori Miniature Roses, 1994

'Crystal White', HT, w, 1965; flowers clear white, large, dbl., exhibition form, moderate fragrance; foliage leathery; moderate, bushy growth; [Princess White X White Queen]; Boerner; J&P

Crystalline™, *see* 'AROBIPY'

Crystal's Double Dark Red, HT, dr

Csárdás, F, op, 1965; flowers luminous orange-pink, large, dbl.; Berger, W.; (Sangerhausen)

'Csl Cerveny Kriz', Pol, mr, 1928; flowers small, dbl.; Böhm, J.; (Sangerhausen)

'Csl Legie', Cl HT, dr, 1933; flowers large, dbl.; Böhm, J.; (Sangerhausen)

'Cuba', HT, rb, 1926; flowers cardinal-red, tinted yellow, fading quickly, large, semi-dbl., globular, intense fragrance; foliage dark, bronze; very vigorous growth; Pernet-Ducher

Cuddle Up, *see* 'TINCUDDLE'

'Cuddles', Min, op, 1978; bud ovoid; flowers deep coral-pink, dbl., 55–60 petals, 1–1.5 in., exhibition form, slight fragrance; compact growth; AOE, 1979; [Zorina X Seedling]; Schwartz, Ernest W.; Nor'East Min. Roses

Cuidad de Oviedo, C, mp

'Cuisse de Emue Nymphe', A, lp; flowers whitish rose; Dumont de Courset, 1802

Cuisse de Nymphe, *see* **'Great Maiden's Blush'**

'Culverbrae', S, dr, 1973; flowers crimson-purple, dbl., 58 petals, 3.5–4 in., some repeat bloom, intense fragrance; foliage light; vigorous growth; [Scabrosa X Francine]; Gobbee, W.D.

Cumba Meillandina®, *see* 'MEINEYTA'

'Cumberland Belle', *see* Le Poilu

'Cunningham Lady Banks', HBank, ly, 1900

'Cunningham West Climber', N, ly, 1900

Cunosa, *see* **'My Girl'**

Cup Final, *see* 'MACSINGAP'

Cupcake™, *see* 'SPICUP'

'Cupid', Cl HT, lp, 1915; flowers glowing flesh tinted peach, large, single; large, orange fruit; vigorous, pillar growth; Cant, B. R.

'Cupido', Min, lp; flowers shell-pink; height 6 in.; Maarse, G.

'Cupidon', F, dr, 1966; flowers brilliant crimson, small, semi-dbl., cupped, slight fragrance; foliage leathery; vigorous, bushy growth; [Chanteclerc X Red Favorite]; Gaujard

'Cupid's Beauty', Min, ob, 1978; bud long, pointed; flowers light orange and cream, small, dbl., 40–45 petals, 1.5 in., exhibition form, moderate fragrance; foliage small, dark; compact, spreading growth; [Seedling X Over the Rainbow]; Williams, Ernest D.; Mini-Roses

'Cupid's Charm', F, pb, 1964; bud pointed; flowers salmon-pink, medium, dbl., 22 petals, moderate fragrance; vigorous, bushy growth; [Little Darling X First Love.]; Fuller; Wyant

Cupid's Heart, HT, rb, 1997

Cupie Doll, *see* 'TINCUPIE'

'CURBAL', Min, pb, 1988; (**Little Ballerina**); flowers white flushed pink, urn-shaped, medium, borne singly, very dbl., 34–40 petals, exhibition form, slight fragrance; no fruit; no prickles; foliage medium, medium green, semi-glossy; bushy, medium growth; [Little Darling X Little Pioneer]; Curtis, Thad; Oregon Miniature Roses, 1988

Curiosa, F, op, 1964; flowers salmon-orange, large, dbl.; deRuiter; (Sangerhausen)

Curiosity, *see* 'COCTY'

'CURLEM', Min, my, 1985; (**Lemon Fluff**); flowers small blooms borne singly and in clusters, dbl., 20 petals, exhibition form, no fragrance; foliage small, medium green, matt; upright, bushy growth; [(Unnamed seedling X Rise 'n' Shine) X Summer Butter]; Curtis, Thad; Hortico Roses

'Curly Locks', Min, lp, 1954; flowers soft pink; dwarf, compact growth; Robinson, T.

'Curly Pink', HT, mp, 1948; bud long, pointed, rose-red; dbl., 3.5–5 in., moderate fragrance; foliage glossy, dark; vigorous, compact growth; [Pink Princess X Crimson Glory]; Brownell, H.C.

'CURMIST', Min, mr, 1986; (**Mister Otis**); flowers small blooms borne singly, dbl., 35 petals, exhibition form, no fragrance; foliage small, medium green, matt; upright, bushy growth; [Rise 'n' Shine X Fire Princess]; Curtis, Thad; Hortico Roses

'CURNEER', Min, mr, 1986; (**Little Pioneer**); flowers gold stamens, small blooms borne singly and in clusters, dbl., 35 petals, exhibition form, moderate fragrance; foliage small to medium, medium green, matt; upright, bushy growth; [(Rise 'n' Shine X Summer Butter) X Sheri Anne]; Curtis, Thad; Hortico Roses

'Current Affair', Min, dr, 1991; flowers red, reverse red with yellow, ages dark pink, medium, very dbl., 80 petals, cupped, no fragrance; foliage medium, dark green, semi-glossy; upright, medium growth; [Red Ace X Seedling]; Gruenbauer, Richard, 1984; Flowers 'n' Friends Miniature Roses, 1993

'Curtain Call', HT, dp, 1977; bud pointed; flowers cherry-red, dbl., 32–35 petals, 4–5 in., exhibition form, slight fragrance; foliage dark, leathery; vigorous, growth; [First Prize X Seedling]; Weeks; Weeks Roses

'Curtis Yellow', HT, my, 1973; flowers clear yellow, large, dbl., exhibition form, moderate fragrance; foliage light, leathery; vigorous, upright growth; [Golden Scepter X Miss Hillcrest]; Curtis, E.C.; Kimbrew

'CURTOY', Min, rb, 1985; (**Toy Soldier**); flowers red-white blend, medium, dbl., 20 petals, no fragrance; foliage medium, medium green, semi-glossy; bushy growth; [Pink Parfait X Over the Rainbow]; Curtis, Thad; Hortico Roses

'Cuthbert Grant', S, dr, 1967; bud ovoid; flowers deep purplish red, large, semi-dbl., cupped, intermittent bloom, slight fragrance; foliage glossy; vigorous, bushy growth; [(Crimson Glory X Assiniboine) X Assiniboine]; Marshall, H.H.; Canada Dept. of Agriculture

'Cutie', Min, mp, 1952; bud pointed; flowers clear pink, base white, semi-dbl., 16 petals, 1 in., flat, slight fragrance; very few prickles; foliage small, glossy, bright green; dwarf (10 in.), bushy growth; [Dancing Doll X Oakington Ruby]; Moore, Ralph S.; Sequoia Nursery

Cutie Pie, *see* 'RENPIE'

Cuwaert, HT, 1992; Delforge, H.; (Cavriglia)

Cybele, HT, 1962; Hauser, V.; (Cavriglia)

Cyclamen, *see* 'DELBRE'

Cyclamen La Sevillana, *see* ORAuna

Cyclamen Meillandecor, *see* 'MEIPELTA'

Cygne Blanc, *see* **'WHITE SWAN'**

Cymbeline, *see* 'AUSlean'

'CYNOSURE', HT, rb, 1973; bud ovoid; flowers red, striped pink, dbl., 50 petals, 4.5 in., intense fragrance; foliage glossy, soft; vigorous, upright growth; [Scarlet Knight X Festival Beauty]; Hardikar, Dr M.N.

'CYNTHIA', HT, mr, 1934; flowers rich oriental red, dbl.; Verschuren-Pechtold; Dreer;, H&S

Cynthia, *see* 'WARDROSA'

'CYNTHIA ANN PARKER', HT, w, 1929; flowers white tinged yellow or cream, dbl., moderate fragrance; Vestal

'CYNTHIA BROOKE', HT, yb, 1943; flowers empire yellow, reverse light salmon, dbl., 45 petals, 4 in., globular, moderate, fruity fragrance; foliage leathery, dark; moderate, compact, bushy growth; [Le Progres X (Mme Mélanie Soupert X Le Progres)]; McGredy; J&P

'CYNTHIA E. HOLLIS', HMult, lp; flowers pale pink, dbl.; foliage small, glossy; Dawson; Eastern Nursery

Cynthia Westcott, S, mp, 1997; Williams, J. Benjamin

'CYNTHIE', HGal, lp; flowers pale rose, circumference almost blush, large, dbl., cupped; erect, moderate growth; Descemet, M. (?), prior to 1833

'CYPRIENNE', Pol, mr, 1969; bud ovoid; flowers light brilliant red, full, large, dbl., abundant, continuous bloom, slight fragrance; foliage glossy, light; moderate, bushy growth; [Sumatra X Fashion]; Delforge

'CYRANO', HT, rb, 1954; flowers bright red shaded purple, large, dbl., moderate fragrance; long stems; very vigorous, upright growth; [(Opera X Seedling) X Seedling]; Gaujard

Cyril Fletcher, *see* 'BEERIL'

'CZARDAS', HT, rb, 1956; flowers red, becoming pink; vigorous growth; [Tango X Seedling]; Delforge

D

'D. ANA GUEDES', HT, ob, 1938; bud ovoid; flowers orange and salmon-pink, veined yellow, large, dbl., cupped, slight fragrance; foliage soft; vigorous, bushy growth; [Angèle Pernet X Mme Méha Sabatier]; da Silva, Moreira

'D. ANGELICA PEREIRA DA ROSA', HT, pb, 1936; flowers pink, shaded golden orange, reverse red, open, very large, dbl.; foliage light; [Angèle Pernet X Edith Nellie Perkins]; da Silva, Moreira

D. D. Ruaux, *see* DORyeco

D. H. Lawrence, *see* 'ROSLAW'

'D. LAURA PINTO D'AZEVEDO', HT, pb, 1936; flowers shrimp-pink, center coral-red, base orange-yellow, very large, dbl., exhibition form; foliage light, soft; [Pink Pearl X Constance]; da Silva, Moreira

D. Malvina Loureiro, F; Moreira da Silva, A.; (Cavriglia)

'D. MARIA ANTONIA PACHECO', HT, mp, 1935; flowers deep carmine-pink, well formed, large, dbl.; foliage rich green, glossy; vigorous growth; [Mme Butterfly X Johanniszauber]; da Silva, Moreira

'D. MARIA DO CARMO DE FRAGOSO CARMONA', HT, pb, 1939; bud long, pointed; flowers flesh-pink, edged yellow, large, dbl., exhibition form; foliage glossy; long stems; vigorous growth; [Charles P. Kilham X Souv. de Claudius Pernet]; da Silva, Moreira

'D. MARIA JOSÉ DE MELO', HT, op; flowers salmon-pink; da Silva, Moreira

'D. MARIA NAVARRO', HT, rb, 1962; flowers carmine, [First Love X Paramount]; da Silva, Moreira

'D. SILVIA FERREIRA', HT, ob; flowers salmon and red, reverse old-gold; da Silva, Moreira

'D. T. POULSEN', F, mr, 1930; flowers bright blood-red, open, semi-dbl., slight fragrance; foliage dark, leathery; bushy growth; (21); [Orléans Rose X Vesuvius]; Poulsen, S.

'D. T. POULSEN IMPROVED', F, mr, 1940; bud small, globular, blood-red; flowers solid cherry-red, open, dbl., 30–35 petals; foliage leathery, wrinkled, dark; short, strong stems; vigorous, compact, bushy growth; Van der Vis; C-P

'DAB', HT, ab, 1984; flowers apricot center, pink reverse, petals tipped deep pink, large, dbl., 35 petals, exhibition form, no fragrance; foliage large, dark, glossy; bushy growth; [Lady X X Flaming Beauty]; Bridges, Dennis A.; Bridges Roses, 1985

'DACAPO', F, op, 1960; flowers deep salmon-pink, blooms in clusters, dbl., 28 petals, 3–4 in., slight fragrance; vigorous, compact, bushy growth; [Fashion X Floribunda seedling]; deRuiter; Horstmann

'DADDIES GIRL', HT, yb, 1963; flowers milky white flushed pink, base golden, dbl., 40 petals, 5–6 in., moderate fragrance; foliage dark; vigorous, spreading growth; [McGredy's Ivory X Peace]; McTeer, Gilbert

Daddy's Pink, A, lp

Dad's Pink Hill, S, mp, 1994

Dady®, *see* 'GAUZINE'

Dafne, *see* BARafne

'DAGENHAM SHOW', F, ob, 1976; flowers salmon-orange, dbl., 25 petals, 3.5 in., intense fragrance; foliage matt; bushy growth; [Elizabeth of Glamis X Seedling]; Warley Rose Gardens

Dagmar, F, lp; Urban, J., 1991; (Czech Rosa Club)

'DAGMAR SPÄTH', F, w, 1936; (Blanc Lafayette, White Lafayette); flowers white, edge flushed pink, fading pure white; [Lafayette sport]; Wirtz & Eicke; Spath

'DAGMAR SPÄTH, CLIMBING', Cl F, w, 1940; Buisman, G. A. H.

'DAGMAR SPÄTH, CLIMBING', Cl F, w, 1943; Howard Rose Co.

'DAGMAR SPÄTH, CLIMBING', Cl F, w, 1961; Huber

'D'AGUESSEAU', HGal, mr; flowers fiery crimson, occasionally shaded dark purple, large, dbl.; compact, erect growth; Vibert (?), 1836

'DAHLILA', F, yb, 1962; flowers yellow, becoming red; [Golden Perfume X Peace]; Leenders, J.

'DAIDALA'®, F, m, 1975; bud ovoid; dbl., 35 petals, 4 in., exhibition form, slight fragrance; foliage dark, soft; vigorous, upright, bushy growth; [Seedling X Silver Star]; Kordes; Willemse

Daily Express, *see* FRYchambi

'DAILY HERALD', HT, yb, 1942; bud pointed; flowers yellowish orange, dbl., 4–5 in., cupped; foliage glossy, dark; vigorous, upright growth; Robinson, T.; J&P

Daily Mail Rose, *see* **'MME EDOUARD HERRIOT'**

Daily Mail Rose, Climbing, *see* **'MME EDOUARD HERRIOT, CLIMBING'**

'DAILY MAIL SCENTED ROSE', HT, rb, 1927; flowers crimson, shaded maroon and vermilion, reverse dark crimson, intense, damask fragrance; Cup for Best New Scented Seedling, Daily Mail, 1927; [Château de Clos Vougeot X K. of K.]; Archer

'DAILY MAIL SCENTED ROSE, CLIMBING', Cl HT, rb, 1930; Archer

Daily Post, *see* 'FRYTROOPER'

Daily Sketch, *see* 'MACAI'

Daily Telegraph, *see* 'PEAHIGH'

'DAIMONJI', HT, or, 1981; flowers large blooms borne usually singly, dbl., 48 petals, exhibition form, slight fragrance; foliage medium, dark; medium, bushy growth; [Seedling X (Miss Ireland X Polynesian Sunset)]; Shibata, T.; K. Hirakata Nursery

'DAINTY', HT, pb, 1921; flowers rosy apricot, tinted cherry-pink, edges and reverse deeper pink, dbl., moderate fragrance; Dickson, H.

'DAINTY', Pol, op, 1931; flowers salmon-pink, cupped, deRuiter

'DAINTY BESS', HT, lp, 1925; flowers soft rose-pink, very distinct maroon stamens, medium, single, 4–11 petals, 5 in., borne in clusters of 5 to 9 single blooms, moderate, tea fragrance,; foliage leathery, dark green; vigorous growth; GM, NRS, 1925; [Ophelia X K. of K.]; Archer

'DAINTY BESS, CLIMBING', Cl HT, lp, 1935; van Barneveld; California Roses

Dainty Bouquet, *see* 'LAVFUHR'

'DAINTY DAWN', Pol, mp, 1931; flowers cerise-pink to mauve, blooms in clusters, semi-dbl., cupped, moderate fragrance; foliage bronze; [Amaury Fonseca X Anchen Müller]; Knight, G.

'DAINTY DELIGHT', Cl HT, mp, 1949; bud ovoid; flowers darker pink than Dainty Bess, medium, borne in clusters, semi-dbl., globular, slight fragrance; foliage glossy; vigorous (6–8 ft) growth; [Ednah Thomas X Dainty Bess]; Duehrsen; California Roses

Dainty Dinah, *see* 'COCAMOND'

'DAINTY LADY', HT, mp, 1959; bud long, pointed; flowers pink, medium, dbl., exhibition form, moderate fragrance; foliage leathery; vigorous, upright, compact growth; [(Girona X Pres. Herbert Hoover) X Michèle Meilland]; Peden, G.H.

'DAINTY LADY', F, op, 1963; flowers coppery salmon-pink, well formed, borne in clusters; moderate growth; [Baby Sylvia X Seedling]; Fryers Nursery, Ltd.

'DAINTY MAID', F, pb, 1940; bud pointed, cerise; flowers silvery pink, reverse carmine, blooms in clusters, single; foliage leathery, dark; vigorous, compact, bushy growth; (28); GM, Portland, 1941; [D.T. Poulsen X Seedling]; LeGrice; C-P

'DAINTY SUPERIOR', Pol, op; flowers have more lasting color; [Dainty sport]; deRuiter

'DAIRY MAID', F, ly, 1957; bud yellow, splashed carmine; flowers cream, fading white, blooms in large clusters, single, 5 petals, 3.5 in.; foliage glossy; vigorous growth; [(Poulsen's Pink X Ellinor LeGrice) X Mrs Pierre S. duPont]; LeGrice

'DAISY', HT, ob, 1923; flowers orange-flamed; [Mme Edouard Herriot sport]; Hicks

'DAISY BRASILEIR', HMult, rb, 1918; flowers bright red and purple-red, anthers yellow, borne in clusters, single, slight fragrance; Turbat

'DAISY BUD', HT, pb, 1933; bud well shaped; flowers rosy pink, shaded carmine and silver, large, moderate fragrance; Dickson, A.

'DAISY DOLL', Min, pb, 1977; bud long, pointed; flowers rose-pink, open, small, dbl., 20–25 petals, 1.5 in., slight fragrance; foliage tiny; compact, upright growth; [Little Amy X Seedling]; Lyon; L. Lyon Greenhouses

'DAISY DUMAS', HT, rb, 1962; flowers bright red, reverse gold edged red, medium, dbl., slight fragrance; foliage dark; bushy growth; [Seedling X (Tzigane X Seedling)]; White; A. Ross & Son

'DAISY HILL', S, mp, 1906; flowers large, single, moderate fragrance; abundant fruit; height 8 ft; [R. X waitziana macrantha hybrid]; Kordes

'DAISY HILLARY', F, op, 1963; bud pointed; flowers salmon-pink, large, borne in clusters, dbl., exhibition form; foliage leathery; vigorous, bushy growth; [Spartan sport]; Mell

'DAISY MAE', HT, dy, 1987; flowers deep yellow, reverse medium yellow, large, borne singly, single, 6 petals, flat; round, ovoid, russet-yellow fruit; large, red down-curved prickles, fading gray-tan; foliage large, dark green, semi-glossy; upright, tall growth; [Golden Showers X Golden Sun]; Stoddard, Louis

'DAISY MAY ROGERS', HMsk, w, 1996; single, 4–7 petals, 1 in., borne in large clusters, slight fragrance; small, moderate prickles; foliage small, medium green, semi-glossy; compact, spreading, medium (4 ft x 5 ft) growth; [Mozart X Unknown]; Muia, Charlotte R.; My Green Thumb, 1992

Daisy Rose, *see* 'KORDAISY'

'DAKAR', HT, pb, 1932; flowers silvery pink, striped rose-pink, large, intense fragrance; long, strong stems; [Julien Potin X Seedling]; Gaujard

Dakota, HT, dr; flowers exhibition form, dbl., borne one per stem; foliage medium, dark green, matte

'DALE FARM', F, or, 1973; flowers vermilion, dbl., 25 petals, 3 in., moderate fragrance; foliage dark; vigorous growth; Smith, E.; Wheatcroft & Sons

Dale's Pink Glow, HMsk, mp, 1994

Dale's Pink Mini Climber, Cl Min, mp

Dale's Ragged Pink, Misc OGR, mp

Dale's Sunrise®, *see* 'KINSUN'

'DALILA', HT, rb, 1943; bud long, pointed; flowers brilliant coppery red, large, semi-dbl., moderate fragrance; foliage leathery; vigorous, erect growth; [Souv. de Claudius Pernet X Seedling]; Gaujard

'DALILA', F, ob, 1958; flowers orange, dbl.; Buyl Frères

Dallas, *see* 'KORLIMIT'

'DALLAS', HT, rb, 1963; flowers crimson-carmine, base primrose-yellow, very large, dbl., 40 petals; foliage dark, glossy; vigorous growth; [Peace sport]; Hunter; Waterhouse Nursery, 1963

'DALLAS GOLD', HT, yb, 1987; flowers large, heavy blooms, dbl., 25 petals, slight fragrance; foliage medium, dark green, glossy; bushy growth; Silver, ARC TG, 1984; [Seedling X Flaming Beauty]; Winchel, Joseph F.; Kimbrew Walter Roses, 1987

Dalli Dalli®, *see* 'TANLILIDA'

'DALVEY', HT, dp, 1971; flowers deep pink, 5.5 in., exhibition form, intense fragrance; foliage medium, light green, matt; vigorous growth; [Peeping Tom X Seedling]; MacLeod

Dama diCuori, *see* **'DAME DE COEUR'**

Damas de Yuste, *see* **'FREIHEITSGLOCKE'**

'DAMAS FRANKLIN', D, pb, 1853; flowers flesh-pink, shading to silver; Robert, ca. 1856

Damask Rose, *see* **'SUMMER DAMASK'**

'DAME BLANCHE', HWich, w, 1923; flowers greenish white, stamens yellow, borne in clusters, single, non-recurrent bloom; vigorous growth; Turbat

'DAME BLANCHE', HT, w, 1927; (White Lady); flowers white tinged green, dbl., moderate fragrance; [Stadtrat Glaser seedling]; Mühle

Dame Cath, *see* 'MACHILVER'

'DAME CATHERINE', HT, dy, 1937; bud ovoid; flowers golden yellow, large, dbl., exhibition form; foliage glossy; long stems; vigorous, bushy growth; Cant, B. R.

'DAME DE COEUR', HT, mr, 1958; (Dama diCuori, Herz-Dame, Queen of Hearts); flowers cherry-red, large, dbl., moderate fragrance; foliage dark, glossy; vigorous growth; [Peace X Independence]; Lens

'DAME DE COEUR, CLIMBING', Cl HT, mr, 1984; [Dame de Coeur sport]; Mungia, Fred A., Sr.

Dame de L'Etoile®, *see* ADAburi

'DAME EDITH HELEN', HT, mp, 1926; flowers glowing pink, very large, dbl., cupped, not very free bloom, moderate fragrance; foliage leathery; long, strong stems; vigorous, bushy growth; GM, NRS, 1926; Dickson, A.

'DAME EDITH HELEN, CLIMBING', Cl HT, mp, 1930; H&S

Dame Joyce Frankland, *see* 'HORNEWGRAM'

'DAME OF SARK', F, ob, 1976; flowers orange flushed red, reverse yellow, dbl., 33 petals, 4.5 in., slight fragrance; foliage large, dark; [(Pink Parfait X Masquerade) X Tablers' Choice]; Harkness

'DAME PRUDENCE', S, lp, 1969; flowers soft pink, reverse lighter, medium, dbl., 65 petals, flat, intense fragrance; [Ivory Fashion X (Constance Spry X Ma Perkins)]; Austin, David

Dame Vera, *see* 'PEAMAX'

Dame Vera Lynn, *see* 'PEAMAX'

Dame Wendy, *see* CANson

Dames Patronesses d'Orleans, HP, dr; flowers crimson red; Vigneron, 1877

Damien's Amulet, T, lp

'DAMON RUNYON', HT, mr, 1955; bud ovoid; flowers crimson, dbl., 50 petals, 4–5.5 in., exhibition form, moderate fragrance; foliage glossy, coppery green; vigorous, upright, bushy growth; [Major Shelley X Heart's Desire]; Duehrsen; H&S

'DANA', HT, mp, 1982; bud long, pointed; flowers spiraled, formal blooms borne singly, sometimes 3 per cluster, dbl., 38 petals, slight, carnation fragrance; long, straight prickles; foliage large, semi-glossy; tall, upright, bushy growth; [White Satin X Bewitched]; Swim, H.C. & Ellis, A.E.; Armstrong Nursery

'DANAË', HMsk, ly, 1913; flowers pale buff-yellow, fading white, in clusters, recurrent bloom; height 6 ft; (14); [Reputedly Trier X Gloire de Chedane-Guinoiseau]; Pemberton

'DANCE OF JOY', F, mr, 1931; flowers vivid scarlet-crimson, large, dbl., moderate fragrance; foliage dark; vigorous growth; (28); GM, Bagatelle, 1931; [Paul's Scarlet Climber X Seedling]; Sauvageot, H.

Dance of Joy 95, *see* SAUdero

'DANCING DOLL', Cl F, dp, 1952; bud small, pointed; flowers deep rose-pink, borne in clusters, semi-dbl., 10–14 petals, cupped, profuse, repeated bloom, intense fragrance; foliage leathery, glossy; vigorous, climbing or spreading (10 ft) growth; [Étoile Luisante X Seedling]; Moore, Ralph S.; Marsh's Nursery

Dancing in the Wind, *see* CLEdan

Dancing Pink, *see* 'HENDAN'

'DANCING SILK', HT, op, 1966; flowers coral-pink, reflexed, 5 in., slight fragrance; foliage light green; vigorous growth; [Ena Harkness X McGredy's Yellow]; Barter

'DANDEE', Min, dr, 1983; bud globular; dbl., 35 petals, flat, borne singly, moderate fragrance; foliage medium, dark, semi-glossy; upright growth; [Seedling X Libby]; Meredith, E.A. & Rovinski, M.E.; Casa de Rosa Domingo

Dandenong, MinFl, mr

'Dandy', Pol, op, 1945; flowers light orange-pink, small, semi-dbl., cupped, profuse, non-recurrent bloom, slight fragrance; foliage wrinkled, glossy, dark; very vigorous growth; [Gloria Mundi X Unidentified Sp rose]; Wiseman

'Dandy Dick', F, mp, 1967; flowers well-formed, large blooms in clusters, dbl., 25 petals, moderate, spicy fragrance; foliage light; [Pink Parfait X Red Dandy]; Harkness

'Dandy Lyon', Min, dy, 1978; bud long, pointed; flowers buttercup-yellow, dbl., 30 petals, 2 in., moderate fragrance; foliage small, dark, glossy; compact, bushy growth; [Seedling X Sunspot]; Lyon

Dania, *see* 'COCgold'

Danica, *see* NOArda

'Daniel', HT, rb, 1943; bud long, pointed; flowers capucine-red on golden yellow base, dbl., cupped; foliage dark, glossy; vigorous, bushy growth; Mallerin, C.; A. Meilland

Daniel, HT, dr, 1995; flowers large, exhibition form, intense fragrance; Williams, J. Benjamin

Daniel Boone, HT, dr, 1967; Morey, Dr Dennison

Daniel Gelin, *see* 'HARquince'

'Daniel Lacombe', HMult, yb, 1885; flowers yellow washed pink, medium, dbl., flat; vigorous growth; [R. multiflora X Général Jacqueminot]; Allard

'Daniel Philip', F, mr, 1999; semi-dbl., 12–16 petals, 1.2–1.5 in., borne in small clusters, slight fragrance; prickles moderate; foliage small, medium green, semi-glossy, disease resistant; upright, medium (6 ft) growth; [(Morden Fireglow X Unknown) X seedling]; Fleming, Joyce L.; Hortico, Inc., 1999

Daniela®, *see* KORlunta

Daniele, Min, 1991; Kordes, R.; (Cavriglia)

Danielle®, HT, w, 1992; Cocker

Danielle, *see* 'BROdan'

'Danielle Darrieux', HT, pb, 1948; bud long, pointed; flowers salmon suffused yellow, reverse salmon-orange-pink, very large, dbl., moderate fragrance; very vigorous, bushy growth; Gaujard

'Danielle Robyn', HT, rb, 1971; bud globular; flowers pinky red and creamy white bicolor, large, dbl., exhibition form, intermittent bloom, slight fragrance; foliage glossy, dark; bushy growth; [Grand Gala X Western Sun]; Hastie

Danina, *see* **'Cara Mia'**

'Daniphyl', HT, mr, 1978; flowers mandarin-red, full, dbl., 45–50 petals, 4–5 in., moderate fragrance; foliage semi-glossy, light; vigorous, upright growth; [Puccini X Chopin]; Ellick; Excelsior Roses

'Danish Gold', F, yb, 1949; flowers yellow fading to creamy white, single, 5–9 petals, 2.5–3 in., moderate, spicy fragrance; foliage glossy; vigorous, compact growth; GM, NRS, 1949; [(Golden Salmon X Souv. de Claudius Pernet) X Julien Potin]; Poulsen, S.; McGredy

'Danish Pink', F, dp, 1965; flowers deep pink, almost medium borne in large clusters, single, 2–2.5 in., moderate, fruity fragrance; vigorous, tall growth; Soenderhousen; Hoersholm Nursery

'Danmark', HT, mp, 1891; flowers large, dbl., intense fragrance; Zeiner-Larsen; (Sangerhausen)

'Dannenberg', HT, op, 1916; flowers coral pink, large, dbl., moderate fragrance; Kiese; (Sangerhausen)

'Dannie's Smile', HT, lp, 1994; dbl., 26–40 petals, 1.5–2.75 in., borne mostly singly, slight fragrance; some prickles; foliage medium, dark green, matt; medium, spreading growth; [Elizabeth Taylor sport]; Maxheimer, Joanne; Giles Rose Nursery, 1993

'Danny Boy', LCl, or, 1969; flowers well-formed, recurrent bloom, intense fragrance; foliage dark; [Uncle Walter X Milord (HT)]; McGredy, Sam IV; McGredy

'Danny Thomas', Min, mr, 1980; flowers yellow stamens, borne usually singly, dbl., 35 petals, moderate fragrance; straight prickles; foliage dark; compact growth; [Rose Hills Red X Self]; Wells, V. W., Jr; Lou McGuire

'DANorang', HT, or, 1972; (**Orange Tango**); flowers large, very dbl., exhibition form, slight fragrance; foliage bronze; vigorous growth; [From unnamed seedlings sport]; McDaniel, Earl; Carlton Rose Nurseries

'Danse Azteque', Min, yb, 1985; (Rigobec 3); flowers yellow-pink blend, medium, semi-dbl., flat, slight fragrance; foliage medium, medium green, semi-glossy; vigorous, upright growth; [Baby Masquerade seedling]; Gailloux, Gilles

Danse des Étoiles, *see* 'GODsensor'

Danse des Sylphes®, *see* 'MALcair'

Danse du Feu, *see* **'Spectacular'**

'Danubio Azul', HT, m, 1957; flowers lilac, medium, dbl., cupped, slight fragrance; upright growth; [Tristesse X Independence]; Camprubi, C.

'Dany Robin', F, op, 1958; bud pointed; flowers salmon-pink, open, medium, borne in clusters, dbl., slight fragrance; foliage leathery; vigorous, compact growth; [Goldilocks X Fashion]; Meilland, F.; URS

Danyrose, HT, 1975; Croix, P.; (Cavriglia)

'Danzig', HT, dr, 1940; flowers shining dark red, medium, borne in clusters, dbl.; upright growth; [Hadley X Kardinal]; Tantau

Danzille, *see* **'Mme Bravy'**

'Daphne', HMsk, lp, 1912; flowers blush-pink, blooms in clusters, semi-dbl., moderate fragrance; vigorous growth; Pemberton

'Daphne', HT, lp, 1925; flowers soft pink, flushed rose, well-formed, dbl., moderate fragrance; Dobbie

'Daphne', HGal, mr; flowers bright red, cupped; Vibert, 1819

'Daphné', HGal, dp, 1819; flowers medium, dbl.; Vibert; (Sangerhausen)

Daphne Claire Jones, *see* 'KIRjack'

'Daphne Gandy', F, mr, 1952; flowers blood-red, blooms in large trusses, 3 in., slight fragrance; foliage dark; vigorous growth; [Farida X Crimson Glory]; Leenders, M.; Gandy Roses, Ltd.

'Daphnis', HT, mp, 1974; flowers brilliant pink, very large; upright growth; [Marylène X Mignonne]; Gaujard

Dapple Dawn, *see* AUSapple

'Dara', HT, ob, 1991; dbl., 30 petals, exhibition form, borne usually singly, intense fragrance; foliage medium, medium green, semi-glossy; upright growth; [Olympiad X Just Joey]; Wambach, Alex A.

Darby, *see* 'DEVcico'

Darby O'Gill, *see* 'SEAdarby'

Darcelle, *see* 'SPOdarc'

'd'Arcet', M, dr, 1851; flowers large, dbl.; Robert; (Sangerhausen)

'Dardanelle', HT, mr, 1926; flowers cherry-rose, dbl., moderate fragrance; foliage wrinkled, dark; bushy growth; [Premier X Ophelia]; Vestal

'Darius', HGal, mr; flowers vivid red, large, dbl.

'Dark Boy', HT, dr, 1965; bud ovoid; flowers velvety dark maroon-red, medium, dbl., slight fragrance; foliage soft; moderate, upright, open growth; [Nigrette X Seedling]; Pal, Dr B.P.; Indian Agric. Research Inst.

Dark Mirage, *see* 'SEAdark'

'Dark Secret', HT, dr, 1937; flowers large, dbl., globular, intense fragrance; foliage leathery; short stems; dwarf growth; [Radiance X Hollywood]; Amling Co.

Darling, *see* 'SUNcredel'

'Darling', HT, lp, 1958; bud ovoid; flowers large, dbl.,moderate, fruity fragrance; foliage dark, glossy; vigorous, bushy growth; [Pink Princess X Charlotte Armstrong]; Taylor, C.A.; California Nursery Co.

Darling, *see* KORspatax

Darling, HT, op, 1985; Noack, Werner

'Darling Annabelle', HT, w, 1992; flowers white with faint pink center, excellent blooms, dbl., 26–40 petals, 3 in., exhibition form, moderate fragrance; foliage large, medium green, matt; upright (170 cms) growth; [South Seas X Peace]; Perry, Astor; Hortico Roses, 1993

Darling Diane, *see* 'BOScherry'

Darling Flame, *see* 'MEIlucca'

'Darling Jenny', HT, pb, 1996; flowers deep pink shaded peach, large blooms borne mostly single, dbl., 26–40 petals, slight fragrance; some prickles; foliage medium, dark green, semi-glossy; bushy, medium growth; [Solitaire X Gavotte]; Poole, Lionel; Poole, 1997

Darlow's Enigma, HMsk, w; Darlow

'D'Artagnan', D, dr, 1969; flowers wine-red, blooms in clusters, dbl., cupped, non-recurrent, intense, damask fragrance; foliage dark, leathery, wrinkled; vigorous, upright (6 ft) growth; [Ma Perkins X York and Lancaster]; Fankhauser

D'Artagnan, *see* LAPniro

Darthuizer Orange Fire, *see* 'INTerfire'

Dart's Dash, HRg, dr

Dart's Defender, S, m, 1971; Darthuis

Dart's Red Dot®, *see* INTermunder

'Das Goldene Prag', *see* **'Zlatá Praha'**

Dasher, *see* 'SPOdash'

'D'Assas', HP, pb, 1850; flowers dark pink, tinged crimson, petals somewhat fringed, medium, dbl.; vigorous, straggling growth; Vibert

Daughter Margaret, *see* 'NOBam'

'Dauntless', HT, pb, 1949; bud long, pointed; flowers pink, reverse yellow, large, dbl., 60 petals, exhibition form, moderate fragrance; foliage leathery, dark; upright growth; [Crimson Glory X Feu Pernet-Ducher]; Davis

'Dauphine', F, op, 1955; bud ovoid; flowers pink shaded salmon, open, large, borne in clusters, dbl., moderate fragrance; very vigorous growth; [Seedling X Opera seedling]; Gaujard

'Dauphine, Climbing', Cl F, op, 1959; Gaujard

'Dave Davis', HT, dr, 1964; bud long, pointed; flowers dark velvety red, large, dbl., 60 petals, exhibition form, intense fragrance; foliage leathery; moderate growth; [Seedling X Charles Mallerin]; Davis; Wyant

Dave Hessayon, *see* 'DRIscobruce'

'David Arnot', HT, rb; flowers bright scarlet, reverse old-gold; vigorous growth

David Charles Armstrong, *see* 'DRIscogeorge'

David Dot, Min, 1978; Dot, Simon; (Cavriglia)

'David Gilmore', HT, mr, 1923; flowers brilliant scarlet; Dickson, H.

'David Gold', HT, rb, 1957; flowers cherry-cerise tinted golden yellow, 6 in., exhibition form, intense fragrance; foliage dark, glossy; very vigorous growth; [Shot Silk X Peace]; Robinson, H.

David Leek, F, dp, 1996; Fleming, Joyce L.

'David McKee', HT, mr, 1933; flowers carmine-red, large, dbl., exhibition form, intense fragrance; foliage leathery; vigorous, bushy growth; Dickson, A.

'David O. Dodd', HT, mr, 1926; flowers rich crimson flushed scarlet, large, dbl., intense fragrance; foliage glossy; bushy growth; Vestal

'David O. Dodd, Climbing', Cl HT, mr, 1937; Howard Rose Co.

David Ruston, HT, pb

'David Thompson', HRg, mr, 1979; bud ovoid; flowers yellow stamens, dbl., 25 petals, 2.5 in., intense fragrance; upright growth; [(Schneezwerg X Frau Dagmar Hartopp) X Seedling]; Svedja, Felicitas; Canada Dept. of Agric.

David Whitfield, *see* GANa

Davidoff, HT, op

Davina Jane, HT, ob, 1997; flowers full, medium, very dbl., 26–40 petals, borne mostly singly; foliage medium, medium green, semi-glossy; upright, medium (3 ft) growth; [Sexy Rexy X Corso]; Rawlins, R.

Davit Dot, S, mr

'DAVona', F, dr, 1982; (**Red Splendour**); flowers deeper red; [Europeana sport]; Davies, Gareth, 1979

'Davy Crockett', F, dr, 1956; flowers large, dbl.; vigorous growth; [Étoile de Hollande X Floribunda seedling]; deRuiter; Gandy Roses, Ltd.

'Dawn', HT, op, 1953; bud long, pointed; flowers salmon-pink, base yellow, dbl., 28–36 petals, 5–6. in., exhibition form, moderate, spicy fragrance; vigorous, upright growth; Jelly; E.G. Hill Co

Dawn Chorus, *see* 'DICquasar'

Dawn Creeper™, *see* 'WILdawn'

'Dawn Fragrance', LCl, op, 1969; flowers salmon-flesh, flushed rose-pink, well formed, cupped, intermittent bloom, intense fragrance; foliage dark, leathery; vigorous growth; [Blossomtime X Blossomtime seedling]; Mason, P.G.

Dawn Haggle, *see* KORboden

'Dawn Mist', F, pb, 1962; bud ovoid; flowers reddish pink, base yellow, dbl., 30–35 petals, 3 in., cupped, intense, spicy fragrance; foliage leathery, glossy; vigorous, bushy, compact growth; [(Goldilocks X Pinocchio) X Vogue]; Boerner; Home Nursery Products Corp.

'Dawn Pink', F, mp, 1962; bud apricot; flowers pink, rosette form, borne in clusters, dbl., 45–50 petals, 3.5 in., moderate fragrance; foliage light green; vigorous, upright growth; [Magenta X Ma Perkins]; Anderson's Rose Nurseries

Dawn Sunsation, *see* KORfeining

Dawn Weller, *see* POUlsum

'Dawnglow', HT, pb, 1937; flowers flesh-pink, dbl., 3.5 in., exhibition form, moderate fragrance; Burbank

'Dawning, Climbing', LCl, op, 1956; flowers salmon-pink, moderate fragrance; foliage glossy; vigorous growth; [New Dawn X Margaret McGredy]; Bennett, H.; Pedigree Nursery

'Dawnlight', F, pb, 1958; bud ovoid; flowers soft pink, dbl., 30–35 petals, 2–2.5 in., flat, slight fragrance; thornless; foliage leathery, light green; vigorous, bushy growth; [Summer Snow, Climbing X Summer Snow]; Motose; G.B. Hart, Inc.

'Dawns Early Light', HT, w, 1972; bud long, palest pink; flowers white, faintly edged pale pink, pointed, bloom lasts well, single, 6 petals, 3.5–4 in., intense fragrance; foliage light; tall growth; [Vesuvius X Vesuvius]; Linscott

'Dawson', HMult, mp, 1888; flowers bright rose-pink, small blooms in clusters of 10–20, dbl.; vigorous (10–25 ft) growth; [R. multiflora X Général Jacqueminot]; Dawson; Strong

Dawson's Climber, Cl HT, lp; Dawson, 1960; (Weatherly, L.)

Dawson's Delight, HT, pb, 1978; Dawson

Dawson's Yellow Climber, LCl, my

Day Break, *see* **'Daybreak'**

'Day Dream', HT, dp, 1969; bud long, pointed; flowers deep pink, large, dbl.; foliage glossy, leathery; vigorous, upright, bushy growth; [Helen Traubel X Tiffany]; Armstrong, D.L.; Armstrong Nursery

Day Glow, *see* 'JACrink'

'Day Is Done', HT, or, 1985; flowers large, dbl., 35 petals, exhibition form, slight fragrance; foliage large, dark, semi-glossy; [Die Welt X Rosalynn Carter]; Schneider, Peter

Day Light®, *see* 'INTerlight'

'Day of Triumph', HT, mp, 1953; (Rendezvous); bud ovoid; flowers pink edged lighter, dbl., 50–65 petals, 5.5 in., cupped, moderate fragrance; foliage leathery; vigorous, upright, bushy growth; [Peace X Europa]; Meilland, F.; Breedlove Nursery, 1955;, URS, 1953

'Daybreak', HMult, op, 1909; flowers deep salmon-pink, borne in clusters, late; [R. wichurana X R. chinensis]; Dawson; Eastern Nursery

'DAYBREAK', HMsk, my, 1918; (Day Break); flowers golden yellow, single, recurrent bloom; foliage dark; vigorous, bushy growth; [Trier X Liberty]; Pemberton

'DAYBREAK', HT, pb, 1935; bud large, long, pointed; flowers salmon-pink and yellow, very dbl., exhibition form, slight fragrance; foliage leathery; long stems; very vigorous growth; [Violet Simpson X Ivy May]; Laxton Bros.

'DAYBREAK', S, dp, 1960; flowers deep pink, dbl., 25–30 petals; [Hansa X R. macounii]; Erskine

Daybreak, *see* **'RASSVET'**

Daydream, *see* 'DICJEEP'

'DAYDREAM', Cl HT, lp; waterlily form, blush pink, semi-dbl.; vigorousd bush or pillar; [Souvenir de Gustave Prat X Rosy Morn or Gwen Nash]; Clark, A., 1925; Hazelwood Bros., 1925, AUS

Daylight, *see* 'INTERLIGHT'

'DAYLIGHT', HT, pb, 1939; flowers creamy blush-pink, base yellow, very dbl., 4 in., exhibition form, moderate fragrance; foliage soft; vigorous growth; RULED EXTINCT 4/92 ARM; [Grange Colombe X Los Angeles]; Hansen, N. J.; B&A

'DAYLIGHT KATY', F, my, 1987; flowers bright yellow, fading slightly paler, urn-shaped, medium, dbl., 22 petals, borne in sprays of 3–15, slight fragrance; round, small, red fruit; pointed, medium, reddish prickles; foliage medium, bronze-green, semi-glossy; bushy, low growth; [Bright Smile X (Princess Michael of Kent X Party Girl)]; Schneider, Peter

Daytona, *see* OLIjzouc

'DAZLA', HT, rb, 1930; bud long, pointed; flowers orange-scarlet, base and reverse golden yellow, wavy petals, semi-dbl., 6 in.; foliage dark; vigorous growth; Cant, B. R.

'DAZLA, CLIMBING', Cl HT, rb, 1950; Cant, B. R.

Dazzler, *see* **'LEONIE'**

Dazzler®, *see* 'KELDAZ'

Dazzler, *see* GENpat

Dazzler, Min, pb, 1992; Kasturi

'DE CANDOLLE', M, lp, 1857; flowers soft pink or rose-tinted, large, dbl.; Portemer fils

'DE CANDOLLE', HWich, op, 1913; Robichon

'DE GREEFF'S JUBILEE', F, dr, 1980; (Fellowship); flowers large, semi-dbl., slight fragrance; foliage medium, dark, semi-glossy; bushy growth; [Diablotin X Gisselfeld]; Verschuren, Ted; Verschuren & Sons

De Kat, HT, w, 1995

'DE LA GRIFFERAIE', HMult, dp, 1845; flowers carmine to pink, medium, dbl.; robust growth; (28); Vibert

'DE LUXE', HT, rb, 1931; flowers bright velvety scarlet, reverse red, semi-dbl., slight fragrance; [Premier sport]; White Bros.; Liggit

De Meaux, *see* **'ROSE DE MEAUX'**

De Meaux White, *see* **'ROSE DE MEAUX WHITE'**

De Montarville, S, mp, 1998; flat, 19 petals, slight; L'Assomption, 1982

De Montarville, *see* 'AC DE MONTARVILLE'

'DE SCHELFHOUT', HGal, lp, 1840; flowers medium, dbl.; Parmentier; (Sangerhausen)

'DEAN COLLINS', Gr, dp, 1953; bud ovoid, red; flowers deep pink, dbl., 53 petals, 4.5–5 in., slight fragrance; foliage dark, glossy, leathery; vigorous growth; [Charlotte Armstrong X Floradora]; Lammerts, Dr Walter; Roseway Nursery

'DEAN HOLE', HT, pb, 1904; flowers silvery carmine, shaded salmon, dbl., moderate fragrance; Dickson, A.

Deantoni, HT, rb

Dear Eleanor, *see* 'GELELE'

Dear One, *see* **'KARA'**

Dear Prudence, *see* 'TALDEARPRU'

'DEAREST', F, pb, 1960; flowers rosy salmon-pink, gold stamens, well-formed, blooms in clusters, dbl., 30 petals, 3.5 in., moderate fragrance; foliage dark, glossy; vigorous, bushy growth; GM, NRS, 1961; [Seedling X Spartan]; Dickson, A.

Dearest, Cl F, op, 1970; Ruston, D.

Dearest One, *see* **'CARA MIA'**

Dearest, Climbing, Cl F, lp; Ruston, D., 1970; (Weatherly, L.)

'DEBBIE', Min, yb, 1966; flowers yellow, edges becoming pink, small, dbl., moderate fragrance; foliage small, leathery; bushy, low, sometimes semi-climbing growth; [Little Darling X Zee]; Moore, Ralph S.; Sequoia Nursery

Debbie, S, dp; Poulsen

Debbie Dawn, F, lp, 1997; flowers double, medium, dbl., 15–25 petals, borne in small clusters, no fragrance; foliage medium, dark green, semi-glossy; upright (2 ft 6 in.) growth; [Margaret Merril X Melrose]; Rawlins, R.

'DEBBIE LYNN', HT, dr, 1992; flowers urn-shaped, blooms borne usually singly, dbl., 36 petals, 4.75 in., exhibition form, moderate fragrance; foliage medium, dark green, semi-glossy; upright, bushy, medium growth; Jerabek, Paul E., 1986

'DEBBIE THOMAS', HT, rb, 1996; flowers crimson, lighter reverse, large, very dbl., exhibition form, slight fragrance; numerous prickles; foliage medium, medium green, semi-glossy; upright, tall growth; [City of Gloucester X My Joy]; Thomas, D.; F. Haynes & Partners, 1992

'DEBBIE-KAREN', HT, mp, 1983; dbl., 4.5–5.5 in., exhibition form, borne singly, slight fragrance; foliage large, light green, semi-glossy; vigorous, upright growth; [Seedling X Seedling]; Burdett, H. J., 1984

Debidue™, *see* 'MICDEB'

'DEBONAIR', HT, my, 1946; bud ovoid; flowers primrose-yellow, dbl., 28–35 petals, 3–4.5 in., exhibition form, moderate fragrance; foliage leathery, glossy, dark; very vigorous, upright, bushy growth; [Golden Rapture X Seedling]; Lammerts, Dr Walter; Armstrong Nursery

Deborah®, *see* MEInoiral

Deborah Devonshire, *see* 'BOSCHERRYDRIFT'

Deborah Beggs Moncrief,

Deborah Moncrief™, *see* 'WILDEB'

Deborah Moncrief™,

Debra Gaye, *see* 'TINDEB'

'DEBRAD', F, mr, 1982; (**Royal Baby**); flowers medium, dbl., 20 petals, slight fragrance; foliage small, medium green, semi-glossy; upright growth; [Generosa X Baby Darling]; Bracegirdle, Derek T.; Arthur Higgs Roses

'DEBRAF', HT, yb, 1986; (**Wendy Pease**); flowers medium, dbl., 35 petals, slight fragrance; foliage medium, medium green, semi-glossy; upright growth; [John Waterer X Tenerife]; Bracegirdle, Derek T.

'DEBRAH', F, mr, 1986; (**Mary Sheffield**); flowers moderately medium, dbl., 15–25 petals, moderate fragrance; foliage small, medium green, glossy; spreading growth; [Doris Tysterman X Admiral Rodney]; Bracegirdle, Derek T., 1987

'DEBRARO', HT, dy, 1984; (**William Walker**); flowers medium, dbl., 20 petals, no fragrance; foliage medium, medium green, semi-glossy; upright growth; [Western Sun X Circus]; Bracegirdle, Derek T.

Deb's Delight, *see* 'LEGSWEET'

Debut™, *see* 'MEIBARKE'

'DEBUTANTE', HMult, lp, 1902; flowers rose-pink, fading to cameo-pink, borne in clusters of 4–6, dbl., non-recurrent, moderate, sweetbriar fragrance; foliage dark, glossy; short stems; height 6–8 ft; [R. wichurana X Baroness Rothschild]; Walsh

Debutante, *see* 'JACINAL'

Deccan Delight, F, yb, 1985; Chiplunkar

Deccan Deluxe, HT, m, 1988; Chiplunkar

'DECEA ANN', F, ab, 1975; flowers flushed peach-pink, full, dbl., 26 petals, 3.5–4 in., intense fragrance; foliage dark, leathery; upright growth; [Queen Elizabeth X Elizabeth of Glamis]; Horsfield

'DECEPTION', HT, dp, 1923; flowers deep rose, large, moderate fragrance; GM, NRS, 1923; Beckwith

Declic, S, m, 1988; Croix

'DÉCOR', LCl, or, 1951; (Record); flowers bright scarlet, blooms in clusters, semi-dbl.; foliage leathery; vigorous growth; [(Love X Paul's Scarlet Climber) X Demain]; Mallerin, C.; URS

Decor Arlequin, *see* 'MEIZOURAYOR'

Decor Rose, *see* 'MEITURAPHAR'

Decora, C, mp; flowers small, very dbl., slight fragrance; Roseraie de l'Hay; (Sangerhausen)

Decorat, *see* 'DEKORAT'

'DECORATOR', HT, rb, 1936; flowers brilliant carmine, base orange-yellow; foliage glossy; very vigorous growth; Dickson, A.

Decosilrose®, HT, pb

Decosilthe®, HT, my

'DEDICATION', F, ly, 1968; flowers creamy ivory, borne in trusses, dbl., slight fragrance; foliage glossy; [Pink Parfait X Circus]; Harkness

Dee Bennett™, *see* 'SAVADEE'

'DEEP PURPLE', F, m, 1980; bud ovoid, pointed; flowers mauve-pink, imbricated, dbl., 30–45 petals, 3–4 in., moderate fragrance; foliage glossy, dark; vigorous, upright, bushy growth; [Zorina X Silver Star]; Kordes, R.; Armstrong Nursery

'DEEP SECRET', HT, dr, 1977; (Mildred Scheel); flowers deep crimson, dbl., 40 petals, 4 in., intense fragrance; foliage glossy, dark; vigorous, upright growth; ADR, 1978; Tantau, Math.; Wheatcroft, 1979

'DEEP SECRET', HT, 1946; flowers rich red becoming darker, compact, intense fragrance; RULED EXTINCT 4/77 ARM; [Matchless X Seedling]; Hildebrandt

'DEEP VELVET'™, Min, dr, 1981; flowers urn-shaped blooms borne singly and 2–3 per cluster, dbl., 33 petals, exhibition form, slight fragrance; straight prickles; foliage tiny, medium green; bushy, compact growth; [(Grand Opera X Jimmy Greaves) X Baby Katie]; Jolly, Betty J.; Rosehill Farm

'DEEPIKA', F, rb, 1975; bud pointed; semi-dbl., 15 petals, 2 in., free bloom, fairly lasting; foliage glossy, dark; vigorous, upright, open growth; [Shepherd's Delight X Seedling]; IARI

'DEEPSHIKHA', F, mr, 1975; bud pointed; flowers open, dbl., 35 petals, 2.5 in., slight, tea fragrance; foliage glossy, light; vigorous, compact, bushy growth; [Sea Pearl X Shola]; IARI

'DÉESSE', HT, rb, 1957; (Goddess); bud long, pointed, white; flowers red spreading from outside to center petals, ending crimson, dbl.; vigorous, upright growth; [Peace X Seedling]; Gaujard

'DEFIANCE', HT, dr, 1914; flowers dark velvety red, dbl., 5–6. in., intense fragrance; foliage dark; vigorous growth; [Gruss an Teplitz X Étoile de France]; Kress

Degenhard, *see* **'Doc'**

DeGrazia's Pink, *see* 'MINHCO'

'DÉJÀ VOUS', Min, ab, 1987; bud pointed; flowers light apricot, aging lighter, large blooms borne usually singly, dbl., 50 petals, exhibition form, slight fragrance; no fruit; no prickles; foliage medium, medium green, semi-glossy; medium, upright, spreading growth; [Anita Charles X Sheri Anne]; Jolly, Nelson & Marie; Rosehill Farm

'DEKORAT', HT, or, 1977; (Decorat, **Freude**®); flowers vermilion and gold blend, moderate fragrance; foliage dark; vigorous, bushy growth; ADR, 1975; [Fragrant Cloud X Peer Gynt]; Kordes, W. Söhne; Mattock, 1975

Del Mar Fair, *see* 'TINDEL'

DELaby, LCl, 1995; (**Papi Delbard**); Delbard; (Cavriglia)

'DELADEL', HT, dr, 1982; (Madame Delbard®, **Mme Georges Delbard**®); flowers large, dbl., 40 petals, exhibition form, no fragrance; foliage large, medium green, semi-glossy; upright growth; PP004391; [(Tropicana X Samourai) X (Tropicana X (Rome Glory X Impeccable))]; Delbard, Georges; Roseraies Delbard, 1980

'DELAMBRE', P, dp, 1863; flowers carmine; Moreau et Robert

'DELANCHE', HT, w, 1981; (**Perle Blanche**); flowers large, dbl., 35 petals, no fragrance; foliage medium, medium green, semi-glossy; upright, bushy growth; [(Virgo X Peace) X (Goldilocks X Virgo)]; Delbard, Georges

Delany Sisters™, *see* 'WILDELS'

DELapo, HT, rb, 1997; (**La Passionate**); Delbard

DELarle, S, pb, 1998; (**Rose des Cisterciens**); Delbard

'DELATUR', HT, rb, 1982; (**Allelulia**®, Hallelujah); flowers velvety deep red, silver reverse, large blooms of heavy substance, dbl., 30 petals; foliage deep green, glossy; [((Impeccable X Papa Meilland) X (Gloire de Rome X Impeccable)) X Corrida]; Delbard, Georges, 1980

'DELAVAL', F, mr, 1986; (**Avalanche Rose**®); flowers large, dbl., no fragrance; vigorous, bushy growth; [(François et Joseph Guy X Sultane seedling) X (Alain X Étoile de Hollande)]; Delbard, 1977

'DELBAF', HT, ob, 1966; (**Apogée**®, 'DELBAL', **'DELbat'**); bud ovoid; flowers coppery, large, dbl., cupped, slight fragrance; foliage bronze, glossy; vigorous, upright growth; [(Queen Elizabeth X Provence) X (Sultane seedling X Mme Joseph Perraud)]; Delbard-Chabert

'DELBARA', F, pb, 1986; (**Bordure Rose**®, Roslyne, Strawberry Ice); flowers medium pink blending to white at base, dbl., 25 petals, 3–4 in., cupped, no fragrance; foliage thick and bright; low, compact growth; GM, Baden-Baden, 1973 Gold Star of the South Pacific, Palmerston North, NZ, 1978; [((Goldilocks X Virgo) X (Orange Triumph X Yvonne Rabier)) X Fashion]; Delbard, 1975

Delbard, F, or; flowers light orange-red, large, dbl., slight fragrance; Delbard-Chabert; (Sangerhausen)

Delbard's Orange Climber, *see* 'DELPAR'

'DELBÉCHIR', HRg, ab, 1979; (**Bernadette Chirac**®); flowers apricot, yellow and orange blend, dbl., 23 petals, 3.5 in., cupped, slight fragrance; foliage rugose; vigorous, bushy growth; [R. rugosa X (First Edition X Floradora)]; Delbard-Chabert

'DELBIR', F, mp, 1965; (**Milrose**®); flowers rose pink, blooms in clusters of 5–15, semi-dbl., cupped, slight fragrance; foliage light green, glossy; vigorous, bushy growth; GM, Baden-Baden, 1964; [Orléans Rose X (Francais X Lafayette)]; Delbard-Chabert

'DELBLAN', F, w, 1988; (**Clos Fleuri Blanc**®); flowers large, dbl., 40 petals, slight fragrance; foliage bright; semi-climbing, vigorous growth; [(Milrose X Legion d'Honneur) X Candeur]; Delbard & Chabert, 1990

DELblatine, S, w, 1993; (**Bouquet Vanille**®); Delbard

'DELBLUE', HT, m, 1984; (**Mamy Blue**®); flowers large, dbl., 35 petals, exhibition form; foliage medium, dark; bushy growth; [((Holstein X Bayadere) X (Prelude X St. Exupery)) X Unnamed seedling]; Delbard, Georges, 1991

DELbobla, F, w, 1997; (**Bordure Blanche**); Delbard

'DELBOIP', HT, dp, 1986; (**Lancéme**®); flowers deep pink, large, dbl., 28 petals, exhibition form, no fragrance; vigorous, upright, bushy growth; [(Dr Albert Schweitzer X (Michele Meilland X Bayadere)) X (MEImet X Present Filial)]; Delbard, 1973

DELbojaune, F, my; (**Bordure d'Or**®); Delbard

'DELBOURBO', Pol, rb, 1987; ('DELBOURDO', **Gloire du Bourdonnais**®); flowers center cream, margin carmine, opening turns purple, large, dbl., 35–40 petals, no fragrance; foliage bright; good, dwarf growth; [(Milrose X Legion d'Honneur) X (Zambra X Sensation)]; Delbard-Chabert, 1988

DELbover, F, rb, 1990; (**Bordure Vermillon**®); Delbard

DELboviv, T, mp, 1985; (**Bordure Vive**®); Delbard

'DELBRAT', HT, dp, 1969; (**Mondovision**); flowers cyclamen-pink, dbl., 35–45 petals, 4–5 in., moderate fragrance; [Dr Albert Schweitzer X (Bayadere X Mme Rene Cassin)]; Schweitzer; Trioreau

'DELBRE', F, mp, 1959; (**Cyclamen**); bud long; semi-dbl., 3 in., blooms in sprays of 4–6; vigorous, bushy growth; [(Frau Karl Druschki seedling X Orange Triumph seedling) X (Orange Triumph seedling X Tonnerre)]; Delbard-Chabert

'DELBRO', HT, pb, 1967; (**Parthenon**®); flowers carmine-pink, reverse soft yellow, large, dbl., cupped; foliage bronze, glossy; vigorous, upright, bushy growth; [Chic Parisien X (Bayadere X Rome Glory)]; Delbard-Chabert

'DELBUT', F, dy, 1965; ('DELCRO', **Jean de la Lune**, Moon Magic, YelloGlo); dbl., cupped, blooms in clusters, slight fragrance; foliage matt; low growth; [(Orléans Rose X Goldilocks) X (Fashion X Henri Mallerin seedling)]; Delbard-Chabert; Cuthbert

'DELCAMA', HT, or, 1978; (**Camara**®); flowers orange-vermilion, recurved, dbl., 33 petals, 4–5 in., slight fragrance; PP004332; [((Chic Parisien X Tropicana) X (Gloire de Rome X Impeccable)) X (Tropicana X Samourai)]; Delbard

'DELCANDE', F, w, 1978; (**Candeur**®); dbl., 30 petals, 3.5 in., exhibition form, no fragrance; low, upright growth; GM, Baden-Baden, 1978; [((Robin Hood X Virgo) X (Frau Karl Druschki X (Queen Elizabeth X Provence)) X (Virgo X Peace)]; Delbard

'DELCAP', LCl, mr, 1963; (**Ténor**); flowers velvety red, blooms in clusters of 6–20, semi-dbl., early; very vigorous growth; Delbard-Chabert

DELcart, HT, dr, 1973; (**Le Rouge et Le Noir**); Delbard

'DELCASCOR', HT, my, 1979; (**Casque d'Or**®); flowers well-shaped, dbl., 30 petals, 4 in., no fragrance; vigorous, upright, bushy growth; [(Zambra X Jean de la Lune) X (Michèle Meilland X Tahiti]; Delbard

'DELCELI', F, ob, 1986; (**Céline Delbard**®, DELcélit, DELCET); flowers salmon, silver reverse, large, dbl., 23 petals, cupped, no fragrance; bushy growth; GM, Monza; [Seedling X (Milrose X Legion d'Honneur)]; Delbard-Chabert, 1983

'DELCHINE', F, pb, 1988; (**Shantung**®); flowers mottled pink, cream and red, large, dbl., 22 petals, flat, slight fragrance; foliage matt; vigorous growth; [(Orléans Rose X Goldilocks) X Unnamed Bordure Rose seedling]; Delbard & Chabert

DELcifra, S, w, 1998; (**Citron-Fraise**); Delbard

DELciste, S, w, 1996; (**Libre Ingenue**); Delbard

'DELCLOPEL', HT, or, 1987; (**Clos de la Pellerie**); flowers medium vermilion red, long, large, dbl., 25–35 petals, slight fragrance; foliage large; good, vigorous, bushy growth; [(Spartan X Baccará) X Seedling]; Delbard-Chabert, 1988

'DELCOLB', HT, pb, 1991; (**Colbert**®); flowers white cream, delicately shaded porcelain pink, cup shaped, large, dbl., 27–35 petals, slight fragrance; foliage dark green, flat; bushy (80–100 cms) growth; [((Peace X Bettina) X (President Herbert Hoover X Tropicana)) X Chateau de Versailles]; Delbard-Chabert, 1990

DELcouro, LCl, lp, 1995; (**Rose Cascade**); Delbard; (Cavriglia)

'DELCOURT', F, ob, 1984; (**Courtoisie**®); flowers orange, reverse orange blended with yellow, large, dbl., 20 petals, moderate fragrance; foliage medium, medium green; bushy growth; [Avalanche Rose X Fashion seedling]; Delbard, Georges; Delbard-Chabert

DELcoussi, F, mp, 1993; (**Bordure Rose #2**); Delbard

'DELCRAF', HT, ab, 1967; (DELCRAFT, Louksor, **Louqsor**®); flowers medium, dbl., globular, slight fragrance; foliage glossy; vigorous, bushy growth; [Dr Albert Schweitzer X Provence]; Delbard-Chabert

DELcreme, F, pb, 1990; (**Regine Crespin**); Delbard

'DELCRIP', Pol, mp, 1967; (**Mme Dimitriu**®); bud globular; flowers pink tinted lighter, large, borne in clusters of 5–10, dbl.; foliage bronze, glossy; vigorous, bushy growth; GM, Geneva, 1967 GM, Rome, 1967; [Chic Parisien X Provence]; Delbard-Chabert

'DELCRO', F, dy, 1965; ('DELBUT', **Jean de la Lune**, Moon Magic, YelloGlo); dbl., cupped, blooms in clusters, slight fragrance; foliage matt; low growth; [(Orléans Rose X Goldilocks) X (Fashion X Henri Mallerin seedling)]; Delbard-Chabert; Cuthbert

'DELCROUF', Min, ly, 1973; (Bordure de Nacrée, **Bordure Nacrée**®); flowers creamy light yellow, small, dbl., 38 petals, no fragrance; low, bushy growth; GM, Baden-Baden, 1972; [(Orléans Rose X Francois et Joseph Guy) X (Goldilocks X DELtorche)]; Delbard

'DELCUS', HT, op, 1966; (**Agéna**®); bud long, pointed; flowers salmon-pink, large, dbl., moderate fragrance; foliage glossy, leathery; vigorous, bushy growth; [Chic Parisien X (Michèle Meilland X Mme Joseph Perraud)]; Delbard-Chabert

'DELDAL', HT, op, 1986; (**Dolce Vita**®, Niagara Pride); flowers rosy salmon, large, dbl., 37 petals, exhibition form, slight fragrance; vigorous, upright, bushy growth; [Voeux de Bonheur X (Chic Parisien X (Michele Meilland X Mme Joseph Perraud))]; Delbard, 1971

'DELDE', F, dr, 1957; (**Walko**); flowers dark crimson, blooms in clusters of 6–8, dbl., 23 petals, moderate fragrance; [(Incendie X Holstein) X Rouge Chabert]; Delbard-Chabert

DELdido, HT, pb, 1994; (Fetes Galantes®, **Fêtes Galantes**); Delbard

'DELDIORE', Gr, m, 1984; (**Dioressence**®); flowers lavender, well-formed, large, dbl., 35 petals, intense fragrance; foliage large, medium green, semi-glossy; bushy growth; [(Holstein X Bayadère) X (Prelude) X Seedling]; Delbard, Georges; Delbard-Chabert

DELdog, S, dy, 1997; (**Pimpernelle**); Delbard

'DELDRA', HT, my, 1966; (**Venusic**®); flowers saffron-yellow, medium, dbl., cupped, moderate fragrance; foliage dark, glossy; vigorous, upright, bushy growth; [(Queen Elizabeth X Provence) X (Mme Joseph Perraud X Bayadere)]; Delbard-Chabert

'DELDROP', HT, mr, 1966; (**Beauté Spatiale**®); flowers velvety red, medium, semi-dbl., exhibition form, slight fragrance; foliage bronze, soft; moderate, bushy growth; [Walko X Impeccable]; Delbard-Chabert

'DELECLA', F, mr, 1988; (**Clos Fleuri Rouge**®, Sparkler); flowers medium, dbl., 20 petals, flat, slight fragrance; foliage bright; bushy, vigorous growth; [Orléans Rose X Queen Elizabeth]; Delbard & Chabert, 1990

'DELEGRAN', HT, pb, 1986; (**Grand Siècle**®, Great Century); flowers creamy pink blend, well-formed, large, dbl., 33 petals, cupped, slight fragrance; foliage large; vigorous, bushy, branching growth; [((Queen Elizabeth X Provence) X (Michele Meilland X Bayadere)) X ((Voeux de Bonheur X MEImet) X (Peace X Dr Debat))]; Delbard, 1987

'DELÉPI', HT, dy, 1981; (**Epidor**®); flowers large, dbl., 35 petals, slight fragrance; foliage large, medium green, matt; bushy growth; GM, Rome; [(Peace X Marcelle Gret) X (Velizy X Jean de la Lune)]; Delbard, Georges

'DELFAT', HT, ab, 1967; (**Louvre**®); bud long, pointed; flowers rosy apricot, medium, dbl., slight fragrance; foliage bronze, glossy; moderate, upright growth; [Souv. de J. Chabert X (Walko X Souv. de J. Chabert)]; Delbard-Chabert

'DELFESROU', F, dr, 1980; (**Festival Rouge**®, Red Festival); flowers well-formed, blooms in clusters, semi-dbl.; foliage dark; [Walko X (Happiness X Sonia)]; Delbard, Georges; Delbard Roses

'DELFIB', F, my, 1965; (**Bebe Lune**®); Delbard

'DELFLIP', HT, mr, 1967; (**Escurial**®); flowers velvety cardinal-red, medium, semi-dbl., exhibition form, slight fragrance; foliage dark, glossy; vigorous, bushy growth; [Gay Paris X Impeccable]; Delbard-Chabert

'DELFLORI', F, mr, 1977; (**Fluorescent®**); flowers medium, dbl., 33 petals, cupped, no fragrance; vigorous, bushy, branching growth; [Zambra X ((DELtorche X Tropicana) X (Alain X Souv. de J. Chabert))]; Delbard

DELfloro, F, mr, 1992; (**Centenaire de Lourdes Rouge®**); Delbard

DELfrei, S, m, 1993; (**Coeur Farouche®**); Delbard

'DELFRI', HT, pb, 1965; (**Tourmaline®**); flowers creamy white, widely edged carmine, large, dbl., 28 petals, exhibition form; foliage light green; moderate growth; GM, Madrid, 1965; [Michele Meilland X Chic Parisien]; Delbard-Chabert; Cuthbert

'DELFROR', HT, pb, 1966; (**Rose du Ciel®**); flowers cream-white broadly edged carmine-pink, large, dbl., globular; foliage dark, glossy; vigorous, bushy growth; [Chic Parisien X (Michele Meilland X Bayadere)]; Delbard-Chabert

'DELGAP', HT, yb, 1968; (**Lido di Roma**); bud long, pointed; flowers deep yellow, shaded red, large, dbl., exhibition form, slight fragrance; foliage glossy, leathery; vigorous, upright, bushy growth; GM, Japan, 1968; [(Chic Parisien X Michele Meilland) X (Sultane X Mme Joseph Perraud)]; Delbard-Chabert

'DELGE', F, mp, 1958; (**Centenaire de Lourdes®**, Centennaire de Lourdes, Mrs Jones); flowers soft rose, blooms in clusters of 5–10, semi-dbl., 3.5–4 in.; vigorous, bushy growth; [(Frau Karl Druschki X Seedling) X Seedling]; Delbard-Chabert

'DELGEOT', HT, dr, 1986; (Isobel Champion, **La Marseillaise®**); bud large, ovoid; flowers well-formed, large, dbl., 40 petals, moderate fragrance; vigorous, bushy growth; [((Gloire de Rome X Impeccable) X (Rouge Meilland X Soraya)) X (MEIsar X Walko)]; Delbard, 1976

'DELGLAP', Cl HT, my, 1971; (**Puerta del Sol®**); flowers medium golden yellow, large, dbl., 28 petals, slight fragrance; vigorous, climbing (to 9 ft) growth; [(Queen Elizabeth X Provence) X (Michele Meilland X Bayadere)]; Delbard

'DELGO', LCl, or, 1961; (**Phare®**); flowers bright orange-red, dbl., 33 petals, 3 in., foliage glossy, dark; vigorous (to 10 ft) growth; [Spectacular X (Floradora X Unnamed seedling)]; Delbard-Chabert

DELgold, HT, dy, 1980; (**Lord Gold®**); Delbard

'DELGORG', HT, rb, 1986; (Chateau de Versailles®, 'DELRICOS', Gorgeous George, **Guy Laroche**, La Tour d'Argent™); flowers brilliant red, silver reverse, large blooms borne usually singly, dbl., 30 petals, exhibition form, slight fragrance; foliage medium, medium green, matt; upright, bushy growth; [Unnamed seedling X (Michele Meilland X Carla)]; Delbard, Georges; Armstrong Nursery, 1985

DELgribla, LCl, w, 1995; (**Blanche Colombell**); Delbard

DELgrim, LCl, my, 1993; (**Dune®**); Delbard

'DELGRORD', HT, w, 1986; (**Grand Nord®**, Great Nord®, Great North); flowers large, dbl., 28 petals, exhibition form, slight fragrance; vigorous, bushy growth; GM, Paris, 1970 GM, Rome, 1973; [((Queen Elizabeth X Provence) X (Virgo X Carina)) X ((Voeux de Bonheur X Virgo) X (Virgo X Peace))]; Delbard, 1975

'DELGUS', Pol, or, 1968; (**Prince Tango®**); bud pointed; flowers mandarin color, open, small, semi-dbl.; foliage bronze, glossy; moderate, bushy growth; GM, Madrid, 1968; [(Orléans Rose X Goldilocks) X (Orange Triumph seedling X Floradora)]; Delbard-Chabert

'DELHI APRICOT', HT, ab, 1964; bud pointed; flowers apricot-yellow, medium, dbl., slight fragrance; foliage light green, soft; moderate, bushy growth; Pal, Dr B.P.; Indian Agric. Research Inst.

'DELHI BRIGHTNESS', F, op, 1963; flowers open, medium, semi-dbl.; foliage glossy; vigorous, upright growth; Pal, Dr B.P.; Indian Agric. Research Inst.

'DELHI DAINTINESS', F, lp, 1963; bud pointed; flowers light pink, reverse darker, medium, semi-dbl.; foliage glossy; vigorous, upright, compact growth; Pal, Dr B.P.; Indian Agric. Research Inst.

'DELHI MAID', F, ob, 1963; bud pointed; flowers flame-orange, base gold, open, medium, single; foliage dark, glossy; vigorous, upright, compact growth; Pal, Dr B.P.; Indian Agric. Research Inst.

'DELHI PINK PEARL', HMult, lp, 1962; bud ovoid; flowers pearly pink, open, medium, semi-dbl.; foliage glossy, light green; very vigorous, compact growth; [Echo sport]; Pal, Dr B.P.; Indian Agric. Research Inst.

'DELHI PRINCE', F, dp, 1963; bud pointed; flowers glowing deep pink, open, medium, semi-dbl., slight fragrance; foliage glossy; vigorous, bushy growth; Pal, Dr B.P.; Indian Agric. Research Inst.

'DELHI PRINCESS', F, mp, 1963; bud ovoid, cerise-red; flowers deep pink, open, large, borne in clusters, semi-dbl., slight fragrance; foliage glossy, bronze; very vigorous, compact growth; Pal, Dr B.P.; Indian Agric. Research Inst.

'DELHI ROSETTE', F, ob, 1965; bud ovoid; flowers bright orange-scarlet, open, medium, dbl., 28 petals; foliage dark, glossy; vigorous, compact growth; Pal, Dr B.P.; Indian Agric. Research Inst.

'DELHI SHERBET', F, mp, 1963; bud ovoid; flowers deep rose-pink, medium, dbl., intense fragrance; foliage glossy; vigorous, bushy, compact growth; [Gruss an Teplitz X Seedling]; Pal, Dr B.P.; Indian Agric. Research Inst.

'DELHI STARLET', Min, ly, 1963; bud pointed; flowers open, small, semi-dbl., slight, musk fragrance; foliage small, glossy; dwarf, compact growth; [Goudvlinder X Seedling]; Pal, Dr B.P.; Indian Agric. Research Inst.

'DELHI SUNSHINE', HT, yb, 1963; bud pointed; flowers deep cream, reverse flushed pink, medium, dbl.; foliage light green; bushy, open growth; [Mme Charles Sauvage X Seedling]; Pal, Dr B.P.; Indian Agric. Research Inst.

'DELHI WHITE PEARL', LCl, w, 1963; bud ovoid; flowers pearly white, blooms in clusters, dbl., repeat bloom, slight fragrance; foliage glossy; vigorous growth; [Prosperity X Seedling]; Pal, Dr B.P.; Indian Agric. Research Inst.

'DELIC', HT, lp, 1978; (**Éterna®**); bud long; flowers light carmine pink, dbl., 28–32 petals, 4–5 in., slight fragrance; vigorous, upright growth; [((Michèle Meilland X Carla) X (Dr Schweitzer X Tropicana)) x (Queen Elizabeth X Provence)]; Delbard-Chabert

'DELICADO', HT, op, 1954; flowers shell-pink shaded peach, well shaped, 6 in.; moderate growth; Lowe

'DELICATA', HRg, lp, 1898; flowers soft lilac-pink, large, semi-dbl., recurrent bloom; vigorous growth; Cooling

Delicate Beauty, *see* KORlarkon

Delicia, *see* 'KORGATUM'

'DELICIOUS', Min, mp, 1995; flowers small blooms borne in small clusters, dbl., 26–40 petals, intense fragrance; few prickles; foliage small, medium green, semi-glossy; tall (65–70 cms), upright, compact growth; [Avandel X [(Seedling X Friesia) X Seedling]]; Welsh, Eric; Rose Hill Roses, 1994

Delicious Weldel, Min, lp; [Avandel X Freesia]; Welsh, Eric, 1991; Rose & Fruit Tree; (Weatherly, L.)

'DÉLIE COMMUNAUDAT', HT, yb, 1933; flowers naples yellow, shaded and edged carmine, dbl., cupped, moderate fragrance; foliage leathery; very vigorous, bushy growth; [Mme Charles Detreaux X Mme Edouard Herriot]; Buatois

'DELIFIC', F, mr, 1983; (**Clos Vougeot®**, Red Prolific, Rouge Prolific); flowers medium, dbl., 28 petals, no fragrance; bushy growth; [(Alain X Charles Mallerin) X (Lafayette X Walko)]; Delbard-Chabert

'DELIGHT', HMult, rb, 1904; flowers bright carmine, base white, stamens yellow, medium to large, semi-dbl., cupped; foliage glossy; long stems; very vigorous, climbing (15–20 ft) growth; Walsh

'DELIGHTFUL', HT, pb, 1931; flowers rose, base yellow, reverse amber-yellow, large, dbl., exhibition form, moderate fragrance; foliage glossy; vigorous growth; [George Dickson seedling]; McGredy

'DELIGHTFUL', HT, yb, 1956; bud pointed; flowers straw-yellow, base shaded red, dbl., 35–50 petals, 4–5 in., exhibition form, moderate fragrance; upright, compact growth; [Curly Pink X Shades of Autumn]; Brownell, H.C.

'DELIGHTFUL KIWI', HT, lp, 1989; bud tapering; flowers blush pink aging to creamy pink, urn-shaped, medium, dbl., 45 petals, borne singly, slight fragrance; prickles light brown, very few; foliage light green, large, shiny; upright, branching growth; [Silent Night X (Prima Ballerina X Irish Mist)]; Cattermole, R.F.; South Pacific Rose Nursery, 1988

'DELIGHTFUL LADY', HT, lp, 1982; flowers large, borne singly in spring, 3–5 per cluster in autumn, dbl., 30–35 petals, exhibition form, slight fragrance; brown prickles; foliage semi-glossy; medium-tall growth; [Pascali X Merry Widow]; Attfield, B.B.

'DELIGHTFUL PINK', F, mp, 1958; bud ovoid; flowers pink, medium, borne in clusters, dbl., 40–45 petals, cupped, moderate fragrance; vigorous, upright growth; [Chic X Demure]; Boerner; J&P

DELilac, S, lp; (**Souv de Louis Amade**); Delbard, 1998

Delilah®, *see* 'SUNDEL'

'DELILLE', M, w, 1852; flowers blush-white, nicely mossed, semi-dbl., may repeat; Robert

'DELJABER', HT, dr, 1990; (**Professeur Jean Bernard®**); bud cupped; flowers very large bloom, 25–30 petals, slight fragrance; foliage dark green, abundant; bushy growth; [(Charles Mallerin X Divine) X [Tropicana X (Rome Glory X Impeccable)]]; Delbard & Chabert, 1989

'DELJACQ', HT, or, 1988; (Joconde, **La Joconde**; flowers well-shaped, large, dbl., 40 petals, no fragrance; upright, vigorous growth; [(Tropicana X (Rome Glory X Impeccable)) X (Spartan X MEIger)]; Delbard & Chabert, 1989

DELjapal, HT, lp; (**Folio Courtisane**); Delbard, 1996

'DELJAUNE', F, my, 1988; (**Clos Fleuri Jaune®**); flowers yellow, shaded ochre, opening to yellow-amber, large, dbl., 18 petals, slight fragrance; foliage bright; raised, semi-climbing, vigorous growth; [(Orléans Rose X Goldilocks) X Parure d'Or]; Delbard & Chabert, 1990

'DELJAUNOR', HT, ob, 1990; (**France Libre®**); flowers nasturtium orange and coppery, yellow and gold reverse, large, dbl., 25–30 petals, cupped, slight fragrance; foliage dark green, glossy; upright growth; [(Zambra X Orange Sensation) X (Unnamed seedling X Unnamed seedling)]; Delbard & Chabert, 1981; Delbard & Chabert

DELjavert, S, w, 1997; (**Pur Caprice**); Delbard

'DELJIS', LCl, or, 1956; (**Tropique**); dbl.; foliage dark; Delbard-Chabert

'DELJOFEM', HT, pb, 1982; (**Tendresse**); flowers light pink, reverse apricot-pink, large, dbl., exhibition form, moderate fragrance; foliage dense; [(Michèle Meilland X Bayadère) X (Grace de Monaco X Present Filial)]; Delbard, Georges; Delbard Roses, 1980

'DELJONQ', F, dy, 1982; (**Jonquille**); flowers large, dbl., 35 petals, no fragrance; foliage medium, medium green, matt; bushy growth; [(Peace X Marcelle Gret) X (Velizy X Jean de la Lune)]; Delbard, Georges

'DELJULI', F, ab, 1986; (**Julie Delbard®**); flowers apricot with yellow and orange hues, ht-form, large, dbl., 28 petals, no fragrance; vigorous, bushy growth; GM, Madrid, 1976; [(Zambra X (Orange Triumph X Floradora)) X ((Orléans Rose X Goldilocks) X (Bettina X Henri Mallerin))]; Delbard, 1976

DELki, Min, ob, 1998; (**Pinocchio**); Delbard

'DELKORT', HT, dr, 1966; (**Atlas®**); bud long, pointed; flowers magenta-red edged darker, large, dbl., cupped; foliage dark, glossy; vigorous, upright, bushy growth; [Chic Parisien X Provence]; Delbard-Chabert

'DELKRI', HT, mr, 1969; (**France Inter**); flowers magenta-red, ovoid, dbl., 35–45 petals, 4–5 in., slight fragrance; [(Rome Glory X La Vaudoise) X Divine]; Delbard; Trioreau

Della Balfour, *see* HARblend

DELmagsa, F, lp; (**The Prince Imperial**); Delbard, 1995

'DELMANCHE', HT, pb, 1980; (**Maman Chérie**); dbl., blooms in clusters; [Gay Paris X (Baccará X Impeccable)]; Delbard, Georges

'DELMAT', HT, yb, 1970; ('DELNAT', **Mitsouko®**); flowers yellow, petals edged red, large, dbl., 50 petals, moderate, fruity fragrance; bronze-red prickles; foliage medium, clear green; dense, bushy growth; [(Michele Meilland X Chic Parisien) X Peace]; Delbard, Georges

'DELMATOR', HT, ob, 1986; (**Beverly Hills**, Malicorne); flowers dark orange, greenhouse variety, blooms in clusters; PP005673; [(Zambra X Orange Sensation) X (Zambra X (Orange Triumph X Floradora))]; Delbard, Georges, 1982

'DELMIR', LCl, yb, 1968; (**Parure d'Or®**); flowers golden yellow edged orange, medium, semi-dbl., repeat bloom, foliage dark, glossy; vigorous, climbing growth; GM, Bagatelle, 1968; [(Queen Elizabeth X Provence) X (Sultane seedling X Mme Joseph Perraud)]; Delbard-Chabert, 1970

DELmistri, Min, rb, 1998; (**Bambi**); Delbard

'DELMOT', LCl, op, 1967; (**Obélisque®**); flowers coppery orange-pink, medium, semi-dbl., globular, abundant, intermittent bloom; foliage bronze, glossy; vigorous, climbing growth; GM, Geneva, 1967; [Spectacular X (Orange Triumph seedling X Floradora)]; Delbard-Chabert

'DELMOUN', Cl HT, mr, 1986; (Sensass, **Sensass Delbard®**); flowers bright velvety red, large, dbl., 28 petals, cupped, no fragrance; vigorous, climbing (to 10 ft) growth; [(Danse du Feu X (Orange Triumph X Floradora)) X Tenor seedling, unnamed]; Delbard, 1973

'DELMUR', LCl, mr, 1966; (**Altissimo®**, Altus); flowers blood-red, single, 7 petals, 4–5 in., cupped, repeat bloom, slight fragrance; foliage dark, serrated; tall growth; [Ténor X Seedling]; Delbard-Chabert; Cuthbert

'DELNAT', HT, yb, 1970; ('DELMAT' **Mitsouko®**); flowers yellow, petals edged red, large, dbl., 50 petals, moderate, fruity fragrance; bronze-red prickles; foliage medium, clear green; dense, bushy growth; [(Michele Meilland X Chic Parisien) X Peace]; Delbard, Georges

'DELNIBLE', HT, m, 1981; (**Blue Nile®**, Nil Bleu); bud ovoid, pointed; flowers deep lavender, dbl., 28–30 petals, 5 in., exhibition form, borne singly or 2–3 per cluster, intense, fruity fragrance; short prickles, hooked downward; foliage large, olive green; tall, upright, spreading growth; PP4671; GM, Bagatelle; [(Holstein X Bayadere) X (Prelude X Saint-Exupery)]; Delbard, Georges; Armstrong Nursery, 1976

'DELNOLLI', F, pb, 1981; (**Tourbillon**); flowers deep pink, yellow reverse, large, dbl., 20 petals, moderate fragrance; foliage small, medium green, semi-glossy; upright growth; [Zambra X ((Orléans Rose X Goldilocks) X (Spartan X Fashion))]; Delbard, Georges

'DELOBLAN', HT, w, 1980; (**Blanche Comète**, White Comet); dbl., blooms in clusters; [(Virgo X Peace) X (Goldilocks X Virgo)]; Delbard, Georges; Delbard Roses

DELodive, F, mp, 1994; (**Clos Fleuri Rose No. 2**); Delbard

Deloitte & Touche, *see* KORaucher

'DELOP', HT, ab, 1967; (**Vatican®**); bud ovoid; flowers apricot-yellow shaded carmine, medium, dbl., exhibition form, slight fragrance; foliage glossy; moderate, bushy growth; [Grande Premiere X (Sultane X Mme Joseph Perraud)]; Delbard-Chabert

Delore, HT, ly, 1982

'DELORFEU', F, ob, 1982; (**Carnaval de Rio®**); flowers orange, blooms in clusters, dbl.; [(Zambra X Orange Sensation) X (Zambra X Seedling (Orange Triumph X Floradora))]; Delbard, Georges; Delbard Roses

'DELOSOL', HT, my, 1984; (**O Sole Mio®**); flowers well-formed, large, dbl., 35 petals, no fragrance; foliage medium, medium green, glossy; [(Peace X Marcelle Gret) X Velizy seedling]; Delbard, Georges

DELpapy, S, my; (**Souvenirs de Marcel Proust**); Delbard, 1993

'DELPAR', LCl, or, 1966; (**Delbard's Orange Climber**, Grimpant Delbard); bud ovoid; flowers medium, dbl., exhibition form, repeat bloom, slight fragrance; foliage dark, glossy, leathery; vigorous, climbing, well branched growth; [Spectacular X (Rome Glory X La Vaudoise)]; Delbard-Chabert, 1963; Armstrong Nursery, 1963

DELperl, LCl, dr, 1994; (**Salammbo**®); Delbard

'DELPÉTRI', HT, mr, 1988; (**Clochermerle**®); flowers large, dbl., 38 petals, cupped, slight fragrance; foliage matt; vigorous, bushy growth; [Seedling X [(Michele Meilland X Karla) X Seedling)]]; Delbard & Chabert

Delphin, HT, m; flowers violet-pink, large, dbl., intense fragrance; VEG; (Sangerhausen)

'DELPLI', Pol, ob, 1969; (**Rosier d'Or**); flowers orange-yellow, shaded apricot-yellow, dbl., 20–28 petals, 2.5–3.5 in., slight fragrance; [Zambra X (Orléans Rose X Goldilocks)]; Delbard; Trioreau

'DELPO', F, mr, 1961; (**Diablotin**®, Little Devil); semi-dbl., 17 petals, 2–3 in., blooms in small clusters, no fragrance; bushy, compact growth; [Orléans Rose X Fashion]; Delbard-Chabert

'DELPOC', HT, mp, 1966; (**Diapason**®); flowers porcelain pink, medium, dbl., 40 petals, globular, moderate fragrance; foliage bronze, glossy; very vigorous, bushy growth; [Chic Parisien X (Sultane seedling X Mme Joseph Perraud)]; Delbard-Chabert

'DELPOMP', F, mp, 1988; (**Clos Fleuri Rose**®); flowers medium, dbl., 30 petals, slight fragrance; foliage bright; bushy, vigorous growth; [(Zambra X Orange Sensation) X (Robin Hood X Virgo)]; Delbard & Chabert, 1990

'DELPOSAR', Cl F, mr, 1970; (**Diablotin, Climbing**®, Grimpant Diablotin, Little Devil, Climbing); GM, Geneva, 1970 GM, Paris, 1970 GM, Rome, 1970; [Diablotin sport]; Delbard

'DELPOUS', HT, dp, 1967; (Fontaine Blue, **Fontainebleau**®); bud long, pointed; flowers magenta-pink, large, dbl., globular, slight fragrance; foliage dark, glossy, leathery; vigorous, bushy growth; [Dr Albert Schweitzer X (Bayadere X Rome Glory)]; Delbard-Chabert

'DELPRAT', HT, lp, 1970; (**Présence**); flowers pink, lighter reverse, large, dbl., 38 petals, slight, fruity fragrance; foliage medium, medium green, matt; upright, bushy growth; [Dr Albert Schweitzer X (Michele Meilland X Bayadere)]; Delbard, Georges; G. Delbard; Society of Nurseries

'DELPRE', Gr, op, 1960; (Happy Anniversary, **Heureux Anniversaire**®); bud urn-shaped; flowers salmon-orange, dbl., 28 petals, 3 in., slight, spicy fragrance; foliage glossy; very vigorous, bushy growth; [(Incendie X Chic Parisien) X (Floradora X Independence)]; Delbard-Chabert; Stark Bros., 1963

'DELPRIMA'®, HT, op, 1968; ('DELTUF', **Grand Prix**®), flowers coral-pink shaded ochre, large, semi-dbl., slight fragrance; foliage glossy, leathery; vigorous, upright, bushy growth; GM, Belgium, 1968; [Chic Parisien X (Grande Premiere X (Sultane X Mme Joseph Perraud))]; Delbard-Chabert

DELredi, HT, ob; (**Renata Tebaldi**); Delbard

'DELRENO', HT, dy, 1987; (**Mme Georges Renoard**); flowers long, large, dbl., 35–40 petals, moderate fragrance; foliage clear, matt; upright, half-divirgent, vigorous growth; [(Peace X Marcelle Gret) X Unnamed Legion d'Honneur seedling]; Delbard-Chabert, 1988

'DELRICOS', HT, rb, 1986; (Chateau de Versailles®, 'DELGORG', Gorgeous George, **Guy Laroche**, La Tour d'Argent™); flowers brilliant red, silver reverse, large blooms borne usually singly, dbl., 30 petals, exhibition form, slight fragrance; foliage medium, medium green, matt; upright, bushy growth; [Unnamed seedling X (Michele Meilland X Carla)]; Delbard, Georges; Armstrong Nursery, 1985

DELrima, F, rb, 1995; (**Sasa**); Delbard

'DELRIO', HT, m, 1983; (Night Flight, **Vol de Nuit**); flowers deep lilac, large, dbl., 33 petals, intense fragrance; bronze-red prickles; foliage medium, light green, matt; bushy growth; GM, Rome, 1970; [(Holstein X (Bayadere X Prelude)) X Saint-Exupery]; Delbard, Georges, 1970

'DELROBLA', F, w, 1991; (**Carmagnole**®); flowers white cream with soft pink, moderately cup shaped, blooms, dbl., 20–25 petals, slight fragrance; foliage dark green; bushy (80–100 cms) growth; [(Milrose X Legion d'Honneur) X (Zambra X Orange Sensation)]; Delbard & Chabert, 1990

'DELROCELES', Cl HT, lp, 1979; (**Rose Céleste**®); flowers well-formed, large, dbl., 33 petals, cupped, moderate fragrance; vigorous, climbing (to 9 ft) growth; [(Queen Elizabeth X Provence) X (Sultane X Mme J. Perraud)]; Delbard

DELroujan, HT, 1988; Delbard, Georges; (Cavriglia)

'DELROUVEL', HT, rb, 1988; (**Château d'Amboise**®); flowers dark red, opening bright, long, dbl., 23–30 petals, slight fragrance; foliage bright; vigorous, semi-raised growth; [Tropicana seedling X [(Rome Glory X Impeccable) X (Rouge Meilland X Soraya)]]; Delbard & Chabert

DELrovrai, HT, 1995; (**Jardins de Villandry**); Delbard, Georges; (Cavriglia)

DELrugro, LCl, lp; (**Tour Eiffel 2000**); Delbard, 1998

'DELRULA', F, pb, 1988; (**Ville du Perreux**); flowers pink with white and cream, long, large, dbl., 28 petals, slight fragrance; foliage bright; bushy, vigorous growth; [Unnamed seedling X (Milrose X Legion d'Honneur)]; Delbard & Chabert

'DELSAB', HT, pb, 1968; (**Royal Ascot**®); flowers pink, reverse shaded crimson, large, semi-dbl., exhibition form; foliage glossy, leathery; vigorous, bushy growth; [Chic Parisien X (Grande Premiere X (Sultane X Mme Joseph Perraud))]; Delbard-Chabert

'DELSAMAR', F, mr, 1974; (**Légion d'Honneur**®); flowers well-formed, medium, dbl., 55 petals, cupped, slight fragrance; low, bushy growth; GM, Geneva, 1974; [(Souv. de J. Chabert X (Walko X Souv. de J. Chabert)) X ((Tamango X Gay Paris) X (Zambra X Jean de la Lune))]; Delbard

'DELSAMO', HT, dp, 1974; (**Paris 2000**); flowers deep pink, large, dbl., 28 petals, cupped, no fragrance; vigorous, upright, bushy growth; [(DELtorche X (Sultane X Mme Joseph Perraud)) X (Queen Elizabeth X Provence)]; Delbard

'DELSAMOUR', HT, ab, 1973; (**Velizy**®); flowers well-formed, large, dbl., 30 petals, slight fragrance; bushy, branching growth; [((Peace X Marcelle Gret) X (Michele Meilland X Tahiti)) X (Peace X Grand Premiere)]; Delbard

'DELSAP', F, mr, 1965; (**Cappa Magna**®); flowers yellow stamens, wavy petals, blooms in clusters of 20–30, 8–10 petals, 4 in., cupped; foliage dark, glossy, large; upright growth; [Tenor seedling]; Delbard-Chabert; Cuthbert, 1967

'DELSATEL', HT, or, 1982; (**Satellite**); flowers medium, dbl., 28 petals, moderate fragrance; foliage medium, medium green, semi-glossy; [((Tropicana X Samourai) X (Tropicana X (Rome Glory X Impeccable))) X Granada]; Delbard, Georges

'DELSET', HT, lp, 1967; (Castel, **Versailles**®); bud ovoid; flowers soft pink, medium, dbl., cupped, slight fragrance; foliage dark, glossy, leathery; vigorous, upright, bushy growth; GM, Baden-Baden, 1965 GM, Bagatelle, 1966 GM, Geneva, 1966; [(Queen Elizabeth X Provence) X (Michele Meilland X Bayadere)]; Delbard-Chabert

'DELSIRE', Cl HT, dr, 1986; (Grandessa, **Messire Delbard**®); flowers deep crimson red, well-formed, large, dbl., 38 petals, slight fragrance; foliage large; vigorous (to 9 ft) growth; [(Danse du Feu X Guinee) X ((Tenor X Fugue) X (Delbard's Orange Climber X Gloire de Dijon))]; Delbard, 1976

'DELSIRP', Pol, ob, 1971; (**Caid**); bud ovoid; flowers orange, medium, dbl., repeat bloom, slight fragrance; foliage leathery; vigorous, bushy growth; [Orangeade X Seedling]; Delforge

'DELSOB', F, ob, 1975; (**Puerto Rico**, Sable Chaud, Sable Cloud); flowers orange, reverse blended with yellow, medium, dbl., cupped, slight, spicy fragrance; foliage bronze, leathery; vigorous, bushy growth; [Zambra X (Orange Triumph X Floradora)]; Delbard; Armstrong Nursery

DELsorb, S, pb; (**Sorbet Framboise**®); Delbard, 1994

'DELSOUCHE', Gr, dr, 1980; (**Souv. de Maurice Chevalier**); semi-dbl., blooms in clusters, [(Walko X Impeccable) X (Papa Meilland X (Baccará X Michele Meilland))]; Delbard, Georges

DELstrabla, HT, pb, 1995; (**Hei Matisse**); Delbard

DELstricol, F, 1996; (**Camille Pisarro**); Delbard, Georges; (Cavriglia)

DELstror, F, pb, 1997; (**Grimaldi**); Delbard

'DELSULAN', HT, mp, 1988; (**Europe 92**); flowers pink magenta, long, large, dbl., 11 petals, slight fragrance; foliage bright; vigorous, bushy growth; [Seedling X (Michèle Meilland X Karla)]; Delbard & Chabert

Delta, *see* **'BREAK O' DAY'**

Delta Dawn, *see* 'IVEmax'

'DELTA GAMMA', HT, ly, 1977; bud pointed; flowers creamy white, base pale yellow, dbl., 38 petals, 4–4.5 in., exhibition form, moderate, spicy fragrance; foliage dark; vigorous, upright, bushy growth; [Queen Elizabeth X Mount Shasta]; Kimbrew-Walter Roses

Delta Gold™, *see* 'BURDEL'

Delta Queen, HT, mp, 1998

'DELTAF', HT, pb, 1962; (**Mme René Cassin**); flowers cyclamen-pink, reverse silvery, dbl., 24 petals, 5 in., exhibition form, moderate fragrance; foliage dark, glossy; vigorous, upright growth; GM, Bagatelle, 1962 GM, Lyon, 1962; [Mme Armand Souzy X Impeccable]; Delbard-Chabert

'DELTANGA', HT, ob, 1978; (**Tango Rose**); bud pointed; flowers salmon-orange, large, dbl., 35–40 petals; foliage matt green; bushy growth; [(Belle Rouge X (Gloire de Rome X Gratitude)) X ((Dr Schweitzer X Tropicana) X (Ena Harkness X Quebec))]; Delbard

DELtendre, S, mp, 1994; (**Comtesse de Segur**, Tendresse); Delbard

'DELTEP', F, op, 1976; (Arnaud Delbard®, **First Edition**); bud ovoid, pointed; flowers luminous coral, shaded orange, dbl., 28 petals, 2–2.5 in., slight, tea fragrance; foliage glossy, light; upright growth; AARS, 1977; [(Zambra X (Orléans Rose X Goldilocks)) X (Orange Triumph seedling X Floradora)]; Delbard; C-P

'DELTOGO', F, ab, 1961; (**Tobago**®); flowers yellow apricot, outer petals aging pink, large, dbl., 35 petals, moderate fragrance; foliage medium, dark, semi-glossy; upright growth; Golden Rose, Geneva, 1981; [Avalanche X (Zambra X Orange Sensation)]; Delbard, Georges

'DELTOP', HT, mr, 1983; (**Crêpe de Chine**®); flowers large, dbl., 20 petals, slight fragrance; bronze-red prickles; foliage glossy, clear green, dense; vigorous, upright, bushy growth; GM, Madrid, 1970; [Joyeux Noel X (Gloire de Rome X Impeccable)]; Delbard, Georges; Delbard-Chabert, 1970

'DELTOS', HT, m, 1961; ('DELVOR', **Saint-Exupéry**®, Waltz Time); flowers mauve tinted silvery, dbl., 5 in., exhibition form, slight fragrance; vigorous, bushy growth; [(Christopher Stone X Marcelle Gret) X (Holstein X Bayadere)]; Delbard-Chabert

DELtou, F, 1986; Delbard, Georges; (Cavriglia)

'DELTRAC', HT, mr, 1967; (**Angkor**®); flowers carmine-red, large, dbl., globular, slight fragrance; foliage bronze, glossy, leathery; vigorous, bushy growth; [Belle Rouge X (Rome Glory X Gratitude)]; Delbard-Chabert

DELtrap, F, mr, 1994; (**Saint-Vincent**®); Delbard, Georges

'DELTRE', HT, mr, 1959; (**Maurice Chevalier**®); bud long; flowers rich red shaded garnet, large, dbl., 25 petals, moderate fragrance; foliage glossy; vigorous growth; [Incendie X (Floradora seedling X Independence)]; Delbard-Chabert

'DELTROB', Pol, w, 1967; (**Cactus Blanc**®); bud globular; flowers creamy white, small, borne in clusters, dbl., cupped, foliage light green, glossy; moderate, bushy growth; [(Orléans Rose X Orléans Rose) X ((Français X Lafayette) X (Orléans Rose X Goldilocks))]; Delbard-Chabert

'DELTRUT', LCl, dp, 1967; (Campanela, **Campanile**®); bud globular; flowers deep magenta-pink, large, dbl., repeat bloom, moderate fragrance; foliage glossy, leathery, bronze; vigorous, climbing growth; [(Queen Elizabeth X Provence) X (Sultane seedling X Mme Joseph Perraud)]; Delbard-Chabert

'DELTUF', HT, op, 1968; ('DELPRIMA'®, **Grand Prix**®); flowers coral-pink shaded ochre, large, semi-dbl., slight fragrance; foliage glossy, leathery; vigorous, upright, bushy growth; GM, Belgium, 1968; [Chic Parisien X (Grande Premiere X (Sultane X Mme Joseph Perraud))]; Delbard-Chabert

'DELUP', HT, dr, 1969; (**Betsy Ross**, La Passionata); flowers deep red, large, dbl., exhibition form, slight fragrance; vigorous, upright growth; [(Gloire de Rome X La Vaudoise) X Divine]; Delbard; Trioreau & Stark Bros.

'DELURT', HT, dr, 1986; (Black Pearl, **Perle Noire**®); flowers velvety dark red, well-formed, large, dbl., 38 petals, slight fragrance; vigorous, bushy growth; [((Impeccable X Papa Meilland) X (Gloire de Rome X Impeccable)) X ((Charles Mallerin X Gay Paris) X (Rouge Meilland X Soraya))]; Delbard, 1976

'DELVEEN', P, lp; Alderton, 1930; (Weatherly, L.)

DELverjaune, F, dy, 1997; (**Nedbank Rose**); Delbard

DELviola, S, 1996; (**Chartreuse de Parme**); Delbard, Georges; (Cavriglia)

DELvirge, S, lp; (**Vierge Folle**); Delbard, 1998

'DELVOR', HT, yb, 1970; (**Femme**); flowers ivory yellow, tinted pink, large, dbl., 28 petals, slight fragrance; bronze-red prickles; foliage medium, dark, glossy; upright growth; [(Gloire de Rome X Bayadere) X (Queen Elizabeth X Provence)]; Delbard, Georges; Delbard Roses

'DELVOR', HT, m, 1961; ('DELTOS', **Saint-Exupéry**®, Waltz Time); flowers mauve tinted silvery, dbl., 5 in., exhibition form, slight fragrance; vigorous, bushy growth; [(Christopher Stone X Marcelle Gret) X (Holstein X Bayadere)]; Delbard-Chabert

'DELZEN', LCl, or, 1982; (**Zenith**); semi-dbl., blooms in clusters; climbing (to 8 ft) growth; [(Spectacular X Tenor seedling) X (Floradora X Incendie)]; Delbard, Georges; Delbard Roses

DELzinsch, HT, dy, 1998; (**President Armand Zinsch**); Delbard

'DEMAIN', HT, mr, 1945; bud pointed; flowers brilliant cardinal-red, reverse saffron-yellow, semi-dbl., cupped; very vigorous, upright growth; [Mrs Pierre S. duPont X Dr Kirk]; Mallerin, C.; A. Meilland

Dembrosky, *see* **'DEMBROWSKI'**

'DEMBROWSKI', HP, dr, 1849; (Dembrosky); flowers deep crimson-violet; Vibert

'DEMOISELLE', HT, pb, 1960; flowers soft pink becoming darker, open, medium, dbl., 25–30 petals, slight fragrance; foliage bronze; vigorous, bushy growth; [Peace X Opera]; Delforge

'DEMOKRACIE', LCl, dr, 1935; (Blaze Superier); said to be recurrent; Böhm, J.

Demon, *see* 'MEIDOMONAC'

'DÉMONE', HT, dr, 1965; bud ovoid; flowers medium, dbl., intense fragrance; foliage dark; Tantau, Math.

'DEMURE', F, dp, 1952; bud ovoid; flowers rose-pink, borne in clusters, dbl., 52 petals, 2 in., flat, moderate fragrance; foliage leathery; vigorous, compact growth; [Garnette X Seedling]; Boerner; J&P

Deneb, LCl, dy

'DENIS HÉLIE', HP, mr, 1864; (Denis Helye); flowers rosy crimson; vigorous growth; Gautreau

Denis Helye, *see* **'DENIS Hélie'**

'DENISE', HT, my, 1955; bud oval; flowers citron-yellow, open, large, dbl., slight fragrance; foliage glossy, clear green; vigorous, bushy growth; [Peace X Brandywine]; Buyl Frères; Delforge

'DENISE CASSEGRAIN', Pol, w, 1922; flowers snow-white, clusters of 30–40, very dbl., intense fragrance; Grandes Roseraies

'DENISE CHAMBARD', HT, yb, 1940; bud long, carmine-yellow; flowers sulfur-yellow, shaded carmine, large, cupped; foliage bright green; long stems; vigorous, upright growth; Chambard, C.; Orard

'DENISE DEWAR', HT, dr, 1968; flowers crimson, pointed, foliage bronze-green; free growth; [Isabelle de France X Karl Herbst]; Trew, C.

Denise Grey®, *see* MEIxetal

Denise Hale, *see* 'WAMHALE'

Denise Hilling, HMoy, dp

'DENISE LEFEUVRE', HT, rb, 1930; flowers nasturtium-red, center yellow, reverse bright red, large, dbl., cupped; strong stems; very vigorous growth; Chambard, C.

'DENISE MCCLELLAND', HT, dr, 1964; flowers claret-red, large, dbl., 30 petals, moderate fragrance; foliage dark, leathery, glossy; vigorous, upright, bushy growth; [Amy Johnson X New Yorker]; Riethmuller; Akhurst

'DENISE-ANNE', F, op, 1973; flowers blush-pink to orange-apricot, full, dbl., 30–35 petals, 4 in., intense fragrance; foliage small, glossy, dark; vigorous growth; [(Memoriam X Orange Sensation) X (Peace X Memoriam)]; Ellick; Radway Roses

Denk an Mich, F, ab, 1995; Huber

Denman, *see* 'LANDEN'

'DENNY BOY', F, yb, 1949; bud small, ovoid; flowers orange-yellow, becoming redder, reverse sulfur-yellow, very dbl., cupped, slight fragrance; very vigorous, bushy, dwarf growth; [Pinocchio X Mrs Erskine Pembroke Thom]; Marsh; Marsh's Nursery

Denoyel, *see* **'SOUV DE CLAUDIUS DENOYEL'**

Dentelle de Bruges®, S, w, 1991; Lens

Dentelle de Bruxelles®, S, mr, 1988; Lens

Dentelle De Malines®, S, mp, 1986; Lens

Denver's Dream™, *see* 'SAVADEN'

'DER KRAD', HT, dr, 1962; bud long, pointed; flowers maroon-red, striped darker, dbl., 3.5 in., exhibition form, intense fragrance; foliage glossy; vigorous, bushy growth; [Ami Quinard X Crimson Glory]; Wyant

'DERBY', F, or, 1963; bud ovoid; flowers medium, semi-dbl., moderate fragrance; foliage light green, soft; very vigorous, bushy growth; [Miss France X Seedling]; Gaujard

Derdinger Sommer, *see* HELstrau

'DEREHAM PRIDE', Pol, dr, 1932; flowers darkest crimson; vigorous, dwarf growth; [Éblouissant X Orange King]; Norfolk Nursery

Derek Nimmo, *see* 'MACWHENU'

'DERNBURG', HT, pb, 1916; flowers bright rose, shaded coral-red and yellow, large, dbl.; Krüger

'DERUITER'S HERALD', Pol, dr, 1949; (Herald); flowers blood-red, prominent yellow stamens, small, borne in large trusses, single, 6 petals; foliage glossy, dark; vigorous, bushy growth; (21); GM, NRS, 1948; [Orange Triumph X Seedling]; deRuiter; Gregory

DESbet, HT, dp; (**Betsy Murchison**); dbl., 34 petals, exhibition form; [Southern Lady X Seedling]; Desamero, Luis, 1998

Descamet, C, mp

Descanso Dream, *see* 'WEKDESC'

'DESCANSO PILLAR', LCl, pb, 1952; bud urn-shaped, carmine to scarlet-red; flowers begonia-rose to deep rose-pink, inside varying to scarlet, dbl., 33 petals, 4.5–5 in., exhibition form, recurrent bloom, moderate fragrance; foliage dark, glossy; height 6–7 ft; [Crimson Glory X Capt. Thomas]; Lammerts, Dr Walter; Germain's

'DESCHAMPS', N, mr, 1877; flowers cherry-red, large, cupped; vigorous growth; Deschamps

'DESDECHARDO', HT, dp, 1980; flowers deep pink; [Red Lion sport]; Taylor, W. J.

'DESDEMONA', HT, lp, 1912; flowers large, dbl., intense fragrance; Paul, W.; (Sangerhausen)

'DESERT CHARM', Min, dr, 1973; flowers deep red, medium, dbl., exhibition form, slight fragrance; foliage dark, leathery; vigorous, dwarf, bushy growth; [Baccará X Magic Wand]; Moore, Ralph S.; Sequoia Nursery

'DESERT DANCE', F, ob, 1975; bud pointed; flowers orange, reverse gold, semi-dbl., 15–18 petals, 3 in., cupped, slight fragrance; foliage glossy; moderately tall growth; [Impala X Seedling]; Herholdt, J. A.

Desert Dawn, Gr, yb, 1998

'DESERT DREAM', HT, lp, 1955; flowers buff-pink, high pointed, 5 in., slight fragrance; foliage light green; very vigorous growth; [R. M. S. Queen Mary X Mrs Sam McGredy]; McGredy, Sam IV; Kordes

Desert Peace, *see* 'MEINOMAD'

'DESERT SANDS', F, ab, 1976; flowers deep apricot, full, dbl., 30 petals, 4.5 in., intense fragrance; vigorous growth; [Arthur Bell X Elizabeth of Glamis]; Bees

'DESERT SONG', HT, rb, 1948; flowers glowing reddish copper, peony form, dbl., 40–45 petals, 4–5 in., intense fragrance; foliage glossy, bronze; [Mrs Sam McGredy X Golden Dawn]; Fletcher; Tucker

'DESERT STORM', Min, op, 1991; bud ovoid; flowers orange in summer, pink in cold weather, urn-shaped, dbl., borne singly; foliage medium, dark green, semi-glossy; upright, medium growth; [Libby X Rise 'n' Shine]; Gruenbauer, Richard, 1984; Flowers 'n' Friends Miniature Roses, 1991

Desert Storm, S, or, 1997; Williams, J. Benjamin

Desert Sun, *see* DEVsleek

'DESERT SUNSET', Gr, or, 1962; bud ovoid; single, 10 petals, 3–4 in., cupped; foliage glossy; upright growth; [Floradora X Chrysler Imperial]; Booy, H.; Booy Rose Nursery

'DESGACHES', B, lp, 1850; Plantier

Desi, HT, yb, 1964; flowers golden yellow with red stripes, large, dbl.; Rupprecht-Radke; (Sangerhausen)

'DESIGNER'S CHOICE'™, Gr, yb, 1989; (Hi Teen); bud pointed; flowers yellow marked orange, medium, borne usually singly and in sprays, dbl., 30 petals, exhibition form, slight fragrance; ovoid, small, orange fruit; prickles slight recurve, small, brown; foliage medium, dark green, glossy; upright, medium growth; [Prominent X Bengali]; deVor Nursery; Co-Operative Rose Growers, 1989

'DÉSIR', HT, rb, 1945; flowers purplish red, large, dbl., intense fragrance; foliage glossy; very vigorous, bushy growth; Gaujard

'DESIRE', HT, mr, 1953; bud ovoid; flowers cardinal-red, dbl., 35–45 petals, 5 in., exhibition form, moderate fragrance; foliage dark; upright, compact growth; [Pink Delight sport]; Obertello; Amling-DeVor Nursery

'DÉSIRŚ BERGERA', HWich, pb, 1910; flowers coppery rose, center brighter, dbl., seasonal bloom; vigorous growth; [R. wichurana X Aurore]; Barbier

Desirée®, *see* 'TANERISED'

'DÉSIRÉE PARMENTIER', HGal, lp; flowers vivid pink, large, dbl., flat; bushy growth; Parmentier, Prior to 1841

'DESLING', F, ob, 1996; (**Cat's Meow**™); flowers medium orange, petal base medium yellow, bright orange, buff, dbl., 29–31 petals, 3.5 in., slight fragrance; numerous prickles; foliage medium, dark green, semi-glossy; upright, tall (40") growth; [Colour Wonder X Unknown]; Desmet, Paul; Desling Farms

'DESMOND GATWARD', HT, mr, 1966; flowers cerise-crimson, well formed, 4.5 in., slight fragrance; foliage red-bronze; free growth;

[Karl Herbst X Dicksons Red]; Kemp, M. L.

'DESMOND JOHNSTON', HT, mr, 1927; flowers brilliant scarlet, base orange, reverse veined orange, large, dbl., exhibition form, moderate fragrance; foliage rich green, leathery, glossy; short stems; bushy growth; GM, NRS, 1927; McGredy

DESmother, Min, w; (**Sadie**); very dbl., exhibition form; [Irresistible sport]; Desamero, Luis, 1998

'DESPERADO', F, yb, 1968; flowers yellow shaded pink, borne in trusses, semi-dbl., slight fragrance; [Pink Parfait X Masquerade]; Harkness

Desprez à Fleur Jaunes, *see* **'JAUNE DESPREZ'**

Dessert, F, rb; flowers red with white, medium, semi-dbl., VEG; (Sangerhausen)

Dessy, F, lp; Williamson, 1959; (Weatherly, L.)

Destin, HT, 1961; Croix, P.; (Cavriglia)

Destin Cl., Cl HT, mr

'DESTINO', HT, pb; flowers salmon-pink, becoming lilac-pink, well formed, dbl., 45 petals; foliage dark; upright growth; Camprubi, C.

'DESTINY', HT, dr, 1935; bud long, pointed, well shaped; flowers rich crimson-scarlet shaded blackish, dbl.; foliage dark, leathery; vigorous, bushy growth; Beckwith

Destiny, *see* MEIkiji

'DETROITER', HT, dr, 1952; (Brilliant, Schlosser's Brilliant); bud long, pointed; dbl., 23 petals, 5.5 in., exhibition form, moderate fragrance; vigorous, upright, bushy growth; GM, NRS, 1952; [Poinsettia X Crimson Glory]; Kordes; J&P

'DETROITER, CLIMBING', Cl HT, dr, 1960; (Schlösser Brilliant, Climbing); J&P

Detty, HT, or; Adam

'DEUIL DE DR REYNAUD', B, mp, 1862; flowers deep crimson pink, intense fragrance; Pradel

'DEUIL DE DUC D'ORLEANS', B, m, 1845; Lacharme, F.

'DEUIL DE DUNOIS', HP, dp, 1864; flowers carmine-pink, medium, dbl.; Verdier, E.; (Sangerhausen)

'DEUIL DE PAUL FONTAINE', M, m, 1873; (Paul de Fontainne); bud somewhat mossy; flowers purple-red, reverse mahogany, cupped, repeat bloom, very prickly; vigorous growth; Fontaine

Deuil du Colonel Denfert, HP, dr; Margottin Père, 1878

'DEUTSCHE HOFFNUNG', HT, yb, 1920; flowers salmon-yellow to apricot-yellow; [Mme Caroline Testout X Grossherzogin Feodora von Sachsen]; Kiese

Deutsche Welle®, F, m, 1983; Leenders

'DEUTSCHES DANZIG', Pol, dp, 1935; flowers carmine-pink with white, small, single; Lambert, P.; (Sangerhausen)

'DEVABE', HT, w, 1982; (**White Mystery**); [Paul's Pink sport]; deVor, Paul F.; DeVor Nurseries, Inc.

'DEVAMARILLO', Min, ob, 1989; (**Rainbow Yellow™**); bud pointed, urn-shaped; flowers tangerine orange, aging lighter, small, borne singly, dbl., intense, fruity fragrance; prickles declining, reddish-brown; foliage medium, medium green, matt; bushy, medium growth; PP007638; [Unnamed seedling X Amber Flash]; Marciel, Stanley G.; DeVor Nurseries, Inc.

'DEVAURORA', Min, ob, 1989; (**Rainbow Sunrise™**); bud pointed; flowers orange with tinge of red, reverse same, dbl., slight, spicy fragrance; prickles sparse, declining, brown; foliage small, medium green, glossy; bushy, medium growth; PP006583; [Amber Flash X Rumba]; Marciel, Stanley G.; DeVor Nurseries, Inc.

'DEVBILL', Cl HT, op, 1995; (**Sheer Elegance, Climbing**); flowers pink blend, blooms, dbl., 26–40 petals, 3–3.5 in., slight fragrance; few prickles; foliage large, dark green, semi-glossy; tall (315–345 cms), upright, branching growth; [Sheer Elegance sport]; deVor, Bill; DeVor Nursery, 1994;, Youngs' American Rose Nursery, 1994

DEVblush, HT, w, 1996; (**Bridal Shower**); deVor

'DEVCAL', HT, w, 1995; (**Victorian Lace**); dbl., 26–40 petals, 1.5–2.5 in., borne mostly singly, no fragrance; some prickles; foliage medium, medium green, semi-glossy; tall (60 in.), upright growth; Marciel, Stanley G.; DeVor Nurseries, Inc., 1995;, Young's American Rose Nursery, 1995

'DEVCALI', HT, dp, 1989; (**Sizzle Pink™**); bud slender, tapering; flowers deep pink, sweetheart, large, dbl., 25 petals, cupped, borne singly, slight fragrance; prickles declining, olive green; foliage large, dark green, matt; upright, tall growth; PP007562; [Unnamed seedling X Unnamed seedling]; Marciel, Stanley G.; DeVor Nurseries, Inc.

'DEVCARLOS', HT, ly, 1992; (**Brianna**); flowers canary yellow, blooms borne mostly singly, dbl., 26–40 petals, 3–3.5 in., moderate fragrance; few prickles; foliage large, medium green, semi-glossy; upright (152 cms) growth; [Seedling 82249–1 (75062–1 X Excitement) X Seedling (75062–1 X Cocktail)]; Marciel, Stanley G.; DeVor Nurseries, Inc.

'DEVCICO', HT, mp, 1993; (**Darby**); flowers moderately medium blooms borne mostly single, dbl., 15–25 petals, 1.5–2.75 in., no fragrance; some prickles; foliage large, dark green, semi-glossy; tall, upright growth; [Dolores X Cerise Dawn]; Marciel, Stanley G. & Jeanne A.; DeVor Nurseries, Inc., 1993;, Matsui Nursery, 1993

'DEVCLAVEL', Min, dp, 1989; (**Rainbow Cerise™**); bud tapering, slender; flowers deep pink, medium, borne singly, dbl., 45 petals, cupped, slight, fruity fragrance; foliage medium, dark green, glossy; bushy, medium growth; PP007560; [Scarlet Sunblaze X Unnamed seedling]; Marciel, Stanley G.; DeVor Nurseries, Inc.

'DEVDICHA', Min, rb, 1989; (**Rainbow Bliss™**); bud slender, tapering; flowers white with cream inside with red edges, reverse same, aging, dbl., cupped, slight, damask fragrance; prickles declining, brown-orange; foliage small, dark green, glossy; upright, low growth; [Unnamed seedling X Scarlet Sunblaze]; Marciel, Stanley G.; DeVor Nurseries, Inc.

'DEVDORADO', HT, ly, 1992; (**Mariko**); dbl., 26–40 petals, 3–3.5 in., borne mostly singly, moderate fragrance; some prickles; foliage large, dark green, semi-glossy; tall (210 cms), upright growth; [Unnamed seedling X Unnamed seedling]; Marciel, Stanley G. & Jeanne A.; DeVor Nurseries, Inc.

'DEVECLIPSAR', Min, rb, 1989; (**Rainbow Eclipse™**); bud pointed; flowers crimson-pink edges, center very light whitish-pink, fading t, dbl., slight fragrance; prickles declining, rusty-brown; foliage small, dark green, glossy; bushy, medium growth; PP007637; [Scarlet Sunblaze X Unnamed seedling]; Marciel, Stanley G.; DeVor Nurseries, Inc.

'DEVFRAGO', HT, ab, 1992; (**Fragrant Fantasy**); dbl., 15–25 petals, 3–3.5 in., borne mostly singly, intense fragrance; few prickles; foliage medium, dark green, matt; medium (3–4 ft), upright growth; [Unnamed seedling 80227-20 X Unnamed seedling 82249-1]; Marciel, Stanley G. & Jeanne A.; DeVor Nurseries, Inc.

'DEVHAUTE', HT, dp, 1994; (**Matty**); flowers deep pink, blooms borne mostly singly, dbl., 15–25 petals, 3–3.5 in., moderate fragrance; some prickles; foliage large, medium green, matt; tall (97 cms), upright growth; [(Happiness X Seedling) X (Emily Post X Visa)]; Marciel, Stanley G. & Jeanne A.; DeVor Nurseries, Inc., 1993;, Sakai Brothers Rose Co., 1993

Devi Gayatri, F, w, 1992; Sen

'DEVICIO', HT, dp, 1989; (**Touch of Raspberry™**); bud slender, tapering; flowers deep pink, large, dbl., 30 petals, cupped, borne singly, slight, fruity fragrance; prickles declining, light lime green with mauve tinges; foliage large, dark green, semi-glossy; upright, tall growth; PP007548; [Love Affair X Paul's Pink]; Marciel, Stanley G.; DeVor Nurseries, Inc.

'DEVIENNE LAMY', HP, mr, 1868; flowers medium, dbl.; Lévêque; (Sangerhausen)

'DEVIENTE', Min, dp, 1989; (**Rainbow Hot Pink™**); bud long, slightly urn-shaped; flowers deep pink, small, borne singly, dbl.; prickles declining, mauve; foliage small, dark green, semi-glossy; bushy, medium growth; PP007689; [Orange Sunblaze X Unnamed seedling]; Marciel, Stanley G.; DeVor Nurseries, Inc.

'DEVIL DANCER', HT, mr, 1979; bud ovoid; flowers brilliant brick-red, single, 5 petals, 4 in., slight fragrance; foliage dark; bushy growth; [Sonora X Matangi]; Hawken, Una

'DEVILK', HT, ob, 1989; (**Sparkling Orange**™); bud tapering; flowers vermillion, reverse scarlet, aging to rose, sweetheart, large, dbl., borne singly, intense, musk fragrance; prickles variation, reddish-brown; foliage large, dark green, glossy; upright, tall growth; [Sonia X Prominent]; Marciel, Stanley G.; DeVor Nurseries, Inc.

Devil's Dance, F, mr

DEVine, HT, pb; (**Dolores**™); deVor

'DEVITE', S, w, 1982; (**Angel**™); flowers flora-tea, small, dbl., 35 petals, slight fragrance; foliage medium, light green, semi-glossy; upright growth; [(Queen Elizabeth X Seedling) X Jack Frost]; deVor, Paul F.

'DEVLASS', HT, w, 1989; (Spun Glass™, **Spunglass**); bud pointed; flowers very large, borne singly, dbl., 50 petals, cupped, slight, spicy fragrance; prickles declining, pea green; foliage large, dark green, glossy; upright, tall growth; PP007613; [Unnamed seedling X Angel]; Marciel, Stanley G.; DeVor Nurseries, Inc.

DEVlicor, HT, dy, 1995; (**Licorice Twist**); deVor

'DEVLOREN', HT, op, 1989; (**Sophia's Song**™); bud slender, tapering; flowers coral, large, dbl., 32 petals, exhibition form, borne singly, slight, fruity fragrance; prickles declining, copper brown with pea green tinges; foliage medium, dark green, semi-glossy; upright, tall growth; PP007455; [Emily Post X Prominent]; Marciel, Stanley G.; DeVor Nurseries, Inc.

DEVmauve, HT, m, 1995; (**Sonora Sunset**); deVor

'DEVMENTA', HT, rb, 1989; (**Peppermint Swirl**™); bud slender, tapering; flowers currant red, reverse same, aging discolors slightly, large,, dbl., 30 petals, cupped, slight, spicy fragrance; prickles declining, copper brown; foliage medium, dark green, semi-glossy; upright, tall growth; PP007612; [Unnamed seedling X Unnamed seedling]; Marciel, Stanley G.; DeVor Nurseries, Inc.

'DEVMESI', Min, dr, 1989; (**Rainbow Crimson**™); bud slender; flowers deep red, aging discolors slightly, small, borne singly, dbl., cupped, slight, damask fragrance; prickles declining, slightly dark mauve; foliage medium, dark green, semi-glossy; upright, low growth; [Unnamed seedling X Unnamed seedling]; Marciel, Stanley G.; DeVor Nurseries, Inc.

'DEVMOMENTO', HT, dy, 1993; (**Glory**); dbl., 26–40 petals, 3–3.5 in., borne mostly singly, no fragrance; some prickles; foliage large, dark green, semi-glossy; tall (205 cms), upright growth; [Capella X Seedling]; Marciel, Stanley G. & Jeanne A.; DeVor Nurseries, Inc., 1992

'DEVMORADA', HT, m, 1989; (**Royal Amethyst**™); bud pointed; flowers lavender, large, borne singly, dbl., 32 petals, intense, fruity fragrance; globular, tangerine orange fruit; prickles declining, henna; foliage medium, medium green, glossy; upright, tall growth; [Angel Face X Blue Moon]; deVor, Paul F.; DeVor Nurseries, Inc.

'DEVNINA', HT, lp, 1989; (Amanda™, **Amanda Marciel**™); bud slender and tapering; flowers very delicate pink, large, borne singly, dbl., 26 petals, cupped, slight, spicy fragrance; no prickles; foliage medium, dark green, glossy; upright, tall growth; [Seedling X Pink Puff]; Marciel, Stanley G.; DeVor Nurseries, Inc.

'DEVNOVIA', HT, mp, 1989; (Megan, **Megan Dolan**); bud urn-shaped; flowers small, sweetheart, 18 petals, cupped, borne singly, spicy fragrance, slight; prickles declining, copper brown with olive green tinges; foliage medium, dark green; upright, tall growth; PP007520; [Angel X Independence '76]; Marciel, Stanley G.; DeVor Nurseries, Inc.

Devon®, *see* POUlrijk

'DEVON MAID', LCl, pb, 1982; flowers light pink, reverse medium pink, large blooms in clusters of 3–4, dbl., 22 petals, moderate, fruity fragrance; ovoid, orange fruit; large, curved, orange prickles; foliage large, medium green, glossy; needs support; vigorous, spreading, tall growth; [Casino X Elizabeth of Glamis]; Warner, Chris; Warner's Roses

'DEVONIENSIS', T, w, 1838; (Magnolia Rose); flowers creamy white, center sometimes tinged blush, very large, dbl., recurrent bloom, moderate fragrance; very vigorous growth; [Parentage uncertain, perhaps Elinthii X Yellow China]; Foster; Lucombe, Prince & Co., 1841

'DEVONIENSIS, CLIMBING', Cl T, w, 1858; [Devoniensis sport]; Pavitt; Curtis

Devonshire Maid, LCl, mp, 1993

DEVoran, HT, yb, 1989; (**Karen**); deVor

'DEVORO', Min, yb, 1991; (**Rainbow Gold**™); flowers yellow-orange blend, full petals, blooms borne in small clusters, slight fragrance; foliage small, dark green, semi-glossy; upright, medium (35 cms) growth; [Amber Flash X Rhumba]; Marciel, Stanley G.; DeVor Nurseries, Inc.

'DEVOTION', F, pb, 1971; flowers light pink, flushed deeper, dbl., 32 petals, 4.5 in., moderate fragrance; foliage light; [Orange Sensation X Peace]; Harkness

'DEVPAJARO', HT, mr, 1994; (**Red Satin**); dbl., 26–40 petals, 3–3.5 in., borne mostly singly, slight fragrance; few prickles; foliage large, medium green, matt; low (55–60 cms), upright growth; [Jacqueline X Prominent]; Marciel, Stanley G. & Jeanne A.; DeVor Nurseries, Inc., 1993;, Pajaro Valley Greenhouses, 1993

'DEVPRESA', Min, ab, 1989; (**Rainbow Surprise**™); bud slender, tapering; flowers medium coral, reverse light coral, aging pink, small, borne, dbl., slight, fruity fragrance; prickles declining, mauve; foliage small, dark green, glossy; bushy, medium growth; PP007558; [Orange Sunblaze X Unnamed seedling]; Marciel, Stanley G.; DeVor Nurseries, Inc.

'DEVRICO', Min, rb, 1989; (**Rainbow Stanford**™); bud pointed, slender; flowers bright red, reverse same, aging discolors to a bright red-or, semi-dbl., fruity fragrance, slight; prickles declining, reddish-brown; foliage medium, dark green, glossy; bushy, medium growth; PP007559; [Candia X Unnamed seedling]; Marciel, Stanley G.; DeVor Nurseries, Inc.

'DEVRISE', HT, dr, 1989; (**Cerise Dawn**™); bud urn-shaped; flowers magenta, reverse tyrian purple, aging no discoloration, large, dbl., 30 petals, globular, slight, damask fragrance; prickles declining and well spaced apart, pea green; foliage large, dark green, semi-glossy; upright, tall growth; [Carina X Angel Face]; Marciel, Stanley G.; DeVor Nurseries, Inc.

'DEVROJO', Min, mr, 1989; (**Rainbow Red**™); bud pointed; flowers small, dbl., borne singly, slight, spicy fragrance; prickles declining, brown with orange; foliage small, dark green, glossy; bushy, medium growth; PP006584; [Scarlet Sunblaze X Rumba]; Marciel, Stanley G.; DeVor Nurseries, Inc.

'DEVROSADO', Min, dp, 1989; (**Rainbow Pink**™); bud ovoid, pointed; flowers deep pink, small, dbl., borne singly, slight, spicy fragrance; prickles declining, purple; foliage medium, dark green, semi-glossy; bushy, medium growth; PP006875; [Unnamed seedling X Orange Sunblaze]; Marciel, Stanley G.; DeVor Nurseries, Inc.

'DEVRUDI', S, lp, 1998; (**First Light**); bud pointed, dark candy pink; flowers single, light pink with purple stamens, single, 5–7 petals, 3.5–4 in., borne in clusters, moderate, spicy fragrance; foliage dark green; compact, low growth; AARS, 1998; [Bonica X Ballerina]; Stanley/Marciel

'DEVSIEM', HT, dr, 1989; (**Always Mine**™); bud pointed, tapering; flowers deep red, large, dbl., 39 petals, cupped, borne singly, intense, spicy fragrance; prickles declining, pea green with cinnamon tinges; foliage large, dark green, glossy; upright, tall growth; [Visa X Sassy]; Marciel, Stanley G.; DeVor Nurseries, Inc.

DEVsleek, HT, dy, 1996; (**Desert Sun**); deVor

DEVsmooth, F, mr, 1995; (**Bella**); deVor

'DEVSOLEAR', Gr, dy, 1989; (**Goldlite**™); bud urn-shaped; flowers canary yellow, reverse buttercup yellow, aging no discoloration, dbl., slight, musk fragrance; round, average, tangerine orange fruit; prickles declining, red; foliage medium, dark green, semi-glossy; upright, tall growth; PP007462; [Unnamed seedling X Excitement]; Marciel, Stanley G.; DeVor Nurseries, Inc., 1987

'DEVSPILIO', HT, w, 1992; (**Silver Fox**); dbl., 26–40 petals, 3–3.5 in., borne mostly singly, slight fragrance; some prickles; foliage large, dark green, semi-glossy; tall (210 cms), upright growth; [Unnamed seedling X Unnamed seedling]; Marciel, Stanley G. & Jeanne A.; DeVor Nurseries, Inc.

'DEVSTAR', HT, w, 1992; (**Stardust**); dbl., 26–40 petals, 3–3.5 in., borne mostly singly, no fragrance; some prickles; foliage large, medium green, semi-glossy; tall (230 cms), upright growth; [Coquette X Unnamed seedling 64022-39]; Marciel, Stanley G. & Jeanne A.; DeVor Nurseries, Inc.

'DEVSTICA', HT, mr, 1989; (**Mystique**™); bud pointed, slender, tapering; flowers bright red, sweetheart, large, dbl., 28 petals, cupped, borne singly, intense, musk fragrance; prickles declining, pale red with tinges of green; foliage large, medium green, semi-glossy; upright, tall growth; [Samantha X Royalty]; Marciel, Stanley G.; DeVor Nurseries, Inc.

DEVsunset, HT, lp; (**Sundancer**); deVor, 1995

'DEVTINTA', HT, dr, 1990; (**Obsession**); bud high-centered; flowers imbricated, large blooms borne singly, dbl., 42 petals, cupped, moderate fragrance; prickles wing-shaped, reddish tinge; foliage medium, medium green, semi-glossy; upright, tall growth; [Unnamed seedling X Unnamed seedling]; Marciel, Stanley G.; DeVor Nurseries, Inc.

'DEVUNICAN', F, dy, 1992; (**Golden Sprite**); dbl., 15–25 petals, 3–3.5 in., borne mostly singly, slight fragrance; some prickles; foliage medium, dark green, matt; medium, upright growth; [Golden Fantasie X Excitement]; Marciel, Stanley G. & Jeanne A.; DeVor Nurseries, Inc.

Dew Drop, *see* 'ZIPDEW'

'DEWDROP', HT, lp, 1921; flowers pale pink to pale rose, moderate fragrance; RULED EXTINCT 11/91 ARM; McGredy

Dezent, HT, ly; flowers yellowish-white, large, dbl., slight fragrance; VEG; (Sangerhausen)

Diablotin®, *see* 'DELPO'

Diablotin, Climbing®, *see* 'DELPOSAR'

'DIABOLO', F, op, 1958; flowers bright salmon, medium, semi-dbl., cupped, slight fragrance; foliage bronze; short stems; very vigorous, bushy growth; [Jolie Princess X (Alain X Miss France)]; Gaujard

'DIADEM', HT, ob, 1922; flowers orange-crimson suffused salmon and yellow, very large, dbl., exhibition form, moderate fragrance; foliage rich green, leathery, glossy; vigorous, bushy growth; McGredy

Diadem, *see* TANmeda

Diamant®, *see* 'KOREB'

Diamant Rose, HMsk, pb, 1995; Lens

'DIAMANTINA', HT, op, 1949; flowers salmon and rose, large; strong stems; [Julien Potin X Ophelia]; Giacomasso

Diamond Anniversary, *see* 'MORSIXTY'

Diamond Border, *see* POUldiam

Diamond Doll, *see* 'JUDDOLL'

'DIAMOND JEWEL', Min, w, 1958; bud globular; flowers white, overcast blush-pink, dbl., 45–50 petals, .5–.75 in., cupped; compact, low, open growth; [Dick Koster sport X Tom Thumb]; Morey, Dr Dennison; J&P

'DIAMOND JUBILEE', HT, ly, 1947; bud ovoid; flowers buff-yellow, dbl., 28 petals, 5–6 in., cupped, moderate fragrance; foliage leathery; upright, compact growth; AARS, 1948; [Maréchal Niel X Feu Pernet-Ducher]; Boerner; J&P

'DIAN', Min, dp, 1957; flowers soft red, dbl., 45 petals, 1 in., moderate, apple fragrance; foliage small, dark, glossy; vigorous (15 in.), bushy growth; [(R. wichurana X Floradora) X (Oakington Ruby X Floradora)]; Moore, Ralph S.; Sequoia Nursery

Diana®, *see* 'TANDINADI'

'DIANA', HT, mp, 1921; flowers malmaison pink, very large, dbl., globular, moderate fragrance; vigorous growth; [Mrs Frank Workman X Sunburst]; Bees

'DIANA', Pol, op, 1922; flowers bright orange shaded pink, large, semi-dbl.; Spek

'DIANA ALLEN', HT, op, 1939; flowers salmon-pink, small, dbl.; short stems; bushy, compact growth; [Mrs Aaron Ward X Seedling]; Clark, A.; NRS New South Wales

'DIANA ARMSTRONG', HT, dy, 1992; dbl., 26–40 petals, 1.5–2.75 in., borne mostly singly, intense fragrance; some prickles; foliage medium, medium green, semi-glossy; upright (80 cms) growth; [Seedling X Prima Ballerina]; Thompson, Robert; Battersby Roses, 1993

'DIANA CANT', HT, rb, 1928; flowers carmine-red, base flushed orange, dbl., moderate fragrance; [Isobel X Seedling]; Cant, B. R.

Diana Festival, Min, lp, 1996; Laver, Keith G.

Diana H. Gupta, *see* 'JAYGUP'

'DIANA MAXWELL', HT, ob, 1957; flowers orange-cerise, high pointed, dbl., 45 petals, 5 in., intense fragrance; foliage bronze; vigorous growth; [Ena Harkness X Sam McGredy]; Kemp, M.L.

'DIANA MENUHIN', HT, dy, 1963; flowers deep buttercup-yellow, dbl., 30 petals, 5.5 in., globular; foliage dark, glossy; vigorous, upright growth; [(Golden Masterpiece X Ellinor LeGrice) X Forward]; LeGrice

'DIANA ROWDEN', LCl, op, 1976; flowers deep copper-salmon to rose-pink, dbl., 30 petals, 5–6. in., profuse, continuous bloom, intense fragrance; foliage large, copper to green; [Mrs Sam McGredy, Climbing X Red Dandy]; Hawker; Harkness

Diana, Princess of Wales™, *see* 'JACSHAQ'

'DIANE', HT, yb, 1958; flowers clear yellow, center orange-yellow, well formed, large; vigorous growth; [Peace X (Seedling X Opera)]; Gaujard

Diane, HT, lp; Dawson, George, 1976; (Weatherly, L.)

'DIANE DE BROGLIE', HT, ob, 1929; flowers coral-orange, very large, dbl., cupped, moderate fragrance; foliage dark; short stems; very vigorous growth; Chambard, C.

Diane de Poltiers, HT, dp

'DIANE D'URFÉ', HT, w, 1958; flowers white, becoming red-edged; vigorous growth; [Peace X Incendie]; Croix, A.; Minier

'DIANNA KAY', Min, or, 1981; flowers small, dbl., 20 petals, no fragrance; foliage small, medium green, semi-glossy; upright growth; [Anytime X Sheri Anne]; Dobbs, Annette E.

'DIANNE FEINSTEIN', HT, yb, 1979; bud ovoid; dbl., 23 petals, exhibition form, borne singly, moderate fragrance; triangular prickles; foliage dark; vigorous, spreading growth; [McGredy's Yellow X Sutter's Gold]; Fong, William P.

Diantheflora, *see* **'FIMBRIATA'**

Dianthiflora, *see* **'FIMBRIATA'**

'DIANY BINNY', LCl, w, 1976; flowers small, single, 5 petals, 2 in., intense fragrance; foliage purplish; [Kiftsgate X R. rubrifolia (?)]; Binny

Diapason®, *see* 'DELPOC'

Diavoletta, 1966; Mansuino, Q.; (Cavriglia)

Dic Della, HT, dp, 1996; Chiplunkar

'DICALOW', F, dy, 1977; (**Yellow Ribbon**); flowers deep golden yellow, medium, dbl., 23 petals, cupped, slight fragrance; foliage medium green, semi-glossy; vigorous, compact, bushy growth; [Illumination X Stroller]; Dickson, Patrick; A. Dickson

'DICAM', HT, mr, 1970; (Coeur d'Amour, **Red Devil**); bud ovoid; flowers medium red, reverse lighter, large, dbl., 72 petals, exhibition form, moderate fragrance; foliage glossy; vigorous growth; GM, Belfast, 1969 GM, Japan, 1967 GM, Portland, 1970; [Silver Lining X Prima Ballerina]; Dickson, A.; J&P

'DICBAND', F, ob, 1978; (**Springfields**); flowers orange, red, gold, dbl., 48 petals,

3 in., cupped, slight fragrance; [Eurorose X Anabell]; Dickson, A.

'DICBAR', F, rb, 1978; (**Memento**®); bud globular; flowers salmon-red, dbl., 22 petals, 3 in., cupped; foliage sage green; bushy growth; GM, Belfast, 1980; [Bangor X Anabell]; Dickson, Patrick; Dickson Nurseries, Ltd.

'DICBEE', F, or, 1978; (**High Summer**); bud ovoid; flowers vermilion, dbl., 26 petals, 3 in., cupped; foliage large; bushy growth; [Zorina X Ernest H. Morse]; Dickson, Patrick; Dickson Nurseries, Ltd.

DICblender, HT, rb, 1997; (**Acapulco**); Dickson, Patrick

'DICBO', HT, mp, 1968; (**Bonsoir**); bud ovoid; flowers peach-pink, dbl., 6 in., intense fragrance; foliage glossy; Dickson, A.

'DICDANCE', F, my, 1980; (**Bright Smile**®); bud pointed; flowers empire yellow, patio, convex blooms borne 15 per cluster, semi-dbl., 15 petals, flat, slight fragrance; concave, purple prickles; foliage dense, mid-green; bushy, medium growth; GM, Belfast, 1982; [Eurorose X Seedling]; Dickson, Patrick, 1981

'DICDIP', HT, pb, 1974; flowers large, dbl., 32 petals, moderate fragrance; foliage medium, purple when young; upright, bushy growth; [Eurorose X Typhoon]; Dickson, Patrick

'DICDIVINE', HT, my, 1980; (**Pot o' Gold**); bud pointed; dbl., 32 petals, flat, borne 7 per cluster, intense fragrance; brown prickles; foliage mid-green with strong purple veins; bushy, medium growth; [Eurorose X Whisky Mac]; Dickson, Patrick

'DICDRUM', F, op, 1982; (**Shona**); flowers medium coral pink, blooms in clusters, dbl., 23 petals, slight fragrance; foliage medium, medium green, semi-glossy; bushy growth; [Bangor X Anabell]; Dickson, Patrick

'DICEL', F, or, 1963; (**Scarlet Queen Elizabeth**®); flowers flame-scarlet, medium, dbl., globular, slight fragrance; foliage dark; vigorous, tall growth; GM, The Hague, 1973 Golden Rose, The Hague, 1973; [(Korona X Seedling) X Queen Elizabeth]; Dickson, Patrick; A. Dickson

'DICFATE', F, mp, 1975; flowers medium, 18 petals, slight fragrance; foliage medium, medium green; upright, bushy growth; [Futura X (Pye Colour X Prominent)]; Dickson, Patrick

'DICFIRE', F, or, 1983; (**Beautiful Britain**); flowers orange-red, reverse deeper, medium, in clusters, dbl., 20 petals, slight fragrance; foliage medium, medium green, semi-glossy; upright, bushy growth; ROTY, 1983; [Red Planet X Eurorose]; Dickson, Patrick; Dickson Nurseries, Ltd.

'DICGROW', Min, ob, 1981; (**Brass Ring**, Peek a Boo); bud pointed; flowers coppery orange, fading to rose pink, patio, blooms in large clusters, dbl., flat; foliage small, pointed, glossy; upright, arching growth; [Memento X Nozomi]; Dickson, Patrick; J&P

'DICINFRA', F, or, 1984; (**Disco Dancer**®); flowers orange scarlet, medium, semi-dbl., slight fragrance; foliage medium, medium green, glossy; bushy growth; GM, The Hague, 1982; [Cathedral X Memento]; Dickson, Patrick, 1983

'DICJANA', HT, ly, 1984; (**Elina**®, Peaudouce); flowers pale yellow to ivory, large, luminous, very dbl., 30–35 petals, 5–5.5 in., exhibition form, borne singly, slight fragrance; foliage large, dark, glossy; long stems; vigorous, tall growth; ADR, 1987 Gold Star of the South Pacific, Palmerston North, NZ, 1987; [Nana Mouskouri X Lolita]; Dickson, Patrick, 1983

'DICJEEP', F, rb, 1984; (Daydream, **Len Turner**); flowers ivory petals flushed and edged carmine, large, dbl., 35 petals, slight fragrance; foliage medium, medium green, glossy; bushy, very compact growth; [Electron X Eyepaint]; Dickson, Patrick

'DICJEM', HT, dy, 1984; (**Freedom**®); flowers chrome yellow, large, dbl., 35 petals, exhibition form, moderate fragrance; foliage medium, medium green, glossy; bushy growth; GM, RNRS, 1983; [(Eurorose X Typhoon) X Bright Smile]; Dickson, Patrick; Dickson Nurseries, Ltd.

'DICJOON', HT, rb, 1984; (**Leslie's Dream**); flowers medium red, yellow reverse, large, borne singly and in truss, dbl., 20 petals, no fragrance; foliage medium, dark, glossy; bushy growth; [Bonfire X Typhoon]; Dickson, Patrick

'DICJOY', F, my, 1984; (**Ards Beauty**); bud large; flowers large, dbl., 20 petals, exhibition form, moderate fragrance; foliage medium, medium green, glossy; bushy growth; GM, RNRS, 1983; [(Eurorose X Whisky Mac) X Bright Smile]; Dickson, Patrick, 1986

'DICJUBELL', HT, mp, 1986; (Dickson's Jubilee, **Lovely Lady**); flowers large, dbl., 35 petals, moderate fragrance; foliage medium, mid-green, glossy; bushy growth; GM, Belfast, 1988; [Silver Jubilee X (Eurorose X Anabell)]; Dickson, Patrick

'DICK KOSTER', Pol, dp, 1929; flowers deep pink; [Anneke Koster sport]; Koster, D. A.

'DICK KOSTER FULGENS', Pol, dp, 1940; flowers light red, borne in clusters, semi-dbl.; low, compact growth; Koster, M.

'DICK KOSTER SUPERIOR', Pol, mr, 1955; flowers rosy red; [Dick Koster sport]; Koster, D. A.

Dick Lindner, HT, mp, 1996

'DICK WILCOX', HT, dr, 1949; bud long, pointed to ovoid; flowers rose-red, dbl., 50–60 petals, 4–5.5 in., exhibition form, moderate fragrance; foliage dark; vigorous growth; [Pink Princess X Crimson Glory]; Brownell, H. C.

'DICKERFUFFLE', F, mp, 1985; (Georgie Girl, **Wishing**); flowers medium peachy pink, large, dbl., 35 petals, slight fragrance; foliage medium, medium green, semi-glossy; bushy growth; [Silver Jubilee X Bright Smile]; Dickson, Patrick; Dickson Nurseries, Ltd., 1984

'DICKERRY', F, pb, 1986; (**Laughter Lines**); flowers large, semi-dbl., 6–14 petals, slight fragrance; foliage small, medium green, semi-glossy; bushy growth; GM, RNRS, 1984; [(Pyecolour X Sunday Times) X Eyepaint]; Dickson, Patrick

'DICKIMONO', F, op, 1984; (Anisley Dickson, **Dicky**®, München Kindl); flowers reddish salmon-pink, reverse lighter, large, dbl., 35 petals, slight fragrance; foliage medium, medium green, glossy; bushy growth; GM, RNRS, 1984 PIT, RNRS, 1984; [Cathedral X Memento]; Dickson, Patrick, 1983

DICkindlel, F, ly, 1986; (**Lemon Honey**); Dickson, Patrick

DICkisser, F, or, 1995; (**Hi Doll**); Dickson, Patrick

DICkitty, F, ob, 1987; (**Old Flame**); Dickson, Patrick

DICknowall, F, my; (**Collegiate 110**); Dickson, Patrick

'DICKOOKY', F, my, 1984; (**Tall Story**®); flowers medium, semi-dbl., slight fragrance; foliage medium, light green, glossy; spreading growth; [Sunsprite X Yesterday]; Dickson, Patrick; Dickson Nurseries, Ltd.

Dick's Apricot Rambler, HWich, ab

Dick's Delight™, *see* 'DICWHISTLE'

'DICKSON'S BOUQUET', HT, ab, 1938; flowers salmon, carmine and apricot, blended saffron, dbl., moderate fragrance; long, wiry stems; vigorous growth; Dickson, A.

'DICKSON'S CENTENNIAL', HT, mr, 1936; bud pointed; flowers crimson to scarlet, loosely formed, very large, moderate fragrance; foliage bronze; long, strong stems; vigorous, bushy growth; Dickson, A.; Dreer, 1937; J&P

'DICKSON'S DELIGHT', HT, ob, 1938; flowers vivid orange, heavily shaded scarlet-orange, moderate fragrance; foliage bronze-green; vigorous growth; Dickson, A.

'DICKSON'S FLAME', F, or, 1958; flowers scarlet-flame, blooms in trusses, dbl., 3.5 in., slight fragrance; vigorous growth; GM, NRS, 1958 PIT, NRS, 1958; [Independence seedling X Nymph]; Dickson, A.

Dickson's Jubilee, *see* 'DICJUBELL'

'DICKSON'S PERFECTION', HT, pb, 1937; flowers shrimp-pink, base orange-yellow, large, dbl., intense fragrance; very vigorous growth; Dickson, A.; Port Stockton Nursery

'DICKSON'S RED', HT, dr, 1938; (Dr F.G. Chandler); flowers velvety crimson-scarlet, large, semi-dbl., 18 petals, cupped, intense, spicy fragrance; foliage leathery, dark; vigorous, bushy growth; AARS, 1940 GM, NRS, 1939, GM, Portland, 1941; Dickson, A.

Dickson's Wonder, HT, ob

Dicky®, *see* 'DICKIMONO'

'DICLADIDA', Min, ob, 1987; (**Cider Cup**); flowers medium, dbl., 15–25 petals, slight fragrance; foliage medium, medium green, glossy; patio growth; [Memento X (Liverpool Echo X Woman's Own)]; Dickson, Patrick, 1988

DIClady, F, or, 1989; (**Feu Follet**®); Dickson, Patrick

'DICLITTLE', Min, pb, 1986; (**Little Woman**); dbl., 15–25 petals, moderate fragrance; foliage small, medium green, semi-glossy; patio; bushy growth; [Memento X (Liverpool Echo X Woman's Own)]; Dickson, Patrick

'DICLULU', Min, lp, 1986; (**Gentle Touch**); flowers moderately small, dbl., 15–25 petals, slight fragrance; foliage small, medium green, semi-glossy; patio; bushy growth; ROTY, 1986; [(Liverpool Echo X Woman's Own) X Memento]; Dickson, Patrick

'DICMADDER', F, rb, 1987; (**Star Child**®, Tangeglow); flowers small, semi-dbl., 6–14 petals, slight fragrance; foliage small, medium green, glossy; strong, straight stems; bushy, prolific growth; PP007312; [Eyepaint X (Liverpool Echo X Woman's Own)]; Dickson, Patrick, 1988

'DICMAGIC', Min, ob, 1986; (**Sweet Magic**); flowers moderately small, dbl., 15–25 petals, no fragrance; foliage small, medium green, glossy; patio; bushy growth; ROTY, 1987; [Peek A Boo X Bright Smile]; Dickson, Patrick

'DICMICKEY', Min, or, 1986; (**Buttons**); flowers medium, dbl., 15–25 petals, slight fragrance; foliage small, medium green, glossy; patio; bushy growth; [(Liverpool Echo X Woman's Own) X Memento]; Dickson, Patrick, 1987

'DICMOPPET', S, my, 1987; (Goldfächer, Mini Lights, **Minilights**); flowers small, semi-dbl., 6–14 petals, slight fragrance; foliage small, dark green, glossy; spreading, compact growth; [White Spray X Bright Smile]; Dickson, Patrick, 1988

'DICNAME', HT, yb, 1986; (**Peter Goldman**); flowers medium, dbl., 26–40 petals, slight fragrance; foliage medium, medium green, glossy; bushy growth; [Silver Jubilee X Bright Smile]; Dickson, Patrick

DICnifty, HT, lp, 1992; (**Empress Michiko**); Dickson, Patrick

'DICNORTH', F, ob, 1989; (**Harvest Fayre**); flowers medium, dbl., 15–24 petals, slight fragrance; foliage medium, medium green, glossy; bushy growth; ROTY, 1990; [Unnamed seedling X Bright Smile]; Dickson, Patrick; Dickson Nurseries, Ltd., 1990

'DICOBEY', HT, rb, 1988; (Beaulieu, **Tequila Sunrise**); dbl., 40 petals, slight fragrance; foliage medium, medium green, glossy; bushy growth; GM, Belfast, 1991 GM, RNRS, 1988; [Bonfire Night X Freedom]; Dickson, Patrick; Dickson Nurseries, Ltd., 1989

'DICODOUR', HT, ab, 1988; (**Fragrant Dream**); flowers apricot blended orange, large, dbl., 20 petals, intense fragrance; foliage large, medium green, glossy; upright growth; [(Eurorose X Typhoon) X Bonfire]; Dickson, Patrick; Dickson Nurseries, Ltd., 1989

'DICOGLE', F, mp, 1989; (**Valentine Heart**); flowers large, dbl., 15–25 petals, intense fragrance; foliage large, medium green, glossy, purple when young; bushy growth; [Shona X Pot o' Gold]; Dickson, Patrick; Dickson Nurseries, Ltd., 1990

'DICOMO', Min, w, 1988; (**Tear Drop**, Teardrop); flowers small, semi-dbl., 6–14 petals, flat, slight fragrance; foliage small, medium green, glossy; patio; bushy growth; [Pink Spray X Bright Smile]; Dickson, Patrick; Dickson Nurseries, Ltd., 1989

'DICOR', F, yb, 1971; (**Redgold**, Rouge et Or); bud ovoid; flowers gold edged deep pink, blooms in large clusters, dbl., slight fragrance; vigorous, upright growth; AARS, 1971 GM, Portland, 1969; [((Karl Herbst X Masquerade) X Faust) X Piccadilly]; Dickson, A.; J&P

'DICORSAR', Cl F, yb, 1984; (Grimpant Rouge et Or, **Redgold, Climbing**, Rouge et Or, Climbing); Pekmez, Paul

'DICPAINT', HT, rb, 1989; (**Painted Moon**); flowers large, dbl., 40 petals, cupped, slight fragrance; foliage medium, medium green, semi-glossy; upright, bushy, stocky growth; GM, Belfast, 1992; [Bonfire X Silver Jubilee]; Dickson, Patrick; Dickson Nurseries, Ltd., 1990

'DICPARTY', F, dp, 1999; (**Party Trick**™); flowers cerise pink, single, 5–11 petals, 2 in., borne in small clusters, no fragrance; prickles moderate; foliage small, medium green, glossy; compact, low (18 in.) growth; [Robin Redbreast X seedling]; Dickson, Colin; Dickson Nurseries, Ltd., 1999

'DICPE', F, ob, 1968; (**Copper Pot**); flowers orange-yellow, deeper reverse, large blooms in trusses, semi-dbl., 15 petals, moderate fragrance; foliage glossy, bronze; tall growth; [Seedling X Golden Scepter]; Dickson, A.

'DICPERHAPS', Gr, op, 1991; (**Quaker Star**); flowers orange with silver reverse, aging to salmon with orange petals, very dbl., 3 in., no fragrance; few prickles, straight, small; foliage medium, dark green, glossy; upright, medium (120 cms) growth; [Anisley Dickson X Unnamed seedling]; Dickson, Colin; Roses by Fred Edmunds, 1992

DICplay, F, ob, 1991; (**New Horizon**); GM, Belfast, 1993; Dickson, Patrick

'DICPLEASANT', F, or, 1992; (**Ruth Woodward**); dbl., 26–40 petals, 3–3.5 in., borne in small clusters, some prickles; foliage medium, medium green, semi-glossy; medium, bushy growth; [Wishing X Unknown seedling]; Dickson Nurseries

'DICQUARREL', Gr, my, 1994; (**Benita**®); *flowers saffron yellow*, blooms, dbl., 26–40 petals, 3 in., moderate fragrance; foliage medium, medium green, semi-glossy; upright (90 cms) to bushy growth; Dickson, Colin; Dickson Nurseries, Ltd., 1994

'DICQUASAR', HT, ob, 1994; (**Dawn Chorus**); flowers orpiment orange with buttercup yellow at base, blooms borne, dbl., 15–25 petals, 1.5–2.75 in., slight fragrance; some prickles; foliage medium, medium to dark green, glossy; medium (74 cms), upright to bushy growth; GM, Dublin, 1991; Roty, 1993; [Wishing X Peer Gynt]; Dickson, Colin; Dickson Nurseries, Ltd., 1993

'DICQUEEN', F, or, 1990; (**Melody Maker**); flowers very full petals, large blooms, slight fragrance; foliage medium, dark green, semi-glossy; bushy growth; ROTY, 1991; [Anisley Dickson X Wishing]; Dickson, Patrick; Dickson Nurseries, Ltd., 1991

'DICQUEUE', HT, ob, 1999; (**Brandy Snap**™, Brandysnap); flowers orange/bold, dbl., 26–40 petals, 3.5 in., borne mostly singly, intense fragrance; prickles moderate; foliage medium, medium green, glossy; upright, medium (30 in.) growth; [seedling X seedling]; Dickson, Colin; Dickson Nurseries Ltd., 1999

'DICQUIET', S, yb, 1994; (**Gypsy Dancer**); bud small, pointed; flowers hand-painted light yellow with orange and light yellow reverse, semi-dbl., 6–14 petals, 1.5–2.75 in., borne in clusters, slight, citrus fragrance; few prickles; foliage medium, dark green, glossy; medium (100–110 cms), bushy growth; PP8900; [Sweet Magic X Little Artist]; Dickson, Patrick; Bear Creek Gardens, 1994

'DICRACER', F, or, 1994; (**Duchess of York**, Sarah, Duchess of York, Sunseeker); flowers mandarin red suffused with sulfur yellow, blooms borne in small clusters, dbl., 26–40 petals, 1.5 in., slight fragrance; some prickles; foliage medium, medium green, semi-glossy; patio; low (56 cms), bushy growth; [Little Prince X Gentle Touch]; Dickson, Colin; Dickson Nurseries, Ltd., 1992

'DICREASON', S, mr, 1994; (**Our Molly**); flowers currant red with a silvery white eye, blooms borne in large, single, 5 petals, 1.5–2.75 in., no fragrance; orange fruit; some prickles; foliage medium, medium green; tall (90 cms), spreading growth; GM, Glasgow, 1996; Dickson, Colin; Dickson Nurseries, Ltd., 1994

DICrelax, F, yb, 1993; (**Flair**); GM, Glasgow, 1995; Dickson, Patrick

'DICREVIVAL', S, dp, 1992; (**Boy Crazy**); flowers deep pink, petal base has cream colored 'half moon', heavy petal substance, dbl., 15–25 petals, 1.5–2.75 in.; some prickles; foliage medium, dark green, glossy; patio; medium (90 cms), upright, bushy growth; [Sweet Magic X DICmerlin]; Dickson, Patrick; Bear Creek Gardens, 1992

'DICROBOT', HT, dy, 1990; (**Belfast Belle**); flowers large, dbl., 26–40 petals, slight fragrance; foliage medium, medium green, semi-glossy; upright, bushy growth; [Seedling X Pot o' Gold]; Dickson, Patrick; Dickson Nurseries, Ltd., 1991

'DICROCKY', S, lp, 1990; (**Space Invader**); flowers large blooms, dbl., 26–40 petals, moderate fragrance; foliage medium, medium green, semi-glossy; spreading growth; [Seedling X Temple Bells]; Dickson, Patrick; Dickson Nurseries; Ltd.

DICrocky, S, lp, 1992; (**Explorer**); Dickson, Patrick

'DICROYAL', HT, ab, 1992; (**Princess Royal**); dbl., 26–40 petals, 3–3.5 in., borne mostly singly, slight fragrance; many prickles; foliage large, medium green, semi-glossy; very stiff, thorny stems; medium (88 cms), bushy growth; [Tequila Sunrise X Unnamed seedling]; Dickson, Colin; Dickson Nurseries, Ltd.

'DICSILVING', Cl HT, pb, 1997; (**Silver Lining, Climbing**); flowers full, large, very dbl., 26–40 petals, intense fragrance; many prickles; foliage medium, medium green, glossy; tall, spreading growth; has 10–12 canes ranging in size of 10–20 ft; [Silver Lining sport]; Dickson, Alex

DICsun, MinFl, lp; (**Mr J C B**); Dickson, Patrick

'DICTALENT', Min, op, 1994; (Just Happy, **Shine On**™); flowers nasturtium red, reverse azalea pink, 2.5 in., borne in small clusters, slight fragrance; prickles moderate; foliage small, medium green, semi-glossy; compact, low (22 in.) growth; [Sweet Magic X seedling]; Dickson, Colin; Dickson Nurseries, Ltd., 1999

'DICTATOR', HT, mp, 1999; (**Pure Bliss**™); flowers pale pink, reverse mid-pink, dbl., 26–40 petals, 2.25–2.5 in., borne mostly singly, moderate fragrance; prickles moderate; foliage medium, medium green, semi-glossy; patio; upright, medium growth; GM, Belfast, 1997; [Elina X seedling]; Dickson, Alex, 1994; Dickson Nurseries Ltd., 1994; (Cavriglia)

'DICUMPTEEN', S, lp, 1999; (**Pretty in Pink**™); flowers pale pink, very dbl., 41 petals, 2 in., borne in small clusters, moderate fragrance; few prickles; foliage medium, medium green, glossy; groundcover; spreading, low (24 in.) growth; [seedling X Grouse]; Dickson, Colin, 1994; Dickson Nurseries Ltd., 1994; (Cavriglia)

'DICUNCLE', S, dp, 1999; (**Wine and Dine**™); flowers rose red, single, 5–11 petals, 2.25 in., moderate fragrance; many prickles; foliage small, medium green, glossy; groundcover; spreading, medium (3 ft) growth; [seedling X seedling]; Dickson, Colin; Dickson Nurseries, Ltd., 1999

'DICUNIFORM', S, mr, 1999; (**Boy O Boy**™); semi-dbl., 12–16 petals, 2.75–3 in., slight fragrance; prickles moderate; foliage small, medium green, glossy; flori-shrub; bushy to spreading (30 in.) growth; [Little Prince X Eye Opener]; Dickson, Colin, 1996; Dickson Nurseries, Ltd., 1996

'DICUPTIGHT', HT, or, 1999; (**Tintinara**™); flowers poppy red, reverse geranium lake, dbl., 26–40 petals, 5 in., borne in small clusters, slight fragrance; prickles moderate; foliage large, medium green, glossy; upright, tall (40 in.) growth; [Melody Maker X seedling]; Dickson, Colin; Dickson Nurseries, Ltd., 1999

'DICVANILLA', F, pb, 1999; (**Happy Ever After**™); flowers pale pink/lemon, reverse-pale pink, semi-dbl., 12–16 petals, 2 in., borne in large clusters, moderate fragrance; prickles moderate; foliage medium, light green, semi-glossy; bushy (33 in.) growth; [The Fairy X seedling]; Dickson, Colin, 1997; Dickson Nurseries, Ltd., 1999

DICvintage, MinFl, yb, 1993; (**Roche Centenary**); Dickson, Patrick; (Cavriglia)

'DICVOOD', S, mr, 1999; (**Glenshane**™); semi-dbl., 12–16 petals, 1.5 in., borne in large clusters, no fragrance; few prickles; foliage medium, medium green, semi-glossy; flori-shrub; spreading to bushy, medium (30 in.) growth; [seedling X Star Child]; Dickson, Colin, 1997

'DICWAFFLE', HT, w, 1999; (**Racy Lady**™); flowers creamy white, reverse cream, dbl., 26–40 petals, 5 in., borne mostly singly, moderate fragrance; many prickles; foliage medium, dark green, glossy; upright, medium (3 ft) growth; [Solitaire X Elina]; Dickson, Colin

'DICWHISTLE', Min, dp, 1999; (**Dick's Delight**™); dbl., 26–40 petals, 1.2 in., borne in large clusters, slight fragrance; prickles moderate; foliage small, dark green, glossy; patio carpet; compact to spreading, low (24 in.) growth; [seedling X The Fairy]; Dickson, Colin; Dickson Nurseries, 1999

DICwillynilly, F, ob, 1999; (**Old John**™); flowers mid-orange, reverse orange-red, dbl., 17–25 petals, 2.5 in., borne in large clusters, moderate fragrance; prickles moderate; foliage medium, dark green, glossy; upright, medium (40 in.) growth; [Sunseeker X New Horizon]; Dickson, Colin; Dickson Nurseries, Ltd, 1999

'DICWITNESS', F, yb, 1999; (**Irish Eyes**™); flowers mid-yellow/red, reverse mid-red/yellow, dbl., 26–40 petals, 2.5 in., borne in large clusters, slight fragrance; prickles moderate; foliage medium, medium green, semi-glossy; bushy (30 in.) growth; [Mr J. C. B. X Gypsy Dancer]; Dickson, Patrick; Dickson Nurseries, Ltd., 2000

'DICWONDER', Min, mp, 1999; (**Marry Me**™); dbl., 26–40 petals, 2 in., borne in small clusters, slight fragrance; prickles moderate; foliage medium, dark green, glossy; patio; upright, medium (28 in.) growth; [seedling X Cider Cup]; Dickson, Patrick; Dickson Nurseries, Ltd., 1998

'DICXPLOSION', Min, rb, 1999; (**Rainbow Magic**™); flowers cerise red, reverse yellow, semi-dbl., 12–16 petals, 1.75 in., borne in large clusters, slight fragrance; prickles moderate; foliage small, medium green, glossy; patio; spreading, low (2 ft × 3 ft) growth; [Sunseeker X seedling]; Dickson, Colin; Dickson Nurseries, Ltd., 1999

DICxtol, S, rb, 1998; (**Red Cottage**); Dickson, Patrick

Die Krone, HT, dy

'DIE MUTTER VON ROSA', HT, lp, 1906; flowers medium, dbl., moderate fragrance; Verschuren; (Sangerhausen)

'DIE PRÄSIDENTIN', HT, w, 1928; flowers marble white, center soft yellow, semi-dbl., slight fragrance; [Harry Kirk seedling]; Mühle

Die Rheinpfalz, *see* HELrobu

Die Schonste, HT, w, 1998

'DIE SPREE', HT, lp, 1907; flowers large, dbl., moderate fragrance; Nauke; (Sangerhausen)

Die Welt®, *see* 'DIEKOR'

'DIEKOR', HT, ob, 1976; (**Die Welt**®, The World); bud long, pointed; flowers orange, red and yellow blend, dbl., 25 petals, 4.5 in.; exhibition form, slight fragrance; foliage glossy; vigorous, upright, very tall, bushy growth; [Seedling X Peer Gynt]; Kordes

'DIENER'S BLUE', HMult, m, 1926; flowers violet, large, heavy clusters, dbl., 2 in.; almost thornless; vigorous growth; Diener

'DIENER'S ROSE UNDERSTOCK', HMult, rb, 1932; bud long; flowers rose-red shaded purple, stamens yellowish, borne in clusters, single, 10 petals, 2 in., foliage small, notched; climbing or trailing growth; producing much new wood in a season; [Veilchenblau X Veilchenblau sport]; Diener

'DIETER WOLF', Pol, or, 1969; flowers salmon-orange, semi-dbl.; foliage glossy, dark; vigorous, compact growth; [Tropicana X Jiminy Cricket]; Buisman, G.A.H.

'DIGNITY', HT, w, 1940; bud long, pointed; flowers creamy white, dbl., intense fragrance; foliage leathery; vigorous, bushy, compact growth; LeGrice

Diletta, HT, op, 1985; Barni, V.

Dil-Ki-Rani, HT, lp, 1985; Pal, Dr B.P.

Dilly Dilly, *see* 'TINDILLY'

'DILLY'S WIEDERKEHR', HFt, 1925; flowers large, dbl.; Schwartzbach; (Sangerhausen)

'DILYS ALLEN', HT, ob, 1952; bud long, pointed, ovoid; flowers orange-red, base saffron, 4 in., slight fragrance; foliage glossy, dark bluish green; vigorous, bushy growth; [Mrs Sam McGredy X Seedling]; Norman; Harkness

'DIMITY', HT, pb, 1956; bud ovoid, pointed; flowers ivory to pure white, edged pink, large, dbl., cupped, moderate fragrance; foliage dark, leathery; upright, bushy growth; [Peace X Seedling]; Taylor, C.A.

'DIMPLES', F, ly, 1968; flowers canary-yellow to ivory, borne in trusses, semi-dbl., slight fragrance; foliage glossy; LeGrice

'DINAH', HT, dr, 1920; flowers deep crimson, shaded darker; Paul, W.

'DINAH SHORE', HT, dp, 1942; flowers cerise-pink, dbl., 65 petals, 5 in., globular, intense fragrance; foliage glossy, dark; vigorous, upright growth; [Jewel sport]; Grillo

'DINKY', Min, or, 1986; flowers orange-red, reverse orange, urn-shaped, blooms borne usually singly, dbl., 20 petals, slight fragrance; long, light red prickles; foliage medium, medium green, semi-glossy; upright growth; [Sheri Anne X Seedling]; Bridges, Dennis A.

Dinsmore, *see* **'MME CHARLES WOOD'**

'DINY HAGE', HT, dr, 1956; flowers crimson-red, large, dbl., intense fragrance; vigorous growth; [Ambassadeur Nemry X Crimson Glory]; Leenders, M.

'DIORAMA', HT, yb, 1965; flowers apricot-yellow, dbl., 4.5 in., exhibition form, moderate fragrance; vigorous, upright growth; [Peace X Beauté]; deRuiter

Dioressence®, *see* 'DELDIORE'

Diorette, *see* **'CANCAN'**

'DIPLOMAT', HT, dr, 1962; bud ovoid; flowers currant red edged blood-red, dbl., 50–55 petals, 3.5–4 in., cupped, moderate fragrance; foliage leathery, dark; vigorous, upright growth; [(Poinsettia X Tawny Gold) X Detroiter]; Boerner; Home Nursery Products Corp.

Diplomatka, F, mr, 1967; flowers blood red, large, dbl.; Klimenko, V. N.; (Sangerhausen)

'DIPUTACION DE TARRAGONA', HT, ob, 1967; (Tarragona); bud pointed; flowers orange-coral, large, dbl., 35 petals, exhibition form, moderate fragrance; foliage glossy, bronze; upright, compact growth; [Baccará X (Chrysler Imperial X Soraya)]; Dot, Pedro; Rosas Dot

'DIRECTEUR ALPHAND', HP, m, 1883; flowers blackish purple, large, dbl.; Lévêque

'DIRECTEUR CONSTANT BERNARD', HT, m, 1886; Soupert & Notting

'DIRECTEUR DONATIEN LELIEVRE', F, ob, 1959; flowers coppery orange, medium, dbl.; Privat

'DIRECTEUR GUÉRIN', HT, ob, 1935; flowers orange-yellow, center coppery, overlarge, dbl.; foliage light; long stems; very vigorous growth; Gaujard

'DIRECTEUR N. JANSEN', HP, m, 1883; flowers purple/pink, large, dbl.; Verdier, E.; (Sangerhausen)

Director Plumecock, *see* **'PRÉSIDENT PLUMECOCQ'**

'DIRECTOR RUBIÓ', HT, rb, 1929; flowers magenta-red, very large, semi-dbl., moderate fragrance; stiff stems; dwarf, bushy growth; [O. Junyent X Jean C.N. Forestier]; Dot, Pedro; C-P

Direktör Benschop, *see* **'CITY OF YORK'**

'DIREKTOR HJELM', Pol, rb, 1927; flowers red; [Prasident Hindenburg sport]; Koster, D. A.

'DIREKTOR REBHUHN', HT, ob, 1929; flowers orange, center reddish, dbl., intense fragrance; [Mme Butterfly X Angèle Pernet]; Kordes

'DIREKTOR RIKALA', F, mp, 1934; [Lafayette sport]; Koster, D. A.

Direktör Rikala, *see* **'FRAU ASTRID SPÄTH'**

'DIREKTOR STRUVE', Pol, w, 1924; [Echo sport]; van Nes

'DIRIGENT'®, S, mr, 1956; (The Conductor); bud pointed; flowers blood-red, blooms in clusters of up to 28, semi-dbl., recurrent bloom, slight fragrance; foliage leathery; vigorous (4 ft) growth; ADR, 1958; [Fanal X Karl Weinhausen]; Tantau, Math.

'DISCO', HT, rb, 1980; bud medium to long, pointed; flowers medium red, reverse cream, blooms borne singly and 2–4 per cluster, dbl., 30 petals, exhibition form, slight, spicy fragrance; prickles hooked downward; foliage leathery, dark; tall, upright growth; GM, Baden-Baden; [Sunrise-Sunset X Seedling]; Weeks, O. L.

Disco®, F, pb, 1980; Harkness

Disco Dancer®, *see* 'DICINFRA'

'DISCOVERY', Gr, pb, 1958; flowers soft pink shaded apricot, dbl., 5–6 in., intense fragrance; vigorous growth; [(Peace X Christopher Stone) X Floribunda seedling]; deRuiter; Blaby Rose Gardens

'DISCRETION', HT, pb, 1952; flowers salmon-pink shaded copper, large, dbl., 28 petals, moderate fragrance; foliage glossy; [Peace X Seedling]; Gaujard

'DISPLAY', F, pb, 1956; flowers salmon-pink, becoming cherry-pink, well formed, borne in large clusters, semi-dbl., 13 petals, 2.5 in., slight fragrance; foliage glossy, bronze-green; very vigorous growth; [Orange Triumph X Golden Scepter]; Arnot; Croll

Disraeli, *see* 'ADADISRES'

'DISTANT DRUMS', S, m, 1985; bud ovoid, pointed; flowers rose-purple, imbricated, large blooms borne 1–10 per cluster, dbl., 40 petals, repeat bloom, intense, myrrhy fragrance; awl-like, brown prickles; foliage medium-large, dark, leathery; vigorous, erect, bushy growth; [September Song X The Yeoman]; Buck, Dr Griffith J.; Iowa State University

Distant Sounds, *see* WEBpriace

'DISTINCT', F, rb, 1953; flowers spectrum-red with white eye, imbricated, small, rounded clusters, dbl., 25–30 petals, cupped, moderate fragrance; foliage glossy; vigorous, compact growth; [Triomphe Orléanais X Mrs Pierre S. duPont]; Boerner; J&P

'DISTINCTION', F, dp, 1927; flowers deep rose-pink, center brighter; [Lafayette sport]; Turbat

'DISTINCTION, CLIMBING', Cl F, dp, 1935; Lens

Ditto, *see* 'LYODIT'

'DIVA', HT, dr, 1976; flowers dark velvety red, dbl., 30 petals, 6 in., slight fragrance; foliage dark; spreading growth; [Sonia X Gisselfeld]; Poulsen

Diva, HT, dy, 1995; Cocker

Diversity, HT, rb, 1995; K&S

'DIVIDEND', HT, dy, 1931; flowers rich yellow, dbl., globular, slight fragrance; foliage dark; dwarf growth; [Franz Deegen X Seedling]; Clark, A.; NRS Victoria

'DIVINE', HT, mr, 1964; bud dark purplish; flowers cardinal-red, well formed, large, dbl., 45 petals; foliage bright green; strong stems; vigorous, upright growth; GM, Geneva, 1964; Delbard-Chabert

'DIVINE LADY', F, op, 1965; flowers salmon-pink suffused brownish, dbl.; very vigorous, dense growth; [Circus X Queen Elizabeth]; Lens

'DIXIE', HT, op, 1925; flowers salmon-pink, more large, dbl., cupped, intense fragrance; [Radiance sport]; Gray, W.R.

'DIXIE BELLE', HT, lp, 1963; bud ovoid, rose-pink; dbl., 38 petals, 5–5.5 in., cupped, moderate fragrance; foliage leathery; vigorous, upright growth; [Golden Masterpiece X Seedling]; Boerner; J&P

'DIXIE CLIMBER', Cl HT, ob, 1935; flowers salmon and gold; [Gov. Alfred E. Smith sport]; Watkins, A.F.; Dixie Rose Nursery; J&P

Dixie Dazzle™, *see* 'KINDIXIE'

Dixie Dream, *see* **'FESTIVAL'**

'DIXIE HOLIDAY', HT, mr, 1968; bud long, pointed; flowers medium, dbl., exhibition form, slight fragrance; foliage bronze, leathery; very vigorous growth; [Étoile de Hollande sport]; Garrison; Kimbrew

Dixieland, *see* 'BRIDIXIE'

Dixieland Linda, *see* 'BEADIX'

Dizzy, F, 1974; Delforge, H.; (Cavriglia)

Dizzy Heights, *see* 'FRYBLISSFUL'

'Doc', Pol, mp, 1954; (Degenhard); flowers phlox-pink, small blooms in large trusses, semi-dbl., 15 petals; compact growth; [Robin Hood X Polyantha seedling]; de-Ruiter; Gregory & Willicher Baumschulen, 1954

Docile, F, 1995; Eve, A.; (Cavriglia)

Docter Dorothy, F, pb, 1997; flowers very full, medium, very dbl., 41 petals, borne in small clusters, intense fragrance; some prickles; foliage medium, medium green, semi-glossy; upright, medium (5 ft) growth; [Jubilee X Little Darling]; Jones, L.J.

Docteur F. Debat, *see* **'DR DEBAT'**

'DOCTEUR LOUIS ESCARRAS', HT, rb, 1922; flowers dark salmon-red shaded carmine-pink, very dbl., 120 petals; [Constance X Seedling]; Nabonnand, C.

'DOCTEUR MARJOLIN', M, mp, 1860; flowers shell-pink, Robert et Moreau

'DOCTEUR MOREL', HT, m, 1946; flowers carmine with chrome-yellow reflections, large, dbl., 40–45 petals; foliage dark; vigorous, upright growth; [Edith Nellie Perkins X Pres. Herbert Hoover]; Laperrière

'DOCTEUR REYMONT', HMult, w, 1907; flowers pure white on pale green base, pyramidal cluster, dbl.; Mermet

'DOCTEUR ROBERT SALMONT', HT, yb, 1946; bud pointed; flowers capucine and yellow, reverse tinted chrome, base coppery, foliage dark; vigorous growth; Gaujard

'DOCTEUR VALOIS', HT, rb, 1950; flowers geranium shaded vermilion, reverse yellow, semi-dbl., 4 in., moderate fragrance; foliage dark, glossy; vigorous, bushy growth; [(Annie Drevet X Condesa de Sástago) X Vive la France]; Mallerin, C.; URS

Doctor Abrahams, HT, ab, 1991; Hallows

Doctor Behring, *see* 'DOTEMIBE'

Doctor Dick, *see* 'COCBADEN'

'DOCTOR ELDON LYLE', Gr, dr, 1968; bud pointed; flowers medium, dbl., exhibition form, moderate fragrance; foliage soft, bronze; vigorous, compact growth; [Pres. Eisenhower X Suspense]; Mackay; Texas Rose Research Foundation

Doctor Faust, F, yb, 1957; flowers dark yellow with orange and pink tints, medium, dbl., slight fragrance; Kordes, W. Söhne; (Sangerhausen)

Doctor Goldberg, *see* 'GANGO'

Doctor Jackson, *see* 'AUSDOCTOR'

Doctor Robert Korns, *see* LETrob

'DOCTOR'S WIFE', HT, op, 1967; bud long, pointed; flowers salmon-pink, large, dbl., exhibition form, slight fragrance; foliage dark, glossy, leathery; vigorous, upright growth; Von Abrams; Edmunds Roses

Dog Rose, *see* **R. CANINA**

'DOHHAWK', Min, mr, 1988; (Night Hawk, **Nighthawk**™); bud globular, pointed; flowers medium, dbl., 22 petals, exhibition form, borne singly and in sprays of 3–5, intense, damask fragrance; prickles straight, slanted down, medium, reddish-brown; foliage medium, medium green; upright, bushy, medium growth; PP007417; AOE, 1989; [Quinella X Poker Chip]; Hardgrove, Donald L.; Nor'East Min. Roses, 1989

Dolce Vita®, *see* 'DELDAL'

Dolcezza®, S, mp

'DOLLAR-ROSE', HT, mr, 1936; flowers carmine-red, medium, dbl.; Tantau, Math.; (Sangerhausen)

Dollie B, *see* 'TROBEE'

Dolly®, *see* 'POULVISION'

'DOLLY BROWNELL', F, lp, 1926; flowers color same as Dr W. van Fleet; [Dr W. Van Fleet seedling]; Brownell, H.C.

'DOLLY DARLING', HT, mp, 1949; bud long, pointed, red; flowers lustrous pink, open, semi-dbl., 20 petals, 4–5 in., moderate fragrance; foliage glossy; vigorous, compact growth; [Pink Princess X Crimson Glory]; Brownell, H.C.

Dolly Dot, MinFl, dy, 1998; J&P

'DOLLY MADISON', HT, dy, 1935; (Super-Dupont); flowers golden yellow, vigorous growth; [Mrs Pierre S. duPont climbing sport seedling sport]; Hillock

'DOLLY PARTON', HT, or, 1984; flowers luminous orange-red, large, dbl., 35 petals, borne mostly singly, intense fragrance; foliage large, medium green, semi-glossy; upright growth; PP005608; Bronze, ARC TG, 1982; [Fragrant Cloud X Oklahoma]; Winchel, Joseph F.; C-P, 1983

'DOLLY VARDEN', HRg, ab, 1914; flowers light apricot-pink, base yellow, large, recurrent bloom, vigorous growth; Paul

'DOLLY VARDEN', Pol, mp, 1930; flowers clear pink, dbl.; deRuiter

'DOLLY'S SISTER', HT, op, 1989; flowers medium coral-pink; [Dolly Parton sport]; Taylor, Thomas E.; Michigan Mini Roses, 1989

Dolly's Sister, Gr, or, 1999; Williams, J. Benjamin

'DOLOMITI', HT, ab, 1933; bud pointed; flowers flesh, with yellow reflex, very large, dbl., moderate fragrance; foliage dark; strong stems; vigorous growth; Ingegnoli

Dolores™, *see* DEVine

'DOMAINE DE CAPUIS', S, m, 1901; Roseraie de l'Hay

Domaine de Courso, *see* MEIdrimy

Dometille Beccard, HGal; (Sangerhausen)

Domila®, *see* LAPnat

'DOMINA', HT, op, 1943; flowers salmon-pink, large, dbl., slight fragrance; Heizmann, E.; (Sangerhausen)

'DOMINANT', HT, op, 1964; bud ovoid; flowers salmon-pink, medium, borne in clusters, dbl., slight fragrance; foliage dark; [Golden Masterpiece X Spartan]; Boerner; Spek

'DOMINATOR', F, mp, 1961; semi-dbl., 3 in., blooms in clusters, moderate fragrance; vigorous, upright growth; [New Yorker X The Optimist]; deRuiter

'DOMINIE SAMPSON', HSpn, lp; flowers soft pink, semi-dbl., non-recurrent; glossy, black fruit; foliage finely divided; dense, shrubby (3–4 ft) growth

'DOMINIQUE', Min, lp, 1981; bud ovoid; flowers light peachy pink, dbl., 30 petals, exhibition form, intense, apple fragrance; curved prickles; foliage medium green, arrow-shaped; upright growth; [Electron X Little Chief]; Bennett, Cecilia 'Dee'; Tiny Petals Nursery

'DOMINO', HT, dr, 1956; bud long, pointed; flowers dark crimson, medium, slight fragrance; foliage dark; [Peace X Seedling]; Gaujard

Domino, *see* TANdomo

'DOMKAPITULAR DR LAGER', HT, pb, 1903; flowers rose and carmine, moderate fragrance; [Mme Caroline Testout X Princesse de Bassaraba de Brancovan]; Lambert, P.

Domstadt Fulda®, *see* KORtanken

'DOMUS AUREA', HT, my, 1940; flowers pure yellow; foliage dark, glossy; strong stems; very vigorous growth; [Julien Potin X Yellow seedling]; Aicardi, D.; Giacomasso

Don Bosco, HT, 1976; Dorieux, Francois; (Cavriglia)

Don Bosco, *see* LAPduf

'DON BRADMAN', HT, rb, 1938; bud long, shapely; flowers coppery claret, fading to silvery pink, dbl., 40–50 petals; Wheatcroft Bros.

'DON CHARLTON', HT, pb, 1991; flowers deep rose pink with silver reverse, large blooms, very dbl., moderate fragrance; foliage large, dark green, glossy; upright growth; [Silver Jubilee X (Chicago Peace X Doris Tysterman seedling)]; Thompson, Robert; Battersby Roses, 1992

'DON DON', Min, dr, 1976; flowers red, reverse blending near white at base, dbl., 60 petals, 1–1.5 in., moderate fragrance; foliage small, glossy, bronze; upright, bushy growth; [Seedling X Over the Rainbow]; Williams, Ernest D.; Mini-Roses

'DON JOSÉ', HT, op, 1922; flowers salmon-pink, semi-dbl., slight fragrance; [Archiduc Joseph X Seedling]; Clark, A.; NRS Victoria

'DON JUAN', LCl, dr, 1958; bud ovoid; flowers velvety crimson red, very dbl., 30–35 petals, 5 in., cupped, recurrent bloom, intense fragrance; foliage dark green, glossy, leathery; climbing, height 12–14 ft; [New Dawn seedling X New Yorker]; Malandrone; J&P

Don Marshall, *see* 'MORBLACK'

Don Pedro, lp; flowers medium, dbl.; Roseraie de l'Hay; (Sangerhausen)

'DON QUICHOTTE', F, rb, 1964; flowers cherry-red, base yellow, well formed, large, borne in clusters, dbl., slight fragrance; foliage glossy, leathery; vigorous, upright growth; [Charles Gregory X Marcelle Auclair]; Robichon; Ilgenfritz Nursery

'DON ROSE', HT, pb, 1943; bud long, pointed, carmine-red; flowers coppery pink, open, large, dbl., 40 petals, cupped, slight fragrance; foliage leathery, bluish green; vigorous, upright, bushy, rather compact growth; [Souer Thérèse X Seedling]; Mallerin, C.; C-P

'DOÑA CLARA', HT, m, 1965; bud ovoid; flowers purplish pink, large, dbl., 50 petals, exhibition form, vigorous growth; Camprubi, C.

Donald Davies, *see* 'CHEWBEAUT'

'DONALD MACDONALD', HT, ob, 1916; flowers orange-carmine, borne in clusters, semi-dbl., moderate fragrance; dwarf growth; GM, NRS, 1916; Dickson, A.

'DONALD PRIOR', F, mr, 1938; bud ovoid; flowers bright scarlet flushed crimson, blooms in large clusters, semi-dbl., 11 petals, 3 in., cupped, moderate fragrance; foliage leathery, dark; vigorous, bushy growth; (28); [Seedling X D.T. Poulsen]; Prior; J&P

'DONALD PRIOR, CLIMBING', Cl F, mr; Farr

Donald Thomas Heald, *see* 'KIRSMILE'

'DONALDO', HT, rb, 1979; bud ovoid; flowers red to pink, shapely, dbl., 35 petals, 4 in.; foliage large, glossy, dark; tall growth; [Honey Favorite X Rose Gaujard]; Murray, Nola

Donatella, *see* **'GRANADA'**

'DONAU', LCl, m, 1913; Praskac

Donauprinzessin®, F, mp, 1994; Noack, Werner

Donauwelle, S, m, 1991; Weihrauch

Donella, *see* JAYdon

'DONNA CLARA', HT, pb; flowers buff, reverse strawberry-pink; Leenders, M.

Donna Darlin', *see* 'WINDONNA'

'DONNA FANNY CAVALIERI', HT, 1953; San Remo Exp. Sta.; (Cavriglia)

'DONNA FAYE', Min, lp, 1976; bud pointed; dbl., 27 petals, 1 in., exhibition form, moderate fragrance; upright growth; [Ma Perkins X Baby Betsy McCall]; Schwartz, Ernest W.; Nor'East Min. Roses

Donna Jean, *see* 'TALDON'

Donna Marella Agnelli®, *see* BARdon

Donna Maria, *see* **'DONNA MARIE'**

'DONNA MARIE', HSem, w, 1830; (Donna Maria); flowers pure white, small, very dbl.; Vibert

Donna Silva Carmine, F, 1974; Cazzaniga, F. G.; (Cavriglia)

Doorenbos Selection, HSpn, dr; Doorenbos

'DOORYARD DELIGHT', HT, pb, 1940; bud short, pointed, spiraled; flowers light pink, reverse rose-pink, petals sharply pointed, large, dbl., 2.5 in., recurrent bloom, slight fragrance; foliage leathery; vigorous, bushy growth; [R. setigera X Lady Alice Stanley]; Horvath; Wyant

'DOPEY', Pol, mr, 1954; (Eberwein); flowers crimson-red, small blooms in trusses, semi-dbl.; compact growth; [Robin Hood X Polyantha seedling]; deRuiter; Gregory & Willicher Baumschulen, 1954

'DORA', HT, or, 1975; bud long; flowers brilliant orange-red, dbl.; foliage bronze; [Tanagra X Rubens]; Gaujard

Dora Delle, *see* 'TALDOR'

'DORA HANSEN', HT, mp, 1908; bud long, pointed; flowers thulite-pink, open, large, dbl., slight fragrance; Jacobs

'DORA STOBER', HT, w, 1925; flowers white shaded yellow, dbl., moderate fragrance; Leenders, M.

Dorabella, HT, 1986; Barni, V.; (Cavriglia)

Dorada, *see* BARdord

DORalp, F, w, 1993; (**Anne-Aymone Giscard d'Estaing®**); Dorieux

Doralta, HT, 1979; Dorieux, Francois; (Cavriglia)

'DORAPRI', HT, ob, 1988; (**Mango®**); Dorieux

DORastri, S, pb, 1996; (**Hugues Aufray**); Dorieux

'DORCAS', S, pb, 1984; bud ovoid, pointed; flowers light pink, pale yellow blend, flecked deeper pink, dbl., 40 petals, cupped, repeat bloom, slight fragrance; awl-like, tan prickles; foliage dark, leathery; erect, bushy growth; hardy.; [Minigold X Freckle Face]; Buck, Dr Griffith J.; Iowa State University

'DORCAS', HWich, pb, 1922; flowers deep rose-pink to coral-pink, base yellow, borne in large clusters, dbl.; vigorous growth; RULED EXTINCT 1/85; English

DORcast, F, dy, 1997; (**Soleillade**); Dorieux

DORcroix, HT, ab, 1995; (**Chateau la Croix**); Dorieux

DORdeli, HT, pb; (**Starion®**); Dorieux, 1987

'DOREEN', HT, ob, 1951; flowers deep golden orange flushed scarlet, well-formed, moderate fragrance; foliage dark; vigorous growth; [Lydia X McGredy's Sunset]; Robinson, H.; Baker's Nursery

'DOREEN JOHNSON', HT, lp, 1977; bud long, pointed; flowers pale pink, dbl., slight fragrance; foliage large, light; vigorous, bushy growth; [(Great Venture X Fort Vancouver) X Memoriam]; Dawson, George; Australian Roses

'DOREEN THORN', HT, pb, 1934; flowers deep pink, base yellow, well shaped, large, dbl., moderate fragrance; vigorous growth; Cant, F.

'DOREEN WELLS', F, or, 1970; flowers orange-scarlet, dbl., 25 petals, 3 in., flat, moderate fragrance; foliage glossy, dark; low, bushy growth; [Soraya X Circus]; Watkins Roses

'DORELL', HMult, dp, 1887; Geschwind, R.

'DOREN', HT, 1951; Robinson, H.; (Cavriglia)

DORfaut, HT, ab, 1997; (**Comtesse Brigitte Chandon-Moet**); Dorieux

DORflo, F, ob, 1974; (**Bibiché®**); Dorieux, Francois; (Cavriglia)

DORfuri, HT, 1991; (**Roseraie de Blois**); Dorieux; (Cavriglia)

DORgold, F, dy, 1982; (**Coucou**); Dorieux

DORgran, ab, 1998; (**Barbara Hendricks**); Dorieux

'DORIC', F, yb, 1963; flowers golden salmon, large, borne in well-spaced clusters, dbl., 40 petals; foliage glossy; vigorous, compact growth; [Masquerade X Korona]; LeGrice

'DORIENNE', HT, mr, 1958; bud short; flowers large, dbl., 34 petals, cupped; bushy, spreading growth; [Mrs Nieminen X Seedling]; Buyl Frères

'DORIENT', Gr, m, 1998; (Melodie Parfumee, **Melody Parfumee™**); flowers dark lavender-plum, lighter reverse, very dbl., 26–40 petals, 4.5 in., borne in large clusters, intense, damask fragrance; prickles moderate, straight; foliage medium, dark green, semi-glossy; bushy, upright, 5 ft growth; [Dioressence X Stephens' Big Purple]; Dorieux, Francois; Bear Creek Gardens, Inc., 1995

'DORINA NEAVE', HT, pb, 1926; flowers silvery pink, large, dbl., globular, intense fragrance; stiff stems; compact growth; Pemberton

Dorinda, S, lp; Peden, R., 1998; (Weatherly, L.)

'DORIS', HT, rb, 1939; flowers cerise striped white; [Briarcliff sport]; Spandikow

'DORIS ANN', Min, dr, 1986; flowers small, dbl., 26–40 petals; foliage small, medium green, matt; [Black Jade X Tiki]; Wambach, Alex A.

'DORIS ARCHER', F, rb, 1962; flowers yellow, bronze and red, well formed, borne in clusters, dbl., 30–35 petals, 4 in., moderate fragrance; foliage glossy; vigorous, compact growth; [Circus X Seedling]; Fryers Nursery, Ltd.

'DORIS DICKSON', HT, ob, 1924; flowers orange-cream, veined cherry-red, moderate fragrance; foliage very dark; stiff, wiry stems; vigorous growth; Dickson, S.

Doris Dowman, *see* 'HORSILKARE'

'DORIS DOWNES', LCl, pb, 1932; flowers pink, shaded red, overlarge, semi-dbl., cupped, early, intense fragrance; climbing growth; Clark, A.; NRS Victoria

'DORIS FINDLATER', HT, ab, 1936; flowers light apricot, reverse flushed reddish salmon and carmine, dbl.; vigorous growth; Dickson, A.

'DORIS GRACE ROBINSON', HT, w, 1943; bud pointed; flowers creamy white, well shaped, large, foliage olive-green; vigorous, upright growth; Bees

'DORIS HOWARD', F, mr, 1957; flowers blood-red, borne in large clusters; vigorous, bushy growth; Wheatcroft Bros.

Doris J. Robertson, HT, lp; flowers whitish-pink, large, dbl., slight fragrance; (Sangerhausen)

'DORIS NORMAN', F, or, 1958; flowers bright orange, to open, borne in small clusters, dbl., 30 petals, 2 in., exhibition form; foliage purplish to dull green; vigorous, bushy growth; [Paul's Scarlet Climber X Mary]; Norman; Harkness

'DORIS OSBORNE', HT, mr, 1937; bud pointed; flowers ruby-cerise, semi-dbl.; bushy growth; [Mme Abel Chatenay X Seedling]; Clark, A.; NRS Victoria

'DORIS PLEASANCE', HT, w, 1978; flowers blush-pink to white; [Queen Elizabeth sport]; Brewer

Doris Reese, *see* 'RESGOLD'

'DORIS RYKER', Pol, ob, 1942; (Dorus Rijkers); flowers salmon-pink, blooms in clusters, dbl., recurrent bloom, moderate fragrance; foliage light green; vigorous, upright growth; Leenders, M.; Klyn

'DORIS TRAYLER', HT, yb, 1924; bud pointed, orange; flowers yellow, reverse flushed crimson and orange, large, dbl., exhibition form, moderate fragrance; foliage light green, leathery, glossy; bushy, dwarf, compact growth; McGredy

'DORIS TYSTERMAN'®, HT, ob, 1975; flowers tangerine and gold, dbl., 28 petals, 4–5 in., slight fragrance; foliage glossy; upright growth; [Peer Gynt X Seedling]; Wisbech Plant Co.

'DORISTEL', F, w, 1986; (**Artiste**®); Dorieux

'DORJURE', HT, rb, 1992; (**Portland Rose Festival**); flowers strawberry red, white reverse, exceptional form with clear s, dbl., 26–40 petals, 3–3.5 in., moderate fragrance; some prickles; foliage large, dark green, glossy; tall (150 cms), upright growth; [Osiria X Pharaon]; Dorieux, Francois; Roses by Fred Edmunds, 1992

DORkade, HT, dr, 1996; (**Georges Truffaut**); Dorieux

DORlain, HT, w, 1997; (**Marie Laforet**); Dorieux

DORlina, HT, w, 1996; (**Provence**); Dorieux

'DORMA', HT, op, 1966; (**Majesté**®); bud oval; flowers salmon, large, dbl., moderate fragrance; foliage glossy, leathery; very vigorous, upright growth; [Radar X Eclipse]; Dorieux; Vilmorin

DORmagi, HT, rb, 1992; (**Imagine**); Dorieux

'DORMAL', HT, mr, 1967; (**Rouge Dorieux**®); bud pointed; flowers cherry-red, open, over large, dbl., slight fragrance; foliage dark, glossy, leathery; vigorous, upright growth; [Seedling X Ena Harkness]; Dorieux; Vilmorin

DORnapa, HT, rb, 1988; (**Flushing Meadow**®); Dorieux

DORneye, F, ob, 1995; (**Copacabana**); Dorieux

DORnice, F, rb, 1997; (**Orange d'Éte**); Dorieux

'DORNRÖSCHEN', S, pb, 1960; bud well-shaped; flowers salmon to deep pink, reverse yellow, large blooms in clusters, dbl., recurrent bloom, moderate fragrance; upright, well-branched growth; [Pike's Peak X Ballet]; Kordes, R.

Dornroschenschloss Sababurg®, *see* KORtensei

Dorola®, *see* 'MACSHANA'

Dorothe®, *see* 'LEGGA'

Dorothea Furrer®, HT, dr

'DOROTHEA HOWARD', HT, pb, 1978; flowers light pink, deeper pink reverse, well-formed, large, dbl., 30 petals, borne singly, moderate, tea fragrance; many hooked, brown prickles; foliage medium green, glossy; [First Prize X Roundelay]; Barclay, Hilary M.

Dorothee Heidorn, Bslt, mp, 1995

'DOROTHY A. GOLIK', HT, ob, 1973; bud ovoid; flowers orange to flaming red, high pointed, dbl., 35 petals, 4 in., moderate, spicy fragrance; foliage glossy; moderate growth; [Tropicana X Peace]; Golik; Dynarose

'DOROTHY ANDERSON', HT, lp, 1949; flowers large, dbl., 33 petals, exhibition form, slight fragrance; free growth; [Sam McGredy X George Dickson]; McGredy

'DOROTHY ANNE', HT, pb, 1985; flowers white blending to deep pink at edges, large, dbl., 35 petals, no fragrance; foliage medium, dark, semi-glossy; upright growth; PP006100; [First Prize X Lady X]; Winchel, Joseph F.; Kimbrew-Walter Roses

'DOROTHY BROSTER', HT, mp, 1978; flowers azalea-pink, full, dbl., 45 petals, 5 in., moderate fragrance; foliage dark; very vigorous growth; [Blue Moon X Karl Herbst]; Ellick; Excelsior Roses

'DOROTHY DENNISON', HWich, lp, 1909; flowers pale pink; [Dorothy Perkins sport]; Dennison

'DOROTHY DIX', Pol, mp, 1923; flowers rose-pink, borne in clusters; Hicks

Dorothy Dix, *see* **'MRS E.M. GILMER'**

Dorothy Donnelly, HT, mr, 1997; flowers full, large, very dbl., 26–40 petals, borne mostly singly, intense fragrance; foliage medium, dark green, semi-glossy; upright, bushy, medium growth; [Adrienne Berman X (Royal William X Gabi)]; Poole, Lionel

'DOROTHY DOUGLAS', HT, rb, 1924; flowers vivid cerise-pink, Dobbie

'DOROTHY DROWNE', HWich, pb, 1924; flowers white to pink, center crimson and scarlet, [Sodenia seedling]; Brownell, H.C.

'DOROTHY FOWLER', HRg, mp, 1938; flowers clear pink, well formed, semi-dbl., 3–3.5 in., non-recurrent, intense fragrance; height 3 ft; [R. rugosa X (R. acicularis X R. spinosissima)]; Skinner

'DOROTHY GOODWIN', HT, yb, 1954; (Perfect Peace); flowers yellow tipped cerise-pink, well formed, dbl., 32 petals, 4 in., moderate fragrance; foliage dark holly-green; vigorous growth; [Peace sport]; Goodwin; Gregory

'DOROTHY GRACE', Cl Min, pb, 1986; flowers yellow with pink petal edges, small blooms in sprays of 3–5, exhibition form, 25 petals, exhibition form, none; very few, brown prickles, hooked downward; foliage small, medium green, semi-glossy; upright, climbing (to 6 ft) growth; [Little Darling X Rise 'n' Shine sport]; Dobbs, Annette E.; Port Stockton Nursery

'DOROTHY HODGSON', HT, ob, 1930; flowers orange-cerise, veined darker, well formed, large, slight fragrance; vigorous growth; Cant, F.

'DOROTHY HOWARTH', Pol, op, 1921; flowers coral-pink, tinted salmon, open, borne in clusters, dbl., moderate fragrance; foliage dark; bushy growth; [Léonie Lamesch X Annchen Müller]; Bees

'DOROTHY JAMES', HT, pb, 1939; flowers peach-pink reverse deep rose; good habit growth; [Golden Dawn sport]; C-P

'DOROTHY KING', HT, rb, 1924; flowers scarlet-crimson and maroon, semi-dbl.; King

'DOROTHY LEE', HT, pb, 1929; flowers silvery shell-pink, base golden yellow, dbl.; Morse

Dorothy Lloyd, *see* 'SHERIGREY'

'DOROTHY MARIE', HT, dp, 1935; [Talisman sport]; Scittine; Lainson

'DOROTHY MAY COOPER', Min, w, 1975; flowers pure white, full, dbl., 25–30 petals, 1–2 in.; foliage dark; very vigorous growth; [R. roulettii X Memoriam]; Ellick

'DOROTHY MCGREDY', HT, rb, 1936; flowers deep vermilion, base and reverse yellow, well shaped, moderate fragrance; foliage cedar-green; strong stems; vigorous growth; McGredy

'DOROTHY MOLLISON', HT, dr, 1930; flowers dark crimson; [Mrs R.C. Bell X Seedling]; Clark, A.; NRS Victoria

'DOROTHY PAGE-ROBERTS', HT, pb, 1907; flowers coppery pink, suffused yellow, open, very large, dbl., slight fragrance; vigorous growth; Dickson, A.

'DOROTHY PEACH', HT, yb, 1957; flowers deep yellow flushed pink, dbl., 37 petals, 5 in., exhibition form, moderate fragrance; foliage dark, glossy; vigorous growth; GM, NRS, 1959; [Lydia X Peace]; Robinson, H.

'DOROTHY PEACH, CLIMBING', Cl HT, yb, 1963; Watkins Roses

'DOROTHY PERKINS', HWich, lp, 1901; flowers rose-pink, dbl., moderate fragrance; foliage dark, glossy; very vigorous (10–20 ft) growth; [R. wichurana X Mme Gabriel Luizet]; J&P

'DOROTHY RATCLIFFE', HT, rb, 1910; flowers coral-red shaded fawn-yellow; vigorous growth; McGredy

Dorothy Rose, *see* 'JONDOROSE'

Dorothy Virginia, *see* 'JUDVIRG'

'DOROTHY WHEATCROFT', F, mr, 1960; flowers oriental red shaded darker, blooms in clusters of 13, semi-dbl., 18 petals, 3.5 in., slight fragrance; foliage bright green; vigorous, bushy growth; GM, NRS, 1961; Tantau, Math.; Wheatcroft Bros.

Dorothy Whitney Wood, HT, op, 1992; Fryer, Gareth

Dorothy Wilson, F, ob, 1995; Beales, Peter

'DOROTHY, CLIMBING', Cl HT, pb, 1935; [Dorothy Page-Roberts sport]; Bostick

Dorothy's Front Porch, Misc OGR, r

'DOROTHY'S REGAL RED', HT, mr, 1984; flowers large blooms in small clusters, dbl., 58 petals, intense fragrance; foliage medium dark, glossy; bushy growth; Jerabek, Paul E.

'DOROTKA DARLING', Min, pb, 1991; flowers pink and cream blend, well-formed, blooms borne mostly singly, dbl., 26–40 petals, 1.5 in., slight fragrance; few prickles; foliage medium, dark green, semi-glossy, clean and abundant; tall (60–80 cms), upright growth; [Orange Darling X Seedling]; Sudol, Julia, 1992

DORouvi, HT, 1992; (Perle du Lac, **Perle du Lac Annecy**); Dorieux; (Cavriglia)

DORpark, F, rb, 1996; (**Le Cid**); Dorieux

DORpurp, HT, rb, 1990; (**Parc des Princes**®); Dorieux

Dorris Lee, *see* 'WELLEE'

'DORRIT', F, ob, 1968; flowers orange-yellow, large blooms in trusses, dbl., 30–40 petals, flat; [Seedling X Folie d'Espagne]; Soenderhousen

Dorrit, *see* **'SUNSONG'**

DORsafr, HT, yb, 1990; (**Open d'Australie**®); Dorieux

DORsand, HT, ab, 1994; (**Gaby Morlay**); Dorieux

'D'ORSAY ROSE', Misc OGR, dp; bud deep pink; flowers deep pink, outer petals fading to pale pink receptacle wide, dbl., summer bloom, moderate fragrance; prickles paired below each leaf; foliage leaflets 5–7, leaden green; erect growth to 5 ft

DORster, HT, pb, 1996; (**Ressins Etienne Gautier**); Dorieux

DORtan, F, ab; (**Vieux Chateau Certan**); Dorieux, 1994

DORtiche, HT, dr; (**Marguerite d'Autricho**); Dorieux

DORtive, HT, op, 1991; (**Pierre Troisgros**®); Dorieux

'DORTMUND'®, HKor, mr, 1955; bud long, pointed; flowers red, white eye, large blooms in large clusters, single, recurrent bloom, moderate fragrance; foliage dark, very glossy; vigorous, climbing growth; ADR, 1954 GM, Portland, 1971; [Seedling X R. kordesii]; Kordes

Dortmunder Kaiserhain, S, lp, 1994; flowers medium-large, dbl.; Noack, Werner; (Sangerhausen)

DORto, HT, lp; (**Sika**); Dorieux, 1974

DORtomyl, Pol, rb, 1992; (**Pleins Feux 92**); Dorieux

Dorus Rijkers, *see* **'DORIS RYKER'**

DORvizo, HT, dp, 1993; (**Rina Hugo**); Dorieux, 1993

DORwest, S, dy; (**Ellen Hamlyn**); Dorieux

DORyeco, HT, ob, 1998; (**D. D. Ruaux**); Dorieux

'DO-SI-DO', S, mp, 1984; flowers medium lavender-pink, large blooms borne 3–10 per cluster, cupped, repeat bloom, moderate fragrance; awl-like, tan prickles; foliage dark olive green, leathery, glossy; vigorous, erect, bushy growth; hardy; [(Autumn Dusk X Solitude) X Wanderin' Wind]; Buck, Dr Griffith J.; Iowa State University

'DOSPEARL', Min, op, 1991; (**Our Coral Pearl**); flowers coral-salmon-pink, blooms borne mostly singly, dbl., 26–40 quilling petals, 1.5–2.75 in., slight fragrance; few prickles, tan, straight and sharp; foliage medium, medium green, semi-glossy; medium growth; [Minnie Pearl sport]; Osburn, Dr William, 1992

'DOTEMIBE', HT, m, 1978; (**Doctor Behring**); bud pointed; flowers red-purple, dbl., 35 petals, 4 in., cupped, moderate fragrance; foliage dark; tall, upright growth; [Amanecer X Tanya]; Dot, Simon

'DOTHAN', HT, mp, 1983; flowers large, dbl., 35 petals, exhibition form, slight fragrance; foliage large, medium green, matt; upright growth; [Koppies X King of Hearts]; Perry, Astor; Perry Roses, 1984

'DOTRAMES', HT, or, 1983; (**Coral Fiesta**®, Maria Teresa de Esteban, Mme Teresa Estaban); flowers large, dbl., 45 petals, cupped, no fragrance; light yellow prickles; foliage large, dark, matt; bushy growth; [Seedling X Seedling]; Dot, Simon; Rose Barni-Pistoia

DOTraner, HT, rb, 1969; (**Jardinero Ortiz**®); Dot

'DOTSUBEBE', HT, or, 1978; (**Galaty**); dbl., 32 petals, 4 in., cupped, intense fragrance; foliage dark; [Tropicana X Lola Montes]; Dot, Simon

'DOTSURODO', HT, or, 1978; (**Simon Dot**); dbl., 35 petals, 5 in., cupped, intense fragrance; foliage dark, upright growth; [Pharaoh X Rose Dot]; Dot, Simon

'DOTTY', HT, dy, 1931; flowers bronze-yellow, large, semi-dbl., globular, moderate fragrance; foliage glossy; very vigorous growth; [Souv. de Claudius Pernet X R. foetida bicolor seedling]; Towill

'DOTTY BASS', HT, dr, 1968; bud long, pointed; flowers medium, dbl.; foliage dark, leathery; vigorous, upright growth; Bass; DeVor Nurseries, Inc.

Double Blush, S, lp; flowers medium, dbl.; (Sangerhausen)

'DOUBLE BLUSH BURNET', HSpn, pb; flowers center blush, fading at edges, reverse white

'DOUBLE BRIQUE', HGal, pb; flowers rosy pink, shading silver toward outside, dbl.

Double Cherokee, *see* **'FORTUNIANA'**

Double Cinnamon, *see* **R. CINNAMOMEA PLENA**

'DOUBLE DARK MARBLED', HSpn, rb; (Petite Red Scotch); flowers red mottled purple, small, semi-dbl., early

Double Date, *see* 'LAVFLUSH'

Double Delight™, *see* 'ANDeli'

Double Delight Supreme, HT, rb, 1993; Chiplunkar

Double Delight, Climbing, *see* 'AROclidd'

Double Else Poulsen, *see* **'Else's Rival'**

'Double Feature', Gr, m, 1976; (Lakeland's Pride); bud pointed; flowers reddish purple, reverse yellow, dbl., 28 petals, 4 in., exhibition form, slight, damask fragrance; foliage large, dark; vigorous, upright growth; [Angel Face X Granada]; Williams, J. Benjamin; Lakeland Nursery Sales

Double French Rose, *see* **R. gallica officinalis**

Double Happy, *see* 'SUNdoha'

'Double Joy', Min, op, 1979; bud long, pointed; dbl., 35 petals, 1.5 in., moderate fragrance; foliage small, matt, green; bushy growth; [Little Darling X New Penny]; Moore, Ralph S.; Sequoia Nursery

Double Mme Butterfly, *see* **'Annie Laurie'**

'Double Ophelia', HT, lp, 1916; [Ophelia X Seedling]; Hill, E.G., Co.

'Double Orléans', Pol, mr, 1924; flowers rosy crimson, center white, [Orléans Rose sport]; Hicks

Double Perfection™, *see* 'BURwin'

'Double Pink Killarney', HT, mp, 1910; flowers large, very dbl., moderate fragrance; Scott; (Sangerhausen)

Double Plum, HCh, m

'Double Scarlet', HEg, mr; flowers bright rosy red, dbl., intense fragrance; weak growth

'Double Star', Min, w, 1978; bud ovoid; 10 petals, borne 2–5 per cluster, slight fragrance; no prickles; foliage small, firm, disease-resistant; vigorous growth; [Fairy Moss X Fairy Moss]; Dobbs, Annette E.; Small World Min. Roses

'Double Talk', F, rb, 1980; bud ovoid, pointed; flowers medium red, creamy white reverse, petals rolled loosely outward, dbl., 48 petals, cupped, slight, spicy fragrance; long prickles, hooked downward; foliage glossy, slightly wrinkled, dark; compact growth; [Plain Talk X Suspense]; Weeks, O.L.

Double Time, HT, rb

Double Treat, *see* 'MORtreat'

'Double White', HEg, w;, flowers flesh-white, dbl., vigorous growth

Double White, *see* **'Elegans'**

'Double White Burnet', HSpn, w; intense fragrance; vigorous growth

Double White Noisette, *see* **'Plena'**

Double Yellow Scots Rose, *see* **'William's Double Yellow'**

'Doubloons', LCl, my, 1934; bud ovoid, deep saffron-yellow; flowers rich gold, large blooms in clusters, dbl., cupped, intermittent bloom, moderate fragrance; foliage glossy; vigorous growth; David Fuerstenberg Prize, ARS, 1936; [R. setigera hybrid X R. foetida bicolor hybrid]; Horvath; J&P

Douce Symphonie, *see* 'MEIbarke'

Douceur Normande®, *see* 'MEIpopul'

Douchka, *see* **'Mary DeVor'**

'Doué Rambler', HWich, mp, 1921; flowers bright pink, borne in well-filled clusters, Begault-Pigné

'Douglas MacArthur', HT, pb, 1943; bud long, pointed; flowers delft rose, base slightly bronze, dbl., 24–30 petals, 4–4.5 in., exhibition form; foliage leathery; vigorous, upright, bushy, compact growth; [Mrs J.D. Eisele X Glowing Sunset]; Howard, F.H.; H&S

'Douglas MacArthur, Climbing', Cl HT, pb, 1949; Howard, F. H.; H&S

'Douglass', Ch, dr, 1848; flowers crimson, medium; vigorous growth; Verdier, V.

'Doulce France', F, pb, 1964; bud round; flowers clear pink flushed apricot, large, dbl.; vigorous, upright growth; [(Peace X Seedling) X Lady Sylvia]; Mondial Roses

'Dourada', HT, dy, 1957; flowers well formed, dbl.; very vigorous growth; [Mme Marie Curie X Julien Potin]; da Silva, Moreira

Doutz, S, lp

Doux Parfum, *see* **'Typhoo Tea'**

Dove, *see* 'AUSdove'

Dove Dale, *see* 'MATdove'

'Dovedale', HT, rb, 1975; flowers cream, petals edged carmine, dbl., 42 petals, 5 in.; foliage dark; low, bushy growth; [Fragrant Cloud X Stella]; Moorhouse & Thornley

'Downland Cherry', HT, mr, 1954; flowers light cerise shaded scarlet, intense, spicy fragrance; foliage dark, leathery, dull green; vigorous growth; [Vanessa X Shot Silk]; Ratcliffe

'Downland Lustre', HT, yb, 1955; bud bronzy gold; flowers maize-yellow, reverse orange, medium, dbl., intense, spicy fragrance; compact, bushy growth; [Vanessa X Shot Silk]; Ratcliffe

Downunder, S, lp; Peden, R., 2000; (Weatherly, L.)

'Dr A. Hermans', HT, ly, 1906; flowers yellowish white; Verschuren

'Dr A. I. Petyt', HT, rb, 1924; flowers maroon-crimson shaded scarlet, large, dbl., exhibition form; vigorous, bushy growth; [George Dickson seedling X Edward Mawley]; Burrell

'Dr A. J. Verhage', HT, dy, 1963; (Golden Wave); flowers large, dbl., 22–30 wavy petals, intense fragrance; foliage dark, glossy; vigorous, bushy growth; [Tawny Gold X (Baccará X Seedling)]; Verbeek; Carlton Rose Nurseries

'Dr A. J. Verhage, Climbing', Cl HT, dy, 1968; Blaby Rose Gardens

'Dr A. S. Thomas', HT, rb, 1951; bud long, pointed; flowers dark crimson shaded darker, large, dbl., 60 petals, exhibition form, intense fragrance; foliage leathery, dark; vigorous, fairly compact growth; Clark, A.; NRS Victoria

'Dr A. Svehla', HT, rb, 1935; flowers dark carmine, very large, dbl.; bushy growth; [Col. Leclerc X Gen. MacArthur]; Böhm, J.

'Dr A. von Erlach', HT, pb, 1932; bud nankeen yellow and salmon; flowers pink and straw-yellow, stamens yellow, semi-dbl., cupped, intense fragrance; vigorous growth; [Prince de Bulgarie X Mrs S.K. Rindge]; Soupert & Notting

Dr Adam Christman™, *see* 'WILdac'

'Dr Albert Schweitzer', HT, pb, 1961; flowers opal-pink, reverse rose-red, well-formed, dbl., 30–35 petals, 5–6 in., moderate fragrance; foliage leathery, glossy; vigorous, upright bushy growth; [Chic Parisien X Michele Meilland]; Delbard-Chabert

'Dr Andrew Carnegie', HT, pb, 1927; flowers light silvery pink, base yellowish; [Mrs Henry Morse sport]; Ferguson, R. C.; Dreer, 1930

'Dr Andry', HP, mr, 1864; flowers rosy crimson, semi-, medium, dbl., 45 petals, cupped; foliage glossy; vigorous, upright growth; Verdier, E.

'Dr Antonin Joly', HP, op, 1886; flowers salmon-pink, large, very dbl.; Besson; (Sangerhausen)

'Dr Auguste Krell', HP, 1877; Verdier, E.; (Cavriglia)

'Dr Augustin Wibbelt', HT, yb, 1928; flowers golden yellow, shaded orange, semi-dbl., moderate fragrance; [Los Angeles sport]; Leenders, M.

'Dr Augusto de Castro', HT, rb, 1954; flowers bright red, reverse yellow, [Sultane X Peace]; da Silva, Moreira

Dr B. Benacerraf, *see* 'MILben'

'Dr B.P. Pal', HT, m, 1980; bud long, pointed; flowers solferino purple, blooms borne singly, dbl., 70 petals, exhibition form, slight fragrance; straight, brown prickles; foliage dark, leathery; upright growth; [Seedling X Seedling]; Division of Vegetable Crops and Floriculture

'Dr Barnardo', F, dr, 1968; flowers crimson, large blooms in trusses, dbl., 30 petals, slight fragrance; upright, bushy growth; [Vera Dalton X Red Dandy]; Harkness

'Dr Belville', Cl HT, ob, 1931; flowers orange-crimson, base yellow, open, large, semi-dbl., profuse spring bloom, then scattering, moderate fragrance; very vigorous (12 ft) growth; [Barbara X Sunstar]; Thomas; H&S

Dr Benjamin Pal, HT, mp, 1993; IARI

Dr Bob Harvey, *see* 'WINbob'

'Dr Bradas Rosa Druschki', HP, mp, 1934; flowers very large, dbl.; Brada, Dr; (Sangerhausen)

'Dr Branscom', HT, ab, 1947; bud pointed; flowers peach-blossom flushed apricot-pink, large, dbl., moderate, spicy fragrance; upright growth; [Pink Dawn sport]; Danegger; J.T. Lovett

'Dr Bretonneau', HP, m, 1858; flowers violet-red, medium, dbl.; Trouillard; (Sangerhausen)

'Dr Briere', B, pb, 1860; Vigneron

'Dr Brownell', HT, yb, 1964; bud long, pointed; flowers buff, center chrome-yellow, dbl., 34 petals, 5.5 in., exhibition form, intense fragrance; foliage glossy, dark; vigorous, upright growth; [Helen Hayes X Peace]; Brownell, H.C.; Brownell

'Dr Burt', LCl, rb, 1942; bud long, pointed; flowers deep red to pink flushed orange, large, dbl., 45 petals, non-recurrent, foliage glossy, light; very vigorous, climbing (to 20 ft), branching growth; [Coral Creeper X Seedling]; Brownell, H.C.

Dr Carabare, HT, m; flowers purple/pink, large, dbl., intense fragrance; (Sangerhausen)

'Dr Carbonaro', HT, pb, 1958; flowers rose, reverse silver, [Happiness X Grand'mere Jenny]; da Silva, Moreira

'Dr Carneiro Pacheco', HT, rb, 1938; flowers carmine, open, large, dbl.; foliage glossy, light; vigorous, bushy growth; [Mev. G. A. van Rossem X Sir David Davis]; da Silva, Moreira

'Dr Cathrall', HT, pb, 1966; flowers deep pink, reverse lighter, 4.5–5 in., intense fragrance; foliage dark, leathery; vigorous growth; [Hector Deane sport]; Hills

'Dr Cazeneuve', HT, dr, 1899; flowers medium, dbl.; Dubreuil; (Sangerhausen)

Dr Charles T. Beaird, *see* 'WAMbeaird'

'Dr D. F. Malan', HT, dr, 1960; bud pointed; flowers very dark maroon, large, dbl., 45–50 petals, moderate fragrance; upright growth; [Happiness X Mirandy]; Herholdt, J.A.; Herholdt's Nursery

Dr Darley, *see* 'HARposter'

'Dr Debat', HT, pb, 1952; (Docteur F. Debat, Dr F. Debat, La Rosée); bud ovoid, pointed; flowers bright pink tinted coral, dbl., 25–30 petals, 5–6 in., exhibition form, moderate fragrance; foliage leathery, dark; vigorous, upright growth; GM, NRS, 1950; [Peace X Mrs John Laing]; Meilland, F.; C-P

'Dr Debat, Climbing', Cl HT, pb, 1955; Barni, V.; URS

Dr Dick, *see* 'COCbaden'

'Dr Domingos Pereira', Cl T, pb, 1925; flowers lilac-rose, center yellow, dbl., moderate fragrance; de Magalhaes

'Dr E. M. Mills', S, yb, 1926; flowers primrose suffused pink, becoming darker, semi-dbl., 2–2.5 in., globular, early, foliage small, dark; vigorous growth; [R. hugonis X Radiance]; Van Fleet; American Rose Society

'Dr Eckener', HRg, pb, 1930; flowers coppery rose on yellow ground, aging soft pink, large, semi-dbl., cupped, repeat bloom, moderate fragrance; vigorous (5–6 ft) growth; [Golden Emblem X Hybrid Rugosa]; Berger, V.; C-P, 1931; Teschendorff, 1930

'Dr Edvard Benes', HT, rb, 1935; flowers red with many white streaks, very large, dbl., moderate fragrance; bushy growth; [Étoile de France sport]; Böhm, J.

'Dr Edward Deacon', HT, ob, 1926; flowers deep salmon-orange to shrimp-pink, large, dbl., globular, moderate fragrance; vigorous, bushy growth; [Mme Edouard Herriot X Gladys Holland]; Morse

'Dr Ernst Mühle', HT, pb, 1928; flowers rose-pink, with salmon-white reflex, very large, very dbl.; [Mme Edmée Metz seedling]; Mühle

Dr F. Debat, *see* **'Dr Debat'**

Dr F. L. Skinner, HSpn, lp; Simonet

'Dr F. Weigand', HT, mr, 1930; flowers cherry-red, dbl., slight fragrance; [Mme Caroline Testout X Hadley]; Weigand, C.

Dr F. G. Chandler, *see* **'Dickson's Red'**

Dr Faust, *see* **'Faust'**

'Dr Felix Guyou', T, dy, 1901; Mari

'Dr Ferrandiz', HT, ob; flowers deep orange-red, vigorous growth; Camprubi, C.

'Dr Fleming', HT, pb, 1960; flowers soft pink, flushed crimson, well-formed, dbl., 40 petals; vigorous growth; [Queen Elizabeth X Baleares]; Dot, M.

Dr Franco Nogueira, HT; Moreira da Silva, A.; (Cavriglia)

'Dr G. Krüger', HT, dr, 1913; flowers crimson; [Mme Victor Verdier X Seedling]; Ulbrich; Kiese

'Dr Gallwey', LCl, w, 1937; flowers snow-white, borne in large clusters, single, 2 in., profuse bloom; very vigorous growth; Reiter

Dr Gentil, HT; Moreira da Silva, A.; (Cavriglia)

'Dr Georges Leger', HT, mr, 1935; flowers blood red, large, dbl., slight fragrance; Ketten, Gebrüder; (Sangerhausen)

'Dr Georges Martin', HP, pb, 1908; flowers carmine-pink, very large, very dbl.; Vilin; (Sangerhausen)

Dr Goldberg, HT, 1988; Gandy, Douglas L.; (Cavriglia)

'Dr Grandvilliers', T, pb, 1893; flowers yellowish-pink with darker pink, medium, dbl., Perny; (Sangerhausen)

'Dr Grill', T, op, 1886; flowers rose shaded coppery, moderate fragrance; [Ophirie X Souv. de Victor Hugo]; Bonnaire

'Dr Guarnero', HT, dp, 1958; flowers deep rose, well formed, large, dbl., intense fragrance; vigorous, bushy growth; [Happiness X Grand'mere Jenny]; da Silva, Moreira

'Dr Guilherme Pereira da Rosa', HT, mr, 1955; flowers cherry-red, well formed, moderate growth; [Charles Mallerin X Lisboa]; da Silva, Moreira

'Dr H. E. Rumble', Min, mr, 1981; flowers scarlet red, small, dbl., 30 petals, cupped, no fragrance; small, brown prickles; foliage large, light green; vigorous, upright growth; [Born Free X Westmont]; Hooper, John C.

'Dr H. I. Gallagher', HT, yb, 1989; bud ovoid; flowers yellow with bright pink on petal tips, reverse same, dbl., 35–50 petals, exhibition form, intense, spicy fragrance; prickles medium, bronze-green; foliage medium, bronze-medium green, semi-glossy heavy, dise; bushy, medium to tall growth; Bronze, ARC TG, 1989; [Spellbinder X Irish Gold]; Anderson, Mrs. Etta S.

'Dr Harry Upshall', S, pb, 1993; dbl., 26–40 petals, 1.5–2.75 in., moderate fragrance; foliage small, medium green, matt; bushy, spreading (2 m) growth; [Liverpool Echo X R. foetida persiana]; Fleming, Joyce L.; Hortico Roses, 1993

'Dr Heinrich Lumpe', HT, pb, 1928; flowers light rose-pink, base yellow, large, dbl., exhibition form, moderate fragrance; strong stems; very vigorous, bushy growth; [Constance X Admiral Ward]; Berger, V.; A. Berger

'Dr Helfferich', HT, pb, 1919; flowers rose, center yellowish orange, edged silvery, dbl., moderate fragrance; [Gustav Grunerwald X Mrs Aaron Ward]; Lambert, P.

'Dr Henri Neuprez', HWich, ly, 1913; flowers canary-yellow to sulfur-white, [R. wichurana X Mme Barthelémy Levet]; Tanne

Dr Herbert Gray, *see* 'AUSfar'

'Dr Herbert Hawkesworth', HT, dr, 1927; flowers deep crimson, center almost black, moderate fragrance; Bees

'Dr Hess von Wichdorf', HT, rb, 1936; flowers red, shaded rose-lilac, large, dbl., exhibition form, intense fragrance; vigorous, bushy growth; [Frank W. Dunlop sport]; Vogel, M.; Heinemann

'Dr Homi Bhabha', HT, w, 1968; bud long, pointed; flowers white, center sometimes tinted cream, large, very dbl., exhibition form, slight fragrance; foliage leathery; vigorous, upright growth; [Virgo X Seedling]; Pal, Dr B.P.; Indian Agric. Research Inst.

'Dr Homi Bhabha, Climbing', Cl HT, w, 1976; IARI

'Dr Hoog', HP, 1880; Laxton; (Cavriglia)

'Dr Huey', LCl, dr, 1914; (Shafter); flowers crimson-maroon, anthers light yellow, blooms in clusters of, semi-dbl., 15 petals, 2 in., slight fragrance; foliage rich green; Gertrude M. Hubbard, ARS, 1924; [Ethel X Gruss an Teplitz]; Thomas; A.N. Pierson, 1920; B&A, 1920

'Dr Ing. H. Blohm', HP, dr, 1919; flowers dark carmine-red, large, dbl., intense fragrance; Lambert, P.; (Sangerhausen)

'Dr J. G. Fraser', HT, ab, 1926; flowers salmon-apricot, suffused vermilion-pink; vigorous growth; [St Helena X Muriel Dickson]; Easlea

'Dr J. H. Nicolas', LCl, mp, 1940; flowers rose-pink, blooms in clusters of 3 or 4, dbl., 50 petals, 5 in., globular, recurrent bloom, moderate fragrance; foliage dark, leathery; vigorous, pillar (8 ft) growth; [Charles P. Kilham X Georg Arends]; Nicolas; J&P

Dr Jack Bender, *see* 'BRIben'

'Dr Jaime Lopes Dias', HT, mp, 1961; flowers large, [Confidence X Juno]; da Silva, Moreira

'Dr John Snow', HT, w, 1979; flowers creamy white, dbl., 35 petals, 5 in., exhibition form, moderate fragrance; foliage light green; tall growth; [Helen Traubel X Seedling]; Gandy, Douglas L.

'Dr Joseph Drew', HT, yb, 1918; flowers salmon-yellow, suffused pink, dbl., moderate fragrance; [Mme Mélanie Soupert seedling X Comtesse Icy Hardegg]; Page; Easlea

Dr Jules Bouche, LCl, w

Dr K. C. Chan, *see* 'TINchan'

'Dr Karel Kramár', HT, dr, 1937; flowers large, dbl., moderate fragrance; Böhm, J.; (Sangerhausen)

'Dr Kater', Pol, dr

'Dr Kidwai', HT, pb, 1999; flowers light pink, edges blended magenta, reverse light pink, dbl., 26–40 petals, 3–4 in., borne mostly singly, slight fragrance; prickles moderate; foliage medium, medium green, semi-glossy; upright, tall (4–5 ft) growth; [Paradise X (Paradise X Oklahoma)]; Chiplunkar, C. R.; KSG Roses, 1998

'Dr Kirk', HT, op, 1940; bud long, pointed; flowers coral, shaded nasturtium-yellow, very large, dbl., 35 petals, exhibition form, slight fragrance; vigorous growth; [Charles P. Kilham X R. foetida bicolor hybrid]; Mallerin, C.; A. Meilland

'Dr Lande', T, 1902; flowers large, dbl., moderate fragrance; Berger, V.; (Sangerhausen)

Dr Lopez Diaz, HT; Moreira da Silva, A.; (Cavriglia)

'Dr M. Euwe', HT, pb, 1936; bud pointed; flowers salmon tinted yellow and pink, dbl., intense fragrance; foliage leathery, bronze; bushy growth; Buisman, G.A.H.

Dr M. S. Randhawa, HT, pb, 1989; Pal, Dr B.P.

'Dr Manuel Alves de Castro', HT, rb; flowers red, reverse golden yellow, da Silva, Moreira

'Dr Margaretha', F, dr, 1960; flowers velvety dark red, medium, in large clusters, dbl.; vigorous, bushy growth; [Red Pinocchio seedling X Alain]; Maarse, G.

'Dr Marx', HP, mr, 1842; flowers crimson red, Laffay, M.

"Dr Maximo de Carvalho", HT, dr, 1960; flowers crimson-red, [Crimson Glory X Charles Mallerin]; da Silva, Moreira

'Dr Mazaryk', HP, lp, 1930; flowers large, dbl., slight fragrance; Böhm, J.; (Sangerhausen)

Dr McAlpine, *see* 'PEAfirst'

'Dr Mendes Correia', HT, mr, 1938; bud pointed; flowers bright red, very large, dbl., intense fragrance; foliage soft; vigorous, bushy growth; [Frau Margarete Oppenheim X Hortulanus Budde]; da Silva, Moreira

'Dr Mengelberg', F, dr, 1952; flowers deep blood-red, large, semi-dbl.; very vigorous growth; Leenders, M.

'Dr Merkeley', (similar to R. spinosissima), dp, 1924; flowers deep pink, dbl., nonrecurrent, moderate fragrance; low to medium growth; (discovered in eastern Siberia; named in honor of Dr Merkeley, who first grew it in Canada)

'Dr Miroslav Tyrs', HP, rb, 1932; flowers crimson, shaded darker, very large, moderate fragrance; [Anna de Diesbach sport]; Böhm, J.

Dr Morse, HT, m

'Dr Müller', HFt, op, 1905; flowers salmon-pink with touches of red, medium, semi-dbl.; Müller, Dr F.; (Sangerhausen)

'Dr Müllers's Rote', HT, m, 1920; flowers purple/pink, medium, dbl., intense fragrance; Müller, Dr F.; (Sangerhausen)

'Dr Nicolas Welter', HT, op, 1912; flowers large, dbl., moderate fragrance; Soupert & Notting; (Sangerhausen)

'Dr Noshir Wadia', HT, rb, 1999; flowers bright red, white stripes, reverse light red with stripes, dbl., 26–40 petals, 3–4 in., borne mostly singly, slight fragrance; many prickles; foliage large, dark green, glossy; compact, medium (4–5 ft) growth; [Norma sport]; Chiplunkar; KSG Roses, 1992

'Dr O'Donel Browne', HT, pb, 1908; flowers carmine-rose, large, well formed, dbl., intense fragrance; vigorous growth; Dickson, A.

'Dr Oliveira Salazar', HT, ob, 1955; flowers salmon and yellow shaded carmine, large, very dbl., intense fragrance; [Mme Marie Curie X Peace]; da Silva, Moreira

Dr Paul Menzel, HT, mp, 1980; flowers carmine-pink, large, dbl., moderate fragrance; Lucke, G.; (Sangerhausen)

Dr R. Maag, *see* 'MEIrigalu'

'Dr Rafael Duque', HT, rb, 1938; flowers velvety purplish red, large, moderate fragrance; [Frau Margarete Oppenheim X Hortulanus Budde]; da Silva, Moreira

Dr Reiner Klimke, F, lp, 1988; Noack, Werner

'Dr Renata Tyrsová', LCl, op, 1937; flowers salmon-pink, medium, semi-dbl.; Böhm, J.; (Sangerhausen)

'Dr Reymont', HMult, w, 1907; flowers large, dbl.; Mermet; (Sangerhausen)

'Dr Ricaud', Pol, lp, 1907; flowers small, dbl., moderate fragrance; Corboeuf; (Sangerhausen)

'Dr Richard Legler', HT, pb; flowers shrimp-pink changing to old-rose and orange, moderate, fruity fragrance; moderate growth

'Dr Rouges', Cl T, rb; flowers large, dbl., moderate fragrance; Schwartz, 1893; (Sangerhausen)

Dr S. S. Bhatnagar, F, dr, 1994; IARI

'Dr Scheiner', HT, dr, 1929; flowers large, dbl., moderate fragrance; Böhm, J.; (Sangerhausen)

'Dr Scott', (strain of R. multiflora), w; mildew-resistant; used as understock

Dr Selma Lagerlof, HRg, mp

Dr Sybil Johnson, *see* 'BATsybil'

Dr Tomin, HT, 1990; Tagashira, Kazuso; (Sangerhausen)

'Dr Trigo de Negreiros', HT, dr, 1954; bud long, pointed; flowers deep red, large, very vigorous growth; [Charles Mallerin X Lisboa]; da Silva, Moreira

'Dr Valois', HT, 1950; Mallerin, C.; (Cavriglia)

'Dr van de Plassche', Pol, mp, 1968; bud ovoid; flowers pink, medium, semi-dbl.; foliage dark; [Heureux Anniversaire X Allotria]; Buisman, G. A. H.

'Dr van Rijn', HT, my, 1952; bud ovoid; flowers lemon-yellow, large, dbl., exhibition form, moderate fragrance; foliage light green, glossy; vigorous, bushy growth; Leenders, M.

'DR VAZQUEZ', HT, op, 1935; flowers salmon, open, medium, semi-dbl.; foliage glossy; upright growth; [Duchess of Atholl X Margaret McGredy]; Camprubi, C.

'DR W.E. HADDEN', HT, rb, 1934; flowers raspberry-red, flushed yellow, deepening at base, well formed, foliage dark; long, strong stems; vigorous growth; McGredy

'DR W. VAN FLEET', LCl, lp, 1910; bud pointed; flowers cameo-pink fading flesh-white, large, dbl., non-recurrent, moderate fragrance; foliage dark, glossy; vigorous, climbing (15–20 ft) growth; (21); [(R. wichurana X Safrano) X Souv. du Prés. Carnot]; Van Fleet; P. Henderson

Dr Wolfgang Pöschl, *see* **'CANADIAN WHITE STAR'**®

'DR ZAMENHOF', HWich, rb, 1935; flowers crimson-red, base yellow, very large, borne in clusters, intense fragrance; very vigorous growth; [R. wichurana X Seedling]; Brada, Dr; Böhm

Dr Zumel, HT; Kordes, R.; (Cavriglia)

Dr Robert Korns

DRAdiwu, S, w, 1996; (**Anita Pereire**); Orard

Dragon's Eye, *see* 'CLEDRAG'

Dragon's Fire™, *see* 'MINIFIRE'

'DRAMBUIE', HT, rb, 1973; flowers orange-red, reverse red, full, high pointed, dbl., 28–30 petals, 5 in., intense fragrance; foliage glossy; vigorous, bushy growth; [Whisky Mac sport]; Anderson's Rose Nurseries

'DREAM', HT, op, 1938; flowers geranium-pink, dbl., 50–60 petals, 4.5 in., moderate fragrance; [Better Times sport]; Dramm

Dream, *see* KORmiller

Dream Baby, *see* 'RENBABY'

Dream Blush, *see* 'TWOAT'

Dream Boat, *see* **'DREAMBOAT'**

Dream Cloud, *see* 'AROPICLU'

Dream Dolly, *see* JACpin

'DREAM DUST', F, mp, 1969; bud pointed; flowers small, borne in clusters, dbl., slight fragrance; foliage leathery; very vigorous, spreading growth; [Lavender Girl seedling X Little Darling seedling]; Gardner, B.C.

'DREAM GIRL', LCl, pb, 1944; (Dreamgirl); flowers salmon-pink overlaid apricot, dbl., 55–65 petals, 3.5 in., recurrent bloom, intense fragrance; foliage glossy; good pillar rose growth; (28); [Dr W. Van Fleet X Senora Gari]; Jacobus; B&A

Dream Lover, *see* 'RENLOVER'

Dream Lover, *see* PEAyetti

Dream Orange, *see* 'TWOAEBI'

'DREAM PARADE', HT, op, 1938; flowers amber in spring, seashell-pink in hot weather, burnt-orange, dbl., moderate fragrance; vigorous growth; [Condesa de Sástago sport]; Hillock

Dream Pink, *see* 'TWOJOAN'

Dream Pink, HT, mp; Twomey, Jerry, 1999

Dream Red, *see* 'TWOPAUL'

'DREAM TIME', HT, mp, 1977; (Dreamtime); dbl., 38 petals, 5 in., exhibition form, intense fragrance; foliage light green; moderately vigorous growth; [Kordes' Perfecta X Prima Ballerina]; Bees

'DREAM WALTZ', F, dr, 1969; flowers large blooms in trusses, dbl.; foliage glossy; Tantau, Math.

Dream Weaver™, *see* 'JACPICL'

Dream Yellow, *see* 'TWOYEL'

'DREAMBOAT', Min, my, 1981; (Dream Boat); flowers medium, dbl., 60–70 petals, exhibition form, slight fragrance; bushy, spreading growth; [Rise 'n' Shine X Grand Opera]; Jolly, Betty J.; Rosehill Farm, 1982

Dreamboat, HT, lp, 1995; Dawson

Dreamcatcher, *see* 'SEACATCH'

Dreamcoat, *see* 'SEACOAT'

Dreamer™, *see* 'SAVADREAM'

Dreamgirl, *see* **'DREAM GIRL'**

'DREAMGLO', Min, rb, 1978; bud long, pointed; flowers white, tipped and blended red, dbl., 50 petals, 1 in., exhibition form, slight fragrance; foliage small, dark; upright growth; [Little Darling X Little Chief]; Williams, Ernest D.; Mini-Roses

Dreaming, *see* 'REIKOR'

Dreaming Parade, Min, op, 1995; Poulsen

'DREAMING SPIRES', LCl, dy, 1973; flowers bright golden yellow, dbl., 25 petals, 3 in., exhibition form, repeat bloom, intense fragrance; foliage dark; GM, Belfast, 1977; [Buccaneer X Arthur Bell]; Mattock

Dreamland, *see* 'RESLAND'

Dreamland, *see* **'Traumland'**

Dreamrider, *see* 'MINRID'

Dreams Come True, *see* 'MEIVESTAL'

Dreamsicle, *see* 'TALDRE'

Dreamtime, *see* **'DREAM TIME'**

Dreamward, *see* 'BURAPWARD'

Dreamy, *see* 'TINDREAM'

Drei Gleichen, HFt, ly; flowers medium, semi-dbl., slight fragrance; (Sangerhausen)

Dreienbrunnen, Pol, mr, 1957; flowers small, dbl.; Berger, W.; (Sangerhausen)

'DRESDEN', HT, w, 1961; (Mathé Altéry); bud ovoid; flowers white lightly suffused pink, dbl., 60 petals, 4–5 in., exhibition form, intense fragrance; foliage leathery, dark; very vigorous, upright growth; [Ophelia X Cathrine Kordes]; Robichon; Ilgenfritz Nursery

'DRESDEN DOLL', Min, lp, 1975; bud mossy; flowers soft pink, mini-moss, small, semi-dbl., 18 petals, 1.5 in., cupped, moderate fragrance; foliage glossy, leathery; low, bushy, compact growth; [Fairy Moss X Moss seedling]; Moore, Ralph S.; Sequoia Nursery

Dresdner Gelbe No. 79311, F, dy; flowers medium-large, semi-dbl., slight fragrance; VEG; (Sangerhausen)

Dresselhuys, *see* 'MACJOCEL'

'DRIES VERSCHUREN', HT, my, 1961; (Blaby Jubilee); bud pointed; flowers buttercup-yellow, well formed, large, dbl., 25 petals, moderate fragrance; foliage glossy, bronze; vigorous, upright, bushy growth; [Golden Rapture X Seedling]; Verschuren, A.; Blaby Rose Gardens

'DRIFTER'S ESCAPE', F, or, 1971; flowers vermilion, semi-dbl., 10 petals, 3–3.5 in.; foliage glossy, reddish when young; [Orangeade X Orange Sensation]; Greenway

'DRIscobruce', HT, mp, 1990; (**Dave Hessayon**); dbl., 26–40 petals, moderate fragrance; foliage medium, medium green, glossy; bushy growth; [Silver Jubilee X Pink Favorite]; Driscoll, W.E.; Rosemary Roses, 1989

'DRIscogeorge', HT, mr, 1992; (**David Charles Armstrong**); flowers coral red, blooms borne mostly singly, dbl., 26–40 petals, 3–3.5 in., slight fragrance; some prickles; foliage medium, medium green, semi-glossy; tall (106 cms), upright growth; [Silver Jubilee X Fragrant Cloud]; Driscoll, W.E., 1993

'DRIsconun', F, ly, 1988; (**Maud Nunn**); bud pointed; flowers creamy yellow, aging paler lemon yellow, loose, small, semi-dbl., 16 petals, exhibition form, slight fragrance; prickles green; foliage medium green, semi-glossy; bushy, medium growth; [Rise 'n' Shine X Rise 'n' Shine]; Driscoll, W.E.

'DRIscored', Min, op, 1999; (**Just William**); flowers salmon pink, reverse pale rose, dbl., 26–40 petals, 2.5 in., no fragrance; few prickles; foliage medium, light green, semi-glossy; upright, tall (40 in.) growth; [Perestroika X Pink Petticoat]; Driscoll, W.E.

'DRIscoroy', HT, dr, 1996; (**Menaka Durga Roy**); flowers deep red, blooms borne mostly single, dbl., 26–40 petals, moderate fragrance; few prickles; foliage medium, medium green, lighter underside, semi-glossy; upright, tall (3 ft) growth; [Silver Jubilee X (Red Planet X Blessings) X (Parkdirector Riggers X Honey Favorite)]; Driscoll, W.E.

'DRONNING ALEXANDRINE', HT, dp, 1926; flowers deep pink, urn-shaped, medium, dbl., no fragrance; foliage medium green, semi-glossy; medium, bushy growth; Poulsen, S.; Poulsen

Dronning Margrethe, *see* 'POULSKOV'

'DROPMORE YELLOW', HFt, my; [R. foetida X R. spinosissima altaica]; Skinner

'DROUJBA', F, yb, 1975; flowers yellow, shaded red, small blooms in clusters of 5–30 per stem, dbl., 42 petals, cupped, moderate, tea fragrance; foliage dark, glossy; bushy growth; [Masquerade X Rumba]; Staikov, Prof. Dr V.; Kalaydjiev and Chorbadjiiski

'DRU', Min, dy, 1985; flowers small, dbl., 35 petals, exhibition form, slight fragrance; foliage medium, medium green, semi-glossy; upright, bushy growth; Hunt, W. Henry, 1986

'DRUMMER BOY', F, mp, 1965; bud ovoid; flowers soft carmine-rose, small, borne in large clusters, dbl., slight fragrance; foliage leathery; vigorous, tall growth; [Pinocchio X Queen Elizabeth]; Lammerts, Dr Walter; Germain's

Drummer Boy, *see* 'HARVACITY'

'DRUSCHKA', HP, dp, 1932; flowers carmine-pink, large, dbl., slight fragrance; Kordes, W.; (Sangerhausen)

'DRUSCHKI RUBRA', HP, mr; Lambert, P., 1929

'DUARTE DE OLIVERA', N, op, 1879; Brassac

'DUBLIN', HT, mr, 1982; flowers large, smoky red, dbl., 35–40 petals, 5 in., exhibition form, borne mostly singly, intense, raspberry fragrance; foliage large, medium green, matt; long straight stems; upright growth; [(Seedling X Mister Lincoln) X Ann Letts]; Perry, Astor; Perry Roses, 1983

Dublin Bay®, *see* 'MACDUB'

'DUBONNET', Pol, dr, 1958; bud small; flowers cardinal-red, dbl., slight fragrance; foliage leathery; vigorous growth; [Stoplite sport]; Jelly; E. G. Hill Co.

'DUC D'ANJOU', HP, dr, 1862; flowers deep red, Boyau

Duc d'Aremberg, HGal, m

'DUC D'AUDIFFRET-PAQUIER', HP, 1887; Verdier, E.; (Cavriglia)

Duc de Angoulême, *see* **'DUCHESSE D'ANGOULÊME'**

'DUC DE BORDEAUX', HGal, pb, 1820; flowers rosy lilac, large, dbl., Vibert

Duc de Brabant, C, mr; flowers medium, dbl.; (Sangerhausen)

'DUC DE CAMBRIDGE', D, m; flowers deep purplish rose, large, dbl.; foliage dark, edged reddish brown when young

'DUC DE CAZES', HP, m, 1861; flowers velvety purple, dbl., cupped, moderate fragrance; [Général Jacqueminot X Seedling]; Touvais

'DUC DE CONSTANTINE', Ayr, mp, 1857; Soupert & Notting

'DUC DE CRILLON', B, mr, 1858; Moreau et Robert

'DUC DE FITZJAMES', HGal, dr; flowers very dark crimson, shaded purple

'DUC DE GUICHE', HGal, m, 1835; flowers light reddish-violet, large, dbl., Prévost, before 1810

'DUC DE MAGENTA', T, ab; flowers flesh shaded fawn, large, dbl.; Margottin, 1859

'DUC DE MARLBOROUGH', HP, dr; Lévêque, 1884

'DUC DE MONTPENSIER', HP, dr, 1875; flowers large, dbl., moderate fragrance; Lévêque; (Sangerhausen)

Duc de Rohan, *see* **'DUCHESSE DE ROHAN'**

'DUC DE SUSSEX', D, lp, 1856; flowers yellowish-white with pink, large, dbl.; Roseraie de l'Hay; (Sangerhausen)

'DUC DE VALMY', HGal, m; flowers light purplish rose, marbled purple, large, dbl., cupped

Duc de Wellington, *see* **'DUKE OF WELLINGTON'**

'DUC D'HARCOURT', HP, mr, 1863; flowers large, dbl.; Moreau et Robert; (Sangerhausen)

'DUC D'ORLÉANS', HGal, 1830; flowers purple/red, medium, dbl.; Vibert, before 1848; (Sangerhausen)

Duc Meillandina, *see* 'MEIPINJID'

'DUCHER', Ch, w, 1869; flowers pure white, small to medium, dbl., flat; vigorous growth; Ducher

'DUCHESS', HT, mp, 1976; bud ovoid; flowers cameo-pink shaded deeper, dbl., 4–4.5 in., exhibition form, moderate fragrance; foliage glossy, leathery; bushy, upright growth; [White Satin X Seedling]; Van Veen; Carlton Rose Nurseries

Duchess, HT, lp; greenhouse variety

'DUCHESS OF ABERCORN', HT, pb, 1919; flowers creamy white edged bright rose, dbl.; Dickson, H.

'DUCHESS OF ALBANY', HT, dp, 1888; (Red La France); flowers deep pink, [La France sport]; Paul, W.

'DUCHESS OF ATHOLL', HT, ob, 1928; flowers vivid orange, flushed old-rose, large, dbl., cupped, intense fragrance; foliage bronze, leathery; vigorous growth; Dobbie

'DUCHESS OF ATHOLL, CLIMBING', Cl HT, ob, 1933; Howard Rose Co.

'DUCHESS OF BEDFORD', HP, mr, 1879; flowers bright medium red, large, dbl., globular; Postans, R. B.; W. Paul & Sons

'DUCHESS OF CONNAUGHT', HT, pb, 1879; flowers deep silvery pink, large, globular, intense fragrance; dwarf growth; [Adam X Duchesse de Vallombrosa]; Bennett

'DUCHESS OF EDINBURGH', T, dr, 1874; (Prince Wasiltchikoff); flowers crimson, becoming lighter, large, dbl.; moderate growth; [Souv. de David d'Angers seedling]; Nabonnand, G.; Veitch

'DUCHESS OF FIFE', HP, lp, 1893; flowers large, dbl., intense fragrance; Cocker; (Sangerhausen)

'DUCHESS OF KENT', F, dp, 1968; flowers rose-neyron-red, blooms in trusses, cupped; low, bushy growth; [Katharine Worsley sport]; Waterhouse, W.P.; Waterhouse Nursery

Duchess of Kent, *see* **'HOMÈRE'**

'DUCHESS OF MARLBOROUGH', HT, pb, 1922; flowers brilliant lilac-rose, reverse carmine-crimson, dbl.; [Jonkheer J.L. Mock seedling X Beauté de Lyon]; Nabonnand, P.

'DUCHESS OF MONTROSE', HT, rb, 1929; flowers vermilion-crimson, large; Dobbie

'DUCHESS OF NORMANDY', HT, pb, 1912; flowers soft salmon-flesh, overlaid yellow, large, dbl., exhibition form; vigorous, branching growth; [Dean Hole sport]; Le Cornu

'DUCHESS OF PADUCAH', HT, w, 1971; bud ovoid; flowers white, edge flushed red, large, dbl., exhibition form, intense fragrance; foliage large, glossy, dark, leathery; vigorous, upright growth; [Kordes' Perfecta X Peace]; Williams, J. Benjamin

'DUCHESS OF PORTLAND', P, mr; (Duchesse de Portland, Portland Rose); flowers bright scarlet

'DUCHESS OF RUTLAND', F, dp, 1956; flowers rich carmine-pink, small, borne on large truss, semi-dbl.; vigorous growth; deRuiter; Gandy Roses, Ltd.

'DUCHESS OF SUTHERLAND', HP, lp, 1839; flowers rosy pink, large, dbl., moderate fragrance; vigorous growth; Laffay, M.

'DUCHESS OF SUTHERLAND', HT, pb, 1912; flowers rose-pink shaded lemon on white base, large, dbl., exhibition form, sweetbriar fragrance, moderate; foliage glossy, olive-green; vigorous growth; Dickson, A.

'DUCHESS OF WELLINGTON', HT, ly, 1909; bud pointed; flowers buff-yellow, deeper toward center, open, large, semi-dbl., 17 petals, moderate fragrance; foliage leathery; long, strong stems; bushy growth; Dickson, A.

'DUCHESS OF WELLINGTON, CLIMBING', Cl HT, ly, 1924; Howard Rose Co.

'DUCHESS OF WESTMINSTER', HT, pb, 1879; flowers pink shaded carmine, large, dbl., slight fragrance; moderate growth; [Adam X Marquise de Castellane]; Bennett

'DUCHESS OF WESTMINSTER', HT, mp, 1911; flowers clear rose-pink, dbl., moderate fragrance; Dickson, A.

Duchess of Windsor, *see* **'PERMANENT WAVE'**

Duchess of York, *see* 'DICRACER'

'DUCHESS OF YORK', HT, yb, 1925; flowers deep golden yellow, center tangerine, very well formed, large, dbl.; vigorous growth; Dickson, S.

'DUCHESSE D'ABRANTES', M, lp; flowers light pink with dark pink shading, large, dbl.; Robert, 1851; (Sangerhausen)

'DUCHESSE D'ALBE', T, yb, 1903; flowers yellowish salmon, shaded coppery purple rose, base golden yellow, dbl., globular; Lévêque

'DUCHESSE D'ANGOULÊME', HGal, lp; (Duc de Angoulême, Wax Rose); flowers blush, center pink, medium, dbl., cupped; moderate, upright growth; [Probably Gallica X Centifolia hybrid]; Vibert, 1821

Duchesse d'Anjou, *see* 'GODZOTY'

'DUCHESSE D'AOSTE', HP, mr, 1867; flowers large, very dbl., moderate fragrance; Margottin; (Sangerhausen)

'DUCHESSE D'ASSUMA', HP, 1877; Jamain, H.; (Cavriglia)

'DUCHESSE D'AUERSTÄDT', N, my, 1888; flowers golden yellow, [Reve d'Or sport]; Bernaix, A.

'DUCHESSE DE BERRY', HGal, pb; Vibert, 1820

'DUCHESSE DE BRABANT', T, lp, 1857; (Comtesse de Labarathe, Comtesse Ouwaroff); flowers soft rosy pink, large, dbl., 45 petals, cupped, intense fragrance; vigorous, spreading growth; Bernède

'DUCHESSE DE BRAGANCE', HP, mp, 1886; flowers very large, dbl.; Verdier; (Sangerhausen)

'DUCHESSE DE BUCCLEUGH', HGal, rb; flowers lively crimson, edges tinged lavender, large, dbl., cupped; vigorous growth; Vibert, 1837

'DUCHESSE DE CAMBACÉRÈS', HP, m, 1854; flowers lilac-rose, dbl.; vigorous growth; Fontaine

'DUCHESSE DE CAYLUS', HP, dp, 1864; flowers brilliant carmine-pink, well formed, large, globular, moderate fragrance; moderate growth; [Alfred Colomb seedling]; Verdier, C.

'DUCHESSE DE DINO', HP, dr, 1889; flowers very large, dbl.; Lévêque; (Sangerhausen)

'DUCHESSE DE GALLIERA', HP, pb; flowers bright rose shaded flesh, large, dbl., cupped; Laffay, M., 1847; 1847

'DUCHESSE DE GRAMMONT', N, w; flowers flesh

Duchesse de Guermantes, M, lp; Morley, Dr B., 1988; (Weatherly, L.)

'DUCHESSE DE LA MOTHE-HOUDANCOURT', HT, dp, 1907; flowers large, dbl., moderate fragrance; Toussaint Mille Fils; (Sangerhausen)

Duchesse de la Tremoille, M, lp; Morley, Dr B., 1988; (Weatherly, L.)

'DUCHESSE DE MONTEBELLO', HGal, lp; flowers rosy pink, changing to flesh-pink, medium, dbl., moderate fragrance; erect, compact growth; Laffay, M., 1824 or 1825

'DUCHESSE DE MORNY', HP, mp, 1863; flowers large, dbl.; Verdier, E.; (Sangerhausen)

Duchesse de Portland, *see* **'DUCHESS OF PORTLAND'**

'DUCHESSE DE ROHAN', HP, pb; (Duc de Rohan); flowers rosy crimson, margined with lilac, compact, large, dbl.; (France; possibly Lévêque), before 1880

Duchesse de Savoie®, *see* LAPbel

'DUCHESSE DE TALLEYRAND', HT, yb, 1944; bud pointed; flowers egg-yolk-yellow to chrome-yellow, dbl., moderate fragrance; well branched growth; [Mme Joseph Perraud X Fred Edmunds]; Meilland, F.

'DUCHESSE DE THURINGE', B, w, 1847; flowers white tinted lilac, Guillot Père

'DUCHESSE DE VALLOMBROSA', HP, ab; flowers flesh shaded rose, large, dbl.; [Jules Margottin seedling]; Schwartz, J., 1875

'DUCHESSE DE VENDOME', HT, rb, 1924; flowers crimson, coppery reflexes, reverse yellow, dbl., moderate fragrance; [Souv. de Gilbert Nabonnand X Juliet]; Nabonnand, P.

'DUCHESSE DE VERNEUIL', M, pb, 1856; bud heavily mossed; flowers flesh-pink deepening to salmon-pink, camellia form, Portemer fils

'DUCHESSE D'ISTRIE', M, lp; Laffay, M., 1855

'DUCHESSE D'ORLEANS', HP, 1851; Quétier; (Cavriglia)

'DUCHESSE D'ORLEANS', HGal, mp, 1821; Laffay, M.

'DUCHESSE MARIE SALVIATI', T, ob; flowers orange-yellow tinted pink; [Mme Lombard X Mme Maurice Kuppenheim]; Soupert & Notting, 1889

DUCmar, Min, op; (**Marquisette®**); Ducher

Dudley Cross, *see* **'MRS DUDLEY CROSS'**

'DUET', HT, mp, 1960; bud ovoid; flowers light pink, reverse dark pink, ruffled, dbl., 25–30 petals, 4 in., exhibition form, borne singly and in small clusters, slight, tea fragrance; foliage leathery, dark green, glossy; vigorous, upright growth; AARS, 1961; GM, Baden-Baden, 1959; [Fandango X Roundelay]; Swim, H.C.; Armstrong Nursery

'DUET SUPREME', HT, pb, 1990; bud rounded; flowers light to medium pink blend, medium pink reverse, urn-shaped, dbl., 35 petals, slight fragrance; foliage medium, dark green, semi-glossy; [Duet sport]; Patterson, William, 1988; Roses Unlimited, 1990

Duett®, F, or, 1982; Noack, Werner

'DUFFEY'S DELIGHT', HT, dr, 1976; bud long, pointed; flowers velvety dark red, very dbl., 60 petals, 5–6 in., exhibition form, slight fragrance; foliage dark; very vigorous, upright growth; [Norman Hartnell sport]; Duffey

'DUFTBELLA'®, F, dr, 1973; bud ovoid; flowers dark velvet red, center lighter, large, dbl., intense fragrance; vigorous, bushy growth; [Fragrant Cloud X (Monique X Mardi Gras)]; Hetzel; GAWA

Duftes Berlin®, HT, ob, 1988; Cocker

Duftgold, *see* 'TANDUGOFT'

Duftparadies, F, dr, 1966

Duftrausch®, *see* 'TANSCHAUBUD'

Duftrausch, *see* 'TANRAUSCH'

'DUFTSTAR', HT, dr, 1974; bud long, pointed; dbl., 24 petals, 4 in., exhibition form, intense fragrance; foliage dark, soft; vigorous, upright growth; [Seedling X Papa Meilland]; Kordes; Dehner & Co.

Duftstern®, HT, op, 1973; Noack, Werner

Duftwolke, *see* 'TANELLIS'

'DUFTWUNDER', F, ob, 1972; flowers yellowish orange, large, very dbl., intense fragrance; moderate, upright growth; [Fragrant Cloud X Goldmarie]; Hetzel

Duftzauber, *see* 'KORDU'

Duftzauber '84, *see* 'KORZAUN'

'DUHAMEL DUMONCEAU', HP, mr, 1872; flowers large, dbl., moderate fragrance; Villin-Hugues; (Sangerhausen)

'DUILIU ZAMFIRESA', HT, mr, 1938; flowers large, dbl., intense fragrance; Palocsay, R.; (Sangerhausen)

'DUISBURG', HT, dp, 1908; flowers carmine-pink, large, dbl., moderate fragrance; Hinner, W.; (Sangerhausen)

'DUKAT', LCl, my, 1955; flowers golden yellow, large, dbl., moderate fragrance; foliage glossy, leathery; vigorous (10–16 ft), upright growth; [Mrs Pierre S. duPont X Golden Glow]; Tantau, Math.

Dukat, HT, my; Brabec, 1980; (Czech Rosa Club)

Duke Meillandina, *see* 'MEIPINJID'

Duke of Argyll, HSpn, pb

'DUKE OF CONNAUGHT', HP, rb, 1875; flowers dark velvety crimson flushed brighter, large, dbl., intense fragrance; vigorous growth; Paul

'DUKE OF CONNAUGHT', HT, mr, 1879; flowers rose-crimson, very large, dbl.; moderate growth; [Adam X Louis van Houtte]; Bennett

'DUKE OF EDINBURGH', HP, dr, 1868; flowers deep red, large, dbl., moderate fragrance; vigorous, erect growth; [Général Jacqueminot X Seedling]; Paul

'DUKE OF NORMANDY', HT, lp, 1921; flowers silvery pink, [St Helena X George Dickson]; Jersey Nursery

'DUKE OF PADUCAH', HT, rb, 1970; bud ovoid; flowers dark velvety crimson, full, large, dbl., slight fragrance; foliage large, glossy, dark, leathery; vigorous, bushy growth; [Grand Gala X Josephine Bruce]; Williams, J. Benjamin

'DUKE OF TECK', HP, dp, 1880; flowers deep pink, dbl., 40 petals, globular, moderate fragrance; vigorous growth; [Duke of Edinburgh X Seedling]; Paul

'DUKE OF WELLINGTON', HP, dr, 1864; (Duc de Wellington); flowers velvety crimson-red, large, dbl., cupped, intense fragrance; vigorous growth; Granger

'DUKE OF WINDSOR', HT, ob, 1969; (Herzog von Windsor); bud pointed; flowers orange, well-formed, large, dbl., 27 petals, intense fragrance; foliage dark, glossy; very vigorous, upright growth; ADR, 1970; Edland fragrance Medal, ARS, 1968; Tantau, Math.

'DUKE OF YORK', HCh, pb, 1894; flowers rosy-pink and white to crimson, variable, Paul & Son

Duke Wayne, *see* **'BIG DUKE'**

'DULCINEA', HT, mr, 1963; flowers oriental red, edged darker, reverse yellow, large, dbl., 50–55 petals, foliage dark, glossy; upright, bushy growth; [Condesa de Sástago X Seedling]; Verschuren, A.; Stassen

'DUMBO', F, mr, 1956; flowers bright cherry-red, very large, semi-dbl.; foliage dark; very vigorous growth; [Mme G. Forest-Colcombet X Independence]; Combe

'DUMORTIER', HGal, pb; flowers light red with silvery reflex, medium, very dbl., flat

'DUNDEE', HSpn, w; flowers white blotched pink, reverse pure white, dbl., cupped; moderate growth; Austin, R., prior to 1832

'DUNDEE RAMBLER', Ayr, w; flowers compact, blooms in large clusters; [Thought to be R. arvensis X Noisette]; Martin, prior to 1837

Dune®, *see* DELgrim

Dunkelrote Ellen Poulsen, *see* **'RED ELLEN POULSEN'**

'DUNKELROTE TAUSENDSCHÖN', HMult, dr, 1942; flowers medium, dbl., slight fragrance; Vogel, M.; (Sangerhausen)

'DUNKERQUE', HT, op, 1940; bud pointed, ovoid; flowers bright pink, slightly coppery, very large, dbl., 30 petals, moderate fragrance; foliage clear olive-green; long, strong stems; vigorus, upright, branching growth; GM, Bagatelle, 1940; [Charles P. Kilham X Seedling]; Laperrière

'DUNKIRK', HT, mr, 1947; flowers rose-red, dbl., 36 petals, 4–5 in., moderate fragrance; foliage glossy; Dickson, A.

'DUNTON GOLD', HT, ob, 1966; flowers deep golden yellow-orange and pink tipped, 6 in., exhibition form, moderate fragrance; foliage glossy; vigorous growth; [Tzigane sport]; Dunton Nursery

Dunwich Rose, HSpn, w

'DUO', F, or, 1955; flowers coppery orange, open, petals fringed, medium, single, moderate fragrance; foliage dark; very vigorous growth; [Peace X Seedling]; Gaujard

'DUPLEX', Misc OGR, mp; (R. pomifera duplex, Wolley-Dod's Rose); flowers clear pink, semi-dbl.; heavy prickles; foliage downy, gray-green; [Chance garden hybrid of R. pomifera X Unidentified garden rose]; Vibert, prior to 1838

'DUPONTII', Misc OGR, w, 1817; (R. freudiana, R. X dupontii); (28); [Perhaps descended from R. gallica X R. moschata hybrid]

'DUPUY JAMAIN', HP, mr, 1868; flowers cerise-red, well-formed, dbl., 30 petals, moderate fragrance; vigorous growth; Jamain, H.

'DUQUESA DE PEÑARANDA', HT, ob, 1931; (Morning Blush); bud pointed; flowers shades of orange, large, dbl., 35 petals, cupped, moderate fragrance; foliage rich green, glossy; vigorous growth; GM, Portland, 1933; [Souv. de Claudius Pernet X Rosella (P. Dot)]; Dot, Pedro; C-P

'DUQUESA DE PEÑARANDA, CLIMBING', Cl HT, ob, 1940; Germain's

Dura, S, w

Durban July, *see* 'KORDURBAN'

Durbankor, *see* 'KORDURBAN'

Durgapur Delight, HT, lp, 1980; Gupta

'DURHAM PILLAR', HMult, mr, 1958; bud globular; flowers rose-red, small, borne in clusters, single, cupped, free, recurrent boom, slight fragrance; foliage dark, leathery, glossy; moderate climbing or trailing growth; [Chevy Chase X Seedling]; Risley

'DURHAM PRINCE BISHOPS', HT, op, 1990; flowers orange, flushed pink, medium, dbl., 26–40 petals, moderate fragrance; foliage large, dark green, glossy; bushy growth; [Silver Jubilee X Doris Tysterman]; Thompson, Robert; Battersby Roses, 1989

Dusky Dancer, *see* FRYtrooper

'DUSKY MAIDEN', F, dr, 1947; flowers deep crimson scarlet, blooms in trusses, single, 3 in., moderate fragrance; foliage dark; vigorous growth; GM, NRS, 1948; [(Daily Mail Scented Rose X Étoile de Hollande) X Else Poulsen]; LeGrice

'DUSKY RED', HT, dr, 1972; flowers medium red, veined, large, dbl., exhibition form, intermittent bloom, intense fragrance; foliage dark, leathery; vigorous, upright, bushy growth; [Karl Herbst X Big Red]; Wyant

'DÜSTERLOHE', Ayr, dp, 1931; flowers rose-red, single, 3 in., non-recurrent bloom, slight fragrance; vigorous, climbing growth; [Venusta Pendula X Miss C. E. van Rossem]; Kordes

'DUSTY PINK', F, lp, 1961; bud ovoid, long, pointed; dbl., 50–60 petals, 3–3.5 in., exhibition form, moderate fragrance; vigorous, upright growth; [Garnette X Garnette seedling]; Jelly; E.G. Hill Co.

'DUSTY ROSE', Min, m, 1974; flowers reddish purple, small, dbl., 40–50 petals, 1.5 in., exhibition form, moderate, spicy fragrance; foliage dark; upright growth; [Amy Vanderbilt X Cécile Brunner]; Morey, Dr Dennison; Pixie Treasures Min. Roses

Dutch Fork China, Ch, rb; (found rose)

'DUTCH GOLD'®, HT, my, 1978; flowers golden yellow, dbl., 32–34 petals, 6 in., moderate fragrance; foliage glossy, dark; vigorous growth; [Peer Gynt X Whisky Mac]; Wisbech Plant Co.

'DUTCH HEDGE', HRg, lp, 1958; flowers small, single, orange-red fruit; compact, upright growth; [(R. rugosa rubra X R. cinnamomea) X R. nitida]; Nyveldt

'DUTCH MISS', Min, pb, 1986; flowers light pink veining to white base, fading slightly, medium, dbl., 22 petals, exhibition form, borne singly, slight fragrance; straight, pointed, medium, tan prickles; foliage medium, dark green, glossy; upright, medium growth; [Summer Spice X Seedling]; Bridges, Dennis A., 1987

'DUTCH PROVENCE', C, mp; flowers rose, large, resembling 'cabbage rose', but blooms are larger, globular

Dwarf Fairy, *see* 'KORFEE'

Dwarf King, *see* **'DWARFKING'**

Dwarf Pavement, S, lp; (Rosazwerg)

Dwarf Queen, *see* **'QUEEN OF THE DWARFS'**

Dwarf Queen '82, *see* **'KORWERK'**

'DWARFKING', Min, mr, 1957; (Dwarf King, Zwergkönig); flowers carmine, small blooms borne singly and in clusters, dbl., 25 petals, cupped, slight fragrance; foliage glossy; compact (8–10 in.) growth; [World's Fair X Tom Thumb]; Kordes; J&P

'DYKNINI', Min, rb, 1995; (**Hanini**); flowers burgundy red stripes serrated by reddish pink, white throat, single, 5 petals, 2 in., moderate, fruity fragrance; foliage medium, medium green, semi-glossy; medium (2 ft), spreading (30 in.) growth; RR; [Sarabande X Hurdy Gurdy]; Dykstra, Dr A. Michael; Justice Miniature Roses, 1995

'DYKTICK', F, pb, 1999; (**Tickles**™); flowers pink and dream stripe, reverse same, semi-dbl., 8–14 petals, 2.5 in., borne in small clusters, slight fragrance; few prickles; [(Pristine X Typhoo Tea) X Hurdy Gurdy]; Dykstra, Dr A. Michael; Certified Roses, 2000

Dyna, *see* **'BINGO'**

Dynamite™, *see* 'JACSAT'

Dynastie, S, 1993; Meilland, Alain A.; (Cavriglia)

Dynasty™, *see* 'JACYO'

Dynasty, HT, rb, 1992; Keisel

'DZAMBUL', HP, mr, 1938; flowers medium, dbl., moderate fragrance; Kosteckij; (Sangerhausen)

E

'E. E. SASKAVÁ', HT, dy, 1933; flowers large, dbl.; Böhm, J.

'E. G. HILL', HT, mr, 1929; bud ovoid; flowers dazzling scarlet, well formed, very large, dbl., intense, damask fragrance; vigorous growth; Hill, E.G., Co.

'E. G. HILL, CLIMBING', Cl HT, mr, 1942; Marlin

'E. GODFREY BROWN', HT, dr, 1919; flowers deep reddish crimson, dbl., moderate fragrance; Dickson, H.

'E. I. FARRINGTON', F, mr, 1953; flowers cardinal-red to blood-red, turning almost crimson, dbl., 50–60 petals, 3.5–4 in., exhibition form, moderate fragrance; vigorous, spreading growth; [Queen o' the Lakes X Seedling]; Brownell, H.C.

'E. J. BALDWIN', HT, my, 1952; flowers rich golden yellow, high pointed, well formed, large, dbl., 30–40 petals; foliage dark; vigorous, upright, branching growth; [Phyllis Gold X Seedling]; Robinson, H.; Baker's Nursery

'E. J. LUDDING', HT, pb, 1931; flowers carmine-pink shaded coral-red and salmon, open, large, dbl., slight fragrance; bushy growth; [Ophelia X Hill's America]; Van Rossem; C-P, Prior

'E. J. MOLLER', HT, dr, 1924; flowers intense red, deepening toward black, dbl.; [George Dickson seedling]; Moller

'E. P. H. KINGMA', HT, ab, 1919; flowers apricot and orange-yellow, dbl.; [Mme Edouard Herriot X Duchess of Wellington]; Verschuren

'E. PEMBERTON BARNES', HT, pb, 1928; flowers light pink, shaded cerise, Pemberton

'E. V. LUCAS', HT, dr, 1934; flowers dark velvety crimson, large, semi-dbl., borne in sprays, slight fragrance; foliage dark; vigorous, upright, branching growth; McGredy

'E. VEYRAT HERMANOS', Cl T, pb, 1895; (E. Veyrath Hermanos, Pillar of Gold); flowers apricot and carmine-pink, reflexes violet-rose, dbl., intense fragrance; vigorous growth; Bernaix, A.

E. Veyrath Hermanos, *see* **'E. VEYRAT HERMANOS'**

'E. Y. TEAS', HP, mp, 1874; (Mons. E.Y. Teas); flowers bright red, large, globular, intense fragrance; [Alfred Colomb seedling]; Verdier, E.

'E.N. WARD', F, lp; Kershaw, 1919

'EADS', F, pb, 1991; flowers pink and yellow, blooms borne in small clusters, semi-dbl., slight fragrance; foliage medium, medium green, semi-glossy; low, compact growth; [(Seedling X Pinocchio) X Seedling]; Burks, Larry; Specialty Roses, 1991

Eagle®, *see* 'HAVEAL'

Eagle Wings®, *see* 'LENSIM'

'EARL BEATTY', HT, dr, 1923; flowers deep crimson, large, dbl., cupped; bushy growth; [Hoosier Beauty X George Dickson]; Chaplin Bros.

'EARL GODARD BENTINCK', HT, mr, 1931; flowers red, base orange, large, dbl., slight fragrance; very vigorous growth; [Pharisaer X Covent Garden]; Buisman, G. A. H.

'EARL HAIG', HT, mr, 1921; flowers brick-red, large, dbl., moderate fragrance; few thorns; vigorous, bushy growth; GM, NRS, 1920; Dickson, A.

'EARL OF DUFFERIN', HP, dr, 1887; flowers velvety crimson, shaded chestnut-red, very large, dbl., 53 petals, globular, moderate fragrance; vigorous growth; Dickson, A.

'EARL OF ELDON', N, ob, 1872; flowers coppery orange; Eldon-Coppin

'EARL OF GOSFORD', HT, dr, 1912; flowers large, dbl., intense fragrance; McGredy

'EARL OF PEMBROKE', HP, mr, 1882; flowers carmine-red, large, dbl., moderate fragrance; Bennett

'EARL OF WARWICK', HT, lp, 1904; flowers pale pinkish buff, reverse livid pink, large, dbl., moderate fragrance; [Souv. de S.A. Prince X Mrs W.J. Grant]; Paul, W.

'EARLDOMENSIS', S, my, 1934; flowers bright yellow; height 6 ft; [R. hugonis X R. sericea pteracantha]; Page

'EARLY BIRD', F, mp, 1965; flowers rose-opal, well formed, dbl., 4 in., moderate fragrance; free growth; [Circus X Fritz Thiedemann]; Dickson, Patrick; A. Dickson

'**Early Blush**', F, lp; [R. rugosa X Unknown]; Wheen, G., 1992; Gretchen's

'EARLY MIST', Min, w, 1971; flowers cream, dbl., 25 petals, 2 in., globular, slight fragrance; foliage dull, light; vigorous growth; de Yssel, Van; Warmerdam

'EARLY MORN', HT, lp, 1944; flowers shell-pink, large, dbl., exhibition form, moderate fragrance; foliage glossy; long stems; vigorous growth; [(Dr W. Van Fleet X Général Jacqueminot) X Break o' Day]; Brownell, H.C.

Early Peace, *see* **'MOLODOST MIRA'**

'**Early Red**', HSpn, rb; flowers carmine-red with gold, medium-large, semi-dbl.; [R. pimpinellifolia X Claudius Denoyel]

'EARTH SONG', Gr, dp, 1975; bud long, pointed to urn-shaped; flowers tyrian red to tyrian rose, blooms, dbl., 25–30 petals, 4–4.5 in., cupped, moderate fragrance; foliage glossy, dark, leathery; upright, bushy growth; [Music Maker X Prairie Star]; Buck, Dr Griffith J.; Iowa State University

Earthquake™, *see* 'MORQUAKE'

Earthquake, Climbing, *see* 'MORSHOOK'

'EASLEA'S GOLDEN RAMBLER', LCl, yb, 1932; (Golden Rambler); flowers rich buff-yellow marked crimson, blooms in clusters, dbl., 35 petals, 4 in., non-recurrent, moderate fragrance; foliage leathery, rich olive-green; vigorous, climbing growth; GM, NRS, 1932; Easlea; Totty

'EAST ANGLIA', HT, mp, 1939; flowers aurora-pink; [Golden Dawn sport]; Morse

'**East Europe**', F, lp; Urban, J.

Easter Bonnet, *see* 'BURSTEIN'

Easter Bunny, *see* 'RENBUN'

Easter Morn, *see* **'EASTER MORNING'**

'EASTER MORNING', Min, w, 1960; (Easter Morn); bud pointed; flowers ivory-white, small, dbl., 60–70 petals, 1.5 in.; foliage leathery, glossy; vigorous, dwarf (12–16 in.) growth; [Golden Glow (Brownell) X Zee]; Moore, Ralph S.; Sequoia Nursery

'EASTER PARADE', F, yb, 1951; bud ovoid, golden yellow; flowers salmon-pink and cerise, reverse yellow, becoming light carmine, dbl., 50–55 petals, 2.5–3.5 in., slight fragrance; foliage dark, glossy, bronze; vigorous, bushy growth; [Sunshine X Herrenhausen]; Whisler; Germain's

'EASTER ROSE', (Rubus coronarius, Rubus rosaefolius coronarius)

Easy, *see* 'ZIPEASY'

Easy Cover, *see* POUleas

Easy Going™, *see* 'HARFLOW'

'EASY ORANGE', F, ob, 1995; flowers bright orange, full (26–40 petals), medium blooms borne in small clusters, fragrant; numerous prickles; foliage medium, dark green, glossy; upright, medium growth; [(Valerie Jeanne X Eyepaint) X Laureate]; Jobson, Daniel, J.; Jobson, 1995

'**Easy to Cut**', HMsk, 1997; Lens, Louis

Easy Vibes, *see* 'GELEASY'

'EBB TIDE', HT, ab, 1961; bud long, pointed; flowers light yellowish pink, dbl., 28 petals, 5 in., exhibition form, slight fragrance; foliage glossy; vigorous, upright, compact growth; [(Sutter's Gold X Seedling) X Peace]; Von Abrams; Peterson & Dering

'EBERHARD JUNG', Pol, mr, 1930; flowers small, semi-dbl., Schmitt, L.

Eberwein, *see* **'DOPEY'**

'ÉBLOUISSANT', Pol, dr, 1918; flowers dazzling deep red, very dbl., globular, blooms in clusters, slight fragrance; foliage bronze, glossy; bushy growth; [Seedling X Cramoisi Superieur]; Turbat

'EBONY', Gr, dr, 1960; bud ovoid; flowers velvety dark red, dbl., 20–30 petals, 3–4 in., exhibition form, slight fragrance; foliage glossy; vigorous, upright growth; [Carrousel X Charles Mallerin]; Von Abrams; Peterson & Dering

'ÉBORACUM', LCl, w, 1977; flowers creamy white, base yellow, full, small, dbl., 30–35 petals, 2.5–3 in., repeat bloom, moderate fragrance; foliage glossy; strong growth; [Casino X Ice White]; Powell

'ÉCARLATE', HT, mr, 1907; flowers brilliant scarlet, somewhat like Gruss Antepliz, open, small, semi-dbl., no fragrance; foliage rich green, glossy; vigorous, bushy growth; [Camoens X Seedling]; Boytard

'Echizo', HT; Camprubi, C.

'ECHO', HMult, pb, 1914; (Baby Tausendschön); flowers varying (like Tausendschon) from dark pink to almost white, semi-dbl., cupped; bushy growth; [Tausendschön sport]; Lambert, P.

'Echo', S, op, 1970; Lens

'ÉCLAIR', HP, dr, 1883; flowers very dark red shaded blackish, well-shaped, small, dbl., moderate fragrance; tall growth; [Général Jacqueminot X Seedling]; Lacharme, F.

'Eclaireur', F, 1971; Godin, M.

'ÉCLAIREUR', HP, dr, 1895; flowers large, dbl., moderate fragrance; Vigneron

'ECLIPSE', HT, ly, 1935; bud remarkably long, pointed, deep gold, with long, narrow; flowers golden yellow, loose blooms, dbl., 28 petals, moderate fragrance; foliage leathery, dark; vigorous, bushy growth; David Fuerstenberg Prize, ARS, 1938 GM, Bagatelle, 1936 GM, Portland, 1935 GM, Rome, 1935; [Joanna Hill X Federico Casas]; Nicolas; J&P

Eclipse, *see* 'SAUdora'

'Eclipse, Climbing', Cl HT, ly

Ecole d'Ecully, *see* 'LAPbu'

'ECSTASY', HT, yb, 1935; flowers pale yellow shaded bronze and cerise, dbl., intense fragrance; erect, branching growth; Dickson, A.

Ecstasy, *see* 'KORazerka'

'ED STEER', HT, mr, 1999; dbl., 26–40 petals, 5 in., borne mostly singly, slight fragrance; prickles moderate; foliage medium, dark green, semi-glossy; upright, medium (30 in.) growth; [(Royal William X Gabi) X Adrienne Berman]; Poole, Lionel

'EDDA', S, mr, 1969; flowers clear rose-red, open, medium, borne in clusters, semi-dbl., 16 petals, profuse, repeated bloom; foliage dark, glossy, leathery; vigorous growth; [Lichterloh X Scharlachglut]; Lundstad

'EDDIE'S ADVENT', HT, lp, 1938; flowers pale buff, tipped pink, fading almost white, large, dbl., exhibition form, slight fragrance; foliage leathery; vigorous growth; [Mrs Sam McGredy X Edith Krause]; Eddie

'EDDIE'S CREAM', F, w, 1956; flowers cream, large, borne in clusters, dbl., apricot fragrance; vigorous growth; [Golden Rapture X Lavender Pinocchio]; Eddie; Harkness

'EDDIE'S CRIMSON', HMoy (S), mr, 1956; flowers blood-red, semi-dbl., 4–5 in., non-recurrent; large, globular fruit; vigorous (9–10 ft) growth; [Donald Prior X R. moyesii hybrid]; Eddie

'EDDIE'S JEWEL', HMoy (S), mr, 1962; flowers fiery red, recurrent bloom; few prickles; bark red; vigorous (8–9 ft) growth; [Donald Prior X R. moyesii hybrid]; Eddie

'EDEL', HT, w, 1919; flowers ivory-white, passing to pure white, well formed, very large, dbl., moderate fragrance; vigorous growth; [Frau Karl Druschki X Niphetos]; McGredy

Edelweiss, *see* **'SNOWLINE'**®

Eden, *see* 'MEIviolin'

Eden Climber™, *see* 'MEIviolin'

'EDEN ELLEN', F, ab, 1984; flowers medium, dbl., 35 petals, slight fragrance; foliage medium, medium green, matt; bushy growth; [Seedling X Seedling]; Schneider, Peter

'EDEN ROSE'®, HT, dp, 1950; bud ovoid; flowers tyrian rose, dbl., 50–60 petals, 4.5 in., cupped, intense fragrance; foliage glossy, bright dark green; vigorous, upright growth; GM, NRS, 1950; [Peace X Signora]; Meilland, F.; C-P, 1953, URS, 1950

Eden Rose 88, *see* 'MEIviolin'

'EDEN ROSE, CLIMBING'®, Cl HT, dp, 1962; Meilland, Alain A.; URS

'EDEN SUNGOLD', F, my, 1976; flowers pure yellow, dbl., 30–35 petals, 3 in., slight fragrance; semi-dwarf growth; [Seedling X Seedling]; Herholdt, J.A.

'EDGAR ANDREU', LCl, mr, 1912; flowers bright blood-red, large, borne in clusters of 7–15, dbl.; foliage dark, glossy; [R. wichurana X Cramoisi Supérieur]; Barbier

'EDGAR BLANCHARD', HT, lp, 1912; flowers pink shading to white, large, dbl., Duron

'EDGAR M. BURNETT', HT, pb, 1914; flowers flesh-pink, center dark pink, dbl., moderate fragrance; GM, NRS, 1913; McGredy

'Edie Anne', HT, mp, 1997

'EDINA', HT, w, 1934; flowers white, occasionally flushed pink, well formed, intense fragrance; foliage bronze red passing to green; vigorous growth; Dobbie

'Edinger White HP', HP, w

'EDITH BELLENDEN', E, mp, 1895; flowers pale rose, single; foliage fragrant; vigorous growth; very hardy. Penzance

'EDITH CAVELL', HT, ly, 1918; flowers pale lemon-white, dbl.; Chaplin Bros.

Edith Cavell, *see* **'MISS EDITH CAVELL'**

'EDITH CLARK', HT, mr, 1928; flowers fiery red, dbl., globular, slight fragrance; foliage rich green; dwarf growth; [Mme Abel Chatenay X Seedling]; Clark, A.; Hackett

'EDITH DE MARTINELLI'®, F, op, 1958; flowers salmon-pink, well-formed; vigorous growth; GM, Geneva, 1958; [(Gruss an Teplitz X Independence) X Floradora]; Arles; Roses-France

Edith de Martinelli, Climbing®, *see* 'ORAdit'

'EDITH DE MURAT', B, w, 1858; Ducher

'EDITH DENNETT', F, op, 1973; flowers salmon-pink, rosette form, dbl., 24–30 petals, 2–3 in., slight fragrance; foliage glossy; free growth; Holmes, R.; Fryer's Nursery, Ltd.

'EDITH FELBERG', HT, w, 1931; flowers cream, center slightly darker, rather, dbl., cupped; foliage leathery; [Seedling X Souv. de H.A. Verschuren]; Felberg-Leclerc

'EDITH HAYWARD', Cl HT, lp, 1967; flowers pastel pink, large, dbl., exhibition form, recurrent bloom, moderate fragrance; foliage glossy, dark; vigorous growth; [Fontanelle X Gen. MacArthur]; Hayward

'EDITH HAZELRIGG', HT, ob, 1953; flowers orange-cerise, pointed, dbl., 25 petals, moderate fragrance; foliage dark; vigorous growth; Cant, F.

Edith Holden, *see* R. filipes 'Kiftsgate'

'EDITH KRAUSE', HT, w, 1930; flowers greenish white, large, dbl., 30 petals, exhibition form, moderate fragrance; very vigorous growth; [Mrs Charles Lamplough X Souv. de H.A. Verschuren]; Krause; J&P

'EDITH MARY MEE', HT, or, 1936; flowers vivid orient red, flushed orange, base yellow, dbl., moderate fragrance; foliage dark, leathery; vigorous, bushy, compact growth; Mee; Beckwith

'EDITH NELLIE PERKINS', HT, op, 1928; flowers salmon-pink, flushed orange, reverse orange-red, shaded orange, dbl., 35–40 petals, moderate fragrance; few prickles; vigorous, bushy growth; Dickson, A.

'EDITH NELLIE PERKINS, CLIMBING', Cl HT, op, 1936; H&S; Howard Rose Co.

'EDITH OLIVER', HT, lp, 1980; flowers soft light pink, large blooms in clusters of 4, dbl., 35 petals, moderate fragrance; brownish-red prickles; foliage medium, medium green, semi-glossy; bushy growth; [Pink Parfait X Seedling]; Singleton, C.H.

'**EDITH PART**', HT, pb, 1913; flowers rich red, suffused deep salmon and coppery yellow, moderate fragrance; McGredy

'**EDITH PIAF**', HT, m, 1964; bud ovoid; flowers purple-red, large, dbl., slight fragrance; foliage dark; [Poinsettia X (Baccará X Seedling)]; Verbeek

Edith Piaf, *see* 'MEInical'

'**EDITH ROBERTS**', HT, ab, 1969; 5 in., exhibition form, moderate fragrance; foliage glossy, dark; [Dorothy Peach sport]; Roberts, P.D.

'**EDITH SCHURR**', S, yb, 1976; bud globular, sulfur-yellow; flowers light yellow, center pink, dbl., 60 petals, 5 in., recurrent bloom, intense, damask fragrance; foliage glossy; spreading growth; [(Wendy Cussons X Gavotte) X Leverkusen]; Stanard; Edmunds Roses

'**Edith Southgate**', Cl Min, pb, 1997; flowers double, medium, dbl., 15–25 petals, borne mostly singly, moderate fragrance; some prickles; foliage medium, medium green, semi-glossy; upright, medium (5 ft) growth; [Laura Ford X Admiral Rodney]; Barker, S.J.L.

'**EDITH WILLKIE**', HT, pb, 1943; bud long, pointed; flowers livid pink, base lemon-chrome, dbl., 25–30 petals, 4.5–5 in., slight fragrance; foliage leathery, dark; vigorous, upright, compact growth; [Joanna Hill X R.M.S. Queen Mary]; Hill, Joseph H., Co., 1943; H&S, 1946, Wayside Gardens Co., 1946

'**EDITH YORKE**', HMult, lp, 1953; flowers light almond-pink, rosette form, borne in trusses, semi-dbl., intense fragrance; foliage light green; very vigorous growth; [Havering Rambler X Seedling]; Miller, A.I.; Jackman

'**EDITOR MCFARLAND**', HT, mp, 1931; flowers glowing pink, slightly suffused yellow, large, dbl., 30 petals, intense fragrance; vigorous, bushy growth; [Pharisaer X Lallita]; Mallerin, C.; C-P

'**EDITOR MCFARLAND, CLIMBING**', Cl HT, mp, 1948; Roseglen Nursery

'**EDITOR STEWART**', HT, mr, 1939; flowers deep cherry-red, open, large, semi-dbl.; foliage bronze; long stems; vigorous pillar or large bush growth; Clark, A.; NRS Victoria

'**EDITOR TOMMY CAIRNS**', HT, pb, 1991; flowers bright pink, light pink reverse, blooms borne mostly singly, dbl., 32–40 petals, slight; foliage medium, medium green, semi-glossy; upright, medium growth; [Seedling X Seedling]; Winchel, Joseph F., 1992

'**EDMÉ ET ROGER**', HT, 1903; Ketten Bros.

'**Edmond Charles-Roux**', HT, 1990; Dorieux, Francois

'**EDMOND DESCHAYES**', HT, ly, 1901; flowers yellowish-white, large, dbl.; Bernaix, A.

'**EDMOND PROUST**', LCl, mr, 1903; flowers pale rose and carmine, large, borne in clusters of 3–6, very dbl., sparse seasonal bloom, moderate fragrance; foliage glossy; height 5–8 ft; [R. wichurana X Souv. de Catherine Guillot]; Barbier

'**EDMUND M. MILLS**', HT, ob, 1927; flowers rosy flame, base deep gold, open, large, semi-dbl., intense fragrance; foliage dark, leathery; very vigorous, upright growth; [Red Radiance X Padre]; Hieatt

Edmund Rice, *see* 'WELPIN'

'**EDNA KAYE**', HT, lp, 1959; flowers light pink tinted buff, dbl., 56 petals, 5 in.; foliage bronze; free growth; [Directeur Guerin X Mirandy]; Kemp, M.L.

Edna Marie, *see* 'MORED'

'**EDNA WALLING**', HMult, w, 1940; Clark

Edna Wilson, *see* 'GRIFED'

'**EDNA-CHRIS**', F, w, 1977; bud pointed; flowers off-white, dbl., 30 petals, 3.5 in., slight fragrance; foliage matt, green; tall, upright growth; [Gene Boerner sport]; Ogden

'**EDNAH THOMAS**', Cl HT, op, 1931; (Bloomfield Improvement); flowers salmon-rose, large, dbl., recurrent bloom, moderate fragrance; strong stems; vigorous, climbing growth; [Seedling climber X Bloomfield Progress]; Thomas; H&S

'**EDO BERGSMA**', HT, ly, 1932; bud pointed; flowers bright flesh and peach, large, intense fragrance; [Capt. F.S. Harvey-Cant X Étoile de Hollande]; Buisman, G. A. H.

'**EDOUARD ANDRÉ**', HP, 1880; Verdier, E.

'**EDOUARD HERVE**', HP, 1884; Verdier, E.

'**EDOUARD MIGNOT**', HT, m, 1927; flowers purplish garnet-red, reverse amaranth, dbl., moderate fragrance; bushy growth; Sauvageot, H.; F. Gillot

'**EDOUARD RENARD**', HT, mr, 1933; flowers carmine, base yellow; long, stiff stems; Dot, Pedro

'**EDUARD SCHILL**', HT, or, 1931; flowers brick-red shaded nasturtium-yellow, very large, semi-dbl., cupped, slight fragrance; foliage glossy; vigorous growth; [Charles P. Kilham X Mev. G.A. van Rossem]; Kordes

'**EDUARDO TODA**', HT, my, 1947; bud pointed; flowers sunflower-yellow, slight fragrance; [Ophelia X Julien Potin]; Dot, Pedro

'**EDWARD BEHRENS**', HT, dr, 1921; flowers very dark velvety crimson, very large, dbl., moderate fragrance; [Richmond X Admiral Ward]; Kordes

Edward Colston, *see* 'SANCOL'

'**EDWARD MAWLEY**', HT, dr, 1911; bud almost black; flowers dark crimson, large, semi-dbl., 18 petals, exhibition form, moderate fragrance; bushy growth; GM, NRS, 1910; McGredy

'**EDWARD MORREN**', HP, mr, 1868; flowers deep cherry-rose, large, dbl., flat, vigorous growth; [Jules Margottin seedling]; Granger

'**EDWARD VII**', Pol, mp, 1911; flowers clear pink, small; [Mme Norbert Levavasseur sport]; Low

Edwardian Lady, *see* R. filipes 'Kiftsgate'

'**EDWIN MARKHAM**', HT, mp, 1923; flowers bright rose-pink suffused silvery, dbl.; [Ophelia X Hoosier Beauty]; Clarke Bros.

'**EDWIN T. MEREDITH**', HT, ob, 1979; bud ovoid, pointed; flowers coral-pink, dbl., 30 petals, 5 in., flat, slight fragrance; bushy, upright growth; [Futura X First Prize]; Warriner, William A.; J&P

'**EFFECTIVE**', LCl, mr, 1913; bud long; flowers crimson, cupped, moderate fragrance; very vigorous growth; [Gen. MacArthur seedling X Paul's Carmine Pillar]; Hobbies

'**EFFEKT**', HT, rb, 1935; flowers scarlet-red, reverse flushed golden yellow, large, dbl., cupped, slight fragrance; vigorous, bushy growth; [I Zingari X Seedling]; Krause

'**Effekt**', F, dr, 1975; Berger

'**EGALITÉ**', F, m, 1946; bud pointed; flowers pale lilac-rose, borne in trusses, dbl., 25 petals, 4 in., moderate fragrance; foliage bronze; vigorous, branching growth; [Irene X Seedling]; Leenders, M.; Longley

'**Egas Monitz**', HT; Moreira da Silva, A.

'**EGERIA**', HP, 1877; Schwartz, J.

'**EGESKOV**'®, F, mp, 1982; flowers bright medium pink, blooms in large clusters, dbl., 20 petals, cupped, slight fragrance; foliage medium, light green, glossy; bushy growth; [Tornado X Matangi]; Olesen, Pernille & Mogens N.; D.T. Poulsen

'**EGLANTINE**', Pol, dp, 1930; flowers carmine, center white, many yellow stamens, single, moderate fragrance; small vigorous, dwarf, bushy growth; [Amaury Fonseca X Rodhatte]; Soupert & Notting

Eglantine, *see* '**R. EGLANTERIA**'

Eglantyne, *see* 'AUSMAK'

Eglantyne Jebb, *see* 'AUSMAK'

Egoli, *see* 'KORameget'

Egon Schiele®, S, lp

'**EGYPTIAN TREASURE**', S, ob, 1973; flowers orange, semi-dbl., 16 petals, 4 in., slight fragrance; foliage dark, glossy; [(Coup de Foudre X S'Agaro) X Vagabonde]; Gandy, Douglas L.

'**EICHSFELDIA**', HMult, ly, 1925; flowers yellowish-white, small, single; Bruder Alfons

'**Eidth Piaf**', HT, dr, 1999; Meilland

'**EIFFEL TOWER**', HT, mp, 1963; (Eiffelturm, Tour Eiffel); bud long, urn-shaped; dbl., 35 petals, 3.5–5 in., exhibition form, intense fragrance; foliage leathery, semi-glossy; vigorous, upright growth; GM, Geneva, 1963 GM, Rome, 1963; [First Love X Seedling]; Armstrong, D.L. & Swim; Armstrong Nursery

'EIFFEL TOWER, CLIMBING', Cl HT, mp, 1967; Laveena Roses

Eiffelturm, *see* **'EIFFEL TOWER'**

'EIKO', HT, yb, 1978; bud pointed; flowers yellow and scarlet, pointed, dbl., 30–35 petals, 5–6. in., slight fragrance; foliage large, glossy, light green; vigorous growth; [(Peace X Charleston) X Kagayaki]; Suzuki, Seizo; Keisei Rose Nursery

'EILEEN BOXALL', HT, dr, 1948; flowers cerise, semi-dbl., 18 petals, 5 in., moderate fragrance; foliage dark; vigorous growth; [Betty Uprichard sport]; Boxall

'EILEEN DOROTHEA', HT, mr, 1931; bud pointed; flowers crimson-scarlet, edged darker, base yellow, dbl., exhibition form, intense fragrance; foliage deeply serrated; vigorous growth; Dickson, A.

'EILEEN LOUISE', HT, mp, 1985; [Admiral Rodney sport]; Brown, Harry G.L.S.

'EILEEN LOW', Pol, mp, 1911; flowers small, dbl.; Levavasseur

'EISENACH', HWich, mr, 1910; flowers bright red, borne in clusters, single; very vigorous growth; Kiese

'Eisprinzessin', HT, pb

'EKTA', G, mp, 1927; single, non-recurrent; [Alika X American Beauty]; Hansen, N.E.

'EL CAPITAN', Gr, mr, 1959; flowers cherry to rose-red, blooms in small clusters, dbl., 30 petals, 3.5–4.5 in., exhibition form, slight fragrance; foliage dark, glossy; vigorous, upright, bushy growth; GM, Portland, 1959; [Charlotte Armstrong X Floradora]; Swim, H.C.; Armstrong Nursery

'EL CAPITAN, CLIMBING', Cl Gr, mr, 1963; Armstrong, D.L.

'EL CATALÁ', Gr, rb, 1981; bud ovoid, pointed; flowers medium red, reverse light pink, slightly blooms borne 1–8 per cluster, dbl., 35 petals, cupped, slight fragrance; awl-like prickles; foliage large, glossy; erect, slightly bushy growth; [Wanderin' Wind X ((Dornroschen X Peace) X Brasilia)]; Buck, Dr Griffith J.; Iowa State University

'EL CID', HT, or, 1969; bud ovoid; flowers large, dbl., exhibition form; foliage soft; vigorous, upright, bushy growth; GM, Rome, 1969; [Fandango X Roundelay]; Armstrong, D.L.; Armstrong Nursery

'EL DORADO', HT, yb, 1972; flowers golden yellow, edged reddish, open, large, dbl., exhibition form, intense fragrance; foliage large, glossy, leathery; vigorous, upright growth; [Manitou X Summer Sunshine]; Armstrong, D.L.; Armstrong Nursery

'El Dorado', HT, my

'El Dorado Cemetary Short Moss', M, mp

'EL PASO', HT, yb, 1988; flowers light yellow on outer petals, deep yellow on inner petals, reverse light yellow, dbl., exhibition form, borne usually singly, slight, spicy fragrance; prickles straight, medium, greenish-brown; foliage medium, medium green, semi-glossy; bushy, medium, floriferous growth; Bronze, ARC TG, 1988; [First Prize X Arlene Francis]; Ohlson, John

El Toro, *see* **'UNCLE JOE'**

Elaina, *see* 'ZIPELA'

'ELAINE', HT, mp, 1950; flowers rose-pink, large, very dbl., exhibition form, moderate fragrance; [Mrs A.R. Barraclough X Lady Sylvia]; Robinson, H.; Baker's Nursery

'ELAINE', HT, ab, 1951; bud pointed; dbl., 35–45 petals, 4.5 in., exhibition form, slight fragrance; foliage leathery; very vigorous, upright growth; [Eclipse X R.M.S. Queen Mary]; Boerner; J&P

'Elaine Frawley', F, lp; [Cornelia X Unknown]; Weatherly, Lila, 1996; Prophyl Pty, Ltd.

'ELAINE HOLMAN', HT, pb, 1975; bud long, pointed; dbl., 50 petals, 3.5–4 in., exhibition form, moderate fragrance; foliage soft; vigorous growth; [Red Devil X Avon]; Watson

'ELAINE STUART', HT, w, 1932; flowers cream, center yellow; long stems; [Antoine Rivoire X Lillian Moore]; Edwards

'ELAINE WHITE', F, w, 1959; bud pointed; flowers cream, white and pink, open, small, borne in clusters, semi-dbl., moderate fragrance; foliage leathery; very vigorous, upright growth; [Gartendirektor Otto Linne X Seedling]; Riethmuller

'ELAINE'S CHOICE', F, pb, 1972; flowers pale pink-apricot, yellow blend, full, dbl., 40 petals, 5 in., intense fragrance; foliage semi-glossy; vigorous growth; [Orange Sensation X Peace]; Ellick

'Élan', F, 1960; Croix, P.

'ELATION', HT, dy, 1974; bud long, pointed; flowers large, dbl., slight fragrance; foliage large, glossy, dark, leathery; vigorous, upright growth; [Buccaneer X Seedling]; Warriner, William A.; Spek

'ELBA', HT, dr, 1963; flowers deep red, reverse gold, [Confidence X Crimson Glory]; da Silva, Moreira

'Elbefreude', F, mr, 1983; flowers vermilion, large, dbl.; Schmadlak, Dr.

'Elbeglut', Pol, dr, 1978; flowers large, semi-dbl.; Schmadlak, Dr.

'Elbegold', Pol, dy, 1973; flowers golden yellow, medium, dbl.; Schmadlak, Dr.

'ELDORA HARVEY', HT, pb, 1930; flowers pink, center tinted lavender, reverse dark pink, dbl., moderate fragrance; [Red Radiance X Maman Cochet]; Harvey

'ELDORADO', HT, ob, 1923; flowers copper, suffused orange and salmon, very large, dbl., moderate fragrance; bushy growth; [Seedling X Mme Edouard Herriot]; H&S

Eldorado, *see* 'JACspri'

'ELEANOR', Min, op, 1960; bud long, pointed; flowers coral-pink, aging darker, dbl., 20–30 petals, 1 in.; foliage leathery, glossy; upright, bushy (12 in.), growth; [(R. wichurana X Floradora) X (Seedling X Zee)]; Moore, Ralph S.; Sequoia Nursery

'ELEANOR FRANCES', HT, dy, 1971; flowers deep aureolin-yellow, full, dbl., 28 petals, 5 in., moderate fragrance; foliage glossy, leathery; vigorous, tall, upright growth; [Vienna Charm X Seedling]; Green Acres Rose Nursery

'ELEANOR HENNING', HT, op, 1920; bud pointed; flowers salmon-pink, Easlea

'ELEANOR PERENYI', Gr, yb, 1985; flowers yellow flushed apricot, reverse yellow flushed salmon, loose, dbl., 25 petals, moderate, fruity fragrance; small, globular, orange-red fruit; very few, hooked, small, red prickles; foliage medium, dark, semi-glossy; medium, upright growth; [America X Sunsong]; French, Richard

'ELECTRA', HT, my, 1970; flowers open, large, dbl., slight fragrance; foliage large, glossy, leathery; vigorous, upright growth; [Eclipse X Seedling]; Boerner; J&P

'ELECTRA', HMult, ly, 1900; flowers yellow, fading white, dbl., globular, slight fragrance; foliage rich green, glossy; very vigorous, climbing growth; [R. multiflora X William Allen Richardson]; Veitch

'ELECTRON'®, HT, dp, 1970; (Mullard Jubilee); flowers shocking pink, classic form, dbl., 32–40 petals, 5 in., exhibition form, borne mostly singly, intense fragrance; foliage large, dark green, glossy; fully branched; upright, medium, bushy growth; AARS, 1973 GM, Belfast, 1972 GM, Portland, 1973 GM, RNRS, 1969 GM, The Hague, 1970; [Paddy McGredy X Prima Ballerina]; McGredy, Sam IV; McGredy

'ELEGANCE', LCl, my, 1937; flowers yellow, fading white at edges, blooms, dbl., 48 petals, 6 in., moderate fragrance; foliage dark, glossy; vigorous growth; [Glenn Dale X (Mary Wallace X Miss Lolita Armour)]; Brownell, H.C.

Elegance, *see* 'MEIbicmarj'

'Elegance', Min, m

'ELÉGANCE', HT, pb, 1955; bud globular; flowers rose-copper, large, dbl., moderate fragrance; foliage dark, leathery; moderate growth; Frères, Buyl

'ELEGANS', Ayr, w; (Double White); semi-dbl., borne in large clusters; very vigorous growth

'Elegant', G, mp

Elegant Beauty, *see* 'KORGATUM'

Elegant Design, *see* 'MORDARCREST'

Elegant Pearl®, *see* 'INTERGANT'

'Elegant touch', HT, lp

'ELEGANTA', Min, mr

'ÉLÉGANTE', HT, ly, 1918; bud pointed; flowers creamy yellow, large, dbl., globular; branching growth; Pernet-Ducher

'ÉLÉGANTE, CLIMBING', LCl, w; bud long, slender; flowers white, center creamy yellow; foliage dull; LeGrice Roses

'ELEGHYA', Gr, dr, 1975; flowers deep blackish-red, blooms in clusters of 2–5, very dbl., 80 petals, cupped; foliage dark; bushy growth; [Spectacular X Seedling]; Staikov, Prof. Dr V.; Kalaydjiev and Chorbadjiiski

Elegy, *see* 'MEILUCRE'

'Elektra', HT, ly, 1964; flowers yellowish-white, large, dbl., moderate fragrance; Rupprecht-Radke

'ELENA CASTELLO', HT, pb, 1932; flowers apricot-yellow to rose, semi-dbl.; vigorous growth; [Mme Butterfly X Angèle Pernet]; Munné, B.

Eleonora®, *see* 'BARleo'

'Eleonore', F, 1973; Noack, Werner

'ELEONORE BERKELEY', HMult, m, 1900

'ELETA', HT, mp, 1934; bud pointed; flowers clear rose-pink, large, dbl., 65 petals, exhibition form; vigorous growth; [Sensation X Seedling]; Dahlgren; Kemble-Smith Co.

'ELETTRA', HT, op, 1940; flowers copper-pink suffused reddish yellow, large, dbl., exhibition form, moderate fragrance; foliage leathery; very vigorous growth; [Julien Potin X Sensation]; Aicardi, D.; Giacomasso

'Eleusine', HT, m

'Elf', Min, dr

Elfe®, *see* 'TANELFE'

'ELFE', F, w, 1951; bud pointed; flowers white tinted rose, shell shaped, large, borne in clusters, intense fragrance; foliage glossy, dark; vigorous, upright, bushy growth; RULED EXTINCT 4/85; [Swantje X Hamburg]; Tantau

Elfe, *see* 'TANcreif'

'ELFE SUPREME', F, mp; flowers rose-pink; [Rosenelfe sport]

'ELFENREIGEN', HMacrantha(S), dp, 1939; flowers deep rose-pink, center brighter, petals shell shaped large, single, profuse, non-recurrent bloom; foliage reddish orange, later gray-green; very vigorous (5 ft) growth; [Daisy Hill X Seedling]; Krause

'ELFIE'E JOY', HT, mp, 1995; flowers medium hot pink, darker edges, lighter reverse, full (26–40 petals), medium (5″) blooms borne mostly singly; fragrant; numerous prickles; foliage large, medium green, semi-glossy; upright, medium growth; [Sheer Elegance sport]; Dry, Elfie; Consolidated Nurseries, Inc.

Elfin, *see* 'POULFI'

'ELFIN', F, or, 1939; flowers cherry-rose shaded orange-salmon, borne in clusters, dbl., 4.5 in.; low growing growth; RULED EXTINCT 12/85; Archer

'ELFIN CHARM', Min, pb, 1974; bud short, pointed; flowers phlox-pink, dbl., 65 petals, 1 in., moderate fragrance; foliage small, glossy, leathery; bushy, compact growth; [(R. wichurana X Floradora) X Fiesta Gold]; Moore, Ralph S.; Sequoia Nursery

'ELFINESQUE', Min, op, 1974; bud pointed; flowers coral-orange to bright pink, small, semi-dbl., slight fragrance; foliage small, glossy, leathery; vigorous, dwarf, upright, bushy growth; [Little Darling seedling X Yellow Bantam]; Morey, Dr Dennison; Pixie Treasures Min. Roses, 1973

'ELFINGLO', Min, m, 1977; bud ovoid; flowers red-purple, micro-mini, dbl., 33 petals, .5 in., cupped, moderate fragrance; foliage small, glossy; compact growth; [Little Chief X Little Chief]; Williams, Ernest D.; Mini-Roses

Elgin Festival, *see* 'AUSpoly'

'ELIANE', HT, ab, 1954; flowers bright salmon, well formed, large, moderate fragrance; foliage dark; vigorous growth; [Mme Joseph Perraud X R. foetida bicolor hybrid]; Gaujard; Gandy Roses, Ltd., 1959

Eliane Gillet, *see* 'MASelgi'

'ELIDA', HT, or, 1966; flowers vermilion, large, dbl., 30 petals, exhibition form, moderate fragrance; foliage dark, glossy; vigorous, branching growth; Tantau, Math.

'ELIE BEAUVILLAIN', Cl T, mp, 1887; flowers coppery pink, large, dbl.; vigorous growth; [Gloire de Dijon X Ophirie]; Beauvillain

Elina®, *see* 'DICJANA'

'ELISA'®, HT, mp, 1981; flowers large, dbl., 40 petals, cupped, no fragrance; foliage large, light green, matt; upright growth; [Seedling X Blessings]; Rose Barni-Pistoia

'ELISA BOËLLE', HP, w, 1869; (Elise Boelle); flowers white, tinted with rose; vigorous growth; Guillot Père

'ELISA ROBICHON', HWich, mp, 1901; flowers salmon-pink, fading pinkish buff, open, large, borne in clusters, semi-dbl., free seasonal bloom, slight fragrance; short, strong stems; vigorous, climbing (10 ft), or trailing growth; [R. wichurana X L'Ideal]; Barbier

'ELISABETH', HMult, lp, 1926; flowers light, creamy pink, small, dbl., moderate fragrance; Bruder Alfons

'ELISABETH DIDDEN', HT, mr, 1918; flowers glowing carmine-red and scarlet, semi-dbl., moderate fragrance; [Mme Caroline Testout X Gen. MacArthur]; Leenders, M.

'ELISABETH FAURAX', HT, w, 1937; bud pointed; flowers white, lightly shaded ivory, large, very dbl.; long stems; upright growth; [Caledonia X Mme Jules Bouche]; Meilland, F.

'ELISABETH FAURAX, CLIMBING', Cl HT, 1937; Meilland, F.

'Elisabeth Tschudin', F, mp, 1981; Hetzel

'ELISE', HT, w, 1969; [Prima Ballerina sport]; Edmunds

Elise Boelle, *see* **'ELISA BOËLLE'**

'Elise Flory', Ch, mr

'Elise Lemaire', HP; Roseraie de l'Hay

'ELISE NOELLE', LCl, mr, 1991; dbl., 15–25 petals, 1.5–2.75 in., borne in small clusters, slight fragrance; foliage medium, medium green, semi-glossy; climbing (300 cms) growth; [Dublin Bay X Burgund]; Alde, Robert O.

'ELISE ROVELLA', G, mp; flowers rosy pink, medium, dbl.; tall growth; Roseraie de l'Hay, before 1842

'ELISE SAUVAGE', T, yb; flowers orange to yellow, sometimes yellow to white, Miellez, 1818

'ELISKA KRÁSNOHORSKÁ', HP, mp, 1932; flowers brilliant pink, large, semi-dbl., exhibition form, moderate fragrance; foliage soft, bronze; bushy growth; [Capt. Hayward X Una Wallace]; Böhm, J.

'ELITE', HT, yb, 1936; bud pointed, red; flowers salmon-pink and yellow blend, very large, dbl., exhibition form; foliage leathery, light; upright growth; [Charles P. Kilham X Pres. Herbert Hoover]; Tantau; J&P, 1941

'ELIZA', S, mp, 1961; flowers clear pale rose, borne in clusters, non-recurrent; foliage dark, glossy; bushy, erect (3 ft) growth; Skinner

Eliza, *see* 'KORlis'

'ELIZA BALCOMBE', HP, pb, 1842; flowers pale flesh; foliage gray-green; Laffay, M.

'Eliza Dorothy', HT, my, 1984; Allender, Robert William

'Eliza Wren', Pol, lp, 1997; Williamson

'ELIZABETH', Pol, mp, 1937; flowers rich salmon, borne in large clusters, semi-dbl.; Letts

Elizabeth Abler, *see* 'TINABLER'

Elizabeth Ann, *see* 'KIRBRONZE'

'ELIZABETH ARDEN', HT, w, 1929; flowers pure white, dbl., intense fragrance; GM, NRS, 1929; [Edith Part X Mrs Herbert Stevens]; Prince

'ELIZABETH BROW', M, 1883; Stamler

'ELIZABETH CONE', HT, lp, 1954; flowers flesh-pink, well shaped; vigorous, low growth; [Picture sport]; Cone; Roger

'ELIZABETH CULLEN', HT, mr, 1921; flowers rich scarlet-crimson, semi-dbl., intense fragrance; GM, NRS, 1917; Dickson, A.

'Elizabeth Fankhauser', HT, op, 1965

'ELIZABETH HAMLIN', LCl, pb, 1987; flowers blush pink, fading to white, small, single, 5 petals, cupped, borne in mass clusters, slight fragrance; few prickles; foliage pennate, 5–7 leaflets, medium; rampant growth; [Seedling X Seedling]; Nobbs, Kenneth J., 1986

Elizabeth Harbour, *see* 'HABONE'

'ELIZABETH HARKNESS'®, HT, ly, 1969; flowers off-white to creamy buff, often with pastel yellow and pink, dbl., 28 petals, moderate fragrance; foliage dark; upright, bushy growth; [Red Dandy X Piccadilly]; Harkness

'ELIZABETH HARKNESS, CLIMBING'®, Cl HT, w, 1975; Harkness

'Elizabeth Harwood', LCl, lp

'ELIZABETH HASSEFRAS', Pol, mp, 1951; flowers buttercup form, glistening rose-pink with many stamens; B&A

Elizabeth Heather Grierson, *see* 'MATTNOT'

'ELIZABETH LEE', HT, dr, 1935; flowers dark velvety red, well shaped, moderate fragrance; Chaplin Bros.

Elizabeth Munn, *see* 'SEABET'

Elizabeth of Glamis®, *see* 'MACEL'

'ELIZABETH OF YORK', HT, dp, 1928; bud large, pointed, cerise; flowers cerise-pink, dbl., 27 petals, exhibition form, intense fragrance; few thorns; foliage dark, glossy; Dobbie

'ELIZABETH PHILP', F, ab, 1976; flowers creamy peach; [Liverpool Echo sport]; Philp, J.B. & Son

'ELIZABETH ROWE', M, dp; bud large, well mossed, deep pink.

Elizabeth Scholtz™, *see* 'WILSCSO'

'ELIZABETH TAYLOR', HT, dp, 1985; flowers deep pink with smoky edges, dbl., 30–35 petals, 4.5–5 in., exhibition form, borne usually singly, moderate, spicy fragrance; foliage large, dark, semi-glossy; long stems; upright growth; PP6492; [First Prize X Swarthmore]; Weddle, Von C.

'ELIZABETH W. ADAM', HT, dp, 1926; flowers pink veined crimson, base yellow, dbl., 50 petals; Adam & Craigmile

'ELIZABETH ZEIGLER', HWich, dp, 1917; flowers deep rose-pink; [Dorothy Perkins sport]; Pierson, A.N.

'ELKA GAARLANDT', Pol, dp, 1966; bud ovoid; flowers medium, borne in large clusters, dbl.; foliage dark; Gold Star of the South Pacific, Palmerston North, NZ, 1970; [Hobby X Kathleen Ferrier]; Buisman, G. A. H.

'ELLA GUTHRIE', HT, mp, 1937; flowers large, dbl., intense fragrance; vigorous growth; [Premier X Seedling]; Clark, A.; NRS Victoria

'ELLA MCCLATCHY', HMult, mp, 1926; flowers rose, borne in clusters, single, sometimes recurrent bloom; thornless; Diener

'ELLA SCOTT', HWich, dp, 1925; flowers deep rose-pink, approaching red, borne in clusters of 15–20, dbl.; few thorns; vigorous, climbing growth; [Orléans Rose X Chance seedling]; Scott, G.J.; Brundrett (NRS Victoria)

Ellamae™, *see* 'SAVAMAE'

'ELLE', S, dp, 1980; bud pointed; flowers deep pink, borne 3–5 per cluster, semi-dbl., 16 petals, non-recurrent, intense fragrance; curved gray prickles; foliage light green, 5–7 leaflets; vigorous, upright dense growth; [Schneezwerg X Splendens]; Lundstad, Arne; Agricultural University of Norway

Elle, *see* **'NOBILITY'**

Ellen®, *see* 'AUScup'

'ELLEN', HT, dp, 1929; bud pointed; flowers unvarying dark pink, large, very dbl., intense fragrance; bushy growth; RULED EXTINCT 12/85; [Premier sport]; Hinner, P.; Gould

'ELLEN GRIFFIN', Min, lp, 1988; flowers medium, dbl., 15–25 petals, slight fragrance; foliage medium, medium green, semi-glossy; upright, bushy growth; [Uwe Seeler X Party Girl]; Hefner, John; Kimbrew Walter Roses, 1988

Ellen Hamlyn, *see* 'DORWEST'

'ELLEN MARY', HT, dr, 1963; flowers well-formed, dbl., 34 petals, 5 in., moderate fragrance; vigorous, upright growth; [Wellworth X Independence]; LeGrice

'ELLEN POULSEN', Pol, mp, 1911; flowers bright cherry-pink, large blooms in clusters, dbl., recurrent bloom, moderate fragrance; foliage glossy, dark; bushy growth; (14); [Mme Norbert Levavasseur X Dorothy Perkins]; Poulsen, D.T.; Teschendorff

'ELLEN POULSEN LYSROSA', Pol, op, 1938; flowers light salmon-pink, medium, dbl., moderate fragrance; Poulsen

'ELLEN POULSEN MORK', Pol, dr, 1928; [Ellen Poulsen sport]; Poulsen, S.

'ELLEN TERRY', HT, ly, 1925; flowers soft sulfur-cream, outer petals soft peach, well shaped, moderate fragrance; upright, vigorous growth; Chaplin Bros.

'ELLEN WILLMOTT', HT, yb, 1936; flowers creamy lemon, flushed rosy pink, large, single; foliage leathery, dark; vigorous, upright growth; [Dainty Bess X Lady Hillingdon]; Archer

'ELLEN WILLMOTT', HT, lp, 1898; flowers silvery flesh to shell-pink, dbl., cupped; very vigorous growth; Bernaix, A.

'ELLEN ZINNOW', HT, my, 1930; flowers yellow, shaded coppery orange and pink, dbl., moderate fragrance; [Souv. de H.A. Verschuren X Sunstar]; Krause

'ELLEN'S JOY', S, lp, 1989; bud ovoid; flowers light shell pink, aging lighter, medium, borne singly and in small clusters, dbl., 23 petals, cupped, repeat bloom, moderate, fruity fragrance; globular, orange-red fruit; prickles awl-like, rusty-green; foliage medium, medium green, semi-glossy; upright, bushy, spreading, medium, winter hardy growth; [Vera Dalton X [Dornroschen X (Tickled Pink X Applejack)]]; Buck, Dr Griffith J.; Historical Roses, 1991, Kimbrew-Walter Roses, 1991

'ELLESMERE', HT, w, 1927; flowers ivory-white to pure white, moderate fragrance; few thorns; [Ophelia seedling]; Allen

'ELLI HARTMANN', HT, my, 1913; flowers yellowish old-gold, dbl., moderate fragrance; [(Souv. du Pres. Carnot X Mme Mélanie Soupert) X Marechal Niel]; Welter

'ELLI KNAB', HT, pb, 1934; flowers flesh-cream flushed bright rose, veined vermilion, very large, dbl., exhibition form, slight fragrance; foliage leathery; upright, very vigorous growth; [Cathrine Kordes X W.E. Chaplin]; Kordes

'ELLI KNAB, CLIMBING', Cl HT, pb, 1953; Tantau

'ELLINOR LEGRICE', HT, my, 1949; bud ovoid; flowers, dbl., 50 petals, 5–5.5 in., cupped, moderate, fruity fragrance; foliage leathery, glossy, dark; vigorous, upright growth; [Mrs Beatty X Yellowcrest]; LeGrice

'ELLINOR LEGRICE, CLIMBING', Cl HT, my, 1959; LeGrice

'ELLIS WOOD', F, yb, 1983; [Arthur Bell sport]; Gateshead Metro. Borough Council

'ELMHURST', HT, pb, 1985; flowers large, peachy pastel pink with yellow, dbl., 30–35 petals, 5 in., exhibition form, borne mostly singly, moderate, fruity fragrance; foliage medium, medium green, matt; upright growth, moderately tall; [Granada X Helmut Schmidt]; Perry, Astor; Perry Roses

'ELMIRA', HRg (S), mr, 1977; bud ovoid; flowers bright red, open, dbl., 25 petals, 1.5–2 in., intense fragrance; foliage yellow-green; upright, bushy growth; [Schneezwerg X Old Blush]; Svedja, Felicitas; Canada Dept. of Agric.

'ELMSHORN', S, dp, 1951; flowers deep pink, pompom type, blooms in large trusses (to 40), dbl., 20 petals, 1 in., cupped, recurrent bloom, slight fragrance; foliage glossy, wrinkled, light green; (21); ADR, 1950; [Hamburg X Verdun]; Kordes; Morse

Elna, HMult, lp, 1964; Petersen

Elna Noack, Pol, or, 1960; flowers orange-red, medium, semi-dbl.; Schmid, P.

'ELNAR TONNING', HT, dp, 1926; flowers fuller and darker; [Ophelia sport]; Gyllin

Elodea®, *see* FEgima

Eloira, *see* 'VARIORA'

Eloquence, *see* 'JACSIL'

'ELSA ARNOT', HT, pb, 1960; flowers golden yellow shaded pink and cerise, dbl., 32 petals, 4 in., intense fragrance; foliage glossy; vigorous, upright growth; GM, NRS, 1959; [Ena Harkness X Peace]; Arnot; Croll

'ELSA KNOLL', HT, op, 1966; flowers shrimp-pink, large, dbl., 30 petals, exhibition form, intense fragrance; foliage dark, glossy, leathery; vigorous, upright growth; [First Love X Castanet]; Morey, Dr Dennison; Country Garden Nursery

'ELSBETH', F, lp, 1962; flowers soft pink, borne in clusters, dbl.; foliage dark; [Valeta sport]; deRuiter

'ELSBETH MEYER', Pol, dr, 1940; flowers medium, semi-dbl.; Vogel, M.

'ELSE', HT, mr, 1929; flowers medium, dbl.; Vogel, M.

'ELSE CHAPLIN', Pol, dp, 1937; flowers deep rich pink, borne in large trusses, semi-dbl.; vigorous growth; Chaplin Bros.

'ELSE KREIS', Pol, dp, 1913; flowers shining deep pink; Kreis

'ELSE POULSEN', F, mp, 1924; (Joan Anderson); flowers bright rose-pink, blooms in clusters, 10 petals, 2 in., slight fragrance; foliage dark, bronze, glossy; vigorous, bushy growth; (21); [Orléans Rose X Red Star]; Poulsen, S.

'ELSE POULSEN MELDUGSFRI', F, mp, 1937; [Else Poulsen sport]; Poulsen, S.; Poulsen's Roses

'ELSE POULSEN MORKROD', F, dr, 1934; [Else Poulsen sport]; Poulsen, S.

'ELSE POULSEN, CLIMBING', Cl F, mp, 1932; Ley

'ELSE'S RIVAL', F, mr, 1938; (Double Else Poulsen); flowers carmine red, more, dbl.; healthier growth than parent; (21); [Else Poulsen sport]; Boer Bros.

'ELSIE', LCl, lp, 1934; flowers soft pink, single; foliage dark, glossy; very vigorous growth; Chaplin Bros.

'ELSIE ALLEN', HT, lp, 1971; flowers pale pink, medium, dbl., exhibition form, slight fragrance; vigorous, bushy growth; [Montezuma sport]; Allen, L.C.; E.T. Welsh

'ELSIE BECKWITH', HT, mp, 1922; flowers rich rosy pink, center deeper, large, dbl., exhibition form, intense fragrance; foliage dark, shaded red, leathery; upright growth; [Ophelia seedling X Mev. Dora van Tets]; Beckwith

'ELSIE BOLDICK', Min, mr, 1978; bud ovoid, mossy; single, borne 1–5 per cluster, no fragrance; foliage small, soft; height 15 in.; [Fairy Moss X Fairy Moss]; Dobbs, Annette E.; Small World Min. Roses

'ELSIE DEVY', F, m, 1967; bud ovoid; flowers soft lavender-pink, reverse mauve-pink, dbl., exhibition form, slight fragrance; foliage light green, leathery; very vigorous, upright, bushy growth; [Ma Perkins X Detroiter]; Fankhauser

'Elsie May', F, lp; Hamilton, 1992

'ELSIE MELTON', HT, pb, 1991; (Picture Perfect); bud pointed; flowers large blooms borne usually singly, dbl., exhibition form, moderate, fruity fragrance; foliage large, dark green, semi-glossy; upright, tall growth; [Pristine X King of Hearts]; Wambach, Alex A., 1986; Alex A. Wambach, 1990

Elsie Warren, *see* 'MILSWEET'

'ELSIE WRIGHT', F, dr, 1982; bud long, pointed; flowers large blooms borne usually singly, sometimes 3 per cluster, 15 petals, intense, damask fragrance; light brown prickles; foliage medium, medium green, glossy; upright growth; [Crimson Glory X Crimson Glory]; Cattermole, R.F.

'ELSIEMAE', Min, op, 1986; flowers light coral pink, blooms in sprays of 2–4, dbl., 25 petals, exhibition form, slight fragrance; very few straight, light brown prickles; foliage medium, medium green, semi-glossy; tall, bushy growth; [Anne Scranton X Patricia Scranton]; Dobbs, Annette E.; Port Stockton Nursery

'ELSINORE', F, mr, 1957; (Helsingör); flowers bright scarlet, borne in large, open clusters, semi-dbl.; GM, NRS, 1957; [Floradora X Pinocchio]; Lindquist; Poulsen

'ELUSIVE', F, pb, 1993; flowers soft pink/cream blend, blooms borne in small clusters, dbl., 15–25 petals, 1.5–2.75 in., moderate fragrance; some prickles; foliage medium, dark green, semi-glossy; medium, upright, bushy growth; [(Party Girl X Laureate) X Ivory Fashion]; Jobson, Daniel J.; Jobson, 1993

Elveshörn®, *see* 'KORBOTAF'

Elvira, *see* 'VARELVI'

'ELVIRA', E, lp; flowers flesh, medium, semi-dbl.; vigorous growth

'ELVIRA ARAMAYO', HT, mr, 1922; flowers indian red, dbl.; bushy growth; GM, Bagatelle, 1922; [Feu Joseph Looymans X (Leslie Holland X Rayon d'Or)]; Looymans

'ELVIRA ARAMAYO, CLIMBING', Cl HT, mr, 1933; Ingegnoli

'ELVIRE POPESCO', HT, my, 1949; bud long, pointed; flowers golden yellow, large, dbl., 25 petals, moderate fragrance; foliage bronze; very vigorous, upright growth; [Comtesse Vandal X Seedling]; Gaujard

'ELVIS', Min, mr, 1978; bud pointed; flowers medium red, base white, dbl., 65 petals, 1 in., exhibition form, slight fragrance; foliage dark; vigorous growth; [Judy Fischer X Seedling]; Wells, V.W., Jr.

'ELVIS', F, lp; flowers pinkish orange with darker edges; Edwards, Eddie; Johnny Becnel Show Roses, 1999

Elwina®, Min, or

'ELYSE FLORY', B, lp, 1852; Guillot

Elysium®, *see* 'KORUMELST'

'EMALINE ROUGE', HT, dr, 1938; flowers deep red; [Better Times sport]; Hofmann

Emanuel®, *see* 'AUSUEL'

'EMBAJADOR LEQUERICA', HT, pb, 1962; bud pointed; flowers strawberry-pink, reverse indian yellow at base passing to brick red at edge, dbl., 30 petals; vigorous growth; La Florida

'EMBASSY', HT, yb, 1967; flowers light gold veined and edged carmine, pointed, large, dbl., moderate fragrance; foliage glossy; [Gavotte X (Magenta X Golden Scepter)]; Sanday, John, 1969

'EMBASSY REGAL', HT, pb, 1976; flowers cream overlaid peach-pink, dbl., 30 petals, 5 in., moderate fragrance; [(Gavotte X Ethel Sanday) X (Crimson Glory X Seedling)]; Sanday, John

Ember, *see* 'SAVEMBER'

'EMBERGLOW', HT, mp, 1935; flowers rich salmon-pink, dbl., 50 petals, 5 in.; foliage leathery; long stems; vigorous growth; [Souvenir sport]; Grillo

'EMBERS', F, mr, 1953; bud ovoid; flowers scarlet, dbl., 23 petals, 2.5–3 in., exhibition form, blooms in clusters, moderate, spicy fragrance; foliage dark, semi-glossy; vigorous, bushy, compact growth; [World's Fair X Floradora]; Swim, H.C.; Armstrong Nursery

Embers, S, yb; Delbard

Emblem™, *see* 'JACBLEM'

'EMBRACE', Pol, lp, 1972; flowers medium, very dbl., exhibition form, slight fragrance; foliage leathery; vigorous, upright, bushy growth; [Seventeen X Jack Frost]; Byrum; J.H. Hill Co.

Embrasement, *see* 'CHAGIP'

Embruixada, *see* **'VIOLETERA'**

'Emely', HT, w, 1998; Kordes

'EMERALD DREAM', F, w, 1976; bud pointed, light to apple green; flowers white, center green, to ivory, loose cup shape, semi-dbl., 12 petals, 1.5–2 in., flat, slight, fruity fragrance; foliage dull, very dark, leathery; low, compact growth; [Pinafore X Ivory Fashion]; Williams, J. Benjamin; Lakeland Nursery Sales

'EMERAUDE D'OR', HT, yb, 1965; flowers yellow suffused carmine-pink, petals serrated, 5 in.; vigorous growth; [Sultane X Queen Elizabeth]; Delbard-Chabert; Cuthbert

'EMERICKROSE', HMult, lp, 1922; flowers small to medium, dbl., moderate fragrance; Bruder Alfons

'EMIL KRUISIUS', F, my, 1943; bud long, pointed; flowers large, dbl., 25–30 petals, borne in clusters, slight fragrance; foliage glossy, light green; vigorous, bushy growth; [Golden Rapture X (Johanna Tantau X Eugenie Lamesch)]; Tantau

'EMILE BARDIAUX', HP, 1889; Lévêque, P.

'EMILE CHARLES', HT, or, 1922; flowers coral-red, [Mme Edouard Herriot sport]; Bernaix, P.

'EMILE CRAMON', HT, or, 1937; bud pointed; flowers coppery carmine, stamens chrome-yellow, very large; foliage dull green; very vigorous growth; Chambard, C.

'EMILE FORTÉPAULE', HWich, w, 1902; flowers white, flushed salmon, borne in large clusters, dbl.; foliage dark; vigorous growth; [R. wichurana X Souv. de Catherine Guillot]; Barbier

'EMILE HAUSBURG', HP, m, 1868; flowers lilac-rose, large, dbl.; Lévêque

'EMILE J. LE DUC', HT, mr, 1931; flowers scarlet-crimson, larger and stronger than the parent; [Scott's Columbia sport]; Le Duc

'EMILE NÉRINI', HWich, dp, 1911; flowers carmine-pink with white, medium, semi-dbl.; Nonin

'EMILIA PLANTIER', N, 1878; flowers medium large to large, dbl., moderate fragrance; Schwartz

'EMILIE', M, w; Roseraie de l'Hay, Prior to 1906

'EMILIE DUPUY', Cl T, op, 1870; flowers light coppery-pink, large, dbl., moderate fragrance; Levet

Emilo Feliu®, *see* 'FEsodios'

Emily, *see* 'AUSBURTON'

'EMILY', HT, lp, 1949; flowers soft pink, dbl., 40 petals, 5–6 in., moderate fragrance; foliage dark; vigorous, upright growth; [Mme Butterfly X Mrs Henry Bowles]; Baines; F. Cant

'EMILY DODD', HT, w, 1927; flowers milk-white, center cream, large, dbl., slight fragrance; Dickson, A.

'EMILY GRAY', LCl, dy, 1918; flowers deep golden buff, stamens yellow, blooms in clusters, dbl., 25 petals, moderate fragrance; foliage very glossy, dark, bronze; vigorous, climbing growth; (21); GM, NRS, 1916; [Jersey Beauty X Comtesse du Cayla]; Williams, A.; B.R. Cant

Emily Hough, *see* 'HOUEMILY'

'EMILY LAXTON', HP, mr, 1878; flowers large, dbl.; Laxton

Emily Louise, *see* 'HARWILLA'

Emily Post, *see* 'OMEGA'

'EMILY RHODES', LCl, mp, 1937; flowers large, dbl., cupped, moderate fragrance; vigorous, climbing or pillar growth; [Golden Ophelia X Zephirine Drouhin]; Clark, A.; NRS Victoria

Emily Victoria, *see* 'BOSHIPEACON'

'EMIN PASCHA', HT, dp, 1894; flowers large, dbl.; Drögemüller

Éminence®, *see* 'GAXENCE'

'EMIR', HT, rb, 1960; flowers yellow with orange-red, dbl., 45 petals, 6 in.; foliage glossy; free growth; [Seedling X Peace]; Verbeek

'EMJAY SKIBA', Min, mp, 1989; bud pointed; flowers medium pink, outer petals lighter, large, borne usually singly, dbl., 45 petals, exhibition form, moderate, fruity fragrance; round, dark green orange fruit; prickles sharp, pointed slightly downward, light green; foliage large, dark green, edged red, semi-glossy; bushy, tall growth; [Sonia X Pink Petticoat]; Skiba, Norman A., 1990

'EMMA', F, dp, 1980; flowers deep pink, blooms borne 6–10 per cluster, dbl., 70 petals, intense fragrance; large prickles; foliage dark, glossy; upright, branching growth; [Chanelle X Prima Ballerina]; Pearce, C.A.; Limes Rose Nursery

Emma, HT, lp, 1996

Emma Clare, HMult, pb, 1994

'EMMA JANE', F, op, 1970; flowers salmon-pink, base orange, semi-dbl., 16 petals, 3 in.; [Vera Dalton X (Masquerade X Independence seedling)]; Sanday, John

Emma Kate, *see* 'JAYEMM'

Emma May, HT, mp

Emma May, *see* 'SHERISILVER'

Emma Mitchell, *see* 'HORHARPDOS'

'EMMA WRIGHT', HT, op, 1918; flowers orange shaded salmon, semi-dbl., moderate fragrance; foliage rich green, glossy; dwarf growth; McGredy

'EMMA WRIGHT, CLIMBING', Cl HT, op, 1932; Cant, F.

'EMMANUELLA DE MOUCHY', LCl, mp, 1922; flowers delicate transparent rose-pink, semi-dbl., intense fragrance; [R. gigantea X Lady Waterlow.]; Nabonnand, P.

Emmanuelle, *see* 'AUSUEL'

'EMMELINE', HT, my, 1921; bud pure deep yellow; flowers lemon-yellow; Paul, W.

Emmeline, *see* **'MADELINE'**

'EMMELOORD', Pol, or, 1973; bud cupped; flowers round, semi-dbl.; foliage glossy, dark; [Olala X Finale]; Buisman, G. A. H.

'EMMERDALE', F, mp, 1983; flowers medium, dbl., 35 petals, moderate fragrance; foliage medium, medium green, semi-glossy; bushy growth; [Seedling X Pink Parfait]; Greensitt, J.A.; Nostell Priory Rose Gardens

'Emmie Gray', Ch, mr

'EMMIE KOSTER', Pol, dr, 1956; flowers deep red; [Dick Koster sport]; Koster, D.A.

Emmy®, Min, ab, 1993; Barni, V.

'EMOTION', B, pb, 1879; Fontaine

Emotion®, F, or, 1981; deRuiter

Emozione, *see* 'BARemoz'

'EMPEREUR DU MAROC', HP, dr, 1858; flowers crimson, tinged purple, very distinct, small, dbl., 40 petals, moderate fragrance; low, compact growth; [Geant des Batailles seedling]; Guinoiseau, B.; E. Verdier

'EMPEROR', HT, dr, 1958; bud urn shaped; flowers rose-red becoming darker, dbl., 35–45 petals, 3–3.5 in., exhibition form, intense fragrance; foliage glossy; vigorous, compact growth; [Pink Delight sport]; Kuramoto, H.

Emperor, *see* 'JACrette'

'EMPIRE GRANGER', HT, mr, 1970; bud long, pointed; flowers velvety blood-red, full, very large, dbl., moderate fragrance; foliage large, glossy, dark, bronze, leathery; very vigorous, upright, bushy growth; [Rose Bowl X Hallmark]; Morey, Dr Dennison

'EMPIRE QUEEN', HT, mr, 1925; flowers brilliant cerise, large, dbl., slight fragrance; upright growth; [Cherry Page X Vanessa]; Easlea

'EMPIRE STATE', HT, mr, 1934; bud pointed; flowers velvety scarlet, base golden yellow, large, dbl., exhibition form, intense fragrance; foliage leathery; vigorous growth; Nicolas; J&P

'EMPRESS', HT, dr, 1933; flowers dark cerise and red, well formed, moderate fragrance; vigorous, upright growth; [Ophelia seedling X Seedling]; Chaplin Bros.

'EMPRESS JOSEPHINE', Misc. OGR, mp; (Francofurtana, Imperatrice Josephine, R. francofurtana, Souv. de l'Imperatrice Josephine); flowers rich pink veined deeper, loosely shaped, large, wavy petals, semi-dbl.; height 4 ft, well-branched growth

'EMPRESS MARIE OF RUSSIA', T, my; flowers canary-yellow, fading, large, moderate fragrance

Empress Michiko, *see* 'DICnifty'

'EMPRESS OF CHINA', Cl Ch, mp, 1896; Jackson

'EMPRESS OF INDIA', HP, 1876; Laxton

Emsie Girl, *see* 'WILCGIR'

Ena Baxter, *see* 'COCBONNE'

'ENA GLADSTONE', HT, dp, 1936; flowers carmine-pink, base yellow, well shaped, large, moderate fragrance; Chaplin Bros.

'ENA HARKNESS', HT, mr, 1946; flowers large, dbl., exhibition form, intense fragrance; foliage leathery; vigorous, upright growth; GM, NRS, 1945 GM, Portland,

1955; [Crimson Glory X Southport]; Norman; C-P, 1949, Harkness, 1946

'Ena Harkness, Climbing', Cl HT, mr, 1954; (Grimpant Ena Harkness); Gurteen & Ritson & R. Murrell

Enchanted, *see* 'JAChon'

'Enchanted Autumn', Gr, ob, 1976; bud ovoid, pointed; flowers orange, dbl., 33 petals, 4–4.5 in., cupped, intense fragrance; foliage glossy, dark, coppery; upright, bushy growth; [(Queen Elizabeth X Ruth Hewitt) X Whisky]; Buck, Dr Griffith J.; Iowa State University

'Enchanter', HT, dp, 1903; flowers deep pink; [Mme Caroline Testout X Mlle Alice Furon]; Cook, J.W.

'Enchantment', HT, pb, 1946; bud long, pointed; flowers shell-pink, base yellow, dbl.; 35 petals, 6 in., slight fragrance; foliage leathery; vigorous, upright growth; [R.M.S. Queen Mary X Eternal Youth]; Hill, E.G., Co.; J&P

Enchantment, *see* 'POUlskov'

Enchantment, *see* **'Sanka'**

Enchantment, *see* POUlspan

'Enchantress', HRg (S), dr; flowers velvety blood-red, very dbl.; extra strong and hardy

'Enchantress', T, mp, 1904; flowers rose-pink; Cook, J.W.

'Enchantress', F, lp; flowers cream white; Paul & Son, 1896

Encore, *see* 'JACore'

'Encore', F, mp, 1958; bud pointed; flowers creamy pink, reverse rose-pink, large cluster, semi-dbl., 10–14 petals, 3 in., cupped, slight, spicy fragrance; foliage glossy; vigorous, upright, bushy growth; RULED EXTINCT 2/84 ARM; [Else Poulsen X Capt. Thomas]; Von Abrams; Peterson & Dering

'Endearment', HT, op, 1988; bud pointed; flowers creamy pink, reverse coral pink, large, borne usually singly, semi-dbl., 10 petals, slight, sweet fragrance; no fruit; prickles straight, medium, light brown; foliage large, medium green, matt; upright, tall growth; [Gladiator X First Prize]; Taylor, Thomas E.; Michigan Mini Roses, 1989

Endeavour, *see* 'TALend'

Endless Dream(TM), *see* 'TWOdream'

Endless Love, *see* 'LYOss'

Endless Summer, *see* 'RENmer'

'Endless Tale', Cl HT, my, 1956; bud ovoid, deep yellow; flowers amber-yellow, outer petals creamy, dbl., 30–35 petals, 6–7 in., abundant, intermittent bloom, moderate fragrance; foliage leathery; vigorous (20 ft or more) growth; [Lestra Hibberd, Climbing X Lestra Hibbard sport]; Motose

Endora, *see* 'ZIPend'

Enduring Love, HT, yb, 1998; Pallek

'Enemy of War', HT, pb, 1986; flowers open, dbl., 50–60 petals, slight fragrance; beak-shaped, light green to deep brown prickles; foliage large, dark green, glossy, leathery; very vigorous, profuse growth; [Festival Beauty X Cynosure]; Hardikar, Dr M.N.

'Enfant d'Orléans', Pol, m, 1929; flowers neyron rose, tinted purple, fading lighter, borne in clusters; Turbat

'Enfant de France', HP, lp, 1860; flowers silvery pink, very dbl.; Lartay

Enfant de lyon, *see* **'Narcisse'**

Enfield in Bloom, *see* 'BOSanneves'

'Eng. D. José de Mendia', HT, mp; flowers rosy salmon; da Silva, Moreira

'Eng. Duarte Pacheco', HT, dr, 1938; flowers blackish crimson, large, very dbl., cupped, intense fragrance; dwarf growth; [Hadley X Presidente Carmona]; da Silva, Moreira

'Eng. Pereira Caldas', HT, mp, 1954; flowers salmon-pink, base yellow; da Silva, Moreira

'Eng. Pulide Garcia', HT, yb, 1961; flowers yellow stained pink; [Grand'mere Jenny X Michele Meilland]; da Silva, Moreira

'Eng. Vitória Pires', HT, dr, 1954; flowers velvety dark red, moderate fragrance; da Silva, Moreira

'Engagement', Gr, op, 1969; bud globular; flowers coral-pink, large, dbl., exhibition form, slight fragrance; foliage dark, leathery; vigorous, bushy growth; [Ma Perkins X Montezuma]; Patterson; Patterson Roses

Enghien, HT, ob, 1991; RvS-Melle

'England's Glory', HT, lp, 1904; flowers large, dbl.; Wood

'England's Rose', S, my, 1995; Austin, David

'English Elegance'®, S, pb, 1986; Austin, David

'English Estates', HT, yb, 1992; flowers deep yellow edged red, blooms borne mostly singly, dbl., 26–40 petals, 1.5–2.75 in., intense fragrance; some prickles; foliage medium, dark green, glossy; upright (80 cms) growth; [Whisky Mac X Catherine Cookson]; Thompson, Robert; Battersby Roses, 1991

English Garden(TM), *see* 'AUSbuff'

'English Hedge', HRg (S), mp, 1959; flowers pink, small, single; red fruit; compact, upright growth; [(R. rugosa rubra X R. cinnamomea) X R. nitida]; Nyveldt

'English Holiday', F, yb, 1977; flowers yellow, blended with salmon, dbl., 33 petals, 4 in., moderate fragrance; foliage large, glossy; [Bobby Dazzler X Goldbonnet]; Harkness

'English Miss', F, lp, 1977; flowers pale pink, blooms in clusters, dbl., 60 petals, 2.5 in., intense fragrance; foliage dark purple to dark green; [Dearest X The Optimist]; Cants of Colchester, Ltd., 1978

English Perfume(TM), *see* 'JAClewt'

English Porcelain, *see* 'MORporc'

English Sachet(TM), *see* 'JACOlfa'

English Sonnet(TM), *see* 'HARverag'

English Violet, S, dp; Austin, David

Enhance, *see* 'SANdaya'

'Enid', Pol, lp, 1936; flowers pale pink, borne in clusters; foliage light; upright growth; Prior

Enjoy, *see* 'LAVjoy'

'Ennio Morlotti', F, mp, 1973; bud globular; flowers clear pink, dbl., 35 petals, 2.5–3 in., exhibition form, intense fragrance; foliage glossy; vigorous, upright growth; [Fashion X Queen Elizabeth]; Cazzaniga, F. G.

Enric Palau, HT, m; Dot

Ensa de Rennes, *see* 'JADecosill'

'Entente Cordiale', HT, w, 1908; flowers nasturtium-red, base yellow; [Mme Caroline Testout X Soleil d'Or]; Guillot, P.

'Entente Cordiale', HT, mr, 1909; flowers creamy white, tinged carmine at edges; [Mme Abel Chatenay X Kaiserin Auguste Viktoria]; Pernet-Ducher

'Enterprise', F, pb, 1957; flowers deep pink edged peach, medium borne in large clusters, semi-dbl., 20 petals, 2 in., moderate fragrance; foliage dark, glossy, vigorous; upright, bushy growth; [Masquerade X Seedling]; Kordes; Morse

'Enver Pascha', HT, lp, 1916; flowers large, dbl.; Kiese

'Enzo Fumagalli', F, mp, 1966; bud globular; flowers salmon-pink, full, medium, very dbl., abundant, intermittent bloom; foliage glossy; vigorous, bushy growth; [Mount Shasta X Papillon Rose]; Cazzaniga, F. G.

'Eos', S, rb, 1950; bud ovoid; flowers sunset-red becoming brighter, center white, semi-dbl., cupped, borne several together, non-recurrent, slight fragrance; foliage leathery, glossy; shrub or pillar (to 6 ft) growth; [R. moyesii X Magnifica]; Ruys

'Epic', F, mp, 1989; bud tapering; flowers medium, borne in sprays of 3–6, dbl., flat, moderate, spicy fragrance; prickles brown; foliage bronze to dark green, glossy; upright, bushy growth; [Silent Night X Irish Mist]; Cattermole, R.F.; South Pacific Rose Nursery

Epidor®, *see* 'DELépi'

Epinal, F, 1994; Croix, P.

Epoca, *see* 'LEN 2'

Epoca Mondadori, Gr, 1966; Lens, Louis

'**Eponine**', HMsk (S), w; dbl., cupped, moderate fragrance

'**Epos**', Pol, mr, 1971; bud ovoid; flowers medium, semi-dbl., cupped, slight fragrance; foliage bronze, leathery; vigorous, upright growth; [Tommy Bright X Seedling]; Delforge

'**Erato**', LCl, w, 1937; bud pointed; flowers open, bornein clusters, semi-dbl., slight fragrance; foliage glossy; long stems; very vigorous climbing growth; [(Ophelia X R. multiflora) X Florex]; Tantau

'**Erbprinzessin Leopold von Anhalt**', HT, ly, 1933; flowers whitish-yellow, large, dbl., moderate fragrance; Behrens

'**Erfurt**', HMsk (S), pb, 1939; bud long, pointed; flowers medium pink, yellow toward base, large blooms in clusters, semi-dbl., recurrent bloom, intense, musk fragrance; foliage leathery, wrinkled, bronze; vigorous (5–6 ft), trailing, bushy growth; [Eva X Reveil Dijonnais]; Kordes, 1931

'**Eric**', F, mr, 1965; bud ovoid; flowers medium, semi-dbl., cupped; foliage dark, glossy, leathery; vigorous growth; [Alain X Coup de Foudre]; Hémeray-Aubert

'**Eric B. Mee**', HT, mr, 1937; flowers vivid cerise, well shaped, small; Mee; Beckwith

'**Eric Hobbis**', HT, pb, 1966; flowers pink, reverse peach, 4.5 in., exhibition form; low growth; [Gavotte X Peace]; Sanday, John

'**Eric Holroyd**', HT, mr, 1925; flowers bright scarlet, base shaded gold, moderate fragrance; Chaplin Bros.

'**Eric Louw**', HT, mr, 1964; bud pointed; flowers cyclamen-red, well formed, dbl., 35–40 petals; foliage leathery, glossy; strong stems; vigorous, bushy growth; [Queen Elizabeth X Confidence]; Herholdt, J.A.; Herholdt's Nursery

Eric The Red, F, lp; Welsh, 1997

Eric von Melnibonée, S, op, 1983; flowers yellowish salmon-pink, large, dbl., slight fragrance; Weihrauch

'**Erica**', F, or, 1964; flowers orange-scarlet, frilled, borne in large clusters, semi-dbl.; [Seedling X Montezuma]; Herholdt, J.A.; Herholdt's Nursery

Erica, *see* 'INTerop'

Erica Herholdt, F, or, 1964; flowers large, semi-dbl.; Herholdt, J.A.

'**Erich Frahm**', F, mr, 1939; bud long, pointed, yellowish red; flowers carmine-scarlet, center yellow, shell shaped, open, borne in umbels of up to 20; foliage dark, glossy, leathery; vigorous growth, very branching.; [Dance of Joy X Mary Hart]; Kordes; Timm

'**Erie**', S, lp, 1946; flowers pale pink, borne in clusters, single, 5 petals, very free, non-recurrent bloom; bottle shaped, bright red fruit; foliage dark, fragrant (sweet-briar); vigorous, spreading growth; hardy.; Preston; Central Exp. Farm

'**Erie Treasure**', HRg (S), w; flowers blush to white, dbl., recurrent bloom, intense fragrance; foliage wrinkled; vigorous (6 ft), bushy growth; [Souv. de Pierre Leperdrieux X Nova Zembla]; Wedrick

'**Erik Hjelm**', HT, op, 1929; flowers pure salmon-pink, very dbl., moderate fragrance; [Lieut. Chaure X Sachsengruss]; Kordes

Erika, HT, mp

'**Erika My Love**', Min, my, 1993; flowers yellow with white tips, medium, dbl., 25–30 petals, 2–3 in., borne mostly singly, no fragrance; some prickles; foliage medium green, semi-glossy; tall (72 cms), upright growth; [Rise 'n' Shine X Seedling]; Armstrong, James L.

Erika My Love, Min, my; flowers purple-violet, dbl., cupped; [Rise 'n' Shine X Seedling]

'**Erika Pluhar**', HT, dr

'**Erika Teschendorff**', HT, mr, 1949; bud long, pointed; flowers fiery scarlet, open, very large, dbl., globular, slight fragrance; foliage glossy, dark; very vigorous, upright growth; Berger, V.; Teschendorff

'**Erikonig**', HMult, m, 1886; Geschwind, R.

Erin Fleming, LCl, mp, 1997; flowers very full, medium, very dbl., 41 petals, borne singly and in large clusters, up to 15 buds per cluster, moderate fragrance; foliage medium, dark green, glossy; upright, tall growth; [Sunsation X Henry Kelsey]; Fleming, Joyce L.

'**Erinnerung an Brod**', HSet, rb; (Souv. de Brod); flowers cerise through crimson to purple, large, dbl., quartered, moderate fragrance; Geschwind, R., 1886

'**Erinnerung an Schloss Scharfenst**', HT, m, 1892; flowers purple/pink, large, dbl., intense fragrance; Geschwind, R.

Erlkönig, *see* '**Roi des Aunes**'

'**Erna**', HWich, lp, 1929; flowers small, dbl.; Vogel, M.

'**Erna Baltzer**', HT, my, 1954; flowers golden yellow, medium, moderate fragrance; vigorous growth; [Tawny Gold X Gaudia]; Leenders, M.

Erna Doris, *see* 'LENdori'

'**Erna Grootendorst**', F, dr, 1938; flowers deep velvety crimson, large, semi-dbl.; foliage glossy, dark; bushy growth; [Bergers Erfolg X Gloria Mundi]; Grootendorst, R.

'**Erna Teschendorff**', Pol, mr, 1911; flowers strawberry-red, open, small, semi-dbl., slight fragrance; foliage rich green, soft; bushy growth; [Mme Norbert Levavasseur sport]; Teschendorff

'**Ernest H. Morse**', HT, mr, 1964; flowers turkey-red, dbl., 30 petals, 4 in., intense fragrance; foliage leathery; vigorous growth; GM, RNRS, 1965; Kordes; Morse

'**Ernest Laurent**', HT, lp, 1914; flowers large, dbl., moderate fragrance; Viaud-Bruant

'**Ernest Metz**', T, mp, 1888; flowers rose-pink, center darker, large, dbl.; Guillot, J. B.

'**Ernest Morel**', HP, dr, 1898; flowers bright garnet-red; Cochet, P.

'**Ernest's Blue**', HT, m, 1992

'**Ernestine Cosme**', HWich, rb, 1926; flowers brilliant red, with large white eye, borne in clusters of 75, single; many thorns; very vigorous, climbing growth; Turbat

Ernie, *see* 'TINernie'

'**Ernie Pyle**', HT, mp, 1946; bud long, pointed; flowers deep rose-pink, reverse deeper, dbl., 35–40 petals, 4.5–5 in., cupped, intense fragrance; foliage leathery; vigorous, upright, bushy growth; [((Royal Red X Talisman) X Seedling (red)) X (Talisman X Nutneyron]; Boerner; J&P

'**Ernst Dechant**', HWich, w, 1928; flowers white with yellow tints, small, semi-dbl.; Vogel, M.

'**Ernst G. Dörell**'®, HMult, dp; Geschwind, R., 1897

'**Ernst Grandpierre**', HWich, w, 1902; flowers pale cream, base yellow, open, borne in clusters, dbl., sparse bloom, slight fragrance; foliage light, glossy; height 8–10 ft; [R. wichurana X Perle des Jardins]; Weigand, C.

Eroica, *see* '**Erotika**'®

Eroika, *see* '**Erotika**'®

'**Eros**', F, dp, 1955; flowers deep rosy pink shaded brick-red, base yellow; dwarf, compact growth; [Pinocchio seedling]; Maarse, G.

Eros, *see* 'BARsor'

Erotica, *see* '**Erotika**'®

'**Erotika**'®, HT, dr, 1968; (Eroica, Eroika, Erotica); bud ovoid; flowers velvety dark red, well-formed, large, dbl., 33 petals, intense fragrance; foliage dark, glossy; vigorous, upright growth; ADR, 1969; Tantau, Math.

Mothers Day *see* '**Mothersday**'

'**Erskine**', Sp, mp; Hansen, N.E.

'**Eruption**', HT, mr, 1934; flowers fiery scarlet-red, large, semi-dbl.; foliage sea-green; bushy growth; [Red-Letter Day X Columbia]; Van Rossem

'**Erwin Hüttmann**', HWich, dp, 1941; flowers medium, dbl., intense fragrance; Krause

Escada, *see* 'TANadac'

'ESCALADE', Cl HT, mr, 1962; bud pointed; flowers carmine, large, exhibition form; vigorous growth; [Spectacular X Charlotte Armstrong]; Combe; Vilmorin-Andrieux

Escapade®, *see* 'HARPADE'

Esco Rose, F; Gregory, C.

'ESCORT', F, dr, 1963; bud pointed to urn-shaped; flowers small to medium, dbl., 30 petals, exhibition form; foliage dark, leathery; vigorous, bushy growth; [Spartan X Garnette]; Swim & Weeks

'ESCULTOR CLARÁ', HT, m, 1956; bud pointed; flowers purple-garnet, reverse magenta, large, dbl., 30 petals, exhibition form, moderate fragrance; foliage dark, glossy; very vigorous, upright, compact growth; [Lilette Mallerin X Floradora]; Dot, Simon

Escurial®, *see* 'DELFLIP'

'ESKIL', HT, yb, 1939; flowers light yellow overlaid red and orange; [Mrs Franklin D. Roosevelt sport]; Ringdahl

Eskimo®, *see* 'KORcilmo'

'ESMÉ', HT, w, 1920; flowers cream-white, edged rosy carmine, dbl.; [Mme Edouard Herriot X Seedling]; Cant, B. R.

Esme Euvrard, *see* 'KORelgas'

'ESMERALDA', F, dp, 1957; flowers deep rose-pink, reverse lighter, small, borne in very large, dbl., moderate fragrance; vigorous growth; [Gartendirector Otto Linne X Seedling]; Riethmuller

Esmeralda®, *see* 'KORMALDA'

Especially for You, *see* 'FRYworthy'

'ESPERANTO', HT, my, 1932; flowers pure yellow, very large, dbl., 60 petals, globular, intense fragrance; [Miss Lolita Armour sport]; Böhm, J.

Esperanto Jubileo, *see* 'SANROZO'

'ESPÉRANZA', F, mr, 1966; bud ovoid; flowers bright red, large blooms in clusters, dbl.; foliage dark, bronze, leathery, glossy; upright growth; GM, Baden-Baden, 1968 GM, The Hague, 1968; [Donald Prior X Reverence]; Delforge

'ESPLANADE', HT, dr, 1961; dbl., 40 petals, moderate fragrance; foliage glossy; vigorous growth; [Soraya X Seedling]; Verbeek

'ESPOIR', HCh, w, 1984; (Interbec); flowers small, single, 5 petals, slight fragrance; foliage very small, medium green, semi-glossy; upright, spreading growth; [HCh seedling X HCh seedling]; Gailloux, Gilles

'ESPOIR', F, 1958; flowers rich salmon, rosette shape, dbl., 50 petals, 4.5 in., moderate fragrance; vigorous, low, bushy growth; RULED EXTINCT 6/83 ARM; [(Oiseau de Feu X Fashion) X (Independence X Seedling)]; Combe; Japan Rose Society

'ESPOIR', HT, 1947; bud long, pointed; flowers pink, center light salmon-pink, large, dbl., 35 petals, abundant bloom, slight fragrance; foliage soft; bushy growth; RULED EXTINCT 6/83?; [Charles P. Kilham X Neville Chamberlain]; Lens

Esprit®, *see* 'KORHOLST'

'ESSENCE', HT, dr, 1930; bud pointed; flowers rich scarlet-crimson, becoming bluish, outer petals slightly fimbriated, cupped, moderate, damask fragrance; Cant, B. R.

'ESSENCE, CLIMBING', Cl HT, dr, 1938; Western Rose Co.

Essex, *see* 'POUlnoz'

Essie Lee, *see* 'JUDLEE'

'ESTAFETTE', F, dr, 1962; flowers open, borne in clusters, semi-dbl., 2.5–3 in.; foliage dark, glossy; vigorous growth; [Alain X Elmshorn]; Delforge

Estelle, HGal, mp; flowers medium, very dbl., intense fragrance; Roseraie de l'Hay

Estelle, *see* 'OLIjbrau'

Estelle, HT, op; Croix

Estelle Pradel, *see* **'ZÉLIA PRADEL'**

Esterel, *see* 'KORIVER'

'ESTHER', F, ob, 1954; bud pointed, turkey-red; flowers golden orange, reverse lighter, open, large, borne in clusters, single, 7–8 petals; foliage glossy, bright green; long stems; very vigorous, bushy growth; [Cocorico X Canzonetta]; San Remo Exp. Sta.

'ESTHER', G, pb, 1849; Vibert

Esther Geldenhuys, *see* 'KORSKIPEI'

Esther Jasik, *see* 'ZIPEST'

'ESTHER JERABEK', F, mp, 1978; semi-dbl., 18 petals; spreading growth; [The Fairy seedling]; Jerabek, Paul E.

Esther O'Farim, *see* 'KORFARIM'

Esther Ofarim, *see* 'KORFARIM'

Esther Peiro®, *see* 'FEbasa'

'ESTHER RANTZEN', F, ob, 1982; flowers medium, semi-dbl., moderate, fruity fragrance; foliage medium, medium green, semi-glossy; upright, bushy growth; [Spartan X Orangeade]; Dwight, Robert & Sons

Esther's Baby®, *see* 'HARKINDER'

Estima, Min, op

Estralia, *see* **'ESTRELLITA'**

'ESTRELLITA', Min, w; (Estralia, Pixie Pearl); flowers large blooms in clusters, dbl.; Said to have been grown in US since about 1910

Estrellita de Oro, *see* **'BABY GOLD STAR'**

'ESTRU', F, lp, 1975; flowers patio, small, dbl., 50 petals, 1.5 in.; foliage small, glossy, dark; low, compact growth; [Rosy Jewel X Floribunda seedling]; deRuiter

'ETAIN', HWich, op, 1953; flowers salmon-pink, borne in trusses, slight fragrance; foliage glossy, dark, almost evergreen; very vigorous growth; Cant, F.

Été Parfumé, *see* **'TYPHOO TEA'**

'ÉTENDARD', LCl, mr, 1956; (New Dawn Rouge, Red New Dawn); flowers bright red, dbl., blooms in large clusters, recurrent bloom, moderate fragrance; foliage very glossy, leathery; vigorous growth; [New Dawn X Seedling]; Robichon

'ETENDARD DE JEANNE D'ARC', N, w, 1882; Margottin

Éterna®, *see* 'DELIC'

Eterna Giovanezza, *see* **'ETERNAL YOUTH'**

'ETERNAL FLAME', LCl, ob, 1955; (Everblooming Pillar No. 12); flowers light orange, semi-dbl., 12–19 petals, 3–4 in., moderate fragrance; upright, climbing growth; [Seedling X Queen o' the Lakes]; Brownell, H.C.

Eternal Flame, F, ob, 1992; Interplant

'ETERNAL SUN', HT, or, 1966; bud ovoid; flowers vermilion, large, dbl., exhibition form, slight fragrance; foliage dark, leathery; vigorous, upright, bushy growth; [Seedling X Jacqueline (HT)]; Hill, Joseph H., Co.

'ETERNAL YOUTH', HT, lp, 1937; (Eterna Giovanezza, Jeunesse Éternelle); bud long, pointed; flowers suffused orange-salmon, dbl., 50 petals, 4–5 in., cupped, intense fragrance; foliage leathery; vigorous, upright, bushy growth; [Dame Edith Helen X Julien Potin]; Aicardi, D.; J&P

Eternally Yours, *see* 'MACSPEEGO'

'ETERNITÉ', F, mr, 1947; flowers scarlet, dbl., 25 petals, 4–5 in., moderate fragrance; foliage dark; very free growth; [Mme Joseph Perraud X Holstein]; Gaujard

Eternity, *see* 'TWOETERN'

'ETHEL', HWich, lp, 1912; flowers flesh-pink, borne in clusters, semi-dbl.; vigorous growth; [Dorothy Perkins seedling]; Turner

Ethel Austin, *see* 'FRYMESTIN'

'ETHEL CHAPLIN', HT, my, 1926; flowers soft lemon-yellow, dbl., moderate fragrance; Chaplin Bros.

'Ethel Dawson', F, lp; [Daily Sketch X (Stella X Charles Mallerin)]; Dawson, George

Ethel Dawson, F, lp; [Red Meillandina X seedling]; Dawson, George, before 1987

'ETHEL DICKSON', HT, dp, 1917; flowers deep salmon-rose with silvery flesh reflexes; Dickson, H.

'ETHEL JAMES', HT, dp, 1921; flowers softer carmine-red than Isobel, flushed orange-scarlet, center yellow, single, 5 petals; bushy growth; GM, NRS, 1920; McGredy

'ETHEL MALCOLM', HT, w, 1909; flowers ivory-white, very large; vigorous growth; GM, NRS, 1909; McGredy

Ethel Orr™, *see* 'MINAMCO'

'Ethel Sanday', HT, yb, 1954; flowers yellow flushed apricot, well-formed, dbl., 34 petals, 4–5 in., slight fragrance; foliage dark; vigorous, upright growth; GM, NRS, 1953; [Rex Anderson X Audrey Cobden]; Mee; Sanday

'Ethel Sloman', HT, mr, 1966; bud ovoid; flowers crimson, very dbl., slight fragrance; foliage leathery; compact, bushy growth; [Baccará X My Choice]; Fankhauser

'Ethel Somerset', HT, mp, 1921; bud pointed; flowers shrimp-pink large, dbl., exhibition form, intense fragrance; vigorous, branching growth; Dickson, A.

'Ethel Utter', LCl, my; bud cherry-red; flowers, dbl.; Wilber

Ethical, F, rb, 1995; Dawson

'Etienne Dubois', HP, dr, 1873; flowers large, dbl.; Damaizin

'Étienne Levet', HP, mr, 1871; flowers carmine-red, large, dbl., 70 petals, sometimes recurrent bloom; vigorous, erect growth; [Victor Verdier seedling]; Levet Père

'Étienne Rebeillard', HT, pb, 1924; flowers flesh-pink, suffused golden, semi-dbl., slight fragrance; Pernet-Ducher

'Étincelante', HT, dr, 1941; flowers brilliant red, tinted purple, dbl., moderate fragrance; [Gruss an Teplitz X Étoile de France]; Chambard, C.

'Étincelle', HT, or, 1958; flowers bright red tinted orange, medium, dbl., 40–45 petals, intense fragrance; low, bushy growth; [Crimson Glory X Seedling]; Moulin; Vilmorin-Andrieux

'Etna', M, dr, 1845; flowers crimson shaded purple, large, very mossy, moderate fragrance; Laffay, M.

'Etna', HT, dr, 1924; flowers deep crimson-maroon, semi-dbl.; [Red-Letter Day X H.V. Machin]; Looymans; Prior

'Étoile d'Alaï', HMsk (S), mr, 1946; flowers brilliant red, prominent golden stamens, medium, very dbl., repeat bloom; bushy growth; [Skyrocket seedling]; Meilland, F.

'Étoile d'Or', HT, yb, 1931; flowers golden yellow, reverse shaded orange, large, semi-dbl., intense fragrance; upright, bushy growth; Pernet-Ducher; Gaujard

'Étoile de Belgique', HT, mr, 1946; flowers brilliant red, very large, dbl., slight fragrance; foliage bronze; vigorous, bushy growth; [Charles P. Kilham X Étoile de Hollande]; Lens

'Étoile de Belgique', HT, mr, 1956; bud ovoid; flowers geranium-red, large, dbl.; bushy, spreading growth; [Independence X Happiness]; Buyl Frères

'Étoile de Feu', HT, op, 1921; flowers salmon-pink and coral-red, large, dbl., globular; foliage bronze; vigorous, bushy, branching growth; Pernet-Ducher

'Étoile de Feu, Climbing', Cl HT, ob, 1930; H&S

'Étoile de France', HT, dr, 1904; bud pointed; flowers dark rose-red, center cerise, medium to small, dbl., cupped, intense fragrance; bushy growth; [Mme Abel Chatenay X Fisher Holmes]; Pernet-Ducher

'Étoile de France, Climbing', Cl HT, dr, 1915; Howard Rose Co.

'Étoile de Hollande', HT, mr, 1919; flowers bright red, large, dbl., 35–40 petals, cupped, intense, damask fragrance; foliage soft; moderate, open growth; (28); [Gen. MacArthur X Hadley]; Verschuren

'Étoile de Hollande, Climbing', Cl HT, mr, 1931; (Grimpant Étoile de Hollande); Leenders, M.

'Étoile de Lyon', T, my, 1881; flowers golden yellow, dbl., sparse intermittent bloom, moderate fragrance; foliage soft; short, weak stems; bushy growth; Guillot, P.

'Étoile de Mai', Pol, ly, 1893; flowers sulfur-white, small, dbl., intense fragrance; vigorous growth; Gamon

'Étoile de Portugal', LCl, dp, 1898; flowers rose-red, dbl.; Cayeux, H.

'Étoile Luisante', Pol, mr, 1918; (Baby Herriot); bud pointed; flowers cerise-red, shaded coppery, semi-dbl., exhibition form, borne in clusters, slight fragrance; few thorns; foliage bronze, glossy; long stems; bushy growth; Turbat; Michell

Etoile Rouge, Min, rb

'Etty van Best', HT, ly, 1934; flowers white, shaded yellow, dbl.; foliage leathery; vigorous growth; [Pharisaer X Souv. de H.A. Verschuren]; Buisman, G. A. H.

'Étude', LCl, dp, 1965; flowers deep rose-pink, semi-dbl., blooms in clusters, recurrent bloom, moderate fragrance; foliage glossy, light green; [Spectacular X New Dawn]; Gregory

'Eucharis', G, dp; flowers bright rose, edged lighter, large, dbl.; Descemet, M. (?), before 1815

'Eugène Appert', HP, dr, 1860; flowers medium, dbl.; Trouillard

'Eugène Barbier', HP, dy, 1920; flowers brilliant canary-yellow, shaded coppery golden yellow, dbl., globular, moderate fragrance; few thorns; upright growth; [Frau Karl Drushchki X Rayon d'Or]; Barbier

'Eugène Boullet', HT, dr, 1909; flowers crimson-red, dbl.; vigorous growth; Pernet-Ducher

'Eugène de Beauharnais', Ch, m, 1838; (Prince Eugène); flowers purple, large, dbl., moderate fragrance; Hardy

'Eugène de Savoie', M, mr; flowers bright red, shaded, dbl.; Moreau et Robert, 1861

'Eugène E. Marlitt', B, mr, 1900; (Mme Eugène Marlitt); flowers bright carmine shaded scarlet, large, dbl.; few prickles; vigorous growth; Geschwind, R.

'Eugène Fürst', HP, dr, 1875; flowers crimson-red, shaded purple, large, dbl., globular, recurrent bloom, moderate fragrance; [Baron de Bonstetten X Seedling]; Soupert & Notting

'Eugène Jacquet', HWich, mr, 1916; flowers cherry-red, borne in clusters of 25–30, dbl., very early bloom, moderate fragrance; foliage bright green; vigorous, symmetrical growth; [wichurana hybrid (red) X Multiflora hybrid (pink)]; Turbat

'Eugène Janvier', G, dp; flowers dark pink, paling to lilac, medium, dbl.

'Eugène Picard', HT, op, 1938; flowers light coppery-pink with red tints, large, very dbl., moderate fragrance; Sauvageot, H.

'Eugène Transon', LCl, or, 1926; flowers orange and copper, reverse orange-red, shaded, borne in clusters; vigorous, climbing growth; [Mme Berard X Constance]; Barbier

'Eugène Verdier', M, dp, 1872; flowers crimson or light red, center deeper, well-formed, very dbl., moderate fragrance; Verdier, E.

'Eugenia', HT, rb, 1920; flowers coral-red to prawn-red, flecked or striped yellow, dbl., moderate fragrance; [Mme Edouard Herriot sport]; Collier

Eugénie de Guinoiseau, *see* **'Eugénie Guinoiseau'**

'Eugénie Guinoiseau', M, mr, 1864; (Eugénie de Guinoiseau); flowers reddish cerise, changing to reddish violet, large, dbl.; vigorous growth; Guinoiseau, B.

'Eugénie Lamesch', Pol, yb, 1899; flowers ochre-yellow and bright yellow, shaded pink, blooms in clusters, dbl., moderate fragrance; foliage glossy; dwarf, compact growth; [Aglaia X William Allen Richardson]; Lambert, P.

'Eugenie Verdier', HP, pb, 1869; Guillot

'Eugenio d'Ors', HT, dr, 1946; flowers oxblood-red, large, dbl., intense fragrance; [Sensation X Margaret McGredy]; Camprubi, C.

'Eugenio Fojo', HT, mr, 1953; bud pointed; flowers vermilion-red, well formed, large, dbl., 35 petals, moderate fragrance; vigorous, bushy growth; [Texas Centennial X Carlos Fargas]; Dot, Pedro

'EUGhien', HT, ab; (**Pearly Shores**); (EUGhien), 1995

'Eulalia', HT, mp, 1934; flowers pink, lighter toward base, large, dbl., intense fragrance; vigorous growth; Verschuren-Pechtold; Dreer, H&S

'Eulalia Berridge', Pol, lp; flowers medium, dbl., slight fragrance

'EULALIA LEBRUN', HGal, w, 1844; flowers white with pink stripes, medium, dbl.; Vibert

'EUMUNDI', F, w, 1953; flowers pure white, very dbl., cupped; foliage light green; vigorous, bushy growth; [Yvonne Rabier X Baby Alberic]; Ulrick, L.W.

'EUMUNDI', F, lp; [Yvonne Rabier X Baby Alberic]; Ulrick, L.W., 1954

Euphoria, *see* 'INTereup'

Euphrates, *see* 'HARUNIQUE'

'EUPHROSYNE', HMult, mp, 1895; flowers pure pink, small, dbl., moderate fragrance; very vigorous growth; [R. multiflora X Mignonette]; Schmitt

'EUREKA', HT, mp, 1914; flowers bright rose; Hobbies

'EUREKA', S, w, 1956; flowers pure white, small, borne in clusters, semi-dbl.; intended for trial as a hardy understock.; [Probably Betty Bland X Ames 5]; Wright, Percy H.

Euro 92, F, mr, 1989; RvS-Melle

'EuroGROOT', F, lp; (**Souvenir**); Kordes/ Grootendorst, 1991

'EUROPA', HT, mp, 1928; flowers bright pink, better; [Columbia sport]; Keessen; Nieuwesteeg

Europa, *see* 'KORTEXUNG'®

Europas Rosengarten, *see* 'HELeuro'

Europawelle Saar, HT, op, 1985; flowers salmon-pink, large, dbl.; Meilland

Europe 92, *see* 'DELSULAN'

Europe Sensation, F, or

European Touch, HT, ab, 1997; bud elongated, edged with pink; flowers large, dbl., 4–5 in., exhibition form, borne singly, no fragrance; foliage large, dark green, matte; long, straight stems; very tall (5 to 6 ft), upright, vigorous growth; Von Koss

'EUROPEANA'®, F, dr, 1968; flowers dark velvety crimson, rosette shape, dbl., 25–30 petals, 3 in., rosette, blooms in large, heavy clusters, slight, tea fragrance; foliage bronze-green to dark green; vigorous growth; AARS, 1968 GM, Portland, 1970 GM, The Hague, 1962; [Ruth Leuwerik X Rosemary Rose]; deRuiter, 1963; C-P, 1968

Europeana, Climbing, *see* 'BUREURO'

'EUROROSE', F, yb, 1973; flowers yellow-ochre, flushed fire-red, dbl., 25 petals, 3.5 in., globular; [Zorina X Redgold]; Dickson, A.

Eurosong®, *see* 'LENORAN'

Eurostar, *see* POUlreb

'EUROVISION', HT, dr, 1961; dbl., 30 petals, 4 in., moderate fragrance; foliage dark; vigorous, bushy growth; [Miss France X Rosita]; Delforge

Euroway, *see* 'HELway'

Euryanthe®, S, mp

'EURYDICE', HSet, 1886; Geschwind, R.

Eustace, HT, w, 1989; Robinson

Eustacia, *see* 'SEAEUS'

'EUTERPE', LCl, ly, 1937; bud pointed; flowers open, borne in clusters, semi-dbl., slight fragrance; foliage glossy; long stems; very vigorous, climbing growth; [(Ophelia X R. multiflora) X Florex]; Tantau

'EUTIN', F, dr, 1940; (Hoosier Glory); bud globular, pointed; flowers glowing carmine-red, blooms in clusters, dbl., cupped, slight fragrance; foliage leathery, glossy, dark; vigorous growth; [Eva X Solarium]; Kordes

'EUTIN, CLIMBING', Cl F, dr, 1957; Lindquist; Howard Rose Co.

'EVA', HMsk (S), rb, 1933; bud pointed; flowers carmine-red, center white, large blooms in clusters to 75, semi-dbl., intermittent bloom, moderate fragrance; very vigorous growth; (28); [Robin Hood (HMsk) X J.C. Thornton]; Kordes

'EVA DE GROSSOUVRE', HT, mp, 1908; Guillot

'EVA EAKINS', HT, or, 1926; flowers scarlet-carmine, flushed orange, base bright yellow, small, dbl., exhibition form, slight fragrance; foliage leathery; bushy growth; McGredy

Eva Gabor, *see* 'POULTAL'

'EVA KNOTT', HT, ob, 1957; flowers coppery orange, well formed, dbl., 35 petals; vigorous growth; [Ethel Sanday X Mrs Sam McGredy]; Mee

'Eva Simone', F, op, 1992; dbl., intense fragrance; Michler, K. H.

'EVA TESCHENDORFF', HMult, w, 1923; flowers greenish-white; [Echo sport]; Grunewald; Teschendorff

'EVA TESCHENDORFF, CLIMBING', Cl Pol, w, 1926; Opdebeeck

'EVALINE', Pol, lp, 1920; flowers light pink, edged brighter, petals quilled, small, borne in, dbl., moderate fragrance; bushy growth; [Orléans Rose X Rayon d'Or]; Prosser

'EVANGELINE', HWich, pb, 1906; flowers rosy white, veined cameo-pink, blooms in clusters, single, 2 in., late seasonal bloom, moderate fragrance; foliage dark, leathery; long stems; very vigorous, climbimg (12–15 ft) growth; [R. wichurana X Crimson Rambler]; Walsh

'EVANGELINE', T, w, 1951; bud deep pink; flowers creamy white, edged blush-pink, medium, dbl., moderate fragrance; almost thornless; vigorous, spreading growth; [Mrs Dudley Cross sport]; Krider Nursery

'EVANGELINE BRUCE', F, yb, 1971; flowers yellow, flushed pink, well-formed, dbl., 24 petals, 4.5 in., moderate fragrance; foliage light; [Colour Wonder X Sea Pearl]; Dickson, A.

Evasion, F, 1972; Hendricks

'EVE', HT, ob, 1954; bud long, pointed; flowers coral-red shaded yellow, very large, dbl., moderate fragrance; foliage glossy; vigorous growth; Gaujard

'EVE ALLEN', HT, rb, 1964; flowers crimson, reverse and base saffron-yellow, dbl., 26 petals, 5 in., moderate fragrance; foliage dark, glossy; vigorous growth; [Karl Herbst X Gay Crusader]; Allen, E.M.; Sanday

'EVEBRIGHT', F, rb, 1991; (**Cecily Gibson**, 'EVERBRIGHT'); bud ovoid; flowers current red, yellow base, blooms borne in sprays of 3–6, dbl., 35 petals, cupped, slight fragrance; foliage medium to large, dark green, glossy; bushy, medium to tall growth; [Southampton X ((Arthur Bell X Maigold) X Glenfiddich)]; Everitt, Derrick, 1990

'EVEGAYTIME', F, ob, 1999; (**Kantha Selvon**); flowers light orange, reverse cream yellow, dbl., 26–40 petals, 3–3.5 in., borne singly and in small clusters, slight fragrance; prickles moderate; foliage medium, medium green, glossy; upright, medium (2.5 ft) growth; TG, RNRS, 1997; [Pot o' Gold X (Mary Sumner X (Glenfiddich X (Arthur Bell X Maigold)))]; Everitt, Derrick

'EVEJAUNTY', F, ob, 1999; (**Alice Jarrett**); flowers orange, reverse cream yellow, dbl., 26–40 petals, 3–3.5 in., borne in small clusters, slight fragrance; prickles moderate; foliage medium, medium green, semi-glossy; upright, tall (3–3.5 ft) growth; [(Mary Sumner X (L'Oreal Trophy X Edith Holden)) X Friendship]; Everitt, Derrick

'EVELEINE', F, ob, 1993; (**Lady Madeleine**); flowers tangerine, paler reverse, blooms borne singly and in small clusters, dbl., 15–25 petals, 1.5–2.75 in., slight fragrance; some prickles; foliage medium, medium green, semi-glossy; medium (30 cms), upright growth; [(Pristine X Edith Holden) X (L'Oreal Trophy X Edith Holden)]; Everitt, Derrick

Evelien, *see* 'INTERLIEN'

Evelina, *see* 'AMAEVELINA'

'EVELYN', HT, lp, 1918; flowers soft pink, base yellow, dbl., 45–50 petals, moderate fragrance; RULED EXTINCT 1/92; [Ophelia sport]; Pierson, A.N.

'EVELYN', HT, MP, 1918; flowers salmon, shaded and edged rose, base yellow, imbricated, large, dbl.; RULED EXTINCT 1/92; Paul, W.

Evelyn, *see* 'AUSSAUCER'

Evelyn, Min, dp; Agel

'EVELYN BUCHAN', HT, pb, 1959; flowers pink tinted yellow, dbl., 23 petals, 3–4 in., exhibition form, moderate fragrance;

foliage leathery, dark; vigorous, upright growth; [Luis Brinas X Crimson Glory]; Riethmuller

'EVELYN DAUNTESSEY', HT, PB, 1909; flowers salmon stained carmine-rose; moderately vigorous growth; McGredy

'EVELYN ELLICE', F, LP, 1966; flowers light pink, becoming white, pointed, borne in clusters, 3 in.; foliage light green; vigorous growth; [Queen Elizabeth sport]; Ellice

Evelyn Fison, *see* 'MACEV'

Evelyn Grace, *see* 'HORAVME'

Evelyn Hough, HT, lp

'EVELYN MAY', HT, dp, 1932; flowers vermilion-pink, dbl., 65 petals, moderate fragrance; free growth; [Lady Alice Stanley X Edith Part]; Edward

'EVELYN MURLAND', HT, pb, 1923; flowers salmon-pink and carmine, veined yellow, reverse veined pink, dbl., moderate fragrance; Dickson, A.

Evelyn Rogers™, *see* 'MINARCO'

Evelyn Taylor, F, ob

'EVELYN THORNTON', Pol, mp, 1919; flowers shell-pink deepening to salmon and lemon shaded orange, open, dbl., moderate fragrance; foliage leathery, glossy, dark bronze; bushy growth; [Léonie Lamesch X Mrs W.H. Cutbush]; Bees

Evening Fire, *see* 'FROFIRE'

'EVENING GLOW', HT, r, 1959; bud long, pointed; flowers buff, large, dbl., 35 petals, moderate fragrance; foliage leathery; moderate growth; [Charlotte Armstrong X Narzisse]; Armbrust; Langbecker

Evening Light, *see* **'TARDE GRIS'**

'EVENING NEWS', HT, my, 1927; flowers apricot-yellow veined rose, base deep buttercup-yellow, open, dbl., moderate fragrance; foliage glossy, rich green; vigorous, bushy growth; [Mme Edouard Herriot sport]; Letts

Evening Queen, *see* 'HADEVENING'

Evening Sentinel, *see* 'FRYevenest'

Evening Shadows, *see* 'MICEVEN'

'EVENING SKY', HT, ob, 1939; flowers orange, tipped scarlet, base yellow, single, 6–8 ruffled petals, 2.5–3 in., moderate fragrance; foliage bluish green; vigorous growth; [Talisman X Unknown]; Moore, Ralph S.

Evening Star®, *see* 'JACVEN'

'EVENING STAR', HT, my, 1919; flowers golden yellow, shaded apricot, large, dbl., slight fragrance; bushy growth; [Mme Edouard Herriot sport]; Morse

'EVENING TELEGRAPH', HT, dy, 1976; [Whisky Mac sport]; Haynes

'EVENSONG', HT, op, 1963; flowers rosy salmon, well-formed, dbl., 25 petals, 5 in.; foliage dark; vigorous growth; [Ena Harkness X Sutter's Gold]; Arnot; Croll

'EVENTAIL', HT, yb, 1989; bud ovoid; flowers light yellow to pink, medium, borne singly, dbl., 50 petals, slight fragrance; prickles downward-pointed, reddish-purple; foliage medium, semi-glossy, slightly denticulated; upright, tall growth; [Sonia X Miyabi]; Kono, Yoshito

'EVENTIDE', HT, dr, 1948; bud ovoid; flowers dark velvety red, open, medium, dbl., moderate fragrance; foliage wrinkled, soft; moderate growth; [Crimson Glory seedling X Rouge Mallerin]; Toogood

'EVEPILK', F, pb, 1993; (**Mary Pilkington**); flowers peach, quickly paling to light pink, creamy yellow reverse, dbl., 15–25 petals, 1.5–2.75 in., moderate fragrance; some prickles; foliage medium, dark green, semi-glossy; tall (80–90 cms), bushy growth; [Mary Sumner X Remember Me]; Everitt, Derrick

'EVER READY', LCl, mr, 1976; flowers bright crimson, dbl., 32 petals, 3 in., recurrent bloom, moderate fragrance; foliage large, medium green, matte; [Aloha X Étoile de Hollande]; MacLeod; Christie Nursery, Ltd.

'EVERBLOOM', Cl Pol, dp, 1939; flowers deep pink, borne in clusters, single, recurrent bloom; foliage glossy; height 3–6 ft first year, 8–10 ft in about 3 years; [Phyllis Bide seedling]; Archer

Everblooming Climbing Cecile Brunner, Cl Pol, lp, 1996; Heirloom

Everblooming Dr W. Van Fleet, *see* **'NEW DAWN'**

Everblooming Jack Rose, *see* **'RICHMOND'**

'EVERBLOOMING PILLAR NO. 122', LCl, yb, 1954; flowers light yellow and orange, dbl., 90 petals, 3.5–4.5 in., moderate fragrance; growth like a hybrid tea, followed by 4–5 ft, canes; [Seedling X Break o' Day, Climbing]; Brownell, H.C.

'EVERBLOOMING PILLAR NO. 126', LCl, mp, 1955; flowers pink, base yellow, dbl., 35–50 petals, 3.5–4.5 in., moderate fragrance; bushy, upright growth; [Seedling X Queen o' the Lakes]; Brownell, H.C.

'EVERBLOOMING PILLAR NO. 214', LCl, my, 1954; flowers amber-yellow, dbl., 75 petals, 3.5 in.; growth like a hybrid tea, followed by 4–5 ft, canes; [Seedling X Break o' Day, Climbing]; Brownell, H.C.

'EVERBLOOMING PILLAR NO. 340', LCl, pb, 1957; flowers pink and yellow, dbl., 35–40 petals, 4.5–5 in., exhibition form, moderate fragrance; growth like a hybrid tea, followed by longer canes.; [Queen o' the Lakes X Scarlet Sensation]; Brownell, H.C.

Everblooming Pillar No. 83, *see* **'SALMON ARCTIC'**

Everblooming Pillar No. 12, *see* **'ETERNAL FLAME'**

Everblooming Pillar No. 73, *see* **'SCARLET SENSATION'**

'EVERBRIGHT', F, rb, 1991; (**Cecily Gibson** 'EVEBRIGHT'); bud ovoid; flowers current red, yellow base, blooms borne in sprays of 3–6, dbl., 35 petals, cupped, slight fragrance; foliage medium to large, dark green, glossy; bushy, medium to tall growth; [Southampton X ((Arthur Bell X Maigold) X Glenfiddich)]; Everitt, Derrick, 1990

'EVERDREAM', HT, my, 1956; bud ovoid; flowers canary-yellow, dbl., 35–40 petals, 4–5 in., moderate fragrance; bushy growth; [Souv. de Claudius Pernet X Kaiserin Auguste Viktoria]; Motose

'EVEREST', HP, w, 1927; flowers cream-white, center tinted green-lemon, very large, dbl., 38 petals, exhibition form, moderate fragrance; foliage light; low, spreading growth; GM, NRS, 1927; [Candeur Lyonnaise X Mme Caristie Martel]; Easlea

'EVEREST DOUBLE FRAGRANCE', F, lp, 1980; bud pointed; dbl., 25 petals, borne 3–7 per cluster, intense fragrance; large prickles; foliage dark, heavily veined; tall, upright growth; [Dearest X Elizabeth of Glamis]; Beales, Peter, 1979

Everglades, *see* 'POUledge'

Everglo, Min, or, 1998; Laver, Keith G.

Everglow, F, ob

Evergold, LCl, my

'EVERGREEN GEM', HWich, w, 1899; bud buff; dbl., 2–3 in., borne in clusters, moderate, sweetbriar fragrance; foliage almost evergreen; vigorous, climbing or trailing growth; [R. wichurana X Marechal Niel]; Horvath; W.A. Manda

Everlasting Love, *see* 'GRAlove'

'EVERT REGTERSCHOT', F, mr, 1965; bud ovoid; flowers bright red, medium, semi-dbl.; foliage dark; [Korona X Seedling]; Buisman, G. A. H.

'EVERT VAN DYK', HT, mp, 1931; flowers rose-pink tinted salmon, large, dbl., exhibition form, slight fragrance; foliage dark; long stems; bushy growth; [Ophelia X Hill's America]; Van Rossem; H&S

'EVGHENYA', F, op, 1975; flowers coral-orange, darker petal edges, base cream, large, very dbl., 75 petals, borne in clusters of 3–12; foliage dark, glossy; vigorous growth; [Highlight X Masquerade]; Staikov, Prof. Dr V.; Kalaydjiev and Chorbadjiiski

'EVIAN CACHAT', HT, mp, 1939; flowers bright pink, center copper salmon, very large, dbl., cupped; vigorous growth; Chambard, C.

Evita, *see* 'POULVITA'

Evita Bezuidenhout, *see* 'JACblar'

'EVODIA', HMult, lp, 1925; flowers small, dbl., moderate fragrance; Bruder Alfons

Evolution, *see* 'OLIjfaon'

Evona's Gallica, G, pb

Evona's Yellow, HT, yb

'Evrard Ketten', HT, m, 1920; flowers bright unshaded carmine-purple, dbl., intense fragrance; [Farbenkonigin X Ruhm de Gartenwelt]; Ketten Bros.

'Excalibur', F, mr, 1967; flowers scarlet, semi-dbl., 14 petals, 2.5 in., blooms in clusters, slight fragrance; foliage dark, glossy; bushy growth; [Vera Dalton X Woburn Abbey]; Harkness

'Excellenz Kuntze', S, ly, 1909; flowers creamy yellow, small, borne in large clusters, dbl., moderate fragrance; foliage dark, glossy; vigorous, upright growth; [Aglaia X Souv. de Catherine Guillot]; Lambert, P.

'Excellenz M. Schmidt-Metzler', HT, w, 1910; flowers large, dbl., slight fragrance; Lambert, P.

'Excellenz von Schubert', Pol, dp, 1909; flowers dark carmine-rose, small blooms in clusters, dbl., late; foliage dark; vigorous growth; [Mme Norbert Levavasseur X Frau Karl Druschki]; Lambert, P.

'Excelsa', HWich, mr, 1909; (Red Dorothy Perkins); flowers irregular blooms in large clusters, dbl., cupped, non-recurrent; foliage rich green, glossy; vigorous, climbing (12–18 ft) growth; (14); Gertrude M. Hubbard, ARS, 1914; Walsh

'Excelsior', F, mp, 1959; flowers salmon, borne in clusters; foliage dark; vigorous, upright growth; [Pinocchio X Mrs Henri Daendels]; Buisman, G. A. H.

Exception, *see* **'Märchenland'**

Excitement™, *see* 'HILco'

Exodus, *see* 'GODusex'

Exotic®, *see* 'FRYxotic'

Exotic, HT, 1995; McGredy, Sam IV

Exotic Beauty, *see* 'LEOexbeau'

Exotica, *see* 'MEIrecrom'

Exploit®, *see* 'MEIlider'

Explorer, *see* DICrocky

Explorer's Dream, *see* 'MICexplore'

Expo 64, F, 1964; Tschanz, E.

'Exposition de Brie', F, lp; flowers bright red; Granger, 1865

Express, LCl, dp; flowers carmine pink, large, dbl.; VEG

Exquisite, HT, lp, 1979; bud long, pointed; dbl., 30 petals, 5.5–6. in., exhibition form, moderate fragrance; vigorous, upright growth; [Memoriam X ((Blanche Mallerin X Peace) X (Peace X Virgo))]; Leon, Charles F., Sr.

'Exquisite', HT, lp, 1918; flowers creamy yellow, dbl., slight fragrance; Therkildsen

Exquisite, HT, lp

'Extase', HT, dr, 1956; bud long, dark red; vigorous growth; [E.G. Hill X Seedling]; Delforge

'Extasis', LCl, dr, 1963; flowers red, shaded darker, blooms in clusters, single, 5 petals; vigorous growth; [Spectacular X Cocktail]; Dot, Simon

'Extravaganza', F, pb, 1974; bud ovoid; flowers pink, base cream, medium, dbl.; foliage leathery; vigorous, bushy growth; [Stella X (Sabrina X Golden Giant)]; Dawson, George; Neil

'Eydie', Min, w, 1981; flowers white with pink blush, fading to white, small blooms in clusters, dbl., 50 petals, moderate, tea fragrance; beige prickles; foliage medium green, purple, when young; vigorous, bushy growth; [Janna X Seedling]; Hooper, John C.; E.M. Brown

Eye Appeal®, *see* 'INTerpeel'

'Eye Liner', HT, dr, 1966; bud long, pointed; flowers blood red, medium, semi-dbl., exhibition form, slight fragrance; foliage dark, leathery; vigorous, upright growth; [Queen Elizabeth X Montezuma]; Armbrust; Langbecker

Eye of the Dragon, *see* 'CLEdrag'

Eye Opener®, *see* 'INTerop'

Eye Paint, *see* 'MACeye'

'Eyecatcher', F, pb, 1976; flowers pink flushed apricot, reverse silvery cream, dbl., 22 petals, 2.5 in., moderate fragrance; foliage glossy, light; [Arthur Bell X Pernille Poulsen]; Cants of Colchester, Ltd., 1977

Eyeopener, *see* 'INTerop'

Eyepaint®, *see* 'MACeye'

F

'F. Cambó', HT, mr, 1933; flowers carmine, large, dbl., cupped, moderate fragrance; foliage glossy; dwarf growth; [Li Bures X Florence L. Izzard]; Dot, Pedro

F. Cuixart, *see* **'Francesca de Cuixart'**

'F. Ferrer', LCl, dr, 1940; bud long, pointed; flowers dark velvety red, open, large, somewhat recurrent bloom, slight fragrance; foliage leathery; vigorous, climbing growth; Pahissa; J&P

'F. J. Grootendorst', HRg, mr, 1918; (Grootendorst, Grootendorst Red, Nelkenrose); flowers bright red, edges serrated like a carnation, small, dbl., blooms in clusters of up to 20, recurrent bloom, slight fragrance; foliage small, leathery, wrinkled, dark; vigorous, bushy growth; [R. rugosa rubra X Unknown Polyantha]; de Goey; F.J. Grootendorst

F. Katom®, Gr, op, 1982; McGredy, Sam IV

'F. L. de Voogd', HT, yb, 1920; flowers clear reddish yellow, semi-dbl.; [Mme Mélanie Soupert X Mme Jenny Gillemot]; Timmermans

'F. M. Vokes', HT, my, 1927; flowers yellow, passing to cream, semi-dbl., moderate fragrance; [Ophelia seedling]; Hicks

'F. P. Merritt', Cl HT, dr, 1951; flowers bright fiery crimson, dbl., 50–60 petals, 5 in., exhibition form, intense fragrance; foliage glossy; very long stems; very vigorous, climbing (12 ft) growth; [Hoosier Beauty sport]; Merritt

'F. R. Patzer', HT, pb, 1909; flowers creamy buff, reverse warm pink, large, dbl.; branching growth; Dickson, A.

'F. W. Alesworth', HT, dr, 1954; (Fred W. Alesworth); flowers deep crimson, well shaped, large, intense fragrance; foliage dark; vigorous growth; [Poinsettia X Crimson Glory]; Robinson, H.

'F. W. Lowe', HT, ab, 1936; flowers rich orange-yellow, well shaped; foliage glossy; vigorous growth; Lowe

F.K. Druschkii, *see* **'Frau Karl Druschki'**

F.R.M. Undritz, *see* **'Gen. John Pershing'**

F.W. Mee, *see* **'Fred W. Mee'**

Fab, *see* 'BOSconpea'

'Fabergé', F, pb, 1969; bud ovoid; flowers light peach-pink, reverse tinted yellow, large, dbl., exhibition form, slight fragrance; foliage dark, leathery; vigorous, dense, bushy growth; [Seedling X Zorina]; Boerner; J&P

'Fabienne', F, mp, 1958; flowers reddish salmon, small, dbl., 35 petals; foliage clear green; bushy, low growth; [Orange Triumph X (Independence X Floradora)]; Arles; Roses-France

Fabiola, *see* **'Queen Fabiola'**

'Fabvier', Ch, mr, 1832; flowers crimson-scarlet, very showy, medium, semi-dbl., recurrent bloom; Laffay, M.

'Facade', HT, mp, 1970; flowers apricot-pink, very large, very dbl., 60 petals, exhibition form, intense fragrance; foliage glossy, dark, leathery; vigorous, upright growth; [Elizabeth Fankhauser X Royal Highness]; Fankhauser

'Fackel', HT, dr, 1937; flowers crimson-red, shaded blackish, very dbl., cupped, slight fragrance; compact growth; [Vaterland X Barcelona]; Krause

'Faïence', HT, op, 1935; flowers peach and apricot, reverse pure yellow, large, dbl., 45 petals, cupped, slight fragrance; foliage leathery; vigorous, bushy growth; [Charles P. Kilham X Julien Potin]; Van Rossem; C-P, 1937

'Faint Heart', HT, w, 1980; flowers cream edged light pink, blooms 1–3 per cluster, dbl., 25 petals, exhibition form, slight fragrance; straight prickles; foliage mid to dark, leathery, semi-glossy; medium, branching growth; [Hawaii X Unnamed seedling]; Pavlick, Mike

Fair Bianca®, *see* 'AUSca'

Fair Dinkum, *see* 'TINdink'

'Fair Dinkum', F, 1966; flowers peach-salmon, 3–4 in., clusters, moderate fragrance; foliage glossy; RULED EXTINCT 6/83 ARM; [Queen Elizabeth X Circus]; Small

Fair Genie, *see* 'LAVmoth'

'Fair Lady', HT, ab, 1959; bud ovoid to pointed; flowers buff overcast pink, dbl., 50 petals, 4.5 in., exhibition form, intense fragrance; foliage glossy; vigorous, upright growth; [Golden Masterpiece X Tawny Gold]; Boerner; J&P

'Fair Maid', HT, mp, 1940; flowers bright rose to strawberry-pink, changing to deep pink, dbl., 60–75 petals, 4–5 in., intense fragrance; foliage light green; vigorous, upright growth; [Talisman sport]; Peirce

'Fair Marjorie', Pol, lp, 1952; bud ovoid, bright pink; flowers blush-pink, lighter in sun, medium, semi-dbl., cupped, intense fragrance; vigorous, bushy growth; [Katharina Zeimet seedling]; Armstrong, P.M.

Fair Molly, *see* 'MORfairpol'

Fair Opal, *see* **'Fire Opal'**

Fair Play®, *see* 'INTerfair'

Fair Pol, Min, 1985; Moore, R. E.; (Cavriglia)

Fair Princess, S, mr, 1999; Williams, J. Benjamin

'Fairest of Fair', Min, ly, 1982; flowers medium yellow, borne singly or 3–5 per cluster, 15–18 petals, exhibition form, slight, tea fragrance; foliage small, medium green; very compact, low growth; [Sunbonnet X Rise 'n' Shine]; Bennett, Cecilia 'Dee'; Tiny Petals Nursery

Fairfield Blaze, LCl, mr

Fairhope, *see* 'TALfairhope'

'Fairlane', Min, ly, 1980; flowers near white flushed pink and yellow, urn-shaped, dbl., 20 petals, exhibition form, borne 1–5 per cluster, slight fragrance; prickles slanted downward; foliage glossy, medium green, deeply serrated; compact, bushy growth; [Charlie McCarthy X Unnamed seedling]; Schwartz, Ernest W.; Nor'East Min. Roses

'Fairlie Rede', HT, op, 1937; flowers salmon, flushed fawn, large, dbl., moderate fragrance; vigorous growth; [Mrs E. Willis X Seedling]; Clark, A.; NRS Victoria

'Fairlight', F, op, 1964; flowers coppery salmon to flame, well-formed, in clusters, dbl., 3.5 in., moderate fragrance; foliage coppery bronze; [Joybells X Seedling]; Robinson, H.

Fairy, *see* **'The Fairy'**

Fairy Changeling, *see* 'HARnumerous'

'Fairy Cluster', F, mp, 1935; flowers rose-pink, borne in clusters, single, slight fragrance; foliage glossy; long stems; very vigorous growth; [Dainty Bess X Ideal]; Archer

Fairy Damsel®, *see* 'HARneatly'

Fairy Dance®, *see* HARward

'Fairy Dancers', HT, ab, 1969; flowers buff-pink, small, dbl., moderate fragrance; low, spreading growth; [Wendy Cussons X Diamond Jubilee]; Cocker

Fairy Floss, Min, lp; [Sweet Chariot X seedling]; Hannemann, F., 1992; The Rose Paradise; (Weatherly, L.)

Fairy Gold, *see* FRYgoldie

Fairy Hedge, *see* **'Baby Jayne'**

Fairy Like, *see* 'HARnimble'

'Fairy Magic', Min, mp, 1979; bud mossy, long, pointed; flowers medium, semi-dbl., 10–15 petals, 1.5 in., moderate fragrance; foliage small, glossy; bushy, upright growth; [Fairy Moss X Unnamed Min. Moss seedling]; Moore, Ralph S.; Sequoia Nursery

Fairy Maid, *see* 'HARlassie'

Fairy Moon, *see* POUlwarb

'FAIRY MOSS', Min, mp, 1969; bud mossy; semi-dbl.; foliage mini-moss, small, light green, leathery; vigorous, bushy, dwarf growth; [(Pinocchio X William Lobb) X New Penny]; Moore, Ralph S.; Sequoia Nursery

Fairy Pompons, *see* 'TRAPOM'

Fairy Prince®, *see* 'HARNOUGETTE'

'FAIRY PRINCESS', Cl Min, lp, 1955; bud pointed, salmon-apricot; very dbl., 1 in., blooms in clusters; foliage small, fern-like; height to $2\frac{1}{2}$ ft; [Eblouissant X Zee]; Moore, Ralph S.; Sequoia Nursery

Fairy Princess, *see* **'LILIBET'**

'FAIRY QUEEN', F, lp, 1971; bud ovoid; flowers bluish pink, center coral-pink, small, dbl., exhibition form, moderate fragrance; foliage small, glossy, bronze; very vigorous, bushy growth; [The Fairy X Queen Elizabeth]; Williams, J. Benjamin

Fairy Queen, Pol, mr; Vurens

Fairy Queen, *see* SPErien

Fairy Red 92, S, mr, 1992; Liebig

Fairy Ring, *see* 'HARNICELY'

Fairy Snow, *see* 'HARLITTLE'

Fairy Snow, *see* HOLfairy

'FAIRY TALE', S, lp, 1960; bud ovoid; flowers open, small, dbl., slight fragrance; foliage dark, glossy; very vigorous, upright (6 ft) growth; RULED EXTINCT 3/93 ARM; [The Fairy X Goldilocks]; Thomson

Fairy Tale, *see* JACchar

Fairy Tale Queen, *see* 'KOROYNESS'

Fairyland®, *see* 'HARLAYALONG'

Fairy-Tale, *see* KEIren

Faith, *see* 'HORFAIWIL'

Faith, HT, lp; [Amatsu Otome sport]; Grayco, 2000; (Weatherly, L.)

'FAITHFUL', F, rb, 1964; flowers crimson edged white, becoming velvety crimson, well shaped, semi-dbl.; foliage small, light green; vigorous growth; [Dusky Maiden X Tabarin]; Latham

Faithful, *see* HARessay

'FAJA LOBBI', HT, mr, 1963; flowers bright red, vigorous growth; [Queen Elizabeth X Florence Mary Morse]; Leenders, J.

Fakir, *see* 'MEICLOUX'

'FALBALA', HT, mp, 1948; bud long; flowers brilliant salmon, moderate fragrance; foliage glossy, dark; erect growth; Gaujard

Falcon, F, 1989; Dot, Simon; (Cavriglia)

Falcon, LCl, dr, 1998; Kordes

'FALKLAND', HSpn, w; flowers pale pink to nearly white, dbl.; low, bushy growth

Fall Festival™, *see* 'LAVFAL'

Fall Splendor, *see* 'MINIDOR'

'FAMA', HT, rb, 1942; flowers amber and red, large, dbl., 25–30 petals, cupped; foliage soft, light green; short stems; vigorous, upright, bushy growth; Dot, Pedro; C-P

Fama, F, dy, 1986; Cocker

Fame!™, *see* 'JACZOR'

'FAMOSA', HT, mp, 1964; flowers pink, well formed; [Tallyho X Flamingo]; Leenders, J.

Famous Cliff, *see* **'YAROSLAVNA'**

'FAN FARE 81', HT, ob, 1981; bud short, pointed; dbl., 23 petals, exhibition form, borne singly, slight, rose fragrance; straight, short, broad-based prickles; foliage medium to large; vigorous growth; [Cotillion X Hoosier Gold]; Byrum, Roy L.; Joseph H. Hill, Co.

'FAN MAIL', F, dy, 1973; bud long, pointed; flowers open, medium, dbl., moderate fragrance; foliage large, leathery; very vigorous, bushy growth; [Spanish Sun X Unnamed]; Boerner; Spek

'FANAL', F, mr, 1946; flowers large, borne in clusters of 10–15, dbl., 20 petals, moderate fragrance; foliage dark, glossy; upright growth; [(Johanna Tantau X Heidekind) X Hamburg]; Tantau

'FANCY', HT, mr, 1928; flowers peach shaded cherry-red, base yellow, open, semi-dbl., moderate fragrance; [Souv. de Claudius Pernet X Gen. Smuts]; Van Rossem

Fancy Beauty, F, ob, 1999

'FANCY FREE', HT, pb, 1922; flowers pink, center white, semi-dbl., moderate fragrance; dwarf growth; [Gustav Grunerwald X Seedling]; Clark, A.; NRS New South Wales

Fancy Hit, *see* POUlfancy

'FANCY LACE', HT, mp, 1967; bud ovoid; flowers pink tipped silver, large, dbl., exhibition form, moderate fragrance; foliage glossy; very vigorous, upright growth; [Unnamed seedling X Queen Elizabeth]; Patterson; Patterson Roses

Fancy Lady, *see* 'CLELADY'

Fancy Pants™, *see* 'KINFANCY'

Fancy Potluck, *see* 'LAVFAN'

Fancy Princess, *see* **'OTOHIME'**

'FANCY TALK', F, pb, 1965; bud urn-shaped; flowers pink tinted orange, small blooms in clusters, dbl., exhibition form, slight fragrance; foliage leathery; vigorous, bushy, low growth; [Spartan X Garnette]; Swim & Weeks; Weeks Wholesale Rose Growers

'FANCY THAT', Min, pb, 1989; bud pointed; flowers different shades of pink, reverse pink-yellow blend, medium, dbl., 68 petals, exhibition form, slight, spicy fragrance; globular, green-brown fruit; no prickles; foliage medium, medium green, semi-glossy; upright, spreading, medium, vigorous growth; [Rise 'n' Shine X Rainbow's End]; Jolly, Marie; Rosehill Farm

'FANDANGO', HT, mr, 1950; bud ovoid, turkey-red, base yellow; flowers orange-red, open, semi-dbl., 16–25 petals, 3.5–4.5 in., moderate fragrance; foliage leathery, glossy, dark; vigorous, upright, bushy growth; [Charlotte Armstrong X Unnamed seedling]; Swim, H.C.; Armstrong Nursery

Fandango, *see* MEIjadel

'FANELY REVOIL', HT, or, 1962; flowers cerise-red, reverse tinted orange, well formed; long stems; [Michele Meilland X Unnamed seedling]; Orard, Joseph

Fanette®, *see* 'LAPCAL'

'FANFARE', F, pb, 1956; bud urn shaped; flowers orange and salmon to pink, open, borne in large clusters, dbl., 20–30 petals, 3–4 in., cupped, moderate, spicy fragrance; foliage glossy, leathery; very vigorous, spreading growth; GM, Rome, 1955; [Fandango X Pinocchio]; Swim, H.C.; Armstrong Nursery

'FANION', HT, dy, 1961; (Puregold); bud long, pointed; flowers rich yellow, dbl., 30 petals, 3–3.5 in., exhibition form, intense fragrance; foliage leathery, dark, glossy; vigorous, upright growth; [Helen Fox X (Mrs Pierre S. duPont X Joanna Hill)]; Robichon; Ilgenfritz Nursery

'FANNY', HT, mr, 1935; flowers bright red; vigorous growth; [Hadley X Mrs Henry Winnett]; Lens

'FANNY BIAS', HGal, mp; flowers blush, center rosy, dbl., 3 in., moderate fragrance; erect, bushy growth; Vibert, before 1811

'FANNY BLANKERS-KOEN', HT, op, 1949; (Luxembourg); bud long, pointed; flowers orange-yellow, flushed and veined red, large, semi-dbl., 16 petals, intense fragrance; foliage glossy; very vigorous, upright, bushy growth; [Talisman X Unnamed seedling]; Verschuren-Pechtold

'FANNY BULLIVANT', HT, lp; [The Rajah sport]; Clark, A., 1941; (Weatherly, L.)

'FANNY ESSLER', HGal, mp; Vibert, 1835

'FANNY OPPENHEIMER', HT, mr, 1923; flowers brilliant cardinal, shaded gold, moderate fragrance; McGredy

Fanny Pavetot, HGal, mp; flowers large, very dbl., moderate fragrance; (Sangerhausen)

Fanny's La France, HT, dp

Fano, S, mp

'FANTAISIE', HT, r, 1948; flowers coppery-salmon, base yellow, slight fragrance; foliage dark; vigorous growth; Gaujard

Fantan, *see* 'MEIMEX'

'FANTAN', F, or, 1956; flowers coppery orange, small; foliage small; [Eternite X Seedling]; Gaujard

Fantasi™, *see* 'KINFANTA'

Fantasia, *see* 'KORFAN'

'FANTASIA', HT, my, 1943; bud long, pointed; flowers golden to lighter yellow, open, medium, dbl., 30–35 petals, intense fragrance; foliage glossy; strong stems; vigorous, bushy, compact growth; RULED EXTINCT; [Seedling X Lord Lonsdale]; Dickson, A.; J&P

'FANTASIA, CLIMBING', Cl HT, my, 1956; Mell

Fantasque, *see* **'FANTASTIQUE'**

'FANTASTIQUE', HT, yb, 1943; (Fantasque); flowers yellow, heavily edged carmine, dbl., 38 petals, 2.5–3 in., cupped, moderate, spicy fragrance; foliage dark, leathery, glossy; vigorous, compact growth; [Ampere X (Charles P. Kilham X (Charles P. Kilham X Capucine-Chambard))]; Meilland, F.; C-P

'FANTASY', HT, rb, 1945; bud long, pointed; flowers cerise, reverse yellow ashed pink, open, blooms, semi-dbl., 9–11 petals, 4.5 in., slight fragrance; foliage leathery, glossy; slender stems; very vigorous, upright growth; RULED EXTINCT 1/86 ARM; [Soeur Thérèse X Unnamed seedling]; Wyant

Fantasy, *see* JACpow

'FANTIN-LATOUR', C, lp; flowers blush, dbl., flat, moderate fragrance; foliage dark, broad; vigorous, bushy growth

'FAR SIDE', F, ob, 1988; bud ovoid, pointed; flowers orange-vermillion, yellow center, small, borne in sprays of, single, 5 petals, moderate, spicy fragrance; globular, medium, vermillion fruit; prickles straight, medium, stout, pink to brown; foliage large, medium green, glossy, smooth; upright, spreading, tall growth; [[Sunsprite X (Many Moons X Maigold)] X Eyepaint]; Stoddard, Louis, 1990

'FARAH', HT, yb, 1961; flowers coppery yellow, large, dbl., moderate fragrance; foliage glossy; long stems; very vigorous, bushy growth; [Peace X Georges Chesnel]; Gaujard

'FARANDOLE', F, mr, 1959; bud oval; flowers vermilion, open, medium, borne in large clusters, dbl., 25 petals; foliage leathery; vigorous, well branched growth; GM, Rome, 1959; [(Goldilocks X Moulin Rouge) X (Goldilocks X Fashion)]; Meilland, Mrs. Marie-Louise; URS

'FARBENKÖNIGEN', HT, mr, 1902; (Queen of Colors, Reine des Couleurs); flowers rosy carmine, dbl., moderate fragrance; [Grand-Duc Adolphe de Luxembourg X La France]; Hinner, W.

'FARBENSPIEL', F, rb, 1960; flowers pink edged red, dbl., intense fragrance; foliage dark; bushy, compact growth; [Pinocchio X Masquerade]; Verschuren, A.; van Engelen

'FARFADET', F, or, 1955; semi-dbl.; very upright growth; [Méphisto X Incendie]; Combe

Faria, *see* OLIjkroet

'FARIDA', F, dp, 1941; flowers deep rose-pink, semi-dbl., moderate fragrance; [Seedling X Permanent Wave]; Leenders, M.

'FARNY WURLITZER', HT, mr, 1968; bud long, pointed; flowers large, dbl., intense fragrance; foliage dark, glossy; vigorous, upright growth; [Poinsettia X Charlotte Armstrong]; Cadey; Ty-Tex Rose Nursery

'FARQUHAR', HWich, mp, 1903; (The Farquhar Rose); flowers bright clear pink, carnation-like, resembling Lady Gay, dbl., late; vigorous, climbing growth; [R. wichurana X Crimson Rambler]; Dawson; Farquhar

'FASCINATING', HT, yb, 1961; bud long, pointed; flowers rose-opal suffused yellow, dbl., 25 petals, 4.5 in., exhibition form, moderate fragrance; foliage leathery, glossy, dark; upright, bushy growth; [Peace X Orange Nassau]; Fisher, G.; C-P

Fascination, *see* 'JACOYEL'

'FASCINATION', HT, pb, 1927; flowers rosy cerise, shaded yellow, moderate fragrance; foliage dark, glossy; vigorous growth; RULED EXTINCT 2/81; Chaplin Bros.

'FASHION', F, pb, 1949; bud ovoid, deep peach; flowers lively coral-peach, blooms in clusters, dbl., 23 petals, 3–3.5 in.; moderate fragrance; vigorous, bushy growth; (28); AARS, 1950 David Fuerstenberg Prize, ARS, 1950 GM, ARS, 1954 GM, Bagatelle, 1949 GM, NRS, 1948 GM, Portland, 1949; [Pinocchio X Crimson Glory]; Boerner; J&P

'FASHION FLAME', Min, op, 1977; bud ovoid, pointed; flowers coral-orange, dbl., 35 petals, 1–1.5 in., exhibition form, slight fragrance; foliage large, leathery; bushy growth; [Little Darling X Fire Princess]; Moore, Ralph S.; Sequoia Nursery

Fashion Flame, Min, op

Fashion Parade, Min, mp, 1994; Poulsen

'FASHION, CLIMBING', Cl F, op, 1951; Boerner

'FASHION, CLIMBING', Cl F, op, 1955; Mattock

'FASHIONETTE', F, pb, 1958; bud pointed to ovoid, coral; flowers pinkish coral, borne in irregular clusters, dbl., 35–40 petals, 3 in., cupped, moderate fragrance; foliage glossy; vigorous, upright growth; [Goldilocks X Fashion]; Boerner; J&P

'FASHIONETTE, CLIMBING', Cl F, pb, 1962; Noack, Werner

Fassadenzauber, LCl, lp, 1997; Noack, Werner

Fat 'n' Sassy, *see* 'TINSASSY'

Fat Tuesday, *see* 'TALFAT'

'FATA MORGANA', F, ob, 1957; flowers orange-yellow, open, borne in large trusses, dbl., 28 petals, 2.5 in., slight fragrance; foliage leathery, glossy; vigorous, upright growth; [Masquerade X Seedling]; Kordes, R.

Father Christmas, *see* 'TALCHRIS'

Father David's Rose, *see* **R. DAVIDII**

Father Hugo Rose, *see* **R. HUGONIS**

Father Hugo's Rose, *see* **R. HUGONIS**

Father's Day, *see* **'VATERTAG'**®

'FATIMA', HT, ob, 1955; bud long, pointed; flowers orange, reverse bright golden yellow, very large, moderate fragrance; vigorous growth; [Opera seedling X Unnamed seedling]; Gaujard

Fatima '67, HT, 1967; Moreira da Silva, A.; (Cavriglia)

'FATIME', HGal, mp, 1820; flowers pink, dotted and spotted both lighter and darker, medium, Descemet, M.

Fatinitza, HT, op, 1956; flowers yellowish salmon-pink, medium, dbl.; Berger, W.; (Sangerhausen)

'FAUST', F, yb, 1957; (Dr Faust); flowers golden yellow shaded orange-pink, blooms in large clusters, dbl., 25 petals, 2 in., moderate fragrance; foliage dark, glossy; vigorous, bushy growth; GM, NRS, 1956; [Masquerade X Golden Scepter]; Kordes, R.; A. Dickson;, McGredy

Faust, *see* **'SCARLANO'**

'FAUST, CLIMBING', Cl F, yb, 1963; deRuiter

Favori, *see* 'LEN 3'

'FAVORITA'™, HT, op, 1954; bud ovoid, burnt-orange; flowers salmon overcast orange, dbl., 48 petals, 5.5–6. in., moderate fragrance; foliage dark; vigorous growth; GM, Rome, 1952; [Unnamed HT seedling X Serenade]; Boerner; Stark Bros.

Favorite, *see* 'LEN 3'

Favorite, *see* LEBbet

Favorite Dream, *see* TWOfavor

Favourite Rosamini, *see* RUIfarol

'FAY', S, pb, 1997; flowers double, medium, dbl., 15–25 petals, borne mostly singly, moderate fragrance; few prickles; foliage medium, light green, semi-glossy; upright, medium (36 in.) growth; [Cecile Brunner X Zepherin Drouhin]; Robert, David

'FAYANNE', F, pb, 1953; flowers deep rose-pink, reverse lighter, small, dbl., 55 petals, 2 in., globular, slight fragrance; foliage leathery; dwarf, bushy growth; [Garnette sport]; Pinchbeck

'FAYBELL', HT, ob, 1974; (**Rebell**); flowers dark orange, large, dbl., exhibition form, intense fragrance; foliage leathery; vigorous, upright growth; [Brandenburg X Seedling]; Kordes; Fey

Faye Reynolds, *see* 'REYFAYE'

FAZcanne, HT, mr, 1999; (**Red France**)

'**Fear Naught**', F, dp, 1968; flowers deep pink, dbl., slight fragrance; [Queen Elizabeth X Ena Harkness]; Harkness

FEbama, HT, lp; (**Iberflora 95**®); exhibition form, 28 petals, none; [FE-85138 X Dallas]; Roses Noves Ferrer, S L; (Asociacion Espanola de la Rose)

FEbasa, F, lp; (**Esther Peiro**®); exhibition form, 26 petals, none; [Lamada X K-881543-02]; Roses Noves Ferrer, S L, 1996; (Asociacion Espanola de la Rose)

FEbesa, Gr, lp; (**Vanessa Campello**®); exhibition form, 32 petals, none; [Lambada X K-881543-02]; Roses Noves Ferrer, S L, 1996; (Asociacion Espanola de la Rose)

FEcasa, HT, lp; (**Barcelona 95**®); exhibition form, 26 petals, slight; [FE-903021 X FE-89081]; Roses Noves Ferrer, S L, 1995; (Asociacion Espanola de la Rose)

Fecirco, HT, 1987; Ferrer, F.; (Cavriglia)

'**Federation**', LCl, op, 1938; bud pointed; flowers rosy pink, orange undertone, dbl., 24–36 wavy petals, 3.5 in., cupped, intense fragrance; foliage leathery, glossy, dark; long, strong stems; very vigorous climbing (12–14 ft) growth; [(R. setigera X Mrs F.F. Prentiss) X Director Rubió]; Horvath; Wayside Gardens Co.

'**Federico Casas**', HT, op, 1931; flowers coppery pink and orange, open, very large, semi-dbl., intense fragrance; foliage dark; vigorous, bushy growth; [Unnamed variety X Eugene Barbier]; Dot, Pedro; C-P

'**Federico Casas, Climbing**', Cl HT, op, 1937; Stell; Stell Rose Nursery

Federico Garcia Lorca®, *see* FEticos

'**Fedra**', F, mr, 1959; flowers brick-red, dbl.; vigorous growth; [(Fiamma X Independence) X Unnamed seedling]; Giacomasso

Fedugia, HT, 1988; Ferrer, F.; (Cavriglia)

FEdugia, HT, lp; (**Marta Salvador**®); globular, 32 petals, moderate; [Zambra X White Satin]; Viveros Fco. Ferrer, S L, 1987; (Asociacion Espanola de la Rose)

'**Fee**', F, or, 1963; flowers orange, large, dbl.; strong, wiry stems; moderate growth; Kordes, R.

Fee, *see* KORmasyl

'**Fée des Champs**', Pol, op, 1965; flowers salmon-orange, dbl., 20–25 petals, foliage bronze; low growth; GM, Bagatelle, 1965; [Queen Elizabeth X Zambra]; Dot; Minier

Fée des Neiges, *see* 'KORbin'

Feeling®, *see* BARfeel

Feeling®, HMsk, yb, 1994; Lens

Feerie, *see* '**The Fairy**'

'**Féerie**', HT, ob, 1938; flowers coppery red, reverse orange-yellow, open, semi-dbl., moderate fragrance; long stems; vigorous, erect growth; Gaujard

FEgama, HT, lp; (**Adela**®); dbl., 32 petals, slight; [Shocking Blue X Kardinal]; Viveros Fco. Ferrer, S L, 1991; (Asociacion Espanola de la Rose)

FEgima, HT, lp; (**Elodea**®); exhibition form, 22 petals, none; [FE-87518 X K-861242-1]; Roses Noves Ferrer, S L, 1994; (Asociacion Espanola de la Rose)

FEgosa, Min, lp; (**Alejandra Conde**®); rosette, 22 petals, none; [M Litigan X Scarlet Meilladina]; Viveros Fco. Ferrer, S L, 1991; (Asociacion Espanola de la Rose)

FEkusa, HT, lp; (**Charlye Rivel**®); exhibition form, 32 petals, none; [Jack-8301365 X Inedita]; Roses Noves Ferrer, S L, 1995; (Asociacion Espanola de la Rose)

'**Felberg's Rosa Druschki**', HP, mp, 1929; flowers bright rose-pink, large, dbl., 25 petals; [Frau Karl Druschki X Farbenkonigin]; Felberg-Leclerc

'**Felicia**', HMsk, pb, 1928; flowers pink fading to blush and partly white, large, branching panicles, semi-dbl., moderate, musk fragrance; pillar or shrub growth; (21); [Trier X Ophelia]; Pemberton

Felicia Teichmann, *see* KORciate

Felicitas, *see* KORberis

'**Félicité Bohain**', M, dp; (Félicité Bohan); flowers vivid pink or bright rose, large, dbl.; (France), before 1866

Félicité Bohan, *see* '**Félicité Bohain**'

'**Félicité et Perpétue**', HSem, w, 1828; flowers pale flesh changing to white, fairly large blooms in clusters, very dbl., flat; foliage almost evergreen; very vigorous growth; [Thought to be R. sempervirens X Noisette]; Jacques, 1827

'**Félicité Parmentier**', A, lp; flowers soft flesh-pink, and then reflexing, very dbl., flat, intense fragrance; foliage gray-green; vigorous, compact growth

'**Felicity**', HT, mp, 1919; flowers rose-pink suffused silvery, large, dbl., 50–60 petals, moderate fragrance; foliage dark; vigorous, branching growth; [Ophelia X Hoosier Beauty]; Clarke Bros.

Felicity II, *see* '**Buttons 'n' Bows**'

Felicity Kendal, *see* 'LANken'

'**Felix Brix**', HT, pb, 1921; flowers soft rose, suffused yellow, passing to salmon-rose, semi-dbl.; [Natalie Boettner X Old Gold]; Brix; Teschendorff

'**Felix Laporte**', HT, m, 1928; flowers blackish velvety purple tinged garnet, dbl., cupped, intense fragrance; foliage dark, leathery, glossy; vigorous, bushy growth; [Yves Druhen X Mme Edouard Herriot]; Buatois

'**Felix Mousset**', HP, 1878; Verdier, E.; (Cavriglia)

'**Fellemberg**', HCh, mr; flowers bright crimson, cupped, 36 petals, cupped; foliage dark; vigorous, spreading growth; Fellemberg, 1857; Introduced prior to 1835

Fellini, *see* INTerdain

Fellowship, *see* '**De Greeff's Jubilee**'

Fellowship, *see* 'HARwelcome'

FEloma, HT, lp; (**Coral**®); exhibition form, 22 petals, none; [FE-87518 X K-861242-1]; Roses Noves Ferrer, S L, 1994; (Asociacion Espanola de la Rose)

FEmental, Gr, lp; (**Juli de Sala**®); exhibition form, 30 petals, none; [FE-85143 X MT-86173]; Roses Noves Ferrer, S L, 1994; (Asociacion Espanola de la Rose)

FEmiento, HT, lp; (**Ana de Cuevas**); exhibition form, 24 petals, slight; [Bucaneer X Cocktail]; Viveros Fco. Ferrer, S L, 1987; (Asociacion Espanola de la Rose)

'**Femina**', HT, op, 1963; bud long, pointed; flowers salmon-pink, large, dbl., moderate fragrance; foliage leathery; vigorous, upright growth; [Fernand Arles X Mignonne]; Gaujard; Ilgenfritz Nursery, 1966

'**Femina**', HT, lp, 1957; flowers soft pink, large, dbl.; vigorous growth; Poulsen, S.

Femme, *see* 'DELvor'

Femnet, *see* PEKaledon

'**Fen Queen**', F, lp, 1963; flowers pale flesh, large, dbl., 25 petals, slight fragrance; foliage light green; vigorous, upright growth; [Queen Elizabeth X Unknown Hybrid Tea]; Sharman

Fenja, S, mp, 1965; Petersen

Fennet, HT, rb; Laperrière

Fennica, *see* 'RUntru'

FEodase, HT, lp; (**Condesa de Barcelona**®); exhibition form, 26 petals, slight; [Mount Shasta X Helmut Schmidt]; Viveros Fco. Ferrer, S L, 1988; (Asociacion Espanola de la Rose)

FEprogo, HT, lp; (**Lourdes Arroyo**®); exhibition form, 26 petals, slight; [Zambra X Inedita]; Viveros Fco. Ferrer, S L, 1990; (Asociacion Espanola de la Rose)

FEpuma, HT, lp; (**Carolina Daza**®); exhibition form, 28 petals, slight; [Zambra X Kolner Karneval]; Viveros Fco. Ferrer, S L, 1991; (Asociacion Espanola de la Rose)

FErama, HT, lp; (**Catalina Frau**®); exhibition form, 28 petals, none; [FE-85146 X Osiana]; Viveros Fco. Ferrer, S L, 1991; (Asociacion Espanola de la Rose)

Ferdi, *see* 'KEItoli'

'**Ferdinand Chaffolte**', HP, mr, 1879; flowers large, very dbl., moderate fragrance; Pernet Père; (Sangerhausen)

'**Ferdinand de Buck**', HGal, mp; (Feu de Buck); flowers brilliant pink, medium, dbl.

'FERDINAND DE LESSEPS', HP, lp; flowers deep crimson; Verdier, E., 1869

'FERDINAND JAMIN', HP, mr, 1888; flowers vermilion-red; Lévêque

'FERDINAND JAMIN', HT, lp; Pernet-Ducher, 1896

'FERDINAND PICHARD', HP, rb, 1921; flowers streaked (striped) pink and scarlet, yellow stamens, dbl., 25 petals, recurrent bloom, vigorous, tall growth; Tanne

'FERDINAND ROUSSEL', HWich, m, 1902; flowers purple/pink to light red, small, dbl.; Barbier; (Sangerhausen)

Ferdinando Dukei, HT, yb; flowers orange-yellow with red, large, semi-dbl., moderate fragrance; (Sangerhausen)

Ferdy™, *see* 'KEItoli'

FErecha®, HT, lp; (**Iris**); exhibition form, 27 petals, none; [Zambra X Jelcanodir]; Viveros Fco. Ferrer, S L, 1991; (Asociacion Espanola de la Rose)

Fergie, *see* 'GANfer'

Fergus Games, S, op, 1999; flowers light salmon pink, dbl.; Williams, J. Benjamin

FERho, LCl, mp, 1997; (**Childhood Memories**); Fergusson

Feria, *see* 'MEIfrison'

Ferline®, *see* LAPdade

'FERN KEMP', HRg, lp, 1918; flowers delicate pink, semi-dbl., 4 in., intense fragrance; vigorous growth; hardy; [Conrad Ferdinand Meyer X Frau Karl Druschki]; Kemp, J.A.

'FERN ROEHRS', LCl, mp, 1943; flowers same as parent but much more, dbl.; [Paul's Scarlet Climber sport]; Graf; Roehrs

'FERNAND ARLES', HT, op, 1949; (Arles); bud long, pointed; flowers orange-salmon shaded red, very large, dbl., moderate fragrance; foliage bronze; very vigorous, bushy growth; [Mme Joseph Perraud X Unnamed seedling]; Gaujard

Fernand Majorel, F, 1975; Gaujard; (Cavriglia)

'FERNAND POINT', HT, rb, 1964; flowers crimson-red, reverse flesh-pink, base yellow, foliage glossy; [Peace X Unnamed seedling]; Orard, Joseph

'FERNAND RABIER', HWich, dr, 1918; flowers pure deep scarlet, borne in clusters of 40–50, dbl., slight fragrance; vigorous, climbing or trailer growth; [Delight X Unnamed seedling]; Turbat

'FERNAND TANNE', LCl, dy, 1920; flowers deep yellow to cream-yellow, large, dbl., intense fragrance; vigorous growth; Tanne; Turbat

Fernanda, HT; (Brazil); (Cavriglia)

'FERNANDE KRIER', HWich, pb, 1925; flowers peach-pink, occasionally margined red or sometimes entirely, [Excelsa sport]; Walter, L.

'FERNANDE LUMAY', HT, ab, 1922; flowers apricot-nankeen-yellow, edged milk-white, dbl., moderate fragrance; [Mrs Aaron Ward X Unnamed seedling]; Buatois

Ferndale Red China, Ch, dp; (found rose)

'FERNIELEA', HT, ab, 1926; flowers apricot, center deeper, moderate fragrance; Adam & Craigmile

Ferris Wheel, *see* 'AROyumi'

'FERRY PORSCHE', HT, mr, 1971; bud long, pointed; dbl., 34 petals, exhibition form, borne singly and in small clusters, slight fragrance; foliage large, dark, soft; vigorous, upright growth; [Tropicana X Americana]; Kordes, W. Söhne

'FERtry', S, my, 1998; (**The Didgemere Rose**); flowers medium yellow, full, very dbl., 26–40 petals, 2.5–3 in., borne mostly singly, moderate fragrance; prickles small, slightly hooked; foliage medium, medium green, dull; upright, low (3–4 ft) growth; [Sunsprite X Trier]; Ferguson, B.; Burston Nurseries, Inc., 1998

'FERVID', F, or, 1960; flowers scarlet-orange, in clusters, single, 3 in., slight fragrance; foliage glossy, dark; vigorous, upright growth; [Pimpernell X Korona]; LeGrice

FEsaru, HT, lp; (**Neus**®); exhibition form, 25 petals, moderate; [Shocking Blue X Carta Blanca]; Viveros Fco. Ferrer, S L, 1991; (Asociacion Espanola de la Rose)

FEsato, Min, lp; (**Clara**®); rosette, 20 petals, none; [M Litigan X Inedita]; Viveros Fco. Ferrer, S L, 1986; (Asociacion Espanola de la Rose)

FEsodios, HT, lp; (**Emilo Feliu**®); exhibition form, 28 petals, slight; [Zambra X Osiana]; Viveros Fco. Ferrer, S L, 1992; (Asociacion Espanola de la Rose)

'FESTIVAL', HT, mr, 1943; (Dixie Dream); flowers rich red, very dbl., moderate fragrance; entirely prickle free; RULED EXTINCT, 8/93 ARM; [E.G. Hill seedling]; Dixie Rose Nursery; Krider Nursery, 1945

Festival, *see* KORdialo

Festival Beauty, *see* **'KRASAVITZA FESTIVALIA'**

Festival Fanfare, *see* 'BLEstogil'

Festival Pink™, *see* 'RUPfespin'

'FESTIVAL QUEEN', HT, lp, 1968; bud long, pointed; flowers light pink, edged deeper, well-formed, exhibition form, slight fragrance; foliage dark, leathery; upright, compact growth; Lindquist; Edmunds Roses

Festival Rouge®, *see* 'DELfesrou'

'FESTIVAL, CLIMBING', Cl HT, mr, 1945; flowers large, globular, free, intermittent bloom, intense fragrance; thornless; foliage glossy; very vigorous, climbing growth; hardy in South.; [E.G. Hill sport]; Watkins, A.F.; Dixie Rose Nursery

'FESTIVE', Pol, dr, 1975; bud short, pointed; dbl., 30 petals, 3–3.5 in., exhibition form, repeat bloom, slight, sweetbriar fragrance; foliage parsley-green; vigorous, upright growth; [Bacarra X Seedling]; Jelly; E.G. Hill Co.

Festivity, *see* 'LENor'

FEstro, HT, lp; (**Alba Garcia**®); exhibition form, 26 petals, intense; [Shocking Blue X Carta Blanca]; Viveros Fco. Ferrer, S L, 1991; (Asociacion Espanola de la Rose)

FEsuma, HT, lp; (**Gloria Ferrer**®); exhibition form, 30 petals, slight; [Zambra X Osiana]; Viveros Fco. Ferrer, S L, 1992; (Asociacion Espanola de la Rose)

Fête des Mères, *see* **'MOTHERSDAY'**

Fête des Pères, *see* **'VATERTAG'**®

Fêtes Galantes®, *see* DELdido

'FÉTICHE', F, mr, 1962; flowers coral-red, center white, open, borne in clusters, semi-dbl., 12 petals, 2 in.; foliage dark; moderate, bushy growth; [Philippe X Tabarin]; Delforge

FEticos, HT, lp; (**Federico Garcia Lorca**®); globular, 30 petals, slight; [Zambra X Susan Hamshire]; Viveros Fco. Ferrer, S L, 1989; (Asociacion Espanola de la Rose)

FEtrone, HT, lp; (**Rocio Elias**®); flat, 32 petals, moderate; [Carinella X Kardinal]; Viveros Fco. Ferrer, S L, 1988; (Asociacion Espanola de la Rose)

FEtural, Min, lp; (**Lucia Cotarelo**®); rosette, 15 petals, none; [M Litigan X Scarlet Meilladina]; Viveros Fco. Ferrer, S L, 1991; (Asociacion Espanola de la Rose)

'FEU D'ARTIFICE', LCl, yb, 1935; (Fireworks, Climbing); bud long, pointed, nasturtium-red; flowers yellow, tinted nasturtium-red, open, borne in clusters, semi-dbl.; long stems; vigorous, climbing (over 8 ft) growth; [R. foetida hybrid X Colette Clément]; Mallerin, C.; B&A, 1939

Feu d'Artifice, *see* **'FEUERWERK'**®

'FEU DE BENGALE', F, lp, 1951; flowers pearl-pink, large, dbl., cupped, moderate fragrance; foliage glossy; vigorous growth; [Orange Triumph X Unnamed seedling]; Gaujard, R.; G. Truffaut

Feu de Buck, *see* **'FERDINAND DE BUCK'**

Feu de Camp, *see* **'LAGERFEUER'**®

'FEU DE JOIE', F, mr, 1951; flowers carmine-red, large, borne in clusters, dbl.; foliage dark, glossy; vigorous growth; [Orange Triumph X Unnamed seedling]; Gaujard, R.; G. Truffaut

'FEU DE SAINT-JEAN', Pol, dr, 1951; flowers blackish red, small, borne in clusters, semi-dbl.; foliage dark, glossy; vigorous, bushy growth; [Orange Triumph X Unnamed seedling]; Gaujard, R.; G. Truffaut

'FEU D'ENFER', F, or, 1958; flowers large, dbl.; upright, well branched growth;

[Orange Triumph X Unnamed seedling]; Gaujard, R.; G. Truffaut

'FEU DU CIEL', Pol, ob, 1951; flowers clear orange, small, borne in clusters, semi-dbl.; foliage glossy; vigorous, bushy growth; [Orange Triumph X Unnamed seedling]; Gaujard, R.; G. Truffaut

'FEU FOLLET', F, mp, 1953; flowers salmon, cupped, moderate fragrance; vigorous growth; Gaujard, R.

Feu Follet®, *see* DIClady

'FEU JOSEPH LOOYMANS', HT, ob, 1921; flowers orange-yellow, large blooms, dbl., cupped; foliage leathery; weak stems; vigorous growth; [Sunburst X Rayon d'Or]; Looymans

'FEU JOSEPH LOOYMANS, CLIMBING', Cl HT, ob, 1935; Western Rose Co.

'FEU MAGIQUE', F, mr, 1956; flowers cherry-red, semi-dbl.; 18 petals; vigorous growth; [Independence X Signal Red]; Frères, Buyl

'FEU PERNET-DUCHER', HT, my, 1935; flowers bright yellow, center apricot, large, dbl., moderate, fruity fragrance; foliage leathery, dark; vigorous, branching growth; GM, Portland, 1936; [Julien Potin X Margaret McGredy]; Mallerin, C.; A. Meilland; C-P

'FEU ROUGE', F, mr, 1956; bud ovoid; flowers open, large, semi-dbl., slight fragrance; foliage glossy; vigorous growth; [Red Favorite X Fanal]; Tantau, Math.

'FEUDOR', HT, yb, 1963; flowers golden yellow shaded vermilion; foliage bright green; vigorous growth; [Peace X Baccará]; Croix, P.; Minier

'FEUERBALL', HT, mr, 1965; bud globular; flowers red lead color, well formed, large, dbl., 25–30 petals; foliage glossy; very vigorous, bushy growth; Tantau, Math.

Feuerfunken, S, or, 1979; flowers large, dbl.; GPG Bad Langensalza; (Sangerhausen)

'FEUERLAND', F, or, 1977; bud ovoid; dbl., 25 petals, 2.5 in., cupped, slight fragrance; very vigorous, upright growth; [Kathe Duvigneau X Topsi]; Kordes, W. Söhne

Feuermeer, *see* **'SEA OF FIRE'**

'FEUERREITER', F, mr, 1968; (Fire-Rider); bud long, pointed; flowers open, large, borne in clusters, semi-dbl., slight fragrance; foliage dark, leathery; very vigorous, upright growth; [Alain X Oskar Scheerer]; Haenchen, E.; Teschendorff

'FEUERSCHEIN', F, mr, 1930; (Krause's Rote Joseph Guy); flowers brilliant red, not turning blue, dbl.; foliage dark; bushy growth; [Lafayette sport]; Krause

'FEUERSCHEIN, CLIMBING', Cl F, mr, 1936; Krause

Feuersturm, F, or; flowers medium, slight fragrance; Verschuren, after 1900; (Sangerhausen)

Feuertaufe, Pol, 1980; dbl.; GPG Bad Langensalza; (Sangerhausen)

'FEUERWERK'®, S, ob, 1962; (Feu d'Artifice, Magneet); flowers bright orange, blooms in clusters, semi-dbl.; foliage glossy; upright, bushy (to 5 ft) growth; Tantau, Math.

Feuerzauber®, *see* 'KORFEU'

'FEUERZAUBER', Ch, mr, 1913

'FEURIO', F, or, 1956; bud ovoid; flowers scarlet-red, borne in clusters, dbl., 30 petals, 2.5 in., cupped, moderate fragrance; foliage glossy, light green; vigorous, low, bushy growth; [Rudolph Timm X Independence]; Kordes, R.; J&P, 1957

'FEURIO, CLIMBING', Cl F., or, 1963; Kordes

Fever, *see* 'PEAFEVER'

Fewell's Noisette, N, lp

'FIABA', HT, rb, 1963; flowers bright red tipped yellow; foliage dark, glossy; very vigorous growth; [(Fiamma X Sovrana) X Unnamed seedling]; Giacomasso

'FIALAWN', HT, mp, 1985; (**Wood Lawn**); flowers clear medium pink, [Fontainebleau sport]; Fiamingo, Joe

'FIALOPI', F, mp, 1991; (**Loads of Pink**); semi-dbl., 6–14 petals, 1.5–2.75 in., borne in large clusters, moderate fragrance; some prickles; foliage medium green, semi-glossy, disease resistant; tall, upright, very vigorous growth; [Fred Loads sport]; Fiamingo, Joe, 1991

'FIAMETTA', F, dr, 1962; flowers velvety red, center yellow; moderate growth; [Karl Weinhausen X Goldilocks]; Leenders, J.

'FIAMMA', F, or, 1948; flowers vermilion, borne in clusters; vigorous growth; GM, Rome, 1951; [Paul's Scarlet Climber X Talisman seedling]; Aicardi, D.; Giacomasso

Fiamma Nera, HMult, w; flowers medium, semi-dbl., slight fragrance; (Sangerhausen)

'FIAMMETTA', LCl, ab, 1922; flowers warm amber-yellow, streaked yellow, single, intense fragrance; vigorous, climbing growth; [R. gigantea X Margaret Molyneux]; Nabonnand, P.

Ficksburg, F, ob, 1994; GM, Durbanville, 1992; Delbard

Fiddler's Gold™, *see* 'MINZCO'

Fiddletown Pink Noisette, N, mp; (found rose)

Fidélio®, *see* 'MEICHEST'

'FIDÉLIO', F, ab, 1961; bud long, pointed, rosy red; dbl., borne in large clusters, vigorous growth; Horstmann

'FIDELITY', HT, mr, 1962; bud long, pointed; flowers large, dbl., exhibition form, moderate fragrance; foliage glossy; tall growth; RULED EXTINCT 7/80 ARM; [Crimson Glory X Peace]; Abrams, Von

Fidelity, *see* **'SCOOP JACKSON'**

Fides, HT, ly, 1970; flowers large, dbl.; Urban, J.; (Sangerhausen)

Field Rose, *see* **R. ARVENSIS**

Fieldfare, S, rb, 1989; Delbard

Fiery Hit, *see* POUlfiry

Fiery Star, *see* UMSfire

Fiery Sunblaze, *see* 'MEINEYTA'

Fiery Sunsation, *see* 'KORTEMMA'

'FIESTA', HT, rb, 1940; bud ovoid; flowers vermilion, splashed bright yellow, large, dbl., moderate fragrance; foliage glossy, dark; vigorous, bushy, compact growth; [The Queen Alexandra Rose sport]; Hansen, C.B.; Armstrong Nursery

Fiesta, *see* 'MACFIRINLIN'

'FIESTA BRAVA', Min, or, 1959; flowers geranium-red, blooms in clusters, semi-dbl., 14 petals; foliage glossy; vigorous, upright, bushy growth; [Méphisto X Perla de Alcañada]; Dot, M.; Combe

Fiesta Charm, F; Schloen, P.; (Cavriglia)

Fiesta Clown, *see* 'LAVCLO'

'FIESTA FLAME', F, mr, 1978; bud pointed; flowers intense scarlet, semi-dbl., 15 petals, 3 in., slight fragrance; low, bushy growth; [Sarabande X Ena Harkness]; Sanday, John

'FIESTA GOLD', Min, yb, 1970; bud long, pointed; flowers yellow orange, small, dbl., cupped, slight fragrance; foliage small, glossy, light, leathery; vigorous, dwarf, upright, bushy growth; [Golden Glow X Magic Wand]; Moore, Ralph S.; Mini-Roses

'FIESTA RUBY', Min, mr, 1977; bud ovoid, pointed; dbl., 40 petals, 1 in., exhibition form, slight fragrance; foliage dark; bushy, compact growth; [Red Pinocchio X Little Chief]; Moore, Ralph S.; Sequoia Nursery

Fiesta Time™, *see* 'MINGCO'

Fièvre d'Or, *see* 'KORBI'

Fifi, *see* 'HENFIF'

'FIFTH AVENUE', HT, dp, 1948; bud large, ovoid; flowers grenadine-pink, dbl., exhibition form, moderate fragrance; vigorous growth; [Orange Nassau sport]; Johnson, W.E.

'FIGARO', HT, dr, 1954; flowers velvety scarlet, dbl., 26 petals, 4.5 in., intense fragrance; vigorous growth; [(Crimson Glory X Grande Duchesse Charlotte) X New Yorker]; Lens

'FIGMENT', F, m, 1998; flowers purple, overlaid with bright magenta in fresh flower, dark magenta reverse, very dbl., 26–40 petals, 4 in., borne in small clusters, moderate fragrance; prickles moderate; foliage medium, medium green, semi-glossy; upright, medium ($2\frac{1}{2}$ ft) growth; [Len Turner X Remember Me]; Rawlins, R.

'**FIGURINE**', HT, pb, 1952; flowers china pink, dbl., 25 petals, moderate fragrance; foliage bluish green; vigorous growth; RULED EXTINCT 11/91 ARM; [Soeur Thérèse X Unnamed seedling]; Lens

Figurine™, *see* 'BENfig'

'**FIJI**', Gr, or, 1965; bud ovoid; flowers bright orange-red, medium, dbl., cupped, moderate fragrance; vigorous, upright growth; [Queen Elizabeth X Unnamed seedling]; Schwartz, Ernest W.; Wyant

Fil d'Ariadne®, *see* LENfil

'**FILAGREE PILLAR**', LCl, dp, 1962; bud ovoid; flowers tyrian rose, heavily veined, reverse lighter, dbl., 50 petals, moderate fragrance; foliage glossy, bronze; strong stems; vigorous, upright (to 5 ft) growth; [Titian X Sterling]; Riethmuller; Hazelwood Bros.

Filipes Kiftsgate, *see* '**KIFTSGATE**'

'**FILLETTE**', F, mp, 1965; bud ovoid; dbl., blooms in clusters, intense fragrance; foliage dark; [Circus X Papillon Rose]; Lens; Spek

'**FIMBRIATA**', HRg, lp, 1891; (Diantheflora, Dianthiflora, Phoebe's Frilled Pink); flowers petals carnation-like, intense fragrance; [R. rugosa X Mme Alfred Carrière]; Morlet

Finale, *see* 'KORAMI'

Financial Times Centenary, *see* 'AUSFIN'

'**FINE FLARE**', HT, or, 1978; bud pointed; flowers vermilion, rounded, dbl., 40 petals, 4 in., moderate fragrance; foliage dark; moderately vigorous, upright growth; [Fragrant Cloud X Mildred Reynolds]; Bees

Fine Gold, *see* 'WEEGOLD'

Fine Touch, *see* 'LYOFIN'

'**FINESSE**', Min, my, 1983; flowers small blooms borne 1–3 per stem, dbl., exhibition form, no fragrance; foliage small, medium green, semi-glossy; bushy, spreading growth; [Picnic X Rise 'n' Shine]; Hardgrove, Donald L.

Finest Hour, *see* 'MICFINEST'

Fingerpaint, *see* 'MORFING'

Finkenrech, F, mp, 1989; dbl., intense fragrance; Michler, K. H.; (Sangerhausen)

'**FINLANDIA**', Pol, op, 1969; flowers orange-salmon; [Greta Kluis sport]; Kraats; Longley

Finnstar®, *see* 'RUFIN'

Finstar, *see* 'RUFIN'

Fintona, HT, w, 1997; Brundrett

Fiocco Bianco, *see* BARfiob

Fiona®, *see* 'MEIBELUXEN'

Fiona, *see* 'KORFI'

Fiona Ravenscroft, HT, lp; Tebbin, N.H., 1972; (Weatherly, L.)

'**FIONIA**', HWich, lp, 1914; flowers light pink to rose, small blooms borne in clusters; foliage dark, glossy; vigorous growth; [Mme Norbert Levavasseur X Dorothy Perkins]; Poulsen, D.T.; Poulsen

Fiord, *see* 'MEICAUF'

Fiorella, *see* 'MEICAPULA'

Fiorella '82, *see* 'MEICAPULA'

Fiorona, F; Mati; (Cavriglia)

'**FIRE BIRD**', F, rb, 1992; flowers vermillion, pale yellow center, blooms borne in large clusters, semi-dbl., 6–14 petals, 1.5–2.75 in., no fragrance; some prickles; foliage medium, dark green, leathery; medium (70 cms), upright growth; [Avocet X Evelyn Fison]; Strange, J.F.

'**FIRE CHIEF**', HT, mr, 1942; bud long, pointed; flowers flame-red, large, dbl., 24–28 petals, exhibition form, moderate fragrance; foliage glossy; very vigorous, bushy growth; [Crimson Glory X Ami Quinard]; Jacobus; B&A

'**FIRE DANCE**', HT, mr, 1951; bud very long, ovoid; flowers brilliant velvety scarlet, dbl., 25–35 petals, 4–6 in., globular, slight fragrance; foliage glossy, light green; upright, moderately vigorous growth; [(Ulrich Brunner Fils X Westfield Star) X (Chieftain X Better Times)]; Verschuren; Totty

'**FIRE FLAME**', F, or, 1958; flowers scarlet, center orange, single, 4 in.; moderate growth; Morse

Fire King, *see* 'MEIKANS'

Fire King, *see* MEIrolyz

Fire Magic, *see* 'KORFEU'

Fire Meidiland™, *see* 'MEIPSIDUE'

Fire 'n' Ice™, *see* 'AROFIRIC'

'**FIRE OPAL**', F, or, 1955; (Fair Opal); bud ovoid; flowers reddish orange-scarlet, reverse lighter, medium, borne in clusters, dbl., 20–25 petals, cupped, moderate fragrance; foliage glossy; vigorous, open habit growth; [Goldilocks X Unnamed orange Polyantha]; Boerner; McGredy

Fire Pillar, *see* '**BISCHOFSSTADT PADERBORN**'®

'**FIRE PRINCESS**', Min, or, 1969; flowers small, dbl.; foliage small, glossy, leathery; vigorous, bushy growth; [Baccará X Eleanor]; Moore, Ralph S.; Sequoia Nursery

'**FIRE QUEEN**', F, dr, 1963; bud ovoid; flowers bright red, medium, borne in clusters, dbl.; foliage dark, glossy; vigorous, bushy growth; [Fusilier X (Carrousel X Queen o' the Lakes)]; Von Abrams; Peterson & Dering

Fire Robe, Gr, rb

Fire Signal, *see* '**SIGNALFEUER**'®

'**FIRE SKY**', HT, mr, 1952; bud long, pointed; flowers large, cupped; foliage dark, glossy; vigorous growth; [Gen. MacArthur X Étoile de Hollande]; Silva

'**FIREBALL**', Pol, or, 1931; flowers glowing reddish orange; foliage dark, glossy; deRuiter

'**FIREBEAM**', F, rb, 1960; flowers yellow, flame, orange and crimson, borne in clusters, semi-dbl., 14 petals, 2.5 in.; intense fragrance; foliage glossy; vigorous growth; [Masquerade X Unnamed seedling]; Fryers Nursery, Ltd.

'**FIREBIRD**', F, ob, 1960; flowers bright orange, base golden yellow, borne in clusters, semi-dbl., 12–18 petals, 2.5–3 in.; foliage glossy, light green; vigorous, upright growth; RULED EXTINCT 4/92; [Masquerade X Mme Henri Guillot]; Watkins Roses

'**FIREBRAND**', HT, or, 1938; flowers bright scarlet, [Flamingo sport]; Cant, B. R.

'**FIREBRAND, CLIMBING**', Cl HT, or, 1953; Raffel; Port Stockton Nursery

Fireburst, *see* 'TWORIGHT'

'**FIRECRACKER**', F, mr, 1956; flowers scarlet, base yellow, blooms in clusters, semi-dbl., 14 petals, 4.5 in., moderate fragrance; foliage leathery, light green; dwarf, bushy growth; [Pinocchio seedling X Numa Fay seedling]; Boerner; A. Dickson, 1956; J & P, 1959; McGredy, 1956

'**FIRECREST**', F, mr, 1964; dbl., 35 petals, 3 in. in clusters; vigorous, low growth; [(Cinnabar X Marjorie LeGrice) X Pimpernell]; LeGrice

'**FIRECREST, CLIMBING**', Cl F, mr, 1969; LeGrice

'**FIREDRAGON**', HT, mr, 1923; flowers fiery red, dbl., moderate fragrance; Clark, A.

'**FIREFALL**', Cl Min, dr, 1980; (Roy Rumsey); bud short; dbl., 43 petals, 1.5 in., flat, blooms in clusters, foliage small, glossy; trailing, arching growth; [Dortmund X Little Chief]; Moore, Ralph S.; Sequoia Nursery

'**FIREFLAME**', F, mr, 1954; bud ovoid; flowers carmine, borne in pyramidal clusters, dbl., 65–70 petals, 2.5–3 in., moderate fragrance; vigorous growth; [Chatter X Red Pinocchio]; Boerner; Stark Bros.

'**FIREFLASH**', HT, yb, 1960; flowers golden yellow splashed scarlet, well formed, dbl., 25–30 petals, 4.5 in., moderate fragrance; foliage glossy; upright growth; [Marjorie LeGrice X Seedling]; LeGrice; Wayside Gardens Co.

'**FIREFLY**', F, yb, 1975; bud slightly pointed; flowers medium yellow, dbl., 22 petals, 3.5 in., cupped, slight fragrance; foliage dark; upright growth; RULED EXTINCT 4/78 ARM; [Contempo sport]; Joliffe

Firefly, *see* '**HOLSTEIN**'

Firefly, *see* 'MACFRABRO'

'**FIREGLOW**', Pol, or, 1929; flowers brilliant vermilion-red, shaded orange, blooms in clusters, single, slight fragrance; dwarf, compact growth; [Orange King sport]; Wezelenburg

'FIREGLOW, CLIMBING', Cl Pol, or, 1950; Guillot, M.

'FIRELIGHT', HT, or, 1971; flowers large, dbl., exhibition form, moderate fragrance; foliage large, light, leathery; vigorous, upright growth; [Detroiter X Orange Delbard]; Kordes, R.; J&P

Fire-Rider, *see* **'FEUERREITER'**

'FIRESIDE', HT, yb, 1977; bud ovoid; flowers yellow, white, red, imbricated, dbl., 35 petals, 5.5–6. in., slight fragrance; foliage large, glossy, dark; vigorous, bushy growth; [Kordes' Perfecta X Belle Blonde]; Lindquist

Firestar®, LCl, mr

Firestorm, *see* 'CLEFIRE'

Firetail, Cl Min, lp; Peden, G.H., 1994; (Weatherly, L.)

Fireworks™, *see* 'SAVAFIRE'

Fireworks, Climbing *see* **'FEU D'ARTIFICE'**

Firlefanz, S, yb, 1968; flowers golden yellow plus carmine-red, medium, dbl.; GPG Bad Langensalza; (Sangerhausen)

'FIRMAMENT', HT, dr, 1971; flowers vermilion-crimson, large, dbl.; foliage bronze; [Chrysler Imperial X Credo]; Gaujard

First Affair, *see* JACvoo

'FIRST BLUSH', Pol, mp, 1965; flowers phlox-pink, borne in clusters, 3 in., cupped, intermittent bloom; foliage dull, light green; moderate growth; [Francais X Orléans Rose]; Delbard-Chabert; Cuthbert

First Born, *see* 'GILFLEUR'

'FIRST CHOICE', F, or, 1958; flowers fiery orange-scarlet, center yellow, in trusses, single, 7 petals, 5 in., moderate fragrance; tall, spreading growth; [Masquerade X Sultane]; Morse

First Class, *see* 'JACARE'

First Edition, *see* 'DELTEP'

'FIRST FEDERAL', HT, pb, 1965; bud ovoid; flowers geranium-pink tinted scarlet, dbl., 35–40 petals, 5–5.5 in., exhibition form, intense, rose geranium fragrance; strong stems; vigorous, upright growth; [(Radiance X Pageant) X Diamond Jubilee seedling]; Boerner; J&P

'FIRST FEDERAL GOLD', HT, dy, 1967; bud ovoid; flowers gold and yellow, large, dbl., exhibition form, moderate fragrance; foliage glossy, leathery; vigorous, upright growth; [Golden Masterpiece seedling X Golden Masterpiece seedling]; Boerner; J&P

'FIRST FEDERAL'S RENAISSANCE', HT, mp, 1980; bud long, pointed; flowers medium pink, tinted lighter, dbl., 23 petals, 5–7 in., very early bloom, slight fragrance; foliage large; compact growth; [Unnamed variety X First Prize]; Warriner, William A.; J&P

First Hit, *see* POUlrang

First Kiss, *see* 'JACLING'

'FIRST LADY', HT, dp, 1961; bud ovoid; flowers rose madder to phlox-pink, semi-dbl., 18–22 petals, 3.5–4.5 in., cupped, slight fragrance; foliage leathery, dark, semi-glossy; very vigorous, upright growth; [First Love X Roundelay]; Swim, H.C.; C.R. Burr

First Lady, *see* TANrif

'FIRST LADY NANCY', HT, yb, 1981; bud ovoid, long, pointed; flowers light yellow tinged light pink, formal, spiraled, dbl., 36 petals, borne singly, slight, tea fragrance; medium prickles; foliage semi-glossy, medium; medium, upright, bushy growth; [American Heritage X First Prize]; Swim, H.C. & Christensen, J.E.; Armstrong Nursery

'FIRST LADY, CLIMBING', Cl HT, dp, 1964; Burr, C.R.

First Light, *see* 'DEVRUDI'

'FIRST LOVE', HT, lp, 1951; (Premier Amour); bud long, pointed; flowers medium, dbl., 25 petals, 2.5–3.5 in., slight fragrance; foliage leathery, light green; moderately bushy growth; [Charlotte Armstrong X Show Girl]; Swim, H.C.; Armstrong Nursery

'FIRST NATIONAL GOLD', Min, dy, 1976; bud pointed; dbl., 38–42 petals, 1–1.5 in., exhibition form, slight fragrance; foliage small; compact, upright growth; [Rise 'n' Shine X Yellow Jewel]; Saville, F. Harmon; Flora World

'FIRST NATIONAL SILVER', Min, w, 1976; bud ovate; flowers small, very dbl., 50–55 petals, 1.5 in., flat, slight fragrance; foliage small, very glossy; compact, spreading growth; [Charlie McCarthy X Little Chief]; Schwartz, Ernest W.; Flora World

'FIRST OFFERING', F, dr, 1975; bud ovoid; flowers open, semi-dbl., 15 petals, 2.5 in., globular, profuse, very lasting bloom, moderate fragrance; foliage large, glossy, bronze, reddish brown when young; vigorous, dwarf, bushy growth; [(Seedling X ?) X Samba]; Viraraghavan, M.S.

'FIRST PRIZE', HT, pb, 1970; bud long, pointed; flowers rose-pink, center blended with old ivory, very large, dbl., 25–30 petals, 5.5–6. in., exhibition form, borne singly, moderate, tea fragrance; foliage large, dark green, leathery; long, stout stems; vigorous, upright growth; AARS, ARS, 1970 Gertrude M. Hubbard, ARS, 1971 GM, ARS, 1971; [Enchantment seedling X Golden Masterpiece seedling]; Boerner; J&P

First Prize, Climbing, *see* 'JACCLIST'

First Red®, *see* PEKcoujenny

'FIRST ROSE CONVENTION', HT, dr, 1971; bud globular; flowers dark crimson, borne singly, dbl., 48 petals, cupped, no fragrance; hooked, pale cream prickles; foliage red when young, turning dark green; vigorous, upright growth; [Flaming Peace X Helen Traubel]; Hardikar, Dr M.N.; The Bombay Rose Society

Firstar, *see* 'JACARY'

Firstburst, *see* 'TWORIGHT'

Fisher & Holmes, *see* **'FISHER HOLMES'**

'FISHER HOLMES', HP, dr, 1865; (Fisher & Holmes); bud long, pointed; flowers deep red, well-formed, large, dbl., 30 petals, recurrent bloom; upright growth; (28); [Probably Maurice Bernardin self-seedling sport]; Verdier, E.

Fisherman's Friend®, *see* 'AUSCHILD'

FISmelody, Min, lp; (**Sweet Melody**); Fischer, 1998

Five-Colored Rose, *see* **'FORTUNE'S FIVE-COLORED ROSE'**

Five-Roses Rose, *see* 'JACOPPER'

'FLAIR', HT, op, 1951; bud ovoid; flowers coral-blush, dbl., 30–35 petals, 3.5–4.5 in., exhibition form; vigorous growth; [Lady Sylvia X Unnamed seedling]; Verschuren-Pechtold; J&P

'FLAIR', HT, rb, 1989; bud pointed; flowers medium red to dark red, reverse lighter, medium, borne singly, dbl., exhibition form, slight fragrance; obovate, one inch, green with red splotch fruit; prickles curved, hooked, reddish-brown, sparse; foliage medium, dark green, semi-glossy; upright, medium growth; [Pristine X Ink Spots]; Cummings, Peter E., 1991

Flair, *see* DICrelax

'FLAMBEAU', HT, mr, 1940; bud pointed; flowers crimson shaded scarlet, open, large, dbl.; foliage glossy; strong stems; vigorous, bushy, open growth; [Royal Red X Johanniszauber]; Nicolas; J&P

'FLAMBÉE', F, or, 1954; flowers medium, borne in clusters of 5–6, dbl., 25–30 petals; foliage reddish to bronze; upright, bushy growth; GM, Bagatelle, 1952; Mallerin, C.; EFR

Flambo, HT, or, 1993; Kordes

Flamboyance, HT, or

'FLAMBOYANT', Pol, mr, 1931; flowers bright scarlet, aging to crimson-carmine, large blooms in clusters, dbl.; foliage glossy; dwarf growth; Turbat

'FLAMBOYANT', F, mr; flowers bright red; vigorous growth; [Holstein X Incendie]; Croix, P.

'FLAME', HMult, op, 1912; flowers bright salmon-pink, borne in large clusters, semi-dbl.; foliage dark, glossy; very vigorous, compact growth; [Crimson Rambler X ?]; Turner

Flame Bouquet™, *see* 'LAVFIRE'

Flame Dance, *see* 'KORFLATA'

Flame Dancer, *see* 'HARDANCER'

'**FLAME OF FIRE**', HT, ob, 1917; flowers orange-flame, open, large, dbl., moderate fragrance; bushy growth; GM, NRS, 1916; McGredy

'**FLAME OF LOVE**', HT, mr, 1950; Hieatt

Flame of the East, *see* '**PLAMYA VOSTOKA**'

'**FLAMEBURST**', F, dr, 1961; bud pointed; flowers crimson, center yellow, borne in clusters, single, 4–5 petals, 2 in., slight fragrance; foliage dark, glossy; very vigorous growth; [Nearly Wild X Unnamed Hybrid Tea seedling]; Brownell, H.C.

Flameche, F, or

'**FLAMEGLO**', Min, yb, 1981; flowers deep yellow to orange-red, reverse deep yellow, small, dbl., slight fragrance; foliage small, dark, glossy; upright, bushy growth; PP005176; [Starburst X Over the Rainbow]; Williams, Ernest D.; Mini-Roses

'**FLAMENCO**', F, lp, 1960; flowers light salmon-pink, blooms in clusters, dbl., 21 petals, 3.5 in., slight fragrance; foliage dark; [Cinnabar X Spartan]; McGredy, Sam IV; McGredy

Flamenco, *see* POUltika

Flamendr, HT, dp, 1969; flowers large, dbl.; Urban, J.; (Sangerhausen)

Flames 'n' Sparks, *see* LUDsporcoma

Flametta, Min, mp

'**FLAMINA**', F, my, 1964; bud pointed, flushed red; flowers golden yellow, large, dbl.; foliage clear green; very vigorous, upright, bushy growth; [(Faust X Peace) X Unnamed seedling]; Mondial Roses

'**FLAMINAIRE**', HT, ob, 1960; (Montagny); bud globular; flowers orange-flame, medium, semi-dbl., slight fragrance; foliage dark, glossy; vigorous, bushy growth; [Eclipse X Independence]; Dorieux; Pin

'**FLAMING ARROW**', F, or, 1965; bud long, pointed; flowers bright orange-red, medium, dbl., moderate fragrance; foliage glossy; vigorous, upright growth; [Montezuma X Nadine]; Schwartz, Ernest W.; Wyant

'**FLAMING BEAUTY**', HT, rb, 1978; flowers yellow and red-orange, dbl., 35 petals, 4 in., exhibition form, slight fragrance; foliage matt, green; bushy growth; Silver, ARC TG, 1979; [First Prize X Piccadilly]; Winchel, Joseph F.; Kimbrew-Walter Roses

Flaming Heart, *see* UMSflame

'**FLAMING JUNE**', Pol, or, 1931; (Mrs A. Hudig); flowers bright orange-scarlet; vigorous growth; Cutbush

Flaming Peace, *see* 'MACBO'

Flaming Potluck, *see* 'LAVFLAME'

Flaming Rosamini, *see* RUIflami

'**FLAMING RUBY**', F, dr, 1963; flowers deep ruby-red, borne in clusters, dbl., 60 petals, 3 in., foliage dark, reddish green, small; vigorous growth; [(Eva self X Guinee) X Guinee]; Hennessey

Flaming Star, *see* BARflam

'**FLAMING SUNSET**', HT, ob, 1948; flowers deep orange, reverse lighter; foliage light bronze; [McGredy's Sunset sport]; Eddie

'**FLAMING SUNSET, CLIMBING**', Cl HT, ob, 1954; Mattock

Flaming Torch, F, op, 1999

'**FLAMINGO**', HRg, mp, 1956; bud pointed; flowers rich pink, large blooms in clusters, single, 5 petals, cupped, recurrent bloom, foliage glossy, gray-green; vigorous (3 ft) growth; [R. rugosa X White Wings]; Howard, F.H.; Wayside Gardens Co.

Flamingo®, *see* 'KORFLÜG'

'**FLAMINGO**', HT, dp, 1929; flowers bright geranium-red to rosy cerise, spiral shape, dbl., exhibition form, intense fragrance; vigorous growth; GM, NRS, 1927; Dickson, A.

'**FLAMINGO**', F, mp, 1958; flowers rose; very vigorous growth; Frères, Buyl

'**FLAMINGO**', F, op, 1961; bud long, pointed; flowers deep scarlet-pink, open, borne in clusters, dbl., 30 petals; vigorous, upright growth; Horstmann

Flamingo, *see* 'HERFLA'

Flamingo Garden Tea, T, op

Flamingo Meidiland, *see* MEIsolroz

'**FLAMINGO QUEEN**', Gr, dp, 1972; flowers deep pink; [Queen Elizabeth sport]; Chan; Canadian Ornamental Plant Foundation

'**FLAMMÈCHE**', F, ob, 1959; flowers orange, large; vigorous growth; Combe

Flammenmeer, *see* '**SHALOM**'®

'**FLAMMENROSE**', HT, ob, 1921; flowers bright orange-yellow, semi-dbl., 16 petals; [Mrs Joseph Hill X Mme Edouard Herriot]; Türke; Kiese

'**FLAMMENSPIEL**', S, mp, 1974; bud ovoid; flowers salmon-pink, dbl.; foliage large, leathery, dark; vigorous, upright growth; [Peer Gynt X Unnamed seedling]; Kordes; Horstmann

Flammentanz®, *see* 'KORFLATA'

Flammette, *see* '**CRIMSON GEM**'

Flanders Field, *see* 'HORFLAN'

'**FLANDRIA**', Pol, mr, 1966; bud ovoid; flowers open, semi-dbl., moderate fragrance; foliage dark, glossy; bushy growth; Delforge

'**FLAPPER**', F, lp, 1998; flowers light pink, slightly curled petals, single, 5–7 petals, 3 in., borne in small clusters, no fragrance; prickles few, medium sized, straight; foliage medium, dark green, semi-glossy; compact, medium (5 ft) growth; [Permanent Wave sport]; Bennett, Frank David

'**FLASH**', LCl, rb, 1938; bud ovoid, yellow suffused scarlet; flowers orange-scarlet, reverse and center yellow, dbl., cupped, long blooming season, slight fragrance; foliage leathery, glossy, bronze; pillar (6–8 ft), compact growth; AARS, 1940 GM Rome, 1939; [Rosella (Cl HT) X Margaret McGredy]; Hatton; C-P

Flash, *see* 'LENANA'

Flash, *see* '**ROSEMARY GANDY**'

Flash Meidiland, *see* MEIstocko

Flashdance, *see* 'POULFLASH'

Flashfire, *see* 'LEEFIR'

'**FLASHLIGHT**', F, or, 1974; flowers orange-scarlet, dbl., 28 petals, 3–3.5 in., moderate fragrance; foliage large, bronze; tall growth; [Vesper X Unnamed seedling]; LeGrice

Flauce, HT, rb

'**FLAVESCENS**', HSpn, ly; flowers pale lemon, rounded; very hardy. Prior to 1824

Flavia, *see* 'INTERETTE'

'**FLAVIEN BUDILLON**', T, lp; flowers pale flesh, large, globular, intense fragrance

'**FLEET STREET**', HT, dp, 1972; flowers deep rose pink, well-formed, dbl., 40 petals, 6 in., intense fragrance; foliage large, dark, leathery; [Flaming Peace X Prima Ballerina]; McGredy, Sam IV; McGredy

'**FLEETWOOD**', F, ob, 1989; bud pointed; flowers bright orange, reverse lighter, aging darker on outer petals, dbl., 32 petals, exhibition form, slight, fruity fragrance, prickles straight, medium pink; foliage medium, dark green, glossy; bushy, medium, vigorous growth; [Little Darling X Orangeade]; Bridges, Dennis A.; Bridges Roses, 1990

'**FLESH TAITO**', F, ob, 1988; flowers yellow at base to vermilion, reverse orange flushed with pink, dbl., 25–30 petals, cupped, moderate fragrance; prickles ordinary, green; foliage medium, dark green, undulated; bushy, medium growth; [Masquerade X Matador]; Kikuchi, Rikichi, 1989

'**FLEUR COWLES**', F, ly, 1972; flowers cream, center buff, dbl., 35 petals, 3 in.; moderate, spicy fragrance; foliage glossy, dark; [Pink Parfait X ?]; Gregory

Fleur d'Amour, S, ab, 1998; Austin, David

'**FLEUR DE FRANCE**', HT, my, 1944; flowers capucine and yellow, medium, semi-dbl., globular, slight fragrance; foliage glossy; vigorous, dwarf growth; Gaujard

'**FLEUR DE PELTIER**', HGal, pb, 1824; Roseraie de l'Hay

Fleurette®, *see* 'INTERETTE'

Fleurop®, *see* 'KORTEXUNG'®

Fleurs de Pelletier, HGal, dp; flowers medium, very dbl.; Roseraie de l'Hay; (Sangerhausen)

Flicker, F, ab

'**Fliegerheld Boelcke**', HT, my, 1920; flowers nankeen yellow, shaded reddish yellow; [Mme Caroline Testout X Sunburst]; Schmidt, J.C.

'**Fliegerheld Öhring**', HT, or, 1919; flowers medium, dbl.; Kiese; (Sangerhausen)

'**Flighty**', F, m, 1969; bud cherry-red; flowers light mauve, medium, borne in trusses, semi-dbl.; foliage dark; free growth; [Orangeade X Sterling Silver]; Trew, C.; Basildon Rose Gardens

Flimo, F, lp, 1990; Poulsen

'**Flipper**', HT, or, 1973; bud long, pointed; flowers large; foliage large; [Tanagra X John S. Armstrong]; Gaujard

'**Flirt**', F, mr, 1952; bud pointed; flowers bright cherry-red, reverse yellow, medium, borne in clusters, dbl., 35–40 petals, moderate fragrance; foliage glossy, dark; vigorous, upright growth; [Pink Princess X Shades of Autumn]; Brownell, H.C.

Flirt, *see* KORvondra

'**Flirtation**', HT, pb, 1953; bud pointed; flowers begonia-rose, reverse lemon to deep yellow, dbl., 30 petals, 5–6 in., exhibition form, moderate fragrance; foliage dark, leathery; strong stems; vigorous, bushy, compact growth; [Fiesta X Peace]; Shepherd; Bosley Nursery

'**Flocon de Neige**', HCh, w, 1898; Lille

'FLOello', Min, w, 1986; (**Hello There**); flowers white, light yellow tints in center, good petal retention, small, very dbl., 108 petals, cupped, borne usually singly, no fragrance; tan, very few prickles; foliage small, medium green, matt; bushy, low growth; [Care Deeply X Red Can Can]; Florac, Marilyn, 1987

'FLOhih', Min, or, 1985; (**High Hope**); flowers bright orange-red, small, borne singly and in clusters, dbl., 45 petals, cupped, no fragrance; foliage medium, light, glossy; vigorous, upright growth; [Young Love X Little Chief]; Florac, Marilyn; M.B. Farm Min. Roses, Inc.

'FLOmor', Min, yb, 1986; (**Country Morning**); flowers light yellow, blended pink, blooms borne singly, dbl., exhibition form, moderate, spicy fragrance; small, reddish prickles; foliage medium, medium green, semi-glossy; medium, upright growth; [Avandel X Young Love]; Florac, Marilyn; MB Farm

'FLOmyst', Min, ob, 1986; (**Sweet Mystery**); flowers deep orange fading to pink, medium, borne usually singly, dbl., 90 petals, cupped, moderate, damask fragrance; few, green prickles; foliage medium, dark green, semi-glossy; bushy, medium growth; (21); [Care Deeply X Red Can Can]; Florac, Marilyn, 1987

'**Flon**', D, mr, 1845; (Gloire des Perpetuelles, La Mienne); flowers bright red, very free bloom, occasionally repeated; Vibert

'FLOper', Min, op, 1985; (**Per Chance**); flowers orange-red fading bright pink, small blooms borne singly and, single, 5 petals, moderate fragrance; foliage small, dark, matt; bushy growth; [Red Can Can X Care Deeply]; Florac, Marilyn; M.B. Farm Min. Roses, Inc.

'**Flor de Torino 61**', HT, w, 1961; flowers white edged violet-pink; [Monte Carlo X Michele Meilland]; da Silva, Moreira

'**Flora**', HSem, m; (Flore); flowers lilac-pink, center deeper, dbl., seasonal bloom, moderate fragrance; Jacques, 1829

'**Flora**', HT, op, 1957; flowers pink tinted salmon, large, dbl., intense fragrance; vigorous, upright growth; [Independence X Charlotte Armstrong]; Maarse, G.

Flora Bama, *see* 'TALflora'

Flora Danica, *see* POUlrim

'**Flora MacLeod**', LCl, lp, 1971; flowers pale rose-pink, dbl., 50 petals, 4.5 in., non-recurrent, intense fragrance; foliage large, medium green, matt; vigorous growth; [New Dawn X Shot Silk, Climbing]; MacLeod

'**Flora McIvor**', HEg, pb, 1894; flowers rosy pink, white center, yellow stamens, nearly small, single, summer bloom; foliage very fragrant (and flowers); vigorous growth; [R. eglanteria X HP or B]; Penzance; Keynes, Williams & Co.

'**Florabelle**', F, op, 1964; flowers soft salmon-pink, large, dbl., 33 petals, globular, moderate fragrance; foliage soft; vigorous, bushy growth; [Ma Perkins X Unnamed seedling]; Schwartz, Ernest W.; Wyant

'**Floradora**', F, or, 1944; bud globular; flowers cinnabar-red, blooms in sprays of 6–12, dbl., 25 petals, 2 in., cupped, slight fragrance; foliage leathery, glossy; upright, bushy growth; AARS, 1945; [Baby Chateau X R. roxburghii]; Tantau; C-P

'**Floradora, Climbing**', Cl F, or, 1951; Shamburger, P.; Shamburger Rose Nursery

Floraece Delattre, *see* MASflodel

Floral Choice, F, ab, 1982; Fryer, Gareth

'**Floral Dance**', Cl HT, or, 1955; flowers orange-cerise; very vigorous, climbing growth; [Souv. de Mme Boullet, Climbing X Crimson Glory]; Homan; Roseglen Nursery

'**Floralies Valenciennoises**', LCl, mr, 1955; bud long, pointed; flowers currant-red, large, semi-dbl., intense fragrance; vigorous growth; [Matador X Soliel d'Orient]; Dorieux; Pin

Florange, F, mr, 1990; RvS-Melle

Floranje, F, or; dbl., 23 petals, 3 in., flat; foliage dense dark green; strong grower growth; RvS-Melle, 1985

'**Floranne**', Min, my, 1996; flowers medium yellow, fades in sun to white, color better in partial shade, single, 1.5–2.75 in., slight fragrance; some prickles; foliage medium, medium green, semi-glossy; upright (30 cms) growth; [Crazy Dottie sport]; Pratt, Florence

Flore, *see* '**Flora**'

'**Flore Berthelot**', Pol, my, 1921; flowers clear lemon-yellow, passing to white, borne in clusters, Turbat

'**Floréal**', Pol, pb, 1923; flowers flesh and rose-pink, dbl., slight fragrance; [Orléans Rose X Yvonne Rabier]; Turbat

'**Floréal**', HT, op, 1944; bud pointed; flowers coral-pink tinted yellow, medium, dbl., cupped, moderate fragrance; foliage dark, glossy; vigorous growth; Gaujard

'**Florence**', HT, lp, 1921; flowers silvery pink; Paul; W.

'**Florence**', HT, w, 1961; bud long; flowers pure white, large; Dorieux; Pin

Florence, *see* KORnagent

'**Florence Chenoweth**', HT, yb, 1918; (Yellow Herriot); flowers yellow, shaded coral-red; [Mme Edouard Herriot sport]; Chenoweth

Florence Delattre, *see* MASflodel

'**Florence Edith Coulthwaite**', HT, ly, 1908; flowers medium, dbl., moderate fragrance; Dickson, A.; (Sangerhausen)

'**Florence Edna**', HT, lp, 1985; [Princesse sport]; Owen, Fred

'**Florence Forrester**', HT, w, 1914; flowers white tinged lemon, large, dbl.; GM, NRS, 1913; McGredy

'**Florence Haswell Veitch**', Cl HT, dr, 1911; flowers bright scarlet, shaded black, large, dbl., intense fragrance; vigorous growth; [Mme Edmee Metz X Victor Hugo]; Paul, W.

'**Florence L. Izzard**', HT, dy, 1923; bud pointed; flowers bright deep golden yellow, large, dbl., exhibition form, intense fragrance; foliage dark, bronze, leathery, glossy; very vigorous, bushy growth; McGredy

'**Florence Lorraine**', F, ab, 1973; bud ovoid; flowers dark apricot, large, very dbl., exhibition form, moderate fragrance; foliage glossy, dark, leathery, wrinkled; very vigorous, upright, bushy growth; [Royal Highness X (Hawaii X Helen Traubel)]; Middlebrooks; Ellesmere Nursery

'**Florence Lydia**', S, ab, 1999; flowers apricot with tint of pink, reverse same, fading to cream, 2 in., borne in large clusters, intense fragrance; prickles moderate; foliage medium, medium green, glossy; upright, tall (6 ft) growth; [Burma Star X Admiral Rodney]; Jones, L.J.

'**Florence Mary**', HT, dy, 1964; flowers deep yellow-ochre, well formed, large, moderate fragrance; bushy growth; [Doreen sport]; Morse

'FLORENCE MARY MORSE', S, mr, 1951; flowers copper-scarlet, blooms in trusses to 30, semi-dbl., 15 petals, 3 in.; foliage dark, glossy; (28); [Baby Chateau X Magnifica]; Kordes

'FLORENCE MAYER', HT, w, 1998; flowers white and pink blend, reverse white, exhibition, very dbl., 26–40 petals, 4–6 in., exhibition form, borne mostly singly, slight fragrance; prickles moderate amount, large, hooked downward; foliage medium, medium green, semi-glossy; upright, tall growth; [Great Scott X Headliner]; Singer, Steven; Wisconsin Roses, 1998

Florence Nightingale, *see* 'GANFLO'

'FLORENCE PAUL', HP, 1886; Paul, W.; (Cavriglia)

'FLORENCE PEMBERTON', HT, w, 1903; flowers creamy white, suffused pink, dbl., exhibition form; foliage rich green, leathery; vigorous growth; Dickson, A.

Florence Rambler, HWich, lp

'FLORENCE RUSS', HT, lp; Russ, 1923; (Weatherly, L.)

'FLORENTIA', HT, ob, 1941; flowers deep orange, well shaped, large; vigorous, bushy, compact growth; GM, Rome, 1940; [Julien Potin X Unnamed variety]; Giacomasso

'FLORENTINA', HT, dr, 1973; (Kordes' Rose Florentine®); bud large, long, pointed; exhibition form, moderate fragrance; foliage leathery; vigorous growth; ADR, 1974; [Liebeszauber X Brandenberg]; Kordes; Horstmann, 1974

'FLORENTINA', F, pb, 1938; flowers hydrangea-pink, reverse rose-red, large, single; foliage leathery, dark; long stems; very vigorous growth; [Unnamed seedling X Permanent Wave]; Leenders, M.

Florescence, F, ob

Florett, F, op, 1983; flowers large, dbl., slight fragrance; GPG Bad Langensalza; (Sangerhausen)

'FLOREX', HT, op, 1927; bud pointed; flowers deep coral-salmon, suffused orange-carmine, large, dbl., exhibition form, intense fragrance; foliage leathery, glossy, dark; long stems; very vigorous growth; [Mme Butterfly X Premier]; Geiger; A.N. Pierson;, Florex Gardens

Flori®, F, ob

'FLORIADE', Gr, ob, 1963; bud urn shaped; flowers bright orange-scarlet, to open, dbl., 4 in., exhibition form; long, strong stems; vigorous growth; [Montezuma sport]; van der Schilden; Armstrong Nursery

Florian, *see* 'MEILAUR'

Florian, Climbing, *see* 'MEILAURSAR'

'FLORICEL', HT, or, 1957; bud pointed; flowers red and salmon, large, dbl., 25 petals, exhibition form; foliage dark, glossy; strong stems; upright, compact growth; [Carito MacMahon X Luis Brinas]; Dot, Pedro

Florida International, *see* 'FOXFLORIN'

Florida International, Min, ab; [Pierrine sport]

'FLORIDA RED', HT, mr, 1964; flowers rich red, dbl., 60 petals, 5 in.; vigorous growth; Hennessey

Florida Sun, *see* 'GELFLA'

'FLORIDA VON SCHARBEUTZ', F, ob, 1957; flowers orange-yellow shaded coppery, very large blooms in large clusters, dbl., exhibition form, moderate fragrance; foliage dark, glossy; vigorous, bushy growth; [Golden Scepter X (Munchen X Peace)]; Kordes, R.

'FLORIMEL', F, mp, 1958; flowers silvery pink to deep rose-pink, well formed, borne in clusters; foliage glossy; vigorous growth; [Pinocchio X Unnamed seedling]; Fryers Nursery, Ltd.

'FLORINDA NORMAN THOMPSON', HT, pb, 1920; flowers delicate rose on lemon, base deeper, dbl., moderate fragrance; Dickson, A.

Florine, HT, 1981; Gaujard; (Cavriglia)

Florita®, *see* BARflor

Florizel, HT, ab; 36 petals, flat, slight fragrance; strong grower growth; RvS-Melle, 1987

'FLORON', HT, ab, 1948; flowers indian yellow-orange, inside lighter, very large, dbl., exhibition form, slight fragrance; upright growth; [Comtesse Vandal X Pilar Landecho]; Camprubi, C.

'FLORRIE JOYCE', F, dp; flowers cherry-pink, blooms in large clusters, dbl., 43 petals, bushy growth; [Gartendirektor Otto Linne X Borderer]; Riethmuller

'FLOSAR', Min, mp, 1986; (**Fresh Start**); flowers bright medium pink, blooms borne usually singly, dbl., globular, no fragrance; small, red prickles; foliage small, medium green, semi-glossy; low, bushy growth; [Avandel X Little Chief]; Florac, Marilyn; M.B. Farm Min. Roses, Inc.

'FLOSOF', Min, pb, 1986; (**Soft Steps**); flowers creamy, petal edges pink, medium, borne singly, dbl., 33 petals, cupped, no fragrance; small, reddish prickles; foliage medium, medium green, semi-glossy; tall, upright growth; [Avandel X ?]; Florac, Marilyn; M.B. Farm Min. Roses, Inc.

Flossie, *see* 'HORFLOSS'

Flower Basket, *see* 'SEABASKET'

Flower Carpet™, *see* 'NOATRAUM'

Flower Child, *see* 'HORFLOWER'

Flower Girl, *see* **'SEA PEARL'**

Flower Girl™, *see* 'FRYYEOMAN'

'FLOWER OF FAIRFIELD', HMult, mr, 1909;, sometimes recurrent bloom; [Crimson Rambler sport]; Ludorf; H. Schultheis

Flower Power, *see* FRYcassia

Flower Power, *see* KORedan

Flower Power, F, mp, 1994; Kordes

Flower Power, *see* FRYcassia

'FLOWER SHOW', HT, rb, 1980; bud ovoid, pointed; flowers scarlet, yellow reverse, borne 3–4 per cluster, dbl., 25 petals, moderate fragrance; dark red prickles; foliage mid-green, semi-matt; vigorous, upright growth; [Fragrant Cloud X Tropicana]; Bees

'FLOWER WORLD', HT, or, 1980; bud pointed, oval; dbl., 20–30 petals, 4 in., exhibition form; foliage dark, leathery; tall growth; [Baccará X South Seas]; Warriner, William A.; Flower World of America

'FLOWIN', Min, mp, 1986; (**Pink Winks**); flowers urn-shaped, small, borne usually singly, semi-dbl., 13 petals, no fragrance; foliage small, medium green, matt; bushy, low growth; (14); [Baby Betsy McCall X Red Can Can]; Florac, Marilyn, 1987

'FLUFFLY RUFFLES', F, pb, 1935; flowers silver-pink, reverse deeper rose, borne in clusters, semi-dbl., cupped, slight fragrance; foliage leathery; vigorous growth; [Miss Rowena Thom X Seedling]; H&S; Dreer

Fluffy, *see* 'INTERFLU'

Fluorescent®, *see* 'DELFLORI'

'FLUORETTE', F, or, 1971; bud pointed; flowers salmon-orange, dbl., 22 petals, 3–3.5 in., cupped, slight fragrance; vigorous, upright, bushy growth; [(Panache X Soprano) X Coloranja]; Lens

'FLUSH O' DAWN', HT, lp, 1900; flowers light pink changing to white, large, dbl., moderate fragrance; vigorous, upright growth; [Margaret Dickson X Sombreuil]; Walsh

Flushing Creek Poly, Pol, w; (found rose)

Flushing Meadow®, *see* DORnapa

Flutterbye™, *see* 'WEKPLASOL'

Flying Colors™, *see* 'SAVAPAINT'

'FLYING COLOURS', LCl, dp, 1922; flowers light red, large, single, slight fragrance; foliage light, leathery, glossy; very vigorous, compact growth; RULED EXTINCT 8/82 ARM; Clark, A.; Hazlewood Bros.

'FLYING TATA', HT, dr, 1983; bud ovoid; dbl., 45 petals, exhibition form, borne singly, moderate fragrance; foliage medium, dark, glossy; vigorous, upright growth; [Scarlet Knight X Cynosure]; Hardikar, Dr M.N.

Foc de Tabara™, F, lp, 1969; small, ovoid; flowers velvety red, cupped, 33 petals, slight; medium green, leathery, glossy; [Paprika X Coup de Fondre]; St. Wagner,

1969; Res. Stn. f. Horticulture, Cluj, 1970, Romania

Focus, *see* 'LENPAC'

Focus, *see* NOAgut

'FOLIACÉE', C, mp, 1810; (Caroline de Berry); flowers light rose, very large, dbl., globular, Descemet, M.

'FOLIE D'ESPAGNE', F, yb, 1965; flowers yellow, orange and scarlet, blooms in clusters, dbl., 20 petals, 2 in., flat, slight fragrance; foliage dark, glossy; Soenderhousen

'FOLIES-BERGÈRE', HT, yb, 1948; flowers yellow shaded coppery, large, dbl., moderate fragrance; foliage leathery, light green; very vigorous, erect growth; [Souv. de Claudius Pernet X Unnamed seedling]; Gaujard

Folio Courtisane, *see* DELjapal

Folk Dance, *see* 'JUDDANCE'

'FOLK SONG', S, pb, 1964; flowers light pink, reverse darker, medium, dbl., recurrent bloom; foliage glossy; vigorous (3–4 ft), compact growth; Von Abrams; Edmunds Roses

'FOLKESTONE', F, dr, 1936; semi-dbl., blooms in clusters, slight fragrance; foliage dark; bushy, spreading growth; (28); Archer

Folklore®, *see* 'KORLORE'

'FOLKSINGER', S, yb, 1985; flowers yellow flushed with dark peach, slightly, large blooms, dbl., 28 petals, cupped, repeat bloom, moderate fragrance; awl-like, tan prickles; foliage leathery, glossy, coppery mid-green; upright, bushy growth; hardy; [Carefree Beauty X Sunsprite]; Buck, Dr Griffith J.; Iowa State University, 1984

Folle Courtisane, HT, 1996; Delbard, Georges; (Cavriglia)

Folletto, F, 1961; Borgatti, G.; (Cavriglia)

Folletto, Min, 1962; Mansuino, Q.; (Cavriglia)

'FONDANT CASCADE', F, mp, 1996; flowers mid pink fading to pale pink, bright yellow stamens, blooms, dbl., 15–26 petals, 1.5 in., slight fragrance; some prickles; foliage medium, medium green, glossy; spreading, medium (20 in.) growth; patio; Bees of Chester; L.W. Van Geest Farms, Ltd., 1995

Fondly™, *see* 'HILSET'

Fontaine, *see* **'FOUNTAIN'**

Fontaine Blue, *see* 'DELPOUS'

Fontainebleau®, *see* 'DELPOUS'

'FONTANELLE', HT, my, 1927; flowers lemon-yellow, center gold, very large, dbl., moderate fragrance; foliage leathery; vigorous growth; [Souv. de Claudius Pernet X Columbia]; Hill, E.G., Co.

'FONTANELLE, CLIMBING', Cl HT, my, 1935; Johns

'FONTENELLE', HP, mr, 1877; flowers bright red, very large, dbl., intense fragrance; vigorous, growth; Moreau et Robert

Fool's Gold, *see* 'AROGOBI'

Footlights, Pol, w, 1983

Footloose™, *see* 'TANOTAX'

For You, *see* **'PARA TI'**

'For You', F, w, 1999; flowers white, deep pink in petal edge, dbl., 30 petals, 2.5 in., slight fragrance; foliage medium green; compact, low (.7 m) growth; [Majorca X Sweet Memory]; Yasuda, Yuji

For You Dad, *see* 'JONDAD'

Forbidden™, *see* 'MINIBID'

Fordham Rose™, *see* 'WILFORD'

Forest Fire, *see* 'LEOFIRE'

Forest Queen, F, dp, 1997

Forever, HT, dr, 1978; Armstrong

'FOREVER AMBER', F, ob, 1975; flowers golden amber, suffused fiery orange, semi-dbl., 15 petals, 4 in., flat, intense fragrance; foliage dark, leathery; vigorous growth; [Arthur Bell X Elizabeth of Glamis]; Bees

Forever Free, *see* 'HORBRIZESEA'

Forever Friends, *see* 'MEIOFFIC'

Forever Mine, *see* 'RENMINE'

'FOREVER SCARLET', HT, pb, 1976; flowers deep pink, reverse lighter, large blooms borne singly, very dbl., 80 petals, exhibition form, moderate, spicy fragrance; globular fruit; bright red prickles; tall, upright growth; [Wini Edmunds X Mister Lincoln]; Epperson, Richard G.

Forever Young, *see* 'BRIYOUNG'

'FOREVER YOURS', HT, dr, 1964; (Concorde); bud long, pointed; flowers cardinal-red, dbl., 38 petals, 4–5 in., exhibition form, moderate, spicy fragrance; vigorous, upright growth; John Cook Medal, ARS, 1969; [Yuletide X Unnamed seedling]; Jelly; E.G. Hill Co., 1964;, URS, 1967

Forevermore, *see* 'LYOFOR'

'FOREZ ROSE', F, mr, 1964; flowers geranium-red, becoming old-rose, borne in clusters, semi-dbl.; vigorous growth; [Sumatra X Antoine Noailly]; Croix, P.

'FORGOTTEN DREAMS', HT, mr, 1981; bud pointed; flowers cardinal red, borne singly and in trusses of 3–5, dbl., 24 petals, intense fragrance; straight, red-brown prickles; foliage medium green, semi-glossy; vigorous, bushy growth; [Fragrant Cloud X Tenerife]; Bracegirdle, Derek T.; Arthur Higgs Roses

'FORMBY FAVOURITE', HT, or; flowers orange-scarlet on golden yellow, [McGredy's Sunset sport]; Wright, R. & Sons

'FORMBY SHOW', HT, mr, 1986; flowers medium, dbl., 26–40 petals, slight fragrance; foliage medium, dark green, semi-glossy; upright growth; [Fragrant Cloud X Elida]; Dwight, Robert & Sons

'FORNARINA', HGal, pb; flowers deep rose, marbled white; Verdier, 1847

'FORNARINA', M, mp, 1862; flowers deep rose, small, dwarf growth; Moreau et Robert

'FORST', HT, or, 1937; flowers fiery scarlet red, well formed, very large; vigorous growth; [Essence X Fritz Schrodter]; Krause

'FORSTMAISTER HEIM', HSet, 1887; Geschwind, R.; (Cavriglia)

Forsythe, *see* 'VERJO'

'FORT KNOX', HT, dy, 1956; bud ovoid; flowers clear yellow, dbl., 20 petals, 3.5–4.5 in., intense fragrance; foliage dark, leathery; vigorous, upright, open growth; [Unnamed seedling X Ville de Paris]; Howard, A.P.; H&S

'FORT VANCOUVER', HT, mp, 1956; bud long, pointed; flowers well-formed, dbl., 42 petals, 5–6 in., intense, damask fragrance; foliage leathery; vigorous growth; [Charlotte Armstrong X Times Square]; Swim, H.C.; Peterson & Dering

Fortissima®, HMsk, pb, 1994; Lens

Fortissimo, S, mr, 1974; flowers large, dbl.; GPG Bad Langensalza; (Sangerhausen)

'FORTSCHRITT', F, yb, 1933; (Progress); flowers yellow-pink, open, large borne in clusters, semi-dbl., slight fragrance; foliage glossy, light; vigorous, bushy growth; [Mrs Pierre S. duPont X Gloria Mundi]; Kordes

Fortuna®, *see* 'KORTUNA'

'FORTUNA', HT, mp, 1927; flowers rose-pink becoming lighter, many golden anthers, semi-dbl., 25 petals, moderate, fruity fragrance; dwarf, bushy growth; RULED EXTINCT 12/85; GM, NRS, 1927; [Lady Pirrie X Nur Mahal]; Pemberton

'FORTUNE', HT, yb, 1951; flowers gold shaded peach, dbl., 60 petals, 4 in., moderate fragrance; foliage glossy; vigorous growth; RULED EXTINCT 12/85; [Phyllis Gold sport]; Watkins Roses

Fortuné Besson, *see* **'GEORG ARENDS'**

Fortune Cookie, *see* 'SAVACOOK'

Fortune Teller, *see* 'JACHEIR'

Fortuneana, *see* **'FORTUNIANA'**

Fortunella, *see* BARfort

'FORTUNE'S DOUBLE YELLOW', Misc OGR, yb; (Beauty of Glazenwood, Gold of Ophir, R. chinensis pseudindica, R. fortuniana, R. pseudindica, R. X odorata pseudindica, San Rafael Rose); flowers salmon-yellow, outside tinged red, blooms in clusters of 3–4, dbl., moderate, sweet fragrance; (14); 1845

'FORTUNE'S FIVE-COLORED ROSE', T, w, 1844; (Five-Colored Rose, Smith's Parish); bud red-tinged; flowers creamy white tinged with pale blush, fading to white, dbl.; foliage light green; vigorous growth; Fortune

'FORTUNIANA', Misc OGR, w; (Double Cherokee, Fortuneana, R. fortuneana, R. X fortuniana); flowers large, dbl.; climbing growth; (14); [Supposedly R. banksiae X R. laevigata]; 1850

'FORTY-NINER', HT, rb, 1949; bud long, pointed; flowers medium red, reverse yellow, dbl., 33 petals, 3.5–4 in., slight fragrance; foliage leathery, glossy, dark; vigorous, upright, compact growth; AARS, 1949 GM, Portland, 1947; [Contrast X Charlotte Armstrong]; Swim, H.C.; Armstrong Nursery

'FORTY-NINER, CLIMBING', Cl HT, rb, 1952; Moffet; Armstrong Nursery

'FORUM', Pol, mr, 1969; flowers bright red, open, large, dbl.; foliage light green, glossy; vigorous, upright growth; [Veronique X Independence seedling]; Delforge

'FORWARD', HT, my, 1962; flowers clear primrose-yellow, moderately large, dbl.; upright growth; [Ethel Sanday X Peace]; LeGrice

'FORWARD MARCH', HT, mp, 1934; bud pointed, bright old-rose; flowers bright rose-pink, becoming lighter, large, dbl.; foliage dark, bronze, leathery, glossy; long stems; very vigorous, bushy growth; [Better Times sport]; Wolfe

'FOSSE WAY', HT, mp, 1980; bud long; flowers rose pink, paler reverse, dbl., 38 petals, exhibition form, borne singly and several to a cluster, slight fragrance; hooked prickles; foliage matt, light green; vigorous, tall, upright growth; [Colour Wonder X Prima Ballerina]; Langdale, G.W.T.

Foster's Melbourne Cup, *see* 'MACMOUHOO'

Foster's Wellington Cup, *see* 'MACMOUHOO'

'FOUCHIN', Min, rb, 1986; (**Chinese Lantern**); flowers red, yellow and white hand painting, reverse light pink, fading red and white, semi-dbl., 15 petals, slight, spicy fragrance; round, medium, red-orange fruit; foliage medium, medium green, red when young, glossy; bushy, spreading, medium growth; [Avandel X Old Master]; Jacobs, Betty A.; Four Seasons Rose Nursery

'FOUMAD', Min, my, 1987; (**Summer Madness**™); flowers medium golden yellow, gold stamens, fading white, deeper in, dbl., 20–25 reflexed petals, exhibition form, slight, fruity tea fragrance; round, small, orange-red fruit; long, red to light brown prickles; foliage medium medium green matt; bushy, medium growth; [Party Girl X Sun Flare]; Jacobs, Betty A.; Four Seasons Rose Nursery

'FOUMAGIC', Min, m, 1986; (**Winter Magic**™); flowers light lavender-gray, golden stamens, large blooms borne, dbl., 30 petals, cupped, moderate, citrus tea fragrance; medium, globular, orange-red fruit; red, slender, straight, medium prickles; foliage medium, medium green, semi-glossy; medium, upright, bushy growth; [Rise 'n' Shine X Blue Nile]; Jacobs, Betty A.; Four Seasons Rose Nursery

'FOUMANDE', Min, ob, 1989; (**Mandarin Delight**); bud pointed; flowers soft, light mandarin orange, reverse lighter, aging peach in cooler weather, dbl., exhibition form, slight fragrance; no fruit; prickles needle, medium, red to tan; foliage medium, medium green, semi-glossy; bushy, medium growth; [Party Girl X Gingersnap]; Jacobs, Betty A.; Four Seasons Rose Nursery

'FOUMOUSE', Min, r, 1989; (**Church Mouse**); bud pointed; flowers tan-brown, with yellow at base, aging light lavender-brown, dbl., 20 petals, moderate, sweet fragrance; no fruit; prickles slightly declining, small, red to tan; foliage large, medium green, matt; low, bushy, compact growth; [Angel Face X Plum Duffy]; Jacobs, Betty A.; Sequoia Nursery, 1989

Found Fancy, HWich, mp; (found rose)

Founder's Pride, *see* 'MICPRIDE'

'FOUNTAIN', HT, mr, 1970; (Fontaine, Red Prince®); flowers crimson, dbl., 35 petals, 5 in., cupped, intense fragrance; foliage dark, glossy; ADR, 1971 GM, RNRS, 1971 PIT, RNRS, 1971; Tantau, Math.; Wheatcroft Bros.

'FOUNTAIN OF BEAUTY', HT, ab, 1974; bud ovoid; flowers creamy to salmon, very dbl., 100 petals, 3–4 in.; foliage leathery; moderate growth; [Unnamed seedling X Colour Wonder]; Golik; Dynarose

Fountain Square™, *see* 'JACMUR'

'FOUR CHEERS', HT, ob, 1988; bud ovoid; flowers soft orange with yellow center, urn-shaped, medium, borne usually singly, dbl., 30 petals, exhibition form, slight fragrance; globular, medium, orange fruit; prickles falcate, moderate, stout, maroon to ivory; foliage medium, medium green, glossy, smooth, leathery; spreading, medium growth; [Daisy Mae X First Prize]; Stoddard, Louis, 1989

Four Seasons, Quatre Saisons, Rose of Castille, *see* **'AUTUMN DAMASK'**

Fourth of July™, *see* 'WEKROALT'

'FOUSUN', Min, dy, 1991; (**San Jose Sunshine**); bud pointed; flowers deep golden yellow, aging light yellow with orange highlight, dbl., 25–30 petals, exhibition form, slight, fruity to tea fragrance; foliage medium, medium green, matt; bushy, spreading, medium to tall growth; [(Rise 'n' Shine X Redgold) X Summer Madness]; Jacobs, Betty A., 1987; Sequoia Nursery, 1991

'FOUTELL', Min, ob, 1988; (**Show 'n' Tell**); flowers bright, velvety orange-red, white border, reverse white, dbl., 30 petals, exhibition form, slight, spicy fragrance; foliage medium, medium green, semi-glossy, disease-resistant; spreading, tall growth; PP007375; [Rocky X (Matangi X Honey Hill)]; Jacobs, Betty A.; Four Seasons Rose Nursery

'FOUTOUCH', Min, pb, 1989; (**Touch-Up**); bud pointed; flowers medium pink with cream border, reverse cream, hand-painted, dbl., 30 petals, exhibition form, moderate, tea fragrance; round, medium, russet-orange fruit; prickles straight, large, red to tan; foliage medium, medium green, matt; upright, bushy, tall growth; [Scarlet Knight X (Matangi X Honey Hill)]; Jacobs, Betty A.; Four Seasons Rose Nursery

'FOUTUM', Min, ob, 1987; (**Autumn Magic**); flowers bright golden-orange, with red at petal tips, reverse yellow, fading to red, dbl., 35 petals, exhibition form, no fragrance; no fruit; very few prickles; foliage medium, dark green, semi-glossy; bushy, medium growth; [Confetti X Anita Charles]; Jacobs, Betty A.; Four Seasons Rose Nursery

'FOUTWIST', Min, dy, 1988; (**Lemon Twist**); bud pointed; flowers medium, dbl., 25 petals, exhibition form, borne usually singly and in sprays of up to 3, slight, tea fragrance; prickles slightly declining, long, red to brown; foliage medium, medium green, glossy; bushy, medium growth; PP007450; [Gold Badge X Great Day]; Jacobs, Betty A.; Four Seasons Rose Nursery

'FOXFIRE', HT, mr, 1990; bud pointed; dbl., 30 petals, exhibition form, borne singly, no fragrance; prickles straight, green; foliage medium, medium green, glossy; upright, tall growth; [Unnamed seedling X First Prize]; Stoddard, Louis, 1991

'FOXFLORIN', Min, ab, 1997; (**Florida International**); flowers full, high centered, exhibition, medium, very dbl., 30–35 petals, borne mostly singly, slight fragrance; foliage medium, medium green, semi-glossy; upright, spreading, medium to tall (1.5–2 ft) growth; [Pierrine X (Loving Touch X Seedling)]; Fox, Tillie

Foxi, *see* 'UHLATER'

Foxi Pavement, *see* 'UHLATER'

Foxtrot, *see* 'BROFOX'

Foxy, S, lp; Peden, R., 2000; (Weatherly, L.)

Foxy Lady™, *see* 'AROshrim'

'FR. LAD. RIEGER', HT, mr, 1939; flowers large, dbl.; Böhm, J.; (Sangerhausen)

'FRAGEZEICHEN', HWich, mp, 1910; flowers shining pink, large, in few-flowered trusses, dbl., 25 petals, globular; foliage glossy; vigorous growth; very hardy.; [Dorothy Perkins X Marie Baumann]; Boettner

Fragola, HT, 1980; Croix, P.; (Cavriglia)

'FRAGRANCE', HT, dp, 1965; bud long, pointed; flowers carmine to rose-madder, large, dbl., exhibition form, intense fragrance; foliage bronze, leathery; vigorous, tall, compact growth; [Charlotte Armstrong X Merry Widow]; Lammerts, Dr Walter; Germain's

'FRAGRANCE', HT, dr, 1924; flowers deep crimson, high pointed, large, dbl., moderate fragrance; vigorous growth; [Hoosier Beauty X George Dickson]; Chaplin Bros.

'FRAGRANT AIR', F, rb, 1977; flowers red changing to magenta-pink, dbl., 20 petals, 2.5 in., intense fragrance; foliage dark; Pearce, C.A.; Limes Rose Nursery

Fragrant Apricot™, *see* 'JACGRANT'

'FRAGRANT BEAUTY', S, dp, 1950; bud ovoid; flowers carmine, large, dbl., 22 petals, cupped, profuse, repeated bloom, intense, spicy fragrance; foliage glossy; very vigorous (4–5 ft), upright, compact growth; [(Pharisaer X Conrad Ferdinand Meyer) X Crimson Glory]; Jacobus; B&A

Fragrant Beauty, HT, mr, 1998; Ghosh

'FRAGRANT BOUQUET', HT, lp, 1922; flowers shell-pink, base yellow, dbl., 30–35 petals, intense fragrance; H&S

Fragrant Carpet, *see* 'BROcade'

Fragrant Charm, *see* 'KORDU'

Fragrant Cloud, *see* 'TANELLIS'

Fragrant Cloud, Climbing, *see* 'COLFRAGRASAR'

'FRAGRANT DELIGHT'®, F, op, 1978; (Wisbech Rose Fragrant Delight®); flowers light orange-salmon, reverse deeper, dbl., 22 petals, 3 in., intense fragrance; foliage glossy, reddish; Edland Fragrance Medal, ARS, 1976 GM, ARS, 1988 James Mason, ARS, 1988; [Chanelle X Whisky Mac]; Wisbech Plant Co.

Fragrant Dream, *see* 'DICODOUR'

Fragrant Fantasy, *see* 'DEVFRAGO'

'FRAGRANT GLORY', HT, dp, 1950; bud dark red; flowers deep cyclamen-pink, dbl., 36 petals, 6 in., exhibition form, intense fragrance; strong stems; vigorous growth; [Phyllis Gold X Crimson Glory]; Cobley

Fragrant Gold, *see* 'TANDUGOFT'

'FRAGRANT HOUR', HT, op, 1973; flowers bronze-pink, high-pointed, dbl., 35 petals, 4.5 in., intense fragrance; foliage light; GM, Belfast, 1975; [Arthur Bell X (Spartan X Grand Gala)]; McGredy, Sam IV

Fragrant Lace™, *see* 'JACSEE'

'FRAGRANT LADY', HT, mp, 1991; bud ovoid; flowers urn-shaped, blooms borne usually singly, semi-dbl., cupped, intense, fruity fragrance; foliage medium, dark green, semi-glossy; upright, medium growth; [Queen Elizabeth X Broadway]; Perry, Anthony; Co-Operative Rose Growers, 1991

Fragrant Lady, *see* 'MEINIACIN'

Fragrant Lavendar, *see* **'VELVET MIST'**

'FRAGRANT LOVE'®, HT, mr, 1979; bud globular, pointed; flowers medium purplish red, blooms borne 1–3 per cluster, dbl., 45 petals, cupped, intense fragrance; curved reddish prickles; foliage large, deep green, matt; upright growth; [Chrysler Imperial X Unnamed seedling]; Rose Barni-Pistoia

Fragrant Memory, *see* 'JACDIS'

Fragrant Mist, *see* 'SMITSBLANC'

Fragrant Morning, *see* 'RENMORNING'

Fragrant Obsession, *see* 'RENFRAGOBSES'

'FRAGRANT PINK TALISMAN', HT, pb, 1938; flowers pink shades, slightly larger than Talisman, semi-dbl., 18 petals, intense fragrance; [Talisman seedling]; Moore, Ralph S.

Fragrant Plum, *see* 'AROplumi'

Fragrant Queen, HT, op

Fragrant Rhapsody™, *see* 'JACTRIG'

Fragrant Surprise, *see* 'HARVERAG'

'FRAÎCHEUR', HWich, lp, 1921; flowers soft pink, borne in pyramidal clusters, foliage glossy, dark; very vigorous growth; Turbat

'FRAÎCHEUR', HT, lp, 1942; bud long, pointed; flowers soft pink tinted pearl-white, medium, semi-dbl., cupped, slight fragrance; foliage leathery, light green; vigorous, bushy growth; [Joanna Hill X ?]; Meilland, F.; A. Meilland

'FRANBURST', HT, dy, 1999; (**Uppingham School**); flowers golden yellow, reverse yellow, dbl., 26–40 petals, 4 in., borne in large clusters, slight fragrance; prickles moderate; foliage medium, dark green, glossy; upright, medium (to 5 ft) growth; [Ann Harkness X (Diamond Jubilee X Picadilly)]; Cowlishaw, Frank

'FRANCA', F, pb, 1961; flowers pink and yellow, [Confidence X Unnamed seedling]; da Silva, Moreira

'FRANÇAIS', F, op, 1951; flowers bright pink tinted orange, borne in clusters, semi-dbl.; vigorous growth; [Holstein X Orange Triumph]; Mallerin, C.; EFR

France Inter, *see* 'DELKRI'

France Libre®, *see* 'DELJAUNOR'

'FRANCES ASHTON', HT, mp, 1937; bud pointed; flowers carmine, stamens wine-colored, large, single, 5 petals, slight fragrance; foliage leathery; vigorous growth; [Lady Battersea X Hawlmark Crimson]; DePuy; Stocking

'FRANCES BLOXAM', HP, op, 1892; flowers medium, dbl.; Paul, G.; (Sangerhausen)

'FRANCES GAUNT', HT, ab, 1918; flowers apricot to salmon-yellow, large, semi-dbl., cupped, intense fragrance; foliage glossy; vigorous, branching growth; GM, Bagatelle, 1920; Dickson, A.

'FRANCES GAUNT, CLIMBING', Cl HT, ab, 1934; Cazzaniga, F. G.

Frances Neale, *see* 'KIRPICE'

Frances Perry, *see* 'BOSREXCITY'

Frances Phoebe, HT, w, 1979; LeGrice

'FRANCESCA', HMsk, ab, 1922; single, blooms in large sprays, recurrent bloom, slight fragrance; foliage leathery; long stems; vigorous (5–6 ft) growth; (21); [Danaë X Sunburst]; Pemberton

'FRANCESCA DE CUIXART', HT, rb, 1960; (F. Cuixart); bud ovoid; flowers lincoln red, reverse fuchsine-red, open, large, dbl., 40 petals, intense fragrance; foliage glossy, bronze; long stems; vigorous, compact growth; [Baccará X Golden Masterpiece]; Dot, Simon; P. Dot

'FRANCESCH MATHEU', HT, ob, 1940; bud long, pointed; flowers rich golden orange, large, dbl., cupped; foliage glossy, dark; strong stems; very vigorous, upright growth; [Luis Brinas X Catalonia]; Dot, Pedro

'FRANCESCO LA SCOLA', HT, or, 1934; flowers bright orange-red, edged nilsson pink, large, semi-dbl., early, slight fragrance; foliage holly-green; vigorous growth; [Hortulanus Budde X Cuba]; Ketten Bros.

'FRANCETTE GIRAUD', F, op, 1961; flowers bright salmon-pink, borne in clusters, dbl.; foliage dark; vigorous growth; [Aloha X (Gloire du Midi X Edith de Martinelli)]; Arles

Francia, HT, 1982; Dorieux, Francois; (Cavriglia)

'FRANCIBEL', HT, m, 1992; (**Tonia**); flowers light lavender/mauve with deeper tones (silvery lavender) at, dbl., 26–40 petals, 3–3.5 in., intense fragrance; some prickles; foliage medium, medium green, matt; upright (175–180 cms), vigorous growth; [Lagerfeld X Remember Me]; Franklin-Smith, Roger, 1995

'FRANCIE SIMMS', HT, mp, 1926; flowers rose-pink, marked carmine, base buttercup yellow, dbl., moderate fragrance; Dickson, A.

'FRANCINE', HT, rb, 1961; bud pointed; flowers crimson, reverse silvery, dbl., 30–35 petals, 6 in., moderate fragrance; foliage glossy; vigorous, bushy growth; Kriloff, Michel; Cramphorn's Nursery

Francine, HMsk, w

Francine Austin, *see* 'AUSram'

'FRANCINE CONTIER'®, HT, my, 1977; bud long, pointed; flowers canary-yellow, dbl., 40–45 petals, 3.5–4.5 in., exhibition form, slight fragrance; foliage glossy, dark; very vigorous growth; [Peer Gynt X Thalia]; Lens

'FRANCIS', HWich, mp, 1933; flowers rose-pink, dbl.; vigorous growth; [R. wichurana hybrid X Crimson Rambler]; Hauser

'FRANCIS DUBREUIL', T, dr, 1894; bud long; flowers velvety crimson, medium; Dubreuil

'FRANCIS E. LESTER', HMsk, w, 1946; flowers white edged pink, blooms in clusters of 25–30, single, 2 in., recurrent bloom, very vigorous (8–10 ft in a season) growth; very hardy.; [Kathleen X Unnamed variety]; Lester Rose Gardens

Francis King, HT, dy

Francis Moreau, HT, ob

'FRANCIS SCOTT KEY', HT, pb, 1913; flowers deep pink, reverse lighter, very large, dbl., exhibition form, slight fragrance; foliage dark, leathery, glossy; long, strong stems; very vigorous growth; [Radiance X Unnamed seedling]; Cook, J.W.

'FRANCIS SCOTT KEY, CLIMBING', Cl HT, pb

'FRANCISCO CURBERA', HT, pb, 1923; flowers salmon-pink and yellow, well formed, very dbl.; vigorous growth; Dot, Pedro

'FRANCITA', HT, dp, 1965; flowers light red, medium, very dbl., slight fragrance; foliage dark; [((Dame Edith Helen X Baccará) X Baccará) X Comtesse Vandal]; Mondial Roses

Francofurtana, *see* **'EMPRESS JOSEPHINE'**

'FRANÇOIS ALLARD', HT, yb, 1927; flowers creamy yellow, reverse salmon-pink, dbl.; [Mme Mélanie Soupert X Mme Segond Weber]; Felberg-Leclerc

'FRANÇOIS ARAGO', HP, dr, 1858; flowers medium, dbl.; Trouillard; (Sangerhausen)

'FRANÇOIS BOLLEZ', HT, op, 1935; bud shrimp-pink; flowers salmon, base orange, large, dbl.; Gillot, F.

'FRANÇOIS COPPÉE', HP, dr, 1895; flowers dark crimson, large, dbl., moderate fragrance; Lédéchaux

'FRANÇOIS CROUSSE', Cl T, mr, 1900; flowers cerise-crimson shaded darker, dbl., globular, vigorous growth; Guillot, P.

'FRANÇOIS CROUSSE', Cl HT, mr, 1900; flowers large, dbl., moderate fragrance; Guillot; (Sangerhausen)

'FRANÇOIS DE SALIGNAC', M, 1854; Moreau et Robert; (Cavriglia)

Francois Drion, HMsk, mp, 1995; Lens

'FRANÇOIS FOUCARD', HWich, my, 1900; flowers lemon-yellow, borne in clusters, semi-dbl.; very vigorous growth; [R. wichurana X L'Ideal]; Barbier

'FRANÇOIS GUILLOT', HWich, w, 1907; flowers milk-white, dbl., slight fragrance; foliage light green, glossy; vigorous (15–18 ft) growth; [R. wichurana X Mme Laurette Messimy]; Barbier

'FRANÇOIS JURANVILLE', HWich, op, 1906; flowers bright salmon-pink, base yellow, quite distinct, large; very vigorous growth; [R. wichurana X Mme Laurette Messimy]; Barbier

Francois Krige, *see* KORharment

'FRANÇOIS LAPLANCHE', HT, w, 1934; flowers flesh-white on yellow ground, veined and edged carmine, large, dbl., cupped, intense fragrance; foliage leathery; bushy growth; [Mme Charles Detreaux X Mme Edouard Herriot]; Buatois

'FRANÇOIS LEVET', HP, dp, 1880; flowers cherry-rose, well formed, Levet, A.

'FRANÇOIS MICHELON', HP, m, 1871; flowers deep rose tinged lilac, large, dbl., globular, moderate fragrance; foliage somewhat wrinkled; upright growth; [La Reine seedling]; Levet, A.

'FRANÇOIS OLIN', HP, dp, 1881; flowers large, dbl.; Ducher; (Sangerhausen)

'FRANÇOIS POISSON', LCl, ly, 1902; flowers pale sulfur-yellow, center shaded orange, passing to white, dbl.; very vigorous growth; [R. wichurana X William Allen Richardson]; Barbier

'FRANÇOIS PREMIER', HP, 1859; Trouillard; (Cavriglia)

Francois Rabelais™, *see* 'MEInusian'

'FRANÇOISE BLONDEAU', HT, op, 1938; flowers coral, large, dbl.; foliage dark; very vigorous growth; [Charles P. Kilham X Colette Clément]; Mallerin, C.; H. Guillot

'FRANÇOISE DE SALIGNAC', M, mp, 1854; flowers rose-pink, vigorous growth; Robert

'FRANK CHAPMAN', HT, yb, 1937; flowers yellowish-orange, medium, dbl., moderate fragrance; Cant, F.; (Sangerhausen)

Frank Leddy, *see* **'FRANS LEDDY'**

'FRANK MACMILLAN', HT, dr, 1978; flowers crimson-red, full, dbl., 26 petals, 4 in., intense fragrance; foliage semi-glossy; moderately vigorous, spreading growth; [Uncle Walter X (Ena Harkness X Fragrant Cloud)]; Scrivens

'FRANK NAYLOR', S, rb, 1978; flowers dark red, yellowish eye, small, single, 5 petals, 1.5 in., moderate, musk fragrance; foliage small, plum shaded; Harkness, 1977

'FRANK NEAVE', HT, ly, 1928; flowers pale mustard-yellow, dbl., moderate fragrance; Morse

'FRANK PENN', HT, pb, 1970; bud long, pointed; flowers cerise, reverse pink, full, large, dbl.; foliage dark, leathery; very vigorous, upright growth; [Wendy Cussons X Lys Assia]; Clayworth; F. Mason

'FRANK READER', HT, my, 1927; flowers lemon-yellow, center apricot, fairly, large, dbl., exhibition form, moderate fragrance; strong stems, vigorous growth; [Golden Ophelia X Souv. de H.A. Verschuren]; Verschuren; Dreer; H&S

'FRANK SERPA', Gr, op, 1960; bud pointed; flowers pink tinted salmon, large, dbl., cupped, moderate fragrance; foliage leathery, glossy, dark; very vigorous growth; [Pres. Macia X ?]; Serpa

'FRANK W. DUNLOP', HT, dp, 1920; flowers deep bright rose-pink, large, dbl., 45 petals, exhibition form, intense fragrance; [Mrs Charles E. Russell X Mrs George Shawyer]; Dunlop; Totty

'FRANK W. DUNLOP, CLIMBING', Cl HT, dp, 1933; Dixie Rose Nursery

Frankenland®, *see* 'TANkenfram'

'FRANKFORT AGATHÉ', HGal, dp; flowers cerise

Frankfurt, *see* **'SPLENDENS'**

'FRANKFURT AM MAIN', F, mr, 1960; flowers blood-red shaded scarlet, well formed, borne in clusters, dbl., 25 petals, 2.5–3 in., slight fragrance; foliage dark; bushy, upright growth; Boerner; Kordes;, Kordes & Tantau

'FRANKLIN', HT, op, 1918; flowers salmon, shaded yellowish salmon, dbl.; Pernet-Ducher

'FRANKLIN D. ROOSEVELT', HT, mr, 1939; flowers bright red fading to rose-purple, medium, dbl., 26 petals, cupped, intense fragrance; foliage glossy, dark; bush and semi-climbing growth; [Betty Uprichard sport]; McClung

'FRANKLIN ENGELMANN', F, dr, 1970; flowers bright scarlet, pointed, very large blooms in trusses, dbl.; 36 petals; vigorous growth; [Heidelberg X (Detroiter X Seedling)]; Dickson, A.

Frank's Climber, LCl, yb

'FRAnlac', HT, m, 1995; (**Road to Freedom**); flowers lilac, gold stamens, blooms, dbl., 15–25 petals, 1.5–2.75 in., moderate fragrance; foliage medium, medium green, glossy; low (15–20 in.), bushy, compact growth; [Seedling X Lilac Charm]; Cowlishaw, Frank; Rearsby Roses, Ltd., 1996

'FRAnluv', F, ab, 1986; (**Summer Love**); flowers light apricot, large blooms in clusters of 5–10, dbl., 27 petals, exhibition form, moderate, spicy fragrance; foliage large, dark, semi-glossy; upright growth; [Pink Parfait X Cynthia Brooke]; Cowlishaw, Frank

FRAnmilro, MinFl, mp, 1998; (**Lovebird**); Cowlishaw, Frank

Frans Hals, S, mp, 1997; Williams, J. Benjamin

'FRANS LEDDY', Pol, op, 1927; (Frank Leddy); flowers light orange-red, turning pink, small, dbl.; [Kersbergen sport]; Kersbergen; Van Nes

'FRAnshine', HT, ab, 1990; (**Chester Cathedral**); bud pointed; flowers light apricot with cream/very light gold reverse, medium, dbl., exhibition form, borne usually singly, slight fragrance; foliage medium, dark green, glossy; low, bushy growth; [Honey Favorite X Piccadilly]; Cowlishaw, Frank, 1989

'FRAnsmoov', HT, yb, 1995; (**With Thanks**); flowers gold/pink bicolor, blooms borne in clusters, dbl., 26–40 petals, 1.5–2.75 in., moderate fragrance; foliage medium,

medium green, glossy; medium (24 in.), upright growth; [(Southampton X Prominent) X Summer Love]; Cowlishaw, Frank; Rearsby Roses, Ltd., 1996

'FRANTASIA', S, m, 1999; (**Rhapsody in Blue**); flowers dark purple/blue, reverse lighter, semi-dbl., 12–16 petals, 2.5 in., borne in large clusters, intense fragrance; prickles moderate; foliage medium, light green, glossy; upright, semi-climbing, tall (6–7 ft) growth; [Summer Wine X seedling]; Cowlishaw, Frank; Warner's Roses, 2000

FRAntier, Min, lp; (**Maureen Lipman**); Cowlishaw, Frank

'FRANTISEK VALÀSEK', Pol, dr, 1930; flowers small, dbl.; Valàsek; (Sangerhausen)

'FRANZ DEEGEN', HT, my, 1901; (Yellow Kaiserin Auguste Viktoria); flowers soft yellow to golden yellow, dbl.; [Kaiserin Auguste Viktoria X ?]; Hinner, W.

'FRANZ GRÜMMER', HT, or, 1927; flowers coral-red, dbl.; [Mme Abel Chatenay X Château de Clos Vougeot]; Maass

'FRASER MCLAY', HT, mr, 1974; bud ovoid; flowers glossy medium red, medium, dbl.; foliage glossy, leathery; bushy growth; [Grand Gala X Suspense]; Dawson, George; Neil

'FRASER'S PINK MUSK', N, lp, 1818; (Blush Musk); flowers blush, medium, borne in large clusters, semi-dbl., intense fragrance; Fraser

Frasquita®, F, op

'FRATERNITÉ', F, pb, 1946; flowers neyron rose, reverse pale lilac-rose, borne in clusters, dbl., 25 petals, 3 in., globular, slight fragrance; foliage bronze; vigorous growth; [Florentina X World's Fair]; Leenders, M.; Longley

'FRAU A. VON BRAUER', HWich, lp, 1913; flowers small, very dbl., moderate fragrance; Lambert, P.; (Sangerhausen)

'FRAU A. WEIDLING', S, mp, 1930; flowers small, semi-dbl., Vogel, M.; (Sangerhausen)

'FRAU ADOLF ANDERS', HT, 1937; Anders, Adolf; (Sangerhausen)

'FRAU ALBERT HOCHSTRASSER', HWich, lp, 1908; flowers small, dbl., intense fragrance; Weigand, C.; (Sangerhausen)

'FRAU ALEXANDER WEISS', Pol, yb, 1909; flowers light yellow with pink, small, dbl.; Lambert, P.; (Sangerhausen)

'FRAU ANNA LAUTZ', HT, mr, 1911; flowers large, dbl., moderate fragrance; Kiese; (Sangerhausen)

'FRAU ANNA PASQUAY', Pol, dp, 1909; flowers small, dbl.; Walter; (Sangerhausen)

'FRAU ANNY BEAUFAYS', F, mr, 1962; (Mrs Annie Beaufays); bud ovoid; flowers salmon-red, borne in clusters, semi-dbl., slight fragrance; low growth; deRuiter; Beaufays

'FRAU ASTRID SPÄTH', F, dp, 1930; (Astrid Späth, Direktör Rikala); flowers clear carmine-rose, [Lafayette sport]; Spath

'FRAU ASTRID SPÄTH, CLIMBING', Cl F, dp, 1935; Lens

'FRAU BERTA GÜRTLER', HMult, lp, 1913; flowers light silky pink, dbl.; Gurtler; P. Lambert

'FRAU BERTHA KIESE', HT, my, 1914; flowers pure golden yellow, dbl.; [Kaiserin Auguste Viktoria X Undine]; Kiese

'FRAU BÜRGERMEISTER KIRSCHSTEIN', HT, dp, 1906; flowers medium, semi-dbl., intense fragrance; Jacobs; (Sangerhausen)

'FRAU CECILIE WALTER', Pol, ly, 1904; flowers small, dbl., intense fragrance; Lambert, P.; (Sangerhausen)

'FRAU CHARLOTTE GIESELER', Pol, dp, 1939; flowers medium, dbl.; Vogel, M.; (Sangerhausen)

'FRAU DAGMAR HARTOPP', HRg, mp; (Frau Dagmar Hastrup, Fru Dagmar Hastrup); flowers silvery pink, single, 5 petals, foliage crinkled, rich green; low growing growth; (14); ca. 1914

Frau Dagmar Hastrup, *see* 'FRAU DAGMAR HARTOPP'

'FRAU DIREKTOR ANNI HARTMANN', HT, ly, 1933; flowers medium, very dbl., moderate fragrance; Brada, Dr.; (Sangerhausen)

'FRAU DR ERRETH', F, dy, 1915; flowers deep golden yellow, passing to white, well shaped, borne in, dbl.; branching growth; [Gruss an Aachen X Mrs Aaron Ward]; Geduldig

'FRAU DR HOOFTMANN', Cl Pol, lp, 1935; flowers large, dbl.; Buisman, G. A. H.; (Sangerhausen)

'FRAU DR KRÜGER', HT, op, 1919; flowers cream-salmon on golden ground, [Baronne Henriette de Loew X Mme Caroline Testout]; Kiese

'FRAU DR SCHRICKER', HCh, rb, 1927; flowers fiery carmine and coppery-red, large, dbl., intense fragrance; dwarf growth; [Gruss an Teplitz X Souv. de Mme Eugene Verdier]; Felberg-Leclerc

'FRAU E. WEIGAND', HT, yb, 1928; (Frau Elisabeth Weigand); flowers canary-yellow, large, dbl., 70 petals, exhibition form, intense fragrance; foliage dark, leathery; strong stems; [Mme Caroline Testout X Souv. de Claudius Pernet]; Weigand, C.; Weigand & H. Schultheis

'FRAU EDUARD BETHGE', HT, dr, 1930; flowers dark crimson and velvety blood-red, large, dbl., intense fragrance; foliage light; [Hadley X Admiral Ward]; Felberg-Leclerc

'FRAU ELISABETH BALZER', HT, w, 1933; flowers white, base orange-yellow, reverse bright flesh-pink, pointed, vigorous growth; [Mrs Henry Morse sport]; Balzer

'FRAU ELISABETH FISHER', HFt, dy, 1930; flowers medium, dbl., moderate fragrance; Fisher & Schulz; (Sangerhausen)

'FRAU ELISABETH MÜNCH', Pol, mr, 1921; flowers scarlet-cherry-red, with deeper reflexes, [Orléans Rose sport. sport]; Münch & Haufe

'FRAU ELISABETH SPRENGER', HT, dy, 1937; flowers medium, dbl.; Wirtz & Eicke; (Sangerhausen)

Frau Elisabeth Weigand, *see* 'FRAU E. WEIGAND'

'FRAU EMMA SASSE', HT, lp, 1908; flowers large, dbl., moderate fragrance; Plog; (Sangerhausen)

'FRAU EMMY HAMMANN', HT, yb, 1923; flowers reddish lemon-yellow shaded sunflower-yellow, dbl.; [Mme Caroline Testout X Mme Hoste]; Weigand, C.; Hammann

'FRAU EVA SCHUBERT', HWich, mp, 1937; Tepelmann

'FRAU FELBERG-LECLERC', HT, my, 1921; flowers pure golden yellow, dbl., slight fragrance; [Louise Catherine Breslau sport]; Felberg-Leclerc

'FRAU FELIX TONNAR', HT, mp, 1924; flowers bright rose, base coppery orange, semi-dbl.; [Mme Mélanie Soupert X Mme Annette Aynard]; Leenders, M.

'FRAU FRITZ PELZER', HT, or, 1927; flowers reddish crimson-orange, dbl.; [Mme Edouard Herriot X Edward Mawley]; Leenders, M.

'FRAU GEHEIMRAT DR STAUB', Cl HT, dr, 1908; flowers large, dbl., intense fragrance; Lambert, P.; (Sangerhausen)

'FRAU GEHEIMRAT SPÄTH', S, lp, 1941; flowers large, dbl., slight fragrance; Tepelmann; (Sangerhausen)

'FRAU GEORG VON SIMSON', HMult, mp, 1909; flowers rose, vigorous growth; [Helene X Rosel Dach]; Walter

'FRAU HEDWIG KOSCHEL', Pol, w, 1921; flowers white, slightly shaded yellow, edges tinted rose-pink, dbl., moderate fragrance; [Ellen Poulsen sport]; Münch & Haufe

'FRAU HEDWIG WAGNER', HT, mp, 1919; dbl., moderate fragrance; [Enchantress X Mrs W.J. Grant]; Krüger; Kiese

'FRAU HELENE KÜHN', HT, w, 1938; flowers medium, dbl., Vogel, M.; (Sangerhausen)

'FRAU HELENE VIDENZ', HMult, lp, 1904; flowers medium, dbl., Lambert, P.; (Sangerhausen)

'FRAU HUGO LAUSTER', HT, dy, 1932; flowers deep canary-yellow, edged lighter, well formed, dbl., moderate fragrance; vigorous growth; Lauster; Pfitzer & Dreer

'FRAU IDA MÜNCH', HT, ly, 1919; flowers light golden yellow, center deeper, dbl.; [Frau

Karl Druschki X Billard et Barre]; Beschnidt

Frau Jenny Wienke, HT, mr, 1958; flowers large, dbl., intense fragrance; Berger, W.; (Sangerhausen)

'Frau Karl Druschki', HP, w, 1901; (F.K. Druschkii, Reine des Neiges, Snow Queen, White American Beauty); bud pointed, tinged carmine-pink; flowers snow-white, center sometimes blush-pink, large, dbl., 35 petals, no fragrance; foliage dark; vigorous growth; (28); [Merveille de Lyon X Mme Caroline Testout]; Lambert, P.

'Frau Karl Druschki, Climbing', Cl HP, w, 1906; (Grimpant Reine des Neiges); Lawrenson

'Frau Käte Schmid', HMult, mp, 1931; flowers large, semi-dbl., slight fragrance; Vogel, M.; (Sangerhausen)

'Frau Liesel Brauer', HWich, mp, 1938; flowers medium, semi-dbl., slight fragrance; Thönges; (Sangerhausen)

'Frau Lieselotte Weber', Pol, 1938; Vogel, M.; (Sangerhausen)

'Frau Lina Strassheim', HMult, op, 1907; flowers reddish salmon-pink; [Crimson Rambler sport]; Strassheim

'Frau Luise Kiese', HT, ly, 1921; (Luise Kiese); flowers ivory-yellow, sometimes clear yellow, very dbl.; Kiese

'Frau Luise Lindecke', HT, dr, 1928; flowers deep claret-red, sometimes crimson, dbl., intense fragrance; [Columbia sport]; Lindecke; Lindecke & Kordes

'Frau Margarete Oppenheim', HT, rb, 1928; flowers intense carmine-red shaded brick-red and yellow, semi-dbl.; [Hortulanus Budde X Souv. de Claudius Pernet]; Felberg-Leclerc

Frau Maria Rüdt, S, lp, 1994; flowers medium-large, dbl.; Tagashira, Kazuso; (Sangerhausen)

'Frau Marie Bromme', Pol, dr, 1928; flowers bright dark red, [Dr Kater sport]; Wirtz & Eicke

'Frau Marie Weinbach', HWich, w, 1906; flowers small, dbl., moderate fragrance; Weigand, C.; (Sangerhausen)

'Frau Martha Schmidt', HT, mr, 1923; flowers carmine-red, dbl., intense fragrance; [Paula Clegg X Edward Mawley]; Kiese

'Frau Math. Noehl', HT, my, 1913; flowers lemon-yellow, dbl., moderate fragrance; [Kaiserin Auguste Viktoria X Mme Ravary]; Welter

'Frau Mathilde Bätz', HT, w, 1929; flowers pure white, stamens yellow, dbl., intense fragrance; [Unnamed seedling X Ophelia]; Felberg-Leclerc

'Frau Mélanie Niedieck', HT, dy, 1916; flowers vivid lemon-yellow, dbl.; [Mme Jenny Gillemot X Prince de Bulgarie]; Leenders, M.

Frau Minka Rödiger, HT, yb, 1959; flowers yellow to light yellow with orange highlights, large, dbl.; Berger, W.; (Sangerhausen)

'Frau O. Plegg', B, dr, 1909; flowers medium, dbl., intense fragrance; Nabonnand; (Sangerhausen)

'Frau Oberhofgärtner Schulze', Pol, mp, 1909; flowers small, dbl.; Lambert, P.; (Sangerhausen)

'Frau Oberhofgärtner singer', HT, pb, 1908; flowers soft pink edged white, moderate fragrance; [Jules Margottin X Mme Eugenie Boullet]; Lambert, P.

'Frau Oberpräsident von Grothe', HT, op, 1920; flowers rose-orange streaked carmine, [Richmond X Farbenkonigin]; Löbner; P. Lambert

'Frau Peter Lambert', HT, op, 1902; flowers pink shading to salmon, [(Kaiserin Auguste Viktoria X Mme Caroline Testout) X Mme Abel Chatenay]; Welter

'Frau Professor Baranov', Pol, rb, 1947; flowers salmon-red, medium, semi-dbl.; Vogel, M.; (Sangerhausen)

'Frau Professor Gnau', HT, w, 1925; flowers creamy white, large, semi-dbl.; Kiese; (Sangerhausen)

'Frau Professor Grischko', S, mr, 1947; flowers medium, semi-dbl.; Vogel, M.; (Sangerhausen)

'Frau Robert Türke', HT, dr, 1928; flowers dark crimson, very dbl., moderate fragrance; foliage dark, glossy; long stems; vigorous growth; [Hadley X Hugh Dickson]; Türke; Teschendorff

'Frau Rudolf Schmidt', Pol, dr, 1919; flowers dark ruby-red without objectionable blue shades, very dwarf growth; [Jessie sport]; Schmidt, R.

'Frau Sophie Meyerholz', S, mp, 1942; flowers medium, dbl.; Vogel, M.; (Sangerhausen)

'Frau Therese Lang', HT, dp, 1910; flowers large, very dbl., intense fragrance; Welter; (Sangerhausen)

'Fräulein Octavia Hesse', HWich, ly, 1910; flowers yellowish white, center deeper, small, dbl., recurrent bloom, moderate fragrance; vigorous, climbing growth; [R. wichurana X Kaiserin Auguste Viktoria]; Hesse

Frazer's Pink Musk, N, lp

'Frazier Annesley', HT, dp, 1935; bud pointed; flowers carmine, base golden yellow, dbl., exhibition form, slight fragrance; foliage glossy, bronze; very vigorous growth; McGredy

Frechdachs, *see* MEHnan

'Freckle Face', S, pb, 1976; bud ovoid; flowers light spirea-red, striped dark spirea, dbl., 23 petals, 3.5–4 in., moderate, clove fragrance; foliage coppery, leathery; bushy, spreading growth; [(Vera Dalton X Dornroschen) X ((World's Fair X Floradora) X Applejack)]; Buck, Dr Griffith J.; Iowa State University

'Freckles', S, pb, 1976; bud ovoid, pointed to urn-shaped; flowers light scarlet, flushed yellow, dbl., 28 petals, 4–4.5 in., cupped, slight fragrance; foliage dark, coppery, leathery; upright, bushy growth; [Tickled Pink X Country Music]; Buck, Dr Griffith J.; Iowa State University

Fred Birch, HT, my, 1995; Poole, Lionel

'Fred Cramphorn', HT, or, 1961; (Manola, Samoa); flowers geranium-red, dbl., 5–6 in., cupped; foliage dark, glossy; vigorous, upright, bushy growth; [Peace X Baccará]; Kriloff, Michel; Cramphorn's Nursery, 1961; Ilgenfritz Nursery, 1964

'Fred Edmunds', HT, ob, 1943; (L'Arlésienne); bud long, pointed; flowers coppery orange, dbl., 25 petals, 5–5.5 in., cupped, intense, spicy fragrance; foliage leathery, glossy; bushy, open habit growth; AARS, 1944 GM Portland, 1942; [Duquesa de Peñaranda X Marie-Claire]; Meilland, F.; C-P

'Fred Edmunds, Climbing', Cl HT, ob, 1989; [Fred Edmunds sport]; Weeks, O.L.; Weeks Wholesale Rose Growers, 1977

'Fred Fairbrother', HT, dp, 1974; flowers bright cerise, dbl., 40 petals, 4–5 in., moderate fragrance; foliage semi-glossy; [(Gavotte X Tropicana) X Fragrant Cloud]; Sanday, John

'Fred Gibson', HT, ab, 1966; flowers apricot suffused gold, dbl., 30 petals, 5 in., slight fragrance; foliage dark; tall, vigorous growth; [Gavotte X Buccaneer]; Sanday, John, 1968

Fred Hollows Vision, F, w, 1998; [Mary McKillop sport]; Stratford's Roses, 1996

'Fred Howard', HT, yb, 1952; bud long; flowers golden orange shaded pink, dbl., 55 petals, 4 in., exhibition form, slight fragrance; vigorous, upright growth; AARS, 1952; [Pearl Harbor X Unnamed seedling]; Howard, F.H.; H&S

'Fred Howard, Climbing', Cl HT, yb, 1954; Howard, A.P.; H&S

'Fred J. Harrison', HT, mr, 1924; flowers cardinal-red shaded crimson, dbl., moderate fragrance; GM, NRS, 1923; Dickson, A.

'Fred Loads', S, or, 1968; flowers vermilion-orange, blooms in clusters, single, 3 in., moderate; foliage glossy; vigorous, tall growth; GM, RNRS, 1967; [Dorothy Wheatcroft X Orange Sensation]; Holmes, R.A.; Fryer's Nursery, Ltd., 1967

'Fred Owen', HT, dy, 1985; [Jan Guest sport]; Owen, Fred

'Fred Streeter', HT, my, 1955; bud pointed; flowers clear yellow, well-formed, dbl., 48 petals, 4 in., moderate fragrance; foliage dark; vigorous growth; [Luis Brinas X Golden Scepter]; Kordes; Wheatcroft Bros.

'Fred Streeter', HMoy, dp, 1951; flowers deep cerise-pink, medium, borne in clusters (up to 3), single, upright, branching growth; Jackman

Fred W. Alesworth, *see* **'F.W. Alesworth'**

'Fred W. Mee', HT, mr, 1957; (F.W. Mee); flowers scarlet-cerise, dbl., 30 petals, exhibition form, intense fragrance; vigorous, upright growth; [Karl Herbst X The Doctor]; Mee

'Fred Walker', HT, mp, 1935; flowers glowing pink, base coppery orange, large, dbl., exhibition form, slight fragrance; foliage soft, light; vigorous growth; McGredy

Fredagh of Bellinchamp, *see* **'Swany River'**

Freddie Mercury, *see* 'BATmercury'

Freddy, *see* 'PEAproof'

Fredensborg, *see* POUlmax

'Frédéric II de Prusse', HCh, m, 1847; (Frederick the Second); flowers rich crimson-purple, medium, dbl., moderate fragrance; vigorous growth; Verdier, V.; 1847; (Sangerhausen)

'Frédéric Lerr', HT, pb, 1950; flowers carmine-red, reverse lighter, large, dbl., exhibition form, foliage bronze; very vigorous growth; [Crimson Glory X (Mrs Pierre S. duPont X Signora)]; Sauvageot, H.; Sauvageot

Frederic Mistral®, *see* 'MEItebros'

'Frédéric Schneider II', HP, dp, 1885; flowers large, dbl.; Ludovic; (Sangerhausen)

'Frédéric Soulié', rb, 1835; flowers carmine-red with purple striping, large, dbl.; Laffay, M.; (Sangerhausen)

Frederic Worth, HRg, dp

'Frederica', HT, mr; flowers deep crimson, dbl., 30 petals, 5 in., moderate fragrance; foliage dark, glossy; vigorous growth; About 1953 in.England

'Frederick S. Peck', LCl, dp, 1938; flowers deep grenadine-pink, center more yellow, semi-dbl., 4 in.; slightly arched stems; [Hybrid creeper X Mrs Arthur Curtiss James]; Brownell, H.C.

Frederick the Second, *see* **'Frédéric II de Prusse'**

'Fredericksbergrosen', F, dp, 1942; flowers deep pink, medium, single, 5 petals, cupped, no fragrance; foliage medium, medium green, semi-glossy; vigorous, bushy growth; [Orléans Rose X Unnamed seedling]; Poulsen, S.; Poulsen's Roses

Frederik Mey, F, op, 1980; Eve, A.

Frederiksborg, *see* POUldrik

Frederyk Chopin, HT, ly, 1995; flowers large, dbl.; Zyla; (Sangerhausen)

'Fredica', S, w, 1974; bud oval; flowers medium, single, 5 petals, cupped; thornless; very vigorous, bushy, upright growth; PP004463; [Indica Major X Multiflora Inermis]; INRA

Free As Air, *see* 'MEHbronze'

'Free Gold', HT, 1948; bud long, pointed; flowers yellow, open, large, dbl., exhibition form, moderate fragrance; foliage glossy; bushy, dwarf growth; RULED EXTINCT 5/83 ARM; [Pink Princess X Shades of Autumn]; Brownell, H.C.

Free Gold®, *see* 'MACfreego'

Free Spirit, *see* 'PIXiree'

Freedom®, *see* 'DICjem'

'Freedom', LCl, w, 1918; (White American Beauty, Climbing); flowers white, center yellow, open, dbl., 75 petals, 4 in., exhibition form, slight fragrance; foliage dark, bronze, glossy; long, strong stems very vigorous, climbing growth; RULED EXTINCT 3/84 ARM; [Silver Moon X Kaiserin Auguste Viktoria]; Undritz

'Freedom's Ring', LCl, rb, 1994; flowers red and white stripes, blooms borne in small clusters, semi-dbl., 6–14 petals, 1.5 in., slight fragrance; some prickles; foliage medium, medium green, matt; medium, bushy, spreading growth; [Stars 'n' Stripes X Paradise]; Dykstra, Dr A. Michael; Certified Roses, Inc., 1994

Freegold®, *see* 'MACfreego'

'Freeleigh', HT, my, 1956; flowers buttercup-yellow, well shaped, small, dbl., slight fragrance; foliage glossy, light green; vigorous growth; [Kingcup X Golden Scepter]; LeGrice

Freemont, *see* 'BRIfree'

Fregate, HT, lp, 1979; flowers large, dbl., slight fragrance; Delbard; (Sangerhausen)

'Freia', HT, ab, 1936; flowers sun-yellow tinted orange, base orange, open, large, dbl.; foliage leathery, glossy; vigorous, bushy growth; [Ville de Paris X Rev. F. Page-Roberts]; Tantau

Freiamt, HT, mr, 1995; Huber

'Freiburg II', HT, pb, 1917; flowers silver-rose, reverse bright apricot-pink, dbl.; [Dr G. Kruger X Frau Karl Drucshki]; Krüger

'Freiburg II, Climbing', Cl HT, pb, 1953; Lindecke

'Freifrau Ida von Schubert', HT, dr, 1912; flowers dark crimson-red, dbl., moderate fragrance; [Oskar Cordel X Frau Peter Lambert]; Lambert, P.

'Freifrau von Marschall', HWich, mp, 1913; flowers fresh pink, small, borne in immense, loose clusters, dbl., mid-season bloom, vigorous (8–12 ft) growth; [Farquhar X Schneewittchen]; Lambert, P.

'Freiheitsglocke', HT, rb, 1963; (Damas de Yuste, Liberty Bell); flowers claret-rose, reverse light cream, dbl., 50 petals, 5 in., globular, moderate fragrance; foliage leathery; vigorous growth; [Detroiter X Kordes' Perfecta]; Kordes, R.; A. Dickson, 1963;, McGredy, 1963;, Wyant, 1966

Freiheitsglocke, *see* Liberty Bell

'Freiherr von Marschall', T, mr, 1903; bud pointed; flowers large, dbl.; foliage blood-red when young; vigorous growth; [Princesse Alice de Monaco X Rose d'Evian]; Lambert, P.

Freisinger Morgenrote®, *see* KORmarter

'French Can Can', HT, pb, 1956; bud well formed; flowers pink, reverse yellow, large, dbl.; vigorous, bushy growth; Frères, Buyl

French Cancan, *see* **Tourbillon**

French Lace, *see* 'JAClace'

French Liberty, *see* 'MEIjette'

French Perfume™, *see* 'KEIbian'

French Rose, *see* **R. gallica**

'French Vanilla', Gr, w, 1986; flowers white and faint pink, reverse same, fading white, medium, dbl., exhibition form, borne usually singly, moderate fragrance; no fruit; few, moderate, brown prickles; foliage medium, dark green, matt; spreading growth; [Araby X Royal Highness]; Thomson, R.

French Vanilla, HT, w, 1999

Frenesie, LCl, 1965; Combe, M.; (Cavriglia)

'Frensham', F, dr, 1946; flowers deep scarlet, blooms in large trusses, semi-dbl., 15 petals, slight fragrance; vigorous growth; (21); GM, ARS, 1955 GM, NRS, 1943; [Floribunda seedling X Crimson Glory]; Norman; C-P, 1949;, Harkness, 1946

'Frensham, Climbing', Cl F, dr, 1958; Bennett, J.A.; Pedigree Nursery

'Frensham's Companion', F, dp, 1952; flowers cerise, loosely formed, medium, borne in trusses, semi-dbl., 18 petals, slight fragrance; very free growth; [Frensham sport]; Morse

Frenzy, *see* 'MEIhigor'

Fresco®, *see* 'RUIco'

Fresh Cream, HT, w, 1991; Kordes

Fresh Hit, *see* POUlfre

'Fresh Pink', Min, lp, 1964; bud ovoid; flowers light pink tipped salmon, blooms in clusters, dbl., 25 petals, cupped, slight fragrance; foliage leathery, glossy; vigorous, bushy growth; [(R. wichurana X Floradora) X Little Buckaroo]; Moore, Ralph S.; Sequoia Nursery

Fresh Pink, *see* MACpinderal

Fresh Start, *see* 'FLOsar'

Freshie, *see* 'MACsweetwa'

Freude®, *see* 'DeKORat'

'**Freudenfeuer**', Pol, mr, 1918; flowers bright red, moderate growth; Kiese

'**Freudentanz**', F, mr, 1973; bud pointed; flowers bright red, reverse flamed red, medium, intense fragrance; foliage glossy; vigorous, upright, bushy growth; [Fragrant Cloud X Goldmarie]; Hetzel; GAWA

'**Freund Pilz**', HT, mr, 1930; flowers carmine-red, medium, dbl.; Schildmann; (Sangerhausen)

'**Freya**', HT, dp, 1956; flowers carmine-red, large, dbl.; vigorous growth; [Seedling X Étoile de Hollande]; Leenders, M.

'**Freya**', HCan, pb, 1910; Geschwind, R.

Friction Lights, *see* 'HORlights'

Friday's Child, *see* 'HORabi'

Fridolin, Min, yb

'**Frieda Krause**', HT, or, 1935; flowers orange-scarlet-red, large, dbl., exhibition form, foliage leathery, dark; very vigorous, bushy growth; [I Zingari X Unnamed seedling]; Krause

'**Friedlanderiana**', HGal, mp; flowers bright rose-pink, single, non-recurrent, [R. gallica hybrid X R. canina]

'**Friedrich Alfred Krupp**', HT, op, 1903; flowers yellowish salmon-pink, large, very dbl., moderate fragrance; Welter; (Sangerhausen)

'**Friedrich Heyer**', S, ob, 1956; flowers bright orange, in large clusters, 10 petals, 3.5 in., moderate fragrance; foliage dark, glossy, leathery; vigorous, upright growth; Tantau, Math.

'**Friedrich Schwarz**', HT, dr, 1952; flowers crimson, large, dbl., 30 petals, cupped, intense fragrance; foliage dark; very tall, branching growth; [Poinsettia X (Crimson Glory X Lord Charlemont)]; Kordes

'**Friedrich Wörlein**', F, dy, 1963; bud globular; flowers golden yellow, large, dbl., moderate fragrance; foliage dark; vigorous, upright, bushy growth; [Clare Grammerstorf X Golden Masterpiece]; Kordes; Wörlein

'**Friedrichsruh**', HT, dr, 1908; flowers dark crimson, shaded black, turning blue, open, large, dbl., intense fragrance; foliage dark, glossy; bushy, open growth; [Princesse de Bearn X Francis Dubreuil]; Türke

Friend for Life, *see* 'COCnanne'

'**Friend of Heart**', HT, pb, 1985; flowers large blooms borne singly, dbl., 30 petals, exhibition form, slight fragrance; deep brown prickles; foliage medium, dark, glossy; upright growth; [(Festival Beauty X Scarlet Knight) X Self]; Hardikar, Dr M.N.

'**Friend of Peace**', HT, yb, 1986; dbl., 20–22 petals, globular, intense fragrance; crescent, light brown prickles; foliage large, dark green, glossy, leathery; upright, open growth; [(Scarlet Knight X Festival Beauty) X Festival Beauty]; Hardikar, Dr M.N.

Friends of Benalla Gardens, Pol, w, 1996; Sutherland

Friendship™, *see* 'LINrick'

'**Friendship**', HT, dp, 1937; bud pointed; flowers dark red, semi-dbl., exhibition form, intense fragrance; foliage glossy; vigorous growth; RULED EXTINCT 9/77; [Templar X Talisman]; Amling Co.

'**Friendship**', HT, dr, 1938; flowers bright strawberry-red, with bright scarlet undertone, very large, dbl.; very vigorous growth; RULED EXTINCT 9/77; GM, NRS, 1938; Dickson, A.

Friendship, *see* 'HAVipip'

Friesensöhne®, *see* 'TANsenfrie'

Friesia, *see* 'KORresia'

'**Frigg**', S, m, 1969; flowers small, borne in clusters, semi-dbl., 15 petals, flat, free, recurrent boom; foliage small, rich green; low growth; [Schneezwerg X R. nitida]; Lundstad

'**Frileuse**', Gr, mp, 1966; flowers urn-shaped, large, dbl., slight fragrance; foliage large, light green, glossy; tall, upright, bushy growth; [Queen Elizabeth X Baronesse Manon]; Poulsen, Niels D.; Vilmorin-Andrieux

'**Frills**', HT, op, 1950; flowers deep salmon-pink becoming lighter, petals scalloped, large, compact, bushy growth; Moss

Frilly Dilly, *see* 'MURfri'

Frilly Dilly, MinFl, lp, 1986; Cocker

'**Frimousse**', F, op, 1959; flowers orange-pink, base yellow, medium, borne in clusters, single, slight fragrance; vigorous, very bushy growth; [Masquerade X ?]; Vilmorin-Andrieux

Frine®, *see* BARfri

'**Friné**', HT, dr, 1961; flowers crimson suffused strawberry-red, well formed, dbl., 35 petals, moderate fragrance; vigorous growth; [Lila Vidri X (Soraya X Vigoro)]; Dot, Simon

'**Fringette**', Min, dp, 1964; flowers deep pink, white center, small, dbl., 25 petals, low (8 in.), compact growth; [Unnamed seedling X Magic Wand]; Moore, Ralph S.; Sequoia Nursery

Frisco®, *see* 'KORflapei'

'**Frisette**', F, or, 1964; flowers bright scarlet, open, borne in clusters, semi-dbl.; vigorous, low to medium height; [Unnamed seeding X Concerto]; Mondial Roses

'**Frisky**', HT, mr, 1959; flowers velvety red, open, large, dbl., 50 petals, moderate fragrance; foliage dark, glossy; vigorous, bushy growth; [Charlotte Armstrong X Chrysler Imperial]; Wyant

Frisson Frais®, HMsk, w, 1990; Lens

'**Fritz Hegar**', Pol, dr, 1930; flowers medium, semi-dbl.; Schmitt-Eltville; (Sangerhausen)

'**Fritz Höger**', HT, dr, 1934; bud pointed; flowers pure crimson, large, dbl., exhibition form, foliage leathery, dark; very vigorous growth; [(Hadley X Comte G. de Rochemur) X Cathrine Kordes]; Kordes

'**Fritz Maydt**', HT, w, 1925; flowers coppery flesh-white, moderate fragrance; [Mev. C. van Marwijk Kooy X Marquise de Sinéty]; Leenders, M.

'**Fritz Nobis**', S, pb, 1940; bud long, pointed, light red; flowers white, reverse reddish salmon-pink, large blooms in clusters, dbl., exhibition form, non-recurrent, intense fragrance; foliage glossy, leathery; vigorous growth; [Joanna Hill X Magnifica]; Kordes

'**Fritz Schrödter**', HT, dr, 1928; flowers brilliant dark scarlet, large, dbl.; foliage bronze, soft; vigorous, bushy growth; [Hortulanus Budde seedling]; Mühle

'**Fritz Thiedemann**', HT, or, 1959; bud pointed; flowers brick-red, well-shaped, dbl., 36 petals, 4 in., moderate fragrance; foliage dark; bushy growth; [Horstmann's Jubilaumsrose seedling X Alpine Glow seedling]; Tantau, Math.

'**Fritz Thiedemann, Climbing**', Cl HT, or, 1961; Kordes

Fritz Walter, *see* HELfriwa

Fritzi, *see* HELtzi

'**Friuli**', HT, ab, 1933; flowers amber-yellow, large, cupped; vigorous growth; Ingegnoli

'**Frivole**', F, mr, 1958; flowers geranium-red, very dbl.; vigorous, low growth; [Independence X Country Girl]; Frères, Buyl

'**Frivolité**', HT, or, 1956; flowers scarlet shaded orange and salmon, large, dbl., 50 petals, intense fragrance; vigorous, compact growth; [Peace X Catalonia]; Dot, Pedro

Frivolous, *see* 'MINfriv'

'FROcrin', Min, w, 1989; (**Crinkles**); bud ovoid; flowers white, aging with flecks of red, slow to fade, crinkled center, very dbl., 60 petals, exhibition form, slight fragrance; fruit not observed; prickles straight, short, red; foliage medium, dark green, semi-glossy; upright, bushy, medium growth; [Rise 'n' Shine X Seedling]; Frock, Marshall J., 1990

'FROfire', Min, rb, 1995; (**Evening Fire**); flowers red, pink reverse, blooms borne mostly single, dbl., 15–25 petals, 1.5 in., moderate fragrance; some prickles; foliage medium, medium green, semi-glossy; bushy growth; [Black Jade X Pierrine]; Frock, Marshall J.

'**FROHSINN**'®, F, ab, 1961; (Joyfulness); bud pointed; flowers apricot, cream and pink blend, large blooms in large clusters, dbl., 20 petals, cupped, slight fragrance; foliage glossy; vigorous, bushy growth; [Horstmann's Jubilaumsrose X Circus]; Tantau, Math.

Frohsinn '82, *see* 'TANSINNROH'

'FROJEAN', Min, dy, 1989; (**Jeannine Michelle**); bud pointed; flowers deep gold-yellow, aging to pale yellow, medium, borne usually singly, dbl., 30 petals, exhibition form, moderate, fruity fragrance; fruit not observed; prickles straight, tan; foliage medium, medium green, semi-glossy; upright, bushy, medium, hardy growth; [Rise 'n' Shine X Unnamed seedling]; Frock, Marshall J.

'**FROLIC**', F, mp, 1953; flowers bright pink, in large sprays, dbl., 21 petals, 2.5 in., slight fragrance; vigorous, bushy growth; [World's Fair X Pinocchio]; Swim, H.C.; Armstrong Nursery

Frolic, F, pb, 1995; Dawson

'FROMITE', Min, w, 1991; (**White Mite**); flowers small, micro-mini blooms, dbl., 26–40 petals, slight fragrance; foliage small, medium green, semi-glossy; upright growth; [Baby Katie X Unnamed seedling]; Frock, Marshall J.

Frondeuse®, S, pb

'FRONEEN', Min, lp, 1992; (**Jeannine**); flowers light pink aging to white with a few red spots, similar to Royal Highness, dbl., 30 petals, 1.5–1.75 in., exhibition form, intense fragrance; foliage medium, medium green, semi-glossy; medium (40–50 cms), upright growth; [Baby Katie X Unnamed seedling]; Frock, Marshall J.

'FRONORM', Min, pb, 1989; (**Norma Margaret**); bud pointed; flowers pink, blending with copper shading, reverse white with pink, dbl., 30 petals, exhibition form, moderate fragrance; fruit not observed; prickles straight, medium, brown; foliage medium, medium green, semi-glossy; upright, medium growth; [Baby Katie X Unnamed seedling]; Frock, Marshall J., 1990

'**FRONTENAC**', S, dp, 1992; dbl., 15–25 petals, 3–3.5 in., borne in small clusters, slight fragrance; some prickles; foliage medium, dark green, glossy; medium (100 cms), upright growth; very winter hardy; Ogilvie, Ian S.; Agriculture Canada, 1992

'**FRONTIER TWIRL**', S, pb, 1984; flowers pink-yellow blend, large blooms borne 1–8 per cluster, dbl., 25 petals, cupped, repeat bloom, moderate fragrance; awl-like, tan prickles; foliage leathery, medium, bronze green; erect, bushy growth; hardy; [Sevilliana X Just Joey]; Buck, Dr Griffith J.; Iowa State University

'**FROSTFIRE**', Min, mr, 1963; flowers red, sometimes flecked white, dbl., 30 petals, 1 in., foliage dark, glossy; bushy, compact (12–14 in.) growth; [((R. wichurana X Floradora) X Seedling) X Little Buckaroo]; Moore, Ralph S.; Sequoia Nursery

'**FROSTY**', Min, w, 1953; bud ovoid, pale pink; flowers clear white, very small blooms in clusters of 3–10 or more, dbl., 45 petals, moderate, honeysuckle fragrance; foliage glossy; vigorous (12–14 in.), compact, spreading growth; [(R. wichurana X Unnamed seedling) X Self]; Moore, Ralph S.; Sequoia Nursery

Frosty Morning, *see* HORFROST

Frothy, *see* MACfrothy

'**FROUFROU**', HT, yb, 1955; flowers chamois-yellow, center carmine, large, very dbl.; vigorous, bushy growth; [Mme Joseph Perraud X Yvonne Plassat]; Moulin-Epinay

'**FROU-FROU**', HT, rb, 1957; flowers crimson, reverse carmine-pink, large, dbl., 25 petals, moderate growth; [Comtesse Vandal X Unnamed seedling]; Laperrière; EFR

'**FROY**', F, or, 1972; bud ovoid; flowers open, medium, semi-dbl., slight fragrance; foliage glossy, dark; bushy growth; [Traumland X Poulsen's Pink]; Lundstad; Norges Landbruks-hogskole

Fru Dagmar Hastrup, *see* '**FRAU DAGMAR HARTOPP**'

'**FRU GERDA HELMUUS**', HT, op, 1935; flowers light orange-pink, large, semi-dbl., slight fragrance; Poulsen, D.T.; (Sangerhausen)

Fru Inge Poulsen, *see* '**MRS INGE POULSEN**'

'**FRU JOHANNE POULSEN**', HT, mp, 1924; flowers bright pink, well formed, vigorous growth; [Margrethe Moller seedling]; Poulsen, S.

Fru Julie Poulsen, *see* '**POULSEN'S DELIGHT**'

'**FRU XENIA JACOBSEN**', HT, dr, 1925; flowers deep red, well shaped, dbl., moderate fragrance; vigorous growth; [Étoile de France X Richmond]; Poulsen, S.

'**FRÜHLINGSANFANG**', HSpn, w, 1950; bud long, pointed; flowers ivory-white, single, 4 in., intermittent bloom, moderate fragrance; foliage leathery; very vigorous, bushy (9 ft in 6 years) growth; [Joanna Hill X R. spinosissima altaica]; Kordes

'**FRÜHLINGSDUFT**, HSpn, pb, 1949; bud ovoid, golden yellow; flowers lemon-yellow with light pink, very large, dbl., exhibition form, non-recurrent, intense fragrance; foliage large, leathery; very vigorous, upright, bushy growth; [Joanna Hill X R. spinosissima altaica]; Kordes

'**FRÜHLINGSGOLD**'®, HSpn, my, 1937; (Spring Gold); bud pointed, nasturtium-red; flowers creamy yellow, single, 3 in., non-recurrent, intense fragrance; foliage large, light, soft, wrinkled; very vigorous, bushy growth; (28); [Joanna Hill X R. spinosissima hispida]; Kordes; B&A, 1951

'**FRÜHLINGSMORGEN**', HSpn, pb, 1942; (Spring Morning); flowers cherry-pink, center soft yellow, stamens maroon, medium, single, occasionally slightly recurrent bloom, large, red fruit; foliage dark; free growth (6 ft); (28); [(E.G. Hill X Cathrine Kordes) X R. spinosissima altaica]; Kordes

'**FRÜHLINGSSCHNEE**', HSpn, w, 1954; bud ovoid; flowers snow-white, very large, single, slight fragrance; foliage leathery, wrinkled, light green; very vigorous, upright growth; [Golden Glow X R. spinosissima altaica]; Kordes

'**FRUHLINGSTAG**', HSpn, my, 1949; bud ovoid; flowers golden yellow, large, borne in small clusters, semi-dbl., profuse non-recurrent bloom, moderate fragrance; many thorns; foliage leathery; [McGredy's Wonder X Fruhlingsgold]; Kordes

'**FRÜHLINGSZAUBER**', HSpn, mp, 1942; flowers pink, medium, semi-dbl., non-recurrent, large, dark red fruit; foliage dark; vigorous (7 ft) growth; [(E.G. Hill X Cathrine Kordes) X R. spinosissima altaica]; Kordes

Fruit Buffet, F, yb

Fruité, *see* MEIfructoz

'**FRÜLINGSSTUNDE**', S, lp, 1942; flowers medium, semi-dbl., slight fragrance; Kordes, W. Söhne; (Sangerhausen)

FRYaffair, HT, lp, 1997; (**Sight Saver**); Fryer, Gareth

FRYbingo, HT, dp, 1997; (**Awareness**); Fryer, Gareth

'FRYBLISSFUL', LCl, mr, 1999; (**Dizzy Heights**); flowers bright red, reverse same, dbl., 17–25 petals, 5–5.5 in., borne in large clusters, slight fragrance; prickles moderate; foliage medium, dark green, semi-glossy; upright, tall (7–10 ft) growth; Fryer, Gareth; Fryer's Nurseries, Ltd., 1999

FRYbountiful, F, dy, 1998; (**Phab Gold**); Fryer, Gareth

FRYbright, F, ob, 1997; (**Razzle Dazzle**); Fryer, Gareth

FRYbubbly, Min, mp, 1998; (**Bubbles**); Fryer, Gareth

FRYcalm, HT, ob, 1998; (**Britannia**); Fryer, Gareth

FRYcassia, MinFl, 1998; (**Flower Power**); Fryer, Gareth

FRYchambi, HT, ab, 1986; (**Daily Express**); Fryer, Gareth

FRYcharm, HT, dy, 1998; (**Lions International**); Fryer, Gareth

FRYclimbdown, LCl, dr, 1991; (**Crimson Cascade**); Fryer, Gareth

FRYdarkeye, S, rb, 1988; (**Biddulph Grange**); GM, Glasgow, 1991; Fryer, Gareth

'FRYELISE', HT, my, 1999; (**Cheshire**); flowers golden honey yellow, large, dbl., 17–25 petals, borne in small clusters, moderate fragrance; prickles moderate; foliage large, dark green, semi-glossy; upright medium (3 ft) growth; Fryer, Gareth; Fryer's Nurseries, Ltd., 1999

'FRYER'S ORANGE', HT, ab, 1934; flowers orange-yellow; [Mrs Sam McGredy sport]; Fryers Nursery, Ltd.

FRYevenest, F, ob, 1986; (**Evening Sentinel**); Fryer, Gareth

FRYgoldie, Min, dy, 1992; (**Fairy Gold**); Fryer, Gareth

'FRYGRAN', HT, ab, 1983; (**Johnnie Walker**); flowers buff apricot, well-formed, large, dbl., 20 petals, intense fragrance; foliage medium, medium green, matt; vigorous, bushy growth; [Sunblest X (Arthur Bell X Belle Blonde)]; Fryers Nursery, Ltd., 1982

'FRYJAM', F, ab, 1984; (**The Flower Arranger**); bud large; flowers peach, well-formed, semi-dbl., slight fragrance; foliage large, dark, glossy; upright growth; [Seedling X Seedling]; Fryers Nursery, Ltd.

'FRYJASSO', F, pb, 1984; (**Inner Wheel**); flowers carmine edged rose pink, large, dbl., 22 petals, slight fragrance; foliage medium, dark red, matt; bushy growth; [Pink Parfait X Picasso]; Fryers Nursery, Ltd.

'FRYJINGO', HT, yb, 1985; (**The Lady**); flowers honey yellow, petals edged salmon, well-formed, large blooms, dbl., 35 petals, exhibition form, slight fragrance; foliage medium, medium green, semi-glossy; upright growth; GM, Baden-Baden, 1987; [Pink Parfait X Redgold]; Fryer, Gareth; Fryer's Nursery, Ltd.

FRYmartor, HT, mr, 1989; (**Marianne Tudor**); Fryer, Gareth

FRYmaxicot, F, ab; (**Summer Dream**); Fryer, Gareth, 1990

'FRYMESTIN', F, dp, 1984; (**Ethel Austin**); flowers deep pink, dbl., 20 petals, moderate fragrance; foliage large, medium green, semi-glossy; upright growth; [Pink Parfait X Redgold]; Fryers Nursery, Ltd.

'FRYMINICOT', F, ab, 1988; (**Sweet Dream**); flowers peach-apricot, medium, dbl., moderate fragrance; foliage medium, medium green, semi-glossy; mini-flora; bushy growth; 1988; [Unnamed seedling X Unnamed seedling]; Fryers Nursery, Ltd.

'FRYMINILES', Min, op, 1987; (**Lesley Anne**); flowers pale peach-pink, pale yellow base, fading slightly paler, large, dbl., 45 petals, cupped, no fragrance; oval, orange-red fruit; thin, small, pointed, light brown prickles; foliage medium, light green, matt; upright growth; [Unnamed seedling X Unnamed seedling]; Fryer, Gareth

FRYministar, Min, or; (**Top Marks**); Fryer, Gareth, 1992

FRYnblue, F, m; (**The Colwyn Rose**); Cowlishaw, Frank; Fryer, 1992

FRYorst, HT, ob, 1996; (**ARC Angel**); Fryer, Gareth

'FRYPERDEE', HT, dr, 1988; (**Velvet Fragrance**); flowers deep crimson, large, dbl., intense fragrance; foliage large, dark green, semi-glossy; upright growth; Edland Fragrance Medal, ARS, 1987 Fragrance Prize, Baden-Baden 1990; [Unnamed seedling X Unnamed seedling]; Fryers Nursery, Ltd.

FRYprincess, F, ob, 1988; (**Julie Cussons**); Fryer, Gareth

FRYrelax, F, ab, 1989; (**Pensioners Voice**); Fryer, Gareth

FRYrhapsody, F, ab, 1990; (**Langdale Chase**); Fryer, Gareth

FRYromeo, HT, ob, 1998; (**Scent-Sation**); Fryer, Gareth

FRYshrewby, HT, dr; (**Shrewsbury Show**); Fryer, Gareth, 1988

'FRYSTAR', HT, or, 1992; (Beauty Star, **Liverpool Remembers**); flowers vermillion, blooms borne mostly singly, dbl., 26–40 petals, 3–3.5 in.; many prickles; foliage medium, medium green, glossy; tall (240 cms), upright growth; [Corso X Unnamed seedling]; Fryer, Gareth; Roses by Fred Edmunds, 1990

FRYstassi, Min, pb, 1988; (**Rosaletta**); Fryer, Gareth

FRYsweetie, Min, op, 1993; (**Mary Gammon**); Fryer, Gareth

FRYtango, HT, ab; (**The Observer**); Fryer, Gareth, 1991

FRYtranquil, HT, dy, 1991; (**Golden Moments**); Fryer, Gareth

'FRYTROOPER', S, dr, 1999; (Daily Post, Dusky Dancer, Karla, Liverpool Daily Post, **Raven**™); flowers deep velvety red, golden stamens, dbl., 20–25 petals, 1.2–1.5 in., borne in very large clusters, slight fragrance; prickles moderate; foliage small, dark green, glossy; upright, bushy, medium (3–4 ft) growth; PP009211; [Lavaglut X [(Anytime X Liverpool Echo) X (New Penny X Unknown)]]; Fryer, Gareth; Weeks Roses, 1992, U.S.

FRYvivacious, F, op, 1992; (**Julie Andrews**); Fryer, Gareth

'FRYWILREY', HT, rb, 1987; (**Audrey Wilcox**®); flowers cerise red and silver cream, large blooms, dbl., 26–40 petals, intense fragrance; foliage large, dark green, glossy; [Alpine Sunset X Whisky Mac]; Fryers Nursery, Ltd., 1985

FRYwinner, F, my, 1994; (**Atco Royal**); Fryer, Gareth

FRYworld, F, op, 1993; (**Atlantic Star**); Fryer, Gareth

FRYworthy, HT, my, 1996; (**Especially for You**); Fryer, Gareth

'FRYXOTIC', HT, ab, 1999; (Chantoli, Exotic®, **Sunset Celebration**™, Warm Wishes); flowers creamy apricot/amber blend, dbl., 35–40 petals, 4.5–5 in., exhibition form, borne mostly singly, moderate, fruity fragrance; prickles moderate; foliage large, medium green, semi-glossy; upright, bushy, medium, (3.5–4 ft.) growth; PP009718; AARS, U.S., 1998 GM, Belfast, 1996 Golden Rose, The Hague 1997; [Pot O'Gold X seedling]; Fryer, Gareth, 1994; Weeks Roses, 1998

FRYxquisite, Min, mp; (**Sweet Petite**); Fryer, Gareth, 1994

FRYyaboo, HT, r, 1994; (**Belle Époque**); Fryer, Gareth

FRYyat, F, ob, 1995; (**Good Morning**); Fryer, Gareth

FRYyat, HT, lp; (**The Cheshire Regiment**); Fryer, Gareth, 1996

FRYyearn, HT, lp, 1995; (**Bride**); Fryer, Gareth

FRYyeh, F, ob; (**Pomona**); Fryer, Gareth

'FRYYEOMAN', S, lp, 1999; (**Flower Girl**™); bud short, ovoid; flowers soft pink fading to light pink centers, semi-dbl., 8–15 petals, 1–1.5 in., borne in very large clusters, slightly pendulous, slight, apple fragrance; few prickles; foliage medium, light green, matt; long slender arching stems, spreading, bushy, medium (4–5 ft) growth; [Amruda X Fairy Snow]; Fryer, Gareth; Weeks Roses, 2000

FRYyippee, S, m, 1993; (**Rosie Larkin**); Fryer, Gareth

FRYyoung, HT, lp; (**Special Occasion**); Fryer, Gareth, 1995

FRYzebedee, HT, ab; (**Cheshire Regiment**); Fryer, Gareth

FRYzippy, F, ob, 1998; (**Bob Greaves**); Fryer, Gareth

Fuchsia, HGal, m

Fuchsia Meidiland™, *see* 'MEIPELTA'

Fuchsia Meillandecor®, *see* 'MEIPELTA'

Fuchsia Sunblaze, *see* MEIrulex

'FUCHSINE GUY', F, m, 1930; flowers lilac-purple, open, large, borne in clusters, semi-dbl.; foliage rich green; bushy growth; [Lafayette sport]; Leenders, M.

'FUËGO', F, or, 1964; bud pointed; flowers chinese vermilion, open, medium, borne in clusters, dbl., 30 petals, slight fragrance; [Aloha X Gabychette]; Arles; Roses-France

Fuggerstadt Augsburg®, *see* 'KORTREU'

'FUGITIVE', F, yb, 1965; bud pointed; flowers apricot-yellow, becoming lighter, open, borne in clusters, semi-dbl.; foliage glossy, bronze; vigorous, bushy growth; [Mrs Oakley Fisher X ?]; Pal, Dr B.P.; Indian Agric. Research Inst.

Fugue®, *see* 'MEITAM'

'FUJIZAKURA', Min, mp, 1997; flowers single, small, single, 4–7 petals, no fragrance;

some prickles; foliage small, medium green, semi-glossy; spreading, low (20 cms) growth; [Azumino X Azumino]; Ohtsuki, Hironaka

'FUKUYAMA', HT, op, 1988; bud pointed; flowers salmon-pink, aging dark, large, dbl., 35 petals, exhibition form, slight fragrance; round, yellow blend fruit; prickles red to dark green; foliage red aging dark green, matt, ovoid; upright, tall growth; [Pristine X Takao]; Tagashira, Kazuso; Hiroshima Rose Nursery, 1988

'FULGENS', HSpn, m; flowers lilac-pink, semi-dbl., early bloom; glossy, black fruit; height 3–4 ft

Fulgens, *see* **'MALTON'**

Fulgens, HGal, lp; Vibert, before 1830

Fulgurante®, *see* WARoujo

Full Moon, *see* 'SPRINGMOON'

Full Sail, *see* 'MACLANOFLON'

'FULLCREAM', HT, ly, 1959; flowers creamy yellow, well formed, dbl., 28 petals, 5–7 in., intense, honey fragrance; foliage glossy; vigorous, low growth; [Wellworth X Diamond Jubilee]; LeGrice

Fullerton Centennial, *see* 'PIXIFULL'

'FULSTEVE', HT, pb, 1991; (**Steve Silverthorne**); flowers white with pink blushing, large blooms borne mostly singly, dbl., slight fragrance; foliage large, medium green, semi-glossy; medium, upright growth; [Dorothy Anne X Headliner]; Fulgham, Mary, 1992

Fulton MacKay, *see* 'COCDANA'

'FULVIA', HT, dp, 1950; flowers pink tinted carmine, medium, dbl., slight fragrance; foliage leathery, light green; vigorous, bushy growth; [Mme Joseph Perraud X Mme Elie Dupraz]; Gaujard

'FUN JWAN LO', S, w; (Indica Major, Odorata 22449); flowers white, center pale pink, small, dbl.; very vigorous growth; not hardy.; ca. 1924

Funchal, HT; Moreira da Silva, A.; (Cavriglia)

'FUNKENMARIECHEN', F, yb, 1973; flowers yellow, red, dbl., globular, slight fragrance; foliage glossy; upright, bushy growth; [Seedling X Samba]; Kordes; Horstmann

Funkuhr®, *see* 'KORPORT'

'FUNNY FACE', Min, rb, 1981; flowers white, petals edged red, aging red, medium, dbl., 35 petals, slight fragrance; foliage small, medium green, semi-glossy; upright, bushy growth; [Avandel X Zinger]; Jolly, Betty J.; Rosehill Farm

Funny Girl, *see* 'JACFUN'

Funny Girl, *see* BARfunn

Furedaiko, *see* **'FURE-DAIKO'**

'FURE-DAIKO', Cl F, op, 1974; (Furedaiko); bud ovoid; flowers yellow to light orange-red, medium, dbl., cupped, moderate fragrance; foliage large, glossy, dark; vigorous, climbing growth; [(Goldilocks seedling X Sarabande) X Golden Giant seedling]; Suzuki, Seizo; Keisei Rose Nursery

Furia, HT, 1968; Combe, M.; (Cavriglia)

'FURORE', HT, or, 1965; flowers scarlet, medium, very dbl., slight fragrance; foliage dark; [(Baccará X Unnamed seedling) X Miracle]; Verbeek

'FÜRST BISMARCK', T, dy, 1886; flowers large, very dbl.; Drögemüller; (Sangerhausen)

'FÜRST LEOPOLD IV ZU SCHAUMBURG-LIPPE', HP, dr, 1918; flowers large, dbl., moderate fragrance; Kiese; (Sangerhausen)

'FÜRSTIN BISMARCK', T, mr, 1887; flowers large, very dbl.; Drögemüller; (Sangerhausen)

Fürstin Infantin von Hohenzollern, *see* **'FÜRSTIN VON HOHENZOLLERN INFANTIN'**

'FÜRSTIN MARIA HATZFELDT', HT, dr, 1927; flowers bright dark red, dbl., moderate fragrance; [Gen. MacArthur X Unnamed seedling]; Boden

'FÜRSTIN VON HOHENZOLLERN INFANTIN', T, m, 1898; (Fürstin Infantin von Hohenzollern); flowers purple rose, center yellowish salmon, medium, dbl.; Brauer, P.

'FURSTIN VON PLESS', HRg, w, 1911; Lambert, P.

Fushimi, HT, pb, 1987; Keihan

'FUSHINO', HT, lp, 1980; flowers large, dbl., 45 petals, exhibition form, no fragrance; foliage medium green; vigorous, spreading growth; [Utage X Ann Letts]; Ota, Kaichiro

'FUSILIER', F, or, 1957; (Grenadier, Red Soldier); bud globular; flowers orange-scarlet, blooms in heavy clusters, dbl., 40 petals, 3–3.5 in., slight fragrance; foliage dark, glossy, leathery; vigorous growth; AARS, 1958; [Red Pinocchio X Floradora]; Morey, Dr Dennison; J&P

'FUTTAKER SCHLINGROSE', HMult, dr; Geschwind, R., before 1900

'FUTURA', HT, or, 1975; bud long, pointed; flowers vermilion, large, dbl., cupped, slight fragrance; foliage glossy, light; vigorous, upright, bushy growth; [Unnamed seedling X Unnamed seedling]; Warriner, William A.; J&P

Futura, *see* KORoketto

Future Award, *see* JACsto

Fuxiana, HT, 1979; Dot, Simon; (Cavriglia)

'FYFIELD PRINCESS', F, my, 1994; dbl., 15–25 petals, 1.5–2.75 in.; borne in large clusters, moderate fragrance; many prickles; foliage medium, medium green, semi-glossy; medium (4 ft), upright growth; [Southampton sport]; Shuttleworth, F.I.; F. Haynes & Partners, 1994

Fyvie Castle, *see* 'COCBAMBER'

G

'G. Amédée Hammond', HT, ab, 1915; flowers apricot-yellow on ivory yellow, dbl., intense fragrance; GM, NRS, 1913; Dickson, A.

'G. F. Veronica', HMult, ly, 1900; flowers creamy yellow, small, semi-dbl., moderate fragrance; Demitrovisi; (Sangerhausen)

'G. H. Davison', HT, mr, 1988; flowers medium, well-defined red, large, borne in sprays of 1–3, dbl.; 45 petals, exhibition form, slight fragrance; oval, large, green fruit; prickles normal, large, red; foliage large, dark green, semi-glossy; upright growth; [Unnamed seedling X Unnamed seedling]; Davison, G. H.; The Central Nursery, 1990

'G. I. Joe', HT, dp, 1943; bud long, pointed; flowers rose-red to deep rose-pink, large, very dbl., moderate fragrance; foliage leathery, dark; vigorous growth; [Red Better Times sport]; Parmentier, J.

G K Rose, HT, dr, 1989; Chiplunkar

G. Nabonnand, *see* **'Gilbert Nabonnand'**

G. P. & J. Baker, *see* 'HARrango'

'G. W. Peart', HT, mr, 1948; bud ovoid; flowers medium, very dbl.; foliage leathery, dark; moderate, bushy growth; [Guinee seedling X Rouge Mallerin]; Toogood

'G. W. Watkins', HT, lp; Williams, A., 1890; (Weatherly, L.)

'Gabi', Gr, or, 1987; (Gabrielle); flowers very bright orange-red, fading to lighter orange, classic form, dbl., 24 petals, exhibition form, moderate fragrance; rounded, small, light brown-green fruit; prickles fairly flat, small, dark brown; foliage medium, dark green, glossy; bushy, medium growth; [Pink Favorite X Red Dandy]; Poole, Lionel, 1988

'Gabriel Lombart', HT, w, 1932; flowers flesh-white to cream-white, very large, dbl., cupped, moderate fragrance; vigorous, bushy growth; [Dr A. Hermans X Rayon d'Or]; Buatois

'Gabriel Noyelle', M, ab, 1933; (Gabrielle Noyelle); bud ovoid; flowers apricot, dbl., cupped, recurrent bloom, moderate fragrance; foliage leathery; very vigorous growth; [Salet X Souv. de Mme Kreuger]; Buatois

Gabriela Sabatini, *see* 'WILsab'

Gabriella®, *see* 'BERgme'

Gabrielle, *see* **'Gabi'**

Gabrielle, *see* 'BERgme'

Gabrielle Noyelle, *see* **'Gabriel Noyelle'**

'Gabrielle Privat', Pol, mp, 1931; flowers brilliant carmine-pink, blooms in pyramidal corymbs of 30–50, semi-dbl.; bushy growth; Barthelemy-Privat; Turbat

Gabriel's Fire, *see* 'BRIfire'

Gaby Morlay, *see* DORsand

'Gabychette', F, rb, 1960; bud ovoid; flowers reddish salmon, base sulfur-yellow, reverse white to rosy white, dbl., exhibition form, slight fragrance; foliage glossy, light green; low growth; [Floradora X Pioupiou]; Arles; Roses-France

'GActa', HT, or, 1969; (**Tanagra®**); bud long, pointed; flowers large, dbl., moderate fragrance; foliage soft; vigorous, upright growth; [Queen Elizabeth X Tropicana]; Gaujard

'GAegui', F, ob, 1963; ('GAegul', **Guitare®**); bud ovoid; flowers gold and orange-red blend, medium, dbl., intense fragrance; foliage light green, leathery; vigorous, bushy growth; GM, Bagatelle, 1966; [Vendome X Golden Slippers]; Gaujard

'GAegul', F, ob, 1963; ('GAegui', **Guitare®**); bud ovoid; flowers gold and orange-red blend, medium, dbl., intense fragrance; foliage light green, leathery; vigorous, bushy growth; GM, Bagatelle, 1966; [Vendome X Golden Slippers]; Gaujard

Gaia®, *see* BARgai

'Gaiata', HT, yb, 1956; flowers indian yellow shaded pink; [Boudoir X Peace]; da Silva, Moreira

'Gaiety', HT, ob, 1926; bud pointed; flowers orange, indian red and silver, large, dbl., cupped; foliage light, glossy; vigorous, branching growth; [Mme Butterfly X Souv. de Claudius Pernet]; Hill, E. G., Co.; Hill Floral Products Co.

'Gaiety', HT, op; flowers salmon-pink flushed yellow, moderate fragrance; vigorous growth; Archer

'Gaiezza', HT, ob, 1940; flowers orange touched red, center yellow, intense fragrance; [Julien Potin X Mme G. Forest-Colcombet]; Giacomasso

Gail, *see* TINgail

'Gail Borden', HT, pb, 1957; bud ovoid; flowers deep rose-pink, reverse overcast cream, dbl., 53 petals, 5.5 in., exhibition form, moderate fragrance; foliage dark, glossy, leathery; vigorous, upright growth; GM, NRS, 1957; [R.M.S. Queen Mary X Viktoria Adelheid]; Kordes; J&P

'Gail Borden, Climbing', Cl HT, pb, 1960; J&P, 1960

'Gainsborough', Cl HT, lp, 1903; flowers flesh-pink, almost white, large, dbl.; long stems; vigorous growth; [Viscountess Folkestone sport]; Good & Reese

'GAisu', F, dr, 1968; (**Sundra®**); bud pointed; flowers large, semi-dbl., cupped, intense fragrance; foliage glossy; vigorous, bushy growth; [Club X Lilli Marlene]; Gaujard

'Gala', F, pb, 1973; bud short, pointed; flowers light pink, dbl., 28–34 petals, 3–3.5 in., exhibition form, slight, spicy fragrance; vigorous, upright, free growth; [Seedling No. 19-64 ps X Seventeen]; Jelly; E. G. Hill Co.

Gala, *see* 'SAVagala'

Gala Charles Aznavour, *see* MEIsazy

'Gala Day', F, or, 1966; flowers vermilion-scarlet, pointed, 4 in., slight fragrance; foliage light green; free, upright growth; [Queen Elizabeth X Dicksons Flame]; Watkins Roses

Gala Gold, *see* 'LAVchro'

Gala Sunrise, *see* 'GELrise'

'Galah', HT, mp, 1956; flowers carmine-pink, base lighter, large, borne in clusters, semi-dbl., moderate fragrance; Riethmuller

'Galahad', HT, w, 1986; flowers very large, dbl., moderate, anise fragrance; foliage medium green, semi-glossy; [Micaela X Lara]; Kriloff, Michel

'Galatea', HMsk, yb, 1914; flowers stone-color, edged pink, small rosette, borne in clusters, recurrent bloom; Pemberton

Galaty, *see* 'DOTsubebe'

Galaxy™, *see* 'MORgal'

'Galaxy', HWich, dp, 1906; flowers bright carmine; vigorous growth; Walsh

Galaxy, *see* MEIhuterb

Galejade, S, op, 1997; Reuter

Galia®, *see* 'MEItinirol'

'Galia', HT, lp, 1966; flowers light pink, center darker, large, very dbl., slight fragrance; foliage dark, leathery; very vigorous, upright growth; RULED EXTINCT 4/81/ARM; Betzel

Galileo, *see* 'MEIgalil'

Galina, HT, mr, 1979; flowers large, dbl., slight fragrance; Cocker; (Sangerhausen)

Galina, F, ab; flowers apricot and yellow, medium, dbl., slight fragrance; VEG; (Sangerhausen)

'Gallagher', F, rb, 1979; bud pointed; flowers cream, edged crimson, shapely, dbl., 41 petals, 2.5 in., moderate, fruity fragrance; foliage leathery; bushy growth; Murray, Nola

'Gallant', F, mr, 1968; flowers scarlet, dbl., 3–3.5 in., blooms in clusters, moderate fragrance; foliage glossy; [Tropicana X Barbecue]; Dickson, A.

Gallantry, HT, pb; Delbard

'GALLERIA', HRg, mr, 1990; (Big John); bud ovoid; flowers medium watermelon pink, reverse silvery pink, large, semi-dbl., 13 petals, exhibition form, borne in sprays of 3–5, moderate, spicy, fruity fragrance; round, small, green yellow fruit; prickles straight, medium, light green to pink; foliage large, dark green, glossy; bushy, tall (8 ft) growth; [The Duke X Hansa]; Weddle, Von C.; Hortico Roses, 1990

'GALLERIA BORGHESE', HT, w, 1954; flowers flesh streaked coral, very large; foliage glossy; strong stems; [Peace X Crimson Glory]; Giacomasso

Gallica Grandiflora, *see* **'ALIKA'**

'GALLICA MACRANTHA', Misc OGR, w; (R. gallica macrantha, R. macrantha, X R. waitziana macrantha); flowers flushed rose at first, changing to nearly white

Gallica Maheca, *see* **'LA BELLE SULTANE'**

'GALLI-CURCI', HT, my, 1924; flowers golden yellow; [Columbia sport]; Kinsman

'GALLIVARDA', HT, rb, 1977; (Galsar); bud long, pointed; flowers red, yellow reverse, dbl., 34 petals, 4.5 in., exhibition form, slight fragrance; foliage glossy; vigorous, upright growth; [Colour Wonder X Wiener Charme]; Kordes, W. Söhne; Willemse

Galsar, *see* **'GALLIVARDA'**

Galway Bay®, *see* 'MACBA'

Gambler, *see* 'GELlor'

'GAMIN DE PARIS', F, dr, 1964; flowers dark blood-red, large, borne in large clusters, dbl.; vigorous, upright growth; [(Orange Triumph X Paprika) X Unnamed seedling]; Mondial Roses

'GAMINE', F, op, 1961; flowers salmon-pink, medium, dbl., moderate fragrance; vigorous growth; [Eclipse X Baccará]; Kriloff, Michel; Verbeek

'GAMMA', HT, pb, 1972; bud pointed; flowers pink, suffused vermilion, intense fragrance; foliage large; [Jouvencelle X American Heritage]; Gaujard

'GAMUSIN', HT, r, 1960; flowers cinnamon to pale pink, dbl., 25 petals, intense fragrance; vigorous, spreading growth; [Grey Pearl X (Lila Vidri X Prelude)]; Dot, Pedro

GANa, F, mp, 1991; (**David Whitfield**); Gandy

GANcre, LCl, w, 1998; (**Crème de la Crème**); Gandy

'GANDOL', HT, my, 1989; (**Doctor Goldberg**, 'GANGO'); flowers large, dbl., 26–40 petals, intense fragrance; foliage large, dark green, matt; upright growth; [Royal Dane X Dutch Gold]; Gandy, Douglas L.; Gandy Roses, Ltd., 1988

'GANDRI', F, pb, 1982; (**Betty Driver**); flowers pale peach and gold blend, yellow stamens, patio, large blooms, semi-dbl., slight fragrance; foliage medium, light green, semi-glossy; low, bushy growth; [Seedling X Topsi]; Gandy, Douglas L.; Gandy Roses, Ltd.

'GANFER', F, ob, 1987; (**Fergie**); flowers large, dbl., slight fragrance; foliage medium, medium green, semi-glossy, disease-resistant; patio; bushy growth; frost-proof; [Unnamed seedling X Copper Pot]; Gandy, Douglas L., 1988

'GANFLO', S, w, 1989; (**Florence Nightingale**); bud pointed; flowers glowing white, flushed buff, reverse tinged pink, aging to white, dbl., 32 petals, moderate, spicy fragrance; rounded, green fruit; prickles very pointed, fawn; foliage medium, medium green, semi-glossy; spreading, medium growth; [Morgengruss X Unnamed seedling]; Gandy, Douglas L.; Gandy Roses, Ltd.

'GANGA', HT, dy, 1970; flowers deep golden yellow, medium, dbl., exhibition form, profuse, intermittent bloom, moderate, tea fragrance; vigorous, upright growth; [Sabina X Unknown]; Division of Vegetable Crops and Floriculture; Indian Agric. Research Inst.

'GANGO', HT, my, 1989; (**Doctor Goldberg**, 'GANDOL'); flowers large, dbl., 26–40 petals, intense fragrance; foliage large, dark green, matt; upright growth; [Royal Dane X Dutch Gold]; Gandy, Douglas L.; Gandy Roses, Ltd., 1988

'GANHOL', HT, ob, 1984; (**Moriah**); flowers creamy yellow, orange reverse, large, dbl., 35 petals, intense fragrance; foliage large, dark, matt; bushy growth; [Fragrant Cloud X Seedling]; Holtzman, Arnold; Gandy Roses, Ltd.

GANjil, F, op, 1998; (**Jill's Rose**); Gandy

'GANSPA', F, lp, 1994; (**Spangles**); flowers pale pink speckled, blooms borne in large clusters, dbl., 15–25 petals, 1.5–2.75 in., moderate fragrance; many prickles; foliage medium, light green, semi-glossy; medium, upright, compact growth; [Florence Nightingale X Silver Jubilee]; Gandy, Douglas L.; Gandy Roses, Ltd., 1995

'GANYMED', F, dr, 1975; bud pointed; dbl., 32 petals, 2.5 in., cupped, slight fragrance; foliage glossy; vigorous, upright, bushy growth; [Europeana X Seedling]; Kordes

'GARDEJÄGER GRATZFELD', Cl Pol, dp, 1940; flowers carmine-red; vigorous growth; [Rodhatte sport]; Gratzfeld

Garden Blanket™, *see* 'JACGRON'

Garden City, *see* 'HARKOVER'

Garden Club, HT; Mansuino; (Cavriglia)

'GARDEN DELIGHT', F, dp, 1956; flowers deep rose-pink, rosette form, borne in clusters, dbl., 34 petals, 3.5–4 in., moderate fragrance; vigorous, branching growth; Norman; Harkness

'GARDEN GEM', HT, lp, 1930; flowers satiny pink, dbl.; very vigorous growth; [Mrs E.T. Stotesbury X Hill's America]; Dingee & Conard

'GARDEN GLOW', HT, op, 1937; flowers scarlet, base copper, dbl.; foliage glossy, bronze; very vigorous, bushy growth; Cant, B. R.

Garden Magic, *see* **Gartenzauber**®

'GARDEN NEWS', HT, dr, 1962; bud pointed; flowers dark crimson-scarlet, large, dbl., 32 petals, intense fragrance; foliage dull, dark, leathery; strong stems; moderate growth; [New Yorker X Étoile de Hollande]; Verschuren; Blaby Rose Gardens

Garden News, *see* POUlrim

'GARDEN PARTY'®, HT, w, 1959; bud urn-shaped; flowers pale yellow to white, often tinged light pink, dbl., 28 petals, 4–5 in., exhibition form, slight fragrance; foliage large, medium green, semi-glossy; vigorous, bushy, well-branched growth; AARS, 1960 GM, Bagatelle, 1959; [Charlotte Armstrong X Peace]; Swim, H.C.; Armstrong Nursery

Garden Party, *see* KORmollis

'GARDEN PARTY, CLIMBING', Cl HT, w, 1964; Itami Rose Nursery

Garden Pavilion Roos, HT, op, 1998

Garden Pearl, HT, w, 1992; Tejganga

Garden Perfume, *see* KORsenter

'GARDEN PRINCESS', F, my, 1961; flowers yellow, becoming lighter, semi-dbl.; growth moderate; [Goldilocks X Lavender Pinocchio]; Leenders, J.

'GARDEN QUEEN', HT, op, 1960; flowers pink to salmon, dbl., 4 in., intense fragrance; vigorous growth; [Ambassadeur Nemry X Tawny Gold]; Leenders, J.

Garden State, *see* 'MEIGENE'

Garden State II, Gr, or, 1997; Williams, J. Benjamin

'GARDEN SUPREME', F, ob, 1959; flowers orange blend shaded reddish, medium, borne in clusters, dbl., 32 petals; foliage leathery, bronze; low, vigorous growth; Jones; Hennessey

'GARDENER'S SUNDAY', F, my, 1975; flowers, dbl., 20 petals, 3 in., moderate fragrance; foliage bright green; [(Pink Parfait X Masquerade) X Arthur Bell]; Harkness

'GARDENIA', HWich, w, 1899; bud pointed, yellow; flowers creamy white, center yellow, well-formed blooms in small sprays; foliage small, dark, glossy; short, strong stems; very vigorous growth; [R. wichurana X Perle des Jardins]; Manda, W.A.

'GARDENIAEFLORA', HMult, w, 1901; flowers pure white, semi-dbl., early bloom, slight fragrance; Benary

Gardens of the World, *see* 'JACCOEUR'

'GARETH DAVIES', HT, mr, 1997; flowers full, large, very dbl., 26–40 petals, borne mostly singly, moderate fragrance; foliage medium, dark green, semi-glossy; upright, bushy, tall growth; [Crimson Glory X Loving Memory]; Poole, Lionel

Garibaldi, B, dp; flowers rosy lilac; Damaizin, 1859

'GARISENDA', HWich, pb, 1911; flowers clear rose-pink, tinted silvery, blooms in clusters, dbl.; [R. wichurana X Souv. de la Malmaison]; Bonfiglio, A.

'GARNETTE', F, dr, 1951; (Garnette Red, Red Garnette); flowers garnet-red, base light lemon-yellow, small, dbl., 50 petals, slight fragrance; foliage leathery, dark; bushy growth; [(Rosenelfe X Eva) X Heros]; Tantau; J&P

Garnette Apricot, Pol, ab

Garnette Carol, *see* **'CAROL AMLING'**

Garnette Pink, *see* **'CAROL AMLING'**

Garnette Red, *see* **'GARNETTE'**

Garnette Rose, Pol, dp

'GARNETTE SUPREME', F, dp, 1954; bud ovoid; flowers carmine, dbl., 35–40 petals, 2.5–3 in., cupped, moderate fragrance; foliage glossy, bronze; vigorous, upright, compact growth; [Yellow Pinocchio seedling X Garnette]; Boerner; J&P

Garnette White, Pol, w

Garnette Yellow, Pol, my

'GARNETTE, CLIMBING', Cl F, dr, 1954; Soria; Amling-DeVor Nursery

'GARNIA', F, dp, 1970; flowers deep pink, full, dbl., 48 petals, 4.5 in., intense fragrance; foliage matt, green; moderately vigorous growth; [Lady Sylvia X Garnette]; Butter; Wood End Gardens

'GARO', LCl, dr, 1986; flowers blackish-red, large, borne 1–5 per cluster, very dbl., 48 petals, exhibition form, slight fragrance; large, brown prickles; foliage red turning dark green, semi-glossy; upright growth; [Uncle Walter X Unnamed seedling]; Garelja, Anita, 1978

Garry Brown, *see* 'MEHBROWN'

Gartenarchitekt Gunther Schulze®, S, ly

Gartenblut®, HT, dr, 1986; Noack, Werner

'GARTENDIREKTOR GLOCKER', F, or, 1957; bud ovoid; flowers cinnabar-red, large, borne in clusters, very dbl., moderate fragrance; foliage glossy, leathery; vigorous, bushy growth; [Obergärtner Wiebicke X Independence]; Kordes

'GARTENDIREKTOR JULIUS SCHUTZE', HT, pb, 1920; flowers pale rosy pink and peach-blossom-pink; [Mme Jules Gravereaux X Pharisaer]; Kiese

'GARTENDIREKTOR NOSE', HT, dr, 1930; bud pointed; flowers dark crimson, large, dbl., exhibition form, moderate fragrance; foliage dark, glossy; vigorous, bushy growth; [Royal Red X Templar]; Kordes; Dreer; H&S

'GARTENDIREKTOR OTTO LINNE', S, dp, 1934; flowers dark carmine-pink, edged darker, base yellowish-white, dbl., borne in clusters of up to 30; foliage leathery, light green; long strong stems; vigorous, bushy growth; [Robin Hood (HMsk) X Rudolph Kluis]; Lambert, P.

Gartengold, HT, dy, 1984; Noack, Werner

'GARTENSTADT LIEGNITZ', HMult, m, 1910; flowers violet/red, medium, semi-dbl.; Lambert, P.; (Sangerhausen)

'GARTENSTOLZ', F, op, 1945; flowers rose tinted salmon, large, borne in clusters of 12–15, single, 8–10 petals, moderate fragrance; foliage leathery, light green; vigorous, upright, bushy growth; [Swantje X Hamburg]; Tantau

Gartenstolz, HT, op, 1974; Noack, Werner

Gartenzauber®, *see* 'KORNACHO'

'GARTENZAUBER', F, 1961; (Garden Magic, Magie des Jardins); flowers blood-red, tinted cinnabar-red, well formed, large; low growth; RULED EXTINCT 6/81 ARM; Kordes, R.

Gartenzauber '84, *see* 'KORNACHO'

Gärtnerfreude, Pol, or, 1965; flowers medium, dbl.; Kordes, W. Söhne; (Sangerhausen)

'GARVEY', HT, pb, 1961; flowers light geranium, reverse pale red, dbl., 30 petals, 6 in., globular, moderate fragrance; foliage dark, leathery; strong stems; vigorous, upright growth; [McGredy's Yellow X Karl Herbst]; McGredy, Sam IV; McGredy

Gary Lineker, *see* 'PEAROBIN'

'GARY PLAYER', HT, ob, 1968; (Goliath); flowers orange-vermilion, dbl., 35 petals, 4–4.5 in., exhibition form; foliage glossy, dark; vigorous growth; [Jolie Madame X Unnamed seedling]; Herholdt, J.A.

'GARY WERNETT', HT, op, 1985; flowers medium coral pink, large blooms borne usually singly, dbl., exhibition form, moderate fragrance; rare to full term fruit; medium, triangular, light red prickles; foliage medium, medium green, matt; bushy growth; [Helen Traubel X Helen Traubel]; French, Richard

'GASPARD MONGE', C, m, 1854; flowers light violet/red, medium, dbl.; Robert; (Sangerhausen)

'GASTON CHANDON', HT, pb, 1884; flowers light pink with coppery highlights, medium, dbl.; Schwartz; (Sangerhausen)

'GASTON LESIEUR', HWich, dp, 1915; flowers carmine-pink, medium, dbl.; Turbat; (Sangerhausen)

'GATESHEAD FESTIVAL', HT, op, 1989; flowers glowing orange flushed salmon with gold at base of petals, dbl., 26–40 petals, intense fragrance; foliage large, dark green, glossy; bushy growth; [Doris Tysterman X Silver Jubilee]; Thompson, Robert; Battersby Roses, 1989

'GAUBIROC', HT, op, 1986; (**Pénélope**®); flowers medium salmon pink, well-formed, large, dbl., 35 petals, moderate fragrance; foliage large, medium green, semi-glossy; upright growth; [(Americana X ?) X Chenonceaux]; Gaujard, Jean; Roseraies Gaujard, 1980

GAUbor, HT, ab; (**Barbara**®); Gaujard

'GAUCOVA', HT, or, 1978; (**Junon**®); bud full; flowers medium, dbl., 45 petals, 3 in.; foliage large, brownish; bushy growth; [Tanagra X Dora]; Gaujard

'GAUDENGI', HT, m, 1981; ('GAUTARA', **Méduse**®); flowers lavender red, dbl., 35 petals, moderate fragrance; foliage large, dark, semi-glossy; upright growth; [Chenonceaux X Tropicana]; Gaujard, Jean; Roseraies Gaujard, 1980

Gaudi, HT; Dot; (Cavriglia)

'GAUDIA', F, mp, 1946; flowers rose-pink, base gold and salmon, borne in clusters, semi-dbl., 15 petals, 4 in., moderate fragrance; foliage bright green edged red; vigorous, tall growth; [Florentina X Talisman]; Leenders, M.; Longley

'GAUDINO', HT, rb, 1945; (**Renaissance**®); bud pointed; flowers red and gold, medium, dbl., slight fragrance; foliage glossy; low growth; RULED EXTINCT 10/86; Gaujard

GAUesca, HT, dp; (**Scala**); Gaujard

'GAUFRARNER', HT, op, 1977; (**Odeon**); bud globular; flowers coral-pink, dbl., 50 petals, 3 in., globular, moderate fragrance; foliage large, dark; vigorous growth; [(Chateau de Chenonceaux X Mignonne) X Americana]; Gaujard

GAUhari, HT, pb, 1980; (**Amara**); Gaujard

'GAUHTI', HT, yb, 1969; (**Antigone**®); bud pointed; flowers yellow shaded red, large, dbl., moderate fragrance; foliage light green, soft; vigorous, upright growth; GM, Bagatelle, 1967; [Rose Gaujard X Guitare]; Gaujard

Gaujard 985, HT, mr, 1974; flowers large, dbl.; Gaujard; (Sangerhausen)

'GAULIMOR', HT, op, 1981; (**Hermione**); flowers deep salmon, large, dbl., 35 petals, moderate fragrance; foliage large, dark, glossy; upright, bushy growth; [Rose Gaujard X Colour Wonder]; Gaujard, Jean; Roseraies Gaujard

'GAUMO', HT, rb, 1957; (**Rose Gaujard**®); flowers cherry-red, reverse pale pink and silvery white, dbl., 80 petals, 3–4 in., exhibition form, slight fragrance; foliage leathery, glossy; vigorous, bushy growth; GM, NRS, 1958; [Peace X Opera seedling]; Gaujard; Armstrong Nursery, 1964

'GAUMOVA', F, or, 1978; (**Circé**); bud long; flowers brilliant orange-red, semi-dbl., 16 petals, 1.5–2 in.; foliage large, dark;

vigorous, compact growth; [Guitare X Prominent]; Gaujard

Gauntlet, *see* 'BRIGAUNT'

'GAURA', F, ob, 1957; (**Vendéme**); bud long; flowers bright salmon, medium, dbl., moderate fragrance; foliage dark, glossy; bushy growth; [Comtesse Vandal X (Fashion X Vogue)]; Gaujard

GAUrama, F, or; (**Arbelle**); Gaujard

'GAUSECA', HT, pb, 1980; ('GAUZECA', **Montreal**®); bud long; flowers cream and pink blend, large, dbl., 45 petals, moderate fragrance; green prickles; foliage large, dark; [Americana X Dora]; Gaujard, Jean

'GAUTARA', HT, m, 1981; ('GAUDENGI', **Méduse**®); flowers lavender red, dbl., 35 petals, moderate fragrance; foliage large, dark, semi-glossy; upright growth; [Chenonceaux X Tropicana]; Gaujard, Jean; Roseraies Gaujard, 1980

GAUtara, HT, ob, 1981; (**Simone Merieux**®); Gaujard

'GAUTIRA', HT, or, 1981; (**Molitor**®); flowers deep orange-red, large, dbl., 35 petals, moderate fragrance; foliage medium, dark, semi-glossy; [Junon X Tanagra]; Gaujard, Jean

GAUvera, HT, mr; (**SERENITÉ**); Gaujard

GAUvila, HT, 1979; (**Renee Columb**); Gaujard; (Cavriglia)

'GAUVITOR', HT, mr, 1977; (**Jean Gaujard**®); flowers brilliant red, dbl., 40 petals; vigorous growth; [Canasta X Rose Gaujard]; Gaujard

'GAUZECA', HT, pb, 1980; ('GAUSECA', **Montreal**®); bud long; flowers cream and pink blend, large, dbl., 45 petals, moderate fragrance; green prickles; foliage large, dark; [Americana X Dora]; Gaujard, Jean

GAUzimi, HT, mr; (**Lisima**); Gaujard

'GAUZINE', F, mp, 1975; (**Dady**®); intense fragrance; [Mignonne X Seedling]; Gaujard

'GAUZOMI', F, or, 1979; (**Odyssée**); flowers well-formed blooms borne 3–5 per cluster, dbl., 25 petals, moderate fragrance; small, brown prickles; foliage dark; [Pampa X Unnamed seedling]; Gaujard, Jean

'GAVÁ', Cl HT, or, 1934; flowers oriental red shaded rose-pink, base yellow, very large, dbl., cupped, intense fragrance; foliage leathery; very vigorous, climbing growth; [Souv. de Claudius Denoyel X Souv. de Claudius Pernet]; Munné, B.; Camprubi

'GAVAL', F, or, 1959; (**Atlantic**®); flowers medium, semi-dbl., slight fragrance; foliage glossy, dark; [Peace X Seedling]; Gaujard

Gavina, HT, w; Dot

Gavno, *see* 'POULGAV'

Gavolda, *see* 'MEICRI'

'GAVOTTE', HT, pb, 1963; flowers pink, reverse light yellow, dbl., 45 petals, 5 in., moderate fragrance; foliage dark; vigorous, upright growth; [Ethel Sanday X Lady Sylvia]; Sanday, John

'GAVROCHE', F, ob, 1963; flowers orange, center yellow, large; vigorous, bushy growth; Robichon

Gavroche™, *see* 'MEIriental'

'GAXENCE', HT, m, 1962; (**Éminence**®); flowers lavender, large, dbl., 40 petals, intense fragrance; foliage leathery, light green; vigorous, upright growth; [Peace X (Viola X Seedling)]; Gaujard; Ilgenfritz Nursery, 1965

'GAY CRUSADER', HT, rb, 1948; flowers red fading to pink, reverse deep yellow, large, exhibition form, moderate fragrance; foliage dark; [Phyllis Gold X Catalonia]; Robinson, H.; Baker's Nursery

'GAY DAWN', HT, op, 1958; bud pointed; flowers large, dbl., cupped, moderate, spicy fragrance; foliage dark, semi-glossy; very vigorous, upright growth; [Eclipse X Mme Henri Guillot]; Taylor, C. A.; California Nursery Co.

'GAY DEBUTANTE', HT, pb, 1960; flowers light pink, base yellow, very large, dbl., 40–45 petals, cupped, slight fragrance; foliage leathery, glossy; vigorous, upright growth; [Peace sport]; Curtis, R.F.; C.R. Burr

'GAY DICKY', F, ob, 1956; flowers tangerine-orange; moderate growth; Verschuren-Pechtold; Gandy Roses, Ltd.

'GAY GOLD', HT, my, 1973; flowers, dbl., 30 petals, 5.5 in., intense fragrance; foliage glossy; [King's Ransom X Piccadilly]; Lowe

'GAY GORDONS'®, HT, yb, 1969; flowers orange-yellow and red, dbl., slight fragrance; foliage dark, glossy; bushy, rather low growth; [Belle Blonde X Karl Herbst]; Cocker

'GAY GYPSY', HT, dr, 1949; bud long, pointed; flowers oxblood-red shades maroon, open, semi-dbl., 15–20 petals, 4.5–5 in., cupped, slight fragrance; foliage leathery; vigorous, upright, bushy growth; [Charles K. Douglas sport]; Crane; Bosley Nursery

'GAY HEART', F, mp, 1951; bud ovoid; flowers bright pink, large, borne in large clusters, dbl., 25 petals, exhibition form, moderate fragrance; foliage leathery; vigorous, upright growth; [Joanna Hill X World's Fair]; Boerner; J&P

'GAY JEWEL', Min, lp, 1958; bud globular; flowers light rose-pink, dbl., 35–40 petals, .5 in., cupped, moderate fragrance; foliage glossy; compact (6–8 in.) growth; [Dick Koster sport X Tom Thumb]; Morey, Dr Dennison; J&P

'GAY LADY', HT, mr, 1953; bud ovoid; flowers currant-red, open, dbl., 20–28 petals, 3.5–4.5 in., moderate, spicy fragrance; foliage dark, leathery, glossy; very vigorous, upright growth; [Charlotte Armstrong X Piccadilly]; Swim, H. C.; Breedlove Nursery

'GAY LYRIC', HT, mp, 1971; flowers rose-pink, large, very dbl., exhibition form, moderate fragrance; foliage glossy, dark, leathery; vigorous, upright growth; [Royal Highness X Elizabeth Fankhauser]; Fankhauser

'GAY MAID', F, or, 1969; flowers red suffused orange-pink, borne in trusses, dbl., 26 petals, globular; foliage light green; very vigorous growth; [Masquerade X ?]; Gregory

'GAY MOOD', LCl, dp, 1940; bud large, ovoid to urn shaped, rose-red; flowers deep rose-pink, open, semi-dbl., 15–25 petals, profuse, repeated bloom; foliage glossy, dark; very vigorous, climbing growth; [Joanna Hill X Sanguinaire]; Lammerts, Dr Walter; Armstrong Nursery

'GAY NINETIES', Cl F, mr, 1955; bud ovoid; flowers rose-red, blooms in clusters of 5–8, dbl., 65 petals, 2–2.5 in., intense fragrance; foliage leathery, glossy; vigorous, pillar (8 ft) growth; [(New Dawn X Red Ripples) X Red Ripples]; Sima

'GAY PARIS', HT, mr, 1960; bud long; flowers bright crimson, well-formed, large, dbl., moderate fragrance; foliage bright green; vigorous growth; [(Floradora X Barcelona) X (Charles Mallerin X Tonnerre)]; Delbard-Chabert

'GAY PRINCESS', F, lp, 1967; bud ovoid; flowers blush-pink, large blooms in clusters, dbl., cupped, moderate fragrance; foliage leathery; vigorous, upright, bushy growth; AARS, 1967; [Spartan X The Farmer's Wife]; Boerner; J&P

'GAY VISTA', S, lp, 1957; flowers very large blooms in very large clusters, single, repeat bloom; height 3 1/2 ft; Riethmuller

Gayness, HT, lp; Reithmuller, 1955; (Weatherly, L.)

'GAYTIME', F, rb, 1966; bud ovoid, pointed; flowers red and yellow, large, dbl., cupped, slight fragrance; foliage dark, glossy, leathery; vigorous, bushy, compact growth; [Unnamed seedling X Circus]; Armstrong, D. L.; Armstrong Nursery

'GAZELLE', HGal, lp; flowers delicate rose, large

G'DayGidday, *see* 'TINHAT'

Ge Korsten, *see* 'HERGEKO'

'GEAAURA', S, my, 1999; (**Precious Child**); flowers medium, dbl., 17–25 petals, borne in small clusters, slight fragrance; prickles moderate; foliage medium, dark green, semi-glossy; spreading (4 ft) growth; [Hokey Pokey X Patio Princess]; Gear, Ian Robert

'GEAAWARD', S, w, 1999; (**The Wedding Rose**); flowers medium, inner petals tinged lemon to white, opens somewhat quartered, very dbl., 41 petals, borne in small clusters, slight fragrance; prickles moderate; foliage dark blue/green, semi-glossy; upright, compact, bushy, medium (3 ft) growth; [Enchantment X English Miss]; Gear, Ian Robert

'GEAcot', S, ab, 1999; (**Apricot Kisses**); flowers dark apricot, aging to pale apricot, dbl., 26–40 petals, 2.5 in., borne in small clusters, slight, fruity fragrance; prickles moderate; foliage medium, medium green, new growth purple-red, semi-gl; spreading, bushy, medium (4–5 ft) growth; patio climber; [Laura Ford X seedling]; Gear, Ian Robert

'Géant des Batailles', HP, mr, 1846; (Giant of Battles); flowers deep fiery crimson, dbl., 85 petals, intense fragrance; moderately vigorous growth; Nerard; Guillot Pere

'Gedge's Glory', *see* WEBheart

Gee Dee, HT, mr, 1979; Dawson

Gee Gee™, *see* 'BENgee'

'Gee Whiz', S, yb, 1984; flowers yellow tinted orange-red, blooms borne 1–10 per cluster, dbl., 23 petals, cupped, repeat bloom, moderate, sweet fragrance; needle-like, brown prickles; foliage medium, leathery, dark olive green; low, bushy, free-branching growth; hardy; [Gingersnap X Sevilliana]; Buck, Dr Griffith J.; Iowa State University

'GEElove', HT, mr, 1996; (**I Love You**); flowers bright red, large blooms borne mostly single, dbl., 26–40 petals, slight fragrance; some prickles; foliage medium, dark green, semi-glossy; upright, medium growth; Bees of Chester; L.W. Van Geest Farms, Ltd., 1995

'GEEpeop', F, mp, 1997; (**The People's Princess**); flowers full, 28 in., very dbl., 26–40 petals, borne in small clusters, slight fragrance; foliage medium, medium green, semi-glossy; upright, medium growth; [Seedling X Seedling]; Bees; L. W. Van Geest Farms Ltd., 1997

'GEEstraw', Min, yb, 1996; (**Strawberries and Cream**); flowers light yellow, pink and red striped, small blooms borne in small clusters, dbl., 26–40 petals, slight fragrance; some prickles; foliage small, medium green, semi-glossy; compact, medium growth; Bees of Chester; L.W. Van Geest Farms, Ltd., 1995

'Geheimrat Dr Mettweg', S, pb, 1909; flowers rose-red, center yellowish white, large, borne in large clusters, recurrent bloom; foliage dark; vigorous, bushy growth; [(Mme Norbert Levavasseur X Trier) X R. foetida bicolor]; Lambert, P.

Geheimrat Duisberg, *see* **'Golden Rapture'**

'Geheimrat Duisberg, Climbing', Cl HT, dy, 1954; flowers large, dbl., moderate fragrance; Knackfuss; (Sangerhausen)

'Geheimrat Richard Willstätter', HT, ab, 1931; flowers apricot-yellow, veined carmine-red, stamens yellow, large, semi-dbl., slight fragrance; foliage bright, thick; vigorous growth; [Constance X Admiral Ward]; Felberg-Leclerc

'Geisha'®, F, mp, 1964; (Pink Elizabeth Arden); bud long; semi-dbl., 2.5–3 in., blooms in clusters of 1–3; foliage dark; bushy, medium height.; Tantau, Math.

'Geisha', HT, ob, 1920; bud orange, marked coral-red; flowers golden yellow; [Mme Edouard Herriot sport]; Van Rossem

'Geisha', HMult, m, 1913; Geschwind, R.

'Geisha Girl', F, my, 1964; dbl., 25 petals, 3.5 in., blooms in clusters; foliage long, pointed; tall growth; [Gold Cup X McGredy's Yellow]; McGredy, Sam IV; McGredy

'Gela Gnau', HT, ab, 1926; flowers amber-yellow, reverse apricot, dbl., moderate fragrance; Leenders, M.

'Gela Tepelmann', HMult, mp, 1950; flowers medium, dbl.; Tepelmann; (Sangerhausen)

'GELance', F, rb, 1999; (**Turbulance**); flowers large, single, 5–11 petals, borne in small clusters, slight fragrance; prickles moderate; foliage medium, light green, glossy; bushy, medium growth; [Pink Favorite X Scentimental]; Giles, Diann; Giles Rose Nursery, 1999

'GELangel', F, m, 1998; (**Silver Angel**); flowers light silver mauve, much lighter than Angel Face, double, (15–25 petals), medium blooms borne in small clusters; fragrant; foliage medium green, semi-glossy; bushy, medium (3 ft) growth; [Angel Face sport]; Giles, Diann; Giles Rose Nursery, 1996

'GELback', HT, ly, 1995; (**Back Home**); flowers light yellow, full (26–40 petals), medium (4–7 cms) blooms borne mostly single; very fragrant; few prickles; foliage medium, light green, semi-glossy; medium, spreading growth; [Dr A. J. Verhage X Midas Touch]; Giles, Diann; Giles Rose Nursery, 1995

Gelbe Dagmar Hastrup, *see* 'MORyelrug'

Gelbe Florida von Scharbeutz, F; (Cavriglia)

'Gelbe Holstein', F, ly, 1951; (Yellow Holstein); bud long, pointed; flowers yellow paling to lemon, blooms in large clusters, dbl., 20 petals, 3 in., slight fragrance; foliage glossy, light green; vigorous, upright, bushy growth; [(Eva X Viscountess Charlemont) X Sunmist]; Kordes; Wheatcroft Bros.

'Gelbe Pharisäer', HT, my, 1927; flowers clear yellow, center deeper, moderate fragrance; [Pharisaer X Mrs Aaron Ward]; Hinner, W.

Gelber Kobold, MinFl, dy, 1996

'GELbird', LCl, mr, 1997; (**Red Bird**); flowers full, medium, very dbl., 26–40 petals, borne mostly singly, slight fragrance; foliage medium, medium green, semi-glossy; spreading, medium (6–8 ft) growth; [Crepe de Chine X Kardinal]; Giles, Diann

'GELburn', F, ob, 1997; (**Sunburnt**); flowers double, small, dbl., 15–25 petals, borne in small clusters, no fragrance; foliage medium, medium green, semi-glossy; spreading, medium (3.5 ft) growth; [Dalton X Select Pollen]; Giles, Diann

'GELcourt', Min, rb, 1999; (**Courtney**); flowers red, reverse gold, dbl., 17–25 petals, 1.5 in., borne in small clusters, no fragrance; many prickles; foliage medium, dark green, glossy; spreading, low (1.5 ft) growth; [Princess Celest X Rainbow's End]; Giles, Diann; Giles Rose Nursery, 1999

'GELdavis', HT, pb, 1998; (**Ken Davis**); flowers light pink with darker edges, medium size, dbl., 15–25 petals, borne mostly singly, no fragrance; prickles medium; foliage medium, light green, dull; upright, medium (4 ft) growth; [Bride's Dream sport]; Davis, Ken; Giles Rose Nursery, 1995

'GELday', S, mp, 1997; (**Sunday China**); flowers small, 1 in., borne in large clusters, no fragrance; foliage small, medium green, dull; compact, low (2 ft 6 in.) growth; [Champney's Pink Cluster X Seedling]; Giles, Kevin

'GELeasy', F, op, 1999; (**Easy Vibes**); flowers medium, dbl., 17–25 petals, borne in small clusters, slight fragrance; few prickles; foliage medium, dark green, glossy; compact, medium growth; [Sun Flare X Easy Living]; Giles, Diann

'GELele', Min, lp, 1999; (**Dear Eleanor**); dbl., 17–25 petals, borne mostly singly, slight fragrance; few prickles; foliage medium, dark green, dull; upright, medium growth; [Party Girl X Snow Bride]; Giles, Diann; Giles Rose Nursery, 1999

'GELfla', F, my, 1997; (**Florida Sun**); flowers full, small, very dbl., 26–40 petals, borne in small clusters, slight fragrance; foliage small, medium green, glossy; compost (2.5–3 ft) growth; [Sun Flare X Select Pollen]; Giles, Diann

'GELfly', F, my, 1999; (**Butterfly Kisses**); flowers medium, single, 5–11 petals, borne in small clusters, moderate fragrance; few prickles; foliage medium, medium green, semi-glossy; upright, spreading, medium growth; [Sun Flare X Summer Snow]; Giles, Diann; Giles Rose Nursery, 1999

'GELgood', F, mp, 1999; (**Good Vibes**); flowers medium, dbl., 17–25 petals, borne in large clusters, slight fragrance; few prickles; foliage medium, medium green, glossy; compact, medium growth; [Sun Flare X Simplicity]; Giles, Diann; Giles Rose Nursery, 1999

'GELgrape', Min, m, 1999; (**Grape Delight**); flowers small, semi-dbl., 12–16 petals, borne in small clusters, slight fragrance; many prickles; foliage medium, medium green, semi-glossy; compact, low (1.5 ft)

growth; [Lavender Sweetheart X Herbie]; Giles, Diann; Giles Rose Nursery, 1999

'GELJAM', HT, mr, 1999; (**Jamie's Love**); flowers medium, dbl., 26–40 petals, borne mostly singly, slight fragrance; few prickles; foliage medium, dark green, dull; upright, medium growth; [Vera Dalton X Special Merit]; Giles, Diann; Giles Rose Nursery, 1999

'GELKEN', HT, mr, 1996; (**Kenny's Rose**); dbl., 26–40 petals, 3–3.5 in., borne mostly singly, moderate fragrance; some prickles; foliage medium, dark green, semi-glossy; medium, spreading, sprawling growth; [Swarthmore X Mister Lincoln]; Giles, Kenneth; Giles Rose Nursery, 1995

'GELLANDO', Min, rb, 1999; (**Rose Magic**); flowers white and red edge, white reverse, medium, dbl., 26–40 petals, borne in small clusters, slight fragrance; few prickles; foliage medium, dark green, semi-glossy; upright, medium growth; [Little Darling X Kristen]; Giles, Diann; Giles Rose Nursery, 1999

'GELLITE', Min, w, 1998; (**Tiffany Lite**); flowers large, white, double, dbl., 15–25 petals, borne in small clusters, slight fragrance; foliage medium, medium green, semi-glossy; upright, bushy, 3–4 ft growth; [Tiffany Lynn sport]; Giles, Diann; Giles Rose Nursery, 1997

'GELLOV', HT, w, 1995; (**Gambler**); flowers white, full (26–40 petals), large (7 cms) blooms borne mostly single; slight fragrance; few prickles; foliage large, medium green, semi-glossy; tall (5 ft), upright growth; [Grace de Monaco X Seedling]; Gimer, Louis; Giles Rose Nursery, 1995

'GELLYNN', HT, rb, 1999; (**Sierra Lynn**); flowers medium, dbl., 26–40 petals, borne mostly singly, slight fragrance; few prickles; foliage medium, dark green, semi-glossy; upright, medium (4–5 ft) growth; [Vera Dalton X Paradise]; Giles, Diann; Giles Rose Nursery, 1999

'GELMAGIC', F, pb, 1997; (**Magic Baby**); flowers full, medium, very dbl., 26–40 petals, borne mostly singly, no fragrance; foliage medium, medium green, dull; upright, medium (3.5 ft)growth; Giles; Diann

'GELMAN', Min, mr, 1997; (**Chess Man**); flowers full, medium, very dbl., 26–40 petals, borne mostly singly, no fragrance; foliage medium, dark green, dull; upright, medium (3 ft) growth; [Chrysler Imperial X Select Pollen]; Giles, Kevin

'GELMAR', Min, mr, 1995 (**Marvie**); flowers medium red, full (26–40 petals), small (0–4 cms) blooms borne mostly single; slight fragrance; few prickles; foliage small, medium green, semi-glossy; medium, upright growth; [Winsome X Seedling]; Giles, Diann; Giles Rose Nursery, 1995

'GELMAUD', Min, mp, 1999; (**Mary Maud**); flowers medium, dbl., 17–25 petals, borne in small clusters, slight fragrance; prickles moderate; foliage medium, medium green, semi-glossy; upright, medium growth; [Little Darling X Unknown]; Giles, Diann; Giles Rose Nursery, 1999

'GELMISS', HT, dp, 1996; (**Miss Mary**); flowers deep pink, blooms, dbl., 26–40 petals, 3–3.5 in., no fragrance; few prickles; foliage large, medium green, matt; tall (5 ft), upright growth; [Big Ben sport]; Miller, Carol; Giles Rose Nursery, 1995

'GELMOON', S, w, 1998; (**South Moon**); flowers pure white, yellow stamens, full, very dbl., 26–40 petals, borne in large clusters, intense fragrance; foliage medium, medium green, semi-glossy; upright, medium (4 ft) growth; [Belle Story sport]; Giles, Diann; Giles Rose Nursery, 1997

'GELOUT', Min, m, 1999; (**Chill Out**); flowers large, dbl., 17–25 petals, borne mostly singly, no fragrance; few prickles; foliage large, medium green, semi-glossy; upright, tall growth; [Lavender Sweetheart X Herbie]; Giles, Diann; Giles Rose Nursery, 1999

'GELPAM', HT, ob, 1999; (**Pam's Passion**); flowers medium, dbl., 26–40 petals, borne mostly singly, slight fragrance; few prickles; foliage dark green, semi-glossy; upright, medium growth; [Vera Dalton X Unknown]; Giles, Diann; Giles Rose Nursery, 1999

'GELPEACH', HT, op, 1997; (**Just Peachy**); flowers double, small, dbl., 15–25 petals, borne mostly singly, no fragrance; foliage medium, dark green, glossy; upright, medium growth; [Vera Dalton X Select Pollen]; Giles, Diann

'GELPEP', Min, or, 1997; (**Chilli Pepper**); flowers full, medium, very dbl., 26–40 petals, borne in small clusters, no fragrance; foliage large, dark green, glossy; bushy, medium (1.5–2 ft) growth; [Vera Dalton X Select Pollen]; Giles, Diann

'GELPUZZLE', LCl, rb, 1999; (**Chinese Puzzle**); flowers small, dbl., 17–25 petals, borne in large clusters, slight fragrance; many prickles; foliage small, dark green, glossy; climbing, tall growth; [Vera Dalton X Roller Coaster]; Giles, Diann; Giles Rose Nursery, 1999

'GELRED', Min, op, 1995; (**Redneck Girl**); flowers orange pink with white base, full (26–40 petals), few prickles; foliage small, dark green, semi-glossy; spreading growth; [Seedling X Seedling]; Giles, Diann; Giles Rose Nursery, 1995

Gelria, F, 1974; Verbeek; (Cavriglia)

'GELRISE', F, rb, 1997; (**Gala Sunrise**); flowers full, medium, very dbl., 26–40 petals, borne in small clusters, no fragrance; foliage medium, dark green, dull; upright, medium (3.5 ft) growth; [Vera Dalton X Rainbow's end]; Giles, Diann

'GELRITE', F, my, 1996; (**Rite Brite**, Right Bright); flowers medium yellow, full (26–40 petals), small (0–4 cms) blooms; slight fragrance; few prickles; foliage small, medium green, matt; low ($3\frac{1}{2}$ ft), spreading, compact growth; [Sun Flare X Rise 'n' Shine]; Giles, Diann; Giles Rose Nursery, 1995

'GELSHAN', Min, pb, 1997; (**Shannie**); flowers full, medium, very dbl., 26–40 petals, borne mostly singly, no fragrance; foliage medium, medium green, semi-glossy; upright, medium (2.5 ft) growth; [Little Darling X Magic Carrousel]; Giles, Diann

'GELSKIES', Min, ab, 1997; (**Morning Skies**); flowers full, apricot blend, very dbl., 26–40 petals, borne mostly singly, no fragrance; foliage medium, medium green, dull; upright, low (1–1/2 ft) growth; [Little Darling X Rainbow's End]; Giles, Diann; Giles Rose Nursery, 1997

'GELSWEET', F, lp, 1999; (**Sweet Vibes**); flowers medium, dbl., 17–25 petals, borne in small clusters, slight fragrance; few prickles; foliage medium, medium green, glossy; compact, medium growth; [Sun Flare X Simplicity]; Giles, Diann; Giles Rose Nursery, 1999

Geltendorf, HT, lb, 1971; Croix; (Cavriglia)

'GELTWO', HT, ab, 1999; (**Two Thumbs Up**); dbl., 26–40 petals, 5 in., borne mostly singly, slight fragrance; few prickles; foliage medium, medium green, dull; upright, medium (5 ft) growth; [South Seas X St Patrick]; Giles, Diann

'GELWIN', Min, w, 1997; (**Winter Princess**); flowers double, medium, dbl., 15–25 petals, borne in small clusters, no fragrance; foliage medium, medium green, semi-glossy; upright, medium (2.5 ft) growth; [Little Darling X Select Pollen]; Giles, Diann

'GEM', HT, mp, 1960; bud long, pointed; flowers deep soft pink, medium, semi-dbl., exhibition form, moderate fragrance; foliage soft; vigorous, upright growth; [Ena Harkness X Mme Butterfly]; Walker

'GEM OF THE PRAIRIES', HSet, dp, 1865; (Bijou des Prairies); flowers rosy red, occasionally blotched white, large blooms in large clusters, flat, non-recurrent, slight fragrance; vigorous growth; [Believed to be Queen of the Prairies X Mme Laffay]; Burgess, A.

'GEMINI', Pol, ob, 1967; bud pointed; flowers orange, small, dbl., exhibition form, slight fragrance; foliage dark, glossy; vigorous, upright, bushy growth; [Unnamed seedling X Rumba]; Hill, Joseph H., Co.

Gemini(TM), *see* 'JACNEPAL'

Gemma, S, mp

'GEMSTONE', HT, mp, 1978; bud high-centered; dbl., 28 petals, 4.5–5 in., exhibition form, slight fragrance; foliage matt; vigorous, upright growth; [Helen Traubel X Swarthmore]; J&B Roses; Eastern Roses

Gén. Jacqueminot, *see* **'GÉNÉRAL JACQUEMINOT'**

'GEN. JOHN PERSHING', LCl, dp, 1917; (F.R.M. Undritz); flowers large, dbl., 53 petals, moderate fragrance; vigorous, climbing growth; [Dr W. Van Fleet X Mrs W.J. Grant]; Undritz

GENdee, S, lp; (**Pearly King**); Genesis

'GENE BOERNER', F, mp, 1968; bud ovoid; flowers deep pink, medium, dbl., 35 petals, exhibition form; foliage glossy; vigorous, upright growth; AARS, 1969; [Ginger X (Ma Perkins X Garnette Supreme)]; Boerner; J&P

Gene Jones, F, W; (Angel's Blush)

Gene Sandberg, *see* 'RESAND'

'GENERAAL SMUTS', HT, mr, 1922; flowers cherry-red, shaded deep coral-red, dbl., moderate fragrance; [Gen. MacArthur X Mme Edouard Herriot]; Van Rossem

'GENERAAL SNIJDERS', HT, dp, 1917; flowers deep carmine shaded coral-red, dbl., intense fragrance; [Mme Mélanie Soupert X George C. Waud]; Leenders, M.

'GENERAL ALLARD', B, dp, 1835; Laffay, M.

'GÉNÉRAL BARON BERGE', HP, mr, 1892; flowers red, center occasionally striped white, large, dbl., 50 petals; erect, vigorous growth; Pernet Père

'GENERAL BARRAL', HP, 1867; Damaizin; (Cavriglia)

'GÉNÉRAL BEDEAU', HP, pb, 1851; flowers pink and red, large, dbl.; Margottin; (Sangerhausen)

'GÉNÉRAL BERTHELOT', HT, dp, 1926; flowers dark pink, slightly streaked white, dbl.; [J.B. Clark X Farbenkonigin]; Walter, L.

'GENERAL BROWNE', HSpn, w; flowers blush-white, fading pure white, quilled, dbl., moderate fragrance; prickles twiggy, prickly growth; foliage dark

General Cavaignac, Pol, mr

'GÉNÉRAL CLERC', M, 1845; Laffay, M.; (Sangerhausen)

'GÉNÉRAL DE VAULGRENANT', HT, dp, 1926; flowers rose-pink, very dbl.; [Mme Henriette Schissele X Mme Adele Gance]; Walter, L.

'GENERAL DESAIX', HP, 1867; Moreau et Robert; (Cavriglia)

'GENERAL DOMINGOS DE OLIVEIRA', HT, ab, 1939; flowers yellow-apricot tinted flesh-pink, large, dbl., cupped, slight fragrance; foliage glossy; dwarf growth; [Frank Reader X Golden Gleam]; da Silva, Moreira

'GENERAL DON', HT, pb, 1919; flowers strawberry tinted coppery, base golden yellow, dbl., moderate fragrance; [Mme Mélanie Soupert X Louise Catherine Breslau]; Le Cornu

'GÉNÉRAL DONADIEU', HGal, mr; flowers purplish red, compact, very dbl.

'GÉNÉRAL DROUOT', mr, 1854; flowers carmine-red, medium, semi-dbl.; Robert; (Sangerhausen)

'GÉNÉRAL DROUST', M, m, 1847; flowers purplish crimson, medium, very dbl., recurrent bloom; vigorous growth; Vibert

'GÉNÉRAL DUC D'AUMALE', HP, dr, 1875; flowers large, dbl.; Verdier, E.; (Sangerhausen)

'GÉNÉRAL FETTER', HT, m, 1922; flowers carmine-purple, glossy, very dbl.; [Jonkheer J.L. Mock X Luise Lilia]; Walter, L.

'GÉNÉRAL GALLIÉNI', T, rb, 1899; flowers coppery red, cupped; vigorous growth; [Souv. de Therese Levet X Reine Emma des Pays-Bas]; Nabonnand, G.

'GÉNÉRAL GUISAN', HT, dr, 1945; flowers large, dbl.; Heizmann, E.; (Sangerhausen)

General Jack, *see* **'GÉNÉRAL JACQUEMINOT'**

'GÉNÉRAL JACQUEMINOT', HP, rb, 1853; (Gén. Jacqueminot, General Jack, Jack Rose); bud scarlet-crimson; flowers dark red, whitish reverse, dbl., 27 petals, recurrent bloom, intense fragrance; foliage rich green; long, strong stems; vigorous, bushy growth; (28); [Probable seedling of Gloire des Rosomanes sport]; Roussel

'GÉNÉRAL KLÉBER', M, mp, 1856; bud well mossed; flowers pink tinted lilac, Robert

General Labutere, Ch, dp

'GENERAL MACARTHUR', HT, dp, 1905; flowers rose-red, dbl., 20 petals, intense, damask fragrance; foliage leathery; (28); E.G. Hill Co.

'GENERAL MACARTHUR, CLIMBING', Cl HT, dp, 1923; Dickson, H.

'GENERAL ROBERT E. LEE', T, my, 1896; bud deep orange-yellow; flowers canary-yellow; Good & Reese

'GÉNÉRAL SCHABLIKINE', T, op, 1878; flowers coppery; vigorous growth; Nabonnand, G.

'GENERÁL STEFÁNIK', HP, m, 1933; (Krásná Azurea); [La Brillante X Unnamed seedling]; Böhm, J.; J&P

'GÉNÉRAL TARTAS', T, dp; flowers deep rose, large, dbl.; Bernède

'GENERAL TESTARD', HWich, rb, 1918; flowers red, center white, small, borne in large clusters, semi-dbl.; Pajotin-Chédane

'GENERAL TH. PESCHKOFF', HT, lp, 1909; flowers very large, dbl., moderate fragrance; Ketten, Gebrüder; (Sangerhausen)

'GENERAL VON BOTHNIA-ANDREAE', HP, mr, 1900; flowers carmine-red, very large, dbl.; Verschuren; (Sangerhausen)

'GENERAL WASHINGTON', HP, dr, 1861; flowers deep crimson, reflexes maroon, large, very dbl., flat, occasionally recurrent bloom, moderate fragrance; moderate growth; [Triomphe de l'Exposition sport]; Granger

'GÉNÉRALE MARIE RAIEWSKY', HP, pb, 1911; flowers pink with dark yellow, large, dbl.; Ketten, Gebrüder; (Sangerhausen)

'GENERALIN ISENBART', HT, pb, 1915; flowers coppery-pink with dark yellow, large, dbl., moderate fragrance; Lambert, P.; (Sangerhausen)

'GENERALOBERST VON KLUCK', HT, dp, 1917; flowers carmine-pink, large, dbl., intense fragrance; Lambert, P.; (Sangerhausen)

'GENERAL-SUPERIOR ARNOLD JANSSEN', HT, dp, 1912; bud pointed; flowers deep rose-pink, veined darker, reverse much darker, large, dbl., moderate fragrance; [Farbenkonigin X Gen. MacArthur]; Leenders, M.

'GENERAL-SUPERIOR ARNOLD JANSSEN, CLIMBING', Cl HT, dp, 1931; Böhm, J.

Generosa, *see* **'MANSUINO ROSE'**

'GENEROSITY', HT, ob, 1982; flowers cream with orange center, large, dbl., 35 petals, moderate fragrance; foliage large, dark, matt; bushy growth; [Fred Gibson X Lady Elgin]; Northfield, G.

'GENESIS', Min, m, 1991; bud ovoid; flowers lavender, reverse white, blooms borne singly, dbl., 45 petals, 2 in., moderate fragrance; foliage medium, medium green, semi-glossy; upright growth; [Lavender Jade X Angel Face]; Jolly, Marie, 1986; Rosehill Farm, 1992

'GENÈVE', HT, or, 1944; bud long; flowers salmon-carmine and capucine-red, high pointed, dbl., moderate fragrance; vigorous growth; [Charles P. Kilham X Mme Joseph Perraud]; Meilland, F.

Genevieve™, *see* 'SAVAGEN'

Geneviève Genest, *see* **'MRS SAM MCGREDY, CLIMBING'**

'GENEVIÈVE LE GOASTER', HT, w, 1923; flowers white, center pale rose, base salmon-rose; Carrette; Richardier

Genevieve Ros, S, lp; Peden, R., 1998; (Weatherly, L.)

'GENIUS MENDEL', HT, mr, 1935; bud pointed; flowers light fiery red to pure red, large, dbl., exhibition form; foliage glossy; bushy growth; [Mrs Henry Winnett X Sir David Davis]; Böhm, J.

GENpat, MinFl, op, 1995; (**Dazzler**); Genesis

GENpen, S, lp; (**Prince Regent**); Genesis

GENpink, S, dp; (**King William**); Genesis

GENruby, S, lp; (**King Richard**); Genesis

Gentiliana, *see* **'POLYANTHA GRANDIFLORA'**

'GENTLE', F, op, 1960; flowers salmon-pink, well formed, borne in clusters, dbl., 26 petals, 2.5–3 in.; vigorous, compact, bushy growth; [Independence X (Lady Sylvia X Fashion)]; Lens

Gentle Clown, *see* HARtopper

Gentle Cover, *see* 'POULLEN'

Gentle Kiss, *see* HARtbrad

'GENTLE LADY', HT, lp, 1971; bud slender, long, pointed; dbl., 35 petals, 3.5 in., cupped, intense fragrance; foliage matt, dark, leathery; upright, bushy growth; [Tiffany X Michele Meilland]; Fuller; Wyant

Gentle Maid, *see* HARvilac

'GENTLE PERSUASION', S, yb, 1984; flowers yellow tinted orange, medium-large blooms borne 1–5 per cluster, dbl., 28 petals, cupped, repeat bloom, slight fragrance; awl-like, tan prickles; foliage large, leathery, semi-glossy, dark olive green; vigorous, bushy, erect growth; hardy.; [Carefree Beauty X Oregold]; Buck, Dr Griffith J.; Iowa State University

Gentle Touch, *see* 'DICLULU'

Gentleman's Agreement, *see* 'BRIGENTLE'

'GENVAL', HT, mp, 1963; flowers cyclamen-pink, moderate fragrance; foliage bronze, dull; vigorous growth; [Rosita X Margaret]; Delforge

GENwine, S, lp; (**King Henry**); Genesis

'GEOFF BOYCOTT', F, w, 1974; dbl., 35 petals, 3.5 in., slight fragrance; foliage dark; [Ice White X Tip-Top]; McGredy, Sam IV

Geoff Hamilton, *see* 'AUSHAM'

'GEORG ARENDS', HP, mp, 1910; (Fortuné Besson); flowers soft pink, large, dbl., 25 petals, intense fragrance; vigorous growth; [Frau Karl Druschki X La France]; Hinner, W.

'GEORG GEUDER', HT, op, 1931; flowers dark salmon-pink, medium, dbl., moderate fragrance; Schmidt, I. C.; (Sangerhausen)

George Armer, *see* 'BRATOWIN'

'GEORGE BAKER', HP, dp, 1881; flowers cerise, very dbl.; Paul & Son

George Burns™, *see* 'WEKCALROC'

George Burns Centennial, *see* 'WEKCALROC'

'GEORGE C. WAUD', HT, mp, 1908; flowers rose, veined darker, dbl., 4.5 in., intense fragrance; bushy growth; Dickson, A.

George Cuvier, B, dp

'GEORGE DAKIN', HT, pb, 1927; flowers silvery pink, flushed apricot, reverse orange to apricot, dbl., exhibition form, moderate fragrance; foliage glossy, bronze; vigorous, bushy growth; [Ophelia X Mrs Henry Morse]; Burbage Nursery

'GEORGE DICKSON', HT, mr, 1912; flowers large blooms with weak stems, dbl., 36 petals, moderate fragrance; weak stems; GM, NRS, 1911; Dickson, A.

'GEORGE DICKSON, CLIMBING', Cl HT, mr, 1949; Woodward

'GEORGE ELGER', Pol, my, 1912; (Yellow Baby Rambler); bud small, golden yellow; flowers coppery yellow to clear yellow, borne in large clusters, very dbl., slight fragrance; foliage small, dark, soft; bushy, dwarf growth; Turbat

'GEORGE ELLIOT', HT, op, 1970; flowers shrimp-pink, dbl., 35 petals, 4–5 in., intense fragrance; foliage bronze; [Highlight X Dorothy Peach]; Wills

'GEORGE FOX', HT, or, 1939; flowers orange-vermilion, medium, globular; foliage glossy; compact growth; [Charles P. Kilham X Lady Forteviot]; Savage Nursery

'GEORGE GEARY', HT, yb, 1953; flowers golden yellow flushed vermilion, high pointed, 4 in., moderate fragrance; foliage dark, bronze; vigorous growth; [Gwyneth Jones X Unnamed seedling]; Geary; Burbage Nursery

'GEORGE GEUDER', HT, yb, 1931; flowers salmon-pink and bright carmine on yellow ground, moderate fragrance; vigorous growth; Schmidt, J.C.

'GEORGE H. MACKERETH', HT, dr, 1924; flowers crimson shaded velvety maroon, dbl., intense fragrance; Dickson, A.

'GEORGE HEERS', HT, pb, 1961; flowers rich pink, touched apricot and yellow; Langbecker

'GEORGE HOWARTH', HT, dp, 1928; flowers bright carmine, dbl., moderate fragrance; [Gorgeous X The Queen Alexandra Rose]; Bees

George IV, *see* **'RIVERS' GEORGE IV'**

'GEORGE R. HILL', HT, w, 1990; dbl., 45 petals; foliage large, dark green; vigorous growth; [Admiral Rodney sport]; Varney, Eric; Battersby Roses, 1991

George Sand, *see* **'LUNELLE'**

'GEORGE THOMAS', HT, w, 1975; flowers pure white, tinged pink, full, dbl., 40 petals, 6–8 in., moderate fragrance; foliage dark; vigorous growth; [Ena Harkness X Memoriam]; Ellick

George Vancouver, S, mr, 1994; flat, 24 petals, slight; Ogilvie, Ian S.

'GEORGE WILL', HRg, dp, 1939; flowers deep pink, blooms in clusters, dbl., 3 in., flat, all-summer bloom, moderate, clove fragrance; foliage rugose; slender branches; height 3–4 ft; [(R. rugosa X R. acicularis) X ?]; Skinner

Georgeous®, *see* 'INTERGEORGE'

'GEORGES CAIN', HRg, dr, 1909; flowers crimson with purple, large, dbl.; very vigorous growth; [Souv. de Pierre Notting X R. rugosa]; Müller, Dr F.

Georges Cassagne, Gr, 1980; Croix; (Cavriglia)

'GEORGES CHESNEL', HT, dy, 1935; bud pointed; flowers deep golden yellow, veined copper, dbl.; foliage glossy; [Julien Potin seedling X Étoile d'Or]; Pernet-Ducher; Gaujard

'GEORGE'S CHOICE', F, mr, 1979; bud small, ovoid; flowers currant-red, moderately full, dbl., 35–40 petals, moderate fragrance; compact, bushy growth; [Evelyn Fison X Tabarin]; Ellick; Excelsior Roses

'GEORGES CLEMENCEAU', HT, ob, 1919; flowers bright orange, shaded umber and carmine; [Mme Edouard Herriot sport]; Lévêque

'GEORGES DE CADONEL', B, lp, 1904; Schwartz

'GEORGES HAMONIÈRE', HT, dr, 1937; flowers medium, dbl.; Moulin; (Sangerhausen)

'GEORGES PAQUEL', HT, my, 1934; flowers saffron-yellow, large, dbl., slight fragrance; vigorous, bushy growth; [Unnamed seedling X Souv. de Claudius Pernet]; Leenders, M.

'GEORGES PERDOUX', HT, pb, 1927; flowers reddish pink tinted coppery red, dbl., moderate fragrance; Barbier

'GEORGES PERNET', Pol, mp, 1887; flowers bright peach-pink; [Mignonette X ?]; Pernet-Ducher

'GEORGES ROUSSET', HP, 1893; Rousset; (Cavriglia)

'GEORGES SCHWARTZ', HT, my, 1899; flowers canary-yellow; [Kaiserin Auguste Viktoria X Souv. de Mme Levet]; Schwartz, Vve.

'GEORGES SCHWARTZ, CLIMBING', Cl HT, my, 1917; Knight, G.

Georges Truffaut, *see* DORkade

'GEORGES VIBERT', HGal, rb, 1853; flowers purplish red, streaked (striped) white, large, dbl., flat, moderate fragrance; Robert

Georgetown Gateside Yellow, N, ly; (found rose)

Georgetown Tea, T, pb; (found rose)

'GEORGETTE'®, Min, mp, 1981; bud ovoid; flowers medium pink, veined darker, dbl., 30 petals, exhibition form, slight fragrance; straight prickles; foliage medium green, dense; upright, compact, bushy growth; [Electron X Little Chief]; Bennett, Cecilia 'Dee'; Tiny Petals Nursery

Georgette, *see* 'INTERORGE'

Georgette, *see* KORacona

Georgeus, Min, mp, 1998

'GEORGIA', HT, ab, 1980; bud short, pointed; flowers peach-apricot blend, borne singly, very dbl., 53–55 petals, 5.5–6. in., moderate, tea fragrance; long prickles, hooked downward; foliage large, glossy, leathery; tall, upright growth; PP4712; [Arizona X Unnamed seedling]; Weeks, O.L.

'GEORGIANNA DOAN', HT, pb, 1942; bud long, pointed; flowers two-tone pink, medium, dbl., 25–30 petals, exhibition form, intense fragrance; foliage leathery, wrinkled, dark; vigorous, upright, much branched growth; [Ophelia X Unnamed seedling]; Hill, Joseph H., Co.

Georgie Anderson, *see* 'ANDgeo'

Georgie Girl, *see* 'DICkerfuffle'

Georgie Lad, *see* 'HORkorblush'

'**Gerald Hardy**', HT, mr, 1936; bud pointed; flowers bright scarlet-red, spiral, large, dbl., intense fragrance; strong, erect stems; bushy growth; Dickson, A.

Geraldine, *see* 'PEAhaze'

'**Geraldine**', HT, 1924; flowers buff, shaded pink, dbl., moderate fragrance; RULED EXTINCT 11/82 ARM; [Antoine Rivoire X Marie Adélaide]; Chaplin Bros.

'**Geraldine Hicks**', HT, ab, 1950; flowers bronze-yellow; [William Moore sport]; Hicks

'**Geranium**', HMoy, mr, 1938; flowers almost scarlet, single, 2 in., blooms in clusters (up to 5); crimson, crimson fruit; upright (8–10 ft), compact growth; Royal Hort. Soc.

Geranium, *see* **'Independence'**

Geranium Primaplant, Pol, mr, 1964; flowers medium, semi-dbl.; Vlaeminck; (Sangerhausen)

'**Geranium Red**', F, or, 1947; flowers bright geranium-red, blooms in clusters, dbl., 50 petals, 4 in., globular, intense, geranium fragrance; foliage dark, glossy; bushy growth; [Crimson Glory X Seedling]; Boerner; J&P

Gerard ter Borch, S, dp, 1997; Williams, J. Benjamin

Gerbe d'Or, *see* 'MACca'

'**Gerbe Rose**', LCl, lp, 1904; flowers delicate pink, large, dbl., slight fragrance; foliage glossy; vigorous growth; [R. wichurana X Baroness Rothschild]; Fauque; Langue

'**Gerda Henkel**', HT, dr, 1964; flowers deep blood-red, large, dbl.; foliage dark, leathery; strong stems; vigorous, upright growth; [New Yorker X Prima Ballerina]; Tantau, Math.

Gerdo, HT, ab, 1986

'**Germaine**', HT, w, 1926; flowers creamy white, center salmon; [Unnamed seedling X Sunburst]; Chambard, C.

'**Germaine Chenault**', HT, lp, 1910; flowers cream-pink with darker edges, large, dbl., moderate fragrance; Guillot; (Sangerhausen)

'**Germanea**', HT, dp, 1929; flowers deep shining rose-pink, well formed, very large, dbl.; [Columbia sport]; Ravenberg

'Germania', *see* **'Charme'**

Germania-Africana, *see* KORtechna

'**Germanica**', HRg, 1900; Mueller, F.; (Cavriglia)

Germiston Gold, *see* 'KORtake'

Germiston Gold, Climbing, Cl HT, dy, 1992; Malanseuns

Gero, F, rb, 1984; flowers pink and red, medium, dbl., slight fragrance; Scholle, E.; (Sangerhausen)

'**Gert Potgieter**', HT, 1968; foliage light green; vigorous growth; Gowie

Gertrud®, F, dr

'**Gertrud Huck**', HT, dp, 1932; bud pointed; flowers flamingo-red, large, dbl., cupped, moderate fragrance; foliage leathery, bronze; vigorous growth; [Wilhelm Kordes sport]; Huck; C-P

'**Gertrud Schweitzer**', HT, ob, 1973; bud long, pointed; flowers apricot-orange, large, dbl., cupped, moderate fragrance; foliage glossy, dark; [Colour Wonder X Seedling]; Kordes; Horstmann

'**Gertrud Westphal**', F, or, 1951; flowers orange-scarlet, single, 5–7 petals, 3 in., slight fragrance; foliage glossy, dark reddish green; dwarf, bushy, much branched growth; [Baby Chateau X Obergärtner Wiebicke]; Kordes

'**Gertrud Westphal, Climbing**', Cl F, or, 1961; Buisman, G. A. H.

'**Gertrude Gregory**', HT, my, 1957; flowers bright golden yellow; [Lady Belper sport]; Gregory

Gertrude Jekyll®, *see* AUSbord

'**Gertrude Raffel**', F, dp, 1956; flowers pink, center rosy, well formed, borne in large clusters, semi-dbl., 15–20 petals, 2–3 in.; slight fragrance; foliage dark; vigorous, bushy growth; Raffel; Port Stockton Nursery

'**Gertrude Reutener**', F, dp, 1954; flowers crimson-pink; vigorous growth; Leenders, M.

'**Gertrude Shilling**', HT, dy, 1988; flowers bright, deep yellow, aging paler, urn-shaped, decorative, large, very dbl., 52 petals, moderate, fruity fragrance; prickles broad, fairly flat, large, dark brown; foliage large, medium green, matt; upright, tall, vigorous, good basal growth; [Golden Splendour X Peer Gynt]; Poole, Lionel; Rearsby Roses, Ltd., 1989

'**Geschwind's Gilda**', HMult, dr, 1887; Geschwind, R.

'**Geschwind's Gorgeous**', Cl HT, mr, 1916; flowers medium, semi-dbl., slight fragrance; Geschwind, R.; (Sangerhausen)

'**Geschwind's Nordlandrose II**', HSet, mr, 1928; Geschwind, R.

'**Geschwind's Nordlandsrose**', HSet, mp, 1884; Geschwind, R.

'**Geschwind's Orden**', HMult, m, 1886; Geschwind, R.

'**Geschwind's Schonste**', HMult, dr, 1900; Geschwind, R.

Geschwister Scholl, Pol, w, 1974; flowers medium, dbl.; GPG Bad Langensalza; (Sangerhausen)

Gessel, HT; Pironti, N.; (Cavriglia)

'**Ghergana**', Gr, dr, 1974; flowers deep blackish-red, large blooms in clusters of 2–5, dbl., 55 petals, cupped, moderate, tea fragrance; foliage dark; vigorous, upright growth; [Spectacular X Unnamed seedling]; Staikov, Prof. Dr V.; Kalaydjiev and Chorbadjiiski

'**Ghislaine de Féligonde**', HMult, ly, 1916; bud bright yellow; flowers yellowish white tinted flesh, borne in clusters of 10–20; vigorous, climbing (8–10 ft) growth; [Goldfinch X ?]; Turbat

Giana®, F, dr

Gianlauro, 1978; Zandri, R.; (Cavriglia)

Giant of Battles, *see* **'Géant des Batailles'**

Giant Pink®, S, pb, 1990; Lens

'**Gibby**', HT, dp, 1977; bud ovoid, pointed; flowers carmine-pink, dbl., 50 petals, 4.5 in., exhibition form; foliage dull, dark, leathery; vigorous, upright, bushy growth; [Christian Dior sport]; Prof. F. Roses; Ludwigs Roses Pty. Ltd.

Gidday, *see* 'TINhat'

'**Gidget**', Min, op, 1975; bud pointed; flowers coral-pink to coral-red, informal shape, 1 in., slight fragrance; foliage small, glossy; vigorous, bushy growth; [(R. wichurana X Floradora) X Fire Princess]; Moore, Ralph S.; Sequoia Nursery

Giesebrecht, *see* **'Bashful'**

Gift of Life, *see* 'HARelan'

Gigantea Cooperi, *see* **'Cooper's Burmese'**

'**Gigantèsque**', T, dp, 1845; flowers deep pink; Odier

Giggles, *see* 'LYOgi'

Giggles, *see* 'KINgig'

'**Gigi**', HT, mp, 1959; flowers rose-pink, reverse brighter, large, dbl., moderate fragrance; foliage light green; long strong stems; vigorous growth; [The Doctor X Unnamed seedling]; Verschuren; Blaby Rose Gardens

Gigliola, 1969; Mansuino; (Cavriglia)

'**Gigolette**', F, rb, 1953; bud ovoid; flowers yellow and red bicolor, medium, borne in clusters, semi-dbl., moderate fragrance; foliage leathery, light green; vigorous growth; Gaujard

'**Gil Blas**', HGal, pb, 1843; flowers light pink with white striping, large, dbl.; Roseraie de l'Hay; (Sangerhausen)

Gilbert Bécaud, *see* 'MEIridorio'

'**Gilbert F. Levy**', F, mr, 1958; flowers currant-red, dbl.; very vigorous growth; [Moulin Rouge X Oiseau de Feu]; Combe

'**Gilbert Nabonnand**', T, op; (G. Nabonnand); flowers large, semi-dbl., moderate fragrance; Nabonnand, 1883; (Sangerhausen)

'**Gilda**', HT, ab, 1936; bud long, pointed; flowers pure orange-yellow, large, dbl.;

foliage leathery, dark; vigorous growth; RULED EXTINCT 3/87; [Souv. de Claudius Pernet X (Lady Hillingdon X Harry Kirk)]; Towill

Gilda, *see* 'PEAHIGH'

'GILFLEUR', HT, pb, 1993; (**First Born**); flowers pink/peach, holds well for cut and arranging, blooms borne mostly singly, very dbl., 47 petals, moderate fragrance; some prickles; foliage medium, dark green, glossy; tall (80–90 cms), upright growth; [Silver Jubilee X Tahiti]; Gilmore, T.O.; Gilmore, 1993

'GILLIAN', HT, op, 1958; bud long, pointed; flowers soft coral-pink; foliage bronze; [Michele Meilland X Mme Butterfly]; Verschuren; Gandy Roses, Ltd.

'GILLIAN DAWN', HT, mr, 1997; flowers full, medium, very dbl., 26–40 petals, borne mostly singly, slight fragrance; foliage medium, medium green, semi-glossy; spreading, medium (24 in.) growth; Robinson, Kenneth G.

Gilmore, HT, ly; Select Roses, B.V.

Gilt Edged, *see* CHEwdainty

Gin Fizz, *see* MEIromar

Gin no suzu, *see* **'SILVER BELL'**

'GINA', F, dr, 1960; flowers velvety dark crimson, borne in large clusters, single, 6 petals, slight fragrance; foliage glossy; vigorous, upright growth; [Alain X Independence]; Kriloff, Michel; Cramphorn's Nursery

Gina Lollobrigida®, *see* 'MEILIVAR'

Gina Louise, *see* 'TROBGINA'

'GINETTE', HT, op, 1924; flowers salmony maize-yellow, dbl., moderate fragrance; [Paul Monnier X Souv. de Claudius Pernet]; Buatois

'GINGER', F, or, 1962; bud ovoid; flowers orange-vermilion, blooms in irregular clusters, dbl., 28 petals, 4 in., cupped, moderate fragrance; foliage leathery; vigorous, compact, bushy growth; [Garnette seedling X Spartan]; Boerner; J&P

Ginger Hill, *see* HILcap

Ginger Meggs, F, ob, 1962; Tantau

'GINGER ROGERS', HT, op, 1969; (Salmon Charm); flowers salmon, loosely formed, large, dbl., 30 petals, moderate fragrance; foliage light green; very tall growth; [Tropicana X Miss Ireland]; McGredy, Sam IV; McGredy

Ginger Toddler, *see* PEAvesta

Gingerbread Man™, *see* 'POULXAS'

Gingernut, *see* 'COCCRAZY'

Gingersnap, *see* 'AROSNAP'

'GINGIA', HT, w, 1983; flowers large, dbl., 35 petals, intense fragrance; foliage large, light green, glossy; bushy growth; [Unnamed seedling X Unnamed seedling]; Fumagalli, Niso

Ginny, *see* 'BISJEN'

Ginny-Lou, *see* 'TROBINKA'

'GINSKY', F, pb, 1983; flowers light salmon pink opening to pale pink to cream; [Liverpool Echo sport]; Barké; L.E.J. Wood

'GINZA KOMACHI', Cl Min, pb, 1980; bud globular; flowers deep pink, white eye, yellow stamens, small blooms in clusters, single, 5 petals, slight fragrance; many, hooked prickles; foliage medium green, glossy; vigorous growth; [Nozomi X Unnamed seedling]; Kono, Yoshito

Gioia, *see* **'PEACE'**

Gioia, Climbing, *see* **'PEACE, CLIMBING'**

'GIOIELLO', Min, my, 1984; flowers small, dbl., 20 petals, no fragrance; foliage small, dark, matt; [Zorina X Sole Di San Remo]; Bartolomeo, Embriaco

'GION', Cl Min, pb, 1979; bud rounded; single, 5 petals, 1 in., flat, non-recurrent, moderate fragrance; foliage tiny, leathery; bushy, climbing growth; [Nozomi X Unnamed seedling]; Onodera, Toru F.; S. Onodera

'GIOVANE', HT, op, 1965; bud pointed; flowers salmon-orange, large, dbl., 28 petals, exhibition form, moderate fragrance; foliage glossy, bronze; dense growth; [Queen Elizabeth X Orient]; Dot, Simon; Rosas Dot

'GIOVANEZZA', HT, rb, 1933; flowers geranium-red, reverse cream-white, edged lighter; vigorous growth; Ingegnoli

Giovanni Paolo II, HT, 1984; McEntire, J.; (Cavriglia)

'GIPSY', HT, dr, 1931; flowers medium, dbl.; Van Rossem; (Sangerhausen)

'GIPSY BOY', B, dr, 1909; (Zigeunerknabe); flowers dark crimson-red, medium, vigorous (3–5 ft) growth; Lambert, P.

Gipsy Jewel, Min, dp, Moore, 1975

'GIPSY LASS', HT, mr, 1932; (Gypsy Lass); flowers scarlet-crimson shaded blackish, dbl., globular, intense fragrance; long, willowy stems; bushy growth; Dickson, A.

'GIPSY LOVE', HT, or, 1964; flowers orange-vermilion, dbl., 25 petals, 4 in., moderate fragrance; vigorous growth; [Chic Parisien X Fashion]; Delbard-Chabert; Cuthbert

'GIPSY MAID', F, dp, 1955; flowers carmine-scarlet, base golden, borne in small clusters of 3 or more, single, moderate, sweetbriar fragrance; foliage olive-green; LeGrice

Giranu Czecheti, Pol, op, 1955; flowers medium, semi-dbl.; Madarsko; (Sangerhausen)

'GIRASOL', HT, my, 1945; bud oval; flowers sunflower-yellow, dbl., 25–30 petals, slight fragrance; foliage dark, glossy; upright, compact growth; [Joanna Hill X Carito MacMahon]; Dot, Pedro

Giriji, HT, dr, 1988; Friends Rosery

Girl Friend, *see* **'PODRUGA'**

Girl Guide, F, op

'GIRL SCOUT', F, my, 1961; bud ovoid; flowers golden yellow, dbl., 50 petals, 3.5–4 in., cupped, moderate fragrance; foliage leathery, glossy; vigorous, medium tall growth; [Gold Cup X Pigmy Gold]; Boerner; J&P

'GIRLIE', Pol, dr, 1923; flowers bright scarlet-crimson, bushy growth; [Orléans Rose sport]; Wezelenburg

Girls' Brigade, *see* 'HARBANJO'

'GIRONA', HT, pb, 1936; flowers soft red and yellow, well-formed, large, dbl., 30 petals, exhibition form, intense, damask fragrance; foliage bright green; vigorous, spreading growth; [Li Bures X Talisman]; Dot, Pedro; C-P, 1939;, H. Guillot, 1939

Girtnerfreude, *see* KORdisard

'GISELA', F, op, 1961; flowers salmon-pink, base straw-yellow, borne in clusters, dbl., 56 petals; foliage dark, glossy, bronze; upright, bushy, compact growth; [Masquerade X Pinocchio]; Verschuren, A.; van Engelen

'GISÈLE ALDAY', HT, mp, 1933; bud pointed; flowers bright rose-pink tinted flesh, large, semi-dbl., cupped, slight fragrance; foliage glossy, dark; vigorous growth; [Mrs Pierre S. duPont X Lallita]; Mallerin, C.; H. Guillot

Giselle, *see* 'JUSELLE'

'GISSELFELD'®, HT, dr, 1972; 17–20 petals, 4–4.5 in., moderate fragrance; foliage dark, leathery; upright growth; [(Tropicana X Champs-Elysees) X Furore]; Poulsen, Niels D.; Poulsen

Gitane, *see* **'BRIGHT WINGS'**

Gites de France®, *see* MEIwaton

Gitta Grummer, *see* 'KORSEE'

Gitte, *see* 'KORITA'

'GIULETTA', B, lp, 1859; Laurentius

'GIULIANA BORGATTI', HT, w, 1936; flowers white, center shaded rose and salmon, very well formed, large, dbl.; foliage dark ivy-green; vigorous growth; [Ophelia X Ville de Paris]; Borgatti, G.

'GIUSEPPE MOTTA', HT, lp, 1936; flowers large, semi-dbl., intense fragrance; Heizmann, E.; (Sangerhausen)

'GIUSEPPINA PAPANDREA', Min, pb, 1989; flowers cerise, reverse lighter; [Petite Folie sport]; Papandrea, John T.

Givenchy™, *see* 'ARODOUSNA'

Gizmo™, *see* 'WEKCATLART'

'GLACIER', F, w, 1952; bud ovoid; flowers white, slightly overcast yellow, dbl., 28 petals, 4.5 in., cupped, moderate fragrance; foliage glossy, dark; vigorous, upright growth; [Unnamed white HT X Summer Snow]; Boerner; J&P

Glad Tidings, *see* 'TANTIDE'

'GLADIADOR', Cl HT, dp, 1954; bud pointed; flowers carmine, very large, dbl., intense fragrance; vigorous growth; [Texas Centennial X Guinee]; Dot, Pedro

'GLADIATOR', LCl, mr, 1955; bud ovoid; flowers rose-red, dbl., 35 petals, 4.5–5 in., exhibition form, moderate fragrance; foliage dark, leathery; vigorous (10–12 ft) growth; [Charlotte Armstrong X (Pink Delight X New Dawn seedling)]; Malandrone; J&P

'GLADNESS', F, pb, 1959; bud pointed; flowers light pink edged darker, large, borne in clusters, dbl., 25 petals; foliage dark, glossy; upright growth; [Sunny Maid X Cinnabar]; Fletcher

'GLADSOME', HMult, mp, 1937; flowers delicate light pink, single; tall hedge rose; Clark, 1940

'GLADYS BENSKIN', HT, op, 1929; flowers rose-cerise, shaded orange, base deeper orange, large, dbl., exhibition form, moderate fragrance; vigorous growth; GM, NRS, 1929; Dickson, A.; Dreer

'GLADYS HARKNESS', HT, op, 1900; flowers very large, dbl., intense fragrance; Dickson, A.; (Sangerhausen)

'GLADYS HOLLAND', HT, lp, 1917; flowers light pink, shaded buff, dbl., moderate fragrance; GM, NRS, 1916; McGredy

'GLADYS MONCRIEF', HT, yb, 1981; flowers golden yellow to apricot, flushed rose red at petal tips; [Granada sport]; Jack, J.; Girraween Nursery

'GLADYS SAAVEDRA', HT, mp, 1922; flowers rosy peach-blossom-pink, dbl.; [Mme Abel Chatenay X Jonkheer J.L. Mock]; Nabonnand, P.

'GLADYS TWEEDIE', HT, dr, 1950; bud long, pointed; flowers crimson, dbl., 30–35 petals, 5 in., exhibition form, moderate fragrance; foliage wrinkled; very vigorous, bushy growth; [Crimson Glory X William Orr]; Toogood

'GLAIVE', HT, pb, 1951; bud long, pointed; flowers cream, center tipped pink, small, dbl., 25 petals, exhibition form, slight fragrance; foliage glossy; vigorous, bushy, compact growth; Clark, A.; NRS Victoria

Glamis Castle, *see* 'AUSLEVEL'

Glamorgan, *see* 'THOOM'

'GLAMOROUS', Min, or, 1979; bud pointed; flowers orange-red, base yellow, dbl., 35 petals, 1–1.5 in., exhibition form, slight fragrance; foliage small, glossy, bronze-green; bushy, spreading growth; [Starburst X Over the Rainbow]; Williams, Ernest D.; Mini-Roses

'GLAMOUR', HT, op, 1939; bud long, ovoid; flowers salmon-pink, large, dbl., intense fragrance; vigorous growth; [Comtesse Vandal X Pres. Macia]; Leenders, M.; T. Robinson, Ltd.

'GLAMOUR GIRL', HT, op, 1942; bud pointed, light jasper-red; flowers light salmon, open, dbl., 45–50 petals, 4–5 in., foliage dark, leathery; strong stems; very vigorous,upright, much branched growth; [Captain Glisson X Justine]; Hill, Joseph H., Co.

Glamour Girl, *see* CLEamour

'GLAnlin', HT, ob, 1985; (**Lincoln Cathedral**, Sarong); flowers outer petals pink, inner ones orange, yellow reverse, large, dbl., 28 petals, slight fragrance; many, reddish prickles; foliage medium, medium green glossy; bushy growth; GM, RNRS, 1985; [Silver Jubilee X Royal Dane]; Langdale, G.W.T.

'GLAnmusic', F, pb, 1997; (**Sir Neville Marriner**); flowers very full, medium, very dbl., 41 petals, borne mostly singly or in small clusters, slight fragrance; few prickles; foliage medium, medium green, semi-glossy; upright, medium (3 × 2 ft) growth; [Seedling X Painted Doll]; Langdale, G.W.T.

'GLAreabit', HT, yb, 1992; (**Lincolnshire Poacher**); flowers yellow flushed apricot and pink, blooms borne mostly singly, dbl., 26–40 petals, 3–3.5 in., slight fragrance; some prickles; foliage medium, medium green, semi-glossy; medium (80 cms–1 m), bushy growth; [Silver Jubilee X Woman and Home]; Langdale, G. W. T., 1993

'GLARONA', HT, w, 1922; flowers creamy flesh, center rose, Krüger; Kiese

'GLASTONBURY', S, rb, 1981; bud globular; flowers dark crimson to deep purple, borne 1–5 per cluster, dbl., 55 petals, repeat bloom, intense fragrance; hooked, red prickles; foliage medium green, sparse, spreading growth; [The Knight X Unnamed seedling]; Austin, David; David Austin Roses, Ltd.

'GLEAMING', F, dy, 1958; flowers deep lemon-yellow, borne in trusses, single, 6–8 petals, 4 in., intense fragrance; foliage dark; very free growth; [Goldilocks X Golden Scepter]; LeGrice

'GLEN ALMOND', HT, pb, 1973; flowers pale orient pink, dbl., 30 petals, 3.5–4 in., foliage glossy, leathery; free growth; [Pascali X Happy Event]; Wallace

'GLEN ARTNEY', HT, dr, 1973; flowers beetroot-purple, dbl., 40 petals, 3.5–4 in., slight fragrance; vigorous growth; [Baccará X Sterling Silver]; Wallace

'GLENARA', HT, dp, 1951; bud long, pointed; flowers deep rosy pink, very large, semi-dbl., 18 petals, slight fragrance; foliage leathery; vigorous, upright bush or pillar growth; Clark, A.; NRS Victoria

Glendora, *see* **'JOASINE HANET'**

Glendora, *see* KORhuba

'GLENFIDDICH', F, dy, 1976; flowers amber-gold, dbl., 25 petals, 4 in., moderate fragrance; foliage glossy, dark; [Arthur Bell X (Sabine X Circus)]; Cocker

'GLENGARRY', F, or, 1969; flowers vermilion, large, dbl., 32 petals, slight fragrance; foliage semi-glossy; compact, bushy growth; [Evelyn Fison X Wendy Cussons]; Cocker

'GLENITI GOLD', HT, dy, 1973; flowers very deep yellow; [Lady Mandeville sport]; Bone, John, & Son; Trevor Griffiths Ltd.

'GLENN DALE', LCl, ly, 1927; flowers lemon, fading to white, large blooms in clusters (to 20), dbl., 40 petals, slight fragrance; foliage dark, leathery; vigorous (10 ft) growth; GM, Portland, 1920; [Believed to be R. wichurana X Isabella Sprunt]; Van Fleet; American Rose Society

Glenshane(TM), *see* 'DICVOOD'

'GLENYS STEWART', HT, dp, 1968; flowers deep rose-pink, dbl., moderate bloom, slight fragrance; moderate growth; [Montezuma X Pink Favorite]; Kemp, M.L.; G. Stewart

'GLETSCHER', F, m, 1955; bud ovoid; flowers pale lilac, large blooms in large trusses, dbl., exhibition form, intense fragrance; foliage glossy; vigorous, upright, bushy growth; [Seedling X Lavender Pinocchio]; Kordes

Gletscherfee®, *see* KORdomal

'GLIMMER', Min, rb, 1988; bud pointed; flowers bright, medium red, yellow at base, reverse slightly darker, dbl., 24 petals, exhibition form, slight fragrance; prickles slightly downward pointed, medium, red; foliage medium, medium green, semi-glossy; upright, medium growth; [Party Girl X Unnamed seedling]; Bridges, Dennis A.; Bridges Roses, 1989

'GLITTERS', Cl HT, mp, 1934; bud pointed; flowers brilliant pink, base orange, open, very large, dbl., globular, recurrent bloom, moderate fragrance; foliage leathery; long stems; very vigorous growth; [Mrs W.J. Grant X Mrs Sam McGredy]; Smith, J.

'GLOAMING', HT, pb, 1935; bud pointed; flowers luminous pink suffused salmon, reverse lighter, open, very large, dbl., 36 petals, intense fragrance; foliage leathery, dark; vigorous, bushy growth; [Charles P. Kilham X Mrs Pierre S. duPont]; Nicolas; J&P

'GLOBE', F, mr, 1956; flowers blood-red, borne in clusters, semi-dbl., 20 petals, 2 in., cupped, moderate fragrance; foliage dark; vigorous, bushy, compact growth; [Fanal X Red Favorite]; Tantau, Math.

'GLOBE HIP', C, w

Gloira Dei-mutace, HT, lp; Lorenc; (Czech Rosa Club)

'GLOIRE D'ANTIBES', HT, dr, 1938; flowers large, dbl.; Mallerin, C.; (Sangerhausen)

Gloire de Bordeau, *see* **'BELLE DE BORDEAUX'**

Gloire de Bordeaux, *see* **'BELLE DE BORDEAUX'**

'GLOIRE DE BOURG-LA-REINE', HP, mr, 1879; Margottin

'GLOIRE DE BRUXELLES', HP, m, 1889; (Gloire de l'Exposition de Bruxelles); flowers very dark, velvety crimson-purple, large, dbl., 60 petals, moderate fragrance; vigorous, upright growth; [Souv. de William Wood X Lord Macaulay]; Soupert & Notting

'GLOIRE DE CHÉDANE-GUINOISEAU', HP, mr, 1907; flowers bright crimson-red, well-formed, large, dbl., 40 petals, cupped, occasionally recurrent bloom, moderate fragrance; foliage dark, soft; vigorous growth; [Gloire de Ducher X Unnamed variety]; Chedane-Pajotin

'GLOIRE DE CIBEINS', HT, or, 1958; flowers deep vermilion-red, well formed, dbl., 30 petals; vigorous growth; [Mme Méha Sabatier X Léonce Colombier]; Arles; Roses-France

'GLOIRE DE DEVENTER', T, ly, 1897; flowers light yellow with pink tints, large, dbl., intense fragrance; Soupert & Notting; (Sangerhausen)

'GLOIRE DE DIJON', Cl T, op, 1853; flowers rich buff-pink shaded orange toward center, very large, dbl., moderate fragrance; very vigorous, climbing growth; (28); [Thought to be an unknown Tea X Souv. de la Malmaison]; Jacotot

'GLOIRE DE DUCHER', HP, dr, 1865; flowers very large, dbl., occasional recurrent bloom, moderate fragrance; Ducher

'GLOIRE DE FRANCE', HGal, lp; flowers pale pink, dbl.; low growth; Bizard, 1828

'GLOIRE DE FRANCE', HT, or, 1946; flowers orange-red variegated copper, well formed, large, dbl.; vigorous growth; GM, Bagatelle, 1945; Gaujard

'GLOIRE DE GUILAN', D, lp; flowers clear pink, center incurved, dbl., quartered, spring bloom, intense fragrance; foliage light green; sprawling shrub (4–5 ft) growth; Nancy Lindsay/Hilling, 1949

'GLOIRE DE HOLLANDE', HT, dr, 1918; Verschuren

Gloire de Hollande, Climbing, Cl HT, dr; flowers large, dbl., moderate fragrance; (Sangerhausen)

'GLOIRE DE LA BRIE', HT, mr; bud long; flowers bright red, large; Grandes Roseraies

Gloire de l'Exposition de Bruxelles, *see* **'GLOIRE DE BRUXELLES'**

'GLOIRE DE MARGOTTIN', HP, dr, 1887; flowers large, dbl., 60 petals, globular, occasional recurrent bloom, intense fragrance; Margottin

'GLOIRE DE MEZEL', M, mp; flowers pale rose, very large

Gloire de Paris, *see* **'ANNA DE DIESBACH'**

Gloire de Rome, *see* **'ROME GLORY'**

'GLOIRE DE VITRY', HP, mp, 1854; flowers bright pink, large, globular; [La Reine seedling]; Masson

Gloire des Anciens, *see* INTerglo

'GLOIRE DES BELGES', HT, dp, 1916; flowers vivid carmine; Chambard, C.

'GLOIRE DES CHARPENNES', Pol, mr, 1898; flowers carmine-red, small, dbl.; Lille; (Sangerhausen)

'GLOIRE DES JARDINS', HGal, m; Vibert, before 1815

'GLOIRE DES LAWRANCEANAS', Min, dr, 1837; flowers dark crimson; dwarf growth

'GLOIRE DES MOUSSEUSES', M, mp, 1852; (Gloire des Mousseux); bud heavily mossed; flowers clear bright pink, center deeper, petals imbricated, large, dbl., borne in clusters; foliage light green; vigorous growth; Robert

Gloire des Mousseux, *see* **'GLOIRE DES MOUSSEUSES'**

Gloire des Perpetuelles, *see* **'FLON'**

'GLOIRE DES POLYANTHA', Pol, mp, 1887; flowers bright pink, well-shaped, small blooms in large clusters, dbl.; dwarf growth; [Mignonette seedling]; Guillot et Fils

'GLOIRE DES ROSOMANES', HCh, mr, 1825; (Ragged Robin, Red Robin); flowers glowing crimson, very large blooms in large clusters, semi-dbl., repeat bloom, moderate fragrance; vigorous growth; (21); Vibert

'GLOIRE D'ORIENT', M, dr, 1856; flowers deep red, Béluze

'GLOIRE D'ORLÈANS', Pol, mr, 1912; flowers small, dbl.; Levavasseur; (Sangerhausen)

Gloire du Beaujolais, HT, 1990; Delbard, Georges; (Cavriglia)

Gloire du Bourbonnais®, *see* 'DELBOURBO'

'GLOIRE DU MIDI', Pol, or, 1932; flowers brilliant orange-scarlet; [Gloria Mundi sport]; deRuiter; J&P; Sliedrecht & Co.

'GLOIRE DU MIDI SUPERIOR', Pol, or; flowers like parent with more lasting color; [Gloire du Midi sport]; deRuiter

'GLOIRE D'UN ENFANT D'HIRAM', HP, mr, 1899; flowers bright red; Vilin

'GLOIRE LYONNAISE', HP, w, 1885; flowers white with trace of yellow at center, very large, very dbl., 84 petals, cupped, moderate fragrance; foliage leathery; very vigorous, bushy growth; not very hardy; [Baroness Rothschild X Mme Falcot]; Guillot et Fils

'GLORIA', HT, dr, 1922; flowers brilliant scarlet-crimson, dbl.; Paul, W.

'GLORIA D'AUTUNNO', HT, 1952; Cazzaniga, F. G.; (Cavriglia)

'GLORIA DE GRADO', HT, mp, 1950; flowers pink, tinted carmine, globular; foliage bright green; [Mari Dot X Comtesse Vandal]; La Florida

Gloria Dei, *see* **'PEACE'**

Gloria Dei, Climbing, *see* **'PEACE, CLIMBING'**

'GLORIA DEL LLOBREGAT', HT, or, 1940; flowers strawberry-red to vermilion, large, dbl., cupped, moderate fragrance; foliage glossy; very vigorous growth; [Sensation X Margaret McGredy]; Camprubi, C.

Gloria di Milano, HT; Ingegnoli; (Cavriglia)

Gloria di Roma, *see* **'ROME GLORY'**

Gloria di Venezia, HT; Kordes, R.; (Cavriglia)

Gloria Dot, Min, 1986; Dot, Simon; (Cavriglia)

Gloria Ferrer®, *see* FEsuma

'GLORIA MUNDI', Pol, or, 1929; flowers striking orange-scarlet, borne in clusters, dbl.; foliage light, glossy; vigorous, bushy growth; [Superb sport]; de-Ruiter; Sliedrecht & Co.; Teschendorff

'GLORIA MUNDI SUPERIOR', Pol, or; flowers like parent with more lasting color; [Gloria Mundi sport]; deRuiter

'GLORIA MUNDI, CLIMBING', Cl Pol, or, 1934; Lens

'GLORIA MUNDI, CLIMBING', Cl Pol, or, 1943; Howard Rose Co.

'GLORIA SOLIS', HT, my, 1949; flowers well formed, 4–5 in.; foliage dark, glossy; vigorous growth; [Ville de Paris X Max Krause]; Giacomasso

'GLORIANA', HT, my, 1936; flowers intense lemon-yellow in heat, deep gold with cerise markings, dbl., 35 petals, cupped; foliage leathery, glossy, dark; vigorous, compact growth; [Condesa de Sástago (probably self seed)]; Hillock

Gloriana 97, *see* 'CHEWPOPE'

Gloriette, Min, rb

Gloriette, S, op, 1979; Cocker

'GLORIFIED LA FRANCE', HT, lp, 1916; flowers silvery pink, deeper than La France, very dbl., 92 petals, slight fragrance; [Frau Karl Druschki X Mrs Charles E. Russell]; Cook, J.W.

'GLORIGLO', Min, ob, 1976; bud pointed; flowers orange, yellow reverse, dbl., 45 petals, 1 in., exhibition form, slight fragrance; foliage small, glossy, bronze; upright, bushy growth; AOE, 1978; [Seedling X Over the Rainbow]; Williams, Ernest D.; Mini-Roses

Glorimontana, F, or, 1974; flowers medium, dbl.; deRuiter; (Sangerhausen)

'GLORIO', HT, mr, 1923; flowers scarlet-cerise, dbl., moderate fragrance; [Premier X Primrose]; Hill, E.G., Co.; Vestal

'GLORIOSA', HT, w, 1920; flowers ivory-white, base yellow, intense fragrance; [Kaiserin Auguste Viktoria X Pharisaer]; Kiese

Glorious, *see* 'LEOGLO'

'GLORIOUS', F, dp, 1947; bud pointed; flowers salmon scarlet, shaded orange, open, medium, large trusses, semi-dbl., 15–17 petals, slight fragrance; foliage leathery, dark; vigorous, upright growth; RULED EXTINCT 7/84 ARM; [(Betty Uprichard X Heidekind) X Heidekind]; Duehrsen; H&S

'GLORIOUS EASTER', HT, op, 1965; bud ovoid; flowers salmon, medium, dbl., exhibition form; foliage leathery; moderate, bushy growth; [Unnamed seedling X Penelope]; Howard, P.J.

'GLORIOUS PERNET', HT, r, 1928; flowers copper, center orange; [Souv. de Claudius Pernet sport]; Myers & Samtmann

'GLORIOUS SUNSET', Pol, rb, 1931; flowers bronze, suffused red, small, borne in clusters, semi-dbl., slight fragrance; foliage small, thick; vigorous growth; [Mariposa sport]; Allen

Glory, *see* 'DEVMOMENTO'

Glory Be, *see* 'SAVABE'

Glory Days, *see* 'JACCOR'

'GLORY OF CALIFORNIA', LCl, lp, 1935; flowers large, dbl., moderate fragrance; Clark, A.; (Sangerhausen)

'GLORY OF CEYLON', F, op, 1967; flowers orange-yellow blended pink, blooms in clusters, semi-dbl., 14 petals, moderate fragrance; foliage dark, glossy; [Vera Dalton X Masquerade]; Harkness

'GLORY OF CHESHUNT', HP, dr, 1880; flowers rich crimson, cupped; vigorous growth; [Charles Lefebvre seedling]; Paul & Son

'GLORY OF EDSELL', HSpn, lp

'GLORY OF HURST', Pol, mr, 1921; flowers cherry-red, borne in clusters, semi-dbl.; foliage small, leathery, glossy, rich green; dwarf growth; [Orléans Rose X Jessie]; Hicks

Glory of Paris, *see* **'ANNA DE DIESBACH'**

Glory of Rome, *see* **'ROME GLORY'**

'GLORY OF SURREY', HT, my, 1935; flowers golden yellow, semi-dbl., moderate fragrance; fairly vigorous growth; Ley

'GLORY OF WALTHAM', HP, dr, 1865; flowers crimson, very large, very dbl., moderate fragrance; vigorous, climbing or pillar growth; Vigneron; W. Paul

Glow Achievement, *see* TANelliv

'GLOW WORM', HT, or, 1919; flowers scarlet, suffused coppery orange, semi-dbl., moderate fragrance; Easlea

Glowing Amber, *see* 'MANGLOW'

'GLOWING CARMINE', HT, dp, 1936; flowers carmine, large, dbl., globular, slight fragrance; foliage leathery; vigorous, open habit growth; [Miss Rowena Thom X Unnamed seedling]; H&S; Dreer

Glowing Carpet, *see* 'MORPAPPLAY'

Glowing Cushion, *see* INTerreflet

Glowing Embers, *see* 'ANDGLO'

Glowing Peace™, *see* 'MEIZOELO'

Glowing Petals, *see* 'MANPETALS'

'GLOWING SUNSET', HT, ob, 1933; (Wilhelm Breder); bud long, pointed; flowers orange shaded yellow and pink, very large, dbl., exhibition form, intense fragrance; foliage leathery, glossy, dark; vigorous growth; [Fontanelle X Julien Potin]; Kordes; Dreer

'GLOWING VELVET', HT, dr, 1975; flowers deep crimson to scarlet, dbl., 25 petals, 4 in., bloom repeats quickly, intense fragrance; foliage dark; Pasley

Glowry™, *see* 'KINGLOW'

Glucksburg, S, pb, 1988; Jensen

'GLÜCKSKIND', HT, dp, 1935; flowers deep pink, well formed, large, moderate fragrance; Berger

'GLÜCKSKIND', F, dr, 1952; flowers dark crimson, medium, semi-dbl.; very vigorous growth; Leenders, M.

'GLYNDYFRDWY', HT, mp, 1978; flowers neyron rose, full, dbl., 35 petals, 4–5 in., moderate fragrance; foliage large, light; very vigorous growth; [Gavotte X George Thomas]; Ellick; Excelsior Roses

'GLYNIS BRYAN', Min, pb, 1997; flowers double, medium, dbl., 15–25 petals, borne in large clusters, no fragrance; some prickles; foliage small, medium green, semi-glossy; spreading, tall (9 ft) growth; [Party Girl X Sheri Anne]; Jones, L.J.

'GNEISENAU', S, w, 1924; flowers snow-white, stamens yellow, borne in clusters, dbl., non-recurrent; height 5–6 ft; [(Schneelicht X Killarney) X Crimson Rambler]; Lambert, P.

Gnom, F, op, 1957; flowers salmon-orange and pink, medium, dbl.; Berger, W.; (Sangerhausen)

'GNOME', Pol, ly, 1936; flowers cream-yellow, large, dbl., intense fragrance; foliage leathery, light; short stems; bushy, dwarf growth; [Unnamed seedling X Mev. Nathalie Nypels]; Leenders, M.

Gnome World, *see* JACpat

Godavari, HT, or, 1987; Kasturi

Goddess, *see* **'DÉESSE'**

Godescalcus Vulf de Sapprothe, S, lp, 1984; flowers small, semi-dbl.; Scholle, E.; (Sangerhausen)

'GODEURPAN', F, ob, 1973; (**Ibis**); bud ovoid; flowers orange, semi-dbl., 14–16 petals, foliage dark; [Europeana X Orangeade]; Godin, M.

Godewind®, *see* KORbraufa

'GODFREY WINN', HT, m, 1968; flowers purplish, dbl., globular, intense fragrance; Dot; Wheatcroft & Sons

'GODFREY'S RED PETITE', HT, dr, 1966; flowers deep red, medium, semi-dbl., 15 petals, slight fragrance; foliage dark; free growth; [Baccará X Audie Murphy]; Godfrey

'GODIALING', HT, or, 1975; (**Chrystelle**); bud oval; flowers deep, dbl., 30–35 petals, 3.5 in., cupped; foliage dark; [(Lady Zia X Wizo) X Silver Lining]; Godin, M.

'GODILOFTER', HT, dr, 1976; (**Maman Pineau**); bud pointed; flowers dark red-purple, dbl., 26–28 petals, 3–3.5 in., cupped, slight fragrance; foliage bronze; vigorous growth; [Maryse Kriloff X Uncle Walter]; Godin, M.

'GODRACHE', HT, m, 1973; (**Grand Duche**); bud ovoid; dbl., 25 petals, 3–3.5 in., cupped, slight fragrance; foliage dark; vigorous growth; [American Heritage X (Seedling X Blue Girl)]; Godin, M.

'GODSENSOR', F, or, 1973; (**Danse des Étoiles**); bud ovoid; flowers red-orange, deep, semi-dbl., 15–18 petals, 2.5 in., cupped; foliage light green; [Orangeade X Orange Sensation]; Godin, M.

'GODUSEX', Gr, mp, 1974; (**Exodus**); bud ovoid; very dbl., 50–52 petals, 3 in., cupped; foliage glossy, dark; [Kordes' Perfecta X Kalinka]; Godin, M.

'GODZOTY', HT, mr, 1976; (**Duchesse d'Anjou**); bud pointed; flowers crimson-red, 3–3.5 in., cupped, slight fragrance; foliage dark; vigorous growth; [Wizo X Soraya]; Godin, M.

Goedele, HT, lp, 1998; RvS-Melle

'GOETHE', M, m, 1911; flowers magenta, small, single; foliage blue-green, rough; new wood bright red; very vigorous growth; Lambert, P.

'GOLCONDA', HT, ly, 1968; bud ovoid; flowers pale yellow, center deep apricot, large, dbl., cupped, intense fragrance; foliage leathery; moderate, bushy, compact growth; [Mme Charles Sauvage X ?]; Pal, Dr B.P.; Indian Agric. Research Inst.

Gold Badge™, *see* 'MEIGRONURI'

Gold Badge, Climbing, *see* MEIgro

'GOLD BLAZE', Min, yb, 1980; bud ovoid, pointed; flowers yellow, dipped red, blooms borne singly or several together, dbl., 23 petals, intense fragrance; tiny, straight prickles; foliage small, glossy, deep green; compact, upright growth; [Unnamed seedling X Unnamed seedling]; Lyon

Gold Britannia, HT, my

Gold Bunny®, *see* 'MEIGRONURI'

'GOLD COAST', Gr, my, 1958; (Golden Pride); bud ovoid; flowers clear yellow, overcast buff-yellow, dbl., 25–30 petals, 4 in., cupped; foliage leathery, glossy; vigorous, upright growth; [Pinocchio X Peace]; Robinson, H.; J&P

'GOLD COIN', Min, dy, 1967; flowers buttercup-yellow, small, dbl., moderate fragrance; vigorous, bushy growth; [Golden Glow (LCl) X Magic Wand]; Moore, Ralph S.; Sequoia Nursery

Gold Country, *see* 'SEAGOLD'

Gold Crest, *see* **'GOLDEN CREST'**

'GOLD CROWN', HT, dy, 1960; (Corona de Oro, Couronne d'Or, Gold Krone, Goldkrone); flowers golden yellow, well-formed, dbl., 35 petals, 5 in., moderate fragrance; foliage leathery, dark; vigorous, upright growth; [Peace X Golden Scepter]; Kordes, R.; McGredy

'GOLD CUP', F, dy, 1957; (Coupe d'Or); bud pointed; flowers golden yellow, blooms in clusters, dbl., 28 petals, 4 in., moderate fragrance; foliage dark, glossy; bushy growth; AARS, 1958; [Goldilocks seedling X King Midas seedling]; Boerner; J&P

'GOLD DAME', HT, dy, 1929; flowers deep golden yellow, semi-dbl., moderate fragrance; foliage dark, glossy; vigorous, bushy growth; Dobbie

Gold Dollar, *see* 'HERDIO'

'GOLD DOT', HT, my, 1963; flowers large, dbl., 25 petals; vigorous, upright growth; [Queen Elizabeth X Peace]; Dot, Simon

Gold Fantasy, Min, dy, 1998

Gold Fever, *see* 'MACPARLEZ'

Gold Fever, *see* 'MORFEVER'

'GOLD GLOW', HT, dy, 1959; flowers bright yellow, very dbl., 100 petals, 3.5–4 in., moderate fragrance; foliage leathery, dark, glossy; vigorous, upright growth; [Fred Howard X Sutter's Gold]; Perry, Anthony; C.R. Burr

Gold Glow Bronze Sport, HT, yb

'GOLD GLOW, CLIMBING', Cl HT, dy, 1964; Burr, C.R.

Gold Heart, *see* **'BURNABY'**

Gold Heart, *see* 'MACYELKIL'

Gold Krone, *see* **'GOLD CROWN'**

'GOLD MAGIC', F, dy, 1990; flowers golden yellow, medium, borne in large clusters, dbl., 15–25 petals, slight fragrance; foliage medium, dark green, glossy; bushy, medium growth; [Gold Badge X Friesensohne]; Christensen, Jack E.; Vaughan's Seed Co., 1991

Gold Magic Carpet, *see* 'POULURT'

Gold Medal®, *see* 'AROYQUELI'

Gold Mine™, *see* LAVoro

'GOLD MINE', HT, my, 1925; flowers indian yellow paling toward edges, base deep orange, dbl., intense fragrance; RULED EXTINCT 9/84; [Golden Rule X Mrs Aaron Ward]; Hill, Joseph H., Co.; A.N. Pierson;, J.H. Hill Co.

Gold Moon®, *see* 'HAVOON'

'GOLD 'N' FLAME', Min, rb, 1980; bud long, pointed; flowers medium red, deep golden yellow reverse, borne singly, dbl., 33 petals, cupped, slight fragrance; very thin, long, tan prickles, curved down; foliage dark, glossy; upright, bushy growth; [Unnamed seedling X Over the Rainbow]; Williams, Ernest D.; Mini-Roses

'GOLD 'N' HONEY', HT, yb, 1976; bud long, pointed; flowers yellow and peach, edged rose, dbl., 28 petals, 5–6. in.; exhibition form, intense fragrance; vigorous, upright, bushy growth; [Helen Traubel X (Unnamed seedling X Ulster Monarch)]; Leon, Charles F., Sr.; Edmunds Roses

'GOLD NUGGET', F, dy, 1972; flowers bright yellow, medium, dbl., exhibition form, moderate fragrance; foliage glossy, abundant; vigorous upright growth; Patterson; Patterson Roses

Gold of Ophir, *see* **'FORTUNE'S DOUBLE YELLOW'**

Gold Patio, MinFl, dy, 1997

'GOLD PIN', Min, dy, 1974; flowers bright golden yellow, semi-dbl., 18 petals, 1 in., slight fragrance; foliage bronze; Mattock

'GOLD PIQUE', Min, my, 1977; bud pointed; flowers small, dbl., 36 petals, 1.5 in., moderate fragrance; foliage small, dark; compact, bushy growth; [Unnamed seedling X Yellow Jewel]; Lyon

Gold Reef, *see* POUldom

'GOLD RUSH', LCl, yb, 1941; flowers gold, dbl., 24 petals, exhibition form, not dependably recurrent, moderate fragrance; foliage glossy, ivy-green; vigorous, climbing growth; Duehrsen; H&S

Gold Rush, *see* 'KIRSUN'

'GOLD SPRAY', F, my, 1971; flowers full, medium, dbl.; foliage soft; moderate, bushy growth; [Philippe X Spek's Yellow]; Delforge

'GOLD STAR', HT, yb, 1933; bud pointed, orange; flowers golden yellow shaded orange, large, dbl.; foliage glossy, bronze, leathery; very vigorous growth; RULED EXTINCT 6/83 ARM; [Souv. de Claudius Pernet X Talisman]; Vestal

Gold Star, *see* 'CANDIDE'

Gold Star, *see* 'TANTERN'

'GOLD STRIKE', F, my, 1955; bud urn shaped; flowers lemon-yellow, borne in rounded clusters, dbl., 30–35 petals, 2–2.5 in., exhibition form, moderate fragrance; foliage leathery; vigorous, bushy, compact growth; [Goldilocks X Pinocchio]; Swim, H.C.; Armstrong Nursery

Gold Strike®, *see* 'SUNLUCK'

Gold Sweetheart™, *see* 'WILGOSH'

Gold Symphonie®, *see* 'MEITOLEIL'

'GOLD TOP', LCl, dy, 1978; flowers golden yellow, full, dbl., 25 petals, 5.5 in., moderate fragrance; foliage large, light matt green; free growth; Pearce; Limes Rose Nursery

Gold Topaz, *see* 'KORTOSSGO'

Goldbeet®, F, dy, 1974; Noack, Werner

'GOLDBONNET', S, my, 1973; flowers, semi-dbl., 13 petals, 4 in., slight fragrance; foliage large, glossy; [(Ann Elizabeth X Allgold) X Golden Showers]; Harkness

'GOLDBUSCH', S, my, 1954; bud long, pointed; flowers yellow, becoming lighter, large blooms in clusters (up to 20), semi-dbl.; foliage leathery, glossy, light green; very vigorous, upright, bushy growth; Kordes

'GOLDDIGGER', HT, dy, 1963; flowers dark saffron-yellow, large, dbl., 50–55 petals, slight, spicy fragrance; foliage glossy, bronze; upright growth; [Marcelle Gret X Dries Verschuren]; Verschuren, A.; Stassen

Golddorf Seppenrade, S, dy, 1969; flowers large, dbl.; Scholle, E.; (Sangerhausen)

'GOLDDUST', HT, my, 1963; flowers golden yellow; foliage clear green; vigorous growth; [Brandywine X Seedling]; Delforge

'GOLDELSE', HT, ob, 1900; flowers golden orange; [Kaiserin Auguste Viktoria seedling]; Hinner, W.

Golden Afternoon, HT, ab, 1984; Pal, Dr B.P.

'GOLDEN ALTAI', HSpn, ly, 1943; flowers cream to pale yellow, single, non-recurrent; very hardy.; [R. spinosissima altaica X Harison's Yellow]; Wright, Percy H.

'GOLDEN ANGEL', Min, dy, 1975; bud short, pointed; dbl., 65 petals, 1 in., moderate fragrance; foliage matt; bushy, compact growth; [Golden Glow (LCl) X (Little Darling X Seedling)]; Moore, Ralph S.; Sequoia Nursery

Golden Anniversary, *see* 'JACARY'

'GOLDEN ANNIVERSARY', HT, dy, 1948; bud ovoid; dbl., 50–60 petals, 4.5–5 in., exhibition form, intense fragrance; foliage leathery; vigorous, upright, bushy growth; RULED EXTINCT 9/82 ARM; [Good News sport]; Mordigan Evergreen Nursery; C-P

Golden Anniversary, MinFl, my, 1997

Golden Arches®, Min, dy

'GOLDEN ARCTIC', LCl, yb, 1954; flowers yellow to orange, dbl., 38 petals, 3.5–4 in., moderate fragrance; growth like a hybrid tea, followed by 4–5 ft canes; [Unnamed seedling X Free Gold]; Brownell, H. C.

Golden Autumn, HT, dy, 1955; flowers large, dbl., slight fragrance; Klimenko, V. N.; (Sangerhausen)

'GOLDEN BAY', S, yb, 1979; bud ovoid; flowers deep buff-yellow, shapely, dbl.,

40 petals, 3.5 in., slight fragrance; foliage large; spreading, bushy growth; [(Tropicana X Sabine) X Zitronenfalter]; Murray, Nola

'GOLDEN BEAUTY', HT, ab, 1937; bud very long; flowers orange buff-yellow, stamens golden, large, semi-dbl.; foliage clear green, glossy; vigorous, bushy growth; Van Rossem

Golden Beauty, *see* 'CLEBEAU'

Golden Beryl, *see* 'MANBERYL'

Golden Bettina, HT, yb; [Bettina sport]; Ruston, D., 1970

Golden Blush, A, ab, 1988; Sievers

Golden Border, S, my, 1993; Verschuren

Golden Bounty(TM), *see* 'JACSUYEL'

Golden Bouquet, F, dy

'GOLDEN BOY', HT, dy, 1964; flowers deep yellow, very, dbl., 35 petals, 5.5 in., exhibition form; foliage long, pointed; moderate growth; [Golden Masterpiece X Belle Blonde]; McGredy, Sam IV; Spek

'GOLDEN BUTTERFLY', HT, ab, 1920; flowers apricot-yellow, shaded carmine, dbl., moderate fragrance; [Old Gold seedling]; Therkildsen

Golden Butterfly, *see* **'GOUDVLINDER'**

'GOLDEN CALIFORNIA', HT, dy, 1966; flowers golden, large, dbl., cupped, moderate fragrance; foliage bronze, leathery, glossy; tall, bushy growth; [California sport]; Howard, P.J.

'GOLDEN CASCADE', LCl, my, 1962; bud ovoid; flowers chrome-yellow, dbl., 25–30 petals, 4.5–5 in., cupped, moderate, fruity fragrance; foliage leathery; vigorous (10–12 ft) growth; [(Capt. Thomas X Joanna Hill) X Lydia]; Morey, Dr Dennison; J&P

Golden Celebration, *see* 'AUSGOLD'

'GOLDEN CENTURY', Cl Min, ob, 1978; bud pointed; flowers cadmium-orange to nasturtium-red, dbl., 35 petals, 1.5–2 in., intense fragrance; foliage glossy, leathery; moderate climber growth; [(R. wichurana X Floradora) X (Sister Therese X Unnamed Miniature)]; Moore, Ralph S.; Sequoia Nursery

'GOLDEN CHALICE', HT, dy, 1960; bud ovoid; flowers clear yellow, open, dbl., 40–45 petals, 4 in., moderate fragrance; foliage glossy; vigorous, upright growth; [(Starlite X Snow White) X Golden Masterpiece]; Boerner; J&P

'GOLDEN CHARM', HT, dy, 1933; [Talisman sport]; Groshens & Morrison

'GOLDEN CHARM, CLIMBING', Cl HT, dy, 1948; Krider Nursery

Golden Chersonese, *see* 'HILGOLD'

'GOLDEN CHOICE', HT, my, 1967; flowers lemon-yellow; [My Choice sport]; Bardill Nursery; LeGrice Roses

'GOLDEN CITY', HT, my, 1922; bud golden yellow; flowers light buff; [Rayon d'Or X Frau Karl Druschki]; Lippiatt

Golden Climber, *see* **'MRS ARTHUR CURTISS JAMES'**

Golden Coach, *see* 'ZIPGOLD'

Golden Colonel, F, ob

'GOLDEN COMET', HT, yb, 1937; bud long, pointed; flowers yellow and pink, open, large, semi-dbl., intense fragrance; foliage dark, leathery; vigorous growth; Burbank; Stark Bros.

'GOLDEN CORONET', F, my, 1967; flowers medium, dbl., exhibition form, slight fragrance; foliage glossy, leathery; vigorous, compact growth; [(Lydia X Golden Scepter) X Isobel Harkness]; Morey, Dr Dennison; Country Garden Nursery

'GOLDEN CREST', Cl HT, my, 1948; (Gold Crest); flowers pure yellow, dbl., 3–4 in., moderate fragrance; foliage glossy, dark; good pillar (6 ft) growth; Archer

Golden Dance®, *see* BARgold

'GOLDEN DAWN', HT, my, 1929; bud yellow, flushed pink; flowers well-formed, dbl., 45 petals, intense fragrance; low, spreading growth; [Elegante X Ethel Somerset]; Grant; Hazelwood Bros;, Prior

'GOLDEN DAWN, CLIMBING', Cl HT, my, 1937; Knight, G.

'GOLDEN DAWN, CLIMBING', Cl HT, my, 1935; Armstrong, J.A.

'GOLDEN DAWN, CLIMBING', Cl HT, my, 1947; LeGrice

'GOLDEN DAY', HT, my, 1931; flowers bright golden yellow, center deeper, larger and fuller, intense fragrance; vigorous growth; [Independence Day sport]; Bentley; Harkness

Golden Days, *see* 'RUGOLDA'

'GOLDEN DELIGHT', F, my, 1956; flowers canary-yellow, dbl., 58 petals, 3 in., moderate fragrance; foliage dark, glossy; dwarf growth; [Goldilocks X Ellinor LeGrice]; LeGrice

'GOLDEN DIAMOND', HT, dy, 1943; flowers large, dbl., 30 petals; foliage leathery, dark; strong stems; vigorous, upright, compact growth; Verschuren; L.C. Lovett

'GOLDEN DREAM', HRg, my, 1932; (Goldener Traum); bud pointed, streaked red; flowers pure yellow, large, dbl., recurrent bloom, intense fragrance; very vigorous ($6\frac{1}{2}$ ft) growth; [Turkes Rugosa Samling X Constance]; Türke; J.C. Schmidt

'GOLDEN DROP', HT, dy, 1939; flowers rich yellow, small blooms in clusters, semi-dbl.; tall growth; [Mme Mascuraud X Seedling]; Clark, A.

Golden Eagle, Gr, dy, 1997

Golden Earing(TM), *see* 'WILGOLD'

Golden Elegance, HT, lp; long; flowers large, exhibition form, 30 petals, single or in clusters of 3–7 florets, good cut flower, large, dark green, glossy foliage; [Ambassador sport]; St Wagner, 1993; Res. Stn. f. Fruit Growing, Cluj, 1995, Romania

Golden Emblem, *see* 'JACGOLD'

'GOLDEN EMBLEM', HT, my, 1917; bud yellow, splashed and shaded red; flowers canary-yellow, well formed, large, dbl., moderate fragrance; foliage dark, glossy, leathery; vigorous growth; GM, NRS, 1915; [Mme Mélanie Soupert X Constance]; McGredy

'GOLDEN EMBLEM, CLIMBING', Cl HT, my, 1927; Armstrong Nursery

'GOLDEN EMPIRE', HT, ob, 1957; flowers orange, dbl.; foliage leathery, glossy; [Orange Everglow X Golden Emblem]; Silva

'GOLDEN FAIRY', Pol, ly, 1889; flowers clear buff, yellow and white; dwarf growth; Bennett

Golden Fantasie, *see* 'HILGOFAN'

'GOLDEN FICTION', F, my, 1958; free growth; [Yellow Pinocchio X Moonbeam]; Spek

Golden Flame®, *see* LAPam

'GOLDEN FLEECE', F, my, 1955; (Toison d'Or); bud ovoid; flowers buff-yellow, in clusters (to 20), dbl., 38 petals, 4.5 in., cupped, intense fragrance; foliage leathery; vigorous, bushy growth; GM, Bagatelle, 1955; [Diamond Jubilee X Yellow Sweetheart]; Boerner; J&P

Golden Flipper, F, dy

Golden Fox, MinFl, dy, 1997; Warner, Chris

Golden Friendship, *see* HARtellody

'GOLDEN FRILLS', HT, my, 1936; flowers rich golden yellow, dbl., cupped; foliage glossy, wrinkled; [Feu Joseph Looymans sport]; B&A

Golden Future, *see* 'HORANYMOLL'

Golden Galaxy, *see* JACyesp

Golden Gardens, *see* 'MORGOGARD'

'GOLDEN GARNETTE', F, dy, 1960; bud ovoid; flowers golden yellow, edged lighter, blooms in clusters, dbl., 33 petals, 3–4 in., cupped, intense, fruity fragrance; foliage leathery, dark, glossy; vigorous, upright, bushy growth; [(Goldilocks seedling X Seedling) X Tawny Gold]; Boerner; J&P

'GOLDEN GATE', HT, my, 1972; bud ovoid; flowers large, dbl., exhibition form, slight fragrance; foliage large, glossy; vigorous, upright, bushy growth; [South Seas X King's Ransom]; Warriner, William A.; J&P

'GOLDEN GATE', T, w, 1891; bud pointed; flowers cream-white, anthers golden yellow, very large, dbl., cupped, moderate fragrance; foliage bright green; vigorous growth; [Safrano X Cornelie Koch]; Dingee & Conard

'GOLDEN GEM', HT, my, 1916; flowers golden yellow; [Lady Hillingdon X Harry Kirk]; Towill

Golden Giant®, *see* 'KORBI'

'GOLDEN GIANT, CLIMBING', Cl HT, dy, 1967; Laveena Roses

Golden Girl, *see* 'MEIVIRGI'

Golden Girls, *see* 'CLEBEAU'

'GOLDEN GLAMOUR', F, my, 1951; bud pointed; flowers large, dbl., 25 petals, exhibition form, moderate fragrance; foliage glossy; vigorous, upright, bushy growth; [Joanna Hill X (Mrs Pierre S. duPont X Amelia Earhart)]; Boerner; J&P

'GOLDEN GLEAM', HT, my, 1926; flowers buttercup-yellow, outer petals streaked, dbl., 25 petals, slight, fruity fragrance; foliage dark; McGredy; Beckwith

Golden Globe, Min, dy

'GOLDEN GLORY', HT, dy, 1931; flowers deep golden yellow large, dbl., intense fragrance; Dobbie

Golden Gloves, F, dy, 1991; Bear Creek Gardens

'GOLDEN GLOW', LCl, my, 1937; dbl., 3.5–5 in., exhibition form, blooms in clusters, non-recurrent, moderate fragrance; foliage leathery, glossy, dark; very vigorous, climbing (20 ft) growth; [Glenn Dale X (Mary Wallace X HT)]; Brownell, H.C.

'GOLDEN GLOW', HT, ab, 1918; flowers apricot, shaded bronzy orange; [Mme Edouard Herriot sport]; Chaplin Bros.

'GOLDEN GRUSS AN AACHEN', F, ob, 1935; bud pointed, red; flowers golden orange, sometimes shaded reddish, very large, dbl., exhibition form, moderate fragrance; foliage glossy; bushy growth; [Mme Butterfly X Gloria Mundi]; Kordes

Golden Halo™, *see* 'SAVAHALO'

Golden Hands, *see* CHEssupremo

Golden Handshake, *see* 'CHEWSUNFORD'

'GOLDEN HARVEST', HT, dy, 1943; flowers clear yellow, dbl., 35 petals, 4.5 in., exhibition form, moderate fragrance; foliage leathery, glossy, bronze; vigorous growth; [McGredy's Ivory X Unnamed seedling]; Mallerin, C.; C-P

'GOLDEN HAZE', HT, yb, 1965; bud ovoid; flowers light golden yellow, large, dbl., moderate fragrance; foliage glossy; vigorous growth; [Peace X Golden Rapture]; Verschuren, H. A. M.; J&P

Golden Heart, *see* **'BURNABY'**

'GOLDEN HERITAGE', HT, my, 1974; flowers canary-yellow, pointed, dbl., 25 petals, 3.5 in., moderate, tea fragrance; bushy growth; [Golden Masterpiece X Unnamed seedling]; Herholdt, J.A.

Golden Hit®, Min, dy

Golden Holstein®, *see* KORtikel

Golden Hope, *see* 'MEHPIC'

'GOLDEN HOUR', HT, ob, 1952; bud ovoid; flowers golden yellow, reverse orange, dbl., 45–55 petals, 4–5 in., exhibition form, intense fragrance; foliage leathery, glossy; very vigorous, upright growth; [Los Angeles X California]; Howard, P.J.

'GOLDEN IDEAL', HT, dy, 1939; bud long, pointed; flowers brilliant chrome-yellow, large, dbl.; vigorous growth; [Roselandia X Joanna Hill]; Lens

Golden Jet, *see* 'TANTASCH'

Golden Jet, *see* 'TANTASCH'

'GOLDEN JEWEL', *see* TANledolg

Golden Jubilee, *see* 'COCAGOLD'

'GOLDEN JUBILEE', F, my, 1948; bud ovoid; flowers golden yellow, becoming buff and chrome, in clusters, dbl., 3 in., slight fragrance; foliage glossy, dark; vigorous, bushy, compact growth; [(Mary Wallace X Talisman) X Mrs Pierre S. duPont]; Jacobus; B&A

Golden Julia™, *see* 'RUPGOLJUL'

'GOLDEN KING', HRg, ly, 1935; flowers pale yellow; [Dr Eckener sport]; Beckwith

'GOLDEN LACE', F, dy, 1962; bud pointed; dbl., 28 petals, 3 in., exhibition form, blooms in clusters, moderate fragrance; foliage glossy; vigorous, upright growth; [Goldilocks X (Golden Scepter X Encore)]; Von Abrams; Peterson & Dering

'GOLDEN LEADER', HT, ab, 1961; flowers apricot, moderate fragrance; [Tawny Gold X Seedling]; Leenders, J.

'GOLDEN LEOPARD', F, my, 1976; bud pointed; flowers daffodil-yellow, small, dbl., 18–25 petals, 2 in., exhibition form, moderate fragrance; foliage glossy, leathery; low, compact growth; [Golden Slippers seedling X Golden Slippers seedling]; Takatori, Yoshiho; Japan Rose Nursery

'GOLDEN LIGHT', LCl, ob, 1939; bud pointed; flowers orange-apricot to buff, edged pink, open, large, dbl., slight fragrance; foliage glossy, dark; strong stems; vigorous, climbing growth; Nicolas

'GOLDEN LION', HSet, my, 1944; flowers clear golden yellow, open, borne in clusters, cupped, non-recurrent; foliage light, glossy; height 8–10 ft; Horvath

'GOLDEN LUSTRE', HT, ab, 1964; bud ovoid; flowers bronze-apricot overcast yellow, dbl., 60–65 petals, 5 in., cupped, moderate fragrance; foliage leathery; moderate growth; [Kate Smith X Tanya]; Boerner; J&P

'GOLDEN MAIN', HT, dy, 1933; (Golden Romance, Goldenes Mainz, Golmain); bud golden yellow striped red; flowers golden yellow, large, dbl., cupped, intense fragrance; foliage glossy; very vigorous, bushy growth; [Fontanelle X Julien Potin]; Kordes; J&P

'GOLDEN MASTERPIECE', HT, my, 1954; bud long, pointed; flowers golden yellow, very large, dbl., 30–35 petals, exhibition form, borne singly, moderate, licorice fragrance; foliage very glossy; vigorous, upright growth; [Mandalay X Golden Scepter]; Boerner; J&P

'GOLDEN MASTERPIECE, CLIMBING', Cl HT, my, 1957; Valdrez

Golden Medaillon, *see* 'KORIKON'

'GOLDEN MEDALLION', HT, my, 1981; bud long, pointed; flowers medium golden yellow, loosely imbricated, dbl., 40–50 petals, 7–8 in., slight fragrance; foliage large, light; upright growth; [Medallion sport]; Permenter; J&P

Golden Medallion, *see* KORnanze

Golden Meillandina, *see* **'RISE 'N' SHINE'**

Golden Melody, *see* **'IRENE CHURRUCA'**

'GOLDEN MEMORIES', HT, dy, 1965; bud long, pointed, canary-yellow tinted coppery; flowers deep canary-yellow, to open, dbl., 28–32 petals, 5–6 in., exhibition form, moderate, spicy fragrance; foliage glossy; vigorous, upright growth; [Golden Rapture sport]; Ravine; Endres Floral Co.

'GOLDEN MME SEGOND WEBER', HT, op, 1923; flowers salmon, center yellow, dbl.; [Mme Segond Weber X Primerose]; Soupert & Notting

Golden Moments, *see* FRYtranquil

Golden Monica®, *see* TANgolca

'GOLDEN MOSS', M, my, 1932; bud globular, peach-yellow, sepals well mossed; flowers tawny yellow, large blooms in clusters of 3–5, dbl., 37 petals, scanty bloom; no repeat, moderate fragrance; foliage almost rugose; vigorous growth; [Frau Karl Druschki X (Souv. de Claudius Pernet X Blanche Moreau)]; Dot, Pedro; C-P

Golden Mozart, S, my, 1986; Verschuren

Golden Mrs Sam McGredy, *see* **'GOLDEN SAM MCGREDY'**

Golden Nugget™, *see* MEIlsinplox

Golden Oldie, *see* 'MACGOLOLD'

'GOLDEN OPHELIA', HT, my, 1918; flowers golden yellow in center, paling slightly on outer petals; foliage glossy; vigorous growth; GM, NRS, 1918; [Ophelia seedling]; Cant, B. R.

'GOLDEN OPHELIA, CLIMBING', Cl HT, my, 1924; Hage; Prior

Golden Opportunity™, *see* 'ANTOPP'

'GOLDEN ORANGE CLIMBER', LCl, ob, 1937; flowers orange to orange-scarlet, often overlaid golden yellow, semi-dbl., 5 in.; vigorous, climbing growth; [Mrs Arthur Curtiss James sport]; Brownell, H. C.

'GOLDEN PAMELA', HT, yb, 1969; bud ovoid; flowers yellow-apricot and pink, large, dbl., cupped, slight fragrance; foliage large, glossy; vigorous, upright growth; [Wellworth sport]; Wheatley

'GOLDEN PEACE', HT, my, 1961; flowers canary-yellow, large, dbl., 45 petals, exhibition form, moderate fragrance; foliage dark, dull; very vigorous, tall growth; LeGrice

'GOLDEN PEARL', HT, my, 1967; flowers large, single, slight fragrance; foliage dark; free growth; Warmerdam

Golden Penny, *see* 'COOSYN'

'GOLDEN PERFECTION', Pol, my, 1937; flowers golden yellow, small, dbl., intense fragrance; bushy, dwarf growth; Leenders, M.

'GOLDEN PERFUME', F, my, 1959; flowers orange-yellow becoming golden yellow, large, borne in clusters, dbl., 50 petals, moderate fragrance; foliage dark, glossy, leathery; vigorous, bushy growth; [Goldilocks X Fashion]; Leenders, J.

Golden Pernet, *see* **'JULIEN POTIN'**

'GOLDEN PERRAUD', HT, my, 1946; bud long, pointed, well formed; flowers brilliant golden yellow, large, dbl., slight fragrance; foliage glossy; vigorous growth; [Mme Joseph Perraud sport]; Lens

'GOLDEN PHEASANT', F, ob, 1951; (Goldfasan); flowers orange and gold, imbricated, dbl., 40 petals, 4 in., moderate fragrance; foliage glossy; low, compact growth; [Pres. Ferier X Dr Debat]; Kordes; Wheatcroft Bros.

Golden Phoenix, *see* 'NAKBET'

Golden Piccolo, *see* 'POULTEX'

'GOLDEN PICTURE', HT, yb, 1967; flowers light yellow tinged pink, medium, slight fragrance; foliage leathery; upright growth; [Picture seedling X Marcelle Gret]; Handover, P.&R.

Golden Pillar™, LCl, my, 1999; flowers large, full, loose, golden yellow, nonfading, very dbl., 30–35 petals, 4 in., slight fragrance; foliage medium, deep green, glossy; Meilland

'GOLDEN PIRRIE', HT, ly, 1921; flowers yellowish white; [Lady Pirrie sport]; Dobbie

'GOLDEN PIXIE', Min, my, 1985; flowers medium golden yellow, medium, dbl., 60 petals, slight fragrance; foliage small, medium green, semi-glossy; vigorous, upright, bushy growth; [Yellow Pages X Rise 'n' Shine]; Hardgrove, Donald L.; Rose World Originals

Golden Plover, S, dy, 1991; Poulsen; (Cavriglia)

'GOLDEN POLY', Pol, yb, 1931; flowers pure yellow, edged carmine, dbl., globular; dwarf growth; [Angèle Pernet X Orange King]; Pahissa

'GOLDEN POLY', Pol, my, 1935; bud yellow with red lines; flowers golden yellow to yellowish white, open, large, semi-dbl., slight fragrance; foliage light, glossy; bushy, dwarf growth; Leenders, M.

Golden Pride, *see* **'GOLD COAST'**

Golden Prince, *see* 'MEIGOLD'

'GOLDEN PRINCESS', S, yb, 1984; bud ovoid, pointed; flowers yellow, petals edged deep pink, large blooms borne 1–5 per cluster, dbl., 33 petals, cupped, repeat bloom, moderate fragrance; large, hooked, tan prickles; foliage medium, olive green, semi-glossy; upright, bushy growth; hardy; [Hawkeye Belle X (Roundelay X Country Music)]; Buck, Dr Griffith J.; Iowa State University

Golden Promise, *see* 'LAVLUV'

Golden Promise, F, dy, 1972; deRuiter

'GOLDEN PYRAMID', LCl, my, 1939; flowers large, semi-dbl., free seasonal bloom; vigorous pyramid to 5–6 ft growth; Brownell, H.C.

'GOLDEN QUEEN', HT, my, 1937; bud gold, slightly marked carmine; flowers golden yellow, large, dbl.; foliage glossy; very vigorous, bushy growth; Chambard, C.

Golden Queen, *see* KORgitte

'GOLDEN RAIN', F, my, 1951; (Goldregen); flowers golden yellow, well-formed, large, borne in clusters of 10–15, semi-dbl., 20 petals; foliage glossy; upright, bushy growth; [Swantje X G. Bentheim]; Tantau

Golden Rambler, *see* **'ALISTER STELLA GRAY'**

Golden Rambler, *see* **'EASLEA'S GOLDEN RAMBLER'**

'GOLDEN RAPTURE', HT, dy, 1933; (Geheimrat Duisberg); bud pointed; flowers golden yellow, very large, dbl., 40 petals, moderate fragrance; foliage glossy; vigorous growth; [Rapture X Julien Potin]; Kordes; Dreer, 1934; H&S, 1934

Golden Rapture No. 5, HT, my; bud ovoid; flowers clear yellow, dbl., 25 petals, 5 in., exhibition form, intense fragrance; foliage dark, leathery; vigorous growth; [Golden Rapture sport]; Krieter; A.N. Pierson

'GOLDEN RAPTURE, CLIMBING', Cl HT, dy, 1941; Swim, H.C.; Armstrong Nursery

'GOLDEN RAPTURE, CLIMBING', Cl HT, dy, 1954; (Climbing Geheimrat Duisberg); Knackfuss

'GOLDEN REVELRY', HT, my, 1952; flowers golden yellow, dbl., 21 petals, exhibition form, slight fragrance; foliage glossy, bright green; [Phyllis Gold X Blossom]; McGredy, Sam IV

Golden Rider, *see* **'GOLDENER REITER'**

Golden Romance, *see* **'GOLDEN MAIN'**

Golden Rosamini, *see* INTergol

Golden Rose of China, *see* **R. HUGONIS**

'GOLDEN RUFFELS', HT, my, 1954; bud long, pointed; flowers golden yellow, medium, dbl., moderate fragrance; upright growth; [Orange Ruffels sport]; Brownell, H.C.

'GOLDEN RULE', HT, my, 1918; flowers clear yellow, dbl.; [Ophelia seedling X Sunburst]; Hill, E.G., Co.

'GOLDEN SALMON', Pol, ob, 1926; (Goldlachs); flowers pure orange, large blooms in huge trusses; vigorous, bushy growth; [Superb sport]; deRuiter

Golden Salmon Improved, *see* **'GOLDEN SALMON SUPÉRIEUR'**

'GOLDEN SALMON SUPÉRIEUR', Pol, ob, 1929; (Golden Salmon Improved); flowers different shade of orange; [Golden Salmon sport]; deRuiter; Sliedrecht & Co.

'GOLDEN SALUTE', HT, dy, 1963; bud ovoid; flowers golden yellow, dbl., 33 petals, 5–5.5 in., cupped, moderate fragrance; foliage leathery, glossy; vigorous, moderately tall growth; [Diamond Jubilee seedling X Golden Masterpiece]; Boerner; J&P

'GOLDEN SAM MCGREDY', HT, yb, 1935; (Golden Mrs Sam McGredy); flowers chrome-yellow, reverse salmon with chrome; [Mrs Sam McGredy sport]; Lens

'GOLDEN SÁSTAGO', HT, dy, 1938; flowers clear yellow, large, dbl., globular, moderate fragrance; foliage soft; vigorous growth; [Condesa de Sástago sport]; Dot, Pedro; C-P

'GOLDEN SCEPTER', HT, dy, 1950; (Spek's Yellow); bud pointed; flowers deep yellow, dbl., 35 petals, 4.5 in., exhibition form, moderate fragrance; foliage leathery, glossy; vigorous, upright growth; [Golden Rapture X Unnamed seedling]; Verschuren-Pechtold; J&P; Spek

'GOLDEN SCEPTER, CLIMBING', Cl HT, dy, 1956; (Climbing Spek's Yellow); Walters

'GOLDEN SÉVERINE', HT, dy, 1929; flowers deep golden yellow, Morse

'GOLDEN SHEEN', F, my, 1966; bud urn shaped; flowers yellow, edged lighter, borne in clusters, dbl., exhibition form; foliage leathery; moderate, upright, bushy growth; [Ophelia X Circus]; Swim & Weeks; Carlton Rose Nurseries

'GOLDEN SHOT', F, dy, 1973; flowers golden yellow, dbl., 24 petals, 4 in., slight fragrance; foliage dark; [Seedling X Allgold]; Martin, J.; Gandy Roses, Ltd., 1976

'GOLDEN SHOWERS'®, LCl, my, 1956; bud long, pointed; flowers daffodil-yellow, open quickly, loose, dbl., 25–28 petals, 4 in., exhibition form, borne singly and in clusters, recurrent bloom, moderate, sweet fragrance; foliage medium, medium green, glossy; vigorous, pillar or climbing (6–10 ft) growth; AARS, 1956 GM, Portland, 1957; [Charlotte Armstrong X Capt. Thomas]; Lammerts, Dr Walter; Germain's

'GOLDEN SIGNORA', HT, my, 1954; flowers golden veined orange, large, dbl., 30 petals, moderate fragrance; foliage glossy; vigorous growth; [Signora sport]; Lowe

Golden Silence, *see* 'RENSILEN'

'**GOLDEN SLIPPERS**', F, yb, 1961; bud pointed; flowers yellow flushed vermilion, center golden yellow, blooms in clusters, dbl., 23 petals, 3 in., exhibition form, moderate fragrance; foliage leathery, glossy; vigorous, compact, low growth; AARS, 1962 GM, Portland, 1960; [Goldilocks X Unnamed seedling]; Von Abrams; Peterson & Dering

'**GOLDEN SONG**', Cl Min, yb, 1980; bud long, pointed; flowers golden yellow, petals edged pink, blooms borne usually singly, dbl., 35 petals, exhibition form, moderate fragrance; long, thin, tan prickles; foliage small, medium to dark, glossy; upright (to about 5 ft) growth; [Little Darling X Golden Angel]; Williams, Ernest D.; Mini-Roses

Golden Spire, *see* KORamgis

'**GOLDEN SPLENDOR**', HT, dy, 1960; bud long, pointed; flowers golden yellow, dbl., 30 petals, 5 in., exhibition form; foliage glossy; vigorous, tall growth; Jones; Hennessey

'**GOLDEN SPLENDOUR**', HT, my, 1962; flowers clear light yellow, large, dbl., 40 petals, moderate fragrance; [Buccaneer X Golden Sun]; Kordes; Wheatcroft Bros.

'**GOLDEN SPRAY**', HT, my, 1917; flowers clear lemon-yellow, huge mass of prominent anthers, semi-dbl., borne in long, arching sprays; GM, NRS, 1915; Dickson, H.

Golden Sprite, *see* 'DEVUNICAN'

'**GOLDEN STAR**', HT, dy, 1974; flowers full, dbl., 32 petals, 4–4.5 in., slight fragrance; foliage dark; free growth; [Whisky Mac sport]; Lowe

'**GOLDEN STATE**', HT, dy, 1937; flowers golden yellow, large, dbl., cupped; foliage leathery, glossy; vigorous growth; GM, Bagatelle, 1937 GM, Portland, 1937; [Souv. de Claudius Pernet X (Charles P. Kilham X Unnamed seedling)]; Meilland, F.; C-P, 1938

Golden Summers, *see* 'KORPORT'

'**GOLDEN SUN**', HT, my, 1957; (Goldene Sonne); bud long, pointed; flowers golden yellow, dbl., 5 in., exhibition form, moderate fragrance; foliage glossy; upright, bushy growth; [(Walter Bentley X Condesa de Sástago) X Golden Scepter]; Kordes, R.

'Golden Sunblaze'™, *see* 'MEIcupag'

Golden Sunblaze, *see* **'RISE 'N' SHINE'**

'**GOLDEN SUNBURST**', HT, yb, 1967; bud ovoid; flowers cadmium-orange and saffron-yellow, large, dbl., intense fragrance; foliage dark, glossy; vigorous, bushy growth; [Golden Wave sport]; Schneeberg; Carlton Rose Nurseries

'**GOLDEN SUNSET**', LCl, my, 1934; flowers golden yellow, often tipped orange-red, large, dbl., cupped; long stems, foliage glossy; vigorous growth; Burbank; Stark Bros.

'**GOLDEN SUNSHINE**', HT, my, 1964; bud pointed, ovoid, chrome-yellow splashed red; flowers canary-yellow, dbl., 50 petals, 5 in., exhibition form, intense fragrance; vigorous, upright growth; [Helen Hayes X Golden Masterpiece]; Brownell, H.C.; Brownell

'**GOLDEN SURPRISE**', F, my, 1978; [Woburn Abbey sport]; Hamilton

'**GOLDEN TALISMAN**', HT, my, 1931; [Talisman sport]; Hill, E.G., Co.

'**GOLDEN TALISMAN, CLIMBING**', Cl HT, my, 1935; [Talisman sport]; Elmer's Nursery

Golden Thoughts, *see* **'MCGREDY'S ORANGE'**

'**GOLDEN TIMES**', HT, my, 1970; flowers lemon-yellow, dbl., 40 petals, 4–5 in., slight fragrance; foliage glossy; [Fragrant Cloud X Golden Splendour]; Cocker; Wheatcroft & Sons

Golden Times, *see* 'KORTIME'

Golden Token™, *see* 'MINITOK'

Golden Topas, *see* 'MANTOPAS'

Golden Touch, S, my

Golden Treasure, *see* **'GOLDSCHATZ'**

'**GOLDEN TREASURE, CLIMBING**', Cl F, my, 1976; Pearson; Burston Nursery

'**GOLDEN TZIGANE**', HT, ob, 1961; flowers orange; [Tzigane sport]; Gregory

'**GOLDEN UNICORN**', S, yb, 1985; flowers yellow, petals edged orange-red, large, dbl., 28 petals, cupped, borne 1–8 per cluster, repeat bloom, moderate fragrance; awl-like, tan prickles; foliage dark olive green, leathery; vigorous, upright, bushy, spreading growth; hardy; [Paloma Blanca X (Carefree Beauty X Antike)]; Buck, Dr Griffith J.; Iowa State University, 1984

Golden Vale, S, lp; [Primula X Unknown]; Sutherland, P, 2000; Golden Vale Nursery; (Weatherly, L.)

'**GOLDEN VAN ROSSEM**', HT, dy, 1937; flowers chrome yellow; [Mev. G. A. van Rossem sport]; Lens

'**GOLDEN VANDAL**', HT, dy, 1935; flowers chrome yellow; [Comtesse Vandal sport]; Lens

'**GOLDEN VISION**', LCl, my, 1922; flowers Marechal Niel Yellow, fading nearly white, semi-dbl., moderate fragrance; [Marechal Niel X R. gigantea]; Clark, A.; NRS Victoria

Golden Wave, *see* **'DR A. J. VERHAGE'**

'**GOLDEN WEDDING**', HT, my, 1938; bud pointed, yellow tinted crimson; flowers clear yellow, dbl., 5 in.; foliage leathery, dark; vigorous growth; [Souv. de H.A. Verschuren X Yellow seedling]; Krebs

Golden Wedding, *see* AROkris

'**GOLDEN WEST**', HT, my, 1936; bud pointed; flowers golden yellow, open, large, semi-dbl., moderate fragrance; foliage dark, leathery, glossy; vigorous, bushy, spreading growth; [Duchess of York sport]; Stocking

'**GOLDEN WINGS**', S, ly, 1956; bud long, pointed; flowers sulfur-yellow, stamens prominent, single, 4–5 in., recurrent bloom, slight fragrance; vigorous, bushy growth; very hardy.; GM, ARS, 1958; [Soeur Thérèse X (R. spinosissima altaica X Ormiston Roy)]; Shepherd; Bosley Nursery

'**GOLDEN WONDER**', HT, yb, 1973; flowers lemon-yellow, edged red, dbl., 36–40 petals, 4–5 in., exhibition form, moderate fragrance; foliage dark; [Miss Ireland X Princess]; Gandy, Douglas L.; Morse Roses

'**GOLDEN WONDER**', HT, ab, 1936; flowers golden apricot suffused pink, globular, moderate fragrance; foliage olive-green; vigorous growth; Gunn

Golden Wonders 99, *see* 'RAWANA'

Golden Years®, *see* 'HARWEEN'

'**GOLDENDALE**', HT, my, 1956; bud long, pointed; flowers golden yellow, dbl., 50 petals, 5 in., moderate fragrance; [Annabella sport]; Grillo

Goldene Aue, F, dy, 1964; flowers large, dbl.; GPG Bad Langensalza; (Sangerhausen)

'**GOLDENE DRUSCHKI**', HP, my, 1936; bud pointed; flowers golden yellow, edged lighter turning creamy yellow in hot weather, cupped, slight fragrance; foliage leathery, dark; very vigorous growth; [Frau Karl Druschki X Friedrich Harms]; Lambert, P.

'**GOLDENE GRÜSS AN AACHEN**', F, ob, 1935; bud pointed, red; flowers golden orange, sometimes shaded reddish, dbl., high centered, very large; fragrant, foliage glossy; bushy growth; [Mme Butterfly X Gloria Mundi]; Kordes

'**GOLDENE JOHANNA TANTAU**', F, my, 1945; (Wheatcroft's Golden Polyantha); bud ovoid; flowers clear golden yellow, borne in clusters, single, 3.5 in., cupped, moderate fragrance; foliage glossy, dark; compact, bushy growth; (28); [Golden Rapture X (Johanna Tantau X Eugenie Lamesch)]; Tantau

Goldene Sonne, *see* **'GOLDEN SUN'**

'**GOLDENER ADLER**', F, my, 1965; bud ovoid; flowers golden yellow, medium, borne in clusters, semi-dbl.; foliage light green; [Allgold X Seedling]; Verschuren, A.; van Engelen

Goldener Olymp®, *see* 'KORSCHNUPPE'

'**GOLDENER REITER**', F, my, 1969; (Golden Rider); flowers golden yellow, large, dbl., 25 petals, cupped, intense fragrance; foliage dark; vigorous, upright growth;

[Circus X Golden Giant]; Haenchen, E.; Teschendorff

Goldener Sommer 83®, F, my, 1983; flowers large, dbl.; Noack, Werner

Goldener Traum, *see* **'Golden Dream'**

'Goldenes Herz', HT, dy, 1975; bud ovoid; flowers medium, dbl., cupped, moderate fragrance; foliage glossy, dark, leathery; vigorous, upright, bushy growth; [Dr A.J. Verhage X Seedling]; Kordes

Goldenes Mainz, *see* **'Golden Main'**

Goldenes Prag, HT, 1966; dbl., intense fragrance; GPG Bad Langensalza; (Sangerhausen)

Goldfächer, *see* 'DICmoppet'

Goldfasan, *see* **'Golden Pheasant'**

Goldfassade, LCl, my, 1967; Baum

'Goldfinch', HMult, ly, 1907; flowers aging white, small blooms in clusters, semi-dbl., non-recurrent, slight fragrance; foliage small, wrinkled; vigorous, climbing growth; [Helene X ?]; Paul

Goldfinger®, *see* 'PEAroyal'

'Goldgleam', F, my, 1966; semi-dbl., 18 petals, 3.5–4 in., blooms in small clusters, intense fragrance; foliage dark, glossy; growth moderate; [Gleaming X Allgold]; LeGrice

Goldglow, HT, 1959; Perry, Anthony; (Cavriglia)

'Goldie', F, my, 1958; bud ovoid; flowers golden yellow, edged lighter, dbl., 25–30 petals, 4 in., moderate fragrance; foliage leathery; vigorous, upright, bushy growth; [Goldilocks X Pigmy Gold]; Boerner; J&P

Goldie, *see* 'MACarom'

Goldika®, HT, ob, 1986; Cocker

'Goldilocks', F, my, 1945; flowers deep yellow, fading to cream, blooms in clusters, dbl., 45 petals, 3.5 in., globular, moderate fragrance; foliage leathery, glossy; vigorous, bushy growth; (28); John Cook Medal, ARS, 1947; [Unnamed seedling X Doubloons]; Boerner; J&P

Goldilocks, *see* 'SUNlock'

'Goldilocks, Climbing', Cl F, my, 1951; Caluya; J&P

Goldina, *see* 'CANdide'

Goldjuwel, *see* TANledolg

Goldjuwel, *see* TANlewu

Goldkrone, *see* **'Gold Crown'**

Goldlachs, *see* **'Golden Salmon'**

Goldlite™, *see* 'DEVsolear'

Goldmarie, *see* 'KORfalt'

'Goldmarie', F, dy, 1958; flowers orange-gold, very large, cluster, semi-dbl., intense fragrance; foliage glossy; very vigorous, upright, bushy growth; RULED EXTINCT 6/84 ARM; [Masquerade X Golden Main]; Kordes, R.

Goldmarie 82®, *see* 'KORfalt'

Goldmarie Nirp®, *see* 'KORfalt'

Goldmarie, Climbing, *see* KORkuma

'Goldmoss', F, my, 1972; bud long, pointed; flowers clear yellow, medium, dbl., intense fragrance; foliage light, leathery; vigorous, dwarf, bushy growth; [Rumba X Moss hybrid]; Moore, Ralph S.; Sequoia Nursery

Goldpin, Min, dy, 1998; Ilsink

Goldpoint, *see* 'JACpo'

'Goldquelle', S, my, 1965; bud ovoid; flowers pure golden yellow, large, borne in clusters, dbl., recurrent bloom, slight fragrance; foliage leathery; strong stems; upright (4 ft) growth; Tantau, Math.

Goldquelle 88, F, 1988; Tantau, Math.; (Cavriglia)

Goldrausch, *see* 'KORbi'

Goldregen, *see* **'Golden Rain'**

'Goldrush', , dy, 1941; flowers medium, dbl., moderate fragrance; Howard & Smith; (Sangerhausen)

'Goldschatz', F, my, 1964; (Golden Treasure); bud pointed; flowers golden yellow, large blooms in clusters (up to 30), dbl.; foliage dark, glossy; bushy, upright growth; Tantau, Math.

Goldschatz, *see* 'TANtasch'

Goldschmied, F, dy; dbl., slight fragrance; VEG; (Sangerhausen)

Goldsmith, *see* 'KORbelma'

Goldstadt Pforzheim, *see* HELpforz

Goldstar, *see* 'CANdide'

Goldstein®, F, ob, 1983; Cocker

Goldstern®, *see* 'TANtern'

Goldstrike, HT, yb, 1997; Perry, Astor

'Goldstück', F, my, 1961; flowers lemon-yellow, borne in clusters, dbl., 40–55 petals; foliage glossy, dark; vigorous, upright growth; [Goldilocks X Unnamed seedling]; Verschuren, A.; van Engelen

'Goldstück', S, my, 1963; bud long, pointed; flowers golden yellow, large, borne in clusters, abundant, non-recurrent bloom, intense fragrance; vigorous, upright, bushy growth; Tantau, Math.

Goldtopas®, *see* 'KORtossgo'

Goldy, *see* 'KORbeen'

Golestan, *see* 'MEIsadina'

Golf, S, w, 1993; Wageningen

Goliath, *see* **'Gary Player'**

Golmain, *see* **'Golden Main'**

Gomathi, HT, mp, 1987; Kasturi

Gomery, HT, dy, 1996; 28 petals, 4.5 in.; foliage matt; strong grower growth; RvS-Melle, 1999

'Gondul', F, mr, 1969; flowers cardinal-red, open, medium, borne in clusters, semi-dbl., 18 petals; foliage dark, glossy; vigorous growth; [Lichterloh X Lumina]; Lundstad

Gone Fishin'™, *see* 'SAVafish'

'Gonsoli Gaelano', HP, mp, 1874; flowers satiny rose; Pernet

Good as Gold, *see* 'CHEWsunbeam'

'Good Cheer', HT, dp, 1937; bud pointed; flowers cerise, open, dbl.; long, strong stems; very vigorous growth; [Talisman X Templar]; Amling Co.

'Good Companion', F, mr, 1961; flowers rich red, large, borne in trusses, dbl., 30 petals, flat, moderate fragrance; foliage dark; vigorous growth; Dickson, A.

Good Day Sunshine, *see* 'TALgoo'

'Good Life', F, or, 1970; dbl., 30 petals, 2 in., exhibition form, slight fragrance; free growth; [Elizabeth of Glamis X John Church]; McGredy, Sam IV; McGredy

Good Luck, *see* BURspec

'Good Morning', HT, w, 1935; flowers white, reverse faintly tinged pink, very large, dbl.; foliage leathery; long, strong stems; very vigorous, bushy growth; [Premier Supreme sport]; Kaucher; Hill Crest Greenhouses

Good Morning, *see* FRYyat

Good Morning America™, *see* 'SAVagood'

Good Morning Sunshine, *see* 'BOSlaur-amber'

'Good Neighbor', HT, ob, 1958; flowers burnt-orange, reverse golden, large, dbl., intense fragrance; foliage dark; vigorous, upright growth; [Fred Howard X Unnamed seedling]; Warriner, William A.; H&S

'Good News', HT, pb, 1940; (Bonne Nouvelle); flowers silvery pink, center tinged apricot, dbl., 50 petals, 5–6 in., globular, moderate fragrance; vigorous, bushy growth; [(Radiance X Souv. de Claudius Pernet) X (Joanna Hill X Comtesse Vandal)]; Meilland, F.; C-P

Good News, F, my, 1992; Warley Rose Gardens

Good Old Summertime, *see* CLEheat

Good Show, *see* JACmeen

'Good Times', HT, pb, 1977; bud ovoid, pointed; flowers ivory to light pink with deep pink edging, dbl., 56 petals, 5–5.5 in., cupped, slight fragrance; foliage large, glossy; compact, upright growth; [Pink Peace X Peace]; Williams, J. Benjamin; Hershey Nursery

Good Vibes, *see* 'GELgood'

'Gooiland', HT, mp, 1922; flowers clear rose-pink, reverse dark coral-rose, dbl., slight fragrance; [Sunburst seedling X Red-Letter Day]; Van Rossem

'**GOOILAND BEAUTY**', HT, ab, 1924; flowers clear golden orange, open, over large, semi-dbl., slight fragrance; foliage dark, leathery; vigorous, bushy growth; GM, Bagatelle, 1925; [Sunburst X Golden Emblem]; Van Rossem; Van Rossem & Prior

'**GOOILAND GLORY**', HT, or, 1925; flowers cherry-red shaded coral-red, semi-dbl., slight fragrance; [Mme Edouard Herriot X Gen. MacArthur]; Van Rossem; Van Rossem & Prior

Goose Fair, *see* 'MEIPOPUL'

'**GOPIKA**', F, op, 1969; flowers light salmon-pink, medium, dbl., globular, slight fragrance; foliage glossy; vigorous, bushy growth; [Marlena X Open pollination]; Singh; Gopalsinamiengar

'**GORDON DRAKE**', F, dp, 1956; flowers cerise-pink, borne in large clusters, dbl.; very vigorous growth; [Eutin sport]; Williams, G.A.

'**GORDON EDDIE**', HT, ab, 1949; flowers deep apricot, edged lighter, very large, dbl., 40 petals, exhibition form, moderate fragrance; foliage leathery, glossy; very vigorous, bushy growth; GM, NRS, 1950; [Royal Visit X Cynthia Brooke]; Eddie

Gordon's College, *see* 'COCJABBY'

'**GORGEOUS**', HT, dy, 1915; flowers deep orange-yellow, veined copper, well-formed, large, dbl., moderate fragrance; foliage rich, green,soft; bushy, open growth; GM, NRS, 1913; Dickson, H.

'**GORGEOUS**', HT, mp, 1956; flowers rose-pink, semi-dbl., 18–20 petals, 5 in., exhibition form, moderate fragrance; foliage leathery; vigorous growth; [Pink Delight sport]; Franc; Carlton Rose Nurseries

Gorgeous George, *see* 'DELRICOS'

'**GOTENHAFEN**', F, mp, 1940; flowers pure bright rose, medium, semi-dbl., 12–15 petals; vigorous, well branched growth; [Mev. Nathalie Nypels X Kardinal]; Tantau

'**GOTHA**', HT, ab, 1932; (Gotha IV); flowers brownish yellow passing to apricot, moderate fragrance; [(Souv. de H.A. Verschuren X Sunset) X Mev. G.A. van Rossem]; Krause

Gotha IV, *see* **'GOTHA'**

'**GOTTFRIED KELLER**', HFt, ab, 1894; flowers apricot-yellow, [((Mme Berard X R. foetida persiana) X (Pierre Notting X Mme Berard)) X R. foetida persiana]; Müller, Dr F.

'**GOUDVLINDER**', HT, ab, 1926; (Golden Butterfly); flowers orange-yellow, small, semi-dbl., 12 petals, foliage glossy, brownish red; vigorous, bushy growth; [Lady Hillingdon X Souv. de Claudius Pernet]; Van Rossem

'**GOURDAULT**', B, m, 1859; Guillot Père

Gourmet Pheasant, Min, dp, 1995

Gourmet Popcorn, *see* 'WEOPOP'

'**GOVERNADOR BRAGA DA CRUZ**', HT, ly, 1954; very vigorous growth; [Peace X Unnamed seedling]; da Silva, Moreira

'**GOVERNOR ALFRED E. SMITH**', HT, yb, 1933; flowers blend of buff, terracotta, gold and salmon, large, dbl., exhibition form; foliage glossy; vigorous growth; [Souv. de F. Bohé seedling X Unnamed seedling]; Denoyel, Vve.; J&P

'**GOVERNOR MARK HATFIELD**', Gr, dr, 1962; bud pointed; flowers rich red, large, dbl., 40 petals, exhibition form, slight fragrance; foliage leathery; vigorous, upright growth; [Carrousel X Charles Mallerin]; Von Abrams; Peterson & Dering

'**GOVERNOR PHILLIP**', Cl HT, mr, 1939; flowers ruby-red, flushed darker, open, large, very dbl., moderate fragrance; foliage leathery, glossy, dark, bronze; long stems; vigorous, climbing growth; [Ophelia seedling X Black Boy]; Fitzhardinge; Hazlewood Bros.

'**GOVERNOR ROSELLINI**', Gr, mr, 1958; flowers rose-red, dbl., 30 petals, 3–4 in., exhibition form, moderate, raspberry fragrance; foliage dark, leathery; vigorous, upright growth; [Baby Chateau X Tiffany]; Lindquist; Howard Rose Co.

Governor's Lady Gloria, *see* 'AROGLOR'

Gowirichs Traum, Pol, dp

'**GOYA**', F, w, 1976; flowers cream, dbl., 30 petals, 5 in., exhibition form, slight fragrance; foliage dark; vigorous growth; [Mildred Reynolds X Arthur Bell]; Bees

Graaff-Reinet, *see* 'KORMATE'

'**GRACE**', HRg, 1923; flowers amber, center apricot, open, very dbl., moderate fragrance; foliage wrinkled; bushy (5–6 ft) growth; RULED EXTINCT ?; [R. rugosa X R. Xharisonii]; Saunders; Central Exp. Farm

'**GRACE ABOUNDING**', F, w, 1968; flowers ivory, blooms in trusses, semi-dbl., moderate, musk fragrance; foliage glossy; [Pink Parfait X Penelope]; Harkness

'**GRACE DARLING**', T, w; flowers cream-white shaded golden, large, dbl., globular, vigorous growth; Bennett, 1885

Grace de Monaco®, *see* 'MEIMIT'

Grace Donnelly, *see* 'HORLEXSTRIP'

'**GRACE HASLAM**', HT, ob; flowers orange to carmine, [Scarlet Glory X Mrs Sam McGredy]; Fryers Nursery, Ltd.

Grace Kelly, *see* 'MEIMAGARMIC'

'**GRACE KIMMINS**', F, mr, 1973; flowers crimson, dbl., 28 petals, 3 in.; foliage glossy; vigorous, bushy growth; [Dainty Maid X Red Dandy]; Gobbee, W.D.

'**GRACE MOLYNEUX**', HT, ab, 1909; flowers creamy apricot, center flesh, dbl., slight fragrance; Dickson, A.

'**GRACE MOORE**', HT, mr, 1948; (Paul Holtge); bud ovoid; flowers crimson-red, large, dbl., cupped, foliage leathery, dark olive-green; vigorous, bushy growth; [Kardinal X Crimson Glory]; Kordes; C-P

Grace Moore's Noisette, N, w

'**GRACE NOLL CROWELL**', HT, pb, 1929; flowers rose-pink, base slightly shaded cream, large, dbl., exhibition form, moderate fragrance; foliage soft, light; vigorous, bushy growth; Vestal

'**GRACE NOTE**', S, pb, 1984; bud ovoid, pointed; flowers medium pink, freckled red, imbricated, large blooms borne 3–8 per cluster, dbl., 38 petals, repeat bloom, moderate fragrance; awl-like, tan prickles; foliage large, leathery, dark; vigorous, erect, bushy growth; hardy; [(Tiki X Marigold) X Freckle Face]; Buck, Dr Griffith J.; Iowa State University

Grace Seward, *see* 'TINGRACE'

'**GRACE WAYMAN**', Cl HT, mp, 1936; flowers pink, very large, dbl., intense fragrance; foliage leathery; vigorous, climbing (10 ft) growth; Wayman

'**GRACE WOOD**', Min, my, 1993; dbl., 26–40 petals, 1.5 in., borne in small clusters, no fragrance; few prickles; foliage small, medium green, matt; medium, compact growth; [Sun Flare X Rise 'n' Shine]; Giles, Diann; Giles Rose Nursery, 1992

'**GRACEFUL**', S, ly, 1966; flowers yellow fading to cream, well formed, dbl., 4 in., slight fragrance; foliage glossy; vigorous, bushy growth; [Paul's Lemon Pillar X Marcelle Gret]; Smith, W. H.

Graceland, *see* 'JACEL'

Graceland, *see* KIRscot

Gracie Allen, *see* 'WEKCRYREG'

'**GRACIE FIELDS**', HT, dy, 1937; flowers vivid buttercup-yellow, moderate, sweetbriar fragrance; foliage glossy; vigorous growth; Letts

'**GRACILIS**', M, dp; flowers deep pink, well mossed, large, dbl., globular, seasonal bloom, foliage large; vigorous growth; good as a standard; Prévost, prior to 1846

'**GRACILIS**', Bslt, mr, 1830; flowers cherry shaded lilac-blush, semi-dbl., cupped; large, long prickles; foliage dark; vigorous, branching growth; Wood

'**GRACILIS**', C; flowers pale rosy pink, edged lilac-blush, compact, dbl.; branching, low-growing growth; Shailer, 1796

'**GRACILIS**', HSet, mp, 1841; flowers pink to rose, borne in clusters, very dbl.; Prince Nursery

Gracilis, *see* **'SHAILER'S PROVENCE'**

'**GRACIOSA**', HT, w, 1957; flowers white edged rose, well formed; very vigorous growth; [Branca X Peace]; da Silva, Moreira

'GRACIOUS LADY', HT, pb, 1965; bud ovoid; flowers peach-pink, base apricot, large, dbl., exhibition form, moderate fragrance; foliage glossy, dark; vigorous, bushy growth; [(Peace seedling X Gail Borden) X Dorothy Peach seedling]; Robinson, H.; J&P

Graeme Douglas, *see* 'MACPIOPI'

'GRAF FRITZ METTERNICH', HP, dr, 1896; flowers large, dbl., intense fragrance; Soupert & Notting; (Sangerhausen)

'GRAF FRITZ VON HOCHBERG', HT, op, 1905; flowers large, dbl.; Lambert, P.; (Sangerhausen)

Graf Lennart, HT, mr, 1991; Meilland

'GRAF SILVA TAROUCA', HT, dp, 1916; flowers carmine-red, very large, dbl., intense fragrance; very vigorous growth; [Étoile de France X Lady Mary Fitzwilliam]; Lambert, P.

'GRAF ZEPPELIN', HMult, dp, 1909; flowers light red to bright pink, medium, borne in large clusters, semi-dbl., very free bloom; vigorous (6–8 ft) growth; [Non Plus Ultra sport]; Boehm

'GRÄFIN ADA BREDOW', HMult, lp, 1909; flowers small, semi-dbl.; Walter; (Sangerhausen)

Grafin Esterhazy, HCh; Geschwind, R.; (Cavriglia)

'GRÄFIN HARDENBERG', S, dp, 1938; flowers carmine-pink, medium, semi-dbl.; Vogel, M.; (Sangerhausen)

'GRÄFIN MARIE HENRIETTE CHOTEK', HMult, mr, 1911; flowers medium, dbl., moderate fragrance; Lambert, P.; (Sangerhausen)

'GRÄFIN MINNIE SCHAFFGOTSCH', HT, w, 1928; flowers cream-white, center pink, very dbl., slight fragrance; [Clio X Unnamed Hybrid Tea]; Mühle

Grafin Sonja, *see* KORfeimot

'GRAFTON PILLAR', LCl, mr, 1958; bud globular; flowers bright red, small, dbl., borne in clusters, free, recurrent boom, slight fragrance; foliage wrinkled; moderate growth; [Second generation Skinner's Rambler X Gruss an Aachen]; Risley

'GRAHAM', HT, dr, 1961; flowers deep crimson-scarlet, large, dbl., 35–40 petals, exhibition form, moderate fragrance; foliage glossy; long stems; vigorous, upright growth; [Eclipse X Unnamed seedling]; Kriloff, Michel; Cramphorn's Nursery

Graham Thomas®, *see* 'AUSMAS'

Graham Thomas Musk, HMsk, w

'GRALOVE', F, mr, 1998; (**Everlasting Love**, Millennium®); flowers large, scarlet, dbl., 15–25 petals, 2.75 in., borne in small clusters, slight fragrance; some prickles; foliage medium, dark green, glossy, new growth is red; medium (80–100 cm), bushy growth; [Molly McGredy X Satchmo]; Grant, Doug

'GRAN PARADA', F, my, 1967; (Grande Parade); flowers yellow, becoming reddish, open, medium, dbl., 25 petals; foliage glossy; vigorous, upright, compact growth; [Gold Dot X (Queen Elizabeth X Zambra)]; Dot, Simon; Rosas Dot

'GRANADA', HT, 1955; flowers deep red, borne in clusters, [Opera X The Doctor]; Delforge

'GRANADA', HT, rb, 1963; (Donatella); bud urn shaped; flowers blend of rose, nasturtium-red and lemon-yellow, to open, dbl., 25–30 petals, 4–5 in., exhibition form, borne singly, intense, damask and spice fragrance; foliage leathery, crinkled; vigorous, upright growth; AARS, 1964 James Alexander Gamble Fragrance Medal, ARS, 1968; [Tiffany X Cavalcade]; Lindquist; Howard Rose Co.

'GRANADA, CLIMBING', Cl HT, rb, 1964; Swim & Weeks; Comley

'GRANADINA', Min, mr, 1956; (Grenadine); flowers oxblood-red, small, dbl., 30 petals, globular; foliage dark; low, upright growth; [Granate X Coralín]; Dot, Pedro; Kordes

'GRANAT', HT, dr, 1937; flowers blackish red, well formed, very vigorous growth; [Barcelona X Château de Clos Vougeot]; Krause

'GRANATE', Min, dr, 1947; flowers velvety oxblood-red, often streaked white, small; almost no prickles; height 6–8 in.; [Merveille des Rouges X Pompon de Paris]; Dot, Pedro

Granatina, HT, 1964; Cazzaniga, F. G.; (Cavriglia)

'GRAND AMOUR', HT, mr, 1956; (Cote Rotie); flowers bright red, large, moderate fragrance; Delbard-Chabert

'GRAND CANARY', HT, my, 1934; dbl., exhibition form, foliage glossy; vigorous, compact growth; [Token sport]; Lowman; U.S. Cut Flower Co.

'GRAND CANYON', S, dy, 1951; bud long, pointed, chrome-yellow; flowers yellow, salmon and copper, turning to crimson, open, medium, recurrent bloom, slight fragrance; foliage glossy, bronze; very vigorous (6 ft), arching growth; [Herrenhausen X Golden Rapture]; Whisler; Germain's

Grand Chateau, *see* 'TANELORAK'

'GRAND CRAMOISI', HGal, dr; flowers crimson, profuse bloom; Vibert (?), ca 1832

Grand Duche, *see* 'GODRACHE'

Grand Finale™, *see* 'JACPIHI'

'GRAND GALA'™, HT, rb, 1954; bud globular; flowers rose-red, reverse white suffused pink, dbl., 45–60 petals, 4.5–5 in., exhibition form, slight fragrance; foliage leathery; vigorous, bushy growth; [Peace X Independence]; Meilland, F.; C-P, 1956;, URS, 1954

Grand Gala, *see* MEIqualis

'GRAND GALA, CLIMBING', Cl HT, rb, 1961; Yamate; Kakujitsuen

Grand Hotel®, *see* 'MACTEL'

Grand Impression™, *see* 'JACLION'

Grand Ivory, HT, lp

'GRAND LADY', HT, mp, 1968; bud ovoid; flowers pink, center lighter, large, dbl., exhibition form, moderate fragrance; foliage leathery; vigorous, upright, bushy growth; [Ma Perkins X Peace]; Patterson; Patterson Roses

Grand Marshall™, *see* 'AROFUTO'

Grand Masterpiece, *see* 'JACPIE'

'GRAND MOGUL', HT, w, 1965; flowers creamy white, large, dbl., 33 petals, exhibition form, moderate fragrance; growth moderate; [Sultane X Chic Parisien]; Delbard-Chabert; Cuthbert

Grand Nord®, *see* 'DELGRORD'

'GRAND OCCASION', HT, pb, 1970; flowers rosy coral shaded yellow, edged carmine, full, dbl., 30 petals, 3 in., intense fragrance; foliage glossy, light; vigorous growth; [Comtesse Vandal X Mme Henri Guillot]; Delbard; Laxton & Bunyard Nursery

'GRAND OPENING', Min, rb, 1991; bud ovoid; flowers orange-red showing yellow eye, red reverse, ages pale red, dbl., 35 petals, exhibition form, moderate, fruity fragrance; foliage medium, dark green, semi-glossy; upright, medium growth; [Poker Chip X Zinger]; Gruenbauer, Richard, 1984; Flowers 'n' Friends Miniature Roses, 1993

'GRAND OPERA', HT, pb, 1964; bud long, pointed; flowers cream edged pink, becoming pink, dbl., 40 petals, 4–5 in., exhibition form, moderate fragrance; foliage leathery; vigorous, bushy growth; [Masquerade X Peace]; Schwartz, Ernest W.; Wyant

Grand Palace, *see* POUlgrad

Grand Prix®, *see* 'DELTUF'

'GRAND PRIZE', HT, rb, 1935; flowers petals red, white and spotted, very large, dbl., cupped, intense fragrance; foliage leathery; very vigorous growth; [Red Radiance sport]; Kistler

Grand Romance, *see* 'BRIGRAN'

Grand Siècle®, *see* 'DELEGRAN'

'GRAND SLAM', HT, mr, 1963; bud urn-shaped; flowers cherry to rose-red, dbl., 28 petals, 4 in., exhibition form, slight fragrance; foliage leathery, dark, semi-glossy; vigorous, upright, spreading growth; [Charlotte Armstrong X Montezuma]; Armstrong, D.L. & Swim; Armstrong Nursery

Grand Trianon, HT, 1967; Truffaut, G.; (Cavriglia)

'GRANDCHILD', Min, mp, 1986; flowers clear, medium pink, small, borne 3–15 per cluster, dbl., 40 petals, exhibition form, moderate, sweet fragrance; hooked, gray-brown prickles; foliage long, narrow, dense, reddish to medium green; bushy, shrub-like, tall growth; [Unnamed Cécile Brunner seedling X Self]; Garelja, Anita, 1987

'GRAND-DUC ADOLPHE DE LUXEMBOURG', HT, rb, 1892; flowers brick-red, reverse carmine, large, dbl., moderate fragrance; weak stems; moderate growth; [Triomphe de la Terre des Roses X Mme Loeben Sels]; Soupert & Notting

'GRAND-DUC ALEXIS', HP, mr, 1892; flowers large, dbl.; Lévêque; (Sangerhausen)

Grande Amore, HT, dr

'GRANDE CENTFEUILLE DE HOLLANDE', C, lp, 1840; flowers large, very dbl., moderate fragrance; Prévost; (Sangerhausen)

Grande Cuisse de Nymphe, *see* **'GREAT MAIDEN'S BLUSH'**

'GRANDE DUCHESSE CHARLOTTE', HT, mr, 1942; bud long, pointed; flowers tomato-red, shaded geranium-red, cactus form, dbl., 25 petals, 5–5.5 in., slight fragrance; foliage glossy, dark; vigorous, bushy growth; AARS, 1943 GM, Portland, 1941 GM, Rome, 1938; Ketten Bros.; C-P

Grande Duchesse de Luxembourg, *see* **'MARIE ADÉLAÏDE'**

Grande Henriette, *see* **'L'ENCHANTRESSE'**

Grande Parade, *see* **'GRAN PARADA'**

'GRANDE PREMIÈRE', HT, yb, 1959; bud long, pointed; flowers yellow edged pink, large, dbl., 38 petals, moderate fragrance; foliage bright green, glossy; upright, bushy growth; [Comtesse Vandal X Mme Henri Guillot]; Delbard-Chabert

'GRANDE RENONCULE', C, mp; flowers dull pink, shading to violet, medium, dbl.

'GRANDE ROUGE', F, dr, 1973; bud ovoid; flowers medium, dbl., slight fragrance; foliage soft; upright, bushy growth; [Unknown X Unknown]; Tantau, Math.; Ahrens & Sieberz

'GRANDE WALZER'®, HT, rb, 1980; bud ovoid; flowers deep orange-red, reverse deep yellow, blooms borne singly, dbl., 35 petals, cupped, slight fragrance; curved, light yellow prickles; foliage rather small, light green, glossy; upright growth; Kordes, W.; Rose Barni-Pistoia

Grandee, *see* 'MACHOMAI'

Grandessa, *see* 'DELSIRE'

'GRANDEUR', HT, dp, 1954; bud long, pointed; flowers cerise-red, dbl., 70 petals, 4 in., exhibition form, moderate fragrance; foliage leathery; very vigorous, upright growth; [Joyance sport]; Grillo

'GRANDEUR', HT; flowers vermilion-red, medium, dbl., 25 petals; moderate, bushy growth; Mallerin, C.

'GRANDEZZA', HT, mp, 1962; bud spiral, pointed; flowers peach-blossom-pink, well formed, large, dbl.; moderate, bushy growth; [Monique X Radar]; Herholdt, J.A.; Herholdt's Nursery

Grandhotel®, *see* 'MACTEL'

'GRANDIOSO', HT, mr, 1961; flowers cherry-red, dbl., 30–40 petals, 5 in., moderate fragrance; foliage dark; [(Happiness X Satisfaction) X (Poinsettia X Happiness)]; Verbeek

'GRANDISSIMA', HGal, m, 1824; (Louis Philippe); flowers rosy crimson, sometimes purplish, very large and compact, dbl., moderate, branching growth; Hardy, Prior to 1848

Grandma's Baby, Min, lp

Grandma's Lace, *see* 'CLEGRAN'

Grandma's Pink, *see* 'MORBOUQUET'

Grandma's Violet, HT, m

'GRANDMASTER', HMsk, ab, 1954; bud long, pointed; flowers apricot shaded lemon and pink, large blooms in clusters, semi-dbl., 10 petals, recurrent bloom, moderate fragrance; foliage light green; bushy growth; (28); GM, NRS, 1951; [Sangerhausen X Sunmist]; Kordes; Morse

'GRAND'MÈRE JENNY', HT, yb, 1950; (Grem); flowers apricot-yellow, edged and suffused pink, dbl., 30 petals, 4–4.5 in., exhibition form, moderate fragrance; foliage dark, glossy; vigorous growth; GM, NRS, 1950 GM, Rome, 1955; [Peace X (Julien Potin X Sensation)]; Meilland, F.; C-P, 1955; URS, 1950

'GRAND'MÈRE JENNY, CLIMBING', Cl HT, yb, 1958; (Gremsar, Grimpant Grand'mère Jenny); Meilland, F.; URS

Grandmother's Hat, HP, mp

Grandpa Alex, *see* 'RENAL'

Grandpa Dan, *see* 'RESPAD'

Grandpa Dickson, *see* **'IRISH GOLD'**

Grandpa Ray, *see* 'RESPAR'

Grandpa Toni, *see* 'RESPAR'

'GRANDPA'S DELIGHT', F, ob, 1984; flowers orange-red, yellow center and stamens, medium, semi-dbl., slight fragrance; foliage medium, medium green, glossy; bushy growth; [Living Fire X Seedling]; Pawsey, Roger; Rearsby Roses, Ltd.

'GRAND-PÈRE LOTTIN', HWich, op, 1918; flowers salmony flesh-pink, center brighter, borne in clusters, very dbl., [Lady Godiva X Mrs W.H. Cutbush]; Lottin

'GRANGE BRIAR', (strain of R. canina), lp; vigorous; once used as understock

'GRANGE COLOMBE', HT, w, 1912; bud pointed; flowers cream-white, center yellow, large, dbl., cupped, moderate fragrance; vigorous growth; [Mme Caroline Testout X Lady Ashtown]; Guillot, P.

'GRANNIE'S ROSE', S, my; flowers clear pink, borne in clusters of 5–7, semi-dbl., non-recurrent, intense fragrance

Granny Grimmetts, HP, dr; flowers dark purple-red, medium, dbl., moderate fragrance; Hilling, 1955

Granny Savoie, HGal, dr

Granny's Favourite, MinFl, lp

Grape Delight, *see* 'GELGRAPE'

Grapeade, Cl Min, m, 1979

Grasten, *see* POUlfeld

'GRATIA', HT, w, 1934; bud pointed; flowers creamy white, large, semi-dbl., exhibition form, moderate fragrance; foliage leathery, dark; vigorous growth; [Seedling X Pius XI]; Leenders, M.

'GRATITUDE', HT, rb, 1960; bud ovoid; flowers reddish orange, dbl., 25–35 petals, 3–3.5 in., slight fragrance; foliage leathery, dark, glossy; vigorous, upright, bushy growth; [Impeccable X Incendie]; Delbard-Chabert

Graton Noisette, N, lp

Gratulation, F, 1985; GPG Bad Langensalza; (Sangerhausen)

'GRAVIN D'ALCANTARA', HT, dr, 1985; flowers well-formed, large blooms borne 1–7 per cluster, dbl., 28 petals, cupped, slight fragrance; red prickles; foliage matt, dark; upright growth; [Montezuma X Forever Yours]; Rijksstation Voor Sierplantenteelt, 1982

Gravin Michel d'Ursel®, HMsk, ab, 1994; Lens

'GRAZIA', HT, op, 1941; flowers salmon-pink, center darker, very large, foliage bright green; long stems; [Julien Potin X Mme G. Forest-Colcombet]; Giacomasso

'GRAZIELLA', F, ob, 1960; bud globular; flowers orange, open, medium, borne in clusters, dbl., slight fragrance; foliage glossy, light green; vigorous, bushy growth; [Feu Follet X Unnamed seedling]; Gaujard, R.; G. Truffaut

'GRAZIELLA', T, lp, 1893; Dubreuil

Great Century, *see* 'DELEGRAN'

Great Day, *see* 'MINBCO'

Great Double White, *see* **'ALBA MAXIMA'**

Great Expectations, *see* 'LANICAN'

Great Expectations, *see* 'JACDAL'

'GREAT MAIDEN'S BLUSH', A, w; (Cuisse de Nymphe, Grande Cuisse de Nymphe, La Royale, La Séduisante, La Virginale, Maiden's Blush, R. alba incarnata, R. alba rubicanda plena, R. alba rubicunda, R. carnea, R. incarnata, R. rubicans); flowers white, tinged pink, dbl., intense fragrance; (42); Roseraie de l'Hay, Cultivated prior to 1738

'GREAT NEWS', F, m, 1973; flowers plum-purple, reverse silver, dbl., 33 petals,

4 in., intense fragrance; foliage large, olive-green; moderate growth; [Rose Gaujard X City of Hereford]; LeGrice

Great Nord®, *see* 'DELGRORD'

Great North, *see* 'DELGRORD'

Great Ormond Street, F, my, 1991; Beales, Peter

Great Phoebus, *see* 'HINGREAT'

'GREAT SCOTT', HT, mp, 1991; flowers large blooms, very dbl.; prickles slightly pubescent peduncles, light green with some; foliage large, medium green, matt; upright (120 cms), bushy growth; [Cleo sport]; Ballin, Don & Paula

'GREAT VENTURE', HT, yb, 1970; bud long, pointed; flowers orange-yellow, flushed pink, medium, dbl., intense fragrance; foliage large, leathery; vigorous, upright growth; [Daily Sketch X Suspense]; Dawson, George; Brundrett

'GREAT WESTERN', B, m, 1840; flowers purplish maroon, large, dbl., blooms mostly in early summer; vigorous growth; Laffay, M.

'GREATER HASTINGS', HT, mp, 1957; Francis

'GREATHEART', HT, op, 1921; flowers pale flesh, shaded salmon, center deeper, dbl., moderate fragrance; [Mrs Walter Easlea sport]; Rosenbluth

'GREBLUB', HT, w, 1990; (**Blushing Bride**); flowers moderately large blooms, dbl., 15–25 petals, intense fragrance; foliage large, dark green, semi-glossy; upright, bushy growth; [Silverado sport]; Greenwood, Chris, 1991

'GREELIZ', F, lp, 1997; (**Turn of the Century**); flowers double, medium, dbl., 15–25 petals, borne in small clusters, slight fragrance; some prickles; foliage medium, semi-glossy; upright, medium (2–2.5) growth; [Elizabeth of Glamis X Seedling]; Sealand Nurseries, Ltd.

'GREEN BUBBLES', Min, w, 1978; bud ovoid; flowers light green, semi-dbl., 12 petals, 1 in., slight fragrance; foliage tiny; very compact, bushy growth; Lyon

'GREEN DIAMOND', Min, w, 1975; bud pointed, dusty pink; flowers soft green, dbl., 25 petals, .5 in., cupped; foliage small, leathery; upright, bushy growth; [Unnamed Polyantha X Sheri Anne]; Moore, Ralph S.; Sequoia Nursery

'GREEN FIRE', F, dy, 1958; bud ovoid, pointed; semi-dbl., 13 petals, 3 in., flat, blooms in clusters, slight fragrance; foliage semi-glossy; vigorous, bushy growth; [Goldilocks X Unnamed seedling]; Swim, H.C.; Armstrong Nursery

'GREEN ICE', Min, w, 1971; bud pointed; flowers white to soft green, small, dbl.; foliage small, glossy, leathery; vigorous, dwarf, bushy growth; [(R. wichurana X Floradora) X Jet Trail]; Moore, Ralph S.; Sequoia Nursery

'GREEN ROSE', Ch, w; (R. chinensis viridiflora, R. viridiflora); flowers green, often touched with bronze, with narrow leaf-like petals, dbl., 1.5–2 in., recurrent bloom; medium, upright growth; (14); Bambridge & Harrison, Cultivated prior to 1856

Green Snake®, *see* 'LENWIGA'

Greenalls Glory, *see* KIRmac

'GREENMANTLE', HEg, rb, 1895; flowers bright rosy red, white eye, golden stamens, single; foliage richly fragrant; very vigorous, tall growth; Penzance

Greensleeves, *see* 'HARLENTEN'

'GREENTAE', F, pb, 1987; (**Hightae**); flowers clear, medium pink outer petals, deeper in color at the heart, dbl., 45–50 petals, exhibition form, intense fragrance; prickles long, curved slightly downward, red; foliage small, bronze aging to medium green, glossy; low, bushy, free-flowering growth; [Duet X Regensberg]; Greenfield, Mrs. P.L.; Hightae Plant Nurs., 1988

'GREER GARSON', HT, pb, 1943; bud pointed; flowers begonia-rose, dbl., 35 petals, 5 in., exhibition form, slight fragrance; foliage leathery, dark; vigorous, tall, bushy growth; Denoyel, Vve.; J&P

'GREET KOSTER', Pol, op, 1933; flowers deep pink, shaded salmon; [Margo Koster sport]; Koster, D.A.

Greetings, *see* **'GRÜSS AN BERLIN'**

Greetings, *see* JACdreco

Greetings from Alma-Aty, *see* **'PRIVET IZ ALMA-ATY'**

Greetje Hennekens, Pol, mr, 1957; flowers carmine-red, medium, semi-dbl.; Loose; (Sangerhausen)

Greg Chappell, HT, ab, 1984

Greg Moore, *see* 'HORGREG'

'GREGART', HT, mr, 1997; (**Never Forgotten**); flowers full, medium, very dbl., 26–40 petals, borne in small clusters, moderate fragrance; some prickles; foliage medium, dark green; compact, medium (2–21 ft) growth; Sealand Nurseries, Ltd.

'GRÉGOR MENDEL', F, op, 1955; flowers coral-pink shaded yellow and carmine, well shaped, borne in, dbl.; vigorous growth; [Pinocchio seedling]; Maarse, G.

'GREGSIL', HT, lp, 1997; (**April Fools Day**); flowers full, medium, very dbl., 26–40 petals, borne in small clusters, moderate fragrance; some prickles; foliage medium, medium green, semi-glossy; bushy, medium (2.5–3 ft) growth; [Silver Jubilee sport]; Gregory, C.

Grem, *see* **'GRAND'MÈRE JENNY'**

Gremsar, *see* **'GRAND'MÈRE JENNY, CLIMBING'**

Grenadier, *see* **'FUSILIER'**

'GRENADIER', HT, mr, 1930; flowers brilliant currant-red shaded scarlet, dbl., cupped, slight fragrance; foliage rich green, leathery, glossy; vigorous, bushy growth; Dickson, A.

Grenadine, *see* **'GRANADINA'**

'GRENOBLE', HT, mr, 1927; (Ville de Grenoble); flowers clear scarlet, dbl., 30–40 petals, 3.5 in., exhibition form, slight, spicy fragrance; foliage thick; very vigorous growth; [Capt. F. Bald X Mme Van de Voorde]; Mallerin, C.; C-P, 1931

'GRENOBLE, CLIMBING', Cl HT, mr, 1939; Western Rose Co.

Greta, F, w, 1995; Briant

'GRETA FEY', HWich, lp, 1909; flowers small, semi-dbl.; Strassheim; (Sangerhausen)

'GRETA KLUIS', Pol, mr, 1916; flowers carmine-red; [Echo sport]; Kluis & Koning

'GRETA KLUIS SUPERIOR', Pol, mr, 1928; flowers deep carmine-red; [Tausendschön sport]; Kluis

'GRETAG', F, lp, 1991; (**Terrell Anne**); bud ovoid; dbl., 26–40 petals, 1.5–2.75 in., exhibition form, borne in small clusters, slight fragrance; some prickles; foliage medium, medium green, glossy; medium (90 cms), upright growth; [Gene Boerner sport]; Greenwood, Chris

'GRETE BERMBACH', HT, pb, 1925; flowers silvery flesh, center rose, sometimes shaded yellow orange, dbl., moderate fragrance; [Mrs Aaron Ward X Pharisaer]; Leenders Bros.

Grete Schickendanz, HT, 1987; Kordes, R.; (Cavriglia)

'GRETEL GREUL', HT, mr, 1939; [Rote Rapture sport]; Greul

'GRETELEIN', HFt, lp, 1933; flowers large, semi-dbl., slight fragrance; Schmitt, K.; (Sangerhausen)

'GRETHE POULSEN', Pol, dp, 1928; flowers light cherry-red, base yellow, semi-dbl., early; dwarf, well branched growth; [Ellen Poulsen X Mme Laurette Messimy]; Poulsen, S.

'GRETOOTS', F, pb, 1990; (**Tootsie**); flowers deep pink painted white, reverse mostly white, aging with a, semi-dbl., 20 petals, slight fragrance; foliage medium, dark green, glossy, disease resistant; upright, bushy, rounded, medium growth; [Angel Face X Old Master]; Greenwood, Chris, 1985; Armstrong Garden Centers, 1991

Gretta, *see* 'SPOGRET'

'GREVINDE ROSE DANNESKJOLD SAMSÖE', HT, dr, 1914; bud dark velvety red; flowers scarlet; Poulsen, D.T.

'GREVINDE SYLVIA KNUTH', LCl, w, 1913; bud yellow; flowers white, center yellow, small, borne in large clusters; Poulsen, D.T.

'GREY DAWN', F, m, 1975; flowers gray, reverse flushed pink and gold, dbl.,

45 petals, 3–4 in., moderate fragrance; foliage glossy; bushy growth; [Brownie X News]; LeGrice

Grey Lady Pink, HT, mp

'Grey Pearl', HT, m, 1945; (The Mouse); bud ovoid; flowers lavender-gray, shaded olive and tan, dbl., 43 petals, 4–4.5 in., exhibition form, moderate fragrance; foliage glossy; vigorous growth; [(Mrs Charles Lamplough X Seedling) X (Sir David Davis X Southport)]; McGredy; J&P

'Grey Pearl, Climbing', Cl HT, m, 1951; Caluya

'Gribaldo Nicola', T, w, 1891; flowers white with touches of yellow, very large, dbl., moderate fragrance; Soupert & Notting; (Sangerhausen)

Grido, F, 1972; Pironti, N.; (Cavriglia)

'GRIfed', HT, my, 1973; (**Edna Wilson**); flowers large, dbl., moderate fragrance; foliage glossy; upright, vigorous growth; [Beauté sport]; Griffiths, Trevor; T. Griffiths, Ltd.

'Grillodale', HT, lp, 1926; flowers light pink, center deeper, dbl., 50 petals, 4.5 in.; moderate fragrance; foliage dark; [Mme Butterfly sport]; Grillo

Grimaldi, *see* DELstror

Grimbeert, HT, 1995; Delforge, H.; (Cavriglia)

'Grimm', LCl, pb, 1932; flowers apple-blossom-pink, center white, stamens golden, edges fluted, single, non-recurrent; vigorous growth; [(Hiawatha X Altmarker) X (Mme Leon Pain X Marquise de Sinéty)]; Lambert, P.

Grimpant Alain, *see* **'Alain, Climbing'**

Grimpant All Gold®, *see* **'Allgold, Climbing'**

Grimpant Allgold®, *see* **'Allgold, Climbing'**

Grimpant Baronne de Rothschild, *see* 'MEIgrisosar'

Grimpant Bettina®, *see* 'MEpalsar'

Grimpant Carina®, *see* 'MEIchimsar'

Grimpant Chrysler Imperial, *see* **'Chrysler Imperial, Climbing'**

Grimpant Clair Matin®, *see* 'MEImont'

Grimpant Comtesse Vandal, *see* **'Comtesse Vandal, Climbing'**

Grimpant Crimson Glory, *see* **'Crimson Glory, Climbing'**

Grimpant Danse des Sylphes®, *see* 'MALcair'

Grimpant Delbard®, *see* 'DELpar'

Grimpant Diablotin, *see* 'DELposar'

Grimpant Double Delight®, *see* 'AROclidd'

Grimpant Ena Harkness, *see* **'Ena Harkness, Climbing'**

Grimpant Étoile de Hollande, *see* **'Étoile de Hollande, Climbing'**

Grimpant Exploit®, *see* 'MEIlider'

Grimpant Grand'mère Jenny, *see* **'Grand'-mère Jenny, Climbing'**

Grimpant Lilli Marleen®, *see* 'PEKlimasar'

Grimpant Marie-Claire, *see* **'Marie-Claire, Climbing'**

Grimpant Michèle Meilland, *see* **'Michèle Meilland, Climbing'**

Grimpant Mrs Herbert Stevens, *see* **'Mrs Herbert Stevens, Climbing'**

Grimpant Opera, *see* **'Opera, Climbing'**

Grimpant Orange Meillandina®, *see* 'MEIjikatarsar'

Grimpant Papa Meilland®, *see* 'MEIsarsar'

Grimpant Pierre de Ronsard®, *see* 'MEIviolin'

Grimpant Queen Elizabeth, *see* **'Queen Elizabeth, Climbing'**

Grimpant Reine des Neiges, *see* **'Frau Karl Druschki, Climbing'**

Grimpant Rose Gaujard®, *see* **'Rose Gaujard, Climbing'**

Grimpant Rouge et Or, *see* 'DICorsar'

Grimpant Sonia Meilland®, *see* 'MEIhelvetsar'

Grimpant Soraya®, *see* 'MEjenorsar'

Grimpant Spartan, *see* **'Spartan, Climbing'**

Grimpant Sutter's Gold, *see* **'Sutter's Gold, Climbing'**

Grimpant Tiffany, *see* **'Tiffany, Climbing'**

Grisbi, *see* **'Sunlight'**

Grisbi, *see* MEIgrelou

Grisbi, Climbing, *see* **'Sunlight, Climbing'**

'Griseldis', B, mp, 1895; dbl., flat, borne in clusters of 1–6; Geschwind, R.

Griseldis, F, w; Select Roses, B. V.

Grootendorst, *see* **'F. J. Grootendorst'**

Grootendorst Red, *see* **'F. J. Grootendorst'**

'Grootendorst Supreme', HRg, dr, 1936; flowers deeper crimson-red; [F. J. Grootendorst sport]; Grootendorst, F. J.

'Gros Choux d'Hollande', C, lp; (Big Cabbage of Holland); flowers soft rose-pink, dbl., intense fragrance; vigorous growth

'Gros Provins Panache', HGal, m; flowers violet-purple streaked white, dbl.

'Grossherzog Ernst Ludwig', Cl HT, mp, 1888; (Red Marechal Niel); flowers silvery carmine, very dbl., intense fragrance; Müller, Dr F.

'Grossherzog Friedrich von Baden', HT, mp, 1908; flowers medium, dbl.; intense fragrance; Lambert, P.; (Sangerhausen)

'Grossherzog Wilhelm Ernst von Sachsen', HT, dp, 1915; flowers large, dbl., intense fragrance; Welter; (Sangerhausen)

'Grossherzogin Eleonore von Hessen', HMult, m, 1907; flowers dark violet-red, dbl.; Strassheim; (Sangerhausen)

'Grossherzogin Feodora von Sachsen', HT, w, 1914; flowers creamy white, base deep yellow; [Frau Karl Druschki X Kaiserin Auguste Viktoria]; Kiese

Grossherzogin Josefine Schararolle, HT, lp, 1989; flowers medium-large, dbl.; Lens, Louis; (Sangerhausen)

Grossmütterchen, HCh, mp, 1983; flowers large, very dbl., intense fragrance; Weihrauch; (Sangerhausen)

Grouse, *see* 'KORimro'

Grugakind, Pol, op; flowers small, dbl.; (Sangerhausen)

'GRUlit', Min, rb, 1992; (**Littlest Spartan**); flowers medium red and yellow, blooms borne mostly singly, dbl., 22–30 petals, 1 in., no fragrance; some prickles; foliage medium, medium green, semi-glossy; low (20–30 cms), upright, bushy growth; [Red Ace X Unnamed seedling]; Gruenbauer, Richard; Flowers 'n' Friends Miniature Roses

'Grumpy', Pol, mp, 1956; (Burkhard, Burkhardt); flowers pink, small blooms in long trusses, dbl.; deRuiter; Gregory & Willicher Baumschulen

'Gruppenkönigin', F, pb, 1935; flowers deep bicolor pink, very large, dbl.; foliage light, leathery; vigorous, bushy growth; [Gruss an Aachen X Mme Edouard Herriot]; Kordes, H.

'Grüss an Aachen', F, lp, 1909; bud orange-red and yellow; flowers flesh-pink fading to creamy white, very dbl., 40–45 petals, 3–3.5 in., borne in clusters, slight, sweet fragrance; foliage rich green, leathery; slender stems, will nod; dwarf growth; (21); [Frau Karl Druschki X Franz Deegen]; Geduldig

'Grüss an Aachen, Climbing', Cl F, lp, 1937; Kordes

Gruss an Aachen, White, F, w

Gruss an Angeln®, HT, mr, 1986; Clausen

Grüss an Bayern®, *see* 'KORmun'

'Grüss an Berlin', HT, mr, 1963; (Greetings); bud ovoid; flowers pure red, dbl., 40 petals, 5.5 in., exhibition form, slight fragrance; foliage dark, glossy; vigorous, upright, bushy growth; Kordes, R.

'Grüss an Breinegg', HMult, dp, 1925; flowers small, single; Bruder Alfons (Sangerhausen)

'Grüss an Coburg', HT, ab, 1927; flowers apricot-yellow, reverse coppery pink, globular, intense fragrance; foliage bronze; vigorous growth; [Alice Kaempff X Souv. de Claudius Pernet]; Felberg-Leclerc

'GRÜSS AN DRESDEN', HT, mr, 1913; flowers fiery red; [Princesse de Bearn seedling]; Türke; Hoyer & Klemm

'GRÜSS AN FÖHR', HFt, yb, 1930; flowers medium, semi-dbl.; Riewers; (Sangerhausen)

'GRÜSS AN FREUNDORF', HWich, dr, 1913; flowers dark velvety crimson, center whitish, stamens bright yellow, semi-dbl.; [R. wichurana hybrid X Crimson Rambler]; Praskac; Teschendorff

'GRÜSS AN FRIEDBERG', N, dy, 1902; flowers medium, dbl., moderate fragrance; Rogmanns; (Sangerhausen)

'GRÜSS AN GERMERSHAUSEN', HMult, mr, 1926; flowers small, single; Bruder Alfons; (Sangerhausen)

'GRÜSS AN HANNOVER', LCl, op, 1938; flowers large, dbl.; vigorous, upright (10–13 ft) growth; Lahmann

Grüss an Heidelberg, *see* 'KORBE'

'GRÜSS AN KOBLENZ', LCl, mr, 1963; flowers bright scarlet, blooms in clusters (up to 10), 20 petals, 3 in., recurrent bloom, slight fragrance; vigorous growth; Kordes, R.

Grüss an Lorrach, F, 1983; Hetzel, K.; (Cavriglia)

Gruss an Maiengrun, F, or

Grüss an Munchen, F, 1962; Tantau, Math.; (Cavriglia)

'GRÜSS AN NAUMBURG', HT, rb, 1928; flowers medium, dbl., intense fragrance; Muller, J.F.; (Sangerhausen)

Gruss an Oldenburg, S, yb, 1995; Weihrauch

'GRÜSS AN RENGSDORF', S, mp, 1920; flowers medium; Boden; (Sangerhausen)

'GRÜSS AN SANGERHAUSEN', HT, mr, 1904; flowers large, dbl., moderate fragrance; Müller, Dr F.; (Sangerhausen)

'GRÜSS AN STEINFURTH', F, my, 1961; flowers open, dbl., 21 petals, cupped, borne in large clusters (to 15); moderate growth; [Goldilocks X Masquerade]; Leenders, J.

'GRÜSS AN STUTTGART', F, mr, 1976; bud ovoid; flowers velvety red, medium, dbl., slight fragrance; vigorous, bushy growth; [(Carina X Seedling) X Sans Souci]; Hetzel

'GRÜSS AN TEPLITZ', HCh, mr, 1897; (Virginia R. Coxe); bud small, ovoid; flowers light crimson, edges sometimes marked dark garnet, dbl., 33 petals, recurrent bloom, intense, spicy fragrance; foliage dark, young growth bronze-red; short, weak stems; vigorous (6 ft), bushy growth; good for hedges; (28); [((Sir Joseph Paxton X Fellenberg) X Papa Gontier) X Gloire des Rosomanes]; Geschwind, R.; P. Lambert

'GRÜSS AN TEPLITZ, CLIMBING', HCh, mr, 1911; (Virginia R. Coxe, Climbing); Storrs & Harrison Co.

'GRÜSS AN WEIMAR', HP, pb, 1919; flowers pink on yellowish ground; [Frau Karl Druschki X Lyon Rose]; Kiese

Gruss an Worishofen, *see* **'BAD WÖRISHOFEN'**®

'GRÜSS AN ZABERN', HMult, w; [Euphrosine X Mme Ocker Ferencz]; Lambert, P., 1903

'GRÜSS AN ZWEIBRÜCKEN', HT, mr, 1915; flowers large, dbl., moderate fragrance; Lambert, P.; (Sangerhausen)

Grüss aus Alma-aty, HT, 1958; Sushkov, K. L.; (Cavriglia)

'GRÜSS VOM WESTERWALD', HT, op, 1914; flowers medium, dbl., moderate fragrance; Kettenbeil; (Sangerhausen)

Guadalajara, *see* 'MACDEEPO'

Guadalupe's Love, HT, w, 1997; flowers full, ruffled, medium, very dbl., 26–40 petals, borne mostly singly, moderate fragrance; foliage medium, medium green, semi-glossy; upright, medium (2.5 to 3 ft) growth; [Bewitched sport]; Price, Kathleen M.

'GUARDSMAN', HT, or, 1937; flowers bright scarlet, base yellow, large, dbl., slight fragrance; foliage glossy; vigorous, compact growth; [Unnamed seedling X Shot Silk]; Archer

Gudhemsrosen, A, w

Guernsey Gold, *see* 'TROBGUERN'

Guernsey Love, *see* 'TROBLOVE'

'GUERREIRO', Cl F, dr; [Unnamed seedling X Alain]; da Silva, Moreira

'GUESCAN', HT, yb, 1997; (**Cerys Ann**); flowers full, medium, very dbl., 26–40 petals, borne mostly singly, slight fragrance; some prickles; foliage large, dark green, glossy; bushy, medium (30 in.) growth; [Fulton Mackay X Freedom]; Guest, M.M.

'GUESCOLOUR', F, yb, 1999; (**Brushstrokes**); flowers yellow and red striped, reverse paler, dbl., 26–40 petals, 3.5 in., borne in small clusters, slight fragrance; few prickles; foliage medium, dark green, glossy; upright, medium (28 in.) growth; [seedling X Solitaire]; Guest, M.M.; A. J. Palmer & Son, 1999

'GUESDELAY', HT, rb, 1997; (**Birthday Wishes**); flowers full, large, very dbl., 26–40 petals, borne mostly singly, intense fragrance; some prickles; foliage large, dark green, glossy; upright, medium (30 in.) growth; [Fragrant Cloud X Honey Favourite]; Guest, M.M.

GUESGENUS, HT, ly, 1997; (**Ted Allen**); flowers double, medium, dbl., 15–25 petals, borne mostly singly, slight fragrance; some prickles; foliage medium, medium green, dull; upright, medium (28 in.) growth; [Manx Queen X (Manx Queen X Rosa Bella)]; Guest, M.M.

'GUESTALL', Cl HT, ob, 1994; (**Just Reward**); flowers orange, blooms borne mostly single, dbl., 15–25 petals, 3–3.5 in., moderate fragrance; some prickles; foliage medium, medium green, glossy; tall (6–8 ft), upright growth; [Basildon Bond X Alexander]; Guest, M.M.; F. Haynes & Partners, 1994

'GUGLIELMO MARCONI', HT, w, 1934; flowers almost white, tinted flesh; [Ophelia X Elisabeth Faurax]; Giacomasso

Guglielmo Marconi, *see* HARbleep

Guiding Spirit, *see* 'HARWOLAVE'

Guildfordian, HT, w

'GUILLAUME KAEMPFF', HT, dr, 1931; flowers dark crimson-red, edged blackish, large, dbl., intense fragrance; foliage thick; vigorous growth; [Hadley X Admiral Ward]; Felberg-Leclerc

'GUINEA GOLD', HT, ab, 1945; bud long, pointed, buff-yellow; flowers apricot-yellow, open, dbl.; foliage dark, leathery; strong stems; vigorous, upright, much branched growth; [Joanna Hill X Golden Rapture]; Hill, Joseph H., Co.

'GUINÉE', Cl HT, dr, 1938; bud pointed; flowers blackish garnet, sometimes mottled scarlet, large, dbl., intense fragrance; foliage leathery; height $6\frac{1}{2}$–9 ft; [Souv. de Claudius Denoyel X Ami Quinard]; Mallerin, C.; A. Meilland; C-P

'GUINEVERE', HT, mp, 1967; dbl., 40 petals, 4.5 in., slight fragrance; foliage glossy; GM, Baden-Baden; [Red Dandy X Peace]; Harkness

'GUINGUETTE', F, pb, 1958; flowers pink edged darker, well formed, cupped; vigorous growth; [Alain X Feu de Joie]; Gaujard, R.; Hémeray-Aubert

Guirlande D'Amour, LCl, w, 1993; Lens

'GUIRLANDE FLEURIE', LCl, mr, 1968; bud ovoid; flowers bright red, large, borne in clusters, semi-dbl., cupped, slight fragrance; foliage leathery; very vigorous, climbing growth; [Valenciennes X Paul's Scarlet Climber]; Robichon; Ilgenfritz Nursery

'GUISEPPE MOTTA', HT, 1936; flowers flesh-pink, reverse red and yellow, large, semi-dbl., intense fragrance; vigorous growth; Heizmann, E.

Guitare®, *see* 'GAEGUL'

Guitare Cl., Cl F, ob, 1974; Kasturi

'GULAB-E-PAL', HT, m, 1985; flowers mauve, blended with yellow, blooms borne singly, dbl., 70 petals, exhibition form, moderate fragrance; brown prickles; foliage medium, light green; upright growth; [Festival Beauty X (Scarlet Knight X Festival Beauty)]; Hardikar, Dr M.N.

'GULDTOP', HSpn, dy; [R. Spinosissima seedling]

Gulf Breeze, *see* 'TALGUL'

Gulgong Gold, HT, yb

Gulletta, *see* 'RUGUL'

'GULLIVER'S GLOW', S, mr, 1954; flowers bright red, small, borne in very large clusters, dbl.; thornless; bushy growth; very hardy.; [Hiawatha X (R. maximowicziana pilosa X Tausendschon)]; Gulliver; Shenandoah Nursery

'GULNARE', HT, my, 1918; flowers golden yellow; Poulsen, D. T.

Gulzar, HT, dr, 1971; IARI

Gumdrop, *see* 'JACGUM'

'GUNDY', F, dp, 1966; bud ovoid; flowers deep rose-pink, borne in clusters, semi-dbl., cupped, slight fragrance; foliage dark; vigorous, upright growth; Schloen, P.; Ellesmere Nursery

'GUNSEI', Cl F, pb, 1986; flowers white flushed pink on fringe, borne large number in clusters, dbl., 13–15 petals, cupped, slight fragrance; no prickles; foliage 7 leaflet, green; vigorous, upright growth; [Unnamed seedling X Summer Snow]; Kikuchi, Rikichi

'GUNSTON HALL', HT, mr, 1929; flowers scarlet-crimson; [Seedling X Hoosier Beauty]; U.S. Dept. of Agric.; C-P

'GURNEY BENHAM', HT, my, 1935; flowers buttercup-yellow, large, dbl., cupped; foliage glossy, bronze; vigorous, bushy growth; [Lady Forteviot sport]; Cant, B. R.

'GURNEY HILL', HT, mr, 1924; flowers pure red, dbl., moderate fragrance; Hill, E.G., Co.

'GUSSIE', Min, pb, 1979; bud ovoid; flowers medium red, reverse pale pink and silver, blooms borne singly, very dbl., 90 petals, exhibition form, slight fragrance; few prickles; foliage green, leathery; bushy, dwarf growth; [Unnamed seedling X Unnamed seedling]; Lorenzen, Frederick

'GUSTAV FRAHM', F, mr, 1959; flowers crimson-scarlet, blooms in large clusters, dbl., 25 petals, 3 in., flat, slight fragrance; foliage light, glossy; vigorous, upright growth; [Fanal X Ama]; Kordes; Timm

'GUSTAV GRÜNERWALD', HT, pb, 1903; flowers carmine-pink, center yellow; (28); [Safrano X Mme Caroline Testout]; Lambert, P.

Gustave Courbet®, *see* SAUbord

'GUSTAVE PIGANEAU', HP, mr, 1889; flowers bright carmine, very large, dbl., cupped; growth moderate; Pernet-Ducher

'GUSTAVE RÈGIS', HT, ly, 1890; flowers creamy yellow, large, semi-dbl., moderate fragrance; Pernet-Ducher; (Sangerhausen)

'GUSTAVE THIERRY', HP, 1881; Oger; (Cavriglia)

'GUSTEL LÖBNER', HT, w, 1927; flowers large, dbl.; Löbner; (Sangerhausen)

'GUSTEL MAYER', Pol, mr, 1909; flowers small, dbl.; Lambert, P.; (Sangerhausen)

Gutersloh 85®, S, mr

Guy de Maupassant™, *see* 'MEISOCRAT'

'GUY FAWKES', F, yb, 1975; flowers yellow center, reverse shading scarlet, full, semi-dbl., 15 petals, 3.5 in., intense fragrance; foliage glossy; [My Choice X Masquerade]; Cadle's Roses

Guy Laroche, *see* 'DELRICOS'

'GUY-GUY', F, ab, 1966; flowers apricot and pink, edged crimson, small, dbl., moderate fragrance; foliage dark, glossy, leathery; low, compact growth; [(Circus X Circus) X (Circus X Circus)]; Fankhauser

Guyscliffe, *see* 'LINDLIFFE'

Gwen Fagan, *see* POUlgewfa

'GWEN MARIE', F, rb, 1964; flowers dark red, center light cream, open, large, semi-dbl., intense fragrance; foliage soft; tall growth; [Dainty Bess seedling]; Robins

Gwen Mayor, *see* 'COCover'

'GWEN NASH', Cl HT, pb, 1920; flowers rich pink, center white, large, semi-dbl., cupped, slight fragrance; foliage glaucous, wrinkled; vigorous, climbing growth; [Rosy Morn X ?]; Clark, A.; NRS New South Wales

Gwen Swane, *see* 'MACWHAKA'

'GWENDOLINE COLLINS', HT, mr, 1937; flowers cerise shaded cherry, large, dbl.; globular, vigorous, bushy growth; Clark, A.; NRS Victoria

Gwent, *see* 'POULURT'

'GWYNETH', Pol, ly, 1923; flowers pale yellow, tinted lemon, changing to nearly white, open, semi-dbl., moderate, musk fragrance; foliage light, leathery; bushy growth; (21); [(Trier X Rayon d'Or) sport X (Gottfried Keller X Entente Cordiale) sport]; Woosman; Easlea

'GWYNETH', HT, ly, 1928; flowers canary-yellow, without shading; [Willowmere X Mrs Wemyss Quin]; Chaplin Bros.

'GWYNETH JONES', HT, op, 1925; bud pointed; flowers brilliant carmine-orange, open, semi-dbl., slight fragrance; foliage light, leathery; vigorous, bushy growth; GM, NRS, 1925; McGredy

'GWYNNE CARR', HT, lp, 1924; flowers silvery pink shaded lilac-rose, dbl.; Dickson, A.

'GWYNNE CARR, CLIMBING', Cl HT, lp, 1934; Easlea

'GYLDENORANGE', F, ob, 1952; flowers golden orange, fades to light yellow, large, dbl., no fragrance; foliage medium, medium green, semi-glossy; medium growth; [Poulsen's Yellow X Unnamed seedling]; Poulsen, S.; Poulsen

'GYMPIE', F, pb, 1953; flowers white and pink, borne in clusters, very dbl., moderate fragrance; foliage light green; vigorous, bushy growth; [Yvonne Rabier X Tip-Top]; Ulrick, L.W.

'GYMPIE BEAUTY', F, dr, 1962; flowers deep red, borne in clusters; good growth; Dunstan; Langbecker

'GYPSY', HT, or, 1972; bud ovoid; flowers fiery orange-red, large, dbl., slight fragrance; foliage large, glossy, leathery; vigorous, upright, bushy growth; AARS, 1973; [((Happiness X Chrysler Imperial) X (El Capitan) X Comanche]; Swim & Weeks; C-P

Gypsy, *see* **'KIBOH'**

Gypsy Bride, Gr, w

Gypsy Carnival, *see* **'KIBOH'**

Gypsy Dancer, *see* 'DICQUIET'

Gypsy Fire, *see* 'MORGLO'

'GYPSY JEWEL', Min, dp, 1975; flowers deep rose-pink, small, dbl., 50 petals, 1.5 in., exhibition form; foliage dark, leathery; vigorous growth; [Little Darling X Little Buckaroo]; Moore, Ralph S.; Park Seed Co.

Gypsy Jubilee, Gr, yb

Gypsy Lady, Gr, mp, 1995

Gypsy Lass, *see* **'GIPSY LASS'**

'GYPSY MOTH', F, op, 1968; flowers salmon, blooms in clusters, dbl., 35 petals, exhibition form; foliage glossy; Tantau, Math.

'GYPSY QUEEN', HMult, mr, 1929; flowers crimson, small, borne in clusters, dbl., vigorous, climbing growth; [Crimson Rambler X ?]; Moore, Ralph S.

Gypsy Song, HT, pb

Gypsy Sunblaze™, *see* 'MEIMAGUL'

'GYPSY'S WINE CUP', F, mr, 1968; flowers deep crimson, borne in trusses, slight fragrance; foliage dark; low, bushy growth; [Highlight X ?]; Austin, David

'GYRENE', S, mr, 1987; flowers bright, medium red, large, dbl., 15–25 petals, repeat bloom, moderate fragrance; foliage medium, medium green, matt, disease resistant; upright, bushy, branching, vigorous, hardy growth; [Arctic Glow X ((Pink Hat X R. arkansana) X R. arkansana)]; James, John; Historical Roses, 1987

'GZLATY DECH', HT, ob, 1936; bud long, pointed; flowers orange-yellow, brown, red and gold shadings, large, semi-dbl., cupped, moderate fragrance; foliage glossy, light; vigorous growth; [Admiration X Talisman]; Böhm, J.

H

H. C. Andersen®, *see* 'POULANDER'

'**H. C. VALETON**', HT, yb, 1926; flowers golden yellow overspread with rose, large, moderate fragrance; strong stems; vigorous growth; [Golden Ophelia X Aspirant Marcel Rouyer]; Verschuren

'**H. C. YOUNG**', HT, op, 1934; bud pointed; flowers shrimp-pink, deepening to salmon, base yellow, strong stems; vigorous growth; Austin & McAslan

'**H. CHAUBERT**', HT, op, 1928; flowers coppery salmon, open, borne in clusters, semi-dbl., slight fragrance; foliage rich green, glossy; bushy growth; [Mrs Aaron Ward X Unnamed seedling]; Barbier

'**H. D. M. BARTON**', HT, dr, 1917; flowers deep velvety crimson, large, dbl., moderate fragrance; bushy growth; Dickson, A.

'**H. E. RICHARDSON**', HT, mr, 1913; flowers dazzling crimson, dbl., exhibition form; vigorous growth; GM, NRS, 1912; Dickson, A.

'**H. F. ALEXANDER**', LCl, ab, 1952; flowers very large, dbl., 30–40 petals, foliage glossy; very long stems; tall, climbing growth; [Duquesa de Peñaranda X Ruth Alexander]; Wilber; Buckley Nursery Co.

'**H. F. EILERS**', HT, mr, 1914; flowers large, dbl., moderate fragrance; Lambert, P.; (Sangerhausen)

'**H. V. MACHIN**', HT, dr, 1914; flowers very dark scarlet-crimson, very large, dbl., globular, slight fragrance; foliage glaucous beech-green; GM, NRS, 1912; Dickson, A.

'**H. V. MACHIN, CLIMBING**', Cl HT, dr, 1919; Dickson, H.

'**H. V. MACHIN, CLIMBING**', Cl HT, dr, 1922; H&S

H. G. Hastings, *see* '**HARRY G. HASTINGS**'

H. P. Pinkerton, *see* '**CLEVELAND**'

H. R. G. Unknown Pink, HP, dp

'**HAAKSBERGEN**', F, my, 1961; flowers bright yellow, medium, borne in clusters, semi-dbl., foliage dark; moderate growth; [Mrs Pierre S. duPont X King Boreas]; Buisman, G. A. H.

'**HABANERA**', S, dr, 1976; bud ovoid, pointed; flowers dark cardinal-red edged lighter, shallow-cupped, dbl., 33 petals, 4–4.5 in.; cupped, repeat bloom, slight fragrance; foliage leathery; upright, bushy growth; [(Vera Dalton X Dornroschen) X ((World's Fair X Floradora) X Applejack)]; Buck, Dr Griffith J.; Iowa State University

'HABONE', HT, pb, 1985; (**Elizabeth Harbour**); bud pointed; flowers light pink, dark pink reverse, well-formed, blooms borne singly, dbl., 30 petals, 3 in., slight fragrance; foliage medium, medium green, semi-glossy; upright (3 ft), bushy growth; [Elizabeth of Glamis X HT seedling (dark red)]; Harbour, E.R.

'**HACKEBURG**', HMult, pb, 1912; flowers soft lilac-pink, center white, borne in clusters, vigorous, climbing growth; Kiese

'HADANGEL', HT, ab, 1986; (**Smooth Angel**); flowers petals apricot blending to cream at edges, large blooms, dbl., 36 petals, cupped, no fruit; no prickles; foliage medium, medium green, matt; medium, bushy, spreading growth; PP006146; [Smooth Sailing X Royal Flush]; Davidson, Harvey D.; Gurney Seed

'HADANGO', HT, ob, 1996; (**Olde Tango**™); flowers orange, old fashioned, blooms borne mostly singly, very dbl., 3.25–3.5 in., flat, moderate, pine fragrance; moderate prickles; foliage medium, medium green, glossy; bushy (80–90 cms) growth; [Shining Ruby X The World]; Davidson, Harvey D.; L E Cooke Co., 1996

'HADCORAL', HT, pb, 1992; (**Shining Coral**); flowers coral pink, blooms borne in sprays of 1–3, dbl., 22–24 petals, 4 in., cupped, moderate, fruity fragrance; foliage large, dark green, very glossy; medium, bushy growth; [Shining Ruby X [Honey Favorite X (Little Darling X Traviata)]]; Davidson, Harvey D.; C & L Valley Rose Co., 1992

'**HADDEN'S VARIETY**', S, m, 1948; flowers rosy purple, borne several together, single, non-recurrent, foliage small, gray-green; Hilling

Haddington, HGal, m; flowers dark purple/pink, small to medium, semi-dbl.; (Sangerhausen)

'HADEMP', HT, mp, 1991; (**Pink Empress**); flowers clean medium pink, urn-shaped, large blooms borne usually singly, dbl., 32–36 petals, exhibition form, moderate, fruity fragrance; foliage medium, medium green, glossy; bushy, medium growth; [Smooth Sailing X Medallion]; Davidson, Harvey D., 1984; Hortico Roses, 1991

'HADEVENING', HT, m, 1995; (**Evening Queen**); flowers slight red on edge of petals with aging, dbl., 26–40 petals, 3–3.5 in., borne in small clusters, intense fragrance; foliage medium, medium green, matt; medium (3 ft), bushy growth; [Blue Ribbon X Great News]; Davidson, Harvey D.; Hortico Roses, 1994

'HADFLARE', HT, or, 1992; (**Shining Flare**); dbl., 24–26 petals, 4.5 in.; cupped, borne in sprays, slight, fruity fragrance; foliage medium, dark green, very glossy; bushy, medium growth; [Shining Ruby X (Smooth Sailing X Futura)]; Davidson, Harvey D.; C & L Valley Rose Co., 1992

'HADLACE', HT, ly, 1993; (**Olde Lace**); very dbl., 3–3.5 in., moderate fragrance; some prickles; foliage large, dark green, glossy; medium (3–4 ft), bushy growth; [Pink Favorite X (Polly X Peace)]; Davidson, Harvey D.; C & L Valley Rose Co., 1994

'HADLADY', HT, mp, 1986; (**Smooth Lady**); flowers urn-shaped to loose, dbl., 21 petals, blooms in clusters of 1–3, moderate, spicy fragrance; medium, globular, orange fruit; no prickles; foliage large, medium green, very glossy; tall, upright, bushy growth; PP006147; [Smooth Sailing X ((Polly X Peace) X Circus)]; Davidson, Harvey D.; Gurney Seed

'**HADLEY**', HT, mr, 1914; flowers rich crimson, well-formed, very large, dbl., intense fragrance; foliage rich green; vigorous growth; [(Liberty X Richmond) X Gen. MacArthur]; Montgomery Co.; A.N. Pierson

'**HADLEY ELATIOR**', HT, mr, 1927; [Hadley sport]; Teschendorff

'**HADLEY, CLIMBING**', Cl HT, mr, 1927; Teschendorff

'HADMELODY', F, rb, 1990; (**Smooth Melody**); bud ovoid; flowers red to white center, white reverse with red on outer edge, a, dbl., 26 petals, intense, fruity fragrance; round, rarely sets seed; thornless; foliage medium, dark green, semi-glossy; PP007729; [Royal Flush X Smooth Lady]; Davidson, Harvey D., 1979; Hortico & C&L Valley Rose Co., 1990

'HADPERFUME', HT, lp, 1990; (**Smooth Perfume**); bud pointed; flowers light pink to very light mauve on edge, urn-shaped, large, b, dbl., 28–30 petals, exhibition form, intense fragrance; thornless; foliage medium, medium green, semi-glossy; bushy, medium growth; PP007728; [(Smooth Sailing X Medallion) X Blue Moon]; Davidson, Harvey D., 1979; Hortico & C&L Valley Rose Co., 1990

'HADPLEASURE', HT, mp, 1995; (**Pink Treasure**); dbl., 26–40 petals, 4.5 in., borne mostly singly, moderate fragrance; some prickles; foliage large, medium green, semi-glossy; upright, medium (110 cms) growth; [Smooth Sailing X Red Planet]; Davidson, Harvey D.; Hortico Roses, 1995

'HADPRETTIE', HT, dr, 1995; (**Velvet Lady**); flowers very dark, velvet red, single (5 petals), borne in small clusters; fragrant; numerous prickles; foliage medium, dark green, very glossy; spreading, medium growth; [Shining Ruby X Precious Platinum]; Davidson, Harvey; Coiner Nursery, 1995–96

'HADPRINCE', HT, mr, 1990; (**Smooth Prince**); bud ovoid; flowers urn-shaped, large, borne usually singly, dbl., 26–28 petals, exhibition form, slight, fruity fragrance; oblong, rarely sets seed. fruit; thornless; foliage medium, medium green, semi-glossy; [Smooth Sailing X Old Smoothie]; Davidson, Harvey D., 1979; Hortico & C&L Valley Rose Co., 1990

'HADROMANCE', HT, w, 1992; (**Smooth Romance**™); flowers cream with tinge of pink in center, urn shaped, blooms borne, dbl., 42–45 petals, 4.75 in., exhibition form, fast repeat, moderate fragrance; will not set seed; thornless; foliage medium, medium green, glossy; straight, upright stems; upright, bushy, tall growth; [Smooth Sailing X Portrait]; Davidson, Harvey D., 1984; Hortico Roses, 1991

'HADROMEO', HT, dr, 1993; (**Olde Romeo**); dbl., 26–40 petals, 3–3.5 in., borne mostly singly, intense fragrance; many prickles; foliage medium, medium green, matt; medium (4 ft), upright growth; [Smooth Sailing X (Old Smoothie X ((Polly X Peace) X Simon Bolivar))]; Davidson, Harvey D.; C & L Valley Rose Co., 1994

'HADRUBY', HT, mr, 1992; (**Shining Ruby**); flowers medium red, fading to blue, blooms borne in sprays, dbl., 24–26 petals, 5.5 in., cupped, moderate, spicy fragrance; foliage large, dark green, very glossy; bushy, medium growth; [Pink Favorite X Simon Bolivar]; Davidson, Harvey D.; C & L Valley Rose Co., 1992

'HADSATIN', HT, mp, 1993; (**Smooth Satin**); dbl., 26–40 petals, 3–3.5 in., borne mostly singly, moderate fragrance; 95–100% thornless; foliage large, medium green, glossy; medium (1 m), upright, bushy growth; [Smooth Lady X Smooth Sailing]; Davidson, Harvey D.; C & L Valley Rose Co.

'HADSUN', HT, ab, 1989; (**Western Sunlight**); bud pointed; flowers apricot-orange, aging to yellow, large, dbl., 31 petals, exhibition form, borne usually singly, slight, fruity fragrance; inverted, does not set seed readily, orange fruit; prickles hooked downward, light brown; foliage medium, dark green, glossy, serrated; upright, medium growth; PP007442; [[(Honey Favorite X Irish Mist) X (San Francisco X Prima Ballerina)] X Just Joey]; Davidson, Harvey D.; Hortico Roses, 1990

'HADSWEE', HT, lp, 1996; (**Olde Sweetheart**™); flowers light pink, with large, raspberry colored stamen, single, 4–7 petals, 3.5 in., slight fragrance; moderate prickles; foliage medium, medium green, glossy; upright, medium (100 cms) growth; [Smooth Sailing X Red Planet]; Davidson, Harvey D.; L E Cooke Co., 1997

'HADTREASURE', HT, w, 1995; (**White Treasure**); flowers white, slight cast of light yellow in center, dbl., 26–40 petals, 4.75 in., borne mostly singly, moderate fragrance; few prickles; foliage large, dark green, semi-glossy; upright, medium (100 cms) growth; [Smooth Sailing X Sunset Jubilee]; Davidson, Harvey D.; Hortico Roses, 1995

'HADVELVET', HT, dr, 1986; (**Smooth Velvet**); flowers large blooms, dbl., 42 petals, cupped, borne usually singly, slight, damask fragrance; medium, globular, orange fruit; no prickles; foliage light green, matt; tall, upright growth; PP006152; [(Smooth Sailing X ((Polly X Peace) X Circus)) X Red Devil]; Davidson, Harvey D.; Gurney Seed

Haendel®, *see* 'MACHA'

Hafiz®, HT, mp

'**HAGOROMO**', LCl, op, 1970; bud ovoid; flowers silvery coral-pink, large, dbl., exhibition form, free, intermittent bloom, moderate fragrance; foliage dark, leathery; vigorous, climbing growth; [Aztec seedling X New Dawn seedling]; Suzuki, Seizo; Keisei Rose Nursery

'**HAIDEE**', S, pb, 1953; flowers clear pink, center cream, large, dbl., cupped, nonrecurrent, large, dark red fruit; sometimes very prickly; foliage small, dark; wood red height 6 ft; [R. laxa X R. spinosissima seedling]; Skinner

Haiku™, *see* 'AROYEFELT'

'**HAILEYBURY**', HP, mr, 1896; flowers large, very dbl., moderate fragrance; Paul, G.; (Sangerhausen)

Hailstorm™, *see* 'COISTOR'

Haircutter's Pink Climber, LCl, mp; (found rose)

'**HAÏSHA**', HT, yb, 1947; flowers gold, edges suffused carmine, dbl., 60 petals, 6 in., intense fragrance; foliage leathery, glossy, dark; upright growth; [Peace X Fantastique]; Meilland, F.

Haitian Belle, LCl, mp

'**HAKATA KANOKO**', Min, rb, 1999; flowers medium red with white eye, 5 petals, 3.75 in., borne in large clusters, no fragrance; foliage small, medium green, disease-resistant; bushy, very compact (6–10 in.) growth; Yamazaki, Kazuko, 1979; Takii & Co., Ltd, 1995

'**HAKKODA**', HT, pb, 1983; flowers white, pink petal edges, large, dbl., 40 petals, exhibition form, intense fragrance; broad prickles, curved downward; foliage medium green, semi-glossy; bushy growth; [Lady X X Izayoi]; Kodoya, Y.; Kogura Rose Nursery, 1986

'**HAKUHOH**', HT, w, 1999; bud pale cream, turning white; dbl., 40 petals, 6 in.; exhibition form, moderate fragrance; foliage medium green; 4.5 ft growth; [White Prince X Bridal Robe]; Hayashi, Shunzo, 1989

'**HAKUSYU**', HT, w, 1999; flowers white, center coral pink, dbl., 26–40 petals, 5.5–6 in.; some prickles; foliage medium, medium green; upright, bushy, medium (4.5–5 ft) growth; Bronze Medal, Japan Rose Concours, 1998; [Sizunomai X Hoshizukuyo]; Ohtsuki, Hironaka

'**HAKUUN**', F, w, 1962; (White Cloud); bud small; flowers creamy white, patio, semi-dbl., 15 petals, 2 in., slight fragrance; foliage light green; low, compact, bushy growth; [Seedling X (Pinocchio X Pinocchio)]; Poulsen, Niels D.; Poulsen

'**HALALI**', F, dp, 1956; flowers deep pink, large blooms in clusters, semi-dbl.; foliage leathery, dense; vigorous (5–6 ft), spreading growth; [Marchenland X Peace]; Tantau, Math.

Halarious, HT, lp; Clark, A., 1935

Haleakala, *see* 'MANHALE'

'**HALF TIME**', HT, rb, 1976; bud pointed; flowers cherry-red, reverse yellow, dbl., 40 petals, 3.5–4 in., moderate, tea fragrance; foliage dark; upright growth; [((Fandango X Roundelay) X (Happiness X Tiffany)) X Peace]; Weeks; Weeks Wholesale Rose Growers

'**HALKA**', HT, lp, 1987; flowers white blush pink, reverse white-silver, large, borne in sprays, dbl., 30 petals, exhibition form, moderate, damask fragrance; straight, medium, red prickles; foliage medium, medium green, glossy; upright, medium growth; [Red Queen X Peace]; Bracegirdle, Derek T., 1988

Hall of Flowers, *see* 'MORMINT'

Hallandsasen, HRg

Hallelujah, *see* 'DELATUR'

'**HALLEY'S COMET**', F, mr, 1986; flowers medium, dbl., 20 petals, slight fragrance; foliage medium, medium green, semi-glossy; [Tip Top sport]; Rearsby Roses, Ltd.

'**HALLMARK**', HT, mr, 1966; bud ovoid; flowers large, dbl., 28 petals, cupped, moderate fragrance; foliage glossy; [Independence X Chrysler Imperial]; Morey, Dr Dennison; J&P

'**HALLOWEEN**', HT, yb, 1962; flowers deep yellow, tipped scarlet, large, dbl., 65 petals, intense fragrance; foliage glossy, dark, leathery; vigorous, upright growth; [(Peace X Fred Howard) X Unnamed seedling]; Howard, A. P.; Great Western Rose Co.

'**HALO**', HT, w, 1956; bud ovoid, seafoam-green; dbl., 25 petals, 4.5–5 in., exhibition form, moderate fragrance; foliage leathery; vigorous, upright growth; [Lady Sylvia X (Virgo X White Briarcliff)]; Lens; J&P

Halo Dolly, *see* 'MORWATEYE'

Halo® **Fire**, *see* 'MORHALFIRE'

Halo Karol, *see* 'MOREYES'

Halo® **Rainbow**, *see* 'MORRAINBOW'

Halo Star, *see* 'MORANYFACE'

Halo Suise, *see* MORsuise

Halo® **Sunrise**, *see* 'MORSUNRISE'

Halo® Today, *see* 'MORTODAY'

'**HAMBURG**', S, dr, 1935; bud pointed; flowers glowing crimson, very large blooms in clusters, semi-dbl., recurrent bloom, slight fragrance; foliage large, leathery, glossy; vigorous growth; (28); [Eva X Daily Mail Scented Rose]; Kordes

Hamburger Deern, *see* KORpalud

'**HAMBURGER PHOENIX**', HKor, mr, 1954; (Hamburger Phonix®); bud long, pointed; flowers rich red, large blooms in clusters, repeat bloom, slight fragrance; large, orange-red fruit; foliage dark, glossy; vigorous, climbing or trailer growth; [R. kordesii X Seedling]; Kordes

Hamburger Phonix®, *see* '**HAMBURGER PHOENIX**'

'**HAMBURG'S LOVE**', F, dy, 1974; dbl., 28 petals, 3 in., intense fragrance; foliage glossy; compact growth; [Fragrant Cloud X Manx Queen]; Timmerman's Roses

Hamish, *see* 'SIMHAHA'

Hammerberg, S, m

Hampshire, *see* KORhamp

Hampton Palace, *see* POUlgret

Hanabusa, *see* '**HANA-BUSA**'

'**HANA-BUSA**', F, or, 1981; (Hanabusa); bud ovoid; semi-dbl., 18 petals, flat, borne 6–10 per cluster, straight prickles; bushy growth; [Sarabande X (Rumba X Olympic Torch)]; Suzuki, Seizo; Keisei Rose Nursery

Hanae Mory, HT, 1990; Delbard, Georges; (Cavriglia)

'**HANAGASA**', F, or, 1978; bud globular; flowers vermilion, dbl., 23 petals, 4–4.5 in., cupped, moderate fragrance; foliage large, light green; vigorous growth; [(Hawaii X Unnamed seedling) X Miss Ireland]; Suzuki, Seizo; Keisei Rose Nursery, 1979

Hanagasumi, *see* '**HANA-GASUMI**'

'**HANA-GASUMI**', F, w, 1985; (Hanagasumi); flowers soft white, aging pink, blooms borne 6–12 per cluster, semi-dbl., 13 petals, flat, moderate fragrance; small, hooked prickles, slanted downward; foliage dark, semi-glossy; bushy growth; [Europeana X (Myo-joh X Fidélio)]; Suzuki, Seizo; Keisei Rose Nursery, 1984

'**HANAGURUMA**', HT, yb, 1977; bud globular; dbl., 58 petals, 6.5–7 in., exhibition form, slight fragrance; foliage light green; upright growth; [Kordes' Perfecta X (Kordes' Perfecta X American Heritage)]; Teranishi, K.; Itami Bara-en

'**HANAKAGO**', F, or, 1972; bud ovoid; flowers deep salmon-vermilion, medium, dbl., cupped, moderate fragrance; foliage glossy, dark; vigorous, bushy growth; [Sarabande seedling X Rondo seedling]; Suzuki, Seizo; Keisei Rose Nursery

'**HANA-KURENAI**', HT, pb, 1980; bud ovoid; flowers light pink, flushed yellow, reverse deeper, blooms borne 1–3 per cluster, dbl., 33 petals, exhibition form, slight fragrance; few, sickle-shaped prickles; foliage medium, medium green, glossy; vigorous, upright growth; [Big Red X Star Queen]; Ohata, Hatsuo

Hanamigawa, *see* '**HANAMI-GAWA**'

'**HANAMI-GAWA**', Cl Min, op, 1986; (Hanamigawa); flowers soft salmon-pink, shaded orange, small blooms borne 6–10 per cluster, dbl., 23 petals, moderate fragrance; small, curved prickles, slanted downward; foliage dark, semi-glossy; vigorous, very bushy growth; [Unnamed seedling X Petite Folie]; Suzuki, Seizo; Keisei Rose Nursery, 1985

'**HANAMORI**', F, rb, 1977; bud circular; dbl., 20–25 petals, 2.5 in., exhibition form, slight fragrance; foliage glossy, dark; bushy growth; [(Tropicana X Karl Herbst) X Lydia]; Teranishi, K.; Itami Bara-en

'**HANATIRUSATO**', Min, dp, 1999; flowers brilliant rose, very dbl., 70 petals, 1.4 in., flat, borne 2–5 per cluster, slight fragrance; foliage medium, medium green; vigorous, bushy, compact (8 in.) growth; [Seedling X Red Minimo]; Yamazaki, Kazuko, 1991; Takii & Co., Ltd., 1996

'**HANAYUZEN**', Min, dy, 1999; flowers deep golden yellow, sometimes flushed orange, very dbl., 45 petals, 1.5 in., exhibition form, moderate fragrance; foliage medium, dark green; very vigorous, upright, bush, compact (12 in.) growth; [Himetatibana X Hanahotaru]; Yamazaki, Kzauko, 1993; Takii & Co., Ltd., 1996

Hand in Hand, *see* HARaztec

Handel®, *see* 'MACHA'

Handel's Largo, *see* '**LARGO D'HAENDEL**'

Handout, *see* 'BARout'

'**HANDSOM RED**', HT, mr, 1954; flowers spectrum-red, dbl., 45 petals, 4–5 in., exhibition form, moderate fragrance; upright, bushy growth; [(Pink Princess X Mirandy) X Queen o' the Lakes]; Brownell, H. C.

'**HANDY ANDY**', HT, ab, 1965; flowers apricot edged pink, 4 in., slight fragrance; free growth; [Kordes' Perfecta X Piccadilly]; McGredy, Sam IV; Geest Industries

Hanib, *see* **Cecilia 89**

Hanini, *see* 'DYKNINI'

Hanka, HT, ly; flowers creamy yellow, very large, dbl.; (Sangerhausen)

Hanky Panky, *see* 'TINPANKY'

Hannah Gordon, *see* 'KORWEISO'

Hannah Hansen, HWich, m, 1997; Nobbs

'**HANNAH HAUXWELL**', F, op, 1991; flowers deep salmon, small, dbl., 26–40 petals, slight fragrance; foliage small, medium green matt; patio; bushy growth; [Unnamed seedling X Unnamed seedling]; Battersby Roses, 1990

'**HANNE**', HT, mr, 1959; flowers scarlet-crimson, medium to large, dbl., exhibition form, intense fragrance; foliage leathery; upright growth; [Ena Harkness X Peace]; Soenderhousen; Hoersholm Nursery

Hanne Dänomik, HT, mr; flowers large, dbl.; (Sangerhausen)

Hanneli Rupert, *see* KORsebue

Hannes, *see* '**ROSE HANNES**'

Hannover, *see* '**MESSESTADT HANNOVER**'

Hannover's Weisse, *see* NOAhan

'**HANS**', HT, w, 1970; flowers open, large, semi-dbl.; foliage glossy, light; vigorous, upright growth; [Message X Virgo]; IARI; Div. of Vegetable Crops & Flori.

Hans Berger, HT, mr, 1958; flowers carmine-red, large, dbl., intense fragrance; Berger, W.; (Sangerhausen)

'**HANS BILLERT**', HT, mr, 1928; flowers brilliant red, very dbl., moderate fragrance; [Laurent Carle X Richmond]; Billert; Teschendorff

Hans Christian Andersen, *see* 'POULANDER'

Hans Erni, F, op, 1992; Meilland

'**HANS HAUBOLD**', HT, mp, 1942; flowers large, dbl., slight fragrance; Vogel, M.; (Sangerhausen)

'**HANS MACKART**', HP, rb, 1884; flowers bright deep geranium red, outer petals tinted carmine purple, dbl., moderate fragrance; Verdier, E.

'**HANS SCHMID**', HWich, dp, 1934; flowers deep pink, rather well formed, small to medium, borne in clusters, dbl., globular, foliage large; vigorous, climbing growth; [Fragezeichen X American Pillar]; Vogel, M.; P. Lambert & Heinemann

'**HANSA**', HRg, mr, 1905; flowers mauvy-red, large blooms, dbl., recurrent bloom, intense, clove-rose fragrance; large, red fruit; short, weak stems; vigorous growth; hardy.; Schaum & Van Tol

Hansaland®, *see* KORhassi

Hansa-Park®, *see* KORfischer

'**HANSEAT**'®, S, mp, 1961; flowers rose-pink, center lighter, medium, single, 5 petals, cupped, slight fragrance; vigorous (6 ft) growth; Tantau, Math.

Hansen's Red Hedge, S, mr

'**HANSESTADT BREMEN**', F, op, 1958; bud ovoid, crimson; flowers deep salmon and reddish pink, large, borne in clusters (up to 10), dbl., 47 petals, moderate fragrance; foliage leathery; very vigorous, bushy growth; [Ama X Fanal]; Kordes, R.

'**HANSESTADT LÜBECK**', F, mr, 1962; (Lübeck); dbl., 3.5 in., slight fragrance; foliage dark; vigorous, tall growth; Kordes, R.; McGredy

'HANSETTE', S, mr, 1938; flowers red, semi-dbl., non-recurrent, [Hansa X R. rubrifolia]; Wright, Percy H.

Hap Renshaw, *see* 'RENHAP'

'HAPPENSTANCE', HBc, ly, 1950; single; thorns large, hooked; low, prostrate ground cover; up to 15 ft; Buss

'HAPPINESS', HT, mr, 1954; (Rim, Rouge Meilland); bud long, pointed; dbl., 38 petals, 5–6 in.; exhibition form, slight fragrance; upright, vigorous growth; [(Rome Glory X Tassin) X (Charles P. Kilham X (Charles P. Kilham X Capucine Chambard))]; Meilland, F.; C-P, 1951;, URS, 1949

'HAPPINESS, CLIMBING', Cl HT, mr, 1954; (Rouge Meilland, Climbing); Meilland, F.; URS

'HAPPY', Pol, mr, 1954; (Alberich®); flowers currant-red, very small blooms in large trusses, semi-dbl.; foliage dark, glossy; vigorous, compact (12–15 in.) growth; [Robin Hood X Katharina Zeimet seedling]; deRuiter; Gregory & Willicher Baumschulen

Happy Anniversary, *see* 'DELPRE'

Happy Anniversary, HT, lp; [Kardinal sport]; Catt, Graeme Charles, 1997; (Weatherly, L.)

'HAPPY BIRTHDAY', HT, dp, 1964; bud ovoid; flowers deep rose, large, dbl., 25 petals, exhibition form, intense fragrance; foliage leathery; vigorous, upright growth; [Peace X The Doctor]; Howard, P. J.

Happy Birthday, MinFl, w, 1997

'HAPPY BUTT', HT, ab, 1995; flowers open apricot, petals deepen to pink as bloom matures, very full (41+ petals), large (7+ cms) blooms borne mostly singly; fragrant; some prickles; foliage medium, medium green, semi-glossy; upright, tall growth; [Nantucket X Medallion]; Carlson, William

Happy Child, *see* 'AUSCOMP'

Happy Day, *see* 'SIMPALNO'

'HAPPY DAYS', HT, or, 1932; bud pointed; flowers geranium-red, open, large, dbl.; foliage dark; long stems; very vigorous growth; RULED EXTINCT 4/87; [Briarcliff sport]; Amling, M. C.; Amling Bros.

'HAPPY DAYS', HT, dr, 1962; bud pointed; flowers oxblood-red, dbl., 3–3.5 in., exhibition form, long stems; vigorous growth; RULED EXTINCT 4/87; [Exciting X Grand Gala]; Herholdt, J.A.; Herholdt's Nursery

Happy Days, *see* 'MACSEATRI'

'HAPPY EVENT', F, pb, 1964; flowers light chrome-yellow, flushed rose-opal, blooms in clusters, semi-dbl., 12 petals, 3 in.; foliage glossy; growth moderate; [(Karl Herbst X Masquerade) X Rose Gaujard]; Dickson, Patrick; A. Dickson

Happy Ever After™, *see* 'DICVANILLA'

Happy Face™, *see* 'SAVAFACE'

Happy Go Lucky™, *see* 'SAVALUCK'

Happy Go Lucky, *see* 'SUNHAP'

Happy Hour™, *see* 'SAVANHOUR'

'HAPPY RED', F, or, 1960; flowers bright brick-red, borne in clusters, single; foliage glossy; moderate growth; [Red Favorite X Cocorico]; Leenders, J.

'HAPPY TALK', F, mr, 1973; flowers cherry-red, small, dbl., slight fragrance; foliage glossy, dark; vigorous, upright, bushy growth; [Escort X Orange Garnet]; Weeks

'HAPPY THOUGHT', Min, op, 1978; bud pointed; flowers pink blended with coral and yellow, dbl., 40 petals; foliage small, glossy; vigorous, bushy growth; [(R. wichurana X Floradora) X Sheri Anne]; Moore, Ralph S.; Sequoia Nursery

'HAPPY TIME', Cl Min, rb, 1974; bud short, pointed; flowers yellow overlaid red, dbl., 35 petals, 1 in., slight fragrance; foliage small, glossy, leathery; climbing growth; [(R. wichurana X Floradora) X (Golden Glow X Zee)]; Moore, Ralph S.; Sequoia Nursery

Happy Times, MinFl, mp, 1995; Chessum, Paul

Happy Trails, *see* 'JACCASP'

'HAPPY WANDERER'®, F, mr, 1972; flowers scarlet, slight fragrance; [Seedling X Marlene]; McGredy, Sam IV; McGredy, 1974

'HAPPY WEDDING BELLS', HT, w, 1966; bud long, pointed; flowers large, dbl., 52 petals, exhibition form, moderate, spicy fragrance; foliage leathery; vigorous, upright growth; [White Swan X Virgo]; Morey, Dr Dennison; Country Garden Nursery

Har Tabor, HT, pb, 1972; Fischel

HARavis, HT, 1995; (**Tower Bridge**); Harkness; (Cavriglia)

HARaztec, MinFl, or, 1994; (**Hand in Hand**); Harkness

HARazz, HT, ab, 1998; (**Chimene**); Harkness

HARbabble, F; (**Sunset Boulevard**); Golden Prize , City of Glasgow, 1998; Harkness, 1997; (Cavriglia)

HARbadge, S, ab, 1997; (**Louisa Stone**); Harkness

'HARBANJO', F, or, 1993; (**Girls' Brigade**); flowers vermilion, orange-red reverse, aging orange-red, dbl., 28 petals, 2 in., no fragrance; foliage small, dark green, glossy; bushy, low growth; [Sexy Rexy X Anna Ford]; Harkness, 1987; Harkness New Roses, Ltd., 1993

HARbanner, S, lp, 1998; (**Country Fair**); Harkness

HARbar, F, 1995; (**Sporting Duo**); Harkness, R.; (Cavriglia)

HARbaroque, S, m, 1995; (**Baroque**); Harkness

HARbell, F, 1995; (**Cathedral Splendour**); Harkness, R.; (Cavriglia)

'HARBELLA', F, pb, 1994; (Peacekeeper, **United Nations Rose**); flowers orange/pink blend, yellowing with age, double, blooms borne, 30 petals, 3.5 in., slight, spicy fragrance; foliage medium, light green, glossy; bushy, medium growth; [Dame of Sark X Bright Smile]; Harkness; Harkness New Roses, Ltd., 1996

'HARBILBO', F, w, 1994 **(St John)**; dbl., 15–25 petals, 1.5–2.75 in., borne in small clusters, slight fragrance; few prickles; foliage medium, medium green, glossy; low (60 cms), bushy, spreading growth; [Prima X Grace Abounding]; Harkness; Harkness New Roses, Ltd., 1994

'HARBINGER', LCl, lp, 1923; bud pointed; flowers soft pink, large, single, slight fragrance; foliage light; vigorous, climbing growth; Clark, A.; Hackett

HARbinger, HT, or, 1994; (**Julie Y**); Harkness

HARbingo, MinFl, my, 1995; (**House Beautiful**); Harkness

HARbleep, F, op, 1996; (**Guglielmo Marconi**); Harkness

HARblend, S, ab, 1994; (**Della Balfour**); Harkness

'HARBONNY', F, mr, 1993; (**Royal Anniversary**); flowers crimson, yellow base, reverse mid-pink, aging dark red, small, dbl., 24 petals, 2.5 in., slight, fruity fragrance; foliage small, dark green, glossy; bushy, low growth; [Intrigue X Anna Ford]; Harkness, 1987; Harkness New Roses, Ltd., 1992

HARboul, MinFl, mp, 1998; (**Pink Pirouette**); Harkness

'HARBRILL', F, my, 1994; (**Conquest**); dbl., 30 petals, 3.5 in., borne in sprays of 7–10, slight, spicy fragrance; foliage medium, light green, glossy; bushy, medium growth; [Dame of Sark X Bright Smile]; Harkness; Harkness New Roses, Ltd., 1994

'HARBRITE', F, lp, 1993; (**Mrs Iris Clow**); flowers blush pink, light pink reverse, paling to near white, loose, dbl., 28 petals, cupped, moderate, spicy fragrance; foliage large, dark green, glossy; upright, medium growth; [Memento X Princess Alice]; Harkness, 1987; Harkness New Roses, Ltd., 1994

'HARBUSY', F, mp, 1971; (**Busy Lizzie**); flowers pink, medium, semi-dbl., 12 petals, 2 in., slight fragrance; foliage glossy, dark; free growth; [(Pink Parfait X Masquerade) X Dearest]; Harkness

HARcheer, S, pb, 1998; (**Country Music**); Harkness

HARchutzpah, HT, 1995; (**June Whitfield**); Harkness, R.; (Cavriglia)

HARclue, HT, ab, 1998; (**Commonwealth Glory**); Harkness

HARcogent, HT, 1996; (**St Christopher**); GM, Belfast, 1998; Harkness, R.; (Cavriglia)

'HARCOMP', LCl, ly, 1980; (**Highfield®**); [Compassion sport]; Harkness, 1981

HARcross, MinFl, dr, 1995; (**Humanity**); Harkness

'HARDANCER', F, ob, 1976; (**Flame Dancer**); flowers orange, red reverse, medium, dbl., cupped, moderate fragrance; large prickles; foliage dense; medium, bushy growth; [Orange Sensation X Alison Wheatcroft]; Harkness; Mason

HARdeed, HT, lp; (**Royal Philharmonic**); Harkness

HARdeluxe, F, 1996; (**Lilian Baylis**); Harkness, R.; (Cavriglia)

HARdenier, S, 1996; (**Marjorie Marshall**); Harkness; (Cavriglia)

HARdimple, F, 1996; (**World Class**); Harkness; (Cavriglia)

HARdinkum, F, w, 1997; (**Princess of Wales**); Harkness

HARdolly, F, lp; (**Tambourine**); Harkness, 1998

HARdwell, LCl, ab, 1998; (**Penny Lane**); Harkness

HAReast, F, 1995; (**Rising Star**); Harkness, R.; (Cavriglia)

HAReco, S, 1997; (**Nipper**); Harkness, R.; (Cavriglia)

HAReden, F, lp, 1997; (**Constance Finn**); Harkness

'HARELAN', HT, yb, 1999; (**Gift of Life**, Poetry in Motion); flowers mid yellow-pink, reverse yellow, dbl., 26–40 petals, 4 in., borne mostly singly, moderate fragrance; many prickles; foliage large, medium green, glossy; bushy, medium (40 in.) growth; GM, Belfast, 1999; [Dr Darley X Elina]; Harkness; Edmunds' Roses, 1999;, Harkness New Roses, 1996

HARelite, F, 1996; (**Wheel Horse Classic**); Harkness; (Cavriglia)

HARencens, S, ab, 1998; (**Miss Dior**); Harkness

HARencore, HT, mr, 1998; (**Pride of England**); Harkness

HARentrap, F, lp; (**St Piers**); Harkness, 1998

HARessay, F, mr, 1998; (**Faithful**); Harkness

HARette, F, ob, 1998; (**Betty Harkness**); Harkness

HAReverso, F, lp; (**Spirit of Tollcross**); Harkness, 1998

Harewood, *see* TANinaso

HARexclaim, F, ly, 1998; (**Irish Hope**); Harkness

HARextra, S, dr, 1998; (**Madrigal**); Harkness

HARfab, MinFl, lp; (**Suncharm**); Harkness, 1998

HARfling, F, lp, 1998; (**Blushing Bride**); Harkness

'HARFLOW', F, yb, 1999; (**Easy Going™**); flowers deep gold apricot, reverse same, peach overlay on fresh blooms, dbl., 26–30 petals, 3.5–4 in., borne in small clusters, moderate, fruity fragrance; many prickles; foliage large, very bright, light green, glossy; upright, bushy, medium (3–3.5 ft) growth; PP10478; [Sport of Livin' Easy sport]; Harkness; Harkness New Roses, 1999;, Weeks Roses, 1999

HARfracas, F, ab, 1998; (**Calliope**); Harkness

'HARHERO', S, rb, 1978; (**Marjorie Fair®**, Red Ballerina, Red Yesterday); flowers medium red, white eye, blooms in very large clusters, single, 5 petals, 1 in., slight fragrance; foliage small, light green, semi-glossy; dense, bushy growth; GM, Baden-Baden, 1979 GM, Rome, 1977; [Ballerina X Baby Faurax]; Harkness, 1977

'HARHESTER', F, op, 1980; (**Caroline Davison**); flowers medium salmon-pink, miniflora, small blooms in clusters of, semi-dbl., 16 petals, slight fragrance; straight, dark green prickles; foliage small, dark reddish-green; low, bushy growth; [Tip Top X Kim]; Harkness, R., & Co., Ltd.

'HARISON LEMON', HSpn, ly, 1929; flowers clear lemon-yellow, semi-dbl., non-recurrent, moderate fragrance; bushy (5 ft) growth; [Harison's Yellow seedling]; Hamblin

Harison Salmon, *see* **'HARISON'S SALMON'**

Harisonii, *see* **'HARISON'S YELLOW'**

'HARISON'S HARDY', HSpn, ly, 1943; flowers cream, center tinted yellow, semi-dbl., non-recurrent; very hardy.; [R. spinosissima altaica X Harison's Yellow]; Wright, Percy H.

'HARISON'S SALMON', HSpn, op, 1929; (Harison Salmon); flowers salmon, semi-dbl., non-recurrent, moderate fragrance; [Harison's Yellow seedling]; Hamblin

'HARISON'S YELLOW', HFt, dy; (Harisonii, R. foetida harisonii, R. lutea hoggii, R. X harisonii); flowers bright yellow, yellow stamens, semi-dbl.; Eurosa, Pimpinellifolia (28); [Probably Persian Yellow X R. spinosissima]; Harison, 1830; ca. 1830

'HARJAMES', S, yb, 1989; (**Xerxes**); bud pointed; flowers rich yellow with scarlet red eye at base, reverse yellow, single, 5 petals, cupped, slight fragrance; fruit not observed; prickles narrow, small, reddish; foliage small, grayish-green, matt; upright, medium to tall growth; [H. persica X Canary Bird]; Harkness, R., & Co., Ltd.

'HARJOOBILY', F, op, 1978; (**Mr E.E. Greenwell**); flowers rosy salmon, semi-dbl., 18 petals, 3 in., flat; vigorous, bushy, spreading growth; [Jove X City of Leeds]; Harkness

'HARJOSINE', HT, ab, 1980; (**Basildon Bond**); flowers loose form, dbl., 27 petals, borne 1–3 per cluster, moderate fragrance; large, straight prickles; foliage large, medium green, very glossy; medium, upright growth; GM, Belfast, 1982; [(Sabine X Circus) X (Yellow Cushion X Glory of Ceylon)]; Harkness, R., & Co., Ltd.

'HARKANTABIL', HT, yb, 1979; (**Marion Harkness®**); flowers canary-yellow, flushed orange-red, dbl., 24 petals, 3.5 in., slight fragrance; bushy growth; [((Manx Queen X Prima Ballerina) X (Chanelle X Piccadilly)) X Piccadilly]; Harkness

'HARKARAMEL', F, ab, 1979; (**Anne Harkness®**); bud globular; flowers deep apricot, dbl., cupped, blooms in trusses, slight fragrance; foliage medium green, semi-glossy; vigorous, upright, tall growth; [Bobby Dazzler X ((Manx Queen X Prima Ballerina) X (Chanelle x Piccadilly))]; Harkness, 1980

'HARKINDER', Min, mp, 1979; (**Esther's Baby®**); bud pointed; flowers persian rose, patio, medium, flat; foliage small, glossy; low, spreading growth; [(Vera Dalton X (Chanelle X Piccadilly)) X Little Buckaroo]; Harkness

'HARKING', F, yb, 1977; (**Judy Garland**); flowers yellow, petals edged orange-red, medium-large, borne singly, dbl., 35 petals, slight fragrance; foliage semi-glossy; medium, bushy growth; [((Tropicana X Circus) X (Sabine X Circus)) X Pineapple Poll]; Harkness, 1978

'HARKITTEN', Min, or, 1980; (**Crispin-Morwenna**); flowers salmon-red, small blooms borne 3–7 per cluster, dbl., 25 petals, cupped, slight, spicy fragrance; reddish prickles; foliage small, dark, glossy; low, spreading growth; [(Vera Dalton X (Chanelle X Piccadilly)) X Little Buckaroo]; Harkness, R., & Co., Ltd.

Harkness Marigold, *see* 'HARTOFLAX'

'HARKOTUR', F, pb, 1977; (**Softly Softly**); flowers pink and creamy pink, dbl., 35 petals, 5 in., slight fragrance; foliage leathery; [White Cockade X ((Highlight X Colour Wonder) X (Parkdirektor Riggers X Piccadilly))]; Harkness, 1980

'HARKOVER', F, op, 1972; (Garden City, **Letchworth Garden City®**); flowers medium salmon-pink, medium, dbl., 20 petals, 2.5 in., moderate, spicy fragrance; foliage medium green, semi-glossy; vigorous, bushy growth; GM, Monza, 1978; [(Sabine X Pineapple Poll) X (Circus X Mischief)]; Harkness, 1979

'HARKREME', HT, pb, 1980; (**Camphill Glory**); flowers creamy pink, large blooms borne singly, dbl., 54 petals, exhibition form, slight fragrance; many prickles; foliage medium green, matt; vigorous, branching growth; [Elizabeth Harkness X Kordes' Perfecta]; Harkness, R., & Co., Ltd., 1982

'HARkuly', F, w, 1977; (**Margaret Merril®**); flowers blush white, full, dbl., 28 petals, 4 in., exhibition form, intense fragrance; Edland Fragrance Medal, ARS, 1978 GM, Geneva, 1978 GM, Monza, 1978 GM, Rome, 1978 Gold Star of the South Pacific, Palmerston North, NZ, 1982 James Mason Medal, RNRS, 1990; [(Rudolph Timm X Dedication) X Pascali]; Harkness, 1978

'HARkushi', F, ob, 1979; (**Amy Brown®**); bud ovoid; flowers burnt-orange to fire-red, rounded, dbl., 28 petals, 2.5–3 in.; moderate, fruity fragrance; foliage large, dark; low, bushy growth; [Orange Sensation X ((Highlight X Colour Wonder) X (Parkdirektor Riggers X Piccadilly))]; Harkness

'HARlacal', S, yb, 1981; (**Rachel Bowes Lyon**); flowers peach pink, reverse yellow, blooms in large clusters, semi-dbl., 14 petals, flat, moderate fragrance; small prickles; foliage small to medium, medium green; low, bushy growth; [Kim X ((Orange Sensation X Allgold) X R. californica)]; Harkness, R., & Co., Ltd.

'HARlassie', Pol, lp, 1980; (**Fairy Maid**); bud short, plump; flowers light rose-pink, medium, dbl., 20 petals, cupped; foliage glossy; low, bushy growth; [The Fairy X Yesterday]; Harkness

'HARlayalong', Pol, lp, 1980; (**Fairyland®**); bud short, fat; flowers medium, dbl., 24 petals, cupped, moderate fragrance; foliage glossy; spreading growth; [The Fairy X Yesterday]; Harkness

HARlecho, LCl, pb, 1994; (**Morning Light**); Harkness

Harlekijn, F, yb

Harlekin®, *see* KORlupo

'HARlenten', F, w, 1980; (**Greensleeves**); bud pointed; flowers chartreuse-green, large, semi-dbl., 15 petals, flat; foliage dark; vigorous, upright growth; [(Rudolph Timm X Arthur Bell) X ((Pascali X Elizabeth of Glamis) X (Sabine X Violette Dot))]; Harkness

'Harlequin', HWich, pb, 1935; flowers half pale pink and half dark red, borne in clusters; very vigorous, climbing growth; RULED EXTINCT 12/83; [Excelsa sport]; Cant, F.

Harlequin, *see* 'AROvule'

Harlequin™, *see* 'KORlette'

Harlew, F, 1973; Petersen, V.; (Cavriglia)

'HARlex', HT, or, 1972; (**Alexander®**, Alexandra); bud pointed; flowers bright vermilion-red, dbl., 25 petals, 5 in., exhibition form, slight fragrance; foliage glossy; tall, vigorous growth; ADR, 1974 GM, Belfast, 1974 GM, Hambourg, 1973 James Mason Medal, RNRS, 1987; [Tropicana X (Ann Elizabeth X Allgold)]; Harkness

'HARlexis', HT, ob, 1981; (Alexis, **L'Oréal Trophy**); flowers orange; GM, Bagatelle, 1984 GM, Belfast, 1984 Golden Rose, Courtrai, 1986; [Alexander sport]; Harkness, R., & Co., Ltd., 1982

Harley®, *see* LENpaga

'HARlightly', F, my, 1981; (**Princess Michael of Kent®**); flowers large blooms borne 1–3 per cluster, dbl., 38 petals, exhibition form, moderate fragrance; short, thick, red prickles; foliage medium large, mid-green, glossy; low, bushy growth; [Manx Queen X Alexander]; Harkness, R., & Co., Ltd.

'HARlino', F, op, 1980; (**Sue Ryder®**); flowers salmon orange, reverse shaded yellow, blooms in large clusters, dbl., 20 petals, cupped, slight fragrance; small prickles; foliage medium, mid-green, semi-glossy; vigorous, medium, bushy growth; [Southampton X ((Highlight X Colour Wonder) X (Parkdirektor Riggers X Piccadilly))]; Harkness, 1983

'HARlittle', Pol, w, 1980; (**Fairy Snow®**); bud squat; dbl., 22 petals, cupped, borne several per cluster, slight fragrance; slender, dark prickles; foliage small, dark, glossy; short, bushy growth; [The Fairy X Yesterday]; Harkness, R., & Co., Ltd.

'Harlow', HT, op, 1968; flowers salmon, large, dbl., 29 petals; foliage glossy; [Fragrant Cloud X Melrose]; Cocker

Harlow Carr, *see* KIRlyl

Harman, F, 1991; Adam, M.; (Cavriglia)

Harman Inermis®, F, dp; Adam

'HARmantelle', F, my, 1982; (**Mountbatten®**); flowers large blooms borne singly or several together, dbl., 45 petals, cupped, moderate fragrance; many, large prickles; foliage large, leathery, glossy; upright, dense growth; GM, Belfast, 1982 GM, Courtrai, 1986 GM, Orleans, 1982 Golden Rose, The Hague, 1986 ROTY, 1982; [Peer Gynt X ((Anne Cocker X Arthur Bell) X Southampton)]; Harkness, R., & Co., Ltd.

'HARmark', F, yb, 1985; (**Hiroshima's Children**); flowers light yellow, petals edged pink, large, dbl., 35 petals, exhibition form, slight fragrance; foliage medium, medium green, matt; bushy growth; Harkness, R., & Co., Ltd.

Harmonie®, *see* 'KORtember'

'Harmonie', S, pb, 1954; (Kordes' Harmonie); bud ovoid, light red; flowers pink bicolor, large, borne in clusters, moderate fragrance; foliage leathery; very vigorous, upright growth; RULED EXTINCT 5/80; [R. eglanteria hybrid X Peace]; Kordes

Harmonie, *see* KRIlamy

'Harmony', Cl HP, ab, 1933; flowers apricot-pink, very large, semi-dbl., exhibition form, intense fragrance; foliage leathery, dark; strong stems; very vigorous, climbing growth; [Rosella (self)]; Nicolas; C-P

Harmony Parade, *see* POUlming

'HARmusky', F, mp, 1980; (**Radox Bouquet**, Rosika); flowers soft medium pink, blooms borne 1–3 per cluster, dbl., 30 petals, cupped, moderate fragrance; large, dark prickles; foliage large, glossy, medium green; upright, rather open growth; [(Alec's Red X Piccadilly) X (Southampton X (Clare Grammerstorfx Fruhlingsmorgen))]; Harkness, R., & Co., Ltd., 1981

'HARneatly', Pol, dr, 1979; (**Fairy Damsel®**); bud short, plump; flowers oxblood-red, medium, dbl., 24 petals, cupped; foliage glossy; low, spreading growth; [The Fairy X Yesterday]; Harkness

'HARnicely', Pol, mp, 1979; (**Fairy Ring**); bud short, plump; flowers rose-pink, dbl., 20 petals, 2 in., cupped; foliage glossy; low, bushy growth; Harkness

'HARnimble', Pol, lp, 1979; (**Fairy Like**); bud short, plump; flowers light rose-pink, small, dbl., 20 petals, cupped; foliage small, glossy; low, spreading growth; [The Fairy X Yesterday]; Harkness

'HARnoble', HT, dp, 1981; (**Pacemaker**); bud pointed; flowers deep pink, blooms borne singly, sometimes 3 per cluster, dbl., 40 petals, 4 in., intense fragrance; large, broad, dark prickles; foliage large, semi-glossy; vigorous, upright, bushy growth; GM, Belfast, 1983; [Red Planet X Wendy Cussons]; Harkness

'HARnougette', Pol, mr, 1981; (**Fairy Prince®**); bud short; flowers medium, dbl., 25 petals, cupped; foliage glossy; spreading growth; First Prize, Copenhagen; [The Fairy X Yesterday]; Harkness

'HARnumerous', Pol, mp, 1979; (**Fairy Changeling**); bud short, plump; flowers pink, pompon, medium, dbl., 22 petals, cupped, slight fragrance; foliage small, dark; recumbent, spreading growth; [The Fairy X Yesterday]; Harkness, 1981

'HARoffen', S, mp, 1985; ('HARroffen', **Rochester Cathedral**); flowers medium to large blooms in clusters, dbl., 58 petals, cupped, repeat bloom, moderate fragrance; foliage medium, dark, matt; medium, dense, spreading growth; [(Unnamed seedling X ((Orange Sensation X Allgold) X R. californica)) X Frank Naylor]; Harkness, R., & Co., Ltd., 1987

Harold Ickes, *see* **'Crepe Myrtle'**

Harold Macmillan, *see* 'HARwestsun'

'HARpade', F, m, 1967; (**Escapade®**); flowers magenta-rose, center white, blooms in clusters, semi-dbl., 12 petals, 3 in., slight fragrance; foliage glossy, light green; ADR, 1973 GM, Baden-Baden, 1969 GM, Belfast, 1969; [Pink Parfait X Baby Faurax]; Harkness

'HARpiccolo', Min, ob, 1980; (**Anna Ford**®, Anne Ford); flowers deep salmon-orange, yellow eye, patio, small, semi-dbl., 18 petals, blooms in large clusters, slight fragrance; small prickles; foliage small, glossy; low, bushy growth; GM, Genoa, 1987 GM, Glosgow, 1989 GM, RNRS, 1981 PIT, RNRS, 1981; [Southampton X Darling Flame]; Harkness

'HARpillar', F, mr, 1982; (**Olive**); flowers large blooms in clusters, dbl., 36 petals, exhibition form, moderate, spicy fragrance; dark prickles; foliage large, dark, glossy; branching growth; [((Vera Dalton X Highlight) X Unnamed seedling) X Dublin Bay]; Harkness, R., & Co., Ltd.

'HARpippin', LCl, ob, 1984; flowers pale salmon-red, yellow reverse, blooms borne singly, dbl., 22 petals, slight fragrance; foliage medium, semi-glossy; upright (to 7 ft) growth; [Royal Dane sport]; Harkness

'HARpluto', F, ob, 1983; (**Avocet**); bud pointed; flowers orange edged vermilion, large, semi-dbl., blooms in large clusters, moderate fragrance; many, dark prickles; foliage dark, glossy; medium, bushy growth; [Dame of Sark X Seedling]; Harkness, R., & Co., Ltd., 1984

'HARpooh', HT, mp, 1983; (**Ruth Harker**); flowers large blooms borne 1–4 per cluster, dbl., 46 petals, intense fragrance; slightly curved, reddish prickles; foliage large, medium green, matt to semi-glossy; upright, bushy growth; [Fragrant Cloud X Compassion]; Harkness, R., & Co., Ltd.

'HARportly', HT, mp, 1982; (**Neville Gibson**); flowers large, dbl., 40 petals, exhibition form, slight fragrance; foliage large, medium green, semi-glossy; medium, upright growth; Golden Rose, Geneva, 1980; [Red Planet X (Carina X Pascali)]; Harkness, R., & Co., Ltd., 1983

'HARposter', HT, mp, 1980; (**Dr Darley**); flowers rose bengal, blooms borne usually singly, dbl., 45 petals, globular, slight fragrance; narrow, reddish prickles; foliage mid-green, semi-glossy; upright, bushy growth; GM, Munich, 1983; [Red Planet X (Carina X Pascali)]; Harkness, R., & Co., Ltd., 1982

'HARprier', S, yb, 1986; (**Tigris**); flowers yellow, dark red eye, dbl., 1 in., slight fragrance; gooseberry-like prickles; foliage small, light green, variable in shape; compact, rounded growth; [Hulthemia persica X Trier]; Harkness, 1985

'HARprincely', HT, dr, 1983; (**Innoxa Femille**); dbl., 50 petals, borne singly, sometimes 3 per cluster, slight fragrance; dark prickles; foliage large, semi-glossy; medium, bushy growth; [Red Planet X Eroica]; Harkness, R., & Co., Ltd.

'HARprocrustes', Min, ab, 1982; (**Clarissa**®); flowers small blooms in large clusters, dbl., 43 petals, exhibition form, slight fragrance; small prickles; foliage small, dark, glossy; tall, upright growth; Gold Star of the South Pacific, Palmerston North, NZ, 1982; [Southampton X Darling Flame]; Harkness, R., & Co., Ltd.; Harkness New Roses, Ltd., 1983

'HARpurl', HT, mp, 1979; (**Leigh-Lo**); bud pointed; flowers rose-bengal, urn-shaped, large, dbl., 42 petals, slight fragrance; foliage large; vigorous, upright growth; [Elizabeth Harkness X Red Devil]; Harkness, 1981

'HARquaker', F, yb, 1985; (**Volunteer**); flowers large, dbl., 35 petals, slight fragrance; foliage medium, light, glossy; bushy growth; [Dame of Sark X Silver Jubilee]; Harkness, R., & Co., Ltd., 1986

'HARquanne', LCl, ab, 1980; (**Breath of Life**); bud plump; flowers apricot to apricot-pink, borne 1–3 per cluster, dbl., 33 petals, moderate fragrance; large, straight, reddish prickles; foliage semi-glossy; upright (to 8 ft) growth; [Red Dandy X Alexander]; Harkness, R., & Co., Ltd., 1982

'HARquantum', F, m, 1984; (**International Herald Tribune**®, Violetta, Viorita®); flowers violet-purple, mini-flora, blooms in trusses, dbl., 20 petals, cupped, moderate fragrance; foliage medium, medium green, semi-glossy; low, bushy growth; GM, Monza, 1984 GM, Tokyo, 1983 Golden Rose, Geneva, 1983; [Unnamed seedling X ((Orange Sensation X Allgold) X R. californica)]; Harkness, R., & Co., Ltd., 1985

'HARqueterwife', HT, op, 1981; (Heart Throb, **Paul Shirville**); flowers light salmon-pink, medium-large blooms borne singly and in 3', dbl., 30 petals, exhibition form, moderate, sweet fragrance; large, reddish prickles; foliage large, dark, semi-glossy; medium, bushy growth; Edland Fragrance Medal, ARS, 1982; [Compassion X Mischief]; Harkness, R., & Co., Ltd., 1983

'HARquhling', S, ly, 1982; (**Anna Zinkeisen**); bud plump; flowers, dbl., 30 petals, blooms in clusters of 3–7, moderate, musk fragrance; small prickles; foliage small, light green, semi-glossy; medium, spreading, dense growth; [Seedling X Frank Naylor]; Harkness, 1983

'HARquibbler', S, pb, 1989; (**Nigel Hawthorne**); bud pointed; flowers pale salmon-rose, deep scarlet eye at base, reverse same, single, 5 petals, cupped, slight, spicy fragrance; plump, small, infrequent, green fruit; prickles thin, narrow, variable, dark to light; foliage medium, medium green, semi-glossy; spreading, low growth; [H. persica X Harvest Home]; Harkness, R., & Co., Ltd.

'HARquillypond', F, mp, 1986; (Blue Carpet, **Cosette**); dbl., slight fragrance; small, dark prickles; foliage small, medium green,matt; patio; low, spreading growth; [Seedling X Esther's Baby]; Harkness, R., & Co., Ltd., 1983

'HARquince', F, op, 1986; (Daniel Gelin, **Wandering Minstrel**); flowers pink shaded orange, large blooms in clusters, dbl., 28 petals, slight fragrance; foliage dark green, glossy; medium, bushy growth; [Dame of Sark X Silver Jubilee]; Harkness

'HARquisp', Min, dp, 1985; (**John Bradshaw**); flowers light rose red, rosette (starry-shaped) blooms in clusters, dbl., 24 petals, flat, slight fragrance; foliage small, semi-glossy; low, bushy growth; [Unnamed seedling X Esther's Baby]; Harkness; White Rose Nurseries, Ltd.

'HARquito', F, op, 1987; (**Bill Slim**); bud globular; flowers salmon, deep pink reverse, large blooms borne singly or in clusters, dbl., 30 petals, slight fragrance; foliage glossy; bushy, spreading growth; [Seedling X Silver Jubilee]; Harkness, 1989

'HARquorgold', F, dy, 1984; (**Rediffusion Gold**); flowers deep golden yellow, blooms in clusters of 3–7, dbl., 70 petals, slight fragrance; small prickles; foliage small, light green, matt; medium, bushy growth; [(Orange Sensation X Allgold) X Sunsprite]; Harkness, R., & Co., Ltd.; R. Harkness & Co., Ltd.

'HARramin', Min, mp, 1985; (**Hollie Roffey**); flowers rosette form, small blooms in clusters, dbl., 35 petals, slight fragrance; foliage small, pointed, medium green, semi-glossy; spreading growth; [(Tip Top X (Manx Queen X Golden Masterpiece)) X Darling Flame]; Harkness, R., & Co., Ltd., 1986

'HARrango', F, op, 1984; (**G. P. & J. Baker**); flowers salmon orange, reverse lighter, blooms in clusters of 3–11, dbl., 36 petals, flat, slight fragrance; many prickles; foliage dark, glossy; medium, bushy growth; [(Bobby Dazzler X Unnamed seedling) X Marion Harkness]; Harkness, R., & Co., Ltd.

'HARregale', S, m, 1984; (**Cardinal Hume**®); bud pointed; flowers violet-purple, blooms in clusters of 3 to many, dbl., 31 petals, cupped, repeat bloom, moderate, musk fragrance; small prickles; foliage variable shades of green, matt; medium, spreading growth; [((Seedling X (Orange Sensation X Allgold) X R. californica)) X Frank Naylor]; Harkness, R., & Co., Ltd.

'Harriet', Cl HT, my, 1931; flowers golden yellow, edged paler, dbl., recurrent bloom; foliage bronze; vigorous, climbing (12–15 ft) growth; [Golden Ophelia sport]; Moore, Ralph S.

'Harriet A. Easlea', HT, rb, 1922; flowers bright carmine, reverse golden yellow, dbl., moderate fragrance; McGredy

'Harriet Elizabeth', S, mp, 1987; flowers pure medium pink, large, borne singly and in clusters, dbl., 26–40 petals, repeat bloom, intense fragrance; foliage medium, red aging dark green, leathery, disease-resistant; upright, bushy, vigorous, fully hardy growth; [Paula X (Micki X

Northlander)]; James, John; Historical Roses, 1987

'HARRIET MILLER', HT, pb, 1972; bud long, pointed; flowers pink, large, very dbl., globular, moderate fragrance; foliage large, glossy; very vigorous, bushy growth; [Helen Hayes X Traviata]; Brownell, H.C.; Stern's Nursery

'HARRIET NEESE', S, op, 1928; flowers coral blended with yellow, base golden yellow, semi-dbl., abundant non-recurrent bloom, slight fragrance; bushy growth; [Ophelia X R. X harisonii]; Conyers

'HARRIET POULSEN', F, mp, 1912; flowers apple-blossom-pink, single; vigorous growth; [Mme Norbert Levavasseur X Dorothy Perkins]; Poulsen, D.T.

'HARRIET SHEPHERD', Gr, mp, 1982; flowers medium, dbl., 35 petals, no fragrance; foliage medium, medium green, holly-like; upright growth; [Queen Elizabeth X Queen Elizabeth]; Shepherd, David

Harriet Sheppard, HWich, m, 1997; Nobbs

'HARRINY', HT, mp, 1967; flowers clear pink, pointed, large, dbl., 40 petals, intense fragrance; foliage dark; [Pink Favorite X Lively]; LeGrice

'HARROFFEN', S, mp, 1985; ('HAROFFEN', **Rochester Cathedral**); flowers medium to large blooms in clusters, dbl., 58 petals, cupped, repeat bloom, moderate fragrance; foliage medium, dark, matt; medium, dense, spreading growth; [(Unnamed seedling X ((Orange Sensation X Allgold) X R. californica)) X Frank Naylor]; Harkness, R., & Co., Ltd., 1987

'HARRONVER', HT, yb, 1990; (**Anneka**); bud ovoid; flowers high-centered to cupped to reflexed, medium, very dbl., 45 petals, exhibition form, borne usually singly or in sprays of 3–5, moderate, fruity fragrance; ovoid, medium to large, medium green fruit; prickles slightly curved, medium, dark reddish; foliage medium, medium green, glossy; upright, medium growth; [Goldbonnet X Silver Jubilee]; Harkness, R., & Co., Ltd.; R. Harkness & Co., Ltd.

'HARROONY', F, ab, 1983; (**Amber Queen**®); bud plump; flowers apricot gold, large, dbl., 25–30 petals, cupped, blooms in clusters of 3–7, moderate, sweet and spicy fragrance; reddish prickles; foliage large, copper red to medium green, semi-glossy; low, compact, bushy growth; disease resistant; PP5582; AARS, 1988 GM, Genoa, 1986 GM, Orléans, 1987 Gold Star of the South Pacific, Palmerston North, NZ, 1988 ROTY, St Albans, UK, 1984; [Southampton X Typhoon]; Harkness, 1984

'HARROTANG', F, or, 1986; (**City of Bradford**); flowers small to blooms in large clusters, semi-dbl., cupped, slight fragrance; foliage dark, semi-glossy; medium, upright growth; [(Manx Queen X Whisky Mac) X ((Highlight X Colour Wonder) X (Parkdirektor Riggers X Piccadilly))]; Harkness

'HARROWBOND', HT, op, 1985; (**Rosemary Harkness**); flowers orange-salmon, orange-yellow reverse, medium-large, dbl., 35 petals, intense fragrance; foliage large, dark, semi-glossy; bushy growth; Fragrance Award, Belfast, 1987 GM, Belfast, 1987; [Compassion X (Basildon Bond X Irish Gold)]; Harkness, R., & Co., Ltd.

Harry®, *see* **'HARRY WHEATCROFT'**

'HARRY CAMPBELL', F, dp, 1977; bud small, pointed; flowers cerise-pink shading to white, dbl., 45 petals, 2.5 in., intense, wild apple fragrance; foliage dark, leathery; vigorous, bushy growth; Burnet; Benefield's Nursery

'HARRY EDLAND', F, m, 1975; flowers lilac-pink, dbl., 26 petals, 4 in., intense fragrance; foliage dark, glossy; Edland Fragrance Medal, ARS, 1975; [(Lilac Charm X Sterling Silver) X (Blue Moon X (Sterling Silver X Africa Star))]; Harkness, 1978

'HARRY G. HASTINGS', HT, dr, 1965; (H.G. Hastings); flowers large, dbl., slight fragrance; foliage leathery; vigorous growth; [Gov. Mark Hatfield X Helene Schoen]; Von Abrams

'HARRY KIRK', T, ly, 1907; bud pointed; flowers light sulfur-yellow, open, dbl., moderate fragrance; foliage leathery; strong stems; vigorous, bushy growth; Dickson, A.

'HARRY MAASZ', LCl, rb, 1939; bud long, pointed; flowers crimson, center white, very large, single, cupped, slight fragrance; foliage large, leathery, wrinkled, dark; very vigorous, climbing growth; [Barcelona X Daisy Hill]; Kordes

Harry Oppenheimer, *see* KORabmask

'HARRY WHEATCROFT', HT, yb, 1972; (Caribia, Harry®); flowers yellow striped red, reverse yellow; [Piccadilly sport]; Wheatcroft & Sons, 1973

Harry Wheatcroft, Climbing, *see* 'AROMONTELIB'

'HARSAMY', HT, op, 1989; (**Shire County**); bud ovoid; flowers peach on primrose yellow base, reverse salmon rose on primrose, dbl., 33 petals, cupped, moderate fragrance; prickles decurved, medium, reddish; foliage medium, medium green, semi-glossy; bushy, medium growth; [Amy Brown X Bonfire Night]; Harkness, R., & Co., Ltd., 1990

'HARSHERRY', F, yb, 1982; (**Sheila's Perfume**); flowers yellow, petals edged red, large blooms borne singly, dbl., 20 petals, intense fragrance; foliage medium, dark, semi-glossy; medium, bushy growth; Edland Fragrance Medal, ARS, 1991 Gold Star of the South Pacific, Palmerston North, NZ, 1993; [Peer Gynt X (Daily Sketch X (Paddy McGredy X Prima Ballerina))]; Sheridan, John; Harkness New Roses, Ltd., 1985

'HARSPRICE', HT, rb, 1986; (**Potton Heritage**); flowers plum red, reverse straw-yellow, large, dbl., 32 petals, exhibition form, moderate fragrance; foliage large, dark, glossy; bushy, branching growth; [Precious Platinum X Dr A.J. Verhage]; Harkness, 1987

HARsuma, S, mr; (**Suma**); Onodera, Toru F.; Harkness, 1989

'HARTANNA', F, my, 1985; (Brite Lites®, **Princess Alice**, Zonta Rose); flowers butter yellow, ruffled, dbl., 28–32 petals, 4–4.5 in., exhibition form, blooms in large clusters, slight, fruity fragrance; foliage medium, medium green, semi-glossy; uprightgrowth; disease resistant; PP6953; GM, Dublin, 1984; [Judy Garland X Anne Harkness]; Harkness, R., & Co., Ltd.

HARtbrad, Min, ab, 1989; (**Gentle Kiss**); Harkness

HARtellody, HT, my, 1992; (**Golden Friendship**); Harkness

'HARTESIA', HT, yb, 1985; (**Beryl Bach**); flowers yellow, blended pink, large, dbl., 40 petals, moderate fragrance; foliage large, light green, matt; tall, upright growth; [Sunsprite X Silver Jubilee]; Harkness, R., & Co., Ltd.; R. Harkness & Co., Ltd.

'HARTILION', F, or, 1986; (**Seafarer**); flowers large blooms in clusters of 3–5, dbl., 30 petals, cupped, slight fragrance; medium, green fruit; medium, red prickles; foliage medium, dark, glossy; medium, bushy growth; [Amy Brown X Judy Garland]; Harkness

HARtillery, HT, 1995; (**Reconciliation**); Harkness, J.; (Cavriglia)

Hartina, *see* **'KAMMERSANGER TERKAL'**

'HARTOFLAX', F, ob, 1986; (**Harkness Marigold**); flowers well-formed, blooms in clusters, dbl., 35 petals, slight fragrance; foliage medium, medium green, semi-glossy; upright growth; [Judy Garland X Anne Harkness]; Harkness, R., & Co., Ltd.

HARtopper, MinFl, pb, 1994; (**Gentle Clown**); Harkness

'HARTRED', F, mr, 1986; (**Save the Children**); flowers bright red, patio, blooms in clusters, dbl., cupped, slight fragrance; narrow, straight prickles; foliage dark, semi-glossy; low, bushy, compact growth; [Amy Brown X Red Sprite]; Harkness

'HARTSAM', HT, ob, 1987; (**Country Lady**); flowers burnt orange, reverse suffused pale scarlet, fading orange-salmon, dbl., 25 petals, slight, spicy fragrance; ovoid, medium, green fruit; decurved, medium, reddish prickles; foliage medium, medium green, semi-glossy; bushy, medium, high-shouldered growth; [Alexander X Bright Smile]; Harkness, R., & Co., Ltd.

'HARTUBOND', HT, op, 1989; (**Lord Houghton of Sowerby**); bud pointed; flowers warm, reddish, salmon-pink, reverse deeper, large, borne singly, exhibition form, slight, spicy fragrance; fruit not observed; prickles broad based, large, reddish; foliage large, dark green, glossy; upright, medium growth; [Silver Jubilee X Basildon Bond]; Harkness, R., & Co., Ltd., 1990

'HARTWIZ', F, yb, 1986; (**Conqueror's Gold**); flowers yellow, petals edged orange-red, blooms in clusters of up to 7, 18 petals, cupped, slight fragrance; foliage medium, dark semi-glossy; medium, bushy growth; [Amy Brown X Judy Garland]; Harkness, R., & Co., Ltd.

'HARUBASIL', HT, ab, 1990; (**Cordon Bleu**); bud pointed; flowers apricot with begonia pink reverse, deepening of apricot with aging, dbl., 20 petals, cupped, moderate, fruity fragrance; prickles recurving, average, reddish; foliage medium to large, dark green, glossy; upright, medium growth; [Basildon Bond X Silver Jubilee]; Harkness, R., & Co., Ltd.; R. Harkness & Co., Ltd., 1992

'HARUBONDEE', HT, op, 1988; (**Queen Charlotte**); flowers deep salmon-red, yellow base, reverse pink-red, aging paler, dbl., exhibition form, slight fragrance; ovoid, large, green fruit; prickles recurved, medium, reddish-green; foliage large, dark green, semi-glossy; upright, tall growth; [Basildon Bond X Silver Jubilee]; Harkness, R., & Co., Ltd.; R. Harkness & Co., Ltd., 1989

Harukaze, *see* **'HARU-KAZE'**

'HARU-KAZE', LCl, op, 1986; (Harukaze); bud ovoid; flowers salmon yellow to orange-red, blooms borne 6–8 per cluster, dbl., 33 petals; small, curved prickles, slanted downward; foliage dark, glossy; bushy, creeping growth; [Charleston X Dorothy Perkins]; Suzuki, Seizo; Keisei Rose Nursery, 1985

'HARUKFORE', F, lp, 1986; (**City of London®**); flowers light pink, fading to blush, large, dbl., 15–25 petals, intense fragrance; small, reddish, sparse prickles; foliage medium, medium green, glossy, ovate to pointed; bushy growth; GM, LeRoeulx, 1985; [Radox Bouquet X Margaret Merril]; Harkness, R., & Co., Ltd.; The Rose Garden & Mini Rose Nursery, 1988

'HARUNIQUE', S, pb, 1986; (**Euphrates**); flowers pale salmon red, deep pink eye, small blooms in clusters of, single, 5 petals, slight fragrance; prickly; foliage small, variable form (usually long and narrow), ligh; low, spreading growth; [Hulthemia persica X Seedling]; Harkness

Harunomai, F, mp

'HARUSEFUL', S, mp, 1988; (**Armada®**); flowers medium pink, aging slightly paler, medium, dbl., 17 petals, cupped, borne in sprays of up to 12, repeat bloom, moderate fragrance; rounded, medium green fruit; prickles slightly recurved, medium green; foliage medium, medium green, glossy; spreading growth; [New Dawn X Silver Jubilee]; Harkness, R., & Co., Ltd.; R. Harkness & Co., Ltd.

Haruyo, HT, w

'HARVACITY', F, dr, 1987; (**Drummer Boy**); flowers deep, vivid, bright scarlet, fading slightly paler, loose, semi-dbl., 15 petals, cupped, slight, spicy fragrance; ovoid, small, greenish fruit; fairly straight, small, purplish-red prickles; foliage small, medium green, semi-glossy, oval-pointed; patio; spreading, low growth; [(Wee Man X (Southampton X Darling Flame)) X Red Sprite]; Harkness, R., & Co., Ltd.

'HARVALEX', F, or, 1985; (**Christingle**); flowers large, dbl., 35 petals, slight fragrance; foliage medium, dark, semi-glossy; bushy growth; [Bobby Dazzler X Alexander]; Harkness, R., & Co., Ltd., 1987

'HARVANDER', Min, lp, 1989; (**Phoebe**); bud pointed; flowers pale rose pink, reverse same, aging very little, rosette, medium, very dbl., slight fragrance; fruit not a noticeable feature; prickles needle-like, long, decurved, small, dark green; foliage small, medium green, semi-glossy; bushy, low growth; [Clarissa X (Unnamed seedling X Mozart)]; Harkness, R., & Co., Ltd.

'HARVARD', HT, dr, 1926; bud pointed; flowers deep crimson, open, very large, dbl., intense fragrance; foliage soft, bronze; long stems; [Hoosier Beauty X Unnamed seedling]; Vestal

'HARVEE', Min, rb, 1989; (**Phoenix**); bud urn-shaped; flowers blood red, with yellow base, reverse same, aging orange-carm, dbl., slight fragrance; fruit not a noticeable feature; prickles narrow, recurved, small, green; foliage small, medium green, semi-glossy, pointed; bushy, low growth; [Clarissa X [Wee Man X (Southampton X Darling Flame)]]; Harkness, R., & Co., Ltd.

'HARVERAG', F, ab, 1999; (**English Sonnet™**, Fragrant Surprise, Samaritan®); flowers apricot pink/yellow, very dbl., 41 petals, 4 in., intense fragrance; prickles moderate; foliage medium, dark green, glossy; upright, spreading, medium (3 ft) growth; Harkness; Harkness New Roses, 1999;, J&P, 1999

Harvest Fayre, *see* 'DICNORTH'

'HARVEST FESTIVAL', HT, ab, 1980; flowers light apricot-orange, reverse apricot flushed pink, urn-shaped, dbl., 28 petals, slight fragrance; globular, large, orange-yellow fruit; medium, reddish-brown prickles; foliage medium, medium green, semi-glossy; tall, bushy growth; [Blessings X Sunblest]; Law, M.J.

'HARVEST GLOW', LCl, rb, 1941; bud long, pointed, ovoid; flowers red to pink, reverse yellow, large, dbl., 60 petals, exhibition form; foliage light green; long stems; vigorous, climbing growth; [Golden Glow X Mercedes Gallart]; Brownell, H.C.

Harvest Home®, *see* 'HARWESI'

'HARVEST MOON', HT, my, 1976; bud long, pointed; dbl., 35 petals, 4–5 in., slight fragrance; vigorous growth; [Whisky sport]; Mason, A.L.; F. Mason

'HARVEST MOON', HT, ly, 1938; flowers cream, open, large, borne in clusters, single; foliage leathery, dark; long stems; vigorous, bushy, compact growth; Cant, B. R.

Harvest Song, S, pb

Harvest Sun, *see* 'BROSUN'

'HARVEST TIME', Cl HT, ab, 1939; flowers apricot, reverse sometimes pinkish, open, very large, semi-dbl., free, recurrent bloom, moderate fragrance; foliage leathery, dark; very vigorous, climbing (15–20 ft) long stems growth; [Sophie Thomas X Souv. de Claudius Pernet]; Thomas; Armstrong Nursery

'HARVESTAL', Min, lp, 1989; (**Pallas**); bud ovoid; flowers light buff pink, paling to buff white, rosette, medium, born, very dbl., no fragrance; fruit not a noticeable feature; prickles narrow, small, dark green; foliage small, medium green, semi-glossy, pointed, plentiful; bushy, spreading, low growth; [Clarissa X New Penny]; Harkness, R., & Co., Ltd.

'HARVESTER', HT, pb, 1975; flowers carmine, reverse silver, center lilac, dbl., 35 petals, 5 in., intense fragrance; foliage large, matt; [Wendy Cussons X Kordes' Perfecta]; Mayhew

HARvilac, Min, m, 1998; (**Gentle Maid**); Harkness

'HARVINTAGE', HT, lp, 1987; (Integrity, **Savoy Hotel**, Vercors, Violette Niestlé); flowers light phlox pink, reverse shaded deeper, well-formed, large, dbl., 40 petals, exhibition form, slight fragrance; rounded, average, green fruit; small, reddish-green, fairly straight, narrow prickles; foliage medium, dark green, semi-glossy; bushy, medium growth; GM, Dublin, 1988; [Silver Jubilee X Amber Queen]; Harkness, R., & Co., Ltd., 1989

HARvissa, Min, ly, 1990; (**Juliet Ann**); Harkness

'HARVOLUTE', F, ab, 1989; (**By Appointment**); bud ovoid; flowers pale buff apricot, aging paler, urn-shaped becoming, medium, dbl., 22 petals, cupped, slight fragrance; prickles rather narrow, medium, dark reddish-green; foliage medium, dark green, semi-glossy; upright, medium growth; [Anne Harkness X Letchworth Garden City]; Harkness, R., & Co., Ltd., 1990

'HARVOOL', F, mp, 1992; (**Muriel**); dbl., 15–25 petals, 1.5 in., borne in large clusters, slight fragrance; few prickles; foliage small, medium green, semi-glossy; patio; low (30 cms), bushy growth; [Liverpool Echo X Unnamed seedling]; Harkness, R., & Co., Ltd.; Harkness New Roses, Ltd., 1991

'HARWADEROX', F, dp, 1992; (**Rosy Future**); flowers carmine, blooms borne in large clusters, dbl., 15–25 petals, 1.5 in., moderate fragrance; few prickles; foliage small, dark green, semi-glossy; patio; medium (60 cms), upright growth; [Radox Bouquet X Anna Ford]; Harkness, R., & Co., Ltd.; Harkness New Roses, Ltd., 1991

'HARWANNA', S, w, 1988; (**Jacqueline du Pré**®); flowers creamy blush to white, loose, large, borne singly or in sprays, semi-dbl., 15 petals, cupped, repeat bloom, moderate, musk fragrance; oval, medium, green orange fruit; prickles small, dark; foliage medium, dark green, glossy; tall, spreading growth; GM, LeRoeulx, 1988; [Radox Bouquet X Maigold]; Harkness, R., & Co., Ltd.; R. Harkness & Co., Ltd., 1989

'HARWANTED', S, pb, 1988; (Many Happy Returns, **Prima**); flowers blush white, reverse blush pink, aging blush to white, large, semi-dbl., 18 petals, cupped, repeat bloom, moderate, fruity fragrance; round, medium, red fruit; prickles straight, small, red; foliage medium, medium green, semi-glossy; spreading, medium growth; GM, Geneva, 1987; [Herbstfeuer X Pearl Drift]; Harkness, R., & Co., Ltd.; R. Harkness & Co., Ltd., 1990

HARward, Pol, mr, 1980; (**Fairy Dance**®); Harkness

'HARWAZZLE', F, or, 1992; (**Sheer Delight**); dbl., 15–25 petals, 1.5 in., borne in large clusters, no fragrance; few prickles; foliage small, light green, semi-glossy; patio; low (40 cms), bushy growth; [Bobby Dazzler X Little Prince]; Harkness, R., & Co., Ltd.; Harkness New Roses, Ltd., 1991

'HARWEEN', F, my, 1988; (**Golden Years**®); bud ovoid; flowers golden yellow, reverse some bronze tint, large, borne in sprays, dbl., 46 petals, cupped, slight, fruity fragrance; fruit not observed; prickles slightly curved, long, thin, greenish-red; foliage medium, dark green, semi-glossy; bushy, medium growth; GM, Orleans, 1990 Hradec Golden Rose, 1989; [Sunblest X Amber Queen]; Harkness, R., & Co., Ltd.; R. Harkness & Co., Ltd., 1990

'HARWELCOME', F, ob, 1992; (Fellowship, **Livin' Easy**™); bud ovoid, short; flowers frilly, orange-apricot, medium sized, very dbl., 25–30 petals, 4–4.5 in., borne in clusters, moderate, sweet and citrus fragrance; foliage medium, medium green, semi-glossy; rounded, medium growth; PP9161; AARS, 1996 GM, RNRS, 1990; [Southampton X Remember Me]; Harkness

'HARWELLINGTON', HT, or, 1992; (**High Sheriff**); dbl., 15–25 petals, 1.5–2.75 in., borne in small clusters, slight fragrance; few prickles; foliage large, dark green, glossy; tall (95 cms), upright growth; [Seedling X Silver Jubilee]; Harkness, R., & Co., Ltd.; Harkness New Roses, Ltd.

'HARWESI', HRg, mp, 1979; (**Harvest Home**®, Harwest Home); bud pointed; flowers mauve-pink, semi-dbl., 14 petals, 4.5 in., cupped, abundant early bloom, then sporadic, slight fragrance; foliage light green, wrinkled; bushy growth; [R. rugosa scabrosa X ?]; Spicer, Mrs. W.E.; Harkness

Harwest Home, *see* 'HARWESI'

'HARWESTSUN', F, or, 1988; (**Harold Macmillan**); flowers medium, borne in sprays of 3–7, dbl., 18 petals, cupped, slight fragrance; rounded, medium, green fruit; prickles broad, medium, green; foliage medium, medium green, glossy, abundant; bushy, medium growth; [Avocet X Remember Me]; Harkness, R., & Co., Ltd.; R. Harkness & Co., Ltd., 1989

'HARWEX', F, or, 1990; (**Spirit of Pentax**); bud pointed; flowers bright red, bright orange-red reverse, aging deeper, to roun, dbl., 21 petals, exhibition form, slight fragrance; prickles straight, narrow, small to medium, medium green; foliage small to medium, dark green, glossy; upright, medium growth; [Alexander X Remember Me]; Harkness, R., & Co., Ltd.; R. Harkness & Co., Ltd.

'HARWHARRY', HT, mr, 1987; (**Malcolm Sargent**, Natascha); flowers shining, bright crimson, urn shaped, loose, medium, borne usually singly, dbl., 25 petals, slight, spicy fragrance; rounded, medium, green fruit; glossy, pointed, narrow, small, reddish-green prickles; foliage medium, dark green; bushy, medium growth; GM, Belfast, 1990; [Herbstfeuer X Trumpeter]; Harkness, R., & Co., Ltd., 1988

'HARWICKLOW', F, op, 1990; (**The Fisherman's Cot**); bud pointed; flowers light salmon pink, aging to deeper pink, medium, borne in sprays, dbl., 28 petals, cupped, moderate, sweet, slightly pungent fragrance; foliage small to medium, dark green, semi-glossy; [Radox Bouquet X Anna Ford]; Harkness, R., & Co., Ltd.; R. Harkness & Co., Ltd., 1991

HARWIGWAM, Cl Min, op, 1991; (**Indian Summer**); Harkness

HARwilla, MinFl, dy, 1990; (**Emily Louise**); Harkness

'HARWINNER', Min, w, 1989; (**Pandora**); bud ovoid; flowers ivory, reverse same, rosette, medium, borne in sprays of 3–15, very dbl., 100 petals, slight fragrance; fruit not a noticeable feature; prickles thin, small, reddish; foliage small, medium green, semi-glossy; bushy, spreading, low, compact growth; [Clarissa X Darling Flame]; Harkness, R., & Co., Ltd.

HARwoey, HT, ab; (**Yesteryear**); Harkness, 1994

'HARWOLAVE', Min, dp, 1988; (**Guiding Spirit**); bud ovoid; flowers deep pink, reverse lighter, medium, borne in sprays of 3–9, dbl., flat, slight fragrance; prickles needle-like, very small; foliage small, dark green, semi-glossy; low, bushy growth; [(Blue Moon X Unnamed seedling) X Little Prince]; Harkness, R., & Co., Ltd.; R. Harkness & Co., Ltd., 1989

'HARWONDER', F, ob, 1990; (**Pride of Maldon**); bud pointed; flowers bright reddish-orange, light orange-yellow reverse, darkening, semi-dbl., 10 petals, cupped, slight fragrance; prickles straight or slightly curved, medium to small, green; foliage medium, dark green, glossy; bushy, medium growth; [Southampton X Wandering Minstrel]; Harkness, R., & Co., Ltd.; R. Harkness & Co., Ltd., 1991

HARwotnext, F, w, 1988; (**Sheila MacQueen**); flowers chartreuse green with apricot tint at certain seasons, mediu, dbl., 24 petals, cupped, slight, peppery fragrance; rounded, medium, green fruit; prickles broad, straight, green; foliage medium, medium green, semi-glossy; upright, medium growth; [Greensleeves X Letchworth Garden City]; Harkness, R., & Co., Ltd.

HARxaglen, LCl, ab, 1998; (**Brass Monkey**); Harkness

'HARXAMPLE', F, mr, 1992; (**Velvia**); dbl., 26–40 petals, 3–3.5 in., borne in small clusters, slight fragrance; some prickles; foliage medium, dark green, semi-glossy; medium, bushy growth; [Dr Darley X Trumpeter]; Harkness, R., & Co., Ltd.; Harkness New Roses, Ltd., 1991

'HARXAMPTON', F, mr, 1992; (**Remembrance**); dbl., 32 petals, 3.5 in., cupped, borne in sprays of 5–7, slight fragrance; foliage dark green, glossy; bushy growth; Golden Prize, Glasgow, 1995; [Trumpeter X Southampton]; Harkness, R., & Co., Ltd., 1983; R. Harkness & Co., Ltd.

HARxever, F, mp, 1996; (**Joy of Health**); Harkness

'HARYAMBER', Min, ob, 1992; (**Cottage Garden**); flowers deep rich orange, blooms borne in large clusters, dbl., 26–40 petals, 1.5–2.75 in.; some prickles; foliage small, dark green, glossy; low (60 cms), upright growth; [Clarissa X Amber Queen]; Harkness, R., & Co., Ltd.; Harkness New Roses, Ltd.

'HARYEARN', HT, mr, 1990; (**Lady Mitchell**); bud pointed; flowers deep rose-red, reverse to rose-red, paling with age, very dbl., 50 petals, cupped, moderate fragrance; prickles slightly declining, medium, green; foliage medium, medium green, semi-glossy; bushy, low to medium growth; [Dr Darley X Silver Jubilee]; Harkness, R., & Co., Ltd.; R. Harkness & Co., Ltd., 1991

'HARYEN', HT, ab, 1990; (**Mary Jean**, Mary-Jean); bud ovoid to pointed; flowers large blooms borne usually singly, dbl., 37 petals, cupped, moderate, sweet fragrance; prickles slightly decurved, medium, green; foliage medium to large, oval, medium green, semi-glossy; bushy, medium growth; [Dr Darley X Amber Queen]; Harkness, R., & Co., Ltd.; R. Harkness & Co., Ltd., 1991

HARyoricks, F, ob, 1991; (**Pat James**); Harkness

HARyup, LCl, mp, 1994; (**High Hopes**); dbl., 32 petals, 4 in., borne in sprays of 1–3, moderate, spicy fragrance; foliage medium, medium green, semi-glossy; upright, tall, climbing growth; [Compassion X Congratulations]; Harkness; Harkness New Roses, Ltd., 1992

HARzap, F, op, 1998; (**Country Life**); Harkness

HARzart, HT, w, 1994; (**Renaissance**); Harkness

HARzazz, HT, (**Sue Hipkin**, Sweet Revelation); Harkness, 1997; (Cavriglia)

'HARzeal', F, mp, 1993; (**Octavia Hill**); flowers very double, blooms borne in sprays of 3–5, 75 petals, 3.25 in., moderate, damask fragrance; foliage medium, dark green, semi-glossy; medium, bushy growth; [Armada X Compassion]; Harkness; Harkness New Roses, Ltd., 1994

HARzelt, HT, or, 1998; (**Hector Berlioz**); Harkness

'HARzest', F, ob, 1993; (**Saga Holiday**); flowers orange, orange-pink reverse, loose, blooms borne in sprays, dbl., 28 petals, 4 in., exhibition form, slight, spicy fragrance; foliage medium, purplish, glossy; bushy, medium growth; [Seedling X Amber Queen]; Harkness, 1985; Harkness New Roses, Ltd., 1993

'HARzippee', HT, pb, 1999; (**LeAnn Rimes**, Perception); flowers rose pink edged ivory yellow, reverse cream, dbl., 26–40 petals, 5 in., exhibition form, borne mostly singly, intense, citrus and rose fragrance; prickles moderate; foliage large, dark green, glossy; upright, tall (4.5 ft) growth; [Dr Darley X Sweetheart]; Harkness; Edmunds' Roses, 1998;, Harkness New Roses, 1997

'HARzodiac', F, lp, 1993; (**Compassionate Friend**); flowers medium pink, lighter at center, reverse light pink, dbl., 24 petals, 4 in., borne in sprays of 5–7, moderate, fruity fragrance; foliage medium, medium green, semi-glossy; spreading, low growth; [Seedling X Memento]; Harkness, 1985; Harkness New Roses, Ltd., 1993

'HARzola', F, lp, 1999; (L'Aimant, **Victorian Spice**™); flowers medium, soft peach pink, old-fashioned, very dbl., 41–50+ petals, 3.5–4 in., cupped, borne in large clusters, intense, damask fragrance; prickles moderate; foliage medium, dark green, glossy; spreading, mounding, bushy, medium (3.5 ft) growth; Harkness, Robert & Philip; J&P, 1994

'HARzorba', LCl, pb, 1993; (**Victorian Spice**); flowers light rose pink, primrose base, reverse salmon pink stained, semi-dbl., 12 petals, 4.75 in., moderate, fruity fragrance; foliage large, dark green, glossy; upright, tall, climbing growth; [Armada X Compassion]; Harkness, 1985; Harkness New Roses, Ltd., 1993

HARzumber, HT; (**Welwyn Garden Glory**); Harkness, 1996; (Cavriglia)

'**Hassan**', F, mr, 1963; flowers scarlet, dbl., 28 petals, 4 in., moderate fragrance; foliage glossy, light green; vigorous, upright growth; [Tivoli X Independence]; McGredy, Sam IV; Fisons Horticulture

'**Hassi-Messaoud**', LCl, or, 1961; flowers garnet-red shaded orange, borne in clusters of 12–15, abundant, recurrent bloom; Hémeray-Aubert

Hat Pin, *see* 'TINpin'

'**Hat Trick**', HT, pb, 1992; flowers pink, darker pink petal edges and reverse, natural recurve, very dbl., 3–3.5 in., slight fragrance; few prickles on upper half of stems; foliage medium, dark green, semi-glossy; medium (90–120 cms), upright growth; [First Prize X Unnamed seedling]; Lienau, David W.; Trophy Roses, Ltd., 1993

'**Hatakeyama**', HT, dp, 1995; flowers deep pink, full (26–40 petals), large (7+ cms) blooms; slight fragrance; some prickles; foliage medium, dark green, matt; medium, upright growth; [Garden Party X Kolner Karneval]; Shimizu, Junji

Hatchell Brown Tea, T, lp; Clark

Hatsukoi, HT, w

'**Hatuzakura**', HT, ab, 1999; (Orange Chateau); flowers apricot-orange, very dbl., 80 petals, 5.5 in., exhibition form, moderate fragrance; 4 ft growth; Teranishi, K.; Itami Rose Nursery, 1997

'**Hauff**', HMult, m, 1911; flowers reddish violet, borne in clusters, dbl., recurrent when established; foliage dark, broad; vigorous, climbing growth; [Aimee Vibert X Crimson Rambler]; Lambert, P.

'HAUmi', HT, mp, 1981; (**Sweet Mimi**®); flowers large, dbl., 20 petals, moderate fragrance; foliage medium, reddish green, semi-glossy; bushy growth; [Tropicana X Elizabeth Harkness]; Hauser, Victor; Roseraies Hauser

Hauraki, *see* 'MACwynscar'

HAUric, Min, op, 1989; (**Ricky**®); Hauser

'**Hauser**', HT, my, 1975; flowers yellow-cream, dbl.; vigorous growth; [Barbara X Guitare]; Gaujard

Haute Pink, *see* 'JAChop'

'HAVam', F, or, 1972; (**Amsterdam**®); bud ovoid; flowers clear orange-red, semi-dbl., 12–15 petals, 2.5 in.; foliage glossy, brown-red; vigorous growth; GM, The Hague, 1972; [Europeana X Parkdirektor Riggers]; Verschuren, Ted; Verschuren

'**Havana**', HT, pb, 1950; flowers salmon-rose, reverse orange-yellow, dbl., 40 petals, 5–5.5 in., slight fragrance; foliage soft; vigorous, compact growth; [Peace X Orange Nassau]; Fisher, G.; Arnold-Fisher Co.

'HAVaps', HT, dp, 1985; (**Poker**®); flowers deep pink, large blooms in sprays of 6–8, semi-dbl., 11 petals, cupped, intense fragrance; rounded fruit; foliage medium, medium green, semi-glossy; bushy, medium growth; [Red Planet X Sonia]; Verschuren, Ted; H.A. Verschuren

'HAVeal', HT, dr, 1984; (**Eagle**®); flowers large blooms in sprays of 3–5, dbl., 28 petals, cupped, moderate fragrance; foliage large, dark, semi-glossy; bushy, tall growth; [Centurio X Red Planet]; Verschuren, Ted; H.A. Verschuren

'**Havering**', HMsk, mp, 1937; flowers china-pink, large, borne in clusters of 4 or 5, moderate fragrance; vigorous growth; Bentall

'**Havering Rambler**', HMult, mp, 1920; flowers almond-blossom-pink, rosette form, small, dbl., borne in large clusters; long stems; very vigorous growth; Pemberton

'HAVink', Cl HT, lp, 1980; (**Pink Ocean**®); flowers large, dbl., 20 petals, intense fragrance; foliage medium, medium green semi-glossy; upright (to 7 ft) growth; [Pink Showers X Alexander]; Verschuren; H.A. Verschuren

'HAVipip', Min, pb, 1981; (**Friendship**); bud ovoid; flowers pink and white blend, small blooms in sprays of 30–50, single, 5–6 petals, no fragrance; very small fruit; foliage medium, light green, semi-glossy; groundcover; spreading, medium growth; [(Swany X Mozart) X Mozart]; Verschuren, Ted; H.A. Verschuren

'**Havlickova Národni**', HP, rb, 1935; flowers dark red with white, medium, semi-dbl.; Böhm, J.; (Sangerhausen)

'HAVoon', Min, dy, 1984; (**Gold Moon**®); bud ovoid; flowers small blooms borne singly, semi-dbl., 15 petals, slight fragrance; no prickles; foliage medium, medium green, glossy; spreading, low growth; (28); [(Aalsmeer Gold X Unnamed seedling) X (Motrea X Golden Times)]; Verschuren, Ted; H.A. Verschuren

'HAVop', HT, dr, 1974; (**Centurio**®); flowers lasting well; moderate, bushy growth; [Orangeade X Baccará]; Verschuren, Ted; Verschuren

'HAVychi', Min, dy, 1984; (**Sunny Child**®); flowers small, single, 8 petals, slight fragrance; no prickles; foliage medium, light green; spreading, low growth; [Gold Bunny X Unnamed seedling]; Verschuren, Ted; H.A. Verschuren

'**Hawa Mahal**', HT, op, 1976; flowers salmon-pink, dbl., 25 petals, 5–6. in., moderate fragrance; foliage dark; [Fragrant Cloud X Kordes' Perfecta]; Harkness; Anand Roses

'**Hawaii**', HT, or, 1960; bud long, pointed; flowers orange-coral, dbl., 33 petals, 6 in., exhibition form, intense fragrance; foliage leathery; vigorous, upright growth; [Golden Masterpiece X Unnamed seedling]; Boerner; J&P

'**Hawaiian Beauty**', HT, pb, 1993; flowers have salmon pink outer petals, light pink inner petals, very dbl., 5–5.5 in., moderate fragrance; few prickles; foliage medium, medium green, semi-glossy; medium (4½–5 ft), bushy growth; [Thriller X Dothan]; Wyckoff, Gilbert R.; Wyckoff, 1994

'HAWAIIAN BELLE', Min, pb, 1982; flowers medium pink, aging to pink blend, small, dbl., 35 petals, slight fragrance; foliage small, medium green, matt; bushy growth; [Pink Ribbon X Pink Ribbon]; Dobbs, Annette E.

'HAWAIIAN DELIGHT', F, op, 1968; flowers burnt-orange to pink, small, dbl., cupped; foliage dark, leathery; vigorous, bushy growth; [Orange Sensation X Circus]; deRuiter; Carlton Rose Nurseries

'HAWAIIAN FRAGRANCE', HT, or, 1994; flowers coral, orange-red, similar to Fragrant Cloud, dbl., 26–40 petals, 5–6 in., intense fragrance; some prickles; foliage large, medium green, semi-glossy; tall (6+ft), upright growth; [Fragrant Cloud X Captain Harry Stebbings]; Wyckoff, Gilbert R.; Wyckoff, 1994

'HAWAIIAN KING', HT, lp, 1993; very dbl., 3–3.5 in., borne mostly singly, moderate fragrance; some prickles; foliage medium, medium to dark green,semi-glossy; tall (5–6 ft), upright growth; [Peggy Lee X Captain Harry Stebbings]; Wyckoff, Gilbert R.; Wyckoff, 1994

'HAWAIIAN LADY', HT, dp, 1993; flowers deep pink, blooms borne mostly singly, very dbl., 3–3.5 in., moderate fragrance; many prickles; foliage medium, medium green, semi-glossy; medium (5 ft), upright, bushy growth; [Peggy Lee X Captain Harry Stebbings]; Wyckoff, Gilbert R.; Wyckoff, 1994

'HAWAIIAN QUEEN MARTHA', HT, op, 1994; flowers salmon, interspered with light pink on both petal surfaces, very dbl., 5.5–6 in., intense fragrance; some prickles; foliage medium, medium green, semi-glossy; medium (4 ft), upright growth; [Kordes Perfecta X Dothan]; Wyckoff, Gilbert R.; Wyckoff, 1995

'HAWAIIAN SUNRISE', Min, rb, 1981; bud pointed; flowers red and yellow blend, blooms borne usually singly, dbl., 40 petals, exhibition form, slight fragrance; thin, reddish prickles; foliage small, dense, glossy, bronze; upright growth; [Unnamed seedling X Over the Rainbow]; Williams, Ernest D.; Mini-Roses

'HAWAIIAN SUNSET', HT, ob, 1962; bud ovoid; flowers orange edged yellow, open, dbl., 45–50 petals, 4–5.5 in., moderate fragrance; foliage leathery, glossy; vigorous, upright, well branched growth; [Charlotte Armstrong X Signora]; Swim & Weeks; C.R. Burr

'HAWAIIAN THRILL', HT, pb, 1993; flowers pink, white center, moderately blooms borne mostly singly, dbl., 23–25 petals, 5.5 in., intense fragrance; some prickles; foliage medium, dark green, semi-glossy; medium (4½–5 ft), upright growth; [Fragrant Cloud X Thriller]; Wyckoff, Gilbert R.; Wyckoff, 1994

Hawkesbury Wonder, S, lp; [Hebe's Lip X Unknown]; (Weatherly, L.)

'HAWKEYE BELLE', S, w, 1975; bud ovoid, pointed; flowers white, tinted azalea-pink, dbl., 38 petals, 4–4.5 in., exhibition form, intense fragrance; foliage large, dark, leathery; vigorous, erect, bushy growth; [(Queen Elizabeth X Pizzicato) X Prairie Princess]; Buck, Dr Griffith J.; Iowa State University

'HAWLMARK CRIMSON', HT, mr, 1920; bud pointed; flowers crimson-scarlet, semi-dbl., moderate fragrance; bushy growth; Dickson, A.

'HAWLMARK SCARLET', HT, mr, 1923; flowers brilliant velvety scarlet-crimson, moderate fragrance; GM, NRS, 1920; Dickson, A.

Haynesville Pink Cluster, N, pb

'HAZEL ALEXANDER', HT, dr, 1933; flowers deep red, intense fragrance; [Ophelia seedling]; Dicksons of Hawlmark

'HAZEL ROSE', HT, lp, 1992; dbl., 26–40 petals, 3–3.5 in., borne singly, slight fragrance; few prickles; foliage medium, dark green, semi-glossy; medium (120 cms), upright growth; [Queen Esther X Selfridges]; Poole, Lionel, 1993

'HAZELDEAN', HSpn, my, 1948; flowers more open, of better form than Persian Yellow, moderate fragrance; very hardy (to –60); [R. spinosissima altaica X Persian Yellow]; Wright, Percy H.

Head of Rivers, D, mp

'HEADLEYENSIS', S, ly; (R. headleyensis); flowers creamy yellow, single; vigorous growth; [R. hugonis X R. spinosissima altaica]

Headline, HT, yb, 1970; Dawson

Headliner, *see* 'JACTU'

Healdsburg Glory, HWich, m

'HEALING HANDS', HT, yb, 1998; flowers pale yellow, edged pink, exhibition form, very dbl., 26–40 petals, 5.5 in., exhibition form, borne mostly singly, slight fragrance; some prickles; foliage large, dark green, glossy; upright, bushy, medium growth; [Gavotte X (Peer Gynt X Golden Splendour)]; Poole, Lionel; David Lister, Ltd., 1999

Heart O' Gold(TM), *see* 'WEKDYKSTRA'

'HEART OF ENGLAND', Gr, dp, 1978; flowers carmine-rose, dbl., 42 petals, 4 in., exhibition form, moderate fragrance; moderately vigorous growth; [Pink Parfait X ?]; Gregory

'HEART OF GOLD', HMoy, rb, 1926; flowers crimson, center white, stamens yellow, open, borne in clusters, single; foliage rich green; vigorous (10 ft) growth; [R. wichurana X R. moyesii]; Van Fleet; American Rose Society

Heart O' Gold, *see* 'WEKDYKSTRA'

Heart of Gold, *see* 'MACYELKIL'

'HEART OF T. D. K.', HT, op, 1983; bud ovoid; flowers salmon-pink, large blooms borne singly, dbl., 50 petals, slight fragrance; heavy hooked prickles; foliage small, light green; vigorous, upright, tall growth; [Sunblest X Red Devil]; Ogawa, Isamu

Heart Throb, *see* 'LEONHART'

'HEART THROB', F, 1966; bud urn shaped; flowers deep yellow, edges flushed pink, medium, cluster, very dbl., intense fragrance; foliage dark, glossy; vigorous, bushy growth; RULED EXTINCT 4/82 ARM; [(Circus X Circus) X (Circus X Circus)]; Fankhauser

Heart Throb, *see* 'HARQUETERWIFE'

Heartache, *see* KOReledas

'HEARTBEAT', F, op, 1970; flowers deep salmon-orange, dbl., 26 petals, 4.5 in., globular, slight fragrance; foliage small, dull; very free growth; [(Castanet X Castanet) X (Cornelia X Seedling)]; Dickson, Patrick; A. Dickson

'HEARTBEAT', HT, mr, 1989; bud pointed; flowers medium red, darker outer petals, reverse lighter, medium, dbl., 45 petals, exhibition form, borne usually singly, slight, fruity fragrance; prickles straight, large, pink to yellow; foliage medium, medium green, semi-glossy; upright growth; [Thriller X Wild Cherry]; Bridges, Dennis A.; Bridges Roses, 1990

Heartbeat 96, *see* 'COCORONA'

Heartbreaker, *see* 'WEKSYBIL'

'HEARTH GLOW', F, or, 1963; bud ovoid; flowers brick-red, medium, borne in clusters, dbl.; foliage soft, light green; vigorous, upright growth; [Red Pinocchio X (Carrousel X Queen o' the Lakes)]; Von Abrams; Peterson & Dering

Heartland(TM), *see* 'SAVSAY'

Heartlight(TM), *see* 'KINHEART'

Hearts A'Fire, *see* 'BRIHEART'

'HEART'S DELIGHT', HT, op, 1933; flowers apricot-coral-orange, veined red, large, very dbl., exhibition form, intense fragrance; foliage soft; vigorous, open habit growth; [Mrs Beckwith sport]; Hart, L.P.

'HEART'S DESIRE', HT, dr, 1942; bud long, pointed; flowers crimson, dbl., 30 petals, 4.5 in., exhibition form, intense, damask fragrance; foliage leathery, dark; vigorous, upright growth; AARS, 1942 GM, Portland, 1941; [Unnamed variety X Crimson Glory]; Howard, F.H.; H&S

Heart's Desire Pink, HWich, pb

'HEART'S DESIRE, CLIMBING', Cl HT, dr, 1945; Howard, F. H.; H&S

Heartsounds, *see* 'MINSOUNDS'

Heartstrings, *see* 'BROHEART'

'HEAT WAVE', F, or, 1958; (Mme Paula Guisez); bud urn-shaped; flowers orange-scarlet, blooms in clusters, dbl., 30 petals, 3.5–4.5 in., cupped, slight fragrance; foliage dark, semi-glossy, rounded; vigorous, upright, bushy growth; [Unnamed seedling X Roundelay]; Swim, H.C.; Armstrong Nursery

Heather, *see* 'JACLAV'

Heather Austin, *see* 'AUSCOOK'

'HEATHER CLAIRE', HT, ab, 1982; bud long; flowers apricot pink, blooms borne singly, dbl., 30 petals, no fragrance; slightly hooked prickles; foliage dark, red reverse; medium growth; [Diamond Jubilee X Bonsoir]; Allender, Robert William

Heather Honey, *see* 'HORSILBEE'

'HEATHER JENKINS', HT, mp, 1968; bud globular; flowers pink, reverse darker, medium, dbl., exhibition form, slight fragrance; foliage light green, wrinkled; moderate, upright, open growth; [Charlotte Armstrong X Ballet]; Watson

Heather Leigh, *see* 'TALHEATHER'

'HEATHER MUIR', sp, (variety of R. sericea), w, 1957; flowers pure white, single, 3 in., blooms over a long period, moderate fragrance; orange fruit; foliage ferny; Sunningdale Nursery

'HEATHER PATON', HT, mr, 1934; flowers carmine, center darker; vigorous growth; Austin & McAslan

Heather Pudney, HT, mp, 1995; Dawson

Heaven, *see* 'JACFON'

Heaven Bound, F, mp, 1999; flowers bright pink, old-fashioned, ruffled, dbl.; compact (2.5 ft) growth; Williams, J. Benjamin

'HEAVEN SCENT', F, op, 1968; flowers salmon, large blooms in trusses, dbl., 30 petals, intense fragrance; [Pernille Poulsen X Isabel de Ortiz]; Poulsen; McGredy

Heaven Scent Pink, *see* 'JALPINK'

Heavenly Days™, *see* 'SAVAHE'

'HEAVENLY FRAGRANCE', HT, lp, 1963; flowers light pink, reverse darker, well formed, 5 in., intense fragrance; moderate growth; [Tiffany X Mme Gregoire Staechelin]; Hennessey

Heavenly Pink, *see* LENneel

Heavenly Rosalind, *see* 'AUSMASH'

'HEAVENLY SCENT', S, mp, 1999; flowers cerise pink, lighter reverse, large, very dbl., 58 petals, borne mostly singly in and small clusters, moderate fragrance; prickles moderate; foliage medium, dark green, glossy; upright, compact, medium (5 ft) growth; Hamilton, Noel

Heavenly Vision, *see* 'MINIVISION'

'HEBE', F, dp, 1941; flowers deep pink, reverse lighter; Leenders, M.

'HEBE', HT, op, 1949; flowers rosy salmon toned orange and apricot-yellow, high pointed, large, dbl., 27 petals, moderate fragrance; foliage glossy, bronze green; vigorous growth; GM, NRS, 1949; Dickson, A.

Hebe Camargo, HT; (Brazil); (Cavriglia)

'HEBE'S LIP', HEg, w; (R. damascena rubrotincta, Reine Blanche, Rubrotincta); flowers creamy white, petals edged pink, vigorous growth, semi-dbl., cupped, nonrecurrent; [Probably R. damascena X R. eglanteria hybrid.]; Lee, before 1846; Re-int. W. Paul, 1912

Heckenfeuer®, *see* KORrohe

Heckengold, F, dy, 1986; flowers large, dbl., slight fragrance; Kordes, W. Söhne; (Sangerhausen)

Heckenzauber, *see* 'MACREXY'

'HECTOR', HGal, m; flowers purple, faintly striped with white, pompon, small, dbl.; Parmentier, before 1847

Hector Berlioz, *see* HARzelt

'HECTOR DEANE', HT, rb, 1938; bud pointed; flowers orange, carmine and salmon-pink, dbl., exhibition form, intense, fruity fragrance; foliage glossy, dark; vigorous, compact growth; [McGredy's Scarlet X Lesley Dudley]; McGredy; J&P

'HEDDA HOPPER', HT, mp, 1952; bud ovoid; flowers light peach passing to pearly pink, dbl., 40 petals, 3.5–4 in., globular, moderate fragrance; foliage coppery; very vigorous growth; [Radiance X Unnamed seedling]; Howard, A. P.; H&S

'HEDE', HT, my, 1934; flowers pure sunflower-yellow, large, dbl., slight fragrance; foliage dark, leathery; strong stems; vigorous growth; [Prof. Gnau X Mev. G. A. van Rossem]; Tantau

Hedgefire, MinFl, mr, 1983; Kordes

Hedgehog Rose, *see* **R. RUGOSA**

'HEDWIG FULDA', Cl Pol, mr, 1934; flowers clear vermilion-red, well formed, large, borne in large clusters, dbl.; foliage bright, dark; long, strong stems; vigorous growth; [Orléans Rose X Farbenkonigin]; Leenders Bros.

'HEDWIG REICHER', HT, ly, 1912; flowers large, very dbl., moderate fragrance; Hinner, W.; (Sangerhausen)

'HEER', HT, lp, 1969; flowers rose-pink, medium, dbl., exhibition form, slight fragrance; vigorous, upright growth; [Picture X Open pollination]; Singh; Gopalsinamiengar

'HEFQUEEN', Min, pb, 1989; (**Regine**™); flowers soft, light pink, silvery-pink reverse, small, dbl., 30 petals, slight fragrance; foliage medium, medium green, semi-glossy; bushy growth; AOE, 1990; [Little Darling X Party Girl]; Hefner, John; Kimbrew Walter Roses, 1990

Hei Matisse, *see* DELstrabla

Hei W. Perron, HT, or

Heidefee®, Pol, dp, 1990; Noack, Werner

Heidefeuer, *see* NOAfeuer

'HEIDEGRUSS', F, pb, 1937; flowers salmon-flesh, base light yellow, large, borne in clusters, very dbl.; foliage leathery; vigorous, bushy growth; [Heidekind X Ophelia]; Tantau; Münch & Haufe

'HEIDEKIND', HRg, dp, 1931; flowers brilliant pink shaded copper-red, large blooms in clusters, dbl., slight fragrance; foliage thick, rugose; [Mev. Nathalie Nypels X R. rugosa hybrid]; Berger, V.; Münch & Haufe

Heidekönigin®, *see* 'KORDAPT'

Heidelberg, *see* 'KORBE'

Heidelinde®, *see* KORdehei

'HEIDEMARIE', F, mr, 1945; flowers carmine-red, large, borne in clusters of 4–6, single, 5–7 petals, slight fragrance; foliage dark, leathery; vigorous, upright, bushy growth; [Hamburg X (Heros X Heidekind)]; Tantau

'HEIDEMARIE PLÜCKER', HP, mp, 1940; flowers large, very dbl., intense fragrance; Westphal; (Sangerhausen)

'HEIDERÖSLEIN', LCl, yb, 1932; bud pointed, orange-red; flowers bright yellowish salmon-pink, base sulfur-yellow, large blooms, single, flat, recurrent bloom, moderate fragrance; broad, bushy growth; [Chamisso X Amalie de Greiff]; Lambert, P.

Heideroslein Nozomi, *see* **'NOZOMI'**

Heideschnee, *see* 'KORCONTA'

Heidesommer®, *see* 'KORLIRUS'

Heidetraum™, *see* 'NOATRAUM'

'HEIDEZAUBER', F, dr, 1936; flowers large, borne in clusters, very dbl., slight fragrance; foliage dark, leathery; bushy growth; [Heidekind X Johanniszauber]; Tantau; Münch & Haufe

Heidi™, *see* 'ARODI'

Heidi, S, mp, 1987; Noack, Werner

'HEIDI JAYNE', HT, dp, 1986; flowers bright deep pink, large, dbl., 32 petals, moderate fragrance; numerous prickles; foliage large, light green, glossy; medium, upright growth; [(Piccadilly X Queen Elizabeth) X (Fragrant Cloud X Unnamed seedling)]; Esser; Harkness

Heidi Kabel, *see* 'KORDIAM'

Heidi Rossin, S, ly

Heidikind, *see* 'KORIVER'

Heidi's Gold, Cl HT, dy, 1994

Heietta, T, m

Heiich Blanc, LCl, w, 1994; Hetzel

Heike, *see* KORrundum

'HEIMATLOS', HCan, mp, 1931; bud pointed; flowers rose-pink, borne in clusters, single, non-recurrent, intense fragrance; vigorous (5–7 ft), open habit growth; has endured –27; [(R. canina X R. roxburghii) X R. canina]; Lohrberg

'HEIN EVERS', F, mr, 1957; bud pointed; flowers bright blood-red, open, borne in clusters, semi-dbl., slight fragrance; foliage leathery; vigorous, upright growth; [Red Favorite X Fanal]; Tantau, Math.

'HEIN EVERS, CLIMBING', Cl F, mr, 1963; Kordes

'HEIN MÜCK'®, S, dr, 1961; flowers velvety blood-red, blooms in clusters, single, cupped; vigorous (6 ft) growth; Tantau, Math.

'HEINE', S, w, 1912; flowers small, dbl.; Lambert, P.; (Sangerhausen)

Heinfels, *see* 'HELFELS'

'HEINRICH CONRAD SOTH', S, pb, 1919; flowers light rosy red, with white eye, small, borne in pyramidal clusters, single, recurrent bloom, moderate fragrance; foliage large, glossy, dark; long, strong stems; very vigorous, bushy growth; [Geheimrat Dr Mittweg X R. foetida bicolor]; Lambert, P.

'HEINRICH EGGERS', HT, op, 1928; flowers orange-copper, often with lighter outer petals, dbl., moderate fragrance; upright growth; [Mrs Charles E. Russell X Mrs Wemyss Quin]; Kordes

'HEINRICH KARSCH', Pol, m, 1927; flowers violet-rose, semi-dbl.; [Orléans Rose X Joan]; Leenders, M.

'HEINRICH MÜNCH', HP, mp, 1911; flowers soft pink, very large, dbl., 50 petals, occasionally recurrent bloom, moderate fragrance; very vigorous growth; [Frau Karl Druschki X (Mme Caroline Testout X Mrs W.J. Grant)]; Hinner, W.; Münch & Haufe

'HEINRICH SCHULTHEIS', HP, lp, 1882; flowers soft pink, well-formed, very large, dbl., occasionally recurrent bloom, intense fragrance; vigorous growth; [Mabel Morrison X E.Y. Teas]; Bennett

'HEINRICH WENDLAND', HT, mr, 1930; flowers nasturtium-red, reverse deep golden yellow, very large, dbl., exhibition form, intense, fruity fragrance; foliage bronze, leathery, glossy; vigorous growth; [Charles P. Kilham X Mev. G.A. van Rossem]; Kordes; Dreer; H&S

'HEINRICH WENDLAND, CLIMBING', Cl HT, mr, 1937; Stell; Stell Rose Nursery

'HEINSOHN'S RECORD', HCan, lp; Heinsohn-Wedel

'HEINZ ERHARDT'®, F, mr, 1962; dbl., 25 petals, 3 in., blooms in clusters (up to 8), slight fragrance; foliage coppery; vigorous, bushy growth; GM, Baden-Baden, 1961; Kordes, R.

Heinz Treffinger, *see* HELtreff

Heinzelmännchen®, *see* 'KORNUMA'

'HEIRESS', HT, mp, 1959; bud pointed; flowers clear rose-pink, well shaped, medium, intense fragrance; Longsdon

Heirloom, *see* 'JACLOOM'

'HELDENGRUSS', HT, dr, 1920; flowers pure deep blood-red, dbl., moderate fragrance; [Étoile de France X Baron Girod de l'Ain]; Kiese

HELdoro, HWich, mp; (**Super Dorothy®**); Hetzel, 1986

'HELEN', HT, pb, 1930; flowers salmon-pink, base shaded yellow, semi-dbl., moderate fragrance; vigorous growth; RULED EXTINCT 11/91; Ferguson, W.

Helen, *see* 'TINHELEN'

'HELEN ALLEN', HT, dr, 1975; flowers velvety red, dbl., 20 petals, 3.5–4 in., moderate fragrance; upright growth; [Evelyn Fison X Vagabonde]; Clayworth

'HELEN BLAND', S, mp, 1950; flowers rose-pink, center deeper, open, medium, borne several together, semi-dbl., profuse non-recurrent bloom, slight fragrance; thornless; foliage soft; stems red-brown; vigorous (7–8 ft), upright growth; [Betty Bland X R. blanda (St Hilaire clone)]; Wright, Percy H.

Helen Boehm, *see* 'AROPRAWN'

'HELEN CHAMBERLAIN', HT, yb, 1918; flowers creamy yellow to orange-gold, paling on outer petals; Easlea

'HELEN DAVIS', HT, dr, 1925; flowers large, dbl.; (Sangerhausen)

'HELEN FOX', HT, my, 1928; bud pointed, indian yellow; flowers golden yellow, dbl., cupped, slight fragrance; foliage bronze; vigorous, bushy growth; GM, Bagatelle, 1926; [Mme Mélanie Soupert X Souv. de Claudius Pernet]; Buatois

Helen Gambier, *see* **'MLLE HÉLÈNE GAMBIER'**

'HELEN GOOD', T, yb, 1907; bud pointed; flowers delicate yellow suffused pink, edged deeper, dbl.; vigorous growth; [Maman Cochet sport]; Good & Reese

Helen Gould, *see* **'BALDUIN'**

'HELEN GOULD, CLIMBING', Cl HT, pb, 1912; [Balduin sport]; Good & Reese

'HELEN HAYES', HT, yb, 1956; bud long, pointed; flowers yellow splashed orange and pink, dbl., 43 petals, 4–5 in., exhibition form, moderate fragrance; foliage glossy; very vigorous growth; [Unnamed R. wichurana hybrid X Sutter's Gold]; Brownell, H.C.

'HELEN JANE BURN', HT, dp, 1999; flowers high-pointed, classic form, dbl., 26–40 petals, 5.5–6 in., borne mostly singly, slight fragrance; some prickles; foliage medium, dark green, semi-glossy; long stems; upright, bushy, medium (3 ft) growth; [Blue Moon X seedling]; Poole, Lionel

'HELEN KELLER', HP, dp, 1895; flowers rosy cerise, petals large, shell shaped, very free bloom; GM, NRS; Dickson, A.

Helen Keller, *see* BARkel

Helen Knight, S, my, 1970; Knight

'HELEN LEENDERS', S, mp, 1924; flowers hydrangea-pink, open, large, borne in clusters, semi-dbl., sometimes recurrent bloom, moderate fragrance; foliage large, rich green; very vigorous, (5 ft), bushy growth; [Orléans Rose X R. foetida bicolor]; Leenders, M.

Helen Margaret, *see* 'HORBLUSH'

'HELEN M. GREIG', HT, lp; flowers pastel pink; [Mrs A.R. Barraclough X Marmion]; Dobbie

Helen Naude, *see* 'KORDIENA'

'HELEN OF TROY', HT, mp, 1956; flowers rose-pink, very dbl., slight, spicy fragrance; vigorous, bushy growth; [Dame Edith Helen X Mrs Henry Morse]; Stevenson; Waterer

Helen Suzman Rose, F, lp

'HELEN TAYLOR', HT, pb, 1924; flowers rosy salmon, dbl., moderate fragrance; Pemberton

'HELEN TRAUBEL', HT, pb, 1951; bud long, pointed; flowers pink to apricot, dbl., 23–30 petals, 5–6 in., exhibition form, borne singly, moderate, fruity fragrance; foliage leathery, matt, green; weak necks; tall, vigorous growth; AARS, 1952 GM, Rome, 1951; [Charlotte Armstrong X Glowing Sunset]; Swim, H.C.; Armstrong Nursery

Helen Traubel, Climbing, Cl HT, ab, 1970; Treloar

'HELEN WILD', HT, op, 1959; flowers orange-pink veined rose-red, dbl., 35 petals, 5–6. in., exhibition form, moderate fragrance; foliage light green; free growth; [Show Girl X Charlotte Armstrong]; Kemp, M.L.

Helena, *see* POUlna

Helena, HT, mr; Strnad; (Czech Rosa Club)

Helena Renaissance, *see* POUlna

'HELENA VAN VLIET', Pol, lp, 1931; flowers soft pink tinted salmon, borne in large trusses; vigorous growth; [Salmonea sport]; Kersbergen

Helene, HT, mp; flowers large, dbl., intense fragrance; Vecera, L.; (Sangerhausen)

'HÉLÈNE', HMult, pb, 1897; flowers soft violet-rose, base yellowish white, borne in clusters, semi-dbl., slight fragrance; vigorous, climbing (12–15 ft) growth; [Hybrid Tea seedling X (Aglaia X Crimson Rambler)]; Lambert, P.

'HÉLÈNE DAPPLES', HT, mr, 1932; bud pointed, dark; flowers glowing crimson-red, intense fragrance; vigorous growth; [Mrs Henry Winnett X Lady Maureen Stewart]; Heizmann, E.

Helene de Gerlache, HT, w, 1982; Lens

'HÉLÈNE DE MONTBRIAND', HT, mr, 1933; flowers deep carmine-red, shaded vermilion, well formed, large, globular; foliage glossy, dark; vigorous growth; [Reine Marie Henriette X Laurent Carle]; Schwartz, A.

'HÉLÈNE DE ROUMANIE', HT, rb, 1949; flowers red to pink, urn shaped, dbl., 35 petals, 5 in., slight fragrance; upright growth; GM, NRS, 1950; [(Mme Joseph Perraud X Unnamed seedling) X (Unnamed seedling X Pres. Herbert Hoover)]; Meilland, F.

Helene de Savoie, *see* KRInirosy

'HÉLÈNE DUCHÉ', HT, lp, 1921; flowers very large, dbl., moderate fragrance; Buatois; (Sangerhausen)

'HÉLÈNE FRANÇOIS', HT, pb, 1923; flowers salmon-pink shaded coppery red, center salmon-orange tinted, dbl.; [Mme Edouard Herriot X Viscountess Enfield]; Schwartz, A.

'HÉLÈNE GRANGER', HMult, pb, 1910; flowers pink, center copper-yellow, borne in clusters of 15–20; [Tea Rambler X Aglaia]; Granger

Helene Leenders, Pol, mp; flowers medium large, very dbl., moderate fragrance; medium growth; Leenders, M., 1925; (Sangerhausen)

Helene Marechal, S, 1995; Lens, Louis; (Cavriglia)

'HÉLÈNE ROBINET', HT, lp, 1928; flowers salmon-white, shaded rose, base yellow, dbl., slight fragrance; [Unnamed variety X Pres Parmentier]; Sauvageot, H.; F. Gillot

'HELENE SCHOEN', HT, mr, 1963; bud long, pointed; dbl., 60 petals, 6 in., exhibition form, slight fragrance; foliage leathery, glossy; vigorous, upright growth; [Multnomah X Charles Mallerin]; Von Abrams; Peterson & Dering

'HÉLÈNE VACARESCO', HT, pb, 1939; flowers salmon, shaded copper-carmine, large, cupped; foliage dark; Chambard, C.

'HÉLÈNE VALABRÈGUE', HT, lp, 1953; bud pointed; flowers pale rose, overlarge, dbl., cupped, slight fragrance; foliage leathery; strong stems; vigorous, bushy growth; [Lorraine X Michele Meilland]; Meilland, F.; URS

'HÉLÈNE VIDENZ', Pol, pb, 1905; flowers salmon-pink, borne in clusters to 75, dbl.; vigorous growth; [Euphrosyne X Louis Philippe]; Lambert, P.

Helenka, HT, lp; Tesar, 1989; (Czech Rosa Club)

HELeuro, F, mp, 1989; (**Europas Rosengarten**); flowers medium, dbl.; Hetzel; (Sangerhausen)

HELexa, LCl, mr; (**Super Excelsa®**); Hetzel, 1986

'HELFELS', LCl, dr, 1996; (**Super Sparkle**, Heinfels); flowers crimson scarlet, blooms borne in large clusters, dbl., 15–25 petals, 2 in., slight fragrance; foliage medium, dark green, glossy; rambling, climbing (300 cms) growth; Helzel, Karl; Eurosa, 1997

HELfriwa, HT, or, 1981; (**Fritz Walter**); Hetzel

Helga®, *see* HelgRUI

'HELGA', HT, yb, 1926; Weigand, C.

'HELGOLAND', F, mr, 1936; bud pointed, dark; flowers crimson to carmine, open, very large, borne in clusters, semi-dbl., slight fragrance; foliage leathery, wrinkled, dark; vigorous, bushy growth; [Else Poulsen X Hybrid Tea seedling (dark crimson)]; Kordes

'HELGOLAND', F, mr, 1973; bud pointed; flowers copper-red, medium, semi-dbl., slight fragrance; foliage glossy, light; moderate, upright, bushy growth; Tantau, Math.; Horstmann

HELgraf, HT, mp; (**Steffi Graf**); flowers large, dbl., intense fragrance; Hetzel, 1993; (Sangerhausen)

HelgRUI, F, w, 1975; (**Helga®**); deRuiter

HELilast, S, lp; (**Werner von Braun**); Hetzel, 1993

'HÉLIODORE DOBER', HGal, mr; flowers deep red edged crimson, ball shaped, quite large

'HELIOS', HT, dy, 1935; flowers deep sunflower-yellow, open, semi-dbl.; foliage leathery, light; vigorous, bushy growth; Leenders, M.

'HELKLEGER', LCl, or, 1996; (**Super Elfin**, Strombergzauber); dbl., 15–25 petals, 2 in., borne in large clusters, slight fragrance; foliage medium, dark green, glossy; rambling, medium (300 cms) growth; Helzel, Karl; Eurosa, 1997

Hellen Ann, HT, 1989; Tantau, Math.; (Cavriglia)

Hello, *see* 'COCHELLO'

Hello, *see* KRImony

Hello There, *see* 'FLOELLO'

HELlux, HT, or, 1989; (**Comtesse Diana**); Hetzel

Helma, Min, 1986; Delforge, H.; (Cavriglia)

HELmedia, HT, mr, 1993; (**Medialis**); Hetzel

Helmut Kohl, *see* TANtumleh

Helmut Schmidt®, *see* 'KORBELMA'

'HELP THE AGED', HT, mp, 1986; flowers clear pink, reverse slightly darker, satin two-tone effect, dbl., 23 petals, exhibition form, intense, damask fragrance; no fruit; straight, brown, very few prickles; foliage medium, light green, semi-glossy; bushy, medium growth; [Mischief X Fragrant Cloud]; Bracegirdle, A.J.; Rosemary Roses, 1987

HELpforz, HT, dy, 1998; (**Goldstadt Pforzheim**); Hetzel

Helping Hands, *see* CLEhelp

Helpmekaar Roos, *see* KORogesa

HELprinzess, Pol, dp, 1993; (**Rosenprinzessin**); Hetzel

HELreg, Min, dr, 1983; (**Kleine Regina®**); Hetzel

HELreib, F, dp, 1992; (**Carolin Reiberl**); Hetzel

HELrobu, F, ob, 1993; (**Die Rheinpfalz**); Hetzel

HELsabin, F, mp, 1988; (**Sabine Ruf**); Hetzel

Helsingör, *see* **'ELSINORE'**

HELstrau, F, mp, 1991; (**Derdinger Sommer**); Hetzel

'HELSUFAIR', LCl, lp, 1996; (**Super Fairy®**); dbl., 15–25 petals, 1.25 in., borne in large clusters, repeat bloom, moderate fragrance; foliage medium, medium green, glossy; rambling (300–500 cms) growth; Helzel, Karl; Eurosa, 1997

HELtoria, HT, 1986; (**Kronprinzessin Victoria**); Hetzel, K.; (Cavriglia)

HELtreff, HT, mp, 1998; (**Heinz Treffinger**); Hetzel

HELtzi, Pol, mp, 1994; (**Fritzi**); Hetzel

'HELVETIA', HT, dp, 1911; flowers large, dbl., moderate fragrance; Heizmann, E.; (Sangerhausen)

Helvétia, *see* **'MANDALAY'**

HELway, HT, mr, 1993; (**Euroway**); Hetzel

HELzwerg, Min, m, 1998; (**Lila Zwerg**); Hetzel

Hemavathy, F, op, 1975; Kasturi

'HEN KAUFFMANN', F, mp, 1954; flowers rosy pink, dbl., moderate fragrance; vigorous growth; Leenders, M.

'HENDAN', F, dp, 1993; (**Dancing Pink**); flowers deep pink, blooms borne in large clusters, semi-dbl., 6–14 petals, 3–3.5 in., moderate fragrance; some prickles; foliage medium, medium green, semi-glossy; medium (75 cms), bushy growth; [Southampton X Dortmund]; Henson, R. W.; Henson, 1993

'HENFIF', F, ab, 1993; (**Fifi**); flowers cream pale peach, blooms borne in large clusters, semi-dbl., 6–14 petals, 3–3.5 in., intense fragrance; many prickles; foliage large, dark green, glossy; tall (100 cms), upright growth; [Lichterloh X Liverpool Echo]; Henson, R.W.; Henson, 1993

'HENKELL ROYAL'®, HT, mr, 1964; bud long; flowers blood-red, well-formed, large, intense fragrance; vigorous, bushy growth; GM, Baden-Baden, 1964; Kordes, R.

'HENNEV', F, rb, 1993; (**Nevertheless**); flowers orange/red/yellow blend, blooms borne in small clusters, dbl., 26–40 petals, 1.5–2.75 in.; moderate fragrance; few prickles; foliage large, medium green, matt; tall (100 cms), upright growth; [Seedling (Silver Jubilee X Trumpeter) X GM]; Henson, R.W.; Henson, 1993

'Henri Barruet', LCl, pb, 1918; flowers coppery yellow, opening to pink and tinted white, borne in clusters; vigorous, climbing (8 ft) growth; Barbier

'Henri Brichard', HT, w, 1891; flowers large, very dbl.; Bonnaire; (Sangerhausen)

Henri Caillaud, LCl; Moreira da Silva, A.; (Cavriglia)

'Henri Coupé', HP, mp, 1916; dbl., moderate fragrance; [Frau Karl Druschki X Gruss an Teplitz]; Barbier

'Henri Declinand', LCl, mr, 1934; flowers bright magenta-red, quite large, dbl.; foliage dark; Mermet

'Henri Foucquier', HGal, mp; (Henri Fouquier); flowers pure rose-pink, large, dbl., moderate fragrance

Henri Fouquier, *see* **'Henri Foucquier'**

Henri IV, HP, m; Verdier, 1862

'Henri Linger', LCl, yb, 1928; flowers clear yellow-orange, open, semi-dbl., slight fragrance; foliage light, glossy; very vigorous, climbing growth; [R. wichurana X Benedicte Seguin]; Barbier

'Henri Mallerin', HT, yb, 1953; bud ovoid; flowers empire-yellow suffused pink, large, dbl., 55–70 petals, slight fragrance; foliage leathery, glossy; bushy growth; [Soeur Thérèse X Duquesa de Peñaranda]; Mallerin, C.; EFR

Henri Mallerin, *see* **'Rouge Mallerin'**

'Henri Martin', M, mr; (Red Moss); bud sparsely mossed; flowers shining crimson, blooms in clusters of 3–8, semi-dbl.; Laffay, M., 1862

Henri Matisse™, F, 1996; Delbard, Georges; (Cavriglia)

'Henri Pauthier', HT, mr, 1933; flowers bright red, open, large, semi-dbl., moderate fragrance; foliage glossy; bushy growth; [Unnamed seedling X Edouard Mignot]; Sauvageot, H.; C-P

Henrietta, *see* 'POUletta'

'Henrietta', HT, op, 1917; bud pointed, orange-crimson; flowers soft coral-salmon, open, semi-dbl., moderate fragrance; foliage dark; vigorous growth; RULED EXTINCT 12/85; [Alister Stella Gray X Andre Gamon]; Merryweather

'Henriette', HT, or, 1916; flowers coppery orange-red, large, dbl.; Dickson, A.; (Sangerhausen)

'Henriette Chandet', HT, op, 1942; bud oval; flowers orange-coral, large, dbl., slight fragrance; foliage glossy; vigorous, bushy growth; [Rochefort X La Parisienne]; Mallerin, C.; A. Meilland

'Henriette Koster', Pol, mr, 1939; [Dick Koster sport]; Koster, D.A.

'Henriette Pechtold', HT, mr, 1946; bud long, pointed; flowers red, reverse salmon-red, large, dbl., moderate fragrance; foliage soft; vigorous, bushy growth; [Briarcliff X Katharine Pechtold]; Verschuren-Pechtold

'HENros', Cl F, dp, 1995; flowers rose pink, blooms borne in small clusters, dbl., 15–25 petals, 1.5–2.75 in., moderate fragrance; some prickles; foliage medium, medium green, glossy; tall (250 cms), climbing growth; [Glenfiddich X Lichterloh]; Henson, R.W.

'Henry Field', HT, mr, 1948; bud ovoid, long, pointed; flowers crimson-red, dbl., 60 petals, 5 in., exhibition form, moderate fragrance; foliage glossy; vigorous, bushy growth; [Pink Princess X Crimson Glory]; Brownell, H.C.; H. Field

Henry Fonda™, *see* 'JACyes'

'Henry Ford', HT, mp, 1954; bud long; flowers silvery pink, dbl., 30 petals, 4–5 in., exhibition form, moderate fragrance; vigorous, upright growth; [Pink Dawn, Climbing X The Doctor]; Howard, A.P.; H&S

'Henry Ford', HT, my, 1927; flowers yellow edged salmon-orange, opening to lemon-yellow, semi-dbl., slight fragrance; [Mme Edouard Herriot X Golden Emblem]; Deverman; B&A

'Henry Hudson', HRg, w, 1976; bud ovoid; flowers yellow stamens, dbl., 25 petals, 2.5–3 in., recurrent bloom, intense fragrance; low, bushy growth; Svedja, Felicitas; Canada Dept. of Agric.

'Henry Irving', Cl HP, dp, 1907; flowers medium, semi-dbl.; Conard & Jones; (Sangerhausen)

'Henry IV', HP, lp; flowers purple-violet, dbl., cupped

'Henry Kelsey', HKor, mr, 1984; dbl., 28 petals, blooms in clusters of 9–18, moderate, spicy fragrance; foliage glossy; trailing growth; remontant, very winter hardy. [R. kordesii hybrid X Unnamed seedling]; Svedja, Felicitas

'Henry King Stanford', F, mr, 1973; flowers small, semi-dbl., 15 petals, 1.5 in., spreading growth; [Red Pinocchio X Open pollination]; Sheridan, V.V.

Henry Lawson, HT, lp; Allender, Robert William, 1999; (Weatherly, L.)

'Henry Morse', F, dr, 1958; flowers deep blood-red shaded scarlet, borne in large trusses, semi-dbl., 3 in., free growth; Kordes; Morse

'Henry Nevard', HP, dr, 1924; flowers crimson-scarlet, very large, dbl., 30 petals, cupped, recurrent bloom, intense fragrance; foliage dark, leathery; vigorous, bushy growth; Cant, F.

'Henry S. Badgery', HT, lp; Johnson, 1885; (Weatherly, L.)

'Henry V', HCh, rb; flowers crimson, center white, cupped

'Hens Verschuren', HT, mr, 1948; bud long; flowers bright red, very large; [Mary Hart X Seedling]; Verschuren

'Her Majesty', HP, mp, 1885; flowers clear rose, with carmine reflexes toward center, very large, dbl., occasionally recurrent bloom, moderate fragrance; very vigorous growth; [Mabel Morrison X Canary (T)]; Bennett

'Hera', HT, mr, 1924; flowers brilliant carmine shaded blood-red, dbl., 40 petals, intense fragrance; [Gen. MacArthur X Luise Lilia]; Van Rossem

Herald, *see* **'DeRuiter's Herald'**

'Heraldo', Cl HT, m, 1949; bud long, pointed; flowers purple-pink, large, very dbl., intense fragrance; foliage leathery, dark; [Guinee X Texas Centennial]; Dot, M.; P. Dot

'HERani', HT, pb, 1962; (**Rina Herholdt**); bud long, pointed; flowers milky white, flushed deep pink at edges, darkening with age, semi-dbl., 60 petals, 3.5–4 in., cupped, moderate fragrance; foliage leathery, glossy; vigorous, bushy growth; [Peace X Seedling]; Herholdt, J.A.; Herholdt's Nursery

'HERaro', HT, w, 1977; (**Kilimanjaro**); bud pointed; flowers pure white, dbl., 35–38 petals, 4.5–5 in., moderate fragrance; foliage rich green; vigorous growth; [Unnamed seedling X Pascali]; Herholdt, J.A.

Herb Garden China, Ch, lp; (found rose)

Herbalist, *see* 'AUSsemi'

HERbeau, HT, ob; (**Southern Sun**)

'Herbemont's Musk Cluster', N, w; flowers pure white, very large blooms in large clusters, dbl., recurrent bloom; Herbemont, prior to 1836

'Herbert Brunning', HT, mr, 1940; flowers brilliant red; Clark, A.

'Herbert Wilson', F, w, 1967; flowers well-formed, slight fragrance; foliage light green; [White Knight X The Optimist]; Latham

'HERbic', Gr, rb, 1981; (**Herero**); flowers yellow, reverse yellow with red overlay, spreading with age, semi-dbl., no fragrance; foliage large, dark, glossy; upright growth; [Angel Bells X Southern Sun]; Herholdt, J.A.

Herbie, *see* 'TINherb'

Herbiz, Min, op, 1987

'Herbstfeuer', HEg, dr, 1961; (Autumn Fire); flowers large blooms in clusters (up to 5), semi-dbl., repeat bloom, moderate fragrance; large, pear-shaped, reddish yellow fruit; vigorous (6 ft) growth; Kordes; Kern Rose Nursery

HERclov, HT, pb; (**Cloverdene**)

'HERcop', HT, dy, 1982; (**Copper King**); flowers copper-gold, large, dbl., 35 petals, no fragrance; foliage dark, glossy; upright growth; [Vienna Charm X Seedling]; Herholdt, J.A.

'HERcres', HT, dp, 1982; (**Summer Crest**); flowers deep pink, large, dbl., slight fragrance; foliage medium, medium green, semi-glossy; [Miss All-American Beauty X Unnamed seedling]; Herholdt, J.A.

'Hercules', LCl, mp, 1938; flowers Dame Edith Helen pink, over large, dbl., cupped, slight fragrance; foliage large, glossy, dark; long stems; very vigorous, climbing growth; [Doubloons X Charles P. Kilham]; Horvath; Wayside Gardens Co.

'HERdio', HT, dy, 1971; (**Gold Dollar**); flowers large, dbl., 35 petals, slight fragrance; foliage glossy; vigorous growth; [Seedling X Weiner Charme]; Herholdt, J.A.; Herholdt's Nursery

Herero, *see* 'HERbic'

'Here's Colette', S, pb, 1999; flowers dark pink, white eye, reverse dark pink, single, 5–11 petals, 1.5 in., borne in large clusters, slight fragrance; few prickles; foliage small, light green, dull; spreading, low (3 ft) growth; [seedling X Ballerina]; Watson, Thomas L. & Glenda, 1994

Here's Ian, *see* 'JUDian'

'HERfla', HT, pb, 1981; (**Flamingo**); flowers light pink, silvery reverse, large, dbl., 35 petals, cupped, moderate fragrance; foliage medium green, semi-glossy; upright growth; [Unnamed seedling X Unnamed seedling]; Herholdt, J.A.

Herfordia, F, dp, 1993; flowers medium, dbl., slight fragrance; Hempelmann; (Sangerhausen)

'Herfsttooi', HT, dr, 1919; flowers dark crimson, dbl., moderate fragrance; [Gen. MacArthur seedling X Leuchtfeuer (HCh)]; Van Rossem

'HERgale', HT, pb, 1970; (**Nightingale**); flowers rich rose-red, blended lighter, large, dbl., 25 petals, exhibition form, slight fragrance; growth moderate; [Rina Herholdt X Tiffany]; Herholdt, J.A.; Herholdt's Nursery

HERgeko, HT, mr; (Ge Korsten)

Heriflor, *see* SAUbaflor

Heritage®, *see* 'AUSblush'

Herman Steyn, *see* KORfolklori

Hermann Berger, HT, or, 1982; flowers large, dbl., slight fragrance; GPG Bad Langensalza; (Sangerhausen)

'Hermann Eggers', HT, or, 1930; flowers deep orange-scarlet, very large, dbl., exhibition form, intense fragrance; foliage dark, leathery; very vigorous growth; [(Pink Pearl X Templar) X Florex]; Kordes; Dreer; H&S

'Hermann Kegel', M, m; flowers reddish violet, sometimes streaked crimson, medium, dbl.; vigorous growth; Portemer fils, 1848; (Cavriglia)

'Hermann Kiese', HT, yb, 1906; flowers yellow and pink, large, semi-dbl., intense fragrance; Geduldig; (Sangerhausen)

'Hermann Lindecke', HT, lp, 1929; flowers whitish pink, reverse salmon-pink large, dbl., exhibition form; [General-Superior Arnold Janssen sport]; Lindecke

'Hermann Löns', HT, dp, 1931; flowers shining light red, large blooms in clusters, single, cupped, moderate fragrance; foliage glossy; vigorous growth; [Ulrich Brunner Fils X Red-Letter Day]; Tantau

'Hermann Neuhoff', HT, mr, 1923; flowers uniform blood-red, dbl., moderate fragrance; [General-Superior Arnold Janssen sport]; Neuhoff; Kordes

'Hermann Robinow', HT, op, 1918; flowers salmon-orange shaded salmon-rose and dark yellow, large, dbl., vigorous growth; [Frau Karl Druschki X Lyon Rose]; Lambert, P.

'Hermann Robinow, Climbing', Cl HT, mp, 1934; flowers large, dbl., moderate fragrance; Lambert, P.; (Sangerhausen)

Hermann Schmidt, LCl, dp, 1986; Hetzel

'Hermann Schönfeld', HT, lp, 1925; flowers large, dbl.; Dechan; (Sangerhausen)

'Hermann Teschendorff', HT, rb, 1949; bud ovoid; flowers copper-red, reverse old-gold, open, very large, dbl., cupped, moderate fragrance; foliage glossy, dark, bronze; very vigorous, upright growth; Berger, V.; Teschendorff

'HERmawi', HT, rb, 1982; (**Mary Wise**); flowers medium red and gold with red reverse, large, dbl., 40 petals, no fragrance; foliage medium green, semi-glossy; bushy growth; [Madelaine X (Unnamed seedling X Apogee)]; Herholdt, J.A.

'HERmela', HT, rb, 1964; (**Angel Bells**®); flowers ivory, flushed orange and red, well-formed, large, dbl., borne singly, slight fragrance; foliage dark, glossy; bushy growth; [Peace X Rina Herholdt]; Herholdt, J.A.; Herholdt's Nursery

'Hermelia Casas', F, mp, 1956; flowers pearly, reverse carmine, medium, borne in clusters of 3–6, dbl., 30 petals; moderate growth; [Méphisto X Perla de Alcañada]; Dot, Pedro

'Hermen Anglada', HT, w, 1933; bud very large; flowers white tinted pink, single, intense fragrance; very vigorous growth; Dot, Pedro

'Hermes', HT, dy, 1935; flowers large, dbl., moderate fragrance; Teschendorff; (Sangerhausen)

'HERmesy', HT, rb, 1983; (**Mystery**); flowers light orange-red, edged amber, golden reverse, large, dbl., 35 petals, slight fragrance; foliage large, medium green semi-glossy; bushy growth; [(Unnamed seedling X Southern Sun) X Southern Sun]; Herholdt, J.A.

'Hermina'®, S, pb, 1996; flowers red with white reverse aging neyron rose with white reverse, dbl., 15–25 petals, good repeat, slight fragrance; some prickles; foliage medium, dark green, semi-glossy; upright, bushy, medium growth; hardy to 18–24", resistant to blackspot; [[(Tickled Pink X Prairie Princess) X Autumn Dusk] X [{Tickled Pink X {Carrousel X (Morning Stars X Suzanne)}} X Maytime]]; Buck, Dr Griffith J.; Kimbrew Walter Roses, 1997

Hermione, *see* 'GAUlimor'

'Hermosa', HCh, lp; (Armosa, Melanie Lemaire, Mme Neumann); bud pointed; flowers light blush-pink, dbl., 35 petals, exhibition form, recurrent bloom, moderate fragrance; foliage bluish green; small vigorous growth; (21); Marcheseau, before 1837; Prior to 1837

Hero®, *see* 'AUShero'

Hermosa, Climbing, *see* **'Setina'**

'Heroïca', HT, dr, 1960; flowers deep velvety red, becoming lighter; vigorous growth; [Rome Glory X Independence]; Lens

'Heroine', HT, op, 1935; bud pointed; flowers salmon-shrimp-pink, large, dbl., exhibition form; foliage leathery, dark; very vigorous, bushy growth; [Wilhelm Kordes X Mrs Atlee]; Krause

'Heroine de Vaucluse', B, mp, 1863; Moreau et Robert

'Heros', HT, mr, 1933; flowers very large, dbl., cupped, intense fragrance; foliage leathery, dark; bushy growth; [Johanniszauber X Étoile de Hollande]; Tantau

'HERpim', F, op, 1972; (**Impala**); bud ovoid, pointed; flowers coppery, reverse orange, dbl., 30 petals, 2.5–3 in.; cupped; foliage bright green; bushy growth; [Zambra X Unnamed seedling]; Herholdt, J.A.

HERpot, HT, yb, 1990; (**Kristo Pienaar**); Herholdt

'Herrenhausen', HMsk, ly, 1938; bud ovoid, greenish yellow; flowers light yellow, fading white, red tints in sun, large, dbl., cupped, borne in clusters, profuse, intermittent bloom, moderate, pansy fragrance; foliage leathery, glossy, light; long stems; vigorous, bushy growth; [Eva X Golden Rapture]; Kordes

'Herrin von Lieser', HT, ly, 1907; flowers cream-yellow, center reddish yellow, intense fragrance; [Frau Karl Druschki X G. Schwartz]; Lambert, P.

'Herself', F, lp, 1966; semi-dbl., 18 petals, 4 in., borne in clusters, free growth; [The

Optimist X Moulin Rouge]; Vincent; Harkness

Hershey's Red, HT, dr, 1996; Williams, J. Benjamin

Hertfordshire, *see* KORtenay

'HERTFORDSHIRE GLORY', F, yb, 1970; flowers yellow, tinted red, large, dbl., 20 petals, slight fragrance; foliage glossy; [Isobel Harkness X Circus]; Harkness

'HERTIE', HT, w, 1976; (**Bettie Herholdt**, Betty Herholdt, Messagere); bud pointed; flowers ivory-white, dbl., 50 petals, 5.5 in., moderate fragrance; vigorous growth; [(White Swan X Seedling) X Pascali]; Herholdt, J.A.

HERtroci, HT, pb; (**Potch Pearl**); (HERtroci)

'HERZ AS'®, HT, mr, 1963; (As de Coeur); bud long, pointed; flowers pure blood-red, well-formed, large, dbl., slight fragrance; vigorous, upright growth; Tantau, Math.

Herz As®, *see* 'TANSARAS'

'HERZBLÄTTCHEN', Pol, dp, 1889; flowers carmine-pink, small, dbl.; Geschwind, R.; (Sangerhausen)

Herzblut, *see* **'COMMONWEALTH'**

Herz-Dame, *see* **'DAME DE COEUR'**

'HERZENSGRUSS', HT, dr, 1973; bud pointed; flowers dark velvety red, medium, dbl., intense fragrance; foliage glossy; vigorous, upright, bushy growth; [(Fragrant Cloud X Goldmarie) X Red American Beauty]; Hetzel; GAWA

'HERZOG FRIEDRICH II VON ANHALT', HT, lp, 1906; flowers soft pink, moderate fragrance; [Souv. du President Carnot X Mme Jules Grolez]; Welter

Herzog von Windsor, *see* **'DUKE OF WINDSOR'**

'HERZOGIN VIKTORIA ADELHEID VON COBURG-GOTHA', HT, op, 1905; flowers coppery carmine-pink, large, dbl., moderate fragrance; Welter; (Sangerhausen)

'HERZOGIN VON CALABRIEN', HT, ly, 1914; flowers large, dbl., moderate fragrance; Lambert, P.; (Sangerhausen)

'HESSENGRUSS', HT, dp, 1928; flowers deep pink, reverse carmine-rose, dbl., moderate fragrance; [Laurent Carle sport]; Thönges

'HESSENSTAR', HT, or, 1973; bud ovoid; flowers orange-red to geranium-red, medium, dbl., 30–40 petals, moderate fragrance; moderate growth; [Baccará X Prima Ballerina]; Hetzel; GAWA

'HESSIE LOWE', HT, op, 1956; flowers peach-pink, dbl., exhibition form, intense fragrance; foliage glossy; very vigorous growth; Lowe

'HESSOISE', HEg, mp; flowers bright rose, semi-dbl.

Hester Prynne, F, dr, 1993; Robinson

Hestia, F; Arles, F.; (Cavriglia)

'HETKORA', HT, ob, 1983; (**Königin Beatrix**®, Queen Beatrix); flowers orange, large, dbl., 35 petals, exhibition form, intense fragrance; foliage medium, medium green, semi-glossy; upright, bushy growth; [Unnamed seedling X Patricia]; Kordes, W.

'HEURE MAUVE', HT, m, 1962; flowers lilac-mauve tinted blush, well-formed, dbl., 35 petals, 5 in.; foliage glossy, bright green; vigorous growth; [Simone X Prelude]; Laperrière; EFR

Heureka, HT, lp; Urban, J.; (Czech Rosa Club)

Heureux Anniversaire®, *see* 'DELPRE'

'HEXHAM ABBEY', F, op, 1975; flowers salmon-pink, base yellow turning copper, 3.5 in., intense fragrance; foliage leathery; vigorous, low to medium height, upright growth; [Fairlight X Arthur Bell]; Wood

Heywood, *see* 'KIRTYN'

Hi Doll, *see* DICkisser

'HI HO', Cl Min, op, 1964; flowers deep pink, small, in clusters, dbl.; foliage glossy; vigorous, climbimg growth; [Little Darling X Magic Wand]; Moore, Ralph S.; Sequoia Nursery

Hi Society, *see* 'COCQUATION'

Hi Teen, *see* **'DESIGNER'S CHOICE'**™

'HI, NEIGHBOR', Gr, mr, 1981; bud ovoid, pointed; flowers slightly blooms borne 1–6 per cluster, dbl., 43 petals, cupped, moderate fragrance; awl-like prickles; foliage leathery, dark, matt; bushy, erect growth; [(Queen Elizabeth X Prairie Princess) X Portrait]; Buck, Dr Griffith J.; Iowa State University

'HIAWATHA', HMult, rb, 1904; flowers deep crimson, center white, anthers golden, blooms in large, single, cupped; foliage rich green, leathery, glossy; very vigorous, climbing (15–20 ft late bloom) growth; [Crimson Rambler X Paul's Carmine Pillar]; Walsh

'HIAWATHA RECURRENT', Cl Pol, op, 1931; flowers carmine suffused orange, white eye, small, borne in clusters, intermittent bloom, foliage small, glossy; long stems; very vigorous, climbing growth; [Hiawatha X Maman Levavasseur]; Sauvageot, H.; C-P

'HIBERNICA', Misc OGR, lp; (R. X hibernica); flowers pale pink, single, 1 in., foliage glaucous green; low growth; (42); [R. canina X R. spinosissima]; Before 1800 in.Great Britain

Hidalgo®, *see* 'MEITULANDI'

'HIDCOTE GOLD', sp, (variety of R. sericea omeiensis), my, 1948; flowers canary-yellow, single blooms borne several together, non-recurrent; foliage ferny; Hilling

Hidcote Yellow, *see* **'LAWRENCE JOHNSTON'**

Hi-de-hi, *see* 'MACANAT'

'HI-FI', F, or, 1958; flowers bright orange-scarlet, borne in clusters, semi-dbl.; foliage glossy; [Independence X ?]; Gregory

High Cloud, *see* 'TINDICK'

'HIGH ESTEEM', HT, pb, 1961; bud pointed; flowers phlox-pink, reverse silvery, dbl., 43 petals, 6 in., exhibition form, intense, fruity fragrance; foliage leathery, light green; vigorous, upright, compact growth; [(Charlotte Armstrong X Mme Henri Guillot) X (Multnomah X Charles Mallerin)]; Von Abrams; Peterson & Dering

'HIGH FASHION', HT, dp, 1972; bud ovoid; flowers deep pink, medium, dbl., exhibition form, intense fragrance; foliage glossy, soft; vigorous, upright growth; [Queen Elizabeth X Peace]; Patterson; Patterson Roses

High Five, *see* 'SUNHIFI'

High Hope, *see* 'FLOHIH'

High Hopes, *see* 'HARYUP'

High Jinks™, *see* 'SAVAJINKS'

High Life, *see* 'SEALIFE'

'HIGH NOON', Cl HT, my, 1946; flowers lemon-yellow, loosely cupped, dbl., 28 petals, 3–4 in., cupped, moderate, spicy fragrance; foliage leathery, glossy; upright, vigorous, climbing (8 ft) growth; AARS, 1948; [Soeur Thérèse X Capt. Thomas]; Lammerts, Dr Walter; Armstrong Nursery

High Point, *see* 'MOLHIP'

High Sheriff, *see* 'HARWELLINGTON'

'HIGH SOCIETY', HT, mr, 1961; flowers bright red, dbl., 30 petals, 4 in., exhibition form, slight fragrance; bushy growth; Kordes, R.; A. Dickson

High Spirits™, *see* 'SAVASPIR'

High Stepper, *see* 'MORCLIM'

High Style, *see* 'LYOHI'

High Summer, *see* 'DICBEE'

'HIGH TIDE', Min, ob, 1986; flowers orange, yellow reverse, medium, borne usually singly, dbl., 25 petals, cupped, slight fragrance; few prickles; foliage small, dark, semi-glossy; upright, bushy growth; [Unnamed seedling X Unnamed seedling]; McDaniel, Earl; McDaniel's Min. Roses

'HIGH TIME', HT, pb, 1959; bud urn-shaped; flowers claret-rose, reverse gold and pink, dbl., 24 petals, 4–5 in., exhibition form, intense, spicy fragrance; foliage dark, glossy; vigorous, upright growth; [Charlotte Armstrong X Signora]; Swim, H.C.; Roseway Nursery

'HIGHDOWNENSIS', S, mr, 1928; (R. X highdownensis); flowers bright medium red, single blooms in clusters, slight fragrance; orange-scarlet fruit; colorful prickles; foliage dark, coppery; vigorous, bushy (10 × 10 ft) growth; [R. moyesii seedling]; Stern, Sir Frederich

Highfield®, *see* 'HARCOMP'

'**HIGHLAND BEAUTY**', F, mr, 1956; flowers rich red, borne in clusters, semi-dbl., 3 in., slight fragrance; foliage dark, glossy; vigorous growth; [Signal Red X Red Wonder]; deRuiter; Anderson

'**HIGHLAND CHARM**', F, op, 1956; flowers coral-salmon, borne in large clusters, semi-dbl., 3.5 in., slight fragrance; foliage leathery; vigorous, bushy growth; [Duchess of Rutland X Fashion]; deRuiter; Anderson

Highland Dancer, F, rb

'**HIGHLAND FLING**', F, or, 1971; flowers orange-scarlet, veined black, dbl., 22 petals, 4–4.5 in.; moderate fragrance; foliage glossy; [Dearest X Elizabeth of Glamis]; Anderson's Rose Nurseries

'**HIGHLAND GLORY**', F, mr, 1956; flowers crimson-red, borne in large clusters, semi-dbl., 3 in., slight fragrance; foliage coppery; vigorous growth; [Sidney Peabody X Floribunda seedling]; deRuiter; Anderson

Highland Laddie, *see* 'COCFLAG'

Highland Lass, *see* 'ZIPHIGH'

'**HIGHLAND PARK**', HT, op, 1942; bud tawny salmon; flowers salmon-pink, open, large, dbl., 35 petals, slight fragrance; foliage leathery; vigorous, upright, bushy growth; [E. G. Hill X Mme Henri Guillot]; Mallerin, C.; C-P

'**HIGHLAND WEDDING**', HT, w, 1971; flowers white, suffused blush, center light gold, dbl., 30 petals, 4.5 in., slight fragrance; foliage large, dark, semi-glossy; vigorous growth; [Virgo X Rose Gaujard]; MacLeod

'**HIGHLIGHT**', F, ob, 1957; flowers orange-scarlet, blooms in large clusters, dbl., 24 petals, 2.5 in., moderate fragrance; vigorous growth; GM, NRS, 1957; [Unnamed seedling X Independence]; Robinson, H.

Hightae, *see* 'GREENTAE'

Highveld Sun, *see* 'MORPAPPLAY'

Highway 290 Pink Buttons, Min, mp; (found rose)

Highway Rose, S, dp

'**HIGOROMO**', HT, dr, 1999; dbl., 30–35 petals, 4–5 in., flat, slight fragrance; foliage dark green, semi-glossy; vigorous (4–5 ft) growth; GM, Japan Rose Concours, 1990; [Duftzauber X (Josephine Bruce X seedling)]; Suzuki, Seizo, 1988; Keisei Rose Nurseries, 1990

Hikari, F, 1969; Suzuki, Seizo; (Cavriglia)

'HILAROMA', HT, pb, 1992; (**Secret**™); flowers light creamy pink edged with deep pink, dbl., 30–40 petals, 4–4.5 in., exhibition form, borne mostly singly, intense, sweet and spicy fragrance; some prickles; foliage large, medium green, semi-glossy; tall(120–130 cms),bushygrowth;PP8494; AARS, 1994; [Pristine X Friendship]; Tracy, Daniel; The Conard-Pyle Co., 1994

HILcap, HT, w; (**Ginger Hill**); flowers blush pink, petals crêped, (HILcap), 1999

'HILCO', Gr, dy, 1985; (**Excitement**™); dbl., 20 petals, exhibition form, borne usually singly, moderate fragrance; medium, slightly pear-shaped, slightly orange fruit; no prickles; foliage medium, dark, semi-glossy; medium, upright, bushy growth; [Golden Fantasie X Coed]; Jelly, Robert G.; E.G. Hill Co.

'**HILDA**', HT, pb, 1928; flowers salmon-pink, reverse orange-carmine, very large, dbl., globular, slight fragrance; foliage leathery; long stems; vigorous growth; Cant, B. R.

Hilda Heinemann, HT, mr, 1971; flowers large, dbl.; Meilland; (Sangerhausen)

Hilda Murrell®, *see* 'AUSmurr'

'**HILDA PHILLIPS**', HT, dy, 1948; flowers deep golden yellow, well-shaped, medium, moderate fragrance; foliage glossy, bronze; [Aureate seedling X Mrs Sam McGredy]; Bees

'**HILDA SCOTT**', HT, my, 1955; flowers butter-yellow, small; [Lady Hillingdon seedling]; Glassford; Morse

Hilde, *see* 'BENHILE'

'**HILDE APELT**', HT, ob, 1927; flowers saffron-yellow, dbl., moderate fragrance; [Unnamed seedling X Souv. de Claudius Pernet]; Leenders, M.

'**HILDE STEINERT**', HT, rb, 1926; flowers coral-red, reverse reddish salmon and old-gold, semi-dbl., moderate fragrance; Leenders, M.

'**HILDEGARDE**', HT, ob, 1946; bud long, pointed; flowers saffron-rose-pink, dbl., 30–35 petals, 5 in., exhibition form, intense fragrance; foliage leathery; very vigorous, upright growth; [Briarcliff sport]; Boerner; J&P

'**HILDENBRANDSECK**', HRg, mp, 1909; flowers shining clear pink, in clusters, single, recurrent bloom; vigorous growth; [Atropurpurea (HRg) X Frau Karl Druschki]; Lambert, P.

'**HILDEPUTCHEN**', HMult, m, 1922; flowers violet-pink with white, small, semi-dbl., moderate fragrance; Bruder Alfons; (Sangerhausen)

'HILGOFAN', HT, my, 1971; (**Golden Fantasie**, Joan Brickhill®); flowers large, semi-dbl., exhibition form, intense fragrance; foliage large, dark, leathery; vigorous, upright, bushy growth; [Dr A.J. Verhage X Anniversary]; Byrum; J.H. Hill Co.

'HILGOLD', S, my, 1967; (**Golden Chersonese**); single, 1.5–2 in., borne singly at each node, early, moderate fragrance; foliage leaflets 7–9; vigorous growth; [R. ecae X Canary Bird]; Allen, E.F.

'**HILL CREST**', HT, rb, 1948; bud short, pointed, oxblood-red; flowers carmine, dbl., 50–55 petals, 3.5–4 in., exhibition form, moderate fragrance; foliage leathery; very vigorous, upright, compact, tall growth; Hill, Joseph H., Co.

'**HILL TOP**', HT, ab, 1942; bud long, pointed, light coral-red; flowers buff, dbl., 28 petals, 4–5 in., globular, moderate fragrance; foliage leathery, dark, wrinkled; vigorous, upright growth; [Joanna Hill X R.M.S. Queen Mary]; Hill, Joseph H., Co.; H&S, 1946;, Wayside Gardens Co., 1946

Hillary First Lady, *see* LAPcli

'**HILLCREST PILLAR**', S, my, 1930; flowers bright yellow, medium, semi-dbl.; pillar (6 ft) growth; [R. X harisonii seedling]; Hillcrest Gardens

'**HILLIER ROSE**', S, dr; (R. hillieri, R. X pruhoniciana); flowers deep red, 2 in., (35); [R. moyesii X R. multibracteata (?)]

'**HILL'S AMERICA**', HT, mp, 1921; flowers rose-pink, dbl., 44 petals, moderate fragrance; GM, NRS, 1924; [Premier X Hoosier Beauty]; Hill, E.G., Co.

'**HILL'S HILLBILLY**', F, mp, 1947; bud ovoid, red; flowers pink, open, borne in clusters, single, 5–6 petals, 1–2 in., slight fragrance; foliage leathery, dark; vigorous, upright, bushy growth; [Juanita X Mrs R.M. Finch]; H&S;, Wayside Gardens Co.

'**HILL'S VICTORY**', HT, rb, 1942; bud red; flowers rose-red to rose-pink, dbl., 50–55 petals, 5–6 in., slight fragrance; foliage leathery, dark, wrinkled; long, strong stems; vigorous, upright, much branched growth; [Chieftain X Sweet Adeline]; Hill, Joseph H., Co.

Hilltop, LCl, pb; Huxley, Ian, 1989

HILrap, HT, mr; (**Classy**™); Hill, E.G., Co., 1992

HILraz, HT, dp; (**Lovely Dream**); (HILraz), 1995

'HILRED', HT, dr, 1989; (**Challenger**™); bud ovoid; flowers reverse medium red, urn-shaped, medium, borne in sprays of 2–3, dbl., 25 petals, exhibition form, moderate fragrance; pear-shaped, small, orange blend fruit; prickles angle downward,small, orange blend; foliage medium, semi-glossy; upright, medium growth; [Jacqueline X Seedling]; Tracy, Daniel; DeVor Nurseries, Inc., 1987

'HILSET', F, lp, 1985; (**Fondly**™); flowers sweetheart, medium, dbl., 20 petals, exhibition form, borne singly and in clusters of 2–4, moderate, spicy fragrance; medium, ovoid, orange-red fruit; few prickles on peduncles; foliage medium, dark, matt; medium, upright growth; PP004983; [Unnamed seedling X Unnamed seedling]; Jelly, Robert G.; E.G. Hill Co.

'HILTACO', F, dr, 1985; (**Sassy**™); flowers sweetheart, small, dbl., 20 petals, exhibition form, borne usually singly, slight fragrance; small, globular, grayed red fruit; prickles only on peduncle; foliage small, medium green, matt; upright growth;

PP004559; [Little Leaguer X Mary DeVor]; Jelly, Robert G.; E.G. Hill Co.

Hilton Edward, *see* 'STIEDWARD'

Himalayan Musk Rose, *see* **R. MOSCHATA NEPALENSIS**

'HIMANGINI', F, w, 1968; bud ovoid; flowers ivory-white, center light buff, open, medium, dbl.; vigorous, bushy, compact growth; [Saratoga X ?]; IARI

'HIMATSURI', HT, dr, 1977; bud circular; 3.5–4 in., exhibition form, slight fragrance; foliage small; upright growth; [(Tropicana X Karl Herbst) X Mainauperle]; Teranishi, K.

Hime, *see* 'KEIFUPIE'

'HIMMELSAUGE', HSet, m; Geschwind, R., 1894

Himmelsstürmer, LCl, pb, 1990; flowers carmine-pink with white, small, single; Wänninger, Franz; (Sangerhausen)

'HINEMOA', F, rb, 1963; flowers buttercup-yellow shading to vermilion, borne in clusters, semi-dbl., 17–20 petals, 3–3.5 in., cupped, moderate fragrance; foliage bronze, leathery; upright, bushy growth; [Circus X ?]; Mason, P.G.

'HINGREAT', LCl, ab, 1999; (**Great Phoebus**); bud white, spherical; flowers opening to orange and yellow, with slight pink edge, very dbl., 41 petals, 5.5 in., pompon, borne mostly singly, moderate fragrance; prickles moderate; foliage large, dark green, semi-glossy; upright, vigorous, medium (8 ft) growth; [Abraham Darby X Breath of Life]; Hintlian, Nancy Sears, 1999

'HINRICH GAEDE', HT, ob, 1931; bud pointed, nasturtium-red; flowers orange-yellow tinted nasturtium-yellow, very large, dbl., exhibition form, moderate, fruity fragrance; foliage glossy, bronze; vigorous growth; [Lady Margaret Stewart X Charles P. Kilham]; Kordes

'HINRICH GAEDE, CLIMBING', Cl HT, ob, 1935; Armstrong, J.A.

'HINRUBY', F, dr, 1999; (**Ruby Velvet**); flowers dark red, turning purple with age, lasts well, very dbl., 41 petals, 5 in., borne in small clusters, intense fragrance; prickles many, very large; foliage medium, medium green, semi-glossy; spreading, medium (4 ft) growth; [The Dark Lady X Oklahoma]; Hintlian, Nancy Sears, 1999

'HINSWEET', S, m, 1999; (**Sweet Ecstasy**); flowers purple-crimson, turning more purple with age, very dbl., 41 petals, 5 in.; borne mostly singly, intense, lasting fragrance; prickles moderate; foliage large, dark green, semi-glossy; upright, vigorous, medium (5 ft) growth; [The Dark Lady X Oklahoma]; Hintlian, Nancy Sears, 1999

'HINTUSCAN', HGal, m, 1999; (**Tuscan Beauty**); flowers red-lavender, aging to near purple, very dbl., 41 petals, 4–4.5 in., borne mostly singly, intense fragrance; prickles moderate; foliage medium, medium green, semi-glossy; spreading, tall (6 ft) growth; [Superb Tuscan X Big Purple]; Hintlian, Nancy Sears, 1999

'HI-OHGI', HT, or, 1981; flowers deep orange-red, large, dbl., 28 petals, moderate fragrance; prickles slanted downward; foliage dark, semi-glossy; tall, upright growth; [San Francisco X (Montezuma X Peace)]; Suzuki, Seizo; Keisei Rose Nursery

Hipango, *see* 'MURGO'

'HIPÓLITO LÁZARO', LCl, dr; flowers carmine, large, intense fragrance; Pahissa

'HIPPOLYTE', HGal, m; flowers vivid carmine shaded violet, small

'HIPPOLYTE JAMAIN', HP, dp, 1874; flowers carmine-red, well-formed, semi-globular, dbl., 38 petals, moderate fragrance; foliage red, when young; vigorous, erect growth; Lacharme, F.

Hippy, HT, 1971; Delbard-Chabert; (Cavriglia)

Hiroshima, HT, dp

Hiroshima Spirit, HT, yb, 1991; flowers dark yellow with red, large, dbl.; Harada, Toshiyuki; (Sangerhausen)

Hiroshima's Children, *see* 'HARMARK'

'HIS MAJESTY', HT, dr, 1909; flowers dark crimson shaded deeper, very large, dbl., exhibition form, intense fragrance; long, strong stems; very vigorous growth; McGredy

'HISAMI', HT, lp, 1988; flowers light cream, flushed crimson at tip, large, borne usually singly, very dbl., 50 petals, exhibition form, moderate fragrance; prickles downward curved, red to light green; foliage medium, dark green, matt; upright, medium growth; [Kordes' Perfecta X Christian Dior]; Harada, Toshiyuki

'HISPANIA', HT, dr, 1938; bud pointed; flowers velvety red, open, overlarge, dbl., intense fragrance; foliage leathery; long, strong stems; very vigorous growth; Pahissa

'HIT PARADE', F, ob, 1962; flowers red, orange and gold, borne in clusters, semi-dbl., 16 petals, 4 in., foliage dark; low growth; [(Independence X Unnamed seedling) X Brownie]; Dickson, Patrick; A. Dickson

'HO NO-O-NO-NAMI', LCl, or, 1968; (Wave of Flame, Waves of Flame); bud pointed; flowers orange-red, reverse lighter, to open, medium, dbl., exhibition form, slight fragrance; foliage glossy, leathery; vigorous, climbing growth; [Spectacular X Aztec]; Suzuki, Seizo; Country Garden Nursery, 1967

Hoag House Cream, HT, w; (found rose)

Hoagy Carmichael, *see* 'MACTITIR'

Hobby, *see* 'TANOB'

'HOCHSOMMER', F, dp, 1960; dbl., borne in clusters; foliage dark; low growth; [Queen Elizabeth X Unnamed seedling]; Verschuren, A.; van Engelen

'HOCUS-POCUS', Gr, or, 1975; dbl., 30 petals, 4–4.5 in., cupped, slight fragrance; foliage large, dark; vigorous, upright growth; [Fandango X Simon Bolivar]; Armstrong, D.L.; Armstrong Nursery

Hoddy Toddy™, *see* 'KINTODDY'

'HOFFMANN VON FALLERSLEBEN', LCl, rb, 1917; flowers salmon-red, shaded yellow and ochre, borne in clusters of 5–2, sometimes repeats; vigorous, climbing growth; [Geheimrat Dr Mittweg X Tip-Top]; Lambert, P.

Hoffnung, HT, ob, 1995; Huber

'HOFGARTENDIREKTOR GRAEBENER', HT, lp, 1899; flowers creamy pink, medium, dbl., moderate fragrance; Lambert, P.; (Sangerhausen)

'HOFGÄRTNER KALB', HCh, pb; flowers bright carmine-rose, center yellow, outer petals shaded red, dbl., 35 petals, moderate fragrance; vigorous, bushy growth; [Souv. de Mme Eugene Verdier X Gruss an Teplitz]; Felberg-Leclerc, 1913

Hoggar, S; Lens, Louis; (Cavriglia)

'HOH-JUN', HT, pb, 1981; flowers pink flushed rose-red, blooms in clusters of 2–5, dbl., 28 petals, cupped, intense fragrance; large prickles; foliage dark, semi-glossy; compact growth; [Granada X Flaming Peace]; Suzuki, Seizo; Keisei Rose Nursery

'HOHSHUN', HT, dr, 1999; flowers velvet deep red, dbl., 25–27 petals, exhibition form, intense fragrance; foliage dark green, half leathery; 5 ft growth; [Charles Mallerin X Mrs Nieminen]; Ohkawara, Kiyoshi, 1970; Nagashima Rose Nursery, 1971

'HOKEY POKEY', Min, ab, 1980; bud long, pointed; flowers deep apricot, borne singly, dbl., 28 petals, exhibition form, slight, spicy fragrance; straight prickles; foliage finely serrated; compact, bushy growth; [Rise 'n' Shine X Sheri Anne]; Saville, F. Harmon; Nor'East Min. Roses

Hokkaido, HRg, mp, 1988

'HOKU-TO', HT, ly, 1979; flowers soft buff yellow, dbl., 42 petals, exhibition form, moderate fragrance; foliage large, light green; vigorous, upright, spreading growth; [(Myoo-Jo X Chicago Peace) X King's Ransom]; Suzuki, Seizo; Keisei Rose Nursery

Hold Slunci, HP, ly, 1956; flowers medium, dbl., slight fragrance; Blatnà; (Sangerhausen)

Hole in One, *see* HOReagle

HOLfairy, Pol, w, 1991; (**Fairy Snow**); Holmes

'HOLIDAY', HT, mr, 1981; flowers small, dbl., 35 petals, slight fragrance; foliage medium green; upright growth; [Cara Mia X Volare]; Strahle, B. Glen; Carlton Rose Nurseries

'HOLIDAY', F, 1948; bud ovoid, orange-yellow flushed pink; flowers flame-pink, reverse clear yellow, clusters of 3–10, semi-dbl., 3–3.5 in., cupped, moderate, clove fragrance; foliage glossy; vigorous, bushy, compact growth; RULED EXTINCT 4/83 ARM; [McGredy's Pillar X Pinocchio]; Boerner; J&P

Holiday Cheer, *see* 'MORLIDAY'

'HOLJE', F, w, 1983; (**Jerusalem**); flowers cream, large blooms in sprays of 5–8, dbl., 40 petals, exhibition form, moderate fragrance; ovoid, small, green fruit; foliage medium, light green, matt; upright, bushy growth; [(Queen Elizabeth X Unnamed seedling) X Moriah]; Holtzman, Arnold; Gandy Roses, Ltd.

'HOLLAND DOUBLE WHITE ALTAI', sp, (form of R. spinosissima altaica), w; dbl., non-recurrent; P.H. Wright

Holländerin, *see* 'TANSCHWEIGRU'

'HOLLANDIA', Pol, dr; flowers deep red, rosette form, small, dbl.

'HOLLANDIA', HT, mr, 1930; flowers brick-red, shaded copper; [Aspirant Marcel Rouyer sport]; Zijverden

'HOLLANDICA', HRg; thought to have been raised by J. Spek, ca. 1888; selected Dutch clone of R. Rugosa, or perhaps a hybrid with Manetti, widely used as an understock, especially for tree roses

Hollands Rugosa, HRg, mp; flowers medium, single; (Sangerhausen)

Hollie Roffey, *see* 'HARRAMIN'

'HOLLY ROCHELLE', HT, dr, 1974; bud pointed; flowers velvety red, very dbl., 60–70 petals, 4 in., exhibition form, moderate fragrance; very vigorous, upright growth; [Charlotte Armstrong X Scarlet Knight]; Graham; South Forrest Rose Nursery

'HOLLYBANK', F, mr, 1966; flowers vermilion, 3–3.5 in., globular, slight fragrance; foliage coppery; upright growth; [Independence X United Nations]; Hooney

'HOLLYWOOD', HT, rb, 1930; bud pointed; flowers dark rose-red, veined darker, dbl., exhibition form, intense fragrance; foliage dark; long, strong stems; bushy growth; [Premier sport]; Scittine

Hollywood, HT, w, 1998

'HOLLYWOOD BEAUTY', HT, dp, 1929; flowers camellia-red, dbl., moderate fragrance; [Rose Marie sport]; Pacific Rose Co.

'HOLSTEIN', F, mr, 1939; (Firefly); bud pointed, dark crimson; flowers clear crimson, blooms in immense clusters, single, 6 petals, 4 in., cupped, slight fragrance; foliage leathery, dark, bronze; very vigorous, bushy growth; GM, Portland, 1939; [Else Poulsen X Dance of Joy seedling]; Kordes; J&P

Holstein 87, *see* 'KORHOLST'

'HOLSTEIN, CLIMBING', Cl F, mr, 1947; Kordes, P.

Holsteinperle®, *see* 'KORDIAM'

'HOLSTENROSE', HT, mr, 1937; flowers scarlet-red, large, dbl., cupped, moderate fragrance; foliage glossy; vigorous, bushy growth; [Gen. MacArthur X Amulett]; Tantau

Holstentor, *see* MEImuide

'HOLT HEWITT', HT, rb, 1925; flowers rich velvety crimson, flushed and edged scarlet, well shaped, intense fragrance; vigorous, bushy growth; Beckwith

Holtermann's Gold, HT, my, 1989; Swane

Holy Toledo™, *see* 'AROBRI'

Homage, HT, pb, 1986; Pal, Dr B.P.

'HOMBRE', Min, pb, 1983; flowers light apricot pink, reverse light pink, small, dbl., exhibition form, slight fragrance; foliage small, medium green, semi-glossy; bushy growth; PP005552; AOE, 1983; [Humdinger X Rise 'n' Shine]; Jolly, Nelson F.; Rosehill Farm, 1982

Home & Country, *see* 'KRIBATIS'

Home of Time, *see* 'COCQUAMBER'

'HOME RUN', HT, pb, 1956; bud ovoid; flowers rose-bengal, dbl., 40 petals, 5 in., exhibition form, moderate fragrance; foliage leathery; vigorous growth; [Pink Delight sport]; Motose

'HOME SWEET HOME', HT, mp, 1941; flowers rich velvety pink, dbl., moderate, damask fragrance; foliage glossy, dark; vigorous growth; Wood & Ingram; C-P

HomeComing, *see* 'MICHOME'

Home-coming, *see* 'MICHOME'

'HOMELAND', HT, pb, 1951; flowers neyron rose, base tinted orange, well formed, dbl., 50–60 petals, 5.5–6 in., intense fragrance; foliage leathery, dark; very free growth; [Hybrid Tea (red) X Guinee]; LeGrice

'HOMENAGEM EGAS MONIZ', HT, mp, 1959; [Walter X Juno]; da Silva, Moreira

'HOMENAGEM GAGO COUTINHO', HT, rb, 1959; flowers crimson-red, reverse yellow, [Confidence X Unnamed seedling]; da Silva, Moreira

'HOMENAGEM PINTO D'AZEVEDO', HT, op, 1959; flowers bright salmon-pink, dbl.; [Super-Congo X Independence]; da Silva, Moreira

'HOMÈRE', T, pb, 1858; (Duchess of Kent); flowers pink, center flesh-white, dbl., cupped, moderate fragrance; vigorous, bushy growth; (21); Robert et Moreau

Home's Beauty, *see* 'JACED'

Home's Choice, *see* 'JACADE'

Home's Pride, *see* 'JACARY'

Homestead, HCh, dr

Homestead, HP, mp

Hommage d'Anjou, HT, or, 1965; flowers coppery-red, large, dbl., slight fragrance; Minier; (Sangerhausen)

'HON A. NORTON', HT, lp; [Mme Lombard sport]; Williams, A., 1910; (Weatherly, L.)

'HON. CHARLOTTE KNOLLYS', HT, pb, 1926; flowers rose, edged lighter, center creamy yellow, dbl., slight fragrance; GM, NRS, 1926; [Antoine Rivoire X Willowmere]; Bees

'HON. EDITH GIFFORD', T, w, 1882; flowers flesh-white tinted rose, large, dbl.; [Mme Falcot X Perle des Jardins]; Guillot et Fils

'HON. GEORGE BANCROFT', HT, rb, 1879; flowers red shaded violet-crimson, intense fragrance; moderate growth; [Mme de St Joseph X Lord Macaulay]; Bennett

'HON. INA BINGHAM', HP, mp, 1905; flowers pink, stamens golden yellow, large, dbl., 23 petals, cupped, intense fragrance; upright growth; Dickson, A.

'HON. JOAN ACTON', HT, pb, 1950; flowers cream edged pink, pointed, very large, very dbl., 90 petals, moderate fragrance; foliage bronze-green; hardy; [Mrs Sam McGredy X Golden Dawn]; Marshall, J.

'HON. LADY LINDSAY', S, pb, 1939; (Honorine Lady Lindsay); flowers pink, reverse darker, dbl., 35 petals, recurrent bloom; foliage dark; bushy (3 × 3 ft) growth; not dependably hardy.; [New Dawn X Rev. F. Page-Roberts]; Hansen, N.J.; B&A, 1938

'HON. VIOLET DOUGLAS PENNANT', HT, pb, 1927; flowers blend of cream and rose, dbl., moderate fragrance; Bees

'HONDO', HT, yb, 1989; flowers medium yellow with red-purple on tips, aging red-purple, large, dbl., 33 petals, moderate fragrance; foliage large, dark green, matt; upright growth; [Irish Gold X Las Vegas]; Perry, Astor, 1990

Hondo, HRg, dp, 1988

Honest Abe™, *see* 'ARON'

'HONEST RED', HT, mr, 1991; flowers large blooms borne mostly singly, dbl., 26–40 petals, intense fragrance; foliage medium, dark green, semi-glossy; tall, bushy growth; [Unnamed seedling X Sea Pearl]; Wambach, Alex A.

'HONEY', F, my, 1955; flowers coppery yellow, becoming lemon-yellow, small, semi-dbl.; [Smiles sport]; Marsh; Marsh's Nursery

Honey Bear, *see* 'PIXIHON'

Honey Bouquet™, *see* 'JACYIMP'

'HONEY BUN', F, my, 1973; bud ovoid; flowers medium, dbl., cupped, intense fragrance; foliage dark, leathery; very vigorous, upright, bushy growth; [Gold

Strike X Golden Garnette]; Ellis; Armstrong Nursery

'HONEY BUNCH', F, dy, 1971; flowers deep gold, shaded peach, long, pointed, dbl., 20 petals, 3 in., slight fragrance; foliage light; moderate, upright growth; [Circus X Soraya]; Watkins Roses

Honey Bunch®, *see* 'COCGLEN'

Honey Butter, *see* 'PIXHONEY'

Honey Child, F, ab, 1999; Ilsink

'HONEY CHILE', F, lp, 1964; bud pointed; dbl., cupped, blooms in clusters, slight fragrance; foliage leathery; vigorous, upright, bushy growth; [Fashion X Queen Elizabeth]; Thomson; Tillotson

'HONEY FAVORITE', HT, lp, 1962; (Honey Favourite); flowers light yellowish pink, base yellow; [Pink Favorite sport]; Von Abrams; Peterson & Dering

Honey Favourite, *see* **'HONEY FAVORITE'**

'HONEY GOLD', F, my, 1956; bud ovoid; flowers maize-yellow overcast buff-yellow, dbl., 43 petals, 3–3.5 in., moderate fragrance; foliage dark, glossy; vigorous, bushy growth; [Yellow Pinocchio X Fashion]; Boerner; J&P

'HONEY HILL', Min, ob, 1981; bud ovoid, pointed; flowers orange, borne singly or several together, dbl., 48 petals, exhibition form, moderate fragrance; curved, light brown prickles; foliage medium green, semi-glossy; vigorous, bushy, upright growth; [Unnamed seedling X Unnamed seedling]; Lyon

Honey Mini-Delite, *see* 'CLEHON'

Honey Moss, *see* 'ARONEMO'

Honey 'n' Spice™, *see* 'MINAGCO'

'HONEY REA', HT, pb, 1973; bud ovoid; flowers pink, very large, semi-dbl., exhibition form, slight fragrance; foliage large, dark, leathery; very vigorous, bushy growth; Concord Floral Co.

Honey Sweet, *see* **'HONEYSWEET'**

Honeybunch, *see* 'COCGLEN'

'HONEYCOMB', Min, ly, 1974; flowers soft yellow to near white, small, dbl., 30 petals, 1.5 in., exhibition form, moderate fragrance; foliage small, glossy, light, leathery; dwarf, bushy growth; [(R. wichurana X Floradora) X Debbie]; Moore, Ralph S.; Sequoia Nursery

Honeycup®, HT, ab, 1990

'HONEYFLOW', F, pb, 1957; flowers white edged pink, borne in very large clusters, single, moderate fragrance; foliage glossy; vigorous growth; [Spring Song (F) X Gartendirektor Otto Linne]; Riethmuller

'HONEYGLOW', F, yb, 1956; flowers lemon-yellow, reverse shaded orange, pointed, borne in clusters, semi-dbl., 2 in., foliage glossy; vigorous growth; [Goldilocks X Ellinor LeGrice]; LeGrice

Honeymoon, *see* **'HONIGMOND'**

'HONEYPOT', S, my, 1969; flowers sulfur-yellow, dbl., 40–50 petals, 4 in., cupped, intense fragrance; foliage dark, semi-glossy; vigorous growth; [Honigmond X Constance Spry]; Austin, David

'HONEYSWEET', S, op, 1984; (Honey Sweet); bud medium-large, ovoid, pointed; flowers yellow-red-orange blend, large blooms borne 1–8 per cluster, dbl., 28 petals, cupped, repeat bloom; awl-like, brown-tan prickles; foliage leathery, dark with copper tints; bushy, erect growth; hardy; [Serendipity X Wiener Charme]; Buck, Dr Griffith J.; Iowa State University

Honeysweet, F, ab

'HONG-KONG', HT, rb, 1962; flowers citron-yellow edged currant-red, becoming red, dbl.; strong stems; bushy growth; [Soraya X (Henri Mallerin X Peace)]; Dot, Pedro

'HONIGMOND', F, my, 1960; (Honeymoon); flowers canary-yellow, rosette shape, blooms in clusters (up to 5), dbl., 40 petals, slight fragrance; foliage dark, veined; vigorous, upright, bushy growth; [Clare Grammerstorf X Golden Scepter]; Kordes, R.; A. Dickson;, McGredy

Honky Tonk, *see* 'TALHON'

Honor™, *see* 'JACOLITE'

'HONORABLE KEN', F, ob, 1993; flowers orange/yellow reverse, blooms borne in small clusters, dbl., 26–40 petals, 1.5–2.75 in., moderate fragrance; some prickles; foliage medium, medium green, semi-glossy; compact (100 cms) growth; [John Lawrence X Zorina]; Bracegirdle, Derek T.; Bracegirdle, 1994

Honore de Balzac®, *see* 'MEIPARNIN'

'HONORINE DE BRABANT', B, pb; flowers pale lilac-pink, spotted and striped mauve and crimson, dbl., cupped, recurrent bloom, foliage light green; vigorous (to 6 ft) growth

Honorine Lady Lindsay, *see* **'HON. LADY LINDSAY'**

Honour, *see* 'JACOLITE'

'HONOUR BRIGHT', LCl, dr, 1950; flowers brilliant crimson, medium, semi-dbl., recurrent bloom, moderate fragrance; foliage bright green; very vigorous (4–6 ft) growth; [(New Dawn X Allen Chandler) X (Mrs W.J. Grant, Climbing X Richmond, Climbing)]; Eacott

'HOOPEARL', Min, lp, 1991; (**Miss Pearl**™); dbl., 15–25 petals, 1.5–2.75 in., borne mostly singly, intense, spicy fragrance; few prickles; foliage medium, medium green, matt; tall (80 cms), upright growth; [Gene Boerner X Pacesetter]; Hooper, John C.; AGM Miniature Roses

'HOOSIER BEAUTY', HT, rb, 1915; bud pointed; flowers glowing crimson shaded darker, large, dbl., intense fragrance; foliage sparse, rich green, glossy; bushy growth; GM, NRS, 1915; [Richmond X Château de Clos Vougeot]; Dorner

'HOOSIER BEAUTY, CLIMBING', Cl HT, rb, 1925; Gray, W.R.

Hoosier Glory, *see* **'EUTIN'**

'HOOSIER GOLD', F, dy, 1974; bud ovoid; dbl., 30–35 petals, 2.5–3.5 in., slight fragrance; vigorous growth; [Lydia X Golden Wave]; Byrum; J.H. Hill Co.

'HOOSIER HONEY', HT, dy, 1973; bud long, pointed; flowers mimosa-yellow, full, dbl., 25–30 petals, 4–5 in., exhibition form, intense fragrance; foliage dark, leathery; vigorous growth; [Unnamed seedling No. 63–704 X Golden Fantasie]; Byrum; J.H. Hill Co.

'HOOSIER HONEY', HT, my, 1955; bud ovoid; flowers amber-yellow, open, dbl., 45–50 petals, 4.5–5 in., moderate fragrance; vigorous, upright, bushy growth; [Anzac X Golden Rapture]; Hill, Joseph H., Co.

'HOOSIER HYSTERIA', S, dr, 1981; bud ovoid; dbl., 45 petals, exhibition form, borne singly or in sprays of 5–7, slight fragrance; very few, curved, dark green prickles; foliage dark, leathery; vigorous, tall, upright growth; [Karl Herbst X Simone]; Schwartz, Ernest W.; Krider Nursery

Hoot 'n' Holler, *see* 'MORPETTI'

Hoot Owl, *see* 'MORHOOT'

'HOOTEA', Min, dr, 1992; (**Billie Teas**); dbl., 26–40 petals, 1.5 in., borne mostly singly, no fragrance; some prickles; foliage medium, medium green, matt; medium (38–45 cms), upright growth; AOE, 1993; [Seedling X Merrimac]; Hooper, John C.; Kimbrew Walter Roses, 1992

'HOPE', Min, w, 1985; flowers well-formed, small, dbl., 35 petals, slight fragrance; foliage large, dark, glossy; bushy growth; [Rise 'n' Shine X Party Girl]; Bridges, Dennis A.; Bridges Roses

Hope, *see* 'ORACREPS'

Hope 98, *see* 'COCLAMENT'

'HOPE FOR HUMANITY', S, dr, 1996; flowers deep dark red, blooms borne in small clusters, dbl., 15–25 petals, 1.75 in., slight fragrance; moderate prickles; foliage medium, semi-glossy; upright, low (56 cms) growth; [(Prairie Princess X Morden Amorette) X (Morden Cardinette X K1)]; Collicutt, L.M. & Davidson, C.G.; Morden Exp. Farm, 1995

Hopeful, *see* 'RESHOPE'

Hopscotch, *see* 'AROyol'

'HORAARDVARK', HT, yb, 1999; (**Mario Lanza**); flowers medium yellow, edged red, reverse paler yellow, dbl., 26–40 petals, 5 in., borne mostly singly, moderate fragrance; prickles moderate; foliage medium, dark green, glossy; bushy, medium (3 ft) growth; [Elina X Remember

Me]; Horner, Colin P.; Warley Rose Gardens, 2000

'HORABI', HT, mr, 1995; (**Friday's Child**); dbl., 26–40 petals, 5 in., borne mostly singly, intense fragrance; foliage medium, medium green, semi-glossy; tall (120 cms), upright growth; [Spirit of Youth X Seedling]; Horner, Colin P.; Battersby Roses, 1997

'HORACE MCFARLAND', HT, op, 1944; (Président Nomblot); bud mahogany-red; flowers coppery pink, dbl., 43 petals, 4.5–5.5 in., exhibition form, moderate, fruity fragrance; foliage leathery, dark; vigorous, bushy growth; AARS, 1945; [Mme Arthaud X Unnamed seedling]; Mallerin, C.; A. Meilland, 1944; C-P, 1944

'HORACE VERNET', HP, dr, 1866; flowers deep red, large, dbl., 40 petals, exhibition form, repeat bloom, intense fragrance; moderate, erect growth; [Général Jacqueminot X ?]; Guillot et Fils

'HORALCAMSTRIP', F, op, 1999; (**County Girl**); flowers orange, pink cream striped, reverse paler striped, ages more salmon, dbl., 17–25 petals, 4 in., borne in small clusters, slight fragrance; prickles moderate; foliage medium, medium green semi-glossy; bushy, medium (3 ft) growth; [Alexander X (Southampton X ((New Penny X White Pet) X Stars 'n' Stripes))]; Horner, Colin P.; Grange Farm Nursery, 2001

'HORANYME', Gr, mr, 1999; (**Kieran Ross Clark**); flowers bright scarlet red, reverse paler, dbl., 17–25 petals, 3 in.; borne in small clusters, slight fragrance; prickles moderate; foliage medium, dark green, glossy; upright, tall (4 ft) growth; [(Anytime X (Liverpool Echo X (Flamenco X Rosa Bella))) X Remember Me]; Horner, Colin P.

'HORANYMOLL', LCl, my, 1997; (**Golden Future**); flowers full, large, very dbl., 26–40 petals, borne in small clusters, moderate fragrance; foliage large, dark green, glossy; upright, medium (9–12 ft) growth; [((Anytime X Liverpool Echo) X (Flamenco X Rosa Bella)) X (Korressia X Kiskadee)]; Horner, Colin P.

'HORAPSUNMOLBABE', S, dy, 1999; (**The Care Rose**); flowers buttercup yellow, reverse medium yellow, small, semi-dbl., 12–16 petals, borne in small clusters, moderate fragrance; few prickles; foliage small, medium green, glossy; compact, low (2 ft); patio growth; [Apricot Sunrise X (Golden Future X Baby Love)]; Horner, Heather M.

Horatio Nelson, *see* 'BEAHOR'

'HORAVME', F, ob, 1993; (**Evelyn Grace**); flowers orange/yellow bicolor, blooms borne in small clusters, dbl., 15–25 petals, 1.5–2.75 in., slight fragrance; some prickles; foliage medium, light green, glossy; medium (90 cms), compact growth; [Avocet X Remember Me]; Horner, Heather M.; Horner, 1995

'HORBARJEFF', HT, my, 1989; (**Taylors Gold**); bud ovoid; flowers do not fade, medium, borne singly and in sprays of 5, dbl., 48 petals, cupped, slight fragrance; fruit not known; prickles narrow, small, brown; foliage medium, medium green, semi-glossy; bushy, medium growth; [Prominent X Helmut Schmidt]; Horner, Heather M.

HORBATBEAUT, HT, ly, 1997; (**Battersby Beauty**); flowers full, medium, very dbl., 26–40 petals, borne mostly singly, slight fragrance; foliage medium, light green, semi-glossy; bushy, medium (90 cms) growth; [Polar Star X (Elina X Remember Me)]; Horner, Colin P.

'HORBLUSH', HT, w, 1997; (**Helen Margaret**); flowers double, medium, dbl., 15–25 petals, borne in small clusters, moderate fragrance; foliage medium, light green, glossy; compact, medium (80 cms)growth; [Champagne Cocktail X Alpine Sunset]; Horner, Colin P.

'HORBONDARC', HT, op, 1997; (**Caroline Clarke**, The Backpackers Rose); flowers full, large, very dbl., 26–40 petals, borne in small clusters, moderate; some prickles; foliage medium, medium green, semi-glossy; bushy, medium (90 cms) growth; [(Prominent X Southampton) X New Year]; Horner, Heather M.

'HORBONDSMILE', F, mp, 1996; (**Brave Heart**); flowers high-pointed blooms of classic form, borne in small clusters, dbl., 15–25 petals, 4 in., moderate fragrance; some prickles; foliage large, medium green, semi-glossy; medium (100 cms), upright growth; [(Prominent X Southampton) X Bright Smile]; Horner, Heather M.; Battersby Roses, 1998

'HORBRITSEA', S, w, 1997; (**Forever Free**); flowers full, small, very dbl., 26–40 petals, borne in large clusters, slight fragrance; some prickles; foliage small, medium green, glossy; spreading, low (45 cms) growth; [(Anna Ford X Little Darling) X Sea Form]; Horner, Colin P.

'HORCABELLERO', HT, 2000; (**Barry Stephens**); flowers apricot edged red, reverse lighter, dbl., 26–40 petals, 6 in.; borne mostly singly, moderate fragrance; prickles moderate; foliage large, dark green, semi-glossy; upright, medium (3.5 ft) growth; [Typhoon X Benson & Hedges Gold]; Horner, Calvin L.; Battersby Roses, 2000

'HORCOCKALORUM', HT, yb, 1997; (**Rebecca Paul**); flowers double, medium, dbl., 15–25 petals, borne in small clusters, slight fragrance; foliage medium, medium green, semi-glossy; bushy, medium (80 cms)growth; [Silver Jubilee X Isobel Derby]; Horner, Colin P.

'HORCOED', S, mp, 1997; (**Kathleen Jane**); flowers very full, medium, very dbl., 41 petals, borne in large clusters, slight fragrance; foliage medium, light green, glossy; spreading, medium (4 ft), growth; [Bonica X Leverkusen]; Horner, Colin P.

'HORCOEXIST', S, yb, 1997; (**Starry Eyed**); flowers single, small, single, 4–7 petals, borne in large clusters, slight fragrance; few prickles; foliage small, light green, dull; spreading, bushy, low (90 cms) growth; [(Anna Ford X Little Darling) X ((Sea Foam X Little Darling) X (Hamburger Phoenix X Prelude))]; Horner, Colin P.

'HORCOFF', F, rb, 1997; (**Hospitality**); flowers double, medium, dbl., 15–25 petals, borne in small clusters, moderate fragrance; some prickles; foliage medium, medium green, glossy; bushy, medium (100 cms) growth; [(Champagne Cocktail X Alpine Sunset) X (Prominent X Southampton)]; Horner, Colin P.

'HORCOFFITUP', S, dy, 1998; (**Celebration 2000**); flowers deep yellow, non-fading, paler reverse, semi-dbl., 8–14 petals, 1.5–2 in., blooms in large clusters, slight fragrance; prickles moderate; foliage small, medium green, very glossy; bushy, tall growth; [(Anytime X Liverpool Echo) X ((Flamenco X Rosa Bella) X Baby Love)]; Horner, Colin P.; Paul Chessum Rose Specialist, 1999

'HORDEN HALL', LCl, w, 1928; flowers pure white, with long yellow stamens, large, single, slight fragrance; [R. wichurana X Frau Karl Druschki]; Conyers

'HOREAGLE', F, pb, 1997; (**Hole in One**); flowers double, small, dbl., 15–25 petals, borne in small clusters, slight fragrance; some prickles; foliage small, medium green, semi-glossy; spreading, medium (90 cms) growth; [((Southampton X New Penny) X (White Pet X Stars 'n' Stripes)) X ((Vester X seedling) X Edith Holden)]; Horner, Colin P.

'HORETHEL', HT, pb, 1992; (**Isobel Derby**); flowers peach pink, reverse lighter pink, aging deeper pink, urn-shaped, dbl., 5 in., moderate, fruity fragrance; foliage medium, medium green, glossy; bushy, medium growth; [Champagne Cocktail X [(Honey Favorite X Dr A.J. Verhage) X Pot o' Gold]]; Horner, Colin P.; Golden Fields Nursery, 1992

'HORFAIWIL', HT, my, 1998; (**Faith**); flowers medium yellow, lighter reverse, very dbl., 26–40 petals, 4 in., borne mostly singly, moderate fragrance; prickles moderate, medium, curved; foliage medium, medium green, glossy; upright, tall growth; [Golden Future X Solitaire]; Horner, Colin P.; Warley Rose Gardens, Ltd., 2000

'HORFLAME', F, mr, 1997; (**Middlesbrough Football Club**); flowers double, large, dbl., 15–25 petals, borne in small clusters, slight fragrance; some prickles; foliage medium, reddish bronze, glossy; upright, medium (60 cms) growth; [Alpha X (Old Master X Southampton)]; Horner, Heather M.

'HORFLAN', F, dp, 1990; (**Flanders Field**); bud ovoid; flowers light red aging slightly lighter, urn-shaped, loose, medium, semi-dbl., slight fragrance; foliage medium, medium green, semi-glossy; medium to tall, upright growth; [Prominent X Southampton]; Horner, Heather M., 1992

'HORFLASH', F, yb, 1983; (**Champagne Cocktail**); flowers pale yellow, flecked and splashed pink, yellow reverse, medium, dbl., 20 petals, blooms in clusters of 3 to 5, moderate fragrance; foliage medium, medium green, glossy; bushy growth; GM, Glasgow, 1990; [Old Master X Southampton]; Horner, Colin P., 1985

'HORFLASHROB', F, rb, 1999; (**Valeria Sykes**); flowers red splashed white, reverse silver, single, 5–11 petals, 2 in., borne in small clusters, slight fragrance; few prickles; foliage small, medium green, semi-glossy; bushy, low (2 ft) growth; [Champagne Cocktail X Robin Redbreast]; Horner, Colin P.; Battersby Roses, 2000

'HORFLOSS', F, mp, 1997; (**Flossie**); flowers double, small, dbl., 15–25 petals, borne in large clusters, moderate fragrance; prickles some; foliage small, medium green, semi-glossy; upright, medium (3 ft) growth; [Coral Reef X (Seedling X Lichtkonigin Lucia)]; Horner, Colin P.

'HORFLOWER', Min, op, 1995; (**Flower Child**); flowers salmon pink, dbl., 15–25 petals, medium blooms; slight fragrance; foliage small, medium green, semi-glossy; compact, tall (40 cms) growth; [Penelope Keith X Gold Bunny]; Horner, Heather M.; Warley Rose Gardens, 1997

'HORFROST', F, w, 1997; (**Frosty Morning**); flowers double, medium, dbl., 15–25 petals, borne in small clusters, slight fragrance; some prickles; foliage medium, medium green, semi-glossy; bushy, medium (100 cms) growth; [Sexy Rexy X Princess Michael of Kent]; Horner, Colin P.

'HORGREG', S, w, 1996; (**Greg Moore**); flowers cream tinged pink, small, camellia-shaped blooms) borne in l, dbl., 15–25 petals, 2 in., slight fragrance; some prickles; foliage small, grey green, dull; medium (100 cms), spreading growth; [Bonica X (Lichtkonigin Lucia X Seedling)]; Horner, Colin P.

'HORHARPDOS', Min, ob, 1992; (**Emma Mitchell**); flowers orange/vermillion striped white, light orange reverse, aging light orange, semi-dbl., 12 petals, 1.25 in., fruity fragrance, slight; foliage small, medium green, semi-glossy; bushy, low growth; Horner, Colin P.; Battersby Roses, 1993

'HORHARRYPLUS', F, w, 1988; (**Whitley Bay**); bud ovoid; flowers cream with pink splashes, reverse cream, loose, medium, born, cupped, slight fragrance; ovoid, medium, yellow fruit; prickles medium, brown; foliage medium, medium green, semi-glossy; spreading, medium growth; [Champagne Cocktail X [Colour Wonder X (Vera Dalton X Piccadilly)]]; Horner, Colin P.; Battersby Roses, 1990

'HORIZON', F, or, 1956; bud pointed; flowers geranium-red, open, large, borne in large clusters, dbl., moderate fragrance; foliage leathery; vigorous, upright, bushy growth; [Crimson Glory X Cinnabar]; Tantau, Math.

Horizon 2000, HT, 1990; Delbard, Georges; (Cavriglia)

'HORJACK', HT, ob, 1995; (**Jack Collier**); flowers orange blend, dbl., 15–25 petals, 4 in., borne mostly singly or in small clusters, slight fragrance; foliage medium, medium green, semi-glossy; tall (120 cms), upright growth; [Marjorie May X Remember Me]; Horner, Colin P.; Battersby Roses, 1997

'HORJEMMA', F, my, 1995; (**Jemma Giblin**); dbl., 15–25 petals, 2.5 in., borne in small clusters, moderate fragrance; foliage medium, medium green, semi-glossy; medium (50 cms), bushy growth; [Gingernut X Amanda]; Horner, Colin P.; Battersby Roses, 1997

'HORJOY', F, pb, 1997; (**Our Joy**); flowers candy pink, silver reverse, dbl., 15–25 petals, 3 in., borne in small clusters, moderate fragrance; prickles moderate; foliage medium, semi-glossy; upright, 80 cms growth; [Seedling X Seedling]; Horner, Heather M.; Battersby Roses, 1997

'HORKEEPOG', F, w, 1994; (**Susan Jellicoe**); flowers cream/pink blend, blooms borne mostly singly and in small clusters, dbl., 26–40 petals, 3–3.5 in., intense fragrance; some prickles; foliage medium, medium green, semi-glossy; medium (100 cms), upright, bushy growth; [Keepsake X Pot o' Gold]; Horner, Colin P.; Horner, 1993

'HORKORBLUSH', HT, r, 1988; (**Georgie Lad**); bud ovoid, red; flowers mahogany-red, yellow at base, reverse lighter red, medium, dbl., cupped, moderate, fruity fragrance; ovoid, medium, medium fruit; prickles small, light brown; foliage medium, medium green, matt; upright, medium growth; [Prominent X (Champagne Cocktail X Alpine Sunset)]; Horner, Colin P.; Battersby Roses, 1990

'HORLEXSTRIP', HT, op, 1991; (**Grace Donnelly**); bud ovoid; flowers pink, orange, yellow striped, urn-shaped, loose, dbl., borne singly or in sprays of 5–9, slight, fruity fragrance; foliage medium, medium green, semi-glossy; bushy, tall growth; [Alexander X [Southampton X [(New Penny X Little White Pet) X Stars 'n' Stripes]]]; Horner, Colin P., 1984; Battersby Roses, 1992

'HORLIGHTS', F, yb, 1987; (**Friction Lights**); flowers canary-yellow edged cherry-red, large, dbl., 15–25 petals, slight fragrance; foliage medium, medium green, semi-glossy; upright growth; [Alexander X Champagne Cocktail]; Horner, Colin P.; Battersby Roses, 1987

'HORLOVEQUEEN', F, ob, 1992; (**Mary Campbell**); flowers orange, blooms borne in small clusters, semi-dbl., 6–14 petals, 1.5–2.75 in., slight fragrance; some prickles; foliage medium, medium green, semi-glossy; medium (80 cms), bushy growth; [Lovers Meeting X Amber Queen]; Horner, Heather M., 1993

'HORMARTIM', F, dr, 1988; (**Margaret Daintry**); flowers crimson-scarlet, large, borne singly and in clusters, dbl., 15–25 petals, intense fragrance; foliage medium, dark green, glossy; bushy growth; [Red Planet X (Blessings X (Parkdirektor Riggers X Honey Favorite))]; Horner, Colin P.; Rosemary Roses, 1988

'HORMASBRICK', F, r, 1992; (**Whickham Highway**); flowers brick red, blooms borne in large clusters, semi-dbl., 6–14 petals, 1.5–2.75 in., slight fragrance; some prickles; foliage medium, medium green, glossy; tall (120 cms), upright growth; [Mary Sumner X (KORp X Southampton)]; Horner, Heather M., 1993

'HORMISLAC', F, dp, 1992; (**Ted Gore**); flowers deep pink, blooms borne in large clusters, dbl., 26–40 petals, 1.5–2.75 in., slight fragrance; some prickles; foliage medium, medium green, semi-glossy; medium (80 cms), bushy growth; [Miss Ireland X Lilac Charm]; Horner, Colin P., 34394

'HORNDON PINK', HT, dp, 1963; flowers old-rose-pink, dbl., 60 petals, 4.5 in., moderate fragrance; foliage dark; [Lady Elgin X Independence]; Barter

'HORNEWGRAM', HT, my, 1988; (**Dame Joyce Frankland**); bud globular, greenish-yellow; flowers urn-shaped, large, borne singly, dbl., 32 petals, slight fragrance; ovoid, large, yellow fruit; prickles straight, medium, light brown; foliage large, medium green, glossy; bushy growth; [(Honey Favorite X Dr A.J. Verhage) X Pot o' Gold]; Horner, Colin P.

Hornish Plot, HP, rb

'HORODD', F, rb, 1998; (**Oddball**); flowers crimson stripes on buff brown, fading to red stripes on pink, semi-dbl., 8–14 petals, 2.5 in., borne in small clusters, slight fragrance; prickles moderate; foliage medium, medium green, dull; spreading growth; [(Licterloh X (Tall Story X Southampton) X ((New Penny X White Pet) X Stars 'n' Stripes)]; Horner, Colin P.; Paul Chessum Rose Specialist, 1998

'HORPEKBOND', Min, or, 1998; (**Hot Gossip**); flowers luminous orange-red, reverse matt orange, semi-dbl., 8–14 petals, 1.5 in., borne in small clusters, slight fragrance; prickles moderate; foliage medium, medium green, dull; compact, medium (50 cms), dwarf-clustered growth; [Penelope Keith X (Prominent X Southampton)]; Horner, Heather M.; Paul Chessum Rose Specialist, 1998

'HORRIDO'®, F, mr, 1963; (Jockey); bud ovoid; flowers pure blood-red, blooms in large clusters, dbl., cupped; foliage dark, glossy; bushy, low growth; Tantau, Math.

'HORSADDLER', F, ab, 1995; (**Saddler's Gold**); flowers amber gold, blooms borne in small clusters, dbl., 15–25 petals, 2.5 in., slight fragrance; foliage medium, medium green, semi-glossy; medium (50 cms), bushy growth; [Gingernut X Gold Bunny]; Horner, Heather M.; Warley Rose Gardens, 1997

'HORSILBEE', HT, ab, 1988; (**Heather Honey**); bud ovoid, bronze; flowers apricot yellow, reverse apricot urn-shaped, medium, borne usually singly, dbl., 25 petals, moderate, fruity fragrance; globular, medium, yellow fruit; prickles small, greenish-brown; foliage medium, medium green, glossy; bushy, medium growth; [Silver Jubilee X (Honey Favorite X Southampton)]; Horner, Colin P.; LeGrice Roses, 1991

'HORSILKARL', HT, mp, 1997; (**Doris Dowman**); flowers full, large, very dbl., 26–40 petals, borne mostly singly, slight fragrance; some prickles; foliage large, medium green, glossy; compact, medium (3 ft) growth; [Silver Jubilee X (Seedling X Karlsruhe)]; Horner, Colin P.

'HORSTACEY', F, pb, 1996; (**Stacey's Star**); flowers blends of pink, short center petals much darker pink, blooms, dbl., 26–40 petals, 2.5 in., flat, moderate fragrance; some prickles; foliage small, medium green, semi-glossy; low (50 cms), bushy growth; Horner, Colin P.; Battersby Roses, 1998

'HORSTMANN'S BERGFEUER', F, dr, 1954; flowers dark blood-red, well formed, very large, borne in large clusters, dbl., moderate growth; [World's Fair X Independence]; Horstmann

'HORSTMANN'S JUBILÄUMSROSE', F, op, 1954; flowers pink tinted peach, well formed, large, borne in clusters of, dbl.; foliage glossy, leathery; dwarf growth; [Golden Rain X Alpine Glow]; Tantau, Math.; Horstmann

'HORSTMANN'S LEUCHTFEUER', F, dr, 1954; flowers blood-red, large, borne in large clusters, dbl.; moderate growth; [Red Favorite X Karl Weinhausen]; Tantau, Math.; Horstmann

'HORSTMANN'S ROSENRESLI', F, w, 1955; flowers pure white, large blooms in clusters, dbl., moderate fragrance; bushy growth; [Rudolph Timm X Lavender Pinocchio]; Kordes; Horstmann

'HORSTMANN'S SCHÖNE BRÜNETTE', HT, or, 1955; flowers coppery brick-red, large, dbl.; strong stems; [Independence X Hens Verschuren]; Horstmann

'HORSUN', F, yb, 1986; (**Playgroup Rose®**); flowers yellow, petals edged red, reverse pale yellow, blooms in sprays, dbl., 25 petals, cupped, moderate fragrance; small, globular, orange fruit; large, light brown prickles; foliage medium, light green, glossy; medium, bushy growth; [Prominent X Southampton]; Horner, Heather M.

'HORSUNPEGY', F, op, 1993; (**Marjorie May**); flowers orange/pink blend, blooms borne in small clusters, dbl., 26–40 petals, 3–3.5 in., slight fragrance; some prickles; foliage medium, medium green, semi-glossy; medium (100 cms), upright growth; [Playgroup X Peer Gynt]; Horner, Heather M.; Horner, 1995

'HORSUNSMILE', F, yb, 1993; (**Voice of Thousands**); flowers yellow edged cherry red, dbl., 15–25 petals, 1.5–2.75 in., borne in small clusters, slight fragrance; some prickles; foliage medium, medium green, semi-glossy; medium (80 cms), bushy growth; [Playgroup X Bright Smile]; Horner, Heather M.; Horner, 1994

'HORTENSE DE BEAUHARNAIS', HGal, pb; flowers rose edged rosy lilac, dbl.

'HORTENSE VERNET', M, w, 1861; flowers white shaded rose, very dbl.; moderate growth; Moreau et Robert

Horticolor®, *see* LAPbal

'HORTICULTOR VIDAL', HT, op, 1952; bud pointed; flowers salmon-pink, large, dbl., 35 petals, intense fragrance; bushy growth; [Mme Butterfly X Federico Casas]; Dot, Pedro

'HORTIFLORA', Pol, mp, 1974; bud oval; flowers full, dbl., 45 petals, 2 in., slight fragrance; Delforge, S.

'HORTROPIC', HT, yb, 1988; (Sir William Leech, **William Leech**); bud ovoid; flowers yellow edged red, reverse light yellow, large, borne usually, dbl., exhibition form, slight fragrance; ovoid, large, orange fruit; prickles medium, red; foliage medium, medium green, glossy; bushy, medium growth; [Royal Dane X Piccadilly]; Horner, Colin P.; Battersby Roses, 1989

'HORTULANUS ALBERT FIET', HT, ab, 1919; flowers apricot and lilac-rose, center coppery orange, dbl., moderate fragrance; [Mme Mélanie Soupert X Mons. Paul Lédé]; Leenders, M.

'HORTULANUS BUDDE', HT, mr, 1919; bud pointed; flowers large, dbl., moderate fragrance; foliage dark; vigorous growth; [Gen. MacArthur X Mme Edouard Herriot]; Verschuren

'HORTULANUS FIET', HT, w, 1919; bud pointed; flowers deep and light cream, open, very large, dbl., slight fragrance; foliage sparse, glossy, dark; vigorous growth; [Cissie Easlea X Golden Star]; Verschuren

'HORTYARD', F, m, 1992; (**Olga Rippon**); flowers purple/mauve, loose, blooms borne in sprays of 5–9, dbl., 2.5 in., slight, spicy fragrance; foliage medium, medium green, semi-glossy; bushy, low growth; [Sexy Rexy X [(INTermezzo X Baby Faurax) X (Tassin X Seedling)]]; Horner, Colin P.

'HORVALKATH', F, yb, 1997; (**Valerie Kathleen**); flowers double, medium, dbl., 15–25 petals, borne in small clusters, moderate fragrance; some prickles; foliage large, dark green, glossy; bushy, medium (2 ft 6 in.) growth; [(Silver Jubilee X Pink Favorite) X Amber Queen]; Horner, Colin P.

'HORWANDER', S, mp, 1997; (**Wanderer**); flowers semi double, small, semi-dbl., 8–14 petals, borne in large clusters, slight fragrance; foliage small, dark green, semi-glossy; spreading, low (50–60 cms) growth; [Sea Foam X Eyeopener]; Horner, Colin P.

'HORWINGFEL', F, or, 1988; (**Winged Fellowship**); bud ovoid; flowers vermillion orange, reverse lighter, urn-shaped, medium, born, dbl., 28 petals, slight fragrance; globular, medium, red-orange fruit; prickles small, brown; foliage medium, medium green, matt; bushy, medium growth; [Guitare X Prominent]; Horner, Heather M.; Rosemary Roses, 1989

'HOSHIZUKUYO', HT, w, 1990; bud pointed; flowers ivory, large, borne usually singly, dbl., 35–40 petals, exhibition form, intense, fruity fragrance; foliage medium, light green, semi-glossy; medium, upright growth; [(Izayoi X Sodori-Hime) X White Success]; Ohtsuki, Hironaka

Hospitality, *see* 'HORCOFF'

'HOSTESS', F, op, 1960; flowers pink tinted salmon, open, large, borne in clusters of 5–7, semi-dbl.; vigorous, upright growth; [Papillon Rose X (Cinnabar X Alain)]; Lens

'HOSTESS GISELA', F, dp, 1973; bud ovoid; flowers medium, dbl., intense fragrance; foliage very glossy; vigorous, upright, bushy growth; [Sympathie X Dr A.J. Verhage]; Hetzel; GAWA

Hot Chile, *see* 'MACBUNBER'

Hot Chocolate, *see* SIMcho

Hot Fire, *see* INTerbricor

Hot Gossip, *see* 'HORPEKBOND'

Hot Gossip, *see* 'JACati'

Hot Lips, *see* 'TINLIPS'

Hot Lips, F, dp

Hot 'n' Spicy, *see* 'MACSODA'

Hot Pants, *see* 'SIMHOPAN'

'HOT PEWTER', HT, or, 1978; (Crucencia, Crucenia, Cruzenia); flowers brilliant orange-red, borne usually singly, dbl., 41 petals, exhibition form, slight fragrance; broad, dark prickles; foliage large, mid-green, semi-glossy; bushy growth; [Alec's Red X Red Dandy]; Harkness, R., & Co., Ltd.

Hot Pink Pastel, *see* 'WILPINK'

Hot Point, *see* BARhop

Hot Princess, HT, lp; exhibition form; foliage large, dark

'HOT SHOT', Min, or, 1982; (Hotshot); flowers vibrant vermilion, medium, dbl., 28 petals, exhibition form, slight fragrance; foliage small, medium green; upright, bushy growth; AOE, 1984; [Futura X Orange Honey]; Bennett, Cecilia 'Dee'; Tiny Petals Nursery

Hot Spot™, *see* 'CARBANE'

'HOT STUFF', Min, mr, 1978; bud pointed; flowers turkey-red, semi-dbl., 10 petals, 1 in., slight fragrance; foliage tiny; very compact, bushy growth; Lyon

Hot Tamale™, *see* 'JACPOY'

'HOTEL HERSHEY', Gr, or, 1977; bud long, pointed to urn-shaped; flowers salmon orange-red, dbl., 4–4.5 in., exhibition form, slight fragrance; foliage dark, leathery; upright growth; [Queen Elizabeth X Comanche]; Williams, J. Benjamin; Hershey Estates

Hotel Royal, S, mr, 1980; McGredy, Sam IV

'HÔTESSE DE FRANCE', HT, dr, 1962; flowers velvety deep red, medium, cupped; foliage bronze; [Soraya X Unnamed seedling]; Hémeray-Aubert

Hotline™, *see* 'AROMIKEH'

Hotshot, *see* **'HOT SHOT'**

'HOUEMILY', HT, lp, 1991; (**Emily Hough**); flowers white, blushing pink toward center, more pink in cooler weather, dbl., 15–25 petals, 3–3.5 in., exhibition form, slight fragrance; some prickles; foliage large, medium green, semi-glossy; medium, upright growth; [Touch of Class sport]; Hough, Robin, 1993

House Beautiful, *see* HARbingo

House of York, HT, w, 1998; flowers creamy white, large, exhibition form; Williams, J. Benjamin

Houston, *see* **'CATHEDRAL'**

'HOUSTON', HT, dy, 1980; bud ovoid; flowers deep bright yellow, borne singly or 3–4 per cluster, dbl., 38 petals, exhibition form, moderate, tea fragrance; long prickles, hooked downward; foliage leathery, wrinkled, dark; vigorous, upright growth; PP4687; [Summer Sunshine X Unnamed seedling]; Weeks, O.L.

'HOUSTONIAN', Gr, dr, 1962; bud ovoid; dbl., 35 petals, 4 in., moderate fragrance; foliage leathery, glossy, bronze; vigorous, upright, tall growth; [Carrousel X Unnamed seedling]; Patterson; Patterson Roses

'HOVYN DE TRONCHÈRE', T, rb; flowers red with orange depths, bordered silver; Puyravaud, 1899

Howard Florey, *see* 'TOMFLO'

'HOWARD JERABEK', S, mp, 1978; flowers pearl-pink, very dbl., 100 petals, moderate, apple-blossom fragrance; foliage large, glossy; vigorous growth; Jerabek, Paul E.

Howard Morrison, *see* 'MACCRACKLE'

'HUBAR', HT, dr, 1975; (**Black Night**®); bud long; dbl., 26 petals, 4 in., foliage dark, leathery; vigorous, upright growth; [Fragrant Cloud X Pharaoh]; Huber

'HUBICKA', HT, pb, 1935; flowers alabaster-white, sometimes rosy, base yellow, very large, dbl., moderate fragrance; foliage dark; bushy growth; [Grete Bermbach sport]; Böhm, J.

'HUDSON', HT, mp; flowers pale rose, well shaped; foliage glossy; Mallerin, C.

Hudson Crimson, HP, dr

Hudson's Bay Rose, *see* **R. BLANDA**

'HUETTE'S DAINTY FLORRIE', Min, lp, 1979; dbl., 24 petals, moderate fragrance; dark red prickles; foliage dark green edged dark red; upright, bushy growth; [Sweet and Low X Mary Marshall]; Schwartz, Ernest W.; Men's Garden Club of Virginia

'HUGH DICKSON', HP, mr, 1905; flowers very large, dbl., 38 petals, exhibition form, recurrent bloom, intense fragrance; vigorous growth; [Lord Bacon X Gruss an Teplitz]; Dickson, H.

Hugh Ringold, HT; Gregory, C.; (Cavriglia)

'HUGH WATSON', HP, op, 1905; flowers deep pink tinged salmon and silver-pink, very large, dbl., 24 petals, flat; vigorous growth; Dickson, A.

'HUGO MAWEROFF', HMult, dp, 1910; flowers carmine-pink, small, dbl.; Soupert & Notting; (Sangerhausen)

'HUGO PILLER', HT, ab, 1927; flowers flesh-white, center pale ecru, [Ophelia sport]; Leenders, M.

'HUGO ROLLER', T, yb, 1907; flowers lemon-yellow, edged and suffused crimson, well-formed blooms, dbl., moderate fragrance; foliage small, rich green; weak stems; compact, bushy growth; Paul, W.

'HUGO ROLLER, CLIMBING', Cl T, lb; [Hugo Roller sport]; Rogers, 1932; (Weatherly, L.)

'HUGO SCHLÖSSER', HT, op, 1955; bud long, pointed; flowers salmon-pink, very large, dbl., exhibition form, moderate fragrance; foliage leathery; strong stem; very vigorous, upright, bushy growth; [World's Fair X Peace]; Kordes

Hugs 'n' Kisses™, *see* 'JACIRAPS'

Hugues Aufray, *see* DORastri

'HUGUETTE', Min, op; flowers shrimp-pink shaded salmon

'HUGUETTE DESPINEY', HWich, yb, 1911; flowers light buff-yellow, edged red, small, borne in large clusters, very dbl.; vigorous growth; [Marco sport]; Girin

'HUGUETTE DUFLOS', HT, op, 1937; bud pointed, dark pink; flowers satiny pink, touched salmon, large, dbl., 30–40 petals, intense fragrance; foliage dark; vigorous, bushy growth; [Betty Uprichard X Unnamed seedling]; Lille

'HUGUETTE VINCENT', HT, op, 1921; flowers brilliant velvety geranium-red, semi-dbl.; [Mrs Edward Powell seedling X Willowmere]; Chambard, C.

Huilito, B, mp

'HULA GIRL', Min, ob, 1975; bud long, pointed; flowers bright orange, dbl., 45 petals, 1 in., moderate, fruity fragrance; foliage small, glossy, embossed; bushy growth; AOE, 1976; [Miss Hillcrest X Mabel Dot]; Williams, Ernest D.; Mini-Roses; , Sequoia Nursery

'HULA HOOP', HT, pb, 1960; flowers striped bright pink and red, intense fragrance; RULED EXTINCT 11/90; Freud; Horstmann

Hula Hoop, *see* 'MORHOOP'

'HULLABALOU', Min, yb, 1990; bud pointed; flowers light orange on yellow with yellow reverse, aging to dull yellow, dbl., 20 petals, moderate fragrance; prickles straight, sparse, small, red; foliage small, medium green when new, aging to maroon, semi-; bushy, low growth; [Rise 'n' Shine X Unknown]; Stoddard, Louis

Hulthemia berberifolia, *see* **'HULTHEMIA PERSICA'**

'HULTHEMIA HARDII', ((Cels) Rowley), Sp; (Rosa X hardii, X Hardiilist as X H. hardii); flowers yellow with the crimson eye of Hulthemia, 2 in.; foliage 1–7 narrow leaflets; Simplicifoliae; [Hulthemia persica X Rosa clinophylla]; Hardy before 1836

'HULTHEMIA PERSICA', ((Michaux) Bornmueller), Sp, yb; (Hulthemia berberifolia, Lowea berberifolia, Rosa berberifolia, Rosa persica, Rosa simplicifolia); flowers buttercup-yellow with a scarlet eye like a cistus, 1 in.; fruit prickly curious xerophyt; foliage leaves simple, bluish-green; Simplicifoliae, (14); 1790

'HULTHEMOSA', (X Hulthemosa)

Humanity, *see* HARcross

'HUMBOLDT', HT, mp, 1922; flowers bright rose-pink, dbl., exhibition form, moderate fragrance; foliage glossy, bronze; vigorous growth; [Ophelia X Unnamed seedling]; Hill, E.G., Co.; Cottage Gardens Co.

'HUMDINGER', Min, op, 1976; bud pointed; flowers orange-red, dbl., 53 petals, 1 in., exhibition form, slight fragrance; foliage glossy, dark; upright, bushy growth; AOE, 1978; [Gold Coin seedling X Unnamed Miniature seedling]; Schwartz, Ernest W.; Nor'East Min. Roses

'HUME'S BLUSH TEA-SCENTED CHINA', T, lp, 1809; dbl., intense fragrance; Hume; A. Hume

'HUMMINGBIRD', HT, ob, 1934; bud pointed; flowers bronze-orange, very dbl., 3 in., vigorous growth; [Talisman sport]; Scittine; Lainson

'HUMORESKE', Min, lp, 1957; flowers white, tinged pink, small, dbl.; bushy, compact growth; [Midget X Pixie]; Spek; McGredy

'HUMPTY-DUMPTY', Min, lp, 1952; flowers soft carmine-pink, center deeper, blooms in clusters, very dbl.; height 6–8 in.; [(R. multiflora nana X Mrs Pierre S. duPont F2) X Tom Thumb]; deVink; T. Robinson, Ltd.

Hungaria, F, mr

Hunslet Moss, M, dp; re-introduced, 1984

'HUNTER', HRg, mr, 1961; (The Hunter); flowers bright crimson, medium, dbl., 43 petals, 2.5 in., borne in clusters, recurrent bloom, moderate, sweet fragrance; foliage rugose, dark, glossy; vigorous (4–5 ft) growth; [R. rugosa rubra X Independence]; Mattock, R.H.; Mattock

'HUNTER'S MOON', HT, my, 1951; dbl., 30 petals, slight fragrance; foliage glossy, dark; very vigorous growth; [Condesa de Sástago X Gorgeous]; McGredy, Sam IV

Huntingburg, F, lp, 1997; Peters

Huntington Brocade, B, rb, 1973; Robinson

Huntington Pink Tea, T, dp

Huntington Seven Sisters, HMult, mp

Huntington Slater's Crimson China, Ch, mr

Huntington's Hero, S, lp, 1995; Martin

'HUNTSMAN', HT, rb, 1951; flowers spectrum-red, reverse yellow, dbl., 35–40 petals, 5.5 in., moderate fragrance; foliage dark; [The Queen Alexandra Rose seedling X Crimson Glory]; Robinson, H.; J&P

'HURDALSROSEN', S, mp, 1860

Hurdy Gurdy, *see* 'MACPLUTO'

'HURON', S, lp, 1932; flowers white flushed pink, semi-dbl., non-recurrent, moderate fragrance; foliage leathery; vigorous ($2\frac{1}{2}$ ft), compact, bushy growth; a good hedge rose; [Pythagoras X R. cinnamomea]; Central Exp. Farm

Huron Sunset, *see* 'RENHURON'

'HURRA'®, F, or, 1962; bud pointed; dbl., blooms in clusters; foliage glossy; vigorous, bushy growth; Tantau, Math.

Hurrikan, Pol, mr; flowers large, dbl., slight fragrance; VEG; (Sangerhausen)

'HURST CHARM', LCl, pb, 1936; flowers pink, slightly tinted mauve, large; Hicks

'HURST CRIMSON', Pol, dr, 1933; flowers deep crimson, borne in large trusses; [Ideal sport]; Hicks

'HURST DELIGHT', LCl, w, 1936; flowers clear pale cream, moderate fragrance; vigorous growth; Hicks

'HURST FAVOURITE', LCl, w, 1936; flowers pure ivory-white, very large, borne in clusters, semi-dbl., moderate fragrance; very vigorous, climbing growth; Hicks

'HURST GEM', Pol, ob, 1931; flowers brilliant orange-scarlet, small, semi-dbl.; [Orléans Rose sport]; Hicks

'HURST GLORY', HT, ob, 1936; flowers pale salmon-cerise, flushed yellow, well shaped, large, dbl., moderate fragrance; vigorous growth; Hicks

'HURST SCARLET', HT, dr, 1933; flowers deep scarlet, large, dbl., moderate fragrance; Hicks

Husmoderrose, HSpn, mp

'HUTTON VILLAGE', HT, my, 1973; flowers bright yellow, [Whisky Mac sport]; Deamer; Warley Rose Gardens

'HUX 1', S, mp, 1998; (**Lily Freeman**); flowers medium pink, single, single, 5–11 petals, 5.25 in., flat, borne in small clusters, moderate fragrance; prickles numerous; foliage small, medium green, dull; bushy, medium tall growth; [Schneezwerg X Schneezwerg]; Huxley, Ian

Hviezdoslav, *see* **'HWIEZDOSLAV'**

'HVISSINGE-ROSE', F, pb, 1943; flowers pinkish with yellow, borne in clusters, single; vigorous growth; [Orléans Rose X Seedling]; Poulsen, S.

'HWIEZDOSLAV', HT, rb, 1936; (Hviezdoslav); flowers copper-red to orange-copper-red, large, dbl., cupped, intense fragrance; foliage glossy; vigorous, bushy growth; Böhm, J.

'HYBRIDE DI CASTELLO', S, ly, 1920

'HYLO', F, mp, 1961; flowers salmon-pink, medium, borne in clusters, dbl., 28 petals, 2–2.5 in., slight fragrance; bushy growth; [Highlight sport]; Cant, B. R.

Hymenee, S, 1969; Delbard-Chabert; (Cavriglia)

'HYMNE', F, mp, 1964; bud short, pointed; flowers neyron rose, large, borne in large clusters, dbl., 38–52 petals, foliage dark, glossy; vigorous growth; [Miracle X Unnamed seedling]; Verbeek

'HYPACIA', C, pb; (Hypathia); flowers bright rose-pink spotted white, center whitish, large, dbl., cupped, moderate, damask fragrance

Hypathia, *see* **'HYPACIA'**

'HYTHE CLUSTER', Pol, dp, 1935; flowers glowing deep pink, borne in large clusters, semi-dbl., cupped; foliage small, glossy, light; vigorous growth; Archer

I

I Love You, *see* 'GEElove'

'I Love You', F, dp, 1981; flowers deep pink; [Margaret Thatcher sport]; Takatori, Yoshiho; Japan Rose Nursery

I Promise, *see* 'LYOpri'

'I Zingari', HT, ob, 1925; flowers orange-scarlet, corymbs, semi-dbl.; foliage dark; stems claret; Pemberton

'I. X. L.', HMult, dp, 1925; flowers magenta, prickles straight, thornless canes (1 in.iam); very vigorous growth; [Tausendschön X Veilchenblau]; Coolidge

'Ian Brinson', HT, dr, 1945; flowers crimson, compact, dbl., 30 petals, 4–5 in., intense fragrance; foliage dark; [Mrs J.J. Hedley-Willis X J.C. Thornton]; Bees

Iberflora 95®, *see* FEbama

Ibica, *see* **'Ibiza'**

Ibis, *see* 'GODeurpan'

'Ibiza', HT, w, 1938; (Ibica); flowers well formed, moderate fragrance; erect growth; [Mme Butterfly X Frau Karl Druschki]; Dot, Pedro

Ice Angel, *see* JACber

Ice Breaker, *see* 'BRIice'

Ice Cascade, S, lp, 1999; very dbl., borne in sprays; Williams, J. Benjamin

Ice Cream, *see* KORzuri

Ice Crystal, *see* 'CLEice'

Ice Fairy, *see* 'SANmed'

Ice Follies, S, w, 1999; flowers white with delicate apricot tint, borne in large sprays, moderate fragrance; foliage dark green; Williams, J. Benjamin

'Ice Maiden', F, w, 1977; bud pointed; flowers pure white, semi-formal, semi-dbl., 15 petals, 3.5 in., moderate fragrance; foliage narrow, glossy, light green; vigorous, upright growth; [Iceberg X Iceberg]; Garelya

Ice Meidiland™, *see* MEIvahyn

Ice Princess, *see* 'LAVice'

Ice Queen™, *see* 'SAVanice'

'Ice White', F, w, 1966; (Vision Blanc®); dbl., 25 petals, 3 in., blooms in clusters; foliage glossy; GM, Portland, 1970; [Mme Leon Cuny X (Orange Sweetheart X Cinnabar)]; McGredy, Sam IV; McGredy

Ice White, Climbing, *see* MACvisar

Iceberg, *see* 'KORbin'

'Iceberg, Climbing', Cl F, w, 1968; Cant, B. R.

'Iced Ginger', F, ob, 1971; flowers petals palest pink, reverse coppery, dbl., 45 petals, 4.5 in., moderate fragrance; foliage red, veined; [Anne Watkins X ?]; Dickson, A.

'Iced Parfait', F, lp, 1972; flowers pale pink, urn-shaped, medium, borne 6 per cluster, dbl., 40 petals, moderate fragrance; straight, red prickles; foliage light green; bushy, compact growth; [Pink Parfait X Iceberg]; Xavier, Sister M.

'Iceland Queen', LCl, w, 1935; flowers creamy white, large, dbl.; vigorous growth; Horvath; Wyant

Ichalkaranji 100, HT, m, 1993; Chiplunkar

ICI Golden Celebration, *see* 'PEAquant'

'Ico', F, m, 1986; flowers light purple, very dbl., 60–70 petals, globular, slight fragrance; brownish-green prickles curving downward; foliage dark green, glossy; vigorous, upright, bushy growth; [Deep Purple sport]; Patil, B.K.; K.S.G. Son's Roses

Ico Ambassador, HT, my, 1988; Patil, B.K.

'Ico Beauty', HT, pb, 1986; flowers rose pink, reverse flushed white, medium, dbl., 25–30 petals, exhibition form, slight fragrance; pale green prickles curving downward; foliage glossy; upright, bushy growth; [Red Planet sport]; Patil, B.K.; K.S.G. Son's Roses

Ico Delight, HT, w, 1989; Patil, B.K.

Ico Deluxe, HT, pb, 1990; Patil, B.K.

Ico Pearl, F, pb, 1990; Patil, B.K.

Ico Talk, F, pb, 1993; Patil, B.K.

Ico Trimurthi, HT, pb, 1989; Patil, B.K.

Icy Tiding, HT, w

Ida Belle™, *see* 'MINayco'

Ida Elizabeth, *see* 'WELiz'

Ida Hoff, HT, mp; flowers large, dbl.; (Sangerhausen)

'Ida Klemm', HMult, w, 1907; flowers snow-white, immense clusters; long, strong stems; vigorous growth; [Crimson Rambler sport]; Walter, L.

'Ida McCracken', HT, op, 1952; flowers salmon and coral, well formed, dbl., 25–30 petals, 4 in., moderate fragrance; foliage leathery, dark; free growth; [Ethel Somerset X Mrs Sam McGredy]; Norman; Harkness

'Ida Scholten', HT, pb, 1933; flowers pink shaded carmine-red, very large, dbl.; foliage dark, leathery, glossy; bushy, dwarf, compact growth; [Capt. F.S. Harvey-Cant X Gen. MacArthur]; Buisman, G. A. H.

'Ideal', Pol, mr, 1921; [Miss Edith Cavell sport]; Spek

Ideal Home, *see* **'Idylle'**

Idée Fixe®, *see* 'LENdec'

Idole, HT, or

'Idun', F, mp, 1969; flowers neyron rose, large, borne in clusters, dbl., 41 petals, cupped; foliage dark, glossy; bushy growth; [Schneewittchen X Fanal]; Lundstad

'Idylle', HT, pb, 1959; (Ideal Home); flowers carmine-pink, base white, well formed, dbl., 25–30 petals, 5 in., moderate fragrance; vigorous, upright growth; [Monte Carlo X Tonnerre]; Laperrière; EFR

'Iga 63', HT, pb, 1963; flowers pink and red; [Confidence X Seedling]; da Silva, Moreira

Iga 83 Munchen, *see* 'MEIbalbika'®

Iga Erfurt, F, op, 1966; flowers medium, semi-dbl.; GPG Bad Langensalza; (Sangerhausen)

'Igloo', HT, w, 1969; bud ovoid; flowers medium, dbl.; foliage dark; [Seedling X White Knight]; Verbeek

'Ignasi Iglesias', HT, mr, 1934; flowers rose in early season, oriental red in summer, dbl., exhibition form, intense fragrance; foliage wrinkled; vigorous, bushy growth; [Angel Guimera X (Souv. de Claudius Pernet X Mme Butterfly)]; Dot, Pedro

'Ignis', S, mr, 1934; flowers fiery red, very large, non-recurrent, foliage leathery, dark; vigorous (3 1/4–$6\frac{1}{2}$ ft) growth; Chotkové Rosarium; Böhm

'Igor', , pb, 1913; flowers carmine-white, medium, dbl., 36 petals, intense fragrance; foliage dark; spreading growth; Geltsendeger

'Ikaruga', HT, yb, 1975; bud ovoid; dbl., 35 petals, 6 in., exhibition form, slight fragrance; vigorous, upright growth; [McGredy's Ivory X Garden Party]; Ito

Ilam, *see* 'SIMilam'

'Ilaria', F, dp, 1962; flowers coral-red, well formed, semi-dbl.; foliage dark; [Cinnabar X Fashion]; Borgatti, G.; Sgaravatti

'Ildiko', F, mr, 1971; bud ovoid; flowers cherry-red, medium, semi-dbl., cupped, profuse, intermittent bloom; moderate, upright growth; [Mardi Gras X Paprika]; Institute of Ornamental Plant Growing

Ile de France, *see* **'Adoration'**

'Ile de France', HWich, rb, 1922; flowers bright scarlet, center white, open, borne in clusters, semi-dbl., slight fragrance; foliage large, leathery, dark; short, strong stems; very vigorous (15–20 ft) growth; [American Pillar Unnamed seedling]; Nonin

Ile de France, *see* PEKsolred

'Illinois', Gr, or, 1969; bud long, pointed; flowers large, dbl., 25 petals, moderate fragrance; foliage leathery; vigorous, bushy growth; [Soprano X Tropicana]; Morey, Dr Dennison; Country Garden Nursery

Illisca®, *see* LAPem

Illos®, HT, m

'Illumination', F, dy, 1970; flowers deep sulfur-yellow, semi-dbl., 12 petals, 3 in., moderate fragrance; foliage glossy, light; free growth; [Clare Grammerstorf X Happy Event]; Dickson, Patrick; A. Dickson

Illumination, F, ab, 1986; Pallek

'Illusion', HKor, mr, 1961; flowers blood-red to cinnabar, large blooms in large clusters, dbl., moderate fragrance; foliage leathery, glossy, light green; vigorous growth; Kordes, R.

'Illusion', HT, my, 1961; dbl., 50 petals, moderate fragrance; foliage glossy; vigorous growth; [Peace X Unnamed seedling]; Verbeek

Ilmenau, S, mr, 1992; Hetzel

Ilona, *see* 'VARlon'

Ilona™, *see* 'MARilo'

'Ilse Haberland'®, S, mp, 1956; flowers crimson-pink, very large, dbl., exhibition form, moderate fragrance; foliage glossy; vigorous, upright, bushy growth; Kordes

'Ilse Krohn', LCl, w, 1957; flowers pure white, overlarge, very dbl., exhibition form, non-recurrent, slight fragrance; foliage glossy, leathery; very vigorous growth; [Golden Glow X R. kordesii]; Kordes

'Ilse Krohn Superior'®, HKor, w, 1964; flowers pure white, dbl., intense fragrance; foliage dark; vigorous (9 ft) growth; Kordes

Ilseta®, *see* 'TANatesil'

Ilsetta, *see* 'TANatesil'

Iluse, HT, ob, 1968; flowers salmon-orange and yellow, large, dbl., slight fragrance; Urban, J.; (Sangerhausen)

Image d'Epinal, LCl, mr, 1997; Croix

Imagination, *see* 'WEKmar'

Imagine, *see* 'RENimag'

Imagine, *see* DORmagi

'Imatra', F, pb, 1930; flowers pink to white; [Orléans Rose seedling]; Poulsen, S.; Olsson

Imbroglio, *see* BENimbro

Imma, HT, lp;, exhibition form

'Immaculada Galan', F, w, 1968; flowers large, borne in trusses, dbl., 60–70 petals, globular; foliage small, blue-gray; very free growth; LeGrice

Immensee®, *see* 'KORimro'

Immortal Juno, *see* 'AUSjuno'

'Imogen', F; Bidwell, 1845; (Weatherly, L.)

Imogene™, *see* 'MINyco'

'Imp', F, rb, 1971; bud globular; flowers red, reverse silver-pink, small, dbl., slight fragrance; foliage large, dark, leathery; bushy growth; [Daily Sketch X Impeccable]; Dawson, George; Brundrett

Impala, *see* 'HERpim'

Impala, *see* 'RUlimpa'

Impatient, *see* 'JACdew'

'Impeccable', HT, dr, 1955; flowers deep velvety red, well-shaped, dbl., intense fragrance; foliage dark; Delbard-Chabert

Imperator, *see* 'MEIperator'

'Impératrice Eugénie', M, m, 1856; flowers lilac-pink, medium, dbl., moderate fragrance; vigorous growth; Guillot Père

Imperatrice Farah, *see* DELivour

Imperatrice Josephine, *see* **'Empress Josephine'**

Impératrice Rouge, *see* **'Red Empress'**

'Imperial', HT, rb, 1957; flowers cardinal-red, reverse golden yellow, well formed, [Geranium X Opera]; da Silva, Moreira

'Imperial Gold', HT, my, 1962; (Canadiana); bud ovoid; flowers lemon-yellow to indian yellow, dbl., 30–35 petals, 3.5–4.5 in., moderate fragrance; foliage leathery, dark, glossy; vigorous, upright growth; [Charlotte Armstrong X Girona]; Swim, H.C.; C.R. Burr

Imperial Palace, *see* POUlchris

'Imperial Pink', HT, mp, 1942; [Royal Beauty sport]; Coddington

'Imperial Potentate', HT, mp, 1921; flowers large, dbl., 45 petals, exhibition form; foliage dark, leathery; vigorous growth; GM, Portland, 1921; [Ophelia X Hoosier Beauty]; Clarke Bros.

'Imperial Queen', HT, mr, 1962; bud long, pointed; flowers cherry-red, dbl., 21 petals, 4.5–5 in., cupped, moderate fragrance; foliage leathery, glossy; vigorous, compact growth; [Queen Elizabeth X Chrysler Imperial]; Lammerts, Dr Walter; C.R. Burr

Imperial Rose, B, dp

Imposant®, F, mp

'Impress', HT, rb, 1929; bud ovoid, cardinal-red, shaded orange; flowers salmon-cerise, tinted golden, very large, dbl., 40–45 petals, foliage dark, glossy; vigorous growth; Dickson, A.; Dreer; Liggit

'Improved Cécile Brünner', F, op, 1948; (Rosy Morn); bud long, pointed; flowers salmon-pink, medium, borne in clusters, dbl., 30 petals, exhibition form, slight fragrance; foliage leathery, dull green; very vigorous, upright growth; [Dainty Bess X R. gigantea]; Duehrsen; H&S

'Improved Lafayette', F, mr, 1935; semi-dbl., slight fragrance; foliage soft; vigorous, bushy growth; [E.G. Hill X Seedling]; H&S; Dreer

'Improved Peace', HT, yb, 1959; bud ovoid; flowers yellow edged and flushed pink, large, dbl., exhibition form, slight fragrance; foliage leathery, wrinkled; very vigorous, bushy growth; [Peace sport]; Dean

'Improved Premier Bal', HT, pb; flowers pale cream edged deep pink, large, moderate fragrance; foliage leathery; vigorous growth; Wheatcroft Bros.

'Improved Prince Philip', Gr, or, 1964; flowers well formed, [Queen Elizabeth X Prince Philip]; Leenders, J.

'Improved Universal Favorite', HWich, mp, 1901; flowers brilliant pink, Manda, W.A.

'Improved Verdun', Pol, dp, 1946; flowers vivid carmine-red, borne in clusters, dbl.; foliage leathery; bushy growth; Kluis; Klyn

'Impulse', Min, op, 1986; flowers salmon-pink, light yellow reverse, small, dbl., 38 petals, cupped, no fragrance; fruit not observed; long, brownish prickles; foliage small, medium green, semi-glossy; medium, upright growth; [Red Ace X Chris Jolly]; Jolly, Marie; Rosehill Farm

Impulse®, *see* 'BARimp'

In the Mood, *see* 'SEAmood'

'In the Pink', F, mp, 1988; flowers medium, borne in sprays of 4–5, dbl., 60 petals, moderate, musk fragrance; prickles hooked, red; foliage greenish-red, glossy; medium, bushy growth; [Baby Faurax X Unnamed seedling]; Ryan, C.; Melville Nursery, 1988

Ina, *see* 'TALina'

Ina an' Mona®, S, ab, 1992; Jensen

'Inano', HT, dy, 1978; flowers urn-shaped, medium, borne singly, dbl., 40 petals, moderate, fruity fragrance; small, brown prickles; foliage medium, light green; medium, bushy growth; [Doreen X Goldilocks]; Teranishi, K.; Itami Rose Nursery

'Inata', F, op, 1967; flowers pink shaded salmon, open, borne in trusses, semi-dbl.; vigorous growth; [Valeta sport]; deRuiter

'Inca de Mallorca', F, mr, 1958; flowers strawberry-red, dbl., 20 petals; strong stems; compact growth; [Soller X Floradora]; Dot, Pedro

Incandescent, HT, lp; globular and short; flowers velvety bright red vermillion-red, dbl., slight; large, reddish dark green, glossy; [Bond Street X Dame de Coeur]; St. Wagner, 1989; Res. Stn. f. Horticulture, Cluj, 1991, Romania

'Incense', HT, dr, 1968; flowers deep red, pointed, dbl., intense fragrance; vigorous growth; [(Karl Herbst X New Yorker) X Konrad Adenauer]; LeGrice

Incense Rose, *see* **R. primula**

'Inch'Allah', HT, pb, 1944; bud long; flowers bright pink, reverse flesh, stamens yellow, very large, semi-dbl.; vigorous growth; [Pres. Macia X Editor McFarland]; Meilland, F.

Incognito™, *see* 'BRIincog'

'Incomparable', HP, w, 1923; flowers rosy white, large; Giraud, A.

'Incredible', S, yb, 1984; bud ovoid, pointed; flowers yellow freckled and streaked with orange-red, urn-shaped, large, dbl., 28 petals, repeat bloom, moderate fragrance; awl-like, brown prickles; foliage medium large, leathery, dark olive green, copper tin; vigorous, erect growth; hardy; [Gingersnap X Sevilliana]; Buck, Dr Griffith J.; Iowa State University

'Indéfectible', Pol, mr, 1919; flowers bright clear red, semi-dbl., moderate fragrance; [Annchen Muller seedling]; Turbat

'Independence', F, or, 1951; (Geranium, Kordes' Sondermeldung, Reina Elisenda, Sondermeldung); bud urn-shaped; flowers pure scarlet, blooms in clusters (up to 10), dbl., 35 petals, 4.5 in., cupped, moderate fragrance; foliage glossy, dark; growth moderate; (28); GM, Bagatelle, 1943 GM, NRS, 1950 GM, Portland, 1953; [F2 seedling (Baby Chateau X Crimson Glory)]; Kordes; J&P

'Independence '76', F, mr, 1974; bud short, pointed; flowers high-pointed, dbl., 22 petals, 3–4 in., moderate, tea fragrance; vigorous, upright growth; [Cotillion X Suspense]; Byrum; J.H. Hill Co.

'Independence Day', HT, ab, 1919; bud pointed; flowers sunflower-gold, stained flame-color and orange-apricot, dbl., exhibition form, intense fragrance; foliage leathery, glossy, dark; vigorous growth; GM, NRS, 1919; [Mme Edouard Herriot X Souv. de Gustave Prat]; Bees

'Independence Day, Climbing', Cl HT, ab, 1930; Brown, W.&J.; E. Murrell

Independence du Luxembourg, HT, w, 1960; flowers creamy white with orange and pink tones, large, dbl., intense fragrance; Lens, Louis

'Independence, Climbing', Cl F, or, 1960; Balducci & Figli

'Indian Chief', HT, rb, 1967; flowers currant-red shaded orange, pointed, dbl., slight fragrance; foliage dark; very free growth; [Tropicana X ?]; Gregory

'Indian Goddess', Cl Min, pb, 1999; semi-dbl., 12–16 petals, 2.5 in., borne singly and in small clusters, no fragrance; few prickles; foliage dark green, semi-glossy, disease-resistant; bushy, medium (4–5 ft) growth; [Crazy Dottie X seedling]; Sridharan, Dr Lakshmi M.

'Indian Gold', F, yb, 1961; bud ovoid, flushed red; flowers yellow flushed soft pink, borne in clusters, dbl., 30–45 petals, 3.5 in., exhibition form, moderate fragrance; foliage glossy, light green; upright, compact growth; [Goldilocks X Seedling]; Von Abrams; Peterson & Dering

'Indian Maid', HT, ob; bud long, pointed; flowers salmon, reverse bronze-yellow, dbl., exhibition form, intense fragrance; foliage glossy; vigorous growth; [Talisman seedling X Souv. de Claudius Pernet]; Padilla

Indian Meillandina, *see* **'Carol-Jean'**

'Indian Pink', HT, pb, 1975; dbl., 30–36 petals, 5 in., globular, foliage leathery; very vigorous, upright, bushy growth; [Unknown seedling X Orange Tango]; McDaniel, Earl; Carlton Rose Nurseries

Indian Princess, *see* 'PIXiprin'

Indian Princess, *see* **'Princess of India'**

'Indian Red', HT, rb, 1948; bud long, pointed; flowers red shaded deeper, large, dbl., exhibition form, moderate fragrance; foliage small, glossy; vigorous, bushy, upright growth; hardy; [Pink Princess X Crimson Glory]; Brownell, H.C.

Indian Silk®, MinFl, w, 1991; Lens

Indian Song, *see* 'MEIhimper'

'Indian Summer', Cl HT, ob, 1938; bud pointed; flowers orange, streaked red, large, dbl., 25 petals, intense fragrance; foliage dark bronze, leathery, glossy; very vigorous (to 12–18 ft) growth; [Ednah Thomas X Autumn]; Duehrsen; H&S

Indian Summer, *see* HARwigwam

Indian Summer, *see* PEAperfume

Indian Sunblaze, *see* **'Carol-Jean'**

Indian Warrior, HT, dr

'Indiana', HT, pb, 1907; flowers bright pink, faintly suffused orange, dbl.; bushy growth; [Rosalind X Frau Karl Druschki]; Hill, E.G., Co.

Indiana, F, mp

'Indianapolis', HT, mp, 1971; flowers deep yellow-pink, medium, very dbl., cupped, slight fragrance; foliage glossy, leathery; moderate, upright growth; [Coloranja X ?]; Schloen, J.; Ellesmere Nursery

'Indica Alba', HCh, lp, 1802; (White Daily Rose); flowers very light blush, [Old Blush sport]

Indica Major, *see* **'Fun Jwan Lo'**

Indica Purpurea, HCh, m; flowers purple/pink, medium, single; Chenault; (Sangerhausen)

'Indigo', P, m, 1830

Indigoletta, LCl, m

'Indira', HT, pb, 1973; bud ovoid; flowers pink, reverse lighter, medium, slight fragrance; foliage soft; vigorous, upright, bushy growth; [Baccará X Prima Ballerina]; Hetzel; GAWA

'Indispensable', LCl, mp, 1947; bud globular; flowers pink, medium, borne in clusters, dbl., exhibition form, slight fragrance; foliage glossy; moderate, upright, pillar growth; [Roserie sport]; Klyn

'Indra', LCl, dp, 1937; bud pointed; flowers rose-pink, open, borne in clusters, semi-dbl., slight fragrance; foliage glossy; long stems; very vigorous, climbing growth; [(Ophelia X R. multiflora) X Florex]; Tantau

Indraman, F, or, 1990; Chiplunkar

Indu Singhal, Gr, pb, 1995; Singhal

'Indy 500', Gr, or, 1976; bud tapered; flowers brilliant orange-red, medium-high, dbl., 32 petals, 4.5–5 in., flat, intense fragrance; foliage large, glossy, dark, reddish; vigorous, upright growth; [(Aztec X Queen Elizabeth seedling) X (Independence X Scarlet Knight seedling)]; Williams, J. Benjamin; Krider Nursery

Inermis Morletii, *see* **'Morletii'**

'Infantania', F, w, 1953; flowers snow-white, sometimes tinged green, small, borne in large clusters, dbl., intense fragrance; [Baby Alberic X ?]; Heers; Langbecker

'Infante Beatrice', HT, ob, 1930; flowers orange-yellow, tinted reddish gold, base golden, dbl., intense fragrance; vigorous growth; [Marie Adélaide X Unnamed seedling]; Guillot, M.

'Infante Maria Cristina', HT, ob, 1930; bud pointed; flowers coppery, tinged carmine, moderate fragrance; foliage reddish bronze; vigorous growth; Gaujard

Inferno, *see* 'AROkunce'

Infinity, *see* 'RESfini'

'Ingar Olsson', F, mr, 1931; flowers brilliant cerise-red, blooms in clusters, semi-dbl., cupped, slight fragrance; foliage leathery; vigorous, rather compact growth; [Else Poulsen X Ophelia]; Poulsen, S.

'Inge Horstmann', HT, rb, 1964; bud long, red, reverse white tinged pink; flowers cherry-red, exhibition form, intense fragrance; long stems; vigorous, bushy growth; Tantau, Math.

'Inge Pein', Pol, mr, 1939; flowers carmine-red, medium, semi-dbl.; Pein; (Sangerhausen)

Inge Schubert®, *see* MAUlave

Ingrid Bergman®, *see* 'POUlman'

'Ingrid Stenzig', Pol, dp, 1951; (Pink Triumph); flowers rose-pink, buttercup form, small, borne inlarge clusters, [Orange Triumph sport]; Hassefras Bros.; B&A

Ingrid Weibull, *see* 'TANweieke'

'Ink Spots', HT, dr, 1985; flowers medium, dbl., 35 petals, slight fragrance; foliage large, dark, semi-glossy; upright, bushy, spreading growth; PP005855; [Unnamed seedling X Unnamed seedling]; Weeks, O.L.

Inka®, *see* TANakni

Inner Glow, *see* 'PIXINNER'

Inner Wheel, *see* 'FRYJASSO'

'**INNISFREE**', F, yb, 1964; flowers yellow, orange and pink, blooms in clusters, dbl., 22 petals, vigorous, tall growth; [(Karl Herbst X Masquerade) X Circus]; Dickson, Patrick; A. Dickson

'**INNOCENCE**', HT, w, 1921; flowers stamens reddish, slightly waved, borne in clusters, semi-dbl., 12 petals, 5 in., moderate fragrance; foliage dark; vigorous growth; Chaplin Bros.

Innocence, *see* 'SAVINN'

Innocence, HT, w; Croix

Innocence 96, *see* 'COCORAY'

'**INNOCENCE, CLIMBING**', Cl HT, w, 1938; Armstrong, J.A.; Armstrong Nursery

Innocencia®, *see* 'KORENBON'

Innocent Blush, *see* 'RENBLUSH'

Innovation, F, 1973; Combe, M.; (Cavriglia)

Innovation Minijet, *see* 'MEIJETTE'

Innoxa Femille, *see* 'HARPRINCELY'

'**INOA**', F, mr, 1958; flowers bright velvety red; vigorous growth; [Gruss an Teplitz X Pioupiou]; Arles; Roses-France

'**INSEL MAINAU**'®, F, dr, 1959; bud ovoid; flowers deep crimson, large blooms in clusters (up to 5), dbl., slight fragrance; foliage leathery, dark; low, compact growth; ADR, 1960; Kordes, R.

Insolite, *see* 'MORCHARI'

'**INSPECTEUR JAGOURT**', F, m, 1932; bud glowing red; flowers purplish pink to china-rose, large white stamens, large, dbl., 25–30 petals; foliage glossy; vigorous growth; [Mrs Henry Winnett X Eblouissant]; Soupert & Notting

'**INSPECTOR ROSE**', HT, rb, 1969; flowers maroon-red, reverse yellow, long, pointed, dbl., 35 petals; foliage coppery bronze-red; very free growth; [Piccadilly sport]; Fryers Nursery, Ltd.

'**INSPEKTOR BLOHM**', HMsk, w, 1942; very dbl., large corymbs, recurrent bloom, intense fragrance; foliage abundant, gray-green; vigorous, well-branched growth; [Joanna Hill X Eva]; Kordes

'**INSPIRATION**', LCl, mp, 1946; flowers large, semi-dbl., moderate fragrance; foliage large, glossy; moderate growth; [New Dawn X Crimson Glory]; Jacobus; B&A

Inspiration, HT, rb, 1991; Perry

Inspiration, *see* 'JACCOH'

'**INSTITUTEUR SIRDEY**', HT, dy, 1905; Pernet-Ducher

'**INSULINDE**', HT, yb, 1923; flowers clear yellow shaded golden yellow, dbl., slight fragrance; [Mr Joh. M. Jolles X Melody]; Van Rossem

'**INSULINDE**', HT, op, 1923; flowers pink and salmon, dbl., moderate fragrance; [Ophelia X Jonkheer J.L. Mock]; Leenders, M.

INTasprint, HT, dy; (**Sprinter**®); Ilsink, 1989

Integrity, *see* 'HARVINTAGE'

Intel, Min, m, 1996; Spooner, Raymond A.

INTepin, F, mp, 1995; (**Pink Fire**®); Ilsink

'INTERALL', S, lp, 1979; (**Rosy Cushion**®); flowers small blooms in large clusters, single, 7–8 petals, repeat bloom, slight fragrance; many medium prickles; foliage dark, glossy; vigorous (3–4 ft) groundcover growth; [Yesterday X Unnamed seedling]; Ilsink

Interama, *see* 'INTRUMA'

INTeramon, S, w; (**White Diamond**®); Interplant, 1994

'INTERANDER', S, op, 1983; (**Oleander Rose**); flowers salmon-pink, small blooms in large clusters, single, 7 petals, repeat bloom, slight fragrance; very few, small prickles; foliage medium, medium green, semi-glossy; upright growth; [Liverpool Echo X Unnamed seedling]; Interplant

INTerbab, F, lp; (**Yellow Dot**); Ilsink; Interplant, 1995

Interbec, *see* '**ESPOIR**'

INTerbricor, S, mr, 1995; (**Hot Fire**); Interplant

'INTERBRONZI', Min, r, 1992; (**Suntan**™); flowers bronze, little fading, urn-shaped, blooms borne in sprays of, dbl., 30 petals, 1.5 in., slight fragrance; foliage small, dark green, glossy; upright, bushy, low (30 cms) growth; [Seedling X The Fairy]; Ilsink, G.P., 1983; Interplant B.V., 1989

'INTERCARP', S, dp, 1983; (**Rosy Carpet**®); flowers deep pink, blooms in clusters, single, 5 petals, repeat blooming, moderate fragrance; many, medium prickles; foliage medium, dark, glossy; spreading (to 4 ft) growth; [Yesterday X Unnamed seedling]; Interplant, 1984

'INTERCEL', S, dp, 1979; ('INTERCELL', **Red Blanket**®); flowers dull, deep pink, small blooms in small clusters, semi-dbl., repeat bloom, slight fragrance; many medium prickles; foliage dark, glossy; vigorous (to 3–4 ft) groundcover growth; [Yesterday X Unnamed seedling]; Ilsink; Dickson Nurseries, Ltd.

'INTERCHER', S, mr, 1992; (**Chimo**®); flowers red aging dark red, blooms borne singly, single, 5 petals, 1.25 in., cupped, repeat bloom, slight fragrance; foliage medium, medium green, glossy; low (90 cms), spreading growth; [Seedling X Immensee]; Ilsink, Peter; & W. Kordes, 1989; Interplant B. V., 1989

'INTERCHIMP', S, mp, 1992; (**Pink Chimo**®); single, 5 petals, 1–1.5 in., cupped, borne singly, repeat bloom, no fragrance; foliage medium, medium green, semi-glossy; spreading, low (30 cms) growth; [Unknown seedling X Immensee]; Ilsink, Peter; Interplant B.V., 1990

'INTERCLEM', S, yb, 1982; (**New Face**®); flowers yellow edged pink, small blooms in large clusters, single, 5 petals, repeat bloom, slight fragrance; many prickles; foliage medium, medium green, semi-glossy; upright growth; Interplant, 1978

'INTERCREAM', F, w, 1986; (**Ivora**); flowers near white, blooms in clusters, dbl., 35 petals, slight fragrance; foliage medium, light green, matt; upright growth; [AmRUda X Unnamed seedling]; Interplant

INTerdain, F, lp, 1996; (**Fellini**); Interplant

INTerdays, HT, lb; (**Tropical Skies**); Ilsink; Interplant, 1997

'INTERDUST', S, lp, 1998; (Stardust, **Sun Runner**); flowers bright yellow, small, single, 4–11 petals, 1.5 in., borne in large clusters, self cleaning, slight fragrance; prickles moderate; foliage small, dark green, glossy; spreading (3 ft), mounding (18 in.), low, growth; PP10240; [Seedling X Seedling]; Ilsink/Interplant; Bear Creek Gardens, 1992

'INTERETTE', S, lp, 1977; (Flavia, **Fleurette**®); single, 5 petals, 2 in., blooms in clusters, repeat bloom, few, medium prickles; foliage medium green, glossy; vigorous (to 4 ft) growth; groundcover; [Yesterday X Unnamed seedling]; Interplant

INTereup, S, ob, 1997; (**Euphoria**); Interplant

'INTERFAIR', S, m, 1982; (**Fair Play**®); flowers light violet, blooms in large clusters, semi-dbl., 18 petals, repeat bloom, slight fragrance; dark green prickles; foliage medium, dark, matt; groundcover; vigorous growth; [Yesterday X Seedling]; Interplant, 1977

'INTERFIRE', F, ob, 1992; (Darthuizer Orange Fire, **Orange Fire**®); flowers very bright orange, urn-shaped, blooms borne in sprays, dbl., 15–20 petals, 3.25 in., slight fragrance; foliage large, medium green, matt; upright growth; [Orange Wave X Unnamed seedling]; Ilsink, Peter; Interplant B.V., 1988; Kordes, 1988; Pekmez, 1987

Interflora, *see* '**INTERVIEW**'

'INTERFLU', Min, w, 1984; (**Fluffy**, Patio Cloud); flowers white, shaded creamy pink, small blooms in clusters, semi-dbl., 15 petals, slight fragrance; few, medium prickles; foliage dark, glossy; groundcover; spreading growth; [Seedling X Nozomi]; Interplant

'INTERFOUR', Min, mp, 1982; (**Petit Four**®); flowers patio, blooms in clusters, semi-dbl., moderate fragrance; many, small prickles; foliage small, medium green, glossy; bushy growth; [Marlena seedling X Unnamed seedling]; Interplant

'INTERGANT', Min, w, 1983; (**Elegant Pearl**®); flowers creamy white, patio, blooms in large clusters, dbl., slight fragrance; few, medium prickles; foliage small, medium green, glossy; bushy growth; [Seedling X Nozomi]; Interplant

'INTERGEORGE', Min, mp, 1992; (Carmela®, **Georgeous**®); semi-dbl., 6–8 petals, 1.25 in., cupped, borne in sprays of 3–8, slight fragrance; foliage small, dark green, glossy; bushy, low (30–40 cms) growth; [Candy Rose X Eyeopener]; Ilsink, Peter; Interplant B.V., 1990

INTerglo, S, pb, 1987; (**Gloire des Anciens**); Interplant

INTergol, Min, dy, 1990; (**Golden Rosamini**); Ilsink

'INTERGRI', F, lp, 1993; (**Silver Dream**®); flowers silver pink, blooms borne in sprays, dbl., 15–25 petals, 1.5–2.75 in., slight fragrance; few prickles; foliage medium, medium green, semi-glossy; medium (60 cms), upright growth; Ilsink, G.P.; Interplant B.V., 1992

'INTERHEL', F, mr, 1979; (**Stadt den Helder**); flowers medium, dbl., 20 petals, slight fragrance; foliage large, dark, matt; upright growth; [Amsterdam X (Olala X Diablotin)]; Interplant

INTerhyro, S, mr; (**Borsalino**®); Interplant

INTerice, HT, op, 1996; (**Coral Ice**); Interplant

INTerim, S, mr, 1991; (**Red Trail**); Ilsink

'INTERJADA', F, dp, 1986; flowers deep pink, blooms in sprays, dbl., 35 petals, slight fragrance; foliage medium, dark, semi-glossy; upright growth; [Unnamed seedling X Unnamed seedling]; Interplant

'INTERLADA', F, lp, 1984; (**Lady of the Dawn**®); flowers large, ruffled, soft cream edged with pink, semi-dbl., borne in large clusters, moderate, fruity fragrance; foliage large, medium green, matte, leathery; upright growth, arching, 4 ft; PP006068; [INTerdress X Stadt den Helder]; Interplant

'INTERLAV', S, m, 1984; (**Lavender Dream**®); flowers deep lilac pink, semi-dbl., 16 petals, blooms in clusters, repeat bloom, no fragrance; few medium prickles; foliage medium, light green, matte, narrow; bushy, arching canes, (5 ft) growth; PP5916; ADR, 1987; [Yesterday X Nastarana]; Interplant, 1985

'INTERLEER', F, ob, 1982; (**Leersum 700**); flowers light orange-yellow, blooms in clusters, semi-dbl., slight fragrance; many, small prickles; foliage medium, light green, matt; upright growth; [Lichtkonigin Lucia X Marlena]; Interplant, 1979

'INTERLIEN', F, lp, 1985; (**Evelien**); flowers, dbl., 35 petals, blooms in clusters, slight fragrance; foliage medium, medium green, semi-glossy; PP007368; [Seedling X Fresh Pink]; Interplant

'INTERLIGHT', F, ab, 1992; (**Day Light**®, Daylight); flowers apricot-yellow, apricot and light pink reverse, aging to champagne, dbl., 15–25 petals, 3.25 in., no fragrance; foliage medium, dark green, matt; upright, medium (+/– 48 cms) growth; [Seedling X New Year]; Ilsink, G.P., 1985; Interplant B.V., 1991

INTerlis, F, lp, 1995; (**Lydia**); Ilsink

'INTERMEZZO', HT, m, 1963; bud ovoid; flowers deep lavender, medium, dbl., 25 petals, 3.5 in., moderate fragrance; foliage dark, glossy; moderately tall, compact growth; [Grey Pearl X Lila Vidri]; Dot, Simon; McGredy; Minier

'INTERMINER', Min, lp, 1982; (**Minerette**); semi-dbl., blooms in clusters, slight fragrance; many small prickles; foliage small, medium green, glossy; bushy growth; GM, Baden-Baden, 1984; [Marlena seedling X Unnamed seedling]; Interplant

'INTERMOTO', F, dp, 1985; (**Joy**®); flowers deep pink, small blooms in clusters, dbl., 35 petals, no fragrance; foliage medium, medium green, semi-glossy; upright growth; [AmRUda X Unnamed seedling]; Interplant

INTermug, S, my, 1996; (**Butterflies**); Ilsink

INTermunder, S, mr, 1989; (**Dart's Red Dot**®); Interplant

International Herald Tribune®, *see* 'HARQUANTUM'

'INTERNIKI', F, mr, 1993; (**Nikita**®); dbl., 15–25 petals, 1.5 in., borne in sprays, slight fragrance; some prickles; foliage medium, medium green, glossy; medium (45 cms), bushy growth; Ilsink, G.P.; Interplant B.V., 1988

'INTERONLY', HT, dr, 1986; (**Only Love**); flowers large, dbl., 35 petals, no fragrance; foliage medium, dark, semi-glossy; upright growth; PP006698; [Unnamed seedling X Caramba]; Interplant

'INTEROP', S, mr, 1987; (Erica, Eye Opener®, **Eyeopener**, Tapis Rouge); flowers small, semi-dbl., no fragrance; foliage medium, medium green, glossy; spreading growth; [(Unnamed seedling X Eyepaint) X (Unnamed seedling X Dortmund)]; Interplant

'INTERORGE', F, w, 1983; (**Georgette**); flowers large, dbl., 35 petals, slight fragrance; foliage large, medium green, semi-glossy; upright growth; [Unnamed seedling X Bordure Rose]; Interplant

'INTERPEEL', S, dr, 1992; (Ardennes, **Eye Appeal**®); semi-dbl., 8 petals, 2 in., cupped, borne singly, repeat bloom, no fragrance; foliage medium, medium green, glossy; spreading, medium (40–60 cms) growth; [Eyeopener X Unknown]; Ilsink, Peter; Interplant B.V., 1991

'INTERPINK', S, mp, 1978; (**Pink Star**®); flowers medium, semi-dbl., slight fragrance; foliage medium, light, semi-glossy; spreading growth; [Yesterday X Seedling]; Interplant, 1977

'INTERPOOL', HT, rb, 1960; (**Verdi**); bud ovoid; flowers brick-red, reverse veined darker, medium, dbl., moderate fragrance; foliage dark, glossy; very vigorous, bushy growth; [Cafougnette X Independence]; Dorieux; Pin

'INTERPRINCE', F, w, 1993; (**Princess**®); dbl., 26–40 petals, 1.5–2.75 in., borne in large clusters, slight fragrance; some prickles; foliage medium, dark green, glossy; tall to medium (60 cms), upright growth; [Pink Delight X Seedling]; Ilsink, G.P.; Interplant B.V., 1989

'INTERPUR', F, m, 1993; (**Purple Prince**®); flowers dark mauve to light purple, blooms borne in large clusters, dbl., 26–40 petals, 1.5–2.75 in., slight fragrance; some prickles; foliage medium, dark green, semi-glossy; medium (55 cms), upright growth; Ilsink, G.P.; Interplant B.V., 1991

INTerreflet, S, mr; (**Glowing Cushion**); Ilsink, 1995; Interplant, 1996

'INTERRO', Min, or, 1984; (**Red Jewel**); flowers small, dbl., 20 petals, no fragrance; foliage small, medium green, semi-glossy; bushy growth; PP005769; [Amanda X Unnamed seedling]; Interplant

'INTERROB', MinFl, rb, 1983; (**Robin Red Breast**, Robin Redbreast); flowers dark red, white eye, reverse silver, small blooms in clusters, single, no fragrance; many, medium prickles; foliage small, medium green, glossy; bushy growth; [Unnamed seedling X Eyepaint]; Interplant, 1984

'INTERSINA', HT, lp, 1984; flowers large, dbl., 35 petals, no fragrance; foliage large, medium green, semi-glossy; upright growth; PP005836; [Seedling X Red Success]; Interplant

'INTERSIREE', F, lp, 1993; (**Swing**®); flowers cream pink, blooms borne in sprays, dbl., 15–25 petals, 1.5–2.75 in., no fragrance; few prickles; foliage medium, dark green, glossy; medium (60 cms), upright growth; Ilsink, G.P.; Interplant B.V., 1990

'INTERSMART', S, lp, 1979; (**Smarty**®); single, 7 petals, 2 in., blooms in clusters, repeat bloom, many, small prickles; foliage bright green, matt; groundcover; vigorous growth; [Yesterday X Unnamed seedling]; Ilsink; Dickson Nurseries, Ltd.

INTerstreep, S, pb, 1997; (**Arabesque**); Interplant

'INTERSUM', S, lp, 1981; (**Summerrose**); single, 6–8 petals, 2 in.; blooms in large clusters, moderate fragrance; many, long prickles; foliage medium, medium green, matt; groundcover; spreading growth; [Yesterday X Unnamed seedling]; Interplant

'INTERTOR', S, mp, 1987; (**Pink Torch**®, Torche Rose®); flowers small blooms in large pyramidal sprays, single, slight fragrance; foliage medium, medium green, glossy; upright (to 4 ft) growth;

[(Mozart X Seedling) X (Seedling X Eyepaint)]; Interplant

'INTERTRA', F, m, 1993; (**Chess**®); flowers deep purple, blooms borne in sprays, dbl., 26–40 petals, 1.5–2.75 in., no fragrance; few prickles; foliage medium, dark green, glossy; medium (55 cms), upright growth; Ilsink, G.P.; Interplant B.V., 1991

INTertropa, HT, lp; (**Temptation**); Interplant, 1995

'INTERTWIK', Min, ly, 1992; (**Twinkle**®); flowers light yellow, does not fade, urn-shaped, blooms borne singly, dbl., 10 petals, 2 in., slight fragrance; foliage medium, medium green, matt; upright, low (40 cms) growth; [McShane X Unknown seedling]; Ilsink, Peter; Interplant B.V., 1989

'INTERTYN', F, op, 1993; (**Sentyna**®); flowers light salmon pink, blooms borne in sprays, dbl., 15–25 petals, 1.5–2.75 in., slight fragrance; few prickles; foliage medium, medium green, glossy; tall (80 cms), upright growth; Ilsink, G.P.; Interplant B.V., 1990

'INTERVAL', HT, w, 1983; (**Porcelina**); flowers cream with pink tint, dbl., 30 petals, borne singly or in small clusters, moderate fragrance; few prickles; foliage large, dark, glossy; upright growth; PP005648; [Seedling X Golden Times]; Interplant

INTervema, HT, mp, 1995; (**Orlando**); Ilsink

'INTERVIEW', HT, dp, 1968; (Interflora); flowers deep pink, large, dbl., 40 petals, exhibition form, slight fragrance; foliage leathery; vigorous, upright growth; [((Baccará X White Knight) X (Baccará X Jolie Madame)) X (Baccará X Paris-Match)]; Meilland

'INTERVILLES'®, LCl, mr, 1968; bud ovoid; flowers large, semi-dbl., cupped, moderate fragrance; foliage dark, glossy; vigorous, climbing growth; [Etendard X Unnamed seedling]; Robichon; Ilgenfritz Nursery

'INTERWAY', HT, w, 1992; (**Milky Way**); flowers white with hint of pink on reverse/edge, urn-shaped, dbl., 20 petals, 4 in., slight fragrance; foliage large, dark green, glossy; upright, medium (100–120 cms) growth; [Esmeralda X True Love]; Ilsink, Peter; Interplant B.V., 1991

INTerwell, S, ly; (**Yellow Fleurette**®); flowers yellow with yellow stamens, semi-dbl., borne in sprays, no fragrance; foliage medium, dark green, glossy; strong, bushy (2.5 ft) growth; Interplant, 1992

'INTIMITÉ', HT, ob, 1956; bud long; flowers golden orange shaded yellow and chamois, open, large, dbl., slight fragrance; foliage glossy, dark; vigorous, bushy growth; [Beauté X Unnamed seedling]; Delforge

Intrepid™, *see* 'PERPID'

Intrepide, LCl, or, 1972; Combe

Intrigue, *see* 'JACUM'

Intrigue, *see* 'KORLECH'

'INTRUMA', F, dr, 1976; (**Interama**); semi-dbl., 18 petals, 3 in.; foliage large, glossy, dark; bushy growth; [Kohima X (Europeana X Kimona)]; deRuiter

Invention, HT, dp, 1988; Patil, B.K.

Invertilis, HSpn, mr; flowers carmine-red, medium-large, single, moderate fragrance; (Sangerhausen)

Invincible, *see* 'RUNTRU'

'INVITATION', HT, ab, 1961; bud long, pointed; flowers rich salmon-pink, base yellow, dbl., 30 petals, 4.5 in., exhibition form, intense, spicy fragrance; foliage leathery, glossy; vigorous, compact, bushy growth; [Charlotte Armstrong X Signora]; Swim & Weeks; C-P

'IOBELLE', HT, pb, 1962; bud ovoid; flowers ivory-white edged and overspread deep pink, large, exhibition form, moderate, fruity fragrance; foliage dark, glossy; vigorous, upright, compact growth; [Dean Collins X Peace]; Buck, Dr Griffith J.; Iowa State University

'IODE', F, dr, 1974; bud ovoid; flowers large, semi-dbl., cupped, slight fragrance; foliage glossy; vigorous, upright growth; [Lichterloh X Red Pinocchio]; Schloen, J.

'IOLANTHE', HT, rb, 1940; flowers bright red, reverse yellow, large, semi-dbl.; vigorous growth; Gaujard

'ION PHILLIPS', HT, dy, 1934; flowers rich yellow, large, dbl.; vigorous growth; Dickson, A.

'IONA HERDMAN', HT, dy, 1914; flowers brilliant yellow, dbl., slight fragrance; GM, NRS, 1913; McGredy

'IONE', Min, ly, 1989; bud ovoid; flowers white with pale yellow center, medium, borne singly and in small clusters, very dbl., 50 petals, exhibition form, slight fragrance; fruit not observed; prickles very few, very small, light green; foliage medium, medium green, semi-glossy; bushy, medium growth; [Unnamed seedling X Unnamed seedling]; Jerabek, Paul E., 1990

Ipitombi, *see* JACorb

'IPSILANTÉ', HGal, m, 1821; flowers light lilac pink, very large, dbl., quartered, vigorous growth

Iranja®, *see* 'LENIRA'

'IRELAND HAMPTON', HT, pb, 1934; flowers flame-pink suffused gold, base gold, large, dbl., cupped, moderate, spicy fragrance; foliage glossy; vigorous, compact growth; [Étoile de Feu X Unnamed seedling]; Hillock

'IRELAND HAMPTON, CLIMBING', Cl HT, pb, 1936; Hillock

'IRENE', F, pb, 1941; flowers rose-white, reverse pure white, semi-dbl., moderate fragrance; [Seedling X Permanent Wave]; Leenders, M.

Irene, *see* MEIridol

Irene au Danmark, *see* **'IRENE OF DENMARK'**

'IRÈNE BONNET', Cl HT, mp, 1920; flowers hermosa pink, dbl., moderate fragrance; Nabonnand, C.

'IRENE CHURRUCA', HT, ly, 1934; (GOLDEN MELODY); bud pointed, yellow; flowers light buff, fading cream, well-formed, large, intense fragrance; [Mme Butterfly X (Lady Hillingdon X Souv. de Claudius Pernet)]; La Florida

'IRENE CURIE', HT, mr, 1952; bud very long, pointed; flowers scarlet, large, dbl., 20 petals; foliage glossy; very vigorous, bushy growth; [Unnamed seedling X Lawrence Johnston (HT)]; San Remo Exp. Sta.

Irene Jane, HWich, m, 1995; Nobbs

'IRENE OF DENMARK', F, w, 1948; (Irene au Danmark, Irene von Danemark); bud pointed; dbl., 40 petals, 3 in., cupped, moderate fragrance; foliage dark; vigorous, upright, bushy growth; (21); [Orléans Rose X (Mme Plantier X Edina)]; Poulsen, S.; C-P, 1950

Irene of Denmark, Climbing, Cl F, lp; [Irene of Denmark sport]; Ruston, D., 1970; (Weatherly, L.)

Irene Smith, *see* 'BOSREXEYE'

'IRENE THOMPSON', HT, yb, 1921; flowers deep ruddy gold shaded bronze or coppery, dbl., intense fragrance; GM, NRS, 1919; McGredy

Irene Virag, *see* 'ZIPVIR'

Irene von Danemark, *see* **'IRENE OF DENMARK'**

'IRENE WATTS', Ch, w, 1896; bud soft apricot-orange; flowers with a button eye, dbl.; Guillot, P.

'IRENE'S BEAUTY', Min, yb, 1999; flowers very colorful, yellow blend, reverse med. yellow, aging to orange-red, dbl., 17–25 petals, 1.5 in., borne in small clusters, no fragrance; few prickles; foliage medium, dark green, semi-glossy; compact, low (1 ft) growth; [Little Darling X Kristin]; Jolly, Betty J.; Langenbach, 1998

'IRENE'S CHOICE', HT, mp, 1978; flowers azalea-pink, full, dbl., 35–40 petals, 4–5 in., slight fragrance; very free growth; [Karl Herbst X Blue Moon]; Ellick; Excelsior Roses

'IRENE'S DELIGHT', HT, lp, 1982; flowers large, dbl., exhibition form, intense fragrance; foliage medium, dark, semi-glossy; upright growth; [Admiral Rodney X Red Lion]; Varney, E.

Irene's Surprise, *see* 'KIRBEE'

'IRINA', F, dr, 1969; flowers dark crimson-red, large, semi-dbl.; foliage soft, glossy; moderate, bushy growth; Grabczewski

'Iris', HSpn, w; dbl.; shining, black fruit; foliage finely divided; dense, shrubby (3–4 ft) growth

Iris, *see* FErecha®

'Iris Foster', HT, mp, 1998; flowers full, medium pink, very large, very dbl., 26–40 petals, 6 in., exhibition form, borne mostly singly, slight fragrance; some prickles; foliage large, medium green, dull; vigorous, upright, bushy, floriferous growth; [Gavotte X (Solitaire X Mischief)]; Poole, Lionel

Iris Gee, F, lp, 1987; Gee

'Iris Patricia Green', HT, mr, 1928; bud pointed; flowers cherry-red, moderate fragrance; foliage dark; Pemberton

'Iris Squire', F, mp, 1966; flowers soft rose, 4–5 in.; foliage dull; tall, vigorous growth; [Unnamed seedling X Queen Elizabeth]; Bees

Iris Webb, *see* 'CHEWell'

Irischer Regen, *see* **'Irish Mist'**

'Irish Afterglow', HT, ob, 1918; flowers very deep tangerine, passing to crushed strawberry; [Irish Fireflame sport]; Dickson, A.

Irish Beauty, *see* 'MACel'

'Irish Brightness', HT, mr, 1904; flowers medium, single, Dickson, A.; (Sangerhausen)

'Irish Charity', HT, rb, 1927; bud intense fiery scarlet with golden sheen; flowers rosy scarlet, dbl.; McGredy; Dreer; H&S

'Irish Charm', HT, ab, 1927; bud pointed; flowers base golden apricot passing to blush-pink, dbl., exhibition form, moderate fragrance; foliage dark, leathery; vigorous growth; McGredy; H&S

'Irish Courage', HT, op, 1927; bud pointed; flowers soft shrimp-pink to salmon, dbl., exhibition form, moderate fragrance; foliage rich green, leathery, glossy; vigorous growth; McGredy

Irish Creme™, *see* 'PERcreme'

'Irish Elegance', HT, ob, 1905; flowers bronze orange-scarlet, large, single, 5 petals; vigorous growth; (21); Dickson, A.

'Irish Engineer', HT, mr, 1904; flowers dazzling scarlet, large, single; Dickson, A.

'Irish Eyes', F, yb, 1974; bud ovoid; flowers yellow edged red, dbl., 45 petals, 2.5–3 in., exhibition form, slight fragrance; vigorous growth; [Unnamed seedling X Gemini]; Byrum; J.H. Hill Co.

Irish Eyes™, *see* 'DICwitness'

'Irish Fireflame', HT, ob, 1914; flowers orange to old-gold, veined crimson, anthers light fawn, single, 5 petals, 5 in., intense fragrance; foliage dark, glossy; compact, bushy growth; GM, NRS, 1912; Dickson, A.

'Irish Fireflame, Climbing', Cl HT, ob, 1916; Dickson, A.

'Irish Glory', HT, pb, 1900; flowers silvery pink, reverse crimson, large, semi-dbl., 10 petals, moderate fragrance; very vigorous growth; Dickson, A.

'Irish Gold', HT, my, 1966; (Grandpa Dickson); bud ovoid; dbl., 33 petals, 7 in., exhibition form, moderate fragrance; foliage dark, glossy, leathery; vigorous, upright, bushy growth; GM, Belfast, 1968 GM, Portland, 1970 GM, RNRS, 1965 GM, The Hague, 1966 PIT, RNRS, 1965; [(Kordes' Perfecta X Governador Braga da Cruz) X Piccadilly]; Dickson, A.; J&P

Irish Gold, Climbing, Cl HT, 1973; Humphreys; (Cavriglia)

Irish Heartbreaker, *see* 'SEAheart'

'Irish Hope', HT, dr, 1927; bud pointed; flowers rosy crimson shaded maroon, large, dbl., exhibition form, intense fragrance; foliage dark, leathery; vigorous growth; McGredy; Dreer; H&S

Irish Hope, *see* HARexclaim

'Irish Lady', Min, pb, 1991; flowers small, dbl., 26–40 petals, slight fragrance; foliage small, dark green, glossy; bushy growth; [Kathy Robinson sport]; Schmidt, Richard; Michigan Mini Roses, 1991

'Irish Mist', F, op, 1966; (Irischer Regen, Irish Summer®); flowers orange-salmon, well-formed, blooms in clusters, 4.5 in., slight fragrance; foliage dark; dense growth; [Orangeade X Mischief]; McGredy, Sam IV; McGredy

'Irish Modesty', HT, op, 1900; flowers light orange-pink, large, single; Dickson, A.; (Sangerhausen)

'Irish Morn', HT, pb, 1927; flowers pink, center coral, dbl., moderate fragrance; McGredy; Dreer; H&S

Irish Rich Marbled, HSpn, rb

'Irish Rover', HT, op, 1970; flowers salmon-pink, dbl., 36 petals, 4 in., slight fragrance; foliage coppery, dark; vigorous growth; [Violet Carson X Tropicana]; McGredy, Sam IV

Irish Squire, F, dp, 1966; flowers carmine-pink, large, dbl.; Bees of Chester; (Sangerhausen)

Irish Summer®, *see* **'Irish Mist'**

'Irish Sweetness', HT, rb, 1927; bud pointed; flowers crimson suffused scarlet, large, dbl., exhibition form, intense fragrance; foliage dark, leathery; vigorous growth; McGredy; Dreer; H&S

Irish Wonder, *see* 'MACev'

Irmela, F, w, 1983; Hetzel, K.; (Cavriglia)

'Iroquois', S, dp, 1932; flowers deep mauve-pink, semi-dbl., non-recurrent, moderate fragrance; foliage leathery; vigorous, bushy, compact growth; [Pythagoras X R. cinnamomea]; Central Exp. Farm

Irresistible™, *see* 'TINresist'

Irwin Lane, B, dr

'Isa', HT, lp, 1931; flowers light pinkish cream, well shaped, dbl., moderate fragrance; vigorous growth; [Abol seedling]; Evans, F.&L.

'Isa Murdock', HSpn, w, 1953; flowers white, sometimes tinged with pink, dbl., non-recurrent, many prickles; foliage spinosissima type; height 3 ft; [R. spinosissima altaica X Dbl. white spinosissima]; Skinner

'Isabel de Ortiz', HT, pb, 1962; (Isabel Ortiz); flowers deep pink, reverse silvery, well-formed, dbl., 38 petals, 5 in.; moderate fragrance; foliage dark, glossy; vigorous, upright growth; GM, Madrid, 1961 GM, NRS, 1962; [Peace X Kordes' Perfecta]; Kordes, R.; J&P, 1965

'Isabel Llorach', HP, yb, 1929; flowers nankeen yellow, tinted red, semi-dbl., moderate fragrance; [Frau Karl Druschki X Benedicte Seguin]; Dot, Pedro

Isabel Ortiz, *see* **'Isabel de Ortiz'**

Isabel Renaissance, *see* POUlisab

'Isabella', HT, ob, 1964; flowers orange, star shaped; [Queen Elizabeth X Pink Lustre]; Leenders, J.

Isabella, *see* POUlisab

'Isabella', HGal, dp, 1834

'Isabella Gray', N, dy; flowers golden yellow, more fragrant, but otherwise similar to parent; [Chromatella seedling]; Gray, Andrew, 1855

'Isabella Skinner', S, mp; flowers pink, well-formed, dbl., blooms on new wood all summer; bushy growth; [(R. laxa X Tea) X Floribunda]

'Isabella Sprunt', T, my, 1855; flowers sulfur-yellow; [Safrano sport]; Sprunt; Buchanan, 1865

Isabella Sprunt Single Seedling, HMsk, 1980; Worl

'Isabelle de France', HT, or, 1956; bud pointed; flowers vermilion, large, dbl., exhibition form, slight fragrance; vigorous, upright growth; [Peace X (Mme Joseph Perraud X Opera)]; Mallerin, C.; Hémeray-Aubert

'Isabelle Milner', HT, ly, 1908; flowers large, dbl., moderate fragrance; Paul, W.; (Sangerhausen)

'Isabelle Nabonnand', T, 1875; Nabonnand, G.; (Cavriglia)

Isabelle Renaissance, *see* POUlisab

Isabel's Jewel, *see* 'RENijewel'

'Isella', Min, lp, 1973; flowers small, borne singly and several together, dbl., 22 petals, cupped, no fragrance; pale pink prickles; foliage small, green; vigorous, compact, upright growth; [(Baccará X Generosa) X Miss Italia]; Bartolomeo, Embriaco

Isfahan, *see* **'ISPAHAN'**

'ISIS', F, w, 1973; flowers ivory-white, dbl., 40–45 petals, 4–5 in., moderate fragrance; compact growth; [Vera Dalton X Shepherdess]; Mattock

Iskra, *see* 'MEIhati'

Iskra 82, LCl, mr

Island of Dreams, F, dy, 1995; Spek

Isle of Man, *see* **'MANX QUEEN'**

Isle of Roses, *see* 'MORisle'

'ISOBEL', HT, pb, 1916; bud pointed; flowers light rose-pink, shaded apricot, large, single, 5 petals, cupped, slight fragrance; foliage rich green, soft; GM, NRS, 1915; McGredy

Isobel Champion, *see* 'DELgeot'

Isobel Derby, *see* 'HORethel'

'ISOBEL HARKNESS', HT, dy, 1957; flowers bright yellow, dbl., 32 petals, 6 in., moderate fragrance; foliage dark, leathery, semi-glossy; vigorous, upright, bushy growth; [McGredy's Yellow X Phyllis Gold]; Norman; Armstrong Nursery; Harkness

Isolde, *see* 'JUDsolde'

'ISPAHAN', D, mp; (Isfahan, Pompon des Princes); flowers bright pink, loosely, dbl., blooms over long season, intense fragrance; foliage small

Istropoliteana, S, 1990; Chorvath, F.; (Cavriglia)

Ita Buttrose, HT, op, 1984; Armstrong

'ITALIA', HT, dr, 1959; bud ovoid; flowers cardinal-red and cherry-red, to open, large, dbl., cupped, moderate fragrance; foliage leathery, dark; long, strong stems; bushy, upright growth; [Baccará X Poinsettia]; Valentino

'ITALIAN PINK', S, mp, 1959; bud short, pointed; flowers begonia-pink, open, large, borne in clusters, dbl., moderate fragrance; foliage dark; vigorous, upright, well branched growth; [Cocorico X Yellow Holstein]; Leenders, J.

'ITALIE IMPÉRIALE', HT, m, 1936; flowers purplish garnet-red, large, vigorous growth; Capiago

'ITALIENISCHES DOERFCHEN', F, or, 1967; bud ovoid; flowers open, small, borne in clusters, single, slight fragrance; foliage small, leathery; vigorous, bushy, low growth; [Highlight X ?]; Haenchen, E.; Teschendorff

It's Show Time™, *see* 'MORjoybon'

'IVAN MISSON', Pol, lp, 1922; flowers small, dbl.; Soupert & Notting; (Sangerhausen)

'IVANHOE', HT, rb, 1928; bud pointed; flowers brilliant scarlet to rich crimson, large, dbl., exhibition form, intense fragrance; foliage glossy; vigorous growth; Easlea

'IVANY', F, rb, 1975; flowers orange-yellow, shaded pink on petal edges, aging red, dbl., 50 petals, cupped, borne in clusters of 5–30; foliage dark, glossy; vigorous growth; [Masquerade X Rumba]; Staikov, Prof. Dr V.; Kalaydjiev and Chorbadjiiski

'IVEmax', HT, my, 1999; (**Delta Dawn**); dbl., 26–40 petals, 5 in., borne mostly singly, intense fragrance; few prickles; foliage medium, medium green, semi-glossy; upright, medium (5 ft) growth; [Out of Africa sport]; Iverson, Halvor

Ivora, *see* 'INTercream'

'IVORY', T, w, 1901; flowers ivory-white, large, dbl., moderate fragrance; vigorous growth; [Golden Gate sport]; Dingee & Conard

'IVORY ANN', S, w, 1993; flowers creamy white, blooms borne in small clusters, single, 5 petals, 3–3.5 in., moderate fragrance; foliage large, medium green, glossy; medium, bushy, spreading growth; [Ivory Fashion X ((Valerie Jeanne X Eyepaint) X Twilight Trail)]; Jobson, Daniel J.; Jobson, 1993

Ivory Beauty, *see* KORivo

Ivory Buccaneer, HT, w

Ivory Carpet™, *see* 'WILivory'

'IVORY CHARM', Cl F, w, 1968; bud ovoid; flowers open, large, semi-dbl., intense fragrance; foliage glossy, leathery; vigorous, climbing growth; [Ivory Fashion sport]; Earing, F.E.

'IVORY FASHION', F, w, 1958; bud ovoid; flowers ivory-white, well-formed, blooms in clusters, semi-dbl., 17 petals, 4–4.5 in., moderate fragrance; foliage leathery; vigorous, upright growth; AARS, 1959; [Sonata X Fashion]; Boerner; J&P

'IVORY FASHION, CLIMBING', Cl F, w, 1964; Williams, J. Benjamin

Ivory Festival™, *see* 'LAVyip'

'IVORY FLUSH', F, pb, 1996; flowers orange centered with pink blended edges, fading to pale pink, very dbl., 3.5 in., slight fragrance; some prickles; foliage medium, light green, dull; bushy, medium (26 in.patio) growth; Bees of Chester; L W Van Geest Farms, Ltd., 1995

Ivory Grand, *see* POUlmount

Ivory Palace, *see* 'MORivory'

'IVORY QUEEN', HT, w, 1954; bud ovoid; flowers ivory-cream, 5–6 in., moderate fragrance; foliage dark, glossy; vigorous, bushy growth; [Edina X McGredy's Ivory]; Fletcher; Tucker

'IVORY QUEEN', Gr, w, 1965; flowers ivory; [Queen Elizabeth sport]; Delforge

'IVORY SPLENDOR', Min, w, 1991; bud rounded; flowers white, pale yellow center, white reverse, aging white, large, dbl., 35 petals, exhibition form, slight, fruity fragrance; foliage medium, medium green, matt; upright, tall growth; [Rise 'n' Shine X Unnamed seedling]; Gruenbauer, Richard, 1984; Flowers 'n' Friends Miniature Roses, 1993

'IVORY TIP TOP', F, lp, 1976; flowers ivory-pink, full, semi-dbl., 12–16 petals, 2–2.5 in., slight fragrance; low, compact, bushy growth; [Tip Top sport]; Fryers Nursery, Ltd.

'IVORY TOWER', HT, w, 1979; bud very long, pointed; flowers ivory-white, shaded light pink and light yellow, dbl., 35 petals, 5.5 in., exhibition form, moderate fragrance; upright, bushy growth; [Colour Wonder X King's Ransom]; Kordes, R.; Armstrong Nursery

'IVORY TRIUMPH', F, w, 1961; bud pointed; flowers ivory, open, borne in clusters, semi-dbl., 12 petals, 3–4 in., slight fragrance; foliage leathery, light green; upright, compact growth; [Goldilocks X Unnamed seedling]; Von Abrams; Peterson & Dering

'IVRESSE', HT, rb, 1958; flowers clear red, reverse silvery, large; vigorous growth; [Peace X Spectacular]; Combe; Japan Rose Society

'IVY ALICE', HWich, op, 1927; flowers soft pink to blush-salmon, splashed carmine when fading, dbl., cupped, borne in very large clusters, slight fragrance; foliage glossy, light; very vigorous, climbing (6 ft) growth; [Excelsa sport]; Letts

'IVY EVANS', HT, mr, 1926; flowers light cerise, moderate fragrance; [George C. Waud seedling X Gen. MacArthur]; Evans

'IVY MAY', HT, pb, 1925; bud pointed; flowers rose-pink, base and edges amber, dbl., moderate fragrance; foliage dark; vigorous growth; [Mme Butterfly sport]; Beckwith

'IWARA', Misc OGR, w; (R. X iwara, R. yesoensis); flowers small, single; (14); [R. multiflora X R. rugosa]

'IZUMI', HT, m, 1999; flowers lavender blended pale pink, dbl., 40 petals, 5–5.5 in., exhibition form, slight fragrance; foliage bronze dark green; 5 ft growth; [(Intermezzo X Soir d'Automne) X (Intermezzo X Soir d'Automne)]; Ohkawara, Kiyoshi, 1994; Komaba Rose Nursery, 1997

'IZY', HT, pb, 1997; flowers full, large, white stripes frequently appear, very dbl., 26–50 petals, 5–6 in., borne mostly singly, intense fragrance; some prickles; foliage long, narrow, medium, medium green semi-glossy; upright, medium to tall growth; [Sheer Bliss sport]; Ballin, Don & Paula

J

'**J.A. GOMIS**', HT, rb, 1933; flowers crimson-red and yellow, medium, dbl., cupped, slight fragrance; foliage dark, glossy; upright growth; [Sensation X Souv. de Claudius Pernet]; Camprubi, C.

'**J. B. CLARK**', HP, dr, 1905; flowers deep scarlet, shaded blackish crimson, large, dbl., 25 petals, exhibition form, moderate fragrance; very prickly; vigorous (8–10 ft), bushy, almost climbing growth; [Lord Bacon X Gruss an Teplitz]; Dickson, H.

'**J. B. CLARK, CLIMBING**', Cl HT, mr, 1939; flowers carmine-red, very large, dbl., moderate fragrance; Vogel, M.; (Sangerhausen)

'**J. B. M. CAMM**', B, lp, 1900; flowers large, very dbl., moderate fragrance; Paul, G.; (Sangerhausen)

'**J. B. MEILLAND**', HT, ob, 1941; flowers orange, reverse golden yellow, large, very dbl., moderate fragrance; very vigorous growth; [Mme Joseph Perraud X (Charles P. Kilham X Margaret McGredy)]; Meilland, F.

J. Bienfait, *see* '**MR J. BIENFAIT**'

'**J. C. THORNTON**', HT, mr, 1926; bud pointed; flowers glowing crimson-scarlet, dbl., slight fragrance; foliage light olive-green, glossy, leathery; vigorous, branching growth; GM, NRS, 1928; [K. of K. X Red-Letter Day]; Bees

'**J. F. BARRY**', HT, my, 1912; flowers light daffodil-yellow; [Arthur R. Goodwin sport]; Piper

'**J. F. MÜLLER**', F, dr, 1929; flowers large; foliage dark; bushy, dwarf growth; [Rodhatte sport]; Muller, J.F.

'**J. G. GLASSFORD**', HT, dr, 1921; bud pointed; flowers deep crimson, very large, exhibition form, moderate fragrance; very vigorous, branching growth; Dickson, H.

'**J. H. BRUCE**', HT, mr, 1937; flowers crimson-scarlet, over large, dbl., exhibition form; foliage glossy; vigorous, bushy growth; GM, NRS, 1936; [H.V. Machin X Marion Horton]; Bees

'**J. H. PEMBERTON**', HT, mr, 1931; flowers scarlet, moderate, damask fragrance; vigorous growth; Bentall

'**J. H. VAN HEYST**', HT, yb, 1936; bud pointed; flowers yellowish flesh, reverse pink; vigorous growth; [Comtesse Vandal X Edith Nellie Perkins]; Leenders, M.

'**J. K. B. ROOS**', HT, m, 1933; bud pointed; flowers pale reddish lilac, shaded salmon-flesh, large, dbl., intense fragrance; very vigorous growth; Leenders, M.

'**J. K. TYL**', LCl, mp, 1936; flowers bright pink, cactus form, very free bloom; vigorous growth; Brada, Dr.; Böhm

'**J. M. LÓPEZ PICÓ**', HT, dr, 1947; bud long, pointed; flowers crimson, large, dbl., exhibition form, slight fragrance; upright growth; [Editor McFarland X Comtesse Vandal]; Camprubi, C.

'**J. M. LÓPEZ PICÓ, CLIMBING**', Cl HT, dr, 1954; Camprubi, C.

J. Michael™, *see* 'KINMIKE'

'**J. MICHEL**', HT, dy, 1930; flowers dark golden yellow, large, dbl.; foliage leathery; vigorous growth; RULED EXTINCT 1/87; [Unnamed seedling X The Queen Alexandra Rose]; Felberg-Leclerc

'**J. N. HART**', HT, dp, 1924; flowers rose-pink, dbl., moderate fragrance; [George Dickson X Edith Cavell]; Chaplin Bros.

J. Otto Ripoff, HT, lp

'**J. OTTO THILOW**', HT, mp, 1927; bud pointed, well shaped; flowers rich glowing rose-pink, large, dbl., exhibition form; very vigorous growth; [Hadley X Souv. de H.A. Verschuren]; Verschuren; Dreer; H&S

'**J. OTTO THILOW, CLIMBING**', Cl HT, mp, 1933; Howard Rose Co.

'**J. P. CONNELL**', S, my, 1987; flowers pale, medium yellow at inner petals, yellow-white on upper petals, dbl., exhibition form, borne 1–8 per cluster, repeat bloom, intense, tea fragrance; no prickles; foliage abundant, dark yellow-green, wide ovate, double serr; bushy, winter hardy growth; [Arthur Bell X Von Scharnhorst]; Svedja, Felicitas; Agriculture Canada, 1986

J. R. Byfield, Cl HT, lp; deep red flowers with petals edged purple, slight; strong and long; [Sensation seedling sport]; Clark, A., 1941

'**J. S. BAAR**', HT, dr, 1934; flowers pure dark carmine-red, dbl., intense fragrance; foliage leathery; long, strong stems; vigorous, bushy growth; Mikes Böhm, J.; Böhm

'**J. S. FAY**', HP, rb, 1905; flowers dark crimson tipped scarlet, dbl.; vigorous growth; [Prince Camille de Rohan X Souv. de Pierre Notting]; Walsh

'**J. W. FARGO**', sp, (variety of R. arkansana), mp; flowers wild-rose-pink, borne in clusters, non-recurrent; well branched (20 in.) growth

J. C. Flat Pink, HT, lp

J. E. Murphy's Pink Tea, T, lp

'**J. F. BAILEY**', HT, lp; [Frau Karl Druschki X Unknown]; Williams, A., 1911; (Weatherly, L.)

J. F. Quadra, S, mp, 1995; Ogilvie, Ian S.

J. G. Mendel, S, lp; Urban, J., 1989; (Czech Rosa Club)

J. G. Sandberg, *see* '**JONKHEER G. SANDBERG**'

J. H. Pierneef, F, rb, 1994

'JACABLE', HT, ob, 1998; (**Spice Twice**™); flowers orange, full, very dbl., 26–40 petals, 5–6 in., borne in large clusters, moderate fragrance; prickles moderate; foliage medium, dark green, semi-glossy; upright, tall (5 ft) growth; [Spirit of Glasnost X Kardinal]; Zary, Dr Keith W.; Bear Creek Gardens, Inc., 1997

'**JACABY**', HT, or, 1979; flowers large, slight fragrance; foliage large, medium green; upright, bushy growth; [South Seas X Tonight]; Warriner, William A.; J&P

'JACADE', HT, pb, 1984; (Home's Choice, **Olympic Dream**); flowers large, dbl., 35 petals, slight fragrance; foliage large, dark, semi-glossy; upright growth; PP005842; [Unnamed seedling X Unnamed seedling]; J&P; McConnell Nurs., Inc.

JACAEBI, HT, dp; (**Rendez-vous**)

Jacaford, S; (Cavriglia)

'**JACAGE**', HT, yb, 1981; flowers medium, dbl., 35 petals, slight fragrance; foliage medium, dark, leathery; upright, bushy growth; [Unnamed seedling X Spellbinder]; Warriner, William A.; J&P

'JACAIM', HT, ob, 1982; (**Acclaim**); flowers orange, large, dbl., 34 petals, slight fragrance; foliage medium green, semi-glossy; upright, bushy growth; PP005638; [Sunfire X Spellbinder]; Warriner, William A.; J&P, 1986

'JACAIR', S, dp, 1998; (**Watermelon Ice**™); flowers dark lavender pink, lighter reverse, small, single, 5–11 petals, 1.5 in., borne in large clusters, repeats quickly, slight fragrance; prickles moderate, straight; foliage small, dark green, semi-glossy; spreading, bushy, low ($1\frac{1}{2}$ ft) growth; PP10229; [The Fairy X Seedling]; Zary, Dr Keith W.; Bear Creek Gardens, Inc., 1997

'JACAKOR'®, HT, mp, 1985; (**Jacaranda**®, Jackaranda); flowers mauve-pink, large, dbl., 35 petals, intense fragrance; foliage large, medium green; upright growth; PP006316; [(Mercedes X Emily Post) X Unnamed seedling]; Kordes, W.

'**JACAL**', HT, or, 1984; flowers large, dbl., 35 petals, no fragrance; foliage medium, dark, semi-glossy; upright growth; PP005341; [Spellbinder X Futura]; Warriner, William A.; J&P

'JACALE', F, yb, 1986; (Arc de Triomphe, **Summer Fashion**); flowers petals light yellow edged pink, pink spreading with a, dbl., 20 petals, moderate fragrance; foliage large, medium green, semi-glossy;

PP005860; [Precilla X Bridal Pink]; Warriner, William A.; J&P, 1985

'JACALP', HT, or, 1980; flowers brick red, large, dbl., 50 petals, slight fragrance; foliage large, dark, leathery; upright growth; [Unnamed seedling X Medallion]; Warriner, William A.; J&P

'JACAMQUE', F, ab, 1999; (**Amber Waves**™); bud copper/gold, pointed, ovoid; flowers amber yellow, reverse yellow/apricot, opens flat, dbl., 26–40 petals, 3.5–4 in., cupped, borne singly and in small clusters, moderate, fruity fragrance; prickles moderate; foliage medium, dark green, glossy; upright, bushy, medium (3.5 ft) growth; Bronze Medal, Australian National Rose Trials, 1999; [Seedling X Amber Queen]; Zary, Dr Keith W.; Bear Creek Gardens, Inc., 2001

JACams, MinFl, lp; (**Sweet Dreams**); J&P, 1995

'JACANGE', HT, ob, 1984; (**John-Paul II**); flowers very large, dbl., no fragrance; foliage medium, dark, semi-glossy; upright growth; PP005639; [Apricot Parfait X Futura]; J&P; McConnell Nurs., Inc.

'JACANGEL', Gr, m, 1991; (**Spellcaster**); flowers lavender and deep mauve, full with heavy substance, blooms b, very dbl., 26–40 petals, 3–3.5 in., moderate fragrance; some prickles; foliage large, dark green, glossy; tall (145–160 cms), upright, spreading, uniform growth; [Unnamed seedling X Angel Face]; Warriner, William A.; Bear Creek Gardens, 1992

'JACANT', F, r, 1985; (**Topaz**); flowers tan, medium, dbl., 35 petals, slight fragrance; foliage medium, medium green, semi-glossy; upright growth; [Unnamed seedling X Intrigue]; Warriner, William A.; J&P

'JACANTH', HT, mr, 1974; ('JACMANTHA', **Samantha**®); bud ovoid; flowers floratea, medium, dbl., exhibition form, slight fragrance; foliage leathery; vigorous, upright growth; [Bridal Pink X Seedling]; Warriner, William A.; J&P

'JACAP', F, ob, 1992; (**Amazon**); Christensen, Jack E.

'JACAPRI', HT, ab, 1991; (**Lucille Ball**); flowers apricot, amber blend, good petal substance, blooms borne mostly singly, dbl., 26–40 petals, 2.75 in., moderate fragrance; some prickles; foliage medium, medium green, semi-glossy; tall (145–160 cms), upright, bushy growth; [Hello Dolly X Unnamed seedling]; Christensen, Jack E.; Bear Creek Gardens, 1993

'JACARA', HT, pb, 1991; (**Spirit of Glasnost**); flowers coral pink and ivory blend, large blooms borne mostly singly, dbl., 26–40 petals, moderate fragrance; foliage medium, dark green, semi-glossy; upright, spreading growth; [Unnamed seedling X Unnamed seedling]; Warriner, William A.; Bear Creek Gardens

Jacaranda®, *see* 'JACAKOR'®

'JACARCH', LCl, w, 1992; (**Lace Cascade**); bud pointed, ovoid; flowers icy white, floriferous, dbl., 26–40 petals, 3–3.5 in., borne in small and large clusters, moderate, sweet fragrance; some prickles; foliage large, medium green to dark green, semi-glossy, milde; tall (150–160 cms), upright, spreading growth; PP8689; [Iceberg X Prairie Fire]; Warriner, William A.; Bear Creek Gardens

'JACARE', F, w, 1988; (**Class Act**, First Class, White Magic); flowers loose, borne in sprays of 3–6, semi-dbl., flat, slight, fruity fragrance; prickles long, narrow; foliage medium, dark green, semi-glossy; upright, bushy, medium growth; AARS, 1989 GM, Portland, 1989 Gold Star of the South Pacific, Palmerston North, NZ, 1990; [Sun Flare X Seedling]; Warriner, William A.; J&P

'JACARED', HT, dr, 1978; (**American Pride**); bud pointed, ovoid; dbl., 33 petals, 4–5 in., exhibition form, slight fragrance; foliage large, dark; tall, upright growth; PP006515; Warriner, William A.; J&P, 1974

'JACARET', Min, dr, 1983; (**Cabaret**); bud small; flowers small, dbl., 35 petals, moderate fragrance; foliage small, medium green, semi-glossy; upright, bushy growth; PP005576; [(Fire Princess X Mary DeVor) X (Seedling X Caliente)]; Warriner, William A.; J&P

'JACARINA', HT, lp, 1986; (**Amorous**); flowers shell pink, dbl., 30 petals, exhibition form, borne singly, slight, fruity fragrance; foliage large, medium green, glossy; upright, tall growth; [White Masterpiece X Marina]; Warriner, William A.; J&P

'JACARY', HT, dy, 1983; (**Firstar**, Golden Anniversary, Home's Pride, Olympic Spirit); flowers large, dbl., 35 petals, moderate fragrance; foliage medium, medium green, glossy; bushy, spreading growth; [New Day X Oregold]; Warriner, William A.; J&P

JACATI, Min, lp; (**Hot Gossip**); Zary, Dr Keith W., 1997

JACAUTE, HT, mp, 1997; (**Sara**); J&P

'JACAVE', F, m, 1983; (**Blue Chip**); flowers medium, dbl., 20 petals, intense fragrance; foliage medium, medium green, semi-glossy; bushy, upright growth; PP005684; [Heirloom X Angel Face]; Warriner, William A.; J&P

JACAY, F, w, 1991; (**Avalanche**); Warriner, William A.

'JACBED', S, mr, 1986; (**Red Rascal**); flowers patio, small, bright red, dbl., 35 petals, 1.5–2 in., cupped, blooms in sprays of 2–5, self cleaning, repeat bloom, slight fragrance; medium, red to brown prickles, hooked downward; foliage small, medium green, semi-glossy; medium, bushy growth; PP006693; [Unnamed seedling X Unnamed seedling]; Warriner, William A.; J&P

JACBER, min, w, 1997; (**Ice Angel**); Zary

'JACBIP', HT, lp, 1990; (**Wedding Pink**); bud ovoid; flowers light pink, reverse slightly paler pink, urn-shaped, large, very dbl., exhibition form, slight fragrance; prickles long, very narrow, reddish-brown; foliage medium, dark green, matt; upright, tall growth; [Bridal White sport]; Nakashima, Tosh; Bear Creek Gardens, 1990

'JACBITOU', HT, yb, 1999; (**Weight Watchers Success**™); flowers light creamy yellow edged pink, dbl., 17–25 petals, 4.5–5 in., exhibition form, borne mostly singly, intense, sweet and spicy fragrance; few prickles; foliage medium, dark green, semi-glossy; upright spreading, tall (4–4.5 ft) growth; [Henry Fonda X French Perfume]; Zary, Dr Keith W.; Bear Creek Gardens, Inc., 1999

JACblar, HT, m, 1995; (**Evita Bezuidenhout**); J&P

'JACBLEM', HT, my, 1981; (**Emblem**™); dbl., 25 petals, exhibition form, borne singly, no fragrance; straight, long, light green prickles; foliage glossy, dark; upright growth; [Seedling X Sunshine]; Warriner, William A.; J&P

'JACBOW', S, m, 1998; (**Kaleidoscope**™); flowers tan mauve blend, reverse yellow mauve blend, very dbl., 26–40 petals, 3 in., borne in large clusters, slight, fruity fragrance; many prickles; foliage dark green, glossy; spreading, bushy, hedge, $3\frac{1}{2}$ ft growth; AARS, 1999; [Pink Pollyanna X Rainbow's End]; Walden, John K.; Bear Creek Gardens, Inc., 1999

JACboy, HT, ob; (**Tango**); J&P, 1992

'JACBRI', F, mp, 1967; (**Bridal Pink**™); bud ovoid, pointed, long; flowers large, pink blended with cream, dbl., 30–35 petals, exhibition form, borne in small clusters or singly, moderate, spicy fragrance; foliage leathery; vigorous, upright, bushy growth; [Summertime seedling X Spartan seedling]; Boerner; J&P

'JACBUB', Min, w, 1989; (**Tiny Bubbles**™); bud ovoid; flowers ivory, near white, fading to white, medium, borne usually singly, dbl., cupped, no fragrance; prickles straight, slightly angled downward, green-yellow; foliage medium, medium green, semi-glossy; upright, spreading, medium growth; [Zorina X Funny Girl]; Warriner, William A.; Bear Creek Gardens, 1990

'JACBUNCH', Min, dp, 1998; (**Raspberry Punch**™); flowers large, deep pink, full, very dbl., 26–40 petals, 2 in., borne in large clusters, slight fragrance; prickles moderate, straight; foliage medium, dark green, semi-glossy; compact, moderate, border plant growth; [Tournament of Roses X Seedling]; Zary, Dr Keith W.; Bear Creek Gardens, Inc., 1999

'JACBUSH', HT, pb, 1990; (**Barbara Bush**™); flowers salmon pink where exposed to sun, lighter pink to ivory white where unexposed, dbl., 25–30 petals, 5 in., exhibition form, borne mostly singly, moderate, damask fragrance; foliage medium, medium green, glossy; tall, upright, spreading growth; PP007542; [Pristine X Antigua]; Warriner, William A.; Bear Creek Gardens, 1991

'JACBUTE', HT, pb, 1991; (**Abracadabra**); bud pointed, ovoid, deep pink/yellow; flowers purple pink, tan and yellow, large blooms, dbl., 26–40 petals, 4.5–5 in., exhibition form, borne mostly singly, intense fragrance; some prickles; foliage large, dark green, semi-glossy; medium (90–120 cms), upright, bushy growth; PP8590; [White Masterpiece X Tribute]; Warriner, William A.; Bear Creek Gardens, 1993

'JACCART', Min, rb, 1989; (**Cartwheel**); bud ovoid; flowers red and white picotee, small, dbl., 20 petals, cupped, borne singly, slight fragrance; prickles hooked, small, tan; foliage small, medium green, matt; low, bushy growth; PP007508; [Libby X Seedling]; Warriner, William A.; Bear Creek Gardens, 1990

'JACCASP', Min, pb, 1992; (**Happy Trails**); flowers pink with cream center, blooms borne in small clusters, very dbl., 1.5 in., no fragrance; some prickles; foliage small, medium to dark green, glossy; low (20–30 cms), groundcover, spreading 60–90 cms across. growth; [Immensee X Roller Coaster]; Warriner, William A. & Zary, Keith W.; Bear Creek Gardens, 1993

JACchar, Min, lp, 1994; (**Fairy Tale**); J&P

'JACCHARI', S, dp, 1998; (**Candy Mountain**™); flowers deep pink, single, 5–11 petals, 1.5–2 in., borne in large clusters, slight fragrance; prickles moderate; foliage small, medium green, dull; bushy, arching, low, 2 ft growth; PP10616; [Sweet Chariot X Seedling]; Walden, John K.; Bear Creek Gardens, Inc., 1997

'JACCHRY', HT, dp, 1993; (**Breathless**); flowers deep pink, blooms borne mostly single, dbl., 15–25 petals, 3–3.5 in., moderate fragrance; some prickles; foliage large, medium green, purple-red when young, semi-glo; tall (150 cms), upright growth; [Seedling X Chrysler Imperial]; Warriner, William A.; Bear Creek Gardens, 1993; Bear Creek Gardens, Inc., 1993, USA

'JACCINQO', Gr, pb, 1998; (**Candelabra**™); bud long, pointed; flowers coral orange, dbl., 20–25 petals, 4 in., borne in small clusters, slight, tea fragrance; prickles moderate; foliage medium, dark green, glossy; bushy, medium tall growth; AARS, J&P, 1999; [Tournament of Roses X Seedling]; Zary, Dr Keith W.; Bear Creek Gardens, 1999

'JACCLAM', LCl, op, 1976; (**America**™); bud ovoid, pointed; flowers salmon, reverse lighter, imbricated, very dbl., 43 petals, 3.5–4.5 in., borne singly and in small clusters, intense fragrance; foliage large, medium green; climbing (10–12 ft) growth; AARS, 1976; [Fragrant Cloud X Tradition]; Warriner, William A.; J&P

'JACCLIP', Cl Min, mp, 1993; (**Bunny Hop**); flowers moderately full, dbl., 15–25 petals, 1.5 in., borne in large clusters, slight fragrance; some prickles; foliage small, dark green, semi-glossy; upright (180 cms), spreading, climbing growth; [Pink Pollyanna X Seedling]; Zary, Dr Keith W.; Bear Creek Gardens, 1995; Bear Creek Gardens, Inc., 1995, USA

'JACCLIST', Cl HT, pb, 1976; (**First Prize, Climbing**); very vigorous, climbing growth; Reasoner; J&P

'JACCLOP', LCl, dr, 1975; (**Tempo**); bud ovoid; flowers deep red, large, very dbl., slight fragrance; foliage large, glossy, dark; climbing growth; [Ena Harkness, Climbing X ?]; Warriner, William A.; J&P

'JACCO', HT, w, 1976; (**Coquette**); bud long; dbl., 30–35 petals, 4 in., exhibition form, slight fragrance; foliage light green, reddish underneath; upright growth; [Seedling X Seedling]; Warriner, William A.; J&P

'JACCOEUR', HT, pb, 1993; (**Gardens of the World**); bud small, ovoid; flowers magenta pink and cream blend, blooms borne mostly singly, dbl., 26–40 petals, 3–3.5 in., slight fragrance; few prickles; foliage medium, medium green, matt; medium (120 cms), upright growth; [Dame de Coeur X Sunbright]; Christensen, Jack E.; Bear Creek Gardens, 1991

'JACCOFL', F, ab, 1993; (**Brass Band**™); flowers melon orange and yellow bicolor, dbl., 30–35 petals, 3–3.5 in., blooms in small clusters, slight fruity fragrance; some prickles; foliage large, dark green, semi-glossy; medium (100 cms), upright, bushy growth; PP9171; AARS, 1995; [Gold Badge X Seedling]; Christensen, Jack E.; Bear Creek Gardens, 1994; Bear Creek Gardens, Inc., 1994, USA

'JACCOGEL', HT, ob, 1999; (**Orange Passion**™); flowers coral orange, reverse lighter, dbl., 26–40 petals, 4.5–5 in., borne mostly singly, slight fragrance; prickles moderate; foliage large, dark green, upright; upright growth; [Anne Morrow Lindbergh X Angelique]; Zary, Dr Keith W.; Bear Creek Gardens, Inc., 1999

'JACCOH', HT, op; (**Inspiration**)

'JACCOL', HT, mp, 1990; (**Colette**); bud ovoid, pointed; flowers medium soft pink, same reverse, aging slightly paler, urn-shaped, dbl., 25–30 petals, exhibition form, slight, damask fragrance; foliage large, dark green, semi-glossy; upright, bushy, tall growth; [Lorena X Seedling]; Warriner, William A. & Zary, Keith W., 1986; Bear Creek Gardens, 1990

'JACCOR', HT, mp, 1991; (**Glory Days**); flowers coral pink, large, dbl., 26–40 petals, moderate fragrance; foliage medium, medium green, semi-glossy; tall, upright, bushy growth; [Unnamed seedling X Showstopper]; Warriner, William A.; Bear Creek Gardens

JACcream, HT, yb; (**Angela**); J&P

'JACDAL', F, mp, 1993; (**Great Expectations**); J&P

'JACDASH', HT, dy, 1993; (**Sun Goddess**); dbl., 15–25 petals, 3–3.5 in., borne mostly singly, moderate fragrance; some prickles; foliage medium, dark green, matt; tall, upright, spreading growth; PP007659; [Sunbright X Seedling]; Warriner, William A.; Bear Creek Gardens, 1993

'JACDAZ', S, dr, 1992; (**Chorus Line**); flowers red, cream base, blooms borne in small clusters, dbl., 16–20 petals, 3–3.5 in., moderate fragrance; some prickles; foliage large, dark green, semi-glossy; medium (120–135 cms), upright, bushy, very vigorous growth; [Razzle Dazzle X Seedling]; Zary, Keith & Warriner, William; Bear Creek Gardens

JACdebu, HT, m, 1992; (**Paul Gauguin**); J&P

JACdeep, HT, lp, 1994; (**Kiko**); J&P

JACdeli, HT, yb, 1992; (**Paul Cezanne**); J&P

'JACDER', HT, dr, 1974; (**Red Masterpiece**); flowers deep red, large, dbl., exhibition form, intense fragrance; foliage large, dark, leathery; vigorous, upright growth; [(Siren X Chrysler Imperial) X (Carrousel X Chrysler Imperial)]; Warriner, William A.; J&P

JACdesa, HT, rb, 1992; (**Claude Monet**®)

'JACDEW', F, or, 1982; (**Impatient**); flowers medium, orange-red, semi-dbl., 20–25 petals, 4 in., borne in large clusters, slight fragrance; foliage medium, light green, glossy; upright, bushy growth; PP005122; AARS, 1984; [America X Unnamed seedling]; Warriner, William A.; J&P, 1984

'JACDIS', HT, mp, 1974; (Fragrant Memory, **Jadis**); flowers large, dbl., exhibition form, intense fragrance; foliage large, light, leathery; vigorous, upright, bushy growth; [Chrysler Imperial X Virgo]; Warriner, William A.; J&P

'JACDOR', F, rb, 1991; (**Señorita**); flowers orange-red to red on top of petal with a yellow petal base a, dbl., 26–40 petals, 1.5–2.75 in., slight fragrance; many prickles; foliage medium, dark green, semi-glossy; upright (75 cms), bushy growth; [Unnamed seedling X Matador]; Warriner, William A.; Bear Creek Gardens, 1992

JACdreco, F, m, 1999; (**Greetings**); Zary, Dr Keith W.

'JACDRIFT', Min, w, 1991; (**Blizzard**); bud pointed; flowers medium, dbl., 26–40 petals, no fragrance; foliage small, dark green, glossy; low, spreading, compact growth; [Petticoat X Orange Honey]; Warriner, William A.; Bear Creek Gardens

JACdrive, S, mr, 1998; (**Red Delicious**); J&P

'JACECOND', HT, dp, 1998; (**Timeless**™); flowers red to deep pink, deep pink reverse, dbl., 17–25 petals, 4.5 in., borne in small clusters, slight fragrance; prickles moderate; foliage medium, dark green, semi glossy; upright, bushy growth; AARS, J&P, 1997; [Spirit of Glasnost X Kardinal]; Zary, Dr Keith W.; Bear Creek Gardens, Inc., 1997

'JACED', HT, dr, 1984; (Home's Beauty, **Olympic Glory** flowers medium, dbl., 35 petals, slight fragrance; foliage large, medium green, glossy; upright growth; [Unnamed seedling X Unnamed seedling]; J&P; McConnell Nurs., Inc.

'JACEL', HT, my, 1988; (**Graceland**); flowers medium yellow, aging lighter at margins, loose, medium, dbl., 30–35 petals, cupped, borne singly, no fragrance; prickles medium, reddish-green; foliage medium, dark green, matt, smooth; upright growth; PP006069; GM, The Hague, 1988; [New Day X Unnamed seedling]; Warriner, William A.; Bear Creek Gardens, 1989

'JACERN', HT, my, 1985; (**Radiant Gold**); flowers large, dbl., 35 petals; foliage large, dark, glossy; upright growth; [Precilla X Sunshine]; J&P; McConnell Nurs., Inc.

JACess, F, w, 1997; (**Moonstruck**); J&P

'JACETTE', HT, w, 1986; (**Regatta**); flowers large, dbl., 48 petals, exhibition form, borne singly, intense fragrance; no fruit; medium red to brown prickles, hooked downward; foliage medium, medium green, matt; medium, upright growth; PP005896; [Bernadette X Coquette]; Warriner, William A.; J&P

JACeve, F, lp; (**Venus**); J&P, 1997

'JACEZZ', F, dr, 1998; (**Crimson Lace**™); flowers dark red, old-fashioned, dbl., 17–25 petals, 3–3.5 in., borne in small clusters, slight fragrance; prickles moderate; foliage medium, dark green, glossy; compact, medium growth; [Esprit X Razzle Dazzle]; Zary, Dr Keith W.; Bear Creek Gardens, Inc., 1997

'JACFED', HT, mp, 1993; very dbl., 3–3.5 in., borne mostly singly, moderate fragrance; some prickles; foliage large, medium green, semi-glossy; tall (150–180 cms), upright, spreading growth; [Grand Masterpiece X First Federal Renaissance]; Warriner, William A.; Bear Creek Gardens, 1992

'JACFISEG', Min, ab, 1992; ('JACFISEQ', **Paper Doll**); flowers light apricot with a hint of pale pink fading to light amber, dbl., 15–25 petals, 1.5–2.75 in., no fragrance; some prickles; foliage small, dark green, glossy; low (45–60 cms), upright growth; [Fiddler's Gold X Sequoia Gold]; Zary, Dr Keith W.; Bear Creek Gardens

'JACFLARE', Min, dy, 1991; (**Miss Daisy**); flowers small, very dbl., slight fragrance; foliage small, dark green, glossy; low, bushy growth; [Unnamed seedling X Sun Flare]; Warriner, William A.; Bear Creek Gardens

'JACFOG', Min, dy, 1992; (**Charm Bracelet**); flowers dark yellow, aging to pink, red blush on petal tips of outer petals, very dbl., 1.5 in., slight fragrance; some prickles; foliage small, dark green, semi-glossy; low (45 cms), bushy, compact growth; [Fool's Gold X Seedling]; Christensen, Jack E.; Bear Creek Gardens, 1992

'JACFON', HT, w, 1993; (**Heaven**); flowers ivory or cream with a light pink blend, blooms borne mostly, dbl., 26–40 petals, 3–3.5 in., intense fragrance; some prickles; foliage large, dark green, matt; tall (150–160 cms), upright, bushy growth; [Honor X First Prize]; Warriner, William A.; Bear Creek Gardens, 1994

'JACFRUIT', HT, lp, 1993; (**Camille**); flowers light pink, near white, blooms borne mostly singly, very dbl., 3–3.5 in., intense fragrance; some prickles; foliage large, dark green, semi-glossy; medium (120 cms), upright, bushy growth; [Honor X Fragrant Memory]; Warriner, William A.; Bear Creek Gardens, 1995

'JACFUN', Min, lp, 1982; (**Funny Girl**); flowers small, dbl., 20 petals, slight fragrance; foliage small, medium green, matt; upright, bushy growth; PP005249; [Bridal Pink X Fire Princess]; Warriner, William A.; J&P, 1983

'JACGATE', HT, mr, 1991; (**Brandenburg Gate**); flowers red with light reverse, large, dbl., 26–40 petals, moderate fragrance; foliage large, medium green, matt; tall, upright growth; [Seedling X Madras]; Warriner, William A.; Bear Creek Gardens, 1990

'JACGLOW', HT, w, 1989; (**White Delight**); bud ovoid, pointed; flowers white with pink blush, large, dbl., 38 petals, exhibition form, borne usually singly, slight fragrance; prickles hooked, medium, brown; foliage medium, dark green, matt; upright, medium growth; PP006664; [White Masterpiece X Futura]; Warriner, William A.; J&P, 1990

'JACGOLD', HT, dy, 1982; (**Golden Emblem**); flowers large, dbl., 20 petals, slight fragrance; foliage large, medium green, glossy; upright growth; PP005121; [(Bridal Pink X Dr A.J. Verhage) X (Golden Sun X South Seas)]; Warriner, William A.; J&P

'JACGRANT', F, ab, 1998; (**Fragrant Apricot**™); bud pointed, oval, apricot blend; flowers light apricot, copper tinted, light coral reverse, very dbl., 26–40 petals, 4 in., borne in large clusters, slight, musk fragrance; prickles moderate; foliage large, dark green, glossy; compact, medium (3 ft) growth; [Impatient X Amber Queen]; Zary, Dr Keith W.; Bear Creek Gardens, Inc., 1999

'JACGRAY', HT, lp, 1998; (**Billy Graham**™); bud long, pointed; flowers light pink, full, large, dbl., 26–40 petals, 5 in., exhibition form, borne singly, slight, sweet fragrance; prickles moderate; foliage large, dark green, glossy; upright, tall, 5 ft growth; [Honor X Color Magic]; Zary, Dr Keith W.; Bear Creek Gardens, Inc., 1998

'JACGRON', S, dp, 1998; (**Garden Blanket**™); flowers dark pink, white at base, lighter reverse, single, 4–11 petals, 1–1.5 in., borne in large clusters, slight fragrance; prickles moderate; foliage small, dark green, glossy; spreading, bushy, tall (3.5 ft) growth; [Magic Carpet X Happy Trails]; Walden, John K.; Bear Creek Gardens, Inc., 1998

'JACGUM', Min, dr, 1981; (**Gumdrop**); bud fat, pointed; flowers, dbl., 25 petals, borne 3–12 per cluster, foliage small, semi-glossy; spreading growth; PP005152; [(San Fernando X Bridal Pink) X (Fire Princess X Mary DeVor)]; Warriner, William A.; J&P

'JACHAL', Min, dy, 1999; (**Sun Sprinkles**™); flowers deep yellow, dbl., 26–40 petals, 2–2.5 in., borne mostly singly and in small clusters, slight fragrance; prickles moderate; foliage medium, dark green, glossy; compact, low (18–20 in.) growth; AOE, U.S., 2001; [Yellow Jacket X seedling]; Walden, John K.; Bear Creek Gardens, 2001

JAChart, F, lp; (**Starburst**); J&P, 1997

'JACHEIR', HT, m, 1993; (**Fortune Teller**); flowers deep mauve/purple, blooms borne mostly singly, dbl., 26–40 petals, 3–3.5 in., intense fragrance; some prickles; foliage large, dark green, semi-glossy; upright, spreading (150–160 cms) growth; [Seedling X Heirloom]; Warriner, William A. & Zary, Keith W.; Bear Creek Gardens

'JACHILL', Min, mp, 1991; (**Suzy Q**™); bud ovoid; flowers medium shell pink, deeper color at petal margins, medium, very dbl., 40–45 petals, cupped, no fragrance; foliage medium, dark green, glossy; upright, bushy, medium growth; [Rose Hills Red X Baby Ophelia]; Warriner, William A. & Zary, Keith W., 1985; Bear Creek Gardens, 1992

JAChoc, HT, w; (**Suzette van der Merwe**); J&P, 1994

JAChon, HT, mp, 1995; (**Enchanted**); J&P

'JACHOP', HT, dp, 1987; (**Haute Pink**); flowers rose pink fading little, loose, medium, dbl., 25–30 petals, cupped, borne singly, slight fragrance; no fruit; long, narrow, red prickles; foliage medium, medium green, matt; upright, tall growth; PP006653; [Bridal Pink X Grand Masterpiece]; Warriner, William A.; J&P

'JACHOTTA', Min, dp, 1999; (**Bedazzled**™); dbl., 26–40 petals, 2–2.5 in., borne in small clusters, slight fragrance; prickles moderate; foliage dark green, semi-glossy; compact, medium (20–24 in.) growth; [Hot Tamale sport]; Walden, John K.; Bear Creek Gardens, Inc., 2000

'JACHY', HT, dy, 1991; (**Old Faithful**); flowers large, dbl., 26–40 petals, moderate fragrance; foliage large, dark green, semi-glossy; upright, bushy growth; PP006445; [Sunbright X Medallion]; Warriner, William A.; Bear Creek Gardens

'JACHYP', HT, mp, 1991; (**Unforgettable**); flowers, dbl., 26–40 petals, 3–3.5 in., borne mostly singly, intense fragrance; foliage large, dark green, semi-glossy; tall (140 cms), upright, spreading growth; PP006318; [Honor X American Dawn]; Warriner, William A.; Bear Creek Gardens, 1992

'JACIAT', Min, mr, 1988; (**Little Sizzler**™, Patio Jewel); bud ovoid, pointed; flowers large, dbl., 38 petals, cupped, borne usually singly and in sprays of 16–20, prickles hooked downward, reddish-brown; foliage large, dark green, semi-glossy; bushy, medium growth; PP006091; [Unnamed seedling X Funny Girl]; Warriner, William A.; Bear Creek Gardens, 1989

'JACIBE', Min, ab, 1982; (**Caribe**); flowers apricot, semi-dbl., slight fragrance; foliage small, light green, semi-glossy; upright, bushy growth; PP005246; [Bridal Pink X Fire Princess]; Warriner, William A.; J&P, 1983

JACibras, F, dy, 1993; (**Catherine McAuley**); Bear Creek Gardens

'JACICE', HT, w, 1984; (**Moonlight**); flowers medium, dbl., 35 petals, no fragrance; foliage large, medium green, matt; upright growth; PP005743; [Coquette X Unnamed seedling]; Warriner, William A.; J&P

'JACIENT', Gr, mp, 1988; (Berkeley, Poesie, **Tournament of Roses**); flowers light coral pink, reverse deep pink, aging coral pink, large, dbl., exhibition form, no fragrance; large prickles; foliage large, dark green, semi-glossy, disease resistant; upright, bushy, medium growth; PP006725; AARS, 1989; [Impatient X Unnamed seedling]; Warriner, William A.; J&P, 1989

'JACIF', F, ab, 1983; (**Pacifica**®); flowers medium, dbl., 35 petals, slight fragrance; foliage medium, light green, matt; upright, bushy growth; PP005261; [Mercedes X Marina]; Warriner, William A.; J&P

'JACIMIN', Min, w, 1993; (**Small Miracle**); bud plump; flowers white, moderately classic hybrid tea, small, dbl., 15–25 petals, 2 in., exhibition form, borne in small clusters, slight fragrance; few prickles; foliage small, dark green, glossy, resistant to powdery mild; low (50–60 cms), bushy growth; PP8850; [Libby X Sun Flare]; Warriner, William A.; Bear Creek Gardens, 1993

'JACINA', S, dp, 1998; (**Wild Dancer**™); flowers bright magenta-pink, single, 4–11 petals, 1.5 in., borne in large clusters, slight fragrance; prickles moderate; foliage small, medium green, semi double; compact, arching, low, 2.5 x 2.5 ft growth; [China Doll X Sweet Chariot]; Walden, John K.; Bear Creek Gardens, Inc., 1998

'JACINAL', F, mp, 1985; (**Debutante**); flowers small, dbl., 20 petals, slight fragrance; foliage medium, medium green, matt; bushy growth; PP005729; [Bridal Pink X Zorina]; Warriner, William A.; J&P

'JACINK', F, mp, 1978; (**Simplicity**™); bud long, pointed; semi-dbl., 18 petals, 3–4 in., flat, slight fragrance; bushy, upright growth; GM, NZ, 1976; [Iceberg X Unnamed seedling]; Warriner, William A.; J&P, 1979

'JACIRAPS', Min, pb, 1999; (**Hugs 'n' Kisses**™); flowers pink, white, reverse near white, medium, dbl., 17–25 petals, 2.5 in., borne in small to large clusters, moderate fragrance; prickles moderate; foliage large, dark, grey-green, glossy; compact, spreading, medium (16–18 in.) growth; [Small Miracle X Seedling]; Walden, John K.; Bear Creek Gardens, Inc., 1999

'JACIRST', HT, op, 1998; (**Artistry**™); flowers coral orange and creamy coral reverse, very dbl., 30–35 petals, 4.5 in., exhibition form, borne singly, full, best form and color in heat, slight fragrance; prickles moderate; foliage large, dark green, semi glossy; upright, bushy, tall growth; PP10230; AARS, J&P, 1997; [Seedling X Seedling]; Zary, Dr Keith W.; Bear Creek Gardens, Inc., 1997

'JACIS', HT, lp, 1976; (Poesie, **Promise**); bud ovoid; flowers large, dbl., exhibition form, slight fragrance; foliage large, glossy; vigorous growth; GM, Bagatelle, 1976; [South Seas X Peace]; Warriner, William A.; J&P

'JACIV', F, dr, 1974; (**Viva**); bud ovoid; flowers medium, dbl., exhibition form, slight fragrance; foliage glossy, dark; vigorous, upright growth; GM, Portland, 1984; [Unnamed seedling X Unnamed seedling]; Warriner, William A.; J&P

'JACJEL', HT, my, 1984; (**Sunbright**®); bud long, pointed; dbl., 28 petals, 4 in., flat, slight fragrance; upright growth; [Unnamed seedling X New Day]; Warriner, William A.; J&P

'JACJEM', F, my, 1981; (**Sun Flare**, Sunflare); bud pointed; flowers, dbl., 30 petals, flat, borne 3–12 per cluster, no fragrance; reddish prickles; foliage small, glossy; low, compact growth; PP005247; AARS, 1983 GM, Japan, 1981 GM, Portland, 1985; [Sunsprite X Unnamed seedling]; Warriner, William A.; J&P, 1983

'JACJUBEL', F, my, 1984; (**Mirabella**™); flowers medium yellow, very dbl., 41 petals, 4–4.5 in., borne singly and in small clusters, intense fragrance; some prickles; foliage medium, dark green, glossy; medium (80–100 cms), upright,-bushy growth; [Sunsprite X Silver Jubilee]; Zary, Dr Keith W.; Bear Creek Gardens, 1994

Jack Collier, *see* 'HORJACK'

Jack Dayson, *see* 'KORWILMA'

'**JACK FOLLY**', HT, w, 1976; bud ovoid; flowers cream color, opening, dbl., 40 petals, 5.5 in., cupped, foliage dull, leathery; spreading growth; Buss, H., Nursery

'**JACK FROST**', F, w, 1962; bud pointed; flowers white to creamy, sweetheart, medium, dbl., 42 petals, exhibition form, moderate fragrance; foliage dark; vigorous, upright growth; [Garnette X Unnamed seedling]; Jelly; E.G. Hill Co.

'**JACK HORNER**', Min, mp, 1955; flowers bright pink, dbl., 50 petals, slight fragrance; no prickles; height 4–8 in.; [Margo Koster X Tom Thumb]; Robinson, T.

'**JACK MCCANDLESS**', HT, rb, 1935; bud long, pointed, carmine and yellow; flowers amber-yellow, veined red, dbl., exhibition form, slight fragrance; foliage small, glossy, dark; McGredy

'**JACK OF HEARTS**', F, 1968; flowers cardinal red, trusses, semi-dbl., cupped, free bloom; foliage dark, glossy; low, bushy growth; Waterhouse Nursery

'**JACK O'LANTERN**', Gr, yb, 1960; bud ovoid; flowers gold and yellow blend, dbl., 25 petals, 4.5 in., exhibition form, slight fragrance; foliage leathery; vigorous, tall, bushy growth; [Circus X Golden Scepter]; Swim & Weeks

Jack Rose, *see* '**GÉNÉRAL JACQUEMINOT**'

Jack Rose, *see* **R. MAXIMOWICZIANA JACKII**

JACka, HT, pb, 1994; (**Loving**); J&P

Jackaranda, *see* 'JACAKOR'®

JACkari, HT, yb; (**Askari**); J&P

'JACKEE', Min, yb, 1998; (**Bees Knees**™); flowers yellow/pink blend, large, very dbl., 26–40 petals, 2.5 in., borne in large clusters, slight fragrance; prickles moderate; foliage medium, dark green, dull; bushy, tall, 3 ft growth; [Seedling X Haute Pink]; Zary, Dr Keith W.; Bear Creek Gardens, Inc., 1998

'**JACKIE**'™, Min, ly, 1955; flowers straw-yellow changing to white, small, dbl., 60 petals, 1.5 in., exhibition form, moderate fragrance; foliage glossy; vigorous, dwarf (12 in.), bushy, spreading growth; [Golden Glow X Zee]; Moore, Ralph S.; Sequoia Nursery

'**JACKIE CLARK**', HT, dr, 1990; bud pointed; very dbl., 35 petals, 3.5 in., exhibition form, borne singly, intense fragrance; prickles curved down, pink; foliage medium, dark green, matt, disease-resistant; vigorous, upright growth; [White Masterpiece X Red Planet]; Wambach, Alex A., 1985; Alex A. Wambach, 1989

'**JACKIE, CLIMBING**', Cl Min, ly, 1957; flowers soft yellow to creamy white, dbl., 60 petals, 1–1.5 in., moderate fragrance; foliage semi-glossy, leathery; height to 10 ft; [Golden Glow X Zee]; Moore, Ralph S.; Sequoia Nursery

'**JACKMAN'S WHITE**', HT, w; (Brookdale Giant White); flowers creamy white, very large, exhibition form; vigorous growth; very hardy for this type; Brookdale-Kingsway; Bosley Nursery, 1940

'JACKO', F, or, 1974; (**Sunfire**); bud ovoid; flowers mandarin-red, medium, dbl., exhibition form, slight fragrance; foliage large, leathery; vigorous, upright, bushy growth; [Tropicana X Zorina]; Warriner, William A.; J&P

Jackpot, *see* 'MORJACK'

Jack's Fantasy, *see* 'BILFAN'

Jackson Square, *see* 'TALJAC'

'JACLACE', F, w, 1980; (**French Lace**); bud pointed; flowers ivory, pastel apricot to white, dbl., 30–35 petals, 4.5 in., exhibition form, borne in small clusters, slight, fruity fragrance; small prickles; foliage small, dark; bushy growth; PP4848; AARS, 1982 GM, Portland, 1984; [Dr A.J. Verhage X Bridal Pink]; Warriner, William A.; J&P

'JACLAF', HT, m, 1998; (**Moon Shadow**™); bud long, pointed, oval; flowers medium lavender, very dbl., 26–40 petals, 4.5–5 in., borne in small clusters, intense, sweet pea fragrance; few prickles; foliage medium, dark green, semi glossy; upright, tall, bushy, 4.5 ft growth; PP9538; Warriner, William A.; Bear Creek Gardens, Inc., 1998

'JACLAM', F, dr, 1986; (**Carmen**); flowers deep red, very little fading, small, dbl., 30 petals, exhibition form, borne usually singly, no fragrance; foliage medium, dark green, semi-glossy; upright, bushy growth; PP006070; [Unnamed seedling X Samantha]; Warriner, William A.; J&P

'JACLAV', HT, m, 1990; (Heather, **Silver Anniversary**™); flowers medium lavender, reverse medium lavender, spot of yellow at base, dbl., 25–30 petals, exhibition form, intense, damask fragrance; foliage large, very dark green, semi-glossy-glossy; long, straight, sturdy stems; average 24–32 in.; long, tall, upright growth; PP007658; [Crystalline X Shocking Blue]; Christensen, Jack E., 1985; Bear Creek Gardens, 1990

JAClay, HT, my; (**Treasure Trove**); Warriner, William A., 1991

'JACLEM', Cl F, my, 1987; (**Sun Flare, Climbing**); [Sun Flare sport]; Warriner, William A.; J&P

'JACLES', HT, rb, 1983; (**Milestone**); flowers medium red, silvery red reverse, opening to coral-pink, dbl., cupped, slight fragrance; foliage large, medium green, semi-glossy; upright growth; [Sunfire X Spellbinder]; Warriner, William A.; J&P, 1985

'JACLEWT', Gr, m, 1999; (**English Perfume**™); flowers lavender blend, very dbl., 41 petals, 5 in., intense fragrance; prickles moderate; foliage medium, dark green, semi-glossy; upright growth; [Seedling X Carefree Wonder]; Zary, Dr Keith W.; Bear Creek Gardens, Inc., 1999

'JACLIANG', Min, m, 1991; (**Plum Dandy**™); bud ovoid, pointed; flowers medium lavender, lighter near petal base, same reverse, fade, very dbl., 35 petals, cupped, moderate, fruity fragrance; foliage medium, medium green, semi-glossy; bushy, spreading, medium growth; [Unnamed seedling X Angel Face]; Warriner, William A.; Bear Creek Gardens

JAClib, Min, pb, 1993; (**Petite Michelle**); J&P

'JACLIN', HT, dr, 1991; (**Patriot**); dbl., 26–40 petals, 3–3.5 in., borne mostly singly, slight fragrance; foliage large, dark green, semi-glossy to glossy; upright (150 cms), spreading growth; [Showstopper X Mister Lincoln]; Warriner, William A.; Bear Creek Gardens

'JACLING', F, pb, 1991; (**First Kiss**); flowers light pink, light yellow blend at base, moderately large, dbl., 15–25 petals, slight fragrance; foliage medium, medium green, matt; bushy, compact growth; [Sun Flare X Simplicity]; Warriner, William A.; Bear Creek Gardens

'JACLION', HT, ab, 1998; (**Grand Impression**™, Monet); flowers yellow/peach/pink blend, apricot yellow reverse, very dbl., 26–40 petals, 6 in., borne in small clusters, slight fragrance; few prickles; foliage large, dark green, semi glossy; upright, tall, 5 ft growth; [Spirit of Glasnost X Medallion]; Warriner, William A.; Bear Creek Gardens, Inc., 1999

'JACLIP', Min, lp, 1986; (**Le Pink**); flowers crowded, small, dbl., 40 petals, cupped, blooms in sprays of 3–20, slight, damask fragrance; long, thin prickles, numerous on peduncle; foliage small, light green, matt; very dense, bushy, upright, medium growth; PP006011; [Unnamed seedling X Watercolor]; Warriner, William A.; J&P

'JACLIP', Min, mp; (**Pink Cloud**®)

'JACLITE', Min, w, 1982; (**Cloud Nine**); flowers medium, semi-dbl., foliage light green, matt; spreading growth; PP005557; [Bon Bon X Calgold]; Warriner, William A.; J&P

'JACLIV', F, my, 1987; (**Blondie** 'JACLIY'); flowers medium yellow, fading slightly, medium, borne usually singly, dbl., 25 petals, exhibition form, slight, fruity fragrance; no fruit; straight, medium prickles; foliage medium, light green, semi-glossy; upright, bushy, medium growth; [(Bridal Pink X Golden Wave) X Gold Rush]; Warriner, William A.; J&P

JACLOGO, HT, ob, 1997; (**Saturn**); J&P

'JACLOOM', HT, m, 1972; (**Heirloom**); bud long, pointed; flowers deep lilac, medium, semi-dbl., intense fragrance; foliage leathery; vigorous, upright growth; [Seedling X Seedling]; Warriner, William A.; J&P

'JACLOPI', HT, mp, 1985; (**Leading Lady**); flowers large, dbl., 35 petals, slight fragrance; foliage large, light green, matt; upright growth; PP005859; [Unnamed seedling X Unnamed seedling]; Warriner, William A.; J&P

'JACLORNA', HT, dp; (**Sugar Daddy**); J&P, 1992

'JACLOVER', S, m, 1992; (**Magic Carpet**); flowers lavender blooms, small, profuse, semi-dbl., 12 petals, 2 in., borne in clusters, intense, spicy fragrance; foliage medium, dark green, glossy; spreading, low, 18 in., ground cover growth; PP9324; Zary, Keith & Warriner, William

'JACMAD', HT, pb; (**Winner's Choice**); J&P, 1992

'JACMAG', HT, pb, 1978; (**Color Magic**); bud long; flowers pale salmon blushing pink, dbl., 25–30 petals, 5 in., flat, borne singly, intense, fruity fragrance; foliage large, dark green; upright growth; AARS, 1978; [Seedling X Spellbinder]; Warriner, William A.; J&P

'JACMANTHA', HT, mr, 1974; ('JACANTH'**Samantha**®); bud ovoid; flowers flora-tea, medium, dbl., exhibition form, slight fragrance; foliage leathery; vigorous, upright growth; [Bridal Pink X Seedling]; Warriner, William A.; J&P

'JACMAR', F, ob, 1991; (**Kaileen**); flowers bright orange, yellow reverse, blooms borne in small clusters, dbl., 26–40 petals, 1.5–2.75 in., slight fragrance; some prickles, light green to yellow; foliage medium, dark green, glossy; upright (180 cms) growth; [Marina sport]; Nakashima, Tosh; Bear Creek Gardens, 1992

'JACMAS', HT, w, 1969; (**White Masterpiece**); bud long, pointed; flowers very large, large peduncle, dbl., exhibition form, slight fragrance; upright growth; Boerner; J&P

'JACMAT', Min, or, 1981; (**Razzmatazz**); flowers, dbl., exhibition form, in clusters, no fragrance; foliage semi-glossy; upright growth; PP005118; [Zorina X Fire Princess]; Warriner, William A.; J&P

JACmeen, F, op, 1998; (**Good Show**); PPAF; J&P

JACMEM, HT, yb, 1994; (**Mango**); J&P

'JACMENT', F, op, 1998; (**Pure Poetry**); flowers yellow/orange/pink blend, very full, very dbl., 41 petals, 4 in., borne in small clusters, slight fragrance; prickles many, straight; foliage medium, dark green, glossy; compact, 2.5 ft growth; [Seedling X Tournament of Roses]; Zary, Dr Keith W.; Bear Creek Gardens, Inc., 1997

'JACMEY', HT, dy, 1987; (**Sun Glory**); flowers medium, borne singly, dbl., 30 petals, exhibition form, slight, fruity fragrance; no fruit; long, red prickles, pointed slightly downwards; foliage medium, medium green, matt; upright growth; PP006612; [Golden Emblem X Unnamed seedling]; Warriner, William A.; J&P

'JACMINNO', Min, ob, 1992; (**Jitterbug**); flowers orange, slightly lighter on reverse, nicely formed, open, dbl., 15–25 petals, 1.5–2.75 in., slight fragrance; some prickles; foliage medium, dark green, glossy; tall (60–75 cms), upright, bushy growth; [Caribe X Impatient]; Warriner, William A.; Bear Creek Gardens, 1993

'JACMINT', Min, pb; (**Peppermint Stick**); J&P

'JACMIRYL', Min, my, 1999; (**Lemon Gems**™); flowers rich, deep yellow, dbl., 26–40 petals, 2.5–2.75 in., cupped, borne mostly singly, slight fragrance; prickles moderate; foliage medium, dark green, glossy; upright, spreading, low (18–20 in.) growth; [Seedling X Small Miracle]; Walden, John K.; Bear Creek Gardens, 1999

'JACMISS', Min, lp, 1989; (**Miss Perfect**™); bud ovoid, pointed; flowers light pink to near white at edge, reverse lighter, medium, very dbl., cupped, slight fragrance; prickles straight to slightly hooked downward, light yellow; foliage medium, medium green, semi-glossy; spreading, low growth; PP007650; [Over the Rainbow X Lavender Lace]; Warriner, William A.; Bear Creek Gardens

'JACMIY', Min, my, 1988; (**Sunsmile**™); bud ovoid; flowers medium, borne singly and in sprays of 2–3, dbl., 48 petals, exhibition form, slight fragrance; prickles straight to slightly hooked downward, brown; foliage medium, medium green, semi-glossy; bushy, spreading, compact, vigorous growth; [Spanish Sun X Calgold]; Warriner, William A.; Bear Creek Gardens, 1989

'JACMO', Min, ob, 1991; (**Pride 'n' Joy**™); bud ovoid; flowers bright, medium orange, reverse orange and cream, fades to sa, very dbl., 30–35 petals, 2 in., exhibition form, moderate, fruity fragrance; foliage medium, dark green, semi-glossy; bushy, spreading, medium (36 in.) growth; PP8578; AARS, 1992; [Chattem Centennial X Prominent]; Warriner, William A., 1983; Bear Creek Gardens, 1992

'JACMOBLI', Min, mp, 1998; (**Barbie**®); flowers medium pink, long lasting, very dbl., 41 petals, 2 in., borne in small clusters, slight fragrance; prickles moderate; foliage medium, dark green, semi glossy; upright, low, suitable for pots growth; [(Seedling X Watercolor) X Red Minimo]; Walden, John K.; Bear Creek Gardens, Inc., 1999

'JACMOCL', Cl Min, ob, 1989; (**Orange Blossom Special**™); bud ovoid, sepals serrated; flowers coral orange, edges darker, reverse lighter coral pink, agin, very dbl., exhibition form, slight fragrance; prickles straight to hooked down, reddish-green; foliage small, dark green, semi-glossy; upright, bushy, tall growth; [Zorina X Andrea]; Warriner, William A.; Bear Creek Gardens, 1990

'JACMUR', HT, w, 1984; (**Fountain Square**™); flowers clear white, dbl., 25–30 petals, 5 in.–5.5 in., exhibition form, borne singly, slight fragrance; foliage large, dark green, semi-glossy; tall growth; PP006805; [Pristine sport]; Humenick, Muriel F.; Fountain Square, Inc., 1986

'JACNEL', HT, dy, 1992; (**Sundance**); very dbl., 3–3.5 in., borne mostly singly, moderate fragrance; some prickles; foliage large, dark green, semi-glossy; tall (180–120 cms), upright growth; [Unnamed seedling X Emblem]; Warriner, William A.; Bear Creek Gardens, 1991

'JACNEPAL', HT, pb, 1999; (**Gemini**™); bud pointed, ovoid; flowers cream, blushing coral pink, dbl., 25–30 petals, 4.5–5 in., exhibition form, usually borne singly, moderate, sweet fragrance; prickles moderate; foliage large, deep green, glossy; upright, spreading, tall (5.5 ft) growth; AARS, U.S., 2000; Silver Medal-Monza, 1999; [Anne Morrow Lindbergh X New Year]; Zary, Dr Keith W.; Bear Creek Gardens, 2000

'JACNON', HT, 1985; flowers medium, dbl., slight fragrance; foliage large, medium green, matt; upright, bushy growth; PP005318; [Baccará X Unnamed seedling]; Warriner, William A.; J&P

'JACNOR', HT, 1998; (**Signature**®); flowers deep pink and cream blend, light pink and cream reverse, very dbl., 26–40 petals, 5 in., exhibition form, borne singly, moderate fragrance; many prickles; foliage large, dark green, semi glossy; upright, bushy, tall, 4–5 ft growth; [Honor X First Federal Renaissance] PP9539; Warriner, William A.; Bear Creek Gardens, Inc., 1996

JACnu, HT, op, 1993; (**Nuance**); J&P

JACnuel, HT, lp; (**Sterling '95**); J&P, 1995

'JACNUYE', Min, ob, 1998; (**Little Flame**™); flowers brilliant orange, dbl., 17–25 petals, 2 in., borne in large clusters, no fragrance; prickles moderate, straight; foliage medium, dark green, semi-glossy; compact, tall, border (3 ft) growth; AOE, J&P, 1999; [New Year X Seedling]; Walden, John K.; Bear Creek Gardens, Inc., 1998

Jacob van Ruysdael, F, op, 1997; flowers peach orange, semi-dbl.; upright, compact growth; Williams, J. Benjamin

Jacobite Rose, *see* White Rose of York

'JACOLBER', HT, dr, 1998; (**Opening Night**™); flowers bright, deep red, very dbl., 26–40 petals, 4.5 in., exhibition form, borne singly, slight fragrance; prickles moderate, straight; foliage medium, dark green, semi glossy; upright, tall, slightly spreading growth; AARS, 1998; [Olympiad X Ingrid Bergman]; Zary, Dr Keith W.; Bear Creek Gardens, Inc., 1998

'JACOLFA', HT, lp, 1999; (**English Sachet**™); flowers cupped, ruffled, almost quartered, delicate petals, very dbl., 40–50 petals, 4–4.5 in., cupped, borne mostly singly, intense, sweet fragrance; prickles moderate; foliage medium, medium green, matte; upright, spreading, tall (5 ft) growth; [Summer Fashion X Silver Jubilee]; Zary, Dr Keith W.; Bear Creek Gardens, Inc., 1999

'JACOLITE', HT, w, 1980; (**Honor**™, Honour, Michèle Torr®); bud ovoid, pointed; flowers loose, dbl., 23 petals, 5 in., slight fragrance; foliage large, dark; upright growth; AARS, 1980 GM Portland, 1978; Warriner, William A.; J&P

'JACOLPUR', HT, m, 1999; (**Purple Passion**™); bud pointed, ovoid; flowers rich purple/lavender, reverse lavender, dbl., 26–40 petals, 4.5–5 in., exhibition form, borne mostly singly, intense, lemony fragrance; prickles moderate; foliage large, dark green, glossy; upright, spreading, tall (5 ft) growth; [seedling X seedling]; Zary, Dr Keith W.; Bear Creek Gardens, Inc., 1999

'JACOPPER', HT, dr, 1999; (City of Newcastle Bicentennary, Five-Roses Rose, Lady in Red, **Veteran's Honor**™); bud 2 i. furled, pointed; flowers bright red, high centered, large, dbl., 26–40 petals, 5–5.5 in., exhibition form, borne mostly singly, slight, raspberry fragrance; prickles moderate; foliage medium, dark green, semi-glossy; upright, spreading, tall (5 ft) growth; Rose of the Year – 2000, J&P, 2000; [Showstopper X (Seedling X Royalty)]; Zary, Dr Keith W.; Jackson & Perkins, 2000, U.S.

'JACORAL', Min, mp, 1988; (**Coral Sprite**™); bud ovoid, pointed; flowers urn-shaped, medium, borne in sprays of 3–18, very dbl., no fragrance; prickles straight to slightly angled downward, yellow-green; foliage medium, dark green, matt; bushy, spreading, low growth; PP007717; [Merci X Party Girl]; Warriner, William A.; Bear Creek Gardens, 1989

JACorb, HT, ob; (**Ipitombi**); J&P

'JACORCA', Min, op, 1998; (**Tropical Twist**™); flowers coral orange and apricot with yellow reverse, dbl., 17–25 petals, borne mostly singly, floriferous, slight fragrance; prickles moderate; foliage medium, dark green, glossy; compact, tall growth; hedging or border; AOE, ARS, 1997; [Seedling X Pink Pollyanna]; Walden, John K.; Bear Creek Gardens, Inc., 1997

'JACORE', HT, dr, 1984; (**Encore**); flowers large, dbl., 20 petals, no fragrance; foliage medium, dark, semi-glossy; upright growth; PP005658; [Seedling X Samantha]; Warriner, William A.; J&P

'JACORG', Min, ob, 1999; (**Crackling Fire**™); flowers copper orange, reverse deep red orange, dbl., 20–40 petals, 2–2.25 in., borne in small clusters, no fragrance; prickles moderate; foliage medium, dark green, semi-glossy; upright, compact, low (14–18 in.) growth; [Seedling X Rainbow's End]; Walden, John K.; Bear Creek Gardens, Inc., 1999

'JACORMIN', Min, or, 1988; (**Vixen**); bud ovoid, pointed, green with reddish-brown; flowers red-orange with yellow base, aging red-orange to pink orange, dbl., 28 petals, cupped, no fragrance; prickles small, straight, reddish-brown; foliage medium, medium green, semi-glossy, small; bushy, spreading, low growth; [Petticoat X Red Minimo]; Warriner, William A.; Bear Creek Gardens, 1990

JACosch, HT, w, 1997; (**Opulence**); J&P

JACosol, HT, my, 1995; (**Moonlight Lady**); J&P

'JACOTTE', LCl, ab, 1920; bud ovoid, orange and yellow; flowers deep coppery yellow, tinted coppery red, open, semi-dbl., 3 in., cupped, moderate fragrance; foliage leathery, glossy, dark; very vigorous, climbing growth; [R. wichurana X Arthur R. Goodwin., Barbier

'JACOUCH', Min, ab, 1984; (**Soft Touch**); flowers medium, semi-dbl., slight fragrance; foliage medium, medium green, semi-glossy; PP005578; [Bridal Pink X Fire Princess]; Warriner, William A.; J&P

'JACOUT', F, dp, 1991; (**Neon Lights**); flowers hot magenta pink, moderately large blooms borne in small clusters, dbl., 15–25 petals, moderate fragrance; foliage medium, medium green, semi-glossy; medium, bushy growth; [Intrigue X Impatient]; Warriner, William A.; Bear Creek Gardens, 1992

'JACOYEL', HT, op, 1982; (**Fascination**); bud nearly globular; flowers orange and rose blend, blooms borne usually singly, dbl., 55 petals, exhibition form, no fragrance; long-based prickles, hooked down; foliage very large, semi-glossy; upright, heavy branching growth; GM, NZ, 1976; [Unnamed seedling X Spellbinder]; Warriner, William A.; J&P

'JACPAL', HT, pb, 1991; (**Brigadoon**™); bud ovoid, pointed; flowers cream pink near center, coral pink at petal margin, rose pink edges, very dbl., 25–30 petals, 5 in., exhibition form, borne mostly singly, moderate, tea fragrance; foliage medium, dark green, semi-glossy; upright, spreading, tall growth; PP8591; AARS, 1992; [Seedling X Pristine]; Warriner, William A., 1983; Bear Creek Gardens, 1992

JACpan, F, mp; (**Meteor Shower**); J&P

JACpat, F, op; (**Gnome World**); J&P

'JACPET', Min, w, 1981; (**Petticoat**); bud short; flowers white tinted pink, dbl., 55 petals, slight fragrance; foliage dark, small, pointed; very compact, spreading growth; PP005039; [Bon Bon X Lemon Delight]; Warriner, William A.; J&P

'JACPICL', Cl F, op, 1998; (**Dream Weaver**™); bud short, pointed; flowers coral pink, very dbl., 26–40 petals, 3–3.5 in., rosette, borne in large clusters, slight, rose fragrance; prickles moderate; foliage large, dark green, glossy; tall, spreading, arching, a climber growth; PP9492; [Seedling X Lady of the Dawn]; Zary, Dr Keith W.; Bear Creek Gardens, Inc., 1997

'JACPICO', HT, w, 1978; (**Pristine**®); bud long; flowers near white, shaded light pink, imbricated, dbl., 28 petals, 5–6 in., slight fragrance; foliage very large, dark; upright growth; Edland Fragrance Medal, ARS, 1979 GM, Portland, 1979; [White Masterpiece X First Prize]; Warriner, William A.; J&P

'JACPIE', HT, mr, 1978; (**Grand Masterpiece**); bud ovoid, pointed; 5 in., exhibition form, slight fragrance; tall, upright growth; [Seedling X Tonight]; Warriner, William A.; J&P

'JACPIF', F, mp, 1988; (**Pleasure**); bud ovoid, pointed; flowers coral pink, reverse lighter, large, dbl., 33 petals, cupped, borne in sprays of 3–7, slight fragrance; globular fruit; prickles slightly hooked downward, reddish-brown; foliage medium, dark green, semi-glossy; low, compact growth; PP007480; AOE, 1990; [(Merci X Faberge) X Intrigue]; Warriner, William A.; Bear Creek Gardens, 1990

'JACPIHI', HT, w, 1998; (**Grand Finale**™); bud long, ovoid; flowers ivory white, full, very dbl., 26–40 petals, 4–4.5 in., exhibition form, borne singly, slight, honeysuckle fragrance; prickles moderate; foliage large, medium green, semi-glossy; upright, medium (4 ft) growth; Rose of the Year, J&P, 1998; [Honor X Pristine]; Zary, Dr Keith W.; Bear Creek Gardens, Inc., 1998

JACpik, HT, or, 1994; (**African Sunset**); J&P

JACpin, F, op, 1993; (**Dream Dolly**); J&P

'JACPLY', HT, yb, 1993; (**Somersault**); flowers yellow and rose red/orange blend, blooms borne in small or l, dbl., 26–40 petals, 1.5–2.75 in., slight fragrance; some prickles; foliage small, dark green, glossy; low (45–50 cms), upright, bushy growth; Zary, Dr Keith W.; Bear Creek Gardens

'JACPO', Min, my, 1984; (**Goldpoint**); flowers small, dbl., 20 petals, slight fragrance; foliage small, light green, matt; upright, bushy growth; PP005645; [Rise 'n' Shine X (Faberge X Precilla)]; Warriner, William A.; J&P

'JACPOP', HT, yb, 1985; (**Prince of Peace**); flowers yellow edged with pink, medium, dbl., 35 petals, slight fragrance; foliage medium, medium green, glossy; upright growth; PP005901; [Bridal Pink X ?]; J&P; McConnell Nurs., Inc.

JACpow, HT, w, 1992; (**Fantasy**); J&P

'JACPOY', Min, yb, 1993; (**Hot Tamale**™); flowers yellow-orange blend changing to yellow pink, finishing pink, dbl., 26–40 petals, 1.5–2.75 in., exhibition form, borne singly and in small clusters, slight fragrance; some prickles; foliage small, dark green, semi-glossy; low (36 cms), bushy, compact growth; PP9015; AOE, ARS, 1994; Zary, Dr Keith W.; Bear Creek Gardens, Inc. (J&P); Nor' East Miniature Roses

JACpri, Min, ab; (**Joyous Moment**); J&P

'JACPURR', F, m, 1991; (**Purple Tiger**); bud pointed, ovoid; flowers very deep purple with stripes and flecks of white and mauve, dbl., 26–40 petals, 3.5–4 in., borne in small clusters, moderate, damask fragrance; nearly thornless; foliage medium, medium green, glossy; stems very glabrous (shiny); medium (70–90 cms), bushy growth; [Intrigue X Pinstripe]; Christensen, Jack E.; Bear Creek Gardens, 1992

'JACPURSH', S, m, 1998; (**Purple Simplicity**); flowers raspberry purple, semi-double, semi-dbl., 12–16 petals, 3 in., borne in large clusters, slight fragrance; prickles few, straight, thin; foliage medium, medium green, semi-glossy; tall, bushy, for hedges growth; [Seedling X Love Potion]; Zary, Dr Keith W.; Bear Creek Gardens, Inc., 1999

Jacqueline, *see* 'MEIGER'

'JACQUELINE', HT, mr, 1961; bud short, pointed; flowers turkey-red, dbl., 30 petals, 4–4.5 in., exhibition form, moderate fragrance; foliage glossy; vigorous, well-branched growth; [Topper X Unnamed seedling]; J.H. Hill Co.

Jacqueline, *see* KORoranki

Jacqueline du Pré®, *see* 'HARWANNA'

'JACQUELINE DUFIER', HT, dr, 1957; flowers crimson shaded black, high pointed, dbl., 25 petals, 4 in., slight fragrance; foliage bronze; vigorous growth; [Dicksons Red X Ena Harkness]; Kemp, M.L.

Jacqueline Humery, HMsk, lp, 1995; Lens

Jacqueline Nebout®, *see* MEIchoiju

'JACQUELINE STERNOTTE', HT, rb, 1974; bud ovoid; flowers red and white, full, very dbl., 51 petals, 4.5 in., moderate fragrance; Delforge, S.

'JACQUES AMYOT', N, lp; flowers lilac pink, medium, good repeat bloom; tall growth; Varangot, 1844

'JACQUES CARROY', Pol, m, 1929; flowers carmine, slightly tinted purple, center velvety, borne in clusters; Turbat

'JACQUES CARTEAU', HT, ly, 1957; bud long; flowers creamy yellow, medium, foliage glossy; moderate growth; Privat

'Jacques Cartier', P, lp, 1868; flowers clear rose, center darker, very dbl., 50 petals, 3.5–4 in., rosette, borne singly and in small clusters, intense, sweet fragrance; foliage medium green, matt; compact, low, branching growth; Moreau et Robert

Jacques Esterel, HT, m; Croix

'JACQUES HACKENBURG', HT, pb, 1919; flowers deep rose-pink and carmine, opening flesh-white, dbl., intense fragrance;

[Jonkheer J.L. Mock X Marquise de Sinéty]; Leenders, M.

'JACQUES LATOUCHE', HT, pb, 1935; flowers orange-pink in spring, red in summer, reverse yellow, very large, dbl., cupped, slight fragrance; foliage glossy; vigorous growth; [Souv. de Claudius Pernet X Director Rubió]; Mallerin, C.; H. Guillot

'JACQUES PORCHER', HT, rb, 1914; flowers blended tints of carmine, saffron and deep yellow, dbl., moderate fragrance; Guillot, P.

Jacques Prevert®, *see* 'MEIMOUSLIN'

'JACQUES PROUST', Pol, m, 1904; flowers violet-red, small, semi-dbl., moderate fragrance; Robichon; (Sangerhausen)

'JACQUES VINCENT', HT, rb, 1908; flowers coral-red, center golden, vigorous growth; [Mme J.W. Budde X Souv. de Catherine Guillot]; Soupert & Notting

Jacquie Williams, *see* 'MORWHEELS'

'JACQUINT', F, ab, 1998; (**Royal Wedding**™); flowers pink amber, very full, very dbl., 41 petals, 3.5–4 in., borne mostly singly, moderate fragrance; moderate, straight prickles; foliage medium, dark green, semi-glossy; compact, old-fashioned growth; [Impatient X Amber Queen]; Zary, Dr Keith W.; Bear Creek Gardens, Inc., 1998

Jacquot, F, 1973; Dorieux, Francois; (Cavriglia)

JACrain, Min, pb, 1993; (**Kleine Renate**); J&P

JACrare, F, pb; (**Sixth Sense**); J&P

'JACRAW', F, rb, 1992; (**Peppermint Twist**); flowers red/white/pink striped, open slightly, blooms borne in small, very dbl., 3–3.5 in., cupped, slight fragrance; some prickles on peduncle; foliage large, medium green, semi-glossy; medium (90–110 cms), upright, bushy growth; [Pinstripe X Maestro]; Christensen, Jack E.; Bear Creek Gardens, 1992

'JACRAZ', F, rb, 1977; (**Razzle Dazzle**); flowers red, reverse white, dbl., 25 petals, 2.5 in., slight fragrance; foliage dark, leathery; bushy growth; GM, Portland, 1978; Warriner, William A.; J&P

'JACREDEM', Min, dr, 1998; (**Jingle Bells**™); flowers dark red, light red reverse, dbl., 17–25 petals, 2 in., borne in large clusters, slight fragrance; moderate, straight prickles; foliage medium, dark green, semi-glossy; bushy, medium growth, 2 ft, border; PP9325 AOE, J&P, 1995; [Seedling X Seedling]; Zary, Dr Keith W.; Bear Creek Gardens, Inc., 1997

JACredi, HT, mr, 1997; (**Colinda**)

'JACREF', HT, mr, 1984; (**Limited Edition**); flowers large, dbl., no fragrance; foliage medium, medium green, semi-glossy; upright growth; [Unnamed seedling X Unnamed seedling]; J&P; McConnell Nurs., Inc.

'JACREG', F, lp, 1995; (**Jubilation**)

'JACRENEW', Min, mr, 1999; (**Very Cherry**™); flowers fresh, clean lavender, dbl., 17–25 petals, 2–2.5 in., borne in small clusters, slight fragrance; prickles moderate; foliage large, medium green, glossy; upright, compact, globular, medium (20–26 in.) growth; [Seedling X New Year]; Walden, John K.; Bear Creek Gardens, Inc., 1999

JACrette, HT, mr, 1997; (**Emperor**); J&P

'JACRIM', F, rb, 1986; (**Target**); flowers white with red petal edges, dbl., 25 petals, exhibition form, borne usually singly, no fragrance; straight, large and small prickles; foliage medium, medium green, semi-glossy; upright, bushy, medium growth; PP006255; [Prominent X Unnamed seedling]; Warriner, William A.; J&P

'JACRINK', Min, dp, 1988; (**Day Glow**); bud ovoid, pointed; flowers deep pink, fading lighter, medium, borne usually singly, very dbl., 60 petals, exhibition form, slight fragrance; prickles straight, short; foliage medium, dark green, matt; low growth; [Petticoat X Red Jewel]; Warriner, William A.; Bear Creek Gardens, 1989

'JACRIS', S, dp, 1987; (**Contessa**); flowers deep pink with yellow center, reverse deep pink, fading slightly, semi-dbl., 6–10 petals, flat, fast cycle, slight, spicy fragrance; straight, long, green-brown prickles; foliage medium, medium green, semi-glossy; upright, tall growth; PP006577; [Sunsprite X Seedling]; Warriner, William A.; J&P

'JACRITE', HT, yb, 1991; (**Rio Samba**™); flowers medium yellow fading to peach-pink, dbl., 15–25 petals, 2.75–3 in., exhibition form, borne singly and small clusters, slight fragrance; some prickles; foliage medium, dark green, matt; medium (110–120 cms), upright, bushy growth; PP8361; AARS, 1993; [Unnamed seedling X Sunbright]; Warriner, William A.; Bear Creek Gardens, 1993

'JACRO', F, dr, 1987; (**Bravado**); flowers, dbl., 26 petals, exhibition form, borne usually singly, slight fragrance; medium, red to brown prickles, hooked downward; foliage medium, dark, semi-glossy; medium, upright growth; PP005978; [Seedling X Gabriella]; Warriner, William A.; J&P

'JACROM', HT, mp, 1983; (**Romance**); flowers medium salmon-pink, large, dbl., 35 petals, slight fragrance; foliage large, medium green, semi-glossy; upright growth; PP005250; [Unnamed variety X Prominent]; Warriner, William A.; J&P

'JACROSE', HT, dp, 1983; (**Tribute**); bud long, pointed; flowers deep pink, loose form, large blooms borne singly, dbl., 30 petals, slight fragrance; reddish prickles, hooked down; foliage large; upright growth; [Unnamed seedling X Unnamed seedling]; Warriner, William A.; J&P

'JACROVE', HT, dr, 1999; (**Perfectly Red**™); dbl., 26–40 petals, 4–4.5 in., borne mostly singly, slight fragrance; prickles moderate; foliage medium, dark green, semi-glossy; upright (5 ft) growth; [Seedling X Love]; Zary, Dr Keith W.; Bear Creek Gardens, Inc., 1999

'JACRULAV', HRg, m, 1999; (**Wild Berry Breeze**™, Wildberry Breeze); flowers lavender-pink, veined, large petals, single, 5–11 petals, 3.5 in., borne in clusters, intense, spicy clove fragrance; round, large, orange fruit; many prickles; foliage medium, dark green, semi-glossy, deeply rugose; upright, bushy, medium (3.5 ft) growth; [Buffalo Gal X seedling]; Zary, Dr Keith W.; Bear Creek Gardens, Inc., 1999

'JACRUWHI', HRg, w, 1999; (**Wild Spice**); flowers snowy white, delicate, ruffled, single, 5–11 petals, 3.5 in., borne in small clusters, intense, clove fragrance; foliage dark green, semi-glossy, deeply rugose; upright, bushy, medium (3.5 ft) growth; [Buffalo Gal X Seedling]; Zary, Dr Keith W.; Bear Creek Gardens, Inc., 1999

'JACRYBI', Min, rb, 1999; (**Petite Perfection**™); bud long, ovoid; flowers bright red/yellow, reverse deep yellow, dbl., 17–25 petals, 2.25–2.5 in., exhibition form, borne singly and in small clusters, slight fragrance; prickles moderate; foliage medium, dark green, glossy; patio mini; upright, compact, medium (16–22 in.) growth; [Seedling X Seedling]; Walden, John K.; Bear Creek Gardens, Inc., 1999

JACsak, HT, mr, 1994; (**Riviera**); J&P

'JACSAL', F, op, 1980; (**Cherish**); bud short, flat; flowers coral-pink, dbl., 28 petals, 3 in., exhibition form, slight fragrance; foliage large, dark; compact, spreading growth; AARS, 1980; [Bridal Pink X Matador]; Warriner, William A.; J&P

JACsan, Gr, ab, 1994; (**Pearl of Joy**); J&P

'JACSASH', HT, lp, 1999; (**Rose Rhapsody**™); flowers deep, dusty pink, very dbl., 41–50+ petals, 6–6.5 in., borne mostly singly, intense, citrus fragrance; prickles moderate; foliage large, dark green, glossy; upright, spreading, tall (4.5–5 ft) growth; [Fragrant Cloud X Ingrid Bergman]; Zary, Dr Keith W.; Bear Creek Gardens, Inc., 1999

'JACSAT', LCl, dr, 1992; (**Dynamite**™); flowers bright red, dbl., 30 petals, 4–4.5 in., borne in small clusters, slight, citrus fragrance; some prickles; foliage large, dark green, glossy, resistant to powdery mildew; tall (150–185 cms), upright, spreading, arching growth; PP8741; [Seedling X Sympathie]; Warriner, William A.; Bear Creek Gardens

'JACSAY', HT, dr, 1990; (**Avanti**); bud ovoid, pointed; flowers dark red, hint of blue reverse, aging dark, dark red, urn-shaped, large, very dbl., 25–30 petals, exhibition form, slight, fruity fragrance; foliage medium to large, dark green, matt to semi-glossy; upright, bushy, tall growth; [Royalty X Samantha]; Warriner, William A.

& Zary, Keith W., 1986; Bear Creek Gardens, 1992

'JACSEDI', F, m, 1993; (**Love Potion**™, Purple Puff); flowers deep clear lavender, dbl., 26–40 petals, 3–3.5 in., borne in small clusters, raspberry fragrance, intense; some prickles; foliage medium, dark green, glossy; medium (90–100 cms), upright, spreading growth; PP9172; [Seedling X Dilly Dilly]; Christensen, Jack E.; Bear Creek Gardens, 1994

'JACSEE', HT, pb, 1998; (**Fragrant Lace**™); flowers cream and lavender pink with cream and light yellow reverse, full, large, very dbl., 26–40 petals, 5 in.–5.5 in., borne mostly singly, intense fragrance; prickles moderate; foliage large, dark green, glossy; upright growth; [Seedling X Cherry Jubilee]; Zary, Dr Keith W.; Bear Creek Gardens, 1998

'JACSERAW', F, ob, 1991; (**Orange Splash**); flowers bright orange to orange-red with white and lighter orange st, dbl., 15–25 petals, 3–3.5 in., moderate fragrance; some prickles; foliage large, dark green, glossy; medium (75–90 cms), upright, bushy growth; [Unnamed seedling X Unnamed seedling]; Christensen, Jack E.; Bear Creek Gardens, 1992

'JACSHAQ', HT, pb, 1998; (**Diana, Princess of Wales**™, The Work Continues); bud ovoid, pointed; flowers luminous pink and creamy ivory blend, very dbl., 26–40 petals, 4 in., exhibition form, borne in small clusters, moderate fragrance; prickles moderate; foliage medium, dark green, glossy; upright, 4.5–5 ft growth; [Anne Morrow Lindbergh X Sheer elegance]; Zary, Dr Keith W.; Bear Creek Gardens, Inc., 1999

'JACSHE', HT, ab, 1986; (**Summer Dream**); flowers apricot pink, blooms borne singly, dbl., 30 petals, exhibition form, slight, fruity fragrance; foliage medium, medium green, matt; upright, tall growth; [Sunshine X Unnamed seedling]; Warriner, William A.; J&P, 1987

'JACSHIP', S, mp, 1989; (**Pink Polyanna**™); bud ovoid; flowers medium pink, reverse slightly darker, aging lighter, small, dbl., exhibition form, repeat bloom, moderate, fruity fragrance; prickles small, straight, red to brown; foliage small, medium green, glossy; upright, bushy, medium growth; [Zorina X Heidi]; Warriner, William A.; Bear Creek Gardens, 1990

'JACSHO', HT, dr, 1981; (**Showstopper**); bud pointed; flowers deep red, spiral blooms borne singly or 3–4 per cluster, dbl., 33 petals, intense fragrance; long prickles; foliage large; strong, upright, bushy growth; [Unnamed seedling X Samantha]; Warriner, William A.; J&P

'JACSHOK', F, m, 1999; (**Cotillion**™); bud pointed, ovoid; flowers lavender, reverse pale lavender, very dbl., 41 petals, 4 in., rosette, borne in large clusters, intense, sweet fragrance; prickles moderate; foliage medium, dark green, glossy; upright, spreading (3.5 ft) growth; GM, Rome, 1998; [Seedling X Shocking Blue]; Zary, Dr Keith W.; Bear Creek Gardens, Inc., 1999

'JACSHUR', S, dr, 1991; (**King Crimson**); flowers large, semi-dbl., 6–14 petals, no fragrance; foliage large, dark green, semi-glossy; tall, upright, bushy growth; [Razzle Dazzle X Unnamed seedling]; Warriner, William A.; Bear Creek Gardens

'JACSIL', F, m, 1986; (**Eloquence**, Little Silver); flowers lavender, mini-flora, small, dbl., 20 petals, flat, borne singly and in clusters, slight, spicy fragrance; foliage medium, medium green, matt; upright, bushy growth; PP006010; [(Merci X Faberge) X Angel Face]; Warriner, William A.; J&P

'JACSIM', F, mp, 1991; (**Sweet Inspiration**); flowers medium pink with some cream coloration at petal base on both, dbl., 26–40 petals, 3–3.5 in., few prickles; foliage medium, medium green, matt; upright (85–100 cms), bushy growth; [Sun Flare X Simplicity]; Warriner, William A.; Bear Creek Gardens, 1993

'JACSIMPL', S, mr, 1991; (**Red Simplicity**); flowers bright red, a bit of blackening near petal edges, moderately, dbl., 15–25 petals, slight fragrance; foliage medium, medium green, semi-glossy; medium, upright, spreading growth; [Unnamed seedling X Sun Flare]; Warriner, William A. & Zary, Keith W.; Bear Creek Gardens, 1992

'JACSLING', HT, mr, 1993; (**Legacy**™); J&P

'JACSMI', Min, my, 1984; (**Smiles**); flowers medium, dbl., 35 petals, slight fragrance; foliage small, light green, semi-glossy; upright, bushy growth; PP005577; [Spanish Sun X Calgold]; Warriner, William A.; J&P

JACsnow, F, w; (**White Simplicity**®); J&P, 1991

'JACSOD', HT, mp, 1985; (**Rhapsody**); flowers medium, dbl., 20 petals, slight fragrance; foliage medium, medium green, matt; [Unnamed variety X Unnamed variety]; Warriner, William A.; J&P

'JACSOS', S, mr, 1991; (**Cherry Jubilee**); flowers light to medium red, moderately medium, dbl., 15–25 petals, no fragrance; foliage medium, medium green, semi-glossy; upright, bushy growth; [Seedling X Simplicity]; Warriner, William A.; Bear Creek Gardens

JACsowe, HT, dp; (**Sowetan Peace**); J&P, 1994

'JACSPIF', F, ly, 1986; (**Satin Doll**); flowers cream to pale apricot, [Pacifica sport]; Warriner, William A.; J&P

JACspri, F, dy, 1992; (**Eldorado**); Rose of the Year, Auckland, NZ, 1994; Christensen, Jack E.; J&P, 1992

'JACSTAND', S, rb, 1999; (**Raspberry Swirl**™); flowers red, white, pink stripe, reverse same, dbl., 17–25 petals, 2 in., borne in large clusters, apple scent fragrance, slight; prickles moderate; foliage small, dark green, glossy; upright, arching, medium (3–4 ft) growth; [seedling X Plum Dandy]; Walden, John K.; Bear Creek Gardens, Inc., 1999

'JACSTINE', HT, yb, 1992; (**Spirit of Peace**); flowers apricot yellow opening with pink tinge where sun strikes the, dbl., 26–40 petals, 3–3.5 in., slight fragrance; some prickles; foliage medium, dark green, semi-glossy; tall (170–180 cms), upright, spreading growth; [Pristine X Unnamed seedling]; Warriner, William A.; Bear Creek Gardens

JACsto, HT, mr, 1993; (**Future Award**); J&P

'JACSTOP', HT, dr, 1990; (**Salsa**); bud ovoid, pointed; flowers urn-shaped, large, very dbl., 30–35 petals, exhibition form, borne singly, moderate, damask fragrance; few prickles; foliage medium, dark green, semi-glossy; upright, tall growth; PP006971; [Showstopper X Unnamed seedling]; Warriner, William A.; Bear Creek Gardens

'JACSUN', Min, dy, 1981; (**Bojangles**); flowers deep yellow, small, dbl., 20 petals, no fragrance; foliage small, light green, semi-glossy; upright, bushy growth; PP004990; [Spanish Sun X Calgold]; Warriner, William A.; J&P, 1983

JACsunov, F, lp; (**Super Nova**); J&P, 1997

'JACSUR', F, op, 1998; (**Bill Warriner**™); bud pointed, ovoid; flowers coral pink, ruffled edges, medium blooms, very dbl., 26–40 petals, 3.5–4 in., borne in large clusters, sweet fragrance, slight; prickles moderate, straight; foliage medium, dark green, semi-glossy; compact growth (3 ft); PP9494; [Sun Flare X Impatient]; Warriner, William A. & Zary, Keith W.; Bear Creek Gardens, Inc., 1997

'JACSUYEL', S, my, 1999; (**Golden Bounty**™); dbl., 17–25 petals, 2.5 in., borne in large clusters, slight fragrance; few prickles; foliage medium, dark green, semi-glossy to glossy; upright, arching, medium (3–4 ft) growth; [Sun Flare X seedling]; Zary, Dr Keith W.; Bear Creek Gardens, Inc., 2001

'JACTAFL', F, lp, 1998; (**Bridal Shower**™); flowers light pink, medium, very dbl., 26–40 petals, 4 in., borne in small clusters, slight fragrance; prickles moderate; foliage medium, dark green, glossy; upright, bushy, tall (4 ft) growth; [Seedling X Sunflare]; Zary, Dr Keith W.; Bear Creek Gardens, Inc., 1999

'JACTAN', LCl, r, 1986; (**Butterscotch**); flowers tannish-orange, double to loose, dbl., 25 petals, cupped, blooms in clusters of 3–5, slight fragrance; no fruit; foliage medium, medium green, semi-glossy; PP005895; [(Buccaneer X Zorina) X Royal Sunset]; Warriner, William A.; J&P

'JACTIG', HT, ab, 1974; (**Antigua**); bud ovoid; flowers large, dbl., exhibition form, slight fragrance; foliage leathery; vigorous, upright, bushy growth; GM, Geneva, 1972; [South Seas X Golden Masterpiece]; Warriner, William A.; J&P

'JACTIGER', F, ob, 1991; ('JACTANG', **Tiger Tail**); flowers deep orange, white or cream colored stripes, cream colored r, dbl., 26–40 petals, no fragrance; foliage medium, medium green, glossy; medium, bushy growth; [Matangi X Pinstripe]; Christensen, Jack E.; Bear Creek Gardens, 1992

'JACTOP', HT, mr, 1992; (**Legend**); flowers, very dbl., 3–3.5 in., exhibition form, borne mostly singly, intense fragrance; very small prickles on peduncle; foliage medium, dark green, semi-glossy; tall (120–140 cms), upright, bushy growth; PP006092; [Grand Masterpiece X Unnamed seedling]; Warriner, William A.; Bear Creek Gardens

'JACTOU', HT, dy, 1992; (**Midas Touch**(TM)); bud urn-shaped; flowers bright, non-fading yellow, dbl., 15–25 petals, 3.5–4 in., exhibition form, borne mostly singly, moderate, musk fragrance; small prickles on peduncle; foliage large, medium green, matt; tall (150–160 cms), upright, bushy growth; PP8706; [Brandy X Friesensohne]; Christensen, Jack E.; Bear Creek Gardens, 1994

'JACTRED', HT, mr, 1988; (**American Spirit**, Medal of Honor); flowers large blooms, dbl., 35 petals, exhibition form, borne usually singly, no fragrance; foliage medium, medium green, semi-glossy; upright, tall growth; PP006083; [Seedling X American Pride]; Warriner, William A.; J&P

'JACTRIG', HT, m, 1999; (**Fragrant Rhapsody**(TM)); flowers light lavender blend, small, dbl., 20–40 petals, borne mostly singly, intense fragrance; prickles moderate; foliage medium, medium green, glossy; upright, spreading, medium (4 ft) growth; [Intrigue X seedling]; Zary, Dr Keith W.; Bear Creek Gardens, Inc., 1999

'JACTRO', HT, w, 1985; (**Sheer Bliss**); flowers white with pink center, large, dbl., 35 petals, exhibition form, borne singly, moderate, spicy fragrance; medium brown prickles; foliage medium, medium green, matt; medium, upright, bushy growth; PP006282; AARS, 1987 GM Japan, 1984; [White Masterpiece X Grand Masterpiece]; Warriner, William A.; J&P, 1987

'JACTU', HT, pb, 1985; (**Headliner**); flowers petals white, blending to deep pink at edges, large, dbl., 40 petals, exhibition form, slight fragrance; foliage large, medium green, glossy; upright growth; PP005340; [Love X Color Magic]; Warriner, William A.; J&P

'JACTUTTI', Min, yb, 1991; (**Tutti-Frutti**); bud ovoid, pointed; flowers yellow with red-orange stripes, medium, dbl., 25–30 petals, cupped, slight fragrance; foliage medium, dark green, semi-glossy; bushy, spreading, medium growth; [Fool's Gold X Pinstripe]; Christensen, Jack E., 1984; Bear Creek Gardens, 1991

'JACTWIN', Gr, rb, 1980; (**Love**); bud short, pointed; flowers bright scarlet red, reverse silvery white, dbl., 35 petals, 3.5 in., exhibition form, borne mostly singly, rose fragrance, slight; foliage dark green, glossy, new growth is red; long stems; upright growth; AARS, 1980 GM, Portland, 1980; [? X Redgold]; Warriner, William A.; J&P

'JACUM', F, m, 1982; (**Intrigue**); flowers reddish purple, large, dbl., 20 petals, intense fragrance; foliage medium, dark, semi-glossy; PP005002; AARS, 1984; [White Masterpiece X Heirloom]; Warriner, William A.; J&P, 1984

'JACUP', Min, mr, 1984; (**Upstart**); flowers small, semi-dbl., slight fragrance; foliage small, medium green, semi-glossy; upright growth; [Merci X Fire Princess]; Warriner, William A.; J&P

JACurp, F, m, 1997; (**Night Star**); J&P

'JACUTE', Min, mp, 1988; (**Confection**); bud ovoid, pointed; flowers pink with yellow to cream base, medium, very dbl., exhibition form, borne usually singly, slight fragrance; prickles straight to slightly hooked, light green; foliage medium, dark green, matt; micro-mini low, bushy growth; PP006518; [Seedling X Seedling]; Warriner, William A.; Bear Creek Gardens

'JACVAL', HT, lp, 1998; (**Ultimate Pink**(TM)); bud pointed, ovoid; flowers large, full, light pink, very dbl., 26–40 petals, 4.5–5 in., borne mostly singly, sweet fragrance, slight; prickles moderate, straight; foliage large, medium green, dull; tall, upright, 4.5 ft, vigorous growth; Rose of the Year, J&P, 1999; [Seedling X Fragrant Memory]; Zary, Dr Keith W.; Bear Creek Gardens, Inc., 1999

'JACVEN', F, w, 1974; (**Evening Star**®); flowers white, base shading pale yellow, large, dbl., exhibition form, slight fragrance; foliage large, dark, leathery; vigorous, upright, bushy growth; GM, Belfast, 1977 GM, Portland, 1977; [White Masterpiece X Saratoga]; Warriner, William A.; J&P

'JACVEP', HT, pb, 1992; (**Corina**); flowers salmon pink with lighter almost cream reverse, yellow crescent at petal base, dbl., 26–40 petals, 3–3.5 in., exhibition form, moderate fragrance; few prickles; foliage large, dark green, matt; medium (150 cms), upright growth; [Bridal Pink X Kardinal]; Warriner, William A. & Zary, Keith W.; Bear Creek Gardens, 1993

JACvet, HT, dr; (**Velvet Treasure**); J&P, 1992

JACvoo, HT, w, 1996; (**First Affair**); J&P

'JACWADE', Min, w, 1992; (**Snow Shower**); flowers snow white, small, opens flat, very dbl., 1.5 in., flat, borne in small clusters, no fragrance; some prickles; foliage small, medium to dark green, glossy; low (15–20 cms), spreading 60–75 cms across; groundcover; PP9374; [Immensee X Roller Coaster]; Warriner, William A. & Zary, Keith W.; Bear Creek Gardens, 1993

'JACWHIM', Min, w, 1988; (**Shy Girl**); bud ovoid, pointed; flowers medium, very dbl., 80 petals, exhibition form, borne usually singly and in sprays of 2–4, slight fragrance; no fruit; prickles straight, short; foliage medium, dark green, semi-glossy; upright, spreading, low growth; PP006514; [Petticoat X Red Minimo]; Warriner, William A.; Bear Creek Gardens

'JACWHIP', S, w, 1993; (**Snow Angel**); flowers white with a light pink tone and a hint of yellow near petal, very dbl., 1.5–2.75 in., slight fragrance; some prickles; foliage small, dark green, semi-glossy; medium (5 ft), upright, bushy growth; [Pink Pollyanna sport]; Bear Creek Gardens; Bear Creek Gardens, 1993

'JACWHY', F, w, 1970; (**Bridal White**(TM)); flowers ivory-white, [Bridal Pink sport]; Warriner, William A.; J&P

'JACWIG', Min, m, 1998; (**Wild Plum**(TM)); flowers lavender blend, darker reverse, dbl., 17–25 petals, 1.5–2 in., borne in large clusters, slight fragrance; prickles few, straight; foliage medium, dark green, glossy; bushy, tall, hedge type, 3 ft growth; [Winsome X Elegance]; Zary, Dr Keith W.; Bear Creek Gardens, Inc., 1995

JACwiht, HT, w, 1995; (**Cathedral Peak**); J&P

'JACWOTTE', HT, ab, 1999; (**Vanilla Perfume**(TM)); bud plump, ovoid; flowers light cream apricot/pink, reverse light apricot, dbl., 26–40 petals, 4–4.5 in., borne mostly singly and in small clusters, intense, sweet, spicy vanilla fragrance; prickles moderate; foliage medium, dark green, semi-glossy to glossy; upright, tall (5 ft) growth; [Anne Morrow Lindbergh X Seedling]; Zary, Dr Keith W.; Bear Creek Gardens, Inc., 1999

'JACYAP', HT, pb, 1993; (**Anne Morrow Lindbergh**); flowers pink, white, yellow blend, dbl., 26–40 petals, 3–3.5 in., borne mostly singly, moderate, some fragrance; some prickles; foliage medium to large, medium green, semi-glossy; tall (150–160 cms), upright growth; [Seedling X Seedling]; Warriner, William A.; Bear Creek Gardens, 1994

'JACYEF', Gr, dy, 1989; (**Shining Hour**(TM)); bud ovoid; flowers deep, bright yellow, medium, borne usually singly and in sprays, dbl., 33 petals, exhibition form, moderate, fruity fragrance; prickles fairly long, hooked downward, red to yellow; foliage large, dark green, semi-glossy; upright, bushy, medium growth; AARS, 1991; [Sunbright X Sun Flare]; Warriner, William A.; Bear Creek Gardens, 1991

'JACYELSH', S, dy, 1998; (**Yellow Simplicity**); flowers deep yellow, medium-sized blooms, 12–16 petals, 3 in., borne in small clusters, slight fragrance; prickles moderate; foliage medium, dark green, glossy; upright, tall hedge growth; [Seedling X Seedling]; Zary, Dr Keith W.; Bear Creek Gardens, Inc., 1997

'JACYEM', HT, my, 1993; (**Santa Fe**); flowers rich, medium yellow, blooms, 3–3.5 in., moderate fragrance; few prickles; foliage large, dark green, matt; tall (180–210 cms), upright growth; [Emblem X Seedling]; Zary, Dr Keith W.; Bear Creek Gardens, 1993

'JACYEPAT', S, dy, 1992; (**Yellow Jacket**); flowers bright yellow, loose, dbl., 26–40 petals, 1.5–2.75 in., borne mostly singly or small clusters, slight fragrance; some prickles; foliage medium, dark green, glossy, very disease resistant; patio; very vigorous, medium (75–90 cms), upright, bushy growth; [Unnamed seedling X Unnamed seedling]; Christensen, Jack E.; Bear Creek Gardens, 1992

'JACYES', HT, dy, 1998; (**Henry Fonda**™); bud pointed, ovoid; flowers deep yellow, pointed, dbl., 20–25 petals, 4.5–5 in., exhibition form, borne most singly, slight fragrance; foliage medium, dark green, glossy; tall growth; PP9390; Zary, Dr Keith W.

JACyesp, F, my, 1997; (**Golden Galaxy**); J&P

'JACYIM', Min, dy, 1989; (**Sunsplash**™); bud ovoid, pointed; flowers deep yellow, aging pale yellow, dbl., 43 petals, exhibition form, borne in sprays of 18–21, slight fragrance; prickles long, slightly hooked downward, red to green-tan; foliage medium, medium green, very glossy, attractive; tall, upright, spreading growth; PP007127; [Rise 'n' Shine X Sun Flare]; Warriner, William A.; Bear Creek Gardens

'JACYIMP', F, yb, 1999; (**Honey Bouquet**™); bud pointed, ovoid; flowers light yellow, hint of pink, reverse light yellow, dbl., 20–40 petals, 4.5–5 in., borne in small clusters, intense, sweet fragrance; foliage medium, dark green, semi-glossy; upright, spreading, medium, 3.5 ft growth; [Seedling X Amber Queen]; Zary, Dr Keith W.; Bear Creek Gardens, Inc., 1999

'JACYO', HT, ob, 1989; (**Dynasty**™); bud pointed; flowers bright orange with yellow blending at petal base, aging to coral, dbl., 30 petals, cupped, no fragrance; prickles hooked down slightly, red to yellow-green; foliage medium, medium green, semi-glossy; upright, spreading, tall growth; PP006443; [Seedling X Seedling]; Warriner, William A.; Bear Creek Gardens, 1991

'JACZAP', F, ob, 1999; (**Outrageous**™); dbl., 26–40 petals, 3.5–4 in., borne in small clusters, moderate fragrance; prickles moderate; foliage medium, dark green, dull; bushy, medium (3.5 ft) growth; [Summer Fashion X seedling]; Zary, Dr Keith W.; Bear Creek Gardens, Inc., 1999

'JACZOR', Gr, dp, 1999; (**Fame!**™); flowers deep shocking pink – almost light red, very dbl., 30–35 petals, 5 in., exhibition form, borne in small clusters, slight fragrance; foliage large, dark green; shrubby, full bush, tall growth; AARS, 1998; [Tournament of Roses X Zorina]; Zary, Dr Keith W.; Bear Creek Gardens, Inc., 1998

'JACZOTTA', F, pb, 1999; (**Sorbet Bouquet**™); flowers rose pink, reverse light yellow, dbl., 26–40 petals, 4 in., borne in large clusters, slight fragrance; prickles moderate; foliage medium, dark green, glossy; bushy, medium (3.5 ft) growth; [Tournament of Roses X Seedling]; Zary, Dr Keith W.; Bear Creek Gardens, Inc., 2000

JADecosill, HT, dp, 1997; (**Ensa de Rennes**); Adam, M.

Jadis, *see* 'JACDIS'

Jaen, HT, pb

'JÄGERBATAILLON', sp, (strain of R. canina), lp; almost thornless; once popular as an understock; Klinken

Jake McIlroy, HT, mp, 1997; Brundrett

'JALACT', Min, dr, 1993; (**Opening Act**); single, 1.5 in., borne in small clusters, no fragrance; some prickles; foliage medium, dark green, glossy; medium (30–35 cms), upright growth; [Anytime X Black Jade]; Jalbert, Brad; Select Roses, 1994

'JALCORAL', Min, op, 1996; (**Super Cascade Coral**); flowers coral, blooms borne in large clusters, very dbl., 1.5 in., slight fragrance; few prickles; foliage medium, dark green, glossy; spreading, medium (12 in.), cascading growth; [Orange Honey X Sexy Rexy]; Jalbert, Brad; Select Roses, 1997

'JALPINK', Min, lp, 1996; (**Heaven Scent Pink**); flowers wide blooms, dbl., 26–40 petals, 2 in., flat, intense fragrance; few prickles; foliage medium, medium green, dull, reddish; upright, tall (24 in.) growth; [Rise 'n' Shine X Rosemary Harkness]; Jalbert, Brad; Select Roses, 1997

'JALRAY', Min, pb, 1995; (**Reiko**); flowers medium pink, lighter pink reverse, dbl. (15–25 petals), medium (1 in.) blooms borne in small clusters; fragrant; some prickles; foliage medium, medium green, glossy; medium (12 in.), compact, bushy growth; [Winsome X Springwood® Gold]; Jalbert, Brad; Select Roses, 1995

'JALSTAR', Min, pb, 1995; (**Rising Star**); flowers cream with strong pink blend edges, full (26–40 petals), large (1–2 in.) blooms borne mostly singly, slight fragrance; some prickles; foliage medium, medium green, semi-glossy; tall (18 in.) upright, bushy growth; [Sans Souci X Pink Petticoat]; Jalbert, Brad; Select Roses, 1995

'JALSUN', Min, my, 1995; (**Sun Chariot**); flowers medium yellow, bright gold stamens, medium, dbl. (15–25 petals), medium blooms borne in small clusters, fragrant; numerous prickles; foliage medium, dark green, matt; medium, upright, tall growth; [Pink Petticoat X Bright Smile]; Jalbert, Brad; Select Roses, 1996

'JALWED', Min, mp, 1995; (**Summer Wedding**); flowers medium pink, full (26–40 petals), medium blooms borne in small clusters, slight fragrance; few prickles; foliage medium, medium green, semi-glossy; compact, bushy, medium growth; [Maurine Neuberger X Sexy Rexy]; Jalbert, Brad; Select Roses, 1995

Jaltinskii Suvenir, F, ly, 1956; flowers large, dbl., moderate fragrance; Klimenko, V. N.; (Sangerhausen)

'JALWINE', Min, mr, 1993; (**Cherry Wine**); dbl., 26–40 petals, 1.5 in., borne mostly singly, slight fragrance; some prickles; foliage medium, medium green, matt; medium (30–35 cms), spreading growth; [Dee Bennett X Winsome]; Jalbert, Brad; Select Roses, 1994

'JALWOOD', Min, r, 1995; (**Sandalwood**); flowers russet with lighter edges, blooms borne in small clusters, dbl., 15–25 petals, 1 in., no fragrance; some prickles; foliage medium, medium green, glossy; medium (12 in.), compact, bushy growth; [Blushing Blue X Springwood Gold]; Jalbert, Brad; Select Roses

Jam Session, *see* 'ZIPJAM'

'JAMAICA', HT, mr, 1965; bud ovoid; flowers cherry-red, large, semi-dbl., cupped, moderate fragrance; foliage dark, glossy, leathery; vigorous, upright growth; [(Charlotte Armstrong X Floradora) X Nocturne]; Lindquist; Howard Rose Co.

'JAMBOREE', F, rb, 1964; flowers cherry-red, reverse lighter, borne in clusters, dbl., 26 petals, 1.5–2 in., flat; foliage glossy, light green; free growth; [Masquerade X ?]; Gregory

'JAMES APPLEBY', F, ob, 1963; flowers orange-scarlet flecked deep crimson, medium, borne in clusters, semi-dbl.; vigorous growth; [Orangeade sport]; Wood

James Biddle, *see* 'WILJAME'

James Bond, *see* **'JAMES BOND 007'**

'JAMES BOND 007', HT, op, 1966; (James Bond); flowers coral-pink, pointed, foliage glossy; bushy growth; [? X Fragrant Cloud]; Tantau, Math.

James Bougault, *see* **'JAMES BOURGAULT'**

'JAMES BOURGAULT', HP, w, 1887; (James Bougault); flowers white, shaded rose, Renault

'JAMES FERRIS', HT, w, 1927; flowers creamy white, vigorous growth; Hall

'JAMES GIBSON', HT, mr, 1928; bud pointed; flowers crimson-scarlet, large, dbl., exhibition form, slight fragrance; foliage dark, leathery; vigorous, bushy growth; GM, NRS, 1929; McGredy

James Mason, HGal, mr, 1982; Beales, Peter

'**James Mitchell**', M, dp; flowers pink, small, dbl., early bloom; very heavily mossed stems; Verdier, E., 1861; E. Verdier

James Pereire, *see* MEInorep

'**James Rea**', HT, dp, 1930; bud pointed; flowers rich carmine or rose-pink, very large, dbl., exhibition form, intense fragrance; foliage light, leathery; vigorous, bushy growth; GM, NRS, 1929; McGredy

'**James Smile**'®, HT, pb, 1987; flowers china-pink, reverse pink-yellow blend, loose, medium, borne, semi-dbl., 40 petals, cupped, moderate, spicy fragrance; rounded, medium, pink-red fruit; rare, small, brown prickles; foliage medium, medium green, semi-glossy; upright, medium growth; [Sea Pearl sport]; Rogin, Josip

'**James Veitch**', M, m, 1865; flowers violet-slate shaded fiery red, blooms in corymbs, dbl.; moderately vigorous growth; Verdier, E.

'**James Walley**', HT, ab, 1923; flowers apricot and fiery salmon, large, dbl., moderate fragrance; foliage olive-green, leathery; vigorous growth; [Ophelia X Unnamed seedling]; Easlea

Jamestown, Gr, op, 1999; flowers soft peach-pink, semi-dbl., moderate fragrance; Williams, J. Benjamin

Jamie™, *see* 'MOGajam'

Jamie Alexander, S, lp; [Ko's Yellow X Eyepaint]; Hannemann, F., 1995; The Rose Paradise; (Weatherly, L.)

Jamie's Love, *see* 'GELjam'

'**Jan Abbing**', HT, rb, 1933; bud pointed; flowers salmon-red shaded yellow, large, cupped, slight fragrance; foliage leathery, dark; vigorous growth; [Columbia X Étoile de Hollande]; Tantau

'**Jan Böhm**', HP, mr, 1928; flowers velvety fiery red, dbl., 40 petals, recurrent bloom, moderate fragrance; vigorous growth; [Hugh Dickson X King George V]; Böhm, J.

'**Jan Guest**', HT, pb, 1975; flowers carmine-pink with yellow reverse, dbl., 43 petals, 4 in.; foliage glossy; vigorous, upright growth; [Fragrant Cloud X Irish Gold]; Guest

'**Jan H. Meyer**', HT, mr, 1954; flowers cinnabar-red, well formed, large; vigorous growth; [Tawny Gold X Gaudia]; Leenders, M.

'**Jan Hus**', HT, lp, 1933; flowers large, dbl., moderate fragrance; Böhm, J.; (Sangerhausen)

'**Jan Spek**'®, F, dy, 1966; dbl., 44 petals, 3 in., flat, blooms in clusters, slight fragrance; foliage dark, glossy; GM, Belfast, 1968 GM, The Hogue, 1970; [Clare Grammerstorf X Faust]; McGredy, Sam IV; McGredy

'**Jan Steen**', HT, mr, 1923; flowers brilliant scarlet-red, semi-dbl., slight fragrance; [Mev. Dora van Tets X Gruss an Dresden]; Spek

Jan Steen, S, rb, 1997; bud high; flowers ivory and red striped, single; upright growth; Williams, J. Benjamin

'**Jan van Riebeeck**', HT, m, 1952; flowers carmine, large, dbl., moderate fragrance; vigorous growth; Leenders, M.

Jan Vermeer, F, m, 1997; flowers purple/burgundy blend, borne in sprays, Williams, J. Benjamin

'**Jan Wellum**', Min, dr, 1986; (Vancouver Centennial); flowers large blooms borne singly and in sprays of 3–5, dbl., 20 petals, fruity fragrance; small, globular fruit; small, straight, near-white prickles; foliage medium, medium green, matt; spreading growth; [Unnamed Shrub X Dwarfking '78]; Fischer, C.&H.; Alpenflora Gardens

Jana, HT, dy, 1976; flowers large, dbl.; Cocker; (Sangerhausen)

'**Janal**', HT, dp, 1978; bud pointed; flowers deep pink, dbl., 55 petals, exhibition form, intense fragrance; hooked red prickles; foliage bronze-red to dark green; vigorous growth; [Charles Mallerin X (Duet X Kordes' Perfecta)]; Dawson, George; Rainbow Roses

'**Jane**', HT, ob, 1956; flowers coppery orange suffused pink, well formed, dbl., 24 petals, 3.5 in., moderate fragrance; foliage dark; vigorous, upright growth; [Signora X Mrs Edward Laxton]; Mee; Sanday

Jane Asher, *see* 'PEApet'

'**Jane Carrel**', HT, yb, 1940; bud pointed; flowers yellow shaded orange, medium, dbl., moderate fragrance; very vigorous, bushy growth; Gaujard

Jane Eyre, *see* 'MEHpark'

Jane Isobella Linton, T, pb; L'Hay

'**Jane Jackson**', F, ob, 1976; flowers strawberry-orange, dbl.; low, vigorous, bushy growth; [Tip Top sport]; Jackson, J.R.

Jane Lathrop Stanford, HT, lp

'**Jane Lazenby**', F, mp, 1959; flowers rose-pink, dbl., 25 petals, 3.5 in., flat, moderate fragrance; foliage dark; very vigorous, bushy growth; [Alain X Mme Henri Guillot]; McGredy, Sam IV; McGredy

'**Jane Pauley**', HT, ob, 1992; flowers orange, reverse orange, aging orange pink, over large blooms borne, dbl., 35 petals, 6.5 in., exhibition form, moderate fragrance; foliage large, medium green, semi-glossy; upright, medium growth; [Elizabeth Taylor X Fortuna]; Weddle, Von C.; Hortico Roses, 1993

'**Jane Piekarski**', HT, w, 1994; very dbl., 3–3.5 in., borne mostly singly, intense fragrance; many prickles; foliage medium, medium green, matt; medium (24 in.), upright growth; [Honor X John F. Kennedy]; Hoshall, Howard

'**Jane Probyn**', HT, dp, 1978; flowers deep pink; [Red Devil sport]; Anderson, K.

'**Jane Rogers**', HT, ab, 1991; flowers large blooms borne singly, sometimes in small clusters, dbl., 26–40 petals, slight fragrance; foliage medium, medium green, semi-glossy; medium, spreading growth; [Fragrant Cloud X Diamond Jubilee]; Mander, George

'**Jane Thornton**', HT, rb; flowers velvety crimson shaded maroon, intense fragrance; Bees, about 1940

'**Janet**', HT, ab, 1915; flowers golden fawn, shaded copper and rose, fading, dbl., moderate fragrance; GM, NRS, 1916; Dickson, A.

Janet B. Wood, Ayr, w, 1990; Beales, Peter

Janet Bebb, *see* 'MEHgoldie'

Janet Carnochan, HT, w, 1996; Pallek

'**Janet Frazer**', F, op, 1937; flowers shrimp-pink and yellow, blooms in clusters, semi-dbl., 3.5 in., slight fragrance; foliage light green; [Mme Leon Cuny X (Orange Sweetheart X Cinnabar)]; McGredy

'**Janet Morrison**', Cl HT, dp, 1936; flowers deep pink, large blooms, semi-dbl., moderate fragrance; long stems; vigorous growth; [Black Boy X Seedling]; Clark, A.; NRS Victoria

'**Janet's Pride**', HEg, pb, 1892; (Clementine); flowers white, edged carmine-rose, semi-dbl., non-recurrent; vigorous growth; Paul; Paul

'**Janette Murray**', HT, op, 1985; flowers orange pink, large blooms borne singly, dbl.; 35 petals, slight fragrance; foliage large, medium green, semi-glossy; vigorous growth; [Daily Sketch seedling X Montezuma]; Bell, Ronald J.; Brundrett & Sons

'**Janice**', Min, mp, 1971; bud ovoid; flowers small, dbl., foliage small, glossy, leathery; vigorous, dwarf, upright, bushy growth; [(R. wichurana X Floradora) X Eleanor]; Moore, Ralph S.; Sequoia Nursery

Janice Heyes, HT, lp, 1986; Dawson

'**Janice Tellian**', Min, op, 1979; bud pointed; flowers light coral-pink, dbl., 40 petals, 1 in., exhibition form, slight fragrance; foliage small; dwarf, bushy, compact growth; [Fairy Moss X Fire Princess]; Moore, Ralph S.; Sequoia Nursery

'**Janida**', HT, rb, 1955; bud long, pointed, blood-red veined maroon; flowers orange, dbl.; [Crimson Glory X Baby Chateau]; Robichon

Janina®, *see* TANija

'JANINE ASTLE', F, mp, 1971; flowers clear pink, dbl., 22 petals, 3 in., exhibition form, slight fragrance; foliage bluish green; vigorous, upright growth; [Charlotte Elizabeth X Grand Slam]; Hunt

Janine Herholdt, HT, ab, 1989; Herholdt

'JANINE VIAUD-BRUANT', HT, mr, 1910; flowers large, dbl., moderate fragrance; Viaud-Bruant; (Sangerhausen)

'JANNA', Min, pb, 1970; bud pointed; flowers pink, reverse white, small, dbl.; foliage leathery; dwarf, bushy growth; [Little Darling X (Little Darling X (R. wichurana X Miniatureseedling))]; Moore, Ralph S.; Sequoia Nursery

Janos, Pol, or; flowers medium, dbl.; VEG; (Sangerhausen)

Jan's Wedding, *see* 'ADAJAN'

'JANTAR', HT, dr, 1966; bud elongated; flowers dark crimson, shaded darker, large, vigorous, upright growth; Grabczewski

'JANTZEN GIRL', Gr, mr, 1961; bud ovoid; dbl., 4–5 in., exhibition form, blooms in clusters, slight fragrance; foliage glossy; upright growth; [Carrousel X (Chrysler Imperial X Seedling)]; Von Abrams; Peterson & Dering

Japanese Rose, *see* **R. RUGOSA**

'JAPONICA', A strain of R. multiflora, probably indistinguishable from it; popular as an understock

Japonica, *see* **'MOUSSEUX DU JAPON'**

'JAPONICA THORNLESS'; prickle-free strain of R. multiflora

Jaquenetta, S, ab, 1983; Austin, David

Jardinero Ortiz®, *see* DOTraner

Jardins de Bagatelle®, *see* 'MEIMAFRIS'

Jardins de France, *see* MEIzebul

Jardins de Valloires, Pol, dp, 1994; Eve, A.

Jardins de Viels Maisons, *see* MASframb

Jardins de Villandry, *see* DELrovrai

Jaris Hudson, Min, mp

Jarlina, *see* 'POULENA'

'JARVIS BROOK', HT, rb, 1928; flowers carmine, reverse orange, dbl., moderate fragrance; Low

Jasena, HT, dy, 1983; flowers large, dbl., slight fragrance; Urban, J.; (Sangerhausen)

'JASLOK', HT, m, 1999; flowers light mauve with dark mauve stripes, reverse lighter, very dbl., 41 petals, 4–5 in., borne mostly singly, slight fragrance; prickles moderate; foliage large, medium green, semi-glossy; compact, medium (2.5–3.5 ft) growth; [Blue Ocean sport]; Chiplunkar, C. R., 1992; KSG's Roses, 1992

Jasnaja Poljana, HT, pb, 1958; flowers salmon-pink with light yellow, large, very dbl., moderate fragrance; Shtanko, E.E.; (Sangerhausen)

Jason, *see* 'SHERICE'

Jaunatella, HT, yb

'JAUNÂTRE', HSem, ly; flowers yellowish white, moderate fragrance

'JAUNE BICOLOR', S, yb, 1633; flowers yellow streaked red, medium, single, [R. foetida bicolor sport]

'JAUNE DESPREZ', N, yb, 1830; (Desprez à Fleur Jaunes, Noisette Desprez); flowers warm yellow shaded peach and apricot, dbl., flat, moderate fragrance; vigorous (to 20 ft) growth; [Blush Noisette X Parks' Yellow Tea-scented China]; Desprez

'JAUNE D'OR', T, ob; flowers coppery yellow, large, dbl., globular; Oger, 1863

'JAVA', F, ob, 1955; bud ovoid; flowers orange-red, medium, borne in clusters of 4–5, dbl., 40–45 petals, slight fragrance; foliage bronze; vigorous, upright, bushy growth; GM, Geneva, 1954; [Francais X Unnamed seedling]; Mallerin, C.; EFR

'JAWAHAR', HT, w, 1980; flowers creamy white, blooms borne 2–6 per cluster, dbl., 47 petals, exhibition form, intense fragrance; straight, brown prickles; foliage light green, glossy; vigorous, bushy growth; [Sweet Afton X Delhi Princess]; Division of Vegetable Crops and Floriculture

Jawarosa Mawa, HT, or; flowers large, dbl.; (Sangerhausen)

'JAY JAY', HT, w, 1971; flowers white, edged pink, very large, dbl., exhibition form, slight fragrance; foliage large, dark, leathery; very vigorous, bushy growth; [Peace sport]; Kern Rose Nursery

'JAYANC'™, Min, mp, 1997; (**Ancestry**); flowers semi double, medium, semi-dbl., 8–14 petals, borne in small clusters, moderate fragrance; foliage medium, medium green, glossy; spreading, bushy, medium (30 cms) growth; [Seedling X Party Girl]; Jellyman, J.S.

'JAYART'™, Min, pb, 1997; (**Arthur Cox**); flowers double, medium, dbl., 15–25 petals, borne in small clusters, slight fragrance; foliage medium, medium green, semi-glossy; compact, medium growth; [Cotswold Gold X Peaches 'n' Cream]; Jellyman, J.S.

Jayatsen, HT, pb, 1998; K&S

'JAYBO', Min, rb, 1988; flowers very bright red, yellow at base, fading to pink, urn-shaped, semi-dbl., 22 petals, slight, fruity fragrance; prickles long, pointed, pink; foliage medium, dark green, glossy; bushy, medium growth; [Rise 'n' Shine X Unnamed seedling]; Bridges, Dennis A.; Bridges Roses

'JAYCOT', Min, ob, 1992; (**Cotswold Gold**); flowers orange, blooms borne in small clusters, semi-dbl., 6–14 petals, 1.5 in., slight fragrance; few prickles; foliage small, medium green, semi-glossy; low (20 cms), bushy growth; [Tony Jacklin X Judy Fischer]; Jellyman, J.S., 1993

'JAYDON', F, mr, 1998; (**Beverley Stoop**); flowers velvety red, dbl., 15–25 petals, borne in clusters, slight fragrance; foliage medium, medium green, semi-glossy; upright, compact, $2\frac{1}{2}$ ft growth; [Seedling X (Tony Jacklin X Andrea)]; Jellyman, J.S.

'JAYEMM', F, mr, 1992; (**Emma Kate**); flowers light red with lighter reverse, full, blooms borne 4–10+ per cluster, dbl., 35 petals, moderate, fruity fragrance; foliage medium, dark green, glossy; upright, bushy, medium growth; [Tony Jacklin X Cairngorm]; Jellyman, J.S., 34001

'JAYGUP', HT, op, 1999; (**Diana H. Gupta**); flowers orange-pink, reverse pale lemon yellow with faint white striping, dbl., 26–40 petals, 3 in., borne in small clusters, slight fragrance; few prickles; foliage medium, medium green, semi-glossy; upright, bushy, medium (3 ft) growth; [Gavotte X Bill Temple]; Jellyman, J.S.

'JAYKOO', F, or, 1997; (**Kooshti-Bok**™); flowers single, medium, single, 4–7 petals, borne in small clusters, slight fragrance; foliage medium, dark green, semi-glossy; upright, spreading, medium to tall (3.5–4 ft) growth; [(Tony Jacklin X Seedling) X (Tony Jacklin X Andrea)]; Jellyman, J.S.

'JAYMARJ', HT, pb, 1997; (**Marjorie Reid**™); flowers double, small, dbl., 15–25 petals, slight fragrance; foliage small, dark green, glossy; low (25 cms) growth; [Seedling X Wee Barbie]; Jellyman, J.S.

'JAYMATT', HT, pb, 1997; (**The Matthew**™); flowers full, medium, very dbl., 26–40 petals, slight fragrance; few prickles; [Silver Jubilee X Pres. Petts]; Jellyman, J.S.

'JAYMER', f, ab, 1994; (**Merrie**); flowers apricot, pale pink, double, blooms borne 6–20 per cluster, 30 petals, moderate fragrance; foliage medium, dark green, semi-glossy; medium, upright, bushy growth; [Minnie Pearl X Seedling]; Jellyman, J.S.

'JAYMOLL', F, pb, 1997; (**Molly Beckley**™); flowers semi double, small, semi-dbl., 8–14 petals, slight fragrance; foliage small, dark green, glossy; spreading, low (1 ft 6in.) growth; [(Tony Jacklin X Andrea) X Wee Barbie]; Jellyman, J.S.

Jayne Austin, *see* 'AUSBREAK'

'JAYSIL', F, dp, 1997; (**Precious Gift**™); flowers double, medium, dbl., 15–25 petals, borne in small clusters, slight fragrance; few prickles; foliage medium, medium green, semi-glossy; upright, bushy, medium (20 cms) growth; [Silver Jubilee X Seedling]; Jellyman, J.S.

'JAYSUN', F, pb, 1992; (**Cotswold Sunset**); flowers gold, pink edge, pale pink, moderately blooms borne in small clusters, dbl., 15–25 petals, 3–3.5 in., exhibition form, slight fragrance; some prickles; foliage medium, dark green, glossy; medium (80–90 cms), upright, compact growth; [Cairngorm X Seedling (Alexander X Wembley)]; Jellyman, J.S., 1993

'JAZZ', F, ob, 1960; flowers orange-yellow flushed crimson, borne in clusters, dbl., 26 petals, 2 in., slight fragrance; foliage dark, glossy; vigorous growth; [Masquerade X Unknown seedling]; deRuiter; Gregory

Jazz, *see* POUlnorm

Jazz Club, S, ab, 1995; Lowery/Robinson

Jazz Dancer, *see* 'SEATWINKLE'

'JAZZ FEST', F, mr, 1971; bud long, pointed; flowers medium, semi-dbl., slight fragrance; foliage large, leathery; vigorous, upright, bushy growth; [Pink Parfait X Garnette]; Armstrong, D.L.; Armstrong Nursery

Jazz Time™, *see* 'MINACCO'

Jean Adrien Mercer, F, 1991; Dorieux, Francois; (Cavriglia)

'JEAN BACH SISLEY', HCh, pb, 1889; flowers silvery rose, outer petals salmon-rose veined carmine, moderate fragrance; growth moderate; Dubreuil

'JEAN BAKER', HT, w, 1998; flowers white, exhibition blooms, very dbl., 26–40 petals, borne singly, intense fragrance; few prickles; foliage medium, dark green, glossy; upright, tall (4–5 ft) growth; [Crystalline X Classic Touch]; Edwards, Eddie

'JEAN BAPTISTE CASATI', HP, dp, 1886; flowers carmine-pink, large, very dbl., moderate fragrance; Schwartz; (Sangerhausen)

'JEAN BODIN', M, lp; flowers light rose-pink, quartered, not very mossy; vigorous growth; Vibert, 1846

'JEAN BOSTICK', HT, my, 1936; (Yellow Condesa de Sastago); flowers deep yellow, sometimes splotched red, large, dbl., 50 petals, globular, intense fragrance; foliage leathery, glossy; very vigorous growth; [Condesa de Sástago sport]; Bostick

'JEAN BROWN', HT, ly, 1930; flowers large, very dbl., moderate fragrance; Evans, F. David; (Sangerhausen)

'JEAN C. N. FORESTIER', HT, rb, 1919; flowers carmine, slightly tinted orange and yellow, very large, dbl.; foliage glossy, bronze; very vigorous growth; GM, Bagatelle, 1919; [Unnamed variety X Mme Edouard Herriot]; Pernet-Ducher

'JEAN CAMPBELL', HT, pb, 1964; flowers blush-pink suffused apricot, well formed, dbl., 28 petals, 4.5 in., moderate fragrance; foliage dark; upright growth; [Ethel Sanday seedling X Lady Sylvia]; Sanday, John

'JEAN CHERPIN', HP, m, 1865; flowers purple/pink, very large, very dbl.; Liabaud; (Sangerhausen)

'JEAN COTE', HT, yb, 1936; flowers old-gold, center deeper, very large, dbl.; foliage brilliant green; vigorous growth; Gaujard; J&P

Jean de la Lune, *see* 'DELCRO'

'JEAN DU TILLEUX', HT, lp, 1980; bud long; flowers medium lavender pink, dbl., 30 petals, exhibition form, slight fragrance; slightly hooked red prickles; foliage deep green, waxy; medium, vigorous growth; [King of Hearts X Golden Masterpiece]; Winchel, Joseph F.; Kimbrew-Walter Roses

'JEAN DUCHER', T, op; flowers salmon to peachy pink, globular, moderate fragrance; Ducher, 1873

Jean Galbraith, HT, ab; [Abraham Darby sport]; Nieuwesteeg, J., 1999; (Weatherly, L.)

Jean Gaujard®, *see* 'GAUVITOR'

Jean Giono™, *see* 'MEIROKOI'

'JEAN GIRIN', HWich, pb, 1910; flowers bright rose-pink, base rosy white, stamens forming a yellow, dbl., profuse bloom, sometimes repeated, vigorous, climbing growth; Girin

'JEAN GUICHARD', HWich, pb, 1905; bud bronzy crimson; flowers copper-pink; vigorous, climbing growth; [R. wichurana X Souv. de Catherine Guillot]; Barbier

'JEAN KATHRYN', F, pb, 1973; flowers neyron rose, reverse white, very full, dbl., 45 petals, 4 in., intense fragrance; foliage light; very vigorous growth; [Memoriam X Gavotte]; Ellick

Jean Kenneally™, *see* 'TINEALLY'

'JEAN LAFITTE', LCl, mp, 1934; bud pointed; flowers Willowmere pink, large, dbl., cupped, moderate fragrance; foliage leathery; very vigorous, climbing (8–10 ft very hardy; [R. setigera seedling X Willowmere]; Horvath; J&P

'JEAN LAPEYRE', S, my, 1960; flowers well formed, recurrent bloom, foliage bronze; vigorous, bushy growth; Gaujard

'JEAN LELIÈVRE', HP, dr, 1879; flowers large, dbl., moderate fragrance; Oger; (Sangerhausen)

'JEAN LHOSTE', HWich, rb, 1926; flowers rosy carmine, base flesh-white, large, borne in clusters, dbl.; [Alexandre Girault X Gerbe Rose]; Congy; Cochet-Cochet

'JEAN LIABAUD', HP, dr, 1875; flowers crimson-maroon, shaded scarlet, large, dbl., 60 petals, some recurrent bloom, moderate fragrance; vigorous growth; [Baron de Bonstetten X ?]; Liabaud

'JEAN LORTHOIS', HT, pb, 1879; flowers rose-pink, center darker, reverse silvery, well formed, large, Ducher, Vve.

'JEAN MACARTHUR', HT, mr, 1942; bud long, pointed, begonia-red; dbl., 30–40 petals, 3.5–4 in., exhibition form, slight fragrance; foliage leathery, wrinkled, dark; vigorous, upright growth; [Joanna Hill X California]; Hill, Joseph H., Co.

Jean Marc, F, or, 1975; Croix; (Cavriglia)

Jean Marc Rosé, F, 1995; Fineschi, G.; (Cavriglia)

Jean Marmoz, *see* **'JEAN MERMOZ'**

'JEAN MAYCOCK', HT, mr, 1999; flowers crimson pink, reverse lighter, dbl., 26–40 petals, 3 in., borne in small clusters, slight fragrance; prickles moderate; foliage medium, medium green, semi-glossy; compact, medium (36 in.) growth; [Sexy Rexy X Sharifa Asma]; Rawlins, R.

'JEAN MCGREGOR REID', HT, w, 1962; flowers cream, large, moderate fragrance; foliage dark, glossy; vigorous growth; [Peace sport]; Sunter; Ross & Son

'JEAN MERMOZ', Pol, mp, 1937; (Jean Marmoz); flowers ruddy pink, imbricated, small blooms in long clusters, very dbl., slight fragrance; foliage glossy, dark; vigorous growth; [R. wichurana X Hybrid Tea]; Chenault; Hémeray-Aubert

Jean Monnet®, *see* PEKcram

Jean Morrison, HT, or, 1996; Sutherland

'JEAN MURAOUR', F, w, 1935; flowers pure white, center light yellow; [Gruss an Aachen sport]; Vogel, M.

'JEAN NOTÉ', HT, op, 1909; flowers light salmon-pink, large, dbl.; Pernet-Ducher; (Sangerhausen)

'JEAN RAMEAU', B, dp, 1918; flowers iridescent rose, [Mme Isaac Pereire sport]; Darclanne; Turbat

'JEAN RENTON', HT, my, 1940; vigorous growth; Clark, A.

Jean Rex, *see* 'REXJEAN'

Jean Rose, *see* 'TINJEAN'

'JEAN ROSENKRANTZ', HP, or, 1864; flowers very bright coral red, large, dbl., intense fragrance; Portemer fils

'JEAN SISLEY', HT, m, 1879; flowers lilac-rose, large; moderate growth; [Adam X Emilie Hausburg]; Bennett

'JEAN SOUPERT', HP, dr, 1875; flowers crimson-maroon, large, dbl., moderate fragrance; [Charles Lefebvre X Souv. du Baron de Semur]; Lacharme, F.

'JEAN THOMSON HARRIS', F, op, 1976; flowers salmon, shaded orange, dbl., 30 petals, 4 in., [(Fragrant Cloud X Heidelberg) X (Heidelberg X Kingcup)]; Cocker

'JEAN WEBB', HT, rb, 1966; flowers cochineal, reverse light bronze, 4 in., moderate fragrance; foliage dark; vigorous growth; [Bettina sport]; Marks

Jeanette Talbot, *see* 'WEBEYAM'

'JEANIE', HT, w, 1959; flowers cream to pink, dbl.; 66 petals, 4–4.5 in., exhibition form, moderate fragrance; foliage dark; vigorous, spreading growth; [Condesa de Sástago X Mme Edmond Labbe]; Eddie; Wyant

'JEANIE WILLIAMS', Min, rb, 1965; flowers orange-red, reverse yellow, small, dbl., slight fragrance; foliage leathery; vigorous, bushy growth; [Little Darling X Magic Wand]; Moore, Ralph S.; Sequoia Nursery

Jeanine, Gr, op

'JEANINE DEFAUCAMBERGE', Pol, pb, 1931; flowers bright salmon-pink, passing to light pink, peony form, large, very dbl.; foliage slender; [Merveille sport]; Turbat

'JEANINE WEBER', HT, ob, 1954; flowers orange, large, moderate fragrance; vigorous growth; [Soestdijk X Mary Hart]; Leenders, M.

'JEANNE CABANIS', HT, rb, 1922; bud coral-red; flowers bright rose-carmine, reverse silvery, center coppery rose, dbl.; Guillot, P.

'JEANNE CORBOEUF', HT, lp, 1902; flowers large, dbl., Corboeuf; (Sangerhausen)

'JEANNE D'ARC', A, w, 1818; flowers creamy flesh fading to ivory-white, large, dbl., intense fragrance; dense bush (to 5 ft) growth; Vibert

'JEANNE D'ARC', N, w, 1848; flowers creamy white, dbl.; vigorous growth; Verdier, V.

'JEANNE D'ARC', Pol, w, 1909; flowers pure milky white, moderate fragrance; [Mme Norbert Levavasseur sport]; Levavasseur

'JEANNE DE MONTFORT', M, mp, 1851; bud heavily mossed, dark carmine; flowers clear pink, edged silver, blooms in large clusters, semi-dbl., flat, moderate fragrance; foliage emerald-green; tall, vigorous growth; Robert

'JEANNE DRIVON', Pol, w, 1883; flowers white, faintly shaded pink, very dbl., moderate fragrance; Schwartz, J.

'JEANNE EXCOFFIER', HT, pb, 1921; flowers daybreak-pink, inside buff, dbl.; [Mme Philippe Rivoire X Mme Edouard Herriot]; Buatois

'JEANNE HACHETTE', M, m, 1851; flowers slaty violet; Robert

'Jeanne Hachette', HP, lp; flowers carmine-rose; Oger

'JEANNE LAJOIE', Cl Min, mp, 1975; bud long, pointed; dbl., 40 petals, 1 in., exhibition form, slight fragrance; foliage small, glossy, dark, embossed; upright, bushy growth; AOE, 1977; [(Casa Blanca X Independence) X Midget]; Sima; Mini-Roses

'JEANNE LALLEMAND', LCl, pb, 1954; flowers pink, reverse salmon-pink, large, dbl., 40–50 petals, very free, recurrent bloom; very vigorous growth; [Mrs Pierre S. duPont X George Dickson]; Buatois

'JEANNE LASSALLE', LCl, mp, 1936; borne in clusters of 50–60, free, recurrent boom; foliage broad, light; long, stiff stems; vigorous (5 ft) growth; Lassalle; Vially

'JEANNE MASSON', HP, lp, 1891; flowers medium, dbl.; moderate fragrance; Liabaud; (Sangerhausen)

'JEANNE MERMET', Pol, w, 1909; vigorous growth; Mermet

'JEANNE NICOD', HT, w, 1929; flowers white, center tinted cream, dbl.; Schwartz, A.

'JEANNE RICHERT', HWich, w, 1929; flowers cream, center red-brown, borne in large clusters; foliage glossy; very vigorous, climbing growth; [Leontine Gervais X Unnamed seedling]; Walter, L.

'JEANNE SAULTIER', HT, pb, 1927; flowers salmon-rose, reverse reddish pink, base yellow, dbl.; moderate fragrance; [Louise Catherine Breslau X Mme Edouard Herriot]; Laperrière

'JEANNE SURY', HP, dp, 1868; flowers large, dbl., moderate fragrance; Faudon; (Sangerhausen)

'JEANNETTE', HGal, dp; flowers bright light red, fading light rose-pink, dbl.; Descemet, M.

Jeannette, Min, pb, 1997; Spooner, Raymond A.

Jeannette Heller, *see* **'WILLIAM R. SMITH'**

'JEANNIE DEANS', HEg, dr, 1895; flowers scarlet-crimson, semi-dbl., very free seasonal bloom, foliage fragrant; vigorous growth; Penzance

'JEANNIE DICKSON', HP, pb, 1890; flowers rose-pink, edged silvery pink, large, dbl., 45 petals, exhibition form, moderate bloom, vigorous growth; Dickson, A.

'JEANNIE SOUPERT', Pol, w, 1913; flowers soft flesh-white, borne in large clusters; moderately vigorous growth; [Mme Norbert Levavasseur X Petite Léonie]; Soupert & Notting

Jeannine, *see* 'FRONEEN'

Jeannine Michelle, *see* 'FROJEAN'

'JEAN'S DREAM', HT, pb, 1971; flowers orient pink to azalea-pink, dbl., 30 petals, 6 in., exhibition form; foliage large; vigorous, upright, bushy growth; [My Choice X (Seedling X Memoriam)]; Ellick

Jeeper's Creeper, *see* 'KORISSEL'

Jeffrey, *see* 'RESJEFF'

Jehoca, *see* 'TANJEKA'

Jehoka, *see* 'TANJEKA'

Jekyll's Own Rambler, Ayr, w

'JELBAR', Min, w, 1980; (**Wee Barbie**); bud globular; flowers cream-white, blooms borne 10–20 per cluster, dbl., 43 petals, moderate fragrance; foliage dark; bushy growth; [Unnamed seedling X Unnamed seedling]; Jellyman, J.S.

'JELCANODIR', F, dy, 1976; flowers aureolin-yellow, full, dbl., 30 petals, 2.5–3 in., almost continuous bloom in glasshouse, slight fragrance; vigorous, upright growth; [Undisseminated seedling 1-61-ys X Golden Garnette]; Jelly; Universal Plants

'JELDANIRAN', F, lp, 1985; (**Mimi Pink**™, Mimi Rose); flowers light pink, reverse darker, sweetheart, large, dbl., 35 petals, no fragrance; foliage medium, dark, semi-glossy; upright growth; PP005636; [Unnamed seedling X Misty Pink]; Jelly, Robert G.; E.G. Hill Co., 1985; Universal Plants, 1980

'JELEIT', F, w, 1971; dbl., 30–35 petals, 3 in., globular, slight fragrance; foliage large, dull, leathery; vigorous, upright growth; [Seventeen X Jack Frost]; Hill, E.G., Co.; Meilland

Jelena de Belder, S, w, 1996; Lens, Louis; (Cavriglia)

'JELICO', HT, or, 1973; flowers vermilion, reverse crimson, dbl., 35 petals, 5 in., cupped, slight fragrance; foliage large, dark; [Baccará X (Forever Yours X Unnamed seedling)]; Jelly; Universal Plants

'JELLO', F, my, 1969; (**Coed**™); bud short, pointed; flowers mini-flora, medium, dbl., exhibition form, slight fragrance; foliage dark, leathery; vigorous, upright growth; [Golden Garnette X Seedling]; Jelly; E.G. Hill Co.

Jelly Bean™, *see* 'SAVABEAN'

'JELPIROFOR', HT, w, 1985; (**White Success**); flowers large, dbl., 55 petals, exhibition form, borne singly, no fragrance; medium, straight prickles; foliage large, dark, semi-glossy; upright growth; PP005632; [Bridal Pink X Unnamed seedling]; Jelly; E.G. Hill Co.

'JELRANDOLI', F, rb, 1976; flowers vermilion-red, base cardinal-red, full, dbl., 35 petals, 3–3.5 in., slight fragrance; foliage large; vigorous, upright growth; [San Francisco X Little Leaguer]; Jelly; Universal Plants

'JELROGANOR', F, rb, 1977; (**Charisma**, Surprise Party); bud ovoid; flowers scarlet and yellow, medium, very full, very dbl., 40–50 petals, 2–2.5 in., exhibition form, borne in small clusters, slight, fruity fragrance; foliage medium, dark green, glossy, leathery; vigorous, bushy, upright growth; PP004173; AARS, 1978 GM, Portland, 1976; [Gemini X Zorina]; Hill, E.G., Co.; C-P

'JELWHITE', HT, w, 1969; (**Promise Me**); dbl., 45 petals, 4 in., exhibition form, slight, sweetbriar fragrance; foliage matt, dark, leathery; vigorous, upright growth; [Snowsong Supreme X Unnamed seedling]; Jelly; E.G. Hill Co.

'JEMA', HT, ab, 1981; bud ovoid; flowers large blooms borne singly, dbl., 45 petals, moderate fragrance; small, triangular, straw prickles; foliage medium, light green; tall, vigorous growth; [Helen

Traubel X Lolita]; Perry, Astor; Perry Roses, 1982

Jemma Giblin, *see* 'HORJEMMA'

Jemne Losos Interflora, HT, lp; Strnad; (Czech Rosa Club)

Jennie Anne(TM), *see* 'KINJEN'

Jennie June(TM), *see* 'TANJENJU'

Jennie Robinson, *see* 'TROBETTE'

Jennifer(TM), *see* 'BENJEN'

'JENNIFER', HT, 1954; flowers pale flesh pink, 4–5 in., intense fragrance; foliage dull green; vigorous growth; RULED EXTINCT 1/85 ARM; [(Mrs Henry Bowles X Phyllis Gold) X Edina.]; Fletcher; Tucker

'JENNIFER', F, 1959; RULED EXTINCT 1/85 ARM; [Independence X Fashion]; Fryers Nursery, Ltd.

Jennifer Hart, *see* 'AROART'

'JENNIFER JAY', HT, pb, 1977; flowers light to medium pink; [Christian Dior sport]; Thomas, Dr A.S.

Jennifer Joy, *see* 'POULJENJOY'

Jennifer-Betty Kenward, *see* 'LANBET'

Jennifer's Rose, HT, mr, 1997; Dawson

'JENNY BROWN', HT, or, 1974; bud long, pointed; flowers salmon-pink, center paler, open, single, 5 petals, 4 in., intense fragrance; foliage glossy; very vigorous growth; [(Pink Favorite X Dorothy Peach) X Dainty Bess]; Parkes, Mrs M.H.; Rumsey

Jenny Butchart, HT, lp; flowers deep orange pink fading to lighter pink, exhibition form, 30–35 petals, slight; medium sized matt green; [Miss Canada X Fragrant Cloud]; Hepworth, George, 1970

Jenny Charlton, *see* SIMway

'Jenny Duval', HCh; flowers rosy bush; no longer in cultivation; thought to be Président de Seze

'JENNY FAIR', HT, mp, 1967; flowers pink, globular, moderate fragrance; foliage dark; slender, upright growth; [Tropicana X ?]; Gregory

'JENNY JONES', T, lp; Williams, A., 1890; (Weatherly, L.)

'JENNY LIND', M, mp, 1845; Laffay, M.

Jenny Lynn, HT, pb

'JENNY WREN', F, ab, 1957; bud salmon-red; flowers creamy apricot, reverse pale salmon, small blooms in large, dbl., intense fragrance; foliage dark; [Cécile Brunner X Fashion]; Ratcliffe

'JENNY'S DREAM', HT, dp, 1980; flowers deep pink, [Red Devil sport]; Beckett, Ian

Jenny's Rose, *see* CANsit

'JENS MUNK', HRg, mp, 1974; bud ovoid; flowers yellow stamens, dbl., 25 petals, 3 in., intense fragrance; upright, bushy growth; [Schneezwerg X Frau Dagmar Hartopp]; Svedja, Felicitas; Canada Dept. of Agric.

Jericho, HT, 1971; Dorieux, Francois; (Cavriglia)

'JERRY', Gr, pb, 1987; flowers white, flushing carmine red, reverse carmine, grading to white, dbl., cupped, intense, sweet fragrance; small, rarely sets fruit; medium, red-brown prickles, hooked downward; foliage medium, medium green, semi-glossy; bushy, tall growth; Silver, ARC TG, 1986; [Unnamed seedling X Unnamed seedling]; Jerabek, Paul E.

'JERRY DESMONDE', HT, mp, 1959; flowers rose-pink, reverse silvery, well formed, dbl., 50 petals, 5 in.; foliage dark, glossy; vigorous, upright growth; [Lord Rossmore X Karl Herbst]; Norman; Harkness

Jerry-O(TM), *see* 'SAVAJERRY'

'JERSEY BEAUTY', HWich, ly, 1899; flowers pale yellow, fading white, large blooms in clusters, single, non-recurrent, intense fragrance; foliage very glossy; vigorous, climbing growth; [R. wichurana X Perle des Jardins]; Manda, W.A.

Jersey Cream, S, ly, 1996; Sutherland

'JERSEY QUEEN', HT, ob, 1920; flowers flame-orange, edged rose, reverse lemon, dbl.; [Mme Mélanie Soupert X Queen Mary]; Le Cornu; Jersey Nursery

Jerusalem, *see* 'HOLJE'

'JESMOND DENE', F, op, 1975; flowers pastel salmon-pink, dbl., 3.5 in., intense fragrance; vigorous, upright growth; [Arthur Bell X Betty May Wood]; Wood

'Jesse's Jewels', *see* 'BURJES'

JESSICA, *see* **Angle**

'JESSICA', HWich, pb, 1910; flowers cream-white, center light rose, large; Walsh

Jessica, *see* 'TANJEKA'

'JESSIE', Pol, rb, 1909; flowers bright crimson, fading rose-pink, center white, small, semi-dbl., 1.5 in., slight fragrance; foliage small, soft, glossy; bushy growth; [Phyllis sport]; Merryweather

'JESSIE ANDERSON', S, dp; flowers deep rose, well formed, large, dbl., blooms continuously on new wood, [(Old Crimson China X R. canina) X Souv. d'Alphonse Lavallee]

'JESSIE BROWN', Min, mp, 1978; bud mossy; flowers loosely formed, small, semi-dbl., 15 petals, 1.5 in.; foliage small; bushy growth; [Fairy Moss X Fairy Moss]; Dobbs; Small World Min. Roses

'JESSIE CLARK', LCl, mp, 1915; flowers rosy pink, becoming lighter, overlarge, single, early spring bloom, foliage dark, leathery; very vigorous growth; [R. gigantea X Mme Martignier]; Clark, A.; NRS Victoria

Jessie Mathews, *see* 'BEEJES'

'JESSIE SEGRAVE', HT, rb, 1937; flowers scarlet on deep chrome base, with pencil markings on inside, dbl.; vigorous growth; Mee; Beckwith

Jessika(R), *see* 'TANJEKA'

'JET', HT, dr, 1948; bud long, pointed; flowers red to very dark red, large, dbl., moderate fragrance; foliage glossy, dark; vigorous, bushy, compact growth; [Pink Princess X Crimson Glory]; Brownell, H.C.

'JET FIRE', F, or, 1964; bud ovoid; flowers large, dbl., cupped, moderate, spicy fragrance; foliage dark, glossy; vigorous, upright, bushy growth; [Sumatra X Fashion]; Schloen, J.; Ellesmere Nursery

Jet Flame(R), *see* 'LENPEN'

Jet Flame Nirpaysage(R), *see* 'LENPEN'

Jet Spray(R), *see* 'LENCARA'

'JET TRAIL', Min, w, 1964; bud pointed; flowers white, sometimes tinted pale green, small, dbl., 40 petals; bushy (12–14 in.) growth; [Little Darling X Magic Wand]; Moore, Ralph S.; Sequoia Nursery

Jeune Fille, *see* **'BROCADE'**

'JEUNE FILLE', F, op, 1964; flowers bright salmon-pink, medium, borne in clusters, dbl., moderate fragrance; foliage leathery; very vigorous, bushy growth; [Rose Gaujard seedling X Vendome]; Gaujard

Jeune France, *see* **'YOUNG FRANCE'**

'JEUNE HENRY', P, mp; flowers vivid rose; Vibert, prior to 1815

'JEUNESSE', HT, mp, 1959; flowers bright pink, dbl.; moderately bushy growth; [(Independence X Tonnerre) X Michele Meilland]; Laperrière; EFR

Jeunesse Éternelle, *see* **'ETERNAL YOUTH'**

'JEWEL', HT, dr, 1938; flowers velvety red, dbl., 50 petals, 5 in., moderate fragrance; [Better Times sport]; Grillo

Jewel Box, *see* 'MORBOX'

Jewel's Delight(TM), *see* 'MINBACO'

'JEZEBEL', HT, mp, 1964; dbl.; strong stems; [Queen Elizabeth X Pink Lustre]; Leenders, J.

JFK, *see* **'JOHN F. KENNEDY'**

Jhb, Garden Club *see* 'POULFAN'

'JIAN', Min, mr, 1965; bud ovoid; flowers medium red, reverse lighter, very small, dbl.; foliage narrow, leathery; very vigorous, bushy, dwarf growth; [Juliette X Oakington Ruby]; Williams, Ernest D.; Mini-Roses

'JIHOCESKE SLUNCE', HT, w, 1937; flowers medium, dbl.; Böhm, J.; (Sangerhausen)

'JILL', F, mr, 1939; flowers cerise-scarlet, open, borne in clusters, semi-dbl., slight fragrance; long stems; vigorous, bushy growth; [(Else Poulsen X Seedling) X ((Seedling (single red) X Étoile de Hollande) X Daily Mail Scented Rose)]; LeGrice

Jill Carter, *see* 'SHERIJILL'

'JILL DARLING', HT, rb, 1937; flowers rich cerise, reverse cinnamon-yellow, moderate fragrance; foliage glossy; vigorous growth; Austin & McAslan

'JILLIAN LOUISE', Min, w, 1994; flowers white with light pink edge and center, blooms borne in large clusters, dbl., 26–40 petals, 1.5–2.75 in., slight fragrance; some prickles; foliage medium, medium green, semi-glossy; medium to tall, upright growth; [Magic Carrousel sport]; Fairweather, Mrs. P.H.

Jillian McGredy, *see* 'MACARNHE'

Jill's Rose, *see* GANjil

Jilly Cooper, *see* LEGcream

Jilly Jewel, *see* BENmfig

Jim Bowie, S, dp, 1996; flowers dark pink to light pink center, semi-dbl., 4 in., borne in sprays, intense fragrance; arching growth; hardy; Williams, J. Benjamin

Jim Dandy™, *see* 'BENJIM'

'JIM INGALL', HT, mr, 1998; flowers medium red, medium sized, full, very dbl., 26–40 petals, 4 in., borne singly, slight fragrance; prickles moderate; foliage medium, dark green, semi-glossy; bushy, upright, 120–150 cms growth; [Debbie Thomas X Maria Theresa]; Thomas, D.

'JIM LOUNSBERY', S, or, 1995; single, 5–7 petals, 1.5–2.75 in., borne in sprays of 5–25, slight fragrance; foliage medium, medium green, matt; upright (120–150 cms), bushy growth; [Liverpool Echo X R. virginiana]; Fleming, Joyce L.; Hortico Roses, 1994

'JIM TODD', HT, rb, 1940; flowers nasturtium-red, reverse touched yellow, large, semi-dbl., cupped; very vigorous, bushy growth; Mallerin, C.; A. Meilland

'JIMINY CRICKET', F, op, 1954; bud ovoid; flowers coral-orange to pink-coral, blooms in clusters, dbl., 28 petals, 3–4 in., cupped, moderate, rose geranium fragrance; foliage glossy; vigorous, upright, bushy growth; AARS, 1955; [Goldilocks X Geranium Red]; Boerner; J&P

Jimmy, Min, ob, 1993; Chandrakant

'JIMMY GREAVES', HT, m, 1971; flowers red-purple, reverse silver, dbl., 55 petals, 5 in., exhibition form, slight fragrance; foliage large; erect, bushy growth; [Dorothy Peach X Prima Ballerina]; Gandy, Douglas L.

Jimmy Savile, *see* 'PEAPOLLY'

Jimmy Saville, *see* 'PEAPOLLY'

Jim's Fence Corner, Pol, w; (found rose)

Jingle Bells™, *see* 'JACREDEM'

'JINGLES', F, mp, 1956; bud ovoid; flowers pink overcast deep rose-pink, to open, dbl., 35–40 petals, 2.5–3 in., cupped, moderate fragrance; foliage leathery, glossy; vigorous, upright, bushy growth; [Goldilocks X Garnette]; Boerner; J&P

Jisraela Amira, HT, 1980; Nevo, Motke

Jitka, F, lp; Strnad; (Czech Rosa Club)

Jitka, F, lp; Urban, J., 1987; (Czech Rosa Club)

'JITRENKA', HWich, lp, 1933; flowers medium, semi-dbl.; Mikes Böhm, J.; (Sangerhausen)

Jitrenka, Pol, dy, 1978; flowers medium, semi-dbl.; slight fragrance; Urban, J.; (Sangerhausen)

Jitterbug, *see* 'JACMINNO'

'JOAN', HMsk, ob, 1919; bud peach; flowers copper, borne in clusters, semi-dbl.; [Trier X Perle des Jeannes]; Pemberton

'JOAN ALDER', HT, pb, 1950; bud long; flowers salmon-pink tinted mauve; foliage dark; very vigorous growth; Moss

Joan Anderson, *see* **'ELSE POULSEN'**

Joan Austin, *see* 'MORDEB'

Joan Ball, *see* 'TROBALL'

'JOAN BELL', HT, dp, 1985; flowers deep pink, [Portland Trailblazer sport]; Bell, John C.

Joan Brickhill®, *see* 'HILGOFAN'

'JOAN CANT', HT, pb, 1929; bud pointed; flowers salmon-pink, reverse brighter, very large, dbl., intense fragrance; foliage light, leathery; vigorous, bushy growth; Cant, B. R.

'JOAN DAVIS', HT, ab, 1927; flowers salmon-apricot shaded cerise-pink, base yellow, dbl., intense fragrance; [Ophelia seedling]; Allen

'JOAN ELIZABETH', HT, yb, 1949; bud long, pointed; flowers golden yellow, reverse flushed pink, well formed, dbl., 30 petals, 5 in., moderate fragrance; foliage dark, glossy; vigorous growth; Fletcher; Tucker

'JOAN FITTALL', HT, yb, 1946; bud long, pointed; flowers bronze and gold fading to pink, open, medium, semi-dbl., slight fragrance; foliage leathery; vigorous, bushy growth; [Luis Brinas X ?]; Moss; F. Mason

Joan Fontaine™, *see* CLEjoan

'JOAN FRUEH', HT, lp, 1924; flowers shell-pink, dbl., moderate fragrance; [Ophelia X General-Superior Arnold Janssen]; Frueh

'JOAN HOWARTH', HT, pb, 1924; flowers shell-pink shaded carmine, very large, dbl., moderate fragrance; [Lyon Rose X Mme Abel Chatenay]; Bees

'JOAN KNIGHT', Cl HT, dr, 1928; vigorous growth; Knight, J.

Joan Kruger, S, w, 1997; Kordes

Joan Longer™, *see* 'WILKTWO'

'JOAN MARGARET DERRICK', Pol, dr, 1953; flowers carmine-red, small, semi-dbl., 15 petals, vigorous growth; [Golden Salmon sport]; Derrick

'JOAN ROSS', HP, pb, 1933; bud pointed; flowers blush, reverse light pink, very large, dbl., profuse, non-recurrent bloom, slight fragrance; vigorous growth; [Frau Karl Druschki X Paul Neyron]; Nicolas

'JOANNA BRIDGE', HT, yb, 1916; flowers canary-yellow shaded strawberry, borne on large trusses, semi-dbl.; vigorous growth; Hicks

'JOANNA HILL', HT, ly, 1928; bud long, pointed; flowers creamy yellow, base flushed orange, large, dbl., 48 petals, moderate fragrance; foliage leathery; vigorous growth; [Mme Butterfly X Miss Amelia Gude]; J.H. Hill Co.

'JOANNA HILL, CLIMBING', Cl HT, yb, 1935; Howard Rose Co.

'JOANNA LUMLEY', HT, ab, 1994; dbl., 26–40 petals, 3–3.5 in., borne singly, slight fragrance; some prickles; foliage medium, dark green, glossy; upright (100 cms) growth; [Chicago Peace X Joanne]; Poole, Lionel; F. Haynes & Partners, 1995

'JOANNA TROUTMAN', HT, ob, 1929; bud pointed; flowers orange-yellow, open, semi-dbl., moderate fragrance; foliage bronze, glossy; [Mme Alexandre Dreux seedling]; Vestal

'JOANNE', HT, op, 1985; flowers medium shrimp pink, large blooms borne singly, dbl., 43 petals, exhibition form, slight fragrance; large, globular, orange fruit; large, dark brown prickles; foliage large, dark, semi-glossy; medium, upright growth; [Courvoisier X Princesse]; Poole, Lionel

'JOANNES GINET', HT, pb, 1929; flowers white, tinted cream, edged oriental red, [The Queen Alexandra Rose sport]; Gaujard

'JOAO MOREIRA DA SILVA', HT, my, 1959; moderate, damask fragrance; [Mme Marie Curie X Dr Manuel Alves de Castro]; da Silva, Moreira

'JOAO PEREIRA DA ROSA', HT, rb, 1936; flowers brilliant red shading orange and yellow, large, dbl., cupped; foliage light, soft; vigorous growth; [Angèle Pernet X Mme Méha Sabatier]; da Silva, Moreira

'JOAQUIN ALDRUFEU', HT, m, 1897; flowers garnet-purple, reverse violet to magenta, dbl., slight fragrance; foliage light green; moderate growth; Aldrufeu

'JOAQUIN MIR', HT, dy, 1940; flowers golden yellow, large, dbl., cupped; foliage glossy, dark; upright growth; [Mrs Pierre S. duPont X Senora Gari]; Dot, Pedro

Joaquina Munoz, HT, 1980; Dot, Simon; (Cavriglia)

'JOASINE HANET', P, m; (Glendora, Portland from Glendora); flowers deep rose tinged with violet, medium, dbl., quartered, heavy bloomer, moderate fragrance; very hardy; Vibert, prior to 1882

'JOCELYN', F, r, 1970; flowers mahogany aging purplish-brown, dbl., 3 in.; LeGrice

Jockey, *see* **'HORRIDO'®**

Joconde, *see* 'DELJACQ'

'JODRELL BANK', HT, pb, 1967; flowers light pink, reverse rose-pink; [Charles F. Warren sport]; Dale, F.

Joe Gray, S, mp, 1999; flowers pink and lavender; dwarf, creeping, ground cover growth; Williams, J. Benjamin

'JOE LONGTHORNE', HT, op, 1995; flowers orange pink, blooms borne singly, dbl., 26–40 petals, 3–3.5 in., intense fragrance; some prickles; foliage medium, dark green, matt; medium (100 cms), upright growth; [Gavotte X Pot of Gold]; Poole, Lionel; Battersby Roses, 1995

'JOE ROSCOE', HT, mp, 1971; flowers rose-red, dbl., 62 petals, 6 in., moderate fragrance; [Karl Herbst X Tzigane]; Wright & Son

Joe-Joe™, *see* 'KINJOE'

Joella®, *see* 'KRITIBAN'

Joëlle®, *see* 'KRITIBAN'

'JOFITALI', F, rb, 1976; flowers rose-bengal, center cardinal-red, PP004083; [Sonia sport]; DeWitte; Meilland

Jogan, HT, ab, 1988; Bansal, O.P.

Johann Strauss®, *see* 'MEIOFFIC'

'JOHANNA OFMAN', HT, dp, 1962; flowers carmine-pink, large, dbl., intense fragrance; [Pink Sensation sport]; Ofman

'JOHANNA RÖPCKE', HWich, op, 1931; (Johanna Ropke); bud pointed; flowers salmon-pink, resembling Ophelia but smaller, small, dbl., cupped, borne in clusters; foliage dark, bronze; very vigorous, climbing growth; [Dorothy Perkins X Ophelia]; Tantau

Johanna Ropke, *see* **'JOHANNA RÖPCKE'**

'JOHANNA TANTAU', Pol, w, 1928; flowers white, center pinkish yellow, large, borne in clusters, dbl.; foliage dark, leathery; bushy, dwarf growth; [Dorothy Perkins X Ophelia]; Tantau

'JOHANNES BOETTNER', F, mr, 1943; flowers light crimson, very large, borne in clusters, dbl.; exhibition form, slight fragrance; vigorous, bushy, compact growth; [Baby Chateau seedling X Else Poulsen]; Kordes

Johannes XXIII, HT, w

Johannesburg Centennial, HT, dy

Johannesburg Garden Club, *see* 'POULFAN'

Johannesburg Sun, *see* 'KORDOUBT'

Johannisfeuer®, *see* TANeufis

'JOHANNISZAUBER', HT, dr, 1926; bud pointed; flowers dark velvety blood-red, very dbl.; vigorous growth; [Château de Clos Vougeot X Unnamed seedling]; Tantau

'JOHASINE HANET', HP, pb, 1847; flowers pink with red, medium, dbl.; Vibert; (Sangerhausen)

'JOHILLGOLD', Gr, dy, 1992; (**Tropical Sunrise**); flowers deep yellow aging to medium yellow, blooms borne singly and, semi-dbl., 21 petals, 4 in., exhibition form, slight fragrance; foliage medium, medium green, semi-glossy; upright, medium growth; [Golden Fantasie X Unnamed seedling]; Hoy, Lowel L.; DeVor Nurseries, Inc.

'JOHILLGOLF', S, mr, 1992; (**Crimson Delight**); dbl., 2.5 in., borne in sprays of 3–4, no fragrance; foliage medium, medium green, semi-glossy; upright, bushy, low growth; [Volare X Seedling]; Hoy, Lowel L.; DeVor Nurseries, Inc.

'JOHILLSTAR', HT, mr, 1992; (**Classical Velvet**); semi-dbl., 25 petals, 5.25 in., borne in sprays of 4–5, no fragrance; foliage medium, dark green, semi-glossy; upright, bushy, tall growth; [Anniversary X Seedling]; Hoy, Lowel L.; DeVor Nurseries, Inc.

JOHilmar, Gr, op; (**After Glow**)

'JOHN A. ALLISON', Gr, lp, 1973; bud ovoid; flowers large, very dbl., slight fragrance; very vigorous, upright growth; [Queen Elizabeth X Montezuma]; Golik; J. Schloen

John A. Macdonald, Gr, dr

John A. Weall, LCl, dr, 1994

'JOHN ABRAMS', F, op, 1976; flowers vermilion and salmon, semi-dbl., 15 petals, 3 in., [Vera Dalton X Sarabande]; Sanday, John

'JOHN ALLEN' sp, (form of R. suffulta), lp, 1944; recurrent bloom, height 18 in., (28); P.H. Wright; (collected in southern Saskatchewan)

John Andrews, HT, dr, 1997; flowers large, foliage disease-resistant; Williams, J. Benjamin

John Bradshaw, *see* 'HARQUISP'

'JOHN BRIGHT', HP, mr, 1878; flowers bright crimson, medium; Paul & Son

'JOHN C. M. MENSING', HT, mp, 1924; (Pink Ophelia); flowers deep bright rose-pink, open, large, dbl., intense fragrance; vigorous growth; [Ophelia sport]; Eveleens

'JOHN CABOT', HKor, mr, 1978; bud ovoid; flowers fuschia tinted red, open flat, dbl., 40 petals, 2.5 in., blooms in clusters, moderate fragrance; foliage yellow-green; vigorous, upright, medium growth; [R. kordesii X Unnamed seedling]; Svedja, Felicitas; Canada Dept. of Agric.

'JOHN CANT', S, dp, 1895; flowers carmine-pink, small, semi-dbl., slight fragrance; Cant, B. R.; (Sangerhausen)

'JOHN CHURCH', F, or, 1964; flowers orange-scarlet, well-formed, blooms in clusters, dbl., 30 petals, 3.5 in., moderate fragrance; vigorous growth; [Ma Perkins X Red Favorite]; McGredy, Sam IV; McGredy

John Clare, *see* 'AUScent'

'JOHN COOK', HT, pb, 1917; bud dark pink; flowers La France pink, reverse very dark, dbl., moderate fragrance; [La France X Unknown seedling]; Krüger; Ketten Bros

John Cramphorn, F, or, 1980; flowers large, dbl.; Kriloff; (Sangerhausen)

'JOHN CRANSTON', M, m, 1861; flowers crimson, shaded purple, expanded, medium, dbl.; vigorous growth; Verdier, E.

'JOHN CRONIN', HT, dp, 1935; flowers deep pink, large, dbl., globular, moderate fragrance; vigorous growth; Clark, A.; NRS Victoria

'JOHN DAVIS', HKor, mp, 1986; flowers medium pink, yellow at base, blooms in clusters of up to 17, dbl., 40 petals, 3.5 in., recurrent bloom, intense, spicy fragrance; straight prickles; foliage glossy, leathery; trailing growth; [(R. kordesii X Unnamed seedling) X ?]; Svedja, Felicitas; Agriculture Canada

'JOHN DAVISON', HT, dr, 1919; flowers rich velvety crimson, dbl., moderate fragrance; McGredy

'JOHN DIJKSTRA', F, dr, 1965; (Letkis); bud ovoid; flowers medium, borne in clusters, semi-dbl.; foliage dark; [Olala X Paprika]; Buisman, G. A. H.

John Donne, S, mp

'JOHN DOWNIE', HT, ob, 1921; flowers salmon, [Lyon Rose sport]; Dobbie

'JOHN E. SLEATH', HT, rb, 1937; flowers carmine-red, suffused vermilion-orange, vigorous growth; Mee; Beckwith

'JOHN EDWARD REED', HT, dy, 1950; bud long, pointed; flowers buttercup-yellow, dbl., 32 petals, 5.5–6. in., moderate fragrance; vigorous, upright growth; [Talisman sport]; Reed

'JOHN F. KENNEDY', HT, w, 1965; (JFK, President John F. Kennedy); bud ovoid, tinted greenish; dbl., 48 petals, 5–5.5 in., exhibition form, moderate fragrance; foliage leathery; vigorous growth; [Unnamed seedling X White Queen]; Boerner; J&P

'JOHN FRANKLIN', S, mr, 1980; bud ovoid; flowers vibrant red, large, dbl., 25 petals, borne in clusters, floriferous, moderate fragrance; yellow-green prickles with purple hues; foliage rounded; upright, bushy growth, 3–4 ft; [Lilli Marleen X Unnamed seedling]; Svedja, Felicitas; Agriculture Canada

'JOHN FRASER', M, rb; flowers bright red, shaded crimson and purple, large, dbl.; shy growth; Lévêque, 1861

John Gold Veitch, HP, mr

'JOHN GREENWOOD', F, mr, 1976; flowers bright red, full, dbl., 25 petals, 2.5 in., slight fragrance; foliage large, dark;

vigorous, free growth; [Marlena X Fragrant Cloud]; Lea

John Grooms, LCl, pb, 1993; Beales, Peter

'John Grow', M, mp, 1859; flowers clear pink, Laffay, M.

'John H. Ellis', HT, dp, 1948; flowers deep rose-pink, well formed, large, dbl., 48 petals; free, bushy growth; McGredy

John Harris, HT, pb, 1995; Dawson

'John Hart', HT, mp, 1922; flowers cherry-pink, dbl., slight fragrance; Hicks

'John Henry', HT, mr, 1925; bud rosy scarlet; flowers rich pink, slight fragrance; Beckwith

'John Hopper', HP, pb, 1862; flowers bright rose edged lilac, center carmine, semi-globular, large, dbl., 70 petals, occasionally recurrent bloom, intense fragrance; vigorous, upright, bushy growth; [Jules Margottin X Mme Vidot]; Ward

John Hughes, *see* 'SANphyllis'

John Keats, *see* 'MEIroupis'

John Keats, LCl, mp

John Kemp, HWich, w

'John Keynes', HP, dr; flowers red shaded maroon, dbl., 48 petals, intense fragrance; vigorous growth; Verdier, E., 1864

'John Kidman', HT, my, 1969; bud ovoid; flowers lemon-yellow, camellia form, very large, dbl., intense fragrance; foliage leathery; vigorous, tall, compact growth; [Radar X Allgold]; Fankhauser

'John Lawrence', F, my, 1990; bud pointed; flower canary yellow, fading as it ages, shaped, medium, borne in small clusters, dbl., 18 petals, flat, moderate, spicy fragrance; foliage medium, glossy; medium, upright growth; [Unnamed seedling X Sunsprite]; Bracegirdle, Derek T., 1984

'John McNabb', HRg, mp, 1932; dbl., profuse midseason bloom, sometimes continuing later; [R. rugosa kamtchatica X R. beggeriana]; Skinner

'John Moore', HT, yb, 1939; flowers buff shaded gold, well shaped, very large, dbl., 47 petals, exhibition form, moderate fragrance; foliage dark; Gaujard

'John Morley', HT, dp, 1945; bud long, pointed; flowers glowing pink, very large, dbl., exhibition form, moderate fragrance; foliage dark, leathery; vigorous, bushy growth; [Joanna Hill X J.C. Thornton]; Duehrsen; California Roses

John Phillip Sousa, S, rb

John Ruskin, *see* **'Ruskin'**

'John Russell', HT, dr, 1924; flowers glowing crimson flushed deeper, well-shaped, large, very dbl.; vigorous growth; GM, Bagatelle, 1924; Dobbie

'John Russell, Climbing', Cl HT, dr, 1930; Ketten Bros.

'John S. Armstrong', Gr, dr, 1961; bud ovoid to urn-shaped; dbl., 40 petals, 3.5–4 in., exhibition form, slight fragrance; foliage leathery, semi-glossy, dark; tall, bushy growth; AARS, 1962; [Charlotte Armstrong X Seedling]; Swim, H.C.; Armstrong Nursery

'John S. Bloomfield', HT, ab, 1964; bud ovoid; flowers deep apricot flushed pink, open, large, dbl., slight fragrance; compact growth; [Ma Perkins X Burnaby]; Fankhauser

John Snowball, HT, w

'John Square', Cl HT, dy, 1937; flowers sunflower-yellow, center deeper, very large, dbl., cupped; foliage glossy, dark; vigorous, climbing (6–8 ft in season) growth; [Souv. de Claudius Pernet sport]; Square

'John Stuart Mill', HP, mr, 1875; flowers large, dbl.; Turner; (Sangerhausen)

'John Wallace', Pol, dr, 1941; flowers deep red, open, large, dbl.; foliage large, leathery, glossy; bushy growth; [Marianne Kluis Superior sport]; Kluis; Klyn

'John Waterer'®, HT, dr, 1970; dbl., 44 petals, 4 in., exhibition form, moderate fragrance; [King of Hearts X Hanne]; McGredy, Sam IV; McGredy

'JOHnago', HT, pb, 1962; (**Chicago Peace®**); flowers phlox-pink, base canary-yellow, very dbl., 50–60 petals, 5–5.5 in., exhibition form, borne mostly singly, slight fragrance; foliage large, dark green, leathery, glossy; tall growth; GM, Portland, 1961; [Peace sport]; Johnston; C-P; URS

John-John™, *see* 'WEKgibotex'

Johnnie Walker, *see* 'FRYgran'

John-Paul II, *see* 'JACange'

John's Rose, *see* 'MEJoria'

Johnson Street Giant, LCl, mp; (found rose)

'Joia', F, yb, 1962; flowers yellow shaded carmine, [Seedling X Virgo]; da Silva, Moreira

'Joie de Vivre', HT, pb, 1949; flowers pink, base gold, well shaped, very large, dbl., moderate fragrance; foliage bronze green; moderately vigorous growth; Gaujard; Wheatcroft Bros.

Jo-Jo, Min, dy, 1988; Interplant

'Joker', HT, ob, 1958; flowers orange-red, reverse lighter, slight fragrance; foliage glossy; [Peace X Karl Herbst]; Lens

'JOLaft', Min, mp, 1991; (**Afternoon Delight**); dbl., 21 petals, 1.5–2.75 in., borne mostly singly and in small clusters, slight fragrance; few prickles; foliage medium, medium green, matt; tall (45 cms), upright, spreading, vigorous growth; [Party Girl X Fashion Flame]; Jolly, Marie; Rosehill Farm, 1992

'Jolanda', HT, dp, 1959; bud long, pointed; flowers rose, cupped; vigorous, upright, bushy growth; Malandrone

'JOLcher', Min, ob, 1991; (**Cherry Bomb**); flowers orange, small, blooms borne mostly singly, very dbl., 1.5 in., no fragrance; few prickles; foliage small, medium green, semi-glossy; medium, upright growth; [Fashion Flame X Sheri Anne]; Jolly, Marie; Rosehill Farm, 1991

'JOLcol', Min, mp, 1991; (**Color Guard**); dbl., 2–5 cms, borne mostly singly, no fragrance; few prickles; foliage medium, medium green, matt; upright (20 cms), bushy, very hardy growth; [Anita Charles X Poker Chip]; Jolly, Marie; Rosehill Farm, 1993

'Joli Coeur', F, dr, 1963; bud globular; flowers dark crimson, medium, dbl., moderate fragrance; foliage dark; symmetrical growth; [Rose Gaujard X (Seedling X Josephine Bruce)]; Gaujard

Joli Mome, HT, 1969; Ducher, Ch.; (Cavriglia)

Jolie Comtoise®, F, op, 1991

Jolie Demoiselle, Pol, lp

'Jolie Madame', HT, or, 1958; bud ovoid; flowers vermilion-red, dbl., 65 petals, 4–4.5 in., cupped, slight fragrance; foliage leathery, glossy; vigorous, upright, bushy growth; [(Independence X Happiness) X Better Times]; Meilland, F.; C-P, 1960; URS, 1958

'Jolie Princesse', F, pb, 1955; flowers pink, shaded ochre, borne on large trusses, dbl., moderate fragrance; foliage leathery, bronze; very vigorous, bushy growth; [Peace X Independence]; Gaujard

Jolie Rose, HT, 1967; Mondial Roses; (Cavriglia)

Jolisquare®, S, mp

'Jollity Jane', Pol, m, 1993; flowers lilac-cream blend, blooms borne in large clusters, very dbl., 1.5 in., slight fragrance; few prickles; foliage small, medium green, glossy; tall, upright, bushy growth; [Valerie Jeanne X Yesterday]; Jobson, Daniel J.; Jobson, 1993

'JOLLmea', F, rb, 1995; (**Atlantic City**); flowers ivory white with dark red blend, full (26–40 petals), large (7+ cms) blooms borne mostly singly, slight fragrance; few prickles; foliage medium, dark green, semi-glossy; medium (105 cms), compact growth; [Crimson Glory X Orange Honey]; Jolly, Marie; Paramount Nursery, 1996

'Jolly', Pol, m, 1934; flowers carmine-purple, center white, borne in clusters, single, slight fragrance; foliage sparse, dark; dwarf growth; [Miss Edith Cavell X Tip-Top]; Leenders, M.

Jolly, *see* 'LuKOR'

Jolly Cupido, *see* RUIxandra

Jolly Dance, HKor, ob

'JOLLY GOOD', F, pb, 1973; bud ovoid; flowers salmon-pink, dbl., 55 petals, 3.5 in., moderate fragrance; foliage glossy, dark leathery; bushy, compact growth; [Cupid's Charm X Lucky Piece]; Fuller; Wyant

Jolly Good, MinFl, op, 1991; Lens

Jolly Joker, F, w

'JOLLY ROGER', F, or, 1973; bud ovoid, pointed; flowers bright reddish orange, medium, semi-dbl., cupped, slight fragrance; foliage wrinkled; growth moderate, bushy; [Spartan X Angelique]; Armstrong, D.L.; Armstrong Nursery

'JOLSIR', Min, ab, 1992; (**Sir**); semi-dbl., 6–14 petals, 1.5 in., moderate fragrance; few prickles; foliage medium, dark green, semi-glossy; medium growth; [Olympic Gold X Rise 'n' Shine]; Jolly, Marie; Rosehill Farm, 1993

'JOLSON', Min, pb, 1991; (**Sonata in Pink**); flowers medium pink fading to light pink, blooms borne mostly singly, dbl., 24 petals, 1.25 in., slight fragrance; few prickles; foliage medium, medium green, semi-glossy; medium (44 cms), upright growth; [Chris Jolly X Chattem Centennial]; Jolly, Marie; Rosehill Farm, 1992

'JOLTIP', Min, mp, 1987; (**Tipper**™); flowers medium pink, aging lighter, medium, long lasting, dbl., 32 petals, exhibition form, borne singly, slight fragrance; round, orange fruit; prickles bayonet, light brown; foliage medium, medium green, semi-glossy; upright, medium growth; PP007340; AOE, 1989; [Chris Jolly X Chattem Centennial]; Jolly, Marie; Rosehill Farm, 1988

Joly Rose Primaplant, HT, mp, 1963; flowers medium to large, dbl., slight fragrance; Vlaeminck; (Sangerhausen)

'JONDAD', Min, op, 1995; (**For You Dad**); flowers coral, single (5 petals), reflexes to form star, medium blooms borne mostly single; slight fragrance; some prickles; foliage medium, dark green, semi-glossy; upright, tall growth; [Heartbreaker X seedling]; Jones, Steve

'JONDOROSE', Min, ob, 1998; (**Dorothy Rose**); flowers orange and white, turning orange red and red, striped, single, 4–7 petals, 2 in., borne in small clusters, slight fragrance; foliage medium, medium green, semi glossy; climbing, 6 ft growth; [Sarabande X Peggy T]; Jones, Steve

Jone Asher, MinFl, 1987; Pearce, C.A.; (Cavriglia)

'JONETSU', HT, mr, 1978; bud pointed; flowers dark scarlet, dbl., 30–35 petals, 4.5–6. in., exhibition form, moderate fragrance; foliage dark, leathery; vigorous growth; [(Kagayaki X Prima Ballerina) X Kagayaki]; Suzuki, Seizo; Keisei Rose Nursery

'JONKHEER G. SANDBERG', HT, my, 1936; (J.G. Sandberg); flowers clear yellow, dbl.; foliage dark, leathery; vigorous growth; [Christine X Mrs Wemyss Quin.; Buisman, G. A. H.; Armstrong Nursery, 1941

'JONKHEER J. L. MOCK', HT, pb, 1910; bud pointed; flowers silvery rose-white, reverse carmine-pink, bluing slightly, dbl., exhibition form, moderate fragrance; foliage dark, leathery; vigorous growth; GM, Bagatelle, 1911; [(Mme Caroline Testout X Mme Abel Chatenay) X Farbenkonigin]; Leenders, M.

'JONKHEER J. L. MOCK, CLIMBING', Cl HT, pb, 1923; Timmermans

'JONKHEER MR G. RUYS DE BEERENBROUCK', HT, ob, 1919; flowers pure orange-yellow fading clear yellow, dbl.; vigorous growth; [Mme Mélanie Soupert X Joseph Hill]; Timmermans

Jonquille, *see* 'DELJONQ'

'JONWHITE', Min, w, 1998; (**White Sunshine**); flowers white, single, like my sunshine, single, 4–7 petals; foliage medium, medium green, semi-glossy; compact, bushy, medium, 2 ft growth; [Sarabande X Peggy T]; Jones, Steve

Jorianda, HT, m

Joro, HT, ob, 1979; RvS-Melle

José Carréras, *see* POUlnew

'JOSEF ANGENDOHR', F, dp, 1982; flowers deep pink, [Dame de Coeur sport]; Angendohr, Hans-Werner; Baumschulen Angendohr

'JOSEF PETER', HT, m, 1929; flowers pale blush, reverse mauve-rose, dbl., moderate fragrance; [Ruth X Frank W. Dunlop]; Ketten Bros.

'JOSEF ROTHMUND', HEg, ob, 1940; (Joseph Rothmund); bud small, ovoid, orange-red; flowers light red with pinkish yellow, borne in clusters, very dbl., profuse, non-recurrent bloom, intense fragrance; foliage bronze, leathery; very vigorous growth; [Joanna Hill X Magnifica]; Kordes

'JOSEF STRNAD', HT, rb, 1932; flowers dark red, with traces of yellow, orange and rose, very large, dbl., cupped, moderate fragrance; foliage leathery, glossy, dark, bronze; very vigorous, bushy, branching growth; [Aspirant Marcel Rouyer X Toison d'Or]; Böhm, J.; J&P, 1934

Josefa, HT, 1982; Laperriere, J.; (Cavriglia)

'JOSEFINA DE SALGADO', HT, dp, 1963; (Joséphine de Salgado); flowers bright pink, large, dbl., 30 petals, moderate fragrance; somewhat weak stems; very vigorous growth; [Queen Elizabeth X Peace]; Dot, Simon

'JOSEPH ARLES', Cl HT, rb, 1964; flowers vermilion-red, reverse silvery white, moderate fragrance; foliage leathery; vigorous, climbing growth; [Aloha X Gabychette]; Arles; Roses-France

'JOSEPH BAUD', HT, yb, 1919; flowers golden yellow and orange-yellow, dbl., intense fragrance; [Rayon d'Or X Unnamed seedling]; Gillot, F.

'JOSEPH BILLARD', HWich, mr, 1906; flowers carmine-red, medium, single; Barbier; (Sangerhausen)

'JOSEPH CHAPPAZ', HP, m, 1883; flowers violet-pink, large, dbl.; Schmitt; (Sangerhausen)

'JOSEPH COURBIS', HT, pb, 1958; flowers carthamus-pink to orange-red, dbl., 48 petals; foliage dark, glossy; vigorous, upright growth; [Margaret McGredy X Emma Wright]; Arles; Roses-France

'JOSEPH F. LAMB', S, dr, 1988; bud ovoid, pointed; flowers dark red, reverse lighter, aging darker, loose, medium, dbl., 23 petals, cupped, moderate, fruity fragrance; yellow-orange fruit; prickles awl-like, small, tan to reddish brown; foliage medium, medium green, semi-glossy; upright, bushy, low, winter hardy growth; [[Prairie Star X [(Dornroschen X Peace) X Music Maker]] X (Music Maker X Topsi)]; Buck, Dr Griffith J., 1989

Joseph Goudreau, B, pb

Joseph Guy, *see* **'LAFAYETTE'**

'JOSEPH GUY', Pol, dp, 1930; flowers carmine-pink, medium, semi-dbl., slight fragrance; Feldmann; (Sangerhausen)

'JOSEPH GUY', Pol, mp, 1950; flowers large, semi-dbl., slight fragrance; Westhus; (Sangerhausen)

'JOSEPH HILL', HT, pb, 1903; bud pointed; flowers pink shaded salmon, reverse coppery pink, dbl., slight fragrance; Pernet-Ducher

Joseph Klimes, HT, dr, 1985; flowers large, very dbl., slight fragrance; Urban, J.; (Sangerhausen)

'JOSEPH LIGER', HWich, yb, 1909; flowers canary-yellow, edged and washed light pink, reverse cream-white, vigorous, climbing growth; [R. wichurana X Irene Watts]; Barbier

'JOSEPH LOWE', HT, op; flowers salmon-pink; [Mrs W.J. Grant sport]

'JOSEPH PERNET D'ANNEMASSE', HT, op, 1934; bud pointed; flowers salmon, dbl.; foliage glossy, dark, bronze; very vigorous growth; Pernet-Ducher; Gaujard

Joseph Rothmund, *see* **'JOSEF ROTHMUND'**

Joseph Sauvageot®, HT, ob, 1989; Sauvageot

'JOSEPHINE', Min, w, 1969; flowers white or soft pink, micro-mini, small, dbl.; foliage small, glossy; dwarf, bushy growth; [(R. wichurana X Carolyn Dean) X Jet Trail]; Moore, Ralph S.; Sequoia Nursery

Joséphine Baker, *see* 'MEIMAUR'

Joséphine Beauharnais, *see* **'BELLE DE SÉGUR'**

'JOSEPHINE BRUCE', HT, dr, 1949; flowers crimson, dbl., 24 petals, 5–6 in., slight fragrance; foliage dark; vigorous, branching growth; [Crimson Glory X Madge Whipp]; Bees, 1949; Totty, 1953

'JOSEPHINE CARMODY', S, lp, 1995; flowers light pink, double, medium sized, dbl., 15–25 petals, borne in small clusters, slight fragrance; prickles numerous; foliage small, dark green, glossy; bushy, medium growth (36 in.); [Sea Foam sport]; Collins, Frank E. Jr.

Joséphine de Beauharnais, *see* **'BELLE DE SÉGUR'**

Joséphine de Salgado, *see* **'JOSEFINA DE SALGADO'**

Josephine Elizabeth, S, lp; Peden, R., 1999; (Weatherly, L.)

'JOSÉPHINE GUYET', B, dr, 1873; (Mme Joséphine Guyet); flowers deep red, recurrent bloom, Touvais

Joséphine Maltot, *see* **'MME BRAVY'**

'JOSÉPHINE MAROT', HT, lp, 1894; flowers white washed pink, Bonnaire

'JOSEPHINE MOREL', Pol, dp, 1892; flowers carmine-pink, small, dbl.; Alégatière; (Sangerhausen)

'JOSEPHINE RITTER', HMult, mp, 1900; flowers medium, dbl.; Geschwind, R.; (Sangerhausen)

'JOSEPHINE SPIECKER', HT, ob, 1939; flowers deep orange to yellow, very dbl., globular, foliage glossy, dark bronze; vigorous, bushy growth; Verschuren-Pechtold; Bentley; Harkness

'JOSEPHINE THOMAS', HT, ob, 1924; flowers orange-salmon to cream-flesh, very dbl., exhibition form, slight fragrance; foliage leathery; vigorous, bushy growth; H&S; Dreer

'JOSEPHINE VESTAL', HT, lp, 1923; flowers soft pink, dbl., exhibition form, moderate fragrance; very vigorous, bushy growth; [Ophelia X Unnamed seedling]; Hill, E.G., Co.; Vestal

Josephine Wheatcroft, *see* **'ROSINA'**

'JOSEPH'S COAT'®, LCl, rb, 1969; flowers yellow and red, repeats, dbl., 23–28 petals, 3 in., blooms in clusters, recurrent bloom, slight fragrance; foliage dark, glossy; vigorous, pillar growth; GM, Bagatelle, 1964; [Buccaneer X Circus]; Armstrong, D.L. & Swim, 1964; Armstrong Nursery, 1964

Josh, *see* 'TINJOSH'

Joshua™, *see* 'MOGAJOSH'

Joshua Bradley, *see* 'CHRISGOBRO'

'JOSPINK', F, lp, 1998; (**Anne Marie Laing**); flowers light pink, darker reverse, full, camellia form, very dbl., 26–40 petals, 4 in., borne in large clusters, moderate fragrance; few prickles; foliage medium, dark green, dull; upright, tall, 110 cms growth; [(Iceberg X Anytime) X Sexy Rexy]; Shipway, John; F. Haynes Partners, 1998

Josyane, S, m, 1980; flowers violet-red, small, dbl.; Lens, Louis; (Sangerhausen)

'JOSYSIGAL', HT, mr, 1975; flowers large, dbl., 60 petals, cupped, intense fragrance; Delforge, S.

Jour de Fête, HT, w, 1968; Lens

Jour des Pères, *see* **'VATERTAG'**®

'JOUR D'ETÉ', HT, mr, 1964; bud very long; flowers bright red, open; vigorous growth; [Coup de Foudre X Berthe Mallerin]; Combe; Vilmorin-Andrieux

'JOURNEY'S END', HT, ob, 1978; flowers indian orange, pointed, dbl., 37 petals, 6 in., slight fragrance; foliage large, glossy; vigorous, upright growth; [Doreen X Vienna Charm]; Gandy, Douglas L.

'JOUVENCELLE', HT, pb, 1969; flowers salmon-pink suffused red, dbl., 50 petals; foliage reddish; [Prima Ballerina X Helen Traubel]; Gaujard

'JOVE', F, or, 1968; flowers scarlet, semi-dbl., blooms in clusters, slight fragrance; foliage glossy; low growth; [Vera Dalton X Paprika]; Harkness

'JOVITA'®, F, or, 1975; blooms in large clusters, slight fragrance; foliage bright green; medium, bushy growth; [Jove X Tip Top]; Harkness; Hauser

'JOVITA PÉREZ', HT, ob, 1929; flowers coppery salmon, shaded coral, dbl., cupped, moderate fragrance; foliage soft, dark; long, strong stems; vigorous, compact growth; [Mme Butterfly X Souv. de Claudius Pernet]; Munné, B.

Joy®, *see* 'INTERMOTO'

'JOY', HT, pb, 1929; bud tangerine-red; flowers rose-pink suffused tangerine, base yellow, dbl., exhibition form, moderate fragrance; foliage leathery; vigorous, branching growth; Beckwith

Joy Bells, *see* **'JOYBELLS'**

Joy Button, *see* LUDpinbu

'JOY O'BRIEN', Gr, op, 1969; flowers pink shaded orange-salmon, large, dbl., moderate fragrance; foliage dark; vigorous growth; [Queen Elizabeth X Seedling]; Verschuren; Stassen

Joy of Health, *see* HARxever

Joy Owens, *see* 'MACRED'

Joy Pagram, HT, lp, 1995; Dawson

'JOY PARFAIT', F, lp, 1965; [Pink Parfait sport]; McIlroy

'JOYANCE', HT, mr, 1939; flowers velvety red, camellia shape, dbl., 50 petals, 4 in., moderate fragrance; foliage leathery, dark; very vigorous, upright growth; [Regina Elena sport]; Grillo

'JOYBELLS', F, mp, 1961; (Joy Bells); flowers rich pink, camellia-shaped, blooms in clusters, dbl., 30 petals, 3.5 in., moderate fragrance; [Unnamed seedling X Fashion]; Robinson, H.

Joybells, *see* 'KORSORB'

'JOYCE', HT, dr, 1953; flowers dark velvety crimson, pointed, medium, dbl., 24 petals, moderate fragrance; foliage leathery; vigorous growth; [George Dickson X Étoile de Hollande]; Cant, F.

Joyce Barden, *see* 'ARDour'

'JOYCE CLAIRE', F, dp, 1965; bud globular; flowers deep pink, open, small, dbl., slight fragrance; foliage glossy; very vigorous, upright growth; [Queen Elizabeth X ?]; Tonkin

Joyce Edmonds, F, op, 1992; Nieuwesteeg, J.

'JOYCE FAIREY', Cl HT, dp, 1929; flowers soft red; pillar growth; Clark, A.; NRS Victoria

Joyce Hunt, lp; [Scented Bouquet X Stella]; Dawson, George

Joyce Lomax, *see* **'SATAN'**

'JOYCE LONGLEY', HT, ob, 1958; flowers in sunset shades, [Opera sport]; Court

'JOYCE NORTHFIELD', HT, ob, 1977; flowers deep orange, high-pointed, dbl., 3–4 in., slight fragrance; foliage dark; vigorous, upright growth; [Fred Gibson X Vienna Charm]; Northfield

'JOYCE RILEY', F, ob, 1978; bud well formed; flowers vermilion, yellow-salmon, full, dbl., 25 petals, 2.5 in., slight fragrance; foliage dark, leathery; vigorous, upright growth; [Paddy McGredy X Arthur Bell]; Wood

'JOYCE ROBINSON', HT, op, 1945; flowers peach-pink, exhibition form, moderate fragrance; foliage dark, leathery; vigorous, bushy growth; [Rose Berkley sport]; Selwood; Rosecraft Nursery

Joycie™, *see* 'MORJOYC'

'JOYENA', LCl, m, 1964; flowers spirea-red, reverse tyrian purple, borne in small clusters, exhibition form, moderate fragrance; foliage dark, bronze, leathery; vigorous growth; [Blossomtime X ?]; Mason, P.G.

'JOYEUX NOËL', HT, or, 1960; bud long, pointed; flowers well formed, medium, dbl., 30–35 petals, moderate fragrance; foliage bronze, leathery; vigorous, bushy growth; [(Floradora X Independence) X (La Vaudoise X Léonce Colombier)]; Delbard-Chabert

'JOYFUL', HT, pb, 1931; bud pointed; flowers pink, reverse streaked red, base red and orange; vigorous growth; Vestal

Joyfulness, *see* **'FROHSINN'**®

Joyfulness, *see* 'TANSINNROH'

Joyfulness, Climbing, Cl HT, lp; [Joyfulness sport]; Ruston, D., 1988; (Weatherly, L.)

'JOYOUS', F, pb, 1939; flowers rose-pink, reverse slightly darker, vigorous, bushy growth; [Else Poulsen sport]; deRuiter; J&P

'JOYOUS CAVALIER', HT, mr, 1926; bud pointed; flowers brilliant red, open, large, dbl., 25–30 petals, slight fragrance; foliage dark, glossy; very vigorous growth; [Red-Letter Day X Clarice Goodacre]; Archer

Joyous Moment, *see* JACpri

Joyride, *see* 'RENRIDE'

'JUAN MARAGALL', HT, mr, 1960; flowers bright strawberry-red, large, dbl., 35 petals, intense fragrance; long, strong stems; vigorous, compact growth; [Chrysler Imperial X Buccaneer]; Dot, Simon

'JUAN PICH', HT, m, 1921; flowers purplish wine-red, dbl., moderate fragrance; Leenders, M.

'JUAN QUEVEDO', HT, ly, 1921; flowers cream-yellow, dbl., moderate fragrance; [Entente Cordiale X My Maryland]; Leenders, M.

'JUANA DE DARDER', HT, yb, 1947; flowers deep yellow shaded salmon, cupped, slight fragrance; foliage bright green; strong stems; vigorous growth; [Souv. de Claudius Pernet X (Sensation X Souv. de Claudius Pernet)]; Munné, M.

Juane Adam, F, my, 1997; Adam, M.

'JUANITA', HGal, pb; flowers pink edged paler, medium, dbl.

'JUANITA', Ch, pb, 1885; flowers pink spotted white, Robert

Jubilaire de Masaryk, *see* **'MASARYKOVA JUBILEJNI'**

'JUBILANT', F, lp, 1967; flowers flesh pink, blooms in clusters, 2.5 in., moderate fragrance; foliage glossy; [Dearest X Circus]; Dickson, A.

Jubilation, *see* 'JACREG'

'JUBILÄUMSROSE', HFt, w, 1910; flowers creamy white, large, semi-dbl.; Schmidt, I. C.; (Sangerhausen)

'JUBILÄUMSROSE', HFt, mr, 1929; flowers large, dbl.; Schmidt, I. C.; (Sangerhausen)

'JUBILEE', HP, m, 1897; flowers purple, shaded maroon, large, dbl., some recurrent bloom, moderate fragrance; moderate growth; [Victor Hugo X Prince Camille de Rohan]; Walsh

'JUBILEE', HT, w, 1930; (Allen's Jubilee); bud pointed; flowers cream, tinged salmon-pink and indian yellow, center coral-pink, dbl., exhibition form, intense fragrance; vigorous growth; [Paul's Lemon Pillar X Aspirant Marcel Rouyer]; Allen

Jubilee, *see* **'MASARYKOVA JUBILEJNI'**

Jubilee 150, *see* 'MEICLOUX'

'JUBILEE CELEBRATION', F, pb, 1977; flowers pink shaded salmon, dbl., 20 petals, 4 in., intense fragrance; foliage matt, green; growth moderate; [Elizabeth of Glamis X Prima Ballerina]; Smith, E.; Wheatcroft, 1976

Jubilee Sunset, *see* 'TALJUB'

'JUBILEJNAJA', HT, pb, 1940; flowers rose pink with dark yellow, medium, semi-dbl., slight fragrance; Kosteckij; (Sangerhausen)

Jubilejni, HT, lp; Urban, J., 1979; (Czech Rosa Club)

Jubileum 110, HT, lp; Urban, J., 1989; (Czech Rosa Club)

'JUDANN', Cl Min, op, 1996; (**Ann's Rose**); flowers light orange pink, blooms borne in small clusters, dbl., 15–25 petals, 1.5 in.; few prickles; foliage medium, light green, glossy; tall (5 ft), upright, pillar (2 ft) growth; Bell, Judy G.; Michigan Mini Roses, 1996

'JUDBABY', MinFl, dp, 1994; (**Betty's Baby**); flowers deep pink with white center, blooms borne in small clusters, semi-dbl., 6–14 petals, 1.5 in., no fragrance; few prickles; foliage small, medium green, semi-glossy; medium (12–14 in.), upright, bushy growth; [Twilight Trail X Charmglo]; Bell, Judy G.; Michigan Mini Roses, 1995

'JUDBRIGHT', Min, ob, 1993; (**Bright Sight**); flowers light orange, dbl., 26–40 petals, 1.5–2.5 in., borne in large clusters, no fragrance; few prickles; foliage small, medium green, matt; upright (18 in.), bushy growth; [Charmglo X Poker Chip]; Bell, Judy G.; Michigan Mini Roses, 1994; Michigan Miniature Roses, 1994, USA

'JUDDANCE', Min, m, 1993; (**Folk Dance**); dbl., 26–40 petals, 1.5 in., borne mostly singly, slight fragrance; few prickles; foliage small, dark green, glossy; bushy (41 cms) growth; [Dale's Sunrise X Angel Face]; Bell, Judy G.; Michigan Mini Roses, 1994

'JUDDOLL', Min, ly, 1996; (**Diamond Doll**); flowers light yellow, darker yellow center, quilled reflex, blooms b, dbl., 26–40 petals, 1.5 in., intense fragrance; few prickles; foliage medium, medium green, matt; upright growth; medium (18 in. winter hardy; [Rise 'n' Shine X Angel Face]; Bell, Judy G.; Michigan Mini Roses, 1997

Jude the Obscure, *see* 'AUSjo'

'JUDFEST', Min, ab, 1999; (**Bloomfest**™); flowers apricot blend, dbl. (17–25 petals), medium (1 in.) blooms borne in small clusters; no fragrance; few prickles; foliage medium, medium green, dull; compact, low (14 in.) growth; [Rise 'n' Shine X Selected Pollen]; Bell, Judy G.; Michigan Miniature Roses, 1996

Judi Dench, *see* 'PEATHUNDER'

'JUDIAN', Min, mr, 1996; (**Here's Ian**); semi-dbl., 6–14 petals, 1.5 in., borne mostly singly, no fragrance; few prickles; foliage small, medium green, semi-glossy, disease resistant; small, compact, rounded growth; low (12 in. winter hardy; [Dale's Sunrise X Seedling]; Bell, Judy G.; Michigan Mini Roses, 1997

'JUDIE DARLING', Cl Min, pb, 1979; flowers marbled pink, reflexed, dbl., 45–50 petals, 1–1.5 in., moderate fragrance; vigorous, climbing growth; Sudol, Julia

Judit, F, lp; cupped, 35 petals, slight; large, dark green, leathery, glossy; [(Frankfurt am Main X Maria Callas) X Dr Faust]; St. Wagner, 1996; Res. Stn. f. Fruit growing, Cluj, 1997, Romania

'JUDITH', HT, rb, 1938; flowers glowing cerise, reverse golden yellow, dbl., globular, intense, fruity fragrance; foliage glossy, bronze; vigorous, bushy growth; LeGrice

'JUDITH ANN', HT, dp, 1998; flowers deep pink with medium reverse and silvery sheen, very dbl., 41 petals, 6–7 in., borne singly, candelabras in summer, slight fragrance; prickles moderate; foliage medium, dark green, glossy; upright, medium, 4–5 ft growth; [Red Devil sport]; Schamel. Al; Edmunds Roses, 1998

'JUDITH BLACK', HT, dr, 1930; bud pointed; flowers rich dark red flushed fiery red, dbl., globular, moderate fragrance; foliage soft; dwarf growth; Clark, A.; Hazlewood Bros.

'JUDITH I. B. HALL', HT, pb, 1953; flowers pink, base orange, dbl., 32 petals, 5 in., intense fragrance; very vigorous growth; [Crimson Glory X Sterling]; Balcombe Nursery

'JUDLEE', MinFl, ob, 1992; (**Essie Lee**); flowers white with orange picotee down 1/2 of petals, white to light orange reverse, dbl., 26–40 petals, 1.5 in., globular, no fragrance; few prickles; foliage small to medium, dark green, semi-glossy; upright (46 cms), bushy growth; [Tennessee X Tennessee]; Bell, Judy G., 1991; Michigan Mini Roses

'JUDLOV', Min, m, 1994; (**Love in Bloom**); flowers mauve to tan with bright yellow stamens, blooms borne in small clusters, single, 5 petals, 1.5 in., slight fragrance; few prickles; foliage medium, dark green, semi-glossy; bushy, spreading (10–12 in.) growth; [Angel Face X Unknown]; Bell, Judy G.; Michigan Mini Roses, 1995

'JUDMARIE', Min, m, 1995; (**Ashley Marie**); flowers light lavender, moderately full (15–25 petals) small blooms borne mostly singly; no fragrance; few prickles; foliage small, dark green, semi-glossy; low (14 in.), bushy growth; [Dale's Sunrise X Angel Face]; Bell, Judy G.; Michigan Miniature Roses, 1996

'JUDMON', Min, m, 1995; (**Monday's Child**); flowers lavender and white, moderately full (15–25 petals), small, blooms borne in small clusters, no fragrance; few prickles; foliage medium, medium green, glossy; medium (16 in.), upright, bushy growth; [Dale's Sunrise X Angel Face]; Bell, Judy G.; Michigan Miniature Roses, 1996

'JUDMORN', Min, m, 1995; (**Misty Morning**); flowers lavender shaded yellow/tan, full (26–40 petals) small blooms borne mostly singly; slight fragrance; few prickles; foliage medium, medium green, matt; tall (18 in.), upright, bushy growth; [Dale's Sunrise X Angel Face]; Bell, Judy G.; Michigan Miniature Roses, 1996

'JUDNIC', Min, op, 1992; (**Baby Dominic**); flowers light orange-coral pink, micro-mini, dime-sized blooms, dbl., 15–25 petals, exhibition form, borne in small clusters, slight fragrance; few prickles; foliage small, light green, matt; low, compact, bushy growth; [Centergold X Cuddles]; Bell, Judy G.; Michigan Mini Roses

'JUDPRIDE', Min, rb, 1996; (**P.J.'s Pride**); bud medium; flowers white with bright pink to red edging shading down petals to, exhibition form, no fragrance; some prickles; foliage medium, dark green, dull; medium, upright growth; [American Rose Centennial X Selected Pollen]; Bell, Judy G.; Michigan Mini Roses, 1997

'JUDRIV', Min, ob, 1996; (**River City Jubilee**); flowers orange, tight, dbl., 26–40 petals, exhibition form, no fragrance; some prickles; foliage medium, dark green, semi-glossy, very disease-resistant; medium (16 in.), upright growth; [Jean Kenneally X Rainbow's End]; Bell, Judy G.; Michigan Mini Roses, 1996

'JUDSAR', Min, dr, 1996; (**Caesar's Rose**); semi-dbl., 1–1.5 in., borne mostly singly, no fragrance; many dark red, short prickles; foliage small, dark green, matt; upright growth; medium (15–18 in. winter hardy; [Jean Kenneally X Angel Face]; Bell, Judy G.; Michigan Mini Roses, 1997

'JUDSILK', Min, ab, 1996; (**Silk 'n' Satin**); flowers apricot, bright yellow stamens, blooms borne in small clusters, semi-dbl., 8–9 wide petals, 1.5 in., no fragrance; no prickles; foliage small, light green, matt; bushy, spreading growth; low (10 in.winter hardy; [Seedling X Unknown]; Bell, Judy G.; Michigan Mini Roses, 1997

'JUDSOLDE', Min, pb, 1994; (**Isolde**); flowers pink blend (light petals in center shading darker toward edges), dbl., 26–40 petals, 1.5 in., slight fragrance; few prickles; foliage medium, medium green, matt; upright, bushy (14–16 in.) growth; [Dale's Sunrise X Charmglo]; Bell, Judy G.; Michigan Mini Roses, 1995

'JUDSPUNK', Min, mr, 1993; (**Spunky**); dbl., 15–25 petals, 1.5 in., borne in small clusters, no fragrance; many prickles; foliage small, dark green, glossy; medium (16 in.), bushy, upright growth; [Dale's Sunrise X J. Michael]; Bell, Judy G.; Michigan Mini Roses, 1994

'JUDVEE', Min, dp, 1993; (**Vee Marie**); flowers bright deep pink, slight white on reverse, yellow stamens, dbl., 26–40 petals, 1.5 in., borne in small clusters, no fragrance; few prickles; foliage small, dark green, glossy; low (15 in.), bushy, spreading, compact growth; [Dale's Sunrise X Angel Face]; Bell, Judy G.; Michigan Mini Roses, 1993

'JUDVIRG', Min, ly, 1999; (**Dorothy Virginia**); flowers medium, dbl., 17–25 petals, borne mostly singly, no fragrance; few prickles; foliage large, dark green, semi-glossy; upright, medium growth; [Loving Touch X Unknown]; Bell, Judy G.; Michigan Mini Roses, 1999

'JUDY', HT, mr, 1940; flowers cerise-red, dbl., 55 petals, 4 in., moderate fragrance; [Jewel sport]; Grillo

Judy, *see* TANydu

Judy Dench, *see* 'PEATHUNDER'

'JUDY FINNIGAN', HT, op, 1999; flowers orange/peach, classic-shaped, dbl., 26–40 petals, 4–4.5 in., borne in small clusters, moderate fragrance; almost thornless; foliage medium, medium green, semi-glossy, very rain-resista; upright, tall (4 ft) growth; [Hazel Rose X Joe Longthorne]; Poole, Lionel; David Lister, Ltd., 2000

'JUDY FISCHER', Min, mp, 1968; bud pointed; flowers rose-pink, small, dbl.; foliage dark, bronze, leathery; vigorous, bushy, low growth; AOE, 1975; [Little Darling X Magic Wand]; Moore, Ralph S.; Sequoia Nursery

Judy Garland, *see* 'HARKING'

'JUDY HART', HT, mp, 1958; bud ovoid; dbl., 30–40 petals, 4–5 in., intense fragrance; foliage leathery; vigorous, bushy growth; [Pink Delight (HT) sport X (Senator X Florex)]; Motose; G.B. Hart

'JUDY ROBERTSON', Min, my, 1998; flowers medium yellow, very full, 80 petals, 2 in., exhibition form, borne singly and in small clusters, slight fragrance; no prickles; foliage medium, medium green, semi-glossy; long, thin stems; upright, bushy, 3 ft growth; [Little Darling X Rise 'n' Shine]; Warner, A.J.

'JUGOSLAVIE', HT, w, 1936; flowers large, dbl., slight fragrance; Böhm, J.; (Sangerhausen)

Juillet®, *see* BRIletjui

'JUJNOBEREJNAIA', F, dr, 1955; (River's South Bank); flowers velvety red, well shaped, medium, [Independence X Vaterland (?)]; Klimenko, V. N.

JuJu™, *see* 'BISRED'

'JULES', LCl, dr, 1997; flowers dark to medium red, velvety, very full, very dbl., 41 petals, 2.5 in., borne in small clusters, moderate fragrance; foliage medium, light green turning medium green, semi-glossy; spreading, medium growth; [Seedling X Seedling]; Jerabek, Paul E.

'JULES BARIGNY', HP, mr, 1886; flowers large, dbl., intense fragrance; Verdier, E.; (Sangerhausen)

'JULES CLOSEN', HWich, dr, 1935; flowers darker and more, dbl., [Excelsa sport]; Opdebeeck

'JULES FINGER', T, rb, 1879; flowers vivid red fading light red, shaded silvery, very large, dbl.; vigorous growth; [Catherine Mermet X Mme de Tartas]; Ducher, Vve.

'JULES GAUJARD', HT, ob, 1928; flowers bright orange-red flushed carmine, very large, cupped, intense fragrance; foliage bright green; very vigorous growth; [Jean C.N. Forestier X Unnamed seedling]; Pernet-Ducher; Gaujard

'JULES GIRODIT', HT, op, 1900; flowers light orange-pink, large, dbl.; Buatois; (Sangerhausen)

Jules Jurgensen, B, m, 1879; flowers purple-crimson, large; Schwartz

'JULES LEVACHER', HWich, lp, 1908; flowers creamy light rose pink, medium, semi-dbl.; Barbier; (Sangerhausen)

'JULES MARGOTTIN', HP, mp, 1853; flowers carmine-rose, rather, large, dbl., 90 petals, flat, recurrent bloom, slight fragrance; vigorous growth; [Probably La Reine seedling sport]; Margottin

'JULES MARGOTTIN, CLIMBING', Cl HP, mr, 1874; flowers red with purple tints, large, dbl., intense fragrance; Cranston; (Sangerhausen)

'JULES SEURRE', HP, 1869; Liabaud, I.; (Cavriglia)

'JULES TABART', HT, pb, 1920; flowers silvery salmon-pink, center coppery coral-pink, dbl., moderate fragrance; [Unnamed seedling X Mme Edouard Herriot]; Barbier

'JULES TOUSSAINT', HT, dr, 1900; flowers large, dbl., moderate fragrance; Bonnaire; (Sangerhausen)

'JULESCHILDREN', F, lp, 1998; (**Me Too**); flowers pink-purple, white eye, white reverse, small, semi-dbl., 8–14 petals, 5–1.5 in., no fragrance; prickles moderate; foliage small, light green, dull; upright, medium growth; [MACkaukaup X Nickelodeon]; Muha, Julius; Garden Jules, 1999

'JULESDEAR', HT, dp, 1998; (**Thanx Mom**, Thanx Mum); flowers hot fuschia pink, opens decorative, dbl., 26–40 petals, 4–4.5 in., borne in small clusters, intense, raspberry fragrance; prickles moderate; foliage medium, medium green, semi-glossy; bushy, medium, 3 ft growth; [Lancome X Stephen's Big Purple]; Muha, Julius; Hortico, Inc., 1998

'JULESLOVE', HT, lp, 1998; (**Luv Ya**); flowers soft apricot pink, full, very dbl., 26–40 petals, 4–4.5 in., borne mostly singly, moderate fragrance; prickles moderate; foliage medium, medium green, semi-glossy; upright, 3–5 ft growth; Muha, Julius

Juli de Sala®, *see* FEmental

Julia®, *see* PEKcouliane

'JULIA ANN BOSTICK', Pol, pb, 1935; flowers apple-blossom-pink, base white, small, single, cupped, moderate fragrance; dwarf growth; [Ideal sport]; Bostick

'JULIA BARTET', HT, yb, 1920; flowers dark canary-yellow, fading pale straw-yellow, dbl., slight fragrance; [Lyon Rose X Georges Schwartz]; Schwartz, A.

'JULIA CLEMENTS', F, mr, 1957; flowers bright red, borne in clusters, single, 3 in.; foliage dark, glossy; very vigorous growth; Wheatcroft Bros.

'JULIA COUNTESS OF DARTREY', HT, pb, 1927; flowers rose pink with golden yellow, very large, dbl., intense fragrance; Hall; (Sangerhausen)

Julia Ferran, *see* **'LADY TRENT'**

'JULIA MANNERING', HEg, lp, 1895; flowers pearly pink, yellow stamens blooms borne along the cane, semi-dbl., summer bloom, moderate fragrance; vigorous growth; Penzance

Julia Renaissance, *see* POUlheart

'JULIA, COUNTESS OF DARTREY', HT, rb, 1927; bud pointed; flowers tyrian rose, base yellow, very large, dbl., exhibition form, intense fragrance; foliage dark, leathery, glossy; very vigorous growth; GM, NRS, 1925; Hall; McGredy

'JULIANA ROSE', Pol, op, 1920; flowers pale salmon; [Orléans Rose sport]; Den Ouden

'JULIA'S ROSE'®, HT, r, 1976; bud long, pointed; flowers parchment and copper shades, pointed, small, dbl., 22 petals, 2.5 in., slight fragrance; foliage reddish; upright growth; GM, Baden-Baden, 1983; [Blue Moon X Dr A.J. Verhage]; Wisbech Plant Co.

Julia's Rose, Climbing, Cl HT, lp; [Julia's Rose sport]; St. Kilda's, 1994; (Weatherly, L.)

'JULIE', HT, dr, 1970; bud ovoid; flowers large, dbl., cupped, intense fragrance; foliage dark, soft; upright growth; [Seedling X Red American Beauty]; Kordes, R.; Kordes

Julie Andrews, *see* FRYvivacious

Julie Ann™, *see* 'SAVAWEEK'

'JULIE ANNE ASHMORE', HT, yb, 1985; flowers deep yellow, suffused with pink throughout, [Peace sport]; Owen, Fred

Julie Cussons, *see* FRYprincess

'JULIE DE MERSAN', M, mp, 1854; (Julie de Mersent); flowers rose shaded blush, Thomas

Julie de Mersent, *see* **'JULIE DE MERSAN'**

Julie Delbard®, *see* 'DELJULI'

'JULIE D'ÉTANGES', HGal, m; flowers rosy lilac, edged blush, large, dbl., cupped; erect, vigorous growth; Vibert, ca 1834

'JULIE SHARP', F, w, 1976; flowers white, pink edge maturing to scarlet, full, dbl., 25–30 petals, 3.5 in., cupped, slight fragrance; foliage matt green; vigorous, upright growth; [Evelyn Fison sport]; Sharp

'JULIE STRAHL', HT, rb, 1928; flowers nasturtium-red, passing to golden yellow, dbl., intense fragrance; [Lady Greenall X Gorgeous]; Leenders Bros.

Julie Y, *see* HARbinger

Julie's Choice, *see* 'RENJULIE'

'JULIEN POTIN', HT, ly, 1927; (Golden Pernet); bud golden, pointed; flowers primrose-yellow, large, dbl., exhibition form, moderate fragrance; foliage bright green; vigorous growth; GM, Portland, 1929; [Souv. de Claudius Pernet X Unnamed seedling]; Pernet-Ducher; Dreer

'JULIEN POTIN, CLIMBING', Cl HT, dy, 1935; Bostick

Julien Renoard, HT, 1987; Dorieux, Francois; (Cavriglia)

'JULIENNE', HT, mp, 1940; flowers silvery pink, [Jewel sport]; Grillo

Julies Choice, *see* 'RENJULIE'

Julie's Choice, *see* 'RENJULIE'

'JULIET', HP, pb, 1910; bud globular, golden yellow; flowers rich rosy red to deep rose, reverse old-gold, large, dbl., occasionally recurrent bloom, moderate fragrance; foliage curiously curled; vigorous growth; [Capt. Hayward X Soleil d'Or]; Paul, W.

Juliet Ann, *see* HARvissa

'JULIET STAUNTON CLARK', HT, w, 1933; flowers white, center blush-white, turning white, large, very dbl., moderate fragrance; [Juliet sport]; Robichon

'JULIETTE', Min, mr; flowers brilliant crimson-scarlet, dbl., 30 petals; foliage bright red in fall; vigorous (10–12 in.) growth; Lamb Nursery

'JULIETTE', HGal, m; Miellez, before 1828

'JULIETTE E. VAN BEUNINGEN', HT, dp, 1937; (Mme Juliette); bud pointed; flowers bright pink, open, very large, semi-dbl., foliage leathery, dark; [Dame Edith Helen X Mrs Sam McGredy]; Buisman, G. A. H.

Julischka®, *see* 'TANJUKA'

'JULIUS FABIANICS DE MISEFA', T, mr, 1902; flowers crimson, Geschwind, R.

'JULIUS GOFFERJE', HT, pb, 1930; flowers peach-pink on yellow ground, dbl.; foliage bright green; vigorous growth; Schmidt, J.C.

'JULKLAP', S, op, 1940; flowers light salmon-pink, large, semi-dbl., slight fragrance; Krause; (Sangerhausen)

'JULY GLORY', HWich, dp, 1932; flowers rich rose-pink, small, borne in large clusters, dbl.; foliage glossy; vigorous growth; Chaplin Bros.

Jumpin' Jack, F, op, 1997

Jumping Jack Flash, *see* 'TALJUM'

'JUNE', HT, pb, 1937; bud pointed; flowers shell-pink, center darker, well shaped, moderate fragrance; vigorous growth; Archer

'JUNE ABERDEEN', F, ob, 1977; flowers salmon, dbl., 20 petals, 2.5 in., slight fragrance; foliage dark; [Anne Cocker X (Sabine X Circus)]; Cocker

'JUNE BOYD', HT, rb, 1924; flowers salmon-carmine, base yellow, opening to bright peach-blossom, dbl., moderate fragrance; McGredy

'JUNE BRIDE', Gr, w, 1957; bud pointed, greenish white tipped pink; flowers creamy white, blooms in clusters of 3–7, dbl., 30 petals, 4 in., exhibition form, moderate fragrance; foliage leathery, crinkled; vigorous, upright growth; [(Mme Butterfly X New Dawn) X Crimson Glory]; Shepherd; Bosley Nursery

June Bug, *see* 'PIXJUN'

'JUNE FLAME', HT, ob, 1949; bud small, tight; flowers bright orange-flame shaded copper, very early bloom; Fletcher; Tucker

June Laver™, *see* 'LAVJUNE'

'JUNE MORN', LCl, rb, 1939; bud ovoid; flowers carmine-red, reverse touched gold, dbl., 5 in., exhibition form, some recurrent bloom, vigorous, climbing (8 ft) growth; [Mme Gregoire Staechelin X Souv. de Claudius Pernet, Climbing]; Nicolas; J&P

'JUNE OPIE', F, ab, 1958; flowers apricot shaded salmon-pink, borne on trusses, semi-dbl., 3 in., slight fragrance; foliage leathery; very free, upright growth; [Masquerade X Seedling]; Kordes; Morse

'JUNE PARK', HT, dp, 1958; flowers rose-pink, dbl., 40 petals, 4.5–5 in., intense fragrance; foliage dark; vigorous, spreading growth; GM, NRS, 1959; [Peace X Crimson Glory]; Park; Sanday and Nonin

'JUNE PATRICIA', HT, lp, 1966; flowers silvery pink, dbl., 3 in., moderate fragrance; foliage dark, glossy; moderate growth; [Peace X Ena Harkness]; Lens

'JUNE TIME', Min, lp, 1963; flowers light pink, reverse darker, small, in clusters, dbl., 75 petals; foliage glossy; bushy, compact (10–12 in. growth; [(R. wichurana X Floradora) X ((Étoile Luisante seedling X Red Ripples) X Zee)]; Moore, Ralph S.; Sequoia Nursery

'JUNE WAY', HT, mp, 1977; dbl., 33 petals, 4–5 in., cupped, slight fragrance; foliage glossy; spreading growth; [Pink Favorite X Chrysler Imperial]; Atkiss; Wyant

'JUNE WEDDING', HT, w, 1977; bud pointed; flowers white, tinted yellow, dbl., 27 petals, 4 in., exhibition form, slight fragrance; foliage glossy, dark; upright growth; [Bewitched sport]; Graham; South Forrest Rose Nursery

June Whitfield, *see* HARchutzpah

'JUNEEN', HT, my, 1967; exhibition form, slight fragrance; compact growth; [Burnaby X Burnaby seedling]; Mason, P.G.

Junior, HT, 1973; St Zila; (Cavriglia)

'JUNIOR BRIDESMAID', F, mp, 1962; bud short, pointed; flowers sweetheart, dbl., 35 petals, 2 in., slight fragrance; foliage leathery; vigorous, upright growth; [Stoplite X Lovelight]; Jelly; E.G. Hill Co.

'JUNIOR GILBERT', HT, or, 1954;, cupped, vigorous growth; Mallerin, C.; EFR

Junior Miss, *see* **'AMERICA'S JUNIOR MISS'**

'JUNIOR MISS', F, pb, 1943; bud well formed; flowers pink and yellow, medium, borne in clusters, semi-dbl., exhibition form, slight fragrance; foliage glossy; vigorous, bushy growth; [Joanna Hill X Heidekind]; Duehrsen; California Roses

'JUNIOR PROM', F, dr, 1962; bud ovoid; flowers crimson, open, dbl., 30–45 petals, 1.5–2 in., moderate fragrance; vigorous, upright growth; [Orange Sweetheart X Lovelight]; Jelly; E.G. Hill Co.

'JUNIOR VAN FLEET', S, lp, 1923; flowers flesh-pink, dbl., non-recurrent, moderate fragrance; [Dr W. Van Fleet X Frau Karl Druschki]; Kemp, J.A.

'JUNO', HCh, lp, 1847; flowers pale rose, very large, dbl., globular; Laffay, M.

'JUNO', HT, mp, 1950; bud ovoid; flowers soft medium pink, large, dbl., 30 petals, exhibition form, slight fragrance; foliage bright, leathery, wrinkled, glossy; moderate upright, bushy growth; [Duquesa de Peñaranda X Charlotte Armstrong]; Swim, H.C.; Armstrong Nursery

'JUNO', C, lp; flowers blush pink, very dbl., globular; arching shrub growth; L1832

Junon®, *see* 'GAUCOVA'

'JUPITER', sp; flowers bright pink, dbl.; shining, black fruit; foliage finely divided; dense, shrubby (3–4 ft) growth

Jupiter, HT, pb

Jupiter, S, mr, 1987; GPG Bad Langensalza

'JUPITER', HP, lp; [Prince Camille de Rohan X Unknown]; Williams, A., 1900; (Weatherly, L.)

Jupon Rose®, *see* 'LENNIRO'

Jura®, *see* **Berliner Luft®**

Jurassic Pink, HKor, mp

'JUSAMANDA', Min, w, 1991; (**Amanda Kay**); bud ovoid; flowers white to near white, some pink shading, small, dbl., exhibition form, borne singly, spicy fragrance, intense; foliage small, medium green, glossy; bushy, low growth; [Seedling X Seedling]; Justice, Jerry G., 1990; Justice Miniature Roses

'JUSCLARE', Min, w, 1995; (**Nancy Clare**); flowers nearly pure white with pink accent at outer edge, blooms bor, dbl., 18–20 petals, 1.75–2 in., moderate, spicy fragrance; foliage medium, light green, semi-glossy; medium (32 in.), upright growth; RR; [Jennifer X Seedling]; Justice, Jerry G.; Justice Miniature Roses, 1995

'JUSELLE', Min, dp, 1991; (**Giselle**); bud small, pointed with medium green sepals; flowers pink outer edges with very light pink (almost white) at midline, semi-dbl., 18 petals, 1.25 in., no fragrance; foliage small, dark green, glossy, disease-resistant; bushy, low growth; [Crazy Dottie X Unnamed seedling]; Justice, Jerry G.; Justice Miniature Roses, 1992

'JUSHAZE', Min, m, 1996; (**Purple Haze**); flowers soft mauve with light tan shading at the throat, blooms born, dbl., 26–40 petals, 2.75 in., intense fragrance; some prickles; foliage medium, medium green, semi-glossy; bushy, medium growth; PPAF; [Twilight Trail X Seedling]; Justice, Jerry G.; Justice Miniature Roses, 1996

'JUSHONEY', Min, ob, 1991; (**My Honey**); bud pointed; flowers orange with light yellow accent at base, acquires pinkish cast, dbl., no fragrance; foliage medium, dark green, semi-glossy; spreading, medium growth; [Orange Honey sport]; Justice, Jerry G., 1990; Justice Miniature Roses

'JUSLATTE', Min, m, 1995; (**Latte**); flowers russet with lavender edge, blooms borne singly, very dbl., 56 petals, 1.5 in., moderate, spicy fragrance; foliage medium, medium green, semi-glossy; medium, bushy growth; RR; [Twilight Trail X Seedling]; Justice, Jerry G.; Justice Miniature Roses, 1995

'JUSLEMON', Min, my, 1996; (**Lemon Meringue**); flowers medium yellow, with white to ivory tips, reverse becomes lig, dbl., 26–40 highly reflexed petals, 1 in., nearly thornless; foliage medium, dark green, dull; compact (16–20 in.) growth; RR; [Rise 'n' Shine X Seedling]; Justice, Jerry G.; Justice Miniature Roses, 1997

'JUSMAIDEN', Min, w, 1995; (**Snow Maiden**); flowers pure white with ivory to bone white throat, reverse pure white, pink tones in cool weather, full (26–40 petals), petals form starlike cluster medium blooms borne mostly singly, slight fragrance; few prickles; foliage medium, dark green, glossy; upright, medium growth; [Snow Twinkle X seedling]; Justice, Jerry G.; Justice Miniature Roses, 1996

'JUSMICHAEL', Min, dp, 1991; (**Baby Michael**); bud pointed; flowers deep rose pink, reverse medium pink with a touch of silver at base, very dbl., 38–40 petals, exhibition form, slight, spicy fragrance; foliage medium, dark green, semi-glossy; upright, medium growth; [Seedling X Seedling]; Justice, Jerry G., 1990; Justice Miniature Roses

'JUSPROP', Min, pb, 1999; (**Whirlygig**); flowers soft pink and white, reverse darker, single, 5–11 petals, 2 in., slight fragrance; foliage medium, dark green, semi-glossy; upright, medium (20–22 in.) growth; [(Seedling X Magic Carrousel) X Unknown]; Justice, Jerry G.; Justice Miniature Roses, 1999

'JUSREGAL', Cl Min, dp, 1999; (**Regal**); flowers old rose pink, reverse same, very dbl., 41 petals, 1.75 in., borne mostly singly, no fragrance; prickles moderate; foliage medium, medium green, glossy; upright, climbing, medium to tall (6 ft) growth; [Rosanna X Unknown]; Justice, Jerry G.; Justice Miniature Roses, 1999

'JUSSABRA', Min, mp, 1999; (**Sabra**); flowers medium pink, reverse light pink, single, 5–11 petals, 1.75 in., borne mostly singly, no fragrance; foliage large, light green, dull; upright, medium (18–21 in.) growth; [(Fairy Moss X Unknown) X Unknown]; Justice, Jerry G.; Justice Miniature Roses, 1999

Just Buddy™, *see* 'KINBUD'

Just Dreamy, *see* 'PERJUSDREAM'

Just For Fun, *see* 'STRILDEW'

Just For You, *see* 'MORYOU'

Just Happy, *see* 'DICTALENT'

Just Jenny, Min, ab, 1993

'JUST JOEY', HT, ob, 1972; flowers buff-orange, classic form, dbl., 30 petals, 5 in., intense fragrance; foliage glossy, leathery; growth moderate; GM, RNRS, 1986 Hall of Fame, WFRS, 1994 James Mason, RNRS, 1986; [Fragrant Cloud X Dr A.J. Verhage]; Cants of Colchester, Ltd.

'JUST JUDY', HT, lp, 1992; dbl., 26–40 petals, 1.5–2.75 in., borne mostly singly, slight fragrance; some prickles; foliage medium, dark green, semi-glossy; upright (0–8 m) growth; [Mischief X Simba]; Poole, Lionel, 1994

'JUST LUCKY', HT, w, 1984; flowers well-formed, large, dbl., 35 petals, intense fragrance; foliage medium, dark, glossy; bushy growth; [Typhoo Tea X Pascali]; Bridges, Dennis A.; Bridges Roses, 1985

Just Magic, *see* 'TROBIC'

Just Peachy, *see* 'GELPEACH'

Just Reward, *see* 'GUESTALL'

Just Simon, *see* 'BURBRAHMVIDYA'

Just William, *see* 'DRISCORED'

Justa Little Goofy, *see* 'RENGOOFY'

Justin, HT, mr, 1997

'JUSTINA', HT, mp; moderate fragrance

'JUSTINE', C, m, 1822; flowers pale lilac-pink, medium, very dbl.; Vibert

'JUSTINE', HT, ly, 1935; bud orange; flowers creamy yellow, base dark orange, reverse almost white, large, dbl., 30–35 petals; foliage dark, leathery; vigorous growth; [Joanna Hill X Sweet Adeline]; Hill, Joseph H., Co.

'JUSTINE RAMET', C, m, 1845; flowers purplish-rose, medium, dbl.; Vibert

Justine Silva, *see* **'Rita Sammons'**

'Justino Henriques', HT, yb, 1926; flowers yellow tinted orange, stamens carmine, dbl., moderate fragrance; [Louise Catherine Breslau sport]; deFreitas; P. Guillot

'Justizrat Dr Hessert', HT, pb, 1919; bud carmine-red; flowers salmon-pink shaded red and yellow, dbl., moderate fragrance; [Gen. MacArthur X Tip-Top]; Lambert, P.

Jutland, F, dp, 1995

'Jutlandia', HWich, mp, 1913; dbl.; vigorous growth; [Mme Norbert Levavasseur X Dorothy Perkins]; Poulsen, D.T.

Jutta, Pol, lp, 1964; flowers large, dbl.; Rupprecht-Radke; (Sangerhausen)

Jutta, Pol, mr, 1970; flowers medium, semi-dbl.; Scholle, E.; (Sangerhausen)

'Jutul', F, dp, 1982; flowers deep pink, small blooms in clusters of 7–9, semi-dbl., 14 petals, cupped, slight fragrance; curved, red-brown prickles; foliage dark, glossy; vigorous growth; [New Dawn X Moulin Rouge]; Lundstad, Arne; Agricultural University of Norway

'Juwel', HT, ly, 1911; flowers large, dbl., moderate fragrance; Hinner, W.; (Sangerhausen)

K

K S G Centenary, HT, yb, 1995; K&S

K S R, HT, 1987; Keisei Rose Nurseries, Inc.; (Cavriglia)

K. of K., *see* **'Kitchener of Khartoum'**

K.A. Viktoria, *see* **'Kaiserin Auguste Viktoria'**

K.T., *see* 'TALkev'

K.T. Marshall, *see* **'Katherine T. Marshall'**

Kabuki®, *see* 'MEIgold'

'Kagayaki', F, rb, 1970; (Brilliant Light); flowers brilliant scarlet and yellow, large, dbl., exhibition form, slight fragrance; foliage glossy, dark; vigorous, upright growth; [(Aztec seedling X (Spectacular X Aztec)) X Cover Girl seedling]; Suzuki, Seizo; Keisei Rose Nursery

Kaikoura®, *see* 'MACwalla'

Kaileen, *see* 'JACmar'

Kaina, HT, pb, 1985; G&L

'Kaiser Wilhelm', T, pb, 1889; flowers golden pink, large, very dbl., moderate fragrance; Drögemüller; (Sangerhausen)

'Kaiser Wilhelm I', HP, m, 1878; flowers purple/pink, large, dbl., moderate fragrance; Ruschpler; (Sangerhausen)

'Kaiser Wilhelm II', HT, mr, 1909; flowers medium, semi-dbl., moderate fragrance; Welter; (Sangerhausen)

'Kaiserin Auguste Viktoria', HT, w, 1891; (K.A. Viktoria); bud long, pointed; flowers snowy white, center tinted lemon, well-formed, very dbl., 100 petals, intense fragrance; foliage rich green, soft; [Coquette de Lyon X Lady Mary Fitzwilliam]; Lambert, P.

'Kaiserin Auguste Viktoria, Climbing', Cl HT, w, 1897; Dickson, A.

Kaiserin Farah, HT, dr, 1965; flowers very large, very dbl., intense fragrance; Kordes, W. Söhne; (Sangerhausen)

'Kaiserin Friedrich', T, pb, 1890; flowers golden yellow on pink, large, very dbl., intense fragrance; Drögemüller; (Sangerhausen)

Kaiserin Zita, HT, lp

Kaitlyn Ainsley, HRg, m, 1998; Baskerville

Kakadu, S, lp; Peden, R., 2000; (Weatherly, L.)

Kakwa, S, w, 1973; Wallace

'Kalahari', HT, op, 1971; flowers salmon-pink, high-pointed, dbl., 25 petals, 4 in., slight fragrance; foliage glossy, dark; [Uncle Walter X (Hamburger Phoenix X Danse de Feu)]; McGredy, Sam IV

'Kalavalla', F, mp, 1935; dbl., borne in large clusters; vigorous growth; [Else Poulsen X Seedling]; Poulsen, S.

'Kaleidoscope', F, ob, 1972; flowers orange and yellow, dbl., 28 petals, 3 in., slight fragrance; foliage glossy; [Circus X Redgold]; Fryer, Gareth; Fryer's Nursery, Ltd.

Kaleidoscope™, *see* 'JACbow'

Kalinka, *see* 'MEIhartfor'

Kalinka, F, 1970; Meilland, L.; (Cavriglia)

Kalinka, Climbing, *see* 'MEIhartforsar'

'Kalmia', HWich, w, 1911; flowers white, upper half of petals tinged pink, borne in clusters, single; foliage dark, glossy; vigorous, climbing growth; Walsh

Kalyana, HT, ob, 1995; Kasturi

Kamaladevi Chattopadhayay, HT, op, 1989; Pal, Dr B.P.

Kambala, HT, my, 1988; Swane

'Kamchin', HT, mp, 1972; flowers neyron rose, outside rose, dbl., 45 petals, 4–5 in., cupped, slight fragrance; foliage dull, dark; vigorous growth; [Carina sport]; Kammeraad; Meilland

Kamelia, HT, mr, 1969; flowers large, dbl., moderate fragrance; Urban, J.; (Sangerhausen)

'Kamion', Min, or, 1988; flowers bright orange-red, reverse matte finish, clear, golden stamens, semi-dbl., 12 petals, exhibition form, no fragrance; oval, large, light, light orange fruit; prickles curved, small, reddish; foliage large, dark green, glossy; upright, tall, sturdy growth; [Starina X Unnamed seedling]; Schoen-Jones, Helen; Justice Miniature Roses, 1988

'Kammersanger Terkal', F, ob, 1971; (Hartina); bud small, globular; flowers pure orange, medium, dbl., slight fragrance; foliage glossy; dwarf, bushy growth; Tantau, Math.; Ahrens & Sieberz

Kammersingerin Perra, S; (Cavriglia)

Kampai, *see* **'Kan-Pai'**

'Kana', HT, dr, 1982; flowers large blooms borne 1–3 per stem, dbl., 38 petals, exhibition form, no fragrance; small prickles, slanted downward; foliage medium green; tall, bushy growth; [Ginger Rogers X Chiyo]; Ota, Kaichiro

Kanaal, F, op, 1998; RvS-Melle

Kanak, F, yb, 1988; Sunil Jolly

'Kanakangi', HT, ab, 1968; bud globular; flowers gold and apricot, open, medium, semi-dbl., intense fragrance; foliage leathery; moderate, bushy, open growth; [Mme Charles Sauvage X ?]; Pal, Dr B.P.; Indian Agric. Research Inst.

'Kanarie', HT, dy, 1919; flowers clear dark yellow, dbl.; [Golden Star X Melody]; Verschuren

Kanchi, HT, dp, 1976; Viraraghavan, M.S.

'Kanegem', F, or, 1985; flowers large blooms, dbl., 42 petals, exhibition form, borne 1–7 per cluster, no fragrance; foliage dark, glossy; upright growth; [Ludwigshafen am Rhein X Satchmo]; Rijksstation Voor Sierplantenteelt, 1982

'Kan-Pai', HT, dr, 1983; (Kampai); flowers deep red, blooms borne 1–3 per cluster, dbl., 48 petals, exhibition form, moderate fragrance; foliage medium, dark; upright growth; GM, Rome, 1983; [(Yu-ai X (Happiness X American Beauty)) X Pharoah]; Suzuki, Seizo; Keisei Rose Nursery, 1980

Kantha Selvon, *see* 'EVEgaytime'

Kanva, HT, op, 1974; Kasturi

Kanyakumari, LCl, op, 1979; Viraraghavan, M.S.

Kapai®, *see* 'MACgam'

Kapiti, *see* 'MACglemil'

'Kara', Min, mp, 1972; (Dear One); bud long, mossy; flowers light to medium pink, micro-mini, mini-moss, small, single, foliage small, soft; vigorous, dwarf, bushy growth; [Fairy Moss X Fairy Moss]; Moore, Ralph S.; Sequoia Nursery

'Kardinal'®, HT, dr, 1934; flowers scarlet-red, sometimes tipped blackish, large, cupped, vigorous, compact, bushy growth; [Château de Clos Vougeot seedling X Unnamed seedling]; Krause

Kardinal™, *see* 'KORlingo'

Kardinal 85, *see* 'KORlingo'

'Kardinal Piffl', HT, ob, 1925; flowers red-orange, reverse golden yellow, very large, dbl., moderate fragrance; foliage good; GM, Bagatelle, 1926; [(Mme Edouard Herriot X Rayon d'Or) X Mme Charles Lutaud]; Leenders Bros.

'Kardinal Schulte', HT, mr, 1926; flowers brilliant scarlet-red, dbl., moderate fragrance; [(Jonkheer J.L. Mock X Radiance) X Commandeur Jules Gravereaux]; Leenders Bros.

Kardinal Schulte, Climbing, Cl HT, mr; flowers large, dbl., moderate fragrance; (Sangerhausen)

Kardinal, Climbing, Cl HT, lp; [Kardinal sport]; Knight, J., 1999; (Weatherly, L.)

Kardinal's Flame, HT, or

'Karel Hynek Mácha', HT, dr, 1936; flowers velvety red, well shaped; Brada, Dr.; Böhm

'KAREL IV', HT, mr, 1935; flowers large, dbl., moderate fragrance; Brada, Dr.; (Sangerhausen)

'KAREN', S, w; flowers creamy white flecked red, well-shaped, very dbl., non-recurrent, moderate fragrance; bushy, erect (5 ft) growth; [R. primula X R. spinosissima cultivar]

'KAREN', F, my, 1961; flowers straw-yellow shaded ochre-yellow, dbl., 30–35 petals, 3 in.; bushy growth; [Goldilocks X Fashion]; Borgatti, G.; Sgaravatti

Karen, *see* DEVoran

Karen Blixen, *see* 'POULARI'

Karen Julie, HT, or; [Alexander X Weinerwold]; Allender, Robert William, 1979

'KAREN MARIE', HT, yb, 1973; flowers cream and yellow, flushed pink, full, dbl., 40 petals, 4 in., intense fragrance; foliage glossy, dark; free growth; [Kordes' Perfecta sport]; Gates

'KAREN POULSEN', F, mr, 1932; flowers scarlet, single blooms in huge trusses, vigorous growth; (28); GM, NRS, 1933 GM, Portland, 1935; [Kirsten Poulsen X Vesuvius]; Poulsen, S., 1932; J&P, 1933

'KAREN POULSEN, CLIMBING', Cl F, mr; Roger

Karenina, Min, op, 1999

Karen's Cream, HT, lp; [Shocking Blue sport]; Neil, J.; (Weatherly, L.)

'KAREN'S PINK LACE', F, mp, 1998; flowers medium pink, creamy reverse at base, very full, 41 petals, borne mostly singly, moderate fragrance; prickles moderate, medium size; foliage medium, medium green, dull; bushy, medium, 3–4 ft growth; [French Lace sport]; Prevatt, Clarence; Giles Rose Nursery, 1998

Karine, Pol, mp, 1992; Knopf, Ruth

Karine Sauvageot®, *see* SAUkar

Karkulka, F, 1986; Urban, J.; (Cavriglia)

Karkulka, HT, lp; Strnad; (Czech Rosa Club)

Karl Fischer, S, or, 1976; Hetzel

Karl Foerster, *see* **'KARL FÖRSTER'**

'KARL FÖRSTER', HSpn, w; (Karl Foerster); bud pointed; flowers snow-white, large, dbl., exhibition form, repeat bloom, slight fragrance; foliage wrinkled, light; vigorous (7 ft) growth; [Frau Karl Druschki X R. spinosissima altaica]; Kordes, 1930

'KARL HERBST', HT, mr, 1950; (Red Peace); flowers dull dark scarlet, well-shaped, large, dbl., 60 petals, intense fragrance; vigorous growth; GM, NRS, 1950; [Independence X Peace]; Kordes

Karl Höchst, F, dp, 1983; flowers carmine-pink, large, dbl., slight fragrance; Hetzel; (Sangerhausen)

Karl Mayer, HT, 1970; Hetzel, K.; (Cavriglia)

'KARL SCHNEIDER', HMult, lp, 1934; flowers medium, semi-dbl.; Vogel, M.; (Sangerhausen)

'KARL WEINHAUSEN', F, dr, 1942; flowers dark red tinted salmon, rosette form, large, borne in clusters, semi-dbl., 20 petals, slight fragrance; vigorous, upright growth; [Baby Chateau X (Heidekind X Ingar Olsson)]; Tantau

Karla, *see* 'FRYtrooper'

'KARLEA', HT, pb, 1965; flowers pink and salmon, base yellow, full, dbl., slight fragrance; foliage glossy; vigorous, upright growth; O'Brien

Karl-Heinz Hanisch, HT, w, 1986; flowers cream white to light pink, medium, dbl., intense fragrance; Meilland; (Sangerhausen)

'KARLIAN', HT, mr, 1991; dbl., borne mostly singly, intense fragrance; foliage medium, medium green, semi-glossy; medium, upright, compact growth; [Pristine X Burgund]; Alde, Robert O.

'KARLSRUHE', HKor, dp, 1957; bud ovoid; flowers deep rose-pink, large blooms in large clusters, very dbl., cupped, repeat bloom, slight fragrance; foliage glossy; vigorous, climbing growth; Kordes

Karma, *see* 'MACNIC'

Karneol Rose, HT, op, 1964; flowers dark salmon-pink, large, dbl.; Rupprecht-Radke; (Sangerhausen)

Karolina, F, mr, 1994; flowers carmine-red, medium, very dbl.; Hempelmann; (Sangerhausen)

Karoline Reiber, Pol, mr, 1991; flowers carmine-red, small, dbl., intense fragrance; Hetzel; (Sangerhausen)

'KAROLINE SVETLA', Pol, lp, 1937; flowers medium, semi-dbl.; Brada, Dr.; (Sangerhausen)

Karol's Rose, Min, mp

Karolyn, *see* **'CORALIN'**

'KAROO', HT, lp; flowers bright pink, large, exhibition form, vigorous growth; RULED EXTINCT 2/88; Elton Farm Nursery

Karoo, *see* 'POULKARO'

Karoo Rose, *see* 'POULKARO'

Kasachstanskaia Jubilejnaja, HT, dr, 1958; flowers large, very dbl., slight fragrance; Sushkov & Besschetnova; (Sangerhausen)

'KASBAH', HT, dp, 1977; bud pointed; flowers dark pink, reverse lighter, very dbl., 80–110 petals, 3.5 in., cupped, moderate fragrance; foliage leathery; upright growth; [Unnamed seedling X Tropicana]; Takatori, Yoshiho; Japan Rose Nursery

Kasbek, HT, w; flowers medium to large, semi-dbl., slight fragrance; VEG; (Sangerhausen)

'KASHMIR', Cl F, m, 1973; bud ovoid; flowers clear mauve, full, medium, dbl., slight fragrance; foliage glossy, leathery; vigorous, climbing growth; [Magenta X Royal Tan]; Thomson

'KASSEL', LCl, or, 1957; bud ovoid; flowers orange-scarlet, large blooms in clusters to 10, semi-dbl., recurrent bloom, moderate fragrance; foliage dark, glossy; vigorous growth; not dependably hardy.; [Hamburg X Scarlet Else]; Kordes, R.; Morse

Kasteel Van Ooidonk, F, lp; single, 10 petals, flat; foliage matt; good grower growth; RvS-Melle, 1995

Kasteel van Ooidonk, F, 1995; Rijksstation Voor Sierplantenteelt; (Cavriglia)

Kasturi Rangan, HT, m, 1983; Agarwal

'KÄTCHEN VON HEILBRONN', Pol, dr, 1922; flowers very dark red, small, dbl.; [Freudenfeuer X Unnamed variety]; Kiese

'KÄTE BEYER', Pol, op, 1947; flowers salmon-pink, medium, dbl.; Vogel, M.; (Sangerhausen)

'KATE EDWARDS', HT, m, 1963; flowers magenta; [Condessa de Mayalde sport]; Edwards

Kate Emily McCormack, *see* 'KENMACHO'

'KÄTE FELBERG', HT, pb, 1930; bud pointed; flowers creamy white, reverse violet-rose, large, dbl., moderate fragrance; vigorous growth; [Unnamed seedling X Mrs Wemyss Quin.]; Felberg-Leclerc

'KATE MOULTON, CLIMBING', Cl HT, pb, 1928; [Miss Kate Moulton sport]; Opdebeeck

'KATE MULL', HT, op, 1934; flowers crushed strawberry and coppery rose, very dbl.; Easlea

'KATE RAINBOW', HT, pb, 1935; flowers blend of glowing pinks and gold, well formed, very large, dbl., intense fragrance; foliage glossy, leathery; very vigorous growth; Beckwith

'KÄTE SCHMID', HMult, dp, 1931; flowers deep rose-pink, very large, borne in clusters, dbl., slight fragrance; foliage light; long stems; very vigorous, climbing growth; [Fragezeichen X Tausendschon]; Vogel, R., Jr.; Kordes

Kate Sheppard, F, mp

'KATE SMITH', HT, pb, 1954; bud ovoid; flowers apricot overcast grenadine-pink, dbl., 35–40 petals, 4.5–5 in., exhibition form, intense fragrance; foliage glossy; vigorous growth; [(Break o' Day X Golden Rapture) X Ballet]; Boerner; J&P

Katelyn Ann, *see* 'MICKATE'

Katerina, HP, mr, 1955; flowers large, very dbl., moderate fragrance; Kosteckij; (Sangerhausen)

Kateryna, *see* CLEkate

Kate's Rose, Pol, m, 1988; [Baby Faurax X Unknown]; Cox, Mrs K., 1988

Katharina Kündgen, S, mp, 1996; dbl., moderate fragrance; Michler, K. H.; (Sangerhausen)

Katharina Sophia, F, lp, 1991; Karwecki

'**Katharina Zeimet**', Pol, w, 1901; (White Baby Rambler); flowers pure white, small blooms in clusters of 25–50, dbl., moderate fragrance; foliage small, rich green; short stems; dwarf, bushy growth; [Étoile de Mai X Marie Pavie]; Lambert, P.

'**Katharine Pechtold**', HT, ob, 1934; bud old-gold and bronzy orange; flowers coppery orange, flushed rose and gold, semi-dbl., moderate, clove pink fragrance; foliage leathery; vigorous, bushy growth; [Roselandia X Charles P. Kilham]; Verschuren-Pechtold; Dreer; H&S

'**Katharine Worsley**', F, mr, 1962; flowers bright oriental red, borne in large, well-spaced clusters, dbl., 28 petals, 3.5 in.; moderate, bushy growth; Waterhouse Nursery

Käthchen, F, mr, 1985; flowers large, dbl.; Hetzel; (Sangerhausen)

'**Käthe Duvigneau**', S, mr, 1942; flowers glistening red tinted salmon, large blooms in clusters of 12–15, semi-dbl., 15 petals, slight fragrance; foliage leathery, glossy, bright green; vigorous, upright growth; [Baby Chateau X R. roxburghii]; Tantau

'**Kathe von Saalfeld**', T, 1914; Eibel

'**Katherine Cook**', HT, mr, 1927; flowers cherry-red, dbl., moderate fragrance; [Crusader X Unnamed seedling]; Cook, J.W.

'**Katherine Harbour**', HT, mp, 1973; bud ovoid; flowers pink, tinted apricot, open, large, dbl., moderate fragrance; foliage glossy; vigorous, upright growth; [Queen Elizabeth X Comtesse Vandal]; Aloe Vera Nursery

Katherine Loker, *see* 'AROkr'

Katherine Mansfield, *see* 'MEIlanein'

'**Katherine McCarty**', Min, yb, 1991; bud ovoid; flowers yellow center, then white, with coral edges, reverse same, very dbl., 60 petals, exhibition form, moderate, damask fragrance; foliage medium, dark green, matt; bushy, medium growth; [Poker Chip X Rise 'n' Shine]; Gruenbauer, Richard, 1984; Flowers 'n' Friends Miniature Roses, 1993

Katherine McGredy, HT, 1997; McGredy, Sam IV; (Cavriglia)

'**Katherine Mock**', HT, my, 1943; [Pres. Herbert Hoover sport]; Mock

'**Katherine T. Marshall**', HT, mp, 1943; (K.T. Marshall); flowers deep rose-pink, flushed yellow, dbl., 22 petals, 5 in., cupped, slight, spicy fragrance; foliage leathery; vigorous, upright growth; AARS, 1944; [Unnamed seedling X Chieftain., Boerner; J&P

'**Kathleen**', HMsk, lp, 1922; flowers blush-pink, small blooms in large clusters, single, recurrent bloom, slight fragrance; vigorous (6 ft) growth; [Daphne X Perle des Jeannes]; Pemberton

'**Kathleen**', HMult, pb; flowers soft rose with white eye, single; Paul, W., 1908

'**Kathleen**', HT, lp, 1934; bud very long; flowers light yellowish salmon, large, slight fragrance; vigorous, free branching growth; Dickson, A.

'**Kathleen Ferrier**', F, op, 1952; flowers deep salmon-pink, blooms in small clusters, semi-dbl., 18 petals, 2.5 in., moderate fragrance; foliage dark, glossy; vigorous, upright growth; [Gartenstolz X Shot Silk]; Buisman, G. A. H.

'**Kathleen Harrop**', B, lp, 1919; flowers soft shell-pink; (21); [Zephirine Drouhin sport]; Dickson, A.

Kathleen Jane, *see* HORcoed

Kathleen Jermyn, *see* 'LEGneed'

'**Kathleen Joyce**', F, lp, 1970; flowers soft pink, dbl., 30 petals, 4 in., exhibition form, intense fragrance; [Paddy McGredy X Ice White]; McGredy, Sam IV

'**Kathleen Kaye**', HT, mp, 1959; flowers rich rose-pink, dbl., 60 petals, 5 in., intense fragrance; foliage dark; vigorous growth; [Directeur Guerin X Mirandy]; Kemp, M.L.

'**Kathleen Kellehan**', S, rb, 1999; (The Nanango Rose); flowers red blend, full (26–40 petals), large (10 cms) bloom borne in large clusters, very fragrant; moderate prickles; foliage medium, medium green, semi-glossy; upright, medium (4 ft) growth; [Lilian Austin X Oklahoma]; Laving, Peter

'**Kathleen Kennedy**', HT, op, 1939; flowers light salmon-carmine, shaded orange, well-formed, large, moderate fragrance; strong stems; vigorous growth; Dickson, A.

'**Kathleen King**', HT, dp, 1930; flowers carmine-pink, well formed, large, exhibition form, intense fragrance; vigorous growth; Marriott

Kathleen Kirkham, *see* 'KIRkit'

'**Kathleen Mills**', HT, pb, 1934; bud long, pointed; flowers pale pink tinted silvery, reverse deep pink, large, semi-dbl., intense fragrance; foliage leathery; vigorous growth; LeGrice

'**Kathleen Nash**', HT, mr, 1944; bud urn shaped; flowers bright cerise, dbl., 28 petals, 4 in., globular, intense fragrance; foliage leathery, dark; vigorous growth; [Pink Delight sport]; Spera; Rose Farms Corp

'**Kathleen O'Rourke**', HT, op, 1976; flowers soft orange-pink, dbl., 38 petals, 4 in., exhibition form, moderate fragrance; foliage large, matt; [Fragrant Cloud X Red Planet]; Dickson, Patrick; A. Dickson

'**Kathleen Peden**', HT, dr, 1959; bud long pointed; flowers crimson, large, dbl., exhibition form, intense fragrance; foliage leathery, dark; very vigorous, upright growth; [Crimson Glory X Charles Mallerin]; Peden, G.H.

Kathleen Rumble, *see* 'WELrum'

'**Kathleen Wiggin**', Cl HP, w, 1932; bud long pointed, opening one at a time on each cluster; flowers white, sometimes tinged pink, large, borne in clusters, very dbl., slight fragrance; foliage glossy, heavy; long stems; vigorous, climbing (12–15 ft) growth; [Frau Karl Druschki seedling]; Wiggin

Kathleen's Rose, *see* KIRkitt

'**Kathryn**', S, ab, 1986; bud copper; flowers apricot, aging to buff, copper stamens, open form, dbl., 25 petals, 4–5 in., moderate, fruity fragrance; foliage semi-glossy; vigorous growth; [Wind Chimes X Yellow HT]; Eggeman, H.W.

'**Kathryn Bailey**', F, w, 1992; dbl., 26–40 petals, 1.5–2.75 in., slight fragrance; some prickles; foliage light green, glossy; medium tall, upright growth; [Gene Boerner sport]; Bailey, Dr Edwin; Bailey's Plant Farm, 1993

'**Kathryn Gram**', F, ab, 1945; bud urn shaped; flowers large, moderate fragrance; bushy, low growth; [Talisman X ?]; Moore, Ralph S.; Sequoia Nursery

Kathryn McGredy, *see* MACauclad

Kathryn Morley, *see* 'AUSclub'

'**Kathy**', Min, mr, 1970; bud pointed; flowers small, dbl., moderate fragrance; foliage small, leathery; moderate, dwarf, bushy growth; [Little Darling X Magic Wand]; Moore, Ralph S.; Sequoia Nursery

'**Kathy Fiscus**', F, mp, 1950; bud ovoid; flowers deep flesh-pink, open, medium, borne in clusters, dbl., 45 petals, moderate fragrance; foliage leathery; very vigorous, upright, bushy growth; [Unnamed seedling X Baby Chateau]; Duehrsen; Elmer Roses Co.

Kathy Reid, HRg, m, 1987; Sandbrook

'**Kathy Robinson**', Min, pb, 1968; flowers pink, creamy reverse, dbl., 26 petals, 1 in., exhibition form, slight fragrance; foliage small, glossy, dark, embossed; upright, bushy growth; [Little Darling X Over the Rainbow]; Williams, Ernest D.; Mini-Roses

Kathy's Hunter's Pink, HWich, mp

'**Katie**', LCl, mp, 1959; bud long, pointed; flowers reverse darker, large, semi-dbl., 17 petals, cupped, recurrent bloom, intense fragrance; foliage glossy; vigorous growth; [New Dawn X Crimson Glory]; O'Neal; Wyant

Katie Crocker, *see* BURbrindley

Katie Pianto's Rose, LCl, pb

Katie's McNear HP, HP, dp; (found rose)

Katrin, HT, op, 1972; flowers salmon-pink, large, dbl.; GPG Bad Langensalza; (Sangerhausen)

'**KATRINA**', HT, op, 1964; bud pointed; flowers orange-scarlet, medium, dbl., cupped, slight fragrance; foliage glossy, leathery; bushy, vigorous growth; [Baccará sport]; Samtmann Bros.; C-P

Katy Lampkin, *see* 'TALKAT'

Katy Road Pink, S, mp

Kauff, *see* '**KAUTH**'

'**KAUTH**', (strain of R. canina), lp; (Kauff); sometimes used as understock

'KAWAMOBLUE', HT, m, 1995; (**Purple Rain**); flowers mauve, full (26–40 petals), large (7+ cms) blooms borne mostly singly, very fragrant; few prickles; foliage medium, dark green, semi-glossy; upright, medium growth; [Seedling X Blue Moon]; Kawamoto, Hiromoto; Carlton Rose Nurseries, 1994

'**KAY**', F, or, 1971; bud ovoid; flowers vermilion, large, dbl., cupped, slight fragrance; foliage large, bronze; vigorous, upright growth; [Queen Elizabeth X Numero Un]; Delforge

'**KAY BARNARD**', HT, lp, 1974; bud long, pointed; flowers pale pink, exhibition form, moderate fragrance; upright growth; [South Seas X Queen Elizabeth]; Dingle

Kay Denise, *see* 'TINKAY'

Kayla, *see* 'MORKAY'

'**KAZANLIK**', D, dp; (R. damascena trigintipetala, Trigintipetala); flowers deep pink, dbl., 30 petals; (28)

'**KDE DOMOV MUJ**', HMult, dp, 1935; flowers carmine-pink, medium, semi-dbl.; Böhm, J.; (Sangerhausen)

Kean, HGal, m

KEBesa, HT, lp; (**Pretty Woman**); (KEBesa), 1999

'**KEELY**', Min, or, 1988; bud pointed; flowers bright, orange-red, reverse slightly darker, aging color fading, dbl., 20 petals, exhibition form; prickles straight, pointed, medium, red; foliage medium, medium green, semi-glossy; bushy, medium growth; [Party Girl X Unnamed seedling]; Bridges, Dennis A.; Bridges Roses, 1989

Keepit, HT, lp, 1988; Dawson

Keepsake, *see* 'KORMALDA'

'**KEEPSAKE**', HT, dr, 1941; flowers deep red, moderate fragrance; RULED EXTINCT 7/81; [Anne Leygues seedling]; Clark, A.

'**KEES KNOPPERS**', Pol, w, 1930; flowers flesh-white, open, large, borne in clusters, semi-dbl.; foliage rich green; vigorous, bushy growth; [Mev. Nathalie Nypels sport]; Leenders, M.

'**KEGON**', HT, mr, 1976; bud pointed; flowers deep red, dbl., 30 petals, 5 in., exhibition form; foliage leathery; tall growth; [Gruss an Berlin X Christian Dior]; Onodera, Toru F.; S. Onodera

'**KEI**', Min, lp, 1980; bud ovoid, pointed; dbl., 33 petals, borne 1–3 per cluster, slight fragrance; straight prickles; foliage tiny, dark; very compact, bushy growth; [Unnamed seedling X Unnamed seedling]; Lyon, Lyndon

'KEIBELMI', Min, rb, 1991; (**Shortcake**™); bud ovoid, pointed; flowers red, white reverse, red color pales with age, blues slightly, dbl., 30–35 petals, cupped, no fragrance; foliage large, dark green, glossy; upright, bushy, tall growth; [Unnamed seedling X Unnamed seedling]; Keisei Rose Nurseries, Inc., 1981; Bear Creek Gardens, 1991

'KEIBIAN', HT, yb, 1993; (**French Perfume**™); flowers light yellow with rose pink picotee, very dbl., 40–45 petals, 5–6.5 in., borne mostly singly, intense, fruity fragrance; foliage large, medium green, purple-red when new, semi-gloss; tall (110–150 cms), upright, spreading growth; PP8476; [(Todoroki X Montana) X Seedling]; Suzuki, Seizo; Bear Creek Gardens, 1993

'KEIFUPIE', Min, dp, 1999; (**Hime**); flowers bright purplish pink, dbl., 20–25 petals, 1.5–2 in., flat, moderate fragrance; foliage dark green, semi-glossy; 10 in. growth; GM, Japan Rose Concours, 1997; [Ko's Yellow X seedling]; Hirabayashi, Hiroshi, 1995; Keisei Rose Nurseries, 1997

'KEIFUPIRA', F, dp, 1999; (**Urara**); flowers pale purplish pink, dbl., 25–30 petals, 3–4 in., cupped, no fragrance; foliage dark green, leathery; 2–3 ft growth; GM, Japan Rose Concours, 1995; [Mimi X Minuette]; Hirabayashi, Hiroshi, 1993; Keisei Rose Nurseries, 1995

'KEIMIPIA', Min, lp, 1999; (**Sayokyoku**); flowers soft pink, very dbl., 45–50 petals, 2–2.5 in., exhibition form, slight fragrance; foliage dark green, leathery; 1.5–2 ft growth; GM, Japan Rose Concours, 1994; [Sonia X Petite Folie]; Hirabayashi, Hiroshi, 1992; Keisei Rose Nurseries, 1996

KEImove, Gr, m; (**Prelude**); (KEImove)

KEInoumi, F, dr; (**Scarlet Mimi**™); (KEInoumi), 1992

KEIreb, HT, mr; (**Mahalia**™); (KEIreb), 1992

KEIren, Pol, dp, 1995; (**Fairy-Tale**); Keiren

KEIromo, HT, op; (**Silva**); (KEIromo)

KEItaibu, Gr, dp; (**Laser**™); (KEItaibu)

'**KEITH HARDER**', HT, lp; Harder, K., 1954; (Weatherly, L.)

Keith Kirsten, *see* TANsirk

Keithie, F, rb

'KEITOLI', S, dp, 1984; (Ferdi, **Ferdy**™); flowers deep pink, small, dbl., 20 petals, no fragrance; foliage medium, medium green, matt; spreading (to 5 ft) growth; groundcover; [Unnamed climbing seedling X Petite Folie seedling]; Suzuki, Seizo; Keisei Rose Nursery

KEIzoubo, HT, ob; (**Pareo**®); (KEIzoubo)

'KELDAZ', Min, yb, 1997; (**Dazzler**™); flowers double, medium, white with red edge and yellow base, dbl., 15–25 petals, borne mostly singly, slight fragrance; foliage medium, dark green, semi-glossy; upright, medium growth; [Rainbow's End X Kristin]; Kelly, Martin

Kelleriis-Rose, *see* '**POULSEN'S SUPREME**'

'**KELLI ANN**', Min, w, 1981; flowers small, dbl., 35 petals, slight fragrance; foliage small, medium green, matt; bushy growth; [Patricia Scranton X (Patricia Scranton X Fairy Moss)]; Dobbs, Annette E.

Kelly Country, LCl, pb, 1997; Sutherland

Kelly Reynolds, *see* 'REYKELLY'

Kelly-Leigh, *see* 'STEKELLY-LEIGH'

Ken Davis, *see* 'GELDAVIS'

'KENDAD', F, ab, 1996; (**May Graham**); flowers coppery apricot, fading to pink, blooms borne in small clusters, dbl., 26 petals, 1.5–2.75 in., slight fragrance; many prickles; foliage medium, medium green, glossy; medium ($3\frac{1}{2}$ ft), bushy growth; [Prominent X Kiskadee]; Kenny, David

'KENDANQU', F, or, 1993; (**Wiggy**); flowers orange red with cerise, dbl., 26–40 petals, 1.5–2.75 in., borne in small clusters, slight fragrance; some prickles; foliage medium, dark green, semi-glossy; tall (90 cms), upright, bushy growth; [(Prominent X Kiskadee) X (Mary Sumner X Kiskadee)]; Kenny, David; Kenny, 1993

'KENFINE', Min, dp, 1997; (**Riverdance**); flowers double, medium, dbl., 15–25 petals, borne in small clusters, moderate fragrance; foliage small, medium green, glossy; bushy, medium (28in.) growth; [MEIdomonac X Freegold]; Kenny, David

'KENMACHO', F, rb, 1999; (**Kate Emily McCormack**); flowers red, pink to white edges,handpainted with white eye, reverse silver, semi-dbl., 8–14 petals, 3 in., borne in large clusters, moderate fragrance; prickles numerous; foliage large, dark green, glossy; spreading, bushy, tall (4 ft) growth; [((Mary Sumner X Kiskadee) X Bassino) X Little Artist]; Kenny, David

'KENMADO', F, op, 1997; (**Sister Joan**); flowers double, large, dbl., 15–25 petals, borne in small clusters, moderate fragrance; some prickles; foliage large, dark green, semi-glossy; upright, bushy, tall (3.5 × 2.5 ft) growth; [(Mary Sumner X Kiskadee) X Maestro]; Kenny, David

'KENMASNIA', F, yb, 1995; (**Aunty Lil**); flowers yellow edged pink, fades to pink, moderately full (15–25 petals), medium (4–7 cms). Blooms borne in large clusters, slight fragrance; some prickles;

foliage medium, medium green, glossy; medium, upright growth; [(Friesia X Kiskadee) X (Mary Sumner X Regensberg)]; Kenny, David; Kenny, 1994

'KENMOLL', F, dy, 1999; (**Tessa O'Keeffe**); dbl., 15–25 petals, 3 in., borne in small clusters, moderate fragrance; few prickles; foliage medium, dark green, glossy; bushy, medium (30 in.) growth; [Friesia X Kiskadee]; Kenny, David

'KENMORE', HT, dy, 1971; flowers deep yellow, dbl., 36 petals, 4.5 in., slight fragrance; foliage glossy; [Jane Lazenby X Golden Delight]; Macara; Sanday

Kenny's Rose, *see* 'GELKEN'

'KENOBSESS', F, ab, 1999; (**Audrey McCormack**); flowers soft apricot yellow, dbl., 26–40 petals, 3 in., borne in large clusters, moderate fragrance; many prickles; foliage large, medium green, glossy; spreading, bushy, medium (3 ft) growth; [Golden Wedding X Bright Smile]; Kenny, David

'KENORA DUET', F, ab, 1976; bud long, pointed; flowers salmon-pink, open, medium, semi-dbl., slight fragrance; foliage glossy; vigorous, bushy growth; [Duet sport]; Tresise

'KENSINGO', F, op, 1995; (**Mary Swindells**); flowers salmon pink, full (26–40 petals), medium (4–7 cms) blooms borne in small clusters; slight fragrance; many prickles; foliage medium, dark green, glossy; tall, upright, bushy growth; [(Mary Sumner X Kiskadee) X (Mary Sumner X Regensberg)]; Kenny, David; Kenny, 1990

Kent®, *see* 'POULCOV'

'KENTFIELD', Pol, mp, 1922; flowers soft cameo-pink to deeper pink, small, dbl., slight fragrance; [Cécile Brunner X Unnamed seedling]; Diener

Kentucky, HT, w

Kentucky Derby, *see* 'ARODER'

Kern White, HP, w; (found rose)

'KERRY GOLD', F, yb, 1967; flowers canary-yellow, outer petals veined red, blooms in clusters, 3 in., globular; foliage dark; [Circus X Allgold]; Dickson, A.

'KERRY MACNEIL', F, mr, 1967; flowers bright vermilion; free growth; [Orangeade X Anna Wheatcroft]; Vincent

'KERRYMAN'®, F, pb, 1971; flowers salmon and pink, dbl., 24 petals, 4.5 in., exhibition form, slight fragrance; [Paddy McGredy X (Mme Leon Cuny X Columbine)]; McGredy, Sam IV; McGredy

'KERSBERGEN', Pol, mr, 1927; flowers bright currant-red; [Miss Edith Cavell sport]; Kersbergen

'KESRI', F, or, 1969; bud ovoid; flowers orient orange, open, small, dbl., slight fragrance; foliage leathery; vigorous growth; [Orangeade, Climbing X Open pollination]; Singh; Gopalsinamiengar

Kessy, F, my

'KETJE', HT, mr, 1938; bud pointed; flowers brilliant red mixed dark pink, well formed, very dbl., foliage bronze; very vigorous growth; [Mrs Sam McGredy X E.G. Hill]; Lens

Kev, *see* 'TALKEV'

Kevin, HT, dp, 1997; Twomey, Jerry

'KEW BEAUTY', HT, dr, 1918; flowers crimson, dbl., moderate fragrance; Therkildsen

'KEW RAMBLER', LCl, mp, 1913; flowers pink, center paler, single, borne in clusters; foliage gray-green; height to 15 ft; [R. soulieana X Hiawatha]

'KEWAI', HT, yb, 1997; flowers full, large, very dbl., 26–40 petals, slight fragrance; few prickles; foliage medium, medium green, semi-glossy; upright, bushy, medium (180 cms) growth; Ohtsuki, Hironaka

Key Rock Rose, Misc OGR, yb

'KEYSTONE', LCl, dy, 1904; flowers deep lemon-yellow, Dingee & Conard

'KI KI PAQUEL', F, or, 1960; flowers bright brick color; [Super-Congo X Independence]; da Silva, Moreira

'KIA ORA', F, or, 1962; flowers orange-scarlet, large, dbl., 35 petals; foliage glossy, dark; vigorous, bushy, compact growth; [Independence X Independence]; Mason, P.G.

Kia Ora, *see* 'SUNCRAL'

Kibo, *see* **'KIBOH'**

'KIBOH', F, rb, 1986; (Gypsy, Gypsy Carnival, Kibo, Lovita); flowers orange-red, reverse yellowish, large, dbl., 50 petals, exhibition form, blooms in clusters of 3, slight fragrance; prickles slanted downward; foliage dark, semi-glossy; vigorous, upright growth; PP007139; GM, The Hague, 1985; [Liberty Bell X Kagayaki]; Suzuki, Seizo; Keisei Rose Nursery

'KICKAPOO', HT, rb, 1990; bud ovoid; flowers yellow, outer petals red, white reverse aging to light red blend, dbl., 20 petals, slight fragrance; round, green light yellow fruit; prickles straight, green aging to tan; foliage medium, medium green, dull; bushy, medium, slightly spreading growth; [(Daisy Mae X First Prize) X Unnamed seedling]; Stoddard, Louis, 1991

'KIDDY', Pol, mr, 1967; bud pointed; flowers dark red, becoming brighter, medium, single, cupped, abundant, recurrent bloom; foliage glossy; vigorous, upright growth; [Mme Dieudonne X Seedling]; Delforge

'KIDWAI', HT, mp, 1933; (Kidway); bud long, pointed; flowers salmon-rose, lower half of petals golden yellow, large, semi-dbl., slight fragrance; foliage leathery, dark, bronze, glossy; very vigorous, bushy growth; Pernet-Ducher; Dreer; Gaujard; H&S

Kidway, *see* **'KIDWAI'**

Kieran Ross Clark, *see* 'HORANYME'

'KIESE', HCan, mr, 1910; flowers bright red, semi-dbl.; [Général Jacqueminot X R. canina]; Kiese

'KIFTSGATE', (R. filipes form), w, 1954; (Filipes Kiftsgate, R. filipes 'Kiftsgate'); flowers creamy white, borne in large clusters, single, 5 petals, moderate fragrance; very vigorous, sprawling growth; E. Murrell

Kika, *see* 'ORTKEE'

Kiki, HT, 1972; Dorieux, Francois; (Cavriglia)

Kiki Paquel, Gr, 1960; Moreira da Silva, A.; (Cavriglia)

Kiki Rose®, *see* 'POULCAT'

Kiko, *see* JACdeep

Kilimanjaro, *see* 'HERARO'

'KILLARNEY', HT, mp, 1898; bud long, pointed; flowers bright medium pink, loose, large, dbl., intense fragrance; foliage bronze; [Mrs W.J. Grant X Charles J. Grahame]; Dickson, A.

'KILLARNEY BRILLIANT', HT, dp, 1914; bud long, pointed; flowers brilliant pink to rosy carmine, open, dbl., moderate fragrance; foliage rich green, soft; [Killarney sport]; Dickson, A.

'KILLARNEY DOUBLE PINK', HT, mp, 1935; flowers sparkling shell-pink; [Killarney Double White sport]; Vestal

'KILLARNEY DOUBLE WHITE', HT, w, 1912; flowers snowy white; [Killarney sport]; Budlong; A.N. Pierson

'KILLARNEY DOUBLE WHITE, CLIMBING', Cl HT, w, 1935; Howard Rose Co.

'KILLARNEY QUEEN', HT, dp, 1912; flowers tyrian rose, brighter than Killarney, [Killarney sport]; Budlong

'KILLARNEY, CLIMBING', Cl HT, mp, 1908; Reinberg

Kilwinning, HSpn, lp; intense; fern like; [R. spinosissima altaica X Hansen's Yellow]; Wright, Percy H.

'KILWORTH GOLD', HT, dy, 1977; flowers golden yellow, [Whisky Mac sport]; Gandy, Douglas L.

'KILWORTH PRIDE', F, dr, 1955; semi-dbl., blooms in clusters; foliage bronze; dwarf, bushy growth; [Better Times X Floribunda seedling]; deRuiter; Gandy Roses, Ltd.

'KIM', F, my, 1971; flowers mini-flora, dbl., 28 petals, 3 in., slight fragrance; foliage small, light green, matt; dwarf growth; [(Orange Sensation X Allgold) X Elizabeth of Glamis]; Harkness, 1973

'KIM', Pol, mp, 1956; flowers geranium-rose, small, borne in clusters; vigorous growth;

[Independence X Salmon Perfection]; Frères, Buyl

'KIMBERLEY ANNE', HT, w, 1983; flowers large, dbl., 35 petals, intense fragrance; foliage medium, medium green, matt; upright growth; Bronze, ARC TG, 1985; [Virgo X Secret Love]; Evans, F. David

Kimberly, *see* 'MICKIM'

Kimbo, F, 1977; Pineau; (Cavriglia)

Kimmy, F, 1975; Laperriere, L.; (Cavriglia)

'KIMONO'®, F, pb, 1961; flowers salmon-pink, blooms in broad clusters, dbl., 30 petals, 3 in., moderate fragrance; vigorous, bushy growth; [Cocorico X Frau Anny Beaufays]; deRuiter; Horstmann

Kim's Cream™, *see* 'RUPKIMCRM'

'KINASH', Min, yb, 1985; (**Baby Ashley**™); flowers light yellow, petals edged light pink, dbl., 24 petals, borne singly and in cluster of 3–5, no fragrance; straight, small, light brown prickles; foliage small, dark, matt; bushy growth; King, Gene; AGM Miniature Roses

'KINBARB', Min, ab, 1990; (**Barbara Mandrell**™); bud pointed; flowers dark apricot changing to apricot-pink at 3/4 open, yellow base, aging to white, dbl., 28 petals, exhibition form, slight fragrance; foliage medium, medium green, matt; upright, medium growth; [(Seedling X Party Girl) X Party Girl]; King, Gene, 1980; AGM Miniature Roses, 1991

'KINBAY', Min, ob, 1989; (**Tampa Bay**™); bud pointed; flowers orange, edges aging darker, medium, borne singly, dbl., 18 petals, exhibition form, slight, fruity fragrance; ovoid, orange fruit; prickles straight, red; foliage medium, medium green, matt; upright, medium growth; [(Arthur Bell X Orange Honey) X Baby Diana]; King, Gene; BDK Nursery, 1988

'KINBEE', Min, dr, 1986; (**B. C.**™); flowers dark red, reverse darker, dbl., 24 petals, exhibition form, slight fragrance; globular fruit; long, hooked, brown prickles; foliage small, medium green, matt; medium, upright, bushy growth; [Evelyn Fison X Magic Mist]; King, Gene; AGM Miniature Roses

'KINBELLE', Min, dp, 1988; (**Rosa Belle**™); flowers deep pink to yellow-cream at base, medium, borne usually singly, dbl., 21 petals, exhibition form; no fruit; prickles straight with hook, few, dark brown; foliage medium, medium green, matt; bushy, medium growth; [Vera Dalton X Party Girl]; King, Gene; AGM Miniature Roses

'KINBO', Min, dy, 1989; (**Tobo**™); bud pointed; flowers medium, borne usually singly and in sprays of 2–3, dbl., 32 petals, exhibition form, slight, fruity fragrance; ovoid, green fruit; prickles straight, slight crook, red to brown; foliage medium, light green, matt; upright, bushy, medium growth; [Arthur Bell X Rise 'n' Shine]; King, Gene; AGM Miniature Roses

'KINBOSCO', Min, ob, 1990; (**Boscobel**™); flowers orange to burnt orange, moderately small, borne mostly singly, dbl., 15–25 petals, no fragrance; foliage medium, light green, matt; upright, medium growth; [(Cheers X Rainbow's End) X Breezy]; King, Gene; AGM Miniature Roses

'KINBUD', Min, ly, 1985; (**Just Buddy**™); flowers mini-flora, large blooms borne singly, dbl., 45 petals, exhibition form, slight fragrance; no hips; straight, light brown prickles; foliage medium, light green, matt; medium upright, bushy growth; [New Day X Rise 'n' Shine]; King, Gene; AGM Miniature Roses, 1986

'KINBUFF', Min, ab, 1986; (**Buffy**™); flowers light apricot, reverse deeper, darker towards center, medium, dbl., 20 petals, exhibition form, slight fragrance; oval, green fruit; slightly hooked, light brown prickles; foliage medium green, matt; mini-flora; bushy spreading growth; [Vera Dalton X Party Girl]; King, Gene; AGM Miniature Roses, 1987

'KINCAP', Min, or, 1991; (**Captivation**™); dbl., borne mostly singly, slight fragrance; some prickles; foliage medium, light green, matt; upright (60–70 cms), bushy growth; [(Arthur Bell X Little Jackie) X Little Jackie]; King, Gene; AGM Miniature Roses, 1992

'KINCHA', Min, mr, 1984; (**Charlie**™); flowers small, dbl., 35 petals, exhibition form, no fragrance; foliage small, dark, matt; upright, bushy growth; [Seedling X Big John]; King, Gene

'KINCLAIRE', Min, w, 1989; (**Mollie Claire**™); bud pointed; flowers white, pink edge blushing toward center, reverse white tipped, dbl., 28 petals, exhibition form, no fragrance; no fruit; prickles crooked, very few, white to brown; foliage small, medium green, matt; bushy, low growth; [(Evelyn Fison X Magic Mist) X Baby Diana]; King, Gene; AGM Miniature Roses

'KINCOACH', Min, op, 1986; (**Stagecoach**™); flowers medium salmon pink, medium, borne usually singly or in spray, dbl., 18 petals, exhibition form, slight fragrance; oval, green fruit; straight, white prickles; foliage medium, light green, matt; upright, spreading growth; [Vera Dalton X Orange Honey]; King, Gene; AGM Miniature Roses, 1987

Kind Regards, *see* 'PEATIGAR'

Kind Regards, *see* PEAtiger

'KINDIDDY', Min, mp, 1991; (**Pink Diddy**™); bud pointed; flowers pink, medium, dbl., 28 petals, cupped, slight, fruity fragrance; foliage small, medium green, semi-glossy; bushy, spreading, low growth; [[(B.C. X Scamp) X Miss Dovey] X Tudelum]; King, Gene, 1986; AGM Miniature Roses, 1990

'KINDIXIE', Min, ob, 1991; (**Dixie Dazzle**™); bud ovoid; flowers orange-red to yellow, small, dbl., 16 petals, exhibition form, no fragrance; foliage small, medium green, semi-glossy; upright, low growth; [(Rainbow's End X Miss Dovey) X Jennie Anne]; King, Gene, 1986; AGM Miniature Roses, 1990

Kindness, HT, dp, 1994

'KINDOV', Min, ab, 1985; (**Miss Dovey**™, Miss Dovie); flowers deep apricot, blooms borne usually singly, dbl., 21 petals, exhibition form, slight fragrance; sets hips; straight, reddish-brown prickles; foliage medium, medium green, semi-glossy; upright, bushy growth; [Anne Harkness X Rise 'n' Shine]; King, Gene; AGM Miniature Roses

'KINFANCY', Min, rb, 1986; (**Fancy Pants**™); flowers deep pink to golden yellow base, edged red, fading deeper pink, dbl., 40 petals, exhibition form, slight, spicy fragrance; fruit not observed; medium, light red prickles, slightly crooked on end; foliage medium, medium green, matt; upright, bushy, medium growth; [Baby Katie X Rose Window]; King, Gene; AGM Miniature Roses, 1987

'KINFANTA', Min, w, 1985; (**Fantasi**™); flowers cream, reverse cream shaded deep pink, mini-flora, large, dbl., 50 petals, exhibition form, slight fragrance; no fruit; medium, hooked, brown prickles; foliage medium, dark, matt; tall, upright growth; King, Gene; AGM Miniature Roses, 1986

King Alexander I, *see* **'ROI ALEXANDRE'**

'KING ARTHUR', F, op, 1967; flowers salmon-pink, blooms in clusters, dbl., 3.5 in., slight fragrance; foliage glossy; [Pink Parfait X Highlight]; Harkness

'KING BOREAS', F, my, 1941; flowers pure yellow shading to nearly white, recurved, medium, very dbl., 100 petals, moderate fragrance; vigorous growth; [Golden Glow (LCl) X ?]; Brownell, H.C.

King Crimson, *see* 'JACSHUR'

King George IV, *see* **'RIVERS' GEORGE IV'**

'KING GEORGE V', HP, dr, 1912; flowers crimson, large, dbl., 40 petals, exhibition form, sparse, intermittent bloom, moderate fragrance; strong stems; very vigorous, open growth; Dickson, H.

King George's Memorial, *see* **'PAMÁTNIK KRÁLE JIRÍHO'**

King Henry, *see* GENwine

King Hey, S, dr

'KING J.', HRg, w, 1996; flowers yellow stamens, blooms borne in small clusters, dbl., 15–25 petals, good repeat, intense, spicy fragrance; numerous prickles; foliage dark green, glossy, very healthy, disease/insect res; bushy, medium growth; hardy zone 6; [R. rugosa alba X Assiniboine]; Fleming, Joyce L.; Hortico Roses, 1996

'KING MIDAS', LCl, my, 1942; bud long, pointed; flowers clear yellow, blooms in clusters of 4–6, dbl., 20 petals, 4–5 in., cupped, repeat bloom, slight fragrance; foliage large, leathery, dark; vigorous, climbing or pillar (8–10 ft) growth; Nicolas; J&P

'**KING O' KINGS**', HT, mr, 1973; dbl., 35–40 petals, 6 in., exhibition form, intense fragrance; foliage dark; [My Love X Duftwolke]; Anderson's Rose Nurseries

'**KING OF HEARTS**', HT, mr, 1968; bud long, pointed; flowers medium, dbl., exhibition form; foliage dark, leathery; vigorous, bushy growth; [Karl Herbst X Ethel Sanday]; McGredy, Sam IV; Edmunds Roses

'**KING OF SCOTS**', HSpn, dp; flowers deep pink, dbl., profuse, early bloom; glossy, black fruit; foliage finely divided; dense, shrubby (3–4 ft) growth; Lee, prior to 1848

'**KING OF THE PRAIRIES**', HSet, mr, 1843; flowers bright red; Feast

King Richard, *see* GENruby

King Tut, *see* 'CHEWARVEL'

King Tut, *see* 'LAVTREK'

King William, *see* GENpink

'**KINGAROY**', HT, dr, 1979; bud long, pointed; flowers urn-shaped blooms borne singly, dbl., 30 petals, slight, fruity fragrance; small prickles; foliage medium, matt; medium growth; [Unnamed seedling X Red Lion]; Perry, Astor, 1981

'**KINGCUP**', HT, my, 1953; flowers buttercup-yellow, well formed, dbl., 40 petals, 4 in., intense, fruity fragrance; foliage dark, glossy; very free growth; [Mrs Sam McGredy X Ellinor LeGrice]; LeGrice

Kingi, *see* 'MURKI'

'KINGIG', Min, mp, 1987; (**Giggles**); flowers light pink, reverse light to dark pink, fading to creamy pink, dbl., 18 petals, exhibition form, slight fragrance; oval, green fruit; slightly crooked, white prickles; foliage medium, medium green, matt; upright, tall growth; [Vera Dalton X Rose Window]; King, Gene; AGM Miniature Roses

'KINGLOW', Min, ob, 1989; (**Glowry**™); bud pointed; flowers bright orange-yellow bicolor, medium, borne singly, dbl., 24 petals, exhibition form, no fragrance; no fruit; prickles straight, red; foliage small, medium green, matt; bushy, low growth; [(Arthur Bell X Orange Honey) X Baby Diana]; King, Gene; AGM Miniature Roses

'**KING'S RANSOM**'®, HT, dy, 1961; bud ovoid; flowers clear golden yellow, dbl., 38 petals, 5–6. in., exhibition form, moderate fragrance; foliage leathery, glossy; vigorous, upright growth; AARS, 1962; [Golden Masterpiece X Lydia]; Morey, Dr Dennison; J&P

'**KING'S ROW**', S, yb, 1965; flowers yellow becoming rose-red, medium, borne in clusters, dbl., moderate fragrance; foliage bronze, leathery; vigorous (4 ft), compact growth; [Easter Parade X Herrenhausen]; Whisler; Germain's

King's Treasure, F, my, 1995; Williams, J. Benjamin

Kingsmead Heritage, *see* KORdalsilk

Kingswood College, *see* KORloher

'KINGUN', Min, pb, 1990; (**Top Gun**™); bud ovoid; flowers peony pink, yellow base, apricot yellow reverse, medium, bor, dbl., 28 petals, exhibition form, no fragrance; foliage medium, medium green, semi-glossy; tall, upright growth; [(Rainbow's End X Vera Dalton) X Vera Dalton]; King, Gene, 1985; AGM Miniature Roses, 1991

'KINHEART', Min, ob, 1985; (**Heartlight**™); flowers orange-yellow, reverse yellow, mini-flora, large, 16 petals, exhibition form, slight fragrance; straight, light brown prickles; foliage medium, medium green, matt; upright, bushy, medium growth; [Golden Slippers X Rise 'n' Shine]; King, Gene; AGM Miniature Roses

'KINJEN', Min, rb, 1986; (**Jennie Anne**™); flowers red, reverse yellow, fading light yellow with red edge, medium, dbl., 16 petals, exhibition form, no fragrance; oval fruit; straight, small, white prickles; foliage medium, medium green, matt; bushy, medium growth; [Gingersnap X Charmglo]; King, Gene; AGM Miniature Roses, 1987

'KINJOE', Min, ob, 1985; (**Joe-Joe**™); flowers orange-yellow, reverse yellow, blooms borne singly, dbl., 25 petals, cupped, slight fragrance; straight, light yellow to brown prickles; foliage medium, dark, matt; bushy, spreading growth; [Seedling X Rise 'n' Shine]; King, Gene; AGM Miniature Roses

'**KINKAKU**', HT, dy, 1975; flowers golden yellow, large blooms borne usually singly, dbl., 38 petals, exhibition form, slight fragrance; brown prickles; foliage medium, light green; medium, bushy growth; [Unnamed seedling X Peace]; Okamoto, K.; K. Hirakata Nursery

'KINLIV', Min, ob, 1991; (**Midwest Living**®); bud globular; flowers orange, slight ruffle to petal edges, good substance, holds, dbl., 26–40 petals, 1.5–2.75 in., slight fragrance; few prickles; foliage medium, medium green, semi-glossy; bushy, spreading, medium (46 cms), healthy, vigorous growth; [Evelyn Fison X Party Girl]; King, Gene; Michigan Mini Roses, 1992

'KINLU', Min, rb, 1990; (**Li'l Alleluia**™); flowers wine to silver with yellow base, moderately small, borne mos, dbl., 15–25 petals, moderate fragrance; foliage medium, medium green, matt; upright, tall growth; [[(B.C. X Scamp) X Tamango] X Magic Carrousel]; King, Gene; AGM Miniature Roses

'KINLUM', Min, dp, 1986; (**Tudelum**™); flowers deep pink to lighter shades, petal edges darken on opening, dbl., 30 petals, exhibition form, no fragrance; no fruit; straight, medium, light pink prickles; foliage medium, medium green, matt; upright, bushy, medium, vigorous growth; [Baby Katie X Watercolor]; King, Gene; AGM Miniature Roses, 1987

'KINMAC', Min, dr, 1989; (**Merrimac**™); bud pointed; flowers deeper at tips, medium, borne usually singly and in sprays, 18 petals, exhibition form, slight, fruity fragrance; ovoid, orange fruit; prickles straight, small, red; foliage medium, medium green, matt; upright, medium growth; [(Alain X Scamp) X Lilli Marleen]; King, Gene; AGM Miniature Roses

'KINMIKE', Min, or, 1987; (**J. Michael**™); flowers orange-red, aging lighter, large, borne usually singly, dbl., 18 petals, exhibition form, moderate, fruity fragrance; straight, medium, white-green prickles; foliage large, light green, matt; mini-flora; upright, tall growth; [Poker Chip X Watercolor]; King, Gene; AGM Miniature Roses

'KINNIGHT', Min, dr, 1990; (**After Midnight**™); bud ovoid; flowers dark red, tips show darker edge, outer petals darker, medium, dbl., 28 petals, exhibition form, borne singly, no fragrance; foliage medium, medium green, semi-glossy; medium, upright growth; [(B.C. X Scamp) X Black Jade]; King, Gene, 1986; AGM Miniature Roses, 1991

'KINNOR', Min, mr, 1987; (**New Orleans**™); flowers medium to dark red toward base, non-fading, reverse medium red, dbl., 32 petals, exhibition form, slight, spicy fragrance; oval, small, green fruit; straight, small, red prickles; foliage medium, medium green, matt; bushy, spreading, medium growth; [Evelyn Fison X Magic Mist]; King, Gene; AGM Miniature Roses

'KINPIC', Min, pb, 1990; (**Pink Picotee**®); flowers white to cream with dark pink picotee edge, small, borne mos, dbl., 26–40 petals, no fragrance; foliage medium, light green, matt; upright, spreading, medium growth; [(Vera Dalton X Fancy Pants) X Magic Carrousel]; King, Gene; AGM Miniature Roses

'KINPLEAS', Min, pb, 1986; ('KINPLEASE' **My Pleasure**™); flowers lavender pink, reverse light pink, fading lighter pink, medium, dbl., 45 petals, exhibition form, moderate, fruity fragrance; oval, medium green fruit; straight, white prickles with brown tips; foliage medium, dark green, matt; upright, medium growth; [Lavender Pinocchio X Unnamed seedling]; King, Gene, 1981; AGM Miniature Roses, 1987

'KINPLEASE', Min, pb, 1986; ('KINPLEAS' **My Pleasure**™); flowers lavender pink, reverse light pink, fading lighter pink, medium, dbl., 45 petals, exhibition form, moderate, fruity fragrance; oval, medium green fruit; straight, white prickles with brown tips; foliage medium, dark green, matt; upright, medium growth; [Lavender

Pinocchio X Unnamed seedling]; King, Gene, 1981; AGM Miniature Roses, 1987

'KINsandy', Min, pb, 1990; (**Pink Sandy**™); flowers light pink to light apricot center, aging deep pink edges, s, dbl., 26–40 petals, slight fragrance; foliage small, medium green, matt; bushy, low growth; [(Vera Dalton X Rainbow's End) X Fancy Pants]; King, Gene; AGM Miniature Roses

'KINschon', Min, rb, 1986; ('KINschon' **Prairie Schooner**™); flowers red, reverse yellow edged red, dbl., 22 petals, exhibition form, borne singly and in sprays, slight fragrance; no fruit; straight, brown prickles; foliage small, dark, matt; upright growth; PP006748; [Vera Dalton X Sheri Anne]; King, Gene; AGM Miniature Roses

'KINslam', Min, dr, 1990; (**Little Slam**®); flowers deep red, moderately small, borne mostly singly, dbl., 15–25 petals, no fragrance; foliage small, medium green, semi-glossy; bushy, low growth; [[(B.C. X Scamp) X Red Ace] X Scamp]; King, Gene; AGM Miniature Roses, 1989

Kinsmen, HT, rb; Delbard

'KINspeech', Min, ob, 1986; (**Speechless**™); flowers dark to light orange, fading darker, medium, borne usually singly, dbl., 30 petals, exhibition form, no fragrance; oval, medium, green fruit; medium, hooked, red prickles, with brown tips; foliage dark green, semi-glossy; upright, bushy growth; [Unnamed seedling X Watercolor]; King, Gene; AGM Miniature Roses, 1987

'KINsun', Min, yb, 1990; (**Dale's Sunrise**®); flowers yellow tipped medium pink, fading pink, small, borne mostly, dbl., 15–25 petals, slight fragrance; foliage small, medium green, semi-glossy; upright, bushy, medium growth; [[(B.C. X Scamp) X Rainbow's End] X Tobo]; King, Gene; AGM Miniature Roses, 1991

'KINtee', Min, mr, 1988; (**Peggy "T"**™); flowers medium red to white, circular base, reverse white, aging med, single, 5 petals; no fruit; prickles straight with hook, white; foliage medium, medium green, matt; bushy, medium growth; [Poker Chip X Rise 'n' Shine]; King, Gene; AGM Miniature Roses

'KINtenn', Min, op, 1988; (**Tennessee**™); flowers coral to white, reverse light coral to white, aging darker coral, dbl., 18 petals, exhibition form; oval, large, orange fruit; prickles straight, light green; foliage medium, light green, matt; upright, tall growth; [Kiskadee X Orange Honey]; King, Gene; AGM Miniature Roses

'KINtoddy', Min, dr, 1989; (**Hoddy Toddy**™); bud pointed; flowers dark red, petals tipped darker, small, borne usually singly, dbl., 28 petals, cupped, no fragrance; no fruit; prickles straight, very small, red; foliage small, medium green, matt; bushy, low growth; [(Alain X Scamp) X Scamp]; King, Gene; AGM Miniature Roses

'Kinugasa', HT, lp, 1984; flowers large blooms borne usually singly, dbl., 45 petals, exhibition form, moderate, fruity fragrance; brown, hooked prickles; foliage medium, light green; medium, bushy growth; [Michelle Meilland X (Michelle Meilland X Anne Letts)]; Shibata, T.; K. Hirakata Nursery

'Kiora', F, mr, 1989; bud pointed; flowers light vermillion shading to darker petal edges, small, dbl., flat, no fragrance; foliage dark green, shiny, veined; upright, bushy growth; [Liverpool Echo X John Church]; Cattermole, R.F.; South Pacific Rose Nursery

Kir Royal, *see* MEInibur

Kirang, HT, pb, 1995; Tejganga

'KIRbee', F, ab, 1993; (**Irene's Surprise**); dbl., 15–25 petals, 1.5–2.75 in., exhibition form, borne in small clusters, moderate fragrance; foliage medium, medium green, semi-glossy; medium ($2\frac{1}{2}$ ft), upright growth; Kirkham, Gordon Wilson; Kirkham, 1993

'KIRbell', HT, dy, 1995; (**Berenice Neville**); flowers moderately blooms borne in small clusters, dbl., 15–25 petals, moderate fragrance; many prickles; foliage medium, dark green, semi-glossy; medium, upright growth; [Tynwald X Bright Smile]; Kirkham, Gordon Wilson; Kirkham, 1996

KIRbill, F, 1995; (**Margaret's World**); Kirkham, Gordon Wilson; (Cavriglia)

'KIRbronze', F, ob, 1993; (**Elizabeth Ann**); flowers bronze and gold, blooms borne in small clusters, dbl., 15–25 petals, 1.5–2.75 in., slight fragrance; some prickles; foliage medium, medium green, glossy; medium, upright growth; [Kathleens Rose X Eurorose]; Kirkham, Gordon Wilson; Kirkham, 1995

'KIRcloud', HT, dp, 1995; (**Mary Hilda Law**); flowers deep pink, blooms, dbl., 15–25 petals, 3–3.5 in., intense fragrance; foliage large, dark green, semi-glossy; medium, upright growth; [Lady Helen X Fragrant Cloud]; Kirkham, Gordon Wilson; Kirkham, 1997

'KIRdex', F, dp, 1985; (**Roccana Diane**); flowers deep pink, medium, dbl., 20 petals, slight fragrance; foliage large, dark, glossy; upright growth; [Pink Favorite X Attraction]; Kirkham, Gordon Wilson

'KIRhol', HT, w, 1993; (**May Lawlor**); flowers cream and pink, borne mostly singly, dbl., 15–25 petals, 2.75–3 in., intense fragrance; many prickles; foliage large, dark green, semi-glossy; tall, upright growth; [Morgengruss X Mary Sumner]; Kirkham, Gordon Wilson

Kiri, *see* MATtkiri

'KIRjack', LCl, or, 1994; (**Daphne Claire Jones**); dbl., 15–25 petals, 3–3.5 in., moderate fragrance; many prickles; foliage large, dark green, semi-glossy; tall, upright growth; [Parkdirecktor Riggers X Seedling]; Kirkham, Gordon Wilson; Kirkham, 1996

'KIRkit', F, pb, 1985; (**Kathleen Kirkham**); flowers medium, semi-dbl., moderate fragrance; foliage medium, medium green, semi-glossy; upright growth; [Manx Queen X Unnamed seedling]; Kirkham, Gordon Wilson

KIRkitt, F, ab, 1986; (**Kathleen's Rose**); Kirkham, Gordon Wilson

'KIRlon', F, lp, 1993; (**Woman O'th North**); flowers shell pink, blooms borne in small clusters, dbl., 15–25 petals, 1.5–2.75 in., moderate fragrance; few prickles; foliage medium, medium green, semi-glossy; low ($1\frac{1}{2}$ x $1\frac{1}{2}$ ft), bushy growth; Kirkham, Gordon Wilson; Kirkham, 1994

KIRlyl, F, mp, 1997; (**Harlow Carr**); Kirkham, Gordon Wilson

KIRmac, F, w, 1989; (**Greenalls Glory**); Kirkham, Gordon Wilson

KIRmelody, HT, 1996; (**Bangor Cathedral**); Kirkham, Gordon Wilson; (Cavriglia)

'KIRpark', HT, mr, 1983; (**Ted Goves**); flowers medium, dbl., 20 petals, intense fragrance; foliage medium, dark, matt; [Prima Ballerina X Teenager]; Kirkham, Gordon Wilson

'KIRpice', HT, mr, 1994; flowers medium red, moderately full (15–25 petals), large (7+ cms) blooms borne in small clusters; slight fragrance; few prickles; foliage large, dark green, semi-glossy; low, bushy growth; [Invincible X Red Dandy]; Kirkham, Gordon; 1995

KIRpink, F, mp, 1968; (**Owen's Pride**); Kirkham, Gordon Wilson

'KIRrad', Min, ab, 1993; (**Arejay**); flowers moderately full, small blooms, dbl., 15–25 petals, 1.5 in., borne in large clusters, slight fragrance; few prickles; foliage small, medium green, glossy; low, bushy growth; [Seedling X Seedling]; Kirkham, Gordon Wilson; Kirkham, 1995

KIRsan, LCl, my, 1999; (**Chris**); Kirkham, Gordon Wilson

Kirsch, Climbing, *see* 'POUllack'

KIRscot, MinFl, op, 1989; (**Graceland**); Kirkham, Gordon Wilson

KIRshru, S, ly; (**St Dunstan's Rose**); Kirkham, Gordon Wilson, 1991

'Kirsi', HT, rb, 1967; flowers cream-white edged pink, becoming red over white, large, dbl.; very vigorous growth; [Rose Gaujard sport]; Palmer, H.E.; Edmunds Roses

'KIRsmile', F, yb, 1994; (**Donald Thomas Heald**); semi-dbl., 6–14 petals, 1.5 in., borne in small clusters, slight fragrance; many prickles; foliage medium, dark green, semi-glossy; patio; low, compact growth; [Seedling X Bright Smile]; Kirkham, Gordon Wilson; Kirkham, 1996

'KIRSTEIN', HT, ob, 1999; (**Sophie Deborah**); dbl., 15–25 petals, 3 in., borne in small clusters, slight fragrance; prickles moderate; foliage medium, dark green, semi-glossy; upright, medium (2.5 ft) growth; [Prunella X Mary Sumner]; Kirkham, Gordon Wilson

Kirsten, *see* POUlkir

Kirsten Klein, HMsk, mp, 1995; Scarman

'KIRSTEN POULSEN', F, mr, 1924; flowers bright scarlet, borne in clusters, single, slight fragrance; foliage leathery; vigorous growth; (21); [Orléans Rose X Red Star]; Poulsen, S.

'KIRSTEN POULSEN IMPROVED', F, mr, 1938; flowers scarlet, borne on trusses, single, free bloom; foliage dark; free growth; (21); Radmore

Kirsty Jane, *see* 'SIMKAYJAY'

'KIRSUN', S, dy; (**Gold Rush**)

'KIRTYN', S, ab, 1995; (**Heywood**); flowers peach, blooms borne in large clusters, semi-dbl., 6–14 petals, 1.5–2.75 in., slight fragrance; many prickles; foliage medium, medium green, matt; tall, bushy growth; [Tynwald X Bright Smile]; Kirkham, Gordon Wilson

'KIRWIM', HT, mp, 1998; (**Clara's Surprise**); flowers lilac pink, high centered, medium size, dbl., 26–40 petals, exhibition form, borne in small clusters, moderate fragrance; few prickles; foliage medium, dark green, semi-glossy; compact, low growth; [Seaspray X Admiral Rodney]; Kirkham, Gordon Wilson

'KIRWORJACKIE', HT, pb, 1993; (**Worjackie**); flowers full, pink blend, dbl., 26–40 petals, 2.75–3 in., borne mostly singly, moderate fragrance; many prickles; foliage medium, dark green, semi-glossy; medium, upright growth; [Solitaire X Brandy Butter]; Kirkham, Gordon Wilson; Battersby Roses, 1994

Kiska Rose, *see* **R. RUGOSA**

'KISKADEE', F, my, 1973; flowers bright yellow, dbl., 25 petals, 3.5 in., exhibition form, slight fragrance; foliage dark; [Arthur Bell X Cynthia Brooke]; McGredy, Sam IV

Kisme, F, pb, 1960; Raffel

'KISMET', HT, my, 1930; flowers clear yellow, center deeper, large, dbl., cupped, intense fragrance; foliage light, glossy; [Talisman sport]; Nicolas; J&P

'KISS', Min, ob, 1980; bud ovoid, pointed; flowers indian orange, reverse lighter, blooms borne 1–3 per cluster, dbl., 28 petals, exhibition form, moderate, fruity fragrance; straight prickles; foliage small, medium green; strong, upright growth; [Unnamed seedling X Unnamed seedling]; Lyon, Lyndon; Dreer

Kiss, *see* KORikis

Kiss and Tell, *see* SEAris

Kiss 'n' Tell, *see* 'SEAKISS'

Kiss of Fire, HT, 1969; Laveena Roses; (Cavriglia)

Kiss of Fire, F, rb, 1960; Gaujard

Kiss the Bride, *see* 'SEAWHI'

'KISSIN' COUSIN', S, op, 1978; bud ovoid, pointed; flowers pink to coral-pink, dbl., 28 petals, 4–5 in., exhibition form, slight fragrance; foliage large, dark, leathery; vigorous, upright, spreading growth; [((Ophelia X Prairie Princess) X Tiki) X ((Corbeille Royale X American Heritage) X Hawkeye Belle)]; Buck, Dr Griffith J.; Iowa State University

'KITANA', HRg, m, 1927; flowers deep lavender-pink, semi-dbl., non-recurrent, intense fragrance; profuse, red fruit; [Tetonkaha X Rose Apples]; Hansen, N.E.

'KITCHENER OF KHARTOUM', HT, mr, 1917; (K. of K.); flowers dazzling velvety scarlet, medium, 10 petals, 3 in., intense fragrance; vigorous, branching growth; GM, NRS, 1916; Dickson, A.

'KITTY', Pol, mr, 1925; flowers carmine-red, small, dbl.; Koster; (Sangerhausen)

'KITTY BICE', LCl, lp; [Ophelia, Climbing X Lady Waterlow]; Fitzhardinge, before 1932; (Weatherly, L.)

'KITTY HAWK', Min, pb, 1986; flowers deep pink, reverse lighter pink, mini-flora blooms borne singly, dbl., 29 petals, exhibition form, slight fragrance; medium, long, red prickles; foliage large, medium green, semi-glossy; medium, upright growth; [Watercolor X ?]; Bridges, Dennis A.

'KITTY KINGSBURY', HT, mp, 1930; flowers shell-pink, moderate fragrance; vigorous growth; [Abol seedling]; Evans

'KITTY KININMONTH', LCl, dp, 1922; flowers deep pink, golden stamens, very large, semi-dbl., cupped, some recurrent bloom, slight fragrance; few prickles; foliage dark, wrinkled; vigorous, climbing growth; [Unnamed seedling X R. gigantea]; Clark, A.; Hackett

'KIWI', HT, w, 1989; bud pointed; flowers creamy pink opening to creamy white, reflexed, pointed, medium, dbl.; slight fragrance; prickles needle-like, red to light brown; foliage dark green, shiny; tall, upright growth; [Judith Morton X (Pascali X Blue Moon)]; Cattermole, R.F.; South Pacific Rose Nursery

'KIWI BELLE', F, pb, 1983; flowers tan apricot to pink, large blooms in clusters of 3–5, dbl., 40 petals, globular, intense fragrance; brown prickles; foliage bronze green when young, turning light green; upright, spreading growth; [Silent Night X Irish Mist]; Cattermole, R.F.

'KIWI CHARM', HT, pb, 1971; bud ovoid; flowers creamy yellow, edged pink, large, semi-dbl., exhibition form, moderate fragrance; foliage glossy, leathery; vigorous, bushy growth; [Kordes' Perfecta X Champagne]; Lindquist; Bell Roses, Ltd.

'KIWI DELIGHT', HT, yb, 1983; flowers golden yellow, petals edged deep pink, aging pink overall, large, dbl., 40 petals, globular, intense fragrance; gray-brown, very small and large prickles; foliage bronze green turning dark green, glossy; upright, bushy growth; [Peace X (Peer Gynt X Irish Mist)]; Cattermole, R.F.

'KIWI GOLD', HT, yb, 1984; bud pointed; flowers light yellow, reverse edges of petals flushed pink, dbl., 22 petals, exhibition form, intense fragrance; very few, light brown prickles; foliage light green, glossy; upright growth; [(Pink Parfait X Pink Parfait) X Waipounamu]; Cattermole, R.F.

'KIWI QUEEN', HT, yb, 1983; flowers yellow, shaded orange and pink, large, dbl., 44 petals, exhibition form, slight fragrance; light brown prickles; foliage medium green; upright, spreading growth; [Peer Gynt X Command Performance]; Cattermole, R.F.

Kiwi Sunrise, *see* 'MACPAURMAR'

'KIYOSUMI', HT, m, 1979; bud slender; flowers clear light purplish blue, dbl., 25 petals, 3 in., exhibition form, slight fragrance; foliage light green, leathery; upright growth; [Unnamed seedling X Sterling Silver]; Onodera, Toru F.; S. Onodera

'KLAUS GROTH', HSpn, ob, 1951; (Claus Groth); flowers salmon-orange shaded apricot-yellow, large, dbl., intense, spinosissima fragrance; large fruit; foliage dark; vigorous, bushy (5 ft) growth; [R.M.S. Queen Mary X R. spinosissima]; Tantau

'KLAUS STÖRTEBEKER', HT, mr, 1962; flowers well-formed, dbl., 40 petals, 5 in.; foliage dark; low, bushy growth; Kordes, R.

Kleine Ballerina, Pol, 1998; Noack, Werner; (Cavriglia)

Kleine Dortmund®, F, rb, 1992; Noack, Werner; (Cavriglia)

'KLEINE ECHO', Pol, mp, 1925; flowers small, semi-dbl.; Kiese; (Sangerhausen)

Kleine Eva, Pol, mp, 1995; Hetzel

'KLEINE LEO', HT, mr, 1921; flowers brilliant red shaded dark red, dbl.; [Farbenkonigin X Gen. MacArthur]; Timmermans

Kleine Regina®, *see* HELreg

Kleine Renate, *see* JACrain

'KLEINE ROSEL', HWich, m, 1929; flowers dark violet-pink, medium, single; Vogel, M.; (Sangerhausen)

'KLEINER ALFRED', Pol, or, 1904; bud garnet-red; flowers well formed, medium; foliage glossy; dwarf growth; [Anna-Maria de Montravel X Shirley Hibberd]; Lambert, P.

Kleopatra, *see* KORverpea

Klerksdorp Horizon, *see* 'KORMOWE'

'KLETTERNDE RUBY', Cl Pol, mr, 1946; bud globular; flowers scarlet, small, borne in clusters, dbl., slight fragrance; foliage dark, wrinkled; very vigorous, trailing growth; [Ruby sport]; Kordes

Klima(TM), *see* 'SAVaklim'

Klimentina, HT, mp, 1955; flowers pink with silvery reverse, large, dbl.; moderate fragrance; Klimenko, V. N.; (Sangerhausen)

'Klondyke', HWich, my, 1911; flowers soft yellow, center deeper, passing to ivory-white, large, dbl.; very vigorous, climbing growth; Paul

'Klondyke', HT, my, 1934; flowers clear golden yellow; [Lady Forteviot sport]; LeGrice

Kluis Orange, *see* **'Klyn's Orange'**

'Kluis Scarlet', F, mr, 1931; flowers brilliant red; very free growth; [Lafayette sport]; Kluis, R.

'Klyn's Orange', Pol, or; (Kluis Orange); flowers orange-scarlet, small, borne in dense clusters, dbl.; foliage light green; dwarf growth; Kluis

'Klyn's Yellow', HT, my, 1948; bud pointed; flowers clear yellow, open, large, dbl., moderate fragrance; foliage glossy; very vigorous, compact growth; [McGredy's Yellow X ?]; Klyn

Knezna Libuse, HT, lp; Cerveny; (Czech Rosa Club)

Knirps, *see* KORverlandus

Knock Out(TM), *see* 'RADrazz'

'Koa', Min, op, 1978; bud ovoid; flowers coral-pink, dbl., 44 petals, 1 in., exhibition form, slight fragrance; tall, open, leggy growth; [Persian Princess X Gene Boerner]; Rovinski & Meredith; Kingsdown Nursery & Miniature Plant Kingdom

Koala®, *see* MEItapov

Koba, *see* 'MEIroverna'

Kobold, F, op, 1981; flowers large, semi-dbl.; McGredy, Sam IV; (Sangerhausen)

Kocher Red, LCl, mr

Ko's Yellow, *see* 'MACkosyel'

'Ko-Choh', F, ob, 1983; flowers orange, reverse yellow shaded orange, small, borne 3–5 per cluster, dbl., 45 petals, exhibition form, moderate fragrance; small prickles; foliage dark, semi-glossy; low, compact growth; [(Rumba X Olympic Torch) X Allgold]; Suzuki, Seizo; Keisei Rose Nursery

Kogane, F, my

'Koha', HT, rb, 1983; flowers medium red, reverse lighter, large, dbl., 32 petals, exhibition form, slight fragrance; brown prickles; foliage light, glossy, veined; upright, branching growth; [Silent Night X (Josephine Bruce X Irish Mist)]; Cattermole, R.F.

'Koharu-Biyori', F, lp, 1999; flowers pale pink, single, 5 petals, 2 in., moderate fragrance; foliage medium green; 2.5 ft growth; GM, Japan Rose Concours, 1993; [Hanagasumi X seedling]; Yasuda, Yuji, 1994

'Kohima', F, rb, 1972; flowers fire-red to orange, reverse cream, very full, dbl., 45–50 petals, 4 in., intense fragrance; vigorous growth; [Orange Sensation X Mischief]; Ellick; Excelsior Roses

'Koh-I-Noor', HT, pb, 1973; flowers white, center deep pink, very full, dbl., 60 petals, 5 in., intense fragrance; foliage glossy, dark; vigorous growth; [(Memoriam X Peace) X My Choice]; Ellick

Kohsai, *see* **'Mikado'**

Koh-Sai, *see* **'Mikado'**

Koigokoro, HT, dp

Kojack, *see* **'Miss Blanche'**

Kojo, *see* **'Sweet Dreams'**

'Kojo no Tsuki', LCl, dy, 1975; bud circular; flowers deep yellow, dbl., 35 petals, 4.5–5 in., exhibition form, slight fragrance; foliage light; vigorous, climbing growth; [(Souv. de Jacques Verschuren X Thais) X Amarillo]; Teranishi, K.; Itami Bara-en

'KoKo', Min, yb, 1982; bud globular; flowers pale yellow, aging orange-red on petal edges, blooms borne singly, dbl., 55 petals, moderate fragrance; thin, triangular, light brown prickles; foliage medium, medium green, matt; low, bushy growth; Meredith, E.A. & Rovinski, M.E.; Casa de Rosa Domingo

Koksijde, F, pb; 19 petals, flat, foliage matt; strong growth; RvS-Melle, 1998

'Kokulensky', (strain of R. canina), lp; susceptible to rust; once used as understock; Kokulensky

Kokyu, HT, lp; good size and form growth

'Koldinghus', F, dp, 1968; flowers deep rose, open, dbl., 23 petals, 2–2.5 in., slight fragrance; foliage dark, glossy; upright growth; [Pernille Poulsen X Seedling]; Poulsen, Niels D.; Poulsen

'Kolibre', F, mr, 1946; bud long, pointed; semi-dbl., 20 petals, 3 in., small trusses, intense fragrance; foliage bronze; moderately vigorous growth; [Mev. Nathalie Nypels X Unnamed seedling]; Leenders, M.; Longley

'Kolkhoznitsa', HT, m, 1957; flowers lilac-pink, large, dbl., 45 petals, moderate fragrance; foliage dark; very vigorous, compact growth; [Peace X Mirandy]; Sushkov & Besschetnova

'Köln am Rhein', HKor, op, 1956; (Cologne); flowers deep salmon-pink, large blooms in clusters, dbl., recurrent bloom, moderate fragrance; foliage dark, glossy; vigorous, climbing growth; Kordes

Kölner Karneval, *see* 'KORgi'

Kölner Karneval, Climbing, Cl HT, m; (Cavriglia)

Komala, HT, pb, 1989; K&S

Kombination, F, 1982; VEG; (Sangerhausen)

Komet, HT, mr; flowers vermilion red, medium to large, dbl., slight fragrance; VEG; (Sangerhausen)

'Komfort', F, mp, 1967; flowers salmon-pink, large, borne in clusters (to 20), semi-dbl.; foliage glossy; very vigorous, bushy growth; Tantau, Math.

'Kommerzienrat W. Rautenstrauch', HFt, mp, 1909; flowers pure salmon-pink, center yellow, reverse lighter, borne in clusters, semi-dbl., slight fragrance; [Léonie Lamesch X R. foetida bicolor]; Lambert, P.

'Kommodore', F, mr, 1959; (The Commodore); flowers blood-red, well-formed, large blooms in clusters, dbl.; vigorous, low growth; Tantau, Math.

'Konfetti', HT, pb, 1965; bud globular; flowers claret-rose, reverse light cream, well formed, dbl., 40–45 petals, 5 in.; foliage dark, leathery; strong stems; bushy, upright growth; Tantau, Math.

König Friedrich II von Dänemark, HP, dr; flowers medium, dbl., slight fragrance; (Sangerhausen)

'König Laurin', HT, w, 1910; flowers large, dbl., moderate fragrance; Türke; (Sangerhausen)

König Ludwig-Rose, S, 1994; flowers large, dbl.; Tantau; (Sangerhausen)

Königen, HWich, mp; flowers small, dbl.; (Sangerhausen)

Königin Beatrix®, *see* 'HETKORa'

'Königin Carola', HT, pb, 1904; (Königin Carola von Sachsen); bud pointed; flowers satiny rose-pink, reverse silvery white, very large, dbl., slight fragrance; foliage dark, leathery; vigorous growth; [Mme Caroline Testout X Viscountess Folkestone]; Türke

Königin Carola von Sachsen, *see* **'Königin Carola'**

Königin der Rosen, *see* 'KORbico'

'Königin Emma', HT, lp, 1903; flowers large, dbl.; Verschuren; (Sangerhausen)

'Königin Juliana', F, 1948; Buisman, G. A. H.; (Cavriglia)

'Königin Luise', HT, w, 1927; flowers very large, dbl., exhibition form, slight fragrance; foliage dark, leathery, glossy; vigorous, bushy growth; [Frau Karl Druschki X Sunburst]; Weigand, C.

'Königin Maria Therese', HT, dp, 1916; flowers carmine-pink, very large, very dbl., intense fragrance; Lambert, P.; (Sangerhausen)

'Königin Viktoria von Schweden', HT, yb, 1919; flowers light saffron-yellow, aging to pale salmon-pink, dbl.; [Mme Segond

Weber X Mrs Joseph Hill]; Ries; Teschendorff

'KÖNIGIN VON DÄNEMARK', A, mp; (Belle Courtisanne, Queen of Denmark, Reine du Dänemark); flowers flesh-pink, center darker, medium, very dbl., non-recurrent, intense fragrance; vigorous growth; [Probably R. alba X Damask hybrid]; Booth, 1816

Konigin Wilhelmine, F, mr, 1997; deVor

'KÖNIGIN WILHELMINE', Pol, dp, 1925; flowers small, dbl.; Koster; (Sangerhausen)

Konigliche Hoheit, *see* **'ROYAL HIGHNESS'**

Königlicht Hoheit, *see* **'ROYAL HIGHNESS'**

'KÖNIGSBERG', HT, mr, 1940; flowers scarlet-red; [Mrs Henry Winnett X Marechal Petain]; Weigand, L.; Teschendorff

'KONINGIN ASTRID', HT, mr, 1935; (Queen Astrid); bud long, pointed, nasturtium-red; flowers reddish apricot and bronze, very large, dbl.; foliage dark, bronze; vigorous growth; Leenders, M.

'KONINGIN JULIANA', F, mp, 1948; flowers salmon-pink becoming light yellow with salmon, large, borne, dbl., blooms late, moderate fragrance; foliage glossy; [Poulsen's Pink X Mrs Pierre S. duPont]; Buisman, G. A. H.

'KONRAD ADENAUER'®, HT, dr, 1955; (Konrad Adenauer Rose®); bud globular; flowers blood-red, dbl., 35 petals, 4 in., cupped, intense fragrance; foliage light green, glossy; vigorous, upright growth; [Crimson Glory X Hens Verschuren]; Tantau, Math.; J&P, 1954

Konrad Adenauer Rose®, *see* **'KONRAD ADENAUER'**

'KONRAD GLOCKER', F, dr, 1962; dbl., 3 in.; foliage dark; vigorous, bushy, low growth; Kordes, R.

Konrad Henkel®, *see* 'KORJET'

'KONRAD THÖNGES', HFt, op, 1929; flowers salmon-orange/pink, large, semi-dbl.; Thönges; (Sangerhausen)

'KON-TIKI', F, mr, 1971; bud ovoid; flowers bright red, large, semi-dbl., cupped; foliage glossy, dark; vigorous, upright growth; [Aztec X Paprika]; Institute of Ornamental Plant Growing

Kontrast, HT, rb, 1981; Hchen

Kooiana Butterscotch, HT, lp; [Golden Times sport]; Gibson, P., 1995; Sunrise Flowers, Int.; (Weatherly, L.)

Kooiana Daybreak, HT, ab

Kooiana Moonlight, HT, lp; [Gerdo sport]; Gibson, P., 1995; Sunrise Flowers, Int.; (Weatherly, L.)

Kooiana Watermelon, HT, lp; [Kooiana Daybreak sport]; Gibson, P., 1995; Sunrise Flowers, Int.; (Weatherly, L.)

Kookaburra, S, lp

'KOOPMANNS SPORT VON ELSE POULSEN', Pol, mr, 1940; flowers medium, semi-dbl.; Koopmann; (Sangerhausen)

Kooshti-Bok, *see* JAYkoo

'KOOTENAY', HT, lp, 1917; flowers primrose-color; Dickson, A.

Köpenicker Sommer, F, dy, 1968; flowers medium-large, dbl.; Rupprecht-Radke; (Sangerhausen)

'KOPPIES', HT, lp, 1979; bud ovoid; dbl., 35 petals, exhibition form, borne usually singly, moderate, fruity fragrance; short, recurved prickles; foliage medium, matt; vigorous, tall growth; [Tropicana X Wendy Cussons]; Perry, Astor

KORabmask, HT, dy, 1996; (**Harry Oppenheimer**); Kordes

KORaby, HT, mp, 1995; (**Barby**); Kordes

KORacona, F, mp, 1995; (**Georgette**); Kordes

'KORAD', HT, yb, 1970; (Susan, **Susan Massu**); bud ovoid; flowers light yellow and salmon blend, large, dbl., cupped, moderate fragrance; foliage glossy, dark, leathery; vigorous, upright growth; GM, Baden-Baden, 1968; [Colour Wonder X Liberty Bell]; Kordes, R.; Kordes

'KORADES', S, dp, 1993; (**Wild at Heart**); flowers deep pink, large, full, dbl., 15–25 petals, 3–3.5 in., borne in large clusters, slight fragrance; foliage large, medium green, glossy; medium (120–130 cms), bushy, spreading (180 cms) growth; [Bonanza X Seedling]; Kordes, W. Söhne; Bear Creek Gardens, 1994

'KORAL', F, or, 1969; (**Bengali**®); bud ovoid; flowers red-orange, medium, dbl.; foliage dark; [Dacapo X Seedling]; Kordes; Buisman, 1969

Koralie, S, lp

'KORALLE', F, mr, 1942; flowers light red, slight fragrance; [Else Poulsen sport]; Koopman; C-P;, Krause;, Tantau

Koralle, F, op, 1991; Harkness

Korallovyj Sjurpriz, F, op, 1966; flowers luminous coral-red, large, dbl., moderate fragrance; Klimenko, V. N.; (Sangerhausen)

KORalogen, F, ab, 1996; (**Apps Rose**); Kordes

'KORALU', HT, pb, 1957; (**Kordes' Perfecta**, Perfecta); bud urn-shaped; flowers cream tipped and then flushed crimson, suffused yellow, dbl., 68 petals, 4.5–5 in., exhibition form, intense fragrance; foliage dark, leathery, glossy; vigorous, upright growth; GM, NRS, 1957 GM, Portland, 1958 PIT, NRS, 1957; [Golden Scepter X Karl Herbst]; Kordes; A. Dickson, 1957;, J&P, 1958;, McGredy, 1957

KORamator, HT, ab, 1996; (**Marjorie Chase**); Kordes

'KORAMBO', HT, my, 1986; (**Professor Fred Ziady**); flowers clear medium yellow, star-shaped, borne singly, very dbl., 43 petals, moderate fragrance; straight, dark brown prickles; foliage medium green; medium, sturdy growth; [Lusambo X Deep Secret]; Kordes, R.; Ludwigs Roses Pty. Ltd., 1985

KORameget, HT, dy, 1995; (**Egoli**); Kordes

KORamgis, HT, dy, 1996; (**Golden Spire**); Kordes

'KORAMI', F, op, 1964; (Ami des Jardins, **Finale**); flowers salmon-rose, well formed, borne in clusters, dbl., 21 petals, 3.5 in., foliage light green; low, compact growth; [Nordlicht X Meteor (F)]; Kordes, R.; A. Dickson;, McGredy

'KORAMPA', F, w, 1985; (**Antique Silk**, Champagner®, Kordes' Rose Champagner); flowers near white, flora-tea, large, dbl., 20 petals, slight fragrance; foliage medium, medium green, semi-glossy; upright, bushy growth; PP005411; [Anabell seedling X Seedling]; Kordes, 1982

KORampli, HT, dr, 1996; (**Bles Bridges**); Kordes

KORamvis, HT, ly, 1997; (**Lemon Dream**); Kordes

KORanderer, HT, ab, 1996; (**Kupferkönigin**); flowers coppery pink, medium, dbl., Kordes, W. Söhne; (Sangerhausen)

KORandpunk, HT, m, 1996; (**Madiba**); Kordes

'KORANGELI', HT, ob, 1985; (Angelique®, **Ankori**, Kordes' Rose Angelique); flowers orange, well-formed, large, dbl., 40 petals, slight fragrance; foliage medium, medium green, matt; bushy growth; [Mercedes X Seedling]; Kordes, 1980

'KORANTEL', HT, yb, 1985; (**Tarantella**); flowers creamy yellow, petals edged and marked with pink, large blooms, dbl., 54 petals, exhibition form; light brown prickles; foliage medium, medium green, glossy; upright growth; [Colour Wonder X Wiener Charme]; Kordes, W.; Horstmann, 1979

KORaucher, S, ab, 1996; (**Deloitte & Touche**); Kordes

'KORAV', HT, op, 1985; (**Ave Maria**®, Sunburnt Country); flowers orange-salmon, large, dbl., 35 petals, exhibition form, borne singly, moderate fragrance; small prickles; foliage large, medium green, semi-glossy; upright growth; [Uwe Seeler X Sonia]; Kordes, W., 1981

KORazerka, HT, dr, 1994; (**Ecstasy**); Kordes

KORbacol, HT, lp; (**Texas**); Kordes, 1993

'KORBAD', F, mr, 1980; (**Bad Füssing**®); bud large; flowers, dbl., 23 petals, cupped, blooms in clusters, moderate fragrance; few prickles; foliage glossy, dark; vigorous, upright, bushy growth; [Gruss an Bayern X Seedling]; Kordes, W.

KORbalem, Min, my, 1988; (**Little Lemmy**); Kordes

KORbarkeit, S, lp; (**Yorkshire**); Kordes, 1998

KORbasren, S, pb, 1993; (**Pink Bassino**); Kordes

KORbasren, S, lp; (**St Tiggywinkles**); Kordes

KORbasta, HT, my; Kordes

KORbaxand, HT, yb, 1973; (**Alexandra**); Kordes

'KORBE', HKor, mr, 1959; (Grüss an Heidelberg, **Heidelberg**); flowers bright crimson, reverse lighter, blooms in clusters, dbl., 32 petals, 4 in., exhibition form; foliage glossy, leathery; very vigorous, bushy growth; Kordes, R.; McGredy

'KORBEEN', HT, dy, 1981; (**Goldy**); flowers large, dbl., 35 petals, moderate fragrance; foliage medium, medium green, semi-glossy; upright, bushy growth; [Berolina X Unnamed seedling]; Kordes, W.; Kordes Sons

Korbel Canyon Red China, Ch, mr; (found rose)

Korbel Vinca Purple, LCl, m; (found rose)

'KORBELL', F, ob, 1972; (**Anabell**®, Annabelle, Kordes' Rose Anabel); flowers orange and silvery blend, well-formed, dbl., 30 petals, 4 in., moderate fragrance; foliage small; [Zorina X Colour Wonder]; Kordes; Dicksons of Hawlmark

'KORBELMA', HT, my, 1979; (Goldsmith, **Helmut Schmidt**®, Simba); bud large, long, pointed; flowers clear, even yellow, dbl., 30–40 petals, 4.5–5 in., exhibition form, borne 1–3 per cluster, moderate, sweet tea fragrance; foliage dark green, matt; vigorous, upright, bushy growth; GM, Belgium, 1979 GM, Geneva, 1979; [New Day X Unnamed seedling]; Kordes, W.

KORberis, S, mp, 1998; (**Felicitas**); Kordes

KORberuhig, LCl, dp, 1996; (**Manita**); Kordes

'KORBETEILICH', Gr, dr, 1999; (**Crimson Bouquet**™); flowers dark garnet red, reverse shiny dark red, good substance, dbl., 20–25 petals, 4–4.5 in., exhibition form, borne in large clusters, slight, sweet fragrance; prickles large; foliage large, dark green, glossy; upright, medium (4.5 ft) growth; AARS, U.S., 2000; [Bad Füssing X Ingrid Bergman]; Kordes; Bear Creek Gardens, Inc., 2000

'KORBI', HT, dy, 1961; (Fièvre d'Or, **Golden Giant**®, Goldrausch); flowers rich golden yellow, well-formed, dbl., 45 petals, 5 in., moderate fragrance; foliage dark; vigorous, tall growth; GM, NRS, 1960; Kordes, R.; A. Dickson;, McGredy

'KORBICO', HT, ob, 1964; (**Colour Wonder**, Königin der Rosen, Queen of Roses, Reine des Roses); bud ovoid; flowers orange-coral, reverse cream, large, dbl., 50 petals, slight fragrance; many, large prickles; foliage glossy, bronze; vigorous, bushy growth; ADR, 1964 GM, Belfast, 1966; [Kordes' Perfecta X Tropicana]; Kordes, R.; A. Dickson, 1964;, McGredy, 1964;, Wyant, 1966

'KORBIDO', HT, m, 1966; (**Silver Star**®); flowers lavender, well-formed, dbl., 5 in., intense fragrance; foliage dark; [Sterling Silver X Magenta (F) seedling]; Kordes, R.; McGredy

'KORBIN', F, w, 1958; (Fée des Neiges, **Iceberg**, Schneewittchen); bud long, pointed; flowers pure white, large blooms in clusters, dbl., intense fragrance; foliage light green, glossy; vigorous, upright, bushy growth; GM, Baden-Baden, 1958 GM, NRS, 1958 Hall of Fame, WFRS, 1983; [Robin Hood X Virgo]; Kordes, R.

'KORBIRAC', Gr, ab, 1992; (**Caribbean**™); flowers apricot orange/yellow blend, dbl., 30–40 petals, 3–3.5 in., exhibition form, borne in small clusters, moderate fragrance; many prickles; foliage large, dark green, semi-glossy, has pronounced red; medium (110–125 cms), upright, bushy growth; PP8592; AARS, 1994; [Mercedes X (New Day X Seedling)]; Kordes, W.; Bear Creek Gardens, 1994

'KORBISCH', S, ob, 1983; (**Kordes' Brillant**®, Kordes' Brilliant); flowers orange, large, dbl., 35 petals, slight fragrance; foliage medium, medium green, glossy; upright, bushy growth; [Sympathie seedling X Unnamed seedling]; Kordes, W., 1982

'KORBLUE', F, m, 1974; (**Shocking Blue**®); bud very large; flowers lilac-mauve, dbl., exhibition form, intense fragrance; foliage dark, glossy, leathery; vigorous growth; [Seedling X Silver Star]; Kordes, 1985

KORboden, HT, op, 1997; (**Dawn Haggle**); Kordes

KORbolak, HT, lp, 1991; (**Melody**); Kordes

KORbonnet, HT, mr; (**Wessie Roos**); Kordes, 1993

'KORBOTAF', S, mp, 1985; (**Elveshörn**®); flowers medium, dbl., 35 petals, slight fragrance; foliage medium, dark, semi-glossy; bushy, spreading growth; [The Fairy X Seedling]; Kordes, W.

KORbraufa, S, mr, 1992; (**Godewind**®); Kordes

'KORCAPAS', F, yb, 1964; (**Samba**®); bud globular; flowers golden yellow touched red, becoming fiery red, blooms in clusters; foliage glossy; low, bushy growth; Kordes, R.

KORcelin, Min, ob, 1987; (**Mandarin**); Kordes

KORciate, HT, ob, 1991; (**Felicia Teichmann**); Kordes

KORcilmo, F, w, 1991; (**Eskimo**®); Kordes

'KORCONTA', S, w, 1998; (**Moon River**™); flowers white, single, small, single, 4–11 petals, 2 in., borne in large clusters, slight fragrance; prickles moderate; foliage medium, dark green, glossy; medium tall groundcover, $2\frac{1}{2}$ ft, spreading growth; [Yesterday X (Sea Foam X Red Max Graf)]; Kordes, W.; Bear Creek Gardens, 1996

'KORCOON', F, yb, 1981; (**Coon Carnival**®); bud ovoid; flowers yellow, changing to pink and red, large blooms borne 1–7 per cluster, dbl., 56 petals, slight fragrance; straight, brown prickles; foliage matt, green; medium high, bushy growth; [Seedling X Seedling]; Kordes, R.; Ludwigs Roses Pty. Ltd.

KORcountry, HT, ab, 1994; (**Bernalene**); Kordes

KORcremkis, F, w, 1991; (**Madeo**); Kordes

KORcrisett, HT, or, 1994; (**Calibra**); Kordes

KORdaba, HT, op, 1992; (**Lambada**); Kordes

KORdadel, Min, yb; (**The Valois Rose**); Kordes, 1990

'KORDAISY', F, pb, 1982; (**Daisy Rose**); flowers pink with white eye, small blooms in large clusters, single, 5 petals; foliage medium, medium green, semi-glossy; bushy growth; [Robin Hood X Topsi]; Kordes, W.

KORdalen, LCl, pb, 1988; (**Antike 89**®); flowers creamy white with pink, large, very dbl.; Kordes, W. Söhne; (Sangerhausen)

KORdalsilk, F, ab, 1996; (**Kingsmead Heritage**); Kordes

'KORDALU', F, mr, 1976; (**Andalusien**®); bud long, pointed; flowers red, dbl., 34 petals, 3 in., cupped; vigorous, bushy growth; ADR, 1976; [Seedling X Zorina]; Kordes, 1977

'KORDAPT', Min, lp, 1985; (**Heidekönigin**®, Palissade Rose®, Pheasant); flowers large, dbl., 35 petals, slight fragrance; foliage small, medium green, glossy; groundcover; spreading growth; [Zwerkonig '78 X R. wichurana seedling]; Kordes, W.

'KORDAY', F, dp, 1984; (Angela®, **Angelica**); flowers deep pink, dbl., 35 petals, cupped, borne singly and in clusters, no fragrance; foliage medium, medium green, glossy; bushy growth; ADR, 1982; [Yesterday X Peter Frankenfeld]; Kordes

KORdehei, F, m, 1991; (**Heidelinde**®); Kordes

KORdehei, S, lp; (**Malverns**); Kordes

'KORDEHN', F, or, 1973; (**Mireille Mathieu**); bud ovoid; dbl., 27 petals, 3.5 in., exhibition form, slight fragrance; foliage soft; vigorous, upright, bushy growth; [Duftwolke X Peer Gynt]; Kordes; Dehner & Co.

KORdemas, Min, mp, 1990; (**Pink Button**); Kordes

Kordes' Brillant®, *see* 'KORBISCH'

Kordes' Harmonie, *see* **'Harmonie'**

Kordes' Magenta, *see* **'MAGENTA'**

Kordes' Perfecta, *see* 'KORALU'

'KORDES' PERFECTA SUPERIOR', HT, mp, 1963; (Perfecta Superior); flowers bright pink; [Kordes' Perfecta sport]; Kordes, 1963; Wyant, 1966

'KORDES' PERFECTA, CLIMBING', Cl HT, pb, 1962; Japan Rose Soc.

Kordes' Rose Anabel, *see* 'KORBELL'

Kordes' Rose Angelique, *see* 'KORANGELI'

Kordes' Rose Bella Rosa, *see* 'KORWONDIS'

Kordes' Rose Champagner, *see* 'KORAMPA'

Kordes' Rose Delicia, *see* 'KORGATUM'

Kordes' Rose Esmeralda, *see* 'KORMALDA'

Kordes' Rose Florentine®, *see* **'FLORENTINA'**

Kordes' Rose Holstein, *see* 'KORHOLST'

Kordes' Rose Immensee®, *see* 'KORIMRO'

Kordes Rose Kardinal, *see* 'KORLINGO'

Kordes' Rose Lady Rose, *see* 'KORLADY'

Kordes' Rose Pasadena, *see* 'KORLAND'

Kordes' Rose Patricia, *see* 'KORPATRI'

Kordes' Rose Repandia®, *see* 'KORSAMI'

Kordes' Rose Robusta®, *see* 'KORGOSA'

Kordes' Rose Sylt®, *see* 'KORYLT'

Kordes' Rose Sylvia, *see* 'KORLIFT'

Kordes' Rose Weiss Immensee, *see* 'KORWEIRIM'

Kordes' Rose Westfalenpark®, *see* 'KORPLAVI'

Kordes' Sondermeldung, *see* **'INDEPENDENCE'**

KORdialo, MinFl, rb, 1994; (**Festival**); Kordes

'KORDIAM', HT, op, 1987; (Heidi Kabel, **Holsteinperle**®, Testa Rossa); flowers orange pink, large, dbl., no fragrance; foliage medium, medium green, semi-glossy; bushy growth; [Unnamed seedling X Flamingo]; Kordes, W., 1985

KORdibor, S, lp, 1988; (**Rosendorf Sparrieshoop**®); Kordes

'KORDIENA', HT, w, 1996; (**Helen Naude**); flowers white flushed with pink, blooms, very dbl., 4.75 in., moderate prickles; foliage large, medium green, dull; bushy, medium growth; Kordes, W. Söhne; Ludwigs Roses Pty. Ltd.

KORdisard, S, mr, 1999; (**Girtnerfreude**); Kordes

'KORDODO', F, mr, 1987; (**Larissa**®); flowers medium, dbl., 26–40 petals, no fragrance; foliage small, medium green, semi-glossy; spreading growth; [(Unnamed seedling X Marina) X Rumba]; Kordes, W., 1989

KORdomal, S, w, 1991; (**Gletscherfee**®); Kordes

'KORDOSELBLA', HT, ab, 1987; (**Sunbeam**®); flowers large, dbl., 26–40 petals, slight fragrance; foliage large, dark green, semi-glossy; bushy growth; Kordes, W.

'KORDOUBT', HT, dy, 1987; (**Johannesburg Sun**); flowers deep golden yellow, large, borne singly, dbl., 22 petals, moderate fragrance; prickles concave, brown; foliage glossy, deep green; tall, upright growth; [Unnamed seedling X Unnamed seedling]; Kordes, W. Söhne; Ludwigs Roses Pty. Ltd., 1988

KORdreischi, HT, w, 1997; (**Beverley Watson**); Kordes

KORdrekes, 1998; (**Myrna's Dream**); Kordes

KORdreweer, 1998; (**Arena's Dream**); Kordes

'KORDU', HT, mr, 1969; (**Duftzauber**, Fragrant Charm); bud ovoid; flowers rose-red, large, dbl., exhibition form, intense fragrance; foliage light, soft; moderate, upright growth; [Prima Ballerina X Kaiserin Farah]; Kordes, R.; Kordes

'KORDURBAN', F, rb, 1982; (**Durban July**, Durbankor); flowers yellow, orange to red, medium, semi-dbl.; foliage medium, medium green, semi-glossy; Kordes, W.

Kore, S, yb, 1980; flowers yellowish-pink, large, dbl., intense fragrance; Urban, J.; (Sangerhausen)

'KOREB', F, or, 1962; (**Diamant**®); bud ovoid; flowers bright orange-scarlet, well-formed, large blooms in clusters, dbl., 40 petals, slight fragrance; foliage dark, glossy; vigorous, upright growth; Kordes, R.; A. Dickson;, McGredy

KORechtem, HT, w; (**White Spire**); Kordes, 1994

KORedan, F, mp, 1995; (Bluewunder, Bluhwunder®, **Flower Power**); Kordes; (Cavriglia)

KOReibei, F, dp, 1993; (**Bridget**); Kordes

'KOREIPARK', S, mr, 1983; (**Rheinaupark**®); flowers large, dbl., 20 petals, slight fragrance; foliage large, dark, glossy; upright, bushy growth; [(Gruss an Bayern X Seedling) X R. rugosa seedling]; Kordes; Kordes Roses

KOReklia, HT, ab, 1989; (New Valencia, Valeccia, **Valencia**®, Valencia 89); Kordes, 1989

KORelasting, S, mp, 1999; (**Linderof**); Kordes

KOReledas, S, lp, 1996; (**NDR I Radio Niedersachsen**); flowers medium, semi-dbl., slight fragrance; Kordes, W. Söhne; (Sangerhausen)

KOReledas, F, lp; (**Centenary**, Heartache); Rose of The Year, Auckland, NZ, 1999; Kordes

KORelgas, HT, w, 1992; (**Esme Euvrard**); flowers tend to light pink, petals creped, dbl., 4–4.5 in., borne mostly singly, slight fragrance; upright, tall (4 ft) growth; Kordes

'KORENBON', F, w, 1987; (**Innocencia**®); flowers medium, dbl., 26–40 petals, no fragrance; foliage medium, medium green, matt; bushy growth; [Lorena sport]; Kordes, W., 1986

'KORENLO', F, mp, 1984; (**Lorena**®, Lorina); flowers medium salmon-pink, flora-tea, large, dbl., 35 petals, exhibition form, no fragrance; foliage medium, medium green, semi-glossy; upright growth; PP005679; [Angelique X Seedling]; Kordes, W., 1983

KORenon, HT, dp, 1997; (**Neon**); Kordes

'KORENPI', F, mp, 1987; (**Loretta**®); flowers medium, dbl., 26–40 petals, no fragrance; foliage medium, medium green, matt; flora-tea bushy growth; PP007479; [Lorena sport]; Kordes, W., 1986

KORfachrit, HT, lp, 1992; (**Andrea Stelzer**); Kordes

'KORFALT', F, dy, 1984; (**Goldmarie**, Goldmarie 82®, Goldmarie Nirp®); flowers deep yellow, red on reverse of outer petal, large, dbl., 35 petals, slight fragrance; foliage medium, medium green, glossy; bushy growth; [((Arthur Bell X Zorina) X (Honeymoon X Dr A.J. Verhage)) x(Seedling X Sunsprite)]; Kordes, W.; Kordes Sons

'KORFAN', HT, pb, 1974; (**Fantasia**); bud ovoid; flowers lilac-red, flora-tea, globular, moderate fragrance; foliage leathery; vigorous growth; [Silver Star X Tradition]; Kordes; Horstmann, 1977

KORfanto, Min, dp, 1989; (**Rosmarin 89**®); Kordes

'KORFARIM', F, ob, 1972; (Esther Ofarim, Esther O'Farim, **Matador**); bud ovoid; flowers light scarlet and orange, reverse gold, medium, dbl., exhibition form, slight fragrance; foliage large, dark, leathery; vigorous growth; [Colour Wonder X Zorina]; Kordes, R.; J&P, 1970

'KORFEE', Min, or, 1979; (Dwarf Fairy, **Zwergenfee**®); bud globular; dbl., 29 petals, 2 in., cupped, moderate fragrance; foliage small, glossy; dwarf, vigorous, upright growth; [Miniature seedling X Traumerei]; Kordes, W.

KORfeimot, HT, pb, 1994; (**Grafin Sonja**); Kordes

KORfeining, S, pb, 1997; (**Dawn Sunsation**); Kordes

KORfeldwo, S, lp; (**Tudor Sunsation**); Kordes, 1997

KORferse, F, or, 1989; (**Coco**); Kordes

'KORFEU', HT, or, 1973; (**Feuerzauber**®, Fire Magic, Magic de Feu); flowers orange-red, reverse lighter, medium to large, dbl., exhibition form; foliage dark, glossy; vigorous, upright growth; [Fragrant Cloud X Unnamed seedling]; Kordes, R., 1974

'KORFI', F, mr, 1976; (**Fiona**); bud long, pointed; dbl., 24 petals, 3 in., exhibition form, slight fragrance; foliage wrinkled; vigorous, upright growth; RULED EXTINCT 9/82 ARM; [Seedling X Prominent]; Kordes

KORfibi, S, dp, 1997; (**Camilla Sunsation**); Kordes

KORfischer, S, m, 1994; (**Hansa-Park**®); Kordes

KORfisro, S, lp; (**Sweet Sunsation**); Kordes

'KORFLAPEI', F, my, 1987; (**Frisco**®, Pamela); flowers medium, dbl., 26–40 petals, slight fragrance; foliage medium, dark green, semi-glossy; mini-flora bushy growth; PP006695; [((New Day X Minigold) X Banzai) X Antique Silk]; Kordes, W., 1986

'KORFLATA', LCl, mr, 1955; (Flame Dance, **Flammentanz**®); flowers crimson, very large blooms in clusters, dbl., exhibition form, non-recurrent, moderate fragrance; foliage dark, leathery; very vigorous (10 ft) growth; ADR, 1952; [R. eglanteria hybrid X R. kordesii]; Kordes

'KORFLOT', HT, dp, 1958; (**Ballet**®); flowers deep pink, dbl., 52 petals, 5 in., slight fragrance; foliage gray-green; vigorous, bushy growth; [Florex X Karl Herbst]; Kordes, R.; A. Dickson;, McGredy

'KORFLÜG', HT, lp, 1979; (**Flamingo**®, Margaret Thatcher, Porcelain, Veronica, Veronika); bud large, long, pointed; flowers large blooms borne singly, dbl., 24 petals, exhibition form, moderate fragrance; many prickles; foliage matt, green; vigorous, upright, bushy growth; [Unnamed seedling X Lady Like]; Kordes, W.; Kordes, 1978;, Ludwigs Roses, 1983

KORfolga, S, lp; (**Splish Splash**); Kordes, 1996

KORfolklori, HT, rb, 1994; (**Herman Steyn**); Kordes

'KORFRILLA', F, rb, 1987; (**Carnaval**®, Carnival); flowers white with red edges, large, dbl., 26–40 petals, no fragrance; foliage medium, dark green, matt; bushy growth; [Seedling X (Die Krone X Simona)]; Kordes, W., 1986

'KORFULLWIND', S, lp, 1993; (**Baby Blanket**, Oxfordshire, Sommermorgen); flowers moderately full, medium blooms, dbl., 15–25 petals, 1.5–2.75 in., borne in large clusters, slight fragrance; few prickles; foliage small, dark green, glossy; medium (60–75 cms), bushy, spreading (120–150 cms) growth; [Weisse Immensee X Goldmarie]; Kordes, W. Söhne; Bear Creek Gardens; Bear Creek Garden, Inc., 1993, USA

KORfungo, S, mr, 1990; (**Royal Bassino**); Kordes

'KORGANE', HT, ob, 1981; (**Las Vegas**®); bud large, pointed; flowers deep orange, reverse lighter, blooms borne 1–3 per cluster, dbl., 26 petals, moderate fragrance; brown prickles; foliage green, slightly glossy; vigorous, upright, bushy growth; GM, Genoa, 1985 GM, Portland, 1988; [Ludwigshafen am Rhein X Feuerzauber]; Kordes, W.

'KORGATER', HT, yb, 1985; ('KORSUN', **Sun City**); flowers deep yellow with red petal edges, red spreading, star-shaped, dbl., 32 petals, slight fragrance; dark brown prickles; foliage deep green, red when young, leathery; upright, tall growth; [((New Day X Minigold) X Unnamed seedling) X MEItakilor]; Kordes, R.; Ludwigs Roses Pty. Ltd.

KORgatine, S, w, 1996; (**Cream Sunsation**); Kordes

'KORGATUM', HT, ly, 1982; (Delicia, **Elegant Beauty**, Kordes' Rose Delicia); flowers light yellow flushed pink, large, dbl., 20 petals; foliage large, dark, matt; upright, bushy growth; [New Day X Seedling]; Kordes, W.

KORgenda, F, m, 1994; (**Shocking Sky**); Kordes

'KORGERA', F, or, 1976; (**Bagheera**); bud ovoid; dbl., 35 petals, 4 in., exhibition form, slight fragrance; foliage glossy, dark; vigorous, upright growth; [Nordia X Seedling]; Kordes, W.

'KORGI', HT, m, 1964; (Blue Girl, Cologne Carnival, **Kölner Karneval**); flowers silvery lilac-lavender, dbl., 35–40 petals, 5.5 in., exhibition form, borne singly, moderate, lightly fruity fragrance; foliage large, deep green; vigorous, bushy growth; GM, Rome, 1964; Kordes, R.

KORgitte, HT, ob, 1986; (**Golden Queen**); Gold Star of the South Pacific, Palmerston North, NZ, 1984; Kordes

'KORGO', F, my, 1963; (Gold Topaz, **Goldtopas**®, Goldtopaz, 'KORTOSSGO'); bud ovoid; flowers amber-yellow, blooms in clusters (up to 10), dbl., 3.5 in., cupped; foliage glossy; vigorous, bushy growth; ADR, 1963; Kordes, R.

'KORGOLD', HT, my, 1977; (Mabella, **New Day**); bud ovoid, pointed; flowers mimosa-yellow, dbl., 30 petals, 4–5 in., exhibition form, intense fragrance; foliage large, light; upright growth; [Arlene Francis X Roselandia]; Kordes, R.; J&P

'KORGOSA', S, mr, 1979; (Kordes' Rose Robusta®, **Robusta**®); bud long, pointed; flowers medium, single, 5 petals, 2.5 in., moderate fragrance; foliage dark, glossy, leathery; very vigorous, upright, bushy growth; ADR, 1980; [Unnamed seedling X R. rugosa]; Kordes, W.

'KORGRAM', S, w, 1983; (**Weisse Max Graf**®); flowers medium, semi-dbl., intense fragrance; foliage small, dark, glossy; spreading (10 ft) growth; groundcover; [Seedling X R. wichurana seedling]; Kordes, W.; Kordes Roses

KORgrapet, F, my; (**Suncluster**); Kordes

KORgrayel, F, m, 1993; (**Saffex Rose**); Kordes

KORgreyel, F, dy; (**The J.S.E. Rose**); Kordes, 1993

'KORGUND', HT, dr, 1977; (Burgund®); bud long, pointed; dbl., 30 petals, 4 in., exhibition form, intense fragrance; vigorous, bushy growth; [Henkell Royal X Seedling]; Kordes, W.

'KORGUND '81', HT, mr, 1983; (Burgund '81, 'KORGUND '81', **Loving Memory**, Red Cedar); flowers large, dbl., exhibition form, slight fragrance; foliage medium green, semi-glossy; upright, bushy growth; [Seedling X Red Planet seedling]; Kordes, W., 1981

'KORGUST', HT, ab, 1987; (Sunsation, **Veldfire**, Wurzburg); flowers orange, reverse chrome-yellow, large, borne singly, dbl., 38 petals, moderate fragrance; prickles concave, yellow-brown; foliage glossy, medium green; upright, well-branched, free-flowering growth; [Unnamed seedling X Unnamed seedling]; Kordes, W. Söhne; Ludwigs Roses Pty. Ltd., 1988

KORhagon, HT, dp, 1992; (**Pretoria**); Kordes

KORhamp, S, mr, 1989; (**Hampshire**); Kordes

'KORHANBU', HT, op, 1985; (**Belami**, 'KORPRILL'); flowers orange pink, large, dbl., 35 petals, moderate fragrance; foliage medium, dark, glossy; upright, bushy growth; [(Prominent X Carina) X Emily Post]; Kordes, W.

KORharment, S, dr, 1993; (**Francois Krige**); Kordes

KORhassi, HRg, dr, 1993; (**Hansaland**®); Kordes

KORhassi, S, lp; (**Charles Notcutt**); Kordes

KORhaugen, S, rb, 1994; (**Carpet of Color**); Kordes

'KORHEIM', S, rb, 1985; (**Morgenrot**®); flowers medium, single, 5 petals, slight fragrance; foliage small, dark, matt; bushy growth; [(Marlena X Europeana) X ((Tropicana X Carina) X (Clare Grammerstorf X Fruhlingsmorgen))]; Kordes, W.

'KORHITOM', Min, dy, 1987; (Perestroika, **Sonnenkind**®); flowers medium, dbl., slight fragrance; foliage small, medium green, semi-glossy; bushy growth; [Unnamed seedling X Goldmarie]; Kordes, W., 1986

'KORHOLST', S, dr, 1987; (City of Birmingham, **Esprit**®, Holstein 87, Kordes' Rose Holstein, Petit Marquis); flowers deep red, aging darker, medium, semi-dbl., 12 petals, flat, borne in sprays of 5–7, repeat bloom, no fragrance; no fruit; medium, tan prickles, slightly down pointed; foliage small, medium green, semi-glossy; upright, bushy, tall growth; PP006117; [Seedling X Chorus]; Kordes, W.; J&P, 1989

KORhood, HT, mr, 1993; (**Rooi Rose**); Kordes

KORhoro, LCl, my, 1988; (**Morgensonne**); Kordes

KORhota, HT, pb; (**Sarie Marais**); Kordes

KORhuba, HT, ab, 1995; (**Glendora**); Kordes

'KORHUG', HT, my, 1987; (**St Hughs**); flowers creamy yellow, large, dbl., 26–40 petals, moderate fragrance; foliage medium, medium green, semi-glossy; upright, bushy growth; [Unnamed seedling X Unnamed seedling]; Kordes, W.; The Rose Nursery, 1986

KORhungen, S, rb, 1998; (**Cambridgeshire**); flowers lavender pink, borne in large clusters; Kordes

KORhurtlen, LCl, lb, 1991; (**Copper Arch**); Kordes

'KORIANT', S, w, 1982; (**Weisse Repandia®**); flowers small, semi-dbl., slight fragrance; foliage small; dark, glossy; spreading (7 ft) growth; groundcover; [The Fairy X R. wichurana seedling]; Kordes, W.; Kordes Roses

KORichard, F, ab, 1990; (**Plisiedame**, Polisiedame); Kordes

'KORICOLE', F, w, 1985; (**Nicole**); flowers white with pink petal edges, large, dbl., 35 petals, slight fragrance; foliage large, dark, semi-glossy; upright growth; [Seedling X Bordure Rose]; Kordes, W., 1984

KORiganta, HT, ab, 1993; (**Antoinette**); Kordes

KORignale, HT, w; (**Amorosa**); Kordes

KORikis, F, op, 1988; (**Kiss**); Kordes

'KORIKON', HT, ly, 1984; (Golden Medaillon, **Limelight**); flowers large, dbl., 35 petals, exhibition form, intense fragrance; foliage medium, dark, semi-glossy; upright, bushy, spreading growth; [Peach Melba X Seedling]; Kordes, W., 1984

'KORILONA', F, my, 1976; (**Bellona**); bud ovoid; flowers golden yellow, pointed, dbl., 27 petals, 3 in., slight fragrance; foliage light; very vigorous, upright growth; [New Day X Minigold]; Kordes, R.; J&P

'KORIMRO', S, lp, 1983; (Grouse, **Immensee®**, Kordes' Rose Immensee®, Lac Rose®); flowers light pink to near white, small, single, moderate fragrance; foliage small, dark, glossy; spreading (to 13 ft) growth; groundcover; GM, RNRS, 1984; [The Fairy X R. wichurana seedling]; Kordes, W., 1982

KORinor, HT, dr, 1987; (**Mandy**); Kordes

'KORINTER', LCl, op, 1982; (**Rosanna**); flowers orange pink, large, dbl., 35 petals, exhibition form, moderate fragrance; foliage medium, medium green, glossy; upright growth; [Coral Dawn X Unnamed seedling]; Kordes, W.

'KORISSEL', S, w, 1993; (**Jeeper's Creeper**); flowers semi-double, blooms borne in large clusters, semi-dbl., 6–14 petals, 1.5–2.75 in., borne in large clusters, slight fragrance; many prickles; foliage medium, dark green, semi-glossy; low (40–50 cms), spreading growth; [Yesterday X Edelweiss]; Kordes, W. Söhne; Bear Creek Gardens, 1994

'KORITA', HT, ab, 1978; (**Gitte**, Peach Melba); bud long, pointed; flowers apricot-pink blend, dbl., 33 petals, 4 in., exhibition form, intense fragrance; foliage dark; vigorous, upright, bushy growth; [(Fragrant Cloud X Peer Gynt) X ((Dr A.J. Verhage X Colour Wonder) X Zorina)]; Kordes, W. Söhne; Horstmann

KORituscha, F, lp; (**Triodene**); Kordes, 1995

'KORIVER', S, dr, 1985; (Esterel, **Heidikind**); flowers medium, dbl., 20 petals, slight fragrance; foliage small, medium green, glossy; upright, bushy growth; [The Fairy X Seedling]; Kordes, W.

KORivo, F, w, 1985; (**Ivory Beauty**); Kordes

'KORIZONT', LCl, dp, 1985; (**Summer Wine**); flowers deep pink, red stamens, large, single, 5 petals, moderate fragrance; foliage large, medium green, semi-glossy; upright growth; Kordes, W.; John Mattock, Ltd

'KORJET', HT, mr, 1983; (Avenue's Red, **Konrad Henkel®**); flowers large, dbl., 35 petals, exhibition form, moderate fragrance; foliage large, medium green, semi-glossy; upright, bushy growth; [Unnamed seedling X Red Planet]; Kordes, W.

'KORJONI', HT, my, 1987; (**Zitronenjette®**); flowers large, dbl., 15–25 petals, intense fragrance; foliage large, medium green, glossy; spreading growth; [Sutter's Gold X Sunblest]; Kordes, W., 1986

KORkalba, Min, w; (**Zepheline**); Kordes, 1994

KORkandel, S, pb; (**Warwickshire**); Kordes, 1991

'KORKEILICH', Min, dp, 1987; (**Pepita**); flowers deep pink, small, dbl., 26–40 petals, no fragrance; foliage small, medium green, semi-glossy; spreading growth; [LENpi X (Mercedes X Garnette)]; Kordes, W., 1985

KORkeindor, F, op, 1994; (**Nancy Gardiner**); Kordes

KORkeltin, S, mr; (**Tradition 95®**); Kordes, 1995

KORkilt, HT, lp; (**Susan**); Kordes, 1996

KORkister, Min, ob, 1987; (**Orange Juwel®**); Kordes

KORkojotie, F, lp; (**Toorenburg**); Kordes, 1995

KORkompo, F, op, 1993; (**Anja**); Kordes

KORkonig, Min, dr; (**Zwergkonig 78**); Kordes, 1978

KORkragor, HT, mp, 1994; (**Coral Spire**); Kordes

KORkultop, F, m, 1993; (**Blue Bajou®**); Kordes

KORkuma, Cl F, yb, 1998; (**Goldmarie, Climbing**); Martens

KORkunde, F, lp; (**Toscana**); Kordes, 1991

'KORLADY', HT, op, 1979; (Kordes' Rose Lady Rose, **Lady Rose®**); bud long, pointed; dbl., 34 petals, 5 in., exhibition form, moderate fragrance; vigorous, upright, bushy growth; GM, Belfast, 1981; [Seedling X Traumerei]; Kordes, W. Söhne

KORlallal, LCl, dp, 1994; (**Pink Curtain**); Kordes

KORlamber, F, mr, 1993; (**Pixie Hat**); Kordes

'KORLAND', HT, or, 1982; (Kordes' Rose Pasadena, **Pasadena®**); flowers large, dbl., 35 petals, exhibition form, no fragrance; foliage large, medium green, matt; upright growth; [Mercedes X (Sweet Promise X (Miss Ireland X Zorina))]; Kordes, W., 1980

'KORLANUM', S, lp, 1985; (Sommerwind, **Surrey**); flowers medium, semi-dbl., slight fragrance; foliage small, medium green, semi-glossy; bushy growth; [The Fairy X Seedling]; Kordes, W.

KORlaper, F, lp; (**La Perla**); Kordes

KORlarkon, HT, w; (**Delicate Beauty**); Kordes

KORlasche, HT, m, 1993; (**Mary Pope**); Kordes

KORlastine, LCl, lp; (**Summer Breeze**); Kordes

'KORLAWE', S, ab, 1969; ('KORWEST', **Westerland®**); bud ovoid; flowers apricot-orange, cupped, 20 petals, 3 in., cupped, repeat bloom, intense; foliage large, dark, soft; vigorous, upright, climbing growth; ADR, 1974; [Friedrich Worlein X Circus]; Kordes

'KORLECH', F, dr, 1978; (Intrigue, Lavaglow, **Lavaglut®**); dbl., 24 petals, 2.5 in., globular, blooms in clusters, slight fragrance; foliage glossy; vigorous, upright, bushy growth; [Gruss an Bayern X Unnamed seedling]; Kordes, W. Söhne, 1979

'KORLEEN', F, mr, 1984; (Bavarian Girl, **Schöne Münchnerin®**); dbl., 22 petals, cupped, borne 2–3 per cluster, slight fragrance; medium, green prickles; foliage medium, medium green, semi-glossy; bushy growth; [Sympathie X Tornado]; Kordes, W.; Kordes;, Ludwig Roses

'KORLETTE', HT, m, 1998; (**Harlequin™**); flowers lavender pink, white reverse, dbl.,

26–40 petals, 4 in., borne mostly singly, slight fragrance; few prickles; foliage medium, dark green, glossy; bushy, low, 3 ft growth; [Prima Ballerina X Peace]; Kordes, W.; Bear Creek Gardens, 1998

KORlichtung, F, my, 1993; (**Ruth Pennington**); Kordes

'KORLIFT', HT, op, 1979; (**Congratulations**, Kordes' Rose Sylvia, Sylvia); bud long, pointed; flowers medium pink, dbl., 42 petals, 4.5 in., exhibition form, moderate fragrance; vigorous, upright, bushy growth; ADR, 1977; [Carina X Seedling]; Kordes, W. Söhne

'KORLILUB', S, my, 1985; (**Lichtkönigin Lucia**®); semi-dbl., 18 petals, cupped, borne 3–5 per cluster, moderate fragrance; foliage medium, dark, glossy; bushy, tall growth; ADR, 1968; [Zitronenfalter X Clare Grammerstorf]; Kordes, W., 1966

'KORLIMA', F, mr, 1959; (Lili Marlene, **Lilli Marleen**®, Lilli Marlene); bud ovoid; dbl., 25 petals, 3 in., cupped, moderate fragrance; foliage leathery; vigorous growth; ADR, 1960 Golden Rose, The Hague, 1966; [(Our Princess X Rudolph Timm) X Ama]; Kordes, R.; J&P, 1961;, McGredy, 1959

'KORLIMIT', HT, mr, 1987; (**Cora Marie**, Dallas); flowers large, dbl., 15–25 petals, no fragrance; foliage large, dark green, semi-glossy; upright growth; [Ankori X Seedling]; Kordes, W., 1986

KORlinde, HT, ob; (**Taihape Sunset**); Kordes, 1987

'KORLINGO', HT, mr, 1986; (**Kardinal**™, Kardinal 85, Kordes Rose Kardinal); flowers large, bright red, dbl., 30–35 petals, 4.5–5 in., exhibition form, borne singly, long vase life, slight fragrance; foliage medium, dark, semi-glossy; upright growth, medium; PP005846; [Unnamed seedling X Flamingo]; Kordes, W., 1985

'KORLIRUS', F, w, 1985; (Cevennes, **Heidesommer**®); flowers medium, dbl., 20 petals, intense fragrance; foliage small, dark, glossy; upright, bushy growth; [The Fairy X Unnamed seedling]; Kordes, W.

KORlis, HT, mp, 1996; (**Eliza**); Kordes

'KORLITA', HT, ab, 1973; ('LITAKOR', **Lolita**®); flowers golden bronze, dbl., 28 petals, 5 in., moderate fragrance; ADR, 1973; [Colour Wonder X Seedling]; Kordes, R.; Dicksons of Hawlmark, 1972

KORlitze, HT, yb, 1992; (**Lady Mavis Pilkington**); Kordes

KORlodera, S, mp, 1995; (**Pink Spectacle**); Kordes

KORloher, F, mr, 1990; (**Kingswood College**); Kordes

KORlomet, S, lp; (**Vogelpark Walsrode**); Kordes, 1988

'KORLORE', HT, ob, 1977; (**Folklore**®); bud long, pointed; flowers orange, reverse lighter, dbl., 44 petals, 4.5 in., exhibition form, intense fragrance; foliage glossy; very tall and vigorous, upright, bushy growth; [Fragrant Cloud X Seedling]; Kordes; Barni-Kordes

KORlowi, Gr, ob; (**Cathie Irwin**); Kordes

'KORLUDWIG', F, dp, 1975; (Ludwigshafen, **Ludwigshafen am Rhein**®); bud ovoid; flowers deep pink, flora-tea, dbl., 45 petals, 4 in., exhibition form, moderate fragrance; foliage soft; vigorous, upright, bushy growth; ADR, 1973; [Seedling X Pink Puff]; Kordes

KORlunta, Min, lp, 1987; (**Daniela**®); Kordes

KORlupo, LCl, pb, 1986; (**Harlekin**®); Kordes

'KORLYN', HT, w, 1985; (**Lynette**); bud long, pointed; flowers cream blended with coral pink, well-formed, large blooms in, dbl., no fragrance; straight, brown prickles; foliage dark; tall, upright growth; [Clivia X MEltakilor]; Kordes, R.; Ludwigs Roses Pty. Ltd., 1983

KORmador, F, lp; (**Tamara**); Kordes, 1988

'KORMAI', HT, dr, 1969; (**Mainauperle**®); bud ovoid; flowers large, dbl., exhibition form, intense fragrance; foliage large, dark, leathery; vigorous, upright, bushy growth; ADR, 1966; [Seedling X Americana]; Kordes, R.

'KORMALDA', HT, pb, 1981; (Esmeralda®, **Keepsake**, Kordes' Rose Esmeralda); bud ovoid; flowers deep pink blended with lighter pink shades, reflexed, large, dbl., 40 petals, moderate fragrance; large, stout prickles; foliage dark; vigorous, bushy growth; GM, Portland, 1987; Kordes, W. Söhne; John Mattock, Ltd.

KORmarec, S, mr, 1995; (**Sommerabend**®); Kordes

'KORMARIE', S, yb, 1982; (**Bonanza**®, Miss Pam Ayres); flowers yellow tipped red, large, dbl., 20 petals, slight fragrance; foliage medium, dark, glossy; upright growth; ADR, 1984; [Seedling X Arthur Bell]; Kordes, W., 1983

KORmarter, S, ob, 1988; (**Freisinger Morgenrote**®); flowers large, semi-dbl.; Kordes, W. Söhne

KORmarter, LCl, lp; (**Sunrise**); Kordes

KORmasyl, HT, lp, 1987; (**Fee**); Martens

'KORMAT', F, ab, 1980; (**Australian Gold**®, Mona Lisa); bud ovoid; flowers apricot-peach, dbl., 20 petals, borne 5 per cluster, moderate fragrance; red prickles; foliage dark, leathery; bushy growth; Kordes; John Mattock, Ltd, 1985

'KORMATE', HT, ab, 1987; (**Graaff-Reinet**); flowers apricot with orange on petal margin, large, borne in sprays, dbl., 36 petals, moderate fragrance; prickles concave, reddish-brown; foliage dull, medium green; compact, medium, well branched, free flowering growth; [Unnamed seedling X Unnamed seedling]; Kordes, W. Söhne; Ludwigs Roses Pty. Ltd., 1988

'KORMATT', F, or, 1982; (**St Boniface**); flowers patio, medium, dbl., 35 petals, slight fragrance; foliage medium, dark, semi-glossy; [Diablotin X Traumerei]; Kordes, W., 1980

KORmauret, LCl, dr, 1994; (**Red Curtain**); Kordes

'KORMAX', HKor, mr, 1980; (Red Max Graf, **Rote Max Graf**®); bud ovoid; single, 6 petals, blooms in clusters, moderate fragrance; dark brown prickles; foliage small, leathery, matt; vigorous, trailing groundcover growth; GM, Baden-Baden, 1981; [R. kordesii X Seedling]; Kordes, W.

KORmazin, F, my, 1991; (**Sommermond**®); Kordes

KORmeeram, HT, 1997; (**Royal Dream**); Kordes

KORmeita, LCl, mp, 1988; (**Ramira**®); Kordes

KORmeneint, S, mp, 1997; (**Mein Schoner Garten**); Kordes

'KORMETTER', F, op, 1985; (**Anna Livia**, Sandton Smile, Trier 2000); flowers orange pink, large, dbl., 20 petals, slight fragrance; foliage medium, medium green, semi-glossy; bushy growth; GM, Orleans, 1987 Golden Prize, Glasgow, 1997; [(Seedling X Tornado) X Seedling]; Kordes, W.

KORmiach, HT, mr, 1990; (**Liebeszauber**); Kordes

KORmiller, HT, lp, 1979; (**Dream**); Kordes

'KORMIORA', F, or, 1975; (**Luminion**®, Rosi Mittermeier); bud globular; dbl., 34 petals, 3 in., cupped, moderate fragrance; foliage glossy, dark; vigorous, upright, bushy growth; [Hurra X Peer Gynt]; Kordes; Vilmorin

'KORMIXEL', S, mr, 1988; (**Bassino**®); Kordes

KORmollis, F, mr, 1999; (**Garden Party**); Kordes

KORmorlet, LCl, ob, 1987; (**Salita**®); Kordes

'KORMOWE', HT, rb, 1987; (**Klerksdorp Horizon**); flowers tomato-red, reverse golden yellow, large, borne 1–3 per cluster, dbl., 32 petals, no fragrance; prickles concave, reddish-brown; foliage medium green, large; upright, medium, free-flowering growth; [Unnamed seedling X Unnamed seedling]; Kordes, W. Söhne; Ludwigs Roses Pty. Ltd., 1988

'KORMUN', F, mr, 1971; (Baveria, **Grüss an Bayern**®); flowers blood-red, medium, semi-dbl., globular, slight fragrance; foliage dark, leathery; vigorous, upright growth; ADR, 1973; [Messestadt Hannover X Hamburg]; Kordes, R.; Kordes

KORmunde, F, mp, 1975; (**Rosamunde**); Kordes

KORmuse, S, mp; (**Wiltshire**); Kordes, 1993

'KORNACHO', F, mr, 1984; (**Gartenzauber**®, Gartenzauber '84); flowers large, dbl., 35 petals, exhibition form, slight fragrance; foliage medium, dark,semi-glossy; upright growth; [(Unnamed seedling X Tornado) X Chorus]; Kordes, W.; Kordes Sons

KORnagent, F, lp, 1987; (**Florence**); Kordes

KORnanze, HT, my, 1990; (**Golden Medallion**); Kordes

KORnauer, HT, lp, 1989; (**Agnes Bernauer**®); Kordes

'KORNEMARK', HT, op, 1992; (**Louisiana**); Kordes

KORnends, HT, mr; (**Charlotte Searle**); Kordes

'KÖRNER', HMsk, yb, 1914; bud reddish; flowers orange-yellow tinted salmon, borne in clusters, dbl.; [Trier X Eugenie Lamesch]; Lambert, P.

KORnicken, S, op; (**Anjou**); Kordes

KORniebon, LCl, lp, 1998; (**Bonny**); Kordes

KORnieoch, HT, w; (**Table Mountain**); Kordes, 1990

'KORNITA', F, or, 1955; (**Korona**®); flowers orange-scarlet, blooms in large trusses, dbl., 20 petals, 2.5 in., slight fragrance; vigorous, upright growth; GM, NRS, 1954; Kordes; Morse

'KORNITZEL', HT, mr, 1987; (**Boksburg Fantasia**); flowers large, borne in sprays of 1–3, dbl., 23 petals, moderate fragrance; prickles needle point, brown; foliage medium green; tall, well branched growth; [Seedling X Seedling]; Kordes, W. Söhne; Ludwigs Roses Pty. Ltd., 1988

'KORNUMA', F, mr, 1983; (**Heinzelmännchen**®, Red Pixie); flowers large, dbl., 35 petals, slight fragrance; foliage medium, medium green, glossy; bushy growth; [(Satchmo X Unnamed seedling) X (Messestadt Hannover X Hamburg)]; Kordes, W., 1984

'KORO', HT, mr, 1983; flowers scarlet, large, dbl., 32 petals, exhibition form, intense, fruity fragrance; reddish prickles; foliage medium green; upright growth; [Pink Parfait seedling X Red Planet]; Cattermole, R.F.

KORoberfinz, HT, w, 1994; (**Margaret Wasserfall**); Kordes

KORocken, HT, ob; (**Out of Africa**); Kordes

KORofaser, HT, ob, 1994; (**Peach Spire**); Kordes

KORogesa, HT, ab, 1993; (**Helpmekaar Roos**); Kordes

KORoketto, F, m, 1994; (**Futura**); Kordes

'KOROL', HT, ob, 1969; (Coppertone, Old Time, **Oldtimer**); flowers bronze, long, pointed, slight fragrance; Kordes, R.; McGredy

'KOROL', HT, yb, 1968; (**Peer Gynt**®); flowers yellow, outer petals edged red, large, dbl., slight fragrance; vigorous, bushy growth; GM, Belfast, 1970; [Colour Wonder X Golden Giant]; Kordes, R.; McGredy

KORomega, HT, or, 1996; (**Pat's Choice**); Kordes

KORommerla, S, ab, 1997; (**Peach Sunsation**); Kordes

KORomtar, HT, w, 1997; (**Cream Dream**); Kordes

Korona®, *see* 'KORNITA'

'KORONA, CLIMBING', Cl F, or, 1957; Kordes

KORonam, HT, w, 1994; Kordes

'KORONET', HT, ob, 1941; bud globular, lemon shaded orange; flowers orange, reverse deep primrose, dbl., 40–50 petals, 4.5 in., exhibition form; foliage dark, glossy, leathery; vigorous, upright, bushy, open growth; [Julien Potin X Bright Wings]; Mallerin, C.; A. Meilland;, C-P

KORonto, S, m, 1991; (**Burghausen**®); Kordes

KORopas, HRg, m, 1988; (**Sir Henry**®); Kordes

KORoranki, F, ob, 1994; (**Jacqueline**); Kordes

'KOROVO', HT, mp, 1931; bud pointed; flowers peach-blossom-pink and coppery old-rose, large, dbl., 30 petals, moderate fragrance; foliage thick; vigorous growth; [Mrs T. Hillas X Étoile de Hollande]; Leenders, M.

'KOROYNESS', HT, lp, 1985; (**Bride's Dream**, Fairy Tale Queen, Marchenkonigin); bud long, pointed, ovoid; flowers very pale pink, large, full, dbl., 25–30 petals, 5 in., exhibition form, borne singly, slight fragrance; dark brown prickles; foliage large, medium green, matt; tall, upright growth; [Royal Highness X Seedling]; Kordes, R., 1984; Ludwigs Roses Pty. Ltd., 1986

KORozon, HT, pb, 1993; (**Mondiale**®); Kordes

'KORP', Gr, or, 1971; (**Prominent**®); bud long, pointed; dbl., 33 petals, 3.5 in., cupped, slight fragrance; foliage matt; upright growth; AARS, 1977 GM, Portland, 1975; [Colour Wonder X Zorina]; Kordes, R.; Kordes

KORpalmor, F, op, 1997; (**Schloss Balthasar**); Kordes

KORpalud, HT, op, 1997; (**Hamburger Deern**); flowers large, dbl., moderate fragrance; Kordes, W. Söhne; (Sangerhausen)

KORpapie, S, op; (**Salmon Sunsation**); Kordes

KORpapiro, F, ab, 1995; (**Apricot Summer**); Kordes

KORparall, S, mp, 1989; (**Mary Hayley Bell**); Kordes

KORparesni, F, lp, 1997; (**Rosenprofessor Sieber**); flowers medium-large, dbl., slight fragrance; Kordes, W. Söhne; (Sangerhausen)

KORpastato, HT, ly, 1997; (**Sebastian Kneipp**); flowers medium-large, very dbl., intense fragrance; Kordes, W. Söhne; (Sangerhausen)

'KORPATRI', F, ab, 1972; (Kordes' Rose Patricia, **Patricia**); flowers apricot, base shaded gold, GM, Orleans, 1979; [Elizabeth of Glamis sport]; Kordes; Fermor

'KORPEAHN', F, mr, 1986; (Carl PhilipKristian IV, The Times Rose, **Mariandel**®; flowers scarlet crimson-red, medium, semi-dbl., 6–14 petals, slight fragrance; foliage medium, dark green, semi-glossy, disease-resistant; bushy growth; Golden Rose, The Hague, 1990 PIT, RNRS, 1982; [Tornado X Redgold]; Kordes, W.; John Mattock, Ltd, 1985

'KORPEK', HT, mr, 1985; dbl., 20 petals, no fragrance; foliage dark, glossy; upright growth; [Unnamed seedling X Unnamed seedling]; Kordes, W.; Paul Pekmez

'KORPERKI', HT, dr, 1986; (**Trojan Victory**); bud ovoid; flowers deep red, dbl., 50 petals, exhibition form, borne singly, moderate, damask fragrance; medium prickles, hooked downward; foliage medium, medium green, semi-glossy; medium, spreading growth; PP005678; [Unnamed seedling X Uwe Seeler]; Kordes, R.; J&P

'KORPESH', Gr, ob, 1981; (**Shreveport**™); bud ovoid, pointed; flowers orange, dbl., 50 petals, exhibition form, borne 1–3 per cluster, slight, tea fragrance; small prickles, hooked downward; foliage large; tall, upright, vigorous growth; PP005157; AARS, 1982; [Zorina X Uwe Seeler]; Kordes, R.; Armstrong Nursery

KORpinka, S, dp, 1992; (**Sommermarchen**); Kordes

'KORPINROB', S, mp, 1987; (**Pink Robusta**®, The Seckford Rose); flowers large, semi-dbl., 16–14 petals, slight fragrance; foliage large, dark green, glossy; bushy, spreading growth; [(Zitronenfalter X Grammerstorf, Climbing) X Robusta]; Kordes, W., 1986

KORplasina, F, w; (**Vanilla**®); Kordes, 1994

'KORPLAVI', S, ab, 1987; (Chevreuse, Kordes' Rose Westfalenpark®, **Westfalenpark**®); flowers large, dbl., 26–40 petals, moderate fragrance; foliage large, dark green, glossy; bushy, spreading growth; [Unnamed seedling X Las Vegas]; Kordes, W., 1986

'KORPON', F, or, 1970; (**Ponderosa**®); bud globular; flowers red-orange, medium, dbl., cupped, slight fragrance; foliage leathery; vigorous, dwarf, bushy growth; ADR, 1971; [Seedling X Marlena]; Kordes, R.; Kordes

KORpora, HT, dy, 1990; (**Christel von der Post**®); Kordes

'KORPORT', HT, yb, 1984; (**Funkuhr®**, Golden Summers, Laser Beam); flowers yellow, petals edged medium red, aging red, large, dbl., 35 petals, no fragrance; foliage medium, medium green, glossy; upright growth; [Unnamed seedling X Unnamed seedling]; Kordes, W.; Kordes Sons

'KORPRILL', HT, op, 1985; (**Belami**, 'KORHANBU'); flowers orange pink, large, dbl., 35 petals, moderate fragrance; foliage medium, dark, glossy; upright, bushy growth; [(Prominent X Carina) X Emily Post]; Kordes, W.

'KORPRIWA', HT, dy, 1984; (Berolina, **Selfridges**); flowers amber yellow, large, dbl., 35 petals, moderate fragrance; foliage medium, medium green, semi-glossy; ADR, 1986; Kordes, W.; John Mattock, Ltd

KORproa, HT, yb; (**Technikon Pretoria**); Kordes, 1993

KORquermu, F, lp, 1991; (**Queen Mother**); Kordes

KORquick, HT, ob, 1988; (**Oudtshorn Joy**); Kordes

'KORRANTU', F, mr, 1968; (**Travemünde®**); bud ovoid; flowers medium, dbl.; foliage dark; ADR, 1966; [Lilli Marleen X Ama]; Kordes; Buisman

'KORRED', HT, dr, 1983; (Ace of Hearts, **Asso di Cuori®**, Toque Rouge); bud large, ovoid; dbl., 30 petals, cupped, moderate fragrance; dark green prickles; foliage large, dark; bushy growth; Kordes, W.; John Mattock, Ltd, 1981

'KORREI', F, ob, 1974; (Dreaming, 'REIKOR', Reverie, **Träumerei®**); bud long, pointed; flowers orange, medium, dbl., cupped, intense fragrance; foliage leathery; vigorous, upright, bushy growth; [Colour Wonder X Seedling]; Kordes

'KORRESIA', F, dy, 1977; (Friesia, **Sunsprite**); bud ovoid; dbl., 28 petals, 3 in., flat, intense fragrance; foliage light; upright growth; GM, Baden-Baden, 1972 Gold Star of the South Pacific, Palmerston North, NZ, 1975; [Unnamed seedling X Spanish Sun]; Kordes, R.; J&P

'KORRESLI', S, dp, 1987; (**Rosenresli®**); flowers deep pink, large, dbl., 26–40 petals, intense fragrance; foliage medium, dark green, glossy; upright, bushy, possibly climbing growth; ADR, 1984; [(New Dawn X Prima Ballerina) X Unnamed seedling]; Kordes, W., 1986

KORretra, HT, mr, 1989; (**Mother's Value**); Kordes

'KORREV', HT, rb, 1962; (**Neue Revue®**, News Review); flowers yellow-white, touched dark red, well-formed, dbl., 30 petals, 4.5 in., intense fragrance; many large prickles; foliage leathery; upright growth; ADR, 1969; [Colour Wonder X ?]; Kordes

'KORRIGAN', F, lp, 1972; bud globular; flowers medium, dbl., 25 petals, 2–2.5 in., slight fragrance; foliage glossy, dark, leathery; vigorous, bushy growth; [Seedling X (Orléans Rose X Eden Rose)]; Poulsen, Niels D.; Poulsen

KORrohe, F, mr, 1984; (**Heckenfeuer®**); Kordes

KORruge, HRg, yb, 1989; (**Rugelda®**); Kordes

KORrundum, HT, my, 1993; (**Heike**); Kordes

'KORSAKU', HRg, dp, 1998; (Playtime, **Rosalina™**); bud slender, pointed; flowers dark pink with traces of lavender, dark pink reverse, single, 4–11 petals, 3 in., borne in small clusters, slight fragrance; many prickles; foliage medium, medium green, dull; bushy, medium, for hedging, 3–4 ft growth; PP9011; [The Fairy X Seedling]; Kordes, W.; Bear Creek Gardens, 1992

'KORSAMI', S, lp, 1983; (Kordes' Rose Repandia®, **Repandia®**); flowers small, semi-dbl., moderate fragrance; foliage small, dark, glossy; low, spreading (to 5 ft groundcover) growth; ADR, 1986; [The Fairy X R. wichurana seedling]; Kordes, W.; Kordes Roses, 1982

KORsanter, HT, 1988; Kordes, R.; (Cavriglia)

'KORSCHAPRAT', HT, ob, 1963; (Charme de Vienne, Charming Vienne, Vienna Charm, **Wiener Charme®**); bud pointed; flowers coppery, orange, dbl., 27 petals, 6 in., exhibition form, moderate fragrance; foliage dark; vigorous, tall growth; [Chantré X Golden Sun]; Kordes, R.; J&P, 1965;, McGredy, 1963

'KORSCHLOSS', F, or, 1975; (**Schloss Mannheim®**); flowers red-orange, dbl., globular, slight fragrance; foliage dark, leathery; vigorous, upright, bushy growth; ADR, 1972; [Marlena X Europeana]; Kordes

'KORSCHNUPPE', LCl, dy, 1984; (**Goldener Olymp®**, Olympic Gold); flowers large, dbl., 20 petals, moderate fragrance; foliage large, medium green, matt; upright, bushy (to 7 ft) growth; [Seedling X Goldstern]; Kordes, W.; Kordes Sons

KORschwama, HT, dr, 1992; (**Schwarze Madonna®**); Kordes

KORschwama, HT, lp; (**Barry Fearn**); Kordes

KORsebue, HT, ob, 1995; (**Hanneli Rupert**); Kordes

'KORSEE', F, ob, 1970; (Gitta Grummer, Orange Vilmoria, Orange Vilmorin, Rainer Maria Rilke, Reiner Maria Rilke, **Uwe Seeler®**); bud ovoid; flowers salmon-orange, large, semi-dbl., exhibition form, moderate fragrance; foliage large, glossy, bronze, leathery; vigorous, upright, bushy growth; [Queen Elizabeth X Colour Wonder]; Kordes, R.; Kordes

KORsenter, HT, pb; (**Garden Perfume**); Kordes

KORseubel, F, ab; (**City of Pretoria**); Kordes

'KORSHEL', HT, op, 1976; (**Seashell**); bud short, pointed; flowers burnt-orange, imbricated, dbl., 48 petals, 3–4 in., slight fragrance; upright growth; AARS, 1976; [Unnamed seedling X Colour Wonder]; Kordes, R.; J&P

'KORSICHT', HT, m, 1984; (**Blue River®**); bud large; flowers lilac, shaded deeper at petal edges, large, dbl., 35 petals, exhibition form, intense fragrance; foliage medium, medium green, semi-glossy; upright growth; GM, Baden-Baden; [Blue Moon X Zorina]; Kordes, W.

KORsilan, S, mp, 1988; (**Palmengarten Frankfurt®**); Kordes

KORsion, Gr, my; (**Randfontein Gold**); Kordes

KORsisten, F, op, 1995; (**Bienkie**); Kordes

'KORSITA', HT, dr, 1982; (**Pasita**); flowers bright dark red, flora-tea, blooms borne singly, dbl., 25 petals, slight fragrance; straight, light brown prickles; foliage glossy; medium-high, densely branched growth; [Mercedes X Seedling]; Kordes, R.; Ludwigs Roses Pty. Ltd.

'KORSKIPEI', HT, op, 1987; (**Esther Geldenhuys**); flowers light coral pink, clam-shaped, large, borne singly, dbl., 32 petals, moderate fragrance; prickles concave, yellow-brown; foliage glossy, purple to medium green; vigorous, very tall, well-branched growth; [Seedling X Seedling]; Kordes, W. Söhne; Ludwigs Roses Pty. Ltd., 1988

'KORSOMMER', F, w, 1984; (**Roseromantic®**); flowers light pink to white, large, single, 5 petals, slight fragrance; foliage small, dark, glossy; bushy, spreading growth; GM, Baden-Baden, 1982; [Seedling X Tornado]; Kordes, W.

'KORSONN', F, dy, 1977; (**Sonnenröschen®**); bud ovoid; flowers deep yellow, dbl., 40 petals, 4 in., cupped, slight fragrance; foliage glossy; vigorous, bushy growth; [Arthur Bell X Yellow seedling]; Kordes, W. Söhne, 1978

'KORSORB', HT, ab, 1988; (**Joybells**); Kordes

'KORSPARKO', F, ob, 1984; (**Orange Sparkle**); flowers bright orange, yellow stamens, blooms in clusters of 1–5, semi-dbl., moderate fragrance; straight, brown prickles; foliage very glossy; tall, bushy growth; [(Colour Wonder X Zorina) X Uwe Seeler]; Kordes, R.; Ludwigs Roses Pty. Ltd.

KORspatax, HT, mp, 1994; (**Darling**); Kordes

KORstacha, S, w; (**Weisse Wolke®**); Kordes, 1993

KORstacha, LCl, lp; (**White Cloud**); Kordes

KORstatis, S, pb, 1989; (**Rosenstadt Zweibrucken®**); Kordes

KORstesgli, S, mr, 1998; (**Lancashire**); Kordes

'KORSUN', HT, yb, 1985; ('KORGATER', **Sun City**); flowers deep yellow with red petal edges, red spreading, star-shaped, dbl., 32 petals, slight fragrance; dark brown prickles; foliage deep green, red when young, leathery; upright, tall growth; [((New Day X Minigold) X Unnamed seedling) X MEItakilor]; Kordes, R.; Ludwigs Roses Pty. Ltd.

KORtabo, F, mr; (**Little Red Hedge**); GM, Durbanville, 1983; Kordes

KORtabris, F, pb, 1989; (**Raspberrry Ice**, Tabris); Kordes

'KORTAG', HT, ob, 1985; (**Clivia®**); bud ovoid; flowers salmon orange-red blend, blooms borne singly, dbl., 30 petals, exhibition form, moderate fragrance; brown prickles; foliage medium, medium green, matt; upright, bushy growth; [Mercedes X (Sonia X Uwe Seeler)]; Kordes, W., 1979

'KORTAKE', HT, dy, 1987; (**Germiston Gold**); flowers deep golden-yellow, large, borne in sprays of 1–3, dbl., 36 petals, intense fragrance; prickles concave, brown; foliage medium green; medium, well branched, free-flowering growth; [Unnamed seedling X Unnamed seedling]; Kordes, W. Söhne; Ludwigs Roses Pty. Ltd., 1988

KORtaly, S, dy, 1988; (**Lucinde®**); Kordes

KORtanken, F, dr, 1994; (**Domstadt Fulda®**); flowers large, semi-dbl., slight fragrance; Kordes, W. Söhne; (Sangerhausen)

KORtara, F, dp, 1987; (**Schleswig 87®**); Kordes

'KORTAT', HT, dr, 1970; (Rosenthal, **Tatjana®**); bud long, pointed; flowers large, dbl., cupped, intense fragrance; foliage leathery, soft; vigorous, upright growth; [Liebeszauber X Prasident Dr H.C. Schroder]; Kordes, R.; Kordes

KORtechna, HT, ob, 1992; (**Germania-Africana**); Kordes

KORtelin, F, yb; (**Spotted Gold**); Kordes, 1994

'KORTELLO', S, dr, 1981; (**Ulmer Münster®**); flowers large, dbl., 35 petals, slight fragrance; foliage large, dark, glossy; [Sympathie X Unnamed seedling]; Kordes, W., 1982

'KORTEMBER', HT, op, 1981; (**Harmonie®**); bud long, pointed; flowers deep salmon, dbl., 20 petals, exhibition form, intense fragrance; foliage slightly glossy; vigorous, upright, bushy growth; GM, Baden-Baden, 1981; [Fragrant Cloud X Uwe Seeler]; Kordes, W.

'KORTEMMA', S, dr, 1998; (**Red Ribbons**, Roselina™, Fiery Sunsation, Chilterns); flowers bright red, double, 17–25 petals, 3 in., borne in large clusters, slight fragrance; prickles moderate; foliage medium, dark green, glossy; low, spreading, 2 ft x 5 ft growth; [Weisse Max Graf X Waltzertraum]; Kordes, W.; Bear Creek Gardens, 1990

KORtenay, S, dp, 1991; (**Hertfordshire**); Kordes

KORtensei, S, mp, 1993; (**Dornroschenschloss Sababurg®**); Kordes, R.

'KORTENSES', Min, or, 1992; (**Sergeant Pepper**); semi-dbl., 6–14 petals, 1.5–2.75 in., borne in small clusters, no fragrance; some prickles; foliage small, dark green, glossy; low (45–60 cms), upright, bushy, spreading growth; [Unnamed seedling X LAVglut]; Kordes, W. Söhne; Bear Creek Gardens, 1992

'KORTERSEN', LCl, dp, 1977; (Netersen, Rosarium Netersen®, **Rosarium Ueteresen®**); bud ovoid; flowers deep pink, very dbl., 142 petals, 3 in., moderate fragrance; foliage large, glossy; vigorous, climbing growth; [Karlsruhe X Seedling]; Kordes, W.

'KORTEXUNG'®, F, mp, 1987; (**Europa**, Fleurop®); flowers medium, dbl., slight fragrance; foliage medium, medium green, matt; bushy growth; PP006513; [(Unnamed seedling X Banzai) X (Mercedes X Carol)]; Kordes, W., 1985

KORtikel, F, dy, 1989; (**Golden Holstein®**); Kordes

'KORTIME', F, my, 1985; (**Golden Times**); flowers medium, dbl., intense fragrance; foliage medium, dark, semi-glossy; bushy growth; [New Day X Minigold]; Kordes, W., 1976

KORtingle, MinFl, mr, 1993; (**Scarlet Patio**); Kordes

KORtionza, S, dy, 1998; (**Postillion**); Kordes

KORtisching, S, lp; (**Yellow Sunsation**); Kordes, 1997

KORtitu, S, mp, 1991; (**Marondo®**); Kordes

'KORTOR', F, or, 1973; (**Tornado®**); flowers medium, semi-dbl., globular, slight fragrance; foliage leathery; vigorous, bushy growth; ADR, 1972; [Europeana X Marlena]; Kordes, R.; Kordes

'KORTOSSGO', F, my, 1963; (Gold Topaz, **Goldtopas®**, Goldtopaz, 'KORGO'); bud ovoid; flowers amber-yellow, blooms in clusters (up to 10), dbl., 3.5 in., cupped, foliage glossy; vigorous, bushy growth; ADR, 1963; Kordes, R.

KORtradkos, F, mr, 1995; (**Parkwood Scarlet**); Kordes

KORtragfei, S, or, 1989; (**Rote Mozart®**); Kordes

'KORTREU', F, or, 1985; (**Fuggerstadt Augsburg®**); flowers medium, semi-dbl., slight fragrance; foliage medium, dark, glossy; upright, bushy growth; [Cordula X Topsi]; Kordes, W.; Kordes Sons

'KORTRI', F, or, 1972; (**Cordula®**); bud globular; flowers red-orange, medium, dbl., slight fragrance; foliage dark, bronze, leathery; vigorous, dwarf, bushy growth; [Europeana X Marlena]; Kordes, R.; Kordes

KORtrolle, F, ob, 1988; (**Bloemfontein**); Kordes

'KORTUEL', F, w, 1989; (**Bella Weiss®**); Kordes

'KORTUNA', HT, op, 1985; (**Fortuna®**); flowers medium salmon-pink, well-formed, large blooms borne singly, dbl., 30 petals, exhibition form, moderate fragrance; foliage medium, medium green, semi-glossy; medium, upright growth; [Sonia X Unnamed seedling]; Kordes, W., 1977

KORturnus, HT, op, 1993; (**Salmon Spire**); Kordes

'KORUB', S, yb, 1971; (**Charivari**); bud ovoid; flowers golden yellow to salmon, large, dbl., cupped, moderate fragrance; foliage glossy; vigorous, upright, bushy growth; [Königin der Rosen X Goldrausch]; Kordes, R.

KORulas, HT, ly, 1997; (**Limona**); Kordes

'KORUMELST', F, mp, 1961; (**Elysium®**); bud pointed; flowers salmon-pink, well-formed, large, dbl., 35 petals, cupped, moderate fragrance; foliage glossy; vigorous, tall, growth; Kordes, R.

'KORVALUE', HT, pb, 1984; (**Woman's Value**); flowers cream with shades of rose and coral, star-shaped blooms born, dbl., 42 petals, slight fragrance; straight, brown prickles; foliage glossy, deep green; medium, bushy, well-branched growth; [((Sonia X ((Dr A.J. Verhage X Colour Wonder) X Zorina)) X Asso di Cuori]; Kordes, R.; Ludwigs Roses Pty. Ltd.

KORveco, HT, or, 1997; (**Corvette**); Kordes

KORvegata, HT, 1996; (**Rebell 96**); Kordes, R.; (Cavriglia)

'KORVEJOH', F, ob; (**Vera Johns**); flowers purple-violet, dbl., cupped

'KORVERA', Gr, or, 1977; (**Vera Johns®**); bud ovoid, pointed; flowers large, dbl., 40 petals, exhibition form, slight fragrance; foliage glossy, dark, leathery; vigorous, upright growth; [? X Prominent]; Kordes, R.; Ludwigs Roses Pty. Ltd.

KORveril, HT, op, 1990; (**Cadillac**); Kordes

KORverlandus, S, dp, 1997; (**Knirps**); Kordes

KORverpea, HT, rb, 1955; (**'Cleopatra'**, Kleopatra); flowers scarlet, reverse old-gold, well-formed, medium, dbl., 45 petals, moderate fragrance; foliage dark, glossy; vigorous growth; GM, NRS, 1955; [(Walter Bentley X Condesa de Sástago) X Golden Scepter]; Kordes

KORviga, Min, lp; (**Young Mistress**); Kordes, 1988

'KORvila', F, mp, 1983; (**Vierländerin**); flowers medium salmon-pink, large, dbl., 35 petals, moderate fragrance; foliage medium, medium green, matt; upright growth; [(Zorina X Zorina) X Rosenelfe]; Kordes, W.; Kordes Roses, 1982

KORvolomin, HT, ab, 1994; (**Saturday Star**); Kordes

KORvondra, F, mp, 1990; (**Flirt**); Kordes

'KORwalbe', Min, rb, 1984; (**Maidy**®); flowers medium, dbl., 20 petals, no fragrance; foliage small, medium green, semi-glossy; bushy growth; [Regensberg X Seedling]; Kordes, W.

'KORweirim', S, w, 1983; (Kordes' Rose Weiss Immensee, Lac Blanc, Partridge, **Weisse Immensee**®); bud light pink; flowers small, single, 5 petals, intense fragrance; foliage small, dark, glossy; spreading (10 ft) growth; groundcover; [The Fairy X R. wichurana seedling]; Kordes, W.; Kordes Roses, 1982

'KORweiso', F, pb, 1983; (**Hannah Gordon**); flowers white with deep pink petal edges, large, dbl., 35 petals, slight fragrance; foliage large, medium green, semi-glossy; upright, bushy growth; [Unnamed seedling X Bordure]; Kordes, W.; John Mattock, Ltd.

'KORwerk', Min, mp, 1982; (Dwarf Queen '82, Zwergkönigin '82); flowers medium, dbl., 35 petals, slight fragrance; foliage small, medium green, glossy; bushy growth; [KORkonig X Sunday Times]; Kordes, W.

'KORwest', S, ab, 1969; ('KORlawe', **Westerland**®); bud ovoid; flowers apricot-orange, cupped, 20 petals, 3 in., cupped, repeat bloom, intense; foliage large, dark, soft; vigorous, upright, climbing growth; ADR, 1974; [Friedrich Worlein X Circus]; Kordes

'KORwilma', HT, rb, 1989; (Jack Dayson, **Perfect Moment**™); bud pointed; flowers red on outer half of petals, yellow on inner, reverse yellow, dbl., 30–35 petals, 4–4.5 in., exhibition form, borne singly and in small clusters, slight fragrance; prickles broad at base, narrowing, hooked down, red to brown; foliage medium, medium green, semi-glossy; upright, bushy, medium growth; PP8007; AARS, 1991; [New Day X Unnamed seedling]; Kordes, W. Söhne; Bear Creek Gardens, 1991

KORwilpa, HGal, mp, 1987; (**Park Wilhelmshohe**); Kordes

'KORwings', HT, ob, 1978; (**Soaring Wings**); bud ovoid; flowers deep dusky orange, large, dbl., 64 petals, exhibition form, moderate fragrance; foliage medium green, matt; vigorous, upright, bushy growth; [Colour Wonder X ?]; Kordes, W. Söhne; Ludwigs Roses Pty. Ltd.

KORwisco, S, lp; (**The Compass Rose**); Kordes, 1997

'KORwonder', F, mp, 1981; (**Bella Rosa**®, Kordes' Rose Bella Rosa, 'KORwondis', Toynbee Hall); flowers mini-flora, large, dbl., 34 petals, slight fragrance; foliage small, medium green, glossy; bushy growth; GM, Baden-Baden, 1983 GM, Copenhagen, 1981 GM, Durbanville, 1981; [Seedling X Traümerei]; Kordes, W., 1982

KORworm, F, mp, 1994; (**Romantic Hedgerose**); Kordes

KORxenna, HT, mr, 1988; (**Robina**); Kordes

KORyard, HT, mp; (**Yardley English Rose**); Kordes, 1990

'KORylt', HKor, dr, 1981; (Kordes' Rose Sylt®, **Sylt**); flowers medium, semi-dbl., slight fragrance; foliage medium, dark, glossy; spreading growth; [R. kordesii X Unnamed seedling]; Kordes, W.

'KORzaun', HT, dr, 1984; (Duftzauber '84, **Royal William**); flowers large, dbl., 35 petals, exhibition form, moderate fragrance; foliage large, dark, semi-glossy; upright, bushy growth; [Feuerzauber X Unnamed seedling]; Kordes, W.

KORzeito, Min, w, 1991; (**Silk Button**); Kordes

'KORzimko', Min, mr, 1987; (**Charmant**®); flowers small, dbl., 40 petals, no fragrance; foliage small, medium green, semi-glossy; bushy growth; [((Seedling X Tornado) X KORkonig) X Trumpeter]; Kordes, W.

KORzimko, HT, op, 1975; (**Charmante**); flowers large, dbl., slight fragrance; Huber; (Sangerhausen)

KORzinta, HT, pb; (**Taubie Kushlik**); Kordes, 1991

'KORzola', F, rb, 1986; (**Zola Budd**); flowers white with ruby red, borne 1–3 per cluster, single, 5–7 petals, slight fragrance; straight, light brown prickles; foliage deep green, leathery; vigorous, bushy, densely-branched growth; [Mabella X [(Dr A.J. Verhage X Colour Wonder) X Zorina]]; Kordes, R.; Ludwigs Roses Pty. Ltd., 1985

KORzuri, HT, w, 1992; (**Memoire**®); GM, Belfast, 1994; Kordes

KORzuri, HT, lp; (**Ice Cream**); Kordes

Ko's Yellow, *see* 'MACkosyel'

Kosai, F, op

Kosmos, F, dr; flowers medium, dbl.; VEG; (Sangerhausen)

'Koster's Orléans', Pol, mr, 1920; flowers brilliant scarlet-red; [Orléans Rose sport]; Koster, M.

'Koster's Triumph', Pol, or, 1920; flowers small, dbl.; Koster; (Sangerhausen)

'Kostior Arteka', F, or, 1955; (Campfire Arteka); flowers coral-red tinted orange, large, slight fragrance; [Independence X ?]; Klimenko, V. N.

'Koto', HT, dy, 1972; bud ovoid; flowers pure deep yellow, large, dbl., exhibition form, moderate fragrance; foliage glossy, dark, leathery; vigorous, upright growth; [Lydia seedling X Peace seedling]; Suzuki, Seizo; Keisei Rose Nursery

'Kotobuki', HT, yb, 1990; bud pointed; flowers creamy to creamy yellow, creamy yellow reverse with light pink, dbl., 30–35 petals, exhibition form, slight fragrance; foliage dark green, glossy; medium growth; [Souma sport]; Ogura Rose Nurseries, 1985

'Kovalam', HT, w, 1976; bud globular; flowers cream-white, dbl., 20–25 petals, 4 in., cupped, intense fragrance; foliage wrinkled; moderate, bushy growth; [(Amberlight X Traumland) X Western Sun]; Viraraghavan, M.S.; Gopalsinamiengar

'Koyo', F, rb, 1989; bud ovoid; flowers vermillion, reverse orange-yellow, small, borne in sprays of many blooms, dbl., 17–20 petals, cupped, slight fragrance; prickles ordinary; foliage dark green, glossy; spreading growth; [Masquerade X Matador]; Kikuchi, Rikichi, 1990

Koyuki, *see* **'Snow Infant'**

'Koza', HRg, dp, 1927; flowers deep pink, semi-dbl.; vigorous (over 7 ft) growth; extremely hardy.; [(R. rugosa X La France) seedling X La Mélusine]; Hansen, N.E.

'Krakow', HT, dr; bud ovate; very dbl.; Grabczewski

'Kralj Alexander I', HT, dr, 1935; flowers velvety blood-red, reflexes fiery red, very large, dbl., intense fragrance; vigorous growth; [Capt. Kilbee Stuart X Jan Böhm]; Böhm, J.

'Kralj Petar II', HT, pb, 1936; flowers salmon-pink, reverse carmine, with coppery sheen, moderate fragrance; Brada, Dr.

'Kralj Tomislav', HT, mr, 1931; bud long, pointed; flowers solferino-red, open, very large, dbl., exhibition form; vigorous growth; [Dora Stober X Étoile de Hollande]; Leenders, M.

'Kraljica Marija', HT, ly, 1935; flowers creamy yellow to creamy white, large, dbl., intense fragrance; [Frau Karl Druschki X Golden Ophelia]; Brada, Dr.; Böhm

'Kranenburg', F, dp, 1963; flowers very dark pink, borne in clusters, dbl., 32 petals; foliage glossy, dark; bushy growth; [Pinocchio X Ma Perkins]; Verschuren, A.; van Engelen

'Krasavitza Festivalia', HT, yb, 1955; (Festival Beauty); flowers yellow edged raspberry-red, medium, dbl., 28 petals, foliage glossy, light green; spreading growth; [(Peace X Crimson Glory) X Poinsettia]; Klimenko, V. N.

Krásná Azurea, *see* **'Generál Štefánik'**

'KRÁSNÁ USLAVANKA', HT, ob, 1930; flowers orange-rose, reverse dark orange-yellow, moderate fragrance; very vigorous, bushy growth; [Mrs Beckwith X Arthur Cook]; Böhm, J.

'KRASNAIA MOSKVA', HT, dr, 1955; (Red Moscow); flowers dark velvety red, medium, slight fragrance; [Peace X Crimson Glory]; Klimenko, V. N.

Krasni Mak-Sin, F, dr, 1975; flowers dark, velvety red, medium, dbl.; Klimenko, V. N.; (Sangerhausen)

'KRASNOKAMENKA', F, mr, 1955; (Red Stone); flowers crimson-red, medium, semi-dbl., 12 petals; foliage dark, glossy; upright growth; [Independence X Kirsten Poulsen]; Klimenko, V. N.

'KRASNYI MAK', F, mr, 1955; (Red Poppy); flowers scarlet, medium, dbl., 26 petals, slight fragrance; foliage glossy, light green; short stems; upright growth; [Independence X Kirsten Poulsen]; Klimenko, V. N.

Krause's Rote Joseph Guy, *see* **'FEUERSCHEIN'**

'KRIBATIS', HT, yb, 1980; (Comtesse d'Alcantara, **Home & Country**); flowers large, moderate fragrance; foliage medium green glossy; upright growth; [Unnamed seedling X Peace]; Kriloff, Michel; Primavera

'KRICARLO', HT, op, 1984; (Marachal Le Clerc, Maréchal le Clerc, **Touch of Class**™); flowers medium pink, shaded coral and cream, large, dbl., 3–35 petals, 4.5–5.5 in., exhibition form, borne mostly singly, slight fragrance; foliage large, dark, semi-glossy; upright, bushy growth; PP005165; AARS, 1986 GM, Portland, 1988; [Micaela X (Queen Elizabeth X Romantica)]; Kriloff, Michel; Armstrong Nursery, 1984

KRIlamy, HT, op; (**Harmonie**); Kriloff

'KRILEVILLE', HT, rb, 1986; (**St Quentin**); flowers red, silver petal edges; foliage dark, glossy; [Unnamed seedling X Tropicana]; Kriloff, Michel

'KRILEXIS', HT, dp, 1986; (**Resurrection**®); bud red; flowers bright pink, large, moderate, sweet fragrance; foliage dark; [Unnamed seedling X Unnamed seedling]; Kriloff, Michel

KRImasar, Cl HT, or, 1984; (**Maryse Kriloff, Climbing**); flowers large, dbl., moderate fragrance; Kriloff; (Sangerhausen)

KRImony, HT, ob, 1994; (**Hello**); Kriloff

KRInico, HT, lp; (**Tresor**); Kriloff, 1996

KRInirosy, HT, op, 1995; (**Helene de Savoie**); Kriloff

'KRIOGA', F, lp, 1986; dbl., 35 petals, blooms in clusters; foliage dense; bushy growth; [Unnamed seedling X Orange Garnet]; Kriloff, Michel

'KRIPRILE', F, yb, 1986; (**Souviens-Toi**); flowers medium yellow, petals edged and washed carmine, large, dbl., 38 petals, moderate fragrance; foliage matt; Kriloff, Michel

Kristall®, F, w, 1979; GPG Bad Langensalza

'KRISTI', HT, mp, 1976; bud ovoid, pointed; flowers clear medium pink, dbl., 45 petals, 5 in., moderate, spicy fragrance; foliage large; vigorous, upright growth; [White Satin X Bewitched]; Swim, H.C. & Ellis, A.E.; Armstrong Nursery

Kristin™, *see* 'BENMAGIC'

'KRISTINA AV TUNSBERG', F, or, 1972; bud long, pointed; flowers large, very dbl., exhibition form, slight fragrance; foliage glossy, dark; vigorous, bushy growth; [Charleston X Toni Lander]; Lundstad; Norges Landbruks-hogskole

Kristo Pienaar, *see* HERpot

'KRITIBAN', HT, w, 1986; (Joella®, **Joëlle**®); flowers white, aging light pink, well-formed, dbl.; foliage glossy; [Unnamed seedling X Unnamed seedling]; Kriloff, Michel

'KRIWI', HT, ob, 1968; (**Wizo**®); flowers salmon-red, reverse darker, dbl., 40 petals, exhibition form; foliage reddish-green, matt; [Tropicana X (Gamine X Romantica)]; Kriloff, Michel; Domaine Agricole de Cronenbourg

Kronborg, *see* POUltry

Kronenbourg, *see* 'MACBO'

Kronenbourg, Climbing, Cl HT, rb, 1973; Kasturi

Kronjuwel, F, mr, 1997; Noack, Werner

'KRONPRINCESSIN VIKTORIA', B, w, 1887; flowers milk-white, center tinted yellow; [Souv. de la Malmaison sport]; Vollert

'KRONPRINSESSE INGRID', F, mp, 1936; (Kronprinzessin Ingrid); bud long, pointed; flowers deep rose-pink, open, semi-dbl., slight fragrance; foliage dark; vigorous growth; [Else Poulsen X Dainty Bess]; Poulsen, S.; C-P, 1942; Poulsen, 1942

'KRONPRINZESSIN CECILIE', HT, lp, 1907; flowers large, very dbl.; Kiese; (Sangerhausen)

Kronprinzessin Ingrid, *see* **'KRONPRINSESSE INGRID'**

Kronprinzessin Victoria, *see* HELtoria

Krymchanka, F, m, 1955; flowers dark purple-red, medium, dbl.; Klimenko, V. N.; (Sangerhausen)

Krymskaja Notsch, HT, dr, 1955; flowers large, dbl., intense fragrance; Klimenko, V. N.; (Sangerhausen)

'KSEELITE', Min, pb, 1996; (**Bright Lights**); flowers dark pink with bright yellow stripe on some petals, reverse, dbl., 15–25 petals, slight fragrance; numerous prickles; foliage medium, medium green, semi-glossy; upright, medium (24 in.) growth; [Red Delight X Hurdy Gurdy]; Apple, Ken; K&C Roses, 1997

Kühnhilda, S, mp, 1983; flowers medium, single, slight fragrance; Weihrauch; (Sangerhausen)

Kumaradhara, F, pb, 1988; K&S

Kumari, F, mp, 1982; Kasturi

'KUMBAYA', F, my, 1981; bud pointed; flowers bright medium yellow, borne up to 15 per cluster, dbl., 22 petals, slight fragrance; slightly hooked, red prickles; foliage deep green; low, bushy growth; [Chatterbox X Allgold]; Sanday, John; Sanday Roses, Ltd.

Kumiko, HT, dp, 1997; Reuter

'KUNIGUNDE', F, 1960; flowers pink, large, dbl., 35 petals, exhibition form, moderate fragrance; vigorous growth; Horstmann

Kupferkönig, F, op, 1993; flowers coppery pink, medium, dbl.; Kordes, W. Söhne; (Sangerhausen)

Kuroshinju, HT, dr, 1988; Suzuki, Seizo

Kurstadt Baden, HT, pb, 1966; flowers pink and red, large, dbl., slight fragrance; Tantau, Math.; (Sangerhausen)

'KURT SCHOLZ', HT, mr, 1934; bud pointed; flowers blood-red with some crimson, very large, dbl., exhibition form, slight fragrance; foliage leathery; vigorous, bushy growth; [Cathrine Kordes X W.E. Chaplin]; Kordes

Kusum, F, lp, 1978; Padhye

'KUTNO', Pol, m, 1965; flowers lavender-pink, dbl.; growth low; [Margo Koster seedling]; Wituszynski, B.

'KWINANA', F, rb, 1962; bud ovoid; flowers crimson overlaid carmine, open, borne in clusters, single, moderate fragrance; foliage leathery; strong stems; vigorous, tall growth; [Orange Triumph X ?]; Riethmuller; Hazlewood Bros.

'KYNAST', HT, mr, 1917; flowers amaranth-red, Krüger

'KYO-MAIKO', F, op, 1974; bud ovoid; flowers deep bright salmon-orange, small, dbl., cupped, slight fragrance; foliage small, glossy, light; moderate, dwarf, bushy growth; [Sarabande seedling X Ruby Lips seedling]; Suzuki, Seizo; Keisei Rose Nursery

Kyoto 1200, *see* **'ARASHIYAMA'**

'KYRIA', F, lp, 1976; (Pitica); PP004017; [Sonia sport]; Royon

'KYSON', Cl HT, mr, 1940; flowers bright red, open, very large, borne in clusters, single, recurrent bloom, slight fragrance; foliage leathery, glossy, dark; shoot, strong stems; very vigorous, climbing (7 ft or more) growth; [New Dawn X Allen Chandler]; Eacott; R. Murrell

L

L. D. Braithwaite®, *see* 'AUSCRIM'

'**L. E. Longley**', HT, mr, 1949; flowers open, large to medium, semi-dbl., moderate fragrance; foliage glossy, dark, bronze; very vigorous, bushy growth; [Pink Princess X Crimson Glory]; Longley, L.E.; Univ. of Minn.

L. G. Harris, HT, 1972; Gregory, C.; (Cavriglia)

'**L. J. de Hoog**', HT, mr, 1934; flowers scarlet-red, well formed, large, dbl., slight fragrance; vigorous, bushy growth; [Hadley X Hawlmark Scarlet]; Leenders Bros.

'**L. R. May**', HT, dr, 1935; flowers scarlet-crimson, paling to silvery pink, base orange, well formed; vigorous growth; Chaplin Bros.

La Baraka®, HT, or

'**La Bella**', HT, dr, 1976; bud large, ovoid; dbl., globular, slight fragrance; foliage glossy, dark; very vigorous, upright growth; [Liebeszauber X Herz As]; Kordes; Horstmann

'**La Belle Distinguée**', HEg, mr; (La Petite Duchesse, Lee's Duchess, Scarlet Sweet Brier); flowers bright crimson, small, dbl., non-recurrent; foliage dainty, fragrant; compact growth

'**La Belle Irisée**', HT, ob, 1943; bud ovoid, coppery; flowers clear orange-yellow, medium, dbl., globular, slight fragrance; foliage leathery; dwarf growth; [Mme Joseph Perraud X Seedling]; Gaujard

'**La Belle Marie**', T, dp; flowers pink, veined darker, reverse deep rose, center incurved, medium, moderate, fruity fragrance; foliage smooth, pointed; vigorous, tall growth; [Old Blush X Mme Laurette Messimy]; Tillotson

'**La Belle Sultane**', HGal, dr; (Gallica Maheca, Violacea); flowers deep crimson, becoming violet, base white, almost single; Hardy, before 1801; ca. 1795

'**La Belle Suzanne**', T, lp; flowers light pink suffused white, moderate fragrance; foliage smooth, pointed; vigorous, tall growth; [Old Blush X Mme Laurette Messimy]; Tillotson

La Belle Villageoise, *see* '**Village Maid**'

'**La Biche**', N, w, 1832; flowers white, center flesh, large, very dbl., cupped; Trouillet

La Blancheur, *see* '**Sodori-Himé**'

'**La Brillante**', HP, mr; flowers bright crimson, well formed, large; Verdier, V., 1861

'**La Caille**', M, mp, 1857; flowers bright rose-pink; Robert et Moreau

La Canada, Min, dp

'**La Canadienne**', HT, ob, 1967; bud long, pointed; flowers orange to shrimp and cream, large, semi-dbl., intense fragrance; foliage dark, bronze, glossy; vigorous, bushy growth; [Royal Sunset X Sierra Sunset]; Morey, Dr Dennison; Country Garden Nursery

'**La Champagne**', HT, ob, 1919; bud long, pointed; flowers light coppery red, base yellow, edged light pink, large, dbl., 25 petals, globular, moderate fragrance; foliage rich green, leathery; vigorous growth; Barbier

La Coquette de Lyon, *see* '**Coquette de Lyon**'

'**La Couronne Tendre**', HGal, lp; flowers flesh-pink, small, dbl.

'**La Detroite**', HT, lp, 1904; flowers large, dbl., intense fragrance; Hopp & Lemke; (Sangerhausen)

'**La Diaphane**', M, lp, 1900

'**La Duchesse**', HGal, mp, 1838

La Favorita, HT, or

'**La Favorite**', HT, mp, 1900; flowers large, dbl.; Schwartz; (Sangerhausen)

'**La Favorite**', HGal, mr, 1844; Vibert

'**La Fiamma**', HWich, ob, 1909; flowers flame, borne in very large clusters, single; vigorous, climbing (15 ft) growth; Walsh

'**La Florida**', HT, mp, 1932; flowers salmon, well formed; foliage glossy; strong stems; very vigorous growth; La Florida

'**La Follette**', LCl, mp, 1910; flowers pink and carmine, large, dbl., moderate fragrance; vigorous (to 20 ft half-hardy; (14); [R. gigantea X ?]; Busby, about 1910

'**La Fontaine**', F, my, 1961; bud pointed, ovoid; flowers barium-yellow, to, borne in clusters, dbl., 20–25 petals, 3.5–4 in., exhibition form, slight fragrance; foliage leathery, dark; vigorous, bushy growth; [Mme Charles Sauvage X Fashion]; Meilland, Mrs. Marie-Louise; C-P

La Fraicheur, HWich, mp

'**La France**', HT, lp, 1867; bud long, pointed; flowers silvery pink, reverse bright pink, large, dbl., 60 petals, intense fragrance; vigorous growth; Guillot et Fils

'**La France de '89**', HT, rb, 1889; bud long, pointed; flowers bright red, sometimes striped white, very large, dbl., very vigorous growth; [Reine Marie Henriette X La France]; Moreau et Robert

'**La France Striped**', HT, pb, 1956; flowers deep pink to red, and white to blush, moderate fragrance; [La France sport]; Hennessey

'**La France Victorieuse**', HT, dp, 1919; flowers silvery carmine-pink, inside tinted yellow, dbl., moderate fragrance; Gravereaux; M. Guillot

'**La France, Climbing**', Cl HT, lp, 1893; Henderson, P.

'**La Giralda**', HT, mp, 1926; [Frau Karl Druschki X Mme Edouard Herriot]; Dot, Pedro

La Grande Parade, S, mr

'**La Joconde**', HT, my, 1920; flowers pure golden yellow, RULED EXTINCT 2/88; [Arthur R. Goodwin sport]; Croibier

La Joconde, *see* 'DELJACQ'

'**La Jolie**', HT, mr, 1956; bud long, pointed; flowers geranium-red, dbl., 30–35 petals; very vigorous growth; [Independence X Hens Verschuren]; Frères, Buyl

'**La Jolla**', HT, pb, 1954; bud long, pointed; flowers soft pink veined deeper, center cream and gold, dbl., 65 petals, 5 in., exhibition form, moderate fragrance; foliage dark, glossy; upright growth; [Charlotte Armstrong X Contrast]; Swim, H.C.; Armstrong Nursery

'**La Louise**', HGal, m, 1840; flowers purple/pink, medium, semi-dbl.; Parmentier; (Sangerhausen)

'**La Maculèe**', HGal, pb, 1810; flowers rose pink with carmine striping, medium, semi-dbl.; Dupony, Andre; (Sangerhausen)

'**La Madelon de Paris**', F, lp, 1962; flowers bright pink, well formed, medium, borne in clusters of 6–8, dbl.; vigorous growth; [Cécile Brunner X Unnamed seedling]; Robichon

'**La Maréchale Pétain**', HT, mr, 1927; flowers carmine, dbl., moderate fragrance; [Col. Leclerc X Château de Clos Vougeot]; Sauvageot, H.; F. Gillot

'**La Marne**', Pol, pb, 1915; flowers blush white, edged vivid pink, single blooms in large, loose clusters; vigorous growth; [Mme Norbert Levavasseur X Comtesse du Cayla]; Barbier

La Marseillaise®, *see* 'DELGEOT'

La Marylene, F, dr

'**La Mascotte**', Cl HT, dy, 1933; flowers deep saffron-yellow, passing to straw-yellow, slightly tinted salmon, dbl.; foliage dark, glossy; [Reine Marie Henriette X Laurent Carle]; Schwartz, A.

'**La Mélusine**', HRg, dp, 1906; flowers pinkish red, large, borne in large clusters, dbl., intense fragrance; very vigorous growth; Spath

'La Mie au Roy', HT, yb, 1927; flowers yellow, salmon and copper, dbl.; [Duchess of Wellington X Pax Labor]; Bernaix, P.

La Mienne, *see* **'Flon'**

La Minuette, *see* **'Laminuette'**

'La Mortola', (form of R. moschata nepalensis), w, 1954; single, 5 petals, 3 in., blooms in trusses, intense fragrance; foliage downy gray-green; very vigorous, climbing growth; Hanbury; Sunningdale Nursery

La Nationale, HGal, pb; flowers pink and red, medium, dbl.; Roseraie de l'Hay; (Sangerhausen)

'La Neige', M, w, 1905; flowers pure white, medium, dbl.; foliage turns purple; vigorous growth; [Blanche Moreau sport]; Moranville

'La Neige', HGal, w, 1853; Robert

'La Noblesse', C, lp, 1856; flowers soft pink, dbl., later bloom than most centifolias, intense fragrance; bushy growth; Soupert & Notting

La Nymphe, N, mp

'La Paloma'®, F, w, 1959; flowers creamy white, well-formed blooms in clusters (to 30), dbl.; foliage dark, glossy; [Yellow Rambler X Goldene Johanna Tantau]; Tantau, Math.

La Paloma '85, *see* 'TANamola'

La Pâquerette, *see* **'Pâquerette'**

'La Parfaite', HT, op, 1956; bud pointed; flowers bright salmon-pink, large, dbl., 40–50 petals; vigorous, bushy growth; [R.M.S. Queen Mary X Lady Sylvia]; Frères, Buyl

La Parfumee, HT, or

'La Parisienne', HT, or, 1937; bud long, pointed, deep coral-red; flowers orange-coral, open, very large, semi-dbl., slight fragrance; foliage glossy, dark; very vigorous growth; [Lucy Nicolas X Charles P. Kilham]; Mallerin, C.; A. Meilland;, C-P

La Parisienne®, *see* 'SUNpari'

La Passionata, *see* 'DELup'

La Passionata, HT, 1987; Barni, V.; (Cavriglia)

La Passionata, *see* DELapo

La Perla, *see* KORlaper

'La Perle', HWich, ly, 1905; flowers pale yellow to white; Fauque

La Petite Duchesse, *see* **'La Belle Distinguée'**

'La Plus Belle des Ponctuées', HGal, pb; flowers deep rose, spotted pale rose, flat; Hébert, Mme., before 1829

'La Pologne', HT, or, 1938; flowers orange-carmine, very large, cupped, foliage bronze; vigorous growth; Chambard, C.

La Presumida, *see* **'Presumida'**

'La Promise', HT, mr, 1956; bud long, pointed; flowers brick-red, semi-dbl.; tall growth; [Seedling X Betty Uprichard]; Frères, Buyl

'La Pucelle', HGal, m, 1820; Vibert

La Quatre Saisons Continue, P, mp

'La Ramée', F, dr, 1949; flowers crimson, borne in clusters, semi-dbl., 13 petals; foliage dark, bronze; vigorous growth; [Holstein X Alain]; Meilland, F.

'La Reine', HP, mp, 1842; (Reine des Francais, Rose de la Reine); flowers glossy rose-pink, large, dbl., 78 petals, cupped, moderate fragrance; vigorous growth; Laffay, M.

La Reine de Hamburg, P, mp

La Reine Victoria, *see* **'Reine Victoria'**

'La Revenante', HGal, mp, 1825; Miellez

La Roche aux Fees, HT, yb, 1997; Adam, M.

La Rose de Mme Raymond Poincaré, *see* **'Mme Raymond Poincaré'**

La Rose de York, *see* White Rose of York

'La Rose Tatouée', HT, m, 1956; (The Rose Tattoo); flowers salmon-pink tipped and spotted lavender, large, semi-dbl., slight fragrance; foliage leathery; very vigorous, upright growth; [Opera seedling X (Opera X Unnamed seedling)]; Gaujard

La Rosée, *see* **'Dr Debat'**

'La Rosée', Pol, w, 1920; flowers sulfur-white, passing to pure white, then to soft pink; Turbat

'La Rosiere', HP, 1861; Verdier, E.; (Cavriglia)

La Rosière, HP, lp; Damaizin, 1874

La Rosière, Climbing, *see* **'Prince Camille de Rohan'**

La Rossa, *see* BARlar

'La Roxelane', HGal, m, 1828; Vibert

La Royale, *see* **'Great Maiden's Blush'**

La Rubanée, *see* **'Village Maid'**

La Sanguine, *see* **'Sanguinea'**

'La Scala', HT, rb, 1964; flowers reddish orange, large, dbl., cupped, intense fragrance; foliage leathery; upright, open growth; GM, Rome, 1961; [(Mme Henri Guillot X Mirandy) X Peace]; Lindquist

La Séduisante, *see* **'Great Maiden's Blush'**

La Serenissima, HT; (Italia); (Cavriglia)

La Sevillana®, *see* 'MEIgekanu'

La Sevillana, Climbing, *see* MEIgekanusar

'La Sirène', HP, m, 1867; (La Syrène); flowers dark reddish-purple; Soupert & Notting

'La Somme', HT, or, 1919; flowers deep coral-red tinted copper, turning salmon, semi-dbl., moderate fragrance; [Mme Caroline Testout X Rayon d'Or]; Barbier

'La Stupenda', HT, pb, 1966; bud long, pointed; flowers pink, reverse darker, large, very dbl., slight fragrance; foliage glossy; upright, bushy growth; [Aztec X First Love]; Taylor, L.R.

La Superba, Gr, 1970; Mansuino; (Cavriglia)

'La Sylphide', T, m; flowers pink tinted lavender, deeper in fall, semi-dbl.; tall growth; Boyau, 1842

La Syrène, *see* **'La Sirène'**

'La Tendresse', HP, mp, 1864; Oger

'La Tosca', HT, pb, 1901; flowers shell-pink, center and reverse darker, dbl., 4 in., moderate fragrance; foliage rich green, leathery; vigorous, bushy growth; [Josephine Marot X Luciole]; Schwartz, Vve.

La Tour d'Argent™, *see* 'DELricos'

'La Tour d'Auvergne', HGal, or, 1842; flowers deep rosy crimson, flecked carmine, incurved, large, dbl.; Vibert

'La Vaudoise', HT, mr, 1946; bud oval; flowers blood-red touched brilliant scarlet, medium, dbl., moderate fragrance; foliage leathery; vigorous, bushy growth; Heizmann, E.; A. Meilland

'La Vendomoise', HT, pb, 1907; flowers pink and red, large, dbl., moderate fragrance; Moullière; (Sangerhausen)

'La Vie', HT, dr, 1931; flowers crimson to scarlet, base orange to carmine, fading to red, [Talisman sport]; Groshens

La Vie en Rose, *see* **'Vie en Rose'**

'La Ville de Bruxelles', D, dp, 1849; (Ville de Bruxelles); flowers pink, center incurved, large, dbl., quartered, foliage glossy, light green; height 5 ft; Vibert

La Villiageoise, *see* **'Village Maid'**

La Virginale, *see* **'Great Maiden's Blush'**

'La Virginale', HP, w, 1858; (Mme Liabaud); flowers white with flesh center, medium, dbl.; delicate habit growth; Lacharme, F.

'La Virginale', A, w, 1840; Moreau et Robert

'La Voulzie', F, dr, 1953; flowers garnet-red, very large, borne in clusters of 8–15, semi-dbl.; very vigorous growth; [Brise Parfumee X Alain]; Robichon

'Labareda', F, mr; flowers bright red; da Silva, Moreira

'L'Abondance', N, w, 1877; flowers pure white; Moreau et Robert

Labrador Rose, *see* **R. blanda**

Lac Blanc, *see* 'KORweirim'

'Lac La Nonne', HRg, dp, 1950; bud pointed, deep red; flowers very deep pink, semi-dbl., 2–3 in., moderate fragrance; foliage light; vigorous (7–8 ft hardy.;

[R. rugosa plena X R. acicularis]; Bugnet; P.H. Wright

Lac Majeau, HRg, w; Bugnet

Lac Rose®, *see* 'KORIMRO'

Lace Cascade, *see* 'JACARCH'

'LACHS', Pol, or, 1943; flowers glowing orange-red; [Dick Koster sport]; Kordes

Lacination, S, mp, 1997; Twomey, Jerry

'LACRE', F, mr, 1963; flowers bright red, loose, semi-dbl.; [Concerto X Seedling]; da Silva, Moreira

Lada, HT, lp; Urban, J., 1978; (Czech Rosa Club)

'LADDIE', HT, or, 1926; flowers deep carmine, flushed orange and scarlet, base orange; bushy growth; McGredy; Beckwith

'LADIES CHOICE', HT, pb, 1969; flowers cerise, reverse silvery, high pointed, large, dbl., intense fragrance; foliage light green; free growth; [Liberty Bell X Prima Ballerina]; Anderson's Rose Nurseries

'LADIES' CHOICE, CLIMBING', Cl HT, pb, 1975; flowers cerise, reverse silvery, high-pointed, large, dbl., intense fragrance; foliage light green; [Ladies' Choice sport]; Anderson's Rose Nurseries

Ladies Home Journal®, *see* 'BURWINLADHOM'

Ladies' View, *see* 'SEAVIEW'

'L'ADMIRATION', HCh, mp, 1856; flowers medium, single; Robert; (Sangerhausen)

'LADY', HT, mp, 1983; flowers medium, dbl., 35 petals; foliage large, medium green, matt to semi-glossy; upright, compact growth; [Song of Paris X Royal Highness]; Weeks, O.L., 1984

'LADY ALICE STANLEY', HT, pb, 1909; bud pointed; flowers pale flesh-pink, reverse coral-rose, large, dbl., 75 petals, moderate fragrance; foliage rich green, leathery; branching growth; McGredy

'LADY ANDERSON', HT, pb, 1920; flowers coral pink to flesh pink and yellow, dbl.; Hall

'LADY ANN', Min, mp, 1961; bud pointed; flowers rose-pink, dbl., 42 petals, 1.75 in., cupped, moderate fragrance; foliage leathery, glossy, dark; vigorous, bushy, low growth; [(R. wichurana X Floradora) X Little Buckaroo]; Moore, Ralph S.; Sequoia Nursery

'LADY ANN KIDWELL', Pol, dp, 1948; bud pointed; flowers deep pink, star-shaped, medium, dbl., slight fragrance; foliage glossy; vigorous, upright growth; [Cécile Brunner X ?]; Krebs; Marsh's Nursery

'LADY ASHTOWN', HT, pb, 1904; bud pointed; flowers carmine-pink, base yellow, large, dbl., 43 petals, exhibition form, slight fragrance; foliage rich green, soft; vigorous, bushy growth; [Mrs W.J. Grant seedling]; Dickson, A.

'LADY ASHTOWN, CLIMBING', Cl HT, pb, 1909; Bradley

'LADY BAILLIE', HSpn, ly; flowers pale sulfur, semi-dbl., profuse, early bloom; glossy, black fruit; foliage finely divided; dense, shrubby (3–4 ft) growth; Lee, prior to 1848

Lady Banks' Rose, *see* **R. BANKSIAE**

'LADY BARBARA', LCl, ob, 1987; flowers tangerine, medium, dbl., 20 petals, moderate fragrance; foliage medium, medium green, semi-glossy; upright growth; [Red Planet X (Elizabeth of Glamis X (Galway Bay X Sutter's Gold))]; Warner, Chris

Lady Barbara Blossom, *see* 'BOSHOTEARL'

'LADY BARNBY', HT, pb, 1930; bud pointed; flowers glowing pink, shaded red, large, dbl., exhibition form, moderate fragrance; foliage rich green, leathery; bushy, low growth; Dickson, A.

'LADY BARNETT', HT, dr, 1957; flowers crimson, reverse darker, exhibition form, moderate fragrance; vigorous, upright growth; Verschuren; Blaby Rose Gardens

'LADY BATTERSEA', HT, pb, 1901; (Red Niphetos); flowers cherry-blossom, base orange, dbl.; [Mme Abel Chatenay X Liberty]; Paul

Lady Be Good, *see* 'SEAGOOD'

'LADY BEATTY', HT, lp, 1918; flowers blush-pink, well formed, moderate fragrance; vigorous growth; Chaplin Bros.

'LADY BEAUTY', HT, pb, 1984; bud ovoid; flowers light pink flushed yellow, reverse deeper, large, dbl., 33 petals, exhibition form, slight fragrance; few sickle-shaped prickles; foliage medium, medium green, glossy; vigorous, upright growth; [Lady X Princess Takamatsu]; Kono, Yoshito, 1986

'LADY BELPER', HT, ob, 1948; flowers bronze-orange shaded light orange, semi-globular, dbl., 38 petals, 4 in., exhibition form, moderate fragrance; foliage glossy, dark; vigorous growth; [Mev. G.A. van Rossem X Seedling]; Verschuren; Gregory

'LADY BETTY', HT, ab, 1930; bud pointed; flowers apricot-pink, veined red, semi-dbl., exhibition form, slight fragrance; [Sunburst X Mrs Aaron Ward]; Bees

'LADY BEVERLEY', F, lp; flowers clear pink; [Independence sport]; Owens

'LADY BIRD', F, yb, 1966; flowers yellow shaded red at edge, small, borne in clusters, dbl., slight fragrance; foliage leathery; vigorous, bushy growth; [Unnamed seedling X Rumba]; Hill, Joseph H., Co.

'LADY BIRD JOHNSON', HT, or, 1971; bud long, pointed; flowers medium, dbl., moderate fragrance; vigorous, upright growth; [Montezuma X Hawaii]; Curtis, E.C.; Texas Rose Research Foundation

'LADY BISSETT', HT, ob, 1928; flowers bright orange, reverse apricot; Lilley

'LADY BLANCHE', R., w, 1913; flowers snow-white, dbl., borne in large clusters, free bloom, sometimes repeated in fall, intense fragrance; very vigorous, climbing growth; Walsh

'LADY BOUNTIFUL', LCl, mr, 1938; flowers scarlet-rose, center white, open, large, borne in clusters, single; foliage leathery, dark; very vigorous, climbing or trailing growth; [American Pillar seedling]; Tait; B&A

'LADY BRAYE', HT, dp, 1960; flowers deep rose-pink, long, pointed; foliage dark; Verschuren; Gandy Roses, Ltd.

Lady Brisbane, *see* **'CRAMOISI SUPÉRIEUR'**

Lady Brisbane, Climbing, *see* **'CRAMOISI SUPÉRIEUR, CLIMBING'**

'LADY CAHN', HT, ab, 1937; bud long, pointed; flowers rich apricot-yellow, veined darker, large, dbl., 40–50 petals; long, strong stems; vigorous growth; Gaujard

'LADY CANADA', HT, mp, 1927; flowers bright rose, dbl., moderate fragrance; [Mme Butterfly X Premier]; Dale

'LADY CAROLINA', S, lp, 1990; bud ovoid; flowers blush pink, same reverse, aging to white, decorative, small, 35–40 petals, repeat bloom, slight fragrance; sound, very small, very red fruit; prickles small, hooked, reddish-brown; foliage average, dark, glossy; bushy, spreading, hedge-type, medium growth; [Lady Gay sport]; Jeremias, Lephon L.

'LADY CASTLEREAGH', T, yb; flowers rosy yellow, well formed; vigorous growth

'LADY CATHERINE', HT, or, 1973; bud urn shaped; dbl., 32–45 petals, 3.5–4 in., exhibition form, slight fragrance; foliage leathery; vigorous, upright, compact growth; [Montezuma X Rubaiyat]; Von Koss; Kern Rose Nursery

'LADY CHARLES TOWNSHEND', HT, ob, 1931; flowers orange, overlaid salmon, large, dbl., globular; vigorous, bushy growth; [The Queen Alexandra Rose X Shot Silk]; Daniels Bros.

'LADY CHARMION', HT, mr, 1923; flowers bright cherry-carmine, dbl., moderate fragrance; [Lyon Rose X Gen. MacArthur]; Bees

Lady Clonbrock, N, lp; flowers medium, dbl.; (Sangerhausen)

'LADY CRAIG', HT, ly, 1922; flowers cream-yellow, center apricot-yellow, well formed, large, dbl., moderate fragrance; vigorous, free branching growth; Dickson, H.

'LADY CROMWELL', HT, dr, 1956; flowers crimson, base gold, large, dbl.; foliage bronze; vigorous growth; Verschuren; Gandy Roses, Ltd.

'LADY CUNLIFFE OWEN', HT, pb, 1932; flowers salmon and cream, base yellow, flushed carmine-rose, outer petals reflexed, dbl., exhibition form, moderate fragrance; foliage leathery; vigorous growth; [Mrs A.R. Barraclough sport]; Ley

'LADY CURZON', HRg, mp, 1901; flowers large, single, moderate fragrance; prickly; arching stems; vigorous growth; (21); [R. macrantha X R. rugosa rubra]; Turner

Lady D, HT, 1982; Huber, R.; (Cavriglia)

'LADY DALLAS BROOKS', HT, mp, 1955; [Peace sport]; Downes

'LADY DAWSON BATES', HT, yb, 1939; flowers golden yellow, flushed pink, opening, exhibition form, moderate fragrance; vigorous growth; McGredy

Lady Di®, HT, lp, 1982; Huber

'LADY DIANA', HT, lp, 1986; flowers, dbl., 37 petals, exhibition form, blooms in sprays of 3–4, slight fragrance; ovoid, orange fruit; short, hooked prickles; foliage medium, medium green, matt; tall, upright growth; PP005360; [Sonia X Caress]; Hoy, Lowel L.; Joseph H. Hill, Co., 1983

'LADY DIXON', HT, ab, 1919; flowers rich apricot, flushed salmon-pink, dbl., moderate fragrance; Dickson, A.

'LADY DIXON-HARTLAND', HT, pb, 1923; flowers centers deep salmon, outer petals pale pink, exhibition form; vigorous growth; Cant, B. R.

Lady Dumas, HT, lp; (Weatherly, L.)

'LADY DUNCAN', HRg, pb, 1900; flowers rich glowing pink, center and stamens yellow, single, 3 in., non-recurrent; foliage glossy; trailing (6 ft) growth; [R. wichurana X R. rugosa]; Dawson; Eastern Nursery

'LADY DUNLEATH', HT, w, 1913; flowers ivory-white edged yellow, small, dbl., moderate fragrance; Dickson, A.

'LADY EDGEWORTH DAVID', HT, mp, 1939; bud long, pointed; flowers Malmaison rose shaded soft pink, open, large, dbl., moderate fragrance; foliage glossy; vigorous growth; [Unnamed seedling X Betty Uprichard]; Fitzhardinge; Hazlewood Bros.

'LADY ELEANORE', Cl HT, yb, 1923; flowers light yellow-cream splashed rose, center golden to copper; [Gruss an Teplitz X Barbara]; Dreer

Lady Elgin, *see* 'MEIMAJ'

'LADY ELPHINSTONE', HT, yb, 1921; flowers indian yellow to clear rose, semi-dbl.; [Mme Edouard Herriot sport]; Dobbie

'LADY EMILY PEEL', N, w, 1862; flowers white, tinged with blush; [Blanche Lafitte X Sappho]; Lacharme, F.

'LADY ENGLISH', HT, or, 1934; flowers bright cerise, center orange, open, very large, dbl., slight fragrance; foliage glossy, bronze; long stems; vigorous growth; Cant, B. R.

Lady Ethel, S, lp; Sutherland, P, 1998; Golden Vale Nursery; (Weatherly, L.)

'LADY EVE', Min, op, 1978; bud globular; flowers creamy white, edged coral-pink, dbl., 40 petals, 1.5–2 in., exhibition form, slight fragrance; vigorous, upright, tall, spreading growth; [Neue Revue X Sheri Anne]; Rovinski & Meredith; Casa de Rosa Domingo

Lady Eve Price, *see* **'CAPRICE'**

'LADY EVELYN GUINNESS', HT, mp, 1932; [Ophelia sport]; Evans

'LADY FAIRBAIRN', HT, mp, 1929; flowers bright pink, dbl., 40 petals; vigorous, upright growth; [Mme Abel Chatenay X Seedling]; Clark, A.; NRS New South Wales

'LADY FAIRE', HT, mp, 1907; flowers salmon-pink; [Mrs W.J. Grant sport]; Bentley

'LADY FAIRFAX', HT, ob, 1930; flowers rose and orange-cerise, flushed orange to yellow, well formed, moderate fragrance; foliage light green; long stems; vigorous growth; Cant, F.

'LADY FLORENCE STRONGE', HT, lp, 1925; bud pointed; flowers pale flesh, base pink and gold, very large, dbl., exhibition form, slight fragrance; foliage leathery, glossy; vigorous, bushy growth; McGredy

'LADY FORTEVIOT', HT, yb, 1926; flowers golden yellow to deep apricot, large, dbl., exhibition form, intense fragrance; foliage bronze, glossy; vigorous, bushy growth; (28); GM, NRS, 1927; Cant, B. R., 1928

'LADY FORTEVIOT, CLIMBING', Cl HT, yb, 1935; Howard Rose Co.

'LADY FRASER', HT, mr, 1941; flowers rich red; [War Paint seedling]; Clark, A.

'LADY FROST', HT, dp, 1935; flowers deep rose, very large, dbl., intense fragrance; foliage leathery; vigorous, bushy growth; [Lady Alice Stanley X Dr Herbert Hawkesworth]; Bees

'LADY GAY', HWich, op, 1905; dbl., blooms in clusters; foliage small, dark, glossy; vigorous, climbing (12–20 ft) growth; [R. wichurana X Bardou Job]; Walsh

Lady Genevieve, HP, mr

'LADY GEORGIA', HT, pb, 1973; flowers pink, base blending to ivory, full, medium, dbl., moderate fragrance; foliage dark, leathery; very vigorous, bushy growth; [Miss Hillcrest X Peace]; Curtis, E.C.; Kimbrew

Lady Glencora, *see* 'ARORAJU'

'LADY GODIVA', HWich, lp, 1908; flowers cameo-pink; [Dorothy Perkins sport]; Paul

'LADY GOWRIE', Cl HT, my, 1938; bud long, pointed; flowers maize and champagne-yellow, large, very dbl., intermittent bloom, moderate fragrance; foliage leathery, glossy, dark; long stems; very vigorous, climbing growth; [Sunburst, Climbing X Rev. F. Page-Roberts]; Fitzhardinge; Hazlewood Bros.

'LADY GRADE', HT, or, 1982; (Lady Kathleen Grade); flowers vermilion, large, dbl., 35 petals, moderate fragrance; foliage large, medium green, semi-glossy; bushy growth; [Tropicana X Unnamed seedling]; Gregory, C.

'LADY GREENALL', HT, yb, 1911; flowers saffron-yellow, edges tinted shell-pink, dbl., moderate fragrance; Dickson, A.

'LADY GREENALL, CLIMBING', Cl HT, yb, 1923; Lippiatt

'LADY GWENDOLINE COLVIN', Cl HT, pb, 1918; flowers apricot-salmon, shaded chrome-yellow, outer petals stained carmine, dbl., moderate fragrance; height 6–10 ft; Chaplin Bros.

'LADY HAILSHAM', HT, or, 1951; bud pointed; flowers orange flushed red, dbl., 30 petals, 3 in., intense fragrance; foliage glossy; vigorous growth; [McGredy's Sunset sport]; Knight's Nursery

'LADY HAMILTON', HSpn, lp; flowers rosy blush, semi-dbl., profuse, non-recurrent bloom; dwarf growth

Lady Harriet, *see* 'WILHERB'

'LADY HELEN', HT, mp, 1970; flowers soft clear pink, pointed, dbl., 30 petals, 5 in., intense fragrance; foliage glossy, dark; bushy growth; [Margaret X (McGredy's Ivory X Peace)]; McTeer, Gilbert; Waterhouse Nursery

'LADY HELEN MAGLONA', HT, dr, 1926; bud pointed; flowers bright crimson-red to scarlet-red, center deeper, very large, dbl., exhibition form, intense fragrance; foliage leathery; vigorous, bushy growth; GM, NRS, 1926; Dickson, A.

'LADY HELEN STEWART', HP, dr, 1887; flowers bright crimson shaded scarlet, dbl., intense fragrance; vigorous growth; Dickson, A.

'LADY HILLINGDON', T, yb, 1910; bud long, pointed; flowers deep apricot-yellow, semi-dbl., moderate fragrance; foliage bronze; bushy growth; sometimes as hardy as a Hybrid Tea; (21); [Papa Gontier X Mme Hoste]; Lowe & Shawyer

'LADY HILLINGDON, CLIMBING', Cl T, yb, 1917; Hicks

'LADY HUDSON', HT, ab, 1930; flowers deep apricot, large, dbl.; vigorous growth; Chaplin Bros.

'LADY HUNTINGFIELD', HT, my, 1937; flowers rich golden yellow, reverse lighter, large, dbl., globular, moderate fragrance; long stems; vigorous, bushy growth; [Busybody X ?]; Clark, A.; NRS Victoria

'LADY ILIFFE', HT, mr, 1976; flowers tyrian rose, dbl., 38 petals, 5 in., intense fragrance; foliage large; [Saul X Wendy Cussons]; Gandy, Douglas L.

Lady in Red, *see* 'SEALADY'

Lady in Red, *see* 'JACOPPER'

'LADY INCHIQUIN', HT, or, 1922; flowers orange-vermilion, large, very dbl., exhibition form, slight fragrance; foliage leathery, rich glossy green; very vigorous, bushy growth; GM, NRS, 1920; Dickson, A.

'LADY JANE', HT, dy, 1992; dbl., 26–40 petals, 3–3.5 in., borne mostly singly, slight fragrance; some prickles; foliage medium, medium green, semi-glossy; medium (75 cms), upright growth; [Dorothe X Helmut Schimdt]; Poole, Lionel, 1993

'LADY JOHNSTONE', LCl, dp, 1922; bud yellow; flowers reddish pink, turning lilac-rose, stamens yellow, large, single, moderate fragrance; vigorous, climbing growth; [R. gigantea X Beauté Lyonnaise]; Nabonnand, P.

Lady Kathleen Grade, *see* **'LADY GRADE'**

'LADY KATHRYN', Min, m, 1989; bud pointed; flowers lavender, aging brown, urn-shaped, medium, borne usually singly, dbl., 22 petals, exhibition form, moderate, damask fragrance; round, green-brown fruit; no prickles; foliage medium, medium green, semi-glossy, disease-resistant; upright, medium, vigorous growth; [Lavender Jade X Angel Face]; Jolly, Marie; Rosehill Farm, 1991

'LADY LAUDER', HT, yb, 1931; flowers deep canary-yellow, reverse flushed crimson, dbl., cupped; foliage thick, light; long stems; very vigorous growth; Morse

'LADY LAVENDER', HT, m, 1993; dbl., 26–40 petals, 3–3.5 in., borne mostly singly, moderate fragrance; some prickles; foliage medium, dark green, glossy; bushy (40–48 in.) growth; [(Paradise X Seedling) X Swarthmore]; Weeks, O.L.; Estrella Rose Company, 1993

'LADY LAYTON', HT, my, 1932; bud long, pointed; flowers light sunflower-yellow, deepening as it opens, large, moderate fragrance; vigorous growth; [Joanna Hill sport]; Layton

'LADY LECONFIELD', HT, w, 1939; bud long, pointed, cream, flushed pink; flowers cream-white, dbl., 25–30 petals, cupped, intense fragrance; foliage leathery; vigorous, bushy growth; Burbage Nursery; C-P

'LADY LE-RU', HRg, dp, 1963; bud round; flowers deep pink, dbl., 50 petals, 3.5–4 in., cupped, moderate, open ($2\frac{1}{2}$ ft free, non-recurrent bloom, moderate fragrance; thornless; long stems; hardy.; [R. rugosa hybrid X Unknown Hybrid Tea]; Lothrop

'LADY LESLIE', HT, rb, 1929; flowers rosy scarlet to scarlet-carmine suffused saffron-yellow, large, dbl., exhibition form; foliage dark, leathery, glossy; vigorous growth; McGredy

'LADY LIBERTY', HT, w, 1986; flowers clear white, yellow at base, large, dbl., 35 petals, exhibition form, borne usually singly or in small clusters, slight fragrance; globular, medium, medium green fruit; straight, medium, light yellow prickles; foliage medium, medium green, semi-glossy; extremely long stems; upright, profuse growth; PP006142; [Lady Diana sport]; deVor, Tom, 1987

'LADY LIKE', HT, ob, 1971; bud globular; flowers dark orange, large, dbl., slight fragrance; vigorous, upright growth; [Seedling X Tropicana]; Tantau, Math.; Ahrens & Sieberz

Lady Like®, *see* TANekily

'LADY LILFORD', HT, my, 1930; flowers clear yellow, center deep golden yellow, slight fragrance; foliage rich green, glossy; vigorous, bushy, branching, compact growth; [Independence Day sport]; Gregory

'LADY LOCH', HT, lp; [Aspasia sport]; Johnson, 1885; (Weatherly, L.)

'LADY LOU', HT, ob, 1948; bud long, pointed; flowers coral-peach, large to medium, dbl., 50 petals, exhibition form, moderate fragrance; foliage glossy, light; vigorous, dwarf growth; [Pink Princess X Shades of Autumn]; Brownell, H.C.

'LADY LUCK', HT, pb, 1956; bud long, pointed; flowers blends of pale to rich pink, dbl., 38 petals, 4–4.5 in., exhibition form, intense, damask fragrance; foliage dark, leathery; vigorous, upright, bushy growth; [Tom Breneman X Show Girl]; Miller, A.J.; Elmer Roses Co.

'LADY MACGREGOR', HT, lp; [Boule de Neige X a Tea]; Williams, A., 1911; (Weatherly, L.)

Lady MacRobert, *see* 'COCLENT'

Lady Madeleine, *see* 'EVELEINE'

'LADY MANDEVILLE', HT, my, 1941; flowers yellow, flushed amber, well-formed, dbl., 35 petals, 5 in., fruity fragrance; slight; foliage dark, bronze; branching, moderate growth; [Unnamed seedling X Mrs Sam McGredy]; McGredy; J&P

'LADY MANN', HT, pb, 1940; flowers rosy salmon; [Lorraine Lee X ?]; Clark, A.

'LADY MARGARET BOSCAWEN', HT, mp, 1911; flowers large, dbl., intense fragrance; Dickson, A.; (Sangerhausen)

'LADY MARGARET STEWART', HT, ob, 1926; bud long, pointed; flowers golden yellow shaded and streaked orange and red, very large, dbl., exhibition form, moderate fragrance; foliage sage-green, leathery; vigorous, bushy growth; GM, Bagatelle, 1928 GM, NRS, 1926; Dickson, A.

'LADY MARINE', HT, or, 1981; flowers dark orange-red, urn-shaped, borne 1–3 per cluster, dbl., 53 petals, exhibition form, moderate fragrance; medium, reddish prickles; foliage dark, leathery; medium growth; [Unnamed seedling X Tropicana]; DeLashmutt; Roseway Nursery

'LADY MARTHA BRUCE', HT, mp, 1925; flowers pink, outer petals tinged peach-blossom-pink; Ferguson, W.

'LADY MARY', HT, mr, 1999; flowers crimson red, dbl., 40 petals, exhibition form, slight fragrance; foliage medium green; 3 ft growth; [Princess Mikasa X seedling]; Teranishi, K.; Itami Rose Nursery, 1990

'LADY MARY CORRY', T, 1900; Dickson, A.; (Cavriglia)

'LADY MARY ELIZABETH', HT, dp, 1927; bud pointed; flowers brilliant carmine-pink, large, dbl., exhibition form, intense fragrance; vigorous, bushy growth; Dickson, A.

'LADY MARY FITZWILLIAM', HT, lp, 1882; flowers flesh-color, large, globular, intense fragrance; vigorous growth; [Devoniensis X Victor Verdier]; Bennett

'LADY MARY WARD', HT, ob, 1913; flowers orange, shaded deeper, dbl., moderate fragrance; GM, NRS, 1912; McGredy

'LADY MAUREEN STEWART', HT, dr, 1920; flowers velvety blackish scarlet-cerise, reflex orange-maroon, dbl., moderate fragrance; Dickson, A.

Lady Mavis Pilkington, *see* KORlitze

'LADY MAYSIE ROBINSON', HT, pb, 1956; flowers deep pink, center white, large, dbl., 22 petals, cupped, moderate fragrance; foliage dark, glossy; vigorous, upright, bushy growth; [Seedling X Peace]; Kordes

'LADY MEDALLIST', Cl HT, lp, 1912; large centered flowers, pink with lighter reverse; dark green foliage; very vigorous early summer climber; Clark, A.

Lady Meilland, *see* MEIalzonite

Lady Meillandina, *see* 'MEILARCO'

'LADY MILLER', HT, dr, 1940; flowers dark-red, well-formed, moderate fragrance; Clark, A.

Lady Mitchell, *see* 'HARYEARN'

'LADY MOND', HT, w, 1920; flowers deep cream, outer petals shaded rose; Paul, W.

'LADY MOYRA BEAUCLERC', HT, mr, 1901; flowers carmine-red, large, dbl.; Dickson, A.; (Sangerhausen)

'LADY MOYRA CAVENDISH', HT, mr, 1939; bud long, pointed; flowers bright strawberry-red, flushed crimson, dbl., exhibition form, slight fragrance; foliage glossy, dark; bushy growth; McGredy

'LADY NUTTING', HT, mp, 1938; flowers soft salmon-pink, large, exhibition form; foliage leathery, dark; vigorous growth; Wheatcroft Bros.

Lady of Hertford, *see* 'BOSCONPEA'

'LADY OF SKY', HT, or, 1974; dbl., 35 petals, exhibition form, slight fragrance; foliage dark; vigorous, upright growth; [Queen Elizabeth X ?]; Gregory

'LADY OF STIFFORD', F, or, 1982; [Matador sport]; Warley Rose Gardens, 1981

Lady of the Dawn®, *see* 'INTERLADA'

Lady Penelope, *see* 'CHEWDOR'

'LADY PENZANCE', HEg, op, 1894; (R. X penzanceana); flowers coppery salmon-pink with yellow stamens, single, summer bloom; foliage dark, fragrant (apple); very vigorous growth; (42); [R. eglanteria X R. foetida bicolor]; Penzance

'LADY PIRRIE', HT, ab, 1910; (Mme Pirrie); bud pointed; flowers apricot-yellow, reverse coppery, large, dbl., moderate fragrance; vigorous, bushy growth; GM, NRS, 1909; Dickson, H.

'LADY PIRRIE, CLIMBING', Cl HT, ab, 1938; Unknown

'LADY PLYMOUTH', T, w, 1914; bud long, pointed; flowers deep ivory-cream, very faintly flushed, large, dbl., moderate fragrance; foliage rich green, leathery; many canes; vigorous, bushy growth; GM, NRS, 1913; Dickson, A.

'LADY R. VERNEY', HT, mp, 1935; flowers medium, dbl., moderate fragrance; Bees of Chester; (Sangerhausen)

Lady Rachel, *see* CANdoodle

'LADY RACHEL VERNEY', HT, mp, 1935; flowers rose, base lemon, large, dbl., cupped; foliage glossy, bronze; vigorous, bushy growth; [Annie Laurie X Lord Charlemont]; Bees

'LADY READING', Pol, mr, 1921; flowers clear red; [Ellen Poulsen sport]; Van Kleef

'LADY RHODES', HT, lp; Clark, A., 1934; (Weatherly, L.)

'LADY ROBERTS', T, ab, 1902; flowers rich reddish apricot, base coppery red, edges shaded orange, dbl., moderate fragrance; vigorous growth; [Anna Olivier bud sport]; Cant, F.

Lady Romsey, F, w, 1985; Beales, Peter

Lady Rose®, *see* 'KORLADY'

Lady Rose, Climbing, *see* ORAkosar

'LADY ROUNDWAY', HT, ab, 1923; flowers bright apricot-orange, fading to creamy buff, open, semi-dbl., slight fragrance; vigorous, but stubby growth; GM, NRS, 1923; Cant, B. R.

Lady Russon, *see* **'ORANGE DELBARD'**

'LADY SACKVILLE', HT, w, 1933; bud pointed; flowers pure white, very large, dbl., exhibition form, moderate fragrance; foliage leathery, bronze; very vigorous growth; Cant, B. R.

Lady Sackville, *see* **'NIGHT'**

'LADY SETON', HT, lp, 1966; dbl., 35 petals, 4.5 in., intense fragrance; vigorous, tall growth; [Ma Perkins X Mischief]; McGredy, Sam IV; McGredy

'LADY SOMERS', HT, lp, 1930; flowers fresh pink, tinted flesh, dbl., slight fragrance; foliage wrinkled, light; bushy growth; [Comte G. de Rochemur X Scorcher]; Clark, A.; NRS Victoria

'LADY SONIA', S, my, 1961; flowers golden yellow, dbl., 20 petals, 4–4.5 in.; foliage dark; vigorous, upright, branching growth; [Grandmaster X Doreen]; Mattock

'LADY STANLEY', T, dr; bud long; flowers crimson, very large, dbl., moderate fragrance

'LADY STUART', HCh, lp, 1851; flowers flesh pink to blush; foliage leaflets 5–7; Portemer fils

Lady Sunblaze™, *see* 'MEILARCO'

'LADY SUNSHINE', HT, my, 1965; bud ovoid; flowers large, very dbl., intense fragrance; foliage dark; [Belle Étoile X (Michele Meilland X Tawny Gold)]; Lens; Spek

'LADY SUSAN BIRCH', HT, ab, 1934; bud pointed; flowers large, dbl., exhibition form, moderate fragrance; foliage glossy, dark; vigorous, bushy growth; Cant, B. R.

'LADY SUZANNE', HT, w, 1985; flowers creamy white, large, dbl., 32 petals, exhibition form, slight fragrance; foliage large, dark, glossy; bushy growth; [Lady X X Flaming Beauty]; Bridges, Dennis A.; Bridges Roses

'LADY SYDNEY EARDLEY-WILMOT', HT, mp, 1925; flowers coppery reddish salmon, tinted fawn and apricot, semi-dbl., moderate fragrance; Chaplin Bros.

'LADY SYLVIA', HT, lp, 1926; [Mme Butterfly sport]; Stevens, W.

'LADY SYLVIA, CLIMBING', Cl HT, pb, 1933; Stevens, W.; Low

Lady Taylor, *see* **'SMItling'**

'LADY TERVUEREN', Pol, mr, 1969; flowers medium to small, dbl.; foliage dark; [Allotria X Seedling]; Buisman, G. A. H.

'LADY TRENT', HT, ob, 1940; (Julia Ferran); flowers coppery orange, large, dbl., 46 petals, exhibition form, moderate fragrance; foliage dark, glossy; vigorous growth; [Rosieriste Gaston Leveque X Federico Casas]; Dot, Pedro; Wheatcroft Bros.

'LADY URSULA', HT, pb, 1908; flowers pink fading lighter, reverse cameo-pink, base lemon, large, very dbl., exhibition form, slight fragrance; foliage dark, leathery, glossy; vigorous, bushy growth; Dickson, A.

'LADY VENABLES VERNON', HT, lp, 1922; flowers soft flesh-color, overlaid blush, moderate fragrance; [Mrs Amy Hammond X Sir Alexander N. Rochfort]; Jersey Nursery

'LADY VERA', HT, pb, 1974; flowers silvery pink, reverse rose-pink, dbl., slight fragrance; vigorous growth; [Royal Highness X Christian Dior]; Smith, R.W.; Brundrett

'LADY VEREY', HT, mp, 1922; flowers rose-pink, dbl., moderate fragrance; Hicks

'LADY VIOLET ASTOR', HT, dp, 1933; flowers deep rose-pink, over large, dbl., exhibition form, slight fragrance; foliage leathery; vigorous growth; Cant, B. R.

'LADY WAKEFIELD', HT, ab, 1926; flowers bright apricot, moderate fragrance; Cant, B. R.

'LADY WATERLOW', Cl HT, pb, 1903; flowers salmon-pink edged carmine, large, dbl.; vigorous (8–10 ft) growth; [La France de '89 X Mme Marie Lavalley]; Nabonnand, G.

'LADY WENLOCK', HT, lp, 1904; flowers china-pink tinted apricot; Bernaix, P.

'LADY WILLINGDON', HT, lp, 1928; flowers very light pink, large, dbl., moderate fragrance; foliage rich green, glossy; vigorous growth; [Ophelia X Premier]; Dale

'LADY WOODWARD', HT, lp, 1959; bud long, pointed; flowers pink veined, large, dbl., exhibition form, moderate fragrance; foliage dark, glossy; vigorous, upright, bushy growth; [Heinrich Wendland X Elli Knab]; Riethmuller

'LADY WORTHINGTON EVANS', HT, dr, 1926; bud pointed; flowers deep crimson shaded blackish, semi-dbl., exhibition form, moderate fragrance; foliage bronze, leathery; vigorous, bushy growth; GM, NRS, 1926; Dickson, A.

Lady X, *see* 'MEIFIGU'

'LADY X, CLIMBING', Cl HT, m, 1976; Takatori, Yoshiho; Japan Rose Nursery

Lady Yvonne, *see* 'BOSBIGSOUTH'

'LADY ZIA', HT, or, 1959; flowers light orange-scarlet, well-formed, dbl., 50 petals, 5–6. in., moderate fragrance; foliage dark, glossy; vigorous growth; GM, NRS, 1959; [Peace X Independence]; Park; Harkness

Ladybug, *see* 'MORBUG'

'LADYLOVE', HT, mp, 1926; flowers light rose-pink fading hydrangea-pink, flushed apricot, well-formed, dbl., intense fragrance; foliage dark; very vigorous growth; [Opehlia X Unnamed seedling]; McGredy; Beckwith

'LAFAYETTE', F, dp, 1924; (August Kordes, Joseph Guy); flowers bright cherry-crimson, large blooms in clusters (up to 40), semi-dbl., cupped, slight fragrance; foliage rich green, glossy; vigorous, bushy growth; [Rodhatte X Richmond]; Nonin; Dreer;, H&S

Lafayette, Climbing, *see* **'AUGUSTE KORDES'**

'L'AFRICAINE', LCl, dr, 1953; flowers garnet shaded coppery, well formed, large, borne in clusters, abundant early bloom, not recurrent; strong stems; very vigorous growth; [Guinee X Crimson Glory]; Mallerin, C.; EFR

'LAFTER', HT, yb, 1948; bud pointed; flowers salmon-yellow, dbl., 23 petals,

4 in., moderate fragrance; vigorous, upright, branching growth; [(V for Victory X (Général Jacqueminot X Dr W. Van Fleet)) X Pink Princess]; Brownell, H.C.

Lagerfeld™, *see* 'AROLAQUELI'

'**LAGERFEUER**'®, F, mr, 1958; (Feu de Camp); bud pointed; flowers velvety scarlet, large blooms in clusters, dbl., slight fragrance; foliage leathery, dark; vigorous, upright growth; [Red Favorite X Kathe Duvigneau]; Tantau, Math.

'**LAGOON**', F, m, 1970; flowers lilac, reverse darker, gold stamens, single, 7 petals, 2.5 in., moderate fragrance; foliage glossy; [Lilac Charm X Sterling Silver]; Harkness, 1973

'**LAGUNA**', HT, or, 1974; bud large, long, pointed; dbl., cupped, moderate fragrance; foliage glossy; vigorous, upright growth; [Hawaii X Orange Delbard]; Kordes; Horstmann

Laguna, HT, lp; flowers yellow with pink edges

Laguna Palace, *see* POUlagun

Lahar, F, my, 1991; IARI

'**LAILA**', F, ob, 1968; flowers bright orange, well formed, borne on trusses, dbl., slight fragrance; foliage glossy; vigorous growth; [Orangeade seedling]; Abdullah

L'Aimant, *see* 'HARZOLA'

'**LAKE COMO**', F, m, 1968; flowers lilac, blooms in trusses, semi-dbl., intense fragrance; [Lilac Charm X Sterling Silver]; Harkness

'**LAKELAND**', HT, lp, 1976; flowers soft shell-pink, dbl., 36 petals, 5–6. in., slight fragrance; [Fragrant Cloud X Queen Elizabeth]; Fryer, Gareth; Fryer's Nursery, Ltd.

Lakeland Princess, *see* 'POULLACK'

Lakeland's Pride, *see* '**DOUBLE FEATURE**'

'**LAL**', HT, pb, 1933; bud long, pointed; flowers deep salmon pink, suffused yellow, intense fragrance; foliage dark; vigorous growth; [Commonwealth (HT) X Florence L. Izzard]; Easlea

'**LAL, CLIMBING**', Cl HT, mp, 1937; flowers very large, dbl., intense fragrance; Vogel, M.; (Sangerhausen)

'**LALIMA**', HT, mr, 1978; flowers large blooms borne singly, dbl., 50 petals, exhibition form, intense fragrance; foliage large; vigorous, upright growth; [Picture X Jour d'Ete]; Pal, Dr B.P.; K.S.G. Son's Roses

'**LALLITA**', HT, mp, 1929; flowers rose, very large, dbl., slight fragrance; foliage rich green, leathery; long, strong stems; [Pres. Briand X ?]; Mallerin, C.

'**LAMARQUE**', N, w, 1830; flowers pure white, center lemon-yellow, blooms in clusters, dbl., intense fragrance; vigorous, climbing growth (long, trailing shoots).; [Blush Noisette X Parks' Yellow Tea-scented China]; Marechal

'**LAMARTINE**', HT, op, 1943; flowers pearly pink shaded orange, dbl.; Meilland, F.

Lamb Chop, S, lp; Peden, R., 1997; (Weatherly, L.)

Lambada, *see* KORdaba

'**LAMBERT CLOSSE**', S, mp, 1994; very dbl., 53 petals, 3–3.5 in., borne mostly singly, slight; some prickles; foliage medium, light green, glossy; medium (85 cms), upright growth; [Arthur Bell X John Davis]; Ogilvie, Ian S.; Agriculture Canada, 1994

'**L'AMI E. DAUMONT**', HP, mr, 1903; flowers carmine-red, large, very dbl.; Vilin; (Sangerhausen)

'**LAMIA**', HT, or, 1918; bud rich apricot; flowers intense reddish orange, semi-dbl.; vigorous growth; GM, NRS, 1918; Easlea

'**LAMINUETTE**', F, rb, 1969; (La Minuette, Minuette); bud ovoid, pointed; flowers petals ivory-white, tipped red, heart-shaped, medium, dbl., slight fragrance; foliage glossy, dark green; bushy, branching, strong (2 ft) growth; [Peace X Rumba (F)]; Lammerts, Dr Walter; DeVor Nurseries, Inc.

'LAMLAM', HT, yb, 1965; (**American Heritage**®); bud long, pointed; flowers ivory and salmon blend, becoming salmon, large, dbl., exhibition form; foliage dark, leathery; tall growth; AARS, 1966; [Queen Elizabeth X Yellow Perfection]; Lammerts, Dr Walter; Germain's

'**LAMOTTE SANGUIN**', HP, mr, 1869; flowers very large, dbl.; Vigneron; (Sangerhausen)

'**LAMPION**', F, or, 1957; bud pointed; flowers blood-red shaded orange, open, large, borne in large clusters, single, 5 petals, slight fragrance; foliage dark, leathery, glossy; dwarf, bushy growth; [Fanal X Kathe Duvigneau]; Tantau, Math.

'**LAMPLIGHTER**', Cl HT, dy, 1948; flowers blend similar to Talisman and Autumn, large, dbl., globular, intense fragrance; foliage leathery; very vigorous, climbing (10–15 ft) growth; [Talisman X Gold Rush]; Duehrsen; California Roses

'**LAMPLIGHTER**', HT, pb, 1950; flowers salmon-rose, reverse gold, very large, dbl., 37 petals, exhibition form, slight fragrance; foliage bronze; vigorous growth; [Sam McGredy X Unnamed seedling]; McGredy, Sam IV

'**LAMPLIGHTER**', HT, yb, 1959; bud long, pointed; flowers mimosa-yellow, dbl., 55–60 petals, 4–4.5 in., exhibition form, moderate fragrance; foliage dark, semi-glossy, leathery; strong stems; vigorous, upright growth; [Peace X Yellow Perfection]; Hill, Joseph H., Co.

Lampo, F; Cazzaniga, F. G.; (Cavriglia)

'LANBET', HT, or, 1985; (**Jennifer-Betty Kenward**); flowers medium, dbl., 35 petals, slight fragrance; foliage medium, medium green, glossy; spreading growth; [Mildred Reynolds X Whisky Mac]; Sealand Nurseries, Ltd.

'**LANCASHIRE**', HT, mr, 1950; flowers fiery red, intense fragrance; [Christopher Stone X Seedling]; Wright, R. & Sons

Lancashire, *see* KORstesgli

'**LANCASHIRE LASS**', HT, mr, 1939; bud well shaped, crimson; flowers scarlet-cerise, semi-dbl., moderate fragrance; foliage glossy, dark; vigorous growth; Archer

Lancashire Life, *see* 'RUILANCA'

'**LANCASTER**', Cl HT, mr, 1962; bud ovoid; flowers medium rose-red, dbl., 35–40 petals, 5 in., exhibition form, intense fragrance; foliage leathery, dark; vigorous growth; [Pres. Eisenhower sport]; Hicks, S.J.; C-P

'**LANCASTRIAN**', HT, mr, 1965; flowers crimson-scarlet, dbl., 40 petals, 3.5–4 in., intense fragrance; foliage light green, glossy; vigorous, upright growth; [Ena Harkness X ?]; Gregory

'**LANCE HIRD**', HT, mp, 1973; flowers deep pink, reverse lighter, full, dbl., 20–30 petals, 4 in., slight fragrance; free growth; [Seedling X Greetings]; Wood; Wood Roses

'**LANCIER**', HT, dr, 1959; flowers crimson-red; [Karl Herbst X Unnamed seedling]; Mallerin, C.; Vilmorin-Andrieux

Lancème®, *see* 'DELBOIP'

Land of the Long White Cloud, *see* 'MACLANOFLON'

'**LANDEN**', HT, ly, 1988; (**Denman**); flowers creamy yellow, urn-shaped, large, borne in sprays of 2–3, dbl., 35 petals, intense fragrance; prickles medium, red; foliage large, dark green, glossy; upright growth; [Mildred Reynolds X Arthur Bell]; Sealand Nurseries, Ltd.

Lander Gold, *see* 'MICGOLD'

'**LANDIA**', HT, ab, 1988; (Anne Diamond); bud pointed; flowers apricot, reverse pink, aging apricot, urn-shaped, medium, bo, dbl., 38 petals; prickles slightly hooked, brown; foliage medium, dark green, semi-glossy; upright, bushy growth; [Mildred Reynolds X Arthur Bell]; Sealand Nurseries, Ltd.

LANdisney, HT, w, 1987; (**Snow White**); Sealand Nurseries, Ltd.

Landmark Rose, S, mp, 1996

Landora, *see* '**SUNBLEST**'

Landora, Climbing, *see* 'CLINORA'

'**LANDOUR**', Cl HT, w, 1978; bud oval; dbl., 40 petals, 6–6.5 in., exhibition form, intermittent bloom, moderate, fruity fragrance; foliage large, glossy, dark, leathery;

upright growth; [Peace, Climbing self seedling]; Thakur; Doon Valley Roses

Landrover, HRg, mr, 1995

Landscape Splendor, *see* CLEland

Landwirtschaftsrat Hubert Schilling, Pol, or, 1964; flowers medium, semi-dbl.; Pieper; (Sangerhausen)

Lane, *see* **'LANEII'**

'LANEII', M, mr, 1845; (Lane, Lane's Moss); bud large, globular, well-mossed; flowers rosy crimson, occasionally tinted purple, large, dbl., globular; foliage large, bright green, 5 leaflets; robust growth; Laffay, M., 1845; Int. into England by Lane & Son, ca. 1846

Lane's Moss, *see* **'LANEII'**

Langdale Chase, *see* FRYrhapsody

'LANGFORD', LCl, dp; flowers nearly red, dbl., free bloom; needs protection in cold areas; [R. setigera seedling]; Preston; Central Exp. Farm

Langford Light, *see* 'LANNIE'

'LANGLEY', HT, dy, 1942; bud streaked red; flowers clear deep yellow, edges flushed old-gold, dbl., 50 petals, 4–5 in., moderate fragrance; foliage bronze, dark; [Mrs Sam McGredy X Phyllis Gold]; Eacott; Ley

'LANGLEY GEM', F, dp, 1939; flowers scarlet-cerise, open, borne in clusters, single; foliage leathery, bronze; strong stems; vigorous growth; [Karen Poulsen X R. moyesii]; Eacott; R. Murrell

'LANICAN', HT, yb, 1988; (**Great Expectations**); flowers light pink, reverse light yellow, aging fading slightly, well-formed, dbl., 55 petals, exhibition form, moderate fragrance; prickles long, pointed, medium, red; foliage medium, medium green, semi-glossy, clean; upright growth; [Rosenella X Cassandra]; Sealand Nurseries, Ltd.

'LANKEN', HT, op, 1985; (**Felicity Kendal**); flowers salmon-orange, large, dbl., 35 petals; foliage large; bushy growth; [Fragrant Cloud X Mildred Reynolds]; Sealand Nurseries, Ltd.

'LANNIE', Min, w, 1985; (**Langford Light**); flowers bright yellow stamens, small, semi-dbl., moderate fragrance; foliage medium, dark, matt; bushy growth; [Ballerina X Little Flirt]; Sealand Nurseries, Ltd., 1984

'LANPIPE', F, dy, 1985; (**St Bruno**); flowers large, dbl., 35 petals, intense fragrance; foliage medium, medium green, semi-glossy; bushy growth; Edland Fragrance Medal, ARS, 1986; [Arthur Bell X Zambra]; Sealand Nurseries, Ltd.

Lantern, *see* 'MACLANTER'

'LANTOR', HT, pb, 1985; (**Torvill & Dean**); flowers pink, yellow reverse, medium, dbl., 35 petals, slight fragrance; foliage medium, dark, semi-glossy; upright growth; [Irish Gold X Alexander]; Sealand Nurseries, Ltd., 1984

Lanvin™, *see* 'AROLEMO'

'LANWOOL', F, ab, 1985; (**Chelsea Gold**); flowers medium, dbl., 20 petals, slight fragrance; many prickles; foliage medium, medium green, semi-glossy; bushy growth; [Arthur Bell X Elizabeth of Glamis]; Sealand Nurseries, Ltd.

LAPad, F, or, 1971; (**Porthos**®); Laperrière

LAPam, F, my; (**Golden Flame**®); Laperrière

LAParan, F, 1958; (**Chantefleur**); Laperriere, L.; (Cavriglia)

'LAPAV', HT, ob, 1967; (**Mme L. Pradel**®); flowers clear orange, medium, dbl., exhibition form, slight fragrance; foliage bronze, glossy, leathery; vigorous, upright growth; [Magicienne X Seedling]; Laperrière; EFR

LAPbal, HT, yb, 1989; (**Horticolor**®); Laperrière

'LAPBAU', HT, my, 1991; (**Bellissima**®); Laperrière

LAPbel, HT, mr, 1988; (**Duchesse de Savoie**®); Laperriere, L.

LAPbu, HT, op, 1988; (**Ecole d'Ecully**); Laperrière

'LAPCAL', HT, rb, 1966; (**Fanette**®); flowers red, reverse white, medium, dbl., 36 petals, exhibition form, slight fragrance; foliage dark, glossy; vigorous, bushy growth; [Jeunesse X Souv. du President Plumecocq]; Laperrière; EFR

'LAPCI', F, mr, 1958; (**Bel Ami**®); flowers large blooms in clusters of 7–8, dbl.; bushy growth; [Michèle Meilland X Tonnerre]; Laperrière; EFR

LAPcli, F, mr, 1997; (**Hillary First Lady**); Laperrière

LAPdade, HT, mr, 1984; (**Ferline**®); Laperrière

LAPdal, HT, yb, 1990; (**Roxane**®); Golden Rose, Geneva, 1990; Laperrière

'LAPDEF', HT, dy, 1986; (**Banco 86**®); Laperrière

LAPdev, F, or, 1991; (**Plein Soleil**®); Laperrière

'LAPDI', HT, or, 1966; (**Renouveau de Provins**®); bud ovoid; flowers geranium color, open, very large, dbl., 36 petals; foliage glossy; vigorous, bushy growth; [Magicienne X Numero Un]; Laperrière; EFR

LAPdil, HT, op, 1988; (**Blondie**); Laperrière

LAPduf, HT, 1989; (**Don Bosco**); Laperrière, L.; (Cavriglia)

'LAPDUL', HT, pb, 1967; (**Régence**®); bud globular; flowers flesh color edged bright pink, open, dbl.; foliage glossy; vigorous, bushy growth; Laperrière; EFR

LAPed, HT, rb; (**Cabaret**®); Laperrière, 1992; (CRL)

LAPej, HT, rb, 1992; (**République de Genève**); Laperrière

LAPem, HT, dp, 1974; (**Illisca**®); Laperrière

LAPjaminal, F, ob, 1983; (**Magic Fire**®); Laperrière

'LAPLAM', HT, ob, 1969; (**Mme Franck Augis**®); bud ovoid; flowers clear orange, medium, dbl., exhibition form, slight fragrance; foliage glossy; vigorous, bushy growth; [Magicienne X Seedling]; Laperrière; EFR

LAPlical, HT, rb, 1981; (**Mamy Laperrière**); Laperrière; (Cavriglia)

'LAPMAU', HT, m, 1968; (**Ludmilla**®); bud ovoid; flowers medium, semi-dbl., slight fragrance; foliage glossy; vigorous, upright growth; [(Peace X Independence) X Heure Mauve]; Laperrière; EFR

LAPmiravi, HT, mr, 1982; (**Carmosine**); Laperriere, L.

LAPmisal, HT, op; (**Mistraline**®); Laperrière

LAPnat, HT, op, 1971; (**Domila**®); Laperrière

'LAPNEUF', F, or, 1967; (**Magic Fire**®); bud pointed; flowers bright orange-red, small, semi-dbl., exhibition form; foliage glossy; vigorous, bushy growth; Laperrière; EFR

LAPniro, F, ob; (**D'Artagnan**); Laperrière

'LAPON', HT, mr, 1968; (**Ursula**®); bud ovoid; flowers clear red, open, medium, dbl., slight fragrance; foliage glossy; vigorous, upright growth; [Jeunesse X (Peace X Independence)]; Laperrière; EFR

'LAPOUDA', HT, ob, 1978; (**Magicienne 78**®); Laperrière

LAPovi, F, dp; (**Capitole**®); Laperrière

Lapponia®, *see* 'TANNIPOLA'

LAPruni, HT, w, 1974; (**Banquise**); Laperrière

'LAPWON', F, ob, 1965; (**Athos**®); flowers bright orange, large, dbl., 28 petals; foliage dark; bushy growth; [Coup de Foudre X Soleil]; Laperrière; EFR

Laque de Chine, HMsk, 1991; Lens

'LARA', HT, op, 1967; flowers salmon-carmine, well-formed, large, dbl., 38 petals, slight fragrance; foliage dark, glossy; vigorous, tall growth; [Tropicana X Romantica]; Kriloff, Michel; Cramphorn's Nursery

Larado, *see* 'SPOLAR'

L'Argentee, HT, rb

'LARGO D'HAENDEL', HT, dp, 1947; (Handel's Largo); flowers reddish apricot to salmon-carmine; Mallerin, C.; URS

Larissa®, *see* 'KORDODO'

L'Arlésienne, *see* 'FRED EDMUNDS'

'LARRY BURNETT', HSpn, w, 1925; flowers blush-white, center deeper, semi-dbl.,

3.5 in., cupped, profuse bloom, intense fragrance; foliage small, rich green, soft; short stems; vigorous, bushy, spreading growth; very hardy.; [R. acicularis X R. spinosissima]; Skinner

Larry's Surprize, *see* 'RENLARRY'

Las Vegas®, *see* 'KORGANE'

'LAS VEGAS', HT, pb, 1957; bud long, pointed; flowers salmon-pink, reverse darker, dbl., 25–30 petals, 5 in., exhibition form, slight fragrance; foliage leathery; vigorous, bushy growth; RULED EXTINCE 6/80; [Charlotte Armstrong X Mission Bells]; Whisler; Germain's

Las Vegas, HT, 1974; Kordes, R.; (Cavriglia)

'LAS-CASES', B, dp, 1828; Vibert

Laser™, *see* KEltaibu

Laser Beam, *see* 'KORPORT'

'LASKER', HT, pb, 1979; bud very long; dbl., 40 petals, exhibition form, borne singly, moderate fragrance; curved prickles; foliage glossy; tall growth; [South Seas X Oregold]; Perry, Astor

'LASSIE', HT, pb, 1946; bud long; pointed; flowers shell-pink, base canary-yellow, dbl., exhibition form, moderate fragrance; foliage leathery; vigorous, compact growth; [Picture sport]; Tuttle Bros. Nursery

Lasting Impression, *see* 'MICIMP'

Lasting Peace, *see* MEImurge

'LATAN', HT, 1979; bud long; flowers light lavender-tan, dbl., 43 petals, 4.5–5.5 in., exhibition form, intense fragrance; foliage leathery; upright, bushy growth; [Lady X X Bronze Masterpiece]; Williams, J. Benjamin; J.B. Williams & Associates

'LATHOM CHAPEL', F, w, 1973; flowers buff-cream, blended to white, full, dbl., 20–25 petals, 4 in., intense fragrance; vigorous growth; [Orange Sensation X Sutter's Gold]; Ellick

'LATHOM PARK', F, dp, 1972; flowers carmine, full, dbl., 40 petals, 4 in., slight fragrance; foliage glossy, light; vigorous growth; [Orange Sensation X Ballet]; Ellick

'LATHOM SUNRISE', F, ob, 1973; flowers orange-flame, full, dbl., 25–30 petals, 4 in., intense fragrance; vigorous growth; [(Orange Sensation X Ballet) X Sutter's Gold]; Ellick

'LATHOM SUNSET', HT, ob, 1973; flowers orange-yellow, pink and red, full, dbl., 20–25 petals, 4.5 in., moderate fragrance; vigorous growth; [Karl Herbst X Mischief]; Ellick

'LATHOM TWILIGHT', HT, m, 1973; flowers deep purple to indian lake, full, dbl., 28–30 petals, 5 in., intense fragrance; foliage dark; vigorous growth; [Karl Herbst X Mischief]; Ellick

Latte, *see* 'JUSLATTE'

Laughter Lines, *see* 'DICKERRY'

'LAURA', HT, op, 1969; flowers coral-pink, large, dbl., exhibition form, moderate fragrance; foliage leathery; vigorous, upright, bushy growth; [((Happiness X Independence) X Better Times) X (Baccará X White Knight)]; Meilland, Mrs. Marie-Louise; C-P

Laura™, *see* 'MEIVOUPLIX'

Laura®, *see* 'MEIDRAGELAC'

'LAURA', C, mp, 1848

Laura '81, *see* 'MEIDRAGELAC'

Laura Anne, *see* 'COCCLARION'

Laura Ashley, *see* 'CHEWHARLA'

Laura Chantal, *see* 'TOMTAL'

Laura Clements, *see* CLEspirit

Laura Ford®, *see* 'CHEWARVEL'

Laura Jane, HT, lp, 1990; Gandy

Laura Louisa, LCl, lp

Laura Louise, HKor, op, 1995; Riches

'LAURA ROSE ELLIOTT', HT, dp, 1995; flowers deep pink with light reverse, blooms borne mostly singly, very dbl., 4–4.75 in., moderate fragrance; numerous prickles; foliage medium, medium green, dull; medium (60–70 cms), upright growth; [Gavotte X City of Gloucester]; Thomas, D.

'LAURA TOWILL', HT, yb, 1929; flowers copper-yellow, becoming copper-pink, semi-dbl., slight fragrance; [Phantom X Buttercup]; Towill

Laura's Red Koster, Pol, mr, 1992

Laure Charton, F, or

'LAURÉ DAVOUST', HMult, lp, 1834; (Marjorie W. Lester); flowers clear pink, fading to flesh, then white, small blooms in clusters, dbl., cupped, non-recurrent; Laffay, M.

'LAURÉ SOUPERT', HMult, ly, 1927; flowers yellowish white to pure white, small blooms in clusters of 80, dbl., recurrent bloom, intense fragrance; foliage small, glossy; strong stems; vigorous, climbing or trailing growth; [Tausendschön X George Elger]; Soupert & Notting; Notting

'LAURE WATTINNE', HT, mp, 1902; flowers large, dbl., moderate fragrance; Soupert & Notting; (Sangerhausen)

'LAUREATE', F, m, 1989; bud pointed; flowers lavender with pink highlights, aging lighter, urn-shaped, medium, dbl., exhibition form, slight, fruity fragrance; fruit not observed; prickles straight, medium, red; foliage medim,dark green, glossy, disease-resistant; upright, bushy, medium, winter-hardy growth; [Baby Talk X Angel Face]; Jobson, Daniel J.

Laurel Louise, MinFl, ab, 1999

'LAURELLE', F, m, 1966; flowers lavender-pink, base veined lemon, borne in clusters, dbl., moderate fragrance; foliage glossy; vigorous, bushy growth; [Rumba X Lavender Princess]; Harris, J.R.; Paulen Park Nursery

Lauren Amy, HT, or, 1995; Dawson

Lauren Elizabeth, *see* 'ORTLAE'

Laurence Olivier, F, 1990; Meilland, Alain A.; (Cavriglia)

'LAURENT CARLE', HT, mr, 1907; flowers brilliant velvety carmine, open, large, dbl., intense fragrance; foliage rich green, soft; bushy growth; Pernet-Ducher

'LAURENT CARLE, CLIMBING', Cl HT, mr, 1923; Rosen, L.P.

'LAURENT CARLE, CLIMBING', Cl HT, mr, 1924; Mermet

'LAURENT DE RILLÈ', HP, mr, 1885; flowers large, dbl., moderate fragrance; Lévêque; (Sangerhausen)

Laurette, HRg, dp, 1999; Hortico

'LAURIE', HT, lp, 1970; flowers creamy pale pink, tipped salmon, full, dbl., 28 petals, 5 in., moderate fragrance; foliage matt green; vigorous growth; [Princesse sport]; Scott, D.H.

Lausitz, S, mp, 1959; flowers medium, semi-dbl., slight fragrance; Berger, W.; (Sangerhausen)

Lava Gold, F, ob, 1999; Williams, J. Benjamin

'LAVABY', Min, lp, 1982; (**Baby Face**); bud tapered; flowers light pink, very small, dbl., 35 petals, blooms in sprays of many, slight; foliage small, light green, matt; compact growth; [Popcorn X Popcorn]; Laver, Keith G.

'LAVACEK', Min, mr, 1984; ('LAVCALE', 'LAVCALI', **Mountie™**); dbl., 35 petals, exhibition form, no fragrance; foliage small, dark, semi-glossy; bushy growth; PP006054; [Party Girl X Dwarfking '78]; Laver, Keith G.; Springwood Roses, 1985

'LAVACRE', Min, ly, 1987; ('LAVCREAM', **Potluck Cream™**); flowers light yellow center, cream edges, reverse cream with yellow, dbl., 30 petals, exhibition form, no fragrance; no fruit; prickles slender, brown; foliage medium, medium green, matt; bushy, low, very free flowering growth; [Cornsilk X Unnamed seedling]; Laver, Keith G.; Springwood Roses, 1988

Lavaglow, *see* 'KORLECH'

Lavaglut®, *see* 'KORLECH'

'LAVALIER', Min, dp, 1989; bud pointed; flowers deep mauve-pink, reverse medium pink, small, borne singly, dbl., 50 petals, cupped, no fragrance; round, red fruit; prickles straight, small, light brown; foliage small, medium green, matt, disease-resistant; upright, bushy, low, prolific growth; [Loving Touch X (Honest Abe X Unnamed seedling)]; Laver, Keith G.; Springwood Roses

'LAVALOT', Min, mp, 1992; (**Potluck Pink**™); flowers rose pink, blooms borne mostly single, dbl., 26–40 petals, 1.5 in., slight fragrance; few prickles; foliage small, medium green, matt; low (25–30 cms), compact growth; [June Laver X Julie Ann]; Laver, Keith G.; Springwood Roses

'LAVALUCK', Min, dp, 1989; bud pointed; flowers vivid, deep cherry pink, aging fuchsia, urn-shaped, small, b, dbl., 28 petals, no fragrance; ovoid, orange-red fruit; no prickles; foliage small, medium green, matt, long sepals, disease resi; bushy, low growth; [Blueblood X Julie Ann]; Laver, Keith G.; Springwood Roses

'LAVAMAZE', Min, lp, 1989; bud ovoid; flowers light pink with deeper center, reverse light pink, full, small, very dbl., 80 petals, exhibition form, moderate fragrance; round, orange-red fruit; prickles straight, narrow, red; foliage small, medium green to red, disease resistant; bushy, low growth; [Loving Touch X Potluck™]; Laver, Keith G.; Springwood Roses

Lavande, F, m; Rennie, Bruce F.

'LAVARU', Min, dr, 1993; (**Springwood Ruby**); dbl., 26–40 petals, 1.5 in., borne in small clusters, no fragrance; some prickles; foliage medium, medium green, matt; low (20–25 cms), compact growth; [(Breezy X June Laver) X (June Laver X Ontario Celebration)]; Laver, Keith G.; Springwood Roses

'LAVAWAY', Min, w, 1989; (**Snowbound**); bud ovoid; flowers ivory, reverse white, aging white, urn-shaped, small, borne, dbl., 55 petals, slight fragrance; round, yellow-orange fruit; prickles straight, narrow, greenish-brown; foliage small, dark green, matt, disease-resistant; bushy, low, compact growth; [Tabris X June Laver]; Laver, Keith G.; Springwood Roses, 1990

'LAVBERT', Min, dr, 1995; (**Springwood® Red Victor**); flowers dark red, very full (41+ petals), small (0–4 cms) blooms borne 3–10 per clusters, no fragrance; some prickles; foliage medium, medium green, semi-glossy; medium (30–33 cms), upright, compact growth; [Seedling X Seedling]; Laver, Keith G.; Springwood Roses, 1995

'LAVBLU', Min, mr, 1982; (**Blueblood**™); flowers velvety texture, small, dbl., 55 petals; foliage small, deep green, semi-glossy; spreading growth; [Dwarfking '78 X Hokey Pokey]; Laver, Keith G.

'LAVBOUND', Min, op, 1989; bud pointed; flowers coral to orange-pink, outer petals pink-apricot, reverse pink, dbl., 53 petals, exhibition form, no fragrance; foliage medium, medium green, semi-glossy, disease-resistant; upright, bushy, low, prolific growth; [June Laver X Black Jade]; Laver, Keith G.; Springwood Roses

'LAVBRIC', Min, lp, 1995; (**Show Carpet**); Flowers light pink, dbl. (15–25 petals), small, micro-mini blooms borne in large clusters; slight fragrance; foliage small, dark green, glossy; upright growth; [Seedling X Seedling]; Laver, Keith G.; Springwood Roses, 1996

'LAVBRUN', Min, ob, 1995; flowers orange, blooms borne mostly single, dbl., 26–40 petals, no fragrance; some prickles; foliage medium, medium green, semi-glossy; bushy, medium growth; [Seedling X Apricot Doll]; Laver, Keith G.; Springwood Roses, 1995

'LAVCALE', Min, mr, 1984; ('LAVACEK', 'LAVCALI', **Mountie**™); dbl., 35 petals, exhibition form, no fragrance; foliage small, dark, semi-glossy; bushy growth; PP006054; [Party Girl X Dwarfking '78]; Laver, Keith G.; Springwood Roses, 1985

'LAVCALI', Min, mr, 1984; ('LAVACEK', 'LAVCALE', **Mountie**™); dbl., 35 petals, exhibition form, no fragrance; foliage small, dark, semi-glossy; bushy growth; PP006054; [Party Girl X Dwarfking '78]; Laver, Keith G.; Springwood Roses, 1985

'LAVCAP', Min, op, 1997; (**Peach Festival**™); flowers very full, medium, very dbl., 41 petals, borne mostly singly, slight fragrance; few prickles; foliage medium, medium green, glossy; upright, bushy, medium growth; [Seedling X Painted Doll]; Laver, Keith G.

'LAVCHIP', Min, m, 1993; (**Blue Ice**); flowers lavender, reflex upon opening, blooms borne mostly singly, dbl., 26–40 petals, 1.5–2.75 in., moderate fragrance; some prickles; foliage small, medium green, semi-glossy; low (25–30 cms), upright, bushy growth; [June Laver X Apricot Doll]; Laver, Keith G.; Springwood Roses, 1993

'LAVCHRO', Min, dy, 1993; (**Gala Gold**); dbl., 26–40 petals, 1.5 in., borne mostly singly, no fragrance; some prickles; foliage small, medium green, matt; low (24–30 cms), upright, bushy, compact growth; [Golden Promise X (June Laver X Tut's Treasure)]; Laver, Keith G.; Springwood Roses, 1994

'LAVCLASS', Min, mp, 1999; (**Springwood Classic**™); dbl., 26–40 petals, 2 in., borne in small clusters, no fragrance; few prickles; foliage medium, medium green, semi-glossy; bushy, medium (10–12 in.) growth; [seedling X seedling]; Laver, Keith G.; Springwood Roses, 1999

'LAVCLO', Min, rb, 1994; (**Fiesta Clown**); flowers scarlet with orange-yellow reverse, bicolor, pointed, recurved, dbl., 26–40 petals, 1.5–2.75 in., intense fragrance; some prickles; foliage medium, dark green, semi-glossy; medium (24–30 cms), upright growth; [(June Laver X Seedling) X (Painted Doll X June Laver)]; Laver, Keith G.; Springwood Roses, 1994

'LAVCOAT', F, yb, 1991; (**Colour Parade**); flowers yellow turning to red, small, blooms borne in small clusters, dbl., 26–40 petals, 1.5 in., slight fragrance; few prickles; foliage medium, medium green, glossy; low (45 cms), bushy growth; [Breezy X Julie Ann]; Laver, Keith G.; Springwood Roses, 1992

'LAVCOM', Min, ob, 1993; (**Silken Laumann**); flowers luminous orange, yellow reverse, blooms borne in large cluster, dbl., 26–40 nicely recurved petals, 1.5–2.75 in., exhibition form, slight fragrance; few prickles; foliage medium, medium green, matt; medium (45 cms), bushy growth; [June Laver X Potluck Red]; Laver, Keith G.; Springwood Roses, 1994

'LAVCREAM', Min, ly, 1987; ('LAVACRE', **Potluck Cream**™); flowers light yellow center, cream edges, reverse cream with yellow, dbl., 30 petals, exhibition form, no fragrance; no fruit; prickles slender, brown; foliage medium, medium green, matt; bushy, low, very free flowering growth; [Cornsilk X Unnamed seedling]; Laver, Keith G.; Springwood Roses, 1988

'LAVCREME', Min, ly, 1988; (**Crème Glacée**); bud pointed; flowers urn-shaped, small, borne singly, dbl., 23 petals, slight fragrance; globular, orange fruit; prickles pointed and straight out, white-beige; foliage small, medium green, matt; bushy growth; [June Laver X Summer Butter]; Laver, Keith G.; Springwood Roses, 1990

'LAVDANCE', Min, pb, 1993; (**Striped Pet**); flowers variable pink stripes on white, blooms borne mostly singly, dbl., 15–25 petals, 1.5–2.75 in., moderate fragrance; few prickles; foliage medium, dark green, matt; low (25–30 cms), upright, compact growth; [June Laver X (June Laver X Ontario Celebration)]; Laver, Keith G.; Springwood Roses, 1993

'LAVDOLL', Min, ab, 1990; (**Apricot Doll**); bud ovoid; flowers apricot, yellow center, reverse lighter, aging light apricot, dbl., 30–35 petals, borne usually singly and in sprays of 1–4, moderate fragrance; foliage small, medium green, matt; spreading, low growth; [Painted Doll X Painted Doll]; Laver, Keith G., 1986; Springwood Roses, 1991

'LAVDUSK', Min, dp, 1992; (**Springwood Pink**™); flowers deep pink, dbl., 15–25 petals, 1.5 in., borne in small clusters, slight fragrance; some prickles; foliage medium, medium green, matt; suitable for pots; low (30–40 cms), spreading growth; [Maurine Neuberger X (June Laver X Ontario Celebration)]; Laver, Keith G.; Springwood Roses

'LAVEENA', HT, yb, 1969; bud long, pointed; flowers yellow tinged pinkish, large, dbl., exhibition form; foliage glossy; vigorous, upright, compact growth; [Kiss of Fire X ?]; Laveena Roses

'LAVENDALE', Min, m, 1989; bud pointed; flowers deep lavender, medium, borne singly, dbl., 40 petals, exhibition form, intense fragrance; round, greenish-brown fruit; no prickles; foliage medium, medium green, semi-glossy; upright, bushy, medium, vigorous growth; [Lavender Jade X

Angel Face]; Jolly, Marie; Rosehill Farm, 1990

'LAVENDER BIRD', HT, m, 1964; bud long, spiral, pointed; flowers lavender-pink, well formed, large, dbl., moderate, lavender fragrance; vigorous growth; Herholdt, J.A.; Herholdt's Nursery

Lavender Blue, *see* 'LEOLAVBLU'

'LAVENDER CHARM', HT, m, 1964; flowers persian lilac, dbl., 45–50 petals, 4.5–5 in., cupped, intense fragrance; foliage dark, leathery; vigorous, bushy growth; [Brownie X Sterling Silver]; Boerner; J&P

Lavender Cover, *see* POUlrust

Lavender Crystal, *see* ASAlav

Lavender Delight, *see* 'MORORCHERI'

Lavender Dream®, *see* 'INTERLAV'

Lavender Friendship, S, m, 1984; Verschuren

'LAVENDER GARNETTE', F, m, 1958; bud globular; flowers lavender, open, medium, dbl., 35–45 petals, moderate fragrance; foliage leathery; vigorous, bushy growth; [Grey Pearl seedling X Garnette]; Boerner; J&P

'LAVENDER GIRL', F, m, 1958; flowers rosy purple, reverse magenta, changing to lavender, dbl., 35–42 petals, 3.5 in., cupped, moderate; spicy fragrance; dwarf, bushy growth; [Fantastique X (Ampere X (Charles P. Kilham X Capucine Chambard))]; Meilland, F.; C-P

Lavender Jade™, *see* 'BENALAV'

'LAVENDER JEWEL', Min, m, 1978; bud pointed; flowers clear lavender-mauve, dbl., 38 petals, 1 in., exhibition form, slight fragrance; foliage dark; compact, bushy growth; [Little Chief X Angel Face]; Moore, Ralph S.; Sequoia Nursery

'LAVENDER LACE', Min, m, 1968; flowers lavender, small, dbl., exhibition form, moderate fragrance; foliage small, glossy; vigorous, bushy dwarf growth; AOE, 1975; [Ellen Poulsen X Debbie]; Moore, Ralph S.; Sequoia Nursery

'LAVENDER LACE, CLIMBING', Cl Min, m, 1971; Rumsey, R.H.

'LAVENDER LADY', F, m, 1956; flowers pastel mauve, large, borne in clusters, semi-dbl., moderate fragrance; vigorous, upright growth; [Unnamed seedling X Lavender Pinocchio]; LeGrice

Lavender Lady, HT, m, 1996; Williams, J. Benjamin

'LAVENDER LASSIE', HMsk, m, 1960; flowers lilac-pink, blooms in large clusters, dbl., 3 in., intense fragrance; very vigorous, tall growth; Kordes; Morse

'LAVENDER LOVE', F, m, 1964; flowers lavender, medium; low growth; [Fashion X Floradora]; Daugherty; Wyant

Lavender Meillandina, Min, m

Lavender Midinette, Min, m; Moore

Lavender Mist, *see* 'TINVIOLET'

'LAVENDER MIST', LCl, m, 1981; (Mystic Mauve); bud ovoid; dbl., 35 petals, borne mostly 3 per cluster, slight, tea fragrance; medium prickles, hooked downward; foliage large; vigorous growth; long arching canes; [Angel Face X Allspice]; Christensen, Jack E.; Armstrong Nursery

'LAVENDER PEARL', HT, m, 1976; flowers lavender center shading to pearl, full, dbl., 40 petals, 3.5 in., moderate, lemon fragrance; foliage small; moderate, bushy growth; [Blue Moon X Grandpa Dickson]; Shaw

Lavender Pink Parfait, Pol, m

'LAVENDER PINOCCHIO', F, m, 1948; bud ovoid, light chocolate-olive-brown; flowers pink-lavender, blooms in clusters, dbl., 28 petals, 3–3.5 in., moderate fragrance; vigorous, bushy, compact growth; [Pinocchio X Grey Pearl]; Boerner; J&P

'LAVENDER PRINCESS', F, m, 1959; bud ovoid; flowers lavender, lightly overcast purplish lilac, open, borne in large clusters, dbl., 25 petals, 3.5–4 in., moderate, fruity fragrance; foliage leathery; vigorous, upright growth; [World's Fair seedling X Lavender Pinocchio seedling]; Boerner; J&P

'LAVENDER QUEEN', HT, m, 1951; bud pointed, touched red; flowers pinkish, lavender, dbl., 20–35 petals, 5–6 in., cupped, moderate fragrance; vigorous growth; Raffel; Port Stockton Nursery

Lavender Simplex™, *see* 'MINTCO'

Lavender Spoon, *see* 'SPOLAV'

Lavender Star™, *see* 'MINAUCO'

Lavender Sunblaze®, Min, m, 1999; flowers small, lavender-mauve, exhibition, dbl., 25–30 petals, 1–1.5 in., exhibition form, borne singly and in small clusters, slight fragrance; some prickles; foliage small, deep green, glossy; bushy, compact, medium, 18 in. growth; Meilland

Lavender Sweetheart™, *see* 'WILLASH'

'LAVENDULA', F, m, 1965; flowers lavender, blooms in clusters, 4 in., intense fragrance; foliage dark; [Magenta (F) X Sterling Silver]; Kordes, R.; McGredy

Laverne's Greenhouse Climber, LCl, lp

Laverne's Little Red Creeper, Min, rb

'LAVFAL', Min, rb, 1997; (**Fall Festival**™); flowers full, medium, very dbl., 26–40 petals, borne mostly singly, no fragrance; some prickles; foliage medium, dark green, semi-glossy; upright, bushy, medium growth; [Seedling X Springwood Red Victor]; Laver, Keith G.

'LAVFAN', Min, dr, 1998; (**Fancy Potluck**); flowers dark red, full, good substance, 26–40 petals, .5–1 in., borne mostly singly, no fragrance; few prickles; foliage small, dark green, matte; low (15–20 cms), compact growth; [Seedling X Antique Gold]; Laver, Keith G.; Springwood Roses, 1998

'LAVFIRE', Min, or, 1999; (**Flame Bouquet**™); flowers brilliant orange red, very long lasting, dbl., 26–40 petals, 2.5–3 in., borne mostly singly, no fragrance; few prickles; foliage medium, medium green, semi-glossy; upright, medium (12 in.) growth; Laver, Keith G.; Springwood Roses, 1999

'LAVFLAME', Min, or, 1995; (**Flaming Potluck**); flowers flaming orange-scarlet, deeper center, blooms borne mostly singly, dbl., 26–40 petals, 1.5 in., slight fragrance; some prickles; foliage medium, dark green, semi-glossy; medium (30–40 cms), upright, bushy growth; Laver, Keith G.; Springwood Roses, 1994

'LAVFLUSH', Min, op, 1993; (**Double Date**); flowers orange pink, blooms borne mostly singly, very dbl., 1.5 in., slight fragrance; some prickles; foliage small, medium green, semi-glossy; low, upright, bushy growth; [(Breezy X June Laver selfed) X Seedling]; Laver, Keith G.; Springwood Roses

'LAVFUHR', Min, mp, 1994; (**Dainty Bouquet**); flowers clear pink, blooms borne mostly singly, very dbl., 1.5 in., slight fragrance; some prickles; foliage medium, medium green, semi-glossy; medium (20 cms), upright, spreading growth; [Seedling X Pink Bouquet]; Laver, Keith G.; Springwood Roses, 1994

'LAVFUN', Min, pb, 1993; (**Blessed Event**); flowers white suffused pink, blooms borne singly and in small clusters, dbl., 26–40 petals, 1.5–2.75 in., slight fragrance; some prickles; foliage medium, dark green, matt; low (25–30 cms), upright growth; [Enjoy X (June Laver X Party Girl)]; Laver, Keith G.; Springwood Roses, 1993

'LAVGLO', Min, my, 1985; (**Potluck Yellow**™, Spicy Minijet, Yellow Mini-Wonder); dbl., 20 petals, no fragrance; foliage small, medium green, semi-glossy; bushy growth; [Rise 'n' Shine X Lemon Delight]; Laver, Keith G., 1986

'LAVGOLD', Min, my, 1991; (**Potluck Gold**™); flowers small, dbl., 20 petals, flat, borne in sprays of 3–5, no fragrance; foliage small, medium green, semi-glossy; bushy, low, very dwarf growth; [Dorola X Julie Ann]; Laver, Keith G., 1987; Springwood Roses

'LAVHOMME', Min, w, 1986; (**Bonhomme**); [Blueblood sport]; Laver, Keith G.

'LAVICE', Min, lp, 1983; (**Ice Princess**); flowers light pink, light yellow stamens, finishing white, medium, dbl., no fragrance; foliage small, medium green, matt; [Unnamed pink seedling X Lemon Delight]; Laver, Keith G.

'LAVICTOR', Min, op, 1991; (**Royal Victoria**); flowers coral pink, white reverse, small blooms borne mostly singly, very dbl., moderate fragrance; foliage small, medium green, semi-glossy; tall, upright, bushy growth; [(Painted Doll X June Laver) X Mountie]; Laver, Keith G.; Springwood Roses

'LAVINA', HT, mp, 1962; flowers buff-pink, large, dbl., 60 petals, exhibition form, moderate fragrance; foliage dark, leathery; vigorous growth; [Queen Elizabeth X Anne Letts]; Reynolds, W.H.

Lavina, HT, lp; Strnad; (Czech Rosa Club)

Lavinia, *see* 'TANKLEWI'

'LAVINIA HARRISON', HT, mp, 1988; flowers medium, shell pink with light undertones, large, dbl., 30–35 petals, borne in sprays of 1–3, slight fragrance; prickles pyramid, red-brown; foliage red to light green, aging dark green; vigorous growth; [Duet X Unnamed seedling]; Harrison, G.

'LAVISLE', Min, m, 1995; (**Springwood® Mauvette**); flowers deep mauve, full (26–40 petals), medium blooms borne mostly singly, slight fragrance; numerous prickles; foliage medium, dark green, glossy; spreading, bushy growth; [Blue Ice X seedling]; Laver, Keith G.; Springwood Roses, 1996

'LAVJACK', Min, or, 1989; (Orange Mini-Wonder, **Potluck Orange**™); bud ovoid; flowers intense, dark orange, compact, full, small, dbl., 24 petals, borne in sprays, no fragrance; fruit not observed, fruit; prickles straight, pointed, light brown; foliage small, medium green, matt; bushy, low, symmetrical growth; PP007326; [Julie Ann X Potluck™]; Laver, Keith G., 1990

'LAVJOY', Min, pb, 1989; (**Enjoy**); bud pointed; flowers blush pink edged deeper pink, reverse white, small, borne singly, dbl., 22 petals, exhibition form, no fragrance; ovoid, orange fruit; prickles straight, very small, sparse, light brown; foliage small, medium green, disease resistant; upright, low growth; [(Moulin Rouge X Unnamed seedling) X Party Girl]; Laver, Keith G.; Springwood Roses, 1990

'LAVJUNE', Min, dy, 1987; (**June Laver**™); flowers dark yellow, aging cream, large, dbl., 20–25 petals, exhibition form, borne usually singly or in small sprays, no fragrance; rounded, light orange-red fruit; prickles small, short, green; foliage large, dark green, matt; bushy, medium, compact growth; PP006859; [Helmut Schmidt X Gold Mine]; Laver, Keith G., 1985; Springwood Roses, 1988

'LAVKIN', Min, dy, 1993; (**Yellow Bouquet**); flowers deep yellow, dbl., 15–25 petals, 1.5 in., borne in small clusters, slight fragrance; few prickles; foliage small, medium green, matt; low (15–20 cms), compact growth; [June Laver X Potluck Gold]; Laver, Keith G.; Springwood Roses, 1994

'LAVLAD', Min, dp, 1993; (**Rosy Forecast**); flowers deep pink, blooms borne in small clusters, dbl., 15–25 petals, 1.5 in., moderate fragrance; few prickles; foliage small, dark green, matt; low (15–18 cms), compact growth; [(June Laver X Ontario Celebration) X Potluck Purple]; Laver, Keith G.; Springwood Roses, 1994

'LAVLEMO', Min, my, 1989; bud ovoid; flowers lemon yellow, deeper in center, medium, borne usually singly, dbl., 28 petals, exhibition form; ovoid, green fruit; prickles beige; foliage small, medium green, matt; upright, bushy, medium growth; [Dorola X Genevieve]; Laver, Keith G.

'LAVLINC', Min, dr, 1984; (**Small Slam**); flowers small, dbl., slight fragrance; foliage small, dark, semi-glossy; bushy growth; [Nic Noc X Party Girl]; Laver, Keith G.

'LAVLINGER', Min, ly, 1991; (**Living Bouquet**, Yellow Festival); bud ovoid; flowers full, blooms borne in sprays of 3–5, very dbl., 40–50 petals, flat, no fragrance; foliage medium, light green, semi-glossy; bushy, compact, semi-upright growth; [Loving Touch X (Dorola X Genevieve)]; Laver, Keith G., 1986; Springwood Roses

'LAVLOW', Min, ob, 1991; flowers orange yellow, small blooms borne in small clusters, dbl., no fragrance; few prickles; foliage small, medium green, semi-glossy; low, compact growth; [(Painted Doll X June Laver) X Potluck Yellow™]; Laver, Keith G.; Springwood Roses

'LAVLUV', Min, dy, 1992; (**Golden Promise**); dbl., 26–40 petals, 1.5 in., borne mostly singly, no fragrance; few prickles; foliage small, dark green, glossy; low (30 cms), bushy growth; [June Laver X Potluck Gold]; Laver, Keith G.; Springwood Roses, 1993

'LAVMILNE', Min, lp, 1987; (**Alexander Milne**); flowers light pink, reverse cream, small, borne singly, dbl., 35 petals, exhibition form, no fragrance; rounded, orange fruit; straight, light brown prickles; foliage small, medium green, semi-glossy; bushy, low growth; [Corn Silk X Ice Princess]; Laver, Keith G.; Springwood Roses

'LAVMIN', Min, dr, 1988; (**Potluck Red**™); bud pointed; flowers deep red, compact, full, small, borne singly, dbl., 35 petals, no fragrance; globular, orange-red fruit; no prickles; foliage small, dark green, young edged in red, matt; upright, low growth; [Breezy X June Laver]; Laver, Keith G.; Springwood Roses, 1989

'LAVMING', Min, rb, 1999; (**Striped Festival**™); flowers very rich stripe, deeper than most, dbl., 17–25 petals, 2.5 in., exhibition form, borne mostly singly, slight fragrance; prickles moderate; foliage medium, medium green, semi-glossy; easy-growing, upright, medium (12–15 in.) growth; Laver, Keith G.; Springwood Roses, 1999

'LAVMOTH', Min, pb, 1988; (**Fair Genie**); bud pointed; flowers light orange in center with pink outer petals, reverse pink, dbl., 33 petals, exhibition form, moderate fragrance; ovoid, orange fruit; prickles slender, straight, almost white, translucent; foliage small, medium green, glossy; upright, bushy, low, strong growth; [Breezy X June Laver]; Laver, Keith G.; Springwood Roses, 1990

'LAVMOUNT', Min, or, 1983; (**Ontario Celebration**); flowers small, dbl., 35 petals, moderate fragrance; foliage small, deep reddish green, semi-glossy; spreading, compact growth; [Nic-Noc X (Party Girl X Queen of the Dwarfs)]; Laver, Keith G., 1984

'LAVNEW', Min, op, 1995; (**Springwood Beauty**™); flowers orange pink, blooms borne mostly single, dbl., 26–40 petals, no fragrance; few prickles; foliage small, medium green, semi-glossy; bushy, low growth; [Seedling X Apricot Doll]; Laver, Keith G.; Springwood Roses, 1995

'LAVNOVA', Min, m, 1995; (**Potluck Blue**™); flowers lavender, reverse white, double, blooms borne mostly, singly, 15–25 petals, no fragrance; some prickles; foliage medium, medium green, semi-glossy; upright, bushy, medium growth; [Blue Ice X Seedling]; Laver, Keith G.; Springwood Roses, 1995

'LAVONT', Min, op, 1995; (**Touche**); flowers orange-pink, very full (41+ petals), medium blooms borne in small clusters; very fragrant; numerous prickles; foliage medium, medium green, semi-glossy; miniature moss; bushy, medium growth; [Seedling X Seedling]; Laver, Keith G.; Springwood Roses, 1996

'LAVPAINT', Min, ob, 1985; (**Painted Doll**™); flowers orange, reverse yellow, small, dbl., 35 petals, slight fragrance; foliage small, light green, matt; bushy growth; [Party Girl X Dwarfking '78]; Laver, Keith G.

'LAVPERT', F, w, 1991; (**Snow Princess**); dbl., 26–40 petals, 1.5–2.75 in., borne in large clusters, no fragrance; some prickles; foliage large, dark green, glossy; bushy, low (45 cms), compact growth; [Regensberg X June Laver]; Laver, Keith G.; Springwood Roses, 1992

'LAVPET', Min, dy, 1992; ('LAVROY', **Perfect Potluck**™); flowers golden yellow, blooms borne in small clusters, dbl., 15–25 petals, 1.5 in., slight fragrance; few prickles; foliage medium, dark green, glossy; low (30 cms), compact growth; [Showbound X (Party Girl X June Laver)]; Laver, Keith G.; Springwood Roses

'LAVPLAT', Min, m, 1988; (**Platinum Lady**); bud pointed; flowers light lavender, reverse white, borne singly, very slow to open, dbl., 21 petals, exhibition form, no fragrance; ovoid, orange fruit; prickles slender "v" shape, red; foliage small, dark green, young edged in red, semi-glossy; upright, bushy, medium growth; [Unnamed seedling (lavender) X Lavender Jade]; Laver, Keith G.; Springwood Roses, 1989

'LAVPLEASE', Min, mp, 1992; (**Pleasantly Pink**); flowers clear pink, blooms borne in small clusters, very dbl., 1.5–2.75 in., moderate fragrance; some prickles; foliage medium, medium green, matt; medium (30–35 cms), bushy growth; [(June Laver X Rosamini Red) X (Dwarfking X Julie Ann)]; Laver, Keith G.; Springwood Roses

'LAVPOP', Min, w, 1993; (**Party Popcorn**); dbl., 26–40 petals, 1.5 in., borne mostly singly, slight fragrance; few prickles; foliage small, medium green, matt; medium (20–25 cms), bushy growth; [(June Laver X Painted Doll) X Popcorn]; Laver, Keith G.; Springwood Roses, 1993

'LAVPOT', Min, dr, 1984; (**Potluck**); flowers small, dbl., 20 petals, no fragrance; foliage small, medium green, glossy; compact, bushy, tiny growth; [Dwarfking '78 X Little Liza]; Laver, Keith G.

'LAVPUP', Min, dr, 1992; (**Potluck Purple**™); flowers fuchsia, blooms borne in small clusters, dbl., 15–25 petals, 1.5 in., no fragrance; few prickles; foliage small, medium green, matt; micro-mini; low (25–30 cms), compact growth; [(June Laver X Painted Doll) X Springwood Purple]; Laver, Keith G.; Springwood Roses

'LAVPURE', Min, w, 1995; (**Potluck® Frosty**); flowers white, full (26–40 petals), medium (4–7 cms) blooms borne mostly singly; no fragrance, some prickles; foliage small, medium green, semi-glossy; low (20–30 cms), bushy, spreading growth; [Seedling X Potluck Yellow]; Laver, Keith G.; Springwood Roses, 1996

'LAVPURR', Min, dp, 1990; (**Springwood Purple**™); flowers fuchsia, loose, sweetheart, medium to large, dbl., 25–30 petals, borne in sprays of 3, no fragrance; foliage large, medium green, semi-glossy; bushy, spreading, medium, very compact growth; [June Laver X (Small Slam X Mountie)]; Laver, Keith G., 1988; Springwood Roses, 1991

'LAVQUEST', Min, lp, 1990; (**Pink Bouquet**, Pink Festival); bud pointed; flowers white to luminous blush pink, white reverse, aging white, very dbl., 50–60 petals, flat, no fragrance; foliage small, dark green, matt; bushy, spreading, low growth; [Loving Touch X Ontario Celebration]; Laver, Keith G., 1987; Springwood Roses, 1991

'LAVRANGE', Min, ob, 1992; (**By Design** 'LANSIGN'); flowers orange, blooms borne in small clusters, dbl., 26–40 petals, 1.5–2.75 in., exhibition form, slight fragrance; some prickles; foliage medium, medium green, semi-glossy; medium (40–50 cms), bushygrowth; [(Breezy X Julie Ann) X (June Laver X Painted Doll)]; Laver, Keith G.; Springwood Roses

'LAVRED', Min, dr, 1988; (**Springwood Red**™); flowers deep red, aging slightly lighter, urn-shaped, small, borne u, dbl., 40 petals, no fragrance; ovoid, orange fruit; prickles slightly recurved, light yellow-green; foliage small, medium green, matt; bushy, low growth; [Small Slam X Mountie]; Laver, Keith G.; Springwood Roses, 1990

'LAVRIDE', Min, dr, 1991; (Red Rider, **Red Rover**); bud pointed; flowers loose blooms borne usually singly, dbl., exhibition form, no fragrance; foliage small to medium, dark green, semi-glossy to glossy; bushy, low growth; [Mountie sport]; Laver, Keith G., 1988; Springwood Roses, 1990

'LAVROSY', Min, mp, 1995; (**Rosy Potluck**); dbl., 26–40 petals, 1.5 in., borne in small clusters, moderate fragrance; few prickles; foliage small, dark green, semi-glossy; low (20–30 cms), compact growth; Laver, Keith G.; Springwood Roses, 1995

'LAVROY', Min, dy, 1992; ('LAVPET' **Perfect Potluck**™); flowers golden yellow, blooms borne in small clusters, dbl., 15–25 petals, 1.5 in., slight fragrance; few prickles; foliage medium, dark green, glossy; low (30 cms), compact growth; [Showbound X (Party Girl X June Laver)]; Laver, Keith G.; Springwood Roses

'LAVSANS', Min, dp, 1986; (**Sans Souci**); flowers deep fuchsia pink, fading slightly, medium, borne usually singly, dbl., exhibition form, slight fragrance; oblong, narrow, red fruit; many, very fine, light brown prickles; foliage medium, light green, matt; bushy, medium growth; [Rise 'n' Shine X Ontario Celebration]; Laver, Keith G.; Springwood Roses

'LAVSAPH', Min, dr, 1995; (**Black Sapphire**); dbl., 26–40 petals, borne mostly singly; few prickles; foliage medium, medium green, glossy; patio, upright, medium growth; [Royal Victoria X Seedling]; Laver, Keith G.; Springwood Roses, 1995

'LAVSASK', Min, rb, 1995; (**Maverick**); flowers striped red and white, double, blooms borne mostly singly, 15–25 petals, slight fragrance; some prickles; foliage medium, medium green, dull; upright, bushy, medium growth; [Striped Pet X Apricot Doll]; Laver, Keith G.; Springwood Roses, 1995

'LAVSAT', Min, mp, 1997; (**Springwood Pink Satin**®); flowers full, small, very dbl., 26–40 petals, borne in small clusters, intense fragrance; few prickles; foliage medium, medium green, semi-glossy; compact, bushy, low growth; [Pleasantly Pink X Springwood Roses]; Laver, Keith G.

'LAVSCENT', Min, or, 1987; (**Springwood Coral**™); flowers coral-pink, turning slightly white at stem, loose, large, bo, dbl., 23 petals, intense fragrance; round, light orange fruit; prickles curved, red; foliage large, light green, matt; bushy, medium growth; [Helmut Schimdt X PotluckTM]; Laver, Keith G., 1986; Springwood Roses, 1990

'LAVSER', Min, pb, 1998; (**Serene Bouquet**); flowers pink with white reverse, 26–40 petals, 1.5–2 in., borne in sprays of 5, slight fragrance; some prickles; foliage medium, medium green, matt; upright, vigorous growth; [Seedling X Antique Gold]; Laver, Keith G.; Springwood Roses, 1998

'LAVSHO', Min, lp, 1984; flowers small blooms in large clusters, dbl., 20 petals, slight fragrance; foliage small, light green, glossy; spreading growth; [Mighty Mouse X Fairy Rose]; Laver, Keith G.

'LAVSHRIMP', Min, mp, 1987; (**Wilfrid H. Perron**); flowers clear shrimp-pink, aging paler, rosea, large, borne singly, very dbl., 70–80 petals, no fragrance; round, very rare, green-red fruit; prickles straight out, very slender, greenish-red; foliage very small, medium green, matt; bushy, low growth; [(Dwarfking '78 X Baby Katie) X Painted Doll]; Laver, Keith G., 1988

'LAVSIGN', Min, ob, 1992; (**By Design**, 'LAVRANGE'); flowers orange, blooms borne in small clusters, dbl., 26–40 petals, 1.5–2.75 in., exhibition form, slight fragrance; some prickles; foliage medium, medium green, semi-glossy; medium (40–50 cms), bushygrowth; [(Breezy X Julie Ann) X (June Laver X Painted Doll)]; Laver, Keith G.; Springwood Roses

'LAVSNO', Min, w, 1984; dbl., no fragrance; foliage small, medium green, matt; upright, bushy growth; [Ice Princess X Sue Lawley]; Laver, Keith G.

'LAVSNOW', Min, w, 1991; (**Springwood White**™, White Festival); very dbl., 1.5–2.75 in., borne mostly singly, no fragrance; few prickles; foliage small, medium green, matt; low (22 cms), bushy growth; [Loving Touch X June Laver]; Laver, Keith G.; Springwood Roses

'LAVSON', Min, dr, 1997; (**Potluck Crimson**®); flowers very full, medium, very dbl., 41 petals, borne mostly singly and in small clusters, no fragrance; few prickles; foliage medium, medium green, semi-glossy; compact, low growth; [June Laver Sport X Springwood Red Victor]; Laver, Keith G.

'LAVSTAR', Min, m, 1989; (**Sweet Butterfly**); bud rounded; flowers mauve pink, pointed, open, loose, medium, star-like form, bo, semi-dbl., 12–15 petals, flat, intense fragrance; globular, yellow fruit; prickles short, pointed, beige; foliage medium, medium green, matt; bushy, low growth; [(Dwarfking X Baby Katie) X (Small Slam X Mountie)]; Laver, Keith G.; Springwood Roses

'LAVSUM', Min, rb, 1997; (**Summer Festival**™); flowers very full, medium, very dbl., 41 petals, borne singly and in small clusters, moderate fragrance; some prickles; foliage medium, dark green, semi-glossy; upright, compact, bushy, medium growth; [Seedling X Painted Doll]; Laver, Keith G.

'LAVTIKI', Min, my, 1999; (**Lemon Bouquet**™); flowers medium yellow, reverse light yellow, dbl., 26–40 petals, 2–2.5 in., exhibition form, borne mostly singly, slight fragrance; no prickles; foliage medium, dark green, glossy; upright, medium (12–15 in.) growth; Laver, Keith G.; Springwood Roses, 1999

'LAVTIPAWS', Min, ob, 1994; (**Tiger Paws**™); flowers orange with yellow reverse, blooms borne mostly singly, dbl., 26–40 petals, 1.5–2.75 in., no fragrance; foliage medium, medium green, matt; medium (20 cms), compact growth; [Painted Doll X Seedling]; Laver, Keith G.; Springwood Roses, 1994

'LAVTIQUE', Min, yb, 1995; (**Antique Gold**); flowers deep chrome yellow tipped with red, dbl., (15–25 petals), small blooms borne mostly singly; slight fragrance; some prickles; foliage medium, light green, semi-glossy; spreading, medium growth; [Seedling X Seedling]; Laver, Keith G.; Springwood Roses, 1996

'LAVTREK', Min, dy, 1989; (**King Tut**); bud pointed; flowers rich, deep yellow, reverse medium yellow, small, borne singly, dbl., 45 petals, exhibition form, moderate fragrance; ovoid, orange-red fruit; prickles curved down, light brown; foliage small, dark green, disease resistant; upright, low, compact growth; [June Laver X Genevieve]; Laver, Keith G.; Springwood Roses

'LAVTRICE', Min, or, 1994; (**Orange Zest**); very dbl., 1.5 in., borne mostly singly, slight fragrance; few prickles; foliage small, dark green, semi-glossy; upright, bushy (25 cms) growth; [Seedling X Painted Doll]; Laver, Keith G.; Springwood Roses, 1994

'LAVTRU', Min, dy, 1997; (**True Gold**™); flowers very full, medium, very dbl., 41 petals, borne mostly singly, no fragrance; few prickles; foliage medium, medium green, dull; upright, bushy, medium growth; [Antique Gold X Yellow Bouquet]; Laver, Keith G.

'LAVTUCH', Min, ob, 1998; (**Painter's Touch**); flowers luminous orange blend, very dbl., 26–40 petals, 1.5–3 in., borne mostly singly, moderate fragrance; prickles moderate; foliage large, medium green, matt; bushy, medium tall growth; [Seedling X Antique Gold]; Laver, Keith G.; Springwood Roses, 1998

'LAVTYNINE', Min, my, 1989; (**Springwood Gold**™); bud pointed; flowers deep, buttery yellow, reverse lighter, medium, borne usually, dbl., 20 petals, exhibition form, slight fragrance; globular, light orange fruit; prickles very narrow, straight, green; foliage small, medium green, semi-glossy; upright, bushy, medium growth; [Rise 'n' Shine X June Laver]; Laver, Keith G.; Springwood Roses

'LAVWYTE', Min, w, 1985; (**Potluck White**™); dbl., 35 petals, no fragrance; foliage small, light green, glossy; spreading growth; [Baby Katie X Mountie]; Laver, Keith G., 1987

'LAVYIP', Min, w, 1997; (**Ivory Festival**™); flowers full, large, very dbl., 26–40 petals, borne singly and in small clusters, intense fragrance; foliage medium, dark green, glossy; compact, bushy, medium growth; [Seedling X Living Bouquet]; Laver, Keith G.

Lawinia®, *see* 'TANKLEWI'

'LAWRENCE JOHNSTON', LCl, my, 1923; (Hidcote Yellow); flowers large, semi-dbl., blooms in clusters, repeat bloom, moderate fragrance; very vigorous, climbing (to 30 ft) growth; [Mme Eugene Verdier X R. foetida persiana]; Ducher

'LAWRENCE JOHNSTON', HT, yb, 1946; bud pointed; flowers yellow edged red and salmon, large, semi-dbl., moderate fragrance; foliage light green; vigorous, upright, bushy growth; GM, Rome, 1954; [Souv. de Denier van der Gon X Brazier]; San Remo Exp. Sta.

'LAWRENCE OF ARABIA', HT, yb, 1938; flowers indian yellow, flushed coppery rose, large, dbl.; vigorous, bushy growth; Dickson, A.

Laxton's Pink Delight, *see* **'PINK DELIGHT'**

'LAXTON'S STANDARD', HT, mp, 1926; flowers clear cerise-pink; Laxton Bros.

Lazy Daze, *see* 'TALDAZ'

'LE CHAMOIS', F, 1954; Combe, M.; (Cavriglia)

'LE CID', HRg, mr, 1908; flowers dazzling crimson, large; vigorous growth

Le Cid, *see* DORpark

'LE DROIT HUMAIN', HMult, pb, 1907; flowers pink with darker carmine-pink, medium, very dbl.; Vilin; (Sangerhausen)

Le Grande Capitain, B, dr; very dbl.

'LE HAVRE', HP, mr, 1870; flowers vermilion, imbricated, large; vigorous growth; Eudes

Le Jacobin, *see* **'MARCEL BOURGOUIN'**

'LE LOBÈRDE', mp, 1800; flowers medium, very dbl.; (Sangerhausen)

'LE LOIRET', Pol, mp, 1920; flowers very brilliant pink to salmon-rose, borne in clusters of 10–15; Turbat

'LE MEXIQUE', HWich, lp, 1912; flowers pale silvery rose, medium, borne in clusters, semi-dbl., slight fragrance; vigorous, climbing growth; [Dorothy Perkins X Marie Pavie]; Schwartz, A.

'LE NANKIN', T, yb, 1871; flowers yellow shaded coppery, well formed, moderate fragrance; Ducher

'LE PACTOLE', T, ly; flowers pale yellow, large, dbl.; [Lamarque X Yellow Tea]; Miellez, before 1841

Le Petit Orange, Pol, ob; Adam

'LE PHOENIX', HGal, mp, 1843; flowers medium carmine-pink, large, very dbl., moderate fragrance; Vibert; (Sangerhausen)

Le Pink, *see* 'JACLIP'

Le Poilu, Cl M, lp, 1900; (**'CUMBERLAND BELLE'**); flowers silvery pink, well mossed, small, very dbl., intense fragrance; [Princesse Adélaide sport]; Dreer

'LE PONCEAU', Pol, dr, 1912; flowers deep garnet-red, small, semi-dbl.; [Gruss an Teplitz X Mme Norbert Levavasseur]; Hémeray-Aubert

'LE PROGRÈS', HT, my, 1903; very vigorous growth; Pernet-Ducher

'LE RÊVE', LCl, ly, 1923; bud pointed; flowers pale yellow, large, semi-dbl., moderate fragrance; foliage rich green, glossy; vigorous, climbing growth; [Mme Eugene Verdier X R. foetida persiana]; Pernet-Ducher

'LE RIGIDE', HMult, mp, 1920; flowers neyron pink, borne in clusters of 25–30, semi-dbl.; vigorous growth; Turbat

'LE RIRE NIAIS', C, mp, 1810; Dupony, Andre

'LE ROITELET', B, mp, 1869; flowers small, dbl.; Soupert; (Sangerhausen)

Le Rosier du Jardin ou Cecile, S, op

Le Rosier Eveque, HGal, m; flowers dark violet-red, medium, dbl.; (Sangerhausen)

Le Rosier Pompon Blanc, *see* **'ROSE DE MEAUX WHITE'**

Le Rouge et Le Noir, *see* DELcart

'LE VÉSUVE', Ch, pb, 1825; flowers carmine shading to pink, large, very dbl.; vigorous growth; (14); Laffay, M.

Lea, F, lp; Urban, J., 1988; (Czech Rosa Club)

'LÉA', HGal, lp; (Lee); flowers blush, shaded flesh, expanded, large, very dbl., abundant early bloom; Vibert, before 1828

'LEA ANN', Gr, w, 1976; bud long, pointed; flowers clear white, dbl., 25–30 petals, 4 in., exhibition form, moderate fragrance; foliage soft; vigorous growth; [Queen Elizabeth X Ivory Fashion]; Patterson; Patterson Roses

'LEA RUBRA', C, mr, 1856

'LEADER', HT, mr, 1924; [Premier sport]; Hill, E.G., Co.

Leading Lady, *see* 'JACLOPI'

'LEADING LADY', HT, lp, 1935; flowers flesh-pink, flushed peach-blossom, large, dbl., exhibition form, intense fragrance; foliage deep green, leathery; vigorous, bushy growth; RULED EXTINCT 1/85; GM, NRS, 1934; Dickson, A.

'LEAGUE OF NATIONS', HT, m, 1929; flowers reddish lilac, shaded salmon-pink, semi-dbl., moderate fragrance; [Frau Felix Tonnar X Solliden]; Leenders, M.

Leah, Pol, mr, 1970; Holzman

'LEALAND JEWEL', S, dp, 1963; flowers deep pink, fading quickly, semi-dbl.; (14); [Athabasca seedling]; Erskine

Leana, *see* ROElanda

Leander®, *see* 'AUSlea'

LeAnn Rimes, *see* 'HARzippee'

Leaping Salmon, *see* 'PEAmight'

LEBbet, F, dp; (**Favorite**); (LEBbet)

Lebensfreude, HT, or, 1979; flowers large, dbl.; GPG Bad Langensalza; (Sangerhausen)

'LECbelpa', HT, dr, 1995; (**Belvedere Park**); flowers dark red, edging to black, blooms borne mostly singly, dbl., 26–40 petals, 3.5 in., slight fragrance; few prickles; foliage medium, dark green, glossy; medium (4 ft), spreading, bushy growth; [Swarthmore X First Prize]; LeCroy, Jack; K&M Nursery, 1995

'LECblu', F, or, 1996; (**Blushing June**); flowers orange, vivid stamens, large blooms borne mostly singly, semi-dbl., 6–14 petals, slight; some prickles; foliage medium, medium green, semi-glossy; upright, medium growth; [Altissimo X Frances Ashton]; LeCroy, Jack

'Lecocq-Dumesnil', HP, mr, 1882; flowers carmine-red, very large, very dbl., moderate fragrance; Verdier, E.; (Sangerhausen)

'Leda', S, lp, 1960; flowers pale pink, shallow, borne in clusters of 4–5, cupped, intense fragrance; large, apple-like, red fruit; foliage dark; bushy (5 ft) growth; Skinner

'Léda', D, w; (Painted Damask); flowers white to blush, edged crimson, medium, dbl., sometimes recurrent bloom, moderate fragrance; foliage dark; compact (3 ft) growth; hardy.

Lee, *see* **'Léa'**

Leea Rubra, C, lp; flowers medium, dbl., moderate fragrance; (Sangerhausen)

'LEEcal', Min, my, 1992; (**Colleen Little**); dbl., 26–40 petals, 1.5 in., exhibition form, borne usually singly or in sprays of 5–7, slight fragrance; some prickles; foliage small, dark green, semi-glossy, disease resistant; medium (26–30 cms), upright, bushy, growth; [Luis Desamero X Cheyenne]; Little, Lee W.; Oregon Miniature Roses, 1992

Leeds Castle, *see* 'TANrupeza'

'LEEfir', LCl, rb, 1992; (**Flashfire**); flowers brilliant coppery red-orange with yellow eye, blooms borne in clusters, single, 5 petals, 3–3.5 in., slight fragrance; some prickles; foliage large, dark green, glossy, disease resistant; tall (210 cms), upright, spreading growth; [Altissimo X Playboy]; Little, Lee W.; Heirloom Old Garden Roses, 1992

'Leenders' Bergfeuer', F, mr, 1959; bud long; flowers scarlet, large, semi-dbl., moderate fragrance; foliage dark, glossy; [Independence X Fashion]; Leenders, J.

'Leenders' Flamingo', F, ob, 1960; bud ovoid; flowers bright coral-peach, open, large, borne in clusters, semi-dbl., moderate fragrance; vigorous, bushy growth; [Cocorico X Ma Perkins]; Leenders, J.

'Leenders' Pink', F, mp, 1959; flowers bright pink, large, borne in heavy clusters, dbl., slight fragrance; foliage dark, leathery, glossy; [Goldilocks X Mrs Inge Poulsen]; Leenders, J.

'LEEpel', Cl HT, mp, 1992; (**Royal Porcelain**); dbl., 26–40 petals, 3–3.5 in., exhibition form, borne in small clusters on strong upright laterals, slight fragrance; some prickles; foliage large, disease resistant; tall (260 cms), upright, spreading growth; [Pele X Altissimo]; Little, Lee W.; Heirloom Old Garden Roses, 1992

Leerder's Harlequin, *see* **'The Jester'**

Leersum 700, *see* 'INTerleer'

Lee's Crimson Perpetual, *see* **'Rose du Roi'**

Lee's Duchess, *see* **'La Belle Distinguée'**

'LEEsho', LCl, pb, 1992; (**Rainbow Shower**); flowers shrimp pink suffused with yellow and darker pink edge, aging, single, 5 petals, 3–3.5 in., slight fragrance; some prickles; foliage medium, medium green, glossy, disease resistant; tall (210 cms), upright, spreading growth; [Altissimo X Playboy]; Little, Lee W.; Heirloom Old Garden Roses, 1992

'Legacy', HT, pb, 1963; flowers rose-pink streaked white, large, dbl., intense fragrance; foliage leathery; moderate growth; [Mrs Bryce Allan sport]; Hamilton; Junior Legacy Club

Legacy™, *see* 'JACsling'

'Legacy Jubilee', HT, yb, 1974; bud long, pointed; flowers yellow, edged red, large, dbl., intense fragrance; foliage large, glossy; vigorous, upright growth; [Great Venture X Fred Streeter]; Dawson, George; Neil

LEGcream, F, w, 1998; (**Jilly Cooper**); LeGrice

Legend, *see* 'JACtop'

'Legendary', HT, lp, 1962; bud long, pointed; flowers soft pink, dbl., 55 petals, 5.5 in., exhibition form, moderate fragrance; foliage leathery; vigorous, upright growth; Von Abrams; Peterson & Dering

'LEGga', HT, my, 1979; (**Dorothe**®, Midas); bud pointed; dbl., 48 petals, borne singly, slight fragrance; large, curved, light brown prickles; foliage glossy, medium green; vigorous, upright growth; [Irish Gold X Dr A.J. Verhage]; LeGrice, E.B., 1980

'LEGgab', S, w, 1981; (**Pearl Drift**®); flowers white flushed pink, blooms borne several together, semi-dbl., 18 petals, slight fragrance; small, light brown prickles; foliage glossy, reddish to dark green; vigorous, compact, spreading growth; [Mermaid X New Dawn]; LeGrice, 1980

'LEGglow', HT, ob, 1981; (**Can Can**, Can-Can); flowers orange, large, dbl., 24 petals, intense fragrance; foliage large, dark, semi-glossy; bushy growth; [Just Joey X (Superior X Mischief)]; LeGrice, E.B., 1982

'LEGgrey', F, m, 1989; (**Nimbus**); bud pointed; flowers lilac-gray, medium, borne in sprays, very dbl., cupped, slight fragrance; foliage medium, medium green, semi-glossy; bushy, medium growth; [Grey Dawn X Unnamed seedling]; LeGrice, E.B. Roses

'Legion', HT, dp, 1920; (American Legion); flowers deep cerise-red, dbl., moderate fragrance; [Milady seedling X Hadley]; Towill

Légion d'Honneur®, *see* 'DELsamar'

'LEGneed', HT, or, 1988; (**Kathleen Jermyn**); flowers urn-shaped, loose, medium, borne usually singly, dbl., slight fragrance; foliage large, dark green, semi-glossy; bushy, medium growth; [Royal Dane X Alexander]; LeGrice, E.B.; E.B. LeGrice Roses, Ltd.

'LEGnews', F, m, 1968; (**News**®); flowers red-purple, blooms in trusses, semi-dbl., moderate fragrance; foliage olive-green, glossy; GM, RNRS, 1970; [Lilac Charm X Tuscany Superb]; LeGrice

LEGram, LCl, dp, 1990; (**Pink Fountain**); LeGrice

'LEGsweet', F, pb, 1983; (**Deb's Delight**); flowers silvery salmon-pink blend, patio, medium, dbl., 35 petals, moderate fragrance; foliage medium, medium green, semi-glossy; low, bushy growth; [Tip Top X Seedling]; LeGrice, E.B., 1982

Leicester Abbey, *see* 'NOSab'

'Leigh Ann'™, Min, pb, 1986; flowers pink, reverse cream, yellow center, fading to light pink, medium, dbl., 34 petals, exhibition form, slight fragrance; round, red-orange fruit; bayonet-shaped, pinkish brown prickles; foliage medium, medium green, semi-glossy; upright, medium, vigorous growth; [Poker Chip X Rise 'n' Shine]; Jolly, Marie; Rosehill Farm, 1987

Leigh-Lo, *see* 'HARpurl'

Leila, *see* 'BERlei'

'Leila Francis', HT, pb, 1948; bud ovoid; flowers two-toned pink, large, dbl., exhibition form, intense fragrance; foliage leathery; vigorous, bushy growth; [Earl Haig X Crimson Glory]; Francis; F. Mason

Leila Verde, F, m

'Leipzig', HMsk, or, 1939; flowers orange-scarlet, open, borne in clusters, semi-dbl., recurrent bloom; foliage leathery, glossy, wrinkled; long, strong stems; vigorous, bushy growth; [Eva X Mermaid]; Kordes

'Leitrim Glory', HT, dy, 1976; flowers deep yellow, full, dbl., 28–30 petals, 5 in., moderate fragrance; foliage bronze, matt green; [Whisky Mac sport]; Hughes

'LELE', HT, dr, 1939; bud pointed; flowers dark red, veined, very large, dbl., 30 petals, moderate fragrance; foliage bright green; strong stems; very vigorous growth; [Marquise d'Andigne X Pres. Herbert Hoover]; San Remo Exp. Sta.

'LELIA LAIRD', Min, ob, 1979; bud long, pointed; flowers orange-red with yellow eye and reverse, blooms borne 1–4 per cluster, dbl., 38 petals, exhibition form, moderate, tea fragrance; long, thin, red prickles; foliage medium green with red edging; upright growth; [Contempo X Sheri Anne]; Bennett, Cecilia 'Dee'; Tiny Petals Nursery, 1980

'LEMANIA', HT, dr, 1937; bud almost black; flowers velvety blackish red, well formed, very large, dbl., intense fragrance; vigorous growth; Heizmann, E.; A. Meilland

'LEMBUT', Min, w, 1980; (**Buttons**); bud globular; dbl., 35 petals, borne singly, no fragrance; no prickles; foliage very tiny, smooth; compact growth; [Seedling X Seedling]; Lemrow, Dr Maynard W.

'LEMCAR', Min, m, 1985; (**Carl's Rose**); flowers small blooms in sprays, dbl., 35 petals, exhibition form, no fragrance; foliage small, medium green, semi-glossy; low, upright growth; Lemrow, Dr Maynard W.

'LEMFALL', Cl Min, w, 1988; (**Snowfall**™); flowers decorative, medium, dbl., cupped, borne usually singly or in sprays of 2–5, no fragrance; no fruit; prickles short, pointed, small; foliage medium, medium green, semi-glossy; spreading, tall, profuse, hardy growth; PP006706; [Jeanne Lajoie X Unnamed seedling]; Lemrow, Dr Maynard W.; Nor'East Min. Roses, 1988

'LEMLIL', Min, pb, 1984; (**Sealily**); flowers white, petals edged pink, small, dbl., 35 petals, no fragrance; foliage medium, dark, semi-glossy; [Max Colwell X ?]; Lemrow, Dr Maynard W.

'LEMMAR', Min, pb, 1981; (**Maryellen**); flowers deep pink, yellow center, blooms borne 3 or more per cluster, single, 5 petals, no fragrance; triangle-shaped prickles; foliage small, deep green; upright growth; [Darling Flame X Unnamed seedling]; Lemrow, Dr Maynard W.

'LEMON BEAUTY', HT, w, 1932; flowers creamy white, quickly fading to paper-white, base lemon-yellow, moderate, fruity fragrance; foliage light; Cant, B. R.

Lemon Blush, *see* SIElemon

Lemon Bouquet™, *see* 'LAVTIKI'

'LEMON CHIFFON', HT, my, 1954; bud long, pointed; flowers lemon-yellow, to open, dbl., 40 petals, 3–4 in., exhibition form, intense, spicy fragrance; compact, bushy growth; RULED EXTINCT 9/86; [Soeur Thérèse X Golden Dawn]; Swim, H.C.; Arp Nursery Co.

'LEMON DELIGHT', Min, my, 1978; bud mossy, long, pointed; flowers mini-moss, small, semi-dbl., 10 petals, 1.5 in., slight fragrance; bushy, upright growth; [Fairy Moss X Goldmoss]; Moore, Ralph S.; Sequoia Nursery

Lemon Dream, *see* KORamvis

'LEMON DROP', Min, ly, 1954; dbl., 0.75 in.; very prickly; foliage very small; dwarf 6 in. growth; [(R. wichurana X Floradora) X Zee]; Moore, Ralph S.; Sequoia Nursery

Lemon Drop, *see* 'WEKYEGI'

'LEMON ELEGANCE', HT, my, 1960; bud long, pointed; flowers lemon-yellow, well-shaped, dbl., 38 petals, 4.5–5 in., moderate fragrance; foliage leathery; vigorous, tall growth; Jones; Hennessey

Lemon Fluff, *see* 'CURLEM'

Lemon Gems™, *see* 'JACMIRYL'

'LEMON GLOW', HT, my, 1964; bud long, pointed; flowers lemon-yellow, dbl., 55 petals, 6–7 in., exhibition form, moderate fragrance; foliage soft; vigorous, upright growth; [Sunlight X Golden Masterpiece]; Schwartz, Ernest W.; Wyant

Lemon Honey, *see* DICkindlel

'LEMON ICE', HT, my, 1960; bud pointed; flowers lemon-yellow, dbl., 65 petals, 4 in., slight fragrance; foliage leathery, dark; vigorous, bushy, compact growth; [Leonard Barron sport]; Kern Rose Nursery

Lemon Ice, Min, lp; [Oz Gold X Lemon Delight]; Hannemann, F., 1990; The Rose Paradise; (Weatherly, L.)

Lemon Light, LCl, ly

Lemon Meringue, *see* 'JUSLEMON'

'LEMON MIST', Min, w, 1991; bud ovoid; flowers white with yellow center, aging white, medium, very dbl., 50 petals, exhibition form, slight, fruity fragrance; foliage medium, medium green, matt; upright growth; [Rise 'n' Shine X Unnamed seedling]; Gruenbauer, Richard, 1984; Flowers 'n' Friends Miniature Roses, 1993

'LEMON OPHELIA', HT, my, 1922; flowers lemon-yellow, dbl., moderate fragrance; [Ophelia sport]; Leenders, M.

Lemon Pearl, *see* 'MICPEARL'

Lemon Pillar, *see* **'PAUL'S LEMON PILLAR'**

'LEMON QUEEN', Cl HT, w, 1912; flowers creamy white, large, dbl., moderate fragrance; Hobbies; (Sangerhausen)

Lemon Rosamini, Min, ly

'LEMON SHERBET', HT, ly, 1973; bud ovoid; flowers white, center light yellow, dbl., 35 petals, 4 in., exhibition form, slight fragrance; foliage large, leathery; upright growth; [Florence sport]; Kern Rose Nursery

'LEMON SPICE', HT, ly, 1966; bud long, pointed; flowers large, dbl., exhibition form, intense fragrance; foliage dark, leathery; vigorous, spreading growth; [Helen Traubel X Seedling]; Armstrong, D.L.; Armstrong Nursery

'LEMON SURPRISE', F, my, 1978; flowers lemon, semi-dbl., 12 petals, moderate, spicy fragrance; foliage glossy; low, upright growth; [Allgold X Elizabeth of Glamis]; Slack

Lemon Swirl, *see* 'RENSWIRL'

Lemon Time, HT, my, 1985; Kasturi

Lemon Twist, *see* 'FOUTWIST'

'LEMON YELLOW', F, my, 1977; flowers lemon-yellow blooms, dbl., 21 petals, 2 in., slight fragrance; foliage dark; bushy growth; [Orange Sensation X King's Ransom]; Gandy, Douglas L.

'LEMONADE', F, w, 1974; flowers cream, dbl., 25 petals, 4.5 in., slight fragrance; foliage matt; [Nancy West sport]; Haynes Roses

'LEMSEA', Min, mp, 1976; (**Seabreeze**); bud short, pointed; dbl., 35 petals, 1 in., slight fragrance; foliage light to medium green; very vigorous, bushy growth; [White Fairy X ?]; Lemrow, Dr Maynard W.; Nor'East Min. Roses

'LEMSUR', Min, ob, 1984; (**Surf's Up**); flowers orange, yellow reverse, small blooms borne singly, dbl., 35 petals, exhibition form, no fragrance; foliage small, medium green, semi-glossy; upright growth; [Avandel X Unnamed seedling]; Lemrow, Dr Maynard W.

'LEMSWI', Min, rb, 1987; (**Swiss Lass**); flowers red, reverse almost all white, medium, very dbl., slight fragrance; foliage small, medium green, semi-glossy; upright growth; [Unnamed seedling X Unnamed seedling]; Lemrow, Dr Maynard W.

'LEN 1', S, pb, 1985; (**Bouquet Fait**); flowers medium pink, white eye blooms in clusters of 1–5, semi-dbl., 2 in., non-recurrent, intense fragrance; foliage grayish green, hairy; bushy (to 5 ft) growth; [R. mollis X Complicata]; Lens, Louis

'LEN 2', Gr, dr, 1966; (**Epoca**); flowers very dark red, large blooms in sprays of 3–18, dbl., 45 petals, exhibition form, no fragrance; brown-green prickles; foliage very dark; upright, bushy growth; [Seedling X Seedling]; Lens, Louis

'LEN 3', F, op, 1980; (**Favori**, Favorite); flowers light salmon-pink, urn-shaped blooms in clusters of 8–24, dbl., 23 petals, 3 in., intense fragrance; hooked, brownish-green prickles; foliage dark; large, bushy growth; [Unnamed seedling X Unnamed seedling]; Lens, Louis

'LEN 4', Min, m, 1984; (**Jet Spray**®, 'LENCARA'); flowers purple-pink, rosette, small, dbl., 21 petals, slight fragrance; no prickles; foliage very small, dark; bushy, spreading growth; [New Penny X Unnamed seedling]; Lens, Louis

Len Turner, *see* 'DICJEEP'

'LENA', T, ab, 1906; flowers apricot, edged yellow; Dickson, A.

Lena, HT, or, 1981; deRuiter

'LENANA', S, lp, 1984; (**Flash**); dbl., 22 petals, 1.5 in., blooms in clusters of 3–50, slight fragrance; hooked, green prickles; foliage small, dark; bushy, spreading growth; [R. multiflora nana X Unnamed seedling]; Lens, Louis

'LENBLANK', Min, w, 1985; ('LENVIR', **White Dream**®); flowers small blooms in clusters of 3–28, dbl., 36 petals, exhibition form, slight fragrance; hooked, greenish-brown prickles; low, bushy growth; [Unnamed seedling X Unnamed seedling]; Lens, 1982

'LENBRAC', HRg, mp, 1987; (**Pink Surprise**); Lens

'LENBRU', HT, dy, 1980; (**Bruocsella**®); flowers golden yellow, urn-shaped, large, dbl., 35 petals, exhibition form, borne singly or in small clusters, intense fragrance; dark green prickles; upright, bushy growth; [Peace X Golden Garnette]; Lens, Louis

'LENCAN', Min, ly, 1984; (**Petit Canard**®); flowers small blooms in clusters of 12–22, dbl., 20 petals, exhibition form, slight fragrance; few, green prickles; foliage small; bushy growth; GM, Paris, 1984; [Rosina X (Le Vesuve X Belle Étoile)]; Lens

'LENCARA', Min, m, 1984; (**Jet Spray**®, 'LEN 4'); flowers purple-pink, rosette, small, dbl., 21 petals, slight fragrance; no prickles; foliage very small, dark; bushy, spreading growth; [New Penny X Unnamed seedling]; Lens, Louis

'L'ENCHANTRESSE', HGal, mp, 1829; (Grande Henriette); flowers clear pink, fading with age, large, dbl.; François, 1824

'LENDA', Pol, rb, 1980; (**Anda**); flowers dark red, white eye, single, 5 petals, 2 in., blooms in clusters of 18–24, moderate, fruity fragrance; purple prickles; bushy growth; [(Britannia X R. moschata) X (Little Angel X Europeana)]; Lens

'LENDADI', S, m, 1985; (**Violine**); flowers lilac, pink, white blend blooms in clusters of 5–28, dbl., 20 petals, 2 in., moderate, fruity fragrance; few, small, brown prickles; foliage light green; upright, bushy growth; [(Little Angel X Picasso) X Skyrocket]; Lens

'LENDEC', F, or, 1980; (**Idée Fixe**®); flowers light orange-red, blooms in clusters of 3–18, 15 petals, 2 in., slight fragrance; hooked, light red prickles; foliage small, dark; low, bushy growth; [Unnamed seedling X (Unnamed seedling X Floradora)]; Lens, Louis

'LENDORI', F, op, 1985; (**Erna Doris**); flowers medium salmon pink, blooms in clusters of 3–12, dbl., 24 petals, exhibition form, slight fragrance; small, hooked, brown-green prickles; foliage small, medium green; upright, bushy growth; [Little Angel X Elizabeth of Glamis]; Lens, Louis

LENfil, S, lp, 1988; (**Fil d'Ariadne**®); Lens

'LENGRA', S, w, 1985; (**Pleine de Grâce**); flowers creamy white, small blooms in clusters of 12–24, single, 5 petals, nonrecurrent, intense fragrance; foliage yellowish-green; bushy, spreading growth; [Ballerina X R. filipes]; Lens, 1983

'LENI NEUSS', HT, pb, 1933; bud pointed; flowers hydrangea-pink, reverse reddish old-rose, very large, dbl., moderate fragrance; very vigorous, bushy growth; [Lilly Jung X Baronesse M. van Tuyll van Serooskerken]; Leenders, M.; C-P

Lenidora, S, 1990; Lens, Louis; (Cavriglia)

'LENIP', HT, w, 1963; (Blanche Pasca®, **Pascali**®); flowers creamy white, wellformed, medium, dbl., 30 petals; foliage dark; vigorous, bushy growth; All-American Rose Selection, AARS, 1969 GM, Portland, 1967 GM, The Hague, 1963 Hall of Fame, WFRS, 1991; [Queen Elizabeth X White Butterfly]; Lens; A. Dickson, 1963; Armstrong Nursery, 1968

'LENIRA', F, or, 1984; (**Iranja**®); flowers orange-red blooms in clusters of 3–24, dbl., 35 petals, 2 in., no fragrance; hooked, red-green prickles; foliage small, brilliant green; low, bushy growth; [Little Angel X (Floradora X Angelina Louro)]; Lens, Louis

'LENISUR', S, lp, 1985; (**Merle Blanc**); flowers blush pink blooms in clusters of 3–22, dbl., 22 petals, 2 in., recurrent bloom, intense fragrance; hooked, brownish-green prickles; foliage dark; tall, spreading growth; [Ballerina X Surf Rider]; Lens

'LENIVILL', S, dr, 1983; (**Ville d'Ettelbruck**); flowers deep red blooms in clusters of 5–18, dbl., 20 petals, 2 in., moderate, fruity fragrance; hooked, green prickles; foliage dark, leathery; upright, bushy growth; [Satchmo X Skyrocket]; Lens

Lenka, F, lp; Vecera, L., 1968; (Czech Rosa Club)

'LENLIT', F, lp, 1984; (**Little Shrimp**®); flowers light shrimp-pink, rosette blooms in clusters of 3–24, dbl., 40 petals, 2 in., slight fragrance; hooked, brownish-red prickles; foliage small; bushy growth; [Little Angel X (Little Angel X Spartan)]; Lens, Louis

'LENMACRA', S, lp, 1984; (**Maria Teresa**®); dbl., 28 petals, cupped, blooms in clusters of 32, recurrent bloom, intense fragrance; hooked, brownish-green prickles; foliage light green; bushy growth; [Unnamed seedling X R. macrantha]; Lens, Louis

'LENMAR', Min, w, 1980; (**Maria Mathilda**®); flowers white, shaded pink, small blooms in clusters of 3–24, dbl., 24 petals, intense fragrance; hooked, brownish-red prickles; foliage very dark, glossy; upright, bushy growth; Golden Rose, The Hague, 1981, Rose d'or à La Haye, 1981; [Unnamed Min seedling X (New Penny X Jour de Fete)]; Lens, Louis

'LENMAU', F, m, 1984; (**Bluesette**®); flowers lilac, blooms in clusters of 3–18, dbl., 50 petals, 2 in., flat, slight fragrance; dark green prickles; bushy growth; [Little Angel X (Westmauve X Blue Diamond)]; Lens, Louis

'LENMOBAR', S, mr, 1984; (**Vivaldi**); single, 5 petals, 1 in., blooms in clusters of 7–50; very hooked, greenish-brown prickles; foliage small; bushy, spreading growth; [R. multiflora seedling X (Unnamed seedling X Robin Hood)]; Lens

'LENMOBRI', S, pb, 1983; (**Rush**®); flowers pink, white eye blooms in clusters of 3–32, single, 5 petals, 2 in., recurrent bloom, moderate, fruity fragrance; hooked, brownish-green prickles; foliage light green; upright, bushy growth; GM, Bagatelle, 1986 GM, Monza, 1982 GM, Rome, 1982; [(Ballerina X Britannia) X R. multiflora]; Lens

'LENMOL', Min, m, 1980; (**Si Bemol**®); flowers lilac blue and white, small blooms in clusters of 3–24, dbl., 30 petals, slight fragrance; few, green prickles; foliage dark; bushy growth; [(Little Angel X Le Vesuve) X Mr Bluebird]; Lens, 1983

'LENMOR', S, pb, 1984; (**Schubert**®); flowers pink, white-eyeblooms in clusters of 7–60, single, 5 petals, 1 in., recurrent bloom, slight fragrance; very hooked, greenish-brown prickles; foliage small; bushy, spreading growth; [Ballerina X R. multiflora]; Lens

LENneel, HMsk, mp, 1997; (**Heavenly Pink**); Lens

'LENNIRO', F, op, 1982; (**Jupon Rose**®); flowers light salmon pink, blooms in clusters of 3–24, dbl., 28 petals, cupped, slight fragrance; hooked, reddish-green prickles; foliage dark; upright, bushy growth; [Little Angel X Pernille Poulsen]; Lens, Louis

'LENNOP', Min, mr, 1984; (**Little Pim**®); flowers rosette, small blooms in clusters of 3–32, dbl., 24 petals, no fragrance; hooked, greenish-brown prickles; foliage dark; bushy growth; [Unnamed Miniature seedling X Ruth Leuwerick]; Lens, Louis

'LENOBIT', HMsk, pb, 1982; (**Poesie**); flowers white shaded pink blooms in clusters of 5–32, dbl., 20 petals, 2 in., intense fragrance; hooked, light brown prickles; foliage large, leathery, dark; bushy, spreading growth; [Ballerina X Moonlight]; Lens, Louis, 1985

'LENOR', F, ob, 1980; (**Festivity**); flowers orange blooms in clusters of 3–24, dbl., 24 petals, 3 in., slight fragrance; hooked, green prickles; foliage dark; bushy growth; [Unnamed seedling X Unnamed seedling]; Lens, Louis

'LENORAN', HT, lp, 1984; (**Eurosong**®); dbl., 38 petals, exhibition form, borne singly or in threes, slight fragrance; large, brownish-red prickles; foliage dark reddish-green; upright growth; [(Queen Elizabeth X Seedling) X Queen Elizabeth]; Lens, Louis

'LENpac', HMsk, lp, 1984; (**Focus**); flowers small blooms in clusters of 3–50, dbl., 20 petals, recurrent bloom, intense fragrance; hooked, brownish-green prickles; foliage small, dark; spreading growth; [Marie Pavie X Unnamed seedling]; Lens, Louis

LENpaga, S, mr; (**Harley**®); Lens

'LENpas', Min, w, 1984; (**Pascaline**®); flowers urn-shaped, blooms in clusters of 3–24, dbl., 35 petals, exhibition form, no fragrance; slightly hooked, brownish-red prickles; foliage dark gray-green; upright, bushy growth; [Unnamed Miniature seedling X (New Penny X Jour de Fete)]; Lens

LENpaya, S, w; (**White Spray**®); Lens, 1980

'LENpen', Min, m, 1984; (**Jet Flame**®, Jet Flame Nirpaysage®); flowers lavender-purple, rosette, micro-mini, small blooms in clusters, dbl., 30 petals, no fragrance; no prickles; foliage very small, brilliant dark green; bushy, spreading growth; GM, Paris, 1984; [New Penny X Violet Hood]; Lens, Louis

'LENpic', F, op, 1982; (**Tache de Beauté**®); dbl., 22 petals, 2 in., exhibition form, blooms in clusters of 3–24, slight fragrance; hooked, brown prickles; foliage dark; upright, bushy growth; [Little Angel X Picasso]; Lens

LENpil, Min, mp, 1982; (**Pink Delight**®); Lens

'LENplat', S, pb, 1982; (**Tapis Volant**®); flowers white-pink blend blooms in clusters of 7–35, semi-dbl., 2 in., recurrent bloom, moderate, fruity fragrance; hooked, reddish-brown prickles; foliage reddish green; groundcover; spreading growth; GM, Kortrijk, 1987; [(R. luciae X ?) X (R. multiflora X Ballerina)]; Lens

'LENpon', Min, or, 1980; (**Asterix**®); flowers small, dbl., 65 petals, flat, blooms in clusters of 12–22, no fragrance; hooked, brown prickles; bushy growth; [(Little Red X (Little Angel X Robin Hood)) X Idée Fixe]; Lens

'LENrag', F, r, 1980; (**Ragtime**); flowers brownish red, blooms in clusters of 7–18, dbl., 25 petals, exhibition form, slight fragrance; foliage brownish-green; low, bushy growth; [Little Angel X Goldtopas]; Lens, Louis

'LENramp', S, m, 1985; (**Running Maid**®); flowers lilac-red, white eye blooms in clusters of 3–32, single, 5 petals, 2 in., recurrent bloom, intense fragrance; hooked, brown prickles; foliage reddish-green; spreading growth; groundcover; [R. multiflora X (R. wichurana X Violet Hood)]; Lens, 1982

'LENrav', HMsk, pb, 1988; (**Ravel**®); Lens

LENray, S, pb, 1980; (**Pink Spray**®); Lens

'LENsim', F, m, 1982; (**Eagle Wings**®); flowers white, shaded lilac blooms in clusters of 3–24, single, 5 petals, 2 in., moderate, spicy fragrance; hooked, brown prickles; foliage very dark; bushy growth; [Seedling X Picasso]; Lens, Louis

'LENsun', Min, yb, 1983; (**Sunnyside '83**); flowers yellow, spotted red, small blooms in clusters of 3–22, dbl., 20 petals, moderate, fruity fragrance; small, hooked, green prickles; foliage small; bushy growth; [Little Angel X (Rosina X Unnamed seedling)]; Lens, Louis

'LENtini', F, lp, 1985; (**Tineke van Heule**®); 7–12 petals, blooms in clusters of 3–12, slight fragrance; few, reddish-brown prickles; foliage large, leathery, greenish-brown; upright, bushy growth; [Unnamed seedling X City of Belfast]; Lens

'LENvie', Pol, dp, 1980; (**View**®); flowers deep pink, blooms in clusters of 3–32, 7–15 petals, 2 in., no fragrance; hooked, brown prickles; foliage greenish-brown; low, bushy growth; GM, The Hague, 1978; [Britannia X R. multiflora]; Lens

'LENvir', Min, w, 1985; ('LENblank', **White Dream**®); flowers small blooms in clusters of 3–28, dbl., 36 petals, exhibition form, slight fragrance; hooked, greenish-brown prickles; low, bushy growth; [Unnamed seedling X Unnamed seedling]; Lens, 1982

'LENway', HT, yb, 1982; (**Waverland**®); flowers petals deep yellow, edged orange-red, large, borne singly, dbl., 30 petals, exhibition form, intense fragrance; brown prickles; foliage dark, leathery; upright, bushy growth; GM, Monza, 1980; [Peer Gynt X Peace]; Lens

'LENwich', S, w, 1985; (**Green Snake**®, 'LENwiga', Serpent Vert®); flowers pure white, small blooms in clusters of 3–24, single, 5 petals, non-recurrent; hooked, light brownish-green prickles; foliage small, spoon-shaped; groundcover growth; [R. arvensis X R. wichurana]; Lens, Louis, 1987

'LENwiga', S, w, 1985; (**Green Snake**®, 'LENwich', Serpent Vert®); flowers pure white, small blooms in clusters of 3–24, single, 5 petals, non-recurrent; hooked, light brownish-green prickles; foliage small, spoon-shaped; groundcover growth; [R. arvensis X R. wichurana]; Lens, Louis, 1987

Lenzburger Duft, HT, mr, 1985; Schaiber

'LEObretak', HT, dp, 1982; (**Breathtaking**); bud ovoid, long, pointed; flowers deep pink, dbl., 40 petals, exhibition form, intense, damask fragrance; foliage medium to large, medium green, semi-glossy; upright, bushy growth; [Pink Silk seedling X (Pink Parfait X Wendy Cussons)]; Leon, Charles F., Sr.

'LEOcok', Min, pb, 1985; (**Cockle Shells**); flowers pale yellow tinged deep pink, medium, dbl., 35 petals, slight fragrance; foliage medium; bushy, spreading growth; [(Kathy Robinson X Seedling) X (Janna X Seedling)]; Leon, Charles F., Sr.

'LEOcrel', HT, dr, 1990; (**Crimson Elegance**); bud rounded; flowers crimson/scarlet, crimson reverse, aging darker red, dbl., 40 petals, exhibition form, borne usually singly, moderate fragrance; foliage medium, medium green, semi-glossy; upright, bushy, tall growth; [Big Red X Swarthmore]; Leon, Charles F., Sr.; Oregon Grown Roses, 1990

'LEOexbeau', HT, op, 1990; (**Exotic Beauty**); bud rounded; flowers orange blend, reverse pink blend, large blooms borne usually singly, dbl., exhibition form, slight, fruity fragrance; foliage large, dark green, glossy; upright, bushy, medium growth; [Silver Jubilee (surmised) X Undetermined seedling]; Leon, Charles F., Sr.; Oregon Grown Roses, 1990

'LEOfire', Min, or, 1985; (**Forest Fire**); flowers medium, dbl., slight fragrance; foliage medium, medium to dark, semi-glossy; bushy growth; [((Sheri Anne X Starina) X (Persian Princess X Starina)) X ((Sheri Anne X Persian Princess) X Starina)]; Leon, Charles F., Sr.

'LEOglo', HT, ab, 1985; (**Glorious**); flowers medium pink tinted apricot, well-formed, large, dbl., 37 petals, slight fragrance; foliage medium to large, medium green, semi-glossy; upright, bushy growth; [Unnamed seedling X Mirato]; Leon, Charles F., Sr.

'LEOlavblu', HT, m, 1982; (**Lavender Blue**); flowers lavender, darker edges, large, dbl., exhibition form, intense fragrance; foliage medium, dark, semi-glossy; upright, bushy growth; [Silver Star X Kolner Karneval]; Leon, Charles F., Sr.

'LEOloyros', HT, mr, 1990; (**Loyal Rosarian**); bud ovoid; flowers medium red becoming darker red as it ages, large blooms borne usually singly, dbl., 36–42 petals, exhibition form, intense fragrance; large fruit, germinates rapidly; foliage medium, leathery, green; upright, bushy, tall growth; [Red Planet X Red Devil (surmised)]; Leon, Charles F., Sr.; Oregon Grown Roses, 1990

'Léon Chenault', HT, pb, 1931; flowers carmine-rose shaded salmon, base deeper, very large, dbl., intense fragrance; foliage dark; very vigorous growth; Pernet-Ducher; Gaujard

'Léon Delaville', HP, dr, 1885; flowers large, dbl.; Verdier, E.; (Sangerhausen)

Leon Lecomte, D, dp

'Léon Renault', HP, mr, 1878; flowers large, very dbl.; Lédéchaux; (Sangerhausen)

'Léon Robichon', HP, w, 1901; flowers large, dbl.; moderate fragrance; Robichon; (Sangerhausen)

'Leonard Barron', HT, op, 1931; bud pointed; flowers salmon-copper and shell-pink, very large, dbl., moderate fragrance; foliage leathery; bushy growth; David Fuerstenberg Prize, ARS, 1933; [Schoener's Nutkana X Souv. de Mme Boullet]; Nicolas; C-P

Leonard Dudley Braithwaite, *see* 'AUSCRIM'

Leonardo de Vinci®, *see* 'MEIDEAURI'

'LÉONCE COLOMBIER', HT, mr, 1943; bud oval; flowers brilliant geranium-red, stamens yellow, medium, dbl., cupped, slight fragrance; foliage leathery; very vigorous, bushy growth; [Charles P. Kilham X (Charles P. Kilham X Capucine Chambard)]; Meilland, F.; A. Meilland

'LEONHART', HT, mr, 1982; (**Heart Throb**); dbl., 37 petals, exhibition form, borne singly and several together, moderate, damask fragrance; foliage medium to large, medium green, leathery; vigorous, bushy, tall growth; [Norita X ((Norita X Unnamed seedling) X Papa Meilland)]; Leon, Charles F., Sr.

Leonidas, *see* MEIcofum

'LEONIE', F, mp, 1964; (Dazzler); dbl.; vigorous growth; [Queen Elizabeth X Circus]; Leenders, J.

'LÉONIE LAMBERT', HT, lp, 1913; flowers silver-pink, shaded yellow and flesh, well shaped, large, moderate fragrance; very vigorous growth; [Frau Karl Druschki X Prince de Bulgarie]; Lambert, P.

'LÉONIE LAMESCH', Pol, ob, 1899; flowers light coppery red, center yellow, edges flecked darker, semi-dbl., foliage rich green, soft; vigorous, bushy growth; [Aglaia X Kleiner Alfred]; Lambert, P.

Leonie's Appoline, HT, dp

'LÉONOR DE MARCH', HT, dr, 1957; bud long, pointed; flowers deep blood-red, large, dbl., moderate fragrance; foliage glossy; upright growth; GM, Rome, 1958; [J.M. Lopez Pico X Poinsettia]; Camprubi, C.

'LÉONORA', HT, mr, 1921; flowers brilliant velvety red, center brighter, dbl.; Paul, W.

'LEONORO', HT, my, 1982; (**Solid Gold**); flowers golden yellow, large, borne 3–6 per cluster, dbl., 36 petals, intense fragrance; foliage dark, glossy; vigorous growth; [(Royal Gold seedling X Golden Giant) X (Bright Gold X PhyllisGold seedling)]; Leon, Charles F., Sr.

'LÉONTINE CONTENOT', HT, yb, 1935; bud pointed; flowers sunflower-yellow bordered pink and yellow, large, dbl., 45–50 petals, exhibition form; foliage quaker green; vigorous growth; [Joanna Hill X Souv. de Claudius Pernet]; Ketten Bros.

'LÉONTINE GERVAIS', LCl, ab, 1903; bud coppery red; flowers salmon-orange and yellow, large blooms in clusters of 3–10, non-recurrent, moderate fragrance; foliage dark, glossy; very vigorous, climbing growth; [R. wichurana X Souv. de Catherine Guillot]; Barbier

'LÉOPOLD I', HP, dr, 1863; flowers deep red, well formed, large, dbl.; [Général Jacqueminot seedling]; Van Assche

'LÉOPOLD LAMBOTTE', HT, mr, 1944; bud long, pointed; flowers scarlet-red, moderate fragrance; very tall growth; [Grenoble X National Flower Guild]; Meilland, F.

'LEOPOLD RITTER', HMult, mr, 1900; Geschwind, R.

'LÉOPOLDINE D'ORLÉANS', HSem, w, 1828; flowers white shaded rose, medium, dbl.; Jacques

'LEORA STEWART', HT, op;, long continued bloom, intense fragrance; Wilber

'LEOSILMU', HT, m, 1982; (**Silvery Moon**); flowers lilac, large, dbl., 35 petals, intense fragrance; foliage medium, medium green, leathery, matt; upright growth; [Silver Star X Blue Moon]; Leon, Charles F., Sr.

'LEOSTRA', Min, w, 1985; (**Strawberry Sundae**); flowers ivory flushed pink, medium, dbl., 35 petals, slight fragrance; foliage medium, light to medium green, semi-glossy; bushy, spreading growth; [(Kathy Robinson X Seedling) X (Kathy Robinson X Seedling)]; Leon, Charles F., Sr.

'LEOTILDE MINGUEZ', HT, rb, 1961; flowers carmine, shaded vermilion and yellow, large, dbl., 30 petals; upright growth; [Pres. Herbert Hoover X Vicky Marfa]; Dot, Pedro

'LEOTOELG', Min, w, 1988; (**Touch of Elegance**); bud pointed; flowers white with creamy yellow center, large, borne in sprays of 1-, dbl., exhibition form, slight fragrance; foliage medium, medium green, semi-glossy; upright, tall growth; [(Gavotte X Buccaneer) X Unnamed Miniature seedling]; Leon, Charles F., Sr.; John Carrigg, 1987

'LEOVELV', HT, rb, 1988; (**Touch of Velvet**); bud pointed, long; flowers magenta red with lighter tones, borne usually singly and in, dbl., 31 petals, exhibition form, moderate fragrance; foliage medium, dark green, semi-glossy; upright, tall growth; [(First Prize X Gypsy) X Unnamed seedling]; Leon, Charles F., Sr.; John Carrigg, 1987

'LEPRECHAUN', F, yb, 1971; bud ovoid; flowers red, reverse yellow, small, dbl., globular, moderate fragrance; foliage glossy; growth moderate, upright; [(Easter Parade X Masquerade) X Little Darling]; Adams, M.R.

'LES AMIS DE LILLE', HT, pb, 1928; bud lincoln red on pale buff ground; flowers pale buff, edges and reverse suffused salmon-pink; [Golden Emblem X Pres. Bouche]; Ketten Bros.

'LES AMIS DE TROYES', HT, mp, 1935; flowers china-pink, passing to carmine-pink, base ochre-yellow, well-formed, dbl.; vigorous growth; [Feu Joseph Looymans X Unnamed seedling]; Vially

Les Amoureux de Peynet®, *see* MEItobla

'LES ROSATI', HFt, mr, 1906; flowers bright carmine; Gravereaux

Les Saisons d'Italie, HGal, mr; flowers carmine-red, medium, very dbl., intense fragrance; (Sangerhausen)

'LES SJULIN', Gr, pb, 1981; flowers coral pink, reverse light red, urn-shaped to imbricated blooms, dbl., 28 petals, moderate, old rose fragrance; awl-like prickles; foliage medium, dark olive green, leathery; erect, bushy growth; [Country Dancer X ((Dornroschen X Peace) X Pink Peace)]; Buck, Dr Griffith J.; Iowa State University

'LES SYLPHIDES', HT, op, 1960; flowers pink shading to yellow and orange, well shaped, intense fragrance; foliage reddish; long stems; very vigorous growth; [Margaret sport]; Watkins Roses

Lesbos, LCl, lp; Peden, R., 1998; (Weatherly, L.)

Lesdain, HT, ob, 1989; RvS-Melle

'LESJA JEAN', HT, lp, 1990; bud pointed; flowers large blooms borne usually singly, very dbl., 37 petals, exhibition form, moderate fragrance; prickles medium, light; foliage medium, medium green, semi-glossy; upright, medium growth; [Unnamed seedling X Unnamed seedling]; Weddle, Von C., 1989

Lesja Ukrajinka, HT, lp, 1964; flowers medium, dbl., moderate fragrance; Lempickij; (Sangerhausen)

Lesley Anne, *see* 'FRYMINILES'

'LESLEY DUDLEY', HT, op, 1932; (Leslie Dudley); flowers warm carmine-pink shaded orange, well shaped, large, dbl., intense fragrance; foliage dark; vigorous, bushy growth; GM, NRS, 1931; McGredy

'LESLEY JOHNS', HT, dp, 1972; flowers deep pink, high-pointed, dbl., 32 petals, 5–6 in., moderate fragrance; foliage dark, glossy; [Soraya X ?]; Gregory

Leslie, *see* 'TINLESLIE'

Leslie Dudley, *see* **'LESLEY DUDLEY'**

'LESLIE EVANS', HT, dr, 1927; flowers rich dark velvety crimson, semi-dbl.; vigorous growth; [J.B. Clark X Red-Letter Day]; Evans; Beckwith

'LESLIE G. HARRIS', HT, mr, 1970; flowers crimson-scarlet, pointed, dbl., 27 petals, 4 in., intense fragrance; moderate growth; Gregory

'LESLIE HOLLAND', HT, dr, 1911; flowers deep velvety crimson, large, dbl., moderate fragrance; GM, NRS, 1909; Dickson, H.

Leslie Johns, HT, 1972; Gregory, C.; (Cavriglia)

'LESLIE PIDGEON', HT, ob, 1922; flowers orange-buff, suffused terra-cotta, semi-dbl., moderate fragrance; Dickson, H.

'Leslie Wheal', F, yb, 1977; bud ovoid; flowers cream-yellow, small, dbl., 25–35 petals, 1.5 in., cupped; foliage light green; compact growth; [Zorina sport]; Buss; H. Buss Nursery

Leslie's Dream, *see* 'DICjoon'

'Lessing', HMsk, rb, 1914; flowers reddish rose streaked white, center citron-yellow, small, dbl., moderate fragrance; foliage large, light; [Trier X Entente Cordiale]; Lambert, P.

'Lestra Hibberd', HT, dy, 1935; bud dark yellow; flowers amber-yellow to orange-yellow, large, dbl., moderate fragrance; foliage leathery; strong stems; vigorous, compact growth; [Joanna Hill X Sweet Adeline]; Hill, Joseph H., Co.

Letchworth Garden City®, *see* 'HARkover'

'Letitia', HT, dr, 1949; bud long, pointed; flowers crimson, dbl., 35 petals, 5–6 in.; foliage dark; very vigorous growth; [Crimson Glory X Southport]; Bees

Letizia, Gr, 1962; Cazzaniga, F. G.; (Cavriglia)

Letizia, *see* BARitzia

Letizia Bianca, 1963; Mansuino; (Cavriglia)

Letkis, *see* **'John Dijkstra'**

Letkiss, F, mr

LETlyda, S, w, 1994; (**Lyda Rose**); Lettunich

'L'Etna', Ch, mp, 1825; flowers pink, becoming brighter, medium, very dbl.; Laffay M.

LETrob, HMsk, ab, 1996; (**Doctor Robert Korns**); Lettunich

LETsilk, HCh, mp, 1992; (**Mateo's Silk Butterflies**); Lettunich

Lettle Len, Min, ab

Lettle Mermaid, Cl Min, ly

'Letty Coles', T, w, 1876; flowers white, center pink; [Mme Mélanie Willermoz sport]; Keynes

'Leuchtfeuer', HCh, mr, 1909; (President Taft); flowers bright red, moderate fragrance; [Gruss an Teplitz X Cramoisi Superieur]; Türke

'Leuchtstern', HMult, pb, 1899; flowers deep rose pink, center white, blooms in clusters, single, slight fragrance; height 8–10 ft; [Daniel Lacombe X Crimson Rambler]; Schmidt, J.C.

Leventina®, HT, op, 1984; Huber

'Leverkusen', HKor, ly, 1954; bud long, pointed; flowers large blooms on long sprays, dbl., exhibition form, recurrent bloom, slight fragrance; foliage glossy, light green; vigorous, creeper or pillar (to 8 ft) growth; [R. kordesii X Golden Glow]; Kordes

Leverson Gower, *see* **'Leweson Gower'**

Leveson Gower, *see* **'Leweson Gower'**

Levianthan, lp; [Great Venture X Fritz Thiedemann]; Dawson, George

'Leweson Gower', B, op, 1845; (Leverson Gower, Leveson Gower, Red Malmaison, Souvenir de la Malmaison Rouge); flowers rose shaded salmon, very large, dbl., cupped; Béluze

'Lewiston', HT, my, 1981; flowers pointed, large blooms borne singly, dbl., 40 petals, exhibition form, intense, fruity fragrance; foliage large, glossy; vigorous, tall growth; [Red Lion X King's Ransom]; Perry, Astor, 1982

'Lexie', Pol, lp; Miers, A., 1935; (Weatherly, L.)

Lexington, *see* POUlgode

'Ley's Perpetual', Class unknown; flowers pale yellow, dbl., cupped, moderate fragrance; vigorous, climbing (to 15 ft) growth

'Li Burés', HT, rb, 1929; flowers rose-red and yellow mixed (very variable), dbl., cupped, slight fragrance; very vigorous, bushy growth; [Château de Clos Vougeot X Souv. de Claudius Pernet]; Dot, Pedro; C-P

Li Schluter, HT; Moreira da Silva, A.; (Cavriglia)

Lia, Min, or

Liaison, Gr, or, 1996; Williams, J. Benjamin

Liane®, LCl, ob, 1989; Cocker

Lianne Reynolds, *see* 'REYlianne'

'Libby', Min, rb, 1978; bud ovoid; flowers white, edged red, high-pointed, dbl., 20 petals, 1–1.5 in., flat; foliage glossy; upright, compact growth; [Overture X Perla de Alcanada]; Rovinski; Casa de Rosa Domingo

'Libby's Gold', HT, yb, 1977; bud long, slender; dbl., 40 petals, 5 in., slight fragrance; foliage dark; upright, compact growth; [Apollo sport]; Carpenter

'Liberté', F, dr, 1946; (Crimson Masse); flowers brilliant crimson-red, large, dbl.; [Florentina X Seedling]; Leenders, M.

'Liberty', HT, dr, 1900; flowers brilliant velvety crimson, large, dbl., intense fragrance; foliage dark; vigorous growth; [Mrs W.J. Grant X Charles J. Grahame]; Dickson, A.

Liberty Bell, *see* **'Freiheitsglocke'**

'Liberty Bell', HT, pb, 1963; (Freiheitsglocke); flowers deep pink, silver reverse, large, dbl.; Kordes, R.; A. Dickson; McGredy

Liberty Miss, *see* 'ZIPlib'

'Liberty, Climbing', Cl HT, dr, 1908; May

'Libia', HT, w, 1934; flowers milk-white, center creamy yellow, large, very dbl.; vigorous growth; Borgatti, G.

Libre Ingenue, *see* DELciste

'Libretto', HT, dp, 1966; bud ovoid; flowers pink-red, medium, borne in clusters, dbl.; foliage dark; [Elli Knab X Seedling]; Verschuren, A.; van Engelen

Libretto, LCl, w

Lichtblick, S, yb, 1972; flowers yellow and pink, large, dbl.; VEG; (Sangerhausen)

'Lichterloh'®, F, mr, 1955; bud ovoid; flowers velvety blood-red, blooms in clusters, semi-dbl., slight fragrance; foliage leathery, dark, glossy; vigorous, upright (3 ft) growth; [Red Favorite X New Dawn]; Tantau, Math.

Lichtkönigin Lucia®, *see* 'KORlilub'

Licorice Twist, *see* DEVlicor

'Lída Baarová', HT, mp, 1934; ('Lida Paar'); bud pointed; flowers rosy salmon, large, dbl., cupped; foliage glossy; vigorous, bushy growth; [Ophelia seedling]; Böhm, J.

'Lida Paar', *see* **'Lída Baarová'**

Lida O, *see* 'BISyel'

'L'Idéal', N, mr, 1887; flowers geranium to turkey-red, base indian yellow, semi-dbl.; vigorous, climbing growth; Nabonnand, G.

Lidice, F, or, 1961; flowers large, dbl., moderate fragrance; Wheatcroft; (Sangerhausen)

Lidka, HT, lp; Urban, J., 1985; (Czech Rosa Club)

Lidka Böhm, *see* **'Lidka Böhmova'**

'Lidka Böhmova', HT, mp, 1929; (Lidka Böhm); flowers salmon-pink, tips veined reddish, base golden yellow, large, dbl., intense fragrance; foliage soft, bronze; very vigorous, bushy growth; [Una Wallace sport]; Böhm, J.

Lido di Roma, *see* 'DELgap'

'Liebesbote', HT, dr, 1935; bud pointed; flowers velvety dark red, very large, dbl., exhibition form, intense fragrance; foliage soft; very vigorous, bushy growth; [Hadley X Miss C.E. van Rossem]; Weigand, C.; Pfitzer

Liebeserklärung, HT, mr, 1968; flowers large, dbl.; Horstmann; (Sangerhausen)

Liebesglut, *see* **'Crimson King'**

Liebeslied, *see* **'Love Song'**

Liebeslied, F, pb, 1990; Noack, Werner

Liebestraum, *see* **'Red Queen'**

'Liebeszauber', HT, mr, 1959; (Charme d'Amour); flowers velvety red, large, cupped, intense fragrance; foliage dark; vigorous, upright growth; [Detroiter X Crimson King]; Kordes

Liebeszauber, *see* KORmiach

Lied, *see* **'Bouquet'**

Lien Budde, *see* **'Caroline Budde'**

Liesbeth Canneman, F, dy, 1997; Interplant

'LIESBETH VAN ENGELEN', HT, w, 1960; flowers creamy white edged lilac-pink, dbl., moderate fragrance; foliage dark, glossy; upright growth; [Briarcliff X Unnamed seedling]; Verschuren, A.; van Engelen

'LIEUTENANT CHAURÉ', HT, dr, 1909; flowers velvety crimson-red, shaded garnet, very large, dbl., cupped, moderate fragrance; foliage rich green, leathery; vigorous, bushy growth; [Liberty X Étoile de France]; Pernet-Ducher

'LIEUTENANT COLONEL A. FAIRRIE', HT, my, 1930; bud pointed; flowers primrose, base deep yellow large, dbl., exhibition form, moderate, apple fragrance; [Rev. F. Page-Roberts X Mme Ravary]; Bees

'LIEUTENANT COLONEL DESMAIRES', F, ob, 1967; bud ovoid; flowers orange and yellow, borne in clusters, dbl., moderate fragrance; foliage leathery; vigorous, low growth; [Rondo X Fashion seedling]; Boerner; J&P

'LIEVEN GEVAERT', HT, mr, 1974; bud ovoid; flowers full, dbl., 30 petals, 4 in., moderate fragrance; foliage dark; Delforge, S.

'LIFEBOAT JUBILEE', HT, mr, 1974; flowers scarlet, shaded crimson, classic form, dbl., 20 petals, 4 in.; [Karl Herbst X (Karl Herbst X Crimson Glory)]; Sanday, John

Lifestyle, *see* 'MORDARAIN'

Lifestyle(TM), Min, pb

'LIFIRANE', HT, mp, 1976; flowers neyron rose, full, dbl., 20–25 petals, 3 in., exhibition form, slight fragrance; foliage dull; vigorous, upright growth; PP004034; [Sweet Promise sport]; Zwemstra; Meilland

'LIGHT EDITOR MCFARLAND', HT, lp, 1950; [Editor McFarland sport]; Thomasville Nursery

'LIGHT TOUCH', S, pb, 1997; flowers double, medium, dbl., 15–25 petals, borne in large clusters, slight fragrance; foliage medium, light green, very glossy; bushy tall (2m) growth; [Cousin Essie X Open Pollination]; Weatherly, Lila

Lights of Broadway(TM), *see* 'SAVALIGHTS'

Ligia, HT; Moreira da Silva, A.; (Cavriglia)

Lijang Rose, Misc OGR, yb

'LIJNBAANROOS', F, ly, 1961; bud yellow; flowers creamy yellow, borne in clusters, semi-dbl.; foliage dark; vigorous, upright growth; [Schneewittchen X Koningin Juliana]; Buisman, G. A. H.

Li'l Alleluia(TM), *see* 'KINLU'

Li'l Rip(TM), *see* 'MOGARIP'

Li'l Touch(TM), *see* 'MOGATOUCH'

'LILA', Pol, lp; Miers, A., 1935; (Weatherly, L.)

Lila Queen of Bermuda, HT, m, 1970; flowers lilac-pink, large, dbl.; Strnad, J.; (Sangerhausen)

'LILA TAN', HT, m, 1961; flowers violet-cobalt, medium, dbl., 30 petals; vigorous, spreading growth; [Grey Pearl X Simone]; Dot, Pedro

'LILA VIDRI', HT, m, 1958; bud pointed; flowers lilac, dbl., 30 petals, exhibition form, moderate fragrance; strong stems; upright, compact growth; [(Unnamed seedling X Prelude) X Rosa de Friera]; Dot, Simon

Lila Zwerg, *see* HELzwerg

Lilac Airs, *see* 'SANLILAC'

'LILAC CHARM', F, m, 1962; bud pointed; flowers pastel mauve, anthers golden, filaments red, blooms in clusters, single, 5–8 petals, 4 in., flat, moderate fragrance; foliage dark; upright, compact growth; GM, NRS, 1931; LeGrice; Wayside Gardens Co., 1952

Lilac Chrysler Imperial, HT, lp; Strnad; (Czech Rosa Club)

'LILAC DAWN', F, m, 1964; bud pointed; flowers lavender-pink to lilac, blooms in clusters, dbl., 43 petals, 2.5 in., moderate, lilac fragrance; foliage leathery, light green; vigorous, bushy growth; [Lavender Pinocchio X Frolic]; Swim & Weeks; Armstrong Nursery

Lilac Joseph Guy, lp; Strnad; (Czech Rosa Club)

Lilac Minimo, Min, m, 1997

Lilac Mystery, S, m, 1994

Lilac Pink Moss, M, dp

Lilac Queen Elizabeth, Gr, m

Lilac Rose, *see* 'AUSLILAC'

'LILAC ROSE', HT, lp, 1962; flowers pink tinted lilac, dbl., 28 petals, 5 in., moderate fragrance; foliage dark; vigorous, upright growth; [Karl Herbst X Chrysler Imperial]; Sanday, John

Lilac Rose(TM), S, pb

'LILAC SNOW', HT, w, 1985; flowers white, strong lilac tinge to inside petals; [Admiral Rodney sport]; Williams, H.; P. Loaney

'LILAC TIME', Min, m, 1955; flowers lilac-pink to light red, becoming lilac tinted, small, dbl.; dwarf (10 in.) growth; [Violette X Zee]; Moore, Ralph S.; Sequoia Nursery

'LILAC TIME', HT, m, 1956; flowers lilac, dbl., 33 petals, 3.5 in., exhibition form, moderate fragrance; foliage light green; growth moderate; [Golden Dawn X Luis Brinas]; McGredy, Sam IV

Lilet Dot, F, 1987; Dot, Simon; (Cavriglia)

'LILETTE MALLERIN', HT, m, 1937; bud pointed, yellow; flowers mauve-red to mauve pink, deepening in cooler weather, reverse yellow, dbl., cupped, slight fragrance; foliage glossy; vigorous growth; [Charles P. Kilham X Unnamed R. foetida bicolor seedling]; Mallerin, C.; A. Meilland

'LILETTE MALLERIN IMPROVED', HT, m, 1942; bud pointed; flowers mauve-red, reverse yellow, large, dbl., 25 petals, exhibition form; foliage bronze, glossy; vigorous, upright, bushy growth; [Lilette Mallerin sport]; J&P

Lili Marlene, *see* 'KORLIMA'

Lili Marlene, Climbing, *see* 'PEKLIMASAR'

'LILIAN', HT, dy, 1931; flowers golden yellow, very large, dbl., cupped, intense fragrance; foliage bronze, glossy; vigorous, bushy growth; RULED EXTINCT 12/85; Cant, B. R.

Lilian Austin®, *see* 'AUSMOUND'

Lilian Baylis, *see* HARdeluxe

Lilian Baylis, *see* HARdelux

'LILIAN BOOTLE', HT, dp, 1973; flowers cerise, dbl., 35 petals, 4.5–5 in., exhibition form, moderate fragrance; foliage matt green; compact, bushy growth; [Margaret X Seedling (red)]; Bootle

'LILIAN NORDICA', HT, w, 1898; flowers large, dbl., moderate fragrance; vigorous growth; [Margaret Dickson X Mme Hoste]; Walsh

'LILIANA', Gr, w, 1977; flowers large, dbl., 50 petals; foliage light green, leathery; vigorous, tall, upright growth; [Queen Elizabeth X Unnamed seedling]; Staikov, Prof. Dr V.; Kalaydjiev and Chorbadjiiski

'LILIANA', HT, pb, 1956; flowers soft pink, reverse carmine; RULED EXTINCT 12/85; GM, Geneva, 1956; [Edith Krause X Fashion]; Camprubi, C.

Liliana, *see* POUlsyng

'LILIBET', F, lp, 1953; (Fairy Princess); bud ovoid; dbl., 30 petals, 2.5–3.5 in., blooms in clusters, moderate, spicy fragrance; foliage glossy; low, bushy growth; AARS, 1954; [Floradora X Pinocchio]; Lindquist; Howard Rose Co.

'LILLAN', Min, pb, 1958; flowers rose pink, white center, small, semi-dbl.; very dwarf growth; [Ellen Poulsen X Tom Thumb]; deVink

'LILLEBROR', Min, mr, 1978; dbl.; dwarf, bushy growth; Hubner; O. Baum

Lilli Marleen®, *see* 'KORLIMA'

Lilli Marleen, Climbing, *see* 'PEKLIMASAR'

Lilli Marlene, *see* 'KORLIMA'

Lilli Marlene, Climbing, *see* 'PEKLIMASAR'

'LILLI VON POSERN', HT, mp, 1910; flowers large, dbl.; Kiese; (Sangerhausen)

Lillian, F, ob, 1997; Williams, J. Benjamin

'LILLIAN FRENCH', HT, pb, 1985; flowers light to medium pink, reverse deep pink, large blooms borne, very dbl., 50–120 petals, quartered, moderate fragrance; large prickles pointed downward; foliage medium, medium green, semi-glossy; upright, bushy growth; [Miss Hillcrest X Bishop Darlington]; French, Richard

'LILLIAN GIBSON', S, mp, 1938; flowers rose-pink, dbl., 40 petals, 3 in., non-recurrent, intense fragrance; vigorous growth; very hardy; [R. blanda X Red Star]; Hansen, N.E.

'LILLIAN GISH', F, ly, 1985; flowers very pale yellow, large blooms borne singly and in small clusters, dbl., 35 petals, slight fragrance; few, straight prickles; foliage medium, medium green, semi-glossy; upright growth; [Queen Elizabeth X Allgold]; French, Richard

'LILLIAN GOMEZ-MENA', Cl HT, pb, 1927; flowers salmon-cream, reverse carmine, dbl., intense fragrance; Chambard, C.

'LILLIAN MOORE', HT, dy, 1917;, moderate fragrance; Dickson, H.

'LILLIE BELL', HT, dp, 1979; bud ovoid to pointed; flowers deep pink to cerise-red, ruffled, dbl., 34 petals, 4.5–5 in., moderate fragrance; foliage large; upright growth; [Pink Peace X Miss All-American Beauty]; Williams, J. Benjamin; Central Mississippi Rose Soc.

'LILLIE DAWBER', HT, dr, 1952; flowers scarlet overlaid crimson, dbl., 25 petals, 5 in., cupped, moderate fragrance; foliage dark; free growth; Kordes

'LILLIPUT', Cl Pol, mr, 1897; flowers carmine-red, small, dbl.; Paul, G.; (Sangerhausen)

'LILLY JUNG', HT, my, 1925; flowers golden yellow, dbl., moderate fragrance; Leenders, M.

Lily Bertschinger®, F, w, 1980; Hauser

'LILY DE GERLACHE', HT, dp, 1971; bud long, pointed; flowers rose-red, large, dbl., cupped, intense fragrance; foliage glossy, bronze, leathery; growth moderate, upright, bushy; [Kordes' Perfecta X Prima Ballerina]; Institute of Ornamental Plant Growing

Lily Freeman, *see* 'HUX 1'

'LILY KEMP', HT, dp, 1928; flowers deep cherry-cerise, dbl., moderate fragrance; [Mme Butterfly X Capt. Ronald Clerk]; Morse

'LILY MERTSCHERSKY', N, m, 1878; flowers violet-red; Nabonnand

'LILY PONS', HT, ly, 1939; flowers yellow center, shading to white outer petals, large, dbl., 50 petals, exhibition form, moderate fragrance; foliage glossy; [Glenn Dale X Stargold]; Brownell, H.C.

'LILY THE PINK', HT, lp, 1992; very dbl., 3–3.5 in., borne mostly singly, slight fragrance; some prickles; foliage medium, medium green, semi-glossy; tall (1000 cms), upright growth; [Seedling X Seedling]; Scrivens, Len, 1991

'LILY VAN OOST', HT, mp, 1938; flowers very large, dbl.; Verschuren-Pechtold; (Sangerhausen)

'LILY WHITE', HT, w, 1950; (Starbright); bud ovoid; dbl., 40 petals, 4.5 in., exhibition form, slight fragrance; foliage glossy; vigorous, upright growth; [Starlite sport]; Hartgerink; J&P

'LIMBURGIA', HT, rb, 1921; flowers glowing deep carmine, reverse lilac-white, dbl., moderate fragrance; Leenders, M.

Lime Kiln, HWich, w, 1970

Limelight, *see* 'KORIKON'

'LIMELIGHT', F, r, 1959; flowers vermilion splashed golden, borne in clusters, semi-dbl., slight fragrance; vigorous growth; RULED EXTINCT 5/85; [Enterprise sport]; Morse

Limelight, *see* 'WEKAMANDA'

Limerick, *see* 'ZIPLIME'

Limited Edition, *see* 'JACREF'

Limona, *see* KORulas

'LINA', Pol, lp, 1930; flowers creamy light pink, medium, dbl.; Vogel, M.; (Sangerhausen)

'LINA SCHMIDT-MICHEL', HT, yb, 1906; flowers yellowish-pink, large, semi-dbl., Lambert, P.; (Sangerhausen)

'LINA VOGEL', HFt, ob, 1936; flowers orange with golden and light yellow shades, medium, single; Vogel, M.; (Sangerhausen)

Linburn Light, F, mp; Andersons

Lincoln Cathedral, *see* 'GLANLIN'

Lincolnshire Poacher, *see* 'GLAREABIT'

'LINDA', HT, yb, 1997; flowers full, medium, very dbl., 26–40 petals, borne in small clusters, slight fragrance; foliage medium, medium green, semi-glossy; upright, medium (3.5 ft) growth; [Champion X (Pink Favourite X Golden Autumn)]; Bracegirdle, A.J.

'LINDA ANN', Gr, mp, 1996; flowers medium pink, deeper pink than parents, more petals than parent, dbl., 26–40 petals, slight fragrance; moderate prickles; foliage medium, medium green, semi-glossy; upright, tall growth; [Queen Elizabeth X Katherine T. Marshall]; Dobbs, Annette E.

Linda Buford, *see* 'TINBUFORD'

Linda Campbell, *see* 'MORTEN'

'LINDA CHRISTINE', F, pb, 1980; flowers cerise, silver reverse; [Molly McGredy sport]; Taylor, W.J.

'LINDA GUEST', HT, ab, 1984; flowers large, dbl., 20 petals, exhibition form, slight fragrance; foliage medium, medium green, matt; bushy growth; [((Golden Jewel X Mischief) X Red Planet) X Valencia]; Guest, M.M.

'LINDA LOU'™, F, ob, 1987; flowers orange-yellow-red, medium, borne in sprays of 1–8, dbl., 30 petals, exhibition form, moderate fragrance; foliage medium, dark green; medium growth; Silver, ARC TG, 1987; [Golden Slippers X Roman Holiday]; Harvey, R.E.; Kimbrew Walter Roses, 1988

Linda Mary, *see* 'BOSELADNEE'

'LINDA PORTER', HT, op, 1957; (Miguel Aldrufeu); bud ovoid; flowers salmon-pink, dbl., 55 petals, 5–6 in., globular, intense fragrance; foliage leathery; vigorous, upright growth; [Senateur Potie X Poinsettia]; Dot, Pedro; B&A

Linda Thomson, *see* 'TOMTWO'

Linda Vista, F, 1974; de Ruiter, G.; (Cavriglia)

Linda's Lipstick, *see* 'SHERILIP'

'LINDBERGH', Pol, mr, 1927; flowers bright geranium-red; [Orléans Rose sport]; Croibier

Lindee, Pol, w

L'Indéfrisible, *see* **'PERMANENT WAVE'**

Linden Heath, S, mp

Linderof, *see* KORelasting

'LINDLIFFE', F, op, 1984; (**Guyscliffe**); flowers orange-salmon, blooms borne 3–5 per cluster, semi-dbl., 12 petals, cupped, intense fragrance; light brown prickles; foliage light green; tall, densely branched growth; Lindner, Richard; Ludwigs Roses Pty. Ltd.

'LINDSEY', HT, op, 1972; flowers salmon-pink, shaded copper, pointed, 3–4 in., moderate fragrance; foliage large; free growth; [Whisky Mac sport]; Watkins Roses

Line & Twine, HP, lp

'LINETTE', Pol, dp, 1922; bud reddish apricot; flowers shrimp-carmine-pink, passing to soft rose-pink, large; Turbat

'L'INGENUE', HGal, w, 1848

Lingo Musk, N, pb

'LINRICK', HT, dp, 1978; (**Friendship**™); bud ovoid; flowers deep pink, dbl., 28 petals, 5.5–6 in., cupped, intense fragrance; foliage large, dark; vigorous, upright growth; AARS, 1979; [Fragrant Cloud X Miss All-American Beauty]; Lindquist; C-P

'LINRO', F, rb, 1966; (**Roman Holiday**); bud ovoid; flowers orange turning blood-red, base yellow, blooms in clusters, dbl., 28 petals, exhibition form, moderate fragrance; foliage dark, leathery; vigorous, bushy, low growth; AARS, 1967; [(Pinkie X Independence) X Circus]; Lindquist; Howard Rose Co.

'LINVILLE', Min, w, 1989; bud pointed; flowers light pink, aging white, large, borne usually singly, dbl., 28 petals, exhibition form, slight, fruity fragrance; prickles straight, medium, deep pink; foliage medium, medium green, semi-glossy; upright, medium growth; PP007737; [Unnamed seedling X Unknown seedling]; Bridges, Dennis A.; Bridges Roses, 1990

'LIOLÀ', HT, w, 1960; flowers ivory edged crimson, dbl.; very vigorous growth; [Peace X (Baiser X Marguerite Chambard)]; Giacomasso

'LION DES COMBATS', HP, mr, 1850; flowers carmine-red, large, dbl.; Lartay; (Sangerhausen)

'LIONEL BARRYMORE', F, my, 1956; bud urn shaped; flowers yellow, center deeper, very dbl., 3–3.5 in., cupped, intense fragrance; foliage dark; strong stems; bushy growth; [Duchess of Atholl X Orange Everglow]; Silva; Booy Rose Nursery

Lionheart, *see* 'MINHEART'

Lions International, *see* FRYcharm

Lippe-Detmold®, HT, mp, 1985; Noack, Werner

'LIPSIANA', Pol, 1952; Cazzaniga, F. G.; (Cavriglia)

'LIPSTICK', F, rb, 1940; flowers deep cerise shaded salmon, blooms in clusters, semi-dbl., 2 in., cupped, slight fragrance; foliage glossy, dark; vigorous (3 ft), bushy growth; Verschuren; Dreer

Lipstick 'n' Lace, *see* 'CLELIPS'

Lisa, F, 1988; Zandri, R.; (Cavriglia)

Lisa, HSpn, lp; [Single Cherry X Primula]; Sutherland, P, 1996; Golden Vale Nursery; (Weatherly, L.)

'LISA COLFAX', F, ob, 1975; bud ovoid; flowers vermilion, overlaid brown, medium, dbl., exhibition form, intense fragrance; foliage leathery; very vigorous growth; [Mignonne X Sherry]; Parkes, Mrs M.H.; Rumsey

Lisa Kent, HT, dp, 1995; Dawson

'LISA MAREE', HT, dp, 1989; flowers deep pink, reverse lighter, [Esther Geldenhuys sport]; Cowper, Mrs. Maree; Cherry Wood's Nursery, 1989

'LISBETH PRIM', HT, or, 1934; flowers coppery red, fading lighter, large, dbl., intense fragrance; vigorous growth; [Hadley X Lady Inchiquin]; Felberg-Leclerc

'LISBETH STELLMACHER', Pol, or, 1919; bud coppery orange-red; flowers coppery red with golden yellow, striped pink, small, dbl., intense fragrance; [Aglaia X Marie van Houtte]; Lambert, P.

'LISBETH VON KAMECKE', HMult, lp, 1910; flowers small, semi-dbl.; Kiese; (Sangerhausen)

'LISBOA', HT, dr, 1953; flowers red shaded darker, large; very vigorous growth; [Barcelona X Crimson Glory]; da Silva, Moreira

'LISBOA DE 1947', HT, 1947; Moreira da Silva, A.; (Cavriglia)

'LISE CHIAVASSA', HT, mr, 1931; flowers carmine, very large, dbl., cupped, intense fragrance; foliage glossy; vigorous growth; [Mme Philippe Rivoire X Yves Druhen]; Buatois

Lise Palais, *see* **'OPAL'**

Liselle, *see* 'RULIS'

Liselotte Hollweg, F, mr, 1962

Lisette, HT, ly; Select Roses, B.V.

'LISETTE DE BERANGER', HP, dp; Guillot et Fils, 1867

Lisima, *see* GAUzimi

'LISSY HORSTMANN', HT, mr, 1943; flowers brilliant scarlet-crimson, large, dbl., 28 petals, cupped; foliage leathery; vigorous growth; [Hadley X Heros]; Tantau; C-P

'LITAKOR', HT, ab, 1973; ('KORLITA', **Lolita**®); flowers golden bronze, dbl., 28 petals, 5 in., moderate fragrance; ADR, 1973; [Colour Wonder X Seedling]; Kordes, R.; Dicksons of Hawlmark, 1972

Literary Giant, HT, pb, 1991; Clements, John K.

'LITLIN', Min, w, 1983; (**Autumn Frost**); flowers medium, dbl., 35 petals, exhibition form, slight fragrance; foliage medium, medium green, semi-glossy; bushy growth; [Seedling X Seedling]; Rose, Euie; Little Gems Mini Rose Nursery

Little Amigo, *see* 'TINAMIGO'

'LITTLE ANGEL', F, mr, 1961; flowers salmon becoming scarlet, borne in large clusters, dbl., 30 petals; foliage dark, glossy; moderate growth; Verschuren; Blaby Rose Gardens

Little Artist®, *see* 'MACMANLY'

Little Ballerina, *see* 'CURBAL'

'LITTLE BEAUTY', F, mp, 1935; (Crown of Jewels); flowers bright cerise-pink, fading to pink, small, very dbl., slight fragrance; foliage leathery; vigorous, bushy growth; [E.G. Hill seedling X Polyantha]; Howard, F.H.; Dreer

'LITTLE BETTY', S, lp, 1940; flowers soft pink, small, non-recurrent; height 3 ft; [Betty Bland X R. nitida]; Wright, Percy H.

Little Bit o' Sunshine, *see* **'BIT O' SUNSHINE'**

Little Bo-Peep, *see* 'POULLEN'

'LITTLE BREEZE', Min, ob, 1981; bud long, slender; flowers orange-red, fading to pink, loose blooms borne singly, semi-dbl., 17 petals, moderate fragrance; straight, gray prickles; foliage large, dark, glossy; vigorous growth; [Anytime X Elizabeth of Glamis]; McCann, Sean

'LITTLE BRIDESMAID', HT, lp; flowers shrimp-pink, well shaped; Archer

Little Brother, *see* 'PIXBRO'

'LITTLE BUCKAROO', Min, mr, 1956; flowers bright red, small, dbl., 23 petals, moderate, fresh apple fragrance; foliage bronze, glossy, leathery; height 14–16 in.; [(R. wichurana X Floradora) X (Oakington Ruby X Floradora)]; Moore, Ralph S.; Sequoia Nursery

'LITTLE CAMEO', F, pb, 1981; bud medium, pointed; flowers pink blend with cream outer petals, dbl., 26 petals, borne mostly singly, slight fragrance; straight, short, red prickles; foliage medium green, leathery; vigorous, upright growth; PP005381; [Lara X Unnamed seedling]; Strahle, Robert; Carlton Rose Nurseries

Little Carol, *see* 'TINCAROL'

'LITTLE CHAMELEON', Min, pb, 1977; bud pointed; flowers rose-pink to red, small, dbl., 30 petals, 1.5 in., slight fragrance; upright growth; [Little Amy X ?]; Lyon

Little Charm, *see* 'LYOLIT'

'LITTLE CHIEF', Min, dp, 1971; bud long, pointed; flowers deep pink, small, semi-dbl.; foliage small, glossy, leathery; growth moderate, dwarf, bushy; [Cotton Candy X Magic Wand]; Moore, Ralph S.; Sequoia Nursery

'LITTLE COMPTON CREEPER', LCl, dp, 1938; flowers deep rose-pink, borne in open clusters, single, yellow-orange red fruit; foliage glossy, dark; Brownell, H.C.

Little Crimson, *see* 'SAVACRIM'

'LITTLE CURT', Min, dr, 1971; bud long, pointed; flowers deep velvety red, medium, semi-dbl.; foliage dark, leathery; vigorous, upright, bushy growth; [Unnamed red Cl F X Westmont]; Moore, Ralph S.; Sequoia Nursery

'LITTLE DARLING', F, yb, 1956; bud ovoid; flowers blend of yellow and soft salmon-pink, well-formed, dbl., 27 petals, 2.5 in., moderate, spicy fragrance; foliage dark, glossy, leathery; very vigorous, spreading growth; David Fuerstenberg Prize, ARS, 1964 GM, Portland, 1958; [Capt. Thomas X (Baby Chateau X Fashion)]; Duehrsen; Elmer Roses Co.

Little Debbie, MinFl, op

Little Devil, *see* 'DELPO'

Little Devil, Climbing, *see* 'DELPOSAR'

Little Diamond, *see* 'SUNDIA'

Little Diane, Min, op

'LITTLE DICKENS', Min, ob, 1979; bud ovoid; flowers orange-red and yellow, small, dbl., 25 petals, cupped, slight fragrance; foliage small; compact growth; [(Ma Perkins X Sheri Anne) X Over the Rainbow]; Schwartz, Ernest W.; Nor'East Min. Roses

'LITTLE DORRIT', Pol, mp, 1930; flowers glowing pink; [Coral Cluster sport]; Reeves

'LITTLE DORRIT, CLIMBING', Cl Pol, mp, 1935; Letts

'LITTLE DOT', Pol, lp, 1889; flowers soft pink, flaked deeper on the outer petals clusters, very dwarf growth; Bennett

'LITTLE EMBERS', F, rb, 1988; flowers cerise to yellow, reverse yellow with cerise on tips, aging, dbl., 25 petals, exhibition form;

no fruit; prickles long, narrow, small, light green; foliage medium, medium green, matt; upright, low growth; [Rise 'n' Shine X Prominent]; McFarland, John

Little Eskimo, *see* 'MORWIT'

Little Fire, *see* **'OGONIOK'**

'LITTLE FIREBALL', Min, or, 1968; bud ovoid; flowers bright coral-red, small, dbl., moderate fragrance; foliage small, glossy; bushy, compact, low growth; [(R. wichurana X Floradora) X New Penny]; Moore, Ralph S.; Mini-Roses;, Sequoia Nursery

Little Flame™, *see* 'JACNUYE'

'LITTLE FLIRT', Min, rb, 1961; bud pointed; flowers orange-red, reverse yellow, small, dbl., 42 petals, 1.5 in., moderate fragrance; foliage light green; vigorous, bushy (12–14 in.) growth; [(R. wichurana X Floradora) X (Golden Glow (LCl) X Zee)]; Moore, Ralph S.; Sequoia Nursery

'LITTLE GEM', M, dp, 1880; bud heavily mossed; flowers bright deep pink, small blooms in clusters, dbl., intense fragrance; dwarf, compact growth; Paul, W.

Little Gem, *see* 'SUNSALM'

'LITTLE GEM', Pol, lp; Alderton, 1930; (Weatherly, L.)

'LITTLE GIRL', Cl Min, op, 1973; bud long, pointed; flowers coral-salmon-pink, medium, dbl.; foliage glossy; bushy, climbing growth; [Little Darling X Westmont]; Moore, Ralph S.; Sequoia Nursery

Little Green Snake, S, w, 1996; Lens

'LITTLE GUY', Min, mr, 1980; flowers medium red, reverse lighter, small blooms in clusters of 3–7, semi-dbl., no fragrance; no fruit; foliage small, medium green, semi-glossy; low, bushy growth; [Magic Wand X Violette]; Moore, Ralph S.; Moore Min. Roses

Little Huzzy, *see* 'TINHUZZY'

Little Jackie™, *see* 'SAVOR'

Little Jewel, *see* 'COCABEL'

Little Jewel, *see* 'SUNJEW'

'LITTLE JOE', HT, dr, 1921; flowers crimson-red, single; [Red-Letter Day X H.V. Machin]; Looymans; Prior

'LITTLE JOHN', Min, rb, 1959; bud pointed; flowers scarlet, center white, medium, borne in clusters, semi-dbl., exhibition form, slight fragrance; foliage soft; vigorous, bushy (12–15 in.) growth; [Baby Masquerade sport]; Mason, F.

'LITTLE JOKER', Min, pb, 1958; flowers rose-pink, center cream, well-shaped, bushy growth; Spek

'LITTLE JUAN', Min, mr, 1966; flowers medium red, reverse lighter, small, dbl., slight fragrance; foliage small, leathery; vigorous, dwarf growth; [Juliette X Seedling]; Williams, Ernest D.; Mini-Roses

'LITTLE JULIET', HT, yb, 1924; flowers apricot and peach on yellow ground, dbl., slight fragrance; [F.J. Looymans seedling X ?]; Looymans; Prior

'LITTLE LADY', Pol, w, 1967; flowers blush to ice-white, small, dbl., 70 petals, slight fragrance; dwarf growth; [Schneewittchen (F) X Baby Faurax]; Harkness

'LITTLE LEAGUER', F, dr, 1962; bud ovoid; dbl., 28–40 petals, 2–2.5 in., cupped, slight fragrance; vigorous, upright growth; [Garnette seedling X Yuletide]; Jelly; E.G. Hill Co.

Little Lemmy, *see* KORbalem

Little Len, MinFl, ab, 1987; Hatfield & Buckley

Little Lighthouse, *see* 'CLELIGHT'

'LITTLE LINDA', Min, ly, 1976; bud high-pointed; flowers micro-mini, semi-dbl., 17 petals, 1 in., exhibition form, slight fragrance; compact growth; [Gold Coin seedling X Seedling]; Schwartz, Ernest W.; Nor'East Min. Roses

'LITTLE LIZA', Min, mp, 1975; bud mossy; flowers rose-pink, semi-dbl., 15 petals, .5–1 in.; low, very compact growth; [Fairy Moss X Fairy Moss]; Saville, F. Harmon; Nor'East Min. Roses

'LITTLE LOVE', Min, mp, 1977; bud pointed; flowers rose-pink, open, small, dbl., 25–30 petals, 1.5 in., slight fragrance; foliage tiny; very low, compact growth; [Little Amy X Unnamed seedling]; Lyon; L. Lyon Greenhouses

Little Mabel, Min, mr

Little Magician™, *see* 'WILMAG'

Little Marvel, *see* 'RUIGERDAN'

Little Meghan, *see* 'SPOMEG'

'LITTLE MELODY', Min, pb, 1979; bud globular; flowers soft peach-yellow blend, dbl., 38 petals, exhibition form, moderate fragrance; light brown prickles, curved downward; foliage medium green; compact, bushy growth; [Neue Revue X Sheri Anne]; Strawn, Leslie E.; Tiny Petals Nursery

Little Mermaid, *see* 'MORWIMER'

'LITTLE MIKE', Min, dr, 1967; bud ovoid; flowers deep red, small, dbl., exhibition form; foliage dark, glossy, leathery; vigorous, dwarf growth; [((R. wichurana X Floradora) X Seedling) X Little Buckaroo]; Moore, Ralph S.; Sequoia Nursery

Little Miss Muffet, *see* POUlense

'LITTLE MISS MUFFETT', F, pb, 1940; bud pointed, cerise; flowers bright rose-pink, reverse deeper, open, large, borne in clusters, semi-dbl., slight fragrance; strong stems; vigorous, bushy growth; [Else Poulsen X Étoile de Hollande]; LeGrice; C-P

'LITTLE NELL', HT, yb, 1933; flowers deep cream, center apricot, reverse primrose, well shaped, intense fragrance; foliage glossy, dark; vigorous growth; Archer

Little Nugget, *see* 'SUNGOLD'

Little One, *see* 'SANONE'

Little Opal, *see* 'SUNPAT'

Little Paradise™, *see* 'WEKLIPS'

Little Peaces, *see* 'TINPEACES'

Little Pearl, *see* 'SUNPEARL'

Little Pim®, *see* 'LENNOP'

'LITTLE PINK', F, pb, 1965; flowers pink, center creamy pink, small, very dbl., exhibition form, slight fragrance; foliage glossy; vigorous, compact growth; [Little Darling X ?]; Castleberry

Little Pink Hedge, F, pb, 1992

Little Pioneer, *see* 'CURNEER'

'LITTLE POOH', Min, or, 1979; bud slim, pointed; single, 5–6 petals, flat, borne 5–6 per cluster, moderate fragrance; triangular prickles; foliage heavy, thick; upright growth; [Anytime X Unnamed seedling]; Fong, William P.

'LITTLE PRANCER', Cl HT, mr, 1996; flowers red, lighter reverse, blooms borne mostly singly, dbl., 15–25 petals, 3 in., moderate fragrance; moderate prickles; foliage medium, medium green striped with red, dull; upright, tall (7–9 ft) growth; [Don Juan X Joseph's Coat]; Thurman, Robert R.; Thurman's Nursery

Little Prince, *see* 'COCCORD'

'LITTLE PRINCESS', Pol, ob, 1937; flowers pale salmon-coral, well-formed, dbl.; vigorous, bushy growth; Knight, G.; Beckwith

'LITTLE PRINCESS', F, mp, 1955; flowers rose-pink, reverse lighter, medium, dbl., 50–75 petals, 2 in., cupped, moderate fragrance; long stems; vigorous growth; [Garnette sport]; Sodano, J.

Little Princess, *see* **'PIXIE'**

Little Purple, F, m

Little Rambler, *see* 'CHEWRAMB'

'LITTLE RASCAL', Min, mr, 1981; flowers medium red, shading to yellow at base, reverse lighter, dbl., 34 petals, exhibition form, slight fragrance; slightly hooked prickles; foliage tiny, light green; compact, bushy growth; [Sheri Anne X Rise 'n' Shine]; Jolly, Betty J.; Rosehill Farm

Little Rascal, *see* PEAalamo

'LITTLE RED', Min, or, 1975; bud ovoid; flowers red-orange, pompon shape, small, semi-dbl., 18–22 petals, 1.5 in., slight fragrance; foliage glossy; compact growth; [New Penny X Coloranja]; Lens; Spek

Little Red Devil, *see* 'AROVIDIL'

Little Red Hedge, *see* KORtabo

'LITTLE RED MONKEY', F, mr, 1954; flowers bright red, short stiff petals, borne singly and in sprays, very dbl.; bushy growth; [Donald Prior X Seedling]; Ratcliffe

Little Red Run-Around, HCh, dr

Little Russel, *see* 'TROBRIC'

Little Sapphire, *see* 'SUNSAP'

'Little Scotch', Min, ly, 1958; bud long; flowers straw-yellow to white, dbl., 55 petals, 1.5 in., moderate fragrance; foliage leathery; vigorous, bushy (12 in.) growth; [Golden Glow (LCl) X Zee]; Moore, Ralph S.; Sequoia Nursery

'Little Showoff', Cl Min, yb, 1960; bud pointed; flowers bright yellow, sometimes tinted red, dbl., 30 petals, 1–1.5 in., exhibition form, moderate fragrance; upright (to 4 ft) growth; [Golden Glow (LCl) X Zee]; Moore, Ralph S.; Sequoia Nursery

Little Shrimp®, *see* 'LENLIT'

Little Silver, *see* 'JACSIL'

'Little Sir Echo', Min, mp, 1977; bud long, pointed; dbl., 48 petals, 1–1.5 in., exhibition form, moderate fragrance; foliage matt, green; compact, upright growth; [Ma Perkins X Baby Betsy McCall]; Schwartz, Ernest W.; Nor'East Min. Roses

'Little Sir Echo, Climbing', Cl Min, mp, 1985; [Little Sir Echo sport]; Watterberg, Leah

Little Sister, *see* 'PIXSIS'

Little Sizzler™, *see* 'JACIAT'

Little Slam®, *see* 'KINSLAM'

'Little Smiles', Min, yb, 1978; bud ovoid; flowers chinese yellow vermilion, semi-dbl., 12 petals, 1 in., moderate, spicy fragrance; compact growth; [Q17a X Redgold]; Lyon

Little Squirt, *see* 'TINSQUIRT'

Little Star Rose, *see* 'BENSTAR'

'Little Starburst'™, Min, yb, 1991; flowers orange-red washing on golden yellow, very small blooms, dbl., slight fragrance; foliage small, medium green, semi-glossy; low, bushy growth; PP006821; [Prominent X Rise 'n' Shine]; Williams, J. Benjamin, 1988

'Little Stephen', HT, mp, 1971; flowers glowing pink, dbl., 40 petals, 3 in., exhibition form, moderate fragrance; foliage matt; upright growth; [Gavotte X Seedling]; Sanday, John

Little Stripes, *see* 'SHERILOWSTRI'

'Little Sunset', Min, pb, 1967; flowers salmon-pink on yellow, star-shaped blooms in clusters; foliage small, light green; [Seedling X Tom Thumb]; Kordes

'Little Sunshine', Pol, my, 1915; flowers creamy yellow, varying to deep golden yellow, occasionally flecked crimson, dbl., 1.5–2 in.; dwarf growth; [R. multiflora nana X Soleil d'Or]; Cumming; A.N. Pierson

Little 't', *see* 'TRAT'

Little Tease, *see* 'ZIPTEASE'

Little Tiger, *see* 'MORSHAKI'

Little Tommy Tucker™, *see* 'TUCTOMMY'

Little Tyke, *see* 'RENYKE'

Little Vegas, *see* 'SPOVEGAS'

'Little Wallace', F, pb, 1952; bud very long; flowers light pink, center yellow, blooms in clusters, dbl., 25 petals, exhibition form, slight fragrance; foliage leathery, dark; vigorous growth; [Elfe sport]; Beall; Beall Greenhouse Co.

Little White Pet, *see* **'White Pet'**

Little White Spray®, Min, w, 1991; Lens

Little Wings™, Min, ly

Little Woman, *see* 'DICLITTLE'

'Little Wonder', F, or, 1975; bud pointed; dbl., 22–28 petals, 3.5–4 in., moderate, spicy fragrance; foliage small; upright to spreading growth; [Duftwolke X Ena Harkness]; Huber

Little Woods Rose, *see* **R. gymnocarpa**

'Littlest Angel', Min, my, 1976; bud short, pointed; flowers medium to deep yellow, micro-mini, dbl., 28 petals, .5 in., exhibition form, slight fragrance; foliage small; low, compact, bushy growth; [Gold Coin seedling X Unnamed Miniature seedling]; Schwartz, Ernest W.; Nor'East Min. Roses

Littlest Spartan, *see* 'GRULIT'

Live Wire™, *see* 'SAVAWIRE'

'Lively', HT, dp, 1959; flowers rose pink, dbl., 32 petals, 4–6. in., intense fragrance; foliage dark, glossy; vigorous, compact, low growth; [Wellworth X Ena Harkness]; LeGrice

'Lively Lady', F, or, 1969; flowers vermilion, large, dbl., slight fragrance; foliage dark, glossy; [Elizabeth of Glamis X Tropicana]; Cocker

Liverpool®, *see* **'Liverpool Echo'**

Liverpool Daily Post, *see* 'FRYtrooper'

'Liverpool Echo', F, op, 1971; (Liverpool®); flowers salmon, dbl., 23 petals, 4 in., exhibition form, slight fragrance; foliage light; tall growth; GM, Portland, 1979; [(Little Darling X Goldilocks) X Munchen]; McGredy, Sam IV; McGredy

Liverpool Remembers, *see* 'FRYSTAR'

'Liverton Lady', Cl HT, mp, 1978; semi-dbl., 16–20 petals, 3–4 in., slight fragrance; foliage small, glossy; tall, climbing growth; [Bantry Bay X Sympathie]; Warner; Bradley Nursery

Livia, *see* MEIkola

Livin' Easy™, *see* 'HARWELCOME'

'Living', HT, rb, 1957; flowers reddish orange, reverse copper streaked red, dbl., 24 petals, 5–6. in., exhibition form, moderate, spicy fragrance; foliage leathery, semi-glossy; vigorous growth; [Charlotte Armstong X Grande Duchesse Charlotte]; Lammerts, Dr Walter; Consolidated Nursery

Living Bouquet, *see* 'LAVLINGER'

'Living Coral', HT, op, 1974; bud ovoid; flowers soft coral-pink, dbl., 28 petals, 5 in., exhibition form, moderate, fruity fragrance; foliage glossy; moderate growth; [Queen of Bermuda X Tropicana]; Golik; Dynarose

'Living Fire', F, ob, 1972; flowers orange, suffused orange-red, rosette form, dbl., 33 petals, 2.5 in., moderate fragrance; foliage dark; [Tropicana X Unknown seedling]; Gregory, 1973

Living Well, S, mp, 1999

Liz, *see* MEIrebuc

Liza®, *see* BARliz

Lizabeth's Lullabye, *see* 'WIRCUROB'

'Lizzie Molk', S, pb, 1983; flowers light pink, white center, blooms in clusters, dbl., 20 petals, does not repeat, intense fragrance; foliage medium, blue-green, glossy; spreading growth; Rusnock, Ann M.

'Ljuba Rizzoli'®, HT, dr, 1980; (Via Romana); bud pointed; dbl., 25 petals, cupped, borne singly, intense fragrance; purple-green prickles; foliage large, dark, matt; upright, bushy growth; Dot, Simon; Rose Barni-Pistoia

'Lleida', HT, rb, 1936; flowers bright red, reverse yellow, intense fragrance; very vigorous growth; [Edouard Renard X Condesa de Sástago]; Dot, Pedro; A. Meilland

Lloyd Center Supreme, *see* 'TWOLOY'

Lloyds of London, *see* CANlloyd

Lloyds Supreme, HT, mp; [Brion X Silver Jubilee]; Twomey, Jerry, 1995

Loads of Pink, *see* 'FIALOPI'

'Lobo', HT, rb, 1987; flowers red-purple, reverse white, large, dbl., 26–40 petals, exhibition form, intense fragrance; foliage large, dark green, matt; upright growth; [Kordes' Perfecta X Gavotte]; Perry, Astor, 1989

'L'Obscurité', M, dr, 1848; flowers dark garnet-crimson, large, semi-dbl.; Lacharme, F.

'Locarno', Pol, or, 1926; flowers large, borne in huge clusters; vigorous, bushy growth; [Orléans Rose sport]; deRuiter

'Locomotion', HT, yb, 1965; flowers salmon tinted peach, becoming citron-yellow; foliage dark, leathery; vigorous, bushy growth; [R.M.S. Queen Mary X Lady Sylvia]; Verschuren, H.A.M.; Ravensberg

'Lodestar', HT, pb, 1953; bud ovoid to globular; flowers buff-pink, center peach, dbl., 35–40 petals, 4.5–5 in., intense, spicy fragrance; foliage leathery; bushy growth; [Diamond Jubilee seedling X Serenade]; Boerner; J&P

'LODEWIJK OPDEBEEK', HT, dr, 1921; flowers oxblood-red, reverse rose, dbl., moderate fragrance; [Jonkheer J.L. Mock X Mev. Dora van Tets]; Leenders, M.

Lodovico, F, 1960; Cazzaniga, F. G.; (Cavriglia)

'LOETA LIGGETT', HT, pb, 1984; flowers light pink, darker pink reverse, often with salmon center; [Duet sport]; Liggett, Myron T.

'LOHENGRIN', HT, lp, 1903; flowers large, dbl.; Kiese; (Sangerhausen)

Lois, *see* 'TINLOIS'

'LOIS CROUSE', HT, lp, 1937; bud pointed; flowers light pink, suffused salmon, peony form, large, dbl.; foliage dark; vigorous, bushy growth; [Mme Butterfly seedling]; Moore, Ralph S.; Brooks & Son

'LOIS MANEY', LCl, op, 1953; flowers salmon-pink, dbl., 4–5 in., abundant, non-recurrent bloom, slight fragrance; foliage leathery; very vigorous (25 ft) growth; [R. maximowicziana pilosa X Templar]; Maney; Iowa State College

Lois Wilson, HT, pb, 1998

Lola de Porcioles, HT; Camprubi, C.; (Cavriglia)

Lola Montes, HT, mr, 1969; Dot

'LOLA VENDRELL', HT, pb, 1957; flowers soft pink streaked deeper, well formed, large, dbl., 40 petals, moderate fragrance; foliage glossy; long stems; vigorous growth; [Serafina Longa X Mme Kriloff]; Bofill; Toree Blanca

'LOLETTE DUPAIN', Cl Pol, pb, 1918; flowers yellowish rose, reverse silvery rose, borne in clusters, dbl., sometimes recurrent bloom; vigorous growth; [Casimir Moulle X Mme Norbert Levavasseur]; Lottin

'LOLI CREUS', HT, dp, 1953; flowers carmine, well formed, large; foliage glossy; strong stems; very vigorous growth; [Cynthia X Manuelita]; Dot, Pedro

Lolita®, *see* 'LITAKOR'

'LOLITA', HT, w, 1937; flowers white, center cream, overlarge, dbl., globular; foliage leathery; vigorous growth; [Frau Karl Druschki X Unnamed Hybrid Tea]; Croibier

'LOLLIPOP', Min, mr, 1959; flowers bright red, dbl., 35 petals, 1–1.5 in., slight fragrance; foliage glossy; vigorous (14 in.), bushy growth; [(R. wichurana X Floradora) X Little Buckaroo]; Moore, Ralph S.; Sequoia Nursery

'LOLLO', HT, rb, 1949; bud pointed; flowers purplish red, reverse crimson-carmine, well formed, dbl., 32–36 petals, moderate fragrance; foliage dark; strong stems; vigorous, bushy growth; [Lele X Crimson Glory]; San Remo Exp. Sta.

'LONDON PRIDE', HT, op, 1954; flowers deep salmon-pink shaded coral, exhibition form, moderate fragrance; upright growth; Ratcliffe

'LONDON STARLETS', HT, ob, 1959; flowers orange to orange-red, well shaped, large, dbl., 50 petals, moderate fragrance; long, strong stems; vigorous growth; [Mission Bells X Jiminy Cricket seedling]; Maarse, G.

'LONDON TOWN', HT, op, 1955; flowers salmon-pink, base buff, well formed, medium, moderate, fruity fragrance; vigorous growth; [Peace X Charles Gregory]; Letts

'LONE STAR', HT, mr, 1925; flowers velvety cardinal-red, large, semi-dbl., moderate fragrance; vigorous, upright growth; [Étoile de France seedling]; Buller; Hillje

'LONE STAR STATE', HT, pb, 1943; flowers darker than parent, irregularly striped white; [Texas Centennial sport]; Collins

'LONETTE CHENAULT', Pol, lp, 1925; flowers creamy rose pink, medium, dbl.; Chenault; (Sangerhausen)

Long Island®, *see* 'CASLON'

'LONG JOHN SILVER', LCl, w, 1934; bud pointed; flowers silvery white, very large blooms in clusters, dbl., cupped, moderate fragrance; foliage large, leathery; vigorous, climbing growth; [R. setigera seedling X Sunburst]; Horvath; J&P

Long Tall Sally™, *see* 'WEKAJAZOUL'

Long White Cloud, *see* 'MACLANOFLON'

'LONGCHAMP', F, mr, 1960; flowers bright red cerise, medium, borne in clusters, semi-dbl.; foliage glossy; vigorous, symmetrical growth; Laperrière; EFR

Longchamp 80, S, 1980; Laperrière; (Cavriglia)

'LONGFORD', S, dp, 1938; flowers carmine-pink, small to medium, dbl.; Central Exp. Farm; (Sangerhausen)

Longleat, *see* 'MACINCA'

'LONGWOOD', LCl, mp, 1914; semi-dbl.; [American Pillar X Unnamed seedling]; Wintzer

'LONGWORTH RAMBLER', LCl, mr, 1880; flowers light crimson, semi-dbl.; vigorous growth; Liabaud

Looks Like Fun, *see* 'SEAFUN'

Looping®, *see* 'MEIROVONEX'

'LORD ALLENBY', HT, dr, 1923; flowers bright crimson, very large, dbl., exhibition form, slight fragrance; foliage rich green, leathery; dwarf, sturdy growth; Dickson, A.

'LORD BACON', HP, dr, 1883; flowers deep crimson shaded scarlet, large, dbl.; globular, vigorous growth; Paul & Son

'LORD BADEN-POWELL', HT, ob, 1937; bud orange striped red; flowers saffron-yellow, large, very dbl., moderate fragrance; vigorous, bushy, compact growth; Leenders, M.

Lord Byron, *see* 'MEITOSIER'

'LORD CALVERT', HT, dr, 1919; flowers dark velvety red, dbl., intense fragrance; [Radiance X Hoosier Beauty]; Cook, J.W.

'LORD CASTLEREAGH', HT, dr, 1927; bud pointed; flowers dark blackish crimson, open, semi-dbl., intense fragrance; foliage dark, leathery; vigorous, bushy growth; Dickson, A.

'LORD CHARLEMONT', HT, dr, 1922; bud long, pointed; flowers clear deep crimson, well shaped, large, dbl., exhibition form, intense fragrance; foliage dark, leathery; bushy growth; McGredy

'LORD CHARLEMONT, CLIMBING', Cl HT, dr, 1932; Hurcombe

'LORD CHELMSFORD', HT, lp; Williams, H., 1906; (Weatherly, L.)

'LORD CLYDE', HP, dr, 1863; flowers bright crimson, well formed, large; Paul & Son

Lord Don, S, rb, 1993

'LORD FAIRFAX', HT, dp, 1925; flowers cherry-rose-pink, moderate fragrance; foliage leathery; long stems; good habit growth; Gray, W.R.

'LORD FREDERICK CAVENDISH', HP, mr; flowers bright scarlet; Frettingham, 1884

Lord Gold®, *see* DELgold

Lord Houghton of Sowerby, *see* 'HARTUBOND'

'LORD KITCHENER', HT, dp, 1918; flowers bright carmine-rose, dbl., intense fragrance; Chaplin Bros.

'LORD LAMBOURNE', HT, yb, 1925; bud pointed; flowers buttercup-yellow, edged carmine-scarlet, very large, dbl., exhibition form, moderate fragrance; foliage light, leathery, glossy; very vigorous, bushy growth; McGredy

'LORD LONSDALE', HT, ab, 1933; bud pointed; flowers deepest orange-yellow, large, dbl., exhibition form, moderate fragrance; foliage glossy, light; vigorous, bushy growth; GM, NRS, 1931; Dickson, A.; Dreer;, H&S

'LORD LOUIS', HT, dp, 1981; bud pointed; flowers light crimson, blooms borne several together, dbl., 30 petals, moderate fragrance; foliage mid-green, glossy; vigorous growth; [Pink Favorite X Seedling]; Gregory, C.; C. Gregory & Sons

'LORD MACAULAY', HP, dr, 1874; flowers crimson, globular; Paul, W.

'LORD PENZANCE', HEg, yb, 1894; flowers soft rosy yellow, paler at base, yellow stamens, blooms in clusters, single, summer bloom, moderate fragrance; foliage small, dark, fragrant; very vigorous growth; [R. eglanteria X Harison's Yellow]; Penzance

'LORD RAGLAN', HP, dr, 1854; flowers bright velvety crimson, very large, very dbl.; [Geant des Batailles seedling]; Guillot Père

'LORD ROSSMORE', HT, w, 1930; flowers creamy white shaded rose toward edge, large, dbl., exhibition form, moderate fragrance; foliage dark olive-green, leathery; vigorous growth; GM, NRS, 1928; Hall; McGredy

Lord Scarman, HGal, pb, 1996; Scarman

'LORD STAIR', HT, dr, 1930; bud pointed; flowers velvety crimson-scarlet, large, dbl., exhibition form, moderate fragrance; foliage dark, leathery; very vigorous, bushy growth; Smith, T.

'LORD WORTHINGTON', HT, mp, 1928; flowers medium, dbl.; Dickson, A.; (Sangerhausen)

Lordly Oberon, *see* 'AUSRON'

L'Oréal Trophy, *see* 'HARLEXIS'

'LOREE', F, w, 1969; bud long, pointed; flowers white tinted pink, small, dbl., cupped, slight fragrance; foliage leathery; vigorous, bushy, compact growth; [Frolic X ?]; Pal, Dr B.P.; Son

'LORELEI', S, ab, 1947; bud ovoid; flowers peach-pink, center yellow, semi-dbl., 3–4 in., non-recurrent, intense fragrance; foliage leathery, dark; vigorous, upright growth; [Joanna Hill X Harison's Yellow]; Fisher, R.C.

'LORELEY', Pol, lp, 1913; flowers creamy rose pink, medium, dbl.; Kiese; (Sangerhausen)

Loreley 82®, HT, dr, 1982; Noack, Werner

Lorena®, *see* 'KORENLO'

Lorenz Schwamborn, Pol, mr, 1960; flowers medium, semi-dbl.; Schmid, P.; (Sangerhausen)

'LORENZO PAHISSA', HT, op, 1941; flowers coral, very large, dbl., moderate fragrance; foliage abundant; very vigorous, upright growth; GM, Bagatelle, 1941; [Seedling X Mari Dot]; Pahissa

Loretta®, *see* 'KORENPI'

'LORETTO', LCl, pb, 1922; flowers reddish, center white, blooms in clusters, semi-dbl., moderate fragrance; few prickles; vigorous, climbing growth; [Jersey Beauty X Seedling]; Clark, A.; Brundrett

'LORI ANN', F, yb, 1990; flowers bright amber yellow, flushed orange and pink, reverse yellow, dbl., 20 petals, cupped, moderate, fruity fragrance; foliage medium, medium green, semi-glossy, slightly elongate; slightly spreading, medium growth; [Judy Garland X Sutter's Gold]; Schneider, Peter, 1987

'LORI NAN', Min, dp, 1965; bud globular; flowers rose-red, small, dbl.; foliage glossy, leathery; moderate growth; [(R. wichurana X Floradora) X (Seedling X Zee)]; Moore, Ralph S.; Sequoia Nursery

Lorina, *see* 'KORENLO'

'LORNA', HT, op, 1936; bud pointed; flowers salmon, dbl., exhibition form; foliage leathery, glossy, light; vigorous, bushy growth; Cant, B. R.

'LORNA ANDERSON', HT, mr, 1940; flowers well-formed, moderate fragrance; Clark, A.

'LORNA DOONE', F, mr, 1972; dbl., 24 petals, 4 in., slight fragrance; foliage dark, glossy; [Red Dandy X Lilli Marleen]; Harkness

'LORNA MAY', F, rb, 1958; flowers crimson-red, center white, large trusses, single, 2 in.; foliage light green; vigorous growth; [Poulsen's Pink X Kathe Duvigneau]; deRuiter; Blaby Rose Gardens

'LORRAINE', HT, op, 1945; flowers salmon-carmine touched red, large, dbl.; vigorous growth; [Peace X Mme Mallerin]; Meilland, F.

'LORRAINE LEE', T, pb, 1924; bud pointed; flowers rosy apricot-pink, dbl., cupped, moderate fragrance; foliage rich green, leathery, glossy; vigorous growth; (14); [Jessie Clark X Capitaine Millet]; Clark, A.; Hackett

'LORRAINE LEE, CLIMBING', Cl T, pb, 1932; McKay

'LORRAINE STEBBINGS', Min, w, 1985; flowers small blooms borne singly, dbl., 23 petals, slight, honey fragrance; foliage small, glossy; upright, bushy growth; [Cinderella X Popcorn]; Morey, Dr Dennison

'LOS ANGELES', HT, op, 1916; bud pointed; flowers coral-pink, base gold, large, dbl., intense fragrance; foliage leathery; vigorous, spreading growth; GM, Bagatelle, 1918; [Mme Segond Weber X Lyon Rose]; Howard, F.H.; H&S

'LOS ANGELES BEAUTIFUL', Gr, yb, 1967; flowers yellow blended with coral and scarlet, medium, dbl., exhibition form, slight fragrance; foliage dark, leathery; vigorous, upright, compact growth; [Queen Elizabeth X Rumba]; Lammerts, Dr Walter; Germain's

Los Angeles Frost, HT, lp, 1989; Robinson

'LOS ANGELES, CLIMBING', Cl HT, pb, 1925; H&S

'LOS TEJAS', HT, mr, 1968; bud globular; flowers large, dbl., exhibition form, moderate fragrance; foliage leathery; vigorous, upright growth; [Chrysler Imperial X Happiness]; Patterson; Patterson Roses

Lost in Paradise, Min, m, 1997; Spooner, Raymond A.

'LOTTE GÜNTHART', HT, mr, 1964; bud ovoid; flowers bright red, peony form, large, very dbl., 90 petals; foliage leathery; tall, upright, bushy growth; [Queen Elizabeth X Bravo]; Armstrong, D.L.; Armstrong Nursery

'LOTTIE FORSTER', Pol, lp; Matthews, W.J., 1915; (Weatherly, L.)

'LOU-CELINA', S, dr, 1979; flowers large, dbl., 50 petals, repeat bloom, moderate fragrance; foliage medium, dark red-green, semi-glossy; upright growth; [(Venture X ((Cecilia X China Belle) X Suzanne)) X (Paula X Soeur Kristin)]; James, John; Historical Roses, 1986

'L'OUCHE', HCh, pb, 1901; flowers rose shaded yellow; Buatois, 1891

'LOUIS BARBIER', HFt, ob, 1909; flowers coppery orange with dark yellow striping, medium, semi-dbl.; Barbier; (Sangerhausen)

Louis Bernard, HT, or, 1969; Croix

'LOUIS BOURGOIN', HT, lp, 1921; flowers flesh-pink passing to silvery pink, dbl., slight fragrance; [Jonkheer J.L. Mock X Frau Karl Druschki]; Gillot, F.

Louis Bugnet, HRg, w, 1960; Bugnet

'LOUIS CALLA', HP, m, 1885; flowers purple/pink, large, dbl.; Verdier, E.; (Sangerhausen)

Louis de Funès®, *see* 'MEIRESTIF'

Louis d'Or, HT, 1965; Dorieux; (Cavriglia)

'LOUIS FAURAX', HT, ob, 1941; flowers coppery-salmon with dark yellow, large, dbl.; Gaujard; (Sangerhausen)

'LOUIS GIMARD', M, mp, 1877; flowers bright pink, very large blooms, dbl.; prickles well-mossed; long stems; vigorous growth; Pernet Père

'LOUIS JOLLIET', HKor, mp, 1991; flowers medium, dbl., 26–40 petals, slight fragrance; foliage medium, medium green, semi-glossy; spreading, low growth; Ogilvie, Ian S.; Agriculture Canada, 1992

'LOUIS KAHLE', HT, mr, 1922; flowers bright cherry-red, intense fragrance; [Lieutenant Chaure X Étoile de France]; Kiese

'LOUIS PAJOTIN', HT, op, 1940; flowers coral, stamens yellow, large, very dbl.; long, stiff stems; very vigorous growth; [Souv. de Claudius Pernet X Margaret McGredy]; Mallerin, C.; Meilland, A.

'LOUIS PAJOTIN, CLIMBING', Cl HT, op, 1959; Pajotin-Chédane

Louis Philippe, *see* **'GRANDISSIMA'**

'LOUIS PHILIPPE', Ch, rb, 1834; flowers dark crimson with edges of center petals blush, aging crimson, dbl., globular; bushy growth; Guérin

Louis Philippe d'Angers, HGal, lp; Hardy, 1824

'LOUIS PHILIPPE D'ORLEANS', HP, 1884; Verdier, E.; (Cavriglia)

'LOUIS PHILIPPE, CLIMBING', Cl Ch, rb

Louis Rambler, HMult, 1997; Lens, Louis; (Cavriglia)

Louis Riel, *see* 'ZUBLOU'

'LOUIS RÖDIGER', LCl, ob, 1935; bud pointed; flowers orange shaded yellow and red, open, very large, semi-dbl., moderate, fruity fragrance; foliage leathery, wrinkled, dark; very vigorous, climbing trailing growth; [Daisy Hill X (Charles P. Kilham X Mev. G.A. van Rossem)]; Kordes

'LOUIS SAUVAGE', HWich, m, 1914; flowers purple/red, small, dbl.; Turbat; (Sangerhausen)

'LOUIS VAN HOUTTE', HP, dr; flowers crimson-maroon, well-formed, large, dbl., 40 petals, some recurrent bloom, intense fragrance; [Général Jacqueminot X ?]; Lacharme, F., 1869

Louis van Till, HGal, mp; flowers luminous pink, very dbl., intense fragrance; (Sangerhausen)

'LOUIS VAN TYLE', m; flowers light crimson, shaded black or purple, small, semi-dbl.

'LOUIS WALTER', HT, ab, 1938; flowers golden orange-yellow, large, dbl.; foliage clear green, glossy; vigorous growth; [Mrs Pierre S. duPont seedling X Charles P. Kilham]; Mallerin, C.

'LOUIS XII', T, m; flowers violet, cupped

'LOUIS XIV', HP, dr, 1859; flowers dark crimson, medium, dbl., 25 petals, intense fragrance; foliage sparse; moderate growth; Guillot et Fils

'LOUISA JANE', HT, m, 1976; flowers center soft white, shading to mauve-pink; [Baronne Edmond de Rothschild sport]; Ross, A., & Son; A. Ross & Son

Louisa Jane Morris, *see* 'BOSADSON'

'LOUISA SCHULTHEIS', HT, op, 1925; flowers pink and salmon, center darker, dbl., 32–36 petals, slight fragrance; [Golden Ophelia X Ruhm von Steinfurth]; Schultheis, A.

Louisa Stone, *see* HARbadge

'LOUISE', HT, mp, 1924; bud pointed; flowers rose-pink shaded cerise, large, dbl., intense fragrance; foliage dark, leathery; vigorous growth; [Isobel X Unnamed variety]; Prince

'LOUISE ABDY', F, mp, 1964; flowers pink, base yellow, single, 6 petals, 4 in., moderate fragrance; foliage dark; vigorous growth; [Donald Prior X McGredy's Yellow]; Abdy

'LOUISE BALDWIN', HT, ob, 1919; bud pointed; flowers rich orange, tinted soft apricot, dbl., exhibition form, intense fragrance; vigorous growth; McGredy

'LOUISE CATHERINE BRESLAU', HT, op, 1912; flowers shrimp-pink shaded reddish coppery orange, reverse chrome-yellow, dbl., moderate fragrance; foliage dark, bronze, leathery; bushy growth; [Unnamed variety X Soleil d'Or seedling]; Pernet-Ducher

'LOUISE CATHERINE BRESLAU, CLIMBING', Cl HT, op, 1917; Kordes

Louise Clements, *see* CLElou

'LOUISE CRETTÉ', HP, w, 1915; flowers snow-white, center creamy white, well-formed, dbl., 55 petals, 6–7 in., exhibition form, moderate fragrance; foliage dark; vigorous, bushy growth; [Frau Karl Druschki X Kaiserin Auguste Viktoria, Climbing]; Chambard, C.

'LOUISE CRINER', HT, w, 1919; flowers snow-white, center creamy, dbl.; few thorns; [Unknown seedling X Louise Crette]; Chambard, C.

'LOUISE D'ARZENS', N, w; Lacharme, F., 1861

'LOUISE DE VILMORIN', HT, ab, 1944; bud ovoid; flowers orange-yellow, overlarge, very dbl., slight fragrance; foliage glossy, dark; dwarf growth; Gaujard

'LOUISE ESTES', HT, pb, 1991; bud pointed; flowers pink blend, reverse white, aging medium pink, medium, dbl., 35 petals, 4 in., exhibition form, quick repeat, moderate, fruity fragrance; foliage medium, medium green, matt, disease-resistant; upright, medium growth; Silver Medal, ARC TG, 1991; [Unnamed seedling X Miss Canada]; Winchel, Joseph F.; Coiner Nursery, 1992

Louise Gardner, *see* 'MACERUPT'

'LOUISE GAUJARD', HT, op, 1941; flowers coppery pink shaded coral, open, very large, very dbl., cupped, moderate fragrance; foliage light green; vigorous, upright growth; [Mme Joseph Perraud X Seedling]; Gaujard

'LOUISE HOPKINS', HMult, lp, 1923; flowers white, center shell-pink, very dbl., 200–225 petals, non-recurrent; vigorous, climbing (20 ft or more) growth; [Trier sport]; Hopkins

'LOUISE JOLY', HT, or, 1922; flowers coral-red, shaded shrimp-pink, dbl., moderate fragrance; [Unnamed seedling X Mme Edouard Herriot]; Buatois

'LOUISE KRAUSE', HT, ob, 1930; flowers reddish orange, passing to golden yellow, large, dbl., slight fragrance; foliage dark, glossy; vigorous, bushy growth; [Mrs Beckwith X Souv. de H.A. Verschuren]; Krause

'LOUISE LE CARDONNEL', HT, yb, 1939; bud globular; flowers yellow tinted coral, very large, dbl.; stiff stems; very vigorous growth; [Unnamed seedling X Mev. G.A. van Rossem]; Mallerin, C.; Meilland, A.

'LOUISE MACK', Min, dr, 1992; dbl., 30 petals, 1.25 in., exhibition form, borne usually singly, slight fragrance; foliage medium, dark green, semi-glossy; medium, spreading growth; [Unknown X Unknown]; Jerabek, Paul E., 1984

'LOUISE MÉHUL', HGal, pb; flowers light red, spotted white, large, flat; Parmentier

'LOUISE ODIER', B, dp, 1851; (Mme de Stella); flowers bright rose-pink, well-formed, full, dbl., cupped, intense fragrance; prickles very few; foliage medium green, matt; tall growth; (28); Margottin

'LOUISE PERNOT', HT, lp, 1903; flowers large, dbl.; Robichon; (Sangerhausen)

'LOUISE PEYRONY', HP, dp, 1844; flowers deep pink shaded carmine, very large, dbl., moderate fragrance; moderate growth; [La Reine seedling]; Lacharme, F.

'LOUISE PIGNÉ', T, mp, 1905; flowers china-pink, base buff-yellow, petals crinkled, very large, very dbl., moderate fragrance; very vigorous growth; [Mme Eugene Resal X Mme Lombard]; Pigné

'LOUISE VERGER', M, lp, 1860; flowers bright pink, medium, dbl.; Robert et Moreau

'LOUISE WALTER', Pol, pb, 1909; flowers white and flesh-pink, open, dbl.; foliage small, rich green; very dwarf growth; (14); [Tausendschön X Rosel Dach]; Walter, L.; P. Lambert

'LOUISIANA', HT, w, 1969; flowers creamy white, dbl., 38 petals, 3–4 in., exhibition form, slight fragrance; foliage dark, leathery; upright growth; [Unnamed seedling X Unnamed seedling]; Weeks

Louisiana, *see* 'KORNEMARK'

Louisiana Lady™, *see* 'TALLOU'

'LOUISIANA PURCHASE', HT, dp, 1954; bud long, pointed; flowers rich cerise, dbl., 20–25 petals, 4–5 in., cupped, intense, damask fragrance; foliage dark, leathery; very vigorous, upright growth; [Charlotte Armstrong X Piccaninny]; Swim, H.C.; Stark Bros.

Louisiana, Climbing, Cl HT, w, 1985; Weeks

'LOUISVILLE LADY', HT, pb, 1986; flowers bright pink, silver reverse, medium, dbl., 35 petals, exhibition form, intense fragrance; foliage medium, dark, semi-glossy; bushy growth; [Osiria X Unnamed seedling]; Weddle, Von C.

Louita, F, rb

Louksor, *see* 'DELCRAFT'

'LOUMUSKOKA', HT, w, 1999; (**Muskoka Moonlight**); flowers near white, medium, dbl., 26–40 petals; some prickles; foliage medium, medium green, glossy; vigorous, spreading, tall growth; very hardy; [Pristine X Lichtkönigin Lucia]; Lougheed, Larry; Lougheed Hybrid Roses, 1998

Louqsor®, *see* 'DELCRAFT'

'LOURDES', HT, dy, 1959; bud ovoid; flowers golden yellow, dbl., 35–50 petals, 4–5 in., moderate fragrance; foliage leathery, glossy, dark; vigorous growth; hardy for the class; [V for Victory X (New Dawn sport X Copper Glow sport)]; Brownell, H.C.; Stern's Nursery

Lourdes Arroyo®, *see* FEprogo

Lou's Sonoma Rose, HCh, dp; (found rose)

'LOUVIOLA', Gr, pb, 1997; (**Viola Lougheed**); flowers full, large, very dbl., 26–40 petals, borne in large clusters, moderate fragrance; many prickles; foliage medium,

medium green, semi-glossy; tall, spreading growth; [Pristine X Peter Frankenfeld]; Lougheed, Larry

Louvre®, *see* 'DELFAT'

'LOVABLE', HT, lp, 1979; bud long, pointed; dbl., 30 petals, 4.5–6 in., exhibition form, moderate fragrance; vigorous, upright, bushy growth; [(Helen Traubel X Michele Meilland) X ((Blanche Mallerin X Peace) X (Peace X Virgo))]; Leon, Charles F., Sr.

'LOVANIA', S, or, 1978; bud ovoid; flowers bright red-orange, pompon shape, small, semi-dbl., 18–22 petals, 1.5 in., recurrent bloom, moderate, fruity fragrance; foliage dark; vigorous, upright, climbing growth; [Robin Hood X (New Penny X Coloranja)]; Lens, 1977

Love, *see* 'JACTWIN'

'LOVE', Cl HT, or, 1935; flowers scarlet, open, semi-dbl., moderate fragrance; foliage dark; long stems; vigorous, climbing ($6\frac{1}{2}$–10 ft), bushy growth; RULED EXTINCT 1/79; [Hadley X Ami Quinard]; Caron, B.; H. Guillot

'LOVE AFFAIR', HT, dr, 1970; bud short pointed; flowers brilliant red, large, very dbl., exhibition form, moderate fragrance; foliage large, dark, leathery; vigorous, upright growth; [Undisseminated red seedling X Forever Yours]; Jelly; E.G. Hill Co.

'LOVE BUG', Min, ob, 1986; flowers orange, white base, reverse orange veining to white base, fades slightly, semi-dbl., 14 petals, exhibition form, slight fragrance; short, small, pink prickles, hooked slightly downward; foliage medium, medium green, semi-glossy; bushy, medium growth; [Heartland X Unnamed seedling]; Bridges, Dennis A., 1987

'LOVE CALL', HT, pb, 1989; bud ovoid; flowers deep pink, lighter at base, reverse light pink, large, dbl., 50 petals, exhibition form, moderate, damask fragrance; prickles pointed slightly downward, medium, light green; foliage medium, dark green, glossy; upright, tall growth; [Lady X X Wini Edmunds]; Bridges, Dennis A.; Bridges Roses, 1990

Love Dove, *see* 'LYOVE'

Love in Bloom, *see* 'JUDLOV'

Love In Bloom, Min, m

Love Knot, *see* 'CHEWGLORIOUS'

'LOVE LETTER'®, F, w, 1977; flowers creamy white, dbl., 30–35 petals, 3–3.5 in., cupped, intense fragrance; foliage glossy, dark; vigorous, bushy, upright growth; [Pink Parfait X Rosenelfe]; Lens, 1980

Love Magic, HT, dr

Love Me, *see* 'WALLOVE'

Love Note, *see* 'ZIPNOTE'

Love Potion™, *see* 'JACSEDI'

'LOVE SONG', HT, pb, 1955; (Liebeslied); bud ovoid; flowers neyron rose, reverse yellow, dbl., 45 petals, 4.5–5 in., cupped, intense fragrance; foliage dark, glossy; vigorous, upright growth; [Peace X Orange Nassau]; Fisher, G.; C-P

Love Story®, *see* 'TANVERY'

'LOVE TOKEN', F, mp, 1964; flowers peach-pink, well formed, dbl., 28 petals, 2 in.; foliage dark, glossy; vigorous growth; Gregory

Love Torch™, *see* 'MINABECO'

Love 'Ya Dad, *see* 'TINDAD'

Love You, *see* 'NEMLOVE'

Lovebird, *see* FRAnmilro

Loveglo™, *see* 'MINJCO'

'LOVELIEST', HT, lp, 1956; bud ovoid; flowers clear rose-pink, large, dbl., exhibition form, intense fragrance; foliage leathery, light green; vigorous, upright growth; [Charlotte Armstrong X Juno]; Leon, Charles F., Sr.

'LOVELIGHT', F, mp, 1959; bud short, pointed; flowers large, dbl., 33 petals, exhibition form, slight fragrance; vigorous, upright growth; [Garnette X Unnamed seedling]; Jelly; E.G. Hill Co.

'LOVELINESS', HWich, lp, 1935; flowers pale pink, white spot at base, round, large, borne in large clusters, dbl.; foliage light; vigorous, climbing growth; Chaplin Bros.

'LOVELY', HT, dp, 1936; bud pointed; flowers carmine-pink, large, semi-dbl., exhibition form, moderate, violet fragrance; foliage soft; vigorous, compact growth; H&S; Dreer

Lovely Dream, *see* HILraz

Lovely Fairy, *see* SPEvu

Lovely Girl, *see* MEIjasonl

Lovely Jubilee, HT, m, 1992; Interplant

Lovely Lady, *see* 'DICJUBELL'

'LOVELY LADY', HT, mr, 1934; bud long, pointed; flowers pure rose-red, large, very dbl., exhibition form, intense fragrance; foliage leathery; long, strong stems; very vigorous, open hait growth; RULED EXTINCT 10/86; [Better Times sport]; Asmus

Lovely Lorrie, Min, mp, 1998; Moore

Lovely Louise, F, ab

Lovely Pink Hybrid Musk, HMsk, mp; (found rose)

Lovely Rosamini, Min, lp

'LOVERLY', F, ab, 1986; flowers medium, semi-dbl., moderate fragrance; foliage medium, medium green, semi-glossy; bushy growth; Silver Medal, ARC TG, 1985; [Restless Native X Apricot Nectar]; Stoddard, Louis

'LOVERS' MEETING', HT, ob, 1980; bud pointed; flowers bright orange, blooms borne singly and in clusters, dbl., 25 petals, exhibition form, moderate fragrance; short prickles; foliage bronze; strong, upright growth; [Unnamed seedling X Egyptian Treasure]; Gandy, O.L.

Lovers Only, *see* 'SEALOVE'

Love's Spring, *see* 'MACIVY'

Loving, *see* JACka

Loving Memory, *see* 'KORGUND '81'

'LOVING SON', HT, w, 1994; very dbl., 1.5–2.75 in., borne mostly singly, moderate fragrance; some prickles; foliage medium, medium green, semi-glossy; medium, bushy growth; [Honor X Captain Harry Stebbings]; Mengel, Russell R.; Mengel, 1994

'LOVING TOUCH'™, Min, ab, 1983; flowers mini-flora, blooms borne usually singly, exhibition form, 25 petals, exhibition form, slight; globular fruit; foliage medium, medium green, semi-glossy; bushy, spreading growth; AOE, 1985; [Rise 'n' Shine X First Prize]; Jolly, Nelson F.; Rosehill Farm, 1982

Lovita, *see* **'KIBOH'**

'LOVITA', HT, mr, 1965; bud ovoid; flowers bright red, large, dbl.; foliage dark; vigorous, upright growth; [Baccará X (Independence X Peace)]; Meilland; C-P, 1967; Moerheim, 1965

Lovita, F, ab, 1989

Lowburn Rose, Misc OGR, dp

Lowea berberifolia, *see* **'HULTHEMI PERSICA'**

'LOWELL THOMAS', HT, dy, 1943; (Botaniste Abrial); bud long, pointed; flowers rich yellow, dbl., 38 petals, 4–4.5 in., exhibition form; foliage leathery; vigorous, upright, bushy, compact growth; AARS, 1944; GM, Portland, 1944; [Mme Mélanie Soupert X Nonin]; Mallerin, C.; A. Meilland, 1943;, C-P, 1943

'LOWELL THOMAS, CLIMBING', Cl HT, dy, 1954; Armstrong, J.A.; Armstrong Nursery

Loyal Rosarian, *see* 'LEOLOYROS'

'LOYALIST', Misc OGR, lp, 1773; flowers pale rose fading almost white, very dbl., cupped, blooms well in June; repeats little, moderate fragrance; bushy, densely branched growth; [Possibly R. damascena X R. virginiana]

Luarca, HT; Dot, Simon; (Cavriglia)

Lübeck, *see* **'HANSESTADT LÜBECK'**

Lübecker Rotspon®, *see* 'TANTIDE'

'LUBOV CHEVTSOVA', HT, lp, 1956; flowers light pink, base tinted orange, large (5 in.), foliage glossy; strong stems; upright growth; [Cathrine Kordes X Peace]; Sushkov, K. L.

'LUBRA', HT, dr, 1938; bud long, pointed; flowers dark crimson, dbl., exhibition form, intense fragrance; foliage leathery, dark; vigorous growth; [Ophelia, Climbing seedling X Black Boy]; Fitzhardinge; Hazlewood Bros.

Luc Steeno, HT, mp, 1997; RvS-Melle

'Luc Varenne', F, mr, 1959; bud oval; flowers scarlet, open, medium, borne in clusters, semi-dbl.; foliage dark, glossy; vigorous, bushy growth; [Alain X Montrouge]; Delforge

'LUCdod', S, mp, 1981; (**Rendez-vous**); flowers medium, semi-dbl., intense fragrance; foliage medium, medium green, matt; bushy growth; [R. wichurana X Alain Blanchard]; Lucas, C.C.

Luce di Todi™, *see* 'WILluce'

'Lucens Erecta', S, op, 1921; Paul & Son

Lucetta®, *see* 'AUSemi'

Luchian, F, lp; globular; flowers large, velvety vermillion red, dbl., 35 petals, slight; [Paprika X Coup de Foudre]; Palocsay and St. Wagner, 1971; Res. Stn. f. Horticulture, Cluj, 1972, Romania

Lucia®, S, my

Lucia Cotarelo®, *see* FEtural

'Lucia Zuloaga', HT, rb, 1932; flowers velvety brownish scarlet with a golden undertone, open, very large, semi-dbl., moderate, fruity fragrance; foliage glossy, dark; [Duquesa de Peñaranda X F. Cambo]; Dot, Pedro; C-P, 1934

'Lucie Duplessis', M, mp; flowers rosy white; Robert, 1853

'Lucie Fernand-David', HT, w, 1924; flowers white, center slightly tinted cream, dbl., intense fragrance; Chambard, C.

'Lucie Marie', HT, yb, 1930; bud pointed; flowers buttercup-yellow veined apricot-orange and shaded salmon-cerise, dbl., exhibition form, moderate fragrance; foliage dark, leathery, glossy; vigorous, compact, bushy growth; Dickson, A.

'Lucie Nicolas Meyer', HT, dp, 1922; flowers dark pink, edged lighter, dbl., moderate fragrance; [Jonkheer J.L. Mock X Unnamed seedling]; Gillot, F.

'Lucie Petzka', Pol, dr, 1953; flowers medium, dbl.; Petzka; (Sangerhausen)

'Lucien Chaurè', HT, lp, 1913; flowers large, dbl.; Soupert & Notting; (Sangerhausen)

'Lucien de Lemos', HT, lp, 1905; flowers large, dbl., moderate fragrance; Lambert, P.; (Sangerhausen)

'Lucien Duranthon', HP, mr, 1894; flowers large, dbl.; Bonnaire; (Sangerhausen)

Lucie's Dream, *see* 'MEIcitrem'

'Lucifer', Pol, or, 1931; flowers small, dbl.; Easlea; (Sangerhausen)

'Lucile', HWich, lp, 1911; flowers flesh-pink, base tinged rosy salmon, borne in clusters, dbl.; foliage large, rich green, glosy; vigorous, climbing growth; Walsh

'Lucile Barker', HT, ab, 1922; bud pointed; flowers apricot-yellow, semi-dbl., exhibition form, moderate fragrance; foliage bronze; vigorous growth; Hicks

'Lucile Duplessis', HGal, dp, 1836; Vibert

'Lucile Hill', HT, dp, 1939; bud long, pointed, rose-red; flowers spinel-pink, very large, dbl., slight fragrance; foliage leathery; long stems; very vigorous, compact growth; [(Senior X De Luxe) X Sweet Adeline]; Hill, Joseph H., Co.

'Lucile Rand', HT, pb, 1930; bud pointed; flowers brilliant carmine, shaded yellow and orange, very large, dbl., intense fragrance; very vigorous, semi-climbing, bushy growth; Pernet-Ducher; Gaujard

'Lucile Supreme', HT, dp, 1941; flowers rose-red, dbl., 25–35 petals, 6 in.; [Lucile Hill sport]; Hill, Joseph H., Co.

'Lucilla', F, ab, 1992; semi-dbl., 6–14 petals, 1.5 in., borne in small clusters, slight fragrance; some prickles; foliage small, medium green, semi-glossy; low (35 cms), compact growth; [Conservation sport]; Stainthorpe, Eric; Battersby Roses, 1993

Lucille Ball, *see* 'JACapri'

Lucille McWherter, *see* 'WELlsned'

'Lucille Ross', LCl, w, 1940; flowers white, center golden, open, semi-dbl., slight fragrance; vigorous, climbing (20–40 ft) growth; [Dr W. Van Fleet sport]; Ross; Roselawn Gardens

'Lucinda', HT, dp, 1927; [Columbia sport]; Heacock

Lucinde®, *see* KORtaly

'Luciole', T, dp; bud long; flowers carmine-rose, base coppery yellow, large, intense fragrance; Guillot et Fils, 1887

'Luciole', HWich, rb, 1923; flowers bright scarlet, center white, small, single, slight fragrance; vigorous, climbing growth; [Hiawatha X Unknown seedling]; Nonin

Lucious, *see* 'RESjuc'

'Lucky', F, dr, 1962; flowers deep velvety red, semi-dbl., 13 petals, 3 in.; [Goldilocks X Independence]; Leenders, J.

'Lucky Beauty', HT, pb, 1970; bud ovoid; flowers pink, yellow reverse, large, very dbl., 90 petals, exhibition form, moderate fragrance; vigorous, upright, bushy growth; [Kordes' Perfecta X Lucky Piece]; Fuller; Wyant

'Lucky Charm', F, yb, 1961; flowers bright yellow tipped red, borne in clusters of 6–10, semi-dbl., 18 petals, 3.5–4 in., intense fragrance; foliage glossy; vigorous growth; Robinson, H.; Lowe

Lucky Charm, *see* 'MORain'

'Lucky Choice', HT, w, 1985; flowers creamy white, flushed light pink in center, large blooms borne 1–3 per stem, dbl., 35 petals, exhibition form, slight fragrance; few, small, slender prickles; foliage large, medium green; upright growth; [American Heritage X Sodori-Hime]; Ota, Kaichiro

Lucky Lady, *see* 'ARMlu'

Lucky Me, *see* SOCluck

'Lucky Piece', HT, pb, 1962; flowers copper, pink and gold blend; [Peace sport]; Gordon; Wyant

'Lucky Star', HT, yb, 1936; flowers golden yellow, suffused flame-scarlet, [Souvenir sport]; Armacost; Armacost & Royston

Lucretia, F, mr

'Lucy', Cl HT, dp, 1936; flowers brilliant carmine, well formed, large; foliage glossy; very vigorous, climbing growth; Williams, A.

Lucy, *see* BARluc

Lucy, *see* MORlucy

'Lucy Ashton', HEg, w, 1894; flowers pure white, edged pink, single, seasonal bloom; foliage dark, fragrant; vigorous growth; (42); Penzance

'Lucy Bertram', HEg, rb, 1895; flowers dark shining crimson, center white, single, seasonal bloom; foliage dark, fragrant; very vigorous growth; Penzance

'Lucy Constable', HT, pb, 1924; flowers silver pink, reverse deep salmon pink, moderate fragrance; Lilley

'Lucy Cramphorn', HT, or, 1960; (Maryse Kriloff); flowers signal-red, well-formed, dbl., 40 petals, 5 in., moderate fragrance; foliage glossy; vigorous, upright growth; [Peace X Baccará]; Kriloff, Michel; Cramphorn's Nursery

'Lucy Marguerite', HT, yb, 1978; flowers buttercup-yellow, diffused red, full, dbl., 35–40 petals, 4 in., moderate fragrance; foliage small, glossy, dark; [Val De Mosa X Denise-Anne]; Ellick; Excelsior Roses

'Lucy Nicolas', HT, r, 1935; bud pointed; flowers coppery salmon, large, dbl., exhibition form; foliage glossy, bronze; very vigorous growth; [Odette Foussier X Cécile Walter]; Mallerin, C.; C-P;, H. Guillot

'Lucy Thomas', Cl HP, mp, 1924; flowers pink, center brighter, semi-dbl., rarely recurrent bloom; [Ulrich Brunner Fils X Georg Arends]; Nabonnand, P.

'Lucyle', HT, mp, 1933; flowers glowing pink, very large, dbl., intense fragrance; foliage leathery, dark; vigorous, bushy growth; Vestal

'Ludek Pik', HP, mr, 1933; flowers large, very dbl.; Böhm, J.; (Sangerhausen)

Ludmilla®, *see* 'LAPmau'

LUDpinbu, Min, ab, 1993; (**Joy Button**); Taschner

LUDsporcoma, HT, rb, 1995; (**Flames 'n' Sparks**); Taschner

LUDsportiana, HT, dp, 1995; (**Alberton Amor**); Taschner

LUDswenic, HT, ob, 1995; (**Nicolette**®); Taschner

Ludvik Vecera, LCl, mr, 1981; flowers medium, dbl., moderate fragrance; Vecera, L.; (Sangerhausen)

'LUDWIG MÖLLER', HP, ab; flowers bright amber-yellow, fading white; [Frau Karl Druschki X Marechal Niel seedling]; Kiese, 1915

'LUDWIG OPPENHEIMER', HT, dr, 1932; bud pointed; flowers crimson-scarlet, large, dbl., moderate fragrance; foliage dark; vigorous growth; [Villa Pia X Capitaine Georges Dessirier]; Leenders Bros.

Ludwigshafen, *see* 'KORLUDWIG'

Ludwigshafen am Rhein®, *see* 'KORLUDWIG'

'LUIS BRINAS', HT, ob, 1934; bud long, pointed; flowers rose-orange, large, dbl., cupped, moderate fragrance; foliage soft; vigorous growth; GM, Bagatelle, 1932 GM, Portland, 1934; [Mme Butterfly X Federico Casas]; Dot, Pedro; C-P

Luis Desamero, *see* 'TINLUIS'

'LUISA FERNANDA DE SILVA', HT, rb, 1946; flowers reddish passing to purplish and then yellowish red, base yellow, moderate fragrance; foliage dark, glossy; vigorous growth; Dot, Pedro

Luisa Stone, S, 1997; Harkness, R.; (Cavriglia)

Luise Kiese, *see* **'FRAU LUISE KIESE'**

'LUISE KÜHNEL', HT, dy, 1937; flowers medium, dbl.; Kühnel; (Sangerhausen)

'LUISE LILIA', HT, dr, 1912;, intense fragrance; moderate growth; Lambert, P.

Luisella Angelini, HT, 1980; Dot, Simon; (Cavriglia)

'LUKOR', HT, rb, 1973; (Jolly, **Lustige**); bud ovoid; flowers copper-red, reverse yellow, large, dbl., cupped, moderate fragrance; foliage large, glossy, leathery; vigorous, upright growth; [Peace X Brandenburg]; Kordes

'LULLABY', Pol, w, 1953; bud ovoid; flowers white, center flushed pink, blooms in loose clusters, dbl., 75 petals, 1.5–2 in., cupped, slight fragrance; foliage dark, leathery; vigorous, bushy, compact growth; [(R. soulieana X Mrs Joseph Hiess) X Cécile Brunner]; Shepherd; Bosley Nursery

'LULLABY', F, dp, 1957; flowers rich rose-pink, camellia shape, borne in clusters (to 28), semi-dbl., 3 in., flat; vigorous growth; Bishop; Baker's Nursery

'LULU', HT, op, 1919; bud very long, pointed, deep orange-red; flowers salmon-pink, large, single, 8 petals, slight fragrance; foliage glossy; bushy, compact growth; Easlea

'LULU', F, ob, 1973; flowers orange-pink, medium, dbl., exhibition form, slight fragrance; foliage glossy, dark, bronze; vigorous, upright, bushy growth; [Zorina X Seedling]; Kordes

'LULUETTE', Min, mp, 1986; flowers small blooms borne singly and in small sprays, single, no fragrance; small, straight, red prickles; foliage small, medium green, semi-glossy; low, compact, tiny growth; [Fairy Moss X Fairy Moss]; Fischer, C.&H.; Alpenflora Gardens

Lumen®, HT, ab, 1990; Sauvageot

'LUMIÈRE', HT, yb, 1944; flowers golden yellow suffused capucine-red, large; foliage dark, leathery; vigorous growth; Mallerin, C.; A. Meilland

'LUMINA', F, or, 1955; flowers orange-scarlet, rosette shape, blooms in large trusses, dbl., 25 petals, 2.5 in.; foliage dark; vigorous growth; [Fanal X Alpine Glow]; Tantau, Math.

'LUMINATOR', Pol, dr, 1938; flowers scarlet-crimson; vigorous growth; [Lady Reading sport]; Smith, J.; Eddie

Luminion®, *see* 'KORMIORA'

'LUMINOSA', HT, or, 1964; bud long, pointed; flowers vermilion, large, dbl., 52 petals, exhibition form; foliage dark, glossy; strong stems; vigorous, upright, bushy growth; Mondial Roses

'LUMINOUS', Pol, or, 1932; flowers brilliant scarlet tinted orange, small, dbl., cupped, slight fragrance; foliage small, light, wrinkled; dwarf growth; [Gloria Mundi sport]; deRuiter; J&P

'LUM'S DOUBLE WHITE', HT, w, 1930; flowers pure white, overlarge, semi-dbl., exhibition form, moderate fragrance; foliage dark, leathery; vigorous growth; [Killarney Double White sport]; Lum; Totty

'LUNA', HT, ly, 1918; bud pointed; flowers pale yellow, large, dbl., exhibition form, intense fragrance; foliage dark; strong stems; very vigorous, bushy growth; [Harry Kirk X Sunburst]; Poulsen, S.

'LUNA PARK', LCl, or, 1964; flowers red shaded orange, large, recurrent bloom; vigorous, climbing growth; [Gladiator X Unnamed seedling]; Croix, P.

Luna Rossa, HT; Delbard-Chabert; (Cavriglia)

Lund's Jubiläum, *see* **'BABY BLAZE'**

Lundy's Lane Yellow, HT, ob; (found rose)

'LUNELLE', HT, lp, 1955; (George Sand); flowers pale pink, very large, dbl., 50 petals, exhibition form, intense fragrance; vigorous growth; [Young France X Signora]; Meilland, F.

'LURAY', F, mr, 1958; flowers rose-red, medium, dbl., 40–45 petals, 2–3 in., moderate fragrance; foliage leathery, glossy; vigorous, bushy growth; [Patty's Pink sport]; Masek; Carlton Rose Nurseries

Lusambo, *see* 'MEILUSAM'

Lustige, *see* 'LUKOR'

'LUSTRE', HT, mp, 1926; flowers rose-pink, large, semi-dbl., moderate fragrance; [Ophelia X Hoosier Beauty]; Hill, E.G., Co.; Amling Bros.

'LUSTRE D'ENGLISE', HGal, mp, 1813; flowers medium pink with lighter reverse, small, dbl., intense fragrance; Roseraie de l'Hay; (Sangerhausen)

Lustrous, *see* **'CELEBRITY'**

'LUTETIA', HT, yb, 1961; flowers coppery yellow, medium, dbl., 28 petals; vigorous, upright growth; [The Optimist X Tudor]; deRuiter

'LUTHER RUSSELL', F, op, 1956; flowers salmon, borne several together, semi-dbl.; foliage olive-green; [Korona sport]; Morse

Lutin, *see* 'MEIRADIA'

Lutin, *see* MEIsecaso

Luv Ya, *see* 'JULESLOVE'

'LUVVIE', Min, pb, 1979; bud ovoid; flowers soft to deep coral pink, micro-mini, very small, dbl., 38 petals, exhibition form, slight fragrance; straight, thin, red prickles; foliage small, deep green with red peduncle and petiole; low, bushy growth; [Little Darling X Over the Rainbow]; Bennett, Cecilia 'Dee'; Tiny Petals Nursery, 1980

Luxembourg, *see* **'FANNY BLANKERS-KOEN'**

'LUXEMBOURG', M, dr; flowers crimson, not very mossy; vigorous growth

Luxembourg, *see* **'MARIE ADÉLAÏDE'**

'LUXEMBOURG, CLIMBING', Cl HT, op, 1932; flowers pinkish orange, base deep yellow; very vigorous, climbing growth; [Marie Adélaide sport]; Wight

Luxor, HT, mp

'LUXURY', HT, rb, 1968; flowers light red, reverse white, large, dbl., exhibition form, moderate fragrance; foliage soft; moderate growth; [Unnamed seedling X Suspense]; Patterson; Patterson Roses

Lybelle, LCl, mp, 1984; flowers large, dbl.; Noack, Werner; (Sangerhausen)

Lybie, *see* 'MARLYB'

'LYCORIS', M, pb; flowers light rosy red spotted white, large, dbl., flat

Lyda Rose, *see* LETlyda

'LYDIA', HT, dy, 1949; bud long, pointed; flowers intense saffron-yellow, medium, very dbl., exhibition form, moderate fragrance; foliage dark, leathery, glossy; vigorous, bushy growth; [Phyllis Gold X Seedling]; Robinson, H.; Baker's Nursery

'LYDIA'®, S, ob, 1973; (Clubrose Lydia); bud ovoid; flowers deep orange, reverse yellow, medium, semi-dbl., cupped, moderate fragrance; foliage glossy, dark, leathery; very vigorous, upright, climbing growth; [Seedling X Circus]; Kordes

'LYDIA', HT, op, 1933; bud pointed; flowers bright orange-rose, large, very dbl., moderate fragrance; foliage glossy; vigorous, bushy growth; [Briarcliff X Florex]; Verschuren; Dreer;, H&S

Lydia, *see* INTerlis

Lydia, HT, op; Croix

'LYDIA GRIMM', HT, ly, 1907; flowers large, dbl., moderate fragrance; Geduldig; (Sangerhausen)

Lydia Morris, S, m

Lydie, F, ob, 1993; Harkness

Lykke Dazia, HT, ob

Lykke White Pet, Pol, w

'LYKKEFUND', LCl, w, 1930; flowers cream color, blooms in clusters, semi-dbl., intense fragrance; no prickles; foliage dark, glossy; vigorous growth; [Seedling from R. helenae, Barbier's form, possibly crossed with Zephirine Drouhin sport]; Olsen

Lyla Barbour, HT, lp; Watt, Mrs B., 1970; (Weatherly, L.)

Lynda, HT, 1970; Mondial Roses; (Cavriglia)

'LYNDA HURST', HT, mp, 1938; flowers large, dbl.; vigorous growth; [Mme Abel Chatenay X Seedling]; Clark, A.; NRS Victoria

Lyndarajha, HT, 1982; Dorieux; (Cavriglia)

Lynette, *see* 'KORLYN'

Lynn Anderson, *see* 'WEKJOE'

'LYNN ANNE', Min, ob, 1981; bud short, pointed; flowers orange-yellow blend, blooms borne singly and in large sprays, dbl., 38 petals, exhibition form, moderate fragrance; long, thin prickles; foliage medium; vigorous, upright, compact growth; [Rise 'n' Shine X Sheri Anne]; Saville, F. Harmon; Nor'East Min. Roses

Lynne Elizabeth™, *see* 'BYRLYNNE'

Lynne Gold®, *see* 'MORLYN'

'LYOAD', Min, or, 1984; (**Added Touch**); bud small; flowers orange-red, touch of yellow in the center and back of petals, dbl., 20 petals, moderate fragrance; foliage small, medium green, semi-glossy; upright, bushy growth; [Dandy Lyon X Seedling]; Lyon

'LYOCA', Min, ab, 1982; (**Care Deeply**); flowers medium, dbl., 35 petals, moderate fragrance; foliage medium, medium green, semi-glossy; upright, bushy growth; [Honey Hill X Seedling]; Lyon, Lyndon; L. Lyon Greenhouses

'LYOCH', Min, dr, 1982; (**Cherokee Fire**); flowers deep red, medium, semi-dbl., no fragrance; foliage medium green, semi-glossy; upright, bushy growth; [Merry Christmas X Seedling]; Lyon, Lyndon; L. Lyon Greenhouses

'LYOCL', Min, mp, 1983; (**Classic Love**); flowers small blooms in clusters, dbl., 20 petals, no fragrance; foliage small, medium green, semi-glossy; upright, bushy growth; [Baby Betsy McCall X Seedling]; Lyon, Lyndon; L. Lyon Greenhouses

'LYOCO', Min, ab, 1982; (**Coral Fantasy**); flowers medium, dbl., 34 petals, moderate fragrance; foliage medium green, semi-glossy; vigorous, upright, bushy growth; [Dandy Lyon X Seedling]; Lyon, Lyndon; L. Lyon Greenhouses

'LYODIT', Min, dr, 1986; (**Ditto**); flowers very small blooms borne usually singly, semi-dbl., 14 petals, no fragrance; no prickles; foliage small, medium green, matt; low, bushy growth; [Baby Betsy McCall X Seedling]; Lyon, Lyndon; M.B. Farm Min. Roses, Inc.

'LYODRO', Min, my, 1986; (**Sun Drops**); dbl., 35 petals, cupped, borne usually singly, slight fragrance; very small, reddish prickles; foliage medium, medium green, matt; low, bushy growth; [Unnamed seedling X Redgold]; Lyon; M.B. Farm Min. Roses, Inc.

'LYOET', Min, dr, 1982; (**Velvet Dreams**); flowers small, semi-dbl., slight fragrance; foliage small, medium green, semi-glossy; very miniature, upright, bushy growth; [Unnamed seedling X Unnamed seedling]; Lyon

'LYOFI', Min, dr, 1982; (**Night Fire**); flowers deep red, petals often edged black, small, dbl., 20 petals, moderate fragrance; foliage medium, dark, semi-glossy; upright, bushy growth; [Unnamed seedling X Unnamed seedling]; Lyon, Lyndon

'LYOFIN', Min, ab, 1986; (**Fine Touch**); flowers pale apricot, deeper in center, large blooms borne singly, dbl., 37 petals, exhibition form, slight fragrance; globular, medium, orange fruit; few, reddish prickles; foliage medium, medium green, matt; medium, upright growth; [Honey Hill X ?]; Lyon; M.B. Farm Min. Roses, Inc.

'LYOFOR', Min, mr, 1986; (**Forevermore**); flowers large blooms borne singly, dbl., 38 petals, exhibition form, intense, spicy fragrance; globular, medium, orange-red fruit; few, reddish, small prickles; foliage medium, medium green, semi-glossy; medium, upright growth; [Unnamed seedling X ?]; Lyon; M.B. Farm Min. Roses, Inc.

'LYOGI', Min, pb, 1982; (**Giggles**); bud pointed; flowers medium pink, white center, small, semi-dbl., slight fragrance; foliage small, medium green, semi-glossy; very small, upright, bushy growth; [Unnamed seedling X Unnamed seedling]; Lyon, Lyndon; L. Lyon Greenhouses

'LYOHI', Min, mr, 1983; (**High Style**); flowers cardinal red, blooms in clusters, dbl., 35 petals, slight fragrance; foliage medium, medium green, semi-glossy; upright, bushy growth; [Unnamed seedling X Unnamed seedling]; Lyon

'LYOLIT', Min, mp, 1984; (**Little Charm**); flowers small, semi-dbl., no fragrance; foliage small, medium green, semi-glossy; upright, bushy growth; [Unnamed seedling X Unnamed seedling]; Lyon, Lyndon

'LYOLITE', Min, ab, 1986; (**Peach Elite**); flowers peach, to loose form, large blooms borne singly, dbl., 26 petals, exhibition form, no fragrance; globular, medium, orange-red fruit; small prickles; foliage large, medium green, matt; upright growth; [Dandy Lyon X ?]; Lyon; M.B. Farm Min. Roses, Inc.

'LYOMA', Min, dr, 1983; (**Magic Splendor**); flowers small blooms in clusters, dbl., 35 petals, no fragrance; foliage medium, medium green, semi-glossy; upright, bushy growth; [Baby Betsy McCall X Seedling]; Lyon, Lyndon

'LYON RAMBLER', HMult, dp, 1909; flowers bright rose-pink, flushed carmine; [Crimson Rambler seedling]; Dubreuil

'LYON ROSE', HT, op, 1907; flowers shrimp-pink, center coral-red shaded yellow, large, dbl., 44 petals, moderate fragrance; (28); GM, Bagatelle, 1909; [Mme Mélanie Soupert X Soliel d'Or seedling]; Pernet-Ducher

'LYON ROSE, CLIMBING', Cl HT, op, 1924; Ketten Bros.

'LYONFARBIGE DRUSCHKI', HP, yb, 1928; flowers yellowish-pink, large, dbl., slight fragrance; Rosarium Sangerhausen; (Sangerhausen)

'LYONNAIS', HP, mp, 1872; flowers large, dbl.; Lacharme, F.; (Sangerhausen)

'LYOPIN', Min, mp, 1984; (**Pink Charmer**); flowers medium, dbl., 20 petals, moderate fragrance; foliage small, medium green, semi-glossy; upright, bushy growth; [Baby Betsy McCall X Unnamed seedling]; Lyon, Lyndon

'LYOPR', Min, my, 1982; (**Precious Moments**); flowers medium, dbl., 35 petals, moderate fragrance; foliage medium, medium green, semi-glossy; upright, bushy growth; [Dandy Lyon X Unnamed seedling]; Lyon

'LYOPRI', Min, pb, 1986; (**I Promise**); flowers pink, reverse pale yellow, medium, borne usually singly, dbl., 28 petals, exhibition form, moderate, spicy fragrance; globular, very small, red-gold fruit; few, small prickles; foliage medium, medium green, semi-glossy; low, bushy growth; [Unnamed seedling X Unnamed seedling]; Lyon; M.B. Farm Min. Roses, Inc.

'LYOPRO', Min, dr, 1985; (**Promenade**); flowers very dark red, informal, ruffled, small blooms borne singly, dbl., 35 petals, no fragrance; foliage medium, dark, matt; bushy growth; [Red Can Can X Unnamed seedling]; Lyon; M.B. Farm Min. Roses, Inc.

'LYORA', Min, ob, 1985; (**Orange Sherbert**); flowers orange, small blooms borne singly, dbl., 35 petals, cupped, moderate fragrance; foliage medium, medium green, matt; upright growth; [Dandy Lyon X ?]; Lyon; M.B. Farm Min. Roses, Inc.

'LYOREN', Min, mr, 1986; (**Renegade**); dbl., 52 petals, cupped, borne usually singly, no fragrance; small, green prickles; foliage medium, medium green, matt; medium, upright growth; [Unnamed seedling X Unnamed seedling]; Lyon; M.B. Farm Min. Roses, Inc.

'LYOSHY', Min, mp, 1985; (**Shy Beauty**); flowers small blooms borne singly, dbl., 55 petals, slight fragrance; foliage medium, medium green, semi-glossy; bushy growth; [Unnamed seedling X ?]; Lyon; M.B. Farm Min. Roses, Inc.

'LYOSS', Min, mr, 1982; (**Endless Love**); flowers medium, dbl., 35 petals, no fragrance; foliage medium, dark, semi-glossy; upright, bushy growth; [Red Can Can X Seedling]; Lyon, Lyndon

'LYOSUN', Min, yb, 1984; (**Sun Sparkle**); flowers yellow blended red, small, dbl., 20 petals, moderate fragrance; foliage small, medium green, semi-glossy; upright, bushy growth; [Dandy Lyon X Unnamed seedling]; Lyon

'LYOTER', Min, mr, 1984; (**Cheery Chatter**); flowers medium, dbl., 20 petals, moderate fragrance; foliage small, medium green, semi-glossy; upright, bushy growth; [Dandy Lyon X Seedling]; Lyon, Lyndon; L. Lyon Greenhouses

'LYOTO', Min, mr, 1982; (**Twinkle Toes**); flowers small, semi-dbl., no fragrance; foliage small, medium green, semi-glossy; very compact growth; [Merry Christmas X Unnamed seedling]; Lyon

'LYOVE', Min, w, 1982; (**Love Dove**); flowers near white, medium, dbl., 35 petals, exhibition form, intense fragrance; foliage medium, medium green, semi-glossy; low growth; [Unnamed seedling X Unnamed seedling]; Lyon, Lyndon

'LYOWE', Min, mr, 1982; (**Wild One**); flowers medium red, lighter center, medium, semi-dbl., no fragrance; foliage medium, dark, semi-glossy; bushy, upright growth; [Merry Christmas X Unnamed seedling]; Lyon

'LYRIC', S, mp, 1951; bud ovoid; flowers rose-pink, blooms in large clusters, dbl., 28 petals, cupped, moderate fragrance; foliage leathery; vigorous (4 ft), upright growth; good for hedge and border; [Sangerhausen X Seedling]; deRuiter; J&P

'LYRICAL', F, pb, 1989; bud pointed; flowers salmon pink shading to cream at base, medium, semi-dbl., 11 petals, flat, borne in sprays of 5–7, moderate, spicy fragrance; fruit not observed; prickles hooked downward, red; foliage medium, dark green, glossy, disease-resistant; upright, bushy, low, winter-hardy growth; [Snow White X Yesterday]; Jobson, Daniel J.

'LYS ASSIA', F, mr, 1958; flowers deep orange-scarlet, borne in small clusters, semi-dbl., 20 petals, 4 in., exhibition form, slight fragrance; foliage glossy; vigorous, upright, bushy growth; [Spartan X Hens Verschuren]; Kordes, R.

Lysa, HT, w, 1989; RvS-Melle

'LYSBETH-VICTORIA', F, lp, 1978; flowers light shell-pink, semi-dbl., 11 petals, 4.5 in., slight fragrance; foliage matt; growth moderate; [Pink Parfait X Nevada]; Harkness

M

M Boncenne, *see* **'Mons. Boncenne'**

'M.A. Keessen', Pol, dp, 1923; flowers darker; [Ellen Poulsen sport]; Keessen

'M.A. Willett', HT, op, 1959; bud globular; flowers pink to salmon-pink, base coppery, large, dbl., moderate fragrance; foliage leathery; moderate, upright growth; [Picture X Tahiti]; Kernovski, V.R.; Langbecker

'M.B.', HT, rb, 1941; flowers red shading to pink, overlaid orange and yellow, open, large, semi-dbl., 25 petals; long stems; vigorous, compact, upright growth; [Dr. W. Van Fleet hybrid X Frau Karl Druschki hybrid]; Brownell, H.C.

'M. Geier', HT, dp, 1929; flowers fiery dark carmine, shaded darker, dbl.; [Augustus Hartmann X Admiral Ward]; Felberg-Leclerc

'M. H. Walsh', HP, mr, 1905; flowers carmine-red, large, dbl., moderate fragrance; Dickson, A.; (Sangerhausen)

'M. Kockel', Pol, lp, 1926; flowers small, dbl.; Kockel; (Sangerhausen)

'M. S. Hershey', HT, dr, 1941; bud long, pointed; flowers velvety crimson, dbl., 33 petals, 4–4.5 in., cupped, slight fragrance; bushy growth; [Seedling X E.G. Hill]; Coddington; C-P

'Ma Fiancée', HT, dr, 1922; flowers dark crimson, often nearly black, dbl., intense fragrance; [Gen. MacArthur seedling X Red-Letter Day]; Van Rossem

'Ma Fille', HT, ob, 1960; flowers orange, large, dbl., 45 petals, slight fragrance; foliage glossy; vigorous, bushy, symmetrical growth; [Berthe Mallerin X Unnamed seedling]; Mallerin, C.; EFR

'Ma Mie', HT, yb, 1955; flowers yellow, edge tinted pink, well formed, very large, dbl., 50 petals; foliage bright green; very vigorous, upright growth; [Peace X Unnamed seedling]; Laperrière; EFR

Ma Pâqueretta, *see* **'Pâquerette'**

'Ma Perkins', F, pb, 1952; bud ovoid; flowers sparkling salmon shell-pink, dbl., 25 petals, 3.5 in., cupped, moderate fragrance; foliage rich green, glossy; vigorous, bushy growth; AARS, 1953; [Red Radiance X Fashion]; Boerner; J&P

'Ma Ponctuée', M, pb, 1847; (Ponctuée); flowers rose, spotted white, medium, semi-dbl., recurrent bloom; moderate growth; Moreau et Robert

'Ma Surprise', S, w, 1872; flowers white tinged salmon, large, dbl., moderate fragrance; [Probably R. roxburghii X R. odorata]; Guillot et Fils

'Ma Tulipe', HT, mr, 1899; flowers carmine-red, large, semi-dbl., moderate fragrance; Bonnaire; (Sangerhausen)

Maaike(TM), HT, pb

Maaseik 750, HMsk, pb, 1994; Lens

'Mab Grimwade', HT, yb, 1937; flowers rich chrome, shaded yellow, center apricot, dbl.; [Souv. de Gustave Prat X ?]; Clark, A.; NRS Victoria

'Mabel Dot', Min, or, 1966; flowers rose-coral, small blooms in clusters, dbl.; foliage small, bronze; [Orient X Perla de Alcanada]; Dot; Minier

'Mabel Drew', HT, w, 1911; flowers deep cream, center canary-yellow, borne in clusters, dbl., moderate fragrance; GM, Bagatelle, 1913 GM, NRS, 1910; Dickson, A.

'Mabel Francis', HT, my, 1943; flowers rose-pink, dbl., 35–40 petals, 5 in., moderate fragrance; foliage light green; free growth; [Leading Lady X Southport]; Bees

'Mabel Jackson', HT, ab, 1924; flowers apricot and pink, dbl., intense fragrance; [Edith Part X Queen Mary]; Easlea

'Mabel Lynas', HT, dr, 1926; bud pointed; flowers crimson-scarlet, base yellow, large, dbl., moderate fragrance; foliage glossy; strong stems; very vigorous, bushy growth; GM, NRS, 1924; McGredy

'Mabel Morrison', HP, w, 1878; flowers flesh-white, becoming pure white, sometimes tinged pink in autumn, dbl., 30 petals, cupped, seasonal bloom; stout, erect growth; [Baroness Rothschild sport]; Broughton

'Mabel Morse', HT, my, 1922; bud pointed; flowers bright golden yellow, well shaped, large, dbl.; foliage dark, bronze; bushy growth; GM, NRS, 1921; McGredy

'Mabel Morse, Climbing', Cl HT, my, 1931; Moulden

'Mabel Morse, Climbing', Cl HT, my, 1932; Ley

'Mabel Prentice', HT, mp, 1923; flowers clear rose-pink; Lippiatt

'Mabel Stewart', LCl, dr, 1942; bud almost black; flowers velvety crimson, semi-single; Clark

'Mabel Turner', HT, pb, 1923; bud pointed; flowers blush, center and reverse rosy carmine, very large, exhibition form, slight fragrance; foliage olive-green; vigorous growth; Dickson, H.

Mabella, *see* 'KORgold'

'Mabelle Stearns', S, mp, 1938; (Maybelle Stearns); flowers peach-pink with silvery reflex, blooms in clusters, dbl., 55 petals, recurrent bloom, intense fragrance; foliage small, glossy, dark; height 2 ft, spreading (6–8 ft) growth; hardy.; [Mrs F.F. Prentiss X Souv. de Georges Pernet]; Horvath; Wayside Gardens Co.

Mableton Crimson China, Ch, rb

'MACaft', HT, op, 1979; (Margaret Trudeau, **Sweepstakes**); bud long, pointed; flowers coral-orange to salmon-pink, imbricated, dbl., 35 petals, 3.5–5 in., exhibition form, moderate fragrance; upright, spreading, bushy growth; [Prima Ballerina X Ginger Rogers]; McGredy, Sam IV; Armstrong Nursery

'MACai', F, pb, 1961; (**Daily Sketch**); flowers petals silver, edged deep pink, well-formed, dbl., 46 petals, 3.5 in., blooms in clusters, moderate fragrance; foliage dark; vigorous, bushy growth; GM, NRS, 1960; [Ma Perkins X Grand Gala]; McGredy, Sam IV; McGredy

'MACal', F, op, 1977; (**Captain Cook**); flowers orange-salmon, 10 petals, 4 in., moderate fragrance; foliage glossy; [Irish Mist X Seedling]; McGredy, Sam IV; Mattock

'MACamster', HT, ab, 1998; (**Matawhero Magic**); flowers apricot blend, double, semi-dbl., 17–25 petals, 4 in., borne mostly singly, intense fragrance; prickles moderate; foliage large, medium green, glossy; bushy, 120 cms growth; [Spek's Centennial X New Year]; McGredy, Sam IV

'MACanat', Min, mp, 1981; (**Hi-de-hi**); flowers small, dbl., 20 petals, moderate fragrance; foliage small, dark, glossy; bushy growth; [Anytime X Gartendirektor Otto Linne]; McGredy, Sam IV

'MACangel', Min, w, 1981; (**Angelita**®, 'MACangeli', Snowball); flowers small, very dbl., slight fragrance; foliage small, dark, glossy; spreading growth; PP005849; Gold Star of the South Pacific, Palmerston North, NZ, 1983; [Moana X Snow Carpet]; McGredy, Sam IV, 1982

'MACar', HT, rb, 1960; (**Piccadilly**®); flowers scarlet, base and reverse gold, high-pointed, dbl., 28 petals, 4.5–5 in.; foliage dark, glossy; vigorous, upright, branching growth; GM, Madrid, 1960 GM, Rome, 1960; [McGredy's Yellow X Karl Herbst]; McGredy, Sam IV; McGredy

'MACarnhe', F, lp, 1998; (**Jillian McGredy**); flowers light pink, double, semi-dbl., 17–25 petals, 3.5 in., borne in small clusters, moderate fragrance; prickles moderate; foliage medium, medium green, semi-glossy; bushy, 120 cms growth; [Sexy Rexy X Lagerfeld]; McGredy, Sam IV

'MACAROM', HT, ly, 1988; (**Goldie**); flowers pale yellow, large, dbl., 26–40 petals, intense fragrance; foliage large, medium green, glossy; upright growth; [Unnamed seedling X Golden Gate]; McGredy, Sam IV

Macartney Rose, *see* **R. BRACTEATA**

'MACATSAN', S, mr, 1992; ('MACCATSAN', **Phantom**); flowers scarlet red, blooms, semi-dbl., 6–14 petals, 3–3.5 in., slight fragrance; foliage large, medium green, semi-glossy; spreading (80 cms) growth; [Pandemonium X Eyeopener]; McGredy, Sam IV; McGredy Roses International

'MACAUCK', HT, mr, 1982; (**Olympiad**™, Olympiode®); flowers brilliant medium red, large, non-fading, dbl., 30–35 petals, 4.5–5 in., exhibition form, borne mostly singly, slight fragrance; foliage large, medium green, matte; upright, bushy growth, vigorous; PP005519; AARS, 1984 GM, Portland, 1985; [Red Planet X Pharaoh]; McGredy, Sam IV, 1983; Armstrong Nursery, 1984

MACauclad, HT, mp, 1998; (**Kathryn McGredy**); flowers medium pink, double, semi-dbl., 17–25 petals, 4.5 in., borne mostly singly, slight fragrance; slight prickles; foliage medium, medium green, glossy; moderate (110 cms) growth; [City of Auckland X Lady Rose]; McGredy, Sam IV, 1996; McGredy, Sam, 1996, NZ

MACausal, S, pb, 1994; (**Pink Magic Carpet**™); McGredy, Sam IV

'MACBA', LCl, op, 1966; (**Galway Bay**®); flowers salmon-pink, well-formed, blooms in clusters, 3.5 in., slight fragrance; [Heidelberg X Queen Elizabeth]; McGredy, Sam IV

'MACBERN', HT, my, 1975; (Yellow Wonder, **Young Quinn**®); bud small, short; dbl., 28 petals, borne singly, slight fragrance; foliage heavily veined, dark; upright, tall growth; GM, Belfast, 1978; [Peer Gynt X Kiskadee]; McGredy, Sam IV

'MACBETH', HT, dr, 1921; bud pointed; flowers deep crimson, shaded darker, large, dbl., exhibition form, intense fragrance; foliage dark, bronze; vigorous, bushy, compact growth; [Richmond X Admiral Ward]; Bees

'MACBIGMA', F, pb, 1988; (Red Hot, **Redhot**); flowers hand-painted, medium, dbl., 15–25 petals, slight fragrance; foliage small, medium green, semi-glossy; bushy growth; [Eyepaint X Ko's Yellow]; McGredy, Sam IV

'MACBIPI', Min, mp, 1978; (**Moana**®); flowers rose-pink, small, dbl., 35 petals, slight fragrance; foliage glossy, dark; [Seedling X New Penny]; McGredy, Sam IV

'MACBO', HT, rb, 1966; (**Flaming Peace**, Kronenbourg); flowers bright medium red, reverse straw-yellow veined red; [Peace sport]; McGredy, Sam IV; Country Garden Nursery, 1966

'MACBUCPAL', HT, w, 1987; (**Auckland Metro**, Métro); flowers large, dbl., 26–40 petals, intense fragrance; foliage large, dark green, semi-glossy; bushy growth; [Sexy Rexy X (Seedling X Ferry Porsche)]; McGredy, Sam IV; McGredy Roses International, 1988

'MACBUNBER', S, or, 1998; (**Hot Chile**); flowers orange-red, semi-double, semi-dbl., 12–16 petals, 2.75 in., borne in small clusters, slight fragrance; prickles moderate; foliage small, dark green, glossy; spreading, low, bushy, 30 cms growth; [Trumpeter X Eyeopener]; McGredy, Sam IV; McGredy, Sam, 1995, NZ

'MACCA', LCl, ly, 1963; (**Casino**®, Gerbe d'Or); flowers soft yellow, well-formed, large, dbl., recurrent bloom, moderate fragrance; foliage dark, glossy; vigorous (10 ft) growth; GM, NRS, 1963; [Coral Dawn X Buccaneer]; McGredy, Sam IV; McGredy

'MACCARLTO', F, mp, 1991; (Northland, **Sweet Gesture**); dbl., 26–40 petals, 3 in., slight fragrance; foliage large, medium green, semi-glossy; ultra healthy, bushy (100 cms) growth; [(Sexy Rexy X New Year) X West Coast]; McGredy, Sam IV; McGredy Roses International, 1992

'MACCARPE', Min, w, 1980; (Blanche Neige, **Snow Carpet**®); flowers small, dbl., 55 petals, slight fragrance; spreading growth; GM, Baden-Baden, 1982; [New Penny X Temple Bells]; McGredy, Sam IV

'MACCARRIB', F, lp, 1992; (**Rexy's Baby**); dbl., 26–40 petals, 1.5–2.75 in., slight fragrance; foliage small, medium green, glossy; bushy (90 cms) growth; [Sexy Rexy X [Freude X ((Anytime X Eyepaint) X Stars 'n' Stripes)]]; McGredy, Sam IV; McGredy Roses International

'MACCATSAN', S, mr, 1992; ('MACATSAN', **Phantom**); flowers scarlet red, blooms, semi-dbl., 6–14 petals, 3–3.5 in., slight fragrance; foliage large, medium green, semi-glossy; spreading (80 cms) growth; [Pandemonium X Eyeopener]; McGredy, Sam IV; McGredy Roses International

'MACCHE', Gr, ab, 1978; ('MACCHEUP', **Sundowner**); flowers golden orange, dbl., 35 petals, 4 in., exhibition form, intense fragrance; foliage leathery; tall, upright growth; AARS, 1979; [Bond Street X Peer Gynt]; McGredy, Sam IV; Edmunds Roses

'MACCHEUP', Gr, ab, 1978; ('MACCHE', **Sundowner**); flowers golden orange, dbl., 35 petals, 4 in., exhibition form, intense fragrance; foliage leathery; tall, upright growth; AARS, 1979; [Bond Street X Peer Gynt]; McGredy, Sam IV; Edmunds Roses

'MACCHOME', HT, or, 1994; (**Big Daddy**®, Rita MacNeil); flowers scarlet-orange, blooms, dbl., 26–40 petals, 3 in., slight fragrance; foliage large, medium green, semi-glossy; bushy (100 cms) growth; [Howard Morrison X Mme Delbard]; McGredy, Sam IV; McGredy Roses International, 1994

'MACCHOPSU', Gr, or, 1998; (**Painted Desert**); flowers orange red, double, dbl., 17–25 petals, 4 in., borne in small clusters, slight fragrance; prickles moderate; foliage large, medium green, bushy growth; upright, bushy, 110 cms growth; [Louise Gardner X Mme. Delbard]; McGredy, Sam IV, 1997; McGredy, Sam, 1997, NZ

'MACCI', F, or, 1968; (**City of Belfast**®); flowers bright red, blooms in trusses, dbl., cupped; foliage glossy; GM, Belfast, 1970 GM, RNRS, 1967 GM, The Hague, 1976 Gold Star of the South Pacific, Palmerston North, NZ, 1969 PIT, RNRS, 1967; [Evelyn Fison X (Circus X Korona)]; McGredy, Sam IV; McGredy

'MACCLACK', HT, ob, 1991; (**Paddy Stephens**); flowers moderately large blooms, dbl., 15–25 petals, slight fragrance; foliage large, dark green, red when young; bushy (100 cms) growth; [Solitaire X [[(Tombola X (Elizabeth of Glamis X (Circus X Golden Fleece))) X Mary Sumner] X Unknown]]; McGredy, Sam IV; McGredy Roses International

'MACCLOSEUP', HT, ab, 1998; (**Carmel Sunset**); flowers apricot blend, double, dbl., 17–25 petals, 4.5 in., borne mostly singly, moderate fragrance; prickles moderate; foliage large, medium green, dull; bushy, tall (120 cms) growth; [Freude X Silver Jubilee]; McGredy, Sam IV, 1997; McGredy, Sam, 1997, NZ

'MACCOLUMB', HT, mp, 1998; (**Carolyn**); flowers medium pink, full, very dbl., 26–40 petals, 4 in., borne mostly singly, intense fragrance; prickles moderate; foliage large, light green, dull; bushy, medium, 110 cms growth; [((Courvoisier X Arthur Bell) X Traumerei) X (Tojo X Vienna Woods)]; McGredy, Sam IV; McGredy, Sam, 1995, NZ

'MACCOMPU', Gr, ob, 1989; (**Today**™); bud ovoid; flowers light orange blending to yellow, urn-shaped, medium, dbl., 33 petals, cupped, borne usually singly, slight fragrance; globular, medium, orange-red fruit; prickles recurved, medium, brown; foliage medium, dark green, glossy; bushy, medium growth; PP007202; [[Typhoo Tea X (Yellow Pages X Kabuki)] X [(Yellow Pages X Kabuki) X (MACjose X Typhoon)]]; McGredy, Sam IV; Co-Operative Rose Growers, 1990

'MACCOURLOD', Min, pb, 1984; (**Ragtime**); flowers small, dbl., slight fragrance; foliage small, dark, matt; bushy growth; [Mary Sumner X Unnamed seedling]; McGredy, Sam IV; McGredy Roses International, 1982

'MACCRACKLE', HT, dr, 1983; (**Howard Morrison**); flowers large, dbl., 35 petals, slight fragrance; foliage medium, medium green, semi-glossy; upright, bushy growth; [Unnamed seedling X Unnamed seedling]; McGredy, Sam IV, 1982

'MACcricke', Min, dp, 1989; (**Marriotta**); flowers deep pink, small blooms with side buds and sprays, dbl., 20 petals, slight fragrance; foliage small, medium green, semi-glossy, resistant to black; bushy growth; [Seaspray X Little Artist]; McGredy, Sam IV; McGredy Roses International, 1989; Oregon Mini. Roses, 1999

'MACdeepo', HT, my, 1985; (**Guadalajara**); flowers large, dbl., 24 petals, exhibition form, borne singly, moderate fragrance; fruit never observed; large, deltoid, red to brown prickles; foliage large, medium green, semi-glossy; medium, upright growth; (28); PP006263; [New Day X Yellow Bird]; McGredy, Sam IV; Roses by Fred Edmunds, 1984

'MACdub', LCl, mr, 1975; (**Dublin Bay**®); bud ovoid; dbl., 25 petals, 4.5 in., moderate fragrance; climbing growth; Rose of The Year, Auckland, NZ, 1993; [Bantry Bay X Altissimo]; McGredy, Sam IV; McGredy Roses International, 1974

'MACel', F, op, 1964; (**Elizabeth of Glamis**®, Irish Beauty); flowers light orange-salmon, blooms in clusters, dbl., 35 petals, 4 in., flat, intense fragrance; vigorous, compact bushy growth; GM, NRS, 1963 PIT, NRS, 1963; [Spartan X Highlight]; McGredy, Sam IV; Edmunds, 1965;, McGredy, 1964

'MACerupt', HT, yb, 1987; (**Louise Gardner**); flowers large, dbl., 15–25 petals, slight fragrance; foliage medium, medium green, matt; upright, bushy growth; [Freude X Sunblest seedling]; McGredy, Sam IV; McGredy Roses International, 1988

'MACesp', F, rb, 1974; (**Old Master**); flowers carmine, white eye and reverse, semi-dbl., 15 petals, 4.5 in., slight fragrance; foliage semi-glossy, medium green; vigorous, bushy growth; [(Maxi X Evelyn Fison) X (Orange Sweetheart X Fruhlingsmorgen)]; McGredy, Sam IV

'MACev', F, mr, 1962; (**Evelyn Fison**, Irish Wonder); flowers scarlet, blooms in broad clusters, dbl., 3 in., slight fragrance; foliage dark, glossy; compact, bushy growth; GM, NRS, 1963; [Moulin Rouge X Korona]; McGredy, Sam IV; J&P, 1964; McGredy, 1962

'MACeye', F, rb, 1975; (Eye Paint, **Eyepaint**®, Tapis Persan); bud ovoid; flowers bright red, whitish eye, gold stamens, single, 5–6 petals, 2.5 in., slight fragrance; foliage small, dark; tall, bushy growth; GM, Baden-Baden, 1974 GM, Belfast, 1978; [Seedling X Picasso]; McGredy, Sam IV

MacFarlane's Own, F, pb, 1999; Williams, J. Benjamin

'MACfirinlin', F, mr, 1998; (Fiesta, **War Dance**); flowers medium red, handpainted in cool weather, semi-dbl., 12–16 petals, 2.5 in., borne in small clusters, slight fragrance; prickles moderate; foliage small, medium green, semi-glossy; compact, low (30 cms) growth; [Howard Morrison X Sue Lawley]; McGredy, Sam IV; McGredy, 1997

'MACfirwal', S, rb, 1988; (Rock 'n' Roll, Tango, **Stretch Johnson**); flowers orange blend, medium, semi-dbl., 6–14 petals, slight fragrance; foliage large, medium green, semi-glossy; bushy growth; PP007472; GM, RNRS, 1988; [Sexy Rexy X Maestro]; McGredy, Sam IV; McGredy Roses International

'MACfrabro', Min, ob, 1985; (**Firefly**); flowers small, dbl., 20 petals, slight fragrance; foliage small, dark, glossy; bushy growth; [Mary Sumner X Ko's Yellow]; McGredy, Sam IV

'MACfreego', Min, dy, 1983; (Free Gold®, **Freegold**®, Penelope Keith); flowers deep yellow, gold reverse, small, dbl., 20 petals, exhibition form, moderate fragrance; foliage small, light green, semi-glossy; upright growth; [Seaspray X Dorola]; McGredy, Sam IV

MACfrothy, Min, w, 1992; (**Frothy**); McGredy, Sam IV

'MACgam', F, or, 1977; (**Kapai**®); flowers medium, dbl., 30 petals, 3.5 in., intense fragrance; foliage small; low, bushy growth; [Madame Bollinger X Tombola]; McGredy, Sam IV

'MACgem', HT, yb, 1979; (**Benson & Hedges Gold**®); bud ovoid; flowers deep golden yellow flushed coppery red, medium, dbl., 33 petals, moderate fragrance; bushy growth; Gold Star of the South Pacific, Palmerston North, NZ, 1978; [Yellow Pages X (Arthur Bell X Cynthia Brooke)]; McGredy, Sam IV; Mattock

'MACgenev', HT, lp, 1989; (Aotearoa-New Zealand, **New Zealand**); flowers large, soft creamy pink, dbl., 34 petals, 4.5–5 in., exhibition form, borne singly, intense, honeysuckle fragrance; foliage large, medium green, semi-glossy; upright, medium growth; PP8279; GM, Portland, USA, 1995 Rose of The Year, Auckland, NZ, 1990; [Harmonie X Auckland Metro]; McGredy, Sam IV, 1991

MACghovie, S, dr; (**Chateau Canon**); McGredy, Sam IV

'MACglemil', F, mp, 1991; (**Kapiti**); semi-dbl., 6–14 petals, 1.5–2.75 in., slight fragrance; foliage medium, medium green, semi-glossy; spreading growth (to 60 cms); [Sexy Rexy X Eyeopener]; McGredy, Sam IV; McGredy Roses International, 1992

'MACgolold', F, ab, 1997; (**Golden Oldie**); flowers apricot blend, dbl., 17–25 petals, 2.75 in., borne in large clusters, slight fragrance; prickles moderate; foliage medium, dark green, semi-glossy; bushy, low (50 cms) growth; [Maiden Voyage X Orange Honey]; McGredy, Sam IV; McGredy, 1997

'MACgoofy', HT, rb, 1990; (**Papageno**®); flowers red blend (striped), large blooms, dbl., 15–25 petals, slight fragrance; foliage large, light green, matt; upright growth; [Freude X [(Anytime X Eyepaint) X Stars 'n' Stripes]]; McGredy, Sam IV; McGredy Roses International, 1989

MACgremli, HT, rb; (**Tiger**); McGredy, Sam IV, 1990

'MACgutsy', Min, pb, 1983; (**Academy**®); flowers patio, small, dbl., 20 petals, slight fragrance; foliage small, medium green, semi-glossy; bushy growth; [Anytime X Matangi]; McGredy, Sam IV, 1982

'MACha', LCl, rb, 1965; (Haendel®, **Handel**®); flowers cream edged red, dbl., 22–30 petals, 3.5 in., recurrent bloom, slight, honey fragrance; foliage glossy, olive-green; blooms on old and new wood, climber growth; GM, Portland, 1975; [Columbine X Heidelberg]; McGredy, Sam IV; McGredy

'MAChaden', HT, dr, 1986; (**Red Perfection**™); flowers dark red, reverse slightly lighter, fading purple-red, medium, semi-dbl., 50–60 petals, exhibition form, moderate, old rose fragrance; globular, large, dark red fruit; slightly recurved, average, green prickles; foliage large, dark green, semi-glossy; tall growth; PP005426; [Karma X Arturo Toscanini]; McGredy, Sam IV; Co-Operative Rose Growers, 1987

'MAChahei', Gr, or, 1998; (**Reba McEntire**, Spirit of Hope); flowers double, orange red blooms, dbl., 17–25 petals, 4–4.5 in., exhibition form, borne in small clusters, slight fragrance; prickles moderate; foliage large, dark green, glossy; bushy, medium (110 cms) growth; Gold Star of the South Pacific, Palmerston North, NZ, 1994; [(Howard Morrison X Red Perfection) X Maiden Voyage]; McGredy, Sam IV; Edmunds Roses, 1997; McGredy, 1997

MAChana, Min, my, 1983; McGredy, Sam IV

'MAChilver', F, lp, 1998; (**Dame Cath**); flowers small, light pink, semi-dbl., 12–16 petals, 2.75 in., borne in small clusters, slight fragrance; prickles moderate; foliage medium, dark green, dull; bushy, low (50 cm) growth; [Chaumant X (Seaspray X Freegold)]; McGredy, Sam IV; McGredy, 1997

Macho Man, *see* 'PERmach'

'MAChomai', Min, dr, 1984; (**Grandee**); flowers gold stamens, medium, semi-dbl., moderate fragrance; foliage medium, medium green, semi-glossy; upright growth; [Regensberg X Ko's Yellow]; McGredy, Sam IV; Oregon Miniature Roses

'MAChoro', HT, mr, 1990; (Placido Domingo, **Velvet Ruby**, Waikato); flowers moderate, large blooms, 15–25 petals, moderate fragrance; foliage large, medium green, matt; bushy growth; [Candella X Auckland Metro]; McGredy, Sam IV; McGredy Roses International

'MACHOROWHEN', HT, or, 1993; (**Sophileo**); dbl., 15–25 petals, 1.5–2.75 in., slight fragrance; foliage medium, medium green, semi-glossy; medium, bushy growth; McGredy, Sam IV; C & K Jones; Golden Fields Nursery

'MACINCA', Min, or, 1978; (Longleat, **Wanaka**®, Young Cale); flowers small, dbl., 40 petals, slight fragrance; foliage light green; low, bushy growth; [Anytime X Trumpeter]; McGredy, Sam IV; McGredy Roses International

'MACINJU', HT, rb, 1981; ('MACKINJU', **Maestro**®); bud ovoid; flowers medium red, painted white, reverse lighter red and white, dbl., 28 petals, slight fragrance; narrow, red prickles; foliage matt, olive green; upright, bushy growth; [Picasso seedling X Unnamed seedling]; McGredy, Sam IV, 1980

'MACIO', F, op, 1964; (**Violet Carson**); flowers peach-pink, reverse silvery, well-formed, blooms in large clusters, dbl., 35 petals, moderate fragrance; foliage dark, glossy; compact, bushy growth; [Mme Leon Cuny X Spartan]; McGredy, Sam IV; McGredy

'MACIR', HT, or, 1961; (**Miss Ireland**); flowers orange-red, yellow reverse, dbl., 37 petals, 5 in., moderate fragrance; foliage dark; vigorous, bushy growth; [Tzigane X Independence]; McGredy, Sam IV; McGredy

'MACIVY', F, ab, 1994; (Love's Spring, **Singin' in the Rain**, Spek's Centennial); flowers apricot/copper, very dbl., 25–30 petals, 1.5–2.75 in., borne in large clusters, moderate, sweet musk fragrance; some prickles; foliage medium, dark green, glossy; medium, (3–4 ft) upright growth, free branching; PP8362; AARS, 1995; [Sexy Rexy X Pot O'Gold]; McGredy, Sam IV; Edmunds Roses, 1994

'MACJILLI', Gr, op, 1987; (**Profile**); flowers orange blend, large, dbl., 26–40 petals, slight fragrance; foliage large, medium green, semi-glossy; upright growth; [Freude X (Unnamed Arthur Bell seedling X Sunsong)]; McGredy, Sam IV; McGredy Roses International, 1988

'MACJOCEL', F, w, 1987; (Dresselhuys, **Petite Penny**); flowers small, semi-dbl., 6–14 petals, moderate fragrance; foliage small, medium green, semi-glossy; bushy growth; [(R. crepuscule X Unnamed seedling) X Royal Occasion]; McGredy, Sam IV; McGredy Roses International, 1988

'MACJULIAT', HT, r, 1990; (Old Spice, Siegfried Sassoon, **Spiced Coffee**, Vidal Sassoon); flowers pale lavender with brown overtones, large blooms, dbl., 15–25 petals, intense fragrance; foliage medium, medium green, matt; upright growth; [Harmonie X Big Purple]; McGredy, Sam IV; McGredy Roses International, 1991

'MACKATI', F, m, 1990; (**Old Port**); dbl., 26–40 petals, moderate fragrance; foliage medium, medium green, matt; bushy growth; [[(Anytime X Eyepaint) X Purple Splendour] X Big Purple]; McGredy, Sam IV; McGredy Roses International, 1991

'MACKEPA', S, ob, 1979; (Blushing Maid, **Rocky**™); bud ovoid; flowers coral orange, reverse whitish, medium, dbl., 25 petals; vigorous, tall, bushy growth; [Liverpool Echo X (Evelyn Fison X (Orange Sweetheart X Fruhlingsmorgen))]; McGredy, Sam IV; McGredy Roses International

'MACKINJU', HT, rb, 1981; ('MACINJU', **Maestro**®); bud ovoid; flowers medium red, painted white, reverse lighter red and white, dbl., 28 petals, slight fragrance; narrow, red prickles; foliage matt, olive green; upright, bushy growth; [Picasso seedling X Unnamed seedling]; McGredy, Sam IV, 1980

'MACKOSRED', Min, mr, 1982; (**Paragon**); dbl., 35 petals, moderate fragrance; foliage small, dark, glossy; bushy growth; [Ko's Yellow X Little Artist]; McGredy, Sam IV

'MACKOSYEL', Min, yb, 1978; (**Ko's Yellow**); flowers yellow, edges marked red, fading to cream, classic form, medium, dbl., 39 petals; foliage dark, glossy; bushy growth; [(New Penny X Banbridge) X (Border Flame X Manx Queen)]; McGredy, Sam IV; McGredy Roses International

'MACKUNG', HT, my, 1980; (**Austragold**®); McGredy, Sam IV

'MACLANOFLON', HT, w, 1998; (**Full Sail**, Land of the Long White Cloud, Long White Cloud); flowers white, full, large, very dbl., 26–40 petals, 4.5 in., exhibition form, borne mostly singly, intense fragrance; few prickles; foliage large, dark green, glossy; bushy, medium, 110 cm growth; [New Zealand sport]; McGredy, Sam IV; Edmunds' Roses, 1998; McGredy, 1998

'MACLANTER', Gr, ob, 1998; (Lantern, **Octoberfest**™, Oktoberfest); bud long, pointed; flowers blended autmnal colors, cream to red-orange, semi-dbl., 127–25 petals, 5–5.5 in., exhibition form, borne in small clusters, slight, fruity fragrance; prickles moderate; foliage large, dark green, glossy; reddish new growth, tall, upright, 6 ft; [Louis Gardner X New Zealand]; McGredy, Sam IV

'MACLAPAZ', F, r, 1986; (**Beachcomber**); flowers medium, semi-dbl., 6–14 petals, slight fragrance; foliage medium, medium green, semi-glossy; bushy growth; Gold Star of the South Pacific, Palmerston North, NZ, 1985; [Seedling X Colorbreak]; McGredy, Sam IV

'MACLEIDE', F, rb, 1998; (**Wizard**); flowers orange-red with white eye, semi-dbl., 12–16 petals, 2.5 in., borne in large clusters, slight fragrance; prickles moderate; foliage small, medium green, semi-glossy; bushy, low, 30 cm growth; [Seedling X Genesis]; McGredy, Sam IV; McGredy, 1996

'MACLOCKER', F, pb, 1992; (Chaumant, Chaumont, **Maiden Voyage**, Oriola); dbl., 15–25 petals, 1.5–2.75 in., borne in small clusters, slight fragrance; foliage medium, medium green, glossy; medium ($3\frac{1}{2}$'), bushy growth; [Sexy Rexy X New Year]; McGredy, Sam IV; Cooperative Rose Growers, 1992

'MACLOUPRI', LCl, or, 1998; (**Too Hot To Handle**); flowers orange-red, double, dbl., 17–25 petals, 2.75 in., borne in small clusters, slight fragrance; prickles moderate; foliage large, dark green, glossy; upright, tall, 150 cm growth; [Waiheke X Eyeopener]; McGredy, Sam IV; McGredy, 1995

'MACMAN', F, rb, 1974; (**Matangi**®); bud ovoid; flowers orange-red, silver eye and reverse, dbl., 30 petals, 3.5 in., slight fragrance; foliage small; bushy growth; GM, Belfast, 1976 GM, Portland, 1982 GM, Rome, 1974 PIT, RNRS, 1974; [Unnamed seedling X Picasso]; McGredy, Sam IV

'MACMANLEY', Min, rb, 1982; (**Little Artist**®, 'MACMANLY', Top Gear); flowers open with hand-painted marks becoming solid medium red, off-white in base half, semi-dbl., slight fragrance; foliage small, medium green, semi-glossy; upright growth; [Eyepaint X Ko's Yellow]; McGredy, Sam IV

MACmatan, HT, mr; (**Ten Ten CFRB**); McGredy, Sam IV, 1987

'MACMELAN', Min, op, 1988; (**Melanie**); bud ovoid; flowers light salmon-pink, reverse lighter, small, borne in sprays o, dbl., 23 petals, cupped, slight fragrance; prickles straight, red-brown; foliage small, medium green, semi-glossy; bushy, low growth; [Seaspray X Wanaka]; McGredy, Sam IV; Sealand Nursery, 1989

'MACMI', HT, op, 1961; (**Mischief**); flowers salmon-pink, dbl., 28 petals, 4 in., moderate fragrance; foliage light green; vigorous, upright growth; GM, NRS, 1961 GM, Portland, 1965 PIT, NRS, 1961; [Peace X Spartan]; McGredy, Sam IV; McGredy

'MACMIGMOU', F, rb, 1980; (**Mighty Mouse**, Painted Star); flowers scarlet, white eye, blooms borne 20 per cluster, 7–12 petals, slight fragrance; sets hips, orange-red fruit; slightly hooked, red prickles; foliage dark, very long, pointed; tall, spreading growth; [Anytime X Eyepaint]; McGredy, Sam IV

Macmillan Nurse, *see* 'BEAMAC'

'MACMINMO', Min, rb, 1987; (Minnie Mouse, **Roller Coaster**); flowers red blend (striped), semi-dbl., 6–14 petals, slight fragrance; foliage small, medium green, glossy; upright growth; PP007319; [(Anytime X Eyepaint) X Stars 'n' Stripes]; McGredy, Sam IV; McGredy Roses International, 1988

'MACmisech', F, op, 1978; (**Taupo®**); flowers light rose-salmon, flora-tea, dbl., 40 petals, 5 in.; tall growth; [Liverpool Echo X Irish Mist]; McGredy, Sam IV; McGredy

'MACmo', F, rb, 1969; (**Molly McGredy**); flowers medium red, reverse silver, well-formed, large blooms in trusses, dbl., 35 petals, slight fragrance; foliage dark, glossy; GM, Belfast, 1971 GM, Portland, 1971 GM, RNRS, 1968 PIT, RNRS, 1968; [Paddy McGredy X (Mme Leon Cuny X Columbine)]; McGredy, Sam IV; McGredy

'MACmooblu', F, m, 1998; (**Moody Blues**); flowers mauve, full, large, very dbl., 26–40 petals, 2.75 in., moderate fragrance; foliage large, medium green, semi-glossy; bushy, medium, 100 cm growth; [Sexy Rexy X Blue Nile]; McGredy, Sam IV; McGredy, 1991

MACmoodre, F, m, 1996; (**Moody Dream**); McGredy, Sam IV

'MACmosco', Min, rb, 1988; (**Whistle Stop**); flowers red blend (striped), small, semi-dbl., 6–14 petals, slight fragrance; foliage small, medium green, semi-glossy; bushy growth; [Mighty Mouse X Hurdy Gurdy]; McGredy, Sam IV, 1989

'MACmota', HT, mr, 1983; (**Crimson Tide**); flowers crimson, large blooms borne singly, dbl., 35 petals, exhibition form, slight fragrance; foliage large, dark, leathery; upright growth; [Seedling X Seedling]; McGredy, Sam IV; Roses by Fred Edmunds

'MACmouhoo', HT, w, 1988; (Christine Horbiger, Foster's Melbourne Cup, Foster's Wellington Cup, **Mount Hood**); flowers medium, dbl., 40–45 petals, moderate; foliage large, medium green, semi-glossy; bushy growth; Gold Star of the South Pacific, Palmerston North, NZ, 1991; [Sexy Rexy X Pot O'Gold]; PP9095; McGredy, Sam IV; McGredy Roses International, 1991

'MACmu', HT, op, 1968; (**Muria**); bud long, pointed; flowers salmon-orange, classic form, dbl., 40 petals, 4 in., slight fragrance; foliage light green; bushy growth; [Miss Ireland X Tropicana]; McGredy, Sam IV; Spek

'MACnaru', HT, mp, 1988; ('MACnauru', 'MACsatur', **Penthouse**); flowers large, dbl., 15–25 petals, moderate fragrance; foliage large, medium green, matt; bushy growth; [Unnamed seedling X Ferry Porsche]; McGredy, Sam IV

'MACnauru', HT, mp, 1988; ('MACnaru', 'MACsatur', **Penthouse**); flowers large, dbl., 15–25 petals, moderate fragrance; foliage large, medium green, matt; bushy growth; [Unnamed seedling X Ferry Porsche]; McGredy, Sam IV

'MACnauru', HT, mp, 1987; ('MACngauru', Metropolitain, Penthouse, **West Coast®**); flowers large, dbl., 15–25 petals, slight fragrance; foliage large, light green, matt; bushy growth; [((Yellow Pages X Kabuki) X Golden Gate) X (Unnamed Poulsen seedling X Picasso)]; McGredy, Sam IV; McGredy Roses International, 1986

'MACnecta', Min, or, 1978; (**Otago®**); flowers medium, dbl., 35 petals, exhibition form, moderate fragrance; bushy growth; [Anytime X Minuette]; McGredy, Sam IV; McGredy Roses International

'MACnew', Min, pb, 1982; ('MACnewing', **Seaspray**); flowers pale pink flushed red, medium, semi-dbl., moderate fragrance; foliage medium, medium green, matt; bushy growth; [Anytime X Moana]; McGredy, Sam IV; John Mattock, Ltd

'MACneweye', Gr, ob, 1983; (Arcadian, 'MACnewye', **New Year®**); flowers orange and gold blend, medium, dbl., 20 petals, slight fragrance; foliage large, dark, glossy; upright growth; PP005428; AARS, 1987; [Mary Sumner X Unnamed seedling]; McGredy, Sam IV, 1982

'MACnewing', Min, pb, 1982; ('MACnew', **Seaspray**); flowers pale pink flushed red, medium, semi-dbl., moderate fragrance; foliage medium, medium green, matt; bushy growth; [Anytime X Moana]; McGredy, Sam IV; John Mattock, Ltd

'MACnewye', Gr, ob, 1983; (Arcadian, 'MACneweye', **New Year®**); flowers orange and gold blend, medium, dbl., 20 petals, slight fragrance; foliage large, dark, glossy; upright growth; PP005428; AARS, 1987; [Mary Sumner X Unnamed seedling]; McGredy, Sam IV, 1982

'MACngauru', HT, mp, 1987; ('MACnauru', Metropolitain, Penthouse, **West Coast®**); flowers large, dbl., 15–25 petals, slight fragrance; foliage large, light green, matt; bushy growth; [((Yellow Pages X Kabuki) X Golden Gate) X (Unnamed Poulsen seedling X Picasso)]; McGredy, Sam IV; McGredy Roses International, 1986

'MACnic', HT, mr, 1981; (Karma, **Pounder Star**); bud long, pointed; dbl., 20 petals, exhibition form, moderate, old rose to spicy fragrance; small, curved prickles; foliage dark, glossy; upright, medium growth; [John Waterer X Kalahari]; McGredy, Sam IV; Roses by Fred Edmunds, 1982

'MACnickel', Min, rb, 1989; (**Nickelodeon**); flowers small, semi-dbl., 6–14 petals, slight fragrance; foliage small, dark green, semi-glossy; patio; bushy growth; [Roller Coaster X [Freude X [(Anytime X Eyepaint) X Stars 'n' Stripes]]]; McGredy, Sam IV, 1991

'MACnon', HT, mp, 1965; (**Shannon®**); bud ovoid; dbl., 58 petals, 5 in.; foliage dark, rounded; vigorous growth; [Queen Elizabeth X McGredy's Yellow]; McGredy, Sam IV; McGredy

'MACoborn', HT, pb, 1995; (**Maggie Barry**); flowers salmon edged orange, blooms, dbl., 26–40 petals, 3 in., slight fragrance; foliage large, light green, matt; bushy (100 cms) growth; [Louise Gardner X West Coast]; McGredy, Sam IV; McGredy Roses International, 1993

MACoffer, HT, op, 1988; (**City of Christchurch**); McGredy, Sam IV

'MACon', HT, mr, 1963; (**Uncle Walter**); flowers crimson-scarlet, dbl., 30 petals, 5 in., exhibition form; foliage leathery, coppery; vigorous, tall growth; [Detroiter X Heidelberg]; McGredy, Sam IV; McGredy

'MACoranlem', S, ob, 1994; (**Oranges 'n' Lemons**); flowers striking orange and yellow striped, dbl., 26–40 petals, 3–3.5 in., borne in small clusters, slight, fruity fragrance; some prickles; foliage medium to large, dark red when new, turning dark green; tall, spreading, very vigorous growth; PP9191; [New Year X (Freude X Seedling)]; McGredy, Sam IV; Weeks Roses, 1995

'MACpa', F, mp, 1962; (**Paddy McGredy**); bud ovoid; flowers deep rose-pink, blooms in clusters, dbl., 33 petals, 4 in., cupped, moderate fragrance; foliage leathery; vigorous, bushy growth; GM, NRS, 1961; [Spartan X Tzigane]; McGredy, Sam IV; J&P

'MACpandem', F, yb, 1988; (Claire Rayner, **Pandemonium**); flowers yellow and red stripes, dbl., 26–40 petals, slight fragrance; foliage small, medium green, glossy; patio; bushy growth; [New Year X ((Anytime X Eyepaint) X Stars 'n' Stripes)]; McGredy, Sam IV

'MACparlez', Min, ab, 1984; (**Baby Sunrise**, Gold Fever); bud small; flowers copper apricot, small, semi-dbl., slight fragrance; foliage small, medium green, semi-glossy; bushy growth; PP007031; [Dorola X Moana]; McGredy, Sam IV; John Mattock, Ltd

'MACpaurmar', Min, ob, 1993; (**Kiwi Sunrise**, MACtaumaruniu); flowers florescent orange with yellow eye, light yellow to white accents, dbl., 15–25 round petals, 2 in., borne in small clusters, moderate, spicy fragrance; few prickles; foliage medium, dark green, glossy; medium (24–30 in.), bushy growth; [Orange Honey X Pandemonium]; McGredy, Sam IV; Justice Miniature Roses, 1994

'MACpic', F, pb, 1971; (**Picasso**); flowers deep pink, petal edges lighter, white eye and reverse, semi-dbl., 18 petals, 3 in.; foliage small; GM, Belfast, 1973 Gold Star of the South Pacific, Palmerston North, NZ, 1971; [Marlena X (Evelyn Fison X (Frulingsmorgen X Orange Sweetheart))]; McGredy, Sam IV; McGredy

MACpinderal, Min, mp, 1987; (**Fresh Pink**); McGredy, Sam IV

'MACpiopi', HT, mp, 1995; (**Graeme Douglas**); dbl., 26–40 petals, 3 in., slight fragrance; foliage large, medium green, semi-glossy; bushy (100 cms) growth; [Silver Jubilee X (Harmonie X Auckland Metro)]; McGredy, Sam IV; McGredy Roses International, 1994

'MACpluto', Min, rb, 1986; (**Hurdy Gurdy**); flowers dark red with white stripes, small, dbl., 26–40 petals, slight fragrance; foliage small, medium green, semi-glossy; mini-flora; upright growth; [Matangi X Stars 'n' Stripes]; McGredy, Sam IV

'MACPONUI', HT, pb, 1988; (**Pinky**); flowers large, dbl., 26–40 petals, slight fragrance; foliage large, dark green, glossy; upright growth; [Freude X Typhoo Tea]; McGredy, Sam IV

'MACPOW', HT, rb, 1977; (**Chivalry**®); flowers red, yellowish reverse, large, dbl., 35 petals; foliage glossy, dark; [Peer Gynt X Brasilia]; McGredy, Sam IV; Mattock

'MACRAIDA', HT, mp, 1982; (**Rainy Day**®); flowers large, dbl., 20 petals, slight fragrance; foliage large, dark, semi-glossy; bushy growth; [Trumpeter X Typhoon]; McGredy, Sam IV; McGredy Roses International

'MACRAMAR', HT, lp, 1998; (**Tess**); flowers large, light pink, very full, very dbl., 41 petals, 4.5 in., borne mostly singly, slight fragrance; prickles moderate; foliage large, medium green, semi-glossy; bushy, tall (120 cms) growth; [((Sexy Rexy X New Year) X (Freude X Courvoisier X Arthur Bell)) X Dreaming]; McGredy, Sam IV; McGredy, 1997

'MACRAT', F, rb, 1978; (**Priscilla Burton**®); flowers deep carmine pink and white blend (with variable combinations), semi-dbl., 10 petals, 2.5 in., moderate fragrance; foliage glossy, dark; PIT, RNRS, 1976; [Old Master X Unnamed seedling]; McGredy, Sam IV; Mattock

'MACRED', HT, mr, 1977; (**Joy Owens**); bud ovoid; 4 in., exhibition form; foliage very dark; moderate, bushy growth; [Electron X Pharaoh]; McGredy, Sam IV; McGredy Roses International

'MACREDPARAP', S, dr, 1994; (**Trinity**); dbl., 15–25 petals, 1.5 in., slight fragrance; foliage small, dark green, semi-glossy; spreading (90 cms) growth; [Sexy Rexy X Eyeopener]; McGredy, Sam IV; McGredy Roses International, 1994

'MACREDSAUS', S, mr, 1999; (**Red Magic Carpet**™); semi-dbl., 12–16 petals, 3 in., borne in small clusters, moderate fragrance; foliage medium, semi-glossy; spreading, low (12 in.); groundcover growth; [Sexy Rexy X Eyeopener]; McGredy, Sam IV, 1994

'MACRELEA', HT, my, 1984; (Chardonnay, Chardony, 'MACRELEASE', **Nobilo's Chardonnay**®); flowers orange yellow, large, dbl., 35 petals, slight fragrance; foliage small, light green, glossy; bushy growth; [Freude X (Wienerwald X Benson & Hedges Gold)]; McGredy, Sam IV

'MACRELEASE', HT, my, 1984; (Chardonnay, Chardony, 'MACRELEA', **Nobilo's Chardonnay**®); flowers orange yellow, large, dbl., 35 petals, slight fragrance; foliage small, light green, glossy; bushy growth; [Freude X (Wienerwald X Benson & Hedges Gold)]; McGredy, Sam IV

'MACRENO', HT, ob, 1994; (**Cardinal's Rose**, Susan Devoy); flowers orange blend, blooms, dbl., 26–40 petals, 3–3.5 in., slight fragrance; foliage large, medium green, matt; bushy (100 cms) growth; [Waiheke X Las Vegas]; McGredy, Sam IV; McGredy Roses International, 1994

'MACRERO', HT, mr, 1977; (**Bulls Red**); bud ovoid; dbl., 28 petals, 4 in.; foliage dark; tall, upright growth; [Sympathie X Irish Rover]; McGredy, Sam IV; McGredy Roses International

'MACREXY', F, mp, 1984; (Heckenzauber, **Sexy Rexy**®); flowers medium pink to light pink, dbl., blooms in large clusters, slight fragrance; foliage small, light green, glossy; bushy growth; PP006713; Gold Star of the South Pacific, Palmerston North, NZ, 1984 Golden Prize, Glasgow, 1989 Rose of The Year, Auckland, NZ, 1991; [Seaspray X Dreaming]; McGredy, Sam IV

'MACROS', Min, mr, 1976; (Rose Baby, **Royal Salute**); flowers rose-red, small, dbl., 30 petals, 1.5 in.; foliage small, dark; [New Penny X Marlena]; McGredy, Sam IV; Mattock

'MACSATUR', HT, mp, 1988; ('MACNARU', 'MACNAURU' **Penthouse**); flowers large, dbl., 15–25 petals, moderate fragrance; foliage large, medium green, matt; bushy growth; [Unnamed seedling X Ferry Porsche]; McGredy, Sam IV

'MACSEATRI', HT, pb, 1987; (**Happy Days**); flowers medium, dbl., 15–25 petals, slight fragrance; foliage medium, medium green, matt; bushy growth; [(Unnamed Poulsen seedling X Picasso) X Paradise]; McGredy, Sam IV; McGredy Roses International, 1988

'MACSEE', F, dy, 1970; (**Courvoisier**®); dbl., 49 petals, 3.5 in., exhibition form, intense fragrance; foliage glossy, dark; [Elizabeth of Glamis X Casanova]; McGredy, Sam IV; McGredy

'MACSEV', HT, op, 1978; (**Christina Atherton**); flowers salmon-pink, classic form, dbl., 33 petals, 4 in., slight fragrance; free growth; [Tiki X Seedling]; McGredy, Sam IV

'MACSHANA', Min, dy, 1982; (Benson & Hedges Special, **Dorola**®, Parkay); flowers mini-flora, medium, dbl., 26 petals, moderate fragrance; foliage small, medium green, semi-glossy; bushy growth; [Darling Flame X New Day]; McGredy, Sam IV; McGredy Roses International, 1983

'MACSINGAL', Min, op, 1988; (**Singalong**); flowers salmon-orange and yellow, medium, dbl., 26–40 petals, slight fragrance; foliage small, medium green, glossy; patio; bushy growth; [(Anytime X Eyepaint) X New Year]; McGredy, Sam IV

'MACSINGAP', HT, or, 1987; (**Cup Final**); flowers large, dbl., 15–25 petals, slight fragrance; foliage medium, medium green, semi-glossy; upright growth; [Benson & Hedges Gold X (Kalahari X Papa Meilland)]; McGredy, Sam IV; McGredy Roses International, 1988

'MACSODA', F, or, 1991; (**Hot 'n' Spicy**); semi-dbl., 6–14 petals, borne in small clusters, slight fragrance; foliage medium, dark green, glossy; bushy growth; [Mary Sumner X Precious Platinum]; McGredy, Sam IV; Co-Operative Rose Growers, 1990

'MACSPASH', F, rb, 1980; (**Sue Lawley**); flowers medium red, petals edged light pink all around, blooms borne, 19 petals, slight fragrance; small, straight prickles; foliage matt, red when young; bushy growth; Gold Star of the South Pacific, Palmerston North, NZ, 1981; [((Little Darling X Goldilocks) X (Evelyn Fison X (Coryana X Tantau's Triumph)) X (((John Church X Elizabeth of Glamis)) x]; McGredy, Sam IV

'MACSPEEGO', HT, rb, 1990; (**Candella**®, Eternally Yours); flowers moderately large blooms, dbl., 15–25 petals, slight fragrance; foliage large, dark green, glossy; bushy growth; [Howard Morrison X Esmeralda]; McGredy, Sam IV; McGredy Roses International

'MACSPICE', Min, m, 1983; ('MACSPIKE'); flowers small, semi-dbl., slight fragrance; foliage small, medium green, semi-glossy; spreading growth; [Anytime X Gartendirektor Otto Linne]; McGredy, Sam IV

'MACSPIKE', Min, m, 1983; (**'MACSPICE'**); flowers small, semi-dbl., slight fragrance; foliage small, medium green, semi-glossy; spreading growth; [Anytime X Gartendirektor Otto Linne]; McGredy, Sam IV

'MACSTEWAR', F, or, 1986; (**Bradley Craig**®, Bradley Graig); flowers scarlet, medium, dbl., 15–25 petals, slight fragrance; foliage medium, medium green, glossy; bushy growth; [Tojo X Montana]; McGredy, Sam IV

'MACSTRA', F, or, 1976; (**Mary Sumner**); flowers medium, semi-dbl., 15 petals, 3 in., slight fragrance; tall, upright growth; [Seedling X Seedling]; McGredy, Sam IV; McGredy Roses International

'MACSUPBOW', Gr, m, 1998; (**Cologne**, Super Bowl); flowers mauve, double, large, dbl., 17–25 petals, 4.5 in., borne in small clusters, intense fragrance; prickles moderate; foliage large, medium green, matte; bushy, medium, 110 cm growth; [Harmonie X Lagerfeld]; McGredy, Sam IV; McGredy, 1996

'MACSUPCAT', HT, yb, 1989; (**Miriam**); flowers medium, dbl., 20 petals, slight fragrance; foliage medium, dark green, semi-glossy; bushy growth; [Sexy Rexy X Yabadabadoo]; McGredy, Sam IV, 1990

'MACSWEETWA', HT, op, 1987; (Freshie, **Sweetwaters**); flowers medium salmon-pink, large, dbl., 26–40 petals, slight fragrance; foliage large, dark green, glossy; upright, tall, very big growth;

McGredy, Sam IV; McGredy Roses International, 1986

'MACTANE', HT, ob, 1981; (**City of Auckland**); flowers large, dbl., intense fragrance; foliage medium green, semi-glossy; bushy growth; [Benson & Hedges Gold X Whisky Mac]; McGredy, Sam IV; McGredy Roses International, 1982

'MACTAURANG', HT, ob, 1998; (Marvelle, **Tropical Sunset**™); bud big, pointed, yellow with orange-red; flowers yellow and pink blend with stripes, dbl., 17–25 petals, 4 in., borne mostly singly, slight; prickles moderate; foliage large, light green, matte; bushy, tall (4 ft) growth; [Louise Gardner X (Auckland Metro X Stars 'n' Stripes seedling)]; McGredy, Sam IV; McGredy, 1995

'MACTEL', LCl, mr, 1972; (**Grand Hotel**®, Grandhotel®); flowers scarlet, 4 in., exhibition form, repeat bloom; foliage dark; ADR, 1977; McGredy, Sam IV

'MACTEMAIK', F, rb, 1998; (**Michelangelo**, The Painter); flowers red blend with stripes, double, dbl., 17–25 petals, 4 in., borne in small clusters, slight fragrance; prickles moderate; foliage medium, medium green, matte; tall, bushy, 120 cm growth; [Louise Gardner X (Auckland Metro X Stars 'n' Stripes seedling)]; McGredy, Sam IV; McGredy, 1995

'MACTENNI', Min, op, 1978; (**Takapuna**®); flowers light peach-pink, medium, dbl., 37 petals, intense fragrance; tall, spreading growth; [New Penny X ((Clare Grammerstorf X Cavalcade) X Elizabeth of Glamis)]; McGredy, Sam IV; McGredy Roses International

'MACTITIR', HT, mr, 1990; (**Hoagy Carmichael**); flowers large blooms, dbl., 26–40 petals, moderate fragrance; foliage medium, dark green, matt; upright, bushy growth; [(Sir Harry Pilkington X Elegy) X Pounder Star]; McGredy, Sam IV; McGredy Roses International, 1990; Roses by Fred Edmunds, 1990

'MACTRAMPOL', F, w, 1998; (**Remuera**); flowers white, very full, small blooms, very dbl., 26–41 petals, 2.75 in., borne in large clusters, slight fragrance; prickles moderate; foliage small, medium green, semi-glossy; bushy, low, 40 cm growth; Gold Star of the South Pacific, Palmerston North, NZ, 1989; [Seaspray X Sexy Rexy]; McGredy, Sam IV; McGredy, 1991

'MACTRUM', F, or, 1977; (**Trumpeter**®); bud ovoid; dbl., 39 petals, 3.5 in., cupped, slight fragrance; foliage medium green, glossy; bushy, compact growth; GM, NZ, 1977 GM, Portland, 1977 James Mason, RNRS, 1991; [Satchmo X Seedling]; McGredy, Sam IV; McGredy Roses International

'MACTUAL', HT, dr, 1998; (**Romeo**); flowers dark red, full, large, very dbl., 26–40 petals, 4.5 in., borne mostly singly, moderate fragrance; prickles moderate; foliage large, dark green, semi-glossy; bushy, tall (110 cms) growth; [Howard Morrison X Harmonie]; McGredy, Sam IV; McGredy, 1995

'MACTURANG', HT, pb, 1998; (**Maria McGredy**); flowers pink blend, large, full, very dbl., 26–40 petals, 4.5 in., borne mostly singly, slight fragrance; prickles moderate; foliage large, medium green, semi-glossy; bushy, tall (120 cms) growth; [Freude X Remember Me]; McGredy, Sam IV; McGredy, 1997

'MACULTRA', F, r, 1983; (**Brown Velvet**, Colorbreak); flowers orange, tinged brownish, medium, dbl., 35 petals, slight fragrance; foliage medium, dark, glossy; upright growth; Gold Star of the South Pacific, Palmerston North, NZ, 1979; [Mary Sumner X Kapai]; McGredy, Sam IV, 1982

MACvisar, Cl F, w; (**Ice White, Climbing**); McGredy, Sam IV

'MACVOLAR', F, or, 1987; (**Volare**); flowers medium, dbl., 15–25 petals, slight fragrance; foliage large, medium green, glossy; upright growth; [Julischka X Matangi]; McGredy, Sam IV; McGredy Roses International, 1988

'MACWAIKE', Gr, op, 1986; (**Waiheke**™, Waikiki); flowers coral-pink fading lighter, medium, semi-dbl., 30 petals, exhibition form, borne in sprays of 5–9, slight, spicy fragrance; ovoid, small, tan-orange fruit; small, green prickles; foliage medium, dark green, glossy; upright, bushy growth; PP005429; [Tony Jacklin X Young Quinn]; McGredy, Sam IV, 1985; Co-Operative Rose Growers, 1987

'MACWAIRAR', HT, my, 1998; (**Apertif**); flowers medium yellow, large, double, dbl., 17–25 petals, 4 in., borne mostly singly, slight; prickles moderate; foliage large, light green, matte; bushy, medium, 110 cm growth; [Solitaire X Sunbright]; McGredy, Sam IV

'MACWAIWER', Min, or, 1987; (Tara, **Tara Allison**); flowers small, semi-dbl., 6–14 petals, slight fragrance; foliage small, medium green, semi-glossy; bushy growth; [Wanaka X Eyepaint]; McGredy, Sam IV; Justice Miniature Roses, 1986

'MACWALLA', Min, ob, 1978; (**Kaikoura**®); flowers orange, patio, medium, dbl., 27 petals, slight fragrance; foliage glossy, dark; vigorous, bushy growth; [Anytime X Matangi]; McGredy, Sam IV

'MACWEEMAT', Min, rb, 1978; ('Waitemata', **Waitmata**®, Wee Matt); flowers medium, dbl., 42 petals, slight fragrance; foliage light green, glossy; bushy growth; [Wee Man X Matangi]; McGredy, Sam IV; McGredy Roses International, 1980

'MACWHAKA', S, mp, 1987; (**Gwen Swane**); flowers medium, dbl., 26–40 petals, slight fragrance; foliage small, medium green, matt; spreading growth; [MACbroey X Snow Carpet]; McGredy, Sam IV; McGredy Roses International, 1988

'MACWHENU', HT, ob, 1981; (**Derek Nimmo**); bud ovoid; flowers orange-red, silvery reverse, borne 1–3 per cluster, dbl., 30 petals, moderate fragrance; red prickles; foliage medium green; vigorous growth; [Seedling X Seedling]; McGredy, Sam IV; John Mattock, Ltd.

'MACWHITOUT', Min, w, 1988; (**Whiteout**); flowers small, dbl., 20 petals, slight fragrance; foliage small, medium green, semi-glossy; bushy growth; [Sexy Rexy X Popcorn]; McGredy, Sam IV; Justice Miniature Roses, 1990

'MACWOODMA', F, rb, 1988; (**Redwood**); flowers hand-painted, medium, semi-dbl., 14 petals, slight fragrance; foliage large, dark green, glossy; bushy growth; [Old Master X Wienerwald]; McGredy, Sam IV

'MACWOOHERM', F, mr, 1998; (**Red Point**); flowers medium red, double, dbl., 17–25 petals, 3.5 in., borne in small clusters, slight fragrance; prickles moderate; foliage large, medium, glossy; bushy, medium, 110 cm growth; [Maiden Voyage X Eyeopener]; McGredy, Sam IV; McGredy, 1993

'MACWYNPIN', F, mp, 1994; (**Manapouri**); dbl., 15–25 petals, 1.5–2.75 in., slight fragrance; foliage medium, medium green, semi-glossy; patio; bushy (35 cms) growth; [MACredparap X Moody Blues]; McGredy, Sam IV; McGredy Roses International, 1994

'MACWYNSCAR', F, or, 1994; (**Hauraki**); dbl., 15–25 petals, 1.5–2.75 in., slight fragrance; foliage small, medium green, semi-glossy; patio; bushy (30 cms) growth; Rose of The Year, Auckland, NZ, 1998; [Trumpeter X Kapiti]; McGredy, Sam IV; McGredy Roses International, 1994

'MACYABA', HT, dy, 1981; (**Yabadabadoo**®); flowers large, dbl., 20 petals, slight fragrance; foliage medium, medium green, semi-glossy; bushy growth; [Yellow Pages X Bonfire Night]; McGredy, Sam IV

'MACYEFRE', HT, yb, 1987; (Chartreuse, **Solitaire**); flowers yellow tinted pink, reverse yellow, fading without blanching, dbl., 25 petals, cupped, slight fragrance; no fruit; slightly hooked, large, reddish-brown prickles; foliage medium, dark green, semi-glossy; bushy, strong growth; PIT, RNRS, 1985; [Freude X Benson & Hedges Gold]; McGredy, Sam IV, 1978; Sealand Nursery, 1987

'MACYELKIL', HT, dy, 1998; (**Gold Heart**, Heart of Gold); flowers deep yellow, dbl., 17–25 petals, 4 in., borne in small clusters, moderate fragrance; prickles moderate; foliage medium, medium green, glossy; upright, very tall, 160 cm growth; [Solitaire X Remember Me]; McGredy, Sam IV; McGredy, 1994

'MACYOUMIS', F, pb, 1979; (Buffalo Bill, **Regensberg**®); flowers petals pink, edged white, white eye, yellow stamens, dbl., 21 petals, 4.5 in., cupped, moderate fragrance; low, bushy growth; GM, Baden-Baden, 1980; [Geoff Boycott X Old Master]; McGredy, Sam IV; McGredy Roses International

Mada, *see* **'SPECTACULAR'**

'MADAM PRESIDENT', F, pb, 1975; (Madame President); bud long, pointed; flowers blend of pink shades, dbl., 70 petals, 4 in., exhibition form, slight fragrance; bushy growth; [Seedling X Handel]; McGredy, Sam IV; Avenue Nursery

Madam Speaker, *see* MEIzuzes

'MADAME BOLLINGER', F, op, 1972; flowers deep orange-salmon, dbl., 25 petals, 3 in., exhibition form, slight fragrance; free growth; [(Little Darling X Goldilocks) X Bobbie Lucas]; McGredy, Sam IV; McGredy

Madame Delbard®, *see* 'DELADEL'

Madame Gilberte Dubois, HT, op; 30 petals, flat; foliage dense, luxurious; RvS-Melle, 1990

Madame President, *see* **'MADAM PRESIDENT'**

Madame Seneclause, HT, lp

'MADAME STOLZ', D, ly, 1848

'MADAME VERBELEN', HT, dr, 1973; flowers full, dbl., 65 petals, 4.5 in., cupped, intense fragrance; Delforge, S.

'MADAME VIOLET', HT, m, 1981; flowers lavender, dbl., 45 petals, exhibition form, borne singly, no fragrance; medium, reddish light-green prickles; foliage medium, medium green, semi-glossy; tall, upright growth; PP006315; Teranishi, K., 1985; Itama Rose Nursery

'MADAME VIOLET, CLIMBING', Cl HT, m, 1999; flowers pale violet, very dbl., 45 petals, 5.5 in., exhibition form, no fragrance; growth to 9–12 ft; Itami Rose Nursery, 1992

Madames de Villeparisis, M, lp; [Henri Martin sport]; Morley, Dr B., 1988; (Weatherly, L.)

'MADCAP', F, mr, 1954; bud urn-shaped; flowers scarlet-red, dbl., 20–25 petals, 4.5 in., cupped, moderate fragrance; foliage leathery; bushy growth; [Independence X Crimson King]; Kordes; Stark Bros.

'MADCAP', HT, rb, 1955; bud dark red; flowers flame-scarlet striped yellow, or orange without stripe; [Grand Duchesse Charlotte sport]; Ratcliffe

'MADDALENA', HT, op, 1934; bud pointed to ovoid; flowers salmon-pink, reverse begonia-rose, very large, dbl., 30–32 petals, cupped, moderate fragrance; foliage light green, glossy, leathery; long stems; very vigorous, upright, bushy growth; [Julien Potin X J.C. Thornton]; San Remo Exp. Sta.

'MADELEINE FAIVRE', HT, lp, 1902; flowers large, very dbl.; Buatois; (Sangerhausen)

'MADELEINE LEMAIRE', HWich, op, 1923; flowers bright salmon-pink, medium, semi-dbl.; [Mrs F.W. Flight X Unnamed seedling]; Nonin

'MADELEINE MONOD', HT, op, 1939; bud long; flowers salmon-carmine, very large, dbl., moderate fragrance; foliage bronze; vigorous growth; Chambard, C.

'MADELEINE PACAUD', HT, op, 1922; flowers silvery rose, tinted salmon; Chambard, C.

Madeleine Rivoire®, HT, lp, 1988; Orard

'MADELEINE SELTZER', HMult, ly, 1926; (Madeleine Selzer, Yellow Tausendschön); flowers pale lemon, fading white; few prickles; vigorous growth; [Tausendschön X Mrs Aaron Ward]; Walter, L.

Madeleine Selzer, *see* **'MADELEINE SELTZER'**

'MADELEINE WEIDERT', HMult, mp, 1928; flowers rose-pink; [Tausendschön X Rosel Dach]; Walter, L.

'MADELINE', HEg, w; (Emmeline); flowers creamy white edged pink, semi-dbl.

'MADELINE CORREY', HT, dp, 1971; bud globular; flowers cerise, full, medium, dbl., moderate, lemon fragrance; foliage glossy, dark; moderate growth; [Minnie Watson X Sterling Silver]; Watson

Madeline Spezzano, *see* 'TINMAD'

Madelon, *see* 'RUIMEVA'

Madelon de Paris, F, 1961; Robichon, M.; (Cavriglia)

'MADELYN LANG', Cl Min, dp, 1970; bud slightly ovoid; flowers deep pink, dbl., 40 petals, 1 in.; foliage small, glossy, dark,embossed; upright growth; [Little Darling X Little Chief]; Williams, Ernest D.; Mini-Roses

'MADEMOISELLE', F, dp, 1950; bud ovoid; flowers rose-red, dbl., 50–60 petals, 3.5 in.; very vigorous, branching growth; [Goldilocks X Marionette]; Boerner; J&P

Mademoiselle, *see* MEIplovon

'MADEMOISELLE DE DINANT', F, w, 1966; bud globular; flowers creamy white, open, dbl., 22 petals, 3–3.5 in., intense, fruity fragrance; foliage light green; vigorous, spreading growth; [Purpurine X Lavender Pinocchio]; Lens

Mademoiselle Jacqueline, Pol, dp, 1985; Weihrauch

Madeo, *see* KORcremkis

'MADETTE', HT, op, 1922; flowers coppery orange-pink; Guillot, P.

'MADGE ELLIOTT', HT, lp, 1964; flowers light pink, center shaded apricot, medium, dbl., slight fragrance; foliage dark, leathery; very vigorous, upright growth; [Queen Elizabeth X ?]; Darvall; G. Knight

'MADGE PRIOR', F, pb, 1934; flowers brilliant claret, white eye, borne in large clusters, single; foliage dark; vigorous growth; Prior

'MADGE TAYLOR', HT, dp, 1930; flowers deep pink, large, dbl., globular, slight fragrance; foliage light; vigorous growth; [Rhea Reid seedling X ?]; Clark, A.; Hazlewood Bros.

'MADGE WHIPP', HT, or, 1936; flowers bright scarlet, dbl., moderate fragrance; foliage leathery; vigorous, bushy growth; [Lady Charmion X J.C. Thornton]; Bees

'MADGE WILDFIRE', HT, or, 1932; bud pointed; flowers indian red, very large, very dbl., exhibition form, moderate fragrance; foliage leathery; vigorous growth; GM, NRS, 1933; Dobbie

Madhatter, *see* 'TINHAT'

'MADHOSH', HT, rb, 1975; bud globular; flowers deep magenta-red, streaked mauve, dbl., 45 petals, 4.5 in., slight fragrance; foliage leathery; vigorous, bushy, compact growth; IARI

'MADHUMATI', HT, mp, 1973; flowers large blooms borne singly, dbl., 55 petals, exhibition form, intense fragrance; brown prickles; foliage medium, medium green, smooth; vigorous, upright, bushy growth; [General MacArthur X Unnamed seedling]; Pal, Dr B.P.; IARI

'MADHURA', F, yb, 1979; bud pointed; dbl., 70 petals, exhibition form, borne 3–6 per cluster, moderate fragrance; foliage glossy; vigorous, upright, bushy growth; [Kiss of Fire X Goudvlinder]; Pal, Dr B.P.; K.S.G. Son's Roses

Mädi, Pol, or, 1969; flowers medium-large, semi-dbl.; GPG Bad Langensalza; (Sangerhausen)

Madiba, *see* KORandpunk

'MADISON', T, w, 1912; flowers pure white; [(Perle X The Bride) X Meteor (N)]; Brant-Hentz

Madison, S, lp, 1998; flowers very small, borne in large clusters; groundcover (2 ft) growth; Poulsen

Madlenka, F, lp; Vecera, L., 1975; (Czech Rosa Club)

'MADOKA', F, rb, 1977; dbl., 25 petals, 2–2.5 in., exhibition form, slight fragrance; foliage glossy, dark; upright growth; [(Zambra X Peace) X Cherry Brandy]; Teranishi, K.; Itami Bara-en

Madona, *see* MEInopia

'MADONNA', HT, w, 1908; [Mlle Alice Furon X Marie van Houtte]; Cook, J.W.

'MADRAS', HT, pb, 1981; bud ovoid; flowers rose with yellow and light pink reverse, borne singly, dbl., 48 petals, moderate fragrance; prickles hooked downward; foliage large, leathery; medium, spreading growth; [Unnamed seedling X Unnamed seedling]; Warriner, William A.; J&P

'MADRIGAL', HT, dp, 1950; bud long, pointed; flowers brilliant salmon-pink flushed coppery, very large, dbl., moderate, spicy fragrance; vigorous growth; [(Mme Joseph Perraud seedling X Unnamed seedling) X (Mme Joseph Perraud seedling X R. foetida bicolor)]; Gaujard

Madrigal, *see* HARextra

'MADY', Cl HT, ly, 1925; flowers large, dbl.; Gemen & Bourg; (Sangerhausen)

Maestro®, *see* 'MACKINJU'

'MAESTRO', HT, dr, 1957; bud oval; flowers deep velvety red, open, large, dbl., intense fragrance; foliage dark, glossy; vigorous, bushy growth; RULED EXTINCT 2/81; [Crimson Glory X Charles Mallerin]; Delforge

Mafalda di Savoia, *see* BARsav

'MAGALI', HT, dp, 1952; flowers carmine, open, medium, dbl., 35–40 petals, slight fragrance; foliage abundant, leathery; very vigorous, upright, bushy growth; [Charles P. Kilham X Brazier]; Mallerin, C.; Meilland-Richardier

Magali, *see* MEIgoufin

'MAGALI BONNEFON', HT, pb, 1916; flowers pink, reverse bright salmon-pink, semi-dbl., moderate fragrance; [Mme Abel Chatenay sport]; Nabonnand

Maganwilare, HT, pb, 1995; Huber

'MAGDALENA DE NUBIOLA', HT, op, 1932; flowers salmon-rose, semi-dbl., slight fragrance; [Li Bures X Mari Dot]; Dot, Pedro; C-P

Magdeburg, HT, 1988; VEG; (Sangerhausen)

'MAGENTA', Pol, m, 1916; flowers violet-red, semi-dbl.; dwarf growth; Barbier

'MAGENTA', F, m, 1954; (Kordes' Magenta); bud ovoid; flowers rosy magenta to soft deep mauve, large blooms in large clusters, dbl., intense fragrance; foliage dark, leathery; vigorous, upright, bushy growth; [Yellow Floribunda X Lavender Pinocchio]; Kordes

'MAGENTA', HT, mr, 1934; bud pointed; flowers crimson-carmine, large, dbl., intense fragrance; foliage glossy, dark; vigorous, bushy growth; Leenders, M.

Magenta Buttons, LCl, dp

Magenta Diadem, *see* TANdelbel

Maggie, B, mr

Maggie Barry, *see* 'MACOBORN'

Magia Nera®, HT, dr

'MAGIC', HT, pb, 1954; bud globular; flowers silver-pink, open, dbl., 75 petals, 5.5 in., moderate fragrance; foliage leathery; very vigorous, upright growth; RULED EXTINCT 11/88; [Thornless Beauty sport]; Grillo

'MAGIC', HT, mr, 1988; bud pointed; flowers medium, borne usually singly, dbl., 24 petals, exhibition form; pear shaped, medium, orange fruit; prickles slender, straight, medium, light green; foliage medium, medium green, matt; upright, tall growth; [MACvolar X Tonight]; Strahle, Robert; Carlton Rose Nurseries, 1987

Magic Baby, *see* 'GELMAGIC'

Magic Blanket™, *see* 'TANIGINO'

'MAGIC CARPET', LCl, yb, 1941; flowers yellow, splashed orange, scarlet and rose, large, borne in clusters, semi-dbl., exhibition form; foliage small, bronze, leathery, glossy; strong stems; vigorous, climbing or trailing growth; [Coral Creeper X Stargold]; Brownell, H.C.

Magic Carpet, *see* 'JACLOVER'

Magic Carrousel®, *see* 'MORROUSEL'

'MAGIC CHARM', HT, mp, 1966; flowers medium, dbl., exhibition form; vigorous growth; [Mount Shasta X Granada]; Aufill

Magic de Feu, *see* 'KORFEU'

'MAGIC DRAGON', Cl Min, dr, 1969; bud short, pointed; flowers small, dbl.; foliage leathery; very vigorous, upright growth; [((R. wichurana X Floradora) X Seedling) X Little Buckaroo]; Moore, Ralph S.; Sequoia Nursery

Magic East, *see* 'VIRSPLASH'

Magic Fire®, *see* 'LAPNEUF'

Magic Fire®, *see* LAPjaminal

Magic Forez, HT, 1970; Croix; (Cavriglia)

Magic Lantern, *see* 'STANQUELI'

Magic Medley, *see* 'ABHAYA'

Magic Meidiland™, *see* 'MEIBONRIB'

Magic Meillandecor®, *see* 'MEIBONRIB'

'MAGIC MIST', Min, mr, 1980; bud pointed; flowers medium red, veined darker, blooms borne usually singly, dbl., 47 petals, exhibition form, moderate fragrance; thin, tan prickles; foliage small, medium to dark green, glossy; bushy, spreading growth; [Tom Brown X Little Chief]; Williams, Ernest D.; Mini-Roses

'MAGIC MOMENT', HT, mr, 1964; bud pointed; flowers scarlet to geranium-red, large, dbl., exhibition form, moderate fragrance; Frères, Buyl; Cuthbert

Magic Moments, HT, op, 1983; Martin

'MAGIC MOON', Gr, ob, 1970; bud ovoid; flowers deep salmon, reverse silver, medium, very dbl., slight fragrance; foliage large, leathery; vigorous, upright growth; [Little Darling X Golden Scepter]; Schwartz, Ernest W.; Wyant

'MAGIC MOUNTAIN', F, yb, 1973; bud ovoid; flowers large, dbl., exhibition form, slight fragrance; foliage glossy, dark, leathery; vigorous, bushy growth; [Circus X Texan]; Armstrong, D.L.; Armstrong Nursery

'MAGIC RED', F, mr, 1942; flowers large blooms in clusters, dbl., 45 petals, globular, slight fragrance; foliage leathery, glossy; vigorous, bushy growth; [Henri Panthier X Dance of Joy]; Kordes; J&P

Magic Silver, HT, m, 1998; Kordes

Magic Splendor, *see* 'LYOMA'

Magic Sunblaze, *see* SCHanbiran

Magic Sunset, F, ob, 1995; Tantau

'MAGIC TOUCH', HT, lp, 1974; bud ovoid; flowers soft pink, dbl., 32 petals, 5 in., exhibition form, moderate fragrance; foliage glossy; moderate growth; [Tropicana X Queen of Bermuda]; Golik; Dynarose

'MAGIC WAND', Cl Min, dp, 1957; flowers light red, blooms in clusters, semi-dbl., 20 petals, 1 in.; orange fruit; foliage small, dark; arching (to 4 ft) growth; [Eblouissant X Zee]; Moore, Ralph S.; Sequoia Nursery

Magician™, *see* 'MINIMAG'

'MAGICIENNE', HT, or, 1957; flowers geranium-red, well formed, dbl., 30 petals; foliage bronze; dwarf, bushy growth; [Comtesse Vandal X (Peace X Independence]; Laperrière; EFR

Magicienne 78®, *see* 'LAPOUDA'

Magie des Jardins, *see* **Gartenzauber**

Magie d'Orient, HMsk, 1991; Lens

Magitta®, *see* 'TANATTIGAM'

Magma, HT, or, 1998; Kordes

'MAGNA CHARTA', HP, mp, 1876; (Casper); flowers bright pink, suffused carmine, very large, dbl., globular, some recurrent bloom, moderate fragrance; foliage thick, rich green; vigorous, compact growth; Paul, W.

'MAGNAFRANO', HT, mr, 1900; flowers rich crimson-rose, large, dbl., intense fragrance; [Magna Charta X Safrano]; Van Fleet; Conard & Jones

Magneet, *see* **'FEUERWERK'**®

Magnet, F, mr; flowers luminous red, large, dbl., slight fragrance; VEG; (Sangerhausen)

'MAGNIFICA', HEg, m, 1916; (R. eglanteria duplex, R. rubiginosa magnifica); flowers purplish red, semi-dbl.; [Lucy Ashton self seedling]; Hesse

'MAGNIFICENCE', HT, op, 1954; flowers salmon-pink tinted yellow, large, dbl., exhibition form, moderate fragrance; foliage dark; vigorous, bushy growth; [Peace seedling]; Gaujard; McGredy

'MAGNIFIQUE', Pol, lp, 1928; flowers clear pink, open, large, borne in clusters, semi-dbl., cupped; foliage rich green, glossy; vigorous growth; [Orléans Rose sport]; deRuiter; Sliedrecht & Co.

Magnolia Rose, *see* **'DEVONIENSIS'**

Magnolia Springs™, *see* 'TALSPRINGS'

'MAGNOLIJA', HP, w, 1940; flowers creamy white, large, dbl.; Kosteckij; (Sangerhausen)

'MAGRANA', F, or, 1954; flowers well formed, dbl.; vigorous growth; [Méphisto X Alain]; Dot, Pedro

'Mahadev', F, or, 1975; bud long, pointed; flowers full, open, small, dbl., 20 petals, 1.5 in., slight fragrance; foliage glossy; vigorous, tall, bushy growth; [Seedling X (Unknown X Unknown)]; Viraraghavan, M.S.

'Mahagona', HT, pb, 1956; (Mahogany); bud pointed; flowers dull orange-scarlet, open, very large, moderate fragrance; foliage leathery, wrinkled; very vigorous, upright, bushy growth; [Golden Rapture X Hens Verschuren]; Kordes; McGredy

'Mahaja', HT, dp, 1936; bud long, pointed; flowers deep rose-pink, center rose-red, base yellow, dbl., 4 in., exhibition form, moderate fragrance; foliage glossy; long stems; vigorous growth; [Rose Hill sport]; Carbaugh; Johnstown Greenhouses

Mahalaxmi, HT, yb, 1988; Patil, B.K.

Mahalia™, *see* KEIreb

'Maharajah', HP, mr, 1931; flowers velvety crimson, golden anthers, large blooms in clusters, semi-dbl., moderate fragrance; foliage dark, bronze, leathery; Cant, B. R.

Maharani, HT, pb, 1996

Maharishi, HT, or, 1995; Tejganga

Maheka, *see* **'Purpurea'**

'Mahina', HT, ab, 1952; flowers reddish apricot, reverse golden yellow, large, dbl., 35 petals, moderate fragrance; foliage leathery; very vigorous, bushy growth; GM, Bagatelle, 1952; [Peace X Fred Edmunds]; Meilland, F.

Mahogany, *see* **'Mahagona'**

Maia, HT, 1986; Barni, V.; (Cavriglia)

'Maid Marian', HT, pb, 1920; flowers carmine-rose, reverse silvery pink, dbl.; Therkildsen

'Maid Marion', HWich, w, 1909; flowers white, tipped pink, center filled with yellow stamens, slightly incurved, single; foliage large, glossy; vigorous growth; Walsh

'Maid Marion', HMsk, w, 1930; flowers white opening blush, borne in very large clusters, semi-dbl.; vigorous (3–4 ft) growth; Pemberton; Bentall

Maid Marion, *see* **'Red Imp'**

'Maid of Gold', Cl HT, my, 1936; bud globular, reddish; flowers golden yellow, large, very dbl., profuse, intermittent bloom; foliage glossy; vigorous, climbing (12 ft), compact growth; [Golden Emblem, Climbing X ?]; Raffel; Port Stockton Nursery

Maid of Honor, *see* **'Maid of Honour'**

'Maid of Honour', HT, yb, 1986; (Maid of Honor); flowers yellow, light pink center, large blooms borne singly and in, dbl., moderate fragrance; foliage large, dark, semi-glossy; tall, upright growth; [Folklore X Unnamed seedling]; Weddle, Von C.; Hortico Roses, 1984

'Maid of Honour', F, op, 1951; (Schleswig); bud long, pointed; flowers salmon-pink, open, large, borne in trusses, single, moderate fragrance; foliage leathery, glossy, light green; vigorous, upright growth; (28); RULED EXTINCT 1/86; [Crimson Glory X Holstein]; Kordes; Morse

'Maid of Kent', HT, op, 1929; flowers soft salmon-pink, moderate fragrance; [Ophelia X Mrs W.J. Grant]; Archer

'Maid of Orleans', F, ob, 1977; flowers orange-flame blend, full, dbl., 40 petals, 4 in., slight fragrance; foliage glossy, dark; very free growth; [Val De Mosa X Tropicana]; Ellick

Maiden Voyage, *see* 'MACLOCKER'

Maiden's Blush, *see* **'Great Maiden's Blush'**

'Maiden's Blush', HAlba, w, 1797; (Small Maiden's Blush); flowers soft blush, medium, dbl., globular, non-recurrent, moderate fragrance; vigorous (to 8 ft) growth; [Presumed to be a natural hybrid of R. alba X R. centifolia]; Kew

Maiden's Blush, *see* **'William R. Smith'**

Maidi, HT, pb

Maids of Jubilee, *see* 'TALMAID'

Maidy®, *see* 'KORWALBE'

'Maigold', S, dy, 1953; flowers bronze-yellow, semi-dbl., 14 petals, 4 in., cupped, non-recurrent, intense fragrance; foliage glossy; bushy (5 ft), pillar or shrub growth; [Poulsen's Pink X Fruhlingstag]; Kordes

Maile's Double Pink OGR, Misc OGR, mp; (found rose)

Maile's Pink Quill, Misc OGR, lp; (found rose)

Mainauperle®, *see* 'KORMAI'

'Mainz', HT, my, 1930; flowers citron-yellow, very large, dbl., moderate fragrance; [Kardinal Piffl sport]; Leenders Bros.

Mainzer Fastnacht, *see* 'TANSI'

Mainzer Fastnacht, Climbing, *see* **'Blue Moon, Climbing'**

Mainzer Rad, *see* **'Mainzer Wappen'**

'Mainzer Wappen', S, or, 1963; (Mainzer Rad); bud pointed; flowers red tinted orange, blooms in clusters (up to 20), dbl., 25 petals, 3.5 in., moderate fragrance; foliage dark; bushy, upright (5 ft) growth; Kordes, R.

Mainzer Wappen, S, or

'Maisie', HT, op, 1926; flowers salmon-orange and pink, medium, semi-dbl.; Dickson, A.; (Sangerhausen)

'Maisie Gowie', HT, ?, 1968; intense fragrance; foliage dark, leathery; bushy growth; Gowie

Maisieres, F, mr, 1993; 29 petals, flat; foliage dense dark green; RvS-Melle, 1992

'Maison Pernet-Ducher', HT, my, 1934; bud pointed; flowers golden yellow, veined copper, large, dbl.; foliage glossy, dark; very vigorous, bushy growth; Pernet-Ducher; Gaujard

Maitland White, T, w

Maiwunder, S, ly, 1966; Kordes

'Maja', F, mr, 1959; flowers cinnabar-red, borne in broad clusters, dbl., 30 petals, cupped; vigorous, bushy growth; [Independence X Signal Red]; deRuiter; Horstmann

Maja Mauser, *see* **'Cara Mia'**

'Maja Mauser', F, or, 1970; flowers dark orange-red, dbl., 20 petals, 4 in., slight fragrance; foliage glossy, dark; bushy, upright growth; [Evelyn Fison X Seedling]; Poulsen, Niels D.; Poulsen

Maja Oetker, *see* **'Coronation Gold'**

'Majestade', HT, w, 1957; flowers cream-white, base deep yellow, well formed; very vigorous growth; [Mme Marie Curie X Peace]; da Silva, Moreira

Majesté®, *see* 'DORMA'

'Majestic', HT, or, 1955; bud long, purplish; flowers cinnabar-red, open, large, dbl., moderate fragrance; vigorous growth; [Peace X Unnamed seedling]; Gaujard

'Majestueuse', HGal, mp, 1811; Guerrapain

Majeure, *see* **'Tanya'**

Majolika®, *see* TANkeijoli

'Major Frank Hayes', HT, dr, 1934; flowers crimson, center darker, very large, dbl., exhibition form, intense fragrance; foliage leathery, dark; vigorous growth; [Joan Howarth X J.C. Thornton]; Bees

'Major Shelley', HT, mr, 1939; bud pointed; flowers rich crimson-scarlet, dbl., 35 petals, 5–6. in., exhibition form, foliage leathery, dark; very vigorous, bushy growth; [Mrs J.D. Eisele X Crimson Glory]; Howard, F.H.; H&S

'Majorca', HT, mr, 1938; (Majorica); flowers scarlet, large, dbl., cupped; foliage glossy, bronze; vigorous, bushy growth; [Aribau X Angels Mateu]; Dot, Pedro; C-P, 1941

'Majorca', HT, op, 1958; bud short, thick; flowers rose-salmon, dbl., 52 petals; moderate growth; Frères, Buyl

Majorette®, *see* 'MEIDAD'

Majorette®, *see* 'MEIPIESS'

Majorette 86, *see* 'MEIPIESS'

Majorica, *see* **'Majorca'**

Make A Wish, *see* 'MEHPAT'

Make Believe™, *see* 'MORMAKE'

Make Mine Sherry™, Min, rb

'MALA RUBINSTEIN', HT, mp, 1971; flowers high-pointed, dbl., 45 petals, 5.5 in., moderate fragrance; foliage large, matt; Edland Fragrance Medal, ARS, 1972 GM, Belfast, 1973; [Sea Pearl X Colour Wonder]; Dickson, A.

'MALAGA', Cl HT, dp, 1971; flowers reddish pink, classic form, dbl., 36 petals, 4.5 in., intense fragrance; foliage glossy, dark; [(Hamburger Phoenix X Danse de Feu) X Copenhagen]; McGredy, Sam IV

'MALAGUENA', S, mp, 1976; bud ovoid, long, pointed; dbl., 28 petals, 4.5 in., cupped, slight fragrance; foliage large, dark, leathery; erect, bushy growth; [Tickled Pink X Country Music]; Buck, Dr Griffith J.; Iowa State University

Malahat, *see* 'BETAHAT'

Malak, HT, pb, 1988; Kumar

Malakarsiddha, HT, lp, 1987; Patil, B.K.

'Mälar-Ros', HT, dr, 1932; bud pointed; flowers glowing ruby-red indoors, dark blood-red with crimson outdoors, dbl., exhibition form, moderate fragrance; very vigorous, bushy growth; [Hadley X Fragrance]; Kordes

'MALCAIR', LCl, or, 1959; (**Danse des Sylphes**®, Grimpant Danse des Sylphes); flowers rich red suffused geranium-red, blooms in large clusters, globular; foliage glossy; very vigorous growth; [Spectacular X (Peace X Independence)]; Mallerin, C.; URS

'MALCOLM', HT, dp, 1979; flowers carmine, full, dbl., 35–40 petals, 4–5 in., slight fragrance; foliage large, glossy, dark; vigorous, free growth; [Hector Deane X Chopin]; Ellick; Excelsior Roses

Malcolm Sargent, *see* 'HARWHARRY'

Maleica, *see* TANcelia

'MALEK-ADEL', HGal, lp; (Melik El Adel); flowers soft pink, dotted white, large

Malene, HT, 1980; Laperrière; (Cavriglia)

'MALESHERBES', HGal, m, 1834; flowers purple, spotted; Vibert

'MALIA', Min, mp, 1992; very dbl., 1.5 in., borne singly, highly recurrent bloom, no fragrance; some prickles; foliage small, medium green, semi-glossy; bushy (100–150 cms) growth; [Seedling (Pink 1172) X Rosa Maria]; Mansuino, Dr Domenico, 1988

'MALIBU', F, or, 1959; bud pointed, ovoid; flowers coral-orange-red, becoming lighter, dbl., 35 petals, 4–4.5 in., exhibition form, intense fragrance; foliage leathery; vigorous, bushy, upright growth; [Charlotte Armstrong X Independence]; Morey, Dr Dennison; J&P

Malibu Lake Shady Pink, HT, dp, 1994

Malicorne, *see* 'DELMATOR'

Malindall, F, mr, 1996; Noack, Werner

Malindi, F, mr, 1974; flowers medium, semi-dbl.; Noack, Werner; (Sangerhausen)

'MALINOVKA', F, dp, 1956; (Raspberry Wine); flowers raspberry-red, medium, slight fragrance; [Staatsprasident Pats X Independence]; Sushkov, K. L.

'MALJOANNA', HT, mp, 1994; (**Querida**); dbl., 26–40 petals, 3–3.5 in., borne mostly singly, slight fragrance; few prickles; foliage large, dark green, semi-glossy; upright (5 ft) growth; PPAF; [Silver Anniversary X Unknown]; Maltagliati, Mark G.; Meilland-Star Roses, Inc., 1995

'MALKARSIDDHA', HT, lp, 1989; [Century Two sport]; Patil, B.K.; K.S.G. Son's Roses, 1987

MALloeur, LCl, mr; (**Coup de Coeur**)

'MALMESBURY', HT, my, 1980; bud pointed; dbl., 23 petals, borne singly and several together, slight fragrance; slightly hooked prickles; foliage medium green; compact, bushy growth; [Vera Dalton X Parasol]; Sanday, John

'MALMIYA', Gr, ly, 1991; (**Ayako**™); flowers pale yellow cream, large, dbl., 26–40 petals, slight fragrance; foliage medium, medium green, semi-glossy; medium, upright growth; [Sonia sport]; Maltagliati, Mark G.; Nino-miya Nursery Co., 1990

'MALNINO', Gr, w, 1989; (**Saint Louis**); flowers white to ivory; PP006908; [Lifirane sport]; Maltagliati, Mark G.; Conard-Pyle Co., 1989

MALren, HT, yb, 1993; (**Nerene**); Malanseuns

'MALSO', F, or, 1958; (**Soleil**®); flowers bright vermilion, blooms in clusters, dbl., 29 petals, 3 in.; foliage clear green; dwarf, bushy growth; Mallerin, C.; EFR

'MALTON', HCh, mr, 1830; (Fulgens); flowers bright crimson; (28); Guérin

'MALVA', F, m, 1935; flowers mauve, center white, open, large, semi-dbl.; foliage large, glossy, light; very vigorous, bushy growth; Leenders, M.

Malvena, HT, lp; Dawson, George, 1971; (Weatherly, L.)

Malverns, *see* KORdehei

'MALVINA', C, lp, 1841; flowers pale pink, edged lighter, well-formed, large blooms in clusters, very dbl.; Verdier, V.

Malwa 94, HT, pb, 1994; Gokhale

Mama, HT, dr, 1991; flowers large, very dbl., moderate fragrance; Wänninger, Franz; (Sangerhausen)

'MAMA DE MEYER', HT, w, 1931; flowers cream, center salmon, well formed, large, dbl.; vigorous growth; [Duchess of Wellington X Aspirant Marcel Rouyer]; Lens

'MAMA LAMESCH', HT, op, 1922; flowers orange-rose, center deeper, reverse reddish rose, dbl., moderate fragrance; [Frau Oberprasident von Grothe X Mme Edouard Herriot]; Lambert, P.

'MAMA LOOYMANS', HT, dp, 1910; flowers carmine-pink, medium, semi-dbl.; Leenders, M.; (Sangerhausen)

Mama Mia, *see* 'ZIPMIA'

'MAMA PECHTOLD', HT, 1938; bud long, pointed; flowers rosy salmon, dbl.; foliage bronze; long, strong stems; vigorous growth; [Katherine Pechtold X Briarcliff]; Pechtold; op

Mamaia, Pol, or, 1972; semi-dbl.; GPG Bad Langensalza; (Sangerhausen)

'MAMAN', HT, or, 1963; flowers orange-vermilion, well formed, dbl., 30–40 petals, 4–5 in., slight fragrance; foliage bright green; bushy growth; GM, Geneva, 1962; [(Rome Glory X La Vaudoise) X Impeccable]; Delbard-Chabert

Maman Chérie, *see* 'DELMANCHE'

'MAMAN COCHET', T, pb, 1893; bud pointed; flowers pale pink, center deeper, base lemon-yellow, dbl., 4 in., exhibition form, moderate fragrance; foliage dark, leathery; vigorous, bushy growth; [Marie van Houtte X Mme Lombard]; Cochet, S.

'MAMAN COCHET, CLIMBING', Cl T, pb, 1909; Upton, 1909 & H&S, 1915

'MAMAN DENTAL', HT, mp, 1921; flowers pure rose-pink; [Mme Caroline Testout sport]; Dental

'MAMAN GENEVIÈVE', F, or, 1960; flowers red tinted orange; bushy growth; Hémeray-Aubert

'MAMAN LEVAVASSEUR', Pol, dp, 1907; (Baby Dorothy); flowers bright crimson-pink, dbl.; [Mme Norbert Levavasseur sport]; Levavasseur

'MAMAN LYLY', HT, lp, 1911; flowers large, semi-dbl.; Soupert & Notting; (Sangerhausen)

Maman Pineau, *see* 'GODILOFTER'

'MAMAN TURBAT', Pol, pb, 1911; flowers china-rose shaded lilac, reverse almost white, large, borne, semi-dbl.; foliage dark, soft; bushy growth; [Mme Norbert Levavasseur X Katharina Zeimet]; Turbat

'MAMBO', F, op, 1971; bud ovoid; flowers salmon-pink, medium, dbl., slight fragrance; dwarf, bushy growth; [Tropicana X Zorina]; Tantau, Math.

'MAMBO', Gr, mr, 1960; bud ovoid; flowers currant-red to cardinal-red, medium, dbl., exhibition form, slight fragrance; foliage dark, leathery; vigorous, tall, bushy growth; [Charlotte Armstrong X Unnamed seedling]; Swim & Weeks

'MAMIE', HT, pb, 1901; flowers pink and red, large, dbl., moderate fragrance; Dickson, A.; (Sangerhausen)

'MAMIE SERPA', F, w, 1955; bud ovoid, creamy; flowers open, dbl., 50 petals, 2.5 in., slight fragrance; foliage dark, soft;

vigorous, bushy, compact growth; [Goldilocks X Snowbird]; Serpa

'MAMILLE', HT, my, 1976; bud long, pointed; dbl., 35 petals, 4.5 in., exhibition form, moderate fragrance; foliage soft; vigorous, upright growth; [Peer Gynt X Valencia]; Kordes; Vilmorin

'MAMITA', HT, mr, 1958; flowers garnet-red, large, intense fragrance; vigorous growth; [Dicksons Red X Unamed seedling]; Robichon

Mamy Blue®, *see* 'DELBLUE'

Mamy Laperrière, *see* LAPlical

'MANA', F, yb, 1989; flowers creamy yellow, edges flushed pink, hybrid-tea shaped, large, dbl., moderate fragrance; prickles light brown; foliage medium green, glossy, veined; upright growth; [Liverpool Echo X Arthur Bell]; Cattermole, R.F.; South Pacific Rose Nursery

Manaia, *see* 'SUNMANI'

'MANAMSUN', Min, ob, 1996; (**Amber Sunset**); flowers orange with orange and dark yellow reverse, star-shaped, dbl., 26–40 petals, 1.75–2 in., exhibition form, no fragrance; some prickles; foliage medium, dark green, glossy, disease resistant; very vigorous, medium (40–50 cms), bushy growth; [June Laver X Rubies 'n' Pearls]; Mander, George; Select Roses, 1996

Manapouri, *see* 'MACWYNPIN'

Manasi, F, lp, 1991; IARI

'MANBERYL', Min, yb, 1995; (**Golden Beryl**); flowers deep yellow brushed orange inside of petals, blooms borne in, dbl., 15–25 petals, 1.75–2 in., slight fragrance; few prickles; foliage medium, medium green, glossy; medium (35–40 cms), bushy growth; [June Laver X Rubies 'n' Pearls]; Mander, George; Oregon Miniature Roses, 1995; Select Roses, 1995

Manchester Guardian Angel, N, w

Manchu Rose, *see* **R. XANTHINA**

'MANCLASSIC', F, op, 1996; (**Buffy Sainte-Marie**, Valia Balkanska); flowers blend of salmon, orange and pink with yellow center and yell, dbl., 15–25 petals, 3.25–4.25 in., slight fragrance; some prickles; foliage medium, dark green, glossy, disease-resistant; very vigorous, medium (70–90 cms), bushy growth; [June Laver X Rubies 'n' Pearls]; Mander, George; Select Roses, 1996

'MANCORAL', Min, op, 1995; (**Coral 'n' Gold**); flowers coral/salmon/pink blend inside, yellow centre, yellow cream reverse, some yellow stripes inside petals, full (26–40 petals) medium (4–6 cms) blooms borne in small clusters; no fragrance; few prickles; foliage medium, medium green semi-glossy; bushy, medium (40–45 cms) growth; [June Laver X Rubies 'n' Pearls]; Mander, George; Select Roses, 1995–96

'MANDALAY', HT, my, 1942; (Helvétia); flowers clear yellow, open, dbl., 40 petals, 6 in., moderate fragrance; foliage leathery; long stems; very vigorous, upright growth; [Soeur Thérèse X Feu Joseph Looymans]; Mallerin, C.; A. Meilland; J&P

'MANDARIN', F, mr, 1951; bud ovoid; flowers mandarin-red, blooms in large clusters, semi-dbl., 18 petals, 3–4 in., exhibition form, moderate fragrance; foliage leathery, glossy; vigorous ($3\frac{1}{2}$–4 ft), upright growth; [Lilette Mallerin sport X Red F seedling]; Boerner; J&P

'MANDARIN', HT, my, 1946; flowers golden yellow, intense fragrance; foliage glossy; vigorous growth; [Betty Uprichard X Ville de Paris]; Robichon

Mandarin, *see* KORcelin

Mandarin Delight, *see* 'FOUMANDE'

Mandarine, *see* **'MANDRINA'**

'MANDA'S TRIUMPH', HWich, w, 1899; flowers pure white, borne in clusters of 10–12, very dbl.; very vigorous growth; [R. wichurana X Paquerette]; Horvath; W.A. Manda

'MANDRINA', F, or, 1964; (Mandarine); dbl., 30 petals, 2.5–3 in., blooms in clusters; foliage dark; vigorous, bushy growth; [Moulin Rouge X Frau Anny Beaufays]; deRuiter; Carlton Rose Nurseries

Mandryka, S, or, 1992; flowers medium red with orange; Mekdeci-Olsen

'MANDY', F, ly, 1963; flowers creamy yellow to soft peach, large, dbl.; foliage coppery; very free growth; [Pinocchio X Sweet Repose]; Robinson, H.

Mandy, *see* KORinor

'MANDY JO', HT, pb, 1970; flowers pale biscuit, flushed light pink, full, dbl., 22 petals, 4.5 in., slight fragrance; foliage very glossy, bright green; vigorous growth; [Pink Favorite sport]; Abrahams

Mandy Singleton, HT, pb, 1995; Dawson

'MANECA', HT, lp, 1929; dbl., moderate fragrance; Pernet-Ducher; Gaujard

'MANETTE', HGal, rb; flowers rosy crimson marbled with purple, green eye, small, blooms l, very dbl., cupped; vigorous, upright growth; Lecoffé, before 1824; Prior to 1848

'MANETTI', N, lp, 1835; (R. chinensis manettii, R. manettii, R. X noisettiana manettii); 2 in.; red shoots; vigorous shrub growth; Rivers; Rivers, 1835, from S. Manetti, Monza Bot. Gdn., It

'MANGLOW', Min, rb, 1996; (**Glowing Amber**); flowers scarlet red with deep yellow reverse and yellow center, dbl., 26–40 petals, 1.5–2 in., borne mostly singly, slight fragrance; some prickles; foliage medium, dark green, glossy; medium (40–50 cms), bushy growth; [June Laver X Rubies 'n' Pearls]; Mander, George; Select Roses, 1996

Mango®, *see* 'DORAPRI'

Mango, *see* 'JACMEM'

'MANHALE', Min, m, 1996; (**Haleakala**); flowers velvety dark ruby-purple inside with yellow center, dbl., 15–25 petals, 1.75–2 in., slight fragrance; some prickles; foliage medium, dark green, glossy; medium (40–50 cms), bushy growth; [Rubies 'n' Pearls X June Laver]; Mander, George; Select Roses, 1996

'MANHATTAN', HT, or, 1936; (Passion Rose); bud urn shaped; flowers jasper-red to coral-red, base yellow, very large, dbl., cupped, intense fragrance; very vigorous growth; [Souvenir sport]; Asmus

Manhatten Blue, HT, m

'MANIFESTO', HT, op, 1920; flowers flesh-pink, tinged salmon, dbl., moderate fragrance; McGredy

Manilla, HT; Kordes, R.; (Cavriglia)

'MANILLE', HT, yb, 1951; flowers golden yellow edged red, very large, dbl., 40–50 petals, globular, intense fragrance; foliage leathery, bronze; vigorous, upright growth; [Pierre X Lumiere]; Mallerin, C.; Vilmorin-Andrieux

Manit, *see* 'UMSDAD'

Manita, *see* KORberuhig

'MANITOU', HT, rb, 1957; flowers coppery red, reverse golden yellow, well formed, very large, dbl.; upright, bushy growth; Swim, H.C.; Kordes

Manja, *see* **'SIX FLAGS'**

Manja Böhm, *see* **'MÁNJA BÖHMOVÁ'**

'MÁNJA BÖHMOVÁ', HMult, w, 1925; (Manja Böhm); flowers greenish white; [Tausendschön sport]; Böhm, J.

Manja Mourier, *see* **'SIX FLAGS'**

'MANJANA', F, op, 1969; flowers salmon-pink-apricot, exhibition form; free growth; [Orange Sensation X (Pink Parfait X Lavender Pinocchio)]; deRuiter

Manmatha, F, m, 1989; Pal, Dr B.P.

'MANNEQUIN', F, mp, 1961;, slight fragrance; vigorous, upright growth; [(Peace X Cinnabar) X Fashion]; Lens

'MANNHEIM'®, S, dr, 1959; flowers crimson, large blooms in clusters, dbl., recurrent bloom; upright (3 ft), bushy growth; [Rudolph Timm X Fanal]; Kordes, R., 1958

'MANNING'S BLUSH', HEg, w; flowers white faintly flushed pink, very small, dbl.; foliage fragrant (apple); height 4–5 ft

Mannington Cascade, HWich, w, 1998; Walpol & Lowne

Manola, *see* **'FRED CRAMPHORN'**

'MANON', HT, ab, 1924; flowers yellow mixed with apricot, semi-dbl.; [Christine X Mrs Farmer]; Bernaix, P.

Manora®, *see* TANaronam

'MANorsun', Min, ob, 1996; (**Orange Sunset**); flowers salmon-orange inside, some yellow stripes, orange-yellow reverse, dbl., 26–40 petals, 1.5–2 in., slight fragrance; some prickles; foliage medium, dark green, glossy; medium (40–50 cms), bushy growth; [June Laver X Rubies 'n' Pearls]; Mander, George; Select Roses, 1996

'Manosque', HT, mr, 1958; flowers well formed, intense fragrance; moderately vigorous growth; Frères, Buyl

Manou, Gr, 1979; Meilland, L.; (Cavriglia)

Manou Meilland®, *see* 'MEItulimon'

Manou Meilland, Climbing, *see* MEItulimonsar

'MANpetals', Min, op, 1996; (**Glowing Petals**); flowers blend of salmon orange and pink, dark yellow reverse, dbl., 26–40 petals, 2–2.5 in., no fragrance; some prickles; foliage large, dark green, glossy; medium (45–60 cms), bushy growth; [June Laver X Rubies 'n' Pearls]; Mander, George; Select Roses, 1996

'MANprincess', HT, lp, 1991; flowers large blooms borne mostly singly, very dbl., slight fragrance; foliage medium, light green, semi-glossy; tall, upright growth; [(Tiffany X Pascali) X Super Sun]; Mander, George

'MANpurpearl', F, m, 1992; flowers purple/cream bicolor, mauve color intensifies with sun and a, dbl., 26–40 petals, 3–3.5 in., slight fragrance; very few prickles; foliage medium, dark green, semi-glossy; medium (80–100 cms), bushy growth; [Rise 'n' Shine X MANpurple]; Mander, George

'MANpurple', HT, m, 1991; flowers purple/cream bicolor, large blooms borne mostly singly, dbl., 15–25 petals, slight fragrance; foliage medium, dark green, glossy; medium, bushy growth; [Mount Shasta X Super Sun]; Mander, George

'MANpursun', Min, m, 1992; (**Purple Sunset**); flowers purple/cream bicolor, very attractive bicolor combination, b, dbl., 15–25 petals, 1.5 in., no fragrance; few prickles; foliage small, dark green, glossy; low (35–40 cms), upright growth; [Rise 'n' Shine X MANpurple]; Mander, George

'MANrupearl', Min, m, 1992; (**Rubies 'n' Pearl**); flowers purple/cream bicolor, very attractive, very dark purple on i, dbl., 15–25 petals, 1.5 in., slight fragrance; few prickles; foliage small, medium green, semi-glossy; low (40–50 cms), upright, bushy growth; [Rise 'n' Shine X MANpurple]; Mander, George

'MANscarlet', F, rb, 1993; (**Scarlet Pearl**); flowers scarlet red with white eye and white reverse, blooms borne i, single, 5 petals, 1.5–2.75 in., slight fragrance; many prickles on basals, few on laterals; foliage medium, dark green, glossy; medium (90–100 cms), bushy, spreading, very vigorous growth; [Pink Meidiland X Seedling]; Mander, George; Christie Nursery, Ltd., 1993

'MANsopas', Cl F, op, 1991; flowers medium pink shaded coral and cream, blooms borne in small clusters, dbl., 26–40 petals, slight fragrance; foliage medium, dark green, glossy; tall, upright, bushy growth; [Shades of Pink X Pascali]; Mander, George

'MANstar', Min, ob, 1999; (**Amber Star**, Brittany's Glowing Star); flowers amber/orange, reverse golden yellow, dbl., 26–40 petals, 1.75–2.25 in., borne in small clusters, slight fragrance; prickles moderate; foliage medium, dark green, glossy; upright, tall (15–20 in.) growth; [Glowing Amber sport]; Mander, George & Pazdzierski, Jim; Select Roses, 2000; The Mini Rose Garden, 1999

'Mansuino Rose', HT, mr, 1964; (Generosa); flowers crimson to spirea-red, small to blooms, dbl., 35 petals, cupped, slight fragrance; foliage small, dark; thin stems; Mansuino, Q.; Carlton Rose Nurseries

'Manteau Pourpre', HGal, 1823; flowers medium, dbl.; Vibert; (Sangerhausen)

Manteo, *see* 'BRIman'

'MANtopas', Min, dy, 1995; (**Golden Topas**); dbl., 15–25 petals, 1.75–2 in., borne in small clusters, slight fragrance; some prickles; foliage medium, medium green, semi-glossy; medium (35 cms), bushy growth; [June Laver X Rubies 'n' Pearls]; Mander, George; Select Roses, 1995

'Manu Mukerji', HT, my, 1972; [Fragrant Cloud sport]; Friends Rosery

Manuel Canovas, *see* MASpagui

'Manuel P. Azevedo', HT, mr, 1911; flowers carmine-red, large, dbl., intense fragrance; Soupert & Notting; (Sangerhausen)

'Manuel Pinto d'Azevedo', HT, pb, 1954; flowers deep rose-pink, reverse lighter, dbl., 48 petals, 4 in., moderate fragrance; foliage dark, glossy, leathery; vigorous, upright growth; [Unnamed seedling X Peace]; da Silva, Moreira

'Manuela'®, HT, mp, 1968; flowers large, dbl., 30 petals, exhibition form, moderate fragrance; foliage glossy; vigorous, upright, bushy growth; Tantau, Math.

'Manuelita', HT, or, 1947; flowers large; very vigorous growth; [Cynthia X Vive la France]; Dot, Pedro

Manureva, *see* SAUdril

'Manx Queen', F, ob, 1963; (Isle of Man); flowers rich gold flushed bronze red, blooms in large clusters, semi-dbl., 18 petals, moderate fragrance; foliage dark; bushy, compact growth; [Shepherd's Delight X Circus]; Dickson, Patrick; A. Dickson

Many Happy Returns, *see* 'HARwanted'

'Many Moons', F, my, 1985; flowers deep yellow to light yellow, large, dbl., 35 petals, moderate fragrance; foliage medium, medium green, matt; upright, bushy, tall, arching growth; [Chinatown X Maigold]; Stoddard, Louis

'Many Summers', HT, ob, 1975; flowers orange-copper, full, dbl., 30 petals, 6 in., intense fragrance; vigorous growth; [Arthur Bell X Belle Blonde]; Fryer, Gareth; Fryer's Nursery, Ltd.

Many Thanks, Min, ab; [Mary Marshall sport]; Geytenbeek, 1976

Manyo, F, ab

'Maori Doll', Min, yb, 1977; flowers buff, yellow center; [Yellow Doll sport]; Bell Roses; Sequoia Nursery

'Maori Lullaby', F, pb, 1963; flowers carmine-rose, base light yellow to white, borne in clusters, semi-dbl., 10–12 petals, 3–3.5 in.; foliage bronze, leathery; bushy, low, compact growth; [Traumland X Unnamed seedling]; Mason, P.G.

'Maori Moon', F, pb, 1974; bud pointed; flowers pink, center cream, semi-dbl., 10 petals, 3.5 in., cupped, slight fragrance; foliage light, aging darker; upright, compact, bushy growth; [Bengali X (Pink Parfait X King Boreas)]; Clayworth

'Maorilander', HT, mr, 1956; bud ovoid, pointed; flowers crimson, lighter reverse, large, dbl., exhibition form, moderate fragrance; foliage bronze, leathery; bushy growth; [Crimson Glory X Peace]; Mason, P.

Marachal Le Clerc, *see* 'KRIcarlo'

Marama, *see* 'SIMaramam'

'Maranta', F, yb, 1974; bud globular; flowers yellow, red, medium, semi-dbl., slight fragrance; foliage soft; upright, bushy growth; Tantau, Math.; Ahrens & Sieberz

'Marathon', F, op, 1956; flowers salmon-rose, borne in clusters, semi-dbl., 4 in.; moderate, compact growth; [Fashion X Hybrid Tea seedling]; Mondial Roses

'MARbel', HT, lp, 1999; (**Belle**); flowers light pink, reverse medium pink, dbl., 17–25 petals, 4.5 in., borne mostly singly, slight fragrance; prickles moderate; foliage small, dark green, semi-glossy; upright, medium (2.5 ft) growth; [First Prize X Handel]; Hiltner, Martin

'Marbrée', P, rb, 1858; flowers red marbled white, large, dbl.; Robert et Moreau

'Marc Guillot', F, dr, 1955; bud ovoid; flowers dark scarlet-red, medium, dbl., 25–35 petals, cupped, slight fragrance; foliage dull green, leathery; vigorous, upright growth; [Happiness X Demain]; Mallerin, C.; EFR

'Marcel Boivin', HT, op, 1954; flowers coral-pink, base yellow, very large, dbl., 50–60 petals, moderate fragrance; vigorous growth; [Souv. de Claudius Pernet X Château de Clos Vougeot]; Buatois

'MARCEL BOURGOUIN', HGal, m; (Le Jacobin); flowers velvety rich scarlet-purple mottled violet; Corboeuf-Marsault, 1899; Corboeuf, 1899

Marcel Pagnol, *see* 'MEIsoyrio'

Marcel Pajotin, F, 1976; Kriloff, Michel; (Cavriglia)

Marcel Turbat, LCl; Moreira da Silva, A.; (Cavriglia)

'MARCELINE', HT, rb, 1928; flowers crimson, edge and reverse violet-rose, dbl.; [Frau Karl Druschki X Yves Druhen]; Buatois

Marcel Pagnoe, *see* 'MEIsoyrio'

Marcella Baldge, *see* **'SOUN DE MARCELLE BALAGE'**

'MARCELLE AUCLAIR', F, or, 1964; flowers medium, borne in clusters of 5–8, semi-dbl., exhibition form, intense fragrance; foliage glossy, leathery; vigorous, upright growth; [Soleil de Lyon X Seedling]; Robichon; Ilgenfritz Nursery

'MARCELLE GRET', HT, my, 1947; bud long, pointed; flowers saffron-yellow, dbl., 28 petals, 6 in., moderate fragrance; foliage dark; vigorous growth; GM, Geneva, 1948; [Peace X Prinses Beatrix]; Meilland, F.

'MARCELLE GRET, CLIMBING', Cl HT, my, 1957; Frères, Brenier

'MARCELLE PETIT', F, op, 1958; flowers glowing salmon-pink; foliage glossy; vigorous growth; [Pinocchio X Independence]; Arles; Roses-France

'MÄRCHEN', Pol, lp, 1927; flowers light rose pink with white, small, single; Kiese; (Sangerhausen)

Marchenkonigin, *see* 'KORoyness'

'MÄRCHENLAND', F, ob, 1951; (Exception); flowers bright rose tinted salmon, large blooms in clusters of 40, semi-dbl., 18 petals, moderate fragrance; foliage dark; vigorous, upright growth; [Swantje X Hamburg]; Tantau, 1946

'MÄRCHENTAG', HT, lp, 1929; flowers creamy pink, large, dbl.; von Württemberg, Herzogin Elsa; (Sangerhausen)

'MARCHESA BOCCELLA', HP, lp, 1842; (Marquise Boccella, Marquise Bocella); flowers delicate pink, edges almost blush, compact, large; petals smaller than other HP's, dbl.; stiff, erect stems; dwarf, robust habit growth; Desprez

'MARCHIONESS OF DUFFERIN', HP, mp, 1891; flowers very large, dbl.; Dickson, A.; (Sangerhausen)

'MARCHIONESS OF LINLITHGOW', HT, dr, 1929; flowers deep blackish crimson, open, large, dbl.; foliage soft, bronze; vigorous growth; GM, NRS, 1930; Dobbie

'MARCHIONESS OF LONDONDERRY', HP, lp, 1893; flowers pale pink, very large, dbl., 50 petals, exhibition form, moderate fragrance; very vigorous growth; Dickson, A.

'MARCHIONESS OF LORNE', HP, pb, 1889; flowers rich rosy pink shaded darker, large, dbl., cupped, intense fragrance; vigorous growth; Paul, W.

'MARCHIONESS OF ORMONDE', HT, ly, 1918; flowers clear wheat-straw color, center deep honey-yellow, dbl.; Dickson, H.

'MARCHIONESS OF SALISBURY', HT, dr, 1890; Pernet

'MARCIA', HT, op, 1952; flowers pink to coral, base yellow, well shaped, large; foliage glossy; vigorous, upright growth; [Étoile de Hollande X Raffel's Yellow]; Raffel; Port Stockton Nursery

'MARCIA COOLIDGE', HT, pb, 1927; flowers very light pink, reverse darker, stamens dark crimson, semi-dbl., intense fragrance; [Gen. MacArthur seedling]; Coolidge

'MARCIA GANDY', HT, rb, 1957; flowers crimson to rose-red, reverse rose-opal, intense fragrance; vigorous growth; Verschuren; Gandy Roses, Ltd.

'MARCIA STANHOPE', HT, w, 1922; flowers pure white, large, dbl., 25 petals, globular, intense fragrance; foliage leathery; GM, NRS, 1924; [Frau Karl Druschki seedling]; Lilley

'MARCO', HWich, w, 1905; flowers white, center coppery; [R. wichurana X Souv. de Catherine Guillot]; Guillot, P.

'MARCO POLO', HT, lp, 1971; flowers soft dawn-pink, large, dbl., exhibition form, moderate fragrance; foliage glossy; vigorous, upright growth; [Memoriam X Elizabeth Fankhauser]; Fankhauser

Marco Polo, *see* 'MEIpaleo'

'MARDI GRAS', HT, dr, 1953; bud ovoid; flowers deep velvety red, dbl., 33 petals, 5 in., exhibition form, moderate fragrance; foliage leathery; vigorous, upright, bushy growth; GM, Baden-Baden, 1953; [Crimson Glory X Poinsettia]; Jordan, G.L.; J&P

'MARDI GRAS, CLIMBING', Cl HT, dr, 1956; Kordes

'MARECHAL CARMONA', HT, op; flowers salmon-pink with reddish tones; da Silva, Moreira

'MARÉCHAL DAVOUST', M, mp, 1853; flowers bright rose, large, cupped; Robert

'MARÉCHAL DE VILLARS', B, dp; flowers deep rose shaded violet, cupped

Marechal du Palais, B, lp; Béluze, 1845

'MARÉCHAL FOCH', Pol, dp, 1918; (Red Orléans Rose); flowers cherry-red to pink, open, borne in compact clusters, semi-dbl., moderate fragrance; vigorous, bushy growth; [Orléans Rose sport]; Levavasseur

Maréchal le Clerc, *see* 'KRIcarlo'

'MARÉCHAL LYAUTEY', HT, dr, 1931; flowers deep red, large, very dbl., exhibition form, intense fragrance; foliage thick, dark, bronze; very vigorous, bushy growth; [Hadley X Laurent Carle]; Croibier

'MARÉCHAL NIEL', N, my, 1864; bud long, pointed; flowers golden yellow, large blooms, dbl., intense fragrance; foliage rich green; weak stems; very vigorous, climbing growth; (14); Pradel

'MARÉCHAL PÉTAIN', HT, yb, 1926; flowers soft pink on yellow ground; Reymond

Maréchal Stalin, HT, mr, 1965; flowers carmine-red, large, dbl., moderate fragrance; Mélichar; (Sangerhausen)

'MARELLA', HT, pb, 1961; bud globular; flowers pink-red, large, dbl., 35 petals, slight fragrance; foliage glossy, leathery; vigorous growth; [(Happiness X Independence) X Better Times]; Meilland, Mrs. Marie-Louise; URS

Maren, *see* 'WILgmar'

Marfa, HT, w, 1975; dbl., slight fragrance; Shtanko, E.E.; (Sangerhausen)

'MARFIL', HT, w, 1962; flowers ivory-white, large; strong stems; tall growth; [White Knight X Angelis]; Dot, Pedro

Marga, HRg, lp, 1979; flowers medium, semi-dbl., slight fragrance; Rieksta; (Sangerhausen)

'MARGA WEIL', HT, or, 1938; flowers salmon-red, large, dbl., slight fragrance; Weil; (Sangerhausen)

'MARGARET', HT, pb, 1954; flowers bright pink, reverse silvery pink, well-shaped, dbl., 70 petals, moderate fragrance; vigorous growth; GM, NRS, 1954; [May Wettern seedling X Souv. de Denier van der Gon]; Dickson, A.

'MARGARET AMOS', HT, dp, 1952; flowers strawberry-red, large, dbl., 25 petals, exhibition form, slight fragrance; foliage dark reddish green; very free growth; [McGredy's Scarlet seedling]; McGredy, Sam IV

'MARGARET ANDERSON', LCl, w, 1931; (Broomfield Novelty); flowers centers deep cream, outer petals cream-flesh, very large, dbl., recurrent bloom, moderate fragrance; foliage thick, leathery; very vigorous, climbing growth; Thomas; H&S

Margaret Anne, HT, lp; Matthews, W.J., 1977; (Weatherly, L.)

'MARGARET ANNE BAXTER', HT, w, 1927; bud pointed; flowers white, sometimes tinted flesh, large, very dbl., 88 petals, moderate fragrance; foliage thick, leathery, glossy, bronze; vigorous, bushy growth; GM, NRS, 1927; [Harry Kirk seedling]; Smith, T.

'MARGARET BELLE HOUSTON', HT, dr, 1929; flowers velvety crimson, very large, dbl., moderate fragrance; foliage light, leathery; Vestal

Margaret Bushby, S, lp; [R. macrantha X Unknown]; Weatherly, Lila, 1999; Prophyl Pty, Ltd.; (Weatherly, L.)

'**MARGARET CHASE SMITH**', HT, dr, 1966; bud long, pointed; flowers large, dbl., moderate fragrance; vigorous, upright growth; [Red Duchess X Queen Elizabeth]; Brownell, H.C.

'**MARGARET CLARA**', HT, ab, 1999; dbl., 26–40 petals, 2 in., borne mostly singly, moderate fragrance; prickles moderate; foliage medium, dark green, glossy, impervious to rain; upright, medium (3.5 ft) growth; [Solitaire X (Alexander X Remember Me)]; Jones, L.J.

Margaret Daintry, *see* 'HORMARTIM'

'**MARGARET DICKSON**', HP, w; flowers white, center pale flesh, well-formed, large, dbl., 65 petals, cupped, occasionally recurrent bloom, slight fragrance; foliage dark; vigorous growth; [Lady Mary Fitzwilliam X Merveille de Lyon]; Dickson, A., 1892

'**MARGARET DICKSON HAMILL**', HT, yb, 1915; flowers straw-yellow, flushed salmon, petals shell-shaped, large, dbl., slight fragrance; foliage dark, leathery; bushy growth; GM, Bagatelle, 1917 GM, NRS, 1914; Dickson, A.

'**MARGARET EGERTON**', HT, dp, 1931; flowers rosy cerise passing to carmine, base yellow, well formed, moderate fragrance; vigorous growth; Chaplin Bros.

'**MARGARET ELBOGEN**', Pol, w, 1936; flowers pinkish white, intense fragrance; very vigorous growth; Brada, Dr.; Böhm

'**MARGARET FLEMING**', HRg, mp, 1995; flowers very clear pink, no lavender tones, blooms borne up to 5 per cluster, semi-dbl., 6–14 petals, 1.5 in., no fragrance; foliage medium to large, medium green, semi-glossy, somewhat; bushy (up to 120 cms) growth; [R. rugosa alba X Masquerade]; Fleming, Joyce L.; Hortico Roses, 1994

'**MARGARET HERBERT**', HT, lp, 1956; Harrison

'**MARGARET HORTON**', HT, ab, 1921; bud pointed; flowers apricot-yellow, open, large, dbl., exhibition form; foliage leathery, glossy, light; vigorous growth; Hicks

'**MARGARET ISABEL**', HT, w, 1983; flowers oyster white, cream center, delicately flushed pink petal edges, dbl., 45 petals, exhibition form, intense fragrance; red-brown prickles; upright growth; [Strawberry Ice X Redgold]; Summerell, B.L.

'**MARGARET JEAN**', Min, dp, 1978; bud ovoid; flowers deep pink, blooms borne 4–8 per cluster, dbl., 40 petals, slight fragrance; very few prickles; foliage small, dark; bushy, vigorous growth; [Fairy Moss X Fairy Moss]; Dobbs, Annette E.; Small World Min. Roses

'**MARGARET LAW**', F, ob, 1981; flowers luminous orange, pink on outer petals, borne 3–23 per cluster, dbl., 25 petals, exhibition form, intense fragrance; hooked, red prickles; foliage dark, glossy; vigorous growth; [Princess Michiko X Gold Gleam]; McTeer, Gilbert; Bridgemere Nursery

'**MARGARET M. WYLIE**', HT, pb, 1921; (Mrs A.J. Wylie); flowers flesh, edges heavily flushed deep rosy pink, dbl., intense fragrance; GM, NRS, 1920; Dickson, H.

Margaret McDowell, *see* 'SEADOW'

'**MARGARET MCGREDY**', HT, or, 1927; flowers orange-scarlet, large, dbl., 35 petals, exhibition form, moderate fragrance; foliage light, leathery, glossy; vigorous growth; GM, NRS, 1925; McGredy

'**MARGARET MCGREDY, CLIMBING**', Cl HT, or, 1936; Dixie Rose Nursery

'**MARGARET MERCER**', HT, yb, 1977; flowers pale yellow, edged light pink, full, very dbl., 53–65 petals, 4.5 in., moderate fragrance; foliage glossy, dark; vigorous growth; [Pink Favorite X Peace]; Mercer

Margaret Merril®, *see* 'HARKULY'

'**MARGARET MOLYNEUX**', HT, my, 1909; flowers canary-yellow, moderate fragrance; Dickson, A.

'**MARGARET MOORE JACOBS**', HT, mp, 1968; bud ovoid; flowers large, dbl., cupped, intense fragrance; foliage glossy; vigorous, bushy growth; [Tiffany X Pink Masterpiece]; Fuller; Wyant

Margaret O, *see* 'ORTMAR'

'**MARGARET ROBERTS**', F, dp, 1976; flowers cerise, full, dbl., 20–25 petals, 3.5 in., intense fragrance; foliage small, dark; moderate, free growth; [Elizabeth of Glamis X Wendy Cussons]; Wood

'**MARGARET RUTH**', HT, w, 1969; flowers creamy white, center pale pink, large, dbl., exhibition form, intermittent bloom, slight fragrance; foliage glossy, leathery; vigorous, upright, bushy growth; [Anne Letts X Christian Dior]; Taylor, L.R.

Margaret Sharpe, T, mp

'**MARGARET SPAULL**', HT, ob, 1928; flowers variable orange and lilac, dbl., moderate fragrance; [Ophelia X Unnamed seedling]; Cant, B. R.

Margaret Telfer, *see* 'SEATEL'

Margaret Thatcher, *see* 'KORFLÜG'

Margaret Thatcher, *see* 'TAKSUN'

Margaret Trudeau, *see* 'MACAFT'

'**MARGARET TURNBULL**', Cl HT, yb, 1931; flowers soft pink on amber ground, large, dbl., cupped, slight fragrance; foliage wrinkled, light; vigorous, pillar growth; Clark, A.; NRS New South Wales

'**MARGARET VAN ROSSEM**', HT, op, 1946; flowers coppery salmon, center old-gold; moderate growth; Van Rossem

Margaret von Hessen, HT, 1973; Hetzel, K.; (Cavriglia)

Margaret Wasserfall, *see* KORoberfinz

Margaret Watson, HT, pb

'**MARGARETE GNAU**', HT, ob, 1930; flowers creamy white on orange ground, very large, dbl., exhibition form, slight fragrance; foliage leathery; very vigorous growth; [Mrs Charles Lamplough X Souv. de H.A. Verschuren]; Krause

'**MARGARETE HERBST**', Pol, dr, 1934; flowers dark blood-red, dbl., late bloom; recurrent; foliage ruby-red when young; vigorous, bushy growth; Herbst

Margarete Krüger, HT, or; flowers salmon-red, large, semi-dbl., slight fragrance; (Sangerhausen)

'**MARGARETHA MÜHLE**', HT, pb, 1925; flowers clear satiny pink with silvery reflex, dbl., moderate fragrance; [Mme Caroline Testout X Mrs W.J. Grant]; Mühle; F.J. Grootendorst

'**MARGARETHE VAN DE MANDERE**', F, dp, 1952; flowers raspberry-red, large, single; very vigorous growth; Leenders, M.

'**MARGARET'S CHOICE**', Gr, or, 1999; flowers orange-vermillion, reverse whitish, stiff petals, dbl., 26–40 petals, 4.5 in., borne mostly singly, moderate fragrance; prickles moderate; foliage medium, dark green, semi-glossy; upright, tall (3–5 ft) growth; Tough, Ian Murray

Margaret's World, *see* KIRbill

'**MARGARITA RIERA**', HT, op, 1924; flowers brilliant rose-salmon, base yellow, dbl., slight fragrance; [Mme Ravary X Mme Edouard Herriot]; Dot, Pedro

'**MARGE**', S, pb, 1999; flowers med. pink, creamy white at base, deep pink splotches on outer petals, dbl., 26–40 petals, 3 in., borne mostly singly, moderate fragrance; prickles moderate; foliage medium, medium green, dull; spreading, medium to tall (5 ft) growth; Jerabek, Paul E.

'**MARGHERITA CROZE**', HT, m, 1914; flowers carmine-purple changing to purple-rose, base shaded deep rose, dbl., moderate fragrance; [Étoile de France X Earl of Warwick]; Ketten Bros.

Margie™, *see* 'MARMIE'

'**MARGIE BURNS**', HT, mp, 1968; flowers rose-pink, reverse tyrian rose, large, dbl., exhibition form, moderate fragrance; foliage leathery; very vigorous, upright, bushy growth; Carrigg

'**MARGO KOSTER**', Pol, ob, 1931; (Sunbeam); flowers salmon; [Dick Koster sport]; Koster, D.A.

'**MARGO KOSTER SUPERIOR**', Pol, ob, 1956; flowers deep salmon-pink; [Dick Koster sport]; Koster, D.A.

'**MARGO KOSTER, CLIMBING**', Cl Pol, ob, 1962; (Sunbeam, Climbing); Golie; Crombie Nursery

'**MARGO'S BABY**', Pol, yb, 1987; flowers creamy ivory, salmon edges, aging dark edge, small, dbl., 30 petals, cupped, no fragrance; prickles few, medium, light green; foliage small, medium green, glossy; upright, bushy, low, vigorous growth; winter hardy.; [Margo Koster sport]; Partain, Joe L., 1988

'MARGO'S SISTER', Pol, lp, 1954; flowers shell pink; [Margo Koster sport]; Ratcliffe

'MARGOT AMOS', HT, op; bud high, pointed; flowers coral-pink flushed strawberry, large; foliage glossy, bronze

'MARGOT ANSTISS', HT, lp, 1947; flowers glossy satin-pink, dbl., 40–45 petals, 6 in., moderate fragrance; vigorous, branching growth; Norman; Harkness

'MARGOT ASQUITH', HT, mr, 1934; flowers shining cerise-red, well formed, moderate fragrance; very vigorous growth; [Betty Uprichard X K. of K.]; Prince

'MARGOT FONTEYN', HT, op, 1964; flowers salmon-orange, dbl., 40 petals, 4 in., intense fragrance; very free growth; [Independence X Ma Perkins]; McGredy, Sam IV; Fisons Horticulture

'MARGRATEN', F, mr, 1949; flowers currant-red, borne in trusses, semi-dbl., 16 petals, 3–4 in., flat, slight fragrance; foliage light green; vigorous, bushy growth; [Donald Prior X World's Fair]; Leenders, M.

'MARGRETHE MÖLLER', HT, dp, 1914; flowers deep cerise-rose, well formed, dbl., moderate fragrance; weak necks; [Lady Mary Fitzwilliam X Seedling]; Poulsen, D.T.

Margriet Hermans, HT, yb, 1994; RvS-Melle

'MARGUERITE AMIDIEU DE CLOS', HT, my, 1926; flowers buttercup-yellow, dbl., moderate fragrance; [Souv. de Claudius Pernet X Golden Emblem]; Ketten Bros.

Marguerite Anne, *see* 'COCREDWOOD'

'MARGUERITE BRASSAC', HP, dr, 1874; flowers large, dbl.; Brassac; (Sangerhausen)

'MARGUERITE CARELS', Cl HP, mp, 1922; flowers neyron pink, center darker, large, dbl., recurrent bloom; vigorous, climbing growth; [Frau Karl Druschki X Gen. MacArthur]; Nabonnand, P.

'MARGUERITE CHAMBARD', HT, or, 1928; bud pointed; flowers geranium-red to vermilion, very large, exhibition form, slight fragrance; foliage dark; very vigorous, bushy growth; Chambard, C.

Marguerite d'Autricho, *see* DORtiche

'MARGUERITE DE ROMAN', HP, w, 1882; Schwartz

Marguerite Defforey, HT, 1973; Orard, Joseph; (Cavriglia)

'MARGUERITE DESRAYAUX', N, mp, 1906; flowers large, semi-dbl.; Nabonnand; (Sangerhausen)

'MARGUERITE GUILLARD', HP, w, 1915; flowers stamens yellow, semi-dbl., 20 petals, flat; [Frau Karl Druschki sport]; Chambard, C.

'MARGUERITE GUILLOT', HT, ly, 1902; flowers very large, dbl., moderate fragrance; Guillot; (Sangerhausen)

'MARGUERITE HEITZMANN', HT, or, 1930; flowers salmon-pink, stamens golden yellow, very large, dbl., cupped, moderate fragrance; foliage leathery; very vigorous growth; [Frau Karl Druschki X Mme Edouard Herriot]; Buatois

'MARGUERITE HILLING', HMoy, mp, 1959; (Pink Nevada); [Nevada sport]; Hilling

'MARGUERITE JAMAIN', HP, lp, 1873; flowers large, very dbl.; Jamain, H.; (Sangerhausen)

'MARGUERITE MOULIN', HT, pb, 1938; flowers lilac-pink, center salmon, large, very dbl., cupped, moderate fragrance; vigorous, bushy growth; [(Mme Edouard Herriot X Mrs Aaron Ward) X Mme Caroline Testout]; Moulin; Hamonière

'MARGUERITE ROSE', Pol, mp, 1905; flowers medium, dbl.; Robichon; (Sangerhausen)

'MARGY', Pol, mr, 1936; flowers brilliant red, open, semi-dbl., moderate, spicy fragrance; foliage small, soft; bushy growth; Sauvageot, H.; C-P

'MARI CAROLYN', HT, m, 1999; dbl., 17–25 petals, 4–5 in., borne mostly singly, intense fragrance; few prickles; foliage medium, medium green, semi-glossy; bushy, medium (5 ft) growth; [Heirloom X Chrysler Imperial]; Sproul, James A.; Roses By Design, 1999

'MARI DOT', HT, ab, 1927; flowers bright salmon to salmon-pink, base yellow, borne in clusters, moderate fragrance; foliage glossy; long, strong stems; very vigorous growth; [O. Junyent X Jean C.N. Forestier]; Dot, Pedro; C-P

'MARIA', HT, rb, 1974; flowers red, yellow reverse, large blooms borne singly, dbl., 60 petals, moderate, tea fragrance; foliage dark, glossy; vigorous growth; [Rina Herholdt X Unnamed seedling]; Staikov, Prof. Dr V.; Kalaydjiev and Chorbadjiiski

'MARIA', F, or, 1965; flowers orange-scarlet, borne in clusters, single, 10 petals, 3 in., slight fragrance; foliage very large, dark, leathery; vigorous, upright growth; [Unknown seedling X Border Beauty]; Gregory

'MARIA ANTONIA CAMPRUBI', HT, mp, 1956; flowers soft carmine, well formed, large, intense fragrance; vigorous growth; [Peace X Rosa Munne]; Munné, M.

'MARIA BURNETT', Min, yb, 1980; bud globular; flowers light yellow, edged pink, small, dbl., 70 petals, globular, slight fragrance; foliage light green; low growth; [Little Darling X Patricia Scranton]; Miniature Plant Kingdom

Maria Callas, *see* 'MEIDAUD'

Maria Callas, Climbing, *see* 'MEIDAUDSAR'

Maria Carta, HT, 1976; Zandri, R.; (Cavriglia)

'MARIA CHAVARRI DE SALAZAR', HT, lp, 1935; flowers large, dbl.; Dot, Pedro; (Sangerhausen)

'MARIA CINTA', HT, op, 1967; bud long, pointed; flowers coral, large, dbl., 28 petals, moderate fragrance; foliage glossy, bronze; low, compact growth; [Duet X (Soraya X Chrysler Imperial)]; Dot, Simon; Minier

'MARIA DE MELLO', HT, dr, 1935; bud pointed; flowers bright velvety carmine-purple, large, dbl., 40–45 petals, cupped; foliage bronze quaker green; vigorous, branching growth; [Mme Gabriel Hanra X Mrs John Bell]; Ketten Bros.

'MARIA DELFORGE', Pol, mp, 1959; bud oval; flowers pink, rosette form, medium, borne in clusters, dbl., slight fragrance; foliage glossy; vigorous, bushy, low growth; [Orange Triumph X Ma Perkins]; Delforge

'MARIA GRAEBNER', S, mp; (R. X mariae-graebneriae); a few borne all summer; subglobose, red fruit; foliage orange and red in fall; 5 ft growth; (28); [R. palustris (?) X R. virginiana]

'MARIA GUARRO', HT, dp, 1935; bud pointed; flowers pink in spring, blood-red in summer, large, very dbl., exhibition form, slight fragrance; foliage glossy, dark; very vigorous growth; [Château de Clos Vougeot X Li Bures]; Dot, Pedro

Maria Hofker, HT, my, 1993; Interplant

'MARIA ISABEL', HT, mr; flowers deep strawberry-red, base yellow, exhibition form; foliage dark; vigorous growth; Camprubi, C.

'MARIA LEONIDA', HBc, w, 1829; (R. X leonida); flowers flesh, anthers purplish, very dbl., intense, tea fragrance; (14); Lemoyne

'MARIA LIESA', HMult, dp, 1925; flowers carmine-pink, small, single; Bruder Alfons; (Sangerhausen)

'MARIA LISA', HWich, pb, 1936; flowers clear rose, center white, stamens yellow, open, small, single, profuse, non-recurrent bloom; foliage dark, leathery; very vigorous, climbing or trailing growth; Liebau

'MARIA MAASS', HT, w, 1927; flowers very large, very dbl., intense fragrance; Maass; (Sangerhausen)

Maria Mathilda®, *see* 'LENMAR'

Maria McGredy, *see* 'MACTURANG'

'MARIA PERAL', HT, yb, 1941; flowers yellow suffused red, large, dbl.; very vigorous growth; Dot, Pedro

'MARIA REID', HT, dp, 1924; flowers dark rose-pink tinted peach, base yellow, dbl.; [Mme Caroline Testout X George C. Waud]; Ferguson, W.

'MARIA SERRAT', HT, op, 1946; flowers salmon-pink, base yellow, reverse deep yellow; [Mrs Pierre S. duPont X Baronesa de Ovilar]; Munné, M.

'MARIA STERN', HT, ob, 1969; bud pointed; flowers orange, large, dbl., 43 petals, globular, moderate fragrance; vigorous, upright growth; [Tip Toes X Queen Elizabeth]; Brownell, H.C.; Stern's Nursery

Maria Teresa®, *see* 'LENmacra'

'Maria Teresa Bordas', HT, mp, 1953; bud ovoid; flowers rose-pink, very large, very dbl., exhibition form, moderate fragrance; foliage dark, glossy; very vigorous, upright, bushy growth; [Sensation X Peace]; Bordas

Maria Teresa de Esteban, *see* 'DOTrames'

'Mariale', HT, ab, 1956; flowers orange-yellow, semi-dbl., slight fragrance; vigorous growth; [Souv. de Jacques Verschuren sport]; de Boer

'Marian Anderson', HT, dp, 1964; flowers deep pink, large, semi-dbl., exhibition form, moderate, spicy fragrance; foliage glossy; vigorous, tall, compact growth; [Queen Elizabeth X Merry Widow]; Lammerts, Dr Walter; Germain's

'Marian Colthorpe', HT, op, 1946; flowers coral shaded lemon and pink; Wheatcroft Bros.

Mariandel®, *see* 'KORpeahn'

Mariandel 92, F, lp; [Ko's Yellow X Eyepaint]; Hannemann, F., 1992; (Weatherly, L.)

'Marianna Rolfs', Cl HT, lp, 1926; flowers silvery pink, semi-dbl.; Walter, L.

'Marianne', HT, yb, 1933; flowers mixture of copper-yellow, pink and red, large, exhibition form, moderate fragrance; vigorous, bushy growth; [Sybil X Sunstar]; Krause

Marianne, *see* MEIklusy

'Marianne Kluis', Pol, mr, 1942; flowers carmine-red, small, dbl.; Kordes, W. Söhne; (Sangerhausen)

'Marianne Kluis Superior', Pol, mr, 1930; flowers deep violet-red; [Greta Kluis Superior sport]; Kluis & Koning

'Marianne Pfitzer', HT, lp, 1902; flowers very large, dbl.; Jacobs; (Sangerhausen)

'Marianne Powell', HT, dr, 1986; flowers large, very dbl., 50 petals, moderate fragrance; foliage large, dark green, glossy; upright growth; [Kerryman X Red Dandy]; Powell, G.

Marianne Tudor, *see* FRYmartor

'Mariano Vergara', T, mr, 1896; flowers magenta-red with vermilion reflections, large, dbl.; vigorous growth; Aldrufeu

Maria-Theresa, HT, ab, 1994; Fryer, Gareth

Maribel, *see* 'COCdana'

'Maribell', HT, rb, 1987; flowers carmine red, reverse silver white, large, dbl., 40–45 petals, slight fragrance; upright, vigorous growth; [(Sea Pearl X Zorina) X Lovita]; Gressard, J.; Delbard Roses, 1988

'Marica', HT, mp, 1964; flowers bright pink, large, dbl., slight fragrance; strong stems; vigorous, symmetrical growth; Mondial Roses

'Marica', Cl Min, pb, 1989; flowers pink, reverse white; [Ginza Komachi sport]; Kono, Yoshito

Marie, *see* 'RESmar'

'Marie Accarie', N; Guillot et Fils, 1872

'Marie Adélaïde', HT, yb, 1912; (Grande Duchesse de Luxembourg, Luxembourg); bud pointed; flowers coppery yellow, center deeper, large, dbl., exhibition form, slight fragrance; foliage bronze, soft; vigorous, spreading growth; [Mme J.W. Budde X Lyon Rose]; Soupert & Notting

'Marie Antoinette', HGal, m; flowers lilac-rose, large, dbl.; Vibert, 1829

'Marie Antoinette', HT, mp, 1968; bud long, pointed; flowers pink, reverse darker, large, dbl., cupped, slight fragrance; foliage dark, glossy; very vigorous, upright growth; [Queen Elizabeth X Chrysler Imperial]; Armstrong, D.L.; Armstrong Nursery

Marie Antoinette Rety, HT, 1983; Orard; (Cavriglia)

'Marie Baumann', HP, mr, 1863; flowers carmine-red, large, dbl., 55 petals, globular, moderate fragrance; foliage dark; vigorous growth; [Alfred Colomb seedling]; Baumann

'Marie Brissonet', Pol, lp, 1913; flowers flesh-rose borne in pyramidal clusters of 75–100; Turbat

'Marie Bugnet', S, w, 1963; bud long, pointed; flowers snow white, dbl., 3 in., recurrent bloom, intense fragrance; foliage light green, rugose; vigorous (3 ft), bushy, compact growth; [(Therese Bugnet X Seedling) X F.J. Grootendorst]; Bugnet; Skinner

Marie Casant, *see* **'Marytje Cazant'**

Marie Christina, MinFl, lp, 1988

Marie Curie, *see* MEIlomit

'Marie de Blois', M, mp; flowers pink tinted lighter, large; Moreau et Robert, 1852

'Marie de Bourgogne', M, mp, 1853; flowers large, dbl.; Robert

'Marie de Saint Jean', P, w, 1869; (Marie de st Jean); flowers large, dbl.; Damaizin

Marie de st Jean, *see* **'Marie de Saint Jean'**

'Marie Dermar', N, ly, 1889; flowers medium, dbl., moderate fragrance; Geschwind, R.; (Sangerhausen)

'Marie Dietrich', LCl, yb, 1928; flowers yellowish red, passing to white; [Leontine Gervais X Eugenie Lamesch]; Walter, L.

'Marie d'Orléans', T, mp; flowers bright pink shaded darker, large, dbl., flat; vigorous growth; Nabonnand, G., 1884

'Marie Dougherty', F, dr, 1976; bud long, slender; single, 5 petals, 3.5 in., slight, tea fragrance; foliage glossy; moderate, bushy growth; [Sarabande X Sarabande]; Linscott

'Marie Dutour', F, op, 1962; flowers reddish salmon, well formed, large; [Aloha X (Gloire du Midi X Edith de Martinelli)]; Arles; Roses France

'Marie Eads', Pol, mr, 1991; bud rounded; flowers medium red with blue overtones, reverse light red, small blooms, dbl., 25 petals, repeat bloom, no fragrance; foliage medium, medium green, semi-glossy with fringed stipu; bushy, low growth; [Baby Faurax X Verdun]; Eads, C.E.

'Marie Elizabeth', F, yb, 1965; flowers yellow shaded rose-pink, dbl., 28 petals, 3 in., flat, intense fragrance; foliage dark, heavily veined; vigorous growth; [Clare Grammerstorf X Cavalcade]; McGredy, Sam IV; McGredy

'Marie Faist', HT, op, 1925; flowers shell-pink tinted salmon, center darker, with orange, dbl., intense fragrance; [Mme Edmond Rostand X Mrs T. Hillas]; Berger, V.; Faist

'Marie Girard', HT, lp, 1898; flowers large, dbl., moderate fragrance; Buatois; (Sangerhausen)

'Marie Gouchault', HWich, mr, 1927; flowers clear red passing to brilliant salmon-rose, small, borne in, dbl., sometimes recurrent bloom; very vigorous growth; Turbat

'Marie Greene', HT, mr, 1941; flowers rich red, moderate fragrance; Clark, A.

'Marie Guillot', TOGR, w, 1874; flowers white, tinged yellow, large, dbl., intense fragrance; strong stems; vigorous growth; Guillot et Fils

'Marie Guillot', T, lp; Guillot et Fils, 1873

'Marie Guillot, Climbing', Cl T, w, 1898; Dingee & Conard

Marie Henriette Grafin Chotek, HMult, dr

'Marie Henriette Gräfin Chotek', HMult, dr, 1911; flowers dark crimson, large, borne in clusters, dbl.; very vigorous, climbing growth; Lambert, P.

'Marie Henry', HT, ly, 1900; flowers large, dbl., moderate fragrance; Buatois; (Sangerhausen)

Marie Laforet, *see* DORlain

'Marie Lambert', T, w, 1886; (Snowflake, White Hermosa); flowers pure white; [Mme Bravy sport]; Lambert, E.

'Marie Lavier', HT, yb, 1935; bud brownish yellow; flowers reddish nankeen yellow to salmon-yellow, rather large, dbl., moderate fragrance; vigorous, bushy growth; [Souv. de Claudius Pernet X Mme Edouard Herriot]; Buatois

Marie Legonde, T, w

'Marie Leonida', HBr, 1832; Moldenke; (Cavriglia)

'Marie Louise', D, mp; flowers mauve-pink, large, very dbl., intense fragrance; bushy, shrubby (about 4 ft) growth

'Marie Louise Mathian', HT, ly, 1912; flowers large, dbl., intense fragrance; Fugier; (Sangerhausen)

'Marie Lünnemann', HT, my, 1920; flowers clear pink, dbl., moderate fragrance; [Pharisaer X Laurent Carle]; Timmermans

'Marie Maass', HT, w, 1928; flowers pure white to ivory-white, very large, dbl., intense fragrance; vigorous, bushy growth; [Kaiserin Auguste Viktoria X Marechal Niel]; Maass

'Marie Menudel', HP, op, 1927; flowers rose-pink, tinted salmon, large, dbl., moderate fragrance; Barbier

Marie Nordlinger, M, lp; [Henri Martin X Unknown]; Morley, Dr B., 1988; (Weatherly, L.)

Marie Palit, HT, pb, 1988; Datt, Braham

Marie Pavic, *see* **'Marie Pavié'**

'Marie Pavié', Pol, w, 1888; (Marie Pavic); flowers white, center flesh, blooms in clusters, dbl., 2 in.; no prickles; foliage large, rich green; vigorous, bushy growth; Allégatière

'Marie Robert', P, m, 1850; flowers lilac-pink, medium, dbl.; Robert et Moreau; (Sangerhausen)

'Marie Schmitt', HT, lp, 1910; flowers large, dbl.; Schmitt; (Sangerhausen)

Marie Shields™, *see* 'MORmari'

'Marie Tudor', HGal, dp

'Marie van Houtte', T, pb, 1871; (Mlle Marie van Houtte); flowers deep cream, tinged pink, base buff-yellow, large, very dbl., exhibition form, moderate fragrance; foliage rich green, leathery; vigorous, bushy, sprawling growth; (14); [Mme de Tartas X Mme Falcot]; Ducher

'Marie van Houtte, Climbing', Cl T, pb, 1936; Thomasville Nursery

'Marie Verbrugh', F, op, 1954; flowers yellow-salmon, reverse coral, well formed, large, dbl., intense fragrance; bushy growth; [Ambassadeur Nemry X Souv. de Claudius Pernet]; Leenders, M.

'Marie Young', HT, or; flowers brilliant orange-red, well formed

'Marie Zahn', HT, lp, 1887; flowers large, dbl.; Müller, Dr F.; (Sangerhausen)

Marie-Antoinette Rety, HT, op

'Marie-Chantal', F, mp, 1959; flowers bright pink, open, large, semi-dbl., slight fragrance; foliage glossy; vigorous, bushy growth; [Peace X Fernand Arles]; Gaujard

'Marie-Claire', HT, or, 1938; bud deep orange-red; flowers golden coral-red, passing to orange-yellow, large, dbl., moderate fragrance; foliage bronze, glossy; strong stems; very vigorous, compact growth; [(Charles P. Kilham X Duquesa de Peñaranda) X (Charles P. Kilham X Margaret McGredy)]; Meilland, F.

'Marie-Claire, Climbing', Cl HT, or, 1944; (Grimpant Marie-Claire); Meilland, F.

'Marie-France', HT, my, 1957; bud apricot shaded orange; flowers pure yellow, large, semi-dbl.; very vigorous growth; [Feu Pernet-Ducher X Léonce Colombier]; Dorieux; Pin

Marie-Francoise Saignes, HT, mr

'Marie-Jeanne', Pol, w, 1913; flowers pale blush-cream borne in clusters of 40–60; no prickles; height 2–3 ft; Turbat

Marieken, F, ab, 1987; RvS-Melle

'Marielle', F, mp, 1963; flowers deep rosy pink, blooms in large clusters, dbl., 25 petals, 4 in., moderate fragrance; foliage dark; vigorous growth; [Independence X F seedling]; deRuiter

Marie-Louise Marjan, HT, ab, 1999; Kordes

'Marie-Louise Poncet', HT, op, 1929; flowers coppery rose to pale coppery pink, reverse carmine-salmon; Gaujard

'Marie-Louise Sondaz', HT, w, 1967; flowers cream shaded red, very large, dbl., moderate fragrance; foliage dark; very vigorous, upright growth; [Rose Gaujard X Peace]; Gaujard

Marie-Louise Velge, F, lp, 1997; 20 petals, 4 in., flat; foliage dense dark green; GM, The Hague, 1996 Golden Rose, Geneva, 1997; RvS-Melle, 1995

Marie-Louise Velge, F, 1997; Institute of Ornamental Plant Growing; (Cavriglia)

'Marie-Rose', Pol, mp, 1930; flowers ruddy pink, carnation shaped, large, borne in clusters of 30, dbl.; foliage glossy; very vigorous growth; [Marie-Jeanne sport]; Truffaut, T.A.; Turbat

'Marie-Rose Besson', HT, lp, 1939; bud long, pointed, yellow, tinted coral; flowers light pink, tinted coral-orange, large, dbl., slight fragrance; foliage glossy; long stems; vigorous growth; [Souv. de Claudius Pernet X Unnamed seedling]; Mallerin, C.; A. Meilland

'Marie-Rose Toussaint', HT, lp, 1946; flowers satiny pink, very large, moderate fragrance; foliage dark, leathery; strong stems; vigorous, bushy growth; Gaujard

Marietta®, *see* 'TANatiram'

'Marietta Silva Taroucová', HMult, mp, 1925; flowers bright rose, large, borne in clusters; foliage rich green; very vigorous, climbing growth; [Colibri X Crimson Rambler]; Zeman

Marie-Victorin, S, lp, 1999; L'Assomption

Marie-Victorin, *see* 'AC Marie-Victorin'

'Marigold', HT, op, 1955; bud long, pointed; flowers salmon-yellow lightly washed pink, dbl., 52 petals, 6 in., exhibition form, intense fragrance; foliage leathery, glossy, bright green; vigorous, upright growth; [Peace X Mme Joseph Perraud]; Lens

'Marijke Koopman', HT, mp, 1979; bud long, pointed; dbl., 25 petals, borne 3–5 per cluster, moderate fragrance; red prickles; foliage dark, leathery; vigorous, medium-tall, upright growth; GM, The Hague, 1978; Fryer, Gareth; Fryer's Nursery, Ltd.

Mariko, *see* 'DEVdorado'

'MARilo', Min, ab, 1999; (**Ilona**™); flowers orange/apricot striped, reverse lighter, single, 5–11 petals, 1 in., borne in small clusters, no fragrance; prickles moderate; foliage small, dark green, near rugose, disease-resistant; compact, low (18–24 in.) growth; [Altissimo X Roller Coaster]; Martin, Robert B., Jr.

'Marilyn', Min, lp, 1955; flowers light pink, base purplish, small blooms in clusters, dbl., 60 petals; very compact growth; [Perla de Montserrat X Bambino]; Dot, M.

'Marilyn', HT, ab, 1954; bud long, pointed; flowers apricot-pink veined red, intense fragrance; foliage dull, green; vigorous growth; [May Wettern X Phyllis Gold]; Fletcher; Tucker

'Marilyn Gowie', F, ?, 1968; borne in trusses; foliage bronze; bushy growth; Gowie

'Marimba'®, F, mp, 1965; [Garnette sport]; Dekkers; Verbeek

Marime, *see* 'MORmarme'

Marina®, *see* 'RinaKOR'

'Marina Fontcuberta', HT, dp, 1924; flowers brilliant carmine, center rose-carmine, dbl., intense fragrance; [Entente Cordiale X Laurent Carle]; Dot, Pedro

Marina Marini®, *see* BARmar

Marinette, *see* 'AUScam'

Mario Clemente, HT; Moreira da Silva, A.; (Cavriglia)

Mario Lanza, *see* 'HORaardvark'

Mariolina, S, 1989; Embriaco, B.; (Cavriglia)

'Marion', F, op, 1956; (Salmon Sensation); flowers pink tinted salmon, medium, semi-dbl.; bushy growth; [Duchess of Rutland X Fashion]; deRuiter

'Marion Cran', HT, or, 1927; bud buttercup-yellow, flushed cerise; flowers scarlet veined orange and yellow, dbl., exhibition form; foliage bronze, leathery, glossy; very vigorous, bushy growth; McGredy

'Marion Dingee', HT, dr, 1889; flowers crimson, [((Comtesse de Caserta X Général Jacqueminot) X Marechal Niel)x (Pierre Notting X Safrano)]; Cook, J.W.

Marion Foster, *see* 'MEIcapula'

Marion Harkness®, *see* 'HARkantabil'

Marion Hess, HT, dr, 1981; Hetzel

'**MARION HORTON**', HT, my, 1929; flowers primrose-yellow; [Gorgeous X Sunstar]; Bees

'**MARION LAWRIE**', HT, mp, 1976; flowers pink, base gold, very full, very dbl., 75 petals, 4 in., slight fragrance; foliage dark, leathery; [Kordes' Perfecta sport]; Lawrie

Marion Manifold, *see* '**MISS MARION MANIFOLD**'

'**MARION R. HALL**', HT, dp; flowers bright cerise-red, well formed, semi-dbl., 20 petals, 5 in.; vigorous growth; [Crimson Glory X Sterling]; Balcombe Nursery

'**MARION RICH**', HT, yb, 1997; flowers double, medium, dbl., 15–25 petals, borne mostly singly, slight fragrance; foliage medium, dark green, glossy; upright, medium (3–3.5 ft) growth; [Solitaire X Remember Me]; Skinner, A.W.

'**MARIONETTE**', F, w, 1944; bud cream-yellow; flowers small, borne in clusters, dbl., 25–30 petals, 1.5 in., moderate fragrance; vigorous, bushy growth; [Pinocchio sport]; deVor, Paul F.; J&P

'**MARIPOSA**', Pol, or, 1927; flowers deeper orange-red; [Orange King sport]; Allen

Mariposa Gem, *see* 'MORMAGEM'

Marisa™, *see* 'ORTISA'

Mariska, *see* VALkita

'**MARISTA**', HT, rb, 1975; flowers medium red, whitish reverse, large blooms borne singly, dbl., 75 petals, moderate, tea fragrance; foliage dark, glossy; upright growth; [Sarah Arnot X Rina Herholdt]; Staikov, Prof. Dr V.; Kalaydjiev and Chorbadjiiski

'**MARISTELLA**', HT, 1952; Giacomasso; (Cavriglia)

'**MARITA**', F, ob, 1961; flowers copper-orange, heavily veined yellow, medium, borne in trusses, dbl., 30–40 petals; foliage coppery; growth very free, straggly; [Masquerade X Serenade]; Mattock

Maritime Bristol, *see* 'SANTANG'

'**MARITIME HEIR**', S, dp, 1986; flowers lavender pink, carnation-like, large, dbl., occasionally recurrent bloom, intense fragrance; fine prickles; foliage small, light green, disease resistant; upright, bushy, very hardy growth; [Therese Bugnet X R. nitida]; James, John

Marjan, HT, ob; Select Roses, B.V.

'**MARJOLIN**', Ch, dr; flowers deep crimson-purple, cupped

'**MARJOLINE**', HT, rb, 1949; flowers cardinal-red, reverse indian yellow, dbl., 40 petals, 6 in., cupped, slight fragrance; vigorous, upright growth; [Boudoir X Léonce Colombier]; Meilland, F.

'**MARJORIE**', HT, w, 1895; flowers white with pink, large, dbl., moderate fragrance; Dickson, A.; (Sangerhausen)

'**MARJORIE ANDERSON**', F, mp, 1973; dbl., 26 petals, 5.5 in., moderate fragrance; foliage very large, matt; [Fragrant Cloud X Sea Pearl]; Dickson, Patrick; Dicksons of Hawlmark

'**MARJORIE ATHERTON**', HT, my, 1977; bud ovoid; dbl.; foliage light, leathery; vigorous, upright, bushy growth; [Mt. Shasta X Peace]; Bell, Ronald J.; Brundrett

'**MARJORIE BULKELEY**', HT, yb, 1921; flowers buff, flushed rose-pink, passing to silvery pink, dbl., moderate fragrance; GM, NRS, 1920; Dickson, H.

Marjorie Chase, *see* KORamator

'**MARJORIE CONN**', F, lp, 1981; bud ovoid; semi-dbl., 14 petals, blooms in clusters of 5–7, slight fragrance; brown prickles, hooked down; foliage dark; bushy growth; [Bon Bon X Unnamed seedling]; Berry, Howard

'**MARJORIE ELLICK**', F, w, 1978; flowers rowanberry, very full, very dbl., 95 petals, 4–6. in., foliage glossy, light; very vigorous, free growth; [(Spion-Kop X Ena Harkness) X (Sam Ferris X Karl Herbst)]; Ellick; Excelsior Roses

Marjorie Fair®, *see* 'HARHERO'

'**MARJORIE FOSTER**', HWich, dr, 1934; flowers deep blood-red, small, dbl.; vigorous growth; Burbage Nursery

'**MARJORIE LEGRICE**', HT, ob, 1949; flowers orange and yellow, pointed, dbl., 30 petals, 5 in., moderate fragrance; foliage glossy; vigorous growth; [Mrs Sam McGredy X President Plumecocq]; LeGrice

'**MARJORIE LEGRICE, CLIMBING**', Cl HT, ob, 1956; Tantau, Math.

Marjorie Marshall, *see* HARdenier

Marjorie May, *see* 'HORSUNPEGY'

'**MARJORIE PROOPS**', HT, dr, 1969; flowers crimson, dbl., exhibition form, moderate fragrance; [Red Dandy X Ena Harkness]; Harkness

Marjorie W. Lester, *see* '**LAURÉ DAVOUST**'

'**MARJORY PALMER**', Pol, mp, 1936; flowers rich pink, blooms in clusters, dbl., intense fragrance; bushy, compact growth; [Jersey Beauty X ?]; Clark, A.; NRS Victoria

Marjorie Reid™, *see* JAYMARJ

Mark 1, *see* 'SAVAMARK'

Mark One™, *see* 'SAVAMARK'

'**MARK SULLIVAN**', HT, op, 1942; (Président Chaussé); flowers gold flushed and veined rose, dbl., 33 petals, 4–4.5 in., exhibition form, moderate fragrance; foliage dark, leathery, glossy; vigorous, upright growth; [Luis Brinas X Brazier]; Mallerin, C.; A. Meilland; C-P

'**MARK SULLIVAN, CLIMBING**', Cl HT, ob

Markgräfin Wilhelmine, F, lp, 1995; flowers large, dbl.; Tantau; (Sangerhausen)

'**MARLENA**', F, mr, 1964; flowers crimson scarlet, patio, blooms in clusters, semi-dbl., 18 petals, flat; low, compact bushy growth; ADR, 1964 GM, Baden-Baden, 1962 GM, Belfast, 1966; [Gertrud Westphal X Lilli Marleen]; Kordes, R.; A. Dickson, 1964; J&P, 1967; McGredy, 1964

Marlowe Soft Orange, LCl, ab

'MARLYB', HT, yb, 1999; (**Lybie**); flowers yellow blend, reverse medium yellow, dbl., 17–25 petals, 4 in., borne in small clusters, moderate fragrance; few prickles; foliage medium, dark green, semi-glossy; upright, medium (3 ft) growth; [Lynn Anderson X ((Carefree Beauty X Picasso) X seedling)]; Hiltner, Martin

Marlyn, *see* TANliram

Marlyse®, *see* PEKomecli

'**MARMALADE**', HT, ob, 1977; bud long, pointed; flowers bright orange, reverse deep yellow, dbl., 30 petals, 5 in., intense, tea fragrance; foliage large, glossy, dark; upright growth; [Arlene Francis X Bewitched]; Swim, H.C. & Ellis, A.E.; Armstrong Nursery

'**MARMALADE MIST**'™, HT, op, 1990; bud pointed; flowers medium salmon pink, with lighter salmon pink reverse, aging, dbl., 25 petals, cupped, slight, fruity fragrance; deep plum and brown prickles; foliage large, dark green, semi-glossy; upright, medium growth; PP007549; Lammerts, Dr Walter, 1981; DeVor Nurseries, Inc., 1991

Marmalade Skies™, *see* 'MEIMONBLAN'

'MARMIE', Cl Min, rb, 1999; (**Margie**™); flowers dark red with white eye, reverse lighter, dbl., 17–25 petals, 2 in., borne in small clusters, slight fragrance; prickles moderate; foliage medium, dark green, semi-glossy; upright, climbing (5–7 ft); shade tolerant growth; [Roller Coaster X Roller Coaster]; Martin, Robert B., Jr.

'**MARMION**', HT, pb, 1934; flowers pale rose flushed orange, reverse salmon-pink, dbl., moderate fragrance; vigorous, branching growth; Dobbie

Marondo®, *see* KORtitu

Marovilla, F, or, 1992

'**MARQUES DE NARROS**', HT, ob, 1951; flowers salmon-pink, pointed; thornless; La Florida

'**MARQUESA DE AGUILAR**', HT, mr, 1955; bud ovoid; flowers cardinal-red to begonia pink, large, dbl., 60 petals, exhibition form, moderate fragrance; foliage glossy; vigorous growth; [Comtesse Vandal X Caprice]; Bofill; Torre Blanca

'**MARQUESA DE BOLARQUE**', HT, my, 1945; bud long, pointed; flowers lemon-yellow, large, dbl., exhibition form, slight fragrance; foliage dark, glossy; vigorous growth; [Shot Silk X Julien Potin]; Camprubi, C.

'**MARQUESA DE CASA VALDÉS**', HT, or, 1955; bud pointed; flowers scarlet-red slightly shaded orange, large, dbl., 35–40 petals, exhibition form, moderate fragrance; foliage dark; very vigorous, compact growth; [Peace X Poinsettia]; Dot, Pedro

'**MARQUESA DE GOICOERROTEA**', HT, ab, 1947; flowers amber-yellow, well formed, slight fragrance; upright growth; [Eclipse X Joanna Hill]; Dot, Pedro

Marquesa de Urquijo, *see* '**PILAR LANDECHO**'

Marquesa de Urquijo, Climbing, *see* '**PILAR LANDECHO, CLIMBING**'

'**MARQUESA DEL VADILLO**', HT, pb, 1945; (Spanish Main); flowers neyron pink, reverse silvery pink, open, 5 in., moderate fragrance; foliage glossy, dark; upright growth; [Girona X Condesa de Sástago]; Dot, Pedro

Marquise Boccella, *see* '**MARCHESA BOCCELLA**'

Marquise Bocella, *see* '**MARCHESA BOCCELLA**'

'**MARQUISE D'ANDIGNÉ**', HT, dr, 1927; flowers velvety scarlet-crimson, dbl., intense fragrance; [(Lieutenant Chaure X George C. Waud) X Laurent Carle]; Leenders Bros.

'**MARQUISE DE BALBIANO**', B, m, 1855; Lacharme, F.

'**MARQUISE DE BARBENTANE**', HT, ab, 1928; flowers apricot-yellow, shaded orange and sunflower-yellow, dbl.; [Mrs Farmer X Severine]; Fugier

'**MARQUISE DE CASTELLANE**', HP, dp, 1869; flowers dark rose-pink, well formed, large, dbl., moderate fragrance; moderate growth; Pere Pernet

'**MARQUISE DE GANAY**', HT, mp, 1910; flowers silvery rose, dbl., moderate fragrance; [Liberty X La France]; Guillot, P.

'**MARQUISE DE SINÉTY**', HT, yb, 1906; flowers golden yellow, shaded bronzy red, dbl., moderate fragrance; GM, Bagatelle, 1907; Pernet-Ducher

'**MARQUISE DE SINÉTY, CLIMBING**', Cl HT, yb, 1912; Griffon

'**MARQUISE DE VIVENS**', T, dp, 1886; flowers carmine, base yellowish, large, dbl.; Dubreuil

'**MARQUISE LITTA DE BRÉTEUIL**', HT, mr, 1893; Pernet-Ducher

Marquisette®, *see* DUCmar

'**MARR**', Pol, op; flowers coral overcast orange, borne in clusters, 1 in., cupped; thornless; rangy habit (10–12 in.) growth

'**MARRAKECH**', HT, dr, 1945; flowers oxblood-red shaded bright scarlet, well formed, very large, dbl.; upright, vigorous growth; [Rome Glory X Tassin]; Meilland, F.

Marriotta, *see* 'MACCRICKE'

Marry Me™, *see* 'DICWONDER'

'**MARS**', HT, op, 1927; flowers deep coral, moderate fragrance; Chaplin Bros.

'**MARSHALL P. WILDER**', HP, dr, 1885; flowers bright deep red, large, dbl., globular, intense fragrance; vigorous, tall growth; [Général Jacqueminot X ?]; Ellwanger & Barry

Marsyanka, HT, dp, 1958; flowers carmine-pink, medium, semi-dbl.; (Sangerhausen)

'**MARTA**', Min, or, 1983; flowers small, dbl., 20 petals, slight fragrance; foliage small, light green, matt; upright growth; [Persian Princess X Anytime]; Dobbs; Small World Min. Roses

Marta Salvador®, *see* FEdugia

'**MARTHA**', Pol, pb, 1906; flowers coppery rose, blooms in clusters of 7–20, dbl.; dwarf growth; [Thalia X Mme Laurette Messimy]; Lambert, P.

'**MARTHA**', B, pb, 1912; flowers pale coral, dbl., borne in clusters of 1–5, moderate fragrance; few thorns; tall, climbing growth; Knudson

'**MARTHA BUGNET**', HRg, mr, 1959; bud long, pointed; flowers purplish red, open, large, semi-dbl., abundant, recurrent bloom, intense fragrance; very large fruit; foliage dark rugosa type; weak stems; vigorous, bushy (5–6 ft) tall and broad growth; [(R. rugosa kamtchatica X R. amblyotis X R. rugosa plena) X F.J. Grootendorst]; Bugnet

'**MARTHA DREW**', HT, w, 1919; flowers creamy white, center rose, dbl., moderate fragrance; GM, NRS, 1919; McGredy

Martha Ford, HMult, ab, 1987; Nobbs

Martha Gonzales, Ch, mr

'**MARTHA KELLER**', Pol, lp, 1912; flowers medium, very dbl.; Walter; (Sangerhausen)

'**MARTHA KORDES**', F, dp, 1941; flowers light capucine-red, becoming pink, open, medium, borne in clusters, semi-dbl., slight fragrance; vigorous, upright growth; [Hedwig Fulda X Holstein]; Kordes

'**MARTHA LAMBERT**', Pol, rb, 1939; flowers brilliant scarlet with small yellow eye, small, borne in clusters, single, recurrent bloom, slight fragrance; foliage glossy; strong stems; vigorous, bushy growth; [Frans Leddy X Paul's Scarlet Climber]; Lambert, P.; C-P

'**MARTHA'S CHOICE**', HT, dp, 1977; bud high-centered; flowers deep pink, full, dbl., 37 petals, 5 in., slight fragrance; [Gavotte X Prima Ballerina]; Bailey

Martha's Vineyard, *see* 'POULANS'

'**MARTHE ANCEY**', HT, w, 1932; (Mme Martha Ancey); flowers cream, tinted salmon-pink, center with straw-yellow reflections, dbl.; vigorous growth; [Souv. de Claudius Pernet X Mme Mélanie Soupert]; Schwartz, A.

'**MARTHE CAHUZAC**', Pol, ly, 1902; flowers medium, dbl.; Ketten, Gebrüder; (Sangerhausen)

'**MARTIAN GLOW**', F, mr, 1972; flowers red, reverse lighter, blooms in clusters, semi-dbl., 10 petals, 1.5 in., slight fragrance; foliage semi-glossy, medium green; vigorous, spreading growth; [Joseph's Coat X Dorothy Wheatcroft]; Gandy, Douglas L.

'**MARTIAN SUNRISE**', S, or, 1979; bud ovoid; flowers urn-shaped, blooms borne singly or 3–5 per cluster, dbl., 43 petals, exhibition form, repeats well, slight fragrance; semi-hooked prickles; foliage light to medium green, semi-glossy; upright, bushy growth; [Paddy McGredy X Heidelberg]; Taylor, Thomas E.

'**MARTIN FAASSEN**', HT, dp, 1965; bud ovoid; flowers pink-red, medium, borne in clusters, dbl.; foliage dark; [Baccará X Seedling]; Verbeek

'**MARTIN FROBISHER**', HRg, lp, 1968; bud ovoid; flowers light pink, center darker, medium, dbl., intense fragrance; foliage light green; vigorous, tall growth; (14); [Schneezwerg X ?]; Svedja, Felicitas; Canada Dept. of Agric.

'**MARTIN LIEBAU**', HP, mp, 1930; flowers large, dbl., moderate fragrance; Kiese; (Sangerhausen)

Martin Martin, *see* 'SEAMAR'

Martina®, HT, w, 1989; Noack, Werner

Martina, F, lp; Urban, J., 1987; (Czech Rosa Club)

Martine Guillot, *see* MASmabay

'**MARTINE HÉMERAY**', F, or, 1958; flowers china-red, well formed, dbl., slight fragrance; foliage leathery; [(Orange Triumph X Mme Edouard Herriot) X Unnamed seedling]; Gaujard, R.; Hémeray-Aubert

Martini, HT, 1967; Delforge; (Cavriglia)

'**MARTONE**', Min, pb, 1989; bud pointed; flowers medium pink, lighter at base, reverse light pink, medium, dbl., 30 petals, exhibition form, slight, damask fragrance; prickles straight, medium, deep pink; foliage medium, medium green, semi-glossy; upright, medium growth; [Queen City X Unknown seedling]; Bridges, Dennis A.; Bridges Roses, 1990

Marty, *see* 'PEASWEET'

Marty's Triumph™, *see* 'BISMAR'

'**MARUSHKA**', HT, dy, 1984; dbl., 35 petals, exhibition form, borne 1–3 per stem; foliage dark, leathery, glossy; vigorous, bushy growth; [Kabuki X Unnamed seedling]; Staikov, Prof. Dr V.; Kalaydjiev and Chorbadjiiski

Marvelle, *see* 'MACTAURANG'

Marvie, *see* 'GELMAT'

'**MARVLOUS**', HT, mr, 1937; flowers crimson, open, large, dbl., intense fragrance;

foliage leathery; very vigorous growth; Cant, B. R.

'MARY', HT, ob, 1931; flowers buff and orange, moderate fragrance; vigorous growth; Bentall

'MARY', Pol, op, 1947; flowers orange-cerise, borne in small clusters; vigorous growth; [Orange Triumph sport]; Qualm; Spek

'MARY ADAIR', Min, ab, 1966; flowers buffy apricot, small, dbl., moderate fragrance; foliage light green, soft; vigorous, bushy, dwarf growth; [Golden Glow (LCl) X Zee]; Moore, Ralph S.; Sequoia Nursery

'MARY ANN', F, dp, 1959; bud ovoid; flowers rose-red, dbl., 60–78 petals, 2 in., exhibition form, slight fragrance; foliage dark; upright growth; [Garnette sport]; Restani

'MARY BARNARD', F, op, 1978; bud ovoid; flowers deep salmon-pink, semi-dbl., 18 petals, 3.5 in., intense fragrance; foliage dark; low, vigorous growth; [(Karl Herbst X Sarabande) X Ernest H. Morse]; Sanday, John

'MARY BEAUFORT', HT, lp, 1969; flowers light peach-pink, well shaped, small, intense fragrance; compact, low growth; [Gavotte X (Ethel Sanday X Crimson Glory)]; Sanday, John

'MARY BELL', Min, w, 1987; flowers small, dbl., 26–40 petals, moderate fragrance; foliage medium, medium green, semi-glossy, disease resistant; bushy, hardy growth; [Cherish X Rise 'n' Shine]; Bell, Charles E., Jr.; Kimbrew Walter Roses, 1987

'MARY BOSTOCK', HT, lp, 1952; flowers shell-pink tinted white, dbl., 60 petals; strong stems; vigorous growth; Clark, A.

Mary Bradby™, *see* 'BYRBRADBY'

'MARY BRUNI', Pol, lp, 1914; flowers light creamy rose pink, small, dbl.; Gratama; (Sangerhausen)

Mary Burke, Min, lp; Hannemann, F., 1990; The Rose Paradise; (Weatherly, L.)

Mary Campbell, *see* 'HORLOVEQUEEN'

'MARY CARVER', HT, lp, 1950; bud globular; flowers shell-pink, dbl., 80 petals, 5 in., cupped, intense fragrance; foliage leathery, light green; vigorous, upright growth; [Red Radiance sport]; Chick

Mary Casant, *see* **'MARYTJE CAZANT'**

Mary Cave, F, 1993; Harkness; (Cavriglia)

'MARY CLARK', Min, lp, 1981; flowers medium, dbl., 55 petals, slight fragrance; straight, light yellow prickles; foliage mid-green; [Janna X Gene Boerner]; Hooper, John C.

'MARY CLAY', HT, dr, 1951; flowers blood-red, dbl., 40–45 petals, 6 in., intense fragrance; foliage very heavy, dark; very free growth; [Kardinal X Crimson Glory]; Kordes; Morse

'MARY CORELLY', HP, op, 1901; flowers dark salmon-pink, medium, dbl.; Prince; (Sangerhausen)

'MARY DELAHUNTY', HT, dr, 1990; bud pointed; flowers large, borne usually singly, dbl., 35–40 petals, moderate, damask fragrance; foliage dark green, glossy; bushy, tall growth; [(Daily Sketch X Impeccable) X Red Planet]; Bell, Ronald J., 1985; Treloar Roses Pty. Ltd., 1990

'MARY DEVOR', F, mr, 1967; (Douchka); flowers cardinal-red, sweetheart, medium, dbl., 35 petals, moderate fragrance; foliage leathery; vigorous, upright growth; [Christian Dior X Rumba]; Lammerts, Dr Walter; Amling-DeVor Nursery

Mary Donaldson, *see* 'CANANA'

'MARY DUTTON', HT, op, 1949; bud long, pointed; flowers salmon-pink, dbl., 40 petals, 6 in., slight fragrance; foliage glossy; very vigorous growth; [Crimson Glory X Mrs Sam McGredy]; Bees

Mary Edith, *see* 'TALMAR'

'MARY EGERTON', F, rb, 1982; flowers white, petals edged orange-red, spreading with age, reverse, dbl., 35 petals; foliage large, dark, glossy; upright, bushy growth; [Fragrant Cloud X Prominent]; Lea, R.F.G.

'MARY ELIZABETH', Min, lp, 1982; flowers small, dbl., 29 petals, slight fragrance; foliage small, light green, matt; bushy growth; [Fairy Moss X Fairy Moss]; Dobbs; Small World Min. Roses

'MARY FLEMING', F, mp, 1995; bud opening to very pale pink; semi-dbl., 6–14 petals, 1.5–2.75 in., borne 2–15 per cluster, moderate fragrance; foliage medium, medium green, matt; bushy (90–100 cms), spreading growth; [Marchenland X Golden Salmon Superieur]; Fleming, Joyce L.; Hortico Roses, 1994

Mary Gammon, *see* FRYsweetie

Mary Gamon, Min, 1993; Fryer, Gareth; (Cavriglia)

'MARY GUTHRIE', Pol, mp, 1929; flowers rich pink, small blooms in large clusters, single, moderate fragrance; foliage light; bushy ($2\frac{1}{2}$ ft) growth; [Jersey Beauty X Scorcher]; Clark, A.; NRS Victoria

'MARY HART', HT, mr, 1931; [Talisman sport]; Hart, G.B.

'MARY HART, CLIMBING', Cl HT, mr, 1937; Western Rose Co.

'MARY HART, CLIMBING', Cl HT, mr, 1942; Meilland, F.

Mary Hayley Bell, *see* KORparall

'MARY HAYWOOD', Min, mp, 1957; flowers bright pink, base white, dbl., 50 petals, 1 in., moderate fragrance; foliage glossy; very compact (10 in.), bushy growth; [(R. wichurana X Floradora) X Oakington Ruby]; Moore, Ralph S.; Sequoia Nursery

'MARY HELEN TANNER', HT, dp, 1932; flowers carmine, stems white or pinkish; [Templar sport]; Tanner

'MARY HICKS', HMult, dr, 1927; flowers deep scarlet, borne in clusters, semi-dbl., moderate fragrance; foliage light; vigorous, climbing growth; Hicks

Mary Hilda Law, *see* 'KIRCLOUD'

'MARY HILL', HT, w, 1916; flowers cream, center deep orange, dbl.; [Ophelia X Sunburst]; Hill, E.G., Co.

Mary Hill, *see* 'MORHILL'

Mary Jean, *see* 'HARYEN'

'MARY JO', HT, my, 1958; bud short, pointed; flowers maize-yellow, to open, medium, dbl., 30–35 petals, 3.5 in., exhibition form, slight fragrance; foliage leathery; strong stems; vigorous, upright, bushy growth; [Unnamed seedling X Orange Delight]; Hill, Joseph H., Co.

Mary Johnston, *see* 'WAMMARY'

'MARY KATE', Min, dp, 1977; bud mossy; flowers light red, dbl., 29 petals, 1 in., flat, slight fragrance; foliage dark, soft; vigorous, upright growth; [Fairy Moss X Fairy Moss]; Dobbs; Small World Min. Roses

Mary Kay™, *see* 'MINOCO'

'MARY KITTEL', HT, dr, 1975; bud pointed; dbl., 35 petals, 5 in., exhibition form, intense fragrance; foliage large, glossy; vigorous growth; [(Chrysler Imperial X Night 'n' Day) X Night 'n' Day]; Harvey, R.E.; Kimbrew-Walter Roses

'MARY L. EVANS', HRg, dp, 1936; flowers deep wild-rose-pink, non-recurrent; very similar to tekonkaha but growth more spreading; [Hansa X R. macounii]; Wright, Percy H.

'MARY LAWRANCE'S SHELL ROSE', Ch, mr, 1799

Mary Lee Johnson Richards, HT, lp

Mary Louise, *see* 'BUSMALOU'

Mary Louise, *see* 'TAYMAR'

'MARY LOVETT', LCl, w, 1915; flowers snow-white, large, dbl., some repeat bloom, moderate fragrance; vigorous, climbing (to 10–12 ft) growth; [R. wichurana X Kaiserin Auguste Viktoria]; Van Fleet; J.T. Lovett

Mary Lynn, *see* 'RENLYNN'

Mary Lyon™, *see* 'WILMTHO'

Mary MacKillop, HT, pb, 1989; Swane

Mary Magdalene, *see* 'AUSJOLLY'

'MARY MALVA', HT, lp, 1971; bud long, pointed; dbl., 18–24 petals, 3.5–4.5 in., exhibition form, moderate, spicy fragrance; foliage dark, leathery; upright growth; [(Pascali X Charlotte Armstrong) X Lilac Charm]; Lens, 1972

'MARY MANGANO', Min, op, 1989; flowers deep coral pink; [Petite Folie sport]; Papandrea, John T.

Mary Manners, HRg, w, 1970; Leicester

Mary Margaret, *see* 'TALMAG'

'MARY MARGARET MCBRIDE', HT, mp, 1942; bud long, pointed; flowers salmon-pink, dbl., 42 petals, 4–5 in., exhibition form, moderate fragrance; foliage dark, leathery, glossy; vigorous, upright, bushy growth; AARS, 1943 David Fuerstenberg Prize, ARS, 1945; [Sunkist X Olympiad]; Nicolas; J&P

'MARY MARQUES', HT, or, 1955; bud long, pointed; flowers orange-red tinted yellow, medium, dbl., 40 petals, exhibition form, slight fragrance; upright growth; [Mediterranea X Suzanne Balitrand]; Bofill; Torre Blanca

'MARY MARSHALL', Min, ob, 1970; bud long, pointed; flowers orange, base yellow, small, dbl., cupped, moderate fragrance; foliage small, leathery; vigorous, dwarf, bushy growth; AOE, 1975; [Little Darling X Fairy Princess]; Moore, Ralph S.; Sequoia Nursery

Mary Marshall, Climbing, *see* 'MINICO'

Mary Maud, *see* 'GELMAUD'

Mary May, *see* 'SHERISKEP'

'MARY MCHUTCHIN', Cl Pol, mr, 1935; flowers crimson, borne in clusters, semi-dbl., cupped, slight fragrance; foliage large, leathery; vigorous, climbing (6–8 ft) growth; Cant, B. R.

'MARY MERRYWEATHER', HT, dy, 1925; bud pointed; flowers deep golden yellow, semi-dbl., cupped, moderate fragrance; foliage glossy; vigorous, bushy growth; [Marquise de Sinéty X Lady Hillingdon]; Merryweather

'MARY MINE', Gr, op, 1973; flowers salmon-pink to light rose, large, dbl., 27 petals, slight fragrance; [Queen Elizabeth X Buccaneer]; Harkness

'MARY MONRO', HT, pb, 1921; flowers carmine-pink, flushed saffron-yellow, dbl., moderate fragrance; Pemberton

'MARY MULLIGAN', HT, or, 1944; bud long, pointed; flowers flame, dbl., cupped, moderate fragrance; vigorous growth; Mallerin, C.; A. Meilland

'MARY MURRAY', HT, ab, 1930; flowers deep apricot-yellow becoming lighter, dbl., 30 petals; foliage glossy, bronze; vigorous growth; Prior

'MARY NISH', HT, w, 1928; (White Radiance); flowers white, center tinted shell-pink, very large, dbl., moderate fragrance; foliage rich green, soft, glossy; vigorous, bushy growth; [Red Radiance sport]; Pacific Rose Co.

Mary of Penola, *see* 'TOMMOP'

'MARY PAULINE', HT, w, 1998; flowers pale pink, double (15–25 petals), exhibition, large blooms borne mostly single; some prickles; foliage large, medium green, semi-glossy; upright, tall growth; cut flowers; [Queen Elizabeth X Chablis]; McCruiry, Curtis

'MARY PICKFORD', HT, ab, 1923; bud pointed, orange-yellow; flowers pale yellow, center deeper, large, dbl., intense fragrance; foliage bronze; vigorous, bushy growth; [Grange Colombe seedling X Souv. de Claudius Pernet]; H&S

Mary Pilkington, *see* 'EVEPILK'

'MARY POPE', F, yb, 1965; flowers golden yellow suffused pink, edged darker, borne in clusters, dbl., 25 petals, 3 in.; foliage glossy, dark; vigorous growth; [Seedling X Independence]; Sanday, John

Mary Pope, *see* KORlasche

'MARY POPPINS', HT, mp, 1967; [Hallmark sport]; Morey, Dr Dennison; Country Garden Nursery

'MARY QUEEN OF SCOTS', HSpn, w

'MARY RAND', HT, w, 1965; flowers cream, edged rose-pink, well formed, 3.5 in.; foliage glossy, light green; vigorous growth; [Caprice X Scandale]; Latham

'MARY RATCLIFFE', HT, pb, 1958; flowers soft pink, reverse darker, moderate fragrance; Ratcliffe

'MARY ROBERTSON', Pol, mr, 1969; bud ovoid; flowers medium, dbl.; foliage dark; [Paprika X Seedling]; Buisman, G. A. H.

Mary Rose®, *see* 'AUSMARY'

'MARY RUSSELL', HT, mr, 1940; flowers well formed, large; Clark, A.

Mary Sampere de Guanabara, F; Moreira da Silva, A.; (Cavriglia)

Mary Sheffield, *see* 'DEBRAH'

Mary Sumner, *see* 'MACSTRA'

Mary Swindells, *see* 'KENSINGO'

Mary Taylor, HT, lp; Taylor, L.R., 1966; (Weatherly, L.)

Mary Thomson, S, lp; Thomson, G.A.; (Weatherly, L.)

'MARY WALLACE', LCl, mp, 1924; bud long, pointed; flowers warm rose pink, very large, semi-dbl., cupped, seasonal bloom, moderate fragrance; foliage glossy, rich green; vigorous, climbing (8–12 ft) growth; [R. wichurana X Pink Hybrid Tea]; Van Fleet; American Rose Society

Mary Warm, Cl Min, or, 1997

'MARY WARREN', Cl HT, mp, 1931; flowers pink, open, large, semi-dbl., recurrent bloom, slight fragrance; foliage soft, large, dark; vigorous, pillar growth; [Mrs Frank Guthrie X Scorcher]; Clark, A.; NRS New South Wales

'MARY WASHINGTON', N, w; flowers white tinted pink, fading to white, dbl., borne in clusters of 7–11, moderate fragrance; very vigorous growth; Registered by Frank L. Ross, Nashville, Tenn

Mary Webb®, *see* 'AUSWEBB'

'MARY WHEATCROFT', HT, ob, 1945; bud high-pointed; flowers deep copper; foliage bronze; [Mrs Sam McGredy X Princess Marina]; Robinson, H.; Wheatcroft Bros.

Mary Wise, *see* 'HERMAWI'

Mary Woodcock, HMsk, mp, 1989; Stydd

'MARY, COUNTESS OF ILCHESTER', HT, mp, 1909; (Countess Mary of Ilchester, Countess of Ilchester); flowers deep rose-pink, open, large, dbl., moderate fragrance; foliage rich green, leathery, glossy; very vigorous, bushy growth; Dickson, A.

Marybeth™, *see* 'MINIBETH'

'MARYCKA MAGDANOVÁ', Pol, dr, 1938; flowers small, semi-dbl.; Böhm, J.; (Sangerhausen)

Maryellen, *see* 'LEMMAR'

Mary-Jean, *see* 'HARYEN'

'MARYKE-MARIKA', F, ob, 1973; bud medium, ovoid; flowers orange, base yellow, dbl., globular, slight fragrance; foliage glossy; vigorous, upright growth; [Colour Wonder X Zorina]; Kordes

Marylea Johnson Richards, *see* 'WILLEA'

'MARYLÈNE', HT, mp, 1965; bud long, pointed; flowers pearl-pink, medium, dbl., moderate fragrance; foliage dark, glossy; upright growth; [Mignonne X Queen Elizabeth]; Gaujard

Marylka, Min, 1967; Wituszynski, B.; (Cavriglia)

'MARY'S FAVORITE', F, dp, 1994; flowers deep pink, blooms borne in small clusters, dbl., 15–25 petals, 3–3.5 in., slight fragrance; few prickles; foliage large, medium green, semi-glossy, disease resistant; medium (48–54 in.), upright, compact growth; [Pink Favorite X Unknown]; Hemphill, William J.

'MARY'S PINK', HT, mp, 1952; bud long, pointed; flowers neyron rose, dbl., 32 petals, 4–4.5 in., slight fragrance; foliage leathery; vigorous, compact, bushy growth; [Better Times sport]; Spanbauer

'MARYSA', Pol, w, 1936; flowers pure white, moderate, lily-of-the-valley fragrance; vigorous growth; Brada, Dr.; Böhm

Maryse Kriloff, *see* **'LUCY CRAMPHORN'**

Maryse Kriloff, Climbing, *see* KRImasar

'MARYTJE CAZANT', Pol, ab, 1927; (Marie Casant, Mary Casant); bud globular; flowers coral-pink, blooms in large clusters; dwarf growth; [Jessie sport]; van Nes

'MASARYKOVA JUBILEJNI', HT, dr, 1931; (Jubilaire de Masaryk, Jubilee, Masaryk's Jubilee, Masaryk's Jubileums-Rose); flowers velvety red shaded black, very large, single, slight fragrance; foliage dark, glossy; vigorous growth; [Blanta sport]; Böhm, J.

Masaryk's Jubilee, *see* **'MASARYKOVA JUBILEJNI'**

Masaryk's Jubileums-Rose, *see* **'MASARYKOVA JUBILEJNI'**

MAScanta, S, my, 1997; (**Claudia Cardinale**, 'MAScatna'); Guillot-Massad, 1997

Mascara™, *see* MEIdalnu

MAScatna, S, my, 1997; (**Claudia Cardinale**, 'MAScanta'); Guillot-Massad, 1997

'MASCOTTE', HT, mp, 1951; bud pointed; flowers hermosa pink, dbl., 4 in., moderate fragrance; foliage dark; vigorous growth; [Michele Meilland X Pres. Herbert Hoover]; Meilland, F.

Mascotte '77®, *see* 'MEITILOLY'

MASdogui, S, op, 1995; (**Sonia Rykiel**); Guillot-Massad, 1995

MASdomo, S, pb, 1996; (**Belle de Dom**); Guillot-Massad

MASelgi, S, w, 1998; (**Eliane Gillet**); Guillot-Massad

MASflodel, S, m, 1997; (**Florence Delattre**); Guillot-Massad

MASflodel, S, ab, 1997; (**Morabito**); Guillot-Massad

MASframb, S, mp, 1998; (**Jardins de Viels Maisons**); Guillot-Massad

'MASKED BALL', HT, pb, 1966; bud globular; flowers scarlet and gold, large, dbl., moderate fragrance; foliage dark, glossy; vigorous, bushy growth; [Masquerade X Peace]; Schwartz, Ernest W.; Wyant

MASlako, S, dr, 1996; (**Anne Laure**); Guillot-Massad

MASlupau, S, ab, 1997; (**Mme Paule Massad**); Guillot-Massad

MASmabay, S, w, 1996; (**Martine Guillot**); Guillot-Massad

MASmarti, S, mp, 1996; (**Orsola Spinola**); Guillot-Massad

MASpagui, S, w, 1995; (**Manuel Canovas**); Guillot-Massad

MASpaujeu, S, ab, 1997; (**Paul Bocuse**); Guillot-Massad, 1999

'MASQUERADE', F, rb, 1949; bud small, ovoid, yellow; flowers bright yellow turning salmon-pink and then dark red, blooms, semi-dbl., 17 petals, 2.5 in., slight fragrance; foliage leathery, dark; vigorous, bushy, compact growth; (28); GM, NRS, 1952; [Goldilocks X Holiday]; Boerner; J&P

'MASQUERADE, CLIMBING', Cl F, rb, 1958; Dillian; Gregory

'MASSABIELLE', HT, w, 1958; flowers well formed, large; upright growth; Guillot, M.

Massara, F, 1995; Sauvageot; (Cavriglia)

'MASTER DAVID', HT, pb, 1949; flowers pink, reverse carmine, pointed, dbl., 25 petals, 4 in., moderate fragrance; almost thornless; vigorous growth; Cox

'MASTER HUGH', (R. macrophylla seedling), dp, 1970; flowers rich rose-pink, blooms in clusters, single, 5 petals, 2.5–3 in., slight fragrance; [R. macrophylla seedling]; Mason, L.M.; Sunningdale Nursery

'MASTER JOHN', Cl HT, or, 1944; flowers fiery orange-red, base gold, large, dbl., globular, profuse spring bloom, then scattered until fall, intense fragrance; foliage glossy, dark; very vigorous, climbing growth; [Ednah Thomas X Golden Rapture]; Duehrsen; California Roses

MASversi, S, lp; (**Versigny**); Guillot-Massad, 1998

Matador, *see* 'KORFARIM'

'MATADOR', HT, dr, 1935; bud long, pointed; flowers scarlet-crimson, shaded darker, large, very dbl., cupped, moderate fragrance; foliage leathery, dark; vigorous, bushy growth; GM, Portland, 1940; [Charles P. Kilham seedling X Étoile de Hollande]; Van Rossem; J&P

'MATADOR, CLIMBING', Cl HT, dr, 1938; Western Rose Co.

Matangi®, *see* 'MACMAN'

Matangi, Climbing, Cl F, ob, 1985; Chiplunkar

Matawhero Magic, *see* 'MACAMSTER'

Matchball®, HMsk, w, 1990; Lens

'MATCHLESS', HT, dp, 1926; flowers cerise-pink, dbl.; upright, bushy growth; [Premier sport]; Duckham-Pierson Co.

'MATDOVE', S, mp, 1998; (**Dove Dale**); flowers medium pink, very full, very dbl., 41 petals, 3.0 in., borne in small clusters, slight fragrance; prickles moderate, medium, straight; foliage medium, dark green, glossy; spreading, low, 3 ft x 6 ft growth; [Seedling X Seedling]; Mather, Wendy

Mateo's Silk Butterflies, *see* LETsilk

Mater Casta, S; (Cavriglia)

'MATEUS ROSE', HT, pb, 1973; flowers peach, reverse pale pink, dbl., 20 petals, 4 in., slight fragrance; [Pink Parfait X Mme Butterfly]; Winship

Mathé Altéry, *see* **'DRESDEN'**

'MATHILDE KÄRGER', Pol, dp, 1929; flowers carmine-pink, medium, dbl.; Kärger; (Sangerhausen)

'MATI BRADOVÁ', HT, dp, 1934; flowers dark rose-pink, sometimes almost carmine, very large, dbl., moderate fragrance; foliage glossy, dark; very vigorous, bushy growth; [Gorgeous X Gen. MacArthur]; Brada, Dr.; Böhm

Matilda®, *see* 'MEIBEAUSAI'

Matilda™, *see* 'MEISOYRIS'

'MATILDA CAMPBELL', HT, dp, 1952; bud ovoid; flowers bengal rose, very dbl., 5–6.5 in., exhibition form, intense fragrance; foliage glossy; vigorous, bushy growth; Campbell

'MATSON MODESTY', HT, lp, 1946; flowers pale pink, dbl., 60 petals, 4 in., moderate fragrance; vigorous growth; [Mrs Sam McGredy X Heinrich Wendland]; Prosser

'MATSUO-HIME', HT, lp, 1999; flowers pale pink, dbl., 31–33 petals, 5.5 in., exhibition form, slight fragrance; foliage medium green; 4.5 ft growth; SM, Japan Rose Concours, 1995; [Hatakeyama X Yonina]; Shimizu, Junji, 1991

Matsuri, LCl, rb

'MATTCHE', Min, mr, 1982; (**Chelsea Pensioner**); flowers scarlet, patio, small, dbl., 20 petals, slight fragrance; foliage small, dark, semi-glossy; bushy growth; [Gold Pin seedling X Seedling]; Mattock, John, Ltd.; John Mattock, Ltd.

MATtdor, S, w, 1990; (**Northamptonshire**); Mattock

Matterhorn®, *see* 'ARMMA'

'MATTGRO', F, mp, 1983; (**Pink Wave**); flowers soft medium pink, medium, semi-dbl., moderate fragrance; foliage medium, medium green, semi-glossy; groundcover; spreading growth; [Moon Maiden X Eyepaint]; Mattock, John, Ltd.

'MATTHEW DUCKETT', HT, or, 1999; dbl., 17–25 petals, 2 in., borne mostly singly, moderate fragrance; few prickles; foliage medium, medium green, bronze when new, semi-glossy; upright, tall growth; [Solitaire X (Alexander X Remember Me)]; Jones, L.J.

'MATTHEWS SURPRISE', T, lp; Matthews, H., 1889; (Weatherly, L.)

Matthias Meilland®, *see* 'MEIFOLIO'

Mattinata, HT; Voightlander; (Cavriglia)

'MATTJO', HT, w, 1983; (**Snowman**); flowers large, dbl., moderate fragrance; foliage large, dark, semi-glossy; upright, bushy growth; [(Peer Gynt X Isis) X Lady Seton]; Mattock, John, Ltd.

MATtkiri, HT, dp, 1994; (**Kiri**); Matthews

MATtlace, F, lp, 1998; (**Anniversary**); Gold Star of the South Pacific, Palmerston North, NZ, 1997; Matthews

'MATTNOT', LCl, lp, 1987; (**Elizabeth Heather Grierson**); flowers soft pink, medium, dbl., 15–25 petals, moderate fragrance; foliage medium, dark green, semi-glossy; upright growth; [Bonfire Night X Dreaming Spires]; Mattock, John, Ltd.; The Rose Nursery, 1986

'MATTSUN', F, ab, 1979; (**Young Venturer**); bud ovoid; flowers rich apricot, dbl., 30 petals, 4.5 in., intense fragrance; foliage dark, glossy, leathery; vigorous, upright growth; [Arthur Bell X Cynthia Brooke]; Mattock

'MATTWYT', HT, ly, 1979; (**Tynwald**); bud ovoid; flowers cream, center yellow, dbl., 60 petals, 5 in., moderate fragrance; bushy, upright growth; [Peer Gynt X Isis]; Mattock

Matty, *see* 'DEVHAUTE'

'MATURITY', HT, mp, 1973; flowers rose-pink, full, pointed, dbl., 50 petals, 7 in.; foliage large, dark; free growth; [Duftwolke X Lively]; LeGrice

'MAUD', HT, pb, 1921; flowers salmon-pink; Paul, W.

'MAUD BETTERTON', HT, mp, 1960; flowers rose-pink, medium, dbl.; vigorous growth; Gregory

'MAUD COLE', F, m, 1968; flowers mauve-purple, dbl., moderate fragrance; foliage dark, glossy; [Lilac Charm X Africa Star]; Harkness

'MAUD CUMING', HT, op, 1923; bud pointed; flowers coral-pink, shaded peach and orange, very large, dbl., exhibition form, slight fragrance; foliage dark, glossy; vigorous, bushy growth; Dickson, A.

'MAUD DAWSON', HT, or, 1915; flowers large, dbl., intense fragrance; Dickson, A.; (Sangerhausen)

'MAUD E. GLADSTONE', Pol, pb, 1926; bud pointed; flowers malmaison pink, shaded coral and chrome-yellow, small, dbl., globular, moderate fragrance; foliage rich green, leathery, glossy; vigorous, bushy growth; [Orléans Rose X Edward VII]; Bees

'MAUD NASH', HT, mr, 1942; flowers rich red with fire; upright growth; Clark, A.

Maud Nunn, *see* 'DRISCONUN'

Maude Sumner, F, ob, 1994

MAUkeole, F, op; (**Vent des Indes**®)

MAUlave, F, rb; (**Inge Schubert**®); (MAUlave), 1994

'MAUNA LOA', HT, mr, 1937; flowers bright red, large, dbl.; foliage heavy; H&S

'MAUPERTUIS', M, mp, 1868; flowers rosy-pink; Moreau et Robert

Maureen Lipman, *see* FRAntier

'MAUREEN MACNEIL', F, op, 1967; flowers salmon, borne in clusters; moderate growth; [Anna Wheatcroft X Orangeade]; Vincent

'MAUREEN THOMPSON', HT, dr, 1949; bud pointed; dbl., 35 petals, 3 in., moderate fragrance; foliage leathery; very vigorous growth; Cant, B. R.; Bosley Nursery

Maurice, *see* **'MCGREDY'S TRIUMPH'**

'MAURICE BERNARDIN', HP, mr, 1861; bud vermilion; flowers bright crimson, moderately large blooms in clusters, dbl., intense fragrance; vigorous growth; [Général Jacqueminot X ?]; Granger

Maurice Chevalier®, *see* 'DELTRE'

'MAURICE LEPELLETIER', HP, dr; flowers vermilion red; Moreau et Robert, 1868

'MAURICE NOYELLE', HT, 1951; Buatois; (Cavriglia)

'MAURICE VILMORIN', HP, rb, 1868; flowers garnet-crimson to purple-maroon, dbl., profuse bloom, sometimes repeated, moderate fragrance; Lédéchaux

'MAURICETTE SISTAU', Pol, w, 1925; flowers pure white to rosy white, large, borne in clusters of 25–50, dbl.; thornless; Turbat

Maurine, F, mp, 1991; Poulsen

Maurine Neuberger™, *see* 'SPOMAUR'

Mauve Mallerin, *see* **'SIMONE'**

'MAUVE MELODEE'®, HT, m, 1962; bud long, pointed, purple; flowers rose-mauve, semi-dbl., 17–25 petals, 4.5–5 in., moderate fragrance; foliage dark, leathery; vigorous, upright growth; [Sterling Silver X Unnamed seedling]; Raffell; Port Stockton Nursery

Maverick, *see* 'LAVSASK'

Maverick, HT, lp; Edwards, Eddie

'MAVIS CAMPBELL', HT, dp, 1942; bud long, deep pink; vigorous, tall growth; Clark, A.

'MAVOURNEEN', S, rb, 1984; bud medium-large, ovoid, pointed; flowers medium red, white reverse, large blooms borne 1–10 per cluster, dbl., 23 petals, repeat bloom, slight fragrance; awl-like, tan prickles; foliage large, leathery, semi-glossy, dark; erect, bushy growth; [(Tickled Pink X Prairie Princess) X El Catala]; Buck, Dr Griffith J.; Iowa State University

'MAX COLWELL', Min, or, 1969; bud long, pointed; flowers orange-red to red, small, dbl., 25 petals, 1.5 in., slight fragrance; foliage leathery; bushy, spreading growth; [Unnamed red F seedling X (Little Darling X Unnamed Min seedling)]; Moore, Ralph S.; Sequoia Nursery

'MAX GRAF', HRg, pb, 1919; flowers bright pink, center golden, medium, single, nonrecurrent; foliage glossy, rugose; vigorous, bushy, trailing growth; valuable hardy groundcover; (14); [Probable hybrid of R. rugosa X R. wichurana]; Bowditch

'MAX HAUFE', HEg, lp, 1939; bud long, pointed, dark pink; flowers large, semi-dbl., seasonal bloom; foliage large, leathery, light; very vigorous (5–7 ft), trailing growth; [Joanna Hill X R. eglanteria hybrid]; Kordes

'MAX HESDÖRFFER', HT, lp, 1903; flowers large, dbl., moderate fragrance; Jacobs; (Sangerhausen)

'MAX KRAUSE', HT, yb, 1930; flowers reddish orange, opening golden yellow, very large, dbl., moderate fragrance; foliage dark, glossy; vigorous growth; [Mrs Beckwith X Souv. de H.A. Verschuren]; Krause; J&P

'MAX KRAUSE SUPERIOR', HT, op, 1940; flowers coppery pink, very large, very dbl., moderate fragrance; Heizmann, E.; (Sangerhausen)

'MAX KRAUSE, CLIMBING', Cl HT, yb, 1940; da Silva, Moreira

'MAX SCHMELING', F, or, 1973; bud ovoid; flowers large, dbl., slight fragrance; foliage large, glossy; upright, bushy growth; [Unknown X Unknown]; Tantau, Math.; Ahrens & Sieberz

'MAX SINGER', HMult, dr, 1885; flowers medium, dbl.; Lacharme, F.; (Sangerhausen)

'MAX VOGEL', HT, op, 1929; flowers coppery orange, large, very dbl., intense fragrance; foliage bronze; [Fritz Maydt X Lilly Jung]; Leenders, M.

'MAXI', F, rb, 1971; flowers red, white eye, semi-dbl., 12 petals, 3 in., slight fragrance; free growth; [(Evelyn Fison X (Tantau's Triumph X R. macrophylla coryana)) X (Hamburger Phoenix X Danse de Feu)]; McGredy, Sam IV; McGredy

'MAXIM', F, op, 1961; bud ovoid; flowers salmon-pink, large, borne in broad clusters, dbl.; foliage leathery; vigorous, bushy growth; Tantau, Math.

Maxim, *see* TANmixa

Maxima, *see* **'ALBA MAXIMA'**

Maxima Regina, *see* 'SULTANA'

'MAXIME CORBON', HWich, rb, 1918; flowers dark coppery red turning apricot-yellow, borne in clusters, dbl., abundant seasonal bloom, moderate fragrance; foliage rich green, glossy, leathery; vigorous, climbing and trailing growth (8–18 ft); [R. wichurana X Léonie Lamesch]; Barbier

'MAXIMIN CHABUEL', HT, op, 1943; Mallerin, C.; A. Meilland

'MAXINE', F, rb, 1958; bud globular, creamy pink; flowers camellia-red becoming maroon flecked pink and white, rosette, dbl., 37–40 petals, flat, moderate fragrance; foliage leathery, glossy; very vigorous, low, bushy growth; [Pinocchio X Crimson Glory]; Silva; Booy Rose Nursery

Maxine, S, dp, 1999; Williams, J. Benjamin

'MAXISTAR', HT, or, 1975; bud round; very dbl., 70–80 petals, 5.5–6 in., foliage dark, leathery; vigorous, upright growth; [Duftwolke X Pharaon]; Huber

'MAY BANKS', T, ly, 1938; flowers lemon-yellow; [Lady Hillingdon sport]; Banks

May Choice, HT, 1958; LeGrice, E.B.; (Cavriglia)

May Graham, *see* 'KENDAD'

May Lawlor, *see* 'KIRHOL'

May Lawlor,

May Lyon, *see* 'COCBAY'

'MAY MARTIN', HT, ly, 1918; flowers pure canary-yellow, center darker, semi-dbl.; [Ophelia sport]; Martin & Forbes Co.

'MAY MILLER', HT, op, 1911; flowers coppery pink, large, dbl.; Hill, E.G., Co.; (Sangerhausen)

'MAY QUEEN', HWich, mp, 1898; flowers pink, very dbl., quartered, moderate, fruity fragrance; height to 25 ft; [R. wichurana X Champion of the World]; W.A. Manda

'MAY QUEEN', HWich, lp, 1898; flowers lilac-pink, large, semi-dbl., flat, profuse bloom, occasionally repeated, intense fragrance; foliage glossy; vigorous, climbing or groundcover growth

'MAY ROBINSON', F, op; flowers bright salmon-pink

'MAY TAYLOR', HT, dp, 1966; flowers deep rose-pink, large, dbl., exhibition form, slight fragrance; foliage soft; vigorous, bushy growth; [Tassin X Ballet]; Taylor, L.R.

'MAY WETTERN', HT, mp, 1928; bud pointed; flowers rosy pink, large, dbl., exhibition form, moderate fragrance; foliage rich green, leathery; vigorous, bushy growth; GM, NRS, 1928; Dickson, A.

'MAY WOOLLEY', F, ab, 1976; flowers bronze-apricot to peach, full, dbl., 25–25 petals, 2.5–3 in., cupped, intense fragrance; foliage small, glossy; moderate, free growth; [Fairlight X Arthur Bell]; Wood

'MAYA', F, or, 1986; blooms in clusters; foliage dense, bright; low growth; [Zorina X Lara]; Kriloff, Michel

'MAYA LEE', HT, or, 1992; very dbl., 50 petals, 3–3.5 in., borne mostly singly, intense fragrance; many prickles; foliage medium, medium green, semi-glossy; long stems; medium (100+ cms), upright growth; [Unknown X Unknown]; Jerabek, Paul E.

Maybelle Stearns, *see* **'MABELLE STEARNS'**

'MAYDAY', F, lp, 1957; bud ovoid; flowers white overcast pink, borne in clusters, dbl., 25–30 petals, 3 in., cupped, moderate fragrance; foliage rich green, leathery; vigorous, bushy growth; [(Pinocchio seedling X Unnamed Hybrid Tea) X Fashion]; Boerner; J&P

'MAYET', HT, mr, 1951; flowers crimson-red passing to neyron pink, base yellow, well formed, dbl., 30 petals, intense fragrance; foliage olive-green; vigorous growth; [Condesa de Sástago X Mme Henri Guillot]; Dot, Pedro

'MAYFAIR', HT, dp, 1935; flowers deep pink, well formed, large, moderate fragrance; foliage bronze; very vigorous growth; Bentall

'MAYFLOWER', HT, pb, 1958; flowers light cerise, reverse silvery pink, dbl., 30 petals, 4 in., intense fragrance; foliage glossy; vigorous growth; [Eden Rose X ?]; Gregory

'MAYLINA', Cl HT, w, 1916; (Mrs Charles E.F. Gersdorff); flowers silvery white, reverse killarney pink to shell-pink, very large, dbl., cupped, abundant, intermittent bloom, moderate, spicy fragrance; foliage large, soft; vigorous, climbing (15 ft), fairly compact growth; [White climbing rose X Killarney]; Gersdorff

'MAYOR BAKER', HT, or, 1928; flowers terracotta to scarlet, base light orange, semi-dbl., moderate fragrance; vigorous growth; [Mons. Paul Lédé X Hadley]; Thomas

'MAYOR CERMÁK', HT, dr, 1932; flowers very dark red, shaded purple, large; vigorous, branching growth; [Mrs Henry Winnett X Vaterland]; Böhm, J.; J&P, 1934

Mayor of Casterbridge, *see* 'AUSBRID'

Maysa, HT; (Cavriglia)

'MAYTIME', S, pb, 1975; bud ovoid, pointed; flowers carmine-rose, base yellow, shallowly-cupped, single, 6–10 petals, 3.5–4 in., cupped, repeat bloom, moderate fragrance; foliage dark, leathery; upright, bushy growth; [Elegance (HT) X Prairie Princess]; Buck, Dr Griffith J.; Iowa State University

'MAYTIME', LCl, lp, 1953; flowers flesh-pink, reverse rose-pink, borne in clusters of 5, single, 5–6 in., profuse, non-recurrent bloom, slight fragrance; foliage leathery; vigorous growth; (21); [R. maximowicziana pilosa X Betty Uprichard]; Maney; Iowa State College

Maywonder, Pol, or, 1968; Grootendorst

'MAYWOOD', HT, mr, 1924; flowers bright red, dbl., moderate fragrance; [Charles K. Douglas X (Killarney X Ophelia)]; Hill, Joseph H., Co.; Amling Bros.

'MAYWOOD RED', HT, mr, 1923; moderate fragrance; [Premier X Seedling]; Hill, E.G., Co.; J.H. Hill Co.

'MAZEPPA', HGal, rb; flowers red, edged and marbled with white, medium, dbl.

'MAZOWSZE', HT, dp, 1966; bud oblong; flowers deep pink edged lighter, dbl.; foliage leathery; very vigorous growth; [Marella seedling]; Grabczewski

'MAZURKA', F, mp, 1965; bud ovoid; flowers pink, medium, borne in clusters, dbl., slight fragrance; foliage dark; Verbeek

Mazurka, *see* MEltune

'MAZZINI', HT, w, 1925; flowers blush-white suffused pink, dbl., moderate fragrance; [Mme Butterfly X Gladys Holland]; Easlea

McCallun House, T, ly

McCartney Rose, *see* 'MEIZELI'

McClinton Tea, T, mp

McGinnis Tea, T, ab

'MCGREDY'S CORAL', HT, op, 1936; flowers coral-pink, overlaid salmon, shaded copper, large, exhibition form; foliage dark cedar-green; very vigorous, branching growth; McGredy

'MCGREDY'S GEM', HT, lp, 1933; bud pointed; flowers creamy pink, base yellow, deepening to rose-pink edges, dbl., 25–40 petals, cupped, slight fragrance; very vigorous growth; McGredy

'MCGREDY'S IVORY', HT, w, 1930; (Portadown Ivory); bud long, pointed; flowers creamy white, base yellow, large, dbl., 28 petals, exhibition form, moderate, damask fragrance; foliage dark, leathery, glossy; vigorous growth; GM, NRS, 1928; [Mrs Charles Lamplough X Mabel Morse]; McGredy; Dreer

'MCGREDY'S IVORY, CLIMBING', Cl HT, w, 1939; Raffel; Port Stockton Nursery

'MCGREDY'S ORANGE', HT, ob, 1936; (Golden Thoughts, Morning Glory, Sunglow); flowers deep indian yellow, reverse orange, flushed salmon, dbl., exhibition form, moderate fragrance; foliage dark, bronze; vigorous growth; [Mrs Sam McGredy sport]; McGredy

'MCGREDY'S PEACH', HT, op, 1933; bud pointed; flowers creamy yellow, washed salmon, very large, cupped, slight fragrance; foliage glossy, dark; vigorous growth; GM, NRS, 1932; McGredy

'MCGREDY'S PERFECTION', HT, lp, 1936; flowers large, dbl.; McGredy; (Sangerhausen)

'MCGREDY'S PILLAR', HT, or, 1935; flowers terra-cotta, moderate fragrance; McGredy

'MCGREDY'S PINK', HT, lp, 1936; flowers bright rose, outer petals pearly cream and pink, base saffron, intense fragrance; foliage dark; vigorous, branching growth; McGredy; J&P

'MCGREDY'S PRIDE', HT, op, 1936; bud long, pointed; flowers orange and salmon-pink, flushed saffron-yellow, reverse yellow, dbl., moderate fragrance; GM, NRS, 1936; [Angèle Pernet X Mrs Charles Lamplough]; McGredy; J&P

'MCGREDY'S SALMON', HT, ab, 1940; bud pointed; flowers apricot-salmon, dbl., slight fragrance; foliage dark, wrinkled; strong stems; vigorous, compact growth; [Mrs Henry Morse X Unnamed seedling]; McGredy; J&P

'MCGREDY'S SCARLET', HT, mr, 1930; flowers medium, dbl., 35 petals, exhibition form, slight fragrance; foliage leathery, glossy; vigorous growth; McGredy

'MCGREDY'S SUNSET', HT, ob, 1936; bud long, pointed; flowers chrome-yellow shading to scarlet, reverse clear buttercup-yellow, dbl., 40 petals, globular, moderate fragrance; foliage glossy, bronze; vigorous growth; [Margaret McGredy X Mabel Morse]; McGredy; J&P

'MCGREDY'S SUNSET, CLIMBING', Cl HT, ob, 1957; Shamburger, P.

'MCGREDY'S TRIUMPH', HT, dp, 1934; (Maurice); flowers soft rose flushed orange, very large, dbl., exhibition form, moderate fragrance; foliage dark reddish bronze, glossy; strong stems; vigorous, branching growth; (28); GM, NRS, 1932; [Admiration X Seedling]; McGredy; J&P

'McGredy's Triumph, Climbing', Cl HT, dp, 1948; Simmonds Nursery

'McGredy's Wonder', HT, ob, 1934; flowers coppery orange, flushed orange-red, reverse orange-red, large, semi-dbl., cupped, moderate, fruity fragrance; foliage glossy, olive-green; vigorous growth; McGredy

'McGredy's Yellow', HT, my, 1933; bud long, pointed; flowers bright buttercup-yellow, large, dbl., 30 petals, cupped, slight fragrance; foliage glossy, bronze; vigorous growth; GM, NRS, 1930 GM, Portland, 1956; [Mrs Charles Lamplough X (The Queen Alexandra Rose X J.B. Clark)]; McGredy, 1934

'McGredy's Yellow, Climbing', Cl HT, my, 1937; Western Rose Co.

McMillan's Pink, *see* **'Affirm'**

McMillan's Yellow, HT, my

'Me Darling', F, pb, 1971; flowers cream suffused pink, dbl., 37 petals, 3.5–4 in., exhibition form, slight fragrance; foliage glossy, light; [Evelyn Fison X Dearest]; Anderson's Rose Nurseries

Me Too, *see* 'JULeschildren'

'Meadow Dancer', Min, lp, 1991; bud ovoid; flowers medium, dbl., 45 petals, cupped, no fragrance; foliage medium, medium green, semi-glossy; spreading, tall growth; [Judy Fischer sport]; Gruenbauer, Richard, 1984; Flowers 'n' Friends Miniature Roses, 1990

'Meadow Ruby', S, mr, 1978; flowers large blooms borne singly, dbl., 40 petals, exhibition form, repeat bloom, moderate fragrance; long, red prickles; foliage leathery; vigorous, upright growth; [Prairie Princess X (Queen Elizabeth X Borealis)]; James, John

'MEban', HT, w, 1955; (Message, **White Knight**); bud long, pointed; flowers clear white, dbl., 33 petals, 4 in., exhibition form; foliage leathery, light green; vigorous, upright growth; AARS, 1958; [(Virgo X Peace) X Virgo]; Meilland, F.; C-P, 1957; URS, 1955

'Mécène', HGal, pb, 1845; flowers white, striped with rose, compact, medium, dbl.; shoots very smooth; erect, moderate growth; Vibert

'Mechak', HT, dr, 1979; flowers very large, dbl., 20 petals, exhibition form, no fragrance; brown prickles; foliage medium, dark, smooth; upright growth; [Samourai X Unnamed seedling]; Pal, Dr B.P.; Arand Roses

'Mechtilde von Neuerburg', HEg, mp, 1920; semi-dbl., 1 in., blooms in corymbs; large, crimson fruit; foliage typical sweetbriar; vigorous (10′ × 10′) growth; very hardy; Boden

'Meda', LCl, op, 1942; flowers shrimp-pink, open, imbricated, dbl., 40–60 petals, 3.5–4 in., intense fragrance; foliage large, leathery; long stems; very vigorous, climbing (10–12 ft) growth; [(R. setigera X Mme Butterfly) X Golden Dawn]; Horvath; Wayside Gardens Co.

Medal of Honor, *see* 'JACtred'

'Medallion'®, HT, ab, 1973; bud long, pointed; flowers light apricot, very large, dbl., moderate fragrance; foliage large, leathery; vigorous, upright growth; AARS, 1973 GM, Portland, 1972; [South Seas X King's Ransom]; Warriner, William A.; J&P

Medeo, F, w

Medialis, *see* HELmedia

'Mediator', HT, op, 1949; flowers coral-pink shading to salmon base, medium, dbl., 45 petals, exhibition form; foliage soft, dark green; vigorous, bushy growth; Totty

Medima®, HT, ob, 1993; Poulsen

'Medina', HT, w, 1918; [Sunburst sport]; White Bros.; American Bulb Co., 1923; White Bros., 1923

'Mediterranea', HT, pb, 1943; flowers carmine with yellow, passing to pink with white markings, dbl., 40 petals, 5 in., exhibition form, moderate fragrance; upright growth; [Signora sport]; Dot, Pedro

'Medley', HT, lp, 1962; bud pointed; flowers bright salmon-pink, reverse flushed yellow, semi-dbl., 35–40 petals, 5–5.5 in., cupped, moderate fragrance; foliage leathery; vigorous, upright growth; [Unnamed Hybrid Tea seedling X Pageant]; Boerner; J&P

Medusa, S, m, 1996; Noack, Werner

Méduse®, *see* 'GAUtara'

Meerzicht Glory, *see* **'Orange Delight'**

'Meg', Cl HT, ab, 1954; flowers salmon-apricot, stamens red, blooms in large clusters, single, 10 petals, 5.5 in., recurrent bloom, moderate fragrance; foliage dark, glossy; vigorous growth; GM, NRS, 1954; [Probably Paul's Lemon Pillar X Mme Butterfly]; Gosset; Harkness

Meg '81, *see* **'String of Pearls'**

'Meg Merrilies', HEg, dp, 1894; flowers rosy crimson, single, summer bloom, moderate fragrance; foliage very fragrant; very vigorous (10 ft) growth; [R. eglanteria X HP or B]; Penzance; Keynes, Williams & Co.

'Me-Gami', HT, op, 1980; flowers medium salmon-pink, large, dbl., 33 petals, exhibition form, moderate fragrance; foliage dark, leathery; upright growth; [Unnamed seedling X Fragrant Cloud]; Suzuki, Seizo; Keisei Rose Nursery

'Megan', HT, pb, 1981; bud pointed; flowers white with pink petal edges, blooms borne 3–5 per cluster, exhibition form, moderate fragrance; broad, slightly hooked prickles; foliage medium green, glossy; upright, strong growth; Bronze, ARC TG, 1981; [Daily Sketch X Unnamed seedling]; Adams, Dr Neil D.

Megan, *see* 'DEVnovia'

Megan Dolan, *see* 'DEVnovia'

'Megan Louise', HT, pb, 1981; bud ovoid; flowers silvery pink, deep pink petal edges, large blooms borne 1–5, dbl., 48 petals, exhibition form, intense fragrance; red-brown prickles; foliage matt, green, tough; short, bushy growth; [Red Lion X Silver Lining]; Erich Welsh Roses; Australian Rose Society & Roy Rumsey Ltd.

'Megastar', HT, or, 1980; bud globular; dbl., 40 petals, cupped, borne singly, intense, fruity fragrance; curved yellow-brown prickles; foliage light green; Sohne, W. Kordes; Rose Barni-Pistoia

'MEgersar', Cl HT, or, 1965; (**Baccará, Climbing**®); URS

Meggie, Min, lp

Meghan's Arrival, *see* 'BULlmeg'

'Megiddo', F, or, 1970; dbl., 25 petals, 4.5 in., slight fragrance; foliage large, olive-green; upright growth; [Coup de Foudre X S'Agaro]; Gandy, Douglas L.

'MEHamber', F, ab, 1996; (**Amber Nectar**); flowers amber/apricot with amber yellow reverse, blooms borne in large clusters, dbl., 26–40 petals, 2 in., slight fragrance; moderate prickles; foliage medium, medium green, glossy; upright, medium (60 cms) growth; [Alexander X Sweet Magic]; Mehring, Bernhard F.; Henry Street Nurseries, 1997

'MEHbronze', Min, r, 1996; (**Free As Air**); flowers orange red/russet with light orange pink reverse, blooms bor, semi-dbl., 8–14 petals, 1.5 in., no fragrance; few prickles; foliage medium, dark green, glossy; compact, medium (35 cms) growth; PVRO 5448; [Anna Ford X Brown Velvet]; Mehring, Bernhard F.; Mehring, 1996

'MEHbrown', F, mr, 1997; (**Garry Brown**); flowers double, medium, dbl., 15–25 petals, borne in small clusters, intense fragrance; some prickles; foliage large, dark green, glossy; upright, medium (2–6 in.) growth; [Roger Lamberlin X Arthur Bell]; Mehring, Bernhard F.

'Meher', HT, mp, 1998; flowers medium pink, lighter reverse, very dbl., 26–40 petals, 3–3.5 in., borne mostly singly, moderate fragrance; numerous small and large prickles; foliage medium, medium green, semi-glossy; upright, tall, 5 ft growth; [American Heritage X Paradise]; Pavri, Nadir J.

'MEHgoldie', F, ly, 1997; (**Janet Bebb**); flowers full, medium, dbl., 15–25 petals, borne in small clusters, slight fragrance; foliage large, medium green, glossy; compact, medium (100 cms) growth;

[Golden Holstein X Seafarer]; Meahring, Bernhard F.

MEHnan, Min, yb, 1998; (**Frechdachs**); Mehring, Bernhard F.

'MEHPARK', LCl, mp, 1998; (**Jane Eyre**); flowers medium pink, double, dbl., 15–25 petals, 4.5 in., borne in small clusters, slight fragrance; prickles moderate; foliage large, medium green, semi-glossy; upright, climbing, medium, 300 cms growth; [(Westfalenpark Dortmund X (Anna Ford X Frank Naylor)) X (Westerland X ((Casino X Mermaid) X (Vesper X Picasso)))]; Mehring, Bernhard F.; Apuldram Roses, 1998

'MEHPAT', Min, pb, 1996; (**Make A Wish**); flowers salmon pink, light pink reverse, blooms borne in large clusters, very dbl., 2 in., moderate fragrance; moderate prickles; foliage medium, dark green, glossy; bushy, medium (35 cms) growth; [Robin Redbreast X Amber Queen]; Mehring, Bernhard F.; Henry Street Nurseries, 1996

'MEHPIC', F, yb, 1998; (**Golden Hope**); flowers yellow, pink edge, reverse amber with pink edge, dbl., 15–25 petals, 2.75 in., borne in small clusters, slight fragrance; no prickles; foliage large, medium green, semi-glossy; upright, medium, 80 cm growth; [Kronprinzessin Victoria von Schweden X Goldene Holstein]; Mehring, Bernhard F.; Henry Street Nurseries, 1998

'MEHREX', F, rb, 1993; (**Sarah Jo**); flowers red with white eye, salmon reverse, blooms borne in small clusters, dbl., 15–25 petals, 1.5–2.75 in., slight fragrance; some prickles; foliage large, dark green, glossy; medium (60 cms), upright, bushy growth; [Sheri Ann X Dortmund]; Mehring, Bernhard F.; Mehring, 1994

'MEHSHERRY', S, dp, 1995; (**The Holt**); flowers fuchsia pink, dbl. (15–25 petals), small blooms borne in large clusters; slight fragrance; some prickles; foliage medium, dark green, glossy; bushy (100 cms) growth; [Sexy Rexy X Robin Redbreast]; Mehring, B. F.; Mehring, 1994–95

'MEIALATE', S, yb, 1998; (**Mystic Meidiland**®); flowers coppery peach fading to butter cream yellow, semi-dbl., 8–14 petals, 1–1.5 in., borne in small clusters, no fragrance; prickles moderate; foliage medium, dark green, semi-glossy; compact, medium, 3 ft growth; Selection Meilland; Conard-Pyle Co., 1997

'MEIALFI', F, ob, 1961; (**Zambra**®); bud ovoid; flowers orange, reverse yellow, blooms in clusters, semi-dbl., 13 petals, 2.5–3 in., flat, slight, sweetbriar fragrance; foliage leathery, glossy, light green; vigorous, well-branched growth; GM, Bagatelle, 1961 GM, Rome 1961; [(Goldilocks X Fashion) X (Goldilocks X Fashion)]; Meilland, Mrs. Marie-Louise; C-P, 1964; URS, 1961

'MEIALFISAR', Cl F, ob, 1969; (**Zambra, Climbing**®); Meilland; URS

MEIalzonite, HT, op, 1986; (**Lady Meilland**); GM, Durbanville, 1982 Gold Star of the South Pacific, Palmerston North, NZ, 1982; Meilland

'MEIARLO', HT, or, 1962; (**Allegro**®); dbl., 30 petals, 3.5–4.5 in., exhibition form, slight fragrance; foliage leathery, glossy; vigorous, bushy growth; GM, Rome, 1962 GM, The Hague, 1962; [(Happiness X Independence) X Soraya]; Meilland, Alain A.; C-P, 1964; URS, 1962

MEIbalani, S, mp, 1980; (**Rétro**); flowers large, dbl., slight fragrance; Meilland; (Sangerhausen)

'MEIBALBIKA'®, F, mr, 1984; (Iga 83 Munchen, Meilland Rosiga '83, Munchen 83, **Rose Iga**, The Wyevale rose); flowers large, dbl., 20 petals; foliage medium, dark, semi-glossy; PP006281; [MEIgurami X (Cruosa X City of Leeds)]; Meilland, Mrs. Marie-Louise; Meilland & Son, 1981

'MEIBARKE', Min, rb, 1988; (**Debut**™, Douce Symphonie, Sweet Symphony); flowers luminous scarlet blending to cream to yellow at base, aging, dbl., 15–18 petals, exhibition form, no fragrance; ovoid, few, dull orange-red fruit; prickles slender, few, straw; foliage medium, dark green, semi-glossy; bushy growth; PP006791; AARS, 1989; [Coppelia X Magic Carrousel]; Selection Meilland; C-P, 1989

'MEIBEAUSAI', F, w, 1988; (Charles Aznavour, **Matilda**®, Pearl of Bedfordview, Seduction); flowers edged pink, large, dbl., 15–20 petals, no fragrance; foliage medium, dark green, semi-glossy; upright, low, compact, proliferous growth; PP007667; GM, Bagatelle, 1987 GM, Courtrai, 1987; [MEIgurami X Nirvana]; Meilland, Alain A.; SNC Meilland & Cie

'MEIBEKARB', HP, mr; (**Cherry Sunblaze**™); flowers purple-violet, dbl., cupped; [Debut sport]

'MEIBELUXEN', S, dr, 1979; (**Fiona**®); flowers small, dbl., 20 petals, slight fragrance; foliage small, dark, semi-glossy; spreading growth; [Sea Foam X Picasso]; Meilland, Mrs. Marie-Louise; Meilland Et Cie, 1982

MEIbicmarj, HT, pb, 1996; (**Elegance**); Meilland

MEIbigoud, HT, mr, 1995; (**Ambassador**); flowers velvety red, frilly; Meilland

'MEIBIL', HT, mp, 1959; (**Pink Peace**®); flowers dusty pink, dbl., 58 petals, 4.5–6 in., intense fragrance; foliage leathery; vigorous, tall, bushy growth; GM, Geneva, 1959 GM, Rome, 1959; [(Peace X Monique) X (Peace X Mrs John Laing)]; Meilland, F.; C-P; URS

'MEIBILSAR', Cl HT, mp, 1968; (**Pink Peace, Climbing**®); Meilland; URS

'MEIBINOSOR', F, or, 1976; (**Concertino**®); flowers cherry-red, medium, dbl., 20 petals, cupped, slight fragrance; foliage matt, dark; vigorous, bushy growth; [((Fidélio X Fidélio) X (Zambra X Zambra)) X Marlena]; Meilland, Mrs. Marie-Louise; Meilland

'MEIBIRANDA', HT, rb, 1978; (**Candia**™); flowers red and yellow blend, large, dbl., 38 petals, exhibition form, slight fragrance; foliage large, light green; PP004705; [Matador X (Tropicana X Flirt)]; Meilland, Mrs. Marie-Louise; Meilland Et Cie

'MEIBLAM', Min, w, 1984; (**White Meillandina**, Yorkshire Sunblaze); flowers medium, semi-dbl., no fragrance; foliage small, light green, semi-glossy; bushy growth; [Katharina Zeimet X White Gem]; Meilland, Mrs. Marie-Louise; Meilland Et Cie, 1983

'MEIBLERI', HT, w, 1985; (**Alliance**); flowers large, semi-dbl., no fragrance; foliage medium, dark, matt; upright growth; [Rustica X Youki San]; Meilland, Mrs. Marie-Louise, 1981; Meilland Et Cie, 1984

MEIbojat, HT, dy, 1987; (**Aurea**, MEIbosat); Meilland

'MEIBOKARB', Min, mr, 1993; (**Cherry Sunblaze**®); flowers small, bright red, non-fading, very dbl., 40–25 petals, 1–1.5 in., borne in small clusters, no fragrance; some prickles; foliage medium, dark green, semi-glossy; medium (35 cms), bushy growth; PP8448; [Coppelia X Magic Carrousel]; Hutton, R.J.; Conard-Pyle, 1993, USA

MEIbolnay, HT, lp; (**Top Secret**); Meilland, 1996

'MEIBONRIB', S, mp, 1994; (**Magic Meidiland**™, Magic Meillandecor®); flowers red purple, lighter centers, dbl., 17–25 petals, 1.5–2.75 in., flat, borne in small clusters, no fragrance; some prickles; foliage small, dark green, glossy; medium to low, spreading, growth; groundcover; PP9469; [R. sempervirens X (Milrose X MEIdomonac)]; Meilland, Alain A.; SNC Meilland & Cie, 1992

MEIbonver, HT, w; (**White Majesty**); Meilland, 1992

MEIborfil, HT, m; (**Silver Anniversary**); Meilland

MEIbosat, HT, dy, 1987; (**Aurea**, MEIbojat); Meilland

'MEIBRICO', HT, op, 1986; (**Bettina '78**); flowers coral, dbl., 30 petals, 4 in., slight fragrance; foliage dark, leathery; vigorous growth; [(Jolie Madame X Sunlight) X (Lady Elgin X Dr A.J. Verhage)]; Paolino; URS, 1974

'MEIBRINPAY', HT, ab, 1996; (**Abbaye de Cluny**™, Romantic Seranade); dbl., 26–40 petals, 4.25 in., borne mostly singly, slight fragrance; moderate prickles; foliage medium, dark green, semi-glossy; bushy, medium (90 cms) growth; Eurosa, Chinensis; US PPAF; GM, Belfast, 1995; [Just Joey X (MEIrestif X MEInarual)];

Meilland International SA, 1992; The Conard-Pyle Co., 1993 (Hay)

'MEIburenac', Min, w, 1978; (**Swany**®); bud ovoid; flowers pure white, large, very dbl., 95 petals, cupped; foliage glossy, bronze; very vigorous, spreading growth; [R. sempervirens X Mlle Marthe Carron]; Meilland, Mrs. Marie-Louise; Meilland, 1977

'MEIburgana', HT, mr, 1981; (**Miriana**); flowers large, dbl., no fragrance; foliage medium, dark, semi-glossy; upright growth; [((Seedling X Independence) X Suspense) X (((Alain X R. mutabilis) X Caprice) X Pharaoh)]; Meilland, Mrs. Marie-Louise; Meilland Et Cie, 1982

'MEIbyba', Min, or, 1965; (**Baby Baccará**®); flowers orange-scarlet, small, dbl., 1.5 in., slight fragrance; foliage dark; [Callisto X Perla de Alcañada]; Meilland, Alain A.; URS; Wheatcroft Bros.

'MEIbyster', HT, or, 1971; (Atoll, **Clarita**); flowers vermilion, dbl., 30–35 petals, 5 in., exhibition form, slight fragrance; foliage dark; very vigorous, upright growth; GM, Geneva, 1971 GM, Lyon, 1971; [Tropicana X (Zambra X Romantica)]; Meilland

'MEIcapinal', HT, pb, 1982; (Aachener Dom, Panthere Rose, **Pink Panther**®); flowers petals silvery pink, edged deep pink, large, dbl., no fragrance; foliage medium, semi-glossy; upright growth; [MEIgurami X MEInaregi]; Meilland, Mrs. Marie-Louise; Meilland Et Cie, 1981

'MEIcapula', Gr, mp, 1981; (**Fiorella**, Fiorella '82, Marion Foster); flowers large, dbl., 35 petals, slight fragrance; foliage medium, dark, matt; upright growth; [(Queen Elizabeth X Nirvana) X (Tropicana X MEInaregi)]; Meilland, Mrs. Marie-Louise; Meilland Et Cie

'MEIcari', HT, dr, 1957; (**Champs-Elysées**®, 'MEIcarl'); flowers rich crimson-red, large, dbl., 35 petals, cupped, slight fragrance; vigorous, bushy growth; GM, Madrid, 1957; [Monique X Happiness]; Meilland, F.; URS

'MEIcarl', HT, dr, 1957; (**Champs-Elysées**®, 'MEIcarl'); flowers rich crimson-red, large, dbl., 35 petals, cupped, slight fragrance; vigorous, bushy growth; GM, Madrid, 1957; [Monique X Happiness]; Meilland, F.; URS

'MEIcarlsar', Cl HT, dr, 1969; (**Champs-Elysées, Climbing**®); Meilland; URS

'MEIcauf', HT, dr, 1994; (**Amalia**™); dbl., 26–40 petals, 3–3.5 in., borne mostly singly, no fragrance; many prickles; foliage medium, dark green, semi-glossy; medium (100–120 cms), upright growth; PP007718; [(Queen Elizabeth X Karl Herbst) X Papa Meilland]; Meilland, Alain A.; Selection Meilland, 1989; The Conard-Pyle Co., 1990

MEIcelna, HT, mp; (**Tino Rossi**®); Meilland, 1990

'MEIcesar', HT, dr, 1963; ('MEIsar', **Papa Meilland**®); bud pointed; flowers dark velvety crimson, large, dbl., 35 petals, exhibition form, intense fragrance; foliage leathery, glossy, olive-green; vigorous, upright growth; GM, Baden-Baden, 1962 Hall of Fame, WFRS, 1988 James Alexander Gamble Fragrance Medal, ARS, 1974; [Chrysler Imperial X Charles Mallerin]; Meilland, Alain A.; URS

'MEIcham', HT, pb, 1964; (**Silva**®); bud long, pointed; flowers yellowish salmon shaded bright rose, dbl., 38 petals, 5.5 in., exhibition form, slight fragrance; foliage dark, glossy, leathery; vigorous, upright growth; GM, The Hague, 1964; [Peace X Confidence]; Meilland, Alain A.; URS

'MEIchanso', HT, or, 1978; (**Parador**®, Tchin-Tchin®); dbl., 20 petals, 3.5 in., cupped; very vigorous growth; GM, Tokyo, 1978; [((Sarabande X Meikim) X (Alain X Orange Triumph)) X Diablotin]; Paolino; URS

MEIchansosar, Cl F, lp; (**Tchin-Tchin, Climbing**); Meilland, 1995

MEIchavrin, LCl, w, 1998; (**Mon Jardin et Ma Maison**); Meilland

'MEIchest', F, or, 1964; (**Fidélio**®); bud long, pointed; flowers medium, dbl., 35 petals, exhibition form, slight fragrance; foliage leathery; vigorous, upright growth; [(Radar X Caprice) X Fire King]; Meilland, Alain A.; URS

'MEIchim', HT, mp, 1963; (**Carina**®); dbl., 40 petals, 5 in., exhibition form, moderate fragrance; foliage leathery; upright, bushy growth; ADR, 1966; [White Knight X (Happiness X Independence)]; Meilland, Alain A.; C-P, 1963; J.H. Hill Co., 1964; URS, 1963

'MEIchimsar', Cl HT, mp, 1968; (**Carina, Climbing**®, Grimpant Carina®); Meilland; URS

MEIchoiju, F, mp, 1989; (**Jacqueline Nebout**®); Meilland

'MEIchonar', S, w, 1996; (**Pearl Sevillana**™); flowers pearly white, blooms borne in small clusters, semi-dbl., 8–14 petals, slight fragrance; moderate prickles; foliage medium, dark green, dull; bushy, medium (4–5 ft) growth; US PPAF; [(MEIdomonac X Paseali) X Edelweiss]; Meilland International SA; The Conard-Pyle Co., 1996

'MEIcijas', HT, yb, 1991; (**Texas Girl**™); flowers creamy yellow, large, dbl., 26–40 petals, slight fragrance; foliage large, medium green; tall, upright growth; PP007784; [Lovely Girl sport]; Meilland, Alain A.; The Conard-Pyle Co., 1991

'MEIcitrem', Min, my, 1995; (**Lucie's Dream**); very dbl., 1.5–2.75 in., borne in small clusters, moderate, lemony fragrance; some prickles; foliage medium, medium green, semi-glossy; medium (25–30 cms), bushy growth; [(Yellow Pages X Gold Badge) X Lemon Delight]; Meilland, Alain A.; SNC Meilland & Cie, 1994

'MEIcloux', F, ob, 1985; (Chacok, Fakir, Jubilee 150, **Pigalle**®, Pigalle 84); flowers yellow blended with orange and orange-red, large, dbl., 40 petals, no fragrance; foliage medium, medium green, semi-glossy; bushy growth; [Frenzy X ((Zambra X Suspense) X King's Ransom)]; Meilland, Mrs. Marie-Louise; Meilland Et Cie, 1984

MEIcobius, HT, lp; (**Terracotta**); Meilland, 1999

MEIcobuis, HT, lp; (**Terracotta**); Meilland, 1999

MEIcofum, HT, rb, 1995; (**Leonidas**); Meilland

'MEIcoublan', S, w, 1986; (Alba Meidiland™, Blanc Meillandécor, **White Meidiland**™); flowers large, very dbl., no fragrance; foliage medium, dark green, glossy; spreading growth; [Temple Bells X MEIgurami]; Meilland, Mrs. Marie-Louise, 1987

'MEIcri', Min, ob, 1958; (**Cricri**®, Cri-Cri, Gavolda); flowers salmon shaded coral, small, dbl.; foliage leathery; dwarf, very bushy growth; [(Alain X Independence) X Perla de Alcañada]; Meilland, F.; URS

'MEIcubasi', Min, dr, 1982; (Scarlet Meillandina, **Scarlet Sunblaze**™); flowers medium, dbl., 20 petals, no fragrance; foliage medium, dark, matt; bushy growth; PP004681; [Tamango X (Baby Bettina X Duchess of Windsor)]; Meilland, Mrs. Marie-Louise; C-P, 1980

'MEIcupag', Min, my, 1994; (**Golden Sunblaze**™); bud short, plump, ovoid; flowers bright golden yellow, like a zinnia, very dbl., 40–45 petals, 1.5 in., cupped, blooms in small clusters, slight fragrance; few prickles; foliage medium, dark green, glossy; medium (15 in.), upright, bushy growth; PP8493; [(Rise 'n' Shine X Mark One) X Yellow Meillandina]; Selection Meilland; The Conard-Pyle Co., 1993

MEIcurbos, F, op; (**Zambra**®, Zambra 92); GM, Bagatelle, 1992; Meilland, 1992

'MEIdacinu', Min, or, 1977; (**Baby Bettina**); bud ovoid; flowers vermilion, reverse carmine, deep, mini-flora, 15–20 petals, cupped, borne 3–11 per cluster, slight fragrance; foliage matt, dense; vigorous growth; [(Callisto X Perla de Alcañada) X Starina]; Meilland, Mrs. Marie-Louise; Meilland Et Cie

'MEIdad', HT, op, 1967; (**Majorette**®, Minna Lerche Lerchenborg®); bud pointed; flowers coppery salmon, large, dbl., exhibition form, slight fragrance; foliage dark, leathery; vigorous, upright growth; GM, Bagatelle, 1966; [Zambra X Fred Edmunds]; Meilland; URS

'MEIdali', HT, pb, 1968; (**Arianna**®); flowers carmine-rose suffused coral, large, dbl., 35 petals, exhibition form, slight fragrance; foliage dark, leathery; vigorous, upright, open growth; GM, Bagatelle, 1965 GM, Rome, 1965 GM, The Hague, 1965; [Charlotte Armstrong X (Peace X Michèle Meilland)]; Meilland, Mrs. Marie-Louise; URS

MEIdalnu, HT, m, 1992; (**Mascara**™); Meilland

'MEIDANCLAR', Min, dp, 1991; (**Candy Sunblaze**™, Romantique Meillandina); flowers deep pink, blooms, very dbl., slight fragrance; foliage medium, dark green, glossy; tall, upright growth; PP007621; [Lady Sunblaze sport]; Selection Meilland, 1988; The Conard-Pyle Co., 1992

'MEIDANEGO', Min, dr, 1984; (**Belle Meillandina**®, Belle Sunblaze); [Meillandina sport]; Meilland, Mrs. Marie-Louise; Meilland Et Cie, 1980

'MEIDANOVER', Min, ob, 1979; (**Colibre 79**®); Meilland

'MEIDANU', F, dr, 1967; (**Tamango**®); flowers large blooms in large clusters, dbl., 35 petals, slight fragrance; [(Alain X Mutabilis) X (Radar X Caprice)]; Meilland, Mrs. Marie-Louise; Wheatcroft Bros.

MEIdarwet, S, pp, 1993; (**Relax Meillandecor**); Meilland

MEIdarwet, S, lp; (**Relax Meidiland**); Meilland

'MEIDAUD', HT, dp, 1965; (Maria Callas, **Miss All-American Beauty**); bud ovoid; flowers large, dbl., 55 petals, cupped, intense fragrance; foliage leathery; vigorous, bushy growth; AARS, 1968 GM, Portland, 1966; [Chrysler Imperial X Karl Herbst]; Meilland, Mrs. Marie-Louise; C-P, 1967; Wheatcroft Bros, 1965

'MEIDAUDSAR', Cl HT, dp, 1969; (Maria Callas, Climbing, **Miss All-American Beauty, Climbing**); Meilland; URS

'MEIDEAURI', F, lp, 1994; (**Leonardo de Vinci**®); flowers bengal pink, blooms borne in small clusters, very dbl., 3–3.5 in., slight fragrance; some prickles; foliage medium, dark green, glossy; bushy (70–110 cms) growth; [Sommerwind X (Milrose X Rosamunde)]; Meilland, Alain A.; SNC Meilland & Cie, 1993

'MEIDEAURI', mp, 1998; (**Leonardo da Vinci**™); flowers medium pink, rosette blooms, very dbl., 25–30 petals, rosette, borne singly and in small clusters, slight fragrance; foliage medium, medium green; climbing (3–4 ft) growth; Meilland

MEIdenji, LCl, mr, 1989; (**Cassandre**®); Meilland

'MEIDESI', Min, mp, 1965; (**Mimi**); bud ovoid; dbl., 33 petals, 1–1.5 in., cupped, blooms in clusters, slight fragrance; foliage leathery; vigorous, bushy (14 in.) growth; [Moulin Rouge X (Fashion X Perla de Montserrat)]; Meilland, Mrs. Marie-Louise; URS

Meidiland Alba, *see* 'MEIFLOPAN'

'MEIDINRO', HT, rb, 1990; (Altesse, **Anticipation**, Rodin); flowers red with silver/white reverse, medium, semi-dbl., 35 petals, exhibition form, borne singly, slight fragrance; foliage medium, medium green, semi-glossy; bushy growth; [Seedling X Seedling]; Meilland, 1980; Co-Operative Rose Growers, 1991

'MEIDIPSER', Min, dp, 1994; (**Spot Meillandina**®); flowers rose bengal, blooms borne in small clusters, very dbl., 74 petals, 1.5 in., no fragrance; few prickles; foliage medium, dark green, semi-glossy; low (20–25 cms), bushy growth; [(Orange Sunblaze X Pink Symphony) X Red Minimo]; Meilland, Alain A.; SNC Meilland & Cie, 1992

'MEIDIRAPO', HT, dr, 1975; flowers red to purple, full, dbl., 25 petals, 3.5–4 in., cupped, slight fragrance; vigorous growth; [(Queen Elizabeth X (Peace X Michele Meilland)) X (Baccará X Seedling)]; Paolino; Meilland

'MEIDO', Min, or, 1961; (**Scarlet Gem**®, Scarlet Pimpernel); bud ovoid; flowers orange-scarlet, dbl., 58 petals, 1 in., cupped, slight fragrance; foliage dark, glossy, leathery; bushy, dwarf (12–15 in.) growth; [(Moulin Rouge X Fashion) X (Perla de Montserrat X Perla de Alcanada)]; Meilland, Alain A.; C-P

'MEIDOMONAC', S, mp, 1985; (**Bonica**™, Bonica '82, Demon); flowers medium pink center, lighter at edges, medium, dbl., no fragrance; foliage small, dark, semi-glossy; bushy growth; PP005105; AARS, 1987 ADR, 1983; [(R. sempervirens X Mlle Marthe Carron) X Picasso]; Meilland, Mrs. Marie-Louise; C-P, 1987; Meilland Et Cie, 1981

'MEIDONA', HT, w, 1965; (Mme Neige, **Youki San**®); flowers large, dbl., 40 petals, intense fragrance; foliage light green; tall growth; GM, Baden-Baden, 1964; [Lady Sylvia X White Knight]; Meilland; URS; Wheatcroft Bros.

'MEIDONFE', Min, dp, 1998; (**Raspberry Sunblaze**); flowers near red purple (deep pink), double (15–25 petals), medium (2 in.) blooms borne in small clusters; no fragrance; moderate prickles; foliage medium, medium green, semi-glossy; bushy, medium (24 in.) growth, pot forcing; [(Orange Sunblaze X Prince Meillandina) X Red Minimo]; Selection Meilland; The Conard Pyle Co., 1998

MEIdorsun, HT, dy, 1997; (**Concorde**); Meilland

'MEIDRAGELAC', HT, or, 1985; (Laura '81, **Natilda**); flowers orange-red, lighter reverse, large, dbl., 30 petals, slight fragrance; foliage small, dark, semi-glossy; medium growth; GM, Japan, 1981; [(Pharaoh X Colour Wonder) X ((Suspense X Suspense) X King's Ransom)]; Meilland, Mrs. Marie-Louise, 1974; Meilland Et Cie, 1981

'MEIDRAGELAC', HT, ob; (**Laura**®)

MEIdrimy, LCl, lp, 1995; (**Domaine de Courso**); Meilland

'MEIDUJARAN', F, mr, 1980; (**Pimlico**, Pimlico '81); flowers large, dbl., 35 petals, no fragrance; foliage large, dark, glossy; bushy growth; GM, Belfast, 1983; [(Tamango X Fidélio) X (Charleston X Lilli Marleen)]; Meilland, Mrs. Marie-Louise; Meilland Et Cie

'MEIELEC', Gr, mr, 1966; (Samourai, **Scarlet Knight**); bud ovoid; flowers crimson-scarlet, dbl., 4–5 in., cupped, slight fragrance; foliage leathery; vigorous, upright, bushy growth; AARS, 1968 GM, Madrid, 1966; [(Happiness X Independence) X Sutter's Gold]; Meilland, Mrs. Marie-Louise; C-P, 1967; URS, 1966

'MEIELPA', LCl, mr, 1967; (**Benvenuto**®); flowers rose-red, medium, semi-dbl., recurrent bloom, slight fragrance; vigorous, climbing growth; [(Alain X Guinée) X Cocktail]; Meilland; URS

'MEIESTHO', F, or, 1966; (**Sangria**®); semi-dbl., in large trusses; vigorous growth; GM, Geneva, 1966 GM, The Hague, 1966; [Fire King X (Happiness X Independence)]; Meilland, Mrs. Marie-Louise; URS

'MEIFAN', HT, rb, 1960; (**Suspense**); bud ovoid; flowers turkey-red, reverse yellow-ochre, dbl., 58 petals, 4.5–5 in., exhibition form, slight fragrance; foliage leathery, dark, glossy; vigorous, upright, bushy growth; [Henri Mallerin X (Happiness X Floradora)]; Meilland, F.; C-P, 1961; URS, 1960

MEIfarent, HT, my, 1986; (**Solidor**®); Meilland

'MEIFERJAC', Min, or, 1996; (**Autumn Sunblaze**™); flowers bright, non-fading orange, very dbl., 40–45 petals, 1.5 in., borne in small clusters, no fragrance; some prickles; foliage medium, medium green, semi-glossy; medium (24 in.), bushy growth; heat tolerant; PP9562; [(Bonfire Night X MEIninrut) X Orange Jewel]; Selection Meilland; Conard-Pyle Co., 1995

MEIfersi, HT, mr, 1986; (**Michel Lis le Jardinier**®); Meilland

'MEIFIGA', HT, or, 1967; (**Pharaoh**, Pharaon®); bud ovoid; flowers bright orange-red, dbl., 5 in., exhibition form, moderate fragrance; foliage dark, glossy, leathery; vigorous, upright growth; GM, Belfast, 1969 GM, Geneva, 1967 GH, The Hague, 1967; [(Happiness X Independence) X Suspense]; Meilland, Mrs. Marie-Louise; C-P, 1969; URS, 1967

'MEIFIGU', HT, m, 1965; (**Lady X**); bud long, pointed; flowers large, dbl., exhibition form, slight fragrance; foliage leathery; vigorous, upright growth; GM, Portland, 1968; [Seedling X Simone]; Meilland, Mrs. Marie-Louise; C-P, 1966

'MEIFIKALIF', HT, or, 1980; (**Allegeo '80**); flowers large, dbl., 35 petals, no fragrance; foliage medium, medium green, semi-glossy; [(Diorette X Tropicana) X (Unnamed seedling X (Diorette X Tropicana))]; Meilland, Mrs. Marie-Louise; Meilland Et Cie

'MEIFINARO', Min, yb, 1982; (**Air France**, Air France Meillandina, American Independence, Rosy Meillandina); flowers yellow, pink petal edges, small, dbl., no fragrance; foliage small, dark, matt; bushy growth; PP005180; [Minijet X (Darling Flame X Perle de Montserrat)]; Meilland, Mrs. Marie-Louise, 1983

'MEIFLAROL', F, ob, 1986; (**Pareo**®); Meilland

'MEIFLOPAN', S, w, 1987; (**Alba Meidiland**™, Alba Meillandécor, Meidiland Alba); flowers medium, very white, dbl., no fragrance; foliage medium, medium green, glossy; spreading growth, vigorous; PP006891; [R. sempervirens X Marthe Carron]; Meilland, Mrs. Marie-Louise, 1985

'MEIFLOREM', F, or, 1970; (**My Fancy**); flowers dutch vermilion, imbricated, dbl., 25–30 petals, 3 in., slight fragrance; foliage dull, dark; vigorous, upright, bushy growth; [(Dany Robin X Fire King) X Rumba]; Meilland; URS

'MEIFLUNEY', Min, dp, 1984; flowers deep pink, medium, dbl., no fragrance; foliage medium, medium green, matt; bushy growth; [(Alain X Fashion) X (Rumba X (Zambra X Cinderella))]; Meilland, Mrs. Marie-Louise; Meilland Et Cie

'MEIFOLIO', F, mr, 1988; (**Matthias Meilland**®); flowers large, dbl., 15–25 petals, no fragrance; foliage medium, dark green, glossy, disease resistant; upright, floriferous growth; [(Mme Charles Sauvage X Fashion) X (Poppy Flash X Parador)]; Meilland, Mrs. Marie-Louise; SNC Meilland & Cie, 1985

'MEIFOTA', HT, mp, 1981; (**Celica**); flowers large, dbl., 20 petals, slight fragrance; foliage medium, dark, semi-glossy; upright growth; PP005683; [Unnamed seedling X (((Zambra X (Baccará X White Knight)) X Golden Garnette) X Unnamed seedling)]; Meilland, Mrs. Marie-Louise; Meilland Et Cie

'MEIFOUR', HT, dp, 1968; (Anne Marie, **Anne Marie Trechslin**®); bud long, pointed; flowers deep pink, large, dbl., exhibition form, intense fragrance; foliage dark, leathery; vigorous, branching growth; [Sutter's Gold X (Demain X Peace)]; Meilland; URS

'MEIFRISON', HT, op, 1968; (**Feria**); flowers coral suffused pink, large, dbl., globular, moderate fragrance; foliage leathery; upright growth; [(Grand Gala X Premier Bal) X Love Song]; Meilland; URS

MEIfructoz, F, ob, 1984; (**Fruité**); Meilland, L.; (Cavriglia)

'MEIFRUIJE', Min, yb, 1994; (**Apricot Sunblaze**®); flowers yellow edged in bright orange, very dbl., 25–40 petals, 1.5 in., exhibition form, borne mostly singly, no fragrance; few prickles; foliage small, dark green, glossy; medium, bushy growth, compact; PP9033; [(Mark One X Yellow Meillandina) X Gold Badge]; Meilland, Alain A.; The Conard-Pyle Co., 1994

'MEIGABI', Min, or, 1965; ('MEIGALI', **Starina**®); flowers orange-scarlet, small, dbl.; foliage glossy; vigorous, dwarf growth; ADR, Germany, 1971 GM, Japan, 1968 Miniature Rose Hall of Fame, ARS, 1999; [(Dany Robin X Fire King) X Perla de Montserrat]; Meilland, Mrs. Marie-Louise; C-P

'MEIGAFOR', HT, w, 1995; (**Confidence**™); Meilland

'MEIGALI', Min, or, 1965; ('MEIGABI', **Starina**®); flowers orange-scarlet, small, dbl., foliage glossy; vigorous, dwarf growth; ADR, Germany, 1971 GM, Japan, 1968 Miniature Rose Hall of Fame, ARS, 1999; [(Dany Robin X Fire King) X Perla de Montserrat]; Meilland, Mrs. Marie-Louise; C-P

'MEIGALIL', HT, mr, 1971; (**Galileo**); flowers currant-red to cherry-red, dbl., 30 petals, 5 in., globular, foliage large, glossy; vigorous, upright growth; [Ma Fille X Love Song]; Meilland

'MEIGANDOR', HT, ab, 1977; (**Sabrina**); flowers apricot-orange, large, dbl., 35 petals; upright growth; PP004520; [(Sweet Promise X Golden Garnette) X ((Zambra X Suspense) X (King's Ransom X Whisky Mac))]; Meilland, Mrs. Marie-Louise; Meilland Et Cie

'MEIGAVESOL', F, ob, 1982; (**Tequila**®); flowers light yellow, overlaid with orange, stained carmine on outer, dbl., 20 petals, slight fragrance; foliage medium, dark, matt; bushy growth; [Poppy Flash X (Rumba X (Meikim X Fire King))]; Meilland, Mrs. Marie-Louise; Meilland Et Cie

'MEIGEKANU', F, or, 1978; (**La Sevillana**®); bud conical; flowers vermilion, medium, semi-dbl., 13 petals; foliage bronze; vigorous, bushy growth; PP006384; ADR, 1979; [((MEIbrim X Jolie Madame) X (Zambra X Zambra)) X ((Tropicana xTropicana) X (Poppy Flash X Rusticana))]; Meilland, Mrs. Marie-Louise; Meilland, 1982

MEIgekanusar, Cl F, or, 1997; (**La Sevillana, Climbing**); Meilland

'MEIGENE', Gr, mp, 1964; (**Garden State**); bud ovoid, pointed; flowers rose-pink, dbl., 42 petals, 3.5–4 in., moderate fragrance; foliage leathery; vigorous, tall, bushy growth; [(Happiness X Independence) X White Knight]; Meilland, Alain A.; C-P

'MEIGER', HT, or, 1954; (**Baccará**®, Jacqueline); flowers bright orange-red, dbl., 75 petals, 3 in., cupped; foliage dark, leathery; bushy, upright growth; [Happiness X Independence]; Meilland, F.; C-P, 1957; URS, 1954

'MEIGERIUM', HT, or, 1975; flowers light vermilion-red, full, dbl., 35 petals, 4–4.5 in.; foliage glossy; very vigorous growth; [(Romantica X Tropicana) X ((Show Girl X Baccará) X Romantica)]; Paolino; URS

'MEIGEROKA', F, mp, 1984; (**Pink La Sevillana**®, Pink La Sevilliana, Pink Sevilliana, Rosy La Sevilliana); PP007117; GM, Baden-Baden, 1985 GM, Durbanville, 1986; [La Sevillana sport]; Meilland, Mrs. Marie-Louise; Meilland Et Cie, 1985

MEIgerokosar, Cl F, mp, 1993; (**Rosy La Sevillana, Climbing**); Meilland

MEIglassol, Min, lp; (**Tropico Sunblaze**); Meilland, 1995

'MEIGLUSOR', Min, dp, 1988; (**Concertino**™); flowers cardinal-pink, medium, dbl., 26–40 petals, no fragrance; foliage small, medium green, semi-glossy; bushy, compact growth; [(Anytime X Julita) X Lavender Jewel]; Meilland, Alain A.; SNC Meilland & Cie

'MEIGOLD', HT, dy, 1968; (Golden Prince, **Kabuki**®); flowers medium, dbl., 45 petals, exhibition form, moderate fragrance; foliage bronze, glossy, leathery; vigorous, upright growth; [(Monte Carlo X Bettina) X (Peace X Soraya)]; Meilland, Mrs. Marie-Louise; C-P; URS

MEIgosar, Cl F, ob; (**Prince Igor, Climbing**); Meilland

MEIgoufin, F, 1986; (**Magali**); Meilland, Alain A.; (Cavriglia)

MEIgovin, Min, w, 1991; (**Snow Meillandina**®); Meilland

MEIgrelou, F, lp; (**Grisbi**); Meilland

'MEIGRISO', HT, rb, 1968; (Baronne de Rothschild, **Baronne Edmond de Rothschild**®); flowers ruby-red, whitish reverse, large, dbl., 40 petals, exhibition form, intense fragrance; foliage glossy, leathery; vigorous growth; GM, Lyon, 1968 GM, Rome, 1968; [(Baccará X Crimson King) X Peace]; Meilland; URS, 1969

'MEIGRISOSAR', Cl HT, rb, 1974; (Baronne de Rothschild, Climbing, **Baronne Edmond de Rothschild, Climbing**®, Grimpant Baronne de Rothschild); Meilland, Mrs. Marie-Louise

MEIgro, Cl F, my, 1991; (**Gold Badge, Climbing**); Meilland

'MEIGRONURI', F, my, 1978; (**Gold Badge**™, Gold Bunny®, Rimosa 79); bud conical; flowers lemon-yellow, dbl., 38 petals, 3 in., cupped; vigorous growth; PP004625; [Poppy Flash X (Charleston X Allgold)]; Paolino; Meilland

MEIgrouge, HT, mr; (**Preference**); Meilland

MEIgurami, F, op, 1976; (**Coppélia 76**); flowers medium, dbl., slight fragrance; Meilland; (Sangerhausen)

MEIhailoil, S, ab, 1995; (**Concerto**); Meilland

'MEIHAITI', Cl F, mr, 1970; (Iskra, 'MEIHAITI', **Sparkling Scarlet**); flowers scarlet-red, small, semi-dbl., 13 petals, 2.5 in., moderate, fruity fragrance; foliage large; GM, Paris, 1969; [Danse des Sylphes X Zambra]; Meilland; URS

MEIhaitoil, S, lp; (**Romantic Sunrise**); Meilland

'MEIHAND', F, or, 1957; ('MEIRABANDE', **Sarabande**®); flowers light orange-red, stamens yellow, blooms in large trusses, semi-dbl., 13 petals, 2.5 in., cupped, slight fragrance; foliage semi-glossy; low, bushy growth; AARS, 1960 GM, Bagatelle, 1957 GM, Geneva, 1957 GM, Portland, 1958 GM, Rome, 1957; [Cocorico X Moulin Rouge]; Meilland, F.; C-P, 1959; URS, 1957

'MEIHANDSAR', Cl F, or, 1968; (**Sarabande, Climbing**®); GM, Japan, 1968; Meilland; URS

'MEIHARTFOR', F, lp, 1970; (Kalinka, **Pink Wonder**); flowers imbricated, dbl., 28 petals, 3 in., intense fragrance; foliage large, glossy, leathery; vigorous, upright growth; GM, Belfast, 1972 GM, Madrid, 1969; [Zambra X (Sarabande X (Goldilocks X Fashion))]; Meilland; URS

'MEIHARTFORSAR', Cl F, lp, 1976; (Kalinka, Climbing, **Pink Wonder, Climbing**); Meilland, Mrs. Marie-Louise; Meilland

'MEIHATI', Cl F, mr, 1970; (Iskra, 'MEIHAITI', **Sparkling Scarlet**); flowers scarlet-red, small, semi-dbl., 13 petals, 2.5 in., moderate, fruity fragrance; foliage large; GM, Paris, 1969; [Danse des Sylphes X Zambra]; Meilland; URS

'MEIHELVET', Gr, pb, 1974; (**Sonia**, Sonia Meilland, Sweet Promise); bud long; flowers pink suffused coral to yellow, dbl., 30 petals, 4–4.5 in., exhibition form, intense, fruity fragrance; foliage glossy, dark, leathery; [Zambra X (Baccará X White Knight)]; Meilland; C-P

'MEIHELVETSAR', Cl Gr, pb, 1976; (Grimpant Sonia Meilland®, Sonia Meilland, Climbing, **Sonia, Climbing**, Sweet Promise, Climbing); Meilland, Mrs. Marie-Louise; Meilland

'MEIHERODE', Min, ob, 1968; (**Petite Folie**®); flowers vermilion, reverse carmine, small blooms in trusses, dbl., globular, slight, fruity fragrance; foliage leathery; vigorous growth; GM, Japan, 1969; [(Dany Robin X Fire King) X (Cricri X Perla de Montserrat)]; Meilland; URS

'MEIHIGOR', F, rb, 1970; (**Frenzy**, Prince Igor®); flowers nasturtium-red, reverse yellow, rounded, dbl., 25 petals, 2 in., moderate, fruity fragrance; foliage matt; vigorous, bushy growth; [(Sarabande X Dany Robin) X Zambra]; Meilland

'MEIHIMPER', HT, pb, 1971; (**Indian Song**, Preziosa); flowers rose, reverse gold, dbl., 40 petals, 5 in., exhibition form, slight fragrance; foliage glossy, dark; vigorous, upright growth; [(Radar X Karl Herbst) X Sabrina]; Meilland

MEIhirvin, HT, mr, 1988; (**Charlotte Rampling**®); Meilland

'MEIHIVANO', F, mp, 1971; (**Sweet Song**); dbl., 35 petals, 4.5 in., moderate, fruity fragrance; bushy growth; [Fidélio X Bettina]; Meilland

'MEIHOURAG', LCl, ob, 1991; (**Arielle Dombasie**®); Meilland

MEIhuterb, F, yb, 1995; (**Galaxy**); Meilland

MEIjadel, LCl, pb, 1989; (**Fandango**); Meilland

'MEIJALITA', F, or, 1977; (**Chorus**®, 'MEIJULITO', 'MEIMORE'); flowers vermilion-red, dbl., 35 petals, 4 in., slight, fruity fragrance; foliage glossy; vigorous growth; ADR, 1977; [Tamango X (Sarabande X Zambra)]; Paolino; URS, 1975

MEIjasonl, HT, ly; (**Lovely Girl**); Meilland

'MEIJENOR', HT, or, 1955; ('MEJENOR', **Soraya**®); bud pointed; flowers orange-red reverse crimson-red, large, dbl., 30 petals, cupped, slight fragrance; foliage glossy; vigorous, bushy growth; [(Peace X Floradora) X Grand'mere Jenny]; Meilland, F.; URS

'MEIJENORSAR', Cl HT, or, 1960; (Grimpant Soraya®, 'MEJENORSAR', **Soraya, Climbing**®); Barni, V.; URS

'MEIJETTE', Min, my, 1986; (French Liberty, **Innovation Minijet**); flowers medium, dbl., slight fragrance; foliage small, light green, glossy; bushy growth; PP006177; [(Rumba X Carol Jean) X (Zambra X Darling Flame)]; Meilland, Mrs. Marie-Louise; SNC Meilland & Cie, 1987

'MEIJIDIRO', Min, mp, 1982; (**Pink Meillandina**®, Pink Sunblaze™); PP004961; [Orange Sunblaze sport]; Meilland, Mrs. Marie-Louise; Meilland Et Cie, 1980

'MEIJIKATAR', Min, or, 1982; (Orange Meillandina, **Orange Sunblaze**™, Sunblaze); flowers non-fading deep orange, very dbl., 40 petals, 1.5–2 in., cupped, borne 1–3 per cluster, slight fragrance; straw-brown prickles; foliage small, light green, matt; upright, bushy (15 in.) growth; drops spent blooms; PP004682; [Parador X (Baby Bettina X Duchess of Windsor)]; Meilland, Mrs. Marie-Louise; C-P, 1981

'MEIJIKATARSAR', Cl Min, or, 1986; (Grimpant Orange Meillandina®, **Orange Meillandina, Climbing**, Orange Sunblaze, Climbing); flowers medium, dbl., 26–40 petals, no fragrance; foliage small, medium green, matt; upright growth; PP006817; [Orange Meillandina sport]; Meilland, Mrs. Marie-Louise

MEIjulitasar, Cl F, or, 1986; (**Chorus, Climbing**); Meilland

'MEIJULITO', F, or, 1977; (**Chorus**®, 'MEIJALITA', 'MEIMORE'); flowers vermilion-red, dbl., 35 petals, 4 in., slight, fruity fragrance; foliage glossy; vigorous growth; ADR, 1977; [Tamango X (Sarabande X Zambra)]; Paolino; URS, 1975

'MEIKANS', F, or, 1959; (**Fire King**); bud ovoid; flowers fiery scarlet, blooms in clusters, dbl., 48 petals, 2.5 in., exhibition form, moderate, musk fragrance; foliage dark, leathery; vigorous, upright, bushy growth; AARS, 1960; [Moulin Rouge X Fashion]; Meilland; C-P, 1959; URS, 1958

MEIkiji, F, w, 1989; (**Destiny**); Meilland

'MEIKINOSI', HT, yb, 1978; (Parador, **Tchin-Tchin**®); flowers chrome-yellow, dbl., 30 petals, 4.5 in., cupped; vigorous growth; [((Zambra X Suspense) X King's Ransom) X (Kabuki X Dr A.J. Verhage)]; Paolino; URS

MEIklusy, F, rb, 1993; (**Marianne**); Meilland

MEIkola, HT, mp; (**Livia**); Meilland

'MEIKROTAL', S, mr, 1987; (**Scarlet Meidiland**™, Scarlet Meillandécor); flowers light cherry red, reverse dark carmine pink, small, dbl., 15–25 petals, blooms in clusters of 10–15, no fragrance; foliage medium, dark green, glossy; spreading growth; PP006087; [MEItiraca X Clair Matin]; Meilland, Mrs. Marie-Louise; SNC Meilland & Cie, 1985

MEIkruza, HT, w; (**Arianna**); Meilland

MEIkyster, F, dp; (**Trudy Mimi**™); Meilland, 1992

'MEILANEIN', HT, m, 1974; (**Charles de Gaulle**®, Katherine Mansfield); flowers lilac, dbl., 38 petals, 3.5–4 in., cupped, intense fragrance; vigorous growth; [(Sissi X Prelude) X (Kordes' Sondermeldung X Caprice)]; Meilland, Mrs. Marie-Louise; URS

'MEILANODIN', F, or, 1976; (**Paso Doble**); flowers geranium-red, 9 petals, 3 in., slight fragrance; vigorous growth; Paolino; URS

'MEILARCO', Min, lp, 1986; (Lady Meillandina, **Lady Sunblaze**™, Peace Meillandina®, Peace Sunblaze); bud plump, pointed; flowers pale orient pink to light coral pink, medium, very dbl., 40 petals, 1.5 in., exhibition form, borne singly, no fragrance; foliage small, dark green, glossy; bushy growth; PP006170; [(Fashion X Zambra) X Belle Meillandina]; Meilland, Mrs. Marie-Louise, 1987

MEIlasso, F, pb, 1989; (**Cocorico**®); Meilland

'MEILAUR', F, mr, 1971; (Florian, **Tender Night**); dbl., 25 petals, 4 in., slight, fruity fragrance; foliage matt, dark; upright growth; GM, Rome, 1971; [Tamango X (Fire King X Banzai)]; Meilland

'MEILAURSAR', Cl F, mr, 1976; (Florian, Climbing, **Tender Night**, **Climbing**); Meilland, Mrs. Marie-Louise

'MEILAVIO', HT, dr, 1998; (**Traviata**); flowers dark red, very full (41+ petals), large

(10–11 cms) blooms borne mostly single; slight fragrance; moderate prickles; large foliage, dark green, glossy; upright, medium (60–80 cms) growth; [(Porta Nigra X Paola) X William Shakespeare]; Selection Meilland; The Conard Pyle Co., 1997

'MEILÉNA', F, or, 1971; (**Poppy Flash**, Rusticana); flowers vermilion, dbl., 20 petals, 3 in., slight, fruity fragrance; vigorous, bushy growth; GM, Geneva, 1970 GM, Rome, 1972; [(Dany Robin X Fire King) X (Alain X Mutabilis)]; Meilland, 1972

MEIlenangal, HT, my, 1981; (**Mitzi**); Meilland; (Cavriglia)

'MEILÉNASAR', Cl F, or, 1975; (**Poppy Flash, Climbing**, Rusticana, Climbing); Paolino; URS

MEIliaxi, S, lp, 1989; (**Schuss**®); Meilland

MEIlicafal, HT, 1983; (**Cannes Festival 83**); Meilland, Alain A.; (Cavriglia)

'MEILIDER', LCl, dp, 1985; (All In One, **Exploit**®, Grimpant Exploit®); flowers deep pink, medium, dbl., 20 petals, no fragrance; foliage small, medium green, matt; very vigorous, spreading, climbing growth; [Fugue X Sparkling Scarlet]; Meilland, Mrs. Marie-Louise, 1983

'MEILIE', HT, mr, 1958; (**Christian Dior**); bud ovoid, pointed; flowers clear true red, high centered, very dbl., 45–50 petals, 4–4.5 in., exhibition form, borne mostly singly, slight, soft spice fragrance; foliage leathery, glossy; long stems; vigorous, upright, bushy growth; AARS, 1962 GM, Geneva, 1958; [(Independence X Happiness) X (Peace X Happiness)]; Meilland, F.; C-P, 1961; URS, 1958

'MEILIMONA', HT, rb, 1961; (**Banzai**); bud ovoid; flowers red, pink and cream blend, dbl., 30 petals, 3.5 in., exhibition form, slight fragrance; foliage leathery, dark; vigorous, bushy growth; [Radar X Caprice (Meilland)]; Meilland, Mrs. Marie-Louise; C-P

'MEILIMONA', HT, my, 1976; (**Banzai**, Banzai '76); flowers canary-yellow, full, dbl., 35 petals, 4 in., cupped, slight fragrance; vigorous growth; PP004013; [Coed X ((Seedling X Seedling) X Verla)]; Paolino; Meilland

'MEILISIA', HT, mp, 1968; ('MEILISIA', **Princess Margaret of England**, Princesse Margaret d'Angleterre®); flowers phlox-pink, large, dbl., exhibition form, slight fragrance; foliage leathery; vigorous, upright growth; GM, Portland, 1977; [Queen Elizabeth X (Peace X Michele Meilland)]; Meilland, Mrs. Marie-Louise; URS, 1969

'MEILISTA', HT, mp, 1968; ('MEILISIA', **Princess Margaret of England**, Princesse Margaret d'Angleterre®); flowers phlox-pink, large, dbl., exhibition form, slight fragrance; foliage leathery; vigorous, upright growth; GM, Portland, 1977; [Queen Elizabeth X (Peace X Michele Meilland)]; Meilland, Mrs. Marie-Louise; URS, 1969

'MEILIVAR', HT, dy, 1997; (**Gina Lollobrigida**®); flowers full, large, very dbl., 70–90 petals, borne mostly singly, moderate fragrance; some prickles; foliage medium, medium green, semi-glossy; upright, medium (5 ft) growth; [MEIdragelac X MEIkinosi]; Meilland International SA

Meilland Decor Arlequin®, *see* 'MEIZOURAYOR'

Meilland Decor Rose, *see* 'MEITURAPHAR'

Meilland Rosiga '83, *see* 'MEIBALBIKA'®

Meillandina, *see* 'MEIROV'

'MEILMERA', Min, w, 1996; (**Bridal Sunblaze**®); very dbl., 30–40 petals, 2 in., borne in small clusters, slight fragrance; few prickles; foliage large, dark green, dull; compact, medium (18 in.) growth; PP10002; [(MEIringa X Schneewitchen) X (MEIzogrel X MEIlarco)]; Meilland International SA; The Conard-Pyle Co., 1996

MEIloise, HT, op, 1985; (**Reve de Paris**®); Meilland

MEIlomit, F, ob, 1996; (**Marie Curie**); Meilland

MEIlomit, S, lp; (**Romantic Dreams**); Meilland

'MEILONTIG', S, w, 1987; (**Repens Meidiland**); flowers medium, single, 5 petals, no fragrance; foliage medium, light green, glossy; spreading, strong growth; PP006598; [Swany X New Dawn]; Meilland, Mrs. Marie-Louise, 1985

'MEILOTUP', HT, dr, 1986; (**Royal Velvet**™); flowers large blooms borne usually singly, dbl., 28 petals, exhibition form, slight fragrance; globular, small, medium green fruit; large, straw-colored prickles; medium, bushy growth; [(Exciting X Suspense) X Duke of Windsor]; Meilland, Mrs. Marie-Louise; Wayside Gardens Co.

MEIlsinplox, HT, my, 1994; (**Golden Nugget**™); Meilland

'MEILUCCA', Min, or, 1971; (**Darling Flame**, Minuette, Minuetto); flowers mandarin-red to vermillion-red, yellow anthers, small, dbl., 25 petals, 1.5 in., globular, slight, fruity fragrance; foliage glossy, dark; vigorous growth; [(Rimosa X Josephine Wheatcroft) X Zambra]; Meilland

'MEILUCRE', HT, or, 1971; (Arturo Toscanini, **Elegy**); flowers vermilion, dbl., 30 petals, 5.5 in., globular, slight fragrance; foliage semi-matt, dark; vigorous growth; [((Happiness X Independence) X Sutter's Gold) X ((Happiness X Independence) X Suspense))]; Meilland

'MEILUMINAC', HT, dr, 1979; (President L. Senghor®, **Président Leopold Senghor**®); bud conical; flowers large, dbl., 25 petals, cupped; foliage glossy, dark; vigorous, bushy growth; Meilland, Mrs. Marie-Louise; Meilland

MEIluminacsar, Cl HT, dr; (**President Leopold Senghor, Climbing**); Meilland

MEIlupin, HT, lp; (**Starlite**); Meilland, 1989

'MEILUSAM', HT, mr, 1973; (Lusambo, **Red Rock**, Red Rocky); flowers cherry-red, open, imbricated, dbl., 35 petals, 5 in., slight fragrance; very vigorous growth; [(Royal Velvet X Chrysler Imperial) X Pharaoh]; Meilland; URS

'MEILUTIDA', F, or, 1977; (**Muchacha**®); bud conical; flowers brilliant vermilion, shallowly-cupped, medium, semi-dbl., 12 petals, cupped; foliage dark; very vigorous, semi-shrub growth; Golden Rose, The Hague, 1976; [(Frenzy X Frenzy) X (Sangria X Sangria)]; Meilland, Mrs. Marie-Louise; Meilland

'MEIMAFRIS', HT, w, 1986; (**Jardins de Bagatelle**®, Sarah); flowers large, very dbl., intense fragrance; foliage large, medium green, semi-glossy; upright growth; GM, Genoa, 1987 Golden Rose, Geneva, 1984; [(Queen Elizabeth X Eleg) X MEIdragelac]; Meilland, Mrs. Marie-Louise; SNC Meilland & Cie, 1987

'MEIMAGARMIC', HT, w, 1982; (Grace Kelly, Preference, Princess Grace, Princess of Monaco, **Princesse de Monaco**®); flowers petals cream, edged pink, large, dbl., 35 petals, exhibition form, moderate fragrance; foliage large, dark, glossy; upright, bushy growth; PP005067; [Ambassador X Peace]; Meilland, Mrs. Marie-Louise; Meilland Et Cie, 1981

'MEIMAGUL', Min, rb, 1994; (**Gypsy Sunblaze**™); flowers red/yellow bicolor, blooms borne mostly single, dbl., 15–25 petals, 1.5 in., no fragrance; some prickles; foliage small, light green, matt; medium, bushy growth; [(Bonfire Night X MEIalfi) X Tapis Jaune]; Meilland, Alain A.; The Conard-Pyle Co., 1993

'MEIMAINGER', HT, ob, 1987; (Bushveld Dawn, **Cary Grant**™); flowers vivid orange blend, lighter at petal base, very dbl., 35–40 petals, 5 in., exhibition form, borne singly, open slowly, intense, spicy fragrance; ovoid, green red-orange fruit; slightly recurved, light green-straw prickles; foliage medium, dark green, glossy; upright, medium growth; PP006792; [(Pharaoh X Königin der Rosen) X ((Zambra X Suspense) X King's Ransom)]; Meilland, Mrs. Marie-Louise; Conard-Pyle, 1989; Wayside Gardens Co., 1989

'MEIMAJ', HT, yb, 1954; (**Lady Elgin**, Thaïs); bud ovoid; flowers buff-yellow washed pink, large, dbl., 40 petals, cupped, moderate fragrance; foliage dark, leathery; vigorous, upright, bushy growth; [Mme Kriloff X (Peace X Geneve)]; Meilland, F.; C-P, 1957; URS, 1954

'MEIMAL', Min, ob, 1958; (Colibre, **Colibri**); bud ovoid; flowers bright orange-yellow, small blooms in clusters, dbl., slight fragrance; foliage glossy; bushy growth; Golden Rose of The Hague, The Hague,

1962; [Goldilocks X Perla de Montserrat]; Meilland, F.; URS

'MEIMALYNA', HT, mr, 1984; (New Rouge Meilland, **Rouge Meilland**); flowers large, dbl., no fragrance; foliage large, dark, semi-glossy; upright growth; [((Queen Elizabeth X Karl Herbst) X Pharoah) X Antonia Ridge]; Meilland, Mrs. Marie-Louise; Meilland Et Cie, 1982

MEImanoir, LCl, mp, 1994; (**Blossom Magic**); Meilland

'MEIMAUR', HT, dr, 1973; (Joséphine Baker, **Velvet Flame**); dbl., 30 petals, 5–5.5 in., slight fragrance; foliage dark; vigorous growth; [Tropicana X Papa Meilland]; Meilland; URS

'MEIMEX', HT, r, 1959; (**Fantan**); bud urn-shaped; flowers burnt-orange to yellow-ochre, dbl., 48 petals, 3.5 in., cupped, slight fragrance; foliage leathery; moderate growth; [(Pigalle X Prelude) X Self]; Meilland, F.; C-P; URS

'MEIMICK', S, rb, 1961; (**Cocktail**®); bud pointed; flowers geranium red, base primrose yellow, medium, single, 5 petals, 2.5 in., borne in clusters, slight, spicy fragrance; foliage leathery, glossy; vigorous, semi-climbing growth; shrub or hedge; [(Independence X Orange Triumph) X Phyllis Bide]; Meilland, F.; C-P, 1961; URS, 1957

'MEIMIT', HT, lp, 1956; (**Grace de Monaco**®); flowers light rose-pink, well-formed, large, dbl., intense fragrance; foliage leathery; vigorous, bushy growth; [Peace X Michele Meilland]; Meilland, F.; URS

'MEIMODAC', S, mp, 1994; (**Royal Bonica**™); dbl., 15–25 petals, 1.5–2.75 in., borne in small clusters, slight fragrance; some prickles; foliage medium, medium green, semi-glossy; medium, upright growth; [MEIdomonac sport]; Meilland, Alain A.; The Conard-Pyle Co., 1993

'MEIMONBLAN', F, ob, 1999; (**Marmalade Skies**™); flowers medium, dbl., 17–25 petals, borne in large clusters, slight fragrance; prickles moderate; foliage medium, dark green, glossy; bushy, medium (3–4 ft) growth; AARS, U.S., 2001; [(Tamango X Parador) X Patricia]; Selection Meilland; Conard-Pyle, 2001

'MEIMONT', LCl, mp, 1960; (**Clair Matin**®, Grimpant Clair Matin); bud pointed; cupped, 15 petals, 2–3 in., cupped, blooms in rounded clusters, moderate, sweetbriar fragrance; foliage dark, leathery; vigorous (10–12 ft), well-branched growth; GM, Bagatelle, 1960; [Fashion X ((Independence X Orange Triumph) X Phyllis Bide)]; Meilland, Mrs. Marie-Louise; C-P, 1963; URS, 1960

'MEIMORE', F, or, 1977; (**Chorus**®, 'MEIJALITA', 'MEIJULITO'); flowers vermilion-red, dbl., 35 petals, 4 in., slight, fruity fragrance; foliage glossy; vigorous growth; ADR, 1977; [Tamango X (Sarabande X Zambra)]; Paolino; URS, 1975

'MEIMOUSLIN', Gr, mr, 1993; (**Cardinal Song**™); very dbl., 3–3.5 in., borne mostly singly, slight fragrance; some prickles; foliage large, dark green, glossy; medium, semi-erect growth; [Olympiad X (Michel Lis Le Jardinier X Red Lady)]; Selection Meilland; The Conard-Pyle Co., 1992

MEImucas, HT, op, 1995; (**Arioso**); Meilland

MEImuide, F, op, 1990; (**Holstentor**); flowers salmon-pink, large, dbl.; Meilland; (Sangerhausen)

MEImurge, Gr, ob, 1997; (**Lasting Peace**); Meilland

Mein München, F, yb, 1987; flowers yellow and red, large, dbl., slight fragrance; Cocker; (Sangerhausen)

'MEIN RUBIN', HT, dr, 1984; flowers urn-shaped, well-formed, blooms borne usually singly, dbl., 35 petals; medium, lavender prickles; foliage medium, medium green, semi-glossy; medium, bushy growth; [(Helene Schoen X Charles Mallerin) X (Helene Schoen X Charles Mallerin)]; Teranishi, K.; Itami Rose Nursery

Mein Schoner Garten, *see* KORmeneint

'MEINABRON', Min, w, 1988; (**Alba Meillandina**®); flowers medium, very dbl., no fragrance; foliage small, medium green, semi-glossy; bushy growth; PP007347; [MEIdonq X (Darling X Jack Frost)]; Meilland, Mrs. Marie-Louise; SNC Meilland & Cie, 1987

MEInagre, F, rb; (**Wapiti**®); Meilland, 1988

MEInalpir, HT, pb, 1997; (**Blue Bell**); Meilland

MEInartemi, F; Meilland; (Cavriglia)

'MEINARVAL', HT, my, 1974; (**Sun King**, Sun King '74); flowers large, dbl., 30 petals, exhibition form; foliage small; vigorous, upright growth; [(Soroya X Signora) X King's Ransom]; Paolino; URS, 1972

'MEINASTUR', HT, or, 1975; (**Alpha**); flowers bright vermilion, dbl., 20 petals, 4 in.; foliage leathery; [((Show Girl X Baccará) X Romantica) X (Romantica X Tropicana)]; Paolino; URS

'MEINATAC', HT, lp, 1972; (**Susan Hampshire**); flowers light fuchsia-pink, dbl., 40 petals, 5.5 in., globular, intense fragrance; foliage matt; vigorous, upright growth; [(Monique X Symphonie) X Miss All-American Beauty]; Paolino; URS

Meindeert Hobbema, S, rb, 1997; Williams, J. Benjamin

'MEINEBLE', S, rb, 1989; (**Red Meidiland**™, Rouge Meillandécor®); bud conical; flowers red with white eye, medium, single, 5 petals, cupped, borne in sprays of 7–15, repeat bloom, no fragrance; globular, small, red fruit; prickles gray-brown; foliage medium, dark green, glossy, disease resistant; spreading, medium, very winter hardy growth; PP007116; [Sea Foam X (Picasso X Eyepaint)]; Meilland, Alain A.; The Conard-Pyle Co., 1989

MEInececa, F, mr; (**Tempo**); Meilland, 1978

'MEINEREAU', HT, yb, 1996; (**Centennial Star**™); very dbl., 4.75 in., moderate fragrance; foliage large, dark green, semi-glossy; medium growth; [(Peace X Landora) X Kings Ransom]; Meilland International SA; Conard-Pyle Co., 1996

'MEINEYTA', Min, ob, 1994; (**Cumba Meillandina**®, Fiery Sunblaze); very dbl., 1.5–2.75 in., borne in small clusters, no fragrance; some prickles; foliage large, medium green, semi-glossy; medium, bushy growth; [Orange Honey X (Darling Flame X Tapis Jaune)]; Meilland, Alain A.; SNC Meilland & Cie, 1992

'MEINIACIN', HT, mp, 1991; (Fragrant Lady, **Perfume Beauty**™); bud conical; flowers rose bengal, blooms borne singly, very dbl., cupped, moderate fragrance; foliage large, dark green; upright, tall growth; PP007819; [MEIlista X (Carina X Silvia)]; Meilland, Alain A., 1990; The Conard-Pyle Co., 1990

MEInibur, LCl, ab, 1995; (**Kir Royal**); Meilland

MEInical, HT, dr, 1999; (**Edith Piaf**); Meilland

'MEINIMO', HT, lp, 1994; (21 Again!, Penny Coelen, Prestige de Lyon, **Regatta**™); dbl., 26–40 petals, 3–3.5 in., borne mostly singly, moderate fragrance; few prickles; foliage large, dark green, matt; medium, bushy growth; PP8390; Geneva, 1989; [MEIgurami X (MEInaregi X MEIdragelac)]; Meilland, Alain A.; SNC Meilland & Cie, 1992; The Conard-Pyle Co., 1994

'MEININRUT', Min, or, 1994; (**Orange Symphonie**®); flowers orange vermillion, blooms borne mostly single or in small clusters, dbl., 26–40 petals, 1.5–2.75 in., no fragrance; some prickles; foliage medium, dark green, semi-glossy; low (35–45 cms), bushy growth; [(Baby Bettina X Anytime) X Meteor]; Meilland, Alain A.; SNC Meilland & Cie, 1993

MEInirlo, S, mr, 1992; (**Rote Woge**); Meilland

MEInitper, HT, ab, 1986; (**Caramella**®); Meilland

'MEINIVOZ', HT, yb, 1994; (**Paul Ricard**, Paul Richard, Spirit of Peace); flowers amber yellow, blooms, very dbl., 3–3.5 in., intense fragrance; foliage large, medium green, matt; tall (110–120 cms), upright growth; PPAF; GM, Rome, 1991; [(Hidalgo X Mischief) X Ambassador]; Meilland, Alain A.; SNC Meilland & Cie, 1990; The Conard-Pyle Co., 1994

'MEINOFRAI', Min, dr, 1990; (**Red Mini-Wonder**™); bud rounded; flowers currant red, cardinal red reverse, aging to dark red, small, very dbl., 40–43 petals, cupped; prickles small, green to tan; foliage small, medium green, very dense, semi-glossy;

bushy, low growth; PP007384; [(Anytime X Parador) X Mogral]; Selection Meilland, 1987; The Conard-Pyle Co., 1989

MEInoiral, S, mp, 1990; (**Deborah**®); Meilland

'MEINOMAD', HT, yb, 1992; (**Desert Peace**); flowers yellow tinged with red, blooms borne mostly singly, dbl., 15–25 petals, 1.5–2.75 in., slight fragrance; foliage large, dark green, glossy; tall (120–130 cms), upright growth; [(Sonia X Rumba) X (Piccadilly X Chicago Peace)]; Selection Meilland; C-P

MEInopia, HT, pb, 1992; (**Madona**); Meilland

MEInorep, HT, pb, 1995; (**James Pereire**); Meilland

'MEINRONSSE', HT, ob, 1992; (Christoph Colombus, Christophe Colomb, **Christopher Columbus**, Cristobal Colon, Cristoforo Colombo, 'MEIRONSSE'); flowers orange blend/copper, blooms borne mostly singly, dbl., 26–40 petals, 3–3.5 in., slight fragrance; many prickles; foliage large, dark green, semi-glossy; medium (120–140 cms), upright growth; GM, Durbanville, 1990; [MEIgurani X (Ambassador X MEInaregi)]; Selection Meilland; France, 1990; Germany; Italy, 1991; Spain, 1991; The Conard-Pyle

'MEINUSIAN', F, mr, 1998; (**Francois Rabelais**™); flowers bright red, very full, very dbl., 41 petals, 2–2.5 in., borne in small clusters, slight fragrance; prickles moderate; foliage medium, dark green, glossy; bushy, medium, 3 ft growth; [(MEIchanso X MEIfolio) X KORlima]; Selection Meilland; Conard-Pyle Co., 1996

MEInustrel, F; Meilland; (Cavriglia)

'MEINUZETEN', HT, ob, 1979; (**Ambassador**®); bud conical; flowers orange-red, reverse blended with golden yellow (orange-apricot), large, dbl., 33 petals, 4 in., cupped; foliage dark, glossy; PP004224; [Seedling X Whisky Mac]; Meilland; C-P

'MEIOFFIC', F, pb, 1994; ('Johann Straus', **Johann Strauss**®, Sweet Sonata); flowers orient pink slightly suffused with aureolin yellow, blooms b, very dbl., 3–3.5 in., slight fragrance; few prickles; foliage medium, dark green, semi-glossy; low (50–60 cms), bushy, compact growth; [Flamingo X (Pink Wonder X Tip Top)]; Meilland, Alain A.; SNC Meilland & Cie, 1993

MEIonagre, HT, mr, 1995; (**Michaelhouse Centenary**); Meilland

MEIouscki, S, dr, 1996; (**Crimson Meillandecor**); Meilland

'MEIPAL', HT, op, 1953; (**Bettina**®, 'MEpal'); flowers salmon-orange, veined, wellformed, dbl., 37 petals, 4 in., moderate fragrance; foliage dark, glossy, bronze; vigorous growth; [Peace X (Mme Joseph Perraud X Demain)]; Meilland, F.; URS

'MEIPALEO', HT, my, 1994; (**Marco Polo**); dbl., 26–40 petals, 3–3.5 in., borne mostly singly, moderate fragrance; some prickles; foliage medium green, semiglossy; tall, upright growth; [Ambassador X (Sunlight X Oregold)]; Meilland, Alain A.; SNC Meilland & Cie, 1992; The Conard-Pyle Co., 1994

'MEIPARADON', HT, mr, 1976; (**Antonia Ridge**®); flowers cardinal-red, dbl., 30 petals, 4–4.5 in., exhibition form, slight fragrance; vigorous growth; [(Chrysler Imperial X Karl Herbst) X Seedling]; Paolino; URS

'MEIPARNIN', HT, pb, 1998; (**Honore de Balzac**®, Romantic Days); flowers light creamy yellow suffused with carmine, very full, 41 petals, 4.5 in., borne in large clusters, moderate fragrance; prickles moderate; foliage medium, medium green, semi-glossy; upright, bushy, medium, 3 ft growth; [(Marion Foster X KORax) X Lancome]; Selection Meilland; Conard-Pyle Co., 1996

MEIparos, HT, mp, 1999; (**Perfumella**); Meilland

MEIpazdia, HT, w, 1998; (**Poker**); Meilland

'MEIPELTA', S, dp, 1994; (Cyclamen Meillandecor, **Fuchsia Meidiland**™, Fuchsia Meillandecor®); flowers deep pink, open flat, semi-dbl., 6–14 petals, 1.5–2.75 in., borne in large clusters, slight fragrance; few prickles; foliage medium, light green, glossy; low, spreading growth; PP8839; [MEItiraca X Clair Matin]; Meilland, Alain A.; The Conard-Pyle Co., 1993

MEIpeluj, LCl, lp; (**Sorbet**); Meilland, 1993

'MEIPERATOR', F, mr, 1971; (**Crimson Wave**, Imperator); flowers cardinal-red, shaded cherry, dbl., 25–30 petals, 4–5 in., slight, apple fragrance; foliage large, semi-matt, dark; vigorous, upright growth; [Zambra X (Sarabande X (Goldilocks X Fashion))]; Meilland

'MEIPIERAR', HT, w, 1993; (Cameo Cream, **Caroline de Monaco**); flowers cream white, blooms borne mostly single, very dbl., 3–3.5 in., no fragrance; foliage large, dark green, semi-glossy; medium, bushy growth; [Chicago Peace X Tchin-Tchin]; Meilland, Alain A., 1989; Conard-Pyle, 1993, USA; The Conard-Pyle Co., 1993

'MEIPIESS', Min, mr, 1986; (**Majorette**®, Majorette 86); flowers medium cardinalred, moderately medium, dbl., 15–25 petals, no fragrance; fruit produced in autumn; foliage small, medium green, semi-glossy; mini-flora; bushy growth; [Magic Carrousel X (Grumpy X Scarletta)]; Meilland, Mrs. Marie-Louise, 1985

'MEIPINJID', Min, mp, 1985; (**Classic Sunblaze**®, Duc Meillandina, Duke Meillandina); flowers rich pink, very double, big, very dbl., 40 petals, 1.5–2 in., borne mostly singly, slight fragrance; foliage medium, dark green; short, bushy, 15 in. growth; PP005958; [Pink Sunblaze sport]; Meilland, Mrs. Marie-Louise; SNC Meilland & Cie

'MEIPITAC', S, pb, 1990; (**Carefree Wonder**™); bud pointed; flowers medium pink with light pink reverse, aging to medium pink, large, dbl., 26 petals, cupped, borne in sprays of 1–4, slight fragrance; oval, reddish-brown fruit; prickles narrow, reddish; foliage medium, medium green, semi-glossy; bushy, medium growth, suitable for hedges; hardy; PP007783; AARS, 1991; [(Prairie Princess X Nirvana) X (Eyepaint X Rustica)]; Selection Meilland, 1978; The Conard-Pyle Co., 1991

'MEIPLATIN', S, lp, 1989; (**Pearl Meidiland**™, Perle Meillandécor®); bud ovoid; flowers light ochre pink, aging white, medium, dbl., 30 petals, flat, borne in sprays of 3–15, repeat bloom, no fragrance; globular, small dish, reddish fruit; prickles small, reddish-brown; foliage medium, dark green, glossy; spreading, low growth; PP006807; [(Sea Foam X MEIsecaso) X Sea Foam]; Meilland, Alain A.; The Conard-Pyle Co., 1989

MEIplovon, HT, mp; (**Mademoiselle**); Meilland

'MEIPOBIL', HT, dp, 1987; (**Rendez-vous**); Meilland

'MEIPONAL', Min, ob, 1986; (Sunblaze, **Sunny Meillandina**®, Sunny Sunblaze); flowers yellow-orange, medium, dbl., 35 petals, no fragrance; foliage small, dark, matt; bushy growth; PP006810; [(Sarabande X Moulin Rouge) X (Zambra X Meikim)]; Meilland, Mrs. Marie-Louise; Meilland, 1985

'MEIPOPUL', S, mp, 1994; (**Douceur Normande**®, Sandton City, Goose Fair); flowers scarlet pink, blooms borne in small clusters, single, 5 petals, 1.5–2.75 in., no fragrance; many prickles; foliage small, medium green, semi-glossy; medium (70–90 cms), bushy growth; [(Immensee X Green Snake) X (Temple Bells X Red Cascade)]; Meilland, Alain A.; SNC Meilland & Cie

'MEIPOQUE', S, pb, 1984; (**Pink Meidiland**®, Schloss Heidegg); flowers deep pink, white eye, medium, single, 5 petals, no fragrance; foliage small, medium green, semi-glossy; bushy growth; PP005956; ADR, 1987; [Anne de Bretagne X Nirvana]; Meilland, Mrs. Marie-Louise; Meilland & Son, 1985

'MEIPOTAL', S, pb, 1994; (Bingo Meidiland®, Bingo Meillandecor®, **Carefree Delight**™); flowers carmine pink with white eye, blooms borne in large clusters, single, 5 petals, 1.5–2.75 in., no fragrance; many prickles; foliage small, dark green, glossy; medium (70–80 cms), bushy growth; PPAF; AARS, 1996; [(Eyepaint X Nirvana) X Smarty]; Meilland, Alain A.; SNC Meilland & Cie, 1991; The Conard-Pyle Co., 1994

MEIpraserpi, HT, op, 1981; (**Catherine Deneuve**®); Meilland

'MEIPSIDUE', S, mr, 1999; (**Fire Meidiland**™); flowers fire engine red, opens flat, dbl., 17–25 petals, 2.5–3 in., borne in small clusters; foliage dark green, glossy; mounding, ground cover growth; Meilland

MEIpsilon, HT, rb; (**Yakimour**®); Meilland, 1980

'MEIPUMA', F, rb, 1975; (**Scherzo**®); flowers bright scarlet, reverse white and crimson, spiraled, dbl., 40 petals, 3.5 in., exhibition form; foliage dark; vigorous, bushy growth; GM, Belfast, 1975; [Tamango X Frenzy]; Paolino; URS

MEIqualis, HT, dr, 1995; (**Grand Gala**); Meilland

'MEIRABANDE', F, or, 1957; ('MEIHAND', **Sarabande**®); flowers light orange-red, stamens yellow, blooms in large trusses, semi-dbl., 13 petals, 2.5 in., cupped, slight fragrance; foliage semi-glossy; low, bushy growth; AARS, 1960 GM, Bagatelle, 1957 GM, Geneva, 1957 GM, Portland, 1958 GM, Rome, 1957; [Cocorico X Moulin Rouge]; Meilland, F.; C-P, 1959; URS, 1957

'MEIRADIA', Min, mp, 1971; (Lutin, **Rosy Gem**); [Scarlet Gem sport]; Meilland, 1973

'MEIRANDIVAL', HT, yb, 1980; [Meibiranda sport]; Meilland, Mrs. Marie-Louise; Meilland Et Cie

'MEIRANOGA', Min, or, 1975; (**Brilliant Meillandina**); semi-dbl., 15 petals, cupped, blooms in clusters of 3–20, no fragrance; foliage dark, semi-glossy; [Parador X (Baby Bettina X Duchess of Windsor)]; Meilland, Mrs. Marie-Louise; Meilland Et Cie

'MEIRANOVI', S, rb, 1982; (**Candy Rose**®); flowers deep pink, reverse medium red, medium, dbl., 20 petals, no fragrance; foliage small, medium green, semi-glossy; spreading growth; PP006385; [(R. sempervirens X Mlle Marthe Carron) X ((Lilli Marleen X Evelyn Fison) X (Orange Sweetheart X Fruhlingsmorgen))]; Meilland, Mrs. Marie-Louise, 1980; Meilland Et Cie, 1983

MEIrasimac, S, mr, 1983; (**Royal Show**); GM, Durbanville, 1984; Meilland

MEIrebuc, F, w, 1995; (**Liz**); Meilland

MEIrecrom, HT, ob, 1999; (**Exotica**); Meilland

'MEIRED', HT, mr, 1972; (**Visa**); flowers turkey-red, dbl., 38 petals, 5 in., exhibition form, slight fragrance; foliage large, leathery; vigorous, upright growth; [(Baccará X Queen Elizabeth) X Lovita]; Meilland

MEIreibat, F, 1993; (**Concorde**); Meilland, Alain A.; (Cavriglia)

'MEIRESTIF', HT, ob, 1987; (Charleston 88, **Louis de Funès**®); flowers orange capucine reverse cadmium yellow, large, dbl., 15–25 petals, slight fragrance; foliage medium, dark green, glossy; upright, strong growth; [(Ambassador X Whisky Mac) X (Arthur Bell X Kabuki)]; Meilland, Mrs. Marie-Louise, 1983

'MEIREVOLT', HT, my, 1994; (**Toulouse Lautrec**®); very dbl., 3–3.5 in., borne mostly singly, no fragrance; some prickles; foliage medium, medium green, glossy; medium (70–80 cms), bushy growth; [Ambassador X (King's Ransom X Sunblest)]; Meilland, Alain A.; SNC Meilland & Cie, 1992

MEIrgano, HT, mr, 1992; (**Porto Nigra**); Meilland

'MEIRIANOPUR', F, or, 1979; (**Orange Bunny**®); bud pointed; flowers orange-red, reverse darker, cup-shaped, borne 1–25 per cluster, semi-dbl., 13 petals, slight fragrance; foliage bronze, matt, very dense; bushy growth; [Scherzo X (Sarabande X Frenzy)]; Meilland, Mrs. Marie-Louise; Meilland Et Cie

'MEIRIDGE', F, yb, 1963; (**Charleston**); bud pointed; flowers yellow flushed crimson, becoming crimson, blooms in clusters, dbl., 20 petals, 3 in., slight fragrance; foliage dark, leathery, glossy; upright, compact growth; [Masquerade X (Radar X Caprice)]; Meilland, Alain A.; URS; Wheatcroft & Sons; Wheatcroft Bros.

MEIridol, F, dp, 1983; (**Irene**); Meilland

'MEIRIDORIO', HT, yb, 1984; (**Gilbert Bécaud**); flowers orange and yellow blend, large, dbl., 45 petals, slight fragrance; foliage bronze, matt; upight growth; [(Peace X Mrs John Laing) X Bettina]; Meilland, Mrs. Marie-Louise; Meilland Et Cie, 1979

MEIRIENTAL, mr, 1997; (**Paprika**™); flowers bright red, large, dbl., 20–25 petals, borne in clusters, slight fragrance; foliage large, deep green, leathery, glossy; climbing (8–10 ft) growth; PP9537; [Marchenland X Red Favorite]; Meilland

'MEIRIGALU', HT, rb, 1968; (**Colorama**, Colourama, Dr R. Maag); bud ovoid; flowers red and yellow, large, dbl., cupped, moderate fragrance; foliage very glossy; vigorous, upright, bushy growth; [Suspense X Confidence]; Meilland, Mrs. Marie-Louise; C-P

'MEIRILOCRA', HT, ab, 1984; (**Capella**, Zambra '80); flowers large, dbl., 35 petals, slight fragrance; foliage medium, dark, matt; PP004868; [Unnamed seedling X Banzai]; Meilland, Mrs. Marie-Louise; Meilland Et Cie, 1979

'MEIRINGA', HT, w, 1975; (**Carte Blanche**™); flowers flora-tea, dbl., 38 petals, 4 in.; foliage matt; vigorous growth; [(Carina X White Knight) X Jack Frost]; Paolino; URS

'MEIRINLOR', HT, my, 1975; semi-dbl., 15–18 petals, 4 in.; foliage dark; [Golden Garnette X ((Golden Garnette X Bettina) X Dr A.J. Verhage)]; Paolino; URS

'MEIRISOURU', F, lp, 1977; (**Nirvana**®); bud ovoid; flowers large, dbl., 20 petals, cupped; foliage glossy; bushy growth; GM, Geneva, 1975; [(Pink Wonder X Kalinka) X Centenaire de Lourdes]; Meilland, Mrs. Marie-Louise; Meilland

MEIrivoui, Min, mp, 1997; (**Belle Symphonie**); Meilland

'MEIROBIDOR', HT, dy, 1980; (Carte d'Or, **Supra**™); flowers flora-tea, large, dbl., 20 petals, no fragrance; foliage medium, dark, semi-glossy; upright growth; PP005042; [((Zambra X (Baccará X White Knight) X Golden Garnette)) X Unnamed seedling]; Meilland, Mrs. Marie-Louise; Meilland Et Cie

'MEIRODIUM', HT, rb, 1976; (**Red Success**); flowers blood-red, base cardinal-red, very full, dbl., 40–45 petals, 4–4.5 in., slight fragrance; foliage large; vigorous, upright growth; PP004037; [(Tropicana X MEIalto) X ((MEIbrem X Zambra) X Tropicana)]; Paolino; URS

'MEIROKOI', HT, yb, 1998; (**Jean Giono**™, Romantic Moments); flowers sunny yellow, veined and edged in orange/apricot, very dbl., 26–40 petals, 3–4 in., borne in small clusters, moderate fragrance; prickles numerous; foliage medium, medium green, glossy; upright, bushy, medium, 90 cms growth; [(Yakimour X Landora) X Graham Thomas]; Selection Meilland; Conard-Pyle Co., 1996

MEIrolange, HT, mp, 1999; (**Paganini**); Meilland

'MEIROLOUR', F, mr; (**Concerto**™)

MEIrolyz, F, ob, 1999; (**Fire King**); Meilland

MEIromar, HT, w, 1995; (**Gin Fizz**); Meilland

'MEIRONSSE', HT, ob, 1992; (Christoph Colombus, Christophe Colomb, **Christopher Columbus**, Cristobal Colon, Cristoforo Colombo, 'MEINRONSSE'); flowers orange blend/copper, blooms borne mostly singly, dbl., 26–40 petals, 3–3.5 in., slight fragrance; many prickles; foliage large, dark green, semi-glossy; medium (120–140 cms), upright growth; GM, Durbanville, 1990; [MEIgurani X (Ambassador X MEInaregi)]; Selection Meilland; France, 1990; Germany; Italy, 1991; Spain, 1991; The Conard-Pyle

MEIrotego, Min, mp, 1977; (**Minijet**®); Meilland

'MEIROUPIS', LCl, mp, 1996; (**Colette**™, John Keats); very dbl., 3–3.5 in., borne in small clusters, intense fragrance; numerous prickles; foliage small, dark green, glossy; spreading, medium (2 m) growth; US PPAF; [(MEIbeluxen X Friesia) X Prairie Princess]; Meilland International SA; The Conard-Pyle Co., 1995

MEIroupis, HT, lp, 1996; (**Colette**™); flowers pink, medium sized, old-fashioned blooms, dbl., 25 petals, 3 in., cupped, intense fragrance; foliage large, medium green; climbing growth; PP9994; [(Centenaire de Lourdes X Pocasso) X MEIhatiti]; Meilland; Meilland, 1998

MEIrouve, HT, ab, 1997; (**Alphonse Daudet**); Meilland

'MEIROV', Min, mr, 1975; (**Meillandina**); flowers currant-red, then imbricated, small, dbl., 20 petals, 1.5 in., cupped; foliage matt; vigorous growth; [Rumba X (Dany Robin X Fire King)]; Paolino; URS

'MEIROVERNA', HT, or, 1979; (**Koba**); flowers scarlet, medium, dbl., 40 petals, cupped; foliage dark; very vigorous, upright growth; [((Unnamed seedling X Rouge Meilland) X Independence) X Queen Elizabeth]; Meilland, Mrs. Marie-Louise; Meilland

'MEIROVONEX', LCl, ob, 1977; (**Looping**®); bud conical; flowers orange-coral, medium, dbl., 40 petals, cupped; spring bloom, slight fragrance; foliage dark; vigorous, climbing growth; [((Zambra X Zambra) X (Malcair X Danse des Sylphes) X (Cocktail Cocktail)) X Royal Gold]; Meilland, Mrs. Marie-Louise; Meilland

'MEIROZRUG', HRg, mp, 1994; (**Turbo**™, Turbo Meidiland®, Turbo Rugostar); flowers fuchsia pink, blooms, dbl., 15–25 petals, 3–3.5 in., no fragrance; many prickles; foliage medium, light green, semi-glossy; tall (130–140 cms), bushy growth; [(Frau Dagmar Hastrup X Manou Meilland) X Pink Grootendorst]; Meilland, Alain A.; SNC Meilland & Cie, 1993

MEIruchka, F, rb, 1999; (**Bolchoi**); Meilland

MEIrulex, MinFl, lp; (**Fuchsia Sunblaze**); Meilland

'MEIRUMOUR', S, rd; 1995; (Cherry Meidiland, **Cherry Meillandecor**®); flowers medium red with white eye, single (5 petals), medium (4–7 cms) blooms borne in small clusters; no fragrance; many prickles; foliage medium, dark green, semi-glossy; GM Geneva, 1994; [MEIpoque X (Regensberg X Fair Play)]; Meilland, 1994; The Conard-Pyle Co.

'MEIRUTRAL', Min, dr, 1988; (**Prince Meillandina**®, Prince Sunblaze, Red Sunblaze®); flowers dark currant-red, medium, dbl., 15–25 petals, 1.5–2 in., cupped, borne in large clusters, no fragrance; foliage medium, dark green, semi-glossy; bushy growth, branching, 15 in.; PP007021; [Parador X Mogral]; Meilland, Alain A.; SNC Meilland & Cie

'MEIRYPOUX', HT, dr, 1990; (**Rotary Rose**); bud pointed; flowers, dbl., 27 petals, exhibition form, borne usually singly, no fragrance; prickles large, red; foliage medium green; vigorous, upright growth; PP007408; [(Mister Lincoln X Pres. Leopold Senghor) X Karl Herbst]; Meilland, Alain A.; Rotary Rose Co., 1989

'MEISADINA', HT, or, 1975; (**Golestan**); bud tapering; flowers vermilion, small, dbl., 25 petals, exhibition form; foliage glossy, dark; vigorous, upright growth; [(Tropicana X Tropicana) X ((Seedling X Rouge Meilland) X Independence)]; Meilland, Mrs. Marie-Louise; Meilland

'MEISAR', HT, dr, 1963; ('MEICESAR', **Papa Meilland**®); bud pointed; flowers dark velvety crimson, large, dbl., 35 petals, exhibition form, intense fragrance; foliage leathery, glossy, olive-green; vigorous, upright growth; GM, Baden-Baden, 1962 Hall of Fame, WFRS, 1988 James Alexander Gamble Fragrance Medal, ARS, 1974; [Chrysler Imperial X Charles Mallerin]; Meilland, Alain A.; URS

MEIsardan, S, yb, 1993; (**Cesar**™, Romantic Occasion); flowers creamy yellow with pink, very large, dbl., slight fragrance; Meilland; (Sangerhausen)

'MEISARSAR', Cl HT, dr, 1970; (Grimpant Papa Meilland®, **Papa Meilland, Climbing**®); Stratford; Rumsey

MEIsazy, F, lp, 1997; (**Gala Charles Aznavour**); Meilland

MEIsecaso, S, pb; (**Lutin**); Meilland

'MEISELGRA', Min, mp, 1990; (**Pink Mini-Wonder**™); bud rounded; flowers light rose bengal, reverse pale rose bengal, aging pale rose, dbl., 39–42 petals, cupped, no fragrance; prickles small, very few, green, aging tan; foliage medium green, semi-glossy; bushy, medium growth; PP007361; [(Anytime X Parador) X Mogral]; Selection Meilland, 1987; The Conard-Pyle Co., 1989

'MEISHUTO', Min, op, 1996; (**Salmon Sunblaze**™); flowers salmon pink, full, dbl., 35–40 petals, 1.5–2 in., borne in small clusters, slight fragrance; few prickles; foliage small, dark green, semi-glossy; compact, low (12–16 in.) growth; US PPAF; [MEIchanso X (MEIlarco X KORwerk)]; Meilland International SA; The Conard-Pyle Co., 1997

MEIsionver, HT, rb, 1998; (**Caprice de Mielland**); Meilland

'MEISOCRAT', F, mp, 1996; (**Guy de Maupassant**™, Romantic Fragrance); very dbl., 3–3.5 in., borne in small clusters, intense fragrance; moderate prickles; foliage large, dark green, glossy; bushy, tall (70–90 cms) growth; US PPAF; [(MEIturaphar X Mrs John Laing) X Egekov]; Meilland International SA; The Conard-Pyle Co., 1995

'MEISOGREL', Min, w, 1988; ('MEIZOGREL', Spot Minijet, **White Mini-Wonder**™); flowers small, clear white, thin petals, very dbl., 40 petals, cupped, borne in small clusters, no fragrance; foliage small, medium green, matt; short, bushy growth, 12 in.; PP007276; [White Gem X Cinderella]; Meilland, Alain A.; SNC Meilland & Cie

MEIsolroz, S, mp, 1990; (**Flamingo Meidiland**); Meilland

'MEISOYRIS', Gr, dr, 1994; (**Matilda**™, Marcel Pagnol, Velveteen); dbl., 26–40 petals, 1.5–2.75 in., moderate fragrance; some prickles; foliage large, light green, glossy; medium, semi-erect growth; PPAF; [Fragrant Cloud X (Oklahoma X Royal William)]; Meilland, Alain A.; SNC Meilland & Cie, 1993; The Conard-Pyle Co., 1994

MEIspola, HT, w; (**Preview**); Meilland

MEIspreyp, F, my, 1994; (**Mimi Golden**™); Meilland

'**MEISTERSTUCK**', F, mr, 1940; bud ovoid; flowers velvety crimson, large, borne in clusters, dbl., cupped, moderate fragrance; foliage dark, glossy; upright, bushy growth; [Holstein X Kardinal]; Kordes

MEIstocko, S, dp, 1991; (**Flash Meidiland**); flowers medium-large, single; Meilland; (Sangerhausen)

MEIsunaj, F, dp, 1989; (**Mme Fernandel**®); Meilland

'MEITABIFOB', HT, my, 1977; (**Cocktail '80**); bud elongated; flowers, dbl., 20 petals, 5 in., exhibition form, slight fragrance; foliage dark; very vigorous growth; Meilland, Mrs. Marie-Louise; Meilland

'MEITALBAZ', F, my, 1994; (**Anthony Meilland**); dbl., 26–40 petals, 3–3.5 in., borne in small clusters; some prickles; foliage large, medium green, semi-glossy; medium, bushy growth; PP8449; [Sunblest X MEIlenangal]; Meilland, Alain A.; SNC Meilland & Cie, 1990; The Conard-Pyle Co., 1994

'MEITAM', LCl, dr, 1958; (**Fugue**®); bud globular; dbl., 30 petals, blooms in clusters, slight fragrance; foliage leathery, glossy; vigorous growth; GM, Madrid, 1958; [Alain X Guinee]; Meilland, Mrs. Marie-Louise; URS

MEItanet, F, lp; (**Starlite**); Meilland, 1995

MEItapov, LCl, w, 1993; (**Koala**®); Meilland

'MEITARAS', HT, pb, 1963; (**Swarthmore**); flowers shades of pink blended, dbl., 50 petals, 4 in., exhibition form, slight fragrance; foliage dark, leathery; vigorous, tall, bushy growth; [(Independence X Happiness) X Peace]; Meilland, Alain A.; C-P

'MEITEBROS', HT, lp, 1998; (**Frederic Mistral**®); flowers venetian pink, reverse suffused rose, full, very dbl., 26–40 petals, 4.5 in., borne mostly singly, intense fragrance; prickles moderate; foliage large, dark green, semi-glossy; upright, tall, 6 ft growth; [(Perfume Delight X Prima Ballerina) X The McCartney Rose]; Selection Meilland; Conard-Pyle Co., 1995

MEItelov, HT, lp; (**Michelangelo**); Meilland, 1997

'MEITIFRAN', Min, rb, 1987; (**Baron Meillandina**®, Baron Sunblaze); flowers white with red edges, umbrella-shaped, medium, dbl., no fragrance; foliage small, medium green, glossy; bushy growth; PP006818; [Magic Carrousel X ((Alain X R. mutabilis) X (Medar X Caprice))]; Meilland, Mrs. Marie-Louise; Meilland Et Cie, 1989

'MEITILOLY', HT, yb, 1976; (**Mascotte '77**®); flowers yellow, edged cardinal-red, dbl., 40 petals, 4.5 in., slight fragrance; foliage glossy; vigorous growth; GM, Belfast, 1979; [(MEIrendal X (Rim X Peace)) X Peace]; Paolino; URS

'MEITINIROL', HT, or, 1977; (**Galia**®); dbl., 38 petals, cupped, borne singly, no fragrance; foliage matt, dense; vigorous growth; [MEIretni X Elegy]; Meilland, Mrs. Marie-Louise; Meilland Et Cie

MEItinor, F, mp; (**Cindy**); Meilland

MEItixia, HT, mr, 1989; (**Revolution Francaise**®); Meilland

MEItobla, F, dp, 1992; (**Les Amoureux de Peynet**®); Meilland

'MEITOFLAPO', F, mp, 1980; flowers large, dbl., 29 petals, no fragrance; foliage small, dark, semi-glossy; [(Jack Frost X (Zambra X (Baccará X White Knight))) X ((Zambrax (Baccará X White Knight)) X Unnamed seedling)]; Meilland, Mrs. Marie-Louise; Meilland Et Cie

'MEITOIFAR', HT, mp, 1994; (**Auguste Renoir**®); flowers old form, blooms borne mostly single, very dbl., 3–3.5 in., moderate fragrance; many prickles; foliage medium, medium green, semi-dull; medium, bushy growth; [(Versailles X Pierre de Ronsard) X Kimono]; Meilland, Alain A.; SNC Meilland & Cie, 1992

'MEITOLEIL', Min, my, 1994; (**Gold Symphonie**); very dbl., 1.5–2.75 in., borne mostly singly or in small clusters, no fragrance; some prickles; foliage medium, dark green, semi-glossy; medium (40–50 cms), bushy growth; [(Rise 'n' Shine X Yellow Meillandina) X Gold Badge]; Meilland, Alain A.; SNC Meilland & Cie, 1993

MEItomkin, Min, lb; (**Symphonie Lumiere**); Meilland, 1995

'MEITONJE', Min, lp, 1987; (Pink Symphonie®, **Pink Symphony**, Pretty Polly, Sweet Sunblaze®); flowers light cardinal pink, medium, unfading, dbl., 26–40 petals, 1.5–2 in., flat, borne singly and in small clusters, slight fragrance; foliage medium, dark green, glossy, disease resistant; vigorous, bushy growth, 18 in.; PP007277; GM, Glasgow, 1992; [Darling Flame X Air France]; Meilland, Mrs. Marie-Louise

MEItorpo, F, mr, 1992; (**Sangria**); Meilland

'MEITOSIER', LCl, ab, 1996; (Lord Byron, **Polka**™, Polka 91, Scented Dawn); flowers ivory yellow, large blooms, very dbl., 30–35 petals, borne in small clusters, intense fragrance; moderate prickles; foliage large, medium green, semi-glossy; upright, climbing, tall (12 ft) growth; PP009233; [Golden Showers X Lichtkonigin Lucia]; Meilland International SA; The Conard-Pyle Co., 1991

MEItrino, Min, mr; (**Tonic Meillandina**®); Meilland, 1994

'MEITRISICAL', Min, yb, 1982; (**Yellow Meillandina**, Yellow Sunblaze); flowers yellow, petals edged pink, dbl., 20 petals, moderate fragrance; foliage medium, dark, semi-glossy; bushy growth; [(Poppy Flash X (Charleston X Allgold)) X Gold Coin]; Meilland, Mrs. Marie-Louise; Meilland Et Cie, 1980

'MEITROGANA', HT, my, 1977; (**Vabene**); bud conical; flowers medium, dbl., 30 petals, exhibition form, slight fragrance; very vigorous, upright growth; [Arthur Bell X (MEIgold X Kabuki)]; Meilland, Mrs. Marie-Louise; Meilland

'MEITULANDI', HT, mr, 1979; (**Hidalgo**®, Michel Hidalgo); bud conical; flowers currant-red, very large, dbl., 30 petals, cupped, intense fragrance; foliage matt, bronze; vigorous, upright growth; [((Queen Elizabeth X Karl Herbst) X (Lady X X Pharaon)) X (MEIcesar X Papa Meilland)]; Meilland, Mrs. Marie-Louise; Meilland

'MEITULIMON', HT, m, 1979; (**Manou Meilland**®); bud conical; flowers mauve-pink, medium, dbl., 50 petals, cupped, slight fragrance; foliage glossy, dark; vigorous, bushy growth; Gold Star of the South Pacific, Palmerston North, NZ, 1980; [(Baronne Edmond de Rothschild X Baronne Edmond de Rothschild) X (Ma Fille X Love Song)]; Meilland, Mrs. Marie-Louise, 1980

MEItulimonsar, Cl F, dp, 1997; (**Manou Meilland, Climbing**); Meilland

MEItune, F, mp, 1995; (**Mazurka**); Meilland

'MEITURAPHAR', S, dp, 1979; (**Anne de Bretagne**®, Decor Rose, Meilland Decor Rose); bud conical; flowers deep pink, shallow-cupped, dbl., 20 petals, cupped; foliage semi-glossy; vigorous, upright growth; [(Malcair X Danse des Sylphes) X ((Zambra X Zambra) X Centenaire de Lourdes)]; Meilland, Mrs. Marie-Louise, 1976

'MEITURUSA', Min, w, 1976; (**White Gem**®); bud long; flowers soft ivory, shaded pale tan, very dbl., 90 petals, 1.5 in., slight fragrance; foliage large, glossy, dark; upright, bushy growth; [Darling Flame X Jack Frost]; Meilland; C-P

MEItylpic, HT, dy, 1999; (**Solidor**); PPAF; Meilland

MEIvahyn, S, w, 1996; (**Ice Meidiland**™); flowers white, opens flat, dbl., 25–30 petals, 2.5–3 in., pompon, borne in clusters, no fragrance; foliage medium, medium green, glossy; compact, mounding, ground cover growth; Meilland

MEIvaleir, LCl, yb, 1998; (**Michka**); Meilland

'MEIVANAMA', HT, or, 1979; (**Ondella**); bud conical; flowers vermilion, large, dbl., 33 petals; foliage dark; vigorous, upright growth; [(Elegy X Arturo Toscanini) X (Peace X Demain)]; Meilland, Mrs. Marie-Louise; Meilland

'MEIVESTAL', HT, dr, 1988; (Dreams Come True, **Senator Burda**®, Spirit of Youth, Victor Hugo®); flowers brilliant currant-red, large, dbl., 26–40 petals, intense fragrance; foliage large, medium green, semi-glossy; upright, strong, floriferous growth; [(Karl Herbst X (Royal Velvet X Suspense)) X Erotika]; Meilland, Mrs. Marie-Louise; SNC Meilland & Cie

MEIvibroum, HRg, lp, 1996; (**Pinky Rugostar**); Meilland

'MEIVILANIC', F, yb, 1981; (**Rustica**®, Stadt Basel, Ville de Bâle); bud very long; flowers yellow-peach blend, reverse buff yellow-orange, blooms borne, dbl., 35 petals, cupped, slight fragrance; foliage dark, semi-matt; half upright growth; [(Queen Elizabeth X Seedling) X Sweet Promise]; Meilland, Mrs. Marie-Louise; Meilland Et Cie

'MEIVILDO', HT, dp, 1985; (Queen Adelaide, The Royal Brompton Rose, **Yves Piaget**®); flowers deep pink, large, very dbl., globular, intense fragrance; foliage medium, dark, semi-glossy; upright growth; PP006895; Golden Rose, Geneve, 1998; [((Pharaoh X Peace) X (Chrysler Imperial X Charles Mallerin)) X Tamango]; Meilland, Mrs. Marie-Louise; Meilland Et Cie, 1983

MEIviola, HT, dr, 1998; (**Traviata**™, Xaviere); flowers dark red, very full, very dbl., 41 petals, 4–4.5 in., borne mostly singly, slight fragrance; prickles moderate; foliage large, dark green, glossy; upright, medium, 80 cm growth; [(Porta Nigra X Paolo) X William Shakespeare]; Selection Meilland; Conard-Pyle Co., 1997

'MEIVIOLIN', LCl, pb, 1987; (Eden, Eden Climber™, Eden Rose 88, Grimpant Pierre de Ronsard®, **Pierre de Ronsard**®); flowers cream white suffused with carmine pink, old rose shape, large, very dbl., 40–55 petals, 4.5–5 in., borne in clusters, repeat bloom, slight fragrance; foliage medium, deep green, semi-glossy; spreading growth, climbing habit; frost resistant; PP006892; [(Danse des Sylphes X Haender) X Pink Wonder, Climbing]; Meilland, Mrs. Marie-Louise, 1985

'MEIVIRGI', Gr, my, 1959; (**Golden Girl**); bud pointed; flowers golden yellow, dbl., 45 petals, 4–4.5 in., exhibition form, moderate fragrance; foliage leathery, light green; upright, vigorous, bushy growth; [(Joanna Hill X Eclipse) X Michele Meilland]; Meilland; C-P

MEIvostro, Min, pb, 1995; (**Carinita**); Meilland

'MEIVOUPLIX', HT, dy; (**Laura**™); Meilland

MEIvraivou, Min, dr; (**Tilt Symphonie**®); Meilland, 1994

MEIvrita, HT, mp, 1999; (**Seduction**); Meilland

MEIvrofix, F, mp; (**Zurella**™); Meilland

MEIwaton, LCl, dp, 1994; (**Gites de France**®); Meilland

'MEIWONDER', Pol, mr, 1965; bud globular; flowers small, dbl., borne in clusters, [Marianne Kluis Superior sport]; Grootendorst, F.J.

'MEIvamo', HT, dp, 1995; (**Paris de Yves St. Laurent**™); flowers deep pink, full (26–40 petals), large (7+ cms) blooms; slight fragrance; some prickles; foliage large, dark green, glossy; upright growth; [Silva sport]; Meiland, 1992; The Conard Pyle Co.

MEIxerul, Min, ab, 1991; (**Peche Meillandina**®); Meilland

MEIxetal, S, lp, 1988; (**Denise Grey**®); Meilland

MEIyacom, HT, mp, 1997; (**Claudia**); Meilland

MEIzalitaf, LCl, yb; (**Spectra**); Meilland, 1983

MEIzebul, F, mp, 1998; (**Jardins de France**); GM, Bagatelle, 1998 GM, Dublin, 1998 GM, Geneva, 1998; Meilland

'MEIzeli', HT, mp, 1995; (**The McCartney Rose**™, McCartney Rose, Paul McCartney, Sweet Lady, The MacCartney Rose); flowers medium pink, full (26–40 petals), large (7+ cms) blooms; very fragrant; foliage large, medium green, semi-glossy; bushy, medium growth; GM, Geneva, 1988; GM LeRoeix, 1988, GM, Monza, 1988; GM Paris, 1988; [(Nirvana X Papa Meilland) X First Prize]; Meilland, Alain; The Conard Pyle Co.

'MEIzoelo', Gr, yb, 1999; (**Glowing Peace**™); flowers yellow and orange blend, medium, dbl., 26–40 petals, borne in small clusters, slight fragrance; prickles moderate; foliage medium, dark green, glossy; upright, bushy, medium growth; AARS, U.S., 2001; [Sun King X Roxane]; Selection Meilland; Conard-Pyle, 2001

'MEIzogrel', Min, w, 1988; ('MEIsogrel' Spot Minijet, **White Mini-Wonder**™); flowers small, clear white, thin petals, very dbl., 40 petals, cupped, borne in small clusters, no fragrance; foliage small, medium green, matt; short, bushy growth, 12 in.; PP007276; [White Gem X Cinderella]; Meilland, Alain A.; SNC Meilland & Cie

'MEIzourayor', S, rb, 1986; (**Decor Arlequin**, **Meilland Decor Arlequin**®); flowers strawberry-red and yellow, medium, semi-dbl., 18 petals, cupped; foliage dark; very vigorous, upright growth; GM, Rome, 1975; [((Zambra X Zambra) X (Suspense X Suspense)) X Arthur Bell]; Meilland, Mrs. Marie-Louise; Meilland Et Cie, 1977

MEIzuzes, HT, rb, 1996; (**Bolcholi**, Madam Speaker); Meilland

'MEJakka', S, rb, 1997; (**Rosy Wings**); flowers semi double, large, semi-dbl., 8–15 petals, borne in large clusters, slight fragrance; foliage large, medium green, glossy; spreading, tall (6 ft) growth; [Dornroschen X Golden Wings]; Mekdeci, John

'MEJenor', HT, or, 1955; ('MEIjenor', **Soraya**®); bud pointed; flowers orange-red reverse crimson-red, large, dbl., 30 petals, cupped, slight fragrance; foliage glossy; vigorous, bushy growth; [(Peace X Floradora) X Grand'mere Jenny]; Meilland, F.; URS

'MEJenorsar', Cl HT, or, 1960; (Grimpant Soraya®, 'MEIjenorsar', **Soraya, Climbing**®); Barni, V.; URS

'MEJoria', S, ob, 1995; (**John's Rose**); flowers brilliant orange to fluorescent pink, blooms borne in small, dbl., 15–25 petals, 3–3.5 in., no fragrance; some prickles; foliage large, dark green, glossy; tall, upright growth; [[(Tropicana X Queen of the Lakes) X Seedling] X Golden Wings]; Mekdeci, John; Hortico Roses, 1995

'MEKkada', S, ob, 1997; (**Butterflies**); flowers single, medium, single, 4–7 petals, borne in large clusters, moderate fragrance; foliage medium, medium green, semi-glossy; bushy (5 ft) growth; [Dornroschen X Golden Wings]; Mekdeci, Dr Anthony Casimir

Melanie, *see* 'MACmelan'

'MELANIE', S, dr, 1946; flowers deep red, semi-dbl., non-recurrent; foliage reddish; hardy (to about -15); [R. rubrifolia X Gruss an Teplitz]; Wright, Percy H.

'MELANIE', HT, dp, 1958; bud long; flowers carmine-pink, dbl.; vigorous growth; [Sterling X Mme Auguste Chatain]; Combe; Japan Rose Society

'MELANIE DE MONTJOIE', HSem, w; flowers pure white, large; vigorous growth; Jacques

Melanie Lemaire, *see* **'HERMOSA'**

'MÉLANIE SOUPERT', T, w; [Gloire de Dijon seedling]; Nabonnand, G., 1882

'MELANIE WALDOR', M, ab, 1865; Moreau et Robert

'MELBA', F, yb, 1963; flowers soft cream and peach, well formed, small, borne in clusters, dbl.; foliage light green; spreading growth; [Masquerade X Independence seedling]; Sanday, John

Meletta, Min, lp; flowers light orange, exhibition form, Giles, Diann

Melflor, F, lp, 1988; RvS-Melle

'MELGLORY', F, mr, 1982; flowers large blooms in clusters of 4–16, dbl., 23 petals, no fragrance; foliage dark, matt; upright growth; [Lilli Marleen X Patricia]; Rijksstation Voor Sierplantenteelt

'MELGOLD', HT, dy, 1980; flowers large blooms in cluster of 1–6, dbl., 62 petals, cupped, no fragrance; prickles reddish-green; foliage dark, matt; upright growth; [Sunblest X Souv. de Jacques Verschuren]; Rijksstation Voor Sierplantenteelt

Melica, HT, lp; Hicl, 1991; (Czech Rosa Club)

Melik El Adel, *see* **'MALEK-ADEL'**

Meli-Melo, *see* ORAstrip

Melina, *see* 'TANema'

Melinda, *see* 'RUlimpa'

Melinda Claire, *see* 'TAYmel'

Melinda Gainsford, *see* 'JACyap'

'MELINDA MARIE', F, lp, 1972; bud ovoid; flowers clear light pink, dbl., 20–25 petals, 3 in., flat, slight fragrance; foliage dark, leathery; bushy growth; [Sarabande X Sarabande]; Linscott

'MÉLISANDE', HT, lp, 1964; bud long; flowers soft pink, reverse darker, well formed, large, dbl., moderate fragrance; foliage dark; long, strong stems; vigorous, symmetrical growth; Mondial Roses

'MELISSA', HT, mp, 1975; dbl., 27–32 petals, 4 in., globular, slight fragrance; foliage leathery; very vigorous, upright, bushy growth; PP004082; [Seedling 1968–1 X Seedling 167-d]; McDaniel, Earl; Carlton Rose Nurseries

Melissa, F, lp, 1996; Noack, Werner

Melissa Joyce, *see* 'RENhom'

'MELISSA MCCARTNEY', F, dp, 1991; bud ovoid; flowers holds color well, does not fade, blooms borne usually singly, very dbl., 36 petals, 5 in., exhibition form, slight fragrance; foliage medium, dark green, glossy, slight red tinge to new; bushy, low (70 cms) growth; [Cherish sport]; Troyer, Ray & Pat

'MELITA', HWich, mp, 1934; flowers carnation-pink, large, dbl.; foliage glossy, light; vigorous, climbing growth; GM, NRS, 1933; [Thelma sport]; Easlea

'MELLE FISCHER', Pol, mp, 1914; flowers small, dbl.; Pfitzer; (Sangerhausen)

Mellow, HT, ab, 1995; Dawson

'MELLOW GLOW', HT, pb, 1989; bud ovoid; flowers medium pink, reverse creamy yellow, medium, borne usually singly, dbl., 52 petals, exhibition form, intense, damask fragrance; prickles pointed slightly downward, large, yellow; foliage medium, medium green, semi-glossy; bushy growth; [Thriller X Wild Cherry]; Bridges, Dennis A.; Bridges Roses, 1990

'MELLOW YELLOW', HT, my, 1968; flowers sunflower-yellow, edged pink, urn shaped; free growth; [Piccadilly sport]; Waterhouse Nursery

Melmore Terrace, HSem, w

Melodie®, F, mr, 1980; Noack, Werner

Melodie Parfumee™, *see* 'DORient'

'MELODY', HT, pb, 1946; bud urn-shaped; flowers deep pink, edged lighter, large, dbl., 35 petals, moderate fragrance; low, bushy growth; [Joanna Hill X Miss C.E.

van Rossem]; Lammerts, Dr Walter; Armstrong Nursery

Melody, *see* KORbolak

Melody Lane, *see* 'MICLANE'

Melody Maker, *see* 'DICQUEEN'

Melody Marshall, *see* 'MORMELMA'

Melody Parfumee™, *see* 'DORIENT'

Melody Queen, HT, pb, 1998; Ghosh

'Melo-melo-day', Cl F, mp, 1955; bud ovoid; flowers cameo-pink, dbl., 40 petals, 2–2.5 in., cupped, moderate fragrance; climbing (10–15 ft) growth; [Demure sport]; Motose

Melonda, HT, ob, 1974; flowers yellowish salmon-orange, very large, dbl., slight fragrance; J&P; (Sangerhausen)

Melpink, F, mp

'Melrose', HT, rb, 1963; flowers creamy white flushed cherry-red, dbl., 35 petals, 3.5 in., moderate fragrance; foliage dark, leathery; vigorous, bushy growth; [Silver Lining X E.G. Hill]; Dickson, A.

Melrose, F, dp, 1990; 24 petals, 3 in.; GM, Durbanville, 1994 GM, Geneva, 1992; RvS-Melle, 1992

'Melvena', HT, op, 1971; bud ovoid; flowers salmon-pink, full, medium, dbl., moderate fragrance; foliage leathery; vigorous, upright growth; [Daily Sketch X Impeccable]; Dawson, George

'Melvin', F, rb, 1980; bud pointed; flowers ivory, flushed red, aging darker red, blooms borne 1–11 per, dbl., 30 petals, exhibition form, slight fragrance; triangular, hooked prickles; foliage medium green; medium, dense growth; Bronze, ARC TG, 1980; [Unnamed seedling X Unnamed seedling]; Jerabek, Paul E.

'Mémé Buy', HT, or, 1935; flowers coppery coral-red, lightly streaked golden yellow, large, very dbl., cupped; foliage bronze; very vigorous growth; Chambard, C.

'Mémée Arles', HT, dr, 1955; flowers deep red tinted vermilion; foliage glaucous green; [Peace X Emma Wright]; Arles; Roses-France

'Mémée Azy', HT, pb, 1921; flowers pink shaded carmine, bordered whitish, stamens orange-yellow, dbl., slight fragrance; [Étoile de France X Le Progres]; Gillot, F.

'Mémée Chanteur', F, dr, 1959; bud ovoid; flowers deep crimson, large, borne in clusters, dbl., cupped, slight fragrance; foliage dark, leathery; vigorous, upright growth; [Karl Herbst X Pioupiou]; Arles; Roses-France

Memento®, *see* 'DICBAR'

Memoire®, *see* KORzuri

Memorial Day Rose, Misc OGR, mp

Memorial Rose, *see* **R. wichurana**

'Memoriam', HT, lp, 1961; bud long, pointed; flowers pastel pink to nearly white, dbl., 55 petals, 6 in., exhibition form, moderate fragrance; foliage dark, leathery; moderately tall growth; GM, Portland, 1960; [(Blanche Mallerin X Peace) X (Peace X Frau Karl Druschki)]; Von Abrams; Peterson & Dering

'Memories', HT, dy, 1977; bud short, pointed, ovoid; flowers empire-yellow, full, small, dbl., 25–30 petals, 2.5–3 in., exhibition form, continuous bloom in greenhouse, intense, tea fragrance; foliage large, glossy; vigorous, upright, bushy growth; [Spanish Sun X Hoosier Gold]; Byrum; J.H. Hill Co.

Memories, *see* CLEjoy

'Memory', HT, pb, 1932; bud pointed; flowers light pink to deeper pink, base yellow, very large, dbl., exhibition form, intense fragrance; foliage rich green, leathery; vigorous, compact growth; GM, NRS, 1932; Cant, B. R.; J&P

Memory Bells, *see* 'POULMA'

'Memory Lane', Min, lp, 1973; bud ovoid; flowers rose-pink, small, very dbl., slight fragrance; foliage leathery; vigorous, dwarf, bushy growth; [(Pinocchio X William Lobb) X Little Chief]; Moore, Ralph S.; Sequoia Nursery

Memory of D. M. Roy, HT, w, 1998; Ghosh

Memory Lane, *see* 'PEAVOODOO'

Memphis, HT, dp; Select Roses, B.V.

'Memphis Heritage', Gr, pb, 1969; bud long, pointed; flowers pink blended with gold, large, dbl., exhibition form, moderate fragrance; very vigorous, compact growth; [Queen Elizabeth X Happiness]; Patterson; Patterson Roses

Memphis Queen, *see* 'WELQUEEN'

'Ménage', A, w, 1847; flowers flesh, medium, dbl., cupped; Vibert

Menaka Durga Roy, *see* 'DRISCOROY'

'MENAP', HT, my, 1955; (**Belle Blonde**); flowers yellow, center darker, well-formed, moderate fragrance; foliage glossy; bushy growth; [Peace X Lorraine]; Meilland, F.; URS

'Mendel', F, mr, 1946; bud pointed; flowers cherry-red, borne in trusses, semi-dbl., 15 petals, 4 in., moderate fragrance; foliage reddish green; vigorous, bushy growth; [Florentina X Unnamed seedling]; Leenders, M.; Longley

Mendocino Delight, HWich, w; Demits

Mendocino Gothic, HMsk, w; Demits, 1993

Mendocino Headlands, HWich, lp; (found rose)

Mendocino Pale Pink China, HCh, lp

Mendocino Smiles, HWich, pb; (found rose)

Menja, HMsk, mp, 1960; Petersen

'Mennie d'Agnin', F, mr, 1962; flowers vermilion-red, large; [Independence X Fashion]; Orard, Joseph

'Mentor', S, mp, 1959; bud globular; flowers light pink, reverse darker, medium, dbl., recurrent bloom, intense fragrance; foliage dark, glossy; very vigorous (6 ft), bushy, upright growth; [Tallyho X New Dawn]; Wyant

Menuett, HT, yb, 1988; flowers dark yellow with carmine pink, large, dbl.; GPG Bad Langensalza; (Sangerhausen)

Menuett, HT, ob, 1961; flowers orange and yellow, large, dbl., slight fragrance; Hill, S. H.; (Sangerhausen)

'Menut', Min, dp, 1956; flowers carmine, small, semi-dbl., 20 petals, abundant, intermittent bloom; dwarf, bushy, compact growth; [Rouletii X Perla de Alcanada]; Dot, Simon

MEpal, HT, op, 1953; (**Bettina**®, 'MEIPAL'); flowers salmon-orange, veined, well-formed, dbl., 37 petals, 4 in., moderate fragrance; foliage dark, glossy, bronze; vigorous growth; [Peace X (Mme Joseph Perraud X Demain)]; Meilland, F.; URS

'MEPALSAR', Cl HT, op, 1958; (**Bettina, Climbing**®, Grimpant Bettina); GM, Geneva, 1959; Meilland, F.; URS

Mephisto, Pol, or, 1985; flowers large, semi-dbl.; GPG Bad Langensalza; (Sangerhausen)

'Méphisto', F, or, 1951; flowers geranium-red, medium, borne in clusters, semi-dbl.; foliage leathery; vigorous, upright, bushy growth; [Francais X Unnamed seedling]; Mallerin, C.; EFR

'Mercator', HT, dr, 1962; flowers deep red, large, dbl., intense fragrance; foliage dark; [Chrysler Imperial X Tango]; Delforge

Mercedes®, *see* 'MERKOR'

'Mercedes', HGal, lp, 1847; flowers white and lilac, changing to pale pink, large, dbl., Vibert

'Mercedes Gallart', Cl HT, dp, 1932; flowers deep pink, base yellow, very large, dbl., recurrent bloom, intense fragrance; foliage glossy; very vigorous, climbing growth; [Souv. de Claudius Denoyel X Souv. de Claudius Pernet]; Munné, B.; J&P

'Mercedes Juncadella', HT, op, 1933; flowers salmon-orange, [Frau Karl Druschki X Angèle Pernet]; Munné, B.

'Mercedes Mendoza', HT, or, 1962; flowers large, dbl., 30 petals, foliage glossy; vigorous growth; [Asturias X Grand'mere Jenny]; Dot, Simon

Mercedes, Climbing, Cl HT, lp; [Mercedes sport]; Ruston, D., 1982; (Weatherly, L.)

'Merci', F, mr, 1974; bud ovoid, long, pointed; flowers medium, dbl., slight fragrance; foliage dark, leathery; vigorous growth; Warriner, William A.; J&P

'MERCURIUS', LCl, op, 1940; flowers light coral-pink, large, borne in clusters, semi-dbl., cupped, profuse seasonal bloom; foliage glossy; vigorous, climbing growth; [(Doubloons X Damask) X Clio]; Horvath; Wayside Gardens Co.

'MERCURY', F, or, 1967; flowers orange-scarlet edged mahogany, borne in clusters, foliage dark; low, bushy growth; [Independence X Paprika]; Sanday, John

Mercy, *see* SIMmer

Mère de la Patrie, *see* **'MOEDER DES VADERLANDS'**

'MEREDITH', LCl, lp, 1984; flowers light pink to almost white, large blooms, dbl., 20 petals, exhibition form, repeat blooming, moderate fragrance; foliage large, dark, semi-glossy; long stems; upright (to 12 ft) growth; [(Charlotte Armstong X New Dawn) X Araby]; Thomson, Richard

Meredith Anne, *see* 'TINMERE'

'MEREDITH HUGHES', Min, op, 1984; flowers medium coral pink, medium, dbl., exhibition form, slight fragrance; foliage medium, medium green, glossy; bushy growth; [Anne Scranton X Patricia Scranton]; Dobbs, Annette E.

Meridian, HT, mr; flowers luminous red, very large, dbl.; VEG; (Sangerhausen)

Meridiana®, *see* BARmeri

Meriggio, *see* BARafne

Merindah, HMult, mp, 1994; Sutherland

Merindah Red, HMult, mr, 1995; Sutherland

'MERIT', F, mp, 1952; bud short, pointed; flowers brilliant rose, reverse lighter, base white, borne in clusters, dbl., 50–90 petals, 1.5–2 in., flat, intense fragrance; foliage leathery, dark; very vigorous, bushy growth; [Garnette sport]; Domilla; Twin Nursery

Merit, *see* 'SPOMERIT'

Merko, *see* 'MERKOR'

'MERKOR', F, or, 1974; (**Mercedes**®, Merko); bud ovoid; flowers bright scarlet, dbl., 33 petals, exhibition form, slight fragrance; foliage large, leathery; [Anabell X Seedling]; Kordes, R.; J&P, 1975

Merle Blanc, *see* 'LENISUR'

'MERLIN', F, pb, 1967; flowers yellow, pink and red, blooms in clusters, dbl., 2.5 in., slight fragrance; foliage glossy; [Pink Parfait X Circus]; Harkness

Merlin, S, lp, 1991; Poulsen

'MERMAID', HBc, ly, 1918; flowers creamy yellow, amber stamens, single, 5 petals, 5–6 in., dependably recurrent, moderate fragrance; foliage dark, glossy; vigorous, climbing, pillar or trailer (6–9 ft) growth; tender in cold regions; (14); GM, NRS, 1917; [R. bracteata X Double yellow Tea rose]; Paul, W.

Merrie, *see* 'JAYMER'

'MERRIE MISS', HT, mp, 1966; bud long, ovoid; flowers soft rose-pink, dbl., 60 petals, 4 in., exhibition form, intense fragrance; foliage dark, matt; upright, bushy growth; [Pink Favorite X Margaret]; Fuller; Wyant

Merrimac™, *see* 'KINMAC'

'MERRIMENT', Min, yb, 1986; flowers bright yellow tipped red, reverse yellow edged red, 8 petals, slight fragrance; medium, long, pink prickles; foliage large, medium green, semi-glossy; tall upright growth; [Rise 'n' Shine X ?]; Bridges, Dennis A.

Merrouw Daendels, *see* **'MRS HENRI DAENDELS'**

'MERRY CHRISTMAS', Min, mr, 1977; bud pointed; flowers currant-red, open, single, 5 petals, 2 in., slight fragrance; foliage dark; upright, compact, branching growth; [Red Can Can X ?]; Lyon; L. Lyon Greenhouses

'MERRY ENGLAND', HP, mr, 1897; flowers satiny light red, shaded carmine, dbl., 50 petals, intense fragrance; vigorous growth; Harkness

Merry Go Round, F, w

'MERRY HEART', Gr, rb, 1960; bud ovoid; flowers orient red, borne in small clusters, dbl., 30 petals, 3.5–4 in., exhibition form, slight fragrance; foliage glossy, dark; vigorous, upright growth; [El Capitan X Seedling]; Swim, H.C.; C.R. Burr

'MERRY WIDOW', Gr, dr, 1958; bud long, pointed; flowers velvety crimson, dbl., 23 petals, 6 in., cupped, intense, spicy fragrance; foliage dark, glossy; vigorous growth; [Mirandy X Grande Duchesse Charlotte]; Lammerts, Dr Walter; Germain's

Merryglo™, *see* 'MINIMERR'

'MERRY-GO-ROUND', HT, op, 1950; bud pointed, becoming urn shaped, orange; flowers orange and pink blend, dbl., 25 petals, 5 in.; foliage dark, leathery; tall, compact growth; [Talisman X R.M.S. Queen Mary]; Fisher, G.; Arnold-Fisher Co.

'MERRYWEATHER'S CRIMSON', HT, mr, 1958; flowers bright crimson, well formed, moderate fragrance; strong stems; vigorous, bushy growth; Merryweather

Mers du Sud, *see* **'SOUTH SEAS'**

Merveille de la Brie, HWich, mr

'MERVEILLE DE LYON', HP, w, 1882; flowers pure white, tinted satiny rose, dbl., 4 in., cupped, some recurrent bloom, vigorous growth; [Baroness Rothschild sport]; Pernet Père

'MERVEILLE DES BLANCHES', HP, w, 1894; flowers medium-large, dbl., moderate fragrance; Pernet; (Sangerhausen)

'MERVEILLE DES JAUNES', Pol, yb, 1920; flowers bright coppery golden yellow, borne in clusters, dbl.; dwarf growth; Turbat

'MERVEILLE DES ROUGES', Pol, rb, 1911; flowers deep velvety crimson, center whitish, borne in large clusters, semi-dbl., cupped; dwarf growth; Dubreuil

Merveilleuse, HT; Dot, Simon; (Cavriglia)

Mervelle, HT, 1995; McGredy, Sam IV; (Cavriglia)

'MERYL JANE GASKIN', HT, pb, 1948; flowers shell-pink edged deeper pink, well formed, dbl., 48 petals, 5–6 in., intense fragrance; free, branching growth; [Rose Berkeley sport]; Mee; Fryer's Nursery, Ltd.

Message, *see* 'MEBAN'

Message, Climbing, *see* **'WHITE KNIGHT, CLIMBING'**

Messagere, *see* 'HERTIE'

Messara, *see* SAUdive

'MESSESTADT HANNOVER', F, mr, 1962; (Hannover); flowers large blooms in clusters, dbl.; foliage light green, glossy; moderate growth; Kordes, R.

Messidor, HT, dy

'MESSIRE', LCl, mr, 1963; flowers bright red, borne in clusters of 6–7, semi-dbl., 10–15 petals, 3 in., profuse, repeated bloom, foliage bronze; moderately vigorous growth; [Unnamed seedling X Spectacular]; Laperrière; EFR

Messire Delbard®, *see* 'DELSIRE'

Metanoia®, LCl, ob

'METEOR'®, F, or, 1959; flowers orange-scarlet, patio, blooms in clusters (up to 10), dbl., 40 petals, 3 in., cupped; foliage light green; vigorous, bushy, low growth; ADR, 1960; [Feurio X Gertrud Westphal]; Kordes, R.; A. Dickson; McGredy

'METEOR', HT, mr, 1887; flowers crimson-carmine, open, dbl., recurrent bloom, slight fragrance; few thorns; foliage small, soft; dwarf growth; Bennett

'METEOR', N, dp, 1887; flowers deep rose tinted carmine-purple, large, dbl., moderate fragrance; vigorous growth; Geschwind, R.

Meteor, B, dp

Meteor Shower, *see* JACpan

'METEOR, CLIMBING', Cl HT, mr, 1901; Dingee & Conard

'METIS', S, mp, 1967; (Simonet); bud ovoid; flowers soft rose, dbl., 35 petals, 2.5–3 in., flat, spring bloom, slight fragrance; foliage small, glossy; [R. nitida X Therese Bugnet]; Harp; Morden Exp. Farm

Métro, *see* 'MACBUCPAL'

'METROPOLE', HT, mp, 1961; dbl., 45 petals, 4–6 in., globular, moderate fragrance; foliage matt, green; vigorous growth;

[Sidney Peabody X Peace]; deRuiter; Blaby Rose Gardens

Metropolitain, *see* 'MACNGAURU'

METset, F, m; (**Cristian**); (METset)

'**MEVROUW A. DEL COURT VAN KRIMPEN**', HT, lp, 1917; flowers flesh-white and pale pink, tinted copper, dbl., moderate fragrance; [Unnamed seedling X Prince de Bulgarie]; Leenders, M.

'**MEVROUW A. H. DE BEAUFORT**', HT, pb, 1934; flowers clear salmon-pink, large, dbl., moderate fragrance; foliage bronze, glossy; vigorous growth; [Morgenglans X Gooiland Beauty]; Van Rossem

'**MEVROUW AMÉLIE MÜLLER**', HT, yb, 1927; flowers old-gold, shaded orange, dbl., moderate fragrance; [Golden Ophelia X Golden Emblem]; Verschuren

'**MEVROUW BOREEL VAN HOGELANDEN**', HT, w, 1918; flowers flesh-white shaded carmine and pink, dbl., intense fragrance; [Mme Leon Pain X Mme Antoine Mari]; Leenders, M.

'**MEVROUW C. VAN MARWIJK KOOY**', HT, w, 1921; flowers white, center indian yellow, sometimes coppery orange, dbl., moderate fragrance; [Mme Caroline Testout X Mrs Aaron Ward]; Leenders, M.

'**MEVROUW D. A. KOSTER**', Pol, mr, 1934; flowers bright red; [Dick Koster sport]; Koster, D.A.

'**MEVROUW DORA VAN TETS**', HT, mr, 1913; flowers velvety deep crimson, dbl., moderate fragrance; GM, Bagatelle, 1914; [Farbenkonigin X Gen. MacArthur]; Leenders, M.

'**MEVROUW DR L. CROBACH**', HT, pb, 1928; flowers carmine, base salmon, dbl., moderate fragrance; [Pink Pearl X Red Star]; Leenders, M.

'**MEVROUW G. A. VAN ROSSEM**', HT, ob, 1929; (Mrs G.A. van Rossem); flowers orange and apricot on golden yellow, veined red, reverse often dark bronze, dbl., intense fragrance; foliage very large, dark, bronze, leathery; vigorous growth; (28); [Souv. de Claudius Pernet X Gorgeous]; Van Rossem; C-P

'**MEVROUW G. A. VAN ROSSEM, CLIMBING**', Cl HT, ob, 1937; (Mrs G.A. van Rossem, Climbing); Gaujard

'**MEVROUW G. DE JONGE VAN ZWYNSBERGEN**', HT, pb, 1923; flowers pale flesh, center flesh-pink and salmon, dbl.; [Mme Mélanie Soupert X George C. Waud]; Leenders, M.

'**MEVROUW H. CREMER**', HT, mr, 1932; flowers large, dbl., intense fragrance; Buisman, G. A. H.; (Sangerhausen)

'**MEVROUW HENRI DAENDELS**', HT, op, 1931; flowers light salmon, large, very dbl., intense fragrance; Buisman, G. A. H.; (Sangerhausen)

'**MEVROUW L. C. VAN GENDT**', HT, ab, 1925; flowers salmon-apricot on yellow ground, dbl., moderate fragrance; [Unnamed seedling X Golden Emblem]; Van Rossem

'**MEVROUW LALA PHILIPS**', HT, or, 1931; flowers brilliant orange-toned shrimp-red, well formed, large, dbl.; foliage dark; vigorous growth; [Elvira Aramayo sport]; Leenders Bros.

'**MEVROUW NATHALIE NYPELS**', Pol, mp, 1919; (Nathalie Nypels); flowers rose-pink, medium, semi-dbl., intense fragrance; dwarf, spreading growth; (14); [Orléans Rose X (Comtesse du Cayla X R. foetida bicolor)]; Leenders, M.

Mevrouw S. van den Bergh, Jr., *see* '**MRS VANDENBERGH**'

'**MEVROUW SMITS GOMPERTZ**', HT, op, 1917; flowers yellowish salmon and coppery orange shaded lilac, dbl., intense fragrance; [Lady Wenlock X (Mme J.W. Budde X Souv. de Catherine Guillot)]; Leenders, M.

Mevrouw van Straaten van Nes, *see* '**PERMANENT WAVE**'

'**MEVROUW WELMOET VAN HEEK**', HT, dp, 1933; flowers carmine-red, well formed; foliage dark; vigorous, bushy growth; Buisman, G. A. H.

'**MEXICALI ROSE**', F, yb, 1957; bud short, pointed, yellow suffused red; flowers deep yellow turning deep rose-pink, then cerise-red, open, dbl., 65–70 petals, 3 in., slight spicy fragrance; foliage dark; vigorous, upright growth; [(Herrenhausen seedling X Golden Rapture) X Easter Parade]; Whisler; Germain's

Mexican Festival, S, dp, 1996; Dickson, Patrick

'**MEXICANA**', HT, rb, 1966; bud ovoid; flowers red, reverse silvery, large, dbl., 33 petals, exhibition form, moderate fragrance; foliage dark, glossy, leathery; vigorous, upright growth; [Kordes' Perfecta X Unnamed seedling]; Boerner; J&P

'**MEXICO**', F, or, 1944; bud globular, deep carmine; flowers deep scarlet, suffused orange, borne in clusters, semi-dbl., 18 petals, 4 in., cupped, slight fragrance; foliage leathery; strong stems; vigorous, bushy growth; [Baby Chateau X Helgoland]; Krause; J&P

Mexico, Min, rb

'MEYPINK', HT, pb, 1971; (**Portrait**, Stéphanie de Monaco); bud ovoid; flowers medium pink and light pink blend, medium, dbl., moderate fragrance; foliage glossy, dark; upright, bushy growth; AARS, (First amateur winner), 1972; [Pink Parfait X Pink Peace]; Meyer, C.; C-P

'**MHAIRI'S WEDDING**', HT, w, 1983; dbl., 35 petals, flat, moderate fragrance; foliage large, dark, matt; upright growth; [Dalvey X Virgo]; MacLeod, Major C.A.; Christie Nursery, Ltd.

Mia, HT, lp; Urban, J., 1975; (Czech Rosa Club)

'**MIA MAID**', HT, mp, 1953; bud ovoid; flowers phlox-pink, open, dbl., 40–50 petals, 3.5–4 in., moderate fragrance; foliage leathery, glossy; upright, compact growth; [Charlotte Armstrong X Signora]; Swim, H.C.; Mt. Arbor Nursery

'**MIA SNOCK**', HT, ly, 1925; flowers lemon-yellow, dbl.; [Mrs T. Hillas X Mrs Wemyss Quin]; Leenders, M.

'**MIAMI**', HT, yb, 1953; flowers orange, veined, reverse yellow, long, pointed, dbl., 25 petals, 5 in., slight fragrance; foliage dark; very vigorous, branching growth; [Mme Joseph Perraud X Fred Edmunds]; Meilland, F.; C-P

'**MIAMI HOLIDAY**', Min, rb, 1976; bud pointed; flowers red, reverse yellow, dbl., 60 petals, 1–1.5 in., moderate fragrance; foliage small, glossy; upright, bushy growth; [Seedling X Over the Rainbow]; Williams, Ernest D.; Mini-Roses

Miami Playa, HT; Dot; (Cavriglia)

'MICADA', Min, pb, 1992; (**Orlando Sunshine**); flowers pink and yellow blend, blooms borne mostly singly, dbl., 15–25 petals, 1.5–2.75 in., slight fragrance; some prickles; foliage large, medium green, semi-glossy; medium (18–24 in.), bushy growth; [Unnamed Miniature seedling X Unnamed Miniature seedling]; Williams, Michael C.; The Mini Rose Garden

'**MICAELA**', M, lp, 1864; flowers blush, compact, medium, very dbl.; erect growth; Moreau et Robert

'**MICAËLA**', HT, or, 1986; (Trudor); flowers orange-red, fading to dark carmine red, blooms in clusters, vigorous growth; [Manola X Unnamed seedling]; Kriloff, Michel

'MICAM', Min, rb, 1989; (**Carolina Morning**, 'MICAR'); bud pointed; flowers red with yellow center, aging red, small, borne singly and in small clusters, dbl., 20 petals, exhibition form, no fragrance; round, orange fruit; prickles straight, small, red; foliage small, medium green, semi-glossy; bushy, medium growth; [Rise 'n' Shine X Rainbow's End]; Williams, Michael C.; The Rose Garden & Mini Rose Nursery, 1990

'MICANGEL', Min, ab, 1996; (**Angel's Blush**); flowers apricot, darkest color is on the very edge of the petals, bl, dbl., 15–25 petals, 1.5–2.75 in., slight fragrance; few prickles; foliage small, medium green, semi-glossy; medium (50 cms), upright growth; [Seedling X Selected Pollen]; William, Michael C.; The Mini Rose Garden, 1997

'MICAR', Min, rb, 1989; (**Carolina Morning**, 'MICAM'); bud pointed; flowers red with yellow center, aging red, small, borne singly and in small clusters, dbl., 20 petals, exhibition form, no fragrance;

round, orange fruit; prickles straight, small, red; foliage small, medium green, semi-glossy; bushy, medium growth; [Rise 'n' Shine X Rainbow's End]; Williams, Michael C.; The Rose Garden & Mini Rose Nursery, 1990

'MICAROLINE', Min, rb, 1998; (**Sweet Caroline**); flowers white with light to medium red on outer petals, dbl., 17–25 petals, 1.5 in., exhibition form, borne mostly singly, no fragrance; prickles moderate, small; foliage medium, dark green, semi-glossy; tall, 36 in., slightly spreading growth; [Seedling X Select Pollen]; Williams, Michael C.; The Mini Rose Garden, 1999

'MICAUTUMN', MinFl, yb, 1999; (**Autumn Splendor**); exhibition form, 26–40 petals, 2 in., exhibition form, borne mostly singly, slight; foliage large, medium green, semi-glossy; upright, tall (30–40 in.) growth; Williams, Michael C.; The Mini Rose Garden, 1999

'MICCOUNTRY', Min, pb, 1996; (**Old Country Charm**); flowers medium to dark pink with salmon, blooms borne mostly singly, dbl., 26–40 petals, no fragrance; foliage medium, dark green, dull; upright, medium growth; [Seedling X Selected pollen]; Williams, Michael C.; The Mini Rose Garden, 1996

'MICDEB', Min, dp, 1991; (**Debidue**™); flowers magenta, blooms borne mostly singly, dbl., 26–40 petals, 1.5 in., slight fragrance; some prickles; foliage medium, dark green, semi-glossy; medium (12 cms), upright growth; AOE, 1992; [Jazz Fest X Party Girl]; Williams, Michael C.; The Rose Garden & Mini Rose Nursery, 1992

'MICEVEN', Min, pb, 1989; (**Evening Shadows**); bud pointed; flowers medium, borne usually singly, dbl., 20 petals, exhibition form, slight, fruity fragrance; globular, light orange fruit; prickles slight downward curve, small, red; foliage large, medium green, semi-glossy; bushy, medium growth; [Tiki X Party Girl]; Williams, Michael C.; The Rose Garden & Mini Rose Nursery, 1990

'MICEXPLORE', Min, op, 1992; (**Explorer's Dream**); flowers deep orange pink, just a touch of yellow at base of each petal, dbl., 15–25 petals, 1.5 in., no fragrance; some prickles; foliage medium, dark green, semi-glossy; medium (50 cms), upright growth; [Miniature seedling X Homecoming]; Williams, Michael C.; The Rose Garden & Mini Rose Nursery

'MICFINEST', Min, ob, 1996; (**Finest Hour**); flowers very bright orange with yellow at base, creamy orange yellow, very dbl., no fragrance; some prickles; foliage medium, medium green, semi-glossy; upright, medium growth; [Glowry X Selected pollen]; Williams, Michael C.; The Mini Rose Garden, 1996

'MICGOLD', Min, my, 1986; (**Lander Gold**); dbl., 36 petals, exhibition form, borne usually singly, moderate, fruity fragrance; no fruit; very few, small, straight prickles; foliage medium, medium green, semi-glossy; medium, bushy growth; [Rise 'n' Shine X Little Jackie]; Williams, Michael C.; The Rose Garden & Mini Rose Nursery

Michael Bates' Moss, M, mp

Michael Crawford, *see* POUlvue

'**MICHAEL JON HALVORSON**', HT, lp, 1999; dbl., 17–25 petals, 3 in., borne in small clusters, slight fragrance; prickles moderate; foliage medium, medium green, semi-glossy; spreading, tall (6 ft) growth; Adams, Dr Neil D.

'**MICHAEL LEEK**', F, or, 1995; flowers bright orange-red, blooms borne 2–8 per cluster, semi-dbl., 5–8 petals, 1.5–2.75 in., slight fragrance; foliage medium, dark green, glossy; bushy (75–90 cms) growth; [Liverpool Echo X Traumerei]; Fleming, Joyce L.; Hortico Roses, 1994

'**MICHAEL SAUNDERS**', HT, pb, 1879; flowers deep pink and coppery red, well formed, medium, moderate fragrance; moderate growth; [Adam X Mme Victor Verdier]; Bennett

Michaelhouse Centenary, *see* MEIonagre

'**MICHEL BONNET**', B, pb, 1864; flowers bright rose, vigorous growth; Guillot Père

Michel Hidalgo, *see* 'MEITULANDI'

Michel Lis le Jardinier®, *see* MEIfersi

Michelangelo, *see* 'MACTEMAIK'

Michelangelo, *see* MEItelov

'**MICHÈLE**', F, op, 1970; flowers deep salmon-pink, large, dbl., 25 petals, moderate fragrance; foliage light green; bushy growth; [Seedling X Orange Sensation]; deRuiter

Michele Bross, F, 1975; Dorieux; (Cavriglia)

'**MICHÈLE MEILLAND**', HT, lp, 1945; flowers light pink shaded lilac, center salmon, large, dbl.; vigorous growth; [Joanna Hill X Peace]; Meilland, F.

'**MICHÈLE MEILLAND, CLIMBING**', Cl HT, lp, 1951; (Grimpant Michèle Meilland); Meilland, F.

Michèle Torr®, *see* 'JACOLITE'

'**MICHELINE**', HT, dp, 1953; flowers deep salmon-pink, base darker, dbl., 30–35 petals; foliage dark; [Edouard Renard X Luis Brinas]; de Basso, Mata

Michelle, F, rb, 1998

Michelle Joy, *see* AROshrel

Michelle Wright, F, ab, 1998

'**MICHIGAN**', HT, pb, 1948; flowers salmon-carmine, reverse indian yellow, very large, dbl.; bushy growth; [Mme Joseph Perraud X Vive la France]; Mallerin, C.; URS

Michka, *see* MEIvaleir

'MICHOME', Min, mp, 1989; (Homecoming, **Home-Coming**); bud pointed; flowers medium pink with slightly darker petal edges, medium, dbl., 35 petals, exhibition form, no fragrance; no fruit; prickles straight, green; foliage medium, medium green, semi-glossy; upright, tall growth; PP007454; [Tiki X Party Girl]; Williams, Michael C.; The Rose Garden & Mini Rose Nursery

'MICIMP', Min, pb, 1998; (**Lasting Impression**); flowers medium pink with darker pink edges, dbl., 17–25 petals, 1.5 in., borne mostly singly, no fragrance; no prickles; foliage medium, dark green, semi-glossy; compact, medium, 20 in. growth; [Seedling X Select Pollen]; Williams, Michael C.; The Mini Rose Garden, 1998

Mick Micheyl, Gr; (Cavriglia)

'MICKATE', F, mp, 1992; (**Katelyn Ann**); flowers deep pink, reverse pale pink with hint of yellow, aging pale, very dbl., 60 petals, 2.5 in., cupped, intense fragrance; foliage medium, dark green, glossy; medium, upright growth; [Unknown X Unknown]; Williams, Michael C., 1987; Roses Unlimited, 1992

'**MICKI**', HT, or, 1982; flowers light tangerine, dbl., 36 petals, exhibition form, moderate fragrance; small, brown prickles; foliage dark, glossy, leathery; vigorous, upright growth; [Fragrant Cloud X Tropicana]; James, John

'MICKIM', Min, ab, 1986; (**Kimberly**); flowers urn-shaped, blooms borne usually singly, dbl., 24 petals, exhibition form, moderate, fruity fragrance; globular, small, orange fruit; no prickles; foliage medium, medium green, semi-glossy; upright, medium growth; [Party Girl X Sheri Anne]; Williams, Michael C.; The Rose Garden & Mini Rose Nursery

'**MICKY**', HT, op, 1951; bud long, pointed; flowers coral-pink, single, 9 petals, slight fragrance; few thorns; foliage leathery; very vigorous, upright, bushy growth; [Lulu X Vesuvius]; Houghton, D.; Elmer Roses Co.

Micky, Cl Min, mr, 1995; Warner, Chris

'MICLANE', Min, m, 1991; (**Melody Lane**); dbl., 26–40 petals, 1.5 in., borne in small clusters, no fragrance; foliage small, medium green, semi-glossy; low (6 cms) growth; [Lavender Jewel X Party Girl]; Williams, Michael C.; The Rose Garden & Mini Rose Nursery, 1992

'MICLEE', Min, yb, 1991; (**Brenda Lee**); bud pointed; flowers red edge to yellow base, urn-shaped, small, dbl., 20 petals, exhibition form; foliage small, medium green, semi-glossy; bushy, low, compact growth; [Rise 'n' Shine X Rainbow's End]; Williams, Michael C., 1988; The Rose Garden & Mini Rose Nursery

'MICLEE', Min, yb, 1989; (**Virginia Lee**); bud ovoid; flowers creamy yellow with pink border, reverse creamy yellow, aging, dbl., 30 petals, cupped, no fragrance; globular, green orange-yellow fruit; prickles straight, green; foliage medium, dark green, glossy; slightly spreading, medium growth; [Rise 'n' Shine X Baby Katie]; Williams, Michael C.; The Rose Garden & Mini Rose Nursery

'MICMAC', S, w; borne in clusters, non-recurrent; foliage deep purplish red; open habit (4 ft) growth; hardy; [R. rubrifolia X R. rugosa]; Central Exp. Farm

'MICMAG', Min, rb, 1988; ('MICMAGIC', **Mini Magic**); flowers white with red edges, reverse more red color, small, dbl., 35 petals, cupped, no fragrance; prickles straight, light green; foliage small, medium green, semi-glossy; slightly spreading, low growth; [Baby Katie sport X Watercolor]; Williams, Michael C.; The Rose Garden & Mini Rose Nursery

'MICMAGIC', Min, rb, 1988; ('MICMAG', **Mini Magic**); flowers white with red edges, reverse more red color, small, dbl., 35 petals, cupped, no fragrance; prickles straight, light green; foliage small, medium green, semi-glossy; slightly spreading, low growth; [Baby Katie sport X Watercolor]; Williams, Michael C.; The Rose Garden & Mini Rose Nursery

'MICODESSA', Min, m, 1998; (**Odessa**); flowers full, mauve, exhibition blooms, very dbl., 26–40 petals, 1.5 in., borne mostly singly, no fragrance; prickles few, small; foliage medium, dark green, matte; upright, tall (30 in.) growth; [Jean Kenneally X Select Pollen]; Williams, Michael C.; The Mini Rose Garden, 1998

'MICONE', Min, mr, 1989; (**Bill Cone**); bud ovoid; flowers medium, borne usually singly, dbl., 35 petals, cupped, no fragrance; round, green orange-yellow fruit; prickles light green, globular; foliage medium, medium green, semi-glossy; upright, tall growth; [Heartland X Anita Charles]; Williams, Michael C.; The Rose Garden & Mini Rose Nursery

'MICPAL', Min, ob, 1992; (**Palmetto Sunrise**); flowers orange with a yellow base, yellow reverse, opening orange, m, dbl., 15–25 petals, 1.5–2.75 in.; some prickles; foliage medium, medium green, semi-glossy; upright (50 cms) growth; AOE, 1993; [Orange Honey X Unnamed Miniature seedling]; Williams, Michael C.; The Rose Garden & Mini Rose Nursery, 1993

'MICPEARL', Min, ly, 1998; (**Lemon Pearl**); flowers light yellow, very full, very dbl., 41 petals, 1–1.5 in., borne mostly singly, no fragrance; prickles few, small, straight; foliage medium, light green, matte; compact, medium (22 in.) growth; [Seedling X Select Pollen]; Williams, Michael C.; The Mini Rose Garden, 1998

'MICPIE', Min, op, 1988; (**Pierrine**); flowers medium pink, reverse slightly lighter, medium, borne singly, dbl., 40 petals, exhibition form, slight damask fragrance; round, green orange-yellow fruit; prickles curved down slightly, light green; foliage medium, medium green, semi-glossy serrated edges; upright, medium growth; [Tiki X Party Girl]; William, Michael C.; The Rose Garden & Mini Rose Nursery

'MICPRIDE', Min, pb, 1991; (**Founder's Pride**); bud pointed; flowers deep pink, white center, mostly white reverse, aging deep pink, dbl., 24 petals, exhibition form, slight, spicy fragrance; foliage large, dark green, semi-glossy; upright growth; [Seedling X Party Girl]; Williams, Michael C., 1987; The Rose Garden & Mini Rose Nursery

'MICRUGOSA', S, lp; (R. vilmorinii, R. wilsonii, R. X micrugosa); single, 3 in.; orange-red fruit; depressed-globose prickles, about 1 3/4 in. diameter; (14); [R. roxburghii X R. rugosa]

'MICRUGOSA ALBA', S, w; flowers otherwise similar to micrugosa, single; [R. roxburghii X R. rugosa]

'MICSHADE', Min, yb, 1989; ('MICSHADY', **Shady Charmer**); bud ovoid; flowers light yellow base with light pink edges, aging cream, medium, dbl., 43 petals, exhibition form, slight, spicy fragrance; no fruit; prickles straight, very few, small, light-green; foliage medium, dark green, semi-glossy; bushy, medium growth; [Party Girl X Anita Charles]; Williams, Michael C.; The Rose Garden & Mini Rose Nursery, 1990

'MICSHADY', Min, yb, 1989; ('MICSHADE', **Shady Charmer**); bud ovoid; flowers light yellow base with light pink edges, aging cream, medium, dbl., 43 petals, exhibition form, slight, spicy fragrance; no fruit; prickles straight, very few, small, light-green; foliage medium, dark green, semi-glossy; bushy, medium growth; [Party Girl X Anita Charles]; Williams, Michael C.; The Rose Garden & Mini Rose Nursery, 1990

'MICSOX', Min, rb, 1986; (**Sox**); flowers medium red, small white eye blooms borne usually singly, single, 7 petals, 1 in., no fragrance; small, globular, orange fruit; small, straight prickles; foliage medium, dark; medium, upright growth; [Baby Katie X Angel Darling]; Williams, Michael C.; The Rose Garden & Mini Rose Nursery

'MICSTEAL', Min, mp, 1995; (**Stolen Dream**); flowers medium pink, touch of yellow at base of petals, dbl., 15–25 petals, 1.5–2.75 in., no fragrance; foliage small, dark green, semi-glossy; upright (55 cms) growth; Williams, Michael C.; The Mini Rose Garden, 1995

'MICSURF', Min, pb, 1988; (**Surfside**); flowers medium pink to creamy yellow to creamy white, medium, borne, dbl., 24 petals, exhibition form, no fragrance; round, medium, green orange-yellow fruit; prickles curved down, very few, red-green; foliage large, light green, semi-glossy; upright, medium growth; [Tiki X Party Girl]; Williams, Michael C.; The Rose Garden & Mini Rose Nursery

'MICTWIST', Min, ob, 1996; (**Tangerine Twist**); flowers medium orange with yellow, light yellow reverse edged orange, dbl., 15–25 petals, no fragrance; few prickles; foliage medium, medium green, semi-glossy; upright, medium growth; [Pierrine X Selected pollen]; Williams, Michael C.; The Mini Rose Garden, 1996

'MICURIN', LCl, mr, 1936; flowers bright red, very large, borne in large clusters, semi-dbl., globular, abundant seasonal bloom; foliage dark, soft; vigorous, climbing growth; Böhm, J.

'MICVOYAGE', Min, yb, 1996; (**Voyager**); flowers yellow blend, at times touch of pink, blooms borne mostly singly, dbl., 26–40 petals, 1.5–2.75 in., no fragrance; foliage medium, dark green, semi-glossy; medium (50 cms), upright growth; [Seedling X Selected Pollen]; Williams, Michael C.; The Mini Rose Garden, 1997

Midas, *see* 'LEGGA'

Midas Touch(TM), *see* 'JACTOU'

Middlesbrough Football Club, *see* 'HORflame'

Middlesbrough Pride, *see* 'NOSMID'

Middlesex County, *see* 'BOSANNE'

Middo, F, 1971; Laperrière; (Cavriglia)

'MIDGET', Min, mr, 1941; bud pink; flowers carmine-red, micro-mini, dbl., 20 petals, .5 in., slight fragrance; foliage fern-like; dwarf growth; [Ellen Poulsen X Tom Thumb]; deVink; C-P

Midget Gem®, S, ly, 1994; Dickson, Patrick

Midi®, HT, op

'MIDINETTE', HT, mp, 1962; flowers cyclamen-pink, well formed, dbl., moderate fragrance; vigorous growth; [Pink Spiral X Seedling]; Delforge

'MIDJET', Min, 1941; deVink, J.; (Cavriglia)

'MIDNIGHT', HT, dr, 1956; bud urn shaped; flowers currant-red to cardinal-red, dbl., 23–30 petals, 3.5–4.5 in., exhibition form, intense fragrance; foliage dark, glossy; long strong stems; vigorous, compact growth; [Gay Lady X Texas Centennial]; Swim, H.C.; Armstrong Nursery

Midnight Magic(TM), *see* 'WILMNMG'

Midnight Rambler, *see* 'TALMID'

'MIDNIGHT RENDEZVOUS', Min, dr, 1985; flowers small, dbl., 27 petals, slight fragrance; foliage small, dark, semi-glossy; bushy growth; [Scarlet King X Big John]; Hardgrove, Donald L.; Rose World Originals

'MIDNIGHT SUN', HT, dr, 1921; flowers deep crimson flushed velvety black, semi-dbl.; [Star of Queensland X Red-Letter Day]; Grant; Kershaw

'MIDNITE SUN', HT, yb, 1955; flowers buttercup-yellow edged red, becoming lighter, dbl., 40 petals, 5 in., moderate fragrance; bushy growth; [Sutter's Gold X R. wichurana seedling]; Brownell, H.C.

'MIDSUMMER', HT, mp, 1947; semi-dbl., 20 petals, 4–5 in., exhibition form, moderate, fruity fragrance; foliage leathery, dark; [Heinrich Wendland X Lady Sylvia]; Prosser

Midtown Mini-Climber, Cl Min, ob

Midwest Living®, *see* 'KINLIV'

'MIE', Min, mr, 1980; bud long, pointed; semi-dbl., 18 petals, cupped, borne several together, slight fragrance; tiny, curved prickles; foliage very tiny, medium green; very compact growth; [Red Can Can X Unnamed seedling]; Lyon

'MIEN DE JONGE', F, mr, 1969; bud ovoid; flowers scarlet, large, semi-dbl., cupped, moderate fragrance; vigorous, bushy growth; [Sumatra X Seedling]; Verschuren, A.; Stassen

'MIES BOUWMAN', F, op, 1973; bud round; flowers salmon-reverse yellow, slight fragrance; foliage glossy, dark; vigorous, bushy growth; [Lijnbaanroos X Unknown seedling]; Buisman, G. A. H.

'MIESZKO', Pol, lp, 1966; flowers pale pink, small; foliage small, light green; low growth; Grabczewski

'MIEZE', Pol, ob, 1909; flowers orange-yellow, borne in small clusters, vigorous growth; very hardy; [Petite Léonie X R. foetida bicolor]; Lambert, P.

'MIEZE SCHWALBE', Pol, mr, 1927; flowers rose-red; [Frau Rudolf Schmidt sport]; Lohse & Schubert

Mighty Moe, *see* 'MOEMIGHTY'

Mighty Mouse, *see* 'MACMIGMOU'

'MIGNARDISE', Pol, mr, 1971; bud ovoid; flowers open, large, single, slight fragrance; foliage large, leathery; vigorous, upright growth; [Luc Varenne X Seedling]; Delforge

Mignon, *see* **'MLLE CÉCILE BRÜNNER'**

Mignon, Climbing, *see* **'MLLE CÉCILE BRÜNNER, CLIMBING'**

'MIGNONETTE', Pol, lp, 1880; flowers rose, sometimes blush white bordered with wine-red spots, dbl., 1 in., blooms in short, full panicles of 50, resembling the Chinas; hooked, red prickles; foliage dark green above, reddish beneath, glossy, 5–7 leafl; branches dark red; very dwarf, bushy growth; (14); [Dbl. Fld. Multiflora X probably China or Tea]; Guillot et Fils

'MIGNONNE', HT, 1962; bud long, pointed; flowers bright salmon-pink, large, dbl., 80 petals, moderate fragrance; foliage leathery; very vigorous growth; [Mme Butterfly X Fernand Arles]; Gaujard; Ilgenfritz Nursery, 1966

Miguel Aldrufeu, *see* **'LINDA PORTER'**

Mijayima, F, dy, 1992; flowers large, dbl.; Tagashira, Kazuso; (Sangerhausen)

'MIKADO', Ch, dy, 1929; flowers deep golden yellow; vigorous growth; RULED EXTINCT 2/87; Dobbie

'MIKADO'™, HT, rb, 1987; (Kohsai, Koh-Sai); flowers brilliant luminous light scarlet, suffused with yellow at base, dbl., 30–35 petals, exhibition form, slight fragrance; rounded, dull orange-red fruit; slightly recurved, medium green prickles, tinged purple; foliage medium, medium green, glossy; upright, tall growth; PP006470; AARS, 1988; [Duftwolke X Kagayaki]; Suzuki, Seizo, 1976; C-P, 1987; Keisei Rose Nursery, 1987

'MIKAGURA', HT, dp, 1999; flowers bright velvet red, dbl., 30 petals, exhibition form, slight fragrance; foliage medium green; 4.5 ft growth; BM, Japan Rose Concours, 1995; [Christian Dior X seedling]; Hayashi, Shunzo, 1992

Mike Thompson, *see* 'SHERIRED'

'MIKE'S OLD-FASHIONED PINK', S, lp, 1999; dbl., 26–40 petals, 2.5 in., borne in small clusters, slight fragrance; prickles moderate; foliage large, dark green, semi-glossy; bushy, medium (4 ft) growth; [Heritage X William Baffin]; Lowe, Malcolm

Mikheil, F, m, 1998

'MIKULÁS ALES', HT, lp, 1936; flowers rosy salmon-white, intense fragrance; very vigorous growth; Bojan

Miky Tea, *see* 'SHERIMIKY'

'MIL LAMP', HT, w, 1991; flowers white with pink border, moderately medium, dbl., 15–25 petals, intense fragrance; foliage medium, medium green, semi-glossy; bushy growth; [Admiral Rodney sport]; Colclasure, C.E., 1989

'MILADY', HT, dp, 1914; flowers deep pink, dbl., exhibition form, moderate fragrance; foliage small, soft; dwarf, sparse bloom. growth; [Richmond X J.B. Clark]; Towill; A.N. Pierson

'MILAGROS DE FONTCUBERTA', HT, m, 1968; bud pointed; flowers violet-mauve, medium, dbl., 50 petals, moderate fragrance; upright, compact growth; [(Sterling Silver X Intermezzo) X (Sterling Silver X Simone)]; Dot, Simon; Rosas Dot

'MILAN', F, rb, 1986; flowers red, silver or yellow reverse, medium, dbl., 28 petals, slight fragrance; foliage medium, medium green, semi-glossy; upright, bushy growth; [Unnamed seedling X Unnamed seedling]; Jerabek, Paul E.

'MILANO', HWich, op, 1923; bud very long, pointed; flowers nasturtium-pink on indian yellow ground, borne in large clusters, dbl., moderate fragrance; foliage glossy; vigorous, climbing (10 ft) growth; Ingegnoli

'MILBEN', Min, pb, 1988; (**Dr B. Benacerraf**); flowers white to light pink, dark pink borders, opening to light pink, dbl., 15–20 petals, slight fragrance; foliage ovoid, medium green, matt, disease resistant; vigorous, compact growth; [(Double Delight X Simplex) X Magic Carrousel]; Miller, F.

'MILDEWFREE ELSE POULSEN', F, pb, 1937; flowers pink slightly tinged yellow, medium, single; [Else Poulsen X Dainty Bess]; Poulsen, S.

'MILDRED CANT', HT, mr, 1935; bud pointed; flowers bright crimson, very large, dbl., exhibition form, intense fragrance; foliage leathery; vigorous, open growth; Cant, B. R.

'MILDRED GRANT', HT, w, 1901; flowers silvery white, tinted pink at edge, dbl., exhibition form, very large, vigorous growth; GM, NRS, 1898; [Niphetos X Mme Mélanie Willermoz]; Dickson, A.

'MILDRED REYNOLDS', F, mr, 1966; flowers cardinal-red, large, borne in large, compact clusters, 3 in., cupped, slight fragrance; foliage glossy; [Peace X Seedling]; Dorieux; Bees

Mildred Scheel, *see* **'DEEP SECRET'**

Milena, HT, 1971; Vecera, L.; (Cavriglia)

Milestone; *see* 'JACLES'

Milevsko, S, mp, 1980; flowers medium, dbl., Vecera, L.; (Sangerhausen)

Mili de Vega, HT; Moreira da Silva, A.; (Cavriglia)

'MILKANA', HT, rb, 1974; flowers brick-red, outer petals shaded pink, large, dbl., 50 petals, foliage dark, glossy; vigorous, upright, tall growth; [Tallyho X Spartan]; Staikov, Prof. Dr V. Kalaydjiev and Chorbadjiiski

'MILKMAID', N, w, 1925; flowers white tinted fawn, small blooms in clusters, semi-dbl., moderate fragrance; foliage rich green; very vigorous, climbing growth; [Crepuscule X Unnamed variety]; Clark, A.; Brundrett

Milky Way, *see* 'INTERWAY'

'MILKY WAY', HWich, w, 1900; flowers pure white, tips lightly tinged pink, stamens yellow, very large, semi-dbl., seasonal bloom, moderate fragrance; foliage glossy; vigorous, climbing growth; RULED EXTINCT 9/86; Walsh

'MILKYWAY', Gr, w, 1986; flowers yellow stamens, loose, medium, borne singly, semi-dbl., 21 petals, cupped, slight, fruity fragrance; straight, small, greenish-yellow prickles; foliage small, medium green, semi-glossy; spreading, medium growth; [Unnamed seedling X Unnamed seedling]; Ohlson, John

'MILLEDGEVILLE', (natural variation of R. setigera), mp, 1842; flowers carmine

Millefleurs, Min, ob

Millennium™, *see* 'GRALOVE'

Millennium™, *see* 'PERMILL'

'MILLER HAYES', HP, mr, 1873; flowers large, dbl., moderate fragrance; Verdier, E.; (Sangerhausen)

Millgrove, HMult, lp, 1998

'MILLICENT', S, mp; flowers light coral-red fading to flesh-pink, reverse yellowish, non-recurrent, flattened, globe shape, light red fruit; foliage dark green veined red-brown; medium-tall growth; hardy; Central Exp. Farm

'MILLIE', Pol, rb, 1937; flowers light cherry-red, base yellow, borne in clusters, foliage dark; strong stems; Russ

'MILLIE PERKINS', HT, mr, 1960; flowers velvety red, well formed, large, dbl.; long, strong stems; vigorous growth; [Ena Harkness X Chrysler Imperial]; Maarse, G.

Millie Walters™, *see* 'MORMILLI'

Millionaire, *see* PEAzara

'MILORD', F, dr, 1961; flowers large, borne in clusters, dbl.; foliage dark; vigorous growth; [Opera X Ville de Gand]; Gaujard

'MILORD', HT, mr, 1962; flowers crimson-scarlet, well formed, dbl., 35 petals, 5–6 in., intense fragrance; foliage dark; upright growth; [Rubaiyat X Karl Herbst]; McGredy, Sam IV; McGredy

'MILOU', F, dr, 1964; flowers dark vermilion-red, medium, semi-dbl.; vigorous, low growth; Mondial Roses

Milrose®, *see* 'DELBIR'

'MILSWEET', F, ab, 1989; (**Elsie Warren**); flowers apricot with lemon yellow eye, medium, dbl., 15–25 petals, moderate fragrance; foliage large, medium green, semi-glossy; upright growth; [Arthur Bell X Arthur Bell]; Milner, William; Battersby Roses, 1990

Milva®, *see* 'TANAVLIM'

Mimi, *see* 'MEIDESI'

'MIMI COERTSE', Gr, mp, 1963; bud pointed; flowers bright rose-pink, dbl., 4 in., exhibition form; foliage glossy; upright growth; [Queen Elizabeth X Constantia]; Herholdt, J.A.; Herholdt's Nursery

Mimi Golden™, *see* MEIspreyp

Mimi Pearl™, F, w

Mimi Pink™, *see* 'JELDANIRAN'

'MIMI PINSON', Pol, dp, 1919; flowers clear crimson, passing to purplish rose and then to neyron purple; Barbier

Mimi Rose, *see* 'JELDANIRAN'

'MIMOLLET', F, lp, 1975; bud pointed; flowers bright light pink, dbl., 40 petals, 3–4 in., exhibition form, foliage glossy; upright growth; [(Queen Elizabeth X Ethel Sanday) X Zambra]; Ota, Kaichiro

'MIN JURGENSEN', HT, w, 1973; flowers center buff, aging to pink edges, full, medium, dbl., slight fragrance; foliage dark, soft; very vigorous, upright growth; [Pascali sport]; Bauer; Trewallyn Nursery

'MINAACO', Min, dp, 1985; (**Cindy**™); flowers deep pink, well-formed, small, dbl., slight fragrance; foliage small, dark, glossy; [Tom Brown X Over the Rainbow]; Williams, Ernest D.; Mini-Roses

'MINABBCO', Min, mr, 1991; (**Redglo**™); flowers very color fast, velvety knapp, outstanding substance, dbl., 26–40 petals, 1.5 in., exhibition form, slight fragrance; some prickles; foliage small, dark green, semi-glossy; low (36 cms), bushy growth; [Starburst X Over the Rainbow]; Williams, Ernest D.; Mini-Roses, 1992

'MINABCCO', Min, m, 1991; (**Space Walk**™); flowers mauve-tan/yellow, russet reverse, blooms borne mostly singly, dbl., 26–40 petals, 1.5 in., intense fragrance; few prickles; foliage small, dark green, semi-glossy; low (30 cms), bushy growth; [Unnamed seedling X Twilight Trail]; Williams, Ernest D.; Mini-Roses, 1992

'MINABCO', Cl Min, yb, 1985; (**Touch o' Midas**™); flowers deep yellow, petals edged deep pink, small blooms borne singly, dbl., 35 petals, slight fragrance; foliage small, dark, semi-glossy; upright (to 5 ft) growth; [Little Darling X Over the Rainbow]; Williams, Ernest D.; Mini-Roses

'MINABDCO', Min, pb, 1991; (**Pretty Woman**); flowers pink with yellow and orange, deeper at edges, aging darker, dbl., 26–40 petals, 1.5 in., exhibition form, moderate fragrance; few prickles; foliage small, medium green, glossy; low (30 cms), bushy growth; [Tom Brown X Over the Rainbow]; Williams, Ernest D.; Mini-Roses, 1992

'MINABECO', Min, rb, 1992; (**Love Torch**™); bud long, pointed; flowers red, orange and yellowblend, holds color well, heavy substance, dbl., 26–40 petals, 1.5 in., slight fragrance; some prickles; foliage small, dark green, glossy; low (30 cms), upright, bushy growth; hardy; [Starburst X Over the Rainbow]; Williams, Ernest D.

'MINACCO', Cl Min, dp, 1985; (**Jazz Time**™); flowers deep pink, small blooms in clusters, dbl., 35 petals, slight fragrance; foliage small, dark, semi-glossy; upright, bushy (4–5 ft) growth; [Little Darling X Little Chief]; Williams, Ernest D.; Mini-Roses, 1986

'MINACO', Min, m, 1982; (**Angelglo**™); flowers lavender, micro-mini, small, dbl., slight fragrance; foliage small, dark, semi-glossy; bushy growth; PP005356; [Angel Face X (Angel Face X Over the Rainbow)]; Williams, Ernest D.; Mini-Roses

'MINACON', Min, rb, 1995; (**Space Probe**); flowers red edges on white petals, blooms, dbl., 15–25 petals, 1.5 in., moderate fragrance; some prickles; foliage small, medium green, semi-glossy; medium, upright, bushy growth; [Little Darling X Over the Rainbow]; Williams, Ernest D.; K&C Roses, 1995

'MINADCO', Min, m, 1986; (**Royalglo**™); flowers non-fading, small, dbl., 40 petals, exhibition form, borne usually singly, intense, damask, sweet fragrance; no fruit; few, long, light tan prickles; foliage small, dark green, semi-glossy; bushy, medium growth; PP006673; [Angel Face X Anita Charles]; Williams, Ernest D.; Mini-Roses

'MINAECO', Min, ab, 1986; (**Apricot Charm**™); flowers apricot, reverse slightly deeper, blending to yellow at base, dbl., 42 petals, exhibition form, borne usually singly, slight fragrance; no fruit; few, long, thin, short, light tan prickles; foliage small, dark green, glossy; bushy, spreading, medium growth; [Gingersnap X Anita Charles]; Williams, Ernest D.; Mini-Roses

'MINAFCO', Min, pb, 1986; (**Orange Love**™); flowers orange, reverse deeper orange to dark red, fading lighter, dbl., 45 petals, exhibition form, slight fragrance; no fruit; very few, short, light tan prickles; foliage small, dark green, semi-glossy; upright, bushy, medium growth; PP006546; [Tom Brown X Over the Rainbow]; Williams, Ernest D.; Mini-Roses

'MINAGCO', Min, r, 1986; (**Honey 'n' Spice**™); flowers tan with red highlights, reverse deeper tan with more red, small, dbl., 45–49 petals, borne usually singly, moderate, fresh honey fragrance; no fruit; few, short, thin, light tan prickles; foliage small, medium green, semi-glossy; bushy, medium growth; [Tom Brown X Over the Rainbow]; Williams, Ernest D.; Mini-Roses

'MINAHCO', Min, dy, 1986; (**Summer Beauty**™); flowers small, borne usually singly or in sprays of 3–5, dbl., exhibition form, moderate fragrance; no fruit; very few, thin, short, light prickles; foliage small, dark green, semi-glossy; upright, bushy, medium growth; [Unnamed seedling X Unnamed seedling]; Williams, Ernest D.; Mini-Roses

'MINAICO', Cl Min, mr, 1986; (**Red Delight**™); flowers medium red, yellow at base, small, dbl., 33 petals, exhibition form, borne singly and in sprays, intense, spicy fragrance; round, red fruit; few, small, tan prickles, slanted downwards; foliage small, dark green, semi-glossy; tall growth; PP006893; [Golden Song X Magic Mist]; Williams, Ernest D.; Mini-Roses, 1987

'MINAJCO', Min, m, 1987; (**Purple Majesty**™); flowers mauve, edged red, reverse mauve, small, dbl., 40 petals, exhibition form, borne usually singly, fruit not observed; tan, needle-like prickles, dilated at base; foliage small, medium green, semi-glossy; bushy, medium growth; PP006894; [Tom Brown X Black Jack]; Williams, Ernest D.; Mini-Roses

'MINAKCO', Min, lp, 1987; (**Pink Kiss**™); flowers light pink, non-fading, small, borne usually singly or in spr, dbl., 40–45 petals, exhibition form, no fragrance; fruit not observed; tan-red prickles, dilated at base; foliage small, medium green, semi-glossy; bushy, medium growth; [Tom Brown X Anita Charles]; Williams, Ernest D.; Mini-Roses

'MINALCO', Min, ob, 1987; (**Orange Star**™); flowers orange, reverse orange with yellow at base, non-fading, high, dbl., 35 petals, exhibition form, intense, damask fragrance; fruit not observed; tan prickles, dilated at base; foliage small, medium green, semi-glossy; bushy, spreading, medium growth; [Unnamed Miniature seedling X Unnamed Miniature seedling]; Williams, Ernest D.; Mini-Roses

'MINAMCO', Min, mr, 1987; (**Ethel Orr**™); flowers small, borne usually singly or in sprays of 3–5, dbl., 35 petals, exhibition form, slight fragrance; no fruit; few, small, tan prickles; foliage small, medium green, glossy; upright, bushy, medium, profuse growth; [Miniature seedling X Big John]; Williams, Ernest D.; Mini-Roses

'MINANCO', Min, or, 1987; (**Oriental Simplex**™); flowers bright orange-red, reverse creamy yellow, aging deeper, non-, single, 5 petals, flat, no fragrance; fruit not observed; tan prickles, declining, dilated at base; foliage small, medium green, glossy; upright, bushy, medium growth; [(Starburst X Over the Rainbow) X Little Chief]; Williams, Ernest D.; Mini-Roses

'MINAPCO', Min, r, 1988; (**Amberglo**™); flowers small, dbl., 34 petals, intense fragrance; foliage small, dark green, semi-glossy; bushy, sturdy growth; [Tom Brown X Twilight Trail]; Williams, Ernest D.

'MINAQCO', Min, w, 1988; (**White Charm**™); bud long; flowers small, dbl., 33 petals, exhibition form, intense fragrance; foliage small, medium green, glossy; bushy growth; [Tom Brown X Over the Rainbow]; Williams, Ernest D.

'MINARCO', Min, mp, 1988; (**Evelyn Rogers**™); flowers small, dbl., 34 petals, slight fragrance; foliage small, medium green, glossy; upright, bushy growth; [Tom Brown X Over the Rainbow]; Williams, Ernest D.

'MINASCO', Min, dy, 1988; (**The Garden Editor**®); flowers small, dbl., 33 petals, moderate fragrance; foliage small, medium green, glossy; upright, bushy growth; [Gold Badge X Yellow Jewel]; Williams, Ernest D.

'MINATCO', Min, mp, 1988; (**Ta Ta**); flowers small, dbl., 33 petals, no fragrance; foliage small, medium green, semi-glossy, dense; bushy growth; [Tom Brown X Over the Rainbow]; Williams, Ernest D.

'MINAUCO', Min, m, 1988; (**Lavender Star**™); flowers mauve blended lavender-tan, small, single, 5 petals, intense fragrance; foliage small, dark green, semi-glossy; upright, bushy, dense growth; [Unnamed seedling X Lavender Simplex]; Williams, Ernest D., 1989

'MINAVCO', Min, r, 1989; (**Sue Jo**™); flowers russet, outside petals tan-lavender, inside petals golden-am, dbl., 33 petals, intense fragrance; foliage small, medium green, glossy; upright, bushy growth; PP007606; [Tom Brown X Twilight Trail]; Williams, Ernest D.; Mini-Roses

'MINAWCO', Min, r, 1989; (**Suntan Beauty**™); flowers tan with rosy highlights, small, dbl., 33–45 reflexed petals, exhibition form, intense fragrance; foliage small, dark green, glossy, disease-resistant; bushy growth; PP007596; [(Angel Face X Golden Angel) X Yellow Jewel]; Williams, Ernest D.; Mini-Roses

'MINAXCO', Min, m, 1989; (**Smoke Signals**™); flowers smoky, lavender-gray, small, dbl., 33 petals, exhibition form, intense fragrance; foliage small, medium green, glossy; upright, bushy growth; PP007604; [[Tom Brown X (Rise 'n' Shine X Watercolor)] X Twilight Trail]; Williams, Ernest D.; Mini-Roses

'MINAYCO', Min, m, 1989; (**Ida Belle**™); flowers lavender with blends of amber, very long-lasting, small, dbl., 33 petals, exhibition form, intense fragrance; foliage small, medium green, glossy; bushy growth; PP007605; [[Tom Brown X ((Rise 'n' Shine) X Watercolor)] X Twilight Trail]; Williams, Ernest D.

'MINAZCO', Min, r, 1989; (**Sands of Time**™); flowers small, dbl., 32 petals, exhibition form, intense fragrance; foliage small, medium green, glossy; upright, bushy growth; PP007607; [Tom Brown X Twilight Trail]; Williams, Ernest D.; Mini-Roses

'MINBACO', Min, mp, 1989; (**Jewel's Delight**™); flowers small, dbl., 33 petals, moderate fragrance; foliage small, medium green, glossy; bushy growth; [Tom Brown X Twilight Trail]; Williams, Ernest D.

'MINBCO', Min, dy, 1982; (**Great Day**); bud long, pointed; flowers well-formed, small, dbl., 35 petals, moderate fragrance; foliage small, light green, glossy; bushy growth; PP005358; [(Little Darling X Gold Coin) X (Little Darling X Gold Coin)]; Williams, Ernest D.; Mini-Roses

'MINCCO', Min, pb, 1982; (**Peggy's Delight**); flowers deep pink, white reverse, small, dbl., slight fragrance; foliage small, dark, glossy; bushy growth; PP005357; [Little Darling X Over the Rainbow]; Williams, Ernest D.; Mini-Roses

'MINDCO', Min, or, 1982; (**Orange Charm**); flowers well-formed, small, dbl., slight fragrance; foliage small, medium green, glossy; bushy growth; [Starburst X Over the Rainbow]; Williams, Ernest D.; Mini-Roses

'MINDOR', F, yb, 1976; flowers deep yellow, slightly flushed orange, urn shaped, dbl., 30–60 petals, 3 in.; foliage dark; vigorous, low growth; [Dr Faust X Mme Lucky]; Station Exp. de Roses

'MINDREAM', Min, m, 1992; (**Twilight Dream**); flowers good color stability, heavy substance, blooms borne mostly singly, very dbl., 1.5 in., exhibition form, good repeat, intense fragrance; few prickles; foliage small, dark green, glossy; low (30 cms), bushy growth; hardy; [Tom Brown X Twilight Beauty]; Williams, Ernest D.

'MINECO', Min, pb, 1982; (**Spring Beauty**); flowers pastel pink and yellow blend, small, dbl., 35 petals, slight fragrance; foliage small, medium green, glossy; upright growth; [Little Darling X Over the Rainbow]; Williams, Ernest D.; Mini-Roses

Minerette, *see* 'INTERMINER'

Miners Cottage, Misc OGR, dp

'MINERVE', HT, yb, 1947; flowers chrome-yellow shaded red, high pointed, dbl., 36 petals, moderate fragrance; foliage thick, dark; vigorous, branching growth; [Peace X Prinses Beatrix]; Meilland, F.

Minette, A, lp

'MINFCO', Cl Min, pb, 1982; (**Break o' Dawn**); flowers white with deep pink edges, white reverse, small, dbl., 35 petals, slight fragrance; foliage small, dark, glossy; upright, bushy, climbing growth; [Little Darling X Over the Rainbow]; Williams, Ernest D.; Mini-Roses

'MINFIRE', Min, or, 1998; (**Watchfire**); flowers bright coral red with deep yellow base, full, very dbl., 35 petals, 1.25–1.5 in., exhibition form, borne mostly singly, slight fragrance; prickles moderate; foliage medium, medium green, semi-glossy; upright, compact, bushy, medium (18 in.) growth; [Sue Jo X Twilight Trail]; Williams, Ernest D.; Texas Mini Roses, 1997

'MINFRIV', Min, mr, 1998; (**Frivolous**); flowers cherry red, broad petals, good substance, single, 8–10 petals, 1.5 in., borne mostly singly, no fragrance; almost thornless; foliage medium, semi-glossy; upright, medium (18 in.) growth; [Twilight Trail X Seedling]; Williams, Ernest D.; Texas Mini Roses, 1997

'MING TOY', F, dp, 1947; bud globular; flowers deep rose-pink, large, loose trusses, dbl., 50 petals, 2 in., moderate fragrance; foliage leathery, dark; vigorous, upright, bushy growth; Krebs; H&S

'MINGCO', Min, rb, 1983; (**Fiesta Time**™); flowers yellow to orange to red, reverse yellow, small, dbl., slight fragrance; foliage small, dark, glossy; bushy growth; [Starburst X Over the Rainbow]; Williams, Ernest D.; Mini-Roses

Minha Mulher, HT, 1969; Moreira da Silva, A.; (Cavriglia)

'MINHCO', Min, mp, 1983; (**DeGrazia's Pink**); flowers heavy substance, small, dbl., slight fragrance; foliage small, dark, glossy; bushy growth; [(Seedling X Over the Rainbow) X (Seedling X Over the Rainbow)]; Williams, Ernest D.; Mini-Roses

'MINHEART', Min, r, 1998; (**Lionheart**); flowers russet with coral edge, deep yellow reverse, dbl., 25 petals, 1.2 in., borne mostly singly, moderate fragrance; prickles moderate; foliage medium, dark green, semi-glossy; upright, bushy, medium growth; [Seedling X Twilight Trail]; Williams, Ernest D.; Texas Mini Roses, 1997

Mini Lights, *see* 'DICMOPPET'

Mini Magic, *see* 'MICMAGIC'

Mini Magic, Min, lp; Walsh, 1998; (Weatherly, L.)

Mini Mercedes, Min, mr

Mini Metro, *see* 'RUFIN'

Mini Nicole, Min, pb

'MINI PINK MELODIES', Min, mp, 1991; flowers moderately small, dbl., 15–25 petals, slight fragrance; foliage small, medium green, semi-glossy; bushy growth; [Unnamed seedling X Unnamed seedling]; Spooner, Raymond A.

'MINIATURE', Pol, lp, 1884; flowers pink, becoming yellowish white, very small, very dbl., intense fragrance; moderate growth; Alégatière

'MINIBETH', Min, m, 1994; (**Marybeth**(TM)); flowers pastel mauve, very double, blooms borne mostly, single, exhibition form, moderate fragrance; few prickles; foliage small, medium green, semi-glossy; medium (18–20 in.), upright growth; [Seedling X Twilight Trail]; Williams, Ernest D.; Texas Mini Roses, 1994

'MINIBID', Min, m, 1994; (**Forbidden**(TM)); flowers medium lavender with dark lavender on edges, dbl., 26–40 petals, 1.5 in., moderate fragrance; few prickles; foliage small, dark green, semi-glossy; low (14–16 in.), bushy growth; [Seedling X Twilight Trail]; Williams, Ernest D.; Texas Mini Roses, 1994

'MINICIN', Min, r, 1993; (**Cinnamon Delight**(TM)); flowers russet with deep yellow base, dbl., 26–40 petals, 1.5 in., borne mostly single, intense fragrance; few prickles; foliage small, dark green, semi-glossy; medium (40 cms), upright growth; [Seedling X Twilight Trail]; Williams, Ernest D.; Mini Roses of Texas, 1993

'MINICO', Cl Min, ob, 1983; (**Mary Marshall, Climbing**); Williams, Ernest D.; Mini-Roses, 1982

'MINIDOR', Min, yb, 1996; (**Fall Splendor**); flowers rich yellow with orange blush on edges, blooms borne mostly, dbl., 26–40 petals, 1.25–1.5 in., moderate fragrance; some prickles; foliage medium, dark green, semi-glossy; medium (16 in.), compact, bushy growth; [Rise 'n' Shine X Twilight Trail]; Williams, Ernest D.; Texas Mini Roses, 1996

'MINIFIRE', Min, rb, 1993; (**Dragon's Fire**(TM)); flowers bright red with deep yellow at base of petals, yellow reverse, dbl., 26–40 petals, 1.5 in., moderate fragrance; few prickles; foliage small, medium green, semi-glossy; medium (35 cms), upright, compact growth; [Starburst X Twilight Trail]; Williams, Ernest D.; Mini Roses of Texas, 1993

'MINIGOLD', F, my, 1970; flowers pointed, dbl., 35 petals, 2.5 in., moderate fragrance; foliage glossy, dark; [Whisky Mac X Zorina]; Tantau, Math

Minijet®, *see* MEIrotego

Minilights, *see* 'DICMOPPET'

'MINIMAG', Min, m, 1993; (**Magician**(TM)); flowers mauve with deeper petal edges, blooms borne mostly single, very dbl., 1.5–2.75 in., intense fragrance; few prickles; foliage small, medium green, semi-glossy; medium (45 cms), upright, bushy growth; [Seedling X Twilight Trail]; Williams, Ernest D.; Mini Roses of Texas, 1993

'MINIMERR', Min, m, 1995; (**Merryglo**(TM)); flowers medium lavender, blooms borne mostly single, dbl., 26–40 petals, 1.25 in., intense fragrance; some prickles; foliage medium, medium green, semi-glossy; medium (18 in.), bushy growth; [Seedling X Twilight Trail]; Williams, Ernest D.; Texas Mini Roses, 1995

'MINIMOON', Min, mr, 1995; (**Red Moon**(TM) (USA)); dbl., 26–40 petals, 1–1.5 in., borne in small clusters, no fragrance; some prickles; foliage medium, medium green, semi-glossy; upright, bushy, climbing (5 ft) growth; [Red Delight X Twilight Trail]; Williams, Ernest D.; Texas Mini Roses, 1995

'MINIMYS', Min, m, 1995; (**Mysterious**); flowers mauve, exhibition, full (26–40 petals), medium blooms borne mostly single, fragrant; some prickles; foliage medium, medium green, semi-glossy; upright (16–18 in.), bushy growth; [Seedling X Twilight Trail]; Williams, Ernest; Texas Mini Roses, 1995

'MINPUM', Min, r, 1995; (**Pumpkin Frost**(TM)); flowers russet, dbl. (15–25 petals), medium (1–$1\frac{1}{4}$ in.) blooms borne small clutsters; fragrant; some prickles; foliage medium, dark green, semi-glossy; low (12 in.), compact growth; [Seedling X Twilight Trail]; Williams, Ernest; Texas Mini Roses, 1995

'MINI-POUL'®, Min, pb, 1978; flowers yellow and deep pink blend, small, dbl., 25 petals, no fragrance; foliage small, dark, glossy; compact growth; [Darling Flame X Unnamed seedling]; Poulsen, Niels D.; D.T. Poulsen

'MINIRASP', Min, mr, 1994; (**Raspberry Beauty**(TM)); flowers non-fading raspberry red, blooms borne mostly single, dbl., 26–40 petals, 1.5 in., exhibition form, slight fragrance; few prickles; foliage small, medium green, semi-glossy; medium (16–20 in.), bushy growth; [Angel Face X Anita Charles]; Williams, Ernest D.; Texas Mini Roses, 1994

Minirosa, Min, lp; flowers small, semi-dbl., slight fragrance; VEG; (Sangerhausen)

Minirot, Min, m; flowers purple/pink with cream reverse, small, semi-dbl., slight fragrance; VEG; (Sangerhausen)

'MINIRYL', Min, m, 1993; (**Royal Lady**(TM)); flowers lavender pink, blooms borne mostly single, dbl., 26–40 ruffled petals, 1.5 in., intense fragrance; few prickles; foliage small, medium green, semi-glossy; medium (40 cms), bushy growth; [Angel Face X Twilight Beauty]; Williams, Ernest D.; Mini Roses of Texas, 1993

'MINISA', HRg, mr, 1927; flowers deep crimson, semi-dbl., 17 petals, bloom repeats, intense fragrance; very hardy.; [R. rugosa X Prince Camille de Rohan]; Hansen, N.E.

'MINISHOW', Min, dr, 1996; (**Showdown**); dbl., 26–40 petals, 1.25 in., borne mostly singly, no fragrance; some prickles; foliage dark green, semi-glossy; medium (16–18 in.), upright, bushy growth; [Rise 'n' Shine X Twilight Trail]; Williams, Ernest D.; Texas Mini Roses, 1996

'MINISONG', Min, ob, 1996; (**Morning Song**); flowers orange with yellow base, blooms borne mostly single, dbl., 26–40 petals, 1.25 in., slight fragrance; some prickles; foliage medium, medium green, semi-glossy; medium (16–18 in.), upright, bushy growth; [Sue Jo X Twilight Trail]; Williams, Ernest D.; Texas Mini Roses, 1996

'MINISTER AFRITSCH', F, or, 1964; flowers large, dbl.; vigorous, bushy growth; [Unknown seedling X Signalfeuer]; Tantau, Math.; Starkl

'MINISTER LUNS', F, dp, 1968; bud ovoid; flowers pink-red, medium, semi-dbl.; foliage dark; [Marchenland X Florence Mary Morse]; Wijnhoven

Minister Rasín, *see* **'MINISTRE DES FINANCES RASÍN'**

'MINISTORM', Min, m, 1994; (**Stormy Weather**(TM)); flowers medium lavender, blooms borne mostly single, dbl., 26–40 petals, 1.5 in., exhibition form, moderate fragrance; few prickles; foliage small, medium green, semi-glossy; tall (24–30 in.), upright growth; [Seedling X Twilight Trail]; Williams, Ernest D.; Texas Mini Roses, 1994

'MINISTRE DES FINANCES RASÍN', HT, dp, 1930; (Minister Rasín); bud long; flowers carmine-rose, large, dbl.; vigorous growth; [Mme Maurice de Luze X Hadley]; Böhm, J.

'MINITOC', Min, r, 1995; (**Touch O' Cloves**); flowers russet, full (26–40 petals), medium (1–1/4 in.) blooms borne mostly single; fragrant; few prickles; foliage medium, dark green, semi-glossy; upright (18 in.) bushy growth; [Seedling X Twilight Trail]; Williams, Ernest; Texas Mini Roses, 1995

'MINITOK', Min, dy, 1993; (**Golden Token**™); foliage small, medium green, glossy; medium (40 cms), upright, bushy growth; [Flowers deep golden yellow, very full (41+ petals), small (0–4 cms) blooms borne in small clusters; slight fragrance; few pr]; Williams, Ernest D.; Mini Roses of Texas, 1993

'MINIUMBER', Min, r, 1996; (**Umberglo**™); flowers russet with red blush on edges, blooms borne mostly singly, dbl., 26–40 petals, 1–1.25 in., moderate fragrance; few prickles; foliage medium, dark green, semi-glossy; low (12–14 in.), compact, bushy growth; [Seedling X Twilight Trail]; Williams, Ernest D.; Texas Mini Roses, 1996

'MINIVIG', Min, w, 1995; (**Vigilance**™); flowers white, full (26–40 petals), medium ($1\frac{1}{2}$–$1\frac{3}{4}$ in.) blooms borne in small clusters; no fragrance; numerous prickles; foliage medium, medium green, glossy; spreading (6 ft), bushy, climbing growth; [Jeanne Lajoie sport]; Williams, Ernest; Texas Mini Roses, 1995

'MINIVISION', Min, mp, 1996; (**Heavenly Vision**); flowers clear, medium pink, blooms borne mostly single, dbl., 15–25 petals, 1.5 in., slight fragrance; few prickles; foliage medium green, semi-glossy; medium (16–18 in.), upright, bushy growth; [Angel Face X Tom Brown]; Williams, Ernest D.; Texas Mini Roses, 1996

'MINIWACK', Min, pb, 1993; (**Paddywack**™); dbl., 26–40 petals, 1.5 in., borne mostly singly, slight fragrance; few prickles; foliage small, medium green, semi-glossy; medium (50 cms), upright, compact growth; [Tom Brown X Over the Rainbow]; Williams, Ernest D.; Mini Roses of Texas, 1993

'MINIWIN', Min, ly, 1995; (**Winter Wheat**™); flowers light yellow, full (26–40 petals), medium ($1\frac{1}{4}$ in.) blooms borne mostly singly; slight fragrance; some prickles; foliage medium; medium green, semi-glossy; upright (16–18 in.), bushy growth; [Seedling X Twilight Trail]; Williams, Ernest; Texas Mini Roses, 1995

'MINIWON', Min, r, 1994; (**Wonderglo**™); flowers unusual russet with red highlights, very dbl., 26–40 petals, 1.5 in., borne mostly singly, moderate fragrance; few prickles; foliage small, medium green, semi-glossy; medium (18 in.), bushy growth; [Seedling X Twilight Trail]; Williams, Ernest D.; Texas Mini Roses, 1994

'MINJCO', Min, op, 1983; (**Loveglo**™); flowers light coral pink and cream, small, dbl., exhibition form, moderate fragrance; foliage small, dark, glossy; bushy growth; PP005677; [Little Darling X Over the Rainbow]; Williams, Ernest D.; Mini-Roses

'MINKCO', Min, dr, 1983; (**Black Jack**™); flowers deep red, golden stamens, small, dbl., 40 petals, exhibition form, slight fragrance; foliage small, dark, semi-glossy; bushy growth; PP005671; [Tom Brown X Over the Rainbow]; Williams, Ernest D.; Mini-Roses

'MINLCO', Min, w, 1983; (**Arctic Snow**™); flowers small, dbl., exhibition form, slight fragrance; foliage small, dark, glossy; upright, bushy growth; [Miniature seedling X Over the Rainbow]; Williams, Ernest D.; Mini-Roses

'MINMCO', Min, ob, 1983; (**Spring Melody**™); flowers orange, small, dbl., 35 petals, slight fragrance; foliage small, bronze green, semi-glossy; bushy growth; [Little Darling X Over the Rainbow]; Williams, Ernest D.; Mini-Roses

'**MINNA**', HEg, w, 1895; flowers pure white, semi-dbl.; very vigorous growth; Penzance

'**MINNA**', F, mp, 1930; flowers rosy pink, large, borne in clusters, dbl., slight fragrance; foliage rich green, leathery; bushy, dwarf growth; [Gruss an Aachen sport]; Kordes

Minna Kordes, *see* **'WORLD'S FAIR'**

Minna Lerche Lerchenborg®, *see* 'MEIDAD'

'MINNCO', Min, mp, 1983; (**Pink Carpet**™); flowers small, dbl., blooms in large clusters, no fragrance; foliage small, light to medium green, glossy; low, spreading (to 6 ft hanging basket) growth; PP005606; [Red Cascade X Red Cascade]; Williams, Ernest D.; Mini-Roses

'**MINNEHAHA**', HWich, lp, 1905; flowers pink fading white, small blooms in large clusters, dbl., non-recurrent, slight fragrance; foliage small, glossy, dark; climbing (15–20 ft) growth; [R. wichurana X Paul Neyron]; Walsh

Minnesota Mo, M, dr

'**MINNIE**', Min, rb; bud long, pointed; flowers red and yellow blend, dbl., 40 petals, 1 in., exhibition form, moderate fragrance; foliage small, glossy; upright growth; [Starburst X Over the Rainbow]; Williams, Ernest D.; Mini-Roses, 1977

'**MINNIE DAWSON**', HMult, w, 1896; flowers pure white, borne in very large clusters; [Dawson X R. multiflora]; Dawson

'**MINNIE FRANCIS**', T, dp, 1905; flowers deep pink, open; vigorous growth; Griffing Nursery

'**MINNIE MARCUS**', HT, pb, 1984; flowers large, dbl., 35 petals, slight fragrance; foliage large, dark, glossy; upright growth; [Queen Elizabeth X Windsounds]; Rodgers, Shafner R.; Neiman-Marcus

Minnie Mouse, *see* 'MACMINMO'

Minnie Pearl™, *see* 'SAVAHOWDY'

Minnie Pearl White, Min, lp; [Minnie Pearl sport]; Lee, 1993; (Weatherly, L.)

'**MINNIE SAUNDERS**', HT, mr, 1921; flowers bright red, single, vigorous growth; Hicks

'**MINNIE WATSON**', HT, lp, 1965; bud globular; semi-dbl., slight fragrance; foliage glossy; compact, bushy growth; [Dickson's Flame X Dickson's Flame]; Watson

'MINOCO', Min, lp, 1984; (**Mary Kay**™); flowers small, dbl., 35 petals, slight fragrance; foliage small, dark, glossy; upright, bushy growth; PP005631; [Tom Brown X Over the Rainbow]; Williams, Ernest D.; Mini-Roses

Minou, F; Moreira da Silva, A.; (Cavriglia)

Minouchette, LCl, 1971; Croix; (Cavriglia)

'MINPCO', Min, yb, 1984; (**Partyglo**™); flowers yellow with pink petal edges, yellow reverse, small, dbl., 35 petals, exhibition form, slight fragrance; foliage small, dark, semi-glossy; PP005880; [Little Darling X Over the Rainbow]; Williams, Ernest D.; Mini-Roses

'MINPEEP', Cl Min, mr, 1992; (**Peep-Eye**™); flowers very distinct shade of medium red, with large clusters of go, dbl., 25 petals, 1.5 in., slight fragrance; some prickles; foliage small, dark green, glossy; tall (4'+), upright, bushy growth; hardy (both heat & cold); [Red Delight X Unnamed seedling]; Williams, Ernest D.

'MINPROP', Min, mr, 1998; (**Prophecy**); flowers rose red, heat stable, medium, very dbl., 26–40 petals, 1–1.25 in., exhibition form, borne mostly singly, slight fragrance; prickles moderate; foliage medium, medium green, semi-glossy; upright, bushy, medium (18 in.) growth; [Red Delight X Twilight Trail]; Williams, Ernest D.; Texas Mini Roses, 1997

'MINQCO', Min, mr, 1984; (**Red Love**™); flowers spiral form, small, dbl., 35 petals, slight fragrance; foliage small, dark, semi-glossy to glossy; upright, bushy growth; [Tom Brown X Over the Rainbow]; Williams, Ernest D.; Mini Roses

'MINRCO', Min, pb, 1984; (**Cheryl's Delight**™); flowers medium pink, reverse white, small, dbl., 35 petals, slight fragrance; foliage small, dark, semi-glossy; bushy growth; PP005910; [Little Darling X Over the Rainbow]; Williams, Ernest D.; Mini-Roses

'MINRID', Min, yb, 1998; (**Dreamrider**); flowers med. yellow, pink edge, reverse yellow, very dbl., 26–40 petals, 1.75 in., borne mostly singly, slight fragrance; few prickles; foliage medium, medium green, semi-glossy; upright, bushy, medium (18 in.) growth; [Sue Jo X Twilight Trail]; Williams, Ernest D.; Texas Mini Roses, 1997

'MINSCO', Min, mp, 1984; (**Pink Bounty**™); flowers small, dbl., 35 petals, exhibition form, slight fragrance; foliage small, medium green, semi-glossy; very dense, bushy growth; [Tom Brown X Over the Rainbow]; Williams, Ernest D.; Mini-Roses

'MINSCO', Min, m, 1985; ('MINXCO', **Twilight Trail**™); flowers lavender-tan, small, dbl., 35 petals, intense fragrance; foliage small, dark, semi-glossy; upright, bushy growth; PP006198; [Angel Face X Anita Charles]; Williams, Ernest D.; Mini-Roses

'MINsounds', Min, mp, 1998; (**Heartsounds**); flowers clear medium pink, cream reverse, very dbl., 30–35 petals, 1–1.25 in., borne mostly singly, slight fragrance; few prickles; foliage medium, medium green, semi-glossy; long stems; upright, bushy, medium (18 in.) growth; [Rise 'n' Shine X Twilight Trail]; Williams, Ernest D.; Texas Mini Roses, 1997

'Minstrel', F, mr, 1967; flowers scarlet, semi-borne in clusters, single, 2.5 in.; foliage dark, glossy; very free growth; [Independence X Paprika]; Sanday, John

Mint Julep, *see* 'AROgresh'

'MINtco', Min, m, 1984; (**Lavender Simplex**™); flowers lavender, purple stamens, small blooms borne singly, single, 5 petals, moderate fragrance; foliage small, dark, semi-glossy; upright, bushy growth; [Angel Face X Yellow Jewel]; Williams, Ernest D.; Mini-Roses

'MINuco', Min, mr, 1984; (**Royal Carpet**™); flowers small, dbl., blooms in clusters, no fragrance; foliage small, dark, semi-glossy; groundcover spreading growth; PP005909; [Red Cascade X Red Cascade]; Williams, Ernest D.; Mini-Roses

'Minuet', HT, my, 1930; flowers open, large, semi-dbl., slight fragrance; foliage leathery; vigorous growth; [Joanna Hill sport]; Thompson's, J.H., Sons

Minuette, *see* 'MEIlucca'

Minuette, *see* **'Laminuette'**

Minuette, Climbing, Cl F; (Cavriglia)

Minuetto, *see* 'MEIlucca'

'MINvco', Min, dr, 1984; (**Red Pendant**™); flowers small, dbl., 35 petals, no fragrance; foliage small, medium green, very glossy; low, spreading growth; [Red Cascade X Red Cascade]; Williams, Ernest D.; Mini Roses

'MINwco', Min, lp, 1985; (**Beautyglo**™); flowers small, dbl., 35 petals, exhibition form, slight fragrance; foliage small, dark, semi-glossy; bushy growth; PP006199; [Tom Brown X Black Jack]; Williams, Ernest D.; Mini-Roses

'Minx', F, ob, 1955; bud globular; flowers orange-pink, borne in pyramidal clusters, dbl., 75–80 petals, 2.5 in., flat, moderate fragrance; strong stems; vigorous, upright growth; [Pinocchio seedling X Garnette]; Boerner; J&P

'MINxco', Min, m, 1985; ('MINsco', **Twilight Trail**™); flowers lavender-tan, small, dbl., 35 petals, intense fragrance; foliage small, dark, semi-glossy; upright, bushy growth; PP006198; [Angel Face X Anita Charles]; Williams, Ernest D.; Mini-Roses

'MINyco', Min, yb, 1985; (**Imogene**™); flowers yellow, marked red, well-formed, small, dbl., 45 petals, slight fragrance; foliage small, dark, glossy; upright, bushy growth; [Little Darling X Over the Rainbow]; Williams, Ernest D.; Mini-Roses

'MINzco', Min, dy, 1985; (**Fiddler's Gold**™); flowers small, dbl., moderate fragrance; foliage small, dark, glossy; bushy growth; [Tom Brown X Golden Angel]; Williams, Ernest D.; Mini-Roses

'Mio Mac', F, dy, 1973; bud ovoid; flowers coppery yellow, medium, dbl., slight fragrance; foliage glossy; growth moderate, upright, bushy; Tantau, Math.; Horstmann

Mirabella™, *see* 'JACjubel'

Mirabilia, F, 1960; Cazzaniga, F. G.; (Cavriglia)

'Miracle', F, op, 1962; flowers soft coral blooms in large trusses, semi-dbl., 16–35 petals, 3 in., exhibition form; foliage glossy; vigorous growth; GM, Bagatelle, 1958; [Unnamed seedling X Fashion]; Verbeek; Ilgenfritz Nursery

'Miragaia', HT, pb, 1958; flowers lilac-pink, [Peace X Coimbra]; da Silva, Moreira

'Mirage', Gr, op, 1966; bud ovoid; flowers bright salmon shaded red, medium, dbl., intense fragrance; foliage dark; very vigorous, bushy growth; [Peace X Circus]; Gaujard

Mirakel, F, yb; flowers dark yellow with orange and red, medium, dbl., moderate fragrance; VEG; (Sangerhausen)

'Miralda', HCh, m; flowers dark crimson-purple, nearly black, small, very dbl.; vigorous, branching growth; Vibert, ca 1830

Miramar, *see* **'Almirante Américo Tomás'**

'Miramar', F, or, 1956; bud globular; flowers cinnabar shaded coppery, large; foliage dark; very vigorous growth; [Opera seedling X Seedling]; Gaujard

'Miranda', P, mp, 1869; flowers satiny pink, very dbl.; de Sansal

'Miranda Jane', F, op, 1972; flowers salmon, dbl., 20 petals, 4 in., exhibition form, moderate fragrance; [Orange Sensation X Red Dandy]; Cocker

Mirandolina®, Min, w

'Mirandy', HT, dr, 1945; flowers garnet-red, aging darker, dbl., 45 petals, 5–6. in., globular, intense, damask fragrance; foliage leathery; vigorous, upright, bushy growth; AARS, 1945; [Night X Charlotte Armstrong]; Lammerts, Dr Walter; Armstrong Nursery

'Mirandy, Climbing', Cl HT, dr, 1961; Moore

'Mirato', HT, mp, 1974; bud ovoid; dbl., intense fragrance; foliage large, glossy; upright, bushy growth; [Unknown X Unknown]; Tantau, Math.

Mirato, *see* TANotari

'Mireille', HT, mr, 1952; flowers coppery crimson-red, dbl., 26 petals, 4 in., moderate fragrance; foliage rich green; vigorous growth; [Opera X Unnamed seedling]; Gaujard

Mireille Mathieu, *see* 'KORdehn'

Mirella®, *see* BARnec

Miriam, *see* 'MACsupcat'

'Miriam', HT, my, 1919; flowers nasturtium-yellow, dbl., slight fragrance; GM, NRS, 1919; Pemberton

Miriam Wilkins, HP, lp

'Miriam's Climber', LCl, op, 1950; bud ovoid, apricot-pink; flowers peach-pink, lighter at tip, large, bornein clusters, dbl., intense fragrance; foliage dark, soft; very vigorous, climbing (25 ft) growth; [Seedling X Edith Nellie Perkins]; Rosen, H.R.

Miriam's Pink Powderpuff, HP, pb

Miriam's Single White Wichurana, HWich, w; (found rose)

Miriana, *see* 'MEIburgana'

'Mirza', F, ob, 1974; bud ovoid; dbl., 35 petals, 4 in., cupped, moderate fragrance; foliage glossy; vigorous, upright, bushy growth; [Zorina X Samba]; Kordes; Willemse

Mischief, *see* 'MACmi'

'MISiljurika', F, mr, 1995; (Preservation); flowers bright red, blooms borne in small clusters, very dbl., 2.75 in., slight fragrance; some prickles; foliage medium, medium green, glossy; medium, upright growth; [Silver Jubilee X Paprika]; Bossom, W.E.

'Miss Ada', F, lp, 1998; flowers light pink, darker edges, reverse darker pink, single, 4–7 petals, 2 in.–4 in., borne singly and in small clusters, slight fragrance; prickles few, brown; foliage medium, medium green, semi-glossy; upright, bushy, medium growth; [Playgirl sport]; Pawlikowski, Martin & Elaine; Giles Rambli' Roses, 1998

'Miss Agnes C. Sherman', T, rb, 1901; flowers rose, salmon and red, Nabonnand

'Miss Alice de Rothschild', T, ly, 1910; flowers light canary-yellow, center deeper, dbl., moderate fragrance; Dickson, A.

'Miss All Australian Beauty', HT, dp, 1969; bud ovoid; flowers light red, reverse darker, medium, dbl., slight fragrance; upright growth; [Aztec X Impeccable]; Armbrust; Langbecker

Miss All-American Beauty, *see* 'MEIdaud'

Miss All-American Beauty, Climbing, *see* 'MEIdaudsar'

Miss All-American Dream™, *see* 'WILdrem'

'Miss Amelia Gude', HT, dy, 1921; flowers deep yellow center shading to cream, dbl., 35–40 petals, moderate fragrance; foliage dark; vigorous growth; [Columbia X Sunburst]; Lemon

'**Miss America**', HT, pb, 1938; flowers light pink, flushed salmon and gold, dbl., 65 petals, 6 in., moderate fragrance; foliage dark, leathery; vigorous growth; [Joanna Hill X S.M. Gustave V]; Nicolas; J&P

'**Miss Annamarie Bally**', HT, or, 1926; flowers reddish copper, reverse suffused whitish fawn, [Aspirant Marcel Rouyer X Lamia]; Easlea

Miss Annie Crawford, *see* '**Annie Crawford**'

Miss Atwood, T, ab

'**Miss Australia**', HT, mp, 1933; flowers pink, center salmon, very large, dbl., 50 petals, globular, intense fragrance; foliage thick; very vigorous, bushy growth; [Dame Edith Helen X Mme Segond Weber]; Knight, G.

'**Miss Blanche**', HT, w, 1980; (Kojack); bud long; flowers urn-shaped to blooms borne usually singly, dbl., 38 petals, exhibition form, no fragrance; straight, reddish prickles; foliage large, dark, leathery; upright growth; [Evening Star X Coquette]; Warriner, William A.; J&P

'**Miss Brisbane**', HT, lp, 1953; flowers shell-pink, medium, dbl., cupped, moderate fragrance; foliage light green; vigorous growth; [? X The Doctor]; Ulrick, L.W.

'**Miss C. E. van Rossem**', HT, mr, 1919; bud long, pointed; flowers crimson-scarlet shaded carmine and black, open, medium to small, semi-dbl., cupped, slight fragrance; foliage leathery, bronze, dark; vigorous, bushy growth; [Leuchtfeuer (HCh) X Red-Letter Day]; Verschuren

'**Miss California**', HT, dp, 1933; flowers deep glowing pink, [Dame Edith Helen sport]; Smith, J.

'**Miss Canada**', HT, pb, 1963; bud ovoid; flowers rose-madder, reverse silver, large, dbl., exhibition form, slight fragrance; foliage glossy, leathery; vigorous, upright, spreading growth; [Peace X Karl Herbst]; Blakeney; Eddie

'**Miss Caroline**', T, pb, 1997; flowers very full, medium, very dbl., 50–60 petals, borne in small clusters, moderate fragrance; foliage medium, medium green, semi-glossy; bushy, medium (4–5 ft) growth; [Duchesse de Brabant sport]; Knopf, Ruth

'**Miss Clipper**', HT, pb, 1942; bud long, pointed to ovoid; flowers pale salmon-pink shaded yellow, dbl., 25–30 petals, 3.5–4 in., exhibition form, moderate, spicy fragrance; foliage glossy, light; strong stems; vigorous, upright, bushy growth; [Angèle Pernet X Pres. Herbert Hoover]; Lammerts, Dr Walter; Armstrong Nursery

'**Miss Conner**', HT, ly, 1920; flowers canary-yellow on lemon-yellow, dbl., moderate fragrance; Dickson, A.

'**Miss Cynthia Forde**', HT, dp, 1909; flowers deep brilliant rose-pink, reverse lighter, large, dbl., moderate fragrance; long stems; vigorous, bushy growth; GM, NRS, 1909; Dickson, H.

Miss Daisy, *see* 'JACflare'

'**Miss Delightful**', F, my, 1966; flowers bright yellow, rosette form, borne in heavy clusters, dbl., 30 petals, 3 in., moderate fragrance; foliage glossy; vigorous, upright growth; [(Masquerade X Seedling) X Golden Scepter]; Sanday, John

Miss Dior, *see* HARencens

Miss Dovey™, *see* 'KINdov'

Miss Dovie, *see* 'KINdov'

'**Miss Edith Cavell**', Pol, dr, 1917; (Edith Cavell, Nurse Cavell); flowers scarlet-crimson overlaid velvety crimson; [Orléans Rose sport]; deRuiter; Spek

'**Miss England**', HT, w, 1936; flowers creamy, very large, dbl., moderate fragrance; foliage leathery, dark; vigorous, compact growth; Cant, B. R.

'**Miss Ethel Richardson**', HP, w, 1897; flowers white tinged pink, very large, dbl.; Dickson, A.; (Sangerhausen)

'**Miss Evelyn Davey**', T, lp; [Souv de S.A. Prince X Niphetos]; Williams, A., 1905; (Weatherly, L.)

Miss Finland, F, ob

Miss Flippins™, *see* 'TUCkflip'

'**Miss Flora Mitten**', LCl, lp, 1913; flowers soft pink, stamens yellow, single, 3 in.; vigorous growth; [R. wichurana X R. canina]; Lawrenson

'**Miss France**', Gr, or, 1955; (Pretty Girl); flowers bright scarlet, large, dbl., globular, moderate fragrance; foliage bronze; vigorous growth; [Peace X Independence]; Gaujard

'**Miss G. Mesman**', Cl Pol, mr; (Baby Rambler, Climbing); [Mme Norbert Levavasseur climbing sport]

'**Miss G. Radcliffe**', HT, lp; [Pharisaer X Unknown]; Williams, A., 1900; (Weatherly, L.)

'**Miss Georgie**', HT, ab, 1980; dbl., 35 petals, borne usually singly, moderate fragrance; long prickles; foliage large, semi-glossy; upright, compact growth; [South Seas X Unnamed seedling]; Warriner, William A.; J&P

Miss Harp, *see* 'TANolg'

'**Miss Hawaii**', HT, w, 1976; flowers creamy white, dbl., 30–35 petals, 5.5–6 in., exhibition form, intense fragrance; foliage leathery; very vigorous growth; [Hawaii sport]; Payne; Lone Star Nursery

'**Miss Helyett**', LCl, pb, 1909; (Miss Heylett); flowers bright carmine-pink, center yellowish salmon-pink, open, large, dbl., slight fragrance; vigorous, climbing (10–12 ft) growth; [R. wichurana X Ernest Metz]; Fauque

'**Miss Henriette Tersteeg**', HT, pb, 1922; flowers flesh and salmon-pink, dbl., slight fragrance; [Mme Abel Chatenay seedling X Mrs Joseph Hill]; Van Rossem

Miss Heylett, *see* '**Miss Helyett**'

'**Miss Hillcrest**', HT, or, 1969; flowers large, dbl., exhibition form, intense, fruity fragrance; foliage glossy; vigorous, tall growth; [Peace X Hawaii]; Curtis, E.C.; Kimbrew

Miss Hillcrest, *see* 'POUlsalfai'

'**Miss House**', HP, w, 1838; flowers satin white, House

'**Miss Huntington**', HT, ob, 1971; bud ovoid; flowers bright orange, large, dbl., exhibition form, moderate fragrance; foliage leathery; vigorous, upright growth; [Ma Perkins X San Francisco]; Patterson; Patterson Roses

Miss Ireland, *see* 'MACir'

Miss Italia, S, 1963; Mansuino; (Cavriglia)

'**Miss Jekyll**', Ayr, w; flowers light rose pink, dbl.; (14)

'**Miss Joan**', Cl HT, op, 1943; flowers copper-bronze and salmon-pink, large, dbl., globular, intense fragrance; foliage dark, glossy; very vigorous, climbing growth; [Ednah Thomas X Golden Dawn]; Duehrsen; California Roses

'**Miss Kate Moulton**', HT, pb, 1906; flowers rosy pink shaded rosy salmon, dbl., [Mme Caroline Testout X (La France X Mrs W.J. Grant)]; Monson; Minneapolis Floral Co.

'**Miss Kate Sessions**', LCl, pb, 1953; flowers deep rose-pink on white base, reverse shell-pink, open, large, dbl., profuse, intermittent bloom, moderate fragrance; foliage light green, leathery; moderate, climbing growth; [Heart of Gold X Ednah Thomas]; Hieatt

'**Miss Koganei**', Cl Min, rb, 1985; flowers medium red, white eye, yellow stamens, single, 7–10 petals, cupped; few prickles; foliage small; vigorous (to 3 ft) growth; [Nozomi X ?]; Asano, S.

'**Miss Leslie**', HT, w, 1998; flowers white with light pink edges, full, very dbl., 26–40 petals, exhibition form, borne mostly singly, slight fragrance; prickles moderate; foliage medium, medium green, semi-glossy; upright, tall growth; [Rina Hugo sport]; Garrett, Troy O.

Miss Liberté, *see* 'AROvule'

Miss Liberty, *see* 'AROvule'

'**Miss Liberty**', LCl, mp, 1956; bud ovoid; flowers tyrian rose, borne in large clusters, semi-dbl., 15–20 petals, 3.5 in., cupped, repeat bloom, moderate fragrance; foliage dark, leathery; strong stems; vigorous, climbing (10–12 ft) growth; [New Dawn X World's Fair, Climbing]; Boerner; Stuart

'**Miss Lolita Armour**', HT, or, 1919; flowers deep coral-red suffused coppery red, base yellow, very large, dbl., cupped, intense fragrance; GM, Bagatelle, 1921; Howard, F.H.; H&S

'Miss Lolita Armour, Climbing', Cl HT, or; H&S, about 1925

Miss Lowe, *see* **'Miss Lowe's Variety'**

'Miss Lowe's Variety', Ch, mr; (Miss Lowe); flowers bright red, single, recurrent bloom; dwarf growth; (14); [Possibly Slater's Crimson China sport]

'Miss M.J. Spencer', HT, my, 1920; flowers clear bright golden yellow, dbl.; Dickson, H.

'Miss M. Robertson', Pol, lp; [Mlle Cécile Brünner X Unknown]; Williams, A., 1902; (Weatherly, L.)

'Miss M.J.F. Gostling', Pol, lp; [Miss Pollock X Unknown]; Williams, A., 1911; (Weatherly, L.)

'Miss Maine 1999', S, dp, 1999; flowers deep pink, reverse medium pink, dbl., 26–40 petals, 2 in., borne in small clusters, slight fragrance; many prickles; foliage small, medium green, semi-glossy; bushy, medium (4 ft) growth; hardy to zone 4; [Henry Kelsey sport]; Law, Stephen

'Miss Marion Manifold', Cl HT, mr; (Marion Manifold); flowers velvety scarlet, shaded crimson, large, dbl., globular, moderate fragrance; foliage large, leathery; vigorous, climbing (12 ft) growth; Adamson, 1913; Brundrett

Miss Mary, *see* 'GELmiss'

'Miss May Marriott', HT, ab, 1917; flowers glowing apricot, [Mme Edouard Herriot sport]; Robinson, T.

'Miss May Thurlowe', T, lp; Williams, A., 1904; (Weatherly, L.)

'Miss Mayne', HT, lp; [Marie van Houtte X General Jacqueminot]; Williams, A., 1900; (Weatherly, L.)

'Miss Middleton', F, op, 1963; flowers coral-pink, borne in small clusters, dbl., 20 petals, 2.5–3 in., cupped; foliage light green; low, bushy growth; [Independence X Masquerade]; Hill, A.

'Miss M'liss', F, mp, 1958; bud short pointed; flowers phlox-pink, dbl., 25–30 petals, 2.5–3 in., exhibition form, slight fragrance; very vigorous, upright growth; [Garnette X Garnette seedling]; Jelly; E.G. Hill Co.

'Miss Modesto', HT, my, 1934; flowers pure yellow, very large, dbl., exhibition form, moderate fragrance; [Rev. F. Page-Roberts sport]; Brooks, L.L.; Brooks & Son

'Miss Muffett', Min, mp, 1955; bud rather mossy; flowers apple-blossom-pink, compact growth, dbl., 80 petals, 4–6 in.; [Baby Bunting X Tom Thumb]; Robinson, T.

'Miss Murine', F, dp, 1937; flowers deep pink, base yellow, small, dbl.; upright, bushy growth; [Unnamed seedling X Cécile Brunner]; Fitzgerald

'Miss N. Robertson', Pol, lp; [Mlle Cécile Brünner X Unknown]; Williams, A., 1902; (Weatherly, L.)

'Miss P. Williams', Pol, lp; [Perle d'Or X Unknown]; Williams, A., 1902; (Weatherly, L.)

Miss Pam Ayres, *see* 'KORmarie'

Miss Pearl™, *see* 'HOOpearl'

Miss Perfect™, *see* 'JACmiss'

'Miss Personality', F, pb, 1972; bud ovoid; flowers cerise-pink, white eye, open, medium, semi-dbl., slight fragrance; foliage small, dark, leathery; vigorous, bushy growth; [Pink Parfait X (Ophelia X Parkdirektor Riggers)]; Sherwood; F. Mason

'Miss Pollock', P, lp; Williams, A., 1906; (Weatherly, L.)

Miss Prissy, HT, w, 1995; Wells

Miss Reus, HT, op; Dot

Miss Rita, F, lp, 1986; Robertson

Miss Rose, Gr; Dorieux; (Cavriglia)

'Miss Rose Hill', HT, lp; Williams, A., 1913; (Weatherly, L.)

'Miss Rowena Thom', HT, pb, 1927; flowers fiery rose and rosy mauve, center washed gold, dbl., 50 petals, 6 in., intense fragrance; vigorous growth; [Radiance X Los Angeles]; H&S

'Miss Rowena Thom, Climbing', Cl HT, pb, 1937; van Barneveld; California Roses

Miss Schweiz, *see* TANziewsim

'Miss Talmadge', HT, dy, 1927; dbl., moderate fragrance; [Constance sport]; Pacific Rose Co.

'Miss Universe', HT, rb, 1956; bud long, pointed; flowers orange-red, reverse tinted copper, large, moderate fragrance; foliage dark; vigorous growth; [(Peace X Seedling) X Seedling]; Gaujard

'Miss Willmott', HT, ly, 1917; flowers soft sulfur-cream, edges flushed pale pink, well formed, dbl., moderate fragrance; vigorous growth; GM, NRS, 1916; McGredy

'Miss Windsor', HT, mr, 1967; flowers medium, dbl., exhibition form, intense fragrance; foliage light green, leathery; very vigorous, upright growth; [Tropicana sport]; Heron

'Miss Winifred Denham', T, lp; [Marie van Houtte X Unknown]; Williams, A., 1911; (Weatherly, L.)

'Mission Bells', HT, pb, 1949; bud long, pointed; flowers vermilion-pink, dbl., 43 petals, 5 in., exhibition form, moderate fragrance; foliage dark, soft; vigorous, bushy growth; AARS, 1950; [Mrs Sam McGredy X Malar-Ros]; Morris; Germain's

'Mission Supreme', HT, ab, 1981; bud pointed; flowers pale peach pink to apricot, blooms borne singly, dbl., 30 petals, moderate fragrance; straight, red-brown prickles; foliage deep green; vigorous, bushy, medium growth; [City of Gloucester X Unnamed seedling]; Sanday, John

'Mississippi', HT, dr, 1976; bud pointed; flowers deep red, dbl., 38 petals, 4.5–5 in., moderate, damask fragrance; foliage dark; upright growth; [Charlotte Armstrong X Mister Lincoln]; Williams, J. Benjamin

Mississippi, *see* **'Mme Charles Sauvage'**

'Mississippi Rainbow', HT, yb, 1977; flowers full, dbl., 41 petals, 4 in., slight fragrance; foliage thick, glossy; upright growth; Graham; South Forrest Rose Nursery

'Missy', Min, mr, 1978; bud pointed; flowers cardinal-red, small, dbl., 22 petals, 1.5 in., slight fragrance; foliage small; compact, bushy growth; [Seedling X Seedling]; Lyon

'Missy Eleanor', S, pb, 1999; flowers light pink, yellow eye, reverse medium pink, single, 5–11 petals, 2.5 in., borne in large clusters, moderate fragrance; prickles moderate; foliage medium, light green, glossy; bushy, medium (4–6 ft) growth; [Anna Ford X Ballerina]; Watson, Thomas L, & Glenda, 1996

'Mistee', Min, ly, 1979; bud long, pointed; flowers white, tinted yellow, small, dbl., 28 petals, 1.5 in., flat, moderate fragrance; foliage small; bushy, upright growth; [Little Darling X Peachy White]; Moore, Ralph S.; Sequoia Nursery

'Mister America', LCl, dr, 1974; flowers blood-red, very large, dbl., exhibition form, abundant bloom, moderate fragrance; foliage leathery; very vigorous, climbing growth; [(Paul's Scarlet Climber X Golden Climber) X Pinocchio]; Zombory; General Bionomics

Mister Chips, HT, 1970; Dickson, A.; (Cavriglia)

'Mister Lincoln'®, HT, dr, 1964; bud urn-shaped; dbl., 35 petals, 4.5–6 in., exhibition form, intense fragrance; foliage leathery, matt, dark; vigorous growth; AARS, 1965; [Chrysler Imperial X Charles Mallerin]; Swim & Weeks; C-P

'Mister Lincoln, Climbing', Cl HT, dr, 1974; Ram

Mister Otis, *see* 'CURmist'

'Mister Softee', HT, w, 1964; flowers creamy white edged pink, open, dbl., 40 petals, 4.5 in., moderate fragrance; foliage glossy, dark; very free growth; Morton's Rose Nursery

'Mistica', HT, m, 1966; bud ovoid; flowers lilac, very large, dbl., cupped, intense fragrance; foliage dark; vigorous growth; [Sterling Silver X Intermezzo]; Dot; Minier

Mistigri, *see* 'TANnimoll'

Mistraline®, *see* LAPmisal

'Mistress Bosanquet', B, lp, 1832; (Mrs Bosanquet); flowers rosy flesh, very dbl.; vigorous growth; Laffay, M.

'Mistress Pat', Pol, mp, 1928; flowers chatenay pink, moderate fragrance; Lilley

Mistress Quickly, *see* 'AUSky'

'Misty', HT, w, 1965; bud ovoid, pointed; flowers creamy white, to formal, dbl., 35 petals, 4 in., cupped, moderate, tea fragrance; foliage large, leathery; vigorous, upright growth; [Mount Shasta X Matterhorn]; Armstrong, D.L.; Armstrong Nursery

'Misty Dawn', Min, w, 1979; bud ovoid, pointed; flowers pure white, dbl., 33 petals, 1 in., cupped, foliage small, dark; vigorous, compact, spreading growth; [Charlie McCarthy X Unnamed seedling]; Schwartz, Ernest W.; Nor'East Min. Roses

'Misty Delight', HT, mp, 1990; dbl., 26–40 petals, borne mostly singly; foliage large, dark green, glossy; upright growth; [Blue Wonder X Unnamed seedling]; Christensen, Jack E.; Vaughan's Seed Co., 1991

'Misty Gold', F, my, 1954; bud ovoid, pointed; flowers empire-yellow, dbl., 45–50 petals, 3.5–4 in., cupped, moderate fragrance; foliage glossy; vigorous growth; [Unnamed Floribunda (Yellow, semi-dbl., self seedling)]; Boerner; Stark Bros.

'Misty Morn', HT, ly, 1949; flowers pale lemon-yellow, pointed, large, dbl., 45 petals, slight fragrance; foliage dark; [Unnamed seedling X Mrs Charles Lamplough]; McGredy

Misty Morning, *see* 'JUDmorn'

Misty Pink, F, mp

Misty Veil, S, mp, 1999; Williams, J. Benjamin

'Mitcheltonii', S, mp, 1966; (Rosa Xmitcheltonii); flowers pink, small, single, profuse spring bloom; thornless; foliage small, leathery; very vigorous, climbing growth; [R. multiflora X I.X.L.]; Armbrust

Mitsouko®, *see* 'DELnat'

'Mitzi', HT, yb, 1956; flowers pearly tints flushed mauve-rose, well formed, dbl., 35 petals, moderate fragrance; foliage dark; strong stems; upright, bushy growth; GM, Rome, 1956; [(Peace X Mme Joseph Perraud) X (Mrs Pierre S. duPont X Mrs John Laing)]; Meilland, F.; URS

Mitzi, *see* MEIlenangal

Miwaku, HT, w, 1988; Suzuki, Seizo

Mix 'n' Match™, *see* 'CHEwily'

Mixed Marriage, F, lb; [Bridal Pink sport]; Ruston, D., 1987; (Weatherly, L.)

'Miyabi', HT, w, 1977; bud ovoid; flowers near white, dbl., 30 petals, 4–4.5 in., exhibition form, slight fragrance; upright growth; [(Amatsu-Otome X Samba) X (Kordes' Perfecta X American Heritage)]; Teranishi, K.; Itami Bara-en, 1976

'Miyagino', Cl Min, pb, 1978; bud rounded; flowers light pink, small, single, 5 petals, flat; foliage tiny; [Nozomi X Unnamed seedling]; Onodera, Toru F.; S. Onodera

Mizar, *see* BARmiz

Mlada, F, yb, 1983; flowers yellow and red, medium, dbl., slight fragrance; Urban, J.; (Sangerhausen)

'Mlle Alice Furon', HT, ly, 1896; flowers yellowish white, [Lady Mary Fitzwilliam X Mme Chedane-Guinoisseau]; Pernet-Ducher

'Mlle Alice Leroy', m, 1842; flowers light violet-pink, medium, dbl.; Vibert; (Sangerhausen)

'Mlle Alice Rousseau', Pol, lp, 1903; flowers small, dbl.; Vilin; (Sangerhausen)

'Mlle Andree Worth', B, lp, 1890; Lévêque

'Mlle Annie Wood', HP, mr, 1866; (Annie Wood); flowers clear red, large, dbl., recurrent bloom, moderate fragrance; Verdier, E.

'Mlle Augustine Guinoisseau', HT, lp, 1889; flowers large, dbl., Guinoisseau; (Sangerhausen)

'Mlle Bep van Rossem', HT, my, 1926; flowers deep canary-yellow, dbl.; [Unnamed seedling X Souv. de Claudius Pernet]; Van Rossem

Mlle Berthe Leveque, *see* **'Mlle Berthe Levet'**

'Mlle Berthe Levet', HP, mp; (Mlle Berthe Leveque)

'Mlle Blanche Lafitte', B, w, 1851; flowers whitish-rose, dbl., blooms in clusters, repeats in autumn, vigorous growth; Pradel

'Mlle Bonnaire', HP, w, 1859; flowers white, center sometimes pink, dbl.; growth moderate; Pernet Père

'Mlle Cécile Brünner', Pol, lp, 1881; (Cécile Brünner, Mignon, Mme Cécile Brünner, Sweetheart Rose); bud long, pointed; flowers bright pink on yellow ground, small blooms in clusters, dbl., moderate; very few prickles; foliage sparse, soft, dark, 3–5 leaflets; growth like a tea, dwarf; (14); [Said to be Dbl. fld. Multiflora X Souv. d'un Ami]; Ducher, Vve.; Pernet-Ducher

'Mlle Cécile Brünner, Climbing', Cl Pol, lp, 1894; (Cécile Brünner, Climbing, Mignon, Climbing, Mme Cécile Brünner, Climbing, Sweetheart Rose, Climbing); Hosp

Mlle Cécile Brünner, Climbing, Cl Pol, lp, 1904; Ardagh

'Mlle Claire Andruejol', HT, lp, 1920; flowers pale pink tinted carmine, dbl., slight fragrance; [Comte G. de Rochemur X Mme Maurice de Luze]; Schwartz, A.

'Mlle Claire Jacquier', N, ly, 1888; flowers small to medium, dbl., moderate fragrance; Bernaix, A.; (Sangerhausen)

'Mlle Claire Truffaut', B, lp, 1887; flowers medium, dbl.; Verdier, E.; (Sangerhausen)

'Mlle Claudine Perreault', T, mp, 1885; flowers rose flesh, center darker, very large, dbl., free bloom; Lambert, E.

'Mlle Danielle Dumur', HT, lp, 1910; flowers large, dbl., moderate fragrance; Laroulandie; (Sangerhausen)

Mlle de Dinant, F, 1966; Lens, Louis; (Cavriglia)

'Mlle de Meux', HT, dp, 1901; flowers carmine-pink, medium, dbl.; Berland; (Sangerhausen)

'Mlle de Morlaincourt', HT, pb, 1934; flowers pink and yellow, [Cécile Walter X Korovo]; Walter, L.; Amis des Roses

'Mlle de Sombreuil', T, w, 1851; Robert

'Mlle Emilienne Moreau', HT, yb, 1920; flowers dark yellow and creamy pink, medium, semi-dbl.; Verschuren; (Sangerhausen)

'Mlle Eugénie Verdier', HP, mp, 1869; flowers clear silvery pink, reverse silvery white, large, dbl., 40 petals, moderate fragrance; vigorous, upright growth; Guillot et Fils

'Mlle Eugénie Verdier', HP, lp; Schwartz, 1872

'Mlle Fernande Dupuy', Pol, dp, 1899; flowers carmine-pink, small, dbl.; Vigneron; (Sangerhausen)

'Mlle Franziska Kruger', T, op; flowers coppery yellow and pink, center often green, large blooms, very dbl., moderate fragrance; weak stems; hardy for this class; (14); [Catherine Mermet X Gén. Schablikine]; Nabonnand, G., 1879

'Mlle Geneviève Godard', T, yb, 1889; flowers yellow with pink and orange, medium, dbl.; Godard; (Sangerhausen)

'Mlle Hélène Gambier', HT, op, 1895; (Helen Gambier); flowers salmon-pink to coppery rose, large, dbl., intense fragrance; Pernet-Ducher

'Mlle Henriette Martin', HT, w, 1936; flowers white, shaded ivory, edges lightly tinted pale pink, moderate fragrance; Reymond; Vially

'Mlle Honorine Duboc', HP, dp, 1894; flowers very large, dbl.; Duboc; (Sangerhausen)

'Mlle Irene Hennessy', HT, or, 1923; flowers bright vermilion-orange, dbl., moderate fragrance; [George C. Waud X Unnamed seedling]; Guillot, P.

'Mlle Jeanne Lenail', Pol, mr, 1924; flowers bright ruby-red shaded carmine, large, dbl.; [Mrs W.H. Cutbush X Mme Taft]; Schwartz, A.

'Mlle Jeanne Philippe', T, yb, 1898; Godard

'Mlle la Comtesse de Leusee', T, pb, 1878; Nabonnand

Mlle Lenari la Granada, HT, 1982; Dot, Simon; (Cavriglia)

'Mlle Louise Chabier', HP, lp, 1867; flowers large, dbl.; Cochet, S.; (Sangerhausen)

'Mlle Madelaine de la Roche', T, 1890; Corboeuf; (Cavriglia)

'Mlle Madeleine de Vauzolles', B, lp, 1881; Vigneron

'Mlle Marcelle Gaugin', Pol, lp, 1910; flowers creamy pink, small, dbl.; Corboeuf; (Sangerhausen)

'Mlle Marie Dauvesse', HP, mp; Vigneron, 1861

'Mlle Marie Drivon', B, pb, 1887; flowers medium, very dbl.; Schwartz; (Sangerhausen)

'Mlle Marie Gaze', N, pb, 1892; flowers yellowish-pink with dark yellow, medium, dbl.; Godard; (Sangerhausen)

'Mlle Marie Louise Bourgeois', M, lp, 1891; flowers light pink, edged white; Corboeuf

'Mlle Marie Magat', HP, mr, 1889; flowers carmine-red, large, dbl.; Liabaud; (Sangerhausen)

'Mlle Marie Mascuraud', HT, w, 1909; flowers white tinted flesh; Bernaix, P.

'Mlle Marie Moreau', T, w; flowers silver-white flushed crimson, well formed, Nabonnand, G., 1879

Mlle Marie van Houtte, *see* **'Marie van Houtte'**

'Mlle Marthe Carron', HWich, w, 1931; flowers white, slightly tinted pink on opening, borne in clusters of 40–50, vigorous growth; [R. wichurana sport X R. wichurana]; Mermet

'Mlle Marthe Moisset', HT, my, 1935; flowers chrome-yellow on ochre ground, very large, cupped, moderate fragrance; foliage glossy, dark; vigorous, erect, bushy growth; [Mme Henri Queuille X Unnamed seedling]; Ducroz

'Mlle Mathilde Lenaerts', T, lp, 1880; flowers medium, dbl., moderate fragrance; Levet; (Sangerhausen)

'Mlle Maurand', HP, lp; flowers pale flesh

'Mlle Renée Denis', HP, w, 1906; flowers white with light pink, dbl.; Chédane-Guinoisseau; (Sangerhausen)

Mlle Rosa Bonheur, *see* **'Rosa Bonheur'**

'Mlle Simone Beaumez', HT, w, 1907; flowers flesh-white, center sometimes tinted saffron-yellow, dbl., moderate fragrance; Pernet-Ducher

'Mlle Sontag', HGal, dp; flowers deep pink, reverse pale blush

'Mlle Stella Mallerin', HT, w, 1926; bud pointed; flowers white, center slightly shaded cream, very large, dbl., cupped, moderate fragrance; Chambard, C.

'Mlle Thérese Levet', HP, mp, 1864; flowers large, very dbl., moderate fragrance; Levet; (Sangerhausen)

'Mlle Yvette Bouquil', F, yb, 1955; flowers yellow edged orange, bushy growth; Privat

'Mme A. Bouchayer', HT, op, 1927; flowers shrimp-pink, base indian yellow; Siret-Pernet

'Mme A. Chatain', HT, 1940; Mallerin, C.; (Cavriglia)

'Mme A. Galland', HT, pb, 1928; flowers rose-pink, shaded shrimp-pink, dbl., intense fragrance; [Pharisaer X (Constance X Unnamed Hybrid Tea]; Mallerin, C.; Laperrière

'Mme A. Labbey', HP, pb, 1843; flowers pink and lilac, medium or small

'Mme A. Lerche', HT, mp, 1928; flowers china-rose shaded carmine, reverse silvery rose, dbl.; [Mme L. Hot X Mrs Henry Winnett]; Bernaix, P.

Mme A. Meilland, *see* **'Peace'**

Mme A. Meilland, Climbing, *see* **'Peace, Climbing'**

'Mme A. Roure', HT, yb, 1932; flowers brilliant chrome-yellow shaded salmon, to amaranth-red, dbl., moderate fragrance; [Wilhelm Kordes X Mev. G.A. van Rossem]; Lens

'Mme Abel Chatenay', HT, pb, 1895; bud pointed; flowers pale pink, center deeper, reverse carmine-pink, dbl., 3 in., moderate fragrance; foliage bronze when young; [Dr Grill X Victor Verdier]; Pernet-Ducher

'Mme Abel Chatenay, Climbing', Cl HT, pb, 1917; Page; Easlea

'Mme Achille Fould', T, yb, 1903; flowers yellow, shaded carmine rose and salmon, very large, dbl., globular, Lévêque

'Mme Achille Villey', HT, or, 1939; flowers coral-red, tinted yellow, dbl.; foliage dark; very vigorous growth; [Charles P. Kilham X Mrs Aaron Ward]; Colombier

Mme Adelaide Ristori, B, dp; flowers cerise and fawn; Praden, 1861

'Mme Adolphe Dahair', T, ly, 1901; flowers large, dbl., moderate fragrance; Puyravaud; (Sangerhausen)

'Mme Adolphe Lafont', HT, ab, 1921; flowers deep apricot-red tinted buff, semi-dbl., intense fragrance; [Joseph Hill sport]; Croibier

'Mme Agathe Nabonnand', T, lp, 1886; (Agathe Nabonnand); flowers rosy flesh tinted amber, petals shell-like, very large; Nabonnand, G.

'Mme Albert Barbier', HP, ob, 1925; flowers salmon, tinted nankeen yellow, center darker, large, dbl., 50 petals, cupped, recurrent bloom, slight fragrance; vigorous, bushy growth; [Frau Karl Druschki X ?]; Barbier; Dreer

'Mme Albert Bernardin', T, w; flowers white, shaded with carmine, center yellow, Mari, 1905

'Mme Albert Gilles', HT, op, 1934; flowers light coral-pink, very large, very dbl.; [Unnamed variety X Jean C.N. Forestier]; Guillot, H.; Mallerin

'Mme Alégatière', Pol, dp, 1888; flowers carmine-pink, medium, dbl.; Alégatière; (Sangerhausen)

'Mme Alexandre', HT, rb, 1926; flowers crimson-vermilion-red, shaded velvety purple, dbl.; Walter, L.

'Mme Alexandre Charvet', HT, or, 1943; bud long, pointed; flowers orange-red, edged lilac, medium, dbl., cupped, slight fragrance; foliage leathery; moderate growth; [Charles P. Kilham X (Charles P. Kilham X Mme Joseph Perraud)]; Meilland, F.; A. Meilland

'Mme Alexandre Dreux', HT, my, 1921; bud pointed; flowers golden yellow, large, dbl.; [Rayon d'Or X Primerose]; Soupert & Notting

'Mme Alfred Carrière', N, w, 1879; flowers pale pinkish white, large, dbl., globular, recurrent bloom, intense fragrance; vigorous, climbing growth; (21); Schwartz, J.

'Mme Alfred de Rougemont', HP, lp, 1862; flowers white tinted pink, medium; vigorous growth; [Blanche Lafitte X Sappho (Vibert's)]; Lacharme, F.

'Mme Alfred Ponnier', HT, w, 1920; Bernaix, P.

'Mme Alfred Schisselé', HT, m, 1930; flowers lilac-white, center coppery orange, large, dbl., moderate fragrance; foliage bronze; [Frau Felix Tonnar X Angèle Pernet]; Leenders, M.

'Mme Alice Garnier', HWich, pb, 1906; flowers bright rose, center yellow to light pink, small, intense, sweet fragrance; foliage dark, glossy; [R. wichurana X Mme Charles]; Fauque

'Mme Ancelot', HRg, mp, 1901; Müller, Dr F.

'Mme André Charmet', HT, mp, 1921; flowers carnation-pink, dbl., moderate fragrance; [Mme Mélanie Soupert X Mme Maurice de Luze]; Croibier

'Mme André de Halloy', HT, ob, 1929; flowers orange, reverse salmon-pink, very dbl., slight fragrance; [Gloire de Hollande X Benedicte Seguin]; Ketten Bros.

'Mme André Dulin', HT, op, 1959; flowers bright coppery pink, large, dbl.; moderate fragrance; vigorous, bushy growth; [Opera X Ville de Gand]; Gaujard

'Mme André Gillier', HT, rb, 1934; bud pointed; flowers coppery red, shaded yellow, semi-dbl., cupped, slight fragrance; [Padre X Unnamed seedling]; Reymond; Vially

'Mme André Leroy', HP, mp, 1864; flowers salmon-rose, large, dbl.; vigorous growth; Trouillard

'Mme André Saint', HP, w, 1926; flowers milk-white to pure white, center clear chamois, dbl., moderate fragrance; few thorns; stocky growth; [Frau Karl Druschki X Benedicte Seguin]; Barbier

'Mme Angélique Veysset', HT, lp, 1890; flowers pink tinted white, large, dbl.; Veysset

'Mme Annette Aynard', HT, w, 1919; flowers milk-white edged pink, passing to amber-yellow, dbl., moderate fragrance; [Mme Caroline Testout X Prince de Bulgarie]; Leenders, M.

'Mme Anth. Kluis', Pol, op, 1924; flowers salmon-pink tinted orange, large, borne in clusters, semi-dbl.; Kluis; Kluis & Koning

'Mme Antoine Mari', T, pb, 1901; flowers rosy flesh, shaded lilac and rose, large, dbl., moderate fragrance; (14); Mari, A.

'Mme Antoine Montagne', HT, lp, 1930; flowers flesh-pink, with reflexes of old ivory, Richardier

'Mme Antoine Rebe', T, rb, 1900; Laperrière

'Mme Armand Souzy', HT, rb, 1945; flowers geranium-red and saffron-yellow, very large; [Charles P. Kilham X (Charles P. Kilham X Margaret McGredy)]; Meilland, F.

'Mme Arthaud', HT, ob, 1938; flowers deep orange, large; very dbl., slight fragrance; vigorous growth; [Charles P. Kilham X (K. of K. X Mari Dot)]; Mallerin, C.

'Mme Arthur Oger', Cl B, mp, 1899; flowers brilliant pink, very large, dbl.; very vigorous, climbing growth; Oger

'Mme Arthur Robichon', Pol, dp, 1912; flowers carmine-pink, small, dbl.; Robichon; (Sangerhausen)

'Mme Auguste Chatain', HT, op, 1940; bud very long; flowers coral-salmon, well formed, large, dbl.; very vigorous growth; GM, Bagatelle, 1939; Mallerin, C.; A. Meilland

'Mme Auguste Choutet', LCl, ob, 1901; flowers orange-yellow, [William Allen Richardson X Kaiserin Auguste Viktoria]; Godard

'Mme Auguste Nonin', HWich, dp, 1914; flowers deep shell-pink, center white, small, borne in clusters, semi-dbl., cupped, slight fragrance; foliage large, glossy; very vigorous, climbing (15–20 ft), open growth; [Dorothy Perkins X Blush Rambler (?)]; Nonin

'Mme Augustine Hammond', HT, mp, 1897; flowers very large, dbl.; Vigneron; (Sangerhausen)

'Mme Autrand', HT, ob, 1922; flowers coppery orange, dbl., intense fragrance; [Mme Caroline Testout X Prince de Bulgarie]; Leenders, M.

'Mme Azelie Imbert', T, 1870; Levet, F.; (Cavriglia)

'Mme Ballu', HRg, m, 1901; flowers violet-pink, medium, semi-dbl.; Roseraie de l'Hay; (Sangerhausen)

'Mme Bardou Job', HT, ly, 1913; flowers canary-yellow, center chrome-yellow, semi-dbl., slight fragrance; [Prince de Bulgarie sport]; Dubreuil

'Mme Barthélemy Levet', T, ly; flowers canary-yellow, Levet, A., 1879

'Mme Baulot', HP, mp, 1885; flowers large, dbl.; Lévêque; (Sangerhausen)

'Mme Bérard', Cl T, ob; flowers salmon-yellow shaded salmon-rose, large, dbl., cupped, moderate fragrance; [Mme Falcot X Gloire de Dijon]; Levet, F., 1870

'Mme Berkeley', T, pb, 1898; Bernaix, A.

'Mme Berthe de Forge', HT, op, 1935; flowers orange-coral, tinted coppery salmon, very large, cupped; foliage bronze; Chambard, C.

'Mme Blondel', HT, mp, 1899; flowers very large, dbl., moderate fragrance; Veysset; (Sangerhausen)

'Mme Blytha Pearkes', HT, ab, 1968; flowers light apricot flushed yellow and pink, medium, dbl.; bushy, compact growth; [Karl Herbst X Lady Hillingdon]; Blakeney; Eddie

'Mme Boll', HP, dp, 1859; flowers carmine-rose, large, recurrent bloom, foliage 5 leaflets per leaf; vigorous growth; Boll, Daniel; Boyeau

'Mme Bollaert', HT, rb, 1938; bud long, coppery red; flowers carmine, shaded nasturtium-red, very large, dbl., cupped, intense fragrance; strong stems; very vigorous, bushy growth; [Ami F. Mayery X Unnamed seedling]; Chambard, C.

Mme Bollinger, F, rb, 1972; flowers coppery red with dark yellow, medium, semi-dbl.; McGredy, Sam IV; (Sangerhausen)

'Mme Bravy', T, w, 1846; (Adèle Pradel, Alba Rosea, Danzille, Joséphine Maltot, Mme de Sertot, Mme Denis); flowers creamy white shaded blush, dbl., moderate fragrance; Guillot Père

Mme Bruno Coquatrix, *see* PINturnia

'Mme Butterfly', HT, lp, 1918; flowers light creamy pink; [Ophelia sport]; Hill, E.G., Co.

'Mme Butterfly, Climbing', Cl HT, lp, 1926; Smith, E.P.

'Mme Byrne', N, w, 1840; flowers cream color, center rose, large, dbl.; [Lamarque seedling]; Buist

'Mme C. Chambard', HT, lp, 1911; flowers rosy flesh-pink, shaded salmon, base yellow, dbl., 72 petals, moderate fragrance; [Frau Karl Druschki X Prince de Bulgarie]; Chambard, C.

'Mme C. Ligier', T, pb, 1900; flowers pink with darker red, large, very dbl., intense fragrance; Berland; (Sangerhausen)

'Mme C. Richardier', HT, my, 1924; flowers yellow, passing to clear yellow, dbl.; Richardier

'Mme Camille', T, lp, 1871; flowers aurora-pink, veined, with white reflections, well formed, large, dbl., moderate fragrance; Guillot et Fils

'Mme Camille Laurens', HT, or, 1956; flowers crimson-red tinted orange, well formed, moderate fragrance; foliage bright green; strong stems; [Peace X Happiness]; Dorieux; Pin

'Mme Caristie Martel', HT, ly, 1916; flowers pure sulfur-yellow, center deeper, 5–6 in.; GM, Bagatelle, 1917; Pernet-Ducher

'Mme Carnot', N, yb, 1889; flowers golden yellow tinged orange, center darker, large, very dbl., globular, moderate fragrance; vigorous, climbing growth; [William Allen Richardson seedling]; Moreau et Robert

'Mme Caroline Küster', N, yb, 1872; flowers yellow and orange blend, medium, dbl., moderate fragrance; Pernet; (Sangerhausen)

'Mme Caroline Schmitt', N, yb, 1878; (Caroline Schmitt); flowers salmon-yellow, dbl., recurrent bloom, [Solfaterre seedling]; Schmitt

'Mme Caroline Testout', HT, mp, 1890; bud pointed; flowers bright satiny rose, center darker, edged soft carmine-pink, dbl., moderate fragrance; foliage rich green, soft; vigorous, bushy growth; (28); [Mme de Tartas X Lady Mary Fitzwilliam]; Pernet-Ducher

'Mme Caroline Testout, Climbing', Cl HT, mp, 1901; Chauvry

'Mme Ceccaldi', HT, op, 1938; bud long; flowers salmon-carmine-pink, shaded vermilion, very large, dbl.; strong stems; very vigorous growth; [Soeur Thérèse X Unnamed seedling]; Chambard, C.

Mme Cécile Brünner, *see* 'Mlle Cécile Brünner'

Mme Cécile Brünner, Climbing, *see* 'Mlle Cécile Brünner, Climbing'

'Mme Cecile Morand', HP, 1890; Corboeuf; (Cavriglia)

'Mme Céline Noirey', T, lp, 1868; flowers soft pink, reverse purple, large, very dbl., moderate fragrance; Guillot et Fils

'Mme Céline Touvais', HP, mp, 1859; flowers large, dbl.; Touvais; (Sangerhausen)

'Mme César Brunier', HP, mp, 1888; flowers large, very dbl., moderate fragrance; Bernaix, A.; (Sangerhausen)

'Mme Chaban Delmas', HT, mr, 1957; flowers bright red, well formed, large; foliage bright green; Privat

'Mme Chamouton-Murgue', HT, op, 1925; flowers orange-carmine, shaded vermilion, very large, dbl., cupped, moderate fragrance; foliage dark; vigorous, erect, branching growth; [Unnamed seedling X Mrs Edward Powell]; Chambard, C.

'Mme Charles', T, yb, 1864; flowers yellow, center salmon; [Mme Damaizin X ?]; Damaizin

'Mme Charles Allizon', HT, w, 1928; flowers rosy white, edges tinted yellow, dbl.; [Mme Vittoria Gagniere X Lady Pirrie]; Schwartz, A.

'Mme Charles Baltet', B, lp, 1865; Verdier, E.

'Mme Charles Détreaux', B, mr; flowers bright carmine-red, large

'Mme Charles Dubreuil', HT, pb, 1911; flowers salmon-rose, reverse shaded carmine, dbl.; [Pharisaer sport]; Guillot, P.

'Mme Charles Frederic Worth', HRg, dp, 1889; (C.F. Worth); flowers rosy carmine, fading, large, borne in large clusters, semi-dbl., profuse early bloom, but sparse in summer and fall, moderate fragrance; vigorous growth; Schwartz, Vve.

'Mme Charles Guillaud', HT, op, 1943; flowers orange-pink with fiery tints, Mallerin, C.

'Mme Charles Guillot', HT, pb, 1943; flowers carmine-pink and orange, large, dbl.; Mallerin, C.; (Sangerhausen)

'Mme Charles Haas', HT, w, 1930; flowers amber-white, tinted flesh-white, large, dbl., 60–70 petals, exhibition form, moderate fragrance; foliage leathery; vigorous, free branching growth; [Mme Abel Chatenay X Golden Emblem]; Ketten Bros.

'Mme Charles Joly', HT, op, 1942; flowers bright salmon shaded coppery yellow, Chambard, C.

'Mme Charles Lejeune', LCl, lp, 1924; flowers soft pink, [Mme Caroline Testout X Lady Ashtown]; Vandevelde

'Mme Charles Lutaud', HT, yb, 1912; flowers chrome-yellow, blending to rosy scarlet at edge, dbl.; GM, Bagatelle, 1913; [Unnamed seedling X Marquise de Sinéty]; Pernet-Ducher

'Mme Charles Lutaud, Climbing', Cl HT, yb, 1922; Guillot, P.

'Mme Charles Magny', HT, rb, 1941; flowers coppery red and golden yellow, very large, very dbl., cupped, moderate fragrance; vigorous, bushy growth; [Mme Joseph Perraud X Seedling]; Gaujard

'Mme Charles Mallerin', HT, ob, 1939; flowers orange-salmon, large, dbl., cupped, slight fragrance; foliage leathery, dark; vigorous growth; [Lucy Nicolas X Brazier]; Mallerin, C.; A. Meilland; C-P

'Mme Charles Meurice', HP, dr, 1878; flowers velvety dark red, well formed, large, dbl.; bushy growth; Meurice

'Mme Charles Rouveure', HT, dy, 1946; dbl., 36 petals, 4–5 in., moderate, fruity fragrance; foliage dark; free growth; Mallerin, C.; A. Meilland

'Mme Charles Sauvage', HT, yb, 1949; (Mississippi); flowers yellow tinted saffron, center orange-yellow, well-shaped, dbl., 30 petals, 5 in., slight fragrance; bushy growth; [Julien Potin X Orange Nassau]; Mallerin, C.; URS

'Mme Charles Singer', T, dp, 1916; flowers garnet, becoming dark velvety purple-garnet, dbl.; vigorous growth; Nabonnand, C.

'Mme Charles Truffant', HP, lp, 1878; flowers satiny rose, large; vigorous growth; Verdier, E.

'Mme Charles Wood', HP, mr, 1861; (Dinsmore); flowers fiery scarlet, large, dbl., 45 petals, moderate fragrance; moderate growth; Verdier, E.

Mme Chatelaine de Lullier, HT, 1987; Meilland, L.; (Cavriglia)

'Mme Chédane-Guinoiseau', T, ly, 1880; flowers sulphur yellow; [Safrano X ?]; Lévêque

'Mme Cheine Duguy', HT, pb, 1929; flowers cerise-red, shaded scarlet, dbl.; Schwartz, A.

'Mme Chevalier', B, mp, 1886; Pere Pernet

'Mme Chiang Kai-shek', HT, ly, 1942; bud long, pointed; flowers lemon yellow, turning lighter, dbl., 27 petals, 5–5.5 in., exhibition form, moderate fragrance; foliage leathery, glossy, dark; vigorous, upright, compact growth; AARS, 1944; [Joanna Hill X Sir Henry Segrave]; Duehrsen; H&S

'Mme Clara d'Arcis', HT, pb, 1931; (Clara d'Arcis); flowers brilliant rose-pink, base yellow, large, dbl., moderate, spicy fragrance; foliage dark, leathery; [Julien Potin X Unnamed seedling]; Gaujard; C-P

'Mme Claude Olivier', HT, op, 1939; bud long, pointed; flowers coral, tinted nasturtium-yellow, very large, dbl., exhibition form, slight fragrance; foliage leathery; strong stems; very vigorous, bushy growth; [Soeur Thérèse X Unnamed R. foetida bicolor seedling]; Mallerin, C.; A. Meilland

'Mme Clémence Joigneaux', HP, pb, 1861; flowers pink and red, very large, dbl.; Liabaud; (Sangerhausen)

'Mme Clemence Marchix', HT, 1899; Bernaix, P.; (Cavriglia)

'Mme Cochet-Cochet', HT, op, 1934; bud very long; flowers coppery rose-pink, tinted coral, large, dbl., 30 petals, cupped, moderate fragrance; foliage glossy; vigorous growth; GM, Bagatelle, 1932; [Mrs Pierre S. duPont X Cécile Walter]; Mallerin, C.; C-P

'Mme Colette Martinet', HT, yb, 1915; flowers old-gold shaded orange-yellow, dbl., moderate fragrance; Pernet-Ducher

'Mme Constans', HWich, lp, 1900

'Mme Constant Soupert', T, yb, 1905; flowers yellow shaded peach; [Souv. de Pierre Notting X Duchesse Marie Salviati]; Soupert & Notting

'Mme Cordier', HP, m, 1903; flowers bright lilac-pink, large, dbl.; Leroy

'Mme Cornélissen', B, lp, 1865; flowers large, very dbl., moderate fragrance; Cornélissen; (Sangerhausen)

'Mme Couibes', HT, ab, 1938; flowers salmon, center apricot, edged lighter, passing to golden coral, dbl., moderate fragrance; foliage fresh green; long stems; vigorous, bushy growth; [Charles P. Kilham X Rochefort]; Meilland, F.

Mme Crego, HT, w

'Mme Crespin', HP, m, 1862; flowers rose shaded violet; Damaizin

'Mme Creux', T, op, 1890; flowers light salmon-pink, large, dbl., moderate fragrance; Godard; (Sangerhausen)

'Mme Croibier', HT, op, 1935; (Mme J. B. Croibier, Mme Jean Croibier); flowers bright salmon, open, very large, dbl.; foliage leathery; very vigorous growth; GM, Rome, 1936; [Unnamed seedling X Mme Nicolas Aussel]; Gaujard

'Mme Crombez', T, op, 1888; flowers rosy buff tinted bronze, well formed, dbl., moderate fragrance; Nabonnand, G.

'Mme Cunisset Carnot', HT, 1900; Buatois; (Cavriglia)

'Mme Cusin', T, rb; flowers crimson, center yellowish white, well formed, dbl.; Guillot et Fils, 1882

'Mme Dailleux', HT, pb, 1901; flowers salmon pink with dark yellow, large, dbl., intense fragrance; Buatois; (Sangerhausen)

'Mme Damaizin', T, w, 1858; flowers creamy white shaded salmon, poorly formed, very large, dbl.; Damaizin

'Mme d'Arblay', HMsk, w, 1835; flowers soft flesh changing to white, borne in large clusters, semi-dbl., moderate fragrance; [R. multiflora X R. moschata]; Wells

'Mme David', T, lp, 1895; flowers pale flesh, center darker, very dbl., flat, moderate fragrance; Pernet Père

'Mme de Beauvoire', HT, lp, 1922; flowers pinkish white, center pale pink, dbl.; [Mme Vittoria Gagniere X Lady Ashtown]; Schwartz, A.

'Mme de Carbuccia', HT, mr, 1941; bud long, pointed; flowers bright crimson-red, medium, dbl., intense fragrance; foliage leathery; very vigorous, upright growth; [Admiral Ward X Mme Méha Sabatier]; Kriloff; A. Meilland

'Mme de la Rôche-Lambert', M, m, 1851; flowers dark reddish purple, large, dbl., globular; moderate growth; Robert

'Mme de Pompadour', HT, rb, 1945; flowers coppery red and bright yellow, medium, borne, dbl., globular, slight fragrance; foliage dark, glossy; short stems; vigorous, bushy growth; Gaujard

Mme de Sancy de Parabère, *see* **'Mme Sancy de Parabère'**

Mme de Sertot, *see* **'Mme Bravy'**

Mme de Sevigne, S, lp, 1998; Delbard

'Mme de Sévigné', B, pb, 1874; flowers bright rose in center, edges lighter, large blooms in clusters, dbl.; vigorous growth; Moreau et Robert

Mme de Soubeyran, *see* **'Mme Soubeyran'**

'Mme de st Joseph', T, lp; flowers fawn shaded salmon, large, dbl., intense fragrance; moderate growth

Mme de Stella, *see* **'Louise Odier'**

'Mme de Tartas', T, lp, 1859; (Mme de Thartas); flowers flush-pink, large, dbl., cupped; vigorous, sprawling growth; Bernède

Mme de Thartas, *see* **'Mme de Tartas'**

'Mme de Trotter', HP, pb, 1854; flowers pink and red, medium, dbl.; Granger; (Sangerhausen)

'Mme de Vatry', T, dp, 1855; flowers deep pink, center lighter, large, dbl., recurrent bloom, moderate fragrance; vigorous growth; Guérin

Mme de Villars, M, mp; Béluze, 1855

'Mme de Watteville', T, yb, 1883; flowers lemon edged pink, large, dbl.; Guillot et Fils

Mme Delaunay, HT, 1963; Privat, J.; (Cavriglia)

'Mme d'Enfert', B, lp, 1904; Vilin

Mme Denis, *see* **'Mme Bravy'**

Mme Denise Cassegrain, Misc OGR, w

'Mme Denise Gallois', HT, pb, 1941; flowers salmon shaded yellow, Sauvageot, H.

'Mme Désiré Giraud', HP, pb; flowers blush-white striped rose; [Baronne Prevost sport]; Van Houtte, 1854

'Mme Desmars', HT, yb, 1929; bud long; flowers golden yellow, tinted nasturtium-red, very large, dbl., exhibition form, slight fragrance; foliage glossy; vigorous, bushy growth; [Ophelia X Constance]; Mallerin, C.; H. Guillot, 1932

'Mme Desmary', HT, yb, 1950; flowers ochre-yellow tinted orange, dbl., 30–35 petals; foliage bright green; vigorous growth; [Aspirant Marcel Rouyer X Emma Wright]; Moulin-Epinay; Vilmorin-Andrieux

'Mme Desprez', B, mp, 1831; flowers rosy lilac, dbl., cupped; Desprez

'Mme Desprez', Ch, w

'Mme Devoucoux', T, my, 1874; flowers bright yellow, well formed, dbl.; Ducher, Vve.

'Mme d'Hébray', HGal, w, 1820; flowers white with pink stripes, large, dbl.; Pradel; (Sangerhausen)

'Mme Didkowsky', HT, rb, 1943; flowers fiery red, reverse golden yellow; Mallerin, C.; A. Meilland

'Mme Dieudonné', HT, rb, 1949; (Mme L. Dieudonné); flowers rose-red, reverse gold, dbl., 30 petals, 4–5 in., exhibition form, moderate fragrance; foliage dark, glossy; vigorous growth; [(Mme Joseph Perraud X Brazier) X (Charles P. Kilham X Capucine Chambard)]; Meilland, F.

'Mme Dieudonné, Climbing', Cl HT, rb, 1959; Anderson

Mme Dimitriu®, *see* 'DELcrip'

'Mme Doré', B, lp, 1863; dbl., moderate fragrance; medium growth; Fontaine

'Mme Driout', Cl T, pb, 1902; flowers bright rose, striped carmine, large, dbl.; very vigorous, climbing growth; Thirat, J.

'Mme Dubost', B, lp, 1890; flowers flesh, center rose; Pernet Père

'Mme Dubroca', T, pb, 1882; flowers salmon shaded carmine, large, dbl.; Nabonnand, G.

'Mme E. Rocque', HMult, m, 1918; flowers violet, sometimes striped white, passing to amethyst, borne, dbl.; [Veilchenblau X Reine des Violettes]; Lottin

'Mme E. Terracol', HT, yb, 1940; bud globular; flowers pure yellow, base orange, very large, dbl., slight fragrance; foliage leathery; very vigorous, upright growth; [Julien Potin X Soeur Therese]; Meilland, F.; A. Meilland

'Mme Edmée Metz', HT, pb, 1901; flowers rosy carmine shaded salmon, dbl.; [Mme Caroline Testout X Ferdinand Jamin]; Soupert & Notting

'Mme Edmond Fabre', HP, mp, 1884; flowers large, dbl., moderate fragrance; Verdier, E.; (Sangerhausen)

'Mme Edmond Gillet', HT, yb, 1921; flowers reddish nankeen yellow, slightly shaded carmine at tips, dbl.; [Mme Edmond Rostand X Marquise de Sinéty]; Pernet-Ducher

'Mme Edmond Labbé', HT, rb, 1938; bud pointed; flowers orange-red, reverse golden yellow, large, dbl., slight fragrance; long stems; very vigorous growth; GM, Bagatelle, 1938; [Souv. de Claudius Pernet X Unnamed Hybrid Tea seedling]; Mallerin, C.; A. Meilland

'Mme Edmond Raynal', HT, ly, 1927; flowers yellowish cream, center salmon, dbl., intense fragrance; Sauvageot, H.; F. Guillot

'Mme Edmond Rostand', HT, pb, 1912; flowers pale flesh, center shaded salmon and reddish orange-yellow, dbl., moderate fragrance; [Unnamed variety X Prince de Bulgarie]; Pernet-Ducher

'Mme Edouard Estaunié', HT, yb, 1936; bud long; flowers nankeen yellow, center reddish, edges and reverse flesh-pink, dbl., moderate fragrance; long stems; very vigorous growth; [Seabird X Souv. de Claudius Pernet]; Buatois

'Mme Edouard Herriot', HT, ob, 1913; (Daily Mail Rose); bud pointed; flowers coral-red shaded yellow and bright rosy scarlet, passing to, semi-dbl., moderate fragrance; foliage bronze, glossy; vigorous, spreading, branching growth; (28); GM, NRS, 1913; [Mme Caroline Testout X HT]; Pernet-Ducher

'Mme Edouard Herriot, Climbing', Cl HT, ob, 1921; (Daily Mail Rose, Climbing); Ketten Bros.

'Mme Edouard Ory', M, dp, 1854; flowers bright carmine-pink, dbl., globular; Robert

'Mme Elie Dupraz', HT, mr, 1948; bud large; flowers brilliant red, medium, dbl., exhibition form, moderate fragrance; foliage dark, glossy; very vigorous, bushy growth; Gaujard

'Mme Elisa de Vilmorin', HT, mr; (Mme Eliza de Vilmorin); flowers dark carmine, large, dbl., 30 petals, sparse bloom; upright, bushy growth; Lévêque, 1867

Mme Eliza de Vilmorin, *see* **'Mme Elisa de Vilmorin'**

'Mme Emile Daloz', HT, pb, 1934; flowers satiny purplish pink, reverse bright rose-pink, very large, dbl., globular, intense fragrance; foliage leathery, glossy; very vigorous, bushy growth; [Frau Karl Druschki X Souv. de Georges Pernet]; Sauvageot, H.; C-P; Sauvageot

'Mme Emile Mayen', HT, ly, 1924; flowers sulfur-yellow passing to cream, dbl.; Chambard, C.

Mme Emile Seneclauze, HT, 1966; Croix, P.; (Cavriglia)

'Mme Emile Thierrard', HT, yb, 1919; flowers chamois-yellow and pink, stamens pure yellow, dbl., moderate fragrance; [Mrs Aaron Ward X Joseph Hill]; Turbat

'Mme Emilie Charron', T, mp, 1895; flowers china-pink, large, cupped; very vigorous growth; Perrier

'Mme Emilie Lafon', HT, mr, 1905; flowers large, dbl.; Moranville; (Sangerhausen)

'Mme Emilie van der Goes', HT, op, 1925; bud pointed, orange-yellow and rosy shadings; flowers more pink than bud, large, semi-dbl., moderate fragrance;

foliage bronze, leathery; bushy growth; [Columbia X Irish Fireflame]; Verschuren

'**Mme Ernest Calvat**', B, mp, 1888; (Mme Ernst Calvat, Pink Bourbon); flowers pink shaded darker; [Mme Isaac Pereire sport]; Schwartz, Vve.

'**Mme Ernest Charles**', HT, or, 1933; bud large, long; flowers coral-red, reverse shrimp-red, dbl., cupped, moderate fragrance; foliage leathery, glossy, bronze; vigorous, bushy growth; [Mme Edmond Rostand X Severine]; Buatois

'**Mme Ernest Levavasseur**', HP, mr, 1900; flowers carmine-red, very large, very dbl., moderate fragrance; Vigneron; (Sangerhausen)

'**Mme Ernest Perrin**', T, op, 1900; flowers light orange-pink, very large, dbl.; Schwartz; (Sangerhausen)

'**Mme Ernest Piard**', HT, dp, 1888; flowers carmine-pink, very large, dbl.; Bonnaire; (Sangerhausen)

Mme Ernst Calvat, *see* '**Mme Ernest Calvat**'

'**Mme Etienne**', T, mp, 1887; flowers rose, well formed; vigorous growth; Bernaix, A.

'**Mme Etienne Levet**', HP, mr, 1871; flowers carmine-red, large, dbl., moderate fragrance; Levet; (Sangerhausen)

'**Mme Eugène Mallet**', N, yb, 1875; flowers pink and yellow, dbl., moderate fragrance; moderate, climbing growth; Nabonnand, G.

Mme Eugène Marlitt, *see* '**Eugène E. Marlitt**'

'**Mme Eugène Moreau**', HT, my, 1925; dbl.; vigorous growth; Richardier

'**Mme Eugène Picard**', HT, my, 1932; [Ariel sport]; Gillot, F.; C-P

'**Mme Eugène Résal**', Ch, pb, 1894; flowers bright pink shaded reddish orange, base yellow; [Mme Laurette Messimy sport]; Guillot, P.

'**Mme Eugène Verdier**', HP, lp; flowers silvery pink, large, dbl., globular, moderate fragrance; vigorous growth; Verdier, E., before 1866

'**Mme Eugène Boullet**', HT, pb, 1897; flowers pink tinted yellow, moderate fragrance; Pernet-Ducher

'**Mme Eugène Frèmy**', HP, dp, 1885; flowers carmine-pink, large, very dbl., moderate fragrance; Verdier, E.

'**Mme Falcot**', T, my, 1858; flowers nankeen yellow passing to clear yellow, large, dbl., moderate fragrance; Guillot et Fils

'**Mme Fanny de Forest**', N, lp, 1882; flowers very large, dbl.; Schwartz; (Sangerhausen)

'**Mme Fanny Pauwels**', T, 1885; Soupert, C.; (Cavriglia)

'**Mme Faurax-Lille**', HT, mr, 1933; bud pointed; flowers bright vermilion-red passing to geranium-red, large, dbl.; [Cuba X Sir David Davis]; Reymond; Vially

'**Mme Fearnley Sander**', HT, m, 1921; flowers carmine, deepening to purple, base yellow, dbl.; [Gen. MacArthur X Rayon d'Or]; Ketten Bros.

'**Mme Ferdinand Jamain**', HP, lp; bud globular; flowers deep pink, large, dbl., 50 petals, cupped, intense fragrance; Lédéchaux, 1875

'**Mme Fernand Gentin**', HT, op, 1939; bud long, pointed; flowers copper shaded coral, large, semi-dbl., cupped, slight fragrance; vigorous growth; [Unnamed seedling X Brazier]; Mallerin, C.; A. Meilland

'**Mme Fernand Gregh**', HT, yb, 1955; flowers canary-yellow shaded coppery, large, slight fragrance; foliage glossy; vigorous growth; [Padre X Madeleine Pacaud]; Robichon

Mme Fernandel®, *see* MEIsunaj

'**Mme Fillion**', HP, op, 1865; flowers salmon-pink, large, dbl.; Gonod; (Sangerhausen)

'**Mme Fojo**', HT, ob, 1937; flowers orange, well formed, large; strong stems; vigorous growth; Dot, Pedro; H. Guillot

'**Mme Forest**', HT, mp, 1928; [Lieutenant Chaure X Mrs George Shawyer]; Walter, L.

'**Mme Foureau**', N, dy, 1913; flowers medium, dbl.; Viaud-Bruant; (Sangerhausen)

'**Mme Fraculy**', HT, dy, 1929; flowers deep golden yellow, dbl.; foliage dark, glossy; long, strong stems; vigorous growth; [Ophelia X Constance]; Siret-Pernet

'**Mme Francisque Favre**', Pol, 1915; Dubreuil; (Cavriglia)

Mme Franck Augis®, *see* 'LAPlam'

'**Mme François Bollez**', HT, op, 1934; flowers coral-pink, center brighter, tinted orange, very large, dbl., moderate fragrance; foliage dark, bronze; very vigorous growth; Gillot, F.

'**Mme François Graindorge**', Pol, pb, 1922; flowers dark reddish pink shaded magenta, base tinged lilac, large, vigorous growth; Grandes Roseraies

'**Mme François Hot**', HT, pb, 1928; flowers salmon shaded coppery rose, edged lighter, base salmon-yellow; [Lady Pirrie X Mme de Bauvoire]; Schwartz, A.

'**Mme François Pittet**', N, w, 1878; Lacharme, F.

'**Mme François Royet**', HMult, mr, 1926; flowers bright red; [Crimson Rambler X Général Jacqueminot]; Royet

Mme Frank Augis, HT, 1969; Laperrière; (Cavriglia)

'**Mme G. Forest-Colcombet**', HT, dp, 1928; flowers deep carmine, strongly tinted scarlet; [Hadley sport]; Mallerin, C.; Grandes Roseraies

'**Mme G. Hekkens**', HT, dr, 1929; flowers velvety dark red (nearly as dark as Chateau de Clos Vougeot), single, intense fragrance; foliage glossy; [Gloire de Holland X Hawlmark Crimson]; Faassen-Hekkens

'**Mme Gabriel Hanra**', HT, rb, 1929; flowers strawberry-red, shaded carmine-purple, dbl., slight fragrance; [The Adjutant X K. of K.]; Ketten Bros.

'**Mme Gabriel Luizet**', HP, lp, 1877; flowers light silvery pink, edged lighter, large, dbl., 34 petals, cupped, non-recurrent, moderate fragrance; vigorous growth; Liabaud

'**Mme Gaston Annouilh**', N, w, 1899; flowers medium, dbl., intense fragrance; Chauvry; (Sangerhausen)

'**Mme Gaston Doumergue**', Pol, lp, 1934; flowers soft salmon; [Probably Gloria Mundi sport]; Levavasseur

'**Mme Gaston Mestreit**', Pol, lp, 1922; flowers very soft flesh-white, borne in clusters; [Jeanny Soupert X Katharina Zeimet]; Soupert & Notting

'**Mme Gaston Nocton**', Pol, w, 1928; flowers white, center flesh-pink, opening pure white, borne in clusters; dwarf growth; [Amaury Fonseca X Jeanny Soupert]; Soupert & Notting

'**Mme Georges Bénard**', HT, dp, 1900; flowers carmine-pink with silvery reverse, large, dbl.; Corboeuf; (Sangerhausen)

'**Mme Georges Brédif**', HT, rb, 1955; flowers red mottled garnet, well formed, moderate fragrance; strong stems; Privat

'**Mme Georges Bruant**', HRg, w, 1887; bud pointed; flowers loose, large blooms in clusters, recurrent bloom, moderate fragrance; (14); [R. rugosa X Sombreuil]; Bruant

Mme Georges Clenet, HT, 1980; Godin, M.; (Cavriglia)

'**Mme Georges Cozon**', HT, pb, 1929; flowers shrimp-pink, reverse yellow; [Mme Charles Lutaud X Unnamed Hybrid Tea seedling]; Laperrière

Mme Georges Delbard®, *see* 'DELadel'

'**Mme Georges Delbard**', HT, or, 1959; flowers bright red suffused orange, dbl., 35–45 petals, 5 in., moderate fragrance; foliage dull green; long, strong stems; very vigorous growth; RULED EXTINCT 4/85; [Impeccable X Mme Robert Joffet]; Delbard-Chabert

Mme Georges Delbard, HT, 1979; Delbard-Chabert; (Cavriglia)

'**Mme Georges Droin**', HT, op, 1930; flowers orange-shrimp-pink; foliage bronze; very vigorous growth; Gaujard, Jules

'**Mme Georges Landard**', HT, lp, 1925; foliage glossy; [Mme Abel Chatenay X Lyon Rose]; Walter, L.; Lamesch

'MME GEORGES PETIT', HT, mr, 1928; bud pointed; flowers bright purple-red to velvety crimson-red, large, dbl., exhibition form, intense fragrance; very vigorous growth; [Gen. MacArthur X Mme Edouard Herriot]; Ketten Bros.

Mme Georges Renoard, *see* 'DELRENO'

'MME GHYS', HMult, m, 1912; flowers lilac-rose, borne in clusters; height 6–10 ft; [Crimson Rambler X ?]; Ghys; Decault

'MME GILBERTE JANAUD', F, mp, 1957; flowers bright pink tinted salmon, well-formed, very dbl.; dwarf growth; Privat

'MME GILLET LAFOND', HT, pb, 1930; flowers salmon-white, center old-rose, large, semi-dbl., moderate fragrance; foliage light, leathery; vigorous growth; [Frau Felix Tonnar X Angèle Pernet]; Leenders, M.

'MME GINA DEMOUSTIER', HT, mr, 1920; flowers pure garnet-red, dbl., intense fragrance; [Étoile de France X Unnamed variety]; Laperrière

'MME GRÉGOIRE STAECHELIN', LCl, pb, 1927; (Spanish Beauty); flowers delicate pink, reverse stained crimson, ruffled, very large, non-recurrent, moderate fragrance; pear-shaped fruit; foliage heavy, large, dark; vigorous (13–14 ft) growth; GM, Bagatelle, 1927 John Cook Medal, ARS, 1929; [Frau Karl Druschki X Château de Clos Vougeot]; Dot, Pedro; C-P, 1929

'MME GUIMET', HT, dp, 1942; flowers large, dbl., slight fragrance; Laperrière; (Sangerhausen)

'MME GUSTAVE METZ', HT, lp, 1906; flowers large, dbl.; Lamesch; (Sangerhausen)

'MME GUSTAVE SOUPERT', HT, pb, 1928; flowers purplish pink, center brighter, reverse silvery carmine, dbl., moderate fragrance; [Augustus Hartmann X Souv. de Georges Pernet]; Soupert & Notting

'MME HARDY', D, w, 1832; flowers pure white, occasionally tinged flesh-pink, green pip, large, dbl., cupped, borne in clusters, intense fragrance; vigorous growth; Hardy

Mme Hébert, *see* **'PRÉSIDENT DE SÈZE'**

'MME HECTOR LEUILLOT', HT, yb, 1903; bud pointed; flowers golden yellow on carmine ground, dbl., moderate fragrance; Pernet-Ducher

'MME HÉLÈNE DAPPLES', HT, mr, 1932; flowers large, semi-dbl., intense fragrance; Heizmann, E.; (Sangerhausen)

'MME HÉLÈNE DUCHÉ', HT, pb, 1921; flowers soft rose with silvery reflexes, edged carmine, dbl.; [Mme Caroline Testout X Reine Emma des Pays-Bas]; Buatois

'MME HÉLÈNE PARMENTIER', HT, or, 1935; bud long; flowers clear nasturtium-red, shaded orange, passing to pink, semi-dbl., cupped, moderate fragrance; foliage glossy, wrinkled, bronze; long stems; vigorous, bushy growth; [Unnamed seedling X Angèle Pernet]; Sauvageot, H.

'MME HENRI BONNET', HT, pb, 1948; bud ovoid; flowers deep salmon-pink suffused golden orange, large, dbl., exhibition form, moderate fragrance; foliage leathery; vigorous, upright, compact growth; [Elite X Unnamed seedling]; Boerner; J&P

'MME HENRI FONTAINE', HT, lp, 1914; flowers creamy pink, large, dbl., moderate fragrance; Guillot; (Sangerhausen)

'MME HENRI GRAVEREAUX', HT, yb, 1926; flowers coppery yellow tinted bronze-yellow, veined orange, dbl., moderate fragrance; [Mrs Aaron Ward X Unnamed seedling]; Barbier

'MME HENRI GRIMM', HT, lp, 1934; bud long; flowers pinkish white tinged carmine, edged crimson, base yellow, dbl., exhibition form, intense fragrance; foliage leathery, dark; vigorous, bushy growth; [Mme Charles Detreaux X Mme Edouard Herriot]; Buatois

'MME HENRI GUILLOT', HT, rb, 1938; flowers orange-coral-red, large, dbl., 25 petals, slight fragrance; foliage glossy; vigorous, bushy growth; GM, Bagatelle, 1936 GM, Portland, 1939; [Rochefort X Unnamed R. foetida bicolor seedling]; Mallerin, C.; A. Meilland; C-P

'MME HENRI GUILLOT, CLIMBING', Cl HT, rb, 1942; Meilland, F.

'MME HENRI GUILLOT, CLIMBING', Cl HT, rb, 1947; van Barneveld; C-P

'MME HENRI LAFOREST', HT, my, 1942; bud pointed; flowers golden yellow, very large, semi-dbl., moderate fragrance; foliage glossy; very vigorous, upright growth; Gaujard

'MME HENRI LUSTRE', HT, rb, 1924; flowers purplish garnet tinted currant-red, very large, dbl., exhibition form, slight fragrance; foliage leathery; very vigorous, bushy growth; [Mme Edouard Herriot X Yves Druhen]; Buatois

'MME HENRI PATÉ', HT, ly, 1929; bud pointed; flowers sulfur-yellow, very large, semi-dbl., intense fragrance; foliage bronze; very vigorous, bushy growth; [Souv. de Claudius Pernet X Unnamed seedling]; Pernet-Ducher; Gaujard

'MME HENRI PELLEY', HT, ly, 1928; flowers transparent cream-yellow; Richardier

'MME HENRI PEREIRE', HP, mr, 1887; flowers crimson-red, moderate fragrance; Vilin

'MME HENRI QUEUILLE', HT, op, 1928; bud long; flowers bright shrimp-pink, center deeper, reverse coppery gold, semi-dbl., moderate fragrance; foliage bronze; very vigorous growth; Pernet-Ducher; Gaujard

'MME HENRI QUEUILLE', HFt, or, 1928; flowers light orange/red, large, dbl., intense fragrance; Gaujard; (Sangerhausen)

'MME HENRI THIEBAUT', HT, op, 1931; bud pointed; flowers salmon-coral base coral-orange, very large, dbl., intense fragrance; foliage bright green; strong stems; vigorous growth; Chambard, C.

'MME HERMANN HAEFLIGER', S, dr, 1951; flowers large; foliage dark; vigorous growth; [R. foetida bicolor seedling X Charles P. Kilham]; Hauser

'MME HERRIOT PANACHÉE', HT, yb, 1921; flowers coral and golden yellow; [Mme Edouard Herriot sport]; Cassegrain

'MME HIDE', HT, pb, 1997; flowers blush edged pink, moderately full (15–25 petals), large (7 cms) blooms; slight fragrance; some prickles; foliage large, medium green, matt; compact (1.2 m) growth; [Pink Picnica X Garden Party]; Ohta, Kaichiro

'MME HIPPOLYTE DUMAS', HT, pb, 1924; flowers flesh tinted salmon-pink, base yellow, dbl., slight fragrance; Guillot, P.

Mme Hippolyte Jamain, HP, lp; Garçon, 1871

'MME HONORÉ DEFRESNE', Cl T, my, 1886; flowers golden yellow, large, dbl., intense fragrance; vigorous growth; [Mme Falcot seedling sport]; Levet, F.

'MME HOSTE', T, w, 1887; flowers yellowish-white, imbricated, large, globular; vigorous growth; Guillot et Fils

'MME ISAAC PEREIRE', B, dp, 1881; flowers deep rose-pink shaded purple, large, dbl., moderate fragrance; vigorous growth; Garçon

Mme J. B. Croibier, *see* **'MME CROIBIER'**

'MME J. BONNAIRE PIERRE', HP, 1892; Bonnaire; (Cavriglia)

'MME J. P. SOUPERT', HT, w, 1900; flowers very large, very dbl., moderate fragrance; Soupert & Notting; (Sangerhausen)

'MME J. M. FRUCTUS', HT, pb, 1935; flowers satiny carmine, shaded salmon, base yellow, well formed, dbl., cupped; foliage dark; very vigorous growth; Chambard, C.

'MME J. W. BÜDDE', HT, dp, 1906; flowers bright carmine; Soupert & Notting

'MME JACQUES PRIVAT', HT, mr, 1959; bud long; flowers red, slight fragrance; foliage glossy; vigorous growth; Privat, J.

Mme Jean Croibier, *see* **'MME CROIBIER'**

'MME JEAN DEMESHAYNE', HT, op, 1934; flowers yellowish salmon-pink, large, very dbl., slight fragrance; Nicolas; (Sangerhausen)

'MME JEAN DUPUY', T, yb, 1902; bud long, pointed; flowers golden yellow washed pink, large, dbl., moderate fragrance; vigorous growth; Lambert, P.

'MME JEAN EVERAERTS', HP, dr, 1907; flowers large, dbl., moderate fragrance; Geduldig; (Sangerhausen)

'MME JEAN GAUJARD', HT, yb, 1938; bud long, pointed; flowers creamy yellow, reverse shaded orange and carmine-pink, very, dbl.; foliage leathery, light; very vigorous, bushy growth; [Julien Potin X Unnamed seedling]; Gaujard; J&P

'Mme Jean Paquel', HT, ly, 1934; bud long, pointed; flowers yellow, passing to cream; foliage glossy; [Alice Stern X Lilly Jung]; Walter, L.; Amis des Roses

Mme Jean Philippe, T, yb

'Mme Jean Raty', HT, w, 1932; bud pointed; flowers amber-white, edges tinted peach-blossom, well formed, large, dbl., moderate fragrance; [Mme Abel Chatenay X Unnamed seedling]; Ketten Bros.

'Mme Jeannine Joubert', B, dp, 1877; Margottin fils

'Mme Jenny', HMult, mp; flowers satiny rose, small, borne in clusters, cupped, moderate fragrance; foliage dark; vigorous, climbing growth; Nonin, 1925

'Mme Jenny Gillemot', HT, ly, 1905; bud pointed; flowers light saffron-yellow, large, dbl., exhibition form, moderate fragrance; Pernet-Ducher

'Mme Joannes Beurrier', HT, op, 1942; bud pointed; flowers bright orange-pink, reddish reflections, very large, very dbl., moderate fragrance; foliage glossy, bronze; very vigorous, bushy growth; Gaujard

'Mme Joseph Bonnaire', HT, pb, 1891; Bonnaire

'Mme Joseph Combet', HT, pb, 1894; flowers pink with yellow and white, large, very dbl., intense fragrance; Bonnaire; (Sangerhausen)

'Mme Joseph Jullien', HT, pb, 1938; flowers coppery carmine, very large, dbl., cupped, moderate fragrance; foliage slightly bronze; [Ami F. Mayery X Unnamed seedling]; Chambard, C.

'Mme Joseph Perraud', HT, yb, 1934; (Sunburst); flowers yellow, center deeply tinted coppery, large, dbl., 33 petals, moderate fragrance; foliage glossy; GM, Bagatelle, 1934; [Julien Potin X Unnamed seedling]; Gaujard

'Mme Joseph Perraud, Climbing', Cl HT, yb, 1945; Marsh's Nursery

'Mme Joseph Schwartz', T, w, 1880; (White Duchesse de Brabant); flowers white washed flesh-pink, medium, dbl.; vigorous growth; [Probably a Duchesse de Brabant sport]; Schwartz, J.

Mme Joséphine Guyet, *see* **'Joséphine Guyet'**

'Mme Jules Bouché', HT, w, 1911; bud pointed; flowers white, center shaded primrose or pale blush, large, dbl., 34 petals, moderate fragrance; tall growth; [Pharisaer X Unknown seedling]; Croibier

'Mme Jules Bouché, Climbing', Cl HT, w, 1938; California Roses

'Mme Jules Finger', HT, lp, 1893; Guillot

'Mme Jules Fontaine-Lamarche', HT, mr, 1936; flowers velvety scarlet-red, large, dbl.; [Sensation X E.G. Hill]; Soupert & Notting

'Mme Jules Gouchault', Pol, op, 1913; flowers bright pink, tinted coral and orange, borne in clusters, dbl., cupped, moderate fragrance; bushy growth; [Maman Turbat X George Elger]; Turbat; Teschendorff

'Mme Jules Gravereaux', Cl T, ab, 1901; flowers flesh, shaded peach or yellow, very dbl., recurrent bloom, slight fragrance; foliage large, dark, glossy; height 3–4 ft, bushy growth; [Reve d'Or X Viscountess Folkestone]; Soupert & Notting

'Mme Jules Grolez', HT, mp, 1896; bud long, pointed; flowers bright china-rose, dbl., exhibition form, moderate fragrance; bushy growth; [Triomphe de l'Exposition X Mme Falcot]; Guillot, P.

'Mme Jules Guérin', HT, ly, 1931; flowers deep cream, very large, dbl., intense fragrance; foliage bronze; Gaujard

'Mme Jules Thibaud', Pol, op; flowers coral-pink; [Cécile Brunner sport]

'Mme Jules Walthery', HT, my, 1924; flowers outer petals becoming white; Allen

'Mme Julie Daran', HP, mr, 1861; flowers crimson; Touvais

'Mme Julien Potin', HRg, lp, 1913; flowers pure flesh-pink, large, dbl., flat, recurrent bloom; foliage very leathery; vigorous growth; [R. rugosa X Gloire de Dijon]; Gravereaux; Cochet-Cochet

Mme Juliette, *see* **'Juliette E. van Beuningen'**

Mme Juliette Guillot, F, dp

'Mme Kahn', HT, ly, 1939; flowers canary-yellow, reverse slightly reddish; vigorous growth; [Charles P. Kilham X Ville de Paris]; Colombier

'Mme Karen Shadish', F, mp, 1999; dbl., 17–25 petals, 2–3 in., borne in small clusters, slight fragrance; prickles moderate; foliage small, medium green, semi-glossy; upright, medium (2–3 ft) growth; [Pink Petticoat X Mary Rose]; Chaney, William

'Mme Kastler', HT, mr, 1934; bud long, pointed; flowers large, dbl.; very vigorous growth; [Mme Adele Gance X Mme Caroline Testout]; Walter, L.; Amis des Roses

'Mme Klatz', M, dp; flowers deep pink, quite large

'Mme Knorr', P, mp, 1855; dbl., moderate fragrance; Verdier, V.

'Mme Kriloff', HT, yb, 1944; flowers clear saffron-yellow, veined reddish orange, large, dbl., globular, moderate fragrance; foliage leathery; vigorous, bushy growth; GM, Bagatelle, 1944 GM, NRS, 1948; [Peace X Signora]; Meilland, F.; A. Meilland

Mme L. Dieudonné, *see* **'Mme Dieudonné'**

'Mme L. Hot', HT, pb, 1926; flowers reddish salmon, shaded salmon-rose and chrome-yellow, dbl.; [Gorgeous X Rosomane Narcisse Thomas]; Bernaix, P.

Mme L. Ladoire, F, op

'Mme L. Laperriere', HT, 1951; Laperrière; (Cavriglia)

Mme L. Pradel®, *see* 'LAPav'

'Mme la Colonelle Desmaires', F, dr, 1969; flowers large, dbl., slight fragrance; foliage soft, bronze; moderate growth; [Mercator X Alain]; Delforge

'Mme la Générale Ardouin', HT, rb, 1927; flowers coppery carmine shaded chrome-yellow, dbl., moderate fragrance; Chambard, C.

'Mme Laffay', HP, mr, 1839; flowers bright crimson, dbl., intense fragrance; very vigorous growth; Laffay, M.

'Mme Lajotte', HT, pb; flowers bright salmon, reverse yellow, large, dbl.; very vigorous, bushy growth; [Rome Glory X Marie-Rose Toussaint]; Gaujard

Mme Lambard, *see* **'Mme Lombard'**

'Mme Landeau', rb, 1875; flowers light red with white striping, medium, dbl.; Moreau et Robert; (Sangerhausen)

'Mme Lanquetin', HT, ab; flowers apricot; Ofman; Pin

'Mme Lauras', HT, mr, 1954; flowers bright red, medium, dbl., slight fragrance; foliage leathery; very vigorous growth; [Rome Glory X Mme Elie Dupraz seedling]; Gaujard

'Mme Laurent Simons', T, op, 1894; flowers coppery-pink, large, very dbl., moderate fragrance; Lévêque; (Sangerhausen)

'Mme Laurette Messimy', Ch, dp, 1887; bud long; flowers rose-pink, base shaded yellow, large, dbl.; vigorous growth; (14); [Rival de Paestum X Mme Falcot]; Guillot et Fils

'Mme Lauriol de Barny', B, lp, 1868; flowers silvery pink, large, dbl., quartered, rarely repeats, moderate, fruity fragrance; height 5–6 ft; Trouillard

'Mme le Guelinel', HT, mr, 1959; flowers large, intense fragrance; [Mme Kriloff X Marrakech]; Gaujard, R.; G. Truffaut

'Mme Lefèbvre', HP, mp, 1886; flowers large, dbl.; Moreau et Robert; (Sangerhausen)

'Mme Legrand', pb, 1863; flowers pink and red, large, dbl.; Fontaine; (Sangerhausen)

'Mme Legras de st Germain', A, w, 1846; flowers white, center rich cream, large, dbl.; few prickles; foliage gray-green; vigorous (6–7 ft) growth

'**Mme Lemesle**', HP, m, 1892; flowers purple/violet, medium, dbl.; Moreau et Robert; (Sangerhausen)

'**Mme Léon Cuny**', HT, mr, 1955; bud long, pointed; flowers bright red, veined purple, very large, dbl., exhibition form, moderate fragrance; foliage dark, bronze; upright growth; [Peace X Seedling]; Gaujard

'**Mme Léon Février**', T, pb, 1884; flowers silvery rose shaded crimson, well formed, dbl., intense fragrance; Nabonnand, G.

'**Mme Léon Guinotte**', HT, pb, 1924; flowers glistening pink shaded yellow; [Mme Edouard Herriot X Old Gold]; Verschuren

'**Mme Léon Pain**', HT, pb, 1904; bud pointed; flowers silvery flesh-pink, center orange-yellow, reverse salmon-pink, dbl., 45 petals, moderate fragrance; vigorous, bushy growth; [Mme Caroline Testout X Souv. de Catherine Guillot]; Guillot, P.

'**Mme Léon Pin**', HT, ob, 1954; flowers soft orange, medium, dbl., slight fragrance; foliage dark, leathery; upright growth; [Louise de Vilmorin X Capucine seedling]; Gaujard

'**Mme Léon Simon**', HT, pb, 1909; flowers dark pink with cream, large, dbl.; Lambert, P.; (Sangerhausen)

'**Mme Léon Troussier**', HT, op, 1941; flowers coral, base golden yellow; Mallerin, C.; A. Meilland

'**Mme Léon Volterra**', HT, mp, 1958; flowers salmon-pink, well shaped, dbl., intense fragrance; vigorous growth; GM, Bagatelle, 1957; [Ambassadeur Nemry X Tawny Gold]; Leenders, M.

'**Mme Léonce Colombier**', HT, ly, 1926; flowers center straw-yellow passing to white, reverse light rose, dbl.; Richardier

'**Mme Leopold Dupuy**', HT, mp, 1912; flowers large, dbl., intense fragrance; Robichon; (Sangerhausen)

Mme Liabaud, *see* La Virginale

'**Mme Lierval**', HP, pb, 1868; Fontaine

'**Mme Line Renaud**', HT, mr, 1956; flowers velvety red, intense fragrance; foliage bright green; [Crimson Glory X Unnamed seedling]; Mondial Roses

'**Mme Loeben Sels**', HT, w, 1879; flowers silvery white shaded rose, large, dbl., flat; moderate growth; Soupert & Notting

'**Mme Lombard**', T, op, 1878; (Mme Lambard); flowers rosy salmon, center darker, sometimes rosy flesh, large, very dbl., moderate fragrance; vigorous growth; [Mme de Tartas seedling]; Lacharme, F.

Mme Louis Armand, HT, 1974; Croix; (Cavriglia)

Mme Louis Ladoire, F; Arles, F.; (Cavriglia)

'**Mme Louis Laperrière**', HT, mr, 1951; flowers rich scarlet, well-formed, medium, dbl., 48 petals, intense fragrance; foliage dark; upright, bushy growth; GM, Bagatelle, 1950; [Crimson Glory X Seedling]; Laperrière; EFR

Mme Louis Lens, *see* '**White Briarcliff**'

'**Mme Louis Lens, Climbing**', Cl HT, w, 1935; [White Briarcliff sport]; Lens

'**Mme Louis Lévêque**', HP, dp, 1864; flowers carmine-rose, very large, dbl.; moderate growth; Lévêque

'**Mme Louis Lévêque**', M, mp; flowers brilliant salmon-pink, large, dbl., globular, sometimes blooms in fall, moderate fragrance; prickles well mossed; Lévêque, 1898

'**Mme Louis Lévêque**', T, yb, 1892; flowers yellow washed pink, dbl.; moderate growth; Lévêque

'**Mme Louis Ricard**', HP, dp, 1904; flowers carmine-pink, medium, dbl.; Boutigny; (Sangerhausen)

'**Mme Louis Ricart**', B, lp, 1892; Duboc

'**Mme Louisa Cointreau**', HT, mr, 1957; flowers garnet-red, dbl., slight fragrance; vigorous growth; [Crimson Glory X Symbole]; Robichon

Mme Louise Guillot, F, 1955; Guillot, M.; (Cavriglia)

'**Mme Louise Piron**', HP, lp, 1903; Piron-Medard

'**Mme Louise Trémeau**', HT, op, 1931; flowers pink, shaded nasturtium-red, center brighter, open, large, dbl., cupped, slight fragrance; vigorous growth; [(Frau Karl Druschki X Mme Edouard Herriot) X (Mrs Edward Powell X R. foetida bicolor)]; Mallerin, C.; C-P

'**Mme Lucien Chaure**', HP, 1884; Vigneron; (Cavriglia)

'**Mme Lucien Perrier**', HT, rb, 1938; bud long; flowers coppery red, lighter on opening, open, very large, dbl., moderate fragrance; foliage glossy, dark; vigorous growth; Gaujard

'**Mme Lucien Villeminot**', HRg, lp, 1901; flowers pale pink, large, dbl., globular, moderate fragrance; vigorous growth; [Conrad Ferdinand Meyer X Belle Poitevine]; L'Hay

'**Mme Lureau-Escalais**', HP, mp, 1886; flowers large, dbl.; Verdier, E.; (Sangerhausen)

'**Mme Mallerin**', HT, rb, 1924; flowers crimson-scarlet shaded vermilion, intense fragrance; [Mrs Edward Powell X Unnamed seedling]; Chambard, C.

Mme Marcel Astic, F, 1978; Orard, Joseph; (Cavriglia)

'**Mme Marcel Delauney**', HT, pb, 1916; flowers pale pink or soft rose, shaded hydrangea-pink, dbl., moderate fragrance; GM, Bagatelle, 1915; Leenders, M.

'**Mme Marcel Fauneau**', HP, mr, 1886; flowers carmine-red, large, dbl.; Vigneron; (Sangerhausen)

'**Mme Margottin**', T, my, 1866; flowers lemon-yellow, large, dbl.; Guillot et Fils

'**Mme Marguerite Lagières**', HT, w, 1955; flowers cream, base golden yellow, very large, dbl.; foliage glossy; strong stems; vigorous growth; Privat

'**Mme Marguerite Marsault**', HP, m, 1894; flowers reddish-violet, large, dbl.; Corboeuf; (Sangerhausen)

'**Mme Marie Croibier**', HT, dp, 1901; flowers large, dbl.; Croibier; (Sangerhausen)

'**Mme Marie Curie**', HT, dy, 1943; (Québec); flowers clear yellow, dbl., 25 petals, 5 in., exhibition form, slight fragrance; foliage leathery, dark; vigorous, bushy, compact growth; AARS, 1944; Gaujard; J&P

'**Mme Marie Eberlin**', HT, w, 1923; flowers cream, passing to white; base light yellow, dbl.; [Comtesse Melanie de Pourtales X Capt. Christy]; Walter, A.; Bacher

'**Mme Marie Lavalley**', HT, pb, 1881; flowers bright rose tinted white; Nabonnand, G.

Mme Marie Van Houtte, HP, lp

'**Mme Marius Dévigne**', HT, pb, 1930; bud long; flowers salmon-pink, reverse vivid carmine, large, globular; Reymond

Mme Martha Ancey, *see* '**Marthe Ancey**'

'**Mme Martignier**', Cl T, m; flowers red tinted purplish, on yellow ground; Dubreuil, 1903

'**Mme Masson**', HP, dr, 1856; flowers pure crimson-rose, free bloom during long season, moderate fragrance; vigorous, compact growth; Masson

'**Mme Massot**', B, w, 1856; Lacharme, F.

'**Mme Maurice Baudot**', HT, ob, 1941; flowers bright orange, medium, semi-dbl., exhibition form, moderate fragrance; foliage dark; very vigorous, bushy growth; [Mme Joseph Perraud X Seedling]; Gaujard

'**Mme Maurice Capron**', HT, ab, 1914; flowers deep apricot-yellow tinted salmon; Guillot, P.

'**Mme Maurice Cazin**', HT, rb, 1931; bud pointed; flowers dark scarlet, reverse clear reddish crimson, large, dbl.; [Gen. MacArthur X Hadley]; Schwartz, A.

'**Mme Maurice de Luze**', HT, pb, 1907; flowers rose-pink, center carmine; [Mme Abel Chatenay X Eugene Furst]; Pernet-Ducher

'**Mme Maurice Fenaille**', HT, 1904; Boutigny; (Cavriglia)

Mme Maurice Genevoix, F, op, 1966; Hémeray-Aubert

'**Mme Maurice Kuppenheim**', T, ab, 1877; flowers salmon-yellow, large, dbl.; Ducher, Vve.

'**Mme Maurin**', T, ob, 1853; flowers creamy white, shaded with salmon, large, dbl.; Guillot Père

'MME MÉHA SABATIER', HT, rb, 1916; flowers deep red, with white stripes in some petals, dbl.; [Seedling X Château de Clos Vougeot]; Pernet-Ducher

'MME MÉLANIE SOUPERT', HT, yb, 1905; bud pointed; flowers salmon yellow, suffused pink and carmine, large, semi-dbl., moderate fragrance; vigorous growth; Pernet-Ducher

'MME MÉLANIE WILLERMOZ', T, lp, 1849; flowers white tinted pink, moderate fragrance; Lacharme, F.

'MME MERCIER DE MOLIN', HT, rb, 1921; flowers fiery red, tinted crimson, edges slightly tinged rose-pink, dbl., moderate fragrance; [Comte G. de Rochemur X Liberty]; Schwartz, A.

'MME MICHEL DUFAY', HMoy, rb, 1932; flowers maroon, reverse purplish garnet, large, dbl., cupped, non-recurrent, slight fragrance; foliage dark; very vigorous, bushy growth; [George Dickson X R. moyesii]; Sauvageot, H.

'MME MILLERAND', HT, w, 1926; flowers rosy white, shaded salmon, dbl.; [Pharisaer X Mme Henriette Schissele]; Walter, L.

'MME MINIVER', HT, ob, 1947; bud pointed; flowers orange-red tinted apricot, base yellow, large, dbl.; vigorous growth; [Joanna Hill X Charles P. Kilham]; Vilmorin-Andrieux

'MME MOISANS', LCl, mp, 1955; flowers hortensia-pink, dbl., recurrent bloom, moderate fragrance; foliage glossy; [Lady Sylvia X Unnamed seedling]; Robichon; Pin

'MME MORAND ANDRÉE', HT, dr, 1957; flowers blackish red, very large; foliage dark; vigorous growth; Privat

'MME MOREAU', M, pb, 1872; flowers rose, edged white, large, dbl.; Moreau et Robert

'MME MOSER', HT, pb, 1889; flowers pink to rosy white, large, dbl., intense fragrance; Vigneron

'MME MOUSEUR-FONTAINE', HT, yb, 1931; bud pointed; flowers sulfur-yellow, center saffron-yellow, large, dbl., slight fragrance; [Sunburst X Primerose]; Soupert & Notting

Mme Neige, *see* 'MEIDONA'

'MME NERARD', B, pb, 1938; Nerard

Mme Neumann, *see* **'HERMOSA'**

'MME NICOLAS AUSSEL', HT, ob, 1930; bud pointed; flowers salmon shaded carmine and ochre, large, dbl.; foliage dark; vigorous growth; GM, Portland; Pernet-Ducher; Gaujard

'MME NICOLAS BOUDLER', HT, pb, 1934; bud round; flowers rose, reverse shaded yellow, large, dbl.; [Souv. de Georges Pernet X Gloire de Dijon]; Boudler

'MME NOBÈCOURT', B, lp, 1893; flowers very large, dbl., intense fragrance; Moreau et Robert; (Sangerhausen)

'MME NOËL', HT, rb, 1939; flowers vermilion-red, reverse carmine-yellow, very large, dbl., cupped; foliage dark; vigorous growth; Chambard, C.

'MME NOËL LE MIRE', HT, rb, 1934; flowers brilliant crimson-red with yellow reflections, borne in clusters, semi-dbl., cupped, moderate fragrance; strong stems; dwarf growth; [George Dawson X Dance of Joy]; Sauvageot, H.

'MME NORBERT LEVAVASSEUR', Pol, mr, 1903; (Red Baby Rambler); flowers crimson-red, center lighter, bluing badly, small blooms in l, semi-dbl., cupped, slight fragrance; foliage glossy, dark; bushy, dwarf growth; (14); [Crimson Rambler X Gloire des Polyantha]; Levavasseur

'MME NORMAN', HP, w, 1867; flowers medium, dbl.; Guillot; (Sangerhausen)

'MME OCKER FERENCZ', T, yb, 1892; flowers canary and carmine; Bernaix, A.

'MME OFMAN', HT, or, 1954; bud long, pointed; flowers large, very dbl., exhibition form, moderate fragrance; foliage dark, glossy; strong stems; vigorous, upright, bushy growth; Ofman

'MME OLYMPE TERETSCHENKO', B, 1882; Lévêque; (Cavriglia)

'MME ORÈVE', HT, ob, 1926; flowers rose-salmon, center coppery salmon, moderate fragrance; Chambard, M.

'MME P. DOITHIER', HT, pb, 1920; flowers glossy pink, shaded shrimp-pink; Chambard, M.

Mme P. Euler, *see* **'MME PIERRE EULER'**

Mme P. Olivier, *see* **'SHINING STAR'**

'MME PAQUEL', HT, my, 1945; flowers chrome-yellow, well formed, dbl.; foliage wood and reddish; Mallerin, C.; A. Meilland

'MME PAUL BOUJU', HT, ob, 1930; bud long, pointed; flowers carmine-orange, very large, cupped; foliage bronze; Chambard, C.

'MME PAUL DURINGE', HT, mp, 1934; flowers deep coral, large; very vigorous growth; Chambard, C.

Mme Paul Euler, *see* **'MME PIERRE EULER'**

'MME PAUL LACOUTIÈRE', HT, op, 1897; flowers coppery-pink, large, semi-dbl., intense fragrance; Buatois; (Sangerhausen)

'MME PAUL MARCHANDEAU', HT, w, 1928; flowers white, base deep yellow, dbl., moderate fragrance; Barbaras

'MME PAUL OLLIVARY', HT, dp, 1924; bud pointed; flowers coppery salmon, reverse shaded yellow, nearly large, single; [Mme Mélanie Soupert X Emma Wright]; Schwartz, A.

'MME PAUL PARMENTIER', HT, ob, 1919; flowers salmon-yellow shaded flesh, copper and daybreak-pink, dbl.; [Le Progres X Lyon Rose]; Gillot, F.

'MME PAUL ROTTIER', F, rb, 1957; flowers red shaded orange, borne in large clusters; moderate growth; Buisman, G. A. H.

Mme Paula Guisez, *see* **'HEAT WAVE'**

Mme Paule Massad, *see* MASlupau

'MME PAULINE LABONTÉ', T, op, 1852; flowers salmon-pink, large, dbl., moderate fragrance; vigorous growth; Pradel

'MME PERNET-DUCHER', HT, dy, 1891; flowers medium, very dbl., moderate fragrance; Pernet-Ducher; (Sangerhausen)

'MME PH. PLANTAMOUR', HRg, mr, 1900; flowers large, semi-dbl.; (Sangerhausen)

'MME PHILIPPE RIVOIRE', HT, ab, 1908; flowers apricot-yellow, center nankeen yellow, reverse red, large, dbl., globular; vigorous growth; Pernet-Ducher

'MME PIERRE COCHET', N, yb, 1891; flowers saffron-yellow, shaded scarlet, center apricot, dbl., intense fragrance; vigorous, climbing growth; [Reve d'Or seedling]; Cochet, S.

'MME PIERRE EULER', HT, rb, 1907; (Mme P. Euler, Mme Paul Euler); flowers silvery vermilion-pink, dbl., moderate fragrance; [Antoine Rivoire X Killarney]; Guillot, P.

'MME PIERRE FORESTIER', HT, ob, 1933; flowers orange, shaded shrimp-carmine to satiny china-pink, very large, cupped, moderate fragrance; Chambard, C.

'MME PIERRE GUILLOT', HT, 1928; Guillot; (Cavriglia)

'MME PIERRE KOECHLIN', HT, op, 1936; flowers salmon-pink, very large, dbl., exhibition form, slight fragrance; [Unnamed seedling X The Queen Alexandra Rose]; Sauvageot, H.; C-P

'MME PIERRE OGER', B, pb, 1878; flowers blush, reverse tinged rosy lilac; (28); [Reine Victoria sport]; Oger; C. Verdier

'MME PIERRE PERNY', T, ly, 1880; flowers large, semi-dbl.; Nabonnand; (Sangerhausen)

Mme Pirrie, *see* **'LADY PIRRIE'**

'MME PIZAY', HT, op, 1920; flowers light salmon; [Unnamed variety X Mme Mélanie Soupert]; Chambard, C.

'MME PIZAY, CLIMBING', Cl HT, op, 1920; flowers light salmon-pink, very large, dbl.; Chambard, C.; (Sangerhausen)

'MME PLANTIER', HAlba, w, 1835; flowers creamy white changing to pure white, green pip, blooms in cl, very dbl., flat, non-recurrent, moderate fragrance; foliage small, 7 leaflets; vigorous, spreading,-bushy (about 5 ft) growth; (21); [Thought to be R. alba X R. moschata]; Plantier

'MME PLATZ', dp, 1864; flowers carmine-pink, medium, dbl., intense fragrance; Moreau et Robert; (Sangerhausen)

'Mme Plumecocq', HT, dy, 1931; flowers golden yellow, center brighter; [Roselandia X Ville de Paris]; Lens

'Mme Plumecocq', HT, pb, 1954; flowers bright pink, reverse silvery, very large, dbl., exhibition form; foliage bronze; vigorous, upright growth; [Peace X Seedling]; Gaujard

Mme Poincaré, *see* **'Mme Raymond Poincaré'**

'Mme Portier Durel', LCl, 1910; Durel, P.; (Cavriglia)

'Mme Prosper Laugier', HP, mr, 1875; flowers carmine-red, large, dbl., moderate fragrance; Verdier, E.; (Sangerhausen)

'Mme Rambaux', HP, dp, 1881; flowers very large, very dbl., moderate fragrance; Rambaux; (Sangerhausen)

'Mme Raoul Fauran', HT, m, 1934; flowers carmine shaded velvety purple, reverse light purple, very large, semi-dbl., cupped, moderate fragrance; foliage leathery, dark; vigorous, bushy growth; Sauvageot, H.

'Mme Ravary', HT, ob, 1899; flowers orange-yellow, dbl., moderate fragrance; vigorous, bushy growth; Pernet-Ducher

'Mme Raymond Chevalier-Appert', HT, rb, 1917; flowers cerise-red, edged lighter, dbl., intense fragrance; [Gen. MacArthur X Richmond]; Guillot, P.

Mme Raymond Gaujard, *see* **Olympiad**(TM)

'Mme Raymond Poincaré', HT, pb, 1919; (La Rose de Mme Raymond Poincaré, Mme Poincaré); flowers pale pink and salmon, center yellow, dbl., slight fragrance; GM, Bagatelle, 1915; Gravereaux; Kieffer & Sons

'Mme Rémond', T, yb; flowers sulfur-yellow, edged red, very dbl., moderate fragrance; Lambert, E., 1883

'Mme Renahy', HP, dp, 1889; (Mme Renshy); flowers carmine; Guillot et Fils

'Mme Renard', HP, op, 1872; flowers salmon-pink, very large, dbl.; Moreau et Robert; (Sangerhausen)

'Mme René André', HT, lp, 1906; flowers large, dbl., slight fragrance; (Sangerhausen)

Mme René Cassin, *see* 'DELtaf'

'Mme René Collette', HT, pb, 1909; flowers yellowish-pink with red, large, dbl., intense fragrance; Gamon; (Sangerhausen)

'Mme René Coty', HT, rb, 1955; bud globular; flowers persian red, reverse yellow, very large, dbl., moderate fragrance; foliage glossy, leathery; long, strong stems; vigorous, upright growth; GM, Bagatelle, 1954; [Peace X Brazil]; Meilland, F.; URS

'Mme René Lefèvre', HT, yb, 1938; bud long, pointed, sulphur-yellow, edged cerise; flowers golden yellow, flushed carmine, open, large, semi-dbl.; foliage leathery, bronze; [Elizabeth of York X Mme Henri Pate]; Robichon

'Mme René Truchot', HT, 1952; Orard, Joseph; (Cavriglia)

'Mme Renee Gravereaux', HRg, lp, 1906; Gravereaux

'Mme Renée Oberthür', HT, w, 1908; flowers large, dbl., moderate fragrance; Vigneron; (Sangerhausen)

Mme Renshy, *see* **'Mme Renahy'**

'Mme Robert', HT, yb, 1917; flowers nankeen yellow with chamois reflexes, dbl.; Chambard, C.

'Mme Robert Fortin', HT, m, 1935; flowers carmine-purple, large, dbl., cupped, intense fragrance; very vigorous growth; [Mme Caroline Testout X Yves Druhen]; Buatois

'Mme Robert Joffet', F, rb, 1956; flowers salmon to geranium, reverse carmine-pink, well formed, large, semi-dbl.; Delbard-Chabert

'Mme Robert Martin', HT, pb, 1943; bud long, pointed; flowers pink, center coral, very large, very dbl., slight fragrance; foliage leathery; vigorous, bushy growth; [Charles P. Kilham X Mme Joseph Perraud]; Meilland, F.; A. Meilland

'Mme Robert Perrier', HT, ob; flowers coppery, stamens saffron, large, semi-dbl.; Orard, Joseph

'Mme Roberte Huet', HT, rb, 1960; flowers velvety scarlet, reverse raspberry-red, base gold, dbl., slight fragrance; foliage glossy, slightly bronze; strong stems; vigorous growth; Hémeray-Aubert

'Mme Rochefontaine', HP, pb; flowers rosy flesh to clear pink, large, dbl., moderate fragrance

'Mme Roger Douine', HT, rb, 1926; flowers crimson, shaded scarlet, dbl.; [Souv. de Claudius Denoyel X Mme Edouard Herriot]; Reymond

'Mme Roger Verlomme', HT, pb, 1951; flowers ochre edged flesh-pink, very large, very dbl.; very vigorous growth; GM, Geneva, 1951; Mallerin, C.

'Mme Rosa Monnet', HP, dp, 1885; Monnet

'Mme Rose Chérie', mp, 1850; flowers medium, dbl.; Laffay, M.; (Sangerhausen)

'Mme Rose Romarin', T, rb, 1888; flowers light coppery red with salmon, large, semi-dbl.; Nabonnand; (Sangerhausen)

'Mme Roudillon', HP, mr, 1903; flowers very large, very dbl., intense fragrance; Vigneron; (Sangerhausen)

Mme Royet, LCl, m

'Mme Ruau', HFt, pb, 1909; flowers orange/pink with dark yellow, large, dbl.; Gravereaux; (Sangerhausen)

'Mme S. Croza', HT, ab, 1935; bud long; flowers flesh-pink, very large, dbl., exhibition form; foliage leathery; long stems; vigorous growth; [Sunburst seedling]; Laperrière

Mme Sachi, HT, w

'Mme Sancy de Parabère', Bslt, lp, 1874; (Mme de Sancy de Parabère); flowers clear soft pink, outer petals are thinner than inner ones, which tend, 5 in., very early, non-recurrent bloom, slight fragrance; no prickles; very hardy.; (14); Bonnet

'Mme Saportas', HGal, dp; flowers bright rosy red, large, dbl., intense fragrance

'Mme Schmitt', HT, ab, 1922; flowers salmon-pink, shaded peach-blossom-pink, dbl.; Schwartz, A.

'Mme Schultz', N, yb; flowers pale yellow, center darker, medium, dbl., intense fragrance; vigorous growth; Béluze, 1856

'Mme Schwaller', HT, mp, 1886; flowers large; vigorous growth; Bernaix, A.

'Mme Scipion Cochet'(TM), T, pb, 1872; flowers pale pink to white, center yellow, large, dbl., cupped, moderate fragrance; vigorous growth; [Anna Olivier X Duchesse de Brabant]; Bernaix, A.

'Mme Scipion Cochet', HP, m; flowers purplish pink edged soft pink, center petals wrinkled, dbl., cupped; vigorous growth; Cochet, S., 1872

'Mme Segond Weber', HT, op, 1907; flowers clear salmon-pink, open, dbl., moderate fragrance; long, strong stems; bushy growth; GM, Bagatelle, 1909; [Antoine Rivoire X Souv. de Victor Hugo]; Soupert & Notting

'Mme Segond Weber, Climbing', Cl HT, op, 1911; Ardagh

'Mme Segond Weber, Climbing', Cl HT, op, 1929; Reymond

'Mme Segond Weber, Climbing', Cl HT, lp; Ardagh, 1911; (Weatherly, L.)

'Mme Soledad de Ampuera de Leguizamon', HT, pb, 1928; flowers hydrangea-pink, reverse carmine-rose, dbl., intense fragrance; [General-Superior Arnold Janssen X Mrs E.G. Hill]; Soupert & Notting

Mme Solvay, LCl, dp, 1992; Eve, A.

Mme Sophie Charlotte, A, lp, 1986; Weihrauch

'Mme Sophie Froppot', HP, mp, 1876; flowers large, dbl.; Levet; (Sangerhausen)

'Mme Sophie Stern', HP, dp, 1887; flowers large, dbl.; Lévêque; (Sangerhausen)

'Mme Soubeyran', HP, dp; (Mme de Soubeyran); flowers bright rose; Gonod, 1872

'Mme Souchet', B, pb, 1843; Souchet

'Mme Soupert', M, mr, 1851; flowers well mossed, red; Soupert & Notting, 1871

Mme Souveton, P, pb

'MME SPOTTI', HT, m, 1955; flowers pink striped mauve, very dbl.; foliage glossy; very vigorous growth; Privat

'MME STEINBACH', HT, pb, 1934; bud very long; flowers coppery pink tinted coral, very large, dbl.; [Mrs Pierre S. duPont X Cécile Walter]; Caron, B.

'MME SUZANNE HERVÉ', HP, dr, 1936; bud very long, pointed; flowers velvety red, heavily streaked maroon, intense fragrance; vigorous growth; [Baron Girod de l'Ain seedling]; Hervé; Vially

'MME TAFT', Pol, dp, 1909; flowers ruby-pink medium, dbl.; vigorous, bushy growth; [Crimson Rambler X Mme Norbert Levavasseur]; Levavasseur

'MME TAHA HUSSEIN', HT, rb, 1939; flowers indian red, reverse darker, semi-dbl.; vigorous growth; [Charles P. Kilham X Betty Uprichard]; Colombier

Mme Teresa Estaban, *see* 'DOTRAMES'

'MME THÉODORE DELACOURT', HT, mp, 1913; flowers large, very dbl.; Pernet-Ducher; (Sangerhausen)

'MME THÉVENOT', HP, dr, 1878; flowers large, very dbl., slight fragrance; Jamain, H.; (Sangerhausen)

'MME THIBAUT', HP, lp, 1889; flowers large, dbl.; Lévêque; (Sangerhausen)

'MME THIERS', B, pb, 1873; Pradel

'MME TIRET', HRg, mr, 1901; flowers large, semi-dbl.; Roseraie de l'Hay; (Sangerhausen)

'MME TONY BABOUD', HT, my, 1895; flowers medium golden yellow, large, semi-dbl.; Godard; (Sangerhausen)

'MME TRIFLE', N, yb, 1869; Levet

'MME TRIFLE', Cl T, lp; flowers flesh pink, very dbl.; tall, vigorous, climbing growth; [Gloire de Dijon X Unknown]

'MME TROTTIER', HT, yb, 1937; flowers yellowish flesh, large, very dbl.; vigorous, bushy growth; Leenders, M.

Mme Tsiranana, HT, mr, 1971; Croix

'MME VAN DE VOORDE', HT, mr, 1928; bud pointed; flowers brilliant scarlet, large, semi-dbl., cupped, intense fragrance; very vigorous growth; [Mme Méha Sabatier X K. of K.]; Mallerin, C.; C-P

'MME VANNIER', HT, or; flowers coppery red, large, dbl., cupped; Orard, Joseph

Mme Verbelen, HT, 1973; Delforge; (Cavriglia)

'MME VERDIER', HP, ab, 1840; flowers pale flesh; foliage gray-green; Verdier, V.

'MME VERRIER-CACHET', HP, mr, 1895; flowers carmine-red, very large, dbl., intense fragrance; Chédane-Guinoisseau; (Sangerhausen)

'MME VICTOR BOZZOLA', HT, mp, 1935; bud pointed; flowers bright coral-pink, well formed, large; [Kardinal Piffl X Mme Edouard Herriot]; Soupert, C.

'MME VICTOR LOTTIN', HWich, rb, 1921; flowers dark red, shaded crimson; vigorous, climbing growth; Lottin

'MME VICTOR RAULT', HT, w, 1920; flowers white tinted salmon, center yellow, dbl.; [Mme Mélanie Soupert X Lyon Rose]; Croibier

'MME VICTOR VERDIER', HP, mr, 1863; flowers clear light crimson, large, dbl., 75 petals, flat, seasonal bloom, intense fragrance; vigorous growth; [Senateur Vaisse X ?]; Verdier, E.

'MME VIDOT', HP, ab, 1854; flowers flesh-white; Couturier; E. Verdier

'MME VIGER', HT, lp, 1901; flowers creamy pink, very large, dbl., moderate fragrance; Jupeau; (Sangerhausen)

'MME VILLATE', HT, ob, 1936; flowers orange with yellow, dbl.; foliage glossy; [Korovo X Unnamed seedling]; Walter, L.

'MME VINCENT AURIOL', HT, yb, 1951; bud long; flowers golden coral, peony form, large, dbl., slight fragrance; foliage glossy; vigorous growth; GM, Bagatelle, 1948; [Trylon X Unnamed seedling]; Caron, B.; URS

'MME VIRGILIO PIROLA', HT, mr, 1939; bud long, pointed; flowers very bright red, dbl., moderate fragrance; foliage dark; vigorous, bushy growth; GM, Bagatelle, 1939; [Charles P. Kilham X Étoile de Hollande]; Lens

'MME VISSEAUX', HT, op, 1936; bud long; flowers orange-pink, base yellow, dbl., 4–5 in., slight fragrance; [Odette Foussier X Elvira Aramayo]; Mallerin, C.; C-P

'MME VITTORIA GAGNIÈRE', HT, w, 1909; flowers white tinted pink; [Anna Chartron X Mrs W.J. Grant]; Schwartz, Vve.

Mme W. Baumann, HT, 1975; Sauvageot; (Cavriglia)

'MME W. C. WHITNEY', HT, lp, 1894; flowers flesh-pink, intense fragrance

'MME WAGRAM, COMTESSE DE TURENNE', T, pb, 1894; flowers bright satiny rose suffused with darker rose, very large, moderate fragrance; vigorous growth; Bernaix, A.

'MME WALTER BAUMANN', HT, rb, 1934; bud long; flowers carmine, base yellow, open, dbl.; [Mlle Franziska Kruger X Gwynne Carr]; Reymond; Vially

'MME WELCH', T, ly, 1878; flowers pale yellow, well formed, large, very dbl.; [Devoniensis X Souv. d'un Ami]; Ducher, Vve.

'MME WILLIAM PAUL', M, dp; flowers bright rose, large, dbl., cupped, recurrent bloom; Moreau et Robert, 1869

'MME YORK', HP, dr, 1881; flowers large, dbl., intense fragrance; Moreau et Robert; (Sangerhausen)

'MME YVES LATIEULE', HT, my, 1949; (Nankin); flowers primrose-yellow, dbl., 70 petals, 5 in., slight fragrance; foliage glossy, dark; vigorous growth; GM, NRS, 1950; [Mme Joseph Perraud X Léonce Colombier]; Meilland, F.

Mme Yvette Gayraud, HT, ob; Wisbech Plant Co.

Mme Yvonne Chaverot, HT, 1976; Orard, Joseph; (Cavriglia)

'MME ZÖETMANS', D, w, 1830; flowers pale flesh, dbl.; Marest

Mme. Hide, Ht, pb, 1997; flowers double, large dbl., 15–25 petals, slight fragrance; some prickles; foliage large, medium green, matt; compact (1.2 m) growth; [Pink Picnica X Garden Party]; Ohta, Kaichiro

'MME LA GÈNÈRALE PAUL DE BENOIST', Cl T, op, 1901; flowers salmon-pink, very large, very dbl.; Berland; (Sangerhausen)

Moana®, *see* 'MACBIPI'

Moane st Yellow Tea, T, my; (found rose)

Mobile Jubilee, *see* 'TALMOBILE'

'MOCK'S ROSA DRUSCHKI', Cl HP, mp, 1935; flowers large, dbl.; Mock; (Sangerhausen)

'MODEL OF PERFECTION', F, ob, 1977; bud globular; flowers yellow, pink and orange, large, dbl., 28 petals, slight fragrance; [Zorina X Arthur Bell]; Dickson, A.

Modern Art®, *see* 'POULART'

'MODERN TIMES', HT, rb, 1956; flowers red striped pink; [Better Times sport]; Verbeek; Minier

'MODESTY', HT, pb, 1916; bud pointed; flowers white, center rose-pink, large, dbl., exhibition form, intense fragrance; GM, NRS, 1915; McGredy

'MOEALPHA', Min, ob, 1998; (**Alpha Moe**); flowers cream white with orange edges, single, 6–11 petals, 1.5 in., borne mostly singly, slight fragrance; few prickles; foliage medium, medium green, semi-glossy; upright, medium (18 in.) growth; [Pink Petticoat X Seedling]; Moe, Mitchie; Mitchie's Roses & More, 1999

'MOEANNE', Min, r, 1999; flowers pale russet, reverse lighter, yellow base, dbl., 17–25 petals, 1–2 in., exhibition form, slight fragrance; few prickles; foliage medium, medium green, semi-glossy; upright, medium (15 in.) growth; [Fairhope X Wistful]; Moe, Mitchie; Mitchie's Roses & More, 1999

'MOEDER DES VADERLANDS', F, mr, 1956; (Mère de la Patrie, Mother's Country); flowers bright vermilion-red; strong stems; vigorous growth; [Ambassadeur Nemry X Cinnabar]; Leenders, M.

'MOEJO-NETSU', Min, dp, 1998; flowers magenta pink, dbl., 15–25 petals, 1–1.5 in., exhibition form, borne mostly singly, slight fragrance; no prickles; foliage medium, dark green, semi-glossy; upright, tall,

18 in. growth; [Klima X Seedling (mauve)]; Moe, Mitchie; Mitchie's Roses & More, 1999

'MOEmarilyn', Min, ab, 1999; flowers apricot, reverse lighter, dbl., 17–25 petals, 1–2 in., exhibition form, slight fragrance; few prickles; foliage medium, dark green, semi-glossy; upright, tall (18 in.) growth; [Sheri Anne X Wistful]; Moe, Mitchie; Mitchie's Roses & More, 1999

'MOEmighty', Min, ob, 1999; (**Mighty Moe**); flowers orange, reverse lighter, semi-dbl., 12–16 petals, .5 in., exhibition form, slight fragrance; few prickles; foliage small, medium green, dull; compact, medium (12 in.) growth; [Luis Desamero X seedling]; Moe, Mitchie; Mitchie's Roses & More, 1999

'Mogador', P, dr; (Roides Pourpres); flowers deeper in color, repeats sparingly; [Rose du Roi sport]; Descemet, M. (?), ca 1819

'MOGajam', Min, lp, 1992; (**Jamie**™); flowers full, outer petals horizontal at exhibition stage, 26–40 petals, 1.5–2.75 in., exhibition form, borne in small clusters, slight fragrance; some prickles; foliage medium, medium green, semi-glossy; medium (45–50 cms), spreading growth; [Party Girl X Fairlane]; Moglia, Thomas; Gloria Dei Nursery, 1993

'MOGajosh', Min, pb, 1989; (**Joshua**™); bud pointed; flowers clear, deep pink with yellow at base, reverse lighter to white, dbl., 20 petals, exhibition form, slight fragrance; fruit not observed; prickles hooked, small, red; foliage medium, medium green, semi-glossy; long stems; upright, medium growth; [Loving Touch X Rainbow's End]; Moglia, Thomas; Gloria Dei Nursery

'MOGaline', Min, op, 1992; (**Party Line**™); flowers coral, blooms borne in clusters of 6 or more on basal canes, dbl., 15–25 petals, 1.5 in., slight fragrance; few prickles; foliage small, medium green, matt; medium (45–50 cms), compact growth; [Party Girl X Fairlane]; Moglia, Thomas; Gloria Dei Nursery, 1993

'MOGarip', Min, w, 1992; (**Li'l Rip**™); flowers white with yellow center, blooms borne in small clusters, dbl., 26–40 petals, 1.5 in., slight fragrance; few prickles; foliage small, medium green, matt; medium (35–45 cms), narrow, upright growth; [Loving Touch X Cupcake]; Moglia, Thomas; Gloria Dei Nursery, 1993

'MOGatouch', Min, ab, 1992; (**Li'l Touch**™); flowers apricot, yellow stamens showing, moderately open, circular o, dbl., 15–25 petals, 1.5 in., slight fragrance; some prickles; foliage medium, medium green, matt; medium (40–50 cms), upright growth; [Loving Touch X Prima Donna]; Moglia, Thomas; Gloria Dei Nursery, 1993

Mohak, HT, ob, 1993; Chiplunkar

'Mohawk', S, m; flowers brighter than aster-purple, center white, single, profuse, non-recurrent bloom; foliage dull green; rounded, dwarf growth; hardy.; [R. rubrifolia X R. rugosa]; Central Exp. Farm

'Mohican', F, mr, 1961; flowers dark rose-red; [Garnette sport]; Thompson's, J.H., Sons

'Mohican', HT, mr, 1937; bud long, pointed; flowers cherry-red, large, dbl., slight fragrance; foliage leathery, glossy, dark; long stems; very vigorous growth; [Unnamed variety X Briarcliff]; J.H. Thompson's Sons

'Mohini', F, ob, 1970; bud long, pointed; flowers chocolate-brown, base tinged yellow, full, medium, dbl.; foliage glossy, dark; moderate, bushy growth; [Sea Pearl X Shola]; Division of Vegetable Crops and Floriculture

Mohykan, HT, lp; Strnad, 1972; (Czech Rosa Club)

'Moïse', HGal, m, 1828; flowers rosy-carmine, shaded purplish slate, expanded, large, dbl.; moderate upright growth; Parmentier

'Mojave', HT, ob, 1954; bud long; flowers apricot-orange tinted red, prominently veined, dbl., 25 petals, 4–4.5 in., exhibition form, moderate fragrance; foliage glossy; vigorous, upright growth; AARS, 1954 GM, Bagatelle, 1953 GM, Geneva, 1953; [Charlotte Armstrong X Signora]; Swim, H.C.; Armstrong Nursery

'Mojave, Climbing', Cl HT, ob, 1964; Trimper, K.; Ruston

'Moje Hammarberg'®, HRg, m, 1931; flowers reddish violet, large blooms, dbl., recurrent bloom, intense fragrance; large, red fruit; short, weak stems; vigorous growth; hardy.; Hammarberg; Stockholm Stads Gatukontor

Moksha, HT, op; flowers large, intense fragrance; (Sangerhausen)

Molde®, *see* 'TANnimoll'

'MOLhip', Min, w, 1986; (**High Point**); [Helen Boehm sport]; Molder, W.A.; Rose Acres

Molineux™, *see* 'AUSmol'

Molitor®, *see* 'GAUtira'

Mollie Claire™, *see* 'KINclaire'

'Molly Abdy', HT, rb, 1957; bud pointed; flowers deep scarlet to lighter red, medium, dbl., moderate fragrance; vigorous growth; [Southport X ?]; Abdy; Shepperson

Molly Beckley™, *see* 'JAYmoll'

'Molly Bishop', HT, op, 1951; bud long, pointed; flowers large, dbl., exhibition form, intense fragrance; foliage leathery, rich green; vigorous growth; Robinson, H.

'Molly Bligh', HT, pb, 1917; flowers deep pink, base deep orange, moderate fragrance; Dickson, A.

'Molly Darragh', HT, pb, 1930; bud pointed; flowers bright old-rose, base orange-yellow, very large, dbl., exhibition form, moderate fragrance; McGredy

'Molly Doyle', F, pb, 1963; flowers old-rose-pink, base silvery white, single, 6 petals, 5 in., intense fragrance; foliage light green; low, bushy growth; [Étoile de Hollande X Queen Elizabeth]; Barter

'Molly Kirby', F, or, 1980; flowers vermilion; [Matangi sport]; Wilson, George D.

Molly McGredy, *see* 'MACmo'

'Molly Sharman-Crawford', T, w; bud long, pointed; flowers greenish white, becoming whiter, very dbl., exhibition form, moderate fragrance; foliage sparse, rich green; bushy growth; Dickson, A., 1909

'Molodost', HT, lp, 1956; (Youth); dbl., 37 petals, 5 in.; foliage dark; strong stems; vigorous, upright growth; [Staatsprasident Pats X Cathrine Kordes]; Sushkov, K. L.

'Molodost Mira', HT, mr, 1955; (Early Peace); flowers coral-red, well formed, large, dbl., 50 petals, moderate fragrance; foliage dark; vigorous, well branched growth; [Peace X (Crimson Glory X Poinsettia)]; Klimenko, V. N.

'MOLsunlem', HT, lp, 1986; (**Sunday Lemonade**); dbl., 30 petals, exhibition form, borne usually singly, intense, fruity fragrance; medium, amber prickles, hooked downward; foliage medium, dark, matt; upright, medium-tall growth; [Lemon Spice sport]; Molder, W.A.; Rose Acres

'Momo', HT, dp, 1987; flowers deep pink, large, borne usually singly, dbl., 35–40 petals, slight fragrance; medium, copper fruit; prickles pointed, straight; foliage medium, dark green; [(Picnic X Kordes' Perfecta) X Christian Dior]; Ota, Kaichiro, 1988

Momo, LCl, dr, 1995; Noack, Werner

Mom's Fancy, *see* 'RENmom'

Momy, F, dp

Mon Amour, *see* 'COGamo'

'Mon Amour, Climbing', Cl HT, pb, 1966; bud ovoid; flowers phlox-pink, reverse silver-rose, large, dbl., cupped, profuse, intermittent bloom, moderate fragrance; foliage dark, leathery; very vigorous growth; [Peace, Climbing X Caprice]; Coggiatti, Stelvio

Mon Cheri™, *see* 'AROcher'

Mon Jardin et Ma Maison, *see* MEIchavrin

'Mon Pays', Gr, w, 1983; (Rigobec); flowers large, dbl., slight fragrance; foliage medium to large, medium green, glossy; upright growth; [Iceberg X Peace]; Gailloux, Gilles

'Mon Petit', Min, dp, 1947; flowers light red, dbl., 80 petals; foliage pointed; dwarf, compact growth; [Merveille des Rouges X Pompon de Paris]; Dot, Pedro; URS

Mon Tresor, *see* **'Red Imp'**

'MONA', F, mr, 1957; flowers light crimson, center lighter, borne in large clusters, dbl., 2 in., exhibition form, slight fragrance; foliage light green; vigorous, upright, bushy growth; [Rudolph Timm X Fanal]; Kordes, R.; Dehner & Co.

Mona Lisa, *see* 'KORMAT'

'MONA LISA', Cl HT, pb, 1956; bud ovoid; flowers warm pink overcast cameo-pink, borne singly and in clusters, dbl., 35–40 petals, 4–4.5 in., cupped, free, recurrent boom, intense fragrance; foliage dark, leathery; vigorous growth; bushy pillar or climbing (8–10 ft); [Mrs Sam McGredy X (Mrs Sam McGredy X (Seedling X Capt. Thomas))]; Malandrone; Armstrong Nursery

Mona Lisa, F, my, 1979; Kordes

'MONA ROSETTE', F, pb, 1963; bud ovoid; flowers pink and white striped, medium, in clusters, dbl., cupped; foliage dark, leathery; bushy, low growth; [Pink Rosette sport]; Manski; Paulen Park Nursery

'MONA RUTH', Min, mp, 1959; dbl., 30 petals, 1–1.5 in., slight fragrance; foliage leathery; vigorous (12–14 in.), bushy growth; [((Soeur Thérèse X Skyrocket) X (Seedling X Red Ripples)) X Zee]; Moore, Ralph S.; Sequoia Nursery

'MONARCH', HT, mp, 1926; bud pointed; flowers silvery pink, very large, dbl., exhibition form, moderate fragrance; Dobbie

'MÖNCH', F, mp, 1952; flowers pink, large large corymbs, dbl.; foliage dark, leathery; vigorous, upright growth; [Karl Weinhausen seedling]; Tantau

'MONCTON', HRg, lp, 1977; bud ovoid; flowers medium, dbl., 20 petals, 2 in., intense fragrance; foliage gray-green; upright, bushy growth; [Schneezwerg X R. chinensis]; Svedja, Felicitas; Canada Dept. of Agric.

Monday's Child, *see* 'JUDMON'

'MONDIAL PINK', HT, pb, 1965; flowers pink, base yellow, medium, exhibition form, slight fragrance; foliage glossy; Mondial Roses

Mondiale®, *see* KORozon

Mondovision, *see* 'DELBRAT'

Monet, *see* 'JACLION'

Monet Rose, HT, m

'MONETA', F, pb, 1969; bud ovoid; flowers pinkish red, small, very dbl.; foliage dark; [Garnette sport]; Freytag; Santhof

'MONETTE', HWich, w, 1921; flowers small, borne in clusters, very dbl., intense fragrance; vigorous, climbing growth; [R. wichurana X Yvonne Rabier]; Hémeray-Aubert

'MONETTE', Pol, rb, 1922; flowers fiery red, passing to rose, with many white streaks, borne i; [Phyllis X Unnamed seedling]; Turbat

Mongioia, Gr, 1966; Mansuino; (Cavriglia)

Monica®, *see* 'TANAKNOM'

Monica Astmann, F, or; flowers light orange, large, dbl., moderate fragrance; (Sangerhausen)

Monika®, *see* 'TANAKNOM'

'MONIQUE', HT, op, 1949; flowers medium salmon-pink, well-shaped, large, dbl., 25 petals, intense fragrance; vigorous, upright growth; GM, NRS, 1950; [Lady Sylvia X Unnamed seedling]; Paolino; URS

Monique van Honsebrouck, Gr; Delforge, H.; (Cavriglia)

Monna Lisa®, HT, w

Monna Lisa, Climbing®, Cl HT, w

'MONS A. MEILLE', B, dr, 1889; flowers very large, dbl., intense fragrance; Moreau et Robert; (Sangerhausen)

'MONS CHÉDANE-GUINOISSEAU', HRg, lp, 1895; flowers very large, dbl.; Chédane-Guinoisseau; (Sangerhausen)

'MONS CORDEAU', B, dp, 1892; flowers deep red with violet, very dbl., intense fragrance; very thorny; Moreau et Robert

'MONS DE MONTIGNY', HP, dp, 1857; flowers carmine-pink, large, dbl.; Paillet; (Sangerhausen)

'MONS DE MORAND', HP, 1891; Schwartz, J.; (Cavriglia)

'MONS EDOUARD DETAILLE', HP, m, 1893; flowers purple/pink, large, dbl.; Gouchault; (Sangerhausen)

'MONS ERNEST DUPRÉ', HP, mr, 1904; flowers large, dbl.; Boutigny; (Sangerhausen)

'MONS FRAISSENON', HT, dp, 1911; flowers large, dbl., moderate fragrance; Gamon; (Sangerhausen)

'MONS FRANCISQUE RIVE', HP, 1893; Schwartz, J.; (Cavriglia)

'MONS HÉLYE', HRg, lp, 1900; flowers creamy pink, medium, semi-dbl.; Morlet; (Sangerhausen)

'MONS HOSTE', HP, mr, 1884; flowers large, dbl.; Liabaud; (Sangerhausen)

'MONS JULES DEROUDILHE', HP, m, 1886; flowers purple/pink, medium, dbl.; Liabaud; (Sangerhausen)

'MONS JULES LEMAITRE', HP, dp, 1890; flowers carmine-pink with lighter reverse, very large, dbl., moderate fragrance; Vigneron; (Sangerhausen)

'MONS JULES MAQUINANT', HP, mr, 1882; flowers large, dbl.; Vigneron; (Sangerhausen)

'MONS LAURIOL DE BARNY', HP, 1894; Trouillard; (Cavriglia)

'MONS LE CAPITAINE LOUIS FRERE', HP, dr, 1883; Vigneron

'MONS MATHIEU BARON', HP, dr, 1886; flowers dark violet-red, large, dbl., moderate fragrance; Schwartz; (Sangerhausen)

'MONS. BONCENNE', HP, dr, 1864; (M Boncenne); flowers deep crimson, medium, dbl.; moderate growth; Liabaud

Mons. E.Y. Teas, *see* **'E. Y. TEAS'**

'MONS. LOUIS RICARD', HP, m, 1894; flowers blackish purple, shaded vermilion, large, dbl., cupped, moderate fragrance; vigorous growth; [Simon de St. Jean X Abel Carrière]; Boutigny

'MONS. PAUL LÉDÉ', HT, pb, 1902; flowers carmine-pink, shaded yellow, large, dbl., cupped, moderate fragrance; Pernet-Ducher

'MONS. PAUL LÉDÉ, CLIMBING', Cl HT, pb, 1913; Low

Mons. Pélisson, *see* **'PÉLISSON'**

'MONS. ROSIER', T, pb, 1887; flowers rose and yellow; Nabonnand

'MONS. TILLIER', T, op, 1891; flowers rosy flesh, shaded salmon-rose and purple-rose, fairly large, dbl.; vigorous growth; Bernaix, A.

Monsieur Cordeau, B, dp; Moreau et Robert, 1892

Monsieur Lede, *see* **'PAUL LEDE, CLIMBING'**

Monsieur Paul Lede, *see* **'PAUL LEDE, CLIMBING'**

Mont Blanc, HRg, lp, 1986; semi-dbl.; numerous hips; 3 ft growth; very hardy; Baum

'MONT HAMEL', LCl, mp, 1937; semi-dbl., globular, borne in clusters; foliage glossy; very vigorous, climbing growth; Constantin

'MONT-À-CAMP', HT, yb, 1928; flowers pure yellow, reverse shaded orange, very large, dbl.; vigorous, upright growth; Delobel

Montagny, *see* **'FLAMINAIRE'**

Montagut, HT, 1991; Dot, Simon; (Cavriglia)

Montana®, *see* **'ROYAL OCCASION'**

Montauban de Bretagne®, *see* ADAmonbu

Montblanc, HRg, w, 1984; Baum

'MONTE CARLO', HT, ob, 1949; flowers indian yellow suffused russet-orange, well shaped, dbl., 45 petals, 5 in., moderate fragrance; foliage glossy; vigorous growth; GM, NRS, 1950; [Peace X Unnamed seedling]; Meilland, F.; URS

Monte Cassino, HRg, dr, 1987; semi-dbl., moderate fragrance; 3 ft growth; Baum

'MONTE CRISTO', HP, dr, 1861; flowers large, dbl., 50 petals, globular; low growth; Fontaine

'MONTE IGUELDO', HT, dr, 1944; flowers velvety dark red, well formed, open, intense fragrance; upright, open habit growth; [Étoile de Hollande X Majorca]; La Florida

'Monte Nevoso', HT, 1929; Cazzaniga, F. G.; (Cavriglia)

Monte Rosa, HRg, mp, 1984; flowers large, semi-dbl., moderate fragrance; numerous hips; 3 ft growth; Baum

'Monte Toro', HT, pb, 1962; flowers strawberry-pink shaded red, large, dbl., moderate fragrance; [Berthe Mallerin X Grand'mere Jenny]; Dot, Simon

'Montecito', S, w, 1930; single, 5 petals, 6 in.; foliage medium green, disease-resistant; vigorous, climbing (to 50 ft) growth; Franceshi-Fenzi

'Monterey', HT, rb, 1933; flowers light gold flushed rose, veined and edged crimson, reverse lighter, dbl., moderate, fruity fragrance; [The Queen Alexandra Rose sport]; Lester Rose Gardens

'Monterosa', HT, pb, 1952; flowers deep rose, edged lighter, peony form, dbl., 40 petals; very vigorous growth; GM, Rome, 1952; [Elettra X Superba (HT)]; Giacomasso

'Montesquieu', HT, pb, 1959; flowers deep rose shaded crimson, dbl., 45 petals, exhibition form, moderate fragrance; strong stems; very vigorous growth; [Loli Creus X Tahiti]; Dot, Simon

Montesuma Rosice, HT, lp; Strnad; (Czech Rosa Club)

'Montezuma', Gr, op, 1955; bud urn-shaped; dbl., 36 petals, 3.5–4 in., exhibition form, slight fragrance; foliage leathery, semi-glossy; very vigorous, compact growth; GM, Geneva, 1955 GM, NRS, 1956 GM, Portland, 1957; [Fandango X Floradora]; Swim, H.C.; Armstrong Nursery

Montezuma, Climbing, Cl HT, lp; [Montezuma sport]; Langbecker, 1970; (Weatherly, L.)

'Monthly Rambler', LCl, dr, 1926; flowers brilliant crimson-red, large, borne in clusters, semi-dbl., blooms throughout summer and on young wood in autumn, moderate fragrance; [R. wichurana X Old Crimson China]; Laxton Bros.

'Monthyon', HGal, m; (Montigny); flowers slaty pink blotched purple, large, dbl., globular

Montigny, *see* **'Monthyon'**

Montigny-le-Tilleul, S, dr; very dbl., 60 petals, flat; foliage dense dark green; RvS-Melle, 1995

'Montijo', F, dr, 1954; bud ovoid; flowers crimson-red, medium, dbl., 40 petals; very vigorous growth; [Méphisto X Magrana]; Dot, Pedro

'Montmartre', F, m, 1955; bud ovoid; flowers bright purplish red, large, dbl., moderate fragrance; foliage dark, bronze; very vigorous growth; [Peace X Seedling]; Gaujard

'Montparnasse, Climbing', Cl HT, ab, 1980; flowers large, dbl., 34 petals, exhibition form, slight fragrance; foliage coppery green, leathery, semi-glossy; bushy, vigorous growth; [Montparnasse sport]; Kodoya, Y.; Ogura Rose Nursery

Montreal®, *see* 'GAUzeca'

Montresor, *see* **'Red Imp'**

'Montrose', HT, dp, 1916; flowers deep pink; [Unnamed red seedling X Laurent Carle]; Cook, J.W.

Montrose, *see* 'TALmontrose'

Montrose Tea, T, lp

'Montrouge', F, rb, 1956; flowers clear red, center tinted copper, medium, semi-dbl., slight fragrance; foliage glossy; vigorous, bushy growth; [Peace X Alain seedling]; Gaujard

'Montseny', HT, op, 1944; flowers salmon-pink, well formed, dbl., 40 petals, exhibition form, intense fragrance; foliage glossy; upright growth; [Mme Butterfly X Jean C.N. Forestier]; Dot, Pedro

'Montserrat', HT, op, 1954; flowers orange-salmon-pink, large, dbl., cupped, moderate fragrance; foliage dark, glossy; very vigorous, upright growth; [Comtesse Vandal X Angels Mateu]; Camprubi, C.

'Montuna', HP, lp; [Prince Camille de Rohan X Unknown]; Luke, E.R., 1928; (Weatherly, L.)

Monty's White, T, w

Monument, HT, mr, 1969; flowers large, dbl., moderate fragrance; Urban, J.; (Sangerhausen)

'Monviso', HT, pb, 1955; bud pointed; flowers white streaked pink, large, dbl., slight fragrance; strong stems; [Julien Potin X Monterosa]; Giacomasso

'Monymusk', HMsk, my, 1954; flowers clear yellow, large, semi-dbl., recurrent bloom, moderate fragrance; vigorous (3–4 ft) growth; [Pax X Phyllis Gold]; Ratcliffe

Monza, HT, 1965; Leenders; (Cavriglia)

'Mood Music', Min, op, 1977; bud mossy; flowers orange to orange-pink, flat to rounded, mini-moss, dbl., 45 petals, 1 in., slight fragrance; foliage small; upright, bushy growth; [Fairy Moss X Goldmoss]; Moore, Ralph S.; Sequoia Nursery

Moody Blues, *see* 'MACmooblu'

Moody Dream, *see* MACmoodre

'Moon Glow', LCl, pb, 1937; flowers creamy primrose, center soft yellow, dbl., 60–75 petals, 3 in., moderate fragrance; strong stems; vigorous growth; [Glenn Dale X Mrs Arthur Curtiss James]; Brownell, H.C.

Moon Magic, *see* 'DELcro'

'Moon Maiden', F, ly, 1970; flowers creamy yellow, dbl., 55 petals, 3.5–4 in., moderate fragrance; foliage dark; [Fred Streeter X Allgold]; Mattock

Moon Mist™, *see* 'PIXimis'

Moon River™, *see* 'KORconta'

Moon River, *see* BENlavscent

Moon Shadow™, *see* 'JAClaf'

'Moonbeam', HT, dy, 1950; flowers deep golden yellow, large, exhibition form, moderate fragrance; foliage dark, glossy; RULED EXTINCT 4/92; [Unnamed seedling X McGredy's Yellow]; Robinson, H.; Baker's Nursery

Moonbeam, *see* 'AUSbeam'

'Moonbeam, Climbing', Cl HT, dy, 1955; Kordes

Moondance, *see* **'MEInivoz'**

'Moondrops', HT, op, 1965; bud ovoid; flowers rose-amaranth shaded salmon, open, single, intense, spicy fragrance; vigorous growth; [Sunny Boy X Seedling]; Delforge

Moondrops 85, HT, m, 1985; Delforge; (Cavriglia)

Moonlight, *see* 'JACice'

'Moonlight', HMsk, ly, 1913; flowers lemon-white, prominent yellow stamens, blooms in small clusters, single, repeat bloom, moderate fragrance; foliage dark, glossy; vigorous (4–5 ft) bushy growth; GM, NRS, 1913; [Trier X Sulphurea]; Pemberton

Moonlight and Roses, *see* 'BRIrose'

Moonlight Bay, *see* 'RENmobay'

Moonlight Fragrance, LCl, ly

Moonlight Lady, *see* 'SOCalp'

Moonlight Lady, *see* JACosol

'Moonlight Magic', HT, m, 1991; flowers lavender, moderately blooms borne mostly singly, dbl., slight fragrance; foliage medium, medium green, matt; medium, bushy growth; [Unknown seedling X Unknown seedling]; Burks, Larry; Co-Operative Rose Growers

'Moonlight Mist', HT, w, 1965; bud pointed; flowers ivory-white, very large, dbl., slight fragrance; very vigorous growth; [Buccaneer X Golden Harvest]; Armbrust; Langbecker

'Moonlight Niagara'®, HT, or, 1987; flowers orange, reverse pale orange, loose, star-shaped, medium, bor, very dbl., 60 petals, exhibition form, moderate, spicy fragrance; rounded, medium, orange-red fruit; semi-thick, medium, brown-red prickles; foliage large, dark green, matt; upright, tall growth; [Red Queen sport]; Rogin, Josip

'Moonlight Serenade', HT, yb, 1994; flowers yellow with red fringe, blooms, dbl., 26–40 petals, 3–3.5 in., moderate fragrance; many prickles; foliage large, medium green, semi-glossy; medium (120 cms), spreading growth; GM, JRC,

1991; [Garden Party X Jana]; Ohata, Hatsuo

'MOONLIGHT SONATA', HT, ab, 1966; bud ovoid; flowers large, dbl., cupped, moderate fragrance; foliage dark, glossy; vigorous, upright, bushy growth; [Diamond Jubilee seedling X (Goldilocks X Orange Nassau)]; Boerner; J&P

'MOONRAKER'®, F, ly, 1968; flowers pale yellow to white, large blooms in clusters, dbl., slight fragrance; foliage light green; [Pink Parfait X Highlight]; Harkness

'MOONRISE', HT, ly, 1997; flowers full, medium, very dbl., 26–40 petals, borne mostly singly, no fragrance; foliage medium, medium green, semi-glossy; upright, low (2 ft) growth; [(Lady X X Antigua) X Midas Touch]; McMillan, Thomas G.

'MOONSPRITE', F, ly, 1956; bud ovoid; flowers creamy white, center pale gold, blooms in clusters, dbl., 80 petals, 2–2.5 in., cupped, intense fragrance; foliage leathery, semi-glossy; dwarf, bushy growth; GM, Baden-Baden, 1955 GM, Rome, 1956; [Sutter's Gold X Ondine]; Swim, H.C.; Armstrong Nursery

Moonstone™, *see* 'WEKCRYLAND'

Moonstruck, *see* JACess

Moorcap, *see* **'RED CASCADE'**

'MOORCAR', Min, rb, 1972; (**Magic Carrousel**®, 'MORROUSEL'); flowers petals white, edged red, small, dbl., exhibition form, slight fragrance; foliage small, glossy, leathery; vigorous, bushy growth; AOE, ARS, 1975 Miniature Rose Hall of Fame, ARS, 1999; [Little Darling X Westmont]; Moore, Ralph S.; Sequoia Nursery

Moore's Classic Perpetual, *see* 'MORCLASSIC'

Moore's Pink Perpetual, *see* 'MORPINKPER'

Moore's Yellow, Min, my

'MOOSE RANGE', HRg, mr, 1944; [Hansa X Mary L. Evans]; Wright, Percy H.

Morabito, *see* MASflodel

'MORAIN', Min, yb, 1989; (**Lucky Charm**); bud ovoid; flowers yellow, reverse tinting red, aging to pink to red, lightly s, dbl., 50–60 petals, exhibition form, no fragrance; no fruit; prickles straight, small, brown; foliage small, medium green, matt; bushy, spreading, low growth; [Rumba X Pinstripe]; Moore, Ralph S.; Sequoia Nursery

'MORALBUQUE', Min, rb, 1997; (**Albuquerque Enchantment**); flowers double, medium, dbl., 15–25 petals, borne in small clusters, slight fragrance; foliage medium, medium green, semi-glossy; medium (30–45 cms) bushy, spreading growth; [Poker Chip X Cherry Magic]; Moore, Ralph S.

'MORALERT', Min, mr, 1990; (**Red Alert**); bud pointed; flowers medium red, slightly lighter reverse, aging similar, medium, dbl., 35 petals, exhibition form, slight fragrance; foliage medium, medium green, semi-glossy; upright, bushy, medium growth; [Orangeade X Rainbow's End]; Moore, Ralph S., 1986; Sequoia Nursery, 1991

'MORANIUM', Min, or, 1983; (**Autumn Fire**); flowers small, semi-dbl., no fragrance; foliage small, medium green, semi-glossy; bushy, very spreading growth; [Little Chief X Anytime]; Moore, Ralph S., 1982

'MORANYBLAC', Min, dr, 1993; (**Red Cameo**); flowers dark red, does not burn, blooms borne mostly single, semi-dbl., 6–14 petals, 1.5–2.75 in., slight fragrance; few prickles; foliage medium, medium green, semi-glossy; medium (34–38 cms), upright, bushy growth; [Anytime X Black Jade]; Moore, Ralph S.; Sequoia Nursery, 1994

'MORANYFACE', Min, ob, 1992; (**Halo Star**); flowers reddish on outside, orange to pink on inside, reddish lavender, single, 5 petals, 1.5 in., no fragrance; foliage small, medium green, matt; medium (35–40 cms), upright, bushy growth; [Seedling (Anytime X Angel Face) X Seedling (Anytime X Angel Face)]; Moore, Ralph S.; Sequoia Nursery, 1993

'MORART', Cl Min, ob, 1989; (**Work of Art**); bud short; flowers orange blend, reverse slightly more yellow, urn-shaped, medium, dbl., 35 petals, exhibition form, slight fragrance; globular, medium, orange fruit; prickles slender, nearly straight, average, brown, few; foliage medium, bronze aging to medium green, semi-glossy; upright, spreading, tall growth; PP007617; [Unnamed yellow Climbing Miniature seedling X Gold Badge]; Moore, Ralph S.; Sequoia Nursery

Morava, HT, 1970; Strnad; (Cavriglia)

'MORBERG', Min, or, 1981; (**Ann Moore**); bud long, pointed; flowers, dbl., 30 petals, exhibition form, borne usually singly, moderate fragrance; long prickles; foliage leathery, semi-glossy; vigorous, bushy, upright growth; [Little Darling X Fire Princess]; Moore, Ralph S.

'MORBERK', Min, m, 1988; (**Berkeley Beauty**™); flowers picotee white, edged lavender to pink (striped), small, dbl., exhibition form, moderate fragrance; prickles varying green to brown; foliage medium, medium green, semi-glossy, abundant; bushy, medium, neat growth; [Pink Petticoat X Make Believe]; Moore, Ralph S.; Sequoia Nursery, 1987

'MORBLACK', Min, dr, 1982; (**Don Marshall**); flowers medium red, reverse blackish-red, small, dbl., 35 petals, exhibition form, slight fragrance; foliage small, dark, matt; bushy, spreading growth; [Baccará X Little Chief]; Moore, Ralph S.; Moore Min. Roses

'MORBOUQUET', S, mp, 1991; (**Grandma's Pink**); very dbl., borne in small clusters, no fragrance; foliage medium, medium green, matt; medium, upright, bushy growth; [Shakespeare Festival X Marchioness of Londonderry]; Moore, Ralph S.; Sequoia Nursery

'MORBOX', Min, pb, 1984; (**Jewel Box**); flowers light to deep pink blend, reverse lighter, small, dbl., 20 petals, slight fragrance; foliage small, medium green, semi-glossy; bushy growth; [Avandel X Old Master]; Moore, Ralph S.; Moore Min. Roses, 1983

'MORBOY', Min, mr, 1990; (**Billy Boy**); bud pointed, short; flowers small blooms borne usually singly, dbl., 15–18 petals, cupped, slight fragrance; prickles slender, straight, small, brownish; foliage small, medium green, matt, dense; bushy, low to medium, compact growth; [Anytime X Happy Hour]; Moore, Ralph S.; Sequoia Nursery

'MORBRIGHTS', Min, ob, 1999; (**Spotlight**); flowers bright orange blended rose with white star-shaped eye, single, 5 petals, 2–2.5 in., borne mostly singly, no fragrance; foliage medium, medium green, semi-glossy; bushy, compact (12–18 in.) growth; [Orangeade X Little Artist]; Moore, Ralph S.; Sequoia Nursery, 1999

'MORBROWN', Min, ab, 1993; (**Apricot Twist**); flowers moderately full, medium sized blooms, dbl., 15–25 petals, 1.5–2.75 in., borne in small clusters, slight fragrance; few prickles; foliage small, medium green, semi-glossy; low (30–32 cms), bushy, compact growth; [Golden Angel X Sequoia Gold]; Moore, Ralph S.; Sequoia Nursery, 1994

'MORBUG', Min, mr, 1992; (**Ladybug**); semi-dbl., 6–14 petals, 1.5 in., borne in small clusters, fast repeat, slight fragrance; few prickles; foliage small, medium green, semi-glossy; low (30–40 cms), bushy, compact growth; [Sheri Anne X Cherry Magic]; Moore, Ralph S.; Sequoia Nursery, 1993

'MORCADE', Cl Min, mp, 1981; (**Pink Cascade**); flowers small, dbl., 35 petals, slight fragrance; foliage small, medium green, matt to semi-glossy; spreading (5–7 ft; similar to red cascade) growth; [(R. wichurana X Floradora) X Magic Dragon]; Moore, Ralph S.; Moore Min. Roses

'MORCALYN', Min, dp, 1986; (**Coral Cameo**); flowers deep pink, borne singly and in small sprays, dbl., exhibition form, slight fragrance; small, globular, orange fruit; very few prickles; foliage small, medium green, semi-glossy; medium, upright, bushy growth; [Little Darling X Anytime]; Moore, Ralph S.; Moore Min. Roses, 1982

'MORCANA', Min, mp, 1980; (**Antique Rose**™, 'MORCARA'); bud pointed; flowers rose pink, mini-flora, medium, dbl., 38 petals, exhibition form, borne usually singly, slight fragrance; straight, brown prickles; foliage dark, semi-glossy; vigorous, upright growth; [Baccará X Little Chief]; Moore, Ralph S.

'MORcara', Min, mp, 1980; (**Antique Rose**™, 'MORcana'); bud pointed; flowers rose pink, mini-flora, medium, dbl., 38 petals, exhibition form, borne usually singly, slight fragrance; straight, brown prickles; foliage dark, semi-glossy; vigorous, upright growth; [Baccará X Little Chief]; Moore, Ralph S.

'MORcarlet', Min, mr, 1988; (**Scarlet Moss**™); flowers intense, scarlet-red, open, small, semi-dbl., borne in sprays of 3–10, no fragrance; round elongated, orange fruit; prickles slender, various, green to brown; foliage medium, medium green, glossy, leathery; upright, bushy, tall growth; PP007128; [(Dortmund X Unnamed mossed Miniature seedling) X (Dortmund X Unnamed mossed Miniature striped seedling)]; Moore, Ralph S.; Sequoia Nursery

'MORcas', Min, w, 1991; (**White Rain**); dbl., 15–25 petals, 1.5 in., borne in large clusters, moderate fragrance; some prickles; foliage small, light green, semi-glossy; low (22 cms), spreading, groundcover growth; [Papoose X Renae]; Moore, Ralph S.; Sequoia Nursery, 1992

'MORcat', Min, mp, 1986; (**Copy Cat**); [Beauty Secret sport]; Moore, Ralph S.; Moore Min. Roses, 1985

'MORcebru', Min, lp, 1981; (**Baby Cécile Brunner**); flowers soft pink, small, slight fragrance; foliage small, medium green, matt; upright, bushy growth; [Cécile Brunner, Climbing X Fairy Princess]; Moore, Ralph S.; Moore Min. Roses

'MORceedee', S, pb, 1990; (**Cee Dee Moss**); bud pointed; flowers pink with occasional white striped, lighter reverse, ages lighter, semi-dbl., 15–25 petals, 2.5–2.75 in., cupped, slight fragrance; foliage medium, light green, glossy; bushy, spreading, medium growth; [Carolyn Dean X Seedling]; Moore, Ralph S., 1981; Sequoia Nursery

'MORchari', Min, m, 1984; (Insolite, **Sweet Chariot**™); flowers lavender to purple blend, small, dbl., blooms in clusters of 5–20, intense fragrance; foliage small, medium green, matt; upright growth; PP005975; [Little Chief X Violette]; Moore, Ralph S.; Moore Min. Roses

'MORcharlie', Min, rb, 1996; (**Charlie Brown**); flowers red and white stripe, blooms borne mostly single or clusters, semi-dbl., 8–14 petals, 1–1.25 in., slight fragrance; few prickles; foliage small, medium green, semi-glossy; compact, bushy, low (12–15 in.) growth; [Seedling X Pinstripe]; Moore, Ralph S.; Sequoia Nursery, 1997

'MORcheer', Min, dr, 1986; (**Cheerleader**™); flowers small, very dbl., blooms in sprays of 5–10, no fragrance; no fruit; foliage small, medium green, semi-glossy; bushy, spreading growth; PP005977; [Fairy Moss X Orange Honey]; Moore, Ralph S.; Moore Min. Roses, 1985

'MORchermag', Min, dr, 1988; (**Cherry Magic**™); flowers deep red, reverse lighter red with silver sheen, aging lighter, dbl., 25 petals, exhibition form, no fragrance; round, small, orange red fruit; prickles short, small, brown; foliage small, medium green, matt; bushy, spreading, low growth; PP007061; [Anytime X Lavender Jewel]; Moore, Ralph S.; Sequoia Nursery

'MORchick', Min, mp, 1990; (**Chick-a-dee**); bud pointed; flowers medium pink with occasional white stripes, reverse similar, dbl., 40–50 petals, exhibition form, slight fragrance; prickles small, hooked downward, brownish; foliage small, medium green, matt to semi-glossy; bushy, low, compact, rounded growth; [Cécile Brunner X (Dortmund X (Fairy Moss X (Little Darling X Ferdinand Pichard)))]; Moore, Ralph S.; Sequoia Nursery

'MORcingold', Min, my, 1995; (**Cinderella Gold**); flowers medium yellow, moderately full (15–25 petals), small (0–4 cms) blooms borne mostly in small clusters; slight fragrance; no prickles; small foliage, light green, semi-glossy; low (15 cms), bushy growth; [Cal Poly X Cal Poly]; Moore, Ralph S.; Sequoia Nursery, 1996

'MORclassic', S, dp, 1999; (**Moore's Classic Perpetual**); very dbl., 41 petals, 3 in., borne in small clusters, moderate fragrance; few prickles; foliage medium, medium green, semi-glossy; dwarf hybrid perpetual, bushy, compact (16–24 in.) growth; [Anytime X Paul Neyron]; Moore, Ralph S.; Sequoia Nursery, 1999

'MORclilav', Cl Min, m, 1992; (**Orchid Jubilee**); flowers mauve blend, holds color well in heat, blooms borne in small, dbl., 15–25 petals, 1.5–2.75 in., profuse repeat bloom, no fragrance; few prickles; foliage medium, medium green, matt; tall (2 meters), upright, climbing growth; [Seedling (Little Darling X Yellow Magic) X Make Believe]; Moore, Ralph S.; Sequoia Nursery, 1993

'MORclim', Cl Min, yb, 1983; (**High Stepper**); flowers yellow overlaid pink, reverse yellow, small, dbl., slight fragrance; foliage medium, dark, semi-glossy; bushy, spreading (to 5 ft; needs support) growth; [(Little Darling X Yellow Magic) X Magic Wand]; Moore, Ralph S.; Moore Min. Roses

'MORclip', Cl F, mp, 1993; (**Playgirl, Climbing**); single, 5 petals, 3–3.5 in., borne in small clusters, slight fragrance; foliage medium, medium green, semi-glossy; tall (2–3 m), upright, spreading, climbing growth; [Playgirl sport]; Moore, Ralph S.; Sequoia Nursery, 1995

'MORcoat', Min, rb, 1992; (**Tag-a-long**); flowers contrast of reddish lavender against white, blooms, somewhat, semi-dbl., 6–14 petals, 1.5–2.75 in., no fragrance; few prickles; foliage medium, medium green, new bronze, matt; medium (30–40 cms), upright, bushy growth; [Seedling (Little Darling X Yellow Magic) X Make Believe]; Moore, Ralph S.; Sequoia Nursery, 1993

'MORcofair', Cl S, pb, 1995; (**Coral Fairy**); single, 5 petals, 1.5 in., borne in large clusters, slight fragrance; few prickles; foliage small, dark green, semi-glossy; tall (3–4 m), upright, bushy, spreading growth; [(R. wichurana X Floradora) X Hallelujah]; Moore, Ralph S.; Sequoia Nursery, 1995

'MORcojo', Min, pb, 1984; (**Country Joy**); flowers light pink, yellow reverse, small, dbl., slight fragrance; foliage small, medium green, matt; compact growth; [Pinocchio X Yellow Jewel]; Moore, Ralph S.; Moore Min. Roses

'MORdan', Min, mr, 1980; (**Red Wagon**); dbl., 23 petals, borne 1–3 or more per cluster, slight fragrance; brown prickles; foliage glossy; vigorous, bushy, rounded growth; [Little Darling X Little Chief]; Moore, Ralph S.; Moore Min. Roses

'MORdarain', Min, pb, 1992; (**Lifestyle**); flowers color holds well, blooms mostly singly, dbl., 15–25 petals, 1.5–2.75 in., slight fragrance; few prickles; foliage medium, medium green, semi-glossy; medium (35–45 cms), upright, bushy, rounded growth; [Little Darling X Rainbow's End]; Moore, Ralph S.; Sequoia Nursery, 1993

'MORdarcrest', F, pb, 1994; (**Elegant Design**); flowers pink, crested sepals, blooms borne in small clusters, dbl., 15–25 petals, 1.5–2.75 in., slight fragrance; few prickles; foliage medium, medium green, matt; medium to tall (45–60 cms), upright, bushy growth; [Little Darling X Crested Jewel]; Moore, Ralph S.; Sequoia Nursery, 1995

'MORdashin', Min, yb, 1991; (**Southern Delight**); flowers yellow edged with red, aging to pink and yellow, blooms born, dbl., 15–25 petals, 1.5–2.75 in., slight fragrance; few prickles; foliage medium, medium green, semi-glossy; medium to tall (22–24 cms), upright, bushy growth; [Little Darling X Rise 'n' Shine]; Moore, Ralph S.; Sequoia Nursery, 1992

'MORday', Min, yb, 1984; (**Sunday Brunch**); flowers soft creamy yellow, petal tips becoming pink, which spreads, dbl., 20 petals, moderate fragrance; foliage small to medium, medium green, semi-glossy; vigorous, upright, bushy growth; [Rumba X Peachy White]; Moore, Ralph S.; Moore Min. Roses, 1983

'MORdeb', Min, pb, 1981; (**Joan Austin**); bud pointed; flowers light to medium pink, white stripes, dbl., 38 petals, exhibition form, borne singly, sometimes 3 or more per cluster, intense fragrance; foliage small, medium green, semi-glossy to matt; very bushy, compact growth; [Avandel X Unnamed seedling]; Moore, Ralph S.; Moore Min. Roses

'MORDEN AMORETTE', S, dp, 1977; bud pointed; flowers deep pink, dbl., 28 petals, 3 in.; foliage dark; [(Independence X (Donald Prior X R. arkansana)) X (Fire King X (J.W. Fargo X Assiniboine))]; Marshall, H.H.; Agriculture Canada

'MORDEN BLUSH', S, lp, 1988; flowers light pink, fading to ivory, small, flat, dbl., 51 petals, flat, borne in sprays of 1–5, repeat bloom; prickles straight; foliage medium, medium green, matt; bushy, low growth, 3–4 ft; hardy; PP8054; [(Prairie Princess X Morden Amorette) X (Prairie Princess X (White Bouquet X (R. arkansana X Assiniboine)))]; Collicutt, L.M. & Marshall, H.H.; Agriculture Canada

'MORDEN CARDINETTE', S, mr, 1980; bud ovoid; flowers cardinal red, blooms borne 1–5 per cluster, occasionally 15, dbl., 25 petals, slight fragrance; foliage 7 leaflets, globuous, dark; Marshall, H.H.; Agriculture Canada

'MORDEN CENTENNIAL', S, mp, 1980; bud ovoid; flowers medium-pink, shapely, very dbl., 40 petals, borne 1–15 per cluster, repeat bloom, slight fragrance; slightly recurved prickles; foliage 7 leaflets, dark, slightly glossy; typical shrub growth, 3–4 ft; [Prairie Princess X (White Bouquet X (J.W. Fargo X Assiniboine))]; Marshall, H.H.; Agriculture Canada

'MORDEN FIREGLOW', S, or, 1989; bud pointed; flowers brilliant red-orange, reverse red, loose, medium, dbl., 28 petals, 3 in., cupped, borne in sprays, repeat bloom, slight fragrance; globular, reflexed calyx fruit; prickles slight downward curve, tan; foliage medium, medium green, matte; bushy, low, medium growth, 4 ft; PP8060; [Unnamed seedling X Morden Cardinette]; Collicutt, L.M. & Marshall, H.H.; Agriculture Canada, 1991

'MORDEN RUBY', S, pb, 1977; bud ovoid; very dbl., 2.5–3.5 in.; heavy bloom; vigorous, irregular growth; [Fire King X (J.W. Fargo X Assiniboine)]; Marshall, H.H.; Agriculture Canada

Morden Snow Beauty, *see* **'MORDEN SNOWBEAUTY'**

'MORDEN SNOWBEAUTY', S, w, 1998; (Morden Snow Beauty); flowers white, medium, semi-dbl., 8–14 petals, 3 in., borne in small clusters, slight fragrance; prickles small; foliage medium, medium green, semi-glossy; compact, low growth; [(Prairie Princess X Morden Amorette) X (Mount Shasta X (Adelaide Hoodless X R. arkansana))]; Davidson, C.G. & Colicutt, L.M.; Agriculture & Agri-Food Canada, 1997

'MORDORA', Cl Min, mr, 1995; (**Sequoia Ruby**); flowers medium red, moderately full (15–25 petals), medium (2 in.) Blooms borne in small clusters; some prickles; foliage medium, dark green, matt; tall (5 ft), arching shrub, upright, bushy, spreading growth; [(Little Darling X Yellow Magic) X Floradora]; Moore, Ralph S.; Sequoia Nursery, 1996

'MORDORT', Min, mr, 1991; (**Sincerely Yours**); semi-dbl., 6–14 petals, 1.5–2.75 in., borne in small clusters, no fragrance; some prickles; foliage medium, medium green, semi-glossy; medium (20–24 cms), bushy growth; AOE, 1992; [Sheri Anne X Dortmund]; Moore, Ralph S.; Sequoia Nursery, 1992

'MORE VALE PRIDE', Gr, mp, 1957; [Ma Perkins X Overloon]; Ulrick, L.W.

'MORECLI', Min, ly, 1984; (**Baby Eclipse**™); bud small; flowers mini-flora, small, semi-dbl., slight fragrance; foliage small, medium green, matt; bushy, spreading growth; AOE, 1984; [(R. wichurana X Floradora) X Yellow Jewel]; Moore, Ralph S.; Moore Min. Roses

'MORED', Min, lp, 1988; (**Edna Marie**); flowers very soft pink, soft yellow base, aging becomes near white, dbl., 20 petals, exhibition form, slight, fruity fragrance; prickles small, brown; foliage small, light green, semi-glossy; upright, bushy, medium growth; [Pinocchio X Peachy White]; Moore, Ralph S.; Sequoia Nursery

'MOREDFAR', Pol, mr, 1995; (**Red Fairy**); flowers deep pink to red, moderately full (15–25 petals), small (0–4 cms) blooms borne in large clusters of 10–25; no fragrance; few prickles; foliage medium, medium green, semi-glossy; upright, bushy, spreading (45–60 cms) growth; [Simon Robinson X Unknown]; Moore, Ralph S.; Sequoia Nursery, 1996

'MORELFIRE', Min, pb, 1982; (**Pink Elf**); flowers medium pink, blended with yellow, small, semi-dbl., no fragrance; foliage small, medium green, matt to semi-glossy; upright, bushy growth; [Ellen Poulsen X Fire Princess]; Moore, Ralph S.; Moore Min. Roses

Morena®, Min, op, 1983; Cocker

'MORENY', Min, mp, 1989; (**Renny**); bud pointed; flowers medium rose pink, reverse lighter, old fashioned, medium, bo, dbl., 25 petals, moderate fragrance; no fruit; no prickles; foliage medium, medium green, matt; upright, bushy, low growth; [Anytime X Renae]; Moore, Ralph S.; Sequoia Nursery

'MOREYES', Min, pb, 1998; (**Halo Karol**); flowers dark pink-lavendar base creates halo effect, small, single, 6–14 petals, 1–1.5 in., borne singly, no fragrance; few prickles; foliage medium, medium green, semi-glossy; bushy, compact growth; [(Anytime X Angel Face) X (Anytime X Angel Face)]; Moore, Ralph S.; Sequoia Nursery, 1998

Morey's Pink, S, mp, 1994; Morey

Morey's Salmon, HT, pb, 1953; Morey, Dr Dennison

'MORFAIR', Min, lp, 1980; (**Pink Cherub**); bud ovoid; flowers medium to light pink, often lighter at tips, dbl., 43 petals, slight fragrance; straight, small prickles; foliage small, medium green, matt; compact, very bushy growth; [Fairy Moss X Fairy Moss]; Moore, Ralph S.; Moore Min. Roses

'MORFAIRPOL', Pol, lp, 1999; (**Fair Molly**); flowers white blend, semi-dbl., 6–14 petals, 1–1.5 in., borne in small clusters, slight fragrance; few prickles; foliage small, medium green, semi-glossy; bushy, spreading, medium (16–24 in.); landscape shrub growth; [(R. polyantha nana X Unknown) X Fairy Moss]; Moore, Ralph S.; Sequoia Nursery, 1999

'MORFEAT', Min, w, 1980; (**White Feather**); flowers loose form, small blooms in clusters of 3–5, semi-dbl., 15 petals, slight fragrance; no fruit; very few prickles; foliage small, light green, semi-glossy; medium, bushy growth; [(R. wichurana X Floradora) X Peachy White]; Moore, Ralph S.; Moore Min. Roses

'MORFEVER', Min, my, 1990; (**Gold Fever**); bud pointed; flowers medium yellow, aging lighter, dbl., 40–50 petals, exhibition form, borne usually singly or in sprays of 3–5, moderate, spicy fragrance; round, small, orange fruit; prickles slender, straight, medium to long, brownish; foliage medium, medium green, semi-glossy; upright, bushy, medium growth; [Sheri Anne X Gold Badge]; Moore, Ralph S.; Sequoia Nursery

'MORFING', Min, ob, 1990; (**Fingerpaint**); bud short, pointed; flowers orange blend, yellow base, with light yellow reverse, aging, semi-dbl., 12–14 petals, flat, no fragrance; prickles small, brownish, straight; foliage medium, medium green, semi-glossy; bushy, spreading, low-medium growth; [Orangeade X Little Artist]; Moore, Ralph S.; Sequoia Nursery

'MORFIRE', Min, yb, 1986; (**Ring of Fire**™); flowers yellow blended orange, reverse yellow, fading lighter, imbricated, very dbl., 60 petals, slight fragrance; no fruit; slender, sharp pointed, medium, green to brown prickles; foliage medium, medium green, semi-glossy; upright, bushy, vigorous growth; PP006618; AOE, 1987; [Pink Petticoat X Gold Badge]; Moore, Ralph S.

'MORFLASH', F, ob, 1993; (**Blastoff**™); flowers scarlet orange, white reverse, petals ruffled, very dbl., 35–40 petals, 1.5–2 in., quartered, borne in large clusters, slight, slight spice fragrance; many prickles; foliage medium, dark green, semi-glossy; medium (100–120 cms), upright growth; PP9405; [Orangeade X Little Artist]; Moore, Ralph S.; Weeks Roses, 1995

'MORFLASH', HP, ob; flowers purple-violet, dbl., cupped

'MORFREE', Min, w, 1983; (**Softee**); flowers creamy white, small, dbl., 35 petals, slight fragrance; no prickles; foliage small to medium, medium green, matt; bushy, spreading growth; [Unnamed seedling X Unnamed seedling]; Moore, Ralph S.; Moore Min. Roses, 1982

'MORGA', HRg, mp; dbl., blooms in clusters

'MORgal', Min, dr, 1980; (**Galaxy**™); bud long, pointed; flowers deep velvety red, borne 3, sometimes 5–10 per cluster, dbl., 23 petals, exhibition form, slight fragrance; prickles slightly curved; foliage small to medium; vigorous, bushy, upright growth; [Fairy Moss X Fairy Princess]; Moore, Ralph S.; Sequoia Nursery

Morgan Hill Red, HT, dr

Morgans Spring, Ch, lp

'MORgel', Cl Min, op, 1987; (**Angel Pink**, Pink Angel); flowers pink to soft coral pink, holds color, dbl., exhibition form, borne in sprays of 3–7, slight fragrance; no fruit; medium, brown prickles, slightly hooked downwards; foliage medium, light green, semi-glossy; upright, tall, climbing (5–7 ft) growth; [Little Darling X Eleanor]; Moore, Ralph S.

'Morgenglans', HT, op, 1916; flowers salmon-flesh, semi-dbl., slight fragrance; Van Rossem

'Morgengrüss', HKor, op, 1962; (Morning Greeting); bud ovoid; flowers light pink tinted orange-yellow, large blooms in clusters, dbl., intense fragrance; foliage glossy, light green; very vigorous (13–14 ft), bushy growth; Kordes, R.

'Morgenluft', F, pb, 1962; flowers dark pink, reverse lighter, borne in clusters, dbl., 55–60 petals; foliage dark buff-green; upright, bushy growth; [La France X Unnamed seedling]; Verschuren, A.; van Engelen

Morgenrot®, *see* 'KORheim'

'Morgenröte', HT, mp, 1951; RULED EXTINCT 12/85; Burkhard

Morgensen, *see* **'Morgensonne'**

'Morgensonne', S, my, 1954; (Morgensen, Morning Sun); bud ovoid; flowers golden yellow, very large, moderate fragrance; foliage glossy, light green; very vigorous, upright, bushy growth; Kordes

Morgensonne, *see* KORhoro

Morgenstern®, S, ly, 1989; Liebig

'MORglo', Min, or, 1981; (**Gypsy Fire**); flowers small blooms in clusters, semi-dbl., no fragrance; foliage small, medium green, semi-glossy to glossy; upright, bushy growth; [(R. wichurana X Carolyn Dean) X Fire Princess]; Moore, Ralph S.; Moore Min. Roses

'MORgogard', Min, my, 1988; (**Golden Gardens**); bud ovoid; flowers bright, clear medium yellow, reverse slightly lighter, informal, dbl., 28 petals, cupped, no fragrance; no fruit; prickles slender, inclined downward, small, brownish; foliage medium, medium green, semi-glossy; upright, medium growth; [(Little Darling X Yellow Magic) X Gold Badge]; Moore, Ralph S.; Sequoia Nursery, 1989

'MORgoldart', Min, pb, 1993; (**Splish Splash**); dbl., 26–40 petals, 1.5–2.75 in., borne in small clusters, slight fragrance; few prickles; foliage medium, medium green, semi-glossy; medium (30 cms), bushy, spreading growth; [Sequoia Gold X Little Artist]; Moore, Ralph S.; Sequoia Nursery, 1994

'MORgoldpoly', Pol, my, 1999; (**Polly Sunshine**); dbl., 26–40 petals, 2–3 in., borne in small clusters; few prickles; foliage medium, medium green, glossy; bushy (16–24 in.) growth; [Golden Angel X seedling]; Moore, Ralph S.; Sequoia Nursery, 1999

'MORgrapes', S, m, 1999; (**Vineyard Song**); dbl., 15–25 petals, 1–1.5 in., borne in large clusters, moderate fragrance; some prickles; foliage dark green, glossy; medium (12–16 in.) growth; Moore, Ralph S.; Sequoia Nursery, 1999

'MORhalfire', Min, rb, 1995; (**Halo Fire**®); flowers bright orange red with darker red halo, semi-dbl. (6–14 petals), medium (4–7 cms) blooms borne in small clusters; some prickles; foliage small, medium green, matt; medium (35–45 cms), upright, bushy growth; [Orangeade X Seedling]; Moore, Ralph S.; Sequoia Nursery, 1996

'MORhill', Min, pb, 1990; (**Mary Hill**); bud pointed; flowers medium pink, yellow reverse, aging lighter, decorative, medium, dbl., 30–35 petals, exhibition form, moderate, fruity fragrance; prickles slender, straight, small, brownish; foliage medium, medium green, semi-glossy; upright, bushy, medium growth; [Little Darling X Golden Angel]; Moore, Ralph S.; Sequoia Nursery

'MORhoney', Min, my, 1983; (**Sun Honey**); [Orange Honey sport]; Moore, Ralph S.; Moore Min. Roses

'MORhoop', F, pb, 1990; (**Hula Hoop**); bud pointed; flowers white with pink to red edge, similar reverse, aging less intense, semi-dbl., 15 petals, 3–3.5 in., flat, borne in sprays of 5–15, no fragrance; foliage medium, medium green, matt; upright, medium growth; [(Dortmund X Unnamed seedling) X Self]; Moore, Ralph S.; Sequoia Nursery, 1991

'MORhoot', Min, rb, 1990; (**Hoot Owl**); bud pointed; flowers red with white eye, small, borne usually singly or in sprays, single, 5 petals, no fragrance; foliage small, medium green, semi-glossy; bushy, low growth; [Orangeade X Little Artist]; Moore, Ralph S., 1987; Sequoia Nursery, 1991

Moriah, *see* 'GANhol'

Moriah, HT, ob

Morinaye, F, rb

'MORisle', Min, yb, 1993; (**Isle of Roses**); dbl., 26–40 petals, 1.5–2.75 in., borne in small clusters, no fragrance; few prickles; foliage medium, medium green, semi-glossy; medium to tall (18–24 in.), upright, bushy growth; [Pink Petticoat X Gold Badge]; Moore, Ralph S.; Sequoia Nursery, 1994

'MORivory', Min, w, 1990; (**Ivory Palace**); bud ovoid; flowers ivory white, white reverse, aging similar color, medium, very dbl., exhibition form, borne in sprays of 3–8, slight fragrance; foliage medium, medium green, semi-glossy; bushy, medium growth; [Sheri Anne X Pinocchio]; Moore, Ralph S., 1982; Sequoia Nursery, 1991

'MORjack', Min, dy, 1984; (**Jackpot**); flowers small, dbl., moderate fragrance; foliage small, medium green, matt; vigorous, bushy, spreading growth; [Little Darling X Sunspray]; Moore, Ralph S.; Moore Min. Roses, 1985

'MORjoybon', HT, ab, 1996; (**It's Show Time**™); flowers reverse or slightly lighter, blooms borne mostly singly, dbl., 15–25 petals, 4–4.5 in.; foliage medium, dark green, semi-glossy; upright, medium (3–5 ft) growth; [Joycie X Bon Silene]; Moore, Ralph S.; Spring Hill Nurseries Co., 1997

'MORjoyc', Min, ob, 1988; (**Joycie**™); flowers orange-apricot, reverse lighter, small, dbl., exhibition form, borne singly or in small clusters, moderate, fruity fragrance; globular, orange fruit; prickles slender, small, brown; foliage small, medium green, semi-glossy; bushy, medium growth; PP007055; [(Little Darling X Yellow Magic) X Gold Badge]; Moore, Ralph S.; Sequoia Nursery

'MORkay', Min, pb, 1994; (**Kayla**); dbl., 15–25 petals, 3–3.5 in., borne mostly single but some in small clusters, slight fragrance; few prickles; foliage large, medium green, semi-glossy; tall (45–55 cms), upright, bushy growth; [Sheri Ann X Violette]; Moore, Ralph S.; Sequoia Nursery, 1995

'MorKINshine', Cl Min, my, 1990; (**Rise 'n' Shine, Climbing**); bud pointed; flowers medium yellow, aging lighter, small, borne usually singly or, dbl., 35 petals, exhibition form, slight fragrance; round, orange fruit; prickles straight, inclined slightly downward, brown; foliage medium, medium green, semi-glossy; upright, spreading, tall, climbing. growth; [Rise 'n' Shine sport]; King, Gene; Sequoia Nursery, 1990

'MORkita', Min, dr, 1988; (**Chiquita**™); flowers rich, dark red with fluorescent glow, borne singly, semi-dbl., 20 petals, flat, slight fragrance; few to none, fruit; prickles hooked, short, brown; foliage small, medium green, matt; upright, bushy, medium growth; [Anytime X Happy Hour]; Moore, Ralph S.; Sequoia Nursery

'MORlavmag', Min, m, 1994; (**Surprise Surprise**); flowers mauve-lavender with occasional petals being striped red, or, semi-dbl., 6–14 petals, 1.5 in., no fragrance; few prickles; foliage small, dark green, semi-glossy; low (12–16 in.), bushy, compact growth; [Cherry Magic sport]; Moore, Ralph S.; Sequoia Nursery, 1994

'MORLEM', Min, ob, 1983; (Sun Gold, **Sungold**); flowers yellow overlaid with orange on outer half of petals, reverse, dbl., 20 petals, no fragrance; foliage small, medium green, matt to semi-glossy; upright, bushy growth; [Rumba X Lemon Delight]; Moore, Ralph S.; Moore Min. Roses, 1982

'MORLETII', Bslt, m, 1883; (Inermis Morletii); flowers magenta, small to blooms in clusters, slight fragrance; branches plum-colored, without prickles; 5 ft growth; very hardy; (14); Morlet

'MORLIDAY', Min, dr, 1982; (**Holiday Cheer**); flowers small blooms in clusters, dbl., 35 petals, no fragrance; foliage small, dark, matt; upright, bushy growth; [Red Pinocchio X Little Chief]; Moore, Ralph S.; Moore Min. Roses

'MORLIYEL', Min, yb, 1980; (**Sierra Sunrise**); bud medium, pointed; flowers soft yellow, petals tipped pink, borne singly or 3–5 per cluster, dbl., 43 petals, exhibition form, slight fragrance; long, straight prickles; foliage medium, medium green; vigorous, bushy, upright growth; Gold Star of the South Pacific, Palmerston North, NZ, 1984; [Little Darling X Yellow Magic]; Moore, Ralph S.; Sequoia Nursery

'MORLIZ', Min, or, 1986; (**Spanish Dancer**); flowers small blooms in sprays of 5–7, semi-dbl., 18 petals, slight fragrance; no fruit; small, brown prickles; foliage medium, medium green, semi-glossy; medium, bushy, spreading growth; [Sarabande X Little Chief]; Moore, Ralph S.; Moore Min. Roses, 1980

'MORLOGEN', Min, yb, 1984; (**Namib Sunrise**); flowers yellow blended with coral pink, blooms in clusters of 15, dbl., 64 petals, slight fragrance; straight, light brown prickles; foliage light; dense growth; [Rumba X Yellow Jewel]; Moore, Ralph S.; Ludwigs Roses Pty. Ltd.

'MORLU', F, mp, 1992; (**Vintage Visalia**); flowers reverse of outer petals deeper pink than inside surface bea, very dbl., 2.75–3 in., slight fragrance; few prickles; foliage large, medium green, semi-glossy; medium (50–60 cms), upright, bushy growth; [Pink Petticoat X Lulu]; Moore, Ralph S.; Sequoia Nursery, 1993

MORlucy, Min, mp, 1997; (**Lucy**); Moore, Ralph S.

'MORLYN', Min, my, 1983; (**Lynne Gold**™); flowers micro-mini, small, dbl., 20 petals, slight fragrance; foliage small, medium green, semi-glossy; bushy, spreading growth; PP005742; [Ellen Poulsen X Yellow Jewel]; Moore, Ralph S.; Moore Min. Roses

'MORMAGEM', S, rb, 1995; (**Mariposa Gem**); flowers red/yellow blend, blooms borne in small clusters, semi-dbl., 6–14 petals, 1.5–2.75 in., slight fragrance; few prickles; foliage medium, medium green, semi-glossy; upright, bushy, spreading($1\frac{1}{2}$ m) growth with arching canes; [Little Darling X Magic Wand]; Moore, Ralph S.; Sequoia Nursery, 1995

'MORMAKE', Min, m, 1986; (**Make Believe**™); flowers mauve and white blend, reverse red-purple, blooms in sprays, 10 petals, no fragrance; small, globular, orange fruit; very few, hooked, dark brown prickles; foliage medium, dark, semi-glossy; upright, bushy growth; [Anytime X Angel Face]; Moore, Ralph S., 1980; Moore Min. Roses, 1985

'MORMARI', Min, mp, 1988; (**Marie Shields**™); flowers medium pink, reverse pink veined white, small, very dbl., exhibition form, borne usually in sprays or clusters, slight fragrance; prickles slightly hooked, small, brown; foliage small, medium green, semi-glossy; bushy, medium growth; [Avandel X (Rumba X Unnamed Floribunda mossed seedling)]; Moore, Ralph S.; Sequoia Nursery

'MORMARME', Min, mr, 1992; (**Marime**); flowers excellent medium red, good form, blooms borne mostly singly, semi-dbl., 6–14 petals, 1.5–2.75 in., no fragrance; few prickles; foliage medium, medium green, semi-glossy; medium (35–45 cms), upright, bushy growth; [Anytime X Happy Hour]; Moore, Ralph S.; Sequoia Nursery

'MORMELMA', Min, ly, 1993; (**Melody Marshall**); dbl., 15–25 petals, 1.5–2.75 in., borne mostly singly, slight fragrance; no prickles; foliage medium, medium green, semi-glossy; medium (25–35 cms), upright, bushy growth; [Mary Marshall sport]; Moore, Ralph S.; Sequoia Nursery, 1989

'MORMILLI', Min, op, 1983; (**Millie Walters**™); flowers deep coral pink, dbl., 45 petals, slight fragrance; foliage small, medium green, matt; upright, bushy growth; PP005741; [Little Darling X Galaxy]; Moore, Ralph S.; Moore Min. Roses, 1984

'MORMINT', Min, my, 1991; (**Hall of Flowers**); bud pointed; flowers lemon yellow, similar reverse, aging slightly lighter, dbl., exhibition form, slight fragrance; foliage medium, medium green, semi-glossy; upright, bushy, medium growth; [Avandel X Gold Badge]; Moore, Ralph S., 1983; Sequoia Nursery

'MORMITCHELL', Min, ly, 1996; (**Annie R. Mitchell**); flowers light yellow to white, reverse similar, blooms borne in small clusters, dbl., 26–40 petals, 1.5–2 in., moderate fragrance; few prickles; foliage medium, medium green, semi-glossy; spreading, bushy, medium (15–18 in.) growth; [Mary Hill sport]; Moore, Ralph S.; Sequoia Nursery, 1996

'MORMUM', Min, rb, 1986; (**Strange Music**, Striped Meillandina); bud mossy; flowers red, striped white, reverse near white, mini-moss, blooms in, dbl., 50 petals, exhibition form, no fragrance; small to medium, straight, brown prickles; foliage medium, medium green, matt; bushy, spreading growth; [Little Darling X (Fairy Moss X (Little Darling X Ferdinand Pichard))]; Moore, Ralph S.; Moore Min. Roses, 1982

'MORMURI', HBc, lp, 1989; (**Muriel**); bud ovoid; flowers light to medium pink, aging slightly lighter, large, borne s, semi-dbl., 15 petals, flat; short, oval, large, prickly, orange fruit; prickles sharp, pointed, average, brown; foliage large, medium green, semi-glossy; spreading, tall growth; [R. bracteata X Guinee]; Moore, Ralph S.; Sequoia Nursery

'MORMYVAL', Min, dr, 1975; (**My Valentine**™); flowers deep red, dbl., 65 petals, 1 in., exhibition form; foliage small, glossy, tinted bronze; vigorous, bushy growth; [Little Chief X Little Curt]; Moore, Ralph S.; Sequoia Nursery

'MORNET', Min, op, 1991; (**Annette Dobbs**); bud pointed; flowers coral red, lighter reverse, ages slightly lighter, with peta, semi-dbl., 15 petals, flat, borne in sprays of 3–5, no fragrance; foliage medium, medium green, semi-glossy; bushy, medium growth; [Anytime X Playgirl]; Moore, Ralph S., 1986; Sequoia Nursery, 1990

Morning Blush, *see* **'DUQUESA DE PEÑARANDA'**

Morning Blush, A, ly, 1988; Sievers

'MORNING DAWN', LCl, lp, 1955; flowers silvery rose flushed salmon, dbl., 63 petals, 5 in., exhibition form, moderate, spicy fragrance; foliage dark, glossy, leathery; vigorous, pillar (6–8 ft) growth; [New Dawn seedling X R.M.S. Queen Mary]; Boerner; J&P

Morning Glory, *see* **'MCGREDY'S ORANGE'**

'MORNING GLORY', Cl HT, rb, 1937; flowers carmine, reverse sulfur-yellow stained carmine, well formed, dbl., moderate fragrance; very vigorous, compact growth; [Portadown Sally sport]; Beckwith

'MORNING GLOW', Min, ly, 1999; flowers light yellow, slight blushing on top surface in cool weather, dbl., 17–25 petals, 1.2 in., exhibition form, borne mostly singly, no fragrance; few prickles; foliage large, medium green, semi-glossy; upright, tall (24 in.) growth; [Tidewater X Kristin]; Jolly, Betty J.; Langenbach, 1998

Morning Greeting, *see* **'MORGENGRÜSS'**

Morning Has Broken, *see* CLEwedding

Morning in Moscow, *see* **'UTRO MOSKVY'**

'MORNING JEWEL'®, LCl, mp, 1968; flowers large, semi-dbl., recurrent bloom, moderate fragrance; foliage glossy; ADR, 1975; [New Dawn X Red Dandy]; Cocker

'MORNING JOY', HT, ab, 1968; flowers creamy amber, reverse flushed coppery pink, well-formed, dbl., slight fragrance; foliage light green; vigorous, upright growth; [Mischief sport]; Williamson

Morning Light, *see* HARlecho

'MORNING MIST', HT, m, 1950; flowers lavender tinted gray, semi-dbl., 15–20 petals, 3.5 in., exhibition form, moderate fragrance; moderately vigorous, upright growth; Fisher, G.; Arnold-Fisher Co.

Morning Mist, *see* 'AUSFIRE'

Morning Skies, *see* 'GELSKIES'

Morning Song, *see* 'MINISONG'

Morning Star, S, yb, 1998

'MORNING STARS', S, w, 1949; bud ovoid; flowers clusters, dbl., 3 in., cupped, free, recurrent bloom, moderate fragrance; foliage glossy; upright, bushy, compact growth; [(New Dawn X Autumn Bouquet) X (New Dawn X Inspiration)]; Jacobus; B&A

Morning Sun, *see* **'MORGENSONNE'**

Morning Sun, *see* AROsumo

'MORNITA', Min, op, 1981; (**Anita Charles**™); bud pointed; flowers bright pink, reverse lighter pink-yellow blend, dbl., 43 petals, exhibition form, borne singly, sometimes 2–3 per cluster, moderate, tea fragrance; straight prickles; foliage small, leathery, matt; vigorous, upright, spreading growth; [Golden Glow (Brownell) X Over the Rainbow]; Moore, Ralph S.

Moroccan Rose, P, dp

'MOROCCO', HT, mr, 1961; bud pointed; flowers velvety red, dbl., 30 petals, 5 in., exhibition form, moderate fragrance; foliage glossy; vigorous, upright growth; [Carrousel X Charles Mallerin]; Von Abrams; Peterson & Dering

'MOROLÉ', Min, r, 1990; (**Café Olé**); bud pointed; flowers medium to large blooms, borne singly or in sprays of 3–5, very dbl., 40–50+ petals, cupped, moderate, spicy fragrance; round, orangish fruit; prickles slender, hooked downward, brown; foliage medium to large, medium green, dull to semi-glossy; upright, bushy, tall, vigorous growth; [Winter Magic sport]; Moore, Ralph S.; Sequoia Nursery

Morongo, S, lp; Peden, G.H., 1996; (Weatherly, L.)

'MORORCHERI', Min, m, 1993; (**Lavender Delight**); flowers semi double, medium, semi-dbl., 6–14 petals, 1.5–2.75 in., borne in small clusters, slight fragrance; few prickles; foliage medium, medium green, semi-glossy; medium, upright, bushy growth; [Orangeade X Cherry Magic]; Moore, Ralph S.; Sequoia Nursery, 1994

'MOROSE', Min, rb, 1987; (**Rose Gilardi**™); flowers red and pink striped, reverse similar, aging well, informal, semi-dbl., 12–15 petals, slight fragrance; slender, straight, small to medium, brownish prickles; foliage small, medium green, semi-glossy; bushy, spreading, medium growth; [Dortmund X ((Fairy Moss X (Little Darling X Ferdinand Pichard)) X Unnamed seedling)]; Moore, Ralph S., 1986

'MORPALE', Min, rb, 1984; (**Painter's Palette**); bud mossy; flowers creamy white, striped deep pink to deep red, blooms borne 3–5 per cluster, dbl., 72 petals, slight fragrance; needle straight, brown prickles; foliage deep green; shrubby growth; Moore, Ralph S.; Ludwigs Roses Pty. Ltd.

'MORPAPPLAY', S, rb, 1988; (Creepy, Glowing Carpet, Highveld Sun, **Ralph's Creeper**); flowers dark orange-red, bright yellow eye, reverse bright yellow to, semi-dbl., 15–18 petals, repeat bloom, moderate, apple-blossom fragrance; round, orange-red fruit; prickles normal, intermediate, tall; foliage small, dark green, matt; spreading, low growth; PP006548; [Papoose X Playboy]; Moore, Ralph S.; Armstrong Nursery, 1987

'MORPETTI', Min, rb, 1993; (**Hoot 'n' Holler**); flowers pleasing blend of red changing to lavender, blooms borne in, semi-dbl., 6–14 petals, 1.5–2.75 in., no fragrance; few prickles; foliage medium, medium green, matt; upright, bushy growth; [Pink Petticoat X Make Believe]; Moore, Ralph S.; Sequoia Nursery, 1993

Morpho, F; Leenders; (Cavriglia)

'MORPICO', Min, rb, 1991; (**Circus Clown**); semi-dbl., 6–14 petals, 1.5 in., borne mostly singly or in small clusters, slight fragrance; few prickles; foliage small, medium green, semi-glossy; low (18–22 cms), bushy, compact growth; [Pink Petticoat X Make Believe]; Moore, Ralph S.; Sequoia Nursery, 1992

'MORPINKPER', S, pb, 1999; (**Moore's Pink Perpetual**); flowers pink blend, very dbl., 41 petals, 3–3.5 in., borne in small clusters, moderate fragrance; many prickles; foliage medium, medium green, semi-glossy; dwarf hybrid perpetual, bushy, compact (16–24 in.) growth; [seedling X Paul Neyron]; Moore, Ralph S.; Sequoia Nursery, 1999

'MORPINTS', Min, rb, 1985; (**Pinstripe**™); flowers red with white stripes, well-formed, small, dbl., 35 petals, slight fragrance; foliage small, medium green, semi-glossy; low, mounded habit growth; [Pinocchio X Seedling]; Moore, Ralph S.; Armstrong Nursery, 1986

'MORPLAG', F, mp, 1986; (**Playgirl**™); flowers yellow stamens, blooms borne usually singly and in sprays, single, slight fragrance; no fruit; few, brown prickles; foliage medium, medium green, semi-glossy; upright, bushy growth; [Playboy X Angel Face]; Moore, Ralph S.; Moore Min. Roses

'MORPLATI', F, or, 1989; (**Playtime**™); bud pointed; flowers vibrant orange-red, aging slightly darker, medium, borne usu, single, 5 petals, flat, slight fragrance; round, medium, orange-red fruit; prickles straight, slightly hooked, medium, light brown; foliage medium, dark green, semi-glossy; upright, bushy, medium growth; [Playboy X Old Master]; Moore, Ralph S.; Sequoia Nursery, 1990

'MORPLAYGOLD', Min, ob, 1997; (**Playgold**); flowers semi double, small, semi-dbl., 6–14 petals, borne in small clusters, slight fragrance; foliage small, medium, green, glossy; low (30–45 min.) growth; [Playgold X Sequoia Gold]; Moore, Ralph S.

'MORPLUM', Min, lp, 1994; (**Sugar Plum**); flowers lavender purple opening to expose a silver shading on inside, semi-dbl., 6–14 petals, 1.5–2.75 in., no fragrance; no prickles; foliage medium, medium green, semi-glossy; medium (30–40 cms), bushy, spreading growth; [Anytime X Angel Face]; Moore, Ralph S; Sequoia Nursery, 1995

'MORPOLY', Min, my, 1991; (**Cal Poly**); flowers long lasting color, non-fading, blooms borne in small clusters, dbl., 15–25 petals, 1.5–2.75 in., slight fragrance; few prickles; foliage medium, medium green, semi-glossy; medium (24 cms), upright, bushy growth; AOE, 1992; [(Little Darling X Yellow Magic) X Gold Badge]; Moore, Ralph S.; Sequoia Nursery, 1992

'MORPOODLE', Min, pb, 1991; (**Pink Poodle**); flowers pink changing to lighter pink to white, typical china characteristics, very dbl., 1.5–2.75 in., moderate fragrance; few prickles; foliage small, medium green, matt; low (26–30 cms), upright growth; [(Little Darling X Yellow Magic) X Old Blush]; Moore, Ralph S.; Sequoia Nursery, 1992

'MORPORC', Min, lp, 1994; (**English Porcelain**); flowers light pink, moderately full (15–25 petals), medium (4–7 cms) blooms borne mostly single; slight fragrance; no prickles; foliage small, medium green, matt; low to medium (28–34 cms), bushy, spreading, compact growth; [Pink Petticoat X Happy time]; Moore, Ralph S.; Sequoia Nursery, 1995

MORposa, S, rb, 1997; (**Royal Midinette**); Moore

'MORPREPEN', Min, ob, 1994; (**Pretty Penny**); semi-dbl., 6–14 petals, 1.5–2.75 in., borne mostly singly or in small clusters, no fragrance; no prickles; foliage medium, medium green, matt; medium (28–34 cms), bushy, spreading, compact growth; [Seedling X Self]; Moore, Ralph S.; Sequoia Nursery, 1995

'MORPROM', Min, dp, 1989; (**Prom Date**); bud short, pointed; flowers deep pink, aging lighter, medium, borne singly, profuse, dbl., 38 petals, globular, no fragrance; globular, orange fruit; prickles straight, pointed, medium, gray to brown; foliage medium, medium green, matt; upright, bushy, medium growth; [Sheri Anne X (Unnamed seedling X Fairy Moss)]; Moore, Ralph S.; Sequoia Nursery

'MORPUFF', Cl HBc, lp, 1990; (**Pink Powderpuff**); bud pointed; flowers light pink, aging slightly lighter, old garden rose type, large, very dbl., exhibition form, repeat bloom, intense, damask fragrance; foliage large, medium green, semi-glossy; tall, spreading, climbing growth; [Lulu X Muriel]; Moore, Ralph S., 1964

'MORQUAKE', Min, rb, 1983; (**Earthquake**™); flowers striped red and yellow, reverse yellow, small, dbl., no fragrance; foliage small, medium green, semi-glossy; upright, bushy growth; PP005791; [Golden Angel X Seedling]; Moore, Ralph S.; Moore Min. Roses, 1984

'MORRAINBOW', Min, pb, 1994; (**Halo Rainbow**); flowers pink edging with center of each petal creamy white and base, single, 5 petals, 1.5–2.75 in., slight fragrance; no prickles; foliage medium, medium green, semi-glossy; medium (28–35 cms), bushy, spreading growth; [Seedling X Make Believe]; Moore, Ralph S.; Sequoia Nursery, 1995

'MORREDFAR', Pol, mr, 1996; (**Red Fairy**); flowers cherry red, dbl., 24–30 petals, 1.5 in., borne in conical sprays, somewhat pendulous, slight fragrance; some prickles; foliage long, bright medium green, semi-glossy; wiry stems; medium, semi-cascading growth; PP10150; [Simon Robinson X Simon Robinson]; Moore, Ralph S.

'MORROUSEL', Min, rb, 1972; (**Magic Carrousel**®, 'MOORCAR'); flowers petals white, edged red, small, dbl., exhibition form, slight fragrance; foliage small, glossy, leathery; vigorous, bushy growth; AOE, ARS, 1975 Miniature Rose Hall of Fame, ARS, 1999; [Little Darling X Westmont]; Moore, Ralph S.; Sequoia Nursery

'MORRUBI', Min, mr, 1986; ('MORUBY', **Ruby Magic**); flowers small to blooms borne usually singly and in sprays of 3–5, dbl., 20 petals, slight fragrance; usually no fruit; brown prickles; foliage small to medium, medium green, semi-glossy; medium, upright, bushy growth; [Orangeade X Pinstripe]; Moore, Ralph S.; Moore Min. Roses

Morsdag, *see* **'MOTHERSDAY'**

Morsdag Alba, Pol, w

Morsdag Red, Pol, mr

'MORSEA', Min, mp, 1986; (**Chelsea**); flowers small blooms borne in clusters of 5 or more, dbl., 25 petals, cupped, slight fragrance; few, globular, medium, orange fruit; foliage small to medium, medium green, semi-glossy; upright, bushy growth; [(Little Darling X Yellow Magic) X Crested Jewel]; Moore, Ralph S.; Sequoia Nursery

'MORSEGOLD', Min, my, 1986; (**Sequoia Gold**™); flowers medium yellow, fading lighter, medium, dbl., 30 petals, exhibition form, borne usually singly, moderate, fruity fragrance; round, orange fruit; slender, medium, pale green-brown prickles; foliage medium, medium green, glossy; bushy, spreading growth; PP006617; AOE, 1987; Rose of The Year, Auckland, NZ, 1995; [(Little Darling X Lemon Delight) X Gold Badge]; Moore, Ralph S.

'MORSEMURI', S, pb, 1995; (**Star Magic**); flowers pink blend, small, single (5 petals, small (0–4 cms) blooms borne in small clusters; no fragrance; some prickles; foliage small, dark green, glossy; groundcover, medium (3 ft x 9 ft), spreading compact growth; [Sequoia Gold X MORmuri]; Moore, Ralph S.; Sequoia Nursery, 1996

'MORSEWEL', Min, mr, 1989; (**Sequoia Jewel**); bud rounded; flowers medium, borne in sprays of 3–7, dbl., 33 petals, cupped, slight fragrance; round, medium, orange-red fruit; prickles straight, short, brown; foliage medium, medium green, matt; upright, bushy, medium growth; [Sheri Anne X Paul Neyron]; Moore, Ralph S.; Sequoia Nursery, 1990

'MORSHACK', Min, ob, 1987; (**Phyllis Shackelford**™); flowers orange, fading pink, small borne usually singly, dbl., 20 petals, exhibition form, moderate, fruity fragrance; rounded, medium, orange fruit; few, medium, brownish prickles, slightly hooked downwards; foliage small, medium green, semi-glossy; upright, bushy, medium growth; [Anytime X Gold Badge]; Moore, Ralph S.

'MORSHAKI', Min, rb, 1989; (**Little Tiger**); bud short, pointed; flowers red, yellow and white stripes of varying patterns, reverse with more yellow, very dbl., exhibition form, no fragrance; prickles average, slender, inclined downward, brown; foliage small, medium green, matt; bushy, low, rounded growth; [Golden Angel X Pinstripe]; Moore, Ralph S.; Sequoia Nursery

'MORSHAKRUG', S, pb, 1992; (**Yesterday's Garden**); very dbl., 1.5–2.75 in., borne in small clusters, no fragrance; foliage medium, medium green, semi-glossy; tall to medium (over 50 cms), upright, bushy, spreading growth; [Shakespeare Festival X Belle Poitevine]; Moore, Ralph S.; Sequoia Nursery, 1993

'MORSHARON', S, w, 1996; (**Sharon's Delight**); single, 4–7 petals, 3.5 in., borne in small clusters, slight fragrance; few prickles; foliage medium, medium green, semi-glossy; bushy, medium (2–3 ft) growth; [Golden Angel X Safrano]; Moore, Ralph S.; Sequoia Nursery, 1996

'MORSHEFRAN', Min, w, 1989; (**New Adventure**); bud pointed; flowers creamy white, medium, borne singly and in sprays of 3–5, dbl., 25 petals, flat, no fragrance; globular, orange fruit; prickles short, pointed, brownish-gray; foliage small, medium green, matt; upright, bushy, medium growth; [Sheri Anne X Safrano]; Moore, Ralph S.; Sequoia Nursery

'MORSHERI', Min, or, 1973; (**Sheri Anne**); bud long, pointed; flowers orange-red, base yellow, semi-dbl., 17 petals, 1–1.5 in., moderate fragrance; foliage glossy, leathery; upright, bushy growth; AOE, 1975; [Little Darling X New Penny]; Moore, Ralph S.; Sequoia Nursery

'MORSHODOT', Min, dr, 1996; (**Cherry Hi**); flowers dark red, reverse slightly lighter, blooms borne in small clusters, very dbl., 1.5 in., no fragrance; few prickles; [Show 'n' Tell X Sincerely Yours]; Moore, Ralph S.; Sequoia Nursery, 1997

'MORSHOOK', Cl Min, rb, 1990; (**Earthquake, Climbing**); bud rounded; flowers red/yellow stripes, yellow reverse, aging similar, urn-shape, dbl., no fragrance; foliage small, medium green, semi-glossy; upright, tall growth; [Earthquake sport]; Moore, Ralph S.; Sequoia Nursery, 1991

'MORSIXTY', Min, m, 1996; (**Diamond Anniversary**); flowers mauve, reverse mauve or slightly lighter, blooms borne in small clusters, dbl., 26–40 petals, 1.5 in., slight fragrance; few prickles; [Joycie X Cherry Magic]; Moore, Ralph S.; Sequoia Nursery, 1997

'MORSNO', Min, w, 1987; (**Snow Twinkle**™); flowers small, borne usually singly, dbl., exhibition form, slight fragrance; round, orange fruit; small, brown prickles; foliage small, medium green, matt; bushy, medium growth; [(Little Darling X Yellow Magic) X Magic Carrousel]; Moore, Ralph S.

'MORSNOP', Min, my, 1993; (**Buttermint**); flowers moderately full, dbl., 15–25 petals, 1.5–2.75 in., borne mostly singly, slight fragrance; few prickles; foliage medium, medium green, semi-glossy; medium to tall (36–40 cms), upright, bushy growth; [Pink Petticoat X Gold Badge]; Moore, Ralph S.; Sequoia Nursery, 1994

'MORSTAR', HRg, mp, 1990; (**Star Delight**, Starry Eyed); bud pointed; flowers rose pink, white base, silvery pink reverse, medium, borne u, single, 5 petals, flat, repeat bloom, slight fragrance; foliage medium, olive to bluish-green, semi-glossy; upright, bushy, tall growth; [Yellow Jewel X Rugosa Magnifica]; Moore, Ralph S., 1989; Sequoia Nursery

'MORSTOCK', Min, yb, 1999; (**Woodstock**); dbl., 26–40 petals, 1–1.5 in., borne mostly singly, slight fragrance; few prickles; foliage small, medium green, semi-glossy; low (8–12 in.), bushy, compact growth; [seedling X Clytemnestra]; Moore, Ralph S.; Sequoia Nursery, 1999

'MORSTRORT', LCl, pb, 1999; (Cl. Shadow Dancer, **Shadow Dancer**™); flowers striped, swirled two-tone pink, reverse same, dbl., 17–25 petals, 3.5–4 in., borne in large clusters, slight fragrance; prickles moderate; foliage medium, dark green, holly-like; climbing, tall (8–10 ft) growth; [Dortmund X (Dortmund X Unknown)]; Moore, Ralph S.; Weeks Roses, 1998

MORsuise, Min, yb, 1997; (**Halo Suise**); Moore

'MORSUN', Min, my, 1981; (**Sweet Sunshine**); flowers small, dbl., 20 petals, intense fragrance; foliage small, medium green, semi-glossy; upright, bushy growth; [Rumba X Yellow Jewel]; Moore, Ralph S.; Moore Min. Roses

'MORSUNRISE', Min, gls, 1997; (**Halo Sunrise**); Moore

'MORSWEET', LCl, mp, 1988; (**Crested Sweetheart**); flowers medium rose pink, large, borne in sprays of 3–5, very dbl., cupped, intense, damask fragrance; prickles small, gray to brown; foliage large, medium green, matt, rugose; upright, tall growth; [Little Darling X Crested Moss]; Moore, Ralph S.; Wayside Gardens Co., 1988

'MORSWISS', Min, ob, 1990; (**Two-Timer**); bud ovoid; flowers orange-red with lighter shades of orange and white striped, very dbl., exhibition form, no fragrance; foliage medium, medium green, semi-glossy; bushy, spreading, low growth; [Orangeade X Pinstripe]; Moore, Ralph S., 1982; Sequoia Nursery, 1991

MORsycheek, S, op; (**Rosy Cheeks**); Moore

'MORTEN', HRg, mr, 1990; (**Linda Campbell**, Tall Poppy); bud pointed; flowers medium red, blooms slightly lighter reverse, aging medium red, semi-dbl., 25 petals, cupped, fast repeat, no fragrance; foliage large, dark green, semi-glossy; upright, bushy, medium growth; [Anytime X Rugosa Magnifica]; Moore, Ralph S., 1985; Wayside Gardens Co., 1991

'MORTODAY', Min, op, 1994; (**Halo Today**); flowers have distinct pink/lavender area at base of each petal, semi-dbl., 6–14 petals, 1.5–2.75 in., no fragrance; no prickles; foliage medium, medium green, semi-glossy; low (16–18 in.), upright, bushy growth; [(Anytime X Gold Badge) X (Anytime X Lavender Jewel)]; Moore, Ralph S.; Sequoia Nursery, 1994

'MORTOPAZ', Min, ab, 1995; (**Peach Candy**); flowers soft peach, moderately full (15–25 petals), small (0–4 cms) blooms borne in small clusters; slight fragrance; few prickles; foliage small, light green, matt; medium (24–30 cms), upright, bushy growth; [Sheri Anne X Topaz Jewel]; Moore, Ralph, S.; Sequoia Nursery, 1996

'MORTORCH', Min, or, 1986; (**Torch of Liberty**™); flowers orange-red, silver reverse, small, dbl., 20 petals, slight fragrance; foliage small, medium green, semi-glossy; upright, bushy growth; PP006254; [Orangeade X Golden Angel]; Moore, Ralph S.; Moore Min. Roses, 1985

'MORTREAT', Min, yb, 1986; (**Double Treat**); bud mossy; flowers bright red and orange-yellow, striped, mini-moss, dbl., cupped, borne usually singly, slight fragrance; small, globular with numerous spines, orange fruit; slender, brown prickles; foliage small to medium, medium green, semi-glossy; medium, upright, bushy growth; [Arizona X ((Fairy Moss X Fairy Moss) X (Little Darling X Ferdinand Pichard))]; Moore, Ralph S.; Sequoia Nursery, 1985

'MORTRIP', Min, rb, 1980; (**Crazy Quilt**); bud pointed; flowers red and white striped, borne 1–3 or more per cluster, dbl., flat; small, straight prickles; foliage medium green; compact, bushy growth; [Little Darling X Seedling]; Moore, Ralph S.; Sequoia Nursery

'MORTWIST', Min, ob, 1986; (**Orange Twist**); flowers tannish-orange; [Sungold sport]; Moore, Ralph S.; Moore Min. Roses, 1985

'MORTWISTER', Cl Min, rb, 1997; (**Twister**); flowers full, medium, very dbl., 26–40 petals, borne in small clusters, slight fragrance; few prickles; foliage medium, medium green, dull; upright, climbing tall (3–ft) growth; [Little Darling X Little Magic]; Rekdeci, John Jason

'MORUBY', Min, mr, 1986; ('MORRUBI', **Ruby Magic**); flowers small to blooms borne usually singly and in sprays of 3–5, dbl., 20 petals, slight fragrance; usually no fruit; brown prickles; foliage small to medium, medium green, semi-glossy; medium, upright, bushy growth; [Orangeade X Pinstripe]; Moore, Ralph S.; Moore Min. Roses

'MORUNUSUAL', F, pb, 1999; (**Most Unusual Day**); dbl., 15–25 petals, 1.5–2.5 in., borne in small clusters, no fragrance; few prickles; foliage medium, medium green, semi-glossy; medium (16–24 in.), upright, bushy growth; [Show 'n' Tell X Blastoff]; Moore, Ralph S.; Sequoia Nursery, 1999

'MORVANDEL', Min, yb, 1977; (**Avandel**); bud long, pointed; flowers pink-yellow blend, dbl., 23 petals, 1–1.5 in., cupped, moderate, fruity fragrance; foliage medium to dark green, leathery; upright, bushy growth; AOE, 1978; [Little Darling X New Penny]; Moore, Ralph S.; Sequoia Nursery

'MORVI', Min, m, 1991; (**Vi's Violet**); flowers soft lavender, small, dbl., slight fragrance; foliage small, medium green, matt; upright, bushy, compact growth; [Unnamed seedling X Angel Face]; Moore, Ralph S.

'MORWATEYE', Min, pb, 1992; (**Halo Dolly**); flowers bicolor, reddish outside, pink inside with reddish lavender, semi-dbl., 6–14 petals, 1.5 in., no fragrance; few prickles; foliage medium, medium green, semi-glossy; medium (30–45 cms), upright, bushy, rounded growth; [Anytime X Seedling (Anytime X Angel Face)]; Moore, Ralph S.; Sequoia Nursery, 1993

'MORWHEELS', Min, rb, 1997; (**Jacquie Williams**); flowers full, medium, very dbl., 26–40 petals, borne mostly single but some clusters, slight fragrance; foliage medium, medium green, semi-glossy; low to medium (30–45 cms) growth; [(Yellow Jewel X Tamango) X Strawberry Ice]; Moore, Ralph S.

'MORWHIT', Min, w, 1981; (**Little Eskimo**, 'MORWIT'); bud long, pointed; flowers near white, blooms borne 3–7 per cluster, sometimes singly, dbl., 55 petals, slight fragrance; long, slender prickles; foliage small, semi-glossy, leathery; vigorous, bushy, upright growth; [(R. wichurana X Floradora) X Jet Trail]; Moore, Ralph S.; Moore Min. Roses

'MORWHY', Min, rb, 1983; (**Why Not**, Why Not); flowers medium red, yellow eye, small, single, 7 petals, slight fragrance; foliage small, medium green, matt to semi-glossy; upright, bushy growth; PP005676; [Golden Angel X Unnamed seedling]; Moore, Ralph S.; Moore Min. Roses

'MORWIMER', Cl Min, ly, 1995; (**Little Mermaid**); flowers light yellow, single, medium (5 petals), medium (4–7 cms) blooms borne in small clusters, no fragrance; some prickles; foliage small, medium green, semi-glossy; bushy, spreading growth; [Mermaid sport] Moore, Ralph S.; Sequoia Nursery, 1995

'MORWINGS', S, ly, 1989; (**Yellow Butterfly**); bud pointed; flowers light yellow, aging to near white, informal, medium, borne i, single, 5 petals, no fragrance; ovoid, small, yellowish-orange fruit; prickles slender, inclined downward, small, brownish; foliage medium, light green, semi-glossy; bushy, spreading, medium, clean growth; [Ellen Poulsen X Yellow Jewel]; Moore, Ralph S.; Sequoia Nursery

'MORWINST', Min, rb, 1994; (**Secret Recipe**); flowers combination of burgundy, red and white stripes, blooms borne, semi-dbl., 6–14 petals, 1.5–2.75 in., slight fragrance; some prickles; foliage medium, medium green, semi-glossy; upright, bushy (30–40 cms) growth; [Little Darling X Seedling]; Moore, Ralph S.; Sequoia Nursery, 1995

'MORWIT', Min, w, 1981; (**Little Eskimo**, 'MORWHIT'); bud long, pointed; flowers near white, blooms borne 3–7 per cluster, sometimes singly, dbl., 55 petals, slight fragrance; long, slender prickles; foliage small, semi-glossy, leathery; vigorous, bushy, upright growth; [(R. wichurana X Floradora) X Jet Trail]; Moore, Ralph S.; Moore Min. Roses

'MORWOOD', Min, or, 1984; (**Redwood Empire**); flowers small, dbl., 20 petals, slight fragrance; foliage small, medium green, semi-glossy; upright, bushy growth; [Rumba X Sheri Anne]; Moore, Ralph S.; Moore Min. Roses, 1983

'MORYEARS', S, w, 1999; (**Out of Yesteryear**); flowers white blend, very dbl., 41 petals, 3 in., borne in large clusters, moderate fragrance; many prickles; foliage medium, dark green, glossy; upright, spreading (4–6 ft) growth; [Golden Angel X Muriel]; Moore, Ralph S.; Sequoia Nursery, 1999

'MORYELRUG', HRg, my, 1987; (Gelbe Dagmar Hastrup, Rustica 91, **Topaz Jewel**™, Yellow Fru Dagmar Hartopp); flowers medium yellow fading to cream, large, dbl., cupped, borne in sprays of 5–8, repeat bloom, moderate, fruity fragrance; no fruit; slender, fairly long, brown prickles in variable sizes; foliage large, medium green, rugose, matt; upright, bushy, spreading, vigorous growth; PP006793; [Golden Angel X Belle Poitevine]; Moore, Ralph S.; Moore Miniature Roses, 1987; Way-side Gardens, 1987

'MORYETTEM', Min, m, 1986; (**Saint Mary**™); flowers deep red purple, small, dbl., cupped, slight fragrance; no fruit; small, brown prickles; foliage small, medium green, semi-glossy; medium, upright, bushy growth; [Little Chief X Angel Face]; Moore, Ralph S.; Moore Min. Roses

'MORYOU', Min, dp, 1990; (**Just For You**); bud pointed; flowers dark pink to light red, lighter reverse, aging lighter, medium, dbl., 35 petals, exhibition form, slight fragrance; foliage medium, medium green, semi-glossy; bushy, medium growth; AOE, 1991; [Orangeade X Rainbow's End]; Moore, Ralph S., 1986; Sequoia Nursery, 1991

Mosaik, F, rb, 1978; flowers red with dark yellow, medium, dbl., slight fragrance; VEG; (Sangerhausen)

'MOSAÏQUE', F, yb, 1960; flowers light yellowish pink, becoming red, open, semi-dbl.; foliage light green; low, bushy growth; [(Alain X Cinnabar) X Circus]; Lens

'MOSCHATA GRANDIFLORA', HMsk, w; flowers prominent golden stamens, large, single, intense fragrance; very vigorous growth; [R. moschata X R. multiflora]; Bernaix, A., ca. 1867

Moscow Morn, *see* **'UTRO MOSKVY'**

'MOSEL', HMult, m, 1920; flowers bluish violet, center reddish violet, dbl., sometimes recurrent bloom, slight fragrance; [Mme Norbert Levavasseur X Trier]; Lambert, P.

'MOSELLIED', HMsk, m, 1932; flowers purplish red, center white, stamens golden yellow, petals he, single, non-recurrent, intense fragrance; foliage dark; vigorous (about $6\frac{1}{2}$ ft), broad growth; [(Geheimrat Dr Mittweg X Tip-Top) X (Chamisso X Parkzierde)]; Lambert, P.

'MOSS MAGIC', Min, mp, 1977; bud cupped, mossy; flowers medium to dark pink, circular, dbl., 48 petals, 1 in., flat, profuse bloom, repeating well; foliage dark; spreading growth; [Fairy Moss X ?]; Sudol, Julia

Moss Rose, *see* **'CENTIFOLIA MUSCOSA'**

'MOSSMAN', S, mp, 1954; flowers heavily mossed, pale dusty pink, very dbl., 2.5 in., vigorous (4 ft) growth; [(R. acicularis X R. rugosa) X Moss rose]; Skinner

Mossy de Meaux, *see* **'MOSSY ROSE DE MEAUX'**

'MOSSY GEM', Min, mp, 1984; bud small, mossy; flowers medium pink, outer petals fading, small, dbl., slight fragrance; foliage medium, dark, semi-glossy; bushy growth; [Heidi X Violette]; Kelly, Martin

'MOSSY ROSE DE MEAUX', M, mp; (Mossy de Meaux); [Rose de Meaux sport]

Most Unusual Day, *see* 'MORUNUSUAL'

Moth, S, m, 1983; Austin, David

'MOTHER', HT, pb, 1939; flowers white edged bright pink, intense fragrance; [Memory sport]; Cant, B. R.

Mother and Baby®, *see* 'SANBABY'

'MOTHER MARIE', Pol, w, 1954; flowers white, center green, borne in clusters, dbl., 40 petals, 1.5–2 in., flat; foliage light green; [Garnette sport]; Podesta

Mother Theresa, HT, w, 1994; IARI

Mother's Country, *see* **'MOEDER DES VADERLANDS'**

Mothers Day, *see* **'MOTHERSDAY'**

'MOTHER'S DAY', HT, w, 1937; flowers white, at times shaded pink, very large; very vigorous growth; Knight, G.

'MOTHER'S DAY ORANGE SPORT', Pol, ob, 1958; flowers orange-yellow; [Mothersday sport]; Klyn

Mother's Love, *see* 'TINLOVE'

Mother's Value, *see* KORretra

'MOTHERSDAY', Pol, dr, 1949; (Fête des Mères, Morsdag, Mothers Day, Muttertag); flowers deep red, blooms in clusters (up to 20), dbl., globular; foliage glossy, small; dwarf growth; [Dick Koster sport]; Grootendorst, F.J.

'MOTHERSDAY, CLIMBING', Cl Pol, dr, 1956; (Muttertag, Climbing); Kordes

Motiv, HT, mp, 1983; flowers large, dbl.; VEG; (Sangerhausen)

'MOTREA', F, mr, 1968; bud ovoid; flowers light red, small, very dbl.; foliage dark; [Coronet sport]; Mosselman & Terreehorst

'MOTYLEK', HT, w, 1954; (Butterfly Papilio); flowers cream, base yellow, medium, dbl., 40–45 petals, exhibition form, moderate fragrance; moderate growth with thin, strong shoots; [Freiburg II X Golden Dawn]; Shtanko, I.

'MOULIN ROUGE'®, F, mr, 1952; (Sans souci); dbl., 20–25 petals, 2 in., cupped, blooms in clusters, slight fragrance; foliage glossy; upright, very bushy growth; GM, Geneva, 1952 GM, NRS, 1952 PIT, NRS, 1952; [Alain X Orange Triumph]; Meilland, F.; URS

'MOULIN ROUGE, CLIMBING', Cl F, mr, 1957; Mondial Roses; URS

Mouna®, HT, mr

'MOUNT EVEREST', S, mp

'MOUNT HOBBEN', HT, mr, 1981; flowers bright scarlet, large blooms borne 1–3 per stem, dbl., 30 petals, exhibition form, no fragrance; many, small prickles; foliage medium green; tall, upright growth; [Duftzauber X Christian Dior]; Ota, Kaichiro

'MOUNT HOMAN', HT, ab, 1983; bud ovoid; flowers large blooms borne 1–3 per stem, dbl., exhibition form, slight fragrance; large prickles; foliage dark, semi-glossy; tall, upright, bushy growth; [Ginger Rogers X Sunblest]; Ota, Kaichiro

Mount Hood, *see* 'MACMOUHOO'

Mount Nelson, F, w, 1998

'MOUNT SHASTA', Gr, w, 1963; bud long, pointed; dbl., 4.5–5 in., cupped, moderate fragrance; foliage leathery, gray-green; vigorous, upright growth; [Queen Elizabeth X Blanche Mallerin]; Swim & Weeks; C-P

Mount Shasta, Climbing, Cl HT, lp; [Mount Shasta sport]; Knight, G., 1968; (Weatherly, L.)

'MOUNTAIN HAZE', F, m, 1967; flowers lavender, reverse silver, medium, dbl., 55–60 petals, exhibition form, moderate fragrance; foliage dark, leathery; vigorous, low, spreading growth; [Amy Vanderbilt X Lilac Time]; Morey, Dr Dennison; Country Garden Nursery

Mountain Mignonette, Pol, lp

'MOUNTAIN MIST', S, m, 1989; bud pointed; flowers purple-lavender shading to white at base, small, borne in sprays, single, 5 petals, flat, intense, multiflora, vanilla fragrance; globular, red fruit; prickles very few, red; foliage small, light green, semi-glossy; spreading, medium (4 × 6'), prolific growth; [Yesterday X Yesterday]; Jobson, Daniel J.

'MOUNTAIN MUSIC', S, pb, 1984; flowers pink and yellow blend, blooms borne 5–15 per cluster, dbl., 23 petals, cupped, repeat bloom, moderate fragrance; awl-like, tan prickles; foliage leathery, semi-glossy; vigorous, erect, bushy growth; hardy.; [Sevilla X Tom Brown]; Buck, Dr Griffith J.; Iowa State University

Mountain Rose, *see* **'R. WOODSII'**

Mountain Side, S

Mountain Snow, *see* 'AUSSNOW'

'MOUNTAINEER', HT, op, 1963; bud long, pointed, orange-red; flowers orange-pink, reverse creamy yellow, dbl., 35 petals, 5 in., exhibition form, moderate fragrance; foliage glossy; vigorous growth; [Mrs Sam McGredy X Crimson Glory]; Wyant

Mountbatten®, *see* 'HARMANTELLE'

Mountie™, *see* 'LAVCALI'

'MOUNTJOY', (natural variation of R. setigera), lp; flowers blush, center darker; (found in Ohio)

'MOURNE GOLD', HT, yb, 1980; flowers deep yellow, pink flush; [Whisky Mac sport]; Kane Brothers

Mouse, *see* 'SEAMOU'

'MOUSHA', HT, lp, 1974; bud long, pointed; dbl., 24 petals, 4 in., exhibition form, foliage glossy, dark; vigorous, upright growth; [Königin der Rosen X King's Ransom]; Kordes; Willemse

Mousseau Ancien, *see* **'COMMUNIS'**

Mousseline, *see* **'ALFRED DE DALMAS'**

Mousseau Ancien, *see* 'CENTIFOLIA MUSCOSA'

'MOUSSEUX DU JAPON', M, m; (Japonica, Moussu du Japon, Muscosa Japonica); bud and stems heavily mossed; flowers purplish rose, quickly fading to lavender, many stamens, semi-dbl.; growth moderate; (Japan), before 1900

Moussu du Japon, *see* **'MOUSSEUX DU JAPON'**

'MOUSSUE DES QUATRE SAISON BLANC', w, 1835; flowers medium, dbl.; (Sangerhausen)

Moussue Partout, *see* **'ZOÉ'**

Movie Star, *see* TANeivom

'MOZART', HMsk, pb, 1937; flowers deep pink with large white eye, small blooms in clusters, single, repeat bloom, moderate fragrance; vigorous, trailing or bushy growth; [Robin Hood X Rote Pharisaer]; Lambert, P.

'MR BLUEBIRD', Min, m, 1960; bud ovoid; flowers lavender-blue, semi-dbl., 15 petals, 1.25 in.; foliage dark; compact, bushy (10–14 in.) growth; [Old Blush X Old Blush]; Moore, Ralph S.; Sequoia Nursery

'MR CHIPS', HT, yb, 1970; flowers yellow and orange, high-pointed, large, dbl.; foliage glossy; [Irish Gold X Miss Ireland]; Dickson, A.

Mr E.E. Greenwell, *see* 'HARJOOBILY'

'MR FAITHFUL', F, mp, 1968; dbl., blooms in trusses, moderate fragrance; [Pink Parfait X Self]; Harkness

Mr J.C.B., *see* DICsun

'MR J. BIENFAIT', HT, mr, 1923; (J. Bienfait); flowers brick-red; [Mme Leon Pain X Red-Letter Day]; Van Rossem

Mr J.C.B., F, dy, 1993; Dickson, Patrick

'MR JOH. M. JOLLES', HT, ly, 1920; flowers clear creamy yellow, shaded apricot and golden yellow, dbl.; [Frau Karl Druschki X Mrs Joseph Hill]; Van Rossem

Mr Leigh, F, mp

Mr McCawber, *see* 'WHICAW'

Mr Nash, LCl, my

'MR PAT', HT, dr, 1966; bud globular; flowers large, dbl., exhibition form, intense, spicy fragrance; foliage leathery; compact growth; [Red Jacket X Mirandy]; Patterson; Patterson Roses

Mr Pomeroy's Rose, S, w

'MR STANDFAST', HT, w, 1968; flowers cream, large, dbl., moderate fragrance; foliage glossy; [Dr A.J. Verhage X Kordes' Perfecta]; Harkness

'MR TALL', HT, op, 1958; bud long, pointed; flowers salmon-pink, open, single, 7 petals, 4 in., intense, cinnamon fragrance; foliage dark, leathery; vigorous, upright growth; [Vogue X Grande Duchesse Charlotte]; Wyant

'MRIDULA', HT, w, 1975; bud long, pointed; flowers white, center soft pink, dbl., 35 petals, 3.5 in., exhibition form, moderate fragrance; foliage large, light, soft; very vigorous, upright, compact growth; [Queen Elizabeth X Sir Henry Segrave seedling]; IARI

Mrilliant Hit, Min, ob

'MRINALINI', HT, mp, 1973; bud long, pointed; flowers very large blooms borne singly, dbl., exhibition form, moderate fragrance; foliage medium, medium green, soft; vigorous, bushy growth; [Pink Parfait X Christian Dior]; IARI

'MRS A. GORDON', Cl T, lp; Fell, J.B., 1916; (Weatherly, L.)

Mrs A. Hudig, *see* **'FLAMING JUNE'**

'MRS A.J. ALLEN', HT, pb, 1930; flowers rich pink, base lemon-yellow, tipped almost white, large, dbl., slight fragrance; [Richmond X Souv. de Mme Boullet]; Allen

Mrs A.J. Wylie, *see* **'MARGARET M. WYLIE'**

'MRS A.R. BARRACLOUGH', HT, mp, 1926; bud long, pointed; flowers bright carmine pink, large, dbl., exhibition form, slight fragrance; vigorous growth; McGredy

'MRS A.R. BARRACLOUGH, CLIMBING', Cl HT, mp, 1935; Fryers Nursery, Ltd.

'MRS A.W. ATKINSON', HT, w, 1918; flowers ivory-white, dbl.; Chaplin Bros.

'MRS AARON WARD', HT, yb, 1907; bud long, pointed; flowers yellow, occasionally washed salmon (quite variable), dbl., exhibition form, moderate fragrance; dwarf, compact growth; Pernet-Ducher

'MRS AARON WARD, CLIMBING', Cl HT, yb, 1922; Dickson, A.

'MRS ALBERT NASH', HT, dr, 1929; moderate fragrance; Clark, A.; NRS Victoria

'MRS ALFRED TATE', HT, rb, 1909; flowers coppery red shaded fawn, base shaded ochre, moderate fragrance; McGredy

'MRS ALFRED WEST', HT, op, 1922; flowers salmon-pink, dbl., intense fragrance; Cant, F.

Mrs Alston's Rose, Pol, rb, 1940; large trusses of reddish pink, semi-double blooms with lighter eye; Clark

'MRS AMBROSE RICARDO', HT, yb, 1914; flowers deep honey-yellow, overlaid brighter yellow, dbl., moderate fragrance; McGredy

'MRS AMY HAMMOND', HT, pb, 1911; flowers cream and amber, sometimes flushed pink, base apricot, dbl., moderate fragrance; GM, NRS, 1910; [Mme Abel Chatenay seedling]; McGredy

'MRS ANDREW CARNEGIE', HT, w, 1913; flowers white, center lightly tinted lemon-yellow, dbl., slight fragrance; GM, NRS, 1912; [Niphetos X Frau Karl Druschki]; Cocker

'MRS ANNE DAKIN', LCl, pb, 1972; flowers salmon-pink, reverse cream, dbl., 3.5 in., slight fragrance; foliage glossy; moderate, climbing growth; Holmes, R.; Albrighton Roses

Mrs Annie Beaufays, *see* **'FRAU ANNY BEAUFAYS'**

'MRS ANTHONY SPALDING', HT, or, 1934; flowers strawberry-red flushed orange, reverse shaded orange, well s, dbl.; vigorous growth; McGredy

'MRS ANTHONY WATERER', HRg, dr, 1898; flowers deep crimson, semi-dbl., intense fragrance; vigorous growth; hardy; (21); [R. rugosa X Général Jacqueminot]; Waterer

'MRS ARCHIE GRAY', HT, ly, 1914; flowers deep creamy yellow, opening light canary-yellow; GM, NRS, 1913; Dickson, H.

'MRS ARNOLD BURR', HT, w, 1945; flowers pure white, large, very dbl., exhibition form, moderate fragrance; foliage glossy; vigorous, upright growth; [Peace X ?]; Burr, A.

'MRS ARTHUR CURTISS JAMES', LCl, my, 1933; (Golden Climber); bud long, pointed; flowers large, semi-dbl., 18 petals, non-recurrent, moderate fragrance; foliage glossy; vigorous, climbing growth; [Mary Wallace X Seedling]; Brownell, H.C.; J&P

'MRS ARTHUR E. COXHEAD', HT, dr, 1911; flowers claret-red, shaded brighter red, dbl., moderate fragrance; McGredy

'MRS ARTHUR JOHNSON', HT, yb, 1920; flowers rich orange-yellow to chrome-yellow, slight fragrance; McGredy

'MRS ARTHUR MOORE', HT, lp; [Victor Hugo X C.J. Graham]; Moore, A.K., 1911; (Weatherly, L.)

'MRS ARTHUR ROBERT WADDELL', HT, pb, 1909; bud long, pointed; flowers reddish salmon, reverse rosy scarlet, fading, open, large, semi-dbl., slight, apricot fragrance; vigorous growth; Pernet-Ducher

'MRS ATLEE', HT, op, 1926; flowers silvery pink, shaded soft salmon, moderate fragrance; Chaplin Bros.

'MRS B. STORY', T, lp; [Mme Lombard X Unknown]; Williams, A., 1905; (Weatherly, L.)

'MRS B.R. CANT', T, mp, 1901; (Mrs Benjamin R. Cant); flowers silvery rose, base suffused buff, reverse deep rose, dbl., cupped, moderate fragrance; vigorous growth; Cant, B. R.

'MRS B.R. CANT, CLIMBING', Cl T, mp, 1960; Hjort

Mrs Baker, HP, mr; Turner, 1876

'MRS BAYARD THAYER', HT, pb, 1915; flowers clear silvery pink, reverse deep rose; [Mrs Charles E. Russell sport]; Waban Conservatories

'MRS BEATTY', HT, my, 1926; flowers marechal niel yellow, well formed, moderate fragrance; foliage bronze; vigorous growth; GM, NRS, 1925; Cant, B. R.

'MRS BECKWITH', HT, my, 1922; bud long, pointed; flowers buttercup-yellow, edged lighter, open, semi-dbl., slight fragrance; vigorous growth; GM, NRS, 1923; Pernet-Ducher

'MRS BELMONT TIFFANY', HT, yb, 1918; flowers golden yellow, base apricot-orange; [Sunburst sport]; Budlong; A.N. Pierson

Mrs Benjamin R. Cant, *see* **'MRS B.R. CANT'**

'MRS BERTRAM J. WALKER', HT, dp, 1915; flowers bright cerise-pink, dbl.; GM, NRS, 1914; Dickson, H.

Mrs Billy Crick, LCl, mp, 1995; Scarman

'MRS BLAMIRE YOUNG', HT, op, 1932; flowers salmon pink, slightly more petals than parent; [Una Wallace sport]; Young

Mrs Bosanquet, *see* **'MISTRESS BOSANQUET'**

'MRS BREEDLOVE', HT, pb, 1947; bud ovoid; flowers pink, base yellow, very large, very dbl., moderate fragrance; foliage leathery, glossy, dark; vigorous, bushy growth; [Golden Dawn sport]; Breedlove; Breedlove Nursery

'MRS BROWNELL', HT, rb, 1942; flowers red to pink and coppery orange, medium, semi-dbl., moderate, spicy fragrance; Brownell, H.C.

'MRS BRYCE ALLAN', HT, dp, 1916; flowers rose-pink, very dbl., intense fragrance; GM, NRS, 1916; Dickson, A.

'MRS BULLEN', HT, rb, 1917; flowers crimson, shaded yellow, passing to carmine, dbl.; Pernet-Ducher

'MRS C.V. HAWORTH, CLIMBING', Cl HT, ab, 1932; flowers dark apricot with pink, large, semi-dbl., intense fragrance; Cant, F.; (Sangerhausen)

'MRS C.E. BERNAYS', HT, lp; [Mme Lombard X Unknown]; Williams, A., 1905; (Weatherly, L.)

'MRS C.E. PRELL', HT, rb, 1938; bud long, pointed; flowers dark cerise, reverse pink, stamens golden, very large, dbl.; long stems; very vigorous growth; [Gustav Grunerwald X Betty Uprichard]; Fitzhardinge; Hazlewood Bros.

Mrs C.J. Bell, *see* **'MRS CHARLES BELL'**

'MRS C.L. FITZGERALD', F, pb, 1937; bud pointed; flowers deep pink, base apricot-yellow, small, dbl., slight fragrance; foliage glossy; upright growth; [Unnamed seedling X Cécile Brunner]; Fitzgerald

'MRS C.V. HAWORTH', HT, ab, 1919; flowers cinnamon-apricot, passing to buff, semi-dbl., intense fragrance; [Mrs Wemyss Quin X Hugh Dickson]; Dickson, A.

'MRS C.W. DUNBAR-BULLER', HT, dp, 1919; flowers deep rosy carmine, large, dbl., intense fragrance; Dickson, A.

'MRS C.W. EDWARDS', HT, rb, 1924; bud pointed; flowers crimson-carmine, base yellow, reverse veined yellow, over large, dbl., exhibition form, moderate fragrance; foliage dark, glossy; long stems; very vigorous growth; McGredy

'MRS C.W. THOMPSON', HMult, dp, 1920; flowers deep pink, quilled, small, borne in clusters, dbl., good seasonal bloom; vigorous, climbing growth; U.S. Dept. of Agric.; Storrs & Harrison Co.

'MRS CALVIN COOLIDGE', HT, yb, 1924; bud long, pointed; flowers golden yellow, deepening to rich orange, semi-dbl., moderate fragrance; [Ophelia sport]; U.S. Cut Flower Co.; F.R. Pierson

'MRS CAMPBELL HALL', T, ab, 1914; flowers soft creamy buff, edged or suffused rose, center warm salmon, exhibition form, intense fragrance; foliage leathery, dark; vigorous growth; Hall; A. Dickson

Mrs Cecily McMullen, *see* 'BOSEMWINE'

'MRS CHAPLIN', HT, pb, 1918; flowers creamy pink, base shaded yellow; Chaplin Bros.

'MRS CHARLES BELL', HT, op, 1917; (Mrs C.J. Bell, Salmon Radiance, Shell-Pink Radiance); flowers shell-pink, shaded soft salmon; [Red Radiance sport]; Bell, Mrs. C.J.; A.N. Pierson

'MRS CHARLES BELL, CLIMBING', Cl HT, op, 1929; Thomasville Nursery

'MRS CHARLES CURTIS-HARRISON', HT, m, 1910; flowers purple/pink, large, dbl., moderate fragrance; Dickson, A.; (Sangerhausen)

'MRS CHARLES E. PEARSON', HT, ob, 1913; flowers orange-apricot, flushed fawn and yellow, well formed, moderate fragrance; GM, NRS, 1912; McGredy

'MRS CHARLES E. RUSSELL', HT, dp, 1914; bud long, pointed; flowers rosy carmine, large, dbl., globular, moderate fragrance; foliage leathery; long stems; vigorous growth; [Mme Caroline Testout X (Mme Abel Chatenay X Marquise Litta de Breteuil)]; Montgomery, A.; Waban Conservatories

Mrs Charles E.F. Gersdorff, *see* **'MAYLINA'**

'MRS CHARLES H. RIGG', HT, my, 1946; bud long, pointed; flowers lemon-yellow, large, dbl., moderate fragrance; vigorous growth; McGredy

'MRS CHARLES HUNTER', HT, dp, 1912; flowers carmine-pink, large, dbl., intense fragrance; Paul, W.; (Sangerhausen)

'MRS CHARLES LAMPLOUGH', HT, ly, 1920; flowers pale lemon-yellow, very large, dbl., moderate fragrance; vigorous growth; GM, NRS, 1919; [Frau Karl Druschki X ?]; McGredy

'MRS CHARLES REED', HT, w, 1914; flowers pale cream, tinted deep peach, base soft golden yellow, dbl., intense fragrance; Hicks

'MRS CHARLES STEWARD', HT, mp, 1959; flowers bright pink, large, dbl., intense fragrance; foliage dark; vigorous growth; Verschuren; Gandy Roses, Ltd.

'MRS CHARLES TENNANT', HT, yb, 1936; flowers clear primrose, shaded rich canary-yellow, moderate fragrance; foliage bronze; vigorous growth; Cant, F.

'MRS CHARLOTTE GUILFOYLE', HT, P; [M. Berard X Unknown]; Johnson, 1885; (Weatherly, L.)

'MRS CLAUDE AVELING', HT, rb, 1929; flowers scarlet-cerise, tinted orange, base buttercup-yellow, anther, semi-dbl., moderate, fruity fragrance; [The Queen Alexandra Rose X Gorgeous]; Bees

'MRS CLEMENT YATMAN', HT, dr, 1927; flowers deep crimson, dbl., intense fragrance; Hicks

'MRS COCKER', HP, mp, 1899; flowers large, dbl., intense fragrance; Cocker; (Sangerhausen)

'MRS COLVILLE', HSpn, m; flowers bright crimson-purple, white eye, single; vigorous (4 ft) growth; [Probably R. spinosissima X R. pendulina]

'MRS CORNWALLIS WEST', HT, w, 1911; flowers white, center blush, very large, imbricated, globular; vigorous growth; GM, NRS, 1910; Dickson, A.

'MRS COURTNEY PAGE', HT, ob, 1922; bud long, pointed; flowers orange-cerise, shaded carmine, very large, dbl., exhibition form, moderate fragrance; strong stems; GM, NRS, 1922; McGredy

'MRS CURNOCK SAWDAY', HT, mp, 1920; flowers satiny pink, dbl., slight fragrance; Hicks

'MRS D.A. KOSTER', Pol, 1934; Koster, D.A.; (Cavriglia)

'MRS DAN PROSSER', HT, rb, 1946; flowers red shaded gold, dbl., 60 petals, 5.5 in., exhibition form; foliage glossy, dark red when young; [Mrs Sam McGredy X Heinrich Wendland]; Prosser

'MRS DAVID BAILLIE', HT, dp, 1912; flowers carmine, penciled deeper, dbl., moderate fragrance; Dickson, H.

'MRS DAVID MCKEE', HT, ly, 1904; flowers creamy yellow, well formed, large; dwarf, compact growth; [Frau Karl Druschki X Kaiserin Auguste Viktoria]; Dickson, A.

Mrs De Graw, *see* '**Champion of the World**'

Mrs Doreen Pike, *see* 'AUSdor'

'**Mrs Douglas Copland**', HT, mp, 1945; dbl.; vigorous growth; Clark, A.

'**Mrs Dudley Cross**', T, yb; (Dudley Cross); flowers pale yellow, tinted in autumn with crimson, dbl., slight fragrance; no prickles; vigorous growth; Paul, W., 1908

'**Mrs Dudley Fulton**', Pol, w, 1931; flowers silver-white, large, borne in large clusters, single; dwarf growth; (14); [Dorothy Howarth X Perle d'Or]; Thomas; Armstrong Nursery

'**Mrs Dunlop Best**', HT, ab, 1916; (Cleveland II); bud long, pointed; flowers reddish apricot, base coppery yellow, large, dbl., 28 petals, moderate fragrance; foliage leathery, bronze, glossy; vigorous growth; Hicks

'**Mrs Dunlop Best, Climbing**', Cl HT, ab, 1933; Rosen, L.P.

'**Mrs E. Alford**', HT, lp, 1913; flowers large, dbl.; Lowe; (Sangerhausen)

'**Mrs E. Claxton**', HT, pb, 1928; flowers light pink, shaded salmon and carmine; Cant, F.

'**Mrs E. Gallagher**', HT, dr, 1924; flowers dark crimson, moderate fragrance; McGredy

'**Mrs E. Townshend**', HT, 1911; Guillot; (Cavriglia)

'**Mrs E. Willis**', HT, lp; semi-dbl., moderate fragrance; growth moderate; [Mme Segond Weber seedling]; Weightman

'**Mrs E. Willis**', HT, op, 1923; flowers light salmon-pink, large, dbl., moderate fragrance; Poulsen; (Sangerhausen)

'**Mrs E. Willis, Climbing**', Cl HT, lp, 1948; Wilson

'**Mrs E. Wood**', HT, yb, 1934; bud very long; flowers light buff-yellow, tinted yellowish salmon, becoming cream-yellow, intense fragrance; vigorous growth; Dickson, A.

'**Mrs E.G. Hill**', HT, w, 1906; flowers alabaster-white, reverse rose-coral, well shaped, large, dbl., moderate fragrance; vigorous growth; [Mme Caroline Testout X Liberty]; Soupert & Notting

'**Mrs E.J. Hudson**', HT, mp, 1923; flowers bright pink, dbl., slight fragrance; [Mrs W.J. Grant seedling]; Lilley

'**Mrs E.J. Manners**', HT, dr, 1938; flowers deep velvety crimson, becoming darker, well shaped; long, strong stems; vigorous, branching growth; Burbage Nursery

'**Mrs E.M. Gibson**', HT, dr, 1940; [Countess of Stradbrooke X ?]; Clark, A.

'**Mrs E.M. Gilmer**', HT, mr, 1927; (Dorothy Dix); [Seedling X Crusader]; Cook, J.W.

'**Mrs E.T. Stotesbury**', HT, pb, 1918; flowers light cream-pink, reverse dark pink, very dbl.; [(Joseph Hill X My Maryland) X Milady]; Towill

'**Mrs E.W. Sterling**', HT, mp, 1916; flowers rose-pink; [Antoine Rivoire X Unnamed pink seedling]; Cook, J.W.

'**Mrs Edith Stanley**', HT, w, 1919; flowers creamy white, shaded indian yellow; Easlea

'**Mrs Edward J. Holland**', HT, op, 1909; flowers salmon-rose, large, moderate fragrance; moderately vigorous growth; GM, NRS, 1909; McGredy

'**Mrs Edward Laxton**', HT, ob, 1935; flowers flaming orange and old-rose, large, very dbl., exhibition form; foliage leathery, dark; very vigorous growth; [Mrs Henry Bowles X Shot Silk]; Laxton Bros.

'**Mrs Edward Powell**', HT, dr, 1911; flowers velvety crimson, dbl., moderate fragrance; Bernaix, P.

'**Mrs Elisha Hicks**', HT, w, 1919; flowers flesh, nearly white, dbl., intense fragrance; [Frau Karl Druschki X Mme Gabriel Luizet]; Hicks

'**Mrs Erskine Pembroke Thom**', HT, my, 1926; bud long, pointed; flowers clear yellow, well-formed, large, dbl., 40 petals, moderate fragrance; vigorous growth; [Grange Colombe X Souv. de Claudius Pernet]; Howard, F.H.; H&S

'**Mrs Erskine Pembroke Thom, Climbing**', Cl HT, my, 1933; Dixie Rose Nursery

'**Mrs Eveline Gandy**', HT, dr, 1959; flowers dark velvety crimson-scarlet, well-formed, large, dbl., 50 petals, intense fragrance; Verschuren; Gandy Roses, Ltd.

'**Mrs F. Millar**', HT, lp; Williams, A., 1904; (Weatherly, L.)

'**Mrs F.F. Prentiss**', LCl, lp, 1925; flowers pale pink, large, dbl.; vigorous, climbing growth; extremely hardy.; [(R. setigera X R. wichurana) X Lady Alice Stanley]; Horvath

'**Mrs F.J. Jackson**', HT, mr, 1933; bud pointed; flowers cerise, dbl., slight fragrance; foliage leathery, bronze; vigorous growth; LeGrice

'**Mrs F.J. Knight**', HT, mr, 1928; flowers velvety scarlet, moderate fragrance; [Lord Charlemont sport]; Knight, J.

'**Mrs F.R. Pierson**', HT, dr, 1926; bud pointed; flowers crimson, shaded scarlet, very large, dbl., intense fragrance; [Premier sport]; Pierson, F.R.

'**Mrs F.W. Flight**', LCl, dp, 1905; flowers rose-pink, open, borne in clusters, semi-dbl., non-recurrent; foliage large, rich green, soft; pillar (6–8 ft) growth; [Crimson Rambler X ?]; Cutbush

'**Mrs F.W. Sanford**', HP, lp, 1898; flowers pink tinged white; [Mrs John Laing sport]; Curtis

'**Mrs F.W. Woodroffe**', T, lp; Williams, A., 1904; (Weatherly, L.)

'**Mrs Farmer**', HT, yb, 1918; bud pointed; flowers yellow, reverse reddish apricot, dbl., moderate fragrance; Pernet-Ducher

'**Mrs Foley Hobbs**', T, pb, 1910; flowers soft ivory-white, edges tinged clear pink, dbl., moderate fragrance; vigorous, upright growth; Dickson, A.

'**Mrs Forde**', HT, pb, 1913; flowers carmine-rose on soft rose-pink, base chrome-yellow, dbl., moderate fragrance; GM, NRS, 1913; Dickson, A.

'**Mrs Francis King**', HT, ly, 1934; flowers cream-white shaded straw-yellow, large, dbl., 80 petals, exhibition form, slight fragrance; foliage leathery; vigorous, bushy growth; [Lady Lilford X Leonard Barron]; Nicolas; J&P

'**Mrs Frank Guthrie**', HT, lp, 1923; flowers deep flesh in autumn, pale in summer, semi-dbl., moderate fragrance; foliage dark, leathery; [R. gigantea seedling X ?]; Clark, A.; Hazelwood Bros.

'**Mrs Frank J. Usher**', HT, yb, 1920; flowers rich yellow, edged rosy carmine, very dbl.; weak stems growth; [Queen Mary X Unnamed variety]; Dobbie

'**Mrs Frank Schramm**', HT, pb, 1934; flowers bright glowing rose-pink, reverse slightly lighter, dbl., moderate fragrance; foliage leathery, dark; very large, long stems; very vigorous growth; [Briarcliff sport]; Schramm

'**Mrs Frank Serpa**', Pol, dp, 1954; flowers deep pink, borne in clusters; dwarf (18 in.) growth; [Rouletii X China Doll]; Serpa; Port Stockton Nursery

'**Mrs Frank Verdon**', HT, my, 1935; flowers creamy yellow, very large, dbl., slight fragrance; foliage leathery, dark; vigorous growth; [Joan Horton X Marion Horton]; Bees

'**Mrs Frank Workman**', HT, dp, 1911; flowers bright rose-pink; Dickson, H.

'**Mrs Franklin D. Roosevelt**', HT, dy, 1933; bud long, pointed; flowers golden yellow, large, dbl., globular, moderate fragrance; foliage glossy; very vigorous growth; [Talisman sport]; Traendly & Schenck

'**Mrs Franklin Dennison**', HT, w, 1915; bud long, pointed; flowers porcelain-white, veiled primrose-yellow, base ochre, very large, dbl., exhibition form, moderate fragrance; McGredy

'**Mrs Fred Cook**', HT, ob, 1920; flowers light terra-cotta, edged silvery white, dbl.; Easlea

'**Mrs Fred Danks**', HT, m, 1951; bud long, pointed; flowers pink tinted lilac, large, semi-dbl., 15 petals, moderate fragrance; foliage leathery; very vigorous, upright, pillar growth; Clark, A.; NRS Victoria

'**Mrs Fred H. Howard**', HT, ob, 1926; flowers orange-apricot edged straw-yellow, dbl., intense fragrance; Dobbie

'Mrs Fred L. Lainson', HT, pb, 1934; bud long, pointed; flowers deep pink, almost red, base yellow and bronze, reverse orange, dbl., moderate fragrance; foliage leathery, glossy, bronze; vigorous growth; [Talisman sport]; Scittine; Lainson

'Mrs Fred Poulsom', HT, mp, 1920; flowers vivid pink; thorny; vigorous growth; [Edith Part X Unnamed variety]; Therkildsen

'Mrs Fred Searl', HT, lp, 1917; flowers very large, dbl., moderate fragrance; Dickson, A.; (Sangerhausen)

'Mrs Fred Straker', HT, yb, 1910; flowers light orange-yellow, medium, dbl., moderate fragrance; Dickson, A.; (Sangerhausen)

'Mrs Frederick W. Vanderbilt', HT, or, 1912; flowers deep orange-red, shaded apricot, dbl.; GM, NRS, 1913; McGredy

'Mrs G.A. Wheatcroft', HT, lp, 1926; flowers light creamy pink, medium, dbl., moderate fragrance; Wheatcroft; (Sangerhausen)

'Mrs G. Payne', T, lp; Williams, A., 1911; (Weatherly, L.)

Mrs G.A. van Rossem, *see* **'Mevrouw G.A. van Rossem'**

Mrs G.A. van Rossem, Climbing, *see* **'Mevrouw G.A. van Rossem, Climbing'**

'Mrs G.A. Wheatcroft', HT, pb, 1926; flowers coppery pink to silver-rose at tips, reverse soft salmon-pink, dbl., intense fragrance; [Lady Pirrie sport]; Wheatcroft Bros.

'Mrs G.M. Smith', HT, dr, 1935; bud long, pointed; flowers deep crimson, dbl., cupped, slight fragrance; vigorous growth; [Red-Letter Day X Mrs J.J. Hedley-Willis]; Bees

'Mrs George B. Easlea', HT, dp, 1939; flowers sparkling carmine-pink, well formed, very large, exhibition form, intense fragrance; strong stems; Easlea

'Mrs George C. Thomas', HMsk, op, 1925; flowers salmon-pink, center orange, semi-dbl., 2–2.5 in., repeat bloom, moderate fragrance; climbing growth to 10 ft; [Mme Caroline Testout, Climbing X Moonlight]; Thomas; B&A

'Mrs George Geary', HT, ob, 1929; bud pointed; flowers orange-cerise, shaded cardinal, very large, dbl., 35–40 petals, exhibition form, intense fragrance; vigorous growth; [Red-Letter Day X Mrs Wemyss Quin]; Burbage Nursery

'Mrs George Marriott', HT, w, 1918; bud long, pointed; flowers deep cream and pearl, suffused rose, very large, dbl., exhibition form, moderate fragrance; GM, NRS, 1917; McGredy

'Mrs George Shawyer', HT, pb, 1911; flowers rosy pink, reverse pale pink, very large, dbl., slight fragrance; [Mme Hoste X Joseph Lowe]; Lowe & Shawyer

'Mrs George Shawyer, Climbing', Cl HT, pb, 1918; Lindquist, E.J.

'Mrs Georgia Chobe', HT, lp, 1937; bud long; flowers large, dbl., exhibition form; foliage leathery, light; vigorous growth; [Miss Rowena Thom X Renault]; H&S

'Mrs Graham Hart', Cl T, lp; Williams, A., 1900; (Weatherly, L.)

'Mrs H. Cobden Turner', HT, rb, 1948; flowers cherry-cerise flushed orange, base yellow, dbl., 26 petals, 3–4 in., intense fragrance; foliage glossy; moderately vigorous growth; [Ophelia X Unnamed seedling]; Mee; Fryer's Nursery, Ltd.

'Mrs H.J. Hedley-Willis', HT, dr, 1930; flowers medium, semi-dbl., intense fragrance; Bees of Chester; (Sangerhausen)

'Mrs H.D. Greene', HT, rb, 1918; flowers reddish bronze, becoming flame and coppery pink, moderate fragrance; [Joseph Hill sport]; Easlea

'Mrs H.G. Johnstone', HT, pb, 1930; bud long; flowers rose-pink, base and edges rose, very large, dbl., exhibition form, slight fragrance; vigorous growth; [Mme Caroline Testout X Mrs George Shawyer]; Bees

'Mrs H.L. Wettern', HT, mp, 1922; flowers vivid pink, dbl., moderate fragrance; McGredy

'Mrs H.M. Eddie', HT, w, 1932; flowers creamy white, passing to purest white, dbl., 40–45 petals, 5–6 in., exhibition form, slight fragrance; foliage dark, leathery, glossy; vigorous, bushy growth; [Mrs Charles Lamplough X Mev. G.A. van Rossem]; Eddie

'Mrs H.M. Eddie, Climbing', Cl HT, w, 1944; Eddie

'Mrs H.P. Abbot', T, lp; [Mme Lombard X Unknown]; Williams, A., 1901; (Weatherly, L.)

'Mrs H.R. Darlington', HT, ly, 1919; bud long, pointed; flowers clear creamy yellow, well formed, balling in wet weather, moderate fragrance; long stems; vigorous growth; GM, NRS, 1919; McGredy

Mrs Harkness, *see* **'Paul's Early Blush'**

'Mrs Harold Alston', Cl HT, mp, 1940; Clark, A.

'Mrs Harold Bibby', HT, pb, 1936; flowers soft pink veined red, outer petals silvery pink, moderate fragrance; foliage dark; Bees

'Mrs Harold Brocklebank', HT, w, 1907; flowers creamy white, center buff, base soft yellow, dbl., moderate fragrance; Dickson, A.

'Mrs Harold Brookes', HT, mr, 1931; flowers very bright red, large, dbl., cupped, moderate fragrance; foliage light; vigorous, bushy growth; [Frau Oberhofgartner Singer X Firebrand]; Clark, A.; NRS Victoria

'Mrs Henri Daendels', HT, ab, 1931; (Mevrouw Daendels); flowers apricot shaded orange, reverse violet-pink, large, very dbl., intense fragrance; foliage bronze; strong stems; very vigorous growth; [Mrs Henry Bowles X Rev. F. Page-Roberts]; Buisman, G. A. H.

'Mrs Henry Balfour', HT, pb, 1918; flowers ivory-white, base primrose, edge penciled rose, like a picotee, moderate fragrance; McGredy

'Mrs Henry Bowles', HT, mp, 1921; flowers rosy-pink, flushed salmon, well-formed, large, dbl., 50 petals, exhibition form; foliage dark, glossy; vigorous growth; [Lady Pirrie X Gorgeous]; Chaplin Bros.

'Mrs Henry Bowles, Climbing', Cl HT, mp, 1929; Dobbie

'Mrs Henry Morse', HT, pb, 1919; bud long, pointed; flowers cream, tinted rose, marked and veined red, large, dbl., exhibition form, moderate fragrance; dwarf growth; GM, NRS, 1919; McGredy

'Mrs Henry Morse, Climbing', Cl HT, pb, 1929; Chaplin Bros.

'Mrs Henry Winnett', HT, dr, 1917; bud long, pointed; flowers deep rich red, large, dbl., exhibition form, moderate fragrance; foliage leathery; vigorous growth; [Mrs Charles E. Russell X Mrs George Shawyer]; Dunlop

'Mrs Henry Winnett, Climbing', Cl HT, dr, 1930; Bernaix, P.

'Mrs Herbert Carter', HT, ab, 1934; flowers apricot-yellow, center deeper, veined bronze, well formed, l, dbl., moderate fragrance; vigorous growth; Cant, F.

'Mrs Herbert Dowsett', HT, pb, 1928; flowers several shades deeper, otherwise similar to parent; [Los Angeles sport]; Easlea

'Mrs Herbert Hawksworth', T, r, 1912; flowers ecru on milk-white, dbl., moderate fragrance; Dickson, A.

'Mrs Herbert Hoover', HT, mr, 1928; bud long, pointed; flowers rich velvety red, dbl., intense fragrance; foliage dark, leathery; vigorous growth; [Ophelia X Hoosier Beauty]; Coddington

'Mrs Herbert Nash', HT, dr, 1925; flowers scarlet-crimson shaded deep crimson, overlarge, moderate fragrance; long, strong stems; vigorous growth; Chaplin Bros.

'Mrs Herbert Stevens', HT, w, 1910; bud long, pointed; dbl., exhibition form, moderate fragrance; foliage light; vigorous, bushy growth; [Frau Karl Druschki X Niphetos]; McGredy

'Mrs Herbert Stevens, Climbing', Cl HT, w, 1922; (Grimpant Mrs Herbert Stevens, Stevens, Climbing); Pernet-Ducher

'MRS HILTON BROOKS', HT, yb, 1929; flowers saffron-yellow, base deeper, suffused pink and carmine, dbl., moderate fragrance; Cant, F.

Mrs Holz Climber #1, LCl, m

Mrs Holz climber #2, LCl, lp

'MRS HORNBY LEWIS', HT, ob, 1921; flowers orange-yellow, very dbl., moderate fragrance; [Gorgeous X Mme Mélanie Soupert]; Hicks

'MRS HOVEY', HSet, w, 1850; flowers blush to almost white; [R. setigera X ?]; Pierce

'MRS HUGH DETTMANN', Cl HT, ab, 1930; flowers bright apricot-yellow; good pillar growth; Clark, A.; NRS Victoria

'MRS HUGH DICKSON', HT, yb, 1915; flowers deep cream, heavily suffused orange and apricot, dbl., 44 petals; GM, NRS, 1916; Dickson, H.

'MRS INGE POULSEN', F, pb, 1949; (Fru Inge Poulsen); flowers pink with a yellow center, open, medium, semi-dbl., slight fragrance; foliage light, matt; vigorous, compact (80 cms) growth; [Poulsen's Pink X Seedling]; Poulsen, S.; McGredy, 1952; Poulsen, 1949

Mrs Iris Clow, *see* 'HARBRITE'

'MRS J. HEATH', HT, yb, 1924; flowers maize-yellow, tinted peach-red, center yellow, dbl., moderate fragrance; McGredy

'MRS J. WYLIE', HT, lp, 1923; flowers silvery blush-pink; Dickson, H.

'MRS J.C. AINSWORTH', HT, mp, 1918; flowers rose-pink, very dbl., moderate fragrance; [Mrs Charles E. Russell sport]; Clarke Bros.

'MRS J.C. MANIFOLD', HP, lp; [Prince Camille de Rohan X Unknown]; Adamson, 1915; (Weatherly, L.)

'MRS J.D. EISELE', HT, pb, 1933; bud long, pointed; flowers brilliant cherry-pink, center shaded scarlet, very large, dbl., intense fragrance; vigorous growth; [Premier Supreme X McGredy's Scarlet]; Howard, F.H.; H&S

'MRS J.D. RUSSELL', HT, dr, 1930; flowers deep crimson, center maroon, almost black in certain lights, dbl., cupped, intense fragrance; [Prince Camille de Rohan X Mrs Aaron Ward]; Bees

'MRS J.F. REDLY', HT, lp; flowers pale flesh-pink, center tinted salmon, large, dbl., 40 petals, lightly repeats in late summer; vigorous growth

'MRS J.J. HEDLEY-WILLIS', HT, dr, 1929; bud pointed; flowers dark crimson, center almost plum-black, dbl., exhibition form, intense fragrance; [Admiral Ward X Richmond]; Bees

'MRS J.T. MCINTOSH', HT, ab, 1935; flowers creamy apricot, center deeper apricot, base golden yellow, dbl.; vigorous growth; McIntosh; Brundrett

'MRS JAMES GARNER', HT, ab, 1931; flowers buff, base orange, moderate fragrance; Cant, F.

'MRS JAMES LYNAS', HT, pb, 1914; flowers pearly pink, reverse and edges flushed rosy peach, dbl., moderate fragrance; GM, NRS, 1913; Dickson, H.

'MRS JAMES SHEARER', HT, w, 1923; flowers pure white, base yellow, large, dbl., exhibition form, moderate fragrance; [Unnamed seedling X Mme Colette Martinet]; Ferguson, W.

'MRS JAMES WILLIAMSON', HT, mp, 1922; flowers clear pink, dbl., moderate fragrance; Dickson, H.

'MRS JEANNETTE G. LEEDS', HT, rb, 1942; bud globular, jasper-red and apricot-buff; flowers venetian pink, dbl., 50–60 petals, 5–6 in., slight fragrance; foliage dark, leathery; very vigorous, upright growth; [Joanna Hill X R.M.S. Queen Mary]; Hill, Joseph H., Co.

'MRS JENNIE DEVERMAN', HT, pb, 1933; flowers cerise edged silvery, base tinted gold; [Pres. Herbert Hoover sport]; Deverman

'MRS JOHN BELL', HT, dr, 1928; bud long; flowers carmine, large, dbl., cupped, moderate fragrance; Bell, Judy G.; Dobbie

'MRS JOHN COOK', HT, w, 1920; bud long, pointed; flowers white, suffused soft pink, deepening in cool weather, large, dbl., cupped, moderate fragrance; [Ophelia X Unnamed seedling]; Cook, J.W.; A.N. Pierson

'MRS JOHN FOSTER', HT, dr, 1915; flowers large, dbl., moderate fragrance; Hicks; (Sangerhausen)

'MRS JOHN INGLIS', HT, dr, 1920; flowers rich crimson, dbl., intense fragrance; McGredy

'MRS JOHN LAING', HP, mp, 1887; bud pointed; flowers soft pink, large, dbl., 45 petals, recurrent bloom, moderate fragrance; foliage light; vigorous, rather dwarf growth; (28); [Francois Michelon X ?]; Bennett

'MRS JOHN MCNABB', S, w, 1941; bud with very long sepals; very dbl., non-recurrent, moderate fragrance; few prickles; foliage large, dark, slightly rugose, prickly underneath; height 5 ft; [R. beggeriana X R. rugosa]; Skinner

'MRS JOHN R. ALLAN', HT, lp, 1920; flowers soft rosy pink, reverse darker, dbl.; GM, NRS, 1920; Dickson, H.

Mrs Jones, *see* 'DELGE'

'MRS JOSEPH H. WELCH', HT, mp, 1911; flowers brilliant rose-pink, large, dbl., slight fragrance; vigorous growth; GM, NRS, 1910; McGredy

'MRS JOSEPH H. WELCH, CLIMBING', Cl HT, mp, 1922; Perkins, H.S.

'MRS JOSEPH HIESS', Pol, lp, 1943; flowers mary wallace pink, base white, borne in clusters of 3–16, dbl., 40 petals, cupped, slight fragrance; foliage leathery; strong stems; vigorous, upright, bushy, compact growth; [Roserie X ?]; Shepherd; Klyn

Mrs K.B. Sharma, HT, w, 1989; Pal, Dr B.P.

Mrs Keays' Red Damask, HCh, mr

Mrs Keays' The Bishop, HCh, m

'MRS L.B. CODDINGTON', HT, mp, 1931; bud long, pointed; flowers large, dbl., moderate fragrance; foliage leathery; vigorous growth; [Templar X Souv. de Claudius Pernet]; Coddington

'MRS L.B. COPELAND', F, pb, 1934; flowers salmon-pink, veined, base yellow, small, dbl., globular, intense fragrance; foliage dark, glossy; vigorous growth; [Unnamed seedling X Cécile Brunner]; Fitzgerald

'MRS LESLIE MOSS', Cl HT, dp, 1944; flowers carmine-pink, medium, semi-dbl.; foliage leathery; vigorous, climbing growth; Moss; Mason, F.

Mrs Lincoln, HT, dr, 1997; Williams, J. Benjamin

'MRS LOVELL SWISHER', HT, pb, 1926; bud pointed; flowers salmon-pink, edged flesh, well-formed, dbl., moderate fragrance; foliage bronze; very vigorous growth; [Unnamed seedling X Souv. de Claudius Pernet]; Howard, F.H.; H&S

'MRS LOVELL SWISHER, CLIMBING', Cl HT, pb, 1930; H&S

'MRS LUTHER BURBANK', HT, mp, 1954; bud long, pointed; flowers rose-pink, dbl., 34 petals, 4–4.5 in., cupped, intense, spicy fragrance; foliage leathery; vigorous growth; [Christopher Stone X Charlotte Armstrong]; Swim, H.C.; Stark Bros.

'MRS M.H. HORVATH', HT, ly, 1940; bud long, pointed; flowers pale yellow, open, large, dbl., 40 petals; foliage glossy; long stems; very vigorous, upright growth; [(Mme Butterfly X ?) X Souv. de Claudius Pernet]; Horvath; Wayside Gardens Co.

'MRS M.H. WALSH', HWich, w, 1913; flowers pure snow-white, small, borne in clusters, dbl.; foliage large, glossy; very vigorous, trailing growth; Walsh

'MRS M.J. GILLON', HT, dr, 1974; bud ovoid; flowers red-purple, dbl., 37 petals, 4 in., globular, moderate fragrance; foliage glossy, dark; upright growth; [Tropicana X Prima Ballerina]; Rijksstation Voor Sierplantenteelt

'MRS MABEL V. SOCHA', HT, my, 1935; bud pointed; flowers pure lemon-yellow, dbl., intense fragrance; long, strong stems; [Seedling X Souv. de Claudius Pernet]; H&S

'MRS MACDONALD'S ROSE', HRg, dp;, non-recurrent; [Apparently R. rugosa plena X R. acicularis]; Reid; P.H. Wright

'MRS MACKELLAR', HT, my, 1915; flowers deep citron or pure canary, passing to primrose, dbl., moderate fragrance; Dickson, A.

'MRS MARY D. WARD', HT, pb, 1927; flowers shell-pink and gold, reverse ivory-white, moderate fragrance; [Double Ophelia X Souv. de Claudius Pernet]; Ward, F.B.

Mrs Mary Thomson, *see* 'TOMONE'

Mrs Matso's Moss, M, dr

'MRS MATTHEWS', HT, lp; Matthews, W.J., 1923; (Weatherly, L.)

'MRS MAUD ALSTON', F, lp; Clark, A., 1940; (Weatherly, L.)

'MRS MAYNARD SINTON', HT, pb, 1909; flowers silvery white suffused pink, very large, moderate fragrance; moderately vigorous growth; GM, NRS, 1909; McGredy

'MRS MINA LINDELL', (form of R. macounii), lp, 1927; semi-dbl., 10–12 petals, non-recurrent; height to 4 ft; hardy.; N.E. Hansen; (found in S. Dakota)

'MRS MINIVER', HT, mr, 1944; (Souv de Louis Simon); flowers scarlet-crimson, reverse slightly darker, dbl., 20 petals, 5.5–6 in., cupped, moderate fragrance; foliage soft; vigorous, upright, bushy, compact growth; Chambard, C.; J&P

'MRS MONA HUNTING', HT, yb, 1916; flowers chamois-yellow, opening fawn, dbl.; Dickson, H.

'MRS MOORFIELD STOREY', HT, lp, 1915; flowers shell-pink, center deeper, dbl.; [Gen. MacArthur seedling X Joseph Hill]; Waban Conservatories

'MRS MUIR MACKEAN', HT, dr, 1912; flowers carmine-crimson, large, exhibition form, intense fragrance; vigorous growth; McGredy

'MRS MURRAY ALLISON', HT, pb, 1925; flowers rose-pink, base carmine, dbl., slight fragrance; Prior

'MRS MYLES KENNEDY', T, pb, 1906; flowers silvery white tinted buff, center and reverse pink, large, dbl.; vigorous growth; Dickson, A.

'MRS NIEMINEN', HT, rb, 1954; flowers blood-red shaded scarlet, well-formed, large, very dbl., slight fragrance; bushy, compact growth; [Hens Verschuren X Poinsettia]; Frères, Buyl

'MRS NORMAN WATSON', Cl HT, dp, 1930; flowers deep cherry-pink, large; very vigorous, pillar growth; [Radiance X Gwen Hash]; Clark, A.; Geelong Hort. Soc.

'MRS NORRIS M. AGNEW', HT, ob, 1934; bud pointed; flowers orange-cerise, large, dbl., slight fragrance; foliage leathery, light; very vigorous growth; [J.C. Thornton X Florence L. Izzard]; Bees

Mrs O.G. Orpen, HGal, dp

'MRS OAKLEY FISHER', HT, dy, 1921; flowers deep orange-yellow, blooms in clusters, single, moderate fragrance; foliage dark, bronze, glossy; vigorous growth; Cant, B. R.

'MRS OLIVE SACKETT', F, mr, 1931; flowers bright red, well formed, semi-dbl., 2–2.5 in.; foliage bronze in autumn; vigorous growth; [Else Poulsen sport]; Wirtz & Eicke; Spath

'MRS OLIVER AMES', HT, my, 1941; bud long, pointed; flowers lemon-yellow, dbl., 63 petals, 3.5–4 in., globular, intense fragrance; foliage leathery; long stems; vigorous growth; [Max Krause X Julien Potin]; Verschuren; Dreer

'MRS OLIVER MEE', HT, rb, 1948; flowers scarlet shaded gold, dbl., 35 petals, 5–6 in., exhibition form, moderate fragrance; foliage glossy, bronze; very vigorous growth; [Mrs Charles Lamplough X Edith Mary Mee]; Mee; Fryer's Nursery, Ltd.

'MRS OSWALD LEWIS', HT, yb, 1936; flowers soft canary-yellow, outer petals edged flame, well formed, dbl., intense fragrance; long stems; vigorous growth; GM, Bagatelle, 1935; Cant, F.

'MRS OSWALD SMEATON', HT, pb, 1932; flowers ivory-cream, center and petal tips pink, very large, dbl., 50–60 petals, moderate fragrance; long stems; vigorous growth; Easlea

'MRS PAT', Pol, lp, 1928; flowers small, dbl., moderate fragrance; Lilley; (Sangerhausen)

'MRS PAUL', B, lp, 1891; flowers blush-white shaded rosy peach, large; Paul & Son

'MRS PAUL GOUDIE', HT, yb, 1932; (Shining Sun); flowers deep buttercup-yellow, edged carmine-scarlet, very large, dbl., moderate, fruity fragrance; McGredy

Mrs Paul Hvid, B, w

'MRS PAUL J. HOWARD', Cl HT, rb, 1938; bud long, pointed; flowers brilliant crimson, reverse flame-red, dbl., 30 petals, 5 in., moderate, spicy fragrance; foliage large, bronze, leathery; long stems; very vigorous, climbing (12–15 ft) growth; [Miss Rowena Thom X Paul's Lemon Pillar]; Howard, F.H.; H&S

'MRS PAUL M. PIERSON', HT, pb, 1930; bud long, pointed; flowers soft pink, reverse brighter, very large, dbl., exhibition form, moderate fragrance; [Premier sport]; Pierson, P.M.

'MRS PAUL R. BOSLEY', HT, my, 1941; flowers apricot-yellow; [Mme Joseph Perraud sport]; Bosley Nursery

'MRS PERCY V. PENNYBACKER', HT, pb, 1929; flowers peach-pink shaded silver, dbl., cupped, moderate fragrance; [Mme Butterfly seedling]; Vestal

'MRS PETER BLAIR', HT, yb, 1906; flowers lemon-chrome, center golden yellow, well formed, large; GM, NRS, 1906; Dickson, A.

'MRS PHILIP RUSSELL', HT, dr, 1927; bud long, pointed; flowers dark red, shaded black, semi-dbl., slight fragrance; foliage glaucous green; vigorous, semi-climbing growth; pillar or large bush; [Hadley X Red Letter Day]; Clark, A.; Hackett

'MRS PIERCE', HSet, mp, 1850; flowers pink, well formed; Pierce

'MRS PIERRE S. DUPONT', HT, my, 1929; bud long, pointed, reddish gold; flowers golden yellow, becoming lighter, dbl., 40 petals, moderate, fruity fragrance; foliage rich green; moderate growth; GM, Bagatelle, 1929; [(Ophelia X Rayon d'Or) X (Ophelia X (Constance X Souv. de Claudius Pernet))]; Mallerin, C.; C-P

'MRS PIERRE S. DUPONT, CLIMBING', Cl HT, my, 1933; Hillock

'MRS PRENTISS NICHOLS', HT, dp, 1923; flowers brilliant deep pink, dbl., moderate fragrance; [Ophelia X Seedling]; Scott, R.

Mrs Price, Pol, pb

'MRS R.D. MCCLURE', HT, op, 1913; flowers salmon-pink, very large, dbl.; Dickson, H.; (Sangerhausen)

'MRS R.B. MCLENNAN', HT, pb, 1924; flowers satiny rose suffused yellow, very dbl., intense fragrance; [George C. Waud X Mme Caristie Martel]; Easlea

'MRS R.B. MOLONEY', HT, mr, 1925; flowers brilliant carmine-red, dbl., intense fragrance; McGredy

'MRS R.C. BELL', HT, mr, 1920; flowers bright red; [Gen. MacArthur X Château de Clos Vougeot]; Clark, A.

'MRS R.G. SHARMAN-CRAWFORD', HP, pb, 1894; flowers rosy pink, outer petals tinted flesh, large, dbl., 75 petals, cupped, recurrent bloom, moderate fragrance; vigorous growth; Dickson, A.

'MRS R.M. FINCH', Pol, mp, 1923; flowers rosy pink, becoming lighter, medium blooms in large clusters, dbl.; bushy growth; [Orléans Rose X ?]; Finch

'MRS R.M. FINCH, CLIMBING', Cl Pol, mp

'MRS R.M. KING', HT, lp; [Mme Abel Chatenay X W.R. Smith]; Harrison, A., 1927; (Weatherly, L.)

'MRS RAMON DE ESCOFET', HT, dr, 1919; flowers flame-crimson, large, dbl.; Easlea

'MRS REDFORD', HT, ab, 1919; flowers bright apricot-orange, semi-dbl., intense fragrance; foliage holly-like; GM, NRS, 1917; McGredy

'MRS REYNOLDS HOLE', T, pb, 1900; flowers carmine shaded purple rose, reverse carmine, large, dbl.; Nabonnand, P.&C.

'MRS RICHARD TURNBULL', LCl, w, 1945; flowers handsome stamens, very large, single; Clark, A.

'MRS RICHARDS', F, pb, 1968; flowers pink tinged apricot, blooms in trusses, single, moderate fragrance; foliage dark, glossy; [Ann Elizabeth X Circus]; Harkness

'Mrs Robert Bacon', HT, ab, 1934; flowers golden apricot shading to coral, base yellow, large, very dbl., cupped, intense fragrance; [Talisman sport]; Bertanzel

'Mrs Robert Garrett', HT, mp, 1900; [Comtesse de Caserta X Mme Eugene Verdier]; Cook, J.W.

'Mrs Robert Mitchell', HT, pb, 1926; flowers salmon-rose, overlaid coppery pink; [St. Helena X Mrs Redford]; Jersey Nursery

'Mrs Robert Peary', Cl HT, w, 1898; [Kaiserin Auguste Viktoria sport]; De Voecht & De Wilde; Dingee & Conard

'Mrs Rosalie Wrinch', LCl, lp, 1915; flowers shell-pink, large, semi-dbl.; vigorous, pillar (5–8 ft) growth; [Frau Karl Druschki X Hugh Dickson]; Brown, W.&J.

'Mrs Roy Green', HT, dr, 1940; flowers very large; long stems; Clark, A.

'Mrs Russell Grimwade', T, mp, 1938; flowers fuchsia-pink; [Lorraine Lee sport]; Grimwade

'Mrs S. Paton', HT, ob, 1928; bud long, pointed; flowers orange-carmine, base orange, large, dbl., slight fragrance; McGredy

Mrs S. Peters, HT, dp

'Mrs S.K. Rindge', HT, yb, 1919; bud long, pointed; flowers deep golden yellow, suffused soft pink with age, over large, semi-dbl., cupped, moderate fragrance; long stems; [Rayon d'Or X Frau Karl Druschki]; H&S

'Mrs S.T. Wright', T, ob, 1914; flowers old gold, center suffused rose-pink on orange, dbl., moderate fragrance; [Harry Kirk sport]; Dickson, A.

'Mrs S.W. Burgess', HT, ab, 1925; flowers apricot-yellow, base deeper, dbl., moderate fragrance; [Mme Mélanie Soupert X Joseph Hill]; Burgess, S.W.

'Mrs Sam McGredy', HT, op, 1929; bud pointed; flowers scarlet-copper-orange, reverse heavily flushed red, large, dbl., 40 petals, exhibition form, moderate fragrance; foliage glossy, reddish bronze; vigorous growth; GM, NRS, 1929 GM, Portland, 1956; [(Donald Macdonald X Golden Emblem) X (Seedling X The Queen Alexandra Rose)]; McGredy

'Mrs Sam McGredy, Climbing', Cl HT, op, 1937; (Geneviève Genest); Buisman, G. A. H.

'Mrs Sam McGredy, Climbing', ClHT, ob, 1938; (Geneviève Genest); Guillaud

'Mrs Sam McGredy, Climbing', ClHT, ob, 1940; Western Rose Co.

'Mrs Spencer Browne', HT, lp; Harrison, A., 1928; (Weatherly, L.)

'Mrs Stewart Clark', HT, mp, 1906; flowers large, dbl., moderate fragrance; Dickson, A.; (Sangerhausen)

'Mrs T. Hillas', HT, my, 1913; flowers pure chrome-yellow, dbl.; Pernet-Ducher

'Mrs T.B. Doxford', HT, rb, 1932; bud pointed; flowers salmon-carmine to peach-blossom-pink, reverse old-rose, large, dbl., exhibition form, slight fragrance; Dickson, A.

'Mrs T.J. English', HT, ab, 1922; flowers apricot and amber, tinted salmon-flesh, heavily veined, moderate fragrance; English

'Mrs Talbot O'Farrell', HT, rb, 1926; flowers cerise flushed bronze, reverse old-gold, dbl., moderate fragrance; McGredy

'Mrs Theodore Roosevelt', HT, lp, 1903; flowers well shaped, very large, dbl., moderate fragrance; vigorous, bushy growth; [La France seedling]; Hill, E.G., Co.

'Mrs Theodore Salvesen', HT, op, 1922; flowers salmon-pink; Dobbie

'Mrs Theonville van Berkel', HT, pb, 1935; bud long, pointed; flowers pink, reverse flushed yellow, large, dbl., exhibition form, intense fragrance; foliage leathery, dark; vigorous growth; [Briarcliff X Mrs Sam McGredy]; Buisman, G. A. H.

'Mrs Tom Paul', HT, yb, 1920; flowers saffron-yellow, suffused pink, dbl.; Dickson, H.

'Mrs Tom Smith', HT, mr, 1924; flowers glowing cerise, dbl., moderate fragrance; Smith, T.

'Mrs Tom Whitehead', HT, ab, 1938; flowers cream, center apricot-orange, outer petals veined, very large, dbl., exhibition form; very vigorous growth; [Mrs Charles Lamplough X Unnamed seedling]; Whitehead; Beckwith

'Mrs Tresham Gilbey', HT, op, 1923; bud pointed; flowers coral-rose shaded salmon, very large, dbl., moderate fragrance; GM, NRS, 1923; [Waltham Flame X Edith Cavell (HT)]; Chaplin Bros.

'Mrs Tresham Gilbey, Climbing', Cl HT, op, 1938; flowers light salmon-pink, very large, dbl., moderate fragrance; Vogel, M.; (Sangerhausen)

'Mrs U.M. Rose', HT, dp, 1931; bud long, pointed; flowers cerise-pink, semi-dbl., cupped, intense fragrance; Vestal

'Mrs van Beresteyn-Frowein', HT, op, 1935; flowers salmon, large, very dbl., moderate fragrance; foliage leathery, light; vigorous growth; [Souv. de Claudius Pernet X Mrs Henry Bowles]; Buisman, G. A. H.

Mrs Van Nes, *see* **'Permanent Wave'**

'Mrs Vandenbergh', HT, mr, 1938; (Mevrouw S. van den Bergh, Jr.); flowers bright red, semi-dbl., exhibition form; foliage dark; strong stems; vigorous growth; [E.G. Hill X Étoile de Hollande]; Buisman, G. A. H.

Mrs Verschuren, *see* **'R.M.S. Queen Mary'**

Mrs Villotti's Pink, Pol, lp

'Mrs W. Ewart', T, lp; [Souv de Catherine Guillot X Unknown]; Williams, A., 1909; (Weatherly, L.)

'Mrs W.A. Lindsay', HT, pb, 1920; flowers peach-pink, center golden yellow, dbl., moderate fragrance; Dickson, H.

'Mrs W.E. Nickerson', HT, pb, 1927; bud pointed; flowers silvery pink deeply shaded old-gold and salmon, large, dbl., moderate fragrance; McGredy; Dreer; H&S

'Mrs W.H. Cutbush', Pol, dp, 1907; flowers deep pink; [Mme Norbert Levavasseur sport]; Levavasseur

'Mrs W.J. Grant', HT, lp, 1895; (Belle Siebrecht); bud long, pointed; flowers light pink fading purplish, dbl., moderate fragrance; [La France X Lady Mary Fitzwilliam]; Dickson, A.

'Mrs W.J. Grant, Climbing', Cl HT, lp, 1899; (Belle Siebrecht, Climbing); Hill, E.G., Co.

'Mrs W.R. Groves', HT, dr, 1941; flowers deep red; foliage good; Clark, A.

'Mrs Wakefield Christie-Miller', HT, pb, 1909; flowers blush, shaded salmon, reverse vermilion-rose, dbl.; foliage light, leathery; dwarf growth; McGredy

'Mrs Wallace H. Rowe', HT, m, 1912; exhibition form; vigorous growth; McGredy

'Mrs Walter Brace', HT, dp, 1939; flowers vivid cerise-rose-pink, slightly larger, of better shape; [Picture sport]; Beckwith

'Mrs Walter Burns', F, mp, 1978; bud ovoid; flowers patio, very dbl., 2.5 in., flat; foliage fragrant (musky), medium, matt, dark; compact, bushy growth; Harkness

'Mrs Walter Easlea', HT, dr, 1910; flowers crimson, large, dbl., moderate fragrance; Dickson, A.

'Mrs Walter Jones', HT, rb, 1930; flowers brilliant coral-red, shaded orange, dbl.; Cant, B. R.

'Mrs Walter T. Sumner', HT, dp, 1920; flowers carmine to deep rose-pink, large, semi-dbl., 12–18 petals, intense fragrance; [Ophelia X Hadley]; Clarke Bros.

'Mrs Walter T. Sumner, Climbing', Cl HT, dp, 1932; Hazlewood Bros.

'Mrs Warren E. Lenon', HT, dr, 1924; bud long, pointed; flowers crimson, large, dbl., globular, intense fragrance; vigorous growth; [Hoosier Beauty X Premier]; Hill, E.G., Co.; Vestal

'Mrs Warren G. Harding', HT, dp, 1923; [Columbia sport]; Pierson, A.N.

'MRS WEMYSS QUIN', HT, dy, 1914; dbl., moderate fragrance; bushy, branching growth; GM, Bagatelle, 1916; Dickson, A.

'MRS WHITMAN CROSS', Cl HT, ob, 1943; bud long, pointed; flowers orange-apricot, overlaid pinkish, reverse sometimes striped, semi-dbl., moderate fragrance; foliage glossy, soft; height 8–9 ft as climbing or pillar, upright.; [Nanjemoy X Marion Cran]; Cross, C.W.; B&A

'MRS WILLIAM C. EGAN', HT, lp, 1922; bud long, pointed; flowers soft pink, dbl., 35 petals, moderate fragrance; H&S

'MRS WILLIAM C. EGAN, CLIMBING', Cl HT, lp, 1933; Howard Rose Co.

'MRS WILLIAM FIFE', HT, pb, 1926; flowers soft rose-pink, flushed blush-pink, dbl., moderate fragrance; Dobbie

'MRS WILLIAM R. HEARST', HT, dp, 1915; flowers clear dark pink; [My Maryland sport]; Pierson, A.N.

'MRS WILLIAM SERGENT', HT, ab, 1923; flowers apricot and peach, edges flushed rose-pink, dbl.; Dickson, H.

'MRS WILLIAM SPROTT', HT, my, 1938; bud long, pointed; flowers large, dbl., slight fragrance; foliage glossy, bronze; vigorous, compact growth; [Portadown Glory X Mrs Sam McGredy]; McGredy

'MRS WM. G. KONING', Pol, w, 1917; flowers pure white, open, borne in clusters, dbl., slight fragrance; vigorous, bushy growth; [Louise Walter sport]; Kluis & Koning

Mrs Woods' Lavender-Pink Noisette, N, m

Ms Mary, *see* 'TINMARY'

Mt Vernon Pompon, LCl, w

Mt Hood, *see* 'MACMOUHOO'

'MT. RAI', HT, w, 1987; flowers white, pale cream at center, aging pale pink at edges, urn-shaped, dbl., 42 petals, exhibition form, slight fragrance; medium, yellow-green fruit; prickles medium, downward-shaped, slightly dented, green; foliage dark green, semi-glossy; bushy, medium growth; [(Golden Scepter X Narcissus) X Ginger Rogers]; Ota, Kaichiro, 1988

'MT. ST HELENS', Gr, dp, 1981; bud long, pointed; flowers deep pink, urn-shaped blooms borne 1–3 per cluster, dbl., 35 petals, slight, spicy fragrance; bronze prickles; foliage dark, leathery; growth to 4–5 ft; [Unnamed seedling X Queen Elizabeth]; Northwest Rose Growers

'MT. TARA', HT, or, 1987; flowers large, borne usually singly, dbl., 35 petals, exhibition form, slight fragrance; medium, yellow-green fruit; prickles medium, downward-shaped, slightly dented, green; foliage dark green, semi-glossy; bushy, medium growth; [Ginger Rogers X Duftzauber]; Ota, Kaichiro, 1988

'MUCABA', HT, pb, 1962; flowers pink stained carmine, [Marechal Carmona X Suzon Lotthe]; da Silva, Moreira

Muchacha®, *see* 'MEILUTIDA'

Mudgee Red, HT, mr; Dot

'MÜHLE HERMSDORF', HWich, w, 1928; flowers pure white, all open at once, dbl.; [R. wichurana X Gruss an Zabern]; Dechant

'MUHME LÖFFLER', HT, dr, 1940; flowers large, dbl., intense fragrance; Krause; (Sangerhausen)

'MULBARTON', LCl, mr, 1965; dbl., 4.5 in., moderate fragrance; foliage coppery; vigorous growth; [Paul's Lemon Pillar X (Ena Harkness X Richmond)]; Hooney

Mulbry Rose, F, ob, 1998

Mullard Jubilee, *see* **'ELECTRON'®**

Mullem, HT, op, 1997; 29 petals, 4 in., cupped, intense fragrance; foliage dense dark green; RvS-Melle, 1996

Multiflora Japonica, *see* **R. MULTIFLORA**

'MULTIFLORA NANA PERPÉTUELLE', Pol, lp, 1893; flowers small, single; Lille; (Sangerhausen)

Multiflore de Vaumarcus, N, lp, 1875; Menet

'MULTNOMAH', HT, rb, 1948; bud long, pointed; flowers carmine, base gold, open, semi-dbl., 18–22 petals, 4–5 in., slight fragrance; foliage glossy, dark; vigorous, upright, compact growth; hardy in Pacific Northwest.; [Contrast X Charlotte Armstrong]; Swim, H.C.; Peterson & Dering

'MUM MUM', Min, w, 1986; flowers pure white, small blooms in clusters of 3–25, dbl., 55 petals, cupped, no fragrance; short, curved, light green prickles; foliage light green, semi-glossy; low, upright, bushy, spreading growth; [Unnamed seedling X Unnamed seedling]; McDaniel, Earl; McDaniel's Min. Roses

'MÜNCHEN', HMsk, dr, 1940; bud long, pointed; flowers scarlet-crimson, large blooms in clusters, semi-dbl., repeat bloom, slight fragrance; foliage dark, glossy; very vigorous, trailing growth; [Eva X Reveil Dijonnais]; Kordes

Munchen 83, *see* 'MEIBALBIKA'®

München Kindl, *see* 'DICKIMONO'

'MÜNCHENER FASCHING', S, mr, 1963; bud ovoid; flowers bright red, blooms in large clusters (up to 40), dbl., 30 petals; foliage dark, glossy; vigorous, bushy ($6–6\frac{1}{2}$ ft) growth; Kordes, R.

'MUNCHKIN', Min, pb, 1986; flowers pink edging, white center, reverse slight pink edged on white, dbl., 22 petals, exhibition form, slight fragrance; short, pointed, small, tan prickles; foliage medium, medium green, semi-glossy; bushy, low growth; [Watercolor X Unnamed seedling]; Bridges, Dennis A., 1987

Munchner Herz®, F, ab, 1990; Cocker

Münchner Kindl, F, op, 1984; flowers large, dbl.; Dickson, Patrick; (Sangerhausen)

Munot®, Gr, or, 1983; Huber

'MUNRO'S IMPROVED PREMIER', HT, dp, 1927; Munro

'MUNSTER', S, pb, 1958; flowers soft pink shaded deeper, blooms in trusses, dbl., 28 petals, 3.5 in., exhibition form, slight fragrance; foliage light green; Kordes; McGredy

Munsterland®, S, pb, 1986; Noack, Werner

'MURASAKI NO SONO', F, m, 1984; flowers silver lilac, blooms borne 1–5 per cluster, semi-dbl., 13 petals, flat, slight fragrance; prickles slanted downward; foliage light green; vigorous, tall growth; [Tasogare X Unnamed seedling]; Kobayashi, Moriji

'MURCA', HT, dr, 1986; (**Plum Cake**); flowers deep plum red, pointed, borne in sprays of 3–7, dbl., 27 petals, almost everblooming, slight fragrance; pointed, brown prickles; foliage medium, dark green, glossy; bushy growth; [(Chanelle X Sabine) X Pompadour]; Murray, Nola

'MURFRI', F, dp, 1986; (**Frilly Dilly**); flowers light magenta red, pointed, small, borne in sprays of 5–7, dbl., 25 petals, slight fragrance; pointed, brown prickles; foliage large, medium green, flat; upright growth; [Red Lion X Magenta]; Murray, Nola

'MURGO', F, yb, 1983; (**Hipango**); flowers deep yellow, petals edged orange, reverse yellow, shapely, large, dbl., 21 petals, slight fragrance; foliage medium, medium to light green; vigorous growth; [Smiley X Una Hawken]; Murray, Nola

'MURHA', HT, mp, 1986; (**Aroha**); flowers medium, soft pink, elongated blooms, dbl., 43 petals, borne in sprays of 3–5, slight fragrance; pointed, brown prickles; foliage medium, light green; tall growth; [Rifleman X Pascali]; Murray, Nola

Muria, *see* 'MACMU'

Muriel, *see* 'HARVOOL'

Muriel, *see* 'MORMURI'

'MURIEL', HT, dr, 1929; bud long, pointed; flowers brilliant velvety scarlet, semi-dbl.; strong stems; Archer

'MURIEL ARMITAGE', HT, dp, 1972; flowers deep rose-pink, very full, very dbl., 60–70 petals, 5 in., moderate fragrance; foliage large, glossy, dark; vigorous, upright growth; [(Orange Sensation X Ballet) X Mischief]; Ellick; Excelsior Roses

'MURIEL DICKSON', HT, ob, 1915; flowers reddish copper; GM, NRS, 1913; Dickson, H.

Muriel Linton, T, dr; Hay

'MURIEL MOORE', HT, w, 1916; [My Maryland sport]; Moore, F.M.

'MURIEL O'LEARY', HT, pb, 1979; bud long; flowers light and deep pink, well-formed, dbl., 27 petals, 3 in., slight fragrance; foliage large, glossy; vigorous, upright growth; [Honey Favorite X Rose Gaujard]; Murray, Nola

'MURIEL PASQUILL', HT, pb, 1927; flowers strawberry-pink, reverse golden yellow, semi-dbl., intense fragrance; [Padre sport]; Pasquill

'MURIEL WILSON', T, ly, 1923; flowers rich lemon-cream, dbl., slight fragrance; GM, NRS, 1921; Hall; Prince

'MURKI', F, pb, 1983; (**Kingi**); flowers yellow flushed rose pink, large, dbl., 35 petals, slight fragrance; large, red prickles; foliage dense, dark; bushy growth; [Liverpool Echo X Una Hawken]; Murray, Nola

Murmure, LCl, 1971; Croix; (Cavriglia)

Murphy's Law, *see* 'SEAMURP'

'MUR-RAY', HT, pb, 1936; flowers darker pink, tinged salmon; [Briarcliff sport]; Murray

'MURRAY HILL', HT, my, 1939; [Joanna Hill sport]; Coddington

'MURSE', HT, lp, 1986; (**Sedgebrook**); flowers very pale pink, almost white, classic form, dbl., 47 petals, slight fragrance; pointed, brown prickles; foliage large, medium green, flat; low growth; [(Chanelle X Prima Ballerina) X Deep Secret]; Murray, Nola

'MURTA', Min, dp, 1983; (**Tasman**); flowers deep pink, small, dbl., 17 petals, flat, no fragrance; foliage medium, dark, leathery; compact growth; [Smiley X ((Pink Parfait X Una Hawken) X Anytime)]; Murray, Nola

'MUSASHINO', HT, op, 1989; bud ovoid; flowers orange-pink, changing to pink, urn-shaped, large, borne usually singly, dbl., 35 petals, exhibition form; few prickles; foliage medium green, oblong, semi-glossy; upright, medium growth; [Garden Party X Daimonji]; Takahashi, Takeshi

Muscosa Japonica, *see* **'MOUSSEUX DU JAPON'**

Muscosa Rubra, *see* **'OLD RED MOSS'**

'MUSCOSA SIMPLEX', M, mp; (R. centifolia andrewsii, R. muscosa simplex); single; (28)

'MUSETTE', F, dr, 1936; flowers glowing light crimson, open, large, borne in clusters, single, slight fragrance; foliage leathery, wrinkled, dark; strong stems; bushy growth; [Ingar Olsson X Johanniszauber]; Tantau

'MUSIC MAKER', S, lp, 1973; flowers medium, dbl., exhibition form, moderate fragrance; foliage glossy, light, leathery; vigorous, dwarf, upright, bushy growth; Buck, Dr Griffith J.; Iowa State University

Music Man, *see* 'BROMAN'

'MUSICALE', Gr, rb, 1964; bud long, pointed; flowers straw-yellow changing to cherry red, small, dbl., moderate fragrance; foliage glossy, bronze; moderate, bushy growth; [(Bravo X Nellie E. Hillock) X Iobelle]; Buck, Dr Griffith J.; Iowa State University

'MUSICIAN', S, rb, 1953; flowers bicolor, nearer red than yellow, with gray tones, borne in s, semi-dbl., 20 petals, non-recurrent; foliage modified rugose; The first bicolor hardy to -50.; [Hansa X Hazeldean]; Wright, Percy H.

Musikantenland, F, mp, 1993; Hetzel

Musimara®, LCl, mr

Musk Rose, *see* **R. MOSCHATA**

'MUSKETEER', HMsk, w; flowers stamens bright yellow, semi-dbl., spring bloom; vigorous (20–25 ft) growth; Lester Rose Gardens

Musketeer, *see* 'WILFOLK'

Musketeer, Gr, op, 1997; Williams, J. Benjamin

Muskoka Moonlight, *see* 'LOUMUSKOKA'

Mustang, *see* **'COLORANJA'**

'MUTABILIS', Ch, yb; (R. chinensis mutabilis, R. mutabilis, Tipo Ideale); flowers sulfur-yellow, changing to orange, red and finally crimson, 2 in.; (14)

Mutace Caramba, HT; (Cavriglia)

'MUTTER BRADA', HT, ab, 1934; flowers apricot-yellow and red, varying, intense fragrance; vigorous growth; [Lady Craig X Freifrau Ida von Schubert]; Brada, Dr.

'MUTTER BRADA II', HT, dp, 1934; flowers large, dbl., intense fragrance; Brada, Dr.; (Sangerhausen)

'MUTTER BRADA III', HT, lp, 1934; flowers large, dbl., slight fragrance; Brada, Dr.; (Sangerhausen)

Muttergruss®, F, dr

Muttertag, *see* **'MOTHERSDAY'**

Muttertag, Climbing, *see* **'MOTHERSDAY, CLIMBING'**

'MY ANGEL', F, dp, 1987; flowers deep pink, medium, borne singly or in sprays of 6–10, dbl., 15–25 petals, slight fragrance; foliage medium, medium green, matt; upright, hardy growth; [Pink Parfait X Roman Holiday]; Pencil, Paul S., 1988

'MY BABY', Min, dp, 1967; bud ovoid; flowers deep pink, small, very dbl., cupped, slight fragrance; foliage small, glossy, dark; moderate, dwarf growth; [Cinderella X R. rouletti]; Quackenbush

'MY CHOICE', HT, pb, 1958; flowers pink, reverse pale yellow, dbl., 33 petals, 4.5–5 in., intense, damask fragrance; foliage leathery; vigorous, upright growth; GM, NRS, 1958 GM, Portland, 1961; [Wellworth X Ena Harkness]; LeGrice

My Delight, *see* 'TINMYDE'

'MY DREAM', HT, dp, 1970; bud long, pointed; flowers deep pink, medium, dbl., exhibition form, slight fragrance; foliage glossy; upright growth; [Pink Favorite X Karl Herbst]; Winchel, Joseph F.; Country Garden Nursery

'MY FAIR LADY', F, pb, 1959; flowers rose-pink, reverse darker, large, borne in large clusters, semi-dbl.; foliage glossy; tall growth; Wheatcroft Bros.

My Fancy, *see* 'MEIFLOREM'

'MY FRIEND', F, lp, 1956; bud ovoid; flowers apple-blossom-pink, dbl., 30 petals, 2.5 in., flat, slight fragrance; dwarf, bushy growth; [Garnette seedling X Summer Snow sport]; Motose

'MY GAL GALE', F, lp, 1958; flowers soft pink, borne in clusters, very dbl.; dwarf growth; [Pinocchio X Morning Star]; Marsh; Marsh's Nursery

'MY GINA', S, mp, 1973; [Dorothy Wheatcroft sport]; Shortland

'MY GIRL', F, ob, 1964; (Cunosa); flowers deep salmon, to open, large, borne in clusters, dbl., 30 petals, cupped; foliage dark; vigorous growth; GM, The Hague, 1963; [Dacapo X Floribunda seedling]; deRuiter; Gregory

My Granny, S, mp, 1991; Poulsen

'MY GUY', F, or, 1986; flowers bright orange-red, fading darker, medium, borne in sprays of, semi-dbl., 13 petals, cupped, moderate fragrance; round, medium, light, light green fruit; medium, light brown, slightly hooked prickles; foliage medium, medium green, semi-glossy; bushy, medium growth; [Rosalynn Carter X Dorothy Wheatcroft]; Milner, William, 1979

My Honey, *see* 'JUSHONEY'

'MY JOY', HT, mp, 1976; [Red Devil sport]; Wood

'MY LADY', HT, ab, 1956; bud pointed; flowers apricot flushed gold, dbl., 46 petals, 5 in., exhibition form, moderate fragrance; foliage dark, leathery; very vigorous, bushy growth; [Seedling X Peace]; Robinson, H.

My Little Boy, Min, my, 1982; Burrows, Steven

My Love™, *see* 'COGAMO'

'MY LOVE', HT, dr, 1960; flowers deep red, large, dbl., 45 petals, exhibition form, intense fragrance; foliage dark; vigorous growth; [Bayadere X Ena Harkness]; Anderson's Rose Nurseries

My Love, Climbing, Cl HT, dr

'MY LUCKY STARR', HT, mr, 1987; flowers large, dbl., 26–40 petals, no fragrance; foliage large, dark green, semi-glossy, leathery; spreading growth; [Secret Love X Karl Herbst]; Evans, F. David

'MY MARYLAND', HT, pb, 1908; flowers bright salmon-pink, edged paler, large, dbl., intense fragrance; vigorous growth; [Madonna X Enchanter]; Cook, J.W.

My Own, *see* 'COOSYL'

My Pleasure™, *see* 'KINPLEASE'

'MY PRAYER', HT, op, 1953; bud long, pointed; flowers peach-pink, dbl., 25–30

petals, 5 in., exhibition form, intense fragrance; foliage leathery; vigorous, upright growth; [Seedling X The Doctor]; H&S

My Sunshine, *see* 'TINSHINE'

My Valentine™, *see* 'MORMYVAL'

My Way, HT, w, 1988; Simpson

'MYFANWY WELBOURNE', HT, pb, 1999; dbl., 26–40 petals, 4.5 in., borne mostly singly, intense fragrance; prickles moderate; foliage medium, dark green, semi-glossy; upright, bushy, medium (3 ft) growth; [(Hazel Rose X Cardiff Bay) X Darling Jenny]; Poole, Lionel; David Lister Roses, 2000

'MYRA', LCl, w, 1926; flowers creamy white, very large, borne in clusters of 2–3, dbl., slight fragrance; foliage dark, bronze; vigorous, climbing growth; RULED EXTINCT 1/92; [Dr W. Van Fleet X Lady Roberts]; Wilber

Myra, *see* 'BATTOO'

Myra, HT, pb

Myra Stegmann, Gr, op, 1991

Myriam®, *see* 'COCGRAND'

'MYRIANTHES RENONCULE', HSem, pb; flowers pale peach, large, dbl., free bloom; not dependably hardy.

Myrna's Dream, *see* KORdrekes

Myrniong, HT, lp; Bell, Ronald J., 1994; (Weatherly, L.)

Myrrh-scented Rose, *see* **'SPLENDOUS'**

'MYSTÈRE', HT, dp, 1969; bud long; flowers scarlet pink, dbl., 45 petals; RULED EXTINCT 10/86; [Premiere Ballerine X Femina seedling]; Gaujard

Mysterious™, *see* 'MINIMYS'

'MYSTERIUM', F, yb, 1963; bud long, pointed; flowers golden yellow striped scarlet, borne in clusters (up to 20), dbl., 25 petals, 3 in., slight fragrance; foliage glossy; bushy, spreading, low growth; [Masquerade X Kordes' Perfecta]; Kordes, R.

Mystery, *see* 'HERMESY'

Mystic, *see* POUlor

Mystic, HT, m; Dot

'MYSTIC GEM', HT, dp, 1959; bud long, pointed; flowers reddish pink, large to medium, dbl., exhibition form, slight fragrance; foliage leathery; vigorous, upright growth; [Rod Stillman X Bravo]; Armbrust; Langbecker

Mystic Mauve, *see* **'LAVENDER MIST'**

Mystic Meidiland®, *see* 'MEIALATE'

Mystique™, *see* 'DEVSTICA'

Mystique, HT, lp; [Mysty sport]; Neil, J., 1989; (Weatherly, L.)

N

N S J Pink Moss, M, dp

N. L. Chrestensen, Pol, or, 1968; flowers medium, semi-dbl.; GPG Bad Langensalza; (Sangerhausen)

'NAARDEN RED', HT, rb, 1932; bud pointed; flowers crimson-red shaded orange, large, dbl., exhibition form, intense fragrance; [Étoile de Hollande X Charles P. Kilham]; Van Rossem

Naas Botha, *see* TANrosilb

Nabab, F, 1959; Delbard; (Cavriglia)

'NACHA POBEDA', HT, dr; (Our Victory); flowers dark velvety red; Costetske

Nachez, *see* 'POULLEN'

Nachsommer, LCl, mp, 1990; flowers large, dbl.; Wänninger, Franz; (Sangerhausen)

Nachtfalter, HT, dr, 1971; flowers large, dbl., intense fragrance; Baum, Oswald; (Sangerhausen)

'NADIA', HT, dr, 1956; foliage light green; [Mme G. Forest-Colcombet X Seedling]; Delforge

'NADINE', F, dr, 1962; bud long, pointed; flowers maroon-red, dbl., 38 petals, 3 in., cupped, intense fragrance; foliage bronze, soft; vigorous, bushy growth; [Red Pinocchio X Unnamed seedling]; Schwartz, Ernest W.; Wyant

Nadine, HT, ob; Croix

Nadja, HT, mr, 1964; flowers large, dbl.; Rupprecht-Radke; (Sangerhausen)

Naema, LCl, lp, 1998; Delbard

Naia, F; Gaujard; (Cavriglia)

Naina, HT, lp, 1998; Kordes

'NAÏR', HFt, yb, 1936; bud long, yellow and carmine; flowers vermilion-red, reverse yellow, stamens yellow, large, semi-dbl., recurrent bloom; vigorous, bushy growth; [R. foetida bicolor X R. wichurana seedling]; Chambard, C.

'NAKBET', HT, my, 1985; (**Golden Phoenix**); [Bettina sport]; Nakashima, Tosh

'NAMBOUR', F, w, 1953; flowers white flushed pink, medium, borne in clusters, very dbl.; foliage bronze; very vigorous growth; [Yvonne Rabier X Mrs Tom Henderson]; Ulrick, L.W.

'NAMENLOSE SCHÖNE', T, w, 1886; flowers white tinted blush, well formed, large, moderate fragrance; Deegen

Namib Sunrise, *see* 'MORLOGEN'

'NAN ANDERSON', F, op, 1970; flowers deep pink, coral sheen, dbl., 30 petals, 2.5–3 in., slight fragrance; foliage glossy, dark; low, bushy growth; Anderson's Rose Nurseries

'NAN POOLE', Min, pb, 1983; bud globular; flowers yellow and pink blend, giving an overall coral-pink effect, dbl., 25 petals, exhibition form, moderate fragrance; thin prickles; foliage small, dark, semi-glossy; low, spreading growth; [Unnamed seedling X Libby]; Meredith, E.A. & Rovinski, M.E.; Casa de Rosa Domingo

'NANA MOUSKOURI', F, w, 1975; flowers well-formed, dbl., 30 petals, 2.5 in., moderate fragrance; [Redgold X Iced Ginger]; Dickson, A.

'NANCY', F, mr; flowers bright red, very small, single; foliage very glossy

'NANCY', HT, dr, 1930; flowers bright scarlet-crimson, semi-dbl., slight fragrance; Ferguson, W.

'NANCY', HT, my, 1934; flowers chamois-yellow, semi-dbl., cupped, slight fragrance; [Mrs T. Hillas X Souv. de Claudius Pernet]; Mallerin, C.; H. Guillot

Nancy, F, or; Croix

Nancy Bennett, *see* 'BOSPARDON'

'NANCY BERGH', F, pb, 1968; bud ovoid; flowers pink, reverse silver-pink, open, large, semi-dbl., intense fragrance; foliage dark, glossy, leathery; vigorous, open growth; [Independence X Impeccable]; Fankhauser; A. Ross & Son

Nancy Clare, *see* 'JUSCLARE'

'NANCY ELIZABETH', HT, w, 1947; bud long, pointed; flowers cream, large, dbl., exhibition form, moderate fragrance; foliage glossy; vigorous, upright growth; [Korovo X Florinda Norman Thompson]; Mason, F.

Nancy Gardiner, *see* KORkeindor

'NANCY HALL', Min, pb, 1972; flowers peachy; [Mary Adair sport]; Moore, Ralph S.; Mini-Roses

'NANCY HAYWARD', LCl, mr, 1937; flowers rich bright cerise, large, single; very vigorous, climbing growth; [Jessie Clark X ?]; Clark, A.; NRS Victoria

'NANCY LEE', HT, pb, 1879; flowers deep pink, intense fragrance; weak growth; [Mme Bravy X Edward Morren]; Bennett

'NANCY PRETTY', HWich, pb, 1917; flowers pink, reverse lighter, small, borne in clusters, dbl.; [Dorothy Perkins X Ellen Poulsen]; MacLellan

'NANCY REAGAN', HT, or, 1967; flowers orange-scarlet, dbl., 28 petals, 5–5.5 in., exhibition form, intense fragrance; foliage dark, bronze, glossy, leathery; vigorous, tall growth; [Orange Delight X Hawaii]; Morey, Dr Dennison; Country Garden Nursery

Nancy Shaw, *see* 'WIWANCY'

'NANCY STEEN', F, pb, 1976; flowers blush pink, center pale cream, dbl., 30 petals, 3.5 in., flat, moderate fragrance; foliage glossy, dark, bronze, leathery; [Pink Parfait X (Ophelia X Parkdirektor Riggers)]; Sherwood; F. Mason

'NANCY WEST', F, yb, 1970; flowers medium yellow, suffused peach; [Elizabeth of Glamis sport]; Haynes

'NANCY WILSON', HT, mp, 1940; Clark, A.

'NANCY WILSON, CLIMBING', Cl HT, mp, 1959; Campton; Hazlewood Bros.

'NANCY'S KEEPSAKE', HT, lp, 1995; flowers pale pink, blooms borne mostly single, dbl., 26–40 petals, 1.5–2.75 in., slight fragrance; some prickles; foliage medium, medium green, glossy; medium ($2\frac{1}{2}$ ft), upright growth; [Keepsake sport]; Grierson, Mrs. Nancy; Battersby Roses, 1995

'NANDA', HT, mp, 1959; bud ovoid to urn shaped, claret-red; flowers light rose, large; vigorous growth; [Dame Edith Helen X Eternal Youth]; Sartore

'NANDINI', HT, pb, 1983; bud long, pointed; flowers pink with creamy white reverse, blooms borne singly, dbl., 46 petals, exhibition form, intense fragrance; brown-gray, hooked prickles; foliage large, medium green, leathery; compact growth; [Kiss of Fire X ?]; Pal, Dr B.P.; Laveena Roses

'NANETTE', LCl, w, 1926; flowers creamy white, large, borne in clusters, dbl.; Hicks

'NANJEMOY', Cl HT, mp, 1937; bud long, pointed; flowers pink, open, large, semi-dbl., free, intermittent bloom; foliage dark; strong stems; vigorous, climbing growth; [Mme Gregoire Staechelin X Bloomfield Comet]; Cross, C.W.

Nankin, *see* **'MME YVES LATIEULE'**

'NANO NAGLE', HT, dr, 1997; flowers full, medium, very dbl., 26–40 petals, borne mostly singly, slight fragrance; foliage medium, medium green, semi-glossy; medium (114 cms) upright growth; [Papa Meilland X Northern Lights]; Xavier, Sister M.

'NANTUCKET', HT, ab, 1972; flowers peach apricot, large, dbl., exhibition form, slight fragrance; foliage leathery; vigorous, upright growth; [Chantré sport]; Kern Rose Nursery

'NAOMI', HT, ob, 1926; flowers coppery buff, dbl., moderate fragrance; RULED EXTINCT 1/88; Pemberton

'NAOMI', HT, dr, 1988; flowers dark red, shaded very dark in center, fading to red-purple, dbl., 32 petals, exhibition form, slight fragrance; prickles flat, medium, medium green; foliage large, dark green, semi-glossy; tall, spreading, upright growth; [Red Lion X Unnamed seedling]; Poole, Lionel, 1989

Naomi, HT, pb, 1998

'NAOMI REBECCA', HT, pb, 1996; flowers deep pink, reverse pale pink, large blooms borne mostly singly, dbl., 26–40 petals, slight fragrance; some prickles; foliage large, dark green, dull; upright, bushy, tall growth; [Solitaire X Gavotte]; Poole, Lionel

Napa Valley, *see* 'POULINO'

Napoleon, HCh, pb, 1835; Laffay, M.

'NAPOLÉON', HGal, m; flowers bright rose, shaded purple, very large, dbl.; erect, vigorous growth; Hardy, before 1814

'NARCISSE', T, my, 1859; (Enfant de lyon); flowers large, dbl.; Avoux

'NARCISSE DE SALVANDY', HGal, mp; flowers deep rose-pink, prominent yellow stamens, large blooms in clusters, flat; large, spreading growth; Van Houtte, before 1847

Nardy, N, dy, 1888; Nabonnand

'NARITA KORINKAKU', HT, dy, 1978; bud pointed; flowers light orange, dbl., 35–40 petals, 5–6 in., exhibition form, intense fragrance; foliage glossy, dark; upright growth; [Burnaby X Montparnasse]; Kikuchi, Rikichi

'NARMADA LAHARI', HT, pb, 1979; bud ovoid; very dbl., 90 petals, borne singly, moderate fragrance; beak-shaped prickles; foliage small, green; bushy, dwarf growth; [Shree Dayananda sport]; Hardikar, Dr M. N.

'NARRE FRAGRANCE', HT, ob, 1942; flowers golden orange; [Portadown Fragrance sport]; Brundrett

'NARRE PEACE', HT, yb, 1960; [Peace sport]; Brundrett

'NARROW WATER', N, lp; flowers darker pink; Daisy Hill Nursery, 1883

Nartaki, Pol, m, 1977; Kasturi

'NARVIK', HT, op, 1960; bud long, pointed; flowers salmon-pink to coppery, large, dbl., slight fragrance; foliage leathery; long, strong stems; vigorous growth; [Unnamed seedling X Praline]; Robichon

'NARZISSE', HT, my, 1942; bud long, pointed; flowers apricot to maize-yellow, large, dbl., 23 petals, exhibition form, moderate fragrance; foliage dark, leathery; vigorous, upright growth; [Golden Rapture X Golden Glory]; Krause; C-P

'NASCAPEE', S, w, 1946; flowers open, single, 5 petals, free, recurrent boom, slight fragrance; tall, vigorous growth; hardy; [(Ross Rambler X (R. rugosa X R. eglanteria)) X ?]; Preston; Central Exp. Farm

'NASE NÁRODNI', HP, m, 1935; flowers purple/pink with white, medium, dbl., intense fragrance; Böhm, J.; (Sangerhausen)

Naseby Rose, HGal, m

'NASTARANA', N, w, 1879; (Persian Musk Rose, R. moschata nastarana, R. pissartii); flowers white tinged pink, blooms in clusters on new wood, semi-dbl., 2 in., recurrent bloom; very vigorous growth; (14); [Probably an early R. chinensis X R. moschata hybrid]

Natacha, *see* **'CARA MIA'**

Natali®, *see* 'TANROTREILI'

Natali, *see* 'TANNALI'

Natalie, Pol, ob, 1972; flowers orange to orange-red, with dark yellow, medium, semi-dbl.; GPG Bad Langensalza; (Sangerhausen)

'NATALIE BOETTNER', HT, my, 1909; flowers sulfur-yellow, passing to cream-yellow, tinted flesh, dbl.; [Frau Karl Druschki X Goldelse]; Boettner

'NATALIE WARD', F, mp, 1974; bud ht type; flowers pink, base yellow, dbl., 20 petals, 2–2.5 in., bloom repeats quickly, slight fragrance; foliage light; bushy growth; [Pink Parfait X ?]; Thomas, R.; Pattollo's Nursery

'NATALKA', HT, lp; Costetske

Natascha, *see* 'HARWHARRY'

'NATASHA', F, my, 1974; flowers lemon yellow, fading white, large blooms in clusters of 7–13, very dbl., 135 petals, foliage dark, glossy; bushy growth; [Highlight X Masquerade]; Staikov, Prof. Dr V.; Kalaydjiev and Chorbadjiiski

'NATASHA MARIA', F, or, 1995; flowers bright, intense, vibrant orange-red, yellow base, blooms bor, semi-dbl., 6–14 petals, 1.5–2.75 in., moderate fragrance; foliage medium, dark green, glossy; 60–80 cms growth; [Wilfrid H. Perron X Golden Olymp]; Fleming, Joyce L.; Hortico Roses, 1994

'NATASHA MONET', HT, m, 1993; flowers very light lavender, dbl., 26–40 petals, 3–3.5 in., borne mostly singly, slight fragrance; some prickles; foliage medium, medium green, semi-glossy; tall, upright, bushy growth; [Crystalline sport]; Pawlikowski, Martin & Elaine

Natchez, *see* 'POULLEN'

Natchitoches Noisette, N, lp

Nathalie, S, mp

Nathalie Nypels, *see* **'MEVROUW NATHALIE NYPELS'**

Nathalie's High Hills, HCh, dp

Natilda, *see* 'MEIDRAGELAC'

'NATIONAL BEAUTY'™, HT, mr, 1983; [The Alamo sport]; Burks, Joe J.; J.B. Williams & Associates

'NATIONAL EMBLEM', HT, dr, 1915; flowers velvety dark crimson, edged vermilion, dbl., exhibition form, moderate fragrance; McGredy

'NATIONAL FLOWER GUILD', HT, mr, 1927; flowers pure scarlet-red, large, dbl.; long stems; very vigorous growth; [(Capt. F. Bald X K. of K.) X Mme Van de Voorde]; Mallerin, C.; C-P, 1930

'NATIONAL TRUST', HT, dr, 1970; (Bad Nauheim, Bad Naukeim); flowers bright red, classic form, dbl., 53 petals, 4 in.; [Evelyn Fison X King of Hearts]; McGredy, Sam IV; McGredy

National Velvet™, *see* 'BURALP'

Nationalstolz, HT, mr, 1970; flowers large, dbl.; McGredy, Sam IV; (Sangerhausen)

'NATIVE WEDDING', S, lp, 1979; bud short, pointed; semi-dbl., 14 petals, cupped, borne singly, repeat bloom, slight fragrance; straight prickles; foliage broad, waved, medium green, semi-glossy, 5–7 leaflet; upright, medium growth; [Restless Native X (Mount Shasta X R. suffulta)]; Stoddard, Louis

Natural Beauty, *see* 'ROGSCRIV'

Naughty But Nice, *see* 'TINNAUGHTY'

'NAUGHTY NANCY', F, pb, 1970; flowers cream to red, semi-dbl., 12 petals, 2 in., slight fragrance; foliage dull, matt green; moderate growth; Cants of Colchester, Ltd.

Naughty Patricia, *see* 'BILPAT'

'NAUTILUS', F, op, 1960; flowers coral-salmon, open, borne in clusters, dbl., 30 petals, slight fragrance; vigorous growth; [Signal Red X Fashion]; deRuiter

'NAVAJO', HT, dp, 1958; bud long, pointed; flowers dark red to rose-red, dbl., 35–40 petals, 4.5–5 in., exhibition form, intense fragrance; foliage leathery; vigorous, upright growth; [Hortulanus Budde X E.G. Hill]; Malandrone; J&P

Navarro Rambler, LCl, pb

Navarro Ridge, N, lp

'NAVID', F, or, 1985; flowers patio, medium, semi-dbl., slight fragrance; foliage medium, medium green, semi-glossy; bushy growth; [Roydon Hall X Trumpeter]; Payne, A.J.

'NAVIGATOR', LCl, mp, 1925; bud globular; flowers radiance pink, large, dbl., 25 petals, cupped; foliage soft; long stems; very vigorous, climbing growth; [R. soulieana X Radiance]; Verhalen; Verhalen Nursery Co., 1943

'NAV-SADABAHAR', F, pb, 1980; bud pointed; flowers deep pink striped white, blooms borne 15 per cluster, dbl., 20 petals, no fragrance; straight, pink to brown prickles; foliage medium, green; medium, spreading, bushy growth; [Sadabahar sport]; Division of Vegetable Crops and Floriculture

'NAYIKA', HT, pb, 1975; bud pointed; flowers delft-rose, reverse and base darker, very dbl., 45 petals, 4 in., exhibition form, slight, tea fragrance; foliage leathery; moderate, upright, bushy growth; Pal, Dr B.P.; Anand Roses

'NAZNEEN', HT, lp, 1969; bud ovoid; flowers very soft pink, large, very dbl., exhibition form, moderate fragrance; foliage light green, glossy; vigorous, upright growth; [Queen Elizabeth X ?]; Pal, Dr B.P.; Son

'NAZR-E-NAZAR', HT, mp, 1968; bud pointed; flowers light pink edges flushed darker, large, dbl., exhibition form, slight fragrance; foliage leathery; vigorous, upright, compact growth; [Clovelly X ?]; Singh

NDR I Radio Niedersachsen, *see* KOReledas

'NE PLUS ULTRA', N, w; flowers creamy white, moderate fragrance

'NÉALA', HGal, dp, 1822; flowers deep rose edged lighter, medium, dbl.; Vibert, 1845 (?)

Near You, *see* 'SEANEAR'

Nearly Black, HT, dr

'NEARLY WILD', F, mp, 1941; bud small, long, pointed; flowers rose-pink, single, white eye, single, 5 petals, moderate, sweet fragrance; bushy growth, 2–3 ft; [Dr W. Van Fleet X Leuchtstern]; Brownell, H.C.

Nearly Wild Gallica, HGal, dp

'NEARLY WILD, CLIMBING', Cl F, mp, 1962; Burks; Co-Operative Rose Growers

'NÉBULEUSE'®, F, dr, 1971; flowers deep crimson, medium, dbl.; vigorous growth; [Ritz (F) X Lilli Marleen]; Gaujard

Nedbank Rose, *see* DELverjaune

'NEDERLAND', HT, dr, 1919; flowers deep glowing red, dbl., 60 petals; [General-Superior Arnold Janssen X George C. Waud]; Verschuren

Neela, HT, m, 1989; K&S

Neelakanti, F, m, 1993; Chiplunkar

'NEELAMABARI', F, dr, 1975; flowers deep red, large, dbl., 35 petals; foliage dark, glossy; vigorous, compact, bushy growth; [Blue Moon X Africa Star]; IARI

Neena, HT, lp, 1990; Bulsara

'NEERVELT', Cl HT, mr, 1910; flowers carmine-red, large, dbl.; Verschuren; (Sangerhausen)

Nefertiti, *see* 'VIRAMBER'

Negridte, F, dr, 1972; flowers medium, dbl.; Urban, J.; (Sangerhausen)

'NEHRU CENTENARY', HT, dr, 1989; bud pointed; flowers dark red, reverse deeper, large, borne singly, 60 petals, exhibition form, slight fragrance; prickles hooked, brown; foliage very large, dark green, dense; tall, upright growth; [Christian Dior X Avon]; IARI

'NEIGE D'AVRIL', HMult, w, 1908; flowers pure white, stamens yellow, nearly large, borne in pyramidal, dbl., early, seasonal bloom; very vigorous, climbing (to 8 ft) growth; Robichon

Neige de Printemps, HMsk, w, 1991; Lens

Neige d'Ete, HMsk, w, 1991; flowers creamy white, medium-large, semi-dbl., moderate fragrance; Lens, Louis; (Sangerhausen)

Neige d'Ete, *see* 'TANOTAX'

'NEIGE PARFUM', HT, w, 1942; flowers white, sometimes tinted cream, large, dbl., intense fragrance; foliage leathery; vigorous growth; [Joanna Hill X (White Ophelia X Unnamed seedling)]; Mallerin, C.; A. Meilland; J & P

'NEIGE ROSE', LCl, pb, 1955; flowers center deep pink, becoming lighter at petal edges, large, slight fragrance; very vigorous growth; Delbard-Chabert

'NEIGES D'ÉTÉ', Min, w, 1984; (Rigobec 2, Summer's Snow); flowers small, dbl., 37 petals, no fragrance; foliage medium, medium green, glossy; upright growth; [Unnamed Baby Masquerade sport X Baby Masquerade seedling]; Gailloux, Gilles

Neisse, S, pb, 1959; flowers yellowish-pink, medium, dbl.; Berger, W.; (Sangerhausen)

'NEJENKA', F, m, 1955; (Tender One); flowers purplish pink, large, dbl., 57 petals, moderate fragrance; foliage dark, glossy; upright growth; Klimenko, V. N.

Nelie Niel, HT, w

Nelkenrose, *see* **'F. J. GROOTENDORST'**

'NELL GWYN', HT, or, 1968; flowers orange-copper, globular, moderate fragrance; upright growth; [Tzigane sport]; Cobley; Blaby Rose Gardens

Nella Martinetti®, HT, mp, 1990; Meilland

'NELLIE CHARLTON', HT, mp, 1923; flowers silvery pink, reverse salmon-pink, dbl., intense fragrance; [Mme Abel Chatenay seedling]; Lilley

'NELLIE E. HILLOCK', HT, mp, 1934; flowers silvery pink, base deep gold, reverse old-rose, peony form, dbl., 60 petals, cupped, moderate fragrance; foliage leathery, dark; low, spreading growth; [Golden Dawn X Unnamed seedling]; Hillock

'NELLIE E. HILLOCK, CLIMBING', Cl HT, mp, 1948; Buck, Dr Griffith J.; Lester Rose Gardens

'NELLIE MAUD POWELL', HT, yb, 1977; flowers yellow edged red, very dbl., 75–80 petals, 4–5 in., intense fragrance; foliage glossy; vigorous growth; [Columbine sport X Kordes' Perfecta]; Powell

Nellie Niel, HT, w, 1975; Dawson

'NELLIE PARKER', HT, w, 1916; flowers creamy white, center darker, well shaped, large, dbl.; vigorous, upright growth; GM, NRS, 1916; Dickson, H.

'NELLY CUSTIS', HMsk, w; dbl., borne in clusters of 7–9, moderate fragrance; vigorous, upright growth

'NELLY VERSCHUREN', HT, my, 1918; flowers clear yellow, intense fragrance; [Unnamed seedling X Duchess of Wellington]; Verschuren

'NÉMÉSIS', Ch, dr, 1836; flowers purplish crimson, pompon form, dbl.; dwarf growth; (14); Bizard

'NEMLOVE', S, mp, 1999; (**Love You**); .75–1 in., borne in large clusters, slight fragrance; prickles moderate; foliage medium, medium green, glossy; bushy, low (2 ft) growth; Nemko, Martin; Bailey Nurseries, 1999

'NÉNETTE LEYDIER', Pol, mr, 1924; flowers crimson-scarlet, base lighter, Richardier

'NENIKUJAKU', F, pb, 1984; flowers pink, blended with yellow and red, blooms in clusters, semi-dbl., 15 petals, cupped, moderate fragrance; vigorous growth; [Masquerade X Matador]; Kikuchi, Rikichi

'NENITA', F, w, 1962; flowers large, dbl., 22 petals, moderate fragrance; foliage light green; vigorous, low growth; [Seedling X Virgo]; da Silva, Moreira

'NEON', F, ob, 1971; flowers intense orange, dbl., 26 petals, 2 in.; compact growth; Waterhouse Nursery

'NEON', HT, mr, 1936; flowers crimson-scarlet, large, very dbl., moderate fragrance; vigorous, branching growth; Nicolas; Beckwith

Neon, *see* KORenon

Neon Lights, *see* 'JACOUT'

'NEPTUNIA', HT, pb, 1998; flowers pink blending to white, classic, very large, very dbl., 26–40 petals, 6 in., exhibition form, borne mostly singly, slight fragrance; some prickles; foliage large, dark green, semi-glossy; upright, medium growth; [(Precious Platinum X Silver Jubilee) X Seedling]; Poole, Lionel

Nerene, *see* MALren

'NERISSA', HT, ly, 1912; flowers cream-yellow, shaded white, dbl., slight fragrance; Paul, W.

'NÉRON', HGal, rb, 1841; flowers crimson, blotched and marbled violet; Laffay, M.

Nerone, HT; Cazzaniga, F. G.; (Cavriglia)

Nervi, HT; Mansuino; (Cavriglia)

'NESTOR', HGal, mr; flowers crimson, large, dbl.; very vigorous growth; Vibert, 1834

'NESTOR BOLDERDIJK', HT, pb, 1938; bud long, pointed; flowers pale ecru, reverse yellowish salmon, base golden, very large, dbl.; foliage glossy; vigorous growth; [Comtesse Vandal X Pres. Macia]; Leenders, M.

Netersen, *see* 'KORTERSEN'

Netravathy, F, mp, 1975; Kasturi

'NETUJOH', HT, dr, 1999; flowers deep red, dbl., 30–35 petals, 4–5 in., exhibition form, slight fragrance; foliage dark green, semi-glossy; upright (3–4 ft) growth; GM, Japan Rose Concours, 1993; Hirabayashi, Hiroshi, 1991; Keisei Rose Nurseries, Inc., 1993

Netujou, HT, dr

Neue Revue®, *see* 'KORREV'

Neues Europa, *see* **'NOUVELLE EUROPE'**

Neus®, *see* FEsaru

'NEUTRON', (form of R. rugosa), m, 1984; flowers purple, shallow- blooms borne singly or few per cluster, semi-dbl., 10 petals, 3 in., cupped, repeat bloom, intense fragrance; straight, gray prickles; foliage thick, rugose, shining, dark; upright, dense growth; [Grown from neutron-irradiated seed of R. rugosa]; Lundstad, Arne; Agricultural University of Norway; (grown from irradiated seed)

'NEVADA', HMoy, w, 1927; bud ovoid, pink or apricot; flowers white, reverse sometimes splashed carmine, large blooms on s, single, heavy bloom usually repeated, vigorous (7 ft), shrubby growth; (28); Dot, Pedro

Nevertheless, *see* 'HENNEV'

Never Forgotten, *see* 'GREGART'

'NEVILLE CHAMBERLAIN', HT, ob, 1940; bud ovoid; flowers salmon, center orange, dbl., 26 petals, 4 in., exhibition form; foliage bronze; vigorous, tall growth; GM, Portland, 1941; [Charles P. Kilham X Mrs Sam McGredy]; Lens; J&P

Neville Gibson, *see* 'HARPORTLY'

Nevis Moss, M, mp

New Adventure, *see* 'MORSHEFRAN'

New Antique, *see* 'BRONEW'

New Arrival, MinFl, or, 1998

New Ave Maria, HT, ob

New Beginning™, *see* 'SAVABEG'

'NEW CASTLE', Min, rb, 1984; flowers red, white reverse, small, dbl., 20 petals, no fragrance; foliage medium, dark, semi-glossy; upright growth; [Watercolor X Unnamed seedling]; Bridges, Dennis A.

'NEW CENTURY', HRg, pb, 1900; flowers flesh-pink, center light red, edges creamy, dbl., intermittent bloom, moderate fragrance; foliage wrinkled, light, tough; vigorous (4–5 ft), bushy growth; [R. rugosa alba X Clotilde Soupert]; Van Fleet; Conard & Jones

'NEW COLUMBIA', HT, mp, 1924; flowers true pink, deepening to glowing pink; [Columbia sport]; Hill, E.G., Co.

'NEW DAILY MAIL', F, dr, 1972; (Pussta); bud globular; flowers large, semi-dbl.; vigorous, upright, bushy growth; [Letkis X Walzertraum]; Tantau, Math.

'NEW DAILY MAIL, CLIMBING', Cl F, dr, 1989; (Pussta, Climbing); [New Daily Mail sport]; Patil, B.K.; K.S.G. Son's Roses, 1987

'NEW DAWN', LCl, lp, 1930; (Everblooming Dr W. Van Fleet, The New Dawn); flowers same as parent except repeat blooming, very dbl., 35–40 petals, borne singly and in clusters, moderate, sweet fragrance; foliage medium, dark green; climbing to 20 ft growth; Hall of Fame, WFRS, 1997; [Dr W. Van Fleet sport]; Somerset Rose Nursery; Somerset Rose Nursery, 1930

New Dawn Rouge, *see* **'ÉTENDARD'**

New Day, *see* 'KORGOLD'

New Europe, *see* **'NOUVELLE EUROPE'**

New Face®, *see* 'INTERCLEM'

'NEW GOLD', Min, dy, 1977; bud long, pointed; flowers buttercup-yellow, open, dbl., 20 petals, 2–2.5 in., intense fragrance; foliage small, very dark; vigorous, upright growth; [Yellow Jewel X Allgold]; Lyon; L. Lyon Greenhouses

'NEW HAVEN QUEEN', HT, ob, 1939; bud pointed; flowers tangerine, dbl., 40 petals, 3.5 in., moderate fragrance; foliage leathery; vigorous growth; [Token sport]; Grillo

'NEW HOPE', Min, w, 1988; bud pointed; flowers creamy white, slight pink edge, medium, borne usually singly, exhibition form; prickles straight, pointed, medium, medium red; foliage medium, dark green, semi-glossy; bushy, medium growth; [Party Girl X Unnamed seedling]; Bridges, Dennis A.; Bridges Roses, 1989

New Horizon, *see* DICplay

'NEW LOOK', F, rb, 1961; flowers maroon, reverse silver, blooms in clusters, dbl., 50 petals, 3 in., moderate fragrance; foliage glossy, coppery; vigorous, bushy growth; [Charles Gregory seedling X Orange Triumph seedling]; Gaujard; Gandy Roses, Ltd.

'NEW LOVE', HT, rb, 1968; flowers cardinal-red, reverse golden yellow, large, dbl., exhibition form; foliage glossy, bronze; vigorous, bushy growth; [South Seas X Coronado]; Morey, Dr Dennison; Country Garden Nursery

'NEW MEXICO', HT, mr, 1967; flowers medium, dbl., exhibition form, moderate fragrance; foliage bronze, leathery; very vigorous growth; [Mount Shasta X Granada]; Aufill

'NEW MOONLIGHT', HT, my, 1935; flowers very large, dbl.; foliage glossy; long stems; very vigorous growth; [Sun Gold X Joseph Hill]; Elmer's Nursery

New Orange, Min, ob

'NEW ORLEANS', HT, yb, 1966; bud ovoid; flowers yellow, veined light red, large, dbl., globular, moderate fragrance; foliage glossy; vigorous, upright growth; RULED EXTINCT 2/87; [Peace sport]; Tate

New Orleans™, *see* 'KINNOR'

New Peace™, *see* 'ARONEWP'

'NEW PENNY', Min, or, 1962; bud short, pointed; flowers orange-red to coral-pink, small, dbl., 20 petals, 1.5 in., moderate fragrance; foliage leathery, glossy; bushy, dwarf (10 in.) growth; [(R. wichurana X Floradora) X Unnamed seedling]; Moore, Ralph S.; Sequoia Nursery

'NEW PLANET', HT, MP, 1930; flowers bright rose-pink, center light pink, dbl., intense fragrance; [Premier sport]; Cleveland Cut-Flower Co.

New Pristine, *see* 'KORDIENA'

New Rouge Meilland, *see* 'MEIMALYNA'

New Star, HT, my, 1987; Keisei Rose Nurseries, Inc.

'NEW STYLE', HT, mr, 1962; bud oval; flowers crimson flushed brighter, large, dbl., 25 petals, slight fragrance; foliage leathery, glossy; long stems; very vigorous, bushy growth; [(Happiness X Independence) X Peace]; Meilland, Alain A.; URS

New Valencia, *see* KOReklia

'NEW WORLD', F, rb, 1945; flowers velvety red and crimson, reverse lighter, dbl., intense fragrance; foliage soft, glossy; bushy growth; (28); [Crimson Glory X Château de Clos Vougeot]; Jacobus; B&A

New Year®, *see* 'MACNEWYE'

New Year, Climbing, *see* 'BURYEAR'

'NEW YORKER', HT, mr, 1947; flowers velvety bright scarlet, dbl., 35 petals, 4–4.5 in., exhibition form, moderate, fruity fragrance; vigorous, bushy growth; [Flambeau X Seedling]; Boerner; J&P

'NEW YORKER, CLIMBING', Cl HT, mr, 1951; Boerner

New Zealand, *see* 'MACGENEV'

Newbury Angel, *see* 'BARDORG'

Newcomer, *see* **'NOVITCHKOVA'**

Newport, *see* 'POULMA'

'NEWPORT FAIRY', HWich, pb, 1908; (Newport Rambler); flowers very deep rosy pink, white eye, golden stamens, small blooms, single; [R. wichurana X Crimson Rambler]; Gardner; Roehrs

Newport Rambler, *see* **'NEWPORT FAIRY'**

News®, *see* 'LEGNEWS'

News Review, *see* 'KORREV'

'NEWSACE', HT, ab, 1961; bud ovoid; flowers light apricot, center darker, dbl., 70 petals, 4 in., moderate fragrance; foliage glossy; strong stems; vigorous, upright growth; [Horace McFarland X Good News]; Wyant

'NEWSGATE', HT, op, 1964; flowers rose and persimmon-orange, base yellow, reverse darker, dbl., 50–58 petals, 4 in.; foliage dark, glossy; vigorous growth; [Paulien Verbeek X (Jolie Madame X Baccará)]; Verbeek

Ngarla, P, Riethmuller, 1955; (Weatherly, L.)

'NIAGARA', HT, mr, 1952; bud long, pointed; flowers bright red, dbl., exhibition form, intense, fruity fragrance; [Crimson Glory X Seedling]; Davis; Garden Town Nursery

'NIAGARA MIST', HT, pb, 1968; bud ovoid; flowers light pink, base yellow, large, dbl., exhibition form, moderate fragrance; foliage glossy; vigorous, upright growth; [Tiffany X Mrs A.R. Barraclough]; Davis; Wyant

Niagara Pride, *see* 'DELDAL'

'NIAGARA SUNSHINE', HT, dy, 1969; flowers chrome-yellow, large, dbl., globular; foliage glossy, light green; vigorous, upright growth; Davis; Wyant

Niccolo Paganini®, F, mr, 1991; Meilland

Nice Day, *see* 'CHEWSEA'

'NICHOLAS SWEETBRIAR', (strain of R. eglanteria), lp; flowers small, 1.5 in.; abundant, red fruit; prickles very thorny; foliage fragrant; height 6–8 ft; Nicholas; Univ. of Neb., 1959; (probably brought from Virginia to Kansas before 1875)

Nickelodeon, *see* 'MACNICKEL'

'NICKY', F, or, 1970; bud medium, pointed; flowers large blooms in clusters, semi-dbl., cupped, no fragrance; foliage medium, light green, sparse, leathery; vigorous, upright, bushy growth; [Cyclamen X Fire King]; Institute of Ornamental Plant Growing

'NIC-NOC', Min, mr, 1978; flowers medium red, lighter reverse, blooms in clusters, dbl., 20 petals, cupped, slight fragrance; foliage small, dark, semi-glossy; spreading growth; [Anytime X Gruss an Bayern]; Poulsen, Niels D.; D.T. Poulsen

'NICOLA', F, dp, 1980; flowers deep rose pink, blooms borne 6–10 per cluster, single, 8 petals, slight fragrance; green prickles; bushy growth; [Unnamed seedling X Unnamed seedling]; Gandy, Douglas L.; Gandy Roses, Ltd.

Nicolas, S, lp; flat, 18 petals, none; L'Assomption, 1996

Nicole, *see* 'KORICOLE'

'NICOLE', HT, yb, 1931; bud long, pointed; flowers yellow, center coppery, shaded carmine, intense fragrance; very vigorous growth; RULED EXTINCT 11/80; Gaujard; C-P

Nicole, HT, rb, 1998; Barni, V.

'NICOLE DEBROSSE', HT, dr, 1962; flowers dark red, shaded scarlet; vigorous growth; [Unnamed seedling X Baccará]; Croix, P.; Minier

'NICOLE, CLIMBING', Cl HT, yb, 1933; Kordes

'NICOLETTA', HT, mp, 1969; bud ovoid; flowers pink, medium, dbl.; foliage dark; [Carla sport]; deRuiter

Nicolette®, *see* LUDswenic

Nicolina, *see* 'ZIPNIC'

'NIDA SENFF', Pol, mp, 1946; flowers soft rosy pink, borne in large clusters, Kersbergen

Nigel Hawthorne, *see* 'HARQUIBBLER'

'NIGGER BOY', HT, dr, 1933; bud long, pointed; flowers very dark velvety blackish maroon, dbl., 56 petals, exhibition form, intense fragrance; foliage thick, glossy, bronze; low, compact growth; [Hadley X Yves Druhen]; Knight, G.

'NIGHT', HT, dr, 1930; (Lady Sackville); bud long, pointed; flowers deepest blackish crimson, shaded maroon, dbl., exhibition form, intense fragrance; foliage dark, glossy; bushy growth; McGredy

Night Fire, *see* 'LYOFI'

Night Flight, *see* 'DELRIO'

Night Hawk, *see* 'DOHHAWK'

'NIGHT LADY', Min, rb, 1983; flowers red, white reverse, medium, borne singly, dbl., 36 petals, exhibition form, intense fragrance; foliage medium, dark, semi-glossy; upright growth; [Unnamed seedling X Libby]; Meredith, E.A. & Rovinski, M.E.; Casa de Rosa Domingo

Night Light®, *see* 'POULLIGHT'

Night Music, *see* 'ZIPMUSIC'

'NIGHT 'N' DAY', HT, dr, 1968; bud pointed; flowers large, dbl., moderate fragrance; foliage dark, leathery; vigorous, tall, bushy growth; [(World's Fair X Chrysler Imperial) X Happiness]; Swim & Weeks

'NIGHT SONG', S, dr, 1984; flowers large blooms borne 1–10 per cluster, dbl., 33 petals, repeat bloom, moderate fragrance; awl-like, tan prickles; foliage medium–large, dark bronze green, semi-glossy; compact, erect, bushy growth; hardy; [(Rosali X Music Maker) X Meisterstuck]; Buck, Dr Griffith J.; Iowa State University

Night Star, *see* JACurp

'NIGHT TIME', HT, dr, 1975; bud long, pointed; flowers dark black-red, dbl., 39 petals, exhibition form, intense fragrance; foliage dark, leathery; vigorous growth; [Forty-niner X Oklahoma]; Weeks

'NIGHT, CLIMBING', Cl HT, dr, 1936; Armstrong, J.A.; Armstrong Nursery

Nighthawk™, *see* 'DOHHAWK'

Nightingale, *see* 'HERGALE'

Night's Musk, LCl, w

'NIGRETTE', HT, dr, 1934; flowers blackish maroon or plum color, varying with season and weath, dbl.; bushy growth; [Château de Clos Vougeot X Lord Castlereagh]; Krause; C-P

'NIGRITELLA', F, dr, 1953; flowers red shaded darker, dbl.; dwarf, bushy growth; Cazzaniga, F. G.

Nikita®, *see* 'INTERNIKI'

'NIKITSKAJA ROSOWAJA', HP, mp, 1937; flowers medium, very dbl., intense fragrance; Kosteckij; (Sangerhausen)

'NIKKI', F, ob, 1981; flowers vermilion, white eye and reverse, medium, semi-dbl., no fragrance; foliage medium, medium green, semi-glossy; bushy growth; [Dusky Maiden X Eyepaint]; Bracegirdle, A.J.

Nil Bleu, *see* 'DELNIBLE'

'NIL DESPERANDUM', HT, ob, 1979; flowers indian orange, full, dbl., 25–30 petals, 4 in.; foliage large, light matt green; very vigorous growth; [Gavotte X Montezuma]; Ellick; Excelsior Roses

'NILES COCHET', T, rb, 1906; (Red Maman Cochet); flowers cherry-red on outer petals, lighter within; [Maman Cochet sport]; California Nursery Co.

'NILSSON GUY', F, dp, 1930; flowers deep rose-pink, open, large, borne in clusters, semi-dbl.; [Lafayette sport]; Leenders, M.

'Nimble', S, pb, 1995; Flowers handpainted bright pink stippling to white center, white reverse, semi-dbl. (6–14 petals), large (4 in.) blooms borne in small and large clusters; slight fragrance; some prickles; foliage medium, medium green, dull, medium (5 ft × 5 ft), upright, bushy growth; [Ivory Fashion X ((Valerie Jeanne X Eyepaint) X Twilight Trail)]; Jobson, Daniel J.; Jobson, 1995

Nimbus, *see* 'LEGGREY'

'NIMES', F, rb, 1970; flowers vermilion, reverse gold, dbl.; [Pampa X Piccadilly]; Gaujard

'NINA MARSHALL', HT, dp, 1966; flowers cerise, base gold, semi-dbl., cupped, slight fragrance; foliage glossy; moderate growth; [Serenade X Queen o' the Lakes]; Golik; Ellesmere Nursery

'NINA POULSEN', F, mr, 1940; flowers clear red, semi-dbl.; [Grethe Poulsen X Hybrid Tea (red)]; Poulsen, S.; Poulsen

'NINA ROSA', HT, op, 1946; flowers coppery pink shaded yellow, very large, dbl., moderate fragrance; [Frank Reader X Condesa de Sástago]; Robichon

'NINA WEIBULL'®, F, dr, 1962; flowers medium, dbl.; foliage dark; compact, bushy growth; [Fanal X Masquerade]; Poulsen, S.

Nina Weibull, Climbing, Cl F, dr

Ninetta®, *see* 'TANATTENIN'

Ninfea, HT, 1957; Borgatti, G.; (Cavriglia)

'NINIE VANDEVELDE', Pol, mp, 1924; flowers salmon, Vandevelde

'NINON VALLIN', HT, ab, 1936; flowers apricot, reverse fresh yellow, large, dbl.; foliage bright green; very vigorous growth; Gaujard

'NIOBE', F, w, 1942; flowers white, center sometimes flushed light pink; [Rosenelfe sport]; J&P

'NIOUMIYA', Min, w, 1999; flowers ivory white, dbl., 35 petals, 1.5 in., exhibition form, borne 2–5 per cluster, moderate fragrance; foliage medium, dark green; vigorous, bushy, compact (10 in.) growth; [seedling X Hatuzakura]; Yamazaki, Kazuko, 1989; Takii & Co., 1995

'NIPHETOS', T, w, 1843; bud pointed; flowers large, globular, intense fragrance; Bougere

'NIPHETOS, CLIMBING', Cl T, w, 1889; Keynes, Williams & Co.

Nipper, *see* HAReco

'NIPPY', HT, yb, 1932; flowers canary-yellow, reverse splashed red, moderate, fruity fragrance; foliage dark; Cant, B. R.

NIPRventyel, HT, lp; (**Versilia**); NIRP, 1996

Niramol, *see* 'UMSNIRA'

Nirpette Pink & White, Min, pb, 1997

NIRPnufdeu, HT, yb; (**Ambiance**); Carlton, 1999; NIRP, 1995

Nirvana®, *see* 'MEIRISOURU'

'NISETTE', F, dp, 1967; flowers pink-red, borne in trusses, globular; foliage small, dark; moderate growth; [Garnette sport]; van't Kruis; DeRuiters Nieuwe Rozen B.V.

Nishikie, *see* **'NISHIKI-E'**

'NISHIKI-E', F, ob, 1981; (Nishikie); flowers orange-yellow, blooms borne 2–5 per stem, dbl., 38 petals, exhibition form, small prickles slanted downward; foliage dark, semi-glossy; upright growth; [(Sarabande X Amanogawa) X Kagayaki]; Suzuki, Seizo; Keisei Rose Nursery

'NITA', Min, ab, 1987; flowers apricot, lighter apricot reverse, blooms borne singly, dbl., 55 petals, exhibition form, slight fragrance; few, light green prickles; foliage medium, dark, semi-glossy; medium, upright, bushy growth; [Unnamed seedling X Unknown Miniature]; McDaniel, Earl; McDaniel's Min. Roses

'NITOUCHE'®, F, pb, 1974; flowers silvery, deep salmon pink reverse, dbl., 25 petals, 4 in., slight fragrance; foliage glossy, dark; bushy, upright growth; [Seedling X Whisky Mac]; Poulsen, Niels D.; Poulsen

Nivaida, HT, ab, 1997

Nivea, *see* **'AIMÉE VIBERT'**

'NIVEA', HT, w, 1949; bud long, pointed; flowers medium, dbl., exhibition form, intense fragrance; foliage sparse; dwarf growth; [Nuria de Recolons X Blanche Mallerin]; Dot, Pedro

Nivescens, HT, lp; flowers medium, semi-dbl., slight fragrance; (Sangerhausen)

NOAbell, F, mp, 1997; (**Bluhendes Barock**); Noack, Werner

Noack's Uberraschung®, HT, yb, 1985; Noack, Werner

NOAdort, F, mp, 1991; (**Schone Dortmunderin**®); Noack, Werner

NOAfeuer, F, dr, 1995; (**Heidefeuer**); Noack, Werner

NOAgut, Gr, op, 1997; (**Focus**); Noack, Werner

'NOAH', HT, ob, 1985; flowers orange; [Dr A.J. Verhage sport]; Nevo, Motke; Maoz Haim Rose Nursery, 1976

NOAhan, S, w, 1997; (**Hannover's Weisse**); Noack, Werner

NOAmet, F, mp, 1997; (**Rose Minarett**); Noack, Werner

NOArda, S, w, 1998; (**Danica**); Noack, Werner

NOAre, S, mr, 2001; (**Red Flower Carpet**); Noack, Werner

NOAschnee, F, lp; (**White Flower Carpet**); Noack, Werner, 1991

NOAsun, S, dy, 1997; (**Celina**); Noack, Werner

'NOATRAUM', S, dp, 1989; (**Flower Carpet**™, Heidetraum™, Pink Flower Carpet); bud globular; flowers deep pink, reverse lighter, small, profuse, semi-dbl., 15 petals, cupped, borne in sprays, slight fragrance; globular, small, light red fruit; prickles crooked, dark; foliage small, dark green, glossy, disease resistant; vigorous, hardy, low, spreading growth; PP007282; GM, Glasgow, 1993 GM, The Hague, 1990; [Immensee X Amanda]; Noack, Werner; Pan-Am Northwest, Inc., 1991

NOAtwi, S, lp; (**Twilight**); Noack, Werner, 1997

'NOBAL', HMult, w, 1985; (**Thornfree Wonder**); flowers peach pink, fading to white, blooms in clusters (up to 14), semi-dbl., 14 petals, cupped, summer bloom, slight fragrance; no prickles; long canes; spreading growth; Nobbs, Kenneth J.

'NOBAM', S, w, 1985; (**Daughter Margaret**); flowers peach, fading to white, borne in clusters of 2–6, dbl., 48 petals, repeat bloom, slight fragrance; broad, pink prickles; foliage typical china; compact growth; [Mutabilis X Cornelia]; Nobbs, Kenneth J.

'NOBILITY', HT, lp, 1961; (Elle); bud ovoid; flowers ivory lightly overcast pink, center deeper, dbl., 35–40 petals, 5–5.5 in., exhibition form, moderate fragrance; foliage leathery; vigorous, upright growth; [Peace seedling X Peace]; Boerner; J&P

Nobilo's Chardonnay®, *see* 'MACRELEASE'

Noble Antony, *see* 'AUSWAY'

Noble Hit, *see* POUlnola

'NOBLESSE', HT, or, 1969; flowers very large, dbl., 28 petals, slight fragrance; foliage glossy; moderate growth; [Coloranja X Coloranja]; Spek; A. Dickson

Noblesse, *see* TANselbon

'NOCTURNE', HT, dr, 1947; bud long, pointed; dbl., 24 petals, 4.5 in., cupped, moderate, spicy fragrance; foliage leathery, dark; vigorous, upright, bushy growth; AARS, 1948; [Charlotte Armstrong X Night]; Swim, H.C.; Armstrong Nursery

'NOCTURNE', F, dr; dbl.; moderate growth; Archer

'NOCTURNE, CLIMBING', Cl HT, dr, 1955; Armstrong, J.A.; Armstrong Nursery

'NOËLLA NABONNAND', Cl T, dr, 1901; flowers velvety crimson-red, large, semi-dbl., moderate fragrance; (21); [Reine Marie Henriette X Bardou Job]; Nabonnand, G.

'NOELLA VIREBENT', Cl T, lp, 1922; flowers flesh-pink, center brighter, semi-dbl., moderate fragrance; few thorns; foliage dark, glossy; very vigorous growth; [R. gigantea X Archiduc Joseph]; Nabonnand, P.

Noelle Marie, Min, w

'NOGAWA', HT, w, 1989; bud ovoid; flowers cream, fringed with pink, large, borne usually singly, dbl., 30 petals, exhibition form, moderate fragrance; prickles almost right-angled to stem; foliage medium, dark green, semi-glossy; bushy, tall growth; [Garden Party X Kordes' Perfecta]; Takahashi, Takeshi

Noisette Desprez, *see* **'JAUNE DESPREZ'**

Noisette Garden Pink Tea, T, mp

'NOISETTE ROSE', (R. indica noisettiana, R. moschata autumnalis, R. X noisettiana)

'NOKOMIS', HWich, dp, 1918; flowers dark rose-pink, larger than lady gay or dorothy perkins, bor, dbl., intense fragrance; foliage light, glossy; vigorous, climbing growth; [R. wichurana X Comte Raimbaud]; Walsh

Nola Emily, HT, mp; [Peter Benjamin sport]; Allender, Robert William, 1990

'NOLDANYU', F, my, 1997; (**Sunny Sam**); flowers very full, medium, very dbl., 41 petals, borne in small clusters, slight fragrance; foliage medium, medium green, semi-glossy; upright, compact, low (2.5–3 ft) growth; [(Gold Bunny X Gold Bunny) X Seedling]; Nolan, Gordon D.

'NON PLUS ULTRA', HMult, dr, 1904; (Weigand's Crimson Rambler); flowers small blooms in clusters, dbl.; vigorous growth; [Crimson Rambler X Blanche Rebatel]; Weigand, C.

'NONA', HT, op, 1924; bud long, pointed; flowers flame and pink, open, semi-dbl., moderate fragrance; [Mme Edouard Herriot X Constance]; Easlea

'NONIN', HT, yb, 1938; flowers golden yellow, tinted coral-orange, very large, dbl., slight fragrance; foliage glossy; vigorous growth; [Souv. de Claudius Pernet X Unnamed seedling]; Mallerin, C.; A. Meilland

Nonino, F, mr, 1997

Nora, *see* 'BERNORA'

'NORA CUNINGHAM', Cl HT, lp, 1920; flowers flesh-pink, center paler, large, semi-dbl., cupped, free bloom, sometimes recurrent, moderate fragrance; foliage wrinkled, light; vigorous growth, long stems; climbing; [Gustav Grunerwald seedling]; Clark, A.; Hackett

'NORA HENSLOW', HT, dp, 1925; flowers crimson-cerise, single, moderate fragrance; [Mme Mélanie Soupert X Gen. MacArthur]; Evans; Beckwith

'NORA HOOKER', F, mr, 1970; dbl., 22 petals, 4 in., slight fragrance; free growth; [Queen Elizabeth sport]; Hooker; Harkness

'NORA JOHNSON', HMoy, dp, 1957; flowers cerise, small, slight fragrance; small, bright fruit; arching wands growth; [Believed to be R. willmottiae X R. moyesii]; Sunningdale Nursery

'NORA POWER', Cl HT, lp; Dearing, 1919; (Weatherly, L.)

Norah Cruickshank, HT, op

'NORAH LONGLEY', Pol, ob, 1948; flowers flame-orange, borne in trusses; foliage bright green; vigorous, branching growth; [Cameo sport]; Longley

Nordfeuer®, F, mr, 1986; Noack, Werner

'NORDHAUSEN', S, dp, 1940; flowers carmine-pink, large, semi-dbl., slight fragrance; Krause; (Sangerhausen)

Nordia, F, or, 1967; flowers medium, dbl.; Poulsen, D.T.; (Sangerhausen)

'NORDIC CHANT', HT, mp, 1974; bud long, pointed; flowers salmon-pink, dbl., 40 petals, 4.5 in., exhibition form, moderate, spicy fragrance; foliage glossy, light; vigorous growth; [Tropicana X Queen of Bermuda]; Golik; Dynarose

'NORDLICHT', HT, or, 1910; flowers coppery red; [Mme Caroline Testout X Luciole]; Kiese

'NORDLICHT', F, or, 1957; (Northlight); flowers deep cinnabar-red, large, borne in small clusters, dbl., exhibition form, slight fragrance; foliage leathery; very vigorous, low, bushy growth; [Bergfeuer X Gertrud Westphal]; Kordes, R.; R. Schmidt

Nordstern, F, 1964; Kordes, R.; (Cavriglia)

Norfolk, *see* POUlfolk

'NORFOLK HARMONY', HT, dp, 1940; bud long, pointed; flowers rosy cerise, large, dbl., exhibition form, slight fragrance; foliage glossy, dark; very vigorous, tall growth; [Comtesse Vandal X Mrs Sam McGredy]; LeGrice

'NORIDA', F, dp, 1967; flowers light crimson-scarlet, medium, dbl.; [(Pinocchio X Pinocchio) X Elsinore]; Poulsen; DeVor Nursery; McGredy

Norita®, *see* 'COMSOR'

Norita-Schwarze Rose®, *see* 'COMSOR'

'NORMA', HT, mr, 1976; flowers brilliant red, large, dbl., 50 petals; [Clio X Credo]; Gaujard

'NORMA BENNETT', F, dr, 1958; flowers crimson; [Florence Mary Morse X Border Queen]; Bennett, H.; Waikato Rose Soc.

Norma Major, HT, mp

Norma Margaret, *see* 'FRONORM'

'NORMAN', HT, mr, 1934; flowers well formed, bright scarlet-red, moderate fragrance; vigorous growth; Dickson, A.

'NORMAN HARTNELL', HT, mr, 1964; flowers crimson-red, well-formed, large, dbl., 21 petals; foliage dark; very vigorous growth; [Ballet X Detroiter]; Kordes, R.; Wheatcroft Bros.

'NORMAN LAMBERT', HT, ob, 1926; bud long, pointed; flowers deep salmon-orange, suffused bronze and yellow, base lighter, dbl., exhibition form, slight fragrance; GM, NRS, 1924; McGredy

'NORMAN ROGERS', HT, dp, 1933; flowers deep rose-pink, base yellow, large; Chaplin Bros.

Normandie, *see* 'CHEWARVEL'

'NORMANDIE', HWich, mp, 1929; flowers salmon-pink, vigorous growth; Nonin

'NORRIS PRATT'®, HT, my, 1964; flowers bright yellow, large; foliage leathery; growth moderate; [Mrs Pierre S. duPont X Marcelle Gret]; Buisman, G. A. H.

'NORRKÖPING', HT, mr, 1961; bud pointed; flowers scarlet, reverse darker; long stems; very vigorous growth; [Karl Herbst X (Baccará X Golden Sun)]; Poulsen, S.

'NORSEMAN', F, mp, 1963; flowers pink, medium, dbl., intense fragrance; foliage soft; vigorous, upright growth; [Unnamed seedling X Pinocchio]; Von Abrams

'NORTH STAR', HT, lp, 1964; bud ovoid; flowers light silvery pink, dbl., 60 petals, 6 in., intense fragrance; foliage glossy; vigorous, medium growth; [Marcia Stanhope X Peace]; Golik; Ellesmere Nursery

Northamptonshire, *see* MATtdor

'NORTHERN DANCER', HT, ob, 1965; bud ovoid; flowers orange-yellow, edges flushed pink, large, dbl., slight fragrance; foliage dark, glossy, leathery; vigorous, tall, compact growth; [Tzigane sport]; Schloen; Ellesmere Nursery

Northern Gold, HT, my

'NORTHERN LIGHTS'®, HT, yb, 1969; flowers lemon-cream, tinted pink, dbl., 50 petals, 5 in., intense fragrance; [Fragrant Cloud X Kingcup]; Cocker, 1971

Northern Lights, S, mp, 1997; Noack, Werner

'NORTHERN STATES', HSpn, w, 1952; bud long, pointed; flowers white tinged pink and yellow, open, large, single, 5 petals, profuse, non-recurrent bloom, moderate fragrance; foliage leathery; bushy, compact ($2\frac{1}{2}$ ft) growth; makes a good hedge; hardy; [R. spinosissima X Irish Charm]; Shepherd; Kern Rose Nursery

Northern Yellow, HGal, yb

Northland, *see* 'MACCARLTO'

'NORTHLANDER', S, mp, 1985; single, 5 petals, 4 in., borne singly and in clusters of 3, repeat bloom, moderate fragrance; foliage medium, dark, matt; vigorous, upright (to 8 ft) growth; [Baronne Prevost X ((Magnifica X Joanna Hill) X (Blanche Mallerin X R. laxa))]; James, John

Northlight, *see* **'NORDLICHT'**

'NORTHUMBERLAND W.I.', HT, ab, 1988; flowers deep apricot, reverse lighter, medium, dbl., 26–40 petals, slight fragrance; foliage large, dark green, glossy; bushy growth; [Silver Jubilee X Doris Tysterman]; Thompson, Robert; Battersby Roses, 1988

Northwest Sunset, MinFl, op

'NORWICH CASTLE', F, ob, 1980; flowers orange, borne 3–5 per cluster, dbl., 30 petals, slight, fruity fragrance; wedged prickles; foliage medium green, shiny, smooth; vigorous, upright growth; [(Whisky Mac X Arthur Bell) X Unnamed seedling]; Beales, Peter, 1979

Norwich Cathedral, *see* 'BEACATH'

'NORWICH CERISE', HT, mr, 1962; flowers cerise, dbl., 25–30 petals, 4–5 in., moderate fragrance; vigorous growth; [Bettina sport]; Morse

'NORWICH GOLD', S, ob, 1962; bud well-formed; flowers yellow shaded orange, large, dbl., 55 petals, moderate fragrance; vigorous, upright growth; Kordes; Morse

'NORWICH PINK', HKor, dp, 1962; flowers bright cerise, semi-dbl., 16 petals, 4 in., moderate fragrance; vigorous, pillar, well-branched growth; Kordes; Morse

'NORWICH SALMON', HKor, op, 1962; flowers salmon-pink, blooms in clusters, dbl., 30 petals, moderate fragrance; foliage glossy; vigorous, pillar, well-branched growth; Kordes; Morse

'NORWICH UNION', F, my, 1975; flowers, 3 in., cupped, intense fragrance; foliage glossy, leathery; [Arthur Bell X (Unnamed seedling X Allgold)]; Beales, Peter, 1976

'NOSAB', HT, mr, 1986; (**Leicester Abbey**); flowers large, very dbl., moderate fragrance; foliage large, dark green, glossy; bushy growth; [(Gavotte X E.H. Morse seedling) X Erotika]; Greensitt, J.A.; Nostell Priory Rose Gardens

'NOSCHAL', HT, dy, 1986; (**Challis Gold**); flowers medium, dbl., 15–20 petals, slight fragrance; foliage medium, medium green, glossy; spreading growth; [Gold Dot sport]; Greensitt, J.A.; Nostell Priory Rose Gardens

'NOSCOOK', HT, lp, 1986; (**Catherine Cookson**); flowers large, dbl., 26–40 petals, moderate fragrance; foliage large, dark green, glossy; bushy growth; [Gavotte X King's Ransom]; Greensitt, J.A.; Nostell Priory Rose Gardens

'NOSMAN', HT, pb, 1986; (**City of Manchester**); flowers large, dbl., moderate fragrance; foliage large, dark green, semi-glossy; bushy growth; [Gavotte X Red Lion]; Greensitt, J.A.; Nostell Priory Rose Gardens

'NOSMID', HT, pb, 1986; (**Middlesbrough Pride**); flowers medium, dbl., 26–40 petals, moderate fragrance; foliage medium, medium green, semi-glossy; bushy growth; [Prima Ballerina X E.H. Morse]; Greensitt, J.A.; Nostell Priory Rose Gardens, 1984

'NOSSA SENHORA DE FÁTIMA', HT, rb; flowers deep red, reverse golden yellow; da Silva, Moreira

'NOSSHEF', HT, yb, 1986; (**City of Sheffield**); flowers medium, dbl., 15–25 petals, intense fragrance; foliage medium, medium green, matt; spreading growth; [Diorama sport]; Greensitt, J.A.; Nostell Priory Rose Gardens

'NOSSUN', HT, lp, 1986; (**Sunderland Supreme**); flowers medium, dbl., 26–40 petals, slight fragrance; foliage medium, medium green, semi-glossy; upright growth; [Paul Neyron X Royal Highness]; Greensitt, J.A.; Nostell Priory Rose Gardens, 1980

Nostalgia™, *see* 'SAVARITA'

Nostalgie, *see* TANeiglat

'NOSTARN', HT, yb, 1986; (**Arnold Greensitt**); flowers medium, dbl., 26–40 petals, intense fragrance; foliage large, light green, matt; bushy growth; [E.H. Morse X Summer Sunshine]; Greensitt, J.A.; Nostell Priory Rose Gardens

'NOSTPRI', HT, mp, 1986; (**Priory Pride**); flowers medium, dbl., 26–40 petals, moderate fragrance; foliage medium, medium green, semi-glossy; bushy growth; [Pink Peace X Chicago Peace]; Greensitt, J.A.; Nostell Priory Rose Gardens, 1981

'NOSTRAD', HT, w, 1986; (Radiant Super Glaze, **Radiant Superglaze**); flowers near white, large, very dbl., moderate fragrance; foliage large, dark green, glossy; [Gavotte X Erotika]; Greensitt, J.A.; Nostell Priory Rose Gardens, 1982

'NOSTROS', HT, mp, 1986; (**Priory Rose**); flowers medium, dbl., 26–40 petals, intense fragrance; foliage medium, medium green, semi-glossy; upright growth; [Unnamed seedling X Unnamed seedling]; Greensitt, J.A.; Nostell Priory Rose Gardens, 1976

'NOSWAN', HT, dp, 1986; (**Pride of Wansbeck**); flowers light red, medium, dbl., 26–49 petals, moderate fragrance; foliage medium, medium green, semi-glossy; bushy growth; [Christian Dior X Unnamed seedling]; Greensitt, J.A.; Nostell Priory Rose Gardens, 1979

Not, *see* **'TONNERRE'**®

Notre Dame, LCl, lp; Peden, R., 1997; (Weatherly, L.)

Notre Pere, F, mr, 1982; Croix; (Cavriglia)

'NOTTINGHAM', HT, yb, 1938; flowers clear yellow, center tinted orange, moderate fragrance; vigorous growth; Robinson, H.; Wheatcroft Bros.

'NOTTINGHAM FOREST', F, mr, 1971; dbl., 28 petals, 3.5 in.; foliage dark; moderate, bushy growth; [Metropole X Diamant]; deRuiter; Geo. deRuiter

'NOTTURNO'®, HT, dr, 1981; flowers dark purplish red, dbl., 35 petals, intense fragrance; reddish, hooked prickles; foliage large, dark; upright, bushy growth; [Papa Meilland X Unnamed seedling]; Rose Barni-Pistoia

Nouveau Monde, HCh, mp

Nouveau Rouge, HGal, dr, 1813

'NOUVEAU VULCAIN', HGal, m; flowers dark purple, medium, very dbl.

'NOUVELLE ETOILE', HT, yb, 1966; flowers creamy yellow, edged carmine-red, well shaped, dbl., 40–48 petals; free growth; [Chic Parisien X Provence]; Delbard-Chabert; Cuthbert

'NOUVELLE EUROPE', F, or, 1964; (Neues Europa, New Europe); flowers bright orange, medium, dbl., moderate fragrance; foliage dark; vigorous, bushy growth; ADR, 1964; [Miss France X Vendome]; Gaujard

'NOUVELLE PIVOINE', HGal, m; flowers violet tinted, center vivid red, large

'NOUVELLE TRANSPARENTE', HGal, dp, 1835; flowers rosy crimson, large, dbl., intense fragrance; Miellez, ca 1835

'NOVA', F, or, 1968; semi-dbl., blooms in clusters, slight fragrance; foliage dark, glossy; [Anne Elizabeth X Paprika]; Harkness

Nova, S, mp

'NOVA LUX', HT, yb, 1955; flowers chrome-yellow with red reflections; foliage glossy; very vigorous growth; [Julien Potin X Sensation]; Aicardi, D.; Giacomasso

'NOVA RED', Min, mr, 1964; bud pointed; flowers crimson, small blooms in clusters, semi-dbl., 10 petals; low (12 in.) growth; [Unnamed seedling X Little Buckaroo]; Moore, Ralph S.; Sequoia Nursery

'NOVA ZEMBLA', HRg, w, 1907; flowers light pink to white; [Conrad Ferdinand Meyer sport]; Mees

Novaia, *see* ORAgofe

November Rain, HT, lp; Urban, J.; (Czech Rosa Club)

'NOVITCHKOVA', HT, ob; (Newcomer); flowers orange-yellow, medium, slight fragrance; foliage dark, leathery; low growth; Novitchkov

'NOWETA', F, mr, 1960; bud ovoid; flowers rose-red, medium, borne in clusters, dbl., moderate fragrance; foliage leathery; vigorous, upright, bushy growth; [Spice X Garnette seedling]; Boerner; J&P

'NOZOMI', Cl Min, lp, 1968; (Heideroslein Nozomi); flowers pearl-pink, blooms in trusses, single, flat, slight fragrance; foliage small, glossy; trailing growth; [Fairy Princess X Sweet Fairy]; Onodera, Toru F.

Nu Gold, Min, my

Nuage Parfume, *see* 'TANELLIS'

Nuage Parfume, Climbing, *see* 'COLFRAGRASAR'

'NUANCE', Min, or, 1992; dbl., 26–40 petals, 1.5 in., borne mostly singly, slight fragrance; few prickles; foliage small, medium green, matt; medium, upright growth; [Pierrine sport]; White, Al; Giles Rose Nursery, 1991

Nuance, *see* JACnu

'NUBIAN', LCl, dr, 1937; flowers dark velvety red, borne in huge clusters, dbl., exhibition form, sometimes recurrent bloom, slight fragrance; foliage large, leathery; vigorous (6–8 ft) growth; B&A

'NUBIENNE', HCh, dr, 1825; flowers medium, dbl.; Laffay, M.; (Sangerhausen)

'NUGGET', F, my, 1973; bud ovoid; flowers small, very dbl., exhibition form, slight fragrance; foliage large, glossy, dark; vigorous growth; [Yellow Pinocchio X Unnamed seedling]; Warriner, William A.; J&P

'NUGGETS', HT, dy, 1941; bud short, pointed, buff-yellow; flowers pale orange-yellow, open, small, semi-dbl., 15–20 petals, 2–3 in., slight fragrance; foliage small, dark, leathery; [Joanna Hill X Unnamed seedling]; Hill, Joseph H., Co.

Nuit d'Orient, *see* 'STEBIGPU'

'NUITS DE YOUNG', M, dr, 1845; (Old Black); flowers reddish purple shading to dusky violet-maroon, well-mossed; Laffay, M.

'NUMA FAY', HT, pb, 1938; flowers salmon-pink, edged pale pink, well-formed, large, dbl.; vigorous growth; Richard; A. Meilland

'NUMÉRO UN', HT, or, 1961; flowers scarlet-red passing to vermilion-red, dbl., 35–40 petals, 5 in., globular; foliage bronze, glossy; vigorous, bushy, symmetrical growth; Mallerin, C.; EFR

'NUNTIUS PACELLI', HT, w, 1929; flowers white, center cream, large, dbl., intense fragrance; [Mrs David McKee X British Queen]; Leenders Bros.; C-P

'NUNTIUS SCHIOPPA', HT, my, 1931; flowers golden yellow, sometimes washed peach-blossom-pink; [Los Angeles sport]; Leenders Bros.

'NUR MAHAL', HMsk, mr, 1923; flowers bright crimson, blooms in clusters, semi-dbl., recurrent bloom, moderate, musk fragrance; foliage small; strong stems; vigorous bush or pillar growth; (21); [Château de Clos Vougeot X Hybrid Musk seedling]; Pemberton

Nuria de Recolona, *see* **'NURIA DE RECOLONS'**

'NURIA DE RECOLONS', HP, w, 1933; (Nuria de Recolona); flowers well-formed very short peduncle, very dbl., slight fragrance; foliage dense; [Canigo X Frau Karl Druschki]; Dot, Pedro

'NURJEHAN', HT, dp, 1980; bud long, pointed; flowers deep pink, blooms borne singly, dbl., 50 petals, exhibition form, intense fragrance; straight prickles; foliage medium, dark green, coppery when young; medium, spreading growth; [Sweet Afton X Crimson Glory]; Division of Vegetable Crops and Floriculture

Nurse Cavell, *see* **'MISS EDITH CAVELL'**

Nursing Centenary®, HT, w

Nutka Rose, *see* **R. NUTKANA**

'NUTKHUT', F, or, 1969; bud long, pointed; flowers coral-red, small, very dbl., globular, slight fragrance; foliage leathery; very vigorous, bushy, open growth; [Rumba X Cocorico]; Pal, Dr B.P.; Son

'NUTNEYRON', S, mp; semi-dbl., occasionally repeats sparingly, height 4 ft; [Paul Neyron X R. nutkana]; Schoener

'NUTZWEDEL', F, mr, 1937; flowers light crimson; [Else Poulsen sport]; Schmidt, K.; Kordes

'NYMPH', F, op, 1953; flowers coral-salmon, borne in trusses, dbl., 30 petals, 3 in., moderate fragrance; foliage dark, glossy; very free growth; [Fashion X Unnamed seedling]; Dickson, A.

'NYMPHE EGERIA', HMult, 1839; Geschwind, R.; (Cavriglia)

'NYMPHE TEPLA', HSet, 1886; Geschwind, R.; (Cavriglia)

'NYMPHENBURG', HMsk, op, 1954; flowers salmon-pink shaded orange, very large blooms in clusters (up, semi-dbl., flat, recurrent bloom, moderate fragrance;) foliage large, glossy; upright growth; [Sangerhausen X Sunmist]; Kordes; Morse

'NYPELS PERFECTION', Pol, pb, 1930; flowers hydrangea-pink, shaded deep pink, large blooms in clusters, semi-dbl.; vigorous, bushy growth; Leenders, M.

'NYVELDT'S WHITE', HRg, w, 1955; flowers snow-white, large, single; orange-red fruit; [(R. rugosa rubra X R. cinnamomea) X R. nitida]; Nyveldt

O

O Sole Mio®, *see* 'DELOSOL'

'**O. JUNYENT**', HT, mr, 1924; (Olegario Junyent); flowers coral-red, base yellow, semi-dbl.; [Frau Karl Druschki X Mme Edouard Herriot]; Dot, Pedro

'**OAKINGTON RUBY**', Min, mr, 1933; bud deep crimson; flowers ruby-crimson, white-eye, dbl., 1–1.5 in.; dwarf (1 ft or less) growth; (14); Bloom

'**OAKLEY**', HT, pb, 1937; flowers bright rose, base deep red, tipped flesh-pink, large, intense fragrance; vigorous growth; Fairhead

'**OAKMONT**', HP, pb, 1893; flowers deep pink, reverse lighter, blooms in clusters, recurrent bloom; May

Oasis Sunset, *see* 'SUNMANI'

'**OBBO**', HT, lp; [Alfred Colomb sport]; (Weatherly, L.)

Obélisque®, *see* 'DELMOT'

Oberbürgermeister Boock, Pol, dr, 1964; flowers medium, dbl.; GPG Bad Langensalza; (Sangerhausen)

'**OBERBÜRGERMEISTER DR KÜLB**', HT, op, 1931; flowers flame-colored, passing to salmon; [Roselandia sport]; Nauheimer

'**OBERBÜRGERMEISTER DR TROENDLIN**', HT, lp, 1904; flowers large, very dbl.; Kaiser; (Sangerhausen)

'**OBERBÜRGERMEISTER HEIMERICH**', HP, mp, 1929; flowers fresh rose, some petals with reverse lighter, dbl., slight fragrance; [Frau Karl Druschki X Souv. de Claudius Pernet]; Weigand, C.

'**OBERGÄRTNER BURGNER**', HT, mr, 1934; flowers medium, semi-dbl., slight fragrance; Burgner; (Sangerhausen)

'**OBERGÄRTNER WIEBICKE**', F, mr, 1950; bud long, pointed; flowers light red, open, very large, borne in clusters, semi-dbl., moderate fragrance; foliage glossy, light green; vigorous, bushy growth; [Johannes Boetnner X Magnifica]; Kordes

'**OBERHOFGÄRTNER A. SINGER**', HP, mr, 1904; flowers carmine, center darker, dbl., 40 petals; dwarf, compact growth; [Mme Caroline Testout X Marie Baumann]; Lambert, H.

'**OBERLEHRER L. BURKHARDT**', HT, dp, 1939; flowers carmine-pink, large, dbl., moderate fragrance; Burkhardt; (Sangerhausen)

'**OBERLEUTNANT IMMELMANN**', HT, lp, 1936; flowers soft yellowish pink, center deeper, petals incurved, somewhat, globular; Henniger

'**OBERON**', F, ab, 1955; flowers salmon-apricot blooms in trusses, dbl., 38 petals, 2–2.5 in., slight fragrance; bushy growth; [Nymph X Seedling]; Dickson, A.

Obsession, *see* 'DEVTINTA'

Ocarina, *see* 'OCARU'

'OCARU', Min, mp, 1978; (**Angela Rippon**®, Ocarina); flowers salmon-pink, patio, small, moderate fragrance; dwarf, compact growth; [Rosy Jewel X Zorina]; deRuiter; Fryer's Nursery, Ltd.

Ocho, HT, ab, 1983; Suzuki, Seizo

'**OCOOCH MOUNTAIN ROSE**', S, mp, 1981; bud small, pointed; single, 5 petals, borne 3–4 per cluster, intense, spicy fragrance; straight, fine prickles; foliage 9 leaflet, small, slightly rugose, medium green; arching growth; Hall, William W.

Octandre, T, pb

Octavia Hill, *see* 'HARZEAL'

'**OCTAVIE**', HGal, lp; flowers light pink, edged blush, open, medium, dbl.; vigorous, branching growth; Coquerel, before 1829

Octavius Weld, T, pb

'**OCTET**', S, m, 1977; bud narrow, pointed; flowers pale purple, single, 5 petals, 3 in., prolific bloom in summer only, slight fragrance; foliage gray-green; extrememly vigorous growth; [R. rudiuscula X R. subglauca]; Rowley; Royal National Rose Soc.

'**OCTOBER**', HT, ob, 1980; bud long, pointed; flowers rich salmon orange, blooms borne singly and 2–3 per cluster, dbl., 30 petals, exhibition form, moderate fragrance; long, hooked prickles; foliage leathery, dark; tall, upright growth; [Unnamed seedling X Unnamed seedling]; Weeks, O.L.

Octoberfest™, *see* 'MACLANTER'

Oddball, *see* 'HORODD'

Odee Pink, T, lp

Odense City, S, yb, 1996

Odeon, *see* 'GAUFRARNER'

'**ODÉRIC VITAL**', HP, lp, 1858; [Baronne Prevost sport]; Oger

Odessa, *see* 'MICODESSA'

Odette, F, mp, 1972; flowers medium, dbl., intense fragrance; GPG Bad Langensalza; (Sangerhausen)

'**ODETTE CHÈNE**', HT, mp, 1940; flowers pink, base coral; vigorous growth; [Richmond, Climbing X Charles P. Kilham]; Colombier

'**ODETTE FOUSSIER**', HT, mp, 1924; flowers salmon-pink, inside chrome on yellow ground, dbl., intense fragrance; Chambard, C.

'**ODETTE FOUSSIER, CLIMBING**', Cl HT, mp, 1929; Chambard, C.

'**ODETTE JOYEUX**', LCl, op, 1959; bud globular, coral-orange; flowers pink to lilac-pink, large, dbl., cupped, abundant, recurrent bloom, intense fragrance; foliage leathery, glossy; very vigorous growth; [Lady Sylvia X Unnamed seedling]; Robichon

Odine, *see* '**ONDINE**'

'**ODORATA**', HT, pb, 1928; flowers carmine-pink, reverse white edged pink, base golden yellow, dbl., intense fragrance; [Sunburst X Ma Fiancee]; Van Rossem

Odorata 22449, *see* '**FUN JWAN LO**'

Odyssée, *see* 'GAUZOMI'

'**IILLET**', C, lp, 1800; flowers bright pink, medium, dbl., moderate fragrance; vigorous growth; Dupony, Andre

Iillet Double, HGal, m, 1835; Prévost

'**IILLET FLAMAND**', HGal, pb, 1845; flowers pale pink striped white and brighter pink, medium, very dbl., flat, intense fragrance; very vigorous growth; Vibert

'**IILLET PANACHÉE**', M, pb, 1888; (Striped Moss); flowers pale pink striped deep pink, petals quilled, small, flat; Verdier, C.

'**IILLET PARFAIT**', HGal, pb; flowers blush, striped light and dark red, medium, dbl., flat; dwarf growth; Foulard, ca 1841

'**OEKONOMIERAT ECHTERMEYER**', HT, dp, 1913; flowers very large, dbl., moderate fragrance; Lambert, P.; (Sangerhausen)

Officinalis, *see* **R. GALLICA OFFICINALIS**

Offrandé, HT, 1982; Gaujard; (Cavriglia)

Offranville, F, dp, 1993; Eve, A.

'**OGNEWAJA**', HT, mr, 1937; flowers medium, dbl.; Kosteckij; (Sangerhausen)

Ogni Jalty, F, mr, 1955; flowers medium, dbl., slight fragrance; Klimenko, V. N.; (Sangerhausen)

'**OGONIOK**', F, ob, 1955; (Little Fire); flowers fiery orange edged darker, medium, slight fragrance; Sushkov, K. L.

Oh Darlin, *see* 'RENDAR'

Oh La La, *see* '**OLALA**'®

Oh My God, *see* 'WEKJOE'

'**OH MY STARS**', Min, ly, 1996; dbl., 26–40 petals, exhibition form, borne mostly singly, slight fragrance; moderate prickles; foliage medium, medium green, dull; upright, tall (2–2½ ft) growth; [Giggles sport]; Bough, Melvin; Giles Rose Nursery

'**OH-CHOH**', HT, yb, 1983; flowers yellow tinted rose, aging red, large, dbl., 38 petals, exhibition form, moderate fragrance; small prickles, slanted downward; foliage dark, semi-glossy; upright growth; [(Rumba X Olympic Torch) X Wisbech Gold]; Suzuki, Seizo; Keisei Rose Nursery

'OHIO', S, mr, 1949; flowers bright red, semi-dbl., recurrent bloom; height 4 ft; hardy; [R. soulieana X Gruss an Teplitz seedling]; Shepherd

'OHIO BELLE', LCl, lp, 1974; bud globular; flowers medium, dbl., 53 petals, 3 in., repeat bloom, slight fragrance; foliage glossy, dark; [New Dawn X ?]; Jerabek, Paul E.; Wyant

'OHL', HGal, m; flowers violet-purple, center bright red, large, dbl.; vigorous growth; Vibert, before 1843

Ohlala, *see* **'OLALA'**®

Ohshima Rose, *see* 'COCHUSTER'

'OIRASE', HT, mr, 1973; bud ovoid; dbl., 24 petals, 5.5 in., exhibition form, below-average bloom continuity, moderate fragrance; foliage glossy, dark; vigorous, upright growth; [Red Lion X Christian Dior]; Ito

'OISEAU BLEU', HT, m, 1970; flowers mauve-rose, dbl., 30 petals, 4–4.5 in., intense fragrance; foliage large, glossy, dark; vigorous growth; Poulsen, Niels D.; Vilmorin-Andrieux

'OISEAU DE FEU', F, mr, 1956; flowers scarlet-red, to open, medium borne in clusters, dbl., 35 petals, 2 in., cupped, slight fragrance; bushy growth; [Chant Indou X Peace]; Mallerin, C.

'OJIBWAY', LCl, w, 1946; bud pointed; semi-dbl., 12–15 petals, 3 in., borne in clusters, free, non-recurrent bloom, slight fragrance; foliage dark; vigorous, spreading growth; hardy.; [Ross Rambler X ((R. rugosa X R. eglanteria) X ?)]; Preston; Central Exp. Farm

'OKAGA', HRg, dp, 1927; flowers deep pink, semi-dbl., non-recurrent; low, bushy growth; very hardy; [Alika X Tetonkaha]; Hansen, N.E.

'OKLAHOMA', HT, dr, 1964; bud ovoid, long, pointed; flowers very dark red, dbl., 48 petals, 4–5.5 in., exhibition form, intense fragrance; foliage leathery, dark, matt; vigorous, bushy growth; GM, Japan, 1963; [Chrysler Imperial X Charles Mallerin]; Swim & Weeks; Weeks Wholesale Rose Growers

'OKLAHOMA, CLIMBING', Cl HT, dr, 1972; Ross, A., & Son

'OKLAHOMA, CLIMBING', Cl HT, dr, 1968; Swim & Weeks; Weeks Wholesale Rose Growers

'OKRESNI HEITMAN CUBR', HT, mp, 1933; flowers large, dbl.; Böhm, J.; (Sangerhausen)

Oksana, HT, pb, 1995; Singer, Steven

Oktoberfest, *see* 'MACLANTER'

'OLALA'®, F, mr, 1956; (Oh La La, Ohlala); bud pointed; flowers blood-red, center lighter, large blooms in clusters to 25, semi-dbl., slight fragrance; foliage leathery, dark, glossy; vigorous, bushy, upright growth; GM, Baden-Baden, 1955; [Fanal X Crimson Glory]; Tantau, Math.

'OLAVE BADEN-POWELL', HT, mr, 1972; flowers scarlet, 5 in., slight fragrance; foliage dark, leathery; Tantau, Math.; Harry Wheatcroft Gardening

'OLAVUS', M, op, 1932; flowers salmon-pink, medium, semi-dbl.; Nielsen; (Sangerhausen)

Old Black, *see* **'NUITS DE YOUNG'**

'OLD BLUSH', Ch, mp; (Common Blush China, Common Monthly, Old Pink Daily, Old Pink Monthly, Parsons' Pink China); flowers two-tone pink, blooms in loose sprays, semi-dbl., dependably recurrent, slight fragrance; vigorous, upright growth; (14); Parsons, 1752; Int. into Sweden in 1752 and into England before 1

Old Blush Single, Ch, pb

'OLD BLUSH, CLIMBING', Cl Ch, mp

Old Castilian, *see* **'AUTUMN DAMASK'**

Old Country Charm, *see* 'MICCOUNTRY'

Old Crimson China, *see* **'SLATER'S CRIMSON CHINA'**

Old Danish, LCl, mp

Old Faithful, *see* 'JACHY'

'OLD FASHION RED', HT, mr, 1947; flowers spectrum-red fading blush, large, very dbl., exhibition form, moderate fragrance; foliage glossy; vigorous, bushy growth; [Pink Princess X Crimson Glory]; Brownell, H.C.

Old Fashioned Girl, *see* 'TINOLD'

Old Fashioned Lady, S, w

Old Flame, *see* DICkitty

Old Gay Hill Red China, Ch, mr

'OLD GLORY', HT, mp, 1940; flowers clear brilliant pink, large, dbl., moderate fragrance; RULED EXTINCT 2/88; [Briarcliff sport]; Hausermann

Old Glory™, *see* 'BENDAY'

'OLD GOLD', HT, or, 1913; flowers vivid reddish orange, shaded coppery red and apricot, semi-dbl., 10 petals, moderate fragrance; foliage dark; short stems; GM, NRS, 1912; McGredy

Old Homestead, HCh, m

Old John™, *see* DICwillynilly

Old Lavender, Ayr, m

Old Lavender Rambler, HWich, m

Old Lilac, *see* 'AUSLILAC'

Old Master, *see* 'MACESP'

Old Nanaimo Rose, Misc OGR, mp

Old Pink Daily, *see* **'OLD BLUSH'**

Old Pink Monthly, *see* **'OLD BLUSH'**

Old Pink Moss, *see* **Communis**

Old Port, *see* 'MACKATI'

Old Red, HCh, mr

'OLD RED BOURSAULT', Bslt, mr; flowers pale red, boursault type, poorly formed, borne in large clusters, semi-dbl.; nearly round fruit; very vigorous growth

'OLD RED MOSS', M, mr; (Muscosa Rubra); flowers carmine-red, medium, very dbl., heavy bloom; non-recurrent, moderate fragrance; vigorous growth

Old Smokey, F, lp; Murley, J.J., 1976; (Weatherly, L.)

'OLD SMOOTHIE', HT, mr, 1970; flowers large, very dbl., exhibition form, slight fragrance; foliage large, glossy, leathery; vigorous, upright growth; [Night 'n' Day X (First Love seedling X (Queen Elizabeth X Chrysler Imperial))]; Weeks

Old Spanish Rose, *see* **'RUSSELLIANA'**

Old Spice, *see* 'MACJULIAT'

Old Stone School, Misc OGR, mp

Old Time, *see* 'KOROL'

Old Velvet Moss, *see* **'WILLIAM LOBB'**

Old White Moss, M, w

Old Yellow Scotch, HSpn, dy

'OLDCASTLE', HT, mp, 1985; flowers large blooms borne singly, dbl., 55 petals, exhibition form, moderate fragrance; foliage large, dark, glossy; medium, upright growth; [Queen Elizabeth X Charlotte Armstrong]; LeMire, Walter; Roses by Walter LeMire

'OLDE ENGLISH', F, dp, 1974; flowers light red, dbl., 30–35 petals, 3–4 in., slight fragrance; foliage glossy, bronze; free growth; [(Floradora X Independence) X Siren]; Orard, Joseph; Harry Wheatcroft Gardening

Olde Fragrance, *see* 'TANSCHAUBUD'

Olde Lace, *see* 'HADLACE'

Olde Romeo, *see* 'HADROMEO'

Olde Sweetheart™, *see* 'HADSWEE'

Olde Tango™, *see* 'HADANGO'

'OLDEN DAYS', Pol, mr, 1989; (Olden Times); flowers small, borne in sprays of 36, semi-dbl., 10 petals, slight fragrance; prickles brown; foliage small, light green, elliptical, slightly serrated, ma; bushy, compact growth; [Unnamed pink Polyantha Paul Carpel sport]; Vash, Ernest J.; Historical Roses, 1987

Olden Times, *see* **'OLDEN DAYS'**

Oldtimer, *see* 'KOROL'

'OLÉ', Gr, or, 1964; bud well-shaped; flowers medium, dbl., 50 petals, exhibition form, moderate fragrance; foliage glossy; vigorous growth; [Roundelay X El Captain]; Armstrong, D.L.; Armstrong Nursery

Olé, Climbing™, *see* 'AROHAICLO'

Oleander Rose, *see* 'INTERANDER'

Olegario Junyent, *see* **'O. JUNYENT'**

Olga, HT, 1988; Delforge; (Cavriglia)

Olga, HT; Mansuino; (Cavriglia)

Olga Rippon, *see* 'HORTYARD'

Olga Tschechowa, HT, ly, 1978; flowers creamy yellow, very large, very dbl., intense fragrance; Cocker; (Sangerhausen)

OLIjbrau, F, r; (**Estelle**); (OLIjbrau), 1999

OLIjcrem, F, w; (**Sahara**); (OLIjcrem), 1997

OLIjfaon, F, pb; (**Evolution**); (OLIjfaon)

OLIjglu, HT, lp; (**Toscanini**)

OLIjkroet, HT, rb; (**Faria**); (OLIjkroet)

OLIjplam, HT, mr, 1999; (**Red Berlin**)

OLIjsab, HT, ob, 1999; (**Orange Flame**)

OLIjzouc, HT, dp, 1997; (**Daytona**); (OLIjzouc)

OLItel, F, pb; (**Super Disco**)

Olive, *see* 'HARPILLAR'

'OLIVE COOK', HT, w, 1934; flowers white, base faintly tinged lemon-yellow, large, exhibition form, moderate fragrance; foliage glossy; vigorous growth; Cant, F.

Olive Elsie, *see* 'BOSSEXEYE'

'OLIVE MCKENZIE', HT, yb, 1970; bud long, pointed; flowers orange-yellow, marked red, large, dbl., moderate fragrance; foliage large, dark; vigorous, upright, bushy growth; [Daily Sketch X Manitou]; Dawson, George; Brundrett

'OLIVE MOORE', HT, lp, 1927; flowers pale rose-pink, reflexed silver-pink, intense fragrance; Allen

'OLIVE PERCIVAL', HT, rb, 1948; bud long, pointed; flowers intense cherry-red, base gold, semi-dbl., 14–20 petals, 3.5–4 in., cupped, intense fragrance; foliage leathery, bronze; very vigorous, upright, free branching growth; [California X Eternal Youth]; Howard, P.J.

Olive Taylor, *see* 'PEAP'

'OLIVE WHITTAKER', HT, op, 1920; flowers rich coppery rose to cerise and salmon; Easlea

'OLIVER DELHOMME', HP, mr, 1861; flowers carmine-red, medium, dbl., moderate fragrance; Verdier, V.; (Sangerhausen)

'OLIVER MEE', HT, mp, 1927; flowers deep salmon tinted fawn, becoming deep salmon-pink, large, dbl., exhibition form, moderate fragrance; Dickson, S.

Oliver Twist, *see* 'SABBYRON'

'OLIVERS', HT, m, 1963; flowers purple to pink, medium, semi-dbl., moderate fragrance; [Pres. Herbert Hoover X ?]; Oliver, H.

Olivet, HMult, dp, 1892; Vigneron

Olivia®, *see* TANavilo

'OLIVIER METRA', HP, 1885; Verdier, E.; (Cavriglia)

'OLLIE', HT, mr, 1982; bud globular, pointed; dbl., 45 petals, borne singly, intense fragrance; red brown prickles; foliage large, leathery, rounded; vigorous, tall growth; [(Pink Garnette X Pink Hat) X ((Frau Karl Druschki X McGredy'sYellow) X (Baronne Prevost X Gruss an Teplitz))]; James, John

'OLWYN', HT, mp, 1946; bud long, pointed; flowers pink, medium, exhibition form, slight fragrance; foliage leathery; vigorous, bushy growth; Bird; F. Mason

Olympe, HT, 1974; Gaujard; (Cavriglia)

'OLYMPE FRESCENCY', T, w; flowers white, becoming yellow; Damaizin, 1859

'OLYMPIA', HT, mr, 1935; flowers bright red, large, dbl., moderate fragrance; long, strong stems; vigorous growth; [Johanniszauber X Hadley]; Tantau

'OLYMPIA', HT, my, 1955; bud bright yellow, well formed; bushy, semi-upright growth; [Eclipse X Seedling]; Delforge

Olympiad™, *see* 'MACAUCK'

'OLYMPIAD', HT, rb, 1931; (Mme Raymond Gaujard); bud long, pointed; flowers blood-red, shaded copper and yellow, large, dbl., intense fragrance; RULED EXTINCT 11/82; GM, Bagatelle, 1930; Pernet-Ducher; Dreer; Gaujard

'OLYMPIAD, CLIMBING', Cl HT, rb, 1938; Raffel; Port Stockton Nursery

'OLYMPIC CHAMPION', HT, yb, 1995; flowers yellow/pink blend, very dbl., 1.5–2.75 in., slight fragrance; some prickles; foliage medium, dark green, semi-glossy; medium (70 cms), bushy growth; [Champion sport]; Pugh, D. J.; Pugh, 1995

'OLYMPIC CHARM', HT, lp, 1963; flowers silvery, reverse bright pink, well formed, dbl., 65 petals, 3.5 in., intense fragrance; foliage dark, glossy; [Grand'mere Jenny X Claude]; Barter

Olympic Dream, *see* 'JACADE'

'OLYMPIC FLAME', F, or, 1962; bud pointed; flowers orange-vermilion, blooms in clusters, single, 8 petals, 3 in.; foliage glossy, bright green; vigorous, upright, bushy growth; Brett

Olympic Glory, *see* 'JACED'

Olympic Gold, *see* 'KORSCHNUPPE'

'OLYMPIC GOLD', Min, ly, 1983; flowers miniflora, medium, dbl., 31 petals, exhibition form, slight fragrance; foliage medium, medium green, semi-glossy; upright, bushy growth; [Rise 'n' Shine X Bonny]; Jolly, Nelson F.; Rosehill Farm

Olympic Spirit, *see* 'JACARY'

Olympic Spirit, *see* 'PEAPRINCE'

'OLYMPIC STAR', HT, mp, 1959; flowers cerise-pink; vigorous growth; [Picture sport]; Trebbin

'OLYMPIC TORCH', HT, rb, 1966; (Seika, Sei-Ka); bud long, pointed; flowers white and red, becoming all red, medium, dbl., exhibition form; foliage glossy, bronze, leathery; vigorous growth; Gold Star of the South Pacific, Palmerston North, NZ, 1970; [Rose Gaujard X Crimson Glory]; Suzuki, Seizo; Keisei Rose Nursery

'OLYMPIC TRIUMPH', F, rb, 1973; flowers red and yellow, dbl., 4.5 in., globular; foliage sage green; upright growth; [Shiralee X Apricot Nectar]; Dickson, Patrick; Dicksons of Hawlmark

Olympiode®, *see* 'MACAUCK'

Olympisches Feuer®, *see* 'TANOLFEU'

Olympisches Feuer 92, *see* TANolfeu92

Olympus, *see* BARoly

'OMAR KHAYYÁM', D, lp, 1893; flowers center incurved, small, very dbl., quartered, moderate fragrance; foliage small, downy; dense, prickly growth (to 3 ft)

'OMAR PACHA', B, mr, 1863; flowers bright cherry, recurrent bloom, slight fragrance; vigorous growth; Pradel

'OMBRÉE PARFAITE', HGal, m, 1823; flowers variable, light pink to deep purple, often in the same flower, dbl.; Vibert

Ombretta, S, 1991; Embriaco, B.; (Cavriglia)

'OMEGA', HT, mp, 1974; (**Emily Post**); flowers soft medium pink, dbl., 48 petals, 3.5–4 in., exhibition form, moderate fragrance; upright, bushy growth; [Eternal Sun X Carina]; Byrum; J.H. Hill Co.

Omega, HT, 1975; Hill, J.H.; (Cavriglia)

Omi Oswald®, HMsk, ly, 1988; Lens

Omni, S, yb, 1988; Lens

'OMPHALE', HGal, 1845; Vibert, J. P.; (Cavriglia)

Omul, S, dy, 1974; flowers dark golden yellow, large, dbl.; GPG Bad Langensalza; (Sangerhausen)

On Fire, S, lp; Peden, R., 1999; (Weatherly, L.)

Onda Rose®, S, mp

Ondella, *see* 'MEIVANAMA'

'ONDINE', HT, w, 1936; (Odine); bud pointed; flowers creamy white, slightly tinted pink, high pointed, large, dbl., 20–25 petals; foliage dark gray-green; long, strong stems; very vigorous growth; [Louise Criner X Souv. de Claudius Pernet]; Ketten Bros.

'O'NEAL'S BEQUEST', S, yb, 1986; flowers yellow with pink petal edges, medium, semi-dbl., slight fragrance; foliage large, medium green, glossy; upright growth; O'Neal, Conrad; Paul Jerabek

O'Neal's Bequest, S, yb

'O'NEAL'S WHITE', LCl, w, 1961; flowers large; vigorous growth; [Blossomtime X New Dawn]; O'Neal; Wyant

Onex, HGal, dr

Onkaparinga, *see* 'TOMRING'

'ONKEL SVEND', F, rb, 1978; flowers medium red and silvery red blend, large blooms in sprays of, dbl., 23 petals, moderate, spicy fragrance; foliage medium, dark, matt; low, bushy, spreading growth; [Sonia X Ernest H. Morse]; Poulsen, Niels D.; D.T. Poulsen

Only Love, *see* 'INTERONLY'

'ONLY YOU', HT, mr, 1970; flowers rose-carmine; [Carina sport]; Vahldiek-Bissingen; URS

Ontario Celebration, *see* 'LAVMOUNT'

'ONYX FLAMBOYANT', HT, lp, 1965; flowers peach to shell-pink, 4 in., exhibition form, slight fragrance; foliage serrated; vigorous, bushy growth; [Sultane X Queen Elizabeth]; Delbard-Chabert; Cuthbert

'OOMASTON PRIDE', HT, yb, 1954; flowers buttercup-yellow suffused deep carnation-pink, large, dbl.; foliage dark, glossy; very vigorous growth; [Haisha sport]; C.W.S. Ltd. Hort. Dept.; Co-op. Wholesale Soc.

Oonagh, *see* 'SEAOONA'

'OOR WULLIE', Min, mp, 1978; flowers salmon-pink, full, medium, dbl., 24 petals, slight fragrance; foliage glossy, leathery; free growth; [New Penny sport]; Anderson's Rose Nurseries

Opa Pötschke, *see* **'PRECIOUS PLATINUM'**

'OPAL', HT, mp, 1934; (Lise Palais); flowers salmon to opal, dbl., 45 petals, 5.5 in., exhibition form, intense fragrance; vigorous growth; [Julien Potin X R. foetida bicolor seedling]; Gaujard; J&P, 1941

'OPAL BRÜNNER', Cl F, lp; bud very small; flowers blush-pink to pale rose, borne in large clusters, dbl., slight, musk fragrance; tall, pillar (10 ft) growth; Marshall, O.C., about 1948

'OPAL GOLD', HT, ab, 1936; flowers apricot, changing to pink; [Joanna Hill sport]; Bate

'OPAL JEWEL', Min, mp, 1962; bud ovoid; flowers pink, center darker, dbl., 45 petals, 1 in., slight fragrance; foliage leathery; vigorous, compact (8–10 in.) growth; [Mothersday X Rosy Jewel]; Morey, Dr Dennison; J&P

'OPAL OF ARZ', HT, ob, 1938; flowers variable orange shades, dbl., moderate fragrance; [Mary Wallace seedling]; Brownell, H.C.

'OPALINE', HT, pb, 1922; flowers pale pink, shaded carmine and old-gold; [Louise Catherine Breslau X Frau Karl Druschki]; Lippiatt

Open Arms, *see* 'CHEWPIXEL'

Open d'Australie®, *see* DORsafr

Open Secret, *see* 'WEEPAINT'

Opening Act, *see* 'JALACT'

Opening Night™, *see* 'JACOLBER'

'OPÉRA', HT, rb, 1950; bud long, pointed; flowers light scarlet-red, base yellow, dbl., 6 in., moderate fragrance; foliage leathery, light green; vigorous, erect growth; GM, NRS, 1949; [La Belle Irisee X Unnamed seedling]; Gaujard; J&P

'OPERA, CLIMBING', Cl HT, rb, 1956; (Grimpant Opera); Armbrust

Operettenrose®, HT, lp

'OPHELIA', HT, lp, 1912; bud long, pointed; flowers salmon-flesh, center tinted light yellow, dbl., 28 petals, moderate fragrance; foliage leathery; vigorous growth; (28); [Parentage unknown. Perhaps a chance seedling of Antoine Rivoire. sport]; Paul, W.

'OPHELIA QUEEN, CLIMBING', Cl HT, lp, 1923; [Ophelia sport]; Westbury Rose Co.

'OPHELIA SUPREME', HT, lp, 1917; flowers light rose-pink, center darker; [Ophelia sport]; Dailledouze Bros.

'OPHELIA, CLIMBING', Cl HT, lp, 1920; Dickson, A.

'OPHIRIE', N, op, 1841; flowers reddish copper, reverse rosy and fawn, very dbl., cupped, moderate fragrance; vigorous growth; Goubault

'OPORTO', HT, dr, 1930; flowers port-wine-red, petals very leathery, dbl., slight fragrance; [Château de Clos Vougeot X Betty Uprichard]; Allen

Opulence, *see* JACosch

Opus, Pol, or; flowers medium to large, semi-dbl., slight fragrance; VEG; (Sangerhausen)

Or Hatabor, HT, dy, 1989; Fischel

Ora Kingsley, *see* CLEhonor

ORAblan, HT, w; (**Therese de Lisieux®**); Orard, 1992

ORAcreps, HT, pb, 2000; (**Hope**); Orard, P.

'ORADIT', Cl F, op, 1983; (**Edith de Martinelli, Climbing®**); Orard, Joseph; Pekmez

ORAdon, HT; (**Triolet**); Orard, 1995; (Cavriglia)

'ORADOUR', F, dr, 1955; dbl.; [Happiness X Demain]; Mallerin, C.; EFR

Orafe, HT, pb, 1997

ORAgofe, F, rb, 1997; (**Novaia**); Orard

ORAju, HT, 1993; (**Tassili**); Orard; (Cavriglia)

ORAkah, HT, m; (**A Caen la Paix®**); Eurosa, Chinensis; Orard, Joseph, 1994; (Hay)

ORAkosar, Cl HT, or, 1993; (**Lady Rose, Climbing**); Orard

ORAlamar, HT, lp; (**Ville de Moulins**); Orard, 1997

'ORAMONT', F, or, 1983; (**Orangina®**); [Royal Occasion sport]; Orard, Joseph; Paul Pekmez

ORAnat, HT, dr, 1997; (**Abbe Lemire**); Orard

Orange Adam®, F, ob, 1993; Adam, M.

Orange Appeal, *see* 'RENAPP'

'ORANGE BEAUTY', Cl F, op, 1961; flowers orange to salmon, blooms in clusters, dbl., exhibition form, slight fragrance; foliage glossy; vigorous (6–8 ft) growth; [Little Darling X Gertrude Raffel]; Raffel; Port Stockton Nursery

Orange Blossom, S, lp; [R. brunonii X Unknown]; McLeod, J., 1990; (Weatherly, L.)

Orange Blossom Special™, *see* 'JACMOCL'

'ORANGE BOUQUET', F, ob, 1972; flowers coppery orange, turning pink, rosette form, dbl., 25 petals, 2.5 in., moderate fragrance; foliage dark; upright, free growth; [Masquerade X Paris-Match]; Northfield

Orange Bunny®, *see* 'MEIRIANOPUR'

Orange Butterfly, S, or, 1997; Williams, J. Benjamin

'ORANGE CASCADE', Cl Min, ob, 1979; bud pointed; flowers yellow-orange, dbl., 20 petals, 1 in., moderate fragrance; foliage small, fern-like; slender, willowy growth; [Unnamed yellow seedling X Magic Wand]; Moore, Ralph S.; Sequoia Nursery

Orange Charm, *see* 'MINDCO'

Orange Chateau, *see* **'HATUZAKURA'**

'ORANGE CHATEAU', HT, ab, 1999; flowers apricot orange, dbl., 30 petals, 5.5 in., moderate fragrance; 3.5 ft growth; winter hardy; heat tolerant; Teranishi, K.; Itami Rose Nursery, 1997

'ORANGE CHEER', Pol, ob, 1937; flowers clear orange, almost, dbl.; vigorous growth; Letts

'ORANGE CHIFFON', F, ob, 1966; flowers orange-salmon, reverse silvery orange, medium, dbl., cupped, foliage dark, leathery; vigorous, bushy growth; [Little Darling X Goldmarie]; Mease; Wyant

Orange Class, Min, ob

'ORANGE COMBE', HT, ob, 1956; flowers orange, very dbl., globular; vigorous, upright growth; [Charlotte Armstrong X Unnamed seedling]; Combe; Delbard

'ORANGE CUP', F, or, 1965; bud pointed; flowers orange-scarlet, open, medium, borne in clusters, single, intense fragrance; foliage dark, glossy; vigorous, compact growth; [Cocorico X ?]; Pal, Dr B.P.; Indian Agric. Research Inst.

'ORANGE DARLING', Cl Min, ob, 1979; flowers orange, small, dbl., 32 petals, 1.5 in., slight fragrance; foliage dark, leathery; climbing or pillar growth; Sudol, Julia

'ORANGE DAWN', F, ob, 1973; flowers orange, tinted vermilion, full, dbl., 30–35 petals, 4 in., intense fragrance; foliage glossy, dark; vigorous growth; [Orange Sensation X Sutter's Gold]; Ellick

'ORANGE DELBARD', HT, or, 1959; (Lady Russon); flowers bright orange, well formed, large, dbl., moderate fragrance; strong stems; vigorous growth; [Impeccable X Mme Robert Joffet]; Delbard-Chabert

'ORANGE DELIGHT', HT, ob, 1950; (Meerzicht Glory, Queen Juliana); bud urn shaped; flowers orange, reverse veined red, dbl., 30 petals, 5.5–6 in., cupped, intense fragrance; foliage glossy, dark; vigorous, tall growth; Verschuren-Pechtold; J&P

'ORANGE DELIGHT, CLIMBING', Cl HT, ob, 1957; Verschuren-Pechtold

Orange d'Éte, *see* DORnice

'ORANGE DOT', HT, ob, 1963; bud long; flowers carthamus-red, large, dbl., 50 petals, moderate fragrance; bushy growth; [Chrysler Imperial X Soraya]; Dot, Pedro

'ORANGE DROP', Min, ob, 1988; flowers soft orange, white at base, reverse soft orange edging veining, semi-dbl., 20 petals, exhibition form; prickles long, very pointed, medium, light green; foliage medium, medium green, semi-glossy; bushy, medium, vigorous growth; [Heartland X Unnamed seedling]; Bridges, Dennis A.; Bridges Roses

'ORANGE ELF', Cl Min, ob, 1959; bud pointed; flowers orange, fading lighter, small, dbl., 25 petals, slight fragrance; trailer or groundcover; vigorous, climbing growth; [Golden Glow (LCl) X Zee]; Moore, Ralph S.; Sequoia Nursery

'ORANGE ELIZABETH OF GLAMIS', F, ob, 1974; flowers orange-flame, dbl., 30 petals, 3 in., intense fragrance; foliage glossy; compact, bushy growth; [Elizabeth of Glamis sport]; Thames Valley Rose Growers

'ORANGE EVERGLOW', LCl, ob, 1942; flowers orange slightly shaded red and yellow; [Copper Glow sport]; Brownell, H.C.

'ORANGE FESTIVAL', HT, op, 1961; flowers pink to coral-orange, small; [Souv. de Jacques Verschuren X Serenade]; Leenders, J.

Orange Fire®, *see* 'INTERFIRE'

'ORANGE FIRE', Min, op, 1974; bud short, pointed; flowers orange, pink, carmine, rose, dbl., 40 petals, 1 in.; foliage very glossy, leathery; upright growth; [(R. wichurana X Floradora) X Fire Princess]; Moore, Ralph S.; Sequoia Nursery

'ORANGE FLAME', HT, or, 1963; bud ovoid, pointed; dbl., 33 petals, 4.5–5 in., exhibition form, moderate fragrance; foliage leathery, glossy; vigorous growth; [Monte Carlo X Radar]; Meilland, Mrs. Marie-Louise; C-P

Orange Flame, F, ob, 1983; Williams, J. Benjamin

Orange Flame, *see* OLIjsab

Orange Flare, F, ob

'ORANGE GARNET', F, or, 1965; bud ovoid; flowers small, dbl.; foliage dark, leathery; vigorous, upright, bushy growth; [(Garnette X Circus) X Spartan]; Swim & Weeks

'ORANGE GLORY', HT, ob, 1936; flowers orange, open, large, semi-dbl.; foliage leathery, light; vigorous growth; [Unnamed seedling X Charles P. Kilham]; Leenders, M.

'ORANGE GLOW', Pol, ob, 1936; flowers bright golden orange, borne in large trusses; Verschuren; Dreer

'ORANGE GOLIATH', HT, op, 1975; bud long, pointed; flowers copper-orange, dbl., 26 petals, 8 in., moderate fragrance; foliage matt, green; vigorous growth; [Beauté X Serenade]; Gandy, Douglas L.

'ORANGE HONEY', Min, ob, 1979; bud pointed; flowers orange-yellow, dbl., 23 petals, 1.5 in., exhibition form, moderate, fruity fragrance; foliage matt, green; bushy, spreading growth; [Rumba X Over the Rainbow]; Moore, Ralph S.; Sequoia Nursery

'ORANGE ICE', F, or, 1963; bud ovoid; flowers light orange-red, blooms in clusters, dbl., 25 petals, 3 in., cupped, slight fragrance; foliage leathery, dark; vigorous, upright, bushy growth; [Fashion X Sumatra]; Thomson, R.; Tillotson

Orange Ice, F, ob, 1999; Williams, J. Benjamin

Orange Ilseta®, *see* 'TANILVOBA'

'ORANGE JADE', Min, or, 1991; dbl., 26–40 petals, borne mostly singly, slight fragrance; foliage medium, dark green, semi-glossy; upright, medium growth; [Black Jade sport]; Cole, Catherine W.; East Tennessee Miniature Roses, 1990

Orange Juice, *see* 'ARORAJU'

Orange Juwel®, *see* KORkister

'ORANGE KING', Pol, ob, 1922; flowers light coral-red, fading greenish, open, borne in clusters, dbl., 0.75 in.; foliage glossy, dark; bushy, dwarf growth; [Orléans Rose sport]; Cutbush

Orange Kordana, Min, or; Kordes

'ORANGE KORONA', F, or, 1959; flowers orange-scarlet, well-formed, blooms in clusters, dbl., 20 petals, 4 in., slight fragrance; foliage olive-green; vigorous, upright growth; [Bergfeuer X Independence]; Morse

Orange Love™, *see* 'MINAFCO'

'ORANGE MARMALADE', Min, ob, 1991; flowers bronze with yellow blend, small, dbl., borne mostly singly, moderate fragrance; foliage medium, light green, matt; low, upright, bushy growth; PP006820; [Gingersnap X Orange Honey]; Williams, J. Benjamin, 1988

'ORANGE MARVEL', Pol, op, 1928; flowers salmon-orange; [Miss Edith Cavell sport]; Van der Vis

'ORANGE MASTERPIECE', F, or, 1970; flowers small, borne in trusses, dbl.; foliage dark; bushy growth; [Seedling X Orange Sensation]; deRuiter

Orange Meillandina, *see* 'MEIJIKATAR'

Orange Meillandina, Climbing, *see* 'MEIJIKATARSAR'

Orange Mini-Wonder, *see* 'LAVJACK'

'ORANGE MIST', F, ob, 1957; flowers orange-salmon to yellow-orange, dbl., 4 in., moderate fragrance; dwarf, bushy growth; [Ma Perkins X Unnamed seedling]; Boerner; J&P

'ORANGE MORSDAG', Pol, ob, 1956; (Orange Mothersday); flowers deep orange; [Mothersday sport]; Grootendorst, F.J.

Orange Mothersday, *see* **'ORANGE MORSDAG'**

Orange Muttertag, *see* **'VATERTAG'®**

'ORANGE NASSAU', HT, ob, 1941; bud long, pointed; flowers two-toned coppery orange, reverse yellow, dbl., cupped, moderate fragrance; foliage leathery; [Mev. G.A. van Rossem X Unnamed seedling]; Verschuren; Dreer

'ORANGE NYMPH', F, or, 1960; flowers medium, borne in small clusters, dbl.; [Nymph sport]; Tulp; G. Maarse

'ORANGE PARFAIT', HT, ob, 1982; bud ovoid, pointed; flowers orange, yellow reverse, large blooms borne singly, dbl., 45 petals, exhibition form, slight, tea fragrance; long, narrow prickles, hooked downward, reddish at base; foliage medium to large, bronze green, leathery; vigorous growth; Weeks, O.L.; Weeks Wholesale Rose Growers

Orange Parfait, Min, ob, 1998; Moore

Orange Passion™, *see* 'JACCOGEL'

Orange Pastel™, *see* 'WILORPAT'

'ORANGE PERFECTION', Pol, or, 1927; [Ideal sport]; Spek

'ORANGE PIXIE', Min, or, 1978; bud ovoid, pointed; flowers bright orange-red, dbl., 48 petals, 1 in., exhibition form; foliage small, glossy, leathery; bushy, compact, upright growth; [Little Chief X Fire Princess]; Moore, Ralph S.; Sequoia Nursery

'ORANGE QUEEN', Pol, op, 1923; flowers salmon-orange; [Orléans Rose sport]; van Nes

'ORANGE RAPTURE', HT, op, 1935; bud long, pointed; flowers orange with pink, large, dbl., exhibition form, moderate fragrance; [Rapture sport]; Schmidt, K.; Kordes

Orange Rosamini®, *see* 'RUIseto'

'ORANGE ROSETTE', Pol, or, 1941; flowers scarlet-orange, borne in clusters, dbl., 30 petals, 1–1.25 in.; vigorous, compact growth; [Gloire du Midi sport]; deRuiter; J&P

'ORANGE RUFFELS', HT, ob, 1952; bud long, pointed; flowers orange to saffron-yellow, dbl., 4–5 in., exhibition form, moderate fragrance; foliage dark, glossy; vigorous, compact growth; [(Dr W. Van Fleet X Général Jacqueminot) X Lafter]; Brownell, H.C.

'ORANGE RUMBA', F, ob, 1962; flowers orange, [Rumba sport]; Zieger; C-P

Orange Sauvageot, F, ob, 1987; Sauvageot

'ORANGE SCEPTER', HT, yb, 1953; flowers orange-yellow with dark golden yellow, large, dbl.; Verschuren-Pechtold; (Sangerhausen)

Orange Scepter, HT, ob

'ORANGE SCHOON', HT, ob, 1938; flowers pure orange; [Katharine Pechtold sport]; Lens

Orange Seabreeze, MinFl, op, 1996; Williams, J. Benjamin

'ORANGE SENSATION'®, F, or, 1961; dbl., 24 petals, 3 in., blooms in clusters, moderate fragrance; foliage dark; vigorous, bushy growth; GM, NRS, 1961 Golden Rose, The Hague, 1968; deRuiter; Gregory

Orange Sherbert, *see* 'LYOra'

'ORANGE SILK', F, or, 1968; flowers orange-vermilion, shallow-, large blooms in large clusters, cupped, slight fragrance; foliage dark, glossy; [Orangeade X (Ma Perkins X Independence)]; McGredy, Sam IV; Gregory

'ORANGE SMOKE', F, ob, 1964; flowers orange, with blue haze in cool weather, 3–4 in.; very vigorous growth; [Orange Ruffels X (Eva X Guinee)]; Hennessey

Orange Sovereign, Pol, ob

Orange Sparkle, *see* 'KORsparko'

'ORANGE SPARKS', HT, ob, 1969; flowers orange-vermilion blend, striped gold, medium, dbl.; foliage glossy, bronze; vigorous, bushy growth; [Cherry Brandy sport]; Vasishth

'ORANGE SPECIAL', F, or, 1957; flowers orange-scarlet, small, borne in large trusses, dbl., 30–35 petals, 1.5 in., cupped, dwarf, bushy growth; [Salmon Perfection sport]; deRuiter; Blaby Rose Gardens

'ORANGE SPICE', Min, ob, 1980; bud ovoid, pointed; flowers orange, deeper on petal edges, blooms borne singly or severa, dbl., 23 petals, cupped, moderate fragrance; recurved prickles; foliage small, medium green; bushy, upright growth; [Unnamed seedling X Unnamed seedling]; Lyon

Orange Spice, *see* 'SOCotra'

Orange Splash, *see* 'JACseraw'

Orange Splendor, F, ob, 1998; Select Roses, B.V.

Orange Star™, *see* 'MINalco'

'ORANGE STARINA', Min, ob, 1981; flowers light orange, [Starina sport]; Graff, Roy; Mini Roses

Orange Sunblaze™, *see* 'MEIjikatar'

Orange Sunblaze, Climbing, *see* 'MEIjikatarsar'

Orange Sunset, *see* 'MANorsun'

'ORANGE SUNSHINE', Min, ob, 1968; flowers orange, [Bit o' Sunshine sport]; Moore, Ralph S.; Sequoia Nursery

Orange Supreme, HT, ob

Orange Surprise, F, ob, 1998

'ORANGE SWEETHEART', F, op, 1952; bud ovoid; dbl., 20–25 petals, 3–3.5 in., cupped, slight, fruity fragrance; foliage dark; upright, bushy growth; [Pinocchio seedling X Fashion]; Boerner; J&P

Orange Symphonie®, *see* 'MEIninrut'

Orange Tango, *see* 'DANorang'

'ORANGE THÉRÈSE', HT, ob, 1943; flowers orange-yellow, [Soeur Thérèse sport]; Howard Rose Co.

'ORANGE TRIUMPH'®, Pol, mr, 1937; flowers small blooms in clusters, semi-dbl., cupped, slight fragrance; foliage glossy; compact, bushy growth; GM, NRS, 1937; [Eva X Solarium]; Kordes; Dreer

'ORANGE TRIUMPH IMPROVED', Pol, or, 1960; flowers orange, borne in large clusters, dbl.; [Orange Triumph sport]; Cant, F.

'ORANGE TRIUMPH SUPERBA', Pol, ob, 1953; flowers clear orange; [Orange Triumph sport]; Maarse, J.D.

'ORANGE TRIUMPH, CLIMBING', Cl Pol, mr, 1945; Leenders, M., 1945; Koopmann, 1948

Orange Twist, *see* 'MORtwist'

Orange Unique, HT, lp; flowers medium orange, reverse rose pink, exhibition form, long stems

'ORANGE VELVET', LCl, or, 1986; flowers bright orange-red, reverse orange-red to orange pink, fading, dbl., 37 petals, exhibition form, moderate, damask fragrance; medium, rounded, medium bright orange fruit; foliage large, dark green, glossy, dark waxed; upright, tall, vigorous growth; [Tropicana, Climbing X Swarthmore]; Williams, J. Benjamin

Orange Vilmoria, *see* 'KORsee'

Orange Vilmorin, *see* 'KORsee'

Orange Zest, *see* 'LAVtrice'

'ORANGEADE'®, F, or, 1959; flowers bright orange-red, blooms in clusters, semi-dbl., slight fragrance; foliage dark; very vigorous, bushy growth; GM, NRS, 1959 GM, Portland, 1965; [Orange Sweetheart X Independence]; McGredy, Sam IV; McGredy

'ORANGEADE, CLIMBING', Cl F, or, 1964; Waterhouse Nursery

'ORANGE-RED SUPREME', F, or, 1958; bud ovoid; flowers open, borne in clusters, dbl., 18–24 petals, 2.5–3 in., moderate fragrance; foliage glossy, wrinkled; bushy growth; [Spice X Garnette seedling]; Boerner; J&P

Oranges 'n' Lemons, *see* 'MACoranlem'

Orangina®, *see* 'ORAmont'

'ORANIËN', F, or, 1962; flowers orange-red, sometimes lined yellow, base light yellow, large, dbl., 42 petals; foliage dark; upright, compact, symmetrical growth; [Highlight X Unnamed seedling]; Verschuren, A.; van Engelen

Oranzova Garnette, F, lp; Strnad; (Czech Rosa Club)

ORAred, F, mr, 1991; (**Romy Schneider®**); Orard

ORAreg, F, lp, 1997; (**Chateau de Bagnols**); Orard

ORAroro, HT, ly; (**Sourire d'Enfant®**); Orard, 1989

ORAstrip, HT, rb, 1998; (**Meli-Melo**); Orard

'ORATAM', S, yb, 1939; bud globular; flowers pink edged copper-pink, base and reverse yellow, large, dbl., non-recurrent, intense, damask fragrance; foliage leathery, dark, matt, yellow-green; bushy growth; [R. damascena X Souv. de Claudius Pernet]; Jacobus; B&A

Oratia Maid, F, lp, 1986; Garelja, Anita

Oratorio, *see* 'WARdido'

ORAuna, F, pb, 1992; (**Cyclamen La Sevillana**); Orard

ORAval, HT, lp; (**Ville de Bar sur Seine**); Orard, 1996

ORAwest, S, yb, 1999; (**President Wilson**); Orard

Orchid Jubilee, *see* 'MORclilav'

Orchid Lace, *see* BENorchid

'ORCHID MASTERPIECE', HT, m, 1960; bud ovoid, deep orchid; flowers lavender-orchid, large, dbl., 68 petals, moderate, fruity fragrance; foliage leathery, dark; vigorous growth; [Golden Masterpiece X Grey Pearl seedling]; Boerner; J&P

Oregold, *see* 'TANolg'

'OREGON CENTENNIAL', HT, mr, 1959; bud pointed; flowers rose-red, dbl., 30–35 petals, 4–5 in., exhibition form, moderate, fruity fragrance; foliage dark; long, strong stems; vigorous, upright, bushy growth; [Charles Mallerin X (Charles Mallerin X Chrysler Imperial)]; Von Abrams; Peterson & Dering

'OREGON OPHELIA', HT, mp, 1921; flowers salmon, edged pink, base yellow, dbl.; [Ophelia sport]; Clarke Bros.

Oregon Rainbow, *see* 'CLERAIN'

Oreste Sgaravatti, HT; Sgaravatti, A.; (Cavriglia)

Orfeo, LCl, 1963; Leenders, J.; (Cavriglia)

'ORFEO, CLIMBING', LCl, dr, 1963; flowers deep red, large, dbl., moderate fragrance; [Curly Pink X Guinee]; Leenders, J.

Organdi, HT, dr

'ORGUEIL DE LYON', HP, or, 1886; flowers vermilion; Besson

'ORIANA', HT, rb, 1970; flowers cherry-red, reverse white, dbl., 38 petals, 5 in., slight fragrance; foliage glossy, dark; Tantau, Math.; Wheatcroft & Sons

'ORIENT', HT, dr, 1959; bud pointed; flowers crimson becoming vermilion, reverse vermilion with yellow base, dbl., 22 petals; [Queen of Bermuda X Henri Mallerin]; Dot, Simon

'ORIENT EXPRESS', HT, or, 1978; flowers deep orange-red, reverse lighter, dbl., 40 petals, 4–5 in., intense fragrance; foliage bronze; vigorous growth; [Sunblest X Seedling]; Wheatcroft

Orient Silk, *see* **'AHIMSA'**

'ORIENT SPICE', HT, m, 1996; (Rajni); flowers pinkish lilac, reverse same, blooms borne mostly single, dbl., 26–40 petals, 5 in., intense fragrance; moderate prickles; foliage large, medium green, semi-glossy; bushy, medium growth; [Violaine X Margaret Merrill]; Viraraghavan, M.S.; Anand Roses, 1984

'ORIENTAL CHARM', HT, mr, 1960; bud globular; semi-dbl., 11 petals, 3–4 in., slight fragrance; foliage leathery, glossy, dark; vigorous, upright, bushy growth; [(Charlotte Armstrong X Gruss an Teplitz) X (Mme Butterfly X Floradora)]; Duehrsen; Elmer Roses Co.

Oriental Dawn, *see* **'ASAGUMO'**

Oriental Emporer, LCl, or, 1999; Clements, John K.

'ORIENTAL GLAMOUR', F, mr, 1972; flowers orient red, semi-dbl., 15 petals, 4–5 in., exhibition form, slight fragrance; foliage bronze; free growth; [Coup de Foudre X Tropicana]; Gandy, Douglas L.

Oriental Palace, *see* POUlorin

'ORIENTAL QUEEN', Ch, or, 1926; flowers brilliant orange-scarlet, base yellow, fading to vivid carmine, dbl., 26 petals; few thorns; foliage dark; McGredy; Beckwith

Oriental Simplex™, *see* 'MINANCO'

'ORIENTALE', Cl HT, ob, 1946; flowers coral, becoming old-rose, free, recurrent bloom, moderate fragrance; foliage glossy; very vigorous growth; not dependably hardy; [George Dickson X Mrs Pierre S. duPont]; Robichon

'ORIFLAMME', HMult, pb, 1914; flowers deep rose-pink, suffused coppery gold, dbl., 55 petals, 2.5 in., moderate fragrance; vigorous, climbing growth; Paul

Origami™, *see* 'AROCHARM'

'ORIHIME', HT, pb, 1983; flowers light pink flushed rose pink, reverse pearly light pink; [Confidence sport]; Itami Rose Nursery

O'Rilla, *see* 'RESILLA'

'ORIMANDA', HT, dr, 1975; bud long, pointed; dbl., 27 petals, 3.5 in., exhibition form, slight fragrance; foliage glossy, dark; vigorous, upright, bushy growth; [Duftwolke X Seedling]; Kordes; Willemse

'ORINDA', S, ly, 1922; flowers deep cream, medium, dbl., profuse, non-recurrent bloom, foliage soft, dark; vigorous (5 ft), bushy growth; [Harison's Yellow X?]; Central Exp. Farm

Oriola, *see* 'MACLOCKER'

'ORIOLE', HMult, my, 1912; flowers marechal niel yellow, dbl.; vigorous, climbing growth; [Aglaia seedling]; Lambert, P.

'ORION', F, mr, 1968; flowers scarlet, medium, dbl., slight fragrance; foliage glossy; [Pink Parfait X Red Dandy]; Harkness

Orione®, HT, ob

Orlando, *see* INTervema

Orlando Sunshine, *see* 'MICADA'

'ORLÉANS IMPROVED', Pol, mp, 1931; flowers vivid rose-pink, [Orléans Rose sport]; Norfolk Nursery

'ORLÉANS ROSE', Pol, rb, 1909; flowers vivid rosy crimson, center white, semi-dbl., repeat bloom, slight fragrance; foliage glossy; vigorous, bushy growth; (14); [Thought to be Mme Norbert Levavasseur seedling sport]; Levavasseur

'ORLÉANS ROSE, CLIMBING', Cl Pol, rb, 1913; Levavasseur

'ORMISTON ROY', S, dy, 1953; flowers large, single, non-recurrent; bushy (3 ft) growth; [R. spinosissima X R. xanthina]; Doorenbos

Ornella Muti®, HT, pb

'ORNEMENT DE LA NATURE', HGal, m, 1826; Hardy, before 1814

Ornement des Bosquets, HSem, dp; Jamain, H., 1860

'ORNEMENT DU LUXEMBOURG', HP, m, 1840; flowers violet/red, small, dbl., intense fragrance; Hardy; (Sangerhausen)

'ORPHELINE DE JUILLET', HGal, m; flowers crimson-purple, base of petals fiery red with occasional str, very dbl.; erect, moderate growth; Paul, before 1836; Prior to 1837

'ORPHEUS', HT, op, 1963; flowers salmon-orange, medium, dbl., 35–40 petals; foliage glossy, dark; long stems; vigorous growth; [Montezuma X Unnamed seedling]; Verschuren, A.; Stassen

'ORPINGTON GEM', HT, or, 1953; bud long, pointed; flowers coppery orange shaded gold, dbl., 32 petals, 3 in., intense fragrance; foliage dark, leathery; vigorous growth; [Princess Marina X Alamein]; Buckwell

'ORPINGTON JEWEL', HT, op, 1953; bud very long; flowers coral-pink shaded coppery, base orange, medium size, dbl., 30 petals, 3 in., intense fragrance; foliage dark, leathery; vigorous growth; [Princess Marina X Picture]; Buckwell

Orsen's Big Red, HT, mr, 1990

Orsola Spinola, *see* MASmarti

'ORTANA', HT, w, 1994; (**Anna Katherine**); flowers white blend, moderately full, dbl., 15–25 petals, 1.5–2.75 in., borne mostly singly, intense fragrance; some prickles; foliage large, medium green, matt; tall, upright growth; [Osiana X Seedling]; Ortega, Carlos, 1994; Aebi Nursery, 1994

'ORTCAL', HT, rb, 1995; (**California Glory**); very dbl., 3–3.5 in., borne mostly singly, intense fragrance; many prickles; foliage medium, dark green, matt; tall (150 cms), upright growth; [Osiana X Kardinal]; Ortega, Carlos; Aebi Nursery

'ORTISA', HT, mp, 1996; (**Marisa**™); flowers medium pink, lighter pink reverse, classic, urn-shaped bloom, dbl., 15–25 petals, 5 in., no fragrance; moderate prickles; foliage medium, dark green, semi-glossy; upright, tall (5 ft) growth; [Paris d'Yves St. Laurent sport]; Ortega, Carlos; Aebi Nursery, 1996

'ORTKEE', HT, ly, 1993; (**Kika**); dbl., 15–25 petals, 3–3.5 in., borne mostly singly, intense fragrance; no prickles; foliage medium, dark green, semi-glossy; medium (150 cms), upright growth; [JACice X Seedling]; Ortega, Carlos; Aebi Nursery, 1993

'ORTLAE', HT, m, 1991; (**Lauren Elizabeth**); flowers mauve blend, large blooms borne mostly singly, dbl., intense fragrance; foliage large, dark green, matt; tall, upright growth; [Moonlight X Unnamed seedling]; Ortega, Carlos, 1992

'ORTMAR', HT, w, 1995; (**Margaret O**); very dbl., 1.5–2.75 in., borne mostly singly, intense fragrance; some prickles; foliage medium, dark green, matt; tall (180 cms), upright growth; [Pristine X Lauren Elizabeth]; Ortega, Carlos; Aebi Nursery, 1995

'ORTSAR', HT, m, 1997; (**Sterling Star**™); flowers very full, large, very dbl., 41 petals, borne mostly singly, intense fragrance; foliage medium, medium green, dull; upright, tall (8–10 ft) growth; [Lauren

Elizabeth X Sterling Silver]; Ortega, Carlos

'ORTsea', HT, pb, 1997; (**Chelsea**™); flowers double, medium, dbl., 15–25 petals, borne mostly singly, slight fragrance; foliage small, dark green, semi-glossy; upright, medium (4–5 ft) growth; [Lauren Elizabeth X Paul's Pink]; Ortega, Carlos

'ORTvic', HT, w, 1997; (**Victoria's Song**™); flowers double, large, dbl., 15–25 petals, borne mostly singly, intense fragrance; foliage medium, medium green, semi-glossy; upright, tall (6.5–7 ft) growth; [Moonlight X Seedling]; Ortega, Carlos

Oscar, S, pb, 1994; Hauser

'Oscar Chauvry', N, mp, 1900; flowers large, dbl., moderate fragrance; Chauvry; (Sangerhausen)

'Oscar II Roi de Suède', HP, mr, 1890; flowers large, very dbl., moderate fragrance; Soupert & Notting; (Sangerhausen)

'Oscar Leclerc', M, 1853; Moreau et Robert; (Cavriglia)

Osiana, HT, ab, 1988; PP007660; Tantau

'Osiria', HT, rb, 1978; bud long, pointed; flowers dark red, white reverse, blooms, dbl., 50 petals, 4.5 in., exhibition form, intense fragrance; short stems; vigorous, upright, bushy growth; [Snowfire X Unnamed seedling]; Kordes, W.; Willemse

Osjen, HT, dp; flowers carmine-pink, medium, dbl., moderate fragrance; (Sangerhausen)

Oskar Cordel, *see* **'Oskar Kordel'**

'Oskar Kordel', HP, mp, 1897; (Oskar Cordel); flowers carmine, large, dbl., 40 petals, cupped, moderate fragrance; vigorous, compact growth; [Merveille de Lyon X Andre Schwartz]; Lambert, P.

'Oskar Scheerer', S, dr, 1961; flowers velvety dark red, large blooms in large clusters, dbl.; vigorous (6 ft), well-branched growth; Kordes, R.

'Osmunda', Pol, dr, 1923; flowers dark carmine; [Jessie sport]; Holland (?)

'Ostara', F, mr, 1964; flowers bright red, medium, borne in clusters, dbl.; foliage dark; [Highlight X Valeta]; deRuiter

Ostrava, S, lp; Vecera, L., 1971; (Czech Rosa Club)

'Oswald Sieper', HT, w, 1933; bud long, pointed; flowers creamy white, very large, dbl., exhibition form, intense fragrance; foliage glossy; vigorous growth; [Mrs Charles Lamplough X Ville de Paris]; Krause; C-P

Otago®, *see* 'MACnecta'

Othello, *see* 'AUSlo'

'Othello', F, or, 1961; flowers light orange-red, medium, dbl., 22 petals, 2.5 in., cupped; [Ma Perkins X Cocorico]; Leenders, J.

'Othello', HT, mr, 1963; flowers velvety bright red, large, dbl., 48–60 petals, moderate fragrance; foliage dark; vigorous, upright, bushy growth; [New Yorker X Unnamed seedling]; Verschuren, A.; Stassen

'Otohime', HT, or, 1977; (Fancy Princess); bud pointed; dbl., 58 petals, 6–6.5 in., exhibition form, moderate fragrance; foliage large, glossy, dark; vigorous growth; [(Hawaii X Tropicana) X (Tropicana X Peace)]; Keisei Rose Nurseries, Inc.

'Otome Pink', HT, pb, 1967; flowers salmon-pink, reverse buff-yellow; [Amatsu-Otome sport]; Rumsey, R.H.

'Otto Krause', HT, dy, 1931; flowers coppery yellow, large, dbl., exhibition form, slight fragrance; foliage soft, glossy, bronze, dark; vigorous growth; [Mme Caroline Testout X Souv. de Claudius Pernet]; Weigand, C.

'Otto Miller', HT, mr, 1968; bud long, pointed; flowers deep red, dbl., 6 in., exhibition form, slight fragrance; foliage glossy, leathery; vigorous, upright growth; [(Chrysler Imperial X Independence) X (Mrs Charles Russell X Happiness)]; Morey, Dr Dennison; Country Garden Nursery

'Otto von Bismarck', HT, mp, 1908; flowers large, dbl., moderate fragrance; Kiese; (Sangerhausen)

Oudtshorn Joy, *see* KORquick

'Ouma Smuts', F, ly, 1950; flowers straw-yellow shaded rosy flesh, intense fragrance; [Egalite X Vanessa]; Leenders, M.

'Our Allies', Pol, lp; Matthews, W.J., 1915; (Weatherly, L.)

'Our Annie', HT, op, 1937; flowers vivid orange-cerise, large, dbl., globular, moderate fragrance; foliage glossy; very vigorous growth; Letts

'Our Bob', HT, mr, 1928; flowers velvety red, center golden, intense fragrance; Dawes

Our Coral Pearl, *see* 'DOSpearl'

Our Diana, *see* 'TWOex'

Our Indira, HT, w, 1998; Viraraghavan, M.S.

Our Joy, *see* 'HORjoy'

Our Jubilee, *see* 'COCcages'

'Our Lady', F, mr, 1956; flowers carmine, moderate fragrance; vigorous growth; [Seedling X Soestdijk]; Leenders, M.

Our Love, *see* 'ANDour'

Our Molly, *see* 'DICreason'

'Our Pearl', LCL, pb, 1993; flowers white with pink edges, blooms borne in small clusters, dbl., 13–18 petals, 1.5–2.75 in., slight fragrance; some prickles; foliage medium, medium green, matt; medium (7 ft), bushy, spreading growth; Bronze, ARC TG, 1993; Jerabek, Paul E.

'Our Princess', F, dr, 1949; (Crimson Glow); flowers deep velvety crimson, borne in large trusses, semi-dbl., slight fragrance; foliage glossy, dark; vigorous growth; (28); [Donald Prior X Orange Triumph]; Robinson, H.; Baker's Nursery

Our Rodeo, F, rb, 1997; Kordes

'Our Rosamond', HT, pb, 1983; flowers silver and pink blend, large blooms borne singly, dbl., 35 petals, exhibition form, slight fragrance; foliage medium, medium green, glossy; upright growth; [Daily Sketch seedling X Red Planet]; Bell, Ronald J.; Treloar Roses Pty. Ltd.

'Our Shirley', HT, w, 1989; bud pointed; flowers medium, borne singly, dbl., 28 petals, no fragrance; prickles long, slender, red; foliage small, medium, dark green; medium, upright growth; [Judith Morton X Sylvia]; Wilson, George D.

'Our Sweet Ann', F, op, 1975; flowers warm pink, with orange glow, full, dbl., 28 petals, 3.5–4 in., intense fragrance; foliage dark, leathery; vigorous, upright growth; [Queen Elizabeth X Elizabeth of Glamis]; Horsfield

Our Town, *see* 'ZIPtown'

Our Victory, *see* **'Nacha Pobeda'**

Out of Africa, *see* KORocken

Out of Canada, HT, pb, 1998; Williams, J. Benjamin

Out of Yesteryear, *see* 'MORyears'

Outrageous™, *see* 'JACzap'

'Outsider', F, mr, 1956; bud ovoid; flowers bright blood-red, open, medium, borne in clusters, semi-dbl., slight fragrance; foliage glossy; vigorous, upright growth; [Fanal X Red Favorite]; Tantau, Math.

'Ovation', HT, or, 1977; bud ovoid; dbl., 25 petals, 4–4.5 in., exhibition form; foliage dark, leathery; vigorous, upright growth; [First Prize X Unnamed seedling]; Weeks

'Over the Rainbow', Min, rb, 1972; flowers red, yellow reverse, small, dbl., exhibition form, slight fragrance; foliage leathery; vigorous, bushy growth; AOE 1975; [Little Darling X Westmont]; Moore, Ralph S.; Sequoia Nursery

'Over the Rainbow, Climbing', Cl Min, rb, 1974; Rumsey, R.H.

'Overloon', F, dp, 1949; flowers rosy pink, medium, dbl.; vigorous growth; [Irene X Hebe (F)]; Leenders, M.

Overnight Scentsation, *see* 'SAVanight'

'Overton on Dee', HT, pb, 1976; flowers pink, reverse cream, very full, dbl., 40 petals, 4.5 in., slight fragrance; [(Ballet X Gavotte) X Orange Sensation]; Ellick

'Overture', F, m, 1960; flowers lilac-lavender, well formed, semi-dbl., 18 petals, 3–3.5 in., moderate fragrance; foliage dark; vigorous, low growth; [(Seedling X Lavender Pinocchio) X Prelude]; LeGrice

'OVID', HSet, 1890; Geschwind, R.; (Cavriglia)

Owen's Pride, *see* KIRpink

Oxbow, *see* 'BERBOW'

'OXFAM', HT, pb, 1973; flowers deep pink, shaded lilac, dbl., 40 petals, 5 in., intense fragrance; foliage dark; upright growth; [Fragrant Cloud X Blue Moon]; Cobley; Harry Wheatcroft Gardening

'OXFORD', HT, yb, 1930; flowers deep warm peach, tipped orange-gold, well-shaped, large, moderate fragrance; Prince

Oxfordshire, *see* 'KORFULLWIND'

Oyster Pearl, HT, lp; [Bewitched sport]

Oz Baby, Min, lp; [Poker Chip X Oz Gold]; Hannemann, F., 1990; The Rose Paradise; (Weatherly, L.)

Oz Gold, Min, ob, 1981; McGredy, Sam IV

P

P.J.'s Pride, *see* 'JUDPRIDE'

P.M. Leenders, *see* **'PRESIDENT MACIA'**

Pablito, Min, mp

Pablo Diez, HT; Moreira da Silva, A.; (Cavriglia)

Pace Setter, *see* 'SAVAPACE'

Pacemaker, *see* 'HARNOBLE'

Pacesetter, *see* 'SAVAPACE'

Pachinko, *see* 'SUNPACH'

'PACIFIC', HT, lp, 1927; flowers soft pink, semi-dbl.; [Los Angeles sport]; Pacific Rose Co.

'PACIFIC', F, mr, 1958; flowers bright red, semi-dbl.; foliage bright green; [Alain X Chanteclerc]; Gaujard

'PACIFIC', HT, lp; [Catherine Kordes X Rose Marie]; Heers, C W, 1948; (Weatherly, L.)

'PACIFIC BELLE', HT, mp, 1989; bud pointed; flowers light pink, reverse deeper pink, reflexed, pointed, medium, dbl., 31 petals, slight fragrance; prickles pointed, light brown; foliage light green, veined, glossy; bushy, branching, upright growth; [(Unnamed Peer Gynt seedling) X Josephine Bruce]; Cattermole, R.F.

'PACIFIC DARLING', Min, pb, 1991; flowers large blooms borne in large clusters, semi-dbl., 6–14 petals, slight fragrance; foliage large, dark green, glossy; low, upright, bushy, compact growth; [Winifred Coulter X Unnamed seedling]; Sudol, Julia

'PACIFIC PRINCESS', HT, mp, 1989; bud tapering; flowers medium, borne usually singly and in sprays of 3–4, very dbl., 56 petals, globular, intense fragrance; foliage dark green, semi-glossy, veined; upright, branching growth; [Pink Parfait X Red Planet]; Cattermole, R.F.; South Pacific Rose Nursery, 1988

Pacific Serenade™, *see* 'SAVANADE'

'PACIFIC SUNSET', HT, pb, 1977; bud deep pointed; flowers pink to orange-scarlet, semi-dbl., 12–16 petals, 5 in., exhibition form, slight fragrance; foliage dark; upright growth; [Mme Henri Guillot X California]; Fong; United Rose Growers

'PACIFIC TRIUMPH', Pol, mp, 1949; flowers salmon-pink, intense fragrance; growth like parent; [Orange Triumph sport]; Heers; Pacific Nursery

Pacifica®, *see* 'JACIF'

'PACOIMA', HT, lp, 1927; flowers flesh tinted yellow, dbl.; [William F. Dreer sport]; Pacific Rose Co.

Paddy McGredy, *see* 'MACPA'

Paddy Stephens, *see* 'MACCLACK'

Paddywack™, *see* 'MINIWACK'

Padmavathi, HT, pb, 1995; K&S

'PADRE', HT, rb, 1921; flowers bright coppery scarlet, flushed yellow, well-formed, semi-dbl., moderate, fruity fragrance; vigorous growth; Cant, B. R.

'PADRE AMÉRICO', HT, mr, 1956; flowers carmine-red, intense fragrance; [Crimson Glory X Peace]; da Silva, Moreira

'PADRE CRUZ', HT, w, 1956; flowers rosy white edged ruby; [Branca X Peace]; da Silva, Moreira

'PADRE MAÑANET', HT, dr, 1957; flowers bright purple-garnet, reverse crimson-red, large, dbl., 30 petals, exhibition form, moderate fragrance; strong stems; upright growth; [Charles Mallerin X (Satan X Mirandy)]; Dot, Pedro

'PAEONIA', HP, dr, 1855; flowers crimson, large, dbl., intense fragrance; Lacharme, F.

Paeonia, HT, lp; Strnad; (Czech Rosa Club)

Paeonienrose, HGal, mp

'PAGAN BEAUTY', F, pb, 1965; flowers bright vermilion, reverse salmon-pink, dbl., 42 petals, 4–5 in., globular; foliage dark, leathery; vigorous growth; [Montezuma X Seedling]; Verschuren; Blaby Rose Gardens

Paganini, HMsk, rb, 1989; Lens

Paganini, *see* MEIrolange

'PAGEANT', HT, rb, 1953; bud ovoid; flowers red, reverse yellow, dbl., 35–40 petals, 4.5–5 in., exhibition form, moderate fragrance; foliage leathery; vigorous, upright growth; [Unnamed Hybrid Tea seedling X Orange Nassau seedling]; Boerner; J&P

'PAGLIACCI', Gr, yb, 1969; bud pointed; flowers yellow, becoming cerise, medium, semi-dbl., moderate fragrance; foliage glossy, bronze; vigorous, upright growth; Von Abrams; Edmunds Roses

'PAGODA', S, dr, 1995; flowers dark velvet red with showy gold stamens, small, single, (5 petals), borne in large clusters, slight fragrance; no prickles; foliage medium, medium green, semi-glossy; low, spreading, bushy growth; [Anytime X Mountain Mist]; Jobson, Daniel J.; Jobson, 1995

Pahadi Dhun, HT, m, 1981; Pal, Dr B.P.

Paheka, HT, ly, 1980; Dawson

Pailine™, HT, ob; flowers yellow with pink edges

'PAINT BOX', F, yb, 1963; bud ovoid; flowers red and golden yellow, becoming deep red, blooms in clusters, semi-dbl., 3 in., flat, slight fragrance; foliage dark; vigorous, upright growth; [Seedling X St Pauli]; Dickson, Patrick; A. Dickson

'PAINTBRUSH', Min, ly, 1975; bud mossy; flowers soft yellow to white, mini-moss, small, 8–10 petals, 1.5 in.; foliage small, glossy, leathery; vigorous, upright, bushy growth; [Fairy Moss X Goldmoss]; Moore, Ralph S.; Sequoia Nursery

Painted Damask, *see* **'LÉDA'**

'PAINTED DESERT', HT, op, 1965; bud long, pointed; flowers pink and copper, open, dbl., moderate fragrance; foliage leathery; vigorous, upright growth; [Talisman sport]; Lone Star Rose Nursery

Painted Desert, *see* 'MACCHOPSU'

Painted Doll™, *see* 'LAVPAINT'

'PAINTED LADY', HT, mp, 1931; bud long, pointed; flowers bright cerise, base old-gold, large, dbl., 42 petals, [(Crusader X Premier) X Julien Potin]; Ward, F. B.

'PAINTED LADY', HT, yb, 1979; bud pointed; flowers cream and gold, red tipped, dbl., 4 in., slight fragrance; foliage glossy, bronze; bushy growth; Herholdt, J.A.

Painted Melody, HT, yb, 1993; K&S

Painted Moon, *see* 'DICPAINT'

'PAINTED SPAIN', Min, ob, 1995; flowers hand-painted orange stippled with white, white reverse, dbl. (15–25) petals, medium blooms borne in small clusters; slight fragrance; numerous prickles; foliage medium, medium green, semi-glossy; tall (3 ft), almost climbing, upright growth; [Rise 'n' Shine X (Queen Elizabeth X Eyepaint)]; Jobson, Daniel J; Jobson.

Painted Star, *see* 'MACMIGMOU'

Painter's Palette, *see* 'MORPALE'

Painter's Touch, *see* 'LAVTUCH'

Paint-Pot, *see* 'TROBGLOW'

Paisley Anniversary, *see* 'ANDPAI'

Pal, HT, ob

'PALACKY', HT, ob, 1936; flowers orange-yellow, open, large, single, intense fragrance; foliage glossy; vigorous growth; [Mme Mélanie Soupert X Sunburst]; Böhm, J.

'PALADIN', F, or, 1960; flowers open, borne in clusters, semi-dbl., 2.5 in.; moderate growth; [Signal Red X Fashion]; deRuiter

Palais de Laeken, HGal, lp; flowers medium, dbl.; (Sangerhausen)

'PALATINO', F, mr, 1956; dbl., very vigorous growth; Frères, Buyl

'PALE HANDS', HT, ab, 1965; bud pointed; flowers ivory-white to buff and peach, very large, dbl., moderate fragrance; foliage leathery; vigorous, upright growth; [McGredy's Ivory X ?]; Pal, Dr B. P.; Indian Agric. Research Inst.

'PALE MOON', HT, my, 1967; flowers large, dbl., exhibition form, moderate fragrance; foliage glossy; vigorous, compact growth; [Ma Perkins X Peace]; Patterson; Patterson Roses

'PALEFACE', Gr, w, 1959; bud ovoid; flowers nearly white, base naples yellow, semi-dbl., 15–25 petals, 3–5 in., exhibition form, slight fragrance; foliage leathery, semi-glossy; vigorous, upright growth; [Joanna Hill X Unnamed seedling]; Lindquist; Howard Rose Co.

'PALETTE', F, yb, 1960; bud ovoid; flowers yellow to salmon-pink and then red, blooms in clusters, dbl., slight fragrance; foliage glossy; compact growth; [Masquerade X High Noon]; Leenders, J.

Palissade Rose®, *see* 'KORDAPT'

Paljas®, HT, pb

Pallas, *see* 'HARVESTAL'

'PALLAS', HT, dr, 1975; flowers garnet-red, reverse paler, dbl., 40 petals, exhibition form, intense, spicy fragrance; foliage large, dull, dark; tall growth; [Chrysler Imperial X Shannon]; Murray & Hawken; Kennedy

'PALLIDA', Ch, mp, 1789; (Bengal Ordinaire); flowers clear rose; Kerr

'PALM SPRINGS', F, rb, 1965; bud ovoid; flowers oriental red to light bronze, center yellowish pink, dbl., 50 petals, intense fragrance; foliage bronze; vigorous, upright growth; [Oriental Charm X Circus]; Duehrsen; Elmer Roses Co.

Palmengarten Frankfurt®, *see* KORsilan

Palmetto Sunrise, *see* 'MICPAL'

Palmira Bastos, HT; Moreira da Silva, A.; (Cavriglia)

'PALMYRE', P, lp, 1844; flowers pale pink, portland type, often repeats, Laffay, M.

Palocsay Rudolf, HT, lp; long; flowers well shaped suitable for glasshouse as cut flower, exhibition form, 30 petals, moderate; leaves medium, medium green, glossy; semi-vigorous upright; [(Charles P. Kilham X R. harisonii) X Ville de Paris]; St Wagner, 1950; Res. Stn. f. Horticulture, Cluj, 1955, Romania

'PALOK', HR, rb, 1993, (**Brisbane Blush**); flowers cerise with silvery cream base, full (26–40 petals), large (7+ cms) blooms borne in small clusters; fragrant; few prickles; foliage medium, medium green, semi-glossy; medium upright growth; [(Golden Slippers X Lavendule) X Prima Ballerin]; Long, Peter; Queensland Rose Society, 1993

'PALOMA', HT, w, 1968; bud urn-shaped; flowers large, dbl., exhibition form; foliage leathery, olive-green; vigorous, bushy growth; GM, Portland, 1971; [Mount Shasta X White Knight]; Swim & Weeks; Weeks Wholesale Rose Growers

'PALOMA BLANCA', S, w, 1984; flowers ivory-white, blooms borne 3–10 per cluster, dbl., 35 petals, cupped, repeat bloom, moderate fragrance; awl-like, tan prickles; foliage leathery, dark olive green; erect, spreading, bushy, compact growth; hardy.; [Vera Dalton X ((Pink Princess X Lillian Gibson) X (Florence Mary Morse X (Josef Rothmund X R. laxa))]; Buck, Dr Griffith J.; Iowa State University

'PALOMA FALCÓ', HT, op, 1930; flowers coral-salmon, dbl., moderate fragrance; [Li Bures X Château de Clos Vougeot]; Dot, Pedro; C-P

'PAM', F, yb, 1962; bud ovoid; flowers yellow edged red, becoming deep red, small, semi-dbl., cupped, slight fragrance; moderate, bushy growth; [Masquerade X ?]; Annabel

'PAMÁTNIK KOMENSKÉHO', LCl, mp, 1936; flowers salmon-pink, large, free early bloom, moderate fragrance; vigorous, climbing growth; Bojan; Böhm

'PAMÁTNIK KRÁLE JIRÍHO', HT, dr, 1936; (King George's Memorial); bud pointed; flowers deep crimson, marked velvety purple, large, dbl., intense fragrance; foliage glossy, dark; vigorous growth; [Gorgeous X Gen. MacArthur]; Böhm, J.

Pamela, *see* 'KORFLAPEI'

'PAMELA', HT, yb, 1924; flowers canary-yellow and bright blush-pink shaded deep carmine, dbl., moderate fragrance; Therkildsen

Pamela Ann, HT, mr, 1998; Tudor

Pamela Joy, S, lp; [Oz Gold X Eye Paint]; Hannemann, F., 1999; The Rose Paradise; (Weatherly, L.)

'PAMELA TRAVERS', HT, mp, 1966; flowers large, dbl., intense fragrance; foliage leathery; vigorous, upright growth; [Pink Favorite X Queen Elizabeth]; Morey, Dr Dennison; Country Garden Nursery

'PAMELA'S CHOICE', HT, my, 1966; flowers golden yellow, 4.5 in., slight fragrance; foliage bronze; [Piccadilly sport]; Bardill Nursery

Pamina®, HT, dp, 1991; Liebig

'PAMPA', F, mr, 1971; bud pointed; flowers brilliant vermilion-red, dbl.; foliage dark; [Colisee X Atlantic]; Gaujard

Pampa, HT, dy, 1988; K&S

Pam's Passion, *see* 'GELPAM'

Pam's Pink, Ch, pb, 1880

'PAN AMERICA', HT, ob, 1941; flowers deep orange suffused tawny yellow, becoming light gold, open, dbl., 30–40 petals, 5.5 in., moderate fragrance; foliage glossy, bronze; long stems; vigorous, upright, bushy, open, growth; [Heinrich Wendland sport X Max Krause]; Boerner; J&P

Panache, *see* 'POULTOP'

'PANACHÉ', Gr, mp, 1959; flowers light salmon-pink, well formed, large, intense fragrance; vigorous growth; [Rubin (HT) X Cinnabar]; Lens

'PANACHEÉ À FLEURS DOUBLES', HGal, pb, 1839; flowers rose pink with violet striping, medium, dbl., moderate fragrance; Viert; (Sangerhausen)

Panachée a Fleurs Plaines, HGal, m, 1839; Vibert

'PANACHÉE D'ANGERS', P, lp, 1879; flowers anthers golden, semi-dbl.; prickles spiny; vigorous growth; Moreau et Robert

'PANACHEÉ DE BORDEAUX', HP, rb, 1898; flowers red with white striping, large, dbl., Duprat; (Sangerhausen)

'PANACHÉE DE LYON', P, pb, 1895; flowers pink, variegated crimson, [Rose du Roi sport]; Dubreuil

'PANACHÉE D'ORLÉANS', HP, pb, 1854; flowers blush-white striped deep rose, [Baronne Prevost sport]; Dauvesse

Panachee Double, *see* **'VILLAGE MAID'**

Panachée Double, *see* **'PERLE DES PANACHÉES'**

'PANAMA', HT, lp, 1913; flowers flesh edged lighter, dbl., slight fragrance; [Frau Karl Druschki X Unnamed seedling (pink)]; Cook, J.W.

'PANAME', LCl, mp, 1959; flowers bright pink, reverse apricot tinted salmon, large, dbl., free, recurrent bloom, slight fragrance; long, strong stems; well branched growth; [Spectacular X Unnamed seedling]; Delbard-Chabert

Panchganga, HT, pb, 1993; Patil, B.K.

'PANCHU', F, mr, 1966; bud globular; flowers ruby-red, open, medium, semi-dbl., foliage leathery; very vigorous, upright, compact growth; Pal, Dr B.P.; Indian Agric. Research Inst.

Pandemonium, *see* 'MACPANDEM'

Pandora, *see* 'HARWINNER'

'PANDORA', HT, my, 1947; bud ovoid; flowers cream to deep yellow, dbl., 35–55 petals, 5 in., exhibition form, moderate fragrance; foliage dark, semi-glossy; very vigorous, tall growth; [Golden Rapture X R.M.S. Queen Mary]; Barké; Arnold-Fisher Co.

'PANEERA', HT, w, 1983; flowers large, dbl., 35 petals, moderate fragrance; foliage medium, dark, glossy; upright growth; [Unnamed X Unnamed]; Fumagalli, Niso

'PANIA', HT, lp, 1968; flowers well-formed, slight fragrance; foliage leathery; Gold Star of the South Pacific, Palmerston North, NZ, 1969; [Paddy McGredy X (Kordes' Perfecta X Montezuma)]; McGredy, Sam IV

'PANORAMA', HT, mp, 1943; flowers rose-pink, reverse silvery pink, dbl., 30 petals, 5 in., cupped, moderate fragrance; foliage glossy; vigorous, upright, compact growth; [Mrs A.R. Barraclough X Unnamed seedling]; McGredy; J&P

Panorama, *see* **'Panorama Holiday'**

'Panorama Holiday', F, mp, 1973; (Panorama); flowers rose, pointed, dbl., 34 petals, 3.5 in., moderate fragrance; foliage glossy, dark; [Queen Elizabeth X Unknown seedling]; Gregory

Panthere Rose, *see* 'MEIcapinal'

'Pantomime', F, dp, 1965; flowers deep pink, in clusters, dbl., 30 petals, 4 in., exhibition form, moderate, fruity fragrance; [Ma Perkins X Karl Herbst]; McGredy, Sam IV; McGredy

Paola, *see* 'TANaloap'

'Papa Gontier', T, pb, 1883; bud long, pointed; flowers bright pink, reverse carmine-red, large, semi-dbl., intermittent bloom, moderate fragrance; foliage rich green; vigorous, bushy growth; (21); Nabonnand, G.

'Papa Gontier, Climbing', Cl T, pb, 1904; Chevrier

'Papa Gouchault', HMult, dr, 1922; bud long, pointed; flowers pure crimson-red, open, borne in clusters of 10–20, dbl., slight fragrance; foliage large, glossy; long stems; very vigorous, climbing growth; [Rubin X Unnamed seedling]; Turbat

'Papa Hémeray', Ch, rb, 1912; flowers red, center white, blooms in clusters, single; very vigorous growth; Hémeray-Aubert

'Papa Hendrickx', HT, or, 1964; bud long, pointed; flowers vermilion-orange, large, dbl., 45–50 petals, exhibition form; foliage bronze; very vigorous, upright growth; [(Jolie Madame X R. rugosa rubra) X Unnamed seedling]; Mondial Roses

'Papa Joao XXIII', HT, lp, 1963; flowers pearl-pink; [Plaisir de France X La Jolla]; da Silva, Moreira

'Papa Klein', HT, or, 1934; flowers reddish coppery orange, passing to salmon-pink, imbricated, dbl., moderate fragrance; foliage cedar-green; very vigorous growth; [Margaret Spaull X Norman Lambert]; Ketten Bros.

'Papa Lambert', HT, mp, 1899; flowers large, dbl., intense fragrance; Lambert, P.; (Sangerhausen)

Papa Meilland®, *see* 'MEIsar'

Papa Meilland, Climbing®, *see* 'MEIsarsar'

Papa Pirosha, HT, my, 1990; Bulsara

'Papa Rouillard', HWich, mr, 1923; flowers bright carmine, borne in long clusters of 15–25, dbl., abundant seasonal bloom; thornless; foliage rich green, glossy; long stems; very vigorous, climbing growth; [Leontine Grevais X Unnamed seedling]; Turbat

'Papa Schneider', HT, dr, 1961; flowers dark red-purple, large, dbl., moderate fragrance; foliage glossy; vigorous, upright growth; [Crimson Glory X Unnamed seedling]; Kriloff, Michel

Papagena, F, yb; flowers light yellow with carmine-red, medium, semi-dbl., slight fragrance; VEG; (Sangerhausen)

Papageno®, *see* 'MACgoofy'

Paper Doll, *see* 'JACfiseq'

Papi Delbard, *see* DELaby

Papilio, HT, 1955; Aicardi, D.; (Cavriglia)

'Papillon', T, pb, 1881; flowers coppery salmon rose, medium, semi-dbl.; Nabonnand

Papillon, Ch, mr, 1900; Dubourg

Papillon, *see* TANollipa

'Papillon Rose', F, mp, 1956; flowers pink tinted salmon, blooms in clusters, dbl., exhibition form, intense fragrance; vigorous, bushy growth; [White Briarcliff X (Lady Sylvia X Fashion)]; Lens

'Papoose', Cl Min, w, 1955; bud pointed; single, 1 in., blooms in clusters; foliage small, fern-like, semi-glossy; vigorous, spreading (to 3–4 ft) growth; trailer or groundcover; [R. wichurana X Zee]; Moore, Ralph S.; Sequoia Nursery

Paprika®, *see* 'TANprik'

Paprika™, *see* MEIriental™

Papst Johannes XXIII, HT, w, 1963; flowers large, dbl., slight fragrance; Brauner; (Sangerhausen)

'Pâquerette', Pol, w, 1875; (La Pâquerette, Ma Pâqueretta); flowers pure white, with imbricated petals, blooms in broad clusters, very dbl., 1 in., cupped, slight fragrance; very few prickles; foliage glossy, 3–5 leaflets; stems bright green; dwarf (12–15 in.), bushy growth; (14); Guillot et Fils

'Para Ti', Min, w, 1949; (For You, Pour Toi, Wendy); flowers white, base tinted yellow, semi-dbl.; foliage glossy; very bushy (6–8 in.) growth; [Eduardo Toda X Pompon de Paris]; Dot, Pedro; A. Meilland, 1946

'Parade', LCl, dp, 1953; bud ovoid; flowers deep rose-pink, dbl., 33 petals, 3.5 in., cupped, moderate fragrance; foliage glossy; vigorous growth; [New Dawn seedlng X World's Fair, Climbing]; Boerner; J&P

'Parade Marshal', F, mr, 1974; flowers full, dbl., 25–30 petals, 2–2.5 in., exhibition form, slight fragrance; vigorous growth; [Little Leaguer X Gemini]; Byrum; J.H. Hill Co.

'Paradis', HT, mr, 1944; bud pointed; flowers clear red, medium, dbl., cupped, slight fragrance; foliage glossy; vigorous growth; Gaujard

'Paradise', HWich, pb, 1907; flowers rose-pink, center white, tips of petals notched, large bloom, single, non-recurrent, slight fragrance; foliage glossy; vigorous, climbing (10–15 ft) growth; RULED EXTINCT 1/79; Walsh

Paradise™, *see* 'WEZip'

'Paradise Rambler', HWich, pb, 1907; flowers pink with white, large, single, slight fragrance; Walsh; (Sangerhausen)

Paradise, Climbing, Cl HT, m, 1985; Weeks

Parador®, *see* 'MEIchanso'

Parador, *see* 'MEIkinosi'

'Paraglider', S, ob, 1984; bud ovoid, pointed; flowers light pink, reverse orange-red, medium-large blooms borne 1-, dbl., 28 petals, cupped, repeat bloom, slight fragrance; awl-like, tan prickles; foliage medium, dark green tinted copper, leathery; vigorous, bushy, spreading growth; hardy; [(Country Dancer X Carefree Beauty) X Alexander]; Buck, Dr Griffith J.; Iowa State University

Paragon, *see* 'MACkosred'

'Paramount', HT, ab, 1950; bud long, pointed; flowers orange-salmon-buff, becoming, dbl., 30 petals, 4–5 in., exhibition form, slight fragrance; foliage glossy; very vigorous, upright, bushy growth; [Charlotte Armstrong X Glowing Sunset]; Swim, H.C.; Paramount Nursery

'Parasol', HT, dy, 1964; flowers rich yellow, dbl., 28 petals, 5 in., moderate fragrance; foliage dark; compact growth; [Peace X Ethel Sanday seedling]; Sanday, John

Paray, F, mr

Parc des Princes®, *see* DORpurp

'Pardinas Bonet', HT, yb, 1931; flowers deep yellow, reverse red, dbl., intense fragrance; [La Giralda X Souv. de Claudius Pernet]; Dot, Pedro; C-P

Paree Pink, Pol, mp

Paree Red, Pol, mr

Paree Salmon, Pol, op

Paree White, Pol, w

'Parel van Aalsmeer', HT, dr, 1941; (Perle von Aalsmeer); flowers deep red; [Better Times sport]; Verschuren

Pareo®, *see* 'MEIflarol'

Pareo®, *see* KEIzoubo

'Parfait', F, mr, 1975; flowers sweetheart, PP004046; [Minuette sport]; Knight, C.; DeVor Nurseries, Inc.

Parfum d'Armor®, *see* ADAnuamn

Parfum de Franche-Comte, HT, op, 1989; Sauvageot

'Parfum de la Neige', HT, w, 1939; flowers large, dbl., intense fragrance; Mallerin, C.; (Sangerhausen)

Parfum de l'Hay, *see* **'Rose à Parfum de l'Hay'**

Parfum LIffreen, F, m, 1995

'Paris', Pol, mr, 1929; flowers bright red, vigorous growth; deRuiter

Paris 2000, *see* 'DELsamo'

Paris de Yves st Laurent™, *see* 'MEIvamo'

Paris Pink, *see* 'AROdousna'

Paris Red, *see* **'Rouge de Paris'**

'Paris Superior', Pol, mr; flowers have more lasting color than parent; [Paris sport]; deRuiter

'Pariser Charme'®, HT, mp, 1965; bud ovoid; flowers pink, well-formed, blooms in clusters (to 10), dbl., 28 petals, 5 in., intense fragrance; foliage dark, glossy; vigorous, upright growth; ADR, 1966; Tantau, Math.

'Parish Life', Gr, or, 1999; flowers orange, red edged, reverse orange, fades to light pink, dbl., 26–40 petals, 4 in., borne mostly singly, slight fragrance; few prickles; foliage medium, medium green, semi-glossy; upright, tall (5 ft) growth; [Piccadilly X Marijke Koopman]; Wilson, George D.; Holy Family Catholic Church, 2000

'Paris-Match', HT, dp, 1957; flowers carmine to rose, center darker, dbl.; foliage leathery; vigorous growth; GM, Bagatelle, 1956; [Independence X Grand'mere Jenny]; Meilland, F.; URS

'Park Avenue', HT, dr, 1962; bud ovoid; flowers cardinal-red, open, dbl., 38–48 petals, 3–4 in., moderate fragrance; foliage leathery, dark; vigorous, upright growth; [Yuletide X San Fernando]; Jelly; E.G. Hill Co.

Park Jewell, *see* **'Parkjuwel'**

Park Place, *see* AROcruby

'Park Royal', F, ob, 1967; bud ovoid; flowers coral-pink, center white, reverse silver, open, medium, dbl., slight fragrance; foliage light green; vigorous, upright growth; [Tropicana X Shepherd's Delight]; Eddie, J.H.; Sheridan Nursery

Park Wilhelmshohe, *see* KORwilpa

Parkay, *see* 'MACshana'

'Parkdirektor Riggers'®, HKor, dr, 1957; bud long, pointed; flowers velvety crimson, blooms in clusters (up to 50), semi-dbl., recurrent bloom, slight fragrance; foliage dark, glossy, leathery; very vigorous, climbing growth; ADR, 1960; [R. kordesii X Our Princess]; Kordes, R.

'Parkfeuer', HFt, or; flowers bright scarlet, single, non-recurrent, vigorous (6–8 ft) growth; Lambert, P., 1906

Parkjewel, *see* **'Parkjuwel'**

'Parkjuwel', S, lp, 1956; (Park Jewell, Parkjewel); bud ovoid; flowers very large, very dbl., cupped, non-recurrent, intense fragrance; foliage leathery, wrinkled, light green; vigorous (4 ft), bushy growth; [Independence X Red Moss]; Kordes; Morse Roses, 1950

'Parklane', HT, my, 1961; bud ovoid; flowers canary-yellow, dbl., 25–35 petals, 4.5–5 in., exhibition form, slight fragrance; foliage glossy; vigorous, upright growth; [Peace X Dawn]; Jelly; E.G. Hill Co.

Parkprinzessin, HT, mp

Parkrose, S, dr; flowers large, dbl., moderate fragrance; Elvinge, H.; (Sangerhausen)

'Parks' Yellow Tea-scented China', T, my, 1824; (R. indica ochroleuca, R. X odorata ochroleuca); flowers pale yellow, dbl.; Int. into England in 1824

'Parkside Rose', S, dp; flowers deep pink, single, vigorous (6 ft) growth

Parkstone Pride, S; Delforge; (Cavriglia)

Parkstown's Pride, *see* TANdirpkrap

Parkstrauchrose, S, rb, 1975; flowers dark red with golden yellow, small to medium, semi-dbl., slight fragrance; Scholle, E.; (Sangerhausen)

Parkwood Scarlet, *see* KORtradkos

'Parkzauber', S, dr, 1956; bud long, pointed; flowers dark crimson, large, dbl., non-recurrent, moderate fragrance; foliage dark, leathery; very vigorous (4 ft), upright, bushy growth; [Independence X Nuits de Young]; Kordes

'Parkzierde', B, dr, 1909; flowers scarlet crimson, petals shell-shaped, dbl., non-recurrent, moderate fragrance; long stems; very vigorous growth; Lambert, P.

Parly 2, HT, lp; flowers large, dbl., slight fragrance; (Sangerhausen)

Parme, *see* **'Simone'**

'Parmelia', HT, mr, 1957; bud long, pointed; flowers flamingo-red, dbl., 24–30 petals, 5–5.5 in., exhibition form; foliage dark, glossy, leathery; very vigorous, upright, compact growth; [Mme Chiang Kai-shek sport]; Lennard

Parmentier, HP, mp; Guillot et Fils, 1860

'Parsifal', F, op, 1968; flowers coral-salmon, well shaped, dbl., 34 petals; vigorous, bushy growth; [Dacapo X Ballade]; deRuiter

Parsons' Pink China, *see* **'Old Blush'**

Parthenon®, *see* 'DELbro'

Partridge, *see* 'KORweirim'

'Party Doll', F, ab, 1958; bud ovoid; flowers pink, dbl., 40 petals, 2.5 in., cupped, moderate fragrance; foliage leathery, glossy; vigorous, bushy growth; [Goldilocks seedling X Fashion]; Boerner; J&P

'Party Dress', HT, ab, 1961; flowers deep apricot shaded buff-peach, dbl., 25 petals, 5 in., exhibition form, moderate fragrance; foliage glossy; vigorous, bushy, compact growth; [Gay Crusader X Unnamed seedling]; Robinson, H.; Lowe

'Party Girl'™, Min, yb, 1979; bud long, pointed; flowers soft apricot-yellow, high centered, dbl., 23 petals, 1–1.5 in., exhibition form, borne mostly singly, moderate, spicy fragrance; compact, bushy growth; PP4598; AOE, ARS, 1981 Miniature Rose Hall of Fame, ARS, 1999; [Rise 'n' Shine X Sheri Anne]; Saville, F. Harmon

Party Line™, *see* 'MOGaline'

'Party Pink', F, mp, 1956; flowers bright pink, borne in clusters, dbl., 30–50 petals, 1–1.5 in., moderate fragrance; moderate growth; [Unnamed seedling X Pinocchio seedling]; Raffel; Port Stockton Nursery

Party Popcorn, *see* 'LAVpop'

'Party Time', HT, yb, 1986; flowers lemon yellow with pink overlay, reverse lemon yellow, fading, dbl., 45 petals, cupped, moderate, fruity fragrance; medium, reddish prickles, hooked downwards; foliage medium, medium green, semi-glossy; upright, medium growth; PP006457; [Perfume Delight X Half Time]; Weeks, O.L.

Party Trick™, *see* 'DICparty'

Partyglo™, *see* 'MINpco'

'Parure', HT, dp, 1965; flowers carmine-rose, 5 in., cupped, slight fragrance; foliage serrated; tall to moderate growth; [Michele Meilland X Chic Parisien]; Delbard-Chabert; Cuthbert

Parure d'Or®, *see* 'DELmir'

Parvifolia, *see* **'Burgundian Rose'**

'Parwana', F, yb, 1974; bud pointed; flowers golden yellow, edged plum-red, open, dbl., 37 petals, 3 in., cupped, slight fragrance; foliage glossy, dark; vigorous growth; [Unknown X Unknown]; Pal, Dr B.P.

Pasadena®, *see* 'KORland'

'Pasadena', HT, yb, 1927; flowers golden yellow, edged flame, dbl., moderate fragrance; RULED EXTINCT 1/82; [The Queen Alexandra Rose sport]; Coolidge

'Pasadena Tournament', F, mr, 1942; (Red Cécile Brünner); bud long, pointed; flowers velvety red, small, dbl., 36 petals, cupped, moderate fragrance; foliage bronze; long stems; very vigorous, bushy growth; [Cécile Brunner X Unnamed seedling]; Krebs; Marsh's Nursery

'Pasadena Tournament, Climbing', Cl F, mr, 1945; (Red Cécile Brünner, Climbing); Marsh's Nursery

Pascali®, *see* 'LENip'

'Pascali, Climbing', Cl HT, w, 1978; Anderson's Rose Nurseries

Pascaline®, *see* 'LENpas'

Pasita, *see* 'KORsita'

Paso Doble, *see* 'MEILANODIN'

Passaya, HT, w

'PASSION', Gr, mr, 1954; bud long, pointed; flowers scarlet-cerise, dbl., 36 petals, 4 in., moderate fragrance; foliage dark; very vigorous, bushy growth; [Peace X Alain]; Gaujard

Passion de J. Renoard®, HT, m

Passion Rose, *see* **'MANHATTAN'**

Passion's Flame, *see* 'RENPAFLAME'

'PASSPORT', Cl HT, dr, 1940; flowers well-formed, moderate fragrance; vigorous growth; Clark, A.

'PASTEL', HT, ab, 1961; bud long, pointed; flowers creamy yellow and soft pink, dbl., 25 petals, 5 in., exhibition form, moderate fragrance; foliage glossy, wrinkled; strong stems; vigorous, upright growth; [(Sutter's Gold X Unnamed seedling) X Fred Edmunds]; Von Abrams; Peterson & Dering

Pastel Delight, HT, lp, 1984; Kasturi

PASTEL PRINCESS, HT, pb, 1995; flowers two-tone pink, dbl. (15–25 petals), medium blooms borne mostly single; fragrant; some prickles; foliage medium, medium green, semi-glossy; upright, tall growth; [Seedling X Seedling]; Perry, Astor; Certified Roses Inc., 1995

Pastel Tower, HT, lp; [Eiffel Tower sport]; Ruston, D., 1968; (Weatherly, L.)

Pastelina, *see* 'SUNPASTE'

'PASTEUR', HT, pb, 1973; bud long; flowers brilliant pink, flushed red, dbl.; upright growth; [Firmament X Femina]; Gaujard

Pastorale®, *see* 'POURALE'

'PASTORELLA', HT, 1953; Meilland, F.; (Cavriglia)

Pastourelle, *see* 'POURALE'

'PASTOURELLE', HT, op, 1952; flowers salmon-pink, very large, very free bloom; RULED EXTINCT 12/85; GM, Rome, 1953; [Comtesse Vandal X Étoile d'Or]; Robichon

Pasture Rose, *see* **R. CAROLINA**

Pat Austin™, *see* 'AUSMUM'

Pat James, *see* HARyoricks

'PAT NIXON', F, dr, 1972; bud ovoid; flowers large, dbl., moderate fragrance; foliage large, glossy, dark; vigorous, upright, bushy growth; [Tamango X (Fire King X Banzai)]; Meilland; Stuart

'PAT PHOENIX', HT, yb, 1964; flowers cream to yellow, flushed pink, base yellow, dbl., 40 petals, 4.5–5 in., moderate fragrance; foliage dark, leathery; very vigorous growth; [(Wellworth X Clarice Goodacre) X Peace]; Latham

'PAT STEWART', HT, dp, 1976; flowers cerise to deep rose-pink, full, dbl., 26 petals, 5–5.5 in., slight fragrance; foliage dark; vigorous, upright growth; [Red Devil X Honey Favorite]; Stewart, G.

Patchwork, HT, rb

'PATCHWORK QUILT', Min, ob, 1990; bud ovoid; flowers orange-yellow-pink blend, aging light orange, urn-shaped, medium, dbl., 60 petals, slight fragrance; foliage small, medium green, matt; upright, bushy, medium growth; [Rise 'n' Shine X Dandy Lyon]; Jolly, Marie, 1985; Rosehill Farm, 1991

Pathfinder, *see* 'CHEWPOBEY'

'PATIENCE', HT, ob, 1927; bud pointed; flowers scarlet-carmine shaded orange and orange-scarlet, large, dbl., exhibition form, moderate fragrance; GM, NRS, 1926; McGredy

'PATIENCE STRONG', HT, mr, 1969; flowers crimson-scarlet, pointed, large, dbl.; foliage dull, gray-green; free growth; [Basildon Belle X Red Dandy]; Trew, C.; Basildon Rose Gardens

'PATIENCE, CLIMBING', Cl HT, ob, 1935; Shamburger, C.S.

Patio Charm, *see* CHEwapri

Patio Cloud, *see* 'INTERFLU'

Patio Dance™, *see* 'WILPADA'

Patio Delight, F, op

Patio Flame, *see* 'POULCOT'

Patio Gem, *see* 'SUNSALM'

Patio Gold™, *see* 'WILPAGO'

Patio Honey, *see* CHEwpiwitel

Patio Jewel, *see* 'JACIAT'

'PATIO JEWEL', MinFl, m, 1975; bud pointed; flowers purple to clear amethyst, open, single, 5–7 petals, 2.5 in., slight fragrance; foliage leathery; very vigorous growth; [Europeana X Angel Face]; Williams, J. Benjamin

Patio Orange, MinFl, or; Harkness

'PATIO PATTY', MinFl, yb, 1975; bud pointed; flowers yellow, washed peach and orange, semi-dbl., 16 petals, 2 in., globular, intense fragrance; foliage small, reddish-green; [(Circus X The Optimist) X (Little Darling X Starina)]; Williams, J. Benjamin

'PATIO PEARL', MinFl, pb, 1975; flowers light pearl-pink, base deeper, semi-dbl., 18–20 petals, 1–1.5 in., exhibition form, slight fragrance; foliage small, glossy, dark; vigorous growth; [Fairy Queen X The Optimist]; Williams, J. Benjamin

Patio Pearl, *see* POUlpearl

Patio Prince, *see* 'POULCRIM'

Patio Princess, *see* POUlholm

Patio Queen, *see* 'CHEWSEA'

'PATIO RIBBON', MinFl, dr, 1975; bud ovoid; flowers velvety bright dark scarlet, semi-dbl., 16 petals, 2.5–3 in., cupped, moderate fragrance; foliage dark; strong growth; [Europeana X Red Favorite]; Williams, J. Benjamin

Patio Snow™, *see* 'WILPASN'

'PATRICE', HT, w, 1985; flowers large, dbl., 25 petals, exhibition form, slight fragrance; foliage medium, dark, semi-glossy; upright, bushy growth; [(Baccará X Generosa) X Zecchino d'Oro]; Bartolomeo, Embriaco

Patricia, *see* 'KORPATRI'

'PATRICIA', HT, mr, 1932; flowers carmine flecked pink, base orange-yellow, moderate fragrance; foliage glossy, dark; vigorous growth; Chaplin Bros.

'PATRICIA ANNE', HT, dp, 1962; flowers deep pink, reverse silvery pink, 4–5 in., moderate fragrance; foliage dark, glossy; vigorous growth; [Kordes' Perfecta sport]; Buzza

'PATRICIA C. OPPMANN', F, w, 1981; bud pointed; flowers very light yellow, blooms borne 1–7 per cluster, dbl., 43 petals, cupped, slight fragrance; slightly hooked prickles; foliage glossy, medium green; vigorous, upright, dense growth; Bronze, ARC TG, 1984; [Unnamed seedling X Unnamed seedling]; Jerabek, Paul E.

'PATRICIA DAWN', HT, or, 1998; flowers orange-red, lighter reverse, ehibition form, very dbl., 41 petals, 4 in., exhibition form, borne singly, moderate fragrance; prickles moderate; foliage medium, medium green, dull; upright, medium growth; [unknown sport]; Austin, M.L.

'PATRICIA HARKNETT', HT, dp, 1961; flowers deep pink, dbl., 23 petals, 4 in., moderate fragrance; foliage bronze-red; vigorous growth; [Lady Sylvia sport]; Harknett

'PATRICIA HYDE', F, mp, 1969; flowers medium, semi-dbl., [Ann Elizabeth X Red Dandy]; Harkness

'PATRICIA MACOUN', LCl, w, 1945; dbl., slight fragrance; foliage dark, glossy; hardy.; [R. helenae X ?]; Central Exp. Farm

'PATRICIA MILLER', HT, w, 1978; bud ovoid; flowers pure white, open, large, semi-dbl., slight fragrance; foliage leathery; vigorous, bushy growth; [Queen Elizabeth sport X Queen Elizabeth sport]; Miller, J.; Treloar Roses Pty. Ltd.

'PATRICIA NEAL', HT, mr, 1963; flowers bright cerise, large, dbl., exhibition form, foliage dark, leathery; vigorous, tall growth; [Better Times sport]; Macres; Paragon Greenhouses

'PATRICIA PIESSE', HT, lp, 1971; flowers luminous light pink, medium, dbl., 50–60 petals, exhibition form, moderate fragrance; foliage glossy; vigorous, upright growth; [Elizabeth Fankhauser X Memoriam]; Fankhauser

'PATRICIA SCRANTON', Cl Min, yb, 1977; bud pointed; flowers light yellow, streaked red, small, semi-dbl., 17 petals, 1.5 in., slight fragrance; foliage glossy, dark, soft; [Fairy Moss X Fairy Moss]; Dobbs; Small World Min. Roses

'PATRICIA WATKINS', F, lp, 1947; flowers bright pink, medium, borne in trusses, single, 5 petals, slight fragrance; foliage dark; vigorous growth; [Karen Poulsen sport]; Watkins Roses

Patricia Weston, *see* 'REYPAT'

'PATRICIAN', HT, mr, 1977; bud ovoid, pointed; flowers cardinal red, dbl., 28 petals, 4–5 in., exhibition form, intense fragrance; foliage large, dark; very upright growth; [Fragrant Cloud X Proud Land]; Warriner, William A.; J&P

'PATRICK ANDERSON', HT, dp, 1938; bud long, pointed; flowers deep rose-pink, large, dbl., exhibition form, intense fragrance; foliage leathery; vigorous growth; [John Henry X Portadown Fragrance]; McGredy; J&P

'PATRICK VINCENT', HT, dr, 1967; flowers crimson, intense fragrance; foliage dark; free growth; [Mirandy X F.W. Alesworth]; Vincent

Patriot, *see* 'JACLIN'

Patriot Flame, *see* 'CLESCRUB'

Pat's Choice, *see* KORomega

Pat's Delight, *see* 'WILRSS'

'PATSY', HT, w, 1930; bud pointed; flowers pure white, open, very large, dbl., cupped, slight fragrance; Dickson, H.; Morse

Patsy Cline™, *see* 'AROCUMU'

'PATTY LOU', Min, pb, 1953; (Petite); bud ovoid; flowers rose pink, reverse silvery pink, dbl., 55 petals, 1 in., moderate fragrance; foliage small; dwarf (10–12 in.), bushy growth; [Oakington Ruby X Oakington Ruby]; Moore, Ralph S.; Sequoia Nursery

Patty Sue, *see* 'TINPAT'

'PATTY'S PINK', F, lp, 1960; bud long, pointed; flowers rose-opal, reverse camellia-rose, dbl., 40–45 petals, 2.5 in., exhibition form, slight fragrance; foliage leathery, glossy; very vigorous, upright, compact growth; [(Cécile Brunner X Mrs R.M. Finch) X Self]; Spanbauer

'PATTY'S RED', F, mr, 1968; flowers cherry-red; [Patty's Pink sport]; Paulen Park Nursery

'PAUL BIGOT', HWich, dp, 1924; flowers bright rose, shaded vermilion, borne in clusters of 5–10; vigorous, climbing growth; Turbat

Paul Bocuse, *see* MASpaujeu

'PAUL BOUCLAINVILLE', HT, yb, 1930; flowers carmine on yellow ground, reverse pinkish white tinted yellow, semi-dbl., cupped, moderate fragrance; very vigorous growth; [Mme Charles Detreaux X Mme Edouard Herriot]; Buatois

'PAUL BUATOIS', Cl HT, dr, 1931; flowers velvety red, base yellow passing to purplish carmine, very large, dbl., cupped, moderate fragrance; foliage leathery; very vigorous, climbing growth; [Marie Baumann X Mme Edouard Herriot]; Buatois

'PAUL BUNYAN', Gr, dr, 1961; bud long, pointed; flowers deep red, dbl., 55 petals, 5 in., exhibition form, slight fragrance; foliage leathery; very vigorous, upright growth; [Charles Mallerin X Carrousel]; Von Abrams; Peterson & Dering

Paul Cezanne, *see* JACdeli

'PAUL CRAMPEL', Pol, or, 1930; flowers deep orange-scarlet, brighter and larger than gloria mundi, dbl.; (14); Kersbergen

'PAUL CRAMPEL, CLIMBING', Cl Pol, or, 1934; Appleton

'PAUL CRAMPEL, CLIMBING', Cl Pol, or, 1934; Vially

'PAUL CRAMPEL, CLIMBING', Cl Pol, or, 1937; Tantau

'PAUL DAUVESSE', HMult, my, 1933; bud long, golden yellow; flowers bright canary-yellow, large, borne in clusters of 4–8, dbl.; vigorous, climbing growth; Barbier

Paul de Fontainne, *see* **'DEUIL DE PAUL FONTAINE'**

'PAUL DE LA MEILLERAYE', HP, mr, 1863; flowers carmine-red with lighter reverse, very large, dbl.; Guillot; (Sangerhausen)

'PAUL DÉLÉPINE', Pol, mp, 1933; flowers brilliant rose-pink, borne in clusters, dbl., globular; foliage leathery, glossy; vigorous growth; [Yvonne Rabier X Dorothy Perkins]; Délépine; Pajotin-Chédane

'PAUL DUVIVIER', HT, mr, 1932; flowers carmine, base yellow, dbl., intense fragrance; [Constance X Pax Labor]; Laperrière

Paul Fromont, *see* **'BOUDOIR'**

Paul Gauguin, *see* JACdebu

Paul Gold, HT, my, 1998; Poulsen

'PAUL HARRIS', HT, dy, 1993; dbl., 26–40 petals, 3–3.5 in., borne mostly singly, slight fragrance; some prickles; foliage medium, medium green, matt; medium (40–48 in.), bushy growth; [(Summer Sunshine X Georgia) X Seedling]; Weeks, O.L.; Estrella Rose Company, 1993

Paul Holtge, *see* **'GRACE MOORE'**

'PAUL KADOLOZIGUE', HMult, lp, 1912; flowers medium, dbl., Lambert, P.; (Sangerhausen)

'PAUL KRAMPEL', Pol, 1930; Kersbergen; (Cavriglia)

'PAUL LAFONT', HT, ly, 1920; flowers golden yellow to white tinted yellow, dbl., slight fragrance; [Mme Maurice Capron X Unnamed seedling]; Guillot, P.

'PAUL LEDE, CLIMBING', Cl T, ab, 1913; (Monsieur Lede, Monsieur Paul Lede); pink, large and pointed; flowers yellowish buff, center flushed carmine, outer petals salmon buff, exhibition form, intense, lush, fruity fragrance; small dark green semi-glossy; vigorous to 15 ft; Lowe

'PAUL LUCCHINI', HT, dr, 1931; flowers purplish garnet, shaded velvety red, dbl., cupped, intense fragrance; foliage bronze, leathery; vigorous growth; [Rhea Reid X Yves Druhen]; Buatois

Paul McCartney, *see* 'MEIZELI'

'PAUL MONNIER', HT, yb, 1902; flowers light yellow shaded salmon, Buatois

'PAUL NABONNAND', T, mp; flowers satiny rose, large, dbl., cupped; vigorous growth; Nabonnand, G., 1876

'PAUL NEYRON', HP, mp, 1869; flowers clear pink to rose-pink, very large, dbl., 50 petals, cupped, occasionally recurrent bloom, moderate fragrance; foliage large, rich green; vigorous growth; [Victor Verdier X Anna de Diesbach]; Levet, A.

'PAUL NOËL', LCl, pb, 1913; flowers old rose and pale yellow, blooms in clusters of 4–6, 2–3 in.; vigorous, climbing growth; [R. wichurana X Mons. Tillier]; Tanne

'PAUL PERRAS', HP, lp, 1870; (The Crepe Rose); flowers pale rose, compact, large, very dbl.; sets many hips; vigorous, pillar growth; Levet

'PAUL PLOTON', HWich, mr, 1910; flowers bright red, blooms late, [R. wichurana X Mme Norbert Levavasseur]; Barbier

Paul Potter, S, rb, 1997; Williams, J. Benjamin

'PAUL REVERE', HT, mr, 1940; flowers carmine, opening scarlet-crimson, dbl., 24–30 petals, 4–5 in., cupped, moderate fragrance; vigorous, upright growth; [Talisman sport]; Roland

Paul Ricard®, *see* 'MEINIVOZ'

'PAUL RICAULT', C, mp, 1845; flowers rose-pink, dbl., quartered, intense fragrance; height to 5 ft; Portemer fils

Paul Richard, *see* 'MEINIVOZ'

Paul Shirville, *see* 'HARQUETERWIFE'

'PAUL TRANSON', LCl, op, 1900; flowers bright pink, large blooms in clusters of 3–5, dbl., slight fragrance; foliage glossy, dark; short, strong stems; height 10 ft; [R. wichurana X L'Ideal]; Barbier

'PAUL VERDIER', HP, dp, 1866; flowers bright rose; Verdier, C.

'PAUL VOGEL', HT, mp, 1932; flowers medium, dbl.; Vogel, M.; (Sangerhausen)

'PAULA', Gr, or, 1980; flowers dusty salmon, [Radiation induced Queen Elizabeth sport]; James, John

'PAULA ANNE CREASEY', F, rb, 1999; flowers upper 2/3 medium red, lower 1/3 white, reverse light pink, dbl., 26 petals, 2.5 in., borne in small clusters, slight fragrance; prickles moderate; foliage medium,

medium green, dull; compact, low (2.5 ft) growth; [Tamango X Anne Scranton]; Dobbs, Annette E., 1998

'PAULA CLEGG', HT, mr, 1919; flowers bright scarlet, [Kaiserin Auguste Viktoria X ?]; Kiese

'PAULA MAYER', HT, pb, 1929; flowers silvery carmine-pink, reverse yellowish pink, semi-dbl., intense fragrance; [Mme Edmee Metz X Betty Uprichard]; Leenders Bros.

'PAULA MEIDINGER', HT, lp, 1923; flowers large, dbl., slight fragrance; Meidinger; (Sangerhausen)

Paula Scholle, S, mp, 1975; flowers medium, semi-dbl., slight fragrance; Scholle, E.; (Sangerhausen)

Paula Scholle II, F, op, 1980; flowers large, dbl.; Scholle, E.; (Sangerhausen)

'PAULE DELAVEY', Pol, ly, 1957; flowers creamy yellow; vigorous, bushy growth; Privat

'PAULETTE', F, ly, 1934; flowers saffron-yellow to yellowish white, borne in clusters, very dbl., cupped, intense fragrance; foliage leathery, glossy; vigorous growth; [Leontine Gervais X Paul Monnier]; Buatois

'PAULETTE', HT, dp, 1946; flowers bright rosy scarlet, center tinted salmon, well formed, large, very dbl., slight fragrance; foliage rich green; tall growth; [Peace X Signora]; Meilland, F.

'PAULETTE BENTALL', HMult, m, 1916; flowers purple/pink, small, semi-dbl., Bentall; (Sangerhausen)

'PAULETTE BUFFET', HT, lp, 1921; flowers pale flesh-pink, reverse silvery pink, dbl., slight fragrance; [Jonkheer J. L. Mock X Unnamed seedling]; Gillot, F.

'PAULETTE COQUELET', HT, mr, 1947; bud long; flowers salmon-red tinted bright coral, large, dbl.; very vigorous growth; [Unnamed seedling X Daniel]; Mallerin, C.; URS

'PAULIEN VERBEEK', HT, ob, 1958; flowers orange-yellow, large, dbl., 55 petals, very vigorous growth; Verbeek

'PAULII', S, w; (R. rugosa repens alba, R. X paulii); flowers yellow stamens, borne in corymbose, single, 2.5 in.; prostrate shrub growth; (14); [R. arvensis X R. rugosa]

'PAULII ROSEA', S, pb; (R. rugosa repens rosea, R. X paulii rosea); flowers medium pink, white eye, yellow stamens, single, [Possibly a Paulii sport]

'PAULINE DAWSON', HWich, dp, 1916; flowers deep pink, large, single; vigorous, climbing growth; Dawson; Eastern Nursery

'PAULINE LANCEZEUR', HP, dr, 1854; (Pauline Lansezeur); flowers crimson shaded violet, large, dbl., recurrent bloom, Lancezeur

Pauline Lansezeur, *see* **'PAULINE LANCEZEUR'**

'PAUL'S CARMINE PILLAR', Cl HT, mr, 1895; (Carmine Pillar); bud long, pointed; flowers carmine-red, open, large, single, very early bloom, slight fragrance; foliage rich green; vigorous growth; Paul

Paul's Double Musk, S, w

'PAUL'S EARLY BLUSH', HP, lp, 1893; (Mrs Harkness); flowers blush; [Heinrich Schultheis sport]; Paul

'PAUL'S HIMALAYAN MUSK RAMBLER', HMsk, lp; flowers blush-lilac-pink, rosette, blooms in clusters, thread-like stems; growth to 30 ft

'PAUL'S LEMON PILLAR', Cl HT, ly, 1915; (Lemon Pillar); bud pale lemon-yellow; flowers pale sulfur-yellow to almost white, very large blooms, dbl., non-recurrent, intense fragrance; long, strong stems; vigorous growth; not dependably hardy; GM, NRS, 1915; [Frau Karl Druschki X Marechal Niel]; Paul

Paul's Perpetual White, *see* **'PAUL'S SINGLE WHITE PERPETUAL'**

'PAUL'S PINK', HT, lp, 1978; flowers soft pink, small, dbl., 28 petals, intense fragrance; vigorous growth; [Snowsong Supreme X Pink Puff]; deVor, Paul F.; DeVor Nurseries, Inc.

'PAUL'S SCARLET CLIMBER', LCl, mr, 1916; flowers vivid scarlet, shaded bright crimson, blooms in large clusters, semi-dbl., sometimes slightly recurrent, slight fragrance; vigorous, climbing or pillar growth; very hardy; (21); GM, Bagatelle, 1918 GM, NRS, 1915; [Paul's Carmine Pillar X Reve d'Or]; Paul, W.

'PAUL'S SINGLE WHITE PERPETUAL', HP, w, 1883; (Paul's Perpetual White); flowers pure white, solitary or in small clusters, single; foliage light green; vigorous growth; Paul

'PAULSPRIDE', LCl, yb, 1996; flowers yellow with pink edge changing to mostly pink, reverse medium, dbl., 26–40 petals, 3.5 in., moderate fragrance; few prickles; foliage medium, medium green, semi-glossy; spreading, medium (8 ft) growth; Jerabek, Paul E.

Pavane, HT, 1969; Herholdt, J.A.; (Cavriglia)

Pavarotti, *see* RUItulvi

'PAVILLON DE PREGNY', N, pb; Guillot, 1863

'PAVILLON DE PREGNY', N, m, 1863; flowers light violet-pink, medium, dbl., moderate fragrance; Guillot; (Sangerhausen)

'PAW MAW', Gr, m, 1994; flowers mauve turning red at edges of petals, blooms borne 1–5 per cluster, dbl., 25–30 petals, 3–3.5 in., moderate fragrance; some prickles; foliage large, medium green, semi-glossy; tall, upright growth; Jerabek, Paul E.

Paws, *see* 'BEAPAW'

'PAX', HMsk, w, 1918; flowers pure white, prominent golden anthers, blooms in clusters, semi-dbl., 3–4 in., recurrent bloom, intense fragrance; foliage large, leathery; long, strong stems; vigorous (4 ft), bushy growth; GM, NRS, 1918; [Trier X Sunburst]; Pemberton

'PAX', F, w, 1946; flowers white tinted greenish, large, semi-dbl., moderate fragrance; [Irene X Mme Alexandre Dreux]; Leenders, M.

'PAX AMANDA', S, lp, 1937; flowers light pink, turning to white, borne in clusters, semi-dbl., 17 petals, free, non-recurrent bloom, thornless; vigorous (7 ft), growth; very hardy.; [Frau Georg von Simson X R. blanda]; Hansen, N.E.

'PAX APOLLO', S, dp, 1938; flowers deep pink, borne in large clusters, semi-dbl., 14 petals, non-recurrent; thornless; vigorous (7 ft), growth; very hardy.; [R. sempervirens X R. blanda]; Hansen, N.E.

'PAX IOLA', S, lp, 1938; flowers clear shell-pink, passing to nearly white, large, borne in l, semi-dbl., 25 petals, non-recurent bloom, thornless; vigorous, pillar, growth; very hardy; [Anci Bohmova X R. blanda]; Hansen, N.E.

'PAX LABOR', HT, yb, 1918; flowers pale golden yellow, slightly shaded coppery carmine, passing, dbl., globular; vigorous growth; [Beauté de Lyon X Unnamed seedling]; Chambard, C.

'PAX LABOR, CLIMBING', Cl HT, yb, 1929; Gaujard

'PAYABLE', F, rb, 1992; (**Anusheh**); flowers red with yellow reverse, dbl., 26–40 petals, 1.5–2.75 in., borne in large clusters, slight fragrance; many prickles; foliage medium, dark green, glossy; medium, upright growth; [Len Turner X Seedling]; Payne, A.J.

'PAYSAGISTE FAURE-LAURENT', HT, ob, 1947; flowers orange, reverse orange-yellow, large, dbl., slight fragrance; foliage glossy, dark; vigorous growth; Gaujard

'PAZ VILA', HT, mr, 1931; flowers large, dbl., moderate fragrance; foliage glossy; vigorous growth; [Jovita Perez X Jean C.N. Forestier]; Munné, B.

PEAalamo, F, dp, 1998; (**Little Rascal**); Pearce

'PEACAP', F, ab, 1981; (**Ragtime**); flowers apricot pink, blooms borne 10–15 per cluster, dbl., 65 petals, slight fragrance; large, straight, red prickles; foliage small, mid-green, glossy; bushy growth; [Vesper X Aloha]; Pearce, C.A.; Limes Rose Nursery

'PEACE', HT, yb, 1945; (Gioia, Gloria Dei, Mme A. Meilland); bud; flowers golden yellow edged rose-pink, dbl., 43 petals, 6 in., exhibition form, slight fragrance; foliage large, very dark, leathery, glossy; very vigorous, tall, bushy growth; All-American Rose Selection, AARS, 1946 GM, NRS, 1947 GM, Portland, 1944 GM

Certificate, ARS, 1947 Golden Rose, The Hague, 1965 Hall of Fame, WFRS, 1976; [((George Dickson X Souv. de Claudius Pernet) X (Joanna Hill X Charles P. Kilham)) X Margaret McGredy]; Meilland, F.; C-P, 1945, U.S.

Peace, T, ly; Piper, 1902

Peace Maker, F, 1990; Tagashira, Kazuso; (Sangerhausen)

Peace Meillandina®, *see* 'MEIlarco'

Peace Sunblaze, *see* 'MEIlarco'

'Peace, Climbing', Cl HT, yb, 1950; (Gioia, Climbing, Gloria Dei, Climbing, Mme A. Meilland, Climbing); Brandy; C-P

'Peace, Climbing', Cl HT, yb, 1951; Kordes

'Peaceful', HT, pb, 1956; bud globular; flowers deep coral rose pink, reverse lighter, dbl., 50 petals, 5.5–6 in., cupped, moderate fragrance; foliage leathery; vigorous, upright growth; [Unnamed seedling X Peace]; Boerner; J&P

Peacekeeper, *see* 'HARbella'

'Peaceport', HT, op, 1960; flowers deep orange-pink; [Peace sport]; Rokos; Wyant

'Peach Beauty', HT, pb, 1970; bud ovoid; flowers peach-pink, large, dbl., moderate fragrance; foliage large, leathery; vigorous, bushy growth; [Ma Perkins X Polynesian Sunset]; Boerner; Thomasville Nursery

Peach Blossom, *see* 'AUSblossom'

'Peach Blossom', F, dp, 1932; bud orange-red; flowers soft carmine-rose, medium, borne in clusters of 6–12, semi-dbl.; vigorous growth; Chaplin Bros.

'Peach Brandy', Min, ab, 1978; bud pointed; flowers small, dbl., 23 petals, exhibition form, moderate fragrance; compact, bushy growth; Schwartz, Ernest W.; Bountiful Ridge Nursery; Gulf Stream Nursery

Peach Candy, *see* 'MORtopaz'

'Peach Dream', F, op, 1996; bud opening to peach orange center with pink edging; dbl., 15–25 petals, 2.5 in., exhibition form, borne in small clusters, slight fragrance; few prickles; foliage medium, dark green, glossy; upright, medium (26 in.) growth; patio; Bees of Chester; L W Van Geest Farms, Ltd., 1995

Peach Elite, *see* 'LYOlite'

Peach Festival™, *see* 'LAVcap'

Peach Fuzz™, *see* 'WEKhelen'

'Peach Glow', F, pb, 1960; bud ovoid; flowers golden coral, base pink, dbl., 30 petals, 3 in., cupped, moderate, spicy fragrance; foliage leathery; vigorous, upright, compact growth; [Goldilocks X Fashion]; Boerner; J&P

Peach Melba, *see* 'KORita'

'Peach Melba', HT, yb, 1960; flowers yellow marked flame and pink, moderate fragrance; vigorous growth; [Golden Scepter X Hazel Alexander]; Dicksons of Hawlmark

Peach Parfait™, *see* 'CARfait'

'Peach Sherbet', F, pb, 1999; flowers salmon pink, dbl., 30 petals, 4 in., exhibition form, slight fragrance; [Sonia X seedling]; Teranishi, K.; Itami Rose Nursery, 1996

Peach Silks, *see* 'CLEpeach'

Peach Spire, *see* KORofaser

Peach Sunsation, *see* KORommerla

Peach Surprise, *see* 'POUlrise'

'Peach Treat', HT, pb, 1968; bud ovoid; flowers peach-pink, large, very dbl., intense fragrance; foliage leathery; vigorous, bushy growth; [Beauté X Kordes' Perfecta]; Fuller; Wyant

'Peachblow', HT, lp, 1942; bud long, pointed; flowers large, dbl., moderate fragrance; foliage glossy; vigorous growth; [Mme Butterfly X Yellow seedling]; Coddington; C-P

'Peaches and Cream', HT, ob, 1936; flowers salmon shaded gold and rose-pink, very dbl., slight fragrance; bushy growth; RULED EXTINCT 4/77; [Seedling X Miss Rowena Thom]; H&S; Dreer

'Peaches 'n' Cream', Min, pb, 1976; bud tapering; flowers light peach-pink blend, dbl., 52 petals, 1 in., exhibition form, slight fragrance; foliage dark; upright, spreading growth; AOE, 1977; [Little Darling X Magic Wand]; Woolcock; Pixie Treasures Min. Roses

'Peachy', Min, pb, 1964; flowers pink tinted yellow, small, dbl., 50 petals, moderate fragrance; foliage light green, soft; vigorous, bushy (12 in.) growth; [Golden Glow (LCl) X Zee]; Moore, Ralph S.; Sequoia Nursery

'Peachy Keen', Min, ab, 1979; bud long, pointed; flowers soft apricot pink, 18–20 petals, 1 in., slight fragrance; bushy, spreading growth; [Little Darling X Sheri Anne]; Bennett, Cecilia 'Dee'; Tiny Petals Nursery

Peachy Pink Magic Carpet, *see* POUlor

'Peachy White', Min, w, 1976; bud long, pointed; flowers near white, often tinted pink, small, semi-dbl., 18 petals, 1.5 in., moderate fragrance; foliage leathery; upright, bushy growth; AOE, 1976; [Little Darling X Red Germain]; Moore, Ralph S.; Nor'East Min. Roses

'Peacock', HT, op, 1985; bud ovoid, pointed; flowers orange-coral, reverse-cream, blooms borne usually singly, dbl., 5.5 in., exhibition form, moderate, spicy fragrance; large, globular, orange-red fruit; foliage large, dark, semi-glossy; vigorous, upright, bushy growth; [Red Queen sport X Red Queen]; LeMire, Walter

'PEAfever', F, dp, 1982; (**Fever**); flowers deep pink, large, dbl., intense fragrance; foliage large, dark; upright growth; Pearce, C.A.; Limes Rose Nursery

'PEAfirst', F, dp, 1981; (**Dr McAlpine**, Seafirst); flowers deep rose-pink, patio, large blooms borne 1–10 per cluster, dbl., 30 petals, exhibition form, intense fragrance; straight, red prickles; foliage dark; low, bushy growth; Pearce, C.A.; Limes Rose Nursery, 1983

'PEAhaze', F, ob, 1982; (**Geraldine**); flowers orange, medium, dbl., 20 petals, slight fragrance; foliage medium, light green, semi-glossy; upright growth; [Unnamed seedling X Unnamed seedling]; Pearce, C.A.; Limes Rose Nursery, 1984

'PEAhigh', F, lp, 1987; (Daily Telegraph, **Gilda**); flowers pale shell pink, moderately medium, dbl., 15–25 petals, intense fragrance; foliage medium, medium green, matt; upright, spreading growth; [Unnamed seedling X Unnamed seedling]; Pearce, C.A.; The Limes New Roses

'Peak Performance', HT, dr, 1991; flowers large blooms borne mostly singly, dbl., slight fragrance; foliage medium, dark green, matt; tall, bushy, spreading growth; [Unknown seedling X Unknown seedling]; Burks, Larry; Co-Operative Rose Growers

'PEAmax', F, or, 1986; (Dame Vera, **Dame Vera Lynn**); flowers brick red, large, dbl., 20 petals, slight fragrance; foliage medium, dark, semi-glossy; upright growth; [Seedling X Seedling]; Pearce, C.A.; Limes Rose Nursery

'PEAmight', LCl, op, 1983; (**Leaping Salmon**); flowers salmon pink, large, dbl., 20 petals, moderate fragrance; foliage large, medium green, semi-glossy; upright (to 8–10 ft) growth; PP007196; [((Vesper X Aloha) X (Paddy McGredy X Maigold)) X Prima Ballerina]; Pearce, C.A.; Limes Rose Nursery, 1986

'PEAnob', Min, pb, 1983; (**Bush Baby**); flowers small, dbl., 35 petals, slight fragrance; foliage small, medium green, matt; bushy growth; Pearce, C.A.; Limes Rose Nursery

'PEAnut', Min, dr, 1985; (**Pot Black**); flowers small, dbl., no fragrance; foliage small, medium green, matt; bushy growth; [Unnamed seedling X Unnamed seedling]; Pearce, C.A.; Limes Rose Nursery

Peanut Butter & Jelly, *see* 'PIXichip'

'PEAp', Min, rb, 1988; (**Olive Taylor**); flowers vermillion red with yellow eye, small, dbl., 20 petals, slight fragrance; foliage medium, dark green, glossy; bushy growth; [Unnamed seedling X Unnamed seedling]; Pearce, C.A.; The Limes New Roses

PEApatio, MinFl, pb, 1992; (**Royal Flush**); Pearce

PEAperfume, HT, ob, 1991; (**Indian Summer**); Pearce

'PEApet', Min, mr, 1988; (**Jane Asher**); flowers scarlet, aging slightly paler, rounded, small, borne in sprays, very dbl., no fragrance; prickles straight, average, red; foliage small, medium green, semi-glossy; patio; bushy, low growth; [Unnamed seedling X Unnamed seedling]; Pearce, C.A.; The Limes New Roses, 1987

'PEApolly', F, ob, 1988; (**Jimmy Savile**, Jimmy Saville); flowers coppery-orange, aging lighter, loose, medium, borne usually, dbl., 25–30 petals, cupped, moderate, fruity fragrance; foliage medium, medium green, matt; bushy, low growth; [Unnamed seedling X Unnamed seedling]; Pearce, C.A.; Rearsby Roses, Ltd., 1988

'PEApost', F, mr, 1989; (**Carol Ann**); bud rounded; flowers scarlet, loose, medium, borne in sprays of 3–15, dbl., 42 petals, slight fragrance; prickles straight, large, red; foliage large, medium green, semi-glossy; spreading, medium growth; [Geraldine X Unnamed seedling]; Pearce, C.A.; Rearsby Roses, Ltd., 1990

'PEAprince', F, ob, 1988; (**Olympic Spirit**); flowers orange, reverse yellow, aging orange-brown, medium, borne in, semi-dbl., cupped, slight, fruity fragrance; prickles pointed, medium, red; foliage small, dark green, glossy; bushy, low growth; [Unnamed seedling X Unnamed seedling]; Pearce, C.A.; Rearsby Roses, Ltd., 1988

'PEAproof', F, op, 1989; (**Freddy**); bud ovoid; flowers deep coral pink, aging pales slightly, urn-shaped, medium, dbl., 25 petals, borne in sprays of 3–21, slight fragrance; rare fruit; prickles hooked, medium, red; foliage medium, medium green, matt; bushy, even growth; [Unnamed seedling X Unnamed seedling]; Pearce, C.A.; Rearsby Roses, Ltd., 1989

'PEAquant', Gr, dy, 1988; (**ICI Golden Celebration**); flowers golden yellow, large, dbl., 26–40 petals, moderate fragrance; foliage medium, medium green, glossy, mildew resistant; bushy, healthy growth; [Unnamed seedling X Unnamed seedling]; Pearce, C.A.; The Limes New Roses

'PEArich', HT, ob, 1992; (**Alistair Sheridan**); flowers orange, reverse orange, aging pinky orange, urn-shaped, dbl., 4.75 in., borne usually singly, slight fragrance; foliage large, medium green, glossy; upright growth; [F seedling (HO79) X HT seedling (315)]; Pearce, C.A., 1986; The Limes New Roses, 1991

'Pearl', HT, lp, 1879; flowers small, very dbl., moderate fragrance; weak growth; [Adam X Comtesse de Serenye]; Bennett

'Pearl', HT, w, 1933; flowers white, shaded pink, moderate fragrance; Bentall

Pearl, HMult, w; Turner, 1915

Pearl, *see* 'WEKpearl'

'Pearl', P, lp; Matthews, D.W., 1915; (Weatherly, L.)

Pearl Anniversary, *see* WHItsun

'Pearl Costin', HT, yb, 1959; bud long, pointed; flowers light yellow, center pink, large, semi-dbl., cupped, slight fragrance; foliage leathery, wrinkled; vigorous, upright, bushy growth; [Elli Knab X Amy Johnson]; Reithmuller

'Pearl Dawn', Min, mp, 1975; bud short, pointed; flowers micro-mini, dbl., 38 petals, 1 in.; very compact, bushy growth; [(Cécile Brunner X Perla de Montserrat) X Perla de Montserrat]; Saville, F. Harmon; Nor'East Min. Roses

Pearl Drift®, *see* 'LEGgab'

'Pearl Harbor', HT, pb, 1943; bud long, pointed; flowers light pink reverse china-rose, dbl., 45 petals, 3–3.5 in., exhibition form, intense fragrance; foliage leathery, dark; very vigorous, upright growth; [Unnamed variety X Miss Rowena Thom]; Howard, F.H.; H&S

Pearl Meidiland™, *see* 'MEIplatin'

'Pearl of Baltimore', HT, lp, 1925; flowers shell-pink, center deeper, very dbl., moderate fragrance; [Ophelia X Glorified La France]; Cook, J.W.

Pearl of Bedfordview, *see* 'MEIbeausai'

Pearl of Canada, *see* **'Perla de Alcañada'**

Pearl of Joy, *see* JACsan

'Pearl S. Buck', HT, yb, 1940; bud long, pointed, deep orange; flowers golden yellow suffused apricot, dbl., 45 petals, 4.5 in., moderate fragrance; foliage leathery, dark; long stems; vigorous, bushy growth; [Joanna Hill X Étoile d'Or]; Kordes; J&P

Pearl Sevillana™, *see* 'MEIchonar'

'Pearl Wilson Kissel', HT, dp, 1954; flowers bright red, dbl., 30–40 petals, 4.5 in., moderate fragrance; foliage leathery; vigorous, bushy growth; [Red Columbia X Chrysler Imperial]; Kissel and Motose

'Pearlie Mae', Gr, ab, 1981; bud ovoid, pointed; flowers yellow blended with pink, reverse pink, blooms borne 1–8 per, dbl., 35 petals, cupped, moderate fragrance; awl-shaped prickles; foliage leathery, semi-glossy, dark olive green, tinted copp; erect, bushy growth; [Music Maker X (Queen Elizabeth X Country Music)]; Buck, Dr Griffith J.; Iowa State University

Pearly Drift, S, 1981; LeGrice, E.B.; (Cavriglia)

Pearly Gates™, *see* 'WEKmeyer'

Pearly Gates, Climbing, *see* 'WEKmeyer'

Pearly King, *see* GENdee

'Pearly Peace', HT, lp, 1959; flowers soft pearl-pink, [Peace sport]; Fryers Nursery, Ltd.

'Pearly Queen', F, lp, 1963; flowers well formed, dbl., 22 petals, 4 in., moderate fragrance; vigorous growth; [Queen Elizabeth sport]; North Hill Nursery

'Pearly Shell', HT, pb, 1972; flowers shell pink, center cream, large, dbl., 30 petals, exhibition form, moderate fragrance; foliage glossy; vigorous, upright growth; [Pink Parfait X Michele Meilland]; Sherwood

Pearly Shores, *see* EUGhien

'Pearly White', LCl, w, 1942; bud long, pointed; flowers white tinted pearl, open, semi-dbl., 21 petals; long, strong stems; vigorous, climbing (to 20 ft), upright growth; [Glenn Dale X Mrs Arthur Curtiss James self seedling]; Brownell, H.C.

'PEArobin', F, ob, 1991; (**Gary Lineker**); flowers luminous orange, yellow reverse, single to blooms borne in small clusters, semi-dbl., 1.5–2.75 in., slight fragrance; many prickles; foliage medium, medium green, glossy; medium (50–90 cms), upright growth; [Seedling X Seedling]; Pearce, C.A.; Rearsby Roses, Ltd., 1991

'PEAroyal', F, dy, 1992; (**Goldfinger**, William David); dbl., 15–25 petals, 1.5–2.75 in., borne in small clusters, slight fragrance; some prickles; foliage medium, dark green, glossy; low (40 cms), compact growth; Pearce, C.A.; Hewlett-Packard, 1992

'PEAshine', F, pb, 1992; (**Anthea Fortescue**); flowers pink, yellow center, buff reverse, urn-shaped, blooms borne, dbl., 2.5 in., slight fragrance; foliage small, medium green, glossy, immune to powdery mildew; low, spreading growth; [F seedling (P135) X F seedling (N147)]; Pearce, C.A., 1987; The Limes New Roses, 1991

'PEAspecial', Min, lp, 1989; (**Audrey Gardner**); bud rounded; flowers shell pink, reverse slightly darker, aging same, large, born, dbl., 45 petals, cupped, no fragrance; no fruit; prickles pointed, red; foliage small, medium green, semi-glossy; bushy, medium growth; [Unnamed seedling X Unnamed seedling]; Pearce, C.A.; Rearsby Roses, Ltd., 1990

'PEAsweet', F, mr, 1991; (**Marty**); flowers dusty red, blooms borne in small clusters, very dbl., 1.5–2.75 in., exhibition form, slight fragrance; few prickles; foliage small, medium green, semi-glossy; low (40–60 cms), bushy growth; [Seedling X Sweetheart]; Pearce, C.A.; Rearsby Roses, Ltd., 1991

'Peat Fire Flame', HT, op, 1985; flowers pale orange, reverse salmon-pink, medium, dbl., 20 petals, moderate fragrance; foliage medium, medium green, glossy; upright growth; [Red Planet X Bonnie Anne]; MacLeod, Major C.A.

'PEAthunder', F, dr, 1995; (Judy Dench, **Red Velvet**); flowers deep red with a hint of brown, blooms borne in large cluster, dbl., 26–40 petals, 1.5–2.75 in., slight fragrance; foliage medium, dark green, semi-glossy; low (50 cms), bushy, spreading growth; Pearce, C.A.; Limes New Roses, 1996; Rearsby Roses, Ltd., 1996

'PEAtigar', F, mr, 1956; (**Kind Regards**); flowers crimson-red, very large, borne in clusters, dbl., vigorous, bushy growth; Kordes

PEAtiger, F, mr, 1995; (**Kind Regards**); Pearce

PEAtrophy, F, lp; (**Charisma**); Pearce

Peaudouce, *see* 'DICjana'

PEAvandyke, F, lp; (**Stardust**); Pearce, 1995

PEAvenus, S, lp; (**St Katherine's**); Pearce, 1995

PEAvesta, F, ob, 1996; (**Ginger Toddler**); Pearce

'PEAvoodoo', F, ab, 1994; (**Memory Lane**); flowers apricot pink, blooms borne in small clusters, dbl., 26–40 petals, 3–3.5 in., slight fragrance; some prickles; foliage large, dark green, semi-glossy; upright (60 cms) growth; [PEApost X Seedling]; Pearce, C.A.; The Limes New Roses, 1995

PEAwinner, F, dr, 1995; (**Ruby Celebration**); Pearce

PEAxanadu, S, mp, 1996; (**Pink Midinette**); Pearce

PEAxanthous, MinFl, yb, 1997; (**Bronze Baby**); Pearce

PEAxi, Cl HT, dr, 1977; (**'Bright Fire'**); flowers crimson, shaded darker, semi-dbl., 15 petals, 4.5 in., slight fragrance; foliage glossy, dark; free growth; [Parkdirektor Riggers X Guinee]; Pearce

PEAyetti, MinFl, mp, 1998; (**Dream Lover**); Pearce

PEAzara, F, dp, 1998; (**Millionaire**); Pearce

'Pebble Mill', F, rb, 1973; flowers magenta, reverse spirea-red, dbl., 28 petals, 3.5 in., flat, moderate fragrance; foliage dark; [Paddy McGredy X Unknown seedling]; Gregory

Peche Meillandina®, *see* MEIxerul

Pechtold's Flame, *see* **'Tudor'**

'Pechtold's Triumph', F, dr, 1961; flowers oxblood-red, medium, borne in large clusters, semi-dbl.; [Red Favorite X Frensham]; Verschuren-Pechtold

'Pedrálbes', HT, w, 1935; flowers cream to pure white, dbl., 30 petals, 5 in., exhibition form, slight fragrance; foliage dark, glossy; very vigorous growth; [Frau Karl Druschki X Souv. de Mme Boullet]; Camprubi, C.; J&P

'Pedro Veyrat', HT, ab, 1933; bud long, pointed; flowers large, dbl., cupped, moderate fragrance; [Li Bures X Benedicte Seguin]; Dot, Pedro; C-P

'Pedrus Aquarius', HT, w, 1999; flowers white/ivory, dbl., 26–40 petals, 6 in., borne mostly singly, slight fragrance; prickles moderate; foliage medium, dark green, glossy; upright, bushy, medium (3 ft) growth; [Solitaire X Joe Longthorne]; Poole, Lionel; David Lister Roses, 1999

Peek a Boo, *see* 'DICgrow'

'Peep o' Day', F, op, 1972; flowers salmon, shaded orange, large, dbl., 28 petals, slight fragrance; foliage dark; [(Pink Parfait X Highlight) X Orion]; Harkness; Mason, 1973

Peep-Eye™, *see* 'MINpeep'

'Peeping Tom', S, dp, 1968; flowers deep pink, pointed, large, dbl., recurrent bloom, slight fragrance; foliage medium, medium green, matt; vigorous, tall growth; [Kordes' Perfecta X Parade]; MacLeod

Peer Gynt®, *see* 'KORol'

'Peerless', HT, mr, 1935; bud long, pointed; flowers bright velvety scarlet-carmine, large, dbl., 38–40 petals, [Better Times sport]; Hill, Joseph H., Co.

Pegasus, *see* 'AUSmoon'

'Peggy', HT, dp, 1934; bud long, pointed; flowers deep rose, dbl., exhibition form, intense fragrance; foliage glossy, light; vigorous growth; [Ophelia X Red-Letter Day]; Bees

Peggy "T"™, *see* 'KINtee'

'Peggy A. Smith', HT, pb, 1936; flowers pink with yellow tints, medium, dbl.; Smith, A.; (Sangerhausen)

'Peggy Ann Landon', LCl, ob, 1938; flowers yellow-orange, becoming lighter, dbl., 3.5–5 in., exhibition form, moderate fragrance; foliage large, leathery, glossy, dark; long, strong stems; very vigorous, climbing (15–25 ft) growth; [Glenn Dale X ((Unnamed seedling X Unnamed seedling) X Mary Wallace)]; Brownell, H.C.

'Peggy Astbury', HT, ab, 1920; flowers soft amber to light yellow; Easlea

'Peggy Bell', HT, ab, 1929; Clark, A.; NRS New South Wales

'Peggy England', HT, ab, 1923; flowers cream-apricot, sometimes tinted carmine, Lilley

'Peggy Grant', Min, lp, 1954; flowers shell-pink, small, dbl., 25 petals; foliage light green; dwarf (5–6 in.), bushy growth; [(Robinette X Mons. Tillier) X Zee]; Moore, Ralph S.; Sequoia Nursery

Peggy Jane™, *see* 'SAVapeg'

Peggy Joan Reynolds, *see* 'REYpeg'

Peggy Lee, *see* 'ARofeigel'

'Peggy M', S, pb, 1996; flowers pink gradually turning white on outer petals, reverse yellow, very dbl., 75 petals, 2.5–3 in., moderate fragrance; numerous prickles; foliage medium, light green, glossy; upright, medium (5 ft) growth; Jerabek, Paul E.

'Peggy Netherthorpe', HT, mp, 1974; dbl., 35 petals, 5 in., slight fragrance; foliage light; [(Voeux de Bonheur X Chic Parisien) X (Michele Meilland X Mme Joseph Perraud)]; Delbard; Harry Wheatcroft Gardening

'Peggy Newton', F, my, 1957; flowers primrose-yellow, borne in clusters, dbl., 40–50 petals, 2.5–3 in., globular, slight fragrance; foliage small, leathery, glossy; dwarf, spreading growth; [Golden Glow X Goldilocks]; Boerner; Stuart

Peggy Rockefeller™, *see* 'WILace'

Peggy's Delight, *see* 'MINcco'

'Peintre Renoir', HT, mp, 1925; flowers large, dbl.; Roseraie de St Jean; (Sangerhausen)

PEKaledon, HT, dp, 1993; (**Femnet**); Pekmez, Paul

'PEKali', HT, ob; flowers orange, large, dbl., 35 petals, no fragrance; foliage large, glossy; upright growth; [Unnamed seedling X Marina]; Pekmez, Paul

'PEKamecel', F, or, 1984; dbl., 20 petals, slight fragrance; foliage dark; upright growth; [Unnamed seedling X Unnamed seedling]; Pekmez, Paul

'PEKatan', HT, pb, 1984; (**Clo-Clo**); dbl., 20 petals, slight fragrance; foliage dark; upright growth; [Emily Post X Bellona]; Pekmez, Paul

PEKcoucan, HT, or; (**Star 2000**®); Pekmez, Paul

PEKcougel, HT, lp, 1990; (**Anna**); Pekmez

PEKcoujenny, HT, dr, 1988; (**First Red**®); Pekmez, Paul

PEKcouliane, HT, ob; (**Julia**®); (PEIcouliane)

PEKcourofondu, HT, lp; (**Vogue**); Pekmez, Paul, 1997

PEKcram, F, mp, 1988; (**Jean Monnet**®); Pekmez, Paul

'PEKgold', HT, dy, 1984; flowers medium, moderate fragrance; foliage dark, glossy; upright growth; [Unnamed seedling X Unnamed seedling]; Pekmez, Paul

'Pekinois', Pol, dr, 1975; (Tapis Afghan); bud round; semi-dbl., 15 petals, 1–2 in., cupped, dwarf growth; [Marlena X Lampion]; Pekmez, Paul

'PEKlican', HT, dy, 1984; flowers large, dbl., 35 petals, moderate fragrance; foliage medium, light green, matt; bushy growth; [Unnamed seedling X Bellona]; Pekmez, Paul

'PEKlimasar', Cl F, mr, 1983; (Grimpant Lilli Marleen®, Lili Marlene, Climbing, **Lilli Marleen, Climbing**, Lilli Marlene, Climbing); Pekmez, Paul

'PEKLIPINK', F, mp, 1984; (**Claridge**); flowers large, dbl., 35 petals, moderate fragrance; foliage medium, light green, semi-glossy; upright growth; [Unnamed seedling X Unnamed seedling]; Pekmez, Paul

PEKmiwhite, S, lp, 1991; (**Perce Neige**®); Pekmez, Paul

'PEKOMEGIR', F, mp, 1984; (**President Souzy**); dbl., 20 petals, moderate fragrance; upright growth; [Emily Post X Unnamed seedling]; Pekmez, Paul

PEKomecli, HT, pb, 1985; (**Marlyse**®); (PElomecli)

PEKoubo, HT, op, 1997; (**Bo**); Pekmez, Paul

PEKsolred, S, mr, 1992; (**Ile de France**); Pekmez, Paul

PEKtarampe, S, dr, 1992; (**Alsace**); Pekmez, Paul

PEKwhina, HT, lp; (**Virginia**); Pekmez, Paul, 1994

'**PELÉ**', Cl HT, w, 1979; bud ovoid; flowers large blooms borne 1–3 per cluster, dbl., 35 petals, repeat bloom, slight, soft, fruity fragrance; triangular, hooked prickles; foliage medium green; upright growth with long canes; Bronze, ARC TG, 1980; [Unnamed seedling X Unnamed seedling]; Benardella, Frank A.

'**PÉLISSON**', M, dr, 1848; (Mons. Pélisson); flowers velvety red, turning purple; Vibert

'**PELTON LONNEN**', HT, my, 1975; flowers pure yellow, full, dbl., 28–30 petals, cupped, moderate fragrance; foliage light; bushy, compact growth; [Whisky Mac sport]; Wood

'**PEMBERTON'S WHITE RAMBLER**', HMult, w, 1914; flowers rosette form, borne in clusters, dbl., Pemberton

'**PEMBRIDGE**', HT, ob, 1934; flowers richer, deeper orange-yellow, [Roselandia sport]; Stevens, E.

'**PENELOPE**', HMsk, lp, 1924; flowers shell-pink fading to white, center lemon, blooms in clusters, semi-dbl., recurrent bloom, moderate fragrance; foliage dark; shrubby growth; (21); GM, NRS, 1925; [Ophelia X Unnamed seedling or possibly William Allen Richardson or Trier]; Pemberton

'**PENELOPE**', T, rb, 1906; flowers dark red, center creamy white, high pointed, dbl.; Williams, A.

Pénélope®, *see* 'GAUBIROC'

Penelope Keith, *see* 'MACFREEGO'

'**PENELOPE PLUMMER**', F, dp, 1970; bud orange; flowers vivid flamingo-pink, semi-dbl., 16 petals, 4 in., flat, slight fragrance; foliage dark; moderate growth; [Anna Wheatcroft X Dearest]; Beales, Peter; Intwood Lane Nursery, 1971

'**PENELOPE, CLIMBING**', Cl T, rb, 1932; Rosen, L.P.

Penkala Slavoljub, *see* 'ROGSLAV'

'**PENNANT**', LCl, mp, 1941; flowers begonia-pink, dbl., profuse, non-recurrent bloom, [Flying Colours X Lorraine Lee]; Clark, A.

'**PENNSYLVANIA**', HT, pb, 1934; bud long, pointed; flowers salmon-pink, center apricot, outer petals striped dark pink, semi-dbl., exhibition form, intense fragrance; [Joanna Hill sport]; Neuner

'**PENNSYLVANIAN**', HT, ob, 1953; bud pointed; flowers apricot-orange, dbl., 25–30 petals, 4–5 in., moderate fragrance; upright, bushy growth; [Luna X (Mrs Pierre S. duPont X Mrs Sam McGredy)]; Ohlhus; C-P

'**PENNY**', F, rb, 1973; flowers strawberry-red, base orange, rosette form, semi-dbl., 17 petals, 2 in., moderate fragrance; dwarf growth; [Sarabande X Circus]; Sanday, John

Penny Annie, *see* 'BISFRA'

Penny Candy(TM), *see* 'SAVPLENTI'

Penny Coelen, *see* 'MEINIMO'

Penny Heyns, *see* TANpen

Penny Lane, *see* 'TALPEN'

Penny Lane, *see* HARdwell

Pensioners Voice, *see* FRYrelax

Penthouse, *see* 'MACSATUR'

Penthouse, *see* 'MACNGAURU'

Peon, *see* '**TOM THUMB**'

'**PEONY OF FRAGRANCE**', HP, mp, 1933; flowers pink, peony-like, large, recurrent bloom, moderate fragrance; Pahissa

'**PEOPLE**', F, dr, 1956; (The People); flowers crimson shaded pink, large, borne in large trusses, dbl., 26 petals, flat, slight fragrance; foliage light green; vigorous, bushy growth; GM, NRS, 1955; [Cinnabar X (Kathe Duvigneau X Cinnabar)]; Tantau, Math.

People's Princess, *see* 'SUNTICK'

'**PEPE**', HT, rb, 1961; flowers flame, base and reverse gold, dbl., 4 in.; foliage dark, glossy; vigorous growth; [Amor X Sutter's Gold]; deRuiter; Blaby Rose Gardens

Pepino®, *see* TANipep

Pepita, *see* 'KORKEILICH'

'**PÉPITA**', HGal, pb; flowers soft rosy pink striped white; Moreau

Pépite, *see* 'SOUB DE J. CHABERT'

'**PEPPER POT**', F, pb, 1973; flowers rose-pink to red, splashed yellow, dbl., 24 petals, 3 in., exhibition form, slight fragrance; foliage light; free growth; [Circus X Seedling]; Fryers Nursery, Ltd.

'**PEPPERMINT**', F, rb, 1964; bud ovoid; flowers red, reverse cream, medium, dbl., cupped; foliage leathery; vigorous, bushy growth; [Jingles X Jingles]; Boerner; J&P

'**PEPPERMINT CANDY**', Min, w, 1991; flowers ivory with red blend, dbl., 26–40 petals, 1.5 in., borne mostly singly, intense fragrance; foliage small, dark green, glossy; medium (12–28 in.), upright, bushy growth; PP006819; [Rose Parade X Easter Morning]; Williams, J. Benjamin, 1988

Peppermint Delight, LCl, rb

Peppermint Ice, *see* 'BOSGREEN'

'**PEPPERMINT PATTY**', Min, rb, 1990; bud pointed; flowers white with red edging, same reverse, aging darker red, urn-shaped, dbl., 33 petals, slight, fruity fragrance; oblong, orange when ripe fruit; prickles straight, very few, tan; foliage medium, medium green, semi-glossy; very hardy, upright growth; [Libby X Libby]; Gruenbauer, Richard, 1984; Richard Gruenbauer

Peppermint Stick, *see* 'JACmint'

'**PEPPERMINT STRIPE**', Min, rb, 1991; flowers moderately small, dbl., 15–25 petals, intense fragrance; foliage small, medium green, semi-glossy; bushy growth; [Roller Coaster X Unnamed seedling]; Spooner, Raymond A.

Peppermint Swirl(TM), *see* 'DEVMENTA'

Peppermint Twist, *see* 'JACRAW'

Pepperoni®, S, or, 1994; Poulsen

Per Chance, *see* 'FLOPER'

Perce Neige®, *see* PEKmiwhite

Perception, *see* 'HARZIPPEE'

'**PERCHÈ SI?**', HT, rb, 1956; flowers carmine-red, reverse silvery white, well formed, foliage dark, glossy; very vigorous growth; [Peace X Crimson Glory]; Giacomasso

'PERCREME', HT, w, 1999; (**Irish Creme**(TM)); flowers light beige, reverse same, dbl., 26–40 petals, 4 in., borne mostly singly, moderate fragrance; prickles moderate; foliage medium, medium green, dull; upright, medium (5 ft) growth; [Butterscotch X Mandelon]; Perry, Astor; Certified Roses, 2000

Percussion, *see* 'ZIPCUSS'

'**PERCY IZZARD**', HT, yb, 1936; flowers maize-yellow, reverse buff flushed warm rose, large, dbl., exhibition form, slight fragrance; foliage leathery; vigorous growth; [May Wettern X Barbara Richards]; Robinson, H.; Wheatcroft Bros.

'**PERCY PILCHER**', F, op, 1961; flowers salmon-orange, large, borne in large clusters, dbl.; Verschuren

'**PERCY THROWER**', HT, mp, 1964; flowers rose-pink, well-formed, dbl., 28 petals, 4–5 in., moderate fragrance; foliage glossy; vigorous, tall growth; [La Jolla X Karl Herbst]; Lens; Gregory

Percy Thrower, Climbing, Cl HT, mp, 1978; Stoneham

Perdita®, *see* 'AUSPERD'

'PERDITA', F, lp; Bidwell, 1845; (Weatherly, L.)

Perestroika, *see* 'KORHITOM'

Perfect Hit, *see* POUlfect

Perfect Moment(TM), *see* 'KORWILMA'

Perfect Peace, *see* **'DOROTHY GOODWIN'**

Perfect Potluck(TM), *see* 'LAVROY'

Perfecta, *see* 'KORALU'

'PERFECTA', Pol, dr, 1920; flowers scarlet-crimson under glass, dark crimson in the open, large, dbl.; vigorous growth; [Ellen Poulsen X Merveille des Rouges]; Spek

Perfecta Superior, *see* **'KORDES' PERFECTA SUPERIOR'**

'PERFECTION', HT, mp, 1925; [Columbia sport]; Stielow Bros.

'PERFECTION', Pol, op, 1932; flowers softer coral-pink; [Marytje Cazant sport]; Prior

'PERFECTION', F, pb, 1957; flowers pink turning red, reverse white; [Masquerade X Self]; Ulrick, L.W.

'PERFECTION DE MONTPLAISIR', T, 1871; Levet, A.; (Cavriglia)

'PERFECTION DES BLANCHES', N, w, 1873; flowers medium, borne in corymbs, dbl.; vigorous growth; Schwartz, J.

Perfectly Red(TM), *see* 'JACROVE'

'PERFEITA', F, or; flowers orange and carmine, [Cocorico X Vogue]; da Silva, Moreira

'PERFUME', HT, dr, 1929; bud long, pointed; flowers velvety, fiery deep crimson, semi-dbl., 15 petals, intense fragrance; vigorous growth; Marriott; Beckwith

Perfume, HT, lp, 1995; Spek

Perfume Beauty(TM), *see* 'MEINIACIN'

'PERFUME DE MULTIFLORE', HT, m, 1935; flowers purple/pink, large, dbl., moderate fragrance; Pahissa; (Sangerhausen)

'PERFUME DELIGHT', HT, mp, 1973; bud long, pointed; flowers deep pink, large, dbl., cupped, intense fragrance; foliage large, leathery; vigorous, upright, bushy growth; AARS, 1974; [Peace X ((Happiness X Chrysler Imperial) X El Capitan)]; Weeks; C-P

Perfume Simplex, Min, yb

Perfume Tiger(TM), *see* 'WINPTIG'

Perfumed Bride, HT, pb, 1994; Bell

Perfumella, *see* MEIparos

'PERGOLÈSE', P, m, 1860; flowers bright purplish crimson, shading to lilac, medium, very dbl., occasionally recurrent bloom, moderate fragrance; Moreau et Robert

'PERJUSDREAM', HT, yb, 1999; (**Just Dreamy**); flowers yellow-gold, reverse apricot gold, dbl., 17–25 petals, 3.5–4 in., borne mostly singly, slight fragrance; few prickles; foliage medium, medium green, semi-glossy; upright, bushy, medium growth; [Folklore X Golden Gate]; Perry, Astor; Certified Roses, 2000

'PERKY', Min, dp, 1958; bud pointed; dbl., 1 in., intense fragrance; foliage glossy; very bushy (12 in.), compact growth; [(R. wichurana X Floradora) X Oakington Ruby]; Moore, Ralph S.; Sequoia Nursery

'PERLA DE ALCAÑADA', Min, dp, 1944; (Baby Crimson, Pearl of Canada, Perle de Alcañada, Wheatcroft's Baby Crimson); bud small, ovoid; flowers carmine, semi-dbl., 18 petals; foliage dark, glossy; dwarf, very compact (6–10 in.) growth; [Perle des Rouges X Rouletti]; Dot, Pedro; A. Meilland

'PERLA DE ALCAÑADA, CLIMBING', Cl Min, dp; Dot, Pedro

'PERLA DE MONTSERRAT', Min, pb, 1945; bud small; flowers hermosa pink edged pearl, blooms in clusters, semi-dbl., 18 petals, dwarf, very compact growth; [Cécile Brunner X Rouletti]; Dot, Pedro; A. Meilland

Perla d'Oriolo, S, 1992; Embriaco, B.; (Cavriglia)

'PERLA ROSA', Min, mp, 1946; bud well-formed; flowers bright pink, very dbl.; very compact (6–8 in.) growth; [Perle des Rouges X Rouletti]; Dot, Pedro

'PERLA ROSA, CLIMBING', Cl Min, mp, 1947; Dot, Pedro

'PERLE', Pol, w, 1920; Easlea

'PERLE ANGEVINE', Pol, lp, 1920; flowers pale rose, small, borne in clusters, dbl.; few thorns; [Jeanne d'Arc X Mrs W.H. Cutbush]; Délépine

Perle Blanche, *see* 'DELANCHE'

Perle d'Amour, B, lp; Weihrauch

Perle d'Angers, B, lp, 1879; Moreau et Robert

Perle d'Anjou, Min, mp

Perle de Alcañada, *see* **'PERLA DE ALCAÑADA'**

'PERLE DE LYON', T, dy, 1872; flowers apricot yellow, large, dbl., Ducher

Perle de Veissenstein, *see* **'PERLE VON WEISSENSTEIN'**

Perle des Blanches, N, w, 1872; Lacharme, F.

'PERLE DES JARDINS', T, ly, 1874; flowers straw-yellow, large, dbl., globular, intense fragrance; foliage dark; slender growth; [Mme Falcot seedling sport]; Levet, F.

'PERLE DES JARDINS, CLIMBING', Cl T, ly, 1890; Henderson, J.

'PERLE DES JAUNES', T, dy, 1913; flowers golden yellow, large, dbl., moderate fragrance; Reymond; (Sangerhausen)

Perle des Neiges, HMult, w; Dubreuil, 1902

'PERLE DES PANACHÉES', HGal, m, 1845; (Panachée Double); flowers white striped with lilac and violet or rose, Vibert

'PERLE DES ROUGES', Pol, dr, 1896; flowers velvety crimson, Dubreuil

'PERLE D'OR', Pol, yb, 1884; (Yellow Cécile Brünner); flowers golden pink, blooms in clusters, very dbl., intense fragrance; foliage rich green, soft; height 3 ft; (14); [Polyantha X Mme Falcot]; Rambaux; Dubreuil

'PERLE D'OR, CLIMBING', Cl Pol, yb; (Yellow Cécile Brünner, Climbing)

Perle du Lac, *see* DORouvi

Perle du Lac Annecy, *see* DORouvi

Perle Meillandécor(R), *see* 'MEIPLATIN'

Perle Noire(R), *see* 'DELURT'

'PERLE ORLÉANAISE', Pol, dp, 1913; flowers carmine-pink, medium, dbl.; Duveau; (Sangerhausen)

'PERLE VOM WIENERWALD', HMult, pb, 1913; flowers carmine-rose, reverse soft rose-pink, borne in clusters of 5, semi-dbl., 2 in.; vigorous, climbing growth; [Helene X Crimson Rambler]; Praskac; Teschendorff

Perle von Aalsmeer, *see* **'PAREL VAN AALSMEER'**

'PERLE VON BRITZ', HMult, lp, 1910; flowers small, semi-dbl.; Kiese; (Sangerhausen)

'PERLE VON GODESBERG', HT, ly, 1902; flowers cream, shaded lemon; [Kaiserin Auguste Viktoria sport]; Schneider

'PERLE VON HEIDELBERG', HT, mr, 1905; flowers silvery pink, intense fragrance; vigorous growth; [La France sport]; Scheurer

'PERLE VON HOHENSTEIN', Pol, lp, 1923; flowers carmine-red, small, borne in clusters, semi-dbl.; [Freudenfeuer X Unnamed seedling]; Kiese

'PERLE VON REMAGEN', HT, lp, 1957; bud pointed; flowers soft pink, reverse tinted creamy white, large, dbl., exhibition form, moderate fragrance; foliage glossy, leathery; long stems; vigorous, upright growth; [R. M. S. Queen Mary X Peace]; Burkhard; Kordes; Kordes & Tantau

'PERLE VON WEISSENSTEIN', HGal, m; (Perle de Veissenstein); flowers brownish, center purple, Schwartzkopf, 1773

Perl-Ilseta(R), *see* 'TANILSEPO'

Perlmutt-Prinzessin, S, lp, 1990; flowers small, very dbl.; Wänninger, Franz; (Sangerhausen)

'PERMACH', HT, m, 1999; (**Macho Man**); flowers lavender, medium, dbl., 17–25 petals, borne mostly singly, moderate fragrance; few prickles; foliage medium, medium green, dull; bushy growth; Perry, Astor; Certified Roses, 2000

'PERMANENT WAVE', F, mr, 1932; (Duchess of Windsor, L'Indéfrisible, Mevrouw van

Straaten van Nes, Mrs Van Nes, Van Nes); flowers bright carmine, petals wavy, large blooms in clusters, semi-dbl., slight fragrance; foliage glossy, dark; vigorous, bushy growth; GM, Bagatelle, 1933; GM, Rome, 1934; [Else Poulsen sport]; Leenders, M.; J&P, 1935

Permeate Peace, HT, pb, 1990; Dawson

'PERMILL', HT, mr, 1997; (**Millennium**™); flowers full, large, very dbl., 26–40 petals, borne mostly singly, slight fragrance; foliage medium, medium green, semi-glossy; upright, medium (5 ft 6 in.) growth; [Fire Magic X Precious Platinum]; Perry, Astor

Permoser, HT, yb; flowers light yellow with carmine edges, large, dbl., moderate fragrance; Institut für Obstbau Dresden Pillnitz; (Sangerhausen)

'PERNETIANA', Pernet-Ducher

'PERNILLE POULSEN'®, F, mp, 1965; flowers light pink, blooms in clusters, semi-dbl., 18 petals, 3.5 in., moderate fragrance; foliage pointed, light green; [Ma Perkins X Columbine]; Poulsen, Niels D.; McGredy

'PERNILLE POULSEN, CLIMBING', Cl F, mp, 1980; Poulsen, Niels D.; Vilmorin-Andrieux

'PERO D'ALENQUER', F, dr; flowers dark velvety red, center lighter; [Unnamed seedling X Alain]; da Silva, Moreira

'PERPETUAL RED', S, mr, 1955; flowers bright red, open, medium, semi-dbl., recurrent bloom; foliage abundant; very vigorous growth; [Gruss an Teplitz X Unnamed seedling]; Gaujard

Perpetual White Moss, *see* **'QUATRE SAISONS BLANC MOUSSEUX'**

'PERPID', HT, dr, 1998; (**Intrepid**™); flowers dark velvety red, exhibition, medium, very dbl., 26–40 petals, 4–4.5 in., exhibition form, borne singly, moderate fragrance; prickles moderate; foliage medium, dark green, dull; medium, upright, 5 ft growth; [Karl Herbst X Burgundy]; Perry, Astor; Certified Roses Inc., 1999

'PERROQUET', LCl, dy, 1957; flowers large, slight fragrance; foliage glossy; vigorous growth; Robichon

'PERROQUET', F, rb, 1960; flowers red, reverse yellow, becoming dark red, well formed; foliage bronze; vigorous growth; [Peace X (Cinnabar X Circus)]; Lens

Persane, HT, 1984; Dorieux; (Cavriglia)

'PERSEPOLIS', HT, dy, 1986; (Truper); dbl., 25 petals, intense, tea fragrance; foliage clear green, glossy; [Unnamed seedling X Unnamed seedling]; Kriloff, Michel

'PERSIAN CARPET', F, rb, 1995; flowers rich crimson with yellow base of petal and reverse, blooms, semi-dbl., 10 petals, 1.5–2.75 in., slight fragrance; foliage medium, dark green, glossy; upright (up to 70 cms) growth; [Masquerade X Traumerei]; Fleming, Joyce L.; Hortico Roses, 1994

'PERSIAN DELIGHT', F, rb, 1995; flowers rich crimson with yellow base of petal and reverse, semi-dbl. (10 petals), medium (4–7 cms) blooms; slight fragrance; foliage medium, dark green, glossy; upright growth; [Masquerade X Traumerei]; Fleming, Joyce; Hortico Roses, 1994

Persian Musk Rose, *see* **'NASTARANA'**

'PERSIAN PRINCESS', Min, or, 1970; flowers coral-red, small, dbl., moderate fragrance; foliage leathery; vigorous, bushy, dwarf (12–14 in.) growth; [Baccará X Eleanor]; Moore, Ralph S.; Sequoia Nursery

Persian Yellow Rose, *see* **R. FOETIDA PERSIANA**

'PERSONALITY', HT, yb, 1960; bud ovoid; flowers golden yellow splashed red, open, dbl., 35–40 petals, 4.5 in., intense fragrance; foliage leathery, glossy; vigorous, upright growth; [Peace X Sutter's Gold]; Morey, Dr Dennison; J&P

'PERSUADER', HT, mp, 1960; flowers bright pink, large, dbl., 30 petals, exhibition form, intense fragrance; vigorous growth; [Golden Scepter X Southport]; Eacott

'PERSUASION', F, lp, 1972; flowers pale blush-pink, base orange, dbl., 30 petals, 3.5 in., exhibition form, moderate fragrance; foliage glossy; [Vera Dalton X Tropicana]; Sanday, John

'PERSUE DE GOSSART', C, dp; flowers velvety cerise, with garnet reflexes, dbl.

'PERSUS', HT, dy, 1998; (**Colossus**™); flowers golden yellow, full, medium sized, very dbl., 26–40 petals, borne in small clusters, moderate fragrance; foliage medium, medium green, glossy; upright, medium, 5 ft growth; [Folklore X Golden Gate]; Perry, Astor; Certified Roses, Inc., 1999

'PERTWENTYFIRST', HT, dy, 1999; (**Twenty First Century**); flowers large, dbl., 26–40 petals, borne mostly singly, slight fragrance; few prickles; foliage medium, dark green, semi-glossy; upright, tall (5 ft) growth; [Butterscotch X Mandelon]; Perry, Astor

Pet, lp; [(Betty Neuss X) X Silver Lining]; Dawson, George

'PETER BENJAMIN', HT, ab, 1978; flowers light apricot pink, borne 2–3 per cluster, dbl., 40 petals, exhibition form, moderate fragrance; pear-shaped, red prickles; foliage light; medium-tall growth; [Benjamin Franklin X Peter Frankenfeld]; Allender, Robert William

'PETER FRANKENFELD'®, HT, dp, 1966; flowers rose-pink, well-shaped, large, slight fragrance; Kordes, R.; A. Dickson

'PETER FRANKENFELD, CLIMBING', Cl HT, dp, 1975; Allen, L.C.; Welsh

Peter Goldman, *see* 'DICNAME'

'PETER LAMBERT', HMult, dp, 1936; flowers carmine-pink, medium, dbl.; Vogel, M.; (Sangerhausen)

'PETER LAWSON', HP, dr, 1862; flowers deep scarlet shaded purple; Thomas

Peter Long, HT, or

'PETER MAY', HT, mr, 1958; flowers scarlet, large, dbl., intense fragrance; foliage leathery; long stems; very vigorous growth; [The Doctor X New Yorker]; Verschuren; Blaby Rose Gardens

'PETER PAN', HT, dr, 1935; flowers dark crimson, large; vigorous growth; Knight, G.

Peter Pan, *see* **'PRESUMIDA'**

Peter Pan, *see* 'CHEWPAN'

Peter Pan, *see* 'SUNPETE'

'PETER PIPER', HT, or, 1969; flowers chinese orange, urn shaped, large, semi-dbl.; bushy growth; [Piccadilly sport]; Waterhouse Nursery

'PETER ROSEGGER', LCl, op, 1914; flowers coral-pink, rosette form, borne in clusters of 5–15, dbl., repeat bloom; foliage dark; vigorous, climbing growth; [Geheimrat Dr Mittweg X Tip-Top]; Lambert, P.

Peter Wessel, *see* 'TANTIDE'

'PETER'S BRIARCLIFF', HT, mp, 1940; flowers true unshaded rose-pink, long pointed, well formed, intense fragrance; foliage dark; vigorous, free growth; J&P

Pétillante, *see* **'ROTER CHAMPAGNER'**

Petit Canard®, *see* 'LENCAN'

'PETIT CONSTANT', Pol, mr, 1899; flowers nasturtium-red, small, dbl., intense fragrance; vigorous growth; [Mignonette X Luciole]; Soupert & Notting

Petit Four®, *see* 'INTERFOUR'

'PETIT FRANÇOIS', F, or, 1957; semi-dbl.; foliage glossy; very dwarf growth; [Alain X (Brazier X Léonce Colombier)]; Dorieux; Pin

'PETIT JEAN', HT, ob, 1926; flowers deep orange-buff, shaded to yellow, edged peach-pink, well s, dbl., moderate fragrance; [White Killarney X Sunburst]; Vestal

'PETIT LOUIS', HWich, op, 1912; flowers shrimp-pink, small, very dbl., slight fragrance; vigorous, climbing growth; [Dorothy Perkins X Seedling]; Nonin

Petit Marquis, *see* 'KORHOLST'

'PETIT POUCET', F, mr, 1955; flowers bright red, center tinted yellow, petals wavy, single; dwarf growth; [Cocorico X Unnamed seedling]; Combe

'PETIT PRINCE', Pol, mr, 1956; flowers geranium-red, borne in clusters of 8–10; very dwarf growth; Laperrière; EFR

Petit Rat de l'Opera®, S, pb, 1990; Lens

'PETIT RENÉ', HWich, mr, 1925; flowers brilliant red, small, borne in clusters, very dbl., intense fragrance; vigorous, climbing growth; Nonin

Petit Sam®, F, dp; Adam

Petit Serpent, S, w, 1994; Lens

Petite, *see* **'PATTY LOU'**

'PETITE CARROUSEL', Min, w, 1985; flowers ivory, coral pink petal edges, small blooms borne singly and, slight fragrance; foliage small, dark, semi-glossy; upright, tiny growth; [Thought to be Magic Carrousel X ?]; Michelis, Dorothy, 1991

'PETITE DE HOLLANDE', C, mp; (Petite Junon de Holland, R. centifolia minor); flowers rose pink, small blooms in clusters, dbl., moderate fragrance; moderate growth

Petite de Orlanaise, *see* **'PETITE ORLÉANAISE'**

Petite de Terre Franche, HMsk, pb, 1996; Louette

'PETITE ECOSSAISE', HSpn, lp; flowers flesh, small, non-recurrent, moderate growth; Vibert, before 1835

Petite Folie®, *see* 'MEIHERODE'

Petite Francoise, Pol, lp, 1915; Gravereaux

Petite Fredaine®, Min, op, 1991; Lens

'PETITE JEANNE', HWich, mr, 1912; flowers currant-red, [Dorothy Perkins seedling]; Nonin

Petite Junon de Holland, *see* **'PETITE DE HOLLANDE'**

'PETITE LÉONIE', Pol, w, 1893; flowers pinkish white, small; [Mignonette X Duke of Connaught]; Soupert & Notting

'PETITE LISETTE', C, dp, 1817; flowers rich rose, pompon-shaped, 1 in.; foliage matt, green; height 3–4 ft; Vibert

'PETITE LOUISE', LCl, op; flowers salmon-pink

Petite Michelle, *see* JACLIB

'PETITE ODETTE', HWich, lp, 1923; very dbl., borne in clusters, slight fragrance; vigorous, climbing growth; [Lady Godiva X Unnamed seedling]; Nonin

'PETITE ORLÉANAISE', C, mp; (Petite de Orlanaise); flowers pink, pompon-shaped, small, dbl.; vigorous, almost climbing growth

Petite Penny, *see* 'MACJOCEL'

Petite Perfection™, *see* 'JACRYBI'

Petite Pink Scotch, S, mp

Petite Red Scotch, *see* **'DOUBLE DARK MARBLED'**

Petite Reine, *see* **'QUEENIE'**

Petito, F, yb; Agel

'PETRA', F, mr, 1974; bud medium, globular; flowers blood-red, dbl., slight fragrance; foliage dark, leathery; moderate, bushy growth; [Seedling X Taora]; Kordes; Dehner & Co.

'PETRINE', HT, ob, 1921; flowers coral-red, shaded chrome-yellow, dbl., slight fragrance; [Old Gold X Mme Edouard Herriot]; Therkildsen

Petro, F, w

'PETRONELLA', HT, rb, 1980; flowers pointed blooms borne singly, dbl., 49 petals, moderate fragrance; straight, red prickles; foliage dark, glossy; compact growth; [Gail Borden X (Dalvey X Fragrant Cloud)]; MacLeod, Major C. A.; Christie Nursery, Ltd.

Petticoat, *see* 'JACPET'

Petticoat Lane, *see* 'SOCAMP'

'PETULA CLARK', HT, mr, 1963; flowers clear red, well-formed, dbl., exhibition form; foliage bronze; vigorous, bushy growth; [Purpurine X Lavender Pinocchio]; Lens

'PETULA CLARK, CLIMBING', Cl HT, mr, 1967; Lens

Pfaffstädt, HP, ly; flowers large, dbl., moderate fragrance; von Württemberg, Herzogin Elsa; (Sangerhausen)

Pfälzer Gold®, *see* 'TANÄLZERGO'

'PFANDER'S CANINA', (strain of R. canina), lp; Pfander, prior to 1954

'PFUSS PFREE', F, mp, 1988; bud ovoid, pointed; flowers light pink, reverse medium pink, full, small, borne in spray, dbl., 25 petals, cupped, slight fragrance; round, deep orange fruit; prickles straight, tan-brown; foliage medium, medium green, very glossy, blackspot resista; spreading growth; [Sea Foam X (Unnamed Restless Native seedling X Europeana)]; Stoddard, Louis, 1989

Phab Gold, *see* FRYbountiful

'PHAENOMEN', HT, op, 1934; (Phenomenon); flowers slightly rosy salmon, reverse light rosy red, very large, dbl.; foliage dark; very vigorous growth; Chotkové Rosarium; Böhm

Phalaenopsis, F, dp

'PHANTASY', HT, lp, 1927; flowers medium light pink, base yellow, dbl., moderate fragrance; [Lady Alice Stanley X Royal (or Priscilla)]; Dunlop

'PHANTOM', HT, ob, 1920; flowers coppery yellow, center lighter, semi-dbl., moderate fragrance; RULED EXTINCT 4/92; [(Joseph Hill X My Maryland seedling) X Lady Hillingdon]; Towill

Phantom, *see* 'MACCATSAN'

Pharaoh, *see* 'MEIFIGA'

Pharaon®, *see* 'MEIFIGA'

Phare®, *see* 'DELGO'

'PHARISÄER', HT, op, 1903; bud long, pointed; flowers rosy white, shaded salmon, blooms, dbl., exhibition form, moderate fragrance; foliage bronze; long, weak stems; vigorous growth; [Mrs W.J. Grant seedling]; Hinner, W.

Pheasant, *see* 'KORDAPT'

'PHEIDIPPIDES', F, mr, 1979; flowers currant-red, full, dbl., 35 petals, 4 in.; foliage light green; moderately vigorous, low growth; [Sam Ferris X Chopin]; Ellick; Excelsior Roses

'PHELAN'S FLAG', HCh, rb, 1952; bud ovoid, greenish white flushed purplish red; flowers unique virgin blooms (1) red, (2) pink, (3) white, (4) red e, dbl., 75–100 petals, 2.5–3 in., quartered, semi-recurrent bloom, intense fragrance; foliage dark; vigorous (6 ft or more) growth, with tendency to climb; Phelan

'PHÉNICE', HGal, dp, 1843; flowers reddish rose, spotted, medium, dbl.; Vibert

Phenomenon, *see* **'PHAENOMEN'**

Philadelphia, *see* **'PHILADELPHIA RAMBLER'**

'PHILADELPHIA RAMBLER', HMult, dr, 1904; (Philadelphia); flowers scarlet-crimson, center lighter, borne in gigantic clusters, dbl., midseason bloom; vigorous, climbing growth; [Crimson Rambler X Victor Hugo]; Van Fleet; Conard & Jones

'PHILÉMON COCHET', B, mp, 1895; flowers rose, Cochet, Sc.

'PHILIBERT BOUTIGNY', HP, mp; flowers silvery rose, very large; vigorous growth

'PHILIP HARVEY', F, ob, 1975; flowers salmon-red, shaded orange, dbl., 25 petals, 5 in., moderate fragrance; foliage glossy, dark; [Fragrant Cloud X Circus]; Harkness

Philippa Pirette, S, pb

'PHILIPPE', F, op, 1959; (Bel Ami); bud oval; flowers peach-salmon, open, medium, borne in clusters, semi-dbl., 15 petals; foliage glossy; moderate growth; [Cognac X Fashion]; Delforge

Philippe Noiret, *see* MEIZOELE

'PHILIPPE PÉTAIN', HT, dp, 1940; flowers velvety carmine with coppery reflections, not turning blue, cupped, slight fragrance; foliage bright chive-green; Nabonnand, C.

'PHILIPPE RIVOIRE', HT, op, 1941; flowers large, dbl.; Gaujard; (Sangerhausen)

Phillipa, *see* POUlheart

Phillipp Melanchthon, F, dr, 1994; Hetzel

'PHILLIPP PAULIG', HP, dr, 1908; flowers large, dbl., moderate fragrance; Lambert, P.; (Sangerhausen)

Philomene, C, m; Hardy

'PHILOMÈNE POLLAERT', HT, or, 1925; flowers crimson tinted orange, dbl., intense fragrance; [Gen. MacArthur seedling X Old Gold]; Pollaert

Phil's Chromatella, N, dy

Phil's Hot Pink Perpetual Damask, D, mp

Phloxy®, F, rb

Phoebe, *see* 'HARVANDER'

'PHOEBE', HT, w, 1922; flowers cream-white, sometime pure white, dbl., slight fragrance; GM, NRS, 1921; [Ophelia X Verna Mackay]; Cant, B. R.

Phoebe's Choice™, *see* 'BILICE'

Phoebe's Frilled Pink, *see* **'FIMBRIATA'**

'PHOEBUS', HP, mp, 1837; flowers bright pink, shading lighter, large, dbl., moderate fragrance; bushy growth

Phoenix, *see* 'HARVEE'

'PHOENIX', HT, dp, 1973; flowers light cerise, large, dbl., exhibition form, moderate fragrance; foliage large, glossy, leathery; vigorous, upright growth; [Manitou X Grand Slam]; Armstrong, D.L.; Armstrong Nursery

'PHOENIX FIRST', F, dr, 1959; bud ovoid; flowers dark red shaded black, small, borne in clusters, very dbl., slight fragrance; foliage leathery; bushy growth; [Our Princess X Pompon Beauty]; Kernovski, V.R.; Langbecker

'PHOTOGENIC', S, w, 1999; flowers white with pink edge, reverse white with narrow pink edge, single, 10 petals, 2.5–3 in., borne in small clusters, moderate fragrance; foliage large, medium green, semi-glossy; upright, tall (10 ft) growth; Jerabek, Paul E.; Freedom Gardens, 1998

'PHYLLIS', Pol, mr, 1908; flowers bright red, small, borne in large clusters, dbl., recurrent bloom; [Mme Norbert Levavasseur X ?]; Merryweather

'PHYLLIS BIDE', Cl Pol, yb, 1923; flowers pale gold, shaded pink, almost small blooms in long, loose clusters, dbl., dependably recurrent; height 6 ft; (14); GM, NRS, 1924; [Perle d'Or X Gloire de Dijon]; Bide

'PHYLLIS BURDEN', HT, op, 1935; bud long, pointed; flowers shrimp-pink and orange, large, dbl.; foliage glossy, light; very vigorous growth; Cant, B. R.

'PHYLLIS GOLD', HT, my, 1935; flowers butter-yellow, edged lighter, dbl., exhibition form, slight fragrance; foliage rich olive-green; very vigorous, branching growth; GM, NRS, 1933; [Lady Florence Stronge X Julien Potin]; Robinson, H.; Wheatcroft Bros.

'PHYLLIS GOLD, CLIMBING', Cl HT, my, 1949; Fryers Nursery, Ltd.

'PHYLLIS LUCAS', HT, ob, 1961; flowers orange shaded bronze, well shaped, moderate fragrance; [Bettina sport]; Wheatcroft Bros.

'PHYLLIS POYSER', HT, op, 1962; bud long, pointed; flowers ornage-pink, large, dbl., 50 petals, exhibition form, intense, spicy fragrance; foliage light green, soft, elongated; vigorous, compact, bushy growth; [Golden Sun X Spartan]; Fankhauser

Phyllis Shackelford™, *see* 'MORSHACK'

'PHYNELIA', HT, mr, 1928; flowers crimson-cerise, borne in clusters; Reeves

'PIA BERGHOUT', Pol, mp, 1967; flowers pink, medium, borne in clusters, dbl.; foliage dark; [Saskia X Seedling]; Buisman, G. A. H.

'PICADOR', F, mr, 1964; flowers cherry-red, borne in clusters, dbl.; foliage glossy, dark; upright, compact growth; [Oranien sport]; Verschuren, A.; van Engelen

Picaninni, *see* WRIpic

'PICARDY', HT, dp, 1967; flowers rose-bengal, globular, moderate fragrance; foliage dark, glossy; free growth; [Rose Gaujard X Bayadere]; Trew, C.; Willik Bros.

Picasso, *see* 'MACPIC'

'PICAYUNE', Ch, lp; flowers light pink to white, small blooms in clusters, dbl.

Piccadilly®, *see* 'MACAR'

'PICCADILLY SUNSET', HT, ob, 1970; flowers orange flushed apricot, reverse gold, high pointed, dbl., 34 petals, 4–5 in., slight fragrance; foliage glossy, dark; moderate, upright growth; [Piccadilly sport]; Goodwin; The Valley Nursery

'PICCADILLY, CLIMBING', Cl HT, rb, 1963; Mized, 1963 & Sutton, 1972

'PICCANINNY', HT, dr, 1941; bud long, pointed; flowers stamens yellow, large, single, 5–6 petals, intense fragrance; foliage dark, glossy; vigorous growth; [Night X Sanguinaire]; Lammerts, Dr Walter; Armstrong Nursery

'PICCIOLA INA', Pol, op, 1937; flowers pure salmon, borne in clusters of 10–50, vigorous, bushy growth; Giacomasso

Piccola, *see* 'TANOLOKIP'

'PICCOLA ES', S, lp, 1983; flowers medium, dbl., 20 petals, slight fragrance; foliage medium, medium green, semi-glossy; bushy growth; [Unnamed X Unnamed]; Fumagalli, Niso

Piccolo®, *see* 'TANOLOKIP'

'PICCOLO', F, dr, 1957; bud ovoid; flowers velvety dark red, open, borne in clusters, dbl., slight fragrance; foliage dark, leathery, glossy; moderate, bushy growth; RULED EXTINCT 4/85; [Red Favorite X Kathe Duvigneau]; Tantau, Math.

'PICCOLO PETE', S, mr, 1984; 7–10 petals, 4 in., flat, borne 1–10 per cluster, repeat bloom, moderate fragrance; slightly hooked, tan prickles; foliage large, leathery, dark olive green; upright, bushy growth; hardy.; [Carefree Beauty X ((Peace X Dornroschen) X Country Music)]; Buck, Dr Griffith J.; Iowa State University

'PICK ME UP', HT, mr, 1941; flowers well-shaped, Clark, A.

Pickering Baby Faurax, Pol, m

Pickering Four Seasons Rose, P, dp

Pickering Old Red China, Ch, dp

Pickering Red, HP, dr

Pickwick, *see* 'SABCLIVE'

'PICNIC', F, or, 1966; (Pic-Nic); bud short, pointed; 3–3.5 in., exhibition form, slight fragrance; foliage leathery; upright growth; [South Seas X Unnamed seedling]; Warriner, William A.; J&P

Pic-Nic, *see* **'PICNIC'**

'PICO', HT, rb, 1962; bud pointed; flowers salmon-red and canary-yellow, dbl.; foliage glossy; bushy growth; Frères, Buyl

Picobello, F, or, 1998

'PICOTEE', F, rb, 1960; flowers white, edged red, blooms in clusters, dbl., 24 petals, 2.5–3.5 in., cupped, slight fragrance; foliage dark, glossy; vigorous, upright, bushy growth; [Little Darling X Gertrude Raffel]; Raffel; Port Stockton Nursery

Picpa, HT, 1974; Dorieux; (Cavriglia)

'PICTURE', HT, lp, 1932; dbl., exhibition form, slight fragrance; foliage glossy, dark; vigorous growth; McGredy

'PICTURE PAGE', HT, lp, 1953; flowers peach-pink shading to flesh-pink, base yellow, well formed, dbl., 23–25 reflexed petals, 4 in., slight fragrance; foliage dark; very free growth; [Picture X Mme Butterfly]; Jordan, B.L.

Picture Perfect, *see* **'ELSIE MELTON'**

'PICTURE, CLIMBING', Cl HT, lp, 1942; Swim, H.C.; Armstrong Nursery

'PICTURESQUE', HT, lp, 1950; bud long; flowers pale pink with pronounced red veining, large, dbl., 30–35 petals, exhibition form, slight fragrance; foliage leathery; vigorous, upright, bushy growth; [Mrs H.M. Eddie X Mrs Sam McGredy]; Eddie

Pie IX, *see* **'PIUS IX'**

Piece d'Or, S, dy

'PIED PIPER', F, mr, 1969; flowers small, dbl., globular; foliage leathery; moderate, low growth; [Garnette X Moulin Rouge]; Lindquist; Howard Rose Co.

Pierette, HRg, dp, 1987; flowers carmine-pink, large, semi-dbl., intense fragrance; Uhl, J.; (Sangerhausen)

Pierette Pavement, HRg, lp, 1987; semi-dbl., moderate fragrance; spreading; ground cover (2.5 ft) growth; Uhl, J.

'PIERRE', HT, yb, 1945; bud large, ovoid; flowers deep golden yellow edged red, very dbl.; foliage glossy; vigorous, upright, bushy growth; [Soeur Thérèse X Lumiere]; Mallerin, C.; A. Meilland

Pierre, HT, ab

'PIERRE AGUETANT', HT, my, 1938; bud long, pointed; flowers chamois-yellow, open, very large, dbl.; moderate fragrance; foliage leathery; vigorous growth; Gaujard

Pierre B., HT, lp; [Dr A.J. Verhage sport]; Bell, Ronald J., 1982; (Weatherly, L.)

'PIERRE BREDY', HT, rb, 1958; flowers currant-red, reverse silvery; foliage dark;

low growth; [Peace X Scheherazade]; Arles; Roses-France

'PIERRE CARO', HP, dr, 1879; flowers medium, dbl., intense fragrance; Levet; (Sangerhausen)

'PIERRE CORMIER', Pol, or, 1926; flowers brilliant scarlet-red, center lighter, borne in clusters of; dwarf growth; Turbat

Pierre de Ronsard®, *see* 'MEIviolin'

'PIERRE DE ST CYR', B, lp, 1838; flowers glossy pale pink, large, very dbl., cupped; vigorous growth; Plantier

'PIERRE GAUJARD', HT, rb, 1944; bud pointed; flowers fiery shades, very large, dbl., slight fragrance; foliage glossy; vigorous growth; Gaujard

'PIERRE GUILLOT', HT, rb, 1879; flowers red with white, large, dbl., moderate fragrance; Guillot; (Sangerhausen)

'PIERRE NOTTING', HP, dr; flowers large, dbl., 42 petals, globular, not often recurrent, moderate fragrance; upright growth; [Alfred Colomb seedling]; Portemer fils, 1862

Pierre Troisgros®, *see* DORtive

'PIERRE WATTINE', HT, dp, 1901; flowers large, dbl., moderate fragrance; Soupert & Notting; (Sangerhausen)

'PIERRETTE', HT, rb, 1931; flowers blackish red streaked white, passing to steel-blue, small, very dbl., cupped, slight fragrance; foliage small; short stems; dwarf growth; [Felix Laporte sport]; Tantau

'PIERRETTE', HT, or, 1945; flowers bright copper red, flecked brighter, large, dbl., 25 petals, intense fragrance; upright, bushy growth; [Texas Centennial sport]; Tantau

Pierrine, *see* 'MICpie'

'PIERROT', F, rb, 1971; bud ovoid; flowers white marked red, becoming red, dbl., 30 petals, 2.5 in., cupped, moderate, spicy fragrance; foliage leathery; vigorous, compact growth; [Poupee X Fillette]; Lens

'PIERSON'S PINK', HT, dp, 1950; bud long, pointed; flowers rose-pink, dbl., 32 petals, 5 in., exhibition form, slight fragrance; very vigorous, upright growth; [Better Times sport]; Pierson, A.N.

'PIET RETIEF', F, dr, 1950; flowers dark morocco-red; [Irene X Donald Prior]; Leenders, M.

'PIET SAVERYS', F, ob, 1955; flowers orange, semi-dbl.; bushy growth; [Independence X Border King]; Frères, Buyl

Pietermaritzburg, F, mp

'PIGALLE', HT, m, 1951; flowers reddish violet, dbl., 4 in., slight fragrance; foliage bronze; bushy growth; [Fantastique X Boudoir]; Meilland, F.

Pigalle®, *see* 'MEIcloux'

Pigalle 84, *see* 'MEIcloux'

'PIGALLE, CLIMBING', Cl HT, m; Roses-France

'PIGMY GOLD', F, dy, 1953; bud ovoid; flowers golden yellow, small, dbl., moderate fragrance; foliage dark, glossy; dwarf growth; Boerner; J&P

'PIGMY LAVENDER', F, m, 1961; bud ovoid; flowers lavender tinted pink, dbl., 30–35 petals, 2 in., cupped, moderate fragrance; foliage leathery, dark; short, strong stems; vigorous, dwarf growth; [Lavender Pinocchio seedling X Unnamed Hybrid Tea seedling]; Boerner; J&P

'PIGMY RED', F, dr, 1953; bud ovoid; flowers deep red, white eye, small blooms in clusters, dbl., moderate fragrance; foliage glossy; dwarf growth; [Chatter X Red Pinocchio]; Boerner; J&P

'PIKE'S PEAK', S, pb, 1940; bud long, pointed; flowers light bright red, center yellow, fading white, blooms in clusters, semi-dbl., 13 petals, 3.5 in., non-recurrent; foliage light, wrinkled; very vigorous (6 ft), bushy growth; [R. acicularis X Hollywood]; Gunter; B&A

'PILAR DE ARBURUA', HT, op; flowers salmon, large, dbl., exhibition form, slight fragrance; foliage dark reddish green; free growth; [Comtesse Vandal X Fashion]; Camprubi, C.

'PILAR DOT', Min, op, 1964; flowers coral, well-formed, small; vigorous, well-branched growth; [Orient X Perla de Alcanada]; Dot, Pedro

'PILAR LANDECHO', HT, ob, 1940; (Marquesa de Urquijo); bud long, pointed; flowers yellow, reverse dark coral-orange, large, dbl., exhibition form, slight fragrance; foliage dark, leathery; vigorous growth; GM, Bagatelle, 1938; [(Sensation X Julien Potin) X Feu Joseph Looymans]; Camprubi, C.; A. Meilland

'PILAR LANDECHO, CLIMBING', Cl HT, ob, 1954; (Marquesa de Urquijo, Climbing); Folgado, Comes

Pilarcitos, N, w

'PILARÍN VILELLA', HT, or, 1936; flowers lacquer-red, large, dbl., cupped, intense fragrance; foliage dark; vigorous growth; [Mrs Pierre S. duPont X Lucia Zuloaga]; Dot, Pedro

'PILGRIM', HT, dr, 1970; bud ovoid; flowers large, dbl., cupped, moderate fragrance; foliage dark, leathery; vigorous, upright, bushy growth; [Unnamed seedling X Chrysler Imperial]; Armstrong, D.L.; Armstrong Nursery

'PILGRIM', HT, pb, 1920; bud long, pointed; flowers silvery pink reverse clear rose-pink, large, dbl., exhibition form, moderate fragrance; foliage leathery, rich green; vigorous, bushy growth; Montgomery Co.; A.N. Pierson

'PÍLINA MATA', HT, ob, 1934; flowers orange-yellow; [Souv. de Claudius Pernet X Los Angeles]; Munné, B.

Pillar Box, *see* 'CHEwaze'

Pillar BoxWardlip, *see* 'CHEwaze'

'PILLAR OF FIRE', Cl F, or, 1963; bud short, ovoid; flowers coral-red, blooms in cluster, dbl., 33 petals, 2–2.5 in., cupped, slight fragrance; foliage leathery, matt; vigorous growth; [Floradora, Climbing sport]; Shamburger, P.; C-P

Pillar of Gold, *see* **'E. VEYRAT HERMANOS'**

'PILLAR STRATFORD', Cl HT, lp, 1946; flowers silvery pink, base deeper, large, very dbl., intense fragrance; vigorous, upright (8 ft) growth; [Stratford sport]; Watkins, A.F.; Krider Nursery

Pillnitzer Marcellina, Pol, dr; semi-dbl.; Institut für Obstbau Dresden Pillnitz; (Sangerhausen)

Pillow Fight™, *see* 'WEKpipogop'

'PILLOW TALK', F, m, 1980; bud short, ovoid; flowers reddish lavender, blooms borne singly or several together, dbl., exhibition form, moderate, tea fragrance; long prickles, hooked downward; foliage medium, thin, leathery, dark; low to medium, rounded growth; [Plain Talk X Angel Face]; Weeks, O.L.

Pilurett, F, or, 1966; flowers medium, semi-dbl.; Schmadlak, Dr.; (Sangerhausen)

Pimlico, *see* 'MEIdujaran'

Pimlico '81, *see* 'MEIdujaran'

'PIMPANT', F, or, 1963; flowers bright orange-red, borne in clusters of 7–8, dbl., 25–30 petals; moderate, bushy growth; [Unnamed seedling X Soleil]; Laperrière; EFR

'PIMPERNELL', F, or, 1954; flowers turkey-red, open, borne in clusters, semi-dbl.; foliage dark; [Unnamed seedling X (Poulsen's Pink X Golden Dawn)]; LeGrice

Pimpernelle, *see* DELdog

'PINAFORE', Pol, ly, 1959; flowers pale yellow to white, tinged pink, borne in large, rounded clusters, single, 1.5–2 in., flat; foliage glossy; low, bushy, compact growth; [China Doll X Mrs Dudley Fulton]; Swim, H.C.; Roseway Nursery

Pinal, Pol, mr; flowers medium, single, Institut für Obstbau Dresden Pillnitz; (Sangerhausen)

'PIÑATA', LCl, yb, 1978; bud ovoid; flowers yellow overlaid vermilion, dbl., 28 petals, 3 in., exhibition form, slight fragrance; semi-climbing growth; Suzuki, Seizo; J&P

PINault, F, 1982; (**Rêve de Deauville**, Revue de Dauville); Pineau; (Cavriglia)

Pincushion, S, mp, 1996; Kordes

'PINEAPPLE POLL', F, ob, 1970; flowers orange-yellow, flushed red, dbl., 30 petals, 2.5 in., moderate fragrance; foliage glossy; [Orange Sensation X Circus]; Cocker

'PINEHURST', Min, pb, 1988; flowers light pink, fading lighter at base, reverse light pink to cream, very dbl., 60 petals, moderate, fruity fragrance; prickles straight, medium, pink; foliage medium, dark green, semi-glossy; bushy, medium, vigorous, neat growth; [Rise 'n' Shine X Unnamed seedling]; Bridges, Dennis A.; Bridges Roses

Pine-Scented Rose, *see* **R. glutinosa**

'**Pingan**', F, w, 1980; bud ovoid, pointed; flowers white with pink petal edges, blooms borne 4–6 per cluster, semi-dbl., 13 petals, moderate fragrance; long prickles; foliage large, leathery; upright growth; [Unnamed seedling X (Ivory Fashion X Little Darling)]; Fong, William P.; Del Rose Nursery

'**Pink Alicia**', HT, mp, 1968; flowers pink, pointed, moderate fragrance; spreading growth; [Duftwolke X Gavotte]; Dale, F.

Pink Angel, *see* 'MORgel'

'**Pink Angel**', Min, mp, 1982; [Starina sport]; Hunton, Claude B.

Pink Angel, HT, lp; Parkes, Mrs M.H., 1977; (Weatherly, L.)

'**Pink Anne**', F, mp, 1951; borne in trusses, slight fragrance; (28); [Anne Poulsen sport]; Cant, B. R.

Pink Arctic, *see* **'Show Garden'**

Pink Avalanche, *see* 'WILpavl'

Pink Babyflor, *see* TANybab

Pink Bassino, *see* KORbasren

'**Pink Beauty**', HT, mp, 1919; flowers clear pink, large, semi-dbl., cupped, intense fragrance; [Ophelia X My Maryland]; Cook, J.W.

'**Pink Bedder**', HT, mp, 1920; flowers rose-pink, center yellow, borne in clusters; Paul, W.

'**Pink Belle**', HT, dp, 1974; flowers deep rose pink, large, dbl., 28 petals, moderate fragrance; foliage dark; RULED EXTINCT 11/83; [Fragrant Cloud X Elizabeth of Glamis]; Harkness; Morse Roses

Pink Bells®, *see* 'POUlbells'

'**Pink Blush**', HT, lp, 1974; bud long, pointed; dbl., 30 petals, 4–5 in., cupped, slight fragrance; foliage large, leathery; very free growth; [Bridal Pink X Seedling]; Warriner, William A.; J&P

'**Pink Bountiful**', F, mp, 1945; bud short, pointed; dbl., 55 petals, 3 in., blooms in clusters, moderate fragrance; foliage dark, leathery; vigorous, upright, much-branched growth; [Juanita X Mrs R.M. Finch]; Hill, Joseph H., Co.; J&P; J.H. Hill Co.

Pink Bounty™, *see* 'MINsco'

'**Pink Bouquet**', F, mp, 1954; flowers china-rose pink, slightly tinted yellow, to ovoid, dbl., 60–75 petals, 3–4 in., exhibition form, moderate fragrance; foliage leathery; upright, open, compact growth; RULED EXTINCT 11/90; [Curly Pink X Free Gold]; Brownell, H.C.

Pink Bouquet, *see* 'LAVquest'

Pink Bouquet, HWich, pb; Treasure

Pink Bourbon, *see* **'Mme Ernest Calvat'**

'**Pink Brocade**', F, lp, 1977; flowers very full, very dbl., 70 petals, 3 in., slight fragrance; foliage glossy; small, compact growth; [Spartan X Lilli Marlene]; Bees

Pink Bunting, *see* 'TALpin'

'**Pink Butterfly**', HT, dp, 1926; flowers bright cerise, base light buff; [Mme Butterfly sport]; Brown, A.C.

Pink Button, *see* KORdemas

'**Pink Cameo**', Cl Min, mp, 1954; (Cameo, Climbing); flowers rose-pink, center darker, blooms in clusters (to 20), dbl., 23 petals, 1.25 in., slight fragrance; foliage small, glossy, rich green; height 3–5 ft; [(Soeur Thérèse X Skyrocket) X Zee]; Moore, Ralph S.; Sequoia Nursery

Pink Candy, HT, lp; [Candy Stripe sport]; Thomas, before 1992; (Weatherly, L.)

Pink Cardinal Hume, S, mp, 1998; Rupert, Kim L.

Pink Carpet™, *see* 'MINnco'

Pink Cascade, *see* 'MORcade'

'**Pink Cascade**', Pol, lp, 1945; flowers la france pink, small, borne in clusters, very dbl., slight fragrance; foliage glossy; vigorous, bushy growth; RULED EXTINCT 12/81; [Mrs Dudley Fulton X Tom Thumb]; Lammerts, Dr Walter; Univ. of Calif.

'**Pink Cavalcade**', F, pb, 1955; flowers deep pink, reverse light yellow to white, borne in pyramidal, dbl., 28–32 petals, 2.5–3 in., cupped, moderate fragrance; foliage leathery, glossy, bronze; vigorous, bushy growth; [Cavalcade sport]; Shamburger, C.S.; Stuart

'**Pink Chameleon**', Pol, pb, 1945; bud rose-red; flowers venetian pink, darkening to purple, open, small, borne in clusters, single, abundant, recurrent bloom, slight fragrance; foliage dark, glossy; very vigorous, bushy growth; [Mrs Dudley Fulton X R. chinensis mutabilis]; Lammerts, Dr Walter; Univ. of Calif.

'**Pink Champagne**', HT, mp, 1956; bud long, pointed; dbl., 32–48 petals, 5–6 in., exhibition form, moderate fragrance; foliage leathery; vigorous, bushy growth; [Unnamed seedling X Pink Bountiful]; Jelly; E.G. Hill Co.

Pink Champagne, *see* CLEcham

Pink Charles Austin, S, pb, 1992; Austin, David

'**Pink Charm**', F, dp, 1938; flowers deep clear pink, very dbl., slight fragrance; foliage leathery; short stems; bushy growth; Kordes; Dreer

Pink Charmer, *see* 'LYOpin'

'**Pink Charming**', HT, lp, 1953; flowers pale pink, loosely formed, dbl., 36 petals, 5.5 in., moderate fragrance; vigorous growth; Leenders, M.

'**Pink Chateau**', HT, w, 1999; flowers white, center soft pink, dbl., 35 petals, 5 in., exhibition form, no fragrance; 4.5 ft growth; [(Sheer Bliss X seedling) X seedling]; Teranishi, K., 1997; Itami Rose Nursery, 1997

Pink Cherokee, *see* **'Anemone'**

Pink Cherokee, *see* R. X anemonoides

Pink Cherub, *see* 'MORfair'

'**Pink Chiffon**', F, lp, 1956; bud ovoid; dbl., 53 petals, 3.5–4 in., cupped, intense fragrance; foliage glossy; vigorous, bushy growth; [Fashion X Fantasia]; Boerner; J&P

Pink Chimo®, *see* 'INTerchimp'

'**Pink Christian Dior**', HT, dp, 1966; bud ovoid; flowers light red and deep pink, large, dbl., exhibition form, slight fragrance; foliage glossy; very vigorous, upright growth; [Christian Dior sport]; Chang, Chi-Shiang

'**Pink Cloud**', LCl, mp, 1952; bud ovoid; flowers rich pink, large blooms in clusters of 5–20, dbl., 28 petals, cupped, recurrent bloom, moderate fragrance; foliage glossy; vigorous, climbing (6–8 ft) growth; [New Dawn X New Dawn seedling]; Boerner; J&P

Pink Cloud®, *see* 'JACLIP'

'**Pink Clouds**', Cl Min, dp, 1956; flowers deep rose pink, small to blooms in clusters, single, intense fragrance; foliage dark, glossy; vigorous (5–8 ft) growth; [Oakington Ruby X R. multiflora]; Moore, Ralph S.; Sequoia Nursery

'**Pink Cluster**', F, op, 1938; flowers salmon-pink, shaded gold, borne in clusters; Morse

'**Pink Cocktail**', HT, pb, 1963; flowers light pink, reverse darker, dbl., 38 petals, 5 in., slight fragrance; foliage dark, glossy; very free growth; [Queen Elizabeth X Claude]; Barter

Pink Cottage, *see* CHEwily

Pink Cover®, S, mp; Poulsen

Pink Crumble Bar, HT, lp; [Crumble Bar sport]; Thomson, G.A.; (Weatherly, L.)

Pink Crystal, *see* WEKslats

Pink Curtain, *see* KORlallal

Pink Cushion, Pol, mp

Pink Danyland®, S, dp

'**Pink Dawn**', HT, mp, 1935; bud long, pointed; flowers large, dbl., 60 petals, exhibition form, intense fragrance; foliage soft; vigorous growth; [Joanna Hill X Seedling]; H&S; Dreer

'**Pink Dawn, Climbing**', Cl HT, mp, 1941; H&S

'PINK DELIGHT', Pol, dp, 1922; (Laxton's Pink Delight); flowers rose-pink, single, moderate fragrance; Laxton Bros.

'PINK DELIGHT', HT, op, 1936; (Vierlanden); bud long, pointed; flowers deep salmon-pink, large, dbl., exhibition form, intense fragrance; foliage leathery, light green; vigorous growth; [Senator X Florex]; Kordes; J&P

Pink Delight®, *see* LENpil

Pink Devil, HT, mp

Pink Diadem®, *see* TANspolett

'PINK DIAMOND', HT, lp, 1942; bud long pointed; flowers shell-pink, base yellow, dbl., 38 petals, 3.5–4 in., cupped, foliage leathery; vigorous, upright growth; Howard, F.H.; Diamond State Nursery

'PINK DIANE', F, dp, 1959; bud ovoid; flowers deep rose-pink, dbl., 50–55 petals, 3.5–4 in., cupped, intense, fruity fragrance; foliage leathery, glossy, dark; vigorous, upright growth; [Rosenelfe sport]; Sodano, A.

Pink Diddy™, *see* 'KINDIDDY'

'PINK DIËLMA', HT, mp, 1969; bud ovoid; flowers pink, medium, very dbl.; foliage dark; [Furore sport]; Tas

'PINK DON JUAN', LCl, mp, 1996; semi-dbl., 6–14 petals, 1.5–2.75 in., borne in small clusters, slight fragrance; some prickles; foliage medium, medium green, semi-glossy; tall, upright, spreading, climbing growth; [Don Juan sport]; Nelson, Brian; Certified Roses, Inc., 1997

'PINK DREAM', HT, pb, 1951; flowers pink, inside rosy white, dbl., 45 petals, 6–7 in., exhibition form, slight fragrance; foliage dark; vigorous growth; [Mrs Sam McGredy X R.M.S. Queen Mary]; McGredy, Sam IV

Pink Drift, *see* 'POULCAT'

'PINK DRUSCHKI', HT, mp, 1949; flowers bright pink, well formed, dbl., 27 petals, 5 in., slight fragrance; very vigorous growth; Longley

'PINK DUCHESS', HT, dp, 1959; bud ovoid; flowers rose-red, dbl., 38 petals, 5–6 in., cupped, moderate fragrance; foliage glossy; vigorous, upright, bushy growth; [Peace seedling X Unnamed seedling]; Boerner; J&P

'PINK ELEGANCE', HT, lp, 1960; bud long, pointed; flowers bright pink, medium, dbl., exhibition form, slight fragrance; foliage glossy; very vigorous growth; [White Butterfly X Baccará]; Hoefer; Carlton Rose Nurseries

Pink Elf, *see* 'MORELFIRE'

Pink Elizabeth Arden, *see* **'GEISHA'®**

'PINK EMPEROR', HT, mp, 1958; flowers clear pink, dbl., 50–60 petals, 4–5 in.; foliage glossy; vigorous, bushy growth; [Peace X Crimson Glory]; Jones; Hennessey

Pink Empress, *see* 'HADEMP'

'PINK EUTIN', F, mp, 1962; [Eutin sport]; Lindquist; Howard Rose Co.

Pink Fairy, Pol, mp

Pink Fantasy, HT, mp, 1995; K&S

'PINK FAVORITE', HT, mp, 1956; (Pink Favourite); bud pointed; flowers neyron rose, dbl., 25 petals, 3–4 in., cupped, slight fragrance; foliage very glossy, bright green; vigorous, upright, bushy growth; GM, Portland, 1957; [Juno X (Georg Arends X New Dawn)]; Von Abrams; Peterson & Dering

Pink Favourite, *see* **'PINK FAVORITE'**

Pink Festival, *see* 'LAVQUEST'

Pink Fire®, *see* INTepin

'PINK FLAIR', F, mp, 1966; bud urn shaped; flowers small, dbl.; foliage leathery; vigorous, bushy growth; [Verona X Escort]; Swim & Weeks; Carlton Rose Nurseries

'PINK FLAMINGO', HT, pb, 1957; flowers rose-pink tinted lighter, dbl., 50 petals, 2.5–3.5 in., intense fragrance; foliage dark, leathery; dwarf, bushy growth; [Golden Dawn sport]; Kern Rose Nursery

'PINK FLORADORA', F, op, 1951; bud ovoid; flowers shrimp-pink, dbl., 35–40 petals, 3 in., cupped, slight fragrance; foliage glossy, light green; bushy growth; [Floradora sport]; Shamburger, P.

Pink Flower Carpet, *see* 'NOATRAUM'

Pink Flurries, F, lp, 1999

'PINK FORMAL', S, pb, 1978; bud pointed; flowers bright coral-pink, loosely-ruffled, dbl., 23 petals, 3.5–4 in., moderate fragrance; foliage leathery; vigorous, upright growth; [(Queen Elizabeth X Gladiator) X (Aztec X Little Darling)]; Williams, J. Benjamin; Krider Nursery

Pink Fountain, *see* LEGram

'PINK FRAGRANCE', HT, mp, 1956; bud long, pointed; flowers rose-pink, open, borne in pyramidal clusters, dbl., 78–85 petals, 4.5–5 in., moderate, spicy fragrance; foliage glossy, leathery; very vigorous, bushy, upright growth; [(Orange Triumph X Golden Rapture) X Peace]; deRuiter; Ilgenfritz Nursery

Pink Friendship, S, mp, 1986; Verschuren

'PINK FRILLS', F, lp, 1954; [Garnette sport]; Carlton Rose Nursery

Pink Fringe®, F, mp, 1990; Interplant

'PINK FROST', HT, mp, 1954; bud ovoid; flowers rose, dbl., 38–45 petals, 4–5 in., exhibition form, intense fragrance; foliage glossy, leathery; vigorous, bushy growth; [Charlotte Armstrong X Texas Centennial]; Swim, H.C.; Arp Nursery Co.

'PINK FROSTFIRE', Min, lp, 1968; [Frostfire sport]; Moore, Ralph S.; Sequoia Nursery

Pink Frosting, *see* 'RENFROST'

'PINK GARLAND', S, mp, 1935; flowers clear pink, open, semi-dbl., 3–3.5 in., non-recurrent, intense fragrance; upright (3 ft) growth; [R. blanda X R. spinosissima cultivar]; Skinner

'PINK GARNETTE', F, dp, 1950; flowers tyrian rose; [Garnette sport]; Schneeberg

'PINK GARNETTE', F, dp, 1951; flowers deep pink, [Garnette sport]; Boerner; J&P

'PINK GARNETTE SUPREME', F, mp, 1946; flowers rose-pink, borne in clusters, dbl., 35–40 petals, 2.5 in., cupped, moderate fragrance; foliage glossy, leathery; vigorous, upright growth; [Garnette Supreme sport]; Perkins, C.H.; J&P

'PINK GEM', HT, lp, 1949; bud long, pointed; flowers pearl-pink, dbl., 30 petals, 4–5 in., flat, intense fragrance; foliage bluish green; vigorous, tall growth; Fletcher; Tucker

'PINK GEM', Min, lp, 1974; bud ovoid; flowers rose-pink, full, medium, dbl., slight fragrance; foliage soft; moderate, upright, bushy growth; [Scarlet Gem sport]; Meilland; C-P

Pink Gift, Pol, lp

Pink Gin, HT, mp; flowers medium, dbl.; (Sangerhausen)

'PINK GLORY', HT, mp, 1960; bud long, pointed; flowers clear pink, dbl., 25 petals, 5 in., exhibition form, intense fragrance; foliage leathery; vigorous, upright growth; [Ernie Pyle seedling X Peace]; Boerner; J&P

'PINK GLOW', HT, lp, 1951; bud globular; flowers rose-pink, dbl., 40 petals, 5–5.5 in., cupped, moderate fragrance; foliage leathery, glossy; vigorous, upright growth; [((Schoener's Nutkana X Unknown seedling) X Mrs Pierre S. duPont) X Home Sweet Home]; Boerner; J&P

'PINK GOLDEN DAWN', HT, mp, 1938; flowers large, dbl., moderate fragrance; [Golden Dawn sport]; Bostick

'PINK GOWN', HT, lp, 1971; flowers light clear pink, dbl., 20 petals, 4 in.; foliage matt green; tall, very free growth; [Vera Dalton X Tropicana]; Sanday, John

'PINK GROOTENDORST', HRg, mp, 1923; flowers clear pink, carnation style petal edges, very dbl., borne in small clusters; foliage medium, dark green, rugose; tall growth; [F.J. Grootendorst sport]; Grootendorst, F.J.

'PINK GRÜSS AN AACHEN', F, op, 1929; flowers light salmon-pink, [Gruss an Aachen sport]; Kluis & Koning

'PINK HADLEY', HT, mp, 1928; flowers clear rose pink; [Hadley sport]; Knight, G.

'PINK HAT', F, pb, 1980; bud pointed; flowers light pink, center deep pink, blooms borne singly or 3–7 per cluster, dbl., 24 petals, exhibition form, slight fragrance; red prickles; foliage red turning dark green, glossy; vigorous, upright, bushy growth; James, John

'PINK HEATHER', Min, lp, 1959; flowers lavender-pink to white, very small blooms in clusters, dbl., 45 petals, foliage very small, glossy; vigorous (10–12 in.), bushy growth; [(R. wichurana X Floradora) X (Violette X Zee)]; Moore, Ralph S.; Sequoia Nursery

'PINK HEDGE', HRg, mp, 1956; red fruit; foliage small, bronze; [(R. rugosa rubra X R. cinnamomea) X R. nitida]; Nyveldt

Pink Heidelberg, HKor, mp

Pink Hit, *see* 'POULINK'

Pink Hit, *see* POUltipe

'PINK HONEY', Min, mp, 1988; flowers medium honey-pink, yellow at base, reverse light yellow at base, semi-dbl., 20 petals, exhibition form, slight, fruity fragrance; prickles long, pointed, medium, pink; foliage large, medium green, semi-glossy; bushy, tall growth; [Summer Spice X Unnamed seedling]; Bridges, Dennis A.; Bridges Roses

Pink Ice, *see* 'ANDpin'

Pink Iceberg(TM), *see* 'PROBERG'

Pink Ilseta, Gr, 1986; Tantau; (Cavriglia)

Pink Jacqueline, HT; (Cavriglia)

'PINK JENNY', HT, mp, 1961; flowers rose-pink, [Grand'mere Jenny sport]; Ruston, D.

'PINK JEWEL', F, mp, 1940; flowers arbutus-pink, center camellia-pink, semi-dbl., cupped, slight fragrance; foliage leathery; vigorous growth; [Crimson Glory X Holstein]; Kordes; Dreer

'PINK JOY', Min, dp, 1953; flowers deep pink, well-shaped, dbl., 30 petals, 1 in., moderate, sweet violet fragrance; dwarf (12 in.), bushy growth; [Oakington Ruby X Oakington Ruby]; Moore, Ralph S.; Sequoia Nursery

Pink Kardinal, HT, dp; Stratford, 1995

'PINK KAREN POULSEN', F, mp, 1936; [Karen Poulsen sport]; Poulsen, S.

'PINK KEY', HT, mp, 1920; [Francis Scott Key sport]; Pierson, F.R.

Pink Kiss(TM), *see* 'MINAKCO'

'PINK KOSTER', Pol, mp; [Margo Koster sport]; Koster, D.A.

Pink La Sevillana®, *see* 'MEIGEROKA'

Pink La Sevilliana, *see* 'MEIGEROKA'

'PINK LACE', F, pb, 1961; flowers light pink, becoming darker and then red, medium ($2-2\frac{1}{2}$"), dbl., 20 petals, flat, slight fragrance; foliage dark; vigorous growth; [The Optimist X Korona]; Watkins Roses

'PINK LADY', HT, mp, 1947; bud urn-shaped; flowers soft pink, reverse darker, semi-dbl., 3.5 in., slight fragrance; medium, vigorous growth; [Pink Delight sport]; Wiltgen; Premier Rose Gardens

'PINK LAFAYETTE', F, mp, 1925; flowers clear rose-pink, [Lafayette sport]; Griffin

'PINK LÉDA', D, mp; flowers light to medium pink, crimson edging on petals; [Leda sport or vice-versa sport]

Pink Lemonade, *see* 'AROFRICHEE'

Pink Licorice, *see* 'RESINK'

'PINK LORRAINE', HT, w, 1972; flowers pink, paling to white, base cream, very full, dbl., 40 petals, 4.5 in., slight fragrance; foliage matt green; vigorous, upright growth; [Milord X Farah]; Williams, J. Benjamin

Pink Love, HT, mp, 1998

'PINK LUSTRE', HT, lp, 1957; bud ovoid; dbl., 48 petals, 5 in., exhibition form, intense fragrance; foliage dark, glossy, leathery; vigorous, upright growth; [Peace X Dame Edith Helen]; Verschuren; J&P

Pink Macartney Rose, HBc, lp

'PINK MAGIC', HT, dp, 1952; bud long, pointed, spinel-red; flowers phlox-pink, dbl., 40–45 petals, 4–6 in., intense fragrance; foliage leathery; vigorous, upright growth; [Better Times sport (white) X Snow White]; Hill, Joseph H., Co.

Pink Magic, HMsk, pb, 1990; Lens

Pink Magic Carpet(TM), *see* MACausal

'PINK MAIDEN', F, mp, 1965; bud ovoid; flowers large blooms in clusters, dbl., slight fragrance; foliage dark; [Spartan seedling X Queen Elizabeth]; Boerner; Spek

Pink Maman Cochet, T, mp

'PINK MANDY', Min, mp, 1974; bud globular; dbl., 40 petals, 1 in.; foliage very glossy, leathery; low, bushy, spreading growth; [Ellen Poulsen X Little Chief]; Moore, Ralph S.; Sequoia Nursery

'PINK MARVEL', F, dp, 1958; flowers spirea-red, borne in clusters, dbl., 45–55 petals, 2.5 in., flat; foliage leathery; bushy, compact growth; [Rosemary Rose X (Cécile Brunner X Floribunda seedling)]; deRuiter; C-P

Pink Masquerade, S, op; single, 5 petals, moderate; [Suzanne X Red Dawn]; Simonet

'PINK MASTERPIECE', HT, pb, 1962; bud ovoid, pointed; flowers la france pink, tinted shrimp-pink, dbl., 38 petals, 6 in., exhibition form, moderate fragrance; foliage leathery; vigorous, upright growth; [Serenade seedling X Kate Smith]; Boerner; J&P

Pink Meidiland®, *see* 'MEIPOQUE'

Pink Meillandina®, *see* 'MEIJIDIRO'

Pink Melody, HT, op, 1993; K&S

'PINK MERINGUE', S, lp, 1999; very dbl., 41 petals, 3.5 in., borne in small clusters, slight fragrance; prickles numerous; foliage medium, medium green, semi-glossy; upright, tall (6 ft) growth; [Mary Rose sport]; Bennett, Frank David

Pink Mermaid, LCl, mp, 1940

'PINK METEOR', F, mp, 1964; [Meteor sport]; Timmerman's Roses

Pink Midinette, *see* PEAxanadu

Pink Minimo, Min, mp

Pink Mini-Wonder(TM), *see* 'MEISELGRA'

'PINK MIST', HT, mp, 1959; bud pointed, ovoid; flowers phlox-pink, becoming darker, dbl., 25–35 petals, 4–5 in., exhibition form, intense fragrance; foliage dark, leathery; vigorous, bushy growth; [Red Better Times sport]; Hill, Joseph H., Co.

Pink Mist, S, op

'PINK MONTEZUMA', Gr, lp, 1964; flowers light pink, reverse blush-pink, large, dbl., exhibition form, slight fragrance; foliage dark, leathery; vigorous, bushy growth; [Montezuma sport]; Williams, J. Benjamin

Pink Moss, *see* **'COMMUNIS'**

Pink Mothersday, Pol, mp

Pink Mystery, Misc OGR, mp, 1997; Lens

Pink 'n' Pretty, HT, dp

Pink 'n' White®, Min, pb

Pink Nevada, *see* **'MARGUERITE HILLING'**

Pink Nevada, *see* 'SANCHARM'

'PINK NYMPH', F, mp, 1959; flowers clear pink; [Nymph sport]; Koster, D.A.

'PINK NYMPH', Min, mp; dbl.; (14); [R. chinensis minima sport]

Pink Ocean®, *see* 'HAVINK'

Pink Olympiad, HT, mp, 1995; Humenick, Muriel F.

Pink Ophelia, *see* **'JOHN C. M. MENSING'**

'PINK OPHELIA', HT, mp, 1916; flowers rose-pink; [Ophelia sport]; Breitmeyer

Pink Osiana, *see* TANpinaiso

Pink Panther®, *see* 'MEICAPINAL'

Pink Parade®, Min, mp

'PINK PARASOL', HT, mp, 1950; bud long, pointed; flowers clear pink, dbl., 25–30 petals, 6 in., slight fragrance; foliage leathery, dark; very vigorous growth; [Rapture X Rome Glory]; Fisher, G.; Arnold-Fisher Co.

'PINK PARFAIT', Gr, pb, 1960; bud ovoid to urn-shaped; flowers outer petals medium pink, center blended pale orange, dbl., 23 petals, 3.5–4 in., exhibition form, slight fragrance; foliage leathery, semi-glossy; vigorous, upright, bushy growth; AARS, 1961 GM, Baden-Baden, 1959 GM, NRS, 1962 GM, Portland, 1959; [First Love X Pinocchio]; Swim, H.C.; Armstrong Nursery

'PINK PASSION', Min, pb, 1978; bud pointed; flowers shell-pink and ivory, small, dbl., 26 petals, 1.5 in., exhibition form, moderate fragrance; upright, bushy growth; [Sweet and Low X ?]; Schwartz, Ernest W.; Bountiful Ridge Nursery

Pink Patio, MinFl, lp, 1997

Pink Pavement, S, op, 1991; flowers salmon pink, semi-dbl., moderate fragrance; compact, spreading (2.5 ft) growth; Baum

Pink Peace®, *see* 'MEIBIL'

Pink Peace, Climbing®, *see* 'MEIBILSAR'

'PINK PEARL', Cl HT, op, 1913; flowers pink shaded salmon, single, [Irish Elegance X Una]; Hobbies

'PINK PEARL', HT, mp, 1924; flowers neyron pink, base salmon, dbl., intense fragrance; long stems; vigorous growth; [Ophelia X (Gen. MacArthur X Marie van Houtte)]; Leenders, M.

'PINK PEARL, CLIMBING', Cl HT, mp, 1933; Dixie Rose Nursery

'PINK PERFECTION', HT, mp, 1927; flowers clear rose-pink, dbl., 45 petals, moderate fragrance; [Premier X Baroness Rothschild]; Ward, F.B.

Pink Perfection, HT, mp, 1998

'PINK PERFEKTA', HT, pb, 1962; flowers deep pink, often edged red, [Kordes' Perfecta sport]; Ross, A., & Son; A. Ross & Son

Pink Perfume, HT, 1955; Robinson, H.; (Cavriglia)

'PINK PERPÉTUÉ', LCl, mp, 1965; flowers bright rose-pink, semi-globular, blooms in clusters, dbl., 32 petals, recurrent bloom, moderate fragrance; foliage glossy, light green; vigorous growth; [Spectacular X New Dawn]; Gregory

'PINK PET', Ch, mp, 1928; flowers bright pink, dbl.; Lilley

'PINK PETTICOAT', Min, pb, 1979; bud pointed; flowers creamy white, edged coral-pink, mini-flora, dbl., 33 petals, 1.5–2 in., exhibition form, slight fragrance; foliage glossy, dark; tall, upright growth; AOE, 1980; [Neue Revue X Sheri Anne]; Strawn; Pixie Treasures Min. Roses

Pink Picotee®, *see* 'KINPIC'

Pink Pillar, LCl, pb, 1940; Brownell, H.C., 1940

Pink Pirouette, *see* HARboul

Pink Pixie, *see* **'PIXIE ROSE'**

Pink Polyanna™, *see* 'JACSHIP'

Pink Poodle, *see* 'MORPOODLE'

Pink Popcorn™, *see* 'WILPOP'

Pink Porcelain, *see* 'TINPORCE'

Pink Posy, *see* 'COCANELIA'

'PINK POWDER PUFF', F, mp, 1965; bud ovoid; flowers soft pink, open, medium, dbl.; foliage leathery; vigorous, upright, compact growth; Pal, Dr B.P.; Indian Agric. Research Inst.

Pink Powderpuff, *see* 'MORPUFF'

'PINK PRIDE', HT, lp, 1959; flowers silvery pink, dbl., 30–35 petals, 5–6 in., exhibition form, intense fragrance; foliage light green; free growth; [May Wettern X Peace]; Fletcher

'PINK PRINCESS', HT, pb, 1939; flowers deep rose pink to yellowish, large, dbl., exhibition form, intense fragrance; foliage dark, leathery, glossy; vigorous, bushy growth; [(Dr W. Van Fleet X Général Jacqueminot) X Break o' Day]; Brownell, H.C.

Pink Princess, Min, mp

'PINK PROFUSION', HSet, pb, 1938; flowers pale flesh-pink, reverse deep coral-rose, small blooms in clusters, dbl., 80 petals, globular, non-recurrent, slight fragrance; foliage large, glossy; long, strong stems; very vigorous growth; [Mrs F.F. Prentiss X Lady Alice Stanley]; Horvath; Wayside Gardens Co.

'PINK PROSPERITY', HMsk, lp, 1931; flowers light pink, blooms in large trusses, moderate, musk fragrance; very vigorous growth; Bentall

'PINK PUFF', F, lp, 1965; flowers soft pink, dbl., moderate fragrance; foliage leathery; vigorous, upright growth; [Pinocchio seedling X (Red Pinocchio seedling X Garnette)]; Boerner; J&P

Pink Quill, *see* 'WEEQUILL'

Pink Radiance, *see* **'RADIANCE'**

Pink Radiance, Climbing, *see* **'RADIANCE, CLIMBING'**

Pink Reflection®, Min, pb

'PINK REVELATION', F, lp, 1979; [Summer Snow sport]; Schramm, D.

'PINK RHAPSODY', LCl, op, 1973; flowers deep vermilion-pink, full, dbl., 35–40 petals, 4 in., slight fragrance; very vigorous growth; [(Heidelberg X Bonn) X Pink Parfait]; Ellick

'PINK RHYTHM', Min, op, 1990; bud pointed; flowers coral pink, white center, medium pink reverse, aging light p, semi-dbl., 18 petals, exhibition form, moderate, fruity fragrance; foliage medium, medium green, matt; upright, spreading, medium growth; [Party Girl X Fashion Flame]; Jolly, Marie, 1984; Rosehill Farm, 1991

'PINK RIBBON', Min, lp, 1966; flowers soft pink, small, dbl., slight fragrance; foliage glossy, light green; vigorous, bushy, dwarf growth; [(R. wichurana X Floradora) X Magic Wand]; Moore, Ralph S.; Sequoia Nursery

'PINK RIPPLES', F, lp, 1956; flowers pink, open, borne in clusters, semi-dbl., 15–20 wavy petals, 2.5–3 in., slight fragrance; very vigorous, upright, compact growth; [Red Ripples sport]; Sanders, H.T.

'PINK ROAMER', HWich, pb, 1897; flowers pink, center white, single, vigorous growth; [R. wichurana X Cramoisi Supérieur]; Horvath; W.A. Manda

Pink Robin, S, pb, 1992; Lens

Pink Robusta®, *see* 'KORPINROB'

'PINK ROCKET', S, pb, 1949; bud ovoid; flowers deep pink suffused copper, single and open, large, semi-dbl., profuse, non-recurrent bloom, slight fragrance; foliage glossy, bronze, dark; very vigorous, upright growth; hardy; [Skyrocket seedling]; Longley, L.E.; Univ. of Minn.

'PINK ROSETTE', F, lp, 1948; bud small, ovoid; flowers soft pink, rosette form, blooms in clusters, dbl., 50 petals, 2 in., cupped, slight fragrance; foliage leathery, dark; vigorous, dwarf, bushy growth; Krebs; H&S

Pink Roundelay, HT, lp; [Roundelay sport]; Taylor, L.R., 1968; (Weatherly, L.)

'PINK ROVER', Cl HT, 1891; Paul, W.; (Cavriglia)

'PINK ROYAL', HT, mp, 1928; flowers glowing pink, dbl., moderate fragrance; [Columbia X Mme Butterfly]; Vestal

'PINK RUBY', HT, mp, 1966; flowers high pointed, 4–4.5 in., moderate fragrance; foliage light green; vigorous growth; [Rubaiyat sport]; Anstiss

'PINK RUFFLES', F, mp, 1977; bud pointed; flowers rose-pink, ruffled, single, 5–7 petals, 3 in., slight, tea fragrance; foliage very glossy, dark; very bushy growth; [Frolic self seedling X Pinafore]; Ellis; Mansion Nursery

Pink Ruffles, HT, dp, 1996; Williams, J. Benjamin

Pink Sachet, HT, dp

Pink Sandy™, *see* 'KINSANDY'

'PINK SATIN', F, mp, 1945; bud large; flowers clean rose-pink, showy stamens, open, semi-dbl., cupped, moderate fragrance; foliage soft, light green; very vigorous, bushy, compact growth; [Indiana X William F. Dreer]; Cross, Mrs. C.W.; B&A

'PINK SATIN', F, lp, 1974; bud ovoid; dbl., 30–35 petals, 4–5 in., exhibition form, foliage large; vigorous, very free growth; [Seedling X Bridal Pink]; Warriner, William A.; J&P

Pink Scotch, HSpn, mp

Pink Secret, *see* 'TROBINA'

'PINK SEMI', (form of R. laxa), lp; resembles R. blanda

Pink Sensation, HT, mp

Pink Sevilliana, *see* 'MEIGEROKA'

'PINK SHADOW', Min, dp, 1977; bud plump, pointed; flowers dusty pink, reverse darker, dbl., 52 petals, 1 in., moderate fragrance; foliage glossy; bushy, spreading growth; [Over the Rainbow X Over the Rainbow]; Williams, Ernest D.; Mini-Roses

Pink Sheri, *see* 'RENERI'

'**PINK SHOWERS**', Cl HT, lp, 1974; flowers large, dbl.; vigorous, tall, climbing growth; [Carla X Golden Showers]; Verschuren, Ted; Verschuren

Pink Showers, Pol, mp, 1968; Ramson

'**PINK SILK**', HT, mp, 1972; flowers carmine-rose, dbl., 42 petals, 4 in., exhibition form, moderate fragrance; Gold Star of the South Pacific, Palmerston North, NZ, 1974; [Pink Parfait X Unknown seedling]; Gregory

'**PINK SNOW**', Cl F, pb, 1981; flowers light pink shading to white in center, deeper on edges of petals, [Summer Snow, Climbing sport]; Reed, Harry, Jr.

'**PINK SOUPERT**', Pol, mp, 1896; in clusters, vigorous growth; [Clotilde Soupert seedling]; Dingee & Conard

Pink Spectacle, *see* KORlodera

'**PINK SPICE**', HT, lp, 1962; bud long, pointed; flowers light pink flushed yellow, dbl., 30 petals, 5 in., exhibition form, intense fragrance; foliage leathery, light green; vigorous, upright growth; Von Abrams; Peterson & Dering

'**PINK SPIRAL**', HT, dp, 1953; flowers deep china-rose-pink, high pointed, dbl., 4 in.; foliage cedar-green; very vigorous growth; McGredy, Sam IV

Pink Spire, HT, mp

'**PINK SPLENDOUR**', HT, mp, 1951; flowers rose-pink, dbl., 48 petals, 6–7 in., intense fragrance; foliage dark; very free growth; [Sam McGredy X Crimson Glory]; McGredy, Sam IV

Pink Spray®, *see* LENray

Pink Star®, *see* 'INTERPINK'

Pink Strike, Min, lp, 1998; Laver, Keith G.

Pink Sublime, HMsk, pb

Pink Summer Snow, Cl F, lb; [Summer Snow sport]; Hannemann, F., 1987; (Weatherly, L.)

Pink Sunblaze™, *see* 'MEIJIDIRO'

'**PINK SUPREME**', HT, lp, 1964; dbl., 23 petals, 4–5 in., intense fragrance; foliage light green; vigorous, upright, well-branched growth; GM, Belfast, 1967; GM, Geneva, 1965; [Amor X Peace]; deRuiter

'**PINK SURPRISE**', Min, mp, 1980; bud ovoid, pointed; dbl., 48 petals, borne singly or several together, moderate fragrance; curved prickles; foliage tiny, medium green; compact, bushy growth; [Unnamed seedling X Unnamed seedling]; Lyon

Pink Surprise, S, dp, 1975; Ottawa

Pink Surprise, *see* 'LENBRAC'

Pink Sweetheart™, *see* 'WILPINK'

Pink Symphonie®, *see* 'MEITONJE'

Pink Symphony, *see* 'MEITONJE'

'**PINK TALISMAN**', HT, 1943; flowers orange to pink with pink predominating, large, upright growth; [Talisman sport]; Howard Rose Co.

Pink Tapestry, F, dp, 1997; Williams, J. Benjamin

'**PINK TINGLE**', Min, mp, 1978; bud ovoid; flowers dawn-pink, to recurved, dbl., 20 petals, 1 in., cupped, slight fragrance; foliage tiny; compact growth; Lyon

Pink Torch®, *see* 'INTERTOR'

Pink Traumland, F, mp, 1996; flowers luminous pink, medium-large, dbl.; Tantau; (Sangerhausen)

Pink Treasure, *see* 'HADPLEASURE'

Pink Triumph, *see* '**INGRID STENZIG**'

'**PINK TRIUMPH**', Min, mp, 1983; flowers small blooms borne usually singly, dbl., 48 petals, moderate fragrance; foliage small, medium green, semi-glossy; upright, bushy growth; [Operetta X Bonny]; Jolly, Nelson F.; Rosehill Farm

Pink Twister, S, mp

Pink Van Fleet, LCl, mp

'**PINK VOGUE**', F, mp, 1960; bud pointed; flowers pink, medium, borne in clusters, semi-dbl., moderate fragrance; moderate, upright growth; [Vogue sport]; Kelleher; Hazlewood Bros.

Pink Wave, *see* 'MATTGRO'

Pink Wings, HT, lp

Pink Winks, *see* 'FLOWIN'

Pink Wonder, *see* 'MEIHARTFOR'

Pink Wonder, Climbing, *see* 'MEIHARTFORSAR'

'**PINK-A-BOO**', F, mp, 1961; bud ovoid; dbl., 33 petals, 3–3.5 in., blooms in large clusters, moderate fragrance; foliage leathery; vigorous, upright growth; [Spartan X Pink Garnette]; Boerner; J&P

Pink-A-ling, *see* 'ZIPPINK'

'**PINKCES**', LCl, yb, 1962; bud pointed; flowers light yellow to pink, open, dbl., 30 petals, 3.5–4 in., abundant, non-recurrent bloom, moderate fragrance; foliage leathery; vigorous, tall, compact growth; [Doubloons sport]; Schmalz & Limpert

'**PINKERTON**', HT, mp, 1949; bud long, pointed; flowers pink shaded deeper, dbl., 30 petals, 4 in., moderate fragrance; foliage light green; vigorous growth; Eacott

'**PINKIE**', Pol, mp, 1947; flowers neyron rose, blooms in large trusses, semi-dbl., 16 petals, 1.75–2.5 in., cupped, intense fragrance; foliage soft, glossy, mostly 7 leaflet leaves; dwarf, bushy growth; (21); AARS, 1948; [China Doll X ?]; Swim, H.C.; Armstrong Nursery

'**PINKIE, CLIMBING**', Cl Pol, mp, 1952; Dering; Armstrong Nursery

Pinky, *see* 'MACPONUI'

Pinky Rugostar, *see* MEIvibroum

'**PINOCCHIO**', F, op, 1940; (Rosenmärchen); flowers pink suffused salmon, edged deeper, blooms in long sprays, dbl., 30 petals, 2 in., cupped, moderate, fruity fragrance; foliage leathery; vigorous, bushy growth; (28); GM, Portland, 1942; (28); [Eva X Golden Rapture]; Kordes; J&P, 1942

Pinocchio, *see* DELki

Pinocchio, *see* 'SUNPINO'

'**PINOCCHIO, CLIMBING**', Cl F, pb, 1951; (Rosenmärchen, Climbing); Parmentier, J.; J&P

'**PINSON**', LCl, my, 1909; flowers chamois-yellow, tinted rosy white, very large, borne in clusters, semi-dbl., early, vigorous, climbing growth; [R. wichurana X Souv. de Catherine Guillot]; Barbier

Pinstripe™, *see* 'MORPINTS'

'**PINTA**', HT, w, 1973; flowers creamy white, dbl., 23 petals, 3.5 in., moderate, sweetbriar fragrance; foliage dark, matt; [Ena Harkness X Pascali]; Beales, Peter

PINturnia, Gr, 1978; (**Mme Bruno Coquatrix**); Pineau; (Cavriglia)

'**PIN-UP**', F, dp, 1959; flowers deep china-rose, medium, borne in clusters, semi-dbl., 16–18 wavy petals; foliage light green; low, bushy growth; [Else Poulsen X ?]; Fletcher

'**PINWHEEL**', Min, pb, 1977; flowers pink and yellow blend; [Jeanie Williams sport]; Moore, Ralph S.; Sequoia Nursery

'**PIONEER**', HT, mr, 1970; flowers very full, dbl., 50 petals, 4 in., intense fragrance; foliage small, dark; very free growth; LeGrice

'**PIONERKA**', F, ob, 1955; (Woman Pioneer); flowers reddish orange, medium, slight fragrance; vigorous growth; Sushkov, K.L.

'**PIOUPIOU**', F, mr, 1958; flowers cerise-red, large; very vigorous, bushy, low growth; [Alain X (Alain X Unnamed seedling)]; Arles; Roses-France

'**PIP**', Min, mr, 1977; bud pointed; flowers small, semi-dbl., 12–15 petals, cupped, slight fragrance; foliage tiny; very compact, bushy growth; RULED EXTINCT 7/90; [N74 X Unnamed seedling]; Lyon

'**PIP**', HT, w, 1990; bud ovoid; flowers white rimmed pink, white reverse, aging same, medium, borne, dbl., 30 petals, exhibition form, slight fragrance; foliage medium, light to medium green, matt; bushy, medium growth; [(Poker Chip X Helmut Schmidt) X [(Rise 'n' Shine X Granada) X Handel]]; Stoddard, Louis, 1991

'PIPE DREAMS', S, mp, 1984; flowers medium pink, reverse darker, blooms borne 5–10 per cluster, dbl., 28 petals, 5 in., cupped, repeat bloom, intense, clove fragrance; awl-shaped, red-brown prickles; foliage leathery, medium olive green; bushy, compact, upright growth; hardy.; [Countryman X ((Meisterstuck X Prairie Princess) X (Tickled Pink X Prairie Princess))]; Buck, Dr Griffith J.; Iowa State University

'PIPPA'S SONG', S, mp, 1984; flowers imbricated, blooms borne 3–10 per cluster, 3 in., repeat bloom, slight fragrance; awl-shaped, tan prickles; foliage medium, leathery, dark olive green; shrubby, erect, bushy growth; hardy.; [Prairie Princess X (Queen Elizabeth X (Morning Stars X Suzanne))]; Buck, Dr Griffith J.; Iowa State University

Pippy, *see* BARpipp

'PIPSQUEAK', Min, m, 1986; flowers pinkish mauve, quilled, small blooms in sprays of 5–20, dbl., 40 petals, flat, no fragrance; no fruit; very small, brown prickles, hooked downward; foliage small, light green, matt; bushy, spreading growth; [Blue Mist X Snow Magic]; Dobbs; Port Stockton Nursery

'PIRATE GOLD', F, dy, 1972; dbl., globular, slight fragrance; foliage glossy, leathery; vigorous, upright growth; [Golden Wave X Seedling]; deRuiter; Carlton Rose Nurseries

'PIRBRIGHT', F, mp, 1948; semi-dbl., borne in trusses, slight fragrance; foliage dark; vigorous growth; Norman

Piro 3, S, lp, 1995; Stritzke

'PIROJA', HT, m, 1999; flowers deep mauve, reverse same, dbl., 26–40 petals, 2.5–3.5 in., borne in small clusters, moderate fragrance; few prickles; foliage medium, medium green, semi-glossy; upright, medium (3–3.5 ft) growth; [Blue Moon X Heirloom]; Chiplunkar, C. R.; KSG's Roses, 1992

Pirol, *see* TANlorip

'PIRON MÉDARD', HP, mp, 1906; flowers large, dbl.; Piron-Médard; (Sangerhausen)

'PIRONIA', Gr, op, 1977; (**Centenaire du Vesinet**); bud pointed; flowers salmon-orange, dbl., 32 petals, 3 in., slight fragrance; vigorous, upright growth; [Sonia X Prominent]; Pineau; Searn

Pironti Arabian, HT; Pironti, N.; (Cavriglia)

Pironti Blithe, F, 1970; Pironti, N.; (Cavriglia)

Pironti Critow, Cl HT; Pironti, N.; (Cavriglia)

Pironti Konrad, Gr, 1971; Pironti; (Cavriglia)

Pironti Tornedos, F, 1971; Pironti; (Cavriglia)

Pirontina, LCl, 1975; Pironti; (Cavriglia)

Piroschka®, *see* 'TANPIKA'

'PIROUETTE', F, pb, 1968; flowers pink tinged orange-red, open, medium, dbl., moderate fragrance; [Ma Perkins X Radar]; Fankhauser; A. Ross & Son

Pirouette, S, rb, 1984; Lens

Pisen, HT, lp; Brabec, 1980; (Czech Rosa Club)

'PISMI', HT, rb, 1972; bud ovoid; flowers solferino-purple to spirea-red, large, dbl., cupped, moderate fragrance; foliage large, dark, leathery; vigorous, upright bushy growth; [Dame Edith Helen X Rote Rapture]; Pecollo

Pitica, *see* **'KYRIA'**

'PITICA', F, lp, 1976; flowers full, dbl., 25–30 petals, 4 in., exhibition form, slight fragrance; [Sweet Promise sport]; Royon; Universal Plants

'PITTSBURGH', HP, lp, 1929; flowers flesh-pink, base yellow, very large, dbl., 25 petals, globular, slight fragrance; foliage leathery; long stems; very vigorous growth; [(R. gigantea seedling X Frau Karl Druschki) X Mrs John Laing]; Schoener; B&A

'PIUS IX', HP, mp, 1849; (Pie IX, Pius the Ninth, Pope Pius IX); flowers violet-rose, very dbl., flat, intense fragrance; vigorous growth; Vibert

Pius the Ninth, *see* **'PIUS IX'**

'PIUS XI', HT, w, 1925; (Pope Pius XI); bud long, pointed; flowers cream-white, center cream-yellow, very large, dbl., moderate fragrance; [Ophelia X Unnamed seedling]; Leenders, M.

'PIXANAH', Min, ab, 1990; (**Savannah Miss**); bud pointed; flowers dark apricot, edges lighter, medium apricot reverse, fading, dbl., 40–50 petals, exhibition form, moderate fragrance; globular, medium green fruit; prickles needle-shaped, light tan; foliage medium, medium green, semi-glossy; bushy, tall growth; [Pounder Star X Ann Moore]; Chaffin, Lauren M.; Pixie Treasures Min. Roses

'PIXANNE', Min, mp, 1993; (**Rebecca Anne**); very dbl., 1.5–2.75 in., borne mostly singly, slight fragrance; some prickles; foliage medium, medium green, semi-glossy; medium (40–45 cms), upright, bushy, vigorous growth; [Gene Boerner X Pink Petticoat]; Chaffin, Lauren M.; Pixie Treasures Min. Roses, 1993

'PIXARLA', Min, w, 1990; (**Starla**); bud pointed; flowers medium, borne singly, semi-dbl., 20–25 petals, exhibition form, slight, fruity fragrance; prickles needle-shaped, tan; foliage medium, medium green, semi-glossy; bushy, medium growth; [Honor X Rainbow's End]; Chaffin, Lauren M.; Pixie Treasures Min. Roses, 1991

'PIXBRO', Min, mr, 1998; (**Little Brother**); flowers medium red, full, very dbl., 26–40 petals, 2 in., borne mostly singly, slight fragrance; no prickles; foliage medium, medium green, semi-glossy; compact, medium growth; [Pink Petticoat X Happy Hour]; Chaffin, Lauren M.; Pixie Treasures Roses, 1998

'PIXCHEK', Min, mr, 1990; (**Checkers**); bud pointed; flowers medium red, aging darker, blooms borne singly, semi-dbl., 20 petals, exhibition form, slight, spicy fragrance; rarely forms fruit; prickles needle-shaped, tan; foliage small, medium green, semi-glossy; bushy, low growth; [Deep Purple X Happy Hour]; Chaffin, Lauren M.; Pixie Treasures Min. Roses

'PIXHAR', Min, ob, 1998; (**Summer Harvest**); flowers very bright orange blend, gold stamens, single, 5 petals, 2 in., borne in small clusters, slight fragrance; prickles moderate, medium, straight; foliage large, dark green, glossy; bushy, tall (24 in.) growth; [Ann Moore X Sequoia Gold]; Chaffin, Lauren M.; Pixie Treasures Roses, 1998

'PIXHONEY', Min, ly, 1998; (**Honey Butter**); flowers creamy yellow center, near white outer edges, very dbl., 26–40 petals, 2–2.5 in., borne mostly singly, slight fragrance; prickles few, small, straight; foliage medium, medium, semi-glossy; bushy, medium growth; [Deep Purple X Rainbow's End]; Chaffin, Lauren M.; Pixie Treasures Roses, 1998

'PIXICHAR', Min, lp, 1992; (**Real Charmer**); dbl., 26–40 petals, 3–3.5 in., borne in small clusters, moderate fragrance; foliage large, medium green, semi-glossy; tall (45 cms), upright growth; [Gene Boerner X Crissy]; Chaffin, Lauren M.; Pixie Treasures Min. Roses

'PIXICHIP', Min, r, 1991; (Chipmonk, **Chipmunk**, Peanut Butter & Jelly); flowers tannish brown blending with mauve when full blown, dbl., 26–40 petals, 1.5–2.75 in., borne mostly singly, moderate fragrance; few prickles; foliage medium, medium green, semi-glossy; medium (30 cms), bushy, neat, compact growth; [Deep Purple X Rainbow's End]; Chaffin, Lauren M.; Pixie Treasures Min. Roses, 1992

'PIXIE', Min, w, 1940; (Little Princess, Princesita); flowers white, center light pink, small blooms, dbl., 55 petals, slight fragrance; foliage very small, soft; short stems; dwarf, compact growth; [Ellen Poulsen X Tom Thumb]; deVink; C-P

Pixie Delight™, *see* 'WILBENTUR'

'PIXIE GOLD', Min, my, 1961; flowers mimosa-yellow, semi-dbl., 11 petals, 1–1.5 in.; foliage very small, thin, dark; bushy growth; [Perla de Montserrat X (Rosina X Eduardo Toda)]; Dot, Pedro; C-P

Pixie Hat, *see* KORlamber

Pixie Hedge, *see* **'BABY JAYNE'**

Pixie Pearl, *see* **'ESTRELLITA'**

'PIXIE PEARL', Min, w; flowers pearly white, small, borne in clusters, very dbl.; vigorous, compact growth; Lamb Nursery

'PIXIE ROSE', Min, dp, 1961; (Pink Pixie); flowers deep pink, blooms in irregular clusters, dbl., 43 petals, 1 in., exhibition form; foliage very small, dark; dwarf, much-branched growth; [Perla de Montserrat X Coralin]; Dot, Pedro; C-P

'PIXIE, CLIMBING', Cl Min, w, 1964; Ruston, D.

'PIXIFULL', Min, or, 1986; (**Fullerton Centennial**); flowers orange-red with white base, bright, fading darker, medium, semi-dbl., 20–25 petals, exhibition form, borne singly or in sprays of 3–5, slight fragrance; occasionally forms globular, medium green fruit; needle-declining, sparse, light tan prickles; foliage small, medium green, semi-glossy, disease-resistant; upright, bushy, medium, neat, symmetrical growth; [Orange Honey X Rise 'n' Shine]; Chaffin, Lauren M., 1983; Friends of the Fullerton Arboretum, 1987

'PIXIHON', Min, ab, 1987; (**Honey Bear**); flowers deep apricot, yellow base, reverse light apricot, aging cream, very dbl., 50–55 petals, cupped; needle-like, straight, light tan prickles; foliage small, medium green, semi-glossy, disease resistant; bushy, low compact growth; PP006844; [Rise 'n' Shine X Holy Toledo]; Chaffin, Lauren M.; Pixie Treasures Min. Roses

'PIXIMIS', Min, w, 1985; (**Moon Mist**(TM)); flowers ivory tinged pale pink, well-formed, small, dbl., 20 petals, moderate fragrance; foliage small, medium green, semi-glossy; bushy growth; [The Optimist X Darling Flame]; Strawn, Leslie E.; Pixie Treasures Min. Roses

'PIXINK', Min, mp, 1989; (**Tickle Me Pink**); bud pointed; flowers medium pink to soft cream atbase, reverse cream blush, aging, dbl., moderate, spicy fragrance; globular, rare, medium green fruit; prickles needle-like, light tan; foliage small, medium green, semi-glossy; bushy, low growth; [Osiria X Magic Carrousel]; Chaffin, Lauren M.; Pixie Treasures Min. Roses

'PIXINNER', Min, rb, 1990; (**Inner Glow**); bud ovoid; flowers red with golden yellow base, yellow reverse, red veining towards outer edges, dbl., 40 petals, exhibition form, slight fragrance; round, medium green fruit; prickles hooked, tan; foliage medium, medium green, semi-glossy; bushy, medium growth; [Ann Moore X Rainbow's End]; Chaffin, Lauren M.; Pixie Treasures Min. Roses, 1991

'PIXIPRIN', Min, r, 1982; (**Indian Princess**); flowers tan flushed orange, reverse burnt umber, shaded garnet-brown, dbl., 20 petals, intense fragrance; foliage small, dark, semi-glossy; upright growth; [Yellow Jewel X Golden Cougar]; Strawn, Leslie E.; Pixie Treasures Min. Roses

'PIXIREE', Min, ab, 1985; (**Free Spirit**); flowers small, dbl., 20 petals, slight fragrance; foliage small, medium green, semi-glossy; upright growth; PP006188; [Prominent X Gold Pin]; Strawn, Leslie E.; Pixie Treasures Min. Roses, 1984

'PIXISLIP', Min, m, 1991; (**Silver Slippers**); bud slender; flowers unique silvery lavender, blooms borne mostly singly, dbl., 26–40 petals, 1.5–2.75 in., moderate fragrance; foliage medium, medium green, semi-glossy; medium (30 cms), bushy growth; [Deep Purple X Jennifer]; Chaffin, Lauren M.; Pixie Treasures Min. Roses, 1992

'PIXISUN', Min, my, 1985; (**Sunshine Girl**); flowers multi-star shaped, small, dbl., 35 petals, exhibition form, slight fragrance; foliage small, medium green, semi-glossy; upright growth; [Sunsilk X Rise 'n' Shine]; Woolcock, Edward P.; Pixie Treasures Min. Roses

'PIXITER', Min, dy, 1982; (**Tiger Butter**); flowers small, dbl., 20 petals, slight fragrance; foliage small, dark, semi-glossy; upright, bushy growth; [Sunblest X Over the Rainbow]; Strawn, Leslie E.; Pixie Treasures Min. Roses

'PIXIWOW', Min, ob, 1985; (**Wow!**); flowers bright orange, small, dbl., 35 petals, slight fragrance; foliage small, dark, glossy; bushy growth; PP006469; [Ann Cocker X Unnamed seedling]; Strawn, Leslie E.; Pixie Treasures Min. Roses

'PIXJUN', Min, ob, 1994; (**June Bug**); flowers bright medium orange, slightly lighter reverse, yellow base, dbl., 26–40 petals, 1.5 in., slight fragrance; some prickles; foliage medium, dark green, semi-glossy; medium (12 in.), bushy, compact growth; [(Prominent X Orange Honey) X Ann Moore]; Chaffin, Lauren M.; Pixie Treasures Min. Roses, 1994

'PIXKICK', F, ob, 1995; (**Side Kick**); flowers orange red, very full (41+ petals), medium (4–7 cms) blooms borne mostly single, slight fragrance; few prickles; foliage medium, medium green, semi-glossy; medium (30 cms), bushy growth; [Dandenong X Ann Moore]; Chaffin, Lauren M.; Pixie Treasures Miniature Rose Nursery, 1996

'PIXRIC', F, rb, 1996; (**Ric Rac**); flowers white edged red, non-fading color, small, long lasting bloom, dbl., 26–40 petals, 2.5–3 in., fast repeat, slight fragrance; moderate prickles; foliage small, medium green, semi-glossy; upright, low (18–20 in.) growth; [Hannah Gordon X Pink Petticoat]; Chaffin, Lauren M.; Pixie Treasures Min. Roses, 1996

'PIXSAR', F, ab, 1998; (**Sarah Elizabeth**); flowers apricot blend, ogr form, very dbl., 41 petals, 2.5–3 in., rosette, borne in small clusters, slight fragrance; prickles few, medium, straight; foliage medium, dark green, glossy; upright, medium growth; [City of Auckland X Seedling]; Chaffin, Lauren M.; Pixie Treasure Roses, 1998

'PIXSIS', Min, mp, 1995; (**Little Sister**); flowers medium pink, very full (41+ petals), medium (4–7 cms) blooms borne mostly single, occasional side buds, slight fragrance; few prickles; foliage medium, medium green, semi-glossy; medium (30 cms), bushy growth; [Gene Boerner X Crissy]; Chaffin, Lauren M.; Pixie Treasures Miniature Rose Nursery, 1996

'PIXSNO', F, rb, 1996; (**Sno Cone**); flowers white edged red, blooms borne in small clusters upright, wa, dbl., 15–25 petals, 2.5–3 in., slight fragrance; moderate prickles; foliage small, medium green, semi-glossy, disease-resistant; bushy, low (18 in.) growth; [Hannah Gordon X Pink Petticoat]; Chaffin, Lauren M.; Pixie Treasures Min. Roses, 1996

'PIXSTA', Min, w, 1993; (**Christa**); dbl., 26–40 petals, 1.5–2.75 in., exhibition form, borne mostly singly, slight fragrance; some prickles; foliage medium, medium green, semi-glossy, disease-resistant; long stems; tall (45 cms), upright, vigorous growth; [Honor X Rise 'n' Shine]; Chaffin, Lauren M.; Pixie Treasures Min. Roses, 1993

'PIXSUN', F, ob, 1994; (**Sunny Side Up**); flowers blended soft yellow to tangerine, blooms borne in small clusters, semi-dbl., 6–14 petals, 3–3.5 in., moderate fragrance; some prickles; foliage medium, dark green, glossy; medium (2–3 ft), bushy growth; [City of Auckland X Rainbow's End]; Chaffin, Lauren M.; Pixie Treasures Min. Roses, 1994

'PIXZIG', Min, rb, 1993; (**Zig-Zag**); flowers white with dark red picotee edge, blooms borne mostly single, very dbl., 26–40 petals, 1.5–2.75 in., exhibition form, slight fragrance; some prickles; foliage similar to peaches 'n' cream, medium, medium green; low (30 cms), upright, bushy, compact growth; [Osiria X Pink Petticoat]; Chaffin, Lauren M.; Pixie Treasures Min. Roses, 1993

'PIZZAZZ', Min, rb, 1991; bud ovoid; flowers various yellows, pink hues, aging to orange-red, blooms born, dbl., 50 petals, 1.5 in., slight fragrance; foliage medium, medium green, semi-glossy; upright, spreading, medium growth; [Orange Honey X Loving Touch]; Jolly, Marie; Rosehill Farm, 1992

'PIZZICATO', S, lp, 1962; bud ovoid, long, pointed; flowers light salmon-rose, dbl., 40 petals, 3–4 in., cupped, moderate fragrance; foliage leathery, bronze; vigorous, upright (3–4 ft), bushy growth; [Florence Mary Morse X (Josef Rothmund X R. laxa)]; Buck, Dr Griffith J.; Iowa State University

Placerville White Noisette, N, w

'PLACET', S, w, 1981; bud small, ovoid; single, 5 petals, borne 40 or more per cluster, all summer bloom, moderate fragrance; foliage soft green, 7 leaflet; bushy growth; Bevan, Mrs. Ruth M.

'PLACIDA', HT, mr, 1959; bud ovoid; flowers crimson, large, dbl., slight fragrance; foliage glossy, dark; vigorous, bushy, compact growth; [Crimson Glory X Una Wallace]; Cayzer

Placido Domingo, *see* 'MACHORO'

'PLAIN TALK', F, mr, 1964; semi-dbl., blooms in clusters, slight fragrance; foliage dark, leathery; vigorous, bushy, low growth; [Spartan X Garnette]; Swim & Weeks; Weeks Wholesale Rose Growers

'PLAISANTE', F, mp, 1957; flowers bright pink, base and reverse lighter, semi-dbl., moderate fragrance; low growth; [Borderer X Unnamed seedling]; Riethmuller

Plaisanterie, HMsk, 1996; Lens; (Cavriglia)

'PLAISIR DE FRANCE', HT, dp, 1952; flowers bright deep coppery pink, dbl., 35 petals, 6 in., moderate fragrance; foliage glossy, dark; vigorous growth; [Peace X Unnamed seedling]; Gaujard

Plameny, HT, mr, 1970; flowers large, dbl., Urban, J.; (Sangerhausen)

'PLAMYA VOSTOKA', F, or, 1955; (Flame of the East); flowers fiery red tinted darker, medium, dbl., 30 petals, moderate fragrance; medium growth; [Independence X Kirsten Poulsen]; Klimenko, V. N.

Planten un Blomen, *see* **'CHERRIO'**

Platinum Lady, *see* 'LAVPLAT'

'PLATO', HSpn, m; flowers red-violet, dbl., non-recurrent, glossy, black fruit; foliage finely divided; dense, shrubby (3–4 ft) growth; Vibert

Play Rose, F, dp, 1989; flowers carmine-pink, large, dbl., Meilland; (Sangerhausen)

'PLAYBOY'®, F, rb, 1976; (Cheerio); flowers scarlet, gold eye, single, 3.5 in., slight fragrance; foliage glossy, dark; GM, Portland, 1989; [City of Leeds X (Chanelle X Piccadilly)]; Cocker

Player, *see* 'SUNPLAY'

Playfair, *see* 'WIWAIT'

'PLAYFUL', HT, ab, 1970; flowers light apricot, edged pink, [Mischief sport]; Watson

Playgirl™, *see* 'MORPLAG'

Playgirl, Climbing, *see* 'MORCLIP'

Playgold, *see* 'MORPLAYGOLD'

Playgroup Rose®, *see* 'HORSUN'

Playmate, *see* 'WEOKAY'

Playtime™, *see* 'MORPLATI'

Playtime, *see* 'KORSAKU'

Pleasant Hill Cemetery, N, w

Pleasantly Pink, *see* 'LAVPLEASE'

Pleasure, *see* 'JACPIF'

'PLEIN CIEL', HT, mp, 1964; bud pointed; flowers large, dbl., exhibition form, moderate fragrance; foliage dark, glossy; vigorous, well branched growth; [Seedling X Golden Rapture]; Mondial Roses

Plein Soleil®, *see* LAPdev

Pleine de Grâce, *see* 'LENGRA'

Pleins Feux 92, *see* DORtomyl

'PLENA', HSem, lp; (Double White Noisette); flowers flesh, dbl.; Laffay, M.

'PLENTIFUL', F, dp, 1961; bud globular; flowers deep pink, blooms in large clusters, dbl., 75 petals, 3–4 in., flat; foliage light green; vigorous, bushy growth; LeGrice

Plisiedame, *see* KORichard

Plnokvety Super Star, HT, lp; Strnad; (Czech Rosa Club)

'PLOMIN', F, dp, 1951; flowers dark rose, well formed, large, borne in large clusters, semi-dbl., intense fragrance; foliage leathery; bushy, dwarf growth; [(Johanna Tantau X Karen Poulsen) X Stammler]; Tantau

'PLUIE DE FEU', LCl, or, 1964; flowers bright scarlet, borne in clusters, semi-dbl., vigorous growth; Mondial Roses

'PLUKOVNIK SVEC', HT, dr, 1935; (Colonel Svec); bud long (over 3 in.), pointed; flowers blood-red, large, semi-dbl., moderate fragrance; foliage small; vigorous, bushy growth; [Pres. Jac. Smits X K. of K.]; Böhm, J.

Plum Brandy, Min, m, 1974

Plum Cake, *see* 'MURCA'

Plum Crazy, *see* 'AROGRAJU'

Plum Dandy™, *see* 'JACLIANG'

'PLUM DUFFY', Min, m, 1978; bud ovoid; flowers deep plum, small, dbl., 25 petals, 1.5 in., exhibition form, slight fragrance; foliage dark; compact growth; [Magic Carrousel X Magic Carrousel]; Bennett, Cecilia 'Dee'; Tiny Petals Nursery

Plum Pudding™, *see* 'WILPLPD'

'PLUTON', HGal, m, 1843; (Rose Pluton); Vibert, 1843

'PLZEN', HT, dr, 1930; (City of Pilsen, Stadt Pilsen); bud very long; flowers dark blood-red, very large, semi-dbl.; very vigorous growth; [Étoile de Hollande X Macbeth]; Böhm, J.

'POBJEDITEL', HT, dp, 1940; flowers large, dbl., slight fragrance; Kosteckij; (Sangerhausen)

Poblet, HT, ob; Dot

Pocahontas, HT, ob, 1998; Williams, J. Benjamin

'POCO', Min, pb, 1986; flowers pink, white center, fades slightly, urn-shaped, medium, born, dbl., 29 petals, slight fragrance; medium, light green prickles, pointed downwards; foliage medium, medium green, semi-glossy; bushy, medium growth; [Heartland X Unnamed seedling]; Bridges, Dennis A., 1987

'PODRUGA', HT, mp, 1939; (Girl Friend); flowers bright pink, base dark red, oval, irregular shape, low, rather weak growth; [Vaterland X Mme Edouard Herriot]; Costetske

'POËMA', Cl Pol, mp, 1933; flowers bright pink, passing to la france pink, borne in clusters of, recurrent bloom; foliage bright, dark; vigorous (5–10 ft) growth; [Tausendschön X Farbenkonigin]; Brada, Dr.; Böhm

'POENTE', F, ob; flowers orange and carmine; [Pinocchio X Goldilocks]; da Silva, Moreira

Poesie, *see* 'LENOBIT'

Poesie, *see* 'JACIS'

Poesie, *see* 'JACIENT'

Poesie, S, w, 1997

'POÈTE JEAN DU CLOS', HT, op, 1919; flowers pink, shaded salmon, dbl., intense fragrance; [Le Progres X Lyon Rose]; Gillot, F.

Poetry, *see* 'ZIPOET'

Poetry in Motion, *see* 'HARELAN'

Pohadka Maje, HT, lp; Certek, 1978; (Czech Rosa Club)

Poiana, HT, dy, 1972; flowers large, dbl.; GPG Bad Langensalza; (Sangerhausen)

'POINSETTIA', HT, mr, 1938; bud long, pointed; flowers bright scarlet, large, dbl., 28 petals, moderate fragrance; foliage glossy; vigorous, compact growth; GM, Portland, 1940; [(Mrs J.D. Eisele X Vaterland) X J.C. Thorton]; H&S; Dreer

'POINSETTIA, CLIMBING', Cl HT, mr, 1950; Thompson, D.L.; Rosemont Nursery Co.

Point Clear, *see* 'TALPOI'

Point de Avignon, F, or

Point du Jour®, *see* 'CANDIDE'

Poker®, *see* 'HAVAPS'

Poker, *see* MEIpazdia

'POKER CHIP', Min, rb, 1979; bud pointed; flowers red with yellow reverse, medium, dbl., 28 petals, exhibition form, intense fragrance; large fruit; foliage glossy, dark; vigorous, compact growth; [Sheri Anne X (Yellow Jewel X Tamango seedling)]; Saville, F. Harmon

'POKORNYANA', S, lp; (R. scopulosa, R. X pokornyana); 1.75 in.; foliage leaflets doubly serrate; (40); [R. canina X R. glauca]

Pol Robson, F, dr, 1955; flowers medium, very dbl., slight fragrance; Klimenko, V. N.; (Sangerhausen)

'POLAR BEAR', HP, w, 1934; flowers white tinted blush, becoming pure white, large, very dbl., globular, recurrent bloom, intense fragrance; foliage large, leathery, wrinkled; vigorous, bushy growth; [Schoener's Nutkana X New Century]; Nicolas; J&P

Polar Star, *see* 'TANLARPOST'

Polareis, *see* STRonin

'POLARIS', LCl, w, 1939; flowers pure snow-white, open, borne in clusters, dbl., profuse, non-recurrent bloom, intense fragrance; foliage glossy, light; very vigorous, climbing (12–15 ft) growth; [(R. wichurana X R. setigera) X R. foetida bicolor]; Horvath; Wayside Gardens Co.

Polarsonne, *see* STRolon

Polarstern®, *see* 'TANLARPOST'

'PÔLE NORD', HT, w, 1944; bud long, pointed; flowers pure white, base tinted greenish; vigorous growth; Mallerin, C.; A. Meilland

Polestar, *see* **'POLSTJÄRNAN'**

'POLIARCHUS', S, w; flowers cream flushed salmon, fading rapidly, non-recurrent, spreading shrub (4 ft) growth; hardy; [Seedling X R. Xharisonii]; Central Exp. Farm

'POLINA', HT, dp, 1984; flowers deep pink, large, dbl., 38 petals; foliage dark, leathery; bushy growth; [Mistica X Chenon]; Staikov, Prof. Dr V.; Kalaydjiev and Chorbadjiiski

Polisiedame, *see* KORichard

Polka™, *see* 'MEITOSIER'

'POLKA', F, mp, 1959; bud ovoid; dbl., 42 petals, 3–3.5 in., exhibition form, blooms in clusters, moderate fragrance; foliage leathery; vigorous, bushy growth; [Moulin Rouge X Fashion]; Meilland, Mrs. Marie-Louise; URS

Polka, *see* 'MEITOSIER'

Polka 91, *see* 'MEITOSIER'

'POLKA DOT', Min, w, 1956; flowers ivory white, small, dbl.; foliage dark, leathery; vigorous, dwarf (10 in.), bushy growth; [Golden Glow X Zee]; Moore, Ralph S.; Sequoia Nursery

'POLKA TIME', S, op, 1984; bud ovoid, pointed; flowers salmon tinted yellow, veined pink, imbricated, blooms borne, dbl., 38 petals, .5 in., cupped, repeat bloom, moderate fragrance; awl-like, tan prickles; foliage dark, leathery; low, bushy, compact growth; hardy; [Bonfire Night X Countryman]; Buck, Dr Griffith J.; Iowa State University

'POLLENTIA', HT, dp, 1942; bud large, long pointed; flowers satiny strawberry-red, dbl., 40 petals, exhibition form, intense fragrance; foliage dark, glossy; upright, compact growth; Dot, Pedro

'POLLINIANA', Misc OGR, w; (R. X polliniana); flowers white to pink, large, long stems; (21); [R. arvensis X R. gallica]

'POLLMERIANA', (strain of R. canina), lp; almost thornless; very vigorous growth; Pollmer

Polluce®, *see* BARpoll

'POLLY', HT, w, 1927; bud long, pointed; flowers cream, center tinted pink or light orange, fading white, large, dbl., 38 petals, exhibition form, intense fragrance; [Ophelia seedling X Mme Colette Martinet]; Beckwith

'POLLY FLINDERS', Min, w, 1954; flowers cream tinted copper-orange, dbl., 30 petals, 1 in.; foliage veined red; [Little Princess (Pol) X Fashion]; Robinson, T.

'POLLY PERKINS', HT, op, 1967; flowers orange-cerise, pointed, moderate fragrance; foliage dark; very free growth; Gregory

Polly Sunshine, *see* 'MORGOLDPOLY'

Polo, *see* TANolop

Polo Club™, *see* 'AROTIGY'

'POLONAISE', S, dp, 1984; bud ovoid, pointed; flowers deep pink, imbricated blooms borne 5–10 per cluster, dbl., 43 petals, 3 in., repeat bloom; awl-like, red-brown prickles; foliage leathery, dark green with reddish veination; dwarf, upright, bushy growth; hardy.; [San Francisco X Prairie Princess]; Buck, Dr Griffith J.; Iowa State University

'POLONAISE', HT, mr, 1961; flowers bright red, dbl., 40–50 ruffled petals, 6 in., exhibition form, moderate fragrance; foliage glossy; vigorous, upright, compact growth; RULED EXTINCT 1/85; [Carrousel X (Chrysler Imperial X Seedling)]; Von Abrams; Peterson & Dering

'POLSTJÄRNAN', LCl, w, 1937; (Polestar, The Polar Star, The Wasa Star, Wasastiernan, White Rose of Finland, White Star of Finland); flowers pure white, very small, blooms in clusters, non-recurrent, very vigorous (to 18 ft) growth; very hardy.; Wasast-jarna

'POLY PEACE', F, yb, 1959; flowers golden yellow edged rose-pink, large, semi-dbl., exhibition form, slight fragrance; foliage dark, glossy; strong stems; very vigorous, tall, bushy growth; [Masquerade X Peace]; Leenders, J.

'POLY PRIM', F, dy, 1953; flowers well shaped, borne in clusters, dbl., 70 petals, 3 in., moderate fragrance; foliage dark, glossy; very vigorous, bushy growth; GM, NRS, 1954; [Goldilocks X Golden Rapture]; Eddie

'POLYANA', S, dp, 1925; [R. rugosa X Polyantha (Mme Norbert Levavasseur?)]; Skinner

'POLYANTHA GRANDIFLORA', HMult, w; (Gentiliana); single; 1888

'POLYBAG JOSHI', HT, lp, 1999; flowers light pale pink, reverse deeper, very dbl., 41 petals, 4–5 in., borne mostly singly, moderate fragrance; many prickles; foliage large, medium green, dull, leathery; spreading, medium (3–4 ft) growth; [First Prize X (First Prize X Perfume Delight)]; Chiplunkar, C. R.; KSG's Roses, 1995

Polygold®, *see* 'TANLYPOLO'

'POLYNESIAN PEARL', Gr, lp, 1977; bud ovoid; flowers shell-pink, small, dbl., 35–59 petals, 2 in., exhibition form, moderate, fruity fragrance; foliage leathery; upright growth; [Floribunda X Tropicana]; Takatori, Yoshiho; Japan Rose Nursery

'POLYNESIAN SUNSET', HT, op, 1965; bud long; flowers coral-orange, dbl., 6 in., exhibition form, moderate, fruity fragrance; foliage leathery; vigorous, bushy growth; [Diamond Jubilee seedling X Hawaii]; Boerner; J&P

Pomona, *see* FRYyeh

'POMPADOUR', HT, dp, 1979; bud pointed; flowers deep pink, dbl., 43 petals, 5.5 in., exhibition form, intense, fruity fragrance; foliage dark, glossy; bushy, compact growth; [Molly McGredy X Prima Ballerina]; Murray, Nola; Rasmussen's

'POMPADOUR RED', F, dp, 1951; bud globular; flowers rose-red, dbl., 30–35 petals, 3.5 in., cupped, moderate, fruity fragrance; vigorous growth; [Orange Triumph X Anne Poulsen]; deRuiter; J&P

Pompeii, HT, rb

Pompon, *see* **'POMPON PANACHÉE'**

'POMPON', C, w; flowers white, streaked lilac-pink, small, dbl.

'POMPON', HGal, dr, 1835; flowers brilliant crimson; Joly

'POMPON BEAUTY', F, or, 1949; flowers scarlet, borne in large trusses, very dbl., vigorous growth; (28); GM, NRS, 1950; [Polyantha seedling X Hybrid Tea seedling]; deRuiter; Spek

'POMPON BLANC PARFAIT', A, wp; (Blanc Parfait); flowers blush-white, pompon form; almost without prickles; compact growth; Verdier, E., 1876; 1876

Pompon de Bourgogne, *see* **'BURGUNDIAN ROSE'**

Pompon de Bretagne, S, dp, 1997

Pompon de Burgogne, *see* **'BURGUNDIAN ROSE'**

Pompon de Panachée, *see* **'POMPON PANACHÉE'**

'POMPON DE PARIS', Ch, mp, 1839; bud very pointed; flowers bright pink, very small, dbl.

'POMPON DE PARIS, CLIMBING', Cl Ch, mp

'POMPON DE SAINT-FRANÇOIS', C, dp; flowers deep violet-pink, small, dbl.; low, bushy growth

Pompon des Princes, *see* **'ISPAHAN'**

Pompon Diadem, *see* TANmop

'POMPON PANACHÉE', HGal, w; (Pompon, Pompon de Panachée); flowers cream to white, striped dark pink, well-formed, dbl., flat; foliage very small; wiry, erect growth; Robert et Moreau, 1856 (?)

Pompon Perpetual, *see* **'BERNARD'**

Pompon Rose, *see* **'ROSE DE MEAUX'**

'POMPON ROUGE', F, mr, 1971; bud ovoid; flowers brilliant red, full, medium, very dbl.; foliage soft; dwarf, bushy growth; [Reverence X Miracle]; Delforge

Pompon Spong, C, m, 1820; flowers small, very dbl., moderate fragrance; Spong, about 1820; (Sangerhausen)

'PONCHEAU-CAPIAUMONT', HCh, dr; (Rosa Poncheaux, Rose Ponceau, Rose Poncheaux); borne in clusters of 3, foliage curiously stiff, thick, deeply veined

Ponctuée, *see* **'MA PONCTUÉE'**

Ponderosa®, *see* 'KORPON'

Pont d'Avignon, F, or; Dot

'PONTBRIANT', HT, mp, 1948; flowers bright pink, very large, dbl., slight fragrance; foliage leathery; vigorous growth; [Mme Joseph Perraud X Kidwai]; Gaujard

'PONTCARRAL', HT, dp, 1943; bud long, pointed; flowers strawberry-red, reverse dull yellow, medium, semi-dbl., cupped, slight fragrance; foliage leathery; moderate, bushy growth; [Charles P. Kilham X (Charles P. Kilham X Margaret McGredy)]; Meilland, F.; A. Meilland

'PONTE D'ARRABIDA', HT, op, 1963; flowers salmon-pink veined deep pink; [Grand'-mere Jenny X Seedling]; da Silva, Moreira

Ponte Rosa, LCl, mr

Ponte Salazar, F; Moreira da Silva, A.; (Cavriglia)

Pontevedra, HT, 1969; Moreira da Silva, A.; (Cavriglia)

'POORNIMA', HT, ly, 1971; bud long, pointed; flowers large blooms borne singly, dbl., 46 petals, exhibition form, moderate fragrance; foliage medium, medium green, smooth; bushy, tall growth; [Fernand Arles X Unnamed seedling]; Pal, Dr B.P.; K.S.G. Son's Roses

'POPCORN', Min, w, 1973; bud ovoid; flowers pure white, semi-dbl., 13 petals, 1 in., moderate, honey fragrance; foliage glossy; upright growth; [Katharina Zeimet X Diamond Jewel]; Morey, Dr Dennison; Pixie Treasures Min. Roses

Pope Pius IX, *see* **'PIUS IX'**

Pope Pius XI, *see* **'PIUS XI'**

'POPPET', F, mp, 1979; dbl., 60 petals, 3.5 in., cupped, slight fragrance; foliage light green, matt; upright growth; [Spartan X Arthur Bell]; Bees

'POPPING WHITE', Min, w, 1977; bud ovoid; semi-dbl., 12 petals, 1 in.; vigorous, compact growth; Lyon

Poppius, HSpn, mp; Steinberg, 1838

'POPPY', HT, op, 1939; flowers coral-pink, passing to shell-pink, petals serrated, dbl., moderate fragrance; vigorous growth; Archer

'POPPY', F, or, 1960; flowers scarlet, open, borne in clusters, semi-dbl., 2–2.5 in.; very vigorous growth; [Cocorico X Geranium Red]; Soenderhousen; Hoersholm Nursery

Poppy Flash, *see* 'MEILÉNA'

Poppy Flash, Climbing, *see* 'MEILÉNASAR'

Pops, *see* 'TINPOPS'

Popsy, S, lp; Peden, G.H.; (Weatherly, L.)

Porcelain, *see* 'KORFLÜG'

'PORCELAIN', F, lp, 1985; flowers shell pink over cream, medium, semi-dbl., slight fragrance; foliage small, dark, semi-glossy; upright, bushy growth; [Dainty Maid X ((Tropicana X Anna Wheatcroft) X (Tropicana X Anna Wheatcroft))]; Gobbee, W.D.

'PORCELAIN PRINCESS', Min, pb, 1990; bud rounded; flowers cream with pale pink edging, aging light pink, medium, borne, dbl., 35 petals, exhibition form, no fragrance; oblong, orange-red fruit; prickles straight, tan; foliage medium, dark green, semi-glossy; bushy, medium growth; [Libby X Unnamed seedling]; Gruenbauer, Richard, 1984; Richard Gruenbauer

Porcelaine de Chine, S, 1996; Lens; (Cavriglia)

Porcelina, *see* 'INTERVAL'

Port St John, HT, 1974; Herholdt; (Cavriglia)

Porta Nigra, HT, dr, 1992; flowers large, dbl., moderate fragrance; Meilland; (Sangerhausen)

Porta Westfalica®, F, dr

'PORTADOWN', HT, dr, 1928; bud long, pointed; flowers velvety deep crimson, large, dbl., exhibition form, slight fragrance; vigorous, bushy growth; GM, NRS, 1929; McGredy

'PORTADOWN BEDDER', HT, op, 1929; bud long, pointed; flowers scarlet-cerise on orange ground, reverse orange-yellow flush, dbl., exhibition form, moderate fragrance; foliage glossy, dark; vigorous, bushy growth; McGredy

'PORTADOWN FRAGRANCE', HT, op, 1931; bud long, pointed; flowers brilliant orange-salmon-pink, flushed orange-scarlet, dbl., exhibition form, intense fragrance; foliage thick, bronze; low, sprawling growth; GM, NRS, 1928; McGredy

'PORTADOWN GLORY', HT, my, 1932; flowers clear canary-yellow, well formed, large, dbl., moderate fragrance; foliage glossy, bright; vigorous growth; GM, NRS, 1933; McGredy

Portadown Ivory, *see* **'MCGREDY'S IVORY'**

'PORTADOWN SALLY', HT, mr, 1931; bud long, pointed; flowers crimson-carmine, base yellow, reverse sulfur-yellow, semi-dbl., exhibition form, moderate fragrance; foliage thick, glossy; vigorous growth; McGredy

Portail Rouge, LCl, 1973; Combe, C.; (Cavriglia)

'PORTCHESTER PINK', HT, dp, 1978; flowers deep rose-pink, intense fragrance; [Red Devil sport]; Poole

Portho, Climbing, Cl F, or, 1976; Laperrière

Porthos®, *see* LAPad

'PORTIA', HT, lp, 1910; flowers pale rose, center yellow; Paul, W.

'PORTIA', HT, mr, 1921; flowers nasturtium-red, paling to pink, dbl., moderate fragrance; [Bridesmaid X Sunburst]; Bees

'PORTLAND', HT, dp, 1958; flowers rose-madder, large, dbl., moderate fragrance; foliage dark, glossy; vigorous, symmetrical growth; Lowe

Portland Dawn, *see* 'SEATIP'

Portland from Glendora, *see* **'JOASINE HANET'**

'PORTLAND PINK', F, dp, 1957; flowers glowing deep pink, [Pinocchio sport]; Murrel, E.

Portland Rose, *see* **'DUCHESS OF PORTLAND'**

Portland Rose Festival, *see* 'DORJURE'

'PORTLAND TRAILBLAZER', HT, dr, 1975; (Big Chief); flowers crimson, dbl., 28 petals, 5.5 in., exhibition form, moderate fragrance; foliage large, matt, brittle; [Ernest H. Morse X Red Planet]; Dickson, A.; Edmunds Roses, 1978

Portmeirion, S, mp, 1988; Austin, David

'PORTO', HT, dr, 1934; flowers deep garnet tinted bright scarlet, large, dbl., slight fragrance; foliage leathery; vigorous growth; [Capt. F. Bald X Mrs Edward Powell]; Mallerin, C.

Porto Nigra, *see* MEIrgano

'PORTOFINO', F, rb, 1964; flowers red, reverse deep yellow; low growth; [Miramar X Seedling]; Delforge

Portrait, *see* 'MEYPINK'

'PORTRAIT OF JENNY', F, dr, 1951; flowers crimson, borne in trusses, dbl., 30 petals, 2 in.; foliage dark; vigorous, bushy growth; [Donald Prior seedling]; Hope

'PORTUGAL PINK', HT, dp, 1959; bud globular; flowers deep pink, large, dbl., cupped, intense fragrance; foliage leathery; vigorous, upright growth; [Independence X Seedling]; Mondial Roses

Post Office, F, rb, 1997; J&P

Postillion, *see* KORtionza

'POSTILLON', HT, ab, 1962; flowers coppery yellow, large, dbl.; foliage glossy; moderate growth; [Peace X Seedling]; Verbeek

'POSY', F, pb, 1951; flowers clear pink, reverse deeper, blooms in large clusters, dbl., 30 petals, 3 in., slight fragrance; foliage dark; dwarf growth; [Rosenelfe X Dusky Maiden]; LeGrice

Pot Black, *see* 'PEANUT'

Pot o' Gold, *see* 'DICDIVINE'

Potch Pearl, *see* HERtroci

'POTIFAR', HT, ab, 1979; (Wieteke van Dordt); flowers pale peachy orange, urn-shaped, large blooms borne usually singly, dbl., 25 petals, slight fragrance; foliage large, leathery, dark, matt; bushy growth; [Royal Dane X Pjerrot]; Poulsen, Niels D.; A. Grumet, 1979; Jan Spek, 1982

Potluck, *see* 'LAVPOT'

Potluck Blue™, *see* 'LAVNOVA'

Potluck Cream™, *see* 'LAVCREAM'

Potluck Crimson®, *see* 'LAVSON'

Potluck Frosty®, *see* 'LAVPURE'

Potluck Gold™, *see* 'LAVGOLD'

Potluck Orange™, *see* 'LAVJACK'

Potluck Pink™, *see* 'LAVALOT'

Potluck Purple™, *see* 'LAVPUP'

Potluck Red™, *see* 'LAVMIN'

Potluck White™, *see* 'LAVWYTE'

Potluck Yellow™, *see* 'LAVGLO'

Potter & Moore, *see* AUSpot

Potton Heritage, *see* 'HARSPRICE'

'POUBICARBE', F, ab, 1974; (**Chipie**); flowers rose-begonia to apricot, open, dbl., 24 petals, 4 in.; foliage light green; bushy growth; [((Elizabeth of Glamis X (Heidelberg X 8366-2)) X ((Pernille Poulsen X (Danish Gold X Mischief))]; Poulsen, Niels D.; Vilmorin-Andrieux

'POUDEX', S, lp, 1994; (**Newport**); flowers pink, old fashioned, dbl., 26–49 petals, 2–2.5 in., pompon, borne singly and in small clusters, slight fragrance; foliage medium green; compact, bushy growth; PP9261; Olesen, 1994

POUlagun, MinFl, op, 1999; (**Laguna Palace**); Poulsen

POUlalo, S, op, 1998; (**Coral Border**); Poulsen

'POULANDER', F, dr, 1986; (America's Choice, **H.C. Andersen**®, Hans Christian Andersen, Touraine); flowers large, semi-dbl., cupped, borne in sprays of 1–25, slight fragrance; foliage medium, dark green, glossy; bushy, tall growth; PP006265; [Royal Occasion X Unnamed seedling]; Olesen, Pernille & Mogens N., 1979; Poulsen Roser ApS, 1986

'POULANS', S, dp, 1995; (**Martha's Vineyard**); flowers hot pink, blending to white at center, semi-dbl., 6–14 petals, 2 in., borne in small clusters, slight fragrance; few prickles; foliage small, light green, semi-glossy; spreading (36 in.) growth, tall, vigorous; PP9664; Olesen, Pernille & Mogens N.; DeVor Nurseries, Inc., 1995; Young's American Rose Nursery, 1995

'POULAPS', S, ly, 1985; (**Armorique Nirpaysage**); flowers creamy yellow, urn-shaped, large blooms in clusters of 1–5, dbl., 20 petals, moderate fragrance; foliage large, leathery, dark, glossy; tall, upright, bushy growth; [Unnamed seedling X Kalahari]; Olesen, Pernille & Mogens N.; Poulsen's Roses

'POULAPS', S, w; (**White Nights**)

'POULARI', HT, w, 1994; (**Karen Blixen**, Silver Anniversary); very dbl., 3–3.5 in., borne in small clusters, slight fragrance; some prickles; foliage large, medium green, glossy; medium (60–80 cms), upright growth; Poulsen Roser APS; Cants of Colchester Ltd., 1995

'POULART', HT, rb, 1985; (**Modern Art**®, Prince de Monaco); flowers medium red, dark red petal edges, spreading with age, revers, dbl., 25 petals, slight fragrance; foliage medium, dark, matt; upright, bushy growth; GM, Rome, 1984; [Unnamed seedling X Unnamed seedling]; Olesen, Pernille & Mogens N.; D.T. Poulsen, 1983

POUlave, S, ab, 1991; (**Sussex**); groundcover growth; Poulsen

POUlbella, F, dp, 1994; (**Berleburg**, Bewitched); Poulsen; (Cavriglia)

'POULBELLS', Min, dp, 1983; (**Pink Bells**®); flowers deep pink, small, dbl., 35 petals, slight fragrance; foliage small, medium green, semi-glossy; groundcover; spreading growth; [Mini-Poul X Temple Bells]; Poulsen Roser APS; John Mattock, Ltd

POUlberin, S, m, 1997; (**Bering Renaissance**); Poulsen

'POULBERO', Gr, ob, 1991; (**Solitude**™); flowers orange-yellow bicolor, blooms borne in large clusters, dbl., 26–40 petals, 1.5–2.75 in., slight fragrance; foliage medium, medium green, semi-glossy; medium (110 cms), bushy growth; [Selfridges X Unnamed seedling]; Olesen, Pernille & Mogens N.; Conard-Pyle Co., 1992

POUlbico, F, pb, 1992; (**Christopher Columbus**, POUlstripe); Poulsen

POUlbo, LCl, w, 1999; (**Bolero**); Olesen, 1998

POUlbon, Min, dy, 1992; (**Bonbon Hit**®); Poulsen

POUlbright, F, 1985; Poulsen; (Cavriglia)

POUlbufi, S, ob; (**Bushfire**); Poulsen

POUlbut, S, mp, 1998; (**Butterflies**); Poulsen

POUlcape, MinFl; (**Summer Palace**); Poulsen, 1995; (Cavriglia)

'POULCAT', Min, lp, 1985; (**Caterpillar**®, Kiki Rose®, Pink Drift); flowers small blooms in large trusses, semi-dbl., no fragrance; foliage small, dark, glossy; groundcover; low, spreading growth; [Temple Bells X Seedling]; Olesen, Pernille & Mogens N.; D.T. Poulsen, 1984

POUlchris, MinFl, mr, 1996; (**Imperial Palace**); Poulsen

'POULCLAUS', Min, dr, 1991; (**Santa Claus**); bud long, well-formed; flowers velvety dark red, large, dbl., 15–25 petals, 1.5–2.75 in., exhibition form, borne mostly singly, slight fragrance; few prickles; foliage medium, dark green, glossy; medium (50–60 cms), upright, bushy growth; PP9063; [Unnamed Floribunda seedling X Unnamed Miniature seedling]; Olesen, Pernille & Mogens N.; Weeks Roses, 1995

'POULCLIMB', LCl, pb, 1999; (**Berries 'n' Cream**™, Calypso®, Cl. Berries 'n' Cream); flowers pink striped white, reverse same, dbl., 26–40 petals, 4–5 in., borne in large clusters, moderate, apple fragrance; few prickles; foliage large, medium green, glossy; climbing, tall (10–12 ft) growth; PP10639; [Evita X seedling]; Olesen; Weeks Roses, 1998

'POULCOT', Min, ab, 1985; (**Apricot Medinette**, Apricot Midinette, Patio Flame); flowers apricot-orange, well-formed, small, dbl., 20 petals, blooms in clusters, slight fragrance; brown prickles; foliage leathery, glossy; spreading growth; [Mini-Poul X Mary Sumner]; Olesen, Pernille & Mogens N.; Ludwigs Roses Pty. Ltd., 1983

'POULCOV', S, w, 1988; (**Kent**®, Pyrenees, Sparkley, White Cover); flowers clear white, bright yellow stamens, semi-dbl., 15–25 petals, 2 in., borne singly and in clusters, slight fragrance; foliage medium green; arching, compact, spreading, low growth; Certificate of Merit, RNRS, 1992 GM, Baden-Baden, 1990 PIT, RNRS, 1990; Olesen, 1988

POUlcrack, F, ob, 1989; (**Crackerjack**); Olesen

'POULCRIM', Min, dr, 1984; (**Crimson Medinette**, Crimson Midinette, Patio Prince); flowers small blooms in clusters of 3–5, dbl., 33 petals, no fragrance; straight brown prickles; foliage small, leathery; dense growth; [Seedling X Pygmae]; Olesen, Pernille & Mogens N.; Ludwigs Roses Pty. Ltd.

POUlcub, MinFl, lp; (**Tiger Cub**); Poulsen, 1996

POUldace, MinFl, lp; (**Sundance Palace**); Poulsen, 1999

'POULDAVA', S, ab, 1995; (**Augusta**); flowers peach, blooms, dbl., 15–25 petals, 2–3 in., slight fragrance; some prickles; foliage small, medium green, dull; low (24 in.), spreading growth; Olesen, Pernille & Mogens N.; DeVor Nurseries, Inc., 1995; Young's American Rose Nursery, 1995

POUldel, Min, w, 1993; (**Pure Hit**); Poulsen

POUldiam, S, w, 1998; (**Diamond Border**); Poulsen

POUldom, F, dy, 1994; (**Gold Reef**); Poulsen

POUldotage, HT, pb; (**Addo Heritage**); Poulsen

POUldra, S, dp, 1997; (**Princesse Alexandra**); Poulsen

POUldrik, F, op, 1998; (**Frederiksborg**); Poulsen

'POULDRON', S, op, 1999; (**Countess Celeste**™, Coral Palace); flowers coral pink, reverse same, old-fashioned form, very dbl., 41 petals, 2.5–3 in., borne in small clusters, moderate fragrance; prickles moderate; foliage medium, dark green, glossy; compact, rounded low (20–24 in.) growth; [Queen Margrethe X seedling]; Poulsen Roser APS; Weeks Roses, 1998

'POULDUCE', HT, my, 1994; (**Tivoli Gardens**); flowers soft yellow, blooms borne in small clusters, dbl., 26–40 petals, 3–3.5 in., slight fragrance; some prickles; foliage large, dark green, semi-glossy; tall (80–100 cms), spreading growth; Olesen, Pernille & Mogens N.; Cants of Colchester Ltd., 1995

POUlduff, HT, mr, 1997; (**Courage**, Roy Castle Rose); Poulsen

POUleas, Min, 1995; (**Easy Cover**); Poulsen; (Cavriglia)

POUledge, F, op, 1999; (**Everglades**); Poulsen

POUlelap, MinFl, mp, 1997; (**Queen's Palace**); Poulsen

'POULEMB', S, w, 1995; (**Cliffs of Dover**); flowers small, white, single, bright yellow stamens, single, 5–10 petals, 1–1.5 in., borne in clusters, no fragrance; few prickles; foliage small, medium green, glossy; low (36 in.), arching, spreading growth; PP9650; Olesen, Pernille & Mogens N.; DeVor Nurseries, Inc., 1995; Young's American Rose Nursery, 1995

'POULENA', HT, yb, 1976; (**Bellevue**®, Jarlina); flowers dark yellow and apricot, edged red, dbl., 23 petals, 6 in.; foliage glossy, leathery; vigorous, upright growth; [(Tropicana X Piccadilly) X Fru Jarl]; Poulsen, Niels D.; Poulsen

POUlense, Cl Min, ly, 1995; (**Little Miss Muffet**); Olesen

'POULETTA', HT, ab, 1984; (**Henrietta**); flowers urn-shaped, large, borne singly, dbl., 25 petals, slight, fruity fragrance; foliage large, dark, glossy; vigorous, upright, bushy growth; [Unnamed seedling X (Pink Nordia X Sonny Boy)]; Olesen, Pernille & Mogens N.; Poulsen's Roses

'POULFAIR', S, my, 1988; (**Yellow Fairy**®); flowers small, semi-dbl., 6–14 petals, slight fragrance; foliage small, light green, semi-glossy; spreading, compact, hardy, abundant growth; GM, Madrid, 1988; [Texas X The Fairy]; Olesen, Pernille & Mogens N.; Poulsen Roser ApS

'POULFAN', S, lp, 1995; (**Cape Cod**, Jhb Garden Club, Johannesburg Garden Club); flowers soft pink, blooms borne in large clusters, single, 5 petals, 2–3 in., no fragrance; few prickles; foliage small, medium green, semi-glossy; medium (36–42 in.), spreading growth; PP9641; Silver, Bundesgartenshau GER, 1991; Olesen, Pernille & Mogens N.; DeVor Nurseries, Inc., 1995; Young's American Rose Nursery, 1995

POUlfancy, Min, mr; (**Fancy Hit**); Poulsen

POUlfect, Min, ab; (**Perfect Hit**); Poulsen

POUlfeld, F, w; (**Grasten**); Poulsen

'POULFI', Min, dp, 1985; (**Alfi**™, Alfie™, Elfin); flowers soft lavendar to pink, small, very dbl., 40–45 petals, blooms in large clusters and cadelabras, no fragrance; foliage small, light green, glossy; compact growth, free flowering; PP005058; [Mini-Poul X Harriet Poulsen]; Olesen, Pernille & Mogens N.; Poulsen's Roses, 1981

POUlfiry, Min, ob; (**Fiery Hit**); Poulsen

'POULFLASH', HT, my, 1984; (**Flashdance**); flowers urn-shaped, large blooms borne singly, dbl., slight fragrance; foliage large, medium green, semi-glossy; upright growth; [Berolina X Unnamed seedling]; Olesen, Pernille & Mogens N.; Poulsen Roser ApS

POUlfolk, S, my, 1990; (**Norfolk**); Poulsen

POUlfre, Min, dp, 1990; (**Fresh Hit**); Poulsen

POUlgan, MinFl, dy, 1992; (**City Lights**); Poulsen

'POULGAV', F, ob, 1988; (Bucks Fizz, **Gavno**); flowers orange, medium, dbl., 20 petals, no fragrance; foliage medium, dark green, glossy; bushy growth; [Unnamed seedling X Mary Sumner]; Olesen, Pernille & Mogens N.; Poulsen Roser ApS

POUlgewfa, S, lp, 1992; (**Gwen Fagan**); Poulsen

POUlgode, F, my; (**Yellow Cover**); Poulsen, 1993

POUlgode, S, ly; (**Lexington**); Olesen, 1993

POUlgrad, MinFl, dr, 1999; (**Grand Palace**); Poulsen

POUlgren, F, w, 1992; (**Snow Gosling**); Poulsen

POUlgret, MinFl, w, 1996; (**Hampton Palace**); Poulsen

POUlharm, S, or, 1999; (**Charming Cover**); Poulsen

POUlheart, S, lp, 1996; (**Julia Renaissance**, Phillipa); Olesen, 1996

'POULHIT', Min, dr, 1984; (**Red Hit**); flowers small, dbl., 20 petals, no fragrance; foliage small, dark, matt; low, bushy, compact growth; [Mini-Poul X Unnamed seedling]; Olesen, Pernille & Mogens N.; Poulsen's Roses

POUlhit, Min, my; (**Sun Hit**); Poulsen, 1994

POUlholm, S, op, 1989; (**Patio Princess**); Olesen

POUlijill, S, lp; (**Bella**); Poulsen

'POULINK', Min, mp, 1988; (**Pink Hit**); flowers small, semi-dbl., 6–14 petals, no fragrance; foliage small, medium green, semi-glossy; bushy, compact, even, abundant growth; [Unnamed seedling X Unnamed seedling]; Olesen, Pernille & Mogens N.; Poulsen Roser ApS, 1986

'POULINO', S, mr, 1995; (**Napa Valley**); flowers, bright red, semi-double, 2″ semi-dbl., 6–14 petals, 2 in., borne in clusters, slight fragrance; foliage small, dark green, glossy; low (36 in.), spreading growth; PP9554; Olesen, Pernille & Mogens N.; DeVor Nurseries, Inc., 1995; Young's American Rose Nursery, 1995

POUlisab, S, dr; (**Isabel Renaissance**, Isabella); Olesen, 1995

POUlisab, S, dr, 1995; (**Isabelle Renaissance**); Poulsen

'POULIT', Min, mp, 1994; (**Baby Grand**™); flowers clear pink, dbl., 25–40 petals, 1.5 in., quartered, borne in small clusters, slight fragrance; some prickles; foliage small, medium green, matt, very disease resistant; low (25–30 cms), bushy, compact, rounded growth; [Egeskov X Seedling]; Poulsen Roser APS; Weeks Roses, 1995

'POULJENJOY', Min, lp, 1985; (**Jennifer Joy**); flowers mini-flora, blooms in clusters, dbl., 75 petals, no fragrance; foliage light green; compact, low, bushy growth; [Mini-Poul X Unnamed seedling]; Olesen, Pernille & Mogens N.; Ludwigs Roses Pty. Ltd., 1983

POUljill, S, my, 1995; (**Bella Renaissance**); Poulsen

'POULJUB', Min, yb, 1978; (**Poulsen's Jubilaeumsrose**); flowers deep pink and yellow blend, blooms in clusters, dbl., 28 petals, slight fragrance; foliage dark, glossy; upright growth; [Darling Flame X Unnamed seedling]; Poulsen, Niels D.

'POULKARO', F, or, 1988; (**Karoo**, Karoo Rose); flowers orange-red, aging watermelon-pink, medium, borne in sprays, dbl., 32 petals, no fragrance; prickles concave, yellow-brown; foliage medium, medium green, very hardy; neat, medium, densely branched growth; [Unnamed seedling X Unnamed seedling]; Poulsen Roser APS; Ludwigs Roses Pty. Ltd., 1988

POUlkef, HT, 1997

POUlkir, HT, lp, 1985; (**Kirsten**); Poulsen

'POULLACK', Cl HT, rb, 1982; (Kirsch, Climbing, Lakeland Princess, **Vintage Wine**); flowers burgundy red, straw-yellow reverse, large, dbl., 40 petals, exhibition form, slight, old rose fragrance; prickles red when new, then tan; foliage large, moss green; pillar (8–10 ft) growth; [Royal Dane X Arthur Bell]; Poulsen, Niels D.; Roses by Fred Edmunds, 1983

POUlland, S, w; (**White Magic Carpet**™); Poulsen, 1994

'POULLEN', Min, mp, 1994; (**Little Bo-Peep**, Nachez); dbl., 26–40 petals, 1.5 in., borne in small clusters, slight fragrance; few prickles; foliage small, dark green; patio; low (65 cms), compact growth; PP9273; PIT, RNRS, 1991; Olesen, Pernille & Mogens N.; DeVor Nurseries, Inc., 1994; Young's American Rose Nursery, 1994

POUllen, S, lp; (Gentle Cover, Little Bo-Peep, **Natchez**); Olesen, 1994

'POULLIGHT', LCl, dy, 1985; (**Night Light**®); flowers large blooms in sprays of 1–5, dbl., 27 petals, moderate fragrance; large, dark red prickles; foliage large, dark, glossy; bushy growth; [Westerland X Pastorale]; Poulsen, Niels D.; D.T. Poulsen, 1982

'POULLOW', Min, dy, 1988; (**Yellow Hit**); flowers small, dbl., 15–25 petals, slight fragrance; foliage small, dark green, semi-glossy; bushy, compact, even growth; PP006860; GM, International Exhibition, 1986; [Unnamed seedling X Unnamed seedling]; Olesen, Pernille & Mogens N.; Poulsen Roser ApS

POUlly, F, 1991; Poulsen; (Cavriglia)

'POULMA', S, lp, 1994; **'Newport'** very dbl., 3–3.5 in., borne in large clusters, no fragrance; some prickles; foliage medium, medium green, glossy; low (2 ft), spreading growth; Olesen, Pernille & Mogens N.; DeVor Nurseries, Inc., 1994; Young's American Rose Nursery, 1994

'POULMAN', HT, dr, 1984; (**Ingrid Bergman**®); flowers large, clear bright red, very dbl., 35–40 petals, 5 in., exhibition form, borne singly, slight, spice fragrance; foliage medium, dark, semi-glossy; upright growth, vigorous, compact; PP006264; GM, Belfast, 1985 GM, Madrid, 1986 Golden Rose, The Hague, 1987; [Unnamed seedling X Unnamed seedling]; Olesen; John Mattock, Ltd, 1985

POUlmanti, MinFl, dp, 1999; (**Romantic Palace**); Poulsen

POUlmar, Min, ob, 1994; (**Rosy Hit**); Poulsen

POUlmax, F, lp, 1995; (**Fredensborg**); Poulsen

POUlming, Min, lp, 1994; (**Harmony Parade**); Olesen

POUlmo, Min, mr, 1991; (**Scarlet Hit**); Poulsen

POUlmount, Gr, ly, 1996; (**Ivory Grand**); flowers ivory white, large, dbl., exhibition form; Poulsen

POUlmulti, S, w, 1992; (**Avon**, Sunnyside); bud blush pink; flowers pearl white, dbl.; compact, spreading (1 ft × 2.5 ft) growth; Rose of the Year, Auckland, NZ, 1996; Poulsen

POUlna, S, lp, 1997; (Helena, **Helena Renaissance**); Olesen, 1997

POUlnew, HT, lp; (**Jose Carreras**); Golden Rose, Hradec Kralove, Czech Republic, 1999; Olesen, Pernille & Mogens N.

'POULNINO', S, mr, 1995; (**Napa Valley**); flowers medium red, semi-dbl. (6–14 petals), medium (2 in.) blooms; slight fragrance; foliage small, dark green, glossy; low (36 in.), spreading growth; [Seedling X Seedling]; Olesen, Pernille & Mogens N.; DeVor Nursery; Young's American Rose Nursery, 1995

'POULNISH', Min, my, 1988; (**Sunbird**™); flowers medium yellow opening to light yellow, fading to light pastel, dbl., 25–30 petals, exhibition form, no fragrance; ovoid, few, pale orange-red fruit; foliage medium, dark green, matt, convex, straw to light mar; bushy, medium growth; [Mini-Poul X Unnamed seedling]; Poulsen Roser APS; Conard-Pyle Co., 1987

POUlnoeu, MinFl, lp; (**Salmo**); Poulsen

POUlnoev, S, op, 1995; (**Salmo**); Olesen

POUlnola, Min, mr; (**Noble Hit**); Poulsen

POUlnorm, LCl, lp; (**Jazz**); Poulsen

POUlnows, Min, w; (**Snow Hit**); Poulsen

POUlnoz, S, mp, 1988; (**Essex**); single, low, spreading growth; Poulsen

POUloma, F, mp, 1992; (**Bossa Nova**); Poulsen

'POULONI', HT, op, 1977; (Benoni '75, **Vision**); flowers medium salmon red, well-formed, large, dbl., 22 petals, borne 1–3 per stem, slight fragrance; foliage dark, glossy; vigorous, spreading growth; PP006939; Poulsen, Niels D.; Poulsen's Roses, 1990

POUlor, S, m; (**Mystic**, Supreme Cover); Olesen, 1994

POUlor, S, lp, 1994; (**Peachy Pink Magic Carpet**); Poulsen

POUlorin, MinFl, op, 1999; (**Oriental Palace**); Poulsen

POUlosy, S, mp, 1998; (**Rosy Border**); Poulsen

POUlover, LCl, lp, 1994; (**Rosenholm**); Poulsen

POUlpearl, F, lp, 1990; (**Patio Pearl**); Olesen

POUlpol, HT, ob, 1997; Poulsen

POUlprima, Min, m; (**Purple Parade**); Poulsen

'POULPYG', S, ab, 1995; (**Central Park**); flowers peach, bright yellow stamens, fades white, dbl., 15–25 petals, 2 in., borne in small clusters, slight fragrance; few prickles; foliage small, medium green, glossy; low (2 ft), spreading growth; PP9665; Olesen, Pernille & Mogens N.; DeVor Nurseries, Inc., 1995; Young's American Rose Nursery, 1995

POUlrang, Min, ob, 1995; (**First Hit**); Poulsen

POUlreb, F, my, 1995; (**Eurostar**); Olesen

'POULRED', Min, mr, 1983; (**Red Bells**®); flowers small, dbl., 35 petals, slight fragrance; foliage small, medium green, semi-glossy; groundcover; spreading growth; [Mini-Poul X Temple Bells]; Poulsen Roser APS; John Mattock, Ltd.

POUlrek, F, lp, 1995; (**Crystal Palace**); flowers creamy pink, medium-large, dbl., Poulsen Roser APS; (Sangerhausen)

POUlria, S, lp; (**Velvet Cover**); Poulsen, 1995

POUlriber, F, op, 1998; (**Riberhus**); Poulsen

POUlrijk, S, mp, 1993; (**Devon**®); Poulsen

POUlrim, HT, 1996; (**Flora Danica**); Poulsen Roser APS; (Sangerhausen)

POUlrim, HT, lp; (**Garden News**, Spellbound); Poulsen

POUlrine, F, mp; (**Bangsbo**); Poulsen

'POULRISE', HT, yb, 1994; (**Peach Surprise**); flowers peach and cream, blooms borne mostly single, dbl., 26–40 petals, 3–3.5 in., moderate fragrance; many prickles; foliage large, dark green, glossy; spreading (80–100 cms) growth; [Freude sport]; Poulsen Roser APS; Cants of Colchester Ltd., 1995

POUlrohill, F, mp; (**Rose Hill**); Poulsen

POUlrougel, Min, or; (**Absolute Hit**); Poulsen

'POULRUSSET', Min, or, 1983; (**Russet Beauty**); flowers small blooms in clusters of 3–15, dbl., 46 petals, straight, brown prickles; foliage matt; compact, bushy growth; [Mini-Poul X Unnamed seedling]; Olesen, Pernille & Mogens N.; Ludwigs Roses Pty. Ltd.

POUlrust, S, m, 1994; (**Lavender Cover**); Poulsen

POUlsail, MinFl, lp; (**Versailles Palace**); Poulsen, 1999

'POULSALFAI', S, op, 1993; (**Miss Hillcrest**); Poulsen

'POULSEN'S BEDDER', F, lp, 1948; (Poulsen's Grupperose); flowers clear pink, blooms in trusses, semi-dbl., 3 in., slight fragrance; foliage bronze; vigorous, upright growth; GM, Portland; [Orléans Rose X Talisman]; Poulsen, S.; C-P, 1952; McGredy, 1948; Poulsen, 1948

'POULSEN'S COPPER', F, op, 1940; flowers rose-pink to orange, base yellow, borne in clusters, dbl., 25–35 petals, 3–3.5 in., cupped, slight, spicy fragrance; foliage small, light green; vigorous growth; [Grethe Poulsen X Souv. de Claudius Pernet]; Poulsen, S.; C-P, 1940; McGredy, 1946; Poulsen, 1940

'POULSEN'S CRIMSON', F, dr, 1950; single, 6 petals, blooms in clusters, slight fragrance; foliage medium, dark, matt; vigorous, bushy growth; [Orange Triumph X (Betty Uprichard X Johanniszauber)]; Poulsen, S.; McGredy; Poulsen

'POULSEN'S DELIGHT', F, lp, 1948; (Fru Julie Poulsen); flowers apple-blossom-pink, borne in trusses, single, 5–7 petals; foliage dark, glossy; vigorous, upright growth; [Else Poulsen X Seedling]; Poulsen, S.; McGredy

'POULSEN'S FAIRY', F, my, 1940; (Rodovrerosen); single, borne in large clusters; very vigorous, tall growth; [Orléans Rose X Dainty Bess]; Poulsen, S.; Poulsen

Poulsen's Grupperose, *see* **'POULSEN'S BEDDER'**

Poulsen's Jubilaeumsrose, *see* 'POULJUB'

'POULSEN'S PARK ROSE', S, lp, 1953; flowers silvery pink, well-shaped, blooms in trusses, dbl., 4–5 in.; vigorous (6 × 6 ft) growth; [Great Western X Karen Poulsen]; Poulsen

'POULSEN'S PEACH', F, ab, 1948; flowers peach, large blooms in clusters, semi-dbl., cupped, slight fragrance; foliage medium, medium green, semi-glossy; medium, bushy growth; Poulsen, S.

'POULSEN'S PEARL', F, lp, 1949; flowers pearly pink, blooms in trusses, single, 5 petals; foliage light; vigorous growth; GM, RNRS; [Else Poulsen X Seedling]; Poulsen, S.; Poulsen's Roses, 1948

'POULSEN'S PINK', F, lp, 1939; flowers soft pink with yellow base, blooms in clusters, semi-dbl., cupped, slight fragrance; foliage glossy, light green; vigorous growth; (28); [Golden Salmon X Yellow HT]; Poulsen; C-P, 1942; Poulsen, 1942

'POULSEN'S SCARLET', F, dp, 1941; flowers bright rose, blooms in clusters, dbl., 30 petals, 2.5 in., slight fragrance; bushy growth; [D.T. Poulsen X Red HT]; Poulsen, S.

'POULSEN'S SUPREME', F, mp, 1945; (Kelleriis-Rose); semi-dbl., 3 in., borne in trusses, foliage light green; very free growth; (28); [Poulsen's Pink X Seedling]; Poulsen, S.; McGredy, 1953

'POULSEN'S YELLOW', F, my, 1938; bud ovoid; semi-dbl., blooms in clusters, intense fragrance; foliage glossy; (21); GM, NRS, 1937; [Mrs W. H. Cutbush X Gottfried Keller]; Poulsen, S.; C-P, 1939

POUlshine, Min, lp, 1988; (**Rutland**); PP009457; Poulsen

POUlsiana, F, dy, 1999; (**Atlantis Palace**); Poulsen

POUlsint, Min, mr, 1998; (**Red Paillette**); Poulsen

POUlske, F, rb, 1996; (**Arlette**); Poulsen

'POULSKOV', S, lp, 1994; (Enchantment, **Queen Margrethe**™, Dronning Margrethe); flowers pastel pink, old-fashioned form, dbl., 40–50 petals, 1.5–2.75 in., quartered, borne in small clusters, moderate, apple fragrance; few prickles; foliage small, medium green, glossy, clean; low to medium (45–50 cms), bushy, compact, rounded growth; PP9062; Gold Star of the South Pacific, Palmerston North, NZ, 1992; [Seedling X Egeskov]; Olesen; NZ, 1994; Weeks Roses, 1995

POUlsol, HT, dy, 1999; (**Solo Mio**); Poulsen

POUlspan, S, my, 1992; (**Cottage Maid**, Enchantment); Rose of the Year, Auckland, NZ, 1997; Poulsen

POUlstar, MinFl, lp; (**Starlight Parade**); Poulsen, 1991

POUlstripe, F, pb, 1992; (**Christopher Columbus**, POUlbico); Poulsen

POUlsue, S, lp; (**Susan**); Poulsen, 1997

POUlsum, F, op, 1995; (**Dawn Weller**); Poulsen

POUlsyng, S, lp, 1995; (**Clair Renaissance**, Liliana); Olesen, 1995

'POULTAL', HT, dp, 1983; (**Eva Gabor**, Sentimental); flowers deep pink, large, dbl., 40 petals, moderate fragrance; foliage large, medium green, glossy; vigorous, upright, bushy growth; [Seedling X Seedling]; Olesen, Pernille & Mogens N.; Roses by Fred Edmunds

POUlterp, S, w, 1999; (**Brilliant Cover**); Poulsen

'POULTEX', Min, my, 1984; (Golden Piccolo, **Texas**); flowers mini-flora, small, semi-dbl., slight fragrance; foliage small, medium green, matt; PP006261; [Seedling X Seedling]; Poulsen Roser APS; John Mattock, Ltd

POUlthe, MinFl, lp; (**Topkapi Palace**); Poulsen, 1996

POUltika, LCl, mp, 1997; (**Flamenco**); Poulsen

POUltime, LCl, pb, 1997; (**Ragtime**); Poulsen

POUltimes, F, my, 1993; Poulsen

POUltin, MinFl, lp; (**Violet Hit**); Poulsen, 1995

POUltipe, MinFl, lp; (**Pink Hit**); Poulsen

'POULTOP', Min, ob, 1991; (**Carrot Top**, Panache); flowers orange, clear lasting color, moderately well-formed, blooms, dbl., 15–25 petals, 1.5 in., prolific bloom, slight fragrance; few prickles; foliage medium, medium green, matt; low (40–50 cms), bushy, compact growth; [Seedling (F) X Seedling (Min)]; Olesen, Pernille & Mogens N.; Weeks Roses, 1994

'POULTRESS', HT, dp, 1988; (**Cherry Velvet**®); flowers deep pink, large, dbl., 20 petals, moderate fragrance; foliage large, medium green, glossy; upright, vigorous growth; [Vision X Seedling]; Olesen, Pernille & Mogens N.; Poulsen Roser ApS, 1986

POUltry, F, dr, 1995; (**Kronborg**); Poulsen

POUltumb, S, lp; (**Tumbling Waters**); Poulsen

'POULURT', S, my, 1995; (**Aspen**, Gwent, Sun Cover, Gold Magic Carpet); dbl., 15–25 petals, 1–2 in., borne in small clusters, few prickles; foliage medium, dark green, semi-glossy; low (14–16 in.), spreading growth; PP9637; 2nd prize, Copenhagen, 1992 Certificate of Merit, Belgium, 1992; Olesen, Pernille & Mogens N.; DeVor Nurseries, Inc., 1995; Young's American Rose Nursery, 1995

'POULUSA', S, w, 1999; (**Princess Marianna**™); flowers ivory white, old-fashioned form, very dbl., 41 petals, 2.5–3 in., borne in small clusters, moderate fragrance; prickles moderate; foliage medium, dark green, glossy; compact, bushy, rounded, low (20–24 in.) growth; PP10638; [seedling X Queen Margrethe]; Poulsen Roser APS; Weeks Roses, 1998

'POULVANG', HT, dr, 1988; (**Solvang**); flowers large, semi-dbl., 6–14 petals, slight fragrance; prickles average; foliage large, dark green, glossy; bushy, vigorous growth; [Vision X Unnamed seedling]; Olesen, Pernille & Mogens N.; Poulsen Roser ApS, 1987

POUlvic, Min, mr; (**Victory Parade**®); Poulsen, 1999

'POULVISION', F, dp, 1975; (**Dolly**®, Springs 75); dbl., 20 petals, 2.5–3 in.; foliage glossy, dark; bushy growth; ADR, 1987 GM, Baden-Baden, 1973; [(Nordia X Queen Elizabeth) X (Seedling X Mischief)]; Poulsen, Niels D.; Poulsen

'POULVITA', Min, w, 1984; (**Evita**); flowers white with touch of pale pink, small blooms in clusters, dbl., 20 petals, slight fragrance; foliage small, dark, glossy; bushy growth; [Mini-Poul X Seedling]; Olesen, Pernille & Mogens N.; Poulsen

POUlvue, HT, ob; (Michael Crawford, **Victor Borge**); flowers pink and yellow blend, dbl., 50 petals, exhibition form, medium growth; Certificate of Merit, Belfast, 1992; Olesen, 1991

POUlwarb, S, ly, 1993; (**Fairy Moon**); Olesen

'POULWEE', Min, dy, 1990; (**Rainbow Yellow Parade**™, Yellow Parade™); bud cupped, globular; flowers bright yellow, aging slightly, small, borne singly, dbl., 28–30 petals, cupped, slight, damask fragrance; no prickles; foliage small, dark green, semi-glossy; upright, low, compact growth; GM, The Hague, 1988; [Unnamed seedling X Texas]; Olesen, Pernille &

Mogens N., 1985; DeVor Nurseries, Inc., 1990

POUlweet, S, lp; (**Sweet Cover**); Poulsen, 1999

'POULWHITE', Min, w, 1983; (**White Bells**®); dbl., 35 petals, slight fragrance; foliage small, medium green, semi-glossy; groundcover; spreading growth; [Mini-Poul X Temple Bells]; Poulsen Roser APS; John Mattock, Ltd, 1980

'POULXAS', Min, ab, 1994; (**Gingerbread Man**™); flowers deep, long-lasting apricot amber, blooms borne in small clusters, dbl., 25–40 petals, 1.5 in., quartered, slight fragrance; some prickles; foliage small, dark green, semi-glossy; medium (40–45 cms), upright, bushy, vigorous growth; [Seedling X Texas]; Poulsen Roser APS; Weeks Roses, 1995

POUlzazz, MinFl, rb, 1996; (**Pzazz**); Poulsen

POUlzin, MinFl, rb, 1999; (**Prince Palace**); Poulsen

'POUMIDOR', HT, ob, 1971; (**Royal Dane**®, Troika); flowers orange, outer petals red, classic form, 6 in., intense fragrance; foliage large, dark, glossy, leathery; vigorous, upright, bushy growth; [(Tropicana X (Baccará X Princesse Astrid)) X Hanne]; Deutsche Rosen-registerstelle; N. Poulsen

Pounder Star, *see* 'MACNIC'

'POUPÉE', F, lp, 1965; flowers flesh-pink, dbl., 25–30 petals, 3–3.5 in., cupped, slight fragrance; foliage glossy; vigorous, upright growth; Lens

Pour Toi, *see* **'PARA TI'**

'POUR VOUS MADAME', F, dp, 1961; flowers rose-red, reverse lighter, open, dbl., 25 petals, 4 in., slight fragrance; foliage dark; vigorous growth; Gaujard; Gandy Roses, Ltd.

'POURALE', S, pb, 1970; (**Pastorale**®, Pastourelle); flowers deep pink, yellow reverse, urn-shaped, large blooms in spray, dbl., 25 petals, slight fragrance; dark green prickles; foliage large, leathery, dark, glossy; very vigorous, upright, bushy growth; [Unnamed seedling X Royal Dane]; Poulsen, Niels D.

'POURPRE', Ch, m, 1827; flowers deep purple; Vibert

'POURPRE DU LUXEMBOURG', M, m, 1848; flowers pink, opening to lilac-pink, Hardy

Poustinia, F, ab, 1994; Orye

'POWDER PUFF', F, lp, 1959; flowers creamy light pink, becoming darker, open, medium, dbl., slight fragrance; foliage glossy; bushy growth; [Masquerade X Masquerade seedling]; Mason, A.L.; F. Mason

'POYNTZFIELD', HT, ab, 1977; dbl., 35 petals, 5 in., slight fragrance; foliage glossy, reddish-green; vigorous growth; [Bonsoir X Percy Thrower]; Simpson, J. W.

'POZNAN', HT, dr, 1966; bud ovoid; flowers vivid dark crimson, well-formed, large, dbl.; vigorous growth; Grabczewski

'PRAECOX', (form of R. sericea), w

PRAeddy, HT, lp; (**Teddy**); Pratt, Florence, 1994

Prairie Belle, *see* **'QUEEN OF THE PRAIRIES'**

'PRAIRIE BREEZE', S, m, 1978; bud ovoid, pointed; flowers tyrian purple, dbl., 25 petals, 4.5 in., cupped, repeat bloom, moderate, spicy fragrance; foliage olive-green, leathery; upright, bushy, spreading growth; [Dornroschen X (Josef Rothmund X R. laxa)]; Buck, Dr Griffith J.; Iowa State University

'PRAIRIE CHARM', S, op, 1959; flowers bright salmon-coral, semi-dbl., free, non-recurrent bloom, foliage light green; stems arching; height 4 ft; hardy on prairies; [Prairie Youth X Prairie Wren]; Morden Exp. Farm

'PRAIRIE CLOGGER', S, mr, 1984; bud medium-small, ovoid, pointed; 8–10 petals, 2.5 in., cupped, borne 1–10 per cluster, repeat bloom, moderate fragrance; awl-like, tan prickles; foliage leathery, dark olive green; vigorous, erect, bushy growth; [Carefree Beauty X (Marlena X Pippa's Song)]; Buck, Dr Griffith J.; Iowa State University

'PRAIRIE DAWN', S, mp, 1959; flowers glowing pink, dbl., 2–2.5 in., repeat bloom on current season's wood; foliage dark, glossy; upright (5 ft) growth; hardy on prairies; [Prairie Youth X (Ross Rambler X (Dr W. Van Fleet X R. spinosissima altaica))]; Modern Exp. Farm

'PRAIRIE FIRE', S, mr, 1960; bud pointed; flowers bright red, base white, blooms in clusters of 35–50, semi-dbl., 9 petals, 2.5–3 in., recurrent bloom, moderate fragrance; foliage glossy, dark; long stems; very vigorous, tall growth; [Red Rocket X R. arkansana]; Phillips; Univ. of Minn.

'PRAIRIE FLOWER', S, rb, 1975; bud ovoid, pointed; flowers cardinal-red, center white, single, 7 petals, 2–3 in., flat, repeat bloom, slight, old rose fragrance; foliage dark, leathery; erect, bushy growth; [(Rose of Tralee X Queen Elizabeth) X (Morning Stars X Suzanne)]; Buck, Dr Griffith J.; Iowa State University

'PRAIRIE HARVEST', S, ly, 1985; flowers medium yellow, imbricated, blooms borne 1–15 per cluster, dbl., 43 petals, 4–5 in., repeat bloom, moderate fragrance; awl-like, tan prickles; foliage leathery, glossy, medium, dark; upright, bushy growth; hardy.; [Carefree Beauty X Sunsprite]; Buck, Dr Griffith J.; Iowa State University, 1984

'PRAIRIE HERITAGE', S, op, 1978; bud ovoid, pointed; flowers peach to coral-pink, dbl., 50 petals, 4–5 in., cupped, repeat bloom, moderate fragrance; foliage dark, leathery; vigorous, upright, spreading, bushy growth; [(Vera Dalton X Prairie Princess) X (Apricot Nectar X Prairie Princess)]; Buck, Dr Griffith J.; Iowa State University

'PRAIRIE JOY', S, mp, 1990; flowers medium pink, aging light to medium pink, borne singly or in, dbl., 30–40 petals, sparse, repeat bloom, slight fragrance; foliage medium, medium green, matt, high resistance to powde; tall, bushy growth; [Prairie Princess X Morden Cardinette]; Collicutt, Lynn M.; Agriculture Canada, 1990

'PRAIRIE LASS', S, pb, 1978; bud ovoid, pointed; flowers claret-rose to rose-red, dbl., 28 petals, 3.5 in., repeat bloom, moderate, spicy fragrance; foliage dark, leathery; vigorous, upright, spreading, bushy growth; [(Hawkeye Belle X Vera Dalton) X (Dornroschen X (World's Fair X Applejack))]; Buck, Dr Griffith J.; Iowa State University

'PRAIRIE MAID', S, w, 1959; flowers cream, dbl., 25 petals, intermittent bloom, moderate fragrance; compact (4 ft) growth; hardy on prairies; [(Ophelia X Turkes Rugosa Samling) X R. spinosissima altaica]; Morden Exp. Farm

'PRAIRIE MOON', LCl, ly, 1953; bud ovoid, deep yellow; flowers creamy yellow, large, borne in clusters of 4–5, dbl., 30–35 petals, abundant, non-recurrent bloom, moderate fragrance; foliage glossy, dark, leathery; vigorous, climbing (15 ft) growth; [R. maximowicziana pilosa X Autumn]; Maney; Iowa State College

'PRAIRIE PINKIE', S, op, 1958; flowers deep coral-pink, dbl., midsummer bloom, intense fragrance; foliage like r spinosissima; upright ($2\frac{1}{2}$ ft), bushy growth; Skinner

'PRAIRIE PRINCESS', S, op, 1972; bud ovoid, long, pointed; flowers light coral-pink, large, semi-dbl., repeat bloom, slight fragrance; foliage large, dark, leathery; vigorous, upright growth; [Carrousel X (Morning Stars X Suzanne)]; Buck, Dr Griffith J.; Iowa State University

'PRAIRIE RED', Min, rb, 1980; flowers red-yellow blend, blooms borne singly or several together, semi-dbl., 18 petals, moderate fragrance; brownish prickles curved downward; foliage tiny, medium green; bushy, upright growth; [Unnamed seedling X Unnamed seedling]; Lyon

Prairie Rose, *see* **R. SETIGERA**

'PRAIRIE SAILOR', HSpn, yb, 1946; flowers golden yellow deeply edged bright red, single, profuse, non-recurrent bloom, vigorous (6 ft) growth; hardy on prairies; Morden Exp. Farm

Prairie Schooner™, *see* 'KINschoon'

'PRAIRIE SQUIRE', S, mp, 1984; dbl., 20 petals, 4–5 in., cupped, borne 1–10 per cluster, repeat bloom, slight fragrance; awl-like, tan prickles; foliage leathery, dark green, tinted copper; vigorous, erect, bushy, spreading growth; hardy.;

[Countryman X Carefree Beauty]; Buck, Dr Griffith J.; Iowa State University

'**Prairie Star**', S, ly, 1975; bud ovoid, pointed; flowers pale chrome-yellow, tinted pink, dbl., 54 petals, 3.5–4 in., cupped, repeat bloom, moderate, green apple fragrance; foliage dark, leathery; vigorous, erect, bushy growth; [Tickled Pink X Prairie Princess]; Buck, Dr Griffith J.; Iowa State University

'**Prairie Sunset**', S, yb, 1984; flowers deep pink, reverse yellow, urn-shaped, blooms borne 5–10 per, dbl., 38 petals, 4 in., repeat bloom, moderate fragrance; tan prickles; foliage moderately large, leathery, dark; erect, bushy, slightly spreading growth; hardy.; [Bonfire Night X (Music Maker X Athlone)]; Buck, Dr Griffith J.; Iowa State University

'**Prairie Valor**', S, mr, 1984; dbl., 38 petals, 4–5 in., cupped, borne 3–5 per cluster, repeat bloom, moderate, damask fragrance; awl-like, brown prickles; foliage leathery, dark; upright, bushy growth; hardy.; [((Dornroschen X (Josef Rothmund X R. laxa)) X Rose du Roi a Fleurs Pourpres) X (Music Maker X Topsi)]; Buck, Dr Griffith J.; Iowa State University

'**Prairie Wren**', S, mp, 1946; flowers rich pink, large, semi-dbl., heavy, nonrecurrent bloom; very hardy.; Morden Exp. Farm

'**Prairie Youth**', S, op, 1948; flowers pure salmon-pink, blooms in clusters, semi-dbl., repeat bloom, slight fragrance; vigorous (6 ft) growth; completely hardy on prairies.; [((Ross Rambler X Dr W. Van Fleet) X R. suffulta) X ((Dr W. Van Fleet X Turkes Rugosa Samling) X R. spinosissima altaica)]; Morden Exp. Farm

'**Praise of Jiro**', F, or, 1959; bud ovoid; flowers large, borne in clusters, dbl., 30 petals, slight fragrance; foliage leathery; vigorous, upright, bushy growth; [Korona X Spartan]; Kordes, R.

'**Praline**', HT, dr, 1955; flowers carmine-purple to crimson, slight fragrance; foliage glossy; [Camelia X Unnamed seedling]; Robichon

'**Prarie Sunrise**', S, ab, 1997; flowers very full, very dbl., 41 petals, borne in small clusters, intense fragrance; foliage large, medium green, glossy; glossy, compact, bushy, medium growth; [Friesia X Freckle Face]; Buck, Dr Griffith J.

'**Präsent**', F, ab, 1969; (Present); flowers yellow-orange, reverse yellow, large, dbl., 27 petals, exhibition form, moderate fragrance; foliage dark, glossy, leathery; vigorous, upright, bushy growth; [Highlight X Allgold]; Haenchen, E.; Teschendorff

'**Präsident Dr H.C. Schröder**', HT, mr, 1959; flowers velvety deep red, large, dbl., strong stems; [New Yorker X Seedling]; Kordes, R.

'**Präsident Hindenburg**', Pol, mr, 1927; flowers carmine to deep red, base white, dbl., [Greta Kluis X ?]; Bom; P. Lambert

Praterstern®, HT, yb

Prato Rosso, Min, 1984; Motta; (Cavriglia)

'**Prattigosa**', HRg, mp, 1953; bud long, pointed, red; flowers pink, overlarge, single, slight fragrance; foliage leathery, light green; vigorous ($3\frac{1}{2}$ ft), upright, bushy growth; [R. prattii X R. rugosa alba]; Kordes

Pray, Gr, 1987; Delforge; (Cavriglia)

'**Precilla**', HT, dy, 1973; bud ovoid; flowers deep golden yellow, medium, dbl., cupped, moderate fragrance; foliage dark, leathery; vigorous, upright growth; [Peer Gynt X Seedling]; Kordes

Preciosa, HT, pb, 1972; flowers pink and yellow, large, dbl., Meilland; (Sangerhausen)

'**Precious**', HT, pb, 1985; flowers blend of light, medium and deep pink, reverse medium pink, l, dbl., 35 petals, exhibition form, moderate fragrance; very few prickles; foliage medium, medium green, semi-glossy; upright, bushy growth; [(Briarcliff X Carla seedling) X Pink Parfait]; Dawson, Charles P.

Precious Child, *see* 'GEAaura'

Precious Gift™, *see* 'JAYsil'

Precious Michelle, HT, lp

Precious Moments, *see* 'LYOpr'

'**Precious Platinum**', HT, mr, 1974; (Opa Pötschke, Red Star); flowers cardinal-red, medium to large, dbl., exhibition form, slight fragrance; foliage glossy, leathery; [Red Planet X Franklin Englemann]; Dicksons of Hawlmark

'**Précoce**', M, dp, 1843; flowers rosy red, sometimes spotted, medium, dbl., very early bloom; Vibert

'**Prediction**', HT, mp, 1974; flowers luminous pink, dbl., 35 petals, 4 in., exhibition form, slight fragrance; foliage leathery; moderate growth; [Queen of Bermuda X Golden Giant]; Golik; Dynarose

Preference, *see* 'MEImagarmic'

Preference, *see* MEIgrouge

'**Prefet Limbourg**', HP, dr, 1878; flowers large, dbl., slight fragrance; Margottin; (Sangerhausen)

Prekrasnaja Rossijanka, HT, mr, 1966; flowers large, dbl., intense fragrance; Klimenko, V. N.; (Sangerhausen)

Prelud, F, lp; Urban, J., 1989; (Czech Rosa Club)

Prelude, *see* '**Sugar Plum**'

Prelude, *see* KEImove

'**Prélude**', HT, m, 1954; flowers lilac-mauve, well formed, medium, dbl., 25 petals, moderate fragrance; foliage dense; strong stems; vigorous, bushy growth; [Fantastique X (Ampere X (Charles P. Kilham X Capucine Chambard))]; Meilland, F.; URS

Preludium, F, lp; [Poker Chip X Unknown]; Hannemann, F., 1989; The Rose Paradise; (Weatherly, L.)

'**Prema**', F, pb, 1970; flowers soft pink, petals edged deep pink, medium, dbl., exhibition form, foliage leathery; vigorous, upright growth; [Sea Pearl X Shola]; Division of Vegetable Crops and Floriculture

'**Premier**', HT, mr, 1918; bud long, pointed; flowers dark velvety rose-rev, veined darker, reverse lighter, open, dbl., intense fragrance; foliage rich green, leathery; vigorous growth; [Ophelia seedling X Mrs Charles E. Russell]; Hill, E. G., Co.

Premier Amour, *see* '**First Love**'

'**Premier Bal**', HT, w, 1955; flowers ivory edged cyclamen-rose, picotee effect, dbl., 45 petals, intense fragrance; vigorous, bushy growth; [(Fantastique X Caprice) X Peace]; Meilland, F.; URS

'**Premier Bal**', HT, w, 1950; flowers ivory edged carmine, dbl., 30–40 petals, cupped, intense fragrance; NOW DISCARDED; [Fantastique X Caprice]; Meilland, F.

'**Premier Supreme**', HT, dp, 1927; bud long, pointed; flowers deep rose-pink, almost scarlet, very large, dbl., exhibition form, moderate fragrance; foliage dark, leathery; long stems; very vigorous growth; [Premier sport]; Zieger

'**Premier, Climbing**', Cl HT, mr, 1927; Vestal

Première Ballerine, *see* '**Prima Ballerina**'®

Présence, *see* 'DELprat'

Present, *see* '**Präsent**'

'**Présent Filial**', HT, pb, 1956; bud long; flowers old rose tinted coppery yellow, center salmon-pink, reverse, very dbl., 85 petals; foliage dense; [Verschuren's Pink X Unnamed seedling]; Delbard-Chabert

Preservation, *see* 'BOSiljurika'

Preservation, *see* '**MISiljurika**'

President, *see* '**Adam**'

'**President**', T, 1860; Paul; (Cavriglia)

President Armand Zinsch, *see* DELzinsch

'**President Boone**', HT, dr, 1936; bud long, pointed; flowers scarlet-crimson, large, dbl., globular, intense fragrance; foliage leathery; stems sometimes weak; vigorous growth; [Unnamed seedling X Miss Rowena Thom]; Howard, F.H.; H&S

'**Président Bouché**', HT, or, 1917; flowers coral-red, shaded carmine, large to medium; vigorous growth; [Unnamed seedling X Lyon Rose]; Pernet-Ducher

'**Président Bray**', HT, op, 1954; flowers salmon-pink veined pink, strong stems; vigorous growth; Privat

'PRÉSIDENT BRIAND', HP, op, 1929; flowers pink suffused salmon, overlarge, dbl., globular, moderate fragrance; foliage wrinkled; strong stems; vigorous, compact growth; [(Unnamed Frau Karl Druschki seedling X Lyon Rose) X (Frau KarlDruschki X Willowmere)]; Mallerin, C.; C-P

Président Charles Hain, *see* **'AMELIA EARHART'**

Président Chaussé, *see* **'MARK SULLIVAN'**

'PRÉSIDENT CHÉRIOUX', HT, rb, 1923; flowers red and salmon-pink, shaded yellow, dbl., slight fragrance; GM, Bagatelle, 1923; Pernet-Ducher

'PRÉSIDENT COCHET-COCHET', HT, dr, 1937; flowers deep garnet-red, tinted scarlet, very large, dbl., slight fragrance; foliage leathery; very vigorous growth; [Grenoble X Unnamed seedling]; Mallerin, C.

'PRESIDENT COOLIDGE', LCl, dr, 1925; flowers glowing crimson, dbl., very hardy; [(R. setigera X R. wichurana) X Château de Clos Vougeot]; Horvath

President Cosnier, HT, mp, 1994; Orard

'PRÉSIDENT DE LA ROCHTERIE', B, m, 1891; flowers dark purple/red, very large, dbl., intense fragrance; Vigneron; (Sangerhausen)

'PRÉSIDENT DE SÈZE', HGal, m; (Mme Hébert); flowers magenta center, paler at edges, cup-shaped, becoming convex, very dbl.; Hébert, Mme., 1828

'PRÉSIDENT DEVILLE', HT, mr, 1929; dbl., moderate fragrance; GM, Bagatelle, 1929; [Fritz Maydt X Mme J. W. Budde]; Leenders, M.

'PRÉSIDENT DUHEM', Pol, mr, 1930; flowers bright red, dbl.; foliage dark; vigorous growth; Reymond

'PRÉSIDENT DUTAILLY', HGal, m, 1888; flowers reddish purple, dbl., intense fragrance; Dubreuil

'PRESIDENT EISENHOWER', HT, mr, 1953; bud ovoid; flowers rose-red, dbl., 35–40 petals, 4–5 in., exhibition form, intense fragrance; foliage dark, leathery; vigorous, bushy growth; [Unnamed seedling (Captivator X Red Delicious) sport]; Hill, Joseph H., Co.; C-P

'PRESIDENT F.A. DES TOMBE', HT, ab, 1925; flowers apricot on golden yellow ground, reverse peach, dbl., moderate fragrance; [Mr Joh. M. Jolles X Golden Emblem]; Van Rossem

'PRÉSIDENT FÉRIER', HT, op, 1938; flowers reddish coppery pink, base tinted yellow, very dbl., intense fragrance; foliage bright green; very vigorous growth; Gaujard

'PRESIDENT FRANKLIN D. ROOSEVELT', HT, or, 1933; flowers velvety scarlet, large, dbl., 35–40 petals, moderate fragrance; [Templar sport]; Traendly & Schenck; S. Reynolds

President Gaupin, HT, dr

President Gausen, B, mr, 1862; Pradel

'PRÉSIDENT GEORGES FEUILLET', LCl, yb, 1954; flowers sulfur-yellow edged vermilion, reverse saffron, semi-dbl., Vially

President Heldar Aliyev, *see* 'COCosimber'

'PRÉSIDENT HENRI QUEUILLE', HT, mr, 1952; flowers carmine-red, very large, dbl., intense fragrance; very vigorous, upright growth; [Rome Glory X Unnamed seedling]; Gaujard

'PRESIDENT HERBERT HOOVER', HT, pb, 1930; bud long, pointed; flowers orange, rose and gold, reverse lighter, large, dbl., 25 petals, moderate, spicy fragrance; foliage leathery; vigorous, tall growth; Gertrude M. Hubbard, ARS, 1934 John Cook Medal, ARS, 1935; [Sensation X Souv. de Claudius Pernet]; Coddington; Totty

'PRESIDENT HERBERT HOOVER, CLIMBING', Cl HT, pb, 1937; Cant, B. R.

'PRESIDENT HOOVER, CLIMBING', Cl HT, pb, 1931; Dixie Rose Nursery

'PRÉSIDENT J.B. CROIBIER', HT, ob, 1940; bud well shaped, pure orange; flowers dark orange, reverse yellow, foliage light green; very vigorous growth; [Mrs Pierre S. duPont X Talisman]; Colombier

'PRESIDENT JAC. SMITS', HT, dr, 1928; (President Jacob Smith); bud long, pointed; flowers brilliant dark red, large, semi-dbl., moderate fragrance; foliage bronze, leathery; very vigorous growth; [Étoile de Hollande X K. of K.]; Verschuren; Dreer

President Jacob Smith, *see* **'PRESIDENT JAC. SMITS'**

President John F. Kennedy, *see* **'JOHN F. KENNEDY'**

President Kekkenen, F, or, 1983; de Ruiter, G.; (Cavriglia)

President L. Senghor®, *see* 'MEIluminac'

Président Leopold Senghor®, *see* 'MEIluminac'

President Leopold Senghor, Climbing, *see* MEIluminacsar

'PRESIDENT LINCOLN', HP, dr; flowers dark red shaded crimson, large, dbl., Granger, 1863

'PRESIDENT MACIA', HT, lp, 1933; (P. M. Leenders); flowers light pink, darker veining, dbl., 25 petals, 6 in., intense fragrance; foliage dark; vigorous, bushy growth; [(Ophelia X Gloire de Hollande) X (Ophelia X Sensation)]; Leenders, M.

'PRÉSIDENT MAGNAUD', Ch, dr, 1916; flowers small, semi-dbl.; Nabonnand, C.

'PRÉSIDENT MOREL JOURNEL', HT, rb, 1934; flowers scarlet, reverse yellow, large, cupped, moderate fragrance; foliage bronze; vigorous growth; [Mrs Edward Powell X R. foetida bicolor hybrid]; Chambard, C.

Président Nomblot, *see* **'HORACE MCFARLAND'**

'PRÉSIDENT PACAUD', HT, yb, 1946; flowers ochre-yellow shaded dark carmine-red and copper, well formed, dbl.; foliage glossy; [Mme Joseph Perraud X Unnamed seedling]; Sauvageot, H.; Sauvageot

'PRÉSIDENT PARMENTIER', HT, ab, 1926; flowers apricot-pink, dbl., slight fragrance; [Col. Leclerc X Le Progres]; Sauvageot, H.; Cochet-Cochet

President Pats, *see* **'STAATSPRÄSIDENT PÄTS'**

'PRÉSIDENT PAUL MARTIN', HT, my, 1942; flowers purplish carmine-red, large, semi-dbl., vigorous growth; [Charles K. Douglas X Souv. de Georges Pernet]; Moulin-Epinay; Hamonière

'PRÉSIDENT PAULMIER', Pol, dr, 1932; flowers pure blood-red, passing to garnet, well formed, borne in clusters, very dbl.; dwarf growth; Turbat

'PRÉSIDENT PLUMECOCQ', HT, yb, 1931; (Director Plumecock); flowers coppery buff and deep salmon, large, dbl., 34 petals, cupped, moderate, fruity fragrance; Gaujard; C-P

'PRÉSIDENT POINCARÉ', HT, 1920; flowers reddish magenta, center crimson, shaded yellow, reverse brighter, dbl., moderate fragrance; Grandes Roseraies

'PRÉSIDENT SCHLACHTER', HP, m, 1877; flowers purple/violet, large, dbl., Verdier, E.; (Sangerhausen)

'PRÉSIDENT SEIZE', F, dp, 1958; bud long; flowers light red becoming darker, semi-dbl., 10–15 petals, 4 in.; foliage dark; strong stems; upright growth; Delbard-Chabert

President Smith, *see* **'WILLIAM R. SMITH'**

'PRESIDENT SONO', HT, ly, 1972; bud pointed; flowers cream-yellow, dbl., 45–50 petals, exhibition form, intense fragrance; foliage dark; very large (5–6 in.) upright growth; [Burnaby X Montparnasse]; Kikuchi, Rikichi

President Souzy, *see* 'PEKomegir'

President Taft, *see* **'LEUCHTFEUER'**

'PRÉSIDENT VAN OOST', HT, my, 1934; flowers golden yellow, edged deeper, large, dbl.; foliage glossy; long, strong stems; vigorous growth; [Souv. de Claudius Pernet X Ville de Paris]; Lens

President Viard, HT, mr

'PRÉSIDENT VIGNET', HT, dr, 1911; flowers deep carmine-red, large, dbl., Pernet-Ducher

'PRÉSIDENT VIGNET, CLIMBING', Cl HT, dr, 1942; flowers large, dbl., slight fragrance; Vogel, M.; (Sangerhausen)

'PRESIDENT W.H. TAFT', HT, op, 1908; bud long, pointed; flowers salmon-pink, large, dbl., exhibition form, moderate fragrance;

foliage glossy; long, strong stems; vigorous growth; McCullough

President Willermoz, HP, mp; flowers birght rose, Ducher, 1867

'**PRESIDENT WILSON**', HT, op, 1918; flowers shrimp-pink, dbl.; Easlea

President Wilson, *see* ORAwest

President Wm. R. Smith, *see* **'WILLIAM R. SMITH'**

Presidente, HT, or; Croix

'**PRESIDENTE CARMONA**', HT, dr, 1937; flowers blackish crimson, shaded salmon, large, dbl., cupped, intense fragrance; foliage soft; vigorous, bushy growth; [Hortulanus Budde X Château de Clos Vougeot]; da Silva, Moreira

'**PRESIDENTE CRAVEIRO LOPES**', HT, rb, 1954; flowers cherry, reverse yellow, medium, dbl., 33 petals, moderate fragrance; foliage dark; strong stems; very vigorous, upright growth; [Sirena X Peace]; da Silva, Moreira

'**PRESIDENTIAL**', Gr, lp, 1960; bud long, pointed; flowers light crimson, reverse china-rose, dbl., 5.5 in., cupped, slight fragrance; foliage leathery, glossy; vigorous, tall growth; [Charlotte Armstrong X (Charlotte Armstrong X Floradora)]; Lammerts, Dr Walter; Germain's

'**PRESQUE PARTOUT**', M, dp; flowers rose, medium, dbl., cupped; vigorous growth

Press and Journal, *see* COClion

Pressin, HT, 1975; Dorieux; (Cavriglia)

Prestance, HT, 1985; Dorieux; (Cavriglia)

'**PRESTATYN ROVER**', HT, op, 1929; flowers salmon-pink, reverse darker, dbl., slight fragrance; [Alfred Colomb X Mrs Wemyss Quin]; Lavender

'**PRESTIGE**', S, mr, 1957; flowers light crimson, large, semi-dbl., recurrent bloom; foliage dense, dark; vigorous, bushy growth; [Rudolph Timm X Fanal]; Kordes, R.; Morse

Prestige de Bellegarde, F, mr, 1992; Eve, A.

Prestige de Lyon, *see* 'MEINIMO'

Prestige de Seine-et-Marne, F, yb, 1992; Eve, A.

'**PRESUMIDA**', Min, yb, 1948; (Baby Talisman, La Presumida, Peter Pan); flowers pumpkin-yellow to white, center yellowish, small, dbl.; dwarf growth; [Eduardo Toda X Pompon de Paris]; Dot, Pedro

'**PRETORIA**', HT, or, 1953; bud long; flowers copper-red to orange, large; very vigorous growth; GM, Rome, 1953; Moro

Pretoria, *see* KORhagon

'**PRETTY BABY**', Min, pb, 1982; flowers light pink, reverse creamy yellow, medium, dbl., 35 petals, exhibition form, slight fragrance; foliage small, medium green, semi-glossy; bushy, spreading growth; [Baby Katie X Unnamed Miniature seedling]; Jolly, Betty J.; Rosehill Farm

Pretty Celine, *see* 'RENCEL'

Pretty Girl, *see* **'MISS FRANCE'**

Pretty in Pink(TM), *see* 'DICUMPTEEN'

Pretty Jessica, *see* 'AUSJESS'

Pretty Lady, *see* 'SCRIVO'

Pretty 'n' Single, *see* 'UMSPRETTY'

Pretty Penny, *see* 'MORPREPEN'

'**PRETTY PINK**', LCl, dp, 1968; flowers deep pink, open, medium, dbl., recurrent bloom, intense fragrance; foliage glossy; very vigorous, climbing growth; [New Dawn X Spartan]; Patterson; Patterson Roses

Pretty Pink, HMsk, pb, 1992; Lens

Pretty Pink, *see* BARprett

'**PRETTY PLEASE**', Min, lp, 1985; flowers urn-shaped, small blooms borne usually singly, dbl., 25 petals, no fragrance; very thin, long, straight, light red prickles; foliage medium, light green, semi-glossy; medium, bushy growth; [Bride's White X Unnamed Miniature seedling]; Epperson, Richard G.

Pretty Polly, *see* 'MEITONJE'

'**PRETTY POLY**', F, mr, 1954; flowers cherry-red, camellia shaped, borne in clusters, flat; vigorous, upright growth; [Mme Butterfly X Our Princess]; Bishop; Baker's Nursery

Pretty Tiny, *see* 'RENPINY'

Pretty Woman, *see* 'MINABDCO'

Pretty Woman, *see* KEBesa

'**PREUSSEN**', HT, dr, 1920; flowers glowing dark blood-red, dbl., intense fragrance; [Farbenkonigin X Richmond]; Löbner; Kordes

Prevalent, F; Verbeek; (Cavriglia)

Preview, *see* MEIspola

'**PREVUE**', S, w, 1978; flowers medium, semi-dbl., no fragrance; foliage medium, medium green, semi-glossy; bushy growth; [(Tausendschon X (Perle d'Or X Old China)) X Safrano]; James, John

Preziosa, *see* 'MEIHIMPER'

Pride Meidiland, *see* MEIrumour

Pride 'n' Joy(TM), *see* 'JACMO'

'**PRIDE OF CANADA**', HT, dp, 1981; bud long, ovoid; flowers deep pink, blooms borne in pairs, dbl., 34 petals, exhibition form, moderate fragrance; small, brown prickles; foliage dark; tall growth; [Ena Harkness X Charlotte Armstrong]; Collins; Pan American Bulb Co.

Pride of Daylesford, Hmult, lp; Huxley, Ian, 1994; (Weatherly, L.)

Pride of England, *see* HARencore

'**PRIDE OF HURST**', Pol, op, 1926; flowers coral-pink, small, very dbl.; [Coral Cluster sport]; Hicks

Pride of Ichalkaranji, HT, yb, 1989; Patil, B.K.

'**PRIDE OF LEICESTER**', HT, pb, 1960; flowers rose-pink, base yellow, large, dbl., exhibition form, intense fragrance; foliage light green; vigorous growth; [R.M.S. Queen Mary X Seedling]; Verschuren; Blaby Rose Gardens

Pride of Maldon, *see* 'HARWONDER'

Pride of Mountbarker, *see* 'TOMMOUNT'

Pride of Nagpur, HT, dr, 1983; Datt, Braham

'**PRIDE OF NEW CASTLE**', HT, dr, 1930; flowers deep velvety crimson, large, dbl., intense fragrance; vigorous, bushy growth; [Hoosier X Beauty unnamed seedling]; Hill, E.G., Co.; Heller Bros.

'**PRIDE OF NEWARK**', F, lp, 1966; flowers shell-pink, dbl., 4 in., cupped, intense fragrance; foliage glossy, bronze, leathery; very vigorous, upright, compact growth; [Joanna Hill X The Fairy]; Morey, Dr Dennison; Country Garden Nursery

'**PRIDE OF OAKLAND**', F, mp, 1977; bud pointed; flowers rosette form, dbl., 26 petals, 2.5 in., moderate, spicy fragrance; upright, spreading growth; [Pinocchio X China Doll]; Lindquist; Howard of Hemet

'**PRIDE OF PACIFIC**', F, op, 1957; flowers salmon-pink, borne in clusters of 4–7, dbl.; symmetrical growth; [Pinocchio X Maxine]; Silva

Pride of Park, F, ob

'**PRIDE OF REIGATE**', HP, rb, 1884; flowers carmine, striped and mottled white; [Comtesse d'Oxford sport]; Brown, J.

'**PRIDE OF REIGATE, CLIMBING**', Cl HP, rb, 1941; flowers red with white striping, large, dbl., moderate fragrance; Vogel, M.; (Sangerhausen)

'**PRIDE OF RUNCORN**', HT, lp; Williams, A., 1901; (Weatherly, L.)

'**PRIDE OF SUNNYBANK**', Gr, mp, 1957; [Ma Perkins X Charlotte Armstrong]; Ulrick, L.W.

'**PRIDE OF WALTHAM**', HP, mp, 1881; flowers silvery rose, large, dbl., moderate fragrance; vigorous growth; [Comtesse d'Oxford sport]; Paul, W.

Pride of Wansbeck, *see* 'NOSWAN'

'**PRIDE OF WASHINGTON**', HSet, m, 1849; flowers rosy violet, very dbl.; Pierce

Pride of Nagpur, HT, dr

Prima, *see* 'HARWANTED'

'**PRIMA BALLERINA**'®, HT, dp, 1957; (Première Ballerine, Primaballerina®); bud long, pointed; flowers cherry-pink, medium to large, dbl., 20 petals, intense fragrance; foliage leathery, light green; [? X Peace]; Tantau, Math.

Prima Ballerina, *see* **'Solistka Baleta'**

'Prima Donna'™, Gr, dp, 1984; (Toboné); flowers deep fuchsia pink, large, dbl., 27 petals, exhibition form, slight fragrance; foliage large, medium green, semi-glossy; bushy, spreading growth; AARS, 1988; [(Unnamed seedling X Happiness) X Prominent]; Shirakawa, Takeshi; Tosh Nakashima, 1983

'Prima Donna', HT, mp, 1944; bud pointed; flowers rich deep salmon-pink tinted buff, dbl., 45 petals, 5 in., exhibition form, slight fragrance; foliage glossy; vigorous, upright, open growth; RULED EXTINCT 1/85; [Heinrich Wendland X Unnamed seedling]; Dickson, A.; J&P

Primaballerina®, *see* **'Prima Ballerina'**®

'Primarosa', HT, dp, 1950; bud long; flowers carmine, base yellow, streaked orange, very large; strong stems; Giacomasso

'Primavera', HT, op, 1936; flowers salmon-pink, well formed, dbl., moderate, musk fragrance; vigorous growth; [Julien Potin X Sensation]; Aicardi, D.; Robichon

'Prime Time', Min, my, 1984; flowers medium, dbl., 35 petals, exhibition form, slight fragrance; foliage small, medium green, semi-glossy; upright, bushy growth; [Picnic X Rise 'n' Shine]; Hardgrove, Donald & Mary; Rose World Originals

'Primerose', HT, my, 1913; flowers melon-yellow, deeper in autumn, shaded apricot, dbl., moderate fragrance; [Mme Mélanie Soupert X Mrs Peter Blair]; Soupert & Notting

'Primerose Sistau', Pol, pb, 1925; flowers carmine, shaded yellow, medium, borne in clusters of 5–6, cupped, few thorns; half dwarf growth; Turbat

Primevere, LCl, yb

'Primevère', LCl, yb, 1929; (Primrose); flowers primrose-yellow to canary-yellow, large blooms in clusters, dbl., non-recurrent, slight fragrance; foliage rich green, glossy; long stems; very vigorous, climbing and trailer growth; [R. wichurana X Constance]; Barbier; Dreer, 1930

Primo Passo, *see* BARprimo

'Primo Premio', HT, pb, 1964; flowers rose suffused yellow, well formed, medium; Giacomasso

Primo Sole®, *see* BARpris

Primrose, *see* **'Primevère'**

'Primrose Bedder', F, my, 1956; flowers primrose-yellow, borne in large trusses, dbl., slight fragrance; foliage light green; Kordes; Morse

'Primrose Pet', Pol, lp; Williams, A., 1913; (Weatherly, L.)

Primrose Queen, HT, my

Primrose Sistau, Pol, mp

Primula, HT, 1956; Aicardi; (Cavriglia)

Prince Abricot, *see* 'AROsnap'

'Prince Albert', HP, rb, 1841; flowers carmine-rose changing to dark crimson, large, globular, Laffay, M.

Prince Albert, B, lp; Laffay, M., 1852

'Prince Arthur', HP, mr, 1875; (Triomphe de Caen); flowers deep crimson, medium, dbl., 55 petals, intense fragrance; vigorous growth; [Général Jacqueminot X ?]; Cant, B. R.

'Prince Bernhard', HT, mr, 1937; bud pointed; flowers shining red, shaded strawberry-red, well formed, large, dbl., intense fragrance; foliage dark; vigorous growth; GM, Bagatelle, 1937; [Matador seedling]; Van Rossem; J&P, 1941

'Prince Camille de Rohan', HP, dr, 1861; (La Rosière, Climbing); flowers very deep velvety crimson-maroon, well-formed, large blooms, very dbl., cupped, sometimes recurrent bloom, intense fragrance; rather weak stems; vigorous, upright growth; [Possibly Général Jacqueminot X Geant des Batailles hybrid]; Verdier, E.

'Prince Charles', B, m, 1842; flowers deep red-purple, fading to lavender, veined magenta, base of, semi-dbl., non-recurrent, intense fragrance; very few prickles; foliage large, dark; vigorous (to 5 ft) growth

Prince Charles, Gr, dr, 1992

Prince Charles d'Aremberg, HP, dp, 1877; Soupert & Notting

'Prince Charles du Luxembourg', Ch, dp; flowers bright carmine, cupped

'Prince Charlie', HT, op, 1932; bud long, pointed; flowers coral-pink, overlaid salmon, base orange, large, dbl., cupped, moderate fragrance; foliage leathery, bronze tinted; vigorous, bushy growth; Dobbie

'Prince Charming', Min, mr, 1953; flowers bright crimson, dbl., 1 in.; foliage tinted red; dwarf (8–12 in.) growth; [Ellen Poulsen X Tom Thumb]; deVink; T. Robinson, Ltd.

'Prince Charming', S, lp, 1958; flowers pale blush-pink, dbl., prolonged bloom, intense fragrance; upright ($2\frac{1}{2}$ ft), bushy growth; Skinner

'Prince Damask', HT, dr, 1954; flowers maroon, medium, dbl., 30 petals, intense fragrance; foliage small, dark; [Guinee X ?]; LeGrice

'Prince de Bulgarie', HT, lp, 1900; bud long, pointed; flowers silvery flesh, center deeper, shaded salmon and saffron-yellow, dbl., cupped, moderate fragrance; foliage bright green; vigorous growth; Pernet-Ducher

Prince de Monaco, *see* 'POUlart'

'Prince de Porcia', HP, dr, 1865; flowers large, dbl.; Verdier, E.; (Sangerhausen)

'Prince de Vaudement', M, 1854; Moreau et Robert; (Cavriglia)

'Prince Engelbert Charles d'Arenberg', HT, or, 1909; flowers scarlet, dbl., moderate fragrance; [Richmond X Étoile de France]; Soupert & Notting

Prince Eugène, *see* **'Eugène de Beauharnais'**

Prince Eugene de Beauharnais, HP, dr, 1864; Moreau et Robert

'Prince Félix de Luxembourg', HT, mr, 1930; flowers carmine-red, shaded purplish, large, dbl., 30–35 petals, moderate fragrance; foliage dark; very vigorous, bushy growth; [Gen. MacArthur X George C. Waud]; Ketten Bros.

Prince Frédéric, *see* **'Prince Frederick'**

'Prince Frederick', HGal, m, 1840; (Prince Frédéric); flowers large, very dbl.; Parmentier; (Sangerhausen)

'Prince Henri', F, mr, 1954; flowers large, dbl.; Reuter, A.; (Sangerhausen)

'Prince Henri d'Orléans', HP, dp, 1886; flowers large, dbl.; Verdier, E.; (Sangerhausen)

'Prince Henry', F, mp, 1926; bud long, pointed; flowers bright pink, passing to blush-pink, clusters of 18–20; [St Helena X Unnamed seedling]; Easlea

Prince Igor®, *see* 'MEIhigor'

Prince Igor, Climbing, *see* MEIgosar

'Prince Jean de Luxembourg', Pol, w, 1926; flowers pure white, very small, borne in immense clusters of 100–150, dbl.; moderately dwarf growth; [Jeanny Soupert X Miniature]; Soupert & Notting

Prince Meillandina®, *see* 'MEIrutral'

Prince Napoleon, B, pb, 1864; Pernet

'Prince Noir', HP, m, 1854; flowers dark velvety crimson-purple, not, very dbl.; moderate growth; Boyau

'Prince of Denmark', HT, mp, 1964; flowers rose-pink, dbl., 4 in.; free growth; [Queen Elizabeth X Independence]; McGredy, Sam IV; Fisons Horticulture

Prince of Orange, HT, ob; Williams, A., 1900

Prince of Peace, *see* 'JACpop'

'Prince of Wales', HT, mr, 1921; bud long, pointed; flowers brilliant cherry-scarlet, semi-dbl., intense fragrance; Easlea

Prince Orange, F, op

Prince Palace, *see* POUlzin

Prince Philip, *see* **'Tudor Prince'**

Prince Regent, *see* GENpen

Prince Sunblaze, *see* 'MEIrutral'

Prince Tango®, *see* 'DELgus'

Prince Wasiltchikoff, *see* **'DUCHESS OF EDINBURGH'**

'PRINCE YUGALA', HT, dr, 1923; bud long, pointed; flowers deep velvety maroon, dbl., slight fragrance; Cant, F.

'PRINCEPS', LCl, mr, 1942; flowers very large, non-recurrent, moderate fragrance; pillar growth; Clark, A.

Princesa Carmen de Borbon, F, 1979; Dot, Simon; (Cavriglia)

Princesa Sofia, HT, 1972; Dot, Simon; (Cavriglia)

Princesita, *see* **'PIXIE'**

Princess®, *see* 'INTERPRINCE'

'PRINCESS ALEXANDRA', F, yb, 1962; flowers creamy yellow suffused carmine-pink, medium, borne in trusse, dbl., 30–35 petals, moderate fragrance; foliage dark, glossy; very vigorous growth; [Masquerade X Seedling]; Cobley

Princess Alexandra, *see* POUldra

Princess Alice, *see* 'HARTANNA'

'PRINCESS ANGELA', HT, lp, 1991; flowers large blooms borne mostly singly, very dbl., slight fragrance; foliage medium, dark green, semi-glossy; tall, upright, bushy growth; [Pristine X Granada]; Alde, Robert O.

'PRINCESS ANGELINE', HT, mp, 1945; flowers pink, becoming open, dbl., 35–50 petals, 4.5–5.5 in., cupped, intense, damask fragrance; foliage leathery; vigorous, tall growth; [Charlotte Armstrong X Times Square]; Swim, H.C.; Peterson & Dering

'PRINCESS BONNIE', HT, dr; flowers vivid crimson, [Bon Silène X William Francis Bennett]; Dingee & Conard, 1895

'PRINCESS CHICHIBU', F, pb, 1971; dbl., 30 petals, 2.5–3 in., slight fragrance; foliage glossy, dark; [(Vera Dalton X Highlight) X Merlin]; Harkness

'PRINCESS ELIZABETH OF GREECE', HT, my, 1926; flowers golden yellow shaded terracotta, well shaped, exhibition form, moderate fragrance; Chaplin Bros.

'PRINCESS FAIR', HT, ab, 1962; flowers light apricot, large, dbl., cupped, moderate fragrance; foliage glossy; vigorous, upright growth; [Queen Elizabeth X (Crimson Glory, Climbing X Happiness)]; Moren; J&P

Princess Grace, *see* 'MEIMAGARMIC'

'PRINCESS HOHENZOLLERN', T, rb; flowers peach-red to crimson, well formed, very large, dbl.

Princess Margaret, HT, or; Poulsen

Princess Margaret of England, *see* 'MEILISTA'

'PRINCESS MARGARET OF ENGLAND, CLIMBING', Cl HT, mp, 1969; Meilland; URS

'PRINCESS MARGARET ROSE', HT, pb, 1933; bud long, pointed; flowers glowing pink suffused orange, large, dbl., cupped, moderate fragrance; foliage leathery; vigorous growth; [Unnamed seedling of Los Angeles parentage X Unnamed seedling]; Cant, B. R.

Princess Marianna™, *see* 'POULUSA'

'PRINCESS MARIE', HSem, 1829; Jacques; (Cavriglia)

'PRINCESS MARINA', HT, ab, 1938; flowers apricot, shaded salmon and copper, well shaped, dbl., slight fragrance; stiff stems; vigorous growth; Robinson, H.; Port Stockton Nursery

'PRINCESS MAY', HT, mp, 1893; flowers large, dbl., moderate fragrance; Paul, W.; (Sangerhausen)

Princess Michael of Kent®, *see* 'HARLIGHTLY'

'PRINCESS MICHIKO', F, ob, 1966; flowers coppery orange, yellow eye, in clusters, semi-dbl., 15 petals, 3 in., cupped; foliage glossy; bushy growth; [Circus X Spartan]; Dickson, A.

'PRINCESS MIKASA', HT, dp, 1983; flowers deep pink, urn-shaped, blooms borne singly, dbl., 45 petals, no fragrance; small, light green prickles; foliage large, medium green, semi-glossy; tall, bushy growth; [(Red Lion X Samantha) X (Red Lion X Samantha)]; Teranishi, K.; Itami Rose Nursery

'PRINCESS NAGAKA', HT, rb, 1922; flowers fiery red, shaded yellow, small, semi-dbl., slight fragrance; Pemberton

'PRINCESS OF INDIA', HT, rb, 1980; (Indian Princess); flowers outer petals deep red, inner petals carmine, large, dbl., 35 petals, exhibition form, intense fragrance; brown to gray prickles; foliage medium to large, dark, smooth; upright, bushy growth; [Tropicana X Granada]; Pal, Dr B.P.; K.S.G. Son's Roses

Princess of Monaco, *see* 'MEIMAGARMIC'

Princess of Nassau, *see* **'PRINCESSE DE NASSAU'**

Princess of Wales, HP, lp; Laxton, 1871

Princess of Wales, *see* HARdinkum

'Princess of Wales', HP, lp; flowers vivid crimson; Paul, W., 1864

'PRINCESS PEARL', F, w, 1960; flowers white center blush, borne in clusters, dbl., 40 petals, 3 in., slight fragrance; foliage leathery, dark; moderate growth; [Carol Amling sport]; Beldam Bridge Nursery

'PRINCESS ROYAL', HT, mp, 1935; flowers rose-pink to hydrangea-pink, petals shell-shaped, very large, intense fragrance; foliage dark, glossy; vigorous growth; RULED EXTINCT 4/92; Dickson, A.

Princess Royal, *see* 'DICROYAL'

'PRINCESS TAKAMATSU', HT, pb, 1974; bud globular; dbl., 50 petals, 6 in., exhibition form, slight fragrance; vigorous, upright growth; [Bonsoir X Christian Dior]; Kono, Yoshito

'PRINCESS VAN ORANGE', Cl Pol, or, 1935; (Princesse d'Orange, Prinses van Orange); [Gloria Mundi, Climbing sport]; deRuiter; J&P

Princess Verona, *see* **'PRINCESSE VERONA'**

'PRINCESS VICTORIA', HT, or, 1920; flowers glowing scarlet-crimson shaded orange, passing to carmine, b, dbl., intense fragrance; GM, NRS, 1920; McGredy

'PRINCESS WHITE', F, w, 1956; bud pointed; flowers medium, dbl., flat, moderate fragrance; foliage leathery; strong stems; vigorous, upright, bushy growth; [Unnamed orange seedling X Demure]; Boerner; J&P

'PRINCESSE', HT, or, 1964; flowers geranium-red, dbl., 45 petals, 4–5 in., globular, vigorous, upright growth; [(Peace X Magicienne) X (Independence X Radar)]; Laperrière; EFR

'PRINCESSE ADÉLAIDE', M, lp, 1845; flowers pale pink, dbl., moderate fragrance; foliage dark, often variegated; vigorous growth; Laffay, M.

Princesse Alexandra, *see* POUldra

'PRINCESSE ALICE', M, m, 1853; flowers violet-rose, not very mossy; [Luxembourg seedling]; Paul, W.

'PRINCESSE ALICE DE MONACO', T, yb, 1893; flowers cream-yellow edged pink; Weber

'PRINCESSE AMÉDÉE DE BROGLIE', HT, or, 1936; bud very long, fiery red; flowers nasturtium-red, deeper in autumn, well formed, large, foliage dark, glossy; vigorous growth; GM, Bagatelle, 1935; [Charles P. Kilham seedling X Colette Clément]; Mallerin, C.; Meilland

'PRINCESSE BACCHIOCHI', mp, 1866; flowers medium, dbl.; Moreau et Robert; (Sangerhausen)

Princesse Béatrice, *see* **'PRINSES BÉATRIX'**

'PRINCESSE DE BASSARABA DE BRANCOVAN', T, pb, 1900; flowers flesh-pink, shaded copper; Bernaix, A.

'PRINCESSE DE BÉARN', HP, mr, 1885; flowers velvety poppy-red, large, dbl., globular; vigorous growth; Lévêque

Princesse de Joinville, B, mp

Princesse de Lamballe, *see* **'PRINCESSE LAMBALLE'**

Princesse de Monaco®, *see* 'MEIMAGARMIC'

'PRINCESSE DE NASSAU', N, ly, 1835; (Princess of Nassau); bud yellowish, tinted pink; flowers creamy yellow, fading to cream, blooms in large clusters, very dbl., cupped; branches grow in zig-zag fashion growth; Laffay, M.

Princesse de Parme, M, lp; [William Lobb X Unknown]; Morley, Dr B., 1988; (Weatherly, L.)

'PRINCESSE DE SAGAN', Ch, dr, 1887; flowers crimson shaded with purple, medium, dbl., cupped; vigorous growth; Dubreuil

Princesse de Vaudemont, lp; flowers medium, dbl., moderate fragrance; Vibert, about 1825; (Sangerhausen)

Princesse des Roses, *see* **'PRINCIPESSA DELLE ROSE'**

Princesse d'Orange, *see* **'PRINCESS VAN ORANGE'**

Princesse d'Orient, HT; Poulsen; (Cavriglia)

'PRINCESSE ÉTIENNE DE CROY', T, yb, 1898; flowers violet rose, dbl.; Ketten Bros.

'PRINCESSE GHIKA', T, mr, 1922; flowers brilliant red with dark reflexes, large, dbl.; [Gén. Schablikine X Papa Gontier]; Nabonnand, P.

'PRINCESSE ITA', HT, yb, 1943; bud oval; flowers indian yellow edged red, open, medium, semi-dbl., slight fragrance; foliage leathery; vigorous, bushy growth; [Julien Potin X Charles P. Kilham]; Meilland, F.; A. Meilland

'PRINCESSE JAUNE', HT, my, 1945; flowers citron-yellow, large, dbl.; Fessel; A. Meilland

'PRINCESSE JOSÉPHINE-CHARLOTTE', Pol, op, 1945; flowers bright pink suffused apricot-salmon, small, semi-dbl., cupped, slight fragrance; very vigorous, bushy growth; [Orange Triumph sport]; Lens

'PRINCESSE L. TROUBEZKOI', HP, 1878; Lévêque; (Cavriglia)

'PRINCESSE LAMBALLE', A, w; (Princesse de Lamballe); flowers white, sometimes tinted flesh, medium, dbl.; vigorous, compact, branching growth

'PRINCESSE LILIANE', HT, mr, 1954; flowers blood-red, dbl., 25–30 petals, 5 in.; foliage dark; vigorous, upright growth; [Happiness X Hens Verschuren]; Frères, Buyl; Morse

'PRINCESSE LOUISE', HSem, w, 1829; flowers creamy white, back petals shaded with rose, large, dbl., cupped; Jacques

'PRINCESSE LOUISE', HP, w, 1869; flowers blush, medium, globular; vigorous growth; [Mme Vidot X Virginal]; Laxton; G. Paul

'PRINCESSE LOUISE', Cl HT, m, 1924; bud long, pointed; flowers rich purple, center tinted brilliant garnet, large, semi-dbl., moderate fragrance; vigorous, climbing growth; [La France de '89 X Victor Hugo]; Nabonnand, P.

Princesse Margaret d'Angleterre®, *see* 'MEILISTA'

'PRINCESSE MARGARETHA', Pol, lp, 1932; flowers pale pink, blooms in clusters, dbl., no fragrance; foliage medium, medium green, semi-glossy; medium, bushy growth; Poulsen, S.

'PRINCESSE MARIE', HSem, mp, 1829; flowers bright pink, blooms in very large clusters; Jacques

Princesse Marie Adelaide de Luxembourg, Pol, w, 1895; Soupert & Notting

'PRINCESSE MARIE CLOTILDE NAPOLÉON', Pol, w, 1924; flowers white shaded pink, base pale yellow, large, dbl.; vigorous growth; Opdebeeck

'PRINCESSE MARIE DAGMAR', T, 1919; Lévêque; (Cavriglia)

'PRINCESSE MARIE DOLGOROUKY', HP, lp, 1878; flowers light pink, often striped with white, very large, dbl.; Gonod; (Sangerhausen)

'PRINCESSE MARIE JOSÉ', Pol, w, 1924; flowers white shaded rose, base yellow, large, dbl.; vigorous growth; Opdebeeck

'PRINCESSE MARIE JOSÉ', HT, or, 1925; bud long, pointed; flowers orange-scarlet, dbl., moderate, fruity fragrance; Klettenberg-Londes; F.J. Grootendorst

'PRINCESSE MARIE-ASTRID', HT, mp, 1964; bud globular; flowers deep rose to camellia-pink, large, dbl., slight fragrance; moderate, bushy growth; [(Mme Edouard Herriot X R. rugosa rubra) X La Jolie]; Mondial Roses

'PRINCESSE MARIE-CHRISTINE', HT, dr, 1955; bud ovoid to long; flowers blood-red, dbl., 25–30 petals; vigorous, bushy growth; [Poinsettia X Ena Harkness]; Frères, Buyl

'PRINCESSE PAOLI', HT, mp, 1966; flowers bright pink, dbl., 5 in., exhibition form, intense fragrance; foliage glossy; vigorous growth; Mondial Roses

Princesse Queen, Pol, pb

'PRINCESSE STÉPHANIE', T, op, 1880; flowers salmon-yellow, large, dbl., moderate fragrance; very vigorous growth; [Gloire de Dijon seedling]; Levet, A.

'PRINCESSE STÉPHANIE DE BELGIQUE', HT, dp, 1929; bud long, pointed; flowers carmine, center deeper, dbl., intense fragrance; [Gen. MacArthur X Unnamed seedling]; Soupert & Notting; C. Soupert

'PRINCESSE VERA ORBELIONI', HT, op, 1909; flowers light salmon-pink, large, dbl.; Schwartz; (Sangerhausen)

'PRINCESSE VERONA', S, mp, 1984; (Princess Verona); dbl., 30 petals, 4 in., cupped, borne 1–15 per cluster, repeat bloom, slight fragrance; awl-like, tan prickles; foliage leathery, dark olive green; upright, bushy, compact growth; hardy.; [Verona X Prairie Princess]; Buck, Dr Griffith J.; Iowa State University

'PRINCESSE YVONNE GHIKA', HT, w, 1927; flowers white, center salmon, dbl., moderate fragrance; [Stadtrat Glaser seedling]; Mühle

'PRINCIPE DE NAPOLI', HT, my, 1937; bud long; flowers large, dbl., intense fragrance; foliage clear green; vigorous growth; Aicardi Bros.

'PRINCIPE DE PIEMONTE', HT, dr, 1929; flowers crimson-red, large, intense fragrance; [Mrs Edward Powell X Gen. MacArthur]; Giacomasso

'PRINCIPESSA DELLE ROSE', HT, mr, 1953; (Princesse des Roses); bud long, pointed; flowers lilac-rose, large, dbl., 30–40 petals, cupped, moderate fragrance; very vigorous, upright growth; [Julien Potin X Sensation]; Aicardi, D.; V. Asseretto

Prins Claus, *see* 'RUPRINS'

'PRINS HAMLET', HT, lp, 1927; bud long, pointed; flowers light pink, base yellow, open, large, semi-dbl., exhibition form, intense fragrance; foliage light, leathery; vigorous, bushy growth; [Ophelia sport]; Mohr

'PRINS WILLEM-ALEXANDER', F, ob, 1970; flowers coral-vermilion, 2–3 in.; foliage dark, red when young; bushy growth; [Tropicana X Europeana]; Verschuren

'PRINSES BÉATRIX', HT, ob, 1940; (Princesse Béatrice); bud long, pointed; flowers terracotta, shaded light apricot, large, dbl., cupped, intense, fruity fragrance; foliage leathery, bronze; vigorous, bushy growth; [Heinrich Wendland X Max Krause]; Buisman; A. Meilland; Morse

'PRINSES CHRISTINA', F, mr, 1945; bud large; flowers clear carmine-red, borne in clusters, dbl.; very vigorous growth; [Lafayette X Donald Prior]; Buisman, G.A.H.

'PRINSES JULIANA', HT, dr, 1918; bud long, pointed; flowers deep crimson-red, shaded darker, dbl., intense fragrance; foliage dark; vigorous growth; [Gen. MacArthur X Marie van Houtte]; Leenders, M.

Prinses van Orange, *see* **'PRINCESS VAN ORANGE'**

'PRINSESSE ASTRID AF NORGE', F, ob, 1958; flowers bright orange, semi-dbl.; vigorous, upright growth; [Pinocchio X Pinocchio]; Poulsen, S.

'PRINSESSE MARGRETHE', HT, op, 1963; flowers dark salmon-orange, dbl., slight fragrance; strong stems; vigorous, upright growth; [Queen Elizabeth X (Independence X Golden Scepter)]; Poulsen, S.

'PRINTEMPS', HT, pb, 1948; bud long; flowers old-rose tinted light red, reverse yellow, dbl., 4 in., moderate fragrance; foliage glossy, dark; very vigorous growth; [Trylon X Brazier]; Mallerin, C.; URS

'PRINTEMPS FLEURI', HMult, m, 1922; flowers bright purple passing to carmine-pink, stamens yellow, medium, semi-dbl., slight fragrance; [Étoile Luisante X Unnamed seedling]; Turbat

'PRINZ HAMLET', HT, lp, 1927; flowers large, dbl., moderate fragrance; Mohr; (Sangerhausen)

'PRINZ HIRZEPRINZCHEN', HMult, m, 1912; flowers medium, dbl., moderate fragrance; Geschwind, R.; (Sangerhausen)

'PRINZ MAXZU SCHAUMBURG', HT, op, 1934; flowers salmon-pink, large, dbl., intense fragrance; vigorous growth; [Frau Karl Druschki X Lyon Rose]; Schaumburg-Lippe

'PRINZESSIN BATHILDIS ZU SCHAUMBURG-LIPPE', HT, lp, 1929; flowers large, dbl., von Württemberg, Herzogin Elsa; (Sangerhausen)

'PRINZESSIN HILDEGARD', HT, my, 1917; flowers bright yellow, fading to cream-yellow, dbl., moderate fragrance; [Frau Karl Druschki X Friedrich Harms]; Lambert, P.

'PRINZESSIN IRRLIEB', F, mr, 1963; flowers velvety bright red, borne in large clusters, dbl.; moderate, symmetrical growth; Kordes, R.

'PRINZESSIN LUDWIG VON BAYERN', HMult, mp, 1911; flowers medium, dbl.; Brög; (Sangerhausen)

'PRINZESSIN M. VON ARENBERG', HT, w, 1928; flowers rosy white, shaded pale rose-pink, dbl., intense fragrance; [Ophelia X Los Angeles]; Leenders, M.

Prinzessin Maria Teresa, F, lp, 1982; flowers medium, dbl.; Lens, Louis; (Sangerhausen)

'PRINZESSIN TATIANA WASILTCHIKOFF', HT, lp, 1941; flowers flesh-pink, dbl.; long stems; upright growth; Spath

'PRIOR M. OBERTHAU', HMult, dp, 1923; flowers small, dbl.; Bruder Alfons; (Sangerhausen)

Priory Pride, *see* 'NOSTPRI'

Priory Rose, *see* 'NOSTROS'

'PRISCILLA', HT, mp, 1922; bud long, pointed; flowers pink, outer petals rose-pink, very large, dbl., cupped, moderate fragrance; foliage leathery, glossy; vigorous growth; [Unnamed seedling X Ophelia]; Montgomery Co.; A.N. Pierson

Priscilla Burton®, *see* 'MACRAT'

'PRISSY MISSY', Min, mp, 1965; flowers medium pink, reverse lighter, small, very dbl., moderate, spicy fragrance; vigorous, bushy, dwarf growth; [Spring Song X Seedling]; Williams, Ernest D.; Mini-Roses

Pristine®, *see* 'JACPICO'

Pristine Pavement, S, w, 1990; semi-dbl., moderate fragrance; foliage glossy; compact, mounding (3 ft) growth; Baum

Priti, Pol, lp, 1971; Kasturi

Priub, *see* 'UMSPRIUB'

Prive™, Gr, mp

'PRIVET IZ ALMA-ATY', HT, op, 1958; (Greetings from Alma-Aty); flowers pink tinted orange, large, dbl., 60 petals, moderate fragrance; foliage dark, glossy; very vigorous growth; [Independence X Peace]; Sushkov & Besschetnova

Priyadarshini, HT, pb, 1988; IARI

'PRIYATAMA', HT, pb, 1981; (The Beloved); bud ovoid; flowers pink, paler reverse, blooms borne singly, dbl., 35 petals, exhibition form; red prickles; foliage glossy, slightly wrinkled; bushy growth; [Inge Horstmann X Picasso]; Viraraghavan, M.S.; K.S.G. Son's Roses

'PROBERG', F, pb, 1995; (**Pink Iceberg™**); flowers medium pink blend, darkens in cool weather, dbl., 15–25 petals, slight; large, light green, glossy; upright, rounded, medium; [Iceberg sport]; Weatherly, Lila, 1998; Swane's Nursery, 1999, AUS; Week's Roses, 1999, USA

'PROBRIL', F, pb, 1999; (**Brilliant Pink Iceberg™**); flowers deep pink, reverse white, dbl., 17–25 petals, 3.5–4 in., borne in large clusters, moderate fragrance; few prickles; foliage large, light green, glossy; upright, rounded, medium (30–36 in.) growth; [Pink Iceberg sport]; Weatherly, Lila; Swane Bros. Pty. Ltd., 1999

'PROBUZENI', S, lp, 1935; flowers medium, very dbl., moderate fragrance; Böhm, J.; (Sangerhausen)

'PRODANÁ NEVESTA', S, w, 1934; flowers snow-white, center tinted yellow, borne inlarge clusters, non-recurrent, intense fragrance; foliage light, glossy; vigorous growth; Brada, Dr; Böhm

Prodigieuse, HT, 1969; Croix; (Cavriglia)

Prof O. L. Kunz, S, mp, 1980; Hetzel

Prof Reynaldo Santos, HT; Moreira da Silva, A.; (Cavriglia)

'PROF. ALFRED DUFOUR', F, mr, 1969; flowers bright red, medium, semi-dbl., slight fragrance; foliage light green; compact growth; [Paprika X Coup de Foudre]; Cazzaniga, F. G.

'PROF. BENTO CARQUEJA', HT, op, 1936; bud long, pointed; flowers pink and salmon, shaded coral-red, large, dbl., exhibition form; foliage soft; vigorous growth; [Ophelia X Mme Edouard Herriot]; da Silva, Moreira

'PROF. C.S. SARGENT', LCl, my, 1903; flowers yellow, center deeper, fading to cream, dbl., early bloom; foliage small, ornamental; vigorous growth; [R. wichurana X Souv. d'Auguste Metral]; Hoopes, Bro. & Thomas

'PROF. CHRIS BARNARD', HT, mr, 1970; flowers blood-red, pointed, dbl., 40 petals, 4 in., slight fragrance; foliage glossy; upright growth; [Ena Harkness X Karl Herbst]; Fishner, P.; Eden Rose Nursery

'PROF. COSTA LEITE', HT, my, 1955; very vigorous growth; [Peace X Julien Potin]; da Silva, Moreira

'PROF. DR VON BECK', HT, pb, 1927; flowers bright rose-pink, reverse carmine-red, dbl., moderate fragrance; [Mme Abel Chatenay X Farbenkonigin]; Ries

'PROF. LEITE PINTO', HT, mr, 1960; flowers cherry-red, center ochre; [Buccaneer X La Jolla]; da Silva, Moreira

'PROF. N.E. HANSEN', HRg, dr, 1892; flowers rich velvety red, Budd

'PROFESSEUR BÉRARD', HT, dp, 1930; flowers bright purple-carmine, base yellow, petals laciniated, dbl., intense fragrance; foliage leathery, dark; very vigorous growth; [Hadley X The Queen Alexandra Rose]; Laperrière

Professeur Christian Cabrol, HMsk, pb, 1993; Lefebvre

'PROFESSEUR DÉAUX', HT, ly, 1935; flowers light yellow streaked chamois, dbl.; foliage glossy, dark, bronze; long stems; very vigorous growth; Pernet-Ducher; Gaujard

'PROFESSEUR ÉMILE PERROT', D, mp; Brought from Persia by Prof. Perrot, 1930; Turbat

'PROFESSEUR GANIVIAT', T, pb, 1890; flowers salmon carmine, shaded carmine purple, large, very dbl.; Perrier

Professeur Jean Bernard®, *see* 'DELJABER'

'PROFESSOR BARANOV', HT, yb, 1947; flowers large, dbl.; Vogel, M.; (Sangerhausen)

Professor Boesman, F, mp, 1994; dbl.; GM, Courtrai, 1992; RvS-Melle, 1991

'PROFESSOR BORZA', HT, or, 1940; flowers medium, dbl.; Ambrosi; (Sangerhausen)

'PROFESSOR DR HANS MOLISCH', Cl HT, mp, 1923; flowers large, dbl.; Mühle; (Sangerhausen)

Professor Dr Kurth, HT, mr, 1960; flowers large, dbl.; Schmid, P.; (Sangerhausen)

'PROFESSOR ERICH MAURER', Cl HT, mp, 1939; flowers large, dbl.; Tepelmann; (Sangerhausen)

Professor Fred Ziady, *see* 'KORAMBO'

'PROFESSOR GNAU', HT, w, 1928; bud long, pointed; flowers creamy white, intense fragrance; [Oskar Cordel X Unnamed seedling]; Tantau

'PROFESSOR IBRAHIM', LCl, mp, 1937; flowers rose-pink, base yellow, large blooms in clusters of 4–5, intense fragrance; very vigorous, climbing (over 13 ft) growth; [Daisy Hill X Talisman]; Krause

Professor Knöll, HT, mr, 1964; flowers very large, very dbl., intense fragrance; Berger, W.; (Sangerhausen)

Professor Panalla, HT, 1988; Dot, Simon; (Cavriglia)

'PROFESSOR SCHMEIL', HFt, yb, 1925; flowers light orange-yellow, large, semi-dbl.; Kröger; (Sangerhausen)

Profil, HT, op, 1985; flowers coral pink, large, dbl., slight fragrance; VEG; (Sangerhausen)

Profile, *see* 'MACjilli'

Profondo Rosso®, F, dr

Profumo Romano, HT; Zandri, R.; (Cavriglia)

'**Profusion**', HT, dp, 1939; bud long, pointed; flowers carmine, base orange-yellow, dbl., moderate fragrance; vigorous growth; Dickson, A.

'**Profusion**', HT, op, 1944; bud very long; flowers orange-salmon and carmine; foliage glossy; [Mme Henri Guillot X Signora]; Meilland, F.

Progress, *see* '**Fortschritt**'

Prohadka Maje, HT; (Cavriglia)

'**Prolet**', S, pb, 1975; flowers brick, shaded deep pink, medium, very dbl., 85 petals, non-recurrent; foliage light green; vigorous, upright, branching growth; [General Stefanik X Bonn]; Staikov, Prof. Dr V.; Kalaydjiev and Chorbadjiiski

'**Prolifera de Redouté**', C, mp

'**Prolifère**', M, dp; flowers deep-rose, too full to open well at times, large, cupped; vigorous growth; (England), Prior to 1800

Prom Date, *see* 'MORprom'

Prom Night, *see* 'ZIPpro'

Promenade, *see* 'LYOpro'

'**Promethean**', S, rb, 1985; flowers medium red, reverse darker, white eye, medium, semi-dbl., 12 petals, repeat bloom, slight fragrance; foliage medium, light green, smooth, matt; bushy growth; [(Blanche Mallerin X Pink Hat) X R. multibracteata]; James, John

Prominent®, *see* 'KORp'

Promise, *see* 'JACis'

'**Promise**', Cl Ch, op, 1929; (Chin Chin); flowers salmon-pink, stamens bright yellow, borne in large clusters, single, non-recurrent, moderate fragrance; foliage almost evergreen; vigorous, climbing growth; [Mme Eugene Resal sport]; Cant, F.

Promise Me, *see* 'JELwhite'

'**Promotion**', F, mp, 1966; bud ovoid; flowers pink, medium, borne in clusters, dbl., moderate fragrance; foliage dark; [Seedling X Miracle]; Verbeek

Prophecy, *see* 'MINprop'

Prophyta™, HT, ab; de Ruiter, G.

Proserpine, B, m, 1841; Lebourg

'**Prosper Laugier**', HP, mr, 1883; flowers scarlet-red, large, dbl., 30 petals, cupped, recurrent bloom; tall growth; Verdier, E.

'**Prosperity**', HMsk, w, 1919; bud creamy white flushed pale pink; flowers ivory white, small to blooms in large clusters, semi-dbl., recurrent bloom, moderate fragrance; foliage glossy; vigorous, pillar (6–8 ft) (21) growth; [Marie-Jeanne X Perle des Jardins]; Pemberton

Prospero®, *see* 'AUSpero'

'**Proteiformis**', S, w; (R. X heterophylla, R. X protieformis); foliage a curiosity of great botanical interest, the very di; (14); [R. rugosa alba X ?]; 1894

Proud Bride®, S, ab, 1992; Austin, David

'**Proud Heritage**', Min, dr, 1986; flowers patio, large blooms, dbl., 28 petals, exhibition form, blooms in sprays of 2–3, slight fragrance; globular, orange fruit; light green to brown prickles, hooked downward; foliage medium, medium green, matt; upright, bushy growth; PP006561; [Red Beauty X Big John]; Jolly, Nelson F.; Rosehill Farm

'**Proud Land**', HT, dr, 1969; bud urn-shaped; flowers deep red, large, dbl., 60 petals, moderate fragrance; foliage dark, leathery; vigorous, upright growth; [Chrysler Imperial X Unnamed seedling]; Morey, Dr Dennison; J&P

Proud Mary, *see* 'TWOhave'

Proud Titania, *see* 'AUStania'

'**Provence**', HT, ob, 1945; flowers orange, tinted copper and pink, well-formed, moderate fragrance; Paolino; A. Meilland

Provence, *see* DORlina

Provence Bergsoe, C, mp

Provence Pink, C, mp, 1759

Provence Rose, *see* '**Cabbage Rose**'

Province d'Anjou, F, mr, 1967; Croix; (Cavriglia)

Province Panachee, B, pb

'**Provins Renonculé**', HGal, mp; (Renoncule); flowers rose; Dupony, Andre, before 1810

Provins Rose, *see* **R. gallica**

'**Prudence**', Cl HT, op, 1938; flowers salmon-pink, large, semi-dbl., cupped, recurrent bloom, intense fragrance; very vigorous, climbing (10 ft) growth; [Warrawee X Souv. de Claudius Pernet]; Fitzhardinge; Hazlewood Bros.

Prudence Rosser, HP, pb

'**Prudhoe Peach**', HT, ab, 1970; flowers apricot-peach, full, dbl., 25 petals; foliage glossy, dark, leathery; free growth; [Piccadilly sport]; Wood; Homedale Nursery

Pruhonice, S, mp, 1973; flowers large, dbl.; Vecera, L.; (Sangerhausen)

Prunella, F, op

Psyche, HMult, mp, 1899; Paul

'**Pteragonis**', ly, 1937; flowers creamy yellow, medium, single, moderate fragrance; Krause; (Sangerhausen)

'**P'tit Pacha**', F, rb, 1965; bud pointed; flowers geranium-red edged dark red, large, semi-dbl.; moderate growth; GM, Madrid, 1963; [Unnamed seedling X Coup de Foudre]; Combe

PTMR, HT, 1975; Kurowski, L.; (Cavriglia)

'**Puanani**', F, lp, 1997; flowers single, medium, single, 4–7 petals, borne in small clusters, moderate fragrance; foliage medium, medium green, semi-glossy; bushy, medium (3 ft) growth; [Playgirl sport]; Belendez, Kitty

Pubta, Climbing, Cl F, dr, 1987; Patil, B.K.

'**Puccini**', HT, rb, 1968; flowers red, veined yellow, full, dbl., 26–33 petals, 4–6 in., intense fragrance; foliage large, light; very vigorous growth; [Opera X Teenager]; Ellick

Puccini, HMsk, lp, 1984; Lens

'**Puccini's Daughter**', F, ob, 1972; flowers orange-flame, reverse streaked yellow, full, dbl., 20–25 petals, 4 in., intense fragrance; foliage glossy, dark; very vigorous, upright growth; [(Puccini X Peace) X Orange Sensation]; Ellick; Excelsior Roses

Pucelle de Lille, M, dp; Miellez, before 1860

'**Puck**', HT, mr, 1921; bud long, pointed; flowers cherry-crimson, dbl., moderate fragrance; [Lyon Rose X Gen. MacArthur]; Bees

'**Puck**', F, or, 1960; bud globular; flowers cinnabar-red, large, dbl., cupped, slight fragrance; foliage leathery; moderate growth; [Pour Toi X Margaret McGredy]; Leenders, J.

'**Puck Pulling**', Cl Min, lp, 1997; flowers single, small, single, 4–7 petals, borne in large clusters, slight fragrance; foliage medium, medium green, semi-glossy; spreading, tall (8 ft) growth; [Seedling X Seedling]; Pulling-Smith, Mrs Pam

Pucker Up, *see* 'TINpuck'

Pudsey Bear, *see* BEDchild

Puerta del Sol®, *see* 'DELglap'

Puerto Rico, *see* 'DELsob'

Puerto Rico, T, w

Puerto Rico Shell Pink Noisette, N, lp

Pulnor, LCl, mr

Pulsar®, HT, mr

'**Pumila**', Min, dp; bud long, pointed; flowers bright pink to red, dbl.; almost thornless; dwarf (8–10 in.) growth

Pumila, HGal, dp, 1879

Pumphouse Climber, LCl, mr

Pumpkin Frost™, *see* 'MINIpum'

'**Punch**', F, dp, 1960; flowers raspberry-red, open, medium, borne in clusters, semi-dbl.; foliage glossy; moderate, bushy growth; [Luc Varenne X Lafayette]; Delforge

Punkin, *see* 'TINkin'

Puppy Love, *see* 'SAVapup'

Pur Caprice, *see* DELjavert

Pur Sang, *see* **'Wiener Walzer'**

Pure Bliss(TM), *see* 'DICtator'

Pure Hit, *see* POUldel

Pure Love(TM), *see* 'BURlov'

Pure Magic, MinFl, ob, 1999

Pure Poetry, *see* 'JACment'

Puregold, *see* **'Fanion'**

'Purezza', LCl, w, 1961; (R. banksiae 'Purezza'); dbl., 1–2 in., blooms in clusters of 39–50, recurrent bloom, moderate fragrance; foliage leathery; very vigorous growth; GM, Rome, 1960; [Tom Thumb X R. banksiae lutescens]; Mansuino, Q.

'Puritan', HT, w; flowers large, moderate fragrance

Puritas, HT, 1969; Borgatti, G.; (Cavriglia)

'Purity', LCl, w, 1917; flowers pure unshaded white, well-formed, large, semi-dbl., slight fragrance; heavy prickles; foliage light; vigorous, climbing growth; [Unnamed seedling X Mme Caroline Testout]; Hoopes, Bro. & Thomas

'Purple Beauty', HT, m, 1979; flowers red-purple, dbl., 30 petals, 5 in., exhibition form, moderate fragrance; foliage leathery; vigorous, upright growth; [Eminence X Tyrius]; Gandy, Douglas L.

'Purple Bengal', Ch, dr, 1827; (Violet Bengal); flowers maroon shaded darker; (14); Vibert

Purple Boursault from Rockville, Bslt, m

Purple Buttons(TM), S, m, 1993; Rupert, Kim L.

Purple Carpet, S, m, 1999; Clements, John K.

Purple Cloud, HT, m, 1993; Keisei Rose Nurseries, Inc.

Purple Dawn, *see* 'BRIdawn'

'Purple East', HMult, m, 1900; flowers crimson-purple, blooms in clusters, semi-dbl., early; [Crimson Rambler X Beauté Inconstante]; Paul

'Purple Elf', Min, m, 1967; flowers fuchsia-purple, small, dbl., 43 petals; foliage glossy; dwarf, bushy (10 in.) growth; [Violette X Zee]; Moore, Ralph S.; Sequoia Nursery

'Purple Fantasy', Min, m, 1982; flowers deep purple, small, dbl., no fragrance; foliage small, medium green, matt; bushy growth; [Blue Mist X Snow Magic]; Dobbs, Annette E.

Purple Haze, *see* 'JUShaze'

'Purple Heart', HT, m, 1946; bud long, pointed; flowers dahlia-purple to blackish red-purple, large, semi-dbl., cupped, intense fragrance; upright, bushy growth; [Crimson Queen X Self]; Moore, Ralph S.; Sequoia Nursery

Purple Heart(TM), *see* 'WEKbipuhit'

'Purple Imp', Min, m, 1967; bud ovoid; flowers magenta to purple, small, dbl.; foliage small, narrow, glossy; vigorous, very compact growth; [Baby Faurax X Red Imp]; Wiliams, E.D.; Mini-Roses

Purple Majesty(TM), *see* 'MINajco'

Purple Noisette, *see* **'Purpurea'**

Purple Parade, *see* POUlprima

Purple Passion(TM), *see* 'JAColpur'

Purple Pavement, *see* **Rotesmear**

Purple Popcorn(TM), *see* 'WILpurp'

Purple Prince®, *see* 'INTerpur'

Purple Puff, *see* 'JACsedi'

Purple Rain, *see* 'KAWamoblue'

Purple Reign, *see* 'WEKpluroco'

Purple Simplicity, *see* 'JACpursh'

Purple Spires, F, m, 1997; Hortico

'Purple Splendour', F, m, 1976; flowers glowing purple, dbl., 26 petals, 4 in., slight fragrance; foliage dark; upright growth; [News X Overture]; LeGrice

Purple Sunset, *see* 'MANpursun'

Purple Tiger, *see* 'JACpurr'

Purple Times, F, m, 1999

'Purpurea', (R. roxburghii hybrid), m; (Maheka, Purpuria); flowers purple-crimson or purplish-rose; [R. roxburghii seedling]; Buist, prior to 1844

'Purpurea', N, dr, 1822; (Purple Noisette); flowers purple-crimson, poorly formed, semi-dbl.; Laffay, M.

Purpurea di Bologna, B, lp; [Variegata di Bologna sport]; Ruston, D., 1970; (Weatherly, L.)

'Purpurea Rubra', M, m; flowers violet-purple, well mossed, large, dbl., intense fragrance

Purpuria, *see* **'Purpurea'**

'Purpurine', F, m; bud long; flowers fuchsia-purple to rhodamine-purple, dbl., slight fragrance; foliage dark; vigorous, bushy growth; [(Peace X Unnamed seedling) X Fashion]; Lens; Galan

Purpurtraum, HWich, dr

'Pusa Christina', HT, mp, 1976; bud globular; very dbl., 50 petals, 3 in., exhibition form, slight fragrance; foliage soft; vigorous, upright growth; IARI

'Pusa Sonia', HT, dy, 1968; bud long, pointed; flowers golden yellow, large, dbl., 24 petals, moderate fragrance; foliage leathery; vigorous, upright growth; [McGredy's Yellow X ?]; IARI

'Pusa Sonora', HT, pb, 1984; flowers rose pink, deeper reverse, medium, single, 5 petals, no fragrance; foliage medium green; bushy, upright growth; [Queen Elizabeth X First Prize]; Division of Floriculture and Landscaping

Pushkala, Min, w, 1973; Kasturi

Pushkarini, F, w, 1990; Chiplunkar

Pussta, *see* **'New Daily Mail'**

Pussta, Climbing, *see* **'New Daily Mail, Climbing'**

'Pye Colour', F, mr, 1972; flowers turkey-red, ovate, dbl., 30 petals, 2 in., slight fragrance; foliage leathery; free growth; [Marlena X Elizabeth of Glamis]; Dickson, A.

'Pygmae'®, Min, or, 1977; (Pygmy); bud globular; flowers bright orange-red, yellow center, blooms in clusters, semi-dbl., 13 petals, 1 in.; foliage small, glossy, dark; low, compact, spreading, bushy growth; [Anytime X Minuette]; Poulsen, Niels D.; D. T. Poulsen

Pygmy, *see* **'Pygmae'**®

Pyrenees, *see* 'POUlcov'

Pyrenees Rose, *see* **R. pendulina pyrenaica**

'Pythagoras', HSpn, pb; flowers light pink, flecked deep pink, semi-dbl., early bloom; glossy, black fruit; foliage finely divided; dense, shrubby (3–4 ft) growth

Pzazz, *see* POUlzazz

Q

QE2, HT, lp; [Queen Elizabeth sport]; Ruston, D.; (Weatherly, L.)

Quadra, HKor, lp; quartered, 60 petals, slight; L'Assomption, 1981

'Quadroon', S, dr; flowers rich dark red, small, single, non-recurrent; growth rather poor; [Supposedly Hansa X (Hansa X R. nitida)]; Wright, Percy H.

'Quaker Beauty', HT, ab, 1936; bud long, pointed; flowers glowing apricot, large, exhibition form; [Joanna Hill sport]; Brookins

'Quaker Maid', F, or, 1959; bud ovoid; flowers nasturtium-red, borne in clusters, semi-dbl., 18–20 petals, 1.5–2 in., flat, moderate, spicy fragrance; foliage leathery; vigorous, upright, bushy growth; [Orange Sweetheart X Pinocchio]; Hill, Joseph H., Co.

Quaker Star, *see* 'DICperhaps'

'Quantock Star', HT, or, 1968; flowers vermilion shaded pink, small, dbl., moderate fragrance; foliage variegated; free growth; [Tropicana sport]; Heard

'Quatre Saisons Blanc Mousseux', M, w; (Perpetual White Moss, Quatre Saisons Blanche, Rosier de Thionville); bud very mossy; dbl., blooms in large clusters, repeats sparingly in fall; [Autumn Damask sport]; Laffay, M., prior to 1837

Quatre Saisons Blanche, *see* **'Quatre Saisons Blanc Mousseux'**

Quatre Saisons d'Italie, P, dp; Dupony, Andre, before 1815

Quatre Saisons Rose, LCl, mp

Québec, *see* **'Mme Marie Curie'**

Queen Adelaide, *see* 'MEIvildo'

'Queen Alexandra', HMult, mp; flowers rosy white, semi-dbl., large corymbs; Veitch, 1915

Queen Alexandra, *see* **'The Queen Alexandra Rose'**

'Queen Ann', HT, dr, 1949; bud long, pointed; flowers crimson, medium, dbl., exhibition form; strong stems; vigorous growth; [Better Times sport]; Spandikow

Queen Astrid, *see* **'Koningin Astrid'**

'Queen Beatrice', HT, lp, 1909; flowers bright silvery pink, large to medium, very dbl., moderate fragrance; vigorous growth; [Mme Abel Chatenay X Liberty]; Kramer

Queen Beatrix, *see* 'HetKORa'

'Queen Bee', S, dr, 1984; dbl., 35 petals, 4–5 in., exhibition form, borne 5–8 per cluster, repeat bloom, moderate, old rose fragrance; awl-like, brown prickles; foliage large, leathery, dark olive green; erect, bushy, branching growth; hardy.; [(Rosali X Music Maker) X (Square Dancer X Tatjana)]; Buck, Dr Griffith J.; Iowa State University

Queen Charlotte, *see* 'HARubondee'

'Queen City', Min, ob, 1986; flowers orange, yellow base, fading lighter, medium, borne usually s, dbl., 20 petals, exhibition form, slight fragrance; long, straight, medium, light colored prickles; foliage medium, medium green, semi-glossy; bushy, medium growth; [Rise 'n' Shine X Unnamed seedling]; Bridges, Dennis A., 1987

'Queen Dina', Gr, dr, 1964; flowers deep scarlet, large, dbl.; foliage leathery; very vigorous, upright, bushy growth; [Cocorico X Geranium Red]; Soenderhousen; Hoersholm Nursery

'Queen Dorothy Bell', HT, dr, 1940; flowers velvety scarlet, very dbl., globular, intense fragrance; foliage light, leathery; vigorous growth; [Oswald Sieper sport]; Stell; Stell Rose Nursery

'Queen Elizabeth'®, Gr, mp, 1954; (Queen of England, The Queen Elizabeth Rose); bud pointed; dbl., 38 petals, 3.5–4 in., exhibition form, borne singly and in clusters, moderate fragrance; foliage dark, glossy, leathery; very vigorous, upright, bushy growth; AARS, 1955 Gertrude M. Hubbard, ARS, 1957 GM, NRS, 1955 GM, Portland, 1954 Hall of Fame, WFRS, 1978 PIT, NRS, 1955; [Charlotte Armstrong X Floradora]; Lammerts, Dr Walter; Germain's

'Queen Elizabeth, Climbing', Cl Gr, mp, 1957; (Grimpant Queen Elizabeth, The Queen Elizabeth Rose, Climbing); Whisler; Germain's, 1957; Wheatcroft Bros., 1960

'Queen Esther', HT, w, 1984; flowers cream, pale pink petal edges, spiraled, large, dbl., 35 petals, slight fragrance; many prickles; foliage medium, medium green, matt; [Golden Masterpiece X Peer Gynt]; Poole, Lionel

'Queen Fabiola', Gr, or, 1961; (Fabiola); [Montezuma sport]; Hazenberg; Delbard

'Queen Frances Connally', HT, mr, 1939; flowers spectrum-red, base lemon-chrome, reverse yellow edged red; [Katharine Pechtold sport]; Stell; Stell Rose Nursery

'Queen Gertrude Anne Windsor', HT, dp, 1936; flowers darker; [Francis Scott Key sport]; Dixie Rose Nursery

Queen Juliana, *see* **'Orange Delight'**

'Queen Louise Boren', HT, op, 1935; flowers pink suffused salmon, large, dbl., intense fragrance; very vigorous growth; [(Emile Charles X La France) X Marechal Niel]; Nicolas; Dixie Rose Nursery

'Queen Lucia', F, op, 1954; flowers salmon-pink, pompon form, borne in large trusses, dbl.; vigorous growth; [Pinocchio X Tapis Rose]; Maarse, G.

'Queen Mab', Ch, ab, 1896; flowers soft rosy apricot, center shaded orange, reverse tinted rose, dbl.; Paul, W.

'Queen Margaret Hunt', HT, dr, 1936; bud long, pointed, spiral; flowers dark velvety crimson-maroon, stamens golden yellow, large, cupped, intense fragrance; foliage leathery; very vigorous growth; [Templar X Ami Quinard]; Nicolas; Dixie Rose Nursery

Queen Margrethe™, *see* 'POUlskov'

'Queen Marie', HT, pb, 1925; flowers rose-pink, reverse deeper pink, base bronze yellow, dbl., moderate fragrance; [Mme Butterfly X Lamia]; Chervenka

'Queen Marie of Jugoslavia', HT, my, 1935; flowers bright yellow, slightly flushed pink; [Mme Butterfly sport]; Hicks

'Queen Mary', HT, yb, 1913; flowers bright canary yellow, shaded red, moderate fragrance; GM, NRS, 1913; Dickson, A.

'Queen Mary', HT, lp; [Frau Karl Druschki X Unknown]; Williams, A., 1911; (Weatherly, L.)

Queen Mother, *see* KORquermu

Queen Nefertiti®, *see* AUSap

'Queen o' the Lakes', HT, dr, 1949; bud ovoid, long, pointed; flowers large, dbl., exhibition form, moderate fragrance; foliage glossy; vigorous, bushy growth; [Pink Princess X Crimson Glory]; Brownell, H.C.

'Queen o' the Lakes, Climbing', Cl HT, dr, 1965; Brownell, H.C.

'Queen of Bath', HT, dy, 1931; bud long, pointed; flowers deep buttercup-yellow, outer edged chrome-yellow, dbl., exhibition form, intense fragrance; foliage thick, glossy, bronze; vigorous growth; [Souv. de Claudius Pernet X Cleveland]; Bees

Queen of Beauty and Fragrance, *see* **'Sour de la Malmaison'**

'Queen of Bedders', B, dp, 1876; flowers deep carmine, well-shaped; dwarf, compact growth; [Sir Joseph Paxton seedling]; Noble

'Queen of Bermuda', Gr, or, 1956; bud ovoid; flowers orange-vermilion, blooms in small clusters, dbl., 35 petals, 4 in., exhibition form, moderate, fruity fragrance; foliage glossy, bronze; vigorous, bushy growth; [(Independence X Orange Triumph) X Bettina]; Bowie; Bermuda Rose Nursery

'QUEEN OF BOURBONS', B, pb, 1834; (Bourbon Queen, Reine des Iles Bourbon, Souv de la Princesse de Lamballe); flowers fawn and rose, cupped, intense fragrance; Mauget

Queen of Colors, *see* **'FARBENKÖNIGEN'**

Queen of Denmark, *see* **'KÖNIGIN VON DÄNEMARK'**

Queen of England, *see* **'QUEEN ELIZABETH'**®

'QUEEN OF FRAGRANCE', HT, lp, 1915; flowers shell-pink, tipped silver, well shaped, large, dbl., intense fragrance; foliage soft; dwarf growth; Paul, W.

Queen of Hearts, *see* **'DAME DE COEUR'**

'QUEEN OF HEARTS', Cl HT, mp, 1920; bud globular; flowers rich pink, large, dbl., cupped, moderate fragrance; foliage dark; very vigorous, climbing growth; [Gustave Grunerwald X Rosy Morn]; Clark, A.; NRS Victoria

Queen of Pearl, HWich, lp, 1898; Van Fleet

Queen of Roses, *see* 'KORBICO'

'QUEEN OF THE BELGIANS', Ayr, w; flowers small, very dbl.

'QUEEN OF THE BELGIANS', HT, op, 1916; bud long, pointed; flowers salmon-pink, semi-dbl.; GM, NRS, 1915; Hicks

'QUEEN OF THE DWARFS', Min, dp, 1955; (Dwarf Queen, Zwergkönigin); flowers deep pink, dbl.; foliage rather coarse; height 10–12 in.; Kordes

'QUEEN OF THE MUSKS', HMsk, pb, 1913; bud coppery red; flowers deep blush and white, blooms in panicles, recurrent bloom, intense fragrance; foliage dark ivy-green; Paul

Queen of the Night, HT; Sgaravatti, A.; (Cavriglia)

'QUEEN OF THE PRAIRIES', HSet, pb, 1843; (Beauty of the Prairies, Prairie Belle); flowers bright pink, frequently striped white, large blooms in clusters, dbl., globular, moderate fragrance; foliage large; vigorous, climbing growth; hardy.; [Probably R. setigera X China]; Feast

Queen of the Violets, *see* **'REINE DES VIOLETTES'**

'QUEEN THORNLESS, CLIMBING', Cl HT, dp; bud pointed; flowers deep pink, dbl., 20–40 petals, 3–4 in., exhibition form, intermittent bloom, moderate fragrance; thornless; foliage leathery, dull green; strong stems; vigorous, arching growth; Kittle; Lincoln Nursery Co.

'QUEEN VICTORIA', HP, lp, 1850; flowers blush-pink, large, dbl., [La Reine seedling]; Fontaine; A. Paul

'QUEEN WILHELMINA', HT, ob, 1942; flowers brilliant orange, base light orange-yellow; [Hinrich Gaede sport]; Deverman

'QUEENIE', F, lp, 1962; (Petite Reine); bud ovoid; dbl., 33 petals, 4 in., cupped, blooms in clusters, moderate fragrance; foliage leathery; vigorous, upright, bushy growth; [Pinocchio seedling X Spartan]; Boerner; J&P

'QUEENIE ROBINSON', HT, op, 1924; bud long, pointed; flowers orange-cerise to flame-pink, semi-dbl., moderate fragrance; Easlea

'QUEENIE'S LOVE', HT, dp, 1968; bud globular; flowers begonia-rose, medium, dbl., moderate fragrance; foliage leathery; vigorous growth; [Libretto X Mme Butterfly]; Verschuren, A.; Stassen

'QUEEN'S KNIGHT', S, mp, 1979; bud short, conical; 6–10 petals, cupped, borne 3 per cluster, profuse for 6 weeks; non-recurrent, slight fragrance; hooked prickles; foliage semi-glossy, deep green, 7 leaflets; erect arches, self-supporting growth; [Don Juan X R. laxa (Retzius)]; Stoddard, Louis

Queen's Palace, *see* POUlelap

'QUEEN'S SCARLET', Ch, mr, 1880; (Red Hermosa, Rote Hermosa); flowers rich velvety scarlet, small, dbl., moderate fragrance; foliage small; bushy, compact growth; Hallock & Thorpe

'QUEEN'S VISIT', HT, dr, 1955; bud long, pointed; flowers dark velvety red veined darker, medium, dbl., exhibition form, moderate fragrance; foliage leathery; vigorous growth; [Crimson Glory sport X Crimson Glory]; Viney; Wynne

'QUEENSLAND BEAUTY', HT, pb, 1934; flowers coppery pink; [Golden Dawn sport]; Alderton & Williams

'QUEENSTOWN', HT, dr, 1991; bud pointed; flowers deep crimson to ruby red, blooms borne up to 6 per cluster, dbl., 32 petals, 5 in., cupped, slight fragrance; foliage leathery, mid-green, semi-glossy; upright, spreading growth; [Silent Night X Josephine Bruce]; Cattermole, R.F., 1975; South Pacific Rose Nursery, 1991

Querida, *see* 'MALJOANNA'

Quicksilver, *see* 'AROSTAL'

Quiet Reflections, T, dy; Hay

Quiet Time, *see* 'TINQUIET'

Quinella, HT, ob

Quintet, HT, op

Quito, HT, yb, 1999

'QUO VADIS?', HT, pb, 1961; bud oval; flowers silvery pink becoming red, large, dbl., 50–60 petals, slight fragrance; very vigorous growth; [Peace X (Baiser X ?)]; Giacomasso

R

R. abietorum, *see* **R. GYMNOCARPA**

R. abyssinica, *see* **R. MOSCHATA ABYSSINICA**

R. ACICULARIS, (Lindley), Sp, dp, 1805; (Arctic Rose, R. acicularis carelica, R. acicularis taquetii, R. carelica, R. fauriei, R. korsakoviensis, R. sayi, R. stricta); flowers deep rose solitary, 1–1.5 in., spring bloom, moderate fragrance; usually pyriform, 1/2–1 in. long fruit; 3 ft growth; Eurosa, Cinnomomeae (28, 42, 56)

R. acicularis bourgeauiana, *see* **R. ACICULARIS SAYI**

R. acicularis carelica, *see* **R. ACICULARIS**

R. acicularis engelmannii, *see* **R. X ENGELMANNII**

R. ACICULARIS FENNICA, (Lallemant) Sp, dp; (R. acicularis gmelinii, R. gmelinii); (28)

R. acicularis gmelinii, *see* **R. ACICULARIS FENNICA**

R. ACICULARIS NIPPONENSIS, ((Crépin) Koehne), Sp, dp, 1894; (R. nipponensis); 1.5 in.; (14, 28)

R. ACICULARIS SAYI, (Rehder), Sp, dp; (R. acicularis bourgeauiana, R. bourgeauiana, R. sayi); 2.5 in.; usually globular fruit; (42, 56)

R. ACICULARIS SAYI PLENA, (Lewis), Sp, dp

R. acicularis taquetii, *see* **R. ACICULARIS**

R. adenosepala, *see* **R. WOODSII HISPIDA**

R. afzeliana, *see* **R. DUMALIS**

R. AGRESTIS, (Savi) Sp, lp; (R. sepium); flowers pale pink or whitish, small; Eurosa, Caninae (35, 42)

R. agrestis inodora, *see* **R. INODORA**

R. alba Allioni, *see* **R. SEMPERVIRENS**

R. alba incarnata, *see* **'GREAT MAIDEN'S BLUSH'**

R. alba maxima, *see* **'ALBA MAXIMA'**

R. alba rubicanda plena, *see* **'GREAT MAIDEN'S BLUSH'**

R. alba rubicunda, *see* **'GREAT MAIDEN'S BLUSH'**

R. alba semi-plena, *see* 'Alba semi-plena'

R. alba suaveolens, *see* 'Alba suaveolens'

R. ALBERTII, (Regel) Sp, w; 1.5 in.; 3 ft growth; Eurosa, Cinnomomeae; 1877

R. albicans, *see* **R. ELYMAITICA**

R. alpina, *see* **R. PENDULINA**

R. altaica, *see* **R. SPINOSISSIMA ALTAICA**

R. AMBLYOTIS, (C.A. Meyer), Sp, mr; 2 in.; solitary or few; subglobose or pyriform, 1/2–1 in., red fruit; Eurosa, Cinnomomeae (14)

R. amoyensis, *see* **R. CYMOSA**

R. andegavensis, *see* **R. CANINA ANDEGAVENSIS**

R. ANDREAE, Sp; (14)

R. ANEMONEFLORA, (Lange), Sp, w; (R. sempervirens anemoniflora, R. triphylla); flowers corymbose, dbl., 1 in.; foliage leaflets 3, lanceolate; climbing growth; Eurosa, Synstelae (14); 1844

R. angustiarum, *see* **R. ARKANSANA**

R. apennina, *see* **R. SERAFINII**

R. ARKANSANA, (Porter), Sp, mp; (Arkansas Rose, R. angustiarum, R. arkansanoides, R. heliophila, R. pratincola, R. rydbergii, R. suffulta); 1.5 in., in corymbs; 1.5 ft growth; Eurosa, Cinnamomeae (28, 14)

R. ARKANSANA ALBA, ((Rehder) Lewis), Sp, w; (R. arkansanoides alba, R. heliophila alba, R. pratincola alba, R. suffulta alba); 1901

R. ARKANSANA PLENA, (Lewis), Sp, lp

R. arkansanoides, *see* **R. ARKANSANA**

R. arkansanoides alba, *see* **R. ARKANSANA ALBA**

R. ARVENSIS, (Hudson), Sp, w; (Field Rose, R. repens, R. serpens, R. silvestris); 1.5–2 in., in few-flowered corymbs, summer bloom, no fragrance; ovoid fruit; foliage deciduous; creeping growth; Eurosa, Synstylae (14)

R. ARVENSIS AYRESHIREA, (Seringe); (Ayrshire Rose, R. arvensis capreolata, R. capreolata)

R. arvensis capreolata, *see* Ayrshire Rose

R. atrovirens, *see* **R. SEMPERVIRENS**

R. aurantiaca, *see* **R. FOETIDA BICOLOR**

R. austriaca, *see* **R. GALLICA**

R. austriaca pygmaea, *see* **R. GALLICA PUMILA**

R. bakerii, *see* **R. X ENGELMANNII**

R. balearica, *see* **R. SEMPERVIRENS**

R. balsamea, *see* **'TACKHOLMII'**

R. BANKSIAE, (Aiton fil), Sp; (Banks' Rose, Banksian Rose, Lady Banks' Rose, R. banksiana); flowers white or yellow on slender pedicels in many-flowered umbels, 1 in., early spring bloom, slight fragrance; foliage evergreen; climbing (20 ft or more) growth; Eurosa, Banksianae (14)

R. banksiae 'Purezza', *see* **'PUREZZA'**

R. banksiae alba, *see* **R. BANKSIAE BANKSIAE**

R. banksiae alba-plena, *see* **R. BANKSIAE BANKSIAE**

R. BANKSIAE BANKSIAE, Sp, w; (Banksiae Alba, R. banksiae alba, R. banksiae alba-plena, White Banksia, White Lady Banks' Rose); dbl., (14); William Kerr, 1807

R. BANKSIAE LUTEA, (Lindley), Sp, ly; (R. banksiae luteaplena, Yellow Lady Banks' Rose); dbl.; (14); 1824

R. banksiae luteaplena, *see* **R. BANKSIAE LUTEA**

R. BANKSIAE LUTESCENS, (Voss), Sp, my; single; (14); ca. 1870

R. BANKSIAE NORMALIS, (Regel), Sp, w; single, (14); 1796, but not distributed until 1877

R. banksiana, *see* **R. BANKSIAE**

R. BANKSIOPSIS, (Baker), Sp, mr; 1 in., borne in corymbs; Eurosa, Cinnomomeae (14); 1907

R. BEGGERIANA, (Schrenk ex Fischer & Meyer), Sp, w; (R. cinnamomea sewerzowii, R. lehmanniana, R. regelii, R. silverhjelmii); 1.5 in., borne in corymbs, late summer bloom; fruit small, globular, without sepals; dense (5 ft) growth; Eurosa, Cinnomomeae (14); 1868

R. belgica, *see* R. X damascena

R. BELLA, (Rehder & Wilson), Sp, mp; flowers pink, solitary, 1.75–2 in.; ovoid, 3/4 in.long, scarlet fruit; growth to 8 ft; Eurosa, Cinnamomeae (28); 1910

R. BELLA PALLENS, (Rehder & Wilson), Sp, lp; flowers pale pink; 1910

R. bengalensis, *see* **'SLATER'S CRIMSON CHINA'**

R. bicolor, *see* **R. FOETIDA BICOLOR**

R. bifera, *see* **'AUTUMN DAMASK'**

R. bifera semperflorens, *see* **'AUTUMN DAMASK'**

R. BILLOTIANA, (Crépin), Sp; fruit like R. sherardii; similar to R. canina; Eurosa, Caninae (28)

R. BLANDA, (Aiton), Sp, mp; (Hudson's Bay Rose, Labrador Rose, R. fraxinifolia, R. gratiosa, R. solandri, R. subblanda, R. virginiana blanda); flowers pink borne usually several on smooth peduncles, single, 2–2.5 in., spring bloom; globular, sometimes elongated fruit; usually no prickles; 5 ft growth; Eurosa, Cinnomomeae (14, 21, 28); 1773

R. blanda, *see* **R. NITIDA**

R. BLANDA CARPOHISPIDA, ((Schuette) Lewis), Sp, mp; (R. blanda hispida); prickles densely covered with fine bristles; (14)

R. blanda hispida, *see* **R. BLANDA CARPOHISPIDA**

R. BLANDA WILLMOTTIANA, (Baker), Sp, mp; flowers bright coral-pink; stems red

R. blinii, *see* **R. MULTIFLORA CARNEA**

R. blondaeana, *see* **R. CANINA BLONDAEANA**

R. bourgeauiana, *see* **R. ACICULARIS SAYI**

R. boursaultii, *see* **'BOURSAULT ROSE'**

R. BRACTEATA, (Wendland), Sp, w; (Chickasaw Rose, Macartney Rose, R. lucida, R. macartnea); 2–2.75 in., borne 1 or few on short stalks, spring-fall bloom; large, globular, tomentose, orange-red fruit; foliage half-evergreen, bright green, somewhat glossy above, climbing growth; Eurosa, Bracteata (14); 1793

R. bracteata, *see* **R. WICHURANA**

R. bracteata alba odorata, R. microphylla alba odorata, *see* **'ALBA ODORATA'**

R. braunii, *see* **R. X INVOLUTA**

R. BRITZENSIS, (Koehne), Sp, lp; flowers pale pink changing to white, borne 1–2 together, 3–4 in.; ovoid, brown fruit; 6 ft growth; Eurosa, Caninae (35); 1901

R. brownii, *see* **R. MOSCHATA NEPALENSIS**

R. brunonii, *see* **R. MOSCHATA NEPALENSIS**

R. burgundensis, *see* **'BURGUNDIAN ROSE'**

R. burgundica, *see* **'BURGUNDIAN ROSE'**

R. calabrica, *see* **R. GLUTINOSA**

R. calendarum, *see* **'SUMMER DAMASK'**

R. CALIFORNICA, (Chamisso & Schlechtendahl), Sp, mp; (California Wild Rose, R. gratissima); 1.5 in., borne few or several on slender pedicels in dense corymbs, summer bloom; globose-ovoid, usually with a prominent neck fruit; usually has paired infrastipular prickles, which are often c; up to 8 ft growth

R. CALIFORNICA NANA, (Bean), Sp, mp; habit very dwarf growth

R. CALIFORNICA PLENA, (Rehder), Sp, 1894; dbl.; (14); Geschwind

R. californica ultramontana, *see* **R. WOODSII ULTRAMONTANA**

R. calva, *see* **R. MULTIFLORA CALVA**

R. camellia, *see* **R. LAEVIGATA**

R. campanulata, *see* **R. X FRANCOFURTANA**

R. CANINA, (Linnaeus), Sp, lp; (Brier Bush, Dog Rose, R. leucantha, R. pseudoscabrata, R. sphaerica, R. surculosa); flowers white or pinkish blooms solitary or in few-flowered corymbs, 2 in., summer bloom; ovoid, glabrous (see endpapers), orange-red scarlet fruit; 10 ft growth; Eurosa, Caninae (35, 42, 34)

R. CANINA ANDEGAVENSIS, ((Bastard) Desportes), Sp; (R. andegavensis); foliage variation with glabrous, eglandular, but glandular; (35)

R. CANINA ANDERSONII, S, mp, 1912; flowers medium, single; Hillier; (Sangerhausen)

R. CANINA BLONDAEANA, ((Ripart) Rouy), Sp; (R. blondaeana); foliage leaflet teeth glandular, bi-serrate; (42)

R. canina borboniana, *see* **'BOURBON ROSE'**

R. canina coriifolia, *see* **R. CORIIFOLIA**

R. canina dumetorum, *see* **R. CORYMBIFERA**

R. CANINA EXILIS, ((Crépin) Keller), Sp, mp; (R. exilis); 1 in.; low growth

R. CANINA FROEBELII, (Christ), Sp, w; (R. coriifolia froebelii, R. froebelii, R. laxa); flowers small; (35)

R. CANINA INERMIS, (hort.), Sp; almost unarmed; vigorous; dog rose popular as an understock

R. canina insignis, *see* **R. CANINA SPURIA**

R. CANINA LUTETIANA, ((Léman) Baker), Sp; (R. lutetiana); (35)

R. CANINA SPURIA, ((Puget) Wolley-Dod), Sp; (R. canina insignis); large fruit; foliage serrations almost entirely simple; (35)

R. canina tomentella, *see* **R. OBTUSIFOLIA**

R. capreolata, *see* Ayrshire Rose

R. carelica, *see* **R. ACICULARIS**

R. carnea, *see* **'GREAT MAIDEN'S BLUSH'**

R. CAROLINA, (Linnaeus), Sp, mp; (Carolina Rose, Pasture Rose, R. humilis, R. parviflora, R. pensylvanica, R. pratensis, R. virginiana humilis); flowers bright pink, 2 in., summer bloom 3–6 ft growth; Eurosa, Carolinae

R. carolina, *see* **R. PALUSTRIS**

R. CAROLINA ALBA, (Rehder), Sp, w; (R. lyonii alba, R. virginiana alba); (28)

R. CAROLINA GLANDULOSA, ((Crépin) Farwell), Sp, mp; (R. mexicana, R. parviflora glandulosa, R. sarrulata); foliage leaflets glandular-serrate leaf-stalk glandular; (28)

R. carolina grandiflora, *see* **R. LAXA**

R. carolina inermis, *see* **R. PALUSTRIS INERMIS**

R. carolina lyonii, *see* **R. CAROLINA VILLOSA**

R. carolina nuttalliana, *see* **R. PALUSTRIS NUTTALLIANA**

R. CAROLINA PLENA, ((Marshall) Doris Lynes), Sp, mp; (R. pennsylvanica plena); dbl.

R. carolina setigera, *see* **R. NITIDA SPINOSA**

R. CAROLINA TRILOBA, ((Watson) Rehder), Sp, mp; (R. humilis triloba); flowers bright pink, petals 3-lobed

R. CAROLINA VILLOSA, ((Best) Rehder), Sp, mp; (R. carolina lyonii, R. humilis villosa, R. lyonii, R. pusilla); foliage pubescent beneath; (28)

R. caroliniana, *see* **R. PALUSTRIS**

R. caryophyllacea, *see* **R. INODORA**

R. cathayensis, *see* **R. MULTIFLORA CATHAYENSIS**

R. cathayensis platyphylla, *see* **'SEVEN SISTERS'**

R. CAUDATA, (Baker), Sp, mr; 2 in., borne in few-flowered corymbs; ovoid fruit; 12 ft growth; Eurosa, Cinnomomeae (14, 28); ca. 1896

R. centifolia, *see* **'CABBAGE ROSE'**

R. centifolia albo-muscosa, *see* **'SHAILER'S WHITE MOSS'**

R. centifolia andrewsil, *see* **'MUSCOSA SIMPLEX'**

R. centifolia cristata, *see* **'CRESTED MOSS'**

R. centifolia minor, *see* **'PETITE DE HOLLANDE'**

R. centifolia muscosa, *see* **'CENTIFOLIA MUSCOSA'**

R. centifolia muscosa, *see* Communis

R. centifolia muscosa cristata, *see* **'CRESTED MOSS'**

R. centifolia parvifolia, *see* **'BURGUNDIAN ROSE'**

R. centifolia pomponia, *see* **'ROSE DE MEAUX'**

R. centifolia sancta, *see* **St John's Rose**

R. centifolia simplex, *see* **'CIUDAD DE OVIEDO'**

R. centifolia variegata, *see* **'VILLAGE MAID'**

R. CERASOCARPA, (Rolfe), Sp, w; 1–1.25 in.; globose, over 1/3 in., deep red fruit; Eurosa, Synstylae (14)

R. charbonneaui, *see* **R. LONGICUSPIS**

R. CHAVINII, (Rapin ex Reuter), Sp, mp; (R. montana chavinii); 2 in.; ovoid, setose fruit; 6–10 ft growth; Eurosa, Caninae (42); [Perhaps a derivative of R. canina X R. montana]

R. cherokeensis, *see* **R. LAEVIGATA**

R. CHINENSIS, (Jacquin), HCh; (Bengal Rose, China Rose, R. chinensis indica, R. indica, R. indica vulgaris, R. nankinensis, R. sinica); flowers crimson or pink, rarely whitish, borne usually several, less, 2 in., recurrent bloom, no fragrance; obovoid or turbinate, about 3/4 in. long fruit; foliage evergreen or partially so; long stems, rarely short-stemmed; Eurosa, Chinensis (14, 21, 28)

R. chinensis fragrans, *see* **R. X ODORATA**

R. chinensis indica, *see* **R. CHINENSIS**

R. CHINENSIS LONGIFOLIA, ((Willdenow) Rehder), Sp; (R. indica longifolia, R. longifolia); foliage leaflets very long and narrow, like a willow; Int. 1820, but not now in cult.

R. chinensis manettii, *see* **'MANETTI'**

R. CHINENSIS MINIMA, HCh; (Angel Rose, R. indica pumila, R. laurentiae, R. lawranceana, R. semperflorens minima); flowers white, pink or red, petals often pointed, semi-dbl., 1.5 in.; height variable; Eurosa, Chinensis

R. chinensis mutabilis, *see* **'MUTABILIS'**

R. chinensis pseudindica, *see* **'FORTUNE'S DOUBLE YELLOW'**

R. chinensis semperflorens, *see* **'SLATER'S CRIMSON CHINA'**

R. chinensis serratipetala, *see* **'SERRATIPETALA'**

R. CHINENSIS SPONTANEA, (Rehder & Wilson) Sp, dr; flowers deep red or pink, borne usually solitary, single

R. chinensis viridiflora, *see* **'GREEN ROSE'**

R. chlorophylla, *see* **R. FOETIDA**

R. CINNAMOMEA, (Linnaeus), Sp, m; (Cinnamon Rose, R. collincola, R. spinosissima); flowers purple borne solitary or few on short, naked pedicels, 2 in., moderate fragrance; depressed-globular, scarlet fruit; 6 ft growth; Eurosa, Cinnomomeae (14, 28)

R. cinnamomea, *see* **R. PENDULINA**

R. CINNAMOMEA PLENA, (Weston), Sp, m; (Double Cinnamon, R. foecundissima, R. majalis, Rose du Saint Sacrament, Stevens Rose, Whitsuntide Rose); dbl.

R. cinnamomea sewerzowii, *see* **R. BEGGERIANA**

R. CLINOPHYLLA, (Thory), Sp, w; (R. involucrata, R. lindleyana, R. lyellii); tender; Eurosa, Bracteata (14); Int. prior to 1817

R. collina, (Jacquin), Misc OGR, mp; flowers rose-colored, sepals shorter

R. collina, *see* **R. CORYMBIFERA**

R. collincola, *see* **R. CINNAMOMEA**

R. coreana, *see* **R. MAXIMOWICZIANA JACKII**

R. CORIIFOLIA, (Fries), Sp, lp; (R. canina coriifolia, R. frutetorum, R. watsonii); flowers pink, short-pediceled, with large bracts; foliage leaflets gray-green, downy; 5 ft growth; Eurosa, Caninae (35)

R. coriifolia froebelii, *see* **R. CANINA FROEBELII**

R. coronata, *see* **R. X INVOLUTA**

R. coruscans, *see* **R. RUGOSA CHAMISSONIANA**

R. CORYMBIFERA, (Borkhausen), Sp, w; (R. canina dumetorum, R. collina, R. dumetorum, R. saxatilis, R. taurica); flowers light pink to white, borne 1-many, 1.75–2 in.; ovoid to subglobose, 3/4 in. fruit; foliage leaflets downy; Eurosa, Caninae (35, 42)

R. corymbosa, *see* **R. PALUSTRIS**

R. CORYMBULOSA, (Rolfe), Sp, rb; flowers red with a white eye, borne in dense, umbel-like corymbs, .75–1 in.; globose fruit; 6 ft growth; Eurosa, Cinnamomeae (14); 1908

R. CROCACANTHA, (Boulenger), Sp, w; .5 in., borne in many-flowered panicles; globose, 1/3 in., red fruit; prickles curved, yellow; Eurosa, Synstylae (14); 1917

R. cursor, *see* **R. SETIGERA TOMENTOSA**

R. cuspidata, *see* **R. TOMENTOSA**

R. CYMOSA, (Trattinnick), Sp, w; (R. amoyensis, R. esquirolii, R. fragariaeflora, R. indica, R. microcarpa, R. sorbiflora); flowers small, borne in many-flowered corymbs; small, globose., red fruit; Eurosa, Banksianae (14); Int. 1904?

R. dalmatica, *see* **R. GLUTINOSA DALMATICA**

R. damascena bifera, *see* **'AUTUMN DAMASK'**

R. damascena rubrotincta, *see* **'HEBE'S LIP'**

R. damascena semperflorens, *see* **'AUTUMN DAMASK'**

R. damascena trigintipetala, *see* **'KAZANLIK'**

R. damascena variegata, *see* **'YORK AND LANCASTER'**

R. damascena versicolor, *see* **'YORK AND LANCASTER'**

R. DAVIDII, (Crépin), Sp, lp; (Father David's Rose); flowers pink borne in corymbs, 1.5–2 in.; oblong-ovoid or ovoid, long-necked, 1/2–3/4 in. long, scarlet fruit; 10 ft growth; Eurosa, Cinnamomeae (28); 1908

R. DAVIDII ELONGATA, (Rehder & Wilson), Sp, lp; (R. parmentieri); large, to 1 in. long fruit; 1908

R. davidii persotosa, *see* **R. PERSETOSA**

R. DAVURICA, (Pallas), Sp, m; (R. willdenowii); ovate fruit; straight instead of curved prickles and smaller leaflets; Eurosa, Cinnamomeae (14); 1910

R. dawsoniana, *see* **R. MULTIFLORA**

R. deserta, *see* **R. WOODSII**

R. dijoniensis, *see* **'ROSE DE MEAUX'**

R. dimorpha, *see* **R. TOMENTOSA**

R. diversifolia, *see* **'SLATER'S CRIMSON CHINA'**

R. doniana, *see* **R. X INVOLUTA**

R. DUMALIS, (Bechstein), Sp; (R. afzeliana, R. glauca, R. gypsicola, R. reuteri); heavily prickled; foliage bluish green glaucous leaflets; Eurosa, Caninae (35)

R. dumetorum, *see* **R. CORYMBIFERA**

R. ECAE, (Aitchison), Sp, dy; (R. xanthina, R. xanthina ecae); flowers short-stalked, 1–1.25 in., borne solitary, obovoid, 1/3–1/2 in., long fruit; upright (4 ft) growth; Eurosa, Pimpinellifoliae (14); 1880

R. ecae, *see* **R. PRIMULA**

R. EGLANTERIA, (Linnaeus), Sp, lp; (Eglantine, R. rubiginosa, R. suavifolia, R. walpoleana, Sweet Brier Rose, Sweetbriar); flowers pink borne solitary or in few-flowered corymbs, single, 2 in.; many prickles; foliage glandular and fragrant (apple); vigorous, to 8 ft growth; Eurosa, Caninae (35, 42)

R. eglanteria, *see* **R. FOETIDA**

R. eglanteria duplex, *see* **'MAGNIFICA'**

R. eglanteria punicea, *see* **R. FOETIDA BICOLOR**

R. ehrrhartiana, *see* **'BURGUNDIAN ROSE'**

R. ELASMACANTHA, (Trautvetter), Sp, my; (R. pimpinellifolia elasmacantha); flowers light yellow, borne solitary; low glabrous shrub growth; Eurosa, Pimpinellifoliae (28)

R. ELEGANTULA, (Rolfe), Sp, mp; flowers pink, with golden anthers, small, free bloomer; summer, 5–10 ft growth; Eurosa, Cinnamomeae (14)

R. elliptica, *see* **R. INODORA**

R. elongata, *see* **R. PALUSTRIS**

R. ELYMAITICA, (Boissier & Haussknecht), Sp, mp; (R. albicans); flowers pink, small, borne 1–3 together; small, globular fruit; dwarf (3 ft) growth; Eurosa, Cinnamomeae (14, 28); 1900

R. engelmannii, *see* **R. X ENGELMANNII**

R. ernestii, *see* **R. RUBUS**

R. ernestii nudescens, *see* **R. RUBUS NUDESCENS**

R. esquirolii, *see* **R. CYMOSA**

R. exilis, *see* **R. CANINA EXILIS**

R. FARGESII, (Boulenger), Sp, w; flowers paniculate, 1.75 in.; Eurosa, Synstylae

R. fargesii, *see* **R. MOYESII FARGESII**

R. fargesii, *see* **R. MOYESII**

R. FARRERI, (Stapf), Sp, w; flowers pale pink to white, 1.25–1.5 in.; ovoid to ellipsoid, red fruit; Eurosa, Pimpinellifoliae

R. FARRERI PERSETOSA, (Stapf), Sp, mp; (Threepenny Bit Rose); flowers pink, smaller; prickles whole plant finely bristly; 1914

R. faureri, *see* **R. MAXIMOWICZIANA**

R. fauriei, *see* **R. ACICULARIS**

R. FEDTSCHENKOANA, (Regel), Sp, w; 1.5–2 in., borne 1–4 together, Eurosa, Cinnamomeae (28); 1876

R. fendleri, *see* **R. WOODSII FENDLERI**

R. fenestrata, *see* **R. SETIGERA**

R. ferox, *see* **R. GLUTINOSA**

R. ferox, *see* **R. HORRIDA**

R. ferox, *see* **R. RUGOSA**

R. ferruginea, *see* R. glauca

R. FILIPES, (Rehder & Wilson), Sp, w; 1 in., borne in large, loose corymbs, moderate fragrance; globose, 1/3–1/2 in., scarlet fruit; 15 ft growth; Eurosa, Synstylae (14); 1908

R. filipes 'Brenda Colvin', *see* **'BRENDA COLVIN'**

R. filipes 'Kiftsgate', *see* **'KIFTSGATE'**

R. fimbriatula, *see* **R. WOODSII**

R. floribunda, *see* **R. HELENAE**

R. floribunda, *see* **R. MICRANTHA**

R. florida, *see* **R. MULTIFLORA CARNEA**

R. foecundissima, *see* **R. CINNAMOMEA PLENA**

R. FOETIDA, (Herrmann), Sp, my; (Austrian Briar, Austrian Brier Rose, Austrian

Yellow Rose, R. chlorophylla, R. eglanteria, R. lutea); flowers bright yellow borne sometimes several, but without bracts o, single, 2–2.5 in., summer bloom, moderate, sickly sweet fragrance; globular, rarely sets fruit; 10 ft growth; Eurosa, Pimpinellifoliae (28); Possibly int. prior to 1542

R. FOETIDA BICOLOR, ((Jacquin) Willmott), Sp, rb; (Austrian Copper Rose, R. aurantiaca, R. bicolor, R. eglanteria punicea, R. lutea punicea, R. luteabicolor, R. punicea); flowers orange-scarlet within, yellow reverse; (28); [R. foetida sport]; Int. prior to 1590

R. foetida harisonii, *see* **'HARISON'S YELLOW'**

R. FOETIDA PERSIANA, ((Lemaire) Rehder), Sp, my; (Persian Yellow Rose, R. hemisphaerica plena, R. lutea persiana, R. lutea plena); dbl.; (28); 1837

R. FOLIOLOSA, (Nuttall ex Torrey & Gray), Sp, mp; flowers rose borne solitary or in few-flowered clusters, 1.5 in., spring bloom; globose fruit; $1\frac{1}{2}$ ft growth; Eurosa, Carolinae (14)

R. FOLIOLOSA ALBA, ((Bridwell) Rehder), Sp, w; (14)

R. FORRESTIANA, (Boulenger), Sp, dp; flowers rose borne solitary or few, 1 in.; Eurosa, Cinnamomeae

R. fortuneana, *see* **'FORTUNIANA'**

R. fortuniana, *see* **'FORTUNE'S DOUBLE YELLOW'**

R. fragariaeflora, *see* **R. CYMOSA**

R. fragrans, *see* **R. PALUSTRIS**

R. franchetii, *see* **R. LUCIAE**

R. franchetii paniculigera, *see* **R. MULTIFLORA**

R. francofurtana, *see* **'EMPRESS JOSEPHINE'**

R. francofurtensis, *see* **R. X FRANCOFURTANA**

R. fraxinifolia, *see* **R. BLANDA**

R. fraxinifolia, *see* **R. PENDULINA**

R. freudiana, *see* **'DUPONTII'**

R. froebelii, *see* **R. CANINA FROEBELII**

R. frutetorum, *see* **R. CORIIFOLIA**

R. fujisanensis, *see* **R. LUCIAE**

R. GALLICA, (Linnaeus), HGal, dp; (French Rose, Provins Rose, R. austriaca, R. grandiflora, R. olympica, R. rubra, R. sylvatica); flowers deep pink to crimson, on stout pedicels, 2–3 in., summer bloom; subglobose or turbinate, brick-red fruit; 5 ft growth; Eurosa, Gallicanae (28)

R. GALLICA AGATHA, ((Thory) Loiseleur), HGal, lp; (Agatha); flowers inner petals concave, dbl.

R. gallica centifolia, *see* **'CABBAGE ROSE'**

R. GALLICA CONDITORUM, (Dieck), HGal, dr; (Conditorum, Tidbit Rose); flowers dark red-purple, paler reverse, perhaps identical with Parkinson's Hungarian Rose, semi-dbl.; (28)

R. gallica damascena, *see* **'SUMMER DAMASK'**

R. gallica grandiflora, *see* **'ALIKA'**

R. GALLICA HAPLODONTA, ((Borbas) Braun), Sp, dp; foliage wild variety with simply serrate and woolly styles

R. gallica macrantha, *see* **'GALLICA MACRANTHA'**

R. gallica maxima, *see* **R. GALLICA OFFICINALIS**

R. GALLICA OFFICINALIS, Sp, dp; ((The) Apothecary's Rose of Provins, (The) Apothecary's Rose of Provins, Apothecary's Rose, Double French Rose, Officinalis, R. gallica maxima, R. gallica plena, R. officinalis, R. provincialis, R. X centifolia provincialis, Red Rose of Lancaster, Rose of Provins); flowers deep pink, yellow stamens, semi-dbl.; few prickles; foliage dark; branching growth

R. gallica plena, *see* **R. GALLICA OFFICINALIS**

R. GALLICA PUMILA, ((Jacquin) Seringe), Sp, dp; (R. austriaca pygmaea, R. humilis, R. pumila); flowers red, single; dwarf growth; (28)

R. gallica rosa mundi, *see* **R. GALLICA VERSICOLOR**

R. gallica variegata, *see* **R. GALLICA VERSICOLOR**

R. GALLICA VELUTINAEFLORA, Sp; (R. velutinaeflora); flowers described by déséglise as deep velvety red, late spring; pyriform, reddish-orange fruit

R. GALLICA VERSICOLOR, Sp, pb; (R. gallica rosa mundi, R. gallica variegata, R. mundi, Rosa Mundi); flowers striped white, pink and red, yellow stamens, semi-dbl.; (28); [**R. GALLICA OFFICINALIS** sport]

R. gallicaremensis, *see* **'BURGUNDIAN ROSE'**

R. gebleriana, *see* **R. LAXA**

R. gechouitangensis, *see* **R. X ODORATA**

R. GENTILI, Sp, w

R. GENTILIANA, (Léveillé & Vaniot), Sp, w; (R. polyantha grandiflora, R. wilsonii); flowers creamy white, borne in dense clusters, semi-dbl.; red fruit; red glandular branches; vigorous shrub growth; Eurosa, Synstylae (14); 1907

R. gentiliana, *see* **R. HENRYI**

R. gentilis, *see* **R. PENDULINA GENTILIS**

R. germanica, *see* **R. X FRANCOFURTANA**

R. GIGANTEA, (Collett), Sp, w; (R. macrocarpa, R. X odorata gigantea, R. xanthocarpa); flowers creamy white, 4–5 in.; vigorous growth; climbing to 50 ft; Eurosa, Chinensis (14); 1889

R. GIGANTEA ERUBESCENS, (Focke), Sp, mp; (R. X odorata erubescens)

R. GIRALDII, (Crépin), Sp, pb; flowers pink with white center borne solitary or few, 1 in.; ovoid, 1/3–1/2 in., scarlet fruit; 6 ft growth; Eurosa, Cinnamomeae (14)

R. giraldii, *see* **R. GLAUCA**

R. GIRALDII VENULOSA, (Rehder & Wilson), Sp, pb; 1907

R. glabrata, *see* **R. X SPINULIFOLIA**

R. glandulosa, *see* **R. PENDULINA**

R. glauca, *see* **R. DUMALIS**

R. GLAUCA, (Pourret), Sp, mp; (R. ferruginea, R. giraldii, R. ilseana, R. lurida, R. rubrifolia); flowers pink borne 1–3 together, 1.5 in., spring bloom, subglobose, (see endpapers-rubrifolia), scarlet fruit; foliage reddish; 6 ft growth; Eurosa, Caninae (28)

R. glaucodermis, *see* **R. GYMNOCARPA**

R. glaucophylla, *see* **R. HEMISPHAERICA**

R. GLOMERATA, (Rehder & Wilson), Sp, w; 1 in., borne in dense corymbs, moderate fragrance; subglobose, about 1/3 in., orange-red fruit; 20 ft growth; Eurosa, Synstylae; 1908

R. GLUTINOSA, (Sibthorp & Smith), Sp, mp; (Pine-Scented Rose, R. calabrica, R. ferox, R. libanotica, R. pulverulenta, R. pustulosa); flowers pink, small, moderate, pine-scented fragrance; globose, small (see endpapers) fruit; dwarf, glandular growth; Eurosa, Caninae (35, 42); 1821

R. GLUTINOSA DALMATICA, ((Kerner) Borbas), Sp, mp; (R. dalmatica); ellipsoid fruit; (35, 42)

R. gmelinii, *see* **R. ACICULARIS FENNICA**

R. godetii, *see* **R. MARGINATA GODETII**

R. GRACILIFLORA, (Rehder & Wilson), Sp, mp; flowers pale rose borne solitary, but numerous along the stem, 1.5 in.; Eurosa, Pimpinellifoliae; 1908

R. grandiflora, *see* **R. GALLICA**

R. grandiflora, *see* **R. SPINOSISSIMA ALTAICA**

R. granulosa, *see* **R. MAXIMOWICZIANA**

R. gratiosa, *see* **R. BLANDA**

R. gratissima, *see* **R. CALIFORNICA**

R. graveolens, *see* **R. INODORA**

R. grevillii, *see* **R. MULTIFLORA CARNEA**

R. grosseserrata, *see* **R. MACOUNII**

R. GYMNOCARPA, (Nuttall ex Torrey & Gray), Sp, lp; (Little Woods Rose, R. abietorum, R. glaucodermis); flowers pale pink borne solitary, on short lateral branchlets, 1 in., early summer; small, globose, orange-red fruit; 10 ft growth; Eurosa, Cinnamomeae (14); 1893

R. gypsicola, *see* **R. DUMALIS**

R. hackeliana, *see* **R. HECKELIANA**

R. hakonensis, *see* **R. LUCIAE**

R. HAWRANA, (Kmet), Sp, lp; 2 in.; globose, densely bristly fruit; Eurosa, Caninae (28)

R. headleyensis, *see* **'HEADLEYENSIS'**

R. HECKELIANA, (Trattinnick), Sp, lp; (R. hackeliana); flowers pink, small, borne usually solitary; dwarf growth; Eurosa, Caninae

R. HELENAE, (Rehder & Wilson), Sp, w; (R. floribunda, R. moschata helenae, R. moschata micrantha); 1.5 in., borne in many-flowered umbel-like corymbs 3–6 in. diam, late spring, moderate fragrance; ovoid or oblong-obovoid, about 1/2 in., scarlet fruit; 15 ft growth; Eurosa, Synstylae (14); 1907

R. heliophila, *see* **R. ARKANSANA**

R. heliophila alba, *see* **R. ARKANSANA ALBA**

R. HEMISPHAERICA, (Hermann), Sp, my; (R. glaucophylla, R. rapinii, R. sulphurea, Sulphur Rose); flowers sulfur-yellow, nodding, borne usually solitary, dbl., cupped, late spring, foliage bluish-green; Eurosa, Pimpinellifoliae (28); Int. before 1625

R. hemisphaerica plena, *see* **R. FOETIDA PERSIANA**

R. HEMISPHAERICA RAPINII, ((Boissier) Rowley), Sp, my; (R. rapinii); (28); 1933

R. HEMSLEYANA, (Täckholm), Sp, lp; (R. macrophylla, R. setipoda); flowers corymbose, 2 in., late spring; ovoid, about 1 in., fruit; stems long, with a distinct neck; Eurosa, Cinnamomeae (42); 1904

R. HENRYI, (Boulenger), Sp, w; (R. gentiliana); flowers corymbose, 1.5 in., summer bloom, moderate fragrance; globose, about 1/3 in. fruit; Eurosa, Synstylae (14); 1907

R. heterophylla, *see* **R. MOLLIS**

R. hibernica, *see* **'TACKHOLMII'**

R. hillieri, *see* **'HILLIER ROSE'**

R. hirtula, *see* **R. ROXBURGHII HIRTULA**

R. hispanica, *see* **R. POUZINII**

R. hispida, *see* **R. POMIFERA**

R. HISPIDA, (Sims, not others), Sp, my; (R. lutescens, R. spinosissima hispida); flowers sulfur-yellow, rather, 2.5–3 in.; (28, 29); Prior to 1781

R. hoffmeisteri, *see* **R. MACROPHYLLA**

R. HOLODONTA, (Stapf), Sp, lp; (R. moyesii rosea); foliage leaves coarsely serrate; Eurosa, Cinnamomeae (28); 1908

R. hookeriana, *see* **R. MACROPHYLLA**

R. HORRIDA, (Fischer), Sp, w; (R. ferox); 1.5 in., borne solitary or few; subglobose, 1/3–1/2 in. across, dark red fruit; prickly; foliage very small, roundish leaflets; low shrub growth; Eurosa, Caninae (35); 1796

R. hudsoniana, *see* **R. PALUSTRIS**

R. hudsoniana scandens, *see* **R. PALUSTRIS SCANDENS**

R. HUGONIS, (Hemsley), Sp, my; (Father Hugo Rose, Father Hugo's Rose, Golden Rose of China, R. xanthina); single, 2.5 in., borne solitary, on slender glabrous pedicels, early spring; depressed-globose, deep scarlet fruit; 6 ft branches drooping growth; Eurosa, Pimpinellifoliae (14); 1899

R. humilis, *see* **R. CAROLINA**

R. humilis, *see* **R. GALLICA PUMILA**

R. humilis, *see* **R. MARGINATA**

R. humilis grandiflora, *see* **R. LAXA**

R. humilis lucida, *see* **R. VIRGINIANA**

R. humilis triloba, *see* **R. CAROLINA TRILOBA**

R. humilis villosa, *see* **R. CAROLINA VILLOSA**

R. hystrix, *see* **R. LAEVIGATA**

R. illinoensis, *see* **R. SPINOSISSIMA**

R. ilseana, *see* **R. GLAUCA**

R. incarnata, *see* **'GREAT MAIDEN'S BLUSH'**

R. inconsiderata, *see* **R. POUZINII**

R. indica, *see* **R. CHINENSIS**

R. indica, *see* **R. CYMOSA**

R. indica cruenta, *see* **'SANGUINEA'**

R. indica fragrans, *see* **R. X ODORATA**

R. indica longifolia, *see* **R. CHINENSIS LONGIFOLIA**

R. indica noisettiana, *see* **'NOISETTE ROSE'**

R. indica ochroleuca, *see* **'PARKS' YELLOW TEA-SCENTED CHINA'**

R. indica odorata, *see* **R. X ODORATA**

R. indica odoratissima, *see* **R. X ODORATA**

R. indica pumila, *see* **R. CHINENSIS MINIMA**

R. indica semperflorens, *see* **'SLATER'S CRIMSON CHINA'**

R. indica vulgaris, *see* **R. CHINENSIS**

R. inermis, *see* **R. X FRANCOFURTANA**

R. INODORA, (Fries), Sp, w; (R. agrestis inodora, R. caryophyllacea, R. elliptica, R. graveolens, R. klukii); flowers white or pink, 1–1.5 in.; ovoid, bright red fruit; Eurosa, Caninae (35, 42)

R. inodora, *see* **R. OBTUSIFOLIA**

R. intermedia, *see* **R. MULTIFLORA**

R. involucrata, *see* **R. CLINOPHYLLA**

R. irridens, *see* **R. LONGICUSPIS**

R. jackii, *see* **R. MAXIMOWICZIANA JACKII**

R. jackii pilosa, *see* **R. MAXIMOWICZIANA PILOSA**

R. jasminoides, *see* **R. LUCIAE**

R. jundzilli godetii, *see* **R. MARGINATA GODETII**

R. jundzillii, *see* **R. MARGINATA**

R. kamtchatica, *see* **R. RUGOSA KAMTCHATICA**

R. kelleri, *see* **R. MAXIMOWICZIANA JACKII**

R. kentuckensis, *see* **R. SETIGERA TOMENTOSA**

R. klukii, *see* **R. INODORA**

R. KORDESII, (Wulff), (classification as "species" disputed), dp; semi-dbl., cupped; elliptic-ovoid fruit; foliage shining green; (28); 1952; (a variety that arose in cultivation as a result of spontaneous chromosome doubling (see Max Graf))

R. KOREANA, (Komarov), Sp, lp; flowers white, flushed pink, borne solitary, but numerous along the, 1 in.; ovoid, 1/2 in. long, orange-red fruit; Eurosa, Pimpinellifoliae (14); 1917

R. korsakoviensis, *see* **R. ACICULARIS**

R. LAEVIGATA, (Michaux), Sp, w; (Cherokee Rose, R. camellia, R. cherokeensis, R. hystrix, R. nivea, R. sinica, R. ternata, R. triphylla); flowers white, rarely rose, borne solitary, 2.5–3.5 in., early spring, moderate fragrance; large, obovoid, bristly, yellow, later turning brown fruit; foliage glossy, leaflets 3; tall, climbing growth; Eurosa, Laevigatae (14); 1759

R. LATIBRACTEATA, (Boulenger), Sp, mp; in two's or corymbose; foliage bracts broad, leaflets 7, up to 1 in. long; 1936

R. laurentiae, *see* **R. CHINENSIS MINIMA**

R. lawranceana, *see* **R. CHINENSIS MINIMA**

R. LAXA, (Lindley), Sp, mp; (R. carolina grandiflora, R. humilis grandiflora, R. lindleyi, R. obovata); flowers bright pink, 2 in.; foliage larger than r; Eurosa, Carolinae (28)

R. laxa, *see* **R. CANINA FROEBELII**

R. LAXA, (Retzius), Sp, w; (R. gebleriana, R. soongarica); flowers small; oblong-ovoid, small; fruit; foliage small, light green; Eurosa, Cinnamomeae (28)

R. lebrunei, *see* **R. MULTIFLORA CARNEA**

R. lehmanniana, *see* **R. BEGGERIANA**

R. LESCHENAULTII, (Wight & Arnott), Sp, w; (R. moschata leschenaultii); flowers larger, in few-flowered corymbs; Eurosa, Synstylae (14)

R. leucantha, *see* **R. CANINA**

R. l'heritierana, *see* **'BOURSAULT ROSE'**

R. libanotica, *see* **R. GLUTINOSA**

R. lindleyana, *see* **R. CLINOPHYLLA**

R. lindleyi, *see* **R. LAXA**

R. linkii, *see* **R. MULTIFLORA**

R. LONGICUSPIS, (Bertoloni), Sp, w; (R. charbonneaui, R. irridens, R. lucens, R. willmottiana); flowers petals silky outside, borne corymbose, 1.5 in.; ovoid, scarlet or orange-red fruit; foliage half-evergreen; 20 ft growth; tender; Eurosa, Synstylae (14)

R. longifolia, *see* **R. CHINENSIS LONGIFOLIA**

R. lucens, *see* **R. LONGICUSPIS**

R. LUCIAE, (Franchet & Rochebrune), Sp, w; (R. franchetii, R. fujisanensis, R. hakonensis, R. jasminoides, R. luciae fujisanensis); flowers smaller; habit more upright growth; Somewhat less hardy; Eurosa, Synstylae (14); 1880

R. luciae, *see* **R. WICHURANA**

R. luciae fujisanensis, *see* **R. LUCIAE**

R. luciae wichurana, *see* **R. WICHURANA**

R. luciaetaquetiana, *see* **R. WICHURANA**

R. lucida, *see* **R. BRACTEATA**

R. lucida, *see* **R. VIRGINIANA**

R. lucida alba, *see* **R. VIRGINIANA**

R. lucida plena, *see* **R. X RAPA**

R. lurida, *see* **R. GLAUCA**

R. lutea, *see* **R. FOETIDA**

R. lutea hoggii, *see* **'HARISON'S YELLOW'**

R. lutea persiana, *see* **R. FOETIDA PERSIANA**

R. lutea plena, *see* **R. FOETIDA PERSIANA**

R. lutea punicea, *see* **R. FOETIDA BICOLOR**

R. luteabicolor, *see* **R. FOETIDA BICOLOR**

R. lutescens, *see* R. hispida

R. lutetiana, *see* **R. CANINA LUTETIANA**

R. lyellii, *see* **R. CLINOPHYLLA**

R. lyonii, *see* **R. CAROLINA VILLOSA**

R. lyonii alba, *see* **R. CAROLINA ALBA**

R. macartnea, *see* **R. BRACTEATA**

R. macdougalii, *see* **R. NUTKANA HISPIDA**

R. MACOUNII, (Greene), Sp, lp; (R. grosseserrata, R. naiadum, R. subnuda, R. woodsii); flowers pale pink, small; depressed-globose fruit; low growth; Eurosa, Cinnamomeae (14, 21); Int. prior to 1826

R. macounii, *see* **R. WOODSII**

R. macrantha, *see* **'GALLICA MACRANTHA'**

R. MACRANTHA, (hort., not Desportes), Sp, lp; flowers blush-pink, large, single; 3/4 in., subglobose, dull red fruit; lax, trailing shrub growth

R. macrocarpa, *see* **R. GIGANTEA**

R. macrophylla, *see* **R. HEMSLEYANA**

R. MACROPHYLLA, (Lindley), Sp, dp; (R. hoffmeisteri, R. hookeriana); flowers red borne 1–3 together, 2 in.; oblong-ovoid, $1-1\frac{1}{2}$ in. long, red fruit; Eurosa, Cinnamomeae (14, 28); 1818

R. macrophylla acicularis, *see* **R. PERSETOSA**

R. macrophylla crasseaculeata, *see* R. setipoda

R. MACROPHYLLA DONCASTERII (DONCASTERII), Sp, dp, 1930; flowers deep pink; hips flagon-shaped, large, red; foliage purplish-green; stems plum colored; arching growth (6 ft); Hurst, C.C.; Int. E. Doncaster of J. Burrell & Co., ca. 1930

R. macrophylla gracilis, *see* **R. PERSETOSA**

R. macrophylla rubrostaminea, *see* **R. MOYESII**

R. majalis, *see* **R. CINNAMOMEA PLENA**

R. manca, *see* **R. NUTKANA**

R. manettii, *see* **'MANETTI'**

R. MARGINATA, (Wallroth), Sp, lp; (R. humilis, R. jundzillii, R. trachyphylla, R. zagrabiensis); flowers pink borne solitary or few, 2–2.5 in.; globose to ellipsoid, scarlet fruit; Eurosa, Caninae (42)

R. MARGINATA GODETII, ((Grenier) Rehder), Sp, lp; (R. godetii, R. jundzilli godetii); flowers small; low shrub growth

R. MARRETII, (Léveillé), Sp, mp; (R. rubrostipullata); flowers pink borne usually 3–6, 1.5–2 in.; subglobose, 1/3–1/2 in. fruit; stems dark purple; Eurosa, Cinnamomeae (14)

R. maximilianii, *see* **R. WOODSII**

R. MAXIMOWICZIANA, (Regel), Sp, w; (R. faureri, R. granulosa); 1.5 in., borne in many-flowered corymbs; partially climbing growth; Eurosa, Synstylae (14)

R. MAXIMOWICZIANA JACKII, ((Rehder) Rehder), Sp, w; (Jack Rose, R. coreana, R. jackii, R. kelleri); prickles without bristles; (14); 1905

R. MAXIMOWICZIANA PILOSA, ((Nakai) Nakai), Sp, w; (R. jackii pilosa); flowers leaf and stalks pubescent; (14)

R. melina, *see* **R. X ENGELMANNII**

R. mexicana, *see* **R. CAROLINA GLANDULOSA**

R. MICRANTHA, (Smith), Sp, lp; (R. floribunda, R. nemorosa, R. rubiginosa, R. rubiginosa nemoralis); flowers pale pink, small; 6 ft growth; Eurosa, Caninae (35, 42)

R. microcarpa, *see* **R. CYMOSA**

R. microcarpa, *see* **R. MULTIFLORA**

R. microphylla, *see* **R. ROXBURGHII**

R. microphyllahirtula, *see* **R. ROXBURGHII HIRTULA**

R. MINUTIFOLIA, (Englemann), Sp; flowers pink or white short-pediceled, 1 in., early spring; subglobose, hispid fruit; 4 ft growth; Minutifoliae (Hesperhodos), (14)

R. mirifica, *see* **R. STELLATA MIRIFICA**

R. mitissima, *see* **R. SPINOSISSIMA INERMIS**

R. mohavensis, *see* **R. WOODSII**

R. mokanensis, *see* **R. WICHURANA**

R. MOLLIS, (Smith), Sp, dp; (R. heterophylla, R. mollissima, R. villosa, R. villosa mollissima); flowers deep pink, rarely white borne 1–3 together, 1.5–2 in., summer bloom, small fruit; foliage branchlets purple, more or less bloomy; 4 ft growth; Eurosa, Caninae (28); 1818

R. mollissima, *see* **R. MOLLIS**

R. MONTANA, (Chaix), Sp, lp; flowers pink borne solitary or few, 1.5 in.; oblong-ovoid, 3/4 in. fruit; vigorous growth; Eurosa, Caninae (42)

R. montana chavinii, *see* **R. CHAVINII**

R. MONTEZUMAE, (Humboldt & Bonpland ex Thory), Sp, dp; flowers pale red, 1.5 in., june; 3 ft growth; Eurosa, Caninae (35)

R. MOSCHATA, (Herrmann), Sp, w; (Musk Rose, R. ruscinonensis); 1.5–2 in., borne usually in 7-flowered corymbs, summer-fall bloom, moderate, musk fragrance; ovoid, small fruit; Eurosa, Synstylae (14); 1540

R. MOSCHATA ABYSSINICA, ((Lindley) Rehder), Sp, w; (R. abyssinica); prickles more prickly; tender growth; (14)

R. moschata autumnalis, *see* **'NOISETTE ROSE'**

R. moschata flore semipleno, *see* **R. MOSCHATA PLENA**

R. moschata helenae, *see* **R. HELENAE**

R. moschata leschenaultii, *see* **R. LESCHENAULTII**

R. moschata micrantha, *see* **R. HELENAE**

R. moschata nastarana, *see* **'NASTARANA'**

R. MOSCHATA NEPALENSIS, (Lindley), Sp, w; (Himalayan Musk Rose, R. brownii, R. brunonii); foliage like r moschata but dull green, downy; (14); 1822

R. MOSCHATA PLENA, (Weston), Sp, w; (R. moschata flore semipleno); dbl.; (14)

R. MOYESII, (Hemsley & Wilson), Sp, mr; (R. fargesii, R. macrophylla rubrostaminea); flowers deep blood-red through deep rose to light pink borne solitary, 1.75–2.5 in., summer bloom; oblong-ovoid, narrowed into a neck, $2-2\frac{1}{2}$ in. long, deep orange-red fruit; 10 ft growth; Eurosa, Cinnamomeae (42); 1894, Re-int. 1903

R. MOYESII FARGESII, (Rolfe), Sp, mr; (R. fargesii); foliage leaflets smaller, broad-oval to suborbicular; (28)

R. moyesii rosea, *see* **R. HOLODONTA**

R. MULLIGANII, (Boulenger), Sp, w; flowers corymbose, 2 in.; ovoid, 1/2 in. long fruit; Eurosa, Synstylae (14); 1917

R. MULTIBRACTEATA, (Hemsley & Wilson), Sp, lp; (R. reducta); flowers pink corymbose, with numerous crowded bracts, sometimes few, 1.25 in.; ovoid, 1/3–1/2 in. long, orange-red fruit; 6 ft growth; Eurosa, Cinnamomeae (28); 1910

R. MULTIFLORA, (Thunberg), Sp, w; (Multiflora Japonica, R. dawsoniana, R. franchetii paniculigera, R. intermedia, R. linkii, R. microcarpa, R. multiflora thunbergiana, R. polyantha, R. thunbergii, R. thyrsiflora, R. wichurae); flowers usually white, or more, borne in many-flowered pyramidal cor, 0.75 in., summer bloom, small, globular fruit; recurving or climbing branches.

growth; Eurosa, Synstylae (14, 28); ca. 1810

R. MULTIFLORA CALVA, (Franchet & Savatier), Sp, w; (R. calva)

R. MULTIFLORA CARNEA, (Thory), Sp, lp; (R. blinii, R. florida, R. grevillii, R. lebrunei, R. multiflora plena, R. rubeoides); dbl.; (14); 1804

R. MULTIFLORA CATHAYENSIS, (Rehder & Wilson), Sp, lp; (R. cathayensis, R. uchiyamana); flowers pink borne in rather corymbs, 1.5 in., flat; (14); 1907

R. multiflora dawsoniana, *see* **R. MULTIFLORA ROSEIFLORA**

R. MULTIFLORA NANA HORT., (hort.), Pol; flowers white to pink, single to very small, in clusters, dbl., recurrent bloom; bushy, dwarf (1–2 ft) growth

R. multiflora platyphylla, *see* **'SEVEN SISTERS'**

R. multiflora plena, *see* **R. MULTIFLORA CARNEA**

R. MULTIFLORA ROSEIFLORA, ((Focke) Rehder), Sp, lp; (R. multiflora dawsoniana); semi-dbl., 1 in.

R. multiflora thunbergiana, *see* **R. MULTIFLORA**

R. MULTIFLORA WATSONIANA, ((Crépin) Matsumura), Sp, lp, 1870; (Bamboo Rose, R. watsoniana); (14); Int. from a Japanese garden in 1870

R. mundi, *see* **R. GALLICA VERSICOLOR**

R. muriculata, *see* **R. NUTKANA**

R. MURIELAE, (Rehder & Wilson), Sp, w; 1 in., borne in 3–7 flowered corymbs; ellipsoid, 1/2–3/4 in., orange-red fruit; growth to 8 ft; Eurosa, Cinnamomeae (28); 1904

R. muscosa, *see* **'CENTIFOLIA MUSCOSA'**

R. muscosa alba, *see* **'SHAILER'S WHITE MOSS'**

R. muscosa simplex, *see* **'MUSCOSA SIMPLEX'**

R. mutabilis, *see* **'MUTABILIS'**

R. myriacantha, *see* **R. SPINOSISSIMA MYRIACANTHA**

R. myriadenia, *see* **R. YAINACENSIS**

R. naiadum, *see* **R. MACOUNII**

R. nankinensis, *see* **R. CHINENSIS**

R. NANOTHAMNUS, (Boulenger), Sp, mp; Eurosa, Cinnamomeae (14); 1935

R. nemorosa, *see* **R. MICRANTHA**

R. nipponensis, *see* **R. ACICULARIS NIPPONENSIS**

R. NITIDA, (Willdenow), Sp, mp; (R. blanda, R. redutea rubescens, R. rubrispina, Shining Rose); flowers bright pink solitary, or in few-flowered corymbs on slender, 1–2 in., summer bloom; $1\frac{1}{2}$ ft growth; Eurosa, Carolinae (14)

R. NITIDA SPINOSA, (Lewis), Sp, mp; (R. carolina setigera); stems with enlarged prickles new england & e canada

R. nivea, *see* **R. LAEVIGATA**

R. NUTKANA, (Presl), Sp, mp; (Nutka Rose, R. manca, R. muriculata, R. spaldingii); flowers pink usually solitary, 2–2.5 in., summer bloom; globose, without a neck (see endpapers) fruit; 5 ft growth; Eurosa, Cinnamomeae (42); ca. 1876

R. NUTKANA HISPIDA, (Fernald), Sp, mp; (R. macdougalii)

R. nuttalliana, *see* **R. PALUSTRIS NUTTALLIANA**

R. obovata, *see* **R. LAXA**

R. OBTUSIFOLIA, (Desvaux), Sp; (R. canina tomentella, R. inodora, R. tomentella); flowers white or rose, 1–few, 1.25 in.; ovoid, 1/2–3/4 in. fruit; foliage leaflets $1\frac{1}{2}$ in. long, tip rounded, 5–7 long; 12 ft growth; Eurosa, Caninae (35)

R. ochroleuca, *see* **R. SPINOSISSIMA LUTEOLA**

R. officinalis, *see* **R. GALLICA OFFICINALIS**

R. olympica, *see* **R. GALLICA**

R. omeiensis, *see* **R. SERICEA OMEIENSIS**

R. omeiensis chrysocarpa, *see* **R. SERICEA CHRYSOCARPA**

R. omeiensis pteracantha, *see* **R. SERICEA PTERACANTHA**

R. omissa, *see* **R. SHERARDII**

R. orbicularis, *see* **R. SERTATA**

R. oreophila, *see* **R. X ENGELMANNII**

R. ORIENTALIS, (Dupont ex Seringe), Sp, lp; flowers pink, solitary, short-pediceled; dwarf growth; Eurosa, Caninae (35)

R. oulengensis, *see* **R. X ODORATA**

R. oxyacanthos, *see* **R. X KOCHIANA**

R. oxyodon, *see* **R. PENDULINA OXYODON**

R. PALUSTRIS, (Marshall), Sp, mp; (R. carolina, R. caroliniana, R. corymbosa, R. elongata, R. fragrans, R. hudsoniana, R. pensylvanica, R. salicifolia, R. virginiana, Swamp Rose); flowers pink usually corymbose, 2 in., spring-summer bloom, growth to 8 ft; Eurosa, Carolinae (14, 28); 1726

R. PALUSTRIS INERMIS, ((Regel) Lewis), Sp, mp; (R. carolina inermis); no prickles; (14)

R. PALUSTRIS NUTTALLIANA, (Rehder), Sp, mp; (R. carolina nuttalliana, R. nuttalliana); (42)

R. PALUSTRIS SCANDENS, Sp, mp; (R. hudsoniana scandens, Swamp Rose); dbl., spring bloom

R. parmentieri, *see* **R. DAVIDII ELONGATA**

R. parviflora, *see* **R. CAROLINA**

R. parviflora glandulosa, *see* **R. CAROLINA GLANDULOSA**

R. parvifolia, *see* **'BURGUNDIAN ROSE'**

R. PENDULINA, (Linnaeus), Sp, dp; (Alpine Rose, R. alpina, R. cinnamomea, R. fraxinifolia, R. glandulosa); flowers pink usually solitary or 2–5 together, single, 2 in., spring bloom; fruit usually nodding, oblong or ovoid, with an elongated neck, scarlet; 3 ft growth; Eurosa, Cinnamomeae (28)

R. PENDULINA GENTILIS, ((Sternberg) R. Keller), Sp, dp; (R. gentilis); flowers deep pink; low shrub with bristly branches growth

R. PENDULINA OXYODON, ((Boissier) Rehder), Sp, mp; (R. oxyodon); flowers rose, solitary; (28)

R. PENDULINA PYRENAICA, ((Gouan) R. Keller), Sp, mp; (Pyrenees Rose, R. pyrenaica); foliage leaves glandular; dwarf growth; (28)

R. pennsylvanica, *see* R. virginiana

R. pennsylvanica plena, *see* **R. CAROLINA PLENA**

R. pensylvanica, *see* **R. CAROLINA**

R. pensylvanica, *see* **R. PALUSTRIS**

R. PERSETOSA, (Rolfe), Sp, dp; (R. davidii persotosa, R. macrophylla acicularis, R. macrophylla gracilis); flowers deep pink in large panicles, 1 in.; Eurosa, Cinnamomeae (14); 1895

R. phoenica, *see* **R. PHOENICIA**

R. PHOENICIA, (Boissier), Sp, w; (R. phoenica); in many-flowered pyramidal corymbs; not hardy; Eurosa, Synstylae (14); ca. 1885

R. pimpinellifolia, *see* R. spinosissima

R. pimpinellifolia, *see* **R. SPINOSISSIMA PIMPINELLIFOLIA**

R. pimpinellifolia elasmacantha, *see* **R. ELASMACANTHA**

R. pimpinellifolia tuschetica, *see* **R. TUSCHETICA**

R. pimpinellifoliainermis, *see* **R. SPINOSISSIMA INERMIS**

R. PINETORUM, (A. Heller), Sp, dp; flowers deep rose, usually solitary, floral tube usually not glandular, 1.5 in.; foliage leaflets doubly serrate, with gland-tipped teeth; 3 ft growth; Eurosa, Cinnamomeae

R. PISOCARPA, (Gray), Sp, mp; flowers pink on short pedicels in several to many-flowered corymbs, 1 in., summer bloom; globose, with a very short neck (see endpapers). fruit; Eurosa, Cinnamomeae (14, 21); Int. ca. 1882

R. pisocarpa ultramontana, *see* **R. WOODSII ULTRAMONTANA**

R. pissartii, *see* **'NASTARANA'**

R. platyphylla, *see* **'SEVEN SISTERS'**

R. poetica, *see* **R. WOODSII FENDLERI**

R. polyantha, *see* **R. MULTIFLORA**

R. polyantha grandiflora, *see* **R. GENTILIANA**

R. polyantho, *see* **'Summer Damask'**

R. pomifera, (Herrmann), Sp, mp; (Apple rose, R. hispida, R. villosa); flowers pink 1–3 together, 1.5–2 in., summer bloom; ovoid or subglobose, to 1 in. fruit; 6 ft growth; Eurosa, Caninae (28); 1771

R. pomifera duplex, *see* **'Duplex'**

R. pomponia, *see* **'Burgundian Rose'**

R. pomponia, *see* **'Rose de Meaux'**

R. pouzinii, (Trattinnick), Sp, mp; (R. hispanica, R. inconsiderata, R. rubiginosa parvifolia); flowers pink, small, 1–few; ellipsoid, small fruit; 3–6 ft growth; Eurosa, Caninae (42)

R. praelucens, (Byhouwer), Sp, lp; near to r. roxburghii, but less bristly, leaflets larger, fe; Microphyllae (Platyrhodon)

R. pratensis, *see* **R. carolina**

R. pratincola, *see* **R. arkansana**

R. pratincola alba, *see* **R. arkansana alba**

R. prattii, (Hemsley), Sp, mp; flowers pink 1–3 together, 0.75 in.; subglobose to ovoid, 1/4–1/3 in., scarlet fruit; growth to 8 ft; Eurosa, Cinnamomeae (14)

R. primula, (Boulenger), Sp, ly; (Incense Rose, R. ecae, R. sweginzowii, R. xanthina, R. xanthina normalis); flowers yellowish white, very early spring; red prickles; foliage small, stiff, remarkably incense-scented; thin, flexible stems; 2–4 ft growth; Eurosa, Pimpinellifoliae (14); 1910

R. procera, *see* White Rose of York

R. prostrata, *see* **R. sempervirens prostrata**

R. provincialis, *see* **R. gallica officinalis**

R. pseudindica, *see* **'Fortune's Double Yellow'**

R. pseudoscabrata, *see* **R. canina**

R. pubescens, *see* **R. rugosa plena**

R. pulchella, *see* **R. spinosissima**

R. pulchella, *see* **'Rose de Meaux'**

R. pulverulenta, *see* **R. glutinosa**

R. pumila, *see* **R. gallica pumila**

R. punicea, *see* **R. foetida bicolor**

R. pusilla, *see* **R. carolina villosa**

R. pustulosa, *see* **R. glutinosa**

R. pyrenaica, *see* **R. pendulina pyrenaica**

R. pyrifera, *see* **R. woodsii**

R. rapinii, *see* **R. hemisphaerica**

R. rapinii, *see* **R. hemisphaerica rapinii**

R. reclinata, *see* **'Boursault Rose'**

R. reducta, *see* **R. multibracteata**

R. redutea rubescens, *see* **R. nitida**

R. regeliana, *see* **R. rugosa**

R. regelii, *see* **R. beggeriana**

R. remensis, *see* **'Burgundian Rose'**

R. repens, *see* **R. arvensis**

R. reuteri, *see* **R. dumalis**

R. rouletii, *see* **'Rouletii'**

R. roxburghii, (Trattinnick), Sp, mp; (Burr Rose, Chestnut rose, Chinquapin Rose, R. microphylla, R. roxburghii plena); bud prickly, like a chestnut burr; flowers lilac pink borne often solitary, on short pedicels, dbl., 2–2.5 in., spring-summer bloom; depressed-globose, 1–1½ in. fruit; 6 ft growth; Microphyllae (Platyrhodon), (14)

R. roxburghii hirtula, ((Regel) Rehder & Wilson), Sp, mp; (R. hirtula, R. microphyllahirtula); flowers lilac pink; foliage pubescent beneath; Microphyllae (Platyrhodon), (14); Prior to 1880

R. roxburghii normalis, (Rehder & Wilson), Sp, lp; (Single Chestnut Rose); single; Microphyllae (Platyrhodon), (14); 1908

R. roxburghii plena, *see* **R. roxburghii**

R. rubella, *see* **R. X reversa**

R. rubeoides, *see* **R. multiflora carnea**

R. rubicans, *see* **'Great Maiden's Blush'**

R. rubifolia, *see* **R. setigera tomentosa**

R. rubiginosa, *see* **R. eglanteria**

R. rubiginosa, *see* **R. micrantha**

R. rubiginosa magnifica, *see* **'Magnifica'**

R. rubiginosa nemoralis, *see* **R. micrantha**

R. rubiginosa parvifolia, *see* **R. pouzinii**

R. rubra, *see* **R. gallica**

R. rubrifolia, *see* **R. glauca**

R. rubrispina, *see* **R. nitida**

R. rubrostipullata, *see* **R. marretii**

R. rubus, (Léveillé & Vaniot), Sp, w; (Blackberry Rose, R. ernestii); 1–1.33 in., borne in dense corymbs, moderate fragrance; subglobose, about 1/3 in., dark scarlet fruit; climbing, 20 ft growth; Eurosa, Synstylae (14); 1907

R. rubus nudescens, ((Stapf) Rowley), Sp, w; (R. ernestii nudescens)

R. rudiuscula, (Greene), Sp, mp; flowers pink corymbose, few, 2 in.; globose, 1/2 in., red fruit; prickles very bristly; Eurosa, Carolinae (28); 1917

R. rugosa, (Thunberg), Sp, m; (Hedgehog Rose, Japanese Rose, Kiska Rose, R. ferox, R. regeliana, Ramanas Rose, Rugosa Rose, Tomato Rose); flowers purple or white borne solitary or few, 2.5–3.5 in., spring-fall bloom; depressed-globose, to 1 in., brick-red fruit; foliage rugose, shining, darkgreen; 6 ft growth; Eurosa, Cinnamomeae (14)

R. rugosa alba, ((Ware) Rehder), Sp, w; (R. rugosa albiflora); (14)

R. rugosa albiflora, *see* **R. rugosa alba**

R. rugosa albo-plena, (Rehder), Sp, w; dbl.

R. rugosa calocarpa, *see* **'Calocarpa'**

R. rugosa chamissoniana, (C.A. Meyer), Sp; (R. coruscans); almost free of bristles

R. rugosa kamtchatica, ((Ventenat) Regel), Sp; (R. kamtchatica); (14); ca. 1770

R. rugosa plena, (Regel), Sp, m; (R. pubescens, R. rugosa rebro-plena); flowers purple, dbl.; very hardy shrub on the prairies, where it is called Empress of the North; (14)

R. rugosa rebro-plena, *see* **R. rugosa plena**

R. rugosa repens alba, *see* **'Paulii'**

R. rugosa repens rosea, *see* **'Paulii Rosea'**

R. rugosa rosea, (Rehder), Sp, mp; (14)

R. rugosa rubra, (hort., not Rehder), Sp, m; (R. rugosa typica); flowers brilliant magenta-purple the largest single-flowered form; very vigorous growth; (14)

R. rugosa rugosa, Sp, m; (R. rugosa thunbergiana); flowers purple; (14)

R. rugosa scabrosa, *see* **'Scabrosa'**

R. rugosa thunbergiana, *see* **R. rugosa rugosa**

R. rugosa typica, *see* **R. rugosa rubra**

R. ruscinonensis, *see* **R. moschata**

R. rydbergii, *see* **R. arkansana**

R. sabinii, *see* **R. X involuta**

R. salicifolia, *see* **R. palustris**

R. salictorum, (Rydberg), Sp, mp; flowers pink, borne in corymbs, 1.5 in.; globose, about 1/2 in., fruit; prickles nearly spineless; foliage leaflets ovate oblong, mostly 2 in.long, 5–7; growth to about 12 ft; Eurosa, Cinnamomeae (14)

R. sancta, *see* **R. woodsii**

R. sandbergii, *see* **R. woodsii**

R. sarrulata, *see* **R. carolina glandulosa**

R. saturata, (Baker), Sp, dr; flowers dark red, anthers purple borne solitary, 2 in.; globose-ovoid, 3/4 in. long, coral-red fruit; growth to 8 ft; Eurosa, Cinnamomeae (28); 1907

R. saxatilis, *see* **R. corymbifera**

R. sayi, *see* **R. acicularis**

R. sayi, *see* **R. acicularis sayi**

R. scandens, *see* **R. sempervirens scandens**

R. scopulosa, *see* **'Pokornyana'**

R. semperflorens, *see* **'Autumn Damask'**

R. semperflorens, *see* **'Slater's Crimson China'**

R. semperflorens minima, *see* **R. chinensis minima**

R. sempervirens, (Linnaeus), Sp, w; (R. alba Allioni, R. atrovirens, R. balearica); 2 in., borne in few-flowered, rarely many-flowered corymbs, summer bloom, slight fragrance; subglobose or ovoid, orange-red fruit; foliage evergreen; Eurosa, Synstylae (14, 21, 28)

R. sempervirens anemoniflora, *see* **R. ANEMONEFLORA**

R. SEMPERVIRENS PROSTRATA, (Desvaux), Sp, w; (R. prostrata); ovoid fruit

R. SEMPERVIRENS SCANDENS, ((Miller) De Candolle), Sp, w; (R. scandens); subglobose fruit

R. sepium, *see* **R. AGRESTIS**

R. SERAFINII, (Viviani), Sp, lp; (R. apennina, R. seraphinii); flowers pink, solitary, very short-stalked, 1 in.; globose-ovoid fruit; 1–3 ft growth; Eurosa, Caninae (35)

R. seraphinii, *see* **R. SERAFINII**

R. seraphinii, *see* **R. SICULA**

R. SERICEA, (Lindley), Sp, w; (R. tetrapetala, R. wallichii); 4–5 petals, 1–1.5 in., early spring; globose or turbinate fruit; 12 ft growth; Eurosa, Pimpinellifoliae (14); 1822

R. SERICEA CHRYSOCARPA, ((Rehder) Rowley), Sp, w; (R. omeiensis chrysocarpa); bright yellow fruit

R. SERICEA DENUDATA, (Franchet), Sp, w; branches unarmed

R. SERICEA HOOKERI, (Regel), Sp, w; branches glandular

R. SERICEA OMEIENSIS, ((Rolfe) Rowley), Sp, w; (R. omeiensis); single; foliage leaflets more numerous; (14); 1901

R. SERICEA POLYPHYLLA, (Geier), Sp, w; foliage leaflets numerous; (14)

R. SERICEA PTERACANTHA, (Franchet), Sp, w; (R. omeiensis pteracantha); single; stems with large, wing-like prickles, deep red and semi-trans; (14); 1890

R. serpens, *see* **R. ARVENSIS**

R. SERTATA, (Rolfe), Sp, mp; (R. orbicularis, R. webbiana); flowers rose or rose-purple solitary on short branchlets, 2–2.5 in.; 5 ft growth; Eurosa, Cinnamomeae (14); 1904

R. SETIGERA, (Michaux), Sp, dp; (Prairie Rose, R. fenestrata, R. trifoliata); flowers deep rose, fading to whitish in few-flowered corymbs, 2 in., summer bloom, slight fragrance; globular, 1/3 in. fruit; 6 ft branches recurving or climbing growth; Eurosa, Synstylae (14); 1810

R. SETIGERA INERMIS, (Palmer & Steyermark), Sp, dp; no prickles; foliage leaves glabrous; 1923

R. SETIGERA SERENA, (Palmer & Steyermark), Sp, dp; no prickles; foliage leaves pubescent; (14); 1924

R. SETIGERA TOMENTOSA, (Torrey & Gray), Sp, dp; (R. cursor, R. kentuckensis, R. rubifolia); foliage leaves tomentose beneath; (14); 1800

R. setigera variegata, Sp, pb

R. setipoda, *see* **R. HEMSLEYANA**

R. SETIPODA, (Hemsley & Wilson), Sp, lp; (R. macrophylla crasseaculeata); flowers pale pink in loose corymbs, 2 in., summer bloom; oblong-ovoid with a narrow neck, about 1 in. long, deep red fruit; foliage sweetbriar-scented; 10 ft growth; Eurosa, Cinnamomeae (28, 42); 1895

R. SHERARDII, (Davies), Sp, dp; (R. omissa, R. subglobsa); flowers deep pink, 1.5–2 in.; ovoid or pear-shaped, 1/2–3/4 in. across fruit; Eurosa, Caninae (28)

R. SHERARDII PERTHENSIS, (Harrison), Sp, dp; remarkable variety with densely glandular hispid fruit; (35)

R. sibirica, *see* **R. SPINOSISSIMA ALTAICA**

R. SICULA, (Trattinnick), Sp; (R. seraphinii); flowers deep red to whitish solitary, rarely to 4, short-stalked, 1–1.25 in.; small, globose fruit; low, $1\frac{1}{2}$–2 ft growth; Eurosa, Caninae (35)

R. silverhjelmii, *see* **R. BEGGERIANA**

R. silvestris, *see* **R. ARVENSIS**

R. sinica, *see* **R. CHINENSIS**

R. sinica, *see* **R. LAEVIGATA**

R. SINOWILSONII, (Hemsley), Sp, w; 1.5 in., in many-flowered corymbs; subglobose, small, red fruit; vigorous growth; climbing to 50 ft; half hardy; Eurosa, Synstylae (14); 1904

R. solandri, *see* **R. BLANDA**

R. SONOMENSIS, (Greene), Sp, mp; flowers bright pink in dense few-flowered corymbs, 1–1.5 in.; 1 ft growth; Eurosa, Cinnamomeae

R. soongarica, *see* **R. LAXA**

R. sorbiflora, *see* **R. CYMOSA**

R. SOULIEANA, (Crépin), Sp, w; flowers corymbose, 1.5 in.; ovoid or subglobose, 1/3–1/2 in. long, orange-red fruit; 12 ft growth; Eurosa, Synstylae (14); 1896

R. spaldingii, *see* **R. NUTKANA**

R. sphaerica, *see* **R. CANINA**

R. spinosissima, *see* **R. CINNAMOMEA**

R. SPINOSISSIMA, (Linnaeus), Sp, w; (Burnet Rose, R. illinoensis, R. pimpinellifolia, R. pulchella, Scotch Rose); flowers cream, but white, yellow, pink or purple in garden forms so, 1.25–2 in., spring bloom; globular, black fruit; 3–4 ft growth; Eurosa, Pimpinellifoliae (28)

R. SPINOSISSIMA ALTAICA, ((Willdenow) Bean), Sp, w; (Altaica, R. altaica, R. grandiflora, R. sibirica, R. spinossissima baltica); flowers large; (28); ca. 1820

R. SPINOSISSIMA ANDREWSII, (Willmott), Sp, mr; dbl.; (28)

R. SPINOSISSIMA BICOLOR, (Andrews), Sp, pb; flowers pale pink becoming cream flushed with pink, later rotate, semi-dbl., cupped; (28)

R. SPINOSISSIMA FULGENS, (Bean), Sp, mp; flowers bright rose; (28)

R. spinosissima hispida, *see* **R. HISPIDA**

R. SPINOSISSIMA INERMIS, ((De Candolle) Rehder), Sp, lp; (R. mitissima, R. pimpinellifoliainermis, R. spinosissima mitissima); branches almost unarmed

R. SPINOSISSIMA LUTEA, (Bean), Sp, my; flowers bright yellow; (28)

R. SPINOSISSIMA LUTEOLA, (Andrews), Sp, ly; (R. ochroleuca); flowers pale yellow, 2 in.; (28)

R. spinosissima mitissima, *see* **R. SPINOSISSIMA INERMIS**

R. SPINOSISSIMA MYRIACANTHA, ((De Candolle) Koehne), Sp, w; (R. myriacantha); flowers white, blushed, small; (28); Prior to 1820

R. SPINOSISSIMA NANA, (Andrews), Sp, w; semi-dbl., 2 in.; (28)

R. SPINOSISSIMA PIMPINELLIFOLIA, ((Linnaeus) Hooker), Sp, w; (R. pimpinellifolia); flowers smooth pedicels; (28)

R. spinossissima baltica, *see* **R. SPINOSISSIMA ALTAICA**

R. SPITHAMEA, (S. Watson), Sp; (California Ground Rose); flowers usually solitary floral tube often covered with gland-tippe; 1/3–1 ft growth; Eurosa, Cinnamomeae (28); Rarely cult.

R. SPITHAMEA SONOMIENSIS, Sp; foliage leaflets often double serrate, with gland-tipped tee

R. STELLATA, (Wooten), Sp, m; (R. vernonii); flowers deep rose-purple, solitary, 1.75–2.25 in.; turbinate, glabrous or puberulent with scattered short bristles; prickles young floral branches with white to yellow, occasio; foliage leaflets 3–5; 2 ft growth; Minutifoliae (Hesperhodos), (14); 1902

R. STELLATA ERLANSONIAE, (Lewis), Sp, m; Minutifoliae (Hesperhodos)

R. STELLATA MIRIFICA, ((Greene) Cockerell), Sp, m; (R. mirifica, The Gooseberry Rose); prickles floral branches glabrous with many internodal gland; 4 ft growth; Minutifoliae (Hesperhodos), (14); 1916

R. STELLATA MIRIFICA MIRIFICA, (Lewis), Sp, m; Minutifoliae (Hesperhodos)

R. STELLATA STELLATA, (Lewis), Sp, m; few internodal prickles; Minutifoliae (Hesperhodos)

R. stricta, *see* **R. ACICULARIS**

R. STYLOSA, (Desvaux), Sp, w; (R. systyla); flowers white or light pink, 1.5–2 in.; (35, 42, 28, 34)

R. suavifolia, *see* **R. EGLANTERIA**

R. subblanda, *see* **R. BLANDA**

R. subglobsa, *see* **R. SHERARDII**

R. subglosa, *see* **R. TOMENTOSA SUBGLOBOSA**

R. subnuda, *see* **R. MACOUNII**

R. SUBSERRULATA, (Rydberg), Sp, mp; flowers rose solitary, 2 in.; $2\frac{1}{2}$ ft growth; Eurosa, Carolinae (14, 28)

R. suffulta, *see* **R. ARKANSANA**

R. suffulta alba, *see* **R. ARKANSANA ALBA**

R. sulphurea, *see* **R. HEMISPHAERICA**

R. surculosa, *see* **R. CANINA**

R. sweginzowii, *see* **R. PRIMULA**

R. SWEGINZOWII, (Koehne), Sp, mp; flowers pink 1–3 together, 1.75 in.; oblong, 1 in. long fruit; Eurosa, Cinnamomeae (42); 1909

R. sweginzowii inermis, *see* **R. WARDII**

R. SWEGINZOWII MACROCARPA, (hort.), Sp, mp

R. sylvatica, *see* **R. GALLICA**

R. systyla, *see* **R. STYLOSA**

R. taquetii, *see* **R. WICHURANA**

R. taurica, *see* **R. CORYMBIFERA**

R. terebinthinacea, *see* **R. X MARCYANA**

R. ternata, *see* **R. LAEVIGATA**

R. tetrapetala, *see* **R. SERICEA**

R. thea, *see* **R. X ODORATA**

R. thoryi, *see* **'SEVEN SISTERS'**

R. thunbergii, *see* **R. MULTIFLORA**

R. thyrsiflora, *see* **R. MULTIFLORA**

R. tomentella, *see* **R. OBTUSIFOLIA**

R. TOMENTOSA, (Smith), Sp, lp; (R. cuspidata, R. dimorpha); flowers pale pink, on longer pedicels; smaller fruit; foliage grayish green, downy; Eurosa, Caninae (35)

R. TOMENTOSA SUBGLOBOSA, ((Smith) Carion), Sp, lp; (R. subglosa); subglobose fruit; prickles hooked

R. tongtchouanensis, *see* **R. X ODORATA**

R. trachyphylla, *see* **R. MARGINATA**

R. trifoliata, *see* **R. SETIGERA**

R. triphylla, *see* **R. ANEMONEFLORA**

R. triphylla, *see* **R. LAEVIGATA**

R. turbinata, *see* **R. X FRANCOFURTANA**

R. TURKESTANICA, (Regel), Sp, my; subglobose fruit; Eurosa, Pimpinellifoliae

R. TUSCHETICA, (Boissier), Sp, mp; (R. pimpinellifolia tuschetica); ovate (not obovate) fruit; differs from R. glutinosa in the leaflets; Eurosa, Caninae; 1945

R. uchiyamana, *see* **R. MULTIFLORA CATHAYENSIS**

R. ultramontana, *see* **R. WOODSII ULTRAMONTANA**

R. unguicularis, *see* R. webbiana

R. usitatissima, *see* White Rose of York

R. velutinaeflora, *see* **R. GALLICA VELUTINAEFLORA**

R. vernonii, *see* **R. STELLATA**

R. villosa, *see* **R. MOLLIS**

R. villosa, *see* **R. POMIFERA**

R. villosa, *see* **R. X INVOLUTA**

R. villosa mollissima, *see* **R. MOLLIS**

R. vilmorinii, *see* **'MICRUGOSA'**

R. virginiana, *see* **R. PALUSTRIS**

R. VIRGINIANA, (Miller), Sp, mp; (R. humilis lucida, R. lucida, R. lucida alba, R. pennsylvanica, Virginia Rose); flowers bright pink usually few or solitary, 2 in., summer bloom; fruit remaining plump until next spring (see endpapers); 6 ft growth; Eurosa, Carolinae (28); Prior to 1807

R. virginiana alba, *see* **R. CAROLINA ALBA**

R. virginiana blanda, *see* **R. BLANDA**

R. virginiana humilis, *see* **R. CAROLINA**

R. VIRGINIANA LAMPROPHYLLA, (Rehder), Sp, mp; foliage leaves lustrous; 1881

R. virginiana plena, *see* **R. X RAPA**

R. virginiana plena, *see* **'ROSE D'AMOUR'**

R. viridiflora, *see* **'GREEN ROSE'**

R. vorbergii, *see* **'VORBERGII'**

R. wallichii, *see* **R. SERICEA**

R. walpoleana, *see* **R. EGLANTERIA**

R. WARDII, (Mulligan), Sp, w; (R. sweginzowii inermis), up to 3 together; almost unarmed

R. WARDII CULTA, (Mulligan), Sp, w; flowers smaller pedicels often glandular, 1.25 in.; (42)

R. watsoniana, *see* **R. MULTIFLORA WATSONIANA**

R. watsonii, *see* **R. CORIIFOLIA**

R. webbiana, *see* **R. SERTATA**

R. WEBBIANA, (Wallich ex Royle), Sp, mp; (R. unguicularis); flowers large, mostly solitary; ovoid (see endpapers) fruit; Eurosa, Cinnamomeae (14); 1879

R. wichurae, *see* **R. MULTIFLORA**

R. WICHURANA, (Crépin), Sp, w; (Memorial Rose, R. bracteata, R. luciae, R. luciae wichurana, R. luciaetaquetiana, R. mokanensis, R. taquetii); 1.5–2 in., in few-flowered pyramidal corymbs, late summer, moderate fragrance; ovoid, to 1/2 in. fruit; foliage half evergreen; prostrate with creeping branches growth; Eurosa, Synstylae (14); 1891

R. willdenowii, *see* **R. DAVURICA**

R. WILLMOTTIAE, (Hemsley), Sp, m; flowers rose-purple, solitary, short-stalked, on short lateral branch, 1–1.25 in., spring bloom; subglobose, 1/3–1/2 in. long, bright orange-red fruit; foliage fern-like; 5–10 ft growth; Eurosa, Cinnamomeae (14); 1904

R. willmottiana, *see* **R. LONGICUSPIS**

R. wilsonii, *see* **'MICRUGOSA'**

R. wilsonii, *see* **R. GENTILIANA**

R. wilsonii, *see* **R. X INVOLUTA WILSONII**

R. woodsii, *see* **R. MACOUNII**

R. WOODSII, (Lindley 1820), Sp, mp; (Mountain Rose, R. deserta, R. fimbriatula, R. macounii, R. maximilianii, R. mohavensis, R. pyrifera, R. sandbergii, R. woodsii mohavensis); flowers pink, rarely white corymbose or solitary, on very short, sm, 1.5–2 in., summer bloom; globose, with a short neck (see endpapers) fruit; 3 ft growth; Eurosa, Cinnamomeae (14)

R. WOODSII FENDLERI, ((Crépin) Rehder), Sp, mp; (R. fendleri, R. poetica); (14)

R. WOODSII HISPIDA, (Lewis), Sp, dp; (R. adenosepala); flowers deep pink; hips glandular; (14)

R. woodsii mohavensis, *see* **R. WOODSII**

R. WOODSII ULTRAMONTANA, (Lindley (Watson) Taylor & Ma), Sp, mp; (R. californica ultramontana, R. pisocarpa ultramontana, R. ultramontana); flowers clusters of 3–10, 2 in.; subglobose, small fruit; 10 ft growth; Eurosa, Cinnamomeae (14)

R. X ALBA, (Linnaeus), A, w; (Bonnie Prince Charlie's Rose, Jacobite Rose, La Rose de York, R. procera, R. usitatissima, White Rose of York); dbl., borne usually several together, summer bloom, moderate fragrance; ovoid (see endpapers), scarlet fruit; 6 ft growth; (42); [**R. GALLICA X R. CANINA**]

R. X anemonoides, *see* **'ANEMONE'**

R. X aschersoniana, *see* **'ASCHERSONIANA'**

R. X borboniana, *see* **'BOURBON ROSE'**

R. X calocarpa, *see* **'CALOCARPA'**

R. X cantabrigiensis, *see* **'CANTABRIGIENSIS'**

R. X centifolia bullata, *see* **'BULLATA'**

R. X centifolia minima, *see* **'ROULETII'**

R. X centifolia provincialis, *see* **R. GALLICA OFFICINALIS**

R. X cooperi, *see* **'COOPER'S BURMESE'**

R. X coryana, *see* **'CORYANA'**

R. X damascena, *see* **'SUMMER DAMASK'**

R. X DULCISSIMA, (Lunell), S; [R. blanda X R. woodsii]

R. X dupontii, *see* **'DUPONTII'**

R. X ENGELMANNII, (S. Watson), S, dp; (R. acicularis engelmannii, R. bakerii, R. engelmannii, R. melina, R. oreophila); (42); [Naturally-occurring hybrid between R. nutkana and R. acicularis]; 1891

R. X fortuniana, *see* **'FORTUNIANA'**

R. X FRANCOFURTANA, (Muenchhausen), Misc OGR, m; (R. campanulata,

R. francofurtensis, R. germanica, R. inermis, R. turbinata); flowers purple borne 1–3 together, semi-dbl., 2–3 in., summer bloom, slight fragrance; turbinate fruit; 6 ft growth; (21, 28); [Probably R. cinnamomea X R. gallica]; Possibly int. prior to 1629

R. X harisonii, *see* **'HARISON'S YELLOW'**

R. X harisonii vorbergii, *see* **'VORBERGII'**

R. X heterophylla, *see* **'PROTEIFORMIS'**

R. X hibernica, *see* **'HIBERNICA'**

R. X highdownensis, *see* **'HIGHDOWNENSIS'**

R. X INVOLUTA, (Smith), Misc OGR, w; (R. braunii, R. coronata, R. doniana, R. sabinii, R. villosa); flowers whitish; (42); [R. spinosissima X R. tomentosa]

R. X INVOLUTA WILSONII, ((Borrer) Baker), Misc OGR; (R. wilsonii); vigorous, shrubby growth; (42)

R. X iwara, *see* **'IWARA'**

R. X JACKSONII, (Willmott), S, mr; flowers bright crimson, very free bloomer; (14); [R. rugosa X R. wichurana]

R. X KOCHIANA, (Koehne), S, dp; (R. oxyacanthos); flowers deep rose borne solitary or 2–3, 1.5 in.; (14); [Possibly an F2 from the cross R. palustris X R. spinosissima]

R. X KOEHNEANA, S, m; flowers purplish red, larger; (21); [R. carolina X R. rugosa]

R. X leonida, *see* **'MARIA LEONIDA'**

R. X MALYI, (Kerner), S, mr; flowers bright red; foliage similar to spinosissima; (28); [Possibly R. pendulina X R. spinosissima]

R. X MARCYANA, (Bouller), S, m; (R. terebinthinacea); flowers pink to light purple, 2.5–3 in.; long-stalked; low (2 ft, rarely 4 ft) growth; (21); [R. gallica X R. tomentosa]

R. X mariae-graebneriae, *see* **'MARIA GRAEBNER'**

R. X micrugosa, *see* **'MICRUGOSA'**

R. X noisettiana, *see* **'NOISETTE ROSE'**

R. X noisettiana manettii, *see* **'MANETTI'**

R. X ODORATA, (Sweet); (R. chinensis fragrans, R. gechouitangensis, R. indica fragrans, R. indica odorata, R. indica odoratissima, R. oulengensis, R. thea, R. tongtchouanensis); flowers white, light pink or salmon-pink or yellowish singly or 2–3, dbl., 2–3.5 in.; short stems; Eurosa, Chinensis (14, 21, 28)

R. X odorata erubescens, *see* **R. GIGANTEA ERUBESCENS**

R. X odorata gigantea, *see* **R. GIGANTEA**

R. X odorata ochroleuca, *see* **'PARKS' YELLOW TEA-SCENTED CHINA'**

R. X odorata pseudindica, *see* **'FORTUNE'S DOUBLE YELLOW'**

R. X paulii, *see* **'PAULII'**

R. X paulii rosea, *see* **'PAULII ROSEA'**

R. X penzanceana, *see* **'LADY PENZANCE'**

R. X pokornyana, *see* **'POKORNYANA'**

R. X polliniana, *see* **'POLLINIANA'**

R. X protieformis, *see* **'PROTEIFORMIS'**

R. X pruhoniciana, *see* **'HILLIER ROSE'**

R. X PTERAGONIS, (Krause), S, my; flowers rich yellow, numerous, 1.5 in.; (14); [R. hugonis X R. sericea]

R. X pteragonis cantabrigiensis, *see* **'CANTABRIGIENSIS'**

R. X RAPA, (Bosc), Misc OGR, mp; (R. lucida plena, R. virginiana plena); flowers bright pink, dbl.; (21); [R. virginiana X Unknown diploid]; Before 1820

R. X REVERSA, (Waldstein & Kitaibel), Misc OGR, mr; (R. rubella); obovoid, pendulous, scarlet fruit; foliage dark green; (28); [R. pendulina X R. spinosissima]

R. X richardii, *see* **St John's Rose**

R. X RUBROSA, (Preston), S, mp; flowers large; (28); [R. glauca X R. rugosa]

R. X ruga, *see* **'RUGA'**

R. X SALAEVENSIS PERRIERI, ((Verlot) Christ), S; flowers rose-purple, borne 1–3 together; [R. dumalis X R. pendulina]

R. X scharnkeana, *see* **'SCHARNKEANA'**

R. X spaethiana, *see* **'SPAETHIANA'**

R. X SPINULIFOLIA, (Dematra), S, mp; (R. glabrata); 2–2.5 in., summer bloom; (28); [R. pendulina X R. tomentosa(?)]

R. X tackholmii, *see* **'TACKHOLMII'**

R. X waitziana, *see* **'WAITZIANA'**

R. X waitziana macrantha, *see* **'GALLICA MACRANTHA'**

R. X warleyensis, *see* **'WARLEYENSIS'**

R. X wintoniensis, *see* **'WINTONIENSIS'**

R. xanthina, *see* **R. ECAE**

R. xanthina, *see* **R. HUGONIS**

R. xanthina, *see* **R. PRIMULA**

R. XANTHINA, (Lindley), Sp, my; (Manchu Rose, R. xanthinoides); flowers short-stalked, dbl., 1.75 in., solitary, 10 ft growth; Eurosa, Pimpinellifoliae (14); 1906

R. xanthina Allard, *see* **'ALLARD'**

R. xanthina 'Canary Bird', *see* **'CANARY BIRD'**

R. xanthina ecae, *see* **R. ECAE**

R. xanthina normalis, *see* **R. PRIMULA**

R. XANTHINA SPONTANEA, (Rehder), Sp, my; single; (14); 1907

R. xanthinoides, *see* **R. XANTHINA**

R. xanthocarpa, *see* **R. GIGANTEA**

R. yainacencis, *see* **R. YAINACENSIS**

R. YAINACENSIS, (Greene), Sp; (R. myriadenia, R. yainacencis); flowers more numerous; smaller fruit; (28)

R. yesoensis, *see* **'IWARA'**

R. zagrabiensis, *see* **R. MARGINATA**

'R.G. CASSON', HT, pb, 1923; flowers rose and copper, moderate fragrance; Cant, B. R.

'R.M.S. QUEEN MARY', HT, mp, 1937; (Mrs Verschuren); flowers salmon-pink suffused orange, very large, dbl., cupped; foliage leathery; vigorous growth; [Briarcliff X Mrs Sam McGredy]; Verschuren; Dreer; J&P

'R.S. HUDSON', HT, yb, 1939; flowers yellow tinged red, large, well-formed; vigorous, upright growth; Wheatcroft Bros.

'R.W. PROCTOR', F, dp, 1947; flowers deep rose-pink, borne in trusses, semi-dbl., 10–12 petals, 4 in., slight fragrance; foliage glossy; vigorous growth; [Anne Poulsen sport]; Proctor

Raalte's Golden Harvest, S, dy, 1992; Interplant

'RAAT-KI-RANI', HT, mr, 1975; bud pointed; flowers velvety crimson-red, full, dbl., 30 petals, 4.5 in., slight fragrance; foliage glossy; vigorous, upright growth; [Seedling X Samourai seedling (open-pollinated)]; IARI

'RABBIE BURNS', HT, mr, 1959; bud long, pointed; flowers bright light red, semi-dbl., intense fragrance; foliage dark; vigorous, upright growth; [Ena Harkness X Sutter's Gold]; Arnot; Croll

Rachel, *see* 'BOOyol'

'RACHEL', HT, ob, 1929; flowers orange-buff, flushed carmine, large, dbl., exhibition form; foliage dark; very vigorous growth; RULED EXTINCT 5/84; Pemberton

Rachel Bowes Lyon, *see* 'HARlacal'

'RACHEL CRAWSHAY', HT, op, 1977; flowers pink to orange-salmon, dbl., 30 petals, 5 in., slight fragrance; foliage olive-green; [Fragrant Cloud X Mary Mine]; Harkness

'RACHEL TOWNSEND', HT, yb, 1963; flowers golden yellow tipped carmine, dbl., 30 petals, 4–4.5 in.; foliage dark, glossy; free growth; [Sultane sport]; Townsend; Townsend & Son

'RACHELLE', F, mp, 1976; bud long; flowers french rose, nearly, dbl., 30 petals, 2–3 in., flat, slight fragrance; upright growth; [Antigua X Unnamed cultivar]; Warriner, William A.; J&P

Racy Lady®, *see* 'DICwaffle'

'RADA', HT, dr, 1975; flowers large, dbl., 75 petals; foliage medium green, glossy; [Baccará X Unnamed seedling]; Staikov, Prof. Dr V.; Kalaydjiev and Chorbadjiiski

'RADAMES', HT, dr, 1983; flowers large, dbl., 35 petals, intense fragrance; foliage large, medium green, glossy; upright growth; [Unnamed X Unnamed]; Fumagalli, Niso

'**RADAR**', HT, or, 1953; bud long; flowers light geranium-red, well formed, large, dbl., 45 petals, moderate fragrance; vigorous growth; [Charles Mallerin X Independence]; Meilland, F.; URS

'**RADAR ITALIANA**', HT, or, 1944; bud pointed; flowers nasturtium-red edged rose, large, dbl., 32–34 petals, moderate fragrance; foliage bright green; vigorous, bushy growth; [Souv. de Denier van der Gon X Brazier]; San Remo Exp. Sta.

'**RADAR, CLIMBING**', Cl HT, or, 1959; Meilland, Mrs. Marie-Louise; URS

'**RADCLIFFE FLAME**', HT, mr, 1986; flowers medium, dbl., 26–40 petals, no fragrance; foliage medium, dark green, matt; upright growth; [Alec's Red X Grandpa Dickson]; Thomson, Colin, 1987

'**RADIANCE**', HT, lp, 1908; (Pink Radiance); bud globular; flowers rose-pink, reverse lighter, large, dbl., 23 petals, cupped, intense, damask fragrance; foliage leathery; vigorous growth; [Enchanter X Cardinal]; Cook, J.W.; P. Henderson

'**RADIANCE, CLIMBING**', Cl HT, lp, 1926; (Pink Radiance, Climbing); Griffing, W.D.

'**RADIANCE, CLIMBING**', Cl HT, lp, 1928; (Pink Radiance, Climbing); Catt

'**RADIANT**', HT, or, 1962; bud spiral; flowers orange-flame, reverse shaded red, dbl., 30 petals, 4–5 in., slight fragrance; foliage glossy; very free growth; RULED EXTINCT 9/87; [Mrs Sam McGredy X Fantasia]; Fletcher; Tucker

Radiant(TM), *see* 'BENRAD'

'**RADIANT BEAUTY**', HT, mr, 1934; flowers deeper crimson than parent, not turning blue, dbl., moderate fragrance; [Francis Scott Key sport]; Cleveland Cut-Flower Co.

'**RADIANT GLOW**', F, or, 1953; bud pointed; flowers bright orange-salmon tinted peach, medium, borne singly and, dbl., cupped; foliage leathery, light green; vigorous, upright growth; [Pinocchio X ?]; Quinn; Roseglen Nursery

Radiant Gold, *see* 'JACERN'

Radiant Super Glaze, *see* 'NOSTRAD'

Radiant Superglaze, *see* 'NOSTRAD'

'**RADIATION**', F, mr, 1960; flowers open, borne in clusters, semi-dbl., 3–3.5 in.; foliage glossy; vigorous growth; [Poulsen's Pink X Pompadour Red]; deRuiter

'**RADIEUSE**', HT, mr, 1955; bud long; flowers clear red, well formed, large, dbl., 30–35 petals; vigorous, bushy growth; Laperrière; EFR

'**RADIMAN**', F, ly, 1975; dbl., 50 petals, cupped, blooms in clusters of 5–35, moderate, tea fragrance; foliage dark, glossy; vigorous growth; [Highlight X Masquerade]; Staikov, Prof. Dr V.; Kalaydjiev and Chorbadjiiski

'**RADIO**', HT, yb, 1937; flowers yellow slightly tinted pink, striped and marked rose, large, dbl., 50 petals, cupped, moderate, spicy fragrance; foliage wrinkled, light; vigorous growth; [Condesa de Sástago sport]; Dot, Pedro; C-P

'**RADIO LANCASHIRE**', F, or, 1987; flowers orange-red, reverse slightly lighter, rosette, medium, borne, dbl., 38 petals, slight, fruity fragrance; round, orange fruit; triangle-shaped, brown prickles; foliage medium, dark green, glossy; upright, medium growth; [Dusky Maiden X Matangi]; Bracegirdle, A.J.; Rosemary Roses, 1987

Radio Times, *see* 'AUSSAL'

Radiosa, HT, 1956; Aicardi; (Cavriglia)

'**RADIUM**', HT, rb, 1922; flowers carmine, shaded coppery red; [Beauté Lyonnaise X Capt. Hayward]; Lippiatt

Radka, HT, dy, 1975; flowers large, dbl., moderate fragrance; Vecera, L.; (Sangerhausen)

'**RADOME**', F, lp, 1966; flowers pale rose; [Queen Elizabeth sport]; Nicol; Minier

Radox Bouquet, *see* 'HARMUSKY'

'RADRAZZ', S, rb, 1999; (**Knock Out**(TM)); flowers medium, light red to deep pink, single, 5–11 petals, 1.5 in., borne in small clusters, slight, tea fragrance; prickles moderate; foliage large, medium green, semi-glossy, very resistant to; bushy, rounded, medium (3 ft) growth; AARS, U.S., 2000; [(Carefree Beauty X Unknown) X (Razzle Dazzle X Unknown)]; Radler, William; Conard-Pyle, 2000

'**RADWAY CHARM**', HT, dp, 1959; flowers deep pink, base deep yellow, semi-dbl., 20 petals, moderate fragrance; vigorous growth; [Christopher Stone X McGredy's Wonder sport]; Waterhouse Nursery

'**RADWAY GLOW**', F, op, 1960; flowers coral-pink, borne in large, open clusters, semi-dbl., 14 petals, slight fragrance; foliage dull green; vigorous growth; Waterhouse Nursery

'**RADWAY JEWEL**', F, dr, 1960; flowers yellow, becoming orange and deep red, medium, borne in large, dbl., 40 petals; foliage light green; Waterhouse Nursery

'**RADWAY PINK**', F, mp, 1965; flowers rose-pink, borne in clusters, dbl., 42 petals, 3.5 in.; foliage dull green; very free growth; [Margaret X Korona]; Waterhouse Nursery

'**RADWAY SCARLET**', F, mr, 1963; low growth; [Karl Weinhausen X Seedling]; Waterhouse Nursery

'**RADWAY SUNRISE**', S, ob, 1962; flowers yellow, shading through orange to red, blooms in clusters, single, 7 petals, 3.5–4 in., recurrent bloom, slight fragrance; foliage dark, glossy; very vigorous growth; [Masquerade X ?]; Waterhouse Nursery

'**RAE DUNGAN**', HT, yb, 1971; bud long, pointed; flowers creamy yellow, edged deep pink, large, dbl., moderate fragrance; foliage dark; vigorous growth; [Daily Sketch X Fred Streeter]; Dawson, George

Raf Braeckman, F, ob, 1985; RvS-Melle

'**RAFAELA G. DE PEÑA**', Pol, ob, 1938; flowers pure orange, dbl.; foliage bright green, leathery; vigorous, bushy growth; Dot, Pedro

'**RAFFEL'S PRIDE**', HT, mr, 1937; bud small; flowers oriental red, reverse gold-splashed, open, dbl.; vigorous growth; [Talisman X ?]; Raffel; Port Stockton Nursery

'**RAFFEL'S YELLOW**', HT, my, 1942; flowers pure yellow, well shaped, large, dbl., 25–35 petals; foliage bronze turning very dark; vigorous growth; [Probably Mrs Beatty X Maid of Gold]; Raffel; Port Stockton Nursery

'**RAFFLES BRUCE**', HT, ab, 1943; flowers apricot and gold, well shaped, 3–4 in.; foliage dark; compact growth; GM, NRS, 1943; [Mrs Sam McGredy X Aureate]; Bees

Ragamuffin, Min, yb

Ragazzina, F; (Cavriglia)

Ragged Robin, *see* '**GLOIRE DES ROSOMANES**'

'**RAGGEDY ANN**', F, mr, 1956; bud short, pointed; semi-dbl., 13 petals, blooms in clusters, slight fragrance; foliage dark, leathery; vigorous, upright growth; [Garnette X Sister Kenny]; Hill, Joseph H., Co.

Ragtime, *see* 'LENRAG'

Ragtime, *see* 'MACCOURLOD'

Ragtime, *see* 'PEACAP'

Ragtime, *see* POUltime

'**RAINBOW**', T, pb, 1889; flowers pink striped carmine; [Papa Gontier sport]; Sievers

Rainbow, S, ob

Rainbow Bliss(TM), *see* 'DEVDICHA'

Rainbow Cerise(TM), *see* 'DEVCLAVEL'

Rainbow Crimson(TM), *see* 'DEVMESI'

Rainbow Delight, Min, mr, 1998

Rainbow Eclipse(TM), *see* 'DEVECLIPSAR'

Rainbow Gold(TM), *see* 'DEVORO'

Rainbow Hot Pink(TM), *see* 'DEVIENTE'

Rainbow Magic(TM), *see* 'DICXPLOSION'

Rainbow Pink(TM), *see* 'DEVROSADO'

Rainbow Red(TM), *see* 'DEVROJO'

Rainbow Robe, HT, m, 1991; Kordes

Rainbow Shower, *see* 'LEESHO'

Rainbow Stanford(TM), *see* 'DEVRICO'

Rainbow Sunrise(TM), *see* 'DEVAURORA'

Rainbow Surprise(TM), *see* 'DEVPRESA'

Rainbow Warrior, *see* CHRisgood

Rainbow Yellow(TM), *see* 'DEVAMARILLO'

Rainbow Yellow Parade(TM), *see* 'POULWEE'

Rainbow's End, Climbing(TM), *see* 'SAVACLENA'

Rainbow's End(TM), *see* 'SAVALIFE'

Raindrops(TM), *see* 'SAVARAIN'

Raindrops, Min, pb, 1997; Williams, J. Benjamin

Rainer Maria Rilke, *see* 'KORSEE'

Rainy Day®, *see* 'MACRAIDA'

Raissa®, HT, mp, 1990; Cocker

'RAJA OF NALAGARH', HT, or, 1977; (Raja Surendra Singh of Nalagarh); bud pointed; flowers red, dbl., 32 petals, 4 in., exhibition form, slight fragrance; foliage dark, leathery; vigorous, upright growth; [Samourai X Montezuma]; Pal, Dr B.P.; Gopalsinamiengar

Raja Surendra Singh of Nalagarh, *see* **'RAJA OF NALAGARH'**

'RAJBALA', F, pb, 1975; bud pointed; flowers open, semi-dbl., 10 petals, 4 in.; foliage large, glossy, light; very vigorous, upright, compact growth; [Delhi Princess X Seedling]; Pal, Dr B.P.; Indian Agric. Research Inst.

'RAJKUMARI', HT, dp, 1975; bud pointed; flowers deep fuchsine-pink, full, very dbl., 70 petals, 4 in.; foliage glossy, light; vigorous, compact growth; [(Charles Mallerin X Delhi Princess) X Seedling]; IARI

Rajni, *see* **'ORIENT SPICE'**

'RAKETA', HT, op, 1952; (Rocket); flowers golden orange-pink, dbl., 50 petals, 5 in.; foliage reddish bronze; vigorous, upright, compact growth; [Narzisse X Comtesse Vandal]; Shtanko, E.E.

Rakete, HT, mr, 1972; flowers large, dbl.; GPG Bad Langensalza; (Sangerhausen)

'RAKTAGANDHA', HT, or, 1975; bud long, pointed; flowers vermilion, dbl., 35 petals, 3 in., exhibition form, slight fragrance; foliage glossy; vigorous, upright growth; [Christian Dior X Carrousel seedling]; IARI

Rakthima, HT, mr, 1991; IARI

'RALLYE', Pol, r, 1966; bud ovoid; flowers cognac color with pink, large, borne in clusters, dbl., moderate fragrance; foliage dark, glossy; vigorous, upright growth; [Cognac X Fashion]; Delforge

Ralph®, HT, ab, 1993; deVor

'RALPH LEIGHTY', F, w, 1971; bud ovoid; flowers near white, medium, dbl., exhibition form, slight fragrance; foliage light, soft; moderate, upright growth; [Gene Boerner sport]; Leighty; McFadden

Ralph Moore, *see* 'SAVARALPH'

'RALPH TIZARD', F, ob, 1979; (Ralph Tizzard); flowers pure salmon, pointed, dbl., 28 petals, 4 in., moderate fragrance; foliage dark; vigorous growth; [Vera Dalton X Tropicana]; Sanday, John

Ralph Tizzard, *see* **'RALPH TIZARD'**

Ralph's Creeper, *see* 'MORPAPPLAY'

Ramanas Rose, *see* **R. RUGOSA**

Ramapo, *see* 'ZIPRAM'

'RAMAT-GAN', F, dy, 1972; bud urn-shaped; flowers deep lemon-yellow, to urn-shaped, dbl., 25 petals, 1 in., cupped, moderate fragrance; foliage light; moderate growth; [Golden Masterpiece X Zorina]; Holtzman; Holtzman Rose Nursery

'RAMBLING RECTOR', HMult, w; semi-dbl., blooms in large clusters, moderate fragrance; vigorous growth

Ramira®, *see* KORmeita

'RAMÓN BACH', HT, ob, 1938; flowers bright orange, edged lighter, reverse reddish gold, stamens, dbl., 80 petals, globular, intense, fruity fragrance; foliage glossy, dark; vigorous growth; [Luis Brinas X Condesa de Sástago]; Dot, Pedro; C-P

'RAMONA', S, mr, 1913; (Red Cherokee); flowers carmine-crimson, single; [Anemone sport]; Dietrich & Turner

'RAMPA PAL', HT, pb, 1975; bud ovoid; flowers fuchsine-pink, reverse lighter, very dbl., 60 petals, 4.5 in., exhibition form, slight fragrance; foliage glossy; moderate, upright, bushy growth; Pal, Dr B.P.

'RAMPANT', HSem, w, 1830; profuse bloom, sometimes in autumn; Jacques

'RAN', S, mp, 1972; bud globular; flowers pink, open, small, semi-dbl., slight fragrance; foliage small, light, soft; vigorous, upright growth; [R. cinnamomea X R. helenae hybrida]; Lundstad; Norges Landbruks-hogskole

'RANDALL', HGal, mp; flowers an old variety taken by early pioneers to western canada, dbl.; not as hardy as Alika

Randfontein Gold, *see* KORsion

Randilla Jaune®, Min, my

Randilla Rose®, Min, mp

Randilla Rouge®, Min, mr

Rangatarang, HT, pb, 1993; Chiplunkar

Ranger, *see* BARger

Rangila, HT, rb, 1995; Tejganga

Rangitoto, *see* 'SUNTOTO'

'RANGOLI', F, op, 1978; bud tapered; flowers coral-pink, dbl., 25 petals, 3–3.5 in., exhibition form, slight, fruity fragrance; foliage glossy; dwarf, bushy growth; [Golden Slippers sport]; Thakur; Doon Valley Roses

'RANGSHALA', HT, ab, 1969; bud pointed; flowers apricot, shaded peach and amber yellow, full, open, medium, dbl.; foliage glossy; moderate growth; [Margaret Spaull X Open pollination]; IARI

'RANJANA', HT, or, 1975; bud pointed; flowers rose-opal, full, dbl., 38 petals, 4.5 in., intense fragrance; foliage dark, leathery; very vigorous, upright, bushy growth; [Samourai X ?]; Pal, Dr B.P.; Anand Roses

'RANKENDE JOHANNA TANTAU', Cl Pol, 1942; Tantau, Math.; (Sangerhausen)

'RANKENDE MINIATURE', Cl Pol, lp, 1908; flowers very small, very dbl., moderate fragrance; Lambert, P.; (Sangerhausen)

'RANKENDE TESCHENDORFF'S JUBILÄUMSROSE', Cl Pol, dp, 1930; flowers medium, dbl., slight fragrance; Teschendorff; (Sangerhausen)

'RANUNCULUS MUSK CLUSTER', HMsk, w; flowers pure white, very dbl.; very vigorous growth

'RAPHAEL', M, w, 1856; flowers well mossed, pinkish white, slightly recurrent bloom, Robert

Raphaela, *see* TANelephar

'RAPPERSWIL'®, HT, or, 1975; bud globular; dbl., 30 petals, 4 in., cupped, intense fragrance; foliage dark, leathery; [Fragrant Cloud X Ena Harkness]; Huber

Rapsodia®, F, ob

'RAPTURE', HT, pb, 1926; flowers deeper pink, flushed gold at base; (28); [Mme Butterfly sport]; Traendly & Schenck

'RAPTURE, CLIMBING', Cl HT, pb, 1933; Dixie Rose Nursery

'RAQUEL MELLER', HT, lp, 1957; flowers soft pink, large, dbl., cupped, intense fragrance; foliage glossy; GM, Geneva, 1956; [Edith Krause X Fashion]; Camprubi, C.

Rare Edition, F, rb, 1982; Kasturi

Rashmi, Pol, dr, 1977; Kasturi

Raspberry Beauty(TM), *see* 'MINIRASP'

'RASPBERRY DELIGHT', HT, rb, 1979; flowers medium red on outer petals shading to creamy coral on inner, dbl., 30 petals, exhibition form, moderate, raspberry fragrance; hooked prickles; foliage medium green, semi-glossy; upright, bushy growth; [Carrousel X First Prize]; Taylor, Thomas E.

Raspberry Ice, *see* 'ZIPBERRY'

Raspberry Ice, *see* 'KORTABRIS'

Raspberry Punch(TM), *see* 'JACBUNCH'

Raspberry Red, F, mr, 1995

Raspberry Ripple, HSet, pb

Raspberry Ripple, F, rb, 1998

Raspberry Rose, F, mr

Raspberry Ruffles, *see* 'TALRAS'

Raspberry Sunblaze®, *see* 'MEIDONFE'

Raspberry Swirl(TM), *see* 'JACSTAND'

Raspberry Swirl, HT, lb; flowers cream goin to yellow throat, with outer half of petal clear raspberry, exhibition form, moderate fragrance; Edwards, Eddie

Raspberry Wine, *see* **'Malinovka'**

'Rassvet', HT, pb, 1955; (Daybreak); flowers soft pink, base creamy yellow, medium, very dbl., 48 petals; foliage dark; spreading growth; [Peace seedling]; Klimenko, V. N.

Ratgeber Rose, HMult, mr; flowers medium, dbl., slight fragrance; (Sangerhausen)

'Rathernice', HT, ob, 1957; flowers coppery orange; foliage coppery; moderate growth; Bishop; Baker's Nursery

Ratio Times, HP; flowers purple-violet, dbl., cupped

'Raubritter', S, lp, 1936; dbl., globular, in clusters, moderate fragrance; foliage leathery, wrinkled; vigorous, climbing growth; (21); [Daisy Hill X Solarium]; Kordes

Ravel®, *see* 'LENrav'

Ravel, *see* RUIsteenka

Raven™, *see* FRYtrooper

Ravensberg®, S, mr, 1986; Noack, Werner

'Ravenswood', Min, yb, 1992; flowers gold, outer petals flushed pink, blooms borne 3–4 per cluster, semi-dbl., 18–20 petals, .5 in., slight fragrance; foliage healthy; upright, bushy growth; [Rise 'n' Shine X Unknown]; Catt, Graeme Charles, 1989; F.D. Catt Wholesale Nursery

'Ravenswood Village', HT, ly, 1996; flowers light yellow, deeper in autumn, very large, blooms borne most, dbl., 26–40 petals, exhibition form, slight fragrance; some prickles; foliage medium, medium green, glossy; upright, medium growth; [Golden Splendour X Queen Esther]; Poole, Lionel; F. Haynes & Partners, 1997

'Raving Beauty', HT, mp, 1948; bud ovoid, rose-red; flowers tyrian rose, dbl., 35–40 petals, 3–4 in., globular, intense fragrance; foliage leathery, dark; vigorous, upright growth; Hill, Joseph H., Co.

'RAWana', HT, dy, 1999; (**Golden Wonders 99**); dbl., 15–25 petals, 3.5 in., borne in small clusters, slight fragrance; prickles moderate; foliage medium, medium green, semi-glossy; bushy (2.5 ft) growth; [(Gold Bunny X (Baby Love X seedling)) X Golden Celebration]; Rawlins, R.

'Ray Adeline', HT, my, 1989; bud pointed; flowers medium, borne usually singly and in sprays of 2–4, dbl., 40 petals, exhibition form, moderate fragrance; small, light green fruit; prickles straight, hooked down, light green; foliage medium, dark green, semi-glossy; upright growth; [First Prize X King's Ransom]; Bevard, Harry D., 1990

'Ray Bunge', Cl HP, dp, 1927; flowers dark rose, reverse lighter, borne singly or clusters of 7–8, dbl., 30–40 petals, 4–5 in., prolific early spring bloom, repeating later in season, slight fragrance; foliage dark; very long stems; very vigorous ($15\frac{1}{2}$ ft) growth; quite hardy.; [Paul Neyron sport]; Bunge; Andrews Nursery Co., 1959

Ray of Hope, *see* COCnilly

Ray of Sunshine, *see* 'COCclare'

'Raymond', HT, pb, 1917; bud long, pointed; flowers peach-blossom-pink, center salmon-carmine, dbl.; [Rayon d'Or seedling]; Pernet-Ducher

Raymond Carver, S, my, 1999; Horner

'Raymond Chenault', HKor, mr, 1960; flowers bright red, blooms in clusters, semi-dbl., 16 petals, 4 in., moderate fragrance; foliage dark, glossy; vigorous (9–12 ft) growth; [R. kordesii X Montezuma]; Kordes, R.

Raymond Kopa, *see* 'WEKjuvoo'

'Raymond Privat', Pol, m, 1935; flowers violet, borne in clusters, dbl.; vigorous growth; Privat

Raymond's Pearl, F, yb, 1997; flowers very full, medium, very dbl., 41 petals, borne in small clusters, no fragrance; some prickles; foliage medium, medium green, semi-glossy; upright, medium (75 cms) growth; [Amber Queen X Brown Velvet]; Rawlins, R.

'Rayon d'Or', HT, my, 1910; flowers golden yellow, dbl.; [Mme Mélanie Soupert X Soleil d'Or]; Pernet-Ducher

'Rayon d'Or', HT, my, 1962; flowers well formed; Combe

Razzle Dazzle, *see* 'JACraz'

Razzle Dazzle, *see* FRYbright

Razzmatazz, *see* 'JACmat'

'Rea Silvia', HT, or, 1958; flowers fiery red; [Baiser X (Peace X seedling)]; Giacomasso

'REAclibal', LCl, pb, 1997; flowers semi double, medium, semi-dbl., 8–15 petals, borne in large clusters, moderate fragrance; some prickles; foliage medium, light green, semi-glossy; upright (15 ft) growth; [Ballerina sport]

Real Charmer, *see* 'PIXichar'

Rea's Rose, HT, ab, 1997

Reba McEntire, *see* 'MAChahei'

'Rebecca', HT, pb, 1930; bud large, long, pointed; flowers silvery pink, reverse salmon-pink, dbl.; foliage dark, leathery; vigorous growth; Pemberton; Bentall

Rebecca®, *see* 'TANrekta'

Rebecca Anne, *see* 'PIXanne'

'Rebecca Claire', HT, op, 1981; flowers coppery orange edged light coral, blooms borne singly and in, dbl., 28 petals, intense fragrance; medium brown prickles; foliage medium green, semi-glossy; vigorous, bushy growth; Edland Fragrance Medal, ARS, 1980; [Blessings X Redgold]; Law, M.J., 1986

'Rebecca Gue', B, mp, 1982; flowers distinction from parent not described; [Mme Ernst Calvat sport]; Gue, Derek J.

Rebecca Paul, *see* 'HORcockalorum'

'Rebecca's Delight', F, mp, 1974; flowers soft salmon pink, shaded darker, large, dbl., 24 petals, slight fragrance; foliage light green; vigorous, upright growth; [(Pink Parfait X Highlight) X Circus]; Harkness; Morse Roses

Rebekah, HT, pb

Rebell, *see* 'FAYbell'

Rebell 96, *see* KORvegata

Rebellastar, HT, 1970; Hetzel; (Cavriglia)

'Recompense', HT, dp, 1957; flowers deep pink, well shaped, medium, moderate fragrance; moderate growth; [Charles P. Kilham X Polly]; Ratcliffe

Reconciliation, *see* HARtillery

Record, *see* **'Décor'**

'Recuerdo de Angel Peluffo', HT, rb, 1928; flowers cardinal-red, center garnet-red, dbl., moderate fragrance; [Mme Edouard Herriot X Elvira Aramayo]; Soupert & Notting

'Recuerdo de Antonio Peluffo', T, yb, 1910; flowers light yellow, edged pink, dbl.; [Mme Mélanie Soupert X Mme Constant Soupert]; Soupert & Notting

'Recuerdo de Blas Munné', Cl HT, rb, 1948; flowers carmine-red shaded cerise red, slight fragrance; foliage dark; long, strong stems; very vigorous growth; [Maria Serrat X Recuerdo del Doctor Ferran]; Munné, M.

'Recuerdo de Felio Camprubi', HT, rb, 1931; flowers crimson suffused pink, reverse yellow suffused red, large, dbl., intense fragrance; vigorous growth; [Hugh Dickson X Souv. de Claudius Pernet]; Camprubi, C.

'Recuerdo del Doctor Ferrari', HT, rb, 1935; bud long, pointed; flowers scarlet-crimson, shaded fiery red, open, large, very dbl., intense fragrance; foliage dark; vigorous growth; [Sensation X K. of K.]; Munné, B.

Red Ace, *see* **'Amruda'**

'Red Ace', Min, mr, 1980; dbl., 23 petals, exhibition form, borne usually singly, slight fragrance; long, thin, straight prickles; low, compact, bushy growth; [Rise 'n' Shine X Sheri Anne]; Saville, F. Harmon; Nor'East Min. Roses

'Red Admiral', F, mr, 1940; flowers scarlet, borne in clusters; Archer

Red Admiral, HT, mr

Red Alert, *see* 'MORalert'

'Red American Beauty', HT, mr, 1959; bud ovoid; flowers scarlet overcast rose-red, dbl., 30–35 petals, 4.5–5 in., exhibition form, intense fragrance; foliage leathery, dark; long stems; vigorous, upright, bushy growth; [Happiness X San Fernando]; Morey, Dr Dennison; J&P

'Red Arrow', Min, mr, 1962; dbl., 40 petals, 1.25 in., exhibition form, blooms in clusters, slight fragrance; foliage leathery; vigorous (12–18 in.) growth; [(R. wichurana X Floradora) X Seedling]; Moore, Ralph S.; Sequoia Nursery

Red Azteca, *see* 'SELazteca'

Red Baby, S, 1992; Embriaco, B.; (Cavriglia)

Red Baby Rambler, *see* **'Mme Norbert Levavasseur'**

Red Bailer, B, mr

Red Ballerina, *see* 'HARhero'

'Red Ballerina', S, mr, 1976; flowers bright crimson, full, small, semi-dbl., 10 petals, slight fragrance; foliage glossy; [Ballerina X Evelyn Fison]; Fryer, Gareth; Fryer's Nursery, Ltd.

'Red Beauty', Min, dr, 1981; flowers dark red, yellow hinge, small to medium, dbl., 35 petals, exhibition form, slight fragrance; foliage small, dark, glossy; bushy growth; [Starburst X Over the Rainbow]; Williams, Ernest D.; Mini-Roses

'Red Beauty', HT, mr, 1929; [Matchless sport]; Dunlop; Liggit

Red Belinda, HMsk, mr

Red Bells®, *see* 'POUlred'

Red Berlin, *see* OLIjplam

'Red Better Times', HT, mr, 1937; flowers bright clear red; [Better Times sport]; Asmus

'Red Bird', HT, mr, 1957; flowers bright red; [Better Times sport]; Manda, Jr., E.A.; J&P

Red Bird, *see* 'GELbird'

Red Blanket®, *see* 'INTercell'

Red Blush, A, m, 1988; Sievers

'Red Boy', HT, or, 1939; bud long, pointed; flowers fiery orange-red, fading to dominant pink, single and open, semi-dbl., slight fragrance; foliage dark, glossy; vigorous, bushy growth; [Charles K. Douglas X Pres. Herbert Hoover]; Hansen, N.J.; B&A

Red Brigand, *see* 'SANtor'

'Red Button', Min, dr, 1978; bud short, pointed; flowers deep red, small, dbl.; foliage very small, glossy; bushy, compact, spreading growth; [(R. wichurana X Floradora) X Magic Dragon]; Moore, Ralph S.; Sequoia Nursery

'Red Camellia', F, or, 1943; bud small; flowers orange-scarlet, dbl., 20–30 petals, slight fragrance; foliage dark, leathery; vigorous, upright, bushy growth; [Baby Chateau X Folkestone]; Krause

Red Camellia, HT, mr

Red Cameo, *see* 'MORanyblac'

'Red Can Can', Min, mr, 1977; bud pointed; flowers cardinal-red, dbl., 28 petals, 2 in., moderate, fruity fragrance; foliage dark; vigorous, upright growth; [Unnamed seedling X ?]; Lyon

Red Candy, F, lp; short; flowers simple light red with white eye, flat, 8 petals, borne in clusters, slight; medium large, dark green, glossy; [Candy Rose chance seedling sport]; St Wagner, 1993; Res. Stn. f. Fruit Growing, Cluj, 1995, Romania

Red Cap, HT, dr

'Red Carpet', Cl F, dr, 1971; bud ovoid; flowers dark scarlet, overlaid darker, medium, dbl., globular, foliage large, dark, leathery; very vigorous, climbing growth; [Don Juan X Red Favorite]; Williams, J. Benjamin

'Red Carrousel', Min, dp, 1984; flowers deep pink; [Magic Carrousel sport]; Rumsey, R.H.; Roy H. Rumsey, Pty. Ltd.

Red Carrousel, Min, dp

'Red Cascade', Cl Min, dr, 1976; (Moorcap); bud pointed; flowers deep red, dbl., 40 petals, 1 in., cupped, slight fragrance; foliage small, leathery; prostrate, bushy growth; AOE, 1976; [(R. wichurana X Floradora) X Magic Dragon]; Moore, Ralph S.; Sequoia Nursery

Red Cécile Brünner, *see* **'Pasadena Tournament'**

Red Cécile Brünner, Climbing, *see* **'Pasadena Tournament, Climbing'**

Red Cedar, *see* 'KORgund '81'

'Red Chateau', HT, mr, 1999; flowers crimson red, dbl., 35 petals, exhibition form, slight fragrance; 3.5 ft growth; [(Samantha X seedling) X seedling]; Teranishi, K.; Itami Rose Nursery, 1997

'Red Cheer', Min, mr, 1975; flowers cherry-red, full, dbl., 45 petals, 1 in., slight fragrance; foliage very small, dark; compact, bushy growth; [Unnamed seedling X Unnamed seedling]; Lyon

'Red Cheerful', HT, dr, 1951; bud pointed; flowers deep red, dbl., 30–40 petals, 4.5 in., exhibition form, moderate fragrance; foliage glossy, leathery, veined red; vigorous, upright growth; [Better Times sport]; Blixen; Woodlawn Gardens

Red Cherokee, *see* **'Ramona'**

'Red Chief', HT, mr, 1967; flowers large, dbl., 35 petals, exhibition form, intense fragrance; foliage leathery; vigorous, upright, bushy growth; [Seedling X Chrysler Imperial]; Armstrong, D.L.; Armstrong Nursery

'Red Coat', F, mr, 1981; (Redcoat); bud pointed; semi-dbl., 10 petals, borne 1–5 per cluster, slight fragrance; hooked, brown prickles; foliage dark; bushy growth; [Unnamed seedling X Golden Showers]; Austin, David, 1973

'Red Columbia', HT, mr, 1920; flowers rich velvety scarlet, large, dbl., intense fragrance; foliage leathery; very vigorous growth; [Columbia sport]; Hill, Joseph H., Co.

Red Cottage, *see* DICxtol

'Red Cross', HT, ob, 1916; flowers orange-crimson-scarlet, moderate fragrance; Dickson, A.

Red Cross, HT, dr, 1998; Meilland

Red Curtain, *see* KORmauret

'Red Cushion', F, dr, 1966; bud pointed; flowers small blooms in clusters, semi-dbl., slight fragrance; foliage dark, glossy, leathery; vigorous, bushy growth; [Circus X Ruby Lips]; Armstrong, D.L.; Armstrong Nursery

Red Dagmar, *see* SPEruge

Red Damask, HGal, mr; flowers medium, dbl., intense fragrance; (Sangerhausen)

'Red Dandy', F, mr, 1959; flowers cherry-red, dbl., 40 petals, 3 in., moderate fragrance; vigorous, upright growth; [Ena Harkness X Karl Herbst]; Norman; Harkness

'Red Dawn', S, mr, 1957; flowers deep rose-red, well-formed, recurrent bloom; hardy.; [New Dawn X ?]; Simonet; Skinner

Red Ded, Min, dr, 1980; flowers dark scarlet-red, small, very dbl.; Cocker; (Sangerhausen)

'Red Delicious', HT, mr, 1942; bud carmine, shaded oxblood-red; flowers brilliant rose-red, dbl., 30–35 petals, 4–6 in., intense fragrance; foliage leathery, dark; weak necks; vigorous, upright growth; [Rome Glory X Chieftain]; Hill, Joseph H., Co.

Red Delicious, *see* JACdrive

'Red Delight', HT, mr; flowers bright red; RULED EXTINCT 11/86; [Pink Delight sport]; Avansino; Mortensen, 1935

Red Delight™, *see* 'MINaico'

'Red Det', Min, mr, 1978; dbl., flat, blooms in clusters, slight fragrance; foliage small, deep green, glossy; vigorous, bushy growth; [(Marlena X Kim) X Little Buckaroo]; Harkness; Rosen-Union

Red Det 80®, Min, or, 1980; Cocker

Red Devil, *see* 'DICam'

Red Diadem®, *see* TANetorde

Red Dorothy Perkins, *see* **'Excelsa'**

'Red Dragon', F, mr, 1970; single, 5 petals, 3.5–4 in., slight fragrance; vigorous growth; [Anna Wheatcroft X Unnamed seedling]; Cants of Colchester, Ltd.

Red Druschki, *see* **'Ruhm von Steinfurth'**

'Red Duchess', HT, mr, 1942; bud long, pointed; dbl., 35–45 petals, 3.5–5 in., exhibition form, moderate fragrance; foliage glossy, bronze; long, strong stems; vigorous, upright, bushy growth; [Pink Princess X Crimson Glory]; Brownell, H.C.

'Red Duchess, Climbing', Cl HT, mr, 1955; Brownell, H.C.

'Red Echo', Pol, or, 1932; flowers vermilion tinted crimson, dbl., cupped, slight fragrance; foliage dark, wrinkled; dwarf growth; [Echo X Unnamed Hybrid Tea (red)]; Kluis & Koning; J&P

Red Elegance™, *see* BALIgance

'Red Elf', Min, dr, 1949; bud ovoid; flowers dark crimson, dbl., 23 petals, .75–1 in., slight fragrance; foliage soft, tiny; vigorous, bushy, dwarf growth; [Eblouissant X Tom Thumb]; deVink; C-P

'Red Ellen Poulsen', Pol, mr, 1918; (Dunkelrote Ellen Poulsen); [Ellen Poulsen sport]; Poulsen, S.

'Red Ember', HT, or, 1953; flowers flame, well formed, dbl., 30 petals, 4–5 in., moderate fragrance; foliage glossy, dark; free growth; Cant, F.

'Red Emblem', F, dr, 1958; bud ovoid; flowers deep red, open, dbl., 30–35 petals, 2.5 in., moderate fragrance; foliage leathery; upright, bushy growth; [Garnette seedling X Pageant]; Boerner; J&P

Red Emperor, *see* 'BROemp'

'Red Empress', LCl, mr, 1956; (Impératrice Rouge, Robur); bud ovoid; flowers cardinal-red, to loosely, blooms borne singly or 2 per stem, dbl., 33 petals, 3.5 in., exhibition form, recurrent bloom, intense fragrance; foliage leathery; vigorous, climbing growth; [(Holstein X Decor) X Self]; Mallerin, C.; C-P; EFR

'Red Ensign', HT, mr, 1947; flowers crimson, dbl., 40–45 petals, 4–5 in., exhibition form, intense, damask fragrance; foliage dark; very vigorous growth; GM, NRS, 1943; [Crimson Glory X Southport]; Norman; Harkness

'Red Explorer', Cl Pol, dr, 1938; flowers deep brilliant crimson, borne in large clusters, recurrent bloom, vigorous, climbing growth; [Miss Edith Cavell sport]; Penny

Red Fairy, *see* 'MORedfar'

Red Fan, HT, or

Red Favorite, *see* 'TANschweigru'

Red Favorite, Climbing,

Red Festival, *see* 'DELfesrou'

'Red Finch', Pol, mp, 1937; flowers cerise, more, more open, more than parent, borne in clusters, dbl., cupped, [Mrs R.M. Finch sport]; Stielow, F.C.

'Red Flare', LCl, dr, 1954; bud globular; flowers carmine overcast spectrum-red, borne singly and clusters, dbl., 40 petals, 4–4.5 in., cupped, profuse bloom repeated sparingly, moderate fragrance; foliage dark, glossy; height 7–8 ft; [Reine Marie Henriette X Paul's Scarlet Climber seedling]; Mansuino, Dr Andrea; J&P

Red Fleurette, S, mr

Red Flower Carpet, *see* NOAre

'Red Flush', Min, mr, 1978; bud ovoid; flowers small, dbl., 53 petals, 1.5 in., cupped, foliage green, matt; very compact growth; AOE, 1979; Schwartz, Ernest W.; Nor'East Min. Roses

'Red Fountain', LCl, dr, 1975; bud ovoid, pointed; flowers scarlet, medium, dbl., 20–25 petals, cupped, blooms in clusters, intense fragrance; foliage large, dark, leathery; very vigorous, climbing growth; [Don Juan X Blaze]; Williams, J. Benjamin; C-P

Red France, *see* FAZcanne

Red Friendship, S, mr, 1986; Verschuren

'Red Frost', HT, rb, 1993; flowers medium red, suffused white, shaded light yellow at base, blo, dbl., 26–40 petals, 3–3.5 in., exhibition form, slight fragrance; some prickles; foliage medium, dark green, glossy; medium (90–120 cms), bushy growth; [First Prize X Seedling]; Lienau, David W.; Trophy Roses, Ltd., 1993

Red Garnette, *see* **'Garnette'**

Red Garter, *see* 'BROgar'

'Red Germain', Min, rb, 1975; bud long, pointed; flowers red, reverse lighter, dbl., 25 petals, 1 in., flat, slight fragrance; foliage small, leathery; vigorous, bushy growth; [(R. wichurana X Floradora) X (Oakington Ruby X Floradora)]; Moore, Ralph S.; Sequoia Nursery

'Red Globe', F, mr, 1971; bud ovoid; flowers full,large, dbl.; foliage soft; moderate, bushy growth; Delforge

'Red Glory', F, mr, 1958; bud ovoid, pointed; flowers cherry to rose-red, blooms in rounded clusters, semi-dbl., 11 petals, 2.5–3.5 in., cupped, slight fragrance; foliage leathery, semi-glossy; very vigorous, tall, bushy growth; good as fence or hedge; [Gay Lady X (Pinocchio X Floradora)]; Swim, H.C.; Armstrong Nursery

Red Gruss an Coburg, *see* **'Clotaria'**

'Red Guard', HT, rb, 1935; flowers dark blood-red, well formed, very dbl.; very vigorous growth; Verschuren

'Red Halo', F, mr, 1968; flowers crimson, borne in trusses, flat, slight fragrance; foliage glossy; vigorous, bushy growth; [Tabarin X Karl Herbst]; Oliver, W.G.

'Red Hedge', HRg, mr, 1958; small, red fruit; upright, compact growth; [(R. rugosa rubra X R. cinnamomea) X R. nitida]; Nyveldt

Red Hermosa, *see* **'Queen's Scarlet'**

'Red Hill', HT, mr, 1941; flowers well-formed, large; [E.G. Hill seedling]; Clark, A.

Red Hit, *see* 'POUlhit'

'Red Hoover', HT, rb, 1937; flowers brilliant red, center salmon-red; [President Herbert Hoover sport]; Lens

Red Hot, *see* 'MACbigma'

'Red Imp', Min, dr, 1951; (Maid Marion, Mon Tresor, Montresor); bud ovoid; flowers deep crimson, micro-mini, very dbl., 54 petals, .75–1 in., flat, slight fragrance; upright, bushy, dwarf (9 in.) growth; [Ellen Poulsen X Tom Thumb]; deVink; C-P

'Red Jacket', HT, mr, 1950; bud ovoid, pointed; flowers urn-shaped, dbl., 21 petals, 3.75 in., flat, slight fragrance; foliage leathery; upright, bushy growth; [World's Fair X Mirandy]; Swim, H.C.; Stuart

Red Jewel, *see* 'INTerro'

Red Joy, HT, mr, 1997

Red Kelly, LCl, lp; Sutherland, P, 1999; (Weatherly, L.)

'Red Knight', LCl, dr, 1964; flowers maroon, medium, very vigorous growth, semi-dbl., globular, recurrent bloom, [Dr Huey sport]; Booy, P.J.; Booy Rose Nursery

Red La France, *see* **'Duchess of Albany'**

Red Lady®, HT, mr

'Red Lion', HT, mr, 1964; flowers red becoming rose-red, dbl., 38 petals, 5 in., exhibition form, [Kordes' Perfecta X Detroiter]; McGredy, Sam IV; Spek

Red Love™, *see* 'MINqco'

'Red Magic', Min, mr, 1977; bud pointed; flowers deep cherry-red, open, medium, semi-dbl., 10–15 petals, slight fragrance; vigorous, upright growth; [Red Can Can X ?]; Lyon; L. Lyon Greenhouses

Red Magic Carpet™, *see* 'MACredsaus'

'Red Maid', F, or, 1975; flowers coral-red, semi-dbl., 18 petals, 3 in.; [Vera Dalton X Stephen Langdon]; Sanday, John

Red Malmaison, *see* **'Leweson Gower'**

Red Maman Cochet, *see* **'Balduin'**

Red Maman Cochet, *see* **'Niles Cochet'**

Red Marechal Niel, *see* **'Grossherzog Ernst Ludwig'**

'Red Margo Koster', Pol, mr

'Red Martini', HT, dr, 1967; bud ovoid; flowers large, dbl., slight fragrance; foliage bronze; vigorous, bushy growth; [Chrysler Imperial X Seedling]; Delforge

'Red Masquerade', F, mr, 1962; bud pointed; flowers red becoming darker, borne in clusters, single, 10 petals, 2.5–3 in.; vigorous growth; [Masquerade X Independence]; Hill, A.

Red Masterpiece, *see* 'JACDER'

Red Matheson, F, mr

Red Max Graf, *see* 'KORMAX'

Red Meidiland®, *see* 'MEINEBLE'

Red Meillandina®, Min, mr, 1982; Meilland

Red Mercedes, F, mr

Red Minimo®, *see* 'RUIMIRED'

Red Mini-Wonder®, *see* 'MEINOFRAI'

Red Moon®, *see* 'MINIMOON'

Red Moscow, *see* **'KRASNAIA MOSKVA'**

Red Moss, *see* **'HENRI MARTIN'**

'RED MOSS', M, mr; flowers heavily mossed, reddish rose, large

Red Moss Rambler, HWich, mr, 1988; Moore

'RED MOSS, CLIMBING', Cl M, mr; Foote; B&A

Red 'n' White Glory, *see* **'CANDYSTICK'**

'RED NEARLY WILD', F, mr, 1960; bud globular; flowers rose-red, small, borne in clusters, single, cupped, foliage soft; vigorous, bushy growth; [Nearly Wild X Unnamed seedling]; Brownell, H.C.; Brownell

Red Nella, HT, dp, 1991; Allender, Robert William

Red Nelly, S, m

Red New Dawn, *see* **'ÉTENDARD'**

Red Niphetos, *see* **'LADY BATTERSEA'**

'RED OPAL', F, mr, 1968; bud pointed; flowers red, reverse cerise, borne in small clusters, flat, foliage small, dark; free growth; [Karl Herbst X Korona]; Northfield

Red Orléans Rose, *see* **'MARÉCHAL FOCH'**

Red Paillette, *see* POUlsint

'RED PARADE', F, mr, 1972; bud ovoid; flowers carmine-red, medium, dbl., exhibition form, moderate fragrance; foliage glossy, bronze; vigorous, bushy growth; [(Frolic X Peace) X Texan]; Patterson; Patterson Roses

Red Parfum®, Cl F, mr, 1972; Eve, A.

Red Pastel, *see* 'WILRIB'

Red Peace, *see* **'KARL HERBST'**

'RED PEARL', HT, mr, 1970; flowers bright red, pointed, dbl., 40 petals, 5 in., moderate fragrance; foliage large, dark; very vigorous growth; [Josephine Bruce X Kordes' Perfecta]; Watkins Roses

Red Pendant®, *see* 'MINVCO'

Red Perfection®, *see* 'MACHADEN'

'RED PETTICOAT', F, mr, 1967; flowers blood-red, short petaled, borne in trusses, dbl., flat, foliage dark, glossy; very vigorous growth; [Buisman's Triumph X Lilli Marlene]; Watkins Roses

'RED PINOCCHIO', F, dr, 1947; flowers velvety carmine-red, blooms in clusters, dbl., 28 petals, 3 in., cupped, moderate fragrance; vigorous, bushy growth; (28); [Unnamed yellow Pinocchio seedling X Donald Prior]; Boerner; J&P

'RED PISTOLS', Min, dy, 1982; flowers deep golden yellow, small, 15–20 petals, no fragrance; brown, needle-like prickles; foliage medium green, matt; upright, bushy growth; [Rise 'n' Shine X Unnamed seedling]; Ballmer, Gordon W.; Biotika International

Red Pixie, *see* 'KORNUMA'

'RED PLANET', HT, dr, 1970; flowers crimson, dbl., 30 petals, 5.5–6 in., intense fragrance; foliage glossy; GM, RNRS, 1969 PIT, RNRS, 1969; [Red Devil X Seedling]; Dickson, Patrick; A. Dickson

'RED PLUME', F, or, 1966; flowers bright scarlet, rosette form, borne in clusters, dbl., 35 petals, 3 in.; foliage glossy; vigorous, bushy growth; [Masquerade X Independence seedling]; Sanday, John

Red Point, *see* 'MACWOOHERM'

Red Poppy, *see* **'KRASNYI MAK'**

'RED PREMIER', HT, dp, 1924; bud long, pointed; flowers bright carmine, large, dbl., intense fragrance; foliage leathery; vigorous growth; [Premier sport]; Scott, R.

'RED PRIDE', F, mr, 1968; bud ovoid; flowers small, dbl.; Verbeek

Red Prince®, *see* **'FOUNTAIN'**

Red Prolific, *see* 'DELIFIC'

'RED PROVENCE', C, mr; flowers clear crimson-red, large, cupped, intense fragrance; low, spreading growth

'RED QUEEN', HT, mr, 1968; (Liebestraum); bud ovoid; flowers large, dbl.; foliage dark; vigorous, upright growth; [Colour Wonder X Liberty Bell]; Kordes; Buisman; McGredy

'RED RADIANCE', HT, dp, 1916; flowers light crimson; [Radiance sport]; Gude Bros.

'RED RADIANCE, CLIMBING', Cl HT, dp, 1927; Pacific Rose Co.

'RED RADIANCE, CLIMBING', Cl HT, dp, 1929; Catt

Red Rascal, *see* 'JACBED'

Red Recker, HT, dr, 1986; Kasturi

'RED REFLECTION', HT, mr, 1964; bud ovoid, pointed; 4–5 in., exhibition form, slight fragrance; foliage large, leathery; upright growth; [Tropicana X Living]; Warriner, William A.; J&P

Red Ribbon®, *see* 'WILRIB'

Red Ribbons, *see* 'KORTEMMA'

'RED RIDER', Cl F, rb, 1970; flowers red, base yellow, full, small, dbl., 50 petals, 2–3 in.; foliage large, glossy; vigorous growth; [Circus X Danse de Feu]; Guest

Red Rider, *see* 'LAVRIDE'

'RED RIDING HOOD', Min, dr, 1955; flowers brilliant dark red; [Red Imp sport]; Robinson, T.

Red Riding Hood, *see* **'RÖDHÄTTE'**

'RED RIPPLES', F, dr, 1942; (Willi Maass); bud globular; flowers deep red, petals wavy, small blooms in large clusters, semi-dbl., slight fragrance; foliage leathery, glossy, wrinkled; upright, bushy growth; [Hamburg X Anne Poulsen]; Krause; C-P

Red River, F; Kordes; (Cavriglia)

Red Robin, *see* **'GLOIRE DES ROSOMANES'**

'RED ROBIN', HT, rb, 1940; flowers red, tending toward scarlet; vigorous growth; Brownell, H.C.

Red Robin, S, rb, 1992; Lens

Red Rock, *see* 'MEILUSAM'

'RED ROCKET', S, mr, 1949; bud ovoid; flowers crimson, large blooms in clusters, semi-dbl., non-recurrent, slight fragrance; foliage large, dark, glossy, bronze; vigorous, upright growth; [Skyrocket seedling]; Longley, L.E.; Univ. of Minn.

Red Rocky, *see* 'MEILUSAM'

Red Roma, HT, mr; flowers large, dbl.; (Sangerhausen)

Red Rosamini®, *see* 'RUIREDRO'

'RED ROSE MARIE', HT, rb, 1938; bud ovoid; flowers cerise-red, large, dbl., exhibition form, slight fragrance; foliage dark, leathery; vigorous, bushy growth; [Rose Marie sport]; Mordigan Evergreen Nursery

Red Rose of Lancaster, *see* **R. GALLICA OFFICINALIS**

Red Roulette, Min, mr

Red Rover, *see* 'LAVRIDE'

'RED RUFFLES', F, dr, 1960; bud pointed; flowers dull dark red, borne in large clusters, dbl., 2.5–3 in., cupped, slight fragrance; foliage dark, glossy; vigorous, bushy, compact growth; [Improved Lafayette X Carrousel]; Von Abrams; Peterson & Dering

Red Rugostar, HRg, mr, 1995; flowers carmine-red, medium-large, semi-dbl.; Meilland; (Sangerhausen)

'RED RUM', F, mr, 1976; flowers red, shaded scarlet, dbl., 24 petals, 2.5 in., slight fragrance; foliage dark; vigorous growth; [Handel X Arthur Bell]; Bees

'RED SARONG', HT, dr, 1974; bud ovoid; flowers deep red, dbl., 36 petals, 5 in., exhibition form, moderate fragrance; foliage dark, leathery; tall growth; [Baccará X Golden Showers]; Golik; Dynarose

Red Satin, *see* 'DEVPAJARO'

Red Sea®, *see* BARred

Red Shadows(TM), *see* 'SAVMORE'

Red Simplicity, *see* 'JACSIMPL'

'RED SKELTON', HT, or, 1968; bud long, pointed; flowers vermilion, large, dbl., exhibition form, moderate fragrance; foliage bronze, leathery; vigorous, upright growth; [Rose Queen X Charlotte Armstrong]; Whisler, D.; Germain's

Red Smith's Parish, T, rb, 1997; flowers double, small, dbl., 15–25 petals, borne in small clusters, slight fragrance; foliage medium, light green, semi-glossy; bushy, medium (4–6 ft) growth; [Smith's Parish sport]; Manners, Dr Malcolm

Red Soldier, *see* **'FUSILIER'**

'RED SPARKLER', HT, rb, 1967; flowers red, striped pink and white, dbl., 55 petals, 4–4.5 in., cupped, intense, damask fragrance; foliage dark, leathery; upright growth; [Scarlet Royal X Rouge Mallerin]; Buck, Dr Griffith J.; Iowa State University

Red Spectacle, S, dr, 1993; Poulsen

'RED SPICE', F, dr, 1958; bud short, flat topped; dbl., 80–85 petals, 2–2.5 in., cupped, moderate fragrance; foliage dark, leathery; vigorous, bushy growth; [Spice X Garnette]; Boerner; J&P

'RED SPLENDOR', HT, mr, 1954; flowers bright red, dbl., 55 petals, 6 in., moderate fragrance; foliage leathery; vigorous, upright growth; [Joyance sport]; Grillo

Red Splendour, *see* 'DAVONA'

'RED SPRITE', F, mr, 1974; flowers glowing red, dbl., 35 petals, 2–2.5 in., slight fragrance; foliage small, glossy, dark; low growth; LeGrice

Red Star, *see* **'PRECIOUS PLATINUM'**

Red Stone, *see* **'KRASNOKAMENKA'**

Red Storm, Min, dr

Red Success, *see* 'MEIRODIUM'

Red Summer, F, or, 1996; Noack, Werner

'RED SUMMIT', Min, dr, 1981; bud ovoid, pointed; dbl., 33 petals, borne usually singly, slight fragrance; long, thin, straight prickles; low, compact, bushy growth; [Unnamed seedling X Unnamed seedling]; Lyon

Red Sunblaze®, *see* 'MEIRUTRAL'

'RED SWEETHEART', F, rb, 1944; bud pointed; flowers crimson-carmine, dbl., 25 petals, 1–1.5 in., exhibition form, moderate, spicy fragrance; foliage small, glossy; strong stems; vigorous, bushy growth; [Intercrossing of Cécile Brunner seedlings sport]; Krebs; Marsh's Nursery

'RED TAG', Min, rb, 1978; bud ovoid, pointed; flowers medium red, white reverse, dbl., 48 petals, 1 in., exhibition form, moderate fragrance; foliage small, glossy, dark; upright, spreading growth; [Seedling X Over the Rainbow]; Williams, Ernest D.; Mini-Roses

'RED TALISMAN', HT, rb, 1931; flowers deep cerise, base yellow; [Talisman sport]; Amling Bros.

'RED TALISMAN, CLIMBING', Cl HT, rb

'RED TAUSENDSCHÖN', HMult, rb, 1931; flowers vivid red with white eye, but varies considerably; Tausendschön sport]; Walter, L.

'RED TORNADO', HT, mr, 1999; flowers medium red, reverse light red, dbl., 26–40 petals, 5–5.5 in., borne mostly singly, slight fragrance; prickles moderate; foliage medium, dark green, glossy; upright, medium (4 ft) growth; [Dorothy Anne X Classic Touch]; McCall, Sharan

Red Trail, *see* INTerim

'RED TRIUMPH', Pol, or, 1956; [Orange Triumph sport]; Morse

Red Trooper, F, dr

Red Velvet, *see* 'PEATHUNDER'

'RED VELVET', F, mr, 1940; bud ovoid; flowers vivid crimson, semi-dbl., cupped, slight fragrance; vigorous, compact growth; Kordes; Dreer

Red Velvet, *see* 'PEATHUNDER'

Red Velvet, *see* TANorelav

Red Wagon, *see* 'MORDAN'

'RED WAND', Cl Min, mr, 1964; flowers light crimson-red, small, dbl.; vigorous, climbing ($3\frac{1}{2}$–4 ft) growth; [((R. wichurana X Floradora) X Orange Triumph) X Unknown Miniature]; Moore, Ralph S.; Sequoia Nursery

Red Wand, S, mr, 1994; Kordes

'RED WAVE', F, mr, 1964; flowers tulip shaped, small; will bloom in pot; low growth; [Carolyn Dean X Unnamed Floribunda seedling]; Moore, Ralph S.; Sequoia Nursery

'RED WINE', HT, mr, 1970; dbl., 22 petals, 3.5 in., exhibition form, blooms in clusters, slight fragrance; foliage medium green, matt; upright growth; [Lilac Rose X Seedling]; Sanday, John

'RED WING', S, my; flowers bright golden yellow; large, winged red prickles; [Probably R. sericea pteracantha X R. hugonis]

'RED WINGS', F, dr, 1958; bud ovoid; flowers rich dark red, semi-dbl., 20 petals, 3.5 in., cupped, moderate fragrance; foliage dark, leathery; vigorous, upright, bushy growth; [(Improved Lafayette X Herrenhausen) X Lavender Pinocchio]; Boerner; J&P

'RED WONDER', F, dr, 1954; bud globular; dbl., 28 petals, 3–3.5 in., cupped, blooms in large clusters, moderate fragrance; foliage leathery, semi-glossy; vigorous, bushy growth; [Better Times X Floribunda seedling]; deRuiter; C-P

Red Yesterday, *see* 'HARHERO'

Red Zinnia, Min, dr

'REDCAP', F, mr, 1954; bud ovoid; semi-dbl., 18 petals, 2.5 in., exhibition form, blooms in clusters, foliage leathery, semi-glossy; vigorous, upright, bushy growth; [World's Fair X Pinocchio]; Swim, H.C.; Armstrong Nursery

'REDCLIFFE', F, mr, 1975; flowers bright crimson, semi-dbl., 15 petals, 3.5 in., slight fragrance; [Seedling X Sarabande]; Sanday, John

Redcoat, *see* **'RED COAT'**

'REDCRAZE', HT, ob, 1959; bud long, pointed; flowers orange-scarlet, medium, dbl., exhibition form, slight fragrance; foliage glossy, dark; vigorous, bushy growth; [Independence X Independence]; Doley

Reddy Teddy, *see* 'SUNHINO'

Redglo(TM), *see* 'MINABBCO'

Redgold, *see* 'DICOR'

'REDGOLD, CLIMBING', Cl F, yb, 1974; Lynch; J&P

Redgold, Climbing, *see* 'DICORSAR'

'REDHEAD', Cl Min, dr, 1956; flowers blood-red; height $2\frac{1}{2}$ ft; [((Soeur Therese X Skyrocket) X (Seedling X Red Ripples)) X Zee]; Moore, Ralph S.; Sequoia Nursery

Redhot, *see* 'MACBIGMA'

'REDHOTS', Min, dr, 1982; dbl., 22 petals, exhibition form, borne singly, slight fragrance; foliage small, long, medium green, finely serrated; very small, tiny growth; [Unnamed seedling X Darling Flame]; Meredith, E.A. & Rovinski, M.E.; Rosa de Casa Domingo

Rediffusion Gold, *see* 'HARQUORGOLD'

'REDIPUGLIA', HT, w, 1933; bud long, pointed; flowers pinkish white, reverse golden rose, base chrome-yellow, large, dbl.; vigorous growth; Ingegnoli

'REDLAND COURT', F, ab, 1982; flowers soft apricot, small, dbl., 20 petals, slight fragrance; foliage small, medium green, glossy; bushy growth; [Red Maid X Sarabande]; Sanday, John; Sanday Roses, Ltd.

Redlands Century(TM), *see* 'BURCEN'

'RED-LETTER DAY', HT, dp, 1914; flowers velvety rose-red, white streak in center of inner petals, st, semi-dbl., 3.5 in.; foliage glaucous sage-green; short stems; vigorous growth; (28); GM, NRS, 1913; Dickson, A.

Redneck Girl, *see* 'GELRED'

'REDONDA', HT, mr, 1968; bud globular; flowers large, dbl., moderate fragrance; foliage leathery, wrinkled; vigorous, upright growth; [Queen Elizabeth X Happiness]; Patterson; Patterson Roses

Redoute(TM), *see* 'AUSpale'

'REDWAY', HT, mr, 1951; flowers crimson-carmine, exhibition form, moderate fragrance; foliage leathery; moderate, upright growth; [Ena Harkness X Unnamed seedling]; Waterhouse Nursery

Redwood, *see* 'MACwoodma'

Redwood Empire, *see* 'MORwood'

Referenz, F, mp, 1983; dbl., slight fragrance; VEG; (Sangerhausen)

'**Reflection**', HT, yb, 1952; bud long, pointed; flowers amber, reverse streaked scarlet, moderate fragrance; vigorous growth; [Mme Henri Guillot sport]; Ratcliffe

Reflection, HT, lp; Brady, M., 1995; (Weatherly, L.)

'**Reflets**', F, or, 1964; flowers large, single; upright growth; [Gertrud Westphal X Sarabande]; Croix, P.

'**Refresher**', LCl, w, 1929; flowers pure white, blooms in clusters, single, early, vigorous, climbing growth; Clark, A.; Hazlewood Bros.

'**Refulgence**', HEg, mr, 1909; flowers scarlet, aging to crimson, large, semi-dbl.; foliage very fragrant; vigorous growth; (41); Paul, W.

'**Reg Willis**', HT, dp, 1966; flowers deep rose-pink, base yellow, well-formed, 4.5 in., slight fragrance; [Golden Masterpiece X Karl Herbst]; McGredy, Sam IV

Regal, *see* 'JUSregal'

'**Regal Gold**', HT, dy, 1958; bud pointed; flowers clear golden yellow, open, dbl., 40–45 petals, 5 in., moderate fragrance; foliage glossy, leathery; vigorous, upright, bushy growth; [Golden Rapture sport]; Dale; Amling-DeVor Nursery

'**Regal Lady**', Min, pb, 1989; bud pointed; flowers white, edged pink, borne usually singly and in sprays of 3–5, dbl., 27 petals, exhibition form, slight fragrance; round fruit; no prickles; foliage medium, medium green, semi-glossy; upright, spreading, tall growth; [Sassy Lassie X First Prize]; Jolly, Nelson F.; Rosehill Farm, 1991

'**Regal Pink**', HT, mp, 1980; [Red Queen sport]; Thomas, Dr A.S.

Regal Red, *see* COCfoster

'**Regalia**', HT, rb, 1964; flowers cherry-red, reverse silver, well-formed, dbl., 60 petals, 5.5 in.; foliage bronze; [Rose Gaujard X Seedling]; Robinson, H.

Regatta™, *see* 'MEInimo'

Regatta, *see* 'JACette'

Régence®, *see* 'LAPdul'

Regensberg®, *see* 'MACyoumis'

Regente Agricola, HT; Moreira da Silva, A.; (Cavriglia)

'**Regierungsrat Rottenberger**', HMult, mp, 1926; flowers medium, dbl.; Praskac; (Sangerhausen)

'**Regina**', HWich, w, 1916; flowers creamy white tipped pink large panicles, single; foliage large, glossy; vigorous, climbing growth; Walsh

Regina, HT, rb

'**Regina Badet**', HRg, dp; flowers deep pink, large, very dbl., repeat bloom, intense fragrance; foliage dark; bushy growth; [(Général Jacqueminot X Empereur du Maroc) X R. rugosa]

'**Régina de Alvéar**', HT, w, 1922; flowers white, center slightly shaded pink, dbl., slight fragrance; [Mme Mélanie Soupert X Mme Segond Weber]; Sauvageot, H.

'**Regina Elena**', HT, mp, 1938; bud long, pointed; flowers darker rose-pink than briarcliff, dbl., 50 petals, 5 in., moderate fragrance; foliage dark; [Briarcliff sport]; Grillo

Regina Louise, *see* CLEconcert

'**Regina Pacis**', HT, w, 1945; 5 in., exhibition form, intense fragrance; foliage soft; branching growth; [Nuria de Recolons X Ibiza]; Dot, Pedro

'**Reginald Fernyhough**', HT, op, 1949; flowers pink lightly suffused orange, long, pointed, dbl., 35 petals, 5 in., intense fragrance; foliage olive-green; very vigorous growth; [Southport X Unnamed seedling]; Bees

Regine™, *see* 'HEFqueen'

Regine Crespin, *see* DELcreme

'**Regulus**', T, pb, 1860; flowers bright coppery pink, medium, dbl., flat, intense fragrance; Moreau et Robert

'**Reichspräsident von Hindenberg**', HP, pb, 1933; flowers dark pink to carmine, veined lighter, reverse darker, very dbl., 6 in., moderate; foliage broad; very vigorous, bushy growth; [Frau Karl Druschki X Graf Silva Tarouca]; Lambert, P.

Reigen, F, ob, 1984; flowers medium, dbl., slight fragrance; GPG Bad Langensalza; (Sangerhausen)

'**Reiko**', HT, op, 1963; bud long, pointed; flowers bright coral-peach, dbl., 25 petals, 3.5–4 in., exhibition form, slight fragrance; foliage dark, leathery; vigorous, upright growth; [Spartan X Fred Streeter]; Teranishi, K.; Itami Rose Nursery

Reiko, *see* 'JALray'

'ReiKOR', F, ob, 1974; (Dreaming, Reverie, **Träumerei**®); bud long, pointed; flowers orange, medium, dbl., cupped, intense fragrance; foliage leathery; vigorous, upright, bushy growth; [Colour Wonder X Seedling]; Kordes

'**Reims**', HT, yb, 1924; bud long, pointed; flowers bright nankeen yellow shaded fiery red, orange-apricot and c; Barbier

Reina Elisenda, *see* '**Independence**'

'**Reina Maria Cristina**', T, ob, 1894; flowers orange-yellow, very dbl., globular; foliage bronze; very vigorous, compact growth; Aldrufeu

Reinaerdiana, F, 1974; Delforge; (Cavriglia)

Reine Amelia, HGal, mp

'**Reine Astrid**', HT, rb, 1937; bud long, pointed; flowers bright deep coppery red, reverse golden yellow, dbl., cupped, moderate fragrance; foliage glossy; vigorous growth; Gaujard

Reine Blanche, *see* '**Hebe's Lip**'

'**Reine Blanche**', M, w; flowers pure white, large, dbl., flat; Moreau et Robert, 1858

Reine Chabeau®, HMsk, w, 1994; Lens

'**Reine de Castille**', B, mp, 1863; bud rose-white; flowers bright rose, well-formed; vigorous growth; Pernet

Reine de Perse, HGal, lp; flowers light cream-pink, small to medium, very dbl., intense fragrance; Roseraie de l'Hay; (Sangerhausen)

Reine de Perse, HGal, w

'**Reine de Portugal**', T, yb; flowers coppery yellow shaded pink, large, very dbl.; Guillot et Fils, 1867

Reine de Saxe, C, mp; flowers medium, dbl., moderate fragrance; Roseraie de l'Hay; (Sangerhausen)

'**Reine des Amateurs**', HGal, m; flowers clear lilac, edged pale pink, well shaped, very large; Hébert, Mme., before 1832

'**Reine des Belges**', HCh, w, 1867; sparse bloom, vigorous growth; [Globe Hip X R. chinensis]; Cochet

Reine des Bordures, *see* '**Border Queen**'

'**Reine des Centfeuilles**', C, mp, 1824; flowers clear pink, reflexed, very large, dbl.; very vigorous growth

Reine des Couleurs, *see* '**Farbenkönigen**'

Reine des Francais, *see* '**La Reine**'

Reine des Iles Bourbon, *see* '**Queen of Bourbons**'

Reine des Mousseuses, M, lp; Moreau et Robert, 1860

Reine des Neiges, *see* '**Frau Karl Druschki**'

Reine des Roses, *see* 'KORbico'

Reine des Vierges, B, lp; Béluze, 1884

'**Reine des Violettes**', HP, m, 1860; (Queen of the Violets); flowers violet-red, large, dbl., 75 petals, recurrent bloom, intense fragrance; foliage sparse; [Pius IX seedling]; Millet-Malet

'**Reine d'Espagne**', HP, mr, 1861; flowers brilliant red; foliage light; bushy growth; Fontaine

Reine du Danemark, A, pb, 1816; Booth

Reine du Dänemark, *see* '**Königin von Dänemark**'

Reine du Forez, HT, 1969; Croix; (Cavriglia)

'**REINE ELISABETH**', HT, mr, 1955; flowers velvety blood-red; [Princesse Liliane X Ena Harkness]; Frères, Buyl

'**REINE ELIZABETH**', Pol, mr; flowers crimson, pompon shape, borne in clusters; compact, dwarf growth

'**REINE ELIZABETH**', HT, rb, 1925; bud long, pointed; flowers coral-red, changing to prawn-red, tinted yellow, dbl., intense fragrance; Opdebeeck

'**REINE EMMA DES PAYS-BAS**', T, my, 1879; flowers coppery yellow; Nabonnand, G.

Reine France, HT, op; flowers large, dbl., moderate fragrance; Delbard-Chabert; (Sangerhausen)

Reine Marguerite, *see* '**TRICOLORE**'

Reine Marguerite d'Italie, HT, mr, 1904; Soupert & Notting

'**REINE MARIA DE ROUMANIE**', HT, w, 1927; bud long, pointed; flowers marble-white, center yellow, dbl.; [Stadtrat Glaser seedling]; Mühle

'**REINE MARIA PIA**', T, dp, 1880; flowers deep pink, center crimson, large, moderate fragrance; very vigorous growth; [Gloire de Dijon seedling]; Schwartz, J.

'**REINE MARIE HENRIETTE**', Cl HT, mr, 1878; flowers pure cherry-red, large, dbl., moderate fragrance; vigorous, climbing growth; [Mme Berard X Général Jacqueminot]; Levet, F.

'**REINE OLGA DE WURTEMBERG**', N, mr, 1881; flowers bright red, large, dbl., moderate fragrance; very vigorous, pillar growth; Nabonnand, G.

'**REINE VICTORIA**', B, mp, 1872; (La Reine Victoria); flowers rich pink, well-formed, dbl., cupped, repeat bloom; foliage soft green; slender, upright (to 6 ft) growth; Schwartz, J.

Reiner Maria Rilke, *see* 'KORSEE'

'**REINHARD BÄDECKER**', HP, my, 1918; flowers clear golden yellow, dbl., slight fragrance; [Frau Karl Druschki X Rayon d'Or]; Kordes

Reinhard Pusch, HT, 1992; flowers large, dbl.; Tagashira, Kazuso; (Sangerhausen)

'**REJOICE**', Gr, pb, 1985; flowers salmon-pink blended with yellow, large blooms in clusters of, dbl., 40 petals, exhibition form, moderate fragrance; foliage large, medium green, glossy; tall, upright, bushy growth; Gold (First Gold Winner), ARC TG, 1985; [Little Darling X Color Magic]; McMillan, Thomas G.

Rejouissence, HT, 1977; Croix; (Cavriglia)

Rekordblueher, F, 1965; Tantau; (Cavriglia)

'**REKORDBLÜHER**', F, mp, 1965; flowers rose-pink, borne in large trusses, semi-dbl., 2–3 in.; foliage dark, glossy; bushy growth; Tantau, Math.

'**REKTOR FOERSTER**', HT, pb, 1936; flowers solferino-pink, flushed yellow, large, dbl., exhibition form, foliage leathery; vigorous growth; [Golden Ophelia X Mme Caroline Testout]; Weigand, C.; Pfitzer

Relax, F, or, 1979

Relax Meidiland, *see* MEIdarwet

Relax Meillandecor, *see* MEIdarwet

Release®, HT, yb

'**RELIEF**', HT, w, 1919; flowers ivory-white, center yellowish pink; [Kaiserin Auguste Viktoria X Sunburst]; Verschuren

'**REMBRANDT**', P, or, 1883; flowers vermilion; Moreau et Robert

'**REMBRANDT**', HT, ob, 1914; flowers salmon, tinted orange-red, well-formed, large, dbl., 40 petals; foliage leathery; vigorous growth; [Frau Karl Druschki X Lyon Rose]; Van Rossem

Rembrandt van Ryn, S, dr, 1997; Williams, J. Benjamin

Remember Me®, *see* 'COCDESTIN'

'**REMEMBRANCE**', HT, yb, 1953; bud ovoid; flowers soft yellow edged dawn-pink, dbl., 50 petals, 4–5 in., exhibition form, profuse bloom, moderate fragrance; foliage leathery, glossy, coppery green; vigorous, upright growth; RULED EXTINCT 4/92; [Fred Howard X Unnamed seedling]; Howard, A.P.; H&S

Remembrance, *see* 'HARXAMPTON'

Remuera, *see* 'MACTRAMPOL'

Remy Martin, *see* 'STARQUELI'

'RENACK', Min, lp, 1995; (**Wee Ack**); dbl., 15–25 petals, 1 in., borne mostly singly, slight fragrance; few prickles; foliage small, light green, semi-glossy; medium (15–18 in.), upright growth; [Seedling X Pink Sheri]; Rennie, Bruce F.; Rennie Roses International, 1994

'**RENAE**', Cl F, mp, 1954; bud pointed; dbl., 43 petals, 2.5 in., blooms in clusters, recurrent bloom, intense fragrance; foliage small, glossy; vigorous, climbing growth; [Étoile Luisante X Sierra Snowstorm]; Moore, Ralph S.; Armstrong Nursery

Renaissance®, *see* 'GAUDINO'

'**RENAISSANCE**'®, HT, ob, 1986; flowers brilliant orange, dbl., exhibition form, moderate fragrance; rounded fruit; small prickles; foliage large; [Unnamed seedling X Pampa]; Gaujard, Jean

Renaissance, *see* HARzart

'RENAL', Min, w, 1995; (**Grandpa Alex**); dbl., 26–40 petals, 1.5 in., borne in small clusters, moderate fragrance; few prickles; foliage medium, dark green, semi-glossy; medium (15–18 in.), bushy growth; [Hap Renshaw X Summer Scent]; Rennie, Bruce F.; Rennie Roses International, 1993

'RENANGEL', F, m, 1992; (**Blue Angel**); dbl., 15–25 petals, 1.5–2.75 in., borne mostly singly, moderate fragrance; few prickles; foliage medium, dark green, semi-glossy; medium, bushy growth; [Lavonde X Shocking Blue]; Rennie, Bruce F.; Rennie Roses International

'RENAPP', Min, or, 1995; (**Orange Appeal**); very dbl., 2.75 in., borne mostly singly, slight fragrance; some prickles; foliage large, medium green, semi-glossy; tall (24 in.), spreading, bushy growth; [Justa Little Goofy X Seedling]; Rennie, Bruce F.; Rennie Roses International, 1992

Renata Tebaldi, *see* DELredi

'**RENATE**', HT, w, 1925; flowers cream-white, dbl., intense fragrance; [Kaiserin Auguste Viktoria seedling]; Berger, V.

'RENBABY', Min, lp, 1993; (**Dream Baby**); dbl., 15–25 petals, 1.5 in., borne mostly singly, slight fragrance; few prickles; foliage small, medium green, semi-glossy; medium, bushy growth; [Party Girl X Silver Phantom]; Rennie, Bruce F.; Rennie Roses International, 1993

'RENBLONDE', Min, yb, 1993; (**California Blonde**); flowers moderately full, dbl., 15–25 petals, 1.5 in., slight fragrance; few prickles; foliage medium, medium green, semi-glossy; medium, upright growth; [Party Girl X Tooth of Time]; Rennie, Bruce F.; Rennie Roses International, 1995

'RENBLUE', Min, m, 1988; (**Blushing Blue**); flowers lavender with magenta edge, yellow-tan at base, medium, dbl., 30–35 petals, exhibition form, borne usually singly, repeat bloom, no fragrance; rounded, medium, orange fruit; prickles straight, medium, transparent yellow-green; foliage medium, medium green, semi-glossy; upright, bushy growth; [Shocking Blue X Twilight Trail]; Rennie, Bruce F.; Rennie Roses International, 1989

'RENBLUSH', Min, lp, 1989; (**Innocent Blush**); bud ovoid; flowers pale blush pink, reverse white, aging white, medium, dbl., 40 petals, exhibition form, borne usually singly and in sprays of 5–7, no fragrance; fruit not observed; prickles straight, medium, yellow; foliage medium, medium green, matt; upright, medium growth; [Paul Shirville X Party Girl]; Rennie, Bruce F.; Rennie Roses International, 1990

'RENBOY', Min, rb, 1995; (**Augie Boy**); dbl., 15–25 petals, 1–1.5 in., borne mostly singly, moderate fragrance; few prickles; foliage medium, medium green, glossy; low (12–15 in.), compact growth; [Party Girl X California Dreaming]; Rennie, Bruce F.; Rennie Roses International, 1994

'RENBUN', Min, m, 1995; (**Easter Bunny**); single, 5 petals, 2 in., borne mostly singly, intense fragrance; few prickles; foliage large, medium green, semi-glossy; tall (2 ft), upright growth; [Lavonde X Seedling]; Rennie, Bruce F.; Rennie Roses International, 1992

'RENcel', Min, pb, 1995; (**Pretty Celine**); dbl., 15–25 petals, 1.25 in., borne mostly singly, no fragrance; some prickles; foliage medium, medium green, semi-glossy; medium (15 in.), upright, bushy growth; [Strawberry Delight X Strawberry Delight]; Rennie, Bruce F.; Rennie Roses International, 1994

Rendan, F, rb

'RENdar', Min, ab, 1995; (**Oh Darlin**); semi-dbl., 6–14 petals, 1.25 in., borne mostly singly, moderate fragrance; few prickles; foliage small, dark green, glossy; low (10–12 in.), compact growth; [Pink Sheri X Hap Renshaw]; Rennie, Bruce F.; Rennie Roses International, 1992

Rendez Vous 81, HT, 1981; Meilland, L.; (Cavriglia)

Rendez-vous, *see* **'Day of Triumph'**

Rendez-vous, *see* 'LUCdod'

Rendez-vous, *see* 'MEIpobil'

Rendez-vous, *see* 'JACaebi'

'RENdude', Min, m, 1995; (**Cool Dude**); dbl., 15–25 petals, 1 in., borne mostly singly, moderate fragrance; no prickles; foliage small, medium green, semi-glossy; low (10–12 in.), compact growth; [Seedling X Blushing Blue]; Rennie, Bruce F.; Rennie Roses International, 1995

'René André', HWich, ab, 1901; bud coppery; flowers saffron-yellow, becoming pale pink and carmine, blooms in clusters, semi-dbl., 2–2.5 in., moderate fragrance; very vigorous growth; [R. wichurana X L'Ideal]; Barbier

'René Buatois', LCl, mp, 1936; flowers ruddy pink; [Leontine Gervais sport]; Buatois

'René d'Anjou', M, dp, 1853; flowers deep pink, veined, small, dbl., globular, Robert

'René Javey', HT, ab, 1934; flowers clear apricot, reverse salmon-pink shaded yellow, passing to, dbl.; foliage clear bronze green; Gillot, F.

'Renée Brightman', HT, rb, 1936; flowers brilliant scarlet, shaded orange, reverse striped yellow, dbl., moderate fragrance; vigorous growth; [Emma Wright seedling]; Hurran

Renee Columb, *see* GAUvila

'Reneée Danielle', LCl, my, 1913; flowers jonquil-yellow to golden yellow, passing to white, medium, b, sometimes blooms again in autumn, vigorous, climbing growth; Guillot, P.

'Renée Wilmart-Urban', HT, pb, 1908; bud long, pointed; flowers salmon-flesh, edged carmine, dbl., slight fragrance; Pernet-Ducher

'RENeen', Min, r, 1987; (**Caribbean Queen**); flowers apricot-copper, reverse yellow, blooms borne singly, dbl., 33 petals, exhibition form, slight fragrance; ovoid, yellow fruit; medium, orange-red prickles; foliage small to medium, medium green, semi-glossy; vigorous, bushy growth; [Sunday Brunch X Gold Mine]; Rennie, Bruce F.; Rennie Roses International

Renegade, *see* 'LYOren'

Renegade, MinFl, lp, 1995; Spooner, Raymond A.

'RENeri', Min, lp, 1985; (**Pink Sheri**); [Sheri Anne sport]; Rennie, Bruce F.

'RENerry', Min, pb, 1989; (**Strawberry Delight**); bud pointed; flowers white with medium pink freckles, reverse white with pink edge, dbl., 28 petals, exhibition form, no fragrance; fruit not observed; prickles straight, medium, brown; foliage medium, dark green, semi-glossy; spreading, medium growth; [Little Darling X California Dreaming]; Rennie, Bruce F.; Rennie Roses International, 1990

Rene's Rose, HT, pb, 1995; Dawson

'RENfer', Min, or, 1988; (**Surfer Girl**); flowers light orange-red, salmon, reverse lighter, medium, borne usua, dbl., 25–30 petals, exhibition form, moderate, spicy fragrance; round, small, yellow-orange fruit; prickles hooked, medium, transparent yellow-brown; foliage medium, medium green, matt; bushy, medium growth; [Pink Sheri X Paul Shirville]; Rennie, Bruce F.; Rennie Roses International

'RENfire', Min, ob, 1988; (**Touch of Fire**); bud ovoid; flowers orange-yellow, reverse light orange, aging apricot-orange, u, dbl., 25 petals, moderate, spicy fragrance; no fruit; prickles straight, small, orange-red; foliage small, medium green, semi-glossy; bushy growth; [Tangerine Mist X California Girl]; Rennie, Bruce F.; Rennie Roses International, 1989

'RENfragobses', Min, m, 1995; (**Fragrant Obsession**); dbl., 15–25 petals, 1.5 in., borne mostly singly, intense fragrance; few prickles; foliage medium, medium green, semi-glossy; medium (18 in.), upright growth; [Lavonde X Blushing Blue]; Rennie, Bruce F.; Rennie Roses International, 1995

'RENfrost', HT, mp, 1992; (**Pink Frosting**); dbl., 26–40 petals, 3–3.5 in., borne mostly singly, intense fragrance; some prickles; foliage large, medium green, matt; tall, upright growth; [Unnamed seedling X Prima Ballerina]; Rennie, Bruce F.; Rennie Roses International

'RENgoofy', F, or, 1992; (**Justa Little Goofy**); dbl., 26–40 petals, 1.5–2.75 in., borne mostly singly, moderate fragrance; some prickles; foliage medium, medium green, semi-glossy; medium, bushy growth; [Unnamed seedling X Unnamed seedling]; Rennie, Bruce F.; Rennie Roses International

'RENhap', Min, ab, 1991; (**Hap Renshaw**); flowers moderately small blooms borne mostly singly, dbl., slight fragrance; foliage small, dark green, semi-glossy; upright growth; [Party Girl X Lavonde]; Rennie, Bruce F.; Rennie Roses International

'RENheart', Min, dr 1995; (**Young Hearts**); flowers dark red, dbl. (15–25 petals), medium (1.5 in.) blooms borne mostly single; fragrant; some prickles; foliage medium, dark green, glossy; medium (18 in.), upright growth; [California Dreaming X Royal City]; Rennie, B.F.; Rennie Roses International, 1993

'RENhom', Min, m, 1988; (Melissa Joyce, **Silver Phantom**); bud pointed; flowers silver-lavender, large, borne usually singly, dbl., 33 petals, exhibition form, slight, licorice fragrance; globular, medium, yellow-orange fruit; prickles hooked, medium, light tan-red to dark; foliage medium, dark green, semi-glossy; upright, bushy, tall growth; [Shocking Blue X Angelglo]; Rennie, Bruce F.; Rennie Roses International, 1989

'RENhuron', Min, ly, 1990; (**Huron Sunset**); flowers small, borne in small clusters, dbl., 26–40 petals, no fragrance; foliage small, medium green, semi-glossy; bushy growth; [Party Girl X Golden Rule]; Rennie, Bruce F.; Rennie Roses International, 1991

Renica, *see* 'TANrekta'

'RENijewel', Min, ob, 1995; (**Isabel's Jewel**); semi-dbl., 6–14 petals, 1.25 in., borne mostly singly, moderate fragrance; some prickles; foliage medium, medium green, glossy; medium (15–18 in.), spreading, bushy growth; [Forever Mine X Fragrant Morning]; Rennie, Bruce F.; Rennie Roses International, 1993

'RENimag', Min, pb, 1995; (**Imagine**); dbl., 26–40 petals, 1 in., borne in clusters, slight fragrance; few prickles; foliage small, medium green, semi-glossy; low (12–15 in.), compact growth; [Forever Mine X Pink Sheri]; Rennie, Bruce F.; Rennie Roses International, 1995

'RENirl', Min, ab, 1986; (**California Girl**); flowers apricot, reverse yellow blend, blooms borne singly, dbl., 28 petals, exhibition form, moderate, fruity fragrance; orange-red fruit; small prickles; foliage medium to large, medium green, semi-glossy; bushy, spreading growth; [Julie Ann X Red Love]; Rennie, Bruce F.; Rennie Roses International, 1987

'RENist', Min, ob, 1987; (**Tangerine Mist**); flowers orange, medium, borne singly, semi-dbl., 15 petals, exhibition form, moderate, fruity fragrance; round, small, orange fruit; prickles straight, medium, red-brown; foliage medium, light green, glossy; upright growth; [(Avandel X Unnamed seedling) X Gold Mine]; Rennie, Bruce F.; Rennie Roses International, 1988

'RENITY', Min, dr, 1989; (**Royal City**); bud ovoid; flowers urn-shaped, medium, borne singly, dbl., 23 petals, exhibition form, no fragrance; globular, medium, orange-red fruit; prickles hooked, medium, yellow; foliage medium, medium green, semi-glossy; upright, tall growth; [Goldmarie X Pink Sheri]; Rennie, Bruce F.; Rennie Roses International, 1990

'RENJULIE', Min, ly, 1990; (**Julie's Choice**); flowers small, borne in small clusters, dbl., 26–40 petals, slight fragrance; foliage small, medium green, glossy; bushy growth; [Unnamed seedling X Unnamed seedling]; Rennie, Bruce F.; Rennie Roses International, 1991

'RENJULIE', Min, ly, 1990; (**Julies Choice**); Rennie, Bruce F.

'RENLARRY', Min, dy, 1990; (**Larry's Surprize**); flowers small, borne mostly singly, semi-dbl., 6–14 petals, slight fragrance; foliage small, light green, matt; bushy (15 in.) growth; [Golden Rule X Rise 'n' Shine]; Rennie, Bruce F.; Rennie Roses International, 1991

'RENLOVER', Min, lp, 1995; (**Dream Lover**); dbl., 26–40 petals, 1–1.5 in., borne mostly singly, moderate, raspberry fragrance; some prickles; foliage medium, medium green, glossy; medium (18 in.), upright growth; [Pink Sheri X Innocent Blush]; Rennie, Bruce F.; Rennie Roses International, 1995

'RENLYNN', HT, dr, 1987; (**Mary Lynn**); flowers maroon red, reverse lighter, aging darker, medium, borne usually singly, semi-dbl., 15 petals, exhibition form, intense, damask fragrance; oblong, medium, yellow fruit; prickles hooked, medium, red-brown; foliage medium, medium green, semi-glossy; upright, tall growth; [(Electron X Watercolor) X Lavonde]; Rennie, Bruce F.; Rennie Roses International, 1988

'RENMER', Min, op, 1988; (**Endless Summer**); bud pointed; flowers shrimp-pink, reverse light pink, small, borne singly, dbl., 33 petals, exhibition form, slight, spicy fragrance; no fruit; prickles straight, small, transparent to brown; foliage small, dark green, semi-glossy; low, bushy growth; [Paul Shirville X California Dreaming]; Rennie, Bruce F.; Rennie Roses International, 1990

'RENMERCY', Min, mr, 1995; (**Tender Mercy**); dbl., 15–25 petals, 1–1.5 in., borne mostly singly, slight fragrance; few prickles; foliage small, medium green, glossy; low (12–15 in.), compact growth; [California Dreaming X Seedling]; Rennie, Bruce F.; Rennie Roses International, 1992

'RENMINE', Min, ob, 1993; (**Forever Mine**); dbl., 26–40 petals, 1.5 in., slight fragrance; some prickles; foliage small, medium green, matt; medium, bushy growth; [Hap Renshaw X Party Girl]; Rennie, Bruce F.; Rennie Roses International, 1994

'RENMOBAY', F, w, 1995; (**Moonlight Bay**); very dbl., 3.25 in., borne mostly singly, no fragrance; some prickles; foliage small, light green, semi-glossy; medium (3 ft), upright growth; [(Paul Shirville X Lavonde) X Party Girl]; Rennie, Bruce F.; Rennie Roses International, 1991

'RENMOM', Min, lp, 1995; (**Mom's Fancy**); flowers slight moss on buds, blooms borne in large clusters, dbl., 15–25 petals, 1.25 in., slight fragrance; some prickles; foliage small, medium green, semi-glossy; low (12–15 in.), compact growth; [Pink Sheri X Hap Renshaw]; Rennie, Bruce F.; Rennie Roses International, 1994

'RENMORNING', Min, my, 1990; (**Fragrant Morning**); flowers small, borne mostly singly, dbl., 26–40 petals, intense fragrance; foliage small, light green, semi-glossy; upright (18 in.) growth; [Sunsprite X Unnamed Miniature seedling]; Rennie, Bruce F.; Rennie Roses International, 1991

Renny, *see* 'MORENY'

'RENO', Gr, mp, 1957; bud urn-shaped; flowers coral-salmon, wavy petals, blooms in clusters, 16–20 petals, 3–3.5 in., globular, slight fragrance; foliage bronze; vigorous growth; [Mrs Sam McGredy X Mme Henri Guillot]; Silva; Booy Rose Nursery

'RENOIR', Gr, pb, 1982; bud plump, pointed; flowers flesh pink with peach tones toward base, borne 1–4 per cluster, semi-dbl., 18 petals, repeat bloom, moderate, tea fragrance; curved prickles; foliage 7–9 leaflet, dark, smooth; vigorous, upright to arching growth; Hall, William W.

Renoncule, *see* **'PROVINS RENONCULÉ'**

'RENONCULE', Pol, dp, 1913; flowers deep pink tinted lighter, buttercup form; Barbier

'RENONCULE PONCTUÉE', HGal, mr, 1835; flowers small, dbl.; Vibert; (Sangerhausen)

Renouveau de Provins®, *see* 'LAPDI'

'RENOVE', Min, mr, 1987; (**California Dreaming**); flowers medium red, reverse lighter, blooms borne usually singly, dbl., 28 petals, exhibition form, moderate, fruity fragrance; globular, yellow-orange fruit; tiny, maroon prickles; foliage medium, dark, semi-glossy; medium, bushy growth; [Julie Ann X Black Jack]; Rennie, Bruce F.; Rennie Roses International

'RENOWN', HT, or, 1927; flowers glowing orange-cerise, shaded cardinal, dbl., intense fragrance; [Red-Letter Day X Mrs Wemyss Quin]; Burbage Nursery

'RENPAFLAME', Min, dp, 1995; (**Passion's Flame**); flowers deep pink, blooms borne in small clusters, semi-dbl., 6–14 petals, 1.5 in., slight fragrance; some prickles; foliage medium, medium green, semi-glossy; tall (18–24 in.), spreading growth; Rennie, Bruce F.; Rennie Roses International, 1993

'RENPIE', Min, ly, 1988; (**Cutie Pie**); bud ovoid; flowers light to medium yellow, small, dbl., 23 petals, exhibition form, borne usually singly and in small clusters, moderate, fruity fragrance; globular, very small, yellow orange fruit; prickles straight, small, yellow-red; foliage small, dark green, glossy; bushy, low growth; [Tangerine Mist X California Girl]; Rennie, Bruce F.; Rennie Roses International, 1989

'RENPINY', Min, w, 1995; (**Pretty Tiny**); dbl., 15–25 petals, 1 in., borne in small clusters, slight fragrance; few prickles; foliage small, medium green, semi-glossy; low (8–10 in.), compact growth; Rennie, Bruce F.; Rennie Roses International, 1995

'RENRIDE', Min, mp, 1995; (**Joyride**); very dbl., 2.75 in., borne mostly singly, slight fragrance; few prickles; foliage large, medium green, semi-glossy; tall (24 in.), upright growth; [Seedling X Innocent Blush]; Rennie, Bruce F.; Rennie Roses International, 1994

'RENRUBY', Min, dr, 1988; (**Ruby Tuesday**); flowers dark, velvety red, reverse lighter, aging darker, pom-pom, s, dbl., 25–30 petals, cupped, no fragrance; round, small, red-orange fruit; prickles slightly hooked, small, brownish; foliage small, dark green, semi-glossy; bushy, low growth; [Pink Sheri X Black Jack]; Rennie, Bruce F.; Rennie Roses International, 1989

'RENSAND', F, op, 1990; (**Coral Sand**); flowers coral, large, borne singly, dbl., 26–40 petals, intense fragrance; foliage medium, medium green, semi-glossy; bushy growth; [Paul Shirville X Shocking Blue]; Rennie, Bruce F.; Rennie Roses International, 1991

'RENSCENT', F, pb, 1988; (**Soft Scent**); bud ovoid; flowers pink blend, reverse light pink, sweetheart, urn-shaped, medium, dbl., 48 petals, exhibition form, borne usually singly, intense, spicy fragrance; elongated, medium, orange fruit; prickles straight, medium, red; foliage medium, dark green, semi-glossy; bushy, medium growth; [Paul Shirville X California Girl]; Rennie, Bruce F.; Rennie Roses International, 1990

'RENSECOBSES', Min, lp, 1995; (**Secret Obsession**); dbl., 26–40 petals, 3 in., borne mostly singly, slight fragrance; few prickles; foliage large, medium green, semi-glossy; tall (24–30 in.), upright growth; [Seedling X Innocent Blush]; Rennie, Bruce F.; Rennie Roses International, 1994

'RENSILEN', Min, yb, 1995; (**Golden Silence**); dbl., 15–25 petals, 1.5 in., borne mostly singly, slight fragrance; some prickles; foliage medium, dark green, semi-glossy; medium (15–18 in.), upright growth; Rennie, Bruce F.; Rennie Roses International, 1995

'RENSUM', Min, lp, 1989; (**Summer Scent**); bud ovoid; flowers medium pink with white center, reverse lighter, medium, born, dbl., 30 petals, exhibition form, moderate, spicy fragrance; fruit not observed; prickles straight, small, few, brown; foliage medium, dark green, glossy; bushy, low growth; [Paul Shirville X Unnamed seedling]; Rennie, Bruce F.; Rennie Roses International, 1990

'RENSUN', Min, ab, 1988; (**California Sun**); flowers golden apricot, opening to golden center, reverse lighter, dbl., 30–35 petals, exhibition form, no fragrance; rounded, medium, yellow-orange fruit; prickles straight, medium, pinking; foliage large, medium green, matt; patio; upright, tall growth; [Shocking Blue X Rise 'n' Shine]; Rennie, Bruce F.; Rennie Roses International, 1989

'RENSWEET', Min, mp, 1993; (**Sweet Scent**); dbl., 26–40 petals, 1.5 in., moderate fragrance; few prickles; foliage small, medium green, semi-glossy; medium, upright growth; [Party Girl X Silver Phantom]; Rennie, Bruce F.; Rennie Roses International, 1994

'RENSWIRL', Min, ly, 1992; (**Lemon Swirl**); dbl., 15–25 petals, 1.5–2.75 in., borne in small clusters, slight fragrance; few prickles; foliage medium, light green, glossy; tall, upright growth; [Unnamed seedling X Sunsprite]; Rennie, Bruce F.; Rennie Roses International

'RENTDEL', Min, m, 1995; (**Timeless Delight**); dbl., 15–25 petals, 2.25 in., borne mostly singly, moderate fragrance; some prickles; foliage large, medium green, dull; tall (24–36 in.), upright growth; [Silver Phantom X Innocent Blush]; Rennie, Bruce F.; Rennie Roses International, 1996

'RENTIME', Min, w, 1989; (**Tooth of Time**); bud pointed; flowers white with cream, urn-shaped, small, borne usually singly an, dbl., 25 petals, slight, spicy fragrance; fruit not observed; prickles hooked downward, medium, light yellow; foliage small, dark green, semi-glossy; bushy, medium growth; [Party Girl X Paul Shirville]; Rennie, Bruce F.; Rennie Roses International, 1990

'RENUGGLE', Min, mp, 1995; (**Snuggles**); dbl., 15–25 petals, 1.5 in., borne in small clusters, no fragrance; some prickles; foliage medium, medium green, semi-glossy; medium (15–18 in.), bushy growth; Rennie, Bruce F.; Rennie Roses International, 1993

'RENURF', Min, w, 1987; (**California Surf**); flowers creamy white with peachy-pink edging, urn-shaped, small, dbl., 20 petals, borne usually singly, slight fragrance; round, small, yellow-orange fruit; prickles straight, small, reddish-brown; foliage small, medium green, semi-glossy; upright growth; [Seedling X Seedling]; Rennie, Bruce F.; Rennie Roses International, 1988

'RENWINNIE', Min, lp, 1991; (**Winnie Renshaw**); flowers moderately small blooms borne in small clusters, dbl., moderate fragrance; foliage small, medium green, semi-glossy; medium growth; [Lavonde X Party Girl]; Rennie, Bruce F.; Rennie Roses International

'RENYKE', Min, pb, 1986; (**Little Tyke**); flowers medium pink, white reverse, loose form small blooms borne singly, dbl., 25 petals, no fragrance; small fruit; small, reddish prickles; foliage small, medium green, matt; small, bushy, very compact growth; [Julie Ann X Red Love]; Rennie, Bruce F.; Rennie Roses International

'RENYOUNG', Min, w, 1993; (**Young 'n' Innocent**); flowers full, small, white, very dbl., 26–40 petals, 1.5 in., borne in small clusters, moderate fragrance; some prickles; foliage small, medium green, semi-glossy; low, spreading growth; [Hap Renshaw X Tooth of Time]; Rennie, Bruce F.; Rennie Roses International, 1994

'RENYOURS', Min, w, 1993; (**Tenderly Yours**); flowers small, white, full, dbl., 26–40 petals, 1.5 in., slight fragrance; some prickles; foliage small, medium green, semi-glossy; low, spreading growth; [Party Girl X Tooth of Time]; Rennie, Bruce F.; Rennie Roses International, 1993

'RENZIG', Min, r, 1995; (**Ziggy Stardust**); dbl., 15–25 petals, 1.25 in., borne mostly singly, slight fragrance; few prickles; foliage medium, medium green, semi-glossy; medium (15 in.), upright growth; [Silver Phantom X Party Girl]; Rennie, Bruce F.; Rennie Roses International

Repandia®, *see* 'KORSAMI'

Repartee, HSpn, w

Repelsteeltje, *see* **'RUMPELSTILZCHEN'**

Repens Meidiland, *see* 'MEILONTIG'

Rephidy, HT, pb

Republique de Geneve, *see* LAPEJ

'RESAND', HT, lp, 1996; (**Gene Sandberg**); flowers light pink with a dark pink line around outer edge of each p, dbl., 26–40 petals, 4–6. in., slight fragrance; moderate prickles; foliage medium, dark green, dull; upright, medium (2–3 ft) growth; [Touch of Elegance X First Prize]; Sheldon, John, Jennifer & Robyn

'RESBOBIE', HT, pb, 1996; (**Bobbie Vesely**); flowers medium pink blend, reverse medium pink with yellowish tinge, dbl., 26–40 petals, slight fragrance; moderate prickles; foliage medium, medium green, dull; upright, medium growth; [Pristine X Gold Medal]; Sheldon, John, Jennifer & Robyn; Sheldon, 1995

'RESCARD', HT, ly, 1995; (**Christmas Card**); flowers light yellow, fading to white on outer petals, full (26–40 petals), large blooms borne mostly singly; fragrant; foliage medium, medium green, matt; upright, medium growth; [Spirit of Glasnost X Lanvin]; Sheldon, John & Robin; Sheldon Manor, 1995

'RESCLARA', HT, ab, 1995; (**Clara**); flowers unusual apricot, full (26–40 petals), medium blooms borne mostly single; slight fragrance; foliage medium, medium green, matt; upright, medium growth; [Nantucket X Lovely Lady]; Sheldon, John & Robin; Sheldon Manor, 1995

'RESFINI', HT, mp, 1995; (**Infinity**); flowers medium pink full (26–40 petals), medium blooms borne mostly singly, slight fragrance; foliage medium, medium green, matt; upright, tall growth; [Kordes Perfecta X Prima Donna]; Sheldon, John & Robin; Sheldon Manor, 1995

'RESGOLD', HT, my, 1996; (**Doris Reese**); bud urn-shaped with tips of petals folding downward; flowers medium yellow, center of rose darker than outer petals, dbl., 26–40 petals, 4 in., exhibition form, moderate fragrance; moderate prickles; foliage medium, dark green, semi-glossy; upright, medium (3 ft) growth; [Seedling X Lanvin]; Sheldon, John, Jennifer & Robyn

'RESHONOR', HT, w, 1995; (**Sheldon's Honor**); flowers white full (26–40 petals), large blooms borne in small clusters; slight fragrance; foliage medium, medium green, matt; upright, medium growth; [Sheer Bliss X Anastasia]; Sheldon, John & Robin; Sheldon Manor, 1995

'RESHOPE', HT, lp, 1995; (**Hopeful**); flowers light pink, full (26–40 petals), medium blooms borne mostly single; slight fragrance; foliage medium, medium green, matt; upright, medium growth; [Sheer Bliss X Elizabeth Taylor]; Sheldon, John & Robin; Sheldon Manor, 1995

'RESILLA', HT, r, 1996; (**O'Rilla**); dbl., 26–40 petals, 5–7 in., exhibition form, slight fragrance; moderate prickles; foliage medium, dark green, semi-glossy; upright, medium (3–4 ft) growth; [Lanvin X First Prize]; Sheldon, John, Jennifer & Robyn

'RESJEFF', HT, my, 1995; (**Jeffrey**); flowers yellow, full (26–40 petals), medium blooms borne in small clusters; slight fragrance; foliage medium, light green, matt; upright, medium growth; [Polarstern X (Spirit of Glasnost X Lanvin)]; Sheldon, John & Robin; Sheldon Manor, 1995

'RESLAND', HT, ob, 1995; (**Dreamland**); flowers orange blend tipped with darker orange, full (26–40 petals), medium blooms borne mostly single; slight fragrance; numerous prickles; foliage medium, medium green, matt; upright, medium growth; [Summer Dream X Lanvin]; Sheldon, John & Robin; Sheldon Manor, 1995

'RESLINE', HT, w, 1995; (**White House**); flowers white with pink tinge, full (26–40 petals), medium blooms borne mostly single; slight fragrance; foliage medium, medium green, matt; upright, medium growth; [Pristine X (Spirit of Glasnost X

Lanvin)]; Sheldon, John & Robin; Sheldon Manor, 1995

'RESLINK', HT, pb, 1995; (**Pink Licorice**); flowers pink blend, full (26–40 petals), medium blooms borne mostly single; licorice fragrance; foliage medium, medium green, matt; upright, medium growth; [Osiria X Elegant Beauty]; Sheldon, John & Robin; Sheldon Manor, 1995

'RESLUC', HT, mr, 1995; (**Lucious**); flowers intense med. red, double, dbl., 15–25 petals, borne mostly singly, slight fragrance; foliage medium, dark green, semi-glossy; upright growth; Sheldon, John & Robin

'RESMAR', HT, w, 1995; (**Marie**); flowers white full (26–40 petals), medium blooms borne mostly single; slight fragrance; foliage medium, medium green, matt; upright, medium growth; [Sheer Bliss X Headliner]; Sheldon, John & Robin; Sheldon Manor, 1995

'RESOLUT', F, or, 1962; bud oval; dbl., 30 petals, 4–5 in., borne in large sprays, slight fragrance; foliage glossy; upright, bushy growth; Tantau, Math.

'RESONE', HT, m, 1991; (**Violet Dawson**); flowers mauve blend, blooms borne in small clusters, dbl., moderate fragrance; foliage medium, dark green, disease-resistant; medium, bushy growth; winter hardy; [Paradise X Unnamed seedling]; Sheldon, John & Jennifer

'RESPAD', HT, pb, 1995; (**Grandpa Dan**); flowers pink blend, full (26–40 petals), medium blooms borne mostly single; slight fragrance; foliage medium, medium green, matt; upright, extremely vigorous, medium growth; [Pristine X Touch of Class]; Sheldon, John & Robin; Sheldon Manor, 1995

'RESPAR', HT, pb, 1995; (**Granpa Ray**); flowers light pink blend, full (26–40 petals), medium blooms borne mostly single, slight fragrance; foliage medium, medium green, matt; upright, medium growth; [Pristine X Touch of Class]; Sheldon, John & Robin; Sheldon Manor, 1995

'RESPAT', HT, pb, 1995; (**Granpa Toni**); flowers light pink blend, full (26–40 petals), medium blooms borne mostly single; slight fragrance; foliage medium, medium green, matt; upright, medium growth; [Pristine X Touch of Class]; Sheldon, John & Robin; Sheldon Manor, 1995

'RESPLENDA', HT, w, 1974; bud ovoid; dbl., 30 petals, 4 in., moderate, fruity fragrance; foliage glossy; moderate growth; [Queen of Bermuda X Golden Giant]; Golik; Dynarose

'RESPLENDENT', HP, lp; Williams, A., 1896; (Weatherly, L.)

'RESPOND', HT, pb, 1995; (**Respond**); flowers dark pink blend, full (26–40 petals), borne mostly single; slight fragrance; foliage medium, medium green, matt; upright, medium growth; [Sheer Elegance X Seedling]; Sheldon, John & Robin; Sheldon Manor, 1995

Respond, *see* 'RESPOND'

Responso, F, yb, 1981; flowers medium, dbl.; GPG Bad Langensalza; (Sangerhausen)

'RESROB', HT, m; (**Robin**)

'RESRUTH', HT, yb; (**Ruthe**)

'RESsox', HT, mr, 1995; (**Roxie**); flowers medium red, full (26–40 petals), medium blooms borne mostly single; slight fragrance; numerous prickles; foliage medium, medium green, matt; upright, medium growth; [Sheer Bliss X Headliner]; Sheldon, John & Robin; Sheldon Manor, 1995

Ressins Etienne Gautier, *see* DORster

Rest in Peace, *see* BEDswap

'RESTFUL', HT, my, 1995; (**Restful**); flowers yellow, full (26–40 petals) large blooms borne mostly single; slight fragrance; foliage medium, medium green, matt; upright, tall growth; [Miyabi X Lanvin]; Sheldon, John & Robin; Sheldon Manor, 1995

Restful, *see* 'RESTFUL'

Restless, HT, mr, 1938; flowers dark red, semi-dbl.; almost thornless; Clark

'RESTLESS NATIVE', S, or, 1973; bud ovoid; flowers orange-scarlet, base white, 8–12 petals, 2.5 in., cupped; foliage large, dark, leathery; vigorous, compact, bushy growth; [Orangeade X R. carolina]; Stoddard, Louis

'RESTO', HT, pb, 1995; (**Thoughtful**); flowers dark pink blend, full (26–40 petals), large blooms borne mostly single; slight fragrance; foliage medium, dark green, matt; upright, medium growth; [Pristine X Seedling]; Sheldon, John & Robin; Sheldon Manor, 1995

Results, *see* 'RESULTS'

'RESULTS', HT, pb, 1995; (**Results**); flowers red, white, pink, purple hues depending on temperature and sunlight, dbl. (15–25 petals), large blooms borne mostly single; slight fragrance; foliage medium, dark green, matt; upright, tall growth; [Pristine X Seedling]; Sheldon, John & Robin; Sheldon Manor, 1995

'RESUNC', HT, mr, 1995; (**Uncle Steve**); flowers medium red, full (26–40 petals), large blooms borne mostly single; slight fragrance; foliage medium, dark green, matt; upright, medium growth; [Pristine X Touch of Class]; Sheldon, John & Robin; Sheldon Manor, 1995

Resurrection®, *see* 'KRILEXIS'

'RETINA', Min, mr, 1980; flowers medium red, reverse lighter, base gold, blooms borne singly, dbl., 43 petals, slight, fruity fragrance; straight, long, tan prickles; foliage green, matt; bushy, spreading growth; [Unnamed seedling X Over the Rainbow]; Williams, Ernest D.; Kimbrew-Walter Roses

Rétro, *see* MEIbalani

Rétro Blanc, S, w

'REUS', HT, m, 1949; bud long, pointed; flowers magenta, large, dbl., exhibition form, moderate fragrance; [Cynthia X Manuelita]; Dot, Pedro

'REV T.C. COLE', Cl T, lp; [Chromatella X Marechal Neil]; Cole, Rev T.C., 1880; (Weatherly, L.)

'REV. DAVID R. WILLIAMSON', HT, mr, 1904; flowers dark crimson, globular, moderate fragrance; Dickson, A.

'REV. F. PAGE-ROBERTS', HT, yb, 1921; bud long, pointed; flowers yellow shaded red, large, dbl., intense fragrance; GM, NRS, 1920; [Queen Mary X Unnamed seedling]; Cant, B. R.

'REV. F. PAGE-ROBERTS, CLIMBING', Cl HT, yb, 1931; Beverley; W.B. Clarke

Rev. Floris Ferwerda, *see* **'BEN STAD'**

'REV. H. D'OMBRAIN', B, mr, 1863; flowers brilliant red, recurrent bloom; vigorous growth; Margottin

'REV. JAMES SPRUNT', Cl HCh, mr, 1856; flowers crimson-red, larger; [Cramoisi Superieur, Climbing sport]; Sprunt; P. Henderson, 1858

'REV. WILLIAMSON', HT, or, 1921; bud long, pointed; flowers coral-red, shaded carmine, dbl.; Pernet-Ducher

'RÊVE DE CAPRI', HT, op, 1953; flowers salmon-orange and yellow; foliage bronze; compact, low growth; [Pres. Herbert Hoover X Seedling]; Frères, Buyl

Rêve de Deauville, *see* PINault

Reve de Paris®, *see* MEIloise

Reve de Valse®, F, mr

'RÊVE D'HÉLÈNE', HT, mp, 1959; flowers bright pink edged silvery, large; foliage clear green; free growth; [Michele Meilland X Unnamed seedling]; Orard, Joseph

'RÊVE D'OR', N, my, 1869; flowers buff-yellow tinted lighter, dbl., moderate fragrance; foliage rich green; vigorous, climbing growth; [Mme Schultz seedling]; Ducher, Vve.

Rêve d'un Soir, HT, m, 1997; Croix

'RÊVE ROSE', F, mp, 1950; dbl., 60 petals, borne in large clusters; Mallerin, C.; Vilmorin-Andrieux

'RÉVEIL', HT, yb, 1924; flowers golden yellow, reverse striped red, dbl., moderate fragrance; [Mr Joh. M. Jolles X Mrs Wemyss Quinn]; Van Rossem

'RÉVEIL DIJONNAIS', Cl HT, rb, 1931; flowers cerise, large golden yellow center, reverse yellow streaked, semi-dbl., 13 petals, 5 in., cupped, moderate fragrance; foliage thick, glossy, bronze; short stems; vigorous, climbing growth; GM, Portland, 1929; [Eugene Furst X Constance]; Buatois

'**REVEILLE**', HT, pb, 1941; flowers light salmon-buff-pink, center deeper pink, open, large, dbl., 40–50 petals, cupped, slight fragrance; foliage dark, leathery; vigorous, upright, bushy, compact growth; [Kidwai X Golden Main]; Nicolas; J&P

'**REVELATION**', HT, pb, 1938; bud long; flowers thulite-pink to rose-red, dbl., 35 petals, 4.5–5 in., exhibition form, moderate fragrance; foliage dark; very vigorous growth; [Briarcliff sport]; Witter; Evans City Cut Flower Co.

Revelry™, *see* 'SWIshrev'

'**REVENANTE**', HGal, pb; flowers light rose-pink, edged lilac; Miellez

'**RÉVÉRENCE**', F, mr, 1962; flowers geranium-red, carnation form, borne in large clusters, dbl., 26 petals, 2 in.; foliage dark, glossy; vigorous, upright growth; [Orange Triumph X Unnamed seedling]; Delforge

'**REVEREND ALAN CHEALES**', HP, mr, 1898; flowers very large, dbl.; Paul, G.; (Sangerhausen)

'**REVERIE**', HT, ab, 1925; flowers apricot, reverse shrimp-pink veined rose, dbl.; [Mme Mélanie Soupert X Jean C.N. Forestier]; Ketten Bros.

Reverie, *see* 'ReiKOR'

'**REVIEW**', HT, mp, 1951; flowers bright rose, reverse flesh, dbl., 50 petals, 6–7 in., moderate fragrance; foliage dull, green; medium, dwarf growth; [Mrs Henry Bowles X Trigo]; Fletcher; Tucker

'**REVIVAL**', HT, ab, 1980; flowers light apricot; [Folklore sport]; Rose Barni-Pistoia, 1979

Revolution, Pol, or, 1972; flowers medium, semi-dbl.; GPG Bad Langensalza; (Sangerhausen)

Revolution Francaise®, *see* MEltixia

Revue de Dauville, *see* PINault

'**REWARD**', HT, yb, 1934; bud long, pointed; flowers clear yellow shaded peach, medium, moderate fragrance; foliage dark, glossy; vigorous, bushy growth; Dickson, A.

'**REX**', HT, mr, 1959; bud long, pointed; flowers velvety cardinal-red, large, semi-dbl., cupped, slight fragrance; foliage leathery, glossy; vigorous, upright growth; [Senior X Better Times]; Spanbauer

'**REX ANDERSON**', HT, w, 1938; flowers ivory-white, very well-formed, large, dbl., shy bloom, moderate, fruity fragrance; foliage gray-green; vigorous growth; [Florence L. Izzard X Mrs Charles Lamplough]; McGredy; J&P

'REXjean', S, dr, 1999; (**Jean Rex**); flowers dark red (3 shades), center vein darkest, reverse dark red, very dbl., 41 petals, 4–4.5 in., borne in small clusters, intense fragrance; prickles numerous; foliage medium, dark green, matte; bushy, very vigorous, strong, medium (4 ft) growth; [Prospero sport]; Rex, Dr Robert W.; Hortico, Inc., 1995

Rexy's Baby, *see* 'MACcarrib'

'REYcantpri', Min, r, 1992; (**Canterbury Pride**); very dbl., 1.5 in., slight fragrance; foliage small, medium green, semi-glossy; low (28 cms), upright growth; [Hot Chocolate X Miniature seedling]; Reynolds, Ted; Reynolds Roses

'REYcidin', Min, yb, 1995; (**City of Dunedin**); flowers old gold changing to yellow then white with red edge, full (26–40 petals), small blooms borne in small clusters; slight fragrance; few prickles; foliage medium, medium green, glossy; upright, medium growth; [(Rise 'n' Shine X Seedling) X Rainbow's End]; Reynolds, Ted; Ted Reynolds Roses International, 1995

'REYfaye', Cl F, dp, 1992; (**Faye Reynolds**); flowers deep pink, blooms borne in small clusters, dbl., 26–40 petals, 3–3.5 in., moderate fragrance; some prickles; foliage sage green, semi-glossy; medium (6′ 6″), upright growth; [Westerland X Gingersnap]; Reynolds, Ted; Reynolds Roses

'REYinver', HT, dp, 1996; (**City of Invercargill**); bud very dark red; flowers deep pink, large blooms borne mostly single, very dbl., moderate fragrance; few prickles; foliage medium, medium green, glossy, disease-free; upright, medium growth; [Westerland X Perfume Delight]; Reynolds, Ted; Ted Reynolds Roses International, 1996

'REYkelly', HT, pb, 1999; (**Kelly Reynolds**); flowers apricot peach pink blend, large, dbl., 26–40 petals, exhibition form, borne mostly singly, slight fragrance; few prickles; foliage large, medium green, glossy; upright, medium growth; [Perfume Delight X seedling]; Reynolds, Ted; Ted Reynolds Roses International, 1999

'REYlianne', F, lp, 1996; (**Lianne Reynolds**); flowers light pink, china porcelain appearance, blooms borne in small clusters, dbl., 26–40 petals, no fragrance; some prickles; foliage small, medium green, glossy; compact, medium growth; [Gloire de Dijon X Hot Pewter]; Reynolds, Ted; Ted Reynolds Roses International, 1997

Reynolda House, *see* 'WILangil'

'**REYNOLDS HOLE**', HT, mp, 1862; flowers medium, very dbl.; Standisch; (Sangerhausen)

Reynold's Rugosa, *see* 'REYrug'

'REYpat', LCl, lp, 1992; (**Patricia Weston**); dbl., 26–40 petals, 3–3.5 in., borne in small clusters, moderate fragrance; some prickles; foliage large, dark green, glossy; medium (2 m), upright growth; [Westerland X Perfume Delight]; Reynolds, Ted; Reynolds Roses

'REYpeg', S, mp, 1992; (**Peggy Joan Reynolds**); single, 5 petals, 1.5 in., borne in small clusters, very remontant, almost perpetual, slight fragrance; few prickles; foliage medium, light green, semi-glossy, disease-resistant; upright (150 cms) growth; very hardy; (14); [Parentage unknown. (Possibly HCh X R. gigantea)]; Reynolds, Ted, 1993

'REYred', HT, mr, 1992; (**Ted's Red**); dbl., 26–40 petals, 3–3.5 in., borne mostly singly, slight fragrance; few prickles; foliage medium, medium, semi-glossy; medium, upright growth; [Alec's Red X Seedling]; Reynolds, Ted; Reynolds Roses

'REYrug', HRg, mr, 1999; (**Reynold's Rugosa**); [double rugosa X double rugosa]; Reynolds, Ted; Ted Reynolds Roses International, 1999

'REYsel', HMsk, dp, 1992; (**Selina**); flowers deep pink, blooms borne in small clusters, single, 5 petals, 1.5 in., slight fragrance; few prickles; foliage small, medium green, semi-glossy; medium, spreading growth; [Cornelia X Trier]; Reynolds, Ted; Reynolds Roses, 1992

'REYshelley', HT, ab, 1998; (**Shelley Higgins**); flowers apricot-peach blend, double, dbl., 15–25 petals, exhibition form, borne mostly singly, intense, spicy fragrance; some prickles; foliage medium, medium green, semi-glossy; upright, medium growth; [Westerland X Seedling]; Reynolds, Ted; Ted Reynolds Roses International, 1998

'**RHAPSODY**', HT, ob, 1951; bud long, pointed; flowers orange, reverse terra-cotta, medium, 20 petals, exhibition form; foliage leathery; vigorous, bushy growth; RULED EXTINCT 1/85; [Lulu X Cecil]; Houghton, D.; Elmer Roses Co.

Rhapsody, *see* 'JACsod'

Rhapsody in Blue, *see* 'FRAntasia'

'**RHEA REID**', HT, mr, 1908; bud long, pointed; flowers crimson-red, dbl., exhibition form, moderate fragrance; foliage soft; GM, Bagatelle, 1908; [American Beauty seedling X Red seedling]; Hill, E.G., Co.

'**RHEA REID, CLIMBING**', Cl HT, mr, 1914; California Nursery Co.

Rheda-Wiedenbruck, F, mp

Rheinaupark®, *see* 'KOReipark'

'**RHEINGOLD**', HT, my, 1934; flowers golden yellow, large, very dbl., intense fragrance; foliage leathery, light; dwarf, bushy, compact growth; [Mrs T. Hillas X Mabel Morse]; Leenders, M.; J&P

'**RHODE ISLAND RED**', LCl, dr, 1957; dbl., 33 petals, 4–5 in., cupped, moderate fragrance; foliage glossy; [Everblooming Pillar No. 73 X Unnamed seedling]; Brownell, H.C.

Rhodes Rose, HMult, w

'**Rhodologue Jules Gravereaux**', T, 1908; Fontes

'**Rhodophile Gravereaux**', HFt, yb, 1900; flowers dark yellow with pink, large, semi-dbl.; Pernet-Ducher; (Sangerhausen)

'**Rhona**', HT, mp, 1984; flowers large, dbl., 20 petals, intense fragrance; foliage medium, medium green, semi-glossy; bushy growth; [Anne Letts X (Dainty Maid X Pink Favorite)]; Gobbee, W.D.

'**Rhonda**', LCl, mp, 1968; bud globular; flowers carmine-rose, large, dbl., slight fragrance; foliage dark, glossy; vigorous, climbing growth; [New Dawn X Spartan]; Lissemore; C-P

Rhotare, Pol, mr; flowers medium, semi-dbl.; (Sangerhausen)

'**Ria Wenning**', HT, dp, 1932; bud long, pointed; flowers carmine, semi-dbl., moderate fragrance; vigorous growth; [Mme Maurice de Luze X Red Star]; Leenders, M.

Ribambelle, F, 1972; Croix; (Cavriglia)

'**Ribatejo**', F, w, 1962; flowers white, center deep yellow; [Virgo X Seedling]; da Silva, Moreira

Riberhus, *see* POUlriber

Ric Rac, *see* 'PIXric'

Ricarda®, F, op, 1989; Noack, Werner

Riccordo di Fernando Scarlatti, HP, dr; flowers large, dbl., intense fragrance; (Sangerhausen)

'**Riccordo di Geo Chavez**', HT, dp, 1911; flowers large, dbl.; Bonfiglio, A.; (Sangerhausen)

'**Riccordo di Giosue Carducci**', HT, lp, 1909; flowers very large, very dbl.; Bonfiglio, A.; (Sangerhausen)

Riccordo di Giovanni Spotti, HT, mr; flowers large, dbl., slight fragrance; (Sangerhausen)

Rich and Rare, *see* 'SEArich'

'**Richard**', HT, dr, 1979; bud long, pointed; flowers spinel-red, dbl., 25–30 petals, 4–5 in., intense fragrance; foliage light green; bushy, upright growth; [Gavotte X Memoriam]; Ellick; Excelsior Roses

Richard Buckley, *see* 'SMItshort'

'**Richard E. West**', HT, my, 1924; bud long, pointed; flowers large, dbl., intense fragrance; foliage leathery; vigorous growth; Dickson, A.

'**Richard Hayes**', S, mp, 1973; 4 in., moderate fragrance; foliage light; vigorous, tall, upright growth; [Fred Loads sport]; Holmes, R.A.; Wonnacott

Richard Strauss®, S, rb, 1989; Noack, Werner

Richard Tauber®, HT, rb, 1986; Pouw

'**Richardson Wright**', HT, pb, 1931; flowers pearl with carmine dashes and lemon reflexes, large, globular, moderate fragrance; foliage dark, leathery; vigorous growth; [Radiance X Ville de Paris]; C-P

'**Richelieu**', HCh, m, 1845; flowers violet-pink, large, very dbl.; Verdier, V.; (Sangerhausen)

'**Richmond**', HT, mr, 1905; (Everblooming Jack Rose); bud long, pointed; flowers bright scarlet, varying greatly at times, dbl., moderate, damask fragrance; vigorous growth; [Lady Battersea X Liberty]; Hill, E.G., Co.

'**Richmond, Climbing**', Cl HT, mr, 1912; Dickson, A.

Ricky®, *see* HAUric

Ricordo di Leone Sgaravatti, HT; Sgaravatti, A.; (Cavriglia)

'**Ridgeway**', HT, pb, 1953; flowers salmon-pink shaded apricot, well shaped, dbl., moderate fragrance; foliage rather sparse; strong stems; very vigorous growth; [Princess Marina X Vanessa]; Ratcliffe

'**Rieder's Solin**', Cl HT, mr, 1930; flowers large, dbl., moderate fragrance; Rieder; (Sangerhausen)

'**Rifleman**', HT, or, 1979; bud ovoid; flowers light vermilion, dbl., 35–43 petals, 3.5 in., exhibition form, slight fragrance; tall, very vigorous, bushy growth; [Tropicana X Orange Sensation]; Murray, Nola

'**Rigaudon**', F, dr, 1957; flowers dark red tinted geranium, dbl.; very dwarf, dense growth; [(Independence X Unnamed seedling) X Unnamed seedling]; Combe; Japan Rose Society

'**Right Royal**', HT, mp, 1979; flowers silvery pink, dbl., 30 petals, 4 in., slight fragrance; foliage dark; moderate growth; [Scented Air X Anne Letts]; Hawken, Una

Rigobec, *see* **'Mon Pays'**

Rigobec 2, *see* **'Neiges d'été'**

Rigobec 3, *see* **'Danse Azteque'**

'**Rigoletto**', F, ab, 1954; flowers apricot-yellow tinted copper, well formed, large, semi-dbl., intense fragrance; vigorous growth; [Floribunda seedling X Souv. de Claudius Pernet]; Leenders, M.

'**Rijswijk**', F, ob, 1964; bud deep yellow; flowers orange-yellow, borne in clusters, semi-dbl.; foliage dark, glossy; vigorous growth; [Goldmarie X Fata Morgana]; Buisman, G. A. H.

'**Riksbyggerosen**', F, or, 1969; flowers open, semi-dbl., 23 petals, 3 in.; foliage glossy, light green; low, compact growth; [Irish Wonder X Seedling]; Poulsen, Niels D.; Poulsen

Rilla, HT, r

Rim, *see* **'Happiness'**

'**Rima**', HT, lp, 1964; flowers light silvery pink; [Prima Ballerina sport]; Samuels

'**Rimosa**', F, my, 1958; flowers indian yellow to citron-yellow, well-formed, medium, borne i, semi-dbl., 18–25 petals, moderate fragrance; foliage leathery; upright, symmetrical, compact growth; [Goldilocks X Perla de Montserrat]; Meilland, F.; URS

Rimosa 79, *see* 'MEIgronuri'

'**Rina**', HT, w, 1999; flowers cream white blended pink, dbl., 32 petals, 5 in., exhibition form, intense fragrance; foliage medium, medium green, semi-glossy; tall (6 ft) growth; [Royal Highness X Shizunomai]; Yonzda, Kazuo, 1997

Rina Herholdt, *see* 'HERani'

'**Rina Herholdt, Climbing**', Cl HT, pb, 1973; Arora, Bal Raj; The Rosery, India

Rina Hugo, *see* DORvizo

'RinaKOR', F, ob, 1974; (**Marina**®); bud long, pointed; flowers orange, base yellow, dbl., moderate fragrance; foliage glossy, dark, leathery; vigorous, upright growth; AARS, 1981; [Colour Wonder X Seedling]; Kordes, 1975

Rinascimento®, *see* BARinas

Ring of Fire™, *see* 'MORfire'

'**Ringfield**', HT, or, 1977; flowers deep vermilion, full, dbl., 30 petals, 5 in., slight fragrance; vigorous growth; [Ernest H. Morse X Fragrant Cloud]; Plumpton, E.

'**Ringlet**', Cl HT, pb, 1922; flowers white, tipped pink and lilac, small blooms in clusters, single; vigorous, climbing growth; [Ernest Morel X Betty Berkeley]; Clark, A.; Brundrett

'**Rio Grande**', HT, dr, 1973; bud ovoid; flowers velvety dark red, medium, dbl., moderate fragrance; foliage soft; moderate, upright, bushy growth; [Unknown X Unknown]; Tantau, Math.; Ahrens & Sieberz

'**Rio Rita**', HT, w, 1931; bud long, pointed; flowers white tinged pink, large, dbl., intense fragrance; foliage glossy; vigorous growth; [Mme Butterfly X Premier]; Hill, Joseph H., Co.

'**Rio Rita**', Cl HT, w, 1935; flowers velvety scarlet-crimson, very large, dbl., cupped, intense fragrance; foliage glossy; very vigorous growth; [E.G. Hill sport]; Elmer's Nursery

Rio Samba™, *see* 'JACrite'

'**Ripples**', F, m, 1971; flowers lilac-lavender, semi-dbl., 3.5 in., slight fragrance; foliage small, green, matt; [(Tantau's Surprise X Marjorie LeGrice) X (Unnamed seedling X Africa Star)]; LeGrice

'**Rise 'n' Shine**', Min, my, 1977; (Golden Meillandina, Golden Sunblaze); bud long, pointed; flowers rich medium yellow, small, dbl., 35 petals, 1.5 in., exhibition form, moderate fragrance; bushy, upright growth; AOE, ARS, 1978; Miniature Rose Hall of Fame, ARS, 1999; [Little Darling X

Yellow Magic]; Moore, Ralph S.; Sequoia Nursery

Rise 'n' Shine, Climbing, *see* 'MORKINSHINE'

Rising Star, *see* 'JALSTAR'

Rising Star, *see* HAReast

'RISING SUN', HT, rb, 1924; bud long, pointed; flowers rich copper, base old-gold, cactus dahlia form; Hicks

'RISQUÉ', Gr, rb, 1985; flowers medium red, light yellow reverse, medium, dbl., 20 petals, no fragrance; foliage medium, dark, semi-glossy; upright, slightly spreading growth; [Bob Hope X Unnamed seedling]; Weeks, O.L.; Weeks Wholesale Rose Growers

'RITA', F, mp, 1960; flowers rich pink, blooms in large clusters, dbl., moderate fragrance; foliage glossy; vigorous growth; [Karl Herbst X Pinocchio]; Fryers Nursery, Ltd.

Rita Applegate, *see* 'TINRITA'

Rita Bugnet, HRg, w, 1958; Bugnet

'RITA JACKSON', HT, rb, 1964; flowers large, dbl., exhibition form, slight fragrance; upright, bushy growth; [Tzigane sport]; Jackson, F.

Rita Levi Montalcini®, *see* BARlev

Rita MacNeil, *see* 'MACCHOME'

'RITA SAMMONS', Pol, pb, 1925; (Justine Silva); flowers deep rose-pink, opening pink, edged lighter; [Cécile Brunner sport]; Clarke, B.; Clarke Bros.

Ritausma, HRg, lp

Ritchie's Red Climber, Cl HT, dr

Rite Brite, *see* 'GELRITE'

Rittenhouse™, *see* 'WILHKPK'

Ritter Taler, HT, rb

'RITTER VON BARMSTEDE', HKor, mp, 1959; 20 petals, 2 in., blooms in clusters of 30–40, foliage glossy; vigorous (10–15 ft) growth; Kordes

'RITZ', LCl, mr, 1955; flowers velvety red blooms in clusters, semi-dbl., 11 petals, 3 in., recurrent bloom, vigorous (8–9 ft) growth; [R. setigera X ?]; Horvath; Wyant

'RITZ', F, mr, 1961; flowers bright scarlet, semi-dbl., 16 petals, 3.5 in., slight fragrance; foliage dark, glossy; vigorous, well branched growth; Gaujard

'RIVA LIGURE', HT, dp, 1947; bud pointed; flowers velvety carmine, dbl., 24–35 petals, intense fragrance; foliage dark; vigorous, bushy growth; [Unnamed seedling X Crimson Glory]; San Remo Exp. Sta.

'RIVAL', HT, mr, 1954; flowers cherry-scarlet, loosely formed, 4–5 in., flat, foliage bronze; free growth; [Southport X The Rev. W.S. Crawford]; Fletcher; Tucker

'RIVAL DE PAESTUM', T, w; bud tinged pink; flowers white, base blush and ivory, dbl.; foliage dark; moderate growth; Béluze, 1841

Rival de Paestum, Ch, w, 1863; Paul

River City Jubilee, *see* 'JUDRIV'

Riverdance, *see* 'KENFINE'

Riverdance, Min, pb, 1998; Laver, Keith G.

Riverina Sunset, T, mp; Hay

'RIVERS', HP, mr, 1832; flowers bright crimson, large, borne in corymbs, dbl., recurrent bloom, vigorous growth; Laffay, M.

'RIVERS' GEORGE IV', HCh, dr; (George IV, King George IV); flowers vivid crimson, shaded with dark purple, loosely, dbl., cupped, non-recurrent, branching shoots tinged with purple; vigorous, growth; [Thought to be Damask X China]; Rivers, ca. 1817

'RIVERS' MUSK', HMsk, mp; flowers rosy buff, small, intense fragrance; Rivers, prior to 1925

River's South Bank, *see* **'JUJNOBEREJNAIA'**

Riverview Centennial, HT, dr, 1980

'RIVIERA', HT, or, 1939; bud globular, yellow; flowers orange-scarlet, reverse lighter, base yellow, open, large, dbl., cupped, foliage glossy, wrinkled, dark; very vigorous, bushy growth; [Luis Brinas X Catalonia]; Dot, Pedro; J&P, 1940

Riviera, *see* JACsak

Riviere de Diamant, *see* BRIsiness

Road to Freedom, *see* 'FRANLAC'

'ROADMAN', HT, w, 1977; bud pointed; flowers near white, dbl., 40–50 petals, 6 in., exhibition form, foliage leathery; upright growth; [Hawaii X Kordes' Perfecta]; Ota, Kaichiro

'ROAMING', HT, dp, 1970; flowers reddish pink shades, pointed, dbl., 24 petals, 3.5 in.; foliage green, matt; [Vera Dalton X Tropicana]; Sanday, John

Rob in des Bois, *see* **'ROBIN HOOD'**

Rob Roy®, *see* 'COROB'

Robbie Burns, *see* 'AUSburn'

Robe de Neige, S, w, 1995; Lens; (Cavriglia)

Robe de Soie, S, lp, 1996; Lens

'ROBE D'ETÉ', F, or, 1966; flowers salmon-orange, dbl., 22–24 petals, 3–3.5 in., exhibition form, slight fragrance; foliage bronze; bushy, upright growth; [(Chatelaine X Mannequin) X (Montezuma X Floradora)]; Lens

Robe Fleuri, S, 1995; Lens; (Cavriglia)

Robe Rose, S, 1995; Lens; (Cavriglia)

'ROBERT', P, pb, 1856; flowers carmine, marbled white; Robert

Robert Aliano, *see* 'WILALNO'

'ROBERT BETTEN', HT, mr, 1920; flowers clear dark carmine-red, not turning blue, dbl.; [Frau Karl Druschki X Corallina]; Schmidt, J.C.

'ROBERT BLAND', S, dp, 1960; flowers open, small, dbl., non-recurrent, thornless; foliage richgreen; vigorous, bushy growth; quite hardy; [((Hansa X R. macounii) X Betty Bland) X (R. blanda X Betty Bland)]; Wright, Percy H.

'ROBERT COTTON', HT, w, 1968; bud ovoid; flowers white, edges flushed pink, large, dbl., slight fragrance; foliage glossy, serrated, leathery; moderate growth; [Marcia Stanhope X Karl Herbst]; Golik; Ellesmere Nursery

'ROBERT DE BRIE', HP, m, 1860; flowers violet-pink with white stripes, large, dbl.; Granger; (Sangerhausen)

'ROBERT DUBOL', HT, mr, 1946; flowers warm orient red, stamens golden, very dbl., exhibition form, Sauvageot, H.; Sauvageot

'ROBERT DUNCAN', HP, pb, 1897; flowers purplish pink, sometimes flamed brilliant red, well-formed, dbl., 70 petals, repeat bloom, moderate fragrance; vigorous growth; Dickson, A.

'ROBERT F. KENNEDY', HT, mr, 1968; flowers scarlet, exhibition form, intense fragrance; foliage dark; very vigorous, upright growth; [Chrysler Imperial X Ena Harkness]; Takatori, Yoshiho; Parnass Rose Nursery

'ROBERT HUEY', HT, dp, 1911; bud long, pointed; flowers carmine edged lighter, bluing slightly, dbl., moderate fragrance; moderate growth; Dickson, A.

'ROBERT LE DIABLE', HGal, m; flowers scarlet-pink aging to deep purple, center often green, dbl.; low, lax growth

'ROBERT LÉOPOLD', M, pb, 1941; flowers salmon-flesh-pink edged light carmine, large, dbl.; Buatois

'ROBERT PERPÉTUEL', P, m, 1856; flowers violet-pink, medium, dbl.; Robert; (Sangerhausen)

Robert Stolz, F, dr, 1974; flowers medium, semi-dbl.; deRuiter; (Sangerhausen)

'ROBERTA BONDAR', LCl, my, 1993; dbl., 26–40 petals, 3–3.5 in., borne 1–8 per truss, moderate fragrance; foliage large, dark green, semi-glossy; climbing (70 cms) growth; [King's Ransom X Buff Beauty]; Fleming, Joyce L.; Hortico Roses, 1993

'ROBESPIERRE', Pol, pb, 1975; bud full; flowers small, very dbl., 88 petals, 2.5 in., cupped, moderate fragrance; foliage bronze; Delforge, S.

'ROBIN', Min, mr, 1956; bud urn-shaped; flowers rich red, blooms in clusters of 15, dbl., 65 petals, 1.25 in., flat, foliage leathery, green, matt; vigorous, dwarf

(12 in.), bushy growth; [Perla de Montserrat X Perla de Alcanada]; Dot, Pedro; C-P

Robin, *see* 'RESROB'

'**ROBIN HOOD**', HMsk, mr, 1927; (Rob in des Bois); flowers cherry-red, blooms in large clusters, recurrent bloom, vigorous (4–5 ft), dense, compact growth; (14); [Seedling X Miss Edith Cavell]; Pemberton

'**ROBIN HOOD**', HT, mr, 1912; flowers soft bright rosy scarlet, changing to bright scarlet-crimson, dbl., intense fragrance; Hill, E.G., Co.

Robin Red Breast, *see* 'INTERROB'

Robin Redbreast, *see* 'INTERROB'

Robina, *see* KORxenna

'**ROBINETTE**', HMult, rb, 1943; flowers amaranth-red, white eye, open, small, borne in clusters, single, intense fragrance; foliage glossy; height 10–12 ft; [Hiawatha X Hiawatha]; Moore, Ralph S.; Hennessey

Robur, *see* **'RED EMPRESS'**

Robusta®, *see* 'KORGOSA'

Robusta, B, mr, 1877; Soupert & Notting

Robusta, Cl T, mp

Roby, HMult, rb; Guillot, 1912

Roccana Diane, *see* 'KIRDEX'

Roche Centenary, *see* DICvintage

'**ROCHEFORT**', HT, ab, 1936; flowers large, dbl., intense, fruity fragrance; foliage leathery; vigorous growth; GM, Portland, 1935; [Mrs Pierre S. duPont X Charles P. Kilham]; Mallerin, C.; C-P

'**ROCHELLE HUDSON**', HT, rb, 1937; bud long, pointed; flowers carmine, base yellow, orange undertone deepening with age, semi-dbl., slight, fruity fragrance; foliage dark; vigorous growth; [Isobel X Mme Edouard Herriot]; Moore, Ralph S.; Brooks & Son

'**ROCHESTER**', F, ab, 1934; flowers buff, reverse orange-carmine, medium, dbl., moderate fragrance; foliage leathery; vigorous, bushy growth; [Echo X Rev. F. Page-Roberts]; Nicolas; J&P

Rochester Cathedral, *see* 'HARROFFEN'

Rocio Elias®, *see* FEtrone

Rock Hill Peach, T, pb

Rock 'n' Roll, *see* 'MACFIRWAL'

Rocket, *see* **'RAKETA'**

'**ROCKET**', HT, mr, 1935; flowers brilliant scarlet, reverse crimson, large, dbl., exhibition form, intense fragrance; foliage leathery, dark, bronze; very vigorous growth; [Dame Edith Helen X Scorcher]; Nicolas; J&P

Rocketeer, *see* 'CHEWALLOP'

Rockin' Robin™, *see* 'WEKBOROCO'

Rocky™, *see* 'MACKEPA'

'**ROCOCO**', F, mr, 1964; flowers scarlet, borne in clusters, semi-dbl., 15 petals, 3 in.; free growth; [Moulin Rouge X Fire Opal]; McGredy, Sam IV; Spek

'**ROD STILLMAN**', HT, lp, 1948; flowers light pink, base flushed orange, large, dbl., 35 petals, intense fragrance; foliage dark; vigorous growth; [Ophelia X Editor McFarland]; Hamilton

Roddy MacMillan, *see* 'COCARED'

'**RODEO**', F, or, 1960; flowers bright scarlet, borne in clusters (up to 10), dbl., 3 in., slight fragrance; foliage light green; bushy, low growth; [Obergärtner Wiebicke X Spartan]; Kordes, R.; McGredy

Rodeo Drive™, *see* 'AROCORE'

'**RÖDHÄTTE**', F, mr, 1912; (Red Riding Hood); flowers clear cherry-red, large blooms in large clusters, semi-dbl.; foliage rich green; bushy, compact growth; [Mme Norbert Levavasseur X Richmond]; Poulsen, D.T.; Poulsen

'**RÖDHÄTTE, CLIMBING**', Cl F, mr, 1925; Grootendorst, F.J.

Rodin, *see* 'MEIDINRO'

Rodinghausen®, S, mr, 1987; Noack, Werner

Rodovrerosen, *see* **'POULSEN'S FAIRY'**

Rody, *see* TANydor

Roedean, *see* BENcamelia

ROElanda, HT, w, 1993; (**Leana**)

'**ROELOF BUISMAN**', HT, mr, 1964; bud ovoid; flowers bright pure red, well-formed, large; vigorous, upright, bushy growth; Kordes, R.

'**ROGER BOUDOU**', Pol, mr, 1957; flowers very bright red; [Lafayette seedling]; Privat

'**ROGER LAMBELIN**', HP, rb, 1890; flowers bright crimson fading maroon, petals margined white, very di, dbl., recurrent bloom; vigorous growth; (28); [Fisher Holmes sport]; Schwartz, Vve.

'**ROGER LAMBELIN STRIPED**', HP, rb, 1953; flowers deep maroon to pink stripes on white ground, recurrent bloom, [Roger Lambelin sport]; Hennessey

Roger Secretrain, LCl; Moreira da Silva, A.; (Cavriglia)

'ROGSCRIV', HT, pb, 1999; (**Natural Beauty**); flowers pink/apricot blend, reverse pink blend, dbl., 26–40 petals, 3 in., rosette, borne in small clusters, intense fragrance; few prickles; foliage medium, medium green, semi-glossy; compact, medium (3 ft) growth; [Pretty Lady X Silver Jubilee]; Scrivens, Len; John Tooby & Co., Ltd., 1998

'ROGslav', HT, ab, 1995; (**Penkala Slavoljub**); flowers apricot-orange striped, full (26–40 petals), medium (4–7 cms) blooms borne mostly single; few prickles; foliage medium, dark green, semi-glossy; medium (100 cms), bushy growth; [Ambassador sport]; Rogin, Josip; Rogin, 1995

'ROGZAG', HT, pb, 1995; (**Zagreb**, Zagreb 900); flowers pink striped, full (26–40 petals), medium (4–7 cms) blooms borne mostly singly, very fragrant; few prickles; foliage large, medium green, matt; tall (150 cms), upright growth; [Eiffel Tower sport]; Rogin, Josip; Rogin, 1995

'**ROI ALBERT**', HT, pb, 1925; bud long, pointed; flowers bright carmine-rose, center tinted scarlet, intense fragrance; [Laurent Carle seedling X Mme Abel Chatenay]; Klettenberg-Londes

'**ROI ALEXANDRE**', HT, ob, 1937; (King Alexander I, S.M. Alexander I); flowers coppery orange, tinted salmon, over large, dbl., moderate fragrance; foliage leathery, glossy, bronze; very vigorous growth; Gaujard

Roi de Nains, Min, mr

'**ROI DE PAYS-BAS**', C, dp, 1848; (Roi des Pays-Bas); flowers deep pink, large, dbl., cupped

Roi de Siam, Cl T, mr, 1825; Laffay, M.

'**ROI DES AUNES**', S, dp, 1885; (Erlkönig); flowers carmine tinted red, large, dbl., globular, non-recurrent, very vigorous growth; Geschwind, R.

'**ROI DES BENGALES**', F, rb, 1958; flowers grenadine-red; vigorous, low growth; [(Hermosa X Gruss an Teplitz) X Independence]; Arles; Roses-France

Roi des Bordures, *see* **'BORDER KING'**

'**ROI DES CRAMOISIS**', HCh, mr; flowers bright red, cupped

Roi des Pays-Bas, *see* **'ROI DE PAYS-BAS'**

Roi des Rois, F, 1955; Delbard-Chabert; (Cavriglia)

'**ROI SOLEIL**', HT, my, 1962; bud long, pointed; flowers citron-yellow; very vigorous growth; GM, Madrid, 1963; [Peace X Independence seedling]; Dorieux; Le Blévenec

Roides Pourpres, *see* **'MOGADOR'**

'**ROKLEA**'®, HT, ob, 1975; flowers bright orange, large, dbl., 20 petals, exhibition form, moderate fragrance; foliage large, dark, semi-glossy; upright growth; Tantau, Math.

Rokoko®, *see* TANokor

Rokoko, HRg, ob, 1985; Baum

'**ROLAND**', F, ob, 1961; flowers salmon-orange-red; low growth; [Karl Weinhausen X Independence]; Leenders, J.

Roland Garros®, *see* BRIgarold

Roletta, F, op, 1983; flowers salmon-orange/pink, large, dbl.; GPG Bad Langensalza; (Sangerhausen)

Roller Coaster, *see* 'MACMINMO'

Roma, HT, op; [Duet sport]; Spronk, 1970; (Weatherly, L.)

Roma di Notte, HT, 1980; Zandri, R.; (Cavriglia)

Romaggi Plot Bourbon, B, pb

'Román', HT, mp, 1961; flowers nilsson pink, becoming hermosa pink, large, dbl., 35 petals, intense fragrance; [Asturias X Rosa de Friera]; Dot, Simon

'Roman Festival', F, pb, 1968; bud ovoid; flowers coral, base yellow, medium, borne in clusters, exhibition form, slight fragrance; foliage dark, glossy; vigorous, low, compact growth; [Queen Elizabeth X Sumatra]; Williams, J. Benjamin

Roman Holiday, *see* 'LINro'

'Roman Triumph', F, mr, 1977; flowers large, semi-dbl., 13 petals, slight fragrance; foliage glossy; upright, bushy growth; [Jove X City of Leeds]; Harkness

'Romana', HT, rb, 1938; bud long, pointed; flowers rose-red to light purple, open, large, dbl.; foliage leathery, glossy, dark; very vigorous growth; [Better Times sport]; Ringdahl

Romana, HT, lp; Vecera, L., 1975; (Czech Rosa Club)

Romance, *see* 'JACrom'

'Romance', HT, my, 1931; bud long, pointed; flowers golden yellow, shading toward lemon, open, large, dbl., moderate fragrance; foliage thick; very vigorous growth; RULED EXTINCT 11/82; [Souv. de Claudius Pernet X Buttercup seedling]; Towill

'Romance', Cl HT, lp, 1933; flowers shell-pink, fading to pale blush, large, not too full, intense fragrance; vigorous, climbing growth; RULED EXTINCT 11/82; [Isa sport]; Beckwith

Romance, *see* 'TANezamor'

Romanina, F, 1974; Zandri, R.; (Cavriglia)

Romantic Days, *see* 'MEIparnin'

Romantic Dreams, *see* MEIlomit

Romantic Fragrance, *see* 'MEIsocrat'

Romantic Hedgerose, *see* KORworm

Romantic Moments, *see* 'MEIrokoi'

Romantic Occasion, *see* MEIsardan

Romantic Palace, *see* POUlmanti

Romantic Seranade, *see* 'MEIbrinpay'

Romantic Sunrise, *see* MEIhaitoil

'Romantica', HT, mp, 1962; bud oval; flowers phlox-pink, large, dbl., 40 petals, exhibition form, slight fragrance; foliage leathery, glossy; very vigorous, upright growth; [Baccará X White Knight]; Meilland, Mrs. Marie-Louise; URS

Romantica 76®, HT, op

Romantique Meillandina, *see* 'MEIdanclar'

'Romany', F, op, 1965; flowers salmon, well formed, borne in clusters, 3.5 in., slight fragrance; free growth; [Orangeade X Mischief]; McGredy, Sam IV; Geest Industries

Romanze, *see* 'TANezamor'

'Rome Glory', HT, mr, 1937; (Gloire de Rome, Gloria di Roma, Glory of Rome); bud ovoid; flowers scarlet, reverse cerise, dbl., 55 petals, 4–5 in., globular, moderate fragrance; vigorous, bushy growth; [Dame Edith Helen X Sensation]; Aicardi, D.; J&P

Rome Glory, Climbing, Cl HT, mr; (Cavriglia)

'Romeo', HT, yb, 1918; flowers indian yellow, suffused coppery pink, slight fragrance; [Edith Part seedling X Ophelia]; Therkildsen

'Romeo', LCl, dr, 1919; flowers deep red, well formed, borne in small clusters, dbl.; vigorous, climbing growth; Easlea

Romeo, *see* 'MACtual'

Romina, *see* TANanim

Romstar, HT, lp; long; flowers medium red, velvety, high centered, dbl., 34 petals, moderate; medium, medium-green, semi-glossy; [Baccara X Coronado]; St Wagner, 1989; Res. Stn. f. Horticulture, Cluj, 1991, Romania

Romy, HT, w; Select Roses, B.V.

Romy Schneider®, *see* ORAred

'Ron West', HT, pb, 1985; flowers white with deep pink petal edges; [Admiral Rodney sport]; West, Ronald

Ronald George Kent, *see* 'BRAndypink'

'Ronald Healy', HT, pb, 1932; bud long, pointed; flowers old-rose, shaded salmon and yellow, dbl., exhibition form, moderate fragrance; foliage glossy; bushy growth; Dobbie

'Ronald Tooke', HT, dr, 1927; flowers deep blackish crimson; [Col. Oswald Fitzgerald sport]; Morse

Roncalli, S, rb, 1997; Noack, Werner

'Ronde Endiablée', F, or, 1963; flowers geranium-red, edged darker, large, semi-dbl.; foliage dark, glossy; moderate growth; GM, The Hague, 1964; Combe

'Rondo', HT, or, 1955; bud ovoid; semi-dbl., 15–20 petals, 3.5–4 in., exhibition form, moderate, fruity fragrance; foliage dark, leathery; vigorous, upright, compact growth; [Danzig X (Crimson Glory X Floradora sister seedling)]; Tantau, Math.; J&P

'Ronny Temmer', F, mp, 1974; bud ovoid; flowers full, dbl., 44 petals, 3.5 in., moderate fragrance; Delforge, S.

'Ronsard', HT, rb, 1937; bud long, pointed, yellowish edged red; flowers brilliant red, reverse yellow and cream, semi-dbl., cupped, slight fragrance; foliage leathery, dark; bushy growth; [Conrad Ferdinand Meyer X R. foetida bicolor seedling]; Gaujard; J&P

Rooi Rose, *see* KORhood

'Roosendaal', F, mr, 1965; flowers medium, dbl.; foliage dark; [Gartendirektor Glocker X Alpine Glow]; Buisman, G. A. H.

'Roquebrune', HT, ob, 1959; bud oval; flowers ochre-yellow edged orange, medium, dbl., slight fragrance; foliage dark, glossy; strong stems; moderate, bushy growth; Delforge

'Roro', HT, 1954; San Remo Exp. Sta.; (Cavriglia)

'Rory Carlton', HT, dp, 1996; flowers deep pink blooms borne mostly single, dbl., 26–40 petals, slight fragrance; few prickles; foliage medium, dark green, semi-glossy; medium (100–120 cms), upright growth; [Sylvia X Royal Highness]; Macredie, W.R.

'Rosa'

Rosa Belle™, *see* 'KINbelle'

Rosa berberifolia, *see* **'Hulthemia persica'**

'Rosa Bonheur', M, mp; (Mlle Rosa Bonheur); flowers pink or bright rose, large, dbl.; moderate growth; Laffay, M., 1871

'Rosa d'Abril', HT, dp, 1948; flowers carmine, large, very dbl., globular, very vigorous growth; Dot, Pedro

'Rosa de Friera', HT, pb, 1956; flowers violet-pink with carmine reflections, large, dbl., 35 petals, intense fragrance; strong stems; vigorous growth; [Rosa Gallart X Paulette]; Dot, Pedro

'Rosa Gallart', HT, mp, 1935; bud long, pointed; flowers rose-pink, large, dbl., cupped, moderate fragrance; foliage glossy; vigorous growth; [Unnamed variety X (Li Bures X Rose Marie)]; Dot, Pedro

'Rosa Grüss an Aachen', F, yb, 1930; flowers satiny yellowish pink; [Gruss an Aachen sport]; Spek

Rosa Kaiserin, HT, mp; flowers large, dbl., slight fragrance; (Sangerhausen)

'Rosa Mamie', HT, mr, 1956; bud ovoid; flowers bright rose, large, dbl., 35 petals, moderate fragrance; long stems; vigorous growth; [Rome Glory sport]; Asseretto, V.

'Rosa Monnet', HP, 1866; Monnet; (Cavriglia)

Rosa Mundi, *see* **R. gallica versicolor**

'Rosa Munné', HT, rb, 1952; flowers red to saffron-pink, large, dbl., intense fragrance; foliage clear green; very vigorous growth; [Maria Serrat X Paz Vila]; Munné, M.

'Rosa Orange Triumph', Pol, mp, 1943; flowers small, semi-dbl., slight fragrance; Verschuren; (Sangerhausen)

Rosa persica, *see* **'HULthemia persica'**

Rosa Poncheaux, *see* **'Poncheau-Capiaumont'**

Rosa Prominent, Gr, 1982; Sieber; (Cavriglia)

Rosa simplicifolia, *see* **'Hulthemia persica'**

Rosa Sinfonie, F, mp; flowers medium, semi-dbl., slight fragrance; VEG; (Sangerhausen)

Rosa Stern, HT, mp; flowers large, dbl.; (Sangerhausen)

'Rosa Traum', F, mp, 1974; bud pointed; dbl., 32 petals, 2.5 in., exhibition form, slight fragrance; foliage glossy; vigorous, upright, bushy growth; [Duftwolke X Seedling]; Kordes; Dehner & Co.

Rosa Union, HT; Dot; (Cavriglia)

'Rosa Verschuren', HT, 1904; Verschuren; (Cavriglia)

'Rosa Vollendung', F, pb, 1943; bud long, pointed; flowers salmon-pink, reverse capucine-red, large, borne in clusters, dbl., slight fragrance; foliage leathery, wrinkled; vigorous, bushy growth; [Crimson Glory X Else Poulsen]; Kordes

Rosa X hardii, *see* **'HULthemia hardii'**

Rosa Xmitcheltonii, *see* **'Mitcheltonii'**

Rosa Zwerg, HRg, mp, 1984; Baum

'Rosabel Walker', HT, mr, 1922; bud long, pointed; flowers brilliant velvety crimson, dbl., 3.5 in., moderate fragrance; very vigorous, bushy, spreading growth; Cant, F.

Rosabell, *see* 'COCceleste'

'Rosabella', HT, mr, 1941; flowers red with cerise reflections, very large; foliage abundant; long stems; [Mrs J.D. Russell X Julien Potin]; Giacomasso

'Rosabella', F, op, 1955; flowers salmon-pink shaded orange, well shaped, borne in very large, very dbl.; vigorous growth; [Pinocchio seedling]; Maarse, G.

'Rosabelle Barnett', F, op, 1970; flowers coral, dbl., 32 petals, 2.5 in., globular, slight fragrance; foliage dark; very free growth; [Tropicana X ?]; Gregory

Rosabunda, F, lp; cylindrical and short; cupped, 30 petals, flowers in clusters of 3–9 florets, moderate; dark green, medium large, glossy; [Frankfurt am Main X Maria Callas]; St Wagner, 1978; Res. Stn. f. Horticulture, Cluj, 1978, Romania

'Rosada', Min, op, 1950; (Rosata); flowers peach edged pink, dbl., 25 petals, cupped, foliage small, glossy; compact (7–8 in.) growth; [Perla de Alcanada X Rouletii]; Dot, Pedro; URS

Rosadora, F, lp; rosette, 25 petals, slight; leaves small, medium green, glossy; spreading habit; [Candy Rose X Yesterday]; St Wagner, 1995; Res. Stn. f. Fruit Growing, Cluj, 1996, Romania

Rosagold, F, lp; medium long; flowers yellow-apricot, cupped, 35 petals, borne in clusters, intense, very; medium large, light green, glossy; vigorous upright tall growth; [Vigorosa X Allgold]; St Wagner, 1995; Res. Stn. f. Fruit Growing, Cluj, 1996, Romania

'Rosalba', HT, pb, 1934; bud long, pointed, streaked carmine; flowers lilac-pink, center shaded salmon-pink, large, dbl.; vigorous growth; [Souv. de Claudius Pernet X Willowmere]; Borgatti, G.

'Rosaleda', HT, yb, 1958; flowers yellow and white, well formed, large, dbl., moderate fragrance; foliage dark; strong stems; upright, bushy growth; [Monte Carlo X Michele Meilland]; da Silva, Moreira

'Rosaleen', HMsk, dr, 1933; dbl., large clusters, recurrent bloom; Bentall

'Rosaleen Dunn', HT, mr, 1942; flowers crimson-red, dbl., 22 petals, 5.5 in., cupped, intense fragrance; foliage dark; vigorous growth; McGredy; J&P

Rosaletta, *see* FRYstassi

Rosali®, *see* 'TANilasor'

Rosali, *see* 'TANli'

Rosali '83®, *see* 'TANilasor'

'Rosalia', HT, dp, 1954; bud long, pointed, light cerise; long stems; Cant, F.

Rosalie, *see* **'Rosella Sweet'**

Rosalie Coral, *see* 'CHEwallop'

'Rosalie Richardson', HT, lp, 1932; flowers soft pink, intense fragrance; Evans

Rosalina™, *see* 'KORsaku'

'Rosalind', HT, op, 1918; bud bright coral; flowers apricot-pink becoming shell-pink, dbl., moderate fragrance; [Ophelia sport]; Pierson, F.R.

Rosalind, HT, pb, 1993; Bell

'Rosalind Orr English', HT, op, 1905; flowers salmon-pink, large, dbl.; Hill, E.G., Co.; (Sangerhausen)

'Rosalind Russell', HT, mp, 1950; bud long, well formed; flowers bright pink, dbl., 45–50 petals, 5 in., exhibition form, moderate fragrance; foliage dark, leathery; very vigorous, upright growth; [Briarcliff X Regina Elena]; Grillo

'Rosalinda', HT, dp, 1945; flowers carmine, well formed, large, very dbl., moderate fragrance; very vigorous growth; [Editor McFarland X Comtesse Vandal]; Camprubi, C.

Rosalinda, F, lp; flowers medium pink borne in clusters, cupped, 39 petals, slight; medium green, semi-glossy; vigorous; [(Frankfurt am Main X Maria Callas) X Foc de Tabara]; St Wagner, 1992; Res. Stn. f. Horticulture, Cluj, 1994, Romania

'Rosalinde', F, mp, 1944; flowers clear pink, blooms in large clusters, dbl.; Krause

Rosalita, HMsk, 1997; Lens; (Cavriglia)

'Rosalpina', HT, pb, 1953; flowers coppery pink, very large, dbl., 50 petals; foliage glossy; strong stems; very vigorous growth; [Signora X Unnamed variety]; Giacomasso

Rosalynn Carter, *see* 'RUprins'

Rosamaria, S, 1974; Mansuino; (Cavriglia)

'Rosamond', HT, op, 1927; flowers orange-salmon, stamens golden, single; [Red-Letter Day X R. foetida bicolor]; Burbage Nursery

'Rosamunde', F, pb, 1941; flowers salmon-carmine, reverse hydrangea-pink, large, dbl., moderate fragrance; [Seedling X Permanent Wave]; Leenders, M.

Rosamunde, *see* KORmunde

Rosanna, *see* 'KORinter'

'Rosanna', HT, mp, 1959; bud urn shaped; flowers spinel-rose, open, medium to large, slight fragrance; foliage leathery; vigorous, upright, compact growth; RU-LED EXTINCT 4/77; [Baccará X Gruss an Coburg]; Valentino

Rosarama, HT, op, 1988; Martin

Rosario®, *see* TANoras

Rosarito, HT; Moreira da Silva, A.; (Cavriglia)

Rosarium den Blakken, F, lp, 1998; RvS-Melle

Rosarium Dortmund, LCl, pb, 1995; Noack, Werner

Rosarium Netersen®, *see* 'KORtersen'

Rosarium Ueteresen®, *see* 'KORtersen'

Rosary, *see* **'Roserie'**

Rosata, *see* **'Rosada'**

Rosazwerg, *see* **Dwarf Pavement**

Röschen Albrecht, F, or, 1981; flowers medium, semi-dbl., slight fragrance; Tantau; (Sangerhausen)

Rose 2000, *see* 'COCquetrum'

Rose à Feuilles de Laitue, *see* **'Bullata'**

'Rose à Feuilles Luisantes', M, lp, 1843; flowers soft pink edged blush, medium, borne in clusters, dbl., globular, moderate fragrance; foliage glossy; branching growth; Vibert

Rose à Parfum de Bulgarie, D, mp; flowers medium, dbl., intense fragrance; (Sangerhausen)

Rose a Parfum de Grasse, D, mp; flowers medium, dbl., intense fragrance

'Rose à Parfum de l'Hay', HRg, mr, 1901; (Parfum de l'Hay); flowers cherry-carmine-red, turning blue in heat, large, dbl., globular, recurrent bloom, intense fragrance; foliage nottypically rugose; vigorous (5 ft) growth; (21); [(Summer Damask X Général Jacqueminot) X R. rugosa]; Gravereaux

'ROSE AIMÉE', HT, m, 1955; flowers gold, flushed and splashed crimson, well formed, dbl., moderate fragrance; very vigorous, bushy growth; [Peace X Seedling]; Gaujard

Rose Angle, *see* **'ANGLE'**

'ROSE ANGLE', HEg, mp; (Rose Angle Blush); flowers bright lilac rose, cupped, foliage very fragrant; very vigorous growth; Martin, prior to 1848

Rose Angle Blush, *see* Rose Angle

Rose Anil, Gr, m, 1998; Viraraghavan, M.S.

'ROSE ANNE', Cl HT, ob, 1938; flowers orange-apricot, base deeper yellow, semi-dbl., cupped, recurrent bloom, foliage large, glossy; very vigorous, climbing (15 ft) growth; RULED EXTINCT 4/77; [Francesca X Margaret McGredy]; Thomas; Armstrong Nursery

'ROSE APPLES', HRg, dp, 1906; flowers carmine-rose, large, semi-dbl., moderate fragrance; vigorous growth; Paul

Rose Baby, *see* 'MACROS'

'ROSE BAMPTON', HT, dp, 1940; bud carmine; flowers bright china-red, camellia form, dbl., 50 petals, 5.5 in., moderate fragrance; foliage dark; vigorous growth; [Charles P. Kilham X Margaret McGredy]; Van Rossem; J&P

'ROSE BANSAL', HT, yb, 1972; flowers yellow, petal edges sometimes crimson; [Ambossfunken sport]; Friends Rosery

'ROSE BENARY', HT, mp, 1908; flowers large, dbl., moderate fragrance; Lambert, P.; (Sangerhausen)

'ROSE BERKLEY', HT, op, 1928; (Souv. de Rose Berkley); bud long, pointed; flowers deep rosy salmon-pink suffused orange, base orange, large, exhibition form, slight fragrance; foliage rich green, glossy; very vigorous growth; McGredy

Rose Bicolore, HT, ob; Croix

Rose Blanche, HT, w; Croix

Rose Bleue, HT, m

'ROSE BOWL', HT, mr, 1961; bud urn shaped; flowers bright red, large, dbl., 35 petals, exhibition form, intense fragrance; foliage leathery; long, strong stems; vigorous, upright, bushy growth; [Mardi Gras X Chrysler Imperial]; Morey, Dr Dennison; J&P

'ROSE BRADWARDINE', HEg, mp, 1894; flowers clear rose-pink, blooms in graceful clusters, single, seasonal bloom, foliage very fragrant; vigorous growth; Penzance

'ROSE BRUFORD', HT, ab, 1961; flowers creamy peach, shaded rosy bronze; [Soraya sport]; Wheatcroft Bros.

Rose Cascade, *see* DELcouro

Rose Céleste®, *see* 'DELROCELES'

'ROSE CHARM', HT, mp, 1934; flowers sanguineous pink; [Talisman sport]; Scittine; Lainson

'ROSE CHEAL', HT, mp, 1970; flowers rose-pink, spiral form, dbl., 42 petals, 4–5 in., moderate fragrance; foliage glossy, dark; free growth; [New Style X Scarlet Queen Elizabeth]; Herincx

Rose Cherry Pastel, Min, mr

Rose Chou, C, mp

Rose Csárdás, F, mr, 1965; flowers medium, dbl.; Berger, W.; (Sangerhausen)

'ROSE D'AMOUR', Misc OGR, dp; (R. virginiana plena, St Mark's Rose, The St Mark's Rose); bud deep pink with long sepals; flowers deep pink, outer petals fading to pale pink receptacle wide, dbl., summer bloom, moderate fragrance; prickles of mixed sizes, scattered; foliage leaflets 5–7, rich green, marked with red on leaf st; lax growth to 10 ft

'ROSE D'AMOUR', HT, rb, 1936; flowers brown-red, reverse yellow, open, large, dbl.; foliage glossy, dark; vigorous growth; Gaujard

Rose d'Annecy®, S, mp; (ADAfetap); Golden Rose, Geneva, 1995; Adam

'ROSE DAWN', HT, pb, 1924; bud long, pointed; flowers soft shell-pink, base yellow, large, dbl., exhibition form, intense fragrance; foliage light; vigorous growth; [(Joseph Hill X Mrs George Shawyer seedling) X Ophelia]; Towill

'ROSE DE FRANCE', HT, op, 1942; bud ovoid; flowers brilliant salmon tinted orange, very large, dbl., slight fragrance; foliage glossy; dwarf growth; Gaujard

Rose de la Reine, *see* **'LA REINE'**

'ROSE DE LOS ANDES', Gr, op, 1974; bud ovoid; flowers pink-salmon-peach, dbl., 44–46 petals, 4–4.5 in.; vigorous growth; [Duet sport]; Gutierrez

'ROSE DE LYON', HT, ob, 1945; bud pointed; flowers orange-yellow, medium, dbl., moderate fragrance; foliage glossy; Gaujard

'ROSE DE MEAUX', C, mp; (De Meaux, Pompon Rose, R. centifolia pomponia, R. dijoniensis, R. pomponia, R. pulchella); dbl., 1.5 in.; dwarf growth; (28)

'ROSE DE MEAUX WHITE', C, w; (De Meaux White, Le Rosier Pompon Blanc, White de Meaux); flowers white with pink centers

Rose de Montfort, F, dp, 1993

Rose de Puteaux, P, mp

Rose de Rennes®, *see* ADAharos

'ROSE DE RESCHT', P, dp; flowers bright fuchsia-red, fading with lilac tints, rosette form, very dbl., long-season bloom, intense, damask fragrance; foliage dense; vigorous, compact growth; re-introduced, 1950

Rose de Schelfhout, HGal, lp; Parmentier, 1840

Rose des Cisterciens, *see* DELarle

'ROSE DES MAURES', HGal, dr; (Sissinghurst Castle); flowers deep plum-crimson, yellow stamens, semi-dbl., 17 petals, 2.5 in., moderate fragrance; foliage thin; vigorous growth; Sackville-West, 1947; Re-int. 1947

'ROSE DES PEINTRES', C, mp; (Centifolia Major); flowers bright pink, dbl., moderate fragrance; vigorous growth

Rose des Quatre Saisons, *see* **'AUTUMN DAMASK'**

'ROSE D'ESPÉRANCE', HT, dr, 1918; flowers medium-large, dbl.; Verschuren; (Sangerhausen)

'ROSE D'EVIAN', T, pb, 1895; flowers pink, center carmine; Bernaix, A.

'ROSE D'HIVERS', D, w; flowers whitish, center shell-pink, well-shaped

'ROSE D'OR', HT, dy, 1941; bud long, pointed; flowers intense yellow, dbl., 35 petals, 4.5–5 in., slight fragrance; foliage dark, bronze, glossy; very vigorous growth; [Julien Potin X Unnamed seedling]; Gaujard; J&P

Rose d'Or de Montreux, *see* ADAtonysil

Rose d'Orsay, S, mp; flowers medium, dbl., moderate fragrance; (Sangerhausen)

'ROSE DOT', HT, rb, 1962; flowers red, reverse white, large, dbl., 35 petals, intense fragrance; foliage dark; vigorous, upright growth; [Baccará X Peace]; Dot, Simon

'ROSE DU BARRI', HT, pb, 1940; flowers salmon-pink, reverse carmine, large, single, moderate fragrance; vigorous growth; Archer

'ROSE DU BENGALE', HCh, 1759; Jacquin; (Cavriglia)

Rose du Ciel®, *see* 'DELFROR'

'ROSE DU MAÎTRE D'ÉCOLE', HGal, m; flowers soft pink to lilac, very large, dbl., flat, intense fragrance; Miellez, 1840

'ROSE DU PRINCE', HT, mr, 1959; bud long, pointed; flowers strawberry-rose, large, dbl., slight fragrance; foliage leathery, light green; vigorous, upright growth; [Blanche Mallerin X Profusion]; Dorieux; Pin

'ROSE DU ROI', P, mr, 1815; (Lee's Crimson Perpetual); flowers bright red shaded violet, large, semi-dbl., remontant bloom, intense fragrance; foliage clear green, slightly fluted; vigorous growth; Lelieur; Souchet

'ROSE DU ROI À FLEURS POURPRES', P, m; flowers purplish, medium, very dbl., intense fragrance; (28); [Rose du Roi sport]; (France), before 1819; 1819

'ROSE DU ROI PANACHÉ', P, rb; flowers pale flesh striped crimson; [Rose du Roi sport]

Rose du Saint Sacrament, *see* **R. CINNAMOMEA PLENA**

Rose Edouard, B, mp, 1818; Perichon/ Neumann

Rose Elf, *see* **'ROSENELFE'**

Rose Ellen, S, lp, 1995; Lykke

'ROSE EUTIN', F, mp, 1958; flowers rose-pink; [Eutin sport]; Hennessey

Rose Festival, S, mr; [Black Beauty X Candy Cane]; Sutherland, P, 1996; Golden Vale Nursery

'ROSE FUKUOKA', HT, pb, 1983; flowers blend of light, medium and deep pink, (deeper at petal edges, dbl., 27 petals, exhibition form, slight fragrance; foliage small, dark, glossy; compact growth; [(Utage X Kordes' Perfecta) X Miss Ireland]; Ota, Kaichiro

Rose Garnette, Pol, mp

Rose Gaujard®, *see* 'GAUMO'

'ROSE GAUJARD, CLIMBING', Cl HT, rb, 1964; (Grimpant Rose Gaujard®); Nakashima, Tosh

Rose Gilardi™, *see* 'MOROSE'

'ROSE HANNES', HT, w, 1982; (Hannes); flowers large, dbl., moderate fragrance; foliage medium, medium green, glossy; upright growth; [Pascali X Unnamed seedling]; Wheatcroft, Christopher; Timmerman's Roses

'ROSE HILL', HT, dp, 1928; bud long, pointed; flowers darker than briarcliff, very large, dbl., exhibition form, moderate fragrance; [Columbia sport]; Hill, Joseph H., Co.

Rose Hill, *see* POUlrohill

'ROSE HILLS RED', Min, dr, 1978; bud pointed; flowers deep red, small, dbl., 30 petals, 1.5 in.; foliage glossy, leathery; vigorous, upright growth; [(R. wichurana X Floradora) X Westmont]; Moore, Ralph S.; Sequoia Nursery

Rose Iga, *see* 'MEIBALBIKA'

'ROSE IMPÉRIALE', HT, ob, 1942; flowers flame and gold, over large, very dbl., globular, moderate fragrance; foliage dark, glossy; vigorous growth; GM, Bagatelle, 1941; Gaujard

Rose Magic, *see* 'GELLANDO'

'ROSE MARIE', HT, mp, 1918; flowers clear rose-pink, very large, dbl., cupped, moderate fragrance; foliage glossy, dark; vigorous growth; [Hoosier Beauty X Sunburst]; Dorner

Rose Marie, F, m

'ROSE MARIE REID', HT, mp, 1956; bud globular; flowers neyron rose, dbl., 48 petals, 5–6 in., cupped, moderate fragrance; foliage dark, leathery; vigorous growth; [Charlotte Armstrong X Katherine T. Marshall]; Whisler; Germain's

'ROSE MARIE, CLIMBING', Cl HT, mp, 1927; Pacific Rose Co.

Rose Mary, HT, mp

'ROSE MERK', HT, or, 1931; flowers bright geranium-red, moderate fragrance; vigorous growth; Cant, F.

Rose Minarett, *see* NOAmet

'ROSE MOET', HT, my, 1962; flowers golden yellow, large, dbl.; foliage leathery; long, strong stems; Dorieux; Pin

'ROSE NABONNAND', T, pb, 1883; flowers salmon-pink tinted yellow, large, intense fragrance; Nabonnand, G.

'ROSE NOBLE', HT, lp, 1927; bud long, pointed; flowers silvery pink, semi-dbl., moderate fragrance; [Mme Caroline Testout X ?]; Mühle

'ROSE NUGGETS', Min, lp, 1991; bud pointed; flowers light pink, aging lighter, self-cleaning, many petaloids, bl, dbl., 30 petals, 1.5 in., exhibition form, no fragrance; foliage small, dark green, glossy, frequently 7 leaflets; very full, bushy, low growth; [Red Ace sport]; Michelis, Dorothy; Justice Miniature Roses, 1992

Rose O'Bree, F, lp; O'Bree, Nancy, 1997; (Weatherly, L.)

Rose Odyssey 2000, *see* 'BALPARTY'

Rose Oeilletde Saint Arquey, *see* **'SERRATIPETALA'**

'ROSE OF CLIFTON', F, yb, 1978; bud pointed; flowers gold edged peach-pink, dbl., 27 petals, 3.5 in., slight fragrance; foliage dark, matt; vigorous, upright growth; [Vera Dalton X Parasol]; Sanday, John

'ROSE OF FREEDOM', HT, mr, 1948; bud ovoid; dbl., 50 petals, 3.5–4.5 in., intense fragrance; foliage leathery, dark; vigorous, upright, bushy growth; [Charlotte Armstrong X Night]; Swim, H.C.; Mt. Arbor Nursery

'ROSE OF LIDICE', HT, yb, 1961; flowers lemon-yellow shaded poppy red; [Tzigane sport]; Wheatcroft Bros.

Rose of Narromine, *see* 'WEKDYKSTRA'

Rose of Provins, *see* **R. GALLICA OFFICINALIS**

'ROSE OF TORRIDGE', HT, dp, 1961; flowers deep pink, dbl., 36 petals, 6 in., moderate fragrance; foliage glossy; vigorous growth; [Karl Herbst X Pink Charming]; Allen, E.M.

'ROSE OF TRALEE', S, op, 1964; flowers deep pink shaded salmon, blooms in small clusters, dbl., 35 petals, 4 in., slight fragrance; foliage dark; very vigorous, bushy growth; [Leverkusen X Korona]; McGredy, Sam IV

Rose of Wagga Wagga, *see* 'JACDASH'

'ROSE OPAL', HT, dp, 1962; flowers pink-opal, well formed, dbl., 25–30 petals, 4–5 in., moderate fragrance; foliage dark (reddish when young); vigorous, upright, bushy growth; [Wellworth X Independence]; LeGrice; Wayside Gardens Co.

'ROSE PARADE', F, pb, 1974; bud ovoid; flowers coral-peach to pink, large, dbl., cupped, moderate fragrance; foliage large, glossy; vigorous, bushy, compact growth; AARS, 1975; [Sumatra X Queen Elizabeth]; Williams, J. Benjamin; Howard Rose Co.

Rose Pluton, *see* **'PLUTON'**

Rose Ponceau, *see* **'PONCHEAU-CAPIAUMONT'**

Rose Poncheaux, *see* **'PONCHEAU-CAPIAUMONT'**

'ROSE QUEEN', HT, or, 1962; bud globular; dbl., 38 petals, 4–5 in., exhibition form, moderate fragrance; foliage leathery, dark; vigorous, upright, bushy growth; [Chrysler Imperial X Unnamed seedling]; Whisler; Germain's

Rose Rhapsody™, *see* 'JACSASH'

'ROSE SHERBET', F, dp, 1962; bud globular; flowers deep rose-pink, medium, dbl., intense fragrance; foliage glossy; vigorous, upright, open growth; [Gruss an Teplitz X ?]; Pal, Dr B.P.; Indian Agric. Research Inst.

Rose Two Thousand, *see* 'COCQUETRUM'

'ROSE VALMAE', S, mr, 1981; flowers decorative blooms borne 3 per cluster, dbl., 40 petals, slight fragrance; foliage light green, 7 leaflet; upright, medium growth; Watts, Mrs. M.A.

Rose Verte, HT, 1989; Croix; (Cavriglia)

'ROSE WINDOW', Min, ob, 1978; bud ovoid, pointed; flowers orange, yellow, and red blend, 15–20 petals, 1 in., slight fragrance; foliage small, dark; spreading, bushy growth; [Seedling X Over the Rainbow]; Williams, Ernest D.; Mini-Roses

Rosea, F, 1974; Delforge; (Cavriglia)

Rosea Centfeuilles, HP, dp; Touvais

'ROSEANNA', Cl Min, dp, 1976; bud long, pointed; flowers deep pink, dbl., 1–1.5 in., exhibition form; foliage glossy; tall, semi-climbing growth; [Little Darling X Seedling]; Williams, Ernest D.; Mini-Roses

'ROSEATE', Cl HT, mr, 1931; flowers large, semi-dbl., cupped, slight fragrance; foliage wrinkled; vigorous, climbing growth; good as a pillar; Clark, A.; Ivanhoe Hort. Soc.

Rosebud, S, pb, 1997

'ROSECARPE', HT, yb, 1993; flowers cream shaded pink, blooms borne mostly single, dbl., 26–40 petals, 3–3.5 in., slight fragrance; some prickles; foliage large, dark green, semi-glossy; medium (100 cms), upright growth; [Selfridges X Mischief]; Poole, Lionel; Battersby Roses, 1994

Rosée du Matin, *see* **'CHLORIS'**

'ROSEE O'BREE', F, mp, 1998; flowers medium pink, reverse light pink, multicolored, double (15–25 petals), small (2.5 cms) blooms borne in large clusters;

slight fragrance; moderate prickles; foliage medium green, semi-glossy; spreading, medium (7 ft) rambling growth; recurrent; hedge; [Parentage unknown]; O'Bree, Nancy

'ROSEFORD', LCl, lp, 1981; bud ovoid; flowers light pink, slightly darker reverse, blooms borne 6 per cluster, dbl., 28 petals, cupped, repeats well, slight fragrance; straight, red prickles; foliage dark, semi-glossy; bushy (can be grown as shrub) growth; Bronze Medal, ARC TG, 1986; [Unnamed seedling X Unnamed seedling]; Jerabek, Paul E.

'ROSEGLEN BOUQUET', F, pb, 1953; flowers ivory-white edged deep pink, small, very dbl., exhibition form, slight fragrance; foliage dark, glossy, leathery; vigorous, upright growth; [Pinocchio X ?]; Quinn; Roseglen Nursery

'RÖSEL DACH', Pol, mr, 1907; flowers cherry-rose, borne in upright clusters, dbl.; dwarf, bushy growth; Walter, L.

'ROSEL VOGEL', S, m, 1938; flowers small, dbl., slight fragrance; Vogel, M.; (Sangerhausen)

'ROSELAND ROSETTE', F, dp, 1952; bud globular; flowers deep rose-pink, small, dbl.; vigorous, bushy growth; [Crimson Rosette sport]; Houghton, T.B.

'ROSELANDIA', HT, my, 1924; flowers darker and larger; [Golden Ophelia sport]; Stevens, W.; Low

'ROSELANDIA, CLIMBING', Cl HT, my, 1933; Lens

'ROSELETTE', F, mp, 1957; flowers soft salmon-pink; foliage bright green; very vigorous growth; [Orange Triumph X Alain]; Lens

Roselina™, *see* 'KORTEMMA'

Roseline de Kersa, HT, dp, 1997

'ROSELLA', F, ab, 1972; (**Bobby Dazzler**); flowers blush, shaded rich apricot, dbl., 50 petals, 3 in., slight fragrance; foliage small, matt; [(Vera Dalton X Highlight) X (Ann Elizabeth X Circus)]; Harkness, 1973

'ROSELLA', F, pb, 1930; flowers salmon-rose-pink, ruffled, borne in large trusses, semi-dbl.; vigorous growth; [Else Poulsen sport]; Prior

'ROSELLA', Cl HT, pb, 1931; bud long, pointed; flowers velvety carmine, base yellow, orange undertone, borne in clusters, single, 2.75 in., slight fragrance; foliage large, glossy; vigorous (8 ft or more) growth; [Mme Edouard Herriot X Roger Lambelin]; Dot, Pedro; C-P

'ROSELLA GRACE CREMER', F, dp, 1957; bud conical to ovoid; flowers dark tyrian rose, reverse lighter, rosette center, medium, dbl., 35–40 petals, moderate fragrance; very vigorous, bushy growth; [Garnette sport]; Cremer

'ROSELLA SWEET', HT, op, 1930; (Rosalie); flowers nasturtium-yellow suffused salmon-pink, large, intense fragrance; Pernet-Ducher; Dreer

Rosellana®, *see* BARros

'ROSEMARIE HINNER', HT, mp, 1949; bud long, pointed; flowers large, exhibition form, moderate fragrance; foliage leathery, light green; very vigorous, upright, bushy growth; [Frau Karl Druschki X (Ellen X Una Wallace)]; Hinner, P.; Bauské Bros. & Hinner

'ROSE-MARIE VIAUD', HMult, m, 1924; (Rosemary Viaud); flowers same as parent except bluer and more, dbl.; [Veilchenblau seeding]; Igoult; Viaud-Bruant

'ROSEMARY', Cl HT, lp, 1925; bud long, pointed; flowers pink shaded old-gold, very dbl., exhibition form, free, recurrent boom, moderate fragrance; foliage leathery; pillar growth; Cant, F.

'ROSEMARY', Min, mp; flowers apple-blossom-pink

'ROSEMARY', F, dp, 1955; bud ovoid; flowers rose-pink, borne in clusters, dbl., 55–60 petals, 2.5 in., flat, moderate fragrance; foliage leathery; vigorous, bushy growth; [Garnette sport]; Schenkel; Amling-DeVor Nursery

'ROSEMARY CLOONEY', F, ab, 1985; flowers creamy apricot, blooms in clusters of 3–6, dbl., 20 petals, slight fragrance; small, red prickles; foliage small, medium green, matt; upright, bushy growth; [Vera Dalton X Elizabeth of Glamis]; French, Richard

'ROSEMARY DUNCAN', F, my, 1964; flowers yellow, edged lighter, medium, dbl., 40 petals; foliage glossy; vigorous, bushy growth; [Pinocchio sport]; Duncan

'ROSEMARY EDDIE', F, mp, 1956; flowers bright pink, well formed; [New Dawn X Fashion]; Eddie

'ROSEMARY GANDY', F, yb, 1957; (Flash); flowers yellow and coppery, medium, semi-dbl., cupped, moderate fragrance; foliage glossy, bronze; bushy growth; [Tabarin seedling X Jolie Princesse seedling]; Gaujard

Rosemary Harkness, *see* 'HARROWBOND'

'ROSEMARY ROSE', F, dp, 1954; flowers deep pink, camellia-shaped, blooms in large trusses, dbl., moderate fragrance; foliage coppery; vigorous, bushy growth; GM, NRS, 1954 GM, Rome, 1954; [Gruss an Teplitz X Floribunda seedling]; deRuiter; Gregory

'ROSEMARY STONE', HT, mp, 1970; flowers pink, very dbl., exhibition form, slight fragrance; foliage leathery; moderate, upright growth; [Waltzing Matilda sport]; Stone

Rosemary Viaud, *see* **'ROSE-MARIE VIAUD'**

'ROSEMARY, CLIMBING', Cl HT, lp, 1920; flowers very large, dbl.; Dingee & Conard; (Sangerhausen)

Rosenau, F, dr, 1961; Eggert

Rosendel, F, m

Rosendorf Schmitshausen, LCl, dr, 1977; Cocker

Rosendorf Sparrieshoop®, *see* KORdibor

Rosendorf Steinfurth, S, mp, 1993; Schultheis

'ROSENDORF UFHOVEN', HEg, dr, 1949; flowers crimson, very large blooms, very dbl., globular, non-recurrent, slight fragrance; foliage leathery, glossy, dark; weak stems; vigorous, upright, bushy growth; [Gen. MacArthur X Magnifica]; Kordes, W. Söhne

'ROSENELFE', F, mp, 1939; (Rose Elf); bud long, pointed; dbl., 2.5 in., exhibition form, blooms in clusters, moderate fragrance; foliage leathery, glossy, light; vigorous, bushy growth; [Else Poulsen X Sir Basil McFarland]; Kordes; Dreer

'ROSENELLA', HT, mp, 1964; flowers clear pink, reverse brighter, intense fragrance; Mondial Roses

Rosenfee®, F, lp, 1967; Boerner

Rosenfest, S, op, 1981; GPG Bad Langensalza

Rosenholm, *see* POUlover

'ROSENKAVALIER', F, dp, 1963; flowers deep rose, borne in clusters, dbl., 45 petals, intense fragrance; foliage dark; vigorous, upright, bushy growth; Verschuren, A.; Stassen

Rosenkavalier, F, or, 1995; Huber

Rosenkreis Neunkirchen, F, mr, 1995; flowers luminous red, semi-dbl., slight fragrance; Michler, K. H.; (Sangerhausen)

'ROSEN-LAMBERT', LCl, dr, 1937; flowers oxblood-red, very large, dbl.; foliage large, bronze, leathery; vigorous, climbing (9 ft) growth; [Fragezeichen X American Pillar]; Vogel, M.; P. Lambert

Rosenmärchen, *see* **'PINOCCHIO'**

Rosenmärchen, Climbing, *see* **'PINOCCHIO, CLIMBING'**

'ROSEN-MÜLLER', S, mr, 1940; flowers medium, single, slight fragrance; Vogel, M.; (Sangerhausen)

'ROSENPFARRER MEYER', HT, or, 1930; flowers coral-red passing to prawn-red, semi-dbl., slight fragrance; very vigorous growth; [Mme Edouard Herriot X Louise Catherine Breslau]; Soupert & Notting

Rosenprinz, S, dp, 1990; flowers small, dbl., slight fragrance; very low, almost dwarf growth; Wänninger, Franz; (Sangerhausen)

'ROSENPRINZESSIN', HT, mp, 1975; bud ovoid; dbl., 3.5–5 in., moderate fragrance; foliage glossy; upright growth; [Nordia X Sans Souci]; Hetzel

Rosenprinzessin, *see* HELprinzess

Rosenprinzessin Andrea, HT, mr, 1993; Hetzel

Rosenprofessor Sieber, *see* KORparesni

Rosenreigen®, S, lp

Rosenresli®, *see* 'KORresli'

Rosenrot®, HT, mr, 1978; Tantau

Rosenstadt Zweibrucken®, *see* KORstatis

Rosentanz, S, rb, 1989; flowers dark red with white, large, semi-dbl.; Fryer, Gareth; (Sangerhausen)

Rosenthal, *see* 'KORtat'

Rosenwalzer®, S, pb, 1993; Dickson, Patrick

'Rosenwunder', HEg, dp, 1934; bud long, pointed; flowers rose-red, very large, semi-dbl., cupped, non-recurrent, slight fragrance; foliage large, leathery, glossy, wrinkled; very vigorous, bushy growth; (29); [W.E. Chaplin X Magnifica]; Kordes

Rosenzauber, F, lp, 1991; flowers medium, semi-dbl., slight fragrance; Wänninger, Franz; (Sangerhausen)

Roseraie de Blois, *see* DORfuri

'Roseraie de l'Hay', HRg, dr, 1901; flowers crimson-red changing to rosy magenta, large, dbl., recurrent bloom, intense fragrance; foliage rugose; vigorous (4–5 ft) growth; (14); Cochet-Cochet

'Roserie', HMult, mp, 1917; (Rosary); flowers tyrian pink, base white, darkening with age, open, semi-dbl., 3.75 in., slight fragrance; thornless; foliage large, rich green, leathery, glossy; very vigorous, climbing growth; [Tausendschön sport]; Witterstaetter

Roseromantic®, *see* 'KORsommer'

Rosetime, *see* 'BENtem'

'Rosetone', Min, mp, 1977; bud ovoid, pointed; dbl., 60 petals, 1 in., exhibition form, slight fragrance; foliage small, dark, leathery; very bushy growth; [Dream Dust X Little Chief]; Moore, Ralph S.; Sequoia Nursery

Rosetta, *see* 'SPOrose'

'Rosette', Pol, mr, 1926; flowers fuchsia-red, medium, dbl.; semi-dwarf growth; Grandes Roseraies

'Rosette', HT, or, 1934; flowers rose red, shaded orange, base yellow, well formed, large, dbl., moderate fragrance; vigorous, bushy growth; Dickson, A.

'Rosette', F, op; flowers peach-pink, borne in clusters, dbl.; vigorous growth; Archer

'Rosette de la Légion d'Honneur', HT, op, 1896; flowers salmon-pink, medium, semi-dbl., moderate fragrance; Bonnaire; (Sangerhausen)

'Rosette Delizy', T, yb, 1922; flowers yellow, apricot reflexes, outer petals dark carmine, well-formed, dbl.; vigorous growth; [Gen. Gallieni X Comtesse Bardi]; Nabonnand, P.

Roseville College, *see* **The Roseville College Rose**

'Rosewood', HT, dp, 1998; flowers neon pink, double, medium-sized, dbl., 15–25 petals, 4–4.5 in., blooms in small clusters, moderate fragrance; prickles moderate; foliage medium, medium green, semi-glossy; compact, medium growth; [Kardinal sport]; Prescott, Cheryl

Roseworld, *see* 'SIMilam'

Rosi Mittermeier, *see* 'KORmiora'

Rosie™, *see* 'BENros'

Rosie Larkin, *see* FRYyippee

'Rosie O'Bree', F, mp, 1998; flowers medium pink, lighter reverse, multi-colored, dbl., 15–25 petals, .75–2.5 in., borne in large clusters, slight fragrance; prickles moderate; foliage medium green, semi-glossy; spreading, rambling, medium growth; [Seedling sport]; O'Bree, Nancy

Rosie O'Donnell, *see* 'WEKwinwin'

Rosier de Damas, D, mp; flowers medium, dbl., moderate fragrance; (Sangerhausen)

Rosier de Thionville, *see* **'Quatre Saisons Blanc Mousseux'**

Rosier des Parfumeurs, HGal, lp, 1800; Cochet

Rosier d'Or, *see* 'DELpli'

'Rosier Gloriette', Min, op; flowers orange-salmon, rosette form

'Rosiériste Gaston Lévêque', HT, dp, 1932; bud long, pointed; flowers brilliant carmine, large, dbl., cupped, intense fragrance; foliage dark; very vigorous growth; [Mme Butterfly X Jean C.N. Forestier]; Dot, Pedro

'Rosieriste Max Singer', HMult, 1885; Lacharme, F.; (Cavriglia)

'Rosiériste Pajotin-Chédane', Pol, dr, 1934; flowers deep red, with white, thread-like markings, borne in clusters, semi-dbl., cupped, slight fragrance; foliage leathery; strong stems; vigorous, bushy growth; Délépine; Pajotin-Chédane

Rosika, *see* 'HARmusky'

Rosilia, Cl Min, or, 1991; flowers small to medium, semi-dbl.; Warner; (Sangerhausen)

'Rosina', Min, my, 1951; (Josephine Wheatcroft, Yellow Sweetheart); flowers sunflower-yellow, small blooms in clusters, semi-dbl., 16 petals, slight fragrance; foliage glossy, light; dwarf, compact growth; [Eduardo Toda X Rouletii]; Dot, Pedro; URS, 1935

'Rosine', HT, pb, 1935; bud citron-yellow with red; flowers flesh-pink, base clear salmon, very dbl.; foliage bronze; Lens

'Rosita', HT, lp, 1956; bud long; flowers soft pink, large, dbl., moderate fragrance; foliage glossy; vigorous growth; [The Doctor X Unnamed seedling]; Delforge

'Rosita Mauri', HT, 1914; Ketten Bros.; (Cavriglia)

Rosita Missoni, *see* BARmiss

'ROSlaw', HT, yb, 1985; (**D. H. Lawrence**); [Gay Gordons sport]; McCarthy, Mrs. Rosemary; Rosemary Roses

'Roslyn', HT, dy, 1929; bud long, pointed, deep orange; flowers golden yellow, reverse darker orange, large, semi-dbl., slight fragrance; vigorous, compact growth; [Souv. de Claudius Pernet X Buttercup]; Towill

'Roslyn, Climbing', Cl HT, dy, 1937; Vestal

Roslyne, *see* 'DELbara'

'Rosmari', HT, pb, 1962; bud pointed; flowers pink, reverse purplish, dbl., 30 petals; strong stems; vigorous, upright growth; [Vigoro X First Love]; Dot, Simon

'Rosmarin'®, Min, pb, 1965; flowers light pink, reverse light red, small, globular, slight fragrance; foliage glossy, light green; dwarf, bushy growth; [Tom Thumb X Dacapo]; Kordes, R.

Rosmarin 89®, *see* KORfanto

Rosnella, Gr, 1958; Moro, L.; (Cavriglia)

'Rosomane Narcisse Thomas', T, rb, 1908; flowers crimson suffused apricot-yellow; Bernaix, P.

'Rosorum', F, or, 1959; flowers orange-scarlet, borne in large clusters, semi-dbl.; foliage dark, glossy; vigorous, upright growth; [Buisman's Triumph X Alpine Glow]; Buisman, G. A. H.

'Ross Gowie', HT, dr, 1968; flowers conical, large, slight fragrance; vigorous, tall growth; Gowie; Gandy Roses, Ltd.

'Ross Rambler', LCl, w, 1938; blooms all summer, height 9 ft; P.H. Wright

'Rossana', HT, dr, 1958; flowers blood-red, well shaped, large, dbl.; vigorous, spreading growth; [Orange Delight X Tudor]; Frères, Buyl

Rosso Giacomasso, HT, 1967; Giacomasso; (Cavriglia)

'Rostock', HMsk, lp, 1937; bud long, pointed; flowers very large blooms in clusters, dbl., exhibition form, recurrent bloom, slight fragrance; foliage leathery, glossy, dark; very vigorous, bushy growth; [Eva X Louise Catherine Breslau]; Kordes

'Roswytha', F, mp, 1968; bud ovoid; flowers pink, small borne in clusters, very dbl.; foliage dark; [Carol sport]; van der Meyden

Rosy, Min, mp, 1993; Chandrakant

Rosy Ann, Min, op, 1994; Lens

Rosy Border, *see* POUlosy

Rosy Carpet®, *see* 'INTercarp'

'Rosy Cheeks', HT, rb, 1975; flowers red, reverse yellow, dbl., 35 petals, 7 in., intense fragrance; foliage dark, glossy;

[? X Irish Gold]; Anderson's Rose Nurseries, 1976

Rosy Cheeks, *see* MORsycheek

Rosy Cheeks, Climbing, *see* 'ANDros'

Rosy Cushion®, *see* 'INTERALL'

'ROSY DAWN', Min, yb, 1982; bud ovoid; flowers yellow, petals edged deep pink, medium, dbl., 28 petals, exhibition form, slight, tea fragrance; foliage small, medium green, glossy; PP005274; [Magic Carrousel X Magic Carrousel]; Bennett, Cecilia 'Dee'; Tiny Petals Nursery

Rosy Forecast, *see* 'LAVLAD'

Rosy Future, *see* 'HARWADEROX'

Rosy Gem, *see* 'MEIRADIA'

'ROSY GLOW', HT, mp, 1947; bud long, pointed; flowers hermosa pink, dbl., 25–35 petals, 4–5 in., exhibition form, intense fragrance; foliage leathery; vigorous, bushy growth; [Better Times sport]; Hill, Joseph H., Co.

Rosy Hit, *see* POUlmar

'ROSY JEWEL', Min, pb, 1958; bud ovoid; flowers rose-red, reverse lighter, center white, dbl., 25 petals, 1 in., moderate fragrance; low (6–8 in.), compact growth; [Dick Koster sport X Tom Thumb]; Morey, Dr Dennison; J&P

Rosy Koster, Pol, mp

Rosy La Sevillana, Climbing, *see* MEIgerokosar

Rosy La Sevilliana, *see* 'MEIGEROKA'

'ROSY LIFE', HT, mp, 1970; flowers sparkling pink, full, dbl., 25 petals, 3–4 in., slight fragrance; foliage glossy; vigorous growth; [Walko X Souv. de J. Chabert]; Delbard; Laxton & Bunyard Nursery

'ROSY MANTLE', LCl, mp, 1968; flowers large, dbl., moderate fragrance; foliage dark, glossy; [New Dawn X Prima Ballerina]; Cocker

Rosy Meillandina, *see* 'MEIFINARO'

Rosy Minimo, Min, mp

Rosy Morn, *see* **'IMPROVED CÉCILE BRÜNNER'**

'ROSY MORN', LCl, mp, 1914; flowers large; [Frau Karl Druschki seedling]; Clark, A.

'ROSY MORN', HP, op, 1878; flowers soft peach shaded salmon-pink, well formed, very large, slight fragrance; very vigorous growth; Paul, W.

'ROSY MORN', Pol, mp, 1930; flowers rose-pink, large, borne in large clusters, moderate fragrance; vigorous growth; Burbage Nursery

Rosy Outlook, *see* 'WEKTORROC'

Rosy Potluck, *see* 'LAVROSY'

Rosy Purple, HMsk, m, 1995; Lens

Rosy Star, HT, mp

'ROSY WINGS', HT, dp, 1962; flowers pink shaded light red, dbl., 35 petals, 2–2.5 in., globular, moderate fragrance; foliage dark, glossy; vigorous growth; [Pink Spiral X Unnamed seedling]; Delforge

Rosy Wings, *see* 'MEJAKKA'

'ROTARIAN', HT, mr, 1921; bud long, pointed; flowers bright cherry-crimson, dbl., 35–40 petals, moderate fragrance; [Ophelia X ?]; Lemon

'ROTARY JUBILEE', HT, yb, 1971; bud ovoid; flowers yellow blend, creamy yellow, large, dbl., exhibition form, moderate fragrance; foliage large, glossy, dark, leathery; vigorous, bushy growth; [Queen Elizabeth X Peace]; Lindquist; Bell Roses, Ltd.

'ROTARY PRESIDENT', F, or, 1975; flowers deep vermilion, dbl., 4 in., intense fragrance; foliage small, plum color, turning dark green; dwarf, bushy growth; [Fairlight X Summer Holiday]; Wood

Rotary Rose, *see* 'MEIRYPOUX'

'ROTARY-LYON', HT, yb, 1936; bud long, pointed, well formed, golden pink; flowers old-gold shaded carmine, reverse yellow, stamens deep yellow, dbl., cupped, moderate fragrance; foliage dark; very vigorous growth; Chambard, C.

Rote Apart, S, mr, 1991; Uhl, J.

'ROTE BETTER TIMES', HT, mr, 1940; flowers large, dbl., intense fragrance; Noack, Werner; (Sangerhausen)

'ROTE CENTIFOLIE', C, dr, 1938; flowers medium, very dbl., intense fragrance; Krause; (Sangerhausen)

'ROTE ELSE POULSEN', F, dp, 1934; (21); [Else Poulsen sport]; Koopmann; Tantau Roses

Rote Flamme®, LCl, dr, 1967; Kordes

'ROTE GABRIELLE PRIVAT', Pol, mr, 1940; [Gabrielle Privat sport]; Koopmann

Rote Hannover, S, mr, 1998

Rote Hermosa, *see* **'QUEEN'S SCARLET'**

Rote Hiroshima, HT, mr, 1990; flowers carmine-red, large, dbl., moderate fragrance; Tagashira, Kazuso; (Sangerhausen)

Rote Max Graf®, *see* 'KORMAX'

'ROTE MEVROUW G.A. VAN ROSSEM', HT, or, 1934; flowers nasturtium-red, reverse lightly tinted yellow, open, large, dbl., intense fragrance; foliage glossy, dark; vigorous growth; [Mev. G.A. van Rossem sport]; Kordes

Rote Mozart®, *see* KORtragfei

'ROTE PERLE', F, dr, 1962; flowers deep red, medium, dbl., 25 petals, 1.5 in.; foliage dark; vigorous, bushy growth; Tantau, Math.

'ROTE PHARISÄER', HT, mr, 1927; bud very long, pointed; flowers well shaped, large, dbl., intense fragrance; foliage reddish green, leathery; very vigorous growth; [Pharisaer X George C. Waud]; Hinner, W.

'ROTE RAPTURE', HT, mr, 1934; flowers bright cherry-red; [Rapture sport]; Weber, J.

'ROTE TESCHENDORFFS JUBILÄUMSROSE', Pol, dp, 1930; flowers dark crimson-pink to light red, large, borne in large clusters, dbl.; vigorous growth; Grunewald; Teschendorff

Rote Woge, *see* MEInirlo

Rotelfe®, *see* 'TANRECKTOR'

'ROTELFE', HT, dr, 1922; flowers very dark red; RULED EXTINCT 4/85; [Château de Clos Vougeot X Ulrich Brunner Fils]; Tantau

'ROTER CHAMPAGNER', HT, dp, 1963; (Pétillante); bud pointed; flowers red champagne color, large, dbl., slight fragrance; foliage bright green; long, strong stems; vigorous, upright growth; Tantau, Math.

Roter Kobold®, F, dr, 1994; Interplant

'ROTER SCHMETTERLING', HFt, mr, 1935; flowers medium, semi-dbl., slight fragrance; Vogel, M.; (Sangerhausen)

Roter Stern, HT, mr, 1958; Meilland

'ROTES MEER', Pol, or, 1960; dbl., borne in large clusters, foliage glossy, dark; upright, bushy, symmetrical growth; [Orange Triumph X Unnamed seedling]; Verschuren, A.; van Engelen

Rotesmeer, HRg, m, 1986; (**Purple Pavement**); Baum

Rotfassade, LCl, mr, 1997; Noack, Werner

'ROTORUA', F, or, 1963; bud ovoid; flowers orange-scarlet, borne in clusters, dbl., 27 petals, 3.5 in., exhibition form, slight fragrance; foliage leathery, dark; short, strong stems; very vigorous, bushy growth; [Independence X Spartan]; McGredy, Sam IV; Avenue Nursery

'ROTRAUT', Pol, dr, 1931; dbl., borne in clusters, vigorous growth; [Miss Edith Cavell sport]; Grunewald

'ROTROU', M, m, 1848; flowers reddish-violet, medium, dbl.; Vibert; (Sangerhausen)

'ROTTKÄPPCHEN', Pol, mr, 1887; flowers medium-large, dbl.; Geschwind, R.; (Sangerhausen)

'ROUGE', Pol, mr, 1934; flowers brilliant scarlet-crimson, open, semi-dbl.; foliage leathery; strong stems; dwarf growth; Verschuren; J&P

Rouge Adam, HT, dr, 1996; Adam, M.; (Cavriglia)

Rouge Admirable, HGal, dp, 1825; Vibert

'ROUGE ANGEVINE', HP, mr, 1907; flowers large, dbl.; Chedane-Pajotin; (Sangerhausen)

Rouge Baiser™, HT, lp; flowers red to reddish orange, exhibition form, greenhouse variety; Delbard

'ROUGE CHAMPION', HT, mr, 1956; flowers bright red, dbl., 40–50 petals; very vigorous growth; [Happiness X Princesse Liliane]; Frères, Buyl

'ROUGE DE PARIS', F, mr, 1958; (Paris Red); flowers light red, dbl., 35–45 petals, 3 in.; foliage purplish; vigorous, bushy growth; [Floradora seedlings X Orange Triumph seedlings]; Delbard-Chabert; Stark Bros., 1964

'ROUGE DE PARME', HT, pb, 1963; flowers mauve-pink, large, dbl., exhibition form; foliage dark, dull; [(Peace X Fred Edmunds) X Buccaneer]; Dorieux; Le Blévenec

Rouge Dorieux®, *see* 'DORMAL'

'ROUGE DOT', HT, mr, 1962; flowers currant-red, dbl., 40 petals, slight fragrance; vigorous growth; [Baccará X Lydia]; Dot, Simon

Rouge et Or, *see* 'DICOR'

Rouge et Or, Climbing, *see* 'DICORSAR'

Rouge Koster, Pol, mr

'ROUGE MALLERIN', HT, mr, 1934; (Henri Mallerin); bud long, pointed; flowers brilliant red, large, dbl., exhibition form, intense, damask fragrance; compact growth; [Mme Van de Voorde X Lady Maureen Stewart]; Mallerin, C.; C-P

Rouge Marbree, B, rb

Rouge Meilland, *see* **'HAPPINESS'**

Rouge Meilland, *see* 'MEIMALYNA'

Rouge Meilland, Climbing, *see* **'HAPPINESS, CLIMBING'**

Rouge Meillandécor®, *see* 'MEINEBLE'

Rouge Pineau, F, 1980; Pineau; (Cavriglia)

Rouge Prolific, *see* 'DELIFIC'

'ROUGEMOSS', F, or, 1972; bud ovoid; flowers medium, dbl., moderate fragrance; foliage leathery; vigorous, dwarf growth; [Rumba X Moss hybrid]; Moore, Ralph S.; Sequoia Nursery

'ROULETII', HCh, mp; (Centifolia Minima, R. rouletii, R. X centifolia minima); flowers rose-pink, dbl., .5 in., recurrent bloom; dwarf growth; Correvon, 1922

Rouletii, Climbing, Cl HCh, mp

Round Robin, Min, lp; [Avandel X Oz Gold]; Hannemann, F., 1990; The Rose Paradise; (Weatherly, L.)

Roundabout, *see* 'ZIPROUND'

'ROUNDELAY', Gr, dr, 1954; bud ovoid; flowers medium-large, dbl., 38 petals, exhibition form, moderate fragrance; foliage dark; very vigorous growth; GM, Geneva, 1954; [Charlotte Armstrong X Floradora]; Swim, H.C.; Armstrong Nursery

Roundelay, Climbing, Cl Gr, dr, 1970

Roussillon®, F, mr

'ROWENA', HWich, dp, 1912; flowers carmine, changing to mauve-pink, small, borne in clusters, dbl.; vigorous, climbing growth; Paul, W.

'ROXANA', HT, yb, 1933; bud large, long, pointed; flowers orange-yellow and copper, open, semi-dbl., moderate fragrance; foliage glossy, light; vigorous, bushy growth; Dickson, A.

Roxana, HT, lp; Urban, J., 1986; (Czech Rosa Club)

Roxane®, *see* LAPdal

Roxburghe Rose, *see* COCember

Roxie, *see* 'RESSOX'

Roxie Baby™, *see* 'TUCKROX'

Roxie's Romance, Min, or

Roy Black, *see* 'POULARI'

Roy Castle Rose, *see* POUlduff

Roy Rumsey, *see* **'FIREFALL'**

'ROYAL AIR FORCE', HT, m, 1969; flowers light lavender-blue, dbl., 35 petals, 6 in., exhibition form, slight fragrance; foliage dark, matt; Laperrière

'ROYAL ALBERT HALL', HT, rb, 1972; flowers wine-red, reverse gold, dbl., 32 petals, 5 in., intense fragrance; foliage dark; [Fragrant Cloud X Postillon]; Cocker

'ROYAL AMERICA', LCl, w, 1994; flowers ivory, blooms borne in small clusters, semi-dbl., 6–14 petals, 1.5–2.75 in., slight fragrance; foliage medium, light green, matt; tall, bushy, spreading growth; [America sport]; Cooper, Curt; Certified Roses, Inc., 1997

Royal Amethyst™, *see* 'DEVMORADA'

Royal Anniversary, *see* 'HARBONNY'

Royal Ascot®, *see* 'DELSAB'

Royal Baby, *see* 'DEBRAD'

Royal Bassino, *see* KORfungo

'ROYAL BATH & WEST', Gr, lp, 1977; flowers pastel pink, dbl., 20 petals, 4 in., intense fragrance; foliage green, matt; [Unnamed seedling X Prima Ballerina]; Sanday, John

'ROYAL BEAUTY', HT, dr, 1940; flowers dark velvety red, dbl., 25 petals, 5 in., exhibition form, moderate, spicy fragrance; foliage bronze; [Better Times sport]; Coddington

Royal Blush, *see* SIEroyal

Royal Bonica™, *see* 'MEIMODAC'

'ROYAL BRIGHT', HT, my, 1976; bud ovoid; dbl., 40 petals, 6 in., exhibition form, slight fragrance; vigorous, upright growth; [Garden Party X (Bronze Masterpiece X Memoriam)]; Kono, Yoshito

'ROYAL CANADIAN', HT, mr, 1968; bud ovoid; flowers scarlet, large, dbl., 35 petals, cupped, moderate fragrance; foliage glossy, leathery; vigorous, upright growth; [Seedling X Talisman]; Morey, Dr Dennison; J&P

Royal Carpet™, *see* 'MINUCO'

'ROYAL CHINOOK', HT, mr, 1939; flowers brilliant rose-red; [Rapture sport]; Chase

Royal City, *see* 'RENITY'

'ROYAL CLUSTER', HMult, lp, 1899; flowers small, single; Conard & Jones; (Sangerhausen)

Royal Dane®, *see* 'POUMIDOR'

'ROYAL DAWN', LCl, mp, 1962; flowers coral-pink, large, dbl., 35 petals, moderate fragrance; foliage dark, glossy; vigorous, climbing growth; [Royal Sunset X Aloha]; Morey, Dr Dennison; J&P

Royal Delight, *see* 'AROreroy'

Royal Dot, HT, pb

Royal Dream, *see* KORmeeram

'ROYAL EDWARD', S, mp, 1994; dbl., 15–25 petals, 1.5 in., borne in large clusters, slight fragrance; few prickles; foliage small, medium green, glossy; low (45 cms), bushy growth; [R. kordesii X [(R. kordesii X (Red Dawn X Suzanne) o.p. X Zeus]]; Ogilvie, Ian S.; Agriculture Canada, 1994

'ROYAL FLARE', HT, rb, 1988; flowers red and white blend, large, borne singly, exhibition form, slight fragrance; prickles downturned, dark green; foliage medium, dark green, disease resistant; upright, medium growth; [Pristine X Standout]; Wambach, Alex A.

'ROYAL FLUSH', LCl, pb, 1970; bud ovoid; flowers cream, edges blending pink, medium, semi-dbl., cupped, repeat bloom, moderate fragrance; foliage dark, leathery; vigorous, upright, climbing growth; [Little Darling X Suspense]; Fuller; Wyant

Royal Flush, *see* PEApatio

Royal Gem, F, lp; Brooks, M.L., 1959; (Weatherly, L.)

'ROYAL GOLD', LCl, my, 1957; flowers golden yellow, blooms borne singly and in clusters of 3–7, dbl., 35 petals, 4 in., cupped, moderate, fruity fragrance; foliage glossy; vigorous, pillar (5–7 ft) growth; [Goldilocks, Climbing X Lydia]; Morey, Dr Dennison; J&P

'ROYAL HIGHNESS', HT, lp, 1962; (Konigliche Hoheit, Königlicht Hoheit); bud long, pointed; flowers soft light pink, dbl., 43 petals, 5–5.5 in., exhibition form, intense fragrance; foliage dark, glossy, leathery; tender, upright, bushy growth; AARS, 1963, David Fuerstenberg Prize, ARS, 1964 GM, Madrid, 1962 GM, Portland, 1960; [Virgo X Peace]; Swim & Weeks; C-P

Royal Lady™, *see* 'MINIRYL'

'ROYAL LAVENDER', LCl, m, 1961; bud ovoid; flowers lavender tinted gray and pink, blooms in clusters, dbl., 35 petals, 3 in., cupped, intense fragrance; foliage leathery; vigorous (6–9 ft) growth; [Lavender Queen X Amy Vanderbilt]; Morey, Dr Dennison; J&P

'ROYAL LUSTRE', HT, op; flowers orange-buff overlaid salmon, dbl., 45 petals, 6–7 in., slight fragrance; foliage coppery; very free growth; [Mrs Sam McGredy X Crimson Glory]; McGredy

Royal Mail, *see* 'ANDROI'

'ROYAL MARBRÉE', HGal, m; flowers lilac and purple, marbled (striped) pink, medium, very dbl.; Moreau-Robert (?), before 1837

Royal Meillandina, *see* 'SCHOBITET'

Royal Midinette, *see* MORposa

'ROYAL OCCASION', F, or, 1974; (Montana®); bud long, pointed; flowers luminous orange-scarlet, dbl., 20 petals, 3 in., cupped, slight fragrance; foliage glossy, large, leathery; upright, compact, strong (3 ft) growth; ADR, 1974; [Walzertraum X Europeana]; Tantau, Math.

'ROYAL PERFECTION', HT, or, 1964; flowers orange-coral, full, dbl., 20 petals, 4.5 in., slight fragrance; vigorous growth; [Rome Glory X Bayadere]; Delbard; Laxton & Bunyard Nursery

Royal Philarmonic, HT, 1997; Harkness, R.; (Cavriglia)

Royal Philharmonic, *see* HARdeed

Royal Porcelain, *see* 'LEEPEL'

'ROYAL PRINCESS', Min, pb, 1981; bud ovoid; flowers pink and yellow blend, blooms borne 1–3 per cluster, dbl., 65 petals, moderate, fruity fragrance; hooked, bronze prickles; foliage dark, leathery; upright, bushy growth; [Unnamed seedling X Unnamed seedling]; Lyon

'ROYAL QUEEN', Gr, w, 1965; flowers greenish white; [Queen Elizabeth sport]; Verschuren

'ROYAL RED', HT, mr, 1924; bud pointed; flowers intense crimson-scarlet, very large, dbl., moderate fragrance; foliage leathery; vigorous growth; Hill, E.G., Co.

'ROYAL RED', HT, dr, 1956; flowers deep scarlet, well formed; moderate growth; [Happiness sport]; Lowe

'ROYAL ROBE', S, mr, 1946; flowers crimson, almost purple, rather large, semi-dbl., nonrecurrent, vigorous growth; [(R. rugosa X Hybrid Perpetual) X (R. multiflora X R. blanda)]; Wright, Percy H.

Royal Romance, *see* 'RULIS'

'ROYAL ROMANCE', HT, my, 1974; flowers bright lemon-yellow, full, dbl., 30 petals, 5 in., moderate fragrance; foliage leathery; very free growth; [Pink Parfait X Unknown seedling]; Fryer, Gareth; Fryer's Nursery, Ltd.

'ROYAL ROSE', HT, mp, 1999; flowers rose pink, very dbl., 80 petals, exhibition form, slight fragrance; 4.5 ft growth; [Happiness X Peter Frankenfeld]; Teranishi, K.; Itami Rose Nursery, 1987

'ROYAL RUBY', Min, mr, 1972; flowers red, base white, small, dbl., globular, slight fragrance; foliage small, dark, leathery; vigorous, upright growth; [Garnette X (Tom Thumb X Ruby Jewel)]; Morey, Dr Dennison; Pixie Treasures Min. Roses

Royal Salute, *see* 'MACROS'

'ROYAL SCARLET', HT, mr, 1966; bud ovoid; flowers scarlet-red, large, dbl., exhibition form, moderate fragrance; foliage glossy; vigorous, bushy growth; [McGredy's Scarlet X Christian Dior]; Kraus; Wyant

'ROYAL SCARLET HYBRID', LCl, mr, 1926; Chaplin Bros.

'ROYAL SCOT', HT, yb, 1928; flowers golden yellow, edged crimson, open, semi-dbl., moderate fragrance; foliage dark; vigorous growth; Dobbie

'ROYAL SHOW', HT, mr, 1973; flowers pointed, dbl., 27 petals, 4 in., slight fragrance; foliage glossy; vigorous growth; [Queen Elizabeth X ?]; Gregory

Royal Show, *see* MEIrasimac

Royal Smile, HT, w, 1980; Beales, Peter

Royal Success™, *see* 'AROREROY'

Royal Sunblaze®, *see* 'SCHOBITET'

'ROYAL SUNSET', LCl, ab, 1960; bud ovoid; dbl., 20 petals, 4.5–5 in., cupped, moderate, fruity fragrance; foliage leathery; vigorous (6 ft) growth; GM, Portland, 1960; [Sungold X Sutter's Gold]; Morey, Dr Dennison; J&P

'ROYAL TAN', HT, m, 1955; flowers pale purple feathered violet and chocolate, high-pointed, 5 in., slight fragrance; foliage dark; vigorous growth; [Charles P. Kilham X Mrs Sam McGredy]; McGredy, Sam IV

Royal Touch, *see* 'ANDROY'

Royal Velvet™, *see* 'MEILOTUP'

'ROYAL VELVET', HT, dr, 1959; flowers rich, velvety cardinal-red, blooms, dbl., 55–65 petals, 4–5 in.; vigorous, tall growth; RULED EXTINCT 3/86 ARM.; [(Happiness X Independence) X (Happiness X Floradora)]; Meilland, F.; C-P

Royal Victoria, *see* 'LAVICTOR'

'ROYAL VISIT', HT, op, 1939; flowers deep tangerine-orange, reverse coral passing to apricot, dbl., slight fragrance; foliage leathery, glossy, dark; bushy growth; [Picture X Mrs Pierre S. duPont]; Eddie

Royal Volunteer, *see* 'COCDANDY'

Royal Wedding™, *see* 'JACQUINT'

'ROYAL WELCOME', HT, rb, 1955; flowers velvety dark red, reverse deep rosy pink, well formed, intense fragrance; foliage glossy; vigorous growth; [Crimson Glory X Peace]; Homan; G.A. Williams

Royal William, *see* 'KORZAUN'

'ROYAL WILLIAM, CLIMBING', Cl HT, dr, 1995; flowers other growth similar to parent; [Royal William sport]; Newman, F. S.; Abbey Rose Gardens

Royal Worcester®, *see* 'TROBROY'

Royale, *see* **'BELLE ÉPOQUE'**

Royale Perfection, HT, or, 1964; flowers large, dbl., moderate fragrance; Delbard-Chabert; (Sangerhausen)

'ROYALET', HT, mr, 1968; flowers crimson-red, dbl., 35–40 petals, 4.5–5.5 in., slight fragrance; foliage dark; upright, bushy growth; [Unnamed seedling X Chrysler Imperial]; Herholdt, J.A.

Royalglo™, *see* 'MINADCO'

'ROYALIST', HT, mp, 1954; flowers deep rose-pink, large, dbl., 28 petals, exhibition form, intense fragrance; foliage dull green; very free growth; [(Billy Boy X Blossom) X Mrs Redford]; McGredy, Sam IV

'ROYALTY', HT, dr, 1976; bud globular, pointed; flowers deep red, dbl., 20 petals, 4–4.5 in., slight fragrance; foliage leathery; vigorous growth; PP004057; [Forever Yours X Love Affair]; Jelly; E.G. Hill Co.

'ROYDEN', F, my, 1989; flowers bright gold, fading quickly to white, large, borne in sprays, dbl., 40 petals, cupped, intense fragrance; foliage light green, glossy; tall, upright growth; [Liverpool Echo X Arthur Bell]; Cattermole, R.F.; South Pacific Rose Nursery

'ROYDON HALL', F, mr, 1983; flowers large, dbl., 35 petals, no fragrance; foliage medium, medium green, semi-glossy; bushy growth; [City of Leeds X (Paprika X Rose Gaujard)]; Scrivens, Len

'ROZE KONINGIN', HT, mr, 1938; bud long, pointed; flowers medium red, passing to pink, well formed large, dbl., moderate fragrance; vigorous growth; [Lady Sylvia seedling X Étoile de Hollande]; Lens

'ROZENAMATEUR A. BOK', F, mr, 1961; flowers bright red, borne in large clusters, single; foliage dark; vigorous, upright growth; [Prinses Christina X Kathe Duvigneau]; Buisman, G. A. H.

Rozorina, *see* **'ZORINA'**

Roztomila, F, lp; Urban, J., 1988; (Czech Rosa Club)

Ruatara, LCl, dp, 1995; Nobbs

'RUBAIYAT', HT, dp, 1946; bud long, pointed; flowers rose-red, reverse lighter, dbl., 25 petals, 4.5–5 in., exhibition form, intense fragrance; foliage dark, leathery; vigorous, upright growth; AARS, 1947, GM, Portland, 1945; [(McGredy's Scarlet X Mrs Sam McGredy) X (Seedling X Sir Basil McFarland)]; McGredy; J&P

'RUBAN ROUGE', F, dr, 1957; bud globular; flowers deep velvety red, borne in clusters; foliage dark, leathery; vigorous growth; [Alain X Cinnabar]; Lens

'RUBELLA', F, dr, 1972; flowers deep red, large, dbl., 24 petals; bushy growth; [Kimono X Lilli Marleen]; deRuiter

'RUBENS', T, w, 1859; flowers white, shaded with rose, center bronzy yellow, large, dbl., cupped; Robert

'RUBENS', HT, mr, 1972; bud pointed; flowers vermilion red, large, dbl.; vigorous growth; [Rose Gaujard X Miss France]; Gaujard

'RÜBEZAHL', HT, m, 1917; flowers large, dbl., slight fragrance; Krüger; (Sangerhausen)

Rubies 'n' Pearl, *see* 'MANRUPEARL'

'RUBIN', HMult, mr, 1899; flowers crimson to spinel-red, medium, borne in clusters, semi-dbl., moderate fragrance; height 10–12 ft; [Daniel Lacombe X Fellemberg]; Schmidt, J.C.

'RUBIN', HT, or, 1956; flowers orange-red, base yellow, large, semi-dbl., slight fragrance; vigorous growth; [Mme Henri Guillot X Grande Duchesse Charlotte]; Lens

'RUBIN', F, dr, 1962; bud long, pointed; flowers deep red, borne in broad clusters, dbl.; foliage dark; bushy, compact, low growth; Kordes; Dehner & Co.

Rubina®, F, dr, 1989; Dickson, Patrick

'RUBINETTE', F, dr, 1971; flowers deep red, dbl., 28 petals, 3.5 in., slight fragrance; foliage dark; vigorous, upright growth; [(Mandrina X Baccará) X (Mandrina X Baccara)]; deRuiter; Carlton Rose Nurseries

Rubino, S, 1966; Mansuino; (Cavriglia)

'RUBIS', Pol, rb, 1926; flowers bright ruby-red, center white, medium, borne in clusters; dwarf growth; [Merveille des Rouges X Jessie]; Nonin

'RUBIS', Cl HT, mr, 1948; flowers blood-red, well formed, large, intense fragrance; foliage bronze; vigorous growth; [Unnamed climber X Mme G. Forest-Colcombet]; Mallerin, C.; URS

'RUBOR', HT, mp, 1947; flowers neyron pink, dbl., moderate fragrance; foliage dark; very vigorous growth; [Cynthia X Director Rubió]; Dot, Pedro

RUbrispa, Min, ob; (**Bright Spark**); Ru?

Rubrotincta, *see* **'HEBE'S LIP'**

Rubus coronarius, Rubus rosaefolius coronarius, *see* **'EASTER ROSE'**

'RUBY', Pol, mr, 1932; flowers glowing scarlet; deRuiter; Sliedrecht & Co.

Ruby Anniversary, *see* 'HARBONNY'

Ruby Celebration, *see* PEAwinner

'RUBY DEE', Gr, mr, 1968; bud ovoid; flowers large blooms in clusters, dbl., exhibition form, moderate fragrance; foliage dark, glossy; vigorous, upright growth; [Queen Elizabeth X Happiness]; Patterson; Patterson Roses

'RUBY GEM', HT, mr, 1962; flowers bright red, large, dbl., 28 petals, slight fragrance; [Seedling X Red Favorite]; Leenders, J.

'RUBY GLOW', LCl, mr, 1955; bud short, oval; flowers spectrum-red, open, borne in clusters, semi-dbl., 13–15 petals, 3 in., moderate fragrance; foliage glossy; moderate pillar (7 ft) growth; [Dream Girl X New World]; Jacobus; B&A

'RUBY JEWEL', Min, mr, 1958; bud ovoid; flowers ruby-red, reverse lighter, open, dbl., 35–40 petals, .5 in., moderate fragrance; foliage glossy; low (6–8 in.), compact growth; [Dick Koster sport X Tom Thumb]; Morey, Dr Dennison; J&P

'RUBY LIPS', F, mr, 1958; bud ovoid, pointed; flowers bright cardinal-red, loose form, blooms in clusters, semi-dbl., 18 petals, 3 in., slight fragrance; foliage semi-glossy; vigorous, semi-spreading, bushy growth; [World's Fair X Pinocchio]; Swim, H.C.; Armstrong Nursery

Ruby Magic, *see* 'MORUBY'

'RUBY MANWARING', HT, dp, 1932; flowers rich rosy cerise; [Betty Uprichard sport]; Longley

'RUBY PENDANT', Min, m, 1979; bud pointed; flowers red-purple, borne singly, dbl., 28 petals, exhibition form, slight fragrance; needle-shaped prickles; foliage reddish-green; vigorous growth; [(Lotte Gunthart X Salvo) X Baby Betsy McCall]; Strawn, Leslie E.; Pixie Treasures Min. Roses

'RUBY PRINCESS', HT, mr, 1949; flowers velvety red, dbl., 50 petals, 5 in., moderate fragrance; [Jewel sport]; Grillo

'RUBY QUEEN', HWich, dp, 1899; flowers deep rose-pink, reverse lighter, small, borne in clusters of, dbl., non-recurrent, slight fragrance; height 10 ft; [R. wichurana X Queen's Scarlet]; Van Fleet; Conard & Jones

Ruby Rain, Cl Min, mr

'RUBY RING', HWich, rb, 1926; flowers white with ruby edge, borne in sprays, dbl.; Clark, A.; (Sangerhausen)

Ruby Star, *see* 'UMSTAR'

'RUBY SUPERIOR', Pol, mr; flowers have more lasting color; [Ruby sport]; deRuiter

'RUBY TALISMAN', HT, mr, 1935; flowers rich ruby-red, more shapely than parent, with reflexed petal; [Talisman sport]; Eddie

Ruby Treasure, Min, lp; [Oz Gold X Lemon Delight]; Hannemann, F., 1990; The Rose Paradise; (Weatherly, L.)

Ruby Tuesday, *see* 'RENRUBY'

'RUBY ULRICK', F, dp, 1953; flowers deep pink, base white, borne in clusters, dbl.; foliage leathery; very vigorous growth; [Mrs Tom Henderson X Gloria Mundi]; Ulrick, L.W.

Ruby Velvet, *see* 'HINRUBY'

'RUBY WEDDING', HT, dr, 1979; flowers deep red, dbl., 44 petals, 3–3.5 in.; vigorous, spreading growth; [Mayflower X ?]; Gregory

'RUDELSBURG', HMult, dp, 1919; flowers shining carmine-rose, blooms in clusters; no prickles; vigorous, climbing growth; Kiese

'RUDI KORTE', Pol, or, 1929; flowers medium, semi-dbl.; Kersbergen; (Sangerhausen)

'RUDOLA', F, mr, 1973; bud ovoid; flowers light geranium-red, open, large, semi-dbl., slight fragrance; foliage leathery; very vigorous, bushy growth; [Dacapo X Kimono]; deRuiter

'RUDOLF ALEXANDER SCHRÖDER', HT, w, 1930; bud long, pointed; flowers white, center tinted lemon, large, dbl., exhibition form, intense fragrance; foliage leathery, light; vigorous growth; [Mrs Herbert Stevens X Pius XI]; Kordes

'RUDOLF SCHMIDTS JUBILÄUMSROSE', F, my, 1955; flowers golden yellow, large, borne in clusters, semi-dbl., exhibition form, moderate fragrance; foliage light green, glossy; very vigorous, upright, bushy growth; Kordes; R. Schmidt

'RUDOLF SCHOCK', HT, mr, 1968; flowers bright currant-red, large, dbl., moderate fragrance; foliage dark, leathery; vigorous growth; [Josephine Bruce X Seedling]; Verschuren, A.; Stassen

'RUDOLF VON BENNIGSEN', LCl, mp, 1932; bud pointed; flowers rosy-pink, edges fading, large, borne in loose trusses of 5-, recurrent bloom, foliage broad, dark, glossy; vigorous, bushy, semi-climbing growth; [(Geheimrat Dr Mittweg X Souv. de Paul Neyron) X Mrs Joseph Hill]; Lambert, P.

'RUDOLPH KLUIS', Pol, mr, 1922; bud globular; flowers pure vermilion-red, dbl., moderate fragrance; foliage rich green, glossy; bushy growth; [Ellen Poulsen sport]; Kluis & Koning

'RUDOLPH KLUIS SUPERIOR', Pol, mr, 1928; flowers glowing scarlet; more compact growth than Rudolph Kluis.; [Ellen Poulsen sport]; Kluis

'RUDOLPH TIMM', F, rb, 1951; flowers white, reverse red, open, small, borne in trusses to 40, semi-dbl., 15–20 petals, 2 in., moderate fragrance; foliage glossy, light green; very free growth; (28); [(Johannes Boettner X Magnifica) X (Baby Chateau X Else Poulsen)]; Kordes; Wheatcroft Bros.

'RUDOLPH VALENTINO', HT, op, 1929; flowers lively shrimp-pink or coral-red, suffused golden coppery, dbl., moderate fragrance; vigorous growth; Pernet-Ducher; Dreer

'RUFFLES', HT, mp, 1994; dbl., 26–40 petals, 1.5–2.75 in., borne mostly singly, moderate fragrance; some prickles; foliage large, medium green, tall growth; [Seedling X Alec's Red]; Perry, Astor; Certified Roses, Inc., 1997

Ruffles 'n' Flourishes, *see* 'CLERUFF'

'RUFIN', Min, op, 1982; (Finnstar®, **Finstar**, Mini Metro); flowers orange-salmon, small, dbl., 20 petals, slight fragrance; foliage small, medium green, semi-glossy; bushy growth; [Minuette X Seedling]; deRuiter, George; Fryer's Nursery, Ltd., 1980

'RUFUS', Pol, mr, 1925; flowers intense scarlet, borne in clusters, dbl.; compact growth; [Orléans Rose sport]; Allen

'RUGA', Ayr, w; (R. X ruga); flowers flesh changing to creamy white, large blooms in several-flow, dbl., cupped; trailing growth; (14); [Thought to be R. arvensis X China or Tea]

Rugelda®, *see* KORruge

'RUGGELDA', HT, dy, 1982; (**Golden Days**, 'RUGOLDA'); flowers large, dbl., 35 petals, slight fragrance; foliage large, medium green, semi-glossy; bushy growth; [Peer Gynt X Seedling]; deRuiter, George; Fryer's Nursery, Ltd., 1980

'RUGOLDA', HT, dy, 1982; (**Golden Days**, 'RUGGELDA'); flowers large, dbl., 35 petals, slight fragrance; foliage large, medium green, semi-glossy; bushy growth; [Peer Gynt X Seedling]; deRuiter, George; Fryer's Nursery, Ltd., 1980

'RUGOSA COPPER', S, ob, 1955; flowers coppery orange, large, recurrent bloom; vigorous growth; [Conrad Ferdinand Meyer X Unnamed seedling]; Gaujard

'RUGOSA MAGNIFICA', HRg, m, 1905; flowers reddish lavender, gold stamens, dbl., moderate fragrance; orange-red fruit; vigorous, spreading growth; very hardy; Van Fleet

Rugosa Ottawa, HRg, lp; flat, 5 petals, intense; L'Assomption, 1969

Rugosa Rose, *see* **R. RUGOSA**

'RUGOTIDA', HRg, 1950; Darthuis; (Cavriglia)

Rugspin, HRg, dr, 1966; Petersen

'RUGUL', Min, my, 1973; (Gulletta, **Tapis Jaune®**); flowers small, dbl., 20 petals, 1.5 in.; foliage small, glossy, dark; low, compact growth; [Rosy Jewel X Allgold]; deRuiter

'RÜHKOR', HT, w, 1984; (Athena); flowers large, dbl., 35 petals, moderate fragrance; foliage large, medium green, matt; upright, bushy growth; [Unnamed seedling X Helmut Schmidt]; Kordes, W., 1981

'RUHM DER GARTENWELT', HP, dr, 1904; flowers large, dbl.; [American Beauty X Fracis Dubreuil]; Jacobs

'RUHM VON STEINFURTH', HP, mr, 1920; (Red Druschki); bud long, pointed; flowers large, dbl., 34 petals, cupped, intense fragrance; foliage dark, leathery; vigorous growth; [Frau Karl Druschki X Ulrich Brunner Fils]; Weigand, C.; H. Schultheis

'RUIBLUN', Min, m, 1983; (Azulabria, **Blue Peter**, Bluenette); flowers lilac-purple, patio, small, semi-dbl., slight fragrance; foliage small, light green, semi-glossy; bushy growth; [Little Flirt X Seedling]; deRuiter, George; Fryer's Nursery, Ltd.

RUIcharo, Min, ob; (**Charming Rosamini**); deRuiter

'RUIco', F, ob, 1968; (**Fresco®**); flowers orange, reverse golden yellow, well-formed, 3 in., slight fragrance; foliage dark, glossy; vigorous, bushy growth; [Metropole X Orange Sensation]; deRuiter

RUIdriko, HT, w; (**Vivaldi™**); deRuiter, 1988

RUIfarol, Min, mp, 1989; (**Favourite Rosamini**); deRuiter

RUIflami, Min, mr, 1987; (**Flaming Rosamini**); deRuiter

'RUIGERDAN', Min, or, 1988; (**Little Marvel**); flowers bright orange-scarlet, small, semi-dbl., 6–14 petals, slight fragrance; foliage small, medium green, semi-glossy; bushy growth; PP007572; [Unnamed seedling X Unnamed seedling]; deRuiter, George; Fryer's Nursery, Ltd., 1987

'RUILANCA', F, or, 1984; (**Lancashire Life**); flowers scarlet, blooms in clusters, semi-dbl., slight fragrance; foliage medium, medium green, semi-glossy; bushy growth; [Robert Stolz X Diablotin]; deRuiter, George; Fryer's Nursery, Ltd.

'RUIMEVA', HT, or, 1983; (**Madelon**); flowers medium, dbl., 20 petals, slight fragrance; foliage medium, medium green, semi-glossy; upright growth; PP005820; [Varlon X MEIgenon]; deRuiter, George; Fryer's Nursery, Ltd., 1981

'RUIMIRED', Min, dr, 1991; (**Red Minimo™**); flowers moderately small, dbl., 15–25 petals, borne in small clusters, no fragrance; foliage small, dark green, semi-glossy; low, bushy growth; PP005770; [Unnamed seedling X Unnamed seedling]; deRuiter, George; Bear Creek Gardens, 1987

'RUIREDRO', Min, dr, 1988; (**Red Rosamini®**); flowers clear crimson, white at base, dbl., 30–40 petals, 1–1.5 in., cupped, floriferous, borne in small clusters, slight fragrance; few prickles; foliage medium, medium green, matt; bushy, medium growth, 15 in.; PP005976; [Unnamed seedling X Unnamed seedling]; deRuiter, George; Conard-Pyle Co., 1987

RUIrupo, Min, mr, 1989; (**Scarlet Rosamini**); deRuiter

RUIsalro, Min, op; (**Salmon Rosamini**); deRuiter

'RUISETO', Min, or, 1988; (**Orange Rosamini®**); semi-dbl.; medium growth; PP006236; [Unnamed seedling X Red Rosamini]; deRuiter, George; DeRuiters Nieuwe Rozen B.V.

RUIsteenka, HT, m, 1994; (**Ravel**); deRuiter

RUItulvi, HT, pb, 1993; (**Pavarotti**); deRuiter

RUIxandra, Min, rb, 1992; (**Jolly Cupido**); deRuiter

RUIzesac, HT, pb, 1989; (**Astra**); deRuiter

'RUKHSAAR', HT, w, 1969; flowers cream, center shell-pink, medium, dbl., exhibition form, slight fragrance; foliage glossy; moderate growth; [Virgo X Open pollination]; Singh; Gopalsinamiengar

'RULIMPA', HT, rb, 1980; (Impala, **Melinda**); flowers scarlet, gold reverse, large, dbl., 35 petals, moderate fragrance; foliage medium, dark, glossy; upright growth; [Whisky Mac X Criterion]; deRuiter, George; Fryer's Nursery, Ltd.

'RULIS', HT, ob, 1982; (**Liselle**, Royal Romance); flowers orange-peach, well-formed, large, dbl., 35 petals, moderate fragrance; foliage large, dark, semi-glossy; bushy growth; [Whisky Mac X Matador]; deRuiter, George, 1980; Fryer's Nursery, Ltd., 1981

Rum Butter, *see* 'TALRUM'

Rum Candy, *see* 'UMSRUM'

'RUMBA'®, F, rb, 1958; bud ovoid; flowers poppy-red, center yellow, blooms in clusters, dbl., 35 petals, 2–2.5 in., cupped, slight, spicy fragrance; foliage dark, glossy, leathery; vigorous, bushy growth; [Masquerade X (Poulsen's Bedder X Floradora)]; Poulsen, S.; C-P, 1962; McGredy, 1960

'RUMBA', HT, op, 1956; bud long, pointed; flowers salmon-orange, intense fragrance; long stems; [Talisman X ?]; Faassen-Houba

'RUMBA, CLIMBING', Cl F, rb, 1972; Bansal, O.P.; Bansal Roses

'RUMPELSTILZCHEN', S, dr, 1956; (Repelsteeltje); flowers deep red, small, single; dwarf growth; deRuiter; Willicher Baumschulen

Running Maid®, *see* 'LENRAMP'

'RUNTRU', F, dr, 1983; (Fennica, **Invincible**); bud large; flowers large, dbl., 20 petals, slight fragrance; foliage large, medium green, glossy; upright growth; [Rubella X National Trust]; deRuiter; Fryer's Nursery, Ltd., 1982

'RUPERT BROOKE', HT, pb, 1928; flowers fawn-pink to cream, dbl., moderate fragrance; [Miss Cynthia Forde X Mrs Wemyss Quin]; Easlea

'RUPFESPIN', S, mp, 1991; (**Festival Pink™**); bud pointed; flowers clear medium pink, white petal base, golden stamens, single, 5–7 petals, 1.5–2.75 in., borne in small and large clusters, slight fragrance; many prickles; foliage medium, bright green, glossy; tall (120–150 cms), bushy,spreading growth; [Festival Fanfare sport]; Rupert, Kim L., 1992

'RUPGOLJUL', HT, my, 1992; (**Golden Julia™**); bud long, pointed; flowers medium golden mustard, opening medium golden yellow, fading, dbl., 18–22 petals, 3–4 in., flat, slight fragrance; few, yellow prickles; foliage medium, medium green, matt; green wood; medium (75–90 cms), upright, bushy growth; [Julia's Rose sport]; Rupert, Kim L.

'RUPKIMCRM', F, w, 1991; (**Kim's Cream**™); bud ovoid, pointed; flowers cream with coffee and gold tints, opening cream, tint remains in cool weather, dbl., 26–40 petals, 3–3.5 in., flat, intense, spicy fragrance; some prickles; foliage leathery, medium, medium green, semi-glossy; low (45–50 cms), bushy, compact, vigorous growth; [Lavender Pinocchio X Lavender Pinocchio]; Rupert, Kim L., 1992

'RUPRINS', Gr, or, 1979; (Prins Claus, **Rosalynn Carter**); flowers coral-red with orange tones, dbl., 30 petals, 3.5–4 in., exhibition form, moderate, spicy fragrance; tall, vigorous, bushy, upright growth; [Unnamed seedling X Scania]; deRuiter; C-P, 1978

'RURAL RHYTHM', S, lp, 1984; dbl., 30 petals, 4 in., cupped, borne 1–5 per cluster, repeat bloom, moderate, myrrh fragrance; awl-like, tan prickles; foliage dark, leathery; erect, bushy, spreading growth; hardy; [Carefree Beauty X The Yeoman]; Buck, Dr Griffith J.; Iowa State University

'RURITANIA', HT, or, 1972; bud ovoid; flowers full, medium, very dbl., intense fragrance; foliage dark, leathery; vigorous, upright growth; [Miss Hillcrest X Hawaii]; Curtis, E.C.; Kimbrew

'RUSALKA', LCl, dp, 1934; flowers carmine to sunset-rose, base yellow, very large, semi-dbl., exhibition form, profuse, repeated bloom; foliage glossy, light; climbing growth; [Tausendschön X Farbenkonigin]; Brada, Dr.; Böhm

Rush®, *see* 'LENMOBRI'

Rushing Stream, *see* 'AUSTREAM'

'RUSKIN', HRg, dr, 1928; (John Ruskin); bud ovoid; flowers deep crimson, large, dbl., 50 petals, cupped, sparingly recurrent, intense fragrance; foliage large, rich green, leathery; vigorous, bushy (4–5 ft) growth; (21); [Souv. de Pierre Leperdrieux X Victor Hugo]; Van Fleet; American Rose Society

'RUSSELL SUPREME', HT, lp, 1927; [Mrs Charles E. Russell sport]; Pacific Rose Co.

'RUSSELLIANA', HMult, m; (Old Spanish Rose, Russell's Cottage Rose Scarlet Grevillea, Souv de la Bataille de Marengo); flowers magenta-crimson fading to mauve, blooms in clusters, dbl., flat, moderate fragrance; foliage coarse; height to 20 ft

Russell's Cottage Rose Scarlet Grevillea, *see* **'RUSSELLIANA'**

Russet Beauty, *see* 'POULRUSSET'

Rustica®, *see* 'MEIVILANIC'

'RUSTICA', HFt, yb, 1929; flowers straw-yellow and gold, center apricot, reverse citron-yellow, semi-dbl., non-recurrent bloom, moderate fragrance; (28); RULED EXTINCT 4/81; [Mme Edouard Herriot X Harison's Yellow]; Barbier

Rustica 91, *see* 'MORYELRUG'

Rusticana, *see* 'MEILÉNA'

Rusticana, F, lp; short; flowers pale pink, flat, 15 petals, borne in clusters, slight; medium large, dark green, glossy; ground cover type; [Candy Rose X Yesterday]; St Wagner, 1994; Res. Stn. f. Fruit Growing, Cluj, 1995, Romania

Rusticana, Climbing, *see* 'MEILÉNASAR'

Ruston's Blush, LCl, lp; [R. gigantea hybrid sport]; Morley, Dr B., 1985; (Weatherly, L.)

'RUTH', HT, ob, 1921; flowers orange flushed carmine, large, dbl., exhibition form, moderate fragrance; vigorous, bushy growth; Pemberton

'RUTH', HGal, mr, 1947; flowers less bright than alika, more double than alika, dbl., non-recurrent; upright growth (to 7 ft); (21); [Mary L. Evans X Alika]; Wright, Percy H.

'RUTH ALEXANDER', LCl, ob, 1937; bud long, pointed; flowers orange, base yellow, large, semi-dbl., exhibition form, moderate fragrance; foliage leathery, glossy, bronze; vigorous, climbing growth; [Myra X Constance Casson]; Wilber; Bertsch

Ruth Harker, *see* 'HARPOOH'

'RUTH HEWITT', F, w, 1963; dbl., 36 petals, 4 in., cupped, borne in open clusters, slight fragrance; foliage dark; vigorous, compact growth; [Unnamed seedling X Queen Elizabeth]; Norman; Harkness

Ruth Knopf, LCl, mp

Ruth Knopf's Pink Poly, Pol, lp

'RUTH LEUWERIK', F, mr, 1961; flowers bright red, blooms in clusters, dbl., 30 petals, 3 in., moderate fragrance; foliage bronze; vigorous, bushy growth; [Kathe Duvigneau X Rosemary Rose]; deRuiter; Gandy Roses, Ltd.

Ruth Pennington, *see* KORlichtung

'RUTH SHAMBURGER', F, lp, 1934; [Kirsten Poulsen sport]; Shamburger, C.S.

Ruth Staley, *see* 'TINSTALEY'

'RUTH TURNER', Pol, dp, 1941; flowers rose-pink, small, borne in clusters of 5–10, dbl., exhibition form, moderate fragrance; foliage glossy, pointed; vigorous, bushy growth; [Étoile Luisante X Sierra Snowstorm]; Moore, Ralph S.; California Roses

'RUTH VESTAL', Cl T, w, 1908; (Climbing Bride); flowers snow-white, well formed, very large, recurrent bloom, moderate fragrance; vigorous growth; [The Bride climbing sport]; Vestal

'RUTH WARNER', F, m, 1978; flowers deep lavender, medium, borne singly and in clusters up to 5, dbl., 48 petals, exhibition form, intense fragrance; very few prickles; foliage dark, smooth; upright growth; Bronze, ARC TG, 1983; [Little Darling X Angel Face]; Warner, A.J.

Ruth Westwood's Single Pink, Pol, lp

Ruth Woodward, *see* 'DICPLEASANT'

'RUTH, CLIMBING', Cl HT, ob; flowers light orange shaded to red at edges, very large, very dbl., recurrent bloom, intense fragrance; foliage very large, glossy, olive-green; vigorous, climbing growth; not hardy

'RUTHCHEN', Pol, mr, 1937; flowers carmine-red, medium, dbl.; Vogel, M.; (Sangerhausen)

Ruthe, *see* 'RESRUTH'

Ruthie, *see* 'TINRUTH'

Ruth's Bright Pink Eglantine, HEg, dp

Ruth's German Rose, HCh, m

Ruth's Pink Musk, HMsk, lp

Ruth's Steeple Rose, HCh, lp

Ruth's Wagon Wheel Bright Pink, HT, mp

'RUTILANT', HT, mr, 1960; bud ovoid; flowers carthamus-red, open, medium, semi-dbl., slight fragrance; foliage glossy; vigorous, bushy growth; [Gloire de Cibeins X Independence]; Arles; Roses-France

Rutland, *see* POUlshine

'RUTRULO', HT, w, 1979; (True Love, **Yorkshire Bank**); dbl., 36 petals, borne 3–5 per cluster, moderate fragrance; pointed, brown prickles; foliage bright green; vigorous, medium, bushy growth; GM, Geneva, 1979, Gold Star of the South Pacific, Palmerston North, NZ, 1979; [Pascali X Peer Gynt]; deRuiter, George; Fryer's Nursery, Ltd.

Ruyton, HT, mp, 1989; (Ruyton Girl's School); Bell, R.

Ruzova Lavina, F, lp; Strnad, 1976; (Czech Rosa Club)

'RUZYNE', HT, rb, 1936; flowers carmine-red marked rose, large, moderate fragrance; Mikes Böhm, J.

'RYOKKOH', F, ly, 1999; (Chidori); flowers light yellow green turning to green, dbl., 15–20 petals, 1–2 in., no fragrance; foliage dark green, leathery; 3–4 ft growth; GM, Baden-Baden, 1989; [Bridal Pink X (seedling X Tasogare)]; Suzuki, Seizo, 1986; Keisei Rose Nurseries, 1991

Ryokukou, F, ly

S

S. Antonio di Padova, HT, 1960; Sgaravatti, A.; (Cavriglia)

'S.&M. Perrier', HT, pb, 1936; flowers very pale rose-pink, with orange glow, large, dbl., 60 petals, exhibition form, slight fragrance; foliage glossy; vigorous growth; [Magdalena de Nubiola X Pres. Cherioux]; Mallerin, C.; C-P

'S.A. Prince Youssof Kamal', HP, rb, 1922; bud long, pointed; flowers crimson, streaked brilliant scarlet, large, semi-dbl., moderate fragrance; [Souv. de Mme Chedane-Guinoisseau X Ulrich Brunner Fils]; Nabonnand, P.

S.H. Freedom, *see* 'TWOfree'

S.M. Alexander I, *see* **'Roi Alexandre'**

'S.M. Gustave V', HP, dr, 1922; bud long, pointed, brilliant crimson; flowers carmine, reflexes crimson, well formed, very large, dbl., 30 petals, cupped, moderate fragrance; vigorous growth; [Frau Karl Druschki X Avoca]; Nabonnand, P.

'S.S. Pennock', HT, pb, 1922; flowers light rose-pink, with sulfur-yellow sheen (under glass, clea, dbl., moderate fragrance); [Lieutenant Chaure X Mrs George Shawyer]; Kordes

'S.S. Pennock, Climbing', Cl HT, pb, 1932; Lens; J&P

S.W.A.L.K., *see* 'SEAwalk'

'S'Aagaró', HT, mr, 1959; bud long, pointed; flowers geranium-red, large, dbl., 30 petals, exhibition form, slight fragrance; foliage dark, glossy; compact growth; [Angels Mateu X (Radar X Grand'mere Jenny)]; Dot, Simon

'Saarbrücken', S, mr, 1959; flowers scarlet-red, blooms in clusters (up to 20), semi-dbl., 3 in.; foliage dark; vigorous, bushy (5 ft) growth; Kordes

Saarlandwelle, HT, or

'Sabaudia', HT, pb, 1934; flowers pink with gold reflections, base light chrome-yellow, large, dbl., cupped, moderate fragrance; foliage leathery, glossy, dark; vigorous, bushy growth; [R. foetida hybrid X Harison's Yellow]; Cazzaniga, F. G.

'SABbelief', Min, rb, 1996; (**Artful Dodger**); flowers cherry red, cream reverse, dbl., 15–25 petals, 1 in., borne in large clusters, slight fragrance; some prickles; foliage medium, dark green, glossy; tall (45 cms), upright, spreading, bushy growth; [Sheri Anne X Richard Buckley]; Burrows, Steven; Burrows Roses, 1997

'SABbyron', Min, w, 1996; (**Oliver Twist**); flowers creamy white, blooms borne in large clusters, dbl., 15–25 petals, 1 in., slight fragrance; some prickles; foliage medium, dark green, glossy; tall (45–60 cms), upright, spreading, bushy growth; [Sheri Anne X Richard Buckley]; Burrows, Steven; Burrows Roses, 1996

SABchurchill, HT, lp; (**Sweet Bouquet**); Burrows, Steven; Burrows, 1996

'SABclive', Min, w, 1998; (**Pickwick**); flowers cream and white, double, medium, dbl., 15–25 petals, borne in small clusters, slight fragrance; prickles moderate; foliage medium, medium green, semi-glossy; compact, bushy, medium growth; [Buttons X Richard Buckley]; Burrows, Steven; Burrows Roses, 1998

'Sabine', HT, ob, 1958; flowers salmon, well formed, dbl., 40 petals; vigorous growth; [Dame Edith Helen X Seedling]; Frères, Buyl

'Sabine', HT, dp, 1962; (Sabine Sinjen); bud pointed; flowers cherry-red, medium, dbl., 30 petals, exhibition form, intense fragrance; foliage dark, glossy; upright, bushy growth; Tantau, Math.

Sabine, HT, pb; RvS-Melle

Sabine Rancy, HT, 1972; Croix; (Cavriglia)

Sabine Ruf, *see* HELsabin

Sabine Sinjen, *see* **'Sabine'**

'Sabinia', HT, pb, 1940; bud ovoid; flowers dark eglantine-pink passing to reddish yellow, large, dbl., exhibition form, moderate fragrance; foliage dark, leathery; very vigorous, upright growth; GM, Rome, 1939; [Julien Potin X Sensation]; Aicardi, D.; Giacomasso

Sable Chaud, *see* 'DELsob'

Sable Cloud, *see* 'DELsob'

Sabra, *see* 'JUSsabra'

'Sabre Dance', F, rb, 1975; flowers scarlet, reverse gold, dbl., 40 petals, 4 in., exhibition form, slight fragrance; foliage dark; vigorous growth; [Mildred Reynolds X Arthur Bell]; Bees

Sabrina, *see* 'MEIgandor'

'Sabrina', HT, rb, 1960; flowers crimson, reverse amber-yellow marked crimson, dbl., 35 petals, 5 in., exhibition form, intense fragrance; foliage dark, leathery; vigorous, bushy growth; [Grand Gala X Premier Bal]; Meilland, Mrs. Marie-Louise; URS

'Sacajawea', F, mr, 1982; flowers small blooms borne singly, dbl., slight fragrance; foliage medium, medium green, smooth; upright growth; [(Baccará X Garnet) X ?]; Elliott, Charles P.

Sacha, *see* SPEkes

Sachalin, HRg, pb, 1988

Sachet™, *see* 'SAVasach'

Sächs. Lichtenstein, S, lp, 1996; flowers medium-large, semi-dbl.; Noack, Werner; (Sangerhausen)

'Sachsengruss', HP, lp, 1912; (Tendresse); flowers soft flesh, well-formed, very large; vigorous growth; [Frau Karl Druschki X Mme Jules Gravereaux]; Neubert; Hoyer & Klemm

Sachsenrose, Pol, dp; flowers carmine-pink, small, semi-dbl.; (Sangerhausen)

'Saclay', F, pb, 1970; flowers opal, full, dbl., 20–22 petals, 2–3 in., cupped, slight fragrance; foliage dull, sage-green; bushy growth; [Miss France sport]; Inst. National Agronomique; Commissariat a l'Energie Atomique

Sacramento, HT, dr, 1981; flowers dark velvety red, medium, dbl.; GPG Bad Langensalza; (Sangerhausen)

'Sadabahar', F, mp, 1970; bud pointed; flowers pink, open, medium, semi-dbl.; foliage glossy, dark; very vigorous, bushy growth; [Frolic X Open pollination]; IARI

Sadabahar, Climbing, Cl F, mp, 1991; IARI

'Sadaranga', HT, mp, 1977; bud ovoid; very dbl., 60 petals, 4.5 in., globular, slight fragrance; foliage glossy; vigorous, upright, compact growth; [Kronenbourg X Peace]; Hardikar, Dr M.N.

Saddler's Gold, *see* 'HORsaddler'

Sadie, *see* DESmother

Sadie Hawkins, *see* 'SPOhawk'

Sadler, *see* 'BIScof'

Sadlers Wells, S, pb, 1983; Beales, Peter

'Safari', F, my, 1966; flowers large, semi-dbl., 22 petals, cupped, blooms in trusses; vigorous, compact growth; Tantau, Math.

Safari, F, ab, 1998; Kordes

Saffex Rose, *see* KORgrayel

'Saffo', HT, 1934; Aicardi, D.; (Cavriglia)

'Safrano', T, ab, 1839; bud pointed; flowers saffron and apricot-yellow, large, semi-dbl., moderate fragrance; vigorous growth; Beauregard

'Saga', F, w, 1973; flowers cream-white, blooms in clusters, semi-dbl., 12 petals, 2.5 in., slight fragrance; [Rudolph Timm X (Chanelle X Piccadilly)]; Harkness, 1974

Saga Holiday, *see* 'HARzest'

'Sahara', HT, w, 1956; bud golden yellow; flowers creamy buff, well shaped, rather late bloom; very vigorous, upright growth; [Mrs Wemyss Quinn X Mev. G.A. van Rossem]; Stevenson; Waterer

Sahara, *see* OLIjcrem

Sahara, S, yb, 1996; Tantau

'**Saïd**', F, rb, 1964; bud ovoid; flowers cinnabar-red, reverse dark red, medium, borne in clusters, dbl., 28 petals, cupped; short stems; [Aloha X Pioupiou]; Arles; Roses-France

'**Saigon**', HT, yb, 1943; flowers dark yellow with coppery pink tints, large, dbl.; Gaujard; (Sangerhausen)

'**Sailoz Mookherjea**', F, or, 1974; flowers cadmium-orange to orange-vermilion, dbl., 20 petals, 3 in., exhibition form, slight fragrance; very vigorous growth; [Unknown X Unknown]; Pal, Dr B.P.

'**Saint Alban**', F, mr, 1977; semi-dbl., 15 petals, 3 in., slight fragrance; foliage glossy; dwarf growth; [Marlena X ?]; Harkness

'**Saint Fiacre**', Pol, or, 1965; bud ovoid; flowers geranium-red, borne in clusters, single, cupped; vigorous, upright growth; [Reverence X Sumatra]; Delforge

Saint Louis, *see* 'MALnino'

Saint Mary™, *see* 'MORyettem'

Saint Patrick, *see* 'WEKamanda'

Saint Patrick, HT, lp; Strickland, Frank A., 1991

'**Saint Prist de Breuze**', HCh, dp, 1838; flowers medium, dbl.; Desprez; (Sangerhausen)

Sainte-Genevieve, S, w, 1987; Guimont, G.

Saint-Exupéry®, *see* 'DELvor'

Saint-Fiacre d'Orleans, Pol, op, 1997; Eve, A.

Saint-Victor, HT, mr, 1979; Croix

Saint-Vincent®, *see* DELtrap

Saison Bluete, HT, dy

'**Sai-Un**', HT, ob, 1980; flowers yellow-orange shaded deep orange-red, large, dbl., 50 petals, moderate fragrance; prickles slanted downward; foliage dark, glossy; upright growth; [(Miss All-American Beauty X Kagayaki) X Unnamed seedling]; Suzuki, Seizo; Keisei Rose Nursery

Sakura Hime, Min, lp, 1990

'**Sakura-Gasumi**', F, op, 1999; flowers pale pink, turning deeper, dbl., 20–25 petals, 1.5–2 in., slight fragrance; foliage dark green, semi-glossy; dwarf, compact (1 ft) growth; GM, Japan Rose Concours, 1990; Suzuki, Seizo, 1988; Keisei Rose Nursery, 1990

'**Salam Aleik**', HT, mr; flowers pure red, very large, dbl.

'**Salamander**', HP, mr, 1891; flowers bright scarlet-crimson, large, dbl., cupped, non-recurrent; upright growth; Paul, W.

'**Salambo**', HT, mr, 1959; flowers carthamus-red; Kordes, R.; Vilmorin-Andrieux

Salammbo®, *see* DELperl

'**Salazar**', HT, rb; flowers bright velvety red with salmon reflections, petals waved; da Silva, Moreira

'**Salden Monarch**', HT, mp, 1968; flowers large, dbl.; foliage glossy; vigorous growth; [Ballet X ?]; Carrigg

'**Salet**', M, mp, 1854; bud well mossed; flowers rosy pink, very large, dbl., flat, some recurrence, vigorous growth; Lacharme, F.

Salita®, *see* KORmorlet

'**Sallie Lewis**', HT, ab, 1924; bud long, pointed; flowers creamy apricot, center deeper, dbl., intense fragrance; [Mme Charles Lutaud X Gladys Holland]; Morse

'**Sally**', F, ab, 1981; flowers pale apricot, fading to ivory, with pink veins; [Elizabeth of Glamis sport]; Walker, B.; Homeric Rose Nurseries

'**Sally**', HT, mp, 1938; bud long, pointed; flowers pink, dbl., exhibition form, foliage leathery; long stems; vigorous growth; RULED EXTINCT 3/81; [Mrs Charles Russell X Mme Butterfly]; Spandikow; Eddie

'**Sally Alder**', HT, mp, 1944; bud long, pointed; flowers pink, medium, semi-dbl., exhibition form; foliage dark; moderate, bushy growth; [Portadown X Seedling]; Moss; F. Mason

Sally Forth, S, lp; Peden, R., 1999; (Weatherly, L.)

'**Sally Holmes**'®, S, w, 1976; bud pointed, apricot; flowers creamy white, hydrangea-like clusters, single, 5–8 petals, 3.5–4 in., blooms in large clusters, slight fragrance; few prickles; foliage dark, glossy; straight, long stems; vigorous, bushy growth; GM, Baden-Baden, 1980; [Ivory Fashion X Ballerina]; Holmes, R.; Fryer's Nursery, Ltd.

'**Sally Mac**', F, ab, 1981; flowers apricot pink blend, yellow at base, blooms borne 4 per cluster, dbl., 40 petals, moderate fragrance; hooked, red-brown prickles; foliage medium, dark, glossy; vigorous, upright growth; [(Joyfulness X Paddy McGredy) X (Circus X Joyfulness)]; McCann, Sean; Hughes Roses

'**Sally Pigtail**', HT, pb, 1959; flowers cream flushed rose-pink, high pointed, large, dbl., 32 petals, moderate fragrance; free growth; [Wilfred Pickles X Karl Herbst]; Mee

'**Sally Tite**', HT, dr, 1930; flowers glowing crimson, over large, dbl., exhibition form; Dickson, S.; Armstrong Nursery; B&A

Sally's Rose, *see* 'CANrem'

Salmo, *see* POUlnoev

Salmo, *see* POUlnoeu

Salmon, Cl Min, op, 1994

'**Salmon Arctic**', LCl, lp, 1954; (Everblooming Pillar No. 83); flowers dawn-pink, very dbl., 90–100 petals, 3.5–4.5 in.; growth like a hybrid tea, followed by 4–5 ft canes; [Unnamed seedling X Break o' Day, Climbing]; Brownell, H.C.

'**Salmon Beauty**', Pol, op, 1929; flowers peach-pink, passing to soft salmon, medium, borne in clusters, very dbl., slight fragrance; dwarf growth; [Orange King sport]; Wezelenburg

Salmon Button, Min, ob

Salmon Charm, *see* '**Ginger Rogers**'

Salmon Dream, MinFl, op, 1999

Salmon Drops, HT, op, 1999; Williams, J. Benjamin

'**Salmon Glory**', Pol, op, 1937; flowers pinkish salmon; [Gloria Mundi sport]; deRuiter

'**Salmon Glow**', F, op, 1970; flowers salmon, full, dbl., 25 petals, 3 in., slight fragrance; foliage matt green; compact, bushy growth; [Vera Dalton X Seedling]; Sanday, John

'**Salmon Marvel**', F, op, 1958; flowers orange-salmon, rosette shape, borne in trusses, dbl., 44 petals, 2.5–3 in.; foliage dark, glossy, crinkled; bushy, upright growth; [Red Pinocchio X Signal Red]; deRuiter; Blaby Rose Gardens

Salmon Midinette, Cl Min, op, 1996; Moore

'**Salmon Perfection**', F, rb, 1952; flowers scarlet-red shaded orange, medium, borne in trusses, dbl., 25 petals, cupped, foliage dark, leathery; vigorous growth; deRuiter; A. Dickson; McGredy

Salmon Pink Garnette, Pol, op

'**Salmon Queen**', Pol, op, 1923; flowers deep salmon; [Juliana Rose sport]; Den Ouden

Salmon Radiance, *see* '**Mrs Charles Bell**'

Salmon Rosamini, *see* RUIsalro

Salmon Sensation, *see* '**Marion**'

Salmon Sensation, F, 1956; deRuiter, G.; (Cavriglia)

Salmon Sorbet, F, pb

Salmon Spire, *see* KORturnus

Salmon Splash, ob, 1997; K&S

'**Salmon Spray**', F, pb, 1923; flowers light salmon-pink, reverse carmine, borne in clusters, semi-dbl., cupped, moderate fragrance; foliage rich green, leathery; very vigorous, bushy growth; [Orléans Rose, Climbing X Midnight Sun]; Grant; Kershaw

'**Salmon Sprite**', F, op, 1964; flowers salmon suffused strawberry, blooms in clusters to 15, dbl., 40 petals, 3 in., intense fragrance; upright growth; [Unnamed seedling X Jiminy Cricket]; LeGrice

Salmon Sunblaze™, *see* 'MEIshuto'

Salmon Sunsation, *see* KORpapie

Salmon Symphony, HT, op

'SALMONE', Gr, lp, 1968; bud ovoid; flowers medium, borne in clusters; foliage dark; [Queen Elizabeth sport]; Faassen-Houba

'SALMONEA', Pol, op, 1927; flowers salmon-pink; Cutbush

'SALMON-KING', F, rb, 1972; flowers red, inside orange-red, full, high pointed, dbl., 28–30 petals, 5 in., intense fragrance; foliage glossy; vigorous, bushy growth; [Show Girl X Orange Sensation]; Ellick; Excelsior Roses

'SALOMÉ', HT, op, 1945; bud ovoid; flowers salmon-flesh-pink, medium, dbl., moderate fragrance; foliage glossy; very vigorous, upright growth; Gaujard

'SALOMÉ', F, pb, 1958; flowers rose-orange, dbl.; vigorous growth; Frères, Buyl

'SALOU', F, pb, 1971; bud long, pointed; flowers deep pink to flesh-pink, open, large, dbl., moderate fragrance; foliage dark, leathery; very vigorous, upright growth; [Astoria X Tiffany]; Delforge

Salsa, *see* 'JACSTOP'

'SALTAIRE', HT, dr, 1925; bud long, pointed; flowers deep velvety crimson, dbl., intense fragrance; Dickson, A.

'SALTATINA', F, or, 1974; bud ovoid; flowers orange-copper to red, open, dbl., 18–22 petals, 2.5–3 in., moderate, spicy fragrance; foliage dark; bushy, spreading growth; [Zorina X Fillette]; Lens

'SALTWELL PARK', F, mr, 1976; flowers very deep scarlet, full, dbl., 20 petals, 3 in., slight fragrance; foliage dark; moderate, upright growth; [Paddy McGredy X Arabian Nights]; Wood

Salut, HT, op, 1987; flowers salmon-orange, large, dbl.; VEG; (Sangerhausen)

Salut à la Suisse, *see* 'TANSCHWEIGRU'

'SALUTATION', LCl, op, 1972; flowers salmon, full, small, dbl., 38 petals, 2.5 in., intense fragrance; foliage glossy, dark; very vigorous growth; [Salute X Mme Isaac Pereire]; Oliver, W.G.

'SALUTE', F, rb, 1958; flowers cherry and ochre bicolor, small borne in trusses, semi-dbl., 20 petals, 1.5 in.; foliage dark, leathery; free growth; [Masquerade X Lady Sylvia]; McGredy, Sam IV; McGredy

'SALVO', HT, mr, 1959; flowers velvety crimson, dbl., 45 petals, 5 in., moderate fragrance; vigorous, bushy growth; [Happiness X Grand Gala]; Herholdt, J.A.; Herholdt's Nursery

Salza, S, dp, 1956; flowers carmine-pink, medium, dbl.; Berger, W.; (Sangerhausen)

Salzagold, HT, dy

Salzajubiläum, F, or, 1982; flowers salmon-red, large, semi-dbl.; GPG Bad Langensalza; (Sangerhausen)

Salzaperle, HT, mp, 1977; flowers large, dbl., moderate fragrance; GPG Bad Langensalza; (Sangerhausen)

Salzaquelle, HT, m, 1977; flowers light lilac, large, dbl., intense fragrance; GPG Bad Langensalza; (Sangerhausen)

Salzburg, S, op, 1967; flowers medium, dbl.; Kordes, W. Söhne; (Sangerhausen)

Sam, *see* 'STESAM'

Sam Buff, *see* **'AFTERGLOW'**

'SAM FERRIS', F, ob, 1967; flowers orange-scarlet, borne in trusses, dbl., slight fragrance; vigorous growth; [Heidleberg sport]; Ellick

Sam Houston Rose, S, pb, 1996; Antique Rose Emporium

'SAM MCGREDY', HT, yb, 1937; flowers dark cream, base sunflower-yellow, large, exhibition form, slight fragrance; foliage dark, leathery; vigorous growth; GM, NRS, 1935; [Delightful X Mrs Charles Lamplough]; McGredy

Sam Tucker, N, my

'SAMANDI', F, w, 1997; flowers double, dbl., 15–25 petals, borne in small clusters, no fragrance; foliage medium; medium (1 m) growth; [Sexy Rexy X Brownie]; Rawlins, R.

Samantha®, *see* 'JACMANTHA'

Samantha®, HT, dr, 1984; Noack, Werner

Samantha Ruth, *see* 'WELSAM'

Samara®, HT, dy, 1989; Noack, Werner

Samaritan®, *see* 'HARVERAG'

Samba®, *see* 'KORCAPAS'

'SAMMETGLUT', F, dr, 1943; bud long, pointed; flowers crimson, open, very large, borne in clusters, semi-dbl., slight fragrance; foliage dark, leathery; very vigorous, upright, bushy growth; [Holstein X Kardinal]; Kordes

'SAMMY', HMsk, dr, 1921; bud small; flowers carmine, borne in erect clusters, single, recurrent bloom, slight fragrance; foliage glossy, bronze; very vigorous, bushy growth; [(Trier X Watta?) X Gruss an Teplitz]; Pemberton

Samoa, *see* **'FRED CRAMPHORN'**

'SAMOA SUNSET', HT, ob, 1965; flowers chrome-yellow and copper-orange, open, large, dbl., moderate fragrance; foliage glossy, bronze; vigorous, bushy, compact growth; [McGredy's Sunset sport]; Lone Star Rose Nursery

Samourai, *see* 'MEIELEC'

Samourai, Climbing, *see* **'SCARLET KNIGHT, CLIMBING'**

Sampson, *see* 'SAVASAM'

Samptosa, F, dr, 1969; flowers large, dbl.; Institut für Obstbau Dresden Pillnitz; (Sangerhausen)

Samson™, *see* 'SAVASAM'

'SAMUEL MARSDEN', Ch, mp, 1989; [Slater's Crimson China sport]; Nobbs, Kenneth J., 1987

'SAMUEL PEPYS', HT, w, 1934; flowers white, center slightly shaded cream, very large, dbl., globular, moderate fragrance; foliage leathery; bushy growth; Cant, B. R.

'SAN ANTONIO', Gr, or, 1967; bud ovoid; flowers large, dbl.; foliage dark, glossy, leathery; very vigorous, upright, bushy growth; [Roundelay X El Capitan]; Armstrong, D.L.; Armstrong Nursery

'SAN DIEGO', HT, my, 1968; bud ovoid; flowers light yellow, large, dbl., 50 petals, exhibition form, moderate fragrance; foliage leathery; vigorous, upright, bushy, compact growth; [Helen Traubel X Tiffany]; Armstrong, D.L.; Armstrong Nursery

San Diego, HT, pb, 1993; deVor

'SAN FERNANDO', HT, mr, 1948; flowers scarlet, large, dbl., 30 petals, exhibition form, intense fragrance; vigorous, upright growth; AARS, 1948; [Heart's Desire X (Crimson Glory X Poinsettia)]; Morris; Western Rose Co.

'SAN FERNANDO, CLIMBING', Cl HT, mr, 1951; Whisler; Germain's

'SAN FRANCISCO', HT, mr, 1962; bud ovoid; flowers signal-red, dbl., 40 petals, 4–5 in., exhibition form, moderate fragrance; foliage dark, glossy, leathery; vigorous, compact, branching growth; [Dean Collins X Independence]; Lammerts, Dr Walter; Amling-DeVor Nursery; Germain's

'SAN GABRIEL', HT, op, 1947; bud long, pointed; flowers deep salmon-pink, dbl., 45 petals, 5.5 in., exhibition form, moderate fragrance; foliage dark, leathery; vigorous, bushy growth; [Poinsettia X Unnamed variety]; Morris; Germain's

San Gerolamo, HGal, dp

'SAN JOAQUIN', HT, ob, 1939; flowers glowing orange; [Talisman sport]; Moore, Ralph S.

San Jordi, HT; Camprubi, C.; (Cavriglia)

'SAN JOSÉ', HT, ob, 1933; bud ovoid; flowers orange-salmon suffused gold, very large, dbl., moderate fragrance; foliage glossy; very vigorous growth; Denoyel, Vve.; J&P

San Jose Sunshine, *see* 'FOUSUN'

San Juan Settler, B, pb

San Leandro Dark Red HT, HT, dr

'SAN LUIS REY', HT, dy, 1947; bud ovoid; flowers deep saffron-yellow, dbl., 40 petals, 4.5 in., cupped, intense fragrance; foliage dark, glossy; bushy growth; [Lady Forteviot X Pedralbes]; Morris; Germain's

San Rafael Rose, *see* **'FORTUNE'S DOUBLE YELLOW'**

San Valentin, Pol, 1971; Dot, Simon; (Cavriglia)

SANaran, HT, w, 1995; (**Aorangi**); Sandbrook

'SANbaby', HT, dr, 1988; (**Mother and Baby**®); flowers dark red with a velvet sheen, some black marking, reverse de, dbl., 48 petals, quartered; prickles barbed, light brown; foliage medium, medium green, semi-glossy; low growth; [Bristol X (Lilac Rose X (Magenta X Crimson Glory))]; Sanday, John; John Sanday Roses, Ltd.

Sanbi, HT, ab

'SANcharm', HMoy, lp, 1985; (**Pink Nevada**); flowers pale lilac pink; [Nevada sport]; Sanday, John, 1970

'SANcol', F, dr, 1990; (**Edward Colston**); bud rounded; flowers dark red with medium red reverse, aging dark red, very dbl., 40–45 petals, cupped, borne in sprays of 3–5, slight, fruity fragrance; no fruit; prickles barbed, red; foliage medium, medium green, matt; upright, medium growth; [Vera Dalton X Stephen Langdon]; Sanday, John, 1982; John Sanday Roses, Ltd.

'Sanctus', HKor, w, 1978; flowers pure white, large blooms borne singly, dbl., 36 petals, exhibition form, repeat bloom, moderate fragrance; large, straight, rust-brown prickles; foliage large, dark, leathery; vigorous, upright, tall growth; [Borealis X Borealis]; James, John

Sandalwood, *see* 'JALwood'

'Sandar', HT, rb, 1946; flowers carmine, center yellow, large, dbl., 25 petals; foliage dark; very vigorous growth; [Charles P. Kilham X Lleida]; Laperrière

'SANdaya', S, ly, 1992; (**Enhance**); flowers soft apricot yellow, blooms borne in large clusters, dbl., 15–25 petals, 1.5 in., moderate fragrance; some prickles; foliage medium, medium green, glossy; low (25 cms), spreading growth; [Malmesbury X The Fairy]; Sanday, John; John Sanday Roses Ltd.

Sandberg, HT, pb

'Sander's White Rambler', HWich, w, 1912; flowers rosette form, small blooms in large clusters, moderate fragrance; foliage bright green, glossy; vigorous growth; Sander & Sons

'SandKOR', HT, op, 1981; (**Sandra**[Rs]); flowers salmon, large, dbl., 35 petals, exhibition form, slight fragrance; foliage medium, medium green, matt; upright growth; [Mercedes X Unnamed seedling]; Kordes, W.

'SANdokan', F, rb, 1977; (**Sanmez**); dbl., 27–32 petals; Sande; Pekmez

'Sandolina de Major', HT, dp, 1977; bud ovoid; flowers deep pink, very dbl., 57 petals, 4 in., cupped, moderate fragrance; vigorous, upright growth; [Dr A.J. Verhage X Königin der Rosen]; Kordes, W. Söhne; Willemse

Sandra[Rs], *see* 'SandKOR'

Sandra Kim®, S, mp

Sandra May Williamson, HT, mp, 1988; Williamson

Sandrina, S, 1996; Embriaco, B.; (Cavriglia)

Sandrina, HT, yb, 1997; Kordes

Sandrine, F, lp, 1975; Eve, A.

'Sandringham', F, my, 1955; dbl., 30 petals, 2.5 in., blooms in large clusters, moderate fragrance; foliage light green, glossy; vigorous, tall growth; Kordes; Morse

'Sandringham Centenary', HT, op, 1980; (Sandringham Century); bud pointed; flowers deep salmon-pink, blooms borne singly, dbl., 22 petals, moderate fragrance; foliage dark, glossy; vigorous, upright growth; [Queen Elizabeth X Baccará]; Wisbech Plant Co., 1981

Sandringham Century, *see* **'Sandringham Centenary'**

Sands of Time™, *see* 'MINazco'

Sandton City, *see* 'MEIpopul'

Sandton Smile, *see* 'KORmetter'

Sandy, HT, op, 1981; flowers coppery-pink, large, dbl., slight fragrance; Wisbech Plant Co.; (Sangerhausen)

'Sangerhausen', HMsk, dp, 1938; bud long, pointed; flowers deep pink, large blooms in large clusters, semi-dbl., cupped, repeat bloom, slight fragrance; foliage large, leathery, wrinkled; vigorous (4–5 ft), bushy growth; (28); [Ingar Olsson X Eva]; Kordes

Sangria®, *see* 'MEIestho'

Sangria, *see* MEItorpo

'Sanguinaire', HRg, pb, 1933; bud long, pointed; flowers brilliant oxblood-red with orange, stamens yellow, open, semi-dbl., 18 petals, 4 in., recurrent bloom, slight fragrance; foliage glossy; short stems; shrub or pillar ($6\frac{1}{2}$ ft) growth; [Bergers Erfolg X Capt. Ronald Clerk]; Gillot, F.

'Sanguinea', Ch, dr; (Bengal Cramoisi Double, Blood-red ChinaRose, La Sanguine, R. indica cruenta); flowers velvety, vivid purple crimson, petals concave with white bas, very dbl., globular, foliage branches, leaves, and flowered stalks very purple; weak, spreading growth

Sanka, HT, ob, 1986; (Enchantment); Keisei

'Sankt Anton', Pol, mr, 1971; bud long, pointed; flowers brilliant red, open, small, single, profuse, continuous bloom, slight fragrance; foliage leathery; vigorous, upright growth; [Alain X Seedling]; Delforge

Sanktflorian, F, mr, 1971; flowers medium to large, dbl.; Meilland; (Sangerhausen)

'SANlilac', HT, lp, 1988; (**Lilac Airs**); flowers soft, lilac pink, yellow at base, reverse silver at base, large, very dbl., 36 petals, exhibition form, intense, damask fragrance; prickles long, fairly straight, medium brown, matt; foliage large, medium green, semi-glossy; bushy, tall, strong growth; [Fred Gibson X Whisky Mac]; Sanday, John; John Sanday Roses, Ltd.

'SANmar', F, or, 1983; (**Bright Eyes**); flowers yellow stamens, medium, dbl., slight fragrance; foliage medium, dark, matt; bushy growth; [Seedling X Circus]; Sanday, John; John Sanday Roses, Ltd.

'SANmed', S, w, 1984; (**Ice Fairy**); [The Fairy sport]; Sanday, John; Sanday Roses, Ltd.

Sanmez, *see* 'SANdokan'

'SANolence', HT, or, 1985; (**Benevolence**); flowers large, dbl., 35 petals, moderate fragrance; long, narrow prickles; foliage large, dark, semi-glossy; bushy growth; [Vera Dalton X Seedling]; Sanday, John, 1986

'SANone', S, rb, 1988; (**Little One**); flowers red with white eye, borne in sprays of 12–20, single, 5 petals, repeat bloom, no fragrance; prickles average, light brown; foliage medium, medium green, glossy; bushy, low growth; [(Sarabande X Unnamed seedling) X Circus]; Sanday, John; John Sanday Roses, Ltd.

'SANphyllis', F, ab, 1986; (**John Hughes**); flowers soft apricot, medium, semi-dbl., 6–14 petals, slight fragrance; foliage medium, dark green, glossy; bushy growth; [City of Gloucester X Bristol Post]; Sanday, John

'Sanremo', HT, mr, 1962; bud ovoid, pointed; flowers cardinal-red, dbl., 20–30 petals, 5.5–6 in., exhibition form, moderate fragrance; foliage glossy; strong stems; vigorous, upright, bushy growth; [(Pink Delight X Rome Glory) X Baccará]; Mansuino, Q.; Carlton Rose Nurseries

'SANroc', F, or, 1985; (**Corsair**); flowers medium, dbl., 20 petals, slight fragrance; foliage medium, dark, glossy; bushy growth; [(Vera Dalton X Stephen Langdon) X Fiesta Flame]; Sanday, John; Sanday Roses, Ltd.

'SANrozo', HT, pb, 1986; (**Esperanto Jubileo**); flowers cream, edged deep rose pink, yellow suffused at base, large, dbl., slight fragrance; foliage medium, medium green, matt; bushy growth; [Gavotte X Piccadilly]; Sanday, John, 1987

Sans Sépales, lp; flowers medium, dbl.; (Sangerhausen)

Sans Souci, *see* 'LAVsans'

Sans souci, *see* **'Moulin Rouge'**®

Sans Souci, Pol, or, 1960; flowers small, dbl.; Schmid, P.; (Sangerhausen)

Sans Souci, *see* BARsan

'SANspic', HT, ab, 1985; (**Apricot Spice**); flowers medium, dbl., 35 petals, slight fragrance; foliage medium, medium green, matt; bushy growth; [City of Gloucester X Seedling]; Sanday, John

'SANTA ANITA', HT, mp, 1932; bud long, pointed; dbl., 22 petals, 3.5–4 in., exhibition form, moderate fragrance; foliage light; vigorous, bushy growth; [(Mrs J.D. Eisele X Unnamed seedling) X E.G. Hill]; Howard, F.H.; H&S

'SANTA ANITA, CLIMBING', Cl HT, mp, 1946; Howard, F.H.; H&S

'SANTA CATALINA', Cl F, lp, 1970; flowers medium, semi-dbl., 18 petals, 3.5 in., slight fragrance; [Paddy McGredy X Heidelberg]; McGredy, Sam IV; McGredy

Santa Claus, *see* 'POULCLAUS'

Santa Fe, *see* 'JACYEM'

Santa Fe, HT, ly; flowers daffodil yellow, exhibition form

'SANTA FÉ', HT, op, 1967; flowers deep salmon-pink, reverse lighter, large, dbl.; [Mischief X Tropicana]; McGredy, Sam IV; McGredy

'SANTA MARIA', F, mr, 1969; flowers scarlet, blooms in trusses, slight fragrance; foliage small; [Evelyn Fison X (Ma Perkins X Moulin Rouge)]; McGredy, Sam IV; McGredy

'SANTA RITA', HT, op, 1961; flowers salmon-pink, well formed, large; vigorous growth; [Independence X Papillon Rose]; Lens

Santa Rosa, *see* **'BURBANK'**

'SANTA ROSA', F, op, 1954; bud ovoid; flowers salmon-pink tipped brick, small, dbl., globular, intense fragrance; foliage glossy, soft, bronze; vigorous, bushy growth; [Golden Salmon X Pinocchio]; Silva

Santa Rosa, HCh, dp, 1899; Burbank

'SANTA TEREZA D'AVILA', HT, op, 1959; flowers salmon-pink and orange, reverse gold, well-formed, moderate fragrance; foliage glossy; vigorous growth; GM, Madrid, 1959; [Monte Carlo X Michele Meilland]; da Silva, Moreira

Santana®, *see* 'TANKLESANT'

'SANTANG', HT, ob, 1983; (**Maritime Bristol**); flowers tangerine, large, dbl., 35 petals, moderate fragrance; foliage medium, dark, semi-glossy; upright, bushy growth; [City of Gloucester X Unnamed seedling]; Sanday, John; Sanday Roses, Ltd.

Santiago®, *see* ADAmona

'SANTOR', F, or, 1984; (**Red Brigand**); flowers medium, dbl., 20 petals, slight fragrance; short, slightly hooked prickles; foliage large, dark, semi-glossy; bushy growth; [Vera Dalton X Stephen Langdon]; Sanday, John; Sanday Roses, Ltd.

'SANWIFE', HT, rb, 1988; (**Wiltshire Pride**); flowers scarlet with yellow at base, reverse yellow with red edging, dbl., 28 petals, exhibition form, slight fragrance; prickles slightly hooked, dark brown; foliage large, dark green, semi-glossy; upright, tall growth; [Bristol X Piccadilly]; Sanday, John; John Sanday Roses, Ltd., 1989

'SAOHIME', HT, lp, 1999; flowers clean, light pink, dbl., 38 petals, exhibition form, slight fragrance; foliage medium green, half-leathery; 3.5 ft growth; Bronze Medal, Japan Rose Concours, 1992; [Gavotte X Bridal Robe]; Hayashi, Shunzo, 1990

'SAONARA', HT, mr, 1962; flowers geranium-red, dbl.; foliage dark; strong stems; vigorous growth; [Baccará X Peace]; Borgatti, G.; Sgaravatti

Saphir, *see* **'SONG OF PARIS'**

Saphir, *see* TANrikas

'SAPHO', HT, ob, 1933; flowers salmon, tinted coppery, very large, dbl., slight fragrance; foliage leathery, glossy, dark; very vigorous growth; Gaujard

'SAPPHO', T, pb; flowers fawn shaded pink, center yellow; vigorous growth; Paul & Son, 1888

Sappho, A, w, 1817

Sara, *see* JACautel

Sarabande®, *see* 'MEIRABANDE'

Sarabande, Climbing®, *see* 'MEIHANDSAR'

'SARAGAT', HT, rb, 1968; flowers deep strawberry to cream, large, dbl., globular, intense fragrance; foliage leathery; very vigorous growth; [Seedling X Seedling]; Malandrone

Sarah, *see* 'MEIMAFRIS'

Sarah, Min, ab; [Poker Chip X Holy Toledo]; Hannemann, F., 1991; The Rose Paradise

'SARAH ARNOT', HT, mp, 1957; flowers warm rose-pink, dbl., 25 petals, 4.5 in., moderate fragrance; foliage leathery; vigorous, upright growth; GM, NRS, 1958; [Ena Harkness X Peace]; Croll

'SARAH BACHERACH', F, mp, 1965; flowers pink, medium, dbl., intense fragrance; foliage dark; [Harmonie X Buisman's Triumph]; Buisman, G. A. H.

'SARAH BERNHARDT', Cl HT, dr, 1906; flowers scarlet-crimson, very large, semi-dbl.; Dubreuil

'SARAH COVENTRY', F, dr, 1956; bud ovoid; flowers cardinal-red, open, medium, borne in irregular clusters, dbl., 40–45 petals, cupped, moderate fragrance; foliage bright green; vigorous, compact, bushy growth; [Red Pinocchio X Garnette]; Boerner; Stuart

'SARAH DARLEY', HT, dy, 1938; flowers clear deep golden yellow, well shaped; foliage dark; vigorous growth; Wheatcroft Bros.

Sarah Elizabeth, *see* 'PIXSAR'

'SARAH HILL', F, mp, 1956; bud short, pointed; flowers phlox-pink, open, dbl., 25–30 petals, 2.5–3 in., slight, spicy fragrance; foliage dark, leathery; vigorous, upright, bushy growth; [Garnette X Pink Bountiful]; Hill, Joseph H., Co.

'SARAH JANE', HT, pb, 1971; flowers rose-pink, edged deeper, reverse white, full, dbl., 35 petals, 4.5–5 in., slight fragrance; foliage glossy; very vigorous, tall growth; [Rose Gaujard sport]; Heath, W.L.

Sarah Jo, *see* 'MEHREX'

'SARAH LYNN', HT, mp, 1991; dbl., borne mostly singly, slight fragrance; foliage large, dark green, semi-glossy; strong stems; vigorous, medium, compact growth; Bronze, ARC TG, 1991; [Seedling X Prima Ballerina]; Ohlson, John

'SARAH MAUD', HT, dy, 1934; bud long; flowers golden yellow, large; foliage bright, dark; very vigorous growth; Mallerin, C.

Sarah Philp, Min, ab

Sarah Robinson, *see* 'TROBINETTE'

'SARAH VAN FLEET', HRg, mp, 1926; flowers wild-rose-pink, large, semi-dbl., cupped, recurrent bloom, intense fragrance; foliage leathery, rugose; compact (6–8 ft) growth; (14); Van Fleet; American Rose Society

'SARAH WRIGHT', HT, dp, 1927; bud long, pointed; flowers rose-pink, dbl., slight fragrance; [Ophelia X Emma Wright]; Morse

Sarah, Duchess of York, *see* 'DICRACER'

Sarah's Tea, T, pb

'SARAJEAN', Min, pb, 1979; bud pointed; flowers peach-pink, dbl., 50 petals, 1 in., globular, moderate fragrance; foliage small, bronze-green, glossy; upright, bushy growth; [Seedling X Over the Rainbow]; Williams, Ernest D.; Mini-Roses

'SARATOGA', F, w, 1963; bud ovoid; flowers gardenia-shaped, blooms in irregular clusters, dbl., 33 petals, 4 in., intense fragrance; foliage glossy, leathery; vigorous, upright, bushy growth; AARS, 1964; [White Bouquet X Princess White]; Boerner; J&P

'SARDANE', F, rb, 1962; bud long; flowers coral-red tinted silvery, dbl., 35–40 petals, 3 in.; foliage bright green; very bushy, compact growth; Laperrière; EFR

'SARGENT', LCl, pb, 1912; flowers apple-blossom-rose to pale pink, base amber-yellow, borne in clusters, semi-dbl., 3 in.; height 10–12 ft; [(R. wichurana X Crimson Rambler) X Baroness Rothschild]; Dawson; Eastern Nursery

'SARIE', HT, mp, 1977; dbl., 35 petals, 4.5–5 in., slight fragrance; foliage glossy, bright green; vigorous growth; [Pink Favorite X Nightingale]; Herholdt, J.A.

Sarie Marais, *see* KORhota

'SARIE MAREIS', F, mr, 1950; flowers glowing scarlet; [Irene X Donald Prior]; Leenders, M.

Saris Sepals, M, mp, 1839

Saroda, *see* **'SHARADA'**

Saroja, F, pb, 1984; Pal, Dr B.P.

Sarong, *see* 'GLANLIN'

'SAROOR', F, or, 1969; bud ovoid; flowers orange-scarlet, open, large, dbl., slight fragrance; foliage large, leathery; upright growth; [Gertrud Westphal X Open pollination]; Singh; Gopalsinamiengar

Sasa, *see* DELrima

Saskabec, *see* **'TANT MIEUX'**

'SASKIA', F, mp, 1961; single, borne in large clusters, foliage dark; bushy growth; [Pinocchio X Gartenstolz]; Buisman, G. A. H.

Sassy(TM), *see* 'HILTACO'

'SASSY LASSY', Min, yb, 1975; bud pointed; flowers yellow and pink blend, dbl., 28 petals, 1 in., slight fragrance; foliage small, bronze, glossy, leathery; upright, spreading growth; [Seedling X Over the Rainbow]; Williams, Ernest D.; Mini-Roses

'SATAN', HT, dr, 1939; (Joyce Lomax, Satanas); flowers very dark red, reverse lighter, nearly, large, dbl., exhibition form, moderate fragrance; foliage leathery; bushy growth; [(Mme Edouard Herriot X Angèle Pernet) X Mari Dot]; Pahissa; J&P

Satanas, *see* **'SATAN'**

'SATCHMO'®, F, or, 1970; flowers bright scarlet, dbl., 25 petals, 3 in., exhibition form, slight fragrance; GM, The Hague, 1970; [Evelyn Fison X Diamant]; McGredy, Sam IV; McGredy

Satellite, *see* 'DELSATEL'

'SATELLITE'®, HT, dr, 1958; bud pointed; flowers deep crimson, dbl., 30 petals, 4.5 in., exhibition form, intense fragrance; foliage dark, glossy; vigorous, upright, compact growth; [Editor McFarland X William Harvey]; Priestly

Satin, HT, dp, 1996

Satin Doll, *see* 'JACSPIF'

Satina®, *see* TANinat

'SATINETTE', Pol, w, 1971; bud ovoid; flowers large, dbl., cupped, slight fragrance; foliage large, light, leathery; vigorous, upright growth; [Maria Delforge X Irene of Denmark]; Delforge

'SATINGLO', F, mp, 1954; bud globular; flowers glowing coral, borne in clusters, dbl., 45–50 petals, 2.5 in., cupped, moderate fragrance; foliage leathery, glossy; vigorous growth; [Pinocchio seedling X Vogue]; Boerner

'SATISFACTION', HT, dp, 1956; flowers carmine-pink, large, dbl., intense fragrance; [Parfum X Unnamed seedling]; Verbeek

'SATMIR', HT, dr, 1956; flowers carmine-red, dbl.; long, strong stems; very vigorous growth; [Satan X Mirandy]; Dot, Pedro

Saturday®, HT, pb, 1990; McGredy, Sam IV

Saturday Star, *see* KORvolomin

Saturn, *see* JAClogo

'SATURNIA', HT, rb, 1936; bud long, pointed; flowers bright scarlet with gold, large, dbl., 20 petals, cupped, moderate, fruity fragrance; foliage dark, glossy; vigorous growth; GM, Portland, 1938, GM, Rome, 1933; [Julien Potin X Sensation]; Aicardi, D.; J&P; Robichon

Satvika, HT, yb; flowers occasionally striped

SAUbaflor, F, lp, 1990; (**Heriflor**); Sauvageot

SAUbima, F, op, 1994; (**Chateau de Filain**); Sauvageot

SAUblim, HT, mr, 1991; (**Atida**); Sauvageot

SAUbord, HT, dp, 1992; (**Gustave Courbet**®); Sauvageot

'SAUCY SUE', F, mp, 1974; flowers full, dbl., 20–25 petals, 3.5 in., moderate fragrance; free growth; [Pink Parfait X Europeana]; Lowe

'SAUDADE D'ANIBAL DE MORAIS', HT, dr, 1935; bud velvety crimson; flowers salmony crimson-red, dbl.; vigorous growth; [Sir David Davis X Pres. Jac. Smits]; da Silva, Moreira

SAUdero, F, 1995; (**Dance of Joy 95**); Sauvageot, H.; (Cavriglia)

SAUdime, F, rb, 1999; (**Amours de Savernem**); Sauvageot

SAUdive, F, 1995; (**Messara**); Sauvageot; (Cavriglia)

SAUdora, F, pb, 1997; (**Eclipse**); Golden Rose, Geneva, 1996; Sauvageot

SAUdril, F, lp, 1997; (**Manureva**); Sauvageot

SAUkar, F, op, 1989; (**Karine Sauvageot**®); Sauvageot

'SAUL', HT, dp, 1970; flowers light rose-madder, dbl., 28 petals, 4 in., moderate fragrance; foliage large, dark; upright growth; [Tropicana X Sterling Silver]; Gandy, Douglas L.

Saumonia, HT, op, 1998

'SAVABE', Min, dy, 1994; (**Glory Be**); dbl., 15–25 petals, 1.5 in., borne singly or in small clusters, no fragrance; foliage small, dark green, semi-glossy; medium (16–20 in.), upright, bushygrowth; [Party Girl X Sonnenkind]; Saville, F. Harmon; Nor'East Min. Roses, 1995

'SAVABEAN', Min, rb, 1981; (**Jelly Bean**(TM)); bud ovoid, pointed; flowers red-yellow blend, micro-mini blooms borne 1–6 per cluster, dbl., 20 petals, exhibition form, moderate, spicy fragrance; no prickles; foliage small; very compact, tiny growth; [Unnamed seedling X Poker Chip]; Saville, F. Harmon; Nor'East Min. Roses, 1982

'SAVABEAR', Min, r, 1989; (**Teddy Bear**(TM)); bud urn shaped; flowers terra-cotta, reverse lighter, aging to mauve-pink, dbl., 28 petals, exhibition form, borne singly, slight fragrance; ovoid, grayed-orange fruit; prickles slight downward curve, medium, purple; foliage medium, dark green, semi-glossy; upright, bushy, medium, vigorous growth; PP007424; [Sachet X Rainbow's End]; Saville, F. Harmon; Nor'East Min. Roses, 1990

'SAVABEG', Min, ob, 1988; (**New Beginning**(TM)); flowers bright orange-yellow bi-color, decorative, medium, very dbl., 40–50 petals, borne usually singly, no fragrance; no fruit; very few prickles; foliage medium, medium green, semi-glossy; bushy, compact growth; PP006707; AARS, 1989; [Zorina X Unnamed seedling]; Saville, F. Harmon; Nor'East Min. Roses, 1989

'SAVABET', Min, mr, 1985; (**Single's Better**(TM)); bud mossy; flowers medium red, yellow hinge, mini-moss, small, single, 5 petals, slight fragrance; foliage medium, medium green, semi-glossy; bushy growth; [(Yellow Jewel X Tamango) X ((Little Chief X Sarabande) X LemonDelight)]; Saville, F. Harmon; Nor'East Min. Roses

'SAVABINO', Min, or, 1997; (**Bambino**); flowers full, small, micro mini, vibrant orange, very dbl., 26–40 petals, borne in small clusters, no fragrance; foliage small, medium green, semi-glossy changing to matte; compact, bushy, low (to 14 in.) growth; [Sequoia Gold X Sparks]; Saville, F. Harmon

'SAVABREZ', Min, ob, 1984; (**Breezy**(TM)); bud small; flowers bright orange-red, yellow reverse, small blooms in sprays, dbl., 20 petals, exhibition form, moderate fragrance; foliage small, medium green, semi-glossy; upright growth; [Sheri Anne X Seedling]; Saville, F. Harmon; Nor'East Min. Roses

'SAVACALL', Min, or, 1984; (**Total Recall**(TM)); flowers medium, dbl., no fragrance; foliage small, medium green, semi-glossy; bushy growth; [Zorina X Baby Katie]; Saville, F. Harmon; Nor'East Min. Roses

'SAVACAMP', Min, mr, 1984; (**Scamp**(TM)); flowers micro-mini, small, dbl., 35 petals, exhibition form, slight fragrance; foliage small, medium green, semi-glossy; compact, bushy growth; [Baby Katie X (Yellow Jewel X Tamango)]; Saville, F. Harmon; Nor'East Min. Roses

'SAVACENT', Min, dy, 1981; (Atkins Beauty, **Center Gold**(TM)); bud pointed; flowers deep yellow, sometimes near white, borne singly or up to 12, dbl., 60 petals, exhibition form, moderate, spicy fragrance; long, thin prickles, slanted downward; foliage glossy, textured; upright, compact growth; AOE, 1982; [Rise 'n' Shine X Kiskadee]; Saville, F. Harmon; American Rose Foundation

'SAVACHASE', Min, rb, 1988; (**Chasin' Rainbows**™, Chasing Rainbows); bud ovoid; flowers very brilliant yellow, edged red with scarlet becoming more red with age, dbl., 21 petals, exhibition form, slight, spicy fragrance; no fruit; prickles long, thin, angled, light brown; foliage small, dark green, semi-glossy; micro-mini bushy, low growth; PP007058; [Zorina X Rainbow's End]; Saville, F. Harmon; Nor'East Min. Roses, 1990

'SAVACHILD', Min, pb, 1991; (**Child's Play**™); flowers porcelain pink/white bicolor, pink edges, semi-dbl., 1.5 in., exhibition form, borne singly or in sprays of 3+, moderate, sweet fragrance; foliage medium, dark green, matt; upright, medium growth; PP8175; AARS, 1993; [(Yellow Jewel X Tamango) X Party Girl]; Saville, F. Harmon; Nor'East Min. Roses

'SAVACIN', Min, r, 1986; (**Cinnamon Toast**™); flowers russet brown, small, dbl., 28 petals, exhibition form, slight fragrance; small, red prickles; foliage small, medium green, semi-glossy; low, upright, bushy growth; [Zorina X (Sheri Anne X Glenfiddich)]; Saville, F. Harmon; Nor'East Min. Roses

'SAVACLEND', Min, yb, 1999; (**Climbing Rainbow's End**™); flowers yellow blooms suffused with red at edges and bloom ages, gradually turni, dbl., 25–40 petals, 1.5 in., exhibition form, borne in small clusters, slight fragrance; prickles moderate; foliage small, medium green, semi-glossy; short, thick stems; climbing to 10 ft growth; [Sport of Rainbow's End sport]; Saville, F. Harmon

'SAVACLOUD', Min, w, 1988; (**White Cloud**™); flowers white with pale pink; PP006876; [Buttons 'n' Bows sport]; Saville, F. Harmon; Nor'East Min. Roses

'SAVACOOK', Min, ab, 1995; (**Fortune Cookie**™); flowers apricot blend, moderately full (15–26 petals), small blooms borne singly and in small clusters, no fragrance; no prickles; foliage small, medium green, semi-glossy; medium (16–18 in.), upright, compact growth; [Baby Katie X Mazurka]; Saville, F. Harmon; Nor'East Miniature Roses, 1996

'SAVACOP', Min, ob, 1988; (**Copper Sunset**™); bud pointed; flowers coppery-orange, flushed orange-red, reverse medium red, dbl., 21 petals, exhibition form, slight fragrance; ovoid, orange fruit; prickles long, thin, slanted, gray-red; foliage medium, dark green, semi-glossy; upright, medium, angular growth; PP007032; [Acey Deucy X Rainbow's End]; Saville, F. Harmon

'SAVACRIM', Min, dr, 1994; (**Little Crimson**); dbl., 15–25 petals, 1.5 in., borne mostly singly, no fragrance; few prickles; foliage small, dark green, semi-glossy; medium (16–20 in.), upright, bushy growth; [Teddy Bear X (Zorina X Baby Katie)]; Saville, F. Harmon; Nor'East Min. Roses, 1995

'SAVADAY', Min, op, 1988; (**Heavenly Days**™; 'SAVAHE'); flowers glowing indian-orange, reverse lemon yellow, flushed fire-red, dbl., 28–32 petals, cupped, no fragrance; foliage medium, medium green, glossy, underside matt; bushy, medium, compact growth; PP006808; AOE, 1988; [Unnamed Climbing Yellow Miniature seedling X (Sheri Anne X Glenfiddich)]; Saville, F. Harmon; Nor'East Min. Roses

'SAVADEE', Min, ob, 1988; (**Dee Bennett**™); flowers yellow and orange becoming orange-yellow, medium, dbl., 25 petals, exhibition form, borne usually singly, slight, fruity fragrance; no fruit; prickles long, thin, curved, gray-orange; foliage medium, dark green, semi-glossy; bushy, medium growth; PP006951; AOE, 1989; [Zorina X (Sheri Anne X (Yellow Jewel X Tamango))]; Saville, F. Harmon; Nor'East Min. Roses, 1989

'SAVADEN', Min, ob, 1995; (**Denver's Dream**™); flowers copper orange with red reverse, dbl., 15–25 petals, 1.5 in., cupped, borne mostly singly, no fragrance; foliage medium, dark green, semi-glossy; medium (16–20 in.), upright growth; PP9435; [Gingersnap X Klima]; Saville, F. Harmon; Nor'East Min. Roses, 1994

'SAVADI', Min, or, 1986; (**Baby Diana**™); flowers orange-red, yellow reverse, small, dbl., 20 petals, exhibition form, borne usually singly, moderate fragrance; long, thin, hooked, brown prickles; foliage small, medium green, semi-glossy; bushy growth; PP005957; [Zorina X (Sheri Anne X Glenfiddich)]; Saville, F. Harmon; Nor'East Min. Roses

'SAVADOLL', Min, ob, 1979; (**Calico Doll**); bud ovoid, pointed; flowers orange, striped yellow, semi-dbl., 18 petals, 1–1.5 in., cupped, foliage dark; compact growth; [Rise 'n' Shine X Glenfiddich]; Saville, F. Harmon; Nor'East Min. Roses

'SAVADREAM', Min, mp, 1990; (**Dreamer**™); bud ovoid, pointed; flowers dusty pink, medium, borne singly or in sprays of 3–5, dbl., 20 petals, cupped, no fragrance; foliage medium, dark green, semi-glossy; upright, bushy, medium growth; PP007757; [Baby Katie X Shocking Blue]; Saville, F. Harmon; Nor'East Min. Roses, 1991

'SAVAFACE', Min, dp, 1991; (**Happy Face**™); bud ovoid; flowers clear rosy pink, blooms borne usually singly or in sprays of 3–5, very dbl., 35–40 petals, cupped, no fragrance; foliage medium, dark green, glossy; bushy, medium, compact growth; [(Sheri Anne X Rise 'n' Shine) X Mountie]; Saville, F. Harmon, 1986; Nor'East Min. Roses

'SAVAFIRE', Min, rb, 1991; (**Fireworks**™); flowers brilliant orange-yellow bicolor, dbl., 35 petals, 1.5 in., exhibition form, profuse, borne singly or in small clusters, slight fragrance; foliage medium, dark green, semi-glossy; bushy, medium growth; PP8182; [(Rise 'n' Shine X Sheri Anne) X Rainbow's End]; Saville, F. Harmon; Nor'East Min. Roses, 1992

'SAVAFISH', Min, ob, 1992; (**Gone Fishin'**™); flowers bright orange, blooms borne singly and in sprays of 4–10, dbl., 28–35 petals, 1.25 in., cupped, slight fragrance; foliage medium, dark green, glossy, very disease resistant; bushy, medium growth; [Fairlane X Zorina]; Saville, F. Harmon; Nor'East Min. Roses, 1993

'SAVAGALA', Min, dp, 1999; (**Gala**); dbl., 17–25 petals, 1.5 in., borne mostly single and in small clusters, loose sprays, no fragrance; prickles moderate; foliage medium, dark green, semi-glossy; upright, spreading, medium (15–16 in.) growth; [High Jinks X seedling]; Saville, F. Harmon; Nor'East Miniature Roses, 2000

'SAVAGEN', Min, yb, 1982; (**Genevieve**™); flowers yellow, streaked scarlet, scarlet increasing with age, small, dbl., 35 petals, moderate, spicy fragrance; foliage medium, medium green, semi-glossy; upright, bushy growth; [Unnamed Miniature Cl. X Unnamed Miniature seedling]; Saville, F. Harmon; Nor'East Min. Roses, 1983

'SAVAGOOD', Min, my, 1991; (**Good Morning America**™); bud ovoid, urn-shaped; flowers urn-shaped, large blooms, sunshine yellow, slight red edges, very dbl., 45–50 petals, 1.5 in., exhibition form, borne mostly singly, moderate, fruity fragrance; foliage medium, dark green, semi-glossy; long stems; upright, bushy, tall growth; PP007761; AOE, 1991; [Fantasia X Rainbow's End]; Saville, F. Harmon, 1984; Nor'East Min. Roses

'SAVAHALO', Min, my, 1991; (**Golden Halo**™); bud ovoid, pointed; flowers bright yellow, medium, dbl., 24–26 petals, cupped, profuse, borne mostly singly, moderate fragrance; foliage medium, medium green, semi-glossy; upright, bushy growth; PP007761; AOE, 1991; [Arthur Bell X Rainbow's End]; Saville, F. Harmon, 1985; Nor'East Min. Roses

'SAVAHE', Min, op, 1988; (**Heavenly Days**™ 'SAVADAY'); flowers glowing indian-orange, reverse lemon yellow, flushed fire-red, dbl., 28–32 petals, cupped, no fragrance; foliage medium, medium green, glossy, underside matt; bushy, medium, compact growth; PP006808; AOE, 1988; [Unnamed Climbing Yellow Miniature seedling X (Sheri Anne X Glenfiddich)]; Saville, F. Harmon; Nor'East Min. Roses

'SAVAHOLD', Min, my, 1996; (**Behold**™); bud medium; flowers clear, bright medium yellow, reverse lighter, holds color, dbl., 15–25 petals, borne mostly singly, no fragrance; few prickles; foliage medium, medium green, semi-glossy; upright, compact, medium (22–26 in.) growth; [(Rise 'n' Shine X Sheri Ann) X

(Heideroslein X Nozomi)]; Saville, F. Harmon; Nor'East Min. Roses, 1997

'SAVAHOWDY', Min, pb, 1982; (**Minnie Pearl**™); flowers light pink, reverse darker, yellow base, exhibition form, borne mostly singly, slight fragrance; few prickles; foliage small, medium green, semi-glossy; thin, wiry stems; upright growth; PP5097; [(Little Darling X Tiki) X Party Girl]; Saville, F. Harmon; Nor'East Min. Roses

'SAVAJERRY', Min, mr, 1998; (**Jerry-O**™); flowers light or medium red, large blooms, dbl., 25 petals, 1.5 in., exhibition form, borne singly, intense fragrance; Saville, F. Harmon

'SAVAJINKS', Min, pb, 1992; (**High Jinks**™); dbl., 25–32 petals, 1.5 in., exhibition form, borne singly and in sprays of 4–30, slight fragrance; foliage medium, dark green, semi-glossy; upright, bushy, medium to tall growth; [Rise 'n' Shine X Sheri Anne]; Saville, F. Harmon; Nor'East Min. Roses, 1993

'SAVAKLIM', Cl Min, dy, 1993; (**Klima**™); semi-dbl., 6–14 petals, 1.5 in., no fragrance; few prickles; foliage small, medium green, semi-glossy; tall (4–5 ft), upright, climbing growth; [Ferris Wheel X COCagold]; Saville, F. Harmon; Nor'East Min. Roses, 1993

'SAVALAV', Min, m, 1994; (**Vista**™); flowers soft lavender, moderately excellent form, dbl., 15–25 petals, 1.5 in., exhibition form, borne mostly singly, slight fragrance; some prickles; foliage medium, medium green, semi-glossy; medium, compact growth; PP9031; [Sachet X Copper Sunset]; Saville, F. Harmon; Nor'East Min. Roses, 1994

'SAVALIFE', Min, yb, 1984; (**Rainbow's End**™); bud small; flowers deep yellow, red petal edges, aging red all over, dbl., 35 petals, exhibition form, borne in small clusters, no fragrance; foliage small, dark, glossy; upright, bushy growth, 18 in.; PP5482; AOE, 1986; [Rise 'n' Shine X Watercolor]; Saville, F. Harmon; Nor'East Min. Roses

'SAVALIGHTS', Min, rb, 1993; (**Lights of Broadway**™); flowers yellow with broad red edge, yellow reverse, very dbl., 1.5 in., borne mostly singly, no fragrance; some prickles; foliage medium, medium green, semi-glossy; medium (16–20 in.), upright, bushy growth; PP8918; [(Tamango X Yellow Jewel) X Party Girl]; Saville, F. Harmon; Nor'East Min. Roses, 1994

'SAVALOT', Min, ob, 1984; (**Cheers**™); flowers orange-red, cream reverse, dbl., 20 petals, slight fragrance; foliage small, medium green, semi-glossy; compact, bushy growth; [Poker Chip X Zinger]; Saville, F. Harmon; Nor'East Min. Roses

'SAVALUCK', Min, ob, 1987; (**Happy Go Lucky**™); flowers brilliant orange-yellow blend, small, dbl., 17–24 petals, exhibition form, borne usually singly, slight, sweet fragrance; no fruit; long, thin pointed prickles; foliage small, dark green, semi-glossy; bushy, medium growth; PP006506; [Cheers X (Sheri Ann X (Yellow Jewel X Tamango))]; Saville, F. Harmon; Nor'East Min. Roses

'SAVALUTE', Min, yb, 1998; (**Absolutely**™); flowers pale to medium yellow, dusted apricot pink, lighter reverse, borne singly and in small clusters, floriferous, slight fragrance; few prickles; foliage medium, medium green, semi-glossy; upright, compact, spreading, bushy, vigorous growth; [Seedling X Rise 'n' Shine]; Saville, F. Harmon; Nor'East Miniature Roses, 1998

'SAVAMAE', Min, ab, 1986; (**Ellamae**™); flowers, dbl., 35 petals, exhibition form, borne singly and in sprays of 3–5, moderate fragrance; long, thin prickles; foliage medium, dark, glossy; medium, upright, bushy growth; PP006053; [Zorina X (Sheri Anne X Glenfiddich)]; Saville, F. Harmon; Nor'East Min. Roses

'SAVAMARK', Min, or, 1982; (Apricot Sunblaze, Mark 1, **Mark One**™); flowers brilliant orange-red, blooms borne singly and up to 10 per cluster, dbl., 43 petals, cupped, moderate, spicy fragrance; long, thin prickles, soft on peduncles; foliage very glossy; compact, bushy growth; [Sheri Anne X Glenfiddich]; Saville, F. Harmon; Nor'East Min. Roses

'SAVAMIST', Min, ab, 1987; (**Apricot Mist**™); flowers apricot with tones of pink and yellow, medium, very dbl., 40–45 petals, exhibition form, borne singly, slight fragrance; no fruit; long, thin, straight, brown prickles; foliage small, dark green, glossy; bushy, low, profuse, compact, symmetrical growth; PP006507; [Fantasia X Baby Katie]; Saville, F. Harmon; Nor'East Min. Roses

'SAVAMOR', Min, m, 1995; (**Scentsational**™); flowers light mauve, large blooms, dbl., exhibition form, borne singly on long stems, intense fragrance; Saville, F. Harmon

'SAVANADE', Min, dy, 1997; (**Pacific Serenade**™); flowers double, medium, bright yellow, dbl., 15–25 petals, cupped, borne mostly singly, moderate fragrance; no prickles; foliage medium, medium green, semi-glossy; upright, compact, spreading, bushy, vigorous growth; [Cal Poly X New Zealand]; Saville, F. Harmon

'SAVANHOUR', Min, mr, 1983; (**Happy Hour**™); flowers bright medium red, yellow eye, small, dbl., 20 petals, moderate fragrance; foliage small, dark, glossy; bushy, spreading growth; PP005449; [(Tamango X Yellow Jewel) X Zinger]; Saville, F. Harmon; Nor'East Min. Roses

'SAVANICE', Min, w, 1991; (**Ice Queen**™); bud ovoid, pointed; flowers medium, white, fully double, very dbl., 60 petals, exhibition form, borne mostly singly, no fragrance; foliage medium, dark green, semi-glossy; bushy, medium growth; PP007771; [Cupcake sport]; Saville, F. Harmon, 1983; Nor'East Min. Roses

'SAVANIGHT', MinFl, mp, 1997; (**Overnight Scentsation**); flowers very full, large, very dbl., 41 petals, borne mostly singly, intense fragrance; foliage medium, medium to dark green, semi-glossy; upright, medium (2 ft) growth; [Taxi X Lavender Jade]; Saville, F. Harmon

'SAVANIN', Min, rb, 1986; flowers medium red with yellow center, moderately small, dbl., 15–25 petals, slight fragrance; foliage small, medium green, semi-glossy; bushy growth; [Rise 'n' Shine X Zinger]; Saville, F. Harmon; SNC Meilland & Cie, 1986

'SAVANNAH', HT, ab, 1980; bud ovoid, pointed; flowers soft apricot, borne singly or 2–3 per cluster, dbl., 32 petals, cupped, slight, spicy fragrance; long prickles, hooked down; foliage moderately leathery; vigorous, upright growth; [Unnamed seedling X Arizona]; Weeks, O.L.

Savannah Miss, *see* 'PIXANAH'

Savanne, HT, dy, 1985; Noack, Werner

'SAVAPACE', Min, w, 1979; (Pace Setter, **Pacesetter**); bud long, pointed; flowers pure white, dbl., 46 petals, 1.5 in., exhibition form, borne in small clusters, needs disbudding, moderate fragrance; few prickles; foliage dark, matt; vigorous, compact growth, tall; PP4513; AOE, 1981; [Ma Perkins X Magic Carrousel]; Schwartz, Ernest W.; Nor'East Min. Roses

'SAVAPAINT', Min, rb, 1982; (**Flying Colors**™); flowers red and yellow blend, aging to pink and white, micro-mini, semi-dbl., slight fragrance; foliage medium green, semi-glossy; upright, bushy growth; [(Yellow Jewel X Tamango) X Sheri Anne]; Saville, F. Harmon; Nor'East Min. Roses, 1983

'SAVAPEG', Min, lp, 1986; (**Peggy Jane**™); flowers light pink, with lighter petal edges and base; PP005999; [Starina sport]; Utz, Peggy L.; Nor'East Min. Roses

'SAVAPIE', Min, rb, 1991; (**Something Else**™); flowers white with very contrasting red edge, blooms borne usually singly, dbl., cupped, slight fragrance; foliage small, medium green, semi-glossy; upright, bushy, medium growth; [(Yellow Jewel X Tamango) X Party Girl]; Saville, F. Harmon; Nor'East Min. Roses, 1985

'SAVAPIECE', Min, mr, 1984; (**Centerpiece**™, Centre Piece); flowers deep medium red, small, dbl., 35 petals, exhibition form, slight fragrance; foliage small, dark, semi-glossy; bushy growth; PP007769; AOE, 1985; [(Sheri Anne X Tamango) X (Sheri Anne X (Yellow Jewel X Tamango))]; Saville, F. Harmon; Nor'East Min. Roses, 1985

'SAVAPPLE', Min, ab, 1999; (**Applause**); flowers medium coral, reverse lighter, dbl., 17–25 petals, 1.2 in., slight fragrance; prickles moderate; foliage small, medium green, semi-glossy; upright, compact, low growth; AOE, ARS, 2000;

[Sequoia Gold X Sparks]; Saville, F. Harmon; Nor'East Miniature Roses, 2000

'SAVAPUP', Min, ob, 1978; (**Puppy Love**); bud pointed; flowers pink, coral, orange blend, small, dbl., 23 petals, 1.5 in., exhibition form, slight fragrance; foliage matt; upright, compact growth; AOE, 1979; [Zorina X Unnamed seedling]; Schwartz, Ernest W.; Nor'East Min. Roses

'SAVARAIN', Min, m, 1989; (**Raindrops**™); bud ovoid; flowers light mauve-purple, light yellow at base, reverse lighter, dbl., 24 petals, borne in small clusters, slight fragrance; fruit not observed; prickles straight, slanted downward, small, gray-red; foliage small, dark green, semi-glossy; upright, medium growth; PP007748; [Sachet X Rainbow's End]; Saville, F. Harmon; Nor'East Min. Roses, 1990

'SAVARALPH', Min, mr, 1999; (**Ralph Moore**); flowers medium, dbl., 17–25 petals, borne mostly singly, slight fragrance; few prickles; foliage medium, dark green, semi-glossy; upright, compact, well-branched, medium (16 in.) growth; AOE, ARS, 2000; [Sachet X seedling]; Saville, F. Harmon; Nor'East Miniature Roses, 2000

'SAVAREND', Min, dp, 1987; (**Adam's Smile**™, 'SAVASMILE'); flowers deep pink, medium, dbl., 23–27 petals, exhibition form, borne usually singly and in sprays of 3–5, no fragrance; no fruit; prickles long, thin, pointed slightly downward, gray-red; foliage medium, medium green, semi-glossy; upright, bushy, medium growth; PP006570; [(Rise 'n' Shine X Sheri Anne) X Rainbow's End]; Saville, F. Harmon; Nor'East Min. Roses, 1991

Savaria, Pol, 1972; Mark; (Cavriglia)

'SAVARITA', Min, mp, 1989; (**Nostalgia**™); bud ovoid; flowers medium pink, reverse lighter, aging lighter, medium, borne u, dbl., 38 petals, cupped, no fragrance; fruit not observed; prickles thin, straight, medium, gray-purple to brown; foliage medium, medium green, semi-glossy; spreading, low growth; [Rita X (Rise 'n' Shine X Sheri Anne)]; Saville, F. Harmon; Nor'East Min. Roses, 1990

'SAVARS', Min, pb, 1991; (A.R.S. Centennial, **American Rose Centennial**™); bud ovoid; flowers creamy white, edged soft pink, very dbl., 50–55 petals, 1.5 in., exhibition form, borne singly or in sprays of 3–5, slight fragrance; foliage medium, dark green, semi-glossy; bushy, medium growth; PP8181; [High Spirits X Rainbow's End]; Saville, F. Harmon; Nor'East Min. Roses, 1992

'SAVASACH', Min, m, 1986; (**Sachet**™); flowers lavender, yellow stamens, urn-shaped to loose, mini-flora, dbl., 30 petals, intense, damask fragrance; medium, thin prickles; foliage small, dark, semi-glossy; medium upright, bushy growth; PP005967; [Unnamed yellow Miniature X Shocking Blue]; Saville, F. Harmon; Nor'-East Min. Roses

'SAVASAM', Min, mr, 1986; (Sampson, **Samson**™); flowers brilliant scarlet red, urn-shaped, blooms in sprays of 20–60, dbl., 28 petals, slight fragrance; long, thin, prickles; foliage large, dark, semi-glossy; tall, upright growth; [Sheri Anne X (Yellow Jewel X Tamango)]; Saville, F. Harmon; Nor'East Min. Roses

'SAVASCHOOL', Min, dy, 1993; (**School Days**); dbl., 26–40 petals, 1.5 in., borne mostly singly, no fragrance; few prickles; foliage medium, medium green, semi-glossy; medium (14–18 in.), bushy, compact growth; [Klima X Sonnenkind]; Saville, F. Harmon; Nor'East Min. Roses, 1994

'SAVASEAT', Min, pb, 1996; (**Seattle Scent-sation**™); flowers yellow, apricot base and mauve-pink blooms, dbl., 15–25 petals, 1.5–2.75 in., borne singly and in small clusters, intense fragrance; no prickles; foliage small, dark green, semi-glossy; medium (26–30 in.), upright, bushy, vigorous growth; Named in honor of ARS Spring Convention, Seattle, WA, ARS, 1996; [Lavender Jade X New Zealand]; Saville, F. Harmon; Nor'East Min. Roses, 1996

'SAVASHEL', Min, pb, 1988; (**Shelly Renee**™); flowers shrimp pink, reverse peach, aging light pink to white; foliage small, dark green, semi-glossy; PP006952; [Spice Drop sport]; Saville, F. Harmon; Nor'East Min. Roses, 1989

'SAVASILK', Min, ly, 1982; (**Cornsilk**™); flowers light pastel yellow, small, dbl., 40 petals, exhibition form, borne mostly singly, moderate fragrance; foliage medium, medium green, semi-glossy; vigorous, bushy growth; PP5164; AOE, 1983; [Rise 'n' Shine X Sheri Anne]; Saville, F. Harmon; Nor'East Min. Roses

'SAVASMILE', Min, dp, 1987; (**Adam's Smile**™, 'SAVAREND'); flowers deep pink, medium, dbl., 23–27 petals, exhibition form, borne usually singly and in sprays of 3–5, no fragrance; no fruit; prickles long, thin, pointed slightly downward, gray-red; foliage medium, medium green, semi-glossy; upright, bushy, medium growth; PP006570; [(Rise 'n' Shine X Sheri Anne) X Rainbow's End]; Saville, F. Harmon; Nor'East Min. Roses, 1991

'SAVASORC', Min, mr, 1994; (**Sorcerer**™); flowers bright, medium red, dbl., 15–25 petals, 1.5 in., borne mostly singly, floriferous, no fragrance; some prickles; foliage small, medium green, semi-glossy; medium (16–20 in.), upright, bushy growth; [Ginger Snap X Rainbow's End]; Saville, F. Harmon; Nor'East Min. Roses, 1995

'SAVASPARK', Min, mr, 1995; (**Sparks**); flowers medium red, semi-dbl. (6–14 petals), small blooms borne mostly in small to large clusters; no fragrance; few prickles; foliage small, medium green, semi-glossy; medium (14–16 in.) upright, compact bush; [(Zorina X Baby Katie) X Red Minimo]; Saville, F. Harmon; Nor'East Miniature Roses

'SAVASPIR', Min, mr, 1983; (**High Spirits**™); flowers small, dbl., 35 petals, exhibition form, blooms in sprays, slight fragrance; foliage small, dark, semi-glossy; upright growth; PP005450; [Sheri Anne X Tamango]; Saville, F. Harmon; Nor'East Min. Roses

'SAVASUN', Min, dy, 1986; (**Sunny Day**™); flowers bright yellow, fading lighter, slight red blush on edges, dbl., 30 petals, exhibition form, borne in small clusters, slight, spicy fragrance; no fruit; long, thin prickles, angled slightly downwards; foliage small, dark green, semi-glossy; upright, bushy, low growth; PP6213; [Golden Slippers X Rise 'n' Shine]; Saville, F. Harmon; Nor'East Min. Roses, 1987

'SAVASWEET', Min, op, 1982; ('SAVSWET', **Spice Drop**™); flowers light salmon-pink, micro-mini, small, dbl., 35 petals, exhibition form, slight fragrance; foliage small, medium green, glossy; tiny growth; [(Sheri Anne X Glenfiddich) X (Unnamed moss seedling X (Sarabande X Little Chief))]; Saville, F. Harmon; Nor'East Min. Roses

'SAVATHREE', Min, mr, 1982; (**Acey Deucy**™); flowers small, dbl., 20 petals, exhibition form, moderate fragrance; foliage small, medium green, semi-glossy; bushy growth; [(Yellow Jewel X Tamango) X Sheri Anne]; Saville, F. Harmon; Nor'-East Min. Roses

'SAVAVAL', Min, dp, 1980; (**Valerie Jeanne**); bud globular; flowers deep magenta pink, blooms borne 1–20 per cluster, dbl., 58 petals, exhibition form, slight fragrance; long, straight, thin prickles; foliage very glossy; vigorous, upright growth; AOE 1983; [Sheri Anne X Tamango]; Saville, F. Harmon; Nor'East Min. Roses

'SAVAVEL', Min, mr, 1993; (**Velvet Touch**); flowers small, medium red, dbl., 15–25 petals, 1.5 in., borne mostly singly, slight fragrance; some prickles; foliage medium, medium green, semi-glossy; medium, bushy, compact growth; [Rainbow's End X Acey Deucy]; Saville, F. Harmon; Nor'East Min. Roses, 1994

'SAVAWED', Min, mp, 1985; (**Wedded Bliss**™); semi-dbl., blooms in clusters, slight fragrance; foliage small, medium green, glossy; groundcover; very spreading growth; [(Yellow Jewel X Tamango) X Nozomi]; Saville, F. Harmon; Nor'East Min. Roses

'SAVAWEEK', Min, or, 1984; (**Julie Ann**™); flowers small, brilliant vermillion orange, dbl., 20 petals, 1–1.5 in., exhibition form, borne mostly singly, moderate fragrance; prickles moderate; foliage small, medium green, semi-glossy; upright, bushy growth; PP5415; AOE, 1984; [Zorina X Poker Chip]; Saville, F. Harmon; Nor'East Min. Roses

'SAVAWHOOP', Min, rb, 1991; (**Whoopi**™); flowers red and white blend, ages as red edges suffused more, dbl., 28–32 petals, 1.5 in., exhibition form, borne mostly singly or in small clusters, slight, spicy fragrance; foliage medium, dark green, semi-glossy; bushy, medium growth; PP8176; [(Yellow Jewel X Tamango) X Party Girl]; Saville, F. Harmon; Nor'East Min. Roses, 1992

'SAVawin', Min, m, 1984; (**Winsome**™); flowers lilac-lavender with red tinge, miniflora, dbl., exhibition form, borne singly and in sprays, no fragrance; foliage medium, dark, semi-glossy; upright, bushy growth; PP5691; AOE, 1985; [Party Girl X Shocking Blue]; Saville, F. Harmon; Nor'East Min. Roses, 1985

'SAVawire', Min, dp, 1988; (**Live Wire**™); flowers deep cardinal red-pink, reverse darker, aging lighter, loose, semi-dbl., 27–30 petals, cupped, no fragrance; no fruit; prickles thin, straight, dark gray-purple; foliage small, medium green, semi-glossy; micro-mini; bushy, low, compact growth; [(Rise 'n' Shine X Sheri Anne) X Rainbow's End]; Saville, F. Harmon; Nor'East Min. Roses

'SAVawist', Min, m, 1994; (**Wistful**); flowers medium mauve, good form, dbl., 15–25 petals, 1.5 in., borne singly, repeats well, slight fragrance; some prickles; foliage small, dark green, semi-glossy; medium (16–22 in.), upright, bushy growth; [Sachet X Rainbow's End]; Saville, F. Harmon; Nor'East Min. Roses, 1995

Save the Children, *see* 'HARtred'

'SAVember', Min, or, 1994; (**Ember**); dbl., 15–25 petals, 1.5 in., borne mostly singly, no fragrance; few prickles; foliage small, medium green, semi-glossy; medium (16–22 in.), upright, bushy, compact growth; [Copper Sunset X (Zorina X Baby Katie)]; Saville, F. Harmon; Nor'East Min. Roses, 1995

'Saverne', HT, rb, 1937; flowers nasturtium tinted brownish red, reverse tinted yellow, large, dbl.; foliage bright, bronze; Heizmann, E.; A. Meilland

'SAVinn', Min, w, 1997; (**Innocence**); flowers full to very full, medium, pure ivory white, very dbl., 26–41 petals, borne mostly singly, slight fragrance; few prickles; foliage medium, dark green, glossy; upright, spreading, bushy, medium (30 in.) growth; [SAVajinks X SAValite]; Saville, F. Harmon

Savkar, HT, mr, 1988; Patil, B.K.

'SAVmore', Min, dr, 1984; (**Red Shadows**™); flowers small, dbl., exhibition form, no fragrance; foliage small, medium green, semi-glossy; bushy, slightly spreading growth; [Tamango X Sheri Anne]; Saville, F. Harmon; Nor'East Min. Roses

'Savoia', HT, m, 1937; bud pointed, rosy lilac; foliage glossy; vigorous growth; [Julien Potin X Sensation]; Aicardi, D.; Giacomasso

'SAVor', Min, ob, 1982; (**Little Jackie**™); flowers light orange-red, yellow reverse, small, dbl., 20 petals, intense fragrance; foliage medium, medium green, semi-glossy; vigorous growth; AOE, 1984; [(Prominent X Sheri Anne) X Glenfiddich]; Saville, F. Harmon; Nor'East Min. Roses

Savoy Hotel, *see* 'HARvintage'

'SAVplenti', Min, ob, 1981; (**Penny Candy**™); bud ovoid; flowers orange-yellow blend, micro-mini, blooms borne singly or seve, dbl., 30 petals, cupped, slight fragrance; long, thin prickles; foliage small; tiny, compact growth; [Rise 'n' Shine X Sheri Anne]; Saville, F. Harmon; Nor'East Min. Roses

'SAVrojet', Min, dp, 1986; flowers deep pink, moderately small, dbl., 15–25 petals, no fragrance; foliage small, medium green, semi-glossy; bushy growth; [(Tamango X Yellow Jewel) X Watercolor]; Saville, F. Harmon; SNC Meilland & Cie, 1985

'SAVsay', Min, op, 1982; (**Heartland**™); bud short, pointed; flowers orange-red, blooms in clusters, dbl., 38 petals, exhibition form, slight fragrance; long, thin prickles; vigorous, upright growth; [Sheri Anne X Watercolor]; Saville, F. Harmon; Nor'East Min. Roses

'SAVswat', Min, dp, 1982; (**Windjammer**); flowers deep pink, reddened by sun, small, dbl., 35 petals, slight fragrance; foliage medium, medium green, semi-glossy; [Sheri Anne X Watercolor]; Saville, F. Harmon; Nor'East Min. Roses

'SAVswet', Min, op, 1982; ('SAVasweet' **Spice Drop**™); flowers light salmon-pink, micro-mini, small, dbl., 35 petals, exhibition form, slight fragrance; foliage small, medium green, glossy; tiny growth; [(Sheri Anne X Glenfiddich) X (Unnamed moss seedling X (Sarabande X Little Chief))]; Saville, F. Harmon; Nor'East Min. Roses

'SAVyk', Min, dy, 1999; (**Y2K**); flowers deep yellow with coating of coral, reverse medium yellow, dbl., 17–25 petals, 1.5–2 in., borne mostly singly, slight fragrance; few prickles; foliage medium, dark green, glossy; bushy, low (12–15 in.) growth; [Cal Poly X New Zealand]; Saville, F. Harmon; Nor'East Miniature Roses, 2000

Saya, Min, ob

Sayokyoku, *see* 'KEImipia'

'Sayonara', HT, yb, 1959; bud long, pointed; flowers yellow blend tinted pink, dbl., 50 petals, 5 in., moderate fragrance; foliage leathery; vigorous, upright growth; [Sunnymount sport]; Grillo

'Sazanami', Min, lp, 1984; flowers soft pink, small blooms borne 2–5 per cluster, dbl., 60 petals, flat, moderate fragrance; prickles slanted downward; foliage dark, semi-glossy; bushy growth; GM, Japan, 1980; [Yorokobi seedling X Yellow Doll]; Suzuki, Seizo; Keisei Rose Nursery, 1982

Sázava, S, mr, 1964; flowers medium, semi-dbl., slight fragrance; Pajer, J.; (Sangerhausen)

'Scabrata', S, m; flowers rich pinkish purple, large, single; foliage downy; [R. corymbifera (?) X R. gallica]

'Scabrosa', HRg, m, 1950; (R. rugosa scabrosa); flowers mauve-pink, stamens light sulfur, blooms in clusters of 5 or, single, 5 petals, 5 in., recurrent bloom, moderate, carnation fragrance; large, bright red fruit; foliage light, glossy, soft; very bushy (5 ft) growth; Harkness

Scala, *see* **'Clubrose Scala'**

Scala, *see* GAUesca

Scaldia, HT, 1987; Delforge; (Cavriglia)

Scamp™, *see* 'SAVacamp'

'Scandale', HT, pb, 1958; bud pink; flowers crimson shaded coppery, dbl., 25 petals, 4 in., moderate fragrance; foliage glossy, dark; long stems; vigorous, upright growth; [Peace X Opera]; Gaujard

'Scandia', Pol, op, 1951; flowers salmon-orange; [Margo Koster sport]; van de Water; van Nes

Scandia, HT, m, 1996; Cooper

Scandica, F, mr

'Scania', F, dr, 1965; flowers deep red, well-formed, 3.5 in.; foliage matt; [Cocorico X Seedling]; deRuiter

Scar P97, Misc OGR, lp, 1997; Scarman

'Scaramouche', F, mr, 1968; bud long, pointed; flowers coral-red, large, dbl., 30–35 petals, exhibition form, slight fragrance; foliage leathery; vigorous, compact, tall growth; [Ma Perkins X Duet]; Fankhauser

'Scarlano', HSet, mr, 1938; (Faust); flowers cerise-red, open, borne in clusters, semi-dbl., sometimes recurrent bloom, slight fragrance; foliage leathery, dark; short, strong stems; bushy (2½ ft) growth; [(R. setigera X Papoose) X Paul's Scarlet Climber]; Horvath; Wayside Gardens Co.

'Scarlet Adventurer', HT, dr, 1957; flowers scarlet, medium, dbl.; foliage dark; Lowe

'Scarlet Beauty', HT, dr, 1934; flowers crimson-scarlet, very large, dbl., slight fragrance; vigorous growth; [Mme Butterfly X Premier Supreme]; Vestal

'Scarlet Bedder', HT, ob, 1927; flowers rich orange-scarlet; dwarf growth; [Mme Edouard Herriot X Gen. MacArthur]; Henderson, W.H.

'Scarlet Betty Uprichard', HT, dr, 1930; bud long, pointed, shaded black; flowers intense scarlet, semi-dbl., cupped, intense fragrance; foliage thick, light; vigorous growth; [Betty Uprichard seedling]; Allen

'Scarlet Button', Pol, dr, 1932; flowers brilliant scarlet; [Locarno sport]; Dreer

'SCARLET CRAMPEL', Pol, dr; flowers scarlet; [Paul Crampel sport]

'SCARLET ELSE', F, dr; flowers scarlet, borne in trusses, semi-dbl., 10–12 petals, 3–4 in., slight fragrance; foliage leathery; very free growth; [Else Poulsen X Hybrid Tea seedling (red)]; Kordes, about 1925

'SCARLET EMPEROR', HT, mr, 1961; flowers scarlet-red, large, dbl., exhibition form; vigorous, upright growth; [Karl Herbst X Fandango]; LeGrice; Wayside Gardens Co.

Scarlet Fire, *see* **'SCHARLACHGLUT'**

'SCARLET FLAME', HT, mr, 1934; flowers brilliant red, petals recurved and fringed, large, dbl., moderate fragrance; very vigorous growth; Burbank; Stark Bros.

'SCARLET GARNETTE', F, mr, 1971; flowers scarlet, small, dbl., 45 petals, globular, slight fragrance; foliage dark; vigorous growth; [Garnette sport]; Newberry

Scarlet Gem®, *see* 'MEIDO'

'SCARLET GLORY', HT, ob, 1925; bud long, pointed; flowers orange-scarlet, dbl., exhibition form, slight fragrance; foliage leathery, rich green; Dickson, A.

'SCARLET GLOW', HT, dr, 1945; flowers brilliant velvety scarlet; [Briarcliff sport]; Sodano, A.; St Leonards Farms

Scarlet Glow, *see* **'SCHARLACHGLUT'**

Scarlet Hit, *see* POUlmo

Scarlet Knight, *see* 'MEIELEC'

'SCARLET KNIGHT, CLIMBING', Cl Gr, mr, 1972; (Samourai, Climbing); [Scarlet Knight sport]; Jack; A. Ross & Son

'SCARLET LADY', Min, mr, 1990; bud ovoid; flowers medium red, aging light pink, large, borne singly, dbl., 30 petals, exhibition form, no fragrance; foliage medium, dark green, semi-glossy; upright, medium growth; [Anita Charles X Chris Jolly]; Jolly, Nelson F., 1983; Rosehill Farm, 1991

'SCARLET LEADER', Pol, ob, 1927; flowers brilliant orange-scarlet, large, borne in clusters, dbl.; Wezelenburg

'SCARLET MARINER', F, mr, 1972; flowers bright scarlet-red, medium, dbl., exhibition form, moderate fragrance; foliage leathery; vigorous, bushy growth; [Unnamed seedling X Showboat]; Patterson; Patterson Roses

'SCARLET MARVEL', F, mr, 1958; flowers orange-scarlet, borne in clusters, dbl., 45–50 petals, 2.5 in., flat; foliage leathery; vigorous, compact growth; [Alain X Floribunda seedling]; deRuiter; C-P

Scarlet Meidiland™, *see* 'MEIKROTAL'

Scarlet Meillandécor, *see* 'MEIKROTAL'

Scarlet Meillandina, *see* 'MEICUBASI'

Scarlet Mimi™, *see* KEInoumi

Scarlet Moss™, *see* 'MORCARLET'

Scarlet Patio, *see* KORtingle

Scarlet Pavement, S, lp, 1991; bud pink; flowers light red, semi-dbl.; numerous, dark red fruit; spreading (2.5 ft) growth; Uhl, J.

Scarlet Pearl, *see* 'MANSCARLET'

Scarlet Pimpernel, *see* 'MEIDO'

'SCARLET QUEEN', F, dr, 1939; bud long, pointed; flowers pure scarlet, open, large, borne in clusters, dbl., slight fragrance; foliage glossy, leathery, bronze; long stems; vigorous, bushy growth; [Dance of Joy X Crimson Glory]; Kordes; Morse

Scarlet Queen Elizabeth®, *see* 'DICEL'

'SCARLET RIBBON', Cl Min, dr, 1961; bud ovoid; flowers red, sometimes almost maroon, dbl., 50 petals, 1.25 in., exhibition form, slight fragrance; vigorous (3 ft) growth; [((Soeur Therese X Wilhelm) X (Seedling X Red Ripples)) X Zee]; Moore, Ralph S.; Sequoia Nursery

Scarlet Rosamini, *see* RUIrupo

'SCARLET ROYAL', HT, mr, 1963; flowers scarlet, base yellow, well formed; vigorous growth; [Karl Herbst X Independence]; Park; Tantau Roses

'SCARLET RUFFLES', Min, or, 1990; bud pointed; flowers orange-red with yellow eye, aging to light red, loose, large, semi-dbl., 8 petals, intense, spicy fragrance; oblong, gold and orange fruit; prickles straight, slightly hooked downwards, tan; foliage medium, dark green, glossy; spreading, medium growth; [Poker Chip X Zinger]; Gruenbauer, Richard, 1984; Richard Gruenbauer

'SCARLET SENSATION', LCl, dr, 1954; (Everblooming Pillar No.73); bud high pointed, crimson; flowers rose-madder, borne in clusters, dbl., 35 petals, 3.5–4.5 in., free bloom, moderate fragrance; growth like a hybrid tea, followed by 4–5 foot canes; [Unnamed seedling X Queen o' the Lakes]; Brownell, H.C.

Scarlet Showers, LCl, mr

Scarlet Spreader™, *see* 'WILSCARLET'

Scarlet Star, *see* 'WILSTAR'

Scarlet Sunblaze™, *see* 'MEICUBASI'

'SCARLET SUNSET', F, or, 1970; flowers small, borne on trusses, semi-dbl., 12 petals; foliage dark, leathery; moderate bushy growth; [Orange Sensation X Seedling]; Geo. deRuiter

Scarlet Sweet Brier, *see* **'LA BELLE DISTINGUÉE'**

'SCARLET TRIUMPH', F, rb, 1951; bud ovoid; flowers deep scarlet, base yellow, small, borne in clusters, semi-dbl., cupped, slight fragrance; foliage glossy, light green; vigorous, bushy growth; [Orange Triumph sport]; Poulter

'SCARLET WAVES', F, mr, 1961; flowers bright scarlet, ruffled petals; tall growth; [(Florence Mary Morse X Border Queen) X Mrs Inge Poulsen]; Bennett, H.; Pedigree Nursery

'SCARLET WONDER', F, or, 1958; flowers bright orange-scarlet, borne in clusters, semi-dbl., 3 in., flat, slight fragrance; foliage dark, glossy; vigorous growth; [Signal Red X Fashion]; deRuiter; Blaby Rose Gardens

'SCARLETINA', Min, or, 1985; flowers spiraled, blooms borne usually singly, dbl., moderate fragrance; foliage small, medium green, semi-glossy; upright, bushy growth; [Futura X Poker Chip]; Hardgrove, Donald L.; Rose World Originals

'SCARLETT O'HARA', Pol, rb, 1947; bud ovoid; flowers brilliant red, overcast orange, open, large, borne in clusters, dbl., slight fragrance; foliage leathery; vigorous, bushy growth; RULED EXTINCT 2/88; [Gloria Mundi sport]; Klyn

Scarlett O'Hara™, *see* 'ARORESAS'

Scarletta®, Min, mr, 1972; deRuiter

Scarman's Crimson China, HCh, mr, 1995; Scarman

Scentasia, *see* 'SUNSCENT'

'SCENTED AIR', F, op, 1965; flowers salmon-pink, well-formed, blooms in clusters, dbl., 5 in., intense fragrance; foliage very large; vigorous growth; GM, Belfast, 1967 GM, The Hague, 1965; [Spartan seedling X Queen Elizabeth]; Dickson, Patrick; A. Dickson

Scented Bouquet, lp; [My Choice X Great Venture]; Dawson, George

'SCENTED BOWL', HT, dr, 1965; bud ovoid; flowers bright red, medium, dbl., intense fragrance; foliage glossy; open, upright growth; [Gen. MacArthur X ?]; Pal, Dr B.P.; Indian Agric. Research Inst.

Scented Dawn, *see* 'MEITOSIER'

'SCENTED STAR', HT, lp, 1973; flowers coral-pink, classic form, dbl., 30–35 petals, 5–5.5 in., intense fragrance; foliage dark; moderate growth; [Fragrant Cloud X Spek's Yellow]; Lowe

Scentimental™, *see* 'WEKPLAPEP'

Scent-Sation, *see* FRYromeo

Scentsational™, *see* 'SAVAMOR'

Scepter'd Isle, *see* 'AUSLAND'

'SCEPTRE', HT, ob, 1923; flowers bright flame, base shaded orange, reverse dull yellow; low growth; McGredy

Schackenborg, HT, dr, 1997; Poulsen

'SCHAFFNERS ERFOLG', F, mr; dbl., 30 petals, 2 in., borne in trusses, vigorous, bushy growth; [Red Favorite X Fanal]; Tantau

SCHanbiran, Min, rb, 1984; (**Magic Sunblaze**); Schwartz

'SCHARLACHGLUT', S, dr, 1952; (Scarlet Fire, Scarlet Glow); flowers scarlet-crimson, single, 5 in.; foliage dull green; vigorous

(4–5 ft), dense, spreading growth; [Poinsettia X Alika]; Kordes

'SCHARNKEANA', S, m; (R. X scharnkeana); flowers rose-purple, 1–5 together; 3 ft growth; [R. californica X R. nitida]

'SCHÉHÉRAZADE', HT, mr, 1942; bud large, oval; flowers fiery red, dbl., cupped, slight fragrance; foliage dark, glossy; vigorous, bushy growth; GM, Rome, 1940; Mallerin, C.; A. Meilland

Scherzo®, *see* 'MEIPUMA'

'SCHIEHALLION', HT, mr, 1982; dbl., 50 petals, 4.5 in., intense fragrance; foliage medium, medium green, glossy; upright growth; [Red Planet X Bonnie Anne]; MacLeod, Major C.A.

'SCHILLER', HMult, mp, 1913; flowers clear pink, medium, borne in clusters; [Trier X Lady Mary Fitzwilliam]; Lambert, P.

Schleswig, *see* **'MAID OF HONOUR'**

Schleswig 87®, *see* KORtara

'SCHLESWIG-HOLSTEIN', HT, rb, 1921; flowers reddish yellow; [Mme Edouard Herriot sport]; Engelbrecht

Schloss Balthasar, *see* KORpalmor

Schloss Dryburg, Cl HT, dy, 1969; flowers luminous golden yellow, large, dbl.; GPG Bad Langensalza; (Sangerhausen)

'SCHLOSS FRIEDENSTEIN', HMult, m, 1915; flowers reddish-violet, small, dbl., moderate fragrance; Schmidt, I. C.; (Sangerhausen)

Schloss Heidegg, *see* 'MEIPOQUE'

Schloss Mannheim®, *see* 'KORSCHLOSS'

'SCHLOSS MORITZBURG', F, dr, 1967; flowers medium, semi-dbl., cupped, slight fragrance; foliage dark, leathery; vigorous growth; [Donald Prior X ?]; Haenchen, E.; Teschendorff

'SCHLOSS SEUSSLITZ', HSpn, ly; flowers very large, semi-dbl., very early bloom, moderate growth; Dechan, 1933

Schlösser Brilliant, Climbing, *see* **'DETROITER, CLIMBING'**

Schlosser's Brilliant, *see* **'DETROITER'**

Schlossgarten, S, mp, 1989; flowers large, dbl.; GPG Bad Langensalza; (Sangerhausen)

'SCHMETTERLING', HFt, yb, 1905; flowers yellow and red, medium, semi-dbl., moderate fragrance; Müller, Dr F.; (Sangerhausen)

'SCHMID'S IDEAL', (strain of R. canina), lp; thorny; used as an understock; R. Schmidt

'SCHMID'S REKORD', S, 1930; Schmid, R.; (Cavriglia)

'SCHNEEBALL', HWich, w, 1905; flowers medium, semi-dbl., moderate fragrance; Weigand, C.; (Sangerhausen)

Schneeberg, S, w

Schnee-Eule, HRg, w, 1989; flowers large, semi-dbl., intense fragrance; Uhl, J.; (Sangerhausen)

Schneekonigin®, *see* 'TANINGO'

Schneekonigin®, *see* TANOTAX

'SCHNEEKOPF', HP, w, 1903; flowers snow-white or light pink, regular form, recurrent bloom, vigorous growth; [Mignonette X Souv. de Mme Sablayrolles]; Lambert, P.

Schneekoppe, HRg, w, 1984; (**Snow Pavement**); Baum

Schneekusschen®, Min, w, 1993; Kordes

'SCHNEELICHT', HRg, w, 1894; flowers pure white, large blooms in clusters, single; makes an impenetrable, prickly hedge; very vigorous, climbing growth; very hardy; Geschwind, R.

'SCHNEEPRINZESSIN', HT, w, 1946; bud long; flowers large, intense fragrance; vigorous growth; Meilland, F.; Pfitzer

'SCHNEESCHIRM', F, w, 1946; flowers white, center tinted rose-yellow, large blooms in clusters, single, slight fragrance; foliage dark; [Johanna Tantau X (Karen Poulsen X Stammler)]; Tantau

Schneesturm®, *see* TANmurse

Schneewalzer, *see* 'TANSCHNEEWA'

Schneewalzer, *see* TANrazlaw

Schneeweisschen®, *see* TANensch

Schneewittchen, *see* 'KORBIN'

'SCHNEEWITTCHEN', Pol, ly, 1901; flowers yellow fading to white; [Aglaia X (Paquerette X Souv. de Mme Levet)]; Lambert, P.

Schneewolke, S, w, 1996; Noack, Werner

'SCHNEEZWERG', HRg, w, 1912; (Snow Dwarf, Snowdwarf); flowers snow-white, stamens golden yellow, blooms in clusters of 3–1, semi-dbl., flat, recurrent bloom, abundant, small, red fruit; prickles spiny; foliage glossy, rugose; vigorous (3–4 ft), growth; (14); [Possibly R. rugosa X Polyantha hybrid]; Lambert, P.

'SCHOBITET', Min, my, 1987; (Royal Meillandina, **Royal Sunblaze**®); flowers lemon yellow, medium, full, quilled petals, very dbl., rosette, borne in small clusters, slight fragrance; foliage medium, medium green, semi-glossy; mini-flora, bushy (15 in.) growth; PP005690; [Unnamed seedling X Unnamed seedling]; Schwartz, Ernest W.; SNC Meilland & Cie, 1984

'SCHOENER'S MUSK', HMsk, w; flowers milk-white, medium enormous trusses; vigorous, pillar growth; Schoener

'SCHOENER'S NUTKANA', S, mp, 1930; flowers clear rose-pink, blooms in arching canes, single, 4 in., non-recurrent, moderate fragrance; few prickles; vigorous, shrub (4–6 ft) growth; [R. nutkana X Paul Neyron]; Schoener; C-P

Scholle's Golden Moss, M, dy, 1985; flowers amber-yellow, medium, dbl., moderate fragrance; Scholle, E.; (Sangerhausen)

'SCHÖN INGEBORG', HP, lp, 1921; flowers large, dbl.; Kiese; (Sangerhausen)

Schone Berlinerin®, *see* TANrised

Schone Dortmunderin®, *see* NOAdort

Schöne Münchnerin®, *see* 'KORLEEN'

'SCHÖNE VON HOLSTEIN', Pol, mp, 1919; flowers pure hermosa pink, dbl.; [Orléans Rose X ?]; Tantau

'SCHÖNE VON KAISERSLAUTERN', F, op, 1957; bud long, pointed; flowers orange tinted salmon-red, dbl., 30 petals, 4 in., exhibition form, free, intermittent bloom, intense fragrance; foliage dark, leathery; strong stems; very vigorous, upright, bushy growth; [R.M.S. Queen Mary X Obergärtner Wiebicke]; Kordes

'SCHÖNE VON MARQUARDT', HWich, rb, 1928; flowers bright dark red, variegated with white; vigorous, climbing growth; [Sodenia sport]; Clauberg

'SCHONERTS MEISTERKLASSE', HT, or, 1952; flowers salmon and coral-red, well formed, very large, dbl., moderate fragrance; vigorous growth; Leenders, M.

School Days, *see* 'SAVASCHOOL'

'SCHOOLGIRL', LCl, ab, 1964; flowers orange-apricot, well-formed, 4 in., recurrent bloom, [Coral Dawn X Belle Blonde]; McGredy, Sam IV; McGredy

Schubert®, *see* 'LENMOR'

Schultheis' American Beauty, HP, dp

'SCHUPRAK', HT, pb, 1989; (**Susan Elizabeth**); bud ovoid; flowers cream tipped with very deep purple-pink, reverse cream to pa, dbl., 40 petals, exhibition form, slight, fruity fragrance; fruit not observed; prickles slightly recurved, sparse, red; foliage medium, medium green, semi-glossy; upright, medium growth; [Pristine X Akebono]; Schlueter, Barry; Schlueter Rose Culture

Schuss®, *see* MEIliaxi

'SCHWABENLAND', HRg, mp, 1928; flowers amaranth-pink, open, large, dbl., profuse, repeated bloom, moderate fragrance; foliage large, rich green, leathery; vigorous (3 ft) growth; [R. rugosa seedling X Elizabeth Cullen]; Berger, V.; Pfitzer

Schwabenmadel, F, dr

'SCHWÄBISCHE HEIMAT', HT, dr, 1934; flowers deep amaranth-red, dbl.; [Jonkheer J.L. Mock sport]; Pfitzer

Schwanensee, *see* **'SWAN LAKE'**

Schwarzaldmadel, F, 1977; Hetzel, K.; (Cavriglia)

Schwarze Madonna®, *see* KORschwama

Schwarzer Samt, HKor, dr, 1969; flowers dark velvety red, medium, semi-dbl.; VEG; (Sangerhausen)

Schwarzwaldmadel®, F, ob, 1977; Hetzel

'**Schweizer Gold**', HT, ly, 1975; (Swiss Gold); bud ovoid; dbl., exhibition form, moderate fragrance; foliage large, light; vigorous growth; GM, Baden-Baden, 1972; [Peer Gynt X King's Ransom]; Kordes; Horstmann

Schweizer Gold, HT, lp; Urban, J.; (Czech Rosa Club)

Schweizer Grüss, *see* 'TANschweigru'

Schweizer Grüss, Climbing,

Schweizer Woche®, HT, w, 1989; Huber

'**Schwerin**', HMsk, dp, 1937; bud long, pointed; flowers light crimson, open, large clusters, cupped, profuse, intermittent bloom, moderate, musk fragrance; foliage large, leathery, glossy, bronze; long, strong stems; very vigorous, bushy growth; [Eva X D.T. Poulsen]; Kordes

'**Scintillation**', S, lp, 1967; flowers pink, blooms in large clusters, semi-dbl., intense fragrance; foliage dark, matt; vigorous growth; [R. macrantha X Vanity]; Austin, David; Sunningdale Nursery

Scipion Cochet, T, dp

Scipion Cochet, HP, dr

Scipion Cochet, T, pb, 1893

'**Scoop Jackson**', Gr, dr, 1983; (Canterbury, Fidelity); flowers large, dbl., 20 petals, slight fragrance; foliage glossy; upright growth; [Kalahari X John Waterer]; McGredy, Sam IV; Roses by Fred Edmunds

'**Scorcher**', Cl HT, dr, 1922; flowers brilliant scarlet-crimson, semi-dbl., 4 in., non-recurrent, slight fragrance; foliage large, wrinkled; vigorous, climbing or pillar (to 10 ft) growth; [Mme Abel Chatenay X Seedling]; Clark, A.; Hackett

'**Scotch Blend**', HT, pb, 1975; bud long, pointed; dbl., 35 petals, 5–6 in., exhibition form, slight fragrance; foliage dark, leathery; upright growth; [Queen Elizabeth X Peace]; J&B Roses; Eastern Roses

Scotch Heather, HSpn, yb, 1996; Sutherland

Scotch Rose, *see* R. spinosissima

Scotch Yellow, HT, my, 1991; Mercer

Scotland Yard, HT, ab

Scotland's Trust, *see* 'COClands'

Scott William, F, lp; [Oz Gold X Gold Bunny]; Hannemann, F., 1992; The Rose Paradise; (Weatherly, L.)

Scottish Soldier, S, yb, 1999; Williams, J. Benjamin

Scottish Special, *see* 'COCdapple'

Scottish Tartan, F, yb

'**Scott's Columbia**', HT, dp, 1928; flowers clear bright pink; [Columbia sport]; Scott, R.

Scout, *see* '**Boy Scout**'

'**Scrabo**', F, op, 1968; flowers light salmon-pink, large, dbl., exhibition form, moderate fragrance; free growth; [Celebration X Elizabeth of Glamis]; Dickson, A.

'SCRivleo', HT, ly, 1997; (**Albert Weedall**); flowers double, large, dbl., 15–25 petals, borne in large clusters, intense fragrance; some prickles; foliage medium, medium green, semi-glossy; upright, tall (5 ft) growth; [Seedling X Seedling]; Scrivens, Len

'SCRivluv', Min, dy, 1992; (**Baby Love**™); bud small, pointed; flowers buttercup yellow, single, 5 petals, 1.5 in., borne mostly singly, slight, licorice fragrance; some prickles; foliage small, medium green, semi-glossy; low (400 cms), compact growth; [Sweet Magic X Seedling]; Scrivens, Len

'SCRivo', F, lp, 1997; (**Pretty Lady**); flowers semi double, medium, semi-dbl., 8–15 petals; many prickles; foliage medium, dark green, semi-glossy; medium (4 ft) growth; [(seedling X (R. davidii elongata X seedling)) X ((Troika X Alpine Sunset) X Freedom)]; Scrivens, Len

Scudbuster, *see* 'CLEscrub'

'**Sea Foam**'®, S, w, 1964; flowers white to cream, blooms in clusters, dbl., recurrent bloom, slight fragrance; foliage small, glossy, leathery; vigorous, climbing trailer, semi-prostrate growth; David Fuerstenberg Prize, ARS, 1968 GM, Rome 1963; [((White Dawn X Pinocchio) X (White Dawn X Pinocchio)) X (White Dawn X Pinocchio)]; Schwartz, Ernest W.; C-P

Sea Foam Rosa, S, 1971; Zandri, R.; (Cavriglia)

'**Sea Mist**', HT, w, 1960; bud long, pointed; flowers cream, center golden, large, semi-dbl., exhibition form, slight fragrance; foliage leathery; vigorous growth; [Helen Traubel X Golden Harvest]; Armbrust; Langbecker

'**Sea Nymph**', Min, pb, 1986; flowers petals blush pink, edged coral, large blooms borne singly, dbl., 48 petals, exhibition form, no fragrance; medium, curved prickles; foliage medium, medium green, semi-glossy; upright, bushy growth; [Unnamed seedling X Unnamed seedling]; McDaniel, Earl; McDaniel's Min. Roses

'**Sea of Fire**', F, or, 1954; (Feuermeer); flowers orange-scarlet, blooms in small clusters, semi-dbl., 3 in., slight fragrance; foliage dark, leathery; vigorous, upright, bushy growth; [(Baby Chateau X Else Poulsen) X Independence]; Kordes

'**Sea of Tranquility**', HT, ab, 1991; bud classical; flowers delicate light pink in bud, opening to apricot pink blush, l, very dbl., 40–45 petals, 4.75 in., exhibition form, slight, sweet fragrance; foliage light green, small, disease-resistant; upright (up to 5 ft), non-spreading growth; [Sylvia sport]; Keene's Rose Nursery

'**Sea Pearl**', F, pb, 1964; (Flower Girl); bud long, pointed; flowers soft pink, reverse flushed peach and yellow, well-formed, bl, dbl., 24 petals, 4.5 in.; foliage dark; upright, bushy growth; [Kordes' Perfecta X Montezuma]; Dickson, Patrick; A. Dickson

'**Sea Spray**', HMsk, w, 1923; flowers stone-white, flushed pink, borne in clusters, non-recurrent, height 3–5 ft; RULED EXTINCT 12/82; Pemberton

'SEAbasket', Min, rb, 1995; (**Flower Basket**); flowers hand-painted carmine red to pink stripes, fades to white with dark red bud, prominent yellow stamens, dbl., (15–25 tightly quilled petals), large blooms borne mostly single, moderate fragrant; numerous prickles; foliage small, medium green, semi-glossy; bushy, very low growth; [Rose Gilardi X Ain't Misbehavin']; McCann, Sean; Justice Miniature Roses, 1996

'SEAbet', Min, pb, 1993; (**Elizabeth Munn**); flowers mainly pink with lighter reverse, dbl., 15–25 petals, flat, borne in small clusters, some prickles; foliage medium, medium green, semi-glossy; medium, upright growth; McCann, Sean; McCann, 1993

'**Seabird**', HT, my, 1913; flowers primrose-yellow; vigorous growth; Dickson, H.

'SEAbla', Min, dr, 1990; (**Ain't Misbehavin'**); single, 5 petals, slight fragrance; foliage small, green, semi-glossy; bushy growth; [Oonagh X (Pot Black X Black Jade)]; McCann, Sean, 1991

Seabreeze, *see* 'LEMsea'

'SEAcatch', Min, rb, 1995; (**Dreamcatcher**); flowers medium to dark red striped with white, blooms borne singly, dbl., 35–40 petals, 2–2.5 in., moderate, damask fragrance; foliage medium, medium green, glossy; tall, spreading growth; RR; [Lady in Red X [(Rose Gilardi X Wit's End) X (Remember Me X Stars 'n' Stripes)]]; McCann, Sean; Justice Miniature Roses, 1995

'SEAcloak', Min, dr, 1999; (**Velvet Cloak**); flowers burgundy red, very prominent yellow stamens, dbl., 17–25 petals, 2 in., borne mostly singly, slight fragrance; prickles moderate; foliage medium, medium green, semi-glossy; bushy, medium (20–24 in.) growth; [Lady in Red X Ain't Misbehavin]; McCann, Sean; Justice Miniature Roses, 1998

'SEAcoat', Min, yb, 1996; (**Dreamcoat**); flowers medium to dark yellow with scarlet accent, fades to reddish, dbl., 15–25 petals, 2.5 in., slight fragrance; foliage large, dark green, semi-glossy; upright, tall (4–6 ft) growth; RR; [Lady in Red X Joseph's Coat]; McCann, Sean; Justice Miniature Roses, 1997

'SEAdarby', Min, ob, 1996; (**Darby O'Gill**); flowers orange with creamy white striping, blooms borne mostly singly, dbl., 18 petals, 1.5 in., slight fragrance; some prickles; foliage medium, dark green, glossy; upright, medium (24–26 in.) growth; PPAF; [Tattooed Lady X Lady in Red]; McCann, Sean; Justice Miniature Roses, 1996

'SEAdark', Min, dr, 1996; (**Dark Mirage**); flowers velvety dark red with very dark shading at tip, reverse dark, dbl., 26–40 petals, 2 in., exhibition form, intense fragrance; foliage medium, dark green, glossy, highly serrated; long stems; medium (18–20 in.) growth; RR; [Lady in Red X Fountain]; McCann, Sean; Justice Miniature Roses, 1997

'SEAdeep', Min, rb, 1993; (**Tattooed Lady**); flowers red, yellow reverse, yellow stripes and veins, blooms borne, dbl., 15–25 petals, 1.5–2.75 in., slight fragrance; many prickles; foliage medium, dark green, bronze, glossy; medium (12–14 in.), upright growth; [Siobhan X (Picasso X Near You)]; McCann, Sean; Justice Miniature Roses, 1994

'SEAdot', Min, ob, 1988; (**Crazy Dottie**); flowers orange-red with star shaped copper center, small, single, 5 petals, slight fragrance; foliage small, medium green, semi-glossy; bushy growth; [Rise 'n' Shine X (Sheri Anne X Picasso)]; McCann, Sean

'SEAdow', HT, mr, 1992; (**Margaret McDowell**); dbl., 26–40 petals, 1.5–2.75 in., slight fragrance; some prickles; foliage medium, medium green, semi-glossy; medium (36"), upright growth; [[Ruby Wedding X (Seedling X Oonagh)] X (Seedling X Oonagh)]; McCann, Sean, 1993

'SEAeus', F, pb, 1999; (**Eustacia**); flowers handpainted with patches of pink and red, yellow center, dbl., 15–25 petals, 3 in., borne in small clusters, slight fragrance; prickles numerous; foliage medium, medium green, glossy; upright, medium (30 in.) growth; [(Copper Pot X Maxi) X Picasso]; McCann, Sean

'**Seafair**', HT, ab, 1959; bud long, pointed; flowers deep apricot, large, dbl., 45 petals, exhibition form, slight fragrance; foliage glossy; vigorous, upright, symmetrical growth; RULED EXTINCT 4/86; [Charlotte Armstrong X Signora]; Von Abrams; Peterson & Dering

Seafarer, *see* 'HARtilion'

Seafirst, *see* 'PEAfirst'

'SEAfun', Min, pb, 1997; (**Looks Like Fun**); flowers very full, medium, very dbl., 41 petals, borne mostly singly, moderate fragrance; foliage medium, medium green, semi-glossy; spreading, bushy, medium (16 in.)growth; [Lady in Red X (Rose Gilardi X Seedling)]; McCann, Sean

'**Seager Wheeler**', HSpn, lp, 1947; dbl., non-recurrent, height 6 ft; hardy; [R. spinosissima altaica X ?]; Wheeler; P.H. Wright

'SEAgold', Min, my, 1987; (**Gold Country**); flowers small blooms borne singly, dbl., 20 petals, exhibition form, intense fragrance; foliage small, light green, semi-glossy; bushy growth; [Rise 'n' Shine X (Rise 'n' Shine X Casino)]; McCann, Sean

'SEAgood', Min, mp, 1990; (**Lady Be Good**); dbl., 15–25 petals, slight fragrance; foliage small, medium green, semi-glossy; bushy growth; [Kiss 'n' Tell X (Irish Mist X Matangi)]; McCann, Sean, 1991

'SEAgru', Min, w, 1983; (**Blushing Groom**); flowers near white, small, dbl., 35 petals, slight fragrance; foliage small, dark, semi-glossy; upright growth; [Rise 'n' Shine X Karl Herbst]; McCann, Sean

'**Seagull**', HWich, w, 1907; flowers pure white, stamens golden, blooms in large clusters, single; very vigorous growth; Pritchard

'SEAheart', Cl Min, rb, 1990; (**Irish Heartbreaker**); flowers small blooms, dbl., 26–40 petals, slight fragrance; foliage medium, medium green, semi-glossy; upright growth; [Rise 'n' Shine X (Oonagh X Siobhan)]; McCann, Sean, 1991

'SEAkiss', Min, ab, 1985; (**Kiss 'n' Tell**); flowers small, dbl., 35 petals, flat, slight fragrance; foliage small, medium green, semi-glossy; bushy growth; [Rise 'n' Shine X (Sally Mac X New Penny)]; McCann, Sean, 1989

'SEAlady', Min, rb, 1988; (**Lady in Red**); flowers red with touch of white at base of petals, small, dbl., 15–25 petals, slight fragrance; foliage small, medium green, semi-glossy; bushy growth; [Rise 'n' Shine X Siobhan]; McCann, Sean, 1990

'SEAlark', HT, rb, 1998; (**Big Jim Larkin**); flowers red with silver reverse, silvers flecks on top surface, very dbl., 26–40 petals, 4.5–5 in., borne in large clusters, moderate fragrance; foliage large, dark green, semi-glossy; free blooming, spreading, medium (4 ft) growth; [Lady in Red X Old Master]; McCann, Sean; Hughes Roses, 1999

'SEAlife', Min, rb, 1996; (**High Life**); flowers medium red with white at throat, reverse white with red, fad, dbl., 26–40 petals, 2.5 in., exhibition form, slight fragrance; foliage large, medium green, dull; upright, tall (30 in.) growth; RR; [Lady in Red X Oriana]; McCann, Sean; Justice Miniature Roses, 1997

Sealily, *see* 'LEMlil'

'**Sealing Wax**', HMoy, mp; flowers large; abundant, bright red fruit; height 8 ft; Royal Hort. Soc.

'SEAlove', Min, rb, 1989; (**Lovers Only**); flowers cherry red, with straw-yellow bicolor, reverse fades to cream, dbl., 15–25 petals, slight fragrance; prickles pubescent on peduncle; foliage medium, medium green, glossy; upright, compact growth; [Rise 'n' Shine X Siobhan]; McCann, Sean

'SEAmar', HT, w, 1985; (**Martin Martin**); flowers cream, petals tipped red, large, exhibition form, foliage large, dark, semi-glossy; upright growth; [Jimmy Greaves X Irish Gold]; McCann, Sean

'SEAmom', Min, m, 1990; (**Stolen Moment**); flowers mauve blend, small, semi-dbl., 6–14 petals, slight fragrance; foliage small, medium green, semi-glossy; bushy growth; [Kiss 'n' Tell X (Aunty Dora X Charles de Gaulle)]; McCann, Sean

'SEAmood', Min, yb, 1988; (**In the Mood**); flowers yellow streaked pink, small, dbl., 20 petals, slight fragrance; foliage small, medium green, semi-glossy; bushy growth; [Rise 'n' Shine X Unnamed seedling]; McCann, Sean

'SEAmou', Min, pb, 1999; (**Mouse**); flowers pink yellow at base, changes to red-pink, dbl., 15–25 petals, 1 in., borne in small clusters, slight fragrance; few prickles; foliage small, medium green, semi-glossy; bushy, medium (18 in.) growth; [Portland Dawn X Siobhan]; McCann, Sean

'SEAmurp', HT, mp, 1982; (**Murphy's Law**); flowers medium pink, lighter reverse, large, dbl., 35 petals, slight fragrance; foliage medium, dark, matt; bushy growth; [Fragrant Cloud X (Prima Ballerina X Gavotte)]; McCann, Sean

Sean, S, lp; [Mme Segond Weber sport]; Sutherland, P; Golden Vale Nursery, 2000; (Weatherly, L.)

'SEAnear', Min, w, 1990; (**Near You**); flowers moderately blooms, dbl., 15–25 petals, slight fragrance; foliage small, medium green, semi-glossy; spreading growth; [Rise 'n' Shine X (Elina X Royal Gold)]; McCann, Sean

'SEAodd', Min, rb, 1992; (**Alice Faye**); flowers full circle of red with distinct yellow eye from the bottom view, dbl., 22–24 petals, 2.5 in., no fragrance; foliage large, medium green, semi-glossy; upright, tall growth; [Seedling X Seedling]; McCann, Sean; Justice Miniature Roses, 1992

'SEAoona', HT, mr, 1990; (**Oonagh**); flowers large blooms, dbl., 26–40 petals, slight fragrance; foliage medium, medium green, semi-glossy; very spreading growth; [Matangi X Gavotte]; McCann, Sean

Search for Life™, *see* 'WILsfol'

'SEArich', Min, rb, 1987; (**Rich and Rare**); flowers scarlet, reverse white veined red, small blooms borne singly, dbl., 35 petals, exhibition form, slight fragrance; foliage small, dark, semi-glossy; bushy growth; [(Rise 'n' Shine X Siobhan) X Beauty Secret]; McCann, Sean

SEAris, Min, lp; (**Kiss and Tell**); McCann, Sean

'SEArodney', Min, w, 1991; (**Admirable**); bud ovoid, sharply pointed sepals; flowers soft pink with darker accent, reverse ivory with pink accent, dbl., 2.5 in., exhibition form, intense, fruity fragrance; foliage large, medium green, semi-glossy; long, straight stems; upright, tall growth; [Seedling X Admiral Rodney]; McCann, Sean; Justice Miniature Roses, 1992

'SEARONAN', Min, op, 1996; (**Street Wise**); flowers orange and pink, reverse yellow at throat, blooms borne most, very dbl., 60 petals, intense fragrance; numerous prickles; foliage medium, medium green, semi-glossy; upright, medium growth; PPAF; [Rise 'n' Shine X Siobhan]; McCann, Sean; Justice Miniature Roses, 1996

Seashell, *see* 'KORSHEL'

'SEASHELL', HWich, mp, 1916; flowers large, borne in large clusters, semi-dbl.; foliage glossy; Dawson

'SEASIO', F, rb, 1984; (**Siobhan**); flowers red, reverse yellow, medium, dbl., 20 petals, exhibition form, slight fragrance; foliage medium, dark bronze, semi-glossy; upright growth; [Maxi X Copper Pot]; McCann, Sean

SEAsol, Min, pb; (**Solitude**); McCann, Sean

'SEASOON', Min, ly, 1992; (**Someday Soon**); flowers light yellow framed in creamy white outer petals, reverse cream, dbl., 52 petals, 1.5 in., exhibition form, slight, spicy fragrance; foliage medium, light green, matt, highly serrated; upright, medium growth; [Unnamed Miniature seedling X Antique Silk]; McCann, Sean; Justice Miniature Roses, 1993

Seaspray, *see* 'MACNEWING'

'SEASWAN', Min, w, 1988; (Swan Song, **Swansong**); flowers small, dbl., 15–25 petals, intense fragrance; foliage small, medium green, semi-glossy; bushy growth; [(Rise 'n' Shine X Party Girl) X Margaret Merril]; McCann, Sean

'SEATEL', Min, w, 1992; (**Margaret Telfer**); dbl., 15–25 petals, 1.5–2.75 in., borne in small clusters, slight fragrance; some prickles; foliage small, medium green, semi-glossy; low (14"), upright growth; [Kiss the Bride X Margaret Merril]; McCann, Sean

'SEATIP', Min, pb, 1988; (**Portland Dawn**); flowers veined pink and bronze, small, dbl., 20 petals, slight fragrance; foliage small, medium green, semi-glossy; bushy growth; [Rise 'n' Shine X (Copper Pot X Maxi)]; McCann, Sean, 1989

'SEATOFF', Min, ob, 1992; (**Toffee**); flowers bright orange, yellow eye, dark orange stamens, reverse oran, semi-dbl., 18 petals, 1 in., slight, fruity fragrance; foliage medium, medium green, semi-glossy, disease-resistant; bushy, medium growth; [Bloomsday X Unnamed Min seedling]; McCann, Sean; Justice Miniature Roses, 1993

Seattle Scentsation™, *see* 'SAVASEAT'

'SEATWINKLE', Min, ob, 1999; (**Jazz Dancer**); flowers orange and yellow, reverse orange, dbl., 26–40 petals, 1.75 in., borne mostly singly, slight fragrance; prickles moderate; foliage medium, dark green, glossy; compact, medium (14–16 in.) growth; McCann, Sean; Justice Miniature Roses, 1999

'SEAVIEW', Min, yb, 1990; (**Ladies' View**); flowers moderately small, dbl., 15–25 petals, slight fragrance; foliage small, medium green, semi-glossy; bushy growth; [You 'n' Me X Amber Queen]; McCann, Sean

'SEAVIN', Min, mr, 1999; (**True Vintage**); flowers medium red to fuchsia, reverse silver/light red, medium, semi-dbl., 12–16 petals, borne mostly singly, slight fragrance; many prickles; foliage medium, medium green, semi-glossy; compact, medium (16–18 in.) growth; [Rose Gilardi X Stolen Moment]; McCann, Sean; Justice Miniature Roses, 1999

'SEAWALK', Cl Min, mr, 1999; (**S.W.A.L.K.**); flowers barn red, dbl., 26–40 petals, 2.25 in., borne mostly singly, slight fragrance; prickles moderate; foliage medium, dark green, glossy; climbing (6–7 ft) growth; trellis or fence for support; McCann, Sean; Justice Miniature Roses, 1998

'SEAWHI', Min, w, 1987; (**Kiss the Bride**); flowers small blooms borne singly and in clusters of 4–6, dbl., 20 petals, moderate, spicy fragrance; foliage medium, medium green, semi-glossy; bushy growth; [Rise 'n' Shine X White Bouquet]; McCann, Sean, 1990

'SEAWISH', Min, rb, 1996; (**Wishful Thinking**); flowers medium red, large light yellow eye, heavy mass of yellow sta, dbl., 15–25 highly reflexed petals, 1.5 in., intense fragrance; foliage medium, dark green to light orange, glossy; upright, medium (16–20 in.) growth; RR; [Seedling X Wit's End]; McCann, Sean; Justice Miniature Roses, 1997

'SEAWIT', Min, rb, 1988; (**Wit's End**); flowers red with yellow reverse, small, borne in sprays of 3–5, dbl., 20 petals, slight fragrance; foliage small, medium green, semi-glossy; bushy growth; [Rise 'n' Shine X Siobhan]; McCann, Sean, 1989

'SEAYOU', Min, w, 1985; (**You 'n' Me**); flowers white, light apricot center, small, dbl., exhibition form, slight fragrance; foliage small, medium green, semi-glossy; bushy growth; [Avandel X Party Girl]; McCann, Sean, 1987

Sebago, *see* **'TYLER'**™

Sebastian Kneipp, *see* KORpastato

Sebastian Schultheis, Gr, dp

'SECOND CHANCE', HT, pb, 1988; bud ovoid; flowers coral orange, tips orange-red, reverse deep pink, aging medi, dbl., 35 petals, exhibition form, moderate, damask fragrance; foliage medium, medium green to maroon, semi-glossy, a bit r; bushy, slightly spreading growth; [Carefree Beauty X Sonia]; Stoddard, Louis, 1991

Secret™, *see* 'HILAROMA'

Secret Garden Musk Climber, LCl, w

'SECRET LOVE', HT, dr, 1973; bud ovoid; flowers deep red, large, dbl., exhibition form, moderate fragrance; foliage leathery; moderate, upright, bushy growth; [Unnamed seedling X Unnamed seedling]; Armstrong, D.L.; Armstrong Nursery

Secret Obsession, *see* 'RENSECOBSES'

Secret Recipe, *see* 'MORWINST'

'SECRÉTAIRE BELPAIRE', HT, op, 1934; flowers brilliant salmon-pink, large, dbl.; foliage bright bronze; vigorous growth; [Angèle Pernet X Mme Edouard Herriot]; Lens

'SECRETAIRE J. NICOLAS', HP, m, 1883; flowers purple; Schwartz

'SECRETARIS ZWART', HT, dp, 1918; flowers bright rose, reverse silvery rose; [Gen. MacArthur X Lyon Rose]; Van Rossem

Sedgebrook, *see* 'MURSE'

Seduction, *see* 'MEIBEAUSAI'

'SEDUCTION', Pol, mp, 1927; flowers peach-blossom-pink, large, borne in clusters of 50–60, dbl.; few thorns; dwarf growth; Turbat

Seduction, *see* MEIvrita

'SEEFELD', F, rb, 1958; bud oval; flowers red tinted lighter, open, medium, borne in large clusters, dbl.; foliage glossy; very vigorous, bushy growth; [Fashion X Orange Triumph]; Delforge

'SEFTOPOLIS', Gr, dp, 1977; flowers deep pink, large, dbl., 36 petals; foliage glossy, leathery; vigorous growth; [Queen Elizabeth X Unnamed seedling]; Staikov, Prof. Dr V.; Kalaydjiev and Chorbadjiiski

Segovia, HT, 1964; Combe, M.; (Cavriglia)

'SÉGUIER', HGal, m, 1853; flowers purple/violet with white stripes, medium, dbl.; Robert; (Sangerhausen)

Sehnsucht®, F, dr

Seika, *see* **'OLYMPIC TORCH'**

Sei-Ka, *see* **'OLYMPIC TORCH'**

Seiko, HT, dy

'SEION', HT, my, 1987; flowers large, dbl., 35–40 petals, exhibition form, moderate fragrance; medium, pale, pale orange fruit; few prickles; foliage light green; upright, medium growth; [Grandpa Dickson X Sunblest]; Yokota, Kiyoshi, 1988

'SEISHO', HT, pb, 1999; flowers soft pink, deep pink petal edge, dbl., 30–35 petals, 5.5 in., exhibition form, moderate fragrance; foliage dark green; 5 ft growth; [Michele Meilland X Maria Teresa Bordas]; Ohkawara, Kiyoshi, 1990; Komaba Rose Nursery, 1992

'SEKEL', S, yb, 1984; flowers light yellow, bright red blend blooms in clusters of 21, semi-dbl., 11 petals, 3.5 in., slight fragrance; foliage dark, leathery; vigorous, upright growth; [Lichterloh X Zitronenfalter]; Lundstad, Arne; Agricultural University of Norway

'SEKI-YOH', HT, or, 1975; flowers large, dbl., 52 petals, exhibition form, slight fragrance; foliage dark, leathery, semi-glossy; upright, compact growth; [Miss France X Christian Dior]; Suzuki, Seizo; Keisei Rose Nursery

'SELANDIA', LCl, mp, 1913; dbl.; vigorous growth; [Mme Norbert Levavasseur X Dorothy Perkins]; Poulsen, D.T.

'SELAZTECA', HT, mr, 1990; (**Red Azteca**); bud pointed, tapering, slender; flowers bright red, no fading, large, borne singly, dbl., 30–35 petals, cupped, prickles reddish, with yellow tip; foliage large, dark green, glossy; upright, tall growth; [Unnamed seedling X Unnamed seedling]; Select Roses, B.V.; DeVor Nurseries, Inc.

Selena™, HT, lp; deVor

Selfridges, *see* 'KORPRIWA'

Selina, *see* 'REYSEL'

'SELSTAR', HT, my, 1989; (**Christina**); bud pointed; flowers bright lemon-yellow, does not fade, to, large, borne singly, dbl., 43 petals, cupped, no fragrance; prickles declining, light red; foliage large, dark green, glossy; upright, tall growth; [Eliora X Unnamed seedling]; Select Roses, B.V.; DeVor Nurseries, Inc., 1991

'SELVETTA'®, HT, pb, 1982; flowers light pink, petals edged salmon, large, dbl., 40 petals, no fragrance; foliage medium, dark, matt; upright growth; Cazzaniga-Como; Rose Barni-Pistoia

'SELWYN BIRD', HT, or, 1969; flowers salmon-cerise, dbl., 35 petals, exhibition form, moderate fragrance; foliage dark, glossy; vigorous growth; [Fragrant Cloud X Stella]; Cocker

'SELWYN TOOGOOD', Min, mp, 1983; flowers small, dbl., 33 petals, exhibition form, slight fragrance; light moss on stems; foliage small, light green; bushy, upright growth; [Heidi X ?]; Eagle, Barry & Dawn; Southern Cross Nursery

'SEMI', (R. laxa form), w, 1913; flowers small, blooms all summer, bright red fruit; tall (8 ft) growth; very hardy.; (grown from seed collected in Siberia by N.E. Hansen)

Semillante, HT, 1976; Combe, M.; (Cavriglia)

'SEMIRAMIS', HT, yb, 1957; bud pointed; flowers rose, center buff or amber, dbl., 35–45 petals, 5 in., exhibition form, intense, raspberry fragrance; foliage dark, glossy; vigorous growth; [Capistrano X (Peace X Crimson Glory)]; Motose

'SÉNATEUR AMIC', LCl, mr, 1924; bud long, pointed; flowers brilliant carmine, nearly, single; very vigorous growth; [R. gigantea X Gen. MacArthur]; Nabonnand, P.

Senateur Lafollette, S, ob

'SÉNATEUR MASCURAUD', HT, ly, 1909; flowers light yellow, center darker, dbl., moderate fragrance; Pernet-Ducher

'SÉNATEUR POTIÉ', HT, ob, 1937; flowers orange-yellow, large, semi-dbl., cupped, foliage glossy, bronze; vigorous growth; [Mme Butterfly X Carito MacMahon]; Dot, Pedro

'SÉNATEUR VAISSE', HP, rb, 1859; flowers red, shaded darker, large, dbl., 32 petals, occasionally recurrent bloom, intense fragrance; upright growth; [? X Général Jacqueminot]; Guillot Père

'SENATOR', HT, mr, 1926; bud long, pointed; flowers brilliant scarlet, dbl., intense fragrance; [Red Columbia X Premier]; Florex Gardens

Senator Burda®, *see* 'MEIVESTAL'

'SENATOR JOE T. ROBINSON', HT, dr, 1938; bud long, pointed; flowers dark crimson, semi-dbl., cupped, intense fragrance; foliage leathery; vigorous growth; [Harvard X David O. Dodd]; Vestal

'SENECA QUEEN', HT, pb, 1965; flowers apricot-pink, reverse darker, dbl., 50 petals, 6 in., exhibition form, moderate fragrance; foliage leathery; vigorous growth; [(Serenade seedling X Fashion) X Golden Masterpiece]; Boerner; J&P

'SÉNÉGAL', Cl HT, dr, 1944; flowers very dark red; very vigorous growth; Mallerin, C.; A. Meilland

'SENFF', (strain of R. canina), lp; almost thornless; sometimes used as understock; Senff

Sengodea, HT, pb

'SENHORA DA GRAÇA', HT, rb; flowers red with carmine reflections; da Silva, Moreira

'SENIOR', HT, rb, 1932; bud pointed; flowers scarlet-crimson, open, dbl., moderate fragrance; foliage thick; long stems; very vigorous growth; [Richmond X Général Jacqueminot]; Spanbauer; Hill Floral Products Co.

'SENIOR PROM', HT, dp, 1964; bud long, pointed; flowers deep pink, dbl., 38 petals, 4.5 in., exhibition form, foliage dark, glossy; vigorous, upright growth; [Pink Princess X Queen Elizabeth]; Brownell, H.C.; Brownell

'SEÑOR PHILIPPE', LCl, m; flowers lilac-pink, center paler, dbl.; vigorous growth

'SEÑORA DE BORNAS', HT, mr, 1955; flowers vermilion-red, medium, very dbl., cupped, intense fragrance; foliage glossy; vigorous, upright growth; [J.M. Lopez Pico X Concerto]; Camprubi, C.

'SEÑORA DE CARULLA', F, mr, 1961; flowers cerise-red; GM, Madrid, 1961; Torre Blanca

'SEÑORA GARI', HT, yb, 1935; bud long, pointed; flowers deep orange-yellow, very large, dbl., exhibition form, moderate fragrance; sprawling growth; [Mari Dot X Constance]; Dot, Pedro; C-P

'SEÑORA LEON DE AUJURIA', HT, ob, 1935; flowers orange; foliage glossy; vigorous growth; La Florida

Señorita, *see* 'JACDOR'

'SEÑORITA CARMEN SERT', HT, yb, 1917; flowers indian yellow, shaded pale pink, edged bright carmine, dbl.; [Marquise de Sinéty seedling]; Pernet-Ducher

'SEÑORITA DE ALVAREZ', HT, mp, 1931; bud long, pointed; flowers glowing salmon, very large, single, cupped, intense fragrance; foliage leathery, dark; vigorous growth; GM, NRS, 1930; Cant, B. R.

Sensass, *see* 'DELMOUN'

Sensass Delbard®, *see* 'DELMOUN'

'SENSATION', HT, mr, 1922; bud long, pointed; flowers scarlet-crimson, open, dbl., 36 petals, 5 in., moderate fragrance; foliage dark; free, branching growth; [Hoosier Beauty X Premier]; Hill, Joseph H., Co.

Sensation, HT, lp

Sensucht, F, 1978; Noack, Werner; (Cavriglia)

'SENTA SCHMIDT', Pol, ob, 1930; flowers coppery orange, semi-dbl.; foliage small, soft, light; dwarf growth; [Suzanne Turbat sport]; Schmidt, R.

Senteur Royale, *see* 'TANSCHAUBUD'

Sentimental, *see* 'POULTAL'

'SENTINEL', HT, pb, 1934; flowers velvety cerise, reverse silvery cerise, large, dbl., cupped, intense fragrance; foliage glossy; vigorous growth; Clark, A.; Wyant

'SENTRY', HT, dr, 1948; bud long, pointed; flowers clear crimson, dbl., 25 petals, 4–5 in., intense fragrance; dwarf, compact growth; Fletcher; Tucker

Sentyna®, *see* 'INTERTYN'

Seppenrade, S, ob, 1970; flowers salmon-orange, large, dbl., moderate fragrance; Scholle, E.; (Sangerhausen)

Seppenrader Elfe, HKor, ly, 1975; flowers medium, dbl., slight fragrance; Scholle, E.; (Sangerhausen)

'SEPTEMBER DAYS', Min, dy, 1976; bud pointed; flowers reflexed, micro-mini, small, dbl., 40 petals, 1.5 in., exhibition form, moderate fragrance; foliage glossy; upright, compact growth; [Rise 'n' Shine X Yellow Jewel]; Saville, F. Harmon; Days Inn

September Eighteenth, *see* 'WILSEPT'

'SEPTEMBER MORN', HT, pb, 1913; flowers flesh-pink, center deeper, large, moderate fragrance; vigorous growth; [Mme Pierre Euler sport]; Dietrich & Turner

'**SEPTEMBER SONG**', Gr, ab, 1981; bud ovoid; dbl., 28 petals, cupped, borne singly and in clusters of 5–8, moderate, fruity fragrance; thin, awl-like prickles; foliage dark, tinted with copper, semi-glossy, leathery; erect, bushy growth; [(Vera Dalton X Prairie Princess) X (Apricot Nectar X Prairie Princess)]; Buck, Dr Griffith J.; Iowa State University

'**SEPTEMBER WEDDING**', HT, mp, 1964; bud ovoid; flowers deep pink, reverse darker, large, dbl., exhibition form, moderate fragrance; foliage dark, glossy; vigorous, tall, compact growth; [Montezuma sport]; Schloen, J.; Ellesmere Nursery

'**SEQUOIA**', HT, ob, 1939; flowers ripe pumpkin-flesh shaded apricot, dbl., globular, slight fragrance; foliage leathery, bronze; vigorous growth; Verschuren-Pechtold; Dreer

Sequoia Gold(TM), *see* 'MORSEGOLD'

Sequoia Jewel, *see* 'MORSEWEL'

Sequoia Ruby, *see* 'MORDORA'

'**SEQUOIA, CLIMBING**', Cl HT, ob, 1940; Swim, H.C.; Armstrong Nursery

'**SERAFINA LONGA**', HT, dp, 1933; flowers old-rose, heavily veined, well formed, large, moderate fragrance; vigorous growth; [Mme Butterfly X Mme Abel Chatenay]; La Florida

Seraphim, *see* '**SERAPHINE**'

'**SERAPHINE**', HSet, pb, 1840; (Seraphim); flowers soft pink, center darker, very dbl.; Prince Nursery

Serena, HT, 1955; Lens; (Cavriglia)

'**SERENADE**', HT, ob, 1949; bud ovoid; flowers coral-orange, dbl., 28 petals, 4–4.5 in., cupped, slight fragrance; foliage glossy, leathery; vigorous, upright growth; [Sonata X R.M.S. Queen Mary]; Boerner; J&P

Serenata, *see* BARser

'**SERENDIPITY**', S, ob, 1978; bud ovoid, pointed; flowers orange to buttercup-yellow, dbl., 23 petals, 4–5 in., cupped, moderate fragrance; foliage dark, glossy, leathery; vigorous, upright, spreading, bushy growth; [(Western Sun X Carefree Beauty) X (Apricot Nectar X Prairie Princess)]; Buck, Dr Griffith J.; Iowa State University

'**SERENE**', HT, w, 1940; bud long, pointed, light buff; flowers shining silvery white, open, very large, dbl., 30–40 petals; foliage sparse, soft; vigorous, upright growth; Mallerin, C.; C-P

Serene Bouquet, *see* 'LAVSER'

'**SERENELLA**', HT, 1954; Cazzaniga, F. G.; (Cavriglia)

'**SERENISSIMA**', LCl, m, 1980; bud pointed; flowers light lilac, dbl., 38 petals, recurrent bloom, intense fragrance; straight, reddish-green prickles; foliage large, light green, matt; upright growth; Takatori, Yoshiho; Rose Barni-Pistoia

Serenité, *see* GAUvera

'**SÉRÉNITÉ**', HT, my, 1946; bud pointed; flowers chrome-yellow, large; foliage reddish; stiff stems; very vigorous growth; RULED EXTINCT 7/86; Gaujard

'**SÉRÉNITÉ**'®, HT, my, 1986; flowers coppery-yellow, borne singly, dbl., exhibition form, intense fragrance; rounded fruit; large prickles; foliage large, medium green; tall growth; [John Armstrong X Tanagra]; Gaujard, Jean

Serenity, F, lp; [Playgirl sport]; Snetsinger

'**SERGE BASSET**', HT, dr, 1918; flowers brilliant garnet-red, dbl.; Pernet-Ducher

Sergeant Pepper, *see* 'KORTENSES'

'**SERGENT ULMANN**', HT, dr, 1930; bud long, pointed; flowers deep garnet, lightened with scarlet, open, very large, semi-dbl., slight fragrance; foliage leathery, bronze; vigorous growth; [Grenoble X Mme Van de Voorde]; Mallerin, C.

Serpent Rose®, S, pb, 1994; Lens

Serpent Vert®, *see* 'LENWIGA'

'**SERRATIPETALA**', Ch, pb, 1912; (R. chinensis serratipetala, Rose Oeilletde Saint Arquey); Vilfray

Set of Gold, HT, yb, 1994; Delbard

'**SETINA**', Cl Ch, lp, 1879; (Hermosa, Climbing); [Hermosa, Climbing sport]; P. Henderson

'**SETINA**', B, lp; Henderson, 1859

'**SETSUKO**', HT, w, 1999; flowers cream white, blended red in petal edge, dbl., 37 petals, 5 in., slight fragrance; foliage dark green, half leathery; 4.5 ft growth; [Marchenkonigin X Hakuchoh]; Sasaki, Keiji, 1996

'**SEVEN SEAS**', F, m, 1973; flowers lilac, dbl., 26 petals, 4 in., moderate fragrance; foliage large, glossy; [Lilac Charm X Sterling Silver]; Harkness

'**SEVEN SISTERS**', HMult, pb, 1817; (R. cathayensis platyphylla, R. multiflora platyphylla, R. platyphylla, R. thoryi, Seven Sisters Rose); flowers pale rose to crimson, much larger than r; (14); 1817

Seven Sisters Rose, *see* '**SEVEN SISTERS**'

'**SEVENTEEN**', F, pb, 1959; bud ovoid; flowers pink-coral, borne in pyramidal clusters, dbl., 20–25 petals, 3.5 in., cupped, intense fragrance; foliage dark, leathery; vigorous, upright growth; [Pinocchio seedling X Fashion seedling]; Boerner; J&P

'**SEVENTH HEAVEN**', HT, dr, 1966; flowers large, dbl., exhibition form, foliage glossy; upright, bushy growth; [Unnamed seedling X Chrysler Imperial]; Armstrong, D.L. & Swim; Armstrong Nursery

'**SÉVERINE**', HT, rb, 1918; flowers coral-red, passing to shrimp-red, semi-dbl.; foliage bronze; vigorous growth; Pernet-Ducher

'**SEVERN VALE**', HT, ob, 1967; flowers salmon, 4.5 in., exhibition form, free growth; [Beauté sport]; Sanday, John

'**SEVILLIANA**', S, pb, 1976; bud ovoid, pointed; flowers light claret-rose, yellow from base, semi-dbl., 15 petals, 3.5–4 in., cupped, moderate, spicy fragrance; foliage tinted copper, leathery; upright, bushy growth; [(Vera Dalton X Dornroschen) X ((World's Fair X Floradora) X Applejack)]; Buck, Dr Griffith J.; Iowa State University

Sexy Rexy®, *see* 'MACREXY'

'**SFINGE**', HT, rb, 1954; flowers deep red edged rose; long stems; vigorous, upright growth; [Julien Potin X Sensation]; Aicardi, D.; Giacomasso

'**SHABNAM**', F, w, 1976; bud ovoid; flowers white, center pinkish, open, small, very dbl., 85 petals, 2 in.; foliage soft; upright, open growth; [Baby Sylvia X ?]; IARI

'**SHADES OF AUTUMN**', HT, rb, 1943; flowers red to pink with some yellow, center yellow, large, dbl., intense fragrance; foliage glossy, leathery; vigorous, compact, upright, bushy growth; [Golden Glow X Condesa de Sástago]; Brownell, H.C.

'**SHADES OF PINK**'(TM), F, pb, 1985; flowers pink with white eye, imbricated, in clusters of 10–50, dbl., 33 petals, 3.5 in., no fragrance; curved, hooked prickles; foliage medium green, glossy; bushy, upright growth; [Robin Hood X Pascali]; Mander, George

Shadow, HT, dr, 1966; Dawson

'**SHADOW DANCE**', F, rb, 1969; bud ovoid; flowers silver-pink edged red, small, dbl., 30–35 petals, intense fragrance; foliage glossy, leathery; vigorous, low, compact growth; [Pink Parfait X Crimson Glory]; Fankhauser

Shadow Dancer(TM), *see* 'MORSTRORT'

Shady Charmer, *see* 'MICSHADY'

'**SHADY FLAME**', Min, or, 1981; dbl., 35 petals, exhibition form, borne usually singly, slight fragrance; straight prickles; foliage small, light green; compact, bushy growth; [Prominent X Zinger]; Jolly, Betty J.; Rosehill Farm

'**SHADY LADY**', Min, dy, 1981; flowers yellow-orange, blooms borne mostly singly, dbl., 35 petals, exhibition form, moderate fragrance; no prickles; foliage tiny, green; upright, bushy growth; [(Prominent X Zinger) X Puppy Love]; Jolly, Betty J.; Rosehill Farm

'**SHADY LANE**', HT, pb, 1990; bud ovoid; flowers deep pink to white base, aging deeper pink, large, borne usu, dbl., 35 petals, exhibition form, intense, damask fragrance; prickles medium, pointed downwards, light green; foliage medium, dark green, semi-glossy; upright, medium growth; [Thriller X Just Lucky]; Bridges, Dennis A.; Bridges Roses

Shafter, *see* **'DR HUEY'**

'SHAIDA', HT, or, 1976; bud long, pointed; flowers red-orange-salmon, dbl., 28–32 petals, 3.5–4.5 in., exhibition form, slight fragrance; foliage dark, leathery; vigorous, upright growth; [(Fandango X Fillette) X Coloranja]; Lens

'SHAILER'S PROVENCE', C, lp; (Gracilis); flowers lilac-pink, base white, inner petals rolled and wrinkled, bl, dbl., cupped, moderate fragrance; foliage small; vigorous (4–5 ft) growth; Shailer, Prior to 1799

'SHAILER'S WHITE MOSS', M, w; (Blanche Mousseuse, Clifton Moss, R. centifolia albo-muscosa, R. muscosa alba, White Bath); flowers medium, dbl.; (28); Shailer, Cult. 1810

'SHAKESPEARE FESTIVAL', Min, my, 1979; bud pointed; flowers clear yellow, dbl., 45 petals, 1.5 in., exhibition form, moderate, tea fragrance; foliage green, matt; bushy, compact growth; [Golden Angel X Golden Angel]; Moore, Ralph S.; Sequoia Nursery

Shakespeare Garden Eglantine, HEg, w

Shaleen Surtie-Richards, HT, mp, 1996

'SHALIMAR', HWich, yb, 1914; flowers creamy blush, picotee edge of bright rose-pink, borne in imm; vigorous, climbing growth; [Minnehaha sport]; Burrell

'SHALOM'®, F, or, 1973; (Flammenmeer); bud globular; dbl., 23 petals, 3.5–4 in.; foliage dark; vigorous, upright growth; [Korona seedling X Korona seedling]; Poulsen, Niels D.; Poulsen

'SHANDON', HT, dr, 1899; flowers large, dbl., intense fragrance; Dickson, A.; (Sangerhausen)

Shanghai Autumn, HT, yb

'SHANGRI-LA', HT, mp, 1945; bud long, pointed; flowers silvery pink, open, dbl., 35 petals, 3.5–4.5 in., intense fragrance; foliage leathery; long stems; very vigorous, upright, bushy growth; [Mrs J.D. Eisele X Pres. Herbert Hoover]; Howard, F.H.; H&S

Shankar Jaikishan, HT, mr, 1998; Shastri

Shannie, *see* 'GELSHAN'

Shannon®, *see* 'MACNON'

Shantaraj, HT, dr, 1998; K&S

Shanthi Pal, HT, op, 1989; Pal, Dr B.P.

Shantung®, *see* 'DELCHINE'

'SHARADA', Gr, lp, 1983; (Saroda); [Queen Elizabeth sport]; Gupta, Dr M.N., Datta, Dr S.K. & Nath, P.; National Botanical Research Institute

'SHARI', HT, pb, 1992; flowers growth, and foliage same as swarthmore color is lighter, dbl., 26–40 petals, slight fragrance; foliage medium, medium green, matt; upright (185 cms) growth; [Sweetie Pie sport]; Perry, Astor; Hortico Roses, 1993

Sharifa Asma™, *see* 'AUSREEF'

'SHARON', HT, lp, 1962; bud long, pointed; flowers soft pink edged lighter, base light yellow, reverse darker pink, dbl., 35–40 petals, 5.5 in., exhibition form; foliage leathery, dull; vigorous, upright growth; [Golden Rapture X Happiness]; Spandikow

Sharon Anne, *see* 'BOSLORVET'

'SHARON FRANCES', HT, pb, 1999; flowers light pink blending to peach center, dbl., 26–40 petals, 5.5–6 in., borne mostly singly, slight fragrance; some prickles; foliage large, dark green, glossy; upright, tall (3 ft) growth; [Tom Foster X Ravenswood Village]; Poole, Lionel

'SHARON LORRAINE', F, rb, 1978; bud pointed; flowers ivory yellow, aging red, reverse red, blooms borne 30–36 per cluster, dbl., 40 petals; long, red prickles; foliage dark, semi-glossy; medium, upright (2–2½ ft) growth; [Unnamed seedling X Unnamed seedling]; Bossom, W.E.

'SHARON LOUISE', HT, w, 1968; bud ovoid; flowers near white, center pale pink, medium, dbl., slight fragrance; foliage dark, leathery; vigorous, tall, bushy growth; [Queen Elizabeth X Virgo]; Parkes, Mrs M.H.

'SHARON MUXLOW', Min, op, 1980; bud globular; flowers bright coppery orange, dbl., 25 petals, 1 in., flat; foliage leathery; upright growth; [Anytime X Persian Princess]; Dobbs; Min. Plant Kingdom

Sharon's Delight, *see* 'MORSHARON'

Sharon's Love, S, lp, 1998; Lens

'SHASTA', F, w, 1962; bud pointed; dbl., 23 petals, 4 in., blooms in clusters, moderate fragrance; foliage leathery; vigorous, bushy growth; [Paul's Lemon Pillar X Fashion]; Schwartz, Ernest W.; Wyant

Shatadhara, F, mp, 1991; Chiplunkar

'SHAYELA', F, yb, 1999; (**Swonderful**); flowers yellow-orange, reverse lighter, similar to little darling, dbl., 26–40 petals, 1.5 in., exhibition form, borne in small clusters, moderate fragrance; prickles moderate; foliage small, dark green, glossy; upright, medium (4 ft); patio growth; [Pink Petticoat X Redgold]; Schakelford, Grace, 1998

'SHE', F, or, 1962; flowers salmon-opal, base lemon, borne in clusters, semi-dbl., 19 petals, 2.5 in., slight fragrance; moderate, bushy growth; [(Independence X Fashion) X Brownie]; Dickson, Patrick; A. Dickson

'SHEELAGH BAIRD', Pol, pb, 1934; flowers shell-pink, overlaid rich rose pink, base yellow, large bloom, dbl.; vigorous growth; Cant, F.

Sheer Bliss, *see* 'JACTRO'

Sheer Delight, *see* 'HARWAZZLE'

Sheer Elegance™, *see* 'TWOBE'

Sheer Elegance, Climbing, *see* 'DEVBILL'

Sheer Grace, HT, pb, 1995; Patil, B.K.

'SHEER STRIPES', S, pb, 1999; flowers dark pink and white stripes, single, 5–11 petals, 3 in., borne in small clusters, slight fragrance; few prickles; foliage medium, medium green, semi-glossy; upright, tall (6 ft) growth; [Hurdy Gurdy X Heritage]; Lowe, Malcolm

'SHEERWATER', HT, ly, 1977; flowers beige to cream, veined carmine, full, dbl., 35 petals, 5 in., moderate fragrance; foliage dark, matt green; free growth; [My Choice X Premier Bal]; Plumpton, E.

Sheffield Pride, *see* 'WEBCYRENE'

'SHEILA', Pol, op, 1930; orange-salmon flowers; Walsh, J.; Beckwith

'SHEILA BELLAIR', HT, op, 1937; flowers salmon-pink, large, semi-dbl.; bushy growth; [Miss Mocatta X ?]; Clark, A.; NRS Victoria

'SHEILA FLEMING', Gr, mp, 1995; flowers prominent stamens, blooms borne 15–20 per cluster, single, 5 petals, 1.5–2.75 in., moderate fragrance; foliage medium, medium green, matt; upright (120 cms), bushy growth; 1st prize, Unnamed Seedling, Toronto Show, 1992; [Marchenland X Montezuma]; Fleming, Joyce L.; Hortico Roses, 1994

Sheila MacQueen, *see* HARwotnext

'SHEILA MITCHELL', HT, yb, 1998; flowers yellow shading to pink, reverse lemon yellow, very full, very dbl., 41 petals, 5 in., exhibition form, borne mostly singly, moderate fragrance; prickles moderate; foliage large, medium green, semi-glossy; compact, medium (3 ft) growth; [Trumpeter X Grandpa Dickson]; Mitchell, Harold V.; Nicholas Maple, 2000

'SHEILA WILSON', HT, mr, 1910; flowers light scarlet; vigorous growth; Hall; A. Dickson

Sheila's Perfume, *see* 'HARSHERRY'

'SHELBY WALLACE', Cl Pol, op, 1929; flowers light salmon-pink, small, semi-dbl., slight fragrance; [Cécile Brunner, Climbing seedling]; Moore, Ralph S.

Sheldon's Honor, *see* 'RESHONOR'

Shell Beach, *see* 'TROBEACH'

'SHELL QUEEN', Gr, lp, 1961; flowers shell pink, fading white; [Queen Elizabeth sport]; Allen, L.C.

'SHELLBROOK ROSE', (form of R. acicularis), dr; long, bottle-shaped fruit; (found near Shellbrook, Sask., Canada)

Shelley Higgins, *see* 'REYSHELLEY'

Shell-Pink Radiance, *see* **'MRS CHARLES BELL'**

'SHELLY', HT, pb, 1987; flowers pale pink with cyclamen-pink shading, reverse fleck & stripe, dbl., 20–25 petals, slight, sweet fragrance; prickles slightly hooked, beige-cream; foliage dark green, glossy, disease resistant; medium growth; [Francine sport]; Melville Nurseries Pty., Ltd., 1988

Shelly Renee(TM), *see* 'SAVASHEL'

'SHENANDOAH', LCl, dr, 1935; bud long, pointed; flowers crimson, large, semi-dbl., exhibition form, intense fragrance; foliage large, glossy; vigorous, climbing (10 ft) growth; [Étoile de Hollande X Schoener's Nutkana]; Nicolas; C-P

'SHEPHERDESS', F, yb, 1967; flowers yellow flushed salmon, blooms in clusters, dbl., 3.5–4 in., slight fragrance; foliage dark, glossy, leathery; vigorous growth; [Allgold X Peace]; Mattock

'SHEPHERD'S DELIGHT', F, rb, 1956; flowers flame and yellow, blooms in trusses, semi-dbl., 15 petals, 3 in., slight fragrance; foliage dark; vigorous growth; GM, NRS, 1958; [Masquerade seedling X Joanna Hill]; Dickson, A.

Sheri Anne, *see* 'MORSHERI'

'SHERICE', HT, mp, 1999; (**Jason**); flowers deep pink, reverse paler, medium, dbl., 15–25 petals, borne mostly singly, slight fragrance; few prickles; foliage medium, dark green, dull; upright (4 ft) growth; TGC, RNRS, 1989; [Silver Jubilee X (Redgold X Golden Slippers)]; Sheridan, John

Sheridan Pink, F, dp, 1998

'SHERIGREY', HT, yb, 1998; (**Dorothy Lloyd**); flowers grey with brown center, reverse grey, dbl., 15–25 petals, 4 in., borne in small clusters, slight fragrance; prickles few, medium, straight; foliage medium, medium green, semi-glossy; bushy, medium (2.5 ft) growth; [Cream Peach X Paradise]; Sheridan, John

'SHERIJILL', F, yb, 1993; (**Jill Carter**); flowers yellow edged red, blooms borne in small clusters, dbl., 15–25 petals, 1.5 in., moderate fragrance; few prickles; foliage small, dark green, semi-glossy; medium, upright growth; [Sheila's Perfume X Darling Flame]; Sheridan, John; Sheridan Nursery

'SHERIJOE', HT, dp, 1999; (**Smokey Joe's Cafe**); dbl., 15–25 petals, 3.5–4 in., borne in small clusters, moderate fragrance; few prickles; foliage medium, medium green, semi-glossy; upright, medium (2.5 ft) growth; [Red Planet X Pretty Lady]; Sheridan, John

'SHERILIP', F, mr, 1998; (**Linda's Lipstick**); flowers scarlet red, white reverse, hand painted, semi-dbl., 8–14 petals, 3.0 in., borne in large clusters, slight fragrance; prickles moderated, pointed; foliage medium, dark green, semi-glossy; upright, tall growth; [Seedling X Picasso]; Sheridan, John

'SHERILOWSTRI', Min, yb, 1999; (**Little Stripes**); flowers yellow with orange stripes, reverse yellow, dbl., 15–25 petals, 2 in., borne in small clusters, slight fragrance; few prickles; foliage medium, semi-glossy; patio; compact, low (18 in.) growth; [Summer Tan X seedling]; Sheridan, John

'SHERIMIKY', HT, pb, 1999; (**Miky Tea**); flowers pink blend, reverse cream, semi-dbl., 8–14 petals, 4 in., borne in small clusters, intense fragrance; few prickles; foliage medium, light green, semi-glossy; upright, low (2 ft) growth; [Cream Peach X Double Delight]; Sheridan, John

'SHERIRED', HT, mr, 1997; (**Mike Thompson**); flowers full, large, very dbl., 26–40 petals, borne mostly singly, slight fragrance; some prickles; foliage large, dark green, semi-glossy; medium, bushy growth; [(Silver Jubilee X seedling) X (Red Planet X Pharoah)]; Thompson, Mike

'SHERISCENT', Min, rb, 1995; (**Betty Wilson**); flowers red, reverse yellow, double (15–25 petals), medium (2 in.) blooms borne in small clusters, very fragrant; few, small, hooked prickles; foliage medium, light green, glossy; medium (2 ft) growth; patio; [Sheila's Perfume X (Little Darling X seedling)]; Sheridan, John

'SHERISILVER', HT, mp, 1999; (**Emma May**); flowers light pink, reverse deeper pink, dbl., 15–25 petals, 5–6 in., exhibition form, borne mostly singly, slight fragrance; prickles moderate; foliage large, dark green, semi-glossy; upright, medium (3.5 ft) growth; TGC, RNRS, 1989; [Silver Jubilee X Dr A. J. Verhage]; Sheridan, John

'SHERISKEP', F, rb, 1999; (**Mary May**); flowers red on white, reverse white, double (15–26 petals), medium (3 in.) blooms borne in small clusters, very fragrant; moderate prickles; foliage small, dark green, semi-glossy; bushy (2 ft) growth; garden decorative; [Daily Sketch X (Cream Peach X Seedling)]; Sheridan, John

Sherrill Anne(TM), *see* 'BYRSHERRILL'

'SHERRY', F, r, 1960; flowers dark sherry color, blooms in clusters, semi-dbl., 14 petals, 2.5 in., slight fragrance; foliage dark; vigorous growth; [Independence X Orange Sweetheart]; McGredy, Sam IV; McGredy

Shi Tz-mei, *see* **'TURNER'S CRIMSON RAMBLER'**

Shikou, HT, m

'SHIMMERING DAWN', F, mp, 1965; bud rose-pink; flowers blush-pink, borne in clusters, 3 in., moderate fragrance; foliage dark, glossy; very vigorous growth; Verschuren

'SHIMMERING SILK', HT, rb, 1968; flowers cerise tinted silvery pink, large, dbl.; foliage dark; vigorous, upright growth; [Ena Harkness X Molly Doyle]; Barter

Shimsha, HT, pb, 1976; Kasturi

Shine On(TM), *see* DICtalent

Shining Coral, *see* 'HADCORAL'

Shining Flare, *see* 'HADFLARE'

Shining Hour(TM), *see* 'JACYEF'

Shining Rose, *see* **R. NITIDA**

Shining Ruby, *see* 'HADRUBY'

'SHINING STAR', HT, my, 1945; (Mme P. Olivier, Trylon); bud long, pointed; flowers vivid chrome-yellow, large, dbl., moderate, fruity fragrance; foliage dark, leathery; [Soeur Thérèse X Feu Pernet-Ducher]; Mallerin, C.; C-P

Shining Sun, *see* **'MRS PAUL GOUDIE'**

'SHINING SUN', HT, yb, 1932; bud long, pointed, golden yellow splashed scarlet; flowers yellow deepening to reddish center, dbl., intense fragrance; foliage thick, bronze; vigorous growth; [Charles P. Kilham X Julien Potin]; Van Rossem

'SHINJU', HT, lp, 1976; bud large, ovoid; flowers light pink, paler at petal edges, large blooms borne 1–3 per cluster, dbl., 28 petals, exhibition form, moderate fragrance; many, medium prickles, slanted downward; foliage medium green, leathery; vigorous, upright growth; [Royal Highness X Garden Party]; Harada, Toshiyuki

Shinsei, *see* **'SHIN-SEI'**

'SHIN-SEI', HT, dy, 1979; (Shinsei); bud pointed; flowers well-formed, large blooms borne singly, dbl., 38 petals, exhibition form, moderate fragrance; prickles slanted downward; foliage medium large, medium green, glossy; upright, bushy growth; [(Ethel Sanday X Lydia) X Koto]; Suzuki, Seizo; Keisei Rose Nursery, 1978

'SHIN-SETSU', LCl, w, 1972; bud ovoid; flowers white, center soft cream, large, very dbl., exhibition form, moderate fragrance; foliage glossy, dark; very vigorous, climbing growth; [(Blanche Mallerin X Neige Parfum) X New Dawn seedling]; Suzuki, Seizo; Keisei Rose Nursery

'SHIRALEE', HT, yb, 1965; flowers yellow flushed orange, dbl., 36 petals, 5.5 in., exhibition form, moderate fragrance; vigorous, tall growth; GM, Japan, 1964; [Seedling X Kordes' Perfecta]; Dickson, Patrick; A. Dickson

Shire County, *see* 'HARSAMY'

'SHIRLEY', HT, rb, 1933; bud shaded russet; flowers light prawn-red, base yellow, intense fragrance; vigorous growth; Dickson, A.

'SHIRLEY HIBBERD', T, my; flowers small; Levet, F., 1873

'SHIRLEY HOLMES', HT, ob, 1958; bud long, pointed; flowers golden orange, dbl., 30 petals, 4 in., intense fragrance; vigorous, bushy growth; [McGredy's Yellow X Ethel Sanday]; Mee; Edenvale Nursery

'SHIRLEY LAUGHARN', HT, yb, 1974; bud ovoid; flowers creamy yellow, edged pink, dbl., 35 petals, 5 in., moderate, fruity fragrance; foliage dark; very vigorous, upright, slightly spreading growth; [Granada X Garden Party]; Swim, H.C.; Laugharn

'**SHIRLEY ROSE**', HT, pb, 1966; flowers cream and carmine, loose, 5 in., slight fragrance; foliage dark; upright growth; [Eden Rose sport]; Lawrence

Shirley Spain, *see* 'COCHAROD'

'**SHIRLEY TEMPLE**', HT, ly, 1936; flowers light yellow, edged lemon-yellow; [Joanna Hill sport]; Engle; Wyant

Shirpa, LCl, op, 1976; Eve, A.

Shiun, *see* **'SHI-UN'**

'**SHI-UN**', HT, m, 1984; (Shiun); flowers deep lilac purple, reverse deeper, blooms borne singly, dbl., 33 petals, exhibition form, moderate fragrance; prickles slanted downward; foliage dark, leathery, semi-glossy; vigorous, upright, bushy growth; [(Blue Moon X Twilight) X (Red American Beauty X Happiness)]; Suzuki, Seizo; Keisei Rose Nursery

'**SHIZU NO MAI**', HT, lp, 1990; flowers light pastel pink, large, borne usually singly, dbl., 32–35 petals, exhibition form, slight fragrance; foliage medium, dark green, semi-glossy; sturdy, upright growth; [Jana X Madame Violet]; Ohtsuki, Hironaka, 1986

Shleby Belogorsky, HMsk, w, 1996

Shobha, HT, ob, 1988; Friends Rosery

'**SHOCKING**', Pol, rb, 1967; flowers dark red, reverse lighter, borne in clusters, 3 in., exhibition form, slight fragrance; foliage dark, glossy; very low, erect growth; [Red Favorite X Alain]; Hémeray-Aubert; McGredy

Shocking Blue®, *see* 'KORBLUE'

'**SHOCKING PINK**', HT, mp, 1968; flowers large, dbl., cupped, intense fragrance; foliage leathery; vigorous, upright growth; [Pink Sensation sport]; McCannon; DeVor Nurseries, Inc.

Shocking Sky, *see* KORgenda

'**SHOLA**', F, or, 1969; bud pointed; flowers sparkling orient red, full, medium, dbl.; foliage leathery; vigorous, dwarf growth; [Anna Wheatcroft X Open pollination]; IARI

Shona, *see* 'DICDRUM'

'**SHOOTING STAR**', Min, yb, 1972; bud ovoid; flowers yellow, tipped red, small, dbl., cupped, slight fragrance; foliage small, light, soft; vigorous, dwarf growth; [Rumba X (Dany Robin X Perla de Montserrat)]; Meilland; C-P

'**SHOOTOUT**', Min, ab, 1981; bud ovoid; flowers apricot-orange, blooms borne singly, dbl., 28 petals, slight fragrance; long, triangular, light green prickles; foliage medium green; vigorous, bushy growth; [Tiki X Darling Flame]; Borst, Jim; Kimbrew-Walter Roses

Shoreline Tawny, N, ly

Short 'n' Sweet, *see* 'TINSHORT'

Shortcake™, *see* 'KEIBELMI'

'**SHOT SILK**', HT, pb, 1924; flowers cherry-cerise, shading to golden yellow at base, dbl., 27 petals, exhibition form, intense fragrance; foliage glossy, slightly curled; vigorous growth; GM, NRS, 1923; [Hugh Dickson seedling X Sunstar]; Dickson, A.

'**SHOT SILK, CLIMBING**', Cl HT, pb, 1931; Knight, C.

Show, HT, lp, 1995; RvS-Melle

Show Carpet, *see* 'LAVBRIC'

'**SHOW GARDEN**', LCl, mp, 1954; (Pink Arctic); flowers crimson to rose-bengal, then magenta, dbl., 43 petals, 4–5 in.; growth like a hybrid tea, followed by 4–5 ft canes; [Unnamed seedling X Queen o' the Lakes]; Brownell, H.C.

Show Garden, LCl, mp

'**SHOW GIRL**', HT, mp, 1946; bud long, pointed; flowers rose-pink, 15–20 petals, 3.5–4.5 in., exhibition form, moderate fragrance; foliage leathery; vigorous, upright, bushy growth; GM, NRS, 1950; [Joanna Hill X Crimson Glory]; Lammerts, Dr Walter; Armstrong Nursery

Show Girl, HT, mp

'**SHOW GIRL, CLIMBING**', Cl HT, mp, 1949; Chaffin; Armstrong Nursery

Show 'n' Tell, *see* 'FOUTELL'

Show Off, *see* 'AROWAGO'

Showbiz, *see* 'TANWEIEKE'

Showbiz, Climbing, Cl F, mr, 1995; Tejganga

'**SHOWBOAT**', F, yb, 1965; bud ovoid; flowers deep yellow on pink to cream ground, dbl., 30 petals, 2.5 in., moderate fragrance; foliage leathery; moderate, bushy growth; [Carrousel X Unnamed seedling]; Patterson; Patterson Roses

Showbound™, Min, op

'**SHOWCASE**', Min, ab, 1986; flowers small, dbl., 35 petals, slight fragrance; foliage small, light green, semi-glossy; bushy growth; [Rise 'n' Shine X Over the Rainbow]; Stoddard, Louis

Showdown, *see* 'MINISHOW'

'**SHOWER OF GOLD**', HWich, my, 1910; flowers golden yellow, rapidly fading to pale yellow, rosette-shaped, dbl.; foliage glossy, fern-like; very vigorous, climbing growth; [Jersey Beauty X Instituteur Sirdey]; Paul

Showoff, *see* 'AROWAGO'

'**SHOWOFF**', LCl, mr, 1952; flowers scarlet, medium to large, borne in large clusters, very dbl., cupped, profuse, repeated bloom; vigorous growth; [Blaze sport]; Moffet; Earl May Seed Co.

'**SHOWPIECE**', HT, mr, 1958; flowers bright scarlet, moderate fragrance; Ratcliffe

Showqueen, HT, mp, 1997; Williams, J. Benjamin

Showstopper, *see* 'JACSHO'

'**SHOWTIME**', HT, mp, 1969; flowers large, dbl., exhibition form, moderate, fruity fragrance; foliage glossy, leathery; vigorous, bushy growth; [Kordes' Perfecta X Granada]; Lindquist; Howard Rose Co.

Showy Gold, F, dy

Showy Pavement, S, mp, 1990; flowers large, semi-dbl., intense fragrance; low, arching, spreading (2 ft) growth; Baum

'**SHREE DAYANANDA**', HT, dp, 1979; bud ovoid; flowers deep pink, very dbl., 90 petals, moderate fragrance; beak-shaped prickles; foliage small, green; bushy, dwarf growth; [Scarlet Knight X Festival Beauty]; Hardikar, Dr M.N.

Shreveport™, *see* 'KORPESH'

Shrewsbury Show, *see* FRYshrewby

Shreyasi, HT, m, 1992; IARI

Shri Swamy Samarth, LCl, dp, 1992; Patil, B.K.

'**SHROPSHIRE LASS**', S, lp, 1968; flowers blush-pink, single, 5 petals, 5 in., summer bloom, moderate fragrance; [Mme Butterfly X Mme Legras de St Germain]; Austin, David

Shrubby Pink, *see* **'SUNDAY TIMES'**

Shugetsu, *see* **'SHU-GETSU'**

'**SHU-GETSU**', HT, dy, 1982; (Shugetsu); flowers large, dbl., 38 petals, exhibition form, moderate fragrance; large, straight prickles; foliage large, dark, glossy; upright, bushy growth; [Seiko X King's Ransom]; Suzuki, Seizo; Keisei Rose Nursery

Shunpo, HT, lp, 1987; Suzuki, Seizo

'**SHUN'YO**', HT, ob, 1982; flowers light yellow orange, large, dbl., 33 petals, exhibition form, slight fragrance; sickle-shaped prickles; foliage glossy; vigorous, upright growth; [(Golden Sun X Summer Holiday) X (Garden Party X Narzisse)]; Kono, Yoshito

Shuo, *see* **'SHU-OH'**

'**SHU-OH**', HT, or, 1982; (Shuo); flowers medium, cupped, moderate fragrance; prickles slanted downward; foliage dark, semi-glossy; upright growth; [San Francisco X Pharaoh]; Suzuki, Seizo; Keisei Rose Nursery

Shurpee, HT, mr

Shuzao Red, S, mr; (China), 1991

Shy Beauty, *see* 'LYOSHY'

Shy Girl, *see* 'JACWHIM'

Shy Maiden, *see* 'ANDshy'

'SI', Min, w; flowers rosy white, micro-mini, small; dwarf growth; [Perla de Montserrat X (Anny X Tom Thumb)]; Dot, Pedro

Si Bemol®, *see* 'LENMOL'

'SIBELIUS', HT, mr, 1958; flowers velvety crimson, dbl., exhibition form, vigorous growth; [New Yorker X Étoile de Hollande]; Verschuren; Blaby Rose Gardens

Sibelius, HMsk, m, 1984; Lens

Sibilla®, *see* BARsib

Sibylle, F, op

Siddartha, HT, rb, 1973; Kasturi

Side Kick, *see* 'PIXKICK'

'SIDI-BRAHIM', HT, dy, 1953; flowers jonquil-yellow, dbl., slight fragrance; strong stems; [Feu Pernet-Ducher X Mme Rene Lefevre]; Robichon

'SIDNEY PEABODY', Gr, dp, 1955; flowers reddish pink, well formed, borne in clusters, dbl., 36 petals, 3.5 in., slight fragrance; foliage dark, leathery; very vigorous growth; [Rome Glory X Unknown Floribunda seedling]; deRuiter; Gandy Roses, Ltd.

Sidonie, *see* **'SYDONIE'**

Siegeslied, *see* **'BOUQUET'**

'SIEGESPERLE', Pol, ly, 1915; flowers small, semi-dbl.; Kiese; (Sangerhausen)

Siegfried Sassoon, *see* 'MACJULIAT'

Sieguier, HGal, m, 1853; Robert

SIElemon, A, ly, 1988; (**Lemon Blush**); Sievers

SIEroyal, A, lp, 1988; (**Royal Blush**); Sievers

'SIERRA DAWN', HT, pb, 1967; bud long, pointed; flowers bright pink blend, large, dbl., exhibition form, moderate fragrance; foliage dark, bronze, leathery; vigorous, upright, bushy growth; [Helen Traubel X Manitou]; Armstrong, D.L.; Armstrong Nursery

'SIERRA GLOW', HT, pb, 1942; bud ovoid; flowers shrimp-pink, reverse strawberry-pink, dbl., 33 petals, 3.5–4.5 in., exhibition form, moderate fragrance; foliage leathery; vigorous, bushy, spreading growth; [Crimson Glory X Soeur Thérèse]; Lammerts, Dr Walter; Armstrong Nursery

'SIERRA GOLD', HT, my, 1960; bud urn shaped; flowers indian yellow, dbl., 45–55 petals, 4–4.5 in., exhibition form, slight fragrance; foliage leathery, semi-glossy; upright, compact growth; [Queen Elizabeth X Tawny Gold]; Lammerts, Dr Walter; Amling-DeVor Nursery

Sierra Lynn, *see* 'GELLYNN'

'SIERRA SNOWSTORM', S, ly, 1936; bud long, pointed, cream and yellow; flowers open, small borne in clusters, single, 2 in., abundant, recurrent bloom, intense fragrance; foliage large, leathery, glossy, light; vigorous (5–6 ft), bushy, arching growth; [Gloire des Rosomanes X Dorothy Perkins]; Moore, Ralph S.; Henderson's Exp. Gardens and Brooks & Son

Sierra Sun, HT, rb

Sierra Sunrise, *see* 'MORLIYEL'

'SIERRA SUNSET', LCl, yb, 1961; bud pointed; flowers blend of yellow, peach, orange and red, in clusters, dbl., 40 petals, 6 in., exhibition form, intense, fruity fragrance; foliage glossy; vigorous (6–8 ft) growth; [(Capt. Thomas X Joanna Hill) X Mme Kriloff]; Morey, Dr Dennison; J&P

Sieska Fervid, F; Klimenko, V. N.; (Cavriglia)

SIEson, A, dr, 1988; (**Crimson Blush**); Sievers

'SIESTA', F, or, 1966; flowers light vermilion on cream base, large, dbl.; foliage glossy, leathery; very free growth; [Sarabande X Dany Robin]; Meilland; URS

'SIF', LCl, ob, 1969; flowers tangerine-orange, large, dbl., 25 petals, cupped, foliage dark, glossy; vigorous, climbing growth; [Traumland X Royal Gold]; Lundstad

Sight Saver, *see* FRYaffair

Sightsaver, *see* FRYaffair

Siglinde, F, 1972; Noack, Werner; (Cavriglia)

'SIGNAL RED', Pol, mr, 1949; flowers scarlet, rosette-shaped, blooms in large trusses, 2 in.; foliage glossy, bronze; vigorous, bushy growth; (28); [DeRuiter's Herald X Polyantha seedling]; deRuiter; Gandy Roses, Ltd.

Signal Rot, F, 1982; VEG; (Cavriglia)

'SIGNALFEUER'®, F, or, 1959; (Fire Signal); flowers cinnabar to orange, blooms in clusters, dbl.; bushy growth; [Lumina X Cinnabar seedling]; Tantau, Math.

Signature®, *see* 'JACNOR'

'SIGNE RELANDER', HRg, dr, 1928; flowers bright dark red, small blooms in clusters, dbl., recurrent bloom, vigorous ($6\frac{1}{2}$ ft) growth; [R. rugosa hybrid X Orléans Rose]; Poulsen, S.; Poulsen

'SIGNET', HT, dp, 1938; bud long, pointed; flowers deep pure pink, large, dbl., exhibition form, intense fragrance; foliage leathery; very vigorous growth; [Premier X Talisman]; Montgomery Co.

'SIGNORA', HT, ob, 1936; (Signora Piero Puricelli); bud long, pointed; flowers orange-apricot, suffused gold, outer petals magenta-pink, la, dbl., 27 petals, cupped, moderate fragrance; foliage glossy; vigorous growth; GM, Portland, 1937; [Julien Potin X Sensation]; Aicardi, D.; J&P

Signora Maria Sgaravatti, HT; Sgaravatti, A.; (Cavriglia)

Signora Piero Puricelli, *see* **'SIGNORA'**

Sika®, HT, op, 1983; flowers large, dbl.; Cocker; (Sangerhausen)

Sika, *see* DORto

Sila, HT, op, 1983; Cocker

'SILBERLACHS', F, op, 1944; (Silver Salmon); flowers pale salmon, shell shaped, borne in clusters of 3–10, single, 6–8 petals, slight fragrance; foliage light green; upright, bushy growth; [Rosenelfe X Hamburg]; Tantau

Silberzauber, F, mp, 1985; flowers large, dbl.; GPG Bad Langensalza; (Sangerhausen)

'SILENT NIGHT', HT, yb, 1969; flowers creamy yellow suffused pink, well-formed, slight fragrance; GM, Geneva, 1969; [Daily Sketch X Hassan]; McGredy, Sam IV

Silhouette, *see* 'TANOLG'

'SILHOUETTE'®, HT, pb, 1984; (Silver Medal); bud long, pointed; flowers white, petals edged pink, dbl., 25 petals, 4–5 in.; foliage light green; upright growth; Warriner, William A.; J&P, 1980

Silk Button, *see* KORzeito

Silk Hat™, *see* 'AROSILHA'

Silk 'n' Satin, *see* 'JUDSILK'

'SILK SASH', HT, pb, 1971; flowers pastel pink, reverse silvery pink, very large, very dbl., 85 petals, exhibition form, slight fragrance; foliage dark, leathery; vigorous, upright growth; [Memoriam X Elizabeth Fankhauser]; Fankhauser

Silken Laumann, *see* 'LAVCOM'

Silky Mist, *see* SIMsilko

Silva®, *see* 'MEICHAM'

Silva, *see* KEIromo

'SILVA GRAÇA', HT, ob, 1956; flowers salmon and yellow shaded pink, well shaped; [Michele Meilland X Comtesse Vandal]; da Silva, Moreira

'SILVABELLA', HT, pb, 1967; flowers deep pink, reverse carmine-pink, large, dbl., cupped, slight fragrance; foliage leathery; upright, bushy growth; [Dr Debat X Eden Rose]; Guiseppe, M.

Silver Angel, *see* 'GELANGEL'

Silver Anniversary, *see* 'POULARI'

Silver Anniversary™, *see* 'JACLAV'

Silver Anniversary, *see* MEIborfil

'SILVER BEAUTY', F, m, 1956; flowers silvery rose-madder; vigorous, bushy growth; [La France seedling]; Verschuren; Gandy Roses, Ltd.

'SILVER BELL', F, lp, 1976; (Gin no suzu); bud ovoid; flowers soft light pink, small, dbl., 35 petals, 2 in., exhibition form, foliage glossy; vigorous, upright growth; [Gene Boerner X Unnamed seedling]; Takatori, Yoshiho; Japan Rose Nursery

'SILVER CHARM', F, m, 1968; flowers lavender-blue, large blooms in large trusses, semi-dbl., moderate fragrance; foliage dark; vigorous, low growth; [Lilac Charm X Sterling Silver]; LeGrice

Silver Cloud, F, r, 1990; Moore

'SILVER COLUMBIA', HT, pb, 1924; flowers clear silver-pink; [Columbia sport]; Leonard

'SILVER DAWN', LCl, w, 1942; flowers creamy white, center deeper, stamens golden yellow, open, large, very dbl., 90–100 petals; foliage large, leathery, glossy, dark; vigorous (10–12 ft), compact, climbing growth; [Silver Moon X Silver Moon]; Zombory

Silver Dream®, *see* 'INTERGRI'

'SILVER ENCHANTMENT', HT, rb, 1971; flowers rich red, reverse silver, conical, large, dbl., 36 petals, moderate fragrance; foliage dark; very free growth; [Tropicana X ?]; Gregory

Silver Fox, *see* 'DEVSPILIO'

'SILVER GEM', F, m, 1969; flowers silver-mauve, large, borne in trusses, semi-dbl.; foliage dark, leathery; bushy growth; deRuiter

'SILVER JUBILEE'®, HT, pb, 1978; flowers silvery pink, reverse darker, dbl., 33 petals, 5 in., exhibition form, slight fragrance; foliage glossy; vigorous growth; GM, Belfast, 1980 GM, RNRS, 1977 PIT, RNRS, 1977; [((Highlight X Colour Wonder) X (Parkdirektor Riggers X Piccadilly)) X Mischief]; Cocker

'SILVER JUBILEE', HT, yb, 1937; flowers light golden yellow, base chrome, edged canary, very large, dbl., moderate fragrance; foliage very large, glossy; vigorous growth; RULED EXTINCT 4/77; Dickson, A.

'SILVER JUBILEE, CLIMBING', Cl HT, pb, 1983; Cocker

'SILVER LADY', F, m, 1991; flowers pale lavender, with a hint of darker lavender on edges of petals, dbl., 15–25 petals, 3–3.5 in., no fragrance; some prickles; foliage medium, medium green, semi-glossy; medium (90 cms), upright, bushy growth; [Azure Sea X Unnamed seedling]; Taylor, Franklin "Pete" & Kay; Taylor's Roses, 1993

'SILVER LINING', HT, pb, 1958; flowers silvery rose, dbl., 30 petals, 5 in., intense fragrance; vigorous growth; GM, NRS, 1958 GM, Portland, 1964; [Karl Herbst X Eden Rose seedling]; Dickson, A.

Silver Lining, Climbing, *see* 'DICSILVING'

Silver Medal, *see* **'SILHOUETTE'**

'SILVER MOON', LCl, w, 1910; bud long, pointed; flowers creamy white, base amber, stamens darker, blooms in clusters, dbl., 20 petals, 4.5 in., non-recurrent, slight fragrance; foliage large, dark, leathery, glossy; very vigorous (to 20 ft or more) growth; [Reputedly (R. wichurana X Devoniensis) X R. laevigata]; Van Fleet; P. Henderson

Silver Peach, Min, lp; [Silver Jubilee X Oz Gold]; Hannemann, F., 1989; The Rose Paradise; (Weatherly, L.)

Silver Phantom, *see* 'RENHOM'

Silver Pink®, F, pb, 1990; Lens

'SILVER PRINCESS', HT, ob, 1934; bud salmon-yellow to delicate pink; flowers almost ivory-white, tinged pink, very large, semi-dbl.; vigorous growth; Burbank; Stark Bros.

Silver Queen, HP, pb; Paul, W., 1886

Silver Queen, HT, w

Silver River®, S, pb, 1989; Lens

Silver Salmon, *see* **'SILBERLACHS'**

'SILVER SHADOWS', HT, m, 1984; flowers light blue-lavender, large blooms borne 3–5 per cluster, dbl., 33 petals, cupped, repeat bloom, intense fragrance; awl-like, tan prickles; foliage large, dark, leathery; upright, bushy growth; hardy.; [((Soir d'Automne X Music Maker) X Solitude) X ((Blue Moon X Tom Brown) X Autumn Dusk]; Buck, Dr Griffith J.; Iowa State University

Silver Slippers, *see* 'PIXISLIP'

'SILVER SPOON', HT, m, 1985; flowers large, dbl., 35 petals, no fragrance; foliage large, medium green, semi-glossy; upright, bushy, spreading growth; [Louisiana X Unnamed seedling]; Weeks, O.L.; Weeks Wholesale Rose Growers

Silver Star®, *see* 'KORBIDO'

'SILVER TIPS', Min, pb, 1961; bud pointed; flowers pink, reverse and tips silvery, becoming soft lavender, dbl., 50 petals, 1 in., slight fragrance; foliage leathery; vigorous, bushy (10–12 in.) growth; [(R. wichurana X Floradora) X Lilac Time (Min)]; Moore, Ralph S.; Sequoia Nursery

'SILVER WEDDING', HT, w, 1921; flowers almost identical to parent; [Ophelia sport]; Amling Co.

Silver Wedding, HT, w, 1976; Gregory

'SILVER WINGS', HT, w, 1943; bud long, pointed; flowers ivory-white, large, dbl., exhibition form, slight fragrance; foliage glossy; vigorous, upright, bushy growth; McGredy

Silverado, *see* 'AROGREWOD'

'SILVERELDA', HT, pb; flowers veined silvery buff-pink, edged lighter, base yellow, dbl., 35 petals, slight fragrance; foliage glossy; vigorous, compact growth; [Heinrich Wendland X Nancy Wilson]; Riethmuller

Silverhill, *see* 'TALSILVER'

Silvery Moon, *see* 'LEOSILMU'

'SILVIA', HT, yb, 1920; (Yellow Ophelia); flowers sulfur-yellow shading to white; [Ophelia sport]; Pierson, F.R.

'SILVIA LEYVA', HT, mr, 1933; flowers cardinal-red, large, dbl., cupped, moderate fragrance; foliage glossy; vigorous growth; [Mrs C.W. Edwards X Mari Dot]; Dot, Pedro; C-P

'SIMARAMAM', HT, w, 1979; (**Marama**); bud pointed; dbl., 45 petals, exhibition form, intense fragrance; dark brown prickles; foliage large, medium green, semi-glossy; bushy, medium growth; [Lady Helen X Unnamed seedling]; Simpson, J.W.

Simba, *see* 'KORBELMA'

SIMcho, F, rb, 1986; (**Hot Chocolate**); Gold Star of the South Pacific, Palmerston North, NZ, 1986; Simpson

'SIMEROSE', HT, rb, 1939; bud long; flowers nasturtium-red, reverse golden yellow; vigorous growth; [Charles P. Kilham X (Charles P. Kilham X Margaret McGredy)]; Meilland, F.

'SIMETNA', F, r, 1979; (**Tane**); bud pointed; dbl., 38 petals, exhibition form, borne 4–10 per cluster, no fragrance; short, straight, dark brown prickles; foliage small, dark, very glossy; vigorous, upright growth; [(Orangeade X Megiddo) X (Unnamed seedling X Jocelyn)]; Simpson, J.W.

Simfonia, HT, lp; high-centered, exhibition form, 29 petals, mostly single or in clusters; dark green, semi-glossy; vigorous upright; [Mount Shasta X Pascali]; St Wagner, 1977; Res. Stn. f. Horticulture, Cluj, 1977, Romania

'SIMHAHA', HT, yb, 1979; (**Hamish**); bud ovoid; flowers foliage medium green, dbl., 50 petals, exhibition form, dark brown prickles; vigorous, medium, upright to bushy growth; [Fairy Dancers X Diamond Jubilee]; Simpson, J.W.

'SIMHOPAN', HT, or, 1979; (**Hot Pants**); bud high-pointed; dbl., 48 petals, exhibition form, borne usually 3 per cluster, no fragrance; few prickles; foliage medium green; spreading growth; [Gypsy Moth X Princesse]; Simpson, J.W.

'SIMILAM', HT, mr, 1994; (Ilam, **Roseworld**); dbl., 26–40 petals, 3–3.5 in., borne mostly singly, slight fragrance; few prickles; foliage medium, medium green, semi-glossy; medium, upright growth; [Melina X Mme G. Delbard]; Simpson, Nola; Simpson, 1994

Simina, F, lp; flowers clear pink, cupped, 40 petals, borne in clusters, moderate; large, dark green, glossy; vigorous tall upright habit; [Vigorosa X Allgold]; St Wagner, 1996; Res. Stn. f. Fruit Growing, Cluj, 1997, Romania

'SIMJEZBEL', HT, dp, 1980; (**Anne Kercher**); bud long, pointed; flowers deep pink, dbl., 30 petals, exhibition form, borne singly or 3 per cluster, straight prickles; foliage large, dense, medium green; vigorous, very bushy, compact growth; [First Prize X Seedling]; Simpson, J.W.

'SIMkayjay', F, or, 1979; (**Kirsty Jane**); bud ovoid; flowers fluted blooms borne 3–10 per cluster, dbl., 28 petals, no fragrance; straight brown prickles; foliage medium, glossy; bushy, vigorous growth; [Orangeade seedling X Megiddo]; Simpson, J.W.

SIMmer, HT, dr, 1997; (**Mercy**); Simpson

'Simon Bolivar', HT, or, 1966; bud ovoid; flowers bright orange-red, large, dbl., 40 petals, exhibition form, slight fragrance; foliage dark, glossy, leathery; vigorous, upright, bushy growth; [Roundelay X El Capitan]; Armstrong, D.L.; Armstrong Nursery

'Simon de st Jean', HP, m, 1861; flowers velvety purple; Liabaud

Simon Dot, *see* 'DOTsurodo'

'Simon Fraser', S, mp, 1992; semi-dbl., 6–14 petals, 1.5–2.75 in., borne in small clusters, slight fragrance; some prickles; foliage dark green, semi-glossy; low (60 cms), upright growth; extremely winter hardy.; Ogilvie, Ian S.; Agriculture Canada, 1992

Simon Robinson, *see* 'TRObwich'

Simona, *see* 'TANmosina'

'Simone', HT, w, 1924; flowers flesh-white, center deeper pink, passing to creamy white, dbl., moderate fragrance; [Mme Caroline Testout X Paul Monnier]; Buatois

'Simone', HT, m, 1957; (Mauve Mallerin, Parme); bud ovoid; flowers pastel lilac, dbl., 50 petals, 6 in., exhibition form, moderate fragrance; foliage leathery, dark, glossy; vigorous growth; [(Peace X Independence) X Grey Pearl]; Mallerin, C.; Hémeray-Aubert

'Simone Damaury', HT, dr, 1925; bud long, pointed; flowers brilliant crimson, semi-dbl., intense fragrance; [Liberty X Gen. MacArthur]; Soupert & Notting

'Simone de Chevigné', HT, pb, 1924; flowers flesh-pink, shaded yellow, dbl.; Pernet-Ducher

'Simone de Nanteuil', HT, pb, 1925; bud long, pointed; flowers rosy white, tinted carmine-pink, reverse flesh, dbl., moderate fragrance; [Ophelia X Mme Vittoria Gagniere]; Schwartz, A.

'Simone Guérin', HT, my, 1929; flowers coral-yellow, large, semi-dbl., exhibition form, slight fragrance; foliage glossy, dark; vigorous growth; [Constance X Unnamed Hybrid Tea seedling]; Mallerin, C.

'Simone Labbé', HT, ab, 1922; bud long, pointed; flowers apricot-yellow, passing to clear saffron-yellow, dbl., intense fragrance; [Le Progres X Lady Greenall]; Ketten Bros.

'Simone Mayery', HT, yb, 1937; bud long, pointed, cream-yellow, shaded carmine-pink; flowers cream, center dark yellow, very large, cupped, foliage bronze; vigorous growth; Chambard, C.

Simone Merieux®, *see* 'GAUtara'

Simonet, *see* **'Metis'**

'Simonet's Double Pink Rugosa', HRg, mp; flowers pink, smaller than hansa, very dbl., non-recurrent, hardy.; [R. macounii X Mme Georges Bruant (?)]; Simonet; P.H. Wright

'SIMpalno', HT, rb, 1979; (**Happy Day**, Velvet Lustre); bud pointed; dbl., 30 petals, exhibition form, borne singly, moderate fragrance; brown prickles; foliage large, dark, semi-glossy; strong, upright, medium growth; [First Prize X Gypsy Moth]; Simpson, J.W.

Simple Pleasures™, *see* 'TALsimple'

'Simple Simon', Min, mp, 1955; flowers carmine-rose, base yellow, dbl.; [(R. multiflora nana X Mrs Pierre S. duPont)F2 X Tom Thumb]; deVink; T. Robinson, Ltd.

'Simplex', Min, w, 1961; bud long, pointed, apricot; single, 5 petals, 1.5 in., blooms in small clusters, slight fragrance; foliage leathery; vigorous, bushy (12–14 in.) growth; [(R. wichurana X Floradora) X Seedling]; Moore, Ralph S.; Sequoia Nursery

Simplex Multiflora, HMult, pb, 1905

Simplicity™, *see* 'JACink'

'Simplicity', HT, w, 1909; flowers pure white, open, single, slight fragrance; foliage glossy, rich green; Dickson, H.

Simplicity, F, op, 1990; Lens

'Simply Divine', F, rb, 1978; semi-dbl., 18 petals, 3.5 in., slight fragrance; foliage dark; [Elizabeth of Glamis X Evelyn Fison]; Anderson's Rose Nurseries

Simply Elegant, *see* 'BUSimel'

Simply Irresistible, *see* 'BURbrahmvidya'

'Simpson's Red', HT, dr, 1977; flowers medium, dbl., 40 petals, 3.5 in., slight fragrance; [Red Lion X Grande Amore]; Simpson, J.W.; Lower Rangitikei Rose Soc.

SIMsilko, HT, lp, 1989; (**Silky Mist**); Simpson

'SIMteono', HT, pb, 1979; (**Te Moana**); bud pointed; dbl., 38 petals, exhibition form, borne mostly singly, intense fragrance; brown prickles; foliage dark, semi-glossy; medium, bushy, upright growth; Gold Star of the South Pacific, Palmerston North, NZ, 1981; [(Gypsy Moth X Percy Thrower) X First Prize]; Simpson, J.W.

SIMway, HT, w; (**Jenny Charlton**); Simpson

'Sinbad', HT, op, 1964; flowers salmon-pink; [Pink Lustre X Circus]; Leenders, J.

Sincera, *see* **'Amistad Sincera'**

Sincerely Mine, Min

Sincerely Yours, *see* 'MORdort'

'Sincerity', HT, ob, 1940; bud long, pointed; flowers flesh, shaded amber and orange, large, dbl., slight fragrance; foliage dark, leathery; vigorous, compact growth; [Comtesse Vandal seedling X Mrs Sam McGredy]; LeGrice

'Sindoor', F, or, 1980; bud long, pointed; dbl., 23 petals, borne singly or 15 per cluster, no fragrance; straight prickles, bending downward; foliage large, glossy, coppery when young; upright, tall growth; [Sea Pearl X Suryodaya]; Division of Vegetable Crops and Floriculture

Singalong, *see* 'MACsingal'

Singin' in the Rain, *see* 'MACivy'

'Single Bliss', Min, pb, 1980; bud short, pointed; flowers deep pink and white, blooms borne 5–30 per cluster, single, 5 petals, flat, slight fragrance; straight prickles; foliage very small, very glossy; very compact, bushy growth; [Seabreeze X Baby Betsy McCall]; Saville, F. Harmon; Nor'East Min. Roses

Single Charm, *see* 'UMScharm'

Single Cherry, S, pb; flowers purple/pink with dary yellow, medium, single; (Sangerhausen)

Single Chestnut Rose, *see* **R. roxburghii normalis**

Single Pink, Ch, pb

Single's Better™, *see* 'SAVabet'

Sinoia, Pol, ob

Sinsetu, LCl, w

Siobhan, *see* 'SEAsio'

'Sioux Beauty', S, rb, 1927; flowers bright rose, center dark crimson, very dbl., 100 petals, non-recurrent bloom, moderate fragrance; hardy.; [Tetonkaha X American Beauty]; Hansen, N.E.

Sir, *see* 'JOLsir'

'Sir Alexander N. Rochfort', HT, ob, 1917; flowers flesh, center darker; [Lady Alice Stanley X Marquise de Sinéty]; Le Cornu

'Sir Arthur Streeton', HT, mp, 1940; flowers pink, well formed; Clark, A.

'Sir Basil McFarland', HT, ob, 1931; bud long, pointed; flowers orange-salmon-pink, flushed yellow, dbl., exhibition form, moderate fragrance; foliage thick; vigorous growth; McGredy

'Sir Billy Butlin', HT, or, 1969; (Sir William Butlin); flowers red-orange, large, dbl., 30 petals, cupped, slight fragrance; foliage bronze; vigorous, upright growth; [Bettina X Majorca]; Gandy, Douglas L.

Sir C.V. Raman, HT, or, 1989; Pal, Dr B.P.

'Sir Cedric Morris', LCl, w, 1980; bud small, globular; single, 5 petals, borne 20–40 per cluster, summer bloom only in large quantity, slight fragrance; large prickles; foliage elongated, large; very vigorous growth; [R. glauca X Unnamed seedling]; Morris, Sir Cedric; Peter Beales, 1979

Sir Clough, *see* AUSclough

'**Sir Dallas Brooks**', HT, mr, 1987; bud long, pointed; flowers large, dbl., exhibition form, moderate fragrance; foliage leathery; vigorous, upright, open growth; [Ena Harkness X Charles Mallerin]; Smith, R.W.; T.G. Stewart

'**Sir David Davis**', HT, rb, 1926; bud long, pointed; flowers deep glowing crimson, base light yellow, dbl., exhibition form, intense fragrance; foliage dark, leathery; vigorous growth; McGredy

'**Sir David Reid**', HT, mr, 1941; flowers crimson-red, large, dbl., 25 petals, flat, moderate fragrance; Dickson, A.

Sir Edward Elgar, *see* 'AUSprima'

Sir Edward's Rose, HT, mp, 1998; Dawson

Sir Frederick Ashton, HT, w, 1987; Beales, Peter

'**Sir Galahad**', F, dp, 1967; flowers deep pink, medium, dbl., slight fragrance; foliage glossy; [Pink Parfait X Highlight]; Harkness

'**Sir Harry**', HT, my, 1989; bud ovoid; flowers medium, borne usually singly, dbl., 30 petals, exhibition form, slight, fruity fragrance; round, orange fruit; prickles triangle, brown; foliage medium, dark green, glossy; bushy, medium growth; [Irish Gold X (Pink Favorite X Golden Autumn)]; Bracegirdle, A.J.; Gregory's Roses, 1989

Sir Harry Pilkington, *see* 'TANema'

Sir Henry®, *see* KORopas

'**Sir Henry Seagrave**', HT, ly, 1932; bud long, pointed; flowers light primrose-yellow, large, dbl., exhibition form, intense fragrance; foliage leathery; GM, NRS, 1932; Dickson, A.

'**Sir Joseph Paxton**', B, dr, 1852; flowers deep red tinted violet, well-formed, medium; Laffay, M.

'**Sir Lancelot**', F, ab, 1967; flowers apricot-yellow, semi-dbl., 3–4 in., slight fragrance; foliage light green, glossy; [Vera Dalton X Woburn Abbey]; Harkness

'**Sir Matthew Nathan**', HT, lp; [Rhea Reid X Laurent Carle]; Harrison, A., 1927; (Weatherly, L.)

Sir Neville Marriner, *see* 'GLAnmusic'

'**Sir Robert Duff**', Cl T, lp; [Gloire de Dijon X Unknown]; Johnson, 1893; (Weatherly, L.)

'**Sir Rowland Hill**', HP, m, 1888; flowers dark purple/red, large, very dbl., moderate fragrance; Mack; (Sangerhausen)

'**Sir Thomas Lipton**', HRg, w, 1900; bud ovoid; flowers very white, round and cupped, dbl., cupped, recurrent bloom, intense fragrance; foliage dark, leathery; vigorous (6–8 ft), bushy growth; [R. rugosa alba X Clotilde Soupert]; Van Fleet; Conard & Jones

Sir Walter Raleigh™, *see* 'AUSspry'

'**Sir Walter Raleigh**', HT, ob, 1919; bud long, pointed; flowers coppery reddish salmon, reverse deep crimson, dbl., moderate fragrance; [Lady Pirrie X Unnamed seedling]; Le Cornu

Sir Wilfrid Laurier, LCl, lp

Sir William Butlin, *see* '**Sir Billy Butlin**'

Sir William Leech, *see* 'HORtropic'

'**Sir Winston Churchill**', HT, op, 1955; flowers salmon-pink shaded orange, dbl., 48 petals, 5 in., exhibition form, moderate fragrance; foliage dark, glossy; very vigorous growth; GM, NRS, 1955; [Unnamed seedling X Souv. de Denier van der Gon]; Dickson, A.

'**Siren**', F, or, 1953; bud ovoid; flowers bright scarlet-red, blooms in clusters, semi-dbl., 18 petals, 3–3.5 in., cupped, moderate fragrance; foliage leathery; vigorous, compact growth; GM, NRS, 1952; [(Baby Chateau X Else Poulsen) X Independence]; Kordes; J&P

'**Sirena**', HT, rb, 1941; bud ovoid; flowers cardinal-red, center shaded yellow, very large, dbl., cupped, intense fragrance; foliage leathery; very vigorous growth; [Saturnia X Anemone]; Aicardi, D.; Giacomasso

Sirenella, HT; (Cavriglia)

Sirenetta®, Min, ab

'**Sirius**', Cl HT, mr, 1939; bud long, pointed; flowers cherry-red, center lighter, large, dbl., intermittent bloom, intense fragrance; vigorous, climbing (15 ft), open habit growth; [Unnamed seedling X Lubra]; Fitzhardinge; Hazlewood Bros.

'**Sirocco**', F, or, 1971; bud globular; flowers vermilion-red, large, very dbl., cupped, slight fragrance; foliage dark, leathery; vigorous, bushy growth; [Independence X Orangeade]; Delforge

Sis, *see* 'BRIsis'

'**Sisi Ketten**', Pol, dp, 1900; flowers medium, dbl.; Ketten, Gebrüder; (Sangerhausen)

Sissi, *see* 'TANsi'

Sissi, Climbing, *see* '**Blue Moon, Climbing**'

Sissinghurst Castle, *see* '**Rose des Maures**'

Sister Joan, *see* 'KENmado'

'**Sister Kenny**', F, mr, 1954; bud ovoid; flowers scarlet, medium, borne in clusters, semi-dbl., 10–12 petals, flat, slight fragrance; foliage leathery; very vigorous, upright growth; [Baby Chateau X Red Delicious]; Hill, Joseph H., Co.; H&S

'**Sister Susan**', HT, r; bud ovoid, yellow stained crimson; flowers cream tinted flesh, center copper-orange, base yellow, medium, dbl., 48 petals, exhibition form, slight fragrance; foliage leathery, glossy, dark; vigorous, spreading growth; [Mrs Sam McGredy X Soeur Therese]; Oliver, F.

Sister Thérèse, *see* '**Soeur Thérèse**'

'**SITpretty**', F, w, 1999; flowers white with shrimp pink edges, reverse darker, dbl., 17–25 petals, 3.5 in., borne in small clusters, no fragrance; prickles moderate; foliage medium dark green, glossy; bushy, shrub-like, tall (5–6 ft) growth; [Nicole sport]; Sitton, John; Edmunds' Roses, 2000

'**Sitting Bull**', HSet, dp; flowers deep pink, semi-dbl., non-recurrent, height 6–8 ft; not hardy.; Horvath

Sitting Pretty, *see* 'TINsit'

'**Six Flags**', Gr, mr, 1962; (Manja, Manja Mourier); flowers cherry-red, dbl., 25–30 petals, 3.5–4 in., exhibition form, foliage slender, leathery, semi-glossy; vigorous, upright, spreading growth; [First Love X Roundelay]; Swim, H.C.; Five M Nursery; Poulsen

Six Fours Auri, Misc OGR, w

Sixth Sense, *see* JACrare

Sizzle Pink™, *see* 'DEVcali'

'**Sizzler**', Min, or, 1975; dbl., 28 petals, 1–1.5 in., exhibition form, moderate, spicy fragrance; vigorous, upright, spreading growth; [Sheri Anne X Prominent]; Saville, F. Harmon; Nor'East Min. Roses

'**Sjoukji Dijkstra**', Pol, mr, 1966; bud ovoid; flowers scarlet, medium, borne in large clusters, semi-dbl.; compact growth; [Chatter X Paprika]; Buisman, G. A. H.

'**Skaggarak**', F, mr, 1970; 18–20 petals, 3.5–6 in., slight fragrance; [Irish Wonder X Seedling]; Poulsen

'**Skarlagen General MacArthur**', HT, dr, 1930; flowers scarlet, large, dbl., slight fragrance; foliage medium, dark, semi-glossy; upright growth; [General MacArthur X Unnamed seedling]; Poulsen, S.; Poulsen's Roses

'**Skinner's Rambler**', LCl, lp, 1955; bud long, pointed; flowers pale pink, open, small, single, 5 petals, borne in clusters of 10–40, slight fragrance; foliage soft green; very vigorous, climbing (20 ft annually) growth; [R. maximowicziana X ?]; Skinner; Univ. of NH

'**Skogul**', F, dp, 1969; flowers rose-madder, open, borne in clusters, semi-dbl., 18 petals; foliage dark, glossy; vigorous growth; [Lichterloh X Lumina]; Lundstad

Skvost, HT; (Cavriglia)

'**Skylark**', F, mp, 1958; bud urn shaped; flowers pink, reverse darker, borne in clusters of 3–15, dbl., 55–65 petals, 2.5 in., exhibition form, foliage glossy; very vigorous, upright growth; [Carol Amling sport]; deVor, Paul F.; Amling Bros.

Skyline, *see* TANilyks

Skyliner, LCl, 1980; Delforge; (Cavriglia)

'SKYLON', HT, rb, 1952; bud very long, pointed; flowers orange shaded peach, veined red, dbl., 26 petals, 5.5–6 in.; foliage glossy; very free growth; [Mme Henri Guillot sport]; Lowe

'SKYROCKET', HMsk, dr, 1934; (Wilhelm); bud long, pointed; semi-dbl., blooms in clusters of 50, recurrent bloom, moderate fragrance; foliage large, glossy, leathery; vigorous, bushy (6–8 ft) growth; (28); [Robin Hood (HMsk) X J.C. Thornton]; Kordes

'SLATER'S CRIMSON CHINA', Ch, mr; (Chinese Monthly Rose, Crimson China Rose, Old Crimson China, R. bengalensis, R. chinensis semperflorens, R. diversifolia, R. indica semperflorens, R. semperflorens); flowers cherry red, usually solitary; slender stems; (14, 21); ca. 1790

'SLATS', HT, mp, 1999; dbl., 26–40 petals, 6 in., borne mostly singly, moderate fragrance; prickles moderate; foliage large, medium green, semi-glossy; upright, tall (5–6 ft) growth; [Louise Estes X Great Scott]; Edwards, Eddie; Johnny Becnel Show Roses, 1999

Slava®, *see* 'WILSLAV'

'SLÁVA BÖHMOVA', HT, rb, 1930; bud long; flowers salmon-red, base golden yellow, large, dbl.; foliage blood-red, mahonia-like; [Covent Garden X Golden Emblem]; Böhm, J.

'SLAVIA', Cl Pol, w, 1934; flowers rosy white, borne in large clusters, recurrent bloom, intense fragrance; vigorous growth; [Tausendschön X Seedling]; Brada, Dr.; Böhm

'SLÁVUSE', HT, w, 1936; flowers cactus dahlia form; Brada, Dr.

'SLEEPING BEAUTY', HT, w, 1966; bud long, pointed; flowers large, dbl., exhibition form, moderate fragrance; foliage leathery; vigorous, upright growth; [(Frau Karl Druschki X Rex Anderson) X Virgo]; Morey, Dr Dennison; Country Garden Nursery

'SLEEPY', Pol, mp, 1955; (Baldwin); flowers very small blooms in trusses, dbl.; [(Orange Triumph X Golden Rapture) X Polyantha seedling]; deRuiter; Gregory & Willicher Baumschulen

'SLEEPY TIME', Min, op, 1973; bud long, pointed; flowers soft peach to soft salmon-pink, small, dbl., slight fragrance; foliage small, light, leathery; vigorous, dwarf, upright, bushy growth; [Ellen Poulsen X Fairy Princess]; Moore, Ralph S.; Sequoia Nursery

'SLEIGH BELLS', HT, w, 1950; bud ovoid, cream; flowers white, center creamy, open, large, dbl., 40 petals, intense fragrance; foliage leathery, glossy; very vigorous, upright growth; [Capt. Thomas X Eternal Youth]; Howard, P.J.; H&S

Sleigh Bells, Climbing, Cl HT, w, 1995; Wilkins

Slot van Laarne, S, lp, 1993; RvS-Melle

'SMALL FANTASY', Min, lp, 1982; flowers small, dbl., slight fragrance; foliage small, light green, semi-glossy; bushy growth; [Snow Magic X Blue Mist]; Dobbs, Annette E.

Small Maiden's Blush, *see* 'Maiden's Blush'

Small Miracle, *see* 'JACIMIN'

Small Slam, *see* 'LAVLINC'

'SMALL TALK', F, my, 1963; dbl., 33 petals, 2.5 in., exhibition form, blooms in large clusters, slight fragrance; foliage leathery, glossy, dark; compact, low growth; [Yellow Pinocchio X Circus]; Swim & Weeks; Weeks Roses

'SMALL VIRTUE', Min, w, 1986; flowers small blooms borne singly and in sprays of 2–7, dbl., 48 petals, exhibition form, slight fragrance; fruit not observed; medium pink prickles; foliage small, medium green, semi-glossy; low, bushy growth; [Party Girl X Snow Bride]; Jolly, Marie; Rosehill Farm, 1983

'SMALL WONDER', Min, pb, 1983; flowers light pink, darker reverse, small blooms borne singly, dbl., exhibition form, no fragrance; foliage small, medium green, semi-glossy; upright, bushy growth; [Futura X Orange Honey]; Hardgrove, Donald L.; Rose World Originals

'SMALL WORLD', Min, or, 1975; bud pointed; flowers rich orange-red, dbl., 21 petals, 2.5 in., flat, foliage small, glossy, leathery; dwarf, compact growth; [Little Chief X Fire Princess]; Moore, Ralph S.; Sequoia Nursery

Smarty®, *see* 'INTERSMART'

Smerisal, HT, op

'SMILES', F, lp, 1937; flowers light salmon-pink, blooms in clusters, semi-dbl., slight fragrance; foliage leathery; bushy growth; [Echo X Rev. F. Page-Roberts]; Nicolas; J&P

Smiles, *see* 'JACSMI'

'SMILEY', F, ob, 1979; bud pointed; flowers orange, apricot, yellow, shapely, dbl., 21 petals, 2.5 in.; foliage matt green; tall, vigorous, upright growth; [Arthur Bell X Little Darling]; Murray, Nola

'SMILING THROUGH', F, or, 1976; flowers orange to red, dbl., 27 petals, 3 in., slight fragrance; foliage glossy, light; [Orange Sensation X Mme Louis Laperriere]; Anderson's Rose Nurseries

Smiling Wings, F, dp

'SMITFIRST', F, ab, 1986; (**SUMMER SÉRÉNADE**®); flowers apricot to gold, fading to cream, urn-shaped, medium, semi-dbl., 25 petals, slight fragrance; round, small, orange fruit; medium, pointed, brown prickles; foliage medium, dark green, glossy; upright medium growth; GM, Bagatelle, 1986 GM, Geneva, 1986; [(Unnamed seedling X Zambra) X Baby Bio]; Smith, Edward; Wheatcroft Roses, 1987

Smith's Parish, *see* **'FORTUNE'S FIVE-COLORED ROSE'**

'SMITLING', F, or, 1983; (Lady Taylor); flowers vermilion, patio, medium, dbl., 35 petals, slight fragrance; foliage medium, medium green, matt; bushy growth; [Elizabeth of Glamis X Topsi]; Smith, Edward; C. Gregory & Sons, 1984

'SMITS' BRIAR', HRg, ?; [R. rugosa X R. canina]; Smits

'SMITSBLANC', F, w, 1984; (**Fragrant Mist**); flowers well-formed, medium, dbl., 35 petals, intense fragrance; foliage medium, medium green, matt; upright growth; [Elizabeth of Glamis X Jubilee Celebration]; Smith, Edward

'SMITSHORT', F, mp, 1994; (**Richard Buckley**); flowers pale salmon pink, blooms borne in small clusters, dbl., 15–25 petals, 1.5–2.75 in., slight fragrance; some prickles; foliage medium, medium green, matt; low (45–55 cms), bushy, compact growth; [Lady Taylor X Regensberg]; Smith, Edward; Smith, 1992

Smoke Signals™, *see* 'MINAXCO'

Smokey Joe, ob, 1998; Williams, J. Benjamin

Smokey Joe's Cafe, *see* 'SHERIJOE'

'SMOKY', HT, rb, 1968; flowers smoky oxblood-red shaded burgundy, medium, dbl.; vigorous, upright growth; Combe; J&P

'SMOKY MOUNTAIN', Min, m, 1988; bud pointed; flowers deep mauve, medium, borne usually singly, dbl., 24 petals, exhibition form, slight fragrance; prickles straight, pointed, small, tan; foliage medium, dark green, semi-glossy; upright, medium growth; [Black Jade X Unnamed seedling]; Bridges, Dennis A.; Bridges Roses, 1989

Smooth Angel, *see* 'HADANGEL'

Smooth Lady, *see* 'HADLADY'

Smooth Melody, *see* 'HADMELODY'

Smooth Perfume, *see* 'HADPERFUME'

Smooth Prince, *see* 'HADPRINCE'

Smooth Princess, HT, w, 1999; Davidson

Smooth Romance™, *see* 'HADROMANCE'

'SMOOTH SAILING', Gr, w, 1977; flowers cream color, dbl., 30 petals, 4 in., slight fragrance; foliage dark, glossy; tall growth; [Little Darling X Pink Favorite]; Davidson, Harvey D.; Burgess Seed & Plant Co.

Smooth Satin, *see* 'HADSATIN'

'SMOOTH TALK', F, w, 1987; flowers ivory-white, aging bright white, medium, abundant, dbl., 25 petals, exhibition form, borne in clusters, slight fragrance; prickles pointed, small, yellow-brown; foliage medium, medium green, semi-glossy; upright, bushy, low, compact growth; PP007156; [Unnamed seedling X

Unnamed seedling]; Weeks, O.L.; Weeks Wholesale Rose Growers, 1988

Smooth Velvet, *see* 'HADVELVET'

'**SMUTS MEMORY**', HT, dr, 1950; flowers brilliant deep crimson-red; [Sensation X World's Fair]; Leenders, M.

'**SNAPPIE**', HT, rb, 1989; bud pointed; flowers medium red with yellow at base, reverse creamy-yellow, aging, dbl., 52 petals, exhibition form, prickles curved slightly downward, medium, deep pink; foliage medium, dark green, semi-glossy; upright, medium growth; [Lady X X Wini Edmunds]; Bridges, Dennis A.; Bridges Roses, 1990

'**SNEEZY**', Pol, dp, 1955; (Bertram); flowers neyron rose, small, single; deRuiter; Gregory & Willicher Baumschulen

'**SNEHURKA**', Pol, w, 1937; flowers small, dbl.; Böhm, J.; (Sangerhausen)

Snehurka, Min, lp; Urban, J.; (Czech Rosa Club)

Snehurka, Min; lp; Vik, 1950; (Czech Rosa Club)

'**SNEPRINCESSE**', Pol, w, 1953; bud ovoid; flowers pure white, small, borne in clusters, dbl., cupped, slight fragrance; foliage light green, glossy; moderate, upright, bushy growth; [Mothersday sport]; Bang

'**SNEPRINSESSE**'®, Pol, w, 1946; [Dick Koster sport]; Grootendorst, F.J.

'**SNO**', Min, w, 1982; dbl., 48 petals, exhibition form, blooms in clusters of 1–3, intense fragrance; foliage narrow, light green, matt; sprawling growth; [Unnamed seedling X Gold Pin]; Meredith, E.A. & Rovinski, M.E.; Casa de Rosa Domingo

Sno Cone, *see* 'PIXSNO'

Snodoll, *see* 'TRASNO'

Snookie, *see* 'TINSNOOK'

'**SNOOPY**', HT, rb, 1973; flowers carmine, reverse silver, urn shaped, very large, dbl., 28 petals, moderate, spicy fragrance; foliage large, glossy; [Paddy McGredy X Rose Gaujard]; Cadle's Roses

'**SNÖVIT**', Pol, w, 1946; flowers pure white; [Dick Koster sport]; Grootendorst, F.J.

Snow Angel, *see* 'JACWHIP'

'**SNOW BALLET**'®, S, w, 1977; (Claysnow, Snowballet®); flowers pure white, dbl., 45 petals, 4 in., cupped, slight fragrance; foliage small, dark, glossy; GM, Baden-Baden, 1980; [Sea Foam X Iceberg]; Clayworth; Harkness New Roses, Ltd., 1978

Snow Bear, *see* 'ADASNOW'

'**SNOW BRIDE**', Min, w, 1982; (Snowbride); dbl., 20 petals, exhibition form, slight fragrance; foliage medium, medium green, semi-glossy; bushy growth; PP005579; AOE, 1983; [Avandel X Zinger]; Jolly, Betty J.; Rosehill Farm

Snow Carpet®, *see* 'MACCARPE'

Snow Cloud, *see* TANinaso

'**SNOW CREAM**', HT, w, 1986; flowers cream, yellow stamens, blooms in clusters of 4–6, semi-dbl., 11 petals, slight fragrance; medium, red, hooked prickles; foliage medium, dark, glossy; medium, upright growth; [Garden Party X Portrait]; Bridges, Dennis A.; Bridges Roses

'**SNOW CRYSTAL**', HT, w, 1978; flowers large, dbl., 20 petals, slight fragrance; foliage large, light green, semi-glossy; [Sonia X Pascali]; Verschuren, Ted; H.A. Verschuren

Snow Drop, Min, w, 1996; Welsh

Snow Dwarf, *see* '**SCHNEEZWERG**'

'**SNOW FAIRY**', F, w, 1963; bud pointed; semi-dbl., 18 petals, cupped, blooms in clusters, moderate fragrance; foliage dark, glossy, leathery; vigorous, bushy growth; [Virgo X Katharina Zeimet]; Camprubi, C.; C-P

'**SNOW FLURRY**', F, w, 1952; bud small, long, pointed, tinted pink in cool weather; flowers open, borne in clusters, semi-dbl., moderate fragrance; foliage leathery, glossy, dark; vigorous, bushy growth; [Unnamed seedling X Red Ripples]; Moore, Ralph S.; Marsh's Nursery

Snow Goose, *see* AUSPOM

Snow Gosling, *see* POUlgren

'**SNOW HEDGE**', HRg, w, 1963; flowers pure white, medium, single; red fruit; upright, compact growth; [(R. rugosa rubra X R. cinnamomea) X R. nitida]; Nyveldt

Snow Hit, *see* POUlnows

'**SNOW INFANT**', Min, w, 1989; (Koyuki); bud ovoid; flowers light green, reverse white, aging white, small, borne in sprays, dbl., 40 petals, no fragrance; ovoid, small, red fruit; prickles small, light green; foliage small, light green, semi-glossy; upright, medium growth; [Katharina Zeimet X Unnamed seedling]; Yamasaki, Kazuko

Snow Kiss, Min, w, 1997; Kordes

'**SNOW MAGIC**', Min, w, 1976; bud short pointed, light pink; flowers medium, dbl., 40 petals, slight fragrance; foliage small, soft; bushy growth; [From unnamed seedling sport]; Moore, Ralph S.; Sequoia Nursery

Snow Maiden, *see* 'JUSMAIDEN'

Snow Meillandina®, *see* MEIgovin

Snow Owl, *see* UHLensch

Snow Pavement, *see* Schneekoppe

Snow Pavement, *see* Schneekoppe

Snow Princess, *see* 'LAVPERT'

Snow Queen, *see* '**FRAU KARL DRUSCHKI**'

Snow Rambler, HMult, w, 1995; Nieuwesteeg, J.

Snow Ruby, MinFl, rb, 1996

Snow Shower, *see* 'JACWADE'

'**SNOW SPRAY**', F, w, 1957; flowers pure white, stamens yellow, dbl., moderate fragrance; dwarf growth; [Gartendirektor Otto Linne X Gartendirektor Otto Linne]; Riethmuller

Snow Twinkle™, *see* 'MORSNO'

'**SNOW WHITE**', HT, w, 1941; bud long, pointed; flowers large, dbl., exhibition form, moderate fragrance; foliage leathery, dark; vigorous, bushy, compact growth; [Joanna Hill X White Briarcliff]; Hill, Joseph H., Co.

'**SNOW WHITE**', Min, w, 1955; flowers white tinted blush, micro-mini, blooms in clusters, dbl., 1 in.; low (3–6 in.), spreading growth; [Little Princess (Pol) X Baby Bunting]; Robinson, T.

'**SNOW WHITE**', Pol, w; [Dick Koster sport]

'**SNOW WHITE**', HT, w, 1938; flowers well formed, large, intense fragrance; [White Ophelia X Nuria de Recolons]; Dot, Pedro

Snow White, *see* LANdisney

'**SNOW WHITE, CLIMBING**', Cl HT, w; Cant, F.

'**SNOW WONDER**', Min, w, 1980; bud ovoid, pointed; dbl., 48 petals, borne 3–5 per cluster, slight fragrance; tiny, curved prickles; foliage tiny, medium green; vigorous, bushy growth; [Red Can Can X Baby Betsy McCall]; Lyon

Snowball, *see* 'MACANGELI'

'**SNOWBALL**', Pol, w, 1901; dbl.; Walsh

Snowballet®, *see* '**SNOW BALLET**'®

'**SNOWBANK**', F, w, 1937; flowers white tinted blush, changing to white, dbl., 30 petals, 3.5 in., cupped, slight fragrance; foliage leathery, dark; dwarf growth; [Mrs E.P. Thom X Gloria Mundi]; Nicolas; J&P

'**SNOWBELT**', Pol, w, 1997; flowers double, medium, dbl., 15–25 petals, borne in large clusters, slight fragrance; some prickles; foliage medium, light green turning medium green, semi-gloss; medium, bushy growth; [Seedling X Seedling]; Jerabek, Paul E.

'**SNOWBIRD**', HT, w, 1936; bud long, pointed; flowers white, center creamy, dbl., exhibition form, intense fragrance; foliage leathery; vigorous, compact, bushy growth; [Chastity X Louise Crette]; Hatton; C-P

'**SNOWBIRD, CLIMBING**', Cl HT, w, 1949; Weeks

Snowblush, *see* 'CLESNOW'

Snowbound, *see* 'LAVAWAY'

Snowbride, *see* '**SNOW BRIDE**'

'**SNOWDANCE**', F, w, 1971; dbl., 36 petals, 3 in., slight fragrance; moderate, bushy growth; [Orange Sensation X Iceberg]; deRuiter

Snowdon, HRg, w, 1989; Austin, David

'SNOWDRIFT', HWich, w, 1913; flowers pure white, small blooms in clusters of 20–30, dbl.; foliage very large, light; vigorous (8–12 ft) growth; RULED EXTINCT 1/85; Walsh

Snowdrift(TM), *see* 'WILSNOD'

Snowdrop, *see* 'AmoRU'

Snowdwarf, *see* **'SCHNEEZWERG'**

Snowfall(TM), *see* 'LEMFALL'

'SNOWFIRE', HT, rb, 1970; bud ovoid; flowers bright red, reverse white, large, dbl., slight fragrance; foliage large, dark, leathery; vigorous growth; [Detroiter X Liberty Bell]; Kordes, R.; J&P

Snowflake, *see* **'MARIE LAMBERT'**

'SNOWFLAKE', T, w, 1890; Strauss & Co.

'SNOWFLAKE', HWich, w, 1922; flowers pure white, blooms in clusters, dbl., intense fragrance; foliage dark, glossy; vigorous, climbing growth; GM, NRS, 1921; Cant, F.

'SNOWFLAKE', Min, w, 1977; [Chipper sport]; Ludwig Roses

Snowflake, *see* 'SUNSNOFLE'

'SNOWFLAKES', Min, w, 1954; flowers small, dbl.; dwarf (6 in.) growth; [(R. wichurana X Floradora) X Zee]; Moore, Ralph S.; Sequoia Nursery

Snowgoose, *see* 'BARSHIFLO'

'SNOWHITE CLIMBER', LCl, w, 1938; flowers large, dbl.; long, strong stems; vigorous climbing growth; Burbank; Stark Bros.

'SNOWLINE'®, F, w, 1970; (Edelweiss); dbl., 31 petals, 3 in., slight fragrance; ADR, 1970; Poulsen, Niels D.; McGredy

Snowman, *see* 'MATTJO'

'SNOWSONG', HT, w, 1961; bud long, pointed; flowers white, base yellow, semi-dbl., 20 petals, 4.5 in., slight, sweetbriar fragrance; foliage dark, glossy; moderately vigorous, upright growth; [Snow White (HT, Hill) X Snow White seedling]; Jelly; E.G. Hill Co.

'SNOWSONG SUPREME', HT, w, 1965; bud long, pointed; flowers white, base greenish, medium, exhibition form, moderate fragrance; moderate growth; [Snowsong X White Butterfly]; Jelly; E.G. Hill Co.

'SNOWSTORM', HMsk, w, 1907; flowers pure white, small blooms in clusters of 6–20, semi-dbl., recurrent bloom, vigorous growth; [R. moschata X Climber]; Paul

Snuggles, *see* 'RENUGGLE'

So In Love, *see* UMSLOVE

Soaring Flight, LCl, pb, 1996; Clements, John K.

Soaring Wings, *see* 'KORWINGS'

'SOBHAG', F, ob, 1972; bud urn shaped; flowers orange, open, medium, semi-dbl., moderate fragrance; foliage glossy, dark, leathery; vigorous, bushy growth; [Orangeade X ?]; Bansal, O.P.; Bansal Roses

'SOCALP', Min, w, 1986; (**Moonlight Lady**); flowers creamy white, center buff pink, small blooms borne 1–3 per cluster, dbl., 52 petals, exhibition form, slight fragrance; narrow, red prickles; foliage medium, dark, semi-glossy; upright growth; [Pink Petticoat X Pink Petticoat]; Eagle, Barry & Dawn; Southern Cross Nursery, 1987

'SOCAMP', Min, pb, 1985; (**Petticoat Lane**); flowers medium pink, deeper in center, light pink reverse, small blooms, dbl., 28 petals, exhibition form, slight fragrance; foliage small, dark, semi-glossy; upright, very small growth; [Pink Petticoat X Pink Petticoat]; Eagle, Barry & Dawn; Southern Cross Nursery

'SOCAPAN', Min, yb, 1985; (**Calumet**); flowers creamy yellow, petals edged with pink, blooms borne 1–3 per, dbl., 30 petals, exhibition form, moderate fragrance; very few, small, red prickles; foliage medium green, semi-glossy; upright, bushy growth; [Golden Angel X Golden Angel]; Eagle, Barry & Dawn; Southern Cross Nursery

Society Special, *see* 'CHEWSOS'

SOCluck, MinFl, mp, 1997; (**Lucky Me**); Eagle

'SOCOTRA', Min, ob, 1985; (**Orange Spice**); flowers orange, yellow reverse, small blooms in clusters of 1–3, dbl., 25 petals, slight fragrance; very few prickles; foliage small, medium green; short, compact, very bushy growth; [Over the Rainbow X Over the Rainbow]; Eagle, Barry & Dawn; Southern Cross Nursery

'SOCRATES', T, pb, 1858; flowers deep rose tinged fawn, large, dbl.; Moreau et Robert

'SOCSPO', Min, ab; (**Apricot Mist**)

'SODENIA', HWich, rb, 1911; flowers bright carmine changing to deep pink, small, borne in large, dbl.; vigorous, climbing growth; Weigand, C.

'SODORI-HIMÉ', HT, w, 1975; (La Blancheur); dbl., 30 petals, 4.5 in., exhibition form, slight fragrance; foliage dark; bushy growth; [White Knight X White Prince]; Onodera, Toru F.; S. Onodera

'SOESTDIJK', F, ob, 1949; flowers deep orange, base buttercup-yellow, medium, semi-dbl., moderate fragrance; vigorous, compact growth; [Vanessa X Seedling]; Leenders, M.

'SOEUR KRISTIN', HKor, my, 1978; flowers large blooms borne singly, dbl., 20 petals, repeat bloom, slight fragrance; foliage medium, medium green, glossy; semi-spreading growth; [Blanche Mallerin X (R. kordesii X (Van Bergen X Soeur Thérèse))]; James, John

'SOEUR MARIE-ANGE', F, w, 1957; flowers snow-white, very dbl.; bushy, dwarf growth; Privat

Soeur Marthe, M, dp, 1848; Vibert

'SOEUR THÉRÈSE', HT, yb, 1931; (Sister Thérèse); bud long, pointed; flowers golden yellow flushed and edged carmine, large, dbl., 25 petals, cupped, slight fragrance; foliage leathery, bronze; vigorous, bushy growth; (28); [(Général Jacqueminot X Juliet) X Souv. de Claudius Pernet]; Gillot, F.; C-P

'SOEUR THÉRÈSE, CLIMBING', Cl HT, yb, 1953; Shira

Soft Scent, *see* 'RENSCENT'

Soft Steps, *see* 'FLOSOF'

Soft Touch, *see* 'JACOUCH'

Softee, *see* 'MORFREE'

Softee, Climbing, Cl Min, lp; [Softee sport]; Hannemann, F., 1992; The Rose Paradise; (Weatherly, L.)

Softly Softly, *see* 'HARKOTUR'

Softy, *see* TANpretty

Softy, Pol, lp, 1992; flowers small, dbl.; Tantau; (Sangerhausen)

'SOGNO', HT, mr, 1973; bud ovoid, globular; flowers orient red to geranium-like, open, medium, dbl., cupped, slight fragrance; foliage large, dark, leathery; vigorous, upright, bushy growth; Calvino

'SOIR D'AUTOMNE', HT, m, 1966; bud long, pointed; flowers violet, medium, dbl., cupped, moderate fragrance; foliage light green, leathery; vigorous growth; [(Sterling Silver X Intermezzo) X (Sterling Silver X Simone)]; Dot; Minier

Soir de Fete, HT, my; Croix

Soir d'Eté, F, mr, 1977; Croix; (Cavriglia)

Soiree, S, op, 1993; (Soiree de Bonheur); Croix; (Cavriglia)

'SOLAR FLAIR', Min, rb, 1996; flowers bright crimson with yellow reverse, small blooms borne mostl, semi-dbl., 6–14 petals, no fragrance; some prickles; foliage small, dark green, glossy; compact, low growth; [Rise 'n' Shine X Captivation]; Tucker, Robbie

Solar Flash, HT, lp

Solarflare, *see* 'CAPFLARE'

Solarflash, *see* 'CAPFLARE'

'SOLARIA', HT, or, 1976; bud long, pointed; dbl., cupped, foliage dark; upright growth; [(Anabell X Zorina) X Seedling]; Kordes, W. Söhne; Barni

'SOLARIUM', HWich, w, 1925; flowers velvety vermilion-red, stamens yellow, very large, borne in, single; foliage rich glossy green; very vigorous, climbing growth; Turbat

'SOLDIER BOY', LCl, mr, 1953; flowers scarlet, large, single, repeat bloom, vigorous, pillar growth; [Unnamed seedling X Guinee]; LeGrice

'SOLDIER'S PRIDE', HT, dr, 1987; flowers medium, dbl., 26–40 petals, slight fragrance; foliage medium, dark green, semi-glossy; upright growth; [Red Planet X Clare]; MacLeod, Major C.A.

Sole di San Remo, S, 1975; Mansuino; (Cavriglia)

Soleil®, *see* 'MALso'

'Soleil d'Angers', HFt, yb, 1910; flowers orange-yellow, medium, very dbl.; Détriché; (Sangerhausen)

'Soleil de France', HT, yb, 1931; flowers sun-yellow, center reddish, large, dbl., moderate fragrance; foliage thick; vigorous growth; [Souv. de Claudius Pernet X Unnamed seedling]; Mermet; J&P

'Soleil de Lyon', F, op, 1955; flowers salmon-pink; vigorous growth; Robichon; Pin

Soleil de Minut, HT; Delbard-Chabert; (Cavriglia)

'Soleil de Rustica', LCl, dy; flowers golden yellow, large; Cognet

Soleil d'Été, *see* **'Summer Sunshine'**

'Soleil d'Or', HFt, yb, 1900; bud long, pointed; flowers orange-yellow to ruddy gold, shaded nasturtium-red, large, dbl., moderate fragrance; foliage rich green; vigorous growth; (28); [Said to be F2 seedling of Antoine Ducher X R. foetida persiana.]; Pernet-Ducher

'Soleil d'Orient', Cl HT, rb, 1935; bud long, pointed; flowers indian red, shaded yellow, very large, dbl., abundant, non-recurrent bloom, foliage glossy; long stems; vigorous, climbing (6 ft) growth; [Frau Karl Druschki X Mme Edouard Herriot]; Croibier

'Soleil Levant', LCl, ob, 1956; flowers orange, borne in clusters, single; [Spectacular X Unnamed seedling]; Mondial Roses

Soleillade, *see* DORcast

'Solfaterre', N, my, 1843; flowers large, dbl., slight fragrance; vigorous growth; [Lamarque seedling]; Boyau

Solid Gold, *see* 'LEOnoro'

Solidor®, *see* MEIfarent

Solidor, *see* MEItylpic

Soliman, F, 1975; VEG; (Cavriglia)

'Solistka Baleta', HT, yb, 1955; (Prima Ballerina); flowers soft lemon-yellow edged pink, large, moderate fragrance; [(Peace X Crimson Glory) X Poinsettia]; Klimenko, V. N.

Solitaire, *see* 'MACyefre'

'Solitaire', F, mp, 1970; flowers coral-pink, reverse silvery, dbl., 25 petals, 3 in., moderate fragrance; foliage glossy, dark; vigorous growth; [Queen Elizabeth X Elysium]; Cants of Colchester, Ltd.

'Solitaire, Climbing', Cl, F, 1997; flowers double, medium, dbl., 15–25 petals, moderate fragrance; many prickles; foliage medium, medium green, semi-glossy; upright (7 ft) growth; [Solitaire sport]; Earnshaw, Ronald

Solitude™, *see* 'POUlbero'

Solitude, *see* SEAsol

'Soller', HT, my, 1949; bud long, pointed; flowers overlarge, very dbl.; foliage glossy, bronze; bushy growth; [Eduardo Toda X Senateur Potie]; Dot, Pedro

'Solliden', HT, rb, 1924; bud long, pointed; flowers carmine, reverse shaded ochre, open, large, semi-dbl., slight fragrance; foliage dark; vigorous growth; [(Mme Mélanie Soupert X George C. Waud) X Mme Edouard Herriot]; Leenders, M.

'Solo'®, Cl HT, mr, 1956; flowers dark red, well-formed, large blooms in large clusters, dbl., slight fragrance; foliage dark, leathery; vigorous growth; [Crimson Glory X ?]; Tantau, Math.

Solo Mio, *see* **Sophia Renaissance**

Solo Mio, *see* POUlsol

'Solus', S, or, 1967; flowers bright orange-scarlet, blooms in clusters, semi-dbl., 19 petals, 2.5 in., slight fragrance; foliage dark, glossy, leathery; very vigorous growth; [Kathleen Ferrier X Dickson's Flame]; Watkins Roses

Solvang, *see* 'POUlvang'

'Soma', HT, m, 1980; bud long, pointed; flowers mauve blend, blooms borne 8 per cluster, dbl., 40 petals, no fragrance; straight prickles; foliage dark, leathery; bushy growth; [Chandrama X Surekha]; Division of Vegetable Crops and Floriculture

'Sombrero', F, pb, 1962; flowers cream flushed pink, well-formed, dbl., 25 petals, 4 in., slight fragrance; foliage light green; vigorous growth; [Masquerade X Rubaiyat]; McGredy, Sam IV; McGredy

'Sombreuil', Cl T, w, 1850; flowers creamy white, often tinted pink, well-formed, large, dbl., flat, vigorous growth; [Gigantesque seedling]; Robert

Someday Soon, *see* 'SEAsoon'

Somersault, *see* 'JACply'

Something Else™, *see* 'SAVapie'

Something for Judy, *see* 'TINjudy'

Sommerabend®, *see* KORmarec

Sommerduft®, *see* 'TANfudermos'

Sommerfreude®, F, lp

Sommerlachen, F, ob, 1972; flowers coppery orange and dark golden yellow, large, dbl., intense fragrance; GPG Bad Langensalza; (Sangerhausen)

Sommerliebe, F, ob, 1986; flowers coppery-pink, large, dbl., intense fragrance; GPG Bad Langensalza; (Sangerhausen)

Sommerliebe, HT, or, 1975; flowers large, dbl., slight fragrance; Pörschmann; (Sangerhausen)

Sommermarchen, *see* KORpinka

'Sommermärchen', F, dp, 1945; flowers dark rose, shell shaped, large, borne in clusters of 12–18, single, slight fragrance; foliage light green, leathery; upright, vigorous, bushy growth; [Prof. Gnau X Baby Chateau]; Tantau

Sommermelodie®, S, lp

Sommermond®, *see* KORmazin

Sommermorgen, *see* 'KORfullwind'

'Sommernachtstraum', Pol, rb, 1944; flowers carmine-red with white, medium, single; Krause; (Sangerhausen)

Sommerschirm, F, 1993; Tantau; (Cavriglia)

Sommerschnee, F, w; flowers large, dbl.; VEG; (Sangerhausen)

Sommertag, HT, or, 1969; flowers large, very dbl.; Gregory; (Sangerhausen)

'Sommertraum', F, pb, 1960; flowers pink, red and yellow, borne in clusters, semi-dbl.; foliage dark; [Masquerade X Unnamed seedling]; Verschuren, A.; van Engelen

Sommerwind, *see* 'KORlanum'

'Sommet', LCl, or, 1960; flowers well formed, dbl.; foliage bright green; vigorous growth; Mallerin, C.; Hémeray-Aubert

'SOMnip', HT, m, 1996; (**Blackberry Nip**); flowers deep purple, blooms borne mostly single, very dbl., intense fragrance; some prickles; foliage medium, medium green, semi-glossy; upright, medium growth; [Deep Secret X Old Port]; Somerfield, Rob

'Sonata', HT, pb, 1942; bud long, pointed; flowers red becoming lively pink, reverse darker, dbl., 30 petals, 4 in., exhibition form, moderate, fruity fragrance; foliage glossy; vigorous, bushy growth; Van Rossem; J&P

Sonata, *see* 'ZIPson'

Sonata in Pink, *see* 'JOLson'

'Sonatina', F, lp, 1982; flowers small, dbl., 35 petals, slight fragrance; foliage medium, dark, semi-glossy; [Red Maid X Sarabande]; Sanday, John; Sanday Roses, Ltd.

Soncy, T, ly

Sondermeldung, *see* **'Independence'**

'Song Bird', Gr, w, 1978; [Tammy sport]; Ryan, C.; J. Hill Co.

'Song of Paris', HT, m, 1964; (Saphir); flowers silvery lavender, dbl., 4 in., exhibition form, moderate fragrance; foliage leathery; upright growth; [(Holstein X Bayadere) X Prelude]; Delbard-Chabert; Armstrong Nursery

'Song of Songs', HT, dr, 1998; flowers dark red, full, classic shape, very dbl., 26–40 petals, 5.5 in., exhibition form, borne mostly singly, intense fragrance; some prickles; foliage medium, dark green, semi-glossy; upright, bushy, medium (3.5 ft) growth; [Adrienne Berman X (Royal William X Gabi)]; Poole, Lionel

Songfest, *see* 'ZIPson'

Sonia, *see* 'MEIhelvet'

'Sonia', HT, rb, 1938; flowers cherry-red, center orange, camellia form, semi-dbl., slight fragrance; foliage glossy, dark; short stems; vigorous, bushy growth; hardy; [(R. multiflora X R. canina) X Hortulanis Budde]; Horvath; Wayside Gardens Co.

Sonia Meilland, *see* 'MEIhelvet'

Sonia Meilland, Climbing, *see* 'MEIhelvetsar'

Sonia Rykiel, *see* MASdogui

Sonia, Climbing, *see* 'MEIhelvetsar'

Sonja 92, Gr, mp

'Sonja Henie', HT, pb, 1949; bud long, pointed to ovoid; flowers pink, reverse darker, very large, very dbl., exhibition form, foliage leathery, dark; vigorous, upright growth; [Briarcliff Supreme X Rosemarie Hinner]; Hinner, P.; Bauské Bros. & Hinner

Sonja Horstmann, HT, mr

Sonja Supreme, Gr, ab

Sonne der Freundschaft, F, dy, 1978; flowers golden yellow, medium, dbl., slight fragrance; GPG Bad Langensalza; (Sangerhausen)

'Sonnengold', HT, dy, 1936; bud long, pointed; flowers golden yellow, large, dbl., exhibition form, slight fragrance; foliage leathery, glossy, light; bushy growth; [Lilian X Sir Basil McFarland]; Kordes

Sonnenkind®, *see* 'KORhitom'

'Sonnenlicht', HFt, my, 1913; flowers canary-yellow, semi-dbl., non-recurrent, moderate fragrance; vigorous growth; [Lady Mary Fitzwilliam X Harison's Yellow]; Krüger; Kriese

Sonnenröschen®, *see* 'KORsonn'

Sonnenschirm®, *see* TANmirsch

Sonnenuntergang, F, 1970; deRuiter; (Cavriglia)

'Sonnet', HT, lp, 1961; bud ovoid; flowers light salmon-pink, dbl., 58 petals, 5 in., cupped, moderate fragrance; foliage leathery; vigorous, upright growth; [Golden Masterpiece X Spartan]; Boerner; J&P

Sonny, S, lp; Peden, G.H., 1995; (Weatherly, L.)

'Sonnychild', HT, yb, 1949; (Zonnekeild); flowers yellow, reverse edged scarlet, semi-dbl., 20 petals, 6 in., moderate fragrance; foliage glossy; vigorous growth; Lowe

'Sonoma', F, mp, 1973; flowers medium salmon-pink, medium, dbl., exhibition form, moderate fragrance; foliage leathery; vigorous, upright, bushy growth; [Sumatra X Circus]; Armstrong, D.L.; Armstrong Nursery

'Sonora', F, yb, 1962; bud ovoid; flowers buff-yellow flushed to pink, dbl., 30 petals, 3.5–4 in., cupped, moderate fragrance; foliage leathery; vigorous, upright growth; [Orange Mist X Mayday]; Boerner; J&P

Sonora Sunset, *see* DEVmauve

'Sonrisa', HT, dr, 1969; flowers deep crimson-red, medium-large, dbl., 48 petals, exhibition form, intense, damask fragrance; foliage dark, leathery; vigorous, upright growth; [Mister Lincoln X Night 'n' Day]; Swim & Weeks; Weeks Wholesale Rose Growers

'Sophia', HT, pb, 1988; flowers medium pink with persimmon-orange, large, borne usually singly, dbl., 27 petals, exhibition form, moderate, fruity fragrance; no fruit; no prickles; foliage large, dark green, glossy; upright, tall growth; [First Prize X Dolce Vita]; Weddle, Von C.; Hortico Roses, 1987

'Sophia Fleur', F, ly, 1978; flowers creamy yellow; [Elizabeth of Glamis sport]; Timmerman's Roses

'Sophia Loren'®, HT, mr, 1967; flowers well-formed, large, dbl., 33 petals; foliage glossy; vigorous, upright growth; Tantau, Math.

Sophia Renaissance, S, my; (Solo Mio)

'Sophia Smith', HT, m, 1996; flowers pinkish mauve with salmon center, reverse pinkish mauve suff, very dbl., 3.25–4 in., slight fragrance; moderate prickles; foliage medium, dark green; upright, medium (5 ft) growth; [Lady X X (Blue Moon X Great News)]; Linck, Robert G.; Linck, 1996

Sophia's Song™, *see* 'DEVloren'

Sophie, F, lp, 1972; Eve, A.

'Sophie de Bavière', A, mp; flowers clear pink, very regular, medium, dbl.; Vibert

'Sophie de Marsilly', M, pb, 1863; flowers pink with light white stripes, dbl., globular, more prickly than most albas; vigorous, upright growth; Moreau et Robert; Prior to 1848

Sophie Deborah, *see* 'KIRstein'

'Sophie MacKinnon', HT, dr, 1937; flowers deep red, large, dbl.; vigorous growth; [John Cromin X ?]; Clark, A.; NRS Victoria

'Sophie Ortlieb', HP, mp, 1933; bud long, pointed; flowers silvery pink, very large, dbl.; foliage wrinkled; very vigorous, open habit growth; [Georg Arends X Unnamed seedling]; Walter, L.

'Sophie Thomas', Cl HT, my, 1931; (Bloomfield Loveliness); flowers deep yellow, passing to lighter yellow but not cream, large, dbl., moderate fragrance; foliage good; long, strong stems; vigorous growth; [Seedling climbing X Los Angeles]; Thomas; H&S

'Sophie's Perpetual', Ch, pb; flowers pale pink, overlaid with deep pink and cerise red, globular, few prickles; foliage dark green; growth to 8 ft; Paul, W., prior to 1905; Re-int. Humphrey Brooke, 1960

Sophileo, *see* 'MAChorowhen'

'Sophisticated Lady', Gr, lp, 1985; dbl., 25 petals, exhibition form, borne usually singly, slight, fruity fragrance; medium, hooked, dull red prickles; foliage medium, medium green, semi-glossy; upright, tall growth; [Queen Elizabeth X Arlene Francis]; Epperson, Richard G.

'Sophocle', HT, m, 1974; flowers purple-red, large; [Rose Gaujard X Credo]; Gaujard

Sophy's Rose, *see* 'AUSlot'

'Soprano', F, or, 1961; flowers large, dbl., moderate fragrance; vigorous growth; [Mannequin X Aztec]; Lens

Soraya®, *see* 'MEjenor'

Soraya, Climbing®, *see* 'MEjenorsar'

Sorbet, *see* MEIpeluj

Sorbet Bouquet™, *see* 'JACzotta'

Sorbet Framboise®, *see* DELsorb

Sorcerer™, *see* 'SAVasorc'

'Sorcier', LCl, ob, 1958; flowers bright orange, recurrent bloom, foliage bronze; vigorous growth; [Seedling X Spectacular]; Hémeray-Aubert

'Soroptimist', HT, dy, 1958; flowers orange-yellow, well formed, dbl., 40 petals, 4 in.; foliage glossy; vigorous growth; [Golden Scepter X Seedling]; Verbeek

Soroptimist International, *see* 'BENstar'

'Souchet', B, m, 1842; flowers purple/pink, large, dbl., intense fragrance; Souchet; (Sangerhausen)

Soukara-Ibara, *see* **'Turner's Crimson Rambler'**

'Souma', HT, mp, 1977; dbl., 45 petals, 6 in., exhibition form, slight fragrance; foliage glossy, dark; vigorous, upright growth; Souma Rose Soc.

'Soupert et Notting', M, dp, 1874; flowers deep pink, very large, globular, recurrent bloom, intense fragrance; foliage 5 leaflets per leaf; dwarf growth; Pernet Père

'Source d'Or', LCl, yb, 1913; flowers amber-yellow, edged creamy yellow, large, borne in clusters, very dbl., abundant, non-recurrent bloom, intense fragrance; foliage glossy, dark; short stems; height 6–8 ft; Turbat

Sourire d'Antan®, S, dp; Lens, 1988

'Sourire de France', HT, ob, 1940; bud oval; flowers orange, base ochre-yellow, open, medium, dbl., slight fragrance; foliage leathery, glossy; vigorous, bushy growth; [Ampere X (Charles P. Kilham X Capucine Chambard)]; Meilland, F.; A. Meilland

Sourire d'Enfant®, *see* ORAroro

Sourire d'Orchidee, S, lp; Golden Rose, Geneva, 1985; Croix, 1985

Sourire Rosel, HMsk, lp; Lens, 1996

'Sousse', HT, lp, 1942; flowers large, dbl.; Meilland, F.; (Sangerhausen)

Sousyun, F, w

South Moon, *see* 'GELmoon'

'South Orange Perfection', HWich, pb, 1899; flowers blush-pink, turning white, rosette form borne in clusters, moderate fragrance; very hardy; [R. wichurana X Cramoisi Superieur]; Horvath; W.A. Manda

South Pacific™, *see* 'ARONesuf'

'South Seas', HT, op, 1962; (Mers du Sud); bud ovoid; flowers coral-pink, dbl., 48 petals, 6–7 in., cupped, moderate fragrance; foliage leathery; vigorous, upright growth; [Rapture X HT, Climbing seedling]; Morey, Dr Dennison; J&P

'Southampton', F, ab, 1971; (Susan Ann); dbl., 28 petals, 3 in., moderate fragrance; foliage glossy; GM, Belfast, 1974; [(Ann Elizabeth X Allgold) X Yellow Cushion]; Harkness, 1972

'Southend Jubilee', F, dp, 1965; flowers deep pink, borne in clusters, semi-dbl., 2.5 in., flat, very vigorous, bushy growth; McCreadie

'Southern Aurora', F, rb, 1999; flowers red, reverse orange, dbl., 17–25 petals, 3 in.; few prickles; foliage medium, medium green, semi-glossy; upright, medium (4.5 ft) growth; [Redgold sport]; Le Fevre, Ian

'Southern Beauty', HT, pb, 1926; flowers deep rose-pink, edged light pink; [Columbia sport]; Rowe

'Southern Belle', HT, pb, 1981; bud ovoid, long, pointed; flowers deep pink, white reverse, spiraled blooms, dbl., 28 petals, borne singly, slight fragrance; long, narrow prickles; foliage large, semi-glossy; medium, upright, spreading growth; PP005077; [Pink Parfait X Phoenix]; Swim, H.C. & Ellis, A.E.; Armstrong Nursery

Southern Charm, *see* 'BRIcharm'

'Southern Cross', HT, mp, 1931; flowers pink, dbl., globular, moderate fragrance; bushy growth; [Joseph Hill X Gen. MacArthur]; Clark, A.; Ballarat Hort. Soc.

Southern Cross, F, lp; [Redgold sport]; Jack, 1977

Southern Delight, *see* 'MORdashin'

'Southern Lady', HT, pb, 1988; bud pointed; flowers light pink, center flesh tones, reverse slightly darker pink, dbl., 40 petals, exhibition form, moderate fragrance; prickles downward pointed, medium, tan; foliage medium, dark green, semi-glossy; upright, medium growth; [Lady X X Flaming Beauty]; Bridges, Dennis A.; Bridges Roses, 1989

'Southern Spring', Min, m, 1988; bud ovoid; flowers light mauve shaded pink, reverse lighter mauve edges, darker, dbl., 65 petals, moderate fragrance; prickles slightly downward pointed, medium, tan; foliage medium, dark green, matt; bushy, low growth; [Twilight Trail X Unnamed seedling]; Bridges, Dennis A.; Bridges Roses, 1989

Southern Sun, *see* HERbeau

Southern Sunset, HT, ob, 1994; (Tamrabarani); flowers silky orange with yellow reverse and copper overtones, foliage mildew resistant; 2.5 ft growth; Viraraghavan, M.S.

'Southport', HT, mr, 1933; bud long, pointed; flowers bright scarlet, semi-dbl., 18 petals, cupped, moderate fragrance; vigorous growth; GM, NRS, 1931; [(George Dickson X Crimson Queen) X Souv. de George Beckwith]; McGredy

'Southport, Climbing', Cl HT, mr, 1946; Howard Rose Co.

Souv d'Adele Launay, B, lp; Moreau et Robert, 1872

Souv d'Adolphe de Charvoik, HWich, lp; flowers small, semi-dbl.; (Sangerhausen)

'Souv d'Adolphe Thiers', HP, mr, 1877; flowers very large, dbl., moderate fragrance; Moreau et Robert; (Sangerhausen)

'Souv d'Adolphe Turc', Pol, op, 1924; (Souvenir de Adolphe Turc); flowers clear salmon-pink; Turc, 1926

'Souv d'Aimée Terrel des Chenes', HCh, op, 1897; flowers coppery-pink, small, dbl., slight fragrance; Schwartz; (Sangerhausen)

'Souv d'Alexandre Bacot', HT, mr, 1958; flowers geranium-red, large, dbl., 50 petals, slight fragrance; very vigorous growth; [Crimson Glory X Independence]; Arles; Roses-France

'Souv d'Alexandre Bernaix', HT, dr, 1926; flowers crimson-scarlet, shaded darker, very large, dbl., cupped; foliage purplish green; vigorous growth; [Étoile de Hollande X Gen. MacArthur]; Bernaix, P.

'Souv d'Alexandre Hardy', HP, mr, 1898; flowers large, dbl.; Lévêque; (Sangerhausen)

'Souv d'Aline Fontaine', HP, lp, 1879; flowers large, very dbl.; Fontaine; (Sangerhausen)

'Souv d'Alphonse Lavallée', HP, dr, 1884; flowers dark velvety crimson to maroon, moderate fragrance; height to 6 ft; Verdier, C.; C. Verdier

'Souv d'André Raffy', HP, dr, 1899; flowers large, dbl., intense fragrance; Vigneron; (Sangerhausen)

'Souv d'Angèle Opdebeeck', HT, my, 1926; bud long, pointed; flowers canary-yellow, dbl., moderate fragrance; [Golden Ophelia X Golden Emblem]; Verschuren

'Souv d'Anne Frank', F, ob, 1960; flowers orange tinted yellow and salmon, open, medium, borne in clusters, semi-dbl., 17 petals; foliage glossy, dark; moderate, bushy growth; [Reve de Capri X Chantecler]; Delforge

'Souv d'Anne-Marie', HT, op, 1902; flowers yellowish salmon-pink, large, dbl.; Ketten, Gebrüder; (Sangerhausen)

Souv d'Anselme, B, dp; flowers large, dbl.; (Sangerhausen)

'Souv d'Antonin Poncet', HT, dr, 1921; bud long, pointed; flowers carmine, flecked paler, dbl.; [Mme Maurice de Luze X Lady Ashtown]; Schwartz, A.

'Souv d'Arthur de Sansal', HP, mp, 1876; flowers large, dbl., intense fragrance; Guénoux; (Sangerhausen)

'Souv d'Auguste Legros', T, mr, 1890; flowers red-crimson, large; Bonnaire

'Souv d'Auguste Métral', HT, dr, 1895; flowers crimson; Guillot, P.

'Souv de Amand Opdebeeck', HT, yb, 1936; bud long, pointed; flowers yellowish apricot-pink, edged pink, very large, dbl., exhibition form, moderate fragrance; foliage leathery, bronze; very vigorous growth; Belge; Opdebeeck

'Souv de Ben-Hur', HT, mr, 1960; bud pointed; flowers crimson-scarlet, dbl., 5 in., intense fragrance; very vigorous growth; [Ena Harkness X Charles Mallerin]; Verschuren; Blaby Rose Gardens

'Souv de Béranger', HP, mp, 1857; flowers rose, large, dbl.; moderate growth; Bruant

'Souv de Bertrand Guinoisseau', HP, m, 1895; flowers large, very dbl., moderate fragrance; Chédane-Guinoisseau; (Sangerhausen)

Souv de Brod, *see* **'Erinnerung an Brod'**

'Souv de Catherine Fontaine', HT, yb, 1934; bud long, pointed; flowers brownish yellow, center brick-red, reverse coral-red, very large, moderate fragrance; [Souv. de Jean Soupert X Mme Edouard Herriot]; Soupert, C.

'Souv de Catherine Guillot', T, rb, 1895; flowers coppery carmine, center shaded orange, large, dbl., intense fragrance; weak growth; Guillot, R.

'Souv de Charles Gouverneur', HT, pb, 1927; bud long, pointed; flowers flesh-pink, center salmon-orange, very large, moderate fragrance; Chambard, C.

'Souv de Charles Laemmel', HT, yb, 1919; flowers golden yellow, streaked orange and shaded pink, intense fragrance; [Frau Karl Druschki X Soleil d'Or]; Gillot, F.

'Souv de Christophe Cochet', HRg, mp, 1894; flowers large, semi-dbl., moderate fragrance; Cochet-Cochet

'Souv de Claude Vially', HT, rb, 1931; bud long, pointed; flowers light red, tinted pink and aurora; Reymond

'Souv de Claudius Denoyel', Cl HT, dr, 1920; (Denoyel); bud long, pointed; flowers rich crimson-red, tinted scarlet, large, dbl., cupped, sparse, recurrent bloom, moderate fragrance; vigorous, climbing growth; [Château de Clos Vougeot X Commandeur Jules Gravereaux]; Chambard, C.

'Souv de Claudius Pernet', HT, my, 1920; bud long, pointed; flowers pure sunflower-yellow, center deeper, large, dbl., 28 petals, moderate fragrance; foliage glossy, rich green; long, strong stems; vigorous growth; GM, Bagatelle, 1920; [Constance X Unnamed variety]; Pernet-Ducher

'Souv de Claudius Pernet, Climbing', Cl HT, my, 1925; Western Rose Co.

'Souv de Claudius Pernet, Climbing', Cl HT, my, 1932; Schmidt, J.C.

'Souv de Claudius Pernet, Climbing', Cl HT, my, 1933; Gaujard

'Souv de Clermonde', HT, pb, 1925; flowers salmon-rose, center darker, shaded yellow, semi-dbl., moderate fragrance; Pernet-Ducher

'Souv de Denier van der Gon', HT, rb, 1935; bud long, pointed; flowers reddish yellow to golden yellow, large, dbl., moderate fragrance; very vigorous growth; [Roselandia X Souv. de Claudius Pernet]; Verschuren-Pechtold

'Souv de E. Guillard', HT, rb, 1912; flowers reddish yellow shaded coppery carmine, large, dbl., moderate fragrance; very vigorous, branching growth; GM, Bagatelle, 1914; [Beauté Inconstante X Le Progres]; Chambard, C.

'Souv de F. Bohé', HT, op, 1922; bud long, pointed; flowers orange-salmon, dbl., 35 petals; [Willowmere X Unnamed seedling]; Chambard, C.

'Souv de Francis Borges', HT, ob, 1932; flowers flesh, center orange, very large, moderate fragrance; vigorous growth; [Mme Leon Pain X Unnamed seedling]; Chambard, C.

'Souv de François Gaulain', T, rb, 1889; flowers crimson and violet; Guillot

'Souv de François Graindorge', HT, yb, 1928; bud long, pointed; flowers ochre-yellow, base indian yellow, dbl., moderate fragrance; [Benedicte Sequin X Lady Hillingdon]; Grandes Roseraies

'Souv de François Mercier', HT, ob, 1923; bud long, pointed; flowers light coppery rose, edged deeper, dbl., moderate fragrance; [Antoine Rivoire sport]; Laperrière

'Souv de François Richardier', HT, rb, 1923; flowers bright carmine-pink, tinted cherry; Richardier

'Souv de Gabriel Luizet', HT, yb, 1922; flowers sulfur-yellow tinted salmon, passing to deep rich yellow and, dbl.; [Mme Mélanie Soupert seedling X Lyon Rose]; Croibier

'Souv de Gabrielle Drevet', T, op; flowers salmon-pink, base coppery, large, well formed, intense fragrance; Guillot et Fils, 1885

'Souv de Gaston Commagères', HT, ob, 1954; flowers orange-ivory veined yellow, edge veined pink, very dbl.; vigorous, bushy growth; Privat

'Souv de George Beckwith', HT, pb, 1919; bud long, pointed; flowers shrimp-pink, tinted chrome-yellow, base deeper, very large, dbl., 55 petals, globular, moderate fragrance; foliage glossy; very vigorous growth; [Unnamed variety X Lyon Rose]; Pernet-Ducher

'Souv de George Knight', HT, my, 1926; flowers nanking yellow; [Rayon d'Or sport]; Knight, J.

'Souv de Georges Pernet', HT, op, 1921; flowers medium salmon-pink, large, dbl., 31 petals, intense fragrance; foliage dark, bronze; vigorous growth; GM, Bagatelle, 1921; [Unnamed seedling X Mme Edouard Herriot]; Pernet-Ducher

'Souv de Georges Pernet, Climbing', Cl HT, op, 1927; Pernet-Ducher

'Souv de Gilbert Nabonnand', T, pb, 1920; flowers in summer: yellow base edged carmine-pink, dbl., intense fragrance; vigorous growth; (28); Nabonnand, C.

'Souv de Gregoire Bordillon', HP, 1860; Moreau et Robert; (Cavriglia)

'Souv de Gustave Prat', HT, ly, 1909; flowers pure light sulfur-yellow, dbl.; Pernet-Ducher

'Souv de Gustave Schickelé', HT, yb, 1927; bud long, pointed; flowers chrome-yellow, reverse bright rosy scarlet, shaded apricot, dbl.; [Mme Edouard Herriot X Duchess of Wellington]; Ketten Bros.

'Souv de H.A. Verschuren', HT, yb, 1922; bud long, pointed; flowers yellow to orange-yellow, well-formed, large, dbl., 38 petals, moderate fragrance; vigorous growth; [Unnamed seedling X Golden Ophelia]; Verschuren

'Souv de Henri Faassen', HT, pb, 1929; bud long, pointed; flowers deep pink, base orange-yellow, open, large, semi-dbl., intense fragrance; foliage bronze; very vigorous growth; [Betty Uprichard X (Mrs George Shawyer X Los Angeles)]; Faassen-Hekkens

'Souv de Henri Venot', HT, mr, 1931; flowers brilliant red, very dbl., intense fragrance; foliage dark; vigorous growth; [Lord Charlemont X Red Star]; Lens

'Souv de J. Chabert', F, mr, 1956; (Bobbie Robbie, Pépite); flowers well-formed, large blooms in clusters of 3–6, dbl.; foliage dark, bronze; vigorous, low growth; [Francais X Unnamed seedling]; Delbard-Chabert

'Souv de J. Mermet', HWich, dr, 1934; flowers silvery carmine, large, borne in clusters, recurrent bloom; foliage bright, bronze; very vigorous growth; Mermet

Souv de J.B. Guillot, T, rb; Guillot, 1897

'Souv de J.B. Weibel', HT, dr, 1930; flowers carmine, very large, dbl., cupped, slight fragrance; foliage dark; very vigorous, bushy growth; [Unnamed seedling of Mrs Bullen X Edouard Mignot]; Sauvageot, H.; C-P, 1933;, F. Gillot, 1930

'Souv de Jacques Verschuren', HT, op, 1950; bud long, pointed; flowers apricot-salmon, large, dbl., moderate fragrance; foliage leathery, dark; vigorous, bushy growth; [Katharine Pechtold X Orange Delight]; Verschuren-Pechtold

'Souv de Jean Croibier', HT, pb, 1921; flowers bright salmon-pink shaded chamois, center coral-red shaded yellow, dbl., slight fragrance; [Mme Mélanie Soupert X Lyon Rose]; Croibier

'Souv de Jean Ginet', HT, mr, 1935; flowers scarlet-red, base coppery, reverse golden yellow; Brenier, E.C.; Buatois

'Souv de Jean Soupert', HT, my, 1929; bud long, pointed; flowers golden yellow, very large, semi-dbl., cupped, intense fragrance; foliage bronze; vigorous growth; [Ophelia X Feu Joseph Looymans]; Soupert & Notting

'Souv de Jeanne Balandreau', HP, mr, 1899; Vilin

'Souv de John E. Knight', HT, ob, 1928; flowers terra-cotta, salmon-pink and flesh streaked yellow, moderate fragrance; foliage dark; vigorous growth; Knight, J.

'Souv de Josefina Plà', HT, mr, 1929; flowers bright red, large, semi-dbl., moderate fragrance; vigorous growth; [Étoile de Hollande X Mme Butterfly]; Munné, B.

'Souv de Joseph Besson', HT, rb, 1931; flowers reddish orange, base yellow; Brenier, E.C.

'Souv de Jules Nicolas Mathieu Lamarche', Pol, rb, 1934; flowers cardinal-red, base yellow, very small large panicles, single; [Eblouissant X Petit Constant]; Soupert, C.

Souv de la Bataille de Marengo, *see* **'Russelliana'**

'Souv de la Malmaison', B, lp, 1843; (Queen of Beauty and Fragrance); flowers creamy flesh, center rosy shaded, very large, dbl., quartered, repeat bloom, intense, spicy fragrance; dwarf, bushy (2 ft) growth; (21); [Mme Desprez X Tea]; Béluze

'Souv de la Malmaison, Climbing', Cl B, lp, 1893; Bennett

'Souv de la Princesse Amélie des Pays-Bas', HP, m, 1873; flowers large, dbl.; Liabaud; (Sangerhausen)

Souv de la Princesse de Lamballe, *see* **'Queen of Bourbons'**

'Souv de la Reine d'Angleterre', HP, mp, 1855; flowers bright pink, large, dbl., recurrent bloom; very vigorous growth; [La Reine X ?]; Cochet Freres

'Souv de Laffay', HP, mr, 1878; flowers crimson-red; Verdier, E.

'Souv de Laurent Guillot', HT, 1894; Bonnaire; (Cavriglia)

'Souv de l'Aviateur Métivier', HWich, ly, 1913; flowers clear yellow, passing to white, dbl.; vigorous, climbing growth; [R. wichurana X Mme Ravary]; Tanne

'Souv de l'Aviateur Olivier de Montalent', HWich, rb, 1913; flowers dull rose, base salmon, borne in clusters, dbl., profuse bloom, rarely recurrent; foliage very glossy, dark; vigorous, climbing growth; [R. wichurana X Anna Olivier]; Tanne

'Souv de l'Exposition de Bordeaux', HMult, mr, 1906; flowers medium, dbl.; Puyravaud; (Sangerhausen)

'Souv de Lilette', HT, w, 1937; bud very long; flowers snow-white, large, intense fragrance; foliage slightly bronze; compact, bushy growth; Chambard, C.

Souv de l'Imperatrice Josephine, *see* **'Empress Josephine'**

Souv de Louis Amade, *see* DELilac

Souv de Louis Simon, *see* **'Mrs Miniver'**

'Souv de Lucie', N, dp, 1893; flowers medium, dbl.; Schwartz; (Sangerhausen)

'Souv de Lucienne Valayer', HT, pb, 1938; bud long; flowers soft pink, shaded light salmon, very large, cupped, moderate fragrance; vigorous, bushy growth; Chambard, C.

'Souv de Madeleine Rouillon', HT, ob, 1929; bud long, pointed; flowers orange, base yellow, dbl., moderate fragrance; [Manon X Elvira Aramayo]; Bernaix, P.

Souv de Malmedy, HGal, lp; Scarman, 1996

'Souv de Maman Corboeuf', HP, mp, 1900; flowers medium, dbl.; Corboeuf; (Sangerhausen)

Souv de Marcel Proust, HT, 1992; Delbard; (Cavriglia)

'Souv de Marcelle Balage', HT, pb, 1930; (Marcella Baldge); bud long, pointed; flowers satiny flesh-pink, center slightly tinted salmon, very large, dbl., cupped, intense fragrance; foliage dark; [Willowmere X Mme Pizay]; Bernaix, P.

'Souv de Maria Clotilde', HT, ly, 1934; bud long; flowers amber-white, often passing to light yellow, center sometimes, dbl., moderate fragrance; vigorous growth; [Mme Abel Chatenay sport]; Carneiro; Ketten Bros.

'Souv de Maria de Zayas', HT, mr, 1906; flowers large, dbl., moderate fragrance; Soupert & Notting; (Sangerhausen)

'Souv de Marie Finon', HT, ab, 1924; flowers apricot-yellow passing to clear yellow shaded salmon, dbl., intense fragrance; Croibier

'Souv de Marie Thérèse Privat', Pol, mr, 1935; flowers bright vermilion, well formed, borne in clusters of 40–50, very dbl.; foliage bright green; dwarf, good habit growth; Privat

'Souv de Marie-Thérèse', HT, w; flowers ivory-white, slight fragrance; low growth; Chabanat; Roses-France

'Souv de Marques Loureiro', HT, rb, 1912; flowers light red shading to rose, tinted yellow and purple, dbl.; [Mons. Paul Lédé X Mme Hoste]; Ketten Bros.

Souv de Maurice Chevalier, *see* 'DELsouche'

Souv de McKinley, HT, lp; Godard, 1902

'Souv de Mlle Juliet de Bricard', Pol, dp, 1934; flowers pale rosy pink, very dbl., globular, slight fragrance; foliage glossy, dark; vigorous, bushy growth; [Cécile Brunner X Yvonne Rabier]; Délépine; Pajotin-Chédane

'Souv de Mme A. Hess', HT, pb, 1936; bud long, pointed; flowers shrimp-pink, center deep coral, very large, cupped, moderate fragrance; foliage bronze; very vigorous, bushy growth; [Unnamed seedling X Ami F. Mayery]; Chambard, C.

'Souv de Mme Achille van Herreweghe', HT, dr, 1936; flowers carmine-red; vigorous growth; Van Herreweghe-Coppitters

Souv de Mme Auguste Charles, B, mp; Moreau et Robert, 1866

'Souv de Mme Augustine Gillot', HT, op, 1920; bud long, pointed; flowers salmony flesh-pink, base salmon-yellow, reverse silvery flesh, slight fragrance; [Frau Karl Druschki X Lyon Rose]; Gillot, F.

'Souv de Mme Berthier', HP, lp; Berthier/ Liabaud, 1881

'Souv de Mme Boullet', HT, dy, 1921; bud long, pointed; flowers well-formed, large, dbl.; vigorous, spreading growth; [Sunburst X Unnamed variety]; Pernet-Ducher

'Souv de Mme Boullet, Climbing', Cl HT, dy, 1930; H&S

Souv de Mme Breuil, B, dp; 1889

'Souv de Mme C. Chambard', HT, op, 1931; flowers coral-rose-pink, center flushed gold, large, semi-dbl., cupped, moderate fragrance; vigorous growth; Chambard, C.; C-P

'Souv de Mme C. Chambard, Climbing', Cl HT, op, 1935; Armstrong, J.A.

'Souv de Mme Canel', HT, rb, 1932; flowers carmine-orange, large, very dbl.; foliage pointed, bronze; robust, bushy growth; Gillot, F.

'Souv de Mme Chédane-Guinoisseau', HP, mr, 1900; flowers bright geranium red, very large, dbl.; Chédane-Guinoisseau

Souv de Mme de Corval, HP, dy; Gonod, 1867

'Souv de Mme Durand', HT, yb, 1954; flowers naples yellow, base bright yellow, edges veined pink, dbl., globular; foliage glossy; strong stems; Privat

'Souv de Mme Eugène Verdier', HT, w, 1894; flowers white, base saffron-yellow; [Lady Mary Fitzwilliam X Mme Chedane-Guinoisseau]; Pernet-Ducher

'Souv de Mme F. Zurich', HT, mp, 1910; flowers silvery pink, large, dbl.; Puyravaud; (Sangerhausen)

'Souv de Mme Gauthier-Dumont', HT, mr, 1921; bud long, pointed; flowers scarlet; Guillot, P.

'Souv de Mme H. Thuret', HP, pb, 1922; flowers salmon-pink, center shrimp-red, edged chrome, well-formed, cupped; foliage rich green; very vigorous growth; [Frau Karl Druschki X Lyon Rose]; Texier; P. Nabonnand

'Souv de Mme Hennecourt', HP, mp, 1869; flowers large, dbl.; Cochet, S.; (Sangerhausen)

'Souv de Mme Jules Pages', HMult, pb, 1937; flowers deep pink shaded orange and red, open, small, dbl., free, intermittent bloom, slight fragrance; vigorous, climbing or pillar growth; [Phyllis Bide X Eblouissant]; Reiter; Stocking

'Souv de Mme Kreuger', HT, op, 1919; flowers pure salmon-orange, passing to coppery pink, dbl.; [Mme Mélanie Soupert X Willowmere]; Chambard, C.

'Souv de Mme l'Advocat', N, op, 1899; flowers coppery pink, medium, dbl., moderate fragrance; Veysset; (Sangerhausen)

'Souv de Mme Lefèbvre', HT, rb, 1929; flowers oriental red, passing to pink, golden yellow and red; vigorous growth; Richardier

'Souv de Mme Léonie Viennot', Cl T, yb, 1898; flowers yellow shaded pink, very dbl.; vigorous growth; Bernaix, A.

'Souv de Mme Levet', T, ob, 1891; flowers orange-yellow; Levet, F.

'Souv de Mme Louise Cretté', HT, yb, 1924; flowers golden yellow, shaded coral and red, dbl.; [Mme Edouard Herriot seedling]; Cretté; C. Chambard

'Souv de Mme Morin-Latune', HT, dp, 1920; bud long, pointed; flowers cream-rose, dbl.; Bernaix, P.

'Souv de Mme Pidoux', HT, yb, 1926; bud long, pointed; flowers chrome-yellow, reverse pink, cupped; [Unnamed seedling X Mrs Aaron Ward]; Chambard, C.

'Souv de Mme Robert', HP, lp, 1879; flowers large, dbl.; Moreau et Robert; (Sangerhausen)

'Souv de Mme Sablayrolles', T, ab, 1891; flowers apricot-pink; Bonnaire

'Souv de Mme Salati-Mongellaz', HP, dp, 1937; bud long, pointed; flowers satiny rose-pink, very large, dbl.; foliage dark; very vigorous growth; [Frau Karl Druschki X Unnamed seedling]; Croibier

'Souv de Mons Rousseau', HP, mr, 1861; flowers medium, dbl.; Fargeton; (Sangerhausen)

'Souv de Norah Lindsay', LCl, mp; flowers bright pink, stamens golden, very large, single; vigorous, pillar growth

'Souv de Nungesser', HT, dr, 1927; flowers brilliant deep carmine-red, dbl., moderate fragrance; [Mme Maurice de Luze X Laurent Carle]; Croibier

'Souv de Papa Calame', HT, dp, 1922; flowers large, dbl., moderate fragrance; Guillot; (Sangerhausen)

'Souv de Paul Grandclaude', HT, yb, 1923; bud large, long, pointed; flowers yellow, shaded pink and in autumn light brown, semi-dbl.; [Mme Mélanie Soupert X Beauté de Lyon]; Sauvageot, H.

'Souv de Paul Neyron', T, yb, 1871; flowers salmon-yellow edged pink, semi-globular, large, intense fragrance; very vigorous growth; [Ophirie seedling]; Levet, A.

'Souv de Paul Raudnitz', HWich, lp, 1910; flowers small, very dbl.; Cochet-Cochet; (Sangerhausen)

'Souv de Philémon Cochet', HRg, w, 1899; flowers white, center rose; [Blanc Double de Coubert sport]; Cochet-Cochet

Souv de Pierre Dupuy, B, lp; Levet, 1876

'Souv de Pierre Guillot', HT, ob, 1928; bud long, pointed; flowers coral-orange blend, dbl.; [Marie Adélaide X Unnamed seedling]; Guillot, M.

'Souv de Pierre Ketten', HT, rb, 1928; bud long, pointed; flowers bright rose, inside pink, base chrome-yellow, dbl., 30–35 petals, intense fragrance; [Mme Mélanie Soupert X Pilgrim]; Ketten Bros.

'Souv de Pierre Leperdrieux', HRg, mr, 1895; flowers bright red; large fruit; very vigorous growth; Cochet-Cochet

'Souv de Pierre Notting', T, yb, 1902; bud long, pointed; flowers sunflower-yellow tinted apricot and coppery yellow, edged rose, very dbl., slight fragrance; foliage rich green, soft; [Marechal Niel X Maman Cochet]; Soupert & Notting

'Souv de Pierre Vibert', M, rb; flowers dark red, shaded carmine and violet, large, dbl., sometimes recurrent bloom; moderate growth; Moreau et Robert, 1867

'Souv de Prosper Fraissenon', HT, m, 1927; bud long, pointed; flowers geranium-red tinted violet; Richardier

'Souv de R.B. Ferguson', HT, pb, 1922; flowers shell-pink and apricot, sometimes shaded rose-pink and cerise, slight fragrance; [Unnamed seedling X Constance]; Ferguson, W.

'Souv de Raymond Gaujard', HT, yb, 1943; bud ovoid; flowers golden yellow, reverse often veined red, open, medium, dbl., moderate fragrance; foliage dark, glossy; vigorous, upright growth; Gaujard

'Souv de René Grognet', HT, ob, 1921; bud long, pointed; flowers coppery orange-yellow, shaded carmine; Chambard, C.

Souv de Rose Berkley, *see* **'Rose Berkley'**

Souv de Rose-Marie, HMsk, lp; Lens, 1998

'Souv de S.A. Prince', T, w, 1889; (The Queen); [Souv. d'un Ami sport]; Prince

'Souv de St Anne's', B, lp; flowers blush-pink, nearly, single, recurrent bloom; (21); [Souv. de la Malmaison sport]; Hilling, 1950

'Souv de Thérèse Levet', T, dr, 1886; [Adam X ?]; Lovet, A.

'Souv de Victor Hugo', T, pb, 1886; flowers china-pink, center salmon-pink, large, dbl., intense fragrance; [Duchesse de Brabant X Regulus]; Bonnaire

'Souv de Victor Hugo', HP, lp, 1885; flowers very large, very dbl.; Pernet; (Sangerhausen)

'Souv de Victor Landeau', B, mr, 1890; bud rose; flowers vivid red; Moreau et Robert

'Souv de Victor Landeau', HP, pb; flowers pale pink to mauve rose, large, dbl., globular; Moreau et Robert, 1890

'Souv de William Wood', HP, dr, 1864; flowers dark velvety red, large, dbl., 37 petals, cupped, sometimes recurrent bloom, moderate fragrance; vigorous growth; Verdier, E.

'Souv de Yeddo', HRg, lp, 1874; flowers large, dbl.; Morlet; (Sangerhausen)

'Souv de. H.A. Verschuren, Climbing', Cl HT, yb, 1927; H&S

'Souv d'Elise Vardon', T, w, 1855; flowers creamy white, center yellowish, very large, dbl., globular; Marest

'Souv d'Emile Mayrisch', HT, dr, 1932; flowers dark crimson-garnet, well formed, large, dbl., 35–40 petals; stiff stems; vigorous growth; Ketten Bros.

'Souv d'Emmanuel Buatois', HT, rb, 1932; bud long, pointed; flowers coral-red shaded shrimp-red, reverse clear rose, base golden, dbl., moderate fragrance; foliage leathery, bronze; vigorous growth; [Mme Edouard Herriot X Souv. de Claudius Pernet]; Buatois

'Souv d'Ernest Thébault', HWich, dr, 1921; dbl., borne in clusters of 10–20; vigorous, climbing growth; Thebault

Souv di Castagneto, HP, lp; Scarman, 1988; re-introduced by Scarman, 1988

'Souv du Baron de Sémur', HP, m, 1874; flowers deep purple-red shaded fiery red and black, large, dbl.; vigorous growth; [Charles Lefebvre X ?]; Lacharme, F.

'Souv du Capitaine Crémona', HT, rb, 1928; flowers salmon-carmine over yellow ground, dbl., moderate fragrance; [Admiration X Gorgeous]; Bernaix, P.

'Souv du Capitaine Fernand Japy', HT, mr, 1922; dbl., slight fragrance; [Le Progres X Les Rosati]; Sauvageot, H.

'Souv du Capitaine Ferrand', HT, rb, 1939; bud long, well formed; flowers nasturtium-red, reverse golden yellow; erect, vigorous growth; Gaujard

'Souv du Docteur Albert Reverdin', HT, rb, 1930; bud long, pointed; flowers brilliant carmine, shaded vermilion, large, very dbl., moderate fragrance; foliage dark; strong stems; vigorous growth; [George C. Waud X Mrs Edward Powell]; Bernaix, P.

'Souv du Docteur Jamain', HP, dr, 1865; flowers plum shaded deep crimson, dbl., moderate fragrance; moderate growth; [Charles Lefebvre X ?]; Lacharme, F.

'Souv du Lieutenant Bujon', B, mr, 1891; flowers very large, dbl., moderate fragrance; Moreau et Robert; (Sangerhausen)

'Souv du Papa Calame', HT, pb, 1921; bud long, pointed; flowers carmine-pink, reverse silvery pink, stamens salmon, dbl.; [Jonkheer J.L. Mock sport]; Gillot, F.

'Souv du President Carnot', HT, lp, 1894; bud long, pointed; flowers flesh-pink, center shell-pink, moderate fragrance; [Unnamed seedling X Lady Mary Fitzwilliam]; Pernet-Ducher

'Souv du Président Carnot, Climbing', Cl HT, lp, 1926; Grandes Roseraies

'Souv du Président Daurel', HT, dp, 1906; flowers large, dbl.; Chauvry; (Sangerhausen)

Souv du President Lincoln, B, mp; Moreau et Robert, 1865

'Souv du Président Plumecocq', HT, rb, 1958; flowers bright red, reverse marked silvery, large, dbl., moderate fragrance; foliage bronze; very vigorous, well branched growth; [Unnamed seedling X Peace]; Laperrière; EFR

'Souv du Reverend Pére Planque', HT, ob, 1932; flowers orange-chrome-yellow, fading lighter; foliage very glossy; very vigorous growth; [Unnamed seedling X Souv. de Georges Pernet]; Bel

'Souv du Rosiériste Gonod', HP, mr, 1889; flowers large, dbl., intense fragrance; Ducher; (Sangerhausen)

Souv du Rosieriste Rambaux, T, pb; Rambaux, 1883

'**Souv du Sénateur Bazire**', HMult, m, 1918; flowers violet, center violet-rose, borne in clusters of 25–50, semi-dbl.; [Veilchenblau X Bordeaux]; Lottin

'**Souv du Sergent Cretté**', HT, yb, 1921; bud long, pointed; flowers coppery golden yellow, shaded carmine, very large, cupped; foliage bronze; very vigorous growth; [Mme Mélanie Soupert X Unnamed seedling]; Chambard, C.

'**Souv d'un Ami**', T, lp, 1846; flowers pale rose tinged salmon, very large, dbl., cupped, intense fragrance; vigorous growth; Belot-Defougere

'**Souv of Miami**', HT, dr, 1925; Cook, J.W.

'**Souv of Stella Gray**', T, ob, 1907; flowers deep orange splashed apricot, salmon and crimson, small to medium, semi-dbl., moderate fragrance; Dickson, A.

'**Souv of the Old Rose Garden**', HT, pb, 1929; flowers silvery pink, brighter inside, very large, dbl., globular, intense fragrance; foliage glossy, light; strong stems; vigorous growth; GM, NRS, 1928; Cant, B. R.

'**Souv of Wootton**', HT, mr, 1888; flowers rich velvety red, very large, dbl., intense fragrance; foliage dark, leathery; vigorous growth; [Bon Silène X Louis van Houtte]; Cook, J.W.

'**Souv of Wootton, Climbing**', Cl HT, mr, 1899; Butler

'**Souvenance**', HT, m, 1965; bud very long, pointed; flowers deep lavender-lilac, dbl., 25–30 petals, 3.5–4.5 in., exhibition form, intense fragrance; foliage dark; moderately vigorous growth; [Unnamed seedling X Sterling Silver]; Lens

'**Souvenir**', HT, dy, 1930; bud pointed; flowers golden yellow, dbl., 36–42 petals, moderate fragrance; foliage glossy; vigorous growth; [Talisman sport]; Pierson, A.N.

Souvenir, *see* EuroGROOT

Souvenir de Adolphe Turc, *see* '**Souv d'Adolphe Turc**'

Souvenir de la Malmaison Rouge, *see* '**Leweson Gower**'

Souvenirs de Marcel Proust, *see* DELpapy

Souviens-Toi, *see* 'KRIprile'

'**Sovereign**', HT, dy, 1922; flowers deep yellow and old-gold, open, cupped, intense, fruity fragrance; foliage glossy, dark, bronze; vigorous growth; [Queen Mary X Unnamed seedling]; Cant, B. R.

'**Sovrana**', HT, ob, 1930; flowers orange streaked yellow, well formed, large; very vigorous, upright growth; [Julien Potin X Signora seedling (yellow)]; Aicardi, D.; Giacomasso

Sowetan Peace, *see* JACsowe

Sox, *see* 'MICsox'

'**Soyécourt**', HT, rb, 1921; flowers blood-red, overlaid orange; [Gen. MacArthur X George C. Waud]; Jersey Nursery

'**Space Girl**', LCl, dr, 1966; flowers dark crimson, very large, dbl., moderate fragrance; foliage dark; vigorous, climbing growth; [Queen Elizabeth X Étoile de Hollande]; Barter

Space Invader, *see* 'DICrocky'

Space Odyssey(TM), *see* 'WEKsnacare'

Space Probe, *see* 'MINacon'

Space Walk(TM), *see* 'MINabcco'

'**Spaethiana**', S, m; (R. X spaethiana); flowers purple, large, corymbose; (21); [R. palustris X R. rugosa]

Spangles, *see* 'GANspa'

Spanish Beauty, *see* '**Mme Grégoire Staechelin**'

Spanish Dancer, *see* 'MORliz'

Spanish Enchantress, S, lp

'**Spanish Eyes**', F, op, 1981; flowers orange-salmon, reverse yellow; [Prominent sport]; Takatori, Yoshiho; Japan Rose Nursery

'**Spanish Gold**', HT, dy, 1960; flowers straw-yellow; tall, bushy growth; Fletcher; Tucker

Spanish Main, *see* '**Marquesa del Vadillo**'

'**Spanish Orange**', F, ob, 1966; flowers orange, small borne in clusters, very dbl., 1.5 in., moderate fragrance; foliage dark, glossy; very free growth; deRuiter; Gregory

'**Spanish Rhapsody**', S, pb, 1984; flowers deep pink, tinted orange and freckled red, shallow-cupped, large blooms, dbl., 30 petals, cupped, repeat bloom, moderate, raspberry fragrance; awl-like prickles; foliage medium-large, leathery, dark olive green; upright, bushy, branching growth; hardy; [Gingersnap X Sevilliana]; Buck, Dr Griffith J.; Iowa State University

'**Spanish Sun**', F, dy, 1966; flowers large, dbl., intense fragrance; foliage glossy, leathery; vigorous, bushy growth; [Yellow Pinocchio X Golden Garnette climbing seedling]; Boerner; J&P

'**Spanky**', Min, yb, 1986; flowers bright yellow, shaded pink to red, reverse yellow, aging pink, dbl., exhibition form, slight fragrance; few, pink, small prickles; foliage large, medium green, semi-glossy; medium, upright growth; [Rise 'n' Shine X ?]; Bridges, Dennis A.; Bridges Roses

'**Sparkels**', HT, rb, 1956; bud urn shaped; flowers red striped white, medium, dbl., 35–50 petals, exhibition form, intense fragrance; foliage leathery; [Briarcliff sport]; Webber

'**Sparkie**', Min, mr, 1957; flowers bright red, becoming darker, blooms in clusters, single, 6 petals; foliage glossy; vigorous (12–16 in.) growth; [(R. wichurana X Floradora) X Little Buckaroo]; Moore, Ralph S.; Sequoia Nursery

'**Sparkle**', HT, pb, 1949; bud pointed, ovoid; flowers white turning cream, rose-pink and yellow toward center, large, dbl., exhibition form, moderate fragrance; foliage glossy, light green; vigorous, upright, bushy growth; [Pink Princess X Shades of Autumn]; Brownell, H.C.

'**Sparkle Plenty**', F, ob, 1976; bud ovoid; flowers bright orange, to open, dbl., 25 petals, 3 in., exhibition form, moderate fragrance; foliage dark, leathery, wrinkled; vigorous growth; [Ma Perkins X Engagement]; Patterson; Patterson Roses

Sparkler, *see* 'DELecla'

'**Sparkler**', Pol, mr, 1929; [Golden Salmon sport]; deRuiter; Sliedrecht & Co.

Sparkley, *see* 'POUlcov'

'**Sparkling Burgundy**', F, dr, 1965; flowers burgundy-red, open, medium, borne in clusters, dbl., moderate fragrance; foliage dark, leathery, glossy; vigorous, upright growth; [Queen Elizabeth X Carrousel]; Williams, J. Benjamin

Sparkling Orange(TM), *see* 'DEVilk'

Sparkling Scarlet, *see* 'MEIhati'

Sparks(TM), *see* 'SAVaspark'

'**Sparrieshoop**', S, lp, 1953; bud long, pointed; flowers gold stamens, single, 4 in., repeat bloom, intense fragrance; foliage leathery; very vigorous (5 ft), upright, bushy growth; GM, Portland, 1971; [(Baby Chateau X Else Poulsen) X Magnifica]; Kordes

'**Spartan**', F, or, 1955; (Aparte); bud pointed; flowers orange-red to reddish coral, blooms borne singly and in clusters, dbl., 30 petals, 3–3.5 in., exhibition form, intense fragrance; foliage dark, leathery, glossy; vigorous, bushy growth; David Fuerstenberg Prize, ARS, 1957 GM, ARS, 1961 GM, NRS, 1954 GM, Portland, 1955 PIT, NRS, 1954; [Geranium Red X Fashion]; Boerner; J&P

'**Spartan Blaze**', Min, rb, 1991; bud ovoid; flowers red with yellow reverse, blooms borne usually singly, dbl., exhibition form, slight fragrance; foliage medium, medium green, semi-glossy; upright growth; [Poker Chip X Rise 'n' Shine]; Gruenbauer, Richard, 1984; Flowers 'n' Friends Miniature Roses, 1991

'**Spartan Dawn**', Min, yb, 1990; bud rounded; flowers yellow with orange edges, reverse yellow to cream to orange, dbl., 30 petals, no fragrance; round, flat on top, orange-red fruit; few or no prickles; foliage medium, medium green, matt; bushy, medium growth; [Rise 'n' Shine X Hokey Pokey]; Gruenbauer, Richard, 1984; Richard Gruenbauer

Spartan II, F; Moreira da Silva, A.; (Cavriglia)

'**SPARTAN, CLIMBING**', Cl F, or, 1960; (Grimpant Spartan); Kordes

'**SPARTAN, CLIMBING**', Cl F, or, 1964; Martinez; J&P

Späth 250®, *see* '**SPÄTH'S JUBILÄUM**'

'**SPÄTH'S JUBILÄUM**', F, or, 1970; (SPÄTH 250®); bud ovoid; dbl., 22 petals, 2.5 in., exhibition form, slight fragrance; foliage dark, soft; vigorous, upright, bushy growth; [Castanet X Seedling]; Kordes

'**SPEAKER SAM**', HT, yb, 1962; flowers light yellow edged red; [Peace sport]; Dean; Arp Nursery Co.

Spearmint, Min, w

SPEcawijk, HT, mr, 1996; (**Aruba**); Spek

'**SPECIAL ANGEL**', Min, m, 1992; flowers mauve, tipped in pink, blooms borne mostly singly, dbl., 15–25 petals, 1.5 in., exhibition form, slight fragrance; few prickles; foliage medium, dark green, semi-glossy; tall (60–70 cms), upright, bushy growth; [Jean Kenneally X (Rise 'n' Shine X Acey Deucy)]; Stoddard, Louis; Bridges Roses, 1992

Special Effects, F, lp; flowers orange and white stripes, intense fragrance; [Scentimental sport]; Desamero, Luis

'**SPECIAL GUEST**', HT, mr, 1992; flowers crimson, blooms borne mostly singly, very dbl., 3–3.5 in., slight fragrance; many prickles; foliage medium, medium green, semi-glossy; medium (70 cms), bushy growth; [Jan Guest sport]; Guest, M.M., 1990

'**SPECIAL MERIT**', HT, mr, 1990; bud pointed; flowers large, borne singly, very dbl., 30 petals, exhibition form; foliage medium, dark green, semi-glossy; medium, upright, bushy growth; [Seedling X First Prize]; Wambach, Alex A.

Special Occasion, *see* FRYyoung

Speckled Delight, HT, lp; Patel, 1998

'**SPECTABILE**', HSem, m; (Spectabilis); flowers bright rosy lilac, petals 'curiously incised' (notched?), cupped, occasionally repeats in autumn; vigorous, climbing growth; [Possibly R. sempervirens X Noisette hybrid]

Spectabilis, *see* '**SPECTABILE**'

'**SPECTACULAR**', LCl, or, 1953; (Danse du Feu, Mada); bud ovoid; flowers scarlet-red, blooms in clusters, dbl., 33 petals, cupped, recurrent bloom, moderate fragrance; foliage glossy, bronze; vigorous, climbing (8–10 ft) growth; [Paul's Scarlet Climber X Unnamed R. multiflora seedling]; Mallerin, C.; EFR, 1953; J&P, 1956

Spectra, *see* MEIzalitaf

Speechless™, *see* 'KINSPEECH'

Speelwark, HT, lp; Kordes, 1999

SPEhewe, HT, lp; (**Tiamo**); Spek, 1996

SPEkes, HT, mr, 1996; (**Sacha**, Our Sacha); Spek

Spek's Centennial, *see* 'MACIVY'

Spek's Improved, HRg, lp

Spek's Yellow, *see* '**GOLDEN SCEPTER**'

Spellbinder, *see* 'WARDIDO'

'**SPELLBOUND**', HT, dr, 1949; flowers deep velvety red, becoming darker, dbl., 25–30 petals, 5.5–6 in., moderate fragrance; thornless; foliage dark; very vigorous, upright growth; [Better Times sport]; Sodano, J.; Amling-DeVor Nursery

Spellbound, *see* POUlrim

Spellcaster, *see* 'JACANGEL'

'**SPENCER**', HP, lp, 1892; flowers soft pink; [Merveille de Lyon sport]; Paul, W.

'**SPENCER'S DELIGHT**', HT, ab, 1998; flowers coral pink, reverse soft amber, full, very dbl., 26–40 petals, 3.5–4.5 in., exhibition form, borne mostly singly, moderate fragrance; prickles moderate; foliage medium, dark green, semi-glossy; long, straight stems; vigorous, upright, tall (48–62 in.) growth; [Maid of Honor sport]; Spencer, Keith V.

Speragina, T, lp; Hay

SPErien, S, lp; (**Fairy Queen**); Spek

SPEruge, HRg, mr, 1997; (**Red Dagmar**); Spek

Spes, HT, mp, 1970; flowers large, very dbl., intense fragrance; Urban, J.; (Sangerhausen)

SPEvu, Pol, dp, 1990; (**Lovely Fairy**); Vurens-Spek

'**SPHINX**', HT, lp, 1967; flowers bright pink tinted lighter, very large, dbl., moderate fragrance; foliage leathery; vigorous, bushy growth; [Rose Gaujard X Gail Borden]; Gaujard

'**SPICE**', F, or, 1954; bud globular; flowers scarlet-red, blooms in clusters, dbl., 53 petals, 3 in., cupped, moderate fragrance; foliage glossy; vigorous, compact, upright growth; [(Goldilocks X Floradora) X Unnamed Floribunda seedling]; Boerner; J&P

Spice, HCh, lp

Spice Drop™, *see* 'SAVSWET'

Spice 'n' Nice, HT, lp; Twomey, Jerry, 1997

Spice Twice™, *see* 'JACABLE'

Spiced Coffee, *see* 'MACJULIAT'

'SPICUP', Min, mp, 1981; (**Cupcake**™); bud ovoid; flowers clear medium pink, long lasting, dbl., 60 petals, exhibition form, borne 1–5 per cluster, no fragrance; no prickles; foliage glossy; compact, bushy growth; PP4835; AOE, 1983; [Gene Boerner X (Gay Princess X Yellow Jewel)]; Spies, Mark C.; Nor'East Min. Roses

Spicy Minijet, *see* 'LAVGLO'

Spielplatz DRS®, HT, yb; Huber, 1987

'**SPINNING WHEEL**'™, S, rb, 1991; flowers cherry red with ivory striping, large blooms borne in large, dbl., slight fragrance; foliage large, dark green, semi-glossy; tall, upright, spreading growth; [Handel X Love]; Williams, J. Benjamin, 1992

'**SPION-KOP**', F, mr, 1968; flowers signal-red, full, dbl., 30–35 petals, 3–5 in., slight fragrance; foliage light, bronze, matt green; vigorous growth; [Evelyn Fison X Orange Sensation]; Ellick

Spirit, HT, lp

'**SPIRIT OF '76**', Gr, or, 1971; bud ovoid; flowers medium, dbl., moderate fragrance; foliage large, glossy, dark; vigorous, upright growth; [Queen Elizabeth X San Francisco]; Whisler, D.; Gro-Plant Industries

SPIRIT OF CANADA, F, dr, 1995; flowers dark red enriched with flashes of scarlet, prominent stamens, semi-dbl., (6–14 petals), medium blooms borne in small and large clusters, numerous prickles; foliage medium, medium green, semi-glossy; upright, bushy, tall growth; [Bambula X Red Max]; Fleming Joyce; Hortico Roses, 1995

Spirit of Glasnost, *see* 'JACARA'

Spirit of Hope, *see* 'MACHAHEI'

Spirit of Peace, *see* 'MEINIVOZ'

Spirit of Peace, *see* 'JACSTINE'

Spirit of Pentax, *see* 'HARWEX'

Spirit of Tollcross, *see* HARreverso

Spirit of Youth, *see* 'MEIVESTAL'

'**SPITFIRE**', Min, mr, 1986; flowers large blooms borne singly, dbl., 38 petals, exhibition form, no fragrance; thin, light green prickles; foliage medium, medium green, matt; upright, bushy growth; [Unnamed seedling X Unnamed seedling]; McDaniel, Earl; McDaniel's Min. Roses

'**SPITFIRE**', HT, mr, 1943; bud oxblood-red; flowers carmine, open, small, dbl., 25–30 petals, 2–3 in., moderate fragrance; foliage dark, leathery; short stems; vigorous, upright, much branched growth; RULED EXTINCT 6/86; [Better Times X Colleen Moore]; Hill, Joseph H., Co.

'**SPITFIRE IMPROVED**', HT, dr, 1949; bud medium, short-pointed, ovoid; flowers velvety dark carmine, semi-dbl., 4–5 in., globular, intense fragrance; foliage leathery, dark; vigorous, upright, bushy growth; RULED EXTINCT 6/86; [Spitfire sport]; Hill, Joseph H., Co.

Spitzenschleier, S, lp; flowers small, single; VEG; (Sangerhausen)

Splash, HT, lp; flowers white with splashes of pink; [Moonstone sport]; Becnel, Johnny

'SPLENDENS', HGal, mr; (Frankfurt); flowers glistening crimson, semi-dbl.

'SPLENDENS', Gr, mr; (Frankfurt, R. gallica Splendens)

'SPLENDID GARLAND', HSem, lp, 1835; flowers medium, dbl.; Wells; (Sangerhausen)

'SPLENDOR', HT, or, 1933; bud long, pointed; flowers orange-carmine, dbl., cupped, moderate fragrance; foliage leathery, glossy; vigorous growth; [La Marechale Petain X Souv. de Claudius Pernet]; Sauvageot, H.; C-P

'SPLENDOR', HT, mr, 1940; bud long; flowers rose-red, reverse slightly lighter, pointed, dbl., 4–5 in., intense fragrance; foliage dark; [Better Times sport]; Abrams, Von

Splish Splash, *see* 'MORGOLDART'

Splish Splash, *see* KORfolga

'SPOBOOM', Min, rb, 1992; (**Boomerang**); flowers red and white, blooms borne in small clusters, dbl., 26–40 petals, 1.5–2.75 in., no fragrance; some prickles; foliage medium, medium green, semi-glossy; medium, bushy growth; AOE, 1993; [Seedling X Seedling]; Spooner, Raymond A.; Oregon Miniature Roses, 1993

'SPOCHEY', Min, ab, 1985; (**Cheyenne**); flowers golden apricot, blooms borne singly, dbl., 20 petals, intense fragrance; foliage medium, medium green, semi-glossy; upright growth; [Rise 'n' Shine X Center Gold]; Spooner, Raymond A.; Oregon Miniature Roses

'SPODARC', Min, w, 1991; (**Darcelle**); flowers small, blooms borne in sprays of 3–5, single, 5 petals, 1.5 in., slight fragrance; foliage small, medium green, semi-glossy; bushy, low (13 cms), very compact growth; [(Whistle Stop X Popcorn) X Nozomi]; Spooner, Raymond A.; Oregon Miniature Roses

'SPODASH', Min, mr, 1993; (**Dasher**); dbl., 26–40 petals, 1.5–2.75 in., borne mostly singly, slight fragrance; some prickles; foliage medium, medium green, semi-glossy; medium (40 cms), upright, bushy growth; [Maurine Neuberger X Seedling]; Spooner, Raymond A.; Oregon Miniature Roses, 1994

'SPODAY', Min, pb, 1995; (**Alisha**); flowers soft pink with white margin on outer edge, blooms borne singly, very dbl., 1.5 in., no fragrance; few prickles; foliage medium, medium green, matt; medium (20 in.), bushy growth; [Marriotta X Seedling]; Spooner, Raymond A.; Oregon Miniature Roses, 1995

'SPODE', HT, ob, 1972; flowers orange-scarlet, flushed cream, long, pointed, dbl., 35 petals, 6 in., intense fragrance; foliage glossy, dark; [Diorama X Fragrant Cloud]; Fryers Nursery, Ltd.

'SPOGLO', Min, or, 1985; (**Superglo**); flowers small, semi-dbl., slight fragrance; foliage small, medium green, semi-glossy; bushy growth; [((Prominent X Rise 'n' Shine) X Trumpeter) X Chattem Centennial]; Spooner, Raymond A.

'SPOGRET', Min, dp, 1993; (**Gretta**); flowers deep pink, blooms borne mostly single, very dbl., 1.5–2.75 in., moderate fragrance; some prickles; foliage medium, dark green, glossy; bushy (44 cms) growth; Spooner, Raymond A.; Oregon Miniature Roses

'SPOHAWK', Min, pb, 1989; (**Sadie Hawkins**); bud globular; flowers medium pink with white stripes, reverse white, aging lighter, dbl., 16 petals, moderate, fruity fragrance; globular, medium green fruit; no prickles; foliage medium, medium green, matt; bushy, medium growth; [Roller Coaster X Unnamed seedling]; Spooner, Raymond A.; Oregon Miniature Roses, 1990

'SPOLAR', Min, mr, 1991; (**Larado**); flowers bright, clear red with small white eye in center, compact, semi-dbl., 6–14 petals, 1.5 in., no fragrance; foliage small, medium green, semi-glossy; micro-mini; low (15 cms), bushy growth; [Tobo Yellow X Seedling]; Spooner, Raymond A.; Oregon Miniature Roses, 1992

'SPOLAV', Min, m, 1995; (**Lavender Spoon**); single, 5 petals, 1.5–2.75 in., borne in sprays of 3–7, slight fragrance; some prickles; foliage small, dark green, semi-glossy; medium (12–14 in.), bushy growth; [(Black Jade X Blue Peter) X Seedling]; Spooner, Raymond A.; Oregon Miniature Roses, 1995

'SPOMAUR', Min, mr, 1989; (**Maurine Neuberger**™); bud pointed; flowers medium, borne singly, dbl., 30 petals, exhibition form, moderate fragrance; ovoid, light green fruit; prickles needle-like, light brown; foliage medium, medium green, matt; upright, medium growth; PP007772; [(Prominent X Zinger) X Centerpiece]; Spooner, Raymond A.; Oregon Miniature Roses

'SPOMEG', Min, dy, 1993; (**Little Meghan**); flowers bright yellow, blooms borne mostly single, dbl., 15–25 petals, 1.5 in., slight fragrance; no prickles; foliage small, light green, matt; low, compact growth; [Good Morning America X Seedling]; Spooner, Raymond A.; Oregon Miniature Roses, 1994

'SPOMERIT', Min, ob, 1989; (**Merit**); bud pointed; flowers brilliant orange, yellow base, reverse yellow, small, dbl., 17 petals, exhibition form, no fragrance; no fruit; prickles needle-like, brown; foliage small, dark green, semi-glossy; bushy, low growth; [(Prominent X Zinger) X Unnamed Miniature seedling]; Spooner, Raymond A.; Oregon Miniature Roses

'SPONG', C, mp; flowers rose-pink, richer in center, rosette form, larger than rose, early bloom; dwarf, compact (to 4 ft) growth; 1805

'SPOROSE', Min, rb, 1991; (**Rosetta**); flowers white center, red on outer half of petals, blooms borne singly, dbl., 26–40 petals, 1.5 in., no fragrance; foliage small, medium green, semi-glossy, disease-resistant; bushy (34 cms) growth; [Scamp X Seedling]; Spooner, Raymond A.; Oregon Miniature Roses, 1992

Sporting Duo, *see* HARbar

Spot Meillandina®, *see* 'MEIDIPSER'

Spot Minijet, *see* 'MEIZOGREL'

Spot 'o Gold, *see* 'WEESPOT'

'SPOTLIGHT', HT, my, 1969; flowers peach and gold, globular, moderate fragrance; foliage dark; free growth; [Seedling X Piccadilly]; Dickson, A.

Spotlight, *see* 'MORBRIGHTS'

Spotted Gold, *see* KORtelin

'SPOVEGAS', Min, mr, 1985; (**Little Vegas**); flowers small, dbl., 20 petals, exhibition form, slight fragrance; foliage small, medium green, semi-glossy; bushy growth; [(Rise 'n' Shine X Prominent) X Unnamed red seedling]; Spooner, Raymond A.; Oregon Miniature Roses, 1984

'SPRAY CÉCILE BRÜNNER', Pol, pb, 1941; bud long, pointed; flowers bright pink on yellow, edged clear pink, center yellowish, dbl., moderate fragrance; foliage sparse, soft, dark; bushy growth; [Cécile Brunner sport]; Howard Rose Co.

Spreeathen, HT, mr, 1968; flowers large, dbl.; Rupprecht-Radke; (Sangerhausen)

Spreeglut, S, dr; flowers medium, semi-dbl.; Kopenick; Institut für Zierpflanzenbau Berlin, 1985; (Sangerhausen)

Spring, *see* **'WESNIANKA'**

Spring Beauty, *see* 'MINECO'

Spring Bouquet, *see* 'TINSPRING'

Spring Break, *see* 'STRAHEIDI'

Spring Fever, *see* 'AROCANT'

Spring Fragrance, *see* **'VESENII AROMAT'**

'SPRING FROLIC', Min, my, 1978; bud ovoid; flowers pompon-shaped, dbl., 70 petals, 1 in., intense fragrance; low, compact, spreading growth; [(Little Darling X Gold Coin) X Golden Angel]; Williams, Ernest D.; Mini-Roses

'SPRING FROST', Min, w, 1978; flowers pure white, ruffled, semi-dbl., 17 petals, 1 in., slight fragrance; foliage small; low, very compact growth; Schwartz, Ernest W.; Bountiful Ridge Nursery

Spring Gold, *see* **'FRÜHLINGSGOLD'**

Spring Hill's Freedom, *see* 'TWOFREE'

Spring Hill's Pink Freedom, S, lp; Spring Hill, 1997

Spring Melody™, *see* 'MINMCO'

Spring Morning, *see* **'FRÜHLINGSMORGEN'**

'**SPRING SONG**', Min, pb, 1957; flowers pink tinted salmon, small blooms in clusters, dbl.; foliage glossy; bushy growth; [(R. wichurana X Floradora) X Thumbelina]; Moore, Ralph S.; Sequoia Nursery

'**SPRING SONG**', S, dp, 1954; flowers rich carmine-pink, semi-dbl., moderate fragrance; vigorous, tall growth; [Gartendirektor Otto Linne seedling]; Riethmuller

Spring Time(TM), *see* 'WILSPRING'

Springfields, *see* 'DICBAND'

Springhill Freedom, *see* TWOFREE

'SPRINGMOON', Min, ly, 1994; (**Full Moon**); flowers creamy light yellow, blooms opening to old fashion form, dbl., 26–40 petals, 1–1.5 in., slight fragrance; few prickles; foliage medium, medium green, semi-glossy, disease-resistant; medium (20 in.), spreading, rounded growth; [Avandel X Olympiad]; Sproul, James A.

'SPRINGPROMISE', Min, dr, 1994; (**Crimson Promise**); flowers antique, smoky dark red, blooms borne mostly single, dbl., 26–40 petals, 1.25–1.75 in., no fragrance; few prickles; foliage medium, dark green, semi-glossy, disease-resistant; medium (20 in.), bushy growth; [Avandel X Chrysler Imperial]; Sproul, James A.

Springs 75, *see* 'POULVISION'

'**SPRINGTIME**', F, lp, 1935; flowers wild-rose-pink, center white, borne in clusters, semi-dbl., cupped, slight fragrance; foliage leathery; bushy growth; RULED EXTINCT 7/90; [Miss Rowena Thom X Unnamed seedling]; Howard, F.H.; Dreer

'**SPRINGVALE**', HT, w, 1975; flowers white to pink, very dbl., 50 petals, 4 in., exhibition form, slight fragrance; foliage dark; [Mme A. Meilland sport]; Miller, J.

Springwood Beauty(TM), *see* 'LAVNEW'

Springwood Classic(TM), *see* 'LAVCLASS'

Springwood Coral(TM), *see* 'LAVSCENT'

Springwood Gold(TM), *see* 'LAVTYNINE'

Springwood Mauvette®, *see* 'LAVISLE'

Springwood Pink(TM), *see* 'LAVDUSK'

Springwood Pink Satin®, *see* 'LAVSAT'

Springwood Purple(TM), *see* 'LAVPURR'

Springwood Red(TM), *see* 'LAVRED'

Springwood Red Victor®, *see* 'LAVBERT'

Springwood Ruby, *see* 'LAVARU'

Springwood White(TM), *see* 'LAVSNOW'

'**SPRINT**', F, mr, 1961; bud oval; flowers bright red, open, borne in clusters, semi-dbl., 12 petals, 2.5 in., slight fragrance; foliage glossy, dark; vigorous growth; [Jolie Princesse X Chanteclerc]; Gaujard

Sprinter®, *see* INTasprint

Spun Glass(TM), *see* 'DEVLASS'

'**SPUN GOLD**', HT, my, 1941; flowers gold, dbl., 27 petals, 4.5 in., exhibition form; foliage glossy, leathery; vigorous, bushy, fairly compact growth; [Unnamed seedling X Portadown Glory]; McGredy; J&P

Spunglass, *see* 'DEVLASS'

Spunky, *see* 'JUDSPUNK'

'**SPUTNIK**', F, or, 1958; semi-dbl., large trusses of 25–30; low, bushy growth; [Jiminy Cricket X Gloria Mundi]; Maarse, G.

'**SQUARE DANCER**', S, dp, 1973; bud ovoid; flowers deep pink, large, dbl., cupped, repeat bloom, moderate fragrance; foliage large, dark, leathery; vigorous, upright, bushy growth; [Meisterstuck X ((World's Fair X Floradora) X Applejack)]; Buck, Dr Griffith J.; Iowa State University

'**SQUATTER'S DREAM**', S, my, 1923; flowers medium yellow, becoming lighter, semi-dbl., cupped, recurrent bloom, moderate fragrance; no prickles; foliage dark bronze-green; dwarf, bushy growth; [Seedling from R. gigantea seedling sport]; Clark, A.; Hackett

'**SRDCE EUROPY**', HWich, dp, 1937; flowers small, single; Böhm, J.; (Sangerhausen)

'**SREBRA**', F, pb, 1975; flowers pink, shaded mauve, blooms in clusters of 5–20, dbl., moderate, tea fragrance; foliage dark, glossy; bushy growth; [Highlight X Masquerade]; Staikov, Prof. Dr V.; Kalaydjiev and Chorbadjiiski

Srinivasa, HT, rb; Kasturi, 1969

St Alban's Gem, S, lp

St Boniface, *see* 'KORMATT'

St Bruno, *see* 'LANPIPE'

St Catherine, LCl, pb

St Cecilia, *see* 'AUSMIT'

St Christopher, *see* HARcogent

St Clair's Rose, HT, lp

St David's, Ch, rb

St Dunstan's Rose, *see* KIRshru

St George Tucker, Misc OGR, lp

St Helena, *see* 'CANLISH'

'**ST HELENA**', HT, ly, 1912; flowers creamy yellow, center pinkish; RULED EXTINCT 6/83; GM, RNRS, 1912; Cant, B. R.

St Helena Texas Centennial, HT, lp

'**ST HILAIRE**', (R. blanda form), mp; no prickles

St Hildas, HT, lp

St Hughs, *see* 'KORHUG'

'**ST INGEBERT**', HP, w, 1926; bud long, pointed; flowers white, center yellowish and reddish, large, dbl., 62 petals, slight fragrance; vigorous growth; [Frau Karl Druschki X Mme Mélanie Soupert]; Lambert, P.

St John, *see* HARBILBO

St John's College, F, lp; J&P, 1998

'**ST JOHN'S ROSE**', S, lp; (R. centifolia sancta, R. sancta, R. X richardii); flowers rose, 2–2.5 in.; (21); [R. gallica X R. phoenicia]

St Katherine's, *see* PEAvenus

'**ST LUCIA**', HT, mp, 1973; bud ovoid; flowers pink, medium, dbl., moderate fragrance; foliage glossy; moderate, upright, bushy growth; [Unknown X Unknown]; Tantau, Math.; Ahrens & Sieberz

St Margaret, HT, lp; Bell, Ronald J., 1998; (Weatherly, L.)

St Mark's Rose, *see* '**ROSE D'AMOUR**'

'**ST NICHOLAS**', D, dp, 1950; flowers rich pink, semi-dbl.; vigorous, erect growth; Hilling; Hilling, 1950

St Patrick(TM), *see* 'WEKAMANDA'

'**ST PAULI**', F, yb, 1958; flowers golden yellow edged red to pink, becoming fused, open, borne, semi-dbl., 15 petals, 3 in., slight fragrance; foliage dark, glossy; very vigorous, upright, bushy growth; [Masquerade X Karl Herbst]; Kordes, R.

St Piers, *see* HARentrap

'**ST PRIST DE BREUZE**', Ch, rb, 1838; flowers rich deep crimson with rose center, medium, dbl., globular, Desprez

St Quentin, *see* 'KRILEVILLE'

St Swithun, *see* 'AUSWITH'

St Thomas China, Ch, m

St Tiggywinkles, *see* KORbasren

St Victor, HT, 1979; Croix; (Cavriglia)

'**ST WILFRID'S HOSPICE**', F, w, 1994; flowers medium red, blooms borne in small clusters, semi-dbl., 6–14 petals, 1.5–2.75 in., slight fragrance; some prickles; foliage small, medium green, semi-glossy; medium, bushy growth; [Len Turner sport]; Sawday, Mrs. D.R.; Apuldram Roses, 1995

St Wilfrid's Hospice, F, mr

'**STAATSPRÄSIDENT PÄTS**', HT, my, 1937; (President Pats); bud long, pointed; flowers amber-yellow, very varying, very large, dbl., intense fragrance; foliage leathery, dark; vigorous growth; [Ophelia X Souv. de Claudius Pernet]; Weigand, C.; Spath

'**STACCATO**', F, mr, 1959; bud ovoid; flowers vermilion-red, open, medium, semi-dbl., 14–20 petals; moderately vigorous, bushy growth; [Cantate sport]; van de Water; Klyn

'**STACEY**', F, yb, 1965; flowers maize-yellow tinted blush, borne in clusters, dbl., 25–30 petals, 4–5 in., moderate fragrance; foliage dark, glossy, leathery; vigorous growth; Verschuren, A.; Blaby Rose Gardens

'**STACEY SUE**', Min, lp, 1976; bud short, pointed; flowers soft pink, dbl., 60 petals, 1 in., slight fragrance; foliage small, glossy; bushy growth; [Ellen Poulsen X Fairy

Princess]; Moore, Ralph S.; Sequoia Nursery

Stacey's Star, *see* 'HORSTACEY'

'**STAD DARMSTADT**', Pol, mr, 1966; bud ovoid; flowers bright red, small, borne in clusters, semi-dbl.; foliage dark; [Red Favorite X The Doctor]; deRuiter

'**STAD DEN HAAG**', F, mr, 1969; flowers scarlet, open, large, semi-dbl., 20 petals, slight fragrance; foliage glossy; free growth; [Evelyn Fison X (Spartan X Red Favorite)]; McGredy, Sam IV

Stadt Basel, *see* 'MEIVILANIC'

Stadt den Helder, *see* 'INTERHEL'

Stadt Eltville®, *see* TANelliv

'**STADT ESSEN**', Pol, mr, 1937; flowers medium, dbl.; Tantau, Math.; (Sangerhausen)

Stadt Ettelbrück, S, mr, 1981; flowers medium-large, semi-dbl.; Lens, Louis; (Sangerhausen)

Stadt Hockenheim, HMsk, lp; Weihrauch

'**STADT KIEL**'®, S, mr, 1962; flowers cinnabar-red, blooms in large clusters, dbl., 30 petals, slight fragrance; foliage dark; very vigorous, bushy growth; Kordes, R.

Stadt Luzern, F, op; Huber, 1967

Stadt Pilsen, *see* '**PLZEN**'

Stadt Pottrop, F, 1961; deRuiter; (Cavriglia)

'**STADT ROSENHEIM**'®, S, or, 1961; dbl., blooms in clusters (up to 10), moderate fragrance; foliage glossy, light green; vigorous, upright growth; ADR, 1960; Kordes, R.

Stadt Wurzburg, HT, 1987; Kordes, R.; (Cavriglia)

'**STADTRAT F. KÖHLER**', HT, 1906; Geduldig; (Sangerhausen)

'**STADTRAT GLASER**', HT, yb, 1910; flowers sulfur-yellow edged soft red; Kiese

'**STADTRAT MEYN**', Pol, mr, 1919; flowers luminous brick-red, large, dbl.; [Orléans Rose X ?]; Tantau

Stagecoach™, *see* 'KINCOACH'

Stainless Steel™, *see* 'WEKBLUSI'

Stamela, F; (Holland); (Cavriglia)

'**STÄMMLER**', HP, mp, 1933; flowers rose-pink, large, very dbl., exhibition form, intense fragrance; foliage glossy; vigorous growth; [Victor Verdier X Arabella]; Tantau; C-P

'**STANDING**', Pol, dr, 1969; flowers open, medium, dbl., slight fragrance; foliage dark, glossy; vigorous growth; [Atlantic X Seedling]; Delforge

Standing Ovation™, *see* 'TUCSTAND'

'**STANDOUT**', HT, rb, 1977; bud ovoid, pointed; flowers red, white reverse, dbl., 40 petals, 3–4 in., exhibition form, slight, tea fragrance; vigorous, upright growth; [Tiffany X Suspense]; Weeks; Weeks Wholesale Rose Growers

Stanley Duncan, Min, ab

'**STANLEY GIBBONS**', HT, op, 1976; flowers salmon-orange, pointed, dbl., 28 petals, 4 in., moderate fragrance; foliage glossy, dark; [Fragrant Cloud X Papa Meilland]; Gregory

'**STANLEY MATTHEWS**', HT, dr, 1964; flowers crimson-scarlet, dbl., 26 petals, 4.5–5 in., moderate fragrance; foliage glossy, light green; vigorous growth; [(Independence X Crimson Glory) X Happiness]; Latham

'**STANWELL PERPETUAL**', HSpn, w, 1838; flowers blush, medium, dbl., recurrent bloom, slight fragrance; many prickles; foliage very small; moderate, spreading growth; (28); [Thought to be a repeat-blooming Damask X Spinosissima]; Lee

'**STANZA**', F, lp, 1969; bud ovoid; flowers light pink, center darker, open, medium, semi-dbl., slight fragrance; foliage glossy, dark; very vigorous, bushy growth; Pal, Dr B.P.; Gopalsinamiengar

Star 2000®, *see* PEKcoucan

Star Burst, S, 1994; Ilsink, Peter; (Cavriglia)

Star Child®, *see* 'DICMADDER'

Star Delight, *see* 'MORSTAR'

Star Gazer, HT, pb

Star Magic, *see* 'MORSEMURI'

'**STAR OF BETHLEHEM**', HWich, w, 1947; flowers stamens red, single, 5 petals, 3–3.5 in., moderate bloom; foliage glossy; [Innocence X Silver Moon]; Fisher, R.C.

'**STAR OF PERSIA**', HFt, my, 1919; flowers bright yellow, stamens golden, medium, semi-dbl.; vigorous (8–10 ft) growth; [R. foetida X Trier]; Pemberton

'**STAR OF QUEENSLAND**', HT, w, 1909; flowers creamy white, medium, dbl., 40 petals, exhibition form; foliage grayish green; vigorous, bushy growth; [Étoile de France X Earl of Dufferin]; Williams, A.

'**STAR OF THAILAND**', HT, w, 1977; flowers creamy white, medium, dbl., 40 petals, exhibition form; foliage grayish green; vigorous, bushy growth; [Mount Shasta X Pascali]; Chinprayoon; Chavalit Nursery

Star of Tokio, Min, 1984; Dot, Simon; (Cavriglia)

'**STAR OF WALTHAM**', HP, mr, 1875; flowers carmine-crimson, semi-globular, medium, dbl., moderate fragrance; smooth green wood with occasional red prickles; foliage very large; dwarf to medium growth; Paul, W.

Star Performer, *see* 'CHEWPEARL'

'**STAR TWINKLE**', Min, pb, 1978; bud pointed; flowers pink, coral and orange, single, 5 petals, 1 in.; foliage glossy; bushy, compact growth; [Fairy Moss X Fire Princess]; Moore, Ralph S.; Sequoia Nursery

Starbright, *see* '**LILY WHITE**'

'**STARBRIGHT**', HT, w, 1962; bud ovoid; dbl., 43 petals, 4 in., exhibition form, moderate fragrance; foliage glossy, dark; vigorous, upright growth; [Princess White X Hybrid Tea seedling]; Boerner; J&P

'**STARBURST**', Gr, rb, 1969; bud ovoid; flowers red and yellow, medium, dbl., exhibition form, slight fragrance; foliage glossy; vigorous, bushy growth; [Zambra X Suspense]; Meilland, Mrs. Marie-Louise; C-P

Starburst, *see* JAChart

Stardance™, *see* 'WILBLANK'

'**STARDUST**', HT, w, 1939; flowers rose-red; RULED EXTINCT 5/92; [Better Times sport]; Florex Gardens

Stardust, *see* 'DEVSTAR'

Stardust, *see* 'INTERDUST'

Stardust, *see* PEAvandyke

'**STARFIRE**', Gr, mr, 1958; bud urn-shaped; dbl., 28 petals, 5 in., exhibition form, blooms in clusters, moderate fragrance; foliage glossy; vigorous, tall, bushy growth; AARS, 1959; [Charlotte Armstrong X (Charlotte Armstrong X Floradora)]; Lammerts, Dr Walter; Germain's

Starfire, Climbing, Cl F, lp; [Starfire sport]; Jack, V., 1972; (Weatherly, L.)

'**STARGAZER**', F, ob, 1977; flowers orange-red, yellow eye, patio, small to medium blooms in large clusters, single, 9 petals, slight fragrance; foliage medium, medium green, matt; medium, bushy growth; [Marlena X Kim]; Harkness

'**STARGLO**', Min, w, 1973; bud long, pointed; flowers small, dbl., exhibition form, slight fragrance; foliage small, leathery; vigorous, bushy growth; AOE, 1975; [Little Darling X Jet Trail]; Williams, Ernest D.; Mini-Roses

'**STARGOLD**', HT, my, 1936; bud long, pointed; flowers yellow, often splashed red, dbl., 32 petals, exhibition form, intense fragrance; foliage dark, glossy; long, strong stems; branching, upright growth; [Mary Wallace self X (Unnamed seedling X Unnamed seedling)]; Brownell, H.C.; Inter-State Nursery

Starina®, *see* 'MEIGALI'

'**STARINA, CLIMBING**', Cl Min, or, 1981; [Starina sport]; Asami, Hitoshi

Starion®, *see* DORdeli

'**STARK WHITECAP**', F, w, 1959; flowers pure white, borne in clusters, dbl., 30–35 petals, 3 in., cupped, moderate fragrance; foliage glossy; vigorous, upright growth; [Glacier X Garnette]; Boerner; Stark Bros.

'**STARKRIMSON**', HT, dr, 1961; flowers scarlet-red, dbl., 48 petals, 4.5 in., cupped, intense fragrance; foliage leathery; vigorous, upright growth; [Happiness X San

Fernando]; Morey Dr Dennison; Stark Bros.

Starla, *see* 'PIXARLA'

'**STARLET**', F, my, 1957; bud pointed; dbl., 60 petals, 2.5 in., exhibition form, blooms in clusters, slight fragrance; foliage dark, leathery, glossy; vigorous, compact, upright growth; [Goldilocks X Unnamed seedling]; Swim, H.C.; Armstrong Nursery

Starlight, *see* 'AROLAQUELI'

'**STARLIGHT**', HT, my, 1934; flowers orange and buff, moderate fragrance; vigorous growth; RULED EXTINCT 3/87; Wood & Ingram

'**STARLIGHT**', HT, my, 1987; flowers medium, dbl., 26–40 petals, slight fragrance; foliage large, light green, matt; bushy growth; [Devotion X Benson & Hedges Gold]; Guest, M.M.

Starlight Express, *see* TRObstar

Starlight Parade, *see* POUlstar

Starlina, Min, mp

'**STARLITE**', HT, w, 1942; bud long, pointed; flowers clear white, center tinted cream, large, dbl., 40 petals, exhibition form, foliage dark, soft; vigorous, upright, bushy growth; RULED EXTINCT 3/87; [Unknown seedling X White Briarcliff]; Nicolas; J&P

Starlite, *see* MEIlupin

Starlite, *see* MEItanet

'STARQUELI', Gr, ab, 1995; (**Magic Lantern**, Remy Martin®); flowers copper-orange-gold, full (26–40 petals); large (7+ cms) blooms borne in small clusters, slight fragrance; some prickles; foliage medium green, semi-glossy; upright, bushy (150 cms) growth; [Gold Medal sport]; Royon, René; Bear Creek Gardens, Inc., 1994

Starry Bouquet, S, lp; Clements, John K., 1999

Starry Eyed, *see* 'HORCOXIST'

Starry Eyed, *see* 'MORSTAR'

'**STARS 'N' Stripes**', Min, rb, 1975; bud long, pointed; flowers evenly striped red and white, blooms in clusters, dbl., 21 petals, exhibition form, moderate, sweet fragrance; foliage light to medium green; bushy, upright growth; [Little Chief X (Little Darling X Ferdinand Pichard)]; Moore, Ralph S.; Sequoia Nursery, 1976

'**STARSCENT**', HT, pb, 1976; flowers rich pink, base yellow, dbl., 20 petals, 4.5 in., intense fragrance; foliage dull, blue-green; upright growth; [Silver Star X Peter Frankenfeld]; Parkes, Mrs M.H.; Rumsey

'**STARSHINE**', HT, yb, 1971; flowers pink and yellow blend, very large, dbl., globular, slight fragrance; foliage glossy; vigorous, bushy growth; [Peace X Condesa de Sástago]; Hamm; Five M Nursery

Starstruck, *see* 'TINSTAR'

Start, F, or; flowers medium, semi-dbl., slight fragrance; VEG; (Sangerhausen)

'**STARTLER**', HT, or, 1955; flowers orange-scarlet, base rich yellow, well formed; vigorous growth; [Hector Deane X Mary Wheatcroft]; Robinson, H.

'**STATE OF MAINE**', HT, w, 1930; flowers white, center slightly tinged greenish, large, very dbl., exhibition form, slight fragrance; [Lady Ursula sport]; de Bree

'**STATELY**', HT, lp, 1930; bud long, pointed; flowers pale flesh, center deeper, well-shaped, large, dbl., exhibition form, slight fragrance; vigorous growth; [Souv. de Gustave Prat X Seedling]; Clark, A.; Hackett

'**STATESMAN**', HT, dr, 1991; flowers large blooms borne mostly singly, dbl., slight fragrance; foliage medium, medium green, matt; medium, bushy growth; [Unknown seedling X Unknown seedling]; Burks, Larry; Co-Operative Rose Growers

'**STEADFAST**', HT, ob, 1939; flowers amber, flushed yellow and pink, semi-dbl.; vigorous growth; [Mme Auguste Choutet X ?]; Clark, A.; T.G. Stewart

'STEBIGPU', HT, m, 1985; (Big Purple, Nuit d'Orient, **Stephen's Big Purple**, Stephens' Rose Big Purple); flowers purple, large, dbl., 35 petals, intense fragrance; foliage large, medium green, matt; upright growth; PP006262; [Unnamed seedling X Purple Splendour]; Stephens, Pat; McGredy Roses International, 1986

'**STEEPLE ROSE**', C, dp; bud full, globular, large; flowers rich rosy crimson with large green foliage; [Perhaps Prolifera de Redoute sport]

'**STEFANOVITCH**', Cl HT, dr, 1943; flowers large, dbl., recurrent bloom, intense fragrance; vigorous growth; [Lemania sport]; Meilland, F.

Steffi Graf, *see* HELgraf

'**STEINO**', HT, my, 1965; bud ovoid; dbl., slight fragrance; foliage dark; [Golden Rapture sport]; Stein

'STEKELLY-LEIGH', Min, yb, 1981; (**Kelly-Leigh**); flowers white and gold, medium, semi-dbl., no fragrance; foliage medium, medium green, glossy; bushy growth; [Ko's Yellow X Unnamed seedling]; Stephens, Paddy

'**STELLA**'®, Gr, w, 1958; bud ovoid; flowers blush, edged deep pink on outer petals, in clusters, dbl., 36 petals, 4–5 in., exhibition form, slight fragrance; foliage dark, leathery; vigorous, upright growth; GM, NRS, 1960; [Horstmann's Jubilaumsrose X Peace]; Tantau, Math.

'**STELLA DI BOLOGNA**', HT, mr, 1909; flowers large, dbl., moderate fragrance; Bonfiglio, A.; (Sangerhausen)

'**STELLA DUCE**', F, dr, 1956; flowers dark crimson-red, stamens deep gold, moderate fragrance; vigorous growth; Leenders, M.

'**STELLA ELIZABETH**', Min, w, 1983; flowers creamy white, petals edged light pink, small, dbl., 35 petals, slight fragrance; foliage small, medium green, semi-glossy; bushy growth; [Unnamed seedling X Unnamed seedling]; Moore, Ralph S.; Roy H. Rumsey, Pty. Ltd.

'**STELLA MATTUTINA**', HT, lp, 1951; flowers flesh-pink, well formed, large, dbl., 35 petals; foliage glossy; vigorous growth; [Numa Fay X Asso Francesco Baracca]; Giacomasso

'**STELLA PACIS**', HT, ob, 1946; flowers combination of orange, salmon, red and yellow, large; vigorous growth; [Julien Potin X Mme G. Forest-Colcombet]; Giacomasso

'**STELLA POLARIS**', HRg, w, 1890; flowers silvery white, large, single, recurrent bloom; foliage dark; vigorous (4 ft) growth; Jensen

'**STELLMACHER**', S, mr, 1937; flowers bright red, well formed umbels, dbl., abundant, recurrent bloom, moderate fragrance; hardy; [D.T. Poulsen X Stammler]; Tantau

Stelvio Coggiatti, *see* BARstel

'**STEPHAN**', Pol, mp, 1967; bud ovoid; flowers cyclamen-pink, large, borne in clusters, dbl., slight fragrance; foliage glossy; vigorous, upright growth; [Maria Delforge X Arc-en-Ciel]; Delforge

'**STEPHANIE**', Cl HT, lp, 1973; flowers open, large, semi-dbl., profuse, repeated bloom, slight fragrance; foliage glossy; vigorous, climbing growth; [Coral Dawn X Titian]; Gatty, Joseph

'**STEPHANIE ANN**', Min, mp, 1978; bud long, pointed; flowers neyron rose, semi-dbl., 16 petals, 1 in., slight fragrance; foliage tiny; upright, bushy growth; Lyon

Stéphanie de Monaco, *see* 'MEYPINK'

'**STEPHANIE DIANE**', HT, mr, 1971; flowers scarlet-red, dbl., 50 petals, 5 in., exhibition form; moderately vigorous, upright growth; [Fragrant Cloud X Cassandra]; Bees

Stephanie Jo, HT, w; [Sylvia sport]; Allender, Robert William, 1995; (Weatherly, L.)

'**STEPHEN FOSTER**', LCl, lp, 1950; bud ovoid; flowers pink, recurved, semi-dbl., 20 petals, 3.5–4 in.; vigorous, upright growth, spreading to 15 ft or more; [Unnamed seedling X Black Knight]; Rosen, H.R.

'**STEPHEN LANGDON**', F, dr, 1969; flowers deep scarlet, large, semi-dbl., slight fragrance; foliage dark; vigorous, compact growth; [Karl Herbst X Sarabande]; Sanday, John

Stephen's Big Purple, *see* 'STEBIGPU'

Stephens' Rose Big Purple, *see* 'STEBIGPU'

Steppin' Out, *see* 'BRISTEP'

'STEREO', F, or, 1972; bud long, pointed; flowers pure orange, open, dbl., 18–24 petals, 3–3.5 in., slight fragrance; foliage brownish, leathery; bushy growth; [Baccará X (Ole X Independence)]; Lens; Spek

'STERKMANNS', HGal, mr, 1847; flowers large, dbl.; Vibert; (Sangerhausen)

'STERLING', HT, mp, 1933; bud long, pointed; flowers brilliant pink, base yellow, large, dbl., 35 petals, moderate fragrance; foliage glossy; vigorous growth; Gertrude M. Hubbard, ARS, 1939 GM; Portland, 1938; [Mme Butterfly X Unnamed seedling]; Hill, E.G., Co.

Sterling '95, *see* JACnuel

'STERLING SILVER', HT, m, 1957; bud long, pointed; flowers lilac, becoming lighter, dbl., 30 petals, 3.5 in., exhibition form, intense fragrance; foliage dark, glossy; vigorous, upright growth; [Seedling X Peace]; Fisher, G.; J&P

'STERLING SILVER, CLIMBING', Cl HT, m, 1963; Miyawaki

Sterling Star(TM), *see* 'ORTSAR'

'STERN VON PRAG', HRg, dr, 1924; flowers dark blood-red, large, dbl., moderate fragrance; foliage large, dark; very vigorous, bushy growth; [R. rugosa seedling X Edward Mawley]; Berger, V.; Faist

Sternenflor®, S, w; Schultheis, 1989

'STESAM', F, r, 1981; (**Sam**); flowers russet, reverse cream, dbl., 20 petals, slight fragrance; foliage medium, dark, glossy; bushy growth; [Brown Eye X Unnamed seedling]; Stephens, Paddy

'STETEAW', HT, rb, 1985; (**Te Awamutu Centennial**); flowers white-edged medium red, dbl., 35 petals, slight fragrance; foliage dark, glossy; bushy growth; [Strawberry Ice X Unnamed seedling]; Stephens, Pat

Steve Silverthorne, *see* 'FULSTEVE'

Stevens Rose, *see* **R. CINNAMOMEA PLENA**

Stevens, Climbing, *see* **'MRS HERBERT STEVENS, CLIMBING'**

Stevie, *see* 'BUSTEV'

'STEYL RAMBLER', HMult, mr, 1915; flowers medium, dbl.; Leenders, M.; (Sangerhausen)

'STIEDWARD', HT, mp, 1993; (**Hilton Edward**); flowers lavender pink, blooms borne mostly single, dbl., 15–25 petals, 3–3.5 in., intense fragrance; some prickles; foliage large, dark green, semi-glossy; tall, upright growth; [Sweet Afton X News]; Stibbard, Robert H.; Tallisker Nursery Enterprises, 1995

Stifontein Rose, *see* TANnietnof

'STIRLING CASTLE', F, mr, 1978; flowers bright scarlet, 18–20 petals, 3 in., slight fragrance; foliage matt; low, compact growth; [(Anne Cocker X Elizabeth of Glamis) X (Orange Sensation X The Optimist)]; Cocker

'STOCKTON BEAUTY', HT, pb, 1948; bud long, pointed; flowers deep salmon-pink, base yellow, open, large, dbl., moderate fragrance; foliage soft; very vigorous, bushy growth; [Banner sport]; Raffel; Port Stockton Nursery

'STOCKTON RED', Cl HT, mr, 1962; flowers clear red, well formed; very long stems, vigorous growth; Raffel; Port Stockton Nursery

'STOKES', HT, op, 1982; flowers peach, reverse salmon, large, dbl., 35 petals, exhibition form, slight, fruity fragrance; foliage medium green, matt; medium-tall growth; [Susan Massu X Yellow seedling]; Perry, Astor; Perry Roses

Stolen Dream, *see* 'MICSTEAL'

Stolen Moment, *see* 'SEAMOM'

Stonelea Rambler, HMsk, lp; Nieuwesteeg, J., 1994

'STOPLITE', F, mr, 1955; flowers rose-red, medium, dbl., 45–55 petals, moderate fragrance; vigorous, upright growth; [Garnette X Unnamed seedling]; Jelly; E.G. Hill Co.

'STORMLY', HT, op, 1966; flowers coral, reverse red-pink, oval, cupped; foliage dull; tall growth; [(Eclipse X Michele Meilland) X Baccará]; Caranta, M.&H.

Stormy Weather(TM), *see* 'MINISTORM'

'STRAHEIDI', HT, dp, 1993; (**Spring Break**); flowers deep florescent pink, dbl., 15–25 petals, 1.5–3 in., exhibition form, borne mostly singly, no fragrance; some prickles; foliage medium, medium green, matt; medium (100–300 cms), upright growth; ARC Test Gardens – Silver Certificate, 1995; [Secret Love X KORlingo]; Strickland, Frank A.; Strickland, 1994

Strange Music, *see* 'MORMUM'

'STRATFORD', HT, op, 1936; bud long-pointed, ovoid; flowers luminous pink tinted salmon, very dbl., intense fragrance; foliage leathery; long stems; very vigorous, bushy growth; [(Emile Charles X La France) X Marechal Niel]; Nicolas; Dixie Rose Nursery

'STRATOSFÉRA', C, dp, 1934; flowers large, very dbl., moderate fragrance; Böhm, J.; (Sangerhausen)

Strauchmaskerade, S, lp

Strawberries and Cream, *see* 'GEESTRAW'

'STRAWBERRY', HT, mr, 1950; flowers strawberry, reverse deep flesh, dbl., 35–40 petals, 5.5–6 in.; foliage light green; vigorous, compact growth; [McGredy's Pink X Phyllis Gold]; Fletcher; Tucker

'STRAWBERRY BLONDE', Gr, or, 1966; bud urn-shaped; flowers light orange-red, medium, dbl., 25 petals, exhibition form, moderate, spicy fragrance; foliage dark, leathery; very vigorous, upright, bushy growth; [Ma Perkins X Spartan]; Armstrong, D.L.; Armstrong Nursery

'STRAWBERRY CREAM', F, pb, 1972; flowers strawberry-pink, streaked cream, very full, dbl., 45 petals, 3 in., intense fragrance; foliage light; vigorous, low growth; Ellick; Excelsior Roses

'STRAWBERRY CRUSH', F, mr, 1974; dbl., 3 in., slight fragrance; foliage red when young; [Bridal Pink X Franklin Engelmann]; Dickson, A.

Strawberry Delight, *see* 'RENERRY'

'STRAWBERRY FAIR', F, mr, 1966; flowers scarlet, dbl., 2 in., blooms in clusters, slight fragrance; foliage dark; vigorous, bushy growth; [Orangeade X ?]; Gregory

Strawberry Fayre, *see* AROwillip

Strawberry Ice, *see* 'DELBARA'

'STRAWBERRY ICE', F, w, 1975; flower petals cream-white, edged pink, dbl., 22 petals, 3–3.5 in.; foliage dark; compact, bushy growth; GM, Madrid, 1974; [((Goldilocks X Virgo) X (Orange Triumph X Yvonne Rabier)) X Fashion]; Bees

Strawberry Kiss, *see* 'CLEKISS'

'STRAWBERRY SOCIAL', F, mr, 1979; bud ovoid; dbl., 23 petals, exhibition form, borne 3–7 per cluster, slight fragrance; hooked prickles; foliage medium green, semi-glossy; upright to spreading, bushy growth; [Pink Parfait X Chrysler Imperial]; Taylor, Thomas E.

Strawberry Sundae, *see* 'LEOSTRA'

'STRAWBERRY SWIRL', Min, rb, 1978; bud ovoid, pointed; flowers red mixed with white, mini-moss, dbl., 48 petals, 1 in., exhibition form; bushy growth; [Little Darling X Unnamed Miniature seedling]; Moore, Ralph S.; Sequoia Nursery

Strawrose, Pol, lp; Schneider, 1998

Street Wise, *see* 'SEARONAN'

Stretch Johnson, *see* 'MACFIRWAL'

'STRILDEW', HT, lp, 1996; (**Just For You**); flowers light pink, outer petals sometimes white with inner petals l, very dbl., 6 in., slight fragrance; moderate prickles; foliage medium, medium green, dull; very long stems, upright, medium (4–5 ft) growth; [Secret Love X First Prize]; Strickland, Frank A.

Strilli, F, my

'STRING OF PEARLS', F, w, 1985; (Meg '81); flowers near white, medium, semi-dbl., slight fragrance; foliage medium, dark, semi-glossy; dwarf, bushy growth; [Meg, Climbing X Sunsprite]; Gandy's Roses, Ltd.; Rearsby Roses, Ltd.

String of Pearls, Min, op; deRuiter, 1991

String of Pearls, S, lp; Williams, J. Benjamin

Striped Fairy Rose, Min, rb

Striped Festival(TM), *see* 'LAVMING'

Striped Meillandina, *see* 'MORMUM'

Striped Moss, *see* **'ILLET PANACHÉE'**

Striped Pet, *see* 'LAVDANCE'

'STRIPED RADIANCE', HT, pb, 1919; flowers distinctly striped white; [Radiance sport]; Vestal

'STRIPED TEXAS CENTENNIAL', HT, ob, 1950; flowers orange-red with dark yellow striping, large, dbl., moderate fragrance; Heckmann; (Sangerhausen)

'STRIPEZ', HT, pb, 1942; flowers striped and variegated; [Better Times sport]; Janssen; Premier Rose Gardens

Stroke-o-Luck, F, lp

'STROLLER', F, rb, 1968; bud loose; flowers cerise, reverse gold, large blooms in trusses, dbl.; foliage matt; [Manx Queen X Happy Event]; Dickson, A.

STRolon, HRg, mp, 1991; (**Polarsonne**); Strobel

Stromboli, S, dr

STRonin, HRg, w, 1991; (**Polareis**); Strobel

'STRYKER', HT, pb, 1994; flowers pink and white, veining in outer petals, blooms borne mostly, dbl., 26–40 petals, 3–3.5 in., slight fragrance; no prickles; foliage medium, dark green, glossy; tall, bushy, upright growth; [Pristine X Suffolk]; Edwards, Eddie; Testing at J&P and Weeks

'STUART'S QUARRY', HT, pb, 1977; bud high centered; flowers pale bluish pink, reverse silver, full, dbl., 30 petals, 4 in., bloom repeats quickly, moderate fragrance; foliage small, glossy, dark; [Silver Lining X Anne Letts]; Bailey

'STUDIENRAT SCHLENZ', Cl HT, mp, 1926; flowers rose-pink, reverse pink, large, dbl., non-recurrent, moderate fragrance; height $6\frac{1}{2}$–8 ft; [Mrs Aaron Ward X Frau Karl Druschki]; Lambert, P.

'STURDY GERTIE', HT, op, 1949; flowers coral-pink, medium, dbl., 30 petals, moderate, raspberry fragrance; foliage small, dark, glossy, somewhat ribbed; [Pink Princess X Peace]; Taylor, C.A.

'STUTTGART', HT, my, 1928; bud long, pointed; flowers pure yellow, medium, moderate fragrance; [Edith Cavell X Mrs Franklin Dennison]; Berger, V.; Pfitzer

'STYLISH', HT, pb, 1953; flowers rose-pink, base yellow, 5–6 in., exhibition form, moderate fragrance; foliage dark, glossy; vigorous growth; Robinson, H.

'SU EXCELENCIA SEÑORA DE FRANCO', F, mr, 1956; bud long, pointed; flowers medium, dbl., cupped, slight fragrance; foliage dark, glossy; vigorous, upright growth; [Cocorico X Independence]; Camprubi, C.

'SUBARU', F, mr, 1977; bud pointed; flowers small, single, 5 petals, 1.5 in., slight fragrance; foliage dark; low growth; [Masquerade X Permanent Wave]; Kikuchi, Rikichi

'SUBLAEVIS', Sp, mp

'SUBLIME', HT, op, 1931; bud long, pointed; flowers orange-salmon, suffused scarlet, semi-dbl., moderate fragrance; long stems; very vigorous growth; [Talisman sport]; Amling Co.

'SUBSTITUT JACQUES CHAPEL', HT, pb, 1922; flowers peach-blossom-pink, edged rose-pink, base shaded citron-yellow, intense fragrance; [Mme Mélanie Soupert X Lyon Rose]; Bernaix, P.

'SUCCES', Hcan, lp, 1954; flowers pinkish white, blooms in trusses, single; foliage dark, glossy; [R. canina seedling]; deRuiter

'SUCCÈS FOU', HT, mr, 1963; bud long; flowers deep cherry-red, well formed; vigorous growth; [Walko X Souv. de J. Chabert]; Delbard-Chabert

Sue Belle, *see* 'TALSUE'

Sue Betts, *see* 'BETSUE'

Sue Hipkin, *see* HARzazz

Sue Jo™, *see* 'MINAVCO'

Sue Lawley, *see* 'MACSPASH'

'SUE LEAT', HT, pb, 1983; flowers medium pink on outer half of petals, golden yellow on lower, dbl., 20 petals, moderate fragrance; brown prickles; upright growth; [Golden Slippers X Anne Letts]; Summerell, B.L.

Sue Ryder®, *see* 'HARLINO'

'SUFFOLK', HT, w, 1983; flowers white with pink petal tips, urn-shaped, large, dbl., exhibition form, slight fragrance; foliage large, medium green, matt; upright growth; [Garden Party X Unnamed yellow seedling]; Perry, Astor; Perry Roses, 1984

Sugandharaj, HT, lp; Datt, Braham, 1992

'SUGANDHINI', HT, lp, 1969; bud pointed; flowers full, medium, dbl., intense fragrance; foliage glossy, light; bushy growth; [Margaret Spaull X ?]; IARI

'SUGAR BABE', Min, lp, 1978; bud long, pointed; semi-dbl., 16 petals, 1 in., slight fragrance; foliage bronze; compact growth; Lyon

Sugar Baby, *see* TANragus

'SUGAR BEAR', Min, w, 1986; flowers medium, borne singly, dbl., 18 petals, exhibition form, slight fragrance; straight, pointed, medium, light green prickles; foliage medium, medium green, semi-glossy; upright, medium growth; [Heartland X Unnamed seedling]; Bridges, Dennis A., 1987

'SUGAR CANDY', Pol, mp, 1951; flowers pink, very small, dbl., profuse, repeated bloom, compact growth; Proctor

Sugar Daddy, *see* 'JACIorna'

'SUGAR ELF', Cl Min, pb, 1974; bud long, pointed; flowers pink and gold blend, semi-dbl., 15 petals, 1 in., slight fragrance; foliage glossy, leathery; bushy, spreading growth; Moore, Ralph S.; Sequoia Nursery

Sugar Magnolia™, *see* 'TALSUGAR'

Sugar 'n' Spice, *see* 'TINSPICE'

'SUGAR PLUM', HT, mp, 1954; (Prelude); bud ovoid; flowers tyrian rose, dbl., 55 petals, 4–5 in., exhibition form, moderate, spicy fragrance; foliage dark, glossy, leathery; vigorous, upright, bushy growth; [Crimson Glory X Girona]; Swim, H.C.; Breedlove Nursery

Sugar Plum, *see* 'MORPLUM'

Sugar Plum Fairy, Min, lp; [Sweet Chariot X seedling]; Hannemann, F., 1995; The Rose Paradise; (Weatherly, L.)

'SUGAR SWEET', F, pb, 1973; flowers soft pink, reverse yellow and pink, semi-dbl., 13 petals, 3–4 in., exhibition form, moderate fragrance; foliage green, matt; Edland Fragrance Medal, ARS, 1974; [Wendy Cussons X Prima Ballerina]; Sanday, John

Suitor, Pol, mp; light pink flowers; low bush; Clark, A., 1942

'SUKUMARI', F, w, 1983; flowers rosy white; [America's Junior Miss sport]; Gupta, Dr M.N. & Datta, Dr S.K.; National Botanical Research Institute

'SULCOVA KLADENSKA', Pol, lp, 1936; flowers bright pink; vigorous growth; Sulc

Sulphur Rose, *see* **R. HEMISPHAERICA**

'SULPHUREA', T, ly, 1900; flowers sulfur-yellow; Paul, W.

Sulphureux, T, lp; flowers sulphur-yellow; Ducher, 1869

'SULTAN OF ZANZIBAR', HP, m; flowers purplish maroon, medium, globular; Paul & Son, 1876

Sultana, Gr, lp; (Maxima Regina)

'SULTANE', HT, rb, 1946; bud long, pointed; flowers vermilion, reverse gold, dbl., 5 in., slight fragrance; foliage leathery, bronze; [J.B. Meilland X Orange Nassau]; Meilland, F.

Suma, *see* HARsuma

'SUMA NO URA', HT, mr, 1980; flowers crimson, blooms borne usually singly, dbl., 40 petals, exhibition form, no fragrance; large, reddish-brown prickles; foliage small, medium green, semi-glossy; medium, bushy growth; [(Rob Roy X Himatsuri) X Unnamed seedling]; Teranishi, K.; Itami Rose Nursery

'SUMATRA', F, or, 1956; bud ovoid; flowers signal-red, blooms in pyramidal clusters, dbl., 26 petals, 3 in., globular, moderate fragrance; foliage leathery; moderate, upright growth; [Olga X Fashion]; Mallerin, C.; C-P

Summer Beauty®, *see* 'MINAHCO'

'**SUMMER BLOSSOM**', F, mp, 1971; flowers soft geranium-pink, open, 2–3 in.; upright, bushy growth; [Orange Sensation X Kimono]; deRuiter

Summer Blush, A, dp; Sievers, 1988

Summer Breeze, *see* 'TALSUM'

Summer Breeze, *see* KORlastine

Summer Breeze, F, lp; Meilland, 1987

'**SUMMER BUTTER**', Min, dy, 1979; bud ovoid, pointed; dbl., 22 petals, 1–1.5 in., cupped, intense, spicy fragrance; foliage glossy; vigorous, compact growth; [Arthur Bell X Yellow Jewel]; Saville, F. Harmon; Nor'East Min. Roses

Summer Cloud, *see* 'ZIPCLOUD'

Summer Crest, *see* 'HERCRES'

'**SUMMER DAMASK**', D; (Damask Rose, R. belgica, R. calendarum, R. gallica damascena, R. polyantho, R. X damascena); flowers red, pink or white, sometimes striped, borne usually in corymbs, dbl., summer bloom, intense fragrance; obovoid fruit; 5 ft growth; Eurosa, Gallicanae (28, 35) Int. to Eu. from Asia Minor in 16th Century

'**SUMMER DAWN**', Pol, mp, 1950; flowers soft rose-pink; foliage small, dainty; [Margo Koster sport]; Proctor

'**SUMMER DAYS**', HT, ly, 1976; flowers pale yellow, full, dbl., 36 petals, 3.5 in., exhibition form, moderate fragrance; vigorous growth; [Fragrant Cloud X Dr A.J. Verhage]; Bees

Summer Dream, *see* 'JACSHE'

Summer Dream, *see* FRYmaxicot

'**SUMMER EVENING**', Min, ob, 1999; flowers orange and white, reverse white, dbl., 26–40 petals, 1.75 in., exhibition form, borne mostly singly, no fragrance; few prickles; foliage medium, dark green, semi-glossy; upright, medium (14 in.) growth; [Tennessee X Kristin]; Jolly, Betty J.; Langenbach, 1998

Summer Fantasy, *see* 'ZIPFAN'

Summer Fashion, *see* 'JACALE'

Summer Festival®, *see* 'LAVSUM'

Summer Fever, HT, my

'**SUMMER FIELDS**', F, dp, 1971; flowers light scarlet, semi-dbl., 17 petals, 5 in., moderate fragrance; foliage red when young, then green, matt; [Tropicana X Goldmarie]; Mattock

Summer Fragrance, *see* 'TANFUDERMOS'

'**SUMMER FROST**', F, w, 1962; bud ovoid; dbl., 30–35 petals, 4.5 in., cupped, moderate fragrance; foliage leathery, dark; vigorous, bushy, compact growth; [Princess White X Golden Masterpiece]; Boerner; Home Nursery Products Corp.

'**SUMMER GLORY**', HT, my, 1961; flowers lemon, well formed, large; upright growth; [Dr van Rijn X Seedling]; Leenders, J.

Summer Harvest, *see* 'PIXHAR'

'**SUMMER HOLIDAY**'®, HT, or, 1967; flowers vermilion, dbl., 48 petals, exhibition form, moderate fragrance; foliage semi-glossy; very vigorous growth; [Tropicana X ?]; Gregory

'**SUMMER JOY**', HWich, dp, 1911; bud pure white; flowers dark rose-pink, borne in clusters, dbl.; foliage large, glossy; vigorous, climbing (18–20 ft) growth; Walsh

Summer Lady®, *see* TANydal

Summer Love, *see* 'FRANLUV'

Summer Madness®, *see* 'FOUMAD'

'**SUMMER MAGIC**', Min, op, 1989; bud ovoid; flowers light pink blended apricot, reverse cream center, pink edging, very dbl., 150 petals, exhibition form, slight, fruity fragrance; no prickles; foliage small, medium green, matt; bushy, spreading, medium growth; [Fashion Flame X Anita Charles]; Jolly, Marie; Rosehill Farm, 1991

'**SUMMER MEETING**', F, my, 1968; flowers large, dbl., 45 petals, blooms in trusses, slight fragrance; foliage glossy; compact, bushy growth; [Selgold X Circus]; Harkness

Summer Morning, *see* 'KORFULLWIND'

Summer Palace, *see* POUlcape

'**SUMMER PERFUME**', LCl, mr, 1962; flowers vermilion-red, semi-dbl., 19 petals, intense fragrance; [Coral Dawn X Cocorico]; Leenders, J.

'**SUMMER PROMISE**', HT, ly, 1969; flowers very large, dbl., exhibition form, slight fragrance; foliage glossy; vigorous growth; Von Abrams; United Rose Growers

'**SUMMER QUEEN**', HT, w, 1964; flowers ivory-white, center tinted pink; [Queen Elizabeth sport]; Delforge

'**SUMMER RAINBOW**', HT, pb, 1966; bud ovoid; flowers pink, reverse yellow, large, dbl., exhibition form, slight fragrance; foliage glossy, leathery; vigorous, bushy growth; [Peace X Dawn]; Jelly; C-P

Summer Scent, *see* 'RENSUM'

Summer Sérénade®, *see* 'SMITFIRST'

'**SUMMER SNOW**', F, w, 1938; 3.5 in., blooms in large clusters, heavy, recurrent bloom, slight fragrance; foliage light green; bushy growth; [Summer Snow, Climbing sport]; Perkins, C.H.; J&P

Summer Snow, F, lp

'**SUMMER SNOW, CLIMBING**', Cl F, w, 1936; semi-dbl., 2 in., cupped, blooms in large clusters, sparse recurrent bloom, slight fragrance; foliage leathery; vigorous, pillar (8–10 ft) growth; [Tausendschön seedling]; Couteau; J&P

Summer Snowflake, Pol, w

'**SUMMER SONG**', F, ob, 1962; (Chanson d'Été); flowers orange and yellow, large blooms in clusters, semi-dbl., 12 petals, slight fragrance; foliage glossy; low, bushy growth; [Seedling X Masquerade]; Dickson, Patrick; A. Dickson

'**SUMMER SPICE**', Min, ab, 1983; flowers light apricot, medium, dbl., 35 petals, moderate fragrance; foliage medium, medium green, semi-glossy; spreading growth; [Sheri Anne X Unnamed seedling]; Bridges, Dennis A.

Summer Sun, *see* 'ADASUN'

Summer Sunrise, S, mp

Summer Sunset, *see* 'BRISUN'

'**SUMMER SUNSHINE**', HT, dy, 1962; (Soleil d'Été); bud ovoid; dbl., 25 petals, 3.5–5 in., exhibition form, slight fragrance; foliage leathery, dark, semi-glossy; vigorous, upright, well-branched growth; [Buccaneer X Lemon Chiffon]; Swim, H.C.; Armstrong Nursery

'**SUMMER SURPRISE**', Min, mp, 1991; bud rounded; flowers pink with yellow stamens, slightly darker reverse, aging lighter, very dbl., 60 petals, no fragrance; foliage small, medium green, matt; bushy, low growth; [Libby X Unnamed seedling]; Gruenbauer, Richard, 1984; Flowers 'n' Friends Miniature Roses, 1990

Summer Sweet, S, pb

'**SUMMER SYMPHONY**', Min, yb, 1984; flowers large, dbl., exhibition form, intense fragrance; foliage medium, medium green, semi-glossy; bushy, spreading growth; [Lady Eve X Little Darling]; Hardgrove, Donald & Mary; Rose World Originals

'**SUMMER TAN**', F, lp, 1979; flowers flesh-pink; vigorous, compact growth; [Golden Slippers X Picasso]; Sheridan, John

Summer Wedding, *see* 'JALWED'

'**SUMMER WIND**', S, op, 1975; bud ovoid, pointed; single, 3.5–4 in., flat, moderate, spicy clove fragrance; foliage dark, leathery; moderately vigorous, erect, bushy growth; [(Fandango X Florence Mary Morse) X Applejack]; Buck, Dr Griffith J.; Iowa State University

Summer Wine, *see* 'KORIZONT'

Summerdale, *see* 'TALDAL'

Summerrose, *see* 'INTERSUM'

Summer's Snow, *see* **'NEIGES D'ÉTÉ'**

'**SUMMERTIME**', HT, mp, 1957; bud ovoid; flowers cameo-pink overcast rose-pink, borne in irregular clusters, dbl., 65–70 petals, 4–4.5 in., exhibition form, intense fragrance; foliage glossy, olive-green; bushy growth; [Diamond Jubilee X Fashion]; Boerner; J&P

Summerwind, S, 1985; Kordes, R.; (Cavriglia)

'**SUMMERWINE**', HT, mp, 1974; bud ovoid, pointed; dbl., 45 petals, 4–6 in., slight fragrance; foliage leathery; [Tiffany X South Seas]; Warriner, William A.; J&P

'SUN BLUSH', HT, yb, 1981; dbl., 37 petals, exhibition form, borne singly or 4 per cluster, slight fragrance; broad based, dark red prickles; foliage mid-green; upright growth; [Circus X Summer Sunshine]; Anderson's Rose Nurseries

Sun Chariot, *see* 'JALSUN'

Sun City, *see* 'KORSUN'

Sun Cover, *see* 'POULURT'

Sun Drops, *see* 'LYODRO'

Sun Flare, *see* 'JACJEM'

Sun Flare, Climbing, *see* 'JACLEM'

Sun Flare, Climbing™, *see* 'BURYELLOW'

Sun Glory, *see* 'JACMEY'

'SUN GLOW', HT, op, 1934; bud long, pointed; flowers dark coral-pink, large, dbl., exhibition form, moderate fragrance; foliage glossy; long, strong stems; very vigorous growth; [Talisman sport]; Florex Gardens

'SUN GOD', HT, pb, 1930; flowers shrimp-pink, yellow, and orange-copper, very dbl., moderate fragrance; foliage leathery, bronze; short stems; dwarf growth; [Unnamed seedling X Mme Edouard Herriot]; Klyn; Wayside Gardens Co.

Sun Goddess, *see* 'JACDASH'

'SUN GOLD', HT, my, 1935; flowers pure yellow, large, dbl., slight fragrance; foliage glossy, dark; very vigorous growth; RULED EXTINCT 3/83; Elmer's Nursery

Sun Gold, *see* 'MORLEM'

Sun Hit, *see* 'POULHIT'

Sun Honey, *see* 'MORHONEY'

Sun King, *see* 'MEINARVAL'

'SUN KING'®, HT, my, 1954; bud long, pointed; flowers bright lemon-yellow, dbl., 45 petals, 3.5 in., exhibition form, moderate fragrance; foliage dark, glossy, leathery; vigorous, upright growth; [Peace X Duchesse de Talleyrand]; Meilland, F.; C-P

Sun King '74, *see* 'MEINARVAL'

Sun Runner, *see* 'INTERDUST'

Sun Sparkle, *see* 'LYOSUN'

Sun Sprinkles™, *see* 'JACHAL'

'SUN UP', HT, lp, 1951; flowers china-pink, large, dbl., moderate fragrance; vigorous growth; [Break o' Day sport]; Brownell, H.C.

'SUN VALLEY', HT, dy, 1952; (Yukon); bud ovoid; flowers golden yellow, open, dbl., 30–35 petals, 4–5 in., intense, spicy fragrance; foliage dark, leathery; vigorous, upright growth; [Soeur Thérèse X Mark Sullivan]; Whisler; Germain's

'SUNALAD', Min, 1999; (**Aladdin**); dbl., 26–40 petals, 1.75–2 in., borne in large clusters, moderate fragrance; prickles moderate; foliage large, dark green, glossy; spreading, tall growth; [Tinkerbell X Texas]; Schuurman, Frank B.; Franko Roses New Zealand, 1995

'SUNAMBRO', Gr, ab, 1999; (**Ambrosia**); dbl., 26–40 petals, 2.75–3 in., borne in small clusters, moderate fragrance; prickles moderate; foliage large, medium green, semi-glossy; bushy, medium growth; [Sexy Rexy X New Year]; Schuurman, Frank B.; Franko Roses New Zealand, 1996

Sunanda, HT, lp; Ghosh, 1998

'SUNAUCK', HT, ab, 1999; (**Barossa Dream**); flowers apricot, reverse light apricot, dbl., 17–25 petals, 4.5–5 in., borne mostly singly, slight fragrance; prickles moderate; foliage large, medium green, semi-glossy; bushy, tall (3–4 ft) growth; [City of Auckland X New Year]; Schuurman, Frank B.; Franko Roses New Zealand, Ltd., 1995

Sunbeam, *see* **'MARGO KOSTER'**

'SUNBEAM', Min, my, 1957; intense fragrance; dwarf (14–18 in.) growth; [(Tom Thumb X Polly Flinders) X Golden Scepter]; Robinson, T.

Sunbeam®, *see* 'KORDOSELBLA'

Sunbeam, Climbing, *see* **'MARGO KOSTER, CLIMBING'**

Sunbird™, *see* 'POULNISH'

Sunbird, *see* 'JACPOY'

Sunblaze, *see* 'MEIJIKATAR'

Sunblaze, *see* 'MEIPONAL'

'SUNBLEST', HT, dy, 1970; (Landora); dbl., 38 petals, 5 in., slight fragrance; foliage glossy; GM, Japan, 1971 Gold Star of the South Pacific, Palmerston North, NZ, 1971; [Seedling X King's Ransom]; Tantau, Math.

Sunblest, Climbing, Cl HT, dy, 1979; Orard, Joseph; (Cavriglia)

'SUNBONNET', F, dy, 1967; bud pointed; flowers bright greenish yellow, medium, dbl., exhibition form, moderate fragrance; foliage dark, leathery; vigorous, low, bushy growth; [Arlene Francis X (Circus X Sweet Talk)]; Swim & Weeks; Weeks Wholesale Rose Growers

'SUNBONNET SUE', S, yb, 1984; flowers yellow stippled with scarlet, large blooms borne 1–10 per cluster, dbl., 25 petals, repeat bloom, moderate fragrance; small, awl-like, brown prickles; foliage large, leathery, semi-glossy, medium green; upright, bushy growth; hardy; [Gold Dot X Malaguena]; Buck, Dr Griffith J.; Iowa State University

Sunbright®, *see* 'JACJEL'

Sunburnt, *see* 'GELBURN'

Sunburnt Country, *see* 'KORAV'

Sunburst, *see* **'MME JOSEPH PERRAUD'**

'SUNBURST', HT, dy, 1912; bud long, pointed; flowers variable yellow, toward orange, dbl., cupped; GM, NRS, 1912; Pernet-Ducher

'SUNBURST JEWEL', HT, rb, 1979; bud pointed; flowers red, reverse pale gold, dbl., 35 petals; bushy growth; [Muchacha X Unnamed seedling]; Herholdt, J.A.

'SUNBURST, CLIMBING', Cl HT, dy, 1914; Howard Rose Co.

'SUNBURST, CLIMBING', Cl HT, dy, 1915; Low

Suncharm, *see* HARfab

Suncluster, *see* KORgrapet

'SUNCOR', Min, op, 1999; (**Coralitos**); dbl., 17–25 petals, 1 in., borne in small clusters, no fragrance; prickles moderate; foliage small, medium green, semi-glossy; spreading, low (12–16 in.) growth; Schuurman, Frank B.; Franko Roses New Zealand, Ltd., 1996

'SUNCRAL', Gr, op, 1999; (**Kia Ora**); 4 in., borne in large clusters, slight fragrance; few prickles; foliage medium green, glossy; bushy, medium growth; [Innocendi X Firefly]; Schuurman, Frank B.; Franko Roses New Zealand, Ltd., 1994

'SUNCREDEL', Gr, lp, 1983; (**Cream Delight**, Darling); [Sonia sport]; Schuurman, Frank B.

Sundance, *see* 'JACNEL'

'SUNDANCE', F, yb, 1954; flowers orange-yellow changing to bright rose-pink, blooms in trusses, dbl., 22 petals, 2.5 in., slight fragrance; foliage light green, matt, thin; vigorous, tall (4 ft), upright, uneven growth; GM, NRS, 1954; [Poulsen's Supreme X Eugene Furst]; Poulsen, S.; McGredy; Poulsen

Sundance Palace, *see* POUldace

Sundancer, *see* DEVsunset

'SUNDAY BEST', LCl, rb, 1924; bud long, pointed; flowers brilliant red, center white, single blooms in clusters, long seasonal bloom, slight fragrance; foliage wrinkled; vigorous, climbing growth; [Frau Karl Druschki X Unnamed seedling]; Clark, A.; NRS Victoria

Sunday Brunch, *see* 'MORDAY'

Sunday China, *see* 'GELDAY'

Sunday Lemonade, *see* 'MOLSUNLEM'

'SUNDAY PRESS', HT, mr, 1970; dbl., 53 petals, 4.5 in., exhibition form; Kordes; McGredy

'SUNDAY TIMES', F, dp, 1972; (Shrubby Pink); flowers deep rosy pink, medium, dbl., globular, slight fragrance; foliage light; moderate, dwarf growth; McGredy, Sam IV; Kordes

'SUNDEL', HT, m, 1999; (**Delilah**®); dbl., 17–25 petals, 4 in., borne mostly singly, no fragrance; few prickles; foliage large, medium green, glossy; upright, tall (3–4 ft) growth; [seedling X Osiana];

Schuurman, Frank B.; Franko Roses New Zealand, Ltd., 1995

Sunderland Supreme, *see* 'NOSSUN'

'SUNDIA', Min, or, 1999; (**Little Diamond**); dbl., 26–40 petals, 1 in., borne in small clusters, no fragrance; few prickles; foliage medium, dark green, glossy; upright, medium (12–18 in.) growth; [Sexy Rexy X Firefly]; Schuurman, Frank B.; Franko Roses New Zealand, Ltd., 1993

Sundial™, *see* 'AROLYME'

'SUNDOHA', HT, lp, 1999; (**Double Happy**); flowers bi-color red and yellow, dbl., 17–25 petals, 3.5 in., borne mostly singly, no fragrance; prickles moderate; foliage medium, dark green, semi-glossy; upright, medium (30–36 in.) growth; [Louise Gardner X Goldmarie]; Schuurman, Frank B.; Franko Roses New Zealand, Ltd., 1999

'SUNDOWN', HT, mp, 1934; bud globular; flowers pink, medium, dbl., 35–40 petals, 3.5 in., intense fragrance; foliage glossy; very vigorous growth; [Talisman sport]; Scittine; Lainson

Sundowner, *see* 'MACCHEUP'

Sundra®, *see* 'GAISU'

Sundream, F, 1971; Leenders; (Cavriglia)

'SUNDUST', Min, yb, 1977; bud pointed; flowers light apricot, small, dbl., 23 petals, 1.5 in., moderate, fruity fragrance; foliage small, light; low, bushy, compact growth; [Golden Glow X Magic Wand]; Moore, Ralph S.; Sequoia Nursery

Sunfire, *see* 'JACKO'

Sunflare, *see* 'JACJEM'

'SUNGIRL', F, yb, 1981; flowers deep yellow, light red on petal tips, dbl., 25 petals, no fragrance; reddish-green prickles; foliage medium, deep green, glossy; upright growth; [Charleston X Unnamed seedling]; Rose Barni-Pistoia

'SUNGLO', F, or, 1976; bud ovoid; flowers open, dbl., 25 petals, 3 in., flat, slight, fruity fragrance; [Woburn Abbey sport]; Kuramoto, H.; J&P

Sunglow, *see* **'MCGREDY'S ORANGE'**

'SUNGOLD', Min, my, 1992; (**Little Nugget**); flowers golden yellow, patio, small, dbl., 15–25 petals, 1.5 in., slight fragrance; foliage small, medium green, semi-glossy; bushy growth; [Lorena X Firefly]; Schuurman, Frank B.; Riverland Nurseries, Ltd., 1991

'SUNGOLD', Cl HT, my, 1939; bud long, pointed; flowers bright golden yellow, large, dbl.; foliage glossy, dark; vigorous (15–18 ft) growth; RULED EXTINCT 3/83; [Margaret Anderson X Souv. de Claudius Pernet, Climbing]; Thomas; Armstrong Nursery

Sungold, *see* 'MORLEM'

'SUNHAP', Gr, dr, 1999; (**Happy Go Lucky**); flowers dark red, reverse medium red, dbl., 17–25 petals, 3 in., borne in large clusters, slight fragrance; prickles moderate; foliage medium, medium green, semi-glossy; spreading, medium (30–36 in.) growth; [Happy Days X Only Love]; Schuurman, Frank B., 1992; Franko Roses New Zealand, Ltd., 1998

'SUNHIFI', F, lp, 1999; (**High Five**); dbl., 26–40 petals, 3.25 in., borne in large clusters, slight fragrance; prickles moderate; foliage medium, dark green, glossy; spreading, medium (30–36 in.) growth; [Jacaranda X French Lace]; Schuurman, Frank B.

'SUNHINO', Min, lp, 1999; (**Reddy Teddy**); dbl., 17–25 petals, 1.5 in., borne mostly singly, no fragrance; prickles moderate; foliage small, medium green, semi-glossy; compact, low (8–12 in.) growth; [Sexy Rexy X Firefly]; Schuurman, Frank B.

Suni®, *see* BARsus

'SUNIL GAVASKAR', F, m, 1985; dbl., 60 petals, 3 in., cupped, blooms in clusters of 7 or more, no fragrance; light green, needle-like prickles; foliage large, dark, leathery; dwarf, bushy growth; [(First Rose Convention X Scarlet Knight) X Ena Harkness]; Hardikar, Dr M.N.

'SUNJEW', Min, ab, 1999; (**Little Jewel**); dbl., 17–25 petals, 1.5–2 in., borne in small clusters, slight fragrance; few prickles; foliage small, medium green, semi-glossy; upright growth; [Innocenti X Firefly]; Schuurman, Frank B.; Franko Roses New Zealand, Ltd., 1991

Sunking™, F, dy

'SUNKISSED', HT, my, 1980; bud ovoid, pointed; flowers yellow-orange, blooms borne singly, no fragrance; long, light green prickles, hooked down; foliage leathery; upright growth; [Minigold X Precilla]; Tantau, Math.; J&P

'SUNKIST'™, HT, yb, 1932; flowers orange-copper; [Joanna Hill sport]; Hill, E.G., Co.

'SUNLAMPO', F, lp, 1999; (**Bellisima**®); dbl., 26–40 petals, 2.5 in., borne mostly singly, slight fragrance; few prickles; foliage medium, dark green, glossy; upright, medium growth; [Lambada X Pot O' Gold]; Schuurman, Frank B.; Franko Roses New Zealand, Ltd., 1998

'SUNLIGHT', HT, my, 1956; (Grisbi); bud ovoid; dbl., 45 petals, 4 in., exhibition form, moderate fragrance; foliage leathery; vigorous, upright, bushy growth; [(Eclipse X Ophelia) X Monte Carlo]; Meilland, F.; C-P, 1958; URS, 1956

'SUNLIGHT, CLIMBING', Cl HT, my, 1963; (Grisbi, Climbing); Meilland, Mrs. Marie-Louise; URS

'SUNLIT', HT, ab, 1937; flowers rich apricot, dbl., globular; compact growth; Clark, A.; NRS Victoria

'SUNLOCK', Min, lp, 1999; (**Goldilocks**); flowers medium yellow, reverse light yellow, dbl., 26–40 petals, 1.5 in., borne in large clusters, moderate fragrance; prickles moderate; foliage large, medium green, glossy; spreading, tall growth; [Scentasia X seedling]; Schuurman, Frank B.; Franko Roses New Zealand, Ltd., 1998

'SUNLUCK', HT, lp, 1999; (**Gold Strike**®); dbl., 26–40 petals, 4.5 in., borne in small clusters, slight fragrance; prickles moderate; foliage large, dark green, glossy; upright, tall growth; Schuurman, Frank B.; Franko Roses New Zealand, Ltd., 1997

Sunmaid®, Min, ob; Spek, 1972

'SUNMANI', HT, ob, 1999; (**Manaia**, Oasis Sunset); flowers orange, reverse apricot, dbl., 17–25 petals, 4.5–5 in., borne mostly singly, slight fragrance; prickles moderate; foliage large, medium green, semi-glossy; bushy, tall (3.5–4 ft) growth; Schuurman, Frank B.; Franko Roses New Zealand, Ltd., 1994

'SUNMIST', F, ly, 1940; bud pointed; flowers clear light sulfur-yellow, open, semi-dbl., slight fragrance; foliage leathery; vigorous growth; [(Eva X Golden Rapture) X Hede]; Kordes; Dreer

'SUNNINGDALE', HT, mp, 1929; bud long, pointed; flowers reddish carmine passing to cherry-pink, dbl., exhibition form, moderate fragrance; Hicks

'SUNNY', F, my, 1952; bud pointed, yellow overlaid red; flowers lighter yellow, open, small, semi-dbl., moderate fragrance; foliage leathery, glossy; vigorous, upright, bushy growth; [Unnamed seedling X Goldilocks]; Moore, Ralph S.; Marsh's Nursery

Sunny Afternoon, *see* 'TALSUN'

'SUNNY BOY', HT, pb, 1961; flowers buff-pink, tinted gold and lilac, center darker, well formed, moderate fragrance; foliage bronze; [Mme Butterfly X The Queen]; Delforge

'SUNNY CALIFORNIA', HT, my, 1936; bud long, pointed; flowers yellow like Ville de Paris, very large, dbl., cupped; [Feu Joseph Looymans sport]; Hanshaw, E.

Sunny Child®, *see* 'HAVYCHI'

Sunny Day™, *see* 'SAVASUN'

'SUNNY DAYS', HT, my, 1939; bud long, pointed; flowers chrome-yellow at edges, large, dbl., exhibition form, slight fragrance; foliage leathery, dark; vigorous growth; RULED EXTINCT 11/86; Verschuren; Dreer

'SUNNY HONEY', F, yb, 1972; flowers yellow-pink blend, reverse red, dbl., 22 petals, 4 in., moderate fragrance; foliage large, dark; [Happy Event X Elizabeth of Glamis]; Dickson, Patrick

'SUNNY JERSEY', HT, ab, 1928; flowers bronze, apricot, salmon and orange; [Mme Edouard Herriot sport]; Le Cornu

'SUNNY JUNE', S, dy, 1952; bud pointed; flowers deep canary-yellow, blooms in large clusters, single, 5 petals, 3–3.5 in., cupped, slight, spicy fragrance; foliage dark, glossy; upright, compact, pillar or shrub (8 ft) growth; [Crimson Glory X Capt. Thomas]; Lammerts, Dr Walter; Descanso Distributors

'SUNNY MAID', F, my, 1949; flowers bright yellow, borne in clusters, single, 6–8 petals, 4 in.; foliage glossy, light green; very vigorous, upright growth; [(Golden Rapture X Fred Walker) X Unnamed seedling]; Fletcher; Tucker

Sunny Meillandina®, *see* 'MEIPONAL'

'SUNNY MORNING', Min, my, 1974; bud long, pointed; flowers medium to creamy yellow, dbl., 35 petals, 1–1.5 in., flat, moderate fragrance; bushy, upright growth; [Golden Glow (LCl) X Peachy White]; Moore, Ralph S.; Sequoia Nursery

Sunny Sam, *see* 'NOLDANYU'

'SUNNY SAN JOAQUIN', F, rb, 1962; flowers cerise-red, reverse tinted ivory well shaped, borne in large clusters, slight fragrance; tall growth; [Little Darling X Gertrude Raffel]; Raffel; Port Stockton Nursery

Sunny Side Up, *see* 'PIXSUN'

'SUNNY SOUTH', HT, pb, 1918; flowers pink flushed carmine, base yellow, large, semi-dbl., cupped, moderate fragrance; foliage rich green; very vigorous growth; [Gustav Grunerwald X Betty Berkeley]; Clark, A.; NRS Victoria

Sunny Sunblaze, *see* 'MEIPONAL'

'SUNNY TODAY', F, dy, 1970; flowers clear deep yellow, medium, dbl., exhibition form, moderate fragrance; foliage glossy, dark; moderate, bushy growth; [(Summer Sunshine X Gold Cup) X Isobel Harkness]; Whisler, D.; Gro-Plant Industries

Sunnybrook, HT, my

'SUNNYDEW', Min, dy, 1978; bud ovoid; semi-dbl., 18 petals, 1 in., exhibition form, slight fragrance; well-branched, spreading growth; [Yellow Doll X Unnamed seedling]; Schwartz, Ernest W.; Bountiful Ridge Nursery

'SUNNYMOUNT', HT, my, 1936; bud sharp pointed but short; flowers buttercup-yellow, large, dbl., 50 petals, intense fragrance; foliage leathery; vigorous growth; [Joanna Hill sport]; Grillo

'SUNNYSIDE'®, Min, yb, 1963; flowers yellow, becoming pink and then red; vigorous growth; [(Purpurine X Miniature seedling) X Rosina]; Lens

'SUNNYSIDE', HT, op, 1949; bud pointed, orange; flowers pink shaded orange, dbl., 24–30 petals, 4.5–5 in., exhibition form, slight fragrance; very vigorous growth; [Yellow Gloria sport]; Andre; Andre Greenhouses

Sunnyside, *see* POUlmulti

Sunnyside '83, *see* 'LENSUN'

'SUNPACH', F, or, 1999; (**Pachinko**); semi-dbl., 12–16 petals, 3 in., borne in large clusters, slight fragrance; prickles moderate; foliage medium, dark green, glossy; spreading, medium (30–36 in.) growth; [Alexander X Golden Emblem]; Schuurman, Frank B.; Franko Roses New Zealand, Ltd., 1997

'SUNPARI', HT, ab, 1999; (**La Parisienne**); dbl., 26–40 petals, 4.5 in., borne mostly singly, slight fragrance; prickles moderate; foliage large, dark green, glossy; upright, tall growth; [Kia Ora X Texas]; Schuurman, Frank B.; Franko Roses New Zealand, Ltd., 1996

'SUNPASTE', F, w, 1992; (**Pastelina**); flowers near white, blooms, dbl., 26–40 petals, 1.5–2.75 in., slight fragrance; foliage medium, medium green, glossy; upright growth; [White Dream X Freegold]; Schuurman, Frank B.; Riverland Nurseries, Ltd., 1991

'SUNPAT', Min, lp, 1992; (**Little Opal**); flowers patio, dbl., 15–25 petals, 1.5 in., slight fragrance; foliage small, medium green, glossy; upright growth; Rose of the Year, Auckland, NZ, 1992; [White Dream X Dicky Bird]; Schuurman, Frank B.; Riverland Nurseries, Ltd., 1991

'SUNPEARL', Min, lp, 1992; (**Little Pearl**); dbl., 26–40 petals, 1.5 in., slight fragrance; few prickles; foliage small, medium green, glossy; medium (50 cms), upright growth; [Innocent X MACfrabro (Firefly)]; Schuurman, Frank B.; Riverland Nurseries, Ltd.

'SUNPETE', F, lp, 1999; (**Peter Pan**); dbl., 26–40 petals, 2 in., borne in large clusters and sprays, no fragrance; prickles moderate; foliage medium, dark green, glossy; upright, medium growth; [Pink Delight X seedling]; Schuurman, Frank B.; Franko Roses New Zealand, Ltd., 1993

'SUNPINO', Min, lp, 1999; (**Pinocchio**); dbl., 17–25 petals, 2–2.5 in., borne in large clusters, no fragrance; prickles moderate; foliage large, dark green, semi-glossy; patio; upright, medium growth; [Tinkerbell X Chess]; Schuurman, Frank B.; Franko Roses New Zealand, Ltd., 1998

'SUNPLAY', HT, w, 1999; (**Player**); dbl., 17–25 petals, 4.5–5 in., borne mostly singly, no fragrance; many prickles; foliage large, dark green, glossy; upright, tall (4.5–5 ft) growth; [Raspberry Ice X Karen]; Schuurman, Frank B.; Franko Roses New Zealand, Ltd., 1993

'SUN-RAY', HT, my, 1932; flowers golden yellow, semi-dbl., moderate fragrance; vigorous growth; Bentall

'SUNRISE', HT, mp, 1939; bud long, pointed; flowers salmon, open, large, dbl.; foliage glossy, bronze; vigorous, bushy growth; [Pilarin Vilella X Rosa Gallart]; Dot, Pedro

Sunrise, *see* KORmarter

'SUNRISE-SUNSET', HT, pb, 1971; flowers blended pink, cream and lavender, large, dbl., exhibition form, slight fragrance; foliage glossy, light to dark, leathery; vigorous growth; [Tiffany X (Unnamed seedling X Happiness)]; Swim & Weeks; Weeks Wholesale Rose Growers

'SUNSALM', Min, op, 1992; (Little Gem, **Patio Gem**); flowers moderately patio, small, dbl., 15–25 petals, 1.5 in., moderate fragrance; foliage small, medium green, matt; spreading growth; [Sexy Rexy X Firefly]; Schuurman, Frank B.; Riverland Nurseries, Ltd., 1991

'SUNSAP', Min, pb, 1999; (**Little Sapphire**); flowers medium pink, reverse light pink, dbl., 17–25 petals, 2.5–3 in., borne in small clusters, slight fragrance; prickles moderate; foliage medium, medium green, semi-glossy; bushy, medium (16–20 in.) growth; [White Dream X seedling]; Schuurman, Frank B.; Franko Roses New Zealand, Ltd., 1993

Sunsation, *see* 'WEKCELPEP'

Sunsation, *see* 'KORGUST'

'SUNSCENT', Min, ly, 1999; (**Scentasia**); dbl., 26–40 petals, 1.5–2 in., borne in large clusters, moderate fragrance; few prickles; foliage large, medium green, semi-glossy; spreading, tall (20–24 in.) growth; [Tinkerbell X Little Nugget]; Schuurman, Frank B.; Franko Roses New Zealand, Ltd., 1997

Sunseeker, *see* 'DICRACER'

'SUNSET', T, dy, 1883; flowers orange-yellow; [Perle des Jardins sport]; Henderson, P.

'SUNSET', Cl HT, yb, 1947; flowers peach and apricot, reverse yellow, dbl., 45–60 petals, 5 in., cupped, slight fragrance; vigorous growth; [Faience sport]; Marsh; Marsh's Nursery

Sunset Boulevard, *see* HARbabble

Sunset Celebration™, *see* 'FRYXOTIC'

'SUNSET GLORY', HT, ab, 1947; bud ovoid; flowers golden yellow overlaid dusty rose-pink, dbl., 35–40 petals, 4–4.5 in., cupped, moderate, fruity fragrance; vigorous, upright, compact growth; [McGredy's Sunset sport]; Boerner; J&P

Sunset Glow, *see* **'CANARIENVOGEL'**

'SUNSET JUBILEE', HT, pb, 1973; bud ovoid; flowers medium pink, tinted lighter, large, dbl., exhibition form, slight fragrance; foliage large, light, leathery; vigorous, upright, bushy growth; [Kordes' Perfecta X Pink Duchess seedling]; Boerner; J&P

Sunset Song, *see* 'COCASUN'

Sunset Strip, *see* 'AROCORE'

'SUNSHINE', HT, yb, 1918; bud long, pointed; flowers golden yellow, shaded apricot, dbl., moderate fragrance; Chaplin Bros.

'SUNSHINE', Pol, ob, 1927; bud ovoid; flowers golden orange, small blooms in clusters, dbl., moderate fragrance; foliage glossy; dwarf growth; (14); [George

Elger X William Allen Richardson]; Robichon; Cutbush

Sunshine, LCl, lp

Sunshine Girl, *see* 'PIXISUN'

Sunshine Princess, *see* 'ANDSUN'

'SUNSHOWER', F, my, 1974; bud ovoid; flowers bright yellow, ruffled, semi-dbl., 16 petals, 2.5 in.; foliage glossy, light; moderate growth; [Golden Showers X Prairie Fire]; Golik; Dynarose

'SUNSIEN', HT, lp, 1999; (**Citronella**®); dbl., 26–40 petals, 4.5–5 in., borne mostly singly, slight fragrance; prickles moderate; foliage large, medium green, glossy; upright, tall growth; [La Parisienne sport]; Schuurman, Frank B.; Franko Roses New Zealand, Ltd., 1998

'SUNSILK', F, my, 1974; flowers lemon-yellow, dbl., 30 petals, 5 in., slight fragrance; foliage dark; GM, Belfast, 1976; [Pink Parfait X Redgold seedling]; Fryer, Gareth; Fryer's Nursery, Ltd.

Sunsmile™, *see* 'JACMIY'

'SUNSNOFLE', Min, lp, 1999; (**Snowflake**); dbl., 26–40 petals, 1 in., borne in small clusters, no fragrance; prickles moderate; foliage small, light green, semi-glossy; compact, low growth; Schuurman, Frank B.; Franko Roses New Zealand, Ltd., 1996

'SUNSONG', Gr, ob, 1966; (Dorrit); bud globular; flowers orange to coral blend, informal, dbl., 63 petals, 3 in., slight, tea fragrance; foliage glossy, leathery; upright, bushy growth; [Folie d'Espagne X (Zambra X Danish Pink)]; Soenderhousen; Armstrong Nursery

Sunsplash™, *see* 'JACYIM'

'SUNSPOT', F, my, 1965; flowers mimosa-yellow, large, borne in clusters, dbl., cupped; foliage leathery; vigorous, bushy growth; [Golden Anniversary X Masquerade]; Fisher, G.; C-P

Sunspray™, *see* 'ARORASP'

Sunsprite, *see* 'KORRESIA'

'SUNSPRITE, CLIMBING', Cl F, dy, 1989; [Sunsprite sport]; Kroeger, Henry, 1990

'SUNSTAR', HT, or, 1921; bud long, pointed; flowers orange and red, semi-dbl., intense fragrance; foliage light green; vigorous, bushy growth; GM, Bagatelle, 1923 GM, NRS, 1917; [Mrs C.V. Haworth X Hugh Dickson]; Dickson, A.

'SUNSTAR, CLIMBING', Cl HT, or, 1925; Dickson, A.

'SUNSTONE', HT, yb, 1957; bud long; flowers bright yellow splashed red, 6 in., slight fragrance; foliage glossy, bronze; strong stems; vigorous, tall growth; [Bridget X Marcelle Gret]; Fletcher; Tucker

'SUNSTRIKE', F, dy, 1974; bud small, pointed, ovoid; flowers medium, dbl., exhibition form, slight fragrance; foliage large, light, leathery; very vigorous, bushy growth; [Spanish Sun X (Buccaneer X Zorina)]; Warriner, William A.; J&P

'SUNSTRUCK', Min, yb, 1978; bud long, pointed; flowers aureolin-yellow and signal-red, dbl., 20 petals, 2 in., intense fragrance; foliage dark; upright growth; [Redgold X ?]; Lyon

'SUNTAN', HT, dy, 1939; flowers deep orange-yellow, fading slowly, large, dbl., 35 petals; foliage leathery, dark, bronze; strong stems; RULED EXTINCT 4/92; [Nanjemoy X Mrs Pierre S. duPont]; Hansen, N.J.; B&A

'SUNTAN', HT, lp, 1940; bud long, pointed; flowers buff-salmon, base yellow, reverse light coral-red, dbl., 25 petals, 3.5 in., cupped, moderate fragrance; vigorous, upright growth; RULED EXTINCT 4/92; [Mrs Franklin D. Roosevelt sport]; Yoder

Suntan™, *see* 'INTERBRONZI'

Suntan Beauty™, *see* 'MINAWCO'

'SUNTHUMB', Min, ly, 1999; (**Thumbelina**); dbl., 17–25 petals, 1.75 in., borne in small clusters, moderate fragrance; prickles moderate; foliage medium, medium green, glossy; patio; compact, medium growth; [Champagner X Tinkerbell]; Schuurman, Frank B.; Franko Roses New Zealand, Ltd., 1998

'SUNTICK', HT, pb, 1999; (People's Princess, **Tickled Pink**); dbl., 26–40 petals, 4.5–5 in., borne mostly singly, intense fragrance; few prickles; foliage large, medium green, glossy; upright, tall growth; [Jacaranda X Gold Medal]; Schuurman, Frank B.; Franko Roses New Zealand, Ltd., 1997

'SUNTINK', F, lp, 1999; (**Tinkerbell**); dbl., 26–40 petals, 1.5 in., borne in large clusters and sprays, moderate fragrance; prickles moderate; foliage medium, dark green, glossy; upright, medium growth; [White Dream X Evelien]; Schuurman, Frank B.; Franko Roses New Zealand, Ltd., 1993

'SUNTOTO', HT, m, 1992; (**Rangitoto**); flowers medium, very dbl., 1.5–2.75 in., slight fragrance; foliage medium, medium green, glossy; upright growth; [Champagne X Chantilly Lace]; Schuurman, Frank B.; Riverland Nurseries, Ltd., 1990

'SUNwend', F, lp, 1999; (**Wendy**); dbl., 26–40 petals, 1.5 in., borne in large clusters and sprays, no fragrance; many prickles; foliage medium, medium green, glossy; upright, medium growth; Schuurman, Frank B.

'SUPER ARIBAU', HT, dr, 1960; flowers crimson-carmine, reverse silvery crimson, large, dbl., 24 petals, intense fragrance; strong stems; upright, vigorous growth; [Aribau X Director Rubió]; Dot, Pedro

Super Bowl, *see* 'MACSUPBOW'

Super Cascade Coral, *see* 'JALCORAL'

'SUPER CHIEF', HT, mr, 1976; bud globular; flowers bright red, aging darker, very dbl., 60–65 petals, 5 in., exhibition form, slight fragrance; foliage dark; vigorous growth; [Queen Elizabeth X Happiness]; Patterson; Patterson Roses

'SUPER CONGO', HT, dr, 1950; bud ovoid; flowers velvety dark blood-red, medium, dbl., 30 petals, slight fragrance; foliage dull green; upright growth; [Congo X Léonce Colombier]; Meilland, F.

Super Derby, *see* 'BURSTEIN'

Super Disco, *see* OLItel

Super Dorothy®, *see* HELdoro

Super Elfin, *see* 'HELKLEGER'

Super Excelsa®, *see* HELexa

Super Fairy, *see* 'HELSUFAIR'

'SUPER GAMUSIN', HT, m, 1962; flowers chamois to mauve; [Grey Pearl X Tristesse]; Dot, Pedro

Super Gold™, *see* 'WINGOLD'

'SUPER HARRINGTON', HT, mr, 1938; flowers medium, dbl., moderate fragrance; Brown, J.; (Sangerhausen)

Super Nova, *see* JACsunov

Super Pink®, LCl, pb; Lens, 1992

Super Prince Tango, F, 1971; Delbard; (Cavriglia)

Super Sparkle, *see* 'HELHEIN'

Super Star, *see* 'TANORSTAR'

'SUPER STAR SUPREME', HT, or, 1982; flowers deeper color; [Tropicana sport]; U.S. Patent Sales, Inc.

Super Star, Climbing, *see* 'TANGOSTAR'

'SUPER SUN', HT, my, 1967; flowers maize-yellow; [Piccadilly sport]; Bentley

Super Swany®, S, w

'SUPER TABARIN', F, ob, 1963; flowers coppery, reverse orange, semi-dbl.; vigorous growth; [Faust X Tabarin]; Gaujard

Super Tan, Min, op, 1995; flowers coppery orange, small to medium, semi-dbl.; Interplant; (Sangerhausen)

'SUPERB', HT, lp, 1924; flowers pale pink, tinted blush; [Mme Caroline Testout X Willowmere]; Evans

'SUPERB', Pol, dr, 1927; flowers crimson, dbl.; deRuiter

Superb Tuscan, *see* **'TUSCANY SUPERB'**

Superb Tuscany, *see* **'TUSCANY SUPERB'**

'SUPERB, CLIMBING', Cl Pol, dr, 1933; Guillot, P.

'SUPERBA', HSet, lp, 1843; flowers blush, borne in large clusters; [R. setigera seedling]; Feast

'SUPERBA', HMoy, dr; flowers dark crimson, large, dbl.; no fruit; [Charles P. Kilham X R. moyesii]; Van Rossem

'**SUPERBA**', (R. setigera seedling), w

'**SUPERBA**', HT, dp, 1940; flowers intense rose with yellow reflections, overlarge, intense fragrance; long stems; [Julien Potin X Seedling]; Aicardi, L.; Giacomasso

Supercandy, Gr, 1967; Mansuino; (Cavriglia)

Super-Dupont, *see* **'DOLLY MADISON'**

'**SUPERGA**', HT, lp, 1956; flowers rose-pearl, high pointed, slight fragrance; foliage dark, glossy; long stems; very vigorous growth; [Savoia sport]; Giacomasso

Superglo, *see* 'SPOGLO'

'**SUPER-HARRINGTON**', HT, dr, 1938; bud oval; flowers dark crimson, mottled scarlet, medium, dbl., moderate fragrance; foliage leathery; vigorous, upright growth; [Tassin X Victoria Harrington]; Brown, John; A. Meilland

'**SUPERIOR**', F, ab, 1968; flowers apricot-yellow, blooms in trusses, semi-dbl., slight fragrance; foliage dark, glossy; vigorous, low growth; [Masquerade X Amberlight]; LeGrice

Supra™, *see* 'MEIROBIDOR'

'**SUPREME**', HT, my, 1963; flowers lemon-yellow, well formed, dbl., 40–45 petals, 4.5–5 in., moderate fragrance; foliage dark, leathery, glossy; vigorous, tall growth; [Golden Masterpiece X Ethel Sanday]; LeGrice

Supreme Cover, *see* POUlor

'**SUPREME RENDEZVOUS**', HT, pb, 1961; flowers cream to pale pink, edged carmine; [Day of Triumph sport]; Langbecker

'**SUPREME'S SISTER**', F, mp, 1960; bud red; flowers coppery salmon to salmon-pink, medium, dbl., 42 petals; foliage leathery, bronze; low, bushy growth; Jones; Hennessey

Supriya, HT, pb; San, 1982

'**SURABHI**', HT, mp, 1975; bud long, pointed; flowers phlox-pink, very dbl., 80 petals, exhibition form, intense fragrance; foliage glossy, dark; upright, compact growth; [Oklahoma X Delhi Princess]; IARI

'**SURAIN**', HT, mp, 1985; flowers large, dbl., 35 petals, exhibition form, slight fragrance; upright growth; [Unnamed seedling X Forever Yours]; Elliott, Charles P.

Surbrunnsgatan, HP, lp

'**SUREKHA**', HT, or, 1969; bud pointed; flowers coral-red, full, large, dbl.; foliage dark, leathery; very vigorous, upright growth; [Queen Elizabeth X Open pollination]; IARI

'**SURF RIDER**', S, w, 1968; flowers creamy white, blooms in large trusses, dbl., 25 petals, moderate, musk fragrance; foliage bright, glossy; very vigorous, erect growth; [Ivory Fashion X Ballerina]; Holmes, R.A.; Fryer's Nursery, Ltd.

Surfer Girl, *see* 'RENFER'

Surf's Up, *see* 'LEMSUR'

Surfside, *see* 'MICSURF'

'**SURKHAB**', HT, m, 1976; bud globular; flowers tyrian purple, reverse silvery amaranth, very dbl., 76 petals, 4 in., cupped, intense fragrance; foliage leathery; very vigorous, upright, compact growth; Pal, Dr B.P.; Anand Roses

Surkhab, HT, lp; Pal, Dr B.P., 1980

'**SURPASSE TOUT**', HGal, mr; (Cérisette la Jolie); flowers rosy crimson, fading cerise-pink, dbl.; height 3–4 ft; Hardy, before 1823

Surpassing Beauty of Woolverstone, HP, dr; re-introduced, 1980

'**SURPRISE**', HT, mp, 1925; flowers salmon-pink, dbl., cupped, slight fragrance; [Frau Karl Druschki X Mme Edouard Herriot]; Van Rossem

Surprise Party, *see* 'JELROGANOR'

Surprise Surprise, *see* 'MORLAVMAG'

Surrey, *see* 'KORLANUM'

'**SURVILLE**', HT, mr, 1924; flowers indian red, shaded cerise, dbl.; [Mme Edouard Herriot seedling]; Croibier

Survivor, S, dr; Ottawa, 1975

'**SURYA KIRAN**', F, or, 1979; bud pointed; 19 petals, borne 2–14 per cluster, slight fragrance; hooked prickles; foliage large, glossy; tall, upright, vigorous, open growth; [Flamenco X Orangeade]; Pal, Dr B.P.; Friends Rosery

Surya Shikha, HT, lp; Diby

'**SURYODAYA**', F, ob, 1968; bud ovoid; flowers bright orange, medium, semi-dbl.; foliage dark, leathery; very vigorous, upright growth; [Orangeade X ?]; IARI

'**SUSAN**', F, mr, 1955; flowers crimson, well-formed blooms in large clusters, semi-dbl.; foliage dark; bushy, compact growth; [Donald Prior X Our Princess]; Robinson, H.

Susan, *see* 'KORAD'

Susan, *see* POUlsue

Susan, *see* KORkilt

Susan Ann, *see* **'SOUTHAMPTON'**

'**SUSAN BECKWITH**', HT, mp, 1998; bud short, squat; flowers medium pink, apricot base, opens flat, very dbl., 26–40 petals, 4–4.5 in., flat, borne in clusters, intense fragrance; prickles moderate; foliage medium, medium green, semi-glossy; bushy, medium (2.5 ft) growth; [Dr Sybil Johnston X Seedling]; Beckwith, R.; Battersby Roses, 1998

Susan Blixen, *see* POULARI

Susan Devoy, *see* 'MACRENO'

Susan Elizabeth, *see* 'SCHUPRAK'

Susan Hampshire, *see* 'MEINATAC'

'**SUSAN HAYWARD**', Cl HT, lp, 1967; flowers light pink, center and reverse darker, large, dbl., exhibition form; foliage dark, glossy; vigorous, climbing, open habit growth; [Fontanelle X Gen. MacArthur]; Hayward

Susan Irvine, *see* 'TOMSUE'

Susan Irvine Cream, T, lp

Susan Irvine Pink Tea, T, lp

Susan Jellicoe, *see* 'HORKEEPOG'

'**SUSAN LOUISE**', S, lp; bud very long, pointed, deep pink; flowers flesh-pink, semi-dbl., recurrent bloom, slight fragrance; vigorous (4–5 ft), bushy growth; [Belle Portugaise seedling]; Adams, Dr Neil D.; Stocking

Susan Massu, *see* 'KORAD'

'**SUSAN NOEL**', Min, ab, 1986; flowers light apricot, urn-shaped, small blooms borne usually singly, very dbl., 80 petals, slight fragrance; small, globular fruit; small, slightly hooked, gray-orange prickles; foliage small, medium green, semi-glossy; medium, upright growth; [Rise 'n' Shine X Orange Honey]; Jolly, Nelson F.; Rosehill Farm

'**SUSAN SCHNEIDER**', S, lp, 1999; flowers med. yellow, turning orange pink, then red, reverse med. yellow, semi-dbl., 8–14 petals, 3 in., borne in small clusters, slight fragrance; prickles moderate, tan; foliage medium, medium green, glossy; spreading, medium (5 ft) growth; Jerabek, Paul E.; Freedom Gardens, 1998

'**SUSANA MARCHAL**', HT, or, 1953; bud pointed; flowers coral-red, large, semi-dbl., 18 petals, exhibition form; foliage clear green; strong stems; upright growth; [Cynthia X Vive la France]; Dot, Pedro

'**SUSANE DOT**', Cl HT, dr, 1963; flowers crimson, large, dbl., 32 petals, moderate fragrance; vigorous growth; [Queen Elizabeth X Peace, Climbing]; Dot, Simon

'**SUSANNA**', Pol, lp, 1914; flowers small, dbl.; Weigand, C.; (Sangerhausen)

Sushma, F, m; Kasturi, 1975

'**SUSIE**', HT, mp, 1955; flowers pink, base yellow, dbl., 45–50 petals, 4–6 in., exhibition form, slight fragrance; foliage leathery; vigorous, bushy growth; RULED EXTINCT 11/90; [Mme Butterfly X Sally]; Spandikow

'**SU-SPANTU**', F, m, 1962; flowers purplish pink, borne in clusters of 7–8, dbl., 25 petals; upright growth; [Alain X Independence]; Borgatti, G.; Sgaravatti

Suspense, *see* 'MEIFAN'

Sussex, *see* POUlave

'**SUTTER'S GOLD**', HT, ob, 1950; bud orange overlaid indian red; flowers golden orange, often with red on outer petals, dbl., 33 petals, 4–5 in., exhibition form, intense fragrance; foliage dark, leathery; very vigorous, upright growth; AARS,

1950 GM, Bagatelle, 1948 GM, Geneva, 1949 GM, Portland, 1946 James Alexander Gamble Fragrance Medal, ARS, 1966; [Charlotte Armstrong X Signora]; Swim, H.C.; Armstrong Nursery

'SUTTER'S GOLD, CLIMBING', Cl HT, ob, 1950; (Grimpant Sutter's Gold); Weeks; Armstrong Nursery

'SUTTON PLACE'(TM), Gr, op, 1991; flowers light pink with salmon pink edge, large blooms borne mostly, single, intense fragrance; foliage large, dark green, semi-glossy; tall, upright, bushy growth; [Queen Elizabeth X (Carla X Command Performance)]; Williams, J. Benjamin, 1992

'SUZAN BALL', F, ab, 1957; bud conical; flowers salmon-coral, to loosely open, borne in clusters, dbl., 25 petals, 2.5–3.5 in., exhibition form, moderate, spicy fragrance; foliage dark; vigorous, upright growth; [Tom Breneman X Fashion]; Warriner, William A.; H&S

'SUZANNE', HSpn, op, 1950; flowers pale coral-pink, very dbl., recurrent bloom; foliage small, dark; height 4 ft; [Second generation R. laxa X R. spinosissima]; Skinner

'SUZANNE ALBRAND', Pol, mr, 1930; flowers bright red, large, borne in large clusters; foliage glossy; vigorous growth; Turbat

'SUZANNE BALITRAND', HT, or, 1942; flowers coral edged fiery red, stamens yellow; Mallerin, C.; A. Meilland

'SUZANNE BIDARD', Pol, mp, 1914; flowers small, dbl.; Vigneron; (Sangerhausen)

'SUZANNE BLANCHET', T, lp, 1885; flowers flesh-pink, large, intense fragrance; Nabonnand, G.

'SUZANNE CARROL OF CARROLTON', HP, mp, 1924; flowers light satiny rose and salmon, large, semi-dbl.; vigorous growth; [Frau Karl Druschki X Mme Gabriel Luizet]; Nabonnand, P.

Suzanne Dolard, Pol, dr, 1966; flowers medium, semi-dbl.; Gaujard; (Sangerhausen)

'SUZANNE HESTER', Cl HT, w, 1950; bud ovoid; flowers white, base tinted yellow, dbl., 30 petals, exhibition form, intense fragrance; foliage glossy; vigorous growth; [Marie Maass sport]; Hester

'SUZANNE MEYER', Pol, w, 1926; flowers white shaded soft pink, center bright rose-pink, medium; medium growth; [Tausendschön X Rosel Dach]; Walter, L.

'SUZANNE MICHELA', HT, my, 1932; bud long, pointed; flowers pure chrome-yellow, very large, dbl., cupped, moderate fragrance; very vigorous growth; [Mme la Generale Ardouin X Unnamed seedling]; Chambard, C.

'SUZANNE MILLER', Pol, mr, 1927; flowers clear bright cherry-red, medium, borne in clusters, dbl.; low, bushy growth; Wezelenburg

'SUZANNE TURBAT', Pol, ob, 1919; flowers coral-red, shaded shrimp-pink, medium, borne in clusters of, dbl., cupped, slight fragrance; dwarf growth; [Petit Constant seedling X Unnamed seedling]; Turbat

'SUZANNE VILLAIN', HT, pb, 1935; flowers peach-blossom-pink suffused cherry-red, reverse salmon-pink, dbl., 60–65 petals; foliage rich green; vigorous, bushy growth; [(Rev. Williamson X Gorgeous) X Mrs John Bell]; Ketten Bros.

'SUZANNE WOOD', HP, mp, 1869; flowers large, dbl.; Verdier, E.; (Sangerhausen)

'SUZANNE-MARIE RODOCANACHI', HP, mp, 1883; flowers dark rosy cerise, shaded and bordered lighter, well formed, dbl., 45 petals, globular, profuse bloom, somewhat recurrent; [Victor Verdier seedling]; Lévêque

Suzette van der Merwe, *see* JAChoc

Suzon, LCl, pb; Eve, 1994

'SUZON LOTTHÉ', HT, pb, 1951; bud peach; flowers pearl-pink, flushed deeper toward edge, dbl., 60 petals, 4 in., exhibition form, intense, damask fragrance; foliage dark; very vigorous growth; [Peace X (Signora X Mrs John Laing)]; Meilland, F.; C-P

'SUZON LOTTHÉ, CLIMBING', Cl HT, pb, 1964; Trimper, K.; Ruston

Suzy, *see* 'BRIsuzy'

Suzy, LCl, yb; Eve, 1994

Suzy Q(TM), *see* 'JACHILL'

'SVATOPLUK CECH', LCl, dy, 1936; flowers orange-yellow, recurrent bloom, intense fragrance; vigorous growth; Brada, Dr.; Böhm

'SVATY VÁCLAV', HT, w, 1936; flowers white, center yellowish, large, intense fragrance; Berger, A.

Svensk Pimpinelle, HSpn, w

'SVETLA ALBENA', F, yb, 1974; flowers opening yellow orange, aging red, blooms in large clusters, very dbl., 85 petals; foliage dark; vigorous, upright growth; [Highlight X Masquerade]; Staikov, Prof. Dr V.; Kalaydjiev and Chorbadjiiski

Svetlana, F, dy; Urban, J., 1975; (Czech Rosa Club)

Svitani, HT, my; Kavka, 1962; (Czech Rosa Club)

'SVORNOST', Pol, dr, 1935; flowers pure dark red, with fiery streaks, borne in large clusters; [Orléans Rose sport]; Böhm, J.

'SWAGATAM', HT, ab, 1989; flowers light pink, reverse apricot; [Surkhab sport]; Patil, B.K.; K.S.G. Son's Roses, 1987

'SWAMI', HT, lp, 1973; bud long, pointed; very dbl., 100 petals, 5 in., exhibition form, moderate fragrance; very vigorous, upright growth; [Scarlet Knight X Festival Beauty]; Hardikar, Dr M.N.

Swamp Rose, *see* **R. PALUSTRIS**

Swamp Rose, *see* **R. PALUSTRIS SCANDENS**

Swan, *see* 'AUSWHITE'

'SWAN LAKE', LCl, w, 1968; (Schwanensee); flowers white, center tinged pinkish, well-formed, large, dbl., 50 petals, slight fragrance; [Memoriam X Heidelberg]; McGredy, Sam IV

Swan Song, *see* 'SEAswan'

'SWANSDOWN', HT, w, 1928; bud large, long, pointed; flowers creamy white, center deeper cream, spiral form, dbl., slight fragrance; foliage olive-green; fairly vigorous growth; Dickson, A.

Swansong, *see* 'SEAswan'

'SWANTJE', F, w, 1936; flowers snow-white, very large, borne in clusters, dbl.; foliage glossy, light; very vigorous, bushy growth; [Johanna Tantau X (Prof. Gnau X Joanna Hill)]; Tantau

Swany®, *see* 'MEIburenac'

'SWANY RIVER', HT, w, 1981; (Fredagh of Bellinchamp); flowers large, dbl., moderate fragrance; foliage medium, dark, semi-glossy; upright growth; [Pascali X Unnamed seedling]; Herholdt, J.A.

Swarthmore, *see* 'MEItaras'

'SWARTHMORE, CLIMBING', Cl HT, pb, 1973; Thomas, Dr A.S.; A. Ross & Son

Swashbuckle, HT, or; Dawson, before 1990

'SWATI', Pol, w, 1968; bud pointed; flowers white, edged deep pink, open, small, semi-dbl.; foliage dark, leathery; moderate, bushy growth; [Winifred Coulter X ?]; IARI

'SWEDISH DOLL', Min, op, 1976; bud long, pointed; flowers small, dbl., 28 petals, 1.5 in., slight fragrance; foliage glossy, leathery; vigorous, upright, branched growth; [Fire King X Little Buckaroo]; Moore, Ralph S.; Sequoia Nursery

Sweepstakes, *see* 'MACaft'

'SWEET', Min, my, 1980; bud long, pointed; dbl., 50 petals, borne 1–3 per cluster, moderate fragrance; tiny, curved prickles; foliage tiny, glossy, deep green; compact, bushy growth; [Unnamed seedling X Unnamed seedling]; Lyon

'SWEET ADELINE', HT, mp, 1929; bud long, pointed; flowers rose-pink, large, semi-dbl., moderate fragrance; [Rapture X Souv. de Claudius Pernet]; Hill, Joseph H., Co.

Sweet Adeline, S, lp; Erskine

'SWEET AFTON', HT, w, 1964; flowers near white, reverse pale pink, dbl., 4.5–5 in., exhibition form, intense fragrance; foliage leathery; tall, spreading, bushy growth; [(Charlotte Armstrong X Signora) X (Alice Stern X Ondine)]; Armstrong, D.L. & Swim; Armstrong Nursery

'Sweet Allison', MinFl, rb, 1985; flowers orange-red shading to yellow or white, reverse cream, deeper, dbl., 35 petals, slight fragrance; foliage medium, medium green, semi-glossy; upright, bushy growth; [Bonny X (Tiki X Seedling)]; Jolly, Nelson F.; Rosehill Farm

Sweet Amazone, HT, op

'Sweet Amy', HT, lp, 1999; flowers medium pink, reverse lighter, dbl., 26–40 petals, 4–5 in., borne mostly singly, intense fragrance; prickles moderate; foliage medium, dark green, glossy; upright, medium (5 ft) growth; [Crystalline X Stainless Steel]; Edwards, Eddie; Johnny Becnel Show Roses, 1999

'Sweet and Low', F, pb, 1962; bud globular; flowers salmon-pink, center lighter, small, borne in large clusters, dbl., 36 petals, slight fragrance; foliage glossy, bronze; compact, low growth; [Pinocchio X Sweet Fairy]; Schwartz, Ernest W.; Wyant

Sweet Bouquet, *see* SABchurchill

'Sweet Briar Queen', F, mp, 1960; flowers rose-pink, borne in clusters, dbl., 45–50 petals, 2 in., flat, moderate fragrance; foliage maroon when young, leathery; vigorous, branching growth; [Rosemary sport]; Schenkel; H.R. Schenkel, Inc.

Sweet Brier Rose, *see* **R. eglanteria**

Sweet Butterfly, *see* 'LAVstar'

'Sweet Caress', F, mp, 1958; bud ovoid; flowers rose-pink, large, dbl., 55–60 petals, moderate fragrance; foliage leathery, glossy; vigorous, upright growth; [Pigmy Red X Demure]; Boerner; J&P

Sweet Caroline, *see* 'MICaroline'

Sweet Chariot™, *see* 'MORchari'

'Sweet Charity', HT, w, 1967; flowers waxy white, sometimes marked pink, medium, semi-dbl., moderate fragrance; foliage dark, glossy; vigorous, upright growth; Park, F.; Moulton-Jones

'Sweet Cherry', HT, mr, 1960; (Cherry Glow); flowers cherry-red and yellow, large, exhibition form; [Margaret X Karl Herbst]; Dickson, A.

Sweet Cory, Pol, lp

Sweet Cover, *see* POUlweet

Sweet Diana, HT, lp; Hortico, 1999

Sweet Dream, *see* 'FRYminicot'

'Sweet Dreams', HT, pb, 1977; (Kojo); bud long, pointed; flowers light pink, reverse deep pink, dbl., 45–50 petals, 6 in., exhibition form, slight, tea fragrance; foliage dark; vigorous, very upright growth; [Jrn. No. 8 seedling X Royal Highness]; Takatori, Yoshiho; Japan Rose Nursery

Sweet Dreams, *see* JACams

Sweet Ecstasy, *see* 'HINsweet'

'Sweet Fairy', Min, lp, 1946; flowers micro-mini, dbl., 57 petals, 1 in., cupped, moderate fragrance; foliage small, dark; vigorous, dwarf (6–8 in.) growth; [Tom Thumb X Unnamed seedling]; deVink; C-P

Sweet Frederique, S, lp; Interplant, 1997

Sweet Gesture, *see* 'MACcarlto'

Sweet Hannah, Min, lp; Moore, 1998

'Sweet Harmony', F, pb, 1962; flowers canary-yellow edged crimson-pink, borne in clusters, dbl., 36 petals, 3.5 in., moderate fragrance; foliage glossy, light green; vigorous growth; [Peace X Masquerade]; Gaujard; Gandy Roses, Ltd.

'Sweet Home', HT, dp, 1969; flowers deep pink, large, dbl., 35 petals, exhibition form, slight fragrance; foliage leathery; vigorous growth; [(Jolie Madame X Baccará) X (Baccará X Jolie Madame)]; Meilland, Mrs. Marie-Louise; URS

Sweet Home Alabama, *see* 'TALsweet'

Sweet Honesty, Min, w

Sweet India, HT, lp; Tejganga, 1996

Sweet Inspiration, *see* 'JACsim'

'Sweet Interlude', HT, lp, 1989; bud pointed; flowers white center with pink edging, large, borne singly, slow opening, dbl., 32 petals, exhibition form, intense fragrance; prickles moderate, brown; foliage medium green, disease-resistant; upright, bushy, medium growth; [Pristine X Olympiad]; Wambach, Alex A.

Sweet Juliet, *see* 'AUSleap'

'Sweet Keri', HT, lp, 1988; flowers light pale pink, large, borne singly, dbl., 44 petals, exhibition form, no fragrance; prickles triangular, red; foliage medium green, large; medium, upright growth; [Judith Marten X Folklore]; Wilson, George D.

Sweet Lady, *see* 'MEIzeli'

'Sweet Lavender', HMult, pb, 1912; flowers blush, edged mauve, borne in large clusters, single; Paul

'Sweet Lavinia', HT, my, 1967; exhibition form, intense fragrance; foliage glossy; [Golden Scepter X Peace]; McTeer, F.

'Sweet Lelanie', HT, lp, 1967; flowers light pink, base yellow, large, dbl., globular, moderate fragrance; foliage wrinkled, bronze; vigorous, upright growth; [Charlotte Armstrong X Seedling.]; Duehrsen; Elmer Roses Co.

'Sweet Love', HT, mp, 1980; flowers medium-large blooms borne 1–7 per cluster, very dbl., 90 petals, flat, no fragrance; red prickles; foliage matt, dark; upright growth; [Queen Elizabeth X Duke of Windsor]; Rijksstation Voor Sierplantenteelt

Sweet Magic, *see* 'DICmagic'

'Sweet Maid', HT, lp, 1950; bud pointed; flowers porcelain-pink, semi-dbl., intense fragrance; Moss

'Sweet Mandarin', Min, ob, 1978; bud pointed; flowers light orange, small, semi-dbl., 16 petals, flat, moderate fragrance; foliage small, soft; compact growth; [Sweet and Low X Gypsy Moth seedling]; Schwartz, Ernest W.; Bountiful Ridge Nursery

Sweet Marie, Min, ob

Sweet Marvel, Min, lp; deRuiter, 1992

Sweet Melody, F; Leenders; (Cavriglia)

Sweet Melody, *see* FISmelody

'Sweet Memorie', HT, mp, 1937; flowers pink, base yellow, reverse purplish, large, semi-dbl., cupped, intense fragrance; foliage leathery; vigorous growth; [Mrs C.W. Edwards X Rose Marie]; Hieatt

Sweet Memories, *see* WHAmemo

Sweet Memories, MinFl, lp; Whartons Roses, 1995

'Sweet Memory', F, lp, 1991; bud pointed; flowers urn-shaped, dbl., 30 petals, 4 in., exhibition form, moderate fragrance; foliage medium, medium green, matt; bushy, medium growth; [Hana-Gasumi X (Lady X X Paradise)]; Yasuda, Yuji, 1992

Sweet Mimi®, *see* 'HAUmi'

Sweet Mystery, *see* 'FLOmyst'

'Sweet 'n' Pink', HT, dp, 1976; bud ovoid, pointed; flowers deep pink, dbl., 48 petals, 4–4.5 in., exhibition form, intense fragrance; foliage dark; upright, branching growth; [(Prima Ballerina X Seedling) X Unnamed seedling]; Weeks; Weeks Wholesale Rose Growers

Sweet Nell, *see* 'COCavoter'

Sweet Pea, Pol, lp; Antique Rose Emporium, 1998

Sweet Perfume, F, or

Sweet Petite, *see* FRYxquisite

Sweet Pickins, *see* 'TINsweet'

'Sweet Prince', HT, pb, 1988; flowers deep pink, petals edged medium red, blooms borne singly, dbl., 20 petals, 3.5 in., intense fragrance; heavily prickled; foliage large, dark, matt; bushy growth; [Honey Favorite X Granada]; Stoddard, Louis

Sweet Promise, *see* 'MEIhelvet'

Sweet Promise, Climbing, *see* 'MEIhelvetsar'

'Sweet Raspberry', Min, dp, 1984; flowers deep purplish pink, urn-shaped, small, dbl., flat, slight fragrance; foliage small, medium green, matt; bushy growth; [Little Rascal X Cinderella]; Jolly, Nelson F.; Rosehill Farm

Sweet Repose, *see* **'The Optimist'**

Sweet Revelation, *see* HARzazz

Sweet Revenge, *see* 'TINREVENGE'

Sweet Rosamini, Min, lp

Sweet Salmon, F; Leenders; (Cavriglia)

Sweet Scent, *see* 'RENSWEET'

Sweet Sensation, F, lp

'SWEET SEVENTEEN', HT, lp, 1923; semi-dbl., moderate fragrance; foliage light, wrinkled; bushy, dwarf growth; [Frau Karl Druschki X Bardou Job]; Clark, A.; NRS Victoria

'SWEET SEVENTEEN', Gr, mp, 1960; flowers pink, small; vigorous growth; [Fred Howard X Cocorico]; Leenders, J.

'SWEET SHADDOW', HT, mr, 1963; flowers scarlet, edged darker, dbl., 22 petals, 4.5 in., moderate fragrance; foliage dark; vigorous growth; [Queen Elizabeth X Étoile de Hollande]; Barter

Sweet Shirley, *see* 'CLESHIR'

'SWEET SIXTEEN', HT, pb, 1943; bud long, pointed; flowers salmon-pink, base yellow, 16–20 petals, 4–5 in., intense fragrance; vigorous, upright, bushy growth; [Mrs Sam McGredy X Pres. Herbert Hoover]; Lammerts, Dr Walter; Armstrong Nursery

Sweet Sonata, *see* 'MEIOFFIC'

Sweet Song, *see* 'MEIHIVANO'

Sweet Success, Gr, lp; Pottschmidt, 1995

'SWEET SUE', Min, lp, 1979; bud ovoid; flowers very small blooms in clusters of 10–25, semi-dbl., 13 petals, exhibition form; foliage small, dark; bushy growth; [Pink Ribbon X Pink Ribbon]; Bennett, Cecilia 'Dee'; Tiny Petals Nursery

'SWEET SUE', HT, op, 1940; bud long, pointed, flame to blood-red; flowers coral-pink, stamens maroon, single, cupped, moderate, spicy fragrance; vigorous growth; [Joanna Hill X Night]; Lammerts, Dr Walter; Armstrong Nursery

'SWEET SULTAN', Cl HT, dr, 1958; (Swet Sultain); flowers crimson shaded maroon, blooms in trusses, single, 5 petals, 4 in., recurrent bloom, intense fragrance; vigorous, pillar growth; [Independence X Honour Bright]; Eacott; LeGrice Roses

Sweet Sunblaze®, *see* 'MEITONJE'

Sweet Sunsation, *see* KORfisro

Sweet Sunshine, *see* 'MORSUN'

'SWEET SURRENDER', HT, mp, 1983; flowers medium silvery pink, large, dbl., 40 petals, cupped, slight, tea fragrance; foliage dark, leathery; AARS, 1983; [Unnamed seedling X Tiffany]; Weeks, O.L.; Weeks Wholesale Rose Growers

Sweet Symphony, *see* 'MEIBARKE'

'SWEET TALK', F, ly, 1964; flowers lemon to white, blooms in clusters, dbl., moderate fragrance; foliage light green, leathery; low, uniform, bushy growth; [Frolic X Lavender Pinocchio]; Swim & Weeks; Weeks Wholesale Rose Growers

'SWEET THOUGHTS', HT, lp, 1965; 5 in., slight fragrance; vigorous growth; [Kordes' Perfecta X Show Girl]; Cant, F.

'SWEET TWILIGHT', Cl HT, ly, 1954; flowers buff-yellow, center darker, dbl., 25 petals, 4–5.5 in., exhibition form, abundant, recurrent bloom, moderate, fruity fragrance; vigorous (15–30 ft) growth; [Tawny Gold sport]; Motose

'SWEET VELVET', F, mr, 1969; flowers large, dbl., moderate fragrance; foliage small, leathery, glossy; moderate growth; [Leverkusen X S'Agaro]; Martin, J.; Gandy Roses, Ltd.

Sweet Vibes, *see* 'GELSWEET'

'SWEET VIVID', Min, pb; flowers pink, yellow center, well-formed, dbl., moderate fragrance; vigorous growth

'SWEET VIVIEN', F, pb, 1961; bud ovoid; flowers pink, center light yellow, blooms in clusters, semi-dbl., 17 petals, 3 in., slight fragrance; large, pear-shaped fruit; foliage small, dark, glossy; short stems; very compact, bushy growth; [Little Darling X Odorata (HT)]; Raffel; Port Stockton Nursery

Sweet Wonder, MinFl, lp

Sweetbriar, *see* **R. EGLANTERIA**

Sweetcorn, Min, lp; Hannemann, F., 1991; The Rose Paradise; (Weatherly, L.)

Sweetheart, *see* 'COCAPEER'

'SWEETHEART', HWich, w, 1901; bud rose-pink; very dbl., 2.5 in., moderate fragrance; foliage glossy, dark; vigorous, climbing growth; RULED EXTINCT 4/80; [R. wichurana X Bridesmaid]; Walsh

Sweetheart Rose, *see* **'MLLE CÉCILE BRÜNNER'**

Sweetheart Rose, Climbing, *see* **'MLLE CÉCILE BRÜNNER, CLIMBING'**

'SWEETIE PIE', HT, lp, 1971; flowers light pink, outer edges darker; [Swarthmore sport]; Hyde

'SWEETNESS', HT, mp, 1919; flowers intense rose-pink, shaded scarlet, well formed, moderate fragrance; McGredy

'SWEETNESS', HT, ab, 1937; bud long, pointed; flowers apricot-yellow shaded pink, spiral, reflexed petals, large, dbl., moderate fragrance; foliage glossy; very vigorous growth; GM, NRS, 1935; Dickson, A.

Sweetnesse, *see* TANsemil

Sweetwaters, *see* 'MACSWEETWA'

Swet Sultain, *see* **'SWEET SULTAN'**

Swift, S, w, 1992; dbl.; groundcover; low (1–1.5 ft) growth; Poulsen

Swing®, *see* 'INTERSIREE'

'SWINGER', Min, my, 1984; flowers, dbl., 35 petals, exhibition form, borne usually singly, slight fragrance; foliage medium, medium green, semi-glossy; spreading growth; PP006560; [Anita Charles X Orange Honey]; Jolly, Nelson F.; Rosehill Farm

'SWINGTIME', F, rb, 1961; bud pointed; flowers crimson, reverse buff-white, medium, dbl., 40 petals, cupped, moderate, sweetbriar fragrance; vigorous, upright growth; [Little Darling X (Red Pinocchio X Masquerade)]; Morgan; Carlton Rose Nurseries

'SWISHREV', HT, dr, 1998; (**Revelry**™); flowers dark red to purple, full, very dbl., 26–40 petals, 4 in., borne moslty single, moderate fragrance; foliage medium, dark green, semi-glossy; upright, tall growth; [Pristine X National Velvet]; Swisher, Stan; Certified Roses, Inc., 1998

'SWISS BLISS', Min, my, 1988; flowers aging lighter, decorative, small, borne singly or in, dbl., 45–50 petals, slight fragrance; prickles reddish-brown; foliage small, medium green, matt, serrated; upright, bushy, medium growth; [Anytime X Unnamed seedling]; Lemrow, Dr Maynard W.

Swiss Fire®, *see* 'HUBAR'

Swiss Gold, *see* **'SCHWEIZER GOLD'**

Swiss Lass, *see* 'LEMSWI'

Swonderful, *see* 'SHAYELA'

'SWORD OF HOPE', HT, mr, 1964; bud long, pointed; dbl., 45 petals, 5–6 in., exhibition form, moderate fragrance; foliage glossy; vigorous, upright growth; Von Abrams; Peterson & Dering

'SYBIL', HT, mp, 1921; bud long, pointed; flowers silvery salmon-rose, center orange-salmon, very dbl., moderate fragrance; [Sunburst X Mary, Countess of Ilchester]; Bees

Sybil Hipkin, HT, lp; Dawson, before 1979

'SYBIL THORNDIKE', HT, mr, 1975; flowers full, dbl., 25 petals, 5 in., slight fragrance; foliage dark; [Liebeszauber X Seedling]; Kordes; Harry Wheatcroft Gardening

'SYDNEY', HCh, mp, 1977; bud ovoid; flowers phlox-pink, dbl., 20 petals, 3–3.5 in., cupped, intense fragrance; foliage rugose, yellow-green; bushy, spreading growth; [Old Blush X Frau Dagmar Hartopp]; Svedja, Felicitas; Canada Dept. of Agric.

'SYDNEY', HP, lp; [Prince Camille de Rohan sport]; Moore, A.K., 1908; (Weatherly, L.)

'SYDONIE', HP, mp, 1846; (Sidonie); flowers rose, medium, very dbl., quartered; red prickles; vigorous growth; Dorisy

Sylphide®, F, dp

Sylt, *see* 'KORYLT'

Sylvaine, HT, lp

'SYLVAN SUNSET', HT, pb, 1979; bud pointed; dbl., 28 petals, exhibition form, slight fragrance; hooked prickles; foliage medium green, semi-glossy; vigorous, upright, tall, bushy growth; [Swarthmore X First Prize]; Taylor, Thomas E.

Sylvana, HT, lp; Huber, 1997

'SYLVANDER', S, my; flowers clear yellow, large, single; dwarf (2–3 ft) growth; [R. X harisonii open-pollinated seedling]; Central Exp. Farm

Sylvia, *see* 'KORLIFT'

'SYLVIA', HWich, ly, 1911; flowers pale lemon-yellow, passing to white, dbl., moderate fragrance; moderate growth; Paul, W.

'SYLVIA', HT, my, 1949; flowers buttercup-yellow, dbl., 25 petals, moderate, fruity fragrance; RULED EXTINCT 1/79; [Golden Gleam sport]; Foster

'SYLVIA DOT', F, lp, 1965; bud pointed; flowers salmon, medium, dbl., 30 petals, exhibition form, slight fragrance; foliage glossy, bronze; upright, dense growth; [Queen Elizabeth X Orient]; Dot, Simon; Minier

'SYLVIA GROEN', HT, pb, 1935; flowers coral-rose, shaded crimson; vigorous, upright growth; [Pres. Herbert Hoover sport]; Groen

'SYLVIA LOUISE', HT, op, 1973; flowers pink and orange blend; foliage glossy; [Ena Harkness, Climbing seedling]; Shaw, H.C.W.

Sylvie Briant, HT, op

Sylvie Leblanc, F, w

Sylvie Vartan®, Pol, lp; Eve, 1969

Sylvor, HT, 1971; Croix; (Cavriglia)

Symbol®, F, mr

'SYMBOL MIRU', HWich, w, 1937; flowers medium, dbl.; Böhm, J.; (Sangerhausen)

'SYMBOLE', S, mr, 1945; bud thick, pointed; flowers carmine, well shaped, very large, slight fragrance; long, strong stems; vigorous growth; [Mev. G.A. van Rossem X Roseraie de l'Hay]; Robichon; Vilmorin-Andrieux

'SYMPATHIE'®, HKor, mr, 1964; flowers velvety dark red, well-formed, intense fragrance; foliage glossy, dark; very vigorous (9–12 ft) growth; ADR, 1966; Kordes, R.

'SYMPHONETTE', Min, pb, 1973; bud ovoid; flowers light pink, reverse deep pink, small, dbl., moderate fragrance; foliage small, glossy, leathery; vigorous, dwarf, bushy growth; [Cécile Brunner X Cinderella]; Morey, Dr Dennison; Pixie Treasures Min. Roses

'SYMPHONIE', HT, pb, 1951; bud pointed, ovoid to globular; flowers shades of pink, broad petals, veined carmine-pink, dbl., 25 petals, 4.5–5 in., exhibition form; foliage glossy, leathery; vigorous, upright, bushy growth; GM, NRS, 1949; [Peace X (Signora X Mrs John Laing)]; Meilland, F.; C-P

Symphonie Lumiere, *see* MEItomkin

'SYMPHONIE, CLIMBING', Cl HT, pb; Elmer's Nursery, about 1952

Symphony, *see* 'AUSLETT'

'SYMPHONY', HP, lp, 1935; bud long, pointed; flowers large, dbl., exhibition form; foliage leathery, dark; [Frau Karl Druschki X Souv. de Claudius Pernet]; Weigand, C.; P.J. Howard

'SYR', S, dr, 1977; bud pointed, ovoid; dbl., 23 petals, 4 in., exhibition form, slight fragrance; foliage glossy, dark; very vigorous, upright growth; [Stadt Rosenheim X Sangerhausen]; Lundstad; Agricultural University of Norway

'SYRACUSE', HT, mr, 1930; bud crimson; flowers scarlet-crimson, large, dbl., slight fragrance; foliage dark, leathery; vigorous growth; [Mme G. Forest-Colcombet X Aspirant Maumejean]; Mallerin, C.; C-P

Syrikit, HT, 1969; Mondial Roses; (Cavriglia)

'SYRINGA', HWich, w, 1931; flowers small, single; Browning; (Sangerhausen)

T

T.B. McQuesten, F, lp

'**T.F. Crozier**', HT, dy, 1918; flowers deep canary-yellow, fading, dbl.; Dickson, H.

'**T.V. Times**', HT, dr, 1970; flowers crimson, ovate, dbl., 35 petals, 5.5 in., moderate fragrance; foliage very large, dull; very free growth; [Gallant X (Brilliant X Seedling)]; Dickson, A.

'**Ta Nase Pisnicka Ceska**', Pol, mp, 1938; flowers small, dbl.; Böhm, J.; (Sangerhausen)

Ta Ta, *see* 'MINATCO'

'**Tabarin**', F, mp, 1956; bud oval; flowers salmon-pink flushed yellow and copper-orange to red, semi-dbl., 3–4 in., slight fragrance; foliage light green; vigorous, bushy growth; [Opera X Masquerade]; Gaujard

Table Mountain, *see* KORnieoch

'**Tablers' Choice**', F, rb, 1974; flowers deep red, yellow reverse, dbl., 38 petals, 4 in., slight fragrance; foliage glossy; [(Ann Elizabeth X Allgold) X (Tropicana X Piccadilly)]; Harkness

Taboo™, *see* 'TANELORAK'

Tabris, *see* 'KORTABRIS'

'**Tabriz**', HT, lp, 1986; flowers soft pink, large, slight fragrance; few prickles; [Belle Epoque X Unnamed seedling]; Kriloff, Michel

Tache de Beauté®, *see* 'LENPIC'

'**Tackholmii**', S, mp; (R. balsamea, R. hibernica, R. X tackholmii); flowers pink, solitary or few, early flowering; vigorous, erect, shrub growth; (56)

'**Tacoma**', HT, ab, 1921; bud long, pointed; flowers flaming apricot, dbl., moderate fragrance; [Mme Butterfly X Honeymoon]; Chervenka

'**Taconis**'®, F, or, 1968; flowers orange-scarlet, in trusses, slight fragrance; foliage glossy; vigorous growth; GM, Baden-Baden, 1968; [Ruth Leuwerik X City of Nottingham]; deRuiter

'**Taffeta**', HT, pb, 1947; bud urn-shaped; flowers pink-yellow blend, dbl., 20 petals, 3–3.5 in., moderate fragrance; foliage leathery, glossy, bronze-green; vigorous, upright growth; AARS, 1948; [Mrs Sam McGredy X Pres. Herbert Hoover]; Lammerts, Dr Walter; Armstrong Nursery

Taffeta, *see* CHEwecrest

'**Taffeta, Climbing**', Cl HT, pb, 1954; Armstrong, J.A.; Weeks Roses

Taffy, *see* 'BAKTAF'

Tag-a-long, *see* 'MORCOAT'

'**Tagore**', F, my, 1955; flowers yellow-ochre, reverse shaded orange, borne in clusters; compact growth; [Pinocchio seedling]; Maarse, G.

'**Tahiti**', HT, yb, 1947; flowers amber-yellow suffused carmine, 6 in., intense fragrance; foliage glossy, dark; [Peace X Signora]; Meilland, F.

'**Tahore**', HT, mp, 1958; flowers rose, large, dbl., 45–50 petals, exhibition form, moderate fragrance; foliage glossy; strong stems; vigorous, upright growth; [Peace X Seedling]; Frères, Buyl

Taifun, *see* '**Typhoon**'

Taiga®, *see* 'TANTIGA'

'**Tai-Gong**', F, or, 1974; (Apachi); bud long, pointed; flowers orange to red, dbl., 28 petals, 3.5 in., exhibition form, slight fragrance; foliage glossy, dark; vigorous, upright growth; [Klaus Stortebeker X Seedling]; Kordes

Taihape Sunset, *see* KORlinde

Taischa, HT, m; Robinson, 1993

'**Taj Mahal**', HT, dp, 1972; flowers deep pink, large, dbl., exhibition form, moderate fragrance; foliage large, leathery; vigorous, upright, bushy growth; [Manitou X Grand Slam]; Armstrong, D.L.; Armstrong Nursery

'**Takao**', HT, yb, 1975; flowers deep yellow, aging scarlet, blooms borne usually singly, dbl., 33 petals, exhibition form, moderate, damask fragrance; deep brown prickles; foliage medium, medium green; medium, bushy growth; [(Masquerade X Lydia) X (Montezuma X Miss Ireland)]; Okamoto, K.; K. Hirakata Nursery

Takapuna®, *see* 'MACTENNI'

'**Takatori Prominent**', F, op, 1981; flowers orange-salmon; [Prominent sport]; Takatori, Yoshiho; Japan Rose Nursery

'TAKSUN', F, rb, 1983; (**Margaret Thatcher**); flowers striped red and white; [Bridal Pink sport]; Takatori, Yoshiho & Sunao; Japan Rose Nursery

'TALALCONBURY', F, dp, 1996; (**Alconbury**); flowers deep pink with touch of violet, blooms borne single and in s, very dbl., 1.5–2.75 in., exhibition form, slight fragrance; few prickles; foliage medium, dark green, semi-glossy; tall (48 in.), upright, bushy growth; Taylor, Franklin "Pete" & Kay; Taylor's Roses, 1997

'TALALF', Min, op, 1991; (**Alfie Luv**); flowers orange-pink with lavender hue, reverse same, yellow base, dbl., 15–25 petals, 1.5–2.75 in., borne mostly singly, slight fragrance; some prickles; foliage medium, medium green, semi-glossy; upright, medium (60 cms) growth; [Azure Sea X Seedling]; Taylor, Franklin "Pete" & Kay; Taylor's Roses, 1992

'TALALI', Min, pb, 1991; (**Alice Lee**); bud pointed; flowers pink with color getting lighter towards edge of petals, giving a lavende, dbl., 15–18 petals, exhibition form, slight fragrance; foliage medium, medium green, semi-glossy; upright, bushy, medium growth; [Azure Sea X Seedling]; Taylor, Franklin "Pete" & Kay, 1986; Taylor's Roses, 1990

'TALBABY', Min, rb, 1995; (**Baby Cakes**); flowers white, edged red, opening to red and white appearance, full (26–40 petals), small borne mostly single; no fragrance; few prickles; foliage medium, medium green, semi-glossy; upright, bushy, medium; [Party Girl X Admiral Rodney]; Taylor, Pete & Kay; Taylor's Roses, 1996

'TALBER', Min, mr, 1993; (**Berry Berry Red**); flowers medium red, moderately full, medium sized, dbl., 15–25 petals, 1.5–2.75 in., borne mostly singly, slight fragrance; few prickles; foliage medium, medium green, semi-glossy; medium (24 in.), upright, bushy growth; [Party Girl X Papa Meilland]; Taylor, Franklin "Pete" & Kay; Taylor's Roses, 1993, USA

'TALBLO', Min, pb, 1993; (**Bloomtown**); flowers pink petals with slightly darker edges, reverse cream, dbl., 15–25 petals, 1.5–2.75 in., small clusters, no fragrance; some prickles; foliage medium, medium green, semi-glossy; medium (24 in.), upright, spreading growth; [Party Girl X Andrea]; Taylor, Franklin "Pete" & Kay; Taylor's Roses, 1993; Taylors Roses, 1993; USA

'TALBRO', Min, ab, 1995; (**Brown Sugar**); flowers apricot, moderately full (15–25 petals), medium (4–7 cms) blooms borne mostly single; slight fragrance; some prickles; foliage medium, dark green, glossy; upright, bushy, medium (30 in.) growth; [Party Girl X Julia's Rose]; Taylor, Pete & Kay; Taylor's Roses, 1996

'TALCAJ', Min, rb, 1993; (**Cajun Dancer**); flowers red with white eye, light reverse, semi-dbl., 6–14 petals, 1.5–2.75 in., borne singly and in large clusters, slight fragrance; some prickles; foliage medium, dark green, semi-glossy; tall (36 in.), upright, bushy growth; [Party Girl X Unknown]; Taylor, Franklin "Pete" & Kay; Taylors Roses; 1993; Taylor's Roses, 1993, USA

'TALCAN', Min, lp, 1999; (**Candy Corn**); flowers orange-yellow, single, 5–11 petals, 1.2 in., borne mostly singly, moderate fragrance; few prickles; foliage medium, medium green, semi-glossy; upright, bushy, medium (18 in.) growth; Taylor,

Franklin "Pete" & Kay; Taylor's Roses, 1998

'TALCHELSEA', Min, mr, 1991; (**Chelsea Belle**); bud ovoid; flowers glowers medium red with white base, whitish reverse, ages lighter, dbl., 28–30 petals, exhibition form, moderate fragrance; foliage medium, medium green, semi-glossy; upright, bushy, medium growth; [Azure Sea X Party Girl]; Taylor, Franklin "Pete" & Kay, 1987; Taylor's Roses, 1990

'TALCHRIS', Min, rb, 1996; (**Father Christmas**); flowers red with white base and white reverse, blooms borne in small clusters, dbl., 26–40 petals, 1.5 in., slight fragrance; some prickles; foliage small, dark green, semi-glossy; low (18 in.), upright, bushy growth; [Party Girl X Seedling]; Taylor, Franklin "Pete" & Kay; Taylor's Roses, 1997

'TALCLOUSSEAU', HT, mr, 1995; (**Clousseau**); flowers medium red, cream reverse full (26–40 petals), large (7+ cms) blooms borne mostly single, slight fragrance; some prickles; foliage medium, dark green, semi-glossy; upright, bushy, medium (48 in.) growth; [Azure Sea X Marijke Koopman]; Taylor, Pete & Kay; Taylor's Roses, 1996

'TALDAL', Min, dy, 1991; (**Summerdale**); bud pointed; flowers deep yellow, aging lighter, loose, medium, borne usually singly, very dbl., 30–35 petals, cupped, no fragrance; foliage small, medium green, matt; bushy, low growth; [Rise 'n' Shine X Unnamed seedling]; Taylor, Franklin "Pete" & Kay, 1986; Taylor's Roses, 1990

'TALDAZ', Min, m, 1991; (**Lazy Daze**); flowers light lavender, reverse slightly darker, white base, blooms, very dbl., 1.5–2.75 in., blooms sometimes quarters, slight fragrance; some prickles; foliage small, medium green, semi-glossy; low (30 cms), compact growth; [Azure Sea X Unnamed seedling]; Taylor, Franklin "Pete" & Kay; Taylor's Roses, 1992

'TALDEARPRU', F, pb, 1995; (**Dear Prudence**); flowers light pink blushing with darker pink edges, moderately full (15–25 petals), medium (4–7 cms) blooms borne in large clusters; slight fragrance; some prickles; foliage medium, medium green, semi-glossy; upright, bushy, medium; [Seedling X Seedling]; Taylor, Pete & Kay; Taylor's Roses, 1996

'TALDON', F, m, 1991; (**Donna Jean**); flowers mauve with white eye, blooms borne singly and in small clusters, semi-dbl., 6–14 petals, no fragrance; foliage medium, dark green, semi-glossy; medium, upright, bushy growth; [Azure Sea X Party Girl]; Taylor, Franklin "Pete" & Kay; Taylor's Roses

'TALDOR', Min, pb, 1991; (**Dora Delle**); flowers light pink with lavender hue, lighter in center, reverse cream, dbl., exhibition form, moderate fragrance; foliage medium, medium green, semi-glossy; medium, upright growth; [Azure Sea X Jean Kenneally]; Taylor, Franklin "Pete" & Kay; Taylor's Roses, 1991

'TALDRE', Min, ob, 1992; (**Dreamsicle**); flowers creamy white edged with orangish pink edges, reverse same, dbl., 15–25 petals, 1.5–2.75 in., slight fragrance; some prickles; foliage small, medium green, semi-glossy; medium (36 cms), compact growth; [Poker Chip X Party Girl]; Taylor, Franklin "Pete" & Kay; Taylor's Roses, 1993

'TALEND', Min, ab, 1993; (**Endeavour**); flowers apricot with yellow base, reverse slightly darker, color fading, dbl., 26–40 petals, 1.5–2.75 in., slight fragrance; few prickles; foliage medium, medium green, semi-glossy; medium, upright growth; [Party Girl X Azure Sea]; Taylor, Franklin "Pete" & Kay; Taylor's Roses, 1993

'TALFAIRHOPE', Min, ly, 1989; (**Fairhope**); bud pointed; flowers soft, light pastel yellow, reverse same, aging same, color holds well, dbl., 16–28 petals, exhibition form, slight fragrance; round, small, green fruit; prickles straight, medium, red; foliage medium, medium green, semi-glossy; upright, bushy, medium growth; [Azure Sea X Unnamed seedling]; Taylor, Franklin "Pete" & Kay; Taylor's Roses

'TALFAT', Min, m, 1991; (**Fat Tuesday**); flowers lavender with darker edges, blending to lighter center, reverse lighter, dbl., 26–40 petals, 1.5–2.75 in., no fragrance; some prickles; foliage medium, medium green, semi-glossy; tall (76 cms), upright growth; [Azure Sea X Lavender Jewel]; Taylor, Franklin "Pete" & Kay; Taylor's Roses, 1992

'TALFLORA', Min, rb, 1996; (**Flora Bama**); flowers medium red blending to cream center, cream reverse, blooms b, dbl., 26–40 petals, 1.5–2.75 in.; few prickles; foliage medium, dark green, semi-glossy; medium (24 in.), upright, spreading growth; [Party Girl X Poker Chip]; Taylor, Franklin "Pete" & Kay; Taylor's Roses, 1997

'TALGOO', Min, dy, 1992; (**Good Day Sunshine**); flowers bright yellow, reverse same, blooms borne mostly singly, very dbl., 1.5–2.75 in., exhibition form, slight fragrance; few prickles; foliage medium, medium green, semi-glossy; medium (60 cms), upright, bushy, spreading growth; [Party Girl X Elina]; Taylor, Franklin "Pete" & Kay; Taylor's Roses, 1993

'TALGUL', Min, pb, 1991; (**Gulf Breeze**); flowers creamy getting darker pink toward edges, reverse same, very dbl., 1.5–2.75 in., slight fragrance; some prickles; foliage medium, medium green, semi-glossy; upright, medium (35 cms), bushy growth; [Baby Katie X Poker Chip]; Taylor, Franklin "Pete" & Kay; Taylor's Roses, 1992

'TALHEATHER', Min, mp, 1989; (**Heather Leigh**); bud pointed; flowers medium pink, reverse slightly darker, aging lighter, holds color well, dbl., 35–40 petals, exhibition form, borne singly, no fragrance; no fruit; prickles straight, small, reddish-brown; foliage medium, medium green, semi-glossy; upright, bushy, medium growth; [Azure Sea X Unnamed Miniature seedling]; Taylor, Franklin "Pete" & Kay; Taylor's Roses

'TALHON', Min, rb, 1996; (**Honky Tonk**); flowers red with darker edges, white reverse, blooms borne mostly singly, dbl., 15–25 petals, 1.5–2.75 in., slight fragrance; many prickles; foliage medium, dark green, glossy; medium (30 in.), upright, bushy growth; [Party Girl X Seedling]; Taylor, Franklin "Pete" & Kay; Taylor's Roses, 1997

Talia, S, 1990; Barni, V.; (Cavriglia)

'TALINA', Min, w, 1994; (**Ina**); flowers white with pinkish edge, moderately good form, blooms borne, dbl., 15–25 petals, 1.5–2.75 in., slight fragrance; few prickles; foliage medium, medium green, semi-glossy; medium (30 in.), upright, bushy growth; [Party Girl X Fairhope]; Taylor, Franklin "Pete" & Kay; Taylor's Roses, 1993

'TALISMAN', HT, yb, 1929; bud pointed; flowers golden yellow and copper, medium, dbl., 25 petals, flat, moderate fragrance; foliage light green, leathery, glossy; vigorous growth; (28); Gertrude M. Hubbard, ARS, 1929 John Cook Medal, ARS, 1932; [Ophelia X Souv. de Claudius Pernet]; Montgomery Co.

'TALISMAN No. 5', HT, yb; flowers deeper color, free bloom; good growth

'TALISMAN, CLIMBING', Cl HT, yb, 1930; Western Rose Co.

'TALISMAN, CLIMBING', Cl HT, yb, 1932; Dixie Rose Nursery

'TALJAC', Min, dr, 1995; (**Jackson Square**); flowers dark red, full (26–40 petals), medium (4–7 cms) blooms borne mostly single; slight fragrance; some prickles; foliage medium, medium green, semi-glossy; upright, bushy, medium (30 in.) growth; [Jean Kenneally X Black Jade]; Taylor, Pete & Kay; Taylor's Roses, 1996

'TALJUB', Min, ob, 1991; (**Jubilee Sunset**); flowers bright vivid orange, yellow eye, reverse creamy yellow, dbl., 15–25 petals, 1.5–2.75 in., slight fragrance; some prickles; foliage medium, medium green, semi-glossy; low (40 cms), upright, bushy growth; [Baby Katie X Poker Chip]; Taylor, Franklin "Pete" & Kay; Taylor's Roses, 1992

'TALJUM', Min, yb, 1992; (**Jumping Jack Flash**); flowers yellow edged with deep pink to red, cream reverse, tipped with deep pink, dbl., 15–25 petals, 1.5–2.75 in., exhibition form, slight fragrance; few prickles; foliage medium, medium green, semi-glossy; medium (42 cms), upright, bushy growth; [Party Girl X Poker Chip]; Taylor, Franklin "Pete" & Kay; Taylor's Roses, 1993

'TALKAT', Min, mp, 1993; (**Katy Lampkin**); flowers medium pink, cream base, reverse same, dbl., 15–25 petals, 1.5 in., borne singly and in small clusters, slight fragrance; few prickles; foliage small, medium green, semi-glossy; low (18 in.), compact growth; [Winsome X Admiral Rodney]; Taylor, Franklin "Pete" & Kay; Taylor's Roses, 1993

'TALKEV', Min, dy, 1991; (K.T., **Kev**); dbl., borne mostly singly, no fragrance; foliage medium, medium green, semi-glossy; medium, upright growth; [Azure Sea X Party Girl]; Taylor, Franklin "Pete" & Kay; Taylor's Roses

Tall Poppy, *see* 'MORTEN'

Tall Story®, *see* 'DICKOOKY'

'TALLOU', Min, yb, 1997; (**Louisiana Lady**™); flowers full, medium, very dbl., 26–40 petals, borne mostly singly, slight fragrance; foliage medium, medium green, semi-glossy; upright, bushy, medium (24 in.) growth; [Seedling X Seedling]; Taylor, Franklin "Pete" & Kay

Tallulah, *see* 'BEELAH'

'TALLYHO', HT, dp, 1948; bud urn-shaped; flowers deep pink, dbl., 35 petals, 3.5–4 in., exhibition form, intense, spicy fragrance; foliage leathery; vigorous, upright, bushy growth; (28) ; AARS, David Fuerstenberg Prize, ARS; 1951; 1949; [Charlotte Armstrong X Unnamed seedling]; Swim, H.C.; Armstrong Nursery

'TALLYHO, CLIMBING', Cl HT, dp, 1952; Armstrong, J.A.; Armstrong Nursery

'TALMAG', Min, op, 1993; (**Mary Margaret**); flowers coral pink, creamy yellow base, reverse same, blooms borne singly, dbl., 26–40 petals, 1.5 in., slight fragrance; few prickles; foliage medium, medium green, semi-glossy; low (18 in.), spreading growth; [Party Girl X Watercolor]; Taylor, Franklin "Pete" & Kay; Taylor's Roses, 1993

'TALMAID', Min, pb, 1989; (**Maids of Jubilee**); bud pointed; flowers bright deep pink with cream base in center, reverse cream and pink blend, dbl., 25–30 petals, exhibition form, no fragrance; prickles yellowish-green; foliage medium, dark green, semi-glossy; upright, bushy, medium growth; [Azure Sea X Unnamed Miniature seedling]; Taylor, Franklin "Pete" & Kay; Taylor's Roses

'TALMAR', Min, w, 1991; (**Mary Edith**); flowers white, tipped with pink edges, flower turns darker pink, tints with age, dbl., moderate fragrance; foliage medium, medium green, semi-glossy; low, upright, bushy growth; [Azure Sea X Unnamed seedling]; Taylor, Franklin "Pete" & Kay; Taylor's Roses

'TALMID', Cl Min, rb, 1993; (**Midnight Rambler**); flowers red with white eye, blooms borne in small clusters, semi-dbl., 6–14 petals, 1.5–2.75 in., no fragrance; some prickles; foliage medium, medium green, semi-glossy; medium (30 in.), spreading growth; [Party Girl X Andrea]; Taylor, Franklin "Pete" & Kay; Taylor's Roses, 1993

'TALMOBILE', Min, op, 1994; (**Mobile Jubilee**); flowers light salmon blending to darker edges, yellow base, reverse, dbl., 26–40 petals, 1.5–2.75 in., slight fragrance; some prickles; foliage medium, medium green, semi-glossy; tall (90 cms), upright, bushy growth; [Party Girl X Gitte]; Taylor, Franklin "Pete" & Kay; Taylor's Roses, 1994

'TALMONTROSE', Min, mp, 1994; (**Montrose**); flowers bright yellow stamens, blooms borne single and in large clusters, dbl., 15–25 petals, 1.5–2.75 in., moderate fragrance; some prickles; foliage medium, dark green, semi-glossy; medium (60 cms), upright, bushy growth; [Party Girl X Andrea]; Taylor, Franklin "Pete" & Kay; Taylor's Roses, 1994

'TALPEN', Min, lp, 1992; (**Penny Lane**); flowers cream base blending to light pink, reverse same, blooms borne, dbl., 26–40 petals, 1.5 in., exhibition form, no fragrance; few prickles; foliage small, medium green, semi-glossy; medium (38 cms), upright, bushy growth; [Party Girl X Maids of Jubilee]; Taylor, Franklin "Pete" & Kay; Taylor's Roses, 1993

'TALPIN', Min, mp, 1993; (**Pink Bunting**); flowers pink, creamy yellow base blending to pink edges around petal, dbl., 15–25 petals, 1.5 in., no fragrance; few prickles; foliage small, dark green, semi-glossy; low (12 in.), spreading, compact growth; Taylor, Franklin "Pete" & Kay; Taylor's Roses, 1993

'TALPOI', Min, w, 1991; (**Point Clear**); flowers white, reverse same, light yellow base, blooms borne mostly, dbl., 26–40 petals, 1.5 in., slight fragrance; some prickles; foliage small, medium green, semi-glossy; low (40 cms), upright growth; [Azure Sea X Party Girl]; Taylor, Franklin "Pete" & Kay; Taylor's Roses, 1992

'TALRAS', F, dp, 1991; (**Raspberry Ruffles**); bud pointed; flowers deep pink with white eye, deep pink reverse, aging lighter, dbl., 18–20 petals, exhibition form, intense fragrance; foliage medium, medium green, semi-glossy; upright, bushy, medium growth; [Garnette X Unnamed seedling]; Taylor, Franklin "Pete" & Kay, 1987; Taylor's Roses, 1990

'TALRUM', Min, ab, 1991; (**Rum Butter**); bud pointed; flowers apricot with yellow base-bloom gives a lavender cast blended, dbl., 12–16 petals, exhibition form, no fragrance; foliage medium, medium green, semi-glossy; upright, bushy, medium growth; [Azure Sea X Unnamed seedling]; Taylor, Franklin "Pete" & Kay, 1987; Taylor's Roses, 1990

'TALSILVER', Min, pb, 1996; (**Silverhill**); flowers pink with mauve tint, blending to cream center and cream reverse, dbl., 26–40 petals, 1.5–2.75 in., exhibition form, slight fragrance; some prickles; foliage medium, medium green, semi-glossy; medium (36 in.), upright, bushy growth; [Azure Sea X Seedling]; Taylor, Franklin "Pete" & Kay; Taylor's Roses, 1997

'TALSIMPLE', Min, rb, 1999; (**Simple Pleasures**™); flowers red petals, white eye, yellow stamens, single, 5–11 petals, 1.75 in., borne mostly singly, no fragrance; few prickles; foliage medium, medium green, semi-glossy; upright, spreading, medium (18 in.) growth; Taylor, Franklin "Pete" & Kay; Taylor's Roses, 1998

'TALSPI', Min, ob, 1993; (**Cajun Spice**); flowers bright orange, white eye, light reverse, semi-dbl., 6–14 petals, 1.5–2.75 in., borne mostly singly, no fragrance; some prickles; foliage medium, medium green, semi-glossy; medium (24 in.), upright, bushy growth; [Party Girl X Baby Diana]; Taylor, Franklin "Pete" & Kay; Taylor's Roses, 1993; Taylor's Roses, 1993, USA

'TALSPRINGS', Min, lp, 1999; (**Magnolia Springs**™); flowers white with pink edge, greenish guard petals, very dbl., 41 petals, 2 in., borne in large clusters, no fragrance; few prickles; foliage medium, medium green, semi-glossy; upright, bushy, medium (18 in.) growth; Taylor, Franklin "Pete" & Kay; Taylor's Roses, 1998

'TALSUE', Min, ob, 1992; (**Sue Belle**); flowers orange with yellow eye, yellow stamens, aging to pink, semi-dbl., 6–14 petals, 1.5–2.75 in., slight fragrance; some prickles; foliage small, medium green, semi-glossy; medium (40 cms), upright, bushy growth; [Party Girl X Poker Chip]; Taylor, Franklin "Pete" & Kay; Taylor's Roses, 1993

'TALSUGAR', F, lp, 1999; (**Sugar Magnolia**™); flowers white, burgundy pistil and stamens, set against fleshtone petals, dbl., 17–25 petals, 2.75 in., borne in small clusters, no fragrance; few prickles; foliage medium, dark green, semi-glossy; upright, bushy (3 ft) growth; Taylor, Franklin "Pete" & Kay; Taylor's Roses, 1998

'TALSUM', Min, op, 1991; (**Summer Breeze**); flowers light orange-pink, ages lighter, yellow base, reverse slight, dbl., 26–40 petals, 1.5 in., slight fragrance; some prickles; foliage medium, medium green, semi-glossy; low (40 cms), spreading growth; [Baby Katie X Poker Chip]; Taylor, Franklin "Pete" & Kay; Taylor's Roses, 1992

'TALSUN', Min, yb, 1995; (**Sunny Afternoon**); flowers light yellow, outer edges pinkish/apricot tinge, blooms born, dbl., 15–25 petals, 1.5 in., slight fragrance; some prickles; foliage medium, medium green, semi-glossy; medium, upright, bushy growth; [Party Girl X Elina]; Taylor,

Franklin "Pete" & Kay; Taylor's Roses, 1995

'TALsweet', Min, pb, 1995; (**Sweet Home Alabama**); flowers magenta pink with white reverse, white eye, very dbl., 26–40 petals, 1.5–3 in., borne mostly singly, slight fragrance; some prickles; foliage medium, medium green, semi-glossy; upright, bushy, medium growth; [Party Girl X Azure Sea]; Taylor, Franklin "Pete" & Kay; Taylor's Roses, 1996

'TALtrick', Min, ob, 1996; (**Trick Or Treat**); flowers orange blending to yellow base, very dbl., 41 petals, 1.5 in., flat, borne mostly singly, slight fragrance; few prickles; foliage small, medium green, semi-glossy; medium (24 in.), upright, bushy growth; [Seedling X Seedling]; Taylor, Franklin "Pete" & Kay; Taylor's Roses, 1997

'TALuptown', Min, m, 1991; (**Uptown**); flowers lavender, edges darker violet, reverse lighter, yellow base, dbl., 26–40 petals, 1.5–2.75 in., exhibition form, no fragrance; some prickles; foliage medium, medium green, semi-glossy; medium (46 cms), upright, bushy growth; [Azure Sea X Party Girl]; Taylor, Franklin "Pete" & Kay; Taylor's Roses, 1992

'TALwil', Min, yb, 1992; (**Wildest Dreams**); flowers yellow blending to deep pink edges, aging lighter, reverse yellow, dbl., 15–25 petals, 1.5–2.75 in., no fragrance; few prickles; foliage small, medium green, semi-glossy; medium (40 cms), upright, bushy growth; [Poker Chip X Unknown]; Taylor, Franklin "Pete" & Kay; Taylor's Roses, 1993

'Tam o'Shanter', F, yb, 1969; flowers yellow and red, borne in clusters, dbl., slight fragrance; low, compact growth; [Orange Sensation X Circus]; Cocker

'Tama', HT, op, 1989; bud pointed; flowers salmon-pink, fringed with pink, reverse creamy yellow, urn-shaped, dbl., 35 petals, exhibition form, slight fragrance; prickles small, lower part hollow; foliage dark green; upright, tall growth; [Red Lion X Garden Party]; Sato, Kohi

Tamango®, *see* 'MEIdanu'

'Tamara', HT, mr, 1966; flowers geranium-red, dbl., exhibition form; foliage dark; very vigorous growth; [Parel van Aalsmeer X Independence]; Mondial Roses

Tamara, *see* KORmador

'Tamarisk', HT, mp, 1954; flowers clear rose-pink, camellia shaped, intense fragrance; foliage leathery, dull green; very vigorous growth; [Mme Butterfly X Signora]; Ratcliffe

'Tambourine', F, rb, 1958; flowers cherry-red, base yellow, reverse light burnt-orange, borne i, dbl., 25 petals, 3.5 in., slight fragrance; foliage dark, veined; vigorous, tall, bushy growth; [Independence seedling X Karl Herbst]; Dickson, A.

Tambourine, *see* HARdolly

Tambrabarani, *see* **Southern Sunset**

'Tammy', Pol, lp, 1972; flowers medium, dbl., exhibition form, moderate fragrance; foliage large, leathery; vigorous, upright, bushy growth; [Seventeen X Jack Frost]; Byrum

Tamora, *see* 'AUStamora'

Tamouree, F; Combe; (Cavriglia)

Tampa Bay™, *see* 'KINbay'

'Tampico', HT, or, 1976; bud ovoid, long, pointed; flowers coral-pink, large, dbl., slight fragrance; foliage large, leathery; vigorous, upright growth; [South Seas X Hawaii]; Warriner, William A.; J&P

'Tamrabarani', HT, ob, 1997; (**Southern Sunset**); flowers full, large, very dbl., 26–40 petals, borne mostly singly, slight fragrance; many prickles; foliage large, dark green, glossy; upright, compact, medium (2.5 ft) growth; Viraraghavan, M.S.

'Tam-Tam', F, my, 1961; bud ovoid; flowers yellow flushed pink, dbl., 35–40 petals, 3 in., cupped, moderate fragrance; foliage glossy; vigorous, upright growth; RULED EXTINCT 1/88; [Goldilocks seedling X Demure seedling]; Boerner; J&P

'Tam-Tam', Min, lp, 1988; flowers inner petals upright, borne in large clusters, dbl., 18–20 petals, flat, slight fragrance; elongated, orange fruit; foliage small, light green; [(Nozomi X Nozomi) X Bo-Peep]; Onodera, Toru F.

Tan Cho, HT, lp; Keisei, 1994

TANadac, F, mr, 1994; (**Escada**); Tantau

'TANagnarat', F, mr, 1982; (**Taranga**®); flowers medium, semi-dbl., no fragrance; foliage medium, medium green, semi-glossy; bushy growth; Tantau, Math., 1981

Tanagra®, *see* 'GActa'

'TANakinom', HT, pb, 1985; (Monica®, **Monika**®, 'TANaknom'); flowers medium, dbl., 35 petals, slight fragrance; foliage medium, dark, glossy; upright growth; Tantau, Math., 1986

TANakni, HT, op, 1978; (**Inka**®); bud pointed; flowers salmon, large, dbl., slight fragrance; foliage large, glossy; upright, bushy growth; Tantau, Math., 1981

TANakni, HT, dy, 1993; (**Inka**®); Tantau

'TANaknom', HT, pb, 1985; (Monica®, **Monika**®, 'TANakinom'); flowers medium, dbl., 35 petals, slight fragrance; foliage medium, dark, glossy; upright growth; Tantau, Math., 1986

'TANal', F, or, 1958; (**Allotria**®); flowers orange-scarlet, dbl., 3 in., flat, blooms in large clusters; foliage dark, glossy; vigorous growth; [Fanal X Cinnabar seedling]; Tantau, Math.

TANalam, HT, lp; (**Unforgettable**); Tantau, 1996

TANalednev, HT, lp; (**Velvet Star**); Tantau, 1997

TANallepal, HT, mr, 1994; (**Acapella**®); bud long, pointed; flowers cherry red, reverse silver, large, very dbl., intense fragrance; foliage dark green, leathery, glossy; upright, vigorous growth; Tantau, Math.; (Sangerhausen)

'TANallet', S, mr, 1985; (**Castella**®); flowers medium, semi-dbl., no fragrance; foliage large, dark, semi-glossy; upright growth; Tantau, Math., 1984

'TANaloap', HT, mr, 1982; (**Paola**); flowers large, dbl., 20 petals, slight fragrance; foliage large, dark, matt; Tantau, Math., 1981

TANaluma, HT, mr, 1932; (**'Amulett'**); bud ovoid; flowers fiery red, very dbl., exhibition form, moderate fragrance; foliage dark, glossy; vigorous, bushy growth; [Mrs Henry Winnett X Johanniszauber]; Tantau; C-P

'TANälzergo', HT, dy, 1982; (**Pfälzer Gold**®); flowers large, dbl., 20 petals, no fragrance; Tantau, Math., 1981

'TANamola', F, w, 1985; (La Paloma '85, **The Dove**); flowers medium, dbl., 20 petals, no fragrance; foliage medium, light green, glossy; upright growth; Tantau, Math.

TANanairan, F, mr, 1996; (**Austriana**); flowers blood red, medium-large, dbl.; Tantau, Math.; (Sangerhausen)

TANanilov, HT, pb, 1997; (**Violina**®); bud long, pointed, dark pink; dbl., borne mostly singly, intense fragrance; foliage large, medium green, glossy; upright, strong (3 ft) growth; Tantau

TANanim, HT, dp, 1994; (**Romina**); Tantau

TANaronam, HT, mr, 1992; (**Manora**®); Tantau

'TANatesil', F, mp, 1985; (**Ilseta**®, Ilsetta); flowers medium, dbl., 35 petals, no fragrance; foliage medium, medium green, matt; upright growth; PP005707; Tantau, Math., 1983

'TANatiram', F, yb, 1985; (**Marietta**®); flowers medium, dbl., 20 petals, no fragrance; foliage medium green, glossy; upright growth; Tantau, Math.

'TANattenin', F, dp, 1985; (**Ninetta**®); flowers deep pink, medium, dbl., 20 petals, slight fragrance; foliage medium green, semi-glossy; upright growth; Tantau, Math.

'TANattigam', HT, mr, 1982; (**Magitta**®); flowers large, dbl., 20 petals, slight fragrance; foliage medium, medium green, semi-glossy; upright growth; Tantau, Math., 1981

TANavilo, HT, mr, 1993; (**Olivia**®); Tantau

'TANavlim', F, m, 1983; (**Milva®**); flowers medium, dbl., 35 petals, slight fragrance; foliage medium, medium green, semi-glossy; upright growth; Tantau, Math.

'TANba', Min, rb, 1956; (Baby Carnaval, Baby Carnival, Baby Mascarade, Baby Maskarade, Baby Maskerade, **Baby Masquerade®**, 'TANbakede'); bud ovoid; flowers yellow aging red, dbl., 23 petals, 1 in., slight, fruity fragrance; foliage leathery; vigorous, compact (8 in. tall) growth; [Tom Thumb X Masquerade]; Tantau, Math.; J&P, 1955

'TANbakede', Min, rb, 1956; (Baby Carnaval, Baby Carnival, Baby Mascarade, Baby Maskarade, Baby Maskerade, **Baby Masquerade®**, 'TANba'); bud ovoid; flowers yellow aging red, dbl., 23 petals, 1 in., slight, fruity fragrance; foliage leathery; vigorous, compact (8 in. tall) growth; [Tom Thumb X Masquerade]; Tantau, Math.; J&P, 1955

'TANbeedee', F, ob, 1971; (**Belinda**); bud ovoid; flowers copper to orange, medium, dbl., moderate fragrance; vigorous, upright, bushy growth; [Seedling X Zorina]; Tantau, Math.

'TANblady', HT, dr, 1976; (**Black Lady®**); bud globose; flowers blackish-red, medium, dbl., intense fragrance; foliage matt; bushy growth; Tantau, Math.

'TANca', HT, rb, 1966; (**Caramba®**); bud pointed; flowers bright red, reverse white, well-formed, large, dbl., 45 petals; foliage glossy; upright growth; Tantau, Math.

'TANcary', HT, yb, 1976; (**Canary**); flowers yellow, petals marked orange, well-formed, medium, dbl., moderate fragrance; Tantau, Math.; Krussmann, 1972

TANcelia, F, w, 1988; (**Maleica**); Tantau

'TANcofeuma', F, or, 1982; (**Cosima®**); flowers medium, dbl., 20 petals, no fragrance; foliage large, dark, glossy; Tantau, Math., 1981

TANcreif, F, w, 1988; (**Elfe**); Tantau

'TANcressor', HT, mr, 1991; (**Celebrate America™**); dbl., 45–55 petals, 4–4.75 in., exhibition form, borne usually singly, slight fragrance; foliage large, medium green, semi-glossy; upright, spreading, medium growth; [Seedling X Seedling]; Rosen Tantau; Bear Creek Gardens, 1990

TANdalaum, F, mp; (**Traumland®**); Tantau, 1988

TANdelbel, F, m, 1994; (**Magenta Diadem**); Tantau

'TANdinadi', F, my, 1977; (**Diana®**); bud globular; flowers medium, dbl., slight fragrance; foliage medium, glossy; vigorous, upright growth; Tantau, Math.

TANdirpkrap, HT, mr; (**Parkstown's Pride**); Tantau

TANdomo, Min, mr, 1994; (**Domino**); Tantau

'TANduft', HT, dy, 1981; (Duftgold, **Fragrant Gold**, 'TANdugoft'); flowers large, semi-dbl., moderate fragrance; foliage medium, dark, glossy; upright growth; Tantau, Math., 1982

'TANdugoft', HT, dy, 1981; (Duftgold, 'TANduGOFT', **Fragrant Gold**); flowers large, semi-dbl., moderate fragrance; foliage medium, dark, glossy; upright growth; Tantau, Math., 1982

Tane, *see* 'SIMetna'

Tanecnice, F, lp; Strnad, 1976; (Czech Rosa Club)

TANedallab, F, lp, 1991; (**Ballade**); Tantau

TANegnaro, Min, or, 1994; (**Ceres**); Tantau

TANeiglat, HT, 1995; (**Nostalgie**); Tantau; (Sangerhausen)

TANeitber, F, dy, 1987; (**Bernstein-Rose®**); Tantau

TANeivom, HT, op, 1995; (**Movie Star**); Tantau

TANekily, HT, pb, 1989; (**Lady Like®**); Tantau

'TANekily',

'TANelaigib', F, pb, 1988; (**Abigaile®**); Tantau

TANelephar, HT, mr, 1994; (**Raphaela**); Tantau

'TANelfe', HT, lp, 1985; (**Elfe®**); flowers medium, dbl., 20 petals, slight fragrance; foliage medium, medium green, matt; upright growth; Tantau, Math., 1972

TANeliet, HT, dp, 1996; (**African Queen**); Tantau

TANellelog, F, lp; (**Goldquelle**); Tantau, 1988

'TANellis', HT, or, 1967; (Duftwolke, **Fragrant Cloud**, Nuage Parfume); bud ovoid; flowers coral-red becoming geranium-red, well-formed, dbl., 28–35 petals, 5 in., exhibition form, borne mostly singly, intense, damask-fruity-spicy-citrus-sharp fragrance; prickles moderate; foliage dark, glossy; vigorous, upright growth; GM, NRS, 1963 GM, Portland, 1963 Hall of Fame, WFRS, 1981 James Alexander Gamble Fragrance Medal, ARS, 1969 PIT, NRS, 1964; [Seedling X Prima Ballerina]; Tantau, Math., 1963; J&P, 1968

TANelliv, F, mr; (Glow Achievement, **Stadt Eltville®**); GM, Durbanville, 1989; Tantau, 1990

'TANelorak', HT, dr, 1993; (Barkarole, Grand Chateau, **Taboo™**); bud pointed, ovoid, almost black; flowers deepest red, velvety, dbl., 26–40 petals, 4–4.5 in., exhibition form, borne mostly singly, moderate, sweet citrus fragrance; some prickles; foliage large, dark green, semi-glossy; tall (170 cms), upright, bushy growth; PP007665; Evers, Hans & Rosen Tantau; Bear Creek Gardens, 1994

'TANema', HT, mr, 1974; (Melina, **Sir Harry Pilkington**); flowers well-formed, dbl., 30 petals, 4–5 in., slight fragrance; foliage dark; [Inge Horstmann X Sophia Loren]; Harry Wheatcroft Gardening

TANemrac, HT, or, 1986; (**Carmen**); Tantau

TANensch, Min, w, 1992; (**Schneeweisschen®**); Tantau

'TANerised', HT, lp, 1985; (**Desirée®**); flowers large, dbl., 35 petals, slight fragrance; foliage large, dark, semi-glossy; upright growth; Tantau, Math., 1986

'TANessenor', HT, or, 1989; (**Baronesse®**); Tantau

TANetee, HT, lp; (**Tea Time**); Tantau, 1994

TANetorde, F, dr, 1991; (**Red Diadem®**); Tantau

TANettola, HT, mr, 1997; (**Charlotte**); Tantau

TANeufis, S, mr, 1988; (**Johannisfeuer®**); flowers blood red, semi-dbl., borne in sprays; foliage large, dark green; bushy, upright, strong (4.5 ft) growth; Tantau

'TANezamor', S, mp, 1985; (Romance, **Romanze**); flowers medium, dbl., 20 petals, slight fragrance; foliage medium, dark, semi-glossy; bushy growth; ADR, 1986 GM, Baden-Baden, 1985; Tantau, Math.; Tantau Roses, 1984

'TANfifum', HT, m, 1978; (**Blue Parfum®**, Blue Perfume 'TANifume', 'TANtifum'); bud ovoid; flowers mauve-blush, large, dbl., intense fragrance; foliage glossy; bushy, upright growth; Tantau, Math.

'TANfudermos', HT, dr, 1985; (**Sommerduft®**, Summer Fragrance); flowers medium-large, dbl., 20 petals, moderate fragrance; foliage medium, dark, semi-glossy; upright growth; Tantau, Math., 1986

Tangeglow, *see* 'DICmadder'

'Tanger', HT, rb, 1949; bud ovoid; flowers crimson, reverse yellow, large, dbl., 60 petals, moderate fragrance; moderate, compact growth; [Condesa de Sástago X Peace]; Dot, Pedro

'Tangerine', HT, op, 1942; bud globular, coral-red; flowers light salmon-orange, dbl., 30–35 petals, 3.5–4 in., slight fragrance; foliage dark, leathery; strong stems; vigorous, very upright, well branched growth; [Capt. Glisson X R.M.S. Queen Mary]; Hill, Joseph H., Co.

'Tangerine Contempo', F, ob, 1983; flowers tangerine orange; [Contempo sport]; Gupta, Dr M.N. & Datta, Dr S.K.; National Botanical Research Institute

Tangerine Mist, *see* 'RENist'

Tangerine Twist, *see* 'MICtwist'

'Tangier', HT, rb, 1955; bud long, pointed; flowers peach-red to scarlet-copper, medium, dbl., 60 petals, exhibition form, slight fragrance; vigorous, bushy growth;

[Katharine Pechtold X R.M.S. Queen Mary]; Parmentier, J.

'TANGO', HT, mr, 1937; flowers scarlet, reverse bronze, base shaded old-gold; vigorous growth; [Seedling X Talisman]; H&S

'TANGO', HT, mr, 1955; bud long; flowers tango-red, open, large, dbl., moderate fragrance; foliage dark, glossy; vigorous, bushy growth; [The Doctor X Karl Herbst]; Delforge

Tango, *see* JACboy

Tango, *see* 'MACFIRWAL'

Tango Rose, *see* 'DELTANGA'

TANgolca, HT, dy, 1988; (**Golden Monica®**); Tantau

'TANGOSAR', Cl HT, or, 1971; (Super Star, Climbing, 'TANGOSTAR', **Tropicana, Climbing**); Boerner; J&P

'TANGOSTAR', Cl HT, or, 1971; (Super Star, Climbing, 'TANGOSAR' **Tropicana, Climbing**); Boerner; J&P

'TANIA VERSTAK', HT, mr, 1962; flowers rich red, large, dbl., slight fragrance; foliage light green, soft; vigorous, upright growth; [Charlotte Armstrong X Happiness]; Armbrust; Langbecker

'TANIBARA', F, ob, 1986; (**ARABIA®**); Tantau

'TANIFUME', HT, m, 1978; (**Blue Parfum®**, Blue Perfume, 'TANFIFUM', 'TANTIFUM'); bud ovoid; flowers mauve-blush, large, dbl., intense fragrance; foliage glossy; bushy, upright growth; Tantau, Math.

TANija, HT, ob, 1974; (**Janina®**); Tantau

'TANILASOR', F, mp, 1983; (**Rosali®**, Rosali '83®); flowers medium, dbl., 20 petals, no fragrance; foliage medium, medium green, glossy; bushy growth; Tantau, Math.

'TANILSEPO', F, w, 1985; (**Perl-Ilseta®**); flowers pearly white; [Ilseta sport]; Tantau, Math., 1984

'TANILVOBA', F, ob, 1985; (**Orange Ilseta®**); flowers orange; [Ilseta sport]; Tantau, Math.

TANilyks, HT, my, 1991; (**Skyline**); Tantau

TANinaso, S, w, 1993; (**Snow Cloud**); Tantau

TANinaso, MinFl, lp; (**Harewood**); Tantau

TANinat, S, mp, 1992; (**Satina®**); Tantau

TANipep, F, lp, 1987; (**Pepino®**); Tantau

TANireb, HT, mp, 1994; (**Belle of Berlin**); Tantau

TANiripsa, F, w, 1997; (**Aspirin**); flowers medium-large, dbl., moderate fragrance; Tantau, Math.; (Sangerhausen)

'TANITO', F, w, 1974; (**Akito**); bud ovoid; flowers, dbl., blooms in clusters, slight fragrance; foliage medium, medium green; upright, bushy growth; [Zorina X Nordia]; Tantau, Math.

'TANJEKA', HT, op, 1971; (Jehoca, Jehoka; Jessica, **Jessika®**); bud long, pointed; flowers peach-salmon, medium, moderate fragrance; vigorous, upright growth; [Colour Wonder X Piccadilly]; Tantau, Math.

'TANJENJU', HT, mr, 1989; (**Jennie June™**); bud ovoid; flowers medium red, reverse lighter with some bluing, large, dbl., 50 petals, exhibition form, borne usually singly, slight fragrance; prickles short, narrow, hooked down, red-green; foliage large, medium green, semi-glossy; upright, spreading, medium growth; [Unnamed seedling X Unnamed seedling]; Rosen Tantau; Bear Creek Gardens, 1990

'TANJUKA', F, mr, 1974; (**Julischka®**); bud long, pointed; flowers bright red, medium, semi-dbl., slight fragrance; foliage glossy, bronze; Gold Star of the South Pacific, Palmerston North, NZ, 1976; Tantau, Math.; Horstmann

TANkalgic, HT, dr, 1997; (**Black Magic**); Tantau

'TANKANUSCH', F, or, 1978; (**Anuschka®**); bud ovoid; flowers large, dbl., 23 petals, slight fragrance; foliage large; bushy, upright growth; Tantau, Math., 1977

TANkeijoli, F, lp, 1988; (**Majolika®**); Tantau

'TANKENFRAM', F, or, 1982; (**Frankenland®**); bud medium, pointed; flowers brilliant scarlet, patio, borne in clusters of many, dbl., 23 petals, exhibition form, no fragrance; straight, brown-red prickles; foliage large, dark, semi-glossy; bushy, low growth; Tantau, Math.; Tantau Roses, 1978

'TANKLESANT', LCl, mr, 1985; (**Santana®**); flowers medium, dbl., 20 petals, no fragrance; foliage large, medium green, glossy; upright growth; Tantau, Math., 1984

'TANKLEVI', LCl, mp, 1982; (Lavinia, **Lawinia®**, 'TANKLEWI'); flowers large, dbl., 20 petals, cupped, moderate fragrance; foliage large, medium green, semi-glossy; spreading (to 8 ft) growth; Tantau, Math., 1980

TANkobi, MinFl, pb, 1986; (**Bubikopf®**); Tantau, Math.; (Cavriglia)

TANkorab, HT, my, 1990; (**Barock**); Tantau

'TANKY', HT, yb, 1967; (Whisky, **Whisky Mac**); bud ovoid; flowers bronze-yellow, well-formed, large, dbl., 30 petals, intense fragrance; foliage glossy; vigorous, upright, bushy growth; Tantau, Math.

'TANLARPOST', HT, w, 1982; (Polar Star, **Polarstern®**); flowers pale yellow to white, medium, dbl., 35 petals, exhibition form, no fragrance; foliage medium, medium green, matt; bushy growth; ROTY, 1985; Tantau, Math.

TANledolg, MinFl, dy, 1959; (Bijou d'Or, **'GOLDEN JEWEL'**, Goldjuwel); flowers golden yellow, blooms in clusters (up to 10), dbl., 3 in., moderate fragrance; foliage dark, glossy; vigorous, bushy growth; [Goldilocks X Masquerade seedling]; Tantau, Math.

TANledolg, Min, dy, 1995; (Bijou d'Or); Tantau

TANlewu, Min, dy, 1993; (**Goldjuwel**); Tantau

'TANLI', F, mp, 1971; (**Rosali**); bud long, pointed; flowers large, dbl., moderate fragrance; vigorous, upright growth; RULED EXTINCT 4/85; [Seedling X Junior Miss]; Tantau, Math.

'TANLILIDA', F, dr, 1977; (**Dalli Dalli®**); bud ovoid; flowers medium, dbl., cupped, slight fragrance; foliage very glossy; upright, bushy growth; ADR, 1975; Tantau, Math.

TANliram, HT, mp, 1997; (**Marlyn**); Tantau

TANlorip, Pol, dy, 1994; (**Pirol**); flowers small, dbl.; Tantau; (Sangerhausen)

'TANLYPOLO', F, dy, 1982; (**Polygold®**); flowers medium, semi-dbl., no fragrance; foliage medium, light green, glossy; bushy growth; Tantau, Math., 1978

TANmeda, F, mp, 1986; (**Diadem**); GM, Durbanville, 1987; Tantau

TANmirsch, S, my, 1993; (**Sonnenschirm®**); Tantau

TANmirson, S, dy; (**Broadlands**); Tantau

TANmixa, HT, rb, 1993; (**Maxim**); Tantau

TANmop, F, mp; (**Pompon Diadem**); Tantau

'TANMOSINA', HT, ab, 1979; (**Simona**); flowers large, dbl., 20 petals, moderate fragrance; foliage large, medium green, matt; upright growth; Tantau, Math.

TANmurse, S, w, 1990; (**Schneesturm®**); Tantau

TANmurse, S, w; (**Blenheim**); Tantau

'TANNACHT', HT, m, 1965; (Blue Monday, **Blue Moon®**, Mainzer Fastnacht, Sissi 'TANSI'); bud long, pointed; flowers lilac, dbl., 40 petals, 4 in., intense fragrance; vigorous growth; ADR, 1964 GM, Rome, 1964; [Sterling Silver seedling X Seedling]; Tantau, Math., 1964

'TANNALI', F, dr, 1973; (**Natali**); bud ovoid; flowers small, semi-dbl.; foliage glossy; dwarf, upright growth; RULED EXTINCT 4/85; [Unknown X Unknown]; Tantau, Math.; Horstmann

TANnietnof, HT, ob; (**Stifontein Rose**); Tantau, 1992

'TANNIMOLL', F, or, 1964; (Mistigri, **Molde®**); dbl., blooms in clusters; foliage dark, glossy; bushy, low, compact. growth; GM, Baden-Baden, 1964; Tantau, Math.

'TANNIPOLA', F, op, 1978; (**Lapponia®**); bud broadly ovoid; flowers medium salmon-pink, dbl., 25 petals, slight fragrance; bushy, upright growth; Tantau, Math.

'TANNIREB', HT, mp, 1985; flowers large, dbl., 20 petals, slight fragrance; foliage medium, dark, semi-glossy; upright growth; Tantau, Math.

TANnus, HT, ob, 1991; (**Capri Sun**); Tantau

'TANOB', F, op, 1955; (**Hobby**); flowers coral-pink, large blooms in open clusters, dbl., slight fragrance; foliage dark; upright, bushy growth; [Red Favorite X Kathe Duvigneau]; Tantau, Math.

TANogrew, Min, ob, 1997; (**Clementine**); Tantau

TANokor, S, ly, 1987; (**Rokoko®**); flowers light yellow with rose tints, large, dbl.; foliage dark green; wide, bushy, well-branched (4.5 ft) growth; Tantau

'TANOLFEU', F, ob, 1971; (**Olympisches Feuer®**); bud ovoid; flowers orange, medium, dbl., slight fragrance; upright growth; [Ahoi X Signalfeuer]; Tantau, Math.; Dehner & Co.

TANolfeu92, F, or, 1992; (**Olympisches Feuer 92**); Tantau

'TANOLG', HT, dy, 1975; (Anneliesse Rothenberger, Miss Harp, **Oregold**, Silhouette); flowers large, dbl., exhibition form, slight fragrance; foliage large, glossy, dark; vigorous, upright, bushy growth; AARS, 1975; [Piccadilly X Colour Wonder]; Tantau, Math.; J&P, 1970

TANollipa, HT, my, 1997; (**Papillon**); Tantau

'TANOLOKIP', F, or, 1985; (Piccola, **Piccolo®**); flowers patio, medium, dbl., 20 petals, no fragrance; foliage large, dark, glossy; upright growth; Tantau, Math., 1984

TANolop, HT, w, 1997; (**Polo**); Tantau

TANopel, F, op, 1963; (**Tip Top**); flowers salmon-pink, large, dbl., moderate fragrance; Tantau, Math.; (Sangerhausen)

TANoras, S, mp, 1993; (**Rosario®**); Tantau

TANorelav, HT, dr, 1994; (**Red Velvet**); Tantau

'TANORSTAR', HT, or, 1960; (Super Star, **Tropicana**); bud pointed; flowers coral-orange, well-formed, dbl., 33 petals, 5 in., intense, fruity fragrance; foliage dark, glossy, leathery; vigorous, upright growth; AARS, 1963; GM, ARS, 1967 GM, NRS, 1960 GM, Portland, 1961 PIT, NRS, 1960; [(Unnamed X Peace) X (Unnamed X Alpine Glow]; Tantau, Math.; J&P, 1962

TANotari, S, mp, 1990; (**Mirato**); Tantau

'TANOTAX', S, dp, 1998; (**Footloose™**, Mirato, Neige d'Ete, Schneekonigin®); flowers dark pink, medium sized, double, flat, borne in large clusters, no fragrance; prickles moderate, straight; foliage small, dark green, glossy; spreading, bushy, medium growth; [Seedling X Seedling]; Evers, Hans & Rosen Tantau; Bear Creek Gardens, 1990

TANotax, MinFl, dp; (**Chatsworth**); Tantau

TANpen, F, ly, 1997; (**Penny Heyns**); Tantau

'TANPIKA', HT, mp, 1972; (**Piroschka®**); bud long, pointed; flowers large, dbl., moderate fragrance; moderate, upright growth; [Fragrant Cloud X Dr A.J. Verhage]; Tantau, Math.

TANpinaiso, HT, lp, 1997; (**Pink Osiana**); Tantau

TANpretty, HT, lp, 1990; (**Softy**); flowers small, dbl.; Tantau; (Sangerhausen)

'TANPRIK', F, or, 1958; (**Paprika®**); bud long, pointed; flowers brick red, blooms in large clusters, semi-dbl., 3.5 in., slight fragrance; foliage leathery, glossy, olive-green; vigorous, upright growth; GM, NRS, 1959 Golden Rose, The Hague, 1961; [Marchenland X Red Favorite]; Tantau, Math.

TANragus, Min, dp; (**Sugar Baby**); flowers small, dbl.; Tantau, 1997; (Sangerhausen)

'TANRAUSCH', F, mp, 1972; (**Duftrausch**); bud globular; semi-dbl., intense fragrance; foliage soft; upright growth; RULED EXTINCT 4/85 ARM; [Unknown X Unknown]; Tantau, Math.; Horstmann

'TANRAVENS', F, or, 1993; (**Wild West**); flowers orange-red, full, very dbl., 26–40 petals, 1.5–2.75 in., borne in large clusters, slight fragrance; small, soft, glandular prickles on peduncles; foliage large, medium green, glossy, resistant to powdery mi; medium (85–95 cms), bushy growth; [Showbiz X Seedling]; Evers, Hans & Rosen Tantau; Bear Creek Gardens, 1993

TANrazlaw, LCl, w, 1987; (**Schneewalzer**); Tantau

'TANRECKERTOR', F, mr, 1985; (**Rotelfe®**, 'TANRECKTOR'); flowers medium, semi-dbl., no fragrance; foliage small, medium green, semi-glossy; groundcover; spreading growth; Tantau, Math.

'TANREKTA', HT, rb, 1970; (**Rebecca®**, Renica); bud ovoid; flowers red, yellow reverse, well-formed, large, dbl., slight fragrance; vigorous, upright growth; [Konfetti X Piccadilly]; Tantau, Math.

TANrif, HT, mp, 1997; (**First Lady**); Tantau

TANrikas, HT, op, 1989; (**Saphir**); Tantau

TANrised, HT, mp, 1986; (**Schone Berlinerin®**); Tantau

TANrolfy, Min, m, 1992; (**Babyflor®**); Tantau

TANrosilb, F, pb, 1990; (**Naas Botha**); Tantau

'TANROTREILI', F, mp, 1981; (**Natali®**); flowers large, dbl., 20 petals, slight fragrance; foliage medium, medium green, matt; Tantau, Math.

'TANROWISA', HT, pb, 1982; (**Wimi®**); flowers shiny pink, silver reverse, well-formed, large, dbl., 20 petals, intense fragrance; foliage large, dark, semi-glossy; bushy growth; Tantau, Math., 1983

'TANRUPEZA', S, m, 1998; (**Leeds Castle**); flowers carmine purple, reverse medium pink, single (4–7 petals), medium (6 cms) blooms borne in large clusters; no fragrance; moderate prickles; foliage medium, dark green, glossy; spreading, low (30–40 cms) growth; groundcover [Seeding X Seedling]; Rosen Tantau; Eurose, 1998

'TANRYRANDY', HT, ob, 1985; (**Cherry Brandy '85**); flowers orange, large, dbl., 35 petals, no fragrance; foliage large, medium green, glossy; upright growth; Tantau, Math.

'TANSARAS', HT, lp, 1999; (**Century Sunset**, Herz As); dbl., 26–40 petals, 4.5–5 in., borne mostly singly, slight fragrance; prickles moderate; foliage large, dark green, semi-glossy; upright, medium (2–2.5 ft) growth; Tantau

'TANSCHAUBUD', HT, mp, 1985; (**Duftrausch®**, Olde Fragrance, Senteur Royale); flowers large, dbl., intense fragrance; foliage medium, medium green, semi-glossy; upright growth; Tantau, Math., 1986

'TANSCHNEEWA', F, w, 1972; (**Schneewalzer**); bud ovoid; flowers large, dbl.; moderate, upright growth; Tantau, Math.

'TANSCHWEIGRU', F, mr, 1954; (Holländerin, **Red Favorite**, Red Favourite, Salut à la Suisse, Schweizer Grüss); bud ovoid; flowers velvety oxblood-red, blooms in trusses, semi-dbl., 13 petals, 2.5 in., slight fragrance; foliage dark, leathery, glossy; vigorous, bushy growth; ADR, 1950; [Karl Weinhausen X Cinnabar]; Tantau, Math.; C-P

TANselbon, HT, op, 1989; (**Noblesse**); Tantau

TANsemil, HT, lp; (**Sweetnesse**); Tantau, 1997

'TANSENFRIE', S, dy, 1982; (**Friesensöhne®**); flowers medium, semi-dbl., no fragrance; foliage medium, light, glossy; PP005706; Tantau, Math.; Tantau Roses, 1981

TANseus, HT, lp; (**The Robe**); Tantau, 1995

'TANSI', HT, m, 1965; (Blue Monday, **Blue Moon®**, Mainzer Fastnacht, Sissi, 'TANNACHT'); bud long, pointed; flowers lilac, dbl., 40 petals, 4 in., intense fragrance; vigorous growth; ADR, 1964 GM, Rome, 1964; [Sterling Silver seedling X Seedling]; Tantau, Math., 1964

'TANSINNROH', HT, ab, 1984; (Frohsinn '82, **Joyfulness**); flowers apricot and orange blend, large, dbl., 35 petals, slight fragrance; foliage large, dark, glossy; Tantau, Math., 1982

TANsirk, F, or, 1990; (**Keith Kirsten**); Tantau

TANspolett, F, dp, 1994; (**Pink Diadem®**); Tantau

'TANT MIEUX', S, ob, 1984; (Saskabec); flowers tangerine, yellow center, blooms in clusters, single, 5 petals, no fragrance;

foliage medium, medium green, glossy; bushy growth; [R. gallica X Maria Stern]; Wright, Percy H.; Gilles Gailloux

'TANTA', F, or, 1968; (**Taora**®); bud long, pointed; flowers large, dbl.; foliage glossy; very vigorous, bushy growth; ADR, 1969; [Fragrant Cloud X Schweizer Gruss]; Tantau, Math.

TANtaatsyr, HT, ob; (**Vrystaat**); Tantau, 1992

Tantaliser, F, lp

'TANTALLON', HT, mp, 1934; flowers salmon-pink, base yellow, reverse cerise-red touched orange, dbl.; foliage glossy, dark; dwarf, stocky growth; Dobbie

Tantarra, HT, lp; Gardner, 1996

'TANTASCH', F, dy, 1998; (Castle Howard Tercentenary, Golden Jet, **Goldschatz**); flowers deep yellow, dbl., 15–25 petals, 4 in., borne in small clusters, slight fragrance; prickles moderate; foliage large, medium green, glossy; bushy, medium (70 cm) growth; [Seedling X Seedling]; Rosen Tantau; Eurosa, 1999

'TANTAU'S DELIGHT', F, or, 1951; bud long, pointed; flowers open, large, borne in clusters, semi-dbl., slight fragrance; foliage dark, glossy; vigorous, bushy growth; [Cinnabar X Kathe Duvigneau]; Tantau

'TANTAU'S SURPRISE', F, mr, 1951; (Tantau's Ueberraschung); flowers blood-red, large, borne in clusters of 5–8, dbl., 40 petals, slight fragrance; foliage dark, glossy; upright growth; [Bouquet X Hamburg]; Tantau

Tantau's Triumph, *see* **'CINNABAR'**

Tantau's Triumph, Climbing, Cl Pol, mr, 1957; flowers medium, semi-dbl.; Rosarium Sangerhausen; (Sangerhausen)

Tantau's Ueberraschung, *see* **'TANTAU'S SURPRISE'**

Tante Frieda, HT, mr, 1995; flowers large, dbl., intense fragrance; Michler, K. H.; (Sangerhausen)

'TANTERN', HKor, my, 1966; (Gold Star, **Goldstern**®); bud long, pointed; flowers golden yellow, large blooms in clusters, recurrent bloom; foliage glossy; vigorous, bushy (7–8 ft) growth; Tantau, Math.

'TANTIDE', F, dr, 1988; (**Glad Tidings**, Lübecker Rotspon®, Peter Wessel); bud ovoid; flowers bright crimson, medium, dbl., 20 petals, cupped, borne in sprays, no fragrance; foliage medium, medium green, semi-glossy; upright, medium growth (2 ft); GM, Durbanville, 1988 ROTY, St Albans, UK, 1989; [Unnamed seedling X Unnamed seedling]; Tantau, R.; Wheatcroft Ltd., 1989

'TANTIFUM', HT, m, 1978; (**Blue Parfum**®, Blue Perfume, 'TANFIFUM', 'TANIFUME'); bud ovoid; flowers mauve-blush, large, dbl., intense fragrance; foliage glossy; bushy, upright growth; Tantau, Math.

'TANTIGA', F, mp, 1972; (**Taiga**®); bud globular; flowers pink, medium, dbl.; vigorous, upright growth; [Geisha X Junior Miss]; Tantau, Math.

'TANTIVVY', Pol, mr, 1954; flowers crimson-scarlet, rosette form, semi-dbl.; compact, bushy growth; [Gloire du Midi sport]; Ratcliffe

TANttelos, S, dp, 1994; (**Violett Satina**); flowers medium, dbl.; Tantau; (Sangerhausen)

TANtumleh, HT, dr, 1996; (**Helmut Kohl**); flowers very large, semi-dbl., slight fragrance; Tantau, Math.; (Sangerhausen)

'TANVERY', HT, ob, 1974; (**Love Story**®); bud ovoid; flowers orange, large, dbl., cupped, slight fragrance; Tantau, Math.; Horstmann, 1972

'TANWEIEKE', F, mr, 1983; (Bernhard Daneke Rose, Ingrid Weibull, **Showbiz**); flowers bright medium red, medium, dbl., 20 petals, no fragrance; foliage medium, dark, semi-glossy; bushy growth; AARS, 1985; Tantau, Math., 1981

'TANYA', HT, ob, 1959; (Majeure); bud pointed; flowers deep orange to apricot-orange, dbl., 48 petals, 5 in., exhibition form, moderate fragrance; foliage leathery, glossy; vigorous, upright growth; [Peace X (Peace X Orange Nassau)]; Combe; J&P; Vilmorin-Andrieux

'TANYA KIM', HT, my, 1975; flowers clear yellow, full, dbl., 24 petals, 4.5 in., intense fragrance; foliage glossy; [Whisky Mac sport]; Marks

'TANYAB', F, my, 1990; (**Bayerngold**®); Tantau

TANybab, Min, dp, 1993; (**Pink Babyflor**); Tantau

TANydal, HT, pb; (**Summer Lady**®); Tantau, 1991

TANydor, S, mr, 1995; (**Rody**); flowers medium, semi-dbl.; Tantau; (Sangerhausen)

TANydu, HT, my, 1997; (**Judy**); Tantau

'TANYELLDA', F, dy, 1972; (**Yellow Belinda**); bud long, pointed; flowers orange-yellow, medium, dbl.; [Tanbeedee sport]; Tantau, Math.

'TANZAHDE', S, mr, 1998; (**Beaulieu**); flowers medium red, single, 4–7 petals, 2.5 in., borne in large clusters, no fragrance; prickles moderate; foliage medium, dark green, glossy; low, spreading (30 cms) growth; [Seedling X Seedling]; Rosen Tantau; Eurosa, 1998

TANzecon, HT, ob, 1992; (**Constanze**); Tantau

TANziewsim, HT, dr, 1996; (**Miss Schweiz**); flowers very dark red, large, dbl.; Tantau; (Sangerhausen)

Taora®, *see* 'TANTA'

'TAPESTRY', HT, rb, 1958; bud pointed; flowers red, yellow and pink, dbl., 38 petals, 4.5–6 in., exhibition form, moderate, spicy fragrance; foliage glossy; upright, bushy growth; [Peace X Mission Bells]; Fisher, G.; C-P

Tapis, Gr, lp

Tapis Afghan, *see* **'PEKINOIS'**

'TAPIS BLANC', Pol, w, 1927; flowers pure white, center tinted cream, large, borne in clusters, dbl.; bushy, dwarf growth; Turbat

Tapis de Soie, *see* **'WEE MAN'**

Tapis Jaune®, *see* 'RUGUL'

Tapis Persan, *see* 'MACEYE'

'TAPIS ROSE', F, mp, 1950; bud ovoid; flowers rose-pink, borne in clusters, dbl., 40 petals, 2–2.5 in., exhibition form, moderate fragrance; foliage leathery, dark; vigorous, upright, bushy growth; (27); [Pinocchio X Mme Jules Gouchault]; Meilland, F.; C-P

Tapis Rouge, *see* 'INTEROP'

'TAPIS ROUGE', F, mr; more compact and dwarf growth

Tapis Rustic, S, lp

Tapis Volant®, *see* 'LENPLAT'

Tapti, Gr, lp; Kasturi, 1978

Tara, *see* 'MACWAIWER'

Tara Allison, *see* 'MACWAIWER'

'TARA RED', HT, dr, 1974; bud ovoid; flowers carmine-red, becoming very dark, dbl., 38 petals, 6 in., exhibition form, moderate fragrance; foliage leathery; tall growth; [Peace X Charles Mallerin]; Golik; Dynarose

Taranga®, *see* 'TANAGNARAT'

Tarantella, *see* 'KORANTEL'

'TARANTELLA'®, HT, dy, 1936; flowers deep golden yellow, large, dbl., cupped, slight fragrance; foliage light, wrinkled; vigorous, bushy growth; [Charles P. Kilham X Pres. Herbert Hoover]; Tantau

'TARANTELLE', HT, or, 1965; flowers well-formed, dbl., 40 petals; foliage bronze; vigorous, bushy growth; [(Beauté X Ma Fille) X Magicienne]; Laperrière; EFR

'TARDE GRIS', HT, m, 1966; (Evening Light); flowers soft lilac, large, dbl., 25 petals, moderate fragrance; foliage dark; upright growth; [(Sterling Silver X Intermezzo) X (Sterling Silver X Simone)]; Dot, Simon; Roses Dot

Target, *see* 'JACRIM'

'TARO', HT, my, 1977; bud oval; flowers high pointed, dbl., 35 petals, 5.5 in.; foliage dark, leathery; vigorous, upright growth; [Edith Krause X Narzisse]; Ota, Kaichiro

Tarragona, *see* **'DIPUTACION DE TARRAGONA'**

'TARRAGONA', HT, ab, 1945; flowers apricot-yellow, open, dbl., 25–35 petals, exhibition form, slight fragrance; foliage glossy; upright growth; [Duquesa de Peñaranda X Federico Casas]; Dot, Pedro

Tarrawarra, Pol, pb; Nieuwesteeg, J., 1992

'TARTUFE', F, mr, 1956; flowers dark geranium-red, dbl., 28 petals; moderate growth; [Pompadour Red X Independence]; Frères, Buyl

Taryn, Min, op

'TARZAN', LCl, mr, 1955; flowers coppery carmine, dbl., free, recurrent bloom, moderate fragrance; foliage bronze; very vigorous growth; Delbard-Chabert

Tasja, HT, lp; RvS-Melle, 1995

Tasman, *see* 'MURTA'

Tasman Bay, LCl, w; Pratt, Florence, 1994

'TASOGARE', F, m, 1977; bud cupped; semi-dbl., 18 petals, 2.5–4 in., flat, moderate fragrance; foliage glossy; vigorous, spreading growth; [Gletscher X (Sterling Silver X Gletscher)]; Kobayashi, Moriji

'TASSIE PRINCESS', F, lp, 1999; dbl., 15–25 petals, 2–2.5 in., borne in small clusters, no fragrance; few prickles; foliage small, medium green, semi-glossy; bushy, low (36 in.) growth; Elszele, L. H.

Tassili, *see* ORAju

'TASSIN', HT, mr, 1942; bud ovoid; flowers large, dbl., intense fragrance; foliage leathery; vigorous, upright growth; [National Flower Guild X Lemania]; Meilland, F.; A. Meilland

'TASSIN, CLIMBING', Cl HT, mr; da Silva, Moreira

'TASTE OF HONEY', F, ob, 1977; bud ovoid; flowers honey color, aging pink, open, large, dbl., intense fragrance; foliage leathery; vigorous, upright growth; [Elizabeth of Glamis sport]; Young, S.A.

'TATA CENTENARY', HT, rb, 1974; flowers deep carmine, striped pale yellow, reverse buff; [Pigalle sport]; Pradhan, Rauf, Murmu, Ghosh

Tata Centenary, Climbing, Cl HT, lp; K&S, 1995

Tatjana®, *see* 'KORTAT'

Tattletale, *see* 'ZIPTALE'

Tattooed Lady, *see* 'SEADEEP'

Taubie Kushlik, *see* KORzinta

'TAUNUSBLÜMCHEN', HMult, m, 1904; flowers violet-pink, small, semi-dbl.; Weigand, C.; (Sangerhausen)

Taupo®, *see* 'MACMISECH'

'TAURANGA CENTENNIAL', F, or, 1982; flowers orange-red, center yellow, large blooms in cluster of 3, dbl., 22 petals, slight fragrance; light brown prickles; upright growth; [Arthur Bell X (Alexander X Sympathy)]; Wareham, Phyllis e.

'TAURINIA', HT, mp, 1942; bud ovoid; flowers bright salmon-pink, large, very dbl., cupped, moderate fragrance; foliage leathery; very vigorous growth; [Julien Potin X Sensation]; Aicardi; Giacomasso

'TAURO', F, dr, 1958; bud globular; flowers oxblood-red, large, borne in clusters, dbl.; foliage glossy; upright growth; [Poinsettia X Alain]; Bofill; Torre Blanca

'TAURUS', Min, or, 1978; bud pointed, ovoid; flowers vermilion, full, dbl., 35–40 petals, 2–3 in.; foliage dark; compact, upright growth; [Baby Masquerade X (Spion-Kop X Evelyn Fison)]; Ellick; Excelsior Roses

'TAUSENDSCHÖN', HMult, pb, 1906; (Thousand Beauties); flowers deep rose-pink, center white, large blooms in large clusters, dbl., cupped, slight fragrance; no prickles; foliage soft; climbing (8–10 ft) growth; [Daniel Lacombe X Weisser Herumstreicher]; Schmidt, J.C.

'TAWNY GOLD', HT, dy, 1951; bud pointed; flowers tawny gold, well-shaped, dbl., 25 petals, 5 in., exhibition form, intense fragrance; foliage leathery; bushy, compact growth; [Vanessa X Burgemeester van Oppen]; Leenders, M.; J&P

'TAWNY SUPERIOR', HT, my, 1958; flowers golden yellow; Leenders, M.

'TAXI', F, rb, 1969; flowers mandarin red, reverse soft rose, pointed, large, dbl.; foliage dark, matt; RULED EXTINCT 12/85; [Isabelle de France X Basildon Belle]; Trew, C.; Basildon Rose Gardens

'TAXI', HT, dr, 1985; flowers large blooms borne usually singly, dbl., 35 petals, exhibition form, intense fragrance; foliage large, dark, matt; vigorous, upright growth; [Fragrant Cloud X Gisselfeld]; Poulsen, Niels D.; Poulsen's Roses, 1978

'TAY CAITLIN', Min, ab, 1997; flowers double, small, dbl., 15–25 petals, .5–1 in., borne mostly singly, slight fragrance; some prickles; foliage medium, medium green, semi-glossy; upright, medium (16–22 in.) growth; [Queen City sport]; Garrett, Troy O.

'TAYBIRD', HT, yb, 1992; (**Bird of Fire**); flowers yellow blending to reddish edges, reverse yellow, aging lighter, exhibition form, 25–30 petals, 4–4.75 in., exhibition form, moderate; foliage medium, dark green, semi-glossy, leathery, reddish w; upright, bushy, medium growth; [Fragrant Delight X Sunsilk]; Taylor, Thomas E., 1993; National Kirtland's Warbler Recovery Team

Taylors Gold, *see* 'HORBARJEFF'

'TAYMAR', HT, pb, 1996; (**Mary Louise**); flowers light pink, reverse medium pink, form similar to first prize, dbl., 15–25 petals, 5 in., borne mostly singly, slight fragrance; moderate prickles; foliage medium, medium green, glossy, similar to honey favor; compact, bushy, medium (3–4 ft) growth; [Honey Favorite X First Prize]; Taylor, Thomas E.

'TAYMEL', Min, m, 1990; (**Melinda Claire**); bud pointed; flowers reddish magenta, yellow stamens, small to blooms borne usually singly, semi-dbl., 15 petals, exhibition form, slight fragrance; globular, medium, orange fruit; prickles curved downward, small, widely spaced, brownish; foliage medium, medium green, semi-glossy; [Charmglo X Charmglo]; Taylor, Thomas E., 1988; Michigan Mini Roses, 1990

Tchin-Tchin®, *see* 'MEICHANSO'

Tchin-Tchin®, *see* 'MEIKINOSI'

Tchin-Tchin, Climbing, *see* MEIchansosar

Te Awamutu Centennial, *see* 'STETEAW'

Te Kawanata Hou, S, lp; Nobbs, 1995

'TE KOWHAI', F, dr, 1960; flowers dark velvety red, reverse white overlaid red; [(Florence Mary Morse X Border Queen) X Confidence]; Bennett, H.; Pedigree Nursery

Te Moana, *see* 'SIMTEONO'

'TEA PARTY', Min, ab, 1972; flowers orange-apricot to pink, small, dbl.; foliage small, light; vigorous, dwarf, bushy growth; [(R. wichurana X Floradora) X Eleanor]; Moore, Ralph S.; Sequoia Nursery

'TEA RAMBLER', HMult, op, 1904; flowers soft salmon-pink, blooms in clusters, dbl., early, moderate fragrance; vigorous, climbing growth; [Crimson Rambler X Tea]; Paul

Tea Rose, *see* '**R. X ODORATA**'

'TEA TIME', F, mp, 1960; bud ovoid; flowers clear pink, medium, borne in clusters, dbl., 60–65 petals, exhibition form, moderate fragrance; foliage leathery; vigorous, bushy growth; [Floribunda seedling X Demure]; Boerner; J&P

Tea Time, *see* TANetee

'TEACHER'S PET', HT, ab, 1979; bud ovoid, pointed; flowers apricot, dbl., 30 petals, exhibition form, slight fragrance; straight prickles; foliage dark green with red mid-rib and edges; upright, bushy growth; [First Prize X Bonsoir]; Taylor, Thomas E.

Tear Drop, *see* 'DICOMO'

'TEASE', F, ob, 1967; bud ovoid, coppery; flowers golden yellow, dbl., cupped, moderate fragrance; foliage leathery; vigorous, upright growth; [Yellow Pinocchio X Fashion seedling]; Boerner; J&P

Teasing Georgia, *see* 'AUSBAKER'

Technikon Pretoria, *see* KORproa

Ted Allen, *see* 'GUESGENUS'

'TED CARNAC', LCl, 1968; vigorous growth; Gowie

Ted Gore, *see* 'HORMISLAC'

Ted Goves, *see* 'KIRPARK'

'TEDCASTLE', HP, m, 1940; flowers purple, open, very dbl., cupped, slight fragrance; foliage glossy; upright, bushy growth;

[Ulrich Brunner Fils X Paul Neyron]; Tedcastle

Teddy, *see* PRAeddy

Teddy Bear™, *see* 'SAVABEAR'

Teddy's Coat, F, dy

Ted's Red, *see* 'REYRED'

'**TEENAGER**', HT, yb, 1958; flowers yellow shaded pink and carmine, loosely formed, dbl., 25 petals, 4–5 in., intense fragrance; foliage dark, leathery; vigorous, upright growth; [Ena Harkness X Sutter's Gold]; Arnot; Croll

'**TEENAGER**', F, pb, 1958; bud ovoid; flowers la france pink, lightly overcast shrimp-pink, dbl., 55–60 petals, 3.5 in., exhibition form, intense fragrance; foliage leathery; strong stems; vigorous, upright growth; [Demure X Demure]; Boerner; J&P

Teeny Bopper, *see* 'TINBOPPER'

Teeny-Weeny, *see* '**BUTTONS 'N' BOWS**'

'**TEGALA**', S, dp, 1927; flowers deep pink, semi-dbl., non-recurrent; height 5 ft; very hardy; [Tetonkaha X Alika]; Hansen, N.E.

'**TEIDE**', LCl, mr, 1948; flowers carmine, well formed; foliage reddish; [Texas Centennial X Guinee]; Dot, M.

Telford's Promise, *see* 'CHEWOZ'

'**TELSTAR**', F, ob, 1962; flowers orange to orange-buff, blooms in clusters (up to 5), semi-dbl., 14 petals, 3.5 in., moderate fragrance; foliage dark; vigorous, upright growth; [Flash (F) X Masquerade]; Gandy, Douglas L.

'**TEMNO**', HT, m, 1936; flowers dark maroon, fairly large, dbl.; vigorous growth; Böhm, J.; J&P

'**TEMPER TANTRUM**', Min, dr, 1986; flowers white center, medium, borne usually singly or in sprays of 2, dbl., 30 petals, exhibition form, slight fragrance; no fruit; no prickles; foliage small, medium green, semi-glossy; upright, medium growth; [Red Beauty X Libby]; Jolly, Marie; Rosehill Farm, 1988

'**TEMPERAMENT**', F, mr, 1957; (Vivacity); flowers light scarlet, open, borne in clusters, dbl., 24 petals, 2.5 in., slight fragrance; foliage glossy, leathery; vigorous, upright growth; [Fanal X Karl Weinhausen]; Tantau, Math.

'**TEMPEST, CLIMBING**', LCl, rb, 1962; bud pointed; flowers yellow, pink and orange-red, borne in clusters, dbl., 26 petals, 2.5–3 in., intense fragrance; foliage dark, glossy; vigorous growth; [Masquerade X Unnamed seedling]; Verschuren; Blaby Rose Gardens

Tempi Moderni, *see* BARdem

'**TEMPIE LEE**', HT, pb, 1991; dbl., 26–40 petals, 1.5–2.75 in., exhibition form, borne mostly singly, slight fragrance; some prickles; foliage medium, medium green, semi-glossy; problems with blockspot; long canes; vigorous, medium, spreading growth; [Elizabeth Taylor sport]; Whittington, J.O., Sr.; Edmunds' Roses, 1998

'**TEMPLAR**', HT, mr, 1924; flowers bright red, dbl., globular; foliage dark, leathery; vigorous growth; GM, NRS, 1925; [Premier X Seedling]; Montgomery Co.; A.N. Pierson

'**TEMPLE BELLS**', Cl Min, w, 1971; flowers small, single, 7 petals, 2 in., slight fragrance; foliage small, glossy; [R. wichurana X Blushing Jewel]; Morey, Dr Dennison; McGredy

'**TEMPLE FLAME**', F, or, 1965; flowers orange flushed red, open, medium, semi-dbl.; foliage leathery; very vigorous, upright, compact growth; [Orangeade X ?]; Pal, Dr B.P.; Indian Agric. Research Inst.

Tempo, *see* 'JACCLOP'

Tempo, *see* MEInececa

'**TEMPTATION**', LCl, mr, 1950; bud ovoid, carmine-red; flowers rose-red, open, large, dbl., 40 petals, free, recurrent boom, moderate fragrance; foliage large, glossy; vigorous, climbing growth; good pillar; very hardy. [(New Dawn X Crimson Glory) X Dream Girl]; Jacobus; B&A

Temptation, *see* 'TROPAT'

Temptation, *see* INTertropa

'**TEMPTRESS**', HT, mr, 1948; bud long, pointed, crimson; flowers carmine, medium, dbl., 35–40 petals, cupped, moderate fragrance; foliage leathery, dark; vigorous, upright, bushy growth; [(Lucile Hill X Chieftain) X Unnamed seedling]; Hill, Joseph H., Co.

Temptress, *see* 'WAMTRESS'

'**TEMPUS-FUGIT**', HT, mr, 1977; flowers cardinal-red, full, dbl., 40–45 petals, 3–4 in., moderate fragrance; foliage glossy, dark; very free growth; [Chopin X Heidelberg]; Ellick

Ten Sisters, *see* '**TURNER'S CRIMSON RAMBLER**'

Ten Ten CFRB, *see* MACmatan

Tender Blush, A, lp; Sievers, 1988

'**TENDER LOVE**', Min, lp, 1980; bud slender, tapering; dbl., 23 petals, borne mostly singly, sometimes 3–5 per cluster, slight fragrance; prickles dilated at base; foliage glossy, medium green; tall, upright growth; [Doris Ashwell X Sheri Anne]; Strawn, Leslie E.; Pixie Treasures Min. Roses

'**Tender Love**', F, lp, 1997; flowers very full, exhibition, medium, very dbl., 41 petals, 2.75 in., borne in large clusters, no fragrance; some prickles; foliage medium, light green, glossy; upright, medium (90 cms) growth; [Sexy Rexy X Seedling]; Bossom, W.E.

'**TENDER LOVING CARE**', *see* BOSpeabay

Tender Mercy, *see* 'RENMERCY'

Tender Night, *see* 'MEILAUR'

Tender Night, Climbing, *see* 'MEILAURSAR'

Tender One, *see* '**NEJENKA**'

Tenderly, HT, 1984; Delforge; (Cavriglia)

Tenderly Yours, *see* 'RENYOURS'

Tendresse, *see* 'DELJOFEM'

Tendresse, *see* '**SACHSENGRUSS**'

Tendresse, *see* DELtendre

'**TENEREZZA**', HT, rb, 1973; bud ovoid, long, pointed; flowers spinel-rose to cherry-red, medium, dbl., cupped, moderate fragrance; foliage large, dark, leathery; vigorous, upright, bushy growth; [Unnamed seedling X Ninfa Rossa]; Calvino

'**TENERIFE**', HT, op, 1972; flowers deep orange, yellowish reverse, pointed, dbl., 45 petals, 5 in., intense fragrance; foliage glossy; [Fragrant Cloud X Piccadilly]; Timmerman's Roses

Tennessee™, *see* 'KINTENN'

Tennessee, F, ob; Select Roses, 1991

'**TENNESSEE BELLE**', LCl, mp; flowers bright rosy blush, large, borne in clusters, dbl., moderate fragrance; vigorous, climbing growth

Ténor, *see* 'DELCAP'

'**TENTATION**', HT, op, 1968; bud pointed; flowers carmine-pink to orange, large, dbl., slight fragrance; foliage glossy; vigorous, upright growth; [Touggout X Flaminaire]; Dorieux; Vilmorin

Tenth Rose Convention, HT, dy; Patil, B.K., 1990

'**TEODORA**', HT, mr, 1983; flowers large, dbl., 30 petals, cupped, slight fragrance; reddish-green prickles; foliage large, brownish-green, matt; upright growth; [Honey Favorite X Command Performance]; Stoddard, Louis; Rose Barni-Pistoia

Tequila®, *see* 'MEIGAVESOL'

Tequila Sunrise, *see* 'DICOBEY'

Tequilla Sunset™, *see* 'WILSUN'

'**TERESA FINOTTI-MASIERI**', HT, m, 1929; bud long, pointed; flowers brilliant purple-rose, rosy scarlet reflexes, bordered yellow; [George C. Waud X Souv. de Gabriel Luizet]; Ketten Bros.

'**TERESA OZORES**', Cl HT, lp, 1957; bud ovoid; flowers soft pearly pink, large, dbl., cupped, intense fragrance; vigorous growth; [Frau Karl Druschki X Lady Sylvia]; Camprubi, C.

Terra Jubilee, F, lp; Zary, Dr Keith W., 1996

Terracotta, *see* MEIcobius

Terrell Anne, *see* 'GRETAG'

'**TERRY O**', F, rb, 1976; flowers dark red, reverse copper yellow; [Redgold sport]; Hughes; Hughes Roses

Tersicore, S, 1985; Barni, V.; (Cavriglia)

'TESCHENDORFFS JUBILÄUMSROSE', Pol, dr, 1928; flowers vivid rosy crimson, darker than Orleans Rose and more brilliant; [Orléans Rose sport]; Teschendorff

'TESCO BERNSTEIN', F, dy, 1972; bud ovoid; flowers brownish yellow, open, medium, semi-dbl., globular, slight fragrance; foliage glossy, dark, leathery; vigorous, bushy growth; [Cognac X Allgold]; Haenchen, E.; Teschendorff

'TESCO BLICKFANG', F, or, 1972; bud ovoid; flowers dark vermilion-red, large, dbl., slight fragrance; foliage dark, soft; vigorous, upright, bushy growth; [Dicksons Flame X Circus]; Haenchen, E.

'TESCO BRENNPUNKT', F, or, 1972; bud ovoid; flowers bright vermilion-red, open, medium, semi-dbl., slight fragrance; foliage dark, soft; vigorous, upright growth; [Highlight X Dicksons Flame]; Haenchen, E.; Teschendorff

'TESCO GOLDTEPPICH', F, my, 1972; bud ovoid; flowers medium, semi-dbl., cupped, slight fragrance; foliage glossy, dark, leathery; vigorous, bushy growth; [Cognac X Allgold]; Haenchen, E.; Teschendorff

'TESCO LICHTBLICK', S, ob, 1972; flowers pink and orange-pink on yellow ground, large, dbl., exhibition form, abundant, continuous bloom, slight fragrance; foliage glossy, leathery; vigorous, upright growth; [Yellow Holstein X Heidelberg]; Haenchen, E.; Teschendorff

'TESCO ROMANZE', F, pb, 1972; bud globular; flowers pink, center white, open, small, semi-dbl., slight fragrance; foliage small, glossy, dark, leathery; vigorous, dwarf, bushy growth; [Marchenland X ?]; Haenchen, E.; Teschendorff

'TESORINO'®, Min, lp, 1982; flowers small, dbl., 20 petals, no fragrance; light green prickles; foliage small, light green, matt; groundcover; spreading growth; [Unnamed X Unnamed]; Rose Barni-Pistoia

'TESS', HMult, mp, 1939; flowers neyron pink, passing to deeper pink, center brighter, large, dbl., slight fragrance; foliage large, rich green; strong stems; very vigorous growth; RULED EXTINCT 11/91; Beckwith

Tess, *see* 'CLETEST'

Tess, *see* 'MACRAMAR'

Tess of the D'Urbervilles, *see* 'AUSMOVE'

Tessa O'Keeffe, *see* 'KENMOLL'

Testa Rossa, *see* 'KORDIAM'

'TETON BEAUTY', HRg, dp, 1927; flowers rich pink to crimson, dbl., cupped, recurrent bloom, moderate fragrance; foliage rugosa-like; very hardy. [Tetonkaha X American Beauty]; Hansen, N.E.

'TETONKAHA', HRg, dp, 1912; flowers deep rich pink, medium, semi-dbl., profuse, non-recurrent bloom, intense fragrance; vigorous (6 ft tall and wide) growth; hardy. [R. macounii X Rugosa hybrid]; Hansen, N.E.

Teutonia®, HT, dp; Noack, Werner, 1985

'TEWANTIN', Cl HT, dp, 1953; bud globular; flowers deep pink, large, dbl., moderate fragrance; long stems; vigorous (12 ft) growth; [Editor McFarland X Self]; Ulrick, L.W.

'TEXAN', F, mr, 1956; (The Texan); bud ovoid; flowers rose-red, borne singly and in clusters, dbl., 23 petals, 3.5–4 in., exhibition form, moderate fragrance; foliage dark, leathery; vigorous, upright growth; [Improved Lafayette X Peace]; Lindquist; Howard Rose Co.

Texas, *see* 'POULTEX'

Texas, *see* KORbacol

'TEXAS CENTENNIAL', HT, rb, 1935; flowers vermilion-red with some gold, center lighter; GM, Portland, 1935; [Pres. Herbert Hoover sport]; Watkins, A.F.; Dixie Rose Nursery

'TEXAS CENTENNIAL PANACHÉE', HT, ob, 1939; flowers light orange with white stripes, large, dbl., moderate fragrance; Tantau, Math.; (Sangerhausen)

'TEXAS CENTENNIAL, CLIMBING', Cl HT, rb, 1936; Dixie Rose Nursery, 1942

Texas Girl™, *see* 'MEICIJAS'

'TEXAS GOLD', HT, my, 1935; flowers golden yellow occasionally tinged pink, large, dbl., intense fragrance; [Pres. Herbert Hoover sport]; Wolfe

'TEXAS QUEEN', HT, mp, 1965; flowers base cream, shading to light and darker pink, large, very dbl., 100 petals, intense fragrance; almost thornless; foliage light green, soft; vigorous growth; [Tip Toes sport]; Leidy; Wilson Nursery

Texas Sunrise, *see* 'UMSTEX'

'TEXAS WAX', S, lp; [Appears to be a hybrid of R. chinensis X R. multiflora]

Thaïs, *see* 'MEIMAJ'

Thalassa, HT, m; Dorieux, 1978

'THALIA', HMult, w, 1895; (White Rambler); flowers small blooms in clusters, dbl., early; [R. multiflora X Paquerette]; Schmitt; P. Lambert

Thalia Remontant, HMult, w; Lambert, P., 1903

Thalie La Gentille, HGal, dp; Vibert, before 1811

'THANET BALLERINA', HT, w, 1973; flowers pearl-white, blushed pink, moderate fragrance; foliage rich green; vigorous growth; [Prima Ballerina sport]; Court

Thank You, *see* CHEsdeep

'THANKSGIVING', HT, ob, 1965; bud ovoid; flowers bronze shades, reverse orange-red, medium, dbl., 38 petals, 3.5 in., exhibition form; foliage dark, leathery; vigorous growth; [Fred Howard X Seedling]; Warriner, William A.; Great Western Rose Co.

Thanx Mom, *see* 'JULESDEAR'

Thanx Mum, *see* 'JULESDEAR'

Thari®, S, dr

'THE ADJUTANT', HT, mr, 1922; bud long, pointed; flowers bright red, dbl., intense fragrance; Pemberton

The Adjutant, Climbing, Cl HT, mr; flowers large, dbl., intense fragrance; (Sangerhausen)

'THE ALAMO', HT, mr, 1959; bud ovoid; flowers cardinal red, very dbl., 110 petals, 3.5–4.5 in., slight fragrance; foliage leathery; vigorous, upright, bushy growth; [Happiness X Independence]; Meilland, F.; Co-Operative Rose Growers

'THE ALEXANDRA', T, op, 1900; flowers light salmon-pink, large, dbl., slight fragrance; Paul, W.; (Sangerhausen)

The Alexandra Rose, *see* 'AUSDAY'

The Alexandria Rose, *see* 'WILALEX'

'THE ALLIES', Pol, w, 1930; flowers white suffused pale pink, open, small, borne in clusters, dbl., slight fragrance; foliage small, glossy; dwarf, bushy growth; Heers

'THE AUSTRALIAN BICENTENNIAL', HT, mr, 1987; flowers medium to deep red, well-formed, borne usually singly, dbl., 30 petals, intense fragrance; curved, green prickles; foliage medium, medium green, glossy, disease resistant; tall, branching, average growth; [Daily Sketch seedling X Red Planet]; Bell, Ronald J.; Roy H. Rumsey, Pty. Ltd., 1988

The Backpackers Rose, *see* 'HORBONDARC'

'THE BAIRN', HT, dy, 1976; flowers golden yellow, very full, medium, dbl., 40–50 petals, 3.5 in.; foliage dark, leathery; dwarf growth; [Arthur Bell X Unnamed seedling]; Wood

'THE BEACON', HWich, rb, 1922; flowers bright fiery-red with white-eye, as in American Pillar, sing, semi-dbl.; vigorous, climbing (to 8 ft) growth; Paul, W.

The Beloved, *see* **'PRIYATAMA'**

'THE BISHOP', C, m; flowers unusual shade of cerise-magenta, fading bluish purple, rosette, flat, early bloom; slender, upright (4–5 ft) growth

'THE BISHOP', HT, dr, 1937; flowers bright crimson, large, dbl., intense fragrance; foliage dark; vigorous growth; GM, NRS, 1937; Dickson, A.

'THE BRIDE', T, w, 1885; flowers white tinged pink; [Catherine Mermet sport]; May

'THE BRIDE', HT, w, 1963; bud pointed, ivory-white; flowers well formed, dbl., 30 petals; foliage glossy; [White Swan X Unnamed seedling]; Herholdt, J.A.; Herholdt's Nursery

The Cambridge Rose, *see* **'CANTABRIGIENSIS'**

The Care Rose, *see* 'HORAPSUNMOLBABE'

The Cheshire Regiment, *see* FRYyat

'The Chief', HT, ob, 1940; (Chief); bud long, pointed; flowers flame, coral and copper, dbl., 35 petals, 4–6 in., intense fragrance; vigorous, bushy, spreading growth; AARS, 1940; [Charles P. Kilham X Pres. Herbert Hoover]; Lammerts, Dr Walter; Armstrong Nursery

The Clerk, S, lp; Jerabek, Paul E., 1997

The Colwyn Rose, *see* FRYnblue

The Commodore, *see* **'Kommodore'**

The Compass Rose, *see* KORwisco

The Conductor, *see* **'Dirigent'**

The Cottage Rose, *see* 'AUSGLISTEN'

The Countryman®, *see* 'AUSMAN'

The Coxswain, *see* 'COCADILLY'

The Crepe Rose, *see* **'Paul Perras'**

The Daniel™, *see* 'WILDAN'

The Dark Lady, *see* 'AUSBLOOM'

'The Dazzler', Cl HT, mr, 1955; flowers bright red, petals ruffled, single, profuse, repeated bloom; [Dainty Bess, Climbing seedling]; Marsh's Nursery

The Didgemere Rose, *see* 'FERTRY'

'The Doctor', HT, mp, 1936; flowers satiny pink, well-formed, dbl., 25 petals, 6 in., intense fragrance; foliage soft, light; dwarf, bushy growth; GM, NRS, 1938; [Mrs J.D. Eisele X Los Angeles]; Howard, F.H.; Dreer

'The Doctor, Climbing', Cl HT, mp, 1950; Dyess; Reliance Rose Nursery

The Dove, *see* 'TANAMOLA'

'The Dowager Countess of Roden', HT, lp, 1919; flowers bright silvery pink, dbl.; [Viscountess Enfield X George C. Waud]; Paul, W.

The Dragon's Eye, *see* 'CLEDRAG'

'The Duke', HT, rb, 1956; bud globular; flowers carmine-red, reverse gold, blooms in clusters of 3–15, dbl., 55 petals, 5–6 in., exhibition form, slight fragrance; foliage semi-glossy; vigorous, open growth; [Applause X Peace]; Von Abrams; Peterson & Dering

The Ednaston Rose, LCl, lp

The Edwardian Lady, *see* **Edith Holden**

'The Fairy', Pol, lp, 1932; (Fairy, Feerie); flowers pink, small blooms in clusters, dbl., recurrent bloom; foliage glossy, small; compact, spreading growth; hardy. [Paul Crampel X Lady Gay]; Bentall, Ann; J.A. Bentall

'The Farmers Wife', F, lp, 1962; bud ovoid; dbl., 38 petals, 4.5 in., cupped, blooms in clusters, intense fragrance; foliage leathery, glossy; upright, moderate growth; [Queen Elizabeth X Spartan]; Boerner; J&P

The Farquhar Rose, *see* **'Farquhar'**

The Faun®, Pol, mp; Poulsen, 1990

The Fisherman's Cot, *see* 'HARWICKLOW'

The Flower Arranger, *see* 'FRYJAM'

The Fordham Rose, HT, rb; Williams, J. Benjamin, 1996

'The Friar', S, lp, 1969; flowers blush, edged white, medium, 10 petals, repeat bloom, intense fragrance; foliage dark; [Ivory Fashion X Seedling]; Austin, David

The Garden Editor®, *see* 'MINASCO'

'The Garland', Misc OGR, w, 1835; (Wood's Garland); flowers faint yellow, pink, and white, blooms in large, long cluster, semi-dbl., mid-season bloom, moderate, musk fragrance; moderate climbing (8 ft) growth; [R. moschata X R. multiflora]; Wells

'The General', HT, mr, 1920; flowers blood-red, flushed lighter, large, dbl., globular, moderate, damask fragrance; foliage dark, bronze, leathery; vigorous, low growth; Pemberton

'The Gentleman', HT, yb, 1958; bud long; flowers creamy yellow flushed pink, semi-dbl., intense fragrance; [Unnamed seedling X Irish Fireflame]; Hay; Marsh's Nursery

The Gift, Pol, w; Demits, 1981

The Gooseberry Rose, *see* **R. stellata mirifica**

The Herbalist™, *see* 'AUSSEM'

The Holt, *see* 'MEHSHERRY'

The Hon Mrs Cat, D, lp; Scarman, 1995

The Hunter, *see* **'Hunter'**

'The Indian', HT, dr, 1939; flowers very dark red, large, semi-dbl.; vigorous growth; [Sensation X Seedling]; Clark, A.; Brundrett

The J.S.E. Rose, *see* KORgreyel

The Jacobite Rose, *see* **'Alba Maxima'**

'The Jester', Gr, yb, 1960; (Leerder's Harlequin); flowers yellow changing to red, compact, borne in clusters, dbl., slight fragrance; foliage dark; [Masquerade X High Noon]; Leenders, J.; British Hort. Co.

'The Knight', S, dr, 1969; flowers crimson turning purple and mauve, medium, dbl., 80 petals, flat, repeat bloom, moderate fragrance; foliage dark; [Chianti X Seedling]; Austin, David

The Lady, *see* 'FRYJINGO'

The Lady Scarman, HMsk, lp; Scarman, 1995

'The Lion', HMult, dp, 1901; flowers medium, single; Paul, G.; (Sangerhausen)

'The Marquess of Bristol', *see* 'WEBlady'

'The Master Cutler', HT, mr, 1954; flowers cherry-red shaded rose, well formed, very large, dbl., intense fragrance; vigorous growth; [The Doctor sport]; Liberty Hill Nurs.

The Matthew™, *see* 'JAYMATT'™

The McCartney Rose™, *see* 'MEIZELI'

'The Meteor', HT, dr, 1887; flowers medium, semi-dbl., moderate fragrance; Evans; (Sangerhausen)

'The Miller', S, mp, 1981; bud globular; flowers rosette-shaped blooms borne 1–4 per cluster, dbl., 40 petals, repeat bloom, moderate fragrance; hooked, red prickles; foliage medium green; upright, bushy growth; [Baroness Rothschild X Chaucer]; Austin, David, 1970

'The Mountie', F, mr, 1949; flowers cherry-red, medium, borne in huge trusses, semi-dbl., 12–15 petals, moderate fragrance; foliage leathery, glossy, light green; bushy growth; (28); [Springtime X World's Fair]; Eddie

The Mouse, *see* **'Grey Pearl'**

The Nanango Rose, *see* **Kathleen Kellehan**

The New Dawn, *see* **'New Dawn'**

The Nun, *see* 'AUSNUN'

The Observer, *see* FRYtango

'The Old Fashioned Rose', HT, 1930; flowers deep crimson, dbl., moderate fragrance; very vigorous growth; Archer

The Old Velvet Rose, *see* **'Tuscany'**

'The Optimist', F, yb, 1955; (Sweet Repose); bud ovoid; flowers maize-yellow tinged carmine, becoming carmine, blooms in large clusters, dbl., 27 petals, 3 in., exhibition form, moderate fragrance; foliage dense, leathery, parsley-green; vigorous, upright, bushy growth; GM, NRS, 1955; [Golden Rapture X Unnamed Floribunda seedling]; deRuiter; C-P, 1956; Gandy, 1955

The Painter, *see* 'MACTEMAIK'

The Painter, F, 1995; McGredy, Sam IV; (Cavriglia)

The People, *see* **'People'**

The People's Princess, *see* 'GEEPEOP'

The Pilgrim™, *see* 'AUSWALKER'

The Polar Star, *see* **'Polstjärnan'**

The Prince™, *see* 'AUSVELVET'

The Prince Imperial, *see* DELmagsa

'The Princess Elizabeth', HT, ob, 1927; flowers orange-yellow, edged deep cerise, large, dbl., moderate fragrance; [The Queen Alexandra Rose sport]; Wheatcroft Bros.

'The Prioress', S, lp, 1969; flowers blush to white, yellow stamens, medium, dbl., 24 petals, cupped, repeat bloom, moderate fragrance; [La Reine Victoria X Seedling]; Austin, David

The Queen, *see* **'Sour de S.A. Prince'**

'THE QUEEN', HT, ob, 1954; bud pointed; flowers orange suffused salmon, medium, dbl., 26–30 petals, moderate fragrance; very free growth; Lowe

'THE QUEEN ALEXANDRA ROSE', HT, rb, 1918; (Queen Alexandra); flowers bright red, reverse shaded old-gold, base orange, dbl., 5 in., moderate fragrance; foliage glossy, dark; bushy growth; GM, NRS, 1917; McGredy

'THE QUEEN ALEXANDRA ROSE, CLIMBING', Cl HT, rb, 1929; Lindecke; Kordes

'THE QUEEN ALEXANDRA ROSE, CLIMBING', Cl HT, rb, 1931; Harkness

The Queen Elizabeth Rose, *see* **'QUEEN ELIZABETH'**®

The Queen Elizabeth Rose, Climbing, *see* **'QUEEN ELIZABETH, CLIMBING'**

The Queen Mother, *see* **'AUGUST SEEBAUER'**

'THE QUEEN MOTHER', F, mp, 1959; flowers soft rosy pink, borne on large trusses, dbl., 45–50 petals, 3.5 in., moderate fragrance; foliage glossy, light green; vigorous growth; [Nymph sport]; Stedman

'THE QUEEN OF PERSIA', HT, ob, 1940; flowers light orange-yellow, large, dbl., slight fragrance; (Sangerhausen)

'THE RAJAH', HT, dr, 1937; flowers deep red, dbl.; bushy growth; [Red-Letter Day X ?]; Clark, A.; NRS Victoria

The Reeve®, *see* 'AUSREEVE'

The Robe, *see* TANseus

The Rose Tattoo, *see* **'LA ROSE TATOUÉE'**

The Roseville College Rose, HT, yb; (Roseville College, The Roseville College, **The Roseville College Rose**)

The Royal Brompton Rose, *see* 'MEIVILDO'

The Seckford Rose, *see* 'KORPINROB'

'THE SENATOR', HT, mr, 1980; bud ovoid, long; flowers bright medium red, petals loosely rolled outward, blooms borne usually singly, dbl., 46 petals, slight, musk fragrance; long, narrow-based, brown prickles, curved down; foliage large, leathery, wrinkled, dark; tall, upright growth; [Unnamed seedling X Suspense]; Weeks, O.L.

The Service Rose, *see* **'WOMEN IN MILITARY'**

The Squire®, *see* 'AUSIRE'

The St Mark's Rose, *see* **'ROSE D'AMOUR'**

'THE STORK', HT, mp, 1952; bud long, pointed; flowers carmine-rose, dbl., 28 petals, 5–5.5 in., exhibition form, slight fragrance; foliage leathery; bushy growth; [Seedling X The Doctor]; Whisler; Germain's

'THE SUN', F, op, 1972; flowers salmon-orange, semi-dbl., 15 petals, 3.5 in., slight fragrance; GM, Madrid, 1973; [(Little Darling X Goldilocks) X Irish Mist]; McGredy, Sam IV; McGredy

'THE SURGEON', HT, mp, 1956; flowers darker rose-pink, more vigorous than the doctor, dbl.; [The Doctor seedling]; Verschuren; Blaby Rose Gardens

The Temptations™, *see* 'WEKAQ'

The Texan, *see* **'TEXAN'**

The Valois Rose, *see* KORdadel

'THE VENERABLE BEDE', F, yb, 1973; flowers yellow to cherry, full, dbl., 20–30 petals, 3–3.5 in., slight fragrance; foliage small, glossy; moderate, free growth; [John Church X Bobby Shafto]; Wood; Homedale Nursery

'THE WALLFLOWER', LCl, mr, 1901; (Wallflower); flowers bright red, large, borne in large clusters, semi-dbl., late; vigorous growth; [Crimson Rambler X Beauté Inconstante]; Paul

The Wasa Star, *see* **'POLSTJÄRNAN'**

The Wedding Rose, *see* 'GEAAWARD'

The Wife of Bath, *see* 'AUSWIFE'

The Wild One, *see* 'TOMWILD'

The Work Continues, *see* 'JACSHAQ'

The World, *see* 'DIEKOR'

The Wyevale rose, *see* 'MEIBALBIKA'®

'THE YANK', HT, mr, 1942; flowers rose-red, very large, dbl., 55–60 petals, intense fragrance; foliage leathery, dark; long, strong stems; very vigorous, upright, much branched growth; [Chieftain X Lucile Hill]; Hill, Joseph H., Co.

'THE YEOMAN', S, op, 1969; flowers salmon-pink, medium, dbl., 50 petals, flat, recurrent bloom, intense fragrance; [Ivory Fashion X (Constance Spry X Monique)]; Austin, David

'THEA HARRISON', HT, lp; Harrison, A., 1927; (Weatherly, L.)

'THEA RUSSEL', F, mr, 1954; flowers bright red, moderate fragrance; vigorous growth; [Ambassadeur Nemry X Cinnabar]; Leenders, M.

Theano, HWich, lp; Geschwind, R., 1895

'THEDA MANSHOLT', Gr, my, 1966; bud ovoid; flowers medium, dbl., slight fragrance; foliage dark; upright growth; [Peace X Garden Party]; Buisman, G. A. H.

'THELMA', HWich, op, 1927; flowers coral-pink, suffused carmine-red, blooms in clusters of 3–10, semi-dbl., 3 in., slight fragrance; few prickles; vigorous, climbing growth; [R. wichurana X Paul's Scarlet Climber]; Easlea

'THELMA AMLING', Pol, w, 1942; bud globular; flowers small, semi-dbl., slight fragrance; foliage small, leathery; short stems; vigorous, bushy growth; [Mrs R.M. Finch sport]; Amling Bros.; J.H. Hill Co.

'THELMA BADER', HT, ob, 1956; flowers orange-scarlet to salmon-pink, semi-dbl., 20 petals, 3 in., slight fragrance; foliage dark, glossy; moderately vigorous growth; [Unknown seedling X (Independence X Fashion)]; Latham

'THELMA WALTON', HT, pb, 1970; flowers deep pink, lightly edged silver, very large, very dbl., exhibition form, slight fragrance; foliage glossy, dark; vigorous, upright growth; [Red Devil sport]; Walton; Cooke

Themis, F, 1984; Gaujard; (Cavriglia)

'THEODORA', HT, dr, 1984; flowers large, semi-dbl., 16 petals; [Apricot Nectar X Dame de Coeur]; Staikov, Prof. Dr V.; Kalaydjiev and Chorbadjiiski

'THEODORA MILCH', HWich, mp, 1906; flowers small, very dbl., moderate fragrance; Weigand, C.; (Sangerhausen)

'THÉODORE LIBERTON', HP, mr, 1887; flowers large, dbl., moderate fragrance; Soupert & Notting; (Sangerhausen)

Theodore Roosevelt, *see* 'BURRED'

Theone, C, mp; Noisette

'THERESA MORLEY', HT, dp, 1928; flowers brilliant carmine-cerise, dbl., moderate fragrance; [Mme Segond Weber X Lady Battersea]; H&S

Theresa Scarman, HGal, lp; Scarman, 1996

'THERESE BAUER', HSet, mp, 1963; flowers open, large, borne in clusters, semi-dbl., profuse, repeated bloom, slight fragrance; very vigorous, upright growth; [(Hansa X R. setigera) X R. setigera]; Ludwig; Kern Rose Nursery

'THÉRÈSE BONNAVIAT', HT, lp, 1934; bud long, pointed; flowers clear pink, center coppery pink, very large, moderate fragrance; foliage purplish green; very vigorous growth; [Mrs Arthur Robert Waddell X Unnamed seedling]; Chambard, C.

'THÉRÈSE BUGNET', HRg, mp, 1950; bud conical but square-tipped; flowers rose aging pale pink, white center line on top of petals, dbl., 35 petals, 4 in., borne in small clusters, repeat bloom, moderate, spicy fragrance; foliage quilted, grey-green; vigorous, 5–6 ft growth; [((R. acicularis X R. rugosa kamtchatica) X (R. amblyotis X R. rugosa plena)) X Betty Bland]; Bugnet; P.H. Wright

Therese de Lisieux®, *see* ORAblan

'THÉRÈSE SCHOPPER', HT, mr, 1933; bud long, pointed, blood-red; flowers nasturtium-red, reverse yellow, open, large, semi-dbl., exhibition form, moderate fragrance; foliage leathery, dark; vigorous growth; [(Charles P. Kilham X Mev. G.A. van Rossem) X Lady Forteviot]; Kordes

'THERESE ZEIMET-LAMBERT', HT, dp, 1922; bud long, pointed; flowers deep rose, yellow ground, base orange, large, dbl., exhibition form, intense fragrance; foliage glossy, bronze; vigorous growth; [Richmond X Mrs Aaron Ward]; Lambert, P.

'THERESETTE', S, mp, 1986; flowers deep soft pink edged white, carnation-like, medium, borne in, dbl., 26–40 frilly petals, moderate fragrance; foliage small, medium green, semi-glossy, disease resistant; bushy, hardy, compact, dwarf growth; [Arctic Glow X Therese Bugnet]; James, John

'THERMIDOR', S, ly, 1909; flowers medium, dbl., moderate fragrance; Corboeuf; (Sangerhausen)

Think Pink, *see* 'TINTHINK'

'THIRZA', HT, or, 1932; flowers orange-scarlet, moderate fragrance; Bentall

'THISBE', HMsk, ly, 1918; flowers chamois-yellow, rosette-shaped blooms in large clusters, dbl., recurrent bloom, intense fragrance; vigorous, bushy growth; [Daphne sport(?)]; Pemberton

'THOMAS A. EDISON', HT, pb, 1931; bud long, pointed; flowers two-toned pink, very large, dbl., cupped, moderate fragrance; very vigorous growth; Bernaix, P.; C-P

'THOMAS ANDREW ELLIOTT', HT, yb, 1995; flowers golden yellow with yellow and pink reverse, very dbl., 41 petals, 4 in., borne mostly singly, slight fragrance; some prickles; foliage medium, medium green, semi-glossy; medium (50–60 cms), upright growth; [Grandpa Dickson X (City of Gloucester X Seedling X Polestar)]; Thomas, D.

Thomas Garces, HT, 1986; Dot, Simon; (Cavriglia)

'THOMAS MILLS', HP, dp, 1873; flowers rosy crimson, large, dbl., cupped, non-recurrent, moderate fragrance; vigorous growth; Verdier, E.

Thomasville Lamarque, N, lp; (found rose)

Thomasville Old Gold, T, lp; (found rose)

'THOOM', F, op, 1994; (**Glamorgan**); flowers coral-salmon, blooms borne in large clusters, semi-dbl., 14 petals, 1.5–2.75 in., slight fragrance; some prickles; foliage medium, red when young turning to medium green, glos; low to medium (2 ft 6 in.), upright, compact growth; [Old Master X Red Splendor]; Thompson, M.L.; Haynes Roses, 1995

'THOR', LCl, dr, 1940; bud ovoid; flowers crimson, dbl., 58 petals, 4–5 in., slight, damask fragrance; foliage large, leathery, dark; vigorous, climbing (8–10 ft) growth; [(Alpha X R. xanthina) X Pres. Coolidge]; Horvath; Wayside Gardens Co.

'THOR SUPREME', LCl, dr, 1951; flowers dark crimson edged deeper, 4 in., intense, damask fragrance; foliage large, leathery, dark; vigorous, climbing growth; Horvath; Wyant

Thora Hird, *see* 'TONYBRAC'

'THORA MCCREA', HT, mr, 1986; dbl., 20–25 petals, exhibition form, slight fragrance; red-brown prickles; foliage dark green; well-shaped, upright, medium growth; [Pink Parfait X Crimson Glory]; Summerell, B.L.

Thoresbyana, *see* **'BENNETT'S SEEDLING'**

Thornfree Wonder, *see* 'NOBAL'

Thornhem®, HT, w

'THORNLESS BEAUTY', HT, dr, 1938; bud long, pointed, dark crimson-red; flowers light crimson-red, dbl., 50 petals, 4 in., moderate fragrance; thornless; foliage leathery; long, strong stems; vigorous growth; [Better Times sport]; Grillo

Thornless Beauty, F, lp; Patil, B.K., 1990

'THORNLESS BLUSH', HT, lp, 1954; bud globular; flowers blush-pink, dbl., 70 petals, 4.5 in., exhibition form, moderate fragrance; thornless stems; foliage leathery; very vigorous, upright growth; [Rosalind Russell sport]; Grillo

'THORNLESS FRINGEDALE', HT, lp, 1955; dbl., 25 petals, 4 in., moderate fragrance; thornless stems; [Thornless Beauty sport]; Grillo

'THORNLESS GLORY', LCl, lp, 1935; flowers clear pink, semi-dbl., 20 petals, 3 in., early summer bloom, slight fragrance; thornless; long stems; vigorous, climbing growth; Izzo

'THORNLESS MIRAGE', HT, mr, 1955; flowers red suffused blush-pink, dbl., 25 petals, 5 in., moderate fragrance; thornless stems; [Jewel sport]; Grillo

'THORNLESS PREMIER', HT, mp, 1955; bud globular, pointed; flowers rich pink, dbl., 75 petals, 4.5 in., moderate fragrance; thornless stems; [Victory Stripe sport X Jewel]; Grillo

'THORNLESS VICTORY STRIPE', HT, rb, 1955; bud long, pointed; flowers dark red and blush-pink, striped, dbl., 50 petals, 5 in.; thornless stems; [Victory Stripe sport X Jewel]; Grillo

Thoughtful, *see* 'RESTO'

Thousand Beauties, *see* **'TAUSENDSCHÖN'**

Threepenny Bit Rose, *see* **R. FARRERI PERSETOSA**

'THRILLER', HT, pb, 1986; flowers white center, pink edges, reverse slight pink edging, pink h, dbl., 28–30 petals, exhibition form, intense, fruity fragrance; medium, long, light green prickles, pointed downwards; foliage medium, dark green, glossy; upright, medium, strong growth; [Lady X X Flaming Beauty]; Bridges, Dennis A., 1987

'THUMBELINA', Min, rb, 1954; flowers cherry-red, white eye, small, semi-dbl.; foliage dark, glossy; dwarf (6–8 in.), bushy growth; [Eblouissant X Zee]; Moore, Ralph S.; Sequoia Nursery

Thumbelina, *see* 'SUNTHUMB'

'THUNDER CLOUD', Min, or, 1979; bud ovoid; dbl., 70 petals, 1–1.5 in.; foliage glossy, leathery; dwarf, bushy, upright growth; [Little Chief X Fire Princess]; Moore, Ralph S.; Sequoia Nursery

'THUNDERBIRD', F, mr, 1958; bud urn-shaped; flowers rose-red, blooms in clusters of 3–8, dbl., 48 petals, 2.5 in., exhibition form; vigorous, spreading growth; [Skylark sport]; deVor, Paul F.; Amling Bros.

Thunderbolt, *see* 'BRIBOLT'

Thungabhadra, HT, ob; Kasturi, 1985

'THUSNELDA', HRg, mp, 1886; flowers rose, semi-dbl., non-recurrent; vigorous growth; [R. rugosa alba X Gloire de Dijon]; Müller, Dr F.

'THYRA HAMMERICH', HP, lp, 1868; flowers medium, very dbl.; Vilin; (Sangerhausen)

Thyrion, HT, 1969; McGredy, Sam IV; (Cavriglia)

Tiamo, *see* SPEhewe

'TIARA', F, w, 1960; bud ovoid; dbl., 48 petals, 3 in., cupped, blooms in clusters, moderate fragrance; foliage leathery; vigorous, bushy growth; [Chic X Demure seedling]; Boerner; J&P

Tibeert, S, 1991; Delforge; (Cavriglia)

Ticino, F, mr, 1979; flowers vermilion red, large, dbl.; Huber; (Sangerhausen)

Tickle Me Pink, *see* 'PIXINK'

'TICKLED PINK', HT, mp, 1963; bud long, pointed; flowers medium, dbl., exhibition form, moderate, spicy fragrance; upright, bushy growth; [Queen Elizabeth X Fashion]; Lammerts, Dr Walter; Amling-DeVor Nursery; Germain's

Tickled Pink, *see* 'SUNTICK'

Tickles™, *see* 'DYKTICK'

'TICK-TOCK', Min, ob, 1973; flowers salmon-pink, shaded orange, open, semi-dbl., 10 petals, 2 in., slight fragrance; free growth; [New Penny X Elizabeth of Glamis]; McGredy, Sam IV; McGredy

Tidbit Rose, *see* Conditorum

Tidewater, *see* 'BRITIDE'

Tiecelijn, HT, 1990; Delforge; (Cavriglia)

'TIERGARTENDIREKTOR TIMM', F, dp, 1944; bud long, pointed; flowers carmine-pink, open, very large, borne in clusters, semi-dbl., slight fragrance; foliage dark, glossy; very vigorous, upright growth; [Sweetness X Hamburg]; Kordes

'TIFFANY', HT, pb, 1954; bud long, pointed; flowers pink-yellow blend, dbl., 28 petals, 4–5 in., exhibition form, intense fragrance; foliage dark; vigorous, upright growth; AARS, 1955 David Fuerstenberg Prize, ARS, 1957 GM, Portland, 1954 James Alexander Gamble Fragrance Medal, ARS, 1962; [Charlotte Armstrong X Girona]; Lindquist; Howard Rose Co.

Tiffany Lite, *see* 'GELLITE'

'TIFFANY LYNN', MinFl, pb, 1985; flowers light to medium pink at edges, blending to white in center, exhibition form, 21 petals, exhibition form, slight; prickles slanted downward; foliage medium, medium green with red edges, semi-glossy; upright, bushy growth; [(Tiki X Seedling) X Party Girl]; Jolly, Nelson F.; Rosehill Farm

'TIFFANY, CLIMBING', Cl HT, pb, 1958; (Grimpant Tiffany); Lindquist; Howard Rose Co.

'TIFFIE', Min, lp, 1979; bud long, pointed; flowers soft pink, sometimes apricot, blooms borne singly, 15–20 petals, exhibition form, moderate, lilac fragrance; large, red prickles; foliage medium green; upright growth; [Little Darling X Over the Rainbow]; Bennett, Cecilia 'Dee' Tiny Petals Nursery

'TIFTON', HT, yb, 1983; flowers brilliant yellow, petal tips red, red spreading with age, large, dbl., exhibition form, moderate fragrance; foliage large, medium, semi-glossy; bushy growth; [Fire Magic X Oregold]; Perry, Astor; Perry Roses

Tiger, *see* MACgremli

'TIGER BELLE', F, ob, 1962; bud ovoid; flowers light signal-red, dbl., 25–35 petals, 2–2.5 in., cupped, moderate growth; [Orange Sweetheart X Lovelight]; Jelly; E.G. Hill Co.

Tiger Butter, *see* 'PIXITER'

Tiger Cub, *see* POUlcub

Tiger Paws™, *see* 'LAVTIPAWS'

Tiger Stripes, *see* 'CLETIG'

Tiger Tail, *see* 'JACTIGER'

'TIGLIA', S, yb, 1962; flowers creamy yellow tinged pink, dbl., recurrent bloom, moderate fragrance; [Ma Perkins X High Noon]; Leenders, J.

Tigris, *see* 'HARPRIER'

'TIKI', F, pb, 1964; flowers light pink and white blend, well-formed, flora-tea, dbl., 30 petals, 3.5 in.; foliage dark; vigorous growth; [Mme Leon Cuny X Spartan]; McGredy, Sam IV; McGredy

'TILL UHLENSPIEGEL', HEg, rb, 1950; bud pointed; flowers bright red, white eye, blooms in clusters, single, 5 petals, non-recurrent; foliage large, glossy, dark reddish-green; tall (to 10 ft), arching growth; [Holstein X Magnifica]; Kordes, W. Söhne

'TILLICUM', HT, 1924; flowers deep rose-pink shaded orange, dbl., slight fragrance; [Général Jacqueminot X Old Gold]; Wilber

Tilly Aston, HT, lp; [Queen Elizabeth sport]; Nieuwesteeg, J., 1996; (Weatherly, L.)

Tilt Symphonie®, *see* MEIvraivou

'TIM PAGE', HT, 1920; flowers rich daffodil yellow, moderate fragrance; Page; Easlea

Timeless™, *see* 'JACECOND'

Timeless Delight, *see* 'RENTDEL'

'TIMES SQUARE', HT, ob, 1943; bud peach; flowers orange, center golden yellow, large, intense fragrance; foliage leathery, glossy; vigorous growth; [Mrs Sam McGredy X Pres. Herbert Hoover]; Lammerts, Dr Walter; Armstrong Nursery

'TIMMIE ARKLES', F, lp, 1954; dbl., 3 in., flat, blooms in clusters of 30–40, moderate, fruity fragrance; vigorous growth; [Mrs R.M. Finch X (Improved Lafayette seedling X Rochester seedling)]; Boerner; J&P

'TIMM'S JUBILAÜMSROSE', F, or, 1975; bud globular; dbl., 27 petals, 2 in., globular, slight fragrance; foliage soft; vigorous, upright, bushy growth; [Marlena X Europeana]; Kordes; Timm

'TIMOTHY BERLEN', Min, ob, 1986; flowers orange, yellow center, reverse yellow fading dark pink, medium, semi-dbl., 20 petals, exhibition form, slight fragrance; round, green yellow-orange fruit; bayonet-shaped, light pink prickles; foliage small, medium green, semi-glossy; upright, spreading, low growth; [Anita Charles X Poker Chip]; Jolly, Marie; Rosehill Farm, 1988

'TIMOTHY EATON', HT, op, 1968; flowers salmon-pink, well-formed, slight fragrance; [Radar X Mischief]; McGredy, Sam IV

'TINA TURNER', HT, ob, 1992; dbl., 26–40 petals, 1.5–2.75 in., borne mostly singly, moderate fragrance; some prickles; foliage large, dark green, glossy; bushy (60 cms) growth; [Silver Jubilee X Doris Tysterman]; Thompson, Robert; Battersby Roses, 1990

'TINABLER', Min, mr, 1991; (**Elizabeth Abler**); flowers medium red, opening to paler red at center, micro-mini, semi-dbl., 6–14 petals, 1.5 in., exhibition form, borne mostly singly, moderate fragrance; no prickles; foliage small, medium green, semi-glossy; low (20–30 cms), bushy, compact growth; [(Christian Dior X Brian Lee) X Unknown]; Bennett, Cecilia 'Dee' Tiny Petals Nursery, 1992

'TINAMIGO', Min, mr, 1983; (**Little Amigo**); flowers small, dbl., 35 petals, exhibition form, no fragrance; foliage small, medium green, semi-glossy; upright growth; [Futura X Orange Honey]; Bennett, Cecilia 'Dee'; Tiny Petals Nursery

'TINBAB', Min, lp, 1999; (**Big and Beautiful**); flowers deep apricot, dbl., 26–40 petals, 1–1.5 in., borne mostly singly, slight fragrance; prickles moderate; foliage medium, medium green, semi-glossy; upright, bushy, tall (5–6 ft) growth; [Futura X Jean Kenneally]; Bennett, Cecilia 'Dee'; Tiny Petals, 2000

'TINBOPPER', Min, rb, 1989; (**Teeny Bopper**); bud ovoid; flowers white with red blush on outer petals, aging red with white, semi-dbl., cupped, slight, damask fragrance; globular, yellow green orange fruit; no prickles; foliage small, medium green, semi-glossy; micro-mini, bushy growth; PP007449; [Little Squirt X Unnamed seedling]; Bennett, Cecilia 'Dee'; Tiny Petals Nursery, 1988

'TINBUFORD', Min, ly, 1993; (**Linda Buford**); flowers soft light yellow, blooms borne mostly singly, dbl., 26–40 petals, 1.5 in., slight fragrance; some prickles; foliage small, medium green, semi-glossy; medium, bushy growth; [Irish Gold X Jean Kenneally]; Bennett, Cecilia 'Dee'; Tiny Petals Nursery, 1994

'TINBUTT', Min, ob, 1985; (**Bread 'n' Butter**); flowers golden orange blended yellow, medium, dbl., 28 petals, exhibition form, moderate fragrance; foliage medium, medium green, semi-glossy; upright, bushy growth; PP006135; [Arizona X Orange Honey]; Bennett, Cecilia 'Dee'; Tiny Petals Nursery

'TINCAROL', Min, dp, 1989; (**Little Carol**); bud ovoid; flowers deep magenta-pink, reverse slightly lighter, medium, borne singly, dbl., 38 petals, exhibition form, slight, spicy fragrance; globular, green brown fruit; prickles hooked slightly downward, red, few; foliage medium, dark green, semi-glossy, disease-resistant; upright, tall growth; [Sonia X Jean Kenneally]; Bennett, Cecilia 'Dee'; Tiny Petals Nursery

'TINCHAN', Min, my, 1987; (**Dr K. C. Chan**); flowers medium yellow, fading pale yellow, urn-shaped, medium, dbl., 25–30 petals, borne usually singly, slight fragrance; globular, medium, brown fruit; slender, straight, reddish prickles; foliage medium, medium green, semi-glossy; mini-flora; upright, bushy, medium growth; [Irish Gold X Rise 'n' Shine]; Bennett, Cecilia 'Dee'; Tiny Petals Nursery, 1986

'TINCHEER', Min, ob, 1986; (**Cheer Up**); flowers deep orange, urn-shaped, dbl., 28 petals, exhibition form, borne usually singly and in small clusters, slight fragrance; globular, 1/2 in., green and brown fruit; small, red prickles; foliage medium, dark, semi-glossy; medium, upright, bushy growth; PP006456; [Futura X Bread 'n' Butter]; Bennett, Cecilia 'Dee'; Tiny Petals Nursery

'TINCHRIS', Min, mr, 1988; (**Christopher**); flowers medium red, medium, dbl., 30–35 petals, exhibition form, borne usually singly or in sprays of 3–5; globular, green brown fruit; prickles hooked slightly downward, yellow-red; foliage medium, medium green, semi-glossy; upright, medium growth; PP007191; [Futura X Big John]; Bennett, Cecilia 'Dee'; Tiny Petals Nursery

'TINCHULA', Min, dr, 1992; (**Chula Vista**); dbl., 26–40 petals, 1.5 in., exhibition form, borne mostly singly, slight fragrance; some prickles; foliage small, medium green, semi-glossy, disease resistant; long stems; medium (60–80 cms), bushy growth; [Christian Dior X Brian Lee]; Bennett, Cecilia 'Dee'; Tiny Petals Nursery, 1993

'TINCUDDLE', Min, yb, 1991; (**Cuddle Up**); bud ovoid; flowers creamy yellow with coral pink blush, medium, dbl., 25–30 petals, exhibition form, moderate, fruity fragrance; foliage medium, medium green, semi-glossy; bushy, medium growth; [Lagerfeld X My Delight]; Bennett, Cecilia 'Dee'; 1987; Tiny Petals Nursery, 1990

'TINCUPIE', Min, lp, 1983; (**Cupie Doll**); flowers light pink with fine coral edging of petals, heavy substance, dbl., 25 petals, exhibition form, slight fragrance; foliage small, medium green, semi-glossy; upright, bushy growth; [Seedling X Coral Treasure]; Bennett, Cecilia 'Dee'; Tiny Petals Nursery

'TINDAD', Min, mr, 1996; (**Love 'Ya Dad**); flowers does not fade, blooms borne mostly single, long-stemmed, dbl., 26–40 petals, 1.5 in., exhibition form, intense fragrance; moderate prickles; foliage medium, dark green, semi-glossy, disease-resistant; upright, bushy, tall (70–80 cms) vigorous growth; [Carrousel X Big John]; Bennett, Cecilia 'Dee'; Tiny Petals Nursery, 1997

'TINDEB', Min, op, 1985; (**Debra Gaye**); flowers orange-pink, soft yellow reverse, medium, dbl., 38 petals, exhibition form, slight fragrance; foliage medium, medium green, semi-glossy; bushy, upright growth; PP006136; [Futura X Fairest of Fair]; Bennett, Cecilia 'Dee'; Tiny Petals Nursery

'TINDEL', F, dp, 1993; (**Del Mar Fair**); flowers deep pink, blooms borne in large clusters, dbl., 26–40 petals, 3–3.5 in., moderate fragrance; some prickles; foliage medium, medium green, semi-glossy; medium, upright, bushy growth; [Gene Boerner X Pucker Up]; Bennett, Cecilia 'Dee'; Tiny Petals Nursery, 1994

'TINDICK', Min, r, 1997; (**High Cloud**); flowers soft tan with yellow base, lighter reverse, dbl., 26–40 petals, 1.5 in., borne singly or in small clusters, moderate fragrance; foliage medium, medium green, glossy; bushy, tall (60–70 cm) growth; [Lagerfeld X Ernie]; Bennett, Cecilia 'Dee'

'TINDILLY', Min, MinFl, m, 1985; (**Dilly Dilly**); flowers lavender, mini-flora, medium, dbl., 35 petals, exhibition form, moderate fragrance; foliage medium, medium green, semi-glossy; upright growth; PP006141; [Chrysler Imperial X Plum Duffy]; Bennett, Cecilia 'Dee'; Tiny Petals Nursery

'TINDINK', Min, pb, 1983; (**Fair Dinkum**); flowers soft pink, petal margins darker, small, dbl., 25 petals, exhibition form, slight fragrance; foliage small, medium green, semi-glossy; upright, bushy growth; PP005564; [Unnamed seedling X Coral Treasure]; Bennett, Cecilia 'Dee'; Tiny Petals Nursery

'TINDONN', Min, dr, 1991; (**Brian Donn**); bud ovoid; flowers darkest ruby red to maroon, dark ruby red reverse, ages only slightly lighter, very dbl., 35–45 petals, exhibition form, moderate, damask fragrance; foliage medium, medium green, semi-glossy; bushy, medium growth; [Intrique X Big John]; Bennett, Cecilia 'Dee'; 1985; Tiny Petals Nursery, 1990

'TINDREAM', Min, yb, 1987; (**Dreamy**); flowers cream, edges blushed pink, reverse cream with more intense blush, dbl., 20–25 petals, slight fragrance; globular, medium, brown fruit; slender, small, reddish prickles, slanted downward; foliage medium, medium green, semi-glossy; upright, medium growth; [Irish Gold X Party Girl]; Bennett, Cecilia 'Dee'; Tiny Petals Nursery

'TINEALLY', Min, ab, 1984; (**Jean Kenneally**™); flowers pale to medium apricot, small, dbl., 22 petals, exhibition form, borne mostly singly, slight fragrance; foliage medium, medium green, semi-glossy; long stems; upright, bushy, vigorous growth; PP005637; AOE, 1986; [Futura X Party Girl]; Bennett, Cecilia 'Dee'; Tiny Petals Nursery

'TINEKE', HT, w, 1989; bud pointed; flowers creamy white, large, borne singly, dbl., 53 petals, cupped, no fragrance; prickles yellow-green with reddish tip; foliage large, dark green, semi-glossy; upright, tall growth; [Unnamed seedling X Unnamed seedling]; Select Roses, B.V.; DeVor Nurseries, Inc., 1990

Tineke van Heule®, *see* 'LENTINI'

'TINERNIE', Min, m, 1989; (**Ernie**); bud ovoid; flowers light mauve, urn-shaped, medium, dbl., 48 petals, borne occasionally singly and in clusters of 6–12, moderate, fruity fragrance; globular, yellow-brown fruit; prickles hooked slightly downward, pale yellow-brown, few; foliage medium, medium green, semi-glossy; upright, bushy, tall growth; [Blue Nile X Blue Mist]; Bennett, Cecilia 'Dee'; Tiny Petals Nursery

TINgail, F, yb, 1986; (**Gail**); flowers golden yellow with orange blush on petal tips, orange spreading with age, dbl., 38 petals, exhibition form, moderate fragrance; globular, 5/8 in., light green fruit; small reddish prickles; foliage large, medium green, semi-glossy; medium, bushy, spreading growth; [Arizona X Orange Honey]; Bennett, Cecilia 'Dee'; Tiny Petals Nursery

'TINGIRL', Min, op, 1983; (**Bloomer Girl**); flowers soft vermilion to medium pink, small, dbl., 35 petals, slight fragrance; foliage medium, medium green, semi-glossy; upright, bushy growth; PP005523; [Futura X Pink Petticoat]; Bennett, Cecilia 'Dee'; Tiny Petals Nursery

'TINGRACE', Min, w, 1991; (**Grace Seward**); bud ovoid; flowers medium, single, 5 petals, moderate, damask fragrance; foliage medium, medium green, semi-glossy; bushy, tall growth; [Watercolor X Unnamed seedling]; Bennett, Cecilia 'Dee'; 1987; Tiny Petals Nursery, 1990

'TINHAT', Min, my, 1988; (G'Day, Gidday, **Madhatter**); flowers medium yellow, aging paler, medium, borne usually singly or, dbl., 25–30 petals, exhibition form, moderate, damask fragrance; globular, green brown fruit; prickles hooked slightly downward, reddish; foliage medium, medium green, semi-glossy; bushy, medium growth; [Autumn X Avandel]; Bennett, Cecilia 'Dee'; Tiny Petals Nursery

'TINHELEN', Min, mr, 1991; (**Helen**); dbl., 26–40 petals, 1.5 in., exhibition form, borne mostly singly, moderate, fruity fragrance; few prickles; foliage small, medium green, semi-glossy, disease-resistant; long cutting stems; tall (60–80 cms), upright, bushy growth; [Carrousel X Starina]; Bennett, Cecilia 'Dee'; Tiny Petals Nursery, 1992

'TINHERB', Min, m, 1987; (**Herbie**); flowers rich mauve, outer petals deep mauve at margins, urn-shaped, dbl., 25–30 petals, slight fragrance; globular, medium, brown fruit; slender, straight, average, reddish prickles; foliage medium, medium green, semi-glossy; upright, bushy, medium growth; PP006787; [Deep Purple X Dilly Dilly]; Bennett, Cecilia 'Dee'; Tiny Petals Nursery

'TINHUZZY', Min, rb, 1991; (**Little Huzzy**); flowers white with striping of pink to medium red over most upper surface, dbl., 26–40 petals, 1.5–2.75 in., exhibition form, moderate fragrance; some prickles; foliage small, medium green, semi-glossy, disease-resistant; long cutting stems; tall (60–80 cms), upright growth; [Futura X Pucker Up]; Bennett, Cecilia 'Dee'; Tiny Petals Nursery, 1992

'TINJEAN', Min, pb, 1987; (**Jean Rose**); flowers peach-pink with yellow base, reverse soft yellow, fading lighter, dbl., 20–25 petals, slight fragrance; globular, medium, brown fruit; few, slender, small, reddish prickles; foliage medium, medium green, semi-glossy; upright, medium growth; [Electron X Fairest of Fair]; Bennett, Cecilia 'Dee'; Tiny Petals Nursery

'TINJOSH', Min, lp, 1999; (**Josh**); dbl., 26–40 petals, 1–1.5 in., borne singly and in small clusters, slight fragrance; prickles moderate; foliage medium, medium green, semi-glossy; upright, bushy, medium (2–3 ft) growth; [Futura X Rise 'n' Shine]; Bennett, Cecilia 'Dee'; Bridges Roses, 2000; Tiny Petals, 2000

'TINJUDY', MinFl, pb, 1995; (**Something for Judy**); flowers cream with deep pink blush, blooms borne mostly single, dbl., 15–25 petals, 1.5–2.75 in., moderate fragrance; foliage medium, medium green, semi-glossy; tall (2–3 ft), bushy growth; [Angel Face X Big John]; Bennett, Cecilia 'Dee'; Tiny Petals Nursery, 1995

'TINKAY', Min, pb, 1995; (**Kay Denise**); flowers soft creamy pink, pink blush margins, full, very dbl., 26–40 petals, exhibition form, borne mostly singly, moderate fragrance; some prickles; foliage small, medium green, semi-glossy; upright, bushy, tall growth; [Tiki X My Delight]; Bennett, Cecilia 'Dee'; Tiny Petals Nursery, 1995

'TINKER BELL', Min, mp, 1954; bud ovoid; flowers bright rose-pink, small, dbl., 60 petals, 1.5 in., cupped, foliage small, leathery; dwarf (8 in.), bushy growth; [Ellen Poulsen X Tom Thumb]; deVink; C-P

Tinkerbell, *see* 'SUNTINK'

'TINKIN', Min, ob, 1983; (**Punkin**); flowers orange with yellow eye, ages red, blooms borne singly, single, 5 petals, no fragrance; foliage small, medium green, semi-glossy; upright growth; [Orange Honey X Orange Honey]; Bennett, Cecilia 'Dee'; Tiny Petals Nursery

'TINLADY', Min, m, 1988; (**Always A Lady**); flowers pale mauve to lavender, medium, dbl., 25–30 petals, exhibition form, borne usually singly or in sprays of 3–5, moderate, damask fragrance; globular, green brown fruit; prickles hooked slightly downward, pale yellow; foliage medium, medium green, semi-glossy; bushy, medium growth; PP007187; [Deep Purple X Dilly Dilly]; Bennett, Cecilia 'Dee'; Tiny Petals Nursery, 1987

'TINLEE', Min, m, 1986; (**Brian Lee**); flowers dark red, aging mauve, urn-shaped, mini-flora, blooms borne, dbl., 28 petals, exhibition form, moderate fragrance; 5/8 in.; globular, brown-green fruit; small reddish prickles; foliage medium, dark, semi-glossy; medium, upright, bushy growth; PP006789; [Carrousel X Plum Duffy]; Bennett, Cecilia 'Dee'; Tiny Petals Nursery

'TINLESLIE', Min, mp, 1997; (**Leslie**); flowers medium pink, 3 cm., semi-dbl., 15–25 petals, 1.5 in., borne in small clusters; foliage medium, medium green, semi-glossy; bushy, 40–50 cms tall growth; [Deep Purple X Jean Kenneally]; Bennett, Cecilia 'Dee'; Tiny Petals Nursery, 1997

'TINLIPS', Min, op, 1988; (**Hot Lips**); flowers deep coral to orange, aging paler, medium, borne usually singly, dbl., 25–30 petals, exhibition form, slight, fruity fragrance; globular, green yellow-brown fruit; prickles hooked slightly downward, reddish; foliage medium, medium green, semi-glossy; upright, bushy, tall growth; [Futura X Why Not]; Bennett, Cecilia 'Dee'; Tiny Petals Nursery

'TINLOIS', Min, mp, 1985; (**Lois**); flowers medium lilac pink, medium, dbl., exhibition form, moderate fragrance; foliage medium, medium green, semi-glossy; upright, bushy growth; PP006140; [Deep Purple X Plum Duffy]; Bennett, Cecilia 'Dee'; Tiny Petals Nursery

'TINLOVE', Min, pb, 1989; (**Mother's Love**); bud ovoid; flowers pastel pink, blending to soft yellow at base, medium, dbl., 23 petals, exhibition form, slight, fruity fragrance; globular, green yellow-brown fruit; prickles straight, tapering, reddish; foliage medium, medium green, semi-glossy; upright, bushy, medium growth; PP007484; [Futura X Party Girl]; Bennett, Cecilia 'Dee'; Tiny Petals Nursery, 1988

'TINLUIS', Min, ly, 1989; (**Luis Desamero**); bud ovoid; flowers pastel yellow, medium, dbl., 28 petals, exhibition form, borne usually singly and in sprays of 3–5, slight, fruity fragrance; globular, green yellow-orange fruit; prickles straight & tapering, pale yellow, reddish base; foliage medium, medium green, semi-glossy; upright, bushy, tall growth; PP007496; [Tiki X Baby Katie]; Bennett, Cecilia 'Dee'; Tiny Petals Nursery, 1988

'TINMAD', MinFl, mp, 1985; (**Madeline Spezzano**); flowers mini-flora, medium, dbl., 45 petals, exhibition form, slight fragrance; foliage large, medium green, semi-glossy; upright, bushy growth; PP006132; [Sonia X Beauty Secret]; Bennett, Cecilia 'Dee'; Tiny Petals Nursery

'TINMARY', Min, yb, 1999; (**Ms Mary**); flowers pale yellow/apricot, reverse pale yellow, dbl., 26–40 petals, 1 in., borne mostly singly, moderate fragrance; prickles moderate; foliage medium, medium green, semi-glossy; compact, bushy, low (18–24 in.) growth; [Deep Purple X Fairest of Fair]; Bennett, Cecilia 'Dee'; Tiny Petals, 1999

'TINMERE', Min, pb, 1983; (**Meredith Anne**); flowers salmon-pink, soft pink reverse, small blooms in clusters, dbl., 23 petals, exhibition form; brown to pale yellow prickles; vigorous growth; [Sonia X Tea Party]; Bennett, Cecilia 'Dee'; Tiny Petals Nursery

'TINMYDE', Min, lp, 1983; (**My Delight**); flowers small, dbl., 30 petals, exhibition form, slight fragrance; foliage medium, medium green, semi-glossy; upright, spreading growth; [Futura X Avandel]; Bennett, Cecilia 'Dee'; Tiny Petals Nursery, 1984

'TINNAUGHTY', Min, ab, 1991; (**Naughty But Nice**); bud ovoid; flowers soft apricot, medium, dbl., 20–25 petals, exhibition form, slight, damask fragrance; foliage medium, medium green, semi-glossy; bushy, tall growth; [Futura X Why Not]; Bennett, Cecilia 'Dee'; 1983; Tiny Petals Nursery, 1990

Tino Rossi®, *see* MEIcelna

'TINOLD', Min, w, 1992; (**Old Fashioned Girl**); flowers soft lavender to white, opening very full, similar to ogr's, dbl., 26–40 petals, 1.5–2.75 in., exhibition form, intense fragrance; some prickles; foliage small, medium green, semi-glossy; medium (60–80 cms), bushy growth; [Blue Ribbon X Unnamed Miniature seedling]; Bennett, Cecilia 'Dee'; Tiny Petals Nursery, 1993

'TINPANKY', Min, ob, 1991; (**Hanky Panky**); flowers range from orange to golden peach as they open, dbl., 26–40 petals, 1.5–2.75 in., borne mostly singly, moderate, fruity fragrance; some prickles; foliage small, medium green, semi-glossy, disease resistant; medium (40–60 cms), bushy, spreading growth; [Deep Purple X Party Girl]; Bennett, Cecilia 'Dee'; Tiny Petals Nursery, 1992

'TINPAT', Min, mp, 1985; (**Patty Sue**); flowers medium, dbl., 35 petals, exhibition form, slight fragrance; foliage medium, medium green, matt; vigorous, upright growth; [Little Darling X Little Chief]; Bennett, Cecilia 'Dee'; Tiny Petals Nursery

'TINPEACES', Min, yb, 1985; (**Little Peaces**); flowers yellow, edged pink, medium, dbl., 33 petals, exhibition form, slight fragrance; foliage medium, medium green, semi-glossy; compact, bushy growth; [Electron X Fairest of Fair]; Bennett, Cecilia 'Dee'; Tiny Petals Nursery

'TINPIN', Min, m, 1987; (**Hat Pin**); flowers pale lavender, urn-shaped, small, semi-dbl., 12–15 petals, borne usually singly, moderate, fruity fragrance; globular, small, brown fruit; straight, extremely small, pale yellow prickles; foliage small, medium green, semi-glossy; micro-mini upright, bushy, low growth; PP006790; [Angel Face X Angelglo]; Bennett, Cecilia 'Dee'; Tiny Petals Nursery, 1986

'TINPLEASURE', Min, lp, 1999; (**Ultimate Pleasure**); flowers light pink, reverse medium pink, dbl., 26–40 petals, .75–1 in., exhibition form, borne mostly singly or in small clusters, moderate fragrance; prickles moderate; foliage medium, medium green, semi-glossy; extremely vigorous, bushy, upright, tall (24–36 in.) growth; [Gene Boerner X Jean Kenneally]; Bennett, Cecilia 'Dee'; Tiny Petals, 1999

'TINPOPS', Min, dp, 1983; (**Pops**); flowers deep pink, small, dbl., 20 petals, exhibition form, slight fragrance; foliage medium, medium green, semi-glossy; bushy growth; [Sheri Anne X Little Girl]; Bennett, Cecilia 'Dee'; Tiny Petals Nursery

'TINPORCE', Min, lp, 1983; (**Pink Porcelain**); flowers small blooms borne singly and in clusters, dbl., 23 petals, exhibition form, slight fragrance; foliage small, medium green, semi-glossy; upright growth; [Futura X Avandel]; Bennett, Cecilia 'Dee'; Tiny Petals Nursery

'TINPUCK', Min, or, 1984; (**Pucker Up**); bud small; flowers bright orange-red, dbl., 23 petals, exhibition form, slight fragrance; foliage medium, medium green, semi-glossy; upright, bushy growth; PP005407; [Futura X Avandel]; Bennett, Cecilia 'Dee'; Tiny Petals Nursery, 1983

'TINQUEEN', Min, pb, 1996; (**Bingo Queen**); flowers light pink with medium pink inner petals, reverse same, one-, dbl., 26–40 petals, 1.5 in., exhibition form, slight fragrance; moderate prickles; foliage medium, medium green, semi-glossy; bushy, medium (50–60 cms) growth; [Brandy X Jean Kenneally]; Bennett, Cecilia 'Dee'; Tiny Petals Nursery, 1997

'TINQUIET', MinFl, m, 1995; (**Quiet Time**); flowers lavender-pink with inner petals of soft tan, blooms borne mo, dbl., 25–40 petals, 1.5 in., slight fragrance; foliage medium, medium green, semi-glossy; bushy growth; [Lagerfeld X Ernie]; Bennett, Cecilia 'Dee' Tiny Petals Nursery, 1995

'TINRESIST', Min, w, 1989; (**Irresistible**®); bud ovoid; flowers white with pale pink center, medium, greenish in shade, dbl., 43 petals, exhibition form, borne singly and in small clusters, moderate, spicy fragrance; globular, green yellow-brown fruit; prickles straight, yellow with red; foliage medium, medium green, semi-glossy; upright, tall growth; PP7971; [Tiki X Brian Lee]; Bennett, Cecilia 'Dee'; Tiny Petals Nursery, 1990

'TINREVENGE', Min, ob, 1996; (**Sweet Revenge**); flowers soft orange with blush of deep orange, blooms borne mostly singly, dbl., 26–40 petals, 1.5 in., slight fragrance; foliage medium, dark green, semi-glossy; tall (2–3 ft), bushy growth; [Tony Jacklin X Pucker Up]; Bennett, Cecilia 'Dee'; Tiny Petals Nursery, 1995

'TINRITA', MinFl, ly, 1996; (**Rita Applegate**); flowers light yellow to softest gold with heart of deeper yellow/gold, very dbl., 26–40 petals, 2 in., exhibition form, borne mostly singly, moderate fragrance; moderate prickles; foliage medium, medium green, semi-glossy; long-stemmed; upright, bushy, tall (80–90 cms), huge growth; [Pink Porcelain X Unknown]; Bennett, Cecilia 'Dee'; Tiny Petals Nursery, 1997

'TINRUTH', Min, w, 1988; (**Ruthie**); flowers cream-white with pink blush on outer petals, aging paler blush, dbl., 25–30 petals, exhibition form, moderate, damask fragrance; globular, green brown fruit; prickles hooked slightly downward, yellow to reddish; foliage medium, medium green, semi-glossy, aging darker; upright, tall growth; PP007189; [Sonia X Little Melody]; Bennett, Cecilia 'Dee'; Tiny Petals Nursery

'TINSASSY', Min, rb, 1986; (**Fat 'n' Sassy**); flowers white with a reddish border, reverse white, aging reddish blush, dbl., 28 petals, cupped, borne singly, slight fragrance; globular, 5/8 in., green-brown fruit; small, reddish prickles; foliage medium, medium green, semi-glossy; medium, upright, bushy growth; PP006455; [Carrousel X Sheri Anne]; Bennett, Cecilia 'Dee'; Tiny Petals Nursery

Tinseltown, HT, pb; Edwards, Eddie, 1996

'TINSHINE', Min, my, 1986; (**My Sunshine**); flowers medium yellow, aging soft orange, bright yellow stamens, single, moderate fragrance; globular, 1/2 in., green fruit; small, reddish-brown prickles; foliage medium, medium green, semi-glossy; medium, upright, bushy growth; PP006453; [Sunsprite X Fool's Gold]; Bennett, Cecilia 'Dee'; Tiny Petals Nursery

'TINSHORT', Min, dp, 1984; (**Short 'n' Sweet**); flowers deep pink, very small, dbl., 35 petals, no fragrance; foliage medium, medium green, semi-glossy; bushy, spreading growth; [Sheri Anne X Unnamed seedling]; Bennett, Cecilia 'Dee'; Tiny Petals Nursery

'TINSIT', Min, pb, 1986; (**Sitting Pretty**); bud ovoid; flowers apricot pink, yellow base, dbl., 38 petals, exhibition form, borne usually singly and in small clusters, intense, damask fragrance; globular, 1/2 in., yellowish-green fruit; small, red prickles; foliage medium, medium green, semi-glossy; medium, upright, bushy growth; PP006454; [Sonoma X Mabel Dot]; Bennett, Cecilia 'Dee'; Tiny Petals Nursery

'TINSNOOK', Min, ob, 1984; (**Snookie**); flowers deep orange, blushing red with age, very small, dbl., 33 petals, slight fragrance; foliage medium green, semi-glossy; bushy, tiny growth; [Torchy X Orange Honey]; Bennett, Cecilia 'Dee'; Tiny Petals Nursery

'TINSPICE', Min, lp, 1985; (**Sugar 'n' Spice**); flowers light peach-pink, medium, dbl., 28 petals, exhibition form, slight fragrance; foliage medium, medium green, semi-glossy; upright, bushy growth; PP006139; [Futura X Avandel]; Bennett, Cecilia 'Dee'; Tiny Petals Nursery

'TINSPRING', Min, pb, 1985; (**Spring Bouquet**); flowers crimson pink, blended with yellow, medium, exhibition form, moderate fragrance; foliage medium, medium green, semi-glossy; upright, bushy growth; [Portrait X Party Girl]; Bennett, Cecilia 'Dee'; Tiny Petals Nursery

'TINSQUIRT', Min, dy, 1984; (**Little Squirt**); flowers micro-mini, very small, semi-dbl., 14 petals, exhibition form, slight fragrance; foliage small, medium green, semi-glossy; upright, compact growth; Bennett, Cecilia 'Dee'; Tiny Petals Nursery, 1983

'TINSTALEY', Min, mp, 1989; (**Ruth Staley**); bud ovoid; flowers medium shell pink, reverse lighter, medium, borne singly and, dbl., 25 petals, exhibition form, moderate, fruity fragrance; globular, green brown fruit; prickles slender, straight, red; foliage medium, dark green, semi-glossy, disease-resistant; upright, tall growth; [Electron X Peachy Keen]; Bennett, Cecilia 'Dee'; Tiny Petals Nursery, 1991

'TINSTAR', Min, ob, 1986; (**Starstruck**); flowers deep golden orange with petals edged red, reverse medium red, dbl., 32 petals, slight fragrance; globular, 3/8 in., brownish fruit; small, thin, reddish prickles; foliage medium, medium green, semi-glossy; low, spreading growth; [October X Orange Honey]; Bennett, Cecilia 'Dee'; Tiny Petals Nursery

'TINSWEET', Min, lp, 1987; (**Sweet Pickins**); flowers pale pink, urn-shaped, medium, dbl., 35–40 petals, borne usually singly, moderate, fruity fragrance; globular, medium, brown fruit; slender, straight, small, reddish prickles; foliage medium, medium green, semi-glossy; micro-mini upright, bushy, medium growth; PP006788; [Futura X Party Girl]; Bennett, Cecilia 'Dee'; Tiny Petals Nursery

Tintagel, LCl, lp; [R. brunonii seedling? sport]; McLeod, J., 1995; (Weatherly, L.)

'TINTAI', MinFl, yb, 1989; (**Yantai**); bud ovoid; flowers pastel yellow, deeper yellow in center and highlights of pas, dbl., 28 petals, exhibition form, moderate, fruity fragrance; prickles hooked downward, yellow to red; foliage medium, medium green, semi-glossy; bushy, spreading, tall, vigorous growth; [Portrait X Party Girl]; Bennett, Cecilia 'Dee'; Tiny Petals Nursery, 1988

'TINTAIL', Min, w, 1983; (**Cottontail**); flowers small, dbl., 33 petals, blooms in large clusters, slight fragrance; foliage medium, medium green, semi-glossy; upright, bushy growth; PP005436; [Pink Petticoat X Pink Petticoat]; Strawn, Leslie E.; Pixie Treasures Min. Roses

'TINTHINK', Min, mp, 1983; (**Think Pink**); flowers golden stamens, small, dbl., 28 petals, intense, apple fragrance; foliage small, medium green, semi-glossy; bushy growth; [Electron X Little Chief]; Bennett, Cecilia 'Dee'

'TINTIN', F, or, 1964; flowers cinnabar-red, large, borne in large clusters, semi-dbl., cupped, moderate growth; Mondial Roses

Tintinara®, *see* 'DICUPTIGHT'

'TINTWEE', Min, op, 1987; (**Tweedle Dee**); flowers coral pink, reverse lighter, urn-shaped, small, borne usuall, dbl., 35–40 petals, slight fragrance; globular, small, brown fruit; no prickles; foliage small, medium, semi-glossy; micro-mini; upright, bushy, low growth; [Deep Purple X Cupcake]; Bennett, Cecilia 'Dee' Tiny Petals Nursery, 1986

'TINVIOLET', MinEl, m, 1993; (**Violet Mist**); flowers pale soft lavender, blooms borne mostly single, dbl., 26–40 petals, 1.5–2.75 in., slight fragrance; some prickles; foliage small, medium green, semi-glossy; medium, bushy growth; [Lagerfeld X Ernie]; Bennett, Cecilia 'Dee'; Tiny Petals Nursery, 1994

Tinwell Moss, M, lp; (found rose)

'TINWILD', Min, ob, 1992; (**Wild Thing**); flowers bright orange-yellow, blooms borne mostly single, dbl., 15–25 petals, 1.5–2.75 in., moderate, fruity fragrance; some prickles; foliage small, medium green, semi-glossy, disease resistant; medium (60–80 cms), bushy growth; [Gingersnap X Baby Katie]; Bennett, Cecilia 'Dee'; Tiny Petals Nursery, 1993

'TINx', Min, pb, 1993; (**X-Rated**); flowers creamy white, edges blush a soft coral to pink, dbl., 26–40 petals, 1.5 in., borne mostly singly, intense fragrance; some prickles; foliage small, medium green, semi-glossy; medium, bushy growth; [Tiki X Baby Katie]; Bennett, Cecilia 'Dee'; Tiny Petals Nursery, 1994

Tiny Bubbles(TM), *see* 'JACBUB'

Tiny Dancer(TM), *see* 'WILDANCE'

'Tiny Flame', Min, or, 1969; flowers coral-orange-red, micro-mini, very small, dbl.; foliage very small; dwarf (6 in.), bushy growth; [(R. wichurana X Floradora) X New Penny]; Moore, Ralph S.; Sequoia Nursery

'Tiny Jack', Min, mr, 1962; bud pointed; dbl., 28 petals, 1–1.5 in., cupped, slight fragrance; foliage leathery, dark; vigorous, bushy (12–14 in.) growth; [(R. wichurana X Floradora) X (Oakington Ruby X Floradora)]; Moore, Ralph S.; Blue Ribbon Plant Co.

'Tiny Jill', Min, mp, 1962; bud pointed; flowers small, dbl., 45 petals, 1.5 in., exhibition form, moderate fragrance; foliage glossy; vigorous, bushy (12–14 in.) growth; [(R. wichurana X Floradora) X Little Buckaroo]; Moore, Ralph S.; Blue Ribbon Plant Co.

'Tiny Love', Min, dr, 1980; bud ovoid; semi-dbl., 11 petals, cupped, borne 1–3 per cluster, intense fragrance; straight prickles; foliage dark; open, bushy growth; [Unnamed seedling X Unnamed seedling]; Lyon

Tiny Petals, *see* 'TINYPETALS'

Tiny Stars, *see* 'TRASTAR'

'Tiny Tears', Min, mp, 1979; bud ovoid; flowers pink, micro-mini, very small blooms in clusters of 10–25, single, 5 petals; foliage glossy, dark; trailing growth; [Pink Ribbon X Pink Ribbon]; Bennett, Cecilia 'Dee'; Tiny Petals Nursery

'Tiny Tim', F, my, 1943; bud long, pointed; flowers clear yellow, open, small, dbl., 23 petals, exhibition form, foliage glossy; moderately vigorous, compact to open growth; [Golden Glow X Shades of Autumn]; Brownell, H.C.

'Tiny Tot', Min, mp, 1955; flowers deep cerise-pink; very dwarf growth; Robinson, T.

Tiny Tot, *see* BENtintot

'Tiny Visions', Min, rb, 1981; bud long; flowers crimson with white center, slightly blooms borne singly, single, 5 petals, cupped, moderate fragrance; very tiny, hooked, brown prickles; foliage dark, apple-scented; very tiny, bushy growth; [Unnamed seedling X Unnamed seedling]; Lyon

'Tiny Warrior', Min, rb, 1975; bud ovoid; dbl., 34 petals, 1 in., flat, slight fragrance; foliage small, glossy, very dark, embossed; upright, bushy growth; [Starburst X Little Chief]; Williams, Ernest D.; Min-Roses

'TINYPETALS', Min, rb, 1992; (**Tiny Petals**); flowers orange-red with cream reverse, blooms borne mostly single, dbl., 15–25 petals, 1.5 in., exhibition form; some prickles; foliage small, dark green, semi-glossy, disease resistant; long stems; medium (60–80 cms), bushy growth; [San Antonio X Jean Kenneally]; Bennett, Cecilia 'Dee'; Tiny Petals Nursery, 1993

'TINZEL', Min, mp, 1988; (**Zelda Lloyd**); flowers medium pink, reverse deep pink, medium, single, 5 petals, cupped, borne usually singly, slight, fruity fragrance; globular, green yellow-brown fruit; prickles hooked slightly downward, reddish; foliage medium, medium green, semi-glossy; bushy, medium growth; PP007190; [Deep Purple X Blue Mist]; Bennett, Cecilia 'Dee'; Tiny Petals Nursery

'Tip Toes', HT, op, 1948; bud pointed; flowers salmon-pink, shading to yellow at base, large, semi-dbl., exhibition form, moderate fragrance; foliage glossy; vigorous, upright growth; [(Général Jacqueminot X Dr W. Van Fleet) X Anne Venderbilt]; Brownell, H.C.

Tip Top, *see* TANopel

Tipo Ideale, *see* **'Mutabilis'**

Tipper(TM), *see* 'JOLTIP'

'Tipperary', HT, my, 1917; bud long, pointed; flowers golden yellow, semi-dbl., moderate fragrance; [Mrs Aaron Ward seedling]; McGredy

'Tipsy', HT, dp, 1969; flowers soft rose, reverse geranium-red, pointed, large, dbl.; foliage dark, dull; free growth; [Basildon Belle X Seedling]; Trew, C.; Basildon Rose Gardens

Tipsy Imperial Concubine, T, pb; re-introduced, 1989

'Tip-Top'®, Pol, w, 1909; (Baby Doll); flowers white, tipped tyrian rose, aging white and pale yellow, reve, dbl.; bushy, dwarf growth; [Trier X R. foetida bicolor seedling]; Lambert, P.

Tipu's Flame, S, rb; Viraraghavan, M.S., 1990

'Tira-Mi-Su', HT, lp, 1999; flowers pale copper and orange blend, dbl., 28 petals, 5 in., slight fragrance; 4.5 ft growth; [(Julia X seedling) X seedling]; Teranishi, K.; Itami Rose Nursery, 1999

'Titania', Ch, rb, 1915; flowers deep salmon-red, shaded clear yellow, small, dbl.; dwarf, bushy growth; Paul, W.

'Titania', Pol, lp, 1938; flowers salmon-flesh to rosy white, large, very dbl., intense fragrance; foliage leathery, dark; vigorous, bushy growth; [Mev. Nathalie Nypels X Unnamed seedling]; Leenders, M.

'Titi Parisien', F, dr, 1959; flowers crimson, center lighter, stamens ivory, single, 5 petals, 3 in.; bushy, low growth; [Francais seedlings X (Orange Triumph seedlings X Floradora)]; Delbard-Chabert

'Titian', F, dp, 1950; flowers deep pink, large blooms in clusters, dbl.; vigorous, tall growth; Riethmuller

'Titian, Climbing', Cl F, dp, 1964; Kordes

'Tivoli', F, mp, 1955; flowers warm rose-pink, center yellow, well formed, borne in clusters, dbl., 24 petals, 3 in., moderate fragrance; foliage dark, glossy; very vigorous growth; [Poulsen's Supreme X (Souv. de Claudius Denoyel X Hvissinge-Rose)]; Poulsen, S.; McGredy

Tivoli Gardens, *see* 'POULDUCE'

Tiz, *see* **'Tzigane'**

'Tiziana', F, op, 1969; bud pointed; flowers copper-pink, large, semi-dbl., slight fragrance; foliage light green; vigorous, compact growth; [Papillon Rose X Gay Paris]; Cazzaniga, F. G.

Tobago®, *see* 'DELTOGO'

Tobo(TM), *see* 'KINBO'

Toboné, *see* **'Prima Donna'**(TM)

'Toby Jo', Min, yb, 1997; flowers yellow and orange, lighter reverse, pink tips, dbl., 15–25 petals, 1 in., borne singly, no fragrance; few prickles; foliage medium, medium green, semi-glossy; upright, medium (16–20 in.) growth; [Old Glory sport]; Garrett, Troy O.

Toby Tristam, HMult, w

Tocade, *see* 'WERINA'

'Toccata', F, or, 1959; flowers well-formed blooms in large clusters; low growth; [Independence X Cinnabar]; Lens

'Toccata', HT, mp, 1963; flowers light rose edged darker, reverse silvery, becoming redder, dbl., 27 petals, 5 in., moderate fragrance; vigorous growth; [Karl Herbst X Seedling]; Sanday, John

Today(TM), *see* 'MACCOMPU'

'Toddler', F, pb, 1977; flowers pink, reverse deeper, semi-dbl., 19 petals, 2 in.; foliage small, dark; very low, bushy growth; [Seedling X Seedling]; LeGrice

'Todoroki', HT, dr, 1977; bud pointed; dbl., 33 petals, 6.5 in., exhibition form, slight fragrance; foliage dark; vigorous growth; [(Pharaon X Kagayaki) X Yu-Ai]; Keisei Rose Nurseries, Inc.

Toffee, *see* 'SEATOFF'

Togo, HT, 1959; Cazzaniga, F. G.; (Cavriglia)

Toison d'Or, *see* **'Golden Fleece'**

'Toison d'Or', HT, ab, 1921; flowers apricot-yellow, shaded orange-red, dbl.; GM, Bagatelle, 1922; Pernet-Ducher

Tojo®, F, mr; McGredy, Sam IV, 1978

Tokay, *see* 'ZIPTOK'

'TOKEN', HT, yb, 1933; bud ovoid; flowers glowing orange, open, large, dbl., moderate fragrance; foliage glossy; strong stems; vigorous growth; [Mme Butterfly X Premier Supreme]; Montgomery Co.

'TOKEN GLORY', HT, or, 1957; bud pointed; flowers orange, dbl., 40 petals, 4 in., moderate fragrance; foliage dark, leathery; long stems; very vigorous, upright, bushy growth; [Token Supreme sport]; Grillo

'TOKEN SUPREME', HT, ob, 1940; flowers deep orange, dbl., 35 petals, 5 in., moderate fragrance; [Token sport]; Grillo

'TOKYO', Cl F, my, 1972; flowers, single, 5 petals, 3 in., profuse May–June bloom; foliage light; vigorous growth; [Salute X Canary Bird]; Oliver, W.G.

Toledo Gold, HT, my

Toliman®, Min, m

'TOLSTOI', HSet, mp, 1938; flowers large, dbl., intense fragrance; Böhm, J.; (Sangerhausen)

'TOM BARR', HT, op, 1932; bud long, pointed; flowers salmon and scarlet, suffused yellow and orange, large, dbl., exhibition form, moderate fragrance; vigorous growth; McGredy

'TOM BRENEMAN', HT, dp, 1950; bud ovoid; flowers rose-pink, dbl., 30–40 petals, 4–4.5 in., globular, intense fragrance; foliage leathery, dark; very vigorous, upright growth; [Mauna Loa X R.M.S. Queen Mary]; Howard, F.H.; H&S

'TOM BRENEMAN, CLIMBING', Cl HT, dp, 1954; H&S

'TOM BROWN', F, r, 1964; flowers orangy-brown, reverse brownish-red, well-shaped, dbl., 32 petals, 3 in., intense fragrance; foliage leathery, dark; vigorous, bushy growth; [Unnamed seedling X Amberlight]; LeGrice

Tom Foster, HT, my, 1995; flowers medium yellow, full, large, very dbl., 26–40 petals, borne mostly singly, moderate fragrance; some prickles; foliage medium, dark green, semi-glossy; bushy, medium growth; [Gertrude Shilling X Helmut Schmidt]; Poole, Lionel; Battersby Roses, 1996

'TOM MANEY', LCl, mp, 1953; bud ovoid; flowers rose-pink, borne in clusters of 3–4, dbl., 35–40 petals, 4–5 in., cupped, non-recurrent, moderate fragrance; foliage leathery, dark; vigorous, climbing (15–20 ft) growth; quite hardy. [R. maximowicziana pilosa X K. of K.]; Maney; Iowa State College

'TOM PILLIBY', F, or, 1963; semi-dbl., in large clusters, slight fragrance; vigorous growth; Combe

'TOM THUMB', Min, rb, 1936; (Peon); flowers deep crimson, center white, semi-dbl., 1 in.; foliage leathery, light green; very dwarf growth; [Rouletii X Gloria Mundi]; deVink; C-P

'TOM TOM', F, dp, 1957; bud ovoid; flowers deep pink, blooms in clusters, dbl., 25 petals, 3–3.5 in., exhibition form, slight, spicy fragrance; foliage dark; vigorous, upright, bushy growth; [Improved Lafayette X Floradora]; Lindquist; Howard Rose Co.

'TOM TOM, CLIMBING', Cl F, dp, 1962; Lindquist; Howard Rose Co.

'TOM WOOD', HP, mr, 1896; flowers cherry-red, large, dbl., cupped, moderate fragrance; Dickson, A.

'TOMÁS BATÂ', HP, dr, 1932; flowers shaded crimson, passing to maroon, large, dbl., moderate fragrance; foliage leathery, glossy, dark; very vigorous growth; [Fisher Holmes X Prince Camille de Rohan]; Böhm, J.

Tomato Rose, *see* **R. RUGOSA**

'TOMBOLA', F, dp, 1967; flowers deep pink, 4 in., intense fragrance; foliage dark, glossy; vigorous, upright, bushy growth; [Amor X (Ena Harkness X Peace)]; deRuiter

'TOMBOT', F, pb, 1998; (**Botanica**); flowers pink with touch of lilac, medium, very dbl., 26–41 petals, 2 in., borne in large clusters, intense fragrance; prickles moderate; foliage medium, medium green, dull; bushy, medium (3–3.5 ft) growth; [Avandel X Madam President]; Thomson, George L.

Tomboy, *see* 'ZIPBOY'

'TOMDEL', Min, lp, 1999; (**Adelaide Tonight**); flowers pink, reverse lighter, dbl., 15–25 petals, .5–.75 in., borne in large clusters, slight fragrance; prickles moderate; foliage medium, medium green, semi-glossy; spreading, medium (18–24 in.) growth; [Avandel X Little Mike]; Thomson, George L.

'TOMFED', F, lp, 1999; (**Canberra Rose**); flowers medium pink, reverse darker, dbl., 16–40 petals, 3 in., borne in large clusters, intense fragrance; prickles moderate; foliage medium, dark green, glossy; upright, medium (3–4 ft) growth; [Watercolour X Madam President]; Thomson, George L.; Ross Roses, 2000

'TOMFLO', F, ab, 1998; (**Howard Florey**); flowers apricot, lighter reverse, double, dbl., 15–25 petals, 2 in., borne in small clusters, moderate fragrance; prickles moderate; foliage medium, medium green, semi-glossy; bushy, medium (3.5–4.5 ft) growth; [Seduction X Apricot Nectar]; Thomson, George L.

'TOMKINS RED', F, dr, 1943; bud long, pointed; flowers deep velvety crimson turning maroon, open, borne in clusters, semi-dbl., 18 petals, moderate fragrance; foliage dark, glossy; long stems; vigorous, compact, upright growth; [(Dr W. Van Fleet X Général Jacqueminot) X Nigrette]; Brownell, H.C.

'TOMMOP', HT, lp, 1999; (**Mary of Penola**); flowers dark pink, reverse darker, very dbl., 41 petals, 6 in., borne mostly singly, moderate fragrance; prickles moderate; foliage medium, medium green, glossy; upright, medium (3–3.5 ft) growth; [Maria Callas X Elina]; Thomson, George L.

'TOMMOUNT', HT, ab, 1996; (**Pride of Mountbarker**); flowers peach cream tinted rose edge, apricot center, lighter reverse, dbl., 26–40 petals, moderate fragrance; moderate prickles; foliage small, medium green, semi-glossy; upright, medium (5 ft x 2 ft) growth; [Gold Medal X Sylvia]; Thomson, George L.

'TOMMY BRIGHT', F, mr, 1961; bud ovoid; flowers scarlet-red, dbl., 35–40 petals, 3.5 in., cupped, moderate fragrance; foliage leathery; strong stems; very vigorous, upright growth; [Chatter seedling X Garnette Supreme]; Boerner; J&P

'TOMMY THOMPSON', HT, or, 1974; bud ovoid; dbl., 50 petals, 5 in., exhibition form, moderate, spicy fragrance; foliage glossy, very dark; moderate growth; [Queen o' the Lakes X Tropicana]; Golik; Dynarose

'TOMMY TUCKER', Min, lp, 1955; flowers silvery pink, borne in clusters, dbl., 1 in.; [Rouletti X Tom Thumb]; Robinson, T.

'TOMONE', S, pb, 1996; (**Mrs Mary Thomson**); flowers lilac pink, cream center, prominent gold stamens, reverse sl, semi-dbl., 8–14 petals, 5.5–6.5 in., intense fragrance; few prickles; foliage medium, medium green, dull; spreading, medium (5 ft x 4 ft) growth; [Dapple Dawn X Ophelia]; Thomson, George L.

'TOMRING', S, ab, 1977; (**Onkaparinga**); flowers apricot pink, reverse lighter, ages pink, dbl., 41 petals, 4–5 in., borne in small clusters, intense, myrrh fragrance; foliage medium, medium green, semi-glossy; spreading growth; [Cymbaline X Troilus]; Thomson, George L.

Tom's Pink, Pol, mp; Horsfield, 1986

Tom's Smelly Pink, Misc OGR, lp; (found rose)

'TOMSUE', S, ob, 1997; (**Susan Irvine**); flowers rose pink, cream center, prominent gold stamens, semi-dbl., 8–14 petals, 2.5–3.5 in., borne in large clusters, moderate fragrance; foliage medium, medium green, semi-glossy; bushy, medium (5 ft) growth; [Mrs. Mary Thomson X R. gigantea seedling]; Thomson, George L.

'TOMTAL', Min, pb, 1997; (**Laura Chantal**); flowers varying shades of pink, dark reverse, dbl., 26–40 petals, borne in small clusters, intense fragrance; foliage medium, medium green, semi-glossy; upright, medium (30 in.) growth; [Avandel X Madam President]; Thomson, George L.

'TOMTWO', F, dp, 1996; (**Linda Thomson**); flowers dark dusky pink, ruffled edge, reverse darker, blooms borne, dbl., 15–25 petals, 3.5 in., moderate fragrance; foliage medium, dark green, glossy; bushy, medium growth; [Fidelio X Showbiz]; Thomson, George L.

'TOMWILD', S, w, 1998; (**The Wild One**); flowers cream with touch of rose, reverse lighter, stamens gold, single, 4–7 petals, 4.5–5.5 in., borne in small clusters, moderate fragrance; prickles numerous; foliage large, light green, semi-glossy; upright, medium (4–5 ft) growth; [Wild Flower X Ophelia]; Thomson, George L.

'TONEHIME', HT, w, 1983; flowers white, creamy center, large blooms borne usually singly, dbl., 30 petals, exhibition form, slight fragrance; few prickles; foliage small, dark, glossy; vigorous, slender, tall growth; [Mizuho X Sodori-Hime]; Kikuchi, Rikichi

'TONGA', HT, ob, 1955; flowers deep golden orange, outer petals veined bronze and scarlet, dbl., intense fragrance; foliage dark, leathery; vigorous, upright growth; Lowe

'TONI CORSARI', F, dr, 1961; flowers velvety deep red, medium, borne in large clusters, semi-dbl., moderate fragrance; bushy, low growth; [Red Favorite X Seedling]; Delforge

'TONI LANDER', F, ob, 1958; flowers salmon-orange, borne in large clusters, dbl., 22 petals, 3 in., slight fragrance; foliage dark; bushy growth; GM, Madrid, 1960; [Independence X Circus]; Poulsen, S.

Toni Thompson's Musk, HMsk, lp; [R. brunonii X Unknown]; Thompson, T., 1984; (Weatherly, L.)

Tonia, *see* 'FRANCIBEL'

Tonic Meillandina®, *see* MEItrino

'TONIGHT', HT, mr, 1972; bud ovoid; flowers bright red, large, dbl., exhibition form, slight fragrance; foliage leathery; vigorous growth; [Seedling X Forever Yours]; Warriner, William A.

Tonimbuk, HT, mp; Dawson, before 1990

'TONNERRE'®, F, dr, 1953; (Not); flowers deep velvety red, blooms in large clusters, dbl., 24 petals, 3 in.; foliage dark; vigorous, bushy growth; [Holstein X Francais]; Mallerin, C.; EFR, 1952

'TONNER'S FANCY', LCl, w, 1928; flowers white, tinted pink, large, dbl., globular, moderate fragrance; vigorous, climbing growth; [R. gigantea seedling X Seedling]; Clark, A.; Gill & Searle

Tonsina, HT, ob

'TONY JACKLIN', F, op, 1972; flowers orange-salmon, flora-tea, dbl., 30 petals, 4 in., exhibition form, slight fragrance; GM, Madrid, 1972 GM, Portland, 1986; McGredy, Sam IV; McGredy

'TONY PEACE', HT, ab, 1964; flowers buff-yellow to apricot; [Peace sport]; Brundrett

'TONY SPALDING', HT, dr, 1933; flowers brilliant crimson, semi-dbl., exhibition form, intense fragrance; foliage glossy; very vigorous growth; McGredy

'TONYBRAC', F, w, 1988; (**Thora Hird**); flowers white deepen to cream in center with pink veins, loose, medium, very dbl., 40 petals, moderate, sweet fragrance; round, amber fruit; prickles straight, brown; foliage medium, dark green, semi-glossy; bushy, medium growth; [Chinatown X Picasso]; Bracegirdle, A.J.; Rosemary Roses, 1987

Too Hot To Handle, *see* 'MACLOUPRI'

Toorenburg, *see* KORkojotie

Tooth of Time, *see* 'RENTIME'

Tootsie, *see* 'GRETOOTS'

'TOP CHOICE', Min, op, 1985; flowers medium salmon-pink, medium, dbl., 36 petals, slight fragrance; foliage small, medium green, semi-glossy; dense, spreading, bushy growth; [Gingersnap X Baby Katie]; Hardgrove, Donald L.; Rose World Originals

Top Gear, *see* 'MACMANLY'

Top Gun™, *see* 'KINGUN'

'TOP HAT', F, ab, 1968; bud ovoid; flowers apricot-blush, medium, dbl., exhibition form, moderate fragrance; foliage small, dark, glossy; compact, low growth; [Ma Perkins X Garden Party]; Fankhauser; A. Ross & Son

Top Marks, *see* FRYministar

'TOP SECRET', Min, mr, 1971; [Beauty Secret sport]; Moore, Ralph S.; Sequoia Nursery

Top Secret, *see* MEIbolnay

Top Star®, HT, ob; Noack, Werner, 1982

Topaz, *see* 'JACANT'

'TOPAZ', Pol, yb, 1937; flowers lemon-yellow, petals edged cream, reverse lemon and cream, very dbl., exhibition form, slight fragrance; foliage small, leathery; dwarf, spreading growth; [Joanna Tantau X (Prof. Gnau X Julien Potin)]; Tantau; C-P

Topaz Jewel™, *see* 'MORYELRUG'

Topaze, F, mp

'TOPAZE ORIENTALE', HT, yb, 1965; flowers maize-yellow to light pink, dbl., 5–6 in., exhibition form, moderate fragrance; tall growth; [Sultane X Queen Elizabeth]; Delbard-Chabert; Cuthbert

'TOPEKA', F, mr, 1978; bud long, pointed; dbl., 21 petals, 3 in., flat; foliage glossy; vigorous, upright growth; [Vera Dalton X Unnamed seedling]; Wisbech Plant Co.

Topis d' Orient, *see* **'YESTERDAY'**®

Topkapi Palace, *see* POUlthe

'TOPPER', HT, mr, 1959; bud short, pointed; flowers signal-red to crimson, to open, dbl., 25–30 petals, 3–3.5 in., exhibition form, slight fragrance; foliage dark, leathery; long, strong stems; vigorous, upright growth; [Pink Bountiful X Sister Kenny]; Hill, Joseph H., Co.

Toprose, *see* 'COCGOLD'

'TOPSI'®, F, or, 1971; bud ovoid; flowers orange-scarlet, patio, medium, semi-dbl., slight fragrance; moderate, dwarf growth; GM, RNRS, 1972 PIT, RNRS, 1972; [Fragrant Cloud X Signalfeuer]; Tantau, Math.

'TOPSI'S FRIEND', F, lp, 1981; flowers rosette blooms borne 5 per cluster, semi-dbl., 15 petals, no fragrance; green prickles; foliage light green, glossy; upright growth; [Dreamland X Topsi]; Anderson's Rose Nurseries

Toque Rouge, *see* 'KORRED'

'TORCH', Cl Pol, or, 1942; flowers scarlet-orange, white eye, blooms in clusters, semi-dbl., 15 petals, 1.5 in., slight fragrance; foliage glossy, dark; vigorous, climbing (6–8 ft) growth; [Unnamed orange Polyantha seedling, sport climbing sport]; deRuiter

Torch of Liberty™, *see* 'MORTORCH'

'TORCH SONG', HT, or, 1959; bud ovoid; flowers vermilion, dbl., 30–35 petals, 5 in., exhibition form, slight fragrance; foliage dark, leathery; vigorous, upright, bushy growth; [(Peace X Floradora) X Grand'mere Jenny]; Meilland, F.; C-P

Torche Rose®, *see* 'INTERTOR'

'TORCHLIGHT', F, or, 1951; flowers brilliant orange-scarlet shaded deeper, borne in clusters to, single, 5–8 petals, 3.5 in., moderate fragrance; foliage dark; vigorous growth; [Dusky Maiden X Holstein]; LeGrice

'TORCHY', F, or, 1969; bud pointed to urn-shaped; flowers brick-orange, medium, dbl., exhibition form, moderate fragrance; foliage leathery; moderate, bushy growth; [Heat Wave X Spartan]; Armstrong, D.L.; Armstrong Nursery

'TOREADOR', HT, pb, 1919; flowers rosy red, reverse golden yellow, semi-dbl.; Paul, W.

'TORESKY', F, lp, 1931; flowers light pink with white eye, borne in large clusters, dbl., cupped, slight fragrance; foliage wrinkled, light; dwarf growth; [Perle d'Or X Antoine Rivoire]; Padrosa

'TORESKY, CLIMBING', Cl F, lp, 1956; Torre Blanca

'TORINO', F, mp, 1960; flowers spinel-rose, small to medium; vigorous growth; [Unnamed seedling X Cocorico]; Mansuino, Q.

Tornado®, *see* 'KORTOR'

Toro, *see* **'UNCLE JOE'**

Toro de Fuego, S, 1970; Delbard-Chabert; (Cavriglia)

'TORRERO', F, ob, 1961; flowers pink-orange-red, semi-dbl., 18 petals, 3 in., flat; [Ma Perkins X Cocorico]; Leenders, J.

Torvill & Dean, *see* 'LANTOR'

'**TOSCA**', F, or, 1972; bud long, pointed; flowers open, medium, dbl., slight fragrance; foliage large, leathery; very vigorous, upright growth; [Seedling X Ginger]; Warriner, William A.; J&P

'**TOSCANA**', HT, dp, 1954; flowers deep pink, dbl., moderate fragrance; vigorous growth; Cazzaniga, F. G.; Fedi

Toscana, *see* KORkunde

Toscanini, *see* OLIjglu

Total Recall™, *see* 'SAVACALL'

'**TOTOTE GELOS**', HT, w, 1915; bud long, pointed; flowers flesh-white, center shaded chrome-yellow, dbl.; Pernet-Ducher

'**TOTTIE**', F, mr, 1961; flowers purplish red, borne in clusters of 6, dbl., 35–40 petals, 3 in.; vigorous, bushy growth; [Alain X Fashion]; Borgatti, G.; Sgaravatti

'**TOTTY'S RED**', HT, dr, 1926; flowers crimson-scarlet; [Premier sport]; Totty

Touch o' Cloves, *see* 'MINITOC'

Touch o' Midas™, *see* 'MINABCO'

Touch of Class™, *see* 'KRICARLO'

Touch of Elegance, *see* 'LEOTOELG'

Touch of Fire, *see* 'RENFIRE'

'**TOUCH OF KIWI**', HT, yb, 1989; bud tapering; flowers creamy yellow, 1/4" wide orange margin on petals, dbl., slight fragrance; prickles pointed, light brown, varying size; foliage medium green; upright growth; [Kiwi Queen X Command Performance]; Cattermole, R.F.; South Pacific Rose Nursery

'**TOUCH OF MAGIC**', HT, or, 1977; bud globular; dbl., 40–45 petals, 5 in., exhibition form, intense fragrance; very vigorous growth; [San Francisco X Peace]; Patterson; Patterson Roses

Touch of Raspberry™, *see* 'DEVICIO'

Touch of Velvet, *see* 'LEOVELV'

'**TOUCH OF VENUS**', HT, w, 1971; flowers near white, center shaded pink, large, dbl., exhibition form, intense fragrance; foliage large, leathery; vigorous, upright growth; [Garden Party X Sweet Afton]; Armstrong, D.L.; Armstrong Nursery

'**TOUCHDOWN**', Min, mr, 1988; bud ovoid; flowers reverse red with white center, aging bluish, urn-shaped, small, dbl., 40 petals, no fragrance; no fruit; no prickles; foliage small, medium green, matt; low, upright growth; [Sheri Anne X Anita Charles]; Jolly, Nelson F.; Rosehill Farm, 1989

Touché, *see* 'LAVONT'

Touch-Up, *see* 'FOUTOUCH'

'**TOUGGOURT**', F, or, 1959; bud globular; flowers orange-shrimp-pink, open, large, dbl.; foliage dark, leathery; vigorous, bushy growth; [(Gruss an Teplitz X Independence) X (Floradora X Independence)]; Arles; Roses-France

Toulouse Lautrec®, *see* 'MEIREVOLT'

'**TOUR DE FRANCE**', F, OR, 1963; flowers deep orange-scarlet shaded blood-red, large to medium, semi-dbl.; vigorous, bushy growth; [(Alain X Orange Triumph) X unnamed seedling]; Mondial Roses

'**TOUR DE MALAKOFF**', C, m, 1856; flowers mauve-pink shaded purple, heavily veined, center green, moderate fragrance; sprawling (7 ft) growth; Soupert & Notting

Tour Eiffel, *see* '**EIFFEL TOWER**'

Tour Eiffel 2000, *see* DELrugro

Touraine, *see* 'POULANDER'

Tourbillon, *see* 'DELNOLLI'

'**TOURBILLON**', F, rb, 1959; (French Cancan); flowers red, reverse silvery pink, well formed, borne in clusters of, dbl., 30–35 petals, 3–4 in.; very vigorous, bushy growth; RULED EXTINCT 4/85; [Michele Meilland X (Incendie X (Floradora X Orange Triumph))]; Delbard-Chabert

Tourmaline®, *see* 'DELFRI'

Tournament of Roses, *see* 'JACIENT'

Tournee, F, dy; flowers luminous golden yellow, medium, semi-dbl., slight fragrance; VEG; (Sangerhausen)

Tove Pedersen, F, dr

Tower Bridge, *see* HARavis

'**TOWN CRIER**', HT, ly, 1961; bud ovoid; flowers straw-yellow, dbl., 33 petals, 5–6 in., exhibition form, moderate fragrance; foliage dark, glossy; vigorous, upright, well-branched growth; [Peace X Yellow Perfection]; Hill, Joseph H., Co.

'**TOWN TALK**', F, or, 1966; bud ovoid; flowers small, dbl., cupped; foliage dark, leathery; moderate growth; [(Circus X Garnette) X Spartan]; Swim & Weeks; Weeks Wholesale Rose Growers

'**TOWN TALK, CLIMBING**', Cl F, or, 1975; Weeks, V.E.; Weeks Wholesale Rose Growers

'**TOWNSEND**', HSpn, mp; dbl., non-recurrent; (Scotland), before 1900

'**TOWNSWOMAN**', HT, rb, 1973; flowers red-purple, reverse silver, dbl., 35 petals, 5 in., globular, intense fragrance; foliage light; free growth; [Seedling X Piccadilly]; Anderson's Rose Nurseries

'**TOY BALLOON**', Min, dr, 1979; bud ovoid, pointed; dbl., 48 petals, 1.5 in., exhibition form, slight fragrance; foliage dark; bushy, spreading growth; [Fairy Moss X Fire Princess]; Moore, Ralph S.; Sequoia Nursery

'**TOY CLOWN**', Min, rb, 1966; flowers white edged red, small, semi-dbl.; foliage small, leathery; bushy, dwarf growth; AOE, 1975; [Little Darling X Magic Wand]; Moore, Ralph S.; Sequoia Nursery

Toy Soldier, *see* 'CURTOY'

'**TOYLAND**', Min, mr, 1977; bud ovoid; flowers indian red, semi-dbl., 10 petals, 1 in., slight fragrance; foliage tiny; compact, bushy growth; Lyon

Toynbee Hall, *see* 'KORWONDIS'

'TRACAIT', Min, w, 1986; (**Caitlin**); flowers flesh pink fading white, urn-shaped, small, borne in sprays, dbl., 12–15 petals, slight, spicy fragrance; no fruit; straight, tan-brown prickles; foliage small, medium green, semi-glossy; bushy, low growth; [Fairy Moss X Fairy Moss]; Travis, Louis R., 1987

'**TRACEY WICKHAM**', Min, yb, 1984; flowers bright yellow, petals edged bright red, patio, small, in clusters, dbl., 30 petals, exhibition form, slight fragrance; foliage medium, medium green, semi-glossy; upright growth; [Avandel X Redgold]; Welsh, Eric; Rose Hill Roses

'TRACLA', Min, yb, 1987; (**Classy**); flowers yellow with pink overlay, small, borne singly, dbl., 13–18 petals, exhibition form, intense, spicy fragrance; round, green-orange fruit; straight, very few, tan-brown prickles; foliage small, medium green, semi-glossy; bushy, low growth; [Yellow Jewel X Yellow Jewel]; Travis, Louis R.

'**TRADE WINDS**', HT, rb, 1964; bud long, pointed; flowers dark red, reverse silver, dbl., 55 petals, 5 in., exhibition form, intense fragrance; foliage glossy; vigorous, tall growth; [(Multnomah X Seedling) X (Carrousel X Seedling)]; Von Abrams; Edmunds Roses

Tradescant, *see* 'AUSDIR'

'**TRADITION**', HT, mr, 1965; flowers scarlet-crimson, dbl., 35 petals, 4.5 in.; [Detroiter X Don Juan]; Kordes, R.; McGredy

Tradition 95®, *see* KORkeltin

Traditional Home®, *see* 'BURWINTRADHOM'

'**TRAFALGAR**', Cl HT, lp; [Mons Desir X Grüss an Teplitz]; Fell, J.B., 1912; (Weatherly, L.)

'**TRAFALGAR SQUARE**', F, mp, 1965; flowers large, dbl.; vigorous growth; Van den Akker Bros.

'**TRAILBLAZER**', F, or, 1975; bud long, pointed; dbl., 22 petals, 2.5 in., exhibition form, slight fragrance; foliage glossy; [Albert X Orange Sensation]; Harvey, R.E.; Kimbrew-Walter Roses

'**TRAKIIKA**', HT, dr, 1974; flowers large, dbl., 25 petals, cupped, moderate fragrance; foliage dark; vigorous, upright, bushy growth; [Tallyho X Spartan]; Staikov, Prof. Dr V.; Kalaydjiev and Chorbadjiiski

'**TRAMONTO**', HT, 1943; Giacomasso; (Cavriglia)

Tranquil, HT, lp; Dawson, 1977

'**TRANQUILITY**', Cl Pol, op; flowers coral-salmon, rosette form, small, borne in trusses; foliage light green; vigorous growth; RULED EXTINCT 11/82; [Princess van Orange sport]; Radmore

Tranquility, *see* 'BAROUT'

Transit, F, 1984; VEG; (Cavriglia)

'TRAPOM', Min, w, 1986; (**Fairy Pompons**); flowers flesh pink fading white, small, borne usually singly, very dbl., 45–60 petals, exhibition form, slight, spicy fragrance; no fruit; straight, tan-brown prickles; foliage small, medium green, matt; bushy, low growth; [Fairy Moss X Fairy Moss]; Travis, Louis R.

'TRASNO', Min, w, 1986; (**Snodoll**); dbl., 30–40 crinkled, translucent petals, borne singly, slight, fruity fragrance; no fruit; bowed, white prickles; foliage small, medium green, semi-glossy; upright growth; [Yellow Doll X Yellow Doll]; Travis, Louis R.

'TRASTAR', Min, rb, 1986; (**Tiny Stars**); flowers white with red narrow edges, small, borne usually singly and, semi-dbl., 12 petals, cupped, no fragrance; no fruit; almost thornless; foliage small, medium green, matt; micro-mini; bushy, low growth; [Magic Carrousel X Magic Carrousel]; Travis, Louis R.

'TRASUZ', Min, ob, 1986; (**Crepe Suzette**); flowers deep yellow with orange overlay, fading white with orange-red, dbl., 25–30 petals, exhibition form, moderate, fruity fragrance; no fruit; bowed, tan-brown prickles; foliage small, medium green, semi-glossy; bushy, low growth; [Orange Honey X Orange Honey]; Travis, Louis R., 1987

'TRAT', Min, w, 1988; (**Little 't'**); bud pointed; flowers white tinged pale pink, reverse white, small, borne in sprays, dbl., cupped, no fragrance; no fruit; no prickles; foliage small, medium green, matt; bushy growth; [Cinderella X Unnamed seedling]; Travis, Louis R., 1989

Träumerei®, *see* 'REIKOR'

'**TRAUMLAND**', F, lp, 1958; (Dreamland); flowers light peach-pink, well-formed blooms in clusters, dbl., 20 petals, slight fragrance; foliage dark, leathery; upright, bushy growth; [Cinnabar Improved X Fashion]; Tantau, Math.

Traumland®, *see* TANdalaum

Travemünde®, *see* 'KORRANTU'

'**TRAVERSER**', LCl, yb, 1928; flowers yellow and cream, well-shaped blooms in clusters; vigorous, climbing growth; Clark, A.

'**TRAVESTI**', F, yb, 1965; flowers yellow to carmine-red, medium, in clusters, dbl., 38 petals, cupped, moderate fragrance; foliage dark; vigorous, bushy growth; [Orange Sensation X Circus]; deRuiter

'**TRAVESTON**', Cl HT, dr, 1953; bud long, pointed; flowers crimson, medium, semi-dbl.; vigorous growth; [Black Boy X Editor McFarland]; Ulrick, L.W.

Traviata™, *see* 'MEILAVIO'

Traviata™, *see* MEIviola

'**TREASURE**', HT, ab, 1929; bud pointed; flowers apricot-pink, dbl., 6 in.; foliage glossy, dark; very vigorous growth; [Golden Rapture X Fred Walker]; Fletcher; Tucker

'**TREASURE CHEST**', HT, my, 1968; bud ovoid; flowers large, dbl., exhibition form, slight fragrance; foliage glossy; vigorous, upright growth; [Charlotte Armstrong X Fred Howard]; Whisler, D.; Germain's

'**TREASURE GOLD**', HT, my, 1950; bud ovoid, pointed; flowers yellow, some petals splashed red, dbl., 38 petals, 4 in., moderate fragrance; upright, branching growth; [Pink Princess X Free Gold (HT)]; Brownell, H.C.

'**TREASURE ISLAND**', HT, op, 1938; bud long, pointed; flowers light salmon, reverse flaming coppery pink, base orange, large, dbl., exhibition form, moderate fragrance; foliage leathery, bronze; vigorous growth; [Comtesse Vandal X Mme Nicolas Aussel]; Raffel; Port Stockton Nursery

'**TREASURE ISLAND, CLIMBING**', Cl HT, op, 1941; Hennessey

'**TREASURE ISLE**', F, op, 1967; flowers salmon-pink, reverse coppery, small, dbl., moderate fragrance; foliage glossy; vigorous, bushy growth; [Seedling X Treasure Island]; Raffel; Port Stockton Nursery

'**TREASURE TROVE**', LCl, ab, 1977; flowers apricot, mauve-pink and cream, dbl., 23 petals, 1.5–2 in., cupped, intense fragrance; vigorous growth; [Kiftsgate X China (possibly Old Blush)]; Treasure

Treasure Trove, *see* JAClay

Trend, HT, op; flowers large, dbl., slight fragrance; VEG; (Sangerhausen)

Tresor, *see* KRInico

Tresor de Thorigny, Cl HT, w

Trevi Fountain, LCl, ob

Trevor Griffiths™, *see* 'AUSOLD'

Trianon, F, pb, 1990; Lens

Tribute, *see* 'JACROSE'

'**TRICENTENAIRE**', HT, mp, 1949; bud long, pointed; flowers pink, center deeper, medium, dbl., 25 petals, exhibition form, moderate fragrance; foliage leathery, small; vigorous growth; [Charles P. Kilham X Neville Chamberlain]; Lens

'**TRICIA**', LCl, lp, 1942; bud long, pointed; flowers flesh-pink to ivory-pink, dbl., exhibition form, intense fragrance; foliage dark, glossy; vigorous (about 8 ft hardy. [New Dawn seedling]; O'Neal

'**TRICIA'S JOY**', LCl, op, 1973; flowers deep coral pink, large, dbl., 35 petals, moderate fragrance; foliage medium, dark, semi-glossy; vigorous growth (to 18 ft); [Mme Caroline Testout, Climbing X Blessings]; McKirdy, J.M.; John Sanday Roses, Ltd.

Trick Or Treat, *see* 'TALTRICK'

Trickster, *see* 'BRITRICK'

'**TRICOLORE**', HGal, pb, 1827; (Reine Marguerite); flowers lilac-pink, fringed at edges, dotted and mottled white; Lahaye Pere

'**TRICOLORE**', HMult, pb, 1863; flowers lilac-pink, similar to tricolore (g), blooms in small clusters; more vigorous than Gallica of the same name; [R. multiflora X ?]; Robert et Moreau

'**TRICOLORE DE FLANDRE**', HGal, pb, 1846; flowers pale blush, striped bright pink, fading to mauve, small, dbl., moderate fragrance; Van Houtte

'**TRIER**'®, HMult, w, 1904; flowers rosy white, base straw-yellow, semi-dbl., moderate fragrance; height 6–8 ft; (14); [Probably Aglaia self seedling sport]; Lambert, P.

Trier 2000, *see* 'KORMETTER'

Trieste, F, 1960; Mansuino; (Cavriglia)

Trigintipetala, *see* '**KAZANLIK**'

'**TRIGO**', HT, ab, 1931; bud long, pointed; flowers indian yellow, reverse apricot, tinted cerise, large, dbl., exhibition form, intense fragrance; foliage thick, bronze, glossy; vigorous growth; GM, NRS, 1931; Dickson, A.

'**TRILBY**', HT, op, 1927; flowers rich salmon, dbl.; Dobbie

Trinity, *see* 'MACREDPARAP'

'**TRINITY**', T, w; flowers pale pink to white, gold stamens, nodding blooms borne usual, semi-dbl., 4 in.; foliage dark; vigorous, spreading growth

'**TRINITY**', HT, pb, 1976; flowers variegated pink and yellow, medium, dbl., globular, intense fragrance; vigorous, upright, compact growth; [Korovo X ?]; de Freitas

Trinity Rambler, LCl, dp

'**TRINKET**', Min, mp, 1965; flowers phlox-pink, micro-mini, small, dbl.; foliage glossy; vigorous, bushy, dwarf growth; [(R. wichurana X Floradora) X Magic Wand]; Moore, Ralph S.; Sequoia Nursery

'**TRINTAGO**', HT, mr, 1962; flowers cardinal-red, well formed, dbl.; [Souv. de Jacques Verschuren X Charles Mallerin]; Leenders, J.

'**TRIO**', F, pb, 1966; flowers gold and pink, well-formed, large, slight fragrance; [Kordes' Perfecta X Shot Silk]; Dickson, A.

Triodene, *see* KORituscha

Triolet, *see* ORAdon

'TRIOMPHE ANGEVIN', Pol, mr, 1934; flowers cerise-red, borne in clusters, semi-dbl., cupped, slight fragrance; foliage glossy; very vigorous growth; Délépine; Pajotin-Chédane

'TRIOMPHE BRIARD', F, op, 1958; flowers coral, borne in clusters, dbl., moderate fragrance; foliage glossy; moderate growth; [Unnamed seedling X Fashion]; Robichon

'TRIOMPHE D'ALENCON', HP, dp, 1858; flowers large, dbl.; Touvais; (Sangerhausen)

'TRIOMPHE DE BOLWYLLER', HSem, w; flowers white, center tinted yellow, large, globular, moderate fragrance; Baumann, prior to 1837

Triomphe de Caen, *see* **'PRINCE ARTHUR'**

'TRIOMPHE DE FLORE', HGal, lp, 1830; flowers medium, very dbl., slight fragrance; Prévost; (Sangerhausen)

'TRIOMPHE DE FRANCE', HP, mp, 1875; flowers very large, very dbl., moderate fragrance; Garçon; (Sangerhausen)

Triomphe de Guillot Fils, T, pb; Guillot, 1861

Triomphe de la Ducher, B, lp

'TRIOMPHE DE LA GUILLOTIÈRE', S, w, 1872; flowers white tinged pink; [Probably R. roxburghii X R. odorata]; Guillot et Fils

'TRIOMPHE DE LA MALMAISON', HT, dp, 1946; flowers dark satiny pink, very large, petals reflexed, intense fragrance; foliage dark; vigorous growth; Gaujard

'TRIOMPHE DE LA TERRE DES ROSES', HP, mp, 1864; flowers violaceous pink, large, dbl.; moderately vigorous growth; Guillot Père

'TRIOMPHE DE LAFFAY', HCh, w; flowers delicate flesh, changing to white, expanded, large, very dbl.; pendulous growth habit

'TRIOMPHE DE L'EXPOSITION', HP, mr, 1855; flowers cherry-red, large, dbl., 55 petals, recurrent bloom, vigorous, bushy growth; Margottin

'TRIOMPHE DE PERNET PÉRE', HT, m, 1890; flowers reddish-violet, large, very dbl., moderate fragrance; Pernet; (Sangerhausen)

Triomphe des Noisettes, N, lp; Pére Pernet , 1887

'TRIOMPHE DU LUXEMBOURG', T, pb; flowers salmon-buff shaded rose; Hardy, 1840

'TRIOMPHE ORLÉANAIS', Pol, mr, 1912; flowers cherry-red, blooms in large clusters, semi-dbl., 1.5–2 in., slight fragrance; foliage leathery, glossy, bright; vigorous growth; Peauger

'TRIOMPHE ORLÉANAIS, CLIMBING', Cl Pol, mr, 1922; Turbat

'TRISTESSE', HT, m, 1953; (Tristeza); flowers gray-mauve, medium, dbl., 30–40 petals, flat, slight fragrance; foliage glossy; vigorous growth; [Charles P. Kilham X Betty Uprichard]; Camprubi, C.

Tristeza, *see* **'TRISTESSE'**

'TRITON', HT, yb, 1978; flowers white to light yellow, petals edged pink, large, dbl., 47 petals; upright growth; [Colour Wonder X Tzigane]; Dickson, Patrick

'TRIUMPHANT', HSet, w, 1850; flowers white tinted flesh, dbl.; [R. setigera X ?]; Pierce

Trix, Pol, or; flowers medium, dbl.; VEG; (Sangerhausen)

'TROBALL', Min, mp, 1987; (**Joan Ball**); flowers small blooms in sprays of 3–6, dbl., 30 petals, exhibition form, moderate, fruity fragrance; short, thin, red-brown prickles; foliage small, dark, glossy; upright, bushy growth; [Orange Sensation X Unnamed seedling]; Robinson, Thomas, Ltd., 1989

'TROBEACH', Min, w, 1983; (**Shell Beach**); bud soft pink; flowers creamy white, blooms in clusters, dbl., 28 petals, slight fragrance; foliage small, medium green, semi-glossy; bushy growth; [Simon Robinson X Simon Robinson]; Robinson, Thomas, Ltd.

'TROBEE', Min, rb, 1982; (**Dollie B**); flowers medium red, silver reverse, small, dbl., 35 petals, no fragrance; foliage small, dark green, red edges, glossy; bushy growth; [Parkdirektor Riggers X Darling Flame]; Robinson, Thomas, Ltd., 1983

'TROBELLE', Min, my, 1982; (**Calay**); flowers lemon yellow, small, semi-dbl., moderate fragrance; foliage small, dark, glossy; upright, bushy growth; [Rumba X New Penny]; Robinson, Thomas, Ltd.; T. Robinson, Ltd.

'TROBETTE', Min, op, 1983; (**Jennie Robinson**); flowers orange flushed pink, patio, small, dbl., 35 petals, moderate fragrance; foliage small, dark, glossy; bushy growth; [Rumba X Darling Flame]; Robinson, Thomas, Ltd.; Thomas Robinson, Ltd.

'TROBGINA', Min, op, 1986; (**Gina Louise**); flowers bright orange-pink, opening to bright yellow, gold anthers, dbl., 35 petals, exhibition form, intense, damask, fruity fragrance; globular, large, orange fruit; thin, red prickles pointed down; foliage small, dark green, semi-glossy; bushy, low growth; [Orange Sensation X Unnamed seedling]; Robinson, T.

'TROBGLOW', Min, or, 1984; (**Paint-Pot**); flowers medium, dbl., 24 petals, intense fragrance; foliage small, medium green, semi-glossy; bushy growth; [Unnamed seedling X Darling Flame]; Robinson, Thomas, Ltd.; T. Robinson, Ltd.

'TROBGOLD', Min, my, 1984; (**Woodland Sunbeam**); flowers patio, medium, dbl., 23 petals, intense fragrance; few prickles; foliage small, dark, glossy; vigorous, bushy growth; [Orange Sensation X Calay]; Robinson, Thomas, Ltd.

'TROBGUERN', Min, dy, 1992; (**Guernsey Gold**); flowers golden yellow, urn-shaped, blooms borne in small clusters, dbl., 26–40 petals, 1.5–2.75 in., slight fragrance; some prickles; foliage small, medium green, semi-glossy; low (30 cms), upright, bushy growth; [Rise 'n' Shine X Seedling]; Robinson, Thomas, Ltd.; Thomas Robinson, Ltd., 1990

'TROBIC', Min, pb, 1989; (**Just Magic**); bud rounded; flowers cream to deep pink, reverse cream tinged pink, aging often strongly, dbl., 21 petals, cupped, moderate, damask fragrance; spheroid, orange fruit; prickles very thin, pointed, red aging brown; foliage small, dark green, glossy; upright, bushy, medium growth; [(Parkdirektor Riggers X New Penny) X [(Parkdirektor Riggers X New Penny) X Unnamed seedling]]; Robinson, Thomas, Ltd., 1992

'TROBIC', Min, mr, 1982; (**Little Russel**, 'TROBRIC'); flowers small, dbl., 20 petals, no fragrance; foliage small, dark, glossy; bushy growth; [Marlena X New Penny]; Robinson, Thomas, Ltd.; T. Robinson, Ltd., 1983

'TROBINA', Min, mp, 1982; (**Pink Secret**); [Beauty Secret sport]; Robinson, Thomas, Ltd.; T. Robinson, Ltd.

'TROBINETTE', Min, ab, 1982; (**Sarah Robinson**); flowers soft apricot, small, dbl., 20 petals, intense fragrance; foliage small, dark, glossy; compact, bushy growth; [Rumba X Darling Flame]; Robinson, Thomas, Ltd.

'TROBINKA', Min, mr, 1983; (**Ginny-Lou**); flowers bright medium red, blooms in clusters, dbl., no fragrance; foliage small, dark, semi-glossy; bushy growth; [Dollie B. X Unnamed seedling]; Robinson, Thomas, Ltd.; Thomas Robinson, Ltd., 1984

'TROBLOVE', Min, dr, 1986; (**Guernsey Love**); flowers small blooms in sprays of 4–5, dbl., 35 petals, cupped, moderate, fruity fragrance; globular, medium, orange-red fruit; thin, red prickles, curving downward; foliage small, dark, glossy; upright, bushy growth; [Dollie B. X Unnamed seedling]; Robinson, Thomas, Ltd., 1990

'TROBRIC', Min, mr, 1982; (**Little Russel**, 'TROBIC'); flowers small, dbl., 20 petals, no fragrance; foliage small, dark, glossy; bushy growth; [Marlena X New Penny]; Robinson, Thomas, Ltd.; T. Robinson, Ltd., 1983

'TROBROY', S, w, 1992; (**Royal Worcester®**); flowers peachy cream, blooms borne in large clusters of 5–69, semi-dbl., 6–14 petals, 3–3.5 in., moderate fragrance; some prickles; foliage medium, dark green, very glossy; low to medium (60 cms), bushy, compact growth; [Simon Robinson X Gina Louise]; Robinson, Thomas, Ltd.; Thomas Robinson, Ltd., 1991

'TRObsa', Min, mp, 1984; (**Woodlands Lady**); flowers medium salmon-pink, small blooms in large clusters, semi-dbl., 15 petals, slight fragrance; foliage small, dark, glossy; compact, bushy growth; [Unnamed seedling X New Penny]; Robinson, Thomas, Ltd.

'TRObspread', S, yb, 1992; (**Unforgettable**); flowers amber, blooms borne in large clusters, dbl., 15–25 petals, 3–3.5 in., repeat bloom, moderate fragrance; some prickles; foliage small to medium, dark green, glossy; medium (90 cms), bushy, spreading (hemisphere shape) growth; [Snow Carpet X Woodland Sunbeam]; Robinson, Thomas, Ltd.; Thomas Robinson, Ltd., 1994

TRObstar, LCl, lp; (**Starlight Express**); Robinson, T., 1997

'TRObsun', Min, yb, 1982; (**Woodlands Sunrise**); flowers small, dbl., 35 petals, slight fragrance; foliage small, dark, glossy; upright, bushy growth; [Rumba X Darling Flame]; Robinson, Thomas, Ltd.

'TRObwich', Min, mp, 1982; (**Simon Robinson**); flowers small blooms in clusters of 10–60, single, 5 petals, moderate fragrance; foliage small, dark, glossy; compact, low, bushy growth; [R. wichurana X New Penny]; Robinson, Thomas, Ltd.

'Trocadero', HT, mr, 1964; bud pointed, bright red; foliage bronze; vigorous growth; [Karl Herbst X Seedling]; Delforge

Trocadero 'D', HT, 1989; Delforge; (Cavriglia)

Troika, *see* 'POUmidor'

'Troika, Climbing', Cl HT, 1964; Thomas, Dr A.S.; Rumsey

Troilus, *see* 'AUSoil'

'Troja', HT, w, 1927; bud long, pointed, cream-yellow; flowers creamy white, dbl., moderate fragrance; vigorous, bushy growth; [Mrs Herbert Stevens sport]; Mikes Böhm, J.; Böhm

'Trojan', Gr, yb, 1961; bud long, pointed; flowers pastel pink, reverse yellow, dbl., 40 petals, 5 in., exhibition form, moderate fragrance; foliage leathery; upright growth; [Sutter's Gold X (Mme Henri Guillot X Seedling)]; Von Abrams; Peterson x Dering

Trojan Victory, *see* 'KORperki'

Trollhattan, HRg, lp

Trompeter von Sackingen, HMult, m; Geschwind, R., before 1890

'TROpat', F, or, 1992; (**Temptation**); single, 5 petals, 3–3.5 in., borne in small clusters, moderate fragrance; some prickles; foliage medium, medium green, semi-glossy; low (70 cms), dense, bushy, compact growth; [Orange Sensation X Seedling]; Robinson, Thomas, Ltd.; Thomas Robinson, Ltd.

'Trophée', LCl, mr, 1968; foliage glossy; vigorous, climbing growth; [Valenciennes X Etendard]; Robichon; Ilgenfritz Nursery

Tropical, F, 1966; Cazzaniga, F. G.; (Cavriglia)

Tropical Paradise, *see* 'BURsrtpara'

Tropical Passion, *see* CARange

Tropical Skies, *see* INTerdays

Tropical Sunrise, *see* 'JOHillgold'

Tropical Sunset™, *see* 'MACtaurang'

Tropical Twist™, *see* 'JACorca'

Tropicana, *see* 'TANorstar'

Tropicana, Climbing, *see* 'TANgostar'

Tropico Sunblaze, *see* MEIglassol

Tropique, *see* 'DELjis'

'Troubadour', HWich, 1911; flowers bright red, shaded maroon, large clusters, dbl., free seasonal bloom; foliage large, glossy, dark; vigorous, climbing growth; RULED EXTINCT; Walsh

'Troubadour', HT, 1956; flowers rose-copper, large, dbl., free bloom, moderate fragrance; strong stems; moderate growth; RULED EXTINCT; Frères, Buyl

Troubadour of Love, HT, dr

Trpaslik, Min, lp; Strnad; (Czech Rosa Club)

'Truby King', F, dp, 1965; flowers carmine, open, large, borne in large clusters, semi-dbl., intense fragrance; foliage leathery, glossy, bronze; vigorous, tall growth; [Border Queen X Helen Traubel]; Harris, L.M.; Cutler

Trudor, *see* **'Micaëla'**

Trudy Mimi™, *see* MEIkyster

True Gold™, *see* 'LAVtru'

'True Love', HT, dr, 1970; flowers deep red, dbl., 30 petals, 3.5–4 in., slight fragrance; foliage glossy; moderately vigorous growth; Delbard; Laxton & Bunyard Nursery

True Love, *see* 'RUTrulo'

True Vintage, *see* 'SEAvin'

'Truly Fair', F, w, 1953; flowers cream, center apricot, small, borne in sprays of 4–5, semi-dbl., moderate fragrance; bushy growth; Ratcliffe

'Truly Yours', HT, ob, 1972; flowers coral-salmon to orange, dbl., 44 petals, 5 in., globular, intense fragrance; foliage large; vigorous, upright growth; Edland Fragrance Medal, ARS, 1971; [Miss Ireland X Stella]; Robinson, H.

Trumpeter®, *see* 'MACtrum'

Truper, *see* **'Persepolis'**

Trylon, *see* **'Shining Star'**

Tschaika, HT, lp, 1959; flowers large, dbl., moderate fragrance; Klimenko, V. N.; (Sangerhausen)

TSCHaka, HT, ab, 1989; (**Charlie Chaplin**); Tschanz, E.

'Tschernowa Ukraijna', HT, mr, 1940; flowers large, semi-dbl., slight fragrance; Kosteckij; (Sangerhausen)

Tschin-Tschin, F, mr, 1973; flowers luminous red, medium, dbl.; Meilland; (Sangerhausen)

'Tsukiakari', HT, w, 1984; flowers creamy white, large blooms borne 1–3 per stem, dbl., 35 petals, exhibition form, no fragrance; medium prickles; foliage medium green, semi-glossy; bushy growth; [Utage X Anne Letts]; Ota, Kaichiro

Tubantia, HT, op, 1975; flowers reddish-orange and pink, large, dbl.; Verbeek; (Sangerhausen)

'TUCkach', MinFl, w, 1997; (**Cachet**™); flowers large, exhibition form, 26–40 petals, 2 in., exhibition form, borne mostly singly, none; foliage medium, medium green, semi-glossy; upright, tall (3 ft) growth; [Seedling X Seedling]; Tucker, Robbie; Rose Hill Nursery, 1998

'TUCkamy', Min, lp, 1998; (**Amy Grant**™); flowers light pink, varies by temperature, large, dbl., 17–25 petals, 1.5–1.75 in., exhibition form, borne mostly singly, slight fragrance; prickles few, small, slightly hooked downward; foliage medium, dark green, glossy; low, upright, 18–24 in. growth; [Loving Touch X White Masterpiece]; Tucker, Robbie; Nor'East Miniature Roses, 1999

'Tucker's Folly', HT, pb, 1951; flowers glowing cerise overlaid orange, dbl., 35–40 petals, 6–7 in.; foliage bronze; very vigorous growth; Fletcher; Tucker

'Tucker's Yellow', HT, dy; flowers deep golden yellow; [Max Krause sport]; Tucker

'TUCkflip', Min, mr, 1997; (**Miss Flippins**™); flowers medium red, deep pink reverse, dbl., 15–25 petals, 1.5 in., exhibition form, borne mostly singly, no fragrance; prickles moderate; foliage medium, dark green, glossy; bushy, medium (2 ft) growth; [Elizabeth Taylor X Kristin]; Tucker, Robbie; Bridges Roses, 1997

'TUCkrox', HT, op, 1998; (**Roxie Baby**™); flowers orange to pink, reverse lighter, very full, very dbl., 65 petals, 3–4 in., exhibition form, borne mostly singly, no fragrance; foliage dark green, glossy; upright, medium (5 ft) growth; [Elizabeth Taylor X Hoagy Carmichael]; Tucker, Robbie; Rosehill Nursery, 1999

'TUCstand', HT, rb, 1998; (**Standing Ovation**™); flowers white with red edge, very full, very dbl., 41 petals, 4.5 in., exhibition form, borne singly or in small clusters, slight fragrance; prickles moderate; foliage medium, dark green, semi-glossy; upright, medium (4–5 ft) growth;

[Elizabeth Taylor X White Masterpiece]; Tucker, Robbie; Edmunds Roses, 1999

'TUCTOMMY', Min, my, 1998; (**Little Tommy Tucker**™); flowers medium yellow, reverse lighter, long-lasting, dbl., 17–25 petals, 1.5–1.75 in., exhibition form, borne in small clusters, no fragrance; prickles few, small to medium, slightly hooked downward; foliage medium, dark green, glossy; upright, low (14–20 in.) growth; AOE, 1999; [Rise 'n' Shine X Captivation]; Tucker, Robbie; Nor'East Miniature Roses, 1999

Tudelum™, *see* 'KINLUM'

'TUDOR', HT, mr, 1953; (Pechtold's Flame); flowers spectrum-red overcast scarlet-red, dbl., 20–25 petals, 5 in., exhibition form, moderate fragrance; long stems; vigorous growth; [Katharine Pechtold X Crimson Glory]; Verschuren-Pechtold; J&P

'TUDOR PRINCE', Gr, mr, 1959; (Prince Philip); bud long, pointed, chestnut; flowers bright red shaded geranium-red, large, dbl., 30–40 petals; foliage glossy, dark; vigorous, upright growth; [Independence X Buccaneer]; Leenders, J.

Tudor Rose, S, lp; [R. brunonii X Unknown]; Porter, L., 1989; (Weatherly, L.)

Tudor Sunsation, *see* KORfeldwo

'TUDOR VICTORY', HT, dr, 1987; flowers dusky, dark red, reverse medium red, aging purple, medium, b, dbl., 27 petals, exhibition form, slight, damask fragrance; prickles straight, large, red; foliage large, medium green, semi-glossy; upright, tall growth; [Queen Elizabeth X John Waterer]; Bracegirdle, Derek T., 1988

'TUHUA', HMult, ab, 1979; bud small, pointed; 11 petals, borne 3–5 per cluster, early bloomer with some repeat, no fragrance; small, red-brown prickles; foliage small, glossy; vigorous, lax, tall growth; [Iceberg X Tausendschon]; Murray, Nola

Tullamore, S, lp

Tulsa, HT, lp; Perry, Astor, 1996

Tumbarumba Schoolhouse, HMult, lp

Tumbling Waters, *see* POUltumb

Tumulte, HT; Kordes, R.; (Cavriglia)

Tunusblumchen, LCl, m

'TUPPERWARE', HT, pb, 1981; bud ovoid, pointed; flowers carmine pink, reverse silver, borne 1–5 per cluster, dbl., 65 petals, exhibition form, moderate fragrance; medium prickles; foliage large, dark, glossy; vigorous, upright growth; [(Pink Peace X Queen Elizabeth) X (Kordes' Perfecta X Peace)]; Williams, J. Benjamin; Tupperware Home Parties

Turbo, *see* 'MEIROZRUG'

Turbo™, HRg, mp

Turbo Meidiland®, *see* 'MEIROZRUG'

Turbo Rugostar®, *see* 'MEIROZRUG'

Turbulance, *see* 'GELANCE'

'TURENNE', HP, dr, 1861; flowers maroon; vigorous growth; [Général Jacqueminot sport]; Verdier, V.

Turenne, HGal, dp; Vibert (?), ca 1846

'TÜRKES RUGOSA SÄMLING', HRg, pb, 1923; bud long, pointed; flowers peach-pink on yellow ground, large, semi-dbl., intense fragrance; foliage dark, leathery; strong stems; vigorous ($2\frac{1}{2}$–3 ft) growth; [Conrad Ferdinand Meyer X Mrs Aaron Ward]; Türke; Teschendorff

Turlock High, *see* 'CLELOCK'

Turn of the Century, *see* 'GREELIZ'

'TURNER'S CRIMSON RAMBLER', HMult, mr; (Crimson Rambler, Shi Tz-mei, Soukara-Ibara, Ten Sisters); flowers bright crimson, fading toward blue, irregular blooms in larg, dbl., heavy, non-recurrent bloom; foliage light, leathery, disposed to mildew; very vigorous, climbing (15–24 ft) growth; (14); Turner, 1893

'TURNVATER JAHN', HP, pb, 1927; flowers white with pink, very large, dbl.; Müller, Dr F.; (Sangerhausen)

Tuscan Beauty, *see* 'HINTUSCAN'

'TUSCANY', HGal, m; (The Old Velvet Rose); flowers velvety blackish-crimson to deep purple, large blooms, semi-dbl.; vigorous, upright growth

'TUSCANY SUPERB', HGal, m; (Superb Tuscan, Superb Tuscany); [Tuscany seedling]; Rivers, prior to 1837

Tuscia®, *see* BARtusc

Tut's Treasure, Min, dy

Tutta's Pink Noisette, N, lp

Tutti-Frutti, *see* 'JACTUTTI'

'TUTU', Min, pb, 1978; bud ovoid; flowers pink and rose, dbl., 40 petals, 1 in., exhibition form, slight fragrance; upright, bushy growth; [Over the Rainbow X Seedling]; Rovinski & Meredith; Kingsdown Nursery

'TUTU MAUVE', F, m, 1963; flowers magenta shaded mauve and rose, well-formed, dbl., 30 petals, 4 in.; bushy, low growth; GM, Madrid, 1962; Delbard-Chabert

'TUTU PETITE', F, pb, 1967; flowers pink, reverse darker, blooms in clusters, semi-dbl., slight fragrance; foliage glossy, light; compact, bushy growth; [Rudolph Timm X ?]; Samuels

Tuxedo, *see* 'AROBRISP'

Tweedle Dee, *see* 'TINTWEE'

'TWEETIE', Min, lp, 1973; bud ovoid, long, pointed; flowers soft pink, small, dbl., slight fragrance; foliage small, light, leathery; moderate, dwarf, bushy growth; [Perle d'Or X Fairy Princess]; Moore, Ralph S.; Sequoia Nursery

Twenty First Century, *see* 'PERTWENTY-FIRST'

Twenty-Fifth, *see* 'BEATWE'

'TWICE AS NICE', HT, rb, 1984; flowers white blending to red at edges, large, dbl., 40 petals, intense fragrance; foliage large, dark, semi-glossy; [Peace X Mirandy]; Patterson, Randell E.

'TWILIGHT', HT, m, 1955; bud pointed; flowers lavender-lilac, reverse silvery, dbl., 30–35 petals, 4.5 in., exhibition form, moderate fragrance; foliage dull green; bushy, upright growth; [Grey Pearl X Lavender Pinocchio]; Boerner; J&P

Twilight, *see* NOAtwi

'TWILIGHT BEAUTY', Min, m, 1977; bud long, pointed; flowers red-purple, dbl., 40 petals, 1–1.5 in., exhibition form, moderate fragrance; bushy, spreading growth; [Angel Face X Over the Rainbow]; Williams, Ernest D.; Mini-Roses

Twilight Dream, *see* 'MINDREAM'

Twilight Mist, LCl, m; Robinson, 1995

Twilight Time, Min, m

Twilight Trail™, *see* 'MINXCO'

'TWILIGHT ZONE', Min, mr, 1985; flowers well-formed, medium, dbl., 60 petals, no fragrance; foliage small, dark, semi-glossy; bushy, dense growth; [Scarlet Knight X Big John]; Hardgrove, Donald L.; Rose World Originals

Twilight Zone, Gr, lp; Williams, J. Benjamin, 1996

Twin Peaks, *see* 'WEKFRAG'

Twin Pinks, *see* 'CLETWIN'

'TWINKIE', Min, lp, 1974; bud pointed; flowers light clear pink, dbl., 40 petals, 1 in.; foliage small, glossy; upright, very bushy growth; [(R. wichurana X Floradora) X Eleanor]; Moore, Ralph S.; Sequoia Nursery

Twinkle®, *see* 'INTERTWIK'

Twinkle Bright, Min, lp

Twinkle Charm, Min, lp

Twinkle Eyes, Min, lp

Twinkle Pink, Min, lp

Twinkle Star, Min, lp

Twinkle Toes, *see* 'LYOTO'

'TWINKLE TWINKLE', Min, ab, 1981; bud pointed; flowers white with apricot petal edges, aging dark pink, blooms born, dbl., 23 petals, slight, tea fragrance; very fine, curved prickles; foliage medium green, semi-glossy; slender, straight, upright growth; [Contempo X Sheri Anne]; Bennett, Cecilia 'Dee'; Tiny Petals Nursery

Twinkler, Min, pb; Datt, Braham, 1989

'TWINKLES', Min, w, 1954; bud flesh; flowers small, dbl., 43 petals, moderate fragrance;

compact, dwarf (8 in.) growth; [Perla de Montserrat X Unnamed Polyantha seedling]; Spek; J&P

Twins®, HMsk, yb; Lens, 1994

Twirly Whirly, S, lp; Peden, R., 1999; (Weatherly, L.)

Twister, *see* 'MORTWISTER'

Two Sisters™, *see* 'WILTWOS'

Two Thumbs Up, *see* 'GELTWO'

'TWOADMIRE', HT, dr, 1991; (**American Glory**); bud ovoid, pointed; flowers cardinal red, large, cupped, 28 petals, cupped, borne usually singly, slight, damask fragrance; foliage medium, dark green, semi-glossy; upright, bushy, medium growth; PP07973; [Portland Trailblazer X Seedling]; Twomey, Jerry, 1983; DeVor Nurseries, Inc., 1991

'TWOADORE', HT, lp, 1991; (**Audrey Hepburn**); bud pointed; flowers blush pink, fading to lighter pink, large, dbl., 30–32 petals, moderate, fruity fragrance; foliage medium, dark green, glossy; upright, bushy, medium growth; PP07980; [Evening Star X Seedling]; Twomey, Jerry, 1983; DeVor Nurseries, Inc., 1992

'TWOADVANCE', S, op, 1991; (**All That Jazz**™); bud pointed; flowers coral salmon blend, loose, large, semi-dbl., 10–15 petals, 4.5 in., cupped, borne in sprays of 3–5, moderate, damask fragrance; prickles moderate; foliage medium, dark green, glossy; upright, bushy, tall growth; PP7978; AARS, 1992; [Gitte X Seedling]; Twomey, Jerry, 1983; DeVor Nurseries, Inc., 1991

'TWOAEBI', HT, ob, 1999; (**Dream Orange**); dbl., 26–40 petals, 3.5 in., borne in small clusters, slight fragrance; prickles moderate; foliage medium, medium green, semi-glossy; bushy, medium (4.5 ft) growth; [Cherish X (Evening Star X Trumpeter)]; Twomey, Jerry, 1999

TWOangel, HT, pb, 1995; (**Angela Lansbury**); PP10172; [Gitte X Silver Jubilee]; Twomey, Jerry

'TWOAT', HT, lp, 1999; (**Dream Blush**); dbl., 17–25 petals, 5 in., borne in small clusters, slight fragrance; few prickles; foliage large, medium green, semi-glossy; compact, medium (4 ft) growth; [Evening Star X Marijke Koopman]; Twomey, Jerry

'TWOBE', HT, op, 1989; (**Sheer Elegance**™); bud pointed; flowers soft creamy pink with dark pink edges, large, exhibition form, 30–35 petals, 4.5 in., exhibition form, borne singly, moderate, musk fragrance; prickles slightly curved, red with green; foliage large, dark green, glossy; upright, tall growth; PP7901; AARS, 1991; [Pristine X Fortuna]; Twomey, Jerry; DeVor Nurseries, Inc., 1990

TWOcherish, Gr, dp, 1995; (**Colorburst**); Twomey, Jerry

'TWODI', HT, mp, 1998; (**Charming Diana**™); flowers pink, reverse medium, dbl., 15–25 petals, 6 in., borne singly, very large, moderate fragrance; prickles numerous; foliage large, dark green, glossy; bushy, medium growth; Twomey, Jerry

'TWODREAM', HT, mp, 1989; (**Endless Dream**™); bud pointed; flowers medium, soft pink, large, borne singly, dbl., 32 petals, cupped, moderate, musk fragrance; prickles declining, grayish-white with black spots; foliage large, dark green, semi-glossy; upright, medium growth; PP007561; [Emily Post X Seedling]; Twomey, Jerry; DeVor Nurseries, Inc., 1990

'TWOETERN', Gr, rb, 1991; (**Eternity**); flowers red/cream bicolor, blooms borne in large clusters, dbl., moderate; some prickles; foliage medium, dark green; tall (183 cms), upright growth; PP08413; [Gitte X Seedling]; Twomey, Jerry; DeVor Nurseries, Inc., 1991

'TWOEX', HT, mp, 1998; (**Our Diana**™); flowers medium pink, double (15–25 petals) drops clean; very large blooms; fragrant; foliage medium, medium green, semi-glossy; upright, tall growth, fast repeat; [Parentage not listed]; Twomey, Jerry

'TWOFAN', HT, op, 1989; (**American Fantasy**™); bud ovoid; flowers salmon pink, reverse lighter, large, dbl., 32 petals, cupped, borne singly, moderate, fruity fragrance; prickles declining, yellow with red tinge; foliage medium, dark green, glossy; upright, medium growth; [Sonia X Seedling]; Twomey, Jerry; DeVor Nurseries, Inc., 1991

TWOfavor, HT, pb, 1995; (**Favorite Dream**); PP10155; [(Friendship X seedling) X (Emily Post X Royalty)]; Twomey, Jerry

'TWOFREE', S, mr, 1989; (S.H. Freedom, **Spring Hill's Freedom**); bud ovoid; flowers medium, scarlet red, medium, borne singly, dbl., 35 petals, cupped, slight, musk fragrance; prickles slightly curved, red-purple; foliage medium, medium green, semi-glossy; upright, medium growth; [Samantha X Unnamed seedling]; Twomey, Jerry; DeVor Nurseries, Inc., 1990

TWOfree, F, mr; (**Springhill Freedom**); PP07868; [Samantha X Fireburst]; Twomey, Jerry, 1998

'TWOHAVE', Gr, or, 1991; (**Proud Mary**); bud pointed; flowers scarlet red, medium, dbl., 19–21 petals, moderate, damask fragrance; foliage medium, dark green, semi-glossy; upright, bushy, medium growth; PP07981; [Unnamed seedling X Royalty]; Twomey, Jerry, 1983; DeVor Nurseries, Inc., 1991

'TWOHONOR', HT, mp, 1993; (**American Honor**); flowers full, large blooms, exhibition form, 26–40 petals, 3–3.5 in., exhibition form, borne mostly singly, slight; few prickles; foliage medium, dark green, semi-glossy; medium (110 cms), upright growth; PP09443; [Sheer Elegance X Seedling]; Twomey, Jerry, 1993; DeVor Nurseries, Inc., 1993

'TWOJOAN', HT, mp, 1999; (**Dream Pink**); flowers large, dbl., 17–25 petals, borne mostly singly, moderate fragrance; prickles moderate; foliage large, medium green, glossy; upright, medium (4 ft) growth; [White Masterpiece X Silver Jubilee]; Twomey, Jerry, 1999

TWOlad, HT, dp, 1995; (**Aladdins Dream**); PP10149; [Silver Jubilee X Evening Star]; Twomey, Jerry, 1994

'TWOLOY', Gr, pb, 1990; (**Lloyd Center Supreme**); bud pointed; flowers light pink with yellow base, dark pink blending to yellow reverse, exhibition form, 25 petals, exhibition form, moderate, fruity fragrance; foliage medium, dark green, glossy; upright, bushy, medium growth; PP07979; [Brion X Seedling]; Twomey, Jerry, 1988; DeVor Nurseries, Inc., 1990

'TWOMIN', Min, or, 1989; (**Apache Princess**); bud ovoid; flowers bright orange-red, medium, dbl., 38 petals, cupped, borne singly, slight, fruity fragrance; prickles declining, purple; foliage medium, medium green, semi-glossy; upright, medium growth; PP008064; [(Cricket X Christ 78) X Seedling]; Twomey, Jerry; DeVor Nurseries, Inc., 1990

'TWOPAUL', HT, mr, 1999; (**Dream Red**); 3.5 in., borne in small clusters, slight fragrance; prickles moderate; foliage medium, dark green, glossy; bushy, medium (4.5 ft) growth; [Esmeralda X Fireburst]; Twomey, Jerry, 1999

'TWORIGHT', F, or, 1993; (**Fireburst**); very dbl., 3–3.5 in., borne mostly singly, slight fragrance; some prickles; foliage medium, medium green, semi-glossy; medium (111 cms), bushy growth; [Seedling X Evening Star]; Twomey, Jerry; DeVor Nurseries, Inc., 1993

Two-Timer, *see* 'MORSWISS'

'TWOWIN', Gr, ob, 1989; (**Winning Colors**™); bud ovoid; flowers orange-yellow blend, medium, borne singly, cupped, 60 petals, cupped, moderate, musk fragrance; prickles declining, yellow-green; foliage medium, dark green, glossy; upright, medium growth; PP07907; [Gingersnap X Marina]; Twomey, Jerry; DeVor Nurseries, Inc., 1990

'TWOYEL', HT, my, 1999; (**Dream Yellow**); dbl., 17–25 petals, 5 in., borne mostly singly, intense fragrance; prickles moderate; foliage medium, medium green, semi-glossy; upright, medium (3 ft) growth; [(Sonia X Prominent) X Whisky Mac]; Twomey, Jerry, 1999

'TWYFORD', HT, op, 1939; bud long; flowers bright salmon-pink, base gold, reverse deep salmon flushed orange, moderate fragrance; foliage dark, reddish; vigorous growth; Waterer

Tycoon, *see* 'BRITY'

'**TYLER**'®, HT, rb, 1988; (Sebago); flowers large, dbl., 15–25 petals, slight fragrance; foliage medium, dark green, matt; upright growth; [(Tamango X Red Planet) X First Prize]; Poor, Cuyler

Tynwald, *see* 'MATTWYT'

Typ Kassel, C, mp

'**TYPHOO TEA**', HT, rb, 1974; (Doux Parfum, Été Parfumé); flowers medium red, silver reverse, classic form, dbl., 50 petals, 5 in., moderate fragrance; foliage small, glossy; [Fragrant Cloud X Arthur Bell]; McGredy, Sam IV

'**TYPHOON**', HT, ob, 1972; (Taifun); flowers salmon, shaded yellow, dbl., 35 petals, 4 in., intense fragrance; [Dr A.J. Verhage X Colour Wonder]; Kordes, R.; McGredy

'**TYRIANA**', HT, dp, 1963; bud pointed, ovoid; flowers rose-pink, well formed, dbl., 40 petals, 4.5–5 in., moderate fragrance; foliage leathery, dark; vigorous, upright growth; [(Happiness X Independence) X Paris-Match]; Meilland, Alain A.; URS; Wheatcroft Bros.

'**TYRIUS**', HT, m, 1972; flowers tyrian purple, dbl., 20 petals, 5 in., slight fragrance; foliage glossy, bronze; [Bettina X Prima Ballerina]; Gandy, Douglas L.

'**TZIGANE**', HT, rb, 1951; (Tiz); bud ovoid; flowers rose-red, reverse yellow, to cactus-formed, large, dbl., cupped, moderate fragrance; foliage dark, glossy, leathery; upright, bushy growth; [Peace X J.B. Meilland]; Meilland, F.; Hennessey, 1956; URS, 1951

'**TZIGANE, CLIMBING**', Cl HT, rb, 1958; Lagoona Nursery; Roseglen Nursery, 1958; URS, 1960; Wheatcroft Bros., 1958

U

'U.P. Hedrick', HSpn, mp, 1932; flowers pink, open, large, single, profuse, non-recurrent bloom, moderate fragrance; foliage soft, dark; vigorous (6 ft), bushy, compact growth; [R. spinosissima altaica X Betty Bland (probably)]; Central Exp. Farm

'Uetersen'®, S, mr, 1939; (Zenith); flowers glowing red, semi-dbl., recurrent bloom; upright, bushy growth; [K. of K. X Stammler]; Tantau

Ufhofen, S, op, 1964; flowers medium, dbl.; Berger, W.; (Sangerhausen)

'Uhland', HMsk, yb, 1916; flowers reddish yellow, petals fringed, borne in clusters of 3–15; foliage pointed, like Tip-Top; [Geheimrat Dr Mittweg X Tip-Top]; Lambert, P.

'UHLater', HRg, dp, 1989; (**Buffalo Gal**, Foxi, Foxi Pavement); flowers lavender-pink, loose form, dbl., 17–25 petals, 3 in., borne in clusters, intense fragrance; foliage large, light green, wrinkled, shiny; upright, 3–4 ft growth; Uhl, J.

UHLensch, HRg, w, 1989; (**Snow Owl**); Uhl, J.

UHLwe, HRg, dp; (**Yankee Lady**); Uhl, J., 1987

Ukrainian Dawn, *see* **'Ukrainskaia Zorka'**

'Ukrainskaia Zorka', F, mr, 1955; (Ukrainian Dawn); flowers bright cinnamon-red, medium, slight fragrance; [Independence X ?]; Klimenko, V.N.

Ulla Land, Misc OGR, mr

Ulmer Münster®, *see* 'KORtello'

Ulrich Brünner, *see* **'Ulrich Brünner Fils'**

'Ulrich Brünner Fils', HP, dp; (Ulrich Brünner); flowers geranium-red to carmine, large, dbl., 30 petals, cupped, intense fragrance; vigorous growth; (28); Levet, A., 1882

'Ulrick's Buttercup', F, ly, 1953; bud ovoid; flowers medium, borne in clusters, dbl., cupped, intense fragrance; foliage bronze; very vigorous, bushy growth; [Yvonne Rabier X Baby Alberic]; Ulrick, L.W.

'Ulrick's Gem', F, pb; flowers deep pink and white, borne in clusters, very dbl.; foliage glossy; very vigorous, bushy growth; [Mrs Tom Henderson X Self]; Ulrick, L.W.

'Ulrick's Red', HT, lp; Ulrick, L.W., 1954; (Weatherly, L.)

'Ulrick's Smokie', F, m, 1953; bud long, pointed; flowers smoky mauve, background white, large, borne in clusters, very dbl., intense fragrance; foliage light green; bushy growth; [Mrs Tom Henderson X Tip-Top]; Ulrick, L.W.

'Ulrick's Yellow', HT, my, 1953; bud globular; flowers yellow, center darker, medium, dbl., cupped, intense fragrance; foliage light green; bushy growth; [Mrs Pierre S. duPont X Lady Hillingdon]; Ulrick, L.W.

Ulrike®, F, r

'Ulster', HP, mr; flowers salmon-red, large, dbl., slight fragrance; Dickson, A., 1899; (Sangerhausen)

'Ulster Gem', HT, my, 1917; bud long, pointed; flowers canary-yellow, large, single, slight fragrance; GM, NRS, 1916; Dickson, H.

'Ulster Monarch', HT, r, 1951; flowers apricot shaded buff, high-pointed, medium, dbl., 50 petals, slight fragrance; foliage glossy, bright green; upright growth; [Sam McGredy X Mrs Sam McGredy seedling]; McGredy, Sam IV

'Ulster Queen', F, ob, 1960; flowers salmon-orange, well formed, borne in clusters, dbl., 25 petals, 3 in., slight fragrance; vigorous growth; [Cinnabar X Independence]; McGredy, Sam IV; McGredy

'Ulster Volunteer', HT, mr, 1918; flowers brilliant cherry-red, base clear white, single, 5–6 in.; Dickson, H.

Ultimate Pink™, *see* 'JACval'

Ultimate Pleasure, *see* 'TINpleasure'

Uma Rao, HT, pb; Pal, Dr B.P., 1989

Umberglo™, *see* 'MINiumber'

'UMScharm', Min, pb, 1990; (**Single Charm**); flowers hand-painted white with pink edge, small, borne mostly singly, single, 5 petals, no fragrance; foliage small, medium green, matt; tall, upright, bushy growth; [Jennifer X Unnamed seedling]; Umsawasdi, Dr Theera

'UMSdad', HT, dr, 1990; (**Manit**); flowers medium, borne mostly singly, semi-dbl., 6–14 petals, no fragrance; foliage medium, medium green, matt; upright, bushy, tall (150–180 cms) growth; [Unnamed seedling X Olympiad]; Umsawasdi, Dr Theera, 1991

UMSlove, F, w, 1996; (**So In Love**); flowers near white, blooms borne mostly single, dbl., 26–40 petals, 1.5–2.75 in., moderate fragrance; few prickles; foliage medium, medium green, semi-glossy; medium, bushy, spreading growth; [Golden Wings X Unknown]; Umsawasdi, Dr Theera

'UMSnira', Min, ab, 1991; (**Niramol**); flowers moderately blooms borne mostly singly, dbl., no fragrance; foliage medium, medium green, semi-glossy; medium, bushy growth; [Loving Touch X Unnamed seedling]; Umsawasdi, Dr Theera

'UMSpretty', Min, dp, 1990; (**Pretty 'n' Single**); flowers deep pink, small blooms, single, 5 petals, intense fragrance; foliage medium, dark green, semi-glossy; upright, bushy, tall growth; [Nymphenburg X Libby]; Umsawasdi, Dr Theera, 1986; Theera Umsawasdi

'UMSpriub', HT, ly, 1990; (**Priub**); flowers yellowish-cream turning to pure white, large blooms, single, 5 petals, slight fragrance; many prickles; foliage medium, medium green, semi-glossy; vigorous, upright, tall growth; [Unnamed seedling X Unnamed seedling]; Umsawasdi, Dr Theera, 1986; Theera Umsawasdi

'UMSrum', Min, r, 1990; (**Rum Candy**); flowers brownish apricot, light apricot shaded light pink when fully, semi-dbl., 6–14 petals, slight fragrance; foliage small, light green, matt; upright, bushy, low (18 cms) growth; [Twilight Trail X Unknown seedling]; Umsawasdi, Dr Theera, 1991

'UMStar', Min, dr, 1990; (**Ruby Star**); flowers deep red, small, borne mostly singly, single, 5 petals, no fragrance; foliage medium, dark green, semi-glossy; upright, bushy, tall (24 cms) growth; [High Spirits X Unknown seedling]; Umsawasdi, Dr Theera, 1991

'UMStex', Min, yb, 1990; (**Texas Sunrise**); flowers yellow at times with pinkish edge, moderately small, dbl., 15–25 petals, no fragrance; foliage medium, medium green, semi-glossy; upright, bushy growth; [Arizona Sunset sport]; Umsawasdi, Dr Theera, 1991

'Una', HCan, ly, 1900; flowers buff-yellow becoming creamy white, semi-dbl., non-recurrent, vigorous growth; [Tea X R. canina]; Paul

'Una Hawken', F, ly, 1972; bud ovoid; flowers butter-yellow to cream, medium, dbl., cupped, moderate fragrance; foliage glossy, leathery; vigorous, upright, bushy growth; [Arthur Bell X Arthur Bell]; Murray, Nola; Rasmussen's

'Una Wallace', HT, dp, 1921; flowers soft, even-toned cherry-rose, well formed, dbl., moderate fragrance; long, strong stems; vigorous growth; GM, NRS, 1920; McGredy

Uncle Bill, HT, mp; Beales, Peter, 1984

'Uncle Joe', HT, dr, 1972; (El Toro, Toro); flowers very large, very dbl., exhibition form; foliage large, dark, leathery; very vigorous, upright, tall growth; [(Mirandy X Charles Mallerin) X Unnamed seedling]; Kern Rose Nursery, 1971

'UNCLE LOU', HT, dp, 1997; flowers deep pink, medium sized, dbl., 26–40 petals, borne singly, slight fragrance; foliage medium, medium green, semi-glossy; upright, tall (4.5–5 ft) growth; [Queen Elizabeth X Christian Dior]; Roth, Louis A.

'UNCLE SAM', HT, dp, 1965; flowers deep rose-pink, large, exhibition form, moderate fragrance; foliage dark, leathery, glossy; vigorous, tall growth; [Charlotte Armstrong X Heart's Desire]; Warriner, William A.; J&P

Uncle Steve, *see* 'RESUNC'

Uncle Walter, *see* 'MACON'

'UNDINE', HT, ob, 1901; flowers dark orange, medium, moderate fragrance; upright, bushy growth; [L'Ideal X Sunset]; Jacobs

Unermüdliche, Ch, m; Lambert, P., 1904

Unforgettable, *see* 'TROBSPREAD'

Unforgettable, *see* 'JACHYP'

Unforgettable, *see* TANalam

UNICEF, *see* COCjojo

Union Redwood Cemetary, HP, lp; (found rose)

Union-Rose St Helena®, *see* 'CANLISH'

'UNIQUE', Ch, w; flowers white edged pink, compact; Laffay, M.

'UNIQUE', LCl, op, 1928; flowers bright fawn-orange-salmon, recurrent bloom; vigorous growth; [Tip-Top X Hybrid Perpetual]; Evans

'UNIQUE BLANCHE', C, w, 1778; (Blanche Unique, Vièrge de Clery, White Provence); flowers white, sometimes tinged pink, large, dbl., moderate fragrance; Grimwood

Unique de Provence, *see* **'UNIQUE MOSS'**

'UNIQUE MOSS', M, w, 1844; (Unique de Provence); flowers pure white, occasionally tinted pink, well-mossed, large, dbl.; prickles shoots very spiny; [Unique Blanche sport]; Robert

'UNIQUE PANACHÉE', C, w, 1821; flowers white, faintly striped rose and lilac, large, dbl., globular; vigorous growth; [Unique Rouge sport]; Caron

'UNIQUE ROUGE', C, mp; flowers bright pink, medium

Unitarian Cemetary Hybrid China, HCh, lp; (found rose)

'UNITED NATIONS', F, op, 1949; flowers salmon-pink, open, medium, borne in clusters, dbl., 26 petals, moderate fragrance; foliage glossy; very vigorous, bushy growth; [Mev. Nathalie Nypels X Rosamunde]; Leenders, M.

United Nations Rose, *see* 'HARBELLA'

'UNITY', HT, dp, 1956; flowers deep pink, reverse slightly darker, well shaped, dbl., 45 petals, 5.5 in., intense fragrance; foliage dark; vigorous, upright growth; [Red Ensign sport]; Sansum

'UNIVERSAL FAVORITE', HWich, mp, 1898; flowers soft rose, borne in large clusters, dbl., moderate fragrance; vigorous, climbing growth; [R. wichurana X Paquerette]; Horvath; W.A. Manda

'UNIVERSITÉ D'ORLÉANS', Pol, mr, 1966; bud globular; semi-dbl., cupped, borne in large clusters; vigorous growth; [? X Ronde Endiablee]; Hémeray-Aubert

'UNN', F, dp, 1972; bud ovoid; flowers deep pink, open, medium, semi-dbl., slight fragrance; foliage glossy; dwarf, moderate growth; [Rimosa X Fidélio]; Lundstad

'UNSER STOLZ', HT, dr, 1960; flowers bright crimson-scarlet, large, dbl.; foliage leathery, dark, glossy; upright growth; [Ena Harkness X Unnamed seedling]; Verschuren, A.; van Engelen

Uppingham School, *see* 'FRANBURST'

Upstart, *see* 'JACUP'

Uptown, *see* 'TALUPTOWN'

'URANIA', HP, dr, 1906; flowers bright crimson, dbl., recurrent bloom; [American Beauty X Suzanne-Marie Rodocanachi]; Walsh

Urara, *see* 'KEIFUPIRA'

'URDH', HP, mp; bud long, pointed; flowers large, dbl., 45 petals, exhibition form, intense fragrance; vigorous growth; [Victor Verdier X Papa Lambert]; Tantau, 1930; C-P, 1930

'URSEL', Pol, mr, 1938; flowers carmine red, medium, semi-dbl.; Vogel, M.; (Sangerhausen)

'URSEL TGARTH', HFt, yb, 1938; flowers yellow and red, medium, dbl.; Ketten, Gebrüder; (Sangerhausen)

Ursula®, *see* 'LAPON'

'USCHI', F, dp, 1972; flowers full, medium, dbl., slight fragrance; foliage glossy, dark; vigorous, bushy growth; [Marimba sport]; Tantau, Math.

'USHA', F, dr, 1975; bud ovoid; flowers deep red, open, small, dbl., 35–40 petals, 1.5 in.; foliage soft; compact, bushy growth; [Orangeade X ?]; IARI

Usmev, HT, lp; Strnad; (Czech Rosa Club)

Usvit, F, op, 1978; flowers salmon-pink, medium, dbl., slight fragrance; Urban, J.; (Sangerhausen)

'UT TYLER ROSE', Gr, ob, 1984; flowers orange, medium, dbl., 35 petals, slight fragrance; foliage large, medium green, matt; upright growth; [Unnamed seedling X Unnamed seedling]; Weeks, O.L.; University of Texas at Tyler

Uta Maro, HT, mr; flowers red with silvery reverse, large, dbl., moderate fragrance; (Sangerhausen)

'UTAGE', HT, w, 1977; bud pointed; flowers very large, dbl., 30–35 petals, exhibition form; foliage leathery; vigorous, upright growth; [Edith Krause X Bridal Robe]; Ota, Kaichiro

Utage, HT, w

'UTRO MOSKVY', HT, mp, 1952; (Morning in Moscow, Moscow Morn); flowers soft rose tinted carmine, large, dbl., 60 petals, slight fragrance; foliage leathery, grayish; spreading growth; GM, International Exhibition, 1961; [Frau Karl Druschki X Independence]; Shtanko, E.E.

'UTTAM', HT, lp, 1969; bud pointed; flowers pastel pink, full, large, dbl.; foliage glossy; vigorous, upright growth; [Elite X Open pollination]; IARI

Uwe Seeler®, *see* 'KORSEE'

V

'V for Victory', HT, ly, 1941; bud ovoid, long, pointed; flowers yellow faintly tinted orange, dbl., 45 petals, exhibition form, intense fragrance; foliage glossy; vigorous, bushy, compact to open growth; [Golden Glow X Condesa de Sástago]; Brownell, H.C.

V G Glowing HT, HT, lp

V O N Canada, F, lp

V Y F Rose, HT, lp

'V. Viviand Morel', T, dr; flowers rich crimson shaded dark red and carmine, large, dbl.

Vabene, *see* 'MEItrogana'

'Vagabonde', F, ob, 1962; bud long, pointed; flowers salmon-orange, blooms in clusters of 3 or more, dbl., 25 petals, exhibition form, foliage dark, glossy; vigorous, bushy growth; [Mannequin X Fashion]; Lens

'Vahine', HT, or, 1964; flowers dark cardinal-red tinted orange; strong stems; vigorous, upright growth; Combe; Vilmorin-Andrieux

'Vainqueur', HT, m, 1937; bud ovoid, dark red; flowers velvety purple, reverse dark red, very large, very dbl., moderate fragrance; foliage bright green; long, strong stems; very vigorous growth; [Sensation seedling]; Heizmann, E.; A. Meilland

Vaire, *see* **'Château de Vaire'**

Vaishnavi, HT, or; K&S, 1992

Vajont, F, 1966; Galesso, G.; (Cavriglia)

Val Boughey, *see* **'Valerie Boughey'**

Val d'Authion, HT, mp; Minier, 1967

'Val De Mosa', HT, my, 1968; very dbl., 70 petals, moderate fragrance; foliage dark, glossy; vigorous growth; [La Jolla X Cynthia Brooke]; Ellick

Valdemar, HMult, lp; Back, 1980

Valdemossa, HT, 1991; Dot, Simon; (Cavriglia)

Vale of Clwyd, *see* 'BEEval'

Valeccia, *see* KOReklia

'Valence Dubois', HGal, mp, 1880; flowers medium, dbl.; Fontaine; (Sangerhausen)

'Valencia', HT, ab, 1967; flowers apricot-orange, large, dbl., exhibition form, moderate fragrance; foliage glossy, leathery; vigorous, upright growth; [Golden Sun X Chantré]; Kordes, R.; J&P

Valencia®, *see* KOReklia

Valencia 89, *see* KOReklia

'Valenciennes', LCl, dr, 1960; bud long, pointed; flowers medium, borne in clusters, semi-dbl., slight fragrance; foliage dark, glossy; very vigorous growth; [Paul's Scarlet Climber X Seedling]; Robichon

'Valentin Beaulieu', HWich, m, 1902; flowers violet-pink, medium, dbl.; Barbier; (Sangerhausen)

Valentina Borgatti, HT, 1969; Borgatti, G.; (Cavriglia)

Valentina Cortese, F, 1976; Cazzaniga, F. G.; (Cavriglia)

'Valentine', F, mr, 1951; flowers bright red, blooms in large clusters, semi-dbl., 18 petals, 2.5–3 in., slight fragrance; foliage dark olive-green; spreading, bushy, compact growth; [China Doll X World's Fair]; Swim, H.C.; Armstrong Nursery

Valentine Heart, *see* 'DICogle'

Valeria Sykes, *see* 'HORflashrob'

'Valerie', F, w, 1932; bud pointed, yellow; flowers cream, large, borne in large clusters; foliage glossy, dark; bushy growth; GM, NRS, 1931; Chaplin Bros.

Valerie, HT, lp; Select, 1994

'Valerie Boughey', HT, op, 1960; (Val Boughey); flowers coppery salmon-pink, high pointed, dbl., 36 petals, 6 in., moderate fragrance; foliage leathery, glossy; vigorous, upright growth; [Tzigane sport]; Fryers Nursery, Ltd.

Valerie Jeanne, *see* 'SAVaval'

Valerie June, HT, lp; Allender, Robert William, 1982

Valerie Kathleen, *see* 'HORvalkath'

'Valerie Purves', HT, mp, 1940; flowers pink, well formed, moderate fragrance; vigorous growth; Clark, A.

Valerie Swane, *see* 'AROBIPY'

'Valeta', F, or, 1960; flowers red shaded vermilion, open, borne in clusters (up to 20), dbl.; foliage dark; vigorous growth; [Signal Red X Fashion]; deRuiter

Valfleury, F, 1971; Croix; (Cavriglia)

Valia Balkanska, *see* 'MANclassic'

'Valiant', HT, mr, 1948; bud long, pointed; flowers bright red, large, dbl., 30 petals, exhibition form, moderate fragrance; foliage dark, leathery; vigorous, upright, branching growth; [Poinsettia X Satan]; Boerner; J&P

VALkita, F, dr, 1995; (**Mariska**); (VALkita)

'Valldemosa', F, mr, 1956; flowers fiery red, large, dbl., 20 petals, globular, foliage glossy; vigorous, open growth; [Magrana X Radar]; Dot, M.

'Valrose', LCl, w, 1964; flowers white suffused pink at edges, medium, borne in clusters of 3, semi-dbl., recurrent bloom, Mondial Roses

'Valstar', HT, dp, 1962; flowers deep pink, well formed, large, dbl., intense fragrance; foliage leathery; strong stems; Mondial Roses

'Vamp', Min, dr, 1981; single, 5 petals, cupped, borne 3 per cluster, no fragrance; pointed prickles; foliage dark; angular growth habit; [Fairy Moss X Fairy Moss]; Gatty, Joseph

'Van Artevelde', HGal, dp; flowers deep pink, petals imbricated in whorls, large, very dbl.; Parmentier, before 1847

'Van Bergen', S, ly, 1980; bud pointed; flowers light yellow fading to white, blooms borne 3, 5, 7 per clust, dbl., 48 petals, repeat bloom, slight, fruity fragrance; hooked, red prickles; foliage small, glossy, dark; vigorous, compact, bushy growth; [(R. wichurana X Baronne Prevost) X Fun Jwan Lo]; James, John

Van Gogh™, *see* 'WILgogh'

Van Houtten, HP, lp

Van Huyssum, HGal, lp; Parmentier, 1800's

Van Nes, *see* **'Permanent Wave'**

'Van Rossem's Jubilee', HT, op, 1937; flowers bright carmine or coral; foliage glossy, bronze; vigorous growth; Van Rossem

'Vanamali', HT, m, 1978; bud long, pointed; flowers orchid-mauve, full, dbl., 35–40 petals, 6 in., exhibition form, moderate fragrance; foliage dark, leathery; tall, vigorous, bushy growth; [Lady X X ((Gruss an Teplitz X Unknown) X (Lake Como X Angel Face))]; Viraraghavan, M.S.

Vancouver, HT, lp; Twomey, Jerry, 1997

Vancouver Centennial, *see* **'Jan Wellum'**

'Vanda Beauty', HT, dy, 1971; flowers pointed, dbl., 28 petals, 3.5 in., moderate fragrance; foliage glossy, dark; very free growth; [Gertrude Gregory X Unnamed seedling]; Gregory

'Vandael', M, m, 1850; bud well mossed; flowers rich purple, edged lilac, large, dbl.; vigorous growth; Laffay, M.

'Vanessa', HT, pb, 1946; bud ovoid; flowers coral, reverse yellow, dbl., 25 petals, 4.5 in., intense fragrance; foliage bright green; vigorous growth; [Arch. Reventos X Lord Baden-Powell]; Leenders, M.; Longley

Vanessa Belinda, *see* 'BRAndyglow'

Vanessa Campello®, *see* FEbesa

'Vanguard', HRg, op, 1932; flowers orange-salmon, large, dbl., moderate fragrance;

foliage light, very glossy; vigorous (to 10 ft) growth; David Fuerstenberg Prize, ARS, 1934 Dr W. Van Fleet Medal, ARS, 1933; [(R. wichurana X R. rugosa alba) X Eldorado]; Stevens, G.A.; J&P

Vanilla®, *see* KORplasina

Vanilla Perfume™, *see* 'JACWOTTE'

'VANITY', HMsk, dp, 1920; flowers rose-pink, near blooms in very large sprays, single, recurrent bloom, intense fragrance; foliage rich green, leathery; very vigorous (to 8 ft), bushy growth; (21); [Château de Clos Vougeot X Seedling]; Pemberton

Vanity, HT, lp; Halstead, 1901; Select Roses, 1999

'VANITY FAIR', HT, lp, 1942; bud long, pointed; flowers cameo-pink, large, semi-dbl., 19 petals, exhibition form, slight fragrance; foliage soft; long stems; very vigorous, upright growth; [Better Times X Golden Rapture]; Roberts; Totty, 1944

Van's Choice, F, lp; Williams, J. Benjamin, 1999

'VANTO', Pol, dr; [Dick Koster sport]; Vanto

'VARBOLE', HT, mr, 1981; (**Caesar**); bud ovoid; flowers cardinal red, blooms borne 2–3 per cluster, dbl., 33 petals, exhibition form, slight, tea fragrance; red prickles; foliage leathery, medium green, semi-glossy; upright, branched growth; [Ilona X Seedling]; van Veen, Jan; Carlton Rose Nurseries

'VARELVI', HT, ab, 1978; (**Elvira**); flowers large, dbl., 35 petals, no fragrance; foliage medium, medium green, semi-glossy; upright growth; [Zorina X Dr A.J. Verhage]; van Veen, Jan; G. Verbeek

Variant, F, op; flowers medium, dbl.; VEG; (Sangerhausen)

'VARIEGATA DI BOLOGNA', B, rb, 1909; flowers white, striped purplish red, large blooms in clusters of 3–5, dbl., globular, moderate fragrance; vigorous (6–8 ft) growth; Bonfiglio, A.

Variegata di Bologna Rouge, B, dr; Lowe, 1984

Varieté, F, m, 1987; flowers purple/pink with lighter shading, large, dbl., slight fragrance; VEG; (Sangerhausen)

'VARIETY CLUB', F, yb, 1965; flowers yellow marked rose-red, well-formed blooms in clusters, dbl., 48 petals, slight fragrance; foliage dark; [Columbine X Circus]; McGredy, Sam IV; McGredy

'VARIORA', HT, my, 1981; (**Eloira**); flowers large, dbl., 35 petals, no fragrance; foliage medium, medium green, semi-glossy; upright growth; [Elvira X Seedling]; van Veen, Jan; G. Verbeek

'VARLON', HT, mr, 1973; (**Ilona**); dbl., 40 petals, 4–4.5 in., slight fragrance; foliage glossy, leathery; bushy growth; [Miracle X (Romantica X Edith Piaf)]; Verbeek

Varna, S, dy, 1979; flowers large, dbl.; GPG Bad Langensalza; (Sangerhausen)

Varo Iglo, HT, w, 1969; flowers medium, dbl.; Verbeek; (Sangerhausen)

Varo Rania, HT, 1969; Verbeek; (Cavriglia)

Varsha, F, w; K&S, 1992

'VASANT', HT, yb, 1980; bud pointed; flowers yellow edged pink, blooms borne 10 per cluster, exhibition form, slight, spicy fragrance; straight, brown prickles; foliage dark; upright, bushy growth; [Sweet Afton X Delhi Princess]; Division of Vegetable Crops and Floriculture

Vasavi, HT, dr; Kasturi, 1976

'VASCO DA GAMA', F, mr; flowers velvety red; [Pinocchio X Alain]; da Silva, Moreira

'VASSAR CENTENNIAL', HT, pb, 1961; flowers peach to shell-pink, dbl., 30 petals, 4.5–5 in., exhibition form, moderate fragrance; foliage dark, leathery; upright, branching growth; [Helene de Roumanie X Confidence]; Meilland, Mrs. Marie-Louise; C-P

'VATER RHEIN', HT, dr, 1922; flowers very dark red, dbl., intense fragrance; [Kynast X Unnamed seedling]; Kiese

'VATERLAND', HT, dr, 1928; flowers dark red with coppery reflexes, large, dbl., moderate fragrance; foliage bronze, leathery; vigorous growth; [National Emblem X Earl Haig]; Berger, V.; Pfitzer

'VATERTAG'®, Pol, ob, 1959; (Father's Day, Fête des Pères, Jour des Pères, Orange Muttertag); flowers salmon-orange; [Mothersday sport]; Tantau, Math.

Vatican®, *see* 'DELOP'

'VEDETTE', HT, mr, 1951; flowers brilliant red, large, dbl., 28 petals, moderate fragrance; foliage leathery; [(Frau Karl Druschki X George Dickson) X Unnamed seedling]; Gaujard

'VEE BRYAN', F, w, 1998; flowers cream, edged pink maroon, full, dbl., 26–40 petals, 2 in., borne in large clusters, moderate fragrance; [Solitaire X Solitaire]; Jones, L.J.; Haynes Roses, 1997

Vee Marie, *see* 'JUDVEE'

'VEILCHENBLAU', HMult, m, 1909; (Blue Rambler, Blue Rosalie, Violet Blue); flowers violet, petals streaked with white, center white, yellow sta, semi-dbl., 1.25 in., cupped, moderate fragrance; very few prickles; foliage large, pointed, glossy, light; short stems; vigorous, climbing (10–15 ft) growth; [Crimson Rambler X Erinnerung an Brod]; Schmidt, J.C.

Velay Rose, LCl, 1972; Croix; (Cavriglia)

Veldfire, *see* 'KORGUST'

'VELINDRE', HT, mr, 1998; flowers red-pink with peach reverse, full, classic, dbl., 26–40 petals, 5.5 in., borne singly, moderate fragrance; some prickles; foliage large, dark green, semi-glossy; upright, bushy, medium growth; [Solitaire X seedling]; Poole, Lionel

Velingstorprosen, HGal, lp

Velizy®, *see* 'DELSAMOUR'

'VELLUTO', HMoy, dr, 1934; bud long, pointed; flowers velvety dark crimson, stamens red, medium, semi-dbl., intermittent bloom, foliage dark; [R. moyesii X J.C. Thornton]; San Remo Exp. Sta.

'VELOUR', F, mr, 1967; bud ovoid; dbl., flat, moderate fragrance; foliage glossy; vigorous, upright, bushy growth; [Garnette seedling X Hawaii seedling]; Boerner; J&P

Velours Épiscopal, HCh, m; flowers violet purple/pink, large, dbl.; Roseraie de l'Hay; (Sangerhausen)

Velours Pourpre, HP, m; Verdier, 1866

'VELSHEDA', HT, lp, 1936; flowers softest rose-pink, well formed, large, dbl., moderate fragrance; foliage dark; strong, erect stems; vigorous growth; Cant, F.

'VELUTINA', HGal, m, 1810; flowers velvety purple shaded violet, golden stamens, three rows of; Van Eeden

'VELUWEZOOM', HT, dp, 1909; flowers large, dbl., moderate fragrance; Pallandt; (Sangerhausen)

Velvet Arrow, HT, dr

Velvet Beauty, *see* **'BARKHATNAIA KRASAVITSA'**

'VELVET BEAUTY', HT, mr; bud high pointed; flowers currant-red, dbl., 40–55 petals, 5–5.5 in., exhibition form, moderate, clove fragrance; vigorous, upright growth; [Happiness X New Yorker]; Fisher, G.; Arnold-Fisher Co.

Velvet Cloak, *see* 'SEACLOAK'

Velvet Cover, *see* POUlria

Velvet Dreams, *see* 'LYOET'

Velvet Flame, *see* 'MEIMAUR'

Velvet Flame, Climbing, Cl HT, 1981; Orard, Joseph; (Cavriglia)

Velvet Fragrance, *see* 'FRYPERDEE'

'VELVET HOUR', HT, dr, 1978; flowers ox-blood-red, full, dbl., 44 petals, 3 in., moderate fragrance; foliage dark; vigorous, upright growth; LeGrice

Velvet Lady, *see* 'HADPRETTIE'

Velvet Lustre, *see* 'SIMPALNO'

'VELVET MIST'™, HT, m, 1990; (Fragrant Lavendar); flowers deep lavender, borne usually singly, medium, dbl., 25–35 petals, exhibition form, moderate, fruity fragrance; foliage large, medium green, matt, disease-resistant; upright, bushy, medium to tall growth; [Blue Ribbon X Shocking Blue]; Christensen, Jack E., 1984; Flowers of the Month, 1990

'VELVET QUEEN', HT, dr, 1965; flowers blood-red, pointed, 4 in., moderate fragrance;

moderate growth; [Fandango X Seedling]; Herholdt, J.A.

Velvet Robe, *see* **'ATOMBOMBE'**

Velvet Ruby, *see* 'MACHORO'

Velvet Ruby, LCl, lp

Velvet Star, *see* BARvelv

Velvet Star, *see* TANalednev

'VELVET TIMES', HT, dr, 1960; bud pointed; flowers rose-red, dbl., 40–50 petals, 4.5–5 in., exhibition form, intense fragrance; foliage leathery; vigorous, upright growth; [Better Times sport]; Peters; J&P

Velvet Touch, *see* 'SAVAVEL'

Velvet Treasure, *see* JACvet

Velveteen, *see* 'MEISOYRIS'

'VELVETIER', HT, dr, 1946; bud long, pointed; flowers deep velvety red, dbl., 28–35 petals, 4–5 in., exhibition form, moderate fragrance; foliage glossy; vigorous, upright growth; hardy for the class; [Pink Princess X Crimson Glory]; Brownell, H.C.

Velvia, *see* 'HARXAMPLE'

Vendome, LCl, lp

Vendéme, *see* 'GAURA'

Vendulka, F, mp, 1974; flowers large, dbl., moderate fragrance; Vecera, L.; (Sangerhausen)

Venere®, F, mp; Barni, V.

'VENEZUELA', F, op, 1957; flowers salmon, edged dull red; very vigorous growth; [Joanna Hill X Pinocchio]; Silva

'VENISE', HT, rb, 1946; flowers red with silvery white reverse flushed salmon-carmine, large, dbl.; spreading growth; [Joanna Hill X Margaret McGredy]; Meilland, F.

Venise, *see* ADAzombar

'VENLO', F, or, 1960; bud pointed; flowers medium, borne in compact clusters, semi-dbl., flat, slight fragrance; foliage dark; moderate, compact growth; [Cinnabar X Fashion]; Leenders, G.

Venrosa, HT, mr, 1973; flowers large, dbl., intense fragrance; St Zila; (Sangerhausen)

Vent des Indes®, *see* MAUkeole

'VENTURE', S, mp, 1984; semi-dbl., blooms in clusters, moderate fragrance; foliage medium, light green, glossy; upright, slender growth; [(Charlotte Armstrong X (Cecilia X China Belle)) X Prevue]; James, John

'VENUS', T, m; flowers carmine purple; Schmidt, J.C., 1896

'VENUS', HT, dp, 1921; bud long, pointed; flowers carmine, edge flushed cream, very dbl., moderate fragrance; GM, NRS, 1922; [J. Barriot X Sunburst]; Bees

'VENUS', F, w, 1955; flowers pure white, well shaped, borne in large trusses; long stems; vigorous, bushy growth; [Pinocchio seedling]; Maarse, G.

Venus, M, pb; Welter, 1904

Venus, *see* JACeve

Venusic®, *see* 'DELDRA'

'VENUSTA PENDULA', Ayr, w, 1928; bud pink; flowers blush white, fading to creamy white, small blooms in cluster, semi-dbl., early flowering, no fragrance; vigorous (to 15 ft) growth; Re-int. Kordes, 1928

'VENU-VAISHALI', HT, pb, 1970; bud ovoid; flowers light pink, striped white, base yellow, full, large, dbl.; foliage large, soft; very vigorous, upright growth; [Astree sport]; Deshpande

'VERA', HT, op, 1922; flowers deep salmon, shaded coral-red, dbl.; Paul, W.

'VERA ALLEN', HT, op, 1939; flowers salmon-pink, well formed, large, dbl., intense fragrance; compact growth; Dickson, A.

'VERA CRUZ', HT, pb, 1938; bud long, pointed; flowers pink shaded mauve, flushed red and yellow, large, dbl., exhibition form, slight fragrance; foliage soft; vigorous growth; [Frank Reader X Johanniszauber]; da Silva, Moreira

'VERA DALTON', F, mp, 1961; bud pointed; flowers soft pink, blooms in clusters, dbl., 24 petals, 4 in., cupped, moderate fragrance; foliage glossy, dark; vigorous, bushy growth; [(Paul's Scarlet Climber X Paul's Scarlet Climber) X (Mary X Queen Elizabeth)]; Norman; Harkness

Vera Johns®, *see* 'KORVERA'

Vera Johns, *see* 'KORVEJOH'

'VERASTELLA', HT, mp, 1954; flowers rose, center deeper; strong stems; vigorous growth; Giacomasso

Verbeeka, F, lp

Verbesserte Tantau's Triumph, *see* **'CINNABAR IMPROVED'**

Vercors, *see* 'HARVINTAGE'

'VERCORS', HT, or, 1946; flowers brilliant orient red, large, dbl., globular, moderate fragrance; vigorous growth; [(Mme Arthaud X Mme Henri Guillot) X (Comtesse Vandal X Brazier)]; Mallerin, C.; A. Meilland

Verdi, *see* 'INTERPOOL'

Verdi, HMsk, lp; Lens, 1984

'VERDUN', Pol, mr, 1918; flowers vivid carmine-red, large blooms in clusters of 25–50, dbl.; vigorous, dwarf growth; Barbier

'VERDUN SUPERIOR', Pol, 1918; Barbier, A.; (Cavriglia)

'VERENA', F, dr, 1973; bud ovoid; flowers dark velvety red, center lighter, medium, dbl., intense fragrance; foliage glossy, bluish green; upright, bushy growth; [Lucy Cramphorn X Inge Horstmann]; Hetzel; GAWA

Vergenal, HP, lp

'VERJO', HT, dp, 1970; (**Forsythe**); flowers venetian pink, carmine-rose, dbl., 45–50 petals, 4.5–5 in., exhibition form, intense fragrance; foliage glossy, dark, leathery; upright growth; [Miracle X Dr A.J. Verhage]; Verbeek

Verlaine, HT, w

'VERMILLON', HT, or, 1929; flowers scarlet tinged orange, base yellow, semi-dbl., slight fragrance; [Constance X Paul's Scarlet Climber]; Barbier; Dreer

'VERNA MACKAY', HT, ly, 1912; flowers buff to bright lemon-yellow; Dickson, A.

'VERONA', F, lp, 1963; bud long, pointed to urn-shaped; dbl., 42 petals, 2.5–3 in., exhibition form, abundant bloom, foliage leathery, dark; vigorous, bushy growth; [Spartan X Garnette]; Swim & Weeks

Veronica, *see* 'KORFLÜG'

'VERONICA', HT, w, 1950; bud pale yellow; flowers snow-white, imbricated, dbl., 32 petals, 4.5 in., intense fragrance; vigorous growth; Prosser

Veronica 92®, F, mp; Noack, Werner, 1992

Veronica Kay, Min, lp; Welsh, Eric; (Weatherly, L.)

Veronika, *see* 'KORFLÜG'

'VERONIQUE', F, mr, 1961; flowers bright raspberry-red, borne in clusters, single, 9 petals, 4 in.; foliage dark, glossy; vigorous, bushy growth; [Sumatra X Philippe]; Delforge

Verrystata, HT, lp; Kordes

Versailles®, *see* 'DELSET'

Versailles Palace, *see* POUlsail

Verschuren, HT, lp; Verschuren, 1904

'VERSCHUREN'S GLOW', Pol, mr, 1939; flowers medium, semi-dbl.; Verschuren; (Sangerhausen)

'VERSCHUREN'S PINK', HT, op, 1950; flowers salmon-pink with darker reflections, reflexed, dbl., 42 petals, 3–4 in., exhibition form, moderate fragrance; foliage glossy, dark; very vigorous growth; GM, NRS, 1949; [Mme Butterfly X Pink Pearl]; Verschuren; Gregory

Versigny, *see* MASversi

Versilia, *see* NIPRventyel

'VERY BUSY', Min, pb, 1973; bud long, pointed; flowers pink and yellow, small, dbl., slight fragrance; foliage small, leathery; dwarf, bushy growth; [Perle d'Or X Fairy Princess]; Moore, Ralph S.; Sequoia Nursery

Very Cherry™, *see* 'JACRENEW'

Vesely, HT, pb

'VESENII AROMAT', HT, or, 1955; (Spring Fragrance); flowers red tinted orange, base lighter, medium, dbl., 73 petals, slight fragrance; foliage dark, glossy; very vigorous, spreading growth; [Crimson Glory X Peace]; Klimenko, V. N.

Vesna, HT, lp; Vecera, L., 1970; (Czech Rosa Club)

'VESPER', F, ob, 1966; flowers orange, reverse burnt orange, blooms in clusters, dbl., slight fragrance; foliage small, blue-gray; moderate growth; LeGrice

'VESTA', F, mr, 1946; flowers currant-red, semi-dbl., moderate fragrance; [Irene X Donald Prior]; Leenders, M.

'VESTAL'S CORAL GEM', HT, op, 1939; flowers soft salmon-pink, reverse glowing carmine with coppery sheen, dbl.; foliage leathery, light; vigorous growth; [Betty Uprichard seedling]; Vestal

'VESTAL'S RED', HT, mr, 1937; flowers clear red, large, dbl., cupped, slight fragrance; foliage leathery, light; vigorous growth; Vestal

'VESTAL'S TORCHLIGHT', HT, rb, 1939; bud long, pointed; flowers red and gold, open, large, dbl., slight fragrance; foliage leathery, bronze; vigorous growth; [Pres. Herbert Hoover X Unnamed seedling]; Vestal

Vestey's Pink Tea, T, lp; (found rose)

Vestey's Yellow Tea, T, lp; (found rose)

Vesuv, F, mr, 1963; flowers medium, dbl.; Vilmorin; (Sangerhausen)

'VESUVIUS', HT, dr, 1923; bud long, pointed; flowers dark velvety crimson, large, single, 6 petals, moderate fragrance; foliage light, leathery; vigorous growth; McGredy

'VESUVIUS', F, or, 1963; flowers geranium-red, large; very vigorous growth; GM, Bagatelle, 1963; Vilmorin-Andrieux

Veteran's Honor, HP; flowers purple-violet, dbl., cupped

Veteran's Honor™, *see* 'JACOPPER'

'VEVEY', HT, my, 1953; flowers sun-yellow, large, dbl.; Heizmann & Co.

Vi Ambler, HMult, lp

Via Mala®, *see* 'VIAKOR'

Via Romana, *see* **'LJUBA RIZZOLI'**®

'VIAKOR', HT, w, 1977; (**Via Mala**®); bud long, pointed; dbl., 33 petals, 4 in., exhibition form, slight fragrance; foliage glossy, dark, leathery; vigorous, upright, bushy growth; [Silver Star X Peer Gynt]; Kordes, W. Söhne

'VIANDEN', HT, pb, 1932; flowers reddish old-rose and pink, reverse ochre-yellow and raw sienna, very dbl., 90–100 petals, intense fragrance; vigorous, bushy growth; [George C. Waud X Ruth]; Ketten Bros.

Vice-President Curtis, *see* **'AUTUMN QUEEN'**

Viceroy, *see* 'BROROY'

'VICKI KENNEDY', HT, pb, 1976; bud globular; flowers deep rose-pink, center yellow, dbl., 53 petals, 4 in., slight fragrance; foliage large, bronze, upright, bushy growth; [Queen Elizabeth X Red Lion]; Murray & Hawken; Rasmussen's

'VICKIE THORNE', HT, lp, 1972; dbl., 25 petals, 4–4.5 in., intense fragrance; foliage dark; vigorous growth; [Prima Ballerina sport]; Thorne

'VICK'S CAPRICE', HP, pb, 1891; flowers lilac-rose, striped white and carmine, large, dbl., cupped, repeat bloom, moderate fragrance; medium growth; [Archiduchesse Elisabeth d'Autriche sport]; Vick

'VICKY', HT, or, 1972; bud long, pointed; flowers orange-vermilion, moderate fragrance; [Canasta X Peace]; Gaujard

Vicky, F, lp; Noack, Werner, 1994

Vicky Brown, HT, rb; Select, 1994

'VICKY MARFÁ', HT, mp, 1958; bud ovoid; flowers begonia-pink, center yellow, large, dbl., 32 petals, exhibition form, moderate fragrance; strong stems; upright, compact growth; [(Soraya X Ellinor LeGrice) X Henri Mallerin]; Dot, Simon

'VICOMTE MAURICE DE MELLON', HT, ab, 1921; flowers apricot and yellowish salmon with coppery reflexes, washed pink, dbl., moderate fragrance; [Earl of Warwick X Sunburst]; Ketten Bros.

Vicomtesse d'Avesnes, N, mp

'VICOMTESSE DE BERNIS', T, op, 1884; flowers coppery rose to fawn and deep salmon, large, dbl., intense fragrance; Nabonnand, G.

'VICOMTESSE DE CHABANNES', LCl, rb, 1921; flowers purplish crimson, center white, forming a distinct eye, large, semi-dbl.; vigorous, climbing growth; Buatois

'VICOMTESSE DE VEZINS', HP, mp, 1867; flowers large, dbl., moderate fragrance; Gautreau; (Sangerhausen)

'VICOMTESSE PIERRE DU FOU', Cl HT, op, 1923; flowers red aging to deep coral pink, large, dbl., recurrent bloom, intense fragrance; foliage large, glossy, bronze; vigorous, climbing growth; [L'Ideal X Joseph Hill]; Sauvageot, H.

'VICTOR', HT, dp, 1918; bud long, pointed; flowers deep rose, often red, semi-dbl., intense fragrance; [Ophelia seedling X Killarney Brilliant]; Hill, E.G., Co.

Victor Borge, *see* POUlvue

Victor Emmanuel, B, dr; Guillot, 1859

'VICTOR FERRANT', HT, dp, 1933; flowers carmine changing to purplish pink, base indian yellow, very, dbl., 60–70 petals, slight fragrance; vigorous growth; [C.W. Cowan X Pres. Cherioux]; Ketten Bros.

Victor Hugo®, *see* 'MEIVESTAL'

'VICTOR HUGO', HP, dr; flowers carmine-red shaded purple, medium, dbl., 30 petals, globular, moderate fragrance; vigorous growth; [Charles Lefebvre(?) X ?]; Schwartz, J., 1885

'VICTOR LEMOINE', HP, dr, 1888; flowers large, dbl.; Lévêque; (Sangerhausen)

'VICTOR MAGNIN', Pol, mr, 1930; flowers bright red, borne in large clusters, dbl.; vigorous growth; Van Gelderen

'VICTOR MAYER', HT, dr, 1921; bud long, pointed; flowers blood-red, reflexes deeper; Buatois

Victor Parmentier, HGal, mp; flowers medium, dbl., moderate fragrance; Roseraie de l'Hay; (Sangerhausen)

'VICTOR TESCHENDORFF', HT, w, 1920; flowers almost pure white on pale greenish yellow ground, very large, dbl., exhibition form, moderate fragrance; foliage glossy, dark; long stems; vigorous growth; [Frau Karl Druschki X Mrs Aaron Ward]; Ebeling; Teschendorff

Victor Veladin, T, w

'VICTOR VERDIER', HP, dp, 1859; flowers bright rose, center carmine, large, dbl., 50 petals, globular, moderate fragrance; vigorous growth; [Jules Margottin X Safrano]; Lacharme, F.

'VICTOR WADDILOVE', HT, dp, 1923; bud long, pointed; flowers bright carmine-pink, base yellow, large, very dbl.; McGredy

'VICTORIA', HT, dp, 1924; bud long, pointed; flowers deep rose-pink, center darker, dbl., intense fragrance; [Isobel X ?]; Prince

'VICTORIA', F, pb, 1946; flowers carmine, center white, large, semi-dbl., moderate fragrance; [Irene X Donald Prior]; Leenders, M.

'VICTORIA', HT, ly, 1947; flowers pale lemon-yellow, large, moderate fragrance; foliage dark; vigorous growth; [Golden Dawn X Phyllis Gold]; Robinson, H.; Baker's Nursery

'VICTORIA', Cl F, or; flowers clear geranium-red; [Unnamed seedling X Alain]; da Silva, Moreira

Victoria, *see* **'VICTORIA DE LOS ANGELES'**

'VICTORIA DE LOS ANGELES', HT, or, 1952; (Victoria); bud ovoid; flowers velvety geranium-red, medium, dbl., 35 petals, moderate fragrance; vigorous, compact growth; [Cynthia X Manuelita]; Dot, Pedro

Victoria Girls, S, lp; Poulsen, 1997

Victoria Gold, *see* WELgold

'VICTORIA HARRINGTON', HT, rb, 1931; flowers very dark red shaded orange-brown, center lighter, large, dbl., intense, spicy fragrance; foliage leathery, dark; vigorous growth; [Diadem X Hadley]; Thomas; H&S

'VICTORIA HARRINGTON, CLIMBING', Cl HT, rb, 1938; Mordigan Evergreen Nursery

'VICTORIA HYLAND', HT, op, 1973; bud ovoid; flowers red-pink to coral, dbl., 34 petals,

4 in., moderate, fruity fragrance; foliage glossy; moderate, compact growth; [Unnamed seedling X Colour Wonder]; Golik; Dynarose

Victoria Park, HT, lp

'VICTORIA REGINA', HT, yb, 1938; flowers golden yellow, reverse brownish yellow, sometimes blushed peach, dbl., 40 petals; vigorous, compact growth; [Nellie E. Hillock X Golden Dawn]; Hillock

Victorian Lace, *see* 'DEVCAL'

Victorian Memory, LCl, pb

Victorian Spice(TM), *see* 'HARzola'

'VICTORIANA', F, ob, 1977; flowers orange, reverse silver, dbl., 28 petals, 5 in., slight fragrance; foliage dark; LeGrice

Victoria's Song(TM), *see* 'ORTVIC'

'VICTORY', LCl, dp, 1918; flowers deep pink, center darker, large, dbl., moderate fragrance; vigorous, climbing growth; [Dr W. Van Fleet X Mme Jules Grolez]; Undritz

'VICTORY', HT, dr, 1920; flowers scarlet-crimson, dbl., moderate fragrance; GM, NRS, 1919; McGredy

Victory Parade®, *see* POUlvic

'VICTORY RED', HT, dp, 1939; flowers rose-red; [Pink Delight sport]; Elliott

'VICTORY STRIPE', HT, rb, 1942; flowers cerise-red variegated white and light pink, dbl., 50 petals, 5 in., moderate fragrance; [Jewel sport]; Grillo

'VICTORY YEAR', S, mp, 1951; bud ovoid; flowers clear pink, open, medium, semi-dbl., profuse, non-recurrent bloom, slight fragrance; foliage leathery; very vigorous, upright growth; [Betty Bland X ?]; Wright, Percy H.

'VIDA BEGLAN', F, ly, 1993; flowers lemon yellow with very small traces of salmon, dbl., 26–40 petals, 1.5–2.75 in., borne in small clusters, intense fragrance; many prickles; foliage medium, medium green, matt; medium (90 cms), upright growth; [Elizabeth of Glamis sport]; Beglan, M.; Beglan, 1992

Vidal Sassoon, *see* 'MACJULIAT'

'VIDIAGO', HT, mr, 1962; flowers currant-red, reverse geranium-red, large, dbl., 26 petals; very vigorous growth; [Baccará X (S'Agaro X Peace)]; Dot, Simon

'VIDYUT', F, dp, 1983; flowers deep pink; [Europeana sport]; Yadava, U.N.; Tata Electric Co.

'VIE EN ROSE', F, lp, 1994; (La Vie en Rose); flowers bright pink, white reverse, blooms borne in large clusters, semi-dbl., 6–14 petals, 1.5–2.75 in., moderate fragrance; some prickles; foliage medium, medium green, semi-glossy; medium, spreading growth; [Charleston X Friesia]; Kameyama, Yasushi; Kame-yama, 1991

Vienna Charm, *see* 'KORSCHAPRAT'

'VIENNA CHARM, CLIMBING', Cl HT, ob, 1972; Gandy, Douglas L.; Gandy Roses, Ltd.

'VIENNA MAID', F, my, 1957; flowers empire-yellow, borne in large clusters, dbl., 30 petals, 2.5 in.; foliage dark, glossy; moderately bushy growth; deRuiter; Blaby Rose Gardens

Vienna Woods, *see* **'WIENERWALD'**

Vièrge de Clery, *see* **'UNIQUE BLANCHE'**

Vierge Folle, *see* DELvirge

Vierlanden, *see* **'PINK DELIGHT'**

Vierländerin, *see* 'KORVILA'

Vieux Chateau Certan, *see* DORtan

View®, *see* 'LENVIE'

Vif Eclat, HMsk, 1992; flowers small to medium, single; Lens, Louis; (Sangerhausen)

'VIGANE', HT, rb, 1962; flowers red and light yellow bicolor, dbl.; Frères, Buyl

Vigilance(TM), *see* 'MINIVIG'

'VIGILANT', HT, dr, 1941; flowers very dark red; [Night seedling]; Clark, A.

Vigo, F; Moreira da Silva, A.; (Cavriglia)

'VIGORO', HT, op, 1953; flowers salmon-pink, large, dbl., 30 petals, exhibition form, moderate fragrance; foliage clear green; very vigorous, upright, compact growth; [Ophelia X Federico Casas]; Dot, Pedro

Vigorosa, S, lp; short; flowers medium sized, flat, 12 petals, moderate; medium-large, dark green, semi-glossy; [(Frankfurt am Main X Maria Callas) X Dr Faust]; St Wagner, 1992; Res. Stn. f. Horticulture, Cluj, 1994, Romania

'VIKING', HT, dr, 1968; bud ovoid; flowers crimson, dbl., 45–50 petals, 4.5–5.5 in., moderate fragrance; foliage leathery, dark; vigorous, upright growth; [Volcano X Happiness]; Moro; Ball Seed Co.

'VIKING QUEEN', LCl, mp, 1963; flowers medium to deep pink, blooms in large clusters, dbl., 60 petals, 3–4 in., globular, recurrent bloom, intense fragrance; foliage dark, glossy, leathery; vigorous growth; [White Dawn X L.E. Longley]; Phillips; Univ. of Minn.

'VIKTORIA ADELHEID', HT, yb, 1932; flowers golden yellow edged and shaded nasturtium-red, large, dbl., moderate fragrance; foliage leathery, glossy; dwarf growth; [Charles P. Kilham X Mev. G.A. van Rossem]; Kordes

'VILIA', F, op, 1960; flowers bright coral-pink, borne in large clusters, single, 2.5 in., moderate fragrance; foliage dark, glossy; moderate growth; GM, NRS, 1958; Robinson, H.; Gregory

'VILLA DE BILBAO', HT, mr, 1933; flowers cardinal-red, large, dbl., cupped, vigorous growth; [O. Junyent X Margaret McGredy]; La Florida

'VILLA DE MADRID', HT, or, 1965; flowers vermilion-red to poppy-red, large, dbl., 60 petals, moderate, musk fragrance; strong stems; upright growth; GM, Madrid, 1961; [Baccará X Peace]; Dot, Simon; C-P

'VILLA DE SITGES', HT, op, 1930; flowers pink shaded salmon; very vigorous, spreading growth; [Frau Karl Druschki X Mme Edouard Herriot]; Munné, B.

'VILLA DES TYBILLES', HRg, mr, 1899; flowers large, single; L'Hay; (Sangerhausen)

'VILLA PIA', HT, dr, 1926; bud long, pointed; flowers velvety deep red, almost black, dbl., moderate fragrance; [Pres. Vignet X Château de Clos Vougeot]; Leenders Bros.

Village Charm, HMult, lp; McLeod, J., 1990; (Weatherly, L.)

Village de Taradeau, F, lp; RVS, 1995

'VILLAGE MAID', C, pb; (Belle des Jardins, Belle Rubine, Belle Villageoise, Centifolia Variegata, Cottage Maid, La Belle Villageoise, La Rubanée, La Villiageoise, Panachee Double, R. centifolia variegata); flowers white, striped pink and purple (stripes varying in width), dbl., cupped; Vibert, 1845

'VILLANDESSA', HT, ob, 1977; bud long, pointed; flowers orange-blend, dbl., 33 petals, 4.5 in., exhibition form, slight fragrance; vigorous, upright, bushy growth; [Peer Gynt X Seedling]; Kordes, W. Söhne; Willemse

'VILLE D'ANGERS', HT, mr, 1934; bud long, pointed; flowers pure currant-red, large, semi-dbl., cupped, foliage leathery, dark; vigorous, bushy growth; [Souv. de Georges Pernet X Souv. de Claudius Denoyel]; Delaunay

Ville d'Arcis sur Aube®, HT, op; Vially

Ville de Bâle, *see* 'MEIVILANIC'

Ville de Bar sur Seine, *see* ORaval

'VILLE DE BORDEAUX', HT, dr, 1955; bud long; flowers dark scarlet-red, dbl., moderate fragrance; very vigorous growth; Privat

'VILLE DE BREST', HT, or, 1942; flowers fiery orange veined reddish copper, medium, semi-dbl., globular, slight fragrance; foliage bronze, glossy; vigorous growth; Gaujard

Ville de Bruxelles, *see* **'LA VILLE DE BRUXELLES'**

'VILLE DE CHALONS', HT, or, 1938; flowers reddish orange shaded darker, dbl.; Champion

Ville de Chine, *see* **'CHINATOWN'**®

Ville de Doué, F; (Cavriglia)

'VILLE DE GAND', HT, op, 1951; flowers deep salmon, well formed, dbl., 25 petals, 3.5 in., moderate fragrance; foliage bronze; very vigorous, upright growth; GM, Geneva, 1950; [Georges Chesnel seedling X (Mme Joseph Perraud seedling X R. foetida bicolor)]; Gaujard

Ville de Grenoble, *see* **'GRENOBLE'**

Ville de Liffre, HT, mp; Adam

Ville de Londres, HGal, dp; Vibert or Robert, 1850

Ville de Lorgues, F, lp

'VILLE DE MALINES', HT, yb, 1929; bud long, pointed, yellow shaded cherry-red; flowers orange to pink shaded yellow; vigorous growth; Lens

Ville de Moulins, *see* ORAlamar

'VILLE DE NANCY', HT, pb, 1940; bud long, pointed, old-rose shaded gold; flowers buff-pink, edges penciled white, reverse light pink, base yellow, dbl., 50–55 petals, cupped, slight fragrance; foliage leathery, dark; vigorous, upright growth; [Souv. de Claudius Pernet X Federico Casas]; Gillot, F.; C-P

'VILLE DE PARIS', HT, my, 1925; flowers clear bright yellow, large, dbl., globular, slight fragrance; foliage reddish green, glossy; vigorous, growth; GM, Bagatelle, 1925; [Souv. de Claudius Pernet X Unnamed seedling]; Pernet-Ducher

'VILLE DE PARIS, CLIMBING', Cl HT, my, 1935; Armstrong, J.A.

'VILLE DE PARIS, CLIMBING', Cl HT, my; Cognet

Ville de Perreux, S, 1987; Delbard; (Cavriglia)

'VILLE DE PRAGUE', HT, or, 1940; bud long, bright coral-red; flowers scarlet and copper, very large; foliage bright green; bushy growth; Chambard, C.; Orard

Ville de Roanne, HT, 1966; Dorieux; (Cavriglia)

Ville de Romilly, HT, 1995; Orard; (Cavriglia)

Ville de Romilly Sur Seine, HT, 1994; Orard; (Cavriglia)

'VILLE DE SAINT DENIS', HP, 1853; Thomas; (Cavriglia)

'VILLE DE SAVERNE', HT, or, 1937; flowers orange-scarlet, tinted brownish red, reverse tinted yellow, dbl.; Heizmann, E.

'VILLE DE TOULOUSE', HGal, mp, 1876; flowers medium, dbl.; Brassac; (Sangerhausen)

Ville de Troyes®, HT, dr

'VILLE DE VALENCIENNES', HT, or, 1954; flowers orange-copper, well formed, very large, dbl., moderate fragrance; vigorous growth; [Peace X Seedling]; Gaujard

Ville de Villeurbanne, HT, op

'VILLE DE ZURICH'®, F, or, 1967; flowers well-formed, dbl., 25 petals, moderate fragrance; foliage bright green; vigorous, bushy growth; [Miss France X Nouvelle Europe]; Gaujard

Ville d'Ettelbruck, *see* 'LENIVILL'

'VILLE DU HAVRE', HT, w, 1931; flowers cream-white, washed rose-pink, base yellow, very dbl.; foliage dark; very vigorous growth; [K. of K. X Souv. de Claudius Pernet]; Cayeux, H.; Turbat

Ville du Perreux, *see* 'DELRULA'

Ville du Roeuix, F, pb

Ville du Roeulx, F, dp; 19 petals, 6 in.; GM, Geneve, 1989; RvS-Melle, 1989

'VIM', HT, mp, 1963; bud pointed; flowers pink, medium, single, moderate bloom, foliage soft; moderate, bushy growth; [Charlotte Armstrong X Applause]; Wyant

'VIN ROSÉ', HT, mp, 1969; bud long, pointed; flowers light coral-pink, large, dbl., 30–35 petals, exhibition form, slight fragrance; foliage glossy; vigorous, upright growth; [Revelry X Hawaii]; Boerner; J&P

'VINCENT GODSIFF', Ch, mr; flowers deep luminous rosy-red, yellow stamens, semi-dbl., 10 petals, 2 in., cupped, foliage dark; compact, upright (to 3 ft) growth

'VINCENT VAN GOGH', Pol, or, 1969; bud ovoid; flowers medium, dbl.; foliage dark; [Allotria X Hobby]; Buisman, G. A. H.

'VINCENTE PELUFFO', HP, mr, 1902; flowers large, dbl.; Lévêque; (Sangerhausen)

'VINCENZ BERGERS WEISSE', HT, w, 1943; (Vinzens Berger's Weisse); bud long, pointed, sulphur; flowers open, very large, dbl., exhibition form, moderate fragrance; foliage glossy, bronze; very vigorous, upright growth; [Mrs Sam McGredy X Seedling]; Berger, V.; Kordes

Vindonissa®, F, ob; Huber, 1986

Vineyard Song, *see* 'MORGRAPES'

'VINO DELICADO', HT, m, 1972; bud long, pointed; flowers mauve, edged purple-red, well-formed, large, dbl., slight fragrance; foliage large, leathery; upright growth; [Unnamed seedling X Mauve Melodee]; Raffel; Port Stockton Nursery

Vintage Visalia, *see* 'MORLU'

Vintage Wine, *see* 'POULLACK'

Vinzens Berger's Weisse, *see* **'VINCENZ BERGERS WEISSE'**

'VIOLA', HT, m, 1955; flowers lilac-pink, very large, dbl., intense fragrance; foliage leathery; vigorous, upright growth; [Orange Triumph X Peace seedling]; Gaujard

Viola, HT, lp; Ghosh, 1998

Viola Lougheed, *see* 'LOUVIOLA'

Violacea, *see* **'LA BELLE SULTANE'**

'VIOLACÉE', M, m, 1876; flowers purple, shaded violet to grayish pink, large, dbl.; Soupert & Notting

'VIOLAINE', HT, m, 1968; bud long, pointed; flowers large, dbl., exhibition form, intense fragrance; foliage leathery; tall, vigorous growth; [Eminence X Simone]; Gaujard; Ilgenfritz Nursery

Violante, S, 1991; Michler, K. H.; (Sangerhausen)

Violet Bengal, *see* **'PURPLE BENGAL'**

Violet Blue, *see* **'VEILCHENBLAU'**

Violet Carson, *see* 'MACIO'

Violet Dawson, *see* 'RESONE'

'VIOLET FONTAINE', S, m, 1972; bud ovoid; flowers violet-purple, large, dbl., abundant, continuous bloom, slight fragrance; foliage large, soft; vigorous, upright, bushy growth; [Unknown X Unknown]; Tantau, Math.; Ahrens & Sieberz

Violet Hit, *see* POUltin

'VIOLET HOOD', S, m, 1976; bud ovoid; flowers dark violet, pompon shape, semi-dbl., 18 petals, 1 in., recurrent bloom, intense fragrance; foliage ribbed, brownish; very vigorous, overhanging growth; [Robin Hood X Baby Faurax]; Lens, 1975

'VIOLET LIDDELL', HT, pb, 1904; flowers light pink to white, center and reverse coppery salmon; Schwartz, A.

'VIOLET MESSENGER', LCl, lp, 1974; flowers shell-pink, base yellow, reflexed, dbl., 30 petals, 6 in., intense fragrance; foliage large, matt green; [Spek's Yellow, Climbing X Masquerade]; Cadle's Roses

Violet Mist, *see* 'TINVIOLET'

'VIOLET PARNCUTT', HT, yb, 1923; bud small, pointed; flowers brownish gold, semi-dbl.; Easlea

'VIOLET QUEEN', HP, m, 1892; flowers marbled crimson and violet; Paul, G.

'VIOLET QUEEN', HT, m, 1970; flowers deep violet-pink, pointed, dbl., 35–40 petals, 4 in., intense fragrance; foliage dark; upright, free growth; [Seedling X Violette Dot]; Northfield

Violet Ruffles, HT, lp; Davidson, 1997

'VIOLET SIMPSON', HT, op, 1930; flowers vivid prawn-pink, base yellow, dbl., moderate fragrance; foliage purple; Simpson; H&S; Laxton Bros.

'VIOLET WILTON', HT, mp, 1930; bud very long, pointed; flowers bright rose-pink on flesh-white ground, tinted yellow, large, dbl., 35–40 petals, moderate fragrance; vigorous growth; [Gen. MacArthur X Mme Charles Lutaud]; Ketten Bros.

'VIOLETERA', HT, m, 1980; (Embruixada); bud ovoid; flowers reddish-mauve, shallow-cupped blooms borne singly to 3 per cluster, dbl., 30 petals, cupped, intense fragrance; curved, reddish-green prickles; foliage medium, light green, matt; bushy growth; Dot, Simon; Rose Barni-Pistoia

Violett Satina, *see* TANttelos

Violetta, *see* 'HARQUANTUM'

'**VIOLETTA**', HT, m, 1957; intense fragrance; foliage glossy; vigorous growth; [Peace X Guinee]; Croix, A.

Violetta, Pol, lp; Bruant, 1924

'**VIOLETTE**', HMult, m, 1921; flowers pure deep violet, blooms in large clusters, very dbl.; vigorous growth; Turbat

'**VIOLETTE BOUYER**', HP, w, 1881; flowers pinkish white, large, cupped, moderate fragrance; [Jules Margottin X Sombreuil]; Lacharme, F.

'**VIOLETTE DOT**', HT, m, 1960; flowers ageratum-blue, medium, semi-dbl., 20 petals, moderate fragrance; strong stems; spreading growth; [Rosa de Friera X Prelude]; Dot, Simon

Violette Niestlé, *see* 'HARVINTAGE'

Violina®, *see* TANanilov

Violine, *see* 'LENDADI'

'**VIOLINISTA COSTA**', HT, rb, 1936; flowers red to deep purplish red, well-formed, large, moderate fragrance; vigorous growth; (28); [Sensation X Shot Silk]; Camprubi, C.

'**VIOLINISTE EMILE LÉVÊQUE**', HT, lp, 1897; flowers medium, dbl.; Pernet-Ducher; (Sangerhausen)

'**VIOLONCELLISTE ALBERT FOURÈS**', HT, ab, 1920; flowers orange-yellow, shaded buff-yellow, dbl.; [Joseph Hill X Unnamed variety]; Croibier

Viorita®, *see* 'HARQUANTUM'

'**VÍRA**', HRg, mr, 1936; flowers bright red, intense fragrance; very vigorous growth; Böhm, J.

'**VIRAGO**', HSet, lp, 1887; flowers large, dbl.; Geschwind, R.; (Sangerhausen)

'VIRAMBER', HT, ab, 1997; (**Nefertiti**); flowers amber yellow blooms, semi-dbl., 15–25 petals, 5 in., borne mostly singly, slight fragrance; foliage large, medium green, semi glossy; upright, medium (3 ft) growth; [Julien Potin X First Prize]; Viraraghavan, M.S.

'VIRBROWN', F, m, 1997; (**Coffee Country**, Coorg); flowers russet and coffee brown, mauve suffusion, light yellow reverse, dbl., 15–25 petals, 4 in., borne in small clusters, slight fragrance; foliage medium, medium green, semi glossy; spreading, medium (3 ft) growth; [(Zorina X Mlle Cécile Brünner) X (The Fairy X Nordia)]; Viraraghavan, M.S.

'**VIRGEN DE FARNÉS**', HT, mp, 1960; flowers bright rose, reverse lighter, well formed, dbl., 26 petals, intense fragrance; strong stems; vigorous growth; [Queen Elizabeth X Virgo]; Dot, M.

Virgin, Min, w; flowers small, dbl.; Lens, Louis; (Sangerhausen)

Virginale, HP, w; Lacharme, F., 1858

Virginale, C, lp

'**VIRGINIA**', LCl, ob, 1934; flowers brilliant flame, suffused gold, open, large, dbl., non-recurrent, moderate fragrance; foliage large, dark; vigorous, climbing (9 ft) growth; [Magnafrano X Eldorado]; Nicolas; C-P

Virginia, *see* PEKwhina

'**VIRGINIA DARE**', HT, dp, 1934; bud long, pointed; flowers deep cerise-pink, large, dbl., moderate fragrance; foliage leathery, dark; long stems; very vigorous growth; [Joanna Hill X Unnamed seedling (dark red)]; Thompson's, J.H., Sons

Virginia Dare, *see* CLEdare

Virginia Lee, *see* 'MICLEE'

Virginia R. Coxe, *see* '**GRÜSS AN TEPLITZ**'

Virginia R. Coxe, Climbing, *see* '**GRÜSS AN TEPLITZ, CLIMBING**'

'**VIRGINIA REEL**', S, dp, 1975; bud ovoid, pointed; flowers light red, dbl., 40 petals, 4–4.5 in., cupped, moderate fragrance; foliage large, dark, leathery; erect, bushy growth; [Tickled Pink X Prairie Princess]; Buck, Dr Griffith J.; Iowa State Univ.

Virginia Rose, *see* '**R. VIRGINIANA**'

Virginian Rambler, Ayr, lp

'**VIRGO**', HT, w, 1947; ('VIRGO LIBERATIONEM'); flowers white, sometimes blush-pink, dbl., 30 petals, 5 in., exhibition form, slight fragrance; foliage dark, leathery; vigorous growth; GM, NRS, 1949; [Blanche Mallerin X Neige Parfum]; Mallerin, C.; Meilland-Richardier

'VIRGO LIBERATIONEM', *see* '**VIRGO**'

'**VIRGO, CLIMBING**', Cl HT, w, 1957; Mondial Roses

'**VIROLAY**', HT, ob; flowers edges orange-red, base deep yellow, large, exhibition form, intense fragrance; vigorous growth; Camprubi, C.

'VIRSPLASH', S, pb, 1998; (Bodhisatva, **Magic East**); flowers white with pink and white splashes, creamy yellow eye, dbl., 15–25 petals, 5 in., borne singly and in clusters, slight fragrance; prickles moderate, medium; foliage medium, blue green, matt; bushy, medium growth; [Honor X Priyatama]; Viraraghavan, M.S.; Hortico, 1995

Vi's Violet, *see* 'MORVI'

Visa, *see* 'MEIRED'

'**VISCOUNT SOUTHWOOD**', HT, pb, 1949; flowers china-pink shaded creamy peach to copper, dbl., 35–40 petals, 4–5 in., exhibition form, vigorous growth; [Walter Bentley X Aribau]; Cobley; Harkness

'**VISCOUNTESS CHARLEMONT**', HT, mp, 1937; bud salmon-rose; flowers satiny rose-pink, base deep buttercup-yellow, large, dbl., intense fragrance; foliage dark cedar green; branching growth; GM, NRS, 1936; McGredy

'**VISCOUNTESS DEVONPORT**', HT, dy, 1923; flowers rich indian yellow, dbl.; Hicks

'**VISCOUNTESS ENFIELD**', HT, pb, 1910; flowers coppery old-rose, shaded yellow; [? X Soleil d'Or seedling]; Pernet-Ducher

'**VISCOUNTESS FALMOUTH**', HT, pb, 1879; flowers mottled pink, very large, dbl., globular, moderate fragrance; thorny; dwarf growth; [Adam X Soupert et Notting]; Bennett

'**VISCOUNTESS FOLKESTONE**', HT, pb, 1886; flowers creamy silver-pink, center deep salmon-pink, dbl., moderate fragrance; Bennett

Vision, *see* 'POULONI'

'**VISION**'®, HT, pb, 1967; flowers gold and pink, dbl., 5.5 in., slight fragrance; foliage glossy; [Kordes' Perfecta X Peace]; Dickson, A.

Vision Blanc®, *see* '**ICE WHITE**'

Vista™, *see* 'SAVALAV'

Vital, HT, dr; Kordes, 1998

Vital Spark, *see* 'COCACERT'

'**VITTONVILLE-ROSE**', HT, ob, 1945; flowers orange, base yellow, reverse lighter; vigorous growth; Mallerin, C.; A. Meilland

'**VIUDA VERDAGUER**', HT, ob, 1934; flowers orange, open, very large, dbl.; foliage glossy, dark; very vigorous growth; [Shot Silk X Mari Dot]; Dot, Pedro

Viva, *see* 'JACIV'

Viva Romana, HT, rb; Dot

'**VIVACÉ**', F, or, 1974; flowers large, dbl., exhibition form, slight fragrance; foliage large, leathery; very vigorous, upright growth; [Klaus Stortebeker X ?]; Kordes

'**VIVACIOUS**', F, mp, 1971; flowers phlox-pink, dbl., 35 petals, 4 in., moderate fragrance; very free growth; [Tropicana X ?]; Gregory

Vivacious Dianne, *see* 'CHRISJEVANS'

Vivacity, *see* '**TEMPERAMENT**'

Vivaldi, *see* 'LENMOBAR'

Vivaldi™, *see* RUIdriko

Vivarose, HT, mr; Croix

'**VIVASTELLA**', HT, dp; flowers laque de robbie color (possibly carmine), well shaped; [Julien Potin X Sensation]; Aicardi, D.; Giacomasso

'**VIVE LA FRANCE**', HT, rb, 1944; bud pointed, well formed; flowers purplish red, reverse yellow, large, dbl., slight fragrance; foliage glossy; vigorous growth; GM, Bagatelle, 1943; [Shining Star X Mme Arthaud]; Mallerin, C.; A. Meilland

Vive La Suisse, F; Tschanz, E.; (Cavriglia)

'**VIVIAN MORELLE**', T, w; bud cream; flowers snow-white, profuse bloom

'**VIVID**', B, m, 1853; flowers brilliant magenta to magenta-pink, dbl., moderate fragrance;

prickles robust, prickly growth; foliage glossy; height to 6 ft; Paul, A.

'VIVID', F, or, 1951; flowers brilliant orange-scarlet, borne in huge clusters, single, 5–7 petals, 3.5 in.; compact growth; LeGrice

'VIVID MASON', HT, dp, 1934; flowers vivid dark pink, base yellow, large, dbl.; foliage leathery, glossy; very vigorous growth; [Premier X Mme Alexandre Dreux]; Mason, J.A.; McLellan Co.

'VIVIEN', HT, dp, 1922; flowers deep rose-pink, dbl., intense fragrance; Paul, W.

'VIVIEN LEIGH', HT, mr, 1963; flowers crimson, dbl., 35 petals, 5 in., exhibition form, moderate fragrance; foliage dark; very free growth; [Queen Elizabeth X Detroiter]; McGredy, Sam IV; Fisons Horticulture

Vivienne Maire, D, mp

Vivre, HT, 1974; Delbard-Chabert; (Cavriglia)

Vixen, *see* 'JACORMIN'

'VLAM', HT, or, 1956; flowers fiery red, dbl., moderate fragrance; vigorous growth; [Tawny Gold X Sarie Mareis]; Leenders, M.

Vlammenspel, LCl, mr

'VLASTA BURIAN', Pol, m, 1937; flowers purple/pink, small, dbl.; Böhm, J.; (Sangerhausen)

'VLTAVA', LCl, m, 1936; flowers violet, passing to purplish red, large, borne in clusters, dbl., globular, foliage glossy; very vigorous, climbing growth; [Veilchenblau X ?]; Böhm, J.

Voeux de Bonheur, *see* **'BON VOYAGE'**

VOEUX DE BONHEUR, HT, pb, 1960; (Bon Voyage); flowers creamy white, petals edged cerise-pink, reverse white, large, dbl., exhibition form, intense fragrance; foliage dark, glossy; vigorous growth; [Michele Meilland X Chic Parisien]; Delbard-Chabert

Vogelpark Walsrode, *see* KORlomet

'VOGUE', F, pb, 1951; bud ovoid; flowers cherry-coral, blooms in clusters, dbl., 25 petals, 3.5–4.5 in., exhibition form, moderate fragrance; foliage glossy; vigorous, upright, bushy, compact growth; (28); AARS, 1952 GM, Geneva, 1950 GM, Portland, 1950; [Pinocchio X Crimson Glory]; Boerner; J&P

Vogue, *see* PEKcourofondu

Voice of Thousands, *see* 'HORSUNSMILE'

'VOIE LACTÉE', Cl HT, w, 1949; bud globular, creamy white; flowers large, dbl., intense fragrance; foliage glossy; very vigorous, climbing growth; [Frau Karl Druschki X Julien Potin]; Robichon

Voie Romaine, HT, 1977; Dot, Simon; (Cavriglia)

Voila, F, 1969; Croix; (Cavriglia)

Vol de Nuit, *see* 'DELRIO'

Volare, *see* 'MACVOLAR'

'VOLARE', HT, dr, 1976; bud ovoid; flowers bright red, dbl., 38–44 petals, 3.5 in., exhibition form, slight fragrance; foliage leathery; bushy, upright growth; PP004172; McDaniel, Earl; Carlton Rose Nurseries

'VOLCANO', HT, dp, 1950; bud long, pointed; flowers cherry-red, dbl., 25 petals, 6.5 in., cupped, moderate, fruity fragrance; foliage dark; vigorous, upright growth; [Charles P. Kilham X Rome Glory]; Moro; J&P

Volunteer, *see* 'HARQUAKER'

'VON HÖTZENDORF', HP, pb, 1916; flowers golden rose-pink, richly tinted coppery red, dbl., moderate fragrance; vigorous growth; [Frau Karl Druschki X Beauté de Lyon]; Schmidt, J.C.

'VON LILIENCRON', HFt, pb, 1916; flowers yellowish light pink with white, reverse salmon-pink, stamen, dbl., exhibition form, profuse bloom, sometimes recurrent, moderate fragrance; foliage glossy, dark, bronze; vigorous (6 ft) growth; [Geheimrat Dr Mittweg X Mrs Aaron Ward]; Lambert, P.

'VON SCHARNHORST', S, ly, 1921; flowers yellow to yellowish-white, semi-dbl., recurrent bloom, slight fragrance; vigorous (6–8 ft) growth; [Frau Karl Druschki X Gottfried Keller]; Lambert, P.

Vonava Elysium, HT, lp; Strnad; (Czech Rosa Club)

Voodoo™, *see* 'AROMICLEA'

'VOORBURG', F, op, 1959; flowers salmon-pink, borne in large clusters, semi-dbl.; foliage bronze to dark; vigorous growth; [Sangerhausen X Vogue]; Buisman, G. A. H.

'VORBERGII', HFt, ly; (R. vorbergii, R. X harisonii vorbergii); flowers pale yellow, single; (28); [R. foetita X R. spinosissima]

'VOX POPULI', HT, dr, 1945; flowers velvety dark red, semi-dbl.; Mallerin, C.

Voyager, *see* 'MICVOYAGE'

Vrystaat, *see* TANtaatsyr

'VULCAIN', HP, dr, 1861; flowers rich dark crimson, well formed; Verdier, V.

'VULCAIN', Pol, mr, 1921; flowers deep cherry-red, dbl.; vigorous growth; Turbat

'VULCANA', LCl, dr, 1964; flowers blood-red, borne in clusters, semi-dbl.; vigorous growth; Mondial Roses

'VULCANIA', HT, mr, 1948; bud long; flowers purplish red, well formed, overlarge; foliage dark; [Matador X Principe di Piemonte]; Giacomasso

'VULGENS', Pol, mr, 1942; flowers medium, semi-dbl.; Koster; (Sangerhausen)

Vulkan, HT, dr; flowers large, dbl., slight fragrance; VEG; (Sangerhausen)

'VUURBAAK', F, or, 1946; flowers scarlet, open, borne in clusters, semi-dbl., 15 petals, 4 in.; foliage reddish green; vigorous, upright growth; [Florentina X World's Fair]; Leenders, M.; Longley

Vydehi, F, ob; Datt, Braham, 1989

'VYSLANEK KALINA', HP, mr, 1935; flowers large, dbl., intense fragrance; Böhm, J.; (Sangerhausen)

W

W.E. Lippiat, HT, dr; Dickson, Patrick, 1907

'**W. Freedland Kendrick**', HT, pb, 1920; flowers pink with peach tones, large, dbl., moderate fragrance; Thomas; (Sangerhausen)

'**W. Freeland Kendrick**', LCl, lp, 1920; (Bloomfield Endurance); flowers flesh, center peach, large, very dbl., non-recurrent, moderate fragrance; foliage dark, bronze, leathery, glossy; vigorous, semi-climbing growth; [Aviateur Bleriot X Mme Caroline Testout]; Thomas; B&A

'**W.A. Bilney**', HT, ab, 1927; flowers pale apricot suffused cerise, reverse yellow tinted pink, large, slight fragrance; foliage dark, leathery; vigorous growth; Easlea

W.A. Willet, HT, lp; Kernovski, V.R., 1959; (Weatherly, L.)

'**W.C. Gaunt**', HT, dr, 1916; bud long, pointed; flowers velvety crimson-scarlet, reflexed petals tipped scarlet, rev, dbl., moderate fragrance; Dickson, A.

'**W.E. Chaplin**', HT, dr, 1929; flowers crimson, deepening to maroon, large, dbl., exhibition form, vigorous growth; GM, NRS, 1930; Chaplin Bros.

'**W.E. Chaplin, Climbing**', Cl HT, dr, 1936; Heizmann, E.

'**W.E. Wallace**', HT, dy, 1922; flowers deep golden yellow, well formed, large, dbl., moderate fragrance; GM, NRS, 1922; [Gorgeous sport]; Dickson, H.

'**W.G. Pountney**', F, or, 1964; flowers scarlet; tall growth; [Moulin Rouge X Mrs Inge Poulsen]; Bennett, H.; Pedigree Nursery

'**W.H. Cotton**', HT, ab, 1946; flowers orange shaded gold, large; foliage dark; vigorous, upright growth; [Mrs Beatty X Mrs Sam McGredy]; Cobley; Leicester Roses

'**W.H. Dunallan**', HT, mr, 1939; flowers very rich bright red, flushed darker, semi-dbl.; vigorous growth; [Edith Clark X ?]; Clark, A.; NRS Victoria

W.J. Matthews, T, lp; Matthews, W.J., 1892

'**W.R. Hawkins**', HT, op, 1948; bud ovoid; flowers tango-pink, large, very dbl., exhibition form, intense fragrance; foliage glossy; vigorous, upright growth; [Crimson Glory seedling X Silver Jubilee]; Toogood

Waanrode, HT, op; Lens, 1990

'**Wabash Dawn**', S, op, 1978; bud tapered; flowers bright orange-pink, dbl., 34 petals, 4–4.5 in., exhibition form, repeat bloom, intense fragrance; foliage large, glossy; vigorous, upright growth; [(Queen Elizabeth X Gladiator) X (Aztec X Little Darling)]; Williams, J. Benjamin; Krider Nursery

'**Wadei**', HRg, mp; single; foliage rich green; weak, prostrate growth; [R. rugosa X ?]

Wagbi, *see* 'BleBAR'

'**Wageningen**', Pol, my, 1968; bud ovoid; flowers medium, dbl.; foliage dark; [Golden Giant X Peace]; Buisman, G. A. H.

Wagtail, S, lp; Poulsen, 1990

Waiheke™, *see* 'MACwaike'

Waikato, *see* 'MAChoro'

Waikiki, *see* 'MACwaike'

'**Waipounamu**', HT, yb, 1982; flowers creamy yellow, shaded mauve-pink on outer petals, dbl., 50 petals, exhibition form, intense fragrance; foliage bronze green, glossy; upright growth; [Peace X Blue Moon]; Cattermole, R.F.

Wait 'n' See, Min, lp; Geytenbeek, 1978; (Weatherly, L.)

Waitemata, *see* 'MACweemat'

Waitmata®, *see* 'MACweemat'

'**Waitziana**', S, dp; (R. X waitziana); flowers deep rose often solitary, 2.5 in.; [R. canina X R. gallica]

'**Waldfee**', HP, mr, 1960; flowers blood-red, camellia-shaped, blooms in small clusters, dbl., 4 in., recurrent bloom, moderate fragrance; foliage glossy; vigorous, tall (10 ft), dense growth; Kordes, R.

'**Waldtraut Nielsen**', M, dp; flowers clear deep pink, large; Nielsen, 1932

Walferdange, HMsk, pb; Lens, 1992

'WALkiss', Min, mp, 1993; (**Autumn Kiss**); flowers medium salmon pink, dbl., 15–25 petals, 1.5 in., borne in small clusters, slight fragrance; few prickles; foliage small, dark green, glossy; medium, upright growth; [Ginger Meggs X Avandel]; Walsh, Richard Bruce; Fradee Nursery

Walko, *see* 'DELde'

'**Walküre**', HT, w, 1919; bud long, pointed; flowers cream-white, center ochre-yellow, dbl.; [Frau Karl Druschki X Mme Jenny Gillemot]; Ebeling

Walküre, S, lp; Geschwind, R., 1909

'**Walkyrie**', HT, dr, 1959; bud urn shaped or ovoid; flowers low-centered, large; bushy growth; [Happiness X Volcano]; Moro

Wall Street®, *see* BARwast

Wallflower, *see* '**The Wallflower**'

'WALlove', Min, w, 1994; (**Love Me**); flowers creamy white, pink edge, blooms borne in small clusters, dbl., 15–25 petals, 1.5 in., slight fragrance; few prickles; foliage small, medium green, matt; low, upright growth; [Magic Carrousel X Old Master]; Walsh, Richard Bruce; Fradee Nursery, 1992

'**Walsham Gold**', HT, my, 1965; flowers yellow with burnished gold and copper overtones in cool weather; LeGrice; Wayside Gardens Co.

'WALshine', Gr, ly, 1997; (**Wollongong Gold**); flowers light creamy yellow, flushed pink, dbl., 26–40 petals, 1.5–3 in., blooms in small clusters, slight fragrance; foliage medium, light green, semi glossy; low (1 m) bushy growth; [Violet Carson X Arizona]; Walsh, Richard Bruce

Walsh's Rambler, *see* '**America**'

'**Walter**', HT, dp, 1949; bud long, pointed; flowers pink, reverse vivid rosy red, open, very large, semi-dbl., 20 petals, intense fragrance; foliage bronze, leathery; very vigorous, bushy growth; [Charles P. Kilham X Comtesse Vandal]; Lens

'**Walter Bentley**', HT, op, 1938; bud long, pointed; flowers coppery orange shaded pink, very large, dbl., exhibition form, slight fragrance; foliage leathery, glossy, dark, bronze; vigorous growth; GM, NRS, 1937; [Mrs Sam McGredy X Dame Edith Helen]; Robinson, H.; Wheatcroft Bros.

Walter Butt, S, dp; Butt, 1905

'**Walter C. Clark**', HT, dr, 1917; bud long, pointed; flowers deep maroon-crimson, shaded black, large, dbl., exhibition form, intense fragrance; foliage dark, leathery; vigorous growth; Paul, W.

'**Walter Rieger**', S, dp, 1977; bud pointed; flowers pink to reddish, medium, dbl., continuous bloom, slight fragrance; foliage glossy; vigorous growth; [Carina X Molde]; Hetzel

'**Walter Ross**', HT, my, 1970; bud ovoid; flowers large, dbl., exhibition form, moderate fragrance; foliage large, glossy, dark, bronze, leathery; vigorous, upright, bushy growth; [Mme Marie Curie X King's Ransom]; Morey, Dr Dennison; General Bionomics

'**Walter Speed**', HT, dy, 1909; flowers deep lemon-yellow, passing to milk-white, dbl.; Dickson, A.

'**Waltham Bride**', LCl, w, 1905; flowers snow-white, medium, dbl., moderate fragrance; Paul, W.

'**Waltham Climber No. 1**', Cl HT, mr, 1885; flowers rosy crimson, imbricated, large, dbl., recurrent bloom, intense fragrance; vigorous, climbing growth; [Gloire de Dijon seedling]; Paul, W.

'**Waltham Climber No. 2**', Cl HT, or, 1885; flowers flame-red, tinted crimson, large, dbl., recurrent bloom, Paul, W.

'WALTHAM CLIMBER NO. 3', Cl HT, mr, 1885; flowers deep rosy crimson, imbricated, large, dbl., recurrent bloom, intense fragrance; vigorous, climbing growth; Paul, W.

'WALTHAM CRIMSON', HT, dr, 1922; flowers deep crimson; Chaplin Bros.

'WALTHAM CROSS', HT, dr, 1927; flowers glowing crimson-scarlet, semi-dbl.; Chaplin Bros.

'WALTHAM FLAME', HT, 1921; flowers deep terra-cotta, shaded bronzy orange; Chaplin Bros.

'WALTHAM RAMBLER', HMult, dp, 1903; flowers deep rosy pink, center paler, stamens bright yellow, borne i, single, early, vigorous, climbing growth; Paul, W.

'WALTHAM SCARLET', HT, dr, 1914; flowers crimson-scarlet, single; Paul, W.

'WALTRAP', Min, op, 1996; (**Bett's Little Rhapsody**); flowers medium blooms, orange-pink, very dbl., 76 petals, borne mostly singly, foliage medium, dark green with pink edge, glossy; upright, very compact, medium (3–4 ft) growth; [Bett's White Delight X Seedling]; Walters, Betty & Richard

Waltz Time, *see* 'DELVOR'

'WALTZING MATILDA', HT, rb, 1965; flowers red splashed light to dark pink, well shaped, large, dbl.; vigorous growth; [Christian Dior sport]; Jack; Girraween Nursery

'WALZERTRAUM', HT, mr, 1968; dbl.; foliage dark; Tantau, Math.

'WAMBEAIRD', Gr, lp, 1999; (**Dr Charles T. Beaird**); flowers coral, medium, dbl., 17–25 petals, borne mostly singly and in small clusters, slight fragrance; prickles moderate; foliage medium, medium green, semi-glossy; upright, spreading, medium growth; Wambach, Catherine; Certified Roses, Inc., 2000

'WAMHALE', HT, pb, 1996; (**Denise Hale**); flowers light cream pink, darker edge, lighter reverse, dbl., 26–40 petals, 4.5 in., borne singly, slight fragrance; moderate prickles; foliage medium, dark green, semi-glossy; upright, medium growth; [Elsie Melton X Seedling]; Wambach, Catherine; Certified Roses, Inc., 1996

'WAMHOME', HT, pb, 1997; (**1997 Traditional Home**™); flowers light pale pink with ivory shading, dbl., 26–40 petals, slight fragrance; foliage medium, dark green, dull; medium (5 ft) growth; [Bobby Charlton X Touch of Class]; Wambach, Catherine

'WAMMARY', HT, lp, 1997; (**Mary Johnston**); flowers full, light pink, dbl., 26–40 petals, 4 in., borne mostly singly, slight fragrance; foliage medium, dark green, semi glossy; medium (4.5 ft) growth; [Elizabeth Taylor X Touch of Class]; Wambach, Alex A.

'WAMTRESS', HT, w, 1998; (**Temptress**); flowers full, white, large, very dbl., 26–40 petals, 4.5–5 in., exhibition form, borne singly, moderate fragrance; few prickles; foliage medium, medium green, semi-glossy; upright, tall (5.5 ft) growth; [Elsie Melton X Elina]; Wambach, Catherine; Certified Roses, Inc., 1999

Wanaka®, *see* 'MACINCA'

'WANDA', HT, mp, 1953; bud short, pointed; flowers rose, dbl., 28–30 petals, 3.5–4.5 in., exhibition form, intense fragrance; foliage leathery; vigorous, upright growth; [Pink Bountiful X Celebrity]; Hill, Joseph H., Co.

Wanda, HT, lp

Wanderer, *see* 'HORWANDER'

'WANDERIN' WIND', S, lp, 1973; flowers two-toned light pink, medium, dbl., exhibition form, intense fragrance; foliage large, glossy, dark, leathery; very vigorous, upright, bushy growth; [Dornroschen X Andante]; Buck, Dr Griffith J.; Iowa State University

Wandering Minstrel, *see* 'HARQUINCE'

'WAPEX', Pol, my, 1968; bud pointed; flowers medium, semi-dbl., borne in clusters, foliage dark; [Golden Showers X Fata Morgana]; Buisman, G. A. H.

Wapiti®, *see* MEInagre

'WAR DANCE', Gr, or, 1962; bud ovoid; flowers dark orange-red, dbl., 34 petals, 4–4.5 in., exhibition form, foliage leathery; vigorous, bushy growth; [Roundelay X Crimson Glory]; Swim & Weeks; C-P

War Dance, *see* 'MACFIRINLIN'

'WAR PAINT', LCl, mr, 1930; flowers large, dbl., globular, slight fragrance; vigorous, pillar or climbing (8 ft) growth; [Rhea Reid seedling X ?]; Clark, A.; Hackett

'WARANA FESTIVAL', HT, mp, 1962; flowers rich pink, base rich apricot, well formed, dbl.; vigorous growth; [Christian Dior sport]; Jack; Langbecker

Warbler, S, my, 1991; Poulsen

'WARDIDO', HT, pb, 1975; (Oratorio, **Spellbinder**); bud ovoid; flowers ivory to crimson, large, dbl., exhibition form, slight fragrance; foliage large, dark, leathery; vigorous growth; [South Seas X Unnamed seedling]; Warriner, William A.; J&P

'WARDROSA', HT, dp, 1975; (Chanterelle, **Cynthia**); bud long; flowers deep pink, dbl., 35 petals, 5 in., exhibition form, moderate fragrance; foliage large, matt, light; tall, upright growth; [Seedling X Bob Hope]; Warriner, William A.; J&P

Warley Jubilee, F, mp; Warley Rose Gardens, 1986

'WARLEYENSIS', S, mp; (R. X warleyensis); (14); [R. blanda X R. rugosa]

'WARM RAIN', Min, mr, 1985; flowers medium coral red, medium, dbl., 70 petals, exhibition form, slight fragrance; foliage small, medium green, semi-glossy; vigorous, upright, bushy growth; [Fragrant Cloud X Orange Honey]; Hardgrove, Donald L.; Rose World Originals

Warm Welcome, *see* 'CHEWIZZ'

Warm Wishes, *see* 'FRYXOTIC'

WARoujo, HT, mr; (**Fulgurante**®); J&P

'WARRAWEE', HT, mp, 1935; bud long, pointed; flowers flesh-pink shaded rose-pink, large, dbl., 30 petals, exhibition form, moderate fragrance; foliage glossy; vigorous growth; [Padre X Rev. F. Page-Roberts]; Fitzhardinge; C-P

'WARRIOR', F, or, 1977; flowers scarlet-red, blooms in trusses, dbl., 32 petals, 3–4 in., slight fragrance; foliage light green, semi-glossy; [City of Belfast X Ronde Endiablee]; LeGrice

'WARSZAWA', HT, ob, 1957; flowers bright orange, reverse golden yellow; foliage dark, glossy; [Carioca sport]; Grabczewski

'WARTBURG', HMult, mp, 1910; flowers pink to magenta, twisted and reflexed petals, dbl., cupped, moderate fragrance; foliage large, light, soft; strong stems; very vigorous, climbing (15–20 ft) growth; Kiese

Wartburg 77, S, my; Berger, 1977

'WARWHOOP', Min, or, 1973; bud ovoid; flowers brilliant orange-red, small, very dbl., slight fragrance; foliage small, glossy, dark; vigorous, bushy growth; [Baccará X Little Chief]; Williams, Ernest D.; Mini-Roses

Warwick Castle®, *see* 'AUSLIAN'

Warwickshire, *see* KORkandel

'WASAGAMING', HRg, mp, 1939; flowers clear rose, dbl., recurrent bloom, moderate fragrance; vigorous (3 ft) growth; [(R. rugosa X R. acicularis) X Gruss an Teplitz]; Skinner

Wasastiernan, *see* **'POLSTJÄRNAN'**

'WASKASOO', HRg, dr, 1963; flowers not clustered, dbl.; growth typical rugosa; (14); [Little Betty X Hansa]; Erskine

Watchfire, *see* 'MINFIRE'

'WATCHUNG', HT, yb, 1932; flowers yellow tipped pink, medium, dbl., 24–26 petals, 3.5 in.; [Pres. Herbert Hoover sport]; Didato

'WATER MUSIC', LCl, dp, 1982; flowers deep pink, darker on petal edges, medium, dbl., 20 petals, slight fragrance; foliage dark, medium, glossy; spreading, climbing growth; [Handel X ?]; Bell, Ronald J.

'WATERCOLOR', Min, mp, 1975; (Watercolour); bud long, pointed; flowers bright pink, small, dbl., 26 petals, 1.5 in., exhibition form, slight fragrance; foliage small, glossy, leathery; vigorous, upright, bushy growth; [Rumba X (Little Darling X Red

Germain)]; Moore, Ralph S.; Sequoia Nursery

Watercolour, *see* **'WATERCOLOR'**

Waterloo, HMsk, 1996; Lens; (Cavriglia)

Watermelon Ice™, *see* 'JACAIR'

Wave of Flame, *see* **'HO NO-O-NO-NAMI'**

Waverland®, *see* 'LENWAY'

'WAVERLEY', F, dr; flowers crimson-scarlet, borne in trusses, dbl., 22 petals, 3.5 in., slight fragrance; foliage glossy, dark; vigorous, bushy, compact growth; Norman

'WAVERLEY TRIUMPH', Pol, mp, 1951; flowers bright pink, base yellow, small, borne in clusters, semi-dbl., cupped, slight fragrance; foliage glossy, light green; vigorous, bushy growth; [Orange Triumph sport]; Poulter

Waverly Garden Club, lp; [Prima Ballerina X (Scented Bouquet X Extravaganza)]; Dawson, George

'WAVES', HT, mp, 1944; (Admiral); bud ovoid; flowers rose-pink, open, dbl., 35 petals, 4.5 in., cupped, intense fragrance; foliage dark, leathery, wrinkled; vigorous growth; [Seedling (deep salmon) X Lucie Marie]; Dickson, A.; J&P

Waves of Flame, *see* **'HO NO-O-NO-NAMI'**

'WAVRIA', F, or, 1973; flowers orange-crimson; [Europeana sport]; Lens

Wax Rose, *see* **'DUCHESSE D'ANGOULÊME'**

Waxwing, S, mp, 1989; groundcover growth; Poulsen

'WAYSIDE GARNET', Min, dr, 1956; flowers garnet-red, small, dbl.; dwarf, compact growth; [Oakington Ruby sport]; Wayside Gardens Co.

'WE ZAIR', C, pb; flowers two-tone pink

'WEBBREAD', Min, op, 1997; (**Bob's Peach**); flowers double, medium, dbl., 15–25 petals, borne in large clusters, slight fragrance; many prickles; foliage large, medium green, glossy; bushy, tall (28 in.) growth; [Robin Redbreast X Pot O'Gold]; Webster, Robert; Handley Rose Nursery, 1998

'WEBCYRENE', Min, mp, 1994; (**Sheffield Pride**); very dbl., 1.5 in., borne in small clusters, slight fragrance; some prickles; foliage small, dark green, semi-glossy; low, compact growth; [Robin Red Breast X Matangi]; Webster, Robert; Handley Rose Nurseries, 1995

'WEBEYAM', F, op, 1998; (**Jeanette Talbot**); flowers orange-pink, yellow at center, small, dbl., 15–25 petals, 2 in., borne in small clusters, slight fragrance; prickles moderate; foliage small, medium green, glossy; compact, low (18 in.) growth; [(Robin Redbreast X Typhoon) X The Lady]; Webster, Robert; Handley Rose Nurseries, 1998

'WEBHEART', HT, op, 1998; (**Gedge's Glory**); flowers pale orange, salmon pink reverse, very dbl., 26–40 petals, 3–4 in., borne mostly singly, slight fragrance; prickles moderate, medium, straight; foliage medium, dark green, glossy; low, compact growth; [Dave Hesswayon X Remember Me]; Webster, Robert

'WEBLADY', HT, yb, 1998; (**The Marquess of Bristol**); flowers yellow, pink edge, yellow reverse, dbl., 17–25 petals, 4–5 in., borne mostly singly, slight fragrance; prickles moderate, medium, hooked; foliage medium, medium green, glossy; upright, medium growth; [Pristine X Remember Me]; Webster, Robert; Handley Rose Nurseries, 1999

WEBpriace, HT, pb, 1996; (**Distant Sounds**); Webb

Wedded Bliss™, *see* 'SAVAWED'

'WEDDIGEN', HT, lp, 1916; flowers silver-pink, moderate fragrance; Lambert, P.

'WEDDING BELLS', HMult, w, 1906; flowers white, outer half of petals soft pink, borne in clusters, semi-dbl., cupped, foliage light; short stems; very vigorous, climbing (12–15 ft) growth; [Crimson Rambler X ?]; Walsh

'WEDDING DAY', LCl, w, 1950; flowers yellow to white flushed pink, single, blooms in huge clusters, intense fragrance; height to 20 ft (14); [R. sinowilsonii X ?]; Stern, Sir Frederich

Wedding Pink, *see* 'JACBIP'

'WEDDING RING', HT, my, 1956; bud long, pointed; flowers golden yellow, dbl., 25–35 petals, 4–4.5 in., slight fragrance; foliage dark, glossy, leathery; vigorous, upright, compact growth; [Ville de Paris X Mrs Sam McGredy]; Shepherd; Bosley Nursery

'WEDDING SONG', HT, w, 1971; slight fragrance; upright growth; [Virgo X Ivory Fashion]; Whisler, D.; Gro-Plant Industries

Wee Ack, *see* 'RENACK'

Wee Barbie, *see* 'JELBAR'

'WEE BETH', Min, ob, 1981; bud pointed, mossy; flowers salmon-pink, blooms borne 3–20 per cluster, semi-dbl., 11 petals, moderate, sweetbriar fragrance; red prickles; foliage small, dark, bristles on reverse; bushy growth; [Orange Silk X Fairy Moss]; Cherry, R. & Welsh; Roy H. Rumsey, Pty. Ltd.

'WEE BUTTERFLIES', Pol, pb, 1989; bud pointed; flowers medium pink with white eye, small, borne in sprays of 3–35, single, 5 petals, slight fragrance; round, small, red fruit; prickles straight, pink; foliage medium, light green, glossy; bushy, low growth; [The Fairy X Unnamed seedling]; Jerabek, Paul E., 1990

Wee Cracker, *see* 'COCMARRIS'

Wee Jock®, *see* 'COCABEST'

'WEE LASS', Min, mr, 1974; bud pointed; flowers blood-red, semi-dbl., 18 petals, .5–1 in.; foliage small, dark; upright, bushy growth; [Persian Princess X Persian Princess]; Moore, Ralph S.; Sequoia Nursery

'WEE MAN', Min, mr, 1974; (Tapis de Soie); flowers scarlet, medium, semi-dbl., 14 petals, 2 in., slight fragrance; foliage glossy, dark; [Little Flirt X Marlena]; McGredy, Sam IV

Wee Matt, *see* 'MACWEEMAT'

Wee One, Min, lp; Tantau, 1994

Wee Topper, *see* 'ANDWEE'

'WEEBLUSH', Min, pb, 1983; (**Blushing Dawn**); flowers creamy white, peach pink eye, yellow stamens, petals painted peach pink, single, 7–8 petals, moderate fragrance; foliage medium, dark, matt; low, bushy growth; [Jeanne Lajoie X Eyepaint]; Weeks, Michael W.J.

'WEEGOLD', HT, dy, 1981; (**Fine Gold**); flowers deep golden yellow, blooms borne 1–3 per stem, dbl., 26 petals, exhibition form, moderate fragrance; foliage medium, dark, glossy; branching, upright growth; [Seedling X Hawaii sport]; Weeks, Michael W.J.

Week-End, S, lp

'WEEPAINT', F, ab, 1987; (**Open Secret**); flowers peach to salmon-pink, sometimes flecked darker, shading to w, single, 5 petals, moderate fragrance; foliage medium, medium green, semi-glossy; bushy, tall growth; [Eyepaint sport]; Weeks, Michael W.J.

Weeping China Doll, *see* **'CHINA DOLL, CLIMBING'**

Weepy, S, mp, 1996; flowers small; Lens, Louis; (Sangerhausen)

'WEEQUILL', Min, mp, 1992; (**Pink Quill**); flowers medium bright pink, blooms borne in large clusters, semi-dbl., 6–14 petals, 1.5 in., moderate fragrance; some prickles; foliage medium, dark green, bronze when young, semi-glossy; tall (25 cms), spreading growth; [Mr Bluebird X Unknown]; Weeks, Michael W.J.

'WEESPOT', Min, my, 1992; (**Spot 'o Gold**); flowers yellow stamens, blooms borne in small clusters, single, 5 petals, 1.5 in., slight fragrance; some prickles; foliage medium, medium green, semi-glossy; bushy (20 cms) growth; [Rise 'n' Shine X Unknown]; Weeks, Michael W.J.

Weetwood, HWich, lp; Bawden, 1983

Wehrinsel, S, dr, 1959; flowers medium, semi-dbl.; Berger, W.; (Sangerhausen)

Weigand's Crimson Rambler, *see* **'NON PLUS ULTRA'**

Weight Watchers Success™, *see* 'JACBITOU'

'WEIHENSTEPHAN', S, dp, 1964; flowers rich pink, blooms in large clusters, semi-dbl.; vigorous ($4\frac{1}{2}$ ft) growth; Kordes

'WEISSE AUS SPARRIESHOOP', S, w, 1962; (White Sparrieshoop); flowers medium, moderate fragrance; Kordes

'WEISSE BETTER TIMES', HT, w, 1940; flowers large, dbl.; Wirth; (Sangerhausen)

'WEISSE ECHO', Pol, w, 1925; flowers medium, semi-dbl.; Kiese; (Sangerhausen)

'WEISSE GRÜSS AN AACHEN', F, w, 1944; bud ovoid; flowers snow-white, large, borne in clusters, very dbl., moderate fragrance; upright, bushy growth; [Gruss an Aachen sport]; Vogel

Weisse Immensee®, *see* 'KORWEIRIM'

'WEISSE MARGO KOSTER', Pol, w, 1939; flowers small to medium, dbl.; Teschendorff; (Sangerhausen)

Weisse Max Graf®, *see* 'KORGRAM'

Weisse Nelkenrose, HRg, w; Münster, 1966

Weisse New Dawn, LCl, w

Weisse Repandia®, *see* 'KORIANT'

Weisse Wolke®, *see* KORstacha

Weisser Engel, F, w; Verschuren, 1966

'WEISSER HERUMSTREICHER', HMult, w, 1895; flowers pure white, large, borne in clusters, dbl.; vigorous, climbing growth; [Daniel Lacombe X Paquerette]; Schmidt, J.C.

Weisses Meer, F, w; flowers medium, semi-dbl., slight fragrance; VEG; (Sangerhausen)

'WEKAJAZOUL', S, w, 1999; (**Long Tall Sally™**); flowers buff white, reverse same, single, 5–11 petals, 3–3.5 in., borne in large clusters, moderate fragrance; prickles moderate; foliage large, dark green, semi-glossy; upright, very vertical, tall (5–7 ft) growth; [All That Jazz X R. soulieana derivative]; Carruth, Tom; Weeks Roses, 1999

'WEKAMANDA', HT, yb, 1999; (Limelight, Saint Patrick, **St Patrick™**); flowers yellow gold, shaded green in heat, dbl., 30–35 petals, 5 in., exhibition form, borne mostly singly, slight fragrance; prickles moderate; foliage medium, medium green, dull; upright, bushy, medium (3–4 ft) growth; PP9591; AARS, U.S., 1996; [Brandy X Gold Medal]; Strickland, Frank A., 1991; Weeks Roses, 1996

'WEKAQ', HT, pb, 1990; (**The Temptations™**); bud pointed; flowers pink blend, medium pink reverse, aging slightly lighter, dbl., 25–30 petals, 4 in., exhibition form, borne singly and in small clusters, moderate, fruity fragrance; oval, medium, orange fruit; prickles slightly hooked, medium, green; foliage medium, dark green, semi-glossy, disease-resistant; upright, tall growth; PP8516; GM, ARC TG, 1989; [Paradise X Admiral Rodney]; Winchel, Joseph F., 1984; Weeks Wholesale Rose Growers, 1993

'WEKBIPUHIT', F, m, 1999; (**Purple Heart™**); flowers magenta purple, dbl., 30–35 petals, 3.5–4 in., cupped, borne in large clusters, intense, clove and spice fragrance; prickles moderate; foliage medium, medium green, matte; rounded, compact, bushy, medium (3.5 ft) growth; [Stephen's Big Purple X International Herald Tribune]; Carruth, Tom; Weeks Roses, 1999

'WEKBLUSI', HT, m, 1991; (**Stainless Steel™**); flowers clean, silvery gray lavender, well-formed, large blooms born, dbl., 26–40 petals, 5–6. in., exhibition form, borne in small clusters, intense, sweet rose fragrance; some prickles; foliage large, medium green, semi-glossy; tall to medium (130–150 cms), vigorous growth; PP10188; [Blue Nile X Silverado]; Carruth, Tom; Weeks Roses, 1995

'WEKBOROCO', S, rb, 1999; (**Rockin' Robin™**); flowers red, white and pink stripes and splashes, ruffled, very dbl., 40–45 petals, 1.5–2 in., borne in large clusters, slight, apple fragrance; prickles moderate; foliage medium, dark green, glossy; foutainous, rounded, bushy (4 ft) growth; PP10070; [Bonica X Roller Coaster]; Carruth, Tom; Weeks Roses, 1997

'WEKCALROC', F, yb, 1996; (**George Burns™**); flowers yellow striped irregularly with red, cream and pink, dbl., 26–40 petals, 3–3.5 in., borne in small clusters, moderate, fruit and citrus fragrance; numerous prickles; foliage large, clean, dark green, glossy; upright, compact, medium ($3–3\frac{1}{2}$ ft) growth; PP10334; [Calico X Roller Coaster]; Carruth, Tom; Spring Hill Nurseries Co., 1997

'WEKCATLART', Min, ob, 1998; (**Gizmo™**); flowers scarlet orange with white eye, long lasting, single, 4–11 petals, 1.5 in.–3. in., borne in small clusters, slight, apple fragrance; prickles moderate, small; foliage medium, dark green, semi-glossy; rounded, compact, medium (18 in.) growth; [Carrot Top X Little Artist]; Carruth, Tom; Weeks Roses, 2000

'WEKCELPEP', HT, pb, 1996; (**Sunsation**); flowers yellow striped pink, large blooms, dbl., 26–40 petals, 5 in., borne mostly singly, intense fragrance; some prickles; foliage large, light green, dull; upright, bushy, medium growth; PPAF; [Celebrity X Peppermint Twist]; Carruth, Tom; Michigan Bulb Co., 1996

'WEKCRYLAND', HT, w, 1998; (Cadillac DeVille, **Moonstone™**); flowers ivory white with a fine, delicate pink edging, very dbl., 26–40 petals, 4.5 in.–5 in., exhibition form, borne singly, slight fragrance; prickles moderate; foliage large, medium green, dull; upright, medium (120–160 cms) growth; [Crystalline X Lynn Anderson]; Carruth, Tom; Weeks Roses, 1998

'WEKCRYPLAG', F, m, 1999; (**Blueberry Hill™**); bud plump, pointed, dark lilac; flowers clear lilac, curved petals, golden stamens, single, 5–11 petals, 4 in., exhibition form, borne in small clusters, moderate, apple fragrance; prickles moderate; foliage very serrated, large, dark green, glossy; upright, medium, rounded, bushy (3–4 ft) growth; PP10072; [Crystalline X Playgirl]; Carruth, Tom; Weeks Roses, 1997

'WEKCRYREG', F, pb, 1999; (**Gracie Allen**); flowers white with a pink heart, dbl., 25–30 petals, 4.5 in., borne in small clusters, slight, apple fragrance; prickles moderate; foliage medium, dark green, glossy; upright, bushy, medium (3–4 ft) growth; [Crystalline X Regensberg]; Carruth, Tom; Weeks Roses, 1998

'WEKDESC', Min, w, 1995; (**Descanso Dream**); flowers white blend, small, dbl., 15–25 petals, 1.5 in., no fragrance; few prickles; foliage small, dark green, semi-glossy; low (25–30 cms), bushy, compact growth; [Origami X Little Artist]; Carruth, Tom; Weeks Roses, 1994

'WEKDOCLEM', HT, dy, 1992; (**Atlanta**); flowers deep clear yellow, long-lasting color, dbl., 15–25 petals, 3–3.5 in., borne mostly singly, slight fragrance; some prickles; foliage medium, medium green, matt; medium (90–120 cms), upright, bushy, full growth; [Honor X Gingersnap]; Lemrow, Dr Maynard W.; Weeks Roses, 1996

'WEKDYKSTRA', Gr, yb, 1999; (**Heart O' Gold™**, Heart of Gold, Rose of Narromine); flowers deep gold surrounded by cerise pink, dbl., 35–40 petals, 4–4.5 in., exhibition form, borne in large clusters, intense, fruit and rose fragrance; prickles moderate; foliage large, medium green, semi-glossy; upright, tall (5–6 ft) growth; PP10713; [Broadway X Gold Medal]; Dykstra, Dr A. Michael; Weeks Roses, 1997

'WEKFRAG', HT, or, 1993; (**Ain't She Sweet™**); flowers large, terra-cotta orange to deep orange-red, dbl., 30–35 petals, 3–3.5 in., exhibition form, borne mostly singly, intense, spice and rose fragrance; some prickles; foliage large, medium green, matt; new growth is dark red; medium, rounded, bushy growth; PP009071; [Seedling X Dolly Parton]; Winchel, Joseph F.; Weeks Roses, 1994

'WEKGIBOTEX', F, lp, 1999; (**John-John™**); flowers bright yellow, dbl., 17–25 petals, 3–3.5 in., borne in small clusters, slight fragrance; prickles moderate; foliage medium, light green, dull; upright, bushy, medium (3–4 ft) growth; [(seedling X Sunsprite) X (Old Master X Texas)]; Carruth, Tom; Weeks Roses, 2000

'WEKHELEN', Min, ab, 1990; (**Peach Fuzz™**); bud pointed, moss; flowers apricot blend fading to pastel, medium, borne singly or in small clusters, dbl., 22 petals, exhibition form, intense, moss fragrance; prickles straight, yellow brown; foliage medium, dark green, glossy; bushy, rounded, full, medium growth; [Fairy Moss X New Year]; Carruth, Tom; Weeks Wholesale Rose Growers, 1991

'WEKIBERTAZ', HT, or, 1999; (**Cancun**™); flowers dark orange red, dbl., 26–40 petals, 5–6 in., borne mostly singly, slight fragrance; prickles moderate; foliage large, dark green, glossy; upright, medium (3.5–4 ft) growth; [Ingrid Bergman X All That Jazz]; Carruth, Tom; Spring Hill Nurseries, Co., 1999

'WEKJOE', HT, pb, 1993; (**Lynn Anderson**, Oh My God); flowers white edged deep pink, dbl., 26–40 petals, 5–5.5 in., exhibition form, borne mostly singly, slight fragrance; few prickles; foliage very large, medium green, matt; tall (170–180 cms), upright, bushy growth; PP9389; [Seedling X Gold Medal]; Winchel, Joseph F.; Weeks Roses, 1995

'WEKJUVOO', HT, ob, 1996; (**Chris Evert**™, Raymond Kopa); flowers clear orange-yellow blushing red, reverse with less blush, dbl., 26–40 petals, exhibition form, borne singly, moderate, fruity fragrance; moderate prickles; foliage clean, medium, dark green, semi-glossy; upright, compact, medium (3–4 ft) growth; PP10071; [Voodoo X (Katherine Loker X Gingersnap)]; Carruth, Tom; Spring Hill Nurseries Co., 1996

'WEKLIPS', Min, m, 1988; (**Little Paradise**™); flowers deep lavender, blushing purple, reverse deep lavender, aging lighter, dbl., 20 petals, exhibition form, slight fragrance; no fruit; prickles nearly straight, small, yellow-brown; foliage small, dark green, semi-glossy, disease resistant; upright, medium, vigorous, abundant growth; PP007580; [Shocking Blue X Helen Boehm]; Carruth, Tom; Weeks Roses, 1991

'WEKMAR', HT, ab, 1992; (**Imagination**); bud large, shapely; flowers apricot orange, yellow reverse, good distinct bicoloration, dbl., 26–30 petals, 3–3.5 in., slight fragrance; some prickles; foliage large, clean, medium green, semi-glossy; medium (100–130 cms), attractive, compact, upright, bushy growth; PP8414; [Marmalade X seedling]; Winchel, Joseph F.; Weeks Roses, 1993

'WEKMEYER', LCl, mp, 1999; (**Pearly Gates**™, Pearly Gates, Climbing); flowers pastel pearl pink, very dbl., 41 petals, 5 in., borne in small clusters on old and new wood, intense, spicy fragrance; few prickles; foliage large, medium green, dull; climbing, tall (9–12 ft) growth; PP10640; [America sport]; Meyer; Weeks Roses, 1999

'WEKPEARL', HT, lp, 1999; (**Pearl**); flowers pearl white, dbl., 17–25 petals, 4.5–5 in., borne mostly singly, slight fragrance; prickles moderate; foliage large, dark green, glossy; upright, bushy, medium (4–5 ft) growth; Winchel, Joseph F.; Weeks Roses, 1999

'WEKPIPOGOP', S, w, 1999; (**Pillow Fight**™); flowers bright white, small rosettes, very dbl., 17–25 petals, 1.5–2 in., rosette, borne in large clusters, floriferous, intense, honey and rose fragrance; few prickles; foliage medium, dark green, glossy; rounded, bushy, medium (30–36 in.) growth; [Pink Pollyanna X Gourmet Popcorn]; Carruth, Tom; Weeks Roses, 2000

'WEKPLAPEP', F, rb, 1999; (**Scentimental**™); flowers striped burgundy and white, or cream, or red, opening quickly, dbl., 25–30 petals, 4–4.5 in., exhibition form, borne in large clusters, intense, spicy fragrance; prickles moderate; foliage large, quilted, medium green, matt; compact, rounded, medium (3–4 ft) growth; PP10126; AARS, U.S., 1997; [Playboy X Peppermint Twist]; Carruth, Tom; Weeks Roses, 1997

'WEKPLAPIC', F, rb, 1999; (**Betty Boop**™); bud pointed; flowers rosy edged, yellow at base when fresh, fading to white; bright stamens, single, 6–12 petals, 4 in., borne in small cluster of 3 to 5; floriferous, moderate, fruity fragrance; prickles moderate; foliage medium, dark green, glossy, dark red new growth; rounded, bushy, medium (4 ft) growth; PP10126; AARS, U.S., 1999; [Playboy X Picasso]; Carruth, Tom; Weeks Roses, 1999

'WEKPLASOL', F, yb, 1999; (**Flutterbye**™); flowers multicolor yellow, coral, orange, tangerine, pink, single, 5–9 petals, 1.5–2 in., borne in large clusters, moderate, spice fragrance; prickles moderate; foliage medium, dark green, glossy; very large, rounded, can be fountainous growth; PP9715; [Playboy X (R. soulieana derivative X Sunsprite)]; Carruth, Tom; Weeks Roses, 1996

'WEKPLUROCO', S, m, 1997; (**Purple Reign**); flowers irregularly striped magenta and white, semi-dbl., 8–14 petals, 1.5–2.5 in., borne in large clusters, intense fragrance; foliage large, dark green, glossy; spreading, medium (60–70 cms) growth; [Fragrant Plum X Roller Coaster]; Carruth, Tom; Michigan Bulb Co., 1997

'WEKQUANEZE', HT, m, 1999; (**Barbra Streisand**); flowers lavender, blushing darker, slow opening, dbl., 26–40 petals, 3.5–4 in., borne in small clusters, intense fragrance; prickles moderate; foliage large, dark green, glossy; upright, bushy, medium (3–4 ft) growth; [(Blue Nile X (Ivory Tower X Angel Face)) X New Zealand]; Carruth, Tom; Weeks Roses, 2000

'WEKROALT', LCl, rb, 1999; (Crazy For You, **Fourth of July**™); flowers velvety red and white striped, ruffled, single, 10–16 petals, 4–4.5 in., borne in large clusters, moderate, apple fragrance; prickles moderate; foliage large, deep green, glossy; climbing, tall (10 ft) growth; AARS, U.S., 1999; [Roller Coaster X Altissimo]; Carruth, Tom; Weeks Roses, 1999

'WEKSAMSOU', LCl, mr, 1999; (**All Ablaze**™); bud short, ovoid; flowers medium cherry red, ruffled, full, very dbl., 35 petals, 4 in., borne in small clusters, slight, spice fragrance; foliage medium, dark green, semi-glossy; vigorous growth; climbing to 12 ft; [Don Juan X (Rosa soulieana seedling X Trumpeter)]; Carruth, Tom

'WEKSAMSOU', LCl, lp, 1999; (**All Ablaze**™); dbl., 26–40 petals, 2.5–3 in., borne in large clusters, slight fragrance; prickles moderate; foliage large, medium green, dull; spreading, climbing, tall (9–12 ft) growth; hardy; [Don Juan X (R. soulieana derivitive X Trumpeter)]; Carruth, Tom; Weeks Roses, 2000

WEKslats, HT, lp, 1998; (**Pink Crystal**); Wathen; Weeks Roses, 1998

'WEKSNACARE', Min, lp, 1999; (**Space Odyssey**™); flowers red with white eye, reverse white, semi-dbl., 12–16 petals, 1–1.5 in., borne in small clusters, slight fragrance; few prickles; foliage small, dark green, glossy; compact, spreading, low (16–20 in.) growth; [Santa Claus X Times Square]; Carruth, Tom; Weeks Roses, 2000

'WEKSYBIL', Min, pb, 1989; (**Heartbreaker**); bud pointed; flowers deep pink with white base, small, borne in sprays of 3–5, dbl., exhibition form, slight fragrance; globular, small, dark orange fruit; prickles nearly straight, small, dark red-brown; foliage small, dark green, glossy; upright, bushy, medium, vigorous growth; PP007588; [Crystalline X Magic Carrousel]; Carruth, Tom; Weeks Roses, 1990

'WEKTORROC', LCl, lp, 1999; (**Rosy Outlook**); flowers deep pink striped white, dbl., 17–25 petals, 3–3.5 in., borne in large clusters, slight fragrance; prickles moderate; foliage large, dark, green, glossy; spreading, climbing, tall (10–12 ft) growth; hardy; [Tournament of Roses X Roller Coaster]; Carruth, Tom; Weeks Roses, 1999

'WEKUZ', F, dp, 1990; (**Columbus**™); bud ovoid, pointed; flowers deep rose pink, very little fading, large blooms, dbl., 28–35 petals, exhibition form, borne usually singly, slight fragrance; prickles almost straight, slightly hooked, medium, pinkish-br; foliage large, medium green, dull; bushy, medium growth; PP8183; [Seedling X Bridal Pink]; Carruth, Tom; Weeks Roses, 1991

'WEKWINWIN', HT, rb, 1999; (**Rosie O'Donnell**, Win Win); bud urnshaped, long; flowers velvety scarlet red with creamy yellow reverse, dbl., 30–35 petals, 5–5.5 in., exhibition form, borne mostly singly, slight fragrance; foliage large, dark green; long stems; medium, upright, slightly spreading growth; Winchel, Joseph F.; Weeks Roses, 1998

'WEKYEGI', Min, my, 1999; (**Lemon Drop**); flowers clear yellow, dbl., 26–40 petals, 1–1.5 in., borne in small clusters, slight fragrance; few prickles; foliage medium, light green, semi-glossy; upright, compact, medium (20–24 in.) growth; [Gingerbread Man X seedling]; Carruth, Tom; Weeks Roses, 1999

'WELAN', HT, pb, 1973; (**Angel Girl**); flowers peach-pink; [Bel Ange sport]; Wyant

'WELANGEL', Min, pb, 1996; (**Angelica Renae**™); flowers bright medium pink blend, dbl., 26–40 petals, exhibition form, borne mostly singly, slight fragrance; some prickles; foliage medium, dark green, matt; upright, bushy (24–30 in.) growth; AOE, 1996; [Anita Charles X Seedling]; Wells, Verlie W.; Nor'East Min. Roses, 1996

'WELCH', (strain of R. multiflora), w; thornless; sometimes used as understock; Mt. Arbor Nursery

'WELCO', HT, pb, 1990; (**Ashgrove Jubilee**); flowers cream, shading to deep rose at edges, 30 petals, exhibition form, borne singly, slight fragrance; foliage thick, shiny, dark green, obtuse at base; tall, upright growth; [Mascot X ((Seedling X Red Lion) X Silver Lining)]; Welsh, Eric, 1985; Rose Hill Roses, 1990

'WELCOME', HT, mp, 1948; bud ovoid; flowers glistening rose-pink, very large, dbl., 35 petals, exhibition form, slight fragrance; foliage leathery; vigorous, upright, bushy growth; Dickson, A.

'WELCOME GUEST', HT, ab, 1984; flowers apricot with a tinge of pink around edges of petals; [Jan Guest sport]; Cox, Arthur George

Welcome Home, *see* 'ANDWEL'

'WELCOME STRANGER', F, my, 1968; flowers large, single, intense fragrance; foliage extra large, glossy, leathery; very vigorous, compact growth; [Ophelia, Climbing X Allgold]; Fankhauser; A. Ross & Son

'WELCORD', Min, rb, 1997; (**Cherry Cordial**); flowers full, medium, bright red with white reverse, very dbl., 26–40 petals, borne mostly singly, slight fragrance; foliage medium, dark green, semi-glossy; upright, medium growth; [Magic Carrousel X Seedling]; Wells, Verlie W.; Kimbrew-Walter Roses, 1997

WELgold, F, lp; (**Victoria Gold**); Welsh, 1999

WELira, HT, lp; (**Woman's Day**); [Avandel X WELpa]; Welsh, Eric, 1992

'WELIZ', HT, mr, 1987; (**Ida Elizabeth**); flowers large, borne singly, dbl., 30 petals, exhibition form, moderate fragrance; foliage matt; medium, bushy growth; [Red Lion X Mainauperle]; Welsh, Eric; Treloar Roses Pty. Ltd., 1990

'WELLEE', Min, ab, 1994; (**Dorris Lee**); flowers apricot blend, dbl., 26–40 petals, 1.5–2.75 in., borne singly and in small clusters, moderate fragrance; some prickles; foliage dark green, semi-glossy; medium, upright growth; [Seedling X Party Girl]; Wells, Verlie W.; Wells Mid-south Roses, 1994

Weller Bourbon, B, lp

'WELLESLEY', HT, dp, 1905; flowers dark rosy pink, large, dbl., moderate fragrance; [Bridesmaid X Liberty]; Montgomery, A.

'WELLINGTON', HGal, m; flowers crimson-purple, cupped; Int. prior to 1848

Well's Climber

Well's Pink and White Climber,

'WELLSNED', Min, pb, 1997; (**Lucille McWherter**); flowers semi double, medium, pink blend, semi-dbl., 8–14 petals, intense fragrance; foliage medium, dark green, glossy; upright, medium (18–24 in.) growth; [Seedling X Seedling]; Wells, Verlie W.

'WELLWORTH', HT, pb, 1949; flowers peach shaded gold, pointed, dbl., 40 petals, 5 in., intense fragrance; foliage grayish green; vigorous growth; [Leontine Contenot X Golden Dawn]; LeGrice

'WELPIN', HT, dp, 1992; (**Edmund Rice**); bud classic ht form; flowers light red, non-fading, blooms borne 5–7 per cluster and singly, dbl., 45–50 pointed, reflexed petals, 5 in., exhibition form, repeats quickly, moderate fragrance; foliage very red when young, dark green, shiny, leathery; bushy, upright (4′) growth; [Red Lion X Pink Silk]; Welsh, Eric, 1989; Christian Bros. College, 1992

'WELQUEEN', Min, w, 1997; (**Memphis Queen**); flowers very full, medium, very dbl., 41 petals, borne mostly singly, slight fragrance; foliage medium, dark green, glossy; upright, medium (60 cms) growth; [Miss Pearl X Pacesetter]; Wells, Verlie W.; Wells Mid-South Roses, 1997

'WELRUM', Min, lp, 1994; (**Kathleen Rumble**); flowers moderately blooms borne mostly single, dbl., 15–25 petals, slight fragrance; some prickles; foliage medium, dark green, glossy; medium (6 cms), upright growth; [Seedling X Miss Pearl]; Wells, Verlie W.; Wells Midsouth Roses, 1994

'WELSAM', Min, yb, 1994; (**Samantha Ruth**); very dbl., 3–3.5 in., borne mostly singly, slight fragrance; some prickles; foliage medium, dark green, semi-glossy; medium, upright growth; [(Party Girl X Magic Carrousel) X Seedling]; Wells, Verlie W.; Wells Midsouth Roses, 1994

'WELSNED', Min, pb, 1997; (**Lucille McWherter**); flowers pink blend, em-dbl. (8–14 petals), medium ($1\frac{1}{2}$ in.) blooms; very fragrant; foliage medium, dark green, glossy; upright, meidum (18–24 in.) growth; [Seedling X Seedling]; Wells, Verlie; Wells Mid-South Roses

Welwyn Garden Glory, *see* HARzumber

Wembley Stadium, F, dp

'WENCHE', HT, mr, 1997; flowers full, large, very dbl., 26–40 petals, borne mostly singly, slight fragrance; foliage medium, dark green, semi-glossy; upright, medium growth; [Silver Jubilee X Loving Memory]; Poole, Lionel

'WENDELIEN', F, mr, 1946; flowers cardinal-red, borne in clusters, dbl., 22 petals, 4 in., globular, slight fragrance; foliage bright green; vigorous, branching growth; [Donald Prior X Unnamed seedling]; Leenders, M.; Longley

Wendy, *see* **'PARA TI'**

'WENDY', Pol, mp, 1949; flowers pink, open, medium, borne in clusters, dbl., moderate fragrance; foliage soft, light green; moderately vigorous, upright growth; RULED EXTINCT 11/91; [Tip-Top X Dorothy Perkins]; Heers; Pacific Nursery

Wendy, *see* 'ZIPWEN'

Wendy, *see* 'SUNWEND'

Wendy, MinFl, lp; Schuurman, Frank B., 1993

Wendy Ackerman Rose, HT, lp; J&P, 1998

'WENDY BARRIE', Pol, op, 1936; flowers orange-salmon, well formed, dbl.; vigorous, dwarf growth; Beckwith

'WENDY CUSSONS', HT, mr, 1963; bud long, pointed; flowers rose-red, dbl., 30 petals, 5–6 in., exhibition form, intense fragrance; foliage leathery, glossy, dark; vigorous, well-branched growth; GM, NRS, 1959 GM, Portland, 1964 Golden Rose, The Hague, 1964 PIT, NRS, 1959; [Believed to be Independence X Eden Rose]; Gregory; Ilgenfritz Nursery

'WENDY CUSSONS, CLIMBING', Cl HT, mr, 1967; Follen; Gregory

'WENDY DUCKETT', F, lp, 1999; flowers red cream, striking color, dbl., 26–40 petals, 1.2 in., borne in small clusters, slight fragrance; few prickles; foliage small, medium green, glossy, impervious to rain; upright, compact, medium growth; [Sheri Anne X (Whippet X Party Girl)]; Jones, L.J.

Wendy Pease, *see* 'DEBRAF'

Wendy van Wanten, F, lp; RvS-Melle, 1995

Wenlock®, *see* 'AUSWEN'

'WENZEL GESCHWIND', HT, m, 1902; flowers purple/pink, medium to large, dbl., moderate fragrance; Geschwind, R.; (Sangerhausen)

'WENZEL GESCHWIND, CLIMBING', Cl HT, m, 1940; flowers purple/pink, medium to large, dbl., moderate fragrance; Vogel, M.; (Sangerhausen)

'WEOKAY', F, rb, 1986; (**Playmate**); flowers vivid scarlet, yellow center, large, borne usually singly, single, 5 petals, slight fragrance; foliage medium, dark green, glossy; upright, bushy growth; [Playboy X Unnamed seedling]; Fonda, Henry; Wee Ones Miniature Roses, 1986

'WEOPOP', Min, w, 1986; (**Gourmet Popcorn**); flowers pure white, medium, semi-dbl., 6–14 petals, slight fragrance; foliage large, dark green, glossy; upright, bushy growth; PP006809; [Popcorn sport]; Desamero, Luis; Wee Ones Miniature Roses, 1986

'WERINA', Gr, ob, 1975; (**Arizona**, Tocade); bud urn-shaped; flowers golden bronze to orangy yellow, medium, dbl., 25–30 petals, exhibition form, borne singly, intense, sweet fragrance; foliage glossy, dark, leathery; long, cutting length stems; vigorous, upright, bushy growth; AARS, 1975; [((Fred Howard X Golden Scepter) X Golden Rapture) X ((Fred Howard X Golden Scepter) X Golden Rapture)]; Weeks; C-P

'WERNER DIRKS', LCl, w, 1937; bud long, pointed; flowers ivory-white, very large, borne in clusters, dbl., exhibition form, moderate fragrance; foliage large, leathery, wrinkled; long, strong stems; very vigorous, climbing growth; [Mrs Pierre S. duPont X Daisy Hill]; Kordes

'WERNER OTTO', HT, my, 1995; flowers medium yellow, very full, very dbl., 56 petals, 4–4.5 in., borne mostly singly, slight fragrance; few prickles; foliage large, dark green, glossy; medium, upright growth; [Golden Fantasie sport]; Mungia, Larry; Montebello Rose Co., 1994

'WERNER TESCHENDORFF', F, op, 1949; bud ovoid; flowers medium, borne in clusters, dbl., cupped, slight fragrance; foliage glossy; vigorous, upright, bushy growth; [Swantje X Hamburg]; Tantau; Teschendorff

Werner von Braun, *see* HELilast

'WESNIANKA', HT, w, 1941; (Spring); flowers white, base carmine-yellow; [Mme Butterfly X Mrs T. Hillas]; Costetske

Wessie Roos, *see* KORbonnet

West Coast®, *see* 'MACNGAURU'

'WESTBROEKPARK', F, or, 1968; flowers medium, borne in clusters, dbl.; foliage dark; [Orange Sensation X Kimono]; deRuiter

Westerland®, *see* 'KORWEST'

'WESTERN GOLD', HT, my, 1932; flowers clear yellow; [Talisman sport]; Western Rose Co.

'WESTERN SUN'®, HT, dy, 1965; flowers deep yellow, dbl., 40 petals, 5 in.; foliage dark; [Golden Scepter seedling X Golden Sun]; Poulsen, Niels D.

Western Sunlight, *see* 'HADSUN'

'WESTERN SUNSET', F, yb, 1958; bud pointed; flowers yellow tipped pink, becoming pink and then mahogany-red, bor, dbl., 35 petals, 1.5–2 in., cupped, foliage leathery; vigorous, upright growth; [Maxine X Masquerade]; Silva

Westfalen, S, 1969; Scholle, E.; (Sangerhausen)

Westfalengold, F, 1979; Noack, Werner; (Cavriglia)

'WESTFALENGRUSS', F, mr, 1978; bud ovoid; flowers fire-red, medium, dbl.; foliage glossy; low, bushy growth; Hubner; O. Baum

Westfalenpark®, *see* 'KORPLAVI'

'WESTFIELD BEAUTY', HT, ab, 1923; bud long, pointed; flowers deep coppery apricot, tinted golden and salmon-pink, dbl., moderate fragrance; [Lady Pirrie X Mme Edouard Herriot]; Morse

'WESTFIELD FLAME', HT, or, 1925; bud long, pointed; flowers very deep flame, dbl., slight fragrance; [Mme Edouard Herriot X Diadem]; Morse

'WESTFIELD GEM', HT, dr, 1925; flowers dark maroon-crimson, dbl., slight fragrance; [Col. Oswald Fitzgerald sport]; Morse

'WESTFIELD SCARLET', HT, mr, 1931; flowers clear scarlet, open, very large, dbl., slight fragrance; foliage leathery, dark; bushy growth; Gold Medal, NRS, 1932; [Lady Inchiquin sport]; Morse

'WESTFIELD STAR', HT, w, 1922; flowers cream, fading white; [Ophelia sport]; Morse

'WESTMINSTER', HT, rb, 1960; flowers red, gold reverse, loosely-formed, large, dbl., 35 petals, intense fragrance; foliage dark; vigorous, tall growth; GM, NRS, 1961; [Gay Crusader X Peace]; Robinson, H.

Westminster Pink, HT, lp; Fryer, Gareth, 1998

'WESTMONT', Min, mr, 1958; bud pointed; flowers bright red, semi-dbl., 1.5 in.; foliage leathery, semi-glossy; small vigorous (12–18 in.), bushy growth; [(R. wichurana X Floradora) X (Oakington Ruby X Floradora)]; Moore, Ralph S.; Sequoia Nursery

'WESTWARD HO!', HT, rb, 1964; flowers mahogany-red, reverse silver, dbl., 42 petals, 4.5 in., moderate fragrance; foliage very dark; vigorous, upright, compact growth; [Karl Herbst X Pink Charming]; Allen, E.M.; Sanday

Wetteriana, F, lp

'WETTRA', F, mr, 1976; flowers large blooms borne 4–15 per cluster, dbl., 20 petals, cupped, no fragrance; red prickles; foliage dark, matt; upright growth; [Pink Puff X Barcarolle]; Rijksstation Voor Sierplantenteelt

'WEZIP', HT, m, 1978; (Burning Sky, **Paradise**™, 'WEZEIP'); bud long, pointed; flowers silvery lavender shading to ruby-red at edge, well-formed, dbl., 28 petals, 3.5–4.5 in., moderate fragrance; foliage glossy, dark; upright growth; AARS, 1979 GM, Portland, 1979; [Swarthmore X Unnamed seedling]; Weeks; C-P

WHAmemo, MinFl, lp; (**Sweet Memories**)

'WHEATCROFT GIANT', HT, lp, 1962; flowers pearly pink, well-formed, large; vigorous growth; Wheatcroft Bros.

Wheatcroft's Baby Crimson, *see* **'PERLA DE ALCAÑADA'**

Wheatcroft's Golden Polyantha, *see* **'GOLDENE JOHANNA TANTAU'**

Wheato, F, dp; Rearsby Roses, Ltd.

Wheel Horse Classic, *see* HARelite

'WHICAW', F, w, 1988; (**Mr McCawber**); bud pointed, white flushed pink; flowers medium, borne in sprays of 8–10, semi-dbl., 15–17 petals, exhibition form, slight fragrance; globular, orange fruit; prickles hooked, light brown; foliage medium, medium green, semi-glossy, disease-resistant; upright, bushy, tall, prolific growth; [(French Lace X Simplex) X (Pristine X White Angel)]; White, James J.

Whickham Highway, *see* 'HORMASBRICK'

'WHIMSICAL', Min, pb, 1980; bud pointed; flowers either light pink or deeper peach-pink on same plant at same, dbl., 48 petals, exhibition form, no fragrance; no prickles; foliage medium green; upright, bushy growth; [Tiki X Baby Betsy McCall]; Strawn, Leslie E.; Pixie Treasures Min. Roses

'WHIPPED CREAM', Min, w, 1968; bud pointed, ivory; flowers small, dbl.; foliage light green, leathery; vigorous, bushy, dwarf growth; [(R. wichurana X Carolyn Dean) X White King]; Moore, Ralph S.; Sequoia Nursery

'WHIPPET', HT, pb, 1973; flowers salmon-pink, reverse lighter, dbl., 35 petals, 5 in., exhibition form, slight fragrance; moderate growth; Scott, D.H.

Whirlygig, *see* 'JUSPROP'

'WHISKY', F, yb, 1964; flowers yellow shaded orange-bronze, open, large, borne in clusters, dbl., slight fragrance; foliage glossy, light green; very vigorous, upright growth; [Cognac X Arc-en-Ciel]; Delforge

Whisky, *see* 'TANKY'

'WHISKY GILL', HT, ob, 1972; flowers burnt-orange to bright orange; [Whisky Mac sport]; Cobley

Whisky Mac, *see* 'TANKY'

Whisky Mac, Climbing, *see* 'ANDMAC'

Whisky, Climbing, *see* 'ANDMAC'

'WHISPER', F, lp, 1971; flowers pale pink, semi-dbl., 10 petals, 3–3.5 in., slight fragrance; foliage light; vigorous growth; [Queen Elizabeth X Monique]; Cants of Colchester, Ltd.

Whisper Louise, Pol, lp; Robinson, 1994

Whistle Stop, *see* 'MACMOSCO'

'WHITE AACHEN', F, ly, 1937; flowers buff-yellow to pure white; [Gruss an Aachen sport]; Western Rose Co.

'WHITE ALASKA', HT, w, 1960; bud long, pointed; flowers pure white, dbl., 20 petals, 4.5–5 in., exhibition form, slight fragrance; foliage leathery, dark; vigorous, upright growth; [White Ophelia X Unnamed seedling]; Hartgerink; Armacost & Royston

White America, LCl, w

White American Beauty, *see* **'FRAU KARL DRUSCHKI'**

White American Beauty, Climbing, *see* **FREEDOM®**

'WHITE ANGEL', Min, w, 1971; flowers small, dbl., exhibition form, slight fragrance; foliage small, light; vigorous, dwarf, bushy growth; AOE, 1975; [(R. wichurana X Floradora) X (Little Darling X Red Miniature seedling)]; Moore, Ralph S.; Sequoia Nursery

White Aster, Min, w; Moore, 1957

White Avalanche, *see* 'WILWAVL'

White Baby Rambler, *see* **'KATHARINA ZEIMET'**

'WHITE BABY STAR', Min, w, 1965; [Baby Gold Star sport]; Spring Hill Nursery; Wyant

White Banksia, *see* **R. BANKSIAE BANKSIAE**

'WHITE BARONESS', HP, w; [Baroness Rothschild sport]; Paul & Son, 1882

White Bath, *see* **'SHAILER'S WHITE MOSS'**

'WHITE BEAUTY', HT, w, 1965; flowers pure white; [The Doctor sport]; Brooks, M.L.; Texas Rose Research Foundation

White Belle of Portugal, LCl, w

White Bells®, *see* 'POULWHITE'

White Blush, A, w; Sievers, 1988

White Bon Silène, *see* **'BON SILÈNE BLANC'**

White Bonnet, F, w

'WHITE BOUGÈRE', T, w, 1898; flowers pure white; Dunlop

'WHITE BOUQUET', F, w, 1956; bud ovoid; flowers gardenia-shaped, blooms in irregular clusters, dbl., 45 petals, 4.5 in., moderate, spicy fragrance; foliage dark, glossy; bushy growth; AARS, 1957; [Glacier X Pinocchio seedling]; Boerner; J&P

'WHITE BRIARCLIFF', HT, w, 1932; (Mme Louis Lens); bud long, pointed; flowers pure white, large, dbl., exhibition form, moderate fragrance; foliage leathery; vigorous growth; [(Briarcliff X Kaiserin Auguste Viktoria) X (Briarcliff X Mrs Herbert Stevens)]; Lens; J&P

'WHITE BUTTERFLY', HT, w, 1954; bud long, pointed; flowers white, inner petals pale chartreuse, center, dbl., 24 petals, 3.5–4.5 in., cupped, moderate fragrance; foliage leathery; vigorous, compact growth; John Cook Medal, ARS, 1957; [Ophelia X Curly White]; Spanbauer

'WHITE CAP', F, w, 1954; dbl., 60 petals, 3.5–4 in., moderate fragrance; [Unnamed seedling X Break o' Day, Climbing]; Brownell, H.C.

White Captain, HRg, lp; Spek/Captain, 1998

White Cascade™, *see* 'CARCADE'

'WHITE CÉCILE BRÜNNER', Pol, w, 1909; flowers white, sulfur yellow and buff; [Cécile Brunner sport]; Fraque

'WHITE CHARM', F, w, 1958; bud long, pointed; flowers white, base yellow-green, semi-dbl., 17–22 petals, 2.5–3 in., exhibition form, moderate fragrance; foliage leathery; bushy, upright growth; RULED EXTINCT 10/88; [Pinocchio X Virgo]; Swim & Weeks

White Charm™, *see* 'MINAQCO'

'WHITE CHATEAU', HT, lp, 1999; bud ivory; flowers pure white, dbl., 26–40 petals, 5 in., exhibition form, slight fragrance; 4.5 ft growth; [Ophelia X seedling]; Teranishi, K.; Itami Rose Nursery, 1996

'WHITE CHATILLON', Pol, w; (Chatillon White); bud pale pink, fading to white.; [Probably a Chatillon Rose sport]

White Chipper, Min, lp

'WHITE CHRISTMAS', HT, w, 1953; bud long, pointed; flowers pure white, medium, dbl., exhibition form, moderate fragrance; foliage leathery, light green; moderate, upright growth; [Sleigh Bells X Seedling]; H&S

White Cloud, *see* **'HAKUUN'**

White Cloud™, *see* 'SAVACLOUD'

White Cloud, *see* KORstacha

'WHITE CLOUDS', Cl HT, w, 1953; bud long, pointed; dbl., 50 petals, 4.5 in., intense fragrance; foliage leathery, dark; vigorous growth; RULED EXTINCT 8/88; [Frau Karl Druschki X Kaiserin Auguste Viktoria]; Silva

White Cochet, *see* **'WHITE MAMAN COCHET'**

'WHITE COCKADE', LCl, w, 1969; flowers large, dbl., moderate fragrance; foliage glossy; vigorous, low, climbing growth; [New Dawn X Circus]; Cocker

White Colorado, F, lp; Select, 1998

'WHITE COLUMBIA, CLIMBING', Cl HT, w, 1934; Clark's Rose Nursery

White Comet, *see* 'DELOBLAN'

White Cover, *see* 'POULCOV'

'WHITE CROSS', LCl, w, 1950; bud ovoid; flowers white, center tinted lemon, large, very dbl., moderate fragrance; foliage leathery, glossy; very vigorous growth; [Mrs Arthur Curtiss James sport]; Hester

White Crystal®, HMsk, w; Lens, 1992

White Daily Rose, *see* **'INDICA ALBA'**

'WHITE DAWN'®, LCl, w, 1949; flowers gardenia-shaped, pure white, dbl., 30–35 petals, 3–3.5 in., blooms in clusters, recurrent bloom, moderate, sweet fragrance; foliage glossy, dark green; very vigorous, climbing growth; [New Dawn X Lily Pons]; Longley, L.E.; Univ. of Minn.

White de Meaux, *see* **'ROSE DE MEAUX WHITE'**

White Delight, *see* 'JACGLOW'

'WHITE DEMURE', F, w, 1952; bud ovoid; dbl., 30–35 petals, 2–2.5 in., flat, slight fragrance; foliage leathery; vigorous, compact, bushy growth; [Demure sport]; Boerner

White Diamond®, *see* INTeramon

'WHITE DIAN', Min, w, 1965; flowers white, sometimes light pink; [Dian sport]; Moore, Ralph S.; Sequoia Nursery

'WHITE DICK KOSTER', Pol, 1946; Grootendorst, F.J.; (Cavriglia)

White Dog, S, w

'WHITE DOROTHY', HWich, w, 1908; (White Dorothy Perkins); flowers creamy white; [Dorothy Perkins sport]; Cant, B. R.

White Dorothy Perkins, *see* **'WHITE DOROTHY'**

White Dr Grill, T, lp

White Dream®, *see* 'LENVIR'

'WHITE DUCHESS', HT, w, 1964; bud spiral, pointed; flowers snow-white, medium; foliage glossy; strong stems; vigorous growth; Herholdt, J.A.; Herholdt's Nursery

White Duchesse de Brabant, *see* **'MME JOSEPH SCHWARTZ'**

'WHITE ELFE', F, w, 1954; bud pointed; flowers white faintly tinted pink, borne in clusters of 3–7, dbl., 40–45 petals, 2–2.5 in., exhibition form, slight fragrance; foliage dark, leathery; vigorous growth; [Rosenelfe sport]; Holmes, V.E.; Avansino; Mortensen

'WHITE ENSIGN', HT, w, 1925; flowers large, very dbl., slight fragrance; foliage leathery, dark, glossy; vigorous growth; McGredy

'WHITE FAIRY', Min, w, 1952; flowers small blooms in clusters, very dbl.; Sequoia Nursery

White Fairy, Pol, lp

White Feather, *see* 'MORFEAT'

White Festival, *see* 'LAVSNOW'

'WHITE FINCH', Pol, w, 1937; flowers white, tinted pink; [Mrs R.M. Finch sport]; Stielow, F.C.

White Fleurette®, *see* BRIseflu

'WHITE FLIGHT', HMult, w, 1923; flowers medium, dbl.; Koster; (Sangerhausen)

'WHITE FLIGHT', HMult, 1923; Koster, D.A.; (Cavriglia)

White Flower Carpet, *see* NOAschnee

'WHITE FORCER', Pol, w, 1926; flowers larger and trusses than parent; [Jesse sport]; Spek

White Fragrant Mist, HT, w

'WHITE GARNETTE', F, w, 1952; flowers white tinged cream, borne in clusters, dbl., 30–35 petals, 2 in., slight fragrance; foliage rich green; bushy growth; [Pinocchio seedling X Garnette]; Boerner; J&P

White Gem®, *see* 'MEITURUSA'

'WHITE GENE BOERNER', F, w, 1978; [Gene Boerner sport]; Takatori, Yoshiho; Japan Rose Nursery

'WHITE GOLD', LCl, w, 1943; bud long, pointed; flowers white, center yellow, petals reflexed, dbl., exhibition form, foliage dark, glossy; vigorous, climbing (20 ft) growth; [Glenn Dale X Mrs Arthur Curtiss James]; Brownell, H.C.

White Gold, *see* 'COCQUIRIAM'

White Gold 98, *see* 'COCQUIRIAM'

'WHITE GROOTENDORST', HRg, w, 1962; [Pink Grootendorst sport]; Eddy

White Hedge, HRg, w

White Hermosa, *see* **'MARIE LAMBERT'**

White Hero, S, lp; Rex, Dr Robert W., 1996

'WHITE HOUSE', HT, w, 1951; bud high-centered to ovoid; flowers satiny white; foliage thick, light green; [McGredy's Yellow X Frau Karl Druschki]; Silva

White House, *see* 'RESLINE'

'WHITE JEWEL', F, w, 1957; dbl., 33 petals, 4 in., cupped, blooms in clusters, moderate fragrance; vigorous, bushy growth; [Starlite seedling X Glacier seedling]; Boerner; J&P

'WHITE JOY', HT, w, 1952; bud pointed; flowers pure white, dbl., 55–65 petals, 4–5 in., flat, moderate fragrance; foliage leathery; bushy, upright growth; Spanbauer

White Junior Miss, Gr; (Cavriglia)

'WHITE KILLARNEY', HT, w, 1909; moderate fragrance; [Killarney sport]; Waban Conservatories

'WHITE KING', Min, w, 1961; bud pointed, ovoid; flowers cream-white, small, dbl., 45 petals, 1.5 in., moderate fragrance; foliage leathery; bushy (12 in.) growth; [Golden Glow (LCl) X Zee]; Moore, Ralph S.; Sequoia Nursery

White Knight, *see* 'MEBAN'

'WHITE KNIGHT, CLIMBING', Cl HT, w, 1959; Komatsu; Kakujitsuen

'WHITE KNIGHT, CLIMBING', Cl HT, w, 1965; (Message, Climbing); Meilland; URS

White La France, *see* **'AUGUSTINE GUINOISEAU'**

White Lady, *see* **'DAME BLANCHE'**

White Lady Banks' Rose, *see* **R. BANKSIAE BANKSIAE**

White Lafayette, *see* **'DAGMAR SPÄTH'**

White Lightnin'™, *see* 'AROWHIF'

White Lightnin'™, Gr, w

White Look, HT, lp

'WHITE LOVE', HT, w, 1973; bud long; flowers ivory-white, large, intense fragrance; foliage dark; bushy growth; [Frau Karl Druschki X (Peace X Unknown seedling)]; Buisman, G. A. H.

'WHITE MA PERKINS', F, w, 1962; bud long, pointed; dbl., 20–40 petals, 3.5 in., borne in clusters, intense fragrance; foliage light green; very vigorous, upright growth; [Ma Perkins sport]; McDonald; Hennessey

'WHITE MADONNA', Min, w, 1973; bud long, pointed; flowers white to pale pink, dbl., 33 petals, 1 in., slight fragrance; foliage glossy, leathery; upright, bushy growth; [(R. wichurana X Floradora) X (Little Darling X Unnamed red Miniature)]; Moore, Ralph S.; Sequoia Nursery

White Magic, *see* 'JACARE'

White Magic, HMsk, lp; Lens, 1989

White Magic, HT, lp

White Magic Carpet™, *see* POUlland

White Majesty, *see* MEIbonver

'WHITE MAMAN COCHET', T, w, 1896; (White Cochet); flowers white, often flushed pink; [Maman Cochet sport]; Cook, J.W.

'WHITE MAMAN COCHET, CLIMBING', Cl T, w, 1907; Knight, G.; Leedle & Co., 1911

White Masterpiece, *see* 'JACMAS'

White Max Graf, HKor, 1983; Kordes, R.; (Cavriglia)

White Meidiland™, *see* 'MEICOUBLAN'

White Meillandina, *see* 'MEIBLAM'

'WHITE MIMOLLET', F, w, 1984; [Mimollet sport]; Ota, Kaichiro

White Mini-Wonder™, *see* 'MEIZOGREL'

White Mite, *see* 'FROMITE'

White Moss, *see* **'COMTESSE DE MURINAIS'**

'WHITE MOUNTAINS', LCl, w, 1958; bud globular; flowers white, becoming greenish, small, borne in clusters, dbl., profuse, intermittent bloom, foliage glossy; vigorous growth; quite hardy; [Skinner's Rambler X Skinner's Rambler]; Risley

White Mountains, LCl, w

'WHITE MRS FLIGHT', HMult, w, 1916; flowers pure white; [Mrs F.W. Flight sport]; T. Rockford

White Mystery, *see* 'DEVABE'

White New Dawn, LCl, w

White Nights, *see* 'POULAPS'

White Noblesse, HT, w

'WHITE NUN', HT, w, 1968; flowers open, medium, dbl.; foliage glossy; vigorous, upright growth; [Virgo X ?]; Pal, Dr B.P.; Indian Agric. Research Inst.

'WHITE OPHELIA', HT, w, 1920; flowers white, center faintly tinted pink, becoming white, well form, semi-dbl., moderate fragrance; [Ophelia sport]; Cleveland Cut-Flower Co.; Cleveland Cut-Flower Co.; E.G. Hill Co.

'WHITE ORLÉANS', Pol, w, 1920; [Orléans Rose sport]; Van Eyk

White Pacific, HT, w

White Pavement, S, lp, 1991; semi-dbl., moderate fragrance; light red fruit; spreading (2.5 ft) growth; Uhl, J.

'WHITE PEARL', HT, w, 1948; bud long, pointed; flowers glistening white, open, well formed, large, dbl., 45–50 petals, slight fragrance; foliage large, soft; very vigorous, upright growth; RULED EXTINCT 4/85; Totty

White Pearl, Min, lp; [Minnie Pearl sport]; Lee, 1993

White Pearl in Red Dragon's Mouth, Ch, mr

White Penthouse, Gr, 1995; Fineschi, G.; (Cavriglia)

'WHITE PET', Pol, w, 1879; (Little White Pet); flowers small blooms in large clusters, dbl.; very dwarf growth; (14); Henderson, P.

'WHITE PET, CLIMBING', Cl Pol, w, 1894; Corboeuf

'WHITE PILLAR', LCl, w, 1958; very dbl.; pillar (8 ft) growth; [William F. Dreer X ?]; Hay; Marsh's Nursery

'WHITE PINOCCHIO', F, w, 1950; bud ovoid; dbl., 50 petals, 2.5 in., globular, moderate fragrance; vigorous growth; [Mrs R.M. Finch X Pinocchio]; Boerner; J&P

'WHITE PRINCE', HT, w, 1961; bud pointed; flowers creamy white, nearly, dbl., 50–80 petals, 5–6 in., globular, slight fragrance; foliage leathery, glossy; vigorous, upright growth; [(Blanche Mallerin X Peace) X (Peace X Frau Karl Druschki)]; Von Abrams; Peterson & Dering

White Provence, *see* **'UNIQUE BLANCHE'**

'WHITE QUEEN', HT, w, 1958; flowers white, center creamy, dbl., 30 petals, 5 in., cupped, moderate fragrance; foliage leathery; vigorous, upright growth; [Starlite seedling X Glacier seedling]; Boerner; J&P

'WHITE QUEEN ELIZABETH', F, w, 1965; (Blanc Queen Elizabeth); flowers pure white, borne in clusters, 4 in., moderate fragrance; foliage light green, leathery; vigorous growth; [Queen Elizabeth sport]; Banner; North Hill Nursery

White Radiance, *see* **'MARY NISH'**

'WHITE RADOX BOUQUET', S, w, 1988; bud small, rounded; flowers white with pink tinge in center, large, borne in sprays of 5, very dbl., 50 petals, moderate fragrance; prickles hooked, small, beige; foliage dark green, leathery; tall, shrub to semi-climber growth; [Radox Bouquet sport]; Melville Nurseries Pty., Ltd., 1989

White Rain, *see* 'MORCAS'

White Rambler, *see* **'THALIA'**

White Rock, *see* BARwrock

White Rosamini, Min, w

White Rose of Finland, *see* **'POLSTJÄRNAN'**

White Rose of York, A, w; (Bonnie Prince Charlie's Rose, Jacobite Rose, La Rose de York, R. procera, R. usitatissima **R. X ALBA**); dbl., borne usually several together, summer bloom, moderate fragrance; ovoid (see endpapers), scarlet fruit; 6 ft growth; (42); Probably R. corymbifera X R. gallica

'WHITE SATIN', HT, w, 1965; bud urn-shaped; flowers white, center greenish yellow, large, dbl., moderate fragrance; foliage light gray-green, leathery; vigorous, tall growth; [Mount Shasta X White Butterfly]; Swim & Weeks; Carlton Rose Nurseries

'WHITE SEDUCTION', Pol, w; [Seduction sport]

'WHITE SHEEN', F, w, 1959; bud ovoid; dbl., 25–30 petals, 2 in., borne in clusters, intense fragrance; foliage dark, glossy; upright growth; [Ma Perkins seedling X Demure seedling]; Boerner; J&P

White Simplicity®, *see* JACsnow

White Sparrieshoop, *see* **'WEISSE AUS SPARRIESHOOP'**

White Spire, *see* KORechtem

White Sporting Koster, Pol, lp

'WHITE SPRAY', F, w, 1968; flowers pure white, well-formed, small blooms in clusters, moderate fragrance; [Seedling X Iceberg]; LeGrice

White Spray®, *see* LENpaya

'WHITE STAR', HT, w, 1920; flowers ivory-white, base shaded lemon-yellow, intense fragrance; [Ophelia sport]; Morse

White Star of Finland, *see* **'POLSTJÄRNAN'**

White Success, *see* 'JELPIROFOR'

White Sunsation, S, lp; Kordes, 1996

White Sunshine, *see* 'JONWHITE'

White Surprise®, S, w; Lens, 1987

'WHITE SWAN', HT, w, 1951; (Cygne Blanc); bud ovoid; flowers pure white, dbl., 30 petals, 4.5–5 in., exhibition form, moderate fragrance; foliage glossy, dark; vigorous, upright growth; [Kaiserin Auguste Viktoria seedling X White seedling]; Verschuren-Pechtold; J&P

'WHITE SWEETHEART', F, w, 1941; flowers white flushed blush-pink when half open, small, dbl., exhibition form, slight fragrance; foliage leathery; short, strong stems; vigorous, upright, bushy growth; [Rosenelfe sport]; J&P

White Symphonie, HT, w; 59 petals, 4 in.; foliage matt; strong growth; RvS-Melle, 1991

'WHITE TAUSENDSCHÖN', HMult, w, 1913; flowers white, sometimes flaked pink; [Tausendschön sport]; Paul, W.

White Treasure, *see* 'HADTREASURE'

White Valerie, HT, lp

White Wedding, LCl, lp; Scarman, 1995

White Wine, *see* 'RESINC'

'WHITE WINGS', HT, w, 1947; bud long, pointed; anthers chocolate-colored, blooms in large clusters, single, 5 petals, 3.5 in., moderate fragrance; foliage leathery, dark; vigorous, upright, bushy growth; [Dainty Bess X Unnamed seedling]; Krebs; H&S

White Wings, HT, lp, 1979; flowers pure white with crimson stamens, single, intense fragrance; 2.5 ft growth; Poulsen

Whitegold, *see* 'COCQUIRIAM'

Whiteout, *see* 'MACWHITOUT'

Whitley Bay, *see* 'HORHARRYPLUS'

Whitney, *see* 'BRINEY'

WHItsun, MinFl, lp, 1995; (**Pearl Anniversary**); Chessum, Paul

Whitsuntide Rose, *see* **R. CINNAMOMEA PLENA**

Whoopi™, *see* 'SAVAWHOOP'

Why Not, *see* 'MORWHY'

'WHYTEWOLD', S, w, 1961; dbl., cupped, non-recurrent, height 3 ft; Skinner

'WICHMOSS', Cl M, lp, 1911; bud long, pointed, mossed; flowers pale blush-pink fading to white, blooms in clusters of 6–15, semi-dbl., 2 in., slight fragrance; foliage dark, leathery; vigorous, climbing growth; [R. wichurana X Salet]; Barbier

Wickwar, S, lp; Steadman, 1960

'WIENER BLUT', F, or, 1961; bud globular; flowers salmon-red, borne in clusters, dbl., 30 petals, intense fragrance; vigorous, upright growth; Horstmann

Wiener Charme®, *see* 'KORSCHAPRAT'

Wiener Donaupark, F, lp

'WIENER WALZER', F, mr, 1965; (Pur Sang); bud pointed; flowers velvety bright red, large, dbl., 25–30 petals; foliage glossy; very vigorous, upright growth; Tantau, Math.

'WIENERWALD', HT, op, 1974; (Vienna Woods); bud long, pointed; flowers pink to light orange, large, dbl., moderate fragrance; foliage large, dark, leathery; vigorous, upright, bushy growth; [Colour Wonder X Seedling]; Kordes

Wieteke van Dordt, *see* **'POTIFAR'**

Wife of Bath, *see* 'AUSWIFE'

Wiggy, *see* 'KENDANQU'

'WIJHE', F, or, 1961; dbl., borne in clusters, foliage dark; moderate growth; [Unnamed seedling X Alpine Glow]; Buisman, G. A. H.

Wil Pink Supreme, S, op

'WILACE', HT, dr, 1992; (**Peggy Rockefeller**™); flowers bright crimson to cherry red with dark smoky red on edge of petals, dbl., 40 petals, moderate, spicy fragrance; few prickles; foliage large, dark green, semi-glossy; upright, bushy, medium growth; [Queen Elizabeth X Swarthmore]; Williams, J. Benjamin, 1991; New York Botanical Garden, 1991

'WILALEX', S, rb, 1988; (**Alexandria Rose**, City of Alexandria, The Alexandria Rose); flowers ivory to white with light red washing on petal edge, dbl., 26–40 petals; foliage large, dark green, glossy, disease resistant; upright, bushy, vigorous, strong, hardy growth; [(Queen Elizabeth X Kordes' Perfecta) X Mount Shasta]; Williams, J. Benjamin

'WILALNO', HT, dr, 1998; (**Robert Aliano**); flowers deep velvet red, dull reverse, dbl., 15–25 petals, 4.5–5 in., exhibition form, borne mostly singly, moderate fragrance; few prickles; foliage large, dark green, semi-glossy; strong, upright, tall, 4–5 ft growth; [Chrysler Imperial X Mister Lincoln]; Williams, J. Benjamin; J. Benjamin Williams & Associates, 1999

'WILANGIL', HT, pb, 1992; (**Reynolda House**); flowers light pink with ivory and coral blend, blooms borne mostly singly, dbl., 26–40 petals, 1.5–2.75 in., intense fragrance; few prickles; foliage medium, dark green, semi-glossy, disease-resistant; upright, bushy growth; medium (3–4 ft winterhardy; [Royal Highness X Command Performance]; Williams, J. Benjamin

'WILBENTUR', Min, w, 1981; (**Pixie Delight**™); dbl., slight fragrance; foliage small, dark, semi-glossy; upright growth; PP005270; [Ma Perkins X Easter Morning]; Williams, J. Benjamin; C-P

'WILBERFORCE', C, dr; flowers dark crimson, large, dbl.; [Probably R. centifolia X R. gallica]; About 1840

'WILBLANK', MinFl, w, 1982; (**Stardance**™); flowers deep yellow stamens, mini-flora, dbl., 35 petals, slight fragrance; foliage small, medium green, semi-glossy; compact growth; PP005222; [Ma Perkins X (Charlie McCarthy X Easter Morning)]; Williams, J. Benjamin; C-P

'WILBRID', HT, w, 2000; (**All-American Bride**); Williams, J. Benjamin

'WILCAMP', Min, or, 1990; (**Campfire**); bud pointed; flowers fiery orange neon red, reverse same, ages deeper, medium, dbl., 26 petals, exhibition form, borne singly, moderate, damask fragrance; foliage medium, dark green, semi-glossy; upright, bushy growth; [Marina X Starina]; Williams, J. Benjamin; White Rose Nurseries, Ltd., 1990

'WILCGIR', HT, ob, 1999; (Country Girl, **Emsie Girl**); bud pointed; flowers creamy white with bright orange edge, dbl., slight fragrance; [Garden Party X American Heritage]; Williams, J. Benjamin

'WILCHER', Min, mr, 1997; (**Cherry Pastel**™); flowers double, medium, bright cherry red, dbl., 15–25 petals, borne mostly singly, moderate fragrance; [Red Sunblaze X Pink Sweetheart]; Williams, J. Benjamin; Paramount Roses, 1997

'WILCREST', F, dy, 1992; (**Cedar Crest College**); flowers deep golden yellow, blooms borne in small clusters, semi-dbl., 6–14 petals, 1.5–2.75 in., slight fragrance; few prickles; foliage medium, dark green, glossy; medium (3–4 ft), upright, bushy growth; [Ivory Fashion X Sunsprite]; Williams, J. Benjamin, 1993

'WILCROWN', S, op, 1992; (**Copper Crown**); flowers orange pink, blooms, single, 5 petals, 3–3.5 in., slight fragrance; foliage medium, dark green, semi-glossy; upright (4 × 4), bushy growth; [Westerland X Orange Velvet]; Williams, J. Benjamin; Hortico Roses, 1992

Wild Amazone, HT, dp

Wild at Heart, *see* 'KORADES'

Wild Berry Breeze™, *see* 'JACRULAV'

'WILD CHERRY', HT, mr, 1984; flowers large, dbl., moderate fragrance; foliage large, dark, glossy; bushy growth; [King of Hearts X Granada]; Bridges, Dennis A.; Bridges Roses, 1985

'WILD CHILD', S, lp, 1982; bud small, pointed; single, 5 petals, borne 3–15 per cluster, repeat bloom, moderate, tea fragrance; curved, thin prickles; foliage small, light green, 9 leaflet, slightly rugose; vigorous, erect, some arching growth; [R. rugosa rubra X Dortmund seedling]; Hall, William W.

Wild Dancer™, *see* 'JACINA'

'WILD FLAME', HT, or, 1973; bud ovoid; flowers medium, dbl., moderate fragrance; foliage leathery; vigorous, upright growth; [Granada X South Seas]; Meyer, C.; Ball Seed Co.

'WILD GINGER', Gr, ob, 1976; bud ovoid, pointed; dbl., 33 petals, 4–4.5 in., moderate, fruity fragrance; foliage dark, leathery; upright, bushy growth; [(Queen Elizabeth X Ruth Hewitt) X Lady Elgin]; Buck, Dr Griffith J.; Iowa State University

'WILD HONEY', HT, ab, 1977; bud long, pointed; flowers coral and peach, dbl., 45 petals, 3–5 in., intense, spicy fragrance; foliage leathery; tall, vigorous growth; Weeks

Wild One, *see* 'LYOWE'

Wild Plum™, *see* 'JACWIG'

Wild Spice, *see* 'JACRUWHI'

'WILD SPIRIT', S, mr, 1985; flowers bright medium red, small, single, 5 petals, non-recurrent, moderate fragrance; foliage small, dark, matt; vigorous, upright, branched growth; [Alika X R. moyesii]; James, John

Wild Thing, *see* 'TINWILD'

Wild West, *see* 'TANRAVENS'

'WILDAC', Gr, dr, 1986; (**Dr Adam Christman**™); flowers dark crimson red to scarlet, large blooms borne in large sprays, dbl., 26–40 petals, slight fragrance; foliage large, dark green, semi-glossy; upright growth; [Queen Elizabeth X Chrysler Imperial]; Williams, J. Benjamin

'WILDAK', MinFl, ob, 1982; (**Amber Flash**™); flowers medium, 15–25 petals, moderate fragrance; foliage medium, dark, semi-glossy; PP005271; [Zorina X Starina]; Williams, J. Benjamin; C-P

'WILDAN', HT, dr, 1989; (**The Daniel**™); bud ovoid; flowers dark velvety red, reverse burgundy, medium, borne usually singly, dbl., 36 petals, exhibition form, no fragrance; fruit not observed; few prickles; foliage large, dark green, semi-glossy; upright, medium growth; [Chrysler Imperial X Queen Elizabeth]; Williams, J. Benjamin; The Peninsula Nursery, 1990

'WILDANCE', Min, lp, 1989; (**Tiny Dancer**™); flowers bright, light coral pink, small blooms, dbl., 34 petals, exhibition form, borne usually single, slight, spicy fragrance; very few prickles; foliage small, dark green, dwarf; upright, low growth; [(Carla X Sonia) X (Circus X Ma Perkins)]; Williams, J. Benjamin; Young's American Rose Nursery, 1990

'WILDAWN', S, dp, 1995; (**Dawn Creeper**™); flowers deep pink, semi-dbl., 6–14 petals, 1.5 in., borne in small clusters, slight fragrance; some prickles; foliage small, dark green, glossy; low (10–12 in.), spreading growth; [The Fairy X Sea Foam (Self) F1]; Williams, J. Benjamin; J. Benjamin Williams, 1996

Wildberry Breeze, *see* 'JACRULAV'

'WILDEB', HT, pb, 1991; (**Deborah Moncrief**™); flowers pink with ivory blend, large blooms borne mostly singly, dbl., 26–40 petals, intense fragrance; few prickles; foliage large, dark green, matt; medium (12–18 in.), bushy, spreading growth; [Carla X Sonia]; Williams, J. Benjamin, 1988

'WILDELS', Gr, pb, 1997; (**Delany Sisters**™); flowers ruffled petals, large, dbl., 15–25 petals, borne mostly single and in large clusters, intense fragrance; foliage large, dark green, semi-glossy; upright, bushy, tall (4.5–5 ft) growth; [Love X Handel]; Williams, J. Benjamin

Wildenfels Gelb, HFt, w; Dechan, 1929

Wildenfels Rosa, HFt, mp; Dechan, 1928

Wildest Dreams, *see* 'TALWIL'

Wildfang®, S, mp; Noack, Werner, 1989

'WILDFEUER', S, mr, 1953; flowers fiery red, semi-dbl., 4.5 in.; bushy growth; Kordes

'WILDFIRE', F, mr, 1955; bud ovoid; flowers bright red, blooms in clusters, 8–10 petals, 3 in., slight fragrance; foliage leathery; vigorous, bushy, compact growth; [World's Fair X Pinocchio]; Swim, H.C.; Armstrong Nursery

Wildflower, *see* 'AUSWING'

Wildlife®, S, pb; Dickson, Patrick, 1989

'WILDREM', HT, dp, 1997; (**Miss All-American Dream**™); flowers large, deep pink, full, very dbl., 26–40 petals, 5.5–6 in., borne mostly singly, intense fragrance; prickles moderate; foliage large, dark green, semi-glossy; upright, bushy, tall growth; ARS Patron Rose, 1996; [Miss All-American Beauty X Bride's Dream]; Williams, J. Benjamin; J. Benjamin Williams & Associates, 1996

'WILDWOOD', HT, op, 1937; bud long, pointed; flowers gold, salmon and bronze, semi-dbl., intense fragrance; foliage glossy, wrinkled; vigorous growth; H&S

'WILF TAYLOR', HT, ab, 1981; flowers large, dbl., 35 petals, exhibition form, slight fragrance; foliage large, dark, matt; upright growth; [Gavotte X Red Lion]; Bracegirdle, A.J.

'WILFOLK', S, ob, 1998; (**Boulie's Dream**™, Musketeer); flowers orange red inside, yellow reverse, very dbl., 26–40 petals, 4.5–5 in., borne singly and in large clusters, intense fragrance; few prickles; foliage large, dark green, semi-glossy; strong, vigorous, tall, upright, climbing, arched growth; [Folklore X self]; Williams, J. Benjamin; J. Benjamin Williams & Associates, 1999

'WILFORD', HT, dr, 1989; (**Fordham Rose**™); bud pointed; flowers deep maroon-red with deep, black, velvety tones, urn-shaped, dbl., moderate fragrance; fruit not observed; prickles few, ovoid, curved down, medium, tan; foliage large, dark green, semi-glossy, thick, disease-resistant; upright, medium growth; [Chrysler Imperial X Josephine Bruce]; Williams, J. Benjamin; Fordham University, 1990

'WILFRED NORRIS', HT, mp, 1972; flowers medium salmon-pink, large, dbl., 23 petals; foliage matt; [Tropicana X Elizabeth of Glamis]; Harkness

'WILFRED PICKLES', HT, op, 1939; flowers peach shaded gold, well-formed, dbl., 26 petals, 6 in., moderate fragrance; foliage glossy, dark; stems red; very vigorous growth; [Mrs Charles Lamplough X Edith Mary Mee]; Mee; Fryer's Nursery, Ltd.

Wilfrid H. Perron, *see* 'LAVSHRIMP'

'WILGMAR', S, mr, 1988; (**Maren**); flowers deep scarlet red with velvet overtones, large, dbl., 26–40 petals, slight fragrance;

foliage large, dark green, semi-glossy, leathery, disease resistant; upright, spreading, vigorous, free-standing growth; [Red Fountain X Red Fountain]; Gimpel, W.F., 1989; J.B. Williams & Associates, 1989

'WILGOGH', S, rb, 1996; (**Van Gogh**™); flowers red and white striped, reverse pale whitish pink with stripes, dbl., 15–25 petals, 3.5–4 in., exhibition form, borne in large clusters, moderate fragrance; moderate prickles; foliage large, dark green, semi-glossy; bushy, medium (3–31/2 ft) growth; [Spinning Wheel X Handel]; Williams, J. Benjamin; J.B. Williams & Associates, 1997

'WILGOLD', Min, dy, 1997; (**Golden Earing**™); flowers double, large, mini flora, dbl., 15–25 petals, borne singly and in clusters, moderate fragrance; [Hershey Yellow X Copper Kittel]; Williams, J. Benjamin

'WILGOSH', Min, my, 1984; (**Gold Sweetheart**™); flowers deep yellow, small, dbl., 35 petals, slight fragrance; foliage small, medium green, semi-glossy; upright, bushy growth; [Sunsprite X Rise 'n' Shine]; Williams, J. Benjamin; J.B. Williams & Associates

Wilhelm, *see* **'SKYROCKET'**

Wilhelm Breder, *see* **'GLOWING SUNSET'**

'WILHELM HANSMANN', HKor, dr, 1955; (William Hansmann); bud ovoid; flowers deep crimson, large blooms in clusters, dbl., exhibition form, recurrent bloom, slight fragrance; foliage dark, leathery; very vigorous growth; Kordes

'WILHELM KAUTH', HT, ly, 1930; flowers large, very dbl., moderate fragrance; Kauth; (Sangerhausen)

'WILHELM KORDES', HT, op, 1922; bud long, pointed; flowers salmon, copper and golden blend, large, dbl., exhibition form, moderate fragrance; foliage bronze, leathery, glossy; bushy growth; [Gorgeous X Adolf Koschel]; Kordes, H.

'WILHELM KORDES, CLIMBING', Cl HT, op, 1927; Wood & Ingram

'WILHELM MARX', S, mr, 1939; flowers medium, dbl.; Vogel, M.; (Sangerhausen)

Wilhelm May, HT, dr, 1989; flowers velvety dark red, dbl., moderate fragrance; Michler, K. H.; (Sangerhausen)

'WILHELM TEETZMANN', F, mr, 1943; (Advance Guard); flowers intense crimson, open, borne in clusters, semi-dbl., 3.5 in.; very vigorous, upright, bushy growth; (28); [Holstein X Crimson Glory]; Kordes

Wilhelm Tell®, F, mr; Meilland, 1989

'WILHERB', HT, op, 1992; (**Lady Harriet**); flowers coral and peach blend, deep orange washings on petals, long, dbl., 26–40 petals, 3–3.5 in., intense fragrance; few prickles; foliage large, dark green, semi-glossy; tall (4–5 ft), upright, bushy growth; [Carla X Sonia]; Williams, J. Benjamin, 1993

'WILHIRE COUNTRY', F, ob, 1981; (Willhire Country); flowers flat-topped blooms borne 6–8 per cluster, 25 petals, intense fragrance; broad, hooked, brown prickles; foliage glossy, round, dark; tall, upright growth; [Elizabeth of Glamis X Arthur Bell]; Beales, Peter; Beales Roses, 1979

'WILHKPK', HT, ob, 1988; (**Rittenhouse**™); flowers fiery orange to copper blended, large, dbl., 26–40 petals, moderate fragrance; foliage large, dark green, glossy, disease resistant; upright, vigorous, mass-blooming, winter hardy growth; [Queen Elizabeth X Zorina]; Williams, J. Benjamin, 1989

'WILIVORY', S, w, 1995; (**Ivory Carpet**™); flowers pure white, full, small, very dbl., 26–40 petals, .5–1 in., borne in small clusters, slight fragrance; some prickles; foliage small, dark green, glossy; low, spreading, 10–12 in. growth; [(Sea Foam X The Fairy) X Star Dance]; Williams, J. Benjamin; J. Benjamin Williams & Associates, 1996

'WILJAME', HT, mr, 1998; (Burgundy Queen, **James Biddle**); flowers burgundy red, reverse medium red, light silver edges, very dbl., 26–40 petals, 4.5–5 in., exhibition form, borne mostly singly, intense fragrance; few prickles; [Pink Peace X Queen Elizabeth]; Williams, J. Benjamin; J. Benjamin Williams & Associates, 1998

'WILKBSP', S, yb, 1985; (**Brown County Splendor**™); flowers ivory to yellow blended with peach to orange-red, large, dbl., 35 petals, intense fragrance; foliage large, dark, semi-glossy; upright, spreading growth; [Paul's Lemon Pillar X (Garden Party X Command Performance)]; Williams, J. Benjamin; Krider Nursery

'WILKTWO', S, w, 1991; (**Joan Longer**™); bud pointed; flowers blush pink opening to ivory with a hint of coral pink in center, semi-dbl., cupped, moderate, damask fragrance; foliage large, dark green, semi-glossy; upright, bushy, tall growth; [Queen Elizabeth X Ivory Fashion, Climbing]; Williams, J. Benjamin, 1971; The Scott Arboretum of Swarthmore College, 1991

'WILL ALDERMAN', HRg, mp, 1954; flowers clear rose-pink, well-shaped, large, dbl., repeat bloom, intense fragrance; erect (4 ft), bushy growth; [(R. rugosa X R. acicularis) X HT]; Skinner

Will Deb, HT, op

'WILL ROGERS', HT, dr, 1936; flowers velvety maroon-crimson, base almost black, burning in sun, dbl., 65 petals, 3 in., intense, damask fragrance; foliage leathery, light; vigorous growth; [Seedling X (Hadley X Crimson Glory)]; Howard, F.H.; H&S

'WILL ROGERS, CLIMBING', Cl HT, dr, 1940; H&S

'WILL SCARLET', HMsk, mr, 1948; flowers scarlet; [Skyrocket sport]; Hilling; Wayside Gardens Co., 1956

'WILLASH', Min, m, 1984; (**Lavender Sweetheart**™); flowers deep blue lavender, small, borne singly, dbl., 20 petals, exhibition form, intense fragrance; foliage small, dark, semi-glossy; bushy growth; [Double Feature X Unnamed Min seedling]; Williams, J. Benjamin; J.B. Williams & Associates

'WILLAVQ', Gr, m, 1995; (**William Carroll**™); flowers mauve, full, very dbl., 26–40 petals, 1.5–2.75 in., borne mostly single and in small clusters, slight fragrance; some prickles; foliage large, dark green, semi-glossy; tall (4–5 ft), upright growth; [Queen Elizabeth X Angel Face]; Williams, J. Benjamin

'WILLEA', HT, ab, 1992; (**Marylea Johnson Richards**); flowers light pink with peach tones, blooms borne mostly singly and, dbl., 26–40 petals, 1.5–2.75 in., intense fragrance; few prickles; foliage medium, dark green, semi-glossy; medium (3–4 ft), upright, bushy growth; [Royal Highness X Command Performance]; Williams, J. Benjamin

Willhire Country, *see* **'WILHIRE COUNTRY'**

Willi Maass, *see* **'RED RIPPLES'**

'WILLIAM A. BILNEY', HT, pb, 1927; flowers pink with light orange, large, very dbl., moderate fragrance; Easlea; (Sangerhausen)

'WILLIAM ALLEN RICHARDSON', N, yb, 1878; bud pointed; flowers orange-yellow, usually white at petal edges, medium to large, dbl., early bloom, repeating, very vigorous, climbing (12 ft) growth; (14); [Reve d'Or seedling]; Ducher, Vve.

William and Mary, S, rb; Beales, Peter, 1988

'WILLIAM BAFFIN', HKor, dp, 1983; bud ovoid; flowers deep pink, free blooming, flat, dbl., 20 petals, 2.5 in., blooms in clusters of 30, slight fragrance; foliage small, medium green, glossy; vigorous, climbing growth; [R. kordesii seedling]; Svedja, Felicitas; Agriculture Canada

William Booth, S, lp; single, 5 petals, slight; L'Assomption, 1999

'WILLIAM BOWYER', HT, mr, 1924; bud long, pointed; flowers velvety red, intense fragrance; [Hadley X Hoosier Beauty]; Chaplin Bros.

'WILLIAM C. EGAN', LCl, lp, 1900; flowers flesh-pink, large, borne in large clusters, very dbl., non-recurrent, moderate fragrance; foliage bright, glossy; long stems; vigorous, semi-climbing growth; [R. wichurana X Général Jacqueminot]; Dawson; Hoopes, Bro. & Thomas

William Carroll™, *see* 'WILLAVQ'

'WILLIAM CARTER', HT, dp, 1957; flowers deep pink, moderate fragrance; very vigorous growth; [Red Ensign sport]; Carter

'WILLIAM COBBETT', F, dp, 1968; flowers claret-rose, borne in trusses, semi-dbl., slight fragrance; vigorous growth; [Dorothy Wheatcroft sport]; Seale Rose Gardens

William David, *see* 'PEAROYAL'

'WILLIAM E. NICKERSON', HT, or, 1928; flowers glowing orange-cerise, large, dbl., moderate fragrance; foliage light; vigorous growth; [Priscilla X William F. Dreer]; Easlea

'WILLIAM F. DREER', HT, lp, 1920; flowers shell-pink, base yellow, sometimes suffusing the entire flower, dbl., exhibition form, moderate fragrance; [Mme Segond Weber X Lyon Rose]; H&S

'WILLIAM F. EKAS', HT, op, 1935; bud cupped; flowers salmon-orange, suffused pink, dbl., 45–50 petals, 4 in., moderate fragrance; [Souvenir sport]; Cremer

'WILLIAM FRANCIS BENNETT', HT, mr, 1880; bud long; flowers crimson, large, dbl., moderate fragrance; Bennett

'WILLIAM GODFREY', S, lp, 1954; bud white; flowers pale pink, well shaped, large, very dbl., intermittent bloom all season, moderately vigorous, erect growth; [Altalaris X Hybrid Perpetual]; Skinner

William Grant, HGal, dp

William Griffith, HP, pb

'WILLIAM GROW', , dr, 1859; flowers velvety violet/red, medium, dbl.; Laffay, M.; (Sangerhausen)

William Hansmann, *see* **'WILHELM HANSMANN'**

'WILLIAM HARVEY', HT, mr, 1948; flowers scarlet, dbl., 4–6 in., exhibition form, intense fragrance; vigorous, bushy growth; [Crimson Glory X Southport]; Norman; Harkness

'WILLIAM III', HSpn, m; flowers magenta-crimson, changing to rich plum, fading to dark lilac, semi-dbl., moderate, spicy fragrance; foliage tiny, dark

'WILLIAM IV', HSpn, lp; (William the Fourth); very dbl., non-recurrent; very dwarf growth; (England), before 1906

'WILLIAM IV', HSpn, 1906; (Cavriglia)

'WILLIAM JESSE', HCh, mr; flowers red suffused violet, large, dbl.; Laffay, M., 1850

William Leech, *see* 'HORTROPIC'

'WILLIAM LOBB', M, m, 1855; (Old Velvet Moss); flowers dark crimson-purple, reverse lilac-pink, fading to grayish-lilac, semi-dbl.; extremely prickly; vigorous growth; Laffay, M.

William McGown, S, lp; Peden, G.H., 1996

'WILLIAM MOORE', HT, mp, 1935; bud long, pointed; flowers soft, even-toned pink, becoming deeper, large, dbl., cupped, intense fragrance; foliage soft, light; bushy growth; McGredy

William Morris, *see* 'AUSWILL'

'WILLIAM NOTTING', HT, rb, 1904; flowers salmon-red, center brighter, dbl., moderate fragrance; [Mme Abel Chatenay X Antoine Rivoire]; Soupert & Notting

'WILLIAM ORR', HT, mr, 1930; bud long, pointed; flowers crimson-scarlet, large, dbl., exhibition form, intense fragrance; foliage light, glossy, leathery; vigorous growth; McGredy

William Quarrier, *see* 'COCLAGE'

'WILLIAM R. SMITH', T, pb, 1908; (Blush Maman Cochet, Charles Dingee, Jeannette Heller, Maiden's Blush, President Smith, President Wm.R. Smith); bud pointed; flowers center pale pink, outer petals creamy flesh, base citron-yellow, dbl.; foliage rich green, leathery; vigorous growth; [Maman Cochet X Mme Hoste]; Bagg; E.G. Hill Co.; P. Henderson

William Shakespeare®, *see* 'AUSROYAL'

'WILLIAM SHEAN', HT, mp, 1906; bud long, tapering; flowers clear rose-pink, veined darker, very large, dbl.; GM, NRS, 1906; Dickson, A.

'WILLIAM SILVA', HT, 1951; few prickles at base; vigorous growth; [Étoile de Hollande X Radiance]; Silva

William Stubbs, *see* 'BOSNEECH'

William the Fourth, *see* **'WILLIAM IV'**

'WILLIAM THOMSON', Pol, pb, 1921; flowers salmon-carmine and bright rose, small, borne in clusters; compact, bushy growth; [Maman Turbat X Ellen Poulsen]; Leenders, M.

William Walker, *see* 'DEBRARO'

'WILLIAM WRIGHT WALCOTT', HT, lp, 1921; flowers outer petals darker, dbl.; [Richmond sport X Ophelia]; McGorum

William-Booth, *see* 'AC WILLIAM-BOOTH'

'WILLIAM'S DOUBLE YELLOW', HSpn, my; (Double Yellow Scots Rose); flowers pine-yellow with green carpels, small, semi-dbl., 2 in., very free blooming in spring, intense, sweet fragrance; foliage tiny, dark; growth free but spindly; Williams, John, 1828

William's Double Yellow, HSpn, my

'WILLIAMSBURG', HT, lp, 1965; bud long, pointed; flowers rose-pink, medium, dbl., cupped, moderate fragrance; foliage dark, glossy; vigorous, upright growth; [Contrast X Queen Elizabeth]; Howard, A.P.; Great Western Rose Co.

Williamsburg Days, SS, lp; Williams, J. Benjamin, 1999

'WILLIE MAE', Min, mr, 1966; flowers small, dbl., slight fragrance; foliage dark, glossy, leathery; vigorous, bushy, dwarf growth; [(R. wichurana X Carolyn Dean) X Little Buckaroo]; Moore, Ralph S.; Mini-Roses

'WILLIE WINKIE', Min, lp, 1955; flowers light rose-pink, micro-mini, small, dbl., globular, [Katharina Ziemet X Tom Thumb]; deVink; T. Robinson, Ltd.

Willits Crescendo, LCl, lp

'WILLIWAW', HT, m, 1978; bud pointed; flowers red-purple, yellow at base, dbl., 60 petals, cupped, moderate fragrance; hooked prickles; foliage glossy, dark; upright growth; [Silver Lining X Duke of Windsor]; Melle

Will-o'-the-Wisp, *see* CLEmist

'WILLOWMERE', HT, op, 1913; bud long, pointed, coral-red; flowers rich shrimp-pink, center yellow, dbl., cupped; foliage light; spreading growth; [Unnamed seedling X Lyon Rose]; Pernet-Ducher

'WILLOWMERE, CLIMBING', Cl HT, op, 1924; Mermet

'WILLUCE', HT, dr, 1998; (**Luce di Todi**™); flowers dark velvet red, full, very dbl., 26–41 petals, 5 in.–5.5 in., exhibition form, borne singly, intense fragrance; few prickles; foliage large, dark green, semi-glossy; tall, upright, 4–5 ft growth; [Mister Lincoln X Miss All-American Beauty]; Williams, J. Benjamin; J. Benjamin Williams & Associates, 2000

'WILLY CHAPEL', HT, yb, 1930; flowers coppery yellow, shaded salmon and shrimp-pink; vigorous growth; Delhaye

'WILLY DEN OUDEN', Pol, or, 1938; flowers intense orange, rosette form, small, borne on trusses; dwarf, compact growth; Den Ouden

'WILMAG', Min, yb, 1989; (**Little Magician**™); bud ovoid; flowers light yellow with orange-red blend, loose, small, borne usually singly, dbl., 32 petals, exhibition form, slight, damask fragrance; fruit not observed; no prickles; foliage small, medium green, semi-glossy; low, winter-hardy growth; [Circus X Magic Carrousel]; Williams, J. Benjamin; White Rose Nurseries, Ltd., 1990

'WILMNMG', HT, dr, 1984; (**Midnight Magic**™); flowers large, dbl., 35 petals, exhibition form, slight fragrance; foliage large, dark, semi-glossy; [(Chrysler Imperial X Mister Lincoln) X (Christian Dior X Josephine Bruce)]; Williams, J. Benjamin, 1986

'WILMTHO', Gr, w, 1988; (**Mary Lyon**™); flowers pure chalk white, medium, dbl., 26–40 petals, intense fragrance; foliage large, dark green, semi-glossy, thick; disease resistant; upright, hardy growth; [(Mount Shasta X Sonia) X (White Masterpiece X Ivory Fashion)]; Williams, J. Benjamin

'WILORPAT', Min, ob, 1997; (**Orange Pastel**™); flowers double, medium, bright orange-red, reverse dull, dbl.,

15–25 petals, 2.5–3 in., borne in small clusters, slight fragrance; few prickles; foliage medium, dark green, semi-glossy; compact, low (12–18 in.) growth; [Red Sunblaze X Orange Sunblaze]; Williams, J. Benjamin; Paramount Roses, 1997

'WILPADA', MinFl, rb, 1984; (**Patio Dance**™); flowers medium red, white reverse, small, dbl., 20 petals, exhibition form, slight fragrance; foliage medium, dark, glossy; strong, low, bushy growth; [Winifred Coulter X White Gem]; Williams, J. Benjamin

'WILPAGO', MinFl, my, 1984; (**Patio Gold**™); flowers small, dbl., slight fragrance; foliage medium, dark, semi-glossy; low, compact, bushy growth; [Patio Patty X Rise 'n' Shine]; Williams, J. Benjamin

'WILPASN', MinFl, w, 1984; (**Patio Snow**™); flowers small, dbl., 35 petals, moderate fragrance; foliage small, medium green, semi-glossy; groundcover; very compact, spreading growth; [Sea Foam X Charlie McCarthy]; Williams, J. Benjamin

'WILPAVL', S, mp, 1988; (**Pink Avalanche**); flowers coral pink to medium red, small, borne in sprays of 20–30, dbl., 15–25 petals, slight fragrance; foliage medium, medium green, glossy; spreading, compact growth; [Sea Foam X The Fairy]; Williams, J. Benjamin, 1989

'WILPINK', F, mp, 1986; (Hot Pink Pastel, **Pink Sweetheart**™); flowers medium coral pink, well-formed, small blooms, dbl., 15–25 petals, 2–2.5 in., borne singly, moderate fragrance; prickles moderate; foliage medium, medium green, semi-glossy; upright, dwarf growth; [(Carla X Sonia) X (Circus X Ma Perkins)]; Williams, J. Benjamin

'WILPINK', Min, mp, 1997; (**Pink Sweetheart**™); flowers double, medium, dbl., 15–25 petals, borne mostly singly, moderate fragrance; foliage medium, medium green, semi-glossy; compact, low (12–18 in) growth; [(Pink Sweet-heart X Patio Patty) X ?]; Williams, J. Benjamin

'WILPLPD', Min, m, 1984; (**Plum Pudding**™); flowers blue-lavender washed red and purple, small, dbl., 35 petals, moderate fragrance; foliage small, dark, semi-glossy; upright, bushy growth; [Angel Face X Unnamed lavender seedling]; Williams, J. Benjamin

'WILPOP', S, lp, 1991; (**Pink Popcorn**™); flowers light pink to coral, blooms borne in large clusters in small, semi-dbl., 6–14 petals, 1.5 in., slight fragrance; few prickles; foliage small, medium green, matt, disease-resistant; low (12–18 in.), compact, bushy growth; [R. chinensis minima X (Sea Foam X The Fairy) seedling]; Williams, J. Benjamin, 1992

'WILPURP', S, m, 1991; (**Purple Popcorn**™); flowers bluish purple, blooms borne in small, grape-like clusters, semi-dbl., 6–14 petals, 1.5 in., slight fragrance; few prickles; foliage small, medium green, matt, disease resistant; low (12–18 in.), compact, bushy growth; [R. chinensis minima X (Sea Foam X The Fairy) seedling]; Williams, J. Benjamin, 1992

'WILRIB', F, mr, 1997; (Red Pastel, **Red Ribbon**™); flowers double, medium, dbl., 15–25 petals, borne in small clusters, moderate fragrance; foliage medium, medium green, semi-glossy; upright, compact, medium (14–18 in.) growth; [Red Sunblaze X Pink Sweetheart]; Williams, J. Benjamin

'WILRSS', S, mr, 1987; (**Pat's Delight**); flowers crimson to light scarlet, loosely, borne in sprays of 5–7, semi-dbl., 16 petals, cupped, slight fragrance; light green prickles, curving downwards; foliage large, dark green, glossy; upright, strong, good branching growth; [Queen Elizabeth X Chrysler Imperial]; Williams, J. Benjamin

'WILSAB', Min, or, 1992; (**Gabriela Sabatini**); flowers bright fire orange-red, sweetheart, dbl., 15–25 petals, 1.5 in., blooms borne mostly singly, slight fragrance; few prickles; foliage small, dark green, semi-glossy; low (14–20″), upright, compact, dwarf type growth; [Marina X Pink Sweetheart]; Williams, J. Benjamin

'WILSCARLET', S, dr, 1995; (**Scarlet Spreader**™); flowers dark scarlet red, small, semi-dbl., 6–14 petals, 1–1.5 in., borne in small clusters, slight fragrance; some prickles; foliage small, dark green, glossy; dwarf, spreading, low (10–12 in. growth; [Seedling X Red Fountain]; Williams, J. Benjamin; J. Benjamin Williams & Associates, 1996

'WILSCSO', Gr, yb, 1988; (**Elizabeth Scholtz**™); flowers deep yellow with orange washing, reverse yellow with orange, dbl., 26–40 petals, exhibition form, moderate, damask fragrance; foliage large, plum-red to dark green, glossy; upright, bushy growth; [(Granada X Oregold) X (Arizona X Sunblest)]; Williams, J. Benjamin

'WILSEPT', HT, pb, 1991; (**September Eighteenth**); bud ovoid; flowers shell pink, reverse deep coral pink, aging light pink, bloom, dbl., exhibition form, intense, damask fragrance; foliage large, light green, semi-glossy; medium, bushy growth; [Carla X Queen Elizabeth]; Williams, J. Benjamin, 1992

'WILSFOL', HT, mr, 1988; (**Search for Life**™); flowers large, dbl., 26–40 petals, exhibition form, slight fragrance; foliage large, dark green, semi-glossy; vigorous, hardy, abundant growth; [Miss All-American Beauty X Mister Lincoln]; Williams, J. Benjamin

'WILSHIRE', HT, mp, 1991; bud pointed; flowers medium, dbl., 25 petals, exhibition form, moderate fragrance; foliage medium, medium green, glossy; upright, tall growth; Bronze, ARC TG, 1985; [Don Juan sport]; Woodard, Joe, 1983; Kimbrew Walter Roses, 1991

'WILSLAV', HT, rb, 1987; (**Slava**™); flowers scarlet red, reverse ivory white, large, borne singly, dbl., 34 petals, exhibition form, moderate, damask fragrance; rounded, medium, pumpkin-orange fruit; medium, few, yellow-green prickles, hooked downwards; foliage large, dark green, glossy, disease resistant; upright, bushy, tall growth; [Garden Party X Love]; Williams, J. Benjamin

'WILSMA', Min, mp, 1982; (**Astra**™); flowers pink, bright yellow stamens, small, single, 5 petals, slight fragrance; foliage small, medium green, semi-glossy; upright growth; [Pinafore X (Lilibet X Fairy Queen)]; Williams, J. Benjamin; C-P

'WILSNOD', S, w, 1986; (**Snowdrift**™); flowers small blooms in large sprays, dbl., 35 petals, repeat bloom, slight fragrance; foliage medium, dark, semi-glossy; low, spreading growth; [Sea Foam X The Fairy]; Williams, J. Benjamin

'WILSPRING', Min, w, 1990; (**Spring Time**™); bud pointed; flowers ivory with light red edge, small, borne usually singly, dbl., 26 petals, exhibition form, intense, damask fragrance; foliage small, dark green, semi-glossy; bushy, low growth; [Easter Morning (F2) X Toy Clown]; Williams, J. Benjamin; White Rose Nurseries, Ltd., 1990

'WILSTAR', LCl, lp, 1999; (**Scarlet Star**); dbl., 17–25 petals, 2.5–3 in., borne in large clusters, slight fragrance; many prickles; foliage large, medium green, semi-glossy; climbing, upright, bushy, tall (12–15 ft) growth; [Red Fountain X Mister LIncoln]; Williams, J. Benjamin; Conary-Pyle, 1999

'WILSUN', F, ob, 1991; (**Tequilla Sunset**™); bud ovoid; flowers brilliant orange and yellow blend, urn-shaped, blooms borne, dbl., moderate, fruity fragrance; foliage medium, medium green, semi-glossy; upright, bushy, medium growth; [Redgold X Tropicana]; Williams, J. Benjamin, 1976; Paramount Nursery, 1992

Wiltshire, *see* KORmuse

Wiltshire Pride, *see* 'SANWIFE'

'WILTWOS', S, w, 1993; (**Two Sisters**™); flowers light pink to white, full, very, dbl., 26–40 petals, 1.5–2.5 in., borne in large clusters slight fragrance; some prickles; foliage medium, dark green, semi-glossy; bushy, spreading growth; [Sea Foam X The Fairy]; Williams, J. Benjamin; J. Benjamin Williams & Associates, 1993

'WILWAVL', S, w, 1988; (**White Avalanche**); flowers cream, opening to pure white, small, borne in clusters, dbl., 15–25 petals, slight fragrance; foliage medium, medium green, glossy, disease resistant; spreading, compact, uniform, winter hardy growth; [Sea Foam X The Fairy]; Williams, J. Benjamin, 1989

'WILWIND', S, rb, 1992; (**Windmill**); flowers burgundy with ivory white on reverse bicolor, blooms borne singly, single, 3–3.5 in., slight fragrance; few prickles;

foliage medium, medium green, semi-glossy; medium (3–5'), bushy, spreading growth; [Handel X (Love X Double Feature)]; Williams, J. Benjamin, 1994

'WILYELP', Min, dy, 1997; (**Yellow Pastel**™); flowers double, medium, dbl., 15–25 petals, borne mostly singly, slight fragrance; foliage medium, medium green, semi-glossy; upright, compact, medium (13–18 in.) growth; [Sunnydew X Rise 'n' Shine]; Williams, J. Benjamin

Wimi®, *see* 'TANROWISA'

Win Win, *see* 'WEKWINWIN'

'WINBILEE', HT, rb, 1997; (**Better Homes & Gardens Diamond Jubilee**™); flowers full, very dbl., 26–40 petals, slight fragrance; foliage medium, medium green, semi-glossy; upright, medium (5 ft) growth; [Lynn Anderson X Seedling]; Winchel, Joseph F.

'WINBOB', HT, lp, 1993; (**Dr Bob Harvey**); dbl., 26–40 petals, 1.5–2.75 in., no fragrance; foliage medium, medium green, matt; bushy growth; Winchel, Joseph F.; Coiner Nursery

'WINBUR', HT, dr, 1987; (**American Dream**™); flowers deep red, large, dbl., 25 petals, exhibition form, borne singly, slight, damask fragrance; round, average, medium, medium orange fruit; average, dark brown, slightly recurved prickles; foliage medium, medium green, semi-glossy; upright, tall growth; Gold, ARC TG, 1986; [(My Dream X Charles Mallerin) X Seedling]; Winchel, Joseph F.; Co-Operative Rose Growers, 1990

Winchester Cathedral, *see* 'AUSCAT'

'WIND CHIMES', HMsk, mp; flowers rosy pink, blooms in clusters, 1 in., recurrent bloom, intense fragrance; vigorous (15–20 ft) growth; Lester Rose Gardens, prior to 1946; Lester Rose Gardens, 1949

'WIND SONG', HT, ab, 1968; flowers orange, large, dbl., moderate fragrance; foliage leathery; bushy growth; [Royal Sunset X Sierra Sunset]; Morey, Dr Dennison; Country Garden Nursery

'WINDEKIND', F, mr, 1974; bud pointed; dbl., 34 petals, 3 in., cupped; foliage matt, dark; upright growth; [Colour Wonder X Geisha]; Institute of Ornamental Plant Growing

'WINDERMERE', LCl, dp, 1932; flowers carmine-rose, blooms in large clusters, semi-dbl.; vigorous growth; Chaplin Bros.

Windflower, *see* 'AUSCROSS'

Windjammer, *see* 'SAVSWAT'

Windlass, Pol, lp; [Simon Robinson seedling sport]; Sutherland, P, 1995; Golden Vale Nursery, 1995, Australia

Windlass, HCh, lp; Sutherland, 1996

Windmill, *see* 'WILWIND'

Windmill, S, lp; Williams, J. Benjamin, 1999

'WINDONNA', HT, rb, 1993; (**Donna Darlin'**); flowers yellow blending pink to bright red, blooms, dbl., 15–25 petals, 1.5–2.75 in., slight fragrance; some prickles; foliage large, dark green, semi-glossy; medium (4 ft), bushy growth; [Seedling X Double Delight]; Winchel, Joseph F.; Coiner Nursery, 1994

Windrose®, S, lp; Noack, Werner, 1993

Windrush®, *see* 'AUSRUSH'

'WINDSOR', HT, mr, 1929; flowers rich crimson-scarlet, moderate fragrance; vigorous growth; Chaplin Bros.

Windsor, T, lp

'WINDSOR CHARM', HT, lp, 1978; bud long, tulip shaped; flowers light pink to ivory, dbl., 27–30 petals, 5–5.5 in., slight fragrance; foliage leathery; vigorous, bushy growth; [First Prize sport]; LeMire, Walter

'WINDSOUNDS', HT, lp, 1976; [First Prize sport]; Scoggins; American Rose Foundation

'WINDY CITY', Min, dp, 1974; flowers deep pink, reverse lighter, small, dbl., exhibition form, slight fragrance; foliage small, bronze; upright, bushy growth; [Little Darling X (Little Darling X (R. wichurana X Unnamed seedling))]; Moore, Ralph S.; Mini-Roses

Wine and Dine™, *see* 'DICUNCLE'

Wine and Roses, S, lp; PPAF; Clements, John K., 1999

Winecup, F, m

'WINEFRED CLARKE', HT, my, 1964; flowers, dbl., 34 petals, 5.5 in., exhibition form; foliage dark, glossy; [Peace X Lydia]; Robinson, H.

Winged Fellowship, *see* 'HORWINGFEL'

Winghaven's White, N, lp; (found rose)

'WINGOLD', HT, ly, 1999; (**Super Gold**™); dbl., 17–25 petals, 2.5 in., borne mostly singly, slight fragrance; few prickles; foliage large, dark green, glossy; upright, tall (36 in.) growth; [Gold Medal X seedling]; Winchel, Joseph F.; Coiner Nursery, 2000

'WINI EDMUNDS', HT, rb, 1973; flowers red, reverse white, large, dbl., exhibition form, moderate fragrance; foliage dark, leathery; vigorous, upright growth; [Red Lion X Hanne]; McGredy, Sam IV; Edmunds Roses

'WINIFRED', Gr, mp, 1981; bud pointed; flowers bright medium pink, dbl., 23 petals, slight fragrance; slightly hooked, red prickles; foliage reddish when young; vigorous, tall growth; [Queen Elizabeth X Unnamed seedling]; Jerabek, Paul E.

'WINIFRED', HT, yb, 1930; flowers deep yellow, shaded peach to old-gold, large, dbl.; vigorous growth; RULED EXTINCT 4/81; Chaplin Bros.

'WINIFRED COULTER', F, rb, 1962; bud pointed; flowers vermilion-scarlet, reverse silvery, blooms in clusters, dbl., 23 petals, 2.5–3 in., flat, intense fragrance; foliage leathery, glossy, dark; vigorous, bushy growth; David Fuerstenberg Prize, ARS, 1968; [Baby Chateau X Contrast]; Kemple; Van Barneveld

Winifred Coulter, Climbing, Cl F, lp; Jarvis, after 1968

'WINKFIELD CRIMSON EMPEROR', LCl, dr, 1958; flowers dark crimson, borne in clusters, dbl., 27 petals, 3.5 in., recurrent bloom, slight fragrance; foliage leathery; vigorous, pillar growth; [Red Empress X Surprise]; Combe; Winkfield Manor Nursery

'WINNERS ALL', HT, dp, 1990; bud ovoid; flowers carmine pink with a touch of orange in center, loose, large, dbl., cupped, slight, fruity fragrance; foliage medium, medium green, semi-glossy; bushy, medium growth; [Cherry Brandy sport]; Rearsby Roses, Ltd., 1986; Rearsby Roses, Ltd.

Winner's Choice, *see* JACmad

'WINNIE DAVIS', HT, mp, 1902; bud long, pointed; flowers pink with lighter reflexes, semi-dbl., moderate fragrance; [Kaiserin Auguste Viktoria X Mrs W.J. Grant]; Nanz & Neuner

Winnie Renshaw, *see* 'RENWINNIE'

Winning Colors™, *see* 'TWOWIN'

'WINNIPEG PARKS', S, dp, 1990; bud pointed; flowers dark pink-red, dark pink reverse, aging dark pink, medium, semi-dbl., 22 petals, cupped, repeat bloom, slight fragrance; foliage medium, medium green, matt; medium, bushy growth; [(Prairie Princess X Cuthbert Grant) X (Unnamed seedling X Morden Cardinette)]; Collicutt, Lynn M., 1981; Agriculture Canada, 1991

'WINPTIG', F, lp, 1999; (**Perfume Tiger**™); flowers dark red and white striped, very dbl., 41 petals, 1.75 in., borne in large clusters, intense fragrance; prickles moderate; foliage medium, medium green, semi-glossy; spreading, medium (27 in.) growth; [Roller Coaster X seedling]; Winchel, Joseph F.; Coiner Nursery, 2000

Winsome™, *see* 'SAVAWIN'

'WINSOME', HT, dp, 1924; flowers cherry-rose, dbl., slight fragrance; RULED EXTINCT 1/85; [Premier X Hoosier Beauty]; Hill, E.G., Co.; Vestal

'WINSOME', Cl HT, mr, 1931; bud long, pointed; flowers cherry-red, large, dbl., intense fragrance; foliage leathery; very vigorous growth; RULED EXTINCT 1/85; Dobbie

'WINSOME NATIVE', S, w, 1979; bud ovoid, pointed; dbl., 20 petals, cupped, borne singly, sparse, repeat bloom, no fragrance; no prickles; foliage finely serrated, 5 leaflet; rounded, medium growth; [Gene Boerner X R. suffulta]; Stoddard, Louis

'WINTER KING', S, dp, 1985; flowers bright deep pink, blooms borne singly, semi-dbl., 11 petals, repeat bloom, intense fragrance; foliage medium, dark, leathery, matt; compact, branching growth; hardy; [((McGredy's Scarlet X Polly) X Frau Karl Druschki) X Northlander]; James, John

Winter Magic™, *see* 'FOUMAGIC'

Winter Princess, *see* 'GELWIN'

'WINTER SUNSET', S, yb, 1997; flowers full, large, very dbl., 26–40 petals, borne in large clusters, moderate fragrance; foliage large, dark green, glossy; upright, bushy, medium growth; [Serendipity X (Country Dancer X Alexandra)]; Buck, Dr Griffith J.

Winter Wheat™, *see* 'MINIWIN'

Wintonbury Parish, *see* 'BERPAR'

'WINTONIENSIS', HMoy, mp; (R. X wintoniensis); flowers rosy pink, blooms in clusters; foliage with sweetbriar (apple) fragrance; (42); [R. moyesii X R. setipoda]; 1935

'WIRCUROB', HT, yb, 1991; (**Lizabeth's Lullabye**); flowers yellow with pink tips, moderately large, dbl., 35–40 petals, moderate fragrance; foliage large, dark green, glossy; upright, tall growth; [Queen Elizabeth X Song of Paris]; Robbins, William C., 1990

Wirrunna White Fairy, Pol, lp; (Weatherly, L.)

'WISBECH GOLD', HT, yb, 1964; flowers golden yellow edged pinkish, well-formed, dbl., 35 petals, 4 in., cupped, moderate fragrance; vigorous, compact growth; [Piccadilly X Golden Star]; McGredy, Sam IV; McGredy

Wisbech Rose Fragrant Delight®, *see* **'FRAGRANT DELIGHT'**

Wise Portia, *see* 'AUSPORT'

Wishful Thinking, *see* 'SEAWISH'

Wishing, *see* 'DICKERFUFFLE'

Wistful, *see* 'SAVAWIST'

'WITCHCRAFT', F, yb, 1961; bud ovoid; flowers yellow, reverse scarlet, borne in large clusters, dbl., 45–50 petals, 2–2.5 in., moderate fragrance; foliage dark, glossy; vigorous growth; [Masquerade X Seedling]; Verschuren; Blaby Rose Gardens

'WITCHING HOUR', F, dr, 1967; bud ovoid; flowers dark red, almost black, medium, dbl.; vigorous, bushy, compact growth; [Rapture X (F.W. Alesworth X Charles Mallerin)]; Morey, Dr Dennison; Country Garden Nursery

With Love, *see* 'ANDWIT'

With Thanks, *see* 'FRANSMOOV'

Wit's End, *see* 'SEAWIT'

'WIWAIT', F, w, 1991; (**Playfair**); single, borne in small clusters, slight fragrance; foliage medium, medium green, semi-glossy; medium growth; [Playgirl X Unknown seedling]; Wilke, William

'WIWANCY', HT, lp, 1992; (**Nancy Shaw**); dbl., 26–40 petals, 1.5–2.75 in., borne mostly singly, slight fragrance; some prickles; foliage medium, medium green, semi-glossy; upright, bushy, tall (4 ft) growth; [Peggy Lee X Unknown]; Wilke, William

Wizard, *see* 'MACLEIDE'

Wizo®, *see* 'KRIWI'

'WOBURN ABBEY', F, ob, 1962; bud ovoid; flowers orange, blooms in clusters, dbl., 25 petals, 3.5 in., cupped, moderate fragrance; foliage dark, leathery; moderate growth; [Masquerade X Fashion]; Sidey & Cobley; Harkness, 1962; J&P, 1964

Woburn Abbey, Climbing, Cl F, lp; Brundrett, S, 1972

'WOBURN GOLD', F, dy, 1970; flowers gold; [Woburn Abbey sport]; Robinson, H.

Woburn Gold, F, lp; [Woburn Abbey sport]; Robinson, H., 1970

'WODAN', HMult, 1890; Geschwind, R.; (Cavriglia)

Wohler Road White Eglantine, HEg, lp; (found rose)

'WOLFE'S GLORIE', Pol, op, 1943; flowers salmon-pink to deep orange; vigorous growth; [Orléans Rose sport]; Wolf, Van der; Hage

'WOLFGANG VON GOETHE', HP, pb; bud very long; flowers delicate bright yellowish pink, very large, dbl., exhibition form, vigorous growth; [Frau Karl Druschki X Souv. de Claudius Pernet]; Weigand, L., 1932

Wolley-Dod's Rose, *see* **'DUPLEX'**

Wollongong Gold, *see* 'WALSHINE'

'WOMAN AND HOME', HT, ob, 1976; flowers orange, pointed, dbl., 33 petals, 5 in., slight fragrance; foliage glossy, dark; [Apricot Silk X Piccadilly seedling]; Gregory

Woman O'th North, *see* 'KIRLON'

Woman Pioneer, *see* **'PIONERKA'**

'WOMAN'S DAY', HT, or, 1966; bud ovoid; flowers scarlet-orange, medium, semi-dbl., exhibition form; foliage leathery; vigorous, upright growth; [Queen Elizabeth X Karl Herbst]; Armbrust; Langbecker

Woman's Day, *see* WELira

Woman's Day, F, lp; Welsh, 1993

Woman's Hour, F, lp; Beales, Peter, 1997

Woman's Institute, HT, lp

'WOMAN'S OWN'®, Min, mp, 1973; flowers pink, small, dbl., 45 petals, 1.5 in., slight fragrance; foliage dark; [New Penny X Tip-Top]; McGredy, Sam IV; McGredy

'WOMAN'S REALM', HT, mr, 1966; flowers scarlet, 3.5–4 in., slight fragrance; foliage glossy; upright growth; [Chrysler Imperial seedling]; Gregory

Woman's Value, *see* 'KORVALUE'

'WOMEN IN MILITARY', S, mr, 1993; (The Service Rose); flowers moderately full, bright red, dbl., 20–30 petals, 2.75–3 in., borne in small clusters, moderate fragrance; some prickles; foliage medium, medium green, glossy; medium (135 cms), upright growth; Jerabek, Paul E.

'Women O'th North', *see* 'KIRLON'

'WONDER OF WOOLIES', F, ab, 1978; flowers deep apricot, dbl., 30 petals, 4.5 in., intense fragrance; vigorous growth; [Arthur Bell X Elizabeth of Glamis]; Bees

Wonderglo™, *see* 'MINIWON'

Wonderland®, S, yb; Dickson, Patrick, 1994

'WONDERSTAR', F, my, 1960; flowers bright yellow, large, semi-dbl.; low, bushy growth; [Goldilocks X Unnamed seedling]; Mondial Roses

Wonderstripe, *see* CLEwonder

Wood Lawn, *see* 'FIALAWN'

Woodland Sunbeam, *see* 'TROBGOLD'

Woodlands Lady, *see* 'TROBSA'

Woodlands Sunrise, *see* 'TROBSUN'

'WOODROW', (form of R. suffulta), mp; dbl., 60 petals, fall bloom; low growth; Hunt, Miss Alice, about 1925; P.H. Wright, 1936; (collected in Sask., Canada)

Wood's Garland, *see* **'THE GARLAND'**

Woodstock, *see* 'MORSTOCK'

Woody, HMsk, mr; Lens, 1978

Worjackie, *see* 'KIRWORJACKIE'

Work of Art, *see* 'MORART'

World Class, *see* HARdimple

World Peace™, *see* 'BURWORPE'

'WORLD'S FAIR', F, dr, 1939; (Minna Kordes); flowers deep crimson fading to scarlet, blooms in clusters, semi-dbl., 19 petals, 4 in., moderate, spicy fragrance; foliage leathery; vigorous, bushy growth; AARS, 1940; [Dance of Joy X Crimson Glory]; Kordes; J&P

'WORLD'S FAIR SALUTE', HT, mr, 1964; bud ovoid; flowers crimson-red, dbl., 33 petals, 5.5 in., exhibition form, moderate fragrance; foliage leathery; vigorous, upright growth; [Mardi Gras X New Yorker]; Morey, Dr Dennison; J&P

'WORLD'S FAIR, CLIMBING', Cl F, dr, 1941; J&P

Wörlitz, S, yb, 1962; flowers orange and yellow, large, dbl.; Berger, W.; (Sangerhausen)

'WORTHINGTON', (natural variation of R. setigera), lp; flowers clear pink; (found in Ohio)

'WORTHWHILE', HT, ob, 1973; flowers orange, reverse lighter, pointed, dbl., 33 petals, 5–6 in., slight fragrance; foliage matt, olive-green; [Gavotte X Vienna Charm]; LeGrice

Woschod, HT, w, 1973; flowers large, dbl.; Bülow; (Sangerhausen)

Wow!, *see* 'PIXIWOW'

Wrams Gunnarstorpsrosen, B, dp

Wretham Rose, C, lp; Hoellering, 1997

'WRIGHT'S SALMON', HT, op, 1960; flowers salmon, dbl., 32 petals, 5 in., exhibition form; [Opera sport]; Wright & Son

WRIpic, Min, ob, 1991; (**Picaninni**); Wright

Wurzburg, *see* 'KORGUST'

'WYNANDA', F, mp, 1968; bud ovoid; flowers pink, medium, dbl., slight fragrance; foliage dark; [United Nations X Seedling]; Heidemij

Wyrala, HT, dr; [Lemania X Ami Guinard]; Heers, C W, 1949

'WYRALLA', HT, dr, 1949; bud ovoid; flowers large, semi-dbl., exhibition form; vigorous, bushy growth; [Lemania X Ami Quinard]; Heers; Pacific Nursery

X

Xanadu, F, pb

'Xavier Olibo', HP, dr, 1865; flowers velvety crimson shaded purplish; [Général Jacqueminot sport]; Lacharme, F.

Xaviere, *see* 'MEIviola'

Xerxes, *see* 'HARjames'

X-Rated, *see* 'TINx'

Y

Y2K, *see* 'SAVyk'

Yabadabadoo®, *see* 'MACyaba'

'Yabusame', HT, lp, 1999; flowers crimson, dbl., 35 petals, 6 in., exhibition form, intense fragrance; 6 ft growth; [Big Chief X Papa Meilland]; Ohtsuki, Hironaka, 1998

'Yachiyo-Nishiki', F, ob, 1986; flowers orange, reverse yellow, medium, borne 2–5 per stem, dbl., 28 petals, moderate fragrance; prickles curved downwards; foliage dark green, semi-glossy; upright, compact growth; [(Maxim X Myo-Joh) X Duftwolke]; Suzuki, Seizo; Keisei Rose Nursery, 1984

Yakiman, F, m

Yakimour®, *see* MEIpsilon

'Yametsu-Hime', Min, pb, 1961; flowers white edged pink, small, semi-dbl.; foliage glossy, light; dwarf growth; [Tom Thumb X Unnamed seedling]; Hebaru; Itami Rose Nursery

'Yamini Krishnamurti', HT, m, 1969; flowers lilac, medium, dbl., exhibition form, slight fragrance; foliage glossy; moderate growth; [Sterling Silver, Climbing X Open pollination]; Singh; Gopalsinamiengar

'Yanina', F, op, 1964; flowers salmon-pink, large, semi-dbl., 16 petals, moderate fragrance; foliage glossy; vigorous growth; [Fashion X Queen Elizabeth]; Yeoman & Sons

'Yanka', S, mp, 1927; semi-dbl., borne in clusters, non-recurrent, height 3–4 ft; very hardy; Hansen, N.E.

Yankee Doodle®, *see* 'YANKOR'

Yankee Lady, *see* UHLwe

'YANKOR', HT, yb, 1965; (**Yankee Doodle®**); bud urn-shaped; flowers apricot to peachy pink and butter-yellow, imbricated, dbl., 75 petals, 4 in., slight, tea fragrance; foliage glossy, dark; upright, bushy growth; AARS, 1976; [Colour Wonder X King's Ransom]; Kordes, R.; Armstrong Nursery

Yantai, *see* 'TINtai'

Yardley Baroque, *see* 'BEAyar'

Yardley English Rose, *see* KORyard

'Yaroslavna', HT, mp, 1958; (Famous Cliff); flowers large, dbl., 60 petals, moderate fragrance; foliage dark; vigorous growth; [La Parisienne X Peace]; Sushkov & Besschetnova

'Yashwant', HT, rb, 1986; flowers red, striped white and yellow, reverse yellow, striped red, dbl., 55–60 petals, exhibition form, slight fragrance; pale, green prickles, curving downwards; foliage glossy; vigorous, upright, bushy growth; [Suspense sport]; Patil, B.K.; K.S.G. Son's Roses, 1985

'Yasnaya Poliana', HT, op, 1958; (Bright Meadow); flowers salmon-pink, well formed, large, dbl., 45–50 petals, moderate fragrance; foliage dark; vigorous, compact growth; [Independence X Luna]; Shtanko, I.

'Yatkan', S, mp, 1927; dbl., 2.5 in., non-recurrent, very hardy; [Probably Gruss an Teplitz X La Mélusine]; Hansen, N.E.

'Yawa', S, op, 1940; flowers light coral-pink, borne in clusters of 4–12 on 8–12″ stem, dbl., 58 petals, 2 in., free, non-recurrent bloom, moderate fragrance; tall, wide spreading (pillar to 9 ft) growth; very hardy; [Anci Bohmova X R. blanda]; Hansen, N.E.

Year 2000, Min, lp

'Yeller Rose O'Texas', HT, dy, 1986; flowers medium, dbl., 26–40 petals, intense fragrance; foliage medium, medium green, semi-glossy; upright growth; [Arizona sport]; Wambach, Alex A.

'Yello Yo Yo', F, dy, 1980; bud slender, long; dbl., 20 petals, exhibition form, borne singly and in clusters, slight, tea fragrance; foliage medium green; upright growth; [Arlene Francis X Unnamed seedling]; Weeks

YelloGlo, *see* 'DELcro'

'Yellow Altai', HSpn, my, 1950; flowers bright yellow, becoming lighter, small, borne in clusters, profuse, non-recurrent bloom, foliage light green, soft; stems reddish-brown; vigorous (7 ft), upright growth; hardy (to -60); [R. spinosissima altaica X Persian Yellow]; Wright, Percy H.

Yellow Angel, Min, dy

Yellow Baby Rambler, *see* **'George Elger'**

'Yellow Bantam', Min, ly, 1960; bud pointed, yellow; flowers yellow to white, micro-mini, dbl., 25 petals, .5 in., slight fragrance; foliage glossy; bushy (10 in.) growth; [(R. wichurana X Floradora) X Fairy Princess]; Moore, Ralph S.; Sequoia Nursery

'Yellow Bantam, Climbing', Cl Min, ly, 1964; Rumsey, R.H.

'Yellow Beauty', HT, my, 1945; bud longer, more pointed, clear golden yellow; more uniform growth than its parent; [Golden Rapture sport]; Ruzicka's

'Yellow Bedder', HT, ly, 1923; flowers clear yellow to cream, dbl., slight fragrance; [Mr Joh. M. Jolles X Mme Edouard Herriot]; Van Rossem

Yellow Belinda, *see* 'TANyellda'

'Yellow Bird', HT, dy, 1974; bud ovoid; dbl., 42 petals, 3.5 in., slight fragrance; upright growth; [Sunlight X Arthur Bell]; McGredy, Sam IV

Yellow Blaze, *see* 'BURyellow'

Yellow Blush, A, ly; Sievers, 1988

Yellow Bouquet, *see* 'LAVkin'

'Yellow Butterfly', HT, ob, 1927; flowers orange-yellow; RULED EXTINCT 1/89; [Mme Butterfly sport]; Scott, R.

Yellow Butterfly, *see* 'MORwings'

'Yellow Button'®, S, yb, 1975; bud globular; flowers medium yellow, deeper in center, rosette blooms borne 1–15 per cluster, very dbl., 90 petals, flat, repeat bloom, slight fragrance; straight, red prickles; foliage dark; bushy growth; [Wife of Bath X Chinatown]; Austin, David

Yellow Cécile Brünner, *see* **'Perle d'Or'**

Yellow Cécile Brünner, Climbing, *see* **'Perle d'Or, Climbing'**

'Yellow Charles Austin'®, S, ly, 1981; [Charles Austin sport]; Austin, David, 1979

'Yellow Cluster', F, my, 1949; flowers mimosa-yellow; spreading growth; [Golden Rapture X Polyantha seedling]; deRuiter; Spek

Yellow Cochet, *see* **'Alexander Hill Gray'**

Yellow Condesa de Sastago, *see* **'Jean Bostick'**

Yellow Condesa de Sástago, *see* **'Yellow Sástago'**

'Yellow Contempo', F, dy, 1983; [Contempo sport]; Gupta, Dr M.N., Datta, Dr S.K. & Banerji, Shri B.K.; National Botanical Research Institute

Yellow Cover, *see* POUlgode

'Yellow Creeping Everbloom', LCl, my, 1953; recurrent bloom, very vigorous, pillar or groundcover growth; [((Orange Everglow X New Dawn) X Unnamed seedling) X Free Gold]; Brownell, H.C.

'Yellow Curls', HT, my, 1947; flowers open, dbl., moderate fragrance; foliage glossy; vigorous growth; [Golden Glow X Self]; Brownell, H.C.

'Yellow Cushion', F, my, 1966; bud pointed; flowers large, dbl., blooms in small clusters, moderate fragrance; foliage glossy, leathery; vigorous, bushy growth; [Fandango X Pinocchio]; Armstrong, D.L.; Armstrong Nursery, 1962

Yellow Dazzler, *see* **'Yellowhammer'**

Yellow Delight, HT, lp

'Yellow Doll', Min, ly, 1962; bud pointed; flowers yellow to cream, small, dbl., 1.5 in., exhibition form, moderate fragrance; foliage leathery, glossy; vigorous, bushy (12 in.) growth; [Golden Glow (LCl) X Zee]; Moore, Ralph S.; Sequoia Nursery

'Yellow Doll, Climbing', Cl Min, ly, 1976; Kirk; Sequoia Nursery

'Yellow Dot', HT, ob, 1938; bud short, pointed; flowers light orange-yellow, reverse coral-red, small, dbl.; foliage soft, dark; short stems; vigorous growth; [Rapture X Betsy Ross]; Hill, Joseph H., Co.

Yellow Dot, *see* INTerbab

'Yellow Faïence', HT, my, 1942; flowers lemon-yellow; [Faience sport]; C-P

Yellow Fairy®, *see* 'POUlfair'

Yellow Festival, *see* 'LAVlinger'

Yellow Festival, Min, lp

Yellow Fleurette®, *see* INTerwell

'Yellow Fontaine', S, my, 1972; bud ovoid; flowers large, dbl., free, continuous bloom, slight fragrance; foliage large, glossy; vigorous, upright growth; [Unknown X Unknown]; Tantau, Math.; Ahrens & Sieberz

Yellow Fru Dagmar Hartopp, *see* 'MORyelrug'

Yellow Gamette, Pol, lp

'Yellow Gloria', HT, my, 1936; bud long, pointed; flowers golden yellow, base saffron, large, dbl., exhibition form, moderate fragrance; foliage leathery; long, strong stems; vigorous growth; [Talisman sport]; Bertanzel

'Yellow Gold', HT, dy, 1957; bud ovoid; flowers deep golden yellow, dbl., 30–35 petals, 5 in., cupped, intense fragrance; foliage leathery, glossy; strong stems; vigorous growth; Boerner; J&P

Yellow Hammer, *see* **'Yellowhammer'**

Yellow Hammer, Climbing, Cl F, lp; Dorieux, 1976

Yellow Herriot, *see* **'Florence Chenoweth'**

Yellow Hit, *see* 'POUllow'

Yellow Holstein, *see* **'Gelbe Holstein'**

'Yellow Hoover', HT, my, 1933; flowers pure yellow; [Pres. Herbert Hoover sport]; Western Rose Co.

Yellow Hybrid Musk, HMsk, my

Yellow Jacket, *see* 'JACyepat'

'Yellow Jewel', Min, my, 1973; bud long, pointed; flowers clear yellow, red eye, medium, semi-dbl., moderate fragrance; foliage small, glossy, leathery; dwarf, bushy growth; [Golden Glow X (Little Darling X Unnamed seedling)]; Moore, Ralph S.; Sequoia Nursery

'Yellow Joanna Hill', HT, my, 1932; bud long, pointed; flowers well formed, dbl., 28 petals, slight fragrance; foliage light; long stems; vigorous growth; [Joanna Hill sport]; White, C.N.; White Bros.

Yellow Kaiserin Auguste Viktoria, *see* **'Franz Deegen'**

Yellow Lady Banks' Rose, *see* **R. banksiae lutea**

Yellow Light, HMsk, my; Lens, 1995

'Yellow Magic', Min, my, 1970; bud long, pointed; flowers small, semi-dbl., moderate fragrance; foliage small, glossy, dark, leathery; vigorous, dwarf, upright, bushy growth; [Golden Glow X (Little Darling X Seedling)]; Moore, Ralph S.; Sequoia Nursery

Yellow Maman Cochet, *see* **'Alexander Hill Gray'**

Yellow Meillandina, *see* 'MEItrisical'

'Yellow Mellow', Min, my, 1984; flowers small, dbl., 35 petals, slight fragrance; foliage large, dark, glossy; bushy growth; [Rise 'n' Shine X ?]; Bridges, Dennis A.; Bridges Roses, 1985

'Yellow Miniature', Min, my, 1961; flowers bright yellow, small; [Rosina seedling sport]; Wayside Gardens Co.

Yellow Mini-Wonder, *see* 'LAVglo'

'Yellow Mme Albert Barbier', HP, my, 1937; flowers sunflower-yellow; [Mme Albert Barbier sport]; Lens

'Yellow Moss', M, dy, 1932; bud ovoid, slightly mossed, deep yellow; flowers yellow, edges tinted pink, semi-dbl., cupped, free, intermittent bloom, moderate fragrance; foliage thick; vigorous growth; [Old Moss X Hybrid Tea]; Walter, L.; J&P

Yellow Mozart, S, my

'Yellow Necklace', Min, ly, 1965; bud pointed; flowers straw-yellow, small, dbl., moderate fragrance; foliage leathery; vigorous, dwarf growth; [Golden Glow (LCl) X Magic Wand]; Moore, Ralph S.; Sequoia Nursery

Yellow Ocean, LCl, lp

Yellow Ophelia, *see* **'Silvia'**

'Yellow Pages', HT, yb, 1971; flowers classic form, dbl., 50 petals, 4 in., moderate fragrance; foliage small, glossy; [Arthur Bell X Peer Gynt]; McGredy, Sam IV; McGredy, 1972

Yellow Parade™, *see* 'POUlwee'

Yellow Pastel™, *see* 'WILyelp'

Yellow Patio, Min, my

'Yellow Perfection', HT, ly, 1952; bud long, pointed, lemon-yellow; flowers canary-yellow, medium, dbl., 25–30 petals, exhibition form, slight fragrance; foliage leathery, dark; vigorous, upright growth; [Pearl Harbor X Golden Rapture]; Hill, Joseph H., Co.

'Yellow Petals', HT, my, 1971; flowers, dbl., 30 petals, 5.5 in., exhibition form, moderate fragrance; foliage large, light; [King's Ransom X Dorothy Peach]; Robinson, H.

'Yellow Pigmy', F, my, 1964; flowers small, dbl., moderate fragrance; foliage small, glossy; will bloom in pots; dwarf, spreading growth; [Unnamed seedling X (Eblouissant X Goldilocks)]; Moore, Ralph S.; Sequoia Nursery

'Yellow Pinocchio', F, my, 1949; bud ovoid; flowers apricot-yellow, fading to cream, blooms in clusters, dbl., 45 petals, 3 in., cupped, moderate fragrance; foliage dark; vigorous, bushy growth; (28); [Goldilocks X Marionette]; Boerner; J&P

'Yellow Pinocchio, Climbing', Cl F, my, 1959; Palmer

'Yellow Queen Elizabeth', Gr, my, 1964; flowers orange-yellow; [Queen Elizabeth sport]; Vlaeminck; Fryer's Nursery, Ltd.

Yellow Quill, MinFl, lp; Williams, J. Benjamin, 1999

Yellow Rambler, *see* **'Aglaia'**

Yellow Rambler, *see* **'Brownell Yellow Rambler'**

Yellow Ribbon, *see* 'DICalow'

'Yellow River', HT, my, 1972; flowers pointed, dbl., 30 petals, 3.5 in., slight fragrance; foliage glossy; vigorous growth; [Bettina sport]; Timmerman's Roses

Yellow River, F, lp

Yellow Robusta, HRg, 1992; Maarse, J.D.; (Cavriglia)

'Yellow Ruffles', HT, my, 1953; flowers, dbl., 35–50 wavy petals, 4–5 in., moderate fragrance; foliage dark, glossy; compact growth; [Orange Ruffels sport]; Brownell, H.C.

'Yellow Sástago', HT, my, 1939; (Yellow Condesa de Sástago); [Condesa de Sástago sport]; Howard Rose Co.

'Yellow Sea', LCl, my, 1977; bud long, pointed; flowers lemon-yellow, semi-dbl., 18–20 petals, 6 in., exhibition form, intense, honey spice fragrance; [Unnamed seedling X Ivory Fashion]; Fong; United Rose Growers

'Yellow Shot Silk', HT, my, 1944; [Shot Silk sport]; Beckwith

Yellow Simplicity, *see* 'JACyelsh'

Yellow Splendor, F, lp; Select, 1998

'Yellow Star', F, my, 1965; flowers well shaped, medium; foliage glossy; moderate growth; deRuiter

Yellow Sunbeam, Min, my

Yellow Sunblaze, *see* 'MEItrisical'

Yellow Sunsation, *see* KORtisching

Yellow Sunset, HT, lp; Dawson, George, 1990

Yellow Sunset, HT, lp

Yellow Sweetheart, *see* **'Rosina'**

'Yellow Sweetheart', F, my, 1952; bud pointed; flowers lemon-yellow, reverse sulfur-yellow, becoming, dbl., 35 petals, 3.5–4 in., globular, foliage leathery; vigorous, branching growth; [Pinocchio X Seedling]; Boerner; J&P

'Yellow Sweetheart, Climbing', Cl F, ly, 1952; flowers apricot-yellow to creamy-yellow, small, dbl., 26 petals, exhibition form, moderate fragrance; foliage leathery, glossy; vigorous, climbing (12 ft) growth; [Étoile Luisante seedling X Goldilocks]; Moore, Ralph S.; Marsh's Nursery

'Yellow Talisman', HT, ly, 1929; flowers yellow; [Talisman sport]; Amling Co.

'Yellow Talisman', HT, ly, 1935; flowers pale sulfur-yellow; [Talisman sport]; Eddie

'Yellow Talisman, Climbing', Cl HT, ly; Stocking

Yellow Tausendschön, *see* **'Madeleine Seltzer'**

'Yellow Tiffany', HT, yb, 1976; flowers yellow tinted pink; [Tiffany sport]; Floradale Nurs.

Yellow Triumph, HT; Herholdt; (Cavriglia)

Yellow Unique, F, lp

'Yellow Wings', S, my, 1986; flowers light-medium yellow, fading to cream, moderately large, born, dbl., 16 petals, moderate fragrance; straight, brown prickles; foliage medium, light green, glossy, disease resistant; bushy growth; [[(Will Scarlet X Sutter's Gold) X Paula] X Golden Wings]; James, John

Yellow Wonder, *see* 'MACbern'

'Yellowcrest', HT, ly, 1935; bud small, pointed; flowers clear canary-yellow, dbl.; foliage light, glossy, small; vigorous growth; LeGrice

'Yellowhammer', F, my, 1956; (Yellow Dazzler, Yellow Hammer); flowers golden yellow to light yellow, blooms in clusters, dbl., 48 petals, 3 in., moderate fragrance; foliage dark, glossy; vigorous, bushy growth; GM, NRS, 1954; [Poulsen's Yellow X Unnamed seedling]; McGredy, Sam IV

'Yellowstone', HT, my, 1969; bud pointed; flowers soft tawny yellow to lighter yellow, full, dbl., 38–44 petals, 3.5–4 in., moderate, tea fragrance; foliage glossy, dark, leathery; tall, upright growth; [Unnamed seedling X Unnamed seedling]; Weeks

'Yesterday'®, Pol, mp, 1974; (Topis d' Orient); flowers lilac pink, golden stamens, blooms in trusses, semi-dbl., 13 petals, 1.5 in., slight fragrance; foliage small, glossy; bushy growth with polyantha characteristics.; ADR, 1978 GM, Baden-Baden, 1975 GM, Monza, 1974; [(Phyllis Bide X Shepherd's Delight) X Ballerina]; Harkness

'Yesterday Reef', HT, w, 1984; flowers white, faintly tinted pink and yellow, large, dbl., 35 petals, exhibition form, slight fragrance; foliage large, medium green, matt; narrowly upright growth; [(Tiffany X Sunblest) X Irish Gold]; Stoddard, Louis

Yesterday's Garden, *see* 'MORshakrug'

Yesteryear, *see* HARwoey

Yeswant, HT, rb; Patil, B.K., 1985

Yeti, lp; Dawson, George

Yi-Hong Yuan, HT, rb, 1995; flowers yellow, edged salmon, reverse lighter, very dbl., 26–40 petals, 4.5 in., borne mostly singly, slight fragrance; some prickles; foliage medium, dark green, semi-glossy; upright, medium growth; [Osiria X New Bal]; Hong-Quan, Li

YKH 501, HT, w, 1997; (**Kyoto 1200**); flowers very full, large, very dbl., 41 petals, moderate fragrance; some prickles; foliage medium, medium green, glossy; medium (1–2 m) growth; [Garden Party X Pristine]; Kameyama, Yasushi

'Yoimatsuri', F, ob, 1989; bud ovoid; flowers orange and yellow tinged, medium, borne in sprays of many, single, 5 petals, cupped, moderate fragrance; prickles ordinary; foliage small, light green, glossy; spreading, medium growth; [Masquerade X Subaru]; Kikuchi, Rikichi, 1990

'Yolande d'Aragon', P, m, 1843; flowers bright purple-pink, large, very dbl.; Vibert

'Yoriko', Pol, lp, 1997; flowers single,small, light pink, reverse white, single, 4–7 petals, 1.5 in., slight fragrance; oval, 6–7 mm, deep orange fruit; few prickles; foliage small, light green, semi-glossy; spreading (45 cms) growth; [Seedling X Seedling]; Ohmori, Yoriko

'York and Lancaster', D, pb; (R. damascena variegata, R. damascena versicolor, York et Lancastre); flowers petals blush white and light pink, sometimes all one, dbl.; mixed prickles; foliage downy light gray-green; tall growth; (28)

York et Lancastre, *see* **'York and Lancaster'**

Yorkshire, *see* KORbarkeit

Yorkshire Bank, *see* 'RUTrulo'

'Yorkshire Lady', HT, ly, 1992; dbl., 26–40 petals, 1.5–2.75 in., exhibition form, borne mostly singly, moderate fragrance; few prickles; foliage medium, medium green, glossy; upright (95–100 cms) growth; [((Red Lion X Royal Highness) X Yellow Petals) X Piccadilly]; Thompson, Robert; Battersby Roses, 1992

Yorkshire Sunblaze, *see* 'MEIblam'

'Yosemite', HT, ob, 1934; bud long, pointed; flowers scarlet-orange, reverse suffused carmine, dbl., cupped, moderate fragrance; foliage leathery; vigorous growth; [Charles P. Kilham X Mrs Pierre S. duPont]; Nicolas; J&P

'Yoshino', HT, w, 1997; flowers double, large, dbl., 15–25 petals, 5 in., intense fragrance; some prickles; foliage medium, medium green, semi-glossy; tall (200 cms) growth; [Confidence X Garden Party]; Ohtsuki, Hironaka

Yosooi, Min, w

You 'n' Me, *see* 'SEAyou'

Youki San®, *see* 'MEIdona'

'Young America', HT, dr, 1902; flowers medium, dbl., moderate fragrance; Hill, E.G., Co.; (Sangerhausen)

Young at Heart, HT, ab

Young Cale, *see* 'MACinca'

'Young France', HT, lp, 1945; (Jeune France); bud long, pointed; flowers apple-blossom-pink, base flushed orange, tipped silvery pink, exhibition form, moderate fragrance; foliage soft, dark; vigorous, upright growth; [Joanna Hill X Mme Joseph Perraud]; Meilland, F.; C-P

Young Hearts, *see* 'RENheart'

'YOUNG LOVE', Min, ob, 1980; bud medium-long, pointed; flowers orange, yellow-center, borne 1–3 per cluster, semi-dbl., 18 petals, intense fragrance; tiny, curved prickles; foliage tiny, dark; compact, bushy growth; [Unnamed seedling X Unnamed seedling]; Lyon

Young Love, Min, lp; Williams, J. Benjamin, 1995

Young Malmaison, lp; Walsh, R, ca. 1994

Young Mistress, *see* KORviga

Young 'n' Innocent, *see* 'RENYOUNG'

Young Quinn®, *see* 'MACBERN'

Young Venturer, *see* 'MATTSUN'

'YOUPI', Min, my, 1966; bud ovoid; flowers lemon-chrome, semi-dbl., 12–19 petals, 1 in., cupped, moderate, fruity fragrance; foliage light green; vigorous, compact growth; [R. wichurana X Pour Toi]; Lens

'YOUR SMILE', HT, lp, 1999; dbl., 35 petals, 6 in., exhibition form, slight fragrance; foliage medium green; 3 ft growth; GM, Japan Rose Concours, 1995; [Miss Harp X Golden Sun]; Ohata, Hatsuo; Komaba Rose Garden, 1997

'YOURS TRULY', HT, mp, 1945; bud long, pointed; flowers rose-pink, base golden yellow, open, dbl., 48 petals, 4.5 in., intense, spicy fragrance; foliage leathery; very vigorous, upright growth; [Unnamed seedling X Texas Centennial]; Morris; Germain's

Youth, *see* **'MOLODOST'**

'YOUTH', HT, w, 1935; bud ovoid; flowers creamy, suffused pink, very large, dbl., 60 petals, exhibition form, moderate fragrance; foliage dark; long stems; [Souv. de Claudius Pernet X My Maryland]; Cook, J.W.

Youth of the World, HT, mr; flowers large, dbl., intense fragrance; (Sangerhausen)

'YU-AI', HT, rb, 1970; bud ovoid; flowers bright red, base yellow, large, dbl., cupped, slight fragrance; foliage dark, leathery; very vigorous, bushy growth; [Sarabande seedling X Peace seedling]; Suzuki, Seizo; Keisei Rose Nursery

Yugiri, HT, pb; Suzuki, Seizo, 1987

'YUHLA', S, mr, 1927; flowers crimson, semi-dbl., non-recurrent; very hardy; [Wild rose from Lake Oakwood (SD) X Général Jacqueminot]; Hansen, N.E.

'YUKI-MATSURI', HT, w, 1989; bud ovoid; flowers urn-shaped, large, borne usually singly, dbl., 42 petals, exhibition form, moderate fragrance; large, light, light orange fruit; prickles sparse; foliage medium, dark green, glossy; medium growth; [(Dolce Vita X Royal Highness) X Nobility]; Yokota, Kiyoshi

Yukon, *see* **'SUN VALLEY'**

'YULETIDE', HT, mr, 1956; bud ovoid; flowers currant-red, blooms, dbl., 43 petals, 3.5–4.5 in., exhibition form, slight fragrance; foliage leathery, semi-glossy; vigorous, upright, bushy growth; [Hill Crest X Silver Kenny]; Hill, Joseph H., Co.

Yume, HT, ab

'YUMEOTOME', Min, lp, 1989; bud ovoid; flowers pink changing to white, dbl., 38 petals, slight fragrance; prickles lower part hollow, about 15 on a stem; foliage small, dark green; bushy growth; [Miyagino sport]; Tokumatsu, Kazuhisa

'YUSAI', HT, yb, 1989; bud ovoid; flowers bright yellow with reddish fringe, reverse changing to red w, dbl., 30 petals, exhibition form; foliage medium, dark green with bronze tinge, slightly dentate; bushy, spreading growth; [(Unnamed American Heritage seedling X Christian Dior) X Miss Ireland]; Ota, Kaichiro

'YUZEN', Gr, pb, 1982; flowers pink, blending deeper at petal edges, urn-shaped, large, bor, dbl., 50 petals, exhibition form; moderate fragrance; prickles slanted downward; foliage dark, glossy; upright growth; [Unnamed seedling X Confidence]; Suzuki, Seizo; Keisei Rose Nursery

Yvan Meneve, HT, lp; RVS, 1998

'YVAN MISSION', Pol, mp, 1922; flowers peach-blossom-pink, borne in large clusters; [Jeanny Soupert X Katharina Zeimet]; Soupert & Notting

'YVES DRUHEN', HT, dr, 1920; flowers dark velvety red, dbl., intense fragrance; [Gen. MacArthur X Château de Clos Vougeot]; Buatois

Yves Piaget®, *see* 'MEIVILDO'

'YVETTE', HT, op, 1921; flowers salmon-yellow; [Mrs T. Hillas X Mme Edouard Herriot]; Buatois

Yvette, Pol, lp; Scarman, 1996

'YVONNE', HWich, pb, 1921; flowers blush-pink, base deep pink shaded yellow, dbl., exhibition form, moderate fragrance; foliage dark, glossy; very vigorous growth; GM, NRS, 1920; Cant, F.

'YVONNE CARRET', HT, ob, 1961; flowers orange-red, reverse golden yellow, dbl., 25 petals, 4.5 in.; foliage bright green; Gaujard

'YVONNE D'HUART', HT, rb, 1932; bud long, pointed; flowers coral-red to lincoln red, base chrome yellow, large, dbl., slight fragrance; very vigorous growth; Ketten Bros.

'YVONNE MILLOT', HT, ab, 1936; bud pointed; flowers apricot-yellow, large, dbl., exhibition form, slight fragrance; foliage glossy; vigorous growth; [(Pharisaer X Constance) X Feu Joseph Looymans]; Mallerin, C.; C-P

'YVONNE PLASSAT', HT, or, 1941; flowers coppery orange to glowing red, dbl.; Moulin-Epinay; Nonin

'YVONNE PRINTEMPS', HT, or, 1939; bud long, almost red; flowers orange, veined copper, reverse bright yellow; Gaujard

'YVONNE RABIER', Pol, w, 1910; flowers pure white, center slightly tinted sulfur, blooms in clusters, dbl., recurrent bloom; foliage rich green, glossy; vigorous growth; (14); [R. wichurana X Polyantha]; Turbat

'YVONNE VACHEROT', HT, w, 1905; bud long, pointed; flowers porcelain-white, strongly flushed rose-pink, imbricated, large, dbl., exhibition form; vigorous growth; [Antoine Rivoire X Souv. d'un Ami]; Soupert & Notting

'YVONNE VIRLET', HT, my, 1936; flowers golden yellow, large, dbl.; Walter, L.

Z

'ZACHARLY ROSE', HT, pb, 1998; flowers pink, shaded peach at base, very dbl., 26–40 petals, 4.5 in., exhibition form, borne mostly singly, slight fragrance; few prickles; foliage medium, medium green, semi-glossy; long stems; very floriferous, upright, tall (4½ ft) growth; [Hazel Rose X Loving Memory]; Poole, Lionel; David Lister, Ltd, 1998

Zagreb, *see* 'ROGZAG'

Zahov Hatabor, F, yb; Fischel, 1979

'ZAÏD', S, rb, 1937; bud bright yellow and copper-red; flowers coppery flame-red, semi-dbl., non-recurrent; foliage glossy, bright green; vigorous growth; [R. foetida bicolor X R. wichurana seedling]; Chambard, C.

'ZAIDA', HT, op, 1922; flowers pale coral-pink, large, dbl.; Lippiatt

Zaire, M, dr; Vibert, 1849

Zambra®, *see* 'MEIALFI'

Zambra®, *see* MEIcurbos

Zambra '80, *see* 'MEIRILOCRA'

Zambra 92, *see* MEIcurbos

Zambra 93, F, 1993; Meilland, Alain A.; (Cavriglia)

Zambra, Climbing®, *see* 'MEIALFISAR'

Zandrina, HT, 1977; Zandri, R.; (Cavriglia)

'ZANI', S, mr, 1927; flowers dark crimson, with a white streak through center petals, semi-dbl., non-recurrent; height 6–8 ft; very hardy; [(R. rugosa X Anna de Diesbach) X Tetonkaha]; Hansen, N.E.

'ZANSHO', HT, rb, 1975; bud ovoid; flowers scarlet and yellow, dbl., 45 petals, 6 in., exhibition form, slight fragrance; foliage dark, leathery; upright, bushy growth; [Garden Party X Christian Dior]; Takahashi, Takeshi

'ZANZIBAR', F, mr, 1964; flowers bright red; [Paradis X Circus]; Leenders, J.

'ZARA HORE-RUTHVEN', HT, mp, 1932; flowers rich pink, large, dbl., cupped, slight fragrance; [Mme Abel Chatenay X Scorcher]; Clark, A.; NRS South Australia

Zärtlichkeit, S, lp, 1992; flowers small, single; Wänninger, Franz; (Sangerhausen)

'ZAUBERLEHRLING', F, op, 1963; flowers light salmon, well formed, large, dbl.; strong stems; bushy, low growth; Kordes, R.

'ZBOROV', Pol, dr, 1935; flowers blood-red, borne in large clusters, very dbl.; foliage glossy, dark; vigorous growth; [Corrie Koster sport]; Böhm, J.

Zburlici, F, lp; rosette, 15 petals, slight; medium large, light green, glossy; [Bonica '82 X La Sevillana]; St Wagner, 1994; Res. Stn. f. Fruit Growing, Cluj, 1995, Romania

Zebra, *see* BENraar

'ZEE', Cl Min, mp, 1940; flowers pink, very small, recurrent bloom; nearly thornless; foliage very small; height 30 in.; [Carolyn Dean X Tom Thumb]; Moore, Ralph S.

Zeiber House, HT, dr

Zelda Lloyd, *see* 'TINZEL'

'ZÉLIA PRADEL', N, w, 1860; (Estelle Pradel); flowers pure white, large, dbl.; Pradel

'ZENA', HT, dr, 1986; flowers dark crimson, large blooms borne singly or several together, dbl., 25 petals, slight fragrance; foliage medium, medium green, semi-glossy; upright growth; [Unnamed seedling X New Style]; Plumpton, E.

'ZENAITTA', Pol, rb, 1991; bud ovoid; flowers bright medium red with small white area at petal base, small, dbl., cupped, no fragrance; foliage small, medium green, glossy; bushy, medium growth; [Unnamed seedling X Unnamed seedling]; Jerabek, Paul E., 1985

Zenith, *see* 'DELZEN'

Zenith, *see* **'UETERSEN'®**

'ZENOBIA', M, mp, 1892; bud well mossed; flowers satiny pink, large, dbl., moderate fragrance; vigorous growth; Paul, W.

Zepheline, *see* KORkalba

'ZÉPHIRINE DROUHIN', B, mp, 1868; bud long, pointed; flowers cerise pink, base white, large, dbl., 25–30 petals, 3 in., borne mostly singly, recurrent bloom, moderate, damask fragrance; no prickles; foliage soft, light; vigorous, semi-climbing growth; Bizot

'ZEPHYR', F, yb, 1961; flowers yellow tipped pink, borne in clusters, dbl., 64 petals, 2.5–3 in.; foliage glossy, light; very vigorous growth; [Goldilocks X Seedling]; Verschuren; Blaby Rose Gardens

'ZEUS', LCl, my, 1959; bud long, pointed; flowers medium, dbl., 25 petals, exhibition form, slight fragrance; foliage leathery, dark; vigorous (15–20 ft) growth; [Doubloons X Unnamed seedling]; Kern Rose Nursery

'ZIGEUNERBARON', F, dr, 1954; flowers large, dbl.; Baron, W.; (Sangerhausen)

Zigeunerblut, B, dr; Geschwind, R., 1890

Zigeunerknabe, *see* **'GIPSY BOY'**

Ziggy Stardust, *see* 'RENZIG'

Zigri, LCl, m

Zig-Zag, *see* 'PIXZIG'

'ZIKA', S, lp, 1927; flowers shell-pink, semi-dbl., non-recurrent bloom, moderate fragrance; very hardy; [(R. rugosa X Anna de Diesbach) X Tetonkaha]; Hansen, N.E.

'ZINGARO', F, mr, 1964; flowers bright crimson, borne in clusters, semi-dbl., 12 petals, 2 in.; foliage dark; compact, low growth; [Masquerade X Independence seedling]; Sanday, John

'ZINGER', Min, mr, 1978; bud long, pointed; flowers small, semi-dbl., 11 petals, 1.5 in., flat, moderate fragrance; very vigorous, spreading growth; AOE, 1979; [Zorina X Magic Carrousel]; Schwartz, Ernest W.; Nor'East Min. Roses

'ZIPALM', Min, ab, 1991; (**Almond Glory**); flowers deep apricot, dbl., 26–40 petals, 1.5 in., exhibition form, borne in small clusters, moderate fragrance; some prickles; foliage small, dark green, glossy; medium (30 cms), compact growth; [Rise 'n' Shine X High Spirits]; Zipper, Herbert; Magic Moment Miniature Roses, 1992

'ZIPAMB', Min, ab, 1991; (**Amber Ribbon**); bud pointed; flowers deep apricot to deep yellow at base, reverse deep apricot to deep yellow, dbl., 22 petals, 1.75–2 in., exhibition form, slight fragrance; foliage medium, dark green, semi-glossy; upright, tall growth; [Rise 'n' Shine X Pot 'o Gold]; Zipper, Herbert; Magic Moment Miniature Roses, 1992

'ZIPBAN', Min, yb, 1987; (**Banana Split**); flowers creamy yellow blushed with pink, medium, dbl., 15–25 petals, no fragrance; foliage medium, medium green, semi-glossy; upright, bushy growth; [Little Darling X Poker Chip]; Zipper, Herbert; Magic Moment Miniature Roses

'ZIPBELL', F, or, 1988; (**Bellissima**); flowers orange-red, darker along edges, medium, dbl., exhibition form, borne singly, no fragrance; foliage medium, medium green, semi-glossy; upright growth; [Pink Petticoat X Lady Rose]; Zipper, Herbert; Magic Moment Miniature Roses, 1989

'ZIPBERRY', MinFl, rb, 1988; (**Raspberry Ice**); flowers white brushed with red, deeper red at petal edge, medium, dbl., 40 petals, no fragrance; foliage medium, dark green, semi-glossy; mini-flora; upright growth; [High Spirits X Charmglo]; Zipper, Herbert; Magic Moment Miniature Roses, 1989

'ZIPBOY', Min, mp, 1988; (**Tomboy**); flowers small, single, 5 petals, no fragrance; foliage small, dark green, semi-glossy; upright, spreading growth; [Maytime X Poker Chip]; Zipper, Herbert; Magic Moment Miniature Roses, 1989

'ZIPBUB', Min, yb, 1986; (**Bubbles**); flowers medium yellow, shaded pink and coral, dbl., 35 petals, borne singly and in small clusters, moderate fragrance; foliage medium, medium green, semi-glossy; upright, spreading growth; [(Little Darling X (Roundabout X Redgold)) X (Maytime X Poker Chip)]; Zipper, H.; Magic Moment Miniature Roses

'ZIPCAN', Min, pb, 1987; (**Candice**); flowers medium, dbl., slight fragrance; foliage medium, medium green, semi-glossy; upright growth; [Libby X Queen Elizabeth]; Zipper, Herbert; Magic Moment Miniature Roses

'ZIPCLOUD', Min, w, 1989; (**Summer Cloud**); bud ovoid; flowers white with hint of pink, urn-shaped, large, borne singly and, dbl., 65 petals, exhibition form, slight fragrance; globular, medium, brown-orange fruit; prickles straight, small, few, light brown; foliage medium, medium green, semi-glossy; bushy, medium growth; [Roundabout X Erfurt]; Zipper, Herbert; Magic Moment Miniature Roses, 1990

'ZIPCOOL', Min, w, 1987; (**Cool Wave**); flowers white with hints of yellow at petal base, medium, dbl., 15–25 petals, exhibition form, no fragrance; foliage medium, medium green, matt; [Poker Chip X Pink Parfait]; Zipper, Herbert; Magic Moment Miniature Roses

'ZIPCORK', F, rb, 1991; (**Corky**); flowers white with red edges, white reverse, blooms borne in small clusters, dbl., 26–40 petals, 3–3.5 in., slight fragrance; few prickles; foliage large, dark green, semi-glossy; medium (100 cms), bushy growth; [Tamango X Seedling]; Zipper, Herbert; Magic Moment Miniature Roses, 1992

'ZIPCUSS', F, rb, 1989; (**Percussion**); bud ovoid; flowers dark red with touch of yellow at base, reverse white at base, dbl., 25 petals, exhibition form, moderate, sweet fragrance; prickles very small, tan; foliage medium, medium green, semi-glossy; mini-flora; upright, tall growth; [Sheri Anne X Deep Purple]; Zipper, Herbert; Magic Moment Miniature Roses, 1990

'ZIPDEW', Min, rb, 1991; (**Dew Drop**); flowers red with white eye, blooms borne mostly singly, single, 5 petals, 1.5 in., slight fragrance; few prickles; foliage small, dark green, semi-glossy, very disease-resistant; low (30 cms), compact growth; [Sheri Anne X Priscilla Burton]; Zipper, Herbert; Magic Moment Miniature Roses, 1992

'ZIPEASY', Min, rb, 1992; (**Easy**); flowers red and white, blooms borne in small clusters, very dbl., 1.5 in., no fragrance; few prickles; foliage small, dark green, glossy; medium (35 cms), bushy growth; [Pink Petticoat X Banana Split]; Zipper, Herbert; Magic Moment Miniature Roses

'ZIPELA', Min, m, 1991; (**Elaina**); dbl., 26–40 petals, 1.5 in., exhibition form, borne in small clusters, moderate fragrance; few prickles; foliage small, dark green, semi-glossy; medium (35 cms), bushy growth; [Blue Nile X Big John]; Zipper, Herbert; Magic Moment Miniature Roses, 1992

'ZIPEND', Min, pb, 1991; (**Endora**); flowers cream, edged deep pink, blooms borne mostly singly, dbl., 26–40 petals, 1.5–2.75 in., exhibition form, slight fragrance; few prickles; foliage small, medium green, semi-glossy; medium (35 cms), upright growth; [Pristine X High Spirits]; Zipper, Herbert; Magic Moment Miniature Roses, 1992

'ZIPEST', MinFl, w, 1997; (**Esther Jasik**); flowers very full, large, very dbl., 41 petals, borne mostly singly, slight fragrance; few prickles; foliage medium, dark green, dull; upright, tall (3 ft) growth; [Shocking Blue X Pink Petticoat]; Zipper, Herbert

'ZIPFAN', S, lp, 1999; (**Summer Fantasy**); flowers light pink, reverse dark pink, dbl., 17–25 petals, 2–2.5 in., borne in small clusters, no fragrance; few prickles; foliage small, dark green, semi-glossy, resistant to mildew; bushy (3–4 ft) growth; [Maytime X Charmglo]; Zipper, Herbert; Island Roses, 1999

'ZIPGOLD', Min, dy, 1991; (**Golden Coach**); flowers small, blooms borne mostly singly, very dbl., 1.5 in., no fragrance; few prickles; foliage small, dark green, semi-glossy; medium (40 cms), upright, compact growth; [Rise 'n' Shine X Pot O'Gold]; Zipper, Herbert; Magic Moment Miniature Roses, 1992

'ZIPHIGH', Min, yb, 1991; (**Highland Lass**); flowers red and yellow with yellow reverse, blooms borne in small clusters, semi-dbl., 6–14 petals, 1.5 in., no fragrance; few prickles; foliage small, dark green, semi-glossy; low (20 cms), compact growth; [Rise 'n' Shine X High Spirits]; Zipper, Herbert; Magic Moment Miniature Roses, 1992

'ZIPHILL', Min, dp, 1989; (**Castle Hill**); bud pointed; flowers deep pink with mauve undertones, shading lighter towards base, dbl., exhibition form, no fragrance; round, small, light orange fruit; prickles curved, small, tan; foliage medium, medium green, semi-glossy, highly mildew res; bushy, low growth; [Sheri Anne X Red Devil]; Zipper, Herbert; Magic Moment Miniature Roses, 1990

'ZIPJAM', Min, op, 1986; (**Jam Session**); flowers coral pink, yellow reverse, small blooms borne singly or in, dbl., 20 petals, no fragrance; foliage small, medium green, matt; upright growth; [Poker Chip X Anytime]; Zipper, H.; Magic Moment Miniature Roses

'ZIPLIB', Min, pb, 1997; (**Liberty Miss**); flowers double, medium, dbl., 8–14 petals, borne mostly singly, moderate fragrance; few prickles; foliage medium, medium green, semi-glossy; bushy, medium (12–18 in.) growth; [Seedling X Cupcake]; Zipper, Herbert

'ZIPLIME', MinFl, rb, 1991; (**Limerick**); flowers red with white reverse, blooms borne in small clusters, dbl., 15–25 petals, 1.5–2.75 in., exhibition form, slight fragrance; few prickles; foliage medium, dark green, glossy; medium (40 cms), spreading growth; [Tamango X Jennifer]; Zipper, Herbert; Magic Moment Miniature Roses

'ZIPMIA', F, mp, 1986; (**Mama Mia**); flowers mini-flora, blooms borne singly or in 3's, dbl., exhibition form, slight fragrance; foliage medium, medium green, matt; upright, bushy growth; [Sheri Anne X Sparrieshoop]; Zipper, H.; Magic Moment Miniature Roses

'ZIPMUSIC', MinFl, mp, 1988; (**Night Music**); flowers deep pink, medium, borne singly and in sprays, dbl., intense fragrance; foliage large, medium green, semi-glossy; upright growth; [Tamango X Pink Petticoat]; Zipper, Herbert; Magic Moment Miniature Roses, 1989

'ZIPNIC', Min, pb, 1991; (**Nicolina**); bud ovoid; flowers coral pink shading to white, reverse pink edge, shading to c, dbl., 25 petals, 1.25 in., exhibition form, slight fragrance; foliage small, dark green, semi-glossy; upright, bushy, tall growth; [(Dandy Lyon X Razzle Dazzle) X Pink Petticoat]; Zipper, Herbert; Magic Moment Miniature Roses, 1992

'ZIPNOTE', MinFl, pb, 1990; (**Love Note**); flowers deep pink, light red going to cream at base, reverse deep pink, dbl., 35 petals, exhibition form, slight fragrance; round, medium, orange fruit; prickles straight, small, light brown; foliage large, medium green, semi-glossy; bushy, tall growth; [Tamango X Cupcake]; Zipper, Herbert, 1986; Magic Moment Miniature Roses, 1991

'ZIPOET', Min, rb, 1990; (**Poetry**); bud pointed; flowers red with white at base, white reverse with some red at edges, dbl., 30 petals, exhibition form, slight fragrance; foliage small, medium green, semi-glossy; bushy, low growth; [Libby X Deep Purple]; Zipper, Herbert, 1987; Magic Moment Miniature Roses, 1991

'ZIPPINK', Min, w, 1988; (**Pink-A-ling**); flowers white edged deep pink, small, borne singly and in sprays, dbl., slight fragrance; foliage medium, dark green, glossy; upright growth; [Tamango X Avandel]; Zipper, Herbert; Magic Moment Miniature Roses, 1989

'ZIPPRO', F, pb, 1987; (**Prom Night**); flowers creamy yellow, shading to pink with deep pink flushing the edges, very dbl., slight fragrance; foliage large, medium green, matt; upright, bushy, candelabra-formed growth; [Poker Chip X Libby]; Zipper, Herbert; Magic Moment Miniature Roses

'ZIPRAM', Min, mp, 1985; (**Ramapo**); flowers small blooms borne singly, single, 5 petals, no fragrance; foliage small, medium green, matt; bushy growth; [Maytime X Libby]; Zipper, Herbert; Magic Moment Miniature Roses

'ZIPROUND', Min, mr, 1985; (**Roundabout**); flowers small blooms borne usually singly, dbl., 20 petals, no fragrance; foliage medium, medium green, matt; bushy growth; [Double Joy X Libby]; Zipper, Herbert; Magic Moment Miniature Roses

'ZIPSON', Min, mp, 1987; (Sonata, **Songfest**); flowers small, borne singly, very dbl., exhibition form, no fragrance; foliage small, medium green, matt; bushy, compact growth; [Queen Elizabeth X Baby Katie]; Zipper, Herbert; Magic Moment Miniature Roses

'ZIPTALE', Min, mp, 1988; (**Tattletale**); flowers deep pink, small, borne singly and in sprays, dbl., 33 petals, no fragrance; foliage small, dark green, semi-glossy; bushy, compact growth; [High Spirits X Charmglo]; Zipper, Herbert; Magic Moment Miniature Roses, 1989

'ZIPTEASE', Min, pb, 1988; (**Little Tease**); flowers yellow to cream at base, suffused with clear pink, darker at edges, exhibition form, no fragrance; foliage small, medium green, semi-glossy; bushy, compact growth; [High Spirits X Charmglo]; Zipper, Herbert; Magic Moment Miniature Roses, 1989

'ZIPTOK', F, lp, 1999; (**Tokay**); flowers apricot suffused with pink, reverse soft pink to apricot, dbl., 26–40 petals, 3 in., borne mostly singly, slight fragrance; prickles moderate; foliage medium, dark green, semi-glossy; upright, tall (4–5 ft) growth; [Olympic Gold X Spanish Sun]; Zipper, Herbert; Island Roses, 1999

'ZIPTOWN', Min, mp, 1986; (**Our Town**); flowers small blooms in small sprays, semi-dbl., no fragrance; foliage small, medium green, semi-glossy; spreading growth; [Maytime X Sheri Anne]; Zipper, H.; Magic Moment Miniature Roses

'ZIPVIA', Min, w, 1991; (**Bella Via**); dbl., 26–40 petals, 1.5–2.75 in., exhibition form, borne mostly singly, moderate fragrance; few prickles; foliage medium, dark green, matt; medium (40 cms), upright growth; [Rise 'n' Shine X Olympic Gold]; Zipper, Herbert; Magic Moment Miniature Roses, 1992

'ZIPVIR', F, lp, 1999; (**Irene Virag**); flowers red, white eye, reverse streaked red and white, single, 5–11 petals, 2–2.5 in., borne in small clusters, slight fragrance; prickles moderate; small, medium green, semi-glossy, new foliage mahogany red; bushy, tall (3–4 ft) growth; [seedling X Playboy]; Zipper, Herbert; Island Roses, 1999

'ZIPWEN', F, mp, 1991; (**Wendy**); dbl., 15–25 petals, 1.5–2.75 in., exhibition form, borne in small clusters, intense fragrance; few prickles; foliage medium, dark green, semi-glossy; medium (75 cms), upright growth; [Wendy Cussons X High Spirits]; Zipper, Herbert; Magic Moment Miniature Roses, 1992

'ZITA', Pol, mr, 1965; bud ovoid; flowers red shaded dark geranium, open, single; foliage dark, leathery; [Amoureuse X Seedling]; Delforge

'ZITKALA', S, mr, 1942; flowers brilliant velvety red, dbl., 25 petals, 3 in., nonrecurrent, intense fragrance; almost thornless; red wood; very hardy; [R. blanda X Amadis]; Hansen, N.E.

'ZITRONENFALTER'®, S, my, 1956; bud ovoid; flowers golden yellow, large blooms in clusters, dbl., 20 petals, repeat bloom, moderate, fruity fragrance; globular, slightly flattened top fruit; few, curved prickles; foliage dark, leathery; upright, bushy growth; [Marchenland X Peace]; Tantau, Math.

'ZITRONENFALTER', HSpn, ly, 1940; flowers brimstone color; [(R. spinosissima altaica X Star of Persia) X Golden Ophelia]; Berger, V.; Teschendorff

Zitronenjette®, *see* 'KORJONI'

'ZIZI', F, dr, 1963; flowers garnet-red, medium, borne in clusters of 3–10, semi-dbl., 15 petals, moderate fragrance; vigorous, bushy growth; [Walko X Souv. de J. Chabert]; Delbard-Chabert

Zlata, F, lp; Vecera, L., 1975; (Czech Rosa Club)

'ZLATÁ PRAHA', HT, my, 1931; ('DAS GOLDENE PRAG'); flowers golden yellow; [Kardinal Piffl sport]; Böhm, J.

'ZLATÉ JUBILEUM', HT, dy, 1938; flowers large, dbl.; Böhm, J.; (Sangerhausen)

'ZLATY DECH', HT, dy, 1936; flowers orange-yellow, large, semi-dbl., moderate fragrance; Böhm, J.; (Sangerhausen)

Zluta Zorina, HT, lp; Strnad; (Czech Rosa Club)

'ZODIAC', F, mr, 1961; flowers camellia-shape, dbl., 74 petals, 2.25–3 in., borne in clusters; foliage leathery, light green; very vigorous growth; [Red Favorite X Seedling]; Verschuren; Blaby Rose Gardens

'ZODIAC', F, pb, 1963; flowers begonia-pink, base yellow, borne in clusters of 25–30, dbl.; low, bushy growth; [Masquerade X Karl Herbst]; Kordes, R.

'ZOE', HGal, mp, 1840; flowers large, very dbl., moderate fragrance; Miellez; (Sangerhausen)

'ZOÉ', M, mp, 1861; (Moussue Partout); flowers rose-pink, globular; prickles well-mossed; Pradel

'ZOLA', S, dy, 1979; bud pointed; flowers golden yellow, semi-dbl., 18 petals, 2.5 in., intense fragrance; foliage dark; vigorous, upright growth; [Spek's Yellow X Magenta]; Sanday, John

Zola Budd, *see* 'KORZOLA'

Zoltan Kodaly, HT; Mark; (Cavriglia)

Zonnekeild, *see* **'SONNYCHILD'**

'ZONNEKIND', HT, my, 1941; flowers buttercup-yellow, large, dbl., moderate fragrance; [Arch. Reventos X Amalia Jung]; Leenders, M.

Zonta Rose, *see* 'HARTANNA'

Zonta-Rose, F, dy, 1988; flowers medium-large, dbl.; Harkness; (Sangerhausen)

'ZORA', F, yb, 1974; flowers yellow, deep pink petal edges, aging deep pink, blooms in clusters, semi-dbl., 18 petals; foliage dark; vigorous growth; [Masquerade X Rumba]; Staikov, Prof. Dr V.; Kalaydjiev and Chorbadjiiski

'ZORINA', F, or, 1963; (Rozorina); bud ovoid; dbl., 25 petals, 3 in., cupped, moderate fragrance; foliage glossy; vigorous, upright growth; GM, Rome, 1964; [Pinocchio seedling X Spartan]; Boerner; J&P

Zorina, HT, lp; Strnad; (Czech Rosa Club)

Zorka, F, rb, 1974; flowers carmine-red with white; Vecera, L.; (Sangerhausen)

'ZUBLOU', S, w, 1996; (**Louis Riel**); single, 5 petals, slight fragrance; numerous prickles; foliage medium, purplish-green, dull; bushy, tall growth; [R. rubrifolia X R. spinosissima 'Altaica']; Zubrowski, Stanley

'ZUKUNFT', F, or, 1951; flowers scarlet-red; [Lafayette sport]; Verschuren

'ZULU QUEEN', HT, dr, 1939; bud long, pointed; flowers dark maroon, large, dbl., exhibition form, moderate fragrance; foliage leathery; vigorous, bushy growth; [(Cathrine Kordes X E.G. Hill) X Fritz Hoger]; Kordes; J&P

'ZULU QUEEN, CLIMBING', Cl HT, dr, 1961; Moore

'ZULU WARRIOR', F, rb, 1980; flowers burgundy red, silver reverse, blooms borne 3–20 per cluster, semi-dbl., 9 petals, slight fragrance; large, reddish prickles; foliage dark, tough; medium growth; [Arthur Bell X Flaming Peace]; Hawken, Una

Zurella™, *see* MEIvrofix

'ZWEIBRÜCKEN', HKor, dr, 1955; bud ovoid; flowers deep crimson, large blooms in large clusters, dbl., recurrent bloom, slight fragrance; foliage dark, leathery; very vigorous, climbing growth; [R. kordesii X Independence]; Kordes

'ZWEMANIA', Gr, or, 1974; flowers light vermilion, full, dbl., 25–30 petals, exhibition form, moderate, fruity fragrance; foliage dark; very vigorous growth; PP004049; [Sweet Promise sport]; Zwemstra; Meilland

Zwergeliebe, Min, 1986; Delforge; (Cavriglia)

Zwergenfee®, *see* 'KORFEE'

Zwergkönig, *see* **'DWARFKING'**

Zwergkonig 78, *see* KORkonig

Zwergkönigin, *see* **'QUEEN OF THE DWARFS'**

Zwergkönigin '82, *see* **'KORWERK'**

ISBN 0122276205
9 780122 276200